Bacterial phylogeny from a shower curtain biofilm

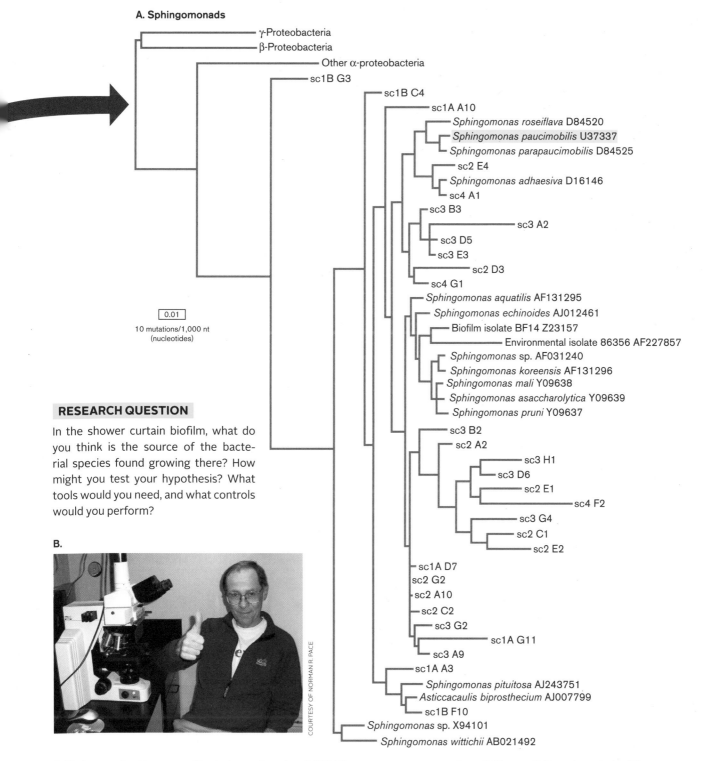

A. Sphingomonads

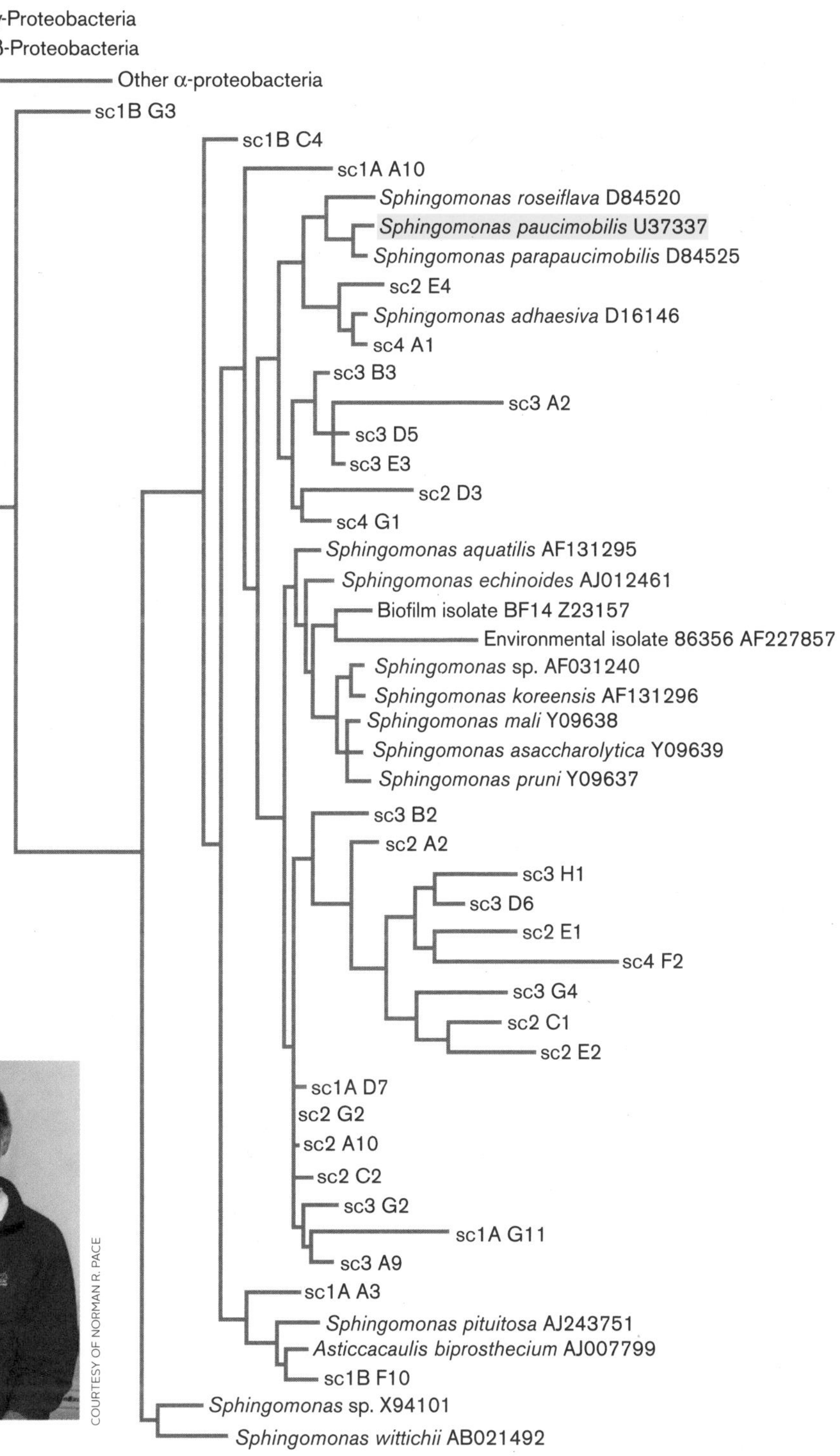

γ-Proteobacteria
β-Proteobacteria
Other α-proteobacteria
sc1B G3
sc1B C4
sc1A A10
Sphingomonas roseiflava D84520
Sphingomonas paucimobilis U37337
Sphingomonas parapaucimobilis D84525
sc2 E4
Sphingomonas adhaesiva D16146
sc4 A1
sc3 B3
sc3 A2
sc3 D5
sc3 E3
sc2 D3
sc4 G1
Sphingomonas aquatilis AF131295
Sphingomonas echinoides AJ012461
Biofilm isolate BF14 Z23157
Environmental isolate 86356 AF227857
Sphingomonas sp. AF031240
Sphingomonas koreensis AF131296
Sphingomonas mali Y09638
Sphingomonas asaccharolytica Y09639
Sphingomonas pruni Y09637
sc3 B2
sc2 A2
sc3 H1
sc3 D6
sc2 E1
sc4 F2
sc3 G4
sc2 C1
sc2 E2
sc1A D7
sc2 G2
sc2 A10
sc2 C2
sc3 G2
sc1A G11
sc3 A9
sc1A A3
Sphingomonas pituitosa AJ243751
Asticcacaulis biprosthecium AJ007799
sc1B F10
Sphingomonas sp. X94101
Sphingomonas wittichii AB021492
0.01
10 mutations/1,000 nt
(nucleotides)

RESEARCH QUESTION

In the shower curtain biofilm, what do you think is the source of the bacterial species found growing there? How might you test your hypothesis? What tools would you need, and what controls would you perform?

B.

COURTESY OF NORMAN R. PACE

A. Phylogeny of sphingomonad bacteria was based on SSU RNA gene sequence comparison. **B.** Norman R. Pace characterized the first sequence phylogeny of thermophiles from high-temperature environments. His laboratory has since characterized the microbial diversity of other environments, such as a shower curtain biofilm.

THIRD EDITION

Microbiology

An Evolving Science

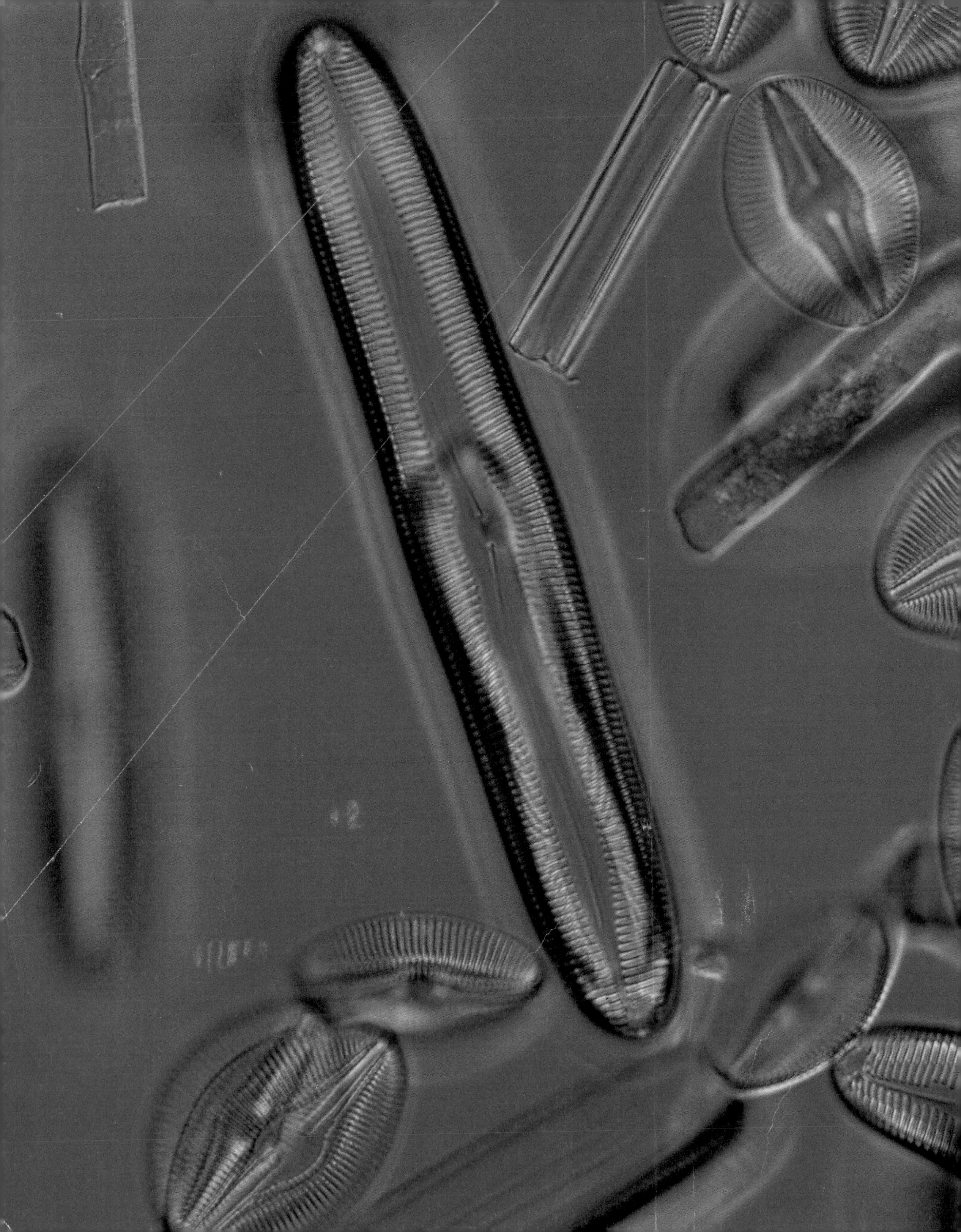

THIRD EDITION

Microbiology
An Evolving Science

Joan L. Slonczewski
Kenyon College

John W. Foster
University of South Alabama

Appendices 1 and 2 by
Kathy M. Gillen
Kenyon College

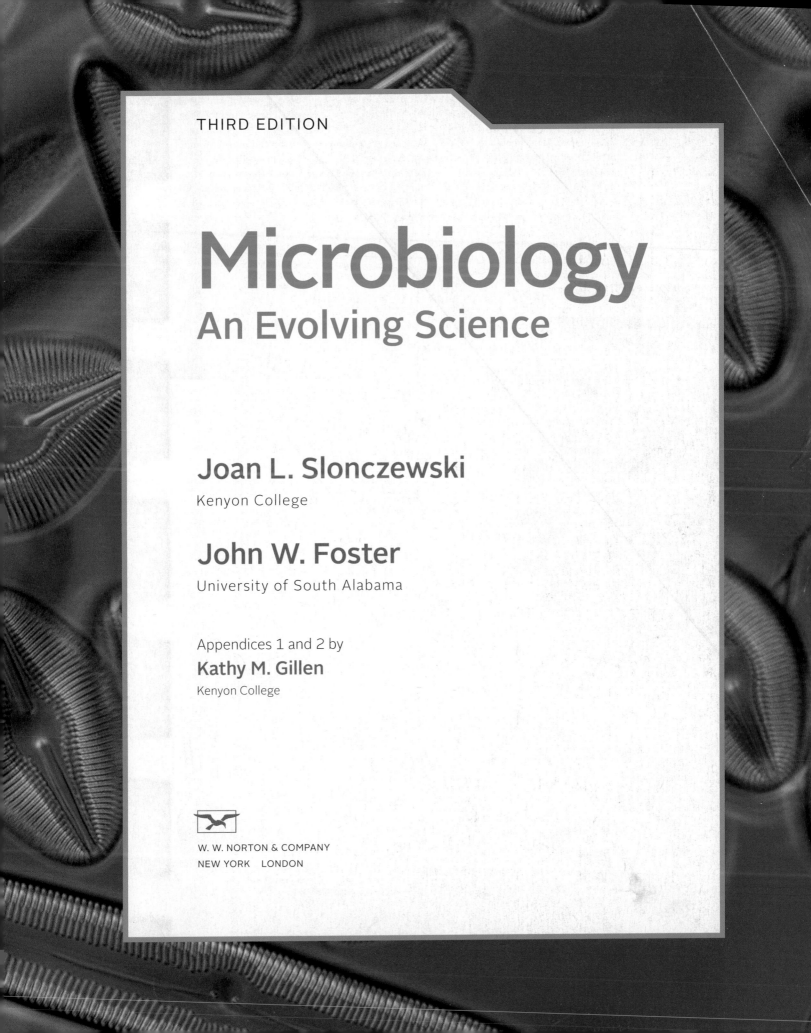

W. W. NORTON & COMPANY
NEW YORK LONDON

W. W. Norton & Company has been independent since its founding in 1923, when William Warder Norton and Mary D. Herter Norton first published lectures delivered at the People's Institute, the adult education division of New York City's Cooper Union. The firm soon expanded its program beyond the Institute, publishing books by celebrated academics from America and abroad. By mid-century, the two major pillars of Norton's publishing program—trade books and college texts—were firmly established. In the 1950s, the Norton family transferred control of the company to its employees, and today—with a staff of four hundred and a comparable number of trade, college, and professional titles published each year—W. W. Norton & Company stands as the largest and oldest publishing house owned wholly by its employees.

Printed in Canada
Third Edition

Art and Composition by Precision Graphics
Project supervisor: Kirsten Dennison
Manufacturing by Transcontinental–Beauceville QC

Editor: Betsy Twitchell
Developmental editors: Michael Zierler and John Murdzek
Senior project editor: Thomas Foley
Copy editor: Stephanie Hiebert
Associate production director: Benjamin Reynolds
Assistant editor: Courtney Shaw
Art director: Rubina Yeh
Designer: Lissi Sigillo
Managing editor, college: Marian Johnson
Science media editors: Patrick Shriner and Robin Kimball
Associate editor, emedia: Callinda Taylor
Marketing manager: Meredith Leo
Photo director: Trish Marx
Photo researcher: Jane Miller
Editorial assistant: Katie Callahan

0-393-91929-5

W. W. Norton & Company, Inc., 500 Fifth Avenue, New York, N.Y. 10110
wwnorton.com

W. W. Norton & Company Ltd., Castle House, 75/76 Wells Street, London W1T 3QT

1 2 3 4 5 6 7 8 9

Dedication

We dedicate this Third Edition to the memory of Lynn Margulis (1938–2011) and Carl Woese (1928–2012), who forever changed our understanding of evolutionary biology. Margulis hypothesized that modern-day eukaryotic cells evolved from symbiotic relationships with bacteria; and Woese used the evidence of gene sequences to propose the now accepted paradigm of three, continually evolving, domains of life: Bacteria, Archaea, and Eukarya. The world will long miss the deep intellect and larger-than-life personalities of these two scientists. We, the authors, were profoundly influenced by their thinking and deeply moved by their passing.

BRIEF CONTENTS

eTopic Contents xix
Preface xxi
About the Authors xxxiii

PART 1
The Microbial Cell 2

1 Microbial Life: Origin and Discovery 5
2 Observing the Microbial Cell 41
3 Cell Structure and Function 79
4 Bacterial Culture, Growth, and Development 119
5 Environmental Influences and Control of Microbial Growth 157
6 Viruses 191

PART 2
Genes and Genomes 234

7 Genomes and Chromosomes 237
8 Transcription, Translation, and Bioinformatics 275
9 Gene Transfer, Mutations, and Genome Evolution 321
10 Molecular Regulation 365
11 Viral Molecular Biology 409
12 Biotechniques and Synthetic Biology 457

PART 3
Metabolism and Biochemistry 488

13 Energetics and Catabolism 491
14 Electron Flow in Organotrophy, Lithotrophy, and Phototrophy 539
15 Biosynthesis 585
16 Food and Industrial Microbiology 623

PART 4
Microbial Diversity and Ecology 662

17 Origins and Evolution 665
18 Bacterial Diversity 711
19 Archaeal Diversity 757
20 Eukaryotic Diversity 797
21 Microbial Ecology 839
22 Microbes in Global Elemental Cycles 889

PART 5
Medicine and Immunology 922

23 Human Microbiota and Innate Immunity 925
24 The Adaptive Immune Response 961
25 Microbial Pathogenesis 1003
26 Microbial Diseases 1051
27 Antimicrobial Therapy 1107
28 Clinical Microbiology and Epidemiology 1147

APPENDIX 1: Biological Molecules A-1
APPENDIX 2: Introductory Cell Biology: Eukaryotic Cells A-23
APPENDIX 3: Laboratory Methods for Microbiology A-43
APPENDIX 4: Taxonomy A-59

Answers to Thought Questions AQ-1
Glossary G-1
Figure Credits F-1
Index I-1

CONTENTS

eTopic Contents xix
Preface xxi
About the Authors xxxiii

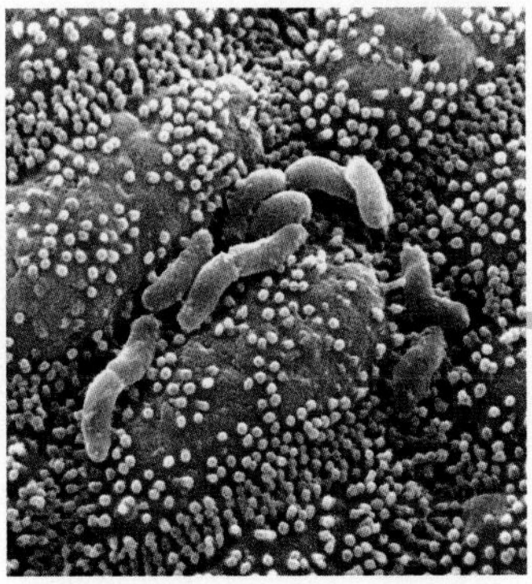

PART 1

The Microbial Cell 2

AN INTERVIEW WITH RICHARD LENSKI:
Evolution in the Lab

CHAPTER 1
Microbial Life: Origin and Discovery 5

1.1 From Germ to Genome: What Is a Microbe? 7
1.2 Microbes Shape Human History 10
Special Topic 1.1: How Did Life Originate? 18
1.3 Medical Microbiology 20
1.4 Microbial Ecology 26
1.5 The Microbial Family Tree 29
1.6 Cell Biology and the DNA Revolution 32

CHAPTER 2
Observing the Microbial Cell 41

2.1 Observing Microbes 42
2.2 Optics and Properties of Light 46
2.3 Bright-Field Microscopy 51
2.4 Fluorescence Microscopy 58
2.5 Dark-Field and Phase-Contrast Microscopy 63
2.6 Electron Microscopy and Tomography 66
2.7 Visualizing Molecules 73
Special Topic 2.1: Molecular "Snapshots": Chemical Imaging 74

CHAPTER 3
Cell Structure and Function 79

3.1 The Bacterial Cell: An Overview 81
3.2 The Cell Membrane and Transport 86
3.3 The Cell Wall and Outer Layers 91
3.4 The Nucleoid, RNA, and Protein Synthesis 101
3.5 Cell Division 104
3.6 Cell Polarity and Aging 107
Special Topic 3.1: Senior Cells Make Drug-Resistant Tuberculosis 108
3.7 Specialized Structures 112

CHAPTER 4
Bacterial Culture, Growth, and Development 119

4.1 Microbial Nutrition 120
4.2 Nutrient Uptake 127
4.3 Culturing Bacteria 133
4.4 Counting Bacteria 136
4.5 The Growth Cycle 139
4.6 Biofilms 145
Special Topic 4.1: Sharks and Biofilms Don't Mix 146
4.7 Cell Differentiation 150

CHAPTER 5
Environmental Influences and Control of Microbial Growth 157

5.1 Environmental Limits on Growth 158
5.2 Adaptation to Temperature 160
Special Topic 5.1: It's Raining Bacteria 163
5.3 Adaptation to Pressure 164
5.4 Water Activity and Salt 166
5.5 Adaptation to pH 167
5.6 Oxygen and Other Electron Acceptors 172
5.7 Nutrient Deprivation and Starvation 176
5.8 Physical, Chemical, and Biological Control of Microbes 178

CHAPTER 6
Viruses 191

6.1 What Is a Virus? 192
6.2 Virus Structure 198
6.3 Viral Genomes and Classification 203
6.4 Bacteriophage Replication 208
6.5 Animal and Plant Virus Replication 213

6.6 Culturing Viruses 221
Special Topic 6.1: The Good Viruses 226
6.7 Viral Ecology 226

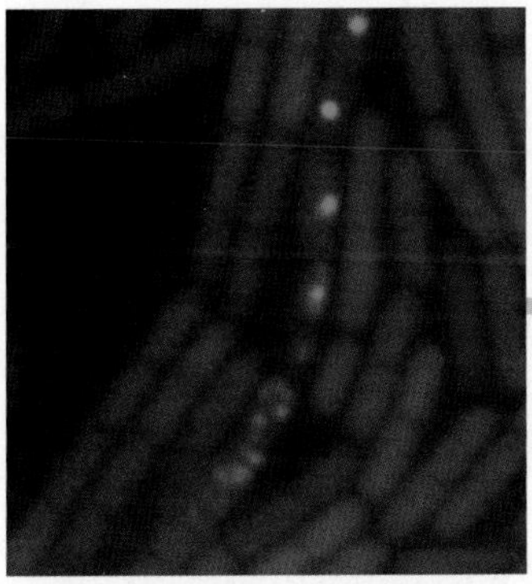

PART 2
Genes and Genomes 234

AN INTERVIEW WITH CHRISTINE JACOBS-WAGNER:
The Thrill of Discovery in Molecular Microbiology

CHAPTER 7
Genomes and Chromosomes........237

7.1 DNA: The Genetic Material 238
7.2 Genome Organization 240
7.3 DNA Replication 248
7.4 Plasmids 258
7.5 Eukaryotic Chromosomes 260
7.6 DNA Sequence Analysis 263
Special Topic 7.1: Where Have All the Bees Gone? Metagenomics, Pyrosequencing, and Nature 268

CHAPTER 8
Transcription, Translation, and Bioinformatics........275

8.1 RNA Polymerases and Sigma Factors 276
8.2 Transcription of DNA to RNA 280
8.3 Translation of RNA to Protein 286
Special Topic 8.1: Stalking the Lone Ribosome 298
8.4 Protein Modification and Folding 303
8.5 Protein Degradation: Cleaning House 304
8.6 Secretion: Protein Traffic Control 306
8.7 Bioinformatics: Mining the Genomes 312

CHAPTER 9
Gene Transfer, Mutations, and Genome Evolution........321

9.1 The Mosaic Nature of Genomes 322
9.2 Gene Transfer 322
Special Topic 9.1: There's a Bacterial Genome Hidden in My Fruit Fly 330

9.3 Recombination 338
9.4 Mutations 341
9.5 DNA Repair 348
9.6 Mobile Genetic Elements 355
9.7 Genome Evolution 358

CHAPTER 10
Molecular Regulation 365

10.1 Regulating Gene Expression 366
10.2 Paradigm of the Lactose Operon 369
10.3 Other Systems of Operon Control 376
10.4 Sigma Factor Regulation 382
10.5 Regulatory RNAs 385
10.6 DNA Rearrangements: Phase Variation by Shifty Pathogens 389
10.7 Integrated Control Circuits 391
10.8 Quorum Sensing: Chemical Conversations 397
10.9 Transcriptomics and Proteomics 401
Special Topic 10.1: Networking with Nanotubes 402

CHAPTER 11
Viral Molecular Biology 409

11.1 Phage T4: The Classic Molecular Model 410
11.2 Hepatitis C: (+) Strand RNA Virus 417
11.3 Influenza Virus: (–) Strand RNA Virus 424
11.4 Human Immunodeficiency Virus (HIV): Retrovirus 432
11.5 Herpes Simplex Virus: DNA Virus 443
Special Topic 11.1: Cytomegalovirus 448
11.6 Gene Therapy with Viruses 450

CHAPTER 12
Biotechniques and Synthetic Biology 457

12.1 Basic Tools of Biotech: A Research Case Study 458
12.2 Genetic Analyses 458
12.3 Classic Molecular Techniques 463
12.4 Viewing the Interactions and Movements of Proteins 471
12.5 Applied Biotechnology 474
12.6 Synthetic Biology: Biology by Design 477
Special Topic 12.1: Bacteria "Learn" to Keep Time and Signal Danger 478

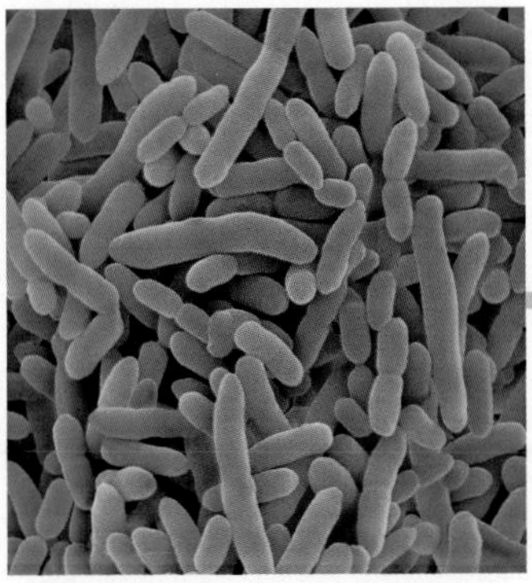

PART 3
Metabolism and Biochemistry 488

AN INTERVIEW WITH DAN WOZNIAK:
Polymer Biosynthesis Makes a Pathogenic Biofilm

CHAPTER 13
Energetics and Catabolism 491

13.1 Energy and Entropy for Life 494
13.2 Energy in Biochemical Reactions 498
13.3 Energy Carriers and Electron Transfer 501
Special Topic 13.1: Microbial Syntrophy Cleans Up Oil 502
13.4 Catabolism: The Microbial Buffet 510
13.5 Glucose Breakdown and Fermentation 516
13.6 The Tricarboxylic Acid (TCA) Cycle 527
13.7 Aromatic Pollutants 532

CHAPTER 14
Electron Flow in Organotrophy, Lithotrophy, and Phototrophy 539

14.1 Electron Transport Systems 541
14.2 The Proton Motive Force 546
Special Topic 14.1: Testing the Chemiosmotic Theory 548
14.3 The Respiratory ETS and ATP Synthase 551
14.4 Anaerobic Respiration in Organotrophs 559
Special Topic 14.2: Bacterial Electric Power 562
14.5 Lithotrophy and Methanogenesis 563
14.6 Phototrophy 571

CHAPTER 15
Biosynthesis 585

15.1 Overview of Biosynthesis 586
15.2 CO_2 Fixation: The Calvin Cycle 589
15.3 CO_2 Fixation: Diverse Pathways 597
15.4 Biosynthesis of Fatty Acids and Polyketides 601
15.5 Nitrogen Fixation 605
Special Topic 15.1: Mining a Bacterial Genome for Peptide Antibiotics 606
15.6 Biosynthesis of Amino Acids and Nitrogenous Bases 612
15.7 Biosynthesis of Tetrapyrroles 618

CHAPTER 16

Food and Industrial Microbiology 623

16.1 Microbes as Food 625
16.2 Fermented Foods: An Overview 627
16.3 Acid- and Alkali-Fermented Foods 629
16.4 Ethanolic Fermentation: Bread and Wine 637
16.5 Food Spoilage and Preservation 641
16.6 Industrial Microbiology 650
Special Topic 16.1: Companies Take On Tuberculosis 652
Special Topic 16.2: Microbial Enzymes Make Money 654

PART 4

Microbial Diversity and Ecology 662

AN INTERVIEW WITH NICOLE DUBILIER:
Marine Animals with Bacterial Symbionts

CHAPTER 17

Origins and Evolution 665

17.1 Origins of Life 667
17.2 Early Metabolism 676
17.3 Microbial Phylogeny and Gene Transfer 681
Special Topic 17.1: Phylogeny of a Shower Curtain Biofilm 686
17.4 Adaptive Evolution 692
17.5 Microbial Species and Taxonomy 696
Special Topic 17.2: Jump-Starting Evolution of a Hyperthermophilic Enzyme 698
17.6 Symbiosis and the Origin of Mitochondria and Chloroplasts 703

CHAPTER 18

Bacterial Diversity 711

18.1 Bacterial Diversity at a Glance 712
18.2 Cyanobacteria: Oxygenic Phototrophs 718
18.3 Firmicutes and Actinobacteria (Gram-Positive) 723
18.4 Proteobacteria (Gram-Negative) 734
Special Topic 18.1: Carbon Monoxide: Food for Bacteria? 738
18.5 Deep-Branching Gram-Negative Phyla 748
18.6 Spirochetes: Sheathed Spiral Cells with Internalized Flagella 750
18.7 Chlamydiae, Planctomycetes, and Verrucomicrobia: Irregular Cells 752

CHAPTER 19
Archaeal Diversity 757

19.1 Archaeal Traits 758
19.2 Crenarchaeota across the Temperature Range 766
19.3 Thaumarchaeota: Symbionts and Ammonia Oxidizers 773
19.4 Methanogens 775
Special Topic 19.1: Eating Ammonia: Thaumarchaeotes 776
19.5 Haloarchaea 784
Special Topic 19.2: Haloarchaea in the Classroom 786
19.6 Thermophilic and Acidophilic Euryarchaeota 790
19.7 Deeply Branching Divisions 793

CHAPTER 20
Eukaryotic Diversity 797

20.1 Phylogeny of Eukaryotes 798
20.2 Fungi 806
Special Topic 20.1: Yeast: A Single-Celled Human Brain? 810
20.3 Algae 818
20.4 Amebas and Slime Molds 824
20.5 Alveolates: Ciliates, Dinoflagellates, and Apicomplexans 827
20.6 Trypanosomes and Metamonads 833
Special Topic 20.2: The Trypanosome: A Shape-Shifting Killer 834

CHAPTER 21
Microbial Ecology 839

21.1 Metagenomes—and Beyond 841
21.2 Functional Ecology 849
21.3 Symbiosis 852
21.4 Marine and Aquatic Microbes 857
Special Topic 21.1: Cleaning Up the *Deepwater* Oil Spill 858
21.5 Soil and Subsurface Microbes 869
21.6 Plant Microbial Communities 876
21.7 Animal Microbial Communities 882

CHAPTER 22
Microbes in Global Elemental Cycles 889

22.1 Biogeochemical Cycles 890
22.2 The Carbon Cycle and Bioremediation 894
22.3 The Hydrologic Cycle and Wastewater Treatment 896
Special Topic 22.1: Bioremediation of Weapons Waste 902
22.4 The Nitrogen Cycle 903
22.5 Sulfur, Phosphorus, and Metals 908
22.6 Astrobiology 916

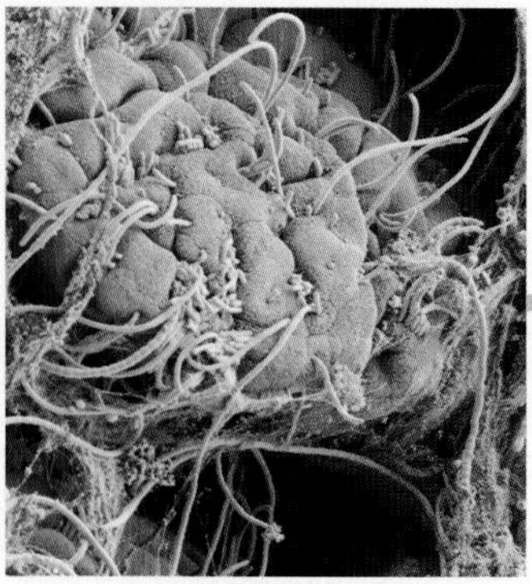

PART 5
Medicine and Immunology 922

AN INTERVIEW WITH FERRIC FANG:
Molecular Microbiology Dissects a Pathogen

CHAPTER 23
Human Microbiota and Innate Immunity 925

23.1 Human Microbiota: Location and Shifting Composition 926
23.2 Risks and Benefits of Microbiota 934
23.3 Overview of the Immune System 935
Special Topic 23.1: Are NETs a Cause of Lupus? 940
23.4 Barbarians at the Gate: Innate Host Defenses 943
23.5 The Acute Inflammatory Response 946
23.6 How Phagocytes Detect and Kill Microbes 950
23.7 Interferon, Natural Killer Cells, and Toll-like Receptors 953
23.8 Complement's Role in Innate Immunity 955
23.9 Fever 958

CHAPTER 24
The Adaptive Immune Response 961

24.1 Overview of Adaptive Immunity 962
24.2 Immunogenicity 965
24.3 Antibody Structure and Diversity 968
24.4 Primary and Secondary Antibody Responses 974
24.5 Genetics of Antibody Production 977
24.6 T Cells Link Antibody and Cellular Immune Systems 982
24.7 Complement as Part of Adaptive Immunity 992
Special Topic 24.1: An Uneasy Peace: Détente at the Microbiota-Intestine Interface 994
24.8 Hypersensitivity and Autoimmunity 995

CHAPTER 25
Microbial Pathogenesis 1003

25.1 Host-Pathogen Interactions 1004
25.2 Virulence Factors and Pathogenicity Islands 1010
25.3 Microbial Attachment: First Contact 1013
25.4 Toxins Subvert Host Function 1018

25.5 Deploying Toxins and Effectors 1028
25.6 Surviving within the Host 1033
Special Topic 25.1: Type VI Secretion: Poison Darts 1034
25.7 Experimental Tools That Define Pathogenesis 1045

CHAPTER 26
Microbial Diseases 1051

26.1 Characterizing and Diagnosing Microbial Diseases 1052
26.2 Skin and Soft-Tissue Infections 1054
26.3 Respiratory Tract Infections 1058
26.4 Gastrointestinal Tract Infections 1066
Special Topic 26.1: Sprouts and an Emerging *Escherichia coli* 1068
26.5 Genitourinary Tract Infections 1075
26.6 Central Nervous System Infections 1083
26.7 Cardiovascular System Infections 1090
26.8 Systemic Infections 1094
26.9 Immunization 1102

CHAPTER 27
Antimicrobial Therapy 1107

27.1 The Golden Age of Antibiotic Discovery 1108
27.2 Fundamentals of Antimicrobial Therapy 1111
27.3 Measuring Drug Susceptibility 1112
27.4 Mechanisms of Action 1116
27.5 Challenges of Drug Resistance 1126
27.6 The Future of Drug Discovery 1134
27.7 Antiviral Agents 1135
Special Topic 27.1: Anti-Quorum Sensing Drug Blocks Pathogen "Control and Command" 1136
Special Topic 27.2: Resurrecting the 1918 Pandemic Flu Virus 1139
27.8 Antifungal Agents 1142

CHAPTER 28
Clinical Microbiology and Epidemiology 1147

28.1 Principles of Clinical Microbiology 1148
28.2 Specimen Collection and Processing 1149
28.3 Conventional Approaches to Pathogen Identification 1152
28.4 Rapid Techniques for Pathogen Identification 1160
28.5 Point-of-Care Rapid Diagnostics 1167
28.6 Biosafety Containment Procedures 1169
28.7 Principles of Epidemiology 1171
Special Topic 28.1: What's Blowing in the Wind? 1178
28.8 Detecting Emerging Microbial Diseases 1179

APPENDIX 1
Biological Molecules A-1
A1.1 Elements, Bonding, and Water A-2
A1.2 Organic Molecules A-6
A1.3 Proteins A-8
A1.4 Carbohydrates A-11
A1.5 Nucleic Acids A-13
A1.6 Lipids A-15
A1.7 Biological Chemistry A-16
Special Topic A1.1: Calculating the Standard Free Energy Change, $\Delta G°$, of Chemical Reactions A-19

APPENDIX 2
Introductory Cell Biology: Eukaryotic Cells A-23
A2.1 The Cell Membrane A-24
A2.2 The Nucleus and Mitosis A-31
A2.3 Problems Faced by Large Cells A-34
A2.4 The Endomembrane System A-34
A2.5 The Cytoskeleton A-38
A2.6 Mitochondria and Chloroplasts A-40

APPENDIX 3
Laboratory Methods for Microbiology A-43
A3.1 Isolating Parts of Cells by Using an Ultracentrifuge A-44
A3.2 Agarose Gel Electrophoresis A-45
A3.3 Protein Identification on 2D Gels with Mass Spectrometry A-47
A3.4 RNA and DNA Identification by Northern and Southern Blots A-49
A3.5 Sanger Method of DNA Sequencing A-51
A3.6 Gene Fusions Identify Regulatory Mutants A-52
A3.7 Primer Extension Identifies Transcriptional Start Sites A-52
A3.8 DNA Microarray A-54
A3.9 Multiplex PCR A-55
A3.10 Fluorescence In Situ Hybridization (FISH) and CARD-FISH A-55
A3.11 Immunoprecipitation Techniques A-57

APPENDIX 4
Taxonomy A-59
A4.1 Viruses A-60
A4.2 Bacteria A-62
A4.3 Archaea A-66
A4.4 Eukarya A-68

Answers to Thought Questions AQ-1
Glossary G-1
Figure Credits F-1
Index I-1

eTOPIC CONTENTS

Access to the eTopics is available through both the ebook and the Norton Coursepack.

1.1 An Interview with Rita Colwell: The Global Impact of Microbiology
1.2 Discovering the Genetic Code
1.3 Clifford W. Houston: From Aquatic Pathogens to Outer Space—An Interview
2.1 Confocal Microscopy
3.1 Isolation and Analysis of the Ribosome
3.2 How Antibiotics Cross the Outer Membrane
3.3 Outer Membrane Proteins: Isolation for Vaccine
3.4 Experiments That Reveal the Bacterial Cytoskeleton
4.1 Transport by Group Translocation: The Phosphotransferase System
4.2 Eukaryotes Transport Nutrients by Endocytosis
4.3 Biofilms, Antibiotics, Garlic, and Disease
5.1 The Arrhenius Equation
5.2 Some Alkaliphilic Enzymes Produce Useful Drug Delivery Systems
5.3 Signaling Virulence
5.4 Oligotrophs
6.1 How Did Viruses Originate?
6.2 West Nile Virus, an Emerging Pathogen
7.1 Genes and Proteins Involved in DNA Replication
7.2 Trapping a Sliding Clamp
7.3 Replication Mechanisms of Bacteriophages
7.4 Plasmid Partitioning and Addiction
7.5 Equilibrium Density Gradient Centrifugation
8.1 Building the Ribosome Machine
8.2 Discovering the mRNA Ribosome-Binding Site
8.3 Ubiquitination: A Ticket to the Proteasome
8.4 What Is the Minimal Genome?
9.1 F Pili and Biofilm Formation
9.2 Mapping Bacterial Chromosome Gene Position by Conjugation
9.3 *Deinococcus* Uses RecA to Repair Fragmented Chromosomes
9.4 Mutation Rate
9.5 The Transposase for a Bacterial Transposon Resembles the Integrase for HIV-1
9.6 Integrons and Gene Capture
10.1 CRP Interactions with RNA Polymerase and CRP-Dependent Promoters
10.2 Glucose Transport Alters cAMP Levels
10.3 Slipped-Strand Mispairing
10.4 Toxin-Antitoxin Modules: Mechanisms for Self-preservation or Altruism?
10.5 The Phage Lambda Lysis/Lysogeny "Decision" Is to Kill or Not to Kill
11.1 Poliovirus: (+) Strand RNA
11.2 The Filamentous Phage M13: Vaccines and Nanowires
11.3 Genetic Resistance to HIV
12.1 Mapping the *E. coli* Interactome
12.2 GFP Proteins Track Cell Movements in Biofilms
12.3 DNA Vaccines
12.4 Gene Therapy and Gene Delivery Systems
12.5 Directed Evolution through Phage Display Technology
12.6 DNA Shuffling Enables In Vitro Evolution
12.7 Site-Directed Mutagenesis Helps Us Probe Protein Function
13.1 Observing Energy Carriers in Living Cells
13.2 Swiss Cheese: A Product of Bacterial Catabolism
13.3 Genomic Analysis of Metabolism
13.4 Pyruvate Dehydrogenase Connects Sugar Catabolism to the TCA Cycle
13.5 Genetic Analysis of Aromatic Catabolism
14.1 Caroline Harwood: A Career in Bacterial Photosynthesis and Biodegradation: An Interview
14.2 Measuring Dy and DpH in Microbes by the Uptake of Molecules
14.3 Environmental Regulation of the ETS
14.4 ATP Synthesis at High pH
15.1 The Discovery of ^{14}C
15.2 Metagenomic Screening for Polyketide Drugs
15.3 Antibiotic Factories: Modular Biosynthesis of Vancomycin
15.4 Riboswitch Regulation
16.1 From Barley and Hops to Beer
16.2 Caterpillar Viruses Produce Commercial Products
17.1 The RNA World: Clues for Modern Medicine
17.2 Horizontal Gene Transfer in *E. coli* O157:H7
17.3 Leaf-cutter Ants with Partner Fungi and Bacteria
18.1 Karl Stetter: Adventures in Microbial Diversity Lead to Products in Industry
20.1 Oomycetes: Lethal Parasites That Resemble Fungi
20.2 A Ciliate Model for Human Aging
21.1 Mapping Bermuda Phytoplankton
21.2 Cold-Seep Ecosystems
22.1 Wetlands: Disappearing Microbial Ecosystems
22.2 Metal Contamination and Bioremediation

23.1 Microbes as Vaccine Delivery Systems
23.2 Do Defensins Help Determine Species Specificity for Infection?
23.3 Cathelicidins
24.1 Factors That Influence Immunogenicity
24.2 ABO Blood Groups: Antigens, Antibodies, and Karl Landsteiner
24.3 Organ Donation and Transplant Rejection
24.4 Case Studies in Hypersensitivity
25.1 Finding Virulence Genes: Signature-Tagged Mutagenesis
25.2 Finding Virulence Genes: In Vivo Expression Technologies
25.3 Caught in the Act: *Streptococcus agalactiae* Evolved through Conjugation
25.4 Pili Tip Proteins Tighten Their Grip
25.5 Normal G-Factor Control of Adenylate Cyclase
25.6 Diphtheria Toxin
25.7 Identifying New Microbial Toxins
25.8 Bacterial Covert Operations: Secreted *Shigella* Effector Proteins Jam Communications between Target Cells and Innate Immunity
26.1 Sequenced Genomes of Pathogens
26.2 Human Papillomavirus
26.3 The Respiratory Tract Pathogen *Bordetella* Binds to Lung Cilia
26.4 The Common Cold versus Influenza
26.5 Intracellular Biofilm Pods Are Reservoirs of Infection
26.6 Human Immunodeficiency Virus: Pathogenesis
26.7 Spongiform Encephalopathies
26.8 Atherosclerosis and Coronary Artery Disease
27.1 Antibiotic Spectrum of Activity
27.2 Antibiotic Biosynthesis Pathways
27.3 Poking Holes with Nanotubes: A New Antibiotic Therapy
27.4 Critical Virulence Factors Found in the 1918 Strain of Influenza Virus
28.1 API Reactions and Generating a Seven-Digit Microbe Identification Code
28.2 DNA-Based Detection Tests
28.3 Microbial Pathogen Detection Gets Wired Up

PREFACE

In the first two editions of *Microbiology: An Evolving Science*, we worked to write the defining core text of our generation—the book that would inspire undergraduate science majors to embrace the microbial world. Our emphasis on genetics and ecology, the use of case histories in the medical section, and the balanced depiction of women and minority scientists, including young researchers, drew—and continues to draw—enthusiastic responses from our more than one hundred adopters. Our focus on evolution, and our modern organization reflecting changes in the field, proved so successful that other textbooks have adjusted their chapter sequence to parallel *An Evolving Science*. In the Third Edition, we maintain this chapter organization to facilitate year-to-year course transitions for instructors. In addition, we incorporate exciting new research advances to ensure that *An Evolving Science* is the most current and engaging microbiology textbook available.

Also in this Third Edition, we maintain our signature balance between cutting-edge ecology and medicine, while adding new research topics and emerging microbial-human partnerships. The book opens with a new Part 1 Interview with Richard Lenski, in which he presents his personal perspective on the groundbreaking bacterial evolution experiment. Experimental evolution now fills a new section in Chapter 17, Origins and Evolution. Other chapters that underwent major revision include Chapter 3, Cell Structure and Function, with a tightened opener and a new section on cell aging; and Chapter 21, Microbial Ecology, which opens with a new section on metagenomics and the culturing of "unculturables."

In many chapters, we relate topics to current events, to keep students interested in and informed on the role of microbiology in the world today. One example is synthetic biology, the construction of microbes with genetic circuits engineered for commercial use (Chapter 12, Biotechnology and Synthetic Biology). Another example is the use of of viral replication cycles to develop lentiviral treatments for cancer and inherited disorders, including the first possible "cure" for pediatric leukemia (presented in Chapter 11, Viral Molecular Biology).

Our Third Edition continues as a community project, drawing on our experience as researchers and educators as well as the input of hundreds of colleagues to create a microbiology text for the twenty-first century. We present the story of molecular microbiology and microbial ecology from its classical history of Koch, Pasteur, and Winogradsky, to twenty-first-century researchers Rita Colwell and Bonnie Bassler. The Third Edition includes many contributions recommended by colleagues from around the world, at institutions such as Washington University, University of California–Davis, University of Wisconsin–Madison, Cornell University, Florida State University, University of Toronto, University of Edinburgh, University of Antwerp, Seoul National University, Chinese University of Hong Kong, and many more. We are grateful to you all.

While we have expanded and developed new topics, we also recognized the need to keep the length and "core" of the book to a size reasonable enough for the undergraduate student. In order to contain length while adding new material, we continue transferring certain topics online as "eTopics." The eTopics are called out in the text, hyperlinked to the ebook, and their key terms are fully indexed in the printed book. Therefore, returning adopters can be confident of keeping access to all of the material they taught from the Second Edition, but now they also have new topics on *Mycobacterium tuberculosis* cell aging and drug resistance (Chapter 3) and on bacteria that convert phage genes into toxin secretion systems (Chapter 25), and many more.

Major Features

Our book targets the science major in biology, microbiology, or biochemistry. Several important features make our book the best text available for undergraduates today:

- **New research on contemporary themes** such as evolution, genomics, metagenomics, molecular genetics, and biotechnology enrich students' understanding of foundational topics and highlight the current state of the field. Every chapter presents numerous current research examples within the up-to-date framework of molecular biology. Examples of current research include measuring the movement of a single translating ribosome; transplanting a whole genome; determining the "pangenome," the overall set of genes available to a species; and the spectroscopic measurement of carbon flux from microbial communities.
- **A comprehensively updated art program with engaging figures that are also dynamic learning tools.** A fresh, contemporary new design and an updated art program presents content in an engaging, visually dynamic manner. New in-figure Thought Questions encourage students to interpret and analyze visuals of important concepts. Figures that pair with a process animation online include a QR code in the text that students can scan using their smartphones to immediately view online.
- **Core concepts are presented in a student friendly way that motivates learning.** Ample Thought Questions throughout every chapter challenge students to think critically about core concepts, the way a scientist would. In addition, scientists pursuing research today are presented alongside the traditional icons. For example, Chapter 1 introduces historical figures such as Koch and Pasteur alongside genome sequencer Claire Fraser-Liggett and young microbial ecologist Kazem Kashefi growing a hyperthermophile in an autoclave, and undergraduate students conducting transcriptomics in *E. coli*. Medical microbiology is presented using the physician-scientist's approach to microbial diseases. Case histories present how a physician-scientist approaches the interplay between the human immune response and microbial diseases.
- **An innovative media package provides powerful tools for instructors and students.** A new Micrograph Database for instructors includes hundreds of micrographs from the book and beyond tagged by easy-to-browse categories as well as by chapter. For students, a new ebook integrates powerful new self-study questions, process animations, quiz questions, weblinks, eTopics, and more to encourage the use of multimedia to enhance their learning of core concepts.

Additional features of the Third Edition include:

- **Genetics and genomics are presented as the foundation of microbiology.** Molecular genetics and genomics are thoroughly integrated with core topics throughout the book. This approach gives students an understanding of how genomes reveal potential metabolic pathways in diverse organisms, and how genomics and metagenomics reveal the character of microbial communities.
- **Microbial ecology and medical microbiology receive equal emphasis,** with particular attention paid to the merging of these fields. Throughout the book, phenomena are presented with examples from both ecology and medicine; for example, when discussing horizontal transfer of "genomic islands" we present symbiosis islands associated with nitrogen fixation, as well as pathogenicity islands associated with disease (Chapter 9).
- **Unlike most microbiology textbooks,** our text provides size scale information for nearly every micrograph.
- **Viruses are presented in molecular detail and in ecological perspective.** For example, in marine ecosystems, viruses play key roles in limiting algal populations while selecting for species diversity (Chapter 6). Similarly, a constellation of bacteriophages influences enteric flora.
- **Microbial diversity that students can grasp.** We present microbial diversity in a manageable framework that enables students to grasp the essentials of the most commonly presented taxa, the continual discovery of organisms ranging from anammox bacteria to emerging pathogenic *Escherichia* strains.
- **Appendices for review and further study.** Our book assumes a sophomore-level understanding of introductory biology and chemistry. For those in need of review, Appendices 1 and 2 summarize the fundamental structure and function of biological molecules and cells. A new Appendix 3 explains many commonly used molecular techniques used to probe microbes, such as fluorescence in situ hybridization (FISH). A new Appendix 4 compiles the taxonomy of commonly studied bacteria, archaea, microbial eukaryotes, and viruses.

Organization

The topics in this book are arranged so that students can progressively develop an understanding of microbiology from key concepts and research tools. The chapters of Part 1 present key foundational topics: history, visualization, the bacterial cell, microbial growth and control, and virology.

The six chapters in Part 1 present many topics that are then developed in further detail throughout Parts 2 through 5. Part 2 presents modern genetics and genomics. Part 3 presents cell metabolism and biochemistry, although the chapters in Part 2 are written in such a way that they can be presented before the genetics material if so desired. Part 4 explores microbial ecology and diversity and discusses the roles of microbial communities in local ecosystems and global cycling. And the chapters of Part 5 (Chapters 23–28) present medical and disease microbiology from an investigative perspective, founded on the principles of genetics, metabolism, and microbial ecology.

What's New in the Third Edition?

The Third Edition of *Microbiology: An Evolving Science* has been thoroughly revised and updated. We have added more up-to-the-minute research and current events as well as incorporating much of the generous feedback that we have received from our many reviewers and adopters. The following list highlights some of the more important content changes for the Third Edition.

Two new part-opening interviews highlight contemporary scientists and their research.

PART 1: Richard Lenski, Professor at Michigan State University, conducts a decades-long evolution experiment in which *Escherichia coli* evolves a new trait of citrate metabolism.

PART 3: Nicole Dubilier, Symbiosis Group Leader at the Max Planck Institute for Marine Microbiology, discovers extraordinary mutualisms between lithotrophic bacteria and deep-ocean invertebrates such as conches and tube worms.

Content changes throughout the text include:

CHAPTER 1: Microbial Life: Origin and Discovery. The chapter opener features biofilm growth of *Vibrio cholerae*, and the Mars landing of Curiosity rover to seek evidence for life.

CHAPTER 2: Observing the Microbial Cell. A new topic presents chemical imaging microscopy—a tool to reveal chemical distributions within a cell, addressing questions about nitrogen fixation.

CHAPTER 3: Cell Structure and Function. A new section presents bacterial cell polarity and aging, contrasting the different models of *Caulobacter crescentis*, *E. coli*, and *Mycobacterium tuberculosis.*

CHAPTER 4: Bacterial Culture, Growth, and Development. The chapter opener highlights an example of bacterial cannibalism in biofilms and a new special topic describes how the unique pattern of shark skin can retard microbial biofilm formation.

CHAPTER 5: Environmental Influences and Control of Microbial Growth. The chapter opener highlights a form of sibling rivalry practiced by *Paenibacillus* in which bacterial colonies spreading toward each other secrete chemical signl molecules that convert the motile rival cells into sessile forms. In addition, new information is included about disinfectants and the growing business of antimicrobial touch surfaces.

CHAPTER 6: Viruses. A new topic explores symbiotic viruses that contribute positive benefits to their hosts, such as the fungal virus that contributes thermal resistance to the host fungus—and to a fungus-associated plant. An expanded section presents CRISPR, a bacterial molecular defense against bacteriophages, with relevance for human gene therapy.

CHAPTER 7: Genomes and Chromosomes. The chapter opener describes the choreography of nucleoid movement during replication. New details about DNA replication have been added, and there is an expanded discussion of deep sequencing technologies and their use in revealing secrets of the many environmental and human microbiomes that exist.

CHAPTER 8: Transcription, Translation, and Bioinformatics. The molecular movements of chaperones as they prepare to refold unfolded proteins are explored in the chapter opener. In addition, the revised chapter includes new details about protein synthesis and the science of bioinformatics, such as how bioinformatic strategies revealed that a eukaryotic pathogen *Trypanosoma brucei* (the cause of African sleeping sickness) employs a bacterial-like enzyme to synthesize a critical membrane lipid.

CHAPTER 9: Gene Transfer, Mutations, and Genome Evolution. This chapter describes the recent discovery of human DNA sequences within the genome of *Neisseria gonorrhoeae* and new topics such as the RNA-based CRISPR defense against phage and its role in preserving biofilms, and the DNA repair mechanism called non-homologous end joining that can fix otherwise catastrophic DNA damage.

CHAPTER 10: Molecular Regulation. The opening research highlight describes a bacterial "lobster trap" engineered to explore quorum sensing parameters of bacterial cell-cell communication. There is also expanded coverage of transcriptomics, proteomics; regulatory RNAs (including cis-antisense RNAs), and of signal molecules such as cyclic di GMP that govern biofilm development. A new special topic describes how cell-to-cell bridge structures called nanotubes allow bacteria to communicate one-on-one.

CHAPTER 11: Viral Molecular Biology. A new section on hepatitis C presents the molecular biology of this emerging virus, and the new CDC recommendations for widespread testing. An expanded section on viral gene therapy presents the growing therapeutic use of lentiviral vectors.

CHAPTER 12: Biotechniques and Synthetic Biology. A new section on synthetic biology presents how bacterial gene circuits are mixed and reassembled like electronic circuits. Next generation synthetic biologists are now building bacterial computer switches that can produce a living computer, and biosensor organisms that "flash" a warning light when toxic materials approach.

CHAPTER 13: Energetics and Catabolism. A new topic on syntrophy explains how pairs of microbial species cooperate to catabolize organic substrates such as sewage and petroleum contaminants.

CHAPTER 14: Electron Flow in Organotrophy, Lithotrophy, and Phototrophy. A new chapter opener highlights nanowires conducting electric current between bacteria of a fuel cell. Electron transport systems are presented for anaerobic ammonia oxidation (anammox) and for iron-oxidizing acidophiles. Coverage of phototrophs is updated.

CHAPTER 15: Biosynthesis. A new topic describes mining a bacterial genome for modules encoding biosynthesis of peptide antibiotics.

CHAPTER 16: Food and Industrial Microbiology. New topics show how bacterial alginate is used to make medical sutures, and how an industrial enzyme company designs a microbial enzyme to brighten laundry.

CHAPTER 17: Origins and Evolution. A new section on experimental evolution shows how evolution in the laboratory reveals bacteria in the act of evolving new traits under natural selection. A new special topic shows how industry applies laboratory evolution to develop a thermophilic enzyme for paper production.

CHAPTER 18: Bacterial Diversity. Additional diverse bacterial species are introduced, such as *Lyngbya*, the giant filamentous cyanobacteria; the biomat-forming sulfur oxidizer *Thioploca*; and the emerging pathogenic mycoplasma *Ureaplasma urealyticum*.

CHAPTER 19: Archaeal Diversity. A new section features the ammonia-oxidizing Thaumarchaeota, including the ecological role of thaumarchaeotes in tidal pools. Coverage of methanogens is updated.

CHAPTER 20: Eukaryotic Diversity. A new topic presents the use of eukaryotic microbes such as yeast to model human biology. The phylogeny of eukaryotic microbes is updated.

CHAPTER 21: Microbial Ecology. A new opening section, "Metagenomics—And Beyond" explains the step by step process of analyzing a metagenome: sampling the target community, high-throughput DNA extraction and sequencing, genome scaffold assembly, and bioinformatic analysis. Featured examples include marine metagenomes, the rumen microbiome, and urban air metagenomes. A new topic presents oil-metabolizing bacteria from habitats contaminated by the Deepwater Horizon oil rig collapse.

CHAPTER 22: Microbes in Global Elemental Cycles. Exanded coverage of bioremediation includes the dechlorination of weapons materials in soil by a bacterial consortium, and the removal of arsenic from groundwater by a bioreactor. Coverage of the nitrogen cycle is updated.

CHAPTER 23: Human Microbiota and Innate Immunity. Expanded coverage of microbiomes includes a description of gut bacteria that are important for immune system development and the role of "fecal transplants" in curing previously untreatable gastrointestinal diseases such as irritable bowel disease. New aspects of innate immunity are discussed such as neutrophil extracellular traps (NETs) "thrown" by neutrophils around nearby pathogens.

CHAPTER 24: The Adaptive Immune Response. The chapter opener describes the finger-like membrane protrusions on immune T cell lymphocytes that probe antigen-presenting cell surfaces for antigens, an early step in the immune response. A new section expands on the relationship between vaccines and different immune system compartments.

CHAPTER 25: Microbial Pathogenesis. This chapter describes how the pathogen *Salmonella enterica* converts a gastrointestinal epithelial cell into a "gateway" M cell that the organism uses to invade the intestine. A new special topic on Type VI secretion shows how Gram negative pathogens capture phage genes and convert them into protein secretion systems that "fire" toxic proteins into host cells. A new section on host mimicry shows how pathogens produce host enzyme mimics that misdirect the host's response to infection, essentially making the pathogen nearly invisible to the immune system.

CHAPTER 26: Microbial Diseases. This chapter highlights recent outbreaks of fungal meningitis in the United States caused by contaminated, injectable steroid solutions, and the spate of hemolytic uremic syndrome cases in Europe caused by the newly emerged *E. coli* O104:H4. Updated sections include those on tuberculosis, *Clostridium difficile*, HIV, and the role of the appendix in reseeding the intestine.

CHAPTER 27: Antimicrobial Therapy. The exciting discovery of a new antibiotic, bedaquiline, is discussed in the opening research highlight, and an expanded section has been added about the success of HIV antiviral therapies. The chapter also describes the emergence of a novel antibiotic resistance enzyme, the New Delhi metallo-beta-lactamase, and the nearly indestructible pathogenic strains of *Klebsiella pneumoniae* that carry it. New information is also presented about the real reason certain classes of antibiotics kill while others only stop growth.

CHAPTER 28: Clinical Microbiology and Epidemiology. The chapter opening highlights heroic efforts of modern day microbe hunters at the CDC to track monkeypox in South America. Revisions reflect how today's diagnostic laboratories use rapid molecular-based technologies, such as PCR, immunochromatography, and even deep sequencing strategies to identify pathogens. A new section describes "point-of-care" diagnostic tests that can be performed not only in a hospital laboratory, but also in a doctor's office or, in some cases, by the patients themselves.

Resources

PRESENTATION TOOLS. Every figure and photograph in the textbook is available in JPEG and PowerPoint format for use in lecture. In order to provide stunning, high-quality visuals, every image has been hand-examined to make sure colors will not fade when projected and to optimize font size and composition for clear, legible viewing even in the back row. Labeled and unlabeled versions are available for download at wwnorton.com/instructors.

MICROGRAPH DATABASE. The NEW Micrograph Database includes searchable access to all of the micrographs in the textbook, tagged by characteristics such as taxonomy, shape, and habitat. The Micrograph Database can be accessed at wwnorton.com/instructors.

PROCESS ANIMATIONS. Sixty process animations depicting key processes of microbiology are offered in Flash, QT, and embedded in PowerPoint files. These animations are all based on the art found in the textbook and were developed under the careful supervision of the textbook authors. Student access to the animations is available in the ebook or via QR codes in the print book, which students can scan with their smartphones. Instructor access to the process animations is available at wwnorton.com/instructors.

Animation Topics Include:

Microscopy
Replisome Movement in a Dividing Cell
Chemotaxis
Phosphotransferase System (PTS) Transport
Dilution Streaking Technique
Biofilm Formation
Endospore Formation
Lysis and Lysogeny
Supercoiling and Topoisomerases
DNA Replication
Rolling Circle Mechanism of Plasmid Replication
PCR
Protein Synthesis
Protein Export
SecA-Dependent General Secretion Pathway
ABC Transporters
Bacterial Conjugation
Recombination
DNA Repair Mechanisms: Methyl Mismatch Repair
DNA Repair Mechanisms: Nucleotide Excision Repair
DNA Repair Mechanisms: Base Excision Repair
Transposition
The *lac* Operon
Transcriptional Attenuation
Chemotaxis: Molecular Events
Quorum Sensing
Influenza Virus Entry into a Cell
Influenza Virus Replication
HIV Replication
Herpes Virus Replication
Construction of a Gene Therapy Vector
Tagging Proteins for Easy Purification
Real-Time PCR
A Bacterial Electron Transport System
ATP Synthase Mechanism
Oxygenic Photosynthesis
Agrobacterium: A Plant Gene Transfer Vector
Phylogenetic Trees
DNA Shuffling
Listeria Infection
Light-Driven Pumps and Sensors
Malaria: A Cycle of Transmission between Mosquito and Human
The Basic Inflammatory Response
Phagocytosis
The Activation of the Humoral and Cell-Mediated Pathways
Cholera Toxin Mode of Action
Process of Type III Secretion
Retrograde Movement of Tetanus Toxin to an Inhibitory Neuron
DNA Sequencing

TEST BANK. Thoroughly revised for the Third Edition and using the Norton Assessment Guidelines, each chapter of the Test Bank consists of five question types classified according to the first five levels of Bloom's taxonomy of knowledge types: Remembering, Understanding, Applying, Analyzing, and Evaluating. Questions are further classified by section and difficulty, making it easy to construct tests and quizzes that are meaningful and diagnostic according to instructors' needs. Questions are multiple-choice and short-answer. The Test Bank is available in *ExamView Assessment Suite*, Word RTF, and PDF formats, downloadable from wwnorton.com/instructors. The Test Bank is also available on a disc. To receive a test bank disc, contact your Norton representative.

COURSEPACKS. At no cost to professors or students, Norton Coursepacks are available in a variety of formats, including all versions of Blackboard and WebCT. With just a simple download, an adopter can bring high-quality Norton digital media into a new or existing online course (no extra student passwords required), and it's theirs to keep forever. Content includes chapter-based assignments, quizzes, and interactive learning tools. Coursepacks can be downloaded at wwnorton.com/instructors.

ENHANCED EBOOK. An affordable and convenient alternative to the print book, Norton Ebooks retain the content and design of the print book and allow students to highlight and take notes, print chapters as needed, and search the text with ease. The enhanced ebook includes:

- **Process animations** based on the text art and developed under the watchful eyes of the textbook authors.
- **Self-study questions** with feedback designed to help students learn as they read. They are included in every section of every chapter as well as for the process animations.
- **Weblinks**, which send students to sites such as the CDC to learn more about concepts and examples discussed in the text.
- **Links to eTopics** written by Joan Slonczewski and John Foster, which supplement and enrich concepts covered in the text.
- **Flashcards** of all the key terms in the book and their definitions.

Acknowledgments

We are very grateful for the help of many people in developing and completing the book, including Norton editors John Byram, Vanessa Drake-Johnson, Mike Wright, and especially Betsy Twitchell, whose heroic efforts assured completion of the Third Edition. Our developmental editors, Michael Zierler and John Murdzek, contributed greatly to the clarity of presentation. Trish Marx and photo researcher Jane Miller did an amazing job of tracking down all kinds of images from sources all over the world. Patrick Shriner and Robin Kimball's coordination of electronic media development has resulted in a superb suite of resources for students and instructors alike. We thank associate editor Callinda Taylor for producing the IM and the Test Bank, as well as contributing in many other ways to the development of the digital resources. Without Thom Foley's incredible attention to detail, the innumerable moving parts of this project would never have become a finished book. Marian Johnson, Norton's managing editor in the college department, helped coordinate the complex process involved in shaping the manuscript over the years. Ben Reynolds ably and calmly managed the production and manufacturing of this book. Assistant editor Courtney Shaw and editorial assistant Katie Callahan coordinated the transfer of many drafts among many people. Finally, we thank Roby Harrington, Drake McFeely, and Julia Reidhead for their support of this book over its many years of development.

For the quality of our illustrations we thank the many artists at Precision Graphics, who developed attractive and accurate representations and showed immense patience in getting the details right.

We thank the numerous colleagues over the years who encouraged us in our project, especially the many attendees at the Microbial Stress Gordon Conferences. We greatly appreciate the insightful reviews and discussions of the manuscript provided by our colleagues, and the many researchers who contributed their micrographs and personal photos. We especially thank the American Society for Microbiology journals for providing many valuable resources. Reviewers Bob Bender, Bob Kadner, and Caroline Harwood offered particularly insightful comments on the metabolism and genetics sections, and Richard Lenski and Zachary Blount provided particularly insightful comments on experimental evolution. We would also like to thank the following reviewers:

Third Edition Reviewers

Emma Allen-Vercoe, University of Guelph
Gregory Anderson, Indiana University–Purdue University Indianapolis
Lisa Antoniacci, Marywood University
Bruce M. Applegate, Purdue University
Dennis Arvidson, Michigan State University
Vicki Auerbuch Stone, University of California, Santa Cruz
Tom Beatty, University of British Columbia
Melody Bell, Vernon College
Prakash Bhuta, Eastern Washington University
Blaise Boles, University of Michigan
Suzanna Bräuer, Appalachian State University
Ginger Brininstool, Lousisiana State University–Baton Rouge Campus
Kathleen L. Campbell, Emory University
Jeff Cardon, Cornell College
Rob Carey, Lebanon Valley College
Maria Castillo, New Mexico State University
Todd Ciche, Michigan State University
Sharron Crane, Rutgers University
Nicola Davies, University of Texas Austin
Angus Dawe, New Mexico State University
Janet Donaldson, Mississippi State
Erastus Dudley, Huntingdon College
Kathleen Dunn, Boston College
Valerie Edwards-Jones, Manchester Metropolitan University
Lehman Ellis, Our Lady of Holy Cross College
David Esteban, Vassar College
Xin Fan, West Chester University
Babu Fathepure, Oklahoma State University
Michael Gadsden, York University
Veronica Godoy-Carter, Northeastern University
Stjepko Golubic, Boston University
Vladislav Gulis, Coastal Carolina University
Ernest Hannig, University of Texas–Dallas
Julian Hurdle, The University of Texas at Arlington
Edward Ishiguro, University of Victoria

Choong-Min Kang, Wayne State University
Bessie Kebaara, Baylor University
John Lee, The City College of The City University of New York
Manuel Llano, University of Texas–El Paso Campus
Aaron Lynne, Sam Houston State University
Ghislaine Mayer, Virginia Commonwealth University
Bob McLean, Texas State University
Sladjana Malic, Manchester Metropolitan University
Gregory Marczynski, McGill University
Naomi Morrissette, University of California, Irvine
Kenneth Murray, Florida International University
Kari Naylor, University of Central Arkansas
Tracy O'Connor, Mount Royal University
Rebecca Parales, University of California, Davis
Roger Pickup, University of Lancaster
Robert Poole, The University of Sheffield
Geert Potters, Antwerp Maritime Academy
Ines Rauschenbach, Rutgers University
Veronica Riha, Madonna University
Marie-Claire Rioux, John Abbott College
Jason A. Rosenzweig, Texas Southern University
Ronald Russell, University of Dublin
Matt Schrenk, East Carolina University
Gary Schultz, Marshall University
Chola Shamputa, Mount Saint Vincent University
Nilesh Sharma, Western Kentucky University
Donald Sheppard, McGill University
Garriet Smith, The University of South Carolina Aiken
Vincent J. Starai, University of Georgia
Lisa Stein, University of Alberta
Karen Sullivan, Louisiana State University
Kapil Tahlan, Memorial University of Newfoundland and Labrador
Liang Tang, The University of Kansas
Tzuen-Rong Jeremy Tzeng, Clemson University
Claire Vieille, Michigan State University
James R. Walker, University of Texas Austin
Susan C. Wang, Washington State University
Chris Weingart, Dennison College
John Zamora, Middle Tennessee State University
Stephanie Zamule, Nazareth College
Fanxiu Zhu, The Florida State University

Second Edition Reviewers

Michael Allen, University of North Texas
Gladys Alexandre, University of Tennessee Knoxville
Hazel Barton, Northern Kentucky University
Suzanne S. Barth, University of Texas at Austin
Barry Beutler, The College of Eastern Utah
Michael J. Bidochka, Brock University
Dwayne Boucaud, Quinnipiac University
Derrick Brazill, Hunter College
Graciela Brelles-Mariño, California State Polytechnic University, Pomona
Jay Brewster, Pepperdine University
Linda Bruslind, Oregon State University
Marion Brodhagen, Western Washington University
Alison Buchan, University of Tennessee Knoxville
Jeffrey Byrd, St. Mary's College of Maryland
Silvia T. Cardona, University of Manitoba
Andrea Castillo, Eastern Washington University
Miguel Cervantes-Cervantes, Rutgers University
Tin-Chun Chu, Seton Hall University
Paul Cobine, Auburn University
Tyrrell Conway, University of Oklahoma
Scott Dawson, University of California–Davis
Jose de Ondarza, SUNY Plattsburgh
Donald W. Deters, Bowling Green State University
Clarissa Dirks, The Evergreen State College
William T. Doerrler, Louisiana State University
Janet R. Donaldson, Mississippi State University
Xin Fan, West Chester University
Babu Z. Fathepure, Oklahoma State University
Clifton Franklund, Ferris State University
Gregory D. Frederick, University of Mary Hardin-Baylor
Christopher French, University of Edinburgh
Jason M. Fritzler, Stephen F. Austin State University
Katrina Forest, University of Wisconsin–Madison
Kimberley Gilbride, Ryerson University
Stjepko Golubic, Boston University
Enid T. Gonzalez, California State University, Sacramento
John E. Gustafson, New Mexico State University
Lynn E. Hancock, Kansas State University
Martina Hausner, Ryerson University
J. D. Hendrix, Kennesaw State University
Michael C. Hudson, University of North Carolina–Charlotte
Jane E. Huffman, East Stroudsburg University
Michael Ibba, Ohio State University
Gilbert H. John, Oklahoma State University
John A. Johnson, University of New Brunswick St. John
Mark C. Johnson, Georgetown College
Carol Ann Jones, University Of California Riverside
Ece Karatan, Appalachian State University
Daniel B. Kearns, Indiana University Bloomington
Robert J. Kearns, University of Dayton
Susan Koval, University of Western Ontario
Deborah Kuzmanovic, University of Michigan
Peter Kennedy, Lewis & Clark College
Greg Kleinheinz, University of Wisconsin Oshkos
Jesse J. Kwiek, Ohio State University
Andrew Lang, Memorial University of Newfoundland
Margaret Liu, University of Michigan
Thomas W. De Lany, Kilgore College
Maia Larios-Sanz, University of St. Thomas
Beth Lazazzera, University of California, Los Angeles
Dr. Lee H. Lee, Montclair State University
Mark Liles, Auburn University
Jun Liu, University of Toronto
Manuel Llano, University of Texas at El Paso
Zhongjing Lu, Kennesaw State University

Aaron Lynne, Sam Houston State University
John C. Makemson, Florida International University
Donna L. Marykwas, California State University Long Beach
Ann G. Matthysse, University of North Carolina at Chapel Hill
Ghislaine Mayer, Virginia Commonwealth University
Robert Maxwell, Georgia State University
William R. McCleary, Brigham Young University
Nancy L. McQueen, California State University, Los Angeles
Scott A. Minnich, University of Idaho
Philip F. Mixter, Washington State University
Christian D. Mohr, University of Minnesota
Craig Moyer, Western Washington University
Scott Mulrooney, Michigan State University
Kari Murad, The College of Saint Rose
William Wiley Navarre, University of Toronto
Ivan J. Oresnik, University of Manitoba
Cleber Costa Ouverney, San Jose State University
Deborah Polayes, George Mason University
Pablo J. Pomposiello, University of Massachusetts Amherst
Joan Press, Brandeis University
Todd P. Primm, Sam Houston State University
Sharon R. Roberts, Auburn University
Michelle Rondon, University of Wisconsin–Madison
Silvia Rossbach, Western Michigan University
Ben Rowley, University of Central Arkansas
Chad R. Sethman, Waynesburg University
Matthew O. Schrenk, East Carolina State University
Anthony Siame, Trinity Western University
Lyle Simmons, University of Michigan
Daniel R. Smith, Seattle University
Garriet W. Smith, University of South Carolina Aiken
Geoffrey B. Smith, New Mexico State University
Ruth Sporer, Rutgers University Camden
Anand Sukhan, Northeastern State University
Karen Sullivan, Louisiana State University
Virginia Stroeher, Bishop's University
Dorothea K. Thompson, Purdue University
Wendy C. Trzyna, Marshall University
Bernard Turcotte, McGill University
Dave Westenberg, Missouri University of Science and Technology
Ann Williams, University of Tampa
Charles F. Wimpee, University of Wisconsin–Milwaukee
Jianping Xu, McMaster University

First Edition Reviewers

Laurie A. Achenbach, Southern Illinois University, Carbondale
Stephen B. Aley, University of Texas, El Paso
Mary E. Allen, Hartwick College
Shivanthi Anandan, Drexel University
Brandi Baros, Allegheny College
Gail Begley, Northeastern University
Robert A. Bender, University of Michigan
Michael J. Benedik, Texas A&M University
George Bennett, Rice University
Kathleen Bobbitt, Wagner College
James Botsford, New Mexico State University
Nancy Boury, Iowa State University of Science and Technology
Jay Brewster, Pepperdine University
James W. Brown, North Carolina State University
Whitney Brown, Kenyon College undergraduate
Alyssa Bumbaugh, Pennsylvania State University, Altoona
Kathleen Campbell, Emory University
Alana Synhoff Canupp, Paxon School for Advanced Studies, Jacksonville, FL
Jeffrey Cardon, Cornell College
Tyrrell Conway, University of Oklahoma
Vaughn Cooper, University of New Hampshire
Marcia L. Cordts, University of Iowa
James B. Courtright, Marquette University
James F. Curran, Wake Forest University
Paul Dunlap, University of Michigan
David Faguy, University of New Mexico
Bentley A. Fane, University of Arizona
Bruce B. Farnham, Metropolitan State College of Denver
Noah Fierer, University of Colorado, Boulder
Linda E. Fisher, late of the University of Michigan, Dearborn
Robert Gennis, University of Illinois, Urbana-Champaign
Charles Hagedorn, Virginia Polytechnic Institute and State University
Caroline Harwood, University of Washington
Chris Heffelfinger, Yale University graduate student
Joan M. Henson, Montana State University
Michael Ibba, Ohio State University
Nicholas J. Jacobs, Dartmouth College
Douglas I. Johnson, University of Vermont
Robert J. Kadner, late of the University of Virginia
Judith Kandel, California State University, Fullerton
Robert J. Kearns, University of Dayton
Madhukar Khetmalas, University of Central Oklahoma
Dennis J. Kitz, Southern Illinois University, Edwardsville
Janice E. Knepper, Villanova University
Jill Kreiling, Brown University
Donald LeBlanc, Pfizer Global Research and Development (retired)
Robert Lausch, University of South Alabama
Petra Levin, Washington University in St. Louis
Elizabeth A. Machunis-Masuoka, University of Virginia
Stanley Maloy, San Diego State University
John Makemson, Florida International University
Scott B. Mulrooney, Michigan State University
Spencer Nyholm, Harvard University
John E. Oakes, University of South Alabama
Oladele Ogunseitan, University of California, Irvine
Anna R. Oller, University of Central Missouri
Rob U. Onyenwoke, Kenyon College
Michael A. Pfaller, University of Iowa
Joseph Pogliano, University of California, San Diego
Martin Polz, Massachusetts Institute of Technology
Robert K. Poole, University of Sheffield
Edith Porter, California State University, Los Angeles
S. N. Rajagopal, University of Wisconsin, La Crosse

James W. Rohrer, University of South Alabama
Michelle Rondon, University of Wisconsin–Madison
Donna Russo, Drexel University
Pratibha Saxena, University of Texas, Austin
Herb E. Schellhorn, McMaster University
Kurt Schesser, University of Miami
Dennis Schneider, University of Texas, Austin
Margaret Ann Scuderi, Kenyon College
Ann C. Smith Stein, University of Maryland, College Park
John F. Stolz, Duquesne University
Marc E. Tischler, University of Arizona
Monica Tischler, Benedictine University
Beth Traxler, University of Washington
Luc Van Kaer, Vanderbilt University
Lorraine Grace Van Waasbergen, The University of Texas, Arlington
Costantino Vetriani, Rutgers University
Amy Cheng Vollmer, Swarthmore College
Andre Walther, Cedar Crest College
Robert Weldon, University of Nebraska, Lincoln
Christine White-Ziegler, Smith College
Jianping Xu, McMaster University

Finally, we offer special thanks to our families for their support. Joan's husband Michael Barich offered unfailing support. John's wife Zarrintaj ("Zari") Aliabadi contributed to the text development, especially the sections on medical microbiology and public health.

To the Reader: Thanks!

We greatly appreciate your selection of this book as your introduction to the science of microbiology. As our textbook continues to evolve, it benefits greatly from the input of its many readers, students as well as professors. We truly welcome your comments, especially if you find text or figures that are in error or unclear. Feel free to contact us at the addresses listed below.

Joan L. Slonczewski
slonczewski@kenyon.edu
John W. Foster
jwfoster@southalabama.edu

ABOUT THE AUTHORS

JOAN L. SLONCZEWSKI received her B.A. from Bryn Mawr College and her Ph.D. in Molecular Biophysics and Biochemistry from Yale University, where she studied bacterial motility with Robert M. Macnab. After postdoctoral work at the University of Pennsylvania, she has since taught undergraduate microbiology in the Department of Biology at Kenyon College, where she earned a Silver Medal in the National Professor of the Year program of the Council for the Advancement and Support of Education. She has published numerous research articles with undergraduate coauthors on bacterial pH regulation, and has published five science fiction novels including *The Highest Frontier* and *A Door into Ocean*, both of which earned the John W. Campbell Memorial Award. She served as At-large Member representing Divisions on the Council Policy Committee of the American Society for Microbiology, and is a member of the Editorial Board of the journal *Applied and Environmental Microbiology.*

JOHN W. FOSTER received his B.S. from the Philadelphia College of Pharmacy and Science (now the University of the Sciences in Philadelphia), and his Ph.D. from Hahnemann University (now Drexel University School of Medicine), also in Philadelphia, where he worked with Albert G. Moat. After postdoctoral work at Georgetown University, he joined the Marshall University School of Medicine in West Virginia; he is currently teaching in the Department of Microbiology and Immunology at the University of South Alabama College of Medicine in Mobile, Alabama. Dr. Foster has coauthored three editions of the textbook *Microbial Physiology* and has published over 100 journal articles describing the physiology and genetics of microbial stress responses. He has served as Chair of the Microbial Physiology and Metabolism division of the American Society for Microbiology and as a member of the editorial advisory board of the journal *Molecular Microbiology.*

THIRD EDITION

Microbiology

An Evolving Science

AN INTERVIEW WITH:

COURTESY OF RICHARD LENSKI

Richard Lenski, Hannah Distinguished Professor, Michigan State University.

Richard Lenski

Evolution in the Lab

Richard Lenski, an evolutionary biologist, has taught for over 20 years as the John Hannah Distinguished Professor at Michigan State University. Since 1988, Lenski and his students have been tracking phenotypic and genetic changes in 12 initially identical populations of bacteria. Their report of *E. coli* bacteria evolving a new trait in the laboratory earned headlines from the *New York Times* and other media around the world. Lenski cofounded BEACON, the National Science Foundation's Center for the Study of Evolution in Action.

How did you decide to make a career in microbial evolution?

I got interested in biology as an undergrad at Oberlin College, and I was especially fascinated by ecology because there were so many unanswered questions. So I went to grad school at the University of North Carolina to study ecology. I began to see the deep connections between ecology and evolution. Many ecologists have lifelong interests in particular organisms—birds, snakes, butterflies, or whatever. But I didn't have any special skills in that respect; I was more interested in the general questions. I remembered the elegance of the genetics experiments with bacteria that I had learned about as an undergraduate. So I decided that, for my postdoctoral work, I should find a lab where I could learn how to work with microbes. I found a superb mentor, Bruce Levin, who was interested in evolution.

Why did you perform your long-term experimental evolution study with *Escherichia coli*? What makes it different from previous studies of evolution?

I started the experiment to ask one main question: How repeatable is evolution? Mutations occur at random, but populations become more fit over time if some of the mutants survive and reproduce better than their ancestors—that's natural selection. In essence, I wanted to know how many different ways there were for the bacteria to adapt to a particular environment.

I set up 12 populations, all started from the same *E. coli* strain, and each one in an identical flask containing a medium where glucose is the source of energy. Every day, someone takes 1% of the volume from each flask and transfers it to a new flask with fresh medium. The bacteria grow and, after some hours, deplete the glucose, so it's a "feast or famine" existence. The dilution and regrowth allows about seven bacterial generations per day. I started the experiment in 1988, and the bacteria have now been evolving for well over 50,000 generations. So many interesting things have happened that I've kept it going all these years. In fact, I hope the experiment will continue even after I'm gone.

This project differs from most research on evolution because we're watching evolution in action. Most evolutionary biologists study fossils or use the comparative approach—that is, quantifying similarities and differences in phenotypes and genomes of living organisms—in order to infer the characteristics of organisms that lived in the past. In this *E. coli* experiment, we can observe changes as the generations go by, and we can directly

COURTESY OF RICHARD LENSKI

One *E. coli* population evolved the novel capacity to consume citrate for energy (clouded flask).

compare the evolving bacteria with their ancestors. We've stored the ancestral strain and samples from every 500 generations in a freezer, and with *E. coli* we can revive the frozen cells. It's like bringing fossils back to life.

Over the 25 years of this experiment, the research has involved dozens of dedicated people. I have an excellent technician, Neerja Hajela, who either does the transfers herself or makes sure someone else does them. Mike Travisano was the first student to base his dissertation research on this experiment, and he's now a professor at the University of Minnesota.

What results have you obtained? Have any results surprised you?

One result is that the average fitness in each population increases over time. We measure fitness by competing bacteria from different generations against the common ancestor. This result is not surprising, since the environment has been constant over time, but it's a concrete demonstration of adaptation by natural selection, the same process that Darwin discovered.

Another finding is that evolution can be quite repeatable; that is, we've seen many examples of parallel changes in the replicate lines. For example, all 12 populations evolved to produce larger individual cells than the ancestor produced. And when we look at their mutations, we see many cases in which some or even all of the lines have mutations in the same genes.

The most dramatic change we've seen happened in only one population. Glucose was the source of energy for the bacteria, but there's been another resource—citrate—in the medium all along. *E. coli* cells can't use the citrate, however, because they're unable to take up citrate in the presence of oxygen. In fact, the inability to grow on citrate is a key feature of *E. coli* as a species. But after about 30,000 generations, a mutant in one population discovered there was something else to eat besides the glucose. At first I thought we had a contaminant—some other species—in this flask, but genetic analyses showed it really was a descendant of the *E. coli* strain used to start the experiment. So here's a case where one population evolved to be very different from all the other populations. Zack Blount, a postdoc in the lab, is analyzing the mutations that allow the bacteria to grow on citrate. Caroline Turner, a grad student, is studying how this new ability changes the ecological interactions between different genotypes in the population.

What new technologies have made it possible to take full advantage of your study?

When I started this experiment in 1988, I couldn't imagine the amazing technologies that would come along and allow us to analyze the evolution that has taken place. The ability to sequence entire genomes is the most important advance. By sequencing the genomes of evolved bacteria and comparing them to the ancestor's genome, we're finding the mutations that led to improved fitness and other phenotypic changes.

Does experimental evolution have industrial applications?

Yes, it does. Humans can apply evolution for practical purposes, just like we use other natural processes, such as gravity and the action of water, to do work via mill wheels and hydroelectric plants. In fact, the selective breeding and domestication of farm animals, crop plants, and even microbes (like baker's yeast) show that our ancestors employed evolution for practical purposes long before the mechanisms of evolution were understood. More recently, scientists have been pursuing genetic engineering to modify microbes for new purposes, like biofuels. Experimental evolution—where scientists construct environments that select for organisms with the desired properties—offers a valuable complement to genetic engineering.

How does your family relate to your work?

I sometimes joke that I have two families: my biological family with my wife and kids, and my lab family, with all the students and postdocs who've been a part of it over the years. As much as I love my work—and I can't imagine a better job than being a biology professor—there's always more research to be done. So I'm grateful that my wife and kids have been supportive of my work. Now I have a granddaughter, and she reminds me just how fortunate I am to see another generation in the great evolutionary tree of life.

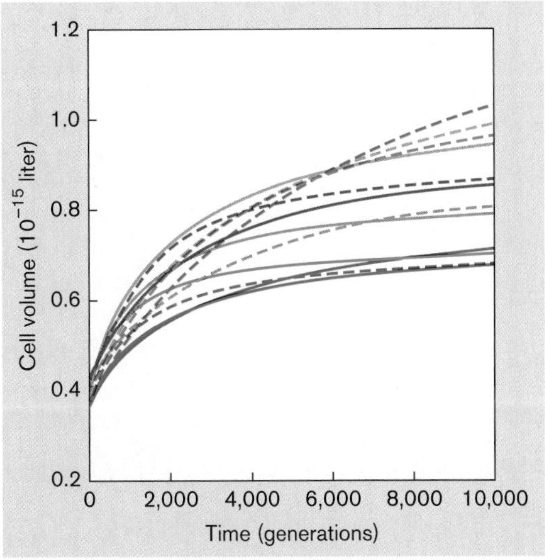
Cell volume (10^{-15} liter)
1.2
1.0
0.8
0.6
0.4
0.2
0
2,000
4,000
6,000
8,000
10,000
Time (generations)

Cell size increased in all 12 evolving populations of bacteria. *Source:* Modified from Richard Lenski and Michael Travisano. Dynamics of adaptation and diversification: a 10,000-generation experiment with bacterial populations. 1994. *PNAS* **91**:6808.

For further details on Lenski's experimental evolution of *E. coli*, see Chapter 17 Origins and Evolutions.

CHAPTER 1

Microbial Life: Origin and Discovery

1.1 From Germ to Genome: What Is a Microbe?

1.2 Microbes Shape Human History

1.3 Medical Microbiology

1.4 Microbial Ecology

1.5 The Microbial Family Tree

1.6 Cell Biology and the DNA Revolution

Life on Earth began early in our planet's history, with microscopic organisms, or microbes. Over the eons, those microbes evolved to shape our atmosphere, our geology, and the energy cycles of all ecosystems. Today microbes are everywhere. A human body contains ten times as many microbes as it does human cells, including numerous bacteria on the skin and in the digestive tract. Throughout history, humans have had a hidden partnership with microbes ranging from food production and preservation to mining for precious minerals.

Yet throughout most of our history, humans had no idea that microbes even existed. To study these unseen organisms required a microscope, first developed in the 1600s. In the nineteenth century—the "golden age" of microbiology—microscopes revealed the tiny organisms at work in our bodies and in our environment. The twentieth century saw the rise of microbes as the engines of biotechnology. Microbial discoveries led to recombinant DNA and revealed the secrets of sequenced genomes. Today, microbes lead the way to synthetic biology—making exciting new products through laboratory evolution.

5 µm

CURRENT RESEARCH **highlight**

Biofilm of cholera bacteria. *Vibrio cholerae* bacteria grow as symbiotic biofilms on marine invertebrates. But when *V. cholerae* enters the human body, it causes the deadly disease cholera. Steven Chu and colleagues at UC Berkeley and the U.S. Department of Energy investigated the spatial organization of a *V. cholerae* biofilm. They used laser confocal microscopy, a technique that reveals layers of cells in 3D. Specific fluorescent labels identify cell cytoplasm (blue), cell surfaces containing a cell-cell adhesion protein (gray), cell-surface sugar chains (red), and substrate adhesion protein (green). These observations will help researchers design antibiotics that target *V. cholerae* growing as a biofilm, a highly resistant state. Biofilm studies also contribute to the design of wastewater treatment, environmental bioremediation, and industrial production systems. *Source:* Veysel Berk et al. 2012. *Science* **337**:236.

In 2012, NASA landed the *Mars Science Laboratory,* or *Curiosity* rover, near the base of a mountain on the planet Mars (**Fig. 1.1**). The car-sized rover had a laser to drill into rock, X-ray and fluorescence analyzers, and camera microscopes. It began its mission to test the Martian soil for water, organic compounds, and other potential evidence of microbial life.

Why do we care whether microbes exist on Mars? The discovery of life beyond Earth would fundamentally change how we see our place in the universe. The observation of Martian life could yield clues as to the origin of our own biosphere and expand our knowledge of how Earth's life evolved. As of this writing, the existence of microbial life on Mars remains unknown, but here on Earth, many terrestrial microbes remain as mysterious as Mars. Barely 0.1% of the microbes in our biosphere can be cultured in the laboratory; even the digestive tract of a newborn infant contains species of bacteria unknown to science. Our "exploration rovers" for Earth's microbiology include, for example, machines that can quickly sequence DNA from unknown microbes, and new kinds of microscopy that scan entire cells in 3D.

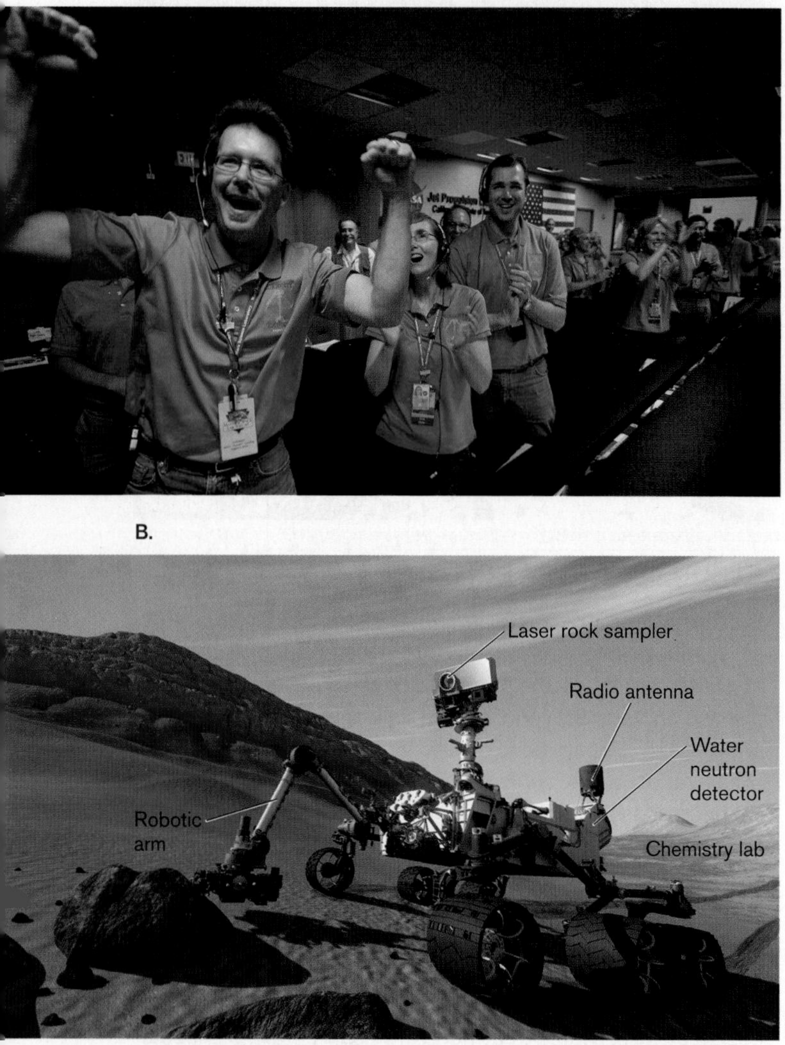
A.
B.
Laser rock sampler
Radio antenna
Water neutron detector
Robotic arm
Chemistry lab

NASA/BILL INGALLS

NASA

FIGURE 1.1 ■ Is there microbial life on Mars? A. NASA scientists celebrate the 2012 landing of the *Mars Science Laboratory,* or *Curiosity* rover, on Mars. **B.** *Curiosity* seeks chemical and physical evidence supporting the possibility that microbial life exists on Mars.

Microbes grow throughout our biosphere, from the superheated black smoker vents at the ocean floor to the subzero ice fields of Antarctica (**Fig. 1.2**). Bacteria such as *Escherichia coli* live in our intestinal tract, while cyanobacteria color wetlands green or orange. Protists are the predators of the microscopic world. Microbes often grow in complex communities such as lichens, which are symbioses between fungi and algae or cyanobacteria (**Fig. 1.2E**). And viruses such as papillomavirus cause disease, as do many bacteria and protists.

Yet before microscopes were developed in the seventeenth century, we humans were unaware of the unseen living organisms that surround us, that float in our air and water, and that inhabit our own bodies. Microbes generate the very air we breathe, including nitrogen gas and much of the oxygen and carbon dioxide. They fix nitrogen for plants, and they make vitamins, such as vitamin B_{12}. In the ocean, microbes produce biomass for the food web that feeds the fish we eat; and microbes consume toxic wastes such as the oil from the *Deepwater Horizon* spill in the Gulf of Mexico in 2010. At the same time, virulent pathogens take our lives—and researchers risk their lives to study them (**Fig. 1.3**). Despite all our advances in medicine and public health, humans continue to die of microbial diseases, such as the avian influenza strain H7N9 that emerged in China in 2013.

Today we discover surprising new kinds of microbes deep underground and in places previously thought uninhabitable, such as the rainless Atacama Desert in Chile, or the glaciers of Antarctica—both considered models for possible life on Mars. Microbes shape our biosphere and provide new tools that impact human society. For example, the use of heat-stable bacterial DNA polymerase (a DNA-replicating enzyme) in a technique called the **polymerase chain reaction** (**PCR**) allows us to detect minute amounts of DNA in traces of blood or fossil bone. Microbial technologies led us from the discovery of the double helix to the sequence of the human genome, the total genetic information that defines our species.

In Chapter 1 we introduce the concept of a microbe, and we survey the history of human discovery. We ask, how do microbes evolve? (See the Part 1 interview with Richard Lenski, on observing evolution in the laboratory.)

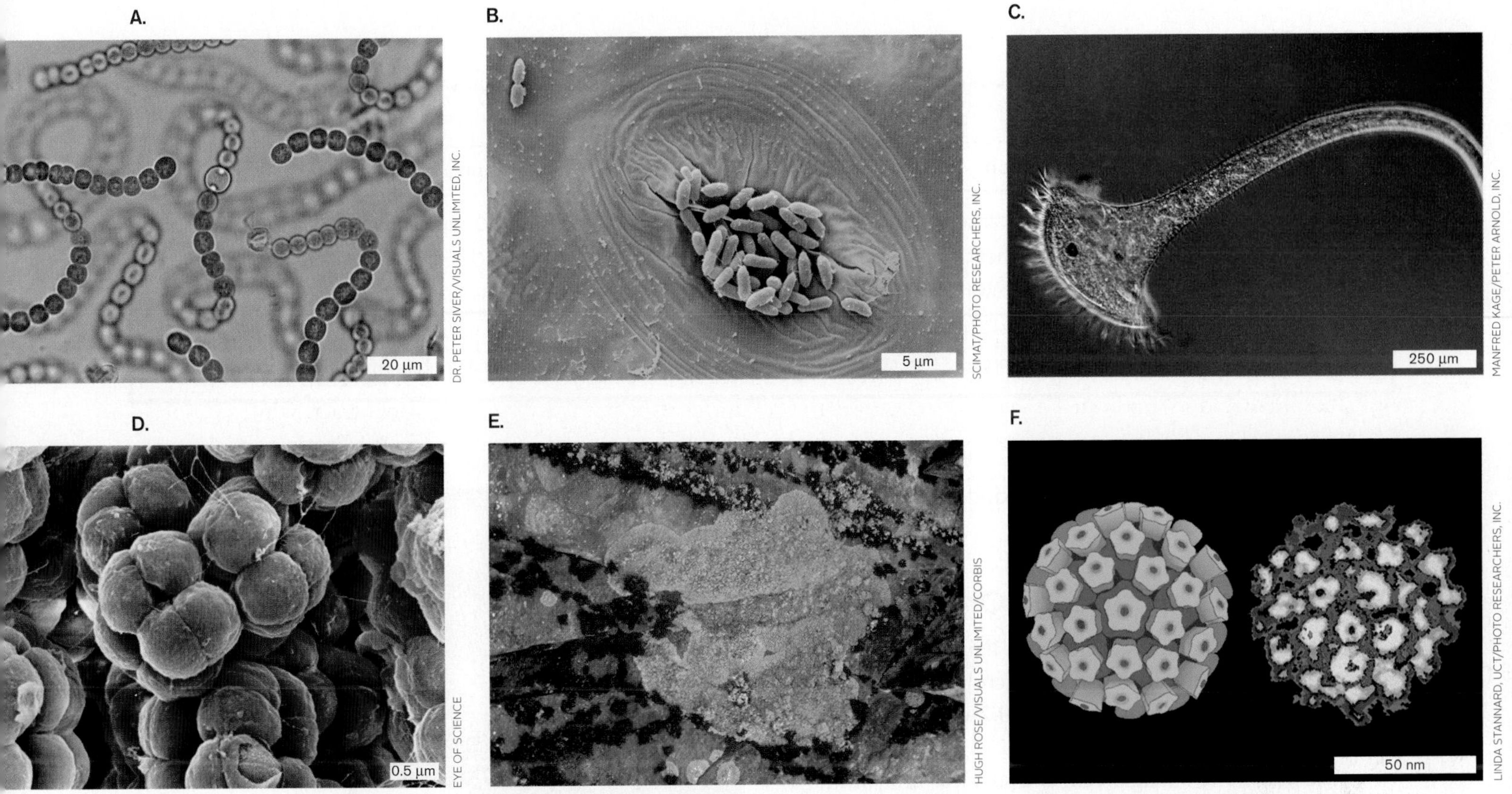
A.
20 µm
DR. PETER SIVER/VISUALS UNLIMITED, INC.
B.
5 µm
SCIMAT/PHOTO RESEARCHERS, INC.
C.
250 µm
MANFRED KAGE/PETER ARNOLD, INC.
D.
0.5 µm
EYE OF SCIENCE
E.
HUGH ROSE/VISUALS UNLIMITED/CORBIS
F.
50 nm
LINDA STANNARD, UCT/PHOTO RESEARCHERS, INC.

FIGURE 1.2 ▪ Representative microbes. A. Filamentous cyanobacteria produce oxygen for planet Earth (bright-field light micrograph). **B.** *Escherichia coli* bacteria colonize a lettuce leaf cell (colorized scanning electron microscopy). **C.** *Stentor* is a protist, a eukaryotic microbe. Cilia beat food into its mouth. **D.** Halophilic archaea, a form of life distinct from bacteria and eukaryotes, grow at extremely high salt concentrations (colorized scanning electron microscopy). **E.** Lichens are mutualistic associations of algae and fungi; these lichens are growing on rock in Antarctica. **F.** Papillomavirus, an infectious agent of disease commonly acquired by young adults, causes cancer and genital warts (model based on electron microscopy).

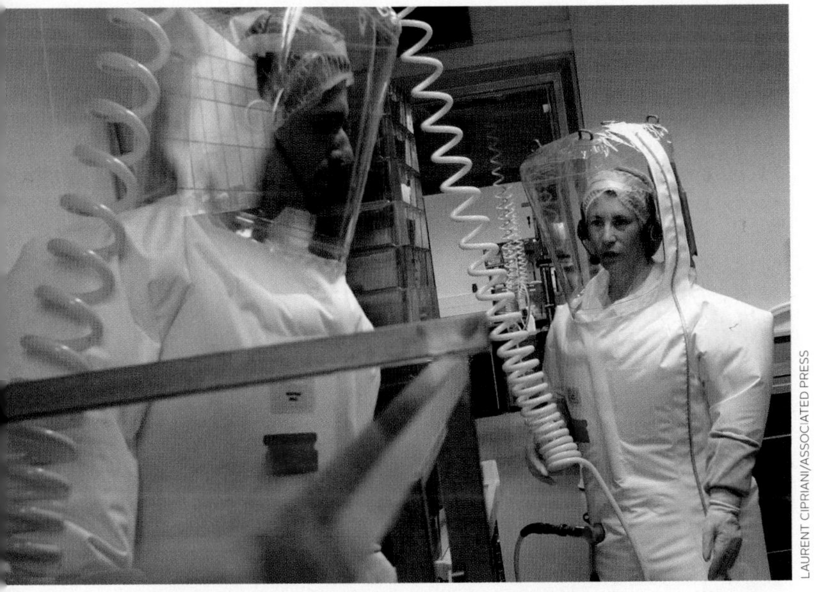
LAURENT CIPRIANI/ASSOCIATED PRESS

FIGURE 1.3 ▪ Researching deadly pathogens. Microbiologists wear protective gear to work in the P4 (protection level 4) high-security laboratory in Lyon, France. This laboratory studies deadly pathogens, including Ebola and influenza virus H1N1.

We explain how to show which pathogen causes a disease. Finally, we address the exciting century of molecular microbiology, in which microbial genetics and genomics have transformed the face of modern biology and medicine.

1.1

From Germ to Genome: What Is a Microbe?

From early childhood, we hear that we are surrounded by microscopic organisms, or "germs," which we cannot see. What are microbes? Our modern concept of a microbe has deepened through the use of two major research tools: advanced microscopy and the sequencing of genomic DNA. Microscopy is covered in Chapter 2, and genome sequencing is presented in Chapter 7. And exciting tools of

TABLE 1.1

Sizes of some microbes.

Microbe	Description	Approximate size
Varicella-zoster virus 1	Virus that causes chickenpox and shingles	100 nanometers (nm) = 10^{-7} meter (m)
Prochlorococcus	Photosynthetic marine bacteria	500 nm = 5×10^{-7} m
Escherichia coli	Bacteria growing within human intestine	1 micrometer (μm) = 10^{-6} m
Spirogyra	Aquatic algae that form long filaments of cells	40 μm = 4×10^{-5} m (cell width)
Pelomyxa	Ameba (a protist) that consumes bacteria in soil or water	5 millimeters (mm)

"synthetic biology," to engineer new kinds of microbes, are described in Chapter 12.

A Microbe Is a Microscopic Organism

A **microbe** is commonly defined as a living organism that requires a microscope to be seen. Microbial cells range in size from millimeters (mm) down to 0.2 micrometer (μm), and viruses may be tenfold smaller (**Table 1.1**). Some microbes consist of a single cell, the smallest unit of life, a membrane-enclosed compartment of water solution containing molecules that carry out metabolism. Each microbe contains a genome used to reproduce its own kind. Microbial cells acquire food, gain energy to build themselves, and respond to environmental change. Microbes evolve at rapid rates—often fast enough to observe in the laboratory (discussed in Chapter 17).

Our simple definition of a microbe, however, leaves us with contradictions.

- **Super-size microbial cells.** Most single-celled organisms require a microscope to render them visible, and thus fit the definition of a microbe. Nevertheless, some species of protists and algae, and even some bacterial cells, are large enough to see with the naked eye. The marine sulfur bacterium *Thiomargarita namibiensis,* called the "sulfur pearl of Namibia," grows as large as the head of a fruit fly (**Fig. 1.4**). Even more surprising, a single-celled plant, the "killer alga" *Caulerpa taxifolia,* spreads through the coastal waters of California. The single cell covers many acres with its leaflike cell parts.
- **Microbial communities.** Many microbes form complex multicellular assemblages, such as mushrooms, kelps, and biofilms. In these structures, cells are differentiated into distinct types that complement each other's function, as in multicellular organisms. And yet, some multicellular worms and arthropods require a microscope for us to see but are not considered microbes.

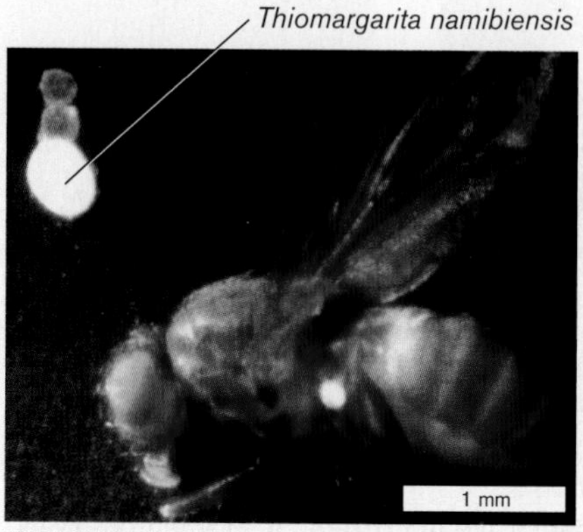
Thiomargarita namibiensis
1 mm
HEIDE N. SCHULZ-VOGT, UNIVERSITY OF HANOVER, GERMANY

FIGURE 1.4 ■ Giant microbial cells. *Thiomargarita namibiensis,* a marine sulfur bacterium, is nearly as large as the head of a fruit fly.
Source: Reprinted with permission from Heide N. Schulz-Vogt et al. 1999. *Science* 284:493. © 2005 AAAS.

- **Viruses.** A **virus** consists of a noncellular particle containing genetic material that takes over the metabolism of a cell to generate more virus particles. Some viruses consist of only a short chromosome packed in protein, whereas others, such as the mimivirus (which infects amebas), show the size and complexity of a cell. Although viruses are not fully functional cells, the genome of a mimivirus shows evidence that it evolved from a cell.

Note: Each section contains Thought Questions that may have various answers. Possible responses are posted at the back of the book.

Thought Questions

1.1 The minimum size of known microbial cells is about 0.2 μm. Could even smaller cells be discovered? What factors may determine the minimum size of a cell?

1.2 If viruses are not functional cells, are they "alive"?

In practice, our definition of a microbe derives from tradition as well as genetic considerations. In this book we consider microbes to include **prokaryotes** (cells lacking a nucleus, including bacteria and archaea) as well as certain

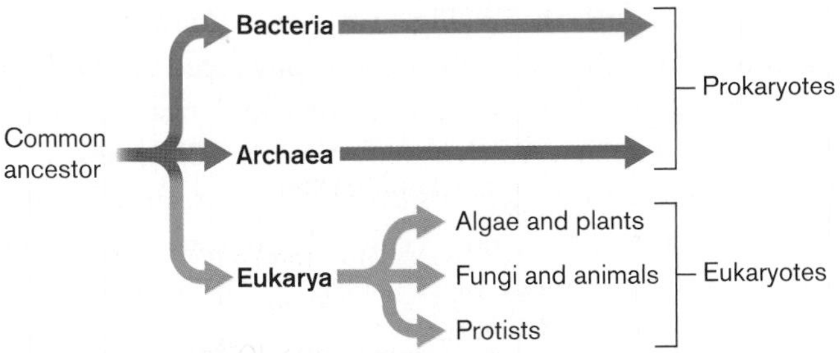
Bacteria
Prokaryotes
Common ancestor
Archaea
Algae and plants
Eukarya
Fungi and animals
Eukaryotes
Protists

FIGURE 1.5 ▪ Three domains of life. Analysis of DNA sequences reveals the ancient divergence of three domains of living organisms: Bacteria and Archaea (both prokaryotes) and Eukarya (eukaryotes). The color code shown here is used throughout this book to indicate the three domains.

classes of **eukaryotes** (cells with a nucleus) that include simple multicellular forms: algae, fungi, and protists (**Fig. 1.5**). The bacteria, archaea, and eukaryotes—known as the three "domains"—evolved from a common ancestral cell. We also discuss viruses and related infectious particles (Chapters 6 and 11).

Note: The formal names of the three domains are **Bacteria**, **Archaea**, and **Eukarya**. Members of these domains are called **bacteria** (singular, **bacterium**), **archaea** (singular, **archaeon**), and **eukaryotes** (singular, **eukaryote**), respectively. The microbiology literature includes alternative spellings for some of these terms, such as "archaean" and "eucaryote."

Microbial Genomes Are Sequenced

How have we learned how microbes work? A key tool is the study of microbial genomes. A **genome** is the total genetic information contained in an organism's chromosomal DNA (**Fig. 1.6**). The genes in a microbe's genome and the sequence of DNA tell us a lot about how that microbe grows and associates with other species. For example, if a microbe's genome includes genes for nitrogenase, a nitrogen-fixing enzyme, that microbe probably can fix nitrogen from the atmosphere into proteins—its own proteins and those of associated plants. And by comparing DNA sequences of different microbes, we can figure out how closely related they are and how they evolved.

The first method of DNA sequencing that was fast enough to sequence large genomes was developed by Fred Sanger (**Fig. 1.7A**) at the University of Cambridge. This achievement—which jump-started the study of molecular biology—earned Sanger the 1980 Nobel Prize in Chemistry, together with Walter Gilbert and Paul Berg. Sanger and colleagues used the new method to sequence DNA containing tens of thousands of base pairs, such as the DNA of the human mitochondrion.

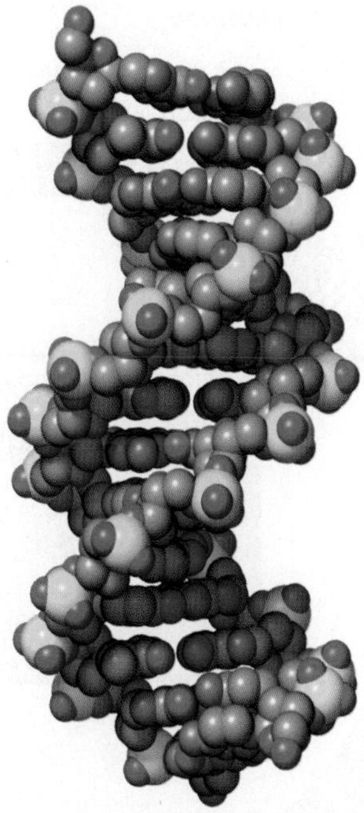

FIGURE 1.6 ▪ DNA. The sequence of base pairs in DNA encodes all the genetic information of an organism.

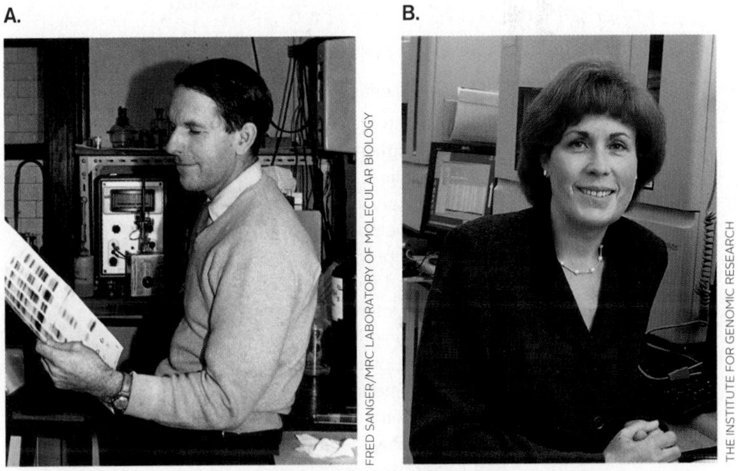
A.
B.
FRED SANGER/MRC LABORATORY OF MOLECULAR BIOLOGY
THE INSTITUTE FOR GENOMIC RESEARCH

FIGURE 1.7 ▪ Microbial genome sequencers. A. Fred Sanger, who shared the 1980 Nobel Prize in Chemistry for devising the method of DNA sequence analysis that is the basis of genome sequencing. Here he is reading sequence data from bands of DNA separated by electrophoresis. **B.** Claire Fraser-Liggett, past president of The Institute for Genomic Research (TIGR), which completed the sequences of *H. influenzae* and many other microbial genomes.

Weblinks The Nobel Prize website presents the lectures and autobiographies of all Nobel Prize winners, including many who were awarded prizes for advances in microbiology. *nobelprize.org*

But most genomes of cells contain millions, or even billions, of base pairs. In 1995, scientists completed the first genome sequence of a cellular microbe, the bacterium

Genome of *Haemophilus influenzae*

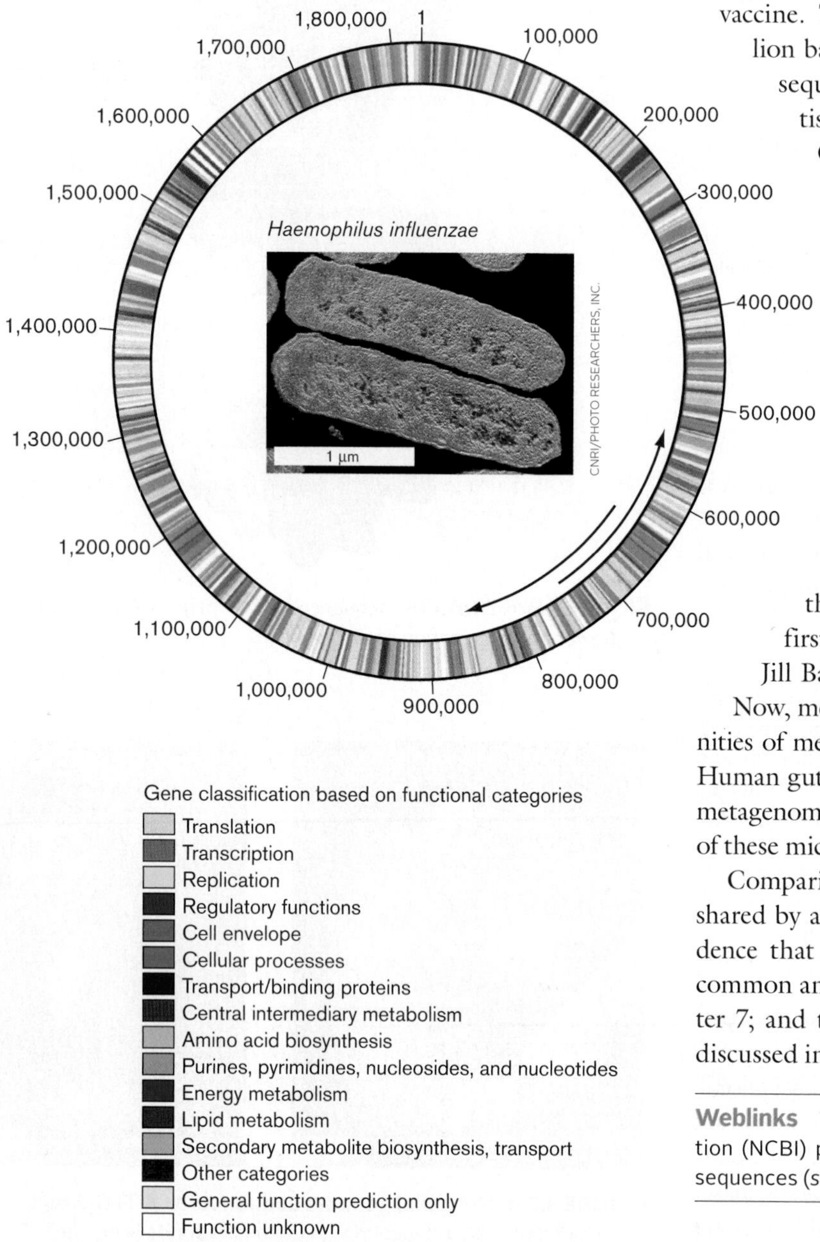
1
100,000
200,000
300,000
400,000
500,000
600,000
700,000
800,000
900,000
1,000,000
1,100,000
1,200,000
1,300,000
1,400,000
1,500,000
1,600,000
1,700,000
1,800,000
Haemophilus influenzae
1 µm
CNRI/PHOTO RESEARCHERS, INC.
Gene classification based on functional categories
Translation
Transcription
Replication
Regulatory functions
Cell envelope
Cellular processes
Transport/binding proteins
Central intermediary metabolism
Amino acid biosynthesis
Purines, pyrimidines, nucleosides, and nucleotides
Energy metabolism
Lipid metabolism
Secondary metabolite biosynthesis, transport
Other categories
General function prediction only
Function unknown

FIGURE 1.8 ▪ The first sequenced genome. The genome of *Haemophilus influenzae,* a bacterium that causes ear infections and meningitis, was the first DNA sequence completed for a cellular organism (inset, colorized electron micrograph). The genome of *H. influenzae* contains nearly 2 million base pairs (denoted by the numbers running clockwise around the map) specifying approximately 1,743 genes, which are expressed to make protein and RNA products. The annotated sequence of the genome appears on the website of the National Center for Biotechnology Information (NCBI). Colored bars indicate gene sequences.

Haemophilus influenzae (**Fig. 1.8**). *H. influenzae* causes meningitis in children, a disease now prevented by the Hib vaccine. The *H. influenzae* genome has nearly 2 million base pairs, which specify about 1,700 genes. The sequence was determined by a large team of scientists led by Craig Venter, Hamilton Smith, and Claire Fraser-Liggett (**Fig. 1.7B**) at The Institute for Genomic Research (TIGR). The TIGR team devised a special computational strategy for assembling large amounts of sequence data—a strategy later used to sequence the human genome.

Today we sequence new bacterial genomes daily. In addition to sequencing individual genomes, computational strategies are used to sequence thousands of genomes of microbes sampled from a natural environment, such as the acid drainage from an iron mine. The collection of sequences taken directly from the environment is called a **metagenome**. The first metagenome of an acid mine was sequenced by Jill Banfield and co-workers at UC Berkeley in 2004. Now, metagenomes are sequenced for microbial communities of medical interest, such as that of the human colon. Human gut microbes contain 100 times more genes in their metagenomes than the human genome contains—and many of these microbial genes contribute to human health!

Comparing genomes has revealed a set of core genes shared by all organisms. These core genes add further evidence that all life on Earth, including humans, shares a common ancestry. Genomes are discussed further in Chapter 7; and the evolution of genomes and metagenomes is discussed in Chapters 17 and 21.

Weblinks The National Center for Biotechnology Information (NCBI) provides free access to all of the published genome sequences (*see ebook*).

1.2

Microbes Shape Human History

Today our knowledge of microbes is enormous, and it keeps growing. Yet throughout most of human history we were unaware of the microbial world and how it shaped human culture. Yeasts and bacteria made foods such as bread and cheese (**Fig. 1.9A**), as well as alcoholic beverages (discussed in Chapter 16). "Rock-eating" bacteria, known as "lithotrophs,"

A.
B.
BETTMANN/CORBIS
JOHNER IMAGES/GETTY IMAGES

FIGURE 1.9 ▪ Production and destruction by microbes. **A.** Roquefort cheeses ripening in France. **B.** Statue decaying from the action of lithotrophic microbes. The process is accelerated by acid rain. (Cathedral of Cologne, Germany.)

leached copper and other metals from ores exposed by mining, enabling ancient human miners to obtain these metals. The lithotrophic oxidation of minerals for energy generates strong acids, which accelerate breakdown of the ore. Today about 20% of the world's copper, as well as some uranium and zinc, is produced by bacterial leaching. Unfortunately, microbial acidification also consumes the stone of ancient monuments (**Fig. 1.9B**)—a process intensified by airborne acidic pollution. Management of microbial corrosion is an important field of applied microbiology.

How did people find out about microbes? (**Table 1.2**, pages 12–13). Microscopists in the seventeenth and eighteenth centuries formulated key concepts about microbes and their existence, including their means of reproduction and death. In the nineteenth century, the "golden age" of microbiology, key principles of disease pathology and microbial ecology were established that scientists still use today. This period laid the foundation for modern biology, in which genetics and molecular biology provide powerful tools for scientists to manipulate microorganisms for medicine, research, and industry.

Microbial Disease Devastates Human Populations

Microbial diseases such as bubonic plague and AIDS have profoundly affected human history (**Fig. 1.10**). The bubonic plague, which wiped out a third of Europe's population in the fourteenth century, was caused by *Yersinia pestis,* a bacterium spread by rat fleas. Ironically, the plague-induced population decline enabled the social transformation that led to the Renaissance, a period of unprecedented

A.
B.
BRIDGEMAN ART LIBRARY
REUTERS/CORBIS

FIGURE 1.10 ▪ Microbial disease in history and culture. **A.** Medieval church procession to ward off the Black Death (bubonic plague). **B.** The AIDS Memorial Quilt spread before the Washington Monument. Each panel of the quilt memorializes an individual who died of AIDS.

TABLE 1.2

Microbes and human history.

Date	Microbial discovery	Discoverer(s)
Microbes impact human culture without detection		
10,000 BCE	Food and drink are produced by microbial fermentation.	Egyptians, Chinese, and others
1500 BCE	Tuberculosis, polio, leprosy, and smallpox are evident in mummies and tomb art.	Egyptians
50 BCE	Copper is recovered from mine water acidified by sulfur-oxidizing bacteria.	Roman metal workers under Julius Caesar
1546 CE	Syphilis and other diseases are observed to be contagious.	Girolamo Fracastoro (Padua)
Early microscopy and the origin of microbes		
1676	Microbes are observed under a microscope.	Antonie van Leeuwenhoek (Netherlands)
1688	Spontaneous generation is disproved for maggots.	Francesco Redi (Italy)
1717	Smallpox is prevented by inoculation of pox material, a rudimentary form of immunization.	Turkish women taught Lady Mary Montagu, who brought the practice to England
1765	Microbe growth in organic material is prevented by boiling in a sealed flask.	Lazzaro Spallanzani (Padua)
1798	Cowpox vaccination prevents smallpox.	Edward Jenner (England)
1835	Fungus causes disease in silkworms (first pathogen to be demonstrated in animals).	Agostino Bassi de Lodi (Italy)
1847	Chlorine as antiseptic wash for doctor's hands decreases pathogens.	Ignaz Semmelweis (Hungary)
1881	Bacterial spores survive boiling but are killed by cyclic boiling and cooling.	John Tyndall (Ireland)
"Golden age" of microbiology: principles and methods established		
1855	Sanitation shows statistical correlation with mortality (Crimean War).	Florence Nightingale (England)
1857	Microbial fermentation produces lactic acid or alcohol.	Louis Pasteur (France)
1864	Microbes fail to appear spontaneously, even in the presence of oxygen.	Louis Pasteur (France)
1866	Microbes are defined as a class distinct from animals and plants.	Ernst Haeckel (Germany)
1867	Antisepsis during surgery prevents patient death.	Joseph Lister (England)
1877	Bacteria are a causative agent of anthrax.	Robert Koch (Germany)
1881	First artificial vaccine is developed (against anthrax).	Louis Pasteur (France)
1882	First pure culture of colonies is grown on solid medium, *Mycobacterium tuberculosis.*	Robert Koch (Germany)
1884	Koch's postulates are published, based on anthrax and tuberculosis.	Robert Koch (Germany)
1884	Gram stain is devised to distinguish bacteria from human cells.	Hans Christian Gram (Netherlands)
1886	Intestinal bacteria include *Escherichia coli,* the future model organism.	Theodor Escherich (Austria)
1889	Bacteria oxidize iron and sulfur (lithotrophy).	Sergei Winogradsky (Russia)
1889	Bacteria isolated from root nodules are proposed to fix nitrogen.	Martinus Beijerinck (Netherlands)
1892, 1899	The concept of a virus is proposed to explain tobacco mosaic disease.	Dmitri Ivanovsky (Russia) and Martinus Beijerinck (Netherlands)
Cell biology, biochemistry, and genetics		
1908	Antibiotic chemicals are synthesized and identified (chemotherapy).	Paul Ehrlich (USA)
1911	Viruses are found to be a cause of cancer in chickens.	Peyton Rous (USA)
1917	Bacteriophages are recognized as viruses that infect bacteria.	Frederick Twort (England) and Félix d'Herelle (France)
1924	The ultracentrifuge is invented and used to measure the size of proteins.	Theodor Svedberg (Sweden)
1928	*Streptococcus pneumoniae* bacteria are transformed by a genetic material from dead cells.	Frederick Griffith (England)
1929	Penicillin, the first widely successful antibiotic, is made by a fungus. The molecule is isolated in 1941.	Alexander Fleming (Scotland), Howard Florey (Australia), and Ernst Chain (Germany)
1933–1945	The transmission electron microscope is invented and used to observe cells.	Ernst Ruska and Max Knoll (Germany)
1937	The tricarboxylic acid cycle is discovered.	Hans Krebs (England)
1938	The microbial "kingdom" is subdivided into prokaryotes (Monera) and eukaryotes.	Herbert Copeland (USA)
1938	*Bacillus thuringiensis* spray is produced as the first bacterial insecticide.	Insecticide manufacturers (France)
1941	One gene encodes one enzyme in *Neurospora.*	George Beadle and Edward Tatum (USA)
1941	Poliovirus is grown in human tissue culture.	John Enders, Thomas Weller, and Frederick Robbins (USA)
1944	DNA is the genetic material that transforms *S. pneumoniae.*	Oswald Avery, Colin MacLeod, and Maclyn McCarty (USA)
1945	The bacteriophage replication mechanism is elucidated.	Salvador Luria (Italy) and Max Delbrück (Germany), working in the USA
1946	Bacteria transfer DNA by conjugation.	Edward Tatum and Joshua Lederberg (USA)
1946–1956	X-ray diffraction crystal structures are obtained for the first complex biological molecules: penicillin and vitamin B_{12}.	Dorothy Hodgkin, John Bernal, and co-workers (England)
1950	Anaerobic culture technique is devised to study anaerobes of the bovine rumen.	Robert Hungate (USA)
1950	Bacteria can carry latent bacteriophages (lysogeny).	André Lwoff (France)
1951	Transposable elements in DNA are discovered in maize and later shown in bacteria.	Barbara McClintock (USA)
1952	DNA is injected into a cell by a bacteriophage.	Martha Chase and Alfred Hershey (USA)

Molecular biology and recombinant DNA		
1953	Overall structure of DNA is identified by X-ray diffraction analysis as a double helix.	Rosalind Franklin and Maurice Wilkins (England)
1953	Double-helical DNA consists of antiparallel chains connected by the hydrogen bonding of AT and GC base pairs.	James Watson (USA) and Francis Crick (England)
1959	Expression of the messenger RNA for the *E. coli* lac operon is regulated by a repressor protein.	Arthur Pardee (England) and François Jacob and Jacques Monod (France)
1960	Radioimmunoassay for detection of biomolecules is developed.	Rosalyn Yalow and Solomon Bernson (USA)
1961	The chemiosmotic theory, which states that biochemical energy is stored in a transmembrane proton gradient, is proposed and tested.	Peter Mitchell and Jennifer Moyle (England)
1966	The genetic code by which DNA information specifies protein sequence is deciphered.	Marshall Nirenberg, H. Gobind Khorana, and others (USA)
1967	Bacteria can grow at temperatures above 80°C in hot springs at Yellowstone National Park.	Thomas Brock (USA)
1968	Serial endosymbiosis is proposed to explain the evolution of mitochondria and chloroplasts.	Lynn Margulis (USA)
1969	Retroviruses contain reverse transcriptase, which copies RNA to make DNA.	Howard Temin, David Baltimore, and Renato Dulbecco (USA)
1972	Inner and outer membranes of Gram-negative bacteria (*Salmonella*) are separated by ultracentrifugation.	Mary Osborn (USA)
1973	A recombinant DNA molecule is made in vitro (in a test tube).	Stanley Cohen, Annie Chang, Robert Helling, and Herbert Boyer (USA)
1974	A rotary motor drives the bacterial flagellum.	Howard Berg, Michael Silverman, and Melvin Simon (USA)
1975	mRNA-rRNA base pairing initiates protein synthesis in *E. coli.*	Joan Steitz and Karen Jakes (USA), and Lynn Dalgarno and John Shine (Australia)
1975	The dangers of recombinant DNA are assessed at the Asilomar Conference.	Paul Berg, Maxine Singer, and others (USA)
1975	Monoclonal antibodies are produced indefinitely in tissue culture by hybridomas, antibody-producing cells fused to cancer cells.	George Kohler and Cesar Milstein (USA)
1977, 1980	A DNA sequencing method is invented and used to sequence the first genome of a virus.	Fred Sanger, Walter Gilbert, and Allan Maxam (USA)
1977	Archaea identified as a third domain of life, the others being eukaryotes and bacteria.	Carl Woese (USA)
1978	The first protein catalog is compiled for *E. coli* based on 2D gels.	Fred Neidhardt, Peter O'Farrell, and colleagues (USA)
1978	Biofilms are a major form of existence of microbes.	William Costerton and others (Canada)
1979	Smallpox is declared eliminated—a global triumph of immunology and public health.	The World Health Organization
Genomics, structural biology, and molecular ecology		
1981	Invention of the polymerase chain reaction (PCR) makes available large quantities of DNA.	Kary Mullis (USA)
1981–1986	Self-splicing RNA is discovered in the protist *Tetrahymena.*	Thomas Cech and Sidney Altman (USA)
1982	Archaea are discovered with optimal growth above 100°C.	Karl Stetter (Germany)
1982	Viable but nonculturable bacteria contribute to ecology and pathology.	Rita Colwell and Norman Pace (USA)
1982	Prions, infectious agents consisting solely of protein, are characterized.	Stanley Prusiner (USA)
1983	Human immunodeficiency virus (HIV) is discovered as the cause of AIDS.	Françoise Barré-Sinoussi and Luc Montagnier (France), and Robert Gallo (USA)
1983	Genes are introduced into plants by use of *Agrobacterium tumefaciens* plasmid vectors.	Eugene Nester, Mary-Dell Chilton, and colleagues (USA)
1984	Acid-resistant *Helicobacter pylori* grow in the stomach, where they cause gastritis.	Barry Marshall and J. Robin Warren (Australia)
1987	*Geobacter* bacteria that can generate electricity are discovered.	Derek Lovley and colleagues (USA)
1988	*Prochlorococcus* is identified as Earth's most abundant marine phototroph.	Sallie Chisholm and colleagues (USA)
1993	A giant bacterium (*Epulopiscium*)—large enough to see—is identified.	Esther Angert and Norman Pace (USA)
1995	The first genome is sequenced for a cellular organism, *Haemophilus influenzae.*	Craig Venter, Hamilton Smith, Claire Fraser-Liggett, and others (USA)
2001	The ribosome structure is obtained at near-atomic level by X-ray diffraction.	Marat Yusupov, Harry Noller, and colleagues (USA)
2004	The mimivirus genome shows that large DNA viruses evolved from cells.	Didier Raoult and colleagues (France)
2006	First metagenomes are sequenced, from Iron Mountain acid mine drainage and from the Sargasso Sea.	Jill Banfield, Craig Venter, and others (USA)
2006	Vaccine prevents genital human papillomavirus (HPV), the most common sexually transmitted infection.	Patented by Georgetown University and other institutions (USA and Australia)
2009	Self-sustained replication of an RNA enzyme shows how life could have originated as an "RNA world."	Gerald Joyce and Tracey Lincoln (USA)
1988–2010	*Escherichia coli* long-term evolution experiment reaches 50,000 generations and continues.	Richard Lenski, Zachary Blount, and colleagues (USA)

cultural advancement. In the nineteenth century, the bacterium *Mycobacterium tuberculosis* stalked overcrowded cities, and tuberculosis became so common that the pallid appearance of tubercular patients became a symbol of tragic youth in European literature. Today, societies throughout the world are devastated by the epidemic of acquired immunodeficiency syndrome (AIDS), caused by the human immunodeficiency virus (HIV). More than 36 million people are living with HIV infection today, and each year 2 million die of AIDS.

Historians traditionally emphasize the role of warfare in shaping human destiny, and the brilliance of leaders or the advantage of new technology, in determining which civilizations rise or fall. Yet the fate of human societies is often determined by microbes. For example, much of the native population of North America was exterminated by smallpox introduced by European invaders. Throughout history, more soldiers have died of microbial infections than of wounds in battle. The significance of disease in warfare was first recognized by the British nurse and statistician Florence Nightingale (1820–1910) (**Fig. 1.11A**).

Better known as the founder of professional nursing, Nightingale also founded the science of medical statistics. She used methods invented by French statisticians to demonstrate the high mortality rate due to disease among British soldiers during the Crimean War. To show the deaths of soldiers due to various causes, she devised the "polar area chart" (**Fig. 1.11B**). In this chart, blue wedges represent deaths due to infectious disease, red wedges represent deaths due to wounds, and black wedges represent all other causes of death. Infectious disease accounts for more than half of all mortality.

Before Nightingale, no one understood the impact of disease on armies, or on other crowded populations, such as in cities. Nightingale's statistics convinced the British government to improve army living conditions and to upgrade the standards of army hospitals. In modern epidemiology, statistical analysis continues to be a crucial tool in determining the causes of disease.

Weblinks The CDC (Centers for Disease Control and Prevention), in Atlanta, is the U.S. agency for medical information and epidemiology. *cdc.gov*

Microscopes Reveal the Microbial World

The seventeenth century was a time of growing inquiry and excitement about the "natural magic" of science and patterns of our world, such as the laws of gravitation and motion formulated by Isaac Newton (1642–1727). Robert Boyle (1627–1691) performed the first controlled experiments on the chemical conversion of matter. Physicians attempted new treatments for disease involving the application of "stone and minerals" (that is, the application of chemicals), what today we would call "chemotherapy." Minds were open to consider the astounding possibility that our surroundings,

A.

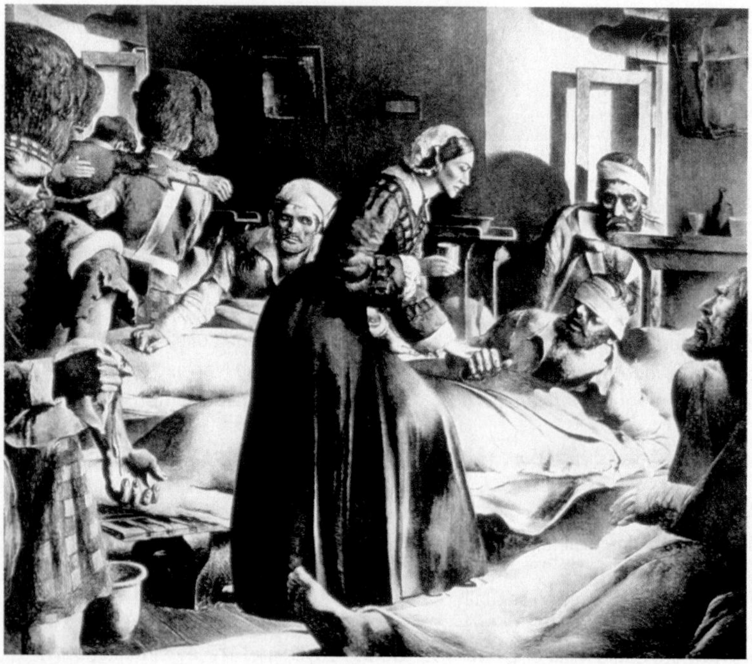

CORBIS

B.

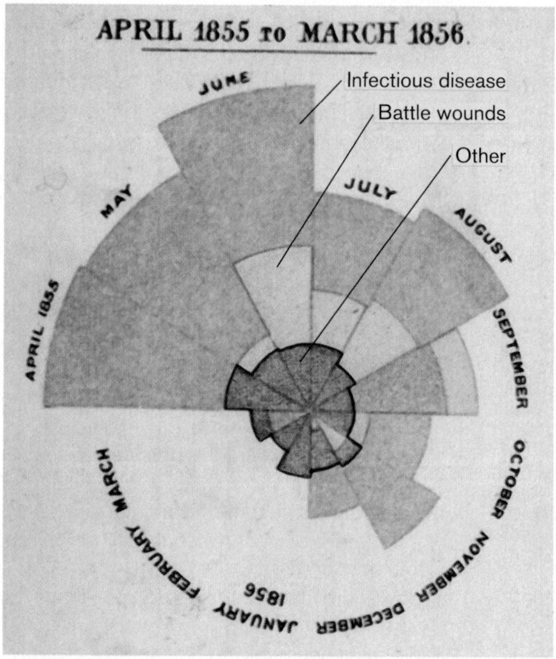
APRIL 1855 TO MARCH 1856
Infectious disease
Battle wounds
Other
JUNE
JULY
AUGUST
SEPTEMBER
OCTOBER
NOVEMBER
DECEMBER
JANUARY
1856
FEBRUARY
MARCH
APRIL 1855
MAY

THE FLORENCE NIGHTINGALE MUSEUM TRUST, LONDON

FIGURE 1.11 ▪ Florence Nightingale, founder of medical statistics. A. Florence Nightingale was the first to use medical statistics to demonstrate the significance of mortality due to disease. **B.** Nightingale's polar area chart of mortality data during the Crimean War.

indeed our very bodies, were inhabited by tiny living beings.

Robert Hooke observes the microscopic world. The first microscopist to publish a systematic study of the world as seen under a microscope was Robert Hooke (1635–1703). As curator of experiments for the Royal Society of London, Hooke built the first compound microscope—a magnifying instrument containing two or more lenses that multiply their magnification in series. With his microscope, Hooke observed biological materials such as nematode "vinegar eels," mites, and mold filaments, drawings of which he published in *Micrographia* (1665), the first publication that illustrated objects observed under a microscope (**Fig. 1.12**).

MILTON S. EISENHOWER LIBRARY

FIGURE 1.12 ▪ Hooke's *Micrographia*. An illustration of mold sporangia, drawn by Hooke in 1665 from his observations of objects using a compound microscope.

Hooke was the first to observe distinct units of living material, which he called "cells." Hooke first named the units cells because the shape of hollow cell walls in a slice of cork reminded him of the shape of monks' cells in a monastery. But his crude lenses achieved at best 30-fold power (30×), so he never observed single-celled organisms.

Antonie van Leeuwenhoek observes bacteria with a single lens. Hooke's *Micrographia* inspired other microscopists, including Antonie van Leeuwenhoek (1632–1723), who became the first individual to observe single-celled microbes (**Fig. 1.13A**). As a young man, Leeuwenhoek lived in the Dutch city of Delft, where he worked as a cloth draper, a profession that introduced him to magnifying glasses. (The magnifying glasses were used to inspect the quality of the cloth, enabling the worker to count the number of threads.) Later in life, he took up the hobby of grinding ever-stronger lenses to see into the world of the unseen.

Leeuwenhoek ground lenses stronger than Hooke's, which he used to build single-lens magnifiers, complete with sample holder and focus adjustment (**Fig. 1.13B**). First he observed insects, including lice and fleas; then the relatively large single cells of protists and algae; then ultimately bacteria. One day he applied his microscope to observe matter extracted from between his teeth. He wrote, "To my great surprise [I] perceived that the aforesaid matter contained very many small living Animals, which moved themselves very extravagantly."

Over the rest of his life, Leeuwenhoek recorded page after page on the movement of microbes, reporting their size and shape so accurately that in many cases we can determine the species he observed (**Fig. 1.13C**). He performed experiments, comparing, for example, the appearance of "small animals" from his teeth before and after drinking hot coffee. The disappearance of microbes from

A.

BETTMANN/CORBIS

B.

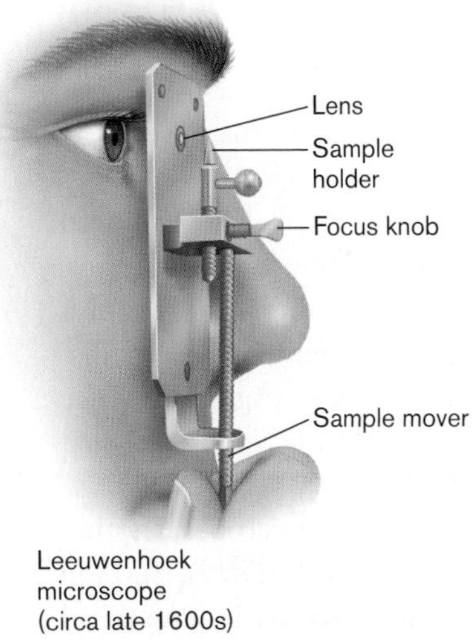
Lens
Sample holder
Focus knob
Sample mover
Leeuwenhoek microscope (circa late 1600s)

C.

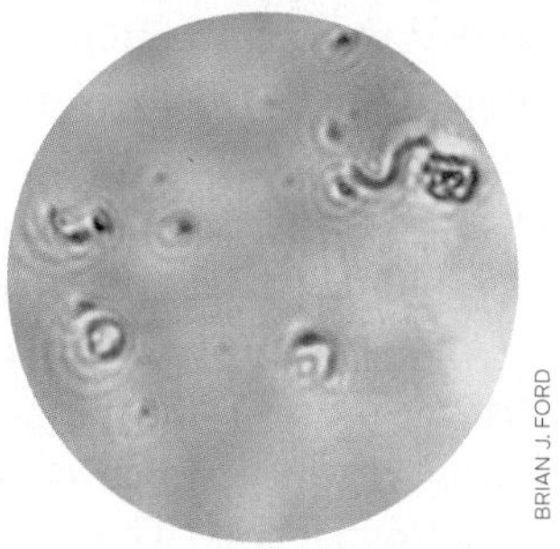
BRIAN J. FORD

FIGURE 1.13 ▪ Antonie van Leeuwenhoek. A. A portrait of Leeuwenhoek, the first person to observe individual microbes. **B.** "Microscope" (magnifying glass) used by Leeuwenhoek. **C.** Spiral bacteria viewed through a replica of Leeuwenhoek's instrument.

his teeth after drinking a hot beverage suggested that heat killed microbes—a profoundly important principle for the study and control of microbes ever since.

Ironically, Leeuwenhoek is believed to have died of a disease contracted from sheep whose bacteria he had observed. Historians have often wondered why it took so many centuries for Leeuwenhoek and his successors to determine the link between microbes and disease. Although observers such as Agostino Bassi de Lodi (1773–1856) noted isolated cases of microbes associated with pathology (see **Table 1.2**), the very ubiquity of microbes—most of them actually harmless—may have obscured their more deadly roles. In addition, it was hard to distinguish between microbes and the single-celled components of the human body, such as blood cells and sperm. It was not until the nineteenth century that human tissues could be distinguished from microbial cells by the application of differential chemical stains (discussed in Chapter 2).

Thought Question

1.3 Why do you think it took so long for humans to connect microbes with infectious disease?

Spontaneous Generation: Do Microbes Have Parents?

The observation of microscopic organisms led priests and philosophers to wonder where they came from. In the eighteenth century, scientists and church leaders intensely debated the question of **spontaneous generation**, the theory that living creatures such as maggots could arise spontaneously, without parental organisms. Chemists of the day tended to support spontaneous generation, as it appeared similar to the changes in matter that could occur when chemicals were mixed. Christian church leaders, however, supported the biblical view that all organisms have "parents" going back to the first week of creation.

The Italian priest Francesco Redi (1626–1697) showed that maggots in decaying meat were the offspring of flies. Meat kept in a sealed container, excluding flies, did not produce maggots. Thus, Redi's experiment argued against spontaneous generation for macroscopic organisms. The meat still putrefied, however, producing microbes that seemed to arise "without parents."

To disprove spontaneous generation of microbes, another Italian priest, Lazzaro Spallanzani (1729–1799), showed that a sealed flask of meat broth sterilized by boiling failed to grow microbes. Spallanzani also noticed that microbes often appeared in pairs. Were these two parental microbes coupling to produce offspring, or did one microbe become two? Through long and tenacious observation, Spallanzani watched a single microbe grow in size until it split in two. Thus he demonstrated cell fission, the process by which cells arise by the splitting of preexisting cells.

Even Spallanzani's experiments, however, did not put the matter to rest. Proponents of spontaneous generation argued that the microbes in the priest's flask lacked access to oxygen and therefore could not grow. The pursuit of this question was left to future microbiologists, including the famous French microbiologist Louis Pasteur (1822–1895) (**Fig. 1.14A**). In addressing spontaneous generation and related questions, Pasteur and his contemporaries laid the foundations for modern microbiology.

Louis Pasteur reveals the biochemical basis of microbial growth. Pasteur began his scientific career as a chemist and wrote his doctoral thesis on the structure of organic crystals. He discovered the fundamental chemical property of chirality, the fact that some organic molecules exist in two forms that differ only by mirror symmetry. In other words,

A.

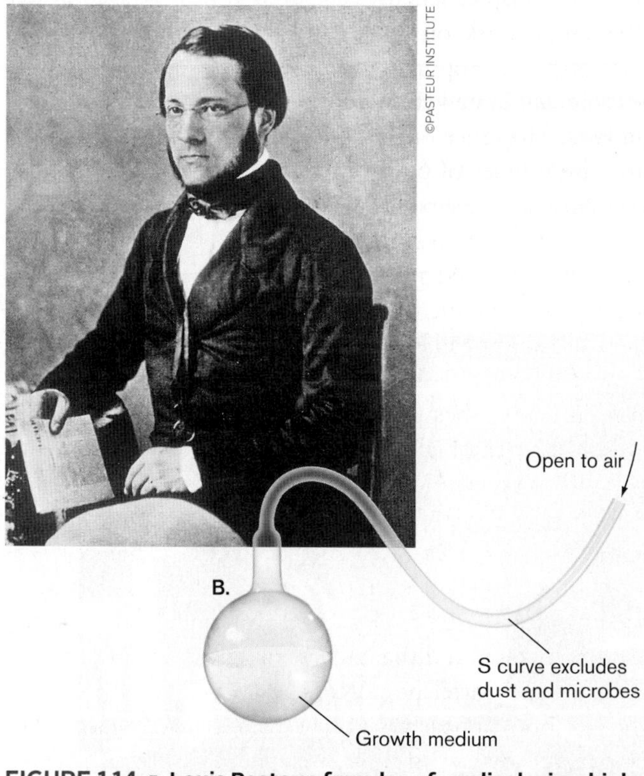
©PASTEUR INSTITUTE
B.
Open to air
S curve excludes dust and microbes
Growth medium

FIGURE 1.14 ■ Louis Pasteur, founder of medical microbiology and immunology. A. Pasteur's contributions to the science of microbiology and immunology earned him lasting fame. **B.** Swan-necked flask. Pasteur showed that, after boiling, the contents in such a flask remain free of microbial growth, despite access to air.

the two structures are mirror images of one another, like the right and left hands. Pasteur found that when microbes were cultured on a nutrient substance containing both mirror forms, only one mirror form was consumed. He concluded that the metabolic preference for one mirror form was a fundamental property of life. Subsequent research has confirmed that most biological molecules, such as DNA and proteins, occur in only one of their mirror forms.

As a chemist, Pasteur was asked to help with a widespread problem encountered by French manufacturers of wine and beer. The production of alcoholic beverages is now known to occur by **fermentation**, a process by which microbes gain energy by converting sugars into alcohol. In the time of Pasteur, however, the conversion of grapes or grain to alcohol was believed to be a spontaneous chemical process. No one could explain why some fermentation mixtures produced vinegar (acetic acid) instead of alcohol. Pasteur discovered that fermentation is actually caused by living yeast, a single-celled fungus. In the absence of oxygen, yeast produces alcohol as a terminal waste product. But when the yeast culture is contaminated with bacteria, the bacteria outgrow the yeast and produce acetic acid instead of alcohol. (Fermentative metabolism is discussed further in Chapter 13.)

Pasteur's work on fermentation led him to test a key claim made by proponents of spontaneous generation. The proponents claimed that Spallanzani's failure to find spontaneous appearance of microbes was due to lack of oxygen. From his studies of yeast fermentation, Pasteur knew that some microbial species do not require oxygen for growth. So he devised an unsealed flask with a long, bent "swan neck" that admitted air but kept the boiled contents free of dust that carried microbes (**Fig. 1.14B**). The famous swan-necked flasks remained free of microbial growth for many years, but when a flask was tilted to enable contact of broth with dust, microbes grew immediately. Thus, Pasteur disproved that lack of oxygen was the reason for the failure of spontaneous generation in Spallanzani's flasks.

But even Pasteur's work did not prove that microbial growth requires preexisting microbes. The Irish scientist John Tyndall (1820–1893) attempted the same experiment as Pasteur, but sometimes found the opposite result. Tyndall found that the broth sometimes gave rise to microbes, no matter how long it was sterilized by boiling. The microbes appear because some kinds of organic matter, particularly hay infusion, are contaminated with a heat-resistant form of bacteria called endospores (or spores). The spore form can be eliminated only by repeated cycles of boiling and resting, in which the spores germinate to the growing, vegetative form that is killed at 100°C.

It was later discovered that endospores could be killed by boiling under pressure, as in a pressure cooker, which generates higher temperatures than can be obtained at atmospheric pressure. The steam pressure device called the **autoclave** became the standard way to sterilize materials for the controlled study of microbes. (Microbial control and antisepsis are discussed further in Chapter 5.)

Although spontaneous generation was discredited as a continual source of microbes, at some point in the past the first living organisms must have originated from nonliving materials. How did the first microbes arise? The earliest fossil evidence of cells in the geological record appears as "microfossils" in sedimentary rock that formed over 2 billion years ago (discussed in Chapter 17).

What was Earth's environment like when life began? Astonishingly, experiments that aim to replicate the proposed chemistry of early Earth yield simple amino acids, as well as adenine, a fundamental base of DNA and RNA. Both biochemical and genetic evidence suggest that life originated in an **"RNA world"** in which RNA performed the functions later adopted by DNA and proteins. **Special Topic 1.1** presents some of the evidence for early biochemistry and the RNA world. **The experimental basis for the origin of life and evolution is discussed in detail in Chapter 17.**

To Summarize

- **Microbes affected human civilization** for centuries before humans guessed at their existence through their contributions to our environment, food and drink production, and infectious diseases.
- **Florence Nightingale** statistically quantified the impact of infectious disease on human populations.
- **Robert Hooke** and **Antonie van Leeuwenhoek** were the first to record observations of microbes through simple microscopes.
- **Spontaneous generation** is the theory that microbes arise spontaneously, without parental organisms. **Lazzaro Spallanzani** showed that microbes arise from preexisting microbes and demonstrated that heat sterilization can prevent microbial growth.
- **Louis Pasteur** discovered the microbial basis of fermentation. He also showed that providing oxygen does not enable spontaneous generation.
- **John Tyndall** showed that repeated cycles of heat were necessary to eliminate spores formed by certain kinds of bacteria.

Special Topic 1.1: How Did Life Originate?

If all life on Earth shares descent from a microbial ancestor, how did the first microbe arise? The earliest fossil evidence of cells in the geological record appears in sedimentary rock that formed as early as 3.8 billion years ago. Although the nature of the earliest reported fossils remains controversial, it is generally accepted that "microfossils" from over 2 billion years ago were formed by living cells. Moreover, the living cells that formed these microfossils looked remarkably similar to bacterial cells today, forming chains of simple rods or spheres (**Fig. 1**).

The exact composition of the first environment for life is controversial. The components of the first living cells may have formed from spontaneous reactions sparked by ultraviolet absorption or electrical discharge. American chemists Stanley Miller (1930–2007) and Harold C. Urey (1893–1981) argued that the environment of early Earth contained mainly reduced compounds—compounds that have a strong tendency to donate electrons, such as ferrous iron, methane, and ammonia. More recent evidence has modified this view, but it is agreed that the strong electron acceptor oxygen gas (O_2) was absent until the first photosynthetic microbes produced it. Today, all our cells are composed of highly reduced molecules that are readily oxidized (lose electrons to O_2). This seemingly hazardous composition may reflect our cellular origin in the chemically reduced environment of early Earth.

In 1953, Miller attempted to simulate the highly reduced conditions of early Earth to test whether ultraviolet absorption or electrical discharge could cause reactions producing the fundamental components of life (**Fig. 2A**). Miller boiled a solution of water containing hydrogen gas, methane, and ammonia and applied an electrical discharge (comparable to a lightning strike). The electrical discharge excites electrons in the molecules and causes them to react. Astonishingly, the reaction produced a number of amino acids, including glycine, alanine, and aspartic acid. A similar experiment in 1961 by Spanish-American researcher Juan Oró (1923–2004) (**Fig. 2B**) combined hydrogen cyanide and ammonia under electrical discharge to obtain adenine, a fundamental component of DNA and of the energy carrier adenosine triphosphate (ATP).

How could early cells have survived the heat and chemically toxic environment of early Earth? Clues may be found

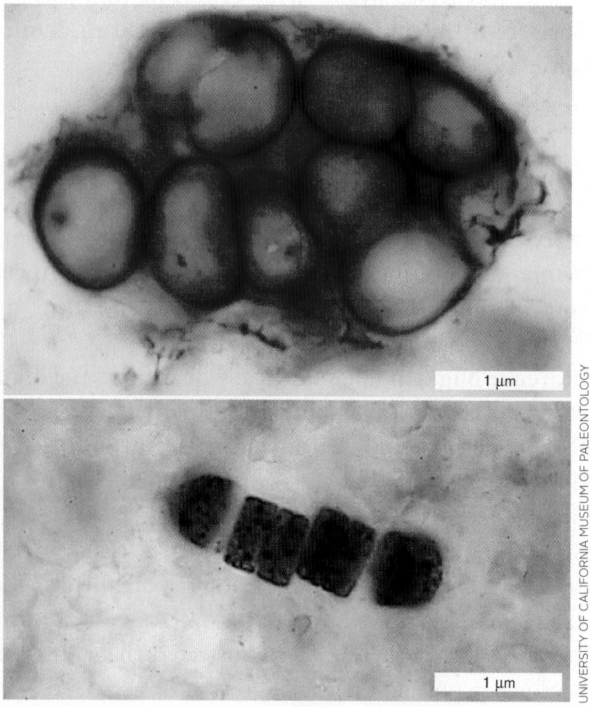
1 µm
1 µm
UNIVERSITY OF CALIFORNIA MUSEUM OF PALEONTOLOGY

FIGURE 1 ■ Evidence of ancient microbial life. Microfossils of ancient cyanobacteria from the Bitter Springs Formation, Australia, about 850 million years old.

A.
B.
JIM SUGAR/CORBIS
UNIVERSITY OF HOUSTON

FIGURE 2 ■ Simulating early Earth's chemistry. A. Stanley Miller with the apparatus of his early-Earth simulation experiment. **B.** Biochemist Juan Oró demonstrated the formation of adenine and other biochemicals from reaction conditions found in comets.

in the survival of archaea that thrive under habitat conditions that we consider extreme, such as solutions of boiling sulfuric acid. The specially adapted structures of such microbes may resemble those of the earliest life-forms.

Research since Miller's day has generated as many questions as answers concerning the origin of life—for example:

- Were the reduced forms of early carbon and nitrogen, such as methane (CH_4) and ammonia (NH_3), supplemented by oxidized forms such as CO_2, spewed out by volcanoes? If oxidized carbon and nitrogen were available, different kinds of early-life chemistry may have occurred.
- How did the origin of informational molecules, such as RNA and DNA, coincide with the origin of metabolism? One possibility is that early metabolism was catalyzed by molecules of RNA instead of protein. RNA molecules capable of catalysis, called ribozymes, were discovered in 1982 by Thomas Cech and Sidney Altman, who earned the Nobel Prize in Chemistry in 1989 (**Fig. 3**). In 2009, researchers made the first self-replicating ribozyme, an RNA that can catalyze reactions and copy itself indefinitely. This achievement supports the theory that early organisms were composed primarily of RNA—the so-called "RNA world."
- Geochemical evidence suggests that cells may have originated on Earth as early as 3.8 billion years ago, when Earth was just barely cool enough to allow the existence of cells. How could cells have formed so quickly? Could the first cells, in fact, have come from somewhere else?

Most of the molecules that spontaneously formed in Miller's experiments are also found in meteorites and comets. This observation led Oró to propose that the first chemicals of life could have come from outer space, perhaps carried by comets. But could life itself have an extraterrestrial origin? Current evidence for the origin and evolution of microbes is discussed in Chapter 17.

RESEARCH QUESTION

The process of DNA synthesis always begins with a short sequence of RNA called a "primer." How does the RNA priming of DNA synthesis in modern cells support the "RNA world" model for early life?

Thomas R. Cech. 2012. The RNA worlds in context. *Cold Spring Harbour Perspectives in Biology* **4**:a006742.

A.

GEOFFERY WHEELER FOR HOWARD HUGHES MEDICAL INSTITUTE

FIGURE 3 ▪ Tom Cech, discoverer of catalytic RNA. **A.** Tom Cech (University of Colorado, Boulder) holding a flask containing *Oxytricha nova,* microbes that make catalytic RNA, the kind of molecule that in early cells may have served both genetic and catalytic functions. **B.** Diagram of a catalytic RNA, where horizontal bars represent bases. The RNA catalyzes cleavage of itself.

B.

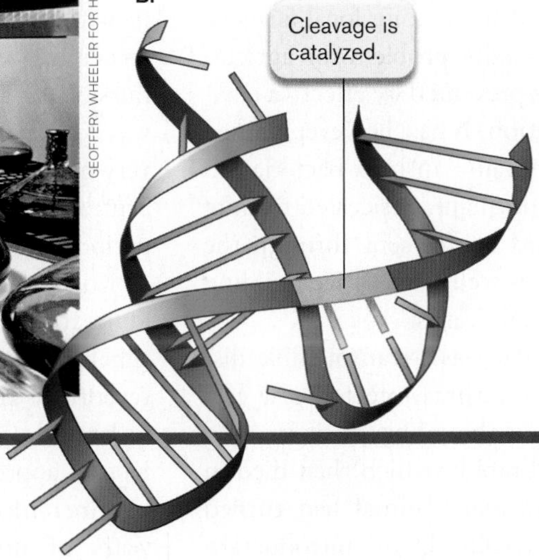
Cleavage is catalyzed.

1.3

Medical Microbiology

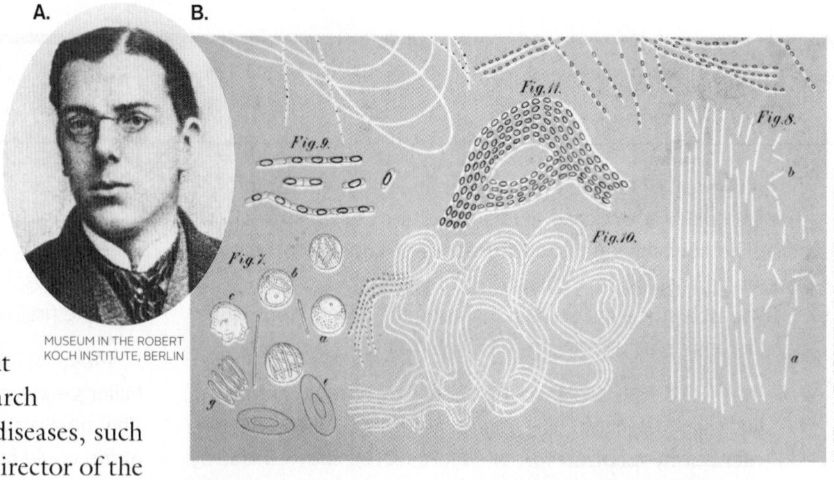
A.
B.
MUSEUM IN THE ROBERT KOCH INSTITUTE, BERLIN
Fig. 9.
Fig. 11.
Fig. 8.
Fig. 10.
Fig. 7.
MUSEUM IN THE ROBERT KOCH INSTITUTE, BERLIN

FIGURE 1.15 ▪ Robert Koch, founder of the scientific method of microbiology. **A.** Robert Koch as a university student. **B.** Koch's sketch of anthrax bacilli in mouse blood.

Over the centuries, thoughtful observers such as Fracastoro and Bassi (see **Table 1.2**) noted a connection between microbes and disease. Ultimately, researchers developed the **germ theory of disease**, the theory that many diseases are caused by microbes. Research today pursues the secrets of many microbial diseases, such as cholera, the focus of Rita Colwell, former director of the National Science Foundation (eTopic 1.1).

The first to establish a scientific basis for determining that a specific microbe causes a specific disease was the German physician Robert Koch (1843–1910) (**Fig. 1.15**). As a college student, Koch conducted biochemical experiments on his own digestive system. Koch's curiosity about the natural world led him to develop principles and techniques crucial to microbial investigation, including the pure-culture technique and the famous Koch's postulates for identifying the causative agent of a disease. He applied his methods to numerous lethal diseases around the world, including anthrax and tuberculosis in Europe, bubonic plague in India, and malaria in New Guinea (**Fig. 1.16**).

MUSEUM IN THE ROBERT KOCH INSTITUTE, BERLIN

FIGURE 1.16 ▪ Robert Koch (second from left) during a visit to New Guinea to investigate malaria.

Growth of Microbes in Pure Culture

Unlike Pasteur, who was a university professor, Koch took up a medical practice in a small Polish-German town. To make space in his home for a laboratory to study anthrax and other deadly diseases, his wife curtained off part of his patients' examining room.

Anthrax interested Koch because its epidemics in sheep and cattle caused economic hardship among local farmers. Today, anthrax is no longer a major problem for agriculture, because its transmission is prevented by effective environmental controls and vaccination. It has, however, gained notoriety as a bioterror agent because anthrax bacteria can survive for long periods in the dormant, desiccated form of an endospore. In 2001, anthrax spores sent through the mail contaminated post offices as well as an office building of the U.S. Senate, causing several deaths.

To investigate whether anthrax was a transmissible disease, Koch used blood from an anthrax-infected cow carcass to inoculate a rabbit. When the rabbit died, he used its blood to inoculate a second rabbit, which then died in turn. The blood of the unfortunate animal had turned black with long, rod-shaped bacilli. Upon introduction of these bacilli into healthy animals, the animals became ill with anthrax. Thus, Koch demonstrated an important principle of epidemiology: the **chain of infection**, or transmission of a disease. In retrospect, his choice of anthrax was fortunate, because anthrax microbes generate disease very quickly, multiply in the blood to an extraordinary concentration, and remain infective outside the body for long periods.

Koch and his colleagues then applied their experimental logic and culture methods to a more challenging disease: tuberculosis. In Koch's day, tuberculosis caused one-seventh of all reported deaths in Europe; today, tuberculosis bacteria continue to infect millions of people worldwide. Koch's approach to anthrax, however, was less applicable to tuberculosis, a disease that develops slowly after many years of dormancy. Furthermore, the causative bacteria, *Mycobacterium tuberculosis,* are small and difficult to distinguish from human tissue or from different bacteria of similar appearance associated with the human body. How could Koch prove that a particular bacterium caused a particular disease?

What was needed was to isolate a **pure culture** of microorganisms, a culture grown from a single "parental" cell. Previous researchers had achieved pure cultures by a laborious process of serially diluting suspended bacteria until a culture tube contained only a single cell. Alternatively, inoculating a solid surface such as a sliced potato could produce isolated **colonies**—distinct populations of bacteria, each grown from a single cell. For *M. tuberculosis,* Koch inoculated serum, which then formed a solid gel after heating. Later he refined the solid-substrate technique by adding gelatin to a defined liquid medium, which could then be chilled to form a solid medium in a glass dish. A covered version called the **petri dish** (also called a "petri plate") was invented by a colleague, Julius Richard Petri (1852–1921). The petri dish is a round dish with vertical walls covered by an inverted dish of slightly larger diameter. Today the petri dish, generally made of disposable plastic, remains an indispensable part of the microbiological laboratory.

Another improvement in solid-substrate culture was the replacement of gelatin with materials that remain solid at higher temperatures, such as the gelling agent **agar** (a polymer of the sugar galactose). The use of agar was recommended by Angelina Hesse (1850–1934), a microscopist and illustrator, to her husband, Walther Hesse (1846–1911), a young medical colleague of Koch (**Fig. 1.17**). Agar comes from red algae (seaweed), which is used by East Indian birds to build nests; it is the main ingredient in the delicacy "bird's nest soup." Dutch colonists used agar to make jellies and preserves, and a Dutch colonist from Java introduced it to Angelina Hesse. The Hesses used agar to develop the first effective growth medium for tuberculosis bacteria. Pure culture and growth conditions are discussed further in Chapters 4 and 5.

Note that some kinds of microbes cannot be grown in pure culture—that is, without other organisms. For example, the marine cyanobacterium *Prochlorococcus,* a major source of Earth's oxygen, requires heterotrophic bacteria to remove toxic oxygen radicals. And all viruses can be cultured only within their host cells (see Chapter 6). The discovery of viruses is explored at the end of this section.

Koch's Postulates

For his successful determination of the bacterium that causes tuberculosis, *Mycobacterium tuberculosis,* Koch was awarded the 1905 Nobel Prize in Physiology or Medicine. Koch for-

A.

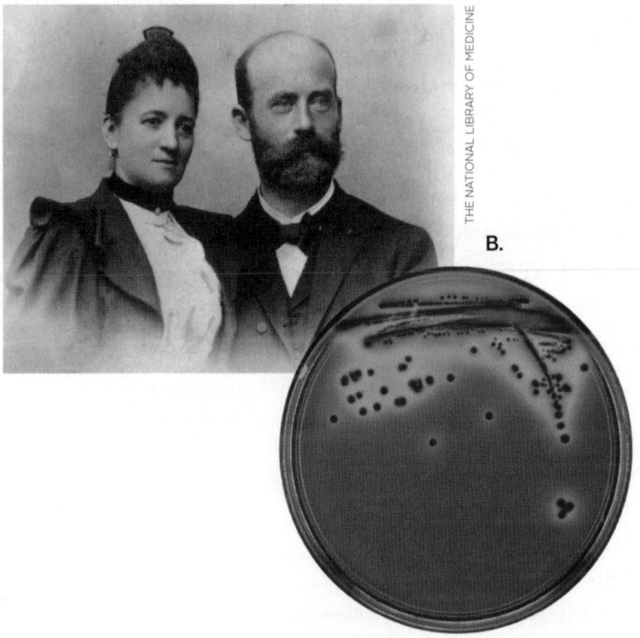
B.
THE NATIONAL LIBRARY OF MEDICINE
CYTOGRAPHICS/VISUALS UNLIMITED

FIGURE 1.17 ■ Angelina and Walther Hesse. A. Portrait of the Hesses, who first used agar to make solid-substrate media for bacterial growth. **B.** Colonies from a streaked agar plate.

mulated his famous set of criteria for establishing a causative link between an infectious agent and a disease (**Fig. 1.18**). These four criteria are known as **Koch's postulates**:

1. The microbe is found in all cases of the disease but is absent from healthy individuals.
2. The microbe is isolated from the diseased host and grown in pure culture.
3. When the microbe is introduced into a healthy, susceptible host (or animal model), the same disease occurs.
4. The same strain of microbe is obtained from the newly diseased host. When cultured, the strain shows the same characteristics as before.

Koch's postulates continue to be used to determine whether a given strain of microbe causes a disease. An example is Lyme disease (borreliosis), a tick-borne infection first described in New England, and shown in 1981 to be caused by the bacterium *Borrelia burgdorferi.* Nevertheless, the postulates remain only a guide; individual diseases and pathogens may confound one or more of the criteria. For example, tuberculosis bacteria are now known to cause symptoms in only 10% of the people infected. If Koch had been able to detect these silent bacilli, they would not have fulfilled his first criterion. In the case of AIDS, the concentration of HIV virus is so low that initially no virus could be detected in patients with fully active symptoms. It took the invention of the polymerase chain reaction (PCR),

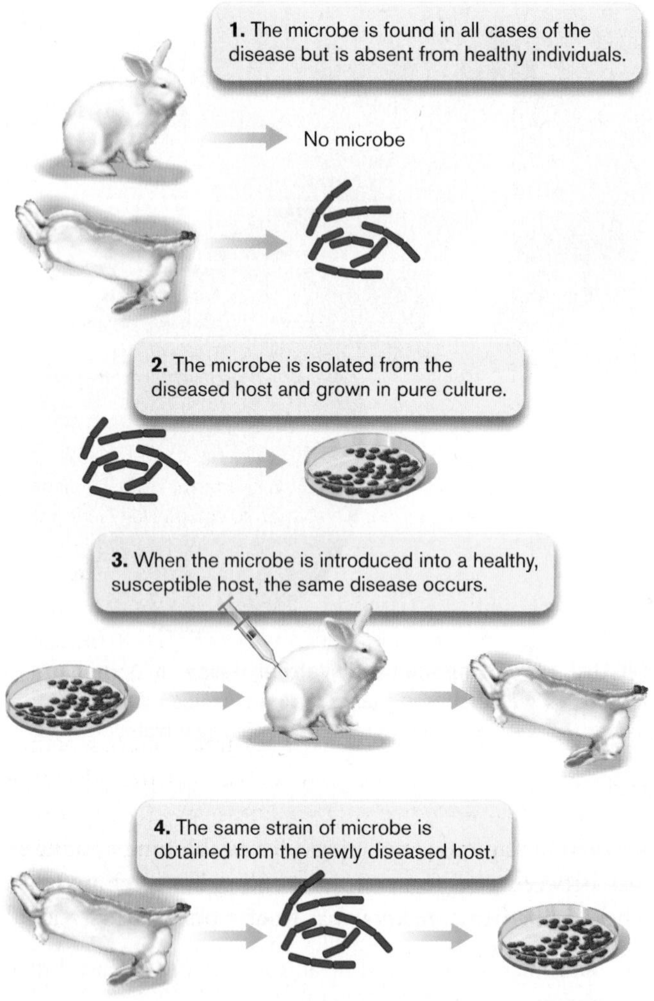
1. The microbe is found in all cases of the disease but is absent from healthy individuals.
No microbe
2. The microbe is isolated from the diseased host and grown in pure culture.
3. When the microbe is introduced into a healthy, susceptible host, the same disease occurs.
4. The same strain of microbe is obtained from the newly diseased host.

FIGURE 1.18 ■ Koch's postulates defining the causative agent of a disease.

a method of producing any number of copies of DNA or RNA sequences, to detect the presence of HIV. Modern research on microbial disease is presented in Chapter 25.

Another difficulty with AIDS and many other human diseases is the absence of an animal host that exhibits the same disease. In the case of AIDS, even chimpanzees (our closest relatives) are not susceptible, although they exhibit a similar disease from a related pathogen: simian immunodeficiency virus (SIV). Experimentation on humans is prohibited, although in rare instances researchers have voluntarily exposed themselves to a proposed pathogen. For example, Australian researcher Barry Marshall ingested *Helicobacter pylori* to convince skeptical colleagues that this organism could colonize the extremely acidic stomach. *H. pylori* turned out to be the causative agent of gastritis and stomach ulcers, conditions that had long been thought to be caused by stress rather than infection. For the discovery of *H. pylori,* Marshall and colleague J. Robin Warren won the 2005 Nobel Prize in Physiology or Medicine.

Thought Question

1.4 How could you use Koch's postulates (**Fig. 1.18**) to demonstrate the causative agent of influenza? What problems not encountered with anthrax would you need to overcome?

Immunization Prevents Disease

Identifying the cause of a disease is, of course, only the first step in developing an effective therapy and preventing further transmission. Early microbiologists achieved some remarkable insights on how to control pathogens (see **Table 1.2**).

The first clue of how to protect an individual from a deadly disease came from the dreaded smallpox. In the eighteenth century, smallpox infected a large fraction of the European population, killing or disfiguring many people. In countries of Asia and Africa, however, the incidence of smallpox was decreased by the practice of deliberately inoculating children with material from smallpox pustules. Inoculated children usually developed a mild case of the disease and were protected from smallpox thereafter.

The practice of smallpox inoculation was introduced from Turkey to Europe in 1717 by Lady Mary Montagu, a smallpox survivor (**Fig. 1.19A**). While traveling in Turkey, Lady Montagu learned that many elderly women there had perfected the art of inoculation: "The old woman comes with a nut-shell full of the matter of the best sort of small-pox, and asks what vein you please to have opened." During a period outside the host, the virus becomes "attenuated"—that is, loses some of its molecular structure required for infection. The attenuated virus stimulates the immune system with much lower mortality than does the fully virulent virus. Lady Montagu arranged for the procedure on her own son and then brought the practice back to England. A similar practice of smallpox inoculation was introduced to the American colonies by a slave, Onesimus, from the Coromantee people of Africa. Onesimus convinced his master, Reverend Dr. Cotton Mather, to promote smallpox inoculation as a defense against an epidemic that was devastating in Boston.

Preventive inoculation with smallpox was dangerous, however, because some infected individuals still contracted serious disease and were contagious. Thus, doctors continued to seek a better method of prevention. In England, milkmaids claimed that they were protected from smallpox

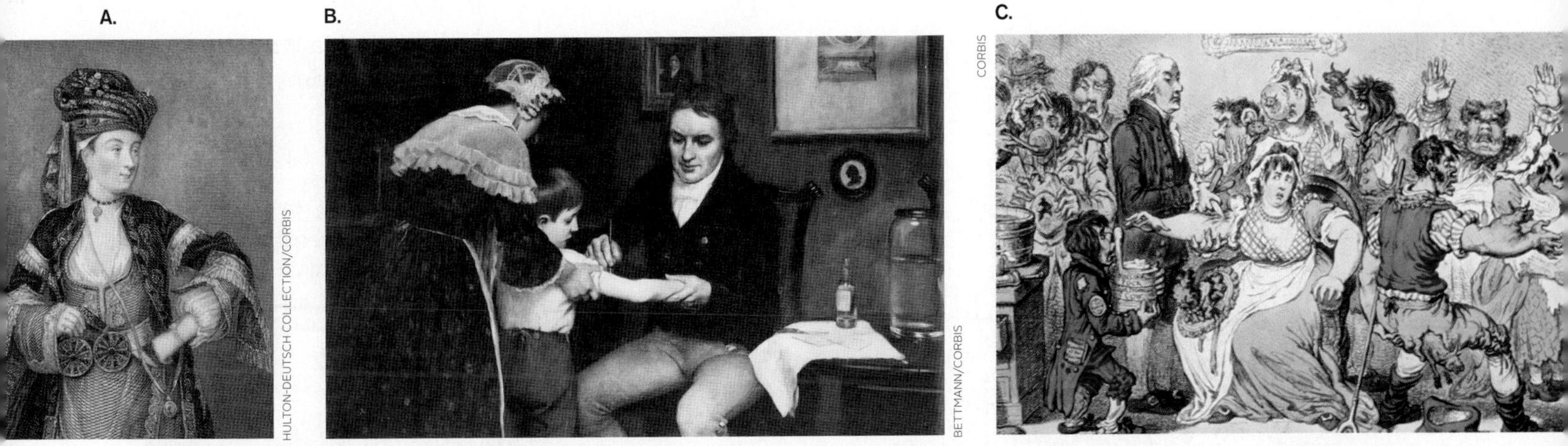
A.
B.
C.
HULTON-DEUTSCH COLLECTION/CORBIS
BETTMANN/CORBIS
CORBIS

FIGURE 1.19 ■ Smallpox vaccination. A. Lady Mary Wortley Montagu, shown in Turkish dress. The artist avoided showing Montagu's facial disfigurement from smallpox. **B.** Dr. Edward Jenner, depicted vaccinating 8-year-old James Phipps with cowpox matter from the hand of milkmaid Sarah Nelmes, who had caught the disease from a cow. **C.** Newspaper cartoon depicting public reaction to cowpox vaccination.

after they contracted cowpox, a related but much milder disease. This claim was confirmed by English physician Edward Jenner (1749–1823), who deliberately infected patients with matter from cowpox lesions (**Fig. 1.19B** and **C**). The practice of cowpox inoculation was called **vaccination**, after the Latin word *vacca,* meaning "cow." To this day, cowpox, or vaccinia virus, remains the basis of the modern smallpox vaccine.

Pasteur was aware of vaccination as he studied the course of various diseases in experimental animals. In the spring of 1879, he was studying fowl cholera, a transmissible disease of chickens with a high death rate. He had isolated and cultured the bacteria that had killed the chickens, but he left his work during the summer for a long vacation. No refrigeration was available to preserve cultures, and when he returned to work, the aged bacteria failed to cause disease in his chickens. Pasteur then obtained fresh bacteria from an outbreak of disease elsewhere, as well as some new chickens. But the fresh bacteria failed to make the original chickens sick (those that had been exposed to the aged bacteria). All of the new chickens, exposed only to the fresh bacteria, contracted the disease. Grasping the clue from his mistake, Pasteur had the insight to recognize that an attenuated (or "weakened") strain of microbe, altered somehow to eliminate its potency to cause disease, could still confer immunity to the virulent disease-causing form.

Pasteur was the first to recognize the significance of attenuation and extend the principle to other pathogens. We now know that the molecular components of pathogens generate **immunity**, the resistance to a specific disease, by stimulating the **immune system**, an organism's exceedingly complex cellular mechanisms of defense (see Chapters 23 and 24). Understanding the immune system awaited the techniques of molecular biology a century later, but nineteenth-century physicians developed several effective examples of **immunization**, the stimulation of an immune response by deliberate inoculation with an attenuated pathogen.

The way to attenuate a strain varies greatly among pathogens. Heat treatment or aging for various periods often turned out to be the most effective approach. The original success of prophylactic smallpox inoculation was due to natural attenuation of the virus during the time between acquisition of smallpox matter from a diseased individual and inoculation of the healthy patient. A far more elaborate treatment was required for the most famous disease for which Pasteur devised a vaccine: rabies.

The rabid dog loomed large in folklore, and rabies was dreaded for its particularly horrible and inevitable course of death. Pasteur's vaccine for rabies required a highly complex series of heat treatments and repeated inoculations. Its success led to his instant fame (**Fig. 1.20**). Grateful survivors of rabies founded the Pasteur Institute for medical research, one of the world's greatest medical research institutions, whose scientists in the twentieth century discovered the virus HIV, which causes AIDS.

Antiseptics and Antibiotics

Before the work of Koch and Pasteur, many patients died of infections transmitted unwittingly by their own doctors. In 1847, Hungarian physician Ignaz Semmelweis (1818–1865) noticed that the death rate of women in childbirth due to puerperal fever was much higher in his own hospital than in a birthing center run by midwives. He guessed that the doctors in his hospital were transmitting pathogens from cadavers that they had dissected. So he ordered

LE BON PASTEUR
Laissez venir à moi les petits enragés.

FIGURE 1.20 ■ Pasteur cures rabies. This cartoon in a French newspaper depicts Pasteur protecting children from rabid dogs.

A.

B.
Penicillium mold

FIGURE 1.21 ■ Alexander Fleming, discoverer of penicillin. **A.** Alexander Fleming in his laboratory. **B.** Fleming's original plate of bacteria with *Penicillium* mold inhibiting the growth of bacterial colonies.

the doctors to wash their hands in chlorine, an **antiseptic** agent (a chemical that kills microbes). The mortality rate fell; but this revelation displeased other doctors, who refused to accept Semmelweis's findings.

In 1865, the British surgeon Joseph Lister (1827–1912) noted that half his amputee patients died of sepsis. Lister knew from Pasteur that microbial contamination might be the cause. So he began experiments to develop the use of antiseptic agents, most successfully carbolic acid, to treat wounds and surgical instruments. After initial resistance, Lister's work, with the support of Pasteur and Koch, drew widespread recognition. In the twentieth century, surgeons developed fully **aseptic** environments for surgery—that is, environments completely free of microbes.

Antibiotics. The problem with most antiseptic chemicals that killed microbes was that if taken internally, they would also kill the patients. Researchers sought a "magic bullet," an **antibiotic** molecule that killed only microbes, leaving their host unharmed.

An important step in the search for antibiotics was the realization that microbes themselves produce antibiotic compounds. This conclusion followed from the famous accidental discovery of penicillin by the English medical researcher Alexander Fleming (1881–1955) (**Fig. 1.21A**). In 1929, Fleming was culturing *Staphylococcus,* which infects wounds. He found that one of his plates of *Staphylococcus* was contaminated with a mold, *Penicillium notatum,* which he noticed was surrounded by a clear region, free of *Staphylococcus* colonies (**Fig. 1.21B**). Following up on this observation, Fleming showed that the mold produced a substance that killed bacteria. We now know this substance as penicillin.

In 1941, British biochemists Howard Florey (1898–1968) and Ernst Chain (1906–1979) purified the antibiotic molecule, which we now know inhibits formation of the bacterial cell wall. Penicillin saved the lives of many Allied troops during World War II, the first war in which an antibiotic became available to soldiers.

The second half of the twentieth century saw the discovery of many new and powerful antibiotics. Most of the new antibiotics, however, were made by little-known bacteria and fungi from endangered ecosystems—a circumstance that focused attention on wilderness preservation. Furthermore, the widespread and often indiscriminate use of antibiotics selects for pathogens to evolve resistance to antibiotics. As a result, antibiotics have lost their effectiveness against certain strains of major pathogens. For example, multidrug-resistant *Mycobacterium tuberculosis* and methicillin-resistant *Staphylococcus aureus* (MRSA) are now serious threats to public health. To combat evolving drug resistance, we continually need to research and develop new antibiotics. Microbial biosynthesis of antibiotics is discussed in Chapter 15, and the medical use of antibiotics is discussed in Chapter 27.

Thought Questions

1.5 Why do you think some pathogens generate immunity readily, whereas others evade the immune system?

1.6 How do you think microbes protect themselves from the antibiotics they produce?

The Discovery of Viruses

Viruses are much smaller than the host cells they infect; most are too small to be seen by a light microscope. So how were they discovered? In 1892, the Russian botanist Dmitri Ivanovsky (1864–1920) studied tobacco mosaic disease, a condition in which the leaves become mottled and the crop yield is decreased or destroyed altogether. Ivanovsky knew that some kind of microbe from the diseased plants transmitted the disease, and he wondered how small it was. He was surprised to find that the agent of transmission could pass through a porcelain filter having a pore size (0.1 μm) that blocked known microbes. Later, the Dutch plant microbiologist Martinus Beijerinck (1851–1931) conducted similar filtration experiments. Beijerinck concluded that because the agent of disease passed through a filter that retained bacteria, it could not be a bacterial cell.

The "filterable agent" of disease was ultimately purified by the American scientist Wendell Stanley (1904–1971), who processed 4,000 kilograms (kg) of infected tobacco leaves. Stanley obtained a sample of infective virus particles pure enough to crystallize, in a 3D array comparable to crystals composed of inert chemicals. The crystal was analyzed by X-ray crystallography (discussed in Chapter 2) to reveal the molecular structure of tobacco mosaic virus (**Fig. 1.22A**)—a feat that earned Stanley the 1946 Nobel Prize in Chemistry. The fact that an object capable of biological reproduction could be stable enough to be crystallized amazed scientists, and ultimately led to a new, more mechanical view of living organisms.

The individual particle of tobacco mosaic virus consists of a helical tube of protein subunits containing its genetic material coiled within (**Fig. 1.22B**). Stanley thought the virus was a catalytic protein, but colleagues later determined that it contained RNA as its genetic material. The structure of the coiled RNA was solved through X-ray crystallography by the British scientist Rosalind Franklin (1920–1958). Other viruses that have RNA genomes include influenza and HIV (AIDS); viruses with DNA genomes include polio and herpes viruses. We now know that all kinds of animals, plants, and microbial cells can be infected by viruses, as well as by even smaller infective particles called viroids and prions. Viral function and disease are discussed in Chapters 6, 11, and 26.

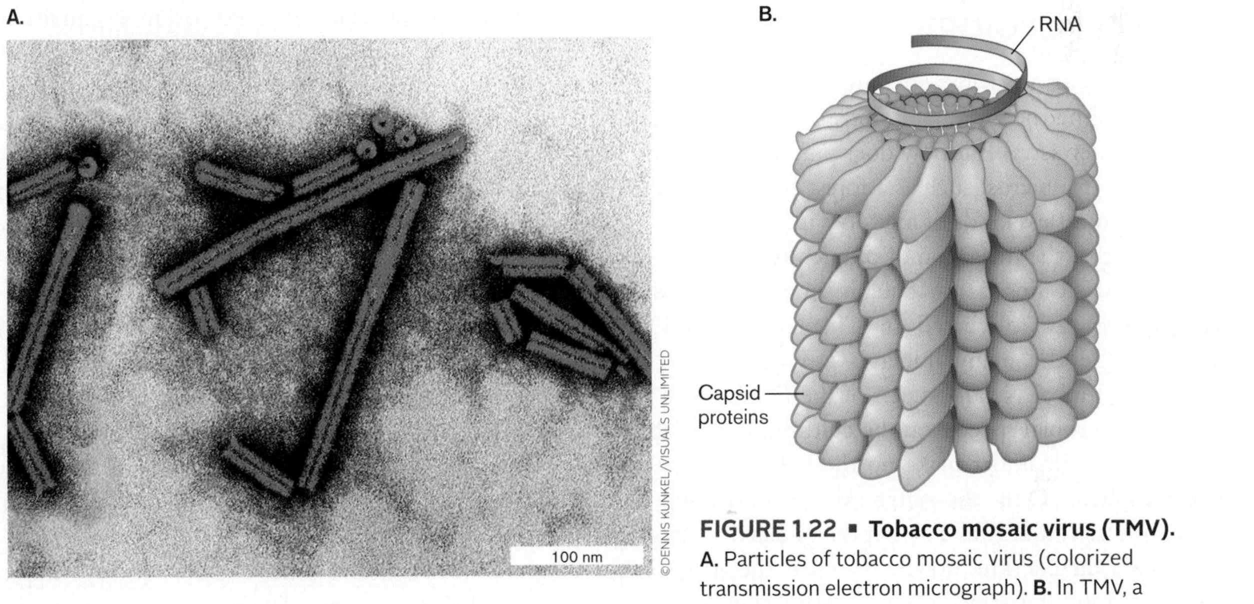
A.
100 nm
©DENNIS KUNKEL/VISUALS UNLIMITED
B.
RNA
Capsid proteins

FIGURE 1.22 ▪ Tobacco mosaic virus (TMV). **A.** Particles of tobacco mosaic virus (colorized transmission electron micrograph). **B.** In TMV, a protein capsid surrounds an RNA chromosome.

To Summarize

- **Robert Koch** devised techniques of pure culture to study a single species of microbe in isolation. A key technique is culture on solid medium using agar, as developed by Angelina and Walther Hesse, in a double-dish container devised by Julius Petri.
- **Koch's postulates** provide a set of criteria to establish a causative link between an infectious agent and a disease.
- **Edward Jenner** established the practice of vaccination, inoculation with cowpox to prevent smallpox. Jenner's discovery was based on earlier observations by Lady Mary Montagu and others that a mild case of smallpox could prevent future cases.
- **Louis Pasteur** developed the first vaccines based on attenuated strains, such as the rabies vaccine.
- **Ignaz Semmelweis** and **Joseph Lister** showed that antiseptics could prevent transmission of pathogens from doctor to patient.
- **Alexander Fleming** discovered that the *Penicillium* mold generated a substance that kills bacteria.
- **Howard Florey** and **Ernst Chain** purified the substance penicillin, the first commercial antibiotic to save human lives.
- **Dmitri Ivanovsky** and **Martinus Beijerinck** discovered viruses as filterable infective particles.

1.4

Microbial Ecology

Koch's growth of microbes in pure culture was a major advance, enabling the systematic study of microbial physiology and biochemistry. But how does pure culture relate to "natural" environments, such as a forest or a human intestine, where countless kinds of microbes interact?

In hindsight, the invention of pure culture in the nineteenth century eclipsed the equally important study of microbial ecology (discussed in Chapters 21 and 22). Microbes cycle the many minerals essential for all life, including all global N_2 and much of the O_2. Yet less than 0.1% of all microbial species can be cultured in the laboratory—and the remainder make up the majority of Earth's entire biosphere. Only the outer skin of Earth supports complex multicellular life. The depths of Earth's crust, to at least 3 kilometers (km) down, as well as the atmosphere 15 km out into the stratosphere, remain the domain of microbes. So, to a first approximation, Earth's ecology is microbial ecology.

Microbes Support Natural Ecosystems

The first microbiologists to culture microbes in the laboratory selected the kinds of nutrients that feed humans, such as beef broth or potatoes. Some of Koch's contemporaries, however, suspected that other kinds of microbes living in soil or wetlands existed on more exotic fare. Soil samples were known to oxidize hydrogen gas, and this activity was eliminated by treatment with heat or acid, suggesting microbial origin. Ammonia in sewage was oxidized by donating electrons to oxygen, forming nitrate. Nitrate formation was eliminated by antibacterial treatment. These findings suggested the existence of microbes that "ate" hydrogen gas or ammonia instead of beef or potatoes, but no one could isolate these microbes in culture.

Among the first to study microbes in natural habitats was the Russian scientist Sergei Winogradsky (1856–1953). Winogradsky waded through marshes to discover microbes with metabolisms quite alien from human digestion. For example, he discovered that species of the bacterium *Beggiatoa* oxidize hydrogen sulfide (H_2S) to sulfuric acid (H_2SO_4). *Beggiatoa* fixes carbon dioxide into biomass without consuming any organic food. Organisms that feed solely on inorganic minerals are known as **chemolithotrophs**, or **lithotrophs** (discussed further in Chapters 4 and 14).

The lithotrophs studied by Winogradsky could not be grown on Koch's plate media containing agar or gelatin. The bacteria that Winogradsky isolated could grow only on inorganic minerals; in fact, some species are actually poisoned by organic food. For example, nitrifiers convert ammonia to nitrate, forming a crucial part of the nitrogen cycle in natural ecosystems. Winogradsky cultured nitrifiers on a totally inorganic solution containing ammonia and silica gel, which supported no other kind of organism. This experiment was an early example of **enrichment culture**, the use of selective growth media that support certain classes of microbial metabolism while excluding others.

Instead of isolating pure colonies, Winogradsky built a model wetland ecosystem containing regions of enrichment for microbes of diverse metabolism. This model is called the **Winogradsky column** (**Fig. 1.23**). The model consists of a glass tube containing mud (a source of wetland bacteria) mixed with shredded newsprint (an organic carbon source) and calcium salts of sulfate and carbonate (an inorganic carbon source for autotrophs). After exposure to light for several weeks, several zones of color develop, full of mineral-metabolizing bacteria. At the top, cyanobacteria conduct

photosynthesis, using light energy to split water and produce molecular oxygen. Below, purple sulfur bacteria use photosynthesis to split hydrogen sulfide, producing sulfur. At the bottom, with O_2 exhausted, bacteria reduce (donate electrons to) alternative electron acceptors such as sulfate. Sulfate-reducing bacteria produce hydrogen sulfide and precipitate iron.

Like a battery cell, the gradient from oxygen-rich conditions at the surface to highly reduced conditions below generates a voltage potential. We now know that the entire Earth's surface acts as a battery—for humans, a potential source of renewable energy. A fuel cell to generate electricity is presented in Chapter 14.

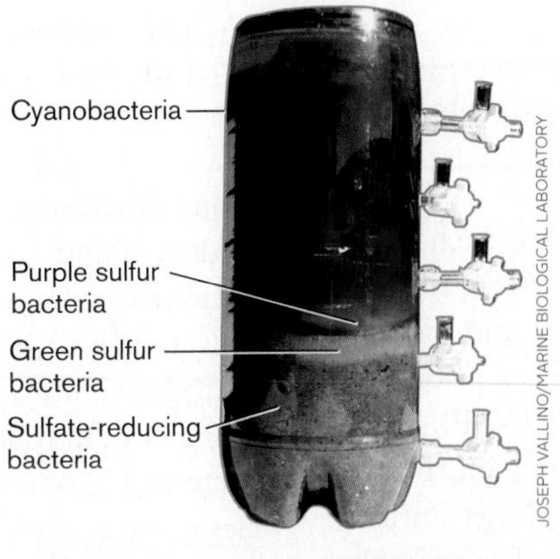
Cyanobacteria
Purple sulfur bacteria
Green sulfur bacteria
Sulfate-reducing bacteria
JOSEPH VALLINO/MARINE BIOLOGICAL LABORATORY

FIGURE 1.23 ▪ Winogradsky column. A wetland model ecosystem designed by Sergei Winogradsky.

Winogradsky and later microbial ecologists showed that bacteria perform unique roles in **geochemical cycling**, the global interconversion of inorganic and organic forms of nitrogen, sulfur, phosphorus, and other minerals. Without these essential conversions (nutrient cycles), no plants or animals could live. Bacteria and archaea fix nitrogen (N_2) by reducing it to ammonia (NH_3), the form of nitrogen assimilated by plants. This process is called **nitrogen fixation** (**Fig. 1.24**). Other bacterial species oxidize ammonium ions (NH_4^+) in several stages back to nitrogen gas. Nitrogen fixation and geochemical cycling are discussed further in Chapters 21 and 22.

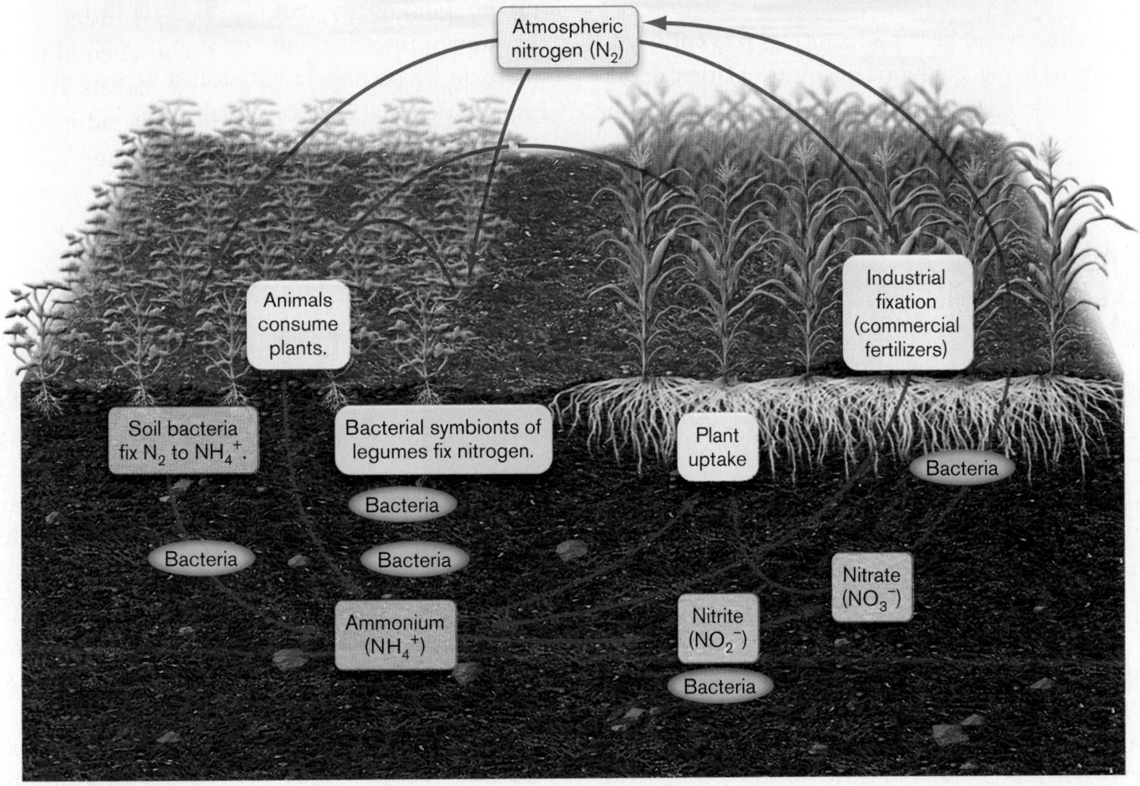
Atmospheric nitrogen (N_2)
Animals consume plants.
Industrial fixation (commercial fertilizers)
Soil bacteria fix N_2 to NH_4^+.
Bacterial symbionts of legumes fix nitrogen.
Plant uptake
Bacteria
Bacteria
Bacteria
Bacteria
Nitrate (NO_3^-)
Ammonium (NH_4^+)
Nitrite (NO_2^-)
Bacteria

FIGURE 1.24 ▪ The global nitrogen cycle. All life depends on these oxidative and reductive conversions of nitrogen—most of which are performed only by microbes.

Thought Question

1.7 Why don't all living organisms fix their own nitrogen?

Today, microbes with unusual properties, such as the ability to digest toxic wastes or withstand extreme temperatures, have valuable applications in industry and bioremediation. For this reason, microbial ecology is a priority for funding by the National Science Foundation (NSF). Microbial ecologist Rita Colwell (**Fig. 1.25A**) directed the NSF from 1998–2004, and founded the Biocomplexity initiative to study complex interactions between microbes and other life in the environment. Such research includes discovery of **extremophiles**, microbes from environments with extreme heat, salinity, acidity, or other factors. One such extremophile microbe is the archaeon *Geogemma*, which reduces rust (iron oxide, Fe_2O_3) to the magnetic mineral magnetite (Fe_3O_4) while growing in an autoclave at 121°C, a temperature high enough to kill all other known organisms (**Fig. 1.25B**). *Geogemma* was discovered by Kazem Kashefi (**Fig. 1.25C**), now at Michigan State University.

Bacterial Endosymbiosis with Plants and Animals

Within plant cells, certain bacteria fix nitrogen as **endosymbionts**, organisms living symbiotically inside a larger organism. Endosymbiotic bacteria known as rhizobia induce the roots of legumes to form special nodules to facilitate bacterial nitrogen fixation. Rhizobial endosymbiosis was first observed by Martinus Beijerinck. Microbial endosymbiosis, in a variety of diverse forms, is widespread in all ecosystems. Many interesting cases involve animal or human hosts. Endosymbiotic microbes make essential nutritional contributions to host animals. Ruminant animals such as cattle, as well as insects such as termites, require digestive bacteria to break down cellulose and other plant polymers. Even humans obtain about 15% of their nutrition from colonic bacteria. Colonic bacteria such as *E. coli* and *Bacteroides* species grow as **biofilms**, organized multispecies communities on a surface. The biofilm shown in **Figure 1.26** is attached to the surface of a digested food particle. Biofilms play major roles in all ecosystems and within parts of the human body (discussed in Chapters 21 and 23).

Many invertebrates, such as hydras and corals, harbor endosymbiotic phototrophs that provide products of photosynthesis in return for protection and nutrients. Other kinds of endosymbiosis involve more than nutrition. In the light organs of squid, luminescent bacteria, such as *Vibrio fischeri,* produce light (bioluminescence) that helps the host evade nocturnal predators. The host squid controls the amount and direction of light that leaves the organ so that it matches the illumination from the moon, rendering the squid nearly invisible. Bacteria that normally inhabit the human intestine and the skin protect our bodies from infection by pathogens. Microbial endosymbiosis has even inspired fictional creations, such as the "breath microbes" of

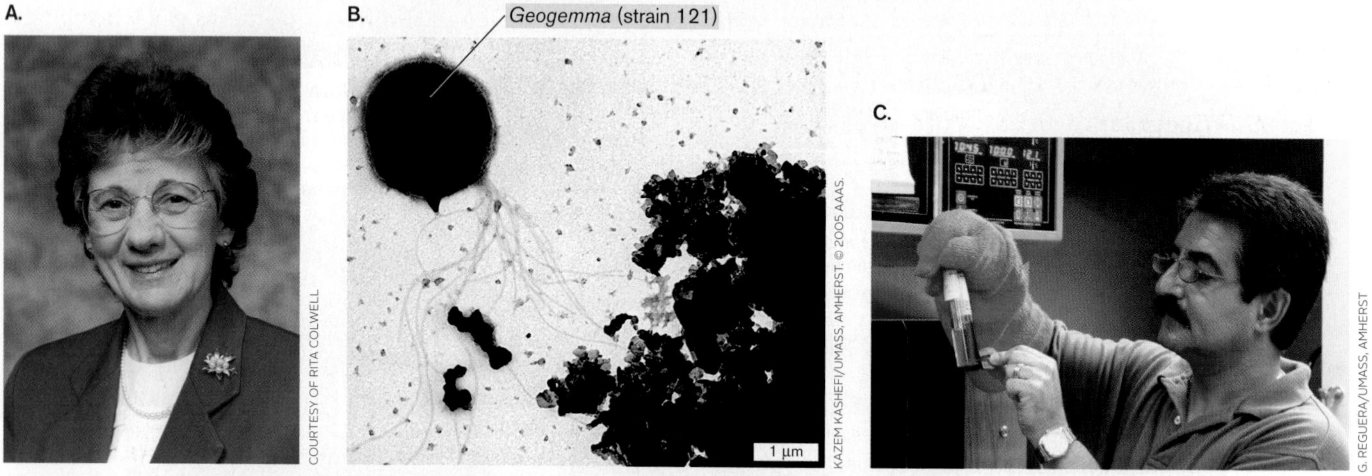
A.
B.
Geogemma (strain 121)
1 μm
C.
COURTESY OF RITA COLWELL
KAZEM KASHEFI/UMASS, AMHERST. © 2005 AAAS.
G. REGUERA/UMASS, AMHERST

FIGURE 1.25 ▪ An extreme thermophile reduces iron oxide to magnetite. **A.** Rita Colwell, director of the National Science Foundation from 1998–2004, promoted the study of environmental microbiology. See **eTopic 1.1** for an interview with Colwell. **B.** *Geogemma* is a round bacterium with a tuft of rotary flagella (transmission electron micrograph). The black material at right is iron oxide. **C.** Kazem Kashefi, now at Michigan State University, pulls a live culture of "strain 121" (*Geogemma*) out of an autoclave generally used to kill all living organisms at 121°C (250°F). The magnet shows that the microbe is converting nonmagnetic iron oxide (rust, Fe_2O_3) to the magnetic mineral magnetite (Fe_3O_4). *Source:* Part B reprinted with permission from Kashefi et al. 2003. *Science* 301(5635):934. © 2005 AAAS.

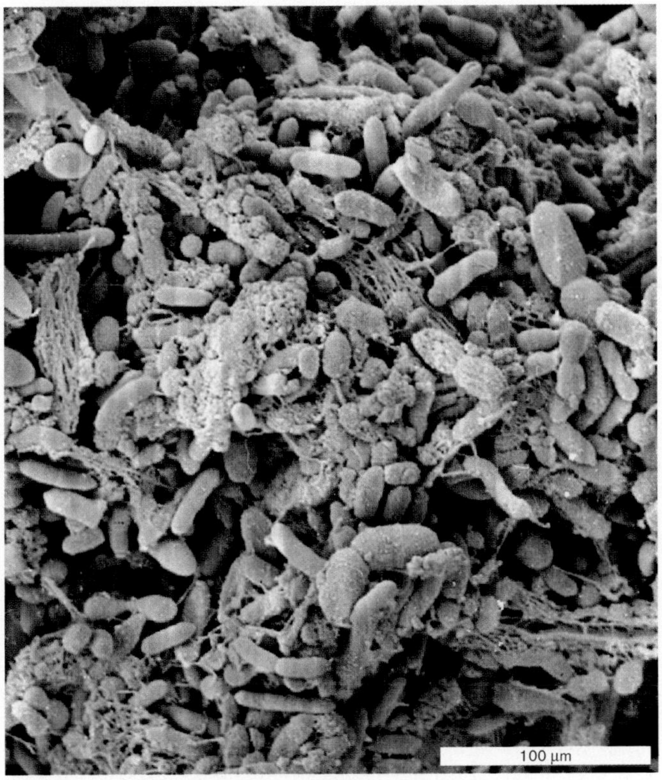
100 μm

FIGURE 1.26 ■ Intestinal biofilm. Bacteria growing on the surface of a residual food particle aid human digestion.
Source: Sandra Macfarlane and George T. Macfarlane. 2006. *Appl. Environ. Microbiol.* 72:6204.

Joan Slonczewski's novel *A Door into Ocean,* which acquire oxygen for human hosts swimming underwater. Microbial endosymbiosis is discussed further in Chapter 21.

To Summarize

- **Sergei Winogradsky** developed the first system of enrichment culture, called the Winogradsky column, to grow microbes from natural environments.
- **Chemolithotrophs (or lithotrophs)** metabolize inorganic minerals, such as ammonia, instead of the organic nutrients used by the microbes isolated by Koch.
- **Geochemical cycling** depends on bacteria and archaea that cycle nitrogen, phosphorus, and other minerals throughout the biosphere.
- **Endosymbionts** are microbes that live within a host organism and provide essential functions for their hosts, such as nitrogen fixation for legume plants.
- **Martinus Beijerinck** first demonstrated that nitrogen-fixing rhizobia grow as endosymbionts within leguminous plants.

1.5

The Microbial Family Tree

The bewildering diversity of microbial life-forms presented nineteenth-century microbiologists with a seemingly impossible task of classification. So little was known about life under the lens that natural scientists despaired of ever learning how to distinguish microbial species. The famous classifier of species, Swedish botanist Carl von Linné (Carolus Linnaeus, 1707–1778), called the microbial world "chaos."

Microbes Are a Challenge to Classify

Early taxonomists faced two challenges as they attempted to classify microbes. First, the resolution of the light microscope visualized little more than the outward shape of microbial cells, and vastly different kinds of microbes looked more or less alike (discussed in Chapter 2). This challenge was overcome as advances in biochemistry and microscopy made it possible to distinguish microbes by metabolism and cell structure, and ultimately by DNA sequence.

Second, microbes do not readily fit the classic definition of a species—that is, a group of organisms that interbreed. Unlike multicellular eukaryotes, microbes generally reproduce asexually. When they do exchange genes, they may do so with related strains or with distantly related species (discussed in Chapter 9). Nevertheless, microbiologists have devised working definitions of microbial species that enable us to usefully describe populations while being flexible enough to accommodate continual revision and change (discussed in Chapter 17). The most useful definitions are based on genetic similarity. For example, two distinct species generally share no more than 95% similarity of DNA sequence.

Note: The names of microbial species are occasionally changed to reflect new understanding of genetic relationships. For example, the causative agent of bubonic plague was formerly called *Bacterium pestis* (1896), *Bacillus pestis* (1900), and *Pasteurella pestis* (1923), but it is now called *Yersinia pestis* (1944). The older names, however, still appear in the literature—a point to remember when carrying out research. Names of bacteria and archaea are compiled at the List of Prokaryotic Names with Standing in Nomenclature (LPSN).

Weblinks List of Prokaryotic Names with Standing in Nomenclature (*see ebook*)

Microbes Include Eukaryotes and Prokaryotes

In the nineteenth century, taxonomists had no DNA information. As they tried to incorporate microbes into the tree of life, they faced a conceptual dilemma because microbes could not be categorized as either animals or plants, which since ancient times had been considered the two "kingdoms" or major categories of life. Taxonomists attempted to apply these categories to microbes—for example, by including algae and fungi with plants. But German naturalist Ernst Haeckel (1834–1919) recognized that microbes differed from both plants and animals in fundamental aspects of their lifestyle, cellular structure, and biochemistry. Haeckel proposed that microscopic organisms constituted a third kind of life—neither animal nor plant—which he called Monera.

In the twentieth century, biochemical studies revealed profound distinctions even within the Monera. In particular, microbes such as protists and algae contain a nucleus enclosed by a nuclear membrane, whereas bacteria do not. Herbert Copeland (1902–1968) proposed a system of classification that divided Monera into two groups: the eukaryotic protists (protozoa and algae) and the prokaryotic bacteria. Copeland's four-kingdom classification (plants, animals, eukaryotic protists, and prokaryotic bacteria) was later modified by Robert Whittaker (1920–1980) to include fungi as another kingdom of eukaryotic microbes. Whittaker's system thus generated five kingdoms: bacteria, protists, fungi, and the multicellular plants and animals.

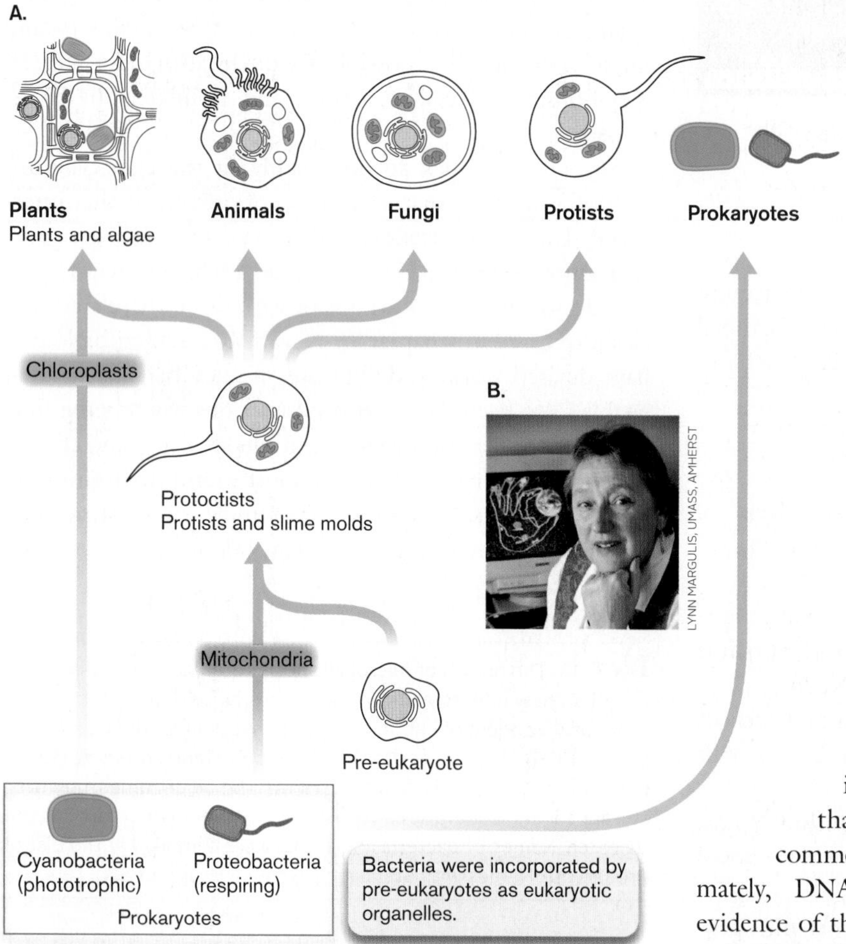
A.
Plants
Plants and algae
Animals
Fungi
Protists
Prokaryotes
Chloroplasts
Protoctists
Protists and slime molds
Mitochondria
Pre-eukaryote
Cyanobacteria
(phototrophic)
Proteobacteria
(respiring)
Prokaryotes
Bacteria were incorporated by pre-eukaryotes as eukaryotic organelles.
B.
LYNN MARGULIS, UMASS, AMHERST

FIGURE 1.27 ■ Lynn Margulis and the serial endosymbiosis theory. A. Five-kingdom scheme, modified by the endosymbiosis theory. **B.** Lynn Margulis (University of Massachusetts, Amherst) proposed that organelles evolve through endosymbiosis.

Eukaryotes Evolved through Endosymbiosis

The five-kingdom system was modified dramatically by Lynn Margulis (1938–2011), at the University of Massachusetts (**Fig. 1.27**). Margulis tried to explain how it is that eukaryotic cells contain mitochondria and chloroplasts, membranous organelles that possess their own chromosomes. She proposed that eukaryotes evolved by merging with bacteria to form composite cells by intracellular endosymbiosis, in which one cell internalizes another that grows within it. The endosymbiosis may ultimately generate a single organism whose formerly independent members are now incapable of independent existence.

Margulis proposed that early in the history of life, respiring bacteria similar to *E. coli* were engulfed by pre-eukaryotic cells, where they evolved into mitochondria, the eukaryote's respiratory organelles. Similarly, she proposed that a phototroph related to cyanobacteria was taken up by a eukaryote, giving rise to the chloroplasts of phototrophic algae and plants.

The endosymbiosis theory was highly controversial because it implied a **polyphyletic**, or multiple, ancestry of living species, inconsistent with the long-held assumption that species evolve only by divergence from a common ancestor (**monophyletic** ancestry). Ultimately, DNA sequence analysis produced compelling evidence of the bacterial origin of mitochondria and chloroplasts. Both of these classes of organelles contain circular molecules of DNA, whose sequences show unmistakable homology (similarity) to those of bacteria. DNA sequences and other evidence established the common ancestry between mitochondria and respiring bacteria, and between chloroplasts and cyanobacteria. The symbiotic origins and evolution of mitochondria and chloroplasts are discussed further in Chapter 17.

Archaea Differ from Bacteria and Eukaryotes

Are there cellular microbes that differ from both bacteria and eukaryotes? Gene sequence analysis led to another startling advance in our understanding of how cells evolved. In 1977, Carl Woese, at the University of Illinois, was studying a group of recently discovered prokaryotes that live in seemingly hostile environments, such as the boiling sulfur springs of Yellowstone, or that exhibit unusual kinds of metabolism, such as production of methane (methanogenesis). Woese used the sequence of the gene for 16S ribosomal RNA (16S rRNA) as a "molecular clock," a gene whose sequence differences can be used to measure the time since the divergence of two species (discussed in Chapter 17). The divergence of rRNA genes showed that the newly discovered prokaryotes were a distinct form of life: **archaea** (**Fig. 1.28**). The archaea resemble bacteria in their relatively simple cell structure, in their lack of a nucleus, and in their ability to grow in a wide range of environments. But some archaea, such as the autoclave-cultured archaeon *Geogemma,* grow in environments more extreme than any that support bacteria. The genetic sequences of archaea differ as much from those of bacteria as from those of eukaryotes; in fact, their gene expression machinery is more similar to that of eukaryotes.

Woese's discovery replaced the classification scheme of five kingdoms with three equally distinct groups, now called the three domains: Bacteria, Archaea, and Eukarya (**Fig. 1.29**). In the three-domain model, the bacterial ancestor of mitochondria derives from ancient proteobacteria (shaded pink in the figure), whereas chloroplasts derive from ancient cyanobacteria (shaded green). The three-domain classification is largely supported by the sequences of microbial genomes, although some horizontal transfer of genes occurs both within and between the domains (discussed in Chapter 17).

Thought Question

1.8 What arguments support the classification of Archaea as a third domain of life? What arguments support the classification of archaea and bacteria together, as prokaryotes, distinct from eukaryotes?

A.

JAMES W. BROWN, NORTH CAROLINA STATE UNIVERSITY

B.

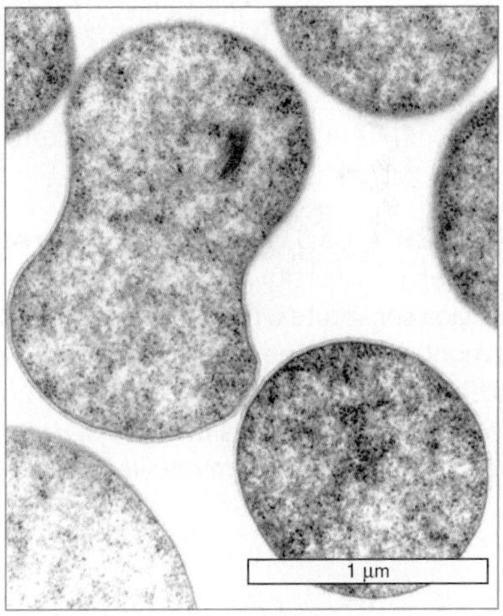
1 μm

HENRY ALDRICH, UNIVERSITY OF FLORIDA

FIGURE 1.28 ■ Archaea, newly discovered life-forms. A. Finding archaea in the hot-spring Obsidian Pool at Yellowstone. **B.** *Pyrococcus furiosus,* an organism that lives at temperatures above 100°C (transmission electron micrograph).

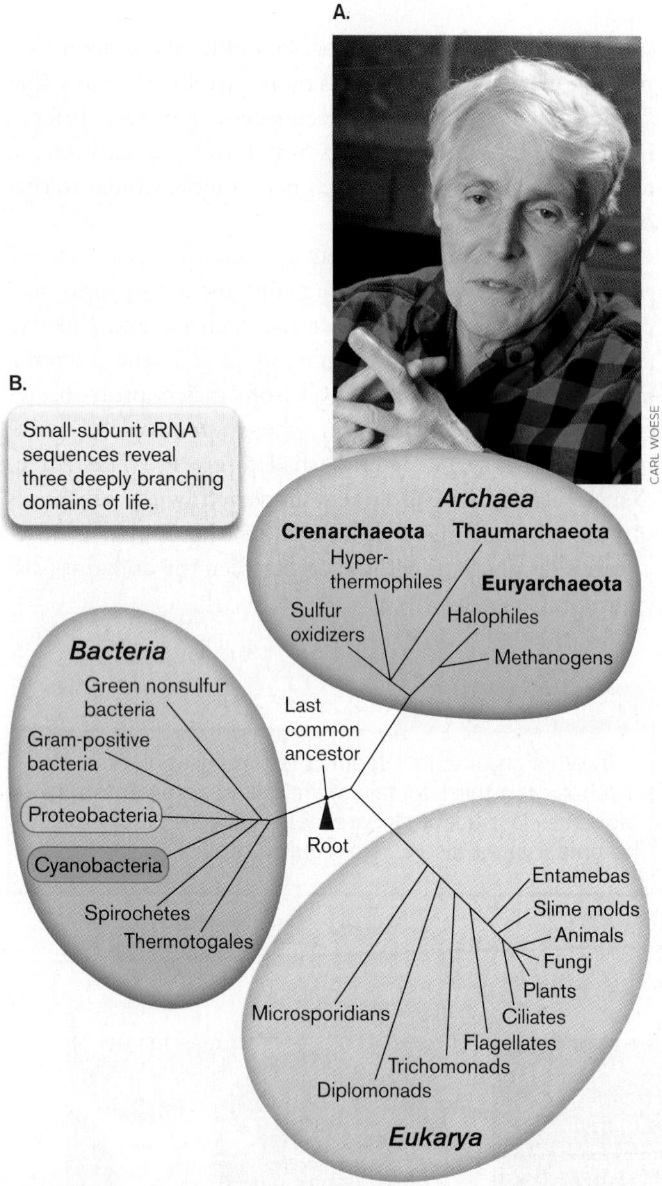
A.
CARL WOESE
B.
Small-subunit rRNA sequences reveal three deeply branching domains of life.
Archaea
Crenarchaeota
Thaumarchaeota
Hyper-thermophiles
Euryarchaeota
Sulfur oxidizers
Halophiles
Methanogens
Bacteria
Green nonsulfur bacteria
Gram-positive bacteria
Proteobacteria
Cyanobacteria
Spirochetes
Thermotogales
Last common ancestor
Root
Entamebas
Slime molds
Animals
Fungi
Plants
Ciliates
Microsporidians
Flagellates
Trichomonads
Diplomonads
Eukarya

FIGURE 1.29 ▪ Carl Woese and the three domains of life. **A.** Carl Woese (University of Illinois at Urbana-Champaign) proposed that archaea constitute a third domain of life. **B.** Three domains form a monophyletic tree based on small-subunit rRNA sequences. Examples of branching groups of organisms are shown. The length of each branch approximates the time of divergence from the last common ancestor. For a more detailed tree, see Chapter 17.

To Summarize

- **Classifying microbes** was a challenge historically because of the difficulties in observing distinguishing characteristics of different categories.
- **Ernst Haeckel** recognized that microbes constitute a form of life distinct from animals and plants.
- **Herbert Copeland** and **Robert Whittaker** classified prokaryotes as a form of microbial life distinct from eukaryotic microbes such as protists.
- **Lynn Margulis** proposed that eukaryotic organelles such as mitochondria and chloroplasts evolved by endosymbiosis from prokaryotic cells engulfed by pre-eukaryotes.
- **Carl Woese** discovered a domain of prokaryotes, Archaea, whose genetic sequences diverge equally from those of bacteria and those of eukaryotes. Archaea grow in a wide range of environments; some species grow under conditions that exclude bacteria and eukaryotes.

1.6

Cell Biology and the DNA Revolution

During the twentieth century, amid world wars and societal transformations, the field of microbiology exploded with new knowledge (see **Table 1.2**). More than 99% of what we know about microbes today was discovered after 1900 by scientists too numerous to cite in this book. Advances in biochemistry and microscopy revealed the fundamental structure and function of cell membranes and proteins. The revelation of the structures of DNA and RNA led to the discovery of the genetic programs of model bacteria, such as *E. coli*, and the bacteriophage lambda. Beyond microbiology, these advances produced the technology of "recombinant DNA," or genetic engineering, the construction of molecules that combine DNA sequences from unrelated species. These microbial tools offered unprecedented applications for human medicine and industry (discussed in Chapters 7–12).

Cell Membranes and Macromolecules

In 1900, the study of cell structure was still limited by the resolution of the light microscope and by the absence of tools that could take apart cells to isolate their components. Both of these limitations were overcome by the invention of powerful instruments. Just as society was being transformed by machines ranging from jet airplanes to vacuum cleaners, the study of microbiology was also being transformed by machines. Two instruments had exceptional impact: The **electron microscope** revealed the internal structure of cells (Chapter 2), and the **ultracentrifuge** enabled isolation of subcellular parts (Chapter 3).

The electron microscope. In the 1920s, at the Technical College in Berlin, student Ernst Ruska (1906–1988) was invited to develop an instrument for focusing rays of electrons. Ruska recalled as a child how his father's microscope could magnify fascinating specimens of plants and animals, but that its resolution was limited by the wavelength of light. He was eager to devise lenses that could focus beams of electrons, with wavelengths far smaller than that of light, to reveal living details never seen before. Ultimately, Ruska built lenses to focus electrons using specially designed electromagnets. Magnetic lenses were used to complete the first electron microscope in 1933 (**Fig. 1.30**). Early transmission electron microscopes achieved about tenfold greater magnification than the light microscope, revealing details such as the ridged shell of a diatom. Further development steadily increased magnification, to as high as a millionfold.

For the first time, cells were seen to be composed of a cytoplasm containing macromolecules and bounded by a phospholipid membrane. For example, the electron micrograph in **Figure 1.31** shows a "thin section" of the photosynthetic bacterium *Chlorobium* sp., including its nucleoid (DNA) and its light-harvesting chlorosomes. Electron microscopy is discussed further in Chapter 2.

Subcellular structures, however, raised many questions about cell function that visualization alone could not answer. Biochemists showed that cell function involves numerous chemical transformations mediated by enzymes. A milestone in the study of metabolism was the elucidation by German biochemist Hans Krebs (1900–1981) of the tricarboxylic acid cycle (TCA cycle, or Krebs cycle), by which the products of sugar digestion are converted to carbon dioxide. The TCA cycle provides energy for many bacteria and for the mitochondria of eukaryotes. But even Krebs understood little of how metabolism is organized within a cell; he and his contemporaries considered the cell a "bag of enzymes." The full understanding of cell structure required experiments on isolated parts of cells.

The ultracentrifuge. Centrifugation can separate whole cells from the fluid in which they are suspended. The first centrifuges spun samples in a rotor with centrifugal force of a few thousand times that of gravity. In the nineteenth century, biochemists proposed that even greater centrifugal forces could separate components of lysed cells, even macromolecules such as proteins. The Swedish chemist Theodor

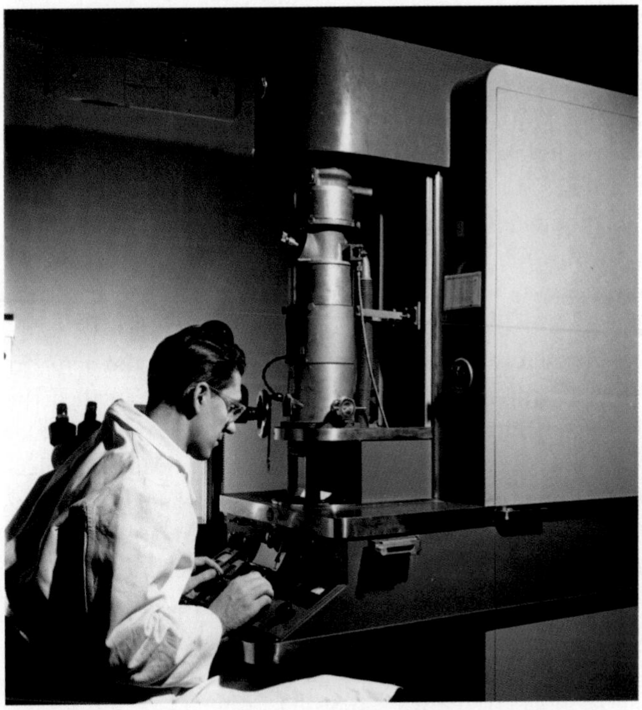

MARINE BIOLOGICAL LABORATORY/WHOI

FIGURE 1.30 ■ Electron microscopy. An early transmission electron microscope. The tall column contains a series of magnetic lenses.

NIELS-ULRIK FRIGAARD ET AL. 2002. *J. BACTERIOL.* **184**:3368.

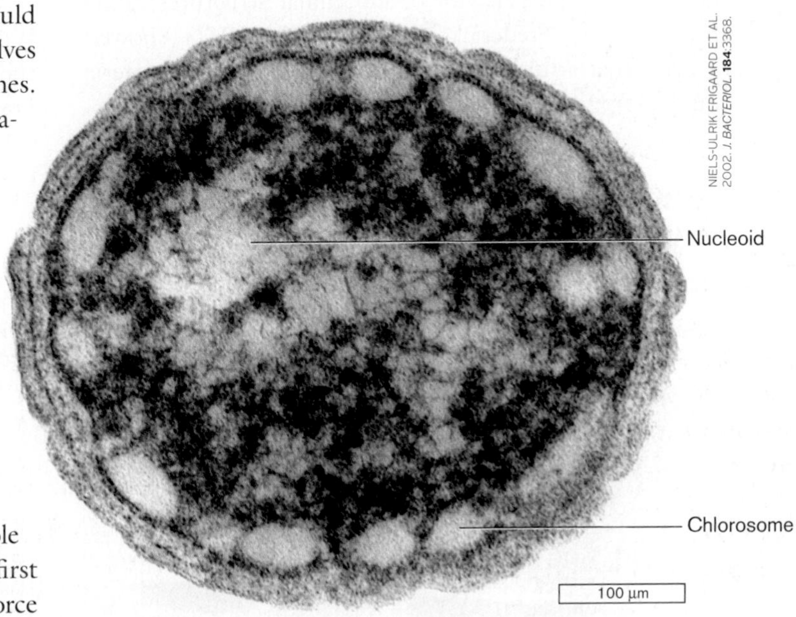
Nucleoid
Chlorosome
100 µm

FIGURE 1.31 ■ Electron micrograph of *Chlorobium* species, a photosynthetic bacterium. The thin section reveals the nucleoid (containing DNA), the light-harvesting chlorosomes, and envelope membranes.

Svedberg (1884–1971), at the University of Uppsala, built such a machine: the ultracentrifuge. By the twentieth century, ultracentrifuges achieved rates so high that they required a vacuum to avoid burning up like a space reentry vehicle. Ultracentrifuges isolated protein complexes such as ribosomes, and DNA molecules such as plasmids.

Experiments combining electron microscopy and ultracentrifugation revealed how membranes govern energy transduction within bacteria and within organelles such as mitochondria and chloroplasts. In the 1960s, English biochemists Peter Mitchell (1920–1992) and Jennifer Moyle proposed and tested a revolutionary idea called the **chemiosmotic theory**. The chemiosmotic theory states that the reduction-oxidation (redox) reactions of the **electron transport system** store energy in the form of a gradient of protons (hydrogen ions) across a membrane, such as the bacterial cell membrane or the inner membrane of mitochondria. The energy stored in the proton gradient, in turn, drives the synthesis of ATP (discussed in Chapter 14).

Microbial Genetics Leads the DNA Revolution

As the form and function of living cells emerged in the early twentieth century, a largely separate line of research revealed patterns of heredity of cell traits. In eukaryotes, the Mendelian rules of inheritance were rediscovered and connected to the behavior of subcellular structures called chromosomes. Frederick Griffith (1879–1941) showed in 1928 that an unknown substance out of dead bacteria could carry genetic information into living cells, transforming harmless bacteria into a strain capable of killing mice—a process called **transformation**. Some kind of "genetic material" must be inherited to direct the expression of inherited traits, but no one knew what that material was or how its information was expressed.

Then, in 1944, Oswald Avery (1877–1955) and colleagues showed that the genetic material for transformation is deoxyribonucleic acid, or DNA. An obscure acidic polymer, DNA had been previously thought too uniform in structure to carry information; its precise structure was unknown. As World War II raged among nations, scientists embarked on an epic struggle: the quest for the structure of DNA.

The discovery of the double helix. The tool of choice to discover the structure of molecules was X-ray crystallography, a method developed by British physicists in the early 1900s. The field of X-ray analysis included an unusual number of women, including Dorothy Hodgkin (1910–1994), who later won a Nobel Prize for the structures of penicillin and vitamin B_{12}. In 1953, crystallographer Rosalind Franklin joined a laboratory at King's College London to study the structure of DNA (**Fig. 1.32A**). As a woman and as a Jew who supported relief work in Palestine, Franklin felt socially isolated at the male-dominated Protestant university; her work was disparagingly called "witchcraft." Nevertheless, her exceptional X-ray micrographs (**Fig. 1.32B**) revealed for the first time that the standard B-form DNA was a double helix.

Without Franklin's knowledge, her colleague Maurice Wilkins showed her data to a competitor, James Watson at the University of Cambridge. The pattern led Watson and Francis Crick (1916–2004) to guess that the four bases of

A. B.

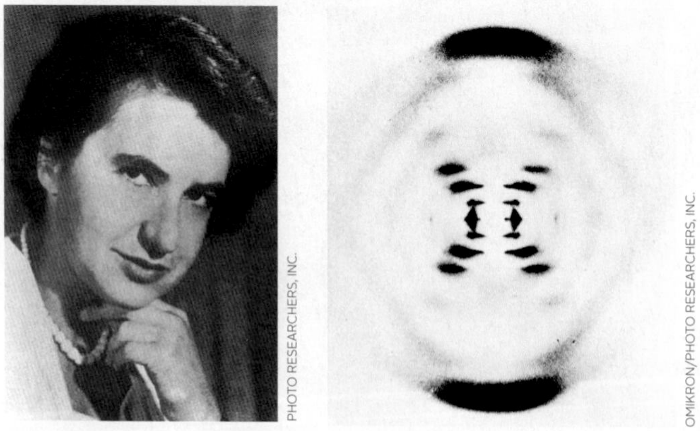
PHOTO RESEARCHERS, INC.
OMIKRON/PHOTO RESEARCHERS, INC.

C.

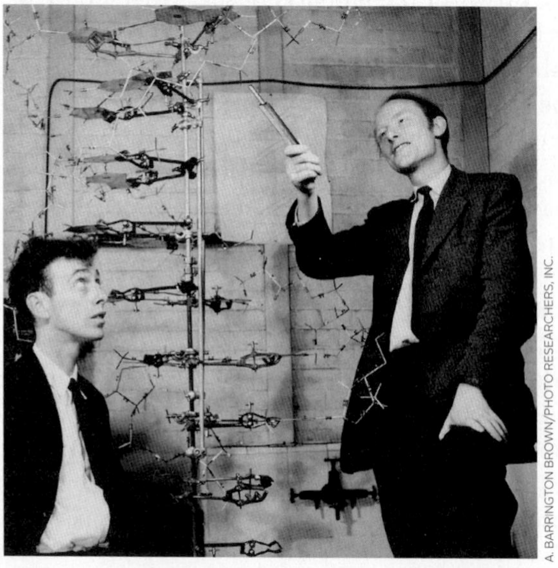
A. BARRINGTON BROWN/PHOTO RESEARCHERS, INC.

FIGURE 1.32 ■ The DNA double helix. A. Rosalind Franklin discovered that DNA forms a double helix. **B.** X-ray diffraction pattern of DNA, obtained by Rosalind Franklin. **C.** James Watson and Francis Crick discovered the complementary pairing between bases of DNA and the antiparallel form of the double helix.

the DNA "alphabet" were paired in the interior of Franklin's double helix (**Fig. 1.32C**). They published their model in the journal *Nature,* while denying that they had used Franklin's data. The discovery of the double helix earned Watson, Crick, and Wilkins the 1962 Nobel Prize in Physiology or Medicine. Franklin died of ovarian cancer before the prize was awarded. Before her death, however, she turned her efforts to the structure of ribonucleic acid (RNA). She determined the helical form of the RNA chromosome within tobacco mosaic virus, the first viral RNA to be characterized.

The complementary pairing of DNA bases led to the development of techniques for **DNA sequencing**, the reading of the sequence of DNA base pairs. **Figure 1.33** shows an example of DNA sequence data, where each color represents one of the four bases and each peak represents a DNA fragment terminating in that particular base. The order of fragment lengths yields the sequence of bases in one strand. Reading the sequence enabled microbiologists to determine the beginning and endpoint of microbial genes, and ultimately entire genomes.

So how does DNA sequence convey information in the cell? What are the "words" and "meanings" of the DNA language? To read DNA information required deciphering the genetic code—how triplets of DNA "letters" specify the amino acid units of proteins. This story is told in eTopic 1.2 and discussed further in Chapter 8.

The DNA revolution began with bacteria. What amazed the world about DNA was that such a simple substance, composed of only four types of subunits, is the genetic material that determines all the different organisms on Earth. The promise of this insight was first fulfilled in bacteria and bacteriophages, whose small genomes and short generation times made key experiments possible (Chapters 7–9). Bacterial tools were later extended to animals and plants; for example:

- **Bacteria readily recombine DNA from unrelated organisms.** The mechanisms of bacterial recombination led to construction of artificially recombinant DNA, or "gene cloning." Recombinant DNA ultimately enabled us to transfer genes between the genomes of virtually all types of organisms.
- **Bacterial DNA polymerases are used for polymerase chain reaction (PCR) amplification of DNA.** A hot spring in Yellowstone National Park yielded the bacterium *Thermus aquaticus,* whose DNA polymerase could survive many rounds of cycling to near-boiling temperature. The Taq polymerase formed the basis of a multibillion-dollar industry of PCR amplification of DNA, with applications ranging from genome sequencing to forensic identification.
- **Gene regulation discovered in bacteria provided models for animals and plants.** The first key discoveries of gene expression were made in bacteria and bacteriophages. Regulatory DNA-binding proteins were discovered in bacteria and then subsequently found in all classes of living organisms.

Public response. In the 1970s, when the DNA revolution began, its implications drew public concern. The use of recombinant DNA to make hybrid organisms—organisms combining DNA from more than one species—seemed "unnatural," although we now know that interspecies gene transfer occurs ubiquitously in nature. Furthermore, recombinant DNA technology raised the specter of placing deadly genes that produce toxins such as botulin into innocuous human-associated bacteria such as *E. coli.*

The unknown consequences of recombinant DNA so concerned molecular biologists that in 1975 they held a conference to assess the dangers and restrict experimentation on recombinant DNA. The conference, led by Paul Berg and Maxine Singer at Asilomar (Pacific Grove, California) was possibly the first time in history that a group of scientists organized and agreed to regulate and restrict their own field.

On the positive side, the emerging world of molecular biology excited students at a time of new ideas and social change. In 1971, the newly discovered process of protein translation by the bacterial ribosome inspired a classic

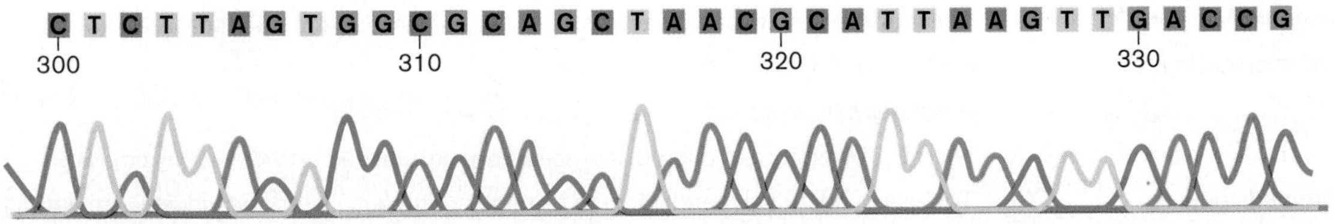
C T C T T A G T G G C G C A G C T A A C G C A T T A A G T T G A C C G
300
310
320
330

FIGURE 1.33 ■ A DNA sequence fluorogram. Sequence obtained using a DNA sequencer from genomic DNA of a bacterium isolated from a discarded beverage container by an undergraduate student at Kenyon College. Each colored trace represents the intensity of fluorescence of one of the four bases terminating a chain of DNA. Units represent DNA bases.

cult film, *Protein Synthesis: An Epic on the Cellular Level,* directed by Gabriel Weiss and choreographed by Jackie Bennington Weiss, America's 1969 Junior Miss (**Fig. 1.34**). Introduced by Nobel laureate Paul Berg, the film depicts dancers on the Stanford University football field forming the shape of the ribosome while "messenger RNA" and "transfer RNAs" meet to assemble a "protein." The dancers sway to "Protein Jive Sutra," performed by a Haight-Ashbury-inspired rock band. Their excitement echoed that of scientists exploring the extraordinary potential of microbial discovery.

Ribosome
Messenger RNA
COURTESY OF GABRIEL WEISS

FIGURE 1.34 ■ "Protein Jive Sutra." At Stanford University, in 1971, rock dancers portrayed the process of protein synthesis in the ribosome. The depiction was immortalized in a film produced by chemist Kent Wilson and directed by Gabriel Weiss, with an introduction by Nobel laureate Paul Berg. *biology.kenyon.edu/protein.mp4*

Thought Question

1.9 Do you think engineered strains of bacteria should be patentable? What about sequenced genes or genomes?

Microbial Discoveries Transform Medicine and Industry

Twentieth-century microbiology transformed the practice of medicine and generated entire new industries of biotechnology and bioremediation. Following the discovery of penicillin, Americans poured millions of dollars of private and public funds into medical research. The March of Dimes campaign for private donations to prevent polio led to the successful development of a vaccine that has nearly eliminated the disease. With the end of World War II, research on microbes and other aspects of biology drew increasing financial support from U.S. government agencies, such as the National Institutes of Health and the National Science Foundation, as well as from governments of other countries, particularly the European nations and Japan. Further support comes from private foundations, such as the Pasteur Institute, the Wellcome Trust, and the Howard Hughes Medical Institute.

Research in microbiology includes fields as diverse as medicine and space science (**Table 1.3**; A scientist whose career combines medicine and space science is profiled in **eTopic 1.3**). These fields all recruit microbiologists (**Fig. 1.35**). Industrial and applied biology (Chapter 16) use bacteria to clone and produce therapeutic proteins, such as insulin for diabetics.

TABLE 1.3

Fields of research in microbiology.

Field	Subject of study
Experimental microbiology	Fundamental questions about microbial form and function, genetics, and ecology
Medical microbiology	The mechanism, diagnosis, and treatment of microbial disease
Epidemiology	Distribution and causes of disease in humans, animals, and plants
Immunology	The immune system and other host defenses against infectious disease
Food microbiology	Fermented foods and food preservation
Industrial microbiology	Production of drugs, cloned gene products, and biofuels
Environmental microbiology	Microbial diversity and microbial processes in natural and artificial environments
Bioremediation	The use of microbial metabolism to remediate human wastes and industrial pollutants
Forensic microbiology	Analysis of microbial strains as evidence in criminal investigations
Astrobiology	The origin of life in the universe and the possibility of life outside Earth

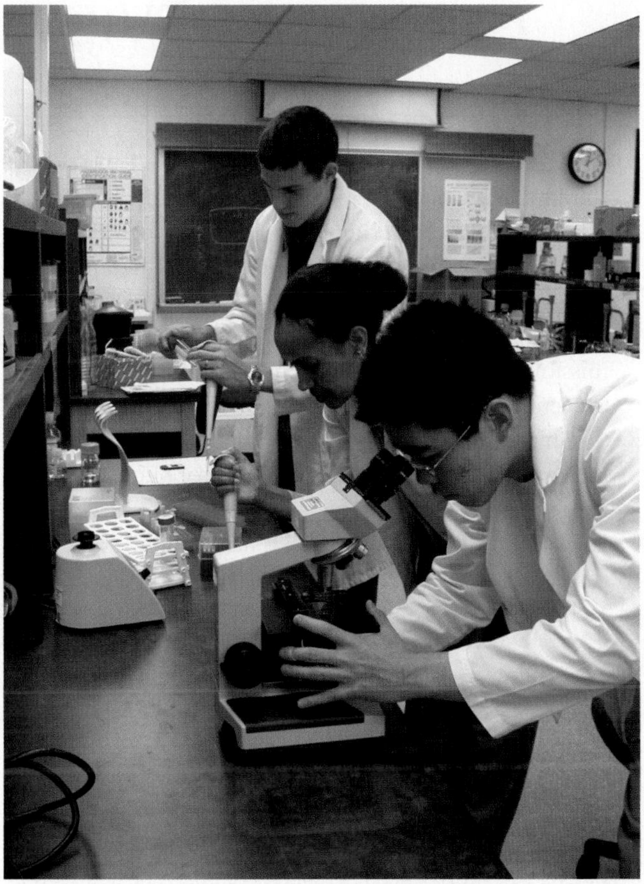
JOAN SLONCZEWSKI, KENYON COLLEGE

FIGURE 1.35 ■ Microbiologists at work. Students at Kenyon College conduct research on bacterial gene expression.

Recombinant viruses make safer vaccines. At the frontiers of science, we use microbes for **synthetic biology**, the construction of novel organisms with useful functions (Chapter 12). For example, synthetic biology may design bacteria with an "on-off" switch to report the presence of arsenic in environmental samples. On a global level, the management of our planet's biosphere, with the challenges of pollution and global warming, increasingly depends on our understanding of microbial populations (Chapter 22).

Weblinks American Society for Microbiology *asm.org*

To Summarize

- **Genetics of bacteria, bacteriophages, and fungi** in the early twentieth century revealed fundamental insights about gene transmission that apply to all organisms.
- **Structure and function of the genetic material, DNA**, were revealed by a series of experiments in the twentieth century.
- **Molecular microbiology generated key advances**, such as the cloning of the first recombinant molecules and the invention of DNA sequencing technology.
- **Genome sequence determination and bioinformatic analysis** became the tools that shape the study of biology in the twenty-first century.
- **Microbial discoveries transformed medicine and industry.** Biotechnology produces new kinds of pharmaceuticals and industrial products. Synthetic biology engineers new kinds of organisms with useful functions.

Concluding Thoughts

Advances in microbial science raise important questions for society. How can medical research control emerging diseases? How do microbes contribute to global cycles of carbon and nitrogen? How can microbial metabolism clean up polluted environments, such as the vast oil spill from the explosion of the *Deepwater Horizon* wellhead in 2010?

This book explores the current explosion of knowledge about microbial cells, genetics, and ecology. We introduce the applications of microbial science to human affairs, from medical microbiology to environmental science. Most important, we discuss research methods—how scientists make the discoveries that will shape tomorrow's view of microbiology. Chapter 2 presents the imaging tools that make possible our increasingly detailed view of the structures of cells (Chapter 3) and viruses (Chapter 6). Chapters 4 and 5 introduce microbial nutrition and growth in diverse habitats, including new information derived from the sequencing of microbial genomes. Throughout, we invite readers to share with us the excitement of discovery in microbiology.

CHAPTER REVIEW

Review Questions

1. Explain the apparent contradictions in defining microbiology as the study of microscopic organisms or as the study of single-celled organisms.
2. What is the genome of an organism? How do genomes of viruses differ from those of cellular microbes?
3. Under what conditions might microbial life have originated? What evidence supports current views of microbial origin?
4. List the ways in which microbes have affected human life throughout history.
5. Summarize the key experiments and insights that shaped the controversy over spontaneous generation. What questions were raised, and how were they answered?
6. Explain how microbes are cultured in liquid and on solid media. Compare and contrast the culture methods of Koch and Winogradsky. How did their different approaches to microbial culture address different questions in microbiology?
7. Explain how a series of observations of disease transmission led to the development of immunization to prevent disease.
8. Summarize key historical developments in our view of microbial taxonomy. What attributes of microbes have made them challenging to classify?
9. Explain how various discoveries in "natural" bacterial genetics were used to develop recombinant DNA technology.

Thought Questions

1. How do Earth's microbes contribute to human health? Include examples of environmental microbes outside the human body, as well as microbes associated with the human body.
2. When space scientists seek evidence for life on Mars, why do you think they expect to find microbes rather than creatures like the "alien monsters" often depicted in science fiction?
3. Why do you think so many environmental microbes cannot be cultured in laboratory broth or agar media?
4. Outline the different contributions to medical microbiology and immunology of Louis Pasteur, Robert Koch, and Florence Nightingale. What methods and assumptions did they have in common, and how did they differ?
5. Outline the different contributions to environmental microbiology of Sergei Winogradsky and Martinus Beijerinck. Why did it take longer for the significance of environmental microbiology to be recognized, as compared with pure-culture microbiology?
6. The Part 1 interview with Richard Lenski describes how he observed *E. coli* bacteria evolving in the laboratory. What historic discoveries and techniques of microbiology made his experiment possible?

Key Terms

agar (21)
antibiotic (24)
antiseptic (24)
Archaea (9, 31)
archaeon (9)
aseptic (24)
autoclave (17)
Bacteria (9)
bacterium (9)
biofilm (28)
chain of infection (20)
chemiosmotic theory (34)
chemolithotroph (26)
colony (21)
DNA sequencing (35)
electron microscope (32)
electron transport system (34)
endosymbiont (28)
enrichment culture (26)
Eukarya (9)
eukaryote (9)
extremophile (28)
fermentation (17)
genome (9)
geochemical cycling (27)
germ theory of disease (20)
immune system (23)
immunity (23)
immunization (23)
Koch's postulates (21)
lithotroph (26)
metagenome (10)
microbe (8)
monophyletic (30)
nitrogen fixation (27)
petri dish (21)
photosynthesis (27)
polymerase chain reaction (PCR) (6)
polyphyletic (30)
prokaryote (8)
pure culture (21)
RNA world (17)
spontaneous generation (16)
synthetic biology (37)
transformation (34)
ultracentrifuge (33)
vaccination (23)
virus (8)
Winogradsky column (26)

Recommended Reading

Angert, Esther, Kendall D. Clements, and Norman R. Pace. 1993. The largest bacterium. *Nature* **362**:239–241.

Blount, Zachary D., Christina Z. Borland, and Richard E. Lenski. 2008. Historical contingency and the evolution of a key innovation in an experimental population of *Escherichia coli*. *Proceedings of the National Academy of Sciences USA* **105**:7899–7906.

Breitbart, Mya, Luke R. Thompson, Curtis A. Suttle, and Matthew B. Sullivan. 2007. Exploring the vast diversity of marine viruses. *Oceanography* **20**:135–139.

Brock, Thomas D. 1999. *Robert Koch: A Life in Medicine and Bacteriology*. ASM Press, Washington, DC.

Dinc, Gulten, and Yesim I. Ulman. 2007. The introduction of variolation "A La Turca" to the West by Lady Mary Montagu and Turkey's contribution to this. *Vaccine* **25**:4261–4265.

Dubos, Rene. 1998. *Pasteur and Modern Science*. Translated by Thomas Brock. ASM Press, Washington, DC.

Fleishmann, Robert D., Mark D. Adams, Owen White, Rebecca A. Clayton, Ewen F. Kirkness, et al. 1995. Whole-genome random sequencing and assembly of *Haemophilus influenzae* Rd. *Science* **269**:496–512.

Gann, Alexander, and Jan Witkowski. 2010. The lost correspondence of Francis Crick. *Nature* **467**:519–524.

Hesse, Wolfgang. 1992. Walther and Angelina Hesse—early contributors to bacteriology. *ASM News* **58**:425–428.

Maddox, Brenda. 2002. *The Dark Lady of DNA*. Harper Collins, New York.

Margulis, Lynn. 1968. Evolutionary criteria in Thallophytes: A radical alternative. *Science* **161**:1020–1022.

Raoult, Didier. 2005. The journey from *Rickettsia* to Mimivirus. *ASM News* **71**:278–284.

Sherman, Irwin W. 2006. *The Power of Plagues*. ASM Press, Washington, DC.

Thomas, Gavin. 2005. Microbes in the air: John Tyndall and the spontaneous generation debate. *Microbiology Today* (Nov. 5): 164–167.

Ward, Naomi, and Claire Fraser. 2005. How genomics has affected the concept of microbiology. *Current Opinion in Microbiology* **8**:564–571.

Woese, Carl R., and George E. Fox. 1977. Phylogenetic structure of the prokaryotic domain: The primary kingdoms. *Proceedings of the National Academy of Sciences USA* **74**:5088–5090.

CHAPTER 2
Observing the Microbial Cell

2.1 Observing Microbes
2.2 Optics and Properties of Light
2.3 Bright-Field Microscopy
2.4 Fluorescence Microscopy
2.5 Dark-Field and Phase-Contrast Microscopy
2.6 Electron Microscopy and Tomography
2.7 Visualizing Molecules

Microscopy reveals the vast realm of microorganisms invisible to the unaided eye. The microbial world spans a wide range of size—over several orders of magnitude. For different size ranges, we use different instruments, from the simple bright-field microscope to the electron microscope. The microscope enables us to count the number of microbes in the human bloodstream or in dilute natural environments such as the ocean. It shows us how microbes swim and respond to signals such as a new food source.

Advanced forms of microscopy reveal microbes in remarkable detail. Fluorescence microscopy even shows how parts function within a living cell. Electron microscopy explores the cell's interior, showing how all the parts of the cell fit together. The electron microscope is used to build models of viruses, and to probe subcellular structures including membranes, ribosomes, and flagellar motors. And new kinds of chemical imaging map the movement of individual molecules within living cells.

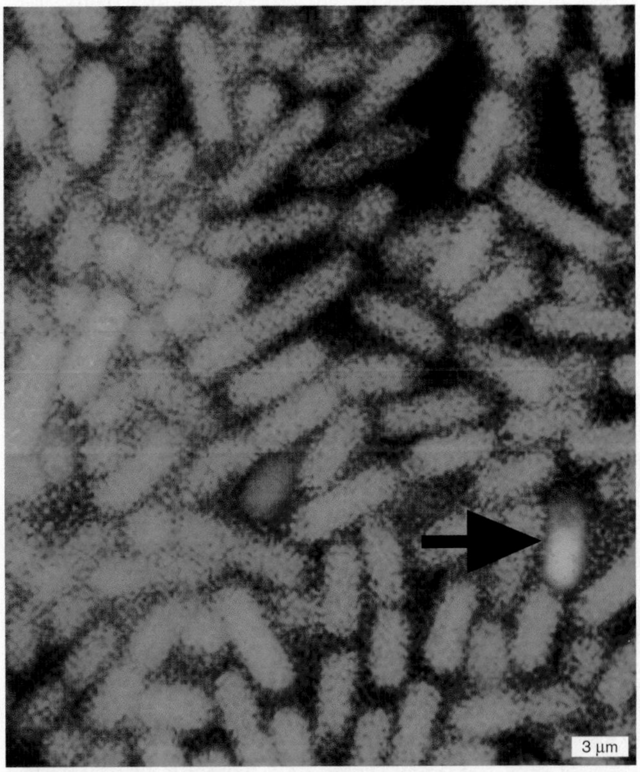
3 µm

CURRENT RESEARCH **highlight**

Observing pH of a single cell. *Escherichia coli* cells in a biofilm were exposed to acid, which they experience while colonizing the human digestive tract. The cells shown contain a gene expressing a form of green fluorescent protein (GFP) whose peak wavelengths of light excitation for fluorescence vary with pH. Colors indicate the ratio between two excitation peaks that increase and decrease with pH. Cells that appear green maintain their cytoplasmic pH at pH 7.5, despite the external acidification. Cells that fail to maintain pH homeostasis appear blue (arrow) as their pH falls to pH 6. In a biofilm, cells show surprisingly diverse responses to acid stress, nutrient starvation, and antibiotics. The fluorescence microscope enables us to map cell responses and develop better ways to control biofilm growth. *Source:* Keith Martinez et al. 2012. *Appl. Environ. Microbiol.* **78**:3706.

How did people first see microbes? When the microscope of Antonie van Leeuwenhoek first revealed the tiny life-forms on his teeth, scientists in England refused to believe it until their own instruments showed the same. Leeuwenhoek's superior lenses were key to his success. Since the time of Leeuwenhoek, microscopists have devised ever-more-powerful instruments to search for microbes in familiar and unexpected habitats.

An example of such a habitat is the human stomach, long believed too acidic (pH 2) to harbor life. In the 1980s, however, Australian scientist Barry Marshall reported finding a new species of bacterium in the stomach. The bacterium, *Helicobacter pylori,* proved difficult to isolate and culture, and for over a decade medical researchers refused to believe Marshall's report. Ultimately, electron microscopy (EM) confirmed the existence of *H. pylori* in the stomach and helped document its role in gastritis and stomach ulcers. The scanning electron micrograph in **Figure 2.1** shows *H. pylori* colonizing the gastric crypt cells. *H. pylori* bacteria are helical rods (colorized green). The actual size of the bacteria can be measured using the scale bar in the lower right of the figure. The cell contours are well resolved by the scanning beam of electrons—an achievement far beyond that possible with a light microscope.

Today, various kinds of microscopes are used in hospitals and veterinary clinics, as well as industrial plants and wastewater treatment facilities. Chapter 2 covers two areas:

- **Light microscopy.** The theory and practice of light microscopy are essential for every student and professional observing microorganisms.
- **Advanced tools for research.** Fluorescence microscopy, electron microscopy, and scanning probe microscopy probe ever farther the frontiers of the unseen.

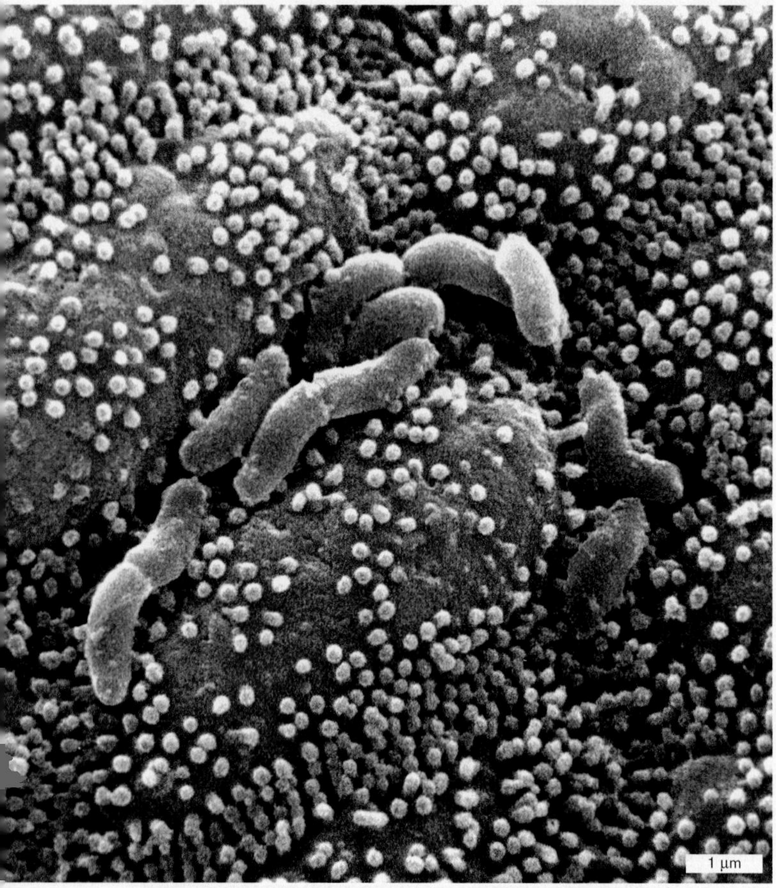
1 µm

VERONIKA BURMEISTER/VISUALS UNLIMITED

FIGURE 2.1 ▪ *Helicobacter pylori* amongst the crypt cells of the stomach lining. Microscopy demonstrated the presence of *H. pylori,* the cause of stomach ulcers, growing on the lining of the human stomach, a location previously believed too acidic to permit microbial growth. Scanning electron micrograph, colorized to indicate bacteria (green).

2.1

Observing Microbes

Most microbes are too small to be seen; that is, they are microscopic, requiring the use of a **microscope** to be seen. But why can't we see microbes without magnification? The answer is surprisingly complex. In fact, our definition of "microscopic" is based on the properties of our eyes. What is microscopic actually lies in the eye of the beholder.

Resolution of Objects by Our Eyes

What determines the smallest object we can see? The size at which objects become visible depends on the eye's ability to resolve detail. **Resolution** is the smallest distance between two objects that allows us to see them as separate objects. The eyes of humans and other animals observe an object by focusing its image on a retina packed with light-absorbing photoreceptor cells (**Fig. 2.2**). The finest resolution is perceived by the fovea, the portion of the retina where the photoreceptors are packed at the highest density. The foveal photoreceptors are cone cells, which detect primary colors (red, green, or blue) and finely resolved detail. A group of cones with their linked neurons forms one unit of detection, comparable to a pixel on a computer screen.

In the human eye, the distance between two foveal "pixels" (groups of cones with neurons) limits resolution to about 150 µm, or one-seventh of a millimeter. So one-seventh of a millimeter is about the smallest object that most of us can see (resolve distinctly) without a magnifier.

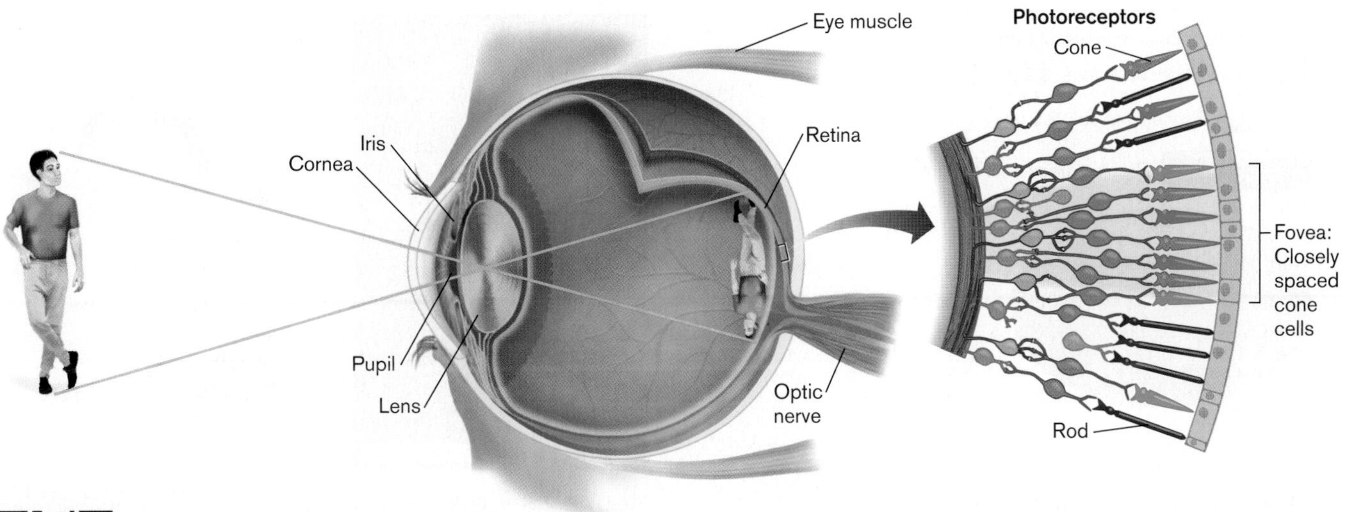
Eye muscle
Photoreceptors
Cone
Iris
Cornea
Retina
Fovea: Closely spaced cone cells
Pupil
Lens
Optic nerve
Rod

FIGURE 2.2 ▪ Defining the microscopic. We define what is visible and what is microscopic in terms of the human eye. Within the human eye, the lens focuses an image on the retina. Adjacent light receptors are sufficiently close to resolve the image of a human being at a distance where its apparent height shrinks to a millimeter. *microbiology2.com/animations*

What if our eyes were formed differently? The retinas of eagles have cones packed more closely than ours, so an eagle can resolve objects eight times as small (or eight times as far away) as a human can; hence, the phrase "eagle-eyed" means "sharp-sighted." On the other hand, insect compound eyes have photoreceptors farther apart than ours; so insect eyes have poorer resolution. The best they can do is resolve objects 100-fold larger than those we can resolve. If a science-fictional giant ameba had eyes with photoreceptors 2 meters apart, it would perceive humans as "microscopic."

Note: In this book, we use standard metric units for size:

1 millimeter (mm)	= one-thousandth of a meter (m)
	= 10^{-3} m
1 micrometer (µm)	= one-thousandth of a millimeter
	= 10^{-6} m
1 nanometer (nm)	= one-thousandth of a micrometer
	= 10^{-9} m
1 picometer (pm)	= one-thousandth of a nanometer
	= 10^{-12} m

Some authors still use the traditional unit angstrom (Å), which equals a tenth of a nanometer, or 10^{-10} meter.

Resolution Differs from Detection

Can we **detect** the presence of objects whose size we cannot resolve? Yes, we can detect their presence as a group. For example, our eyes can detect a large population of microbes, such as a spot of mold on a piece of bread (about a million cells) or a cloudy tube of bacteria in liquid culture (a million cells per milliliter) (**Fig. 2.3A**). **Detection**, the ability to determine the presence of an object, differs from resolution. When the unaided eye detects the presence of mold or bacteria, it cannot resolve distinct cells.

To resolve bacterial cells, such as those of the wetland phototroph *Rhodospirillum rubrum* (**Fig. 2.3B**), our eyes need assistance—that is, **magnification**. In microscopy, magnifying an object means increasing the object's apparent dimensions. As the distance increases between points of detail, our eyes can now resolve the object's shape as a magnified image.

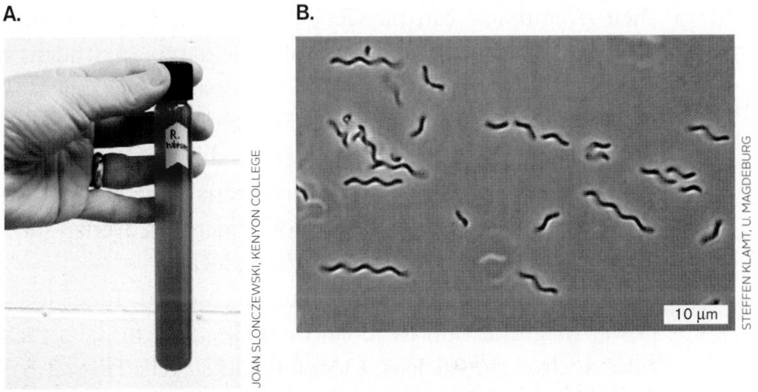
A.
B.
JOAN SLONCZEWSKI, KENYON COLLEGE
10 µm
STEFFEN KLAMT, U. MAGDEBURG

FIGURE 2.3 ▪ Detecting and resolving bacteria. A. A tube of bacterial culture, *Rhodospirillum rubrum.* The presence of bacteria is detected, though individual cells are not resolved. **B.** Individual cells of *R. rubrum* are resolved by light microscopy.

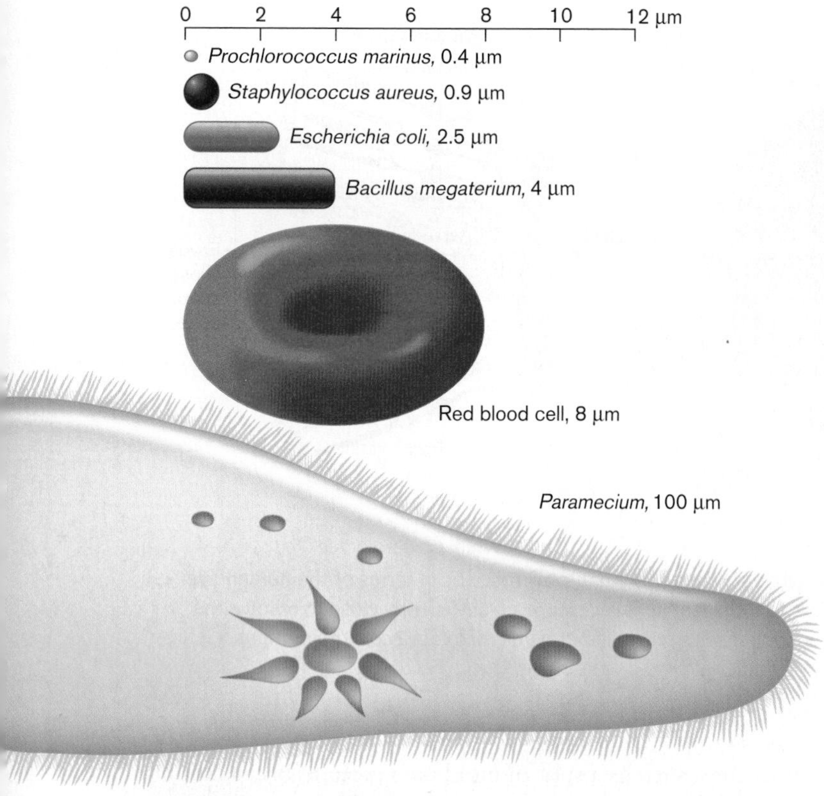
0
2
4
6
8
10
12 µm
Prochlorococcus marinus, 0.4 µm
Staphylococcus aureus, 0.9 µm
Escherichia coli, 2.5 µm
Bacillus megaterium, 4 µm
Red blood cell, 8 µm
Paramecium, 100 µm

FIGURE 2.4 ▪ Relative sizes of different cells. Microbial cells come in various sizes, most of which are below the limit of resolution perceived by the unaided human eye (150 µm). Note that most bacteria are much smaller than the human red blood corpuscle, a small eukaryotic cell.

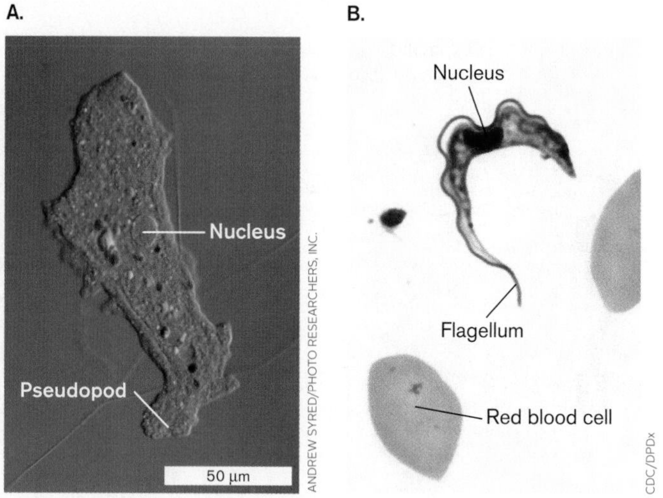
A.
B.
Nucleus
Pseudopod
50 µm
ANDREW SYRED/PHOTO RESEARCHERS, INC.
Nucleus
Flagellum
Red blood cell
CDC/DPDx

FIGURE 2.5 ▪ Eukaryotic microbial cells. Eukaryotic microbes are large enough that details of internal and external organelles can be seen under a light microscope. **A.** *Amoeba proteus.* **B.** *Trypanosoma brucei,* (cause of sleeping sickness) among blood cells. (Length, 14–33 µm.)

Microbial Size and Shape

Different kinds of microbes differ in size, over a range of several orders of magnitude, or powers of ten (**Fig. 2.4**). Eukaryotic microbes are often sufficiently large (10–100 µm) that we can resolve their compartmentalized structure under a light microscope. Prokaryotes (bacteria and archaea) are generally smaller (0.4–10 µm); thus, their overall size can be seen, but most of their internal structures are too small to resolve by a typical student's microscope. Nevertheless, a few bacterial species, such as *Thiomargarita namibiensis,* are large enough to be seen by the unaided eye, while many unculturable marine eukaryotes are as small as the smallest bacteria. The actual size ranges of eukaryotic and prokaryotic microbes overlap substantially (discussed in Chapters 20 and 21).

Many eukaryotic microbes can be resolved under low-power magnification to reveal internal and external structures such as the nucleus, vacuoles, and flagella (**Fig. 2.5**). Protists show complex shapes and appendages. For example, an ameba from an aquatic ecosystem shows a large nucleus and pseudopods to engulf prey (**Fig. 2.5A**). Pseudopods can be seen moving by the streaming of their cytoplasm. Another protist readily observed by light microscopy is *Trypanosoma brucei,* an insect-borne blood parasite that causes African sleeping sickness (**Fig. 2.5B**). In the trypanosome, we observe a nucleus and a flagellum. Eukaryotic flagella propel the cell by a whiplike action. For more on microbial eukaryotes, see Chapter 20.

Prokaryotic cell structures are generally simpler than those of eukaryotes (see Chapter 3). **Figure 2.6** shows representative members of several common cell types, as visualized by light microscopy or by scanning electron microscopy. Note that with bright-field light microscopy, the cell shape is just discernible under the highest power. With scanning electron microscopy, cell shapes appear in greater detail. Subcellular structures are better visualized by fluorescence microscopy, or by transmission electron microscopy (discussed in the next section).

Certain shapes of bacteria are common to many taxonomic groups (**Fig. 2.6**). For example, both bacteria and archaea form similarly shaped **rods**, or **bacilli** (singular, **bacillus**), and **cocci** (spheres; singular, **coccus**). Thus, rods and spherical shapes evolve independently within different taxa. In contrast, a unique bacterial shape that evolved in only one taxon is the **spirochete**, a tightly coiled spiral. Species of spirochetes cause diseases such as syphilis and Lyme disease. The spiral form of the cell is maintained by internal axial filaments and flagella, as well as an outer sheath (for more on spirochetes, see Section 18.6). A different, unrelated spiral form is the "spirillum," a wide, rigid spiral cell that is similar to a rod-shaped bacillus.

Note: The genus name *Bacillus* refers to a specific taxonomic group of bacteria, but the term "bacillus" (plural, bacilli) refers to any rod-shaped bacterium or archaeon.

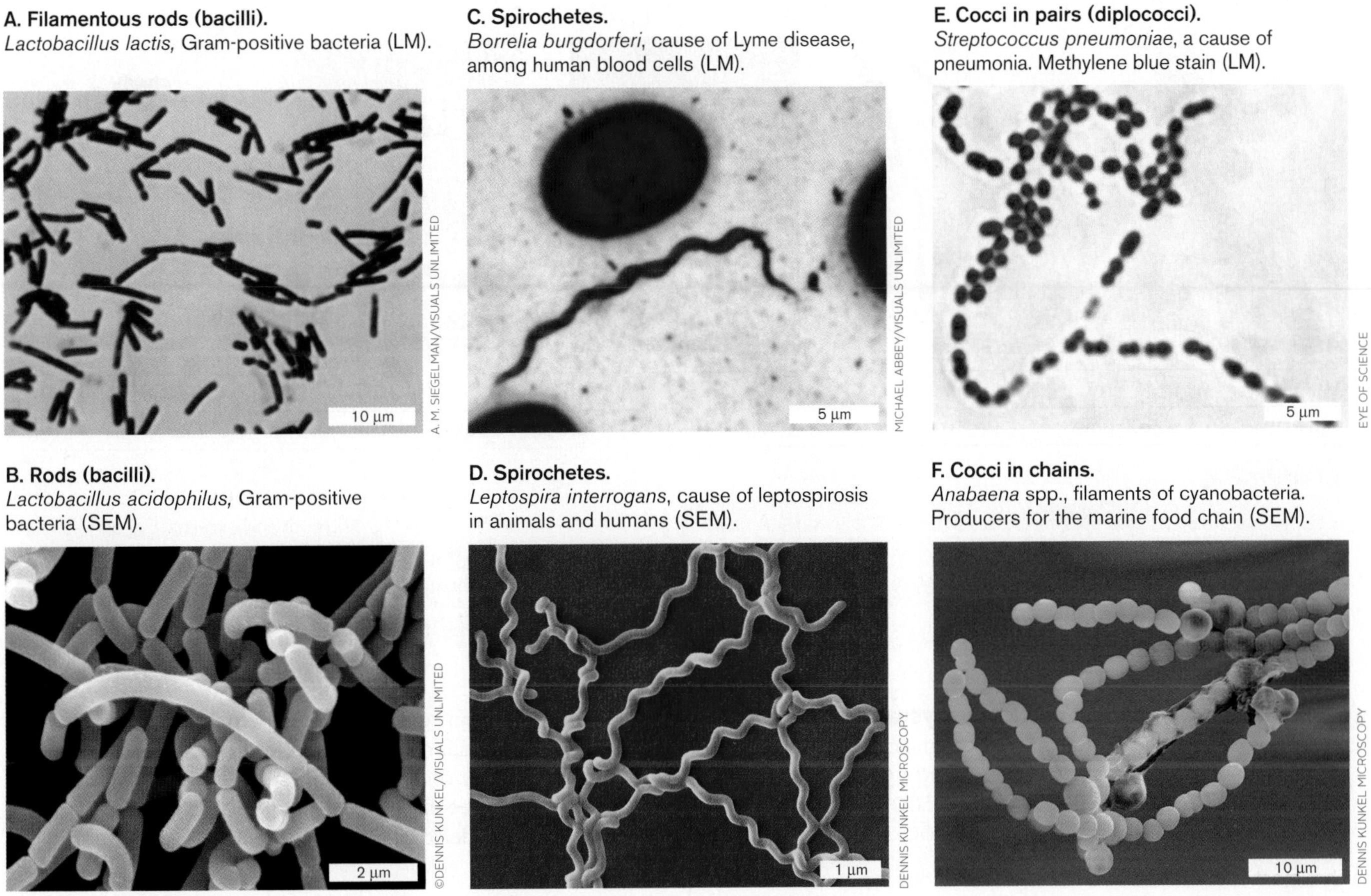
A. Filamentous rods (bacilli).
Lactobacillus lactis, Gram-positive bacteria (LM).
10 μm
A. M. SIEGELMAN/VISUALS UNLIMITED
C. Spirochetes.
Borrelia burgdorferi, cause of Lyme disease, among human blood cells (LM).
5 μm
MICHAEL ABBEY/VISUALS UNLIMITED
E. Cocci in pairs (diplococci).
Streptococcus pneumoniae, a cause of pneumonia. Methylene blue stain (LM).
5 μm
EYE OF SCIENCE
B. Rods (bacilli).
Lactobacillus acidophilus, Gram-positive bacteria (SEM).
2 μm
©DENNIS KUNKEL/VISUALS UNLIMITED
D. Spirochetes.
Leptospira interrogans, cause of leptospirosis in animals and humans (SEM).
1 μm
DENNIS KUNKEL MICROSCOPY
F. Cocci in chains.
Anabaena spp., filaments of cyanobacteria. Producers for the marine food chain (SEM).
10 μm
DENNIS KUNKEL MICROSCOPY

FIGURE 2.6 ■ Common shapes of bacteria. The shapes of most bacterial cells can be discerned with light microscopy (LM) (**A, C, E**), but their subcellular structures and surface details cannot be seen. Surface detail is revealed by scanning electron microscopy (SEM) (**B, D, F**). These SEM images are colorized to enhance clarity.

Microscopy for Different Size Scales

To resolve microbes and microbial structures of different sizes requires different kinds of microscopes. **Figure 2.7** shows the different techniques used to resolve microbes and structures of various sizes. For example, a paramecium can be resolved under a light microscope, but an individual ribosome requires electron microscopy.

- **Light microscopy** resolves images of individual bacteria by their absorption of light. The specimen is commonly viewed as a dark object against a light-filled field, or background; this is called **bright-field microscopy** (seen in **Fig. 2.7A** and **B**). Advanced techniques, based on special properties of light, include fluorescence, dark-field, and phase-contrast microscopy.
- **Electron microscopy (EM)** uses beams of electrons to resolve details several orders of magnitude smaller than those seen under light microscopy. In **scanning electron microscopy (SEM)**, the electron beam is scattered from the metal-coated surface of an object, generating an appearance of 3D depth. In **transmission electron microscopy (TEM)** (**Fig. 2.7C** and **D**), the electron beam travels through the object, where the electrons are absorbed by an electron-dense metal stain.
- **Atomic force microscopy (AFM)** uses intermolecular forces between a probe and an object to map the 3D topography of a cell, or of cell parts.
- **X-ray crystallography** detects the interference pattern of X-rays entering the crystal lattice of a molecule. From the interference pattern, researchers build a computational model of the structure of the individual molecule, such as a protein or a nucleic acid, or even a molecular complex such as a ribosome (**Fig. 2.7E**).

Thought Question

2.1 (refer to **Fig. 2.7**) You have discovered a new kind of microbe, never observed before. What kinds of questions about this microbe might be answered by light microscopy? What questions would be better addressed by electron microscopy?

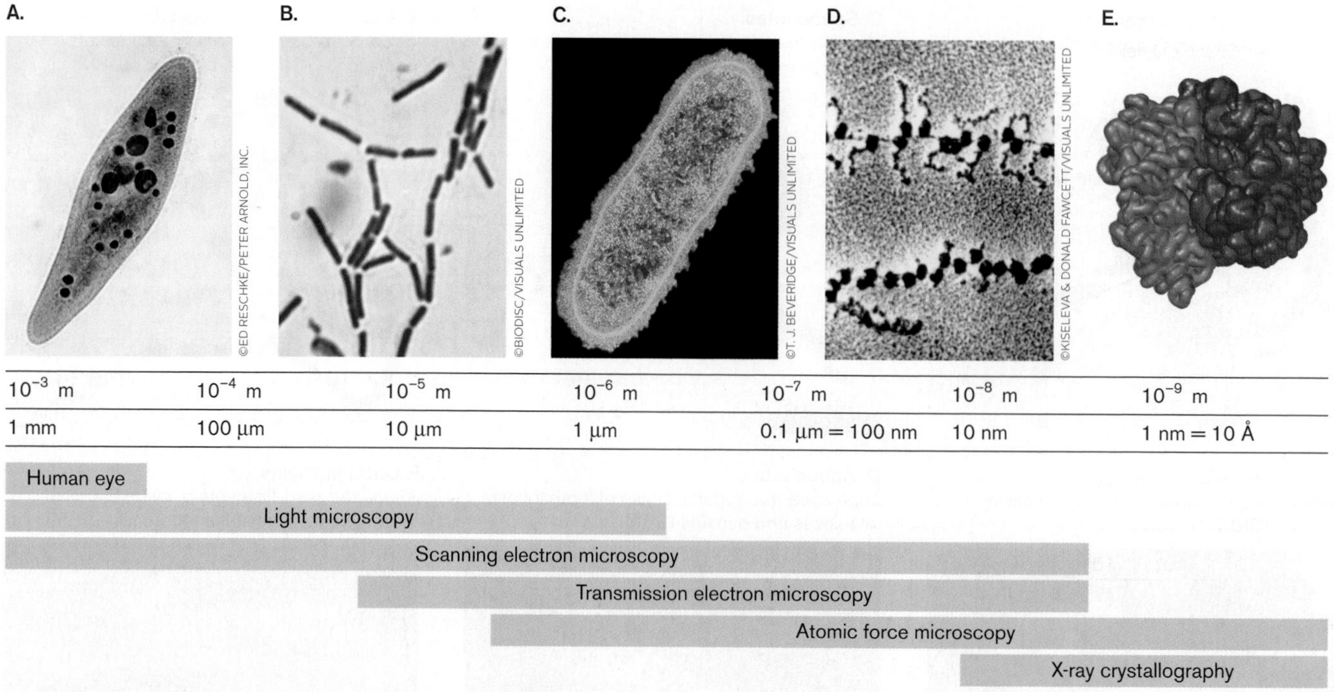
A.
B.
C.
D.
E.
©ED RESCHKE/PETER ARNOLD, INC.
©BIODISC/VISUALS UNLIMITED
©T. J. BEVERIDGE/VISUALS UNLIMITED
©KISELEVA & DONALD FAWCETT/VISUALS UNLIMITED
10^−3 m
10^−4 m
10^−5 m
10^−6 m
10^−7 m
10^−8 m
10^−9 m
1 mm
100 µm
10 µm
1 µm
0.1 µm = 100 nm
10 nm
1 nm = 10 Å
Human eye
Light microscopy
Scanning electron microscopy
Transmission electron microscopy
Atomic force microscopy
X-ray crystallography

FIGURE 2.7 ■ Microscopy and X-ray crystallography, range of resolution. **A.** Light microscopy reveals internal structures of a paramecium (a eukaryote, 100 µm). LM, magnification 100× at 35-mm size. **B.** *Pseudomonas* sp., rod-shaped bacteria (1–2 µm), are resolved. LM, methylene blue stain, magnification 1,200×. **C.** Internal structures of the bacterium *Escherichia coli* (1–2 µm) are revealed by transmission electron microscopy (TEM). Magnification 32,000×. **D.** Individual ribosomes are attached to the messenger RNA molecules that are translated to make peptides. TEM, magnification 150,000×. **E.** A ribosome (diameter, 21 nm) can be modeled with X-ray crystallography.

Chemical Imaging Microscopy

A new class of imaging techniques known as **chemical imaging microscopy** goes beyond size and shape to reveal the chemical composition of the microbial object. These techniques apply mass spectroscopy to a microscopic object. The object, such as a microbe, is bombarded with a beam of ions that vaporizes organic molecules from the object's surface. The molecular fragments that come off are identified through mass analysis. The microbe may be cultured with heavy isotopes of an element such as carbon or nitrogen, whose incorporation into biomass alters the mass spectra. These imaging techniques show where in the cell different nutrients are fixed into biomass (discussed in **Special Topic 2.1**).

To Summarize

- **Detection** is the ability to determine the presence of an object.
- **Resolution** is the smallest distance by which two objects can be separated and still be distinguished.
- **Magnification** means an increase in the apparent size of an image.
- **Eukaryotic microbes may be large enough to resolve subcellular structures** under a light microscope, although some eukaryotes are as small as bacteria.
- **Bacteria and archaea may be too small for subcellular resolution by a light microscope.** Their shapes include characteristic forms such as rods and cocci.
- **Different kinds of microscopy** resolve cells and subcellular structures of different sizes. Chemical imaging microscopy reveals the pattern of chemical composition.

2.2

Optics and Properties of Light

How do light rays magnify an image? Light microscopy directly extends the lens system of our own eyes. Light is part of the spectrum of **electromagnetic radiation** (**Fig. 2.8**), a form of energy that is propagated as waves associated with electrical and magnetic fields. Regions of the electromagnetic spectrum are defined by wavelength, which for visible light is about 400–750 nm. Radiation of longer

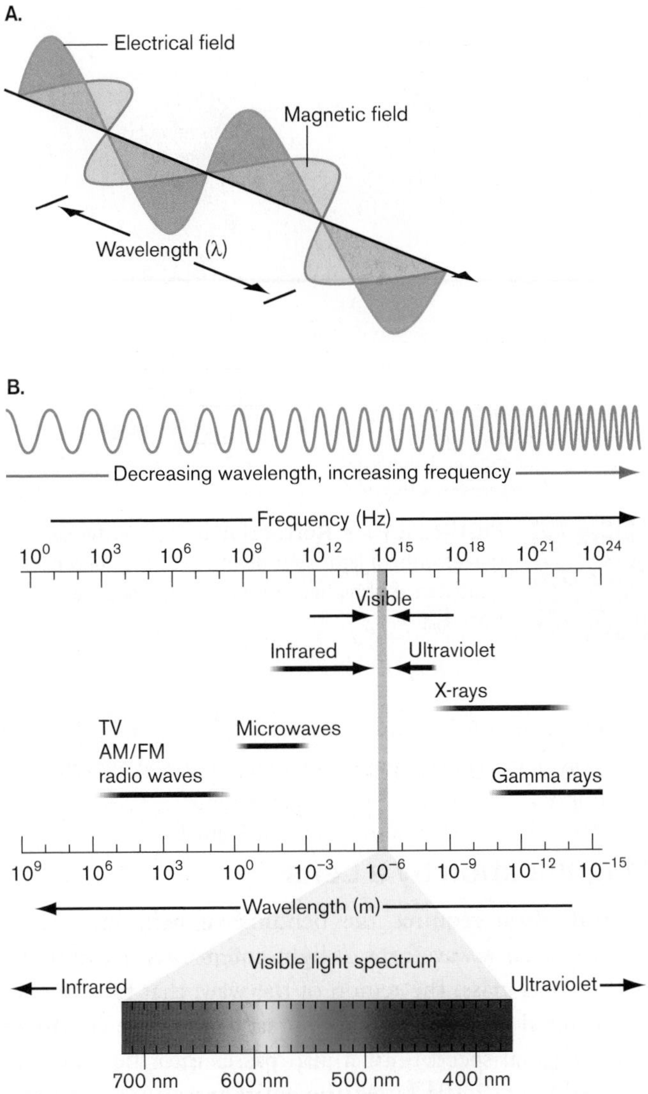
A.
Electrical field
Magnetic field
Wavelength (λ)
B.
Decreasing wavelength, increasing frequency
Frequency (Hz)
10^0 10^3 10^6 10^9 10^12 10^15 10^18 10^21 10^24
Visible
Infrared
Ultraviolet
X-rays
TV
AM/FM
radio waves
Microwaves
Gamma rays
10^9 10^6 10^3 10^0 10^-3 10^-6 10^-9 10^-12 10^-15
Wavelength (m)
Visible light spectrum
Infrared
Ultraviolet
700 nm
600 nm
500 nm
400 nm

FIGURE 2.8 ■ Electromagnetic energy. A. Electromagnetic radiation is composed of electrical and magnetic waves perpendicular to each other. **B.** The electromagnetic spectrum includes the visible range, used in light microscopy.

wavelengths includes infrared and radio waves, whereas shorter wavelengths include ultraviolet rays and X-rays.

Light Carries Information

All forms of electromagnetic radiation carry information from the objects they interact with. The information carried by radiation can be used to detect objects; for example, radar (using radio waves) detects a speeding car. All electromagnetic radiation travels through a vacuum at the same speed: about 3×10^8 meters per second (m/s), the speed of light. The speed of light (c) is equal to the wavelength (λ) of the radiation multiplied by its frequency (ν), the number of wave cycles per unit time:

$$c = \lambda\nu$$

Because c is constant, the longer the wavelength λ is, the lower the frequency ν is. Frequency is usually measured in hertz (Hz), reciprocal seconds (1/s).

For electromagnetic radiation to resolve an object from neighboring objects or from its surrounding medium, certain conditions must exist:

- **Contrast between the object and its surroundings. Contrast** is the difference in light and dark. If an object and its surroundings absorb or reflect radiation equally, then the object will be undetectable. It is hard to observe a cell of transparent cytoplasm floating in water, because the aqueous cytoplasm and the extracellular water tend to transmit light similarly, producing little contrast.
- **Wavelength smaller than the object.** For an object to be resolved, the wavelength of the radiation must be equal to or smaller than the size of the object. If the wavelength of the radiation is larger than the object, then most of the wave's energy will simply pass through it, like an ocean wave passing around a dock post. Thus, radar, with a wavelength of 1–100 centimeters (cm), cannot resolve microbes, though it easily resolves cars and people.
- **Magnification.** The human retina absorbs radiation within a range of wavelengths, 400–750 nm (0.40–0.75 μm), which we define as "visible light." But the smallest distance our retina can resolve is 150 μm, about 300 times the wavelength of light. Thus, our eyes are unable to access all of the information contained in the light that enters. To use more of the information carried within the light, we must spread the light rays apart far enough for our retina to perceive the resolved image.

Light Interacts with an Object

The physical behavior of light resembles in some ways a beam of particles and in other ways a waveform. The particles of light are called photons. Each photon has an associated wavelength that determines how the photon will interact with a given object. The combined properties of particle and wave enable light to interact with an object in several different ways: absorption, reflection, refraction, and scattering.

- **Absorption** means that the absorbing object gains the photon's energy (**Fig. 2.9A**▶). The energy is converted to a different form, usually heat. (That is why a live specimen eventually "cooks" on the slide if observed for too long.) When a microbial specimen absorbs light, it

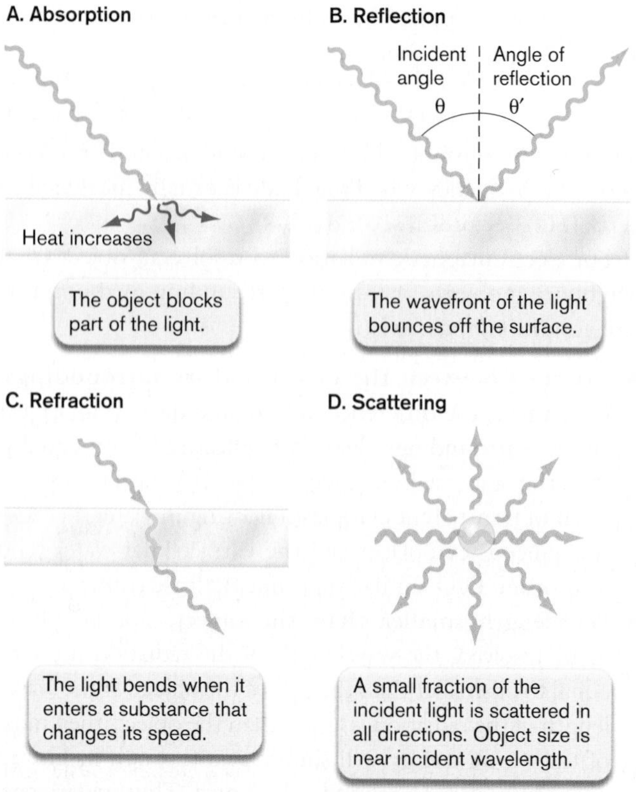
A. Absorption
Heat increases
The object blocks part of the light.
B. Reflection
Incident angle
θ
Angle of reflection
θ′
The wavefront of the light bounces off the surface.
C. Refraction
The light bends when it enters a substance that changes its speed.
D. Scattering
A small fraction of the incident light is scattered in all directions. Object size is near incident wavelength.

FIGURE 2.9 ■ Interaction of light with matter.

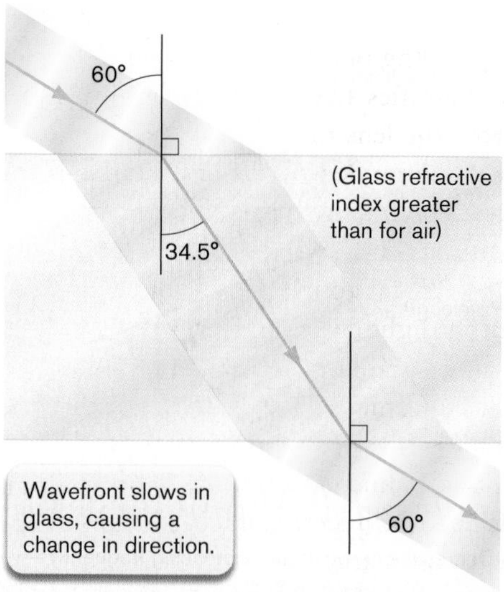
60°
(Glass refractive index greater than for air)
34.5°
Wavefront slows in glass, causing a change in direction.
60°

FIGURE 2.10 ■ Refraction of light waves. Wavefronts of light shift direction as they enter a substance of higher refractive index, such as glass.

can be observed as a dark spot against a bright field, as in bright-field microscopy. Some molecules that absorb light of a specific wavelength reemit energy as light with a longer wavelength; this is called **fluorescence**. Fluorescence microscopy is discussed in Section 2.4.

- **Reflection** means that the wavefront redirects from the surface of an object at an angle equal to its incident angle (**Fig. 2.9B**). Reflection of light waves is analogous to the reflection of water waves. Reflection from a silvered mirror or a glass surface is used in the optics of microscopy.
- **Refraction** is the bending of light as it enters a substance that slows its speed (**Fig. 2.9C**). Such a substance is said to be "refractive" and, by definition, has a higher **refractive index** than air. Refraction is the key property that enables a lens to magnify an image.
- **Scattering** occurs when a portion of the wavefront is converted to a spherical wave originating from the object (**Fig. 2.9D**). If a large number of particles simultaneously scatter light, we see a haze—for example, the haze of bacteria suspended in a culture tube. Special optical arrangements (such as dark-field microscopy, discussed in Section 2.5) can use scattered light to detect (but not resolve) microbial shapes smaller than the wavelength of light.

Magnification by a Lens

Magnification requires the bending of light rays, as in refraction. As a wavefront of light enters a refractive material, such as glass, the region of the wave that first reaches the material is slowed, while the rest of the wave continues at its original speed until it also passes into the refractive material (**Fig. 2.10**). As the entire wavefront continues through the refractive material, its path continues, bent at an angle from its original direction.

Thought Question

2.2 Explain what happens to the refracted light wave as it emerges from a piece of glass of even thickness. How do its new speed and direction compare with its original (incident) speed and direction?

How does refraction accomplish magnification? Refraction magnifies an image when light passes through a refractive material shaped so as to spread its rays. One shape that spreads light rays is a parabolic curve. When light rays enter a **lens** of refractive material with a parabolic surface (**Fig. 2.11**), parallel rays each bend at an angle such that all of the rays meet at a certain point, called the **focal point**. From the focal point (F′) behind the lens, the light rays continue, spreading out with an expanding wavefront. This expansion magnifies the image carried by the wave. The distance from the lens to the focal point (called the "focal distance") is determined by the degree of curvature of the lens, and by the refractive index of its material.

In **Figure 2.11**, the object under observation is placed near the focal point (F) in front of the lens. The lens bends all of the light rays so that they converge at the opposite focal point (F′). As the light rays continue through point F′, they diverge, spreading out. The image carried by the light rays is inverted, and it expands. This expansion, or magnification, increases the distances between points of the image. In a microscope, the details of the magnified image become larger than the spacing of photoreceptor units in our retina. Thus our eyes can perceive the details that the unaided eye could not see.

Thought Question

2.3 Parabolic lenses are generally "biconvex"—that is, curving outward on both sides. What will happen to parallel light rays that pass through a lens that is concave on both sides? Or a lens that is convex on one side and concave with equal curvature on the other?

Resolution of Detail

What limits the effect of magnification? The spreading of light rays does not in itself increase resolution. For example, an image composed of dots does not gain detail when enlarged on a photocopier, nor does an image composed of pixels gain detail when enlarged on a computer screen. In these cases, magnification fails to show details because the individual details of the image expand in proportion to the expansion of the overall image. Magnification without increasing detail is called **empty magnification**.

The resolution of detail in microscopy is limited by the wave nature of light. Light rays actually form wavefronts, which undergo **interference** (**Fig. 2.12A**). In interference, two wavefronts (or two portions of a wavefront) interact with each other by addition (amplitudes in phase) or subtraction (amplitudes out of phase). The result of interference between two waves is a pattern of alternating zones of constructive and destructive interference (brightness and darkness) (**Fig. 2.12B**).

In theory, a perfect lens that focuses all of the light from an object should have no interference as its rays converge toward the focal point. In fact, however, the limited diameter of the wavefronts entering the lens causes interference (**Fig. 2.13**). The converging edges of the

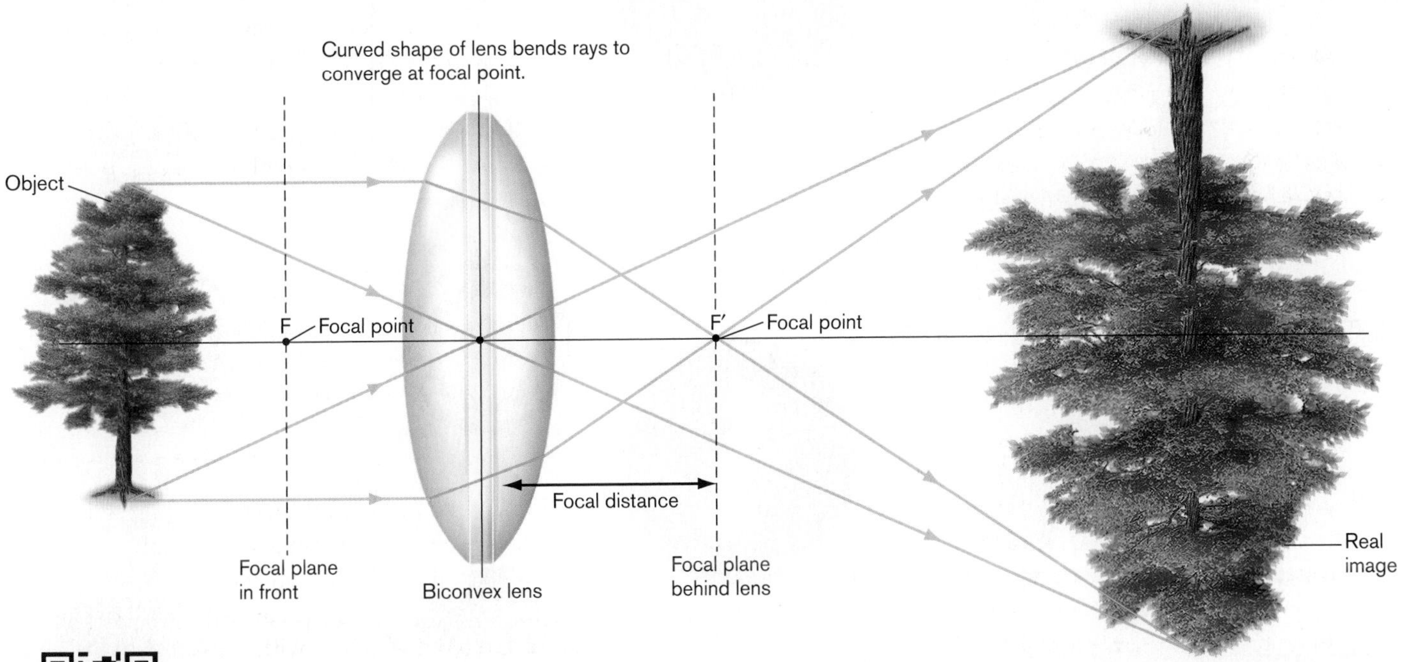
Curved shape of lens bends rays to converge at focal point.
Object
F
Focal point
F′
Focal point
Focal distance
Focal plane in front
Biconvex lens
Focal plane behind lens
Real image

FIGURE 2.11 ▪ A lens magnifies an image. The object is placed near the focal point (F) in front of the lens. All light rays from the object are bent by the lens, converging at the focal point (F′) behind the lens. The light rays continue through, spreading apart. The image of the object is inverted and magnified.

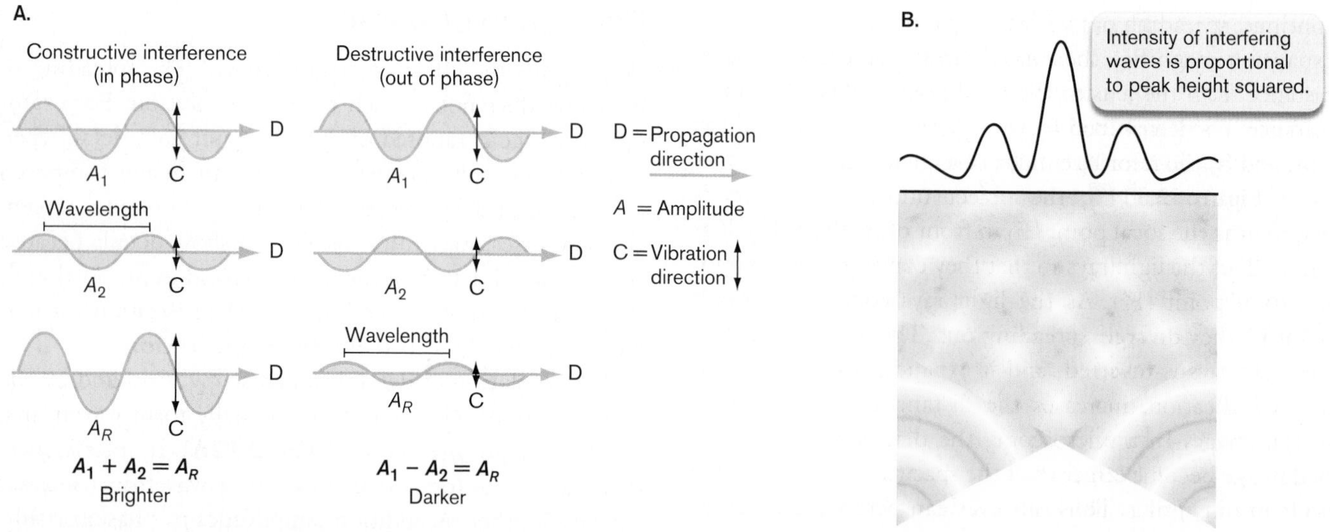
A.
Constructive interference
(in phase)
Destructive interference
(out of phase)
D
A_1
C
Wavelength
A_2
A_R
$A_1 + A_2 = A_R$
Brighter
$A_1 - A_2 = A_R$
Darker
D = Propagation direction
A = Amplitude
C = Vibration direction
B.
Intensity of interfering waves is proportional to peak height squared.

FIGURE 2.12 ■ Constructive (additive) and destructive (subtractive) interference of light waves. A. In constructive interference (left), the peaks of the two wave trains rise together; their amplitudes are additive, forming a wave of greater amplitude. In destructive interference (right), the peaks of the waves are opposite one another, so their amplitudes cancel, forming a wave of lesser amplitude. **B.** Two wavefronts approaching at an angle generate an interference pattern in which intensity alternately increases and decreases.

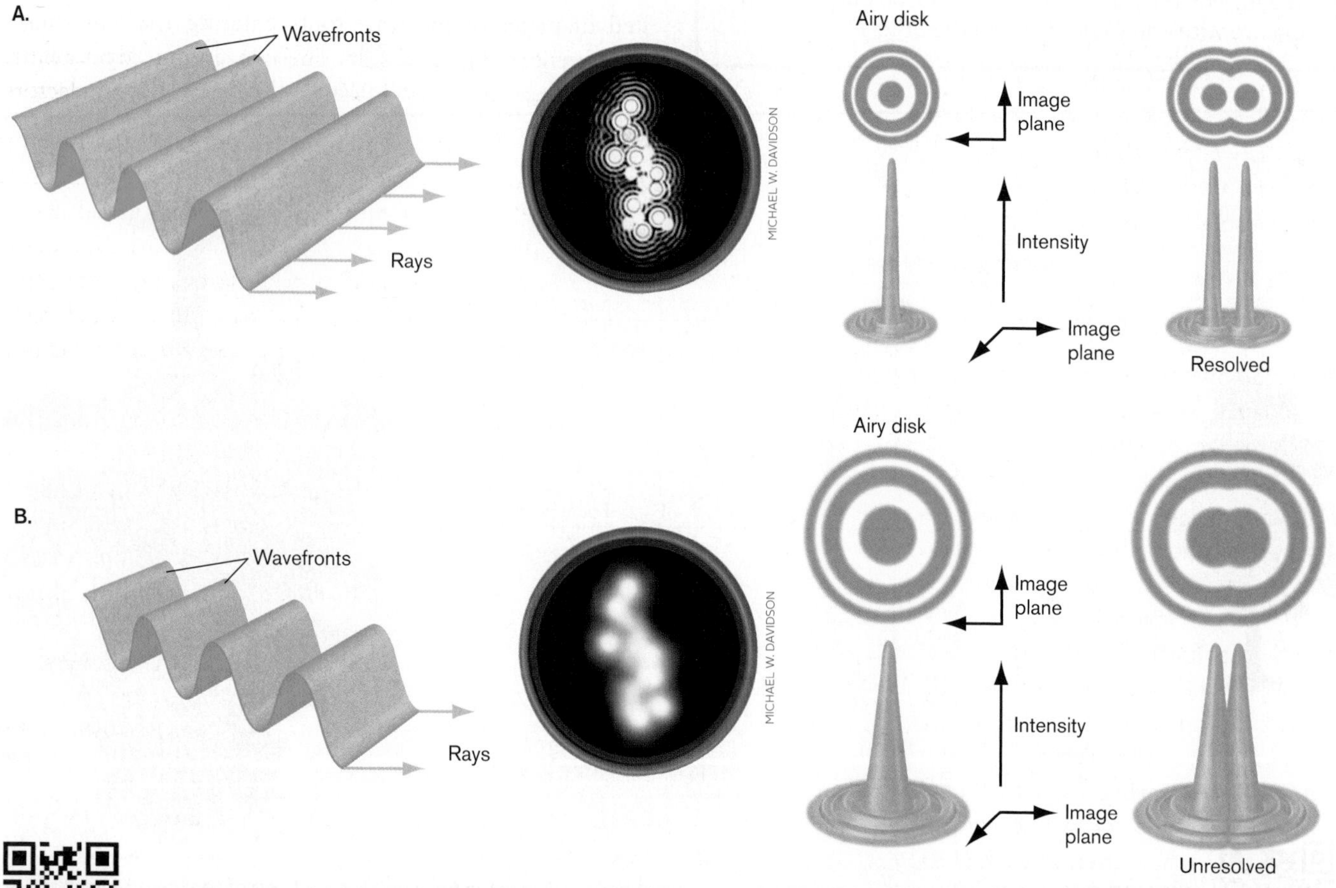
A.
Wavefronts
Rays
MICHAEL W. DAVIDSON
Airy disk
Image plane
Intensity
Image plane
Resolved
B.
Wavefronts
Rays
MICHAEL W. DAVIDSON
Airy disk
Image plane
Intensity
Image plane
Unresolved

FIGURE 2.13 ■ Interference of light waves at the focal point generates an Airy disk. Interference of the wavefront with itself converts each point of detail to an Airy disk: a peak intensity surrounded by rings. **A.** Wide wavefronts generate narrow Airy disks that are resolved. **B.** Narrow wavefronts generate wide airy disks that are unresolved.

wave interfere with each other to form alternating regions of light and dark. Through interference, a point source of light (such as a point of detail in a specimen) forms an image of a disk containing a bright central peak surrounded by rings of light and dark. This disk is called an "Airy disk," named for the astronomer George Airy (1801–1892), who showed that the stars viewed through a telescope could never appear as true point sources, but only as tiny disks of light surrounded by rings. Similarly, under a microscope a well-focused bright object appears as a bright disk surrounded by faint rings arising from wave interference.

Suppose an object consists of a collection of point sources of light. Each point source generates an Airy disk with one central peak of intensity. The width of this central peak will define the resolution, or separation distance, between any two points of the object (**Fig. 2.13**). This resolution determines the degree of detail that can be observed. In practice, any object, such as a stained microbe against a bright field, can be considered a large collection of points of light that act as partly resolved peaks of intensity.

What factors limit resolution of an image? The wavelength of light limits the width of the peak intensity of a point of detail. For this reason, bright-field light microscopy (Section 2.3) resolves only details that are greater than half the wavelength of light, about 200 nm (0.2 µm). The width of the wavefront of light gathered by the lens also limits resolution, as discussed in the next section.

To Summarize

- **Electromagnetic radiation** interacts with an object and acquires information we can use to detect the object. **Contrast** between object and background makes it possible to detect the object and resolve its component parts.
- **The wavelength of the radiation** must be equal to or smaller than the size of the object for a microscope to resolve the object's shape.
- **Absorption** means that the energy from light (or other electromagnetic radiation) is acquired by the object. **Reflection** means that the wavefront bounces off the surface of a particle at an angle equal to its incident angle. **Scattering** occurs when a wavefront interacts with an object of smaller dimension than the wavelength. Light scattering enables the detection of objects whose detail cannot be resolved.
- **Refraction** is the bending of light as it enters a substance that slows its speed. Refraction through a curved lens **magnifies** an image, enlarging its details beyond the spacing between our eye's photoreceptors.
- **Interference** between wavefronts converts a point light source to a peak of intensity surrounded by rings. The width of the peak limits the **resolution** of details of an image.

2.3

Bright-Field Microscopy

The most common kind of light microscopy is called **bright-field microscopy**, in which an object such as a bacterial cell is perceived as a dark silhouette blocking the passage of light (for examples, see **Fig. 2.6A**, **C**, and **E**). Details of the object are defined by the points of light surrounding its edge. Here we explain how a typical student's microscope works, and how to use it to image microbes.

Magnification

How do the optics of a bright-field microscope maximize the observation of detail? We consider the following factors:

- **Wavelength and resolution.** The resolution limit derived from the wavelength of light is 200 nm, while our eye's detector spacing is a thousandfold larger (approximately 150 µm, or 150,000 nm). Thus, the greatest magnification that can improve our perception of detail is about 1,000×. Any greater magnification expands the image size, but with no greater resolution (empty magnification).
- **Light and contrast.** For any given lens system, a balanced amount of light yields the highest contrast between the dark specimen and the light background. High contrast is needed to perceive the full resolution at a given magnification.
- **Lens quality.** All lenses contain inherent **aberrations** that detract from perfect curvature. Optical properties limit the perfection of a single lens, but manufacturers construct microscopes with a series of lenses that multiply each other's magnification and correct for aberrations.

Let's first consider magnification of an image by a single lens. **Figure 2.14** shows an **objective lens**, a lens situated directly above an object or specimen that we wish to observe at high resolution. How can we maximize the resolution of details?

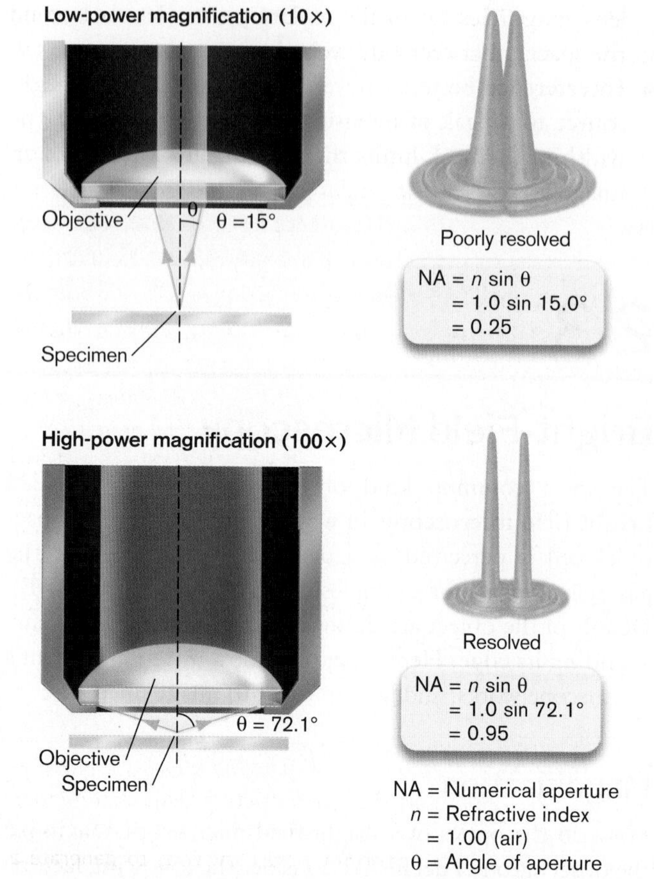
Low-power magnification (10×)
Objective
θ
θ =15°
Specimen
Poorly resolved
NA = n sin θ
= 1.0 sin 15.0°
= 0.25
High-power magnification (100×)
θ = 72.1°
Objective
Specimen
Resolved
NA = n sin θ
= 1.0 sin 72.1°
= 0.95
NA = Numerical aperture
n = Refractive index
= 1.00 (air)
θ = Angle of aperture

FIGURE 2.14 ■ Numerical aperture and resolution. The numerical aperture (NA) equals the refractive index (n) of the medium containing the light cone, multiplied by the sine of the angle of the light cone (θ). Higher NA allows greater resolution.

Thought Question

2.4 (refer to **Fig. 2.14**) What angle θ might offer magnification even greater than 100×? What practical problem would you have in designing such a lens to generate this light cone?

An object at the focal point of a lens sits at the tip of a cone of light formed by rays from the lens converging at the object. The angle of the light cone is determined by the curvature and refractive index of the lens. The lens fills an aperture, or hole, for the passage of light; and for a given lens the light cone is defined by an angle θ (theta) projecting from the midline, known as the **angle of aperture**. As θ increases and the horizontal width of the light cone ($\sin \theta$) increases, a wider cone of light passes through the specimen. The wider the cone of light rays, the less the interference between wavefronts—and the narrower the peak intensities in the image. Thus, a wider light cone allows us to resolve smaller details. The greater the angle of aperture of the lens ($\sin \theta$), the better the resolution.

Resolution also depends on the refractive index of the medium that contains the light cone, which is usually air. The refractive index (n) is the ratio of the speed of light in a vacuum to its speed in another medium. For air, the refractive index (n) is extremely close to 1. For water, $n = 1.33$; for lens material, n ranges from 1.4 to 1.6. As light passing through air or water enters a lens of higher refractive index, the light bends, at angles up to a maximum (θ). The product of the refractive index (n) of the medium multiplied by $\sin \theta$ is the **numerical aperture** (NA):

$$\text{NA} = n \sin \theta$$

In **Figure 2.14** we see the calculation of NA for an objective lens of magnification 10×, and for a lens of magnification 100×. As NA increases, the peak intensities of an image narrow, and the distance between two objects that can be resolved decreases. The minimum resolution distance R varies inversely with NA:

$$R = \frac{\lambda/\text{NA}}{2}$$

where λ represents the wavelength of incident light.

Note that as the lens strength increases and the light cone widens, the lens must come nearer the object. Defects in lens curvature become more of a problem, and focusing becomes more challenging. As θ becomes very wide, too much of the light from the object is lost owing to refraction at the glass-to-air interface. The greater the refractive index of the medium between the object and the objective lens, the more light can be collected and focused.

For the highest-power objective lens, generally 100×, a zone of constant refractive index is maintained by placing **immersion oil** between the object and the lens. Immersion oil has a refractive index comparable to that of the lens ($n = 1.5$) (**Fig. 2.15**). Immersion oil minimizes the loss of light rays at the widest angles and makes it possible to reach 100× magnification with minimal distortion.

The Compound Microscope

A **compound microscope** is a system of multiple lenses designed to correct or compensate for lens aberrations (deviation from perfect curvature). Why do we use a compound microscope instead of a single perfect lens? The manufacture of high-power lenses is difficult because the effects of aberration increase faster than the magnification. Instead, a series of lenses can multiply their magnification with minimal aberration. A typical arrangement of a

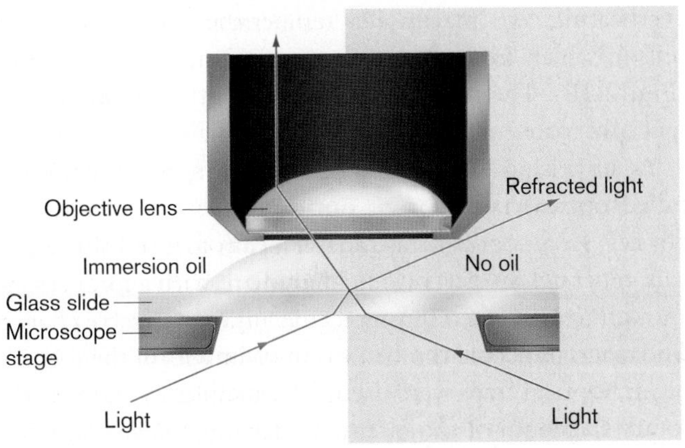
Objective lens
Refracted light
Immersion oil
No oil
Glass slide
Microscope stage
Light
Light

FIGURE 2.15 ■ Use of immersion oil in microscopy. Immersion oil with a refractive index comparable to that of glass (n = 1.5) prevents light rays from bending away from the objective lens. Thus, more light is collected, NA increases, and resolution improves.

compound microscope is shown in **Figure 2.16**. In this arrangement, the light source is placed at the bottom, shining upward through a series of lenses, including the condenser, objective, and ocular lenses.

Between the light source and the condenser sits a **diaphragm**, a device to cut the diameter of the light column. Lower-power lenses require lower light levels because the excess light makes it impossible to observe the darkening effect of absorbance by the specimen (contrast). Higher-power lenses require more light and thus an open diaphragm.

Above the diaphragm, the **condenser** consists of one or more lenses that collect a beam of rays from the light source onto a small area of the slide, where light may be absorbed by the object or specimen. The **objective lens** forms a magnified image (I) of the object (**Fig. 2.16A**). As the image forms, each light ray traces a path toward a position opposite its point of origin; thus, the image is mirror-reversed. Keep this mirror reversal in mind when exploring a field of cells.

The first image (I) is then amplified by a secondary magnification step through the **ocular lens**. The final virtual image (I') includes the total magnification of the object.

The nosepiece of a compound microscope typically holds three or four objective lenses of different magnifying power, such as 4×, 10×, 40×, and 100× (requiring immersion oil). These lenses are arranged so as to rotate in turn into the optical column. A high-quality instrument will have the lenses set at different heights from the slide so as to be **parfocal**. In a parfocal system, when an object is focused with one lens, it remains in focus, or nearly so, when another lens is rotated to replace the first.

Note: Objective lenses can be obtained in several different grades of quality, manufactured with different kinds of correction for aberrations. Lenses should feature at minimum the following corrections: "plan" correction for field curvature, to generate a field that appears flat; and "apochromat" correction for spherical and chromatic aberrations.

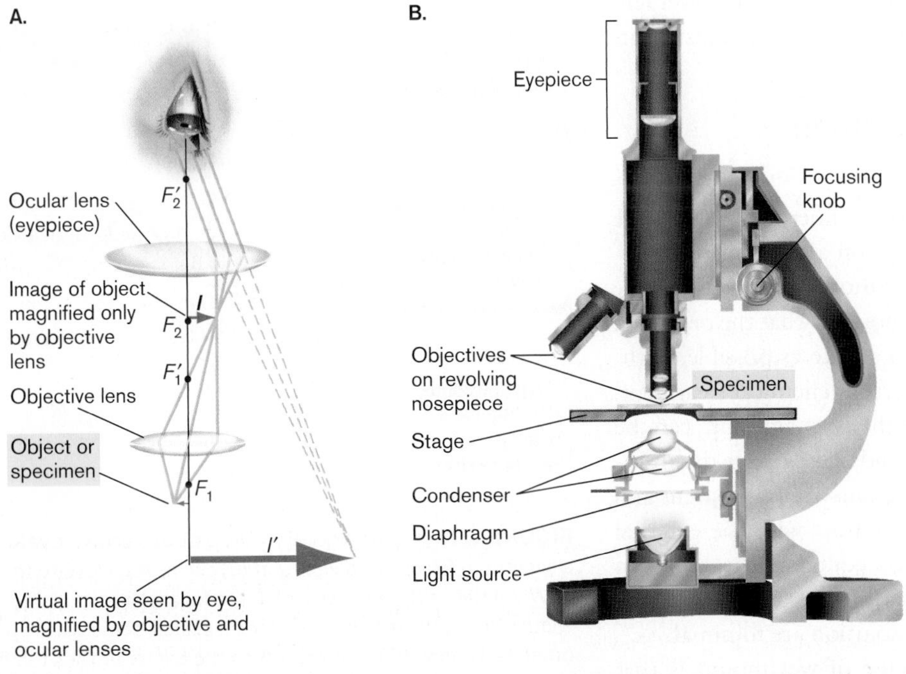
A.
Ocular lens (eyepiece)
F_2'
Image of object magnified only by objective lens
I
F_2
F_1'
Objective lens
Object or specimen
F_1
I'
Virtual image seen by eye, magnified by objective and ocular lenses
B.
Eyepiece
Focusing knob
Objectives on revolving nosepiece
Specimen
Stage
Condenser
Diaphragm
Light source

FIGURE 2.16 ■ Anatomy of a compound microscope. A. Light path through a compound microscope. **B.** Cutaway view of a compound microscope.

The magnification factor of the ocular lens is multiplied by the magnification factor of the objective lens to generate the **total magnification** (power). For example, a 10× ocular multiplied by a 100× objective generates 1,000× total magnification.

Observing a specimen under a compound microscope requires several steps:

- **Position the specimen centrally in the optical column.** Only a small area of a slide can be visualized within the field of view of a given lens. The higher the magnification, the smaller the field of view that will be seen.
- **Optimize the amount of light.** At lower power, too much light will wash out the light absorption of the specimen. At higher power, more light needs to be collected by the condenser. To optimize light, the condenser must be set at the correct vertical position to focus on the specimen, and the diaphragm must be adjusted to transmit the amount of light that produces the best contrast.
- **Focus the objective lens.** The focusing knob permits adjustment of the focal distance between the objective lens and the specimen on the slide so as to bring the specimen into the **focal plane**. Typically, we focus first using a low-power objective, which generates a greater **depth of field**—that is, a planar section along the optical column in which the object appears in reasonable focus. After focusing under low power, we can rotate a higher-power lens into view and then fine-tune the adjustment.

Preparing a Specimen for Microscopy

A simple way to observe microbes is to place them in a drop of water on a slide with a coverslip. This is called a **wet mount** preparation. The advantage of the wet mount is that the organism is viewed in as natural a state as possible, without artifacts resulting from chemical treatment; and live behavior such as swimming can be observed (**Figs. 2.3** and **2.17**). The disadvantage is that most living cells are transparent and therefore show little contrast with the external medium. With limited contrast, the cells can barely be distinguished from background, and both detection and resolution are minimal.

Another disadvantage of wet mount is that the sample rapidly converts absorbed light to heat, thus overheating and drying out. To avoid overheating, we may use a temperature-controlled flow cell, in which fresh medium passes through the specimen (**Fig. 2.18**). The microbe to be observed must adhere to a specially coated slide within the flow cell.

A surprising new method of "holding" a specimen, called **optical tweezers** or **optical trap**, arises from laser physics. In an optical trap, a laser beam focused through a lens generates an electrical field gradient with a force strong enough to attract a macromolecule with polarizable charge. The macromolecule can be held in place within the focused beam. Optical traps were used, for example, to "tether" the rotary flagellum of the marine bacterium *Vibrio alginolyticus,* in order to watch the bacterium rotate and to study its response to stimuli. A different application was to "trap" an isolated ribosome and watch its progress along a messenger RNA molecule (Special Topic 8.1).

Focusing the Object

An object appears in focus (that is, it is situated within the focal plane of the lens) when its edge appears sharp and distinct from the background. The shape of the dark object is

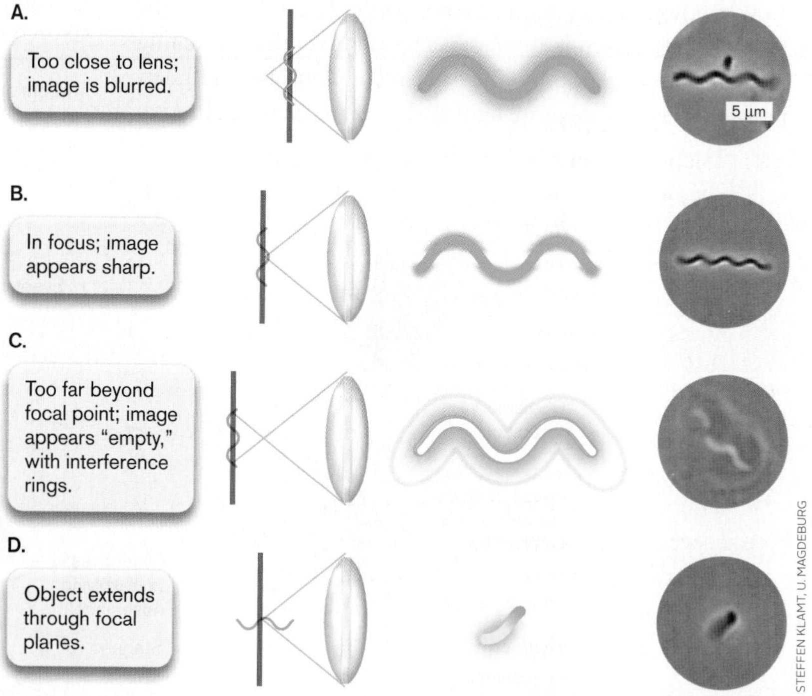
A.
Too close to lens; image is blurred.
5 μm
B.
In focus; image appears sharp.
C.
Too far beyond focal point; image appears "empty," with interference rings.
D.
Object extends through focal planes.
STEFFEN KLAMT, U. MAGDEBURG

FIGURE 2.17 ■ Bacteria observed at different levels of focus. A. When a bacterium (*Rhodospirillum rubrum*) swims too close to the lens, its image blurs. **B.** When the bacterium lies within the focal plane, its image appears sharp. If the width of the cell crosses several focal planes, some parts appear sharp, whereas other parts appear blurred. **C.** When the bacterium lies too far from the lens, its image appears "empty" or "hollow," surrounded by rings of Airy-like interference. **D.** When the cell body extends through the focal plane, different parts show different focal effects (in focus, too near, or too far).

BIOPTECHS FLOW CELL (PHOTO BY J. SLONCZEWSKI)

FIGURE 2.18 ■ Flow cell allows extended observation of living microbes. Culture medium flows through an inlet tube into the slide chamber, and then exits through the outlet.

actually defined by the points of light surrounding its edge. At higher power, as we reach the resolution limit, these points of light are only partly resolved. The partial resolution of these points of light generates interference effects, such as extra rings of light surrounding an object.

In **Figure 2.17**, we see microscopic observation of *Rhodospirillum rubrum,* photosynthetic bacteria that enrich wetlands. As *R. rubrum* cells swim in and out of the focal plane, their appearance changes through optical effects. When a bacterium swims out of the focal plane too close to the lens, resolution declines and the image blurs (**Fig. 2.17A**). When the bacterium swims within the focal plane, its image appears sharp, with a bright line along its edge. A helical cell like *R. rubrum* shows well-focused segments alternating with hazier segments that are too close to the lens (**Fig. 2.17B**). When the cell swims too far past the focal plane, the bright interference lines collapse into the object's silhouette, which now appears bright or "hollow," or surrounded by rings (**Fig. 2.17C**). In fact, the bacterium is not hollow at all; only its image has changed.

When the cell extends across several focal planes, different portions appear out of focus (either too near or too far from the lens). In addition, when the end of a cell points toward the observer, light travels through the length of the cell before reaching the observer, so the cell absorbs more light and appears dark (**Fig. 2.17D**). Motile bacteria swimming in and out of the focal plane present a challenge even to experienced microscopists. The higher the magnification, the narrower the depth of the focal plane; thus, observing swimming organisms requires a trade-off between magnification and depth of field.

Thought Question

2.5 Under starvation conditions, a bacterium such as *Bacillus thuringiensis,* the biological insecticide, packages its cytoplasm into a spore, leaving behind an empty cell wall. Suppose, under a microscope, you observe what appears to be a hollow cell. How can you tell whether the cell is indeed hollow or is simply out of focus?

Fixation and Staining Improve Resolution and Contrast

Detection and resolution of cells under a microscope are enhanced by **fixation** and **staining**, procedures that usually kill the cell. Fixation is a process by which cells are made to adhere to a slide in a fixed position. We can fix cells with methanol or by heat treatment to denature cellular proteins, whose exposed side chains then adhere to the glass. A stain absorbs much of the incident light, usually over a wavelength range that results in a distinctive color. The use of chemical stains was developed in the nineteenth century, when German chemists used organic synthesis to invent new coloring agents for women's clothing. Clothing was made of natural fibers such as cotton or wool, so a substance that dyed clothing would be likely to react with biological specimens.

How do stains work? Most stain molecules contain conjugated double bonds or aromatic rings that absorb visible light (**Fig. 2.19**) and one or more positive charges that bind to negative charges such as phosphoryl groups on membrane phospholipids (discussed in Section 3.2). Different stains vary with respect to the strength of their binding and the degree of binding to different parts of the cell.

Different Kinds of Stains

A **simple stain** adds dark color specifically to cells, but not to the external medium or surrounding tissue (in the case of pathological samples). The most commonly used simple stain is methylene blue, originally used by Robert Koch to stain bacteria (see **Figs. 2.6E** and **2.7B**). A typical procedure for fixation and staining is shown in **Figure 2.20**. First we fix a drop of culture on a slide by treating with methanol or by heating on a slide warmer. Either treatment

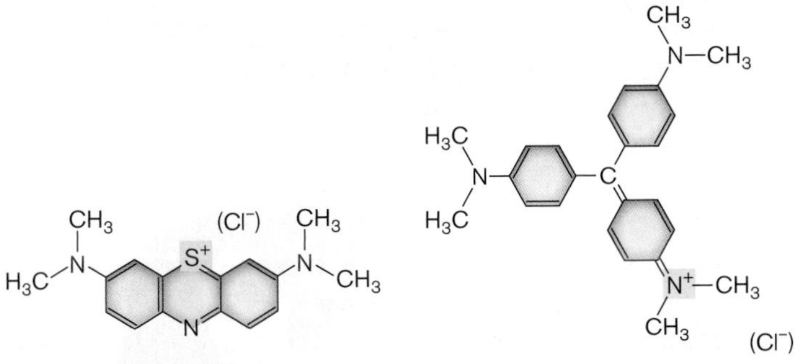

Methylene blue

Crystal violet

FIGURE 2.19 ■ Chemical structure of stains. Methylene blue and crystal violet are cationic (positively charged) dyes. The positively charged groups react with the bacterial cell envelope, which carries mainly negative charge. Chloride (Cl^-) is the counter-ion.

denatures cell proteins, exposing side chains that bind to the glass. We then flood the slide with methylene blue solution. The positively charged molecule binds to the negatively charged cell envelope of fixed bacteria. After excess stain is washed off and the slide has been dried, we observe it under high-power magnification using immersion oil.

A **differential stain** stains one kind of cell but not another. The most famous differential stain is the **Gram stain**, devised in 1884 by the Dutch physician Hans Christian Gram (1853–1938). Gram first used the Gram stain to distinguish pneumococcal (*Streptococcus pneumoniae*) bacteria from human lung tissue. In **Figure 2.21A**, Gram-stained *S. pneumoniae* bacteria appear dark purple amongst human white blood cells. Other species of bacteria such as *Proteus mirabilis* (a cause of urinary infections) fail to retain the purple stain (**Fig. 2.21B**). Different bacterial species

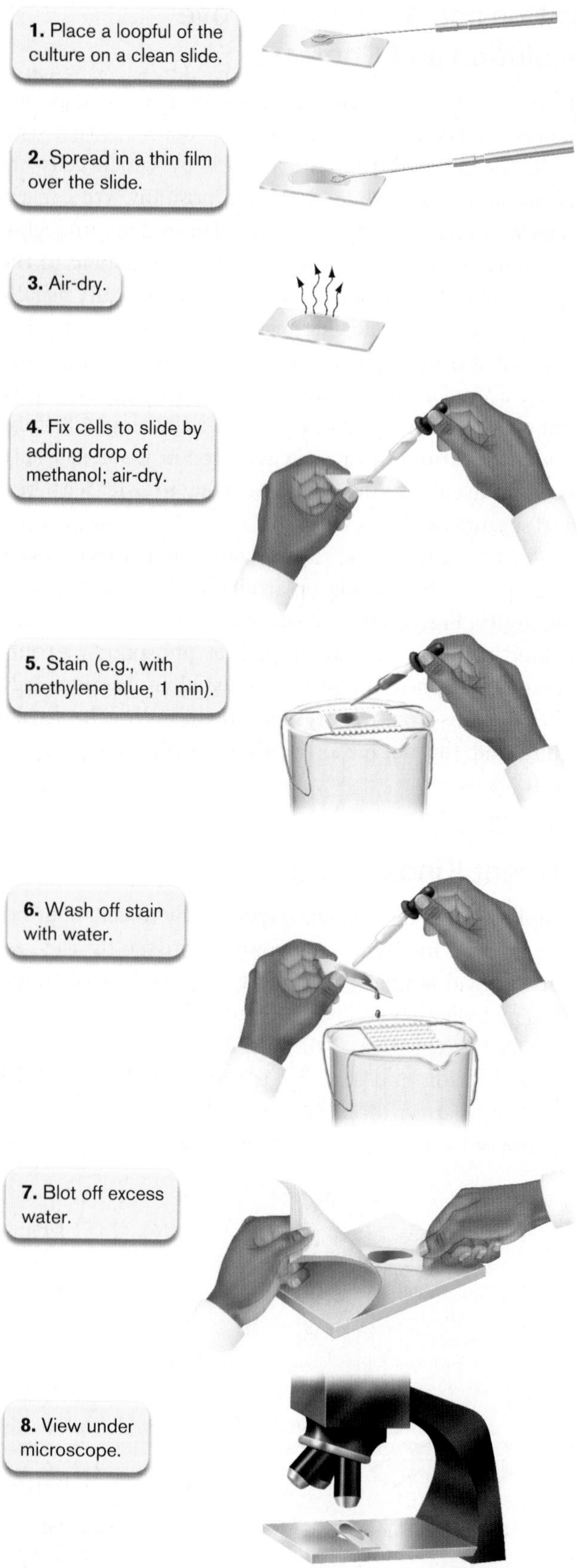
1. Place a loopful of the culture on a clean slide.
2. Spread in a thin film over the slide.
3. Air-dry.
4. Fix cells to slide by adding drop of methanol; air-dry.
5. Stain (e.g., with methylene blue, 1 min).
6. Wash off stain with water.
7. Blot off excess water.
8. View under microscope.

FIGURE 2.20 ■ Procedure for simple staining with methylene blue.

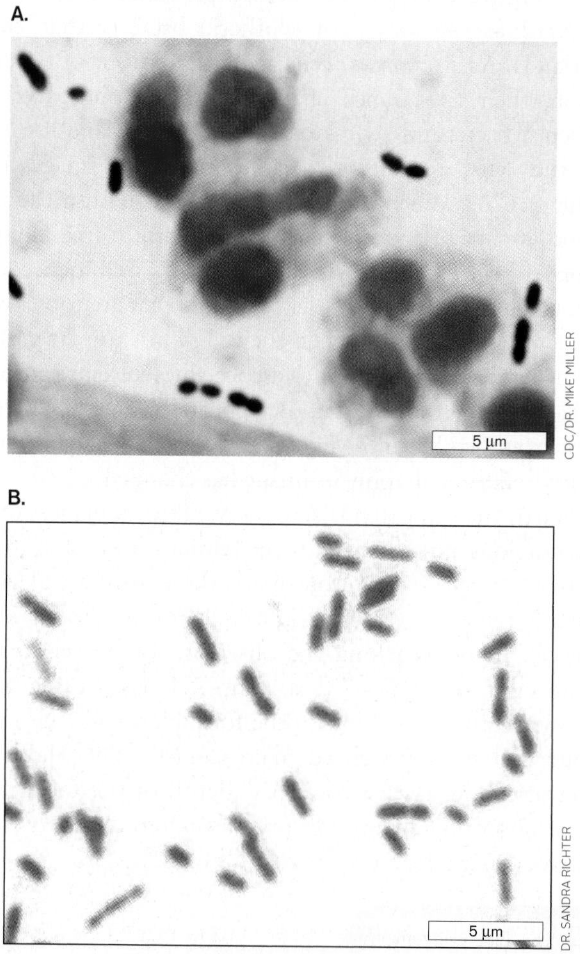
A.
5 µm
CDC/DR. MIKE MILLER
B.
5 µm
DR. SANDRA RICHTER

FIGURE 2.21 ■ Gram staining of bacteria (a type of differential stain). A. Gram stain of a sputum specimen from a patient with pneumonia, containing Gram-positive *Streptococcus pneumoniae* (purple diplococci) among white blood cells in pus. Cell length 05–1.0 µm. **B.** Gram-negative *Proteus mirabilis* (pink rods). Cell length 1–2 µm.

are classified as Gram-positive or Gram-negative, depending on whether they retain the purple stain.

Gram Staining Separates Bacteria into Two Classes

In the Gram stain procedure (**Fig. 2.22**), a dye such as crystal violet binds to the bacteria; it also binds to the surface of human cells, but less strongly. After the excess stain is washed off, a **mordant**, or binding agent, is applied. The mordant used is iodine solution, which contains iodide ions (I^-). The iodide complexes with the positively charged crystal violet molecules trapped inside the cells. The crystal violet–iodide complex is now held more strongly within the cell wall. The thicker the cell wall, the more crystal violet–iodide molecules are held.

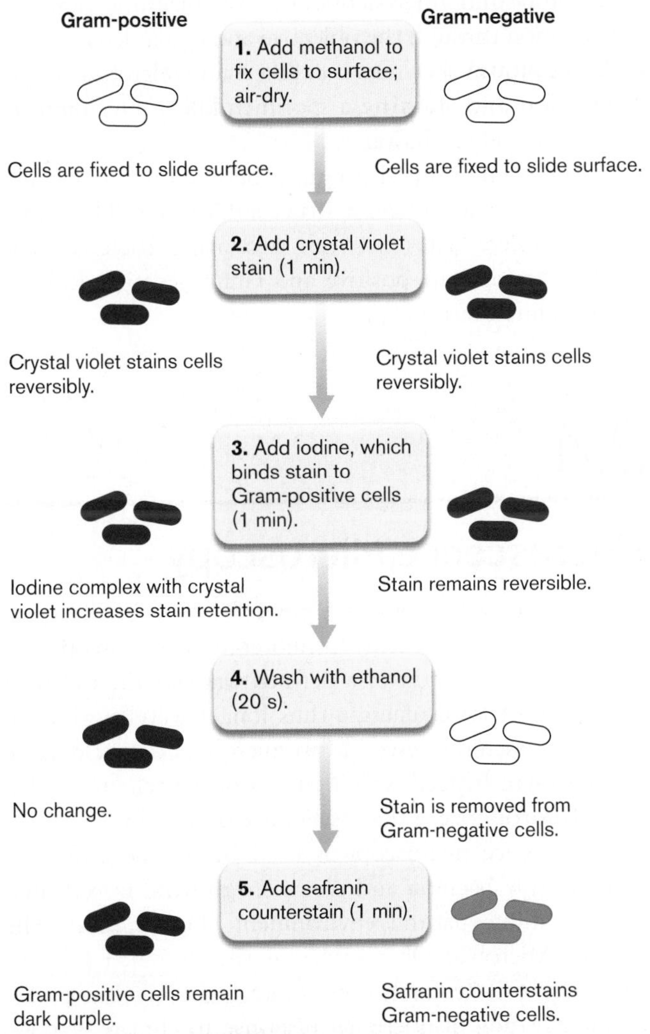
Gram-positive
Gram-negative
1. Add methanol to fix cells to surface; air-dry.
Cells are fixed to slide surface.
Cells are fixed to slide surface.
2. Add crystal violet stain (1 min).
Crystal violet stains cells reversibly.
Crystal violet stains cells reversibly.
3. Add iodine, which binds stain to Gram-positive cells (1 min).
Iodine complex with crystal violet increases stain retention.
Stain remains reversible.
4. Wash with ethanol (20 s).
No change.
Stain is removed from Gram-negative cells.
5. Add safranin counterstain (1 min).
Gram-positive cells remain dark purple.
Safranin counterstains Gram-negative cells.

FIGURE 2.22 ■ The Gram stain procedure. The Gram stain distinguishes between Gram-positive cells, with thick cell walls, which retain the crystal violet stain, and Gram-negative cells, with thinner cell walls, which lose the crystal violet stain but are counterstained by safranin.

Next a decolorizer, ethanol, is added for a precise time interval (typically 20 seconds). The decolorizer removes loosely bound crystal violet–iodide, but Gram-positive cells retain the stain tightly. The **Gram-positive** cells that retain the stain appear dark purple, while the **Gram-negative** cells are colorless. The decolorizer step is critical because if it lasts too long, the Gram-positive cells, too, will release their crystal violet stain. In the final step, a **counterstain**, safranin, is applied. This process allows the visualization of Gram-negative material, which is stained pale pink by the safranin.

It turns out that Gram-negative species of bacteria possess a cell wall thinner and more porous than that of Gram-positive species (discussed in Chapter 3). A Gram-negative cell wall has only one to three layers of peptidoglycan (sugar chains cross-linked by peptides), whereas a Gram-positive cell has five or more layers (**Fig. 2.23**). The multiple layers of peptidoglycan retain enough stain complex that the cell appears purple.

Thus the Gram stain became a key tool for identifying species in the clinical laboratory. Note, however, that still other groups of bacteria and archaea have different kinds of cell walls that may stain either positive or negative and are thus not distinguished by the Gram stain (discussed in Chapters 18 and 19).

Other Differential Stains

Other differential stains applied to various types of prokaryotes are illustrated in **Figure 2.24**. These include:

- **Acid-fast stain** (Ziehl-Neelsen). Carbolfuchsin specifically stains mycolic acids of *Mycobacterium tuberculosis* and *M. leprae,* the causative agents of tuberculosis and leprosy, respectively (**Fig. 2.24A**).
- **Spore stain.** When samples are boiled with malachite green, the stain binds specifically to the endospore coat (**Fig. 2.24B**). It detects spores of *Bacillus* species such as the insecticide *B. thuringiensis* and *B. anthracis,* the cause of anthrax, as well as spores of *Clostridium botulinum,* which produces botulinum toxin.
- **Negative stain.** Some bacteria, such as the soil bacterium *Flavibacterium capsulatum*, synthesize a capsule of extracellular polysaccharide filaments. The capsule protects the cell from predation or from engulfment by white blood cells. The capsule is transparent and invisible in suspended cells. However, a suspension of opaque particles such as India ink or Congo red dye can be added to darken the surrounding medium. The particles are excluded by the thick polysaccharide capsule, which

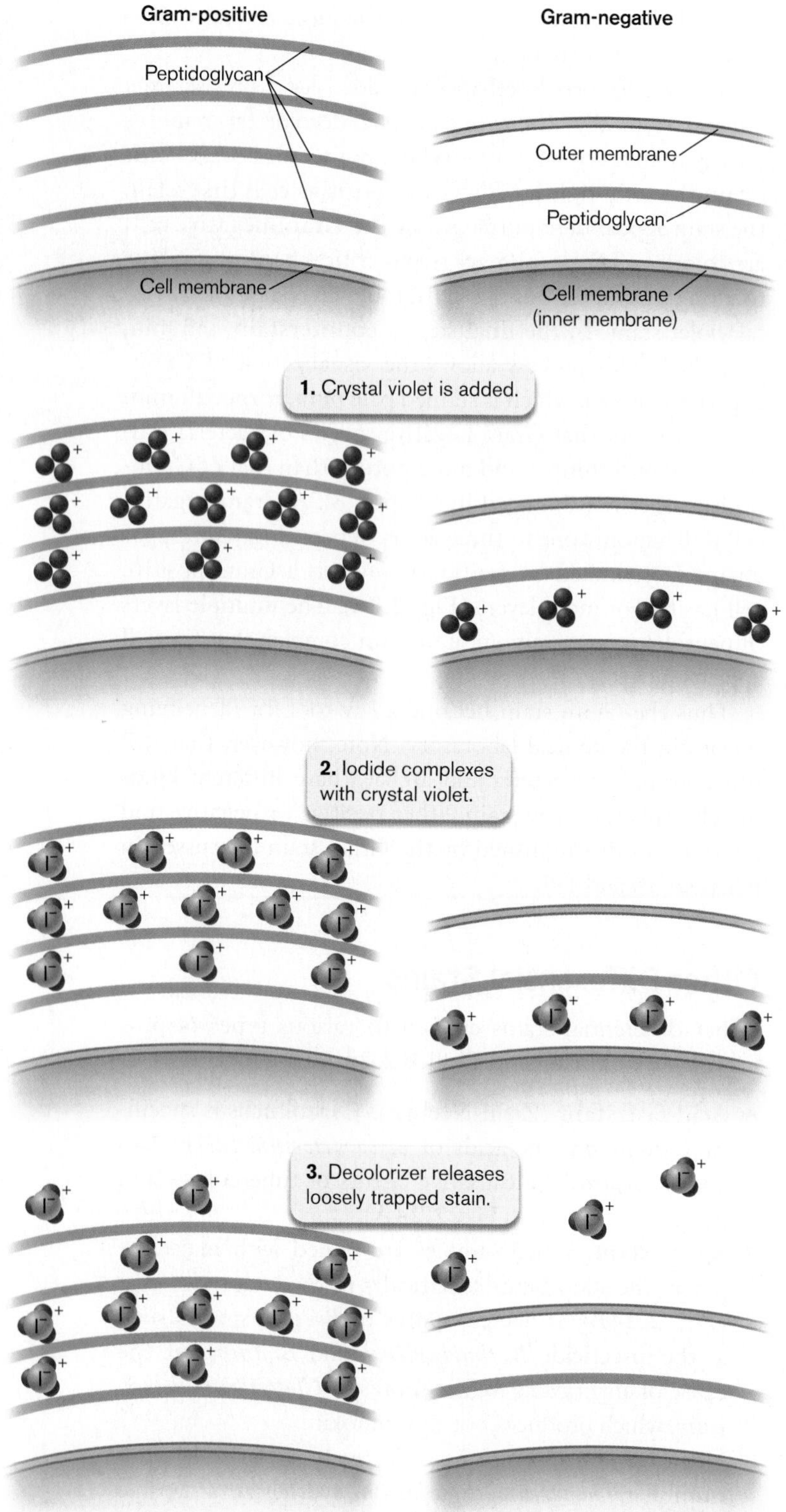
Gram-positive
Gram-negative
Peptidoglycan
Outer membrane
Peptidoglycan
Cell membrane
Cell membrane
(inner membrane)
1. Crystal violet is added.
2. Iodide complexes with crystal violet.
3. Decolorizer releases loosely trapped stain.

FIGURE 2.23 ■ Mechanism of the Gram stain. In a Gram-positive cell, the crystal violet–iodide complex is retained by multiple layers of peptidoglycan. In a Gram-negative cell, the stain leaks out. If the decolorizer is applied for too long, the Gram-positive cell will lose its stain as well.

thus appears clear against the dark background; this is called a negative stain (**Fig. 2.24C**).

- **Antibody stains.** Stains linked to antibodies can identify precise strains of bacteria or even specific molecular components of cells. The antibody (which binds a specific cell protein) is linked to a reactive enzyme for detection, or to a fluorophore (fluorescent molecule) for immunofluorescence microscopy (discussed next, in Section 2.4).

To Summarize

- **In bright-field microscopy**, image quality depends on the wavelength of light; the magnifying power of a lens; and the position of the focal plane, the region where the specimen is "in focus" (that is, where the sharpest image is obtained).
- **A compound microscope** achieves magnification and resolution through the objective and ocular lenses.
- **A wet mount** specimen contains living microbes.
- **Fixation and staining** a specimen kills it but improves contrast and resolution.
- **Differential stains** distinguish between different kinds of bacteria with different structural features. **The Gram stain** differentiates between two major bacterial taxa, which stain Gram-positive and Gram-negative. Eukaryotes stain negative.

2.4

Fluorescence Microscopy

Fluorescence microscopy is a powerful tool for detecting microbes and parts of cells. In fluorescence microscopy, the specimen absorbs light of a defined wavelength, and then emits light of lower energy, thus longer wavelength; that is, the specimen fluoresces. Most microbes do not fluoresce on their own. Instead, specific parts of the cell are labeled with a **fluorophore**, a fluorescent dye or protein.

Fluorescence microscopy is used by marine ecologists to reveal tiny bacteria and plankton growing in seawater, a highly dilute natural environment (**Fig. 2.25A**). The Oceanic Microbial Observatory at the Bermuda Biological Station for Research uses fluorescence microscopy for long-term study of microbial response to climate change. Microbes, including viruses, bacteria, and protists, are detected by means of DNA-specific fluorescence. The advantage of this fluorescence technique is that it detects only live cells whose DNA is intact, distinguishing them from environmental debris.

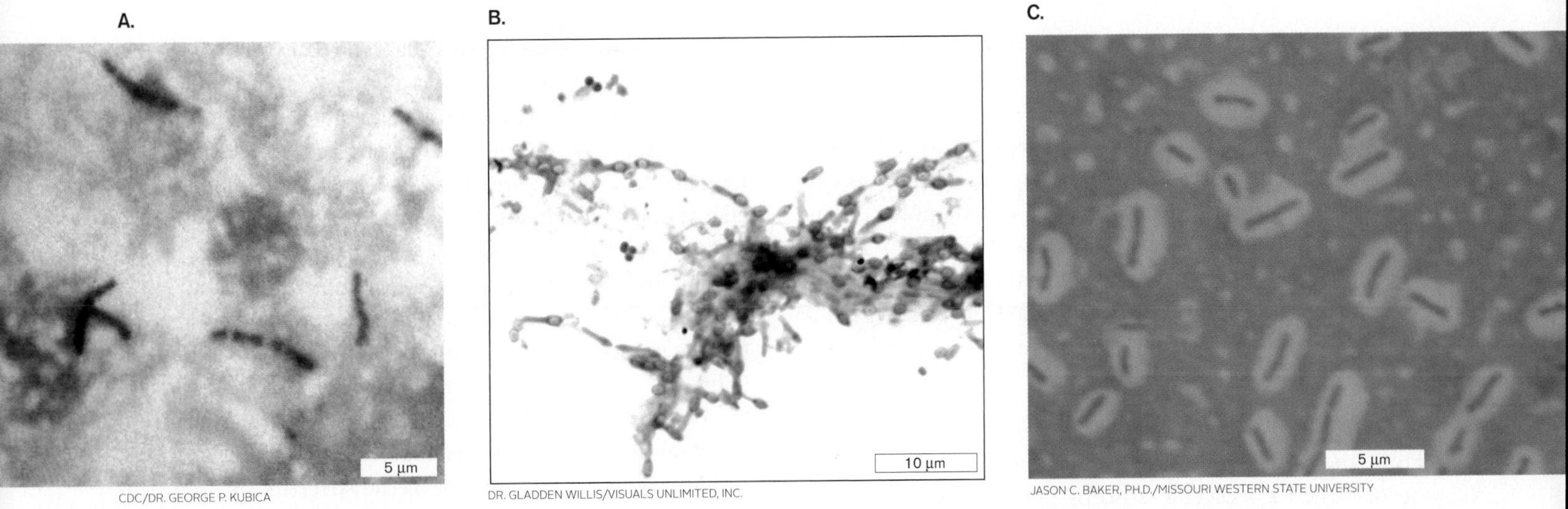
A.
B.
C.
5 µm
10 µm
5 µm
CDC/DR. GEORGE P. KUBICA
DR. GLADDEN WILLIS/VISUALS UNLIMITED, INC.
JASON C. BAKER, PH.D./MISSOURI WESTERN STATE UNIVERSITY

FIGURE 2.24 ▪ Differential stains. A. *Mycobacterium tuberculosis,* acid-fast stain (stained cells are red). **B.** *Clostridium tetani,* endospore stain (stained endospores are blue). **C.** *Flavobacterium capsulatum;* negative stain with Congo red reveals translucent capsule surrounding each bacterium.

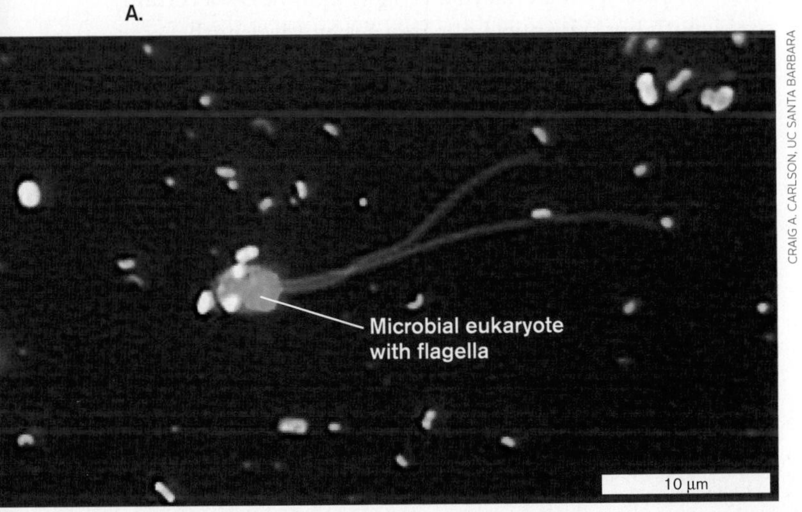
A.
Microbial eukaryote with flagella
10 µm
CRAIG A. CARLSON, UC SANTA BARBARA

FIGURE 2.25 ▪ Fluorescence microscopy. A. DNA-specific fluorophore DAPI reveals bacteria and protists from the Sargasso Sea; vast numbers of these tiny plankton cycle the ocean's carbon. **B.** *Listeria* bacteria (green) invade human cells, propelled by polymerizing "tails" of actin (violet). Two human cell nuclei fluoresce red.

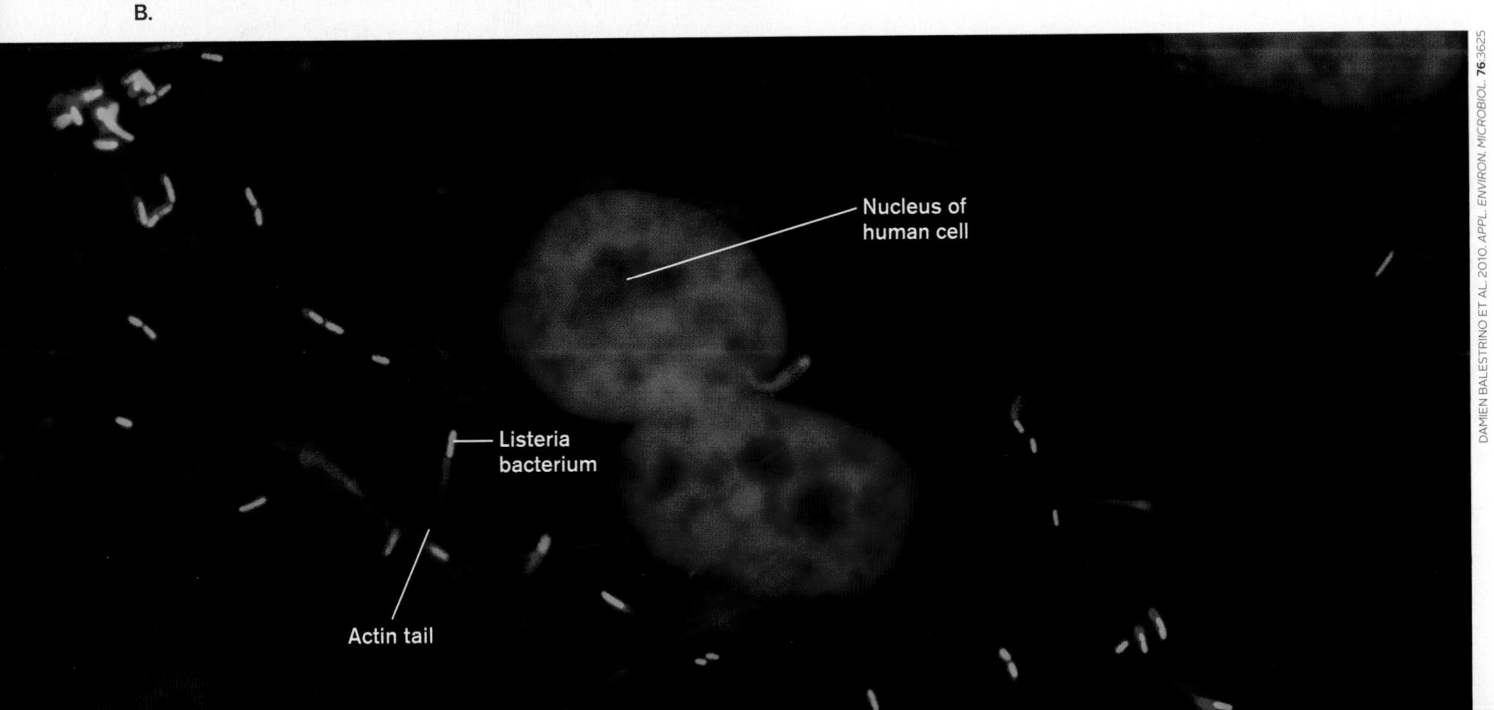
B.
Nucleus of human cell
Listeria bacterium
Actin tail
DAMIEN BALESTRINO ET AL. 2010. APPL. ENVIRON. MICROBIOL. 76:3625

For medicine, fluorescence enables detection of microbial pathogens within the relatively large cells and organ systems of the human body. **Figure 2.25B** shows infection by *Listeria monocytogenes*, the cause of listeriosis, a dangerous foodborne disease. The *Listeria* bacteria (green fluorophore) propel themselves through a human cell by polymerizing "tails" of human cell actin (violet, with an actin-specific fluorophore). The human cell nuclei are labeled red with a DNA-specific fluorophore. Thus, fluorescence can visualize three different components within one microscopic field.

Excitation and Emission

How and when does a molecule fluoresce? Fluorescence occurs when a molecule absorbs light of a specific wavelength (the **excitation wavelength**) that has just the right energy needed to raise an electron to a higher-energy orbital (**Fig. 2.26**). Because this higher-energy electron state is unstable, the electron decays to an orbital of slightly lower energy, while losing some energy as heat. The electron then falls to its original level by emitting a photon of less energy and longer wavelength (the **emission wavelength**). The emitted photon has a longer wavelength (less energy) because part of the electron's energy of absorption was lost as heat.

The optical system for fluorescence microscopy uses filters to limit incident light to the wavelength of excitation and emitted light to the wavelength of emission (**Fig. 2.27**). The wavelengths of excitation and emission are determined by the choice of fluorophore; in this case we show excitation light as blue, and emission as green. Because only a small portion of the spectrum is used, fluorescence requires a high-intensity light source such as a tungsten arc lamp. The light passes through a filter that screens out all but the peak wavelengths of excitation. The excitation light (blue) is then reflected by a dichroic mirror (dichroic filter), a material that reflects light below a certain wavelength but transmits light above that wavelength.

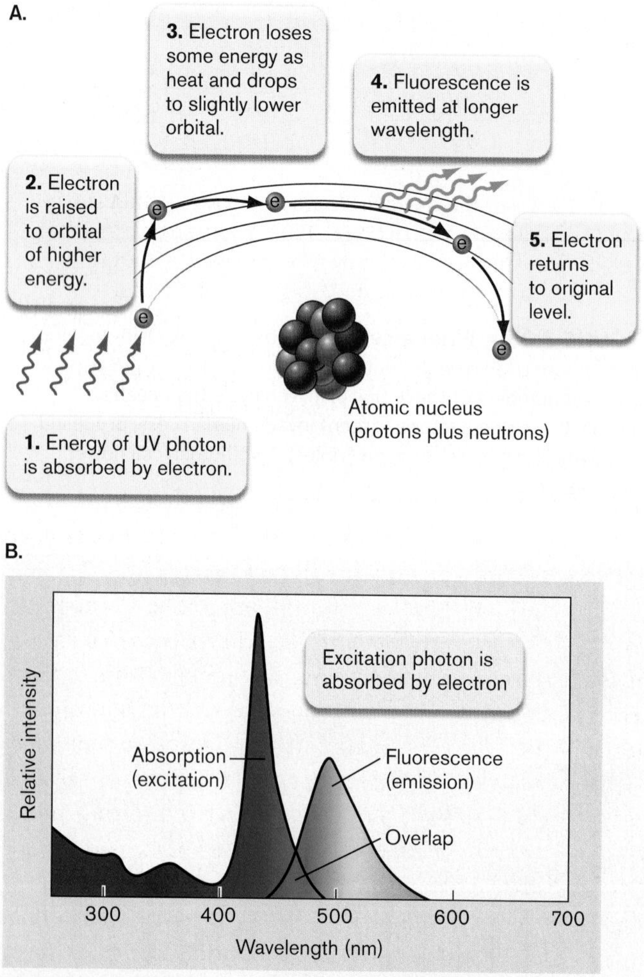
A.
3. Electron loses some energy as heat and drops to slightly lower orbital.
4. Fluorescence is emitted at longer wavelength.
2. Electron is raised to orbital of higher energy.
5. Electron returns to original level.
1. Energy of UV photon is absorbed by electron.
Atomic nucleus (protons plus neutrons)
B.
Excitation photon is absorbed by electron
Relative intensity
Absorption (excitation)
Fluorescence (emission)
Overlap
300
400
500
600
700
Wavelength (nm)

FIGURE 2.26 ■ Fluorescence. Energy gained from UV absorption is released as heat and as a photon of longer wavelength in the visible region. **A.** Fluorescence on the molecular level. **B.** Comparison of absorption and emission spectra for a fluorophore.

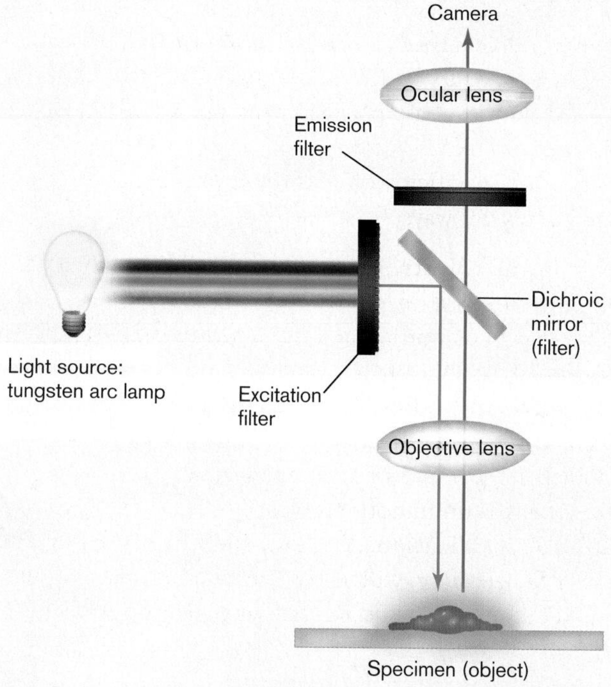
Camera
Ocular lens
Emission filter
Dichroic mirror (filter)
Light source: tungsten arc lamp
Excitation filter
Objective lens
Specimen (object)

FIGURE 2.27 ■ Fluorescence microscopy. A high-intensity light source shines through an excitation filter that transmits only the range of wavelengths to excite the fluorophore (blue in this case). The excitation light is reflected by the dichroic mirror, focused by the objective lens, and absorbed by the specimen. Fluorescence (emitted light, green) is focused by the objective lens and penetrates the dichroic mirror. The emitted light range of wavelengths is narrowed further by the emission filter to eliminate stray light, and then focused by the ocular lens for detection by the CCD (charge-coupled device) camera.

The reflected blue light enters the objective lens, which focuses it onto the specimen, where it excites fluorophores to fluoresce green. The fluorescence emanates in all directions from the specimen (like a point source). A small portion of the fluorescent light (green) returns through the objective lens to reach the dichroic mirror. The green light now has a longer wavelength, above the penetration limit of the mirror, so it continues through to the ocular lens. The ocular lens focuses the fluorescent light onto the photodetectors of a highly sensitive digital camera.

Fluorescence can be observed in live organisms. The organism must, however, be fixed to a substrate, such as a slide coated with positively charged molecules that bind the negatively charged cell-surface molecules of the microbe.

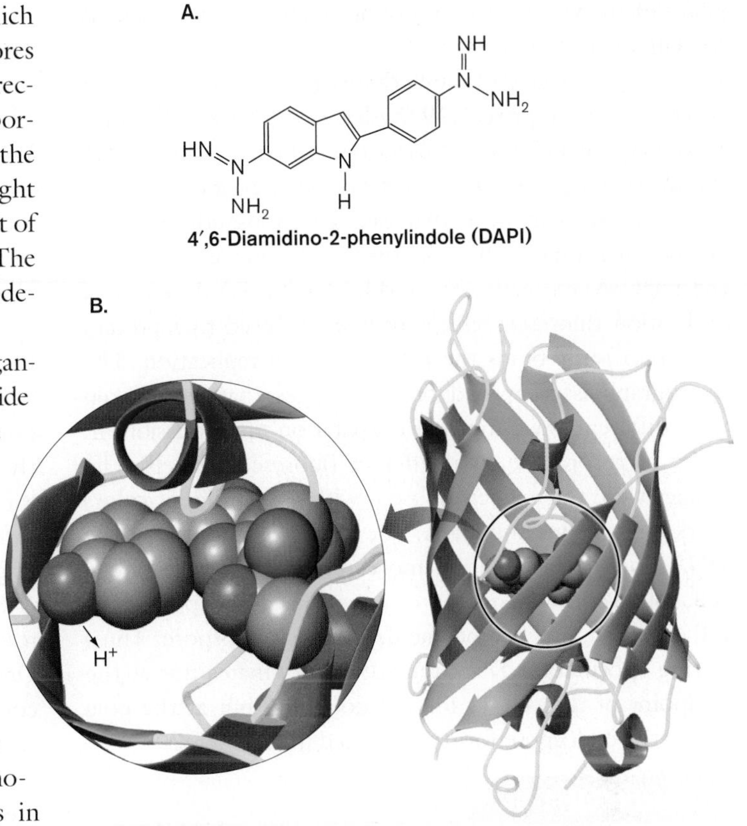
A.
NH
N
NH2
HN
N
NH2
N
H
4′,6-Diamidino-2-phenylindole (DAPI)
B.
H+

FIGURE 2.28 ■ Fluorescent molecules (fluorophores) commonly used in microscopy. The conjugated double bonds in these molecules provide closely spaced molecular orbitals that give rise to fluorescence. **A.** DAPI specifically labels DNA. **B.** Green fluorescent protein (GFP) is expressed endogenously by the cell. **Inset:** Three GFP amino acid residues (serine, tyrosine, and glycine) condense to form the fluorophore. The fluorophore can deprotonate; thus, its fluorescence depends on pH.

Fluorophores for Labeling

What determines the properties of a fluorophore? The molecular structure of each fluorophore determines its peak wavelengths of excitation and emission, as well as its binding properties. For example, the slides for cell counting in the marine ecology study (see **Fig. 2.25**) used the DNA-specific stain 4′,6-diamidino-2-phenylindole (DAPI) (**Fig. 2.28A**). DAPI absorbs in the UV and emits in the blue range. The aromatic rings of DAPI mimic a base pair, enabling intercalation between base pairs of DNA.

The cell specificity of the fluorophore can be determined in several ways:

- **Chemical affinity.** Certain fluorophores have chemical affinity for certain classes of biological molecules. For example, the fluorophore FM4-64 (green excitation, red emission) specifically binds membranes.
- **Labeled antibodies.** Antibodies that specifically bind a cell component are chemically linked to a fluorophore molecule. The use of antibodies linked to fluorophores is known as immunofluorescence.
- **DNA hybridization.** A short sequence of DNA attached to a fluorophore will hybridize to a specific sequence in the genome. Thus we can label one position in the chromosome.

A special category of fluorophores occurs within fluorescent proteins that the living cell makes itself. The most famous of these is green fluorescent protein (GFP) (**Fig. 2.28B**), whose discovery led to the 2008 Nobel Prize in Chemistry for Osamu Shimomura, Martin Chalfie, and Roger Tsien. Originally isolated from a jellyfish, *Aequorea victoria*, GFP can be expressed from a gene spliced into the DNA of any organism; even monkeys have been engineered to "glow" green. Bacteria can be engineered with a **gene fusion**, a fused gene that expresses one of their own bacterial proteins combined with GFP. The GFP fluorescence lets us detect a single protein within the cell.

How does GFP act as a fluorophore? The fluorophore part of GFP consists of three amino acid residues that fuse to form an aromatic ring structure, embedded within a "beta barrel" protein tube. The properties of the fluorophore are modified by the surrounding protein, so mutation of the gene encoding GFP generates numerous variants with different spectral ranges. Another interesting property of GFP is that its fluorophore can ionize (labeled in **Fig. 2.28B**). The ionized and protonated forms have slightly different excitation ranges, thus affording a way to measure hydronium ion concentration (pH) within a cell. This property enables GFP to report on environmental stress responses of bacteria, such as the biofilm shown in the Current Research Highlight. The wide range of "colors" and environmental sensitivities of GFP variants provides an extraordinary set of probes for cell structure, as we will see in Chapter 3.

Another example of multi-fluorophore labeling of a cell is the image in **Figure 2.29** (by Richard Losick at Harvard University), which tracks cells of the soil bacterium *Bacillus subtilis.* Under starvation, *Bacillus* cells develop an endospore—a dormant form that can last thousands of years. In the micrograph, the cell envelope fluoresces red, from the membrane-specific dye FM4-64. The DNA origin of replication fluoresces green, from GFP fused to a protein that specifically binds the DNA origin of replication. The micrograph is composed of two images that were taken with different wavelengths for excitation and emission, in order to record the two different fluorescent labels. The images were then superimposed, with two different colors marking envelope and DNA. The composite shows how the DNA origin of replication moves toward one pole of the cell, followed by formation of a septum, a new portion of cell envelope, just behind the developing endospore. Thus, during sporulation DNA replication originates not at the midpoint of the cell, as for cell doubling, but at the pole where the endospore forms (discussed in Chapter 4).

Thought Question

2.6 What experiment could you devise to determine the actual order of events in DNA movement toward the pole during formation of an endospore?

In **Fig. 2.29**, note how the labeled membranes and DNA origin appear diffuse—that is, unresolved. The emitted light travels in all directions from the "point source" of the object, and its resolution is limited by the wavelength. We can detect the location of a DNA-binding protein within a cell and resolve it as distinct from another fluorescent object located elsewhere. But fluorescence cannot resolve the detailed shape of a protein, nor distinguish two proteins that are close together.

Another concern with the use of GFP fluorescence is that proteins fused to GFP may behave differently from the original non-fused protein. In some cases, the GFP portion causes fusion proteins to form complexes at the cell poles that are absent in non-GFP cells. Thus, in cellular biology, it is always important to confirm data with the results of a different kind of technique; for example, localization of the target protein with a labeled antibody.

Some advanced forms of fluorescence microscopy can resolve subcellular details, and even visualize cells in 3D. One such method is **confocal laser scanning microscopy** (or "confocal microscopy"), in which a microscopic laser light source scans the specimen. **Figure 2.30** shows a biofilm composed of the pathogenic bacterium *Pseudomonas aeruginosa,* treated with the antibiotic tobramycin. The biofilm is treated with fluorophores that reveal live cells (green) beneath the dead cells (red)—a situation that causes problems for medical therapy. The method of confocal microscopy is presented in **eTopic 2.1**.

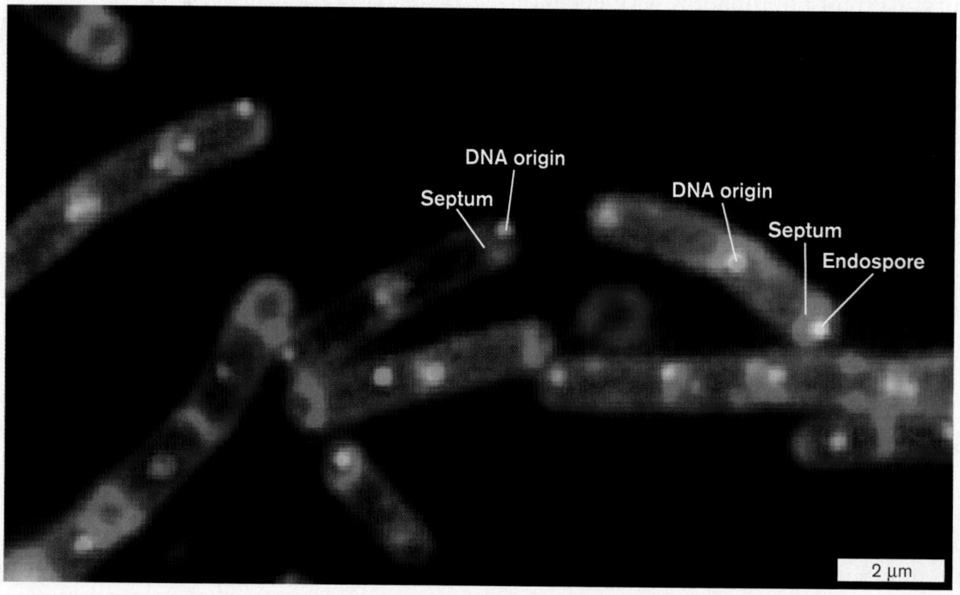
DNA origin
Septum
DNA origin
Septum
Endospore
2 μm

P. L. GRAUMANN AND R. L. LOSICK. 2001. *J. BACTERIOL.* **183**:4052

FIGURE 2.29 ■ Fluorescence micrograph of *Bacillus subtilis* cells during sporulation. Red fluorescence arises from membrane stained with the dye FM4-64. Green fluorescence arises from green fluorescent protein (GFP) fused to a protein that binds the DNA origin of replication. The yellow color occurs where green fluorescence overlaps red.

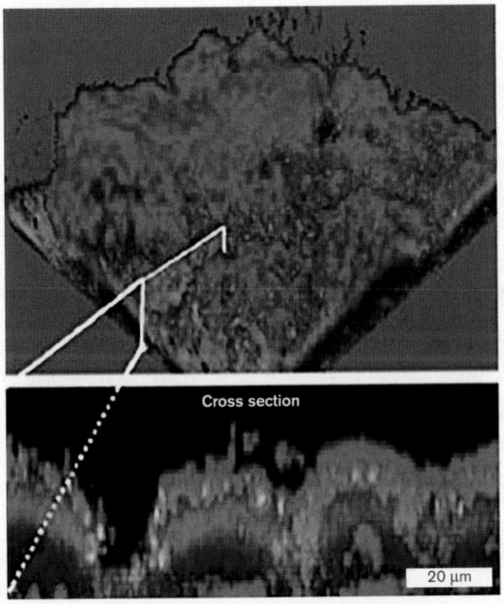
Cross section
20 µm
MORTEN HENTZER AND MICHAEL GIVSKOV. 2003. J. CLIN. INVEST. 112:1300

FIGURE 2.30 ■ Biofilm observed by confocal laser scanning microscopy. *Pseudomonas aeruginosa* cells growing in a biofilm, treated with the antibiotic tobramycin. Live/dead stain fluoresces red (dead cells) or green (live cells).

To Summarize

- **Fluorescence microscopy** involves detection of specific cells or cell parts based on fluorescence by a fluorophore.
- **The specimen absorbs light at one wavelength and then emits light at a longer wavelength.** Color filters allow only light in the excitation range to reach the specimen, and only emitted light to reach the photodetector.
- **A fluorophore can label a cell part by:** chemical affinity for a component such as membrane; attachment to an antibody stain; or attachment to a short nucleic acid that hybridizes to a DNA sequence.
- **Fluorescent proteins such as GFP can be fused to a specific protein expressed by the cell.** Endogenous GFP-type proteins can track intracellular movement of cell parts, and can report environmental stress responses.

2.5

Dark-Field and Phase-Contrast Microscopy

Advanced optical techniques enable us to visualize structures that are difficult or impossible to detect under a bright-field microscope, either because their size is below the limit of resolution of light or because their cytoplasm is transparent. These techniques take advantage of special properties of light waves, such as light scattering (dark field) and phase contrast.

Dark-Field Microscopy Detects Unresolved Objects

Dark-field optics enables microbes to be visualized as halos of bright light against the darkness, like stars are observed against the night sky. A tiny object whose size is well below the wavelength of light, such as a virus particle, can be detected by light scattering. A famous application of **dark-field microscopy** is the detection of *Treponema pallidum*, the cause of syphilis. *T. pallidum* cells are so narrow (0.1 µm) that their shape cannot be fully resolved by light microscopy. Nevertheless, the spiral form of *T. pallidum* can be detected by dark-field microscopy (**Fig. 2.31**).

Light scattering. The wavefront of scattered light is spherical, like a wave emitted by a point source (see **Fig. 2.9D**). The scattered wave has a much smaller amplitude than the incident (incoming) wave has. Therefore, with ordinary bright-field optics, scattered light is washed out. Detection of scattered light requires a modified condenser arrangement that excludes all light that is transmitted directly

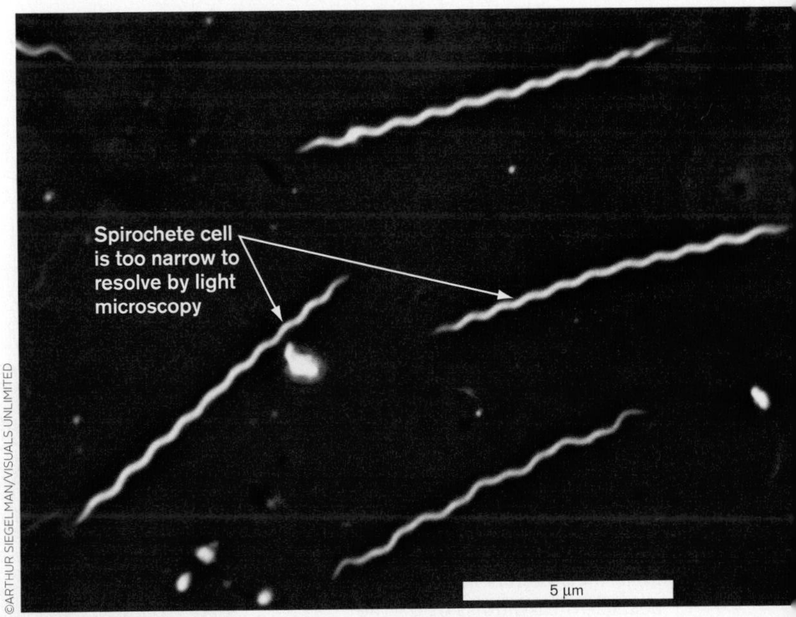
Spirochete cell is too narrow to resolve by light microscopy
5 µm
©ARTHUR SIEGELMAN/VISUALS UNLIMITED

FIGURE 2.31 ■ Dark-field observation of bacteria. *Treponema pallidum* specimen from a patient with syphilis. Note the detection of dust particles. Dark-field light microscopy.

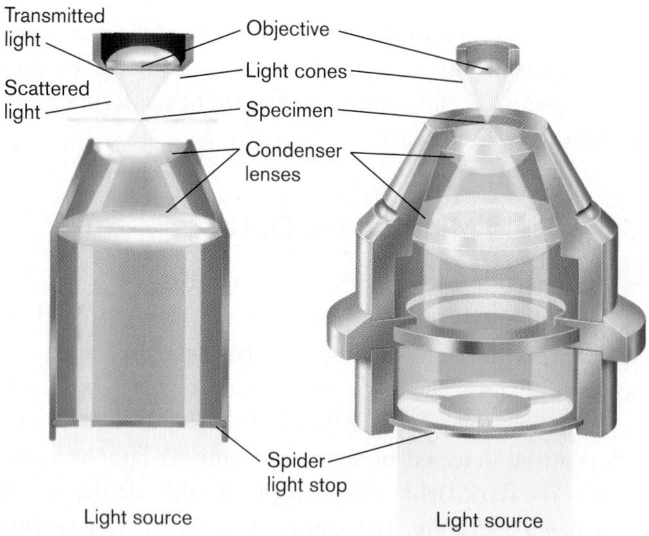
Transmitted light
Scattered light
Objective
Light cones
Specimen
Condenser lenses
Spider light stop
Light source
Light source

FIGURE 2.32 ▪ A dark-field condenser system with a spider light stop. Below the condenser, the spider light stop excludes all but an annular ring of light from the light source. The annular ring converges as a hollow cone of light focused on the specimen. Objects in the specimen scatter light in all directions. The scattered light is collected by the objective lens, but the transmitted light shines outside the range of the objective; thus, in the absence of scattering objects, the field appears dark. Only light scattered by the specimen enters the objective lens.

(**Fig. 2.32**). The condenser contains a "spider light stop," an opaque disk held by three "spider legs" across an open ring. The ring permits only a hollow cone of light to focus on the object. The incident light converges at the object and then generates an inverted hollow cone radiating outward.

The objective lens is positioned in the central region, where it completely misses the directly transmitted light. For this reason, the field appears dark. However, light scattered by the object radiates outward in a spherical wave. A sector of this spherical wave enters the objective lens and is detected as a halo of light.

An intriguing application of dark-field optics is the study of bacterial motility. Motility is important in bacterial diseases such as urethritis, in which the pathogen needs to swim up through the urethra. The bacterial swimming apparatus consists of helical filaments called **flagella** (singular, **flagellum**), which are rotated by a motor device embedded in the bacterial cell wall (for flagellar structure, see Section 3.7). The "swimming strokes" of bacteria were first elucidated by Howard Berg and Robert Macnab (1940–2003) using dark-field optics to view the helical flagella. The flagella are too narrow to resolve, but dark-field optics can detect them (**Fig. 2.33**).

Note, however, that the bacterial cell itself appears "overexposed"; its shape is unresolved owing to the high light intensity. Another disadvantage of dark-field microscopy is that any tiny particle, including specks of dust, can scatter light and interfere with visualization of the specimen. Unless the medium is extremely clear, it can be difficult to distinguish microbes of interest from particulates.

Weblinks Videos of swimming bacteria (*see ebook*)

Thought Questions

2.7 Some early observers claimed that the rotary motions observed in bacterial flagella could not be distinguished from whiplike patterns, comparable to the motion of eukaryotic flagella. Can you imagine an experiment to distinguish the two and prove that the flagella rotate? *Hint:* Bacterial flagella can get "stuck" to the microscope slide or coverslip.

2.8 Compare and contrast fluorescence microscopy with dark-field microscopy. What similar advantage do they provide, and how do they differ?

Phase-Contrast Microscopy

Phase-contrast microscopy exploits differences in refractive index between the cytoplasm and the surrounding

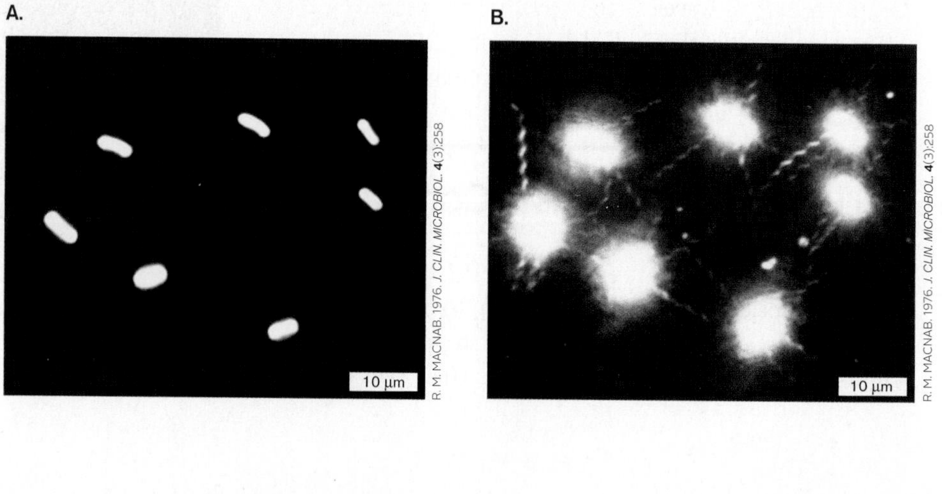
A.
B.
10 μm
10 μm
R. M. MACNAB. 1976. J. CLIN. MICROBIOL. 4(3):258
R. M. MACNAB. 1976. J. CLIN. MICROBIOL. 4(3):258

FIGURE 2.33 ▪ Motile bacteria observed under dark-field microscopy. A. Flagellated *E. coli* observed with low light, which limits scattering. Only cell bodies are detected; no flagella. **B.** The light intensity is increased. Flagella are detected, although their fine structure is not resolved, because their width is below the threshold of resolution by light.

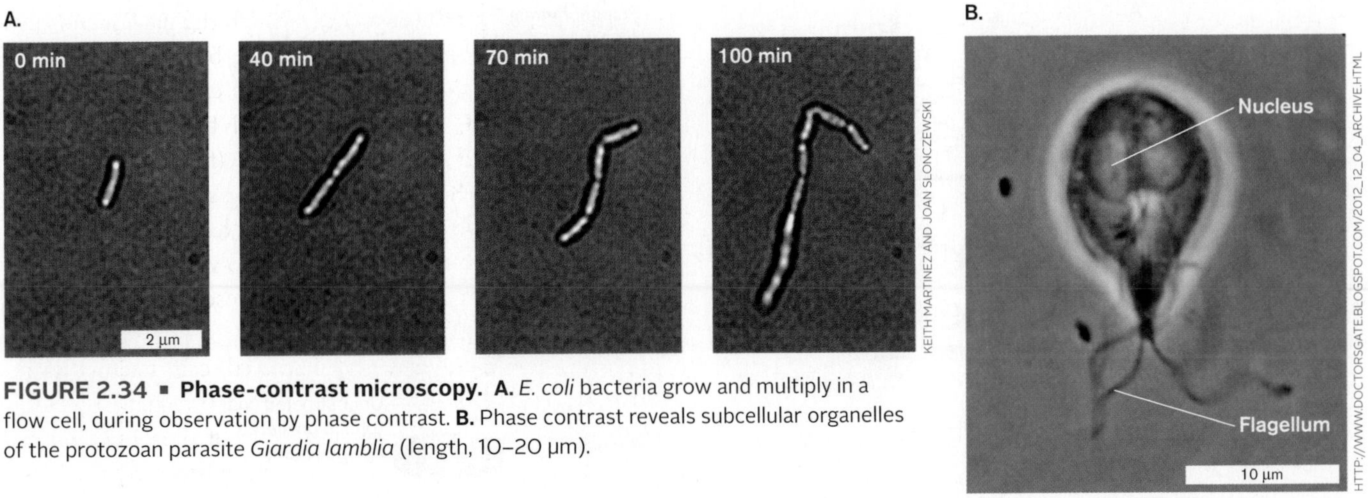
A.
0 min
40 min
70 min
100 min
2 µm
KEITH MARTINEZ AND JOAN SLONCZEWSKI
B.
Nucleus
Flagellum
10 µm
HTTP://WWW.DOCTORSGATE.BLOGSPOT.COM/2012_12_04_ARCHIVE.HTML

FIGURE 2.34 ■ Phase-contrast microscopy. **A.** *E. coli* bacteria grow and multiply in a flow cell, during observation by phase contrast. **B.** Phase contrast reveals subcellular organelles of the protozoan parasite *Giardia lamblia* (length, 10–20 µm).

medium or between different organelles. **Figure 2.34A** shows a live cell of *E. coli* bacteria growing and dividing. The cell outline appears dark because light passes through the envelope, whose refractive index differs from that of cytoplasm. For eukaryotic cells, phase contrast can distinguish many intracellular compartments. For example, **Figure 2.34B** shows a phase-contrast image of *Giardia lamblia,* a water-borne parasite commonly found in wildlife. The nucleus and other intracellular organelles are visible, as are the eukaryotic flagella (whiplike organelles).

The optical system for phase contrast was invented in the 1930s by the Dutch microscopist Frits Zernike (1888–1966), for which he earned the Nobel Prize in Physics. In this system, slight differences in the refractive index of the various cell components are transformed into differences in the intensity of transmitted light. Zernike's scheme makes use of the fact that living cells have relatively high contrast because of their high concentration of solutes. Given the size and refractive index of commonly observed cells, light is retarded by approximately one-quarter of a wavelength when it passes through the cell. In other words, after having passed through a cell, light exits the cell about one-quarter of a wavelength behind the phase of light transmitted directly through the medium.

The Zernike optical system is designed to retard the refracted light by an additional quarter of a wavelength, so that the light refracted through the cell is slowed by a total of half a wavelength compared with the light transmitted through the medium. When two waves are out of phase by half a wavelength, they produce destructive interference, canceling each other's amplitude (see **Fig. 2.12A**). The result is a region of darkness in the image of the specimen.

As in dark-field microscopy, the light transmitted through the medium in phase-contrast microscopy needs to be separated from the light interacting with the object—in this case, light waves slowed by refraction. This separation is performed by a ring-shaped slit, called an "annular ring," similar in function to the spider light stop. The annular ring stops light from passing directly through the center of the lens system, where the specimen is located, and generates a hollow cone of light, which is focused through the specimen and generates an inverted cone above it (**Fig. 2.35**). Light passing through the specimen, however, is not only retarded; it is also refracted and thus bent into the central region within the inverted cone.

Both the refracted light from the specimen and the outer cone of transmitted light enter the phase plate. The phase plate consists of refractive material that is thinner in the region met by the outer (transmitted) light cone. The refracted light passing through the center of the phase plate is retarded by an additional one-quarter wavelength compared with the transmitted light passing through the thinner region on the outside; the overall difference approximates half a wavelength. When the light from the inner and outer regions focuses at the ocular lens, the amplitudes of the waves cancel and produce a region of darkness. In this system, small differences in refractive index can produce dramatic differences in contrast between the offset phases of light.

Other kinds of optical systems use light interference to enhance cytoplasmic contrast. **Differential interference contrast microscopy (DIC)**, also known as "Nomarski microscopy," enhances contrast by superimposing the image of the specimen on a second beam of light that generates interference fringes. The interference patterns are highly sensitive to slight differences in the refractive index of the specimen. They produce an illusion of shadowing across the specimen.

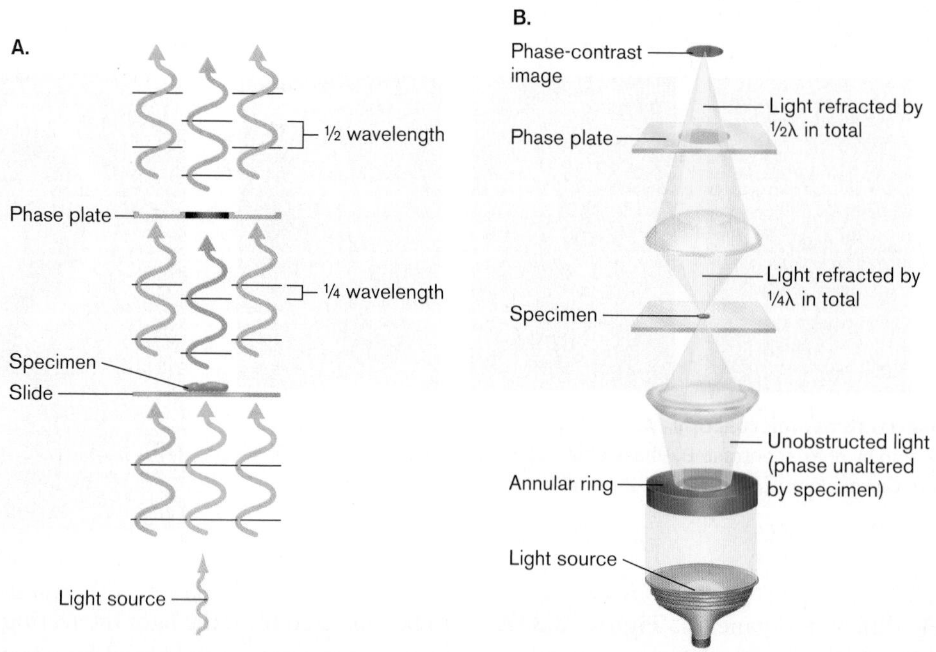
A.
½ wavelength
Phase plate
¼ wavelength
Specimen
Slide
Light source
B.
Phase-contrast image
Light refracted by ½λ in total
Phase plate
Light refracted by ¼λ in total
Specimen
Unobstructed light (phase unaltered by specimen)
Annular ring
Light source

FIGURE 2.35 ▪ Phase-contrast optics. A. The specimen retards light by approximately one-quarter of a wavelength. The phase plate contains a central disk of refractive material that retards light from the specimen by another quarter wavelength, increasing the phase difference to half a wavelength. The light from the specimen and the transmitted light are now fully out of phase; they cancel, making the specimen appear dark. **B.** In the phase-contrast microscope, the annular ring forms a hollow cone of light. As the light cone passes through the refractive material of the specimen, it is delayed by about one-quarter of a wavelength, and its path bends inward to the central region. This refracted light is surrounded by the hollow cone of unrefracted light. The light refracted by the specimen enters the lens through the dense central disk of the phase plate, which retards the wave by another quarter wavelength. When the transmitted and refracted light cones re-join at the focal point, they are out of phase; their amplitudes cancel each other, and that region of the image appears dark against a bright background.

To Summarize

- **Dark-field microscopy** uses scattered light to detect objects too small to be resolved by light rays. Extremely small microbes and thin structures can be detected. The shapes of objects are not resolved.
- **Phase-contrast microscopy** superimposes refracted light and transmitted light shifted out of phase so as to reveal differences in refractive index as patterns of light and dark. Live cells with transparent cytoplasm, and the organelles of eukaryotes, can be observed with high contrast.
- **Differential interference contrast microscopy (DIC)** superimposes interference bands on an image, accentuating small differences in refractive index.

2.6 Electron Microscopy and Tomography

All cells are built of macromolecular structures. The tool of choice for observing the shapes of these macromolecular structures is **electron microscopy (EM)**. In electron microscopy, beams of electrons are focused to generate images of cell membranes, chromosomes, and ribosomes with a resolution a thousand times that of light microscopy. Developed in the 1950s, the electron microscope yielded astonishing views of the subcellular landscape that captured the public imagination. In Michael Crichton's classic film *The Andromeda Strain* (1971), an electron microscope is used to analyze a fictional pathogen from outer space.

Beams of Electrons

How does an electron microscope work? Electrons are ejected from a metal subjected to a voltage potential. Like photons, the electrons travel in a straight line, interact with matter, and carry information about their interaction. And like photons, electrons can exhibit the properties of waves. The wavelength associated with an electron is 100,000 times smaller than that of a photon; for example, an electron accelerated over a voltage of 100 kilovolts (kV) has a wavelength of 0.0037 nm, compared with 400–750 nm for visible light. However, the actual resolution of electrons in microscopy is limited not by the wavelength, but by the aberrations of the lensing systems used to focus electrons. The magnetic lenses used to focus electrons never achieve

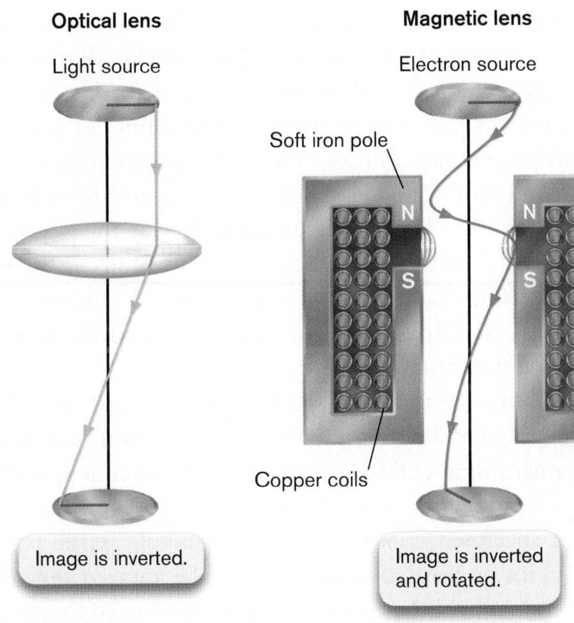
Optical lens
Light source
Image is inverted.
Magnetic lens
Electron source
Soft iron pole
N
S
N
S
Copper coils
Image is inverted and rotated.

FIGURE 2.36 ■ A magnetic lens. The beam of electrons spirals around the magnetic field lines. The U-shaped magnet acts as a lens, focusing the spiraling electrons much as a refractive lens focuses light rays.

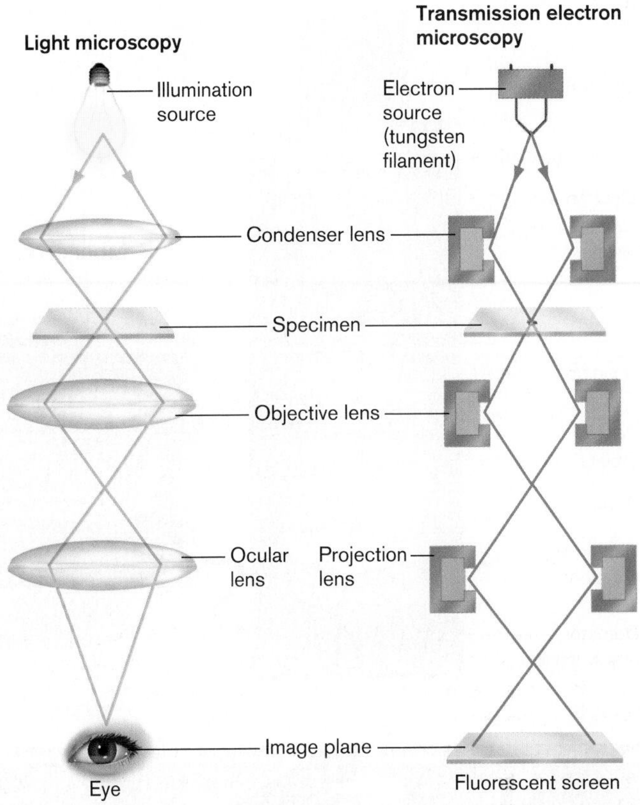
Light microscopy
Transmission electron microscopy
Illumination source
Electron source (tungsten filament)
Condenser lens
Specimen
Objective lens
Ocular lens
Projection lens
Image plane
Eye
Fluorescent screen

FIGURE 2.37 ■ Transmission electron microscopy. The light source is replaced by an electron source consisting of a high-voltage current applied to a tungsten filament, which gives off electrons when heated. Each magnetic lens shown (condenser, objective, projection) actually represents a series of lenses.

the precision required to utilize the full potential resolution of the electron beam.

Electrons are focused by means of a magnetic field directed along the line of travel of the beam (**Fig. 2.36**). As a beam of electrons enters the field, it spirals around the magnetic field lines. The shape of the magnet can be designed to generate field lines that will focus the beam of electrons in a manner analogous to the focusing of photons by a refractive lens. The electron beam, however, forms a spiral because electrons travel around magnetic field lines. Magnetic lenses generate large aberrations; thus, we need a series of corrective magnetic lenses to obtain a resolution of about 0.2 nm. This resolution is a thousand times greater than the 200-nm resolution of light microscopy.

Thought Question

2.9 An electron microscope can be focused at successive powers of magnification, as in a light microscope. At each level, the image rotates at an angle of several degrees. Given the geometry of the electron beam, as shown in **Figure 2.36**, why do you think the image rotates?

Transmission EM and scanning EM. Two major types of electron microscopy are **transmission electron microscopy (TEM)** and **scanning electron microscopy (SEM)**. In TEM, electrons are transmitted through the specimen as in light microscopy to reveal internal structure. In SEM, the electron beams scan across the surface of the specimen and are reflected to reveal the contours of its 3D surface.

The transmission electron microscope closely parallels the design of a bright-field microscope, including a source of electrons (instead of light), a magnetic condenser lens, a specimen, and an objective lens (**Fig. 2.37**). The light source is replaced by an electron source consisting of a high-voltage current applied to a tungsten filament, which gives off electrons when heated. The electron beam is focused onto the specimen by a magnetic condenser lens. The specimen image is then magnified by a magnetic objective lens. The projection lens, analogous to the ocular lens of a light microscope, focuses the image on a fluorescent screen.

The scanning electron microscope is arranged somewhat differently from the TEM in that a series of condenser lenses focuses the electron beam onto the surface of

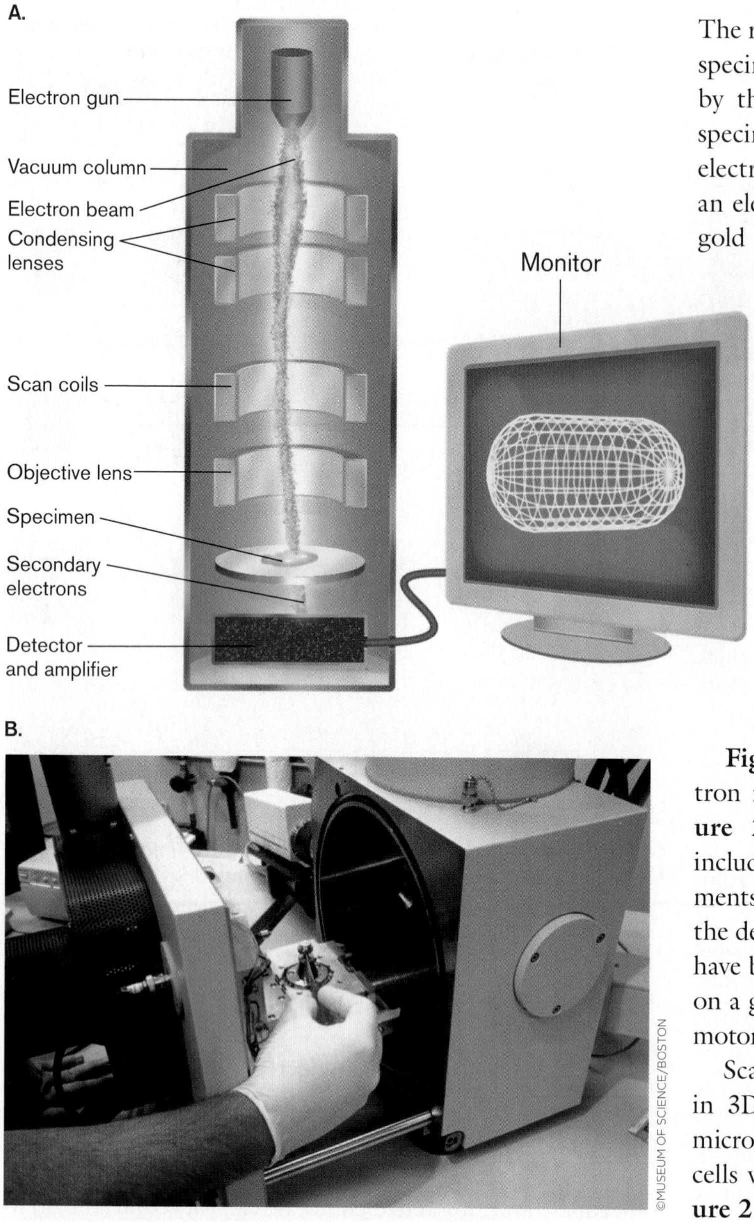
A.
Electron gun
Vacuum column
Electron beam
Condensing lenses
Scan coils
Objective lens
Specimen
Secondary electrons
Detector and amplifier
Monitor
B.

FIGURE 2.38 ▪ Scanning electron microscopy. A. In the scanning electron microscope (SEM), the electron beam is scanned across a specimen coated in gold, which acts as a source of secondary electrons. The incident electron beam ejects secondary electrons toward a detector, generating an image of the surface of the specimen. **B.** Loading a specimen into the vacuum column.

the specimen. Reflected electrons are then picked up by a detector (**Fig. 2.38**).

Sample Preparation

Standard electron microscopy of biological specimens at room temperature poses special problems. The entire optical column must be maintained under vacuum to prevent the electrons from colliding with the gas molecules in air. The requirement for a vacuum precludes the viewing of live specimens, which in any case would be quickly destroyed by the electron beam. Moreover, the structure of most specimens lacks sufficient electron density (ability to scatter electrons) to provide contrast. Thus, the specimen requires an electron-dense stain using salts of heavy metals such as gold or uranium. The heavy atoms collect at the surface of cell structures such as membranes, where their electron scatter reveals the outline of the structure. Staining, however, can be avoided for **cryo-electron microscopy** (next section).

We can prepare a specimen by embedding it in a polymer for thin sections. A special knife called a microtome cuts slices through the specimen, each slice a fraction of a micrometer thick. Alternatively, a specimen consisting of, for example, virus particles or isolated organelles can be sprayed onto a copper grid. In either case, the electron beam penetrates the object as if it were transparent. The electrons are actually absorbed by the heavy-atom stain, which collects at the edge of biological structures.

Figure 2.39 shows examples of transmission electron microscopy. The TEM of *Bacillus anthracis* in **Figure 2.39A** shows a thin section through a bacillus, including a cell wall, membranes, and glycoprotein filaments. The image includes electron density throughout the depth of the section. In **Figure 2.39B**, flagellar motors have been isolated from a bacterial cell envelope and spread on a grid for TEM. The micrograph reveals details of each motor, including the axle and individual rings.

Scanning electron microscopy can show whole cells in 3D view, with much greater resolution than in light microscopy. SEM is particularly effective for visualizing cells within complex communities such as a biofilm. **Figure 2.40** shows *Alcaligenes xylosoxidans* bacilli growing on the interior surface of a catheter inserted in the blood vessel of an immunocompromised patient. The SEM reveals the irregular contours of the bacteria, as well as the fibrin-like filaments deposited among them.

Thought Question

2.10 What kinds of research questions could you investigate using SEM? What questions could you answer using TEM?

An important limitation of traditional electron microscopy, whether TEM or SEM, is that the fixatives and heavy-atom stains can introduce artifacts into the image, especially at finer details of resolution. In some cases, different preparation procedures have led to substantially different

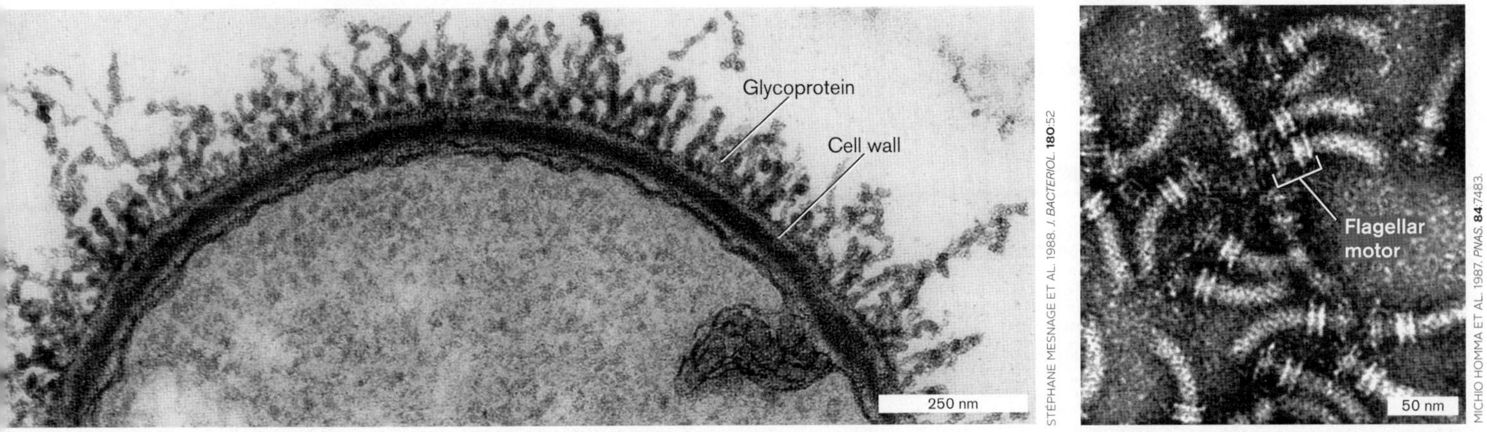
A. TEM of *Bacillus anthracis* showing envelope and cytoplasm
Glycoprotein
Cell wall
250 nm
STÉPHANE MESNAGE ET AL. 1988. *J. BACTERIOL.* **180**:52
B. TEM of flagellar motors from *Salmonella enterica*
Flagellar motor
50 nm
MICHIO HOMMA ET AL. 1987. *PNAS.* **84**:7483.

FIGURE 2.39 ▪ Transmission electron micrographs.

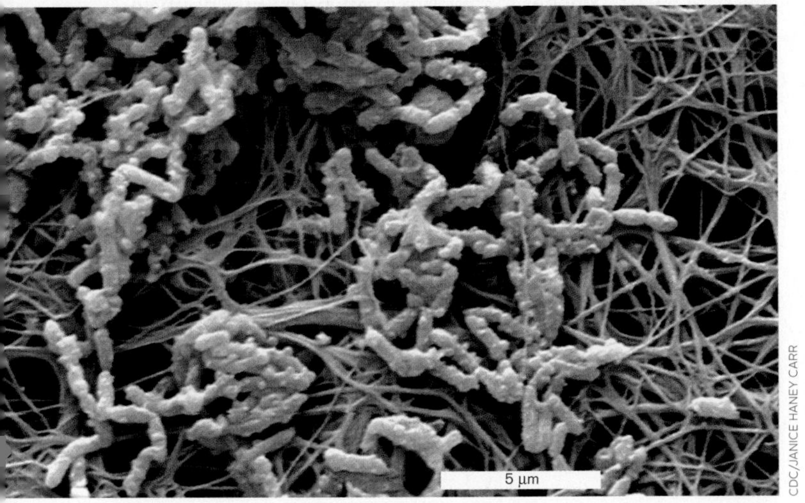
5 µm
CDC/JANICE HANEY CARR

FIGURE 2.40 ▪ Scanning electron micrograph of a catheter biofilm. *Alcaligenes xylosoxidans* bacilli growing amid fibrin-like filaments on the interior surface of a catheter in a patient's blood vessel.

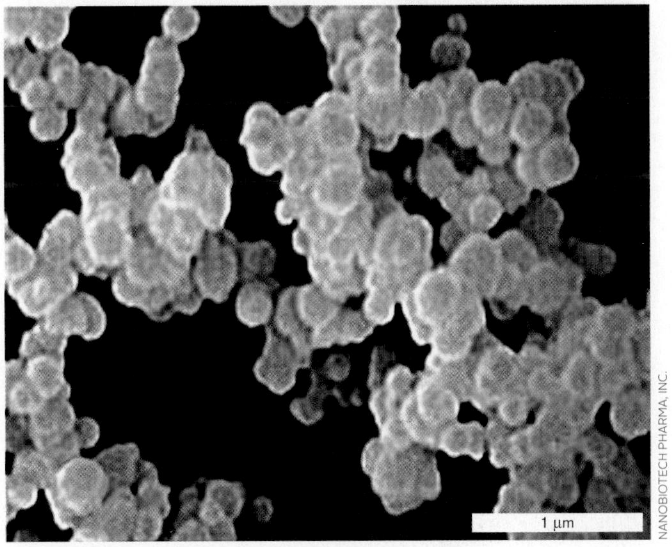
1 µm
NANOBIOTECH PHARMA, INC.

FIGURE 2.41 ▪ Artifacts of microscopy. Objects identified as "nanobacteria" turned out to be mineral deposits. *Source:* E. Olavi Kajander and Neva Cifcioglu, 1998. *PNAS* USA **95**:8274.

interpretations of subcellular structure. For example, an oval that appears hollow might be interpreted as a cell when in fact it represents a deposit of staining material. A microscopic structure that is interpreted incorrectly is termed an **artifact**. Avoiding artifacts is an important concern in microscopy.

In controversial reports, scanning electron microscopy was used to claim the isolation of pathogenic bacteria only 0.2–0.3 µm in diameter (**Fig. 2.41**). These proposed cells were termed "nanobacteria." However, other researchers were unable to confirm these results and determined that the objects observed were actually mineral deposits that had crystallized in the growth medium. The existence of small infective particles in the blood remains a question of interest to medical researchers, but the results published were probably artifacts.

Thought Question

2.11 What kinds of experiments could prove or disprove the interpretations of the images of "nanobacteria" in blood plasma?

Cryo-Electron Microscopy and Tomography

How can electron microscopy achieve finer resolution? High-strength electron beams now permit low-temperature **cryo-electron microscopy (cryo-EM)**, also known as **electron cryomicroscopy**. In cryo-EM, the specimen does not require staining, because the high-intensity electron beams

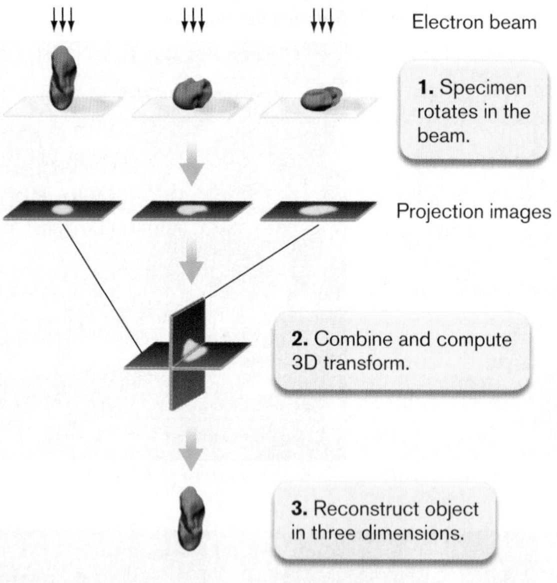
Electron beam
1. Specimen rotates in the beam.
Projection images
2. Combine and compute 3D transform.
3. Reconstruct object in three dimensions.

FIGURE 2.42 ■ 3D image construction in cryo-electron tomography. Cryo-EM images are obtained in multiple focal planes throughout an object. The images are combined through a mathematical transformation to model the entire object in 3D.

can detect smaller signals (contrast in the specimen) than in earlier instruments. The specimen must, however, be flash-frozen—that is, suspended in water and frozen rapidly in a refrigerant of high heat capacity (ability to absorb heat). The rapid freezing avoids ice crystallization, leaving the water solvent in a glass-like amorphous phase. The specimen retains water content and thus closely resembles its living form, although it is still ultimately destroyed by electron bombardment.

Another innovation made possible by cryo-EM is **tomography**, the acquisition of projected images from different angles of a transparent specimen. **Cryo-electron tomography**, or **electron cryotomography**, avoids the need to physically slice the sample for thin-section TEM. The images from tomography are combined digitally to visualize the entire object in 3D. Repeated scans can be summed computationally to obtain an image at high resolution (**Fig. 2.42**). The scans are taken either at different angles or within different focal planes. Each different scan images slightly different parts of the object. The scans are summed computationally to generate a 3D model.

One use of cryo-electron tomography is to generate high-resolution models of virus particles. For a highly symmetrical virus, images of multiple particles can be averaged together. The digitally combined images can achieve high resolution, nearly comparable to that of X-ray crystallography. Wah Chiu at Baylor College of Medicine (**Fig. 2.43A**) pioneered the use of cryo-EM to visualize virus particles at high resolution. An example is rice dwarf virus (**Fig. 2.43B**), one of the world's most economically destructive agricultural pathogens. Other viruses modeled recently include herpes virus and human immunodeficiency virus (HIV). Cryo-EM is especially useful for particles that cannot be crystallized for X-ray diffraction analysis, the most common means of molecular visualization (Sections 2.6 and 2.7).

A.

WAH CHIU, BAYLOR COLLEGE OF MEDICINE

B.

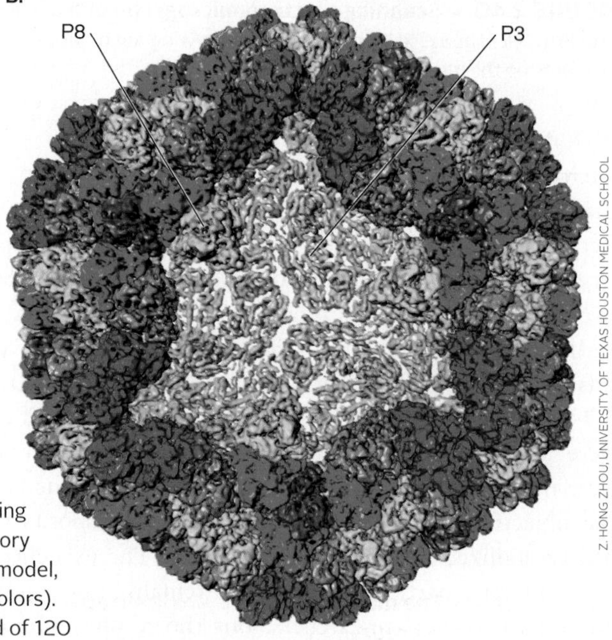
P8
P3
Z. HONG ZHOU, UNIVERSITY OF TEXAS HOUSTON MEDICAL SCHOOL AND GUANGYING LU, PEKING UNIVERSITY

FIGURE 2.43 ■ Cryo-electron tomography reveals virus structure. **A.** Wah Chiu (standing) and Joanita Jakana, at Baylor College of Medicine, using a JEOL 300-kilo-electron-volt (keV) electron cryomicroscope. Chiu's laboratory images virus structures using cryo-electron tomography. **B.** Rice dwarf virus model, at resolution 0.68 nm. The outer shell is composed of 396 subunits (bright colors). A cutaway from the outer shell reveals the inner shell (pale colors) composed of 120 subunits. P3 and P8 are structural proteins of the virus.

Cryo-EM models of a virus are impressive, but can we build a 3D model of an entire cell? Grant Jensen and colleagues at the California Institute of Technology use cryo-electron tomography to visualize an entire flash-frozen bacterium. Such a model includes all the cell's parts and their cytoplasmic connections—and reveals new structures never seen before.

The cell modeled in **Figure 2.44** is *Magnetospirillum magneticum,* a bacterium that can swim along magnetic field lines because the cell contains a string of magnetic particles composed of the mineral magnetite (iron oxide, Fe_3O_4). A cryo-EM section through the bacterium (**Fig. 2.44A**) shows fine details, including the inner membrane (equivalent to the cell membrane), peptidoglycan cell wall, and outer membrane, an outer covering found in Gram-negative bacteria. Four dark magnetosomes (particles of magnetite) appear in a chain, each surrounded by a vesicle of membrane.

Figure 2.44B models the magnetosomes, reconstructed using multiple cryo-EM scans across the volume of the cell. The magnetosomes are colorized red, each surrounded by a membrane vesicle (green). The vesicles are organized within the cell by a series of protein axial filaments, colorized yellow. **Figure 2.44C** shows an expanded view of the magnetosomes viewed from the cell interior. This expansion reveals that the magnetosome vesicles consist of invaginations from the cell membrane. Thus, the 3D model shows how the magnetite particles are fixed in position by invaginated membranes and held in a line by axial filaments.

Chapter 3 presents additional structures visualized by cryo-EM tomography, such as the flagellar motors of spirochete bacteria, and the carbon dioxide–fixing structures of marine cyanobacteria. Most surprisingly, cryo-EM reveals bacterial microtubules—structures never seen before, which offer a promising new target for antibiotics.

Emerging Methods of Microscopy

Can living cells be imaged without waves of light or electrons? New methods of microscopy use physical forces or chemical probes for unique views of single cells and of microbial communities.

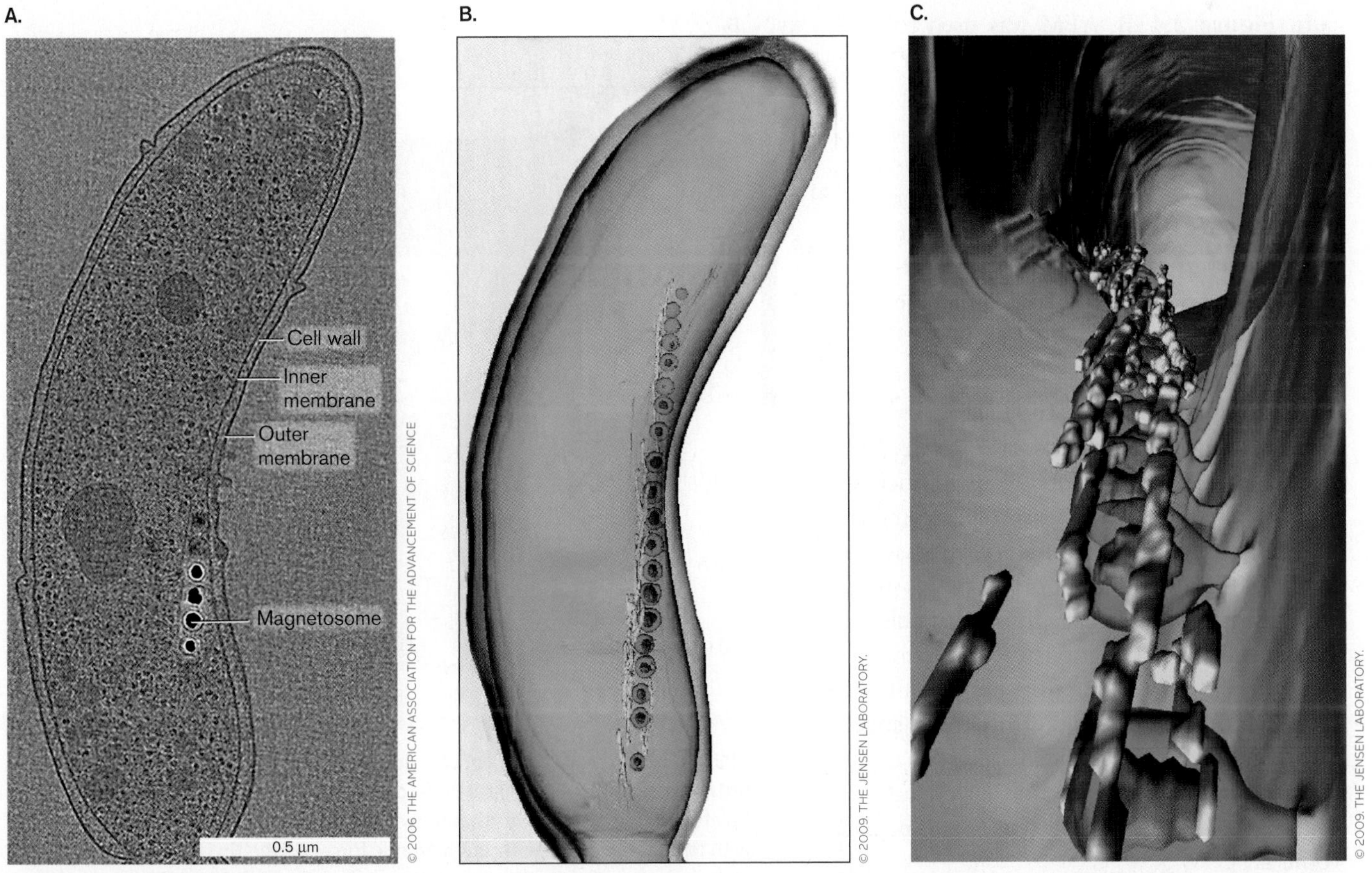
A.
B.
C.
Cell wall
Inner membrane
Outer membrane
Magnetosome
0.5 μm
© 2006 THE AMERICAN ASSOCIATION FOR THE ADVANCEMENT OF SCIENCE
© 2009. THE JENSEN LABORATORY.
© 2009. THE JENSEN LABORATORY.

FIGURE 2.44 ■ The magnetotactic cell visualized by cryo-electron tomography. A. A single cryo-EM scan lengthwise through *Magnetospirillum magneticum.* **B.** 3D model of *M. magneticum* based on multiple scans. **C.** Expanded view of the cell interior. *Source:* Arash Komeili et al. 2006. *Science* **311**:242–245.

Scanning probe microscopy. Scanning probe microscopy (SPM) enables nanoscale observation of cell surfaces. SPM techniques measure a physical interaction, such as the "atomic force" between the sample and a sharp tip. **Atomic force microscopy (AFM)** measures the van der Waals forces between the electron shells of adjacent atoms of the cell surface and the sharp tip. Unlike electron microscopy, AFM can be used to observe live bacteria in water or exposed to air. In AFM, an instrument probes the surface of a sample with a sharp tip a couple of micrometers long and often less than 10 nm in diameter (**Fig. 2.45A**). The tip is located at the free end of a lever that is 100–200 µm long. The lever is deflected by the force between the tip and the sample surface. Deflection of the lever is measured by a laser beam reflected off a cantilever attached to the tip as the sample scans across. The measured deflections allow a computer to map the topography of cells in liquid medium with a resolution below 1 nm.

In **Figure 2.45B**, AFM was used to observe live bacteria collected on a filter, from seawater off the coast of California. Two round bacteria and a helical bacterium can be seen (raised regions, green-white). The cells were observed in water suspension, without stain. This technique can help assess the ecological contributions of marine bacteria that cannot be cultured.

Chemical imaging microscopy. Another class of emerging microscopy methods is called "chemical imaging microscopy." These methods use mass spectrometry (analysis of molecular fragments by mass) to visualize the distribution of chemicals in a biological sample, such as a microbial colony. The chemical distribution can actually be observed within a cell, at 100-nm resolution, using **nanoscale secondary ion mass spectrometry (NanoSIMS)**. Chemical imaging offers extraordinary opportunities to map the structure and function of cells (see **Special Topic 2.1**).

Weblinks Molecular Expressions: Exploring the World of Optics and Microscopy (*see ebook*)

A.

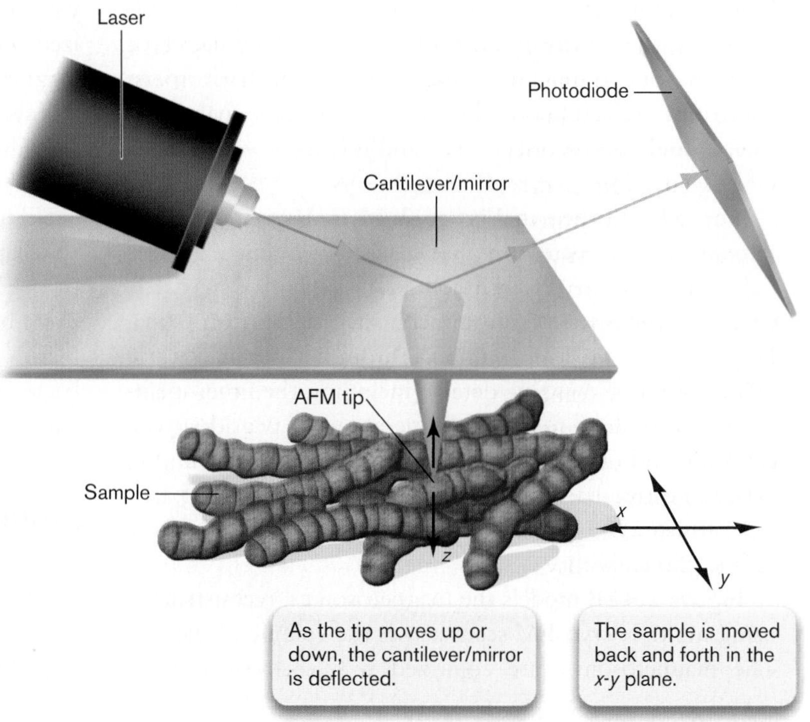
Laser
Photodiode
Cantilever/mirror
AFM tip
Sample
x
y
z
As the tip moves up or down, the cantilever/mirror is deflected.
The sample is moved back and forth in the x-y plane.

B.

µm
137.4
114.5
91.6
68.7
45.8
22.9
0.0
68.7
0.0 0.5 1.0 1.5 2.0 2.5 µm
µm
2.5
2.0
1.5
1.0
0.5
0.0

NICOLE HANSMEIER ET AL. 2006. *MICROBIOLOGY* **152**:923–935

FIGURE 2.45 ■ Atomic force microscopy enables visualization of untreated cells. A. The atomic force microscope (AFM) has a fine-pointed tip attached to a cantilever that moves over a sample. The tip interacts with the sample surface through atomic force. As the tip is pushed away, or pulled into a depression, the cantilever is deflected. The deflection is measured by a laser light beam focused onto the cantilever and reflected into a photodiode detector. **B.** AFM image shows live bacteria collected on a filter, from seawater off the coast of California. Two round bacteria and a helical bacterium can be seen (raised regions, green-white).

To Summarize

- **Electron microscopy (EM)** focuses beams of electrons on an object stained with a heavy-metal salt that scatters electrons. Much higher resolution can be obtained than with light microscopy.
- **Transmission electron microscopy (TEM)** transmits electron beams through a thin section.
- **Scanning electron microscopy (SEM)** involves scanning of a 3D surface with an electron beam.
- **Cryo-electron microscopy (cryo-EM)** involves the observation of samples flash-frozen in water solution. **Tomography** combines multiple images by computation to achieve high resolution.
- **Atomic force microscopy (AFM)** uses intermolecular force measurement to observe cells in water solution.
- **Chemical imaging methods of mass spectrometry such as NanoSIMS** map the distribution of chemicals within a cell.

2.7

Visualizing Molecules

To know a cell, we need to isolate the cell's individual molecules. The major tool used at present to visualize a molecule is **X-ray diffraction analysis**, or **X-ray crystallography**. Alternatively, cryo-EM modeling has also reached near-atomic resolution. Another emerging alternative to X-ray diffraction for analysis of small molecules and proteins is nuclear magnetic resonance (NMR). The advantage of NMR is that it presents a dynamic view of molecules in solution.

Unlike microscopy, X-ray diffraction does not present a direct view of a sample, but generates computational models. Dramatic as the models are, they can only represent particular aspects of electron clouds and electron density that are fundamentally "unseeable." That is why we represent molecular structures in different ways that depend on the context—by electron density maps, as models defined by van der Waals radii, or as stick models. Proteins are frequently presented in a cartoon form that shows alpha helix and beta sheet secondary structures.

X-Ray Diffraction Analysis

For a molecule that can be crystallized, X-ray diffraction makes it possible to fix the position of each individual atom in the molecule. Atomic resolution is possible because the wavelengths of X-rays are much shorter than the wavelengths of visible light and are comparable to the size of atoms. X-ray diffraction is based on the principle of wave interference (see **Fig. 2.12**). The interference pattern is generated when a crystal containing many copies of an isolated molecule is bombarded by a beam of X-rays (**Fig. 2.46A**). The wavefronts associated with the X-rays are diffracted as they pass through the crystal, causing interference patterns. In the crystal, the diffraction pattern is generated by a symmetrical array of many sample molecules (**Fig. 2.46B**). The

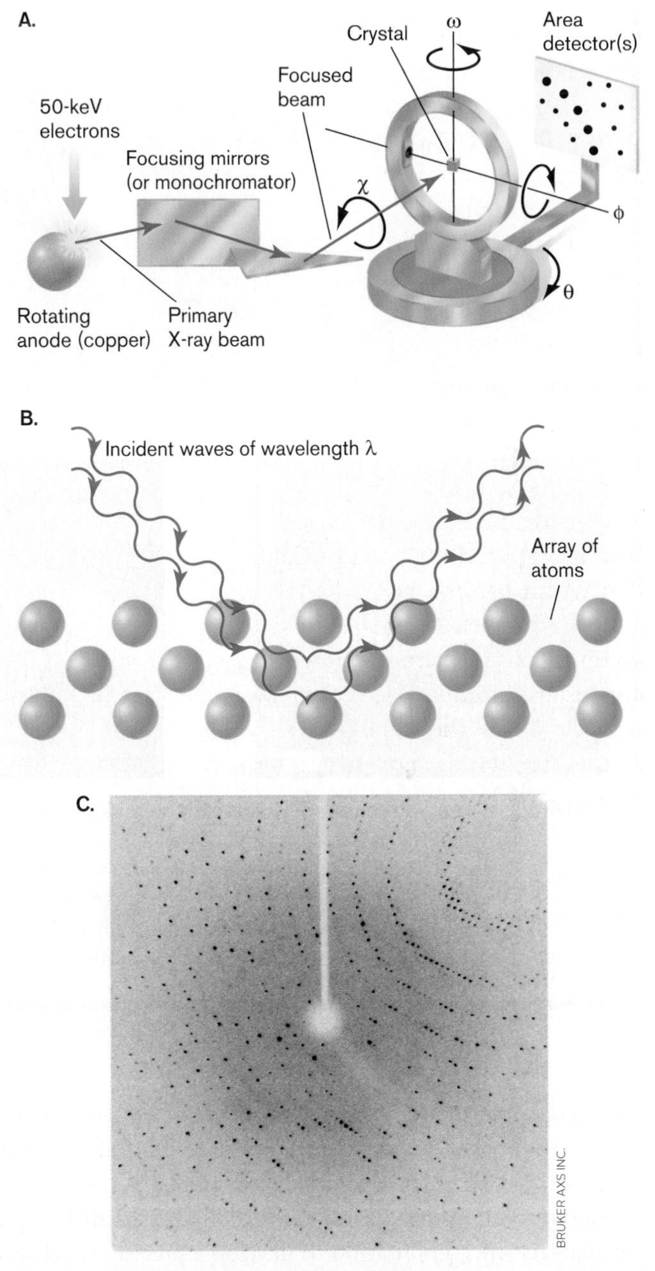
A.
Crystal
ω
Area detector(s)
Focused beam
50-keV electrons
Focusing mirrors (or monochromator)
χ
ϕ
θ
Rotating anode (copper)
Primary X-ray beam
B.
Incident waves of wavelength λ
Array of atoms
C.
BRUKER AXS INC.

FIGURE 2.46 ■ Visualizing molecules by X-ray crystallography. **A.** Modern apparatus for X-ray crystallography. The X-ray beam is focused onto a crystal, which is rotated over all angles to obtain diffraction patterns. The intensity of the diffracted X-rays is recorded on film or with an electronic detector. **B.** X-rays are diffracted by rows of identical molecules in a crystal. The diffraction pattern is analyzed to generate a model of the individual molecules. **C.** Diffraction pattern from a crystal.

Special Topic 2.1: Molecular "Snapshots": Chemical Imaging

Light rays and electron beams offer glimpses into the astonishing world within a cell. But what are all those intriguing knobs and tubules made of? "Chemical imaging microscopy" can yield molecular snapshots of a cell—and of global processes they take part in, such as nitrogen and carbon fixation.

A high-resolution method for chemical imaging is called **nanoscale secondary ion mass spectrometry (NanoSIMS)**. The imaging method starts with an ionizing probe, a source of energy that breaks up the large organic molecules of a sample (**Fig. 1A**). The molecular fragments, called "secondary ions," fly

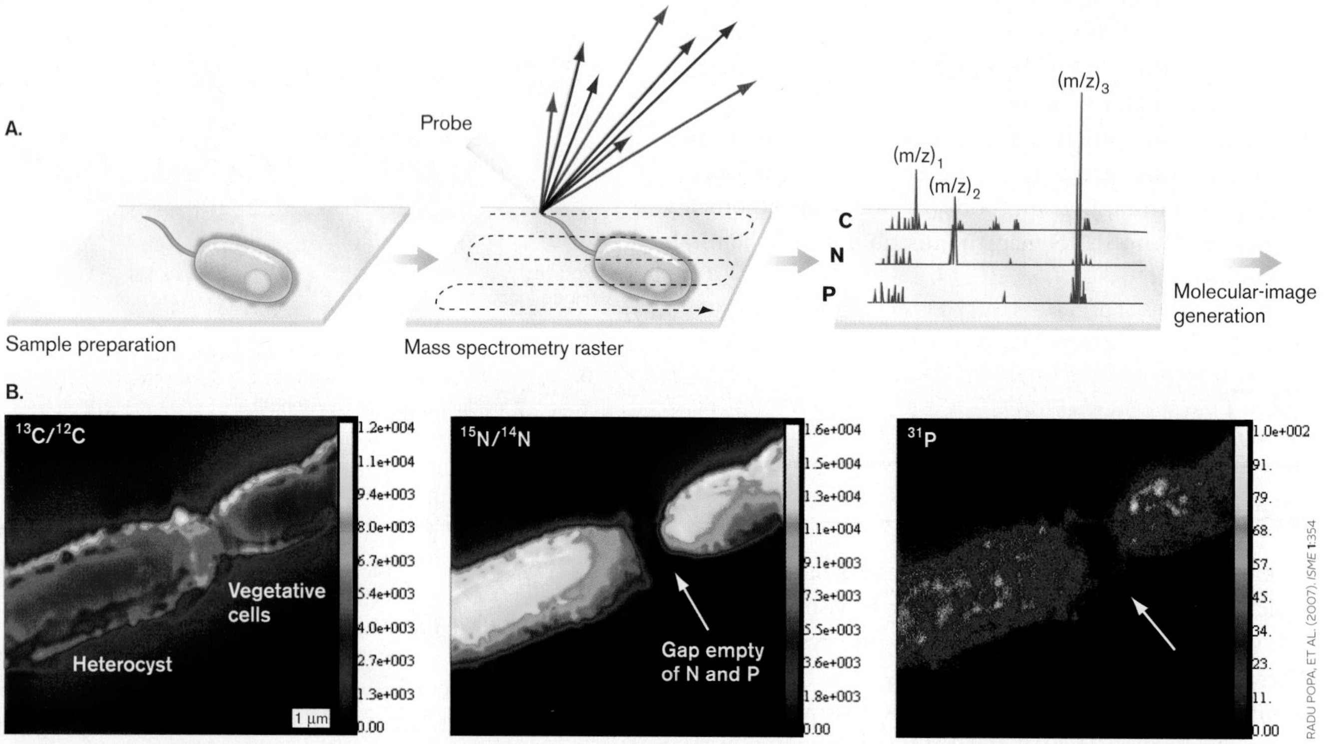
A.
Probe
(m/z)1
(m/z)2
(m/z)3
C
N
P
Sample preparation
Mass spectrometry raster
Molecular-image generation
B.
13C/12C
Vegetative cells
Heterocyst
1 µm
1.2e+004
1.1e+004
9.4e+003
8.0e+003
6.7e+003
5.4e+003
4.0e+003
2.7e+003
1.3e+003
0.00
15N/14N
Gap empty of N and P
1.6e+004
1.5e+004
1.3e+004
1.1e+004
9.1e+003
7.3e+003
5.5e+003
3.6e+003
1.8e+003
0.00
31P
1.0e+002
91.
79.
68.
57.
45.
34.
23.
11.
0.00
RADU POPA, ET AL. (2007). ISME 1:354

FIGURE 1 ■ Imaging mass spectrometry. Mass spectra are obtained from thousands of locations throughout the sample surface. **A.** Molecular fragments are selected for analysis of mass-to-charge ratio (*m*/*z*). Selected isotopes may label specific atoms—for example, C, N, or P. The relative intensities of individual compounds are visualized using false-color gradients. **B.** Nanoscale secondary ion mass spectrometry (NanoSIMS) of *Anabaena* heterocyst and vegetative cells, showing intracellular carbon (C), nitrogen (N), and phosphorus (P). The cells show continuous connection by carbon compounds, but a gap empty of N and P (arrow) separates the nitrogen-fixing heterocyst from the CO_2-fixing vegetative cells.

more copies of the molecule in the array, the narrower the interference pattern and the greater the resolution of atoms within the molecule. Diffraction patterns obtained from the passage of X-rays through a crystal (**Fig. 2.46C**) can be analyzed by computation to develop a precise structural model for the molecule, detailing the position of every atom in the structure.

The application of X-ray crystallography to complex biological molecules was pioneered by the Irish crystallographer John Bernal (1901–1971) (**Fig. 2.47A**). Bernal was particularly supportive of women students and colleagues, including Rosalind Franklin (1920–1958), who made important discoveries about DNA and RNA, and Nobel laureate Dorothy Crowfoot Hodgkin (1910–1994). Hodgkin (**Fig. 2.47B**) won the 1964 Nobel Prize in Chemistry for solving the crystal structures of penicillin and vitamin B_{12} (**Fig. 2.47C**). She later solved one of the first protein structures—that of the hormone insulin.

off from the source and are captured by a mass spectrometer. This instrument measures fragment masses of the secondary ions, generating a mass spectrum. Mass spectra are taken from thousands of locations, scanned across a bacterial cell.

What information does NanoSIMS yield? Different kinds of isotope labeling can focus on specific chemical questions. For example, the cell's uptake of nitrogen can be detected by incorporation of the heavy isotope ^{15}N. The increased weight of ^{15}N compared to the normally predominant ^{14}N (as indicated by the mass ratio $^{15}N/^{14}N$) is detected quantitatively in the molecular fragment masses. Alternatively, we can detect isotopes of carbon or phosphorus.

NanoSIMS can answer questions such as: How do marine cyanobacteria manage the juggling act of fixing nitrogen while fixing carbon at the same time? Nitrogen fixation is poisoned by oxygen, which oxygenic photosynthesis produces while fixing carbon dioxide. Because cyanobacteria manage to do both, they are keystone species in the global cycles of carbon and nitrogen. These questions were addressed in 2007 by microbiologist Radu Popa at Portland State University (**Fig. 2**) working with Douglas Capone and colleagues at UCLA, and Peter Weber and colleagues at Lawrence Livermore National Laboratory.

The cyanobacterium *Anabaena oscillarioides* grows as chains containing two types of cells: vegetative cells, conducting photosynthesis and fixing carbon dioxide; and heterocysts, spaced at approximately ten-cell intervals (**Fig. 1B**). Popa and Capone used NanoSIMS to reveal the uptake patterns of carbon, nitrogen, and phosphorus. The images show that carbon is incorporated into biomass throughout the chain, but nitrogen and phosphorus are tightly compartmentalized, kept out of the linking region between heterocyst and vegetative cells. Timed NanoSIMS exposures reveal that nitrogen levels remain high throughout the carbon-fixing cells; thus, a highly efficient mechanism must pump the nitrogenous biochemicals from the heterocyst to the carbon-fixing vegetative cells. It is thought-provoking to realize that Earth's entire global ecosystem depends on the efficiency of nitrogen transfer in these tiny marine life-forms.

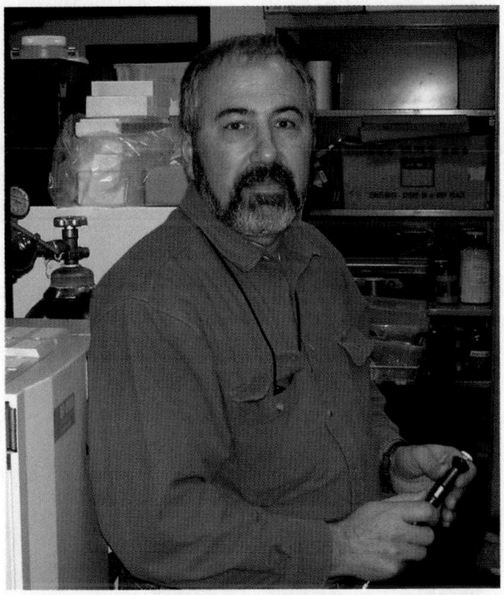
PROVIDED BY POPA

FIGURE 2 ▪ **Radu Popa, at Portland State University, images single cells using NanoSIMS, a form of imaging mass spectrometry.**

RESEARCH QUESTION

Cyanobacteria could separate carbon fixation and nitrogen fixation by fixing carbon dioxide during the day and fixing nitrogen at night. Can you propose an experiment using NanoSIMS to test this model?

Popa, Radu, Peter K. Weber, Jennifer Pett-Ridge, Juliette A. Finzi, Stewart J. Fallon, et al. 2007. Carbon and nitrogen fixation and metabolite exchange in and between individual cells of *Anabaena oscillarioides*. *ISME Journal* **1**:354–360.

Today, X-ray data undergo digital analysis to generate sophisticated molecular models, such as the one seen in **Figure 2.48** of anthrax lethal factor, a toxin produced by *Bacillus anthracis* that kills the infected host cells. The model for anthrax lethal factor was encoded in a Protein Data Bank (PDB) text file that specifies coordinates for all atoms of the structure. The Protein Data Bank is a worldwide database of solved X-ray structures, freely available on the Internet. Visualization software is used to present the structure as a "ribbon" of amino acid residues, color-coded for secondary structure. In **Figure 2.48**, the red coils represent alpha helix structures, whereas the blue arrows represent beta sheets (for a review of protein structures, see Appendix 1).

Weblinks Protein Data Bank: Research Collaboratory for Structural Bioinformatics (RCSB), protein and nucleic acid databases (*see ebook*)

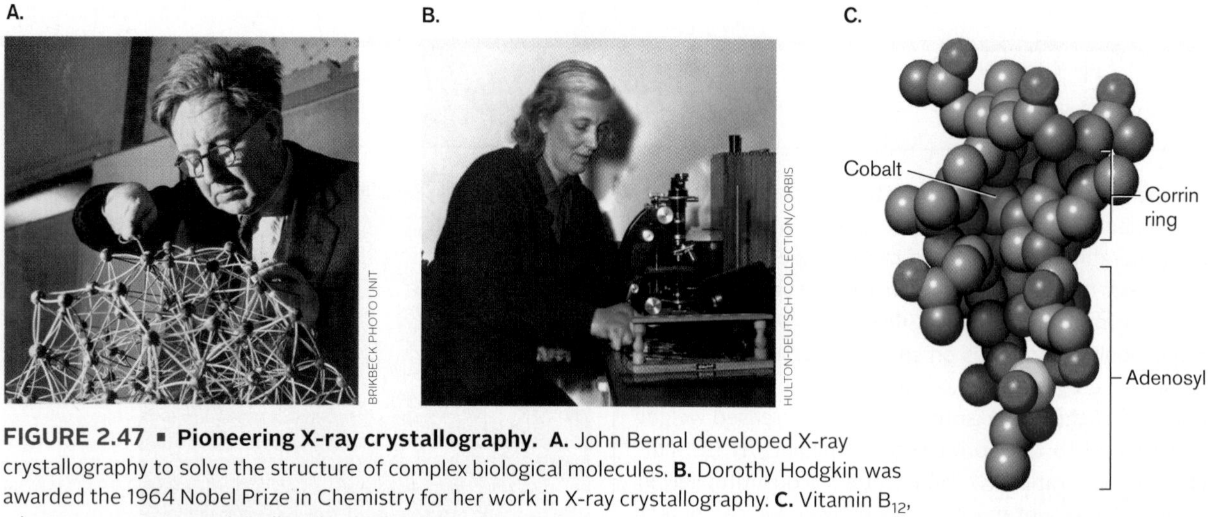
A.
B.
C.
BRIKBECK PHOTO UNIT
HULTON-DEUTSCH COLLECTION/CORBIS
Cobalt
Corrin ring
Adenosyl

FIGURE 2.47 ■ Pioneering X-ray crystallography. A. John Bernal developed X-ray crystallography to solve the structure of complex biological molecules. **B.** Dorothy Hodgkin was awarded the 1964 Nobel Prize in Chemistry for her work in X-ray crystallography. **C.** Vitamin B_{12}, whose structure was originally solved by Hodgkin. The corrin ring structure is built around an atom of cobalt (pink). Carbon atoms are gray; oxygen, red; nitrogen, blue; phosphorus, yellow. Hydrogen atoms are omitted for clarity.

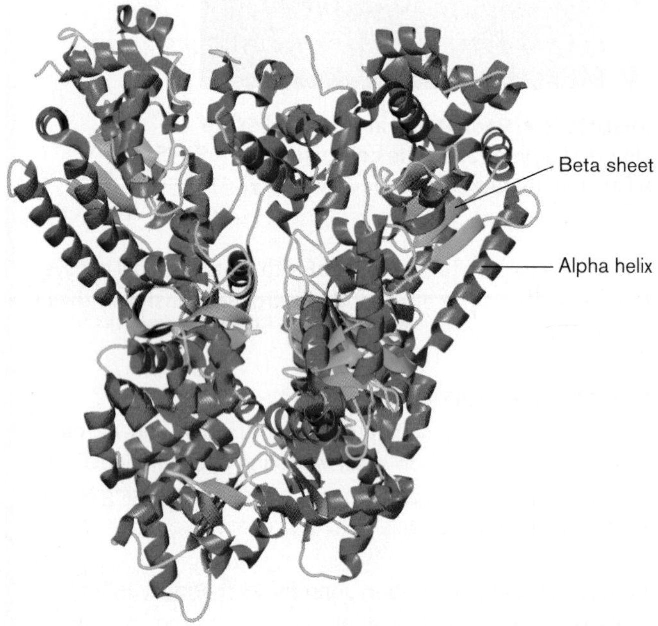
Beta sheet
Alpha helix

FIGURE 2.48 ■ X-ray crystallography of a protein complex, anthrax lethal factor. The toxin consists of a butterfly-shaped dimer of two peptide chains. This cartoon model is based on X-ray-crystallographic data, showing alpha helix (red coils) and beta sheet (blue arrows). (PDB code: 1J7N)

Note: Molecular and cellular biology increasingly rely on visualization in 3D . Many of the molecules illustrated in this book are based on structural models deposited in the Protein Data Bank, as indicated by the PDB file code. You may view these structures in 3D by downloading the PDB file and viewing with a free plug-in such as Jmol.

A limitation of X-ray analysis is the unavoidable deterioration of the specimen under bombardment by X-rays. The earliest X-ray diffraction models of molecular complexes such as the ribosome relied heavily on components from thermophilic bacteria and archaea that grow at high temperatures. Because thermophiles have evolved to grow under higher thermal stress, their macromolecular complexes form more stable crystals than do their homologs in organisms growing at moderate temperatures.

X-ray diffraction analysis of crystals from a wide range of sources was made possible by cryocrystallography. In cryocrystallography, as in cryo-EM, crystals are frozen rapidly to liquid-nitrogen temperature. The frozen crystals have greatly decreased thermal vibrations and diffusion, thus lessening the radiation damage to the molecules. Models based on cryocrystallography can present multisubunit structures such as the bacterial ribosome complexed with transfer RNAs and messenger RNA. Much of our knowledge of microbial genetics (Chapters 7–12) and metabolism (Chapters 13–16) comes from crystal structures of key macromolecules.

To Summarize

- **X-ray diffraction analysis**, or **X-ray crystallography**, uses X-ray diffraction (interference patterns) from crystallized macromolecules to determine structure at atomic resolution.
- **Cryocrystallography** uses frozen crystals with greatly decreased thermal vibrations and diffusion, enabling the determination of structures of large macromolecular complexes, such as the ribosome.
- **Molecular visualization** by crystallography can only model the "appearance" of a molecule at atomic resolution. Different models emphasize different structural features and levels of resolution.

Concluding Thoughts

The tools of microscopy and molecular visualization described in this chapter have shaped our current understanding of microbial cells—how they grow and divide, organize their DNA and cytoplasm, and interact with other cells. Our current models of cell structure and function are explored in Chapter 3. In Chapter 4 we learn how cells use their structures to obtain energy, reproduce, and develop dormant forms that can remain viable for thousands of years.

CHAPTER REVIEW

Review Questions

1. What principle defines an object as "microscopic"?
2. Explain the difference between detection and resolution.
3. How do eukaryotic and prokaryotic cells differ in appearance under the light microscope?
4. Explain how electromagnetic radiation carries information and why different kinds of radiation can resolve different kinds of objects.
5. Describe how light interacts with an object through absorption, reflection, refraction, and scattering.
6. Explain how refraction enables magnification of an image.
7. Explain how magnification increases resolution and why "empty magnification" fails to increase resolution.
8. Explain how angle of aperture and resolution change with increasing lens magnification.
9. Summarize the optical arrangement of a compound microscope.
10. Explain how to focus an object and how to tell when the object is in or out of focus.
11. Explain the relative advantages and limitations of wet mount and stained preparations for observing microbes.
12. Explain the significance (and limitations) of the Gram stain for bacterial taxonomy.
13. Explain the basis of dark-field, phase-contrast, and fluorescence microscopy. Give examples of applications of these advanced techniques.
14. Explain the difference between transmission and scanning electron microscopy, and the different applications of each.

Thought Questions

1. Explain which features of bacteria you can study by (a) light microscopy; (b) fluorescence microscopy; (c) scanning EM; (d) transmission EM.
2. Explain how resolution is increased by magnification. Why can't the details be resolved by your unaided eye? Explain why magnification reaches a limit; why can it not go on resolving greater detail?
3. Explain why artifacts appear, even with the best lenses. Explain how you can tell the difference between an optical artifact and an actual feature of an image.
4. How can "detection without resolution" be useful in microscopy? Explain with specific examples.

Key Terms

aberration (51)
absorption (47)
acid-fast stain (57)
angle of aperture (52)
antibody stain (58)
artifact (69)
atomic force microscopy (AFM) (45, 72)
bacillus (44)
bright-field microscopy (45, 51)
chemical imaging microscopy (46)
coccus (44)
compound microscope (52)
condenser (53)
confocal laser scanning microscopy (62)
contrast (47, 51)
counterstain (57)

cryo-electron microscopy (cryo-EM) (electron cryomicroscopy) (68, 69)
cryo-electron tomography (electron cryotomography) (70)
dark-field microscopy (63)
depth of field (54)
detection (43)
diaphragm (53)
differential interference contrast microscopy (DIC) (65)
differential stain (56)
electromagnetic radiation (46)
electron microscopy (EM) (45, 66)
emission wavelength (60)
empty magnification (49)
excitation wavelength (60)
fixation (55)
flagellum (64)
fluorescence (48)
fluorophore (58)
focal plane (54)
focal point (48)
gene fusion (61)
Gram-negative (57)
Gram-positive (57)
Gram stain (56)
immersion oil (52)
interference (49)
lens (48)
light microscopy (45)
magnification (43, 47)
micrometer (μm) (43)
microscope (42)
millimeter (mm) (43)
mordant (57)
nanometer (nm) (43)
nanoscale secondary ion mass spectrometry (NanoSIMS) (72, 74)
negative stain (57)
numerical aperture (52)
objective lens (51, 53)
ocular lens (53)
optical tweezers (optical trap) (54)
parfocal (53)
phase-contrast microscopy (64)
picometer (pm) (43)
reflection (48)
refraction (48)
refractive index (48)
resolution (42)
rod (44)
scanning electron microscopy (SEM) (45, 67)
scattering (48)
simple stain (55)
spirochete (44)
spore stain (57)
staining (55)
tomography (70)
total magnification (54)
transmission electron microscopy (TEM) (45, 67)
wet mount (54)
X-ray crystallography (X-ray diffraction analysis) (45, 73)

Recommended Reading

Altindal, Tuba, Suddhashil Chattopadhyay, and Xiao-Lun Wu. 2011. Bacterial chemotaxis in an optical trap. *PLoS ONE* **6**:e18231.

Chiu, W., M. L. Baker, W. Jiang, and Z. H. Zhou. 2002. Deriving folds of macromolecular complexes through electron cryomicroscopy and bioinformatics approaches. *Current Opinion in Structural Biology* **12**:263–269.

Graumann, Peter L., and Richard Losick. 2001. Coupling of asymmetric division to polar placement of replication origin regions in *Bacillus subtilis*. *Journal of Bacteriology* **183**:4052–4060.

Jiang, W., J. Chang, J. Jakana, P. Weigele, J. King, et al. 2006. Structure of epsilon15 bacteriophage reveals genome organization and DNA packaging-injection apparatus. *Nature* **439**:612–616.

Komeili, A., Z. Li, D. K. Newman, and G. J. Jensen. 2006. Magnetosomes are cell membrane invaginations organized by the actin-like protein MamK. *Science* **311**:242–245.

Lucic, Vladan, Friedrich Förster, and Wolfgang Baumeister. 2005. Structural studies by electron tomography: From cells to molecules. *Annual Review of Biochemistry* **74**:833–865.

Matias, Valério R. F., Ashruf Al-Amoudi, Jacques Dubochet, and Terry J. Beveridge. 2003. Cryo-transmission electron microscopy of frozen-hydrated sections of *Escherichia coli* and *Pseudomonas aeruginosa*. *Journal of Bacteriology* **185**:6112–6118.

Murphy, Douglas B. 2001. *Fundamentals of Light Microscopy and Electronic Imaging*. Wiley-Liss, Hoboken, NJ.

Popescu, Aurel, and R. J. Doyle. 1996. The Gram stain after more than a century. *Biotechniques in Histochemistry* **71**:145–151.

Tocheva, E., Z. Li, and G. Jensen. 2010. Electron cryotomography, p. 213–232. *In* Lucy Shapiro and Richard M. Losick (eds.), *Cell Biology of Bacteria: A Subject Collection from Cold Spring Harbor Perspectives in Biology*. Cold Spring Harbor Laboratory Press, Cold Spring Harbor, NY.

CHAPTER 3
Cell Structure and Function

3.1 The Bacterial Cell: An Overview

3.2 The Cell Membrane and Transport

3.3 The Cell Wall and Outer Layers

3.4 The Nucleoid, RNA, and Protein Synthesis

3.5 Cell Division

3.6 Cell Polarity and Aging

3.7 Specialized Structures

Early in Earth's history, the first living cells evolved. Ancient cells diverged into three domains that continue today: the bacteria, archaea, and eukaryotes (introduced in Chapter 1). Bacterial cells are remarkable for their small size and efficient growth. With just a few thousand genes in its genome, how does a bacterium grow and reproduce? The bacterial cell coordinates its DNA replication through the DNA replisome and the cell fission ring. Other devices, such as flagellar propellers, enable microbial cells to swim, communicate, and build biofilm communities. Microbial cells face environmental stress, enduring extreme changes in temperature and salinity. And pathogens face the chemical defenses of their hosts.

Cell form and function have exciting applications for medicine and biotechnology. The structures of ribosomes and cell envelope provide targets for new antibiotics. And devices such as the flagellar motor inspire "nanotechnology," the design of molecular machines.

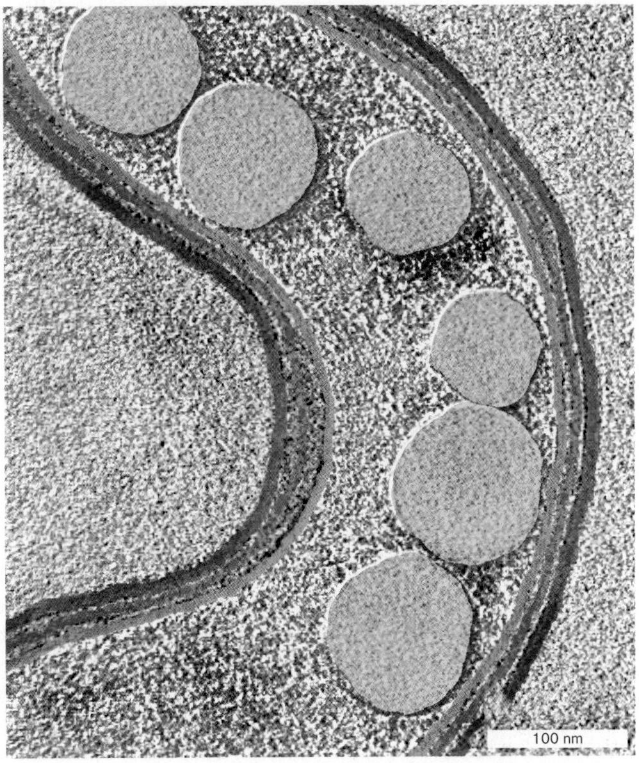
100 nm

CURRENT RESEARCH highlight

Spirochete makes mystery organelles. *Leptospira biflexa* is a spirochete, a type of bacterium shaped like a corkscrew. *Leptospira* bacteria commonly grow in soil and wetlands, but some species infect the kidneys of animals such as dogs and rodents. The disease, leptospirosis, can be transmitted to humans through contaminated water, causing 500,000 cases annually. New drugs to treat leptospirosis may target the spirochete's unique subcellular structures. The colorized cryo-EM reveals the cell's inner membrane (orange), peptidoglycan cell wall (pink), and outer membrane or sheath (purple). Peptidoglycan is the target of antibiotics such as penicillin and vancomycin. The spherical cytoplasmic organelles (yellow) are unique to *Leptospira;* their function remains a mystery. *Source:* Gianmarco Raddi et al. 2012. *J. Bacteriol.* **194**:1299.

The microbial cell has formidable tasks: to obtain nutrients faster than its competitors, to protect itself from toxins and predators, and to reproduce. To accomplish these tasks, cells have evolved an amazing range of molecular parts, such as rotary machines that make ATP, and cell walls with tensile strength comparable to that of steel.

Chapter 3 assumes an elementary knowledge of cell biology (reviewed in Appendices 1 and 2). We begin with a tour of an *Escherichia coli* bacterial cell (**Fig. 3.1**). We explore the structures common to most microbial cells, as well as more specialized devices, such as light-harvesting complexes and magnetosomes. We dissect the cell's complex outer layers and see how they interact with the replicating chromosome to accomplish cell fission. Along the way, we learn how our views of a cell emerge from microscopy, cell fractionation, and genetic analysis. Many of the cell's structures offer targets for antibiotic design, as well as models for biotechnology.

Overall, this chapter focuses on bacteria. Bacterial cells show a wide range of form, whose diversity is explored in Chapter 18. But most bacteria share these traits:

- **Thick, complex outer envelope.** The envelope protects the cell from environmental stress and mediates exchange with the environment.
- **Compact genome.** Prokaryotic genomes are compact, with relatively little noncoding DNA. Small genomes maximize the production of cells from limited resources.

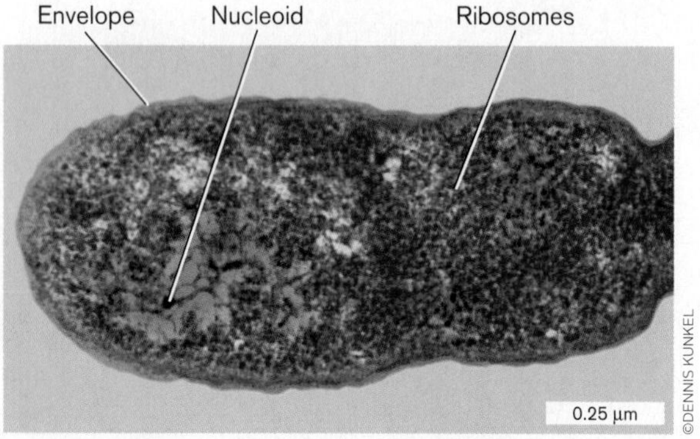
Envelope
Nucleoid
Ribosomes
0.25 µm
©DENNIS KUNKEL

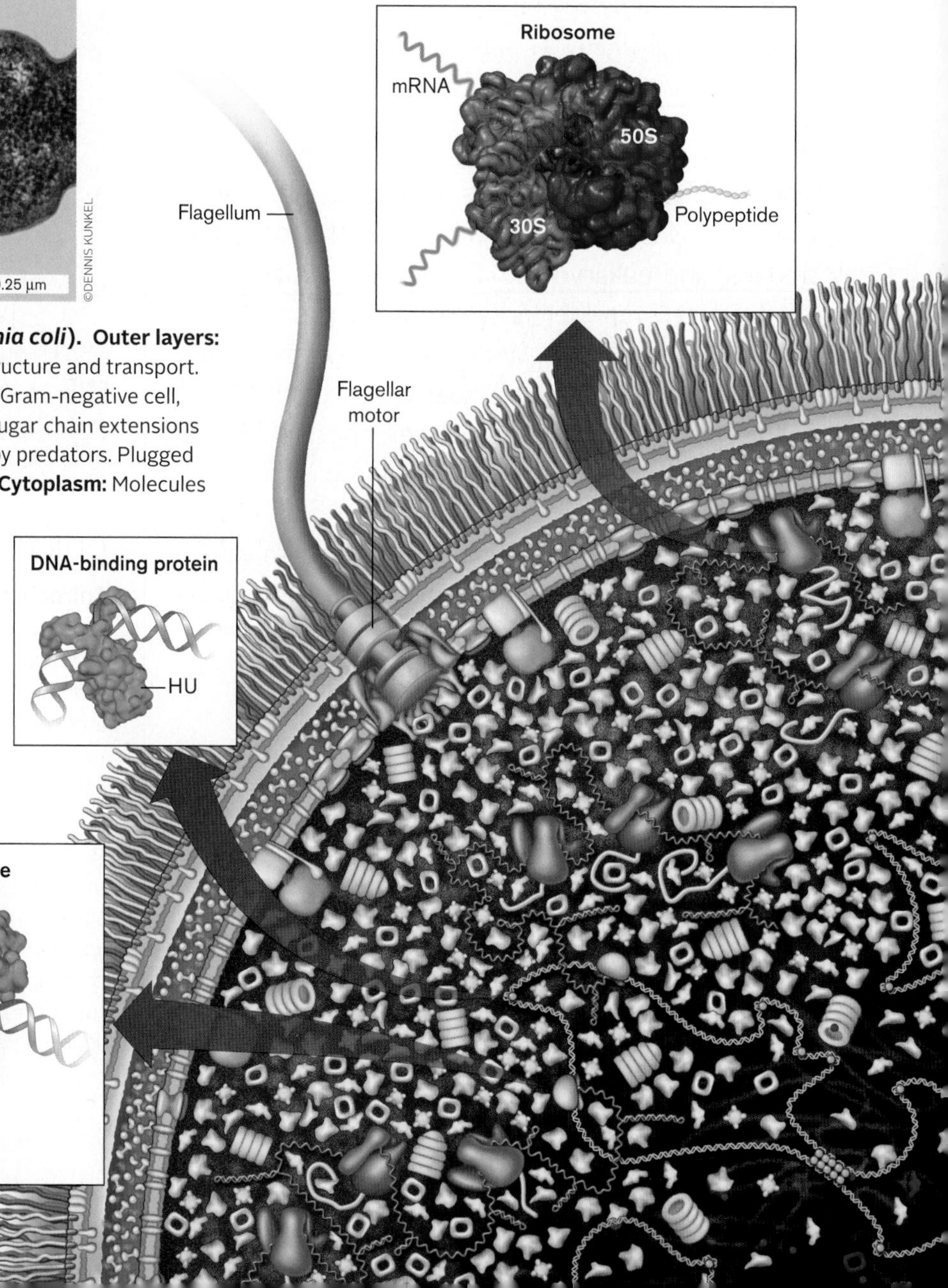
Ribosome
mRNA
50S
30S
Polypeptide
Flagellum
Flagellar motor
DNA-binding protein
HU

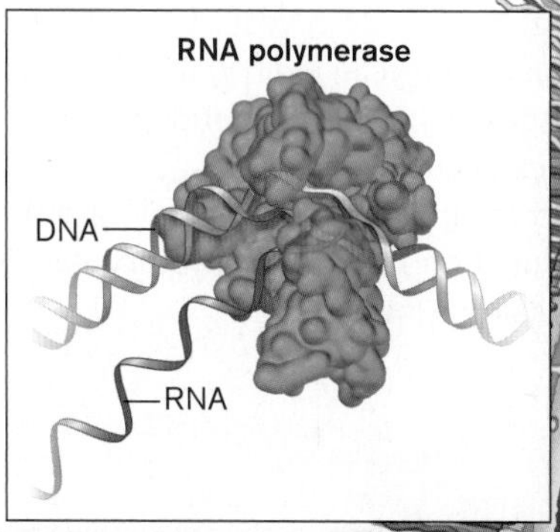
RNA polymerase
DNA
RNA

FIGURE 3.1 ■ Model of a bacterial cell (*Escherichia coli*). Outer layers: The cell membrane contains embedded proteins for structure and transport. The cell membrane is supported by the cell wall. In this Gram-negative cell, the cell wall is coated by the outer membrane, whose sugar chain extensions protect the cell from attack by the immune system or by predators. Plugged into the membranes is the rotary motor of a flagellum. **Cytoplasm:** Molecules of nascent messenger RNA (mRNA) extend out of the nucleoid to the region of the cytoplasm rich in ribosomes. Ribosomes translate the mRNA to make proteins, which are folded by chaperones. **Nucleoid:** The chromosomal DNA is wrapped around binding proteins. Replication by DNA polymerase and transcription by RNA polymerase occur at the same time within the nucleoid. (PDB codes: ribosome, 1GIX, 1GIY; DNA-binding protein, 1P78; RNA polymerase, 1MSW)

- **Tightly coordinated functions.** The cell's parts work together in a highly coordinated mechanism, which may enable a high reproduction rate.

In this chapter we also refer to archaea and eukaryotes for comparisons of cell form and function with those of bacteria.

- **Archaea, like bacteria, are prokaryotes (cells without a nucleus).** Archaea have unique membrane and envelope structures that enable survival in extreme environments. Archaeal diversity is explored in Chapter 19.
- **Eukaryotic cells have a nucleus and extensive membranous organelles.** Organelles such as endoplasmic reticulum and Golgi complex are reviewed in Appendix 2. The mitochondria and chloroplasts of eukaryotic cells evolved by endosymbiosis with engulfed bacteria (Chapter 17). Diverse microbial eukaryotes, such as fungi and protists, are explored in Chapter 20.

3.1

The Bacterial Cell: An Overview

In the early twentieth century, the cell was envisioned as a bag of "soup" full of floating ribosomes and enzymes. But research shows that, in fact, the cell's parts fit together in a structure that is ordered, though flexible. Our model of the bacterial cell (**Fig. 3.1**) offers an interpretation of how the major components of one cell fit together. The model represents *Escherichia coli*, but its general features apply to many kinds of bacteria. Remember that we cannot literally

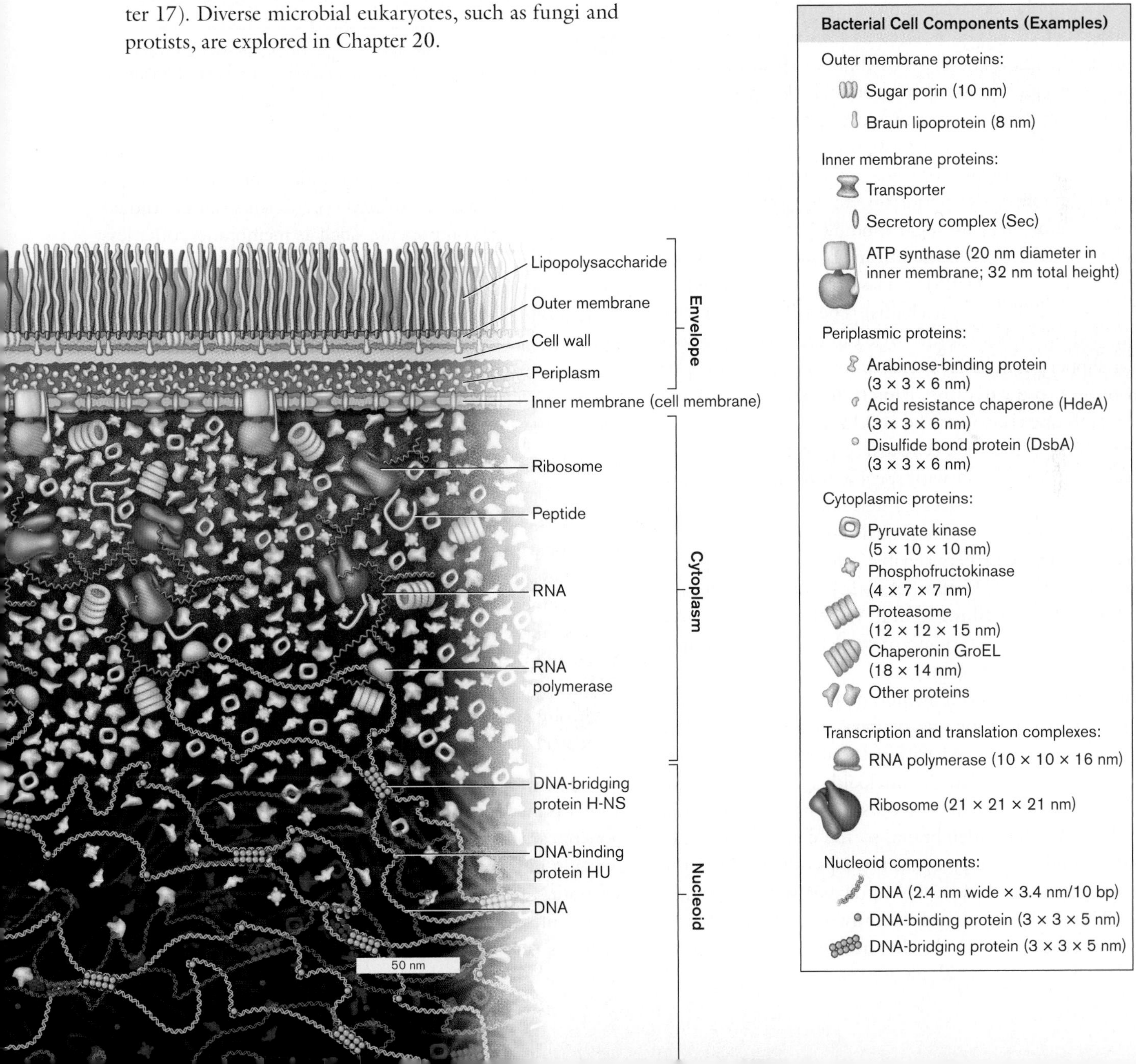

Lipopolysaccharide
Outer membrane
Cell wall
Periplasm
Inner membrane (cell membrane)
Envelope
Ribosome
Peptide
RNA
RNA polymerase
Cytoplasm
DNA-bridging protein H-NS
DNA-binding protein HU
DNA
Nucleoid
50 nm
Bacterial Cell Components (Examples)
Outer membrane proteins:
Sugar porin (10 nm)
Braun lipoprotein (8 nm)
Inner membrane proteins:
Transporter
Secretory complex (Sec)
ATP synthase (20 nm diameter in inner membrane; 32 nm total height)
Periplasmic proteins:
Arabinose-binding protein (3 × 3 × 6 nm)
Acid resistance chaperone (HdeA) (3 × 3 × 6 nm)
Disulfide bond protein (DsbA) (3 × 3 × 6 nm)
Cytoplasmic proteins:
Pyruvate kinase (5 × 10 × 10 nm)
Phosphofructokinase (4 × 7 × 7 nm)
Proteasome (12 × 12 × 15 nm)
Chaperonin GroEL (18 × 14 nm)
Other proteins
Transcription and translation complexes:
RNA polymerase (10 × 10 × 16 nm)
Ribosome (21 × 21 × 21 nm)
Nucleoid components:
DNA (2.4 nm wide × 3.4 nm/10 bp)
DNA-binding protein (3 × 3 × 5 nm)
DNA-bridging protein (3 × 3 × 5 nm)

"see" the molecules within a cell, but microscopy and subcellular analysis generate a remarkably detailed view.

Model of a Bacterial Cell

Within a cell, the cytoplasm consists of a gel-like network composed of proteins and other macromolecules. The cytoplasm is contained by a **cell membrane**. For Gram-negative bacteria (discussed in Section 3.3) the cell membrane is called the "inner membrane," in order to distinguish it from the additional "outer membrane." The inner membrane is composed of phospholipids, transporter proteins, and other molecules. This membrane prevents cytoplasmic proteins from leaking out and maintains gradients of ions and nutrients. Between the inner and outer membranes lies the **cell wall**, a cage-like structure composed of polysaccharides linked covalently by peptides (peptidoglycan). The cell wall forms a single molecule that surrounds the cell. The cell wall limits expansion of the cytoplasm, keeping the cell membrane intact when water flows in; the resulting turgor pressure makes the cell rigid.

The cell wall of *E. coli* consists of one or two layers of peptidoglycan, a polymer of sugars and peptides. The cell wall extends within the periplasm, an aqueous layer containing proteins such as sugar-binding proteins. Outside the cell wall lies the **outer membrane** of phospholipids and **lipopolysaccharides (LPS)**, a class of lipids attached to long polysaccharides (sugar chains). The LPS layer may be surrounded by a thick capsule. The capsule polysaccharides form a slippery mucus layer that inhibits phagocytosis by macrophages (presented in Chapter 26). The cell wall and outer membrane constitute the **envelope**.

The envelope includes cell-surface proteins that enable the bacterium to interact with specific host organisms. For example, *E. coli* cell-surface proteins help the bacterium colonize the human intestinal epithelium, whereas *Sinorhizobium* cell-surface proteins enable colonization of legume plants for nitrogen fixation. Another common external structure is the **flagellum** (plural, **flagella**), a helical protein filament whose rotary motor propels the cell in search of a more favorable environment.

Within the cell, the cell membrane and envelope provide an attachment point for one or more chromosomes. The chromosome is organized within the cytoplasm as a system of looped coils called the **nucleoid**. Unlike the round, compact nucleus of eukaryotic cells, the bacterial nucleoid is not bounded by a membrane, so the coils of DNA can extend throughout the cytoplasm. Loops of DNA from the nucleoid are transcribed by RNA polymerase to form messenger RNA (mRNA), as well as transfer RNA (tRNA) and ribosomal RNA (rRNA). As the mRNA transcripts grow, they bind ribosomes to start synthesizing polypeptide chains. As the polypeptides grow, protein complexes called chaperones help them fold into their functional conformations. The information flow from DNA to RNA to protein is presented in Chapters 7–10.

Observing Cell Parts

How can we know how all the cell parts shown in **Figure 3.1** interact and work together? Electron microscopy largely defines how we "see" the cell's interior as a whole. But smaller parts, such as the ribosomes, appear only as small, densely packed particles. Furthermore, even higher-resolution images from electron microscopy cannot tell us the chemical composition of ribosomes or how they function in the living cell.

Note: Isolation and analysis of the ribosome by cell fractionation, crystallography, and genetic analysis is presented online, in **eTopic 3.1**.

Cell fractionation. One way to study cell parts is by **cell fractionation**, the isolation of cellular components such as ribosomes and flagella. Isolated components can be studied in detail. Other components, such as membranes, offer less information when isolated, because their functions require interactions with other parts of the cell.

To isolate cell components, we first need to **lyse** (break open) the cell. The lysis method must generate enough force to separate the membrane lipids (held together by hydrophobic force) but not enough to disintegrate complexes of protein and RNA. Examples of such methods include:

- **Mild detergent lysis.** Cells can be lysed with a detergent capable of dissolving membranes but not denaturing proteins.
- **Sonication.** Cells can be lysed by intense ultrasonic vibrations that are above the range of human hearing.
- **Enzymes.** Enzymes such as lysozyme can break the cell wall, allowing the cell to be lysed by mild osmotic shock.
- **Mechanical disruption.** Cells can be broken open by the application of high pressure (with a French press) or through beating with microscopic beads (with a bead beater).

A key tool of cell fractionation is the **ultracentrifuge**, a device in which tubes containing solutions of cell components are spun at very high speed (**Fig. 3.2A**). The high rotation rate generates centrifugal forces strong enough to separate subcellular particles (more on ultracentrifugation is presented in Appendix 3). The particles are collected in fractions of sample from the tube, and then the fractions

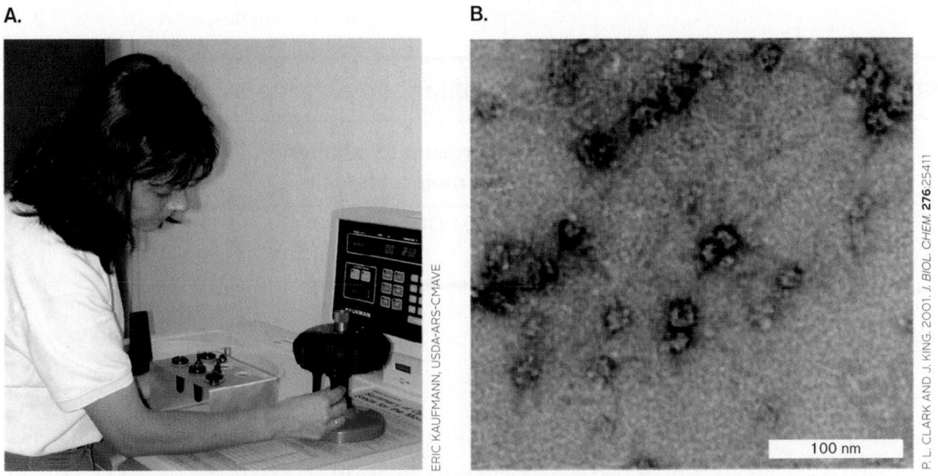
A.
B.
ERIC KAUFMANN, USDA-ARS-CMAVE
100 nm
P. L. CLARK AND J. KING. 2001. J. BIOL. CHEM. 276:25411

FIGURE 3.2 ■ Cell fractionation by ultracentrifugation. **A.** An ultracentrifuge is used to fractionate cell components. **B.** *Salmonella* ribosomes (TEM) isolated from the cytoplasm by ultracentrifugation through a linear sucrose density gradient. A polysome consists of two or more ribosomes attached to mRNA.

are observed by electron microscopy. **Figure 3.2B** shows bacterial ribosomes isolated by centrifugation. In some cases, two or more ribosomes are connected by a strand of messenger RNA (mRNA). This multiple-ribosome structure, called a polysome, provides our first clue as to the intracellular organization of the protein translation apparatus, as seen in **Figure 3.1**, where a number of ribosomes translate each mRNA at the same time. Remarkably, ribosomes isolated by centrifugation can translate messenger RNA in cell-free systems. Experiments in cell-free systems provide the basis for much of our knowledge of protein synthesis (see Chapter 8).

Isolated ribosomes and other components can be crystallized for structure analysis by X-ray diffraction crystallography (discussed in Chapter 2). X-ray-crystallographic data enable us to build a 3D model of the entire ribosome (discussed in Chapter 8). A limitation of crystallographic analysis, however, is that it applies only to isolated particles capable of crystallization. X-ray analysis remains impractical for proteins having flexible, nonrigid structures, and for many membrane-soluble proteins.

A limitation of cell fractionation is that it provides little information about processes that require an intact cell. For example, the role of the transmembrane electrochemical potential, or proton potential, in ATP synthesis was obscured for many years because biochemists were unable to isolate a cytoplasmic complex that generates it. In bacteria, the entire cell membrane of an intact cell supports the proton potential (discussed in Chapter 14).

Genetic analysis. In science, different methods may yield complementary information; that is, different approaches may confirm or extend the conclusions from each. An approach that is complementary to cell fractionation is **genetic analysis**. In genetic analysis, mutant strains are selected for loss of a given function, or a strain can be intentionally mutated so as to lose or alter a gene. The phenotype of the mutant cell may yield clues about the function of the altered part. For example, a cell with a mutation in a gene for the cell division apparatus may form a long filament of incompletely divided daughter cells ("filamentous" mutants, discussed in Section 3.5).

An exciting extension of genetic analysis is the construction of strains with a "reporter gene" fused to a gene encoding a protein of interest. An example of a reporter gene is green fluorescent protein (GFP), which fluoresces at the site of the fused protein. As we saw in Chapter 2, fluorescent reporter genes enable us to observe the function of single proteins within a live cell. In Section 3.5, fluorescent reporter genes show how chromosomes replicate and cells divide. Further methods of genetic analysis are discussed in Chapter 12.

Biochemical Composition of Bacteria

The bacterial cell model in **Figure 3.1** represents the shape and size of cell parts but tells us little about their chemistry—that is, how the parts interact with other molecules of the cell or of the cell's environment. Chemistry explains, for example, why wiping a surface with ethanol kills microbes, whereas water has little effect. Water is a universal constituent of cytoplasm but is excluded by cell membranes. Ethanol, however, dissolves both polar and nonpolar substances; thus, ethanol disintegrates membranes and destroys the folded structure of proteins.

All cells share common chemical components:

- **Water**, the fundamental solvent of life
- **Essential ions**, such as potassium, magnesium, and chloride ions
- **Small organic molecules**, such as lipids and sugars, that are incorporated into cell structures and that provide nutrition by catabolism
- **Macromolecules**, such as nucleic acids and proteins, that contain information, catalyze reactions, and mediate transport, among many other functions

TABLE 3.1

Molecular composition of a bacterial cell, *Escherichia coli,* during balanced exponential growth.[a]

Component	Percentage of total weight[b]	Approximate number of molecules/cell	Number of different kinds
Water	70	20,000,000,000	1
Proteins	16	2,400,000	4,000[c]
RNA:			
rRNA, tRNA, and other small RNA (sRNA) molecules	6	250,000	200
mRNA	0.7	4,000	2,000[c]
Lipids:			
Phospholipids (membrane)	3	25,000,000	50
Lipopolysaccharides (outer membrane)	1	1,400,000	1
DNA	1	2[d]	1
Metabolites and biosynthetic precursors	1.3	50,000,000	1,000
Peptidoglycan (murein sacculus)	0.8	1	1
Inorganic ions	0.1	250,000,000	20
Polyamines (mainly putrescine and spermine)	0.1	6,700,000	2

[a]Values shown are for a hypothetical "average" cell cultured with aeration in glucose medium with minimal salts at 37°C.

[b]The total weight of the cell (including water) is about 10^{-12} gram (g), or 1 picogram (pg).

[c]The number of kinds of mRNA and of proteins is difficult to estimate because some genes are transcribed at extremely low levels and because RNA and proteins include kinds that are rapidly degraded.

[d]In rapidly growing cells, cell fission typically lags approximately one generation behind DNA replication—hence, two identical DNA copies per cell.

Source: Modified from F. Neidhardt and H. E. Umbarger. 1996. Chemical composition of *Escherichia coli,* p. 14. In F. C. Neidhardt (ed.), *Escherichia coli and Salmonella: Cellular and Molecular Biology,* 2nd ed. ASM Press, Washington, DC.

Cell composition varies with species, growth phase, and environmental conditions. **Table 3.1** summarizes the chemical components of a cell for the model bacterium *Escherichia coli* during exponential growth.

Small molecules and ions. The *E. coli* cell consists of about 70% water, the essential solvent required to carry out fundamental metabolic reactions and to stabilize proteins. The water solution contains inorganic ions, predominantly potassium, magnesium, and phosphate. Inorganic ions store energy in the form of transmembrane gradients, and they serve essential roles in enzymes. For example, a magnesium ion is required at the active site of RNA polymerase to help catalyze the incorporation of ribonucleotides into RNA.

The cell also contains many kinds of small charged organic molecules, such as phospholipids and enzyme cofactors. A major class of organic cations is the **polyamines**, molecules with multiple amine groups that are positively charged when the pH is near neutral. Polyamines balance the negative charges of the cell's DNA and stabilize ribosomes during translation.

Macromolecules. Many cells have similar content of water and small molecules, but their specific character is defined by their macromolecules, especially their nucleic acids (DNA and RNA) and their proteins. DNA and RNA molecules can be isolated by size using polyacrylamide gel **electrophoresis**, in which the negatively charged molecules migrate in an electrical field (Appendix 3).

A cell's proteins are specified by the genomic DNA (discussed in Chapters 7–10). But a given cell uses different genes to make different proteins, depending on environmental conditions such as temperature, nutrient levels, and entry into a host organism. Individual proteins are made in very different amounts, from 10 per cell to 10,000 per cell. The proteins expressed by a cell under given conditions are known collectively as a **proteome** (**Fig. 3.3A**). Early attempts to define the proteome of a cell were conducted by Fred Neidhardt (**Fig. 3.3B**) and colleagues at the University of Michigan, who compiled the first protein catalog of *E. coli* using **two-dimensional polyacrylamide gel electrophoresis** (**2D PAGE** or **2D gels**). In 2D PAGE, cell contents are first separated through a gel in a pH gradient, dependent on the sum of positive and negative charges on their amino acid side chains. In the "second dimension," the proteins are coated with negative charge and drawn through a voltage potential that separates them by size. The method of 2D PAGE is described in Appendix 3.

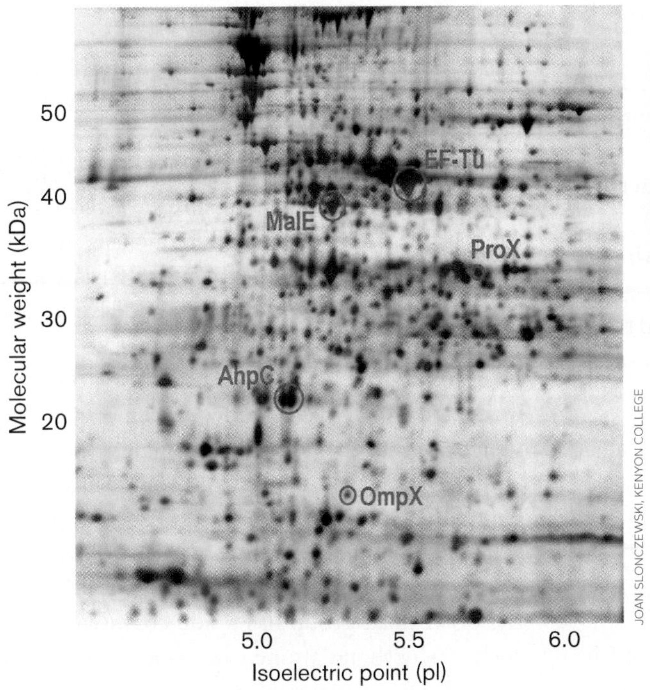
A.
50
40
30
20
Molecular weight (kDa)
EF-Tu
MalE
ProX
AhpC
OmpX
5.0
5.5
6.0
Isoelectric point (pI)
JOAN SLONCZEWSKI, KENYON COLLEGE

B.
FREDERICK C. NEIDHARDT, U. OF MICHIGAN

FIGURE 3.3 ■ **Proteins of *E. coli*.** **A.** 2D gel of proteins of *Escherichia coli* grown aerobically in a casein–yeast extract medium. Proteins were identified by their N-terminal sequence and by mass spectroscopy. EF-Tu is an elongation factor for translation; MalE is an outer membrane maltose-binding protein; ProX and OmpX are periplasmic transporters; and AhpC is an antioxidant stress protein. **B.** Fred Neidhardt at the University of Michigan, Ann Arbor, used 2D gel electrophoresis to complete the first protein catalog of *E. coli*.

In a 2D gel of *E. coli* (**Fig. 3.3A**), a proteome of about 500 different proteins can be distinguished (out of a complete proteome of about 4,300 protein-encoding genes). The most highly expressed proteins include ribosomal proteins and translation factors such as elongation factor Tu (EF-Tu). Outer membrane proteins such as OmpX and ProX are also present in high concentration. Conversely, many proteins of the inner (cell) membrane are too insoluble in water to appear, and some important regulators are present at levels too low to be seen. Nevertheless, 2D gels are useful for identifying changes in protein expression that occur under different environmental conditions or when bacteria are invading host cells.

Protein synthesis is directed by DNA and RNA. The content of nucleic acids in *E. coli* is nearly 8% by weight—much higher than in multicellular eukaryotes. For microbes, the high nucleic acid content is advantageous, allowing the cell to maximize reproduction of its chromosome while minimizing resources for protein-rich cytoplasm. The high level of nucleic acids is actually toxic to human consumers, who lack the enzymes to digest the uric acid waste product of digested nucleotides. That is why most kinds of bacteria cannot be eaten as a major part of our diet.

Other kinds of macromolecules are found in the cell wall and outer membrane. The bacterial cell wall consists of **peptidoglycan**, an organic polymer of peptide-linked sugars that constitutes nearly 1% of the cell mass, approximately the same mass as that of DNA. Peptidoglycan limits the volume of the enclosed cell, so as water rushes in, it generates turgor pressure. This investment of biomass in the cell wall shows the importance (for most species) of maintaining turgor pressure in dilute environments, where water would otherwise enter by osmosis, causing osmotic shock (see Appendix 2 and Section 3.2).

Thought Question

3.1 Which chemicals occur in the greatest number in a bacterial cell? The smallest number? Why does a cell contain 100 times as many lipid molecules as strands of RNA?

To Summarize

- **Bacterial cells are protected by a thick cell envelope.** The envelope includes a cell membrane and a peptidoglycan cell wall. A Gram-negative cell includes an outer membrane, and the cell membrane is called the inner membrane.
- **Bacteria are composed of nucleic acids, proteins, phospholipids, and other organic and inorganic**

chemicals. Proteins in the cell vary, depending on the species and environmental conditions.
- **The bacterial cytoplasm is highly structured.** DNA replication, RNA transcription, and protein synthesis occur coordinately within the cytoplasm.
- **Cell fractionation isolates cell parts for structural and biochemical analysis.**
- **Microscopy can show how cell parts fit within the cell as a whole.**
- **Genetic analysis of mutants shows why cell parts are needed.** Reporter genes can show how cell parts work.

3.2

The Cell Membrane and Transport

How does a cell distinguish "itself" from what is outside? The structure that defines the existence of a cell is the cell membrane (**Fig. 3.4**). The cell membrane consists of a phospholipid bilayer containing lipid-soluble proteins. The overall amounts of lipid and of protein are approximately equal. Overall, the membrane serves two kinds of functions: It contains the cytoplasm within the external medium, mediating transport between the two; and it carries many proteins with specific functions, such as biosynthetic enzymes and environmental signal receptors.

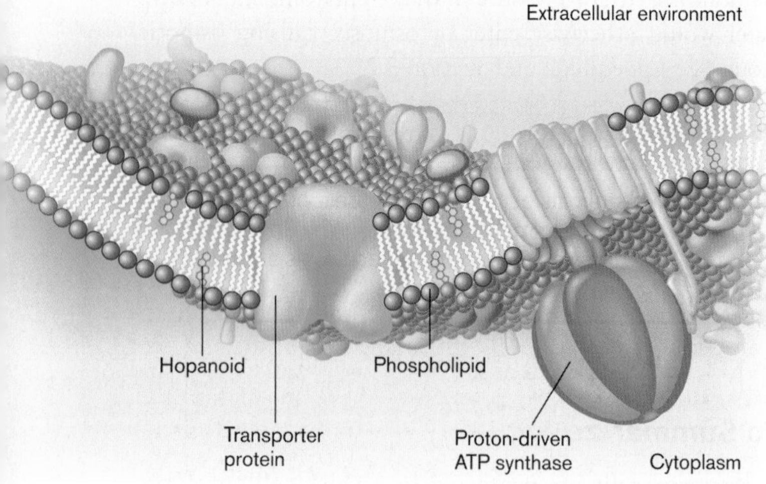

Extracellular environment
Hopanoid
Phospholipid
Transporter protein
Proton-driven ATP synthase
Cytoplasm

FIGURE 3.4 ▪ Bacterial cell membrane. The cell membrane consists of a phospholipid bilayer, with hydrophobic fatty acid chains directed inward, away from water. The bilayer contains stiffening agents such as hopanoids, which serve the same function as cholesterol in eukaryotic membranes. Half the volume of the membrane consists of proteins.

Membrane Lipids

Most membrane lipids are **phospholipids**. A phospholipid possesses a charged phosphate-containing "head" that contacts the water interface, as well as a hydrophobic "tail" of fatty acid packed within the bilayer. Lipid biosynthesis is a key process that is vulnerable to antibiotics. For example, the bacterial enzyme enoyl reductase, which synthesizes fatty acids (discussed in Chapter 15) is the target of triclosan, a common antibacterial additive in detergents and cosmetics.

A typical phospholipid consists of glycerol with ester links to two fatty acids and a phosphoryl polar head group, which at neutral pH is deprotonated (negatively charged) (**Fig. 3.5**). This kind of phospholipid is called a **phosphatidate**. The negatively charged head group of the phosphatidate can contain various organic groups, such as glycerol to form **phosphatidylglycerol** (**Fig. 3.5A**). In other lipids, the polar head group has a side chain with positive charge, such as ethanolamine in **phosphatidylethanolamine** (**Fig. 3.5B**), a major phospholipid of *E. coli.* Phospholipids with positive charge or with mixed charges are concentrated in portions of the membrane that interact with DNA, which has negative charge.

In the bilayer, all phospholipids face each other tail to tail, keeping their hydrophobic side chains away from the water inside and outside the cell. The two layers of phospholipids in the bilayer are called **leaflets**. One leaflet of phospholipids faces the cell interior; the other faces the exterior. As a whole, the phospholipid bilayer imparts fluidity and gives the membrane a consistent thickness (about 8 nm).

Membrane Proteins

The cell membrane can be thought of as a 2D fluid crowded with many different hydrophobic proteins (see **Fig. 3.4**). Membrane proteins serve functions such as transport, communication with the environment, and structural support:

- **Structural support.** Some membrane proteins anchor together different layers of the cell envelope (discussed in Section 3.4). Other proteins attach the membrane to the cytoskeleton, or form the base of structures extending out from the cell, such as flagella.
- **Detection of environmental signals.** In *Vibrio cholerae,* the causative agent of cholera, the membrane protein ToxR detects acidity and elevated temperature—signs in the host digestive tract. The ToxR domain facing the cytoplasm then binds to a DNA sequence activating expression of cholera toxin.
- **Secretion of virulence factors and communication signals.** Membrane protein complexes export toxins

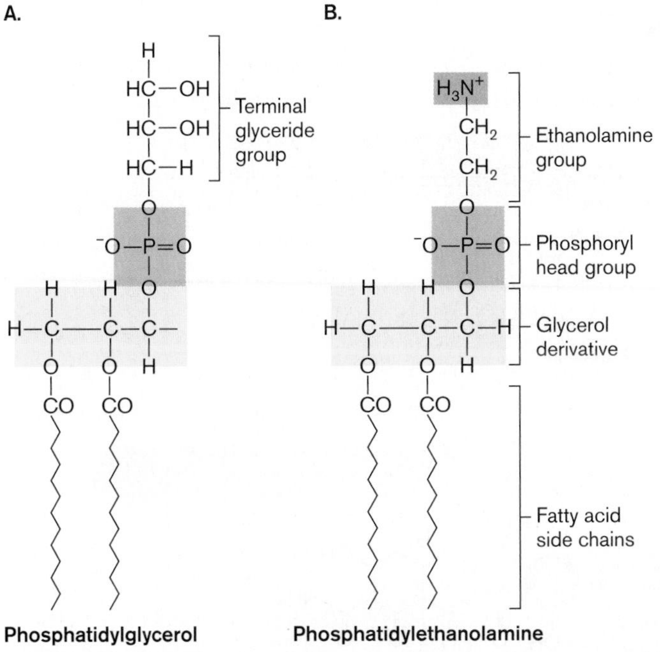
A.
B.
Terminal glyceride group
Ethanolamine group
Phosphoryl head group
Glycerol derivative
Fatty acid side chains
Phosphatidylglycerol
Phosphatidylethanolamine

FIGURE 3.5 ▪ Phospholipids. A. Phosphatidylglycerol consists of glycerol with ester links to two fatty acids, and a phosphoryl group linked to a terminal glycerol. **B.** Phosphatidylethanolamine contains a glycerol linked to two fatty acids, and a phosphoryl group with a terminal ethanolamine. The ethanolamine carries a positive charge.

and cell signals across the envelope. For example, symbiotic nitrogen-fixing rhizobia require membrane proteins NodI and NodJ to transport nodulation signals out to the host plant roots, inducing the plant to form root nodules containing the bacteria.

- **Ion transport and energy storage.** Transport proteins manage ion flux between the cell and the exterior, and store energy in ion gradients. Ion gradients and energy storage are discussed in Chapter 4.

Proteins embedded in a membrane require a hydrophobic portion that is soluble within the membrane. Typically, several hydrophobic alpha helices thread back and forth through the membrane. Other peptide regions extend outside the membrane, containing charged and polar amino acids that interact favorably with water. The combination of hydrophobic and hydrophilic regions effectively locks the protein into the membrane.

Transport across the Cell Membrane

The cell membrane acts as a barrier to keep water-soluble proteins and other cell components within the cytoplasm. But how do nutrients from outside get into the cell—and how do secreted products such as toxins get out? Specific membrane proteins transport molecules across the membrane between the cytoplasm and the outside. Selective transport is essential for cell survival; it means the ability to acquire scarce nutrients, exclude waste, and transmit signals to neighbor cells.

Note Diffusion and osmosis are presented in detail in Appendix 2. Protein transporters are presented in Chapter 4.

Passive diffusion. Small uncharged molecules, such as O_2, CO_2, and water, easily permeate the membrane. Some molecules, such as ethanol, also disrupt the membrane—an action that can make such molecules toxic to cells. By contrast, large, strongly polar molecules such as sugars, and charged molecules such as amino acids, generally cannot penetrate the hydrophobic interior of the membrane and thus require transport by specific proteins. Water molecules permeate the membrane, but their rate of passage is increased by protein channels called aquaporins.

Osmosis. Most cells maintain a concentration of total **solutes** (molecules in solution) that is higher inside the cell than outside. As a result, the internal concentration of water is lower than the concentration outside the cell. Because water can cross the membrane but charged solutes cannot, water tends to diffuse across the membrane into the cell, causing the expansion of cell volume, in a process called **osmosis**. The resulting pressure on the cell membrane is called **osmotic pressure** (reviewed in Appendix 2). Osmotic pressure will cause a cell to burst, or lyse, in the absence of a countering pressure such as that provided by the cell wall. That is how bacteria are killed by penicillin, which disrupts cell wall synthesis.

Membrane-permeant weak acids and bases. A special case of movement across cell membranes is that of **membrane-permeant weak acids** and **weak bases** (**Fig. 3.6**), which exist in equilibrium between charged and uncharged forms:

$$\text{Weak acid: } HA \rightleftharpoons H^+ + A^-$$

$$\text{Weak base: } B + H_2O \rightleftharpoons BH^+ + OH^-$$

Weak acids and weak bases cross the membrane in their uncharged form: HA (weak acid) or B (weak base). On the other side, upon reentering aqueous solution they dissociate (HA to A^- and H^+) or reassociate with H^+ (B to BH^+). In effect, they conduct acid (H^+) or base (OH^-) across the membrane, causing acid or alkali stress. A high proton concentration outside the cell (low pH) will increase the amount of uncharged weak acid that can freely enter the cell. Thus, if the H^+ concentration (acidity) outside the cell is greater than inside, it will drive weak acids into the cell.

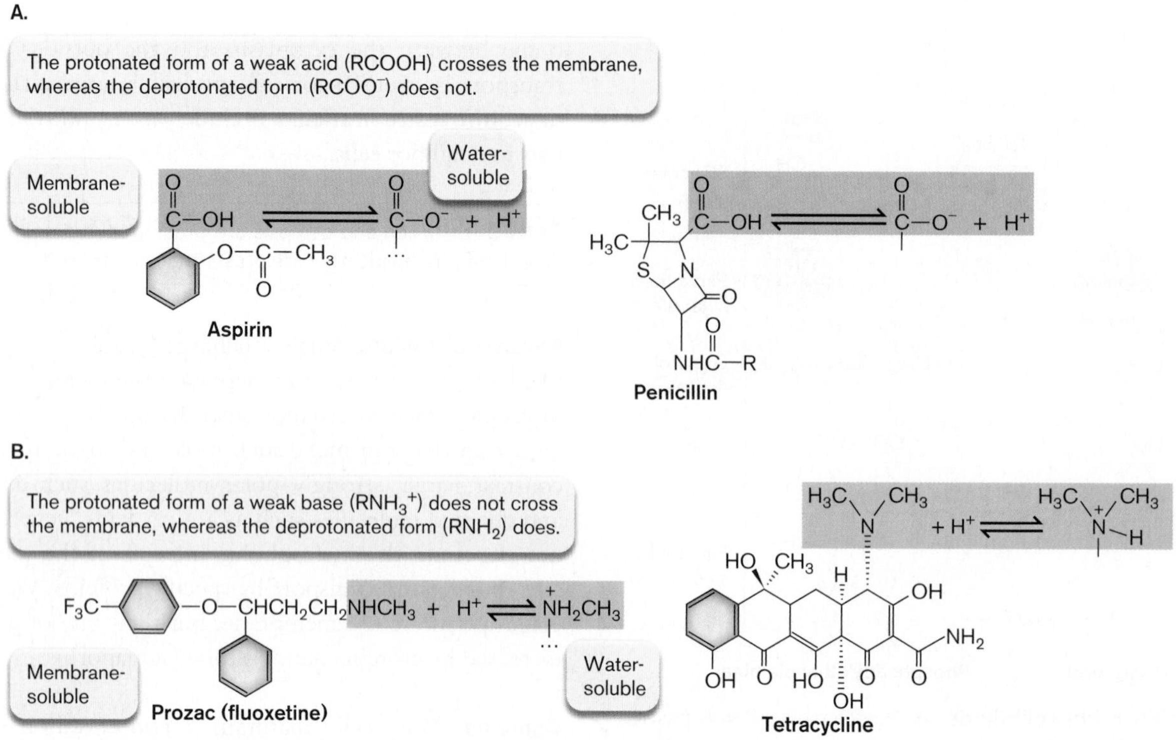
A.
The protonated form of a weak acid (RCOOH) crosses the membrane, whereas the deprotonated form (RCOO⁻) does not.
Membrane-soluble
Water-soluble
Aspirin
Penicillin
B.
The protonated form of a weak base (RNH3+) does not cross the membrane, whereas the deprotonated form (RNH2) does.
Membrane-soluble
Water-soluble
Prozac (fluoxetine)
Tetracycline

FIGURE 3.6 ■ Common drugs are membrane-permeant weak acids and bases. In its charged form, each drug is soluble in the bloodstream. The uncharged form is hydrophobic and penetrates the cell membrane.

Many key substances in cellular metabolism are membrane-permeant weak acids and bases, such as acetic acid. Most pharmaceutical drugs—therapeutic agents delivered to our tissues via the bloodstream—are weak acids or bases whose uncharged forms exist at sufficiently low concentration to cross the membrane without disrupting it. Examples of weak acids that deprotonate (acquiring negative charge) at neutral pH include aspirin (acetylsalicylic acid) and penicillin (**Fig. 3.6A**). Examples of weak bases that protonate (acquiring positive charge) at neutral pH include Prozac (fluoxetine) and tetracycline (**Fig. 3.6B**).

Thought Question

3.2 Amino acids have acidic and basic groups that can dissociate. Why are they not membrane-permeant weak acids or weak bases? Why do they fail to cross the phospholipid bilayer?

Transmembrane ion gradients. Molecules that carry a fixed charge, such as hydrogen and sodium ions (H^+ and Na^+), cannot cross the phospholipid bilayer. Such ions usually exist in very different concentrations inside and outside the cell. An **ion gradient** (ratio of ion concentrations) across the cell membrane can store energy for nutrition, or to drive the transport of other molecules. The role of ion gradients for storing energy is discussed in Section 4.1.

Inorganic ions require transport through specific **transport proteins**, or **transporters**. So, too, do organic molecules that carry a charge at cytoplasmic pH, such as amino acids and vitamins. For example, the vitamin B_{12} transporter in *E. coli* spends 2 ATP for energy to drive vitamin B_{12} across the inner cell membrane into the cytoplasm (**Fig. 3.7**). The vitamin B_{12} transporter is a member of the ABC (ATP-binding cassette) transporter family, discussed further in Section 4.2.

What happens to membrane transport in different environments? Bacteria have evolved different transporters that take up different ions and organic substances under different environmental conditions. Organisms that live in complex, changing environments express numerous transporters. For example, the genome of *Streptomyces coelicolor*, a soil-dwelling actinomycete bacterium that produces several antibiotics, contains genes for over 400 different transporters. These include exporters for toxic ions such as arsenate and chromate, exporters for oligosaccharides and peptides, and multiple-drug pumps. When the substrate for a transporter appears in the environment, it may induce bacterial expression of an uptake transporter (for a nutrient) or an efflux pump (for a toxic drug). Efflux pumps send antibiotics such as tetracycline out of the bacterial cell, enabling harmful bacteria to grow in the presence of antibiotics. Pathogens and cancer cells evolve multidrug transporters that enable them to survive chemotherapy.

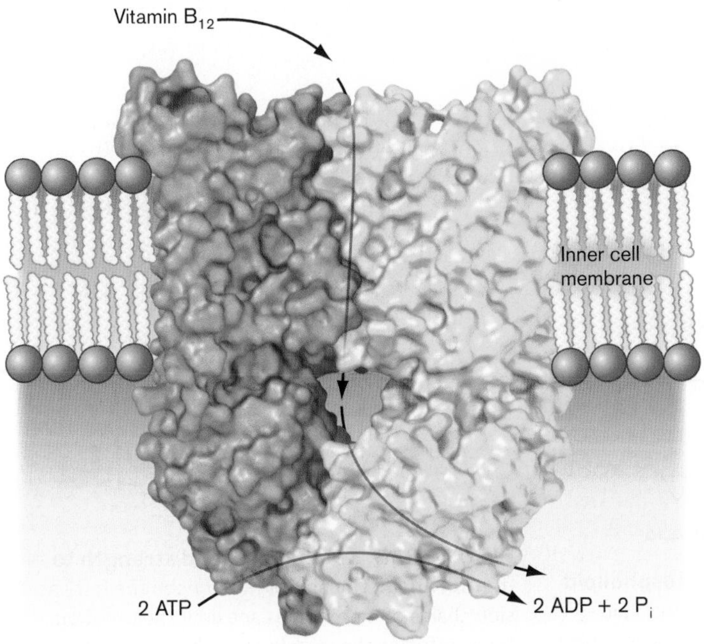

Vitamin B_{12}
Inner cell membrane
2 ATP
2 ADP + 2 P_i

FIGURE 3.7 ▪ A membrane-embedded transport protein: the vitamin B_{12} ABC transporter. The protein complex hydrolyzes ATP to drive transport of vitamin B_{12} across the membrane. (PDB code: 1L7V)

Transport may be passive or active. In **passive transport**, molecules accumulate or dissipate along their concentration gradient. **Active transport**—that is, transport from lower to higher concentration—requires cells to spend energy. The energy for active transport may be obtained by cotransport of another substance down its gradient from higher to lower concentration. For example, many transporters couple uptake of amino acids to uptake of sodium ions. Alternatively, transport may be powered by a coupled chemical reaction that spends energy, such as ATP hydrolysis for ABC transporters (**Fig. 3.7**). Transport protein mechanisms are discussed in Section 4.2.

Membrane Lipid Diversity

Membranes require a uniform thickness and stability to maintain structural integrity and function. So why do individual membrane lipids differ in structure? Membrane lipids help determine whether an organism grows in a Yellowstone hot spring or in the human lungs, where it may cause pneumonia. Phospholipids vary with respect to their phosphoryl head groups and with respect to their hydrocarbon side chains. Some membrane lipids lack phosphate altogether, substituting other polar groups; and some replace mobile side chains with fused rings.

Environmental stress. An interesting phosphatidate is **cardiolipin**, or **diphosphatidylglycerol**, which is actually a double phospholipid linked by a glycerol (**Fig. 3.8A**). Cardiolipin concentration increases in bacteria grown to starvation or stationary phase (discussed in Chapter 4). Within a cell, cardiolipin does not diffuse at random; it concentrates in patches called "domains" near the cell poles. The polar localization of cardiolipin was demonstrated by fluorescence microscopy, in which a cardiolipin-specific fluorophore localized to the poles of *E. coli* (**Fig. 3.8B**). The "wedge" shape of cardiolipin, with its narrow head group and wide fatty acid group, is thought to form concave domains of lipid that stabilize the curve of the polar membrane. At the pole, cardiolipin binds certain environmental stress proteins, such as a protein that transports osmoprotectants when the cell is under osmotic stress. Thus, a phospholipid can have specific functions associated with specific membrane proteins.

The fatty acid component of phospholipids also varies greatly among bacteria. The most common bacterial fatty acids are hydrogenated chains of varying length, typically between 6 and 22 carbons. But soil bacteria and pathogens such as *Mycobacterium tuberculosis* (the cause of tuberculosis) have several hundred different kinds of fatty acids, including chains of 60 carbons, as well as chains containing aromatic groups.

A.

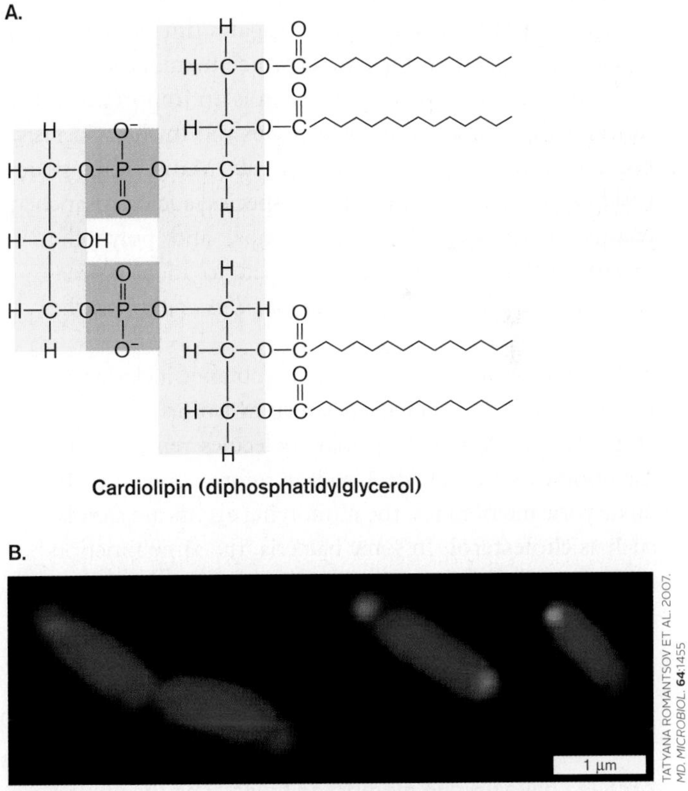

Cardiolipin (diphosphatidylglycerol)
B.
1 µm
TATYANA ROMANTSOV ET AL. 2007. MOL. MICROBIOL. 64:1455

FIGURE 3.8 ▪ Cardiolipin localizes to the poles. A. Cardiolipin is a double phospholipid joined by a third glycerol. **B.** Cardiolipin localizes to the cell poles, as shown by microscopy with a cardiolipin-specific fluorophore (red). The cell's DNA is stained with DAPI (blue).

Some fatty acid chains are partly unsaturated (possess one or more carbon-carbon double bonds). Most unsaturated bonds in membranes are *cis,* meaning that both alkyl chains are on the same side of the bond, so the unsaturated chain has a "kink," as in the cis form of oleic acid (**Fig. 3.9**). Because the kinked chains do not pack as closely as the straight hydrocarbon chains do, the membrane is more "fluid." This is why, at room temperature, unsaturated vegetable oils are fluid, whereas highly saturated butterfat is solid. The enhanced fluidity of a kinked phospholipid improves the function of the membrane at low temperature; hence, bacteria can respond to cold and heat by increasing or decreasing their synthesis of unsaturated phospholipids.

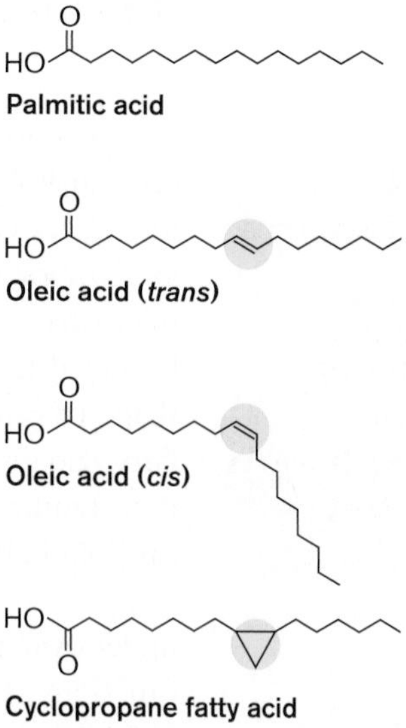

HO
O
Palmitic acid
Oleic acid (*trans*)
Oleic acid (*cis*)
Cyclopropane fatty acid

FIGURE 3.9 ■ Phospholipid side chains. Bacterial lipid side chains include palmitic acid, oleic acid, and cyclopropane fatty acid.

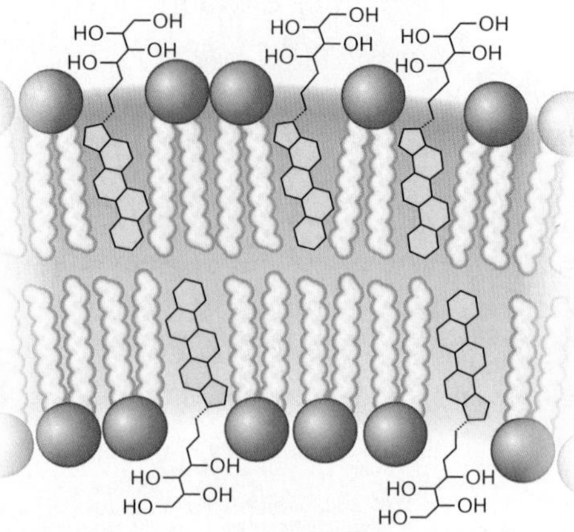

HO
OH

FIGURE 3.10 ■ Hopanoids add strength to membranes. Hopanoids fit between the fatty acid side chains of membranes and limit their motion, thus stiffening the membrane. A hopanoid has five fused rings and an extended hydroxylated tail.

Another interesting structural variation is cyclization of part of the chain to form a stiff planar ring with decreased fluidity. The double bond of unsaturated fatty acids can incorporate a carbon from *S*-adenosyl-L-methionine to form a three-membered ring, generating a cyclopropane fatty acid (see **Fig. 3.9**). Bacteria convert unsaturated fatty acids to cyclopropane during starvation and acid stress, conditions under which membranes require stiffening. Cyclopropane conversion is an important factor in the pathogenesis of *M. tuberculosis* and in the acid resistance of food-borne toxigenic *E. coli.* Many other structural variants are seen in different species, such as branched chains, hydroxyl and sulfate groups, and polycyclic groups. Fatty acid profiles are used to identify certain kinds of pathogens, such as *Bacillus anthracis,* the cause of anthrax.

In addition to phospholipids, membranes include planar molecules that fill gaps between hydrocarbon chains (**Fig. 3.10**). These stiff, planar molecules reinforce the membrane, much as steel rods reinforce concrete. In eukaryotic membranes, the reinforcing agents are sterols, such as **cholesterol**. In some bacteria, the same function is filled by pentacyclic (five-ring) hydrocarbon derivatives called **hopanoids**, or **hopanes**. Hopanoids appear in geological sediments, where they indicate ancient bacterial decomposition; they provide useful data for petroleum exploration.

Archaea have unique membrane lipids. The membrane lipids of archaea differ fundamentally from those of bacteria and of eukaryotes. All archaeal phospholipids replace the ester link between glycerol and fatty acid with an ether link, C–O–C (**Fig. 3.11**). Ethers are much more stable than esters, which hydrolyze easily in water. This is one reason why some archaea can grow at higher temperatures than all other forms of life. Another modification is that archaeal hydrocarbon chains are branched **terpenoids**, polymeric structures derived from isoprene, in which every fourth carbon extends a methyl branch. The branches strengthen the membrane by limiting movement of the hydrocarbon chains.

The most extreme hyperthermophiles, which live beneath the ocean at 110°C, have terpenoid chains linked at the tails, forming a tetraether monolayer. In some species,

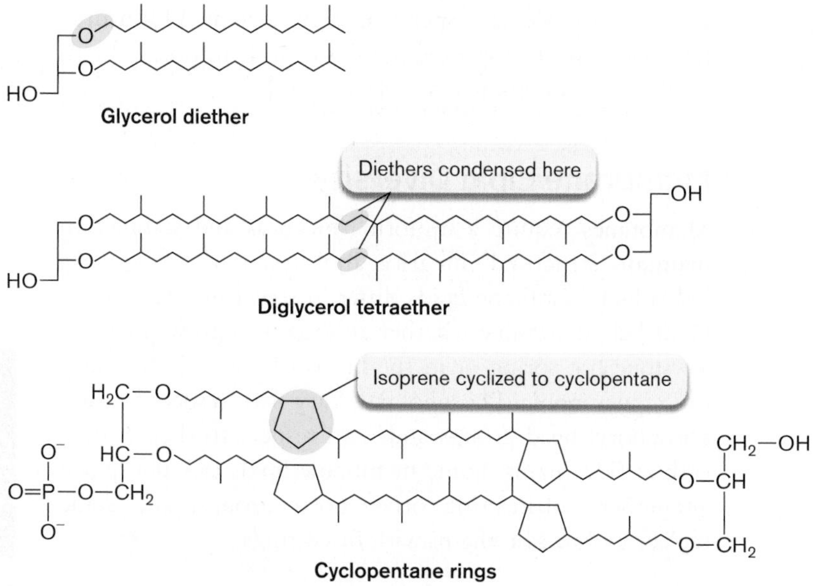

Glycerol diether
Diethers condensed here
Diglycerol tetraether
Isoprene cyclized to cyclopentane
Cyclopentane rings

FIGURE 3.11 ■ Terpene-derived lipids of archaea. In archaea, the hydrocarbon chains are ether-linked to glycerol, and every fourth carbon has a methyl branch. In some archaea, the tails of the two facing lipids of the bilayer are fused, forming tetraethers; thus, the entire membrane consists of a single layer of molecules (a monolayer).

the terpenoids cyclize to form cyclopentane rings. These planar rings stiffen the membrane under stress to an even greater extent than the cyclopropyl chains of bacteria. For more on archaeal cells, see Chapter 19.

Interestingly, both hopanoids of bacteria and the cholesterol of eukaryotes are synthesized from the same precursor molecules as the unique lipids of archaea (**Fig. 3.12**). It may be that hopanoids and cholesterol persist in bacteria and eukaryotes as derivatives of lipids once possessed by a common ancestor of all three domains.

To Summarize

- **The cell membrane** consists of a phospholipid bilayer containing hydrophobic membrane proteins. Membrane proteins serve diverse functions, including transport, cell defense, and cell communication.
- **Small uncharged molecules**, such as oxygen, can penetrate the cell membrane by diffusion.
- **Weak acids and weak bases** exist partly in an uncharged form that can diffuse across the membrane and increase or decrease, respectively, the H^+ concentration within the cell.
- **Polar molecules and charged molecules** require membrane **proteins to mediate** transport. Such facilitated transport can be active or passive.
- **Active transport** requires input of energy from a chemical reaction or from an ion gradient across the membrane. **Ion gradients** generated by membrane pumps store energy for cell functions.
- **Diverse fatty acids** are found in different microbial species and in microbes grown under different environmental conditions.
- **Archaeal membranes have ether-linked terpenoids**, which confer increased stability at high temperature and acidity. Some species have diglycerol tetraethers, which generate a lipid monolayer.

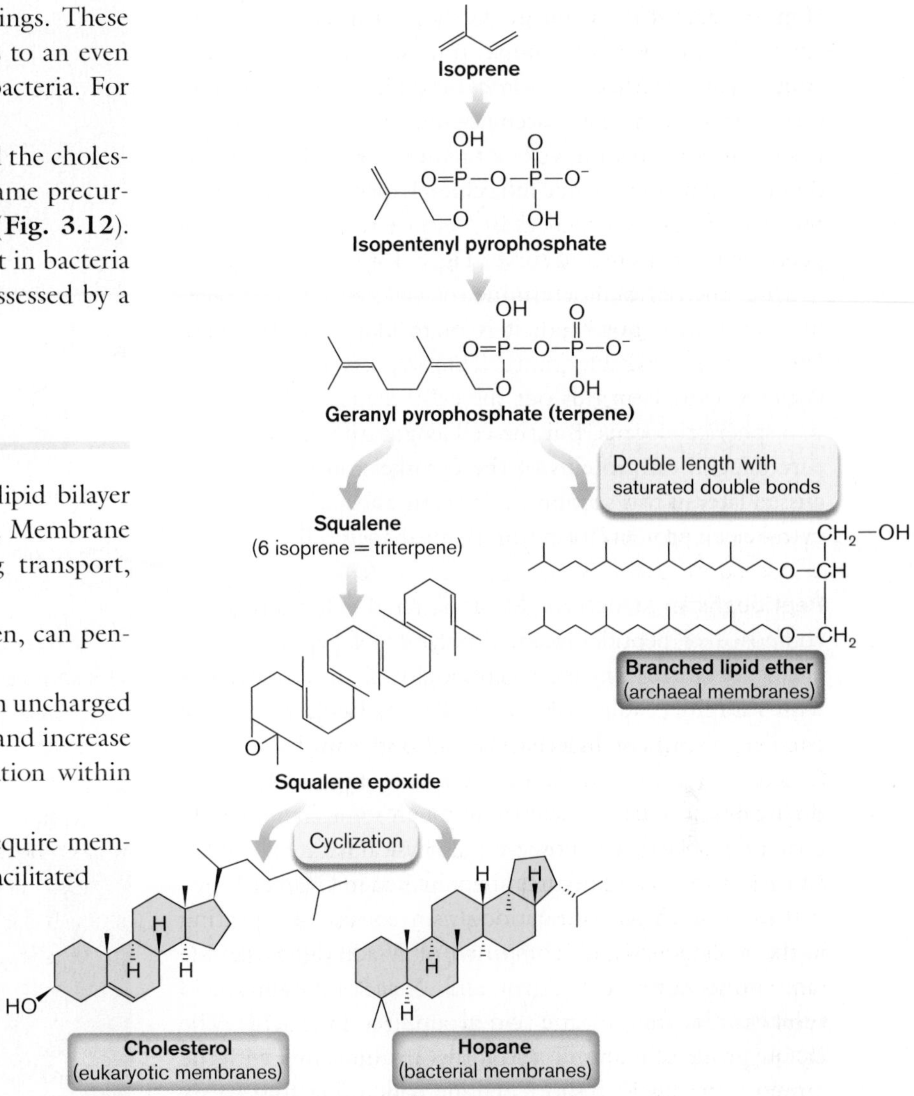
Isoprene
Isopentenyl pyrophosphate
Geranyl pyrophosphate (terpene)
Double length with saturated double bonds
Squalene
(6 isoprene = triterpene)
Branched lipid ether
(archaeal membranes)
Squalene epoxide
Cyclization
Cholesterol
(eukaryotic membranes)
Hopane
(bacterial membranes)

FIGURE 3.12 ■ Synthesis of terpene-derived lipids. Archaeal lipids are synthesized from isoprene chains. Two isoprene units (five carbons in each) link to form a terpene (ten carbons). Terpenoid multimers, such as the triterpene derivative squalene, generate the lipid chains of archaea, in which most of their double bonds have been hydrogenated. In bacteria, squalene cyclizes to form hopanes (hopanoids); and in eukaryotes, squalene is converted to cholesterol.

3.3

The Cell Wall and Outer Layers

How do bacteria and archaea protect their cell membrane? For most species, the cell envelope includes at least one structural supporting layer, like an external skeleton, located outside the cell membrane. The most common structural support is the cell wall (see **Fig. 3.1**). Many species possess additional coverings, such as an outer membrane or an S-layer. Nevertheless, a few prokaryotes, such as the mycoplasmas, have a cell membrane with no outer layers, depending on host fluids for osmotic balance. The archaeon *Ferroplasma* sp. grows in extreme acid, near pH 0, with no support outside its membrane; how this cell survives is unknown.

The Cell Wall Is a Single Molecule

The bacterial cell wall, also known as the **sacculus**, consists of a single interlinked molecule that envelops the cell. The sacculus has been isolated from *E. coli* and visualized by TEM

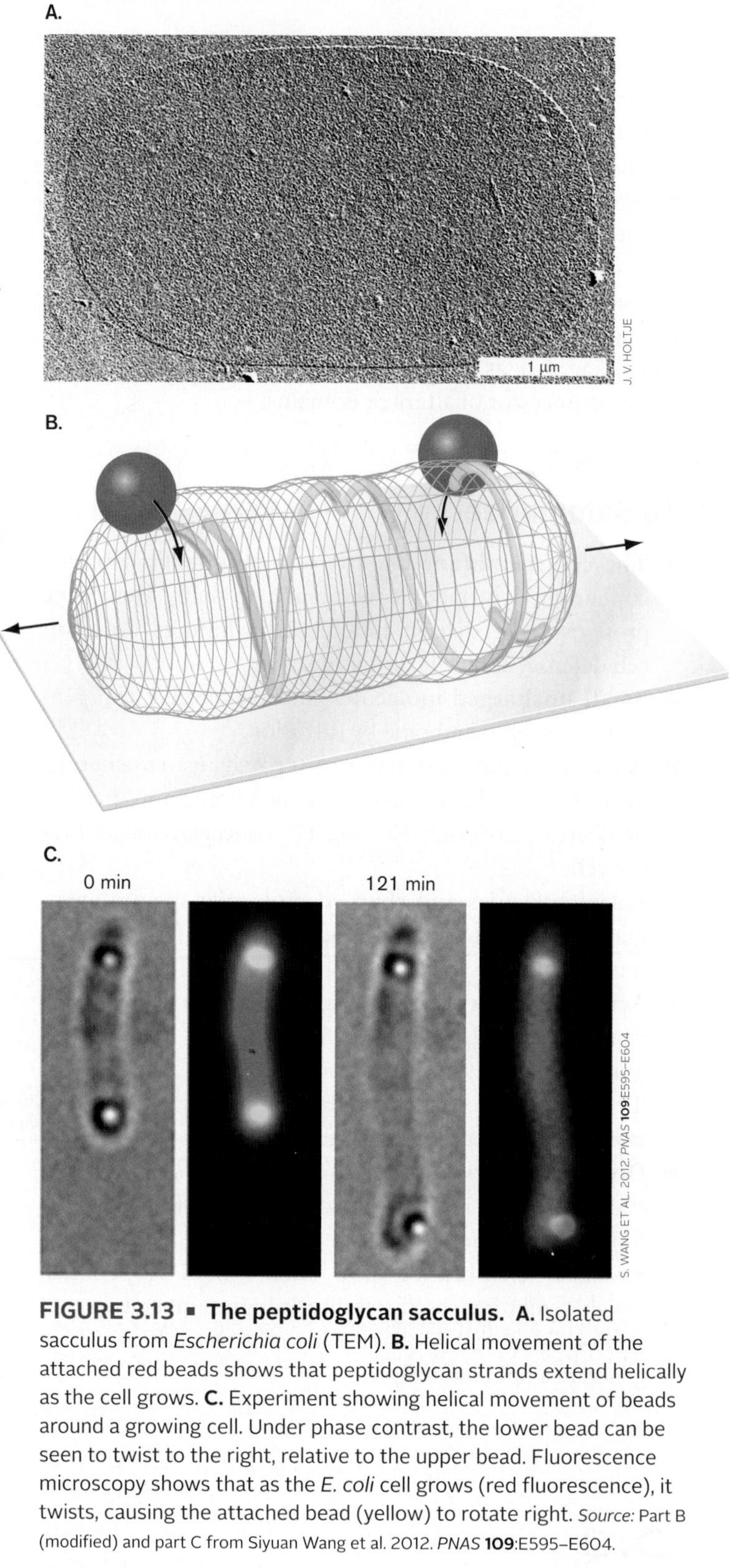
A.
1 µm
J. V. HOLTJE
B.
C.
0 min
121 min
S. WANG ET AL. 2012. PNAS 109:E595–E604

FIGURE 3.13 ▪ The peptidoglycan sacculus. A. Isolated sacculus from *Escherichia coli* (TEM). **B.** Helical movement of the attached red beads shows that peptidoglycan strands extend helically as the cell grows. **C.** Experiment showing helical movement of beads around a growing cell. Under phase contrast, the lower bead can be seen to twist to the right, relative to the upper bead. Fluorescence microscopy shows that as the *E. coli* cell grows (red fluorescence), it twists, causing the attached bead (yellow) to rotate right. *Source:* Part B (modified) and part C from Siyuan Wang et al. 2012. *PNAS* **109**:E595–E604.

(**Fig. 3.13A**). In the image shown, the isolated sacculus appears flattened on the sample grid like a deflated balloon. Its geometrical structure encloses maximal volume with minimal surface area. The sacculus—unlike the membrane—is a single-molecule cage-like structure (**Fig. 3.13B**), highly porous to ions and organic molecules. Experiments show that the sacculus grows with a "twist," incorporating new units of peptidoglycan in a helical curve (**Fig. 3.13C**).

Does the cell wall determine the cell's shape? The cage-like cell wall is not rigid; it is more like a flexible mesh bag with unbreakable joints. Turgor pressure within the enclosed cytoplasm fills out the cell's shape, such as a rod or a spherical coccus. But the cell length and curvature depend on proteins of the cytoskeleton (discussed later in this section). *E. coli* mutants lacking cytoskeletal proteins form round, misshapen cells.

Peptidoglycan structure. Most bacterial cell walls are composed of peptidoglycan, a polymer of peptide-linked chains of amino sugars. "Peptidoglycan" is synonymous with **murein** ("wall molecule"). The molecule consists of parallel polymers of disaccharides called **glycan** chains cross-linked with peptides of four amino acids (**Fig. 3.14**). Peptidoglycan is unique to bacteria, although some archaea build analogous structures whose overall physical nature is similar. (Archaeal cell wall structures are presented in Chapter 19.)

The long chains of peptidoglycan consist of repeating units of a disaccharide composed of *N*-acetylglucosamine (an amino sugar derivative) and *N*-acetylmuramic acid (glucosamine plus a lactic acid group) (see **Fig. 3.14**). The lactate group of muramic acid forms an amide link with the amino terminus of a short peptide containing four to six amino acid residues. The peptide extension can form **cross-bridges** connecting parallel strands of glycan (**Fig. 3.14**).

The peptide contains two amino acids in the unusual D mirror form: D-glutamic acid and D-alanine. The third amino acid, *m*-diaminopimelic acid, has an extra amine group, which forms an amide link to a cross-bridged peptide. The amide link forms with the fourth amino acid of the adjacent peptide, D-alanine (1). The cross-bridge forms by removal of a second D-alanine (2) at the end of the chain. The cross-linked peptides of neighboring glycan strands form the cage of the sacculus.

Note: Amino acids occur in two forms that are "mirror opposites," D and L, of which only the L form is incorporated by ribosomes into protein. The D-form amino acids, however, are used by microbes for many nonprotein structural molecules.

The details of peptidoglycan structure vary among bacterial species. Some Gram-positive species, such as *Staphylococcus aureus* (a cause of toxic shock syndrome), have peptides linked by bridges of pentaglycine instead of the D-alanine link to *m*-diaminopimelic acid. In Gram-negative species, the *m*-diaminopimelic acid is linked to the outer membrane, as discussed shortly.

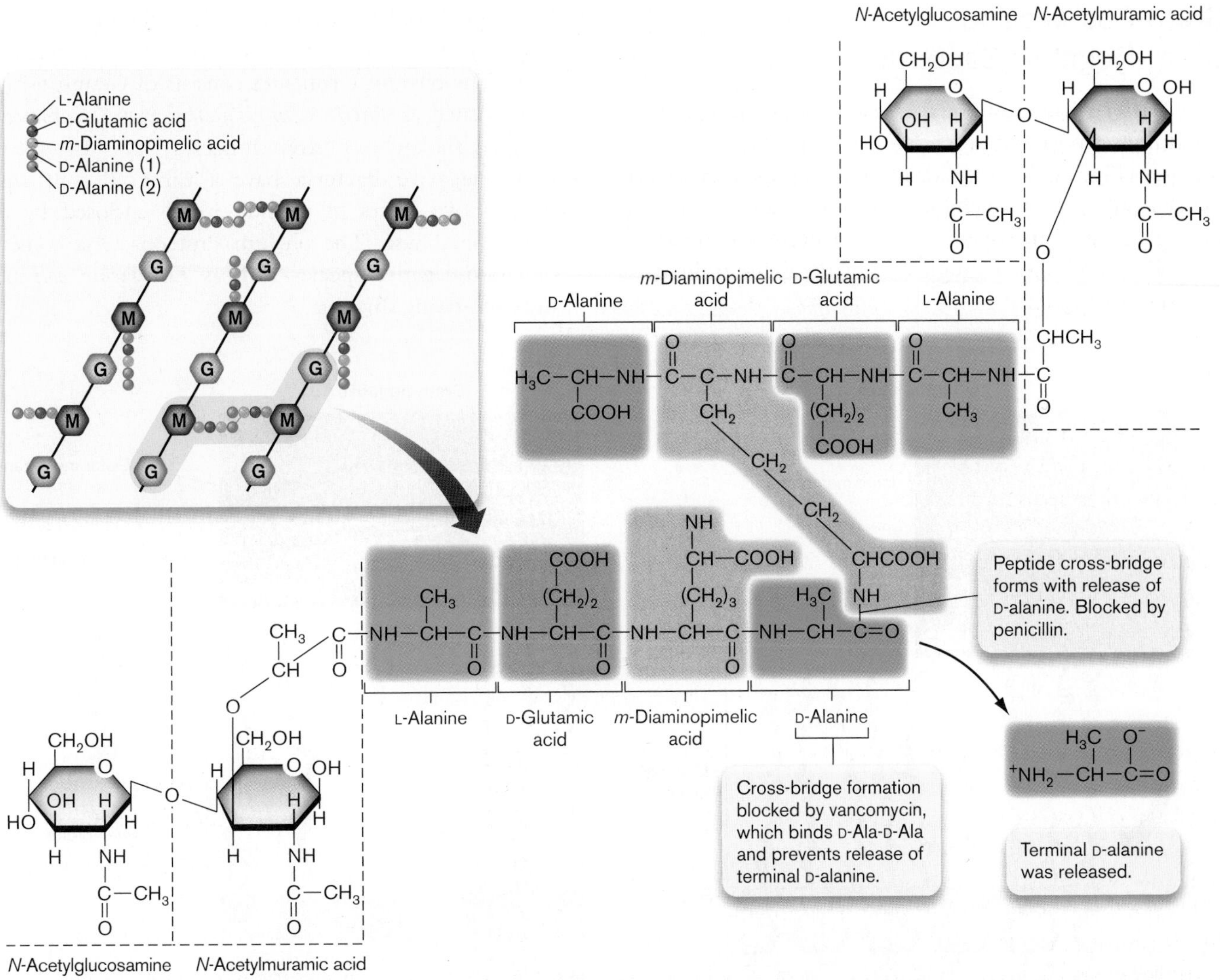

N-Acetylglucosamine
N-Acetylmuramic acid
L-Alanine
D-Glutamic acid
m-Diaminopimelic acid
D-Alanine (1)
D-Alanine (2)
D-Alanine
m-Diaminopimelic acid
D-Glutamic acid
L-Alanine
Peptide cross-bridge forms with release of D-alanine. Blocked by penicillin.
L-Alanine
D-Glutamic acid
m-Diaminopimelic acid
D-Alanine
Cross-bridge formation blocked by vancomycin, which binds D-Ala-D-Ala and prevents release of terminal D-alanine.
Terminal D-alanine was released.
N-Acetylglucosamine
N-Acetylmuramic acid

FIGURE 3.14 ■ Peptidoglycan cross-bridge formation. A disaccharide unit of glycan has an attached peptide of four to six amino acids. The amino terminus of the peptide forms an amide bond with the lactate group of muramic acid. On the peptide, the extra amino group of *m*-diaminopimelic acid can cross-link to the carboxyl terminus of a neighboring peptide. The connection of D-alanine to the peptide is blocked by vancomycin, which binds D-Ala-D-Ala. The cross-bridge formation by transpeptidase is blocked by penicillin.

Peptidoglycan synthesis as a target for antibiotics. Synthesis of peptidoglycan requires many genes encoding enzymes to make the special sugars, build the peptides, and seal the cross-bridges. Because peptidoglycan is unique to bacteria, these biosynthetic enzymes make excellent targets for antibiotics (see **Fig. 3.14**). For example, the transpeptidase that cross-links the peptides is the target of penicillin. Vancomycin, a major defense against *Clostridium difficile* and drug-resistant staphylococci, prevents cross-bridge formation by binding the terminal D-Ala-D-Ala dipeptide, thus preventing release of the terminal D-alanine.

Unfortunately, the widespread use of such antibiotics selects for evolution of resistant strains. One of the most common agents of resistance is the enzyme beta-lactamase, which cleaves the lactam ring of penicillin, rendering it ineffective as an inhibitor of transpeptidase. In a different mechanism, strains resistant to vancomycin, contain an altered enzyme that adds lactic acid to the end of the branch peptides in place of the terminal D-alanine. The altered peptide is no longer blocked by vancomycin. As new forms of drug resistance emerge, researchers continue to seek new antibiotics that target cell wall formation (discussed in Chapter 27).

Thought Question

3.3 If the sacculus (cell wall) consists of a single molecule, how do you think it expands as the cell grows?

Gram-Positive and Gram-Negative Bacteria

Most bacteria have additional envelope layers that provide structural support and protection from predators and host defenses (**Fig. 3.15**). Additional molecules are attached to the cell wall and cell membrane, in some cases threading through them. Envelope composition defines two major categories of bacteria distinguished by the Gram stain (discussed in Chapter 2):

- **Gram-positive bacteria** have a thick cell wall with 3–20 layers of peptidoglycan, interpenetrated by teichoic acids. The phylum Firmicutes consists of Gram-positive species such as *Bacillus thuringiensis* and *Streptococcus pyogenes,* the cause of "strep throat."
- **Gram-negative bacteria** have a thin cell wall with one or two layers of peptidoglycan, enclosed by an outer membrane. The phylum Proteobacteria consists of Gram-negative species such as *Escherichia coli* and nitrogen-fixing rhizobia.

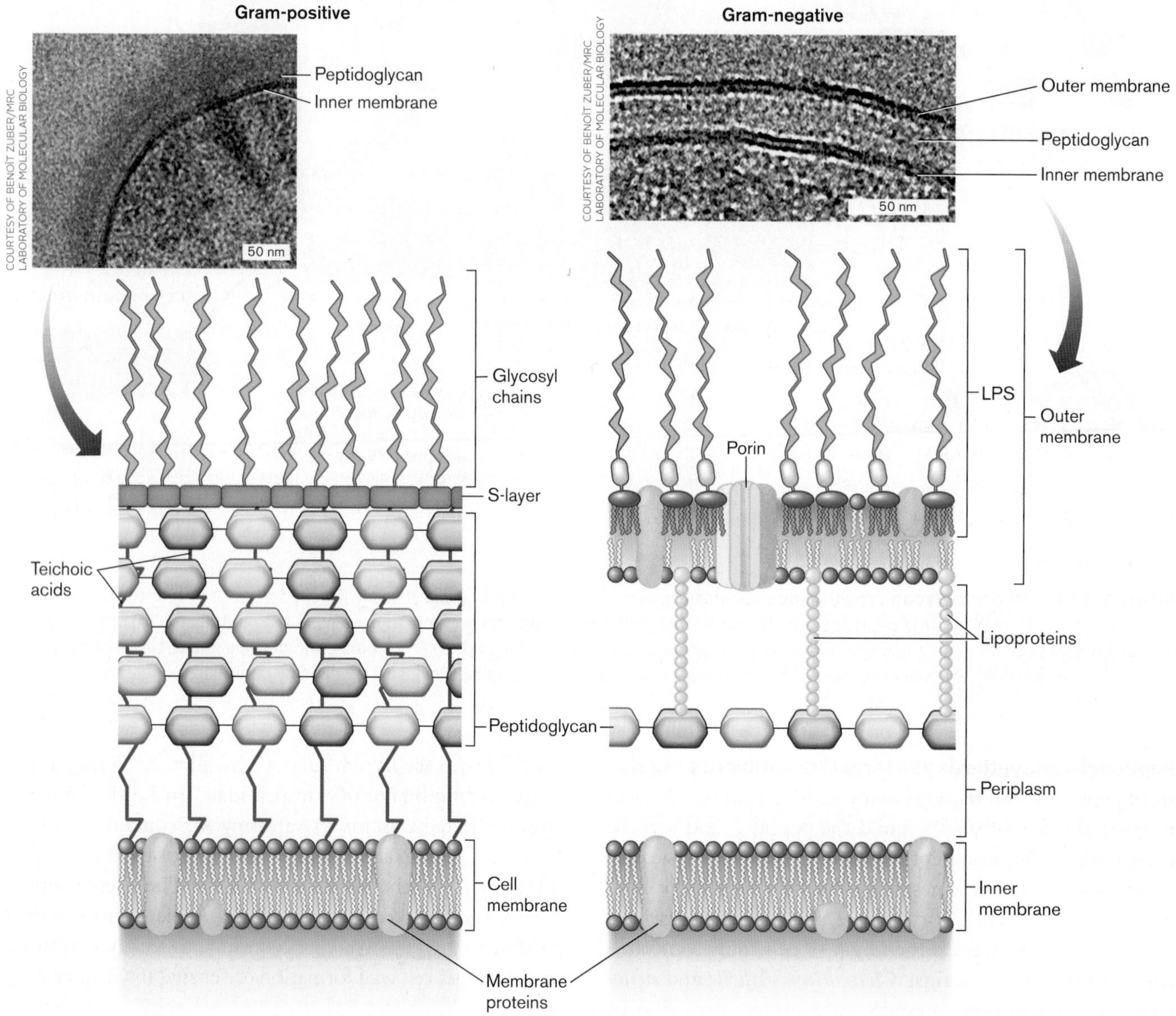
Gram-positive
Peptidoglycan
Inner membrane
50 nm
COURTESY OF BENOÎT ZUBER/MRC LABORATORY OF MOLECULAR BIOLOGY
Gram-negative
Outer membrane
Peptidoglycan
Inner membrane
50 nm
COURTESY OF BENOÎT ZUBER/MRC LABORATORY OF MOLECULAR BIOLOGY
Glycosyl chains
S-layer
Teichoic acids
Peptidoglycan
Cell membrane
Membrane proteins
LPS
Outer membrane
Porin
Lipoproteins
Periplasm
Inner membrane

FIGURE 3.15 ▪ Cell envelope: Gram-positive and Gram-negative. The Gram-positive cell has a thick cell wall with multiple layers of peptidoglycan, threaded by teichoic acids. The cell wall may be covered by an S-layer and a capsule (not shown). The Gram-negative cell has a single layer of peptidoglycan covered by an outer membrane. Some Gram-negative species include an S-layer or a capsule (not shown). The cell membrane of Gram-negative species is called the inner membrane. **Left inset:** Gram-positive envelope of *Bacillus subtilis* (TEM), showing cell membrane, cell wall, and capsule. **Right inset:** Gram-negative envelope of *Pseudomonas aeruginosa* (TEM), showing inner membrane, thin cell wall in the periplasm, and outer membrane.

Thought Question

3.4 **Figure 3.15** highlights the similarities and differences between the cell envelopes of Gram-negative and Gram-positive bacteria. What do you think are the advantages and limitations of a cell's having one layer of peptidoglycan (Gram-negative) versus several layers (Gram-positive)?

Outside these two groups, however, many prokaryotic species cannot be classified according to the Gram stain models. For example, archaeal cell envelopes (see Chapter 19) are highly diverse and cannot be distinguished by Gram stain.

Note: The exact number and arrangement of layers of peptidoglycan in bacteria remains controversial. The question is important because thickening of the cell wall may confer antibiotic resistance. For differing views, see Lu Gan et al., 2008, *PNAS* **105**:18953–18957; and Longzhu Cui et al., 2003, *J. Clin. Microbiol.* **41**:5–14.

The Gram-Positive Cell Envelope

A section of a Gram-positive cell envelope is shown in **Figure 3.15**. The multiple layers of peptidoglycan are reinforced by **teichoic acids** threaded through its multiple layers (**Fig. 3.16**). Teichoic acids are chains of phosphodiester-linked glycerol or ribitol, with sugars or amino acids linked to the middle OH groups. The negatively charged cross-threads of teichoic acids, as well as the overall thickness of the Gram-positive cell wall, help retain the Gram stain.

How does the cell wall attach extracellular structures? Gram-positive bacteria have a type of enzyme called "sortase" that forms a peptide bond from a cell wall cross-bridge to a protein extending from the cell. Such external proteins, for example, can help the cell acquire nutrients, or help attach the cell to a substrate. Sortases are now used in the protein engineering industry.

Outside the cell wall, Gram-positive cells are often encased in a slippery **capsule** consisting of loosely bound polysaccharides. The capsule can be visualized under the light microscope with a negative stain consisting of particles which the capsule excludes (shown in **Fig. 2.24C**). Gram-negative cells may also have a capsule.

The S-layer. An additional protective layer commonly found in free-living bacteria and archaea is the surface layer, or **S-layer**. The S-layer is a crystalline layer of thick subunits consisting of protein or glycoprotein (proteins with attached sugars) (**Fig. 3.17**).

Each subunit of the S-layer contains a pore large enough to admit a wide range of molecules. The subunits form a smooth layer on the cell wall or outer membrane (see **Fig. 3.17**). The proteins are arranged in a highly ordered array, either hexagonally or tetragonally. The S-layer is rigid, but it also flexes and allows substances to pass through it in either direction.

The functions of the S-layer are uncertain, in part because the S-layer is often lost by bacteria after repeated subculturing in the laboratory. The protein subunits serve

FIGURE 3.16 ▪ Teichoic acids. Teichoic acids in the Gram-positive cell wall consist of glycerol or ribitol phosphodiester chains. The middle hydroxyl group of each glycerol or ribitol is typically linked to D-alanine, to D-lysine, or to a sugar such as galactose or *N*-acetylglucosamine.

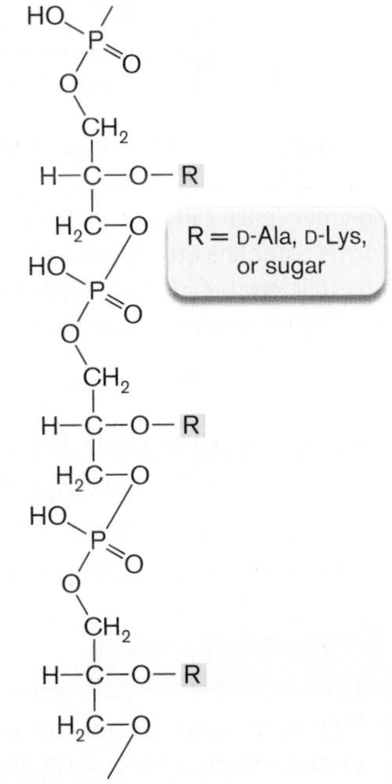
HO
P
O
CH2
H—C—O—R
H2C—O
R = D-Ala, D-Lys, or sugar

Glycerol teichoic acid

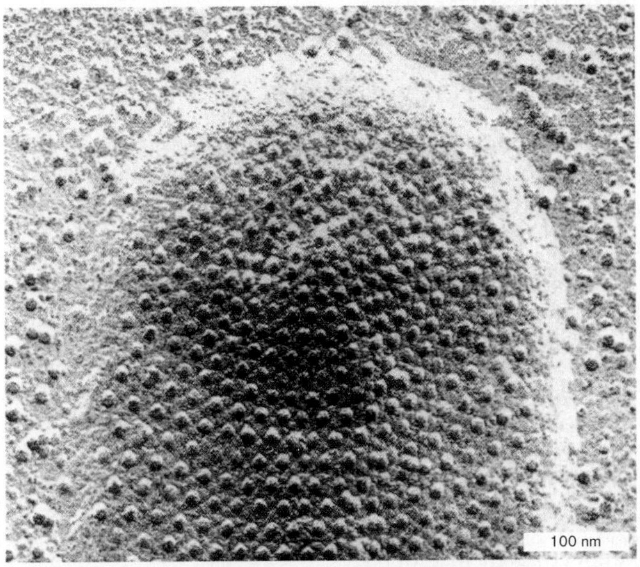
100 nm

PAUL MESSNER ET AL. 1986. *J. BACTERIOL.* **166**:1046

FIGURE 3.17 ▪ The S-layer. The archaeon *Thermoproteus tenax* has a single tetraether membrane encased by an S-layer (SEM). Note the regular pattern of tiled S-layer proteins.

as attachment platforms for polysaccharide filaments. In natural environments, the tough proteins may deter predators such as amebas or host defenses such as phagocytes. The S-layer may contribute to cell shape and help protect the cell from osmotic stress.

Traits diminish in the absence of selective pressure for genes encoding them (discussed in Chapter 17). For example, the mycoplasmas are close relatives of Gram-positive bacteria, yet they have permanently lost their cell walls, as well as the S-layer. Mycoplasmas have no need for cell walls, because they are parasites living in host environments, such as the human lung, where they are protected from osmotic shock.

Mycobacterial cell envelopes. Exceptionally complex cell envelopes are found in actinomycetes, a large and diverse family of soil bacteria that produce antibiotics and other industrially useful products (discussed in Chapter 18). The most complex envelopes known are those of actinomycete-related bacteria, the mycobacteria. Mycobacteria include the famous pathogens *Mycobacterium tuberculosis* (the cause of tuberculosis) and *Mycobacterium leprae* (the cause of leprosy).

Mycobacterial cell envelopes include unusual membrane lipids called mycolic acids (**Fig. 3.18**). Mycolic acids contain a hydroxy acid backbone with two hydrocarbon chains—one comparable in length to typical membrane lipids (about 20 carbons), the other about threefold longer. The long chain includes ketones, methoxyl groups, and cyclopropane rings. Hundreds of different forms are known. The phenolic glycolipids include a phenol group, also linked to sugar chains.

The mycobacterial cell wall includes peptidoglycan linked to chains of galactose, called galactans (**Fig. 3.18**). The galactans are attached to arabinans, polymers of the five-carbon sugar arabinose. The arabinan-galactan polymers are known as arabinogalactans. Arabinogalactan biosynthesis is inhibited by two major classes of anti-tuberculosis drugs: ethambutol and the benzothiazinones. The ends of the arabinan chains form ester links to mycolic acids. Mycolic acids provide the basis for acid-fast staining, in which cells retain the dye carbolfuchsin, an important diagnostic test for mycobacteria and actinomycetes (described in Chapter 2). In *M. tuberculosis* and *M. leprae*, the mycolic acids form a kind of bilayer interleaved with phenolic glycolipids. The extreme hydrophobicity of the phenol derivatives generates a waxy surface that prevents phagocytosis by macrophages.

Overall, the thick, waxy envelope excludes many antibiotics and offers exceptional protection from host defenses, enabling the pathogens of tuberculosis and leprosy to colonize their hosts over long periods. However, the mycobacterial envelope also retards uptake of nutrients. As a result, *M. tuberculosis* and *M. leprae* grow extremely slowly and are a challenge to culture in the laboratory.

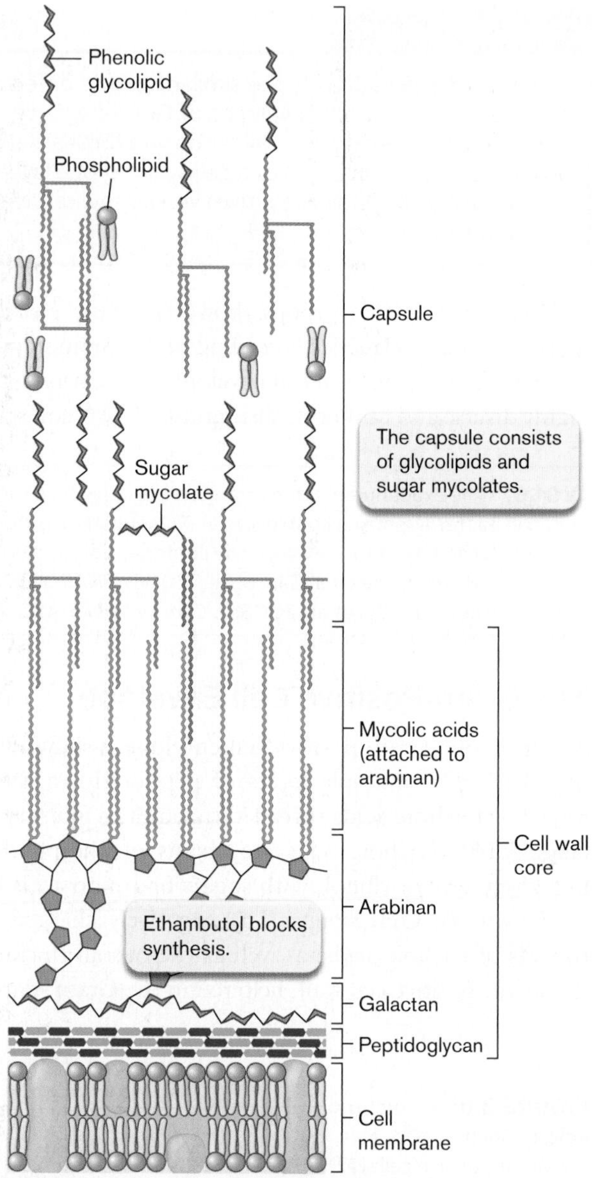
Phenolic glycolipid
Phospholipid
Sugar mycolate
Capsule
The capsule consists of glycolipids and sugar mycolates.
Mycolic acids (attached to arabinan)
Cell wall core
Arabinan
Ethambutol blocks synthesis.
Galactan
Peptidoglycan
Cell membrane

FIGURE 3.18 ■ Mycobacterial envelope structure. A complex cell wall includes a peptidoglycan layer linked to a chain of galactose polymer (galactan) and arabinose polymer (arabinan). Arabinan forms ester links to mycolic acids, which form an outer bilayer with phenolic glycolipids. Outside the outer bilayer is a capsule of loosely associated phospholipids and phenolic glycolipids.

Thought Question

3.5 Why would laboratory culture conditions select for evolution of cells lacking an S-layer?

The Gram-Negative Outer Membrane

A Gram-negative cell envelope is seen in **Figure 3.15C**. Calculations of molecular density show that the thin layer of peptidoglycan consists of one or two sheets. The peptidoglycan is covered by an outer membrane. The Gram-negative outer membrane confers defensive abilities and toxigenic properties on many pathogens, such as *Salmonella* species and enterohemorrhagic *E. coli* (strains that cause hemorrhaging of the colon). Between the outer and inner (cell) membranes is the periplasm.

Lipoprotein and lipopolysaccharide (LPS). The inward-facing leaflet of the outer membrane has a phospholipid composition similar to that of the cell membrane (which in Gram-negative species is called the **inner membrane** or inner cell membrane). The outer membrane's inward-facing leaflet includes lipoproteins that connect the outer membrane to the peptide bridges of the cell wall. The major lipoprotein is called **murein lipoprotein**, also known as "Braun lipoprotein" (**Fig. 3.19A**). Murein lipoprotein consists of a protein with an N-terminal cysteine attached to three fatty acid side chains. The side chains are inserted in the inward-facing leaflet of the outer membrane. The C-terminal lysine forms a peptide bond with the *m*-diaminopimelic acid of peptidoglycan (murein). What happens to a mutant cell that fails to make murein lipoprotein? As the cell grows and divides, it fails to attach its outer membrane to the growing cell wall, causing the outer membrane to balloon out in the region where the daughter cells separate (**Fig. 3.19C**).

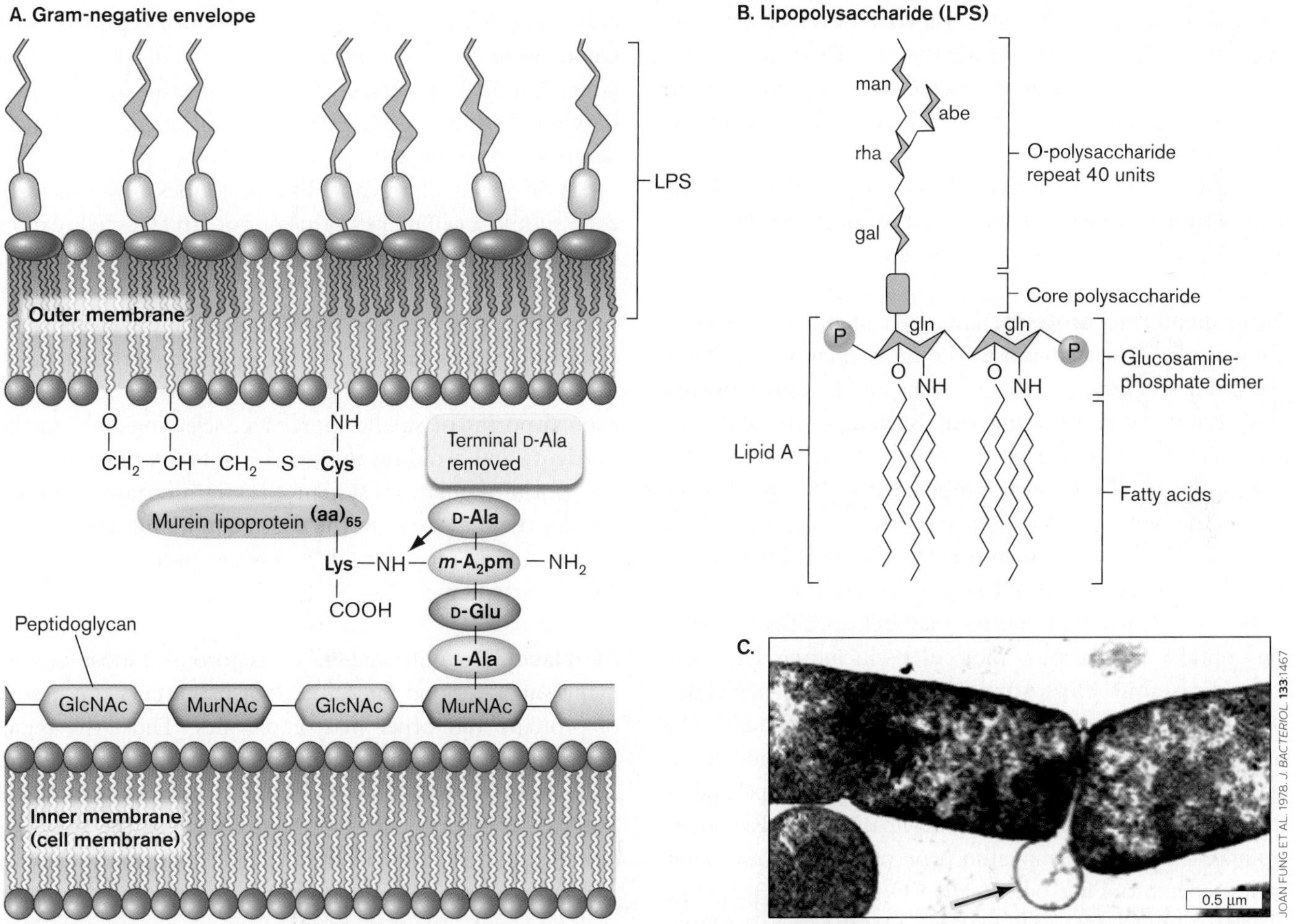
A. Gram-negative envelope
LPS
Outer membrane
O
O
NH
CH2—CH—CH2—S—Cys
Terminal D-Ala removed
Murein lipoprotein (aa)65
D-Ala
Lys—NH—
m-A2pm
—NH2
COOH
D-Glu
L-Ala
Peptidoglycan
GlcNAc
MurNAc
GlcNAc
MurNAc
Inner membrane
(cell membrane)
B. Lipopolysaccharide (LPS)
man
abe
rha
gal
O-polysaccharide repeat 40 units
Core polysaccharide
gln
gln
P
P
O
NH
O
NH
Glucosamine-phosphate dimer
Lipid A
Fatty acids
C.
0.5 µm
JOAN FUNG ET AL. 1978. J. BACTERIOL. 133:1467

FIGURE 3.19 ■ Lipoprotein and lipopolysaccharide. A. Murein lipoprotein has an N-terminal cysteine triglyceride inserted in the inward-facing leaflet of the outer membrane. The C-terminal lysine forms a peptide bond with the *m*-diaminopimelic acid of the peptidoglycan (murein) cell wall. **B.** Lipopolysaccharide (LPS) consists of branched and unbranched short-chain fatty acids linked to a dimer of phosphoglucosamine. One glucosamine is linked to a core polysaccharide extending out from the cell, which is attached to about 40 repeating units of a polysaccharide known as O antigen. **C.** Lack of murein lipoprotein in mutant *Salmonella* bacteria causes the outer membrane to balloon out (arrow), when the cell tries to divide (TEM).

The outward-facing leaflet of the outer membrane has very different lipids from the inner leaflet. The main outward-facing phospholipids are called **lipopolysaccharide** (**Fig. 3.19B**). The LPS lipids have shorter fatty acid chains than those of the inner cell membrane, and some are branched. LPS is of crucial medical importance because it acts as an **endotoxin**. An endotoxin is a cell component that is harmless as long as the pathogen remains intact; but when released by a lysed cell, endotoxin overstimulates host defenses, inducing potentially lethal endotoxic shock. Thus, antibiotic treatment of an LPS-containing pathogen can kill the cells but can also lead to death of the patient.

In LPS the fatty acids are esterified to **glucosamine**, an amino sugar found also in peptidoglycan. Two glucosamine dimers each condense with two fatty acid chains, making four fatty acid chains in all. Like the glycerol of phospholipid, each glucosamine of LPS connects to a phosphate, whose negative charge interacts with water. One of the glucosamines is attached to a **core polysaccharide**, a sugar chain that extends outside the cell. The core polysaccharide consists of about five sugars with side chains such as phosphoethanolamine. It extends to an **O polysaccharide**, a chain of as many as 200 sugars. The O polysaccharide may extend longer than the cell itself. These chains form a layer that helps bacteria resist phagocytosis by white blood cells.

Outer membrane proteins. The outer membrane contains unique proteins not found in the inner membrane. Outer membranes contain a class of transporters called **porins** that permit the entry of nutrients such as sugars and peptides. Outer membrane porins have a distinctive cylinder of beta sheets, also known as a beta barrel. A typical outer membrane porin exists as a trimer of beta barrels, each of which acts as a pore for nutrients. **Figure 3.20** shows a model of the sucrose porin based on a crystal structure.

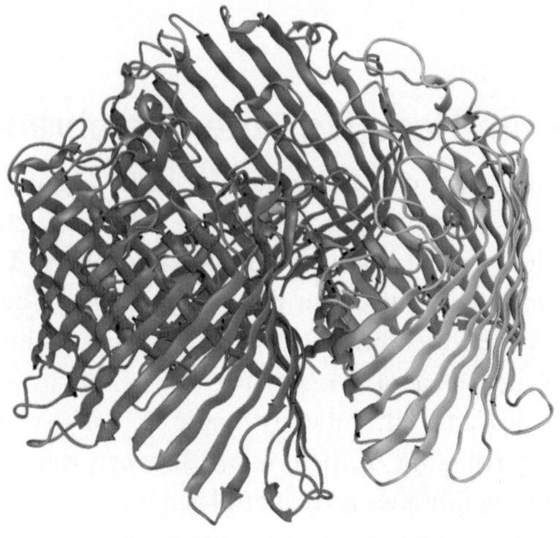

FIGURE 3.20 ■ Sucrose porin. Outer membrane porin (OMP) for sucrose transport in *Salmonella enterica,* based on X-ray crystallography. The porin comprises three beta barrel channels (red, yellow, and blue). Each transports a molecule of sucrose through its channel. (PDB code: 1AOS)

But outer membrane porins may lack specificity, allowing uptake of various molecules—including antibiotics such as ampicillin. Ampicillin is a form of penicillin, which must get through the outer membrane to access the cell wall in order to block the formation of peptide cross-bridges. Ampicillin contains two charged groups and is thus unlikely to diffuse through a lipid bilayer. Researchers have shown that ampicillin penetrates the *E. coli* outer membrane through a porin called OmpF (**Fig. 3.21**). To observe ampicillin passage through OmpF, a single OmpF trimer (with three protein channels) was solubilized within an artificial membrane bilayer. The bilayer was formed across a chamber of solution containing electrodes. A voltage was applied, and the current was measured. Dips in the current indicated transient blockage of a channel by ampicillin (**Fig. 3.21B**). When two of the three channels of the trimer were blocked simultaneously, the dip was twice as large. A model of ampicillin penetration through OmpF is shown in **Figure 3.21C**. Further details of the research are described in eTopic 3.2.

If porins can admit dangerous molecules as well as nutrients, should a cell make porins or not? In fact, cells express different outer membrane porins under different environmental conditions. In a dilute environment, cells express porins of large pore size, maximizing the uptake of nutrients. In a rich environment—for example, within a host—cells down-regulate the expression of large porins and express porins of smaller pore size, selecting only smaller nutrients and avoiding the uptake of toxins. For example, the porin regulation system of Gram-negative bacteria enables them to grow in the colon containing bile salts—a hostile environment for Gram-positive bacteria, which lack an outer membrane.

Periplasm. The outer membrane is porous to most ions and many small organic molecules, but it prevents the passage of proteins and other macromolecules. Thus, the region between the inner and outer membranes of Gram-negative cells, including the cell wall, defines a separate membrane-enclosed compartment of the cell known as the periplasm (see **Fig. 3.15**). The periplasm contains specific enzymes and nutrient transporters not found within the cytoplasm, such as periplasmic transporters for sugars, amino acids, or other nutrients. Periplasmic proteins are subjected to fluctuations in pH and salt concentration, because the outer membrane is porous to ions. Some periplasmic proteins help refold proteins unfolded by oxidizing agents or by acidification.

A.

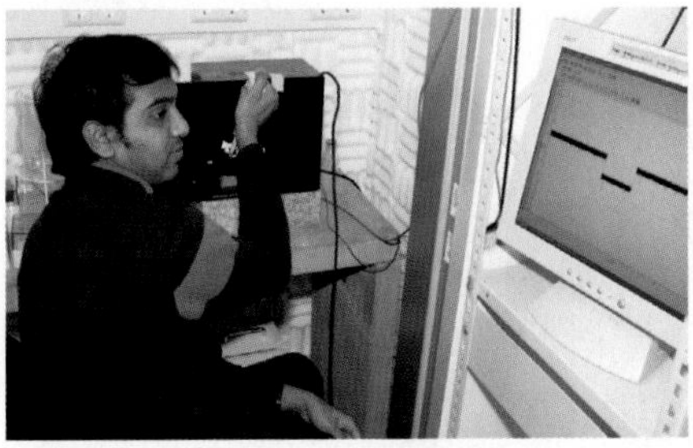

COURTESY OF MATHIAS WINTERHALTER/ JACOBS. UNIV.

FIGURE 3.21 ■ Ampicillin passage rate and pH dependence. A. Graduate student Kozhinjampara R. Mahendran measures voltage across a membrane during ampicillin passage. **B.** Passage events are observed as dips in transmembrane current, caused by ampicillin briefly blocking the OmpF channel. **C.** Model of OmpF blocked by ampicillin, based on X-ray crystallography (PDB code: 2OMF). The blowup shows amino acid residues contacting ampicillin. Ampicillin's positively charged amine group is attracted to the carboxylate of glutamate-117, and its negatively charged carboxylate is attracted to the arginine residues. *Source:* Part B modified from Ekaterina M. Nestorovich et al. 2002. *PNAS* **99**:9789–9794

B.

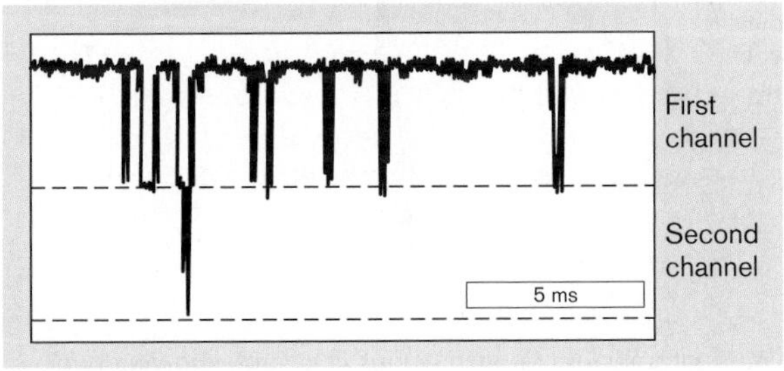
First channel
Second channel
5 ms

C.

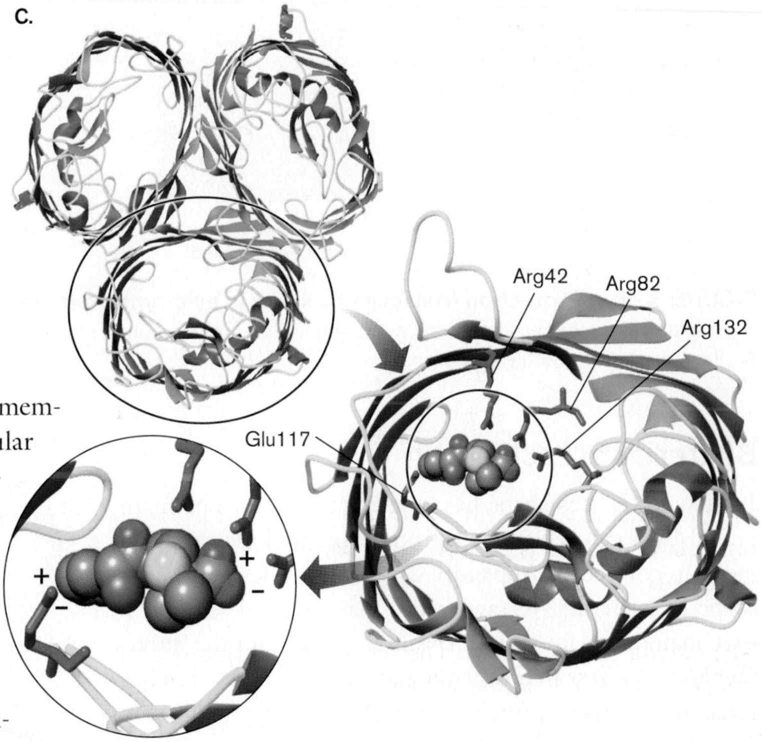
Arg42
Arg82
Arg132
Glu117
+
–
+
–

Overall, the outer membrane, periplasm, inner membrane, and cytoplasm define four different cellular compartments within a Gram-negative cell: two membrane-soluble compartments (outer and inner membrane), and two aqueous compartments (periplasm and cytoplasm). Each type of cellular protein is typically found in only one of these locations. For example, the proton-translocating ATP synthase is found only in the inner membrane fractions, whereas sugar-accepting porins are only in the outer membrane. Cell fractionation techniques can separate the two membranes and the cytoplasm and periplasm, which is the first step in isolating proteins from each region for study. **eTopic 3.3** discusses isolating outer membrane proteins from *Borrelia burgdorferi,* the cause of Lyme disease, for study as candidate vaccines.

Thought Question

3.6 Why would proteins be confined to specific cell locations? Why would a protein not be able to function everywhere in the cell?

Eukaryotic Microbes: Protection from Osmotic Shock

Eukaryotic cells are usually larger and contain more intracellular compartments than do bacteria. But, like bacteria, eukaryotes face environmental stresses such as osmotic shock. How do they adapt? Some osmotic adaptations of eukaryotes resemble the bacterial cell wall. For example, algae form cell walls of cellulose fibers, and fungi form walls of chitin. Diatoms form intricate exoskeletons of silicate.

Other eukaryotic microbes, such as amebas and ciliated protists, have no rigid cell wall; instead, they have a flexible outer coating, the pellicle. The flexible pellicle allows the uptake of large particles by endocytosis and of larger objects, even entire cells, by phagocytosis. Thus, eukaryotic cells have the nutritional option of engulfing prey—an option unavailable to prokaryotes and to eukaryotes with cell walls. Eukaryotic microbes that lack a cell wall possess a **contractile vacuole** to pump water out of the cell, avoiding osmotic shock (**Fig. 3.22**). The vacuole takes up water from the cytoplasm through an elaborate network of intracellular channels, and then expels the water through a pore.

Thought Question

3.7 What do you think are the advantages and disadvantages of a contractile vacuole, compared with a cell wall?

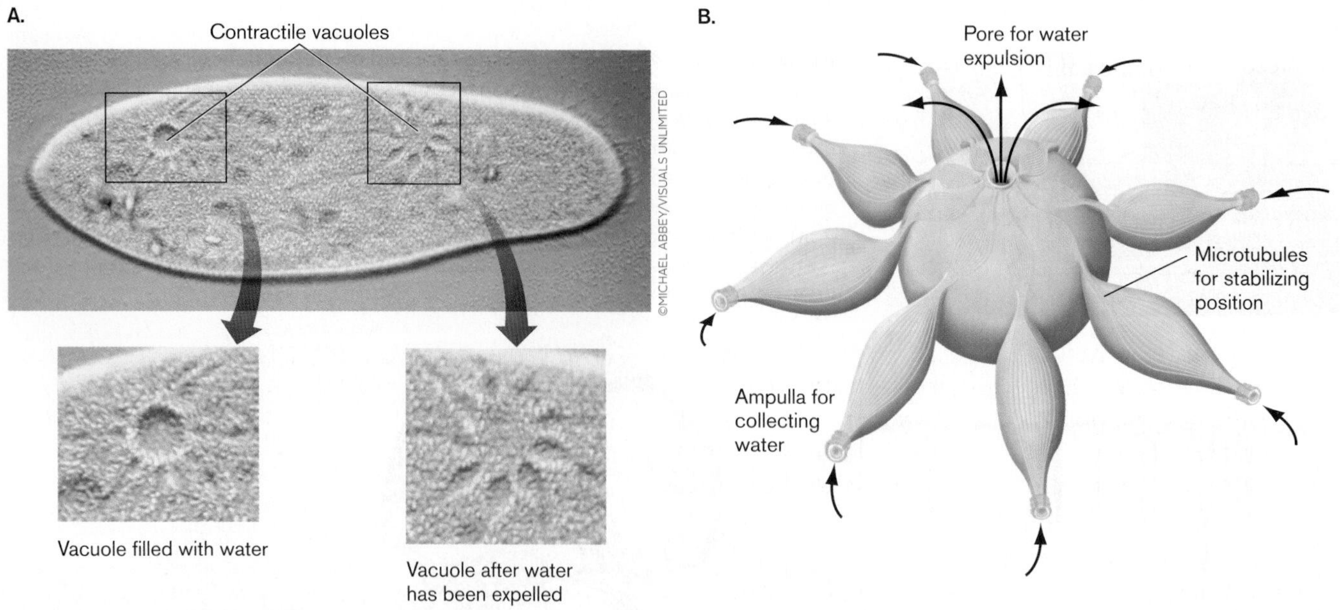
A.
Contractile vacuoles
©MICHAEL ABBEY/VISUALS UNLIMITED
Vacuole filled with water
Vacuole after water has been expelled
B.
Pore for water expulsion
Microtubules for stabilizing position
Ampulla for collecting water

FIGURE 3.22 ▪ Protection from osmotic shock in eukaryotic microbes. **A.** A paramecium (length, about 200 nm), showing two contractile vacuoles (phase-contrast micrograph). The blowups show radiating channels that collect water and expel it from the cytoplasm. **B.** Diagram of a contractile vacuole.

Bacterial Cytoskeleton

In eukaryotes, cell shape has long been known to be maintained by a cytoskeleton of protein microtubules and filaments (reviewed in Appendix 2). But what determines the shape of bacteria? We saw earlier that bacterial shape is in part maintained by the cell wall and the resulting turgor pressure. But research over the past decade shows that bacteria also possess protein cytoskeletal components—and remarkably, some of them are homologous to eukaryotic proteins such as tubulin and actin. The bacterial cytoskeletal proteins were discovered through gene defects that confer a loss of cell shape, and by fluorescent labeling of the corresponding gene products in wild-type cells. Experiments that led to these discoveries are detailed in **eTopic 3.4**.

The first bacterial cytoskeleton protein to be discovered was FtsZ, which polymerizes around the circumference of the cell division plane to form a ring called the Z-ring (**Fig. 3.23A**). In both spherical bacteria (cocci) and rod-shaped bacilli, the Z-ring determines the cell diameter. Elongation of a rod-shaped cell requires polymerization of a second cytoskeletal protein, MreB (**Fig. 3.23B**). MreB localizes in arc-shaped patches, just beneath the cell membrane of the rod-shaped cell. Some models propose a helical coil of MreB around the entire cell, but more recent data indicate arcs of MreB driven by elongating strands of cell wall. If the rod shape is curved (called a vibrio, or crescent shape), a third cytoskeletal protein, CreS (crescentin) polymerizes along the inner curve of the crescent (**Fig. 3.23C**). All three of these cytoskeletal proteins can be visualized in cells by fluorescence microscopy, either by using GFP gene

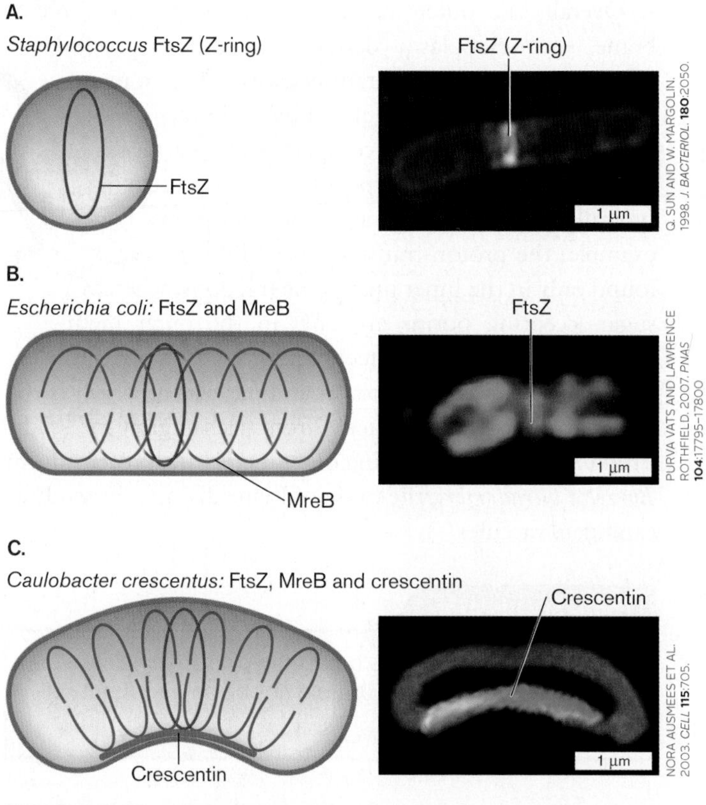
A.
Staphylococcus FtsZ (Z-ring)
FtsZ
FtsZ (Z-ring)
1 µm
Q. SUN AND W. MARGOLIN. 1998. J. BACTERIOL. 180:2050.
B.
Escherichia coli: FtsZ and MreB
MreB
FtsZ
1 µm
PURVA VATS AND LAWRENCE ROTHFIELD. 2007. PNAS 104:17795–17800
C.
Caulobacter crescentus: FtsZ, MreB and crescentin
Crescentin
Crescentin
1 µm
NORA AUSMEES ET AL. 2003. CELL 115:705.

FIGURE 3.23 ▪ Shape-determining proteins. **A.** Cell diameter is maintained by FtsZ polymerization to form the Z-ring. **B.** Elongation of a rod-shaped cell requires Mre proteins. MreB polymerizes around an *E. coli* cell (YFP-MreB fluorescence) along with a Z-ring of FtsZ (fluorescent anti-FtsZ antibody). **C.** Crescent-shaped cells possess a third shape-determining protein, CreS (crescentin), which polymerizes along the inner curve of the crescent. Crescentin protein fused to green fluorescent protein (CreS-GFP) localizes to the inner curve of *C. crescentus.* Membrane-specific stain FM4-64 (red fluorescence) localizes to the membrane around the cell.

fusions or by immunostaining with an antibody attached to a fluorophore.

Note: GFP reporter gene fusions provide valuable data, but in some cases they form polar complexes that are absent in non-GFP cells. This kind of microscopy artifact is discussed in Section 2.4.

To Summarize

- **The cell wall maintains turgor pressure.** The cell wall is porous, but its network of covalent bonds generates turgor pressure that protects the cell from osmotic shock.
- **The Gram-positive cell envelope** has multiple layers of peptidoglycan, threaded by teichoic acids.
- **The S-layer**, composed of proteins, is highly porous but can prevent phagocytosis and protect cells in extreme environments. In archaea, the S-layer serves the structural function of a cell wall.
- **The capsule**, composed of polysaccharide and glycoprotein filaments, protects cells from phagocytosis. Either Gram-positive or Gram-negative cells may possess a capsule.
- **The Gram-negative outer membrane** regulates nutrient uptake and excludes toxins. The outer membrane contains LPS and protein porins of varying selectivity.
- **Eukaryotic microbes** are protected from osmotic shock by polysaccharide cell walls or by a contractile vacuole.
- **The bacterial cytoskeleton** includes proteins that regulate cell size, play a role in determining the rod shape of bacilli, and generate curvature in a vibrio.

3.4

The Nucleoid, RNA, and Protein Synthesis

How are DNA and its expression machinery organized within the cell? Bacteria organize their DNA very differently from eukaryotes. **Figure 3.24** compares the organization of chromosomal material in enteropathogenic *E. coli* cells with that in a cultured human cell that they have colonized. In this thin-section TEM, each bacterium contains a filamentous nucleoid region that extends through the cytoplasm. In contrast, the nucleus of the eukaryotic cell, only a fraction of which is visible in the figure, is many times larger than the entire bacterial cell, and the chromosomes it contains are separated from the cytoplasm by the nuclear membrane.

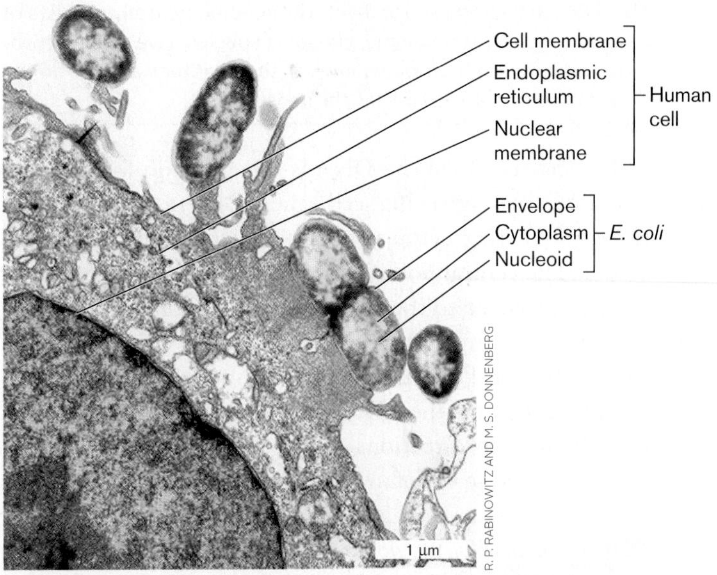
Cell membrane
Endoplasmic reticulum
Nuclear membrane
Human cell
Envelope
Cytoplasm
Nucleoid
E. coli
1 µm
R. P. RABINOWITZ AND M. S. DONNENBERG

FIGURE 3.24 ■ Bacterial nucleoid and eukaryotic nucleus. Enteropathogenic *Escherichia coli* bacteria attaching to the surface of a tissue-cultured human cell (TEM). The human cell has a well-defined nucleus delimited by a nuclear membrane, whereas the *E. coli* nucleoid DNA extends throughout the cell.

DNA Is Organized in the Nucleoid

All living cells on Earth possess chromosomes consisting of DNA. The genetic functions of microbial DNA are discussed in detail in Chapters 7–12. Here we focus on the physical organization of DNA within the nucleoid of bacterial and archaeal cells (**Fig. 3.25**).

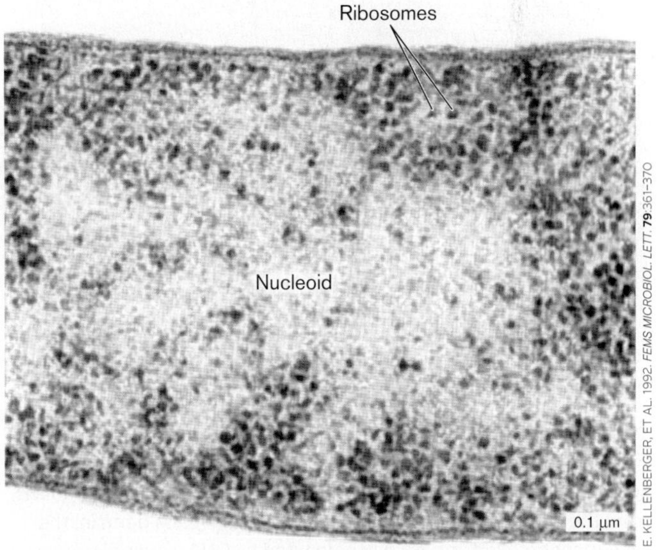
Ribosomes
Nucleoid
0.1 µm
E. KELLENBERGER, ET AL. 1992. FEMS MICROBIOL. LETT. 79:361–370

FIGURE 3.25 ■ The nucleoid. The *E. coli* nucleoid appears as clear regions that exclude the dark-staining ribosomes and contain DNA strands (cryo-TEM).

Note: In bacteria and archaea, the genome typically consists of a single circular chromosome, but some species have a linear chromosome or multiple chromosomes. In this chapter we focus on the simple case of a single circular chromosome.

In a bacterial cell the DNA is organized in loops called domains. In a growing cell, the DNA domains extend throughout the cytoplasm and generally exclude ribosomes. The central point is the **origin of replication (*ori*)**, which is attached to the cell envelope at a point on the cell's "equator," halfway between the two poles (**Fig. 3.26**). At the origin, the DNA double helix is melted open by binding proteins, and then DNA polymerase synthesizes new strands in both directions (bidirectionally). DNA replication is covered in detail in Chapter 7.

Note: In biology, the word "domain" is used in several different ways, each referring to a defined portion of a larger entity.

- **DNA domains** of the nucleoid are distinct loops of DNA that extend from the origin.
- **Protein domains** are distinct functional or structural regions of a protein.
- **Lipid domains** are clusters of one type of lipid within a membrane.
- **Taxonomic domains** are genetically distinct classes of organisms, such as Bacteria, Archaea, and Eukarya.

How does all of the cell's DNA fit neatly into the nucleoid? The DNA is compacted by supercoiling and by DNA-binding proteins.

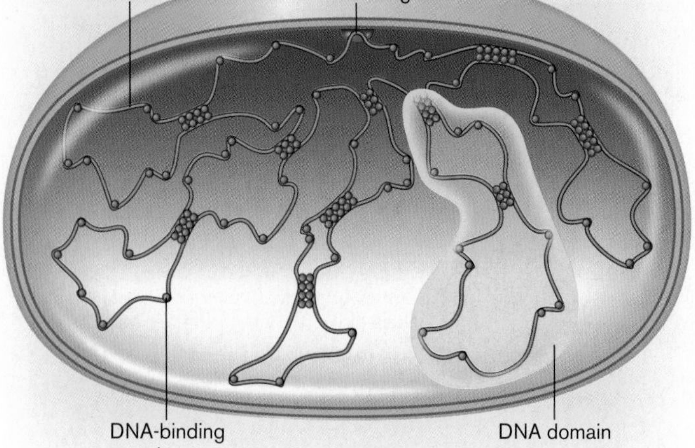
DNA
DNA origin
DNA-binding protein
DNA domain

FIGURE 3.26 ■ Nucleoid organization. The nucleoid forms approximately 50 chromosome loops called domains, which radiate from the center (one domain is shaded). Within each domain, the DNA is supercoiled and partly compacted by DNA-binding proteins (shown in green).

Supercoiling. The chromosome includes a number of extra twists, called **supercoils** or superhelical turns, beyond those inherent in the structure of the DNA duplex (double helix). The supercoiling causes portions of DNA to double back and twist upon itself, resulting in compaction. Supercoiling is generated by enzymes such as gyrase, a major target for antibiotics such as quinolones.

DNA-binding proteins. Prokaryotic DNA is condensed by winding around various classes of binding proteins (see **Fig. 3.26**). Some binding proteins, such as H-NS and HU, also function as regulators of gene expression. Binding proteins can respond to the state of the cell; for example, under starvation conditions, when most RNA synthesis ceases, the binding protein Dps is used to organize the DNA into a protected crystalline structure. Such "biocrystallization" by Dps and related proteins may be a key to the extraordinary ability of microbes to remain viable for long periods in stationary phase or as endospores.

Transcription and Translation

The information encoded in DNA is "read" by the processes of transcription and translation to yield gene products. Within the prokaryotic cell, DNA transcription to RNA is coupled tightly with RNA translation to proteins.

DNA transcription to RNA. The initial product of a gene is a single strand of ribonucleic acid (RNA), produced when DNA is transcribed by RNA polymerase (**Fig. 3.27**). In some cases, the newly made RNA has a function of its own, such as one of the RNA components of a ribosome, the cell's protein-making machine. For most genes, however, the RNA is a messenger RNA (mRNA) that immediately binds to a ribosome for translation.

mRNA translation to protein. A growing bacterial cell typically invests 40% of its energy in the translation of mRNA by ribosomes. The ribosome, with its large number of protein and RNA components, is probably the most complex subcellular machine to be discovered. Its task of converting the four-letter RNA code into 20 amino acids requires the assistance of 27 different transfer RNAs (tRNAs) and an equal number of aminoacyltransferase enzymes, each encoded by a different gene. Each amino acid is brought by a tRNA to fit into the acceptor site within the ribosome, sandwiched between the 30S and 50S subunits. The amino acid is then transferred onto a peptide chain, where the peptide bond forms. The process of translation is discussed further in Chapter 8.

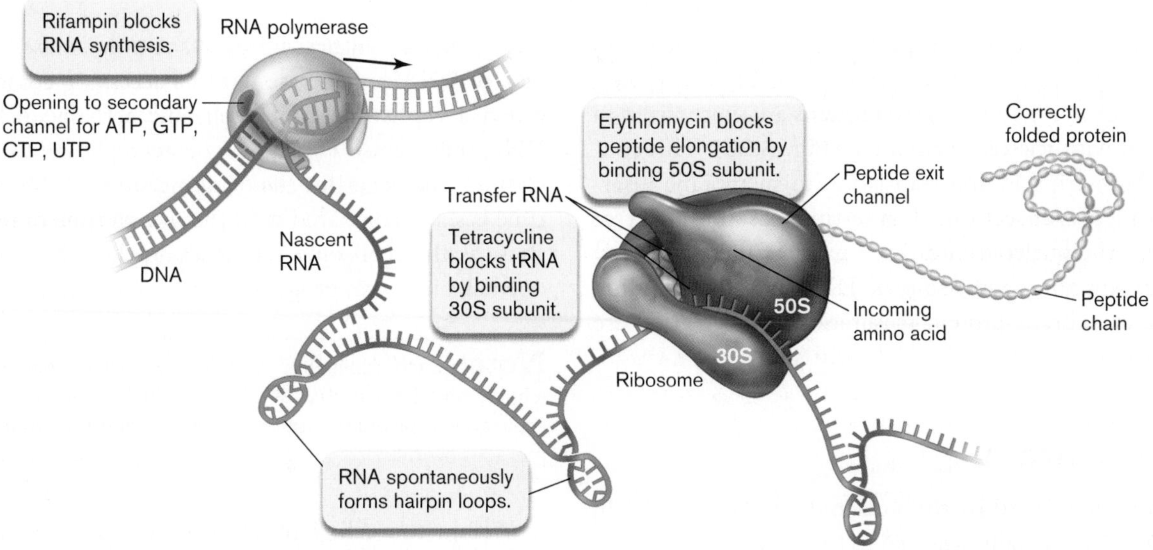
Rifampin blocks RNA synthesis.
RNA polymerase
Opening to secondary channel for ATP, GTP, CTP, UTP
DNA
Nascent RNA
Transfer RNA
Tetracycline blocks tRNA by binding 30S subunit.
Erythromycin blocks peptide elongation by binding 50S subunit.
Peptide exit channel
Correctly folded protein
50S
30S
Ribosome
Incoming amino acid
Peptide chain
RNA spontaneously forms hairpin loops.

FIGURE 3.27 ▪ RNA synthesis is coupled to protein synthesis. As the DNA sequence is transcribed into a strand of messenger RNA, ribosomes already start translating the RNA to synthesize proteins.

Because of its complexity, the ribosome is targeted by many kinds of antibiotics. For example, tetracyclines block aminoacyl-tRNA from arriving at the 30S subunit, and erythromycin blocks peptide elongation at the 50S subunit (see **Fig. 3.27**).

Coupling of transcription and translation. In bacteria and archaea, translation is tightly coupled to transcription; the ribosomes bind to mRNA and begin translation even before the mRNA strand is complete. Thus, a growing bacterial cell is full of mRNA strands dotted with ribosomes. These composite mRNA-ribosome structures are known as **polysomes**. Polysomes can be observed as the mRNA elongates, attached by RNA polymerase to the DNA strand (**Fig. 3.28**).

In rapidly growing bacteria, both transcription and translation occur at top speed while the DNA itself is being replicated. This remarkable coordination of replication, transcription, and translation explains why some bacterial cells can divide in as little as 10 minutes.

By contrast, how do eukaryotes organize DNA replication and gene expression? Eukaryotes replicate their DNA during S phase, while expressing most genes and proteins during G phase of interphase (reviewed in Appendix 2). Within the nucleus, nascent (elongating) mRNA is translated by ribosomes, which check the transcript for errors and target faulty transcripts for destruction. The majority of eukaryotic translation, however, occurs outside the nucleus, in the cytoplasm. Whereas bacteria coordinate their DNA replication and gene expression, eukaryotes separate these processes both spatially (nucleus versus cytoplasm) and temporally (different growth phases).

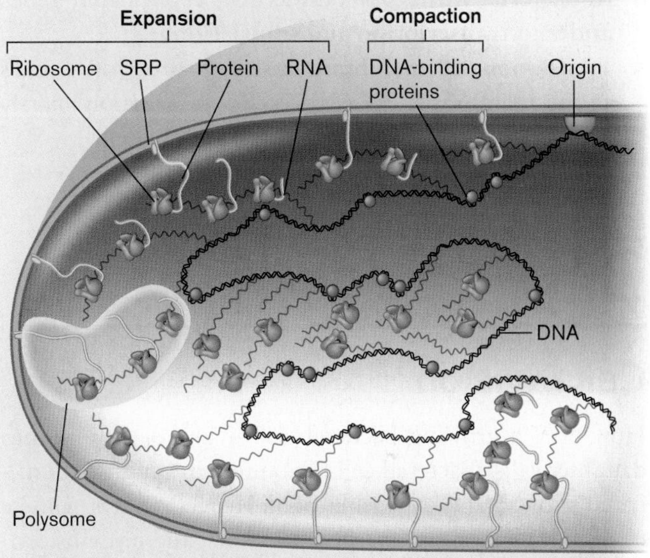
Expansion
Compaction
Ribosome
SRP
Protein
RNA
DNA-binding proteins
Origin
DNA
Polysome

FIGURE 3.28 ▪ Protein synthesis and secretion (transertion). Bacterial transcription of DNA to RNA is coordinated with translation of RNA to make proteins. Growing peptides destined for the membrane bind the signal recognition particle (SRP) for insertion into the membrane. While DNA is compacted by binding proteins (green), it is also pulled outward by the nascent chains of RNA (blue) and membrane-inserted peptides (orange). Each mRNA is translated by multiple ribosomes, which together are called a polysome (highlighted yellow).

Inserting proteins into the membrane (transertion). Some of the newly translated proteins are destined for the membrane or for secretion outside the cell. Proteins destined for membrane insertion must be hydrophobic and hence are poorly soluble in the aqueous cytoplasm where translation occurs. How can proteins that are insoluble in water be folded correctly in the cytoplasm? In prokaryotes, membrane proteins and secreted proteins are synthesized in association with the cell membrane (**Fig. 3.28**). This coupling of transcription and translation to membrane insertion is called **transertion**. Transertion has the effect of expanding the nucleoid into distal parts of the cell, partly counteracting the condensing of DNA by DNA-binding proteins. Membrane protein maturation and secretion are discussed in Chapter 8.

To Summarize

- **DNA is organized in the nucleoid.** In a bacterial cell, the DNA is generally attached to the envelope at the origin of replication, on the cell's equator.
- **DNA is compacted by supercoiling and by DNA-binding proteins.**
- **In bacteria, while DNA undergoes replication, genes undergo transcription and translation.**
- **Translation of membrane-inserted proteins occurs at the membrane.** This process of "transertion" partly expands the nucleoid into distal parts of the cell.

3.5

Cell Division

How does a growing bacterial cell divide, or fission, into daughter cells? Bacterial cell fission requires highly coordinated growth and expansion of all the cell's parts. Unlike eukaryotes, prokaryotes synthesize RNA and proteins continually while the cell's DNA undergoes replication. Bacterial DNA replication is coordinated with the expansion of the cell wall and ultimately the separation of the cell into two daughter cells. Bacteria do not undergo mitosis or meiosis (these eukaryotic processes are reviewed in Appendix 2). Bacterial DNA replication is outlined here as it relates to cell division; the molecular details of the process are discussed in Chapter 7.

DNA Is Replicated Bidirectionally

How does a bacterium begin to replicate a circular chromosome? Replication begins at the origin, or *ori* site (see **Fig. 3.26**), a unique DNA sequence in the chromosome. At the origin sequence, the DNA double helix begins to unzip, forming two replication forks. At each replication fork, DNA is synthesized by DNA polymerase. The complex of DNA polymerase with its accessory components is called a **replisome**. The replisome actually includes two DNA polymerase enzymes: one to replicate the "leading strand," the other for the "lagging strand." The actual lag time is short compared with the overall time of replication; thus, as the replisome travels along the DNA, it converts one helix into two progeny helices almost simultaneously.

Note: Each replisome contains two DNA polymerases (for leading and lagging strands). Each dividing nucleoid requires two replisomes (for bidirectional replication), and thus four DNA polymerases overall.

Within the cell, replication proceeds outward in both directions around the genome. Thus, bidirectional replication requires two replisomes, one for each replicating fork. A long-standing question has been, Do the two replisomes move oppositely around the DNA, or do they stay in the middle while the DNA helices slide through them?

To answer this question, fluorescence microscopy is used to observe the process of DNA replication within a growing cell of *Bacillus subtilis* (**Fig. 3.29**). The DNA origin of replication (*ori*) and the pair of replisomes are labeled by fluorescence. The origin of replication is labeled blue by a hybrid protein fused to a gene encoding cyan fluorescent protein (CFP). CFP binds to a promoter sequence cloned in *B. subtilis* near its origin site. The replisomes are labeled yellow by a hybrid protein expressed from a gene encoding a DNA polymerase subunit fused to a gene encoding yellow fluorescent protein (YFP). The replisomes usually locate together near the center of the cell, but sometimes they separate, visible as two yellow spots.

The fluorescence data are consistent with a model in which two replisomes are located near the middle of the growing cell (**Fig. 3.30**). Each of the two replisomes forms a replicating fork that directs two daughter strands of DNA toward opposite poles. The two copies of the DNA origin of replication (green in the figure), attached to the cell envelope, move apart as the cell expands. The termination site (red) remains in the middle of the cell, where the two replisomes continue replication at both forks. Finally, as the termination site replicates, the two replisomes separate from the DNA. At each new *ori* site, however, two pairs of new replisomes have formed. Replication of the new *ori* sites begins, sometimes even before termination of the previous round of replication.

Note that the contents of the cytoplasm and envelope must expand coordinately with DNA replication for the cell to generate progeny equivalent to the parent. In a rod-shaped cell, the cell envelope and cell wall must elongate

A.

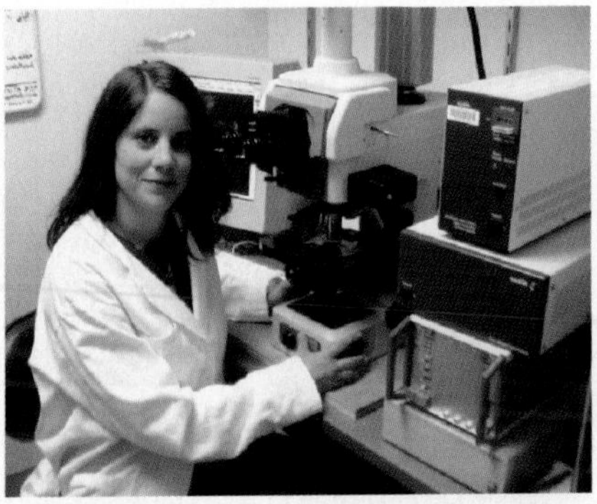
COURTESY OF MELANIE BERKMEN, MIT

B.

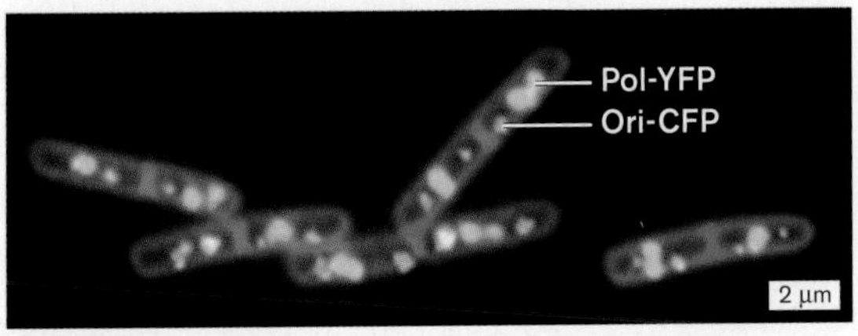
Pol-YFP
Ori-CFP
2 µm
COURTESY OF MELANIE BERKMEN, MIT

FIGURE 3.29 ■ The replisome and the DNA origin of replication. A. Melanie Berkmen, working in the laboratory of Alan Grossman, obtains the fluorescence micrographs shown. **B.** Fluorescence microscopy reveals the DNA origin, labeled blue by a protein fused to cyan fluorescent protein, binding at a sequence near the origin (Ori-CFP). Replisomes are labeled yellow by fusion of a DNA polymerase subunit to yellow fluorescent protein (Pol-YFP) in dividing cells of *Bacillus subtilis.* The cell envelope is labeled red with the membrane stain FM4-64.

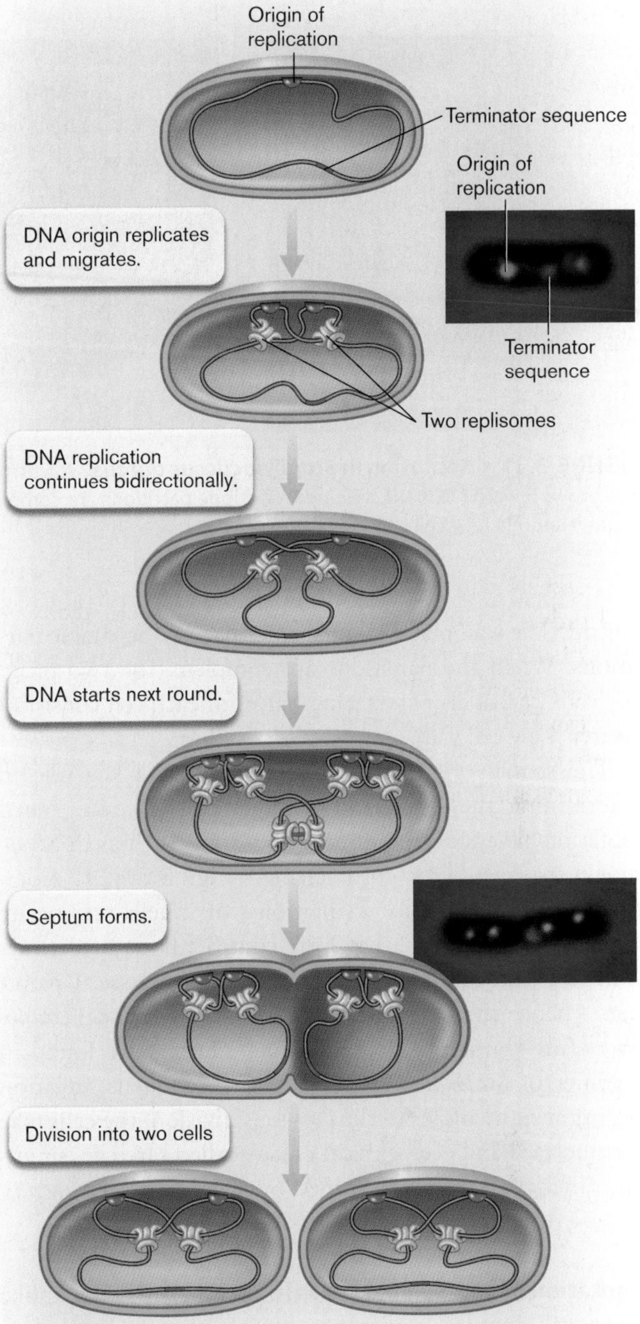
Origin of replication
Terminator sequence
Origin of replication
DNA origin replicates and migrates.
Terminator sequence
Two replisomes
DNA replication continues bidirectionally.
DNA starts next round.
Septum forms.
Division into two cells

FIGURE 3.30 ■ Replisome movement within a dividing cell. The DNA origin-of-replication sites (green) move apart in the expanding cell as the pair of replisomes (yellow) stay near the middle, where they replicate around the entire chromosome, completing the terminator sequence last (red). *Source:* Ivy Lau et al. 2003. *Mol. Microbiol.* **49**:731.

in a process coordinated by cytoskeletal proteins so as to maintain progeny of even girth and length.

Septation Completes Cell Division

For the cell to divide, DNA replication must be complete. Replication of the DNA termination site triggers growth of the dividing partition of the envelope, called the **septum** (plural, **septa**). The septum grows inward from the sides of the cell, at last constricting and sealing off the two daughter cells. This process is called **septation**. Septation and envelope extension require rapid biosynthesis of all envelope components, including membranes and cell wall. Recall how the lack of murein lipoprotein connecting the outer membrane to the cell wall causes deformation at septation (see **Fig. 3.19C**). The biosynthetic enzymes required for cell division are all of great interest as antibiotic targets. Cell wall biosynthesis poses an interesting theoretical problem: How is it possible to expand the covalent network of the sacculus without breaking links to insert new material, thus weakening the wall? The answer remains unclear.

Septation of spherical cells. In spherical cells (cocci), such as *Staphylococcus aureus,* the process of septation generates most of the new cell envelope to enclose the expanding cytoplasm (**Fig. 3.31**). Furrows form in the cell envelope, in a ring all around the cell equator, as new cell wall grows

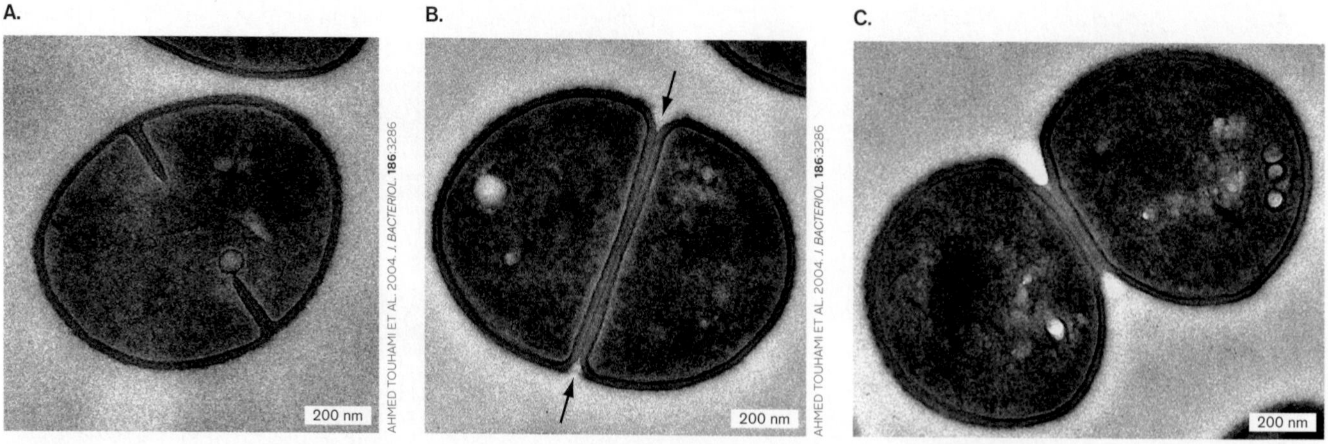
A.
B.
C.
200 nm
200 nm
200 nm

AHMED TOUHAMI ET AL. 2004. *J. BACTERIOL.* **186**:3286

AHMED TOUHAMI ET AL. 2004. *J. BACTERIOL.* **186**:3286

AHMED TOUHAMI ET AL. 2004. *J. BACTERIOL.* **186**:3286

FIGURE 3.31 ■ Septation in *Staphylococcus aureus*. A. Furrows appear in the cell envelope, all around the cell equator, as new cell wall grows inward (TEM). **B.** Two new envelope partitions are complete. **C.** The two daughter cells peel apart. The facing halves of each cell contain entirely new cell wall.

inward. The wall material must compose two separable partitions. When the partitions are complete, the two progeny cells peel apart. The facing halves of each cell consist of entirely new cell wall.

The spatial orientation of septation has a key role in determining the shape and arrangement of cocci. When septation always occurs in parallel planes, such as in *Streptococcus* species, cells form chains (**Fig. 3.31**). If septation occurs in random orientations, or if cells reassociate loosely after septation, they form compact hexagonal arrays similar to the grape clusters portrayed in classical paintings—hence the Greek-derived term **staphylococci** (*staphyle* means "bunch of grapes"). Such clusters are found in colonies of *Staphylococcus aureus.* If subsequent septation occurs at right angles to the previous division, the cells may form tetrads and even cubical octads called sarcinae (singular, sarcina). Tetrads are formed by *Micrococcus tetragenus,* a cause of pulmonary infections (**Fig. 3.32**).

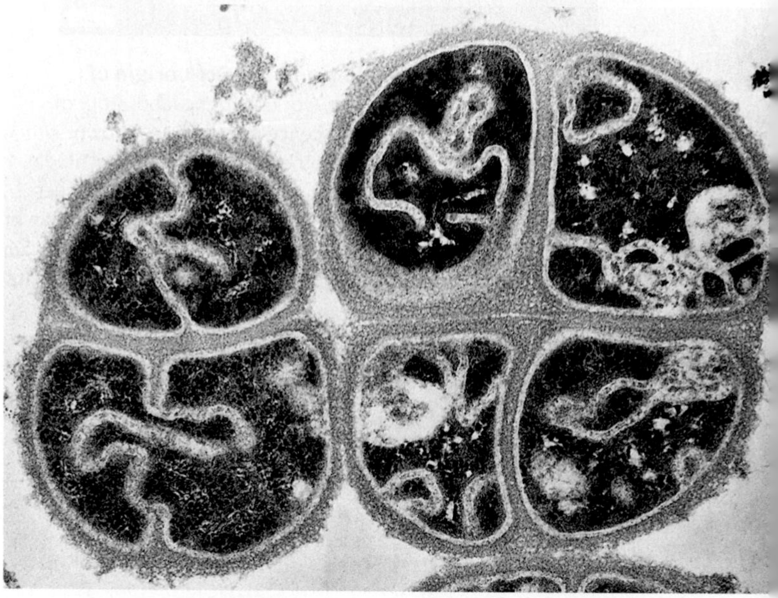

KWANGSHIN KIM/PHOTO RESEARCHERS, INC.

FIGURE 3.32 ■ Septation orientation determines the arrangement of progeny cells. Septation in two planes generates a tetrad (*Micrococcus tetragenus*, 0.5–1.0 μm, negative stain).

Septation of rod-shaped cells. In rod-shaped cells, unlike cocci, cell division requires the envelope to elongate before septation, followed by the formation of a new polar envelope for each progeny cell. The process of septation involves an intricate series of molecular signals that is at the frontier of current research.

Certain mutant strains of *E. coli* form long filaments instead of dividing normally. This filamentation results from a failure to form a septum between cells. The mutant genes causing this behavior were called *fts* for "filamentation temperature sensitive" because the cells divide normally at the **permissive temperature** but fail to septate at the **restrictive temperature**, forming long filaments.

Some of the *fts* genes encode proteins directly involved in formation of the septum. The most dramatic example is the protein FtsZ, which, we noted earlier, assembles to form the "Z-ring," a constriction ring around the equator (**Fig. 3.33**). FtsZ is universally found in bacteria and archaea as a key septation protein. FtsZ is also an ancient homolog of tubulin, which is the major component of the mitotic apparatus in eukaryotes. The discovery of FtsZ is interesting because it implies that the processes of mitosis

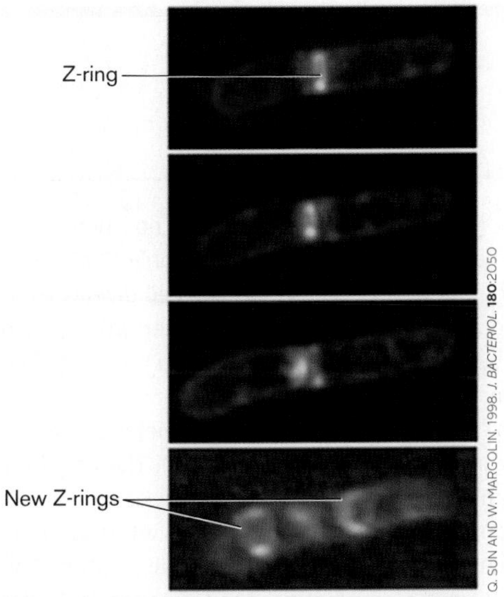
Z-ring
New Z-rings
Q. SUN AND W. MARGOLIN. 1998. J. BACTERIOL. 180:2050

FIGURE 3.33 ▪ Septation and the Z-ring. Fluorescence microscopy of *E. coli* (1–2 μm) based on FtsZ-GFP, a genetic fusion of FtsZ with green fluorescent protein (GFP). The Z-ring of FtsZ subunits forms around the equator of the constricting cell, as the septum grows inward.

in eukaryotes and cell division in prokaryotes might have evolved from a common process in an ancestral cell.

To Summarize

- **The nucleoid** region contains loops of DNA, supercoiled and bound to DNA-binding proteins.
- **DNA is transcribed** in the cytoplasm, often at the same time that it is being replicated.
- **The ribosome translates RNA to make proteins,** which are folded by chaperones and in some cases secreted at the cell membrane.
- **DNA is replicated** bidirectionally by the replisome.
- **Cell expansion and septation** are coordinated with DNA replication.

3.6

Cell Polarity and Aging

Are rod-shaped bacterial cells as symmetrical as they appear? That is, are the two poles equivalent to each other physically and chemically? In fact, bacterial cell poles differ in their origin and "age"—a phenomenon called **polar aging**. Polar aging has surprising consequences for bacterial interactions with their environment and for antibiotic resistance (**Special Topic 3.1**).

In bacteria that appear superficially symmetrical, polar differences may appear at cell division. For example, *Bacillus* species can undergo an asymmetrical cell division to form an endospore at one end. In this process, one daughter nucleoid forms an inert endospore capable of remaining dormant but viable for many years (discussed in Chapter 4). Other bacteria expand their cells by extending one pole only. In the non-spore-former *Corynebacterium diphtheriae,* the cause of diphtheria, polar extension occurs at variable rates and direction, generating irregularly shaped rods. Unipolar extension is common among actinomycetes and mycobacteria.

Other bacteria have poles with different structure and generate two different forms of progeny. The wetland bacterium *Caulobacter crescentus* has one plain pole and a pole with either a flagellum or a cytoplasmic extension called a stalk (**Fig. 3.34**). A flagellated cell swims about freely in an aqueous habitat, such as a pond or a sewage bed. After swimming for about half an hour, if the bacterium finds a place with enough nutrients, the cell sheds its flagellum and replaces it with a stalk. The stalked cell attaches to sediment and then immediately starts to replicate its DNA and divides, producing a flagellated daughter cell, as well as a daughter cell containing the original stalk.

How does *C. crescentus* organize itself to produce two different cell types, each with a different organelle at one pole? The process is a rudimentary form of cell differentiation,

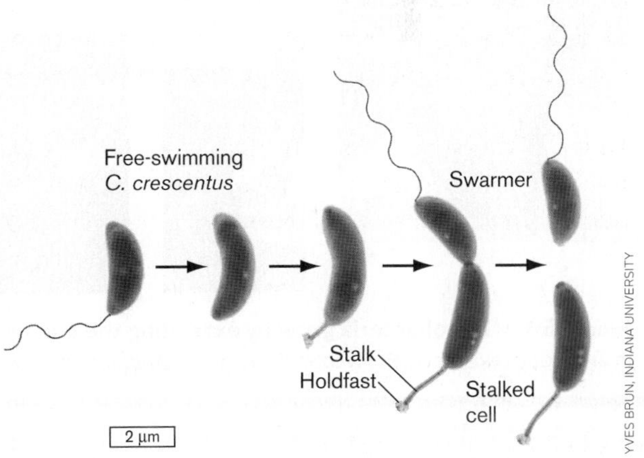
Free-swimming
C. crescentus
Swarmer
Stalk
Holdfast
Stalked
cell
2 μm
YVES BRUN, INDIANA UNIVERSITY

FIGURE 3.34 ▪ Asymmetrical cell division: a model for development. A swarmer cell of *Caulobacter crescentus* loses its flagellum and grows a stalk. The stalked cell divides to produce a swarmer cell (TEM).

Special Topic 3.1: Senior Cells Make Drug-Resistant Tuberculosis

Tuberculosis, caused by *Mycobacterium tuberculosis,* is a global health problem—amplified by the specter of resistance to nearly every known antibiotic. Effective treatment of tuberculosis can require months or years of therapy with multiple antibiotics, and when treatment lapses, resistant strains appear. Why does resistance occur so fast? The secret may lie in the diversity of growth phenotypes that arise from the unusual course of cell division in mycobacteria. The diverse growth states cause different cells to resist different antibiotics.

At the Harvard School of Public Health, microbiologist Sarah Fortune (**Fig. 1A**) and her students designed an experiment to test the hypothesis that different growth states of mycobacteria have different antibiotic sensitivities. First, Fortune established a way to grow the bacteria in a "microfluidic chamber." The chamber contains ridges to keep the growing cells in place during many hours of perfusion with fresh medium. The chamber was used to observe growth starting with a single isolated cell of *M. tuberculosis,* or of *M. smegmatis,* a nonpathogenic model organism with very similar growth traits.

In **Figure 1B**, a single septating cell is labeled with a green fluorophore, while new cell growth appears red. If this cell were *E. coli,* what would happen? The newly septated *E. coli* cells would each expand throughout, like a balloon, and separate again. But the mycobacterial cell does no such thing. Instead, the cell extends at the two old poles only (2 h). Once growth from the old poles is complete, the cell divides (5 h) and the "new poles" start to elongate. In other words, each cell alternates between elongating from an old pole and then elongating from the opposite pole.

The process of alternating pole extension yields classes of cells that differ in their growth traits (**Fig. 2A**). The daughter cell arising from the rapidly extended pole is called "accelerator." The daughter cell left behind at the alternate pole (composed mainly of old material) is called "alternator." The alternator cell delays a generation, then extends its older pole. The alternator grows more slowly than the accelerator. Differing in both growth rate and size, the alternator and accelerator cells were expected to show distinctly different patterns of metabolism. These metabolic states are assigned age classes: "age 1" and "age 2," respectively.

In the next generation, division of the alternator cell again yields cells of the classes "age 1" and "age 2." But division of the accelerator cell yields a super-accelerator that extends even faster than the parent; this metabolic class is called "age 3." **Figure 2B** shows some of the data that demonstrate the statistical differences in growth rate among cells observed in

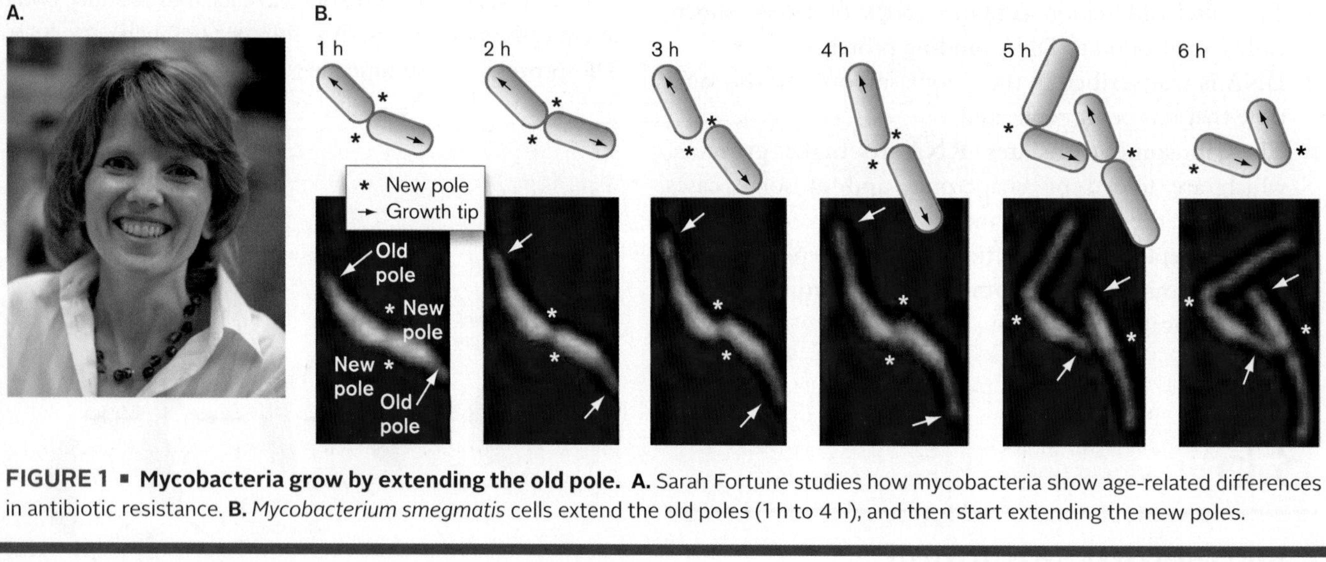
A.
B.
1 h
2 h
3 h
4 h
5 h
6 h
* New pole
→ Growth tip
Old pole
New pole
New pole
Old pole

FIGURE 1 ■ Mycobacteria grow by extending the old pole. A. Sarah Fortune studies how mycobacteria show age-related differences in antibiotic resistance. **B.** *Mycobacterium smegmatis* cells extend the old poles (1 h to 4 h), and then start extending the new poles.

comparable to the differentiation processes that animal cells undergo in the embryo. The *C. crescentus* life cycle is governed by regulators such as TipN, a cell cycle protein studied by students of Christine Jacobs-Wagner at Yale University (see the Part 2 interview). Mutants lacking TipN make serious mistakes in development. Instead of making a single flagellum at the correct cell pole, the cell makes multiple flagella at various locations, even on the stalk (**Fig. 3.35A**).

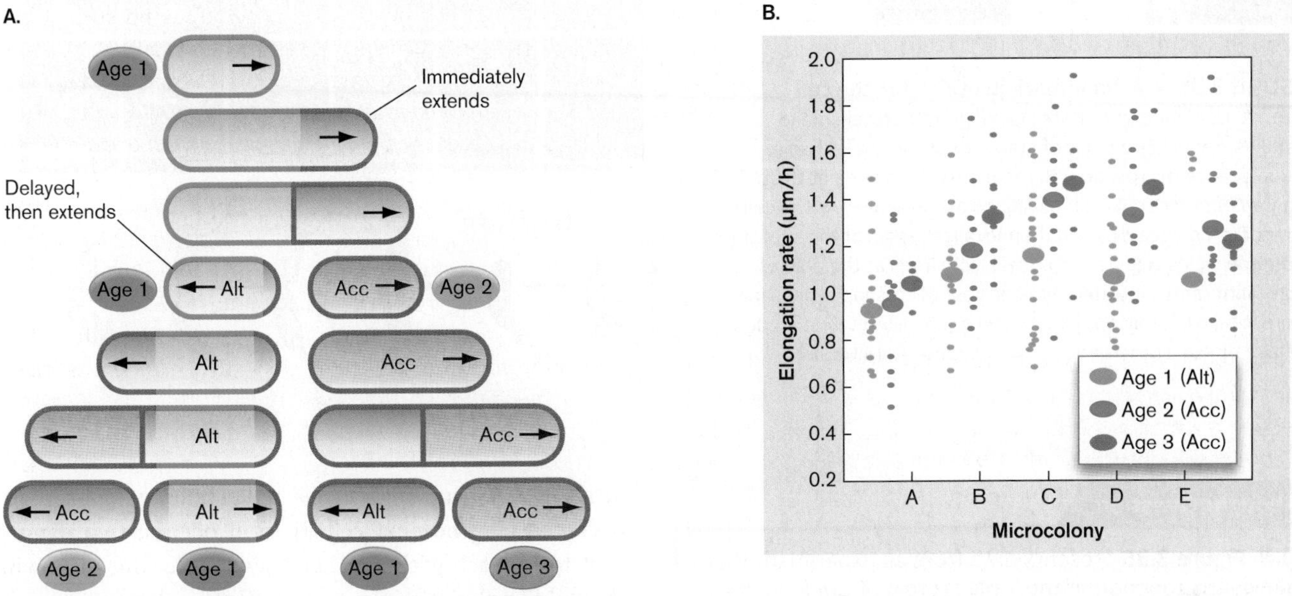
A.
Age 1
Immediately extends
Delayed, then extends
Age 1
Alt
Acc
Age 2
Alt
Acc
Alt
Acc
Acc
Alt
Alt
Acc
Age 2
Age 1
Age 1
Age 3
B.
Elongation rate (μm/h)
2.0
1.8
1.6
1.4
1.2
1.0
0.8
0.6
0.4
0.2
A
B
C
D
E
Microcolony
Age 1 (Alt)
Age 2 (Acc)
Age 3 (Acc)

FIGURE 2 ■ Mycobacteria with older poles grow faster. A. In mycobacteria, cells extend first the older pole [accelerator (Acc) cell] and then the newer pole [alternator (Alt) cell] from the parental cell. **B.** In mycobacteria (unlike *E. coli*) the cells with older poles (age 2, age 3) extend faster than cells with newer poles (age 1). Extension rates also vary greatly within each age class. *Source:* Bree Aldridge et al. 2012. *Science* **335**:100

the three age classes. Subsequent cell divisions of "age 3" cells yield mixed data, Fortune reports; at some point, an unknown mechanism resets the growth rate to that of the ancestral cell. Thus, mycobacterial cells (unlike *E. coli*) can reverse polar aging.

How do these different cell types respond to antibiotics? Fortune compared the survival rates of alternator and accelerator cells. Alternator cells showed greater resistance to the antibiotics cycloserine and meropenem, which inhibit peptidoglycan synthesis. This makes sense because alternator cells make relatively small amounts of cell wall, compared to the rapidly extending accelerator cells. On the other hand, accelerator cells were more resistant to rifampicin, which inhibits RNA polymerase. The rifampicin sensitivity of alternator cells might be explained by the rapid RNA turnover required to initiate extension of the alternate pole.

Overall, we see that the "senior cells" from mycobacterial cell division play distinctive roles in generating antibiotic resistance. And the data scatter in **Figure 2** suggests that other factors besides polar age remain to be discovered. No doubt the mycobacterial cell cycle has further surprises in store for medical researchers developing therapies to cure and curb the spread of tuberculosis.

Research Question

Which age class of *M. tuberculosis* (age 1 or age 3) do you think would be more resistant to quinolones (which block gyrase that supercoils DNA)? Or to TMC207 (which inhibits ATP synthase)? Propose experiments to test your hypotheses.

Aldridge, Bree B., Marta Fernandez-Suarez, Danielle Heller, Vijay Ambravaneswaran, Daniel Irimia, et al. 2012. Asymmetry and aging of mycobacterial cells lead to variable growth and antibiotic susceptibility. *Science* **335**:100–103.

Jacobs-Wagner proposed that TipN is a "landmark protein" that correctly marks the site of a new cell pole, and polar placement of flagella. **Figure 3.35B** shows cells expressing TipN fused to GFP, which is then detected by fluorescence. The fluorescent TipN-GFP localizes to the cell pole opposite the stalk. As the cell prepares to divide, TipN leaves the pole, delocalizing around the cell; but it relocalizes at the septum, where the new poles appear.

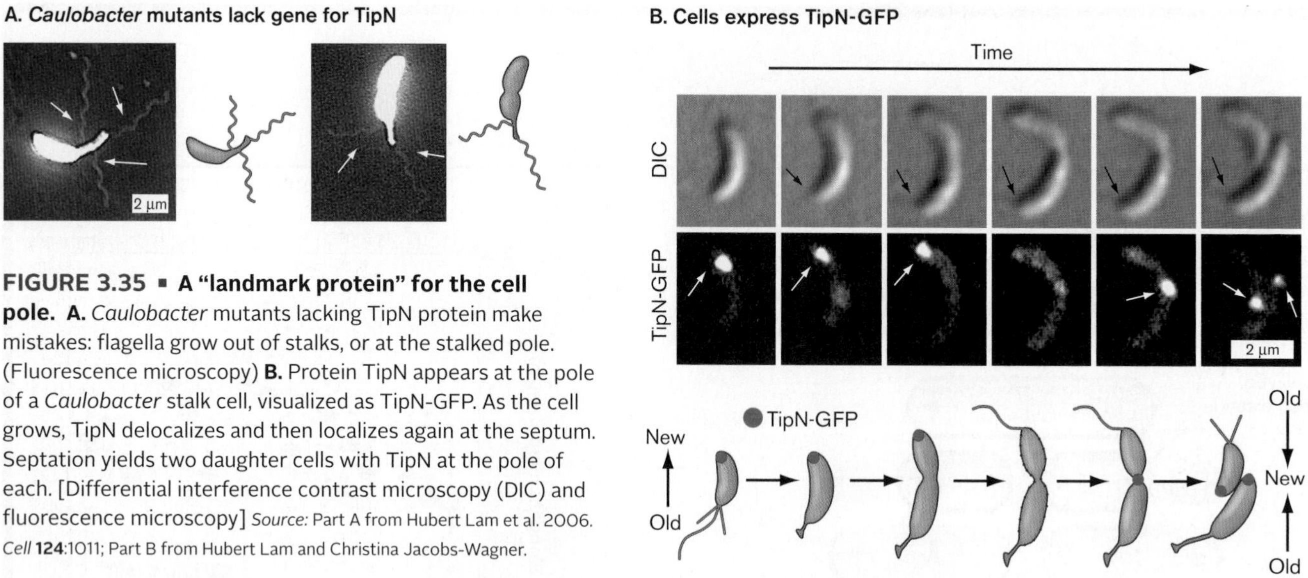

A. *Caulobacter* mutants lack gene for TipN
2 µm
B. Cells express TipN-GFP
Time
DIC
TipN-GFP
2 µm
TipN-GFP
New
Old
Old
New
Old

FIGURE 3.35 ■ A "landmark protein" for the cell pole. A. *Caulobacter* mutants lacking TipN protein make mistakes: flagella grow out of stalks, or at the stalked pole. (Fluorescence microscopy) **B.** Protein TipN appears at the pole of a *Caulobacter* stalk cell, visualized as TipN-GFP. As the cell grows, TipN delocalizes and then localizes again at the septum. Septation yields two daughter cells with TipN at the pole of each. [Differential interference contrast microscopy (DIC) and fluorescence microscopy] *Source:* Part A from Hubert Lam et al. 2006. *Cell* **124**:1011; Part B from Hubert Lam and Christina Jacobs-Wagner.

Thought Question

3.8 Figure 3.35 presents data from an experiment that allows the function of the TipN protein of *Caulobacter* to be visualized by microscopy. Can you propose an experiment with mutant strains of *Caulobacter* to test the hypothesis that one of the proteins shown in the next figure (**Fig. 3.36**) is required for one of the cell changes to occur?

Cell development involves many such proteins working together. **Figure 3.36** shows how TipN interacts with two other polar proteins: the flagellar marker PodJ, and the stalk marker DivJ. Each young cell (at right of cycle) has a new pole containing TipN. The flagellated "swarmer" cell loses its flagellum, replacing it with a stalk to settle. Now PodJ is replaced by the stalk marker DivJ. As the stalked cell grows, TipN delocalizes around the cell; then it localizes again at the middle, where the cell septates and divides. Once division is complete, TipN is evident at both new poles. As PodJ moves to the new pole, that pole grows a flagellum. Overall, throughout the cycle a series of polar proteins localize and delocalize to define the polar functions.

Does an apparently symmetrical cell such as *E. coli* actually show polar differences? In fact, it turns out that not only is a bacterial cell asymmetrical, but the actual process of cell division itself determines that the poles of each daughter cell differ chemically from each other (**Fig. 3.37**). Each cell starts out with one "old" pole (red in the figure) and one "new" pole (blue) where septation occurred. As the next cell division occurs, two daughter cells form, each with another "new" pole. But meanwhile, the "old" poles continue to age. With each generation, the polar cell wall material degrades slightly, increasing the chance of cell lysis. In a population of *E. coli* under environmental stress, at each cell division about 2% of the population dies—of polar old age.

Why does polar aging matter? One consequence of polar aging is that cells of different ages differ in their resistance to antibiotics. This phenomenon could cause problems for antibiotic therapy, as discussed in **Special Topic 3.1**.

To Summarize

- **The two poles of a bacterial cell differ in age.** One pole arises from the septum of the parental cell, whereas the other pole arises from a parental pole.
- **The two poles may differ in function.** *Caulobacter crescentus* has one plain pole, and one pole that has either a flagellum or a stalk. A stalked cell fissions to produce one stalked cell and one flagellar cell.
- **In *E. coli*, successive cell divisions yield progeny with a mixture of polar ages.** Cells with a very old pole cease replication and die.
- **In *Mycobacterium tuberculosis*, cells extend at the poles, which alternate between fast and slow extension.** Mycobacterial cells can thus reverse polar aging.

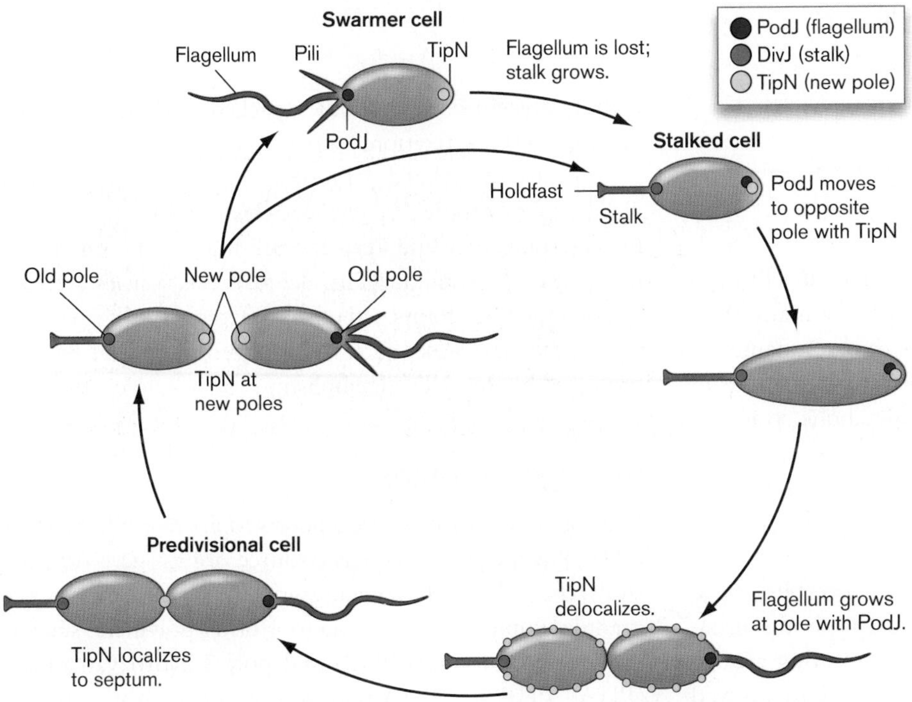
Swarmer cell
Flagellum
Pili
TipN
PodJ
Flagellum is lost; stalk grows.
PodJ (flagellum)
DivJ (stalk)
TipN (new pole)
Stalked cell
Holdfast
Stalk
PodJ moves to opposite pole with TipN
Old pole
New pole
Old pole
TipN at new poles
Predivisional cell
TipN localizes to septum.
TipN delocalizes.
Flagellum grows at pole with PodJ.

FIGURE 3.36 ▪ Cell cycle of *Caulobacter*. A swarmer cell loses its flagellum and grows a stalk. PodJ protein (purple) is at the flagellar pole, while DivJ protein (red) is at the stalk. TipN (yellow) is found at "new" poles (newly septated). TipN delocalizes and then localizes at the cell equator, midway between poles. The pole with PodJ grows a flagellum. The cell septates, forming two new poles, each containing TipN. The stalked cell still has DivJ at the stalk, and the new swarmer cell has PodJ at the flagellum. *Source:* Modified from Melanie Lawler and Yves Brun. 2006. *Cell* **124**:891.

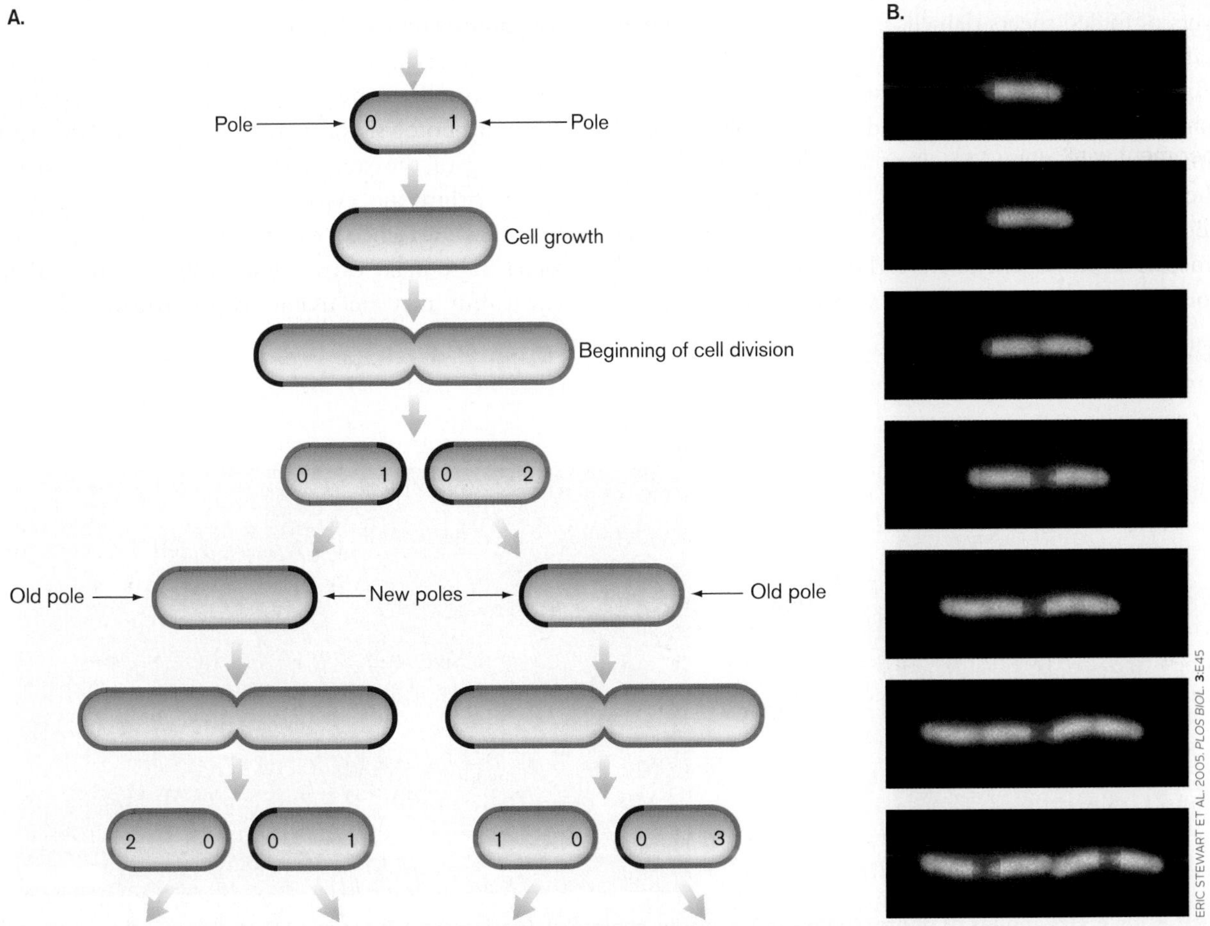
A.
Pole
0
1
Pole
Cell growth
Beginning of cell division
0
1
0
2
Old pole
New poles
Old pole
2
0
0
1
1
0
0
3
B.
ERIC STEWART ET AL. 2005. PLOS BIOL. 3:E45

FIGURE 3.37 ▪ Bacterial cell division generates cells with an old pole and a new pole. A. Succeeding generations have cells with diverse combinations of new poles (blue), old poles (red), and very old poles (two or more generations, also red). **B.** *E. coli* cells during growth, colorized for new and old poles.

3.7

Specialized Structures

We have introduced the major structures that all cells need to contain and organize their contents, maintain their DNA, and synthesize new parts. Besides these fundamental structures, different species have evolved different kinds of specialized devices adapted to diverse metabolic strategies and environments.

Thylakoids and Carboxysomes

Photosynthetic bacteria, also called phototrophs, need to collect as much light as possible to drive photosynthesis (see Section 14.6). To maximize the collecting area of their photosynthetic membranes, phototrophs have evolved specialized systems of extensively folded intracellular membrane called **thylakoids** (**Fig. 3.38A**). Thylakoids consist of layers of folded sheets (lamellae) or tubes of membranes packed with chlorophylls and electron carriers. Cyanobacteria containing thylakoids structurally resemble eukaryotic chloroplasts, which are believed to have evolved from a common ancestor of modern cyanobacteria.

The thylakoids conduct only the "light reactions" of photon absorption and energy storage. The energy obtained is rapidly spent to fix carbon dioxide—a process that occurs within **carboxysomes**. Carboxysomes are polyhedral, protein-covered bodies packed with the enzyme Rubisco for CO_2 fixation.

Another specialized structure, found in aquatic and marine phototrophs, is **gas vesicles**, which are used to increase buoyancy and keep the cell high in the water column, near the sunlight (**Fig. 3.38B**). Gas vesicles are specialized vacuoles composed of specific proteins. The vesicles trap and collect gases such as hydrogen or carbon dioxide produced by the cell's metabolism.

Storage Granules

During times of starvation, phototrophs may digest their thylakoids for energy and as a source of nitrogen. Alternatively, the cell may digest energy-rich materials from storage granules composed of glycogen or other polymers, such as polyhydroxybutyrate (PHB) and poly-3-hydroxyalkanoate (PHA). PHB and PHA polymers are of interest as biodegradable plastics, which bacteria are engineered to produce industrially. Similar storage granules are also produced by nonphototrophic soil bacteria.

Another type of storage device is sulfur—granules of elemental sulfur produced by purple and green phototrophs through photolysis of hydrogen sulfide (H_2S). Instead of disposing of the sulfur, the bacteria store it in granules, either within the cytoplasm (purple phototrophs) or as "globules" attached outside of the cell. Sulfur-reducing bacteria also make extracellular sulfur globules (**Fig. 3.39**). The sulfur may be usable as an oxidant when reduced

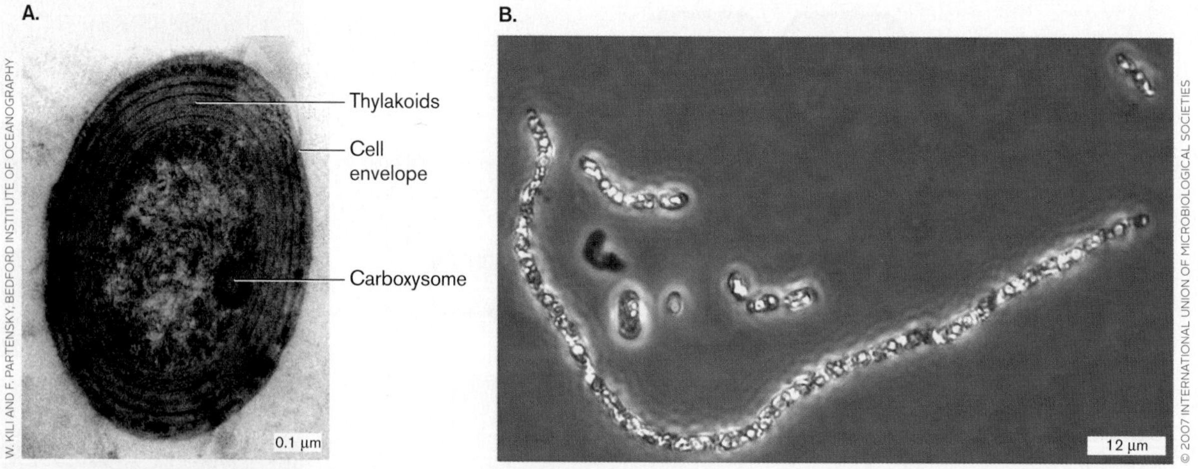
A.
Thylakoids
Cell envelope
Carboxysome
0.1 μm
W. KILI AND F. PARTENSKY, BEDFORD INSTITUTE OF OCEANOGRAPHY
B.
12 μm
© 2007 INTERNATIONAL UNION OF MICROBIOLOGICAL SOCIETIES

FIGURE 3.38 ▪ Organelles of phototrophs. A. The marine phototroph *Prochlorococcus marinus* (TEM). Beneath the envelope lie the photosynthetic double membranes called thylakoids. Carboxysomes are polyhedral, protein-covered bodies packed with the Rubisco enzyme for CO_2 fixation. **B.** Filaments (chains of cells) of the Gram-negative phototroph *Halochromatium roseum.* Gas vesicles (seen as white spots) provide buoyancy, enabling the phototroph to remain at the surface of the water, exposed to light. *Source:* Part B from P. Anil Kumar et al. 2007. *Int. J. Syst. Evol. Microbiol.* **57**:2110–2113.

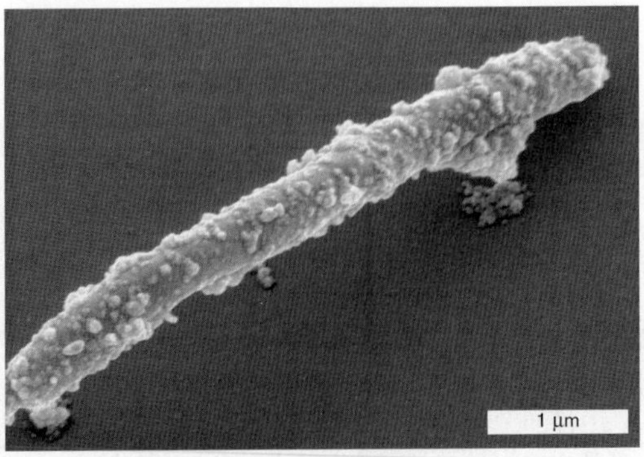
1 µm
COURTESY OF JUERGEN WIEGEL AND MANFRED ROHDE

FIGURE 3.39 ■ External sulfur particles. Sulfur globules dot the surface of *Thermoanaerobacter sulfurigignens*, an anaerobic thermophilic bacterium that gains energy by reducing thiosulfate ($S_2O_3^{2-}$) to elemental sulfur (S^0).

substrates become available. And the presence of potentially toxic sulfur granules may help cells avoid predation.

Magnetosomes

An unusual structure possessed by magnetotactic species of bacteria (bacteria showing magnetically directed motility) is the magnetosome. **Magnetosomes** are microscopic membrane-embedded crystals of the magnetic mineral magnetite, Fe_3O_4 (shown in the previous chapter, Fig. 2.44). They are found in anaerobic pond-dwelling organisms such as *Magnetospirillum gryphiswaldense*. The crystals generate a magnetic dipole moment along the length of a bacterium, constraining it to swim along a magnetic field. This magnetic orientation of swimming is called **magnetotaxis**, the ability to sense and respond to magnetism. Magnetotactic bacteria can be collected by placing a magnet in a jar of pond water; bacteria orienting by the field lines collect nearby.

The natural function of magnetosomes appears to be to orient bacterial swimming toward the bottom of the pond. Magnetotactic organisms are anaerobes, which prefer the lower part of the water column, where oxygen concentration is lowest. Because Earth's magnetic field lines point downward in the northern latitudes, bacteria that are magnetotactic swim "downward" toward magnetic north.

Thought Question

3.9 How would a magnetotactic species have to behave if it was in the Southern Hemisphere instead of the Northern Hemisphere?

Magnetotactic bacteria are being studied for their potential applications in wastewater treatment. Through their anaerobic metabolism, some magnetotactic bacteria accumulate high concentrations of toxic metals from the water. They and the toxic metals they scavenge can then be removed by application of a magnetic field to attract and concentrate the bacteria.

Pili, Stalks, and Nanotubes

In a favorable habitat, such as a running stream full of fresh nutrients or the epithelial surface of a host, it is advantageous for a cell to adhere to a substrate. **Adherence**, the ability to attach to a substrate, requires specific structures. A common adherence structure is the **pilus** (plural, **pili**), also called **fimbria** (plural, **fimbriae**), which is constructed of straight filaments of protein monomers called **pilin**. For example, fimbriae attach the oral pathogen *Porphyromonas gingivalis* to gum epithelium, where it causes periodontal disease (**Fig. 3.40**). A different kind of pili, the **sex pili**, attach a "male" donor cell to a "female" recipient cell for transfer of DNA. This process of DNA transfer is called conjugation. The genetic consequences of conjugation are discussed in Chapter 9.

Another kind of attachment organelle is a membrane-embedded extension of the cytoplasm called a **stalk**, seen earlier in the stalked cell of *Caulobacter*. The tip of the stalk secretes adhesion factors called **holdfasts**, which firmly attach the bacterium in an environment that has proved favorable. The mechanism of stalk and holdfast attachment has been extensively studied in iron-oxidizing bacteria that interfere with mining operations by producing massive biofilms. An example is *Gallionella ferruginea*, an iron-oxidizing species that grows a long stalk (**Fig. 3.41**). The long, twisted stalks of adherent *Gallionella* cells become

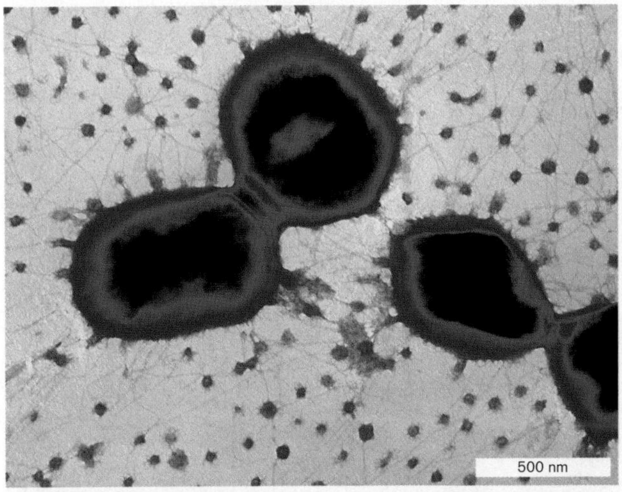
500 nm
TSUTE CHEN AND MARGARET J. DUNCAN, THE FORSYTH INSTITUTE

FIGURE 3.40 ■ Pili: protein filaments for attachment. *Porphyromonas gingivalis*, a causative agent of gum disease or gingivitis. The *P. gingivalis* cells show fimbriae along with vesicles budding from the cell's outer membrane. (TEM)

A.

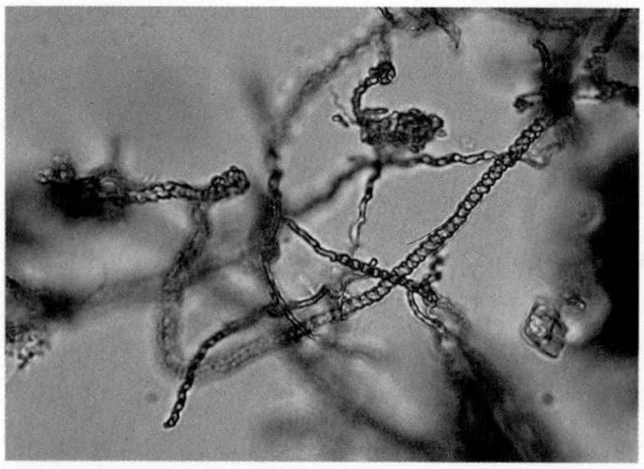

ELEANORA I. ROBBINS, U.S. GEOLOGICAL SURVEY

B.

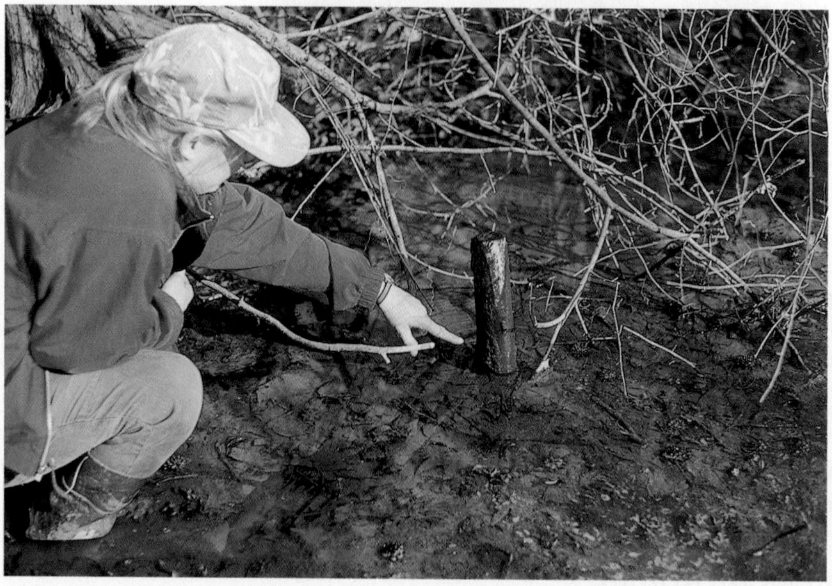

ELEANORA I. ROBBINS, U.S. GEOLOGICAL SURVEY

FIGURE 3.41 ▪ *Gallionella ferruginea:* iron-oxidizing, stalked bacteria. A. The oval cell of *G. ferruginea* generates a long, twisted stalk about 2 μm wide, encrusted with iron oxides. **B.** Karen L. Prestegaard, of the University of Maryland, studies iron-oxidizing bacteria attached to iron surfaces, such as this rod in a stream, where the bacteria promote rusting, coating the surface orange.

coated by iron hydroxides, and *Gallionella* contributes a major part of the process of biomineralization (biological crystallization of minerals) in iron mines.

An exciting recent discovery is the formation of intercellular **nanotubes**. Nanotubes are extensions of cell envelope that connect the cytoplasm or periplasm between two different cells. **Figure 3.42** shows an example of a nanotube connecting two cells of *Bacillus subtilis.* Nanotubes can transmit materials from one cell to another—even between cells of different species. As yet, we know little about how nanotubes form, but we know that they contribute to the structure of biofilms (discussed in Chapter 10).

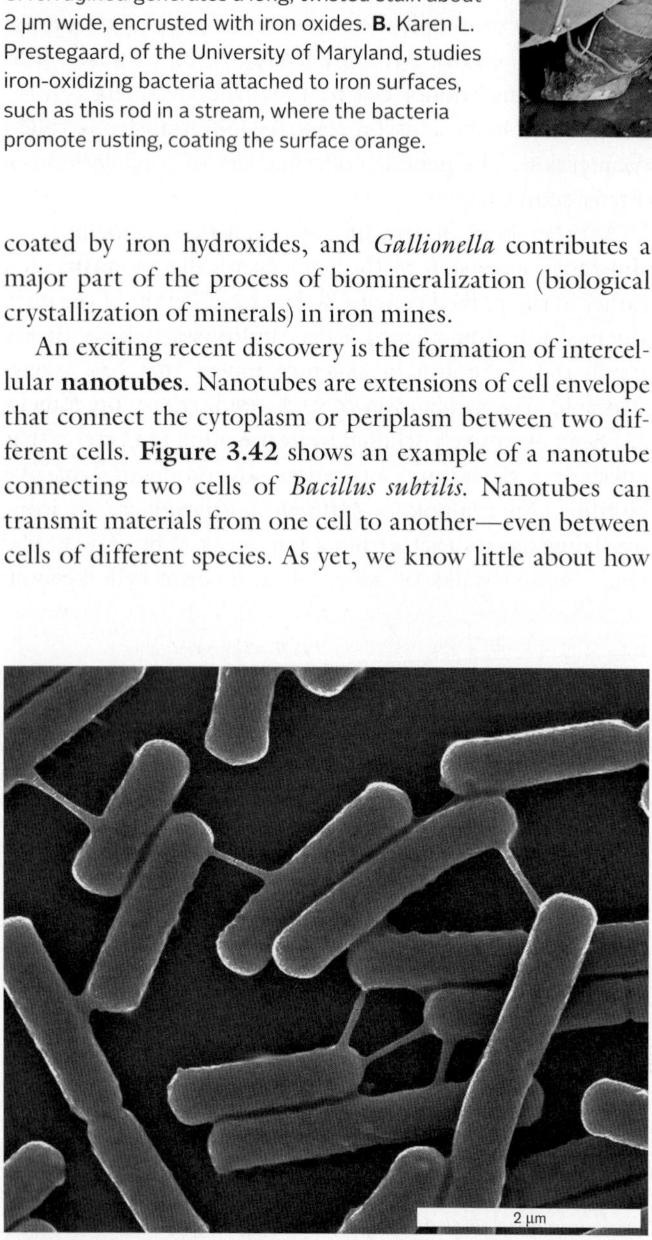
2 μm

G. P. DUBEY AND S. BEN-YEHUDA. 2011. *CELL* **144**:590–600

FIGURE 3.42 ▪ Intercellular nanotubes. *Bacillus subtilis* bacteria with intercellular connections called nanotubes that pass material from one cell to the next.

Rotary Flagella

What happens when the cell's environment runs out of nutrients, or becomes filled with waste? In rapidly changing environments, cell survival requires **motility**, the ability to move and relocate. Many bacteria and archaea can swim by means of rotary **flagella** (singular, **flagellum**). Flagella are helical propellers that drive the cell forward like the motor of a boat. Howard Berg at the California Institute of Technology originally described the bacterial flagellar motor, which was the first rotary device to be discovered in a living organism.

Different bacterial species have different numbers and arrangements of flagella. Peritrichous cells, such as *E. coli* and *Salmonella* species, have flagella randomly distributed around the cell (**Fig. 3.43A**). The flagella rotate together in a bundle behind the swimming cell (**Fig. 3.43B**). Lophotrichous cells, such as *Rhodospirillum rubrum,* have flagella attached at one or both ends. In monotrichous (polar) species, such as the *Caulobacter* swarmer cell (see **Fig. 3.36**), the cell has a single flagellum at one end.

Flagellar rotation. Each flagellum is a spiral filament of protein monomers called flagellin. The filament actually rotates by means of a motor driven by the cell's transmembrane proton current—the same proton potential that drives the membrane-embedded ATP synthase (presented in Chapter 14). The flagellar motor is embedded in the layers of the cell envelope (**Fig. 3.44**). The motor possesses

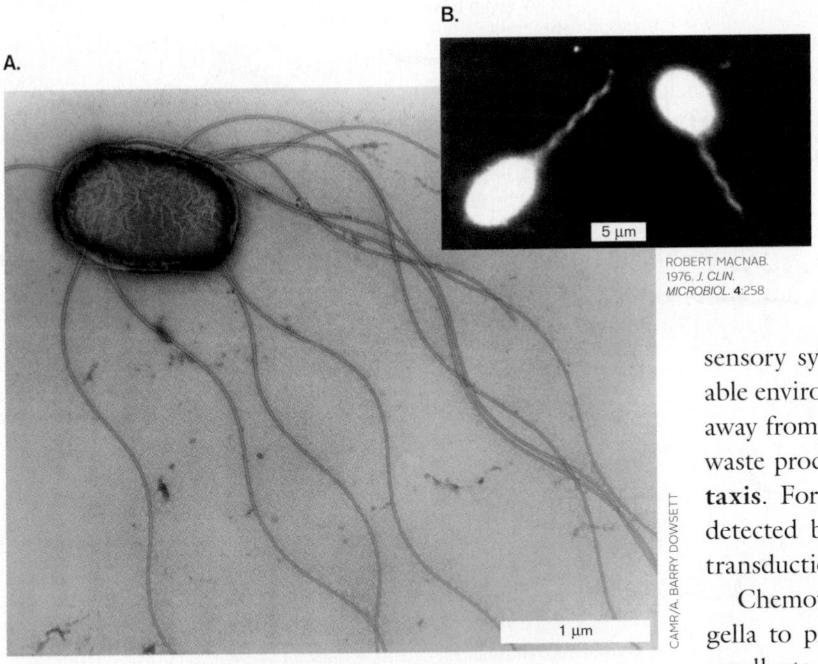
A.
B.
5 µm
1 µm
ROBERT MACNAB. 1976. J. CLIN. MICROBIOL. 4:258
CAMR/A. BARRY DOWSETT

FIGURE 3.43 ■ Flagellated *Salmonella* bacteria. A. *Salmonella enterica* has multiple flagella (colorized TEM). **B.** The flagella collect in a bundle behind a swimming cell. Under dark-field microscopy, the cell body appears overexposed, about five times as large as the actual cell.

an axle and rotary parts, all composed of specific proteins. Much of the structure and function of the motor was elucidated by Scottish microbiologist Robert Macnab (1940–2003) at Yale University.

Flagellar motility benefits the cell by causing the population to disperse, decreasing competition. Surprisingly, flagella have also evolved an alternate function—adherence of cells to a substrate to begin forming a biofilm (discussed in Chapter 4). This is an example of a common natural phenomenon in which an organism evolves a structure that serves one function but later evolves to serve another function.

Chemotaxis. How do cells decide where to swim? Most flagellated cells have an elaborate sensory system that enables them to swim toward favorable environments (attractant signals, such as nutrients) and away from inferior environments (repellent signals, such as waste products). This sensory system is known as **chemotaxis**. For chemotaxis, the attractants and repellents are detected by a polar array of chemoreceptors. The signal transduction mechanism is explained in Chapter 10.

Chemotaxis requires a mechanism for the rotary flagella to propel the cell toward attractants or away from repellents. This movement is accomplished by flagellar rotation either clockwise or counterclockwise relative to the cell (**Fig. 3.45A**▶). When a cell is swimming toward an attractant chemical, the flagella rotate counterclockwise (CCW), enabling the cell to swim smoothly for a long stretch. When the cell veers away from the attractant, receptors send a signal that allows one or more flagella to switch rotation clockwise (CW), against the twist of the helix. This switch in the direction of rotation disrupts the bundle of flagella, causing the cell to tumble briefly, ending up pointed in a random direction. The cell then swims off in the new direction. The resulting pattern of movement generates a "biased random walk" in which the cell tends to migrate toward the attractant (**Fig. 3.45B**).

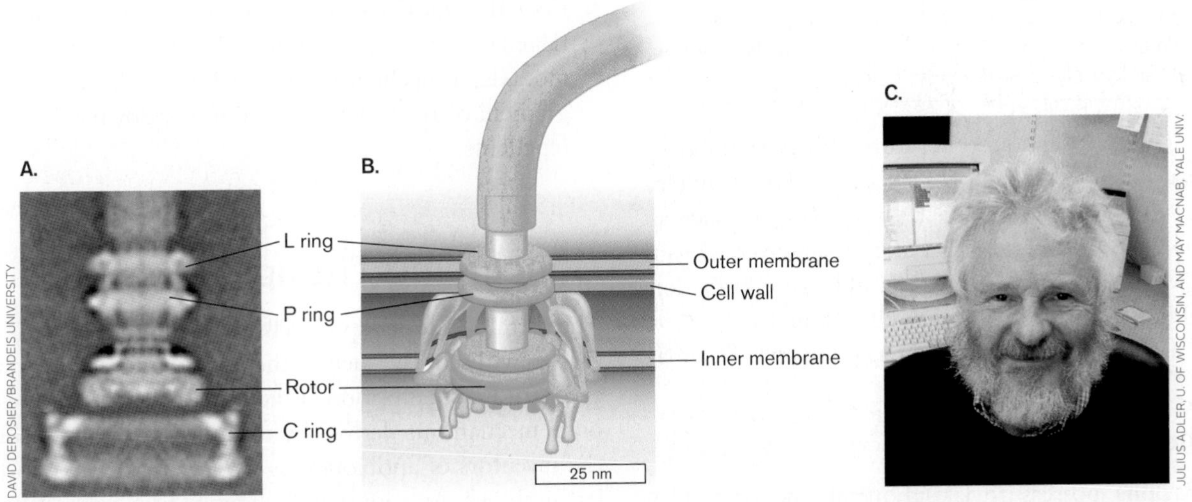
A.
B.
C.
L ring
P ring
Rotor
C ring
Outer membrane
Cell wall
Inner membrane
25 nm
DAVID DEROSIER/BRANDEIS UNIVERSITY
JULIUS ADLER, U. OF WISCONSIN, AND MAY MACNAB, YALE UNIV.

FIGURE 3.44 ■ The flagellar motor. A. The basal body, or motor, of the bacterial flagellum (TEM). The image is based on digital reconstruction, in which electron micrographs of purified hook-basal bodies were rotationally averaged. **B.** Diagram of the flagellar motor, including major protein components. **C.** Robert Macnab (Yale University) identified many components of the motor and chemotaxis signaling.

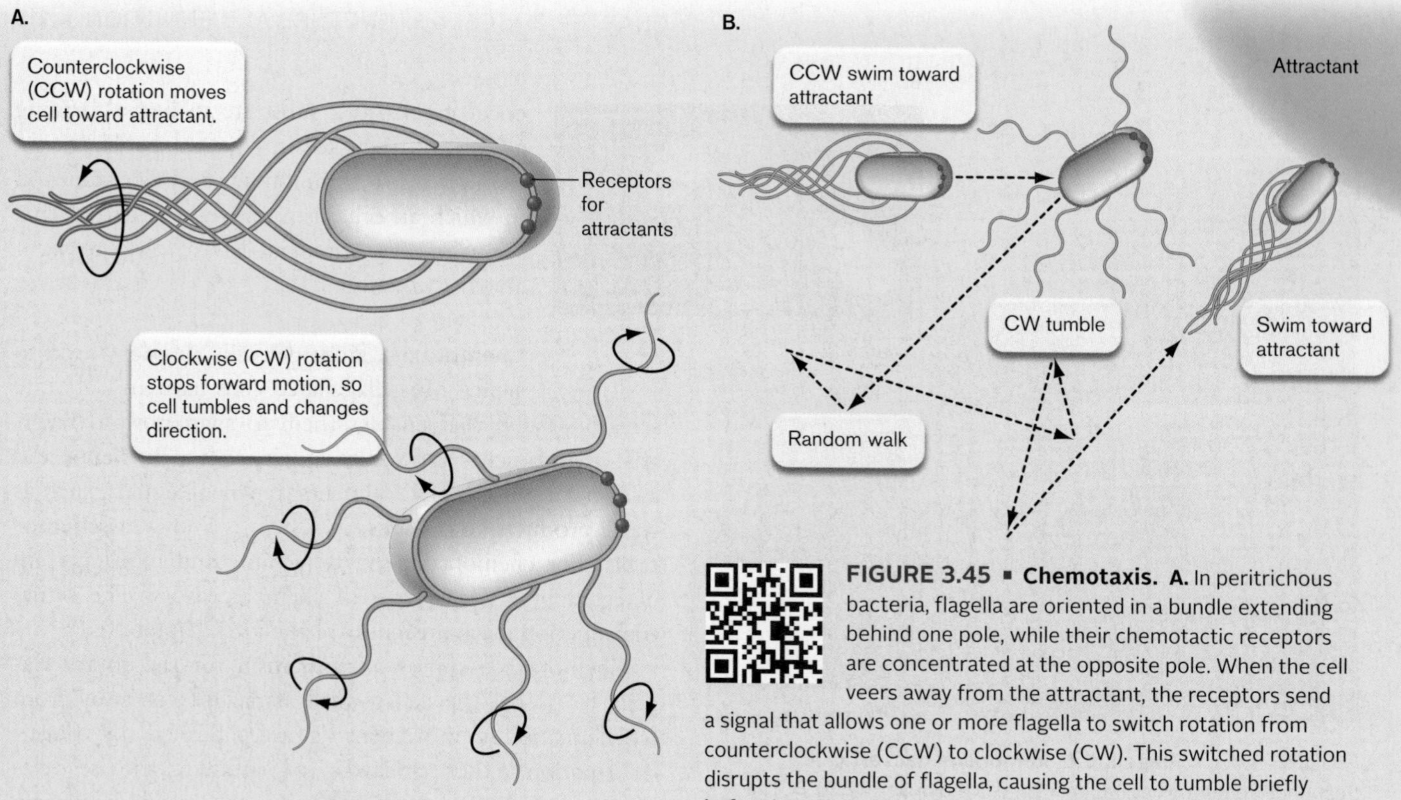
A.
Counterclockwise (CCW) rotation moves cell toward attractant.
Receptors for attractants
Clockwise (CW) rotation stops forward motion, so cell tumbles and changes direction.
B.
CCW swim toward attractant
Attractant
CW tumble
Swim toward attractant
Random walk

FIGURE 3.45 ▪ Chemotaxis. A. In peritrichous bacteria, flagella are oriented in a bundle extending behind one pole, while their chemotactic receptors are concentrated at the opposite pole. When the cell veers away from the attractant, the receptors send a signal that allows one or more flagella to switch rotation from counterclockwise (CCW) to clockwise (CW). This switched rotation disrupts the bundle of flagella, causing the cell to tumble briefly before it swims off in a new direction. **B.** The pattern of movement resulting from alternating swimming and tumbling is a "biased random walk" in which the cell sometimes moves randomly but overall tends to migrate toward the attractant. ▶

Thought Question

3.10 Most laboratory strains of *E. coli* and *Salmonella* commonly used for genetic research lack flagella. Why do you think this is the case? How can a researcher maintain a motile strain?

Note: Bacterial flagella differ completely from the whiplike flagella and cilia of eukaryotes. Eukaryotic flagella are much larger structures containing multiple microtubules enclosed by a membrane (shown in Chapter 20). They move with a whiplike motion, powered by ATP hydrolysis all along the flagellum.

Besides flagellar rotation, other forms of bacterial motility are just beginning to be understood. For example, "twitching motility" is a process by which bacteria such as *Pseudomonas aeruginosa* use pili to drag themselves across a surface. Twitching motility is involved in biofilm formation (discussed in Section 4.6). Another kind of motility, called "gliding," is observed in cyanobacteria and in myxobacteria.

To Summarize

- **Phototrophs** possess thylakoid membrane organelles packed with photosynthetic apparatus and carboxysomes for carbon dioxide fixation. Other subcellular structures may include sulfur granules from H_2S photolysis and gas vesicles for buoyancy in the water column.
- **Storage granules** store polymers for energy.
- **Magnetosomes** orient the swimming of magnetotactic anaerobic bacteria.
- **Adherence structures** enable prokaryotes to remain in an environment with favorable environmental factors. Major adherence structures include pili, or fimbriae (protein filaments), and the holdfast (a cell extension).
- **Flagellar motility** occurs by rotary motion of helical flagella. The direction of motility is determined by chemotaxis, a mechanism by which the cell migrates up a gradient of an attractant substance or down a gradient of repellents.

Concluding Thoughts

The form and function of cells continue to amaze us with each new discovery, such as the polar aging of bacteria and the intricate organization of their chemoreceptors. The elaborate mechanisms derived by microbial evolution challenge the inventors of antibiotics, as well as designers of molecular machines. As a journalist observed in *Science*, "When it comes to nanotechnology, physicists, chemists, and materials scientists can't hold a candle to the simplest bacteria."

CHAPTER REVIEW

Review Questions

1. What are the major features of a bacterial cell, and how do they fit together for cell function as a whole?
2. What fundamental traits do most prokaryotes have in common with eukaryotic microbes? What traits are different?
3. Give examples of how our views of ribosome structure and function have emerged from microscopy, cell fractionation, X-ray crystallography, and genetic analysis. Explain the advantages and limitations of each technique.
4. Outline the structure of the peptidoglycan sacculus, and explain how it expands during growth. Cite two different kinds of experimental data that support our current views of the sacculus.
5. Compare and contrast the structure of Gram-positive and Gram-negative cell envelopes. Explain the strengths and weaknesses of each kind of envelope.
6. Explain how the process of DNA replication is coordinated with cell wall septation.
7. Explain how DNA transcription to RNA is integrated with translation and with protein processing and secretion.
8. What kinds of subcellular structures are found in certain cells with different functions, such as magnetotaxis or photosynthesis?
9. Compare and contrast bacterial structures for attachment and motility. Explain the molecular basis of chemotaxis.

Thought Questions

1. The aquatic bacterium *Caulobacter crescentus* alternates between two cell forms: a cell with a flagellum that swims, and a stalked cell that adheres to particulate matter. The flagellar cell can discard its flagellum to grow a stalk and adhere, and then the stalked cell divides to give one stalked cell and one flagellated cell. What would be the adaptive advantage of this alternating morphology?
2. Suppose that one cell out of a million has a mutant gene blocking S-layer synthesis, and suppose that the mutant strain can grow twice as fast as the S-layered parent. How many generations would it take for the mutant strain to constitute 90% of the population?
3. Explain two different ways that an aquatic phototroph might use its subcellular structures to maximize its access to light. Explain how an aerobe (an organism requiring molecular oxygen for growth) might remain close to the surface, with access to air.
4. How do pathogenic bacteria avoid engulfment by phagocytes of the human bloodstream? How do you think various aspects of the cell structure can prevent phagocytosis?

Key Terms

active transport (89)
adherence (113)
capsule (95)
carboxysome (112)
cardiolipin (89)
cell fractionation (82)
cell membrane (82)
cell wall (82)
chemotaxis (115)
cholesterol (90)
contractile vacuole (99)
core polysaccharide (98)
cross-bridge (92)
diphosphatidylglycerol (89)
electrophoresis (84)
endotoxin (98)
envelope (82)
fimbria (113)
flagellum (82, 114)
gas vesicle (112)
genetic analysis (83)
glucosamine (98)
glycan (92)
holdfast (113)
hopanoid (hopane) (90)
inner membrane (97)
ion gradient (88)
leaflet (86)
lipopolysaccharide (LPS) (82, 98)
lysis (lyse) (82)
magnetosome (113)
magnetotaxis (113)
membrane-permeant weak acid (87)
membrane-permeant weak base (87)
motility (114)
murein (92)
murein lipoprotein (97)
nanotube (114)
nucleoid (82)

O polysaccharide (98)
origin of replication (*ori*) (102)
osmosis (87)
osmotic pressure (87)
outer membrane (82)
passive transport (89)
peptidoglycan (85)
permissive temperature (106)
phosphatidate (86)
phosphatidylethanolamine (86)
phosphatidylglycerol (86)
phospholipid (86)
pilin (113)
pilus (113)
polar aging (107)
polyamine (84)
polysome (103)
porin (98)
proteome (84)
replisome (104)
restrictive temperature (106)
S-layer (95)
sacculus (91)
septation (105)
septum (105)
sex pilus (113)
solute (87)
stalk (113)
staphylococcus (106)
supercoil (102)
teichoic acid (95)
terpenoid (90)
thylakoid (112)
transertion (104)
transport protein (transporter) (88)
two-dimensional polyacrylamide gel electrophoresis (2D PAGE, 2D gels) (84)
ultracentrifuge (82)

Recommended Reading

Aldridge, Bree B., Marta Fernandez-Suarez, Danielle Heller, Vijay Ambravaneswaran, Daniel Irimia, et al. 2012. Asymmetry and aging of mycobacterial cells lead to variable growth and antibiotic susceptibility. *Science* **335**:100–103.

Bardy, Sonia L., and Janine R. Maddock. 2007. Polar explorations: Recent insights into the polarity of bacterial proteins. *Current Opinion in Microbiology* **10**:617–623.

Celler, Katherine, Roman I. Koning, Abraham J. Koster, and Gilles P. van Wezel. 2013. Multidimensional view of the bacterial cytoskeleton. *Journal of Bacteriology* **195**:1627–1636.

Feucht, Andrea, and Jeff Errington. 2005. *ftsZ* mutations affecting cell division frequency, placement and morphology in *Bacillus subtilis. Microbiology* **151**:2053–2064.

Lam, Hubert, Whitman B. Schofield, and Christine Jacobs-Wagner. 2006. A landmark protein essential for establishing and perpetuating the polarity of a bacterial cell. *Cell* **124**:1011–1023.

Lenz, Peter, and Lotte Søgaard-Andersen. 2011. Temporal and spatial oscillations in bacteria. *Nature Reviews. Microbiology* **9**:565–577.

Lele, Uttara N., Ulfat I. Baig, Milind G. Watve. 2011. Phenotypic plasticity and effects of selection on cell division symmetry in *Escherichia coli. PLoS ONE* **6**:e14516.

Libby, Elizabeth A., Manuela Roggiani, and Mark Gouliana. 2012. Membrane protein expression triggers chromosomal locus repositioning in bacteria. *Proceedings of the National Academy of Sciences USA*. [Epub ahead of print.] doi:10.1073/pnas.1109479109.

Nilsen, Trine, Arthur W. Yan, Gregory Gale, and Marcia B. Goldberg. 2005. Presence of multiple sites containing polar material in spherical *Escherichia coli* cells that lack MreB. *Journal of Bacteriology* **187**:6187–6196.

Noji, Hiroyuki, Ryohei Yasuda, Masasuke Yoshida, and Kazuhiko Kinoshita, Jr. 1997. Direct observation of the rotation of F_1-ATPase. *Nature* **386**:299–302.

Pagès, Jean-Marie M., Chloë E. James, and Mathias Winterhalter. 2008. The porin and the permeating antibiotic: A selective diffusion barrier in Gram-negative bacteria. *Nature Reviews. Microbiology* **6**:893–903.

Renner, Lars D., and Douglas B. Weibel. 2011. Cardiolipin microdomains localize to negatively curved regions of *Escherichia coli* membranes. *Proceedings of the National Academy of Sciences USA* **108**:6264–6269.

Ruiz, Natividad, Daniel Kahne, and Thomas J. Silhavy. 2006. Advances in understanding bacterial outer-membrane biogenesis. *Nature Reviews. Microbiology* **4**:57–66.

Saier, Milton H., Jr. 2008. Structure and evolution of prokaryotic cell envelopes. *Microbe* **3**:323–328.

Stewart, Eric J., Richard Madden, Gregory Paul, and François Taddei. 2005. Aging and death in an organism that reproduces by morphologically symmetric division. *PloS Biology* **3**:e45.

CHAPTER 4

Bacterial Culture, Growth, and Development

4.1 Microbial Nutrition

4.2 Nutrient Uptake

4.3 Culturing Bacteria

4.4 Counting Bacteria

4.5 The Growth Cycle

4.6 Biofilms

4.7 Cell Differentiation

The adage "To eat well is to live well" is as true for microbes as it is for humans. Microorganisms constantly struggle to survive in natural habitats because they compete for food, yet a single microbial pathogen can multiply within the course of a day to cause deadly illness, and a few algae can abruptly bloom and cover the entire surface of a lake. In each case, the population explosion results from the sudden availability of food. Over eons, bacteria have evolved ingenious strategies to find, acquire, and metabolize a wide assortment of potential food sources, ranging from glucose to mothballs. This metabolic diversity arose from the need to find new sources of food ignored by competitors.

The remarkable plasticity of microbial genomes enables microorganisms to adapt to new food sources. During the course of DNA replication, gene systems involved in the use of one food naturally undergo duplications and mutations, some of which sculpt new biochemical pathways capable of metabolizing novel substrates. This genomic flexibility raises hopes that we can engineer microbial biochemistry to remediate pollution and produce tomorrow's wonder drugs.

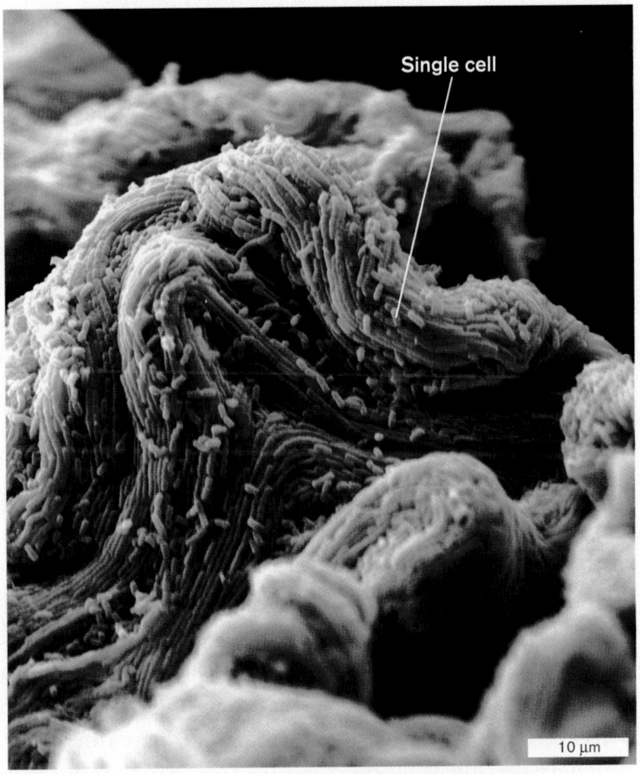

Single cell
10 μm

HTTP://WYSS.HARVARD.EDU/VIEWMEDIA/134/BACTERIAL-BIOFILM-2

CURRENT RESEARCH highlight

Microbial cannibalism. Biofilms of *Bacillus subtilis* are held together by an extracellular matrix that ensnares matrix-producing cells, motile cells, and dormant spores alike. Surprisingly, matrix-producing cells become "cannibals" that kill motile *B. subtilis* or other microbes as a way to release nutrients. Cannibals feed on the nutrients to maintain viability and delay sporulation. *B. subtilis* and related species secrete chemicals that prompt the production of matrix-producing cannibals. Thus, different members of a microbiome, as in soil, influence cannibalization and biofilm formation. *Source*: E. A. Shank et al. 2011. *PNAS* **108**:E1236–E1243. Photo courtesy of Benjamin Hatton and Joanna Aizenberg of the Wyss Institute.

Microorganisms need sources of carbon in order to grow. Understanding how they use food to increase their cell mass and, ultimately, cell number enables us to control their growth and manipulate them to make useful products. Yeast, for example, consume glucose and break it down to ethanol and carbon dioxide gas. These end products are merely waste to the yeast but are extremely important to humans who enjoy beer. Brewers have learned to control the amount of sugar supplied to yeast growing in fermentation vats to produce just the right amount of alcohol and CO_2, which causes beer's bubbling carbonation. Many a home brewer has learned the hard way that providing too much sugar (in malt) will cause yeast to make too much CO_2 gas, turning a homemade fermentor into an unpredictable explosive device.

Learning how bacteria grow provides a window through which to view the core processes of life. As a result of studying bacterial growth and nutrition, we now know how DNA replicates, how RNA is made, and how proteins are assembled. We have also learned that the availability of nutrients influences the evolution of sophisticated microbial processes designed to avoid or survive starvation. For example, bacteria communicate with each other to build elaborate multicellular reproductive structures, such as fruiting bodies, or complex multispecies biofilms, such as those that erode our teeth and rust the surfaces of ocean liners.

In Chapter 4 we consider the microbial biosphere and the basics of bacterial growth. How diverse is this biosphere? Is it true that we have identified only about 0.1% of the bacterial species that inhabit Earth? Where do bacteria get their energy? Why do some bacteria gain energy from light (photosynthesis), whereas others gain energy from oxidizing sulfur? With so many microbes consuming nutrients, how has nature arranged biosystems to avoid depleting Earth of key compounds?

To understand the biochemistry behind metabolic diversity, we need to grow microbes. How do we ensure their growth and measure it? Some bacteria make elaborate, multicellular structures, whereas others form environmentally resistant fortresses called spores. In this chapter, we discuss how microbes obtain energy and nutrients for growth, and how cell populations develop. A more detailed treatment of the mechanisms of energy gain and biosynthesis is presented in Chapters 13–15. Microbial growth in ecosystems is discussed in Chapter 21.

4.1

Microbial Nutrition

Bacterial cells, for all their apparent simplicity, are remarkably complex and efficient replication machines. One cell of *Escherichia coli*, for example, can divide into two cells every 20–30 minutes. At a rate of 30 minutes per division, one cell could potentially multiply to over 1×10^{14} cells in 24 hours—that is, 100 trillion organisms! Although these 100 trillion cells would weigh only about 1 gram, the mass of cells would explode to 10^{14} grams (that is, 10^7 tons) after another 24 hours (two days total) of replicating every 30 minutes. Why, then, are we not buried under mountains of *E. coli*?

Nutrient Supplies Limit Microbial Growth

One factor limiting growth is the finite supply of nutrients. Microbes commonly encounter environments where essential nutrients are scarce. **Essential nutrients** are those compounds a microbe cannot make itself but must gather from its environment if the cell is to grow and divide. Microbial cells obtain all the essential nutrients for growth from their immediate environment, so they stop growing when an environment becomes depleted of one or more essential nutrients. How organisms cope with these periods of starvation until nutrients are restored is a rapidly developing field of microbiology.

All microorganisms require a minimum set of **macronutrients**, nutrients needed in large quantities. Six of these macronutrients—carbon, nitrogen, phosphorus, hydrogen, oxygen, and sulfur—make up the carbohydrates, lipids, nucleic acids, and proteins of the cell. Four other macronutrients are cations whose roles range from serving as enzyme **cofactors** that fit into specific enzymes and aid in the catalytic process (Mg^{2+}, Fe^{2+}, and K^+) to acting as regulatory signal molecules (Ca^{2+}). All cells also require very small amounts of certain trace elements, called **micronutrients**. These include cobalt, copper, manganese, molybdenum, nickel, and zinc, which are ubiquitous contaminants on glassware and in water. As a result, these trace elements are not added to laboratory media unless heroic measures have been taken to first remove the elements from the medium.

Micronutrients are required by cells as essential components of enzymes or cofactors. Cobalt, for example, is part of the cofactor vitamin B_{12}. Some organisms, such as the laboratory "workhorse" bacterium *E. coli*, make all their proteins, nucleic acids, and cell wall and membrane components from this very simple blend of chemical elements and compounds. For many other microbes, this basic set of nutrients is insufficient. *Borrelia burgdorferi,* for example, the cause of Lyme disease, requires an extensive mixture of complex organic supplements to grow.

Microbes Evolved to Grow in Different Environments

As a response to the ecological niche it inhabits, an organism may have evolved to require additional **growth factors** (**Table 4.1**), specific nutrients that are not required by all cells. Why, for example, should *Streptococcus pyogenes* make glutamic acid or alanine if both are readily available in its normal environment (such as the human oral cavity)? Because it never needs to make these compounds, *S. pyogenes,* as a matter of efficiency, has lost the genes whose protein products synthesize glutamic acid and alanine.

A **defined minimal medium** contains only those compounds needed for an organism to grow (**Table 4.2**). In the case of *S. pyogenes,* this medium would include glutamic acid and alanine in addition to the macro- and micronutrients mentioned previously. Other organisms have adapted so well to their natural habitats that we still do not know how to grow them in the laboratory. For example, *Rickettsia prowazekii,* the cause of epidemic typhus fever, grows only within the cytoplasm of eukaryotic cells (**Fig. 4.1**). This obligate intracellular bacterium lost key pathways needed for independent growth because the host cell supplies them (that species also evolved new mechanisms facilitating invasion and intracellular growth). Despite extensive efforts to grow them outside a host cell (**axenic growth**), cells of *R. prowazekii* have proved uncooperative.

In 2009, another supposed obligate intracellular bacterium (*Coxiella burnetii,* the cause of Q fever) was coaxed into dividing axenically. This pathogen normally grows only within acidified phagosome vacuoles. Robert Heinzen and colleagues at Rocky Mountain Laboratories designed an acidified (pH 5) medium containing numerous supplements that would support growth. Finding ways to grow these pathogens axenically will aid studies to identify virulence factors and develop treatments.

Although it will not grow in a defined medium, we can at least grow *Rickettsia* in the laboratory using eggs (the bacteria grow inside endothelial cells comprising the blood vessels formed in fertilized eggs) or animal cell tissue culture. But it is estimated that 99.9% of all the microorganisms on Earth cannot be grown in the laboratory at all. They are currently unculturable. Either these organisms need nutrients we have yet to discover, or they require cooperation with other species in their normal environment. Chapter 21 describes approaches to culturing previously unculturable microbes.

How do we even know that these microbes exist if we cannot grow them? All known microorganisms have a set of genes encoding RNA molecules present in ribosomes. The ribosomal RNA molecules are highly conserved across the phylogenetic tree. A DNA-amplifying procedure called the polymerase chain reaction (PCR, described in Section 7.6) can screen for the presence of these genes in soil and water samples. Comparing the DNA sequences of the PCR products with the DNA sequences of similar genes from known organisms reveals that nature harbors many

TABLE 4.1

Growth factors and natural habitats of organisms associated with disease.

Organism	Diseases	Natural habitats	Growth factors
Shigella	Bloody diarrhea	Humans	Nicotinamide (NAD)[a]
Haemophilus	Meningitis, chancroid	Humans and other animal species, upper respiratory tract	Hemin, NAD
Staphylococcus	Boils, osteomyelitis	Widespread	Complex requirement
Abiotrophia	Osteomyelitis	Humans and other animal species	Vitamin K, cysteine
Legionella	Legionnaires' disease	Soil, refrigeration cooling towers	Cysteine
Bordetella	Whooping cough	Humans and other animal species	Glutamate, proline, cystine
Francisella	Tularemia	Wild deer, rabbits	Complex, cysteine
Mycobacterium	Tuberculosis, leprosy	Humans	Nicotinic acid (NAD),[a] alanine
Streptococcus pyogenes	Pharyngitis, rheumatic fever	Humans	Glutamate, alanine

[a]Both nicotinamide and nicotinic acid are derived from NAD, nicotinamide adenine dinucleotide.

TABLE 4.2

Composition of commonly used media.

Medium	Ingredients per liter		Organisms cultured
Luria Bertani (complex)	Bacto tryptone[a]	10 g	Many Gram-negative and Gram-positive organisms (such as *Escherichia coli* and *Staphylococcus aureus*, respectively)
	Bacto yeast extract	5 g	
	NaCl	10 g	
	pH 7		
M9 medium (defined)	Glucose	2.0 g	Gram-negative organisms such as *E. coli*
	Na_2HPO_4	6.0 g (42 mM)	
	KH_2PO_4	3.0 g (22 mM)	
	NH_4Cl	1.0 g (19 mM)	
	NaCl	0.5 g (9 mM)	
	$MgSO_4$	2.0 mM	
	$CaCl_2$	0.1 mM	
	pH 7		
Sulfur oxidizers (defined)	NH_4Cl	0.52 g	*Acidithiobacillus thiooxidans*
	KH_2PO_4	0.28 g	
	$MgSO_4 \cdot 7H_2O$	0.25 g	
	$CaCl_2$	0.07 g	
	Elemental sulfur	1.56 g	
	CO_2	5%	
	pH 3		

[a]Bacto tryptone is a pancreatic digest of casein (bovine milk protein).

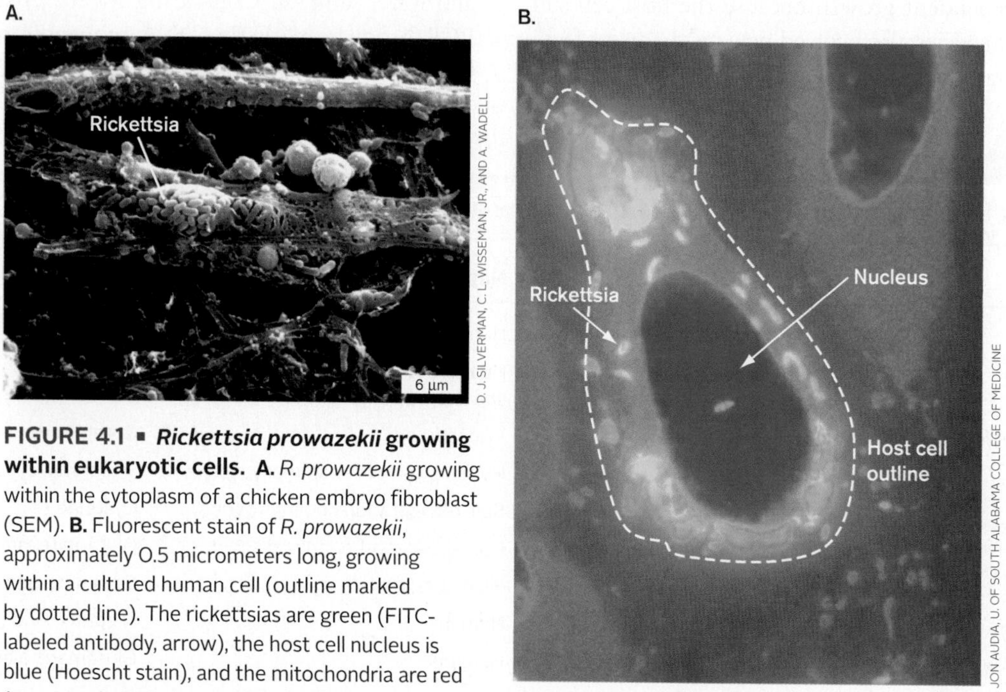
A.
Rickettsia
6 µm
D. J. SILVERMAN, C. L. WISSEMAN, JR., AND A. WADELL
B.
Rickettsia
Nucleus
Host cell outline
JON AUDIA, U. OF SOUTH ALABAMA COLLEGE OF MEDICINE

FIGURE 4.1 ■ *Rickettsia prowazekii* growing within eukaryotic cells. **A.** *R. prowazekii* growing within the cytoplasm of a chicken embryo fibroblast (SEM). **B.** Fluorescent stain of *R. prowazekii*, approximately 0.5 micrometers long, growing within a cultured human cell (outline marked by dotted line). The rickettsias are green (FITC-labeled antibody, arrow), the host cell nucleus is blue (Hoescht stain), and the mitochondria are red (Texas Red MitoTracker). The bacterium grows only in the cytoplasm, not in the nucleus.

microbes hitherto undiscovered because we cannot yet grow them in the laboratory. Even though we don't know the growth and nutritional requirements of these phantom microbes, modern genomic techniques can expose their existence. We can gain remarkable insight into the physiology of these "invisible," uncultured organisms by comparing their gene sequences, mined by PCR, with known gene sequences from culturable organisms that have well-characterized physiologies (see Chapter 8).

Microbes Build Biomass through Autotrophy or Heterotrophy

Maintaining life on this planet is an amazing process. All of Earth's life-forms are based on carbon, which they acquire through delicately choreographed processes that recycle key nutrients. The carbon cycle, a critical part of this process, involves two counterbalancing metabolic groups of organisms: heterotrophs and autotrophs. Their pathways of metabolism are discussed in Chapters 13 and 14.

Heterotrophs (such as *E. coli*) rely on other organisms to form the organic compounds (such as glucose) that they use as carbon sources. During heterotrophic metabolism, organic carbon sources are disassembled to generate energy and then reassembled to make cell constituents, such as proteins and carbohydrates (**Fig. 4.2A**). This process converts a large amount of the organic carbon source to CO_2, which is then released to the atmosphere. Thus, left on their own, heterotrophs would deplete the world of organic carbon sources (converting them to unusable CO_2) and starve to death. For life to continue, CO_2 must be recycled.

Autotrophs (such as cyanobacteria) assimilate CO_2 as a carbon source, reducing it (adding hydrogen atoms; see Chapter 15) to make complex cell constituents made up of C, H, and O (for example, carbohydrates, which have the general formula CH_2O) (**Fig. 4.2B**). These organic compounds can later be used as carbon sources by heterotrophs. Autotrophs are classified as photoautotrophs or chemoautotrophs by how they obtain energy. **Photoautotrophs** use light for photosynthesis, whereas **chemoautotrophs**, also known as **chemolithotrophs** or **lithotrophs** (literally, "rock-eaters"), gain energy by oxidizing inorganic substances such as iron or ammonia. Both photoautotrophs and chemoautotrophs carry out the autotrophic process of CO_2 fixation, which forms part of the carbon cycle between autotrophs and heterotrophs (**Fig. 4.2**). You should also realize that many microorganisms can use both heterotrophy and autotrophy to gain energy (for example, phototropic soil bacteria).

Thought Question

4.1 In a mixed ecosystem of autotrophs and heterotrophs, what happens when a heterotroph allows the autotroph to grow and begin to make excess organic carbon?

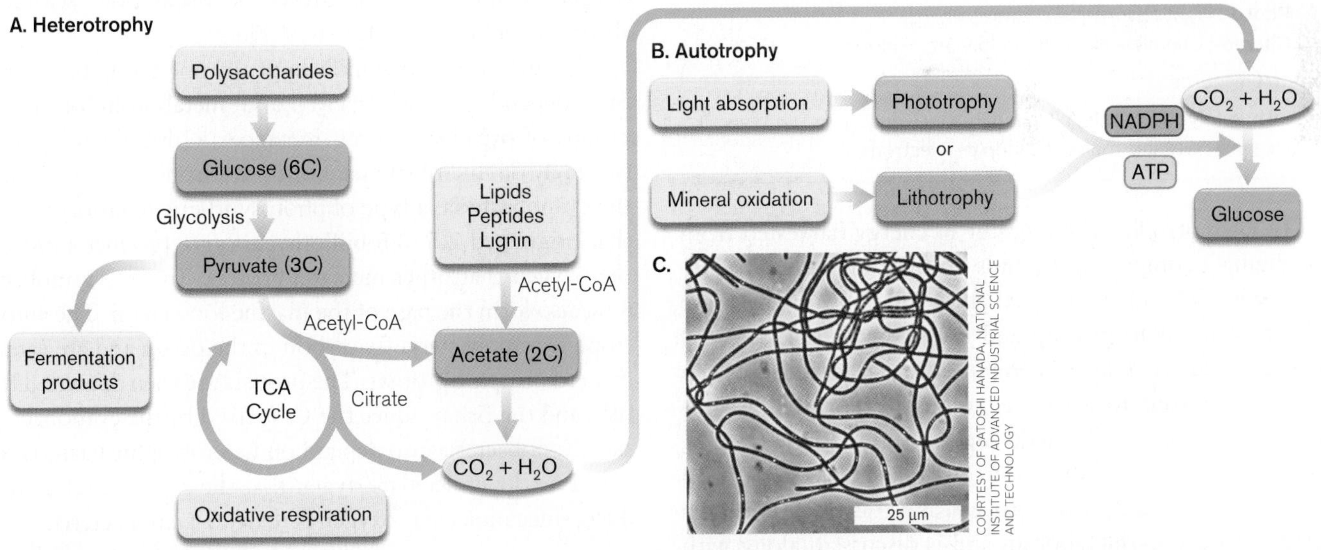
A. Heterotrophy
Polysaccharides
Glucose (6C)
Glycolysis
Pyruvate (3C)
Lipids
Peptides
Lignin
Acetyl-CoA
Acetyl-CoA
Fermentation products
Acetate (2C)
TCA Cycle
Citrate
$CO_2 + H_2O$
Oxidative respiration
B. Autotrophy
Light absorption
Phototrophy
or
Mineral oxidation
Lithotrophy
NADPH
ATP
$CO_2 + H_2O$
Glucose
C.
25 µm
COURTESY OF SATOSHI HANADA, NATIONAL INSTITUTE OF ADVANCED INDUSTRIAL SCIENCE AND TECHNOLOGY

FIGURE 4.2 ■ **The carbon cycle.** The carbon cycle requires both autotrophs and heterotrophs. **A.** Heterotrophs gain energy from degrading complex organic compounds (such as polysaccharides) to smaller compounds (such as glucose and pyruvate). The carbon from pyruvate moves through the tricarboxylic acid (TCA) cycle and is released as CO_2. In the absence of a TCA cycle, the carbon can end up as fermentation products, such as ethanol or acetic acid. **B.** Autotrophs use light energy or energy derived from the oxidation of minerals to capture CO_2 and convert it to complex organic molecules. **C.** *Chloroflexus aggregans,* originally isolated from hot springs in Japan, possesses extraordinary metabolic versatility. It grows anaerobically (without oxygen) as a photoheterotroph and aerobically (with oxygen) as a chemoheterotroph.

Microbes Obtain Energy through Phototrophy or Chemotrophy

Although the macronutrients mentioned earlier (C, N, P, H, O, and S) provide the essential building blocks to make proteins and other cell structures, all synthetic processes require an energy source. Depending on the organism, energy can be obtained from chemical reactions triggered by the absorption of light (**phototrophy**, or photosynthesis) or from oxidation-reduction reactions that transfer electrons from high-energy compounds to make products of lower energy (**chemotrophy**). Chemotrophic organisms fall into two classes that use different sources of electron donors: **lithotrophs** (chemolithotrophs, also described previously as chemoautotrophs) or **organotrophs** (chemoorganotrophs). Lithotrophs oxidize inorganic chemicals (for example, H_2, H_2S, NH_4^+, NO_2^-, and Fe^{2+}) for energy, whereas organotrophs oxidize organic compounds (for example, sugars). Phototrophs are also subcategorized as photolithotrophs or photoorganotrophs depending on their source of electron donors.

Note: The following prefixes for "-trophy" terms help distinguish different forms of biomass-building and energy-yielding metabolism.

Carbon source for biomass:

Auto-: CO_2 is fixed and assembled into organic molecules.
Hetero-: Preformed organic molecules are acquired from outside and assembled.

Energy source:

Photo-: Light absorption captures energy.
Chemo-: Chemical electron donors are oxidized.

Electron source:

Litho-: Inorganic molecules donate electrons.
Organo-: Organic molecules donate electrons.

In chemotrophy, the amount of energy harvested from oxidizing a compound depends on the compound's reduction state. The more reduced the compound is, the more electrons it has to give up and the higher is the potential energy yield. A reduced compound, such as glucose, can donate electrons to a less reduced (more oxidized) compound, such as nicotinamide adenine dinucleotide (NAD), releasing energy (in the form of donated electrons) and becoming oxidized in the process. NAD is a cell molecule critical to energy metabolism and is discussed, along with oxidation-reduction reactions, in Chapter 13.

In short, microbes are classified on the basis of their carbon and energy acquisition as follows:

- **Autotrophs.** Autotrophs build biomass by fixing CO_2 into complex organic molecules. Autotrophs gain energy through one of two general metabolic routes that either use or ignore light:

  Photoautotrophy generates energy through light absorption by the photolysis (light-activated breakdown) of H_2O or H_2S. The energy is used to fix CO_2 into biomass.

  Chemoautotrophy (or lithotrophy) produces energy from oxidizing inorganic molecules such as iron, sulfur, or nitrogen. This energy is used to fix CO_2 into biomass.

- **Heterotrophs.** Heterotrophs break down organic compounds from other organisms to gain energy and to harvest carbon for building their own biomass. Heterotrophic metabolism can be divided into two classes, also based on whether light is involved.

  Photoheterotrophy produces energy through photolysis of organic compounds or via a light-powered proton pump. Organic compounds are broken down and used to build biomass.

  Chemoheterotrophy (or organotrophy) yields energy and carbon for biomass solely from organic compounds. Chemoorganotrophy is commonly called heterotrophy.

Note that many species, particularly free-living soil and aquatic bacteria, can utilize more than one of these strategies, depending on environmental conditions. They do this by having multiple gene systems that are expressed under different conditions and yield products that carry out different functions. For example, *Rhodospirillum rubrum* grows by photoheterotrophy when light is available and oxygen is absent, but it grows by respiration, without absorbing light, when O_2 is available.

The survival and metabolism of any one group of organisms depend on the survival and metabolism of other groups of organisms. Even metazoa (multicellular organisms) rely on microbial metabolism to survive. For example, the cyanobacteria, a type of photosynthetic microorganism that originated 2.5–3.5 billion years ago, produce most of the oxygen that other metazoa breathe. In fact, cyanobacteria also form the base of the marine food chain. The autotrophic cyanobacteria fix carbon in the ocean and are eaten by heterotrophic protists. The protists are then devoured by fish, and the fish produce the CO_2 fixed by the cyanobacteria. Cyanobacteria also depend on heterotrophic bacteria to consume the molecular oxygen that the cyanobacteria produce, since molecular oxygen is toxic to cyanobacteria.

Energy Is Stored for Later Use

Whatever the source, energy, once obtained, must be converted to a form useful to the cell. This form can be chemical energy, such as that contained in the high-energy phosphate bond in adenosine triphosphate (ATP), or it can be electrochemical energy, which is stored in the form of an electrical potential formed between compartments separated by a membrane (see Chapter 14). Energy stored by an electrical potential across the membrane is known as the **membrane potential**. A membrane potential is generated when chemical (or light) energy is used to pump protons (and in some cases Na^+) outside of the cell, so that the proton concentration is greater outside the cell than inside. For example, membrane proteins such as cytochrome oxidases use energy from respiration to pump hydrogen ions across the cell membrane, generating a hydrogen ion gradient. This ion movement produces an electrical gradient across the cell membrane, making the inside of the cell more negatively charged than the outside. The hydrogen ion gradient plus the charge difference (voltage potential) across the membrane forms an **electrochemical potential**. Because this electrochemical potential includes the hydrogen ion gradient, it is called the **proton potential**, or **proton motive force**. The energy stored in the proton motive force can be used directly to move nutrients into the cell via specific transport proteins (see Section 3.3), to drive motors that rotate flagella, and to drive synthesis of ATP by a membrane-embedded **ATP synthase** (**Fig. 4.3A**).

The membrane-embedded ATP synthase, also called F_1F_o ATP synthase, provides most of the ATP for aerobic respiring cells such as *E. coli*, and essentially the same complex mediates ATP generation in our own mitochondria. ATP synthase is composed of many different proteins. The enzyme includes a channel (F_o) that allows H^+ to move across the membrane and drive rotation of the ATP synthase complex (F_1). Rotation of F_1 mediates the formation of ATP. The role of the proton potential in metabolism is discussed in detail in Chapters 13 and 14.

The idea that a living organism could contain rotating parts was highly controversial when such parts were first discovered in bacterial flagella (discussed in Section 3.7). Before this discovery, scientists had believed that living body parts could not rotate and that only humans had "invented the wheel." The discovery of rotary biomolecules has inspired advances in nanotechnology, the engineering of microscopic devices. For example, a "biomolecular motor" was devised using an ATP synthase F_1 complex to drive a metal submicroscopic propeller (**Fig. 4.3B**). In the future, such biomolecular design may be used to build microscopic robots that enter the bloodstream to perform microsurgery.

Note: The name F_1F_o derives from early biochemical studies. F_1 comes from fraction 1 obtained during a purification scheme, whereas F_o comes from this fraction's ability to bind oligomycin.

The Nitrogen Cycle Depends on Bacteria and Archaea

Nitrogen is an essential component of proteins, nucleic acids, and other cellular constituents, and as such it is required in large amounts by living organisms. Nitrogen gas (N_2) makes up nearly 79% of Earth's atmosphere, but most organisms are unable to use nitrogen gas because the triple bond between the two nitrogen atoms is highly stable and requires considerable energy to be broken. For nitrogen to be used for growth, it must first be "fixed," or converted

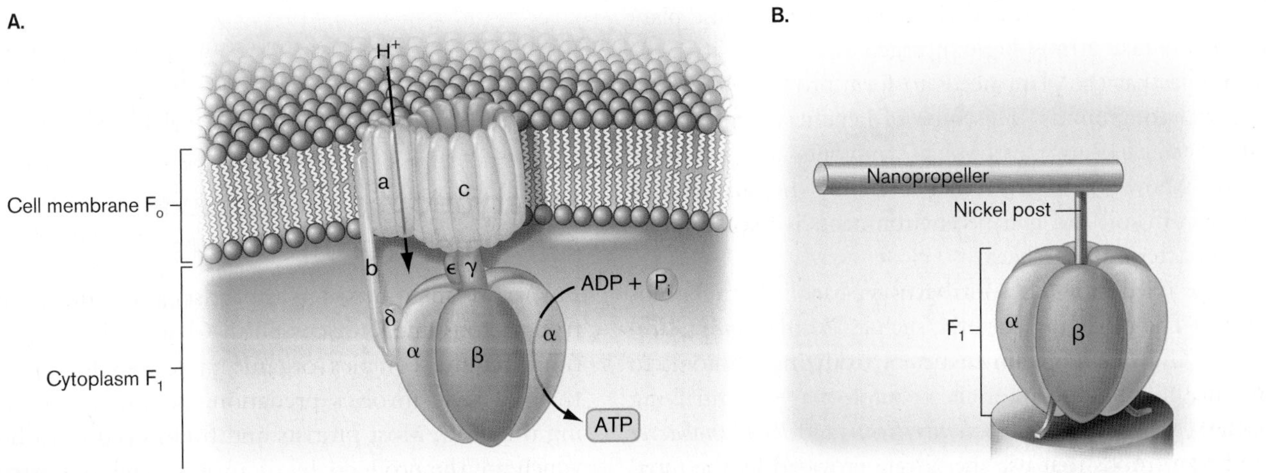
A.
H+
a
c
b
ε γ
δ
α
β
α
ADP + Pi
ATP
Cell membrane Fo
Cytoplasm F1
B.
Nanopropeller
Nickel post
F1
α
β

FIGURE 4.3 ■ Bacterial membrane ATP synthase. A. The F_o portion of the F_1F_o complex of ATP synthase is embedded in the cell membrane. The F_o channel/rotor can admit hydrogen ions (H^+) because the flow of H^+ across the membrane is driven by the concentration and charge differences between inside and outside the cell. The flow of H^+ causes F_1 to rotate, and rotation of F_1 drives the conversion of ADP + P_i to ATP. **B.** An artificial "biomolecular motor" was built from an ATP synthase F_1 unit attached to a nickel post and a nanopropeller.

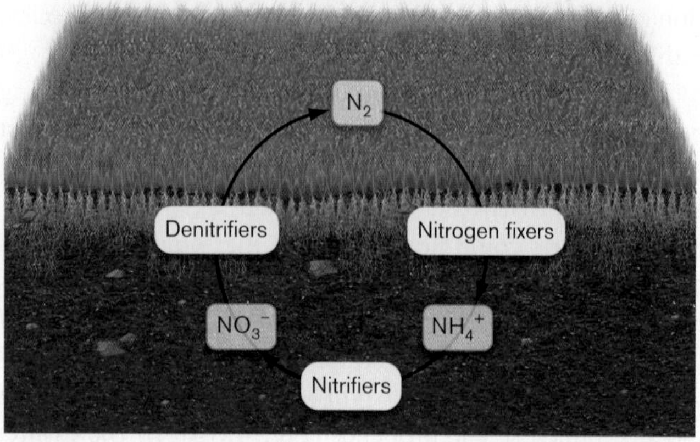
N$_2$
Denitrifiers
Nitrogen fixers
NO$_3^-$
NH$_4^+$
Nitrifiers

FIGURE 4.4 ▪ The nitrogen cycle. Nitrogen gas (N_2) is fixed by species of bacteria (nitrogen fixers) that possess the enzyme nitrogenase. Other bacteria (nitrifiers) oxidize ammonia (as the ammonium ion, NH_4^+) to generate energy. Still others (denitrifiers) use oxidized forms of nitrogen, such as nitrate (NO_3^-), as an alternative electron acceptor in place of O_2.

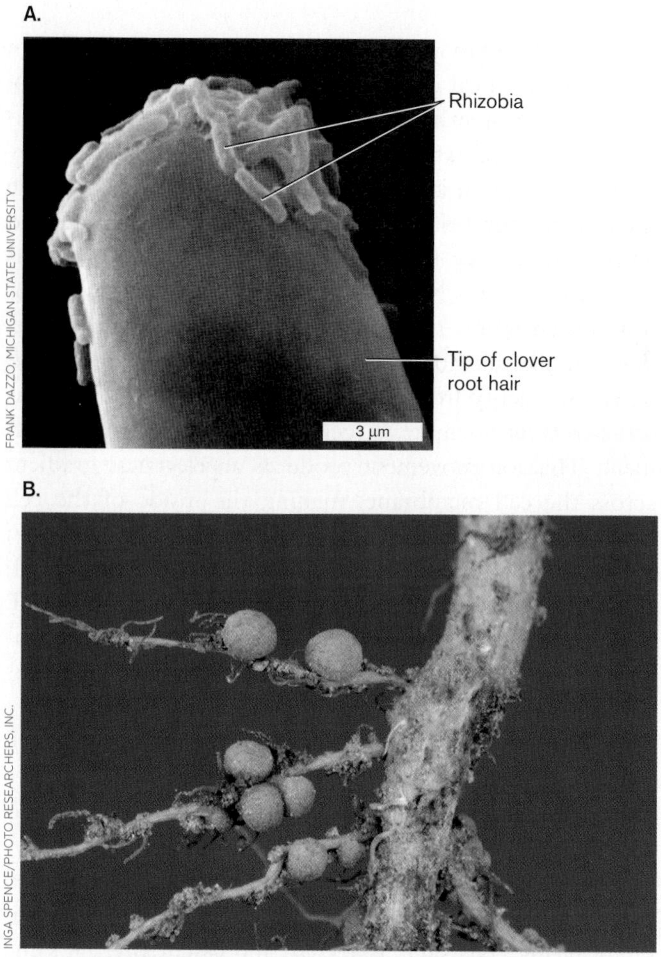
A.
Rhizobia
Tip of clover root hair
3 μm
FRANK DAZZO, MICHIGAN STATE UNIVERSITY
B.
INGA SPENCE/PHOTO RESEARCHERS, INC.

FIGURE 4.5 ▪ *Rhizobium* and a legume. A. Symbiotic *Rhizobium* cells clustered on a clover root tip (SEM). Although the rhizobia (about 0.9 μm × 3 μm) are shown here clustered on the surface of the root, they will soon invade the root and begin a symbiotic partnership that will benefit both organisms. **B.** Root nodules. After the rhizobia invade the plant root, symbiosis between plant and microbe produces nodules.

to ammonium ions (NH_4^+). As with the carbon cycle, various groups of organisms collaborate to interconvert nitrogen gas, ammonium ions, and nitrate ions (NO_3^-) in what is called the nitrogen cycle. One group "fixes" atmospheric nitrogen, while other bacteria do the opposite, transforming ammonia to nitrate (**nitrification**) and then converting nitrate to N_2 (**denitrification**) (**Fig. 4.4**). For the environmental significance of nitrogen metabolism, see Chapter 22.

Nitrogen-fixing bacteria may be free-living in soil or water, or they may form symbiotic associations with plants or other organisms. A **symbiont** is an organism that lives in intimate association with a second organism. *Rhizobium, Sinorhizobium,* and *Bradyrhizobium* species, for example, are nitrogen-fixing symbionts in leguminous plants such as soybeans, chickpeas, and clover (**Fig. 4.5**). These plant symbionts take atmospheric nitrogen and convert it to the ammonia that the plant needs to form proteins and other essential compounds. This type of beneficial symbiosis is called "mutualism." Although symbionts are the most widely known nitrogen-fixing bacteria, the majority of nitrogen in soil and marine environments is fixed by free-living bacteria and archaea.

Once fixed, how does nitrogen get back into the atmosphere? The nitrifying bacteria, such as *Pseudomonas, Alcaligenes,* and *Bacillus,* gain energy by oxidizing ammonia to produce nitrate (a form of chemoautotrophy). Denitrifying bacteria such as *Paracoccus denitrificans* and *Rhodanobacter* are heterotrophs that use the nitrate provided by the nitrifiers to produce a series of nitrogen compounds, ending with gaseous N_2. The activity of denitrifying bacteria results in a substantial loss of nitrogen into the atmosphere that roughly balances the amount of nitrogen fixation occurring each year. Like the carbon cycle, the nitrogen cycle illustrates how nature manages to replenish planet Earth.

Eukaryotic Microbes Include Consumers and Producers

Eukaryotic microbes such as protists and fungi are heterotrophic consumers (discussed in Chapter 21). Protists and fungi require complex organic molecules for growth, so their lifestyle involves predation, parasitism, or scavenging the dead. Most protists and fungi need mitochondria, which are the products of an ancient symbiotic partnering with internalized prokaryotes (discussed in Section 17.6). Heterotrophic fungi possess the exceptional ability to digest complex organic compounds such as lignin, a major component of wood and bark. In marine and aquatic systems, protists form a huge part of the food chain, so eating trout or swordfish means, in effect, that you are consuming trillions of protists.

Just like photosynthetic bacteria, eukaryotic algae are photoautotrophs that produce biomass through photosynthesis. In addition to needing chloroplasts for photosynthesis, however, algae require mitochondria for energy production. (Recall from Section 1.5 that chloroplasts, like mitochondria, evolved from internalized bacteria.) Protist algae, such as single-celled *Euglena,* are **mixotrophic**, capable of utilizing photosynthesis or heterotrophic respiration, depending on environmental conditions.

To Summarize

- **Microorganisms** require certain essential macro- and micronutrients to grow.
- **Microbial genomes** evolve in response to nutrient availability.
- **Obligate intracellular bacteria** lose metabolic pathways provided by their hosts and develop requirements for growth factors supplied by their hosts.
- In the **global carbon cycle**, autotrophs use CO_2 as a carbon source, either through photosynthesis or through lithotrophy.
- **Autotrophs** make complex organic compounds that are consumed by **heterotrophs**.
- **Nitrogen fixers**, **nitrifiers**, and **denitrifiers** contribute to the nitrogen cycle.
- **Bacteria** or **archaea** carry out all of the carbon, nitrogen, and energy reactions just described, whereas eukaryotes carry out only a limited range of heterotrophic and photosynthetic reactions.

4.2 Nutrient Uptake

How do bacteria gather nutrients? Whether a microbe is propelled by flagella toward a favorable habitat or, lacking motility, drifts through its environment, the organism must be able to find nutrients and move them across the membrane into the cytoplasm. The membrane, however, presents a daunting obstacle. Membranes are designed to separate what is outside the cell from what is inside. So, for a cell to gain sustenance from the environment, the membrane must be selectively permeable to nutrients the cell can use. A few compounds, such as oxygen and carbon dioxide, can passively diffuse across the membrane, but most cannot. Selective permeability is achieved in three ways:

- By the use of substrate-specific carrier proteins (**permeases**) in the membrane
- With the aid of dedicated nutrient-binding proteins that patrol the periplasmic space
- Through the action of membrane-spanning protein channels, or pores, that discriminate between substrates

Microbes must also overcome the problem of low nutrient concentrations in the natural environment (for example, lakes or streams). If the intracellular concentration of nutrients were no greater than the extracellular concentration, the cell would remain starved of most nutrients. To solve this dilemma, most organisms have evolved efficient transport systems that concentrate nutrients inside the cell relative to outside. However, moving molecules against a concentration gradient requires some form of energy.

In contrast to environments where nutrients are available but exist at low concentrations (for example, aqueous environments), certain habitats have plenty of nutrients but those nutrients are locked in a form that cannot be transported into the cell. Starch, a large, complex carbohydrate, is but one example. Many microbes unlock these nutrient vaults by secreting digestive proteins that break down complex carbohydrates or other molecules into smaller compounds that are easier to transport. The amazing techniques that cells use to extrude these large digestive proteins through the membrane and into their surrounding environment are discussed in Chapter 8.

Facilitated Diffusion

Although most transport systems use cellular energy to bring compounds into the cell, a few do not. **Facilitated diffusion** uses the concentration gradient of a compound to move that compound across the membrane from a compartment of higher concentration to a compartment of lower concentration. The best these passive systems can do is to equalize the internal and external concentrations of a solute; facilitated transport cannot move a molecule against its gradient. Facilitated diffusion systems are used for compounds either too large or too polar to diffuse on their own.

The most important facilitated diffusion transporters are those of the aquaporin family that transport water and small polar molecules such as glycerol (which is used for energy and for building phospholipids). Glycerol transport is performed by an integral membrane protein in *E. coli*

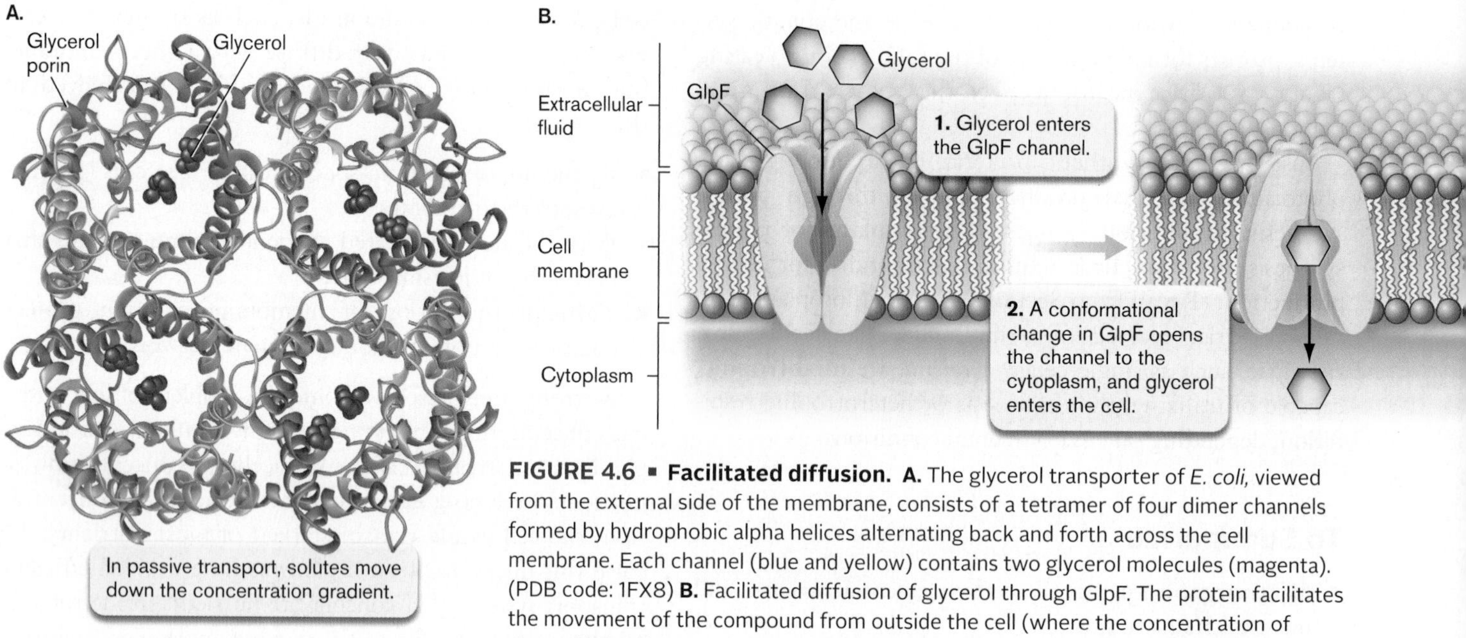
A.
Glycerol porin
Glycerol
In passive transport, solutes move down the concentration gradient.
B.
Extracellular fluid
Cell membrane
Cytoplasm
GlpF
Glycerol
1. Glycerol enters the GlpF channel.
2. A conformational change in GlpF opens the channel to the cytoplasm, and glycerol enters the cell.

FIGURE 4.6 ■ Facilitated diffusion. A. The glycerol transporter of *E. coli,* viewed from the external side of the membrane, consists of a tetramer of four dimer channels formed by hydrophobic alpha helices alternating back and forth across the cell membrane. Each channel (blue and yellow) contains two glycerol molecules (magenta). (PDB code: 1FX8) **B.** Facilitated diffusion of glycerol through GlpF. The protein facilitates the movement of the compound from outside the cell (where the concentration of glycerol is high) to inside the cell (where the concentration of glycerol is low).

called GlpF. The structure of a glycerol channel is shown in **Figure 4.6A**, where the complex is viewed from the outer face of the membrane. The complex is a tetramer of four channels, each of which transports a glycerol.

In the membrane, GlpF reversibly (and randomly) assumes two conformations. One form exposes the glycerol-binding site to the external environment, whereas the second form exposes this site to the cytoplasm. When the concentration of glycerol is greater outside than inside the cell, the form with the binding site exposed to the exterior is more likely to find and bind glycerol. After binding glycerol, GlpF changes shape, closing itself to the exterior and opening to the interior (**Fig. 4.6B**). Bound glycerol is then released (diffuses) into the cytoplasm (influx). Of course, this form of GlpF could also bind a glycerol molecule in the cytoplasm and release it outside the cell. Thus, once the cytoplasmic concentration of glycerol equals the concentration outside the cell, bound glycerol can be released to either compartment. However, facilitated diffusion normally promotes glycerol influx, because the cell consumes the compound as it enters the cytoplasm, keeping cytoplasmic concentrations of glycerol low.

Weblinks GlpF animation (*see ebook*)

Active Transport Requires Energy

Most forms of transport expend energy to take up molecules from outside the cell and concentrate them inside. The ability to import nutrients against their natural concentration gradients is critical in aquatic habitats, where nutrient concentrations are low, and in soil habitats, where competition for nutrients is high.

The simplest way to use energy to move molecules across a membrane is to exchange the energy of one chemical gradient for that of another. The most common chemical gradients used are those of ions, particularly the positively charged ions Na^+ and K^+. These ions are kept at different concentrations on either side of the cell membrane. When an ion moves down its concentration gradient (from high to low), energy is released. Some transport proteins harness that free energy and use it to drive transport of a second molecule up, or against, its concentration gradient in the process called **coupled transport**.

The two types of coupled transport systems are **symport**, in which the two molecules travel in the same direction (**Fig. 4.7A**), and **antiport**, in which the actively transported molecule moves in the direction opposite to the driving ion (**Fig. 4.7B**). An example of a symporter is the lactose permease LacY of *E. coli,* one of the first transport proteins whose function was elucidated. This work was carried out by the pioneering membrane biochemist H. Ronald Kaback (**Fig. 4.7C**). LacY moves lactose inward, powered by a proton that is also moving inward (symport). LacY proton-driven transport is said to be **electrogenic** because an unequal distribution of charge results (for example, symport of a neutral lactose molecule with H^+ results in net movement of positive charge).

An example of **electroneutral** coupled transport, in which there is no net transfer of charge, is that of the Na^+/H^+

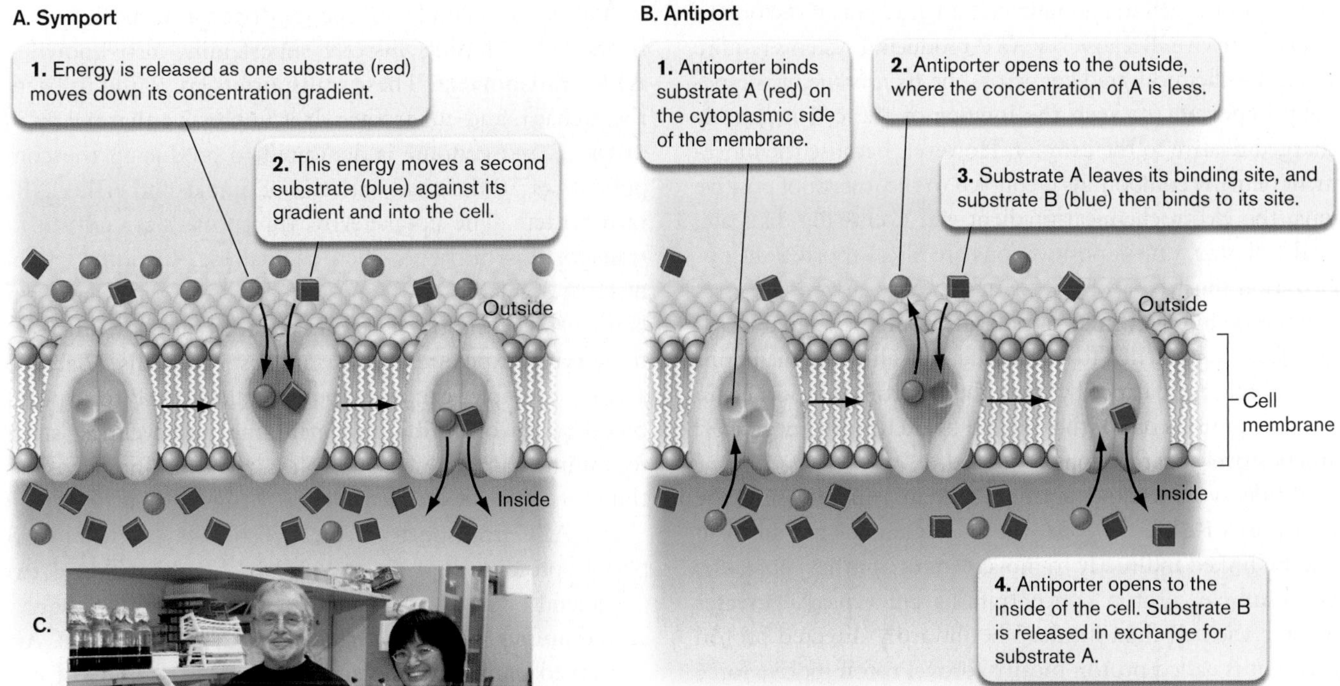
A. Symport
1. Energy is released as one substrate (red) moves down its concentration gradient.
2. This energy moves a second substrate (blue) against its gradient and into the cell.
Outside
Inside
B. Antiport
1. Antiporter binds substrate A (red) on the cytoplasmic side of the membrane.
2. Antiporter opens to the outside, where the concentration of A is less.
3. Substrate A leaves its binding site, and substrate B (blue) then binds to its site.
Outside
Cell membrane
Inside
4. Antiporter opens to the inside of the cell. Substrate B is released in exchange for substrate A.

C.

COURTESY OF RONALD KABACK

D.

COURTESY OF ETANA PADAN

FIGURE 4.7 ■ Coupled transport. **A.** Symport. **B.** Antiport. In both symport and antiport, substrate B (blue) is taken up against its gradient because of the energy released by substrate A (red) traveling down its gradient. **C.** Ronald Kaback (left) of UCLA elucidated the mechanism of transport by the proton-driven lactose symporter LacY. **D.** Etana Padan (second from right) of the Hebrew University of Jerusalem dissected the molecular mechanism of the Na^+/H^+ antiporter NhaA.

antiporter. The Na^+/H^+ antiporter couples the export of Na^+ with the import of a proton (antiport). Because molecules of like charges are merely exchanged, there is no net movement of charge. The mechanism by which Na^+/H^+ antiporters function was elucidated by the Israeli biochemist Etana Padan (**Fig. 4.7D**). Sodium exchange is important for all organisms and is particularly critical for organisms living in a high-salt habitat (see Section 5.4).

Thought Questions

4.2 In what situation would antiport and symport be passive rather than active transport?

4.3 What kind of transporter, other than an antiporter, could produce electroneutral coupled transport?

Symport and antiport transporter proteins function by alternately opening one end or the other of a channel that spans the cell membrane. The channel contains solute-binding sites (**Fig. 4.7A** and **B**). When the channel is open to the high-concentration side of the membrane, the driving ion (solute) attaches to the binding sites. The transport protein then changes shape to open that site to the low-concentration side of the membrane, and the ion leaves. When and where the second (cotransported) solute binds depends on whether the transport protein is an antiporter or a symporter.

Weblinks Movie on antiport (*see ebook*)

With all this ion traffic going on across the membrane, a careful accounting must be kept of how many ions are inside the cell relative to outside. The cell must recirculate ions into and out of the cell to maintain certain gradients if the organism is to survive. Because key ATP-producing systems require an electrochemical gradient across the membrane, it is especially important to keep the interior of the cell negatively charged relative to the exterior. However, because the movement of many compounds is coupled to the import of positive ions, the electrochemical gradient will eventually dissipate, or depolarize, unless positive ions are also exported. Depolarization must be avoided, because a depolarized cell loses membrane integrity and cannot carry out the simple transport functions needed to sustain life. A healthy cell maintains a proper charge balance by using the electron transport chain to move protons out of the cell and by exchanging negatively and positively charged ions as needed.

While symport and antiport systems directly link the transport of different ions, the movement of one ion can also be linked indirectly to movement of another molecule. For example, proton concentrations are typically greater outside the cell than inside. The inwardly directed proton gradient is called proton motive force. Proton motive force can impel the exit of Na^+ through the Na^+/H^+ antiporter. The resulting Na^+ gradient can then drive the symport of amino acids into the cell. In this case, Na^+ moves back into the cell down its gradient, and the energy released is tied to the import of an amino acid against its gradient.

ABC Transporters Are Powered by ATP

As we pointed out in Section 3.3, a major function of proton transport is to form the proton motive force that powers ATP synthesis (for details on the proton motive force, see Chapter 14). The energy stored in ATP, whether generated by the proton current or by cytoplasmic means such as fermentation, can drive membrane transport of nutrients.

The largest family of energy-driven transport systems is the ATP-binding cassette superfamily, also known as **ABC transporters**. These transporters are found in bacteria, archaea, and eukaryotes. It is impressive that nearly 5% of the *E. coli* genome is dedicated to producing the components of 70 different varieties of uptake and efflux ABC transporters. The uptake ABC transporters are critical for transporting nutrients such as maltose, histidine, arabinose, and galactose. The efflux ABC transporters are generally used as multidrug efflux pumps that allow microbes to survive exposures to hazardous chemicals. *Lactococcus,* for example, can use a pump called LmrP to export a broad range of antibiotics, including tetracyclines, streptogramins, quinolones, macrolides, and aminoglycosides, thus conferring resistance to those drugs (see Section 27.5).

An ABC transporter typically consists of two hydrophobic proteins that form a membrane channel and two peripheral cytoplasmic proteins that contain a highly conserved amino acid motif involved with binding ATP. Any conserved amino acid sequence found in a family of proteins is viewed as a "cassette" because the sequence appears to have been inserted into those proteins for specific functions (hence the name "ATP-binding cassette"). The ABC transporter superfamily contains both uptake and efflux transport systems. The uptake systems (but not the efflux systems) possess an additional, extracytoplasmic protein, called a **substrate-binding protein**, that initially binds the substrate (also called solute). In Gram-negative bacteria, these substrate-binding proteins float in the periplasmic space between the inner and outer membranes. In Gram-positive bacteria, which lack an outer membrane, the proteins must be tethered to the cell surface.

ABC transport (**Fig. 4.8**) starts with the substrate-binding protein snagging the appropriate solute, either as it

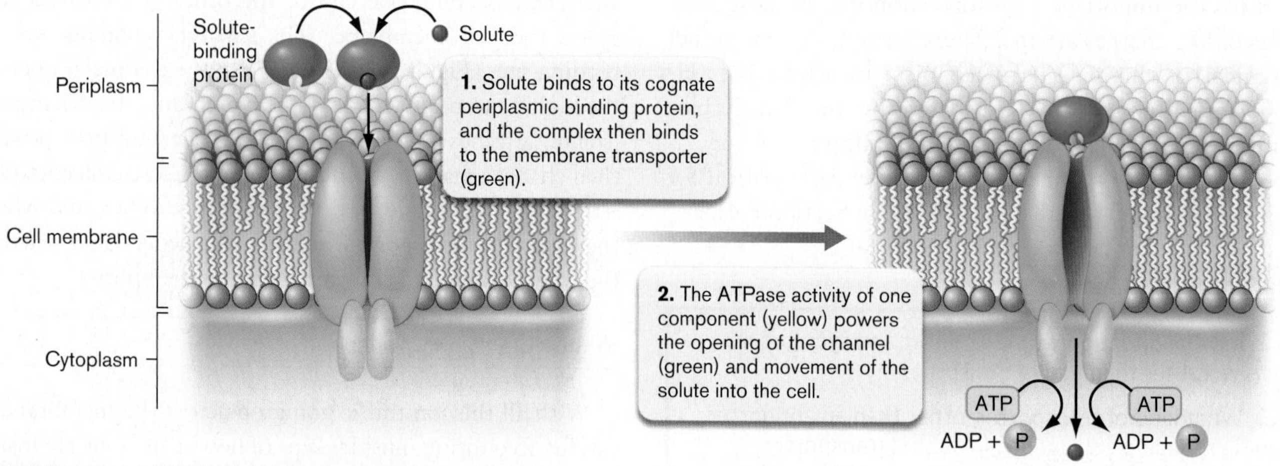
Solute-binding protein
Solute
Periplasm
Cell membrane
Cytoplasm
1. Solute binds to its cognate periplasmic binding protein, and the complex then binds to the membrane transporter (green).
2. The ATPase activity of one component (yellow) powers the opening of the channel (green) and movement of the solute into the cell.
ATP
ADP + P
ATP
ADP + P

FIGURE 4.8 ■ ABC (ATP-binding cassette) transporters.

floats by a Gram-positive microbe or as the molecule enters the periplasm of a Gram-negative cell. Most substrates nonspecifically enter the periplasm of Gram-negative organisms through the outer membrane pores, although some high-molecular-weight substrates, like vitamin B_{12}, require the assistance of a specific outer membrane protein to move the substrate into the periplasm. Because substrate-binding proteins have a high affinity for their cognate (matched) solutes, their use increases the efficiency of transport when concentrations of solute are low.

Once united with its solute, the binding protein binds to the periplasmic face of the channel protein and releases the solute, which now moves to a site on the channel protein. This interaction triggers a structural (or conformational) change in the channel protein (the green "membrane transporter" in **Fig. 4.8**) that is telegraphed to the nucleotide-binding proteins on the cytoplasmic side (yellow). On receiving this signal, the nucleotide-binding proteins start hydrolyzing ATP and send a return conformational change through the channel, signaling the channel to open its cytoplasmic side and allow the solute to enter the cell.

Many different ABC transporters mediate the transport of a wide variety of substrates. All of them appear to have arisen from a common ancestral porter, so they share a considerable amount of amino acid sequence homology.

Siderophores Are Secreted to Scavenge Iron

Iron, an essential nutrient of most cells, is mostly locked up in nature as $Fe(OH)_3$, which is insoluble and unavailable for transport. Many bacteria and fungi have solved this transport dilemma by synthesizing and secreting specialized molecules called **siderophores** (Greek for "iron bearer"), which have a very high affinity for whatever soluble ferric iron is available in the environment. These iron scavenger molecules are produced and sent forth by cells when the intracellular iron concentration is low (**Fig. 4.9**). In most Gram-negative organisms, the siderophore binds iron in the environment, and the siderophore-iron complex then attaches to specific receptors in the outer membrane. At this point, either the iron is released directly and is passed to other transport proteins or the complex is transported across the cytoplasmic membrane by a dedicated ABC transporter. The iron is released intracellularly and reduced from Fe^{3+} to Fe^{2+} for biosynthetic use. Other Gram-negative microorganisms, such as *Neisseria gonorrhoeae* (the causative agent of gonorrhea), do not use siderophores at all but employ receptors on their surface that bind human iron complexes (for example, transferrin or lactoferrin) and wrest the iron from them.

Group Translocation

The uptake transporters we have just considered increase concentrations of solute inside the cell relative to the outside. They move nutrients "uphill" against a concentration gradient. An entirely different system, known as **group translocation**, cleverly accomplishes the same result but

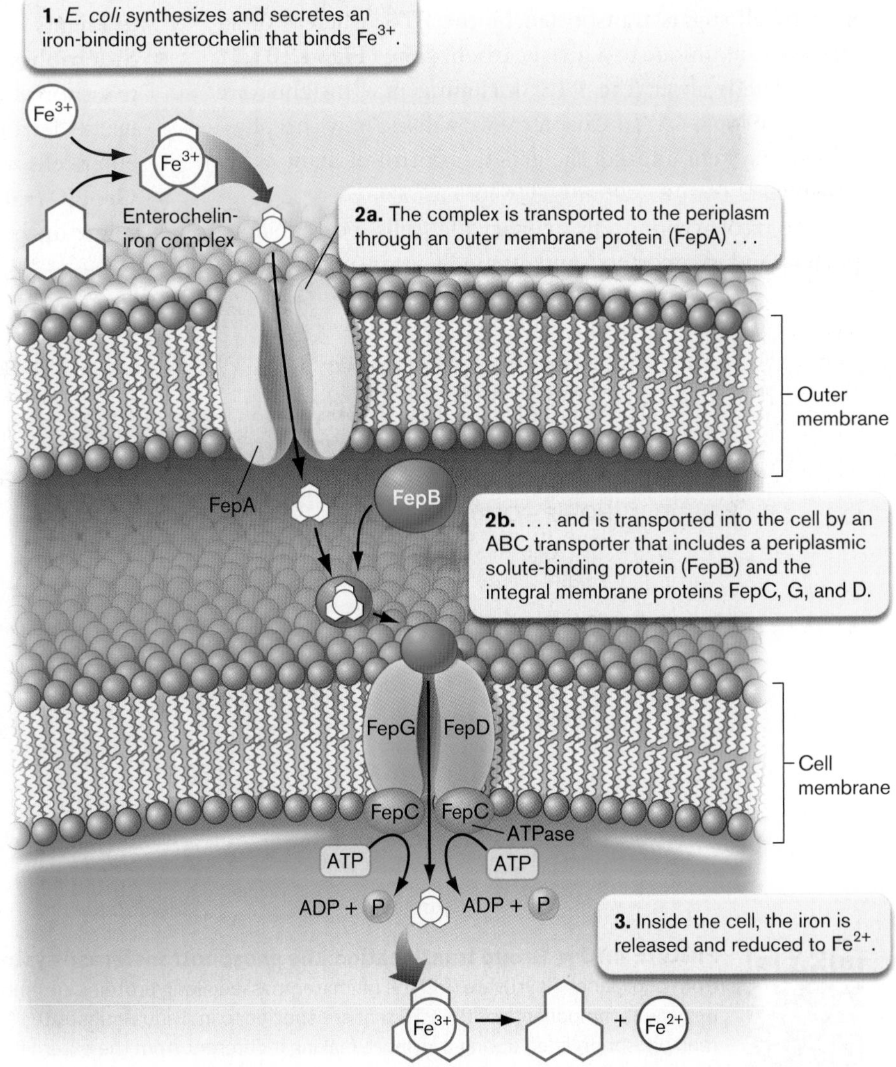
1. E. coli synthesizes and secretes an iron-binding enterochelin that binds Fe3+.
Fe3+
Fe3+
Enterochelin-iron complex
2a. The complex is transported to the periplasm through an outer membrane protein (FepA) . . .
Outer membrane
FepA
FepB
2b. . . . and is transported into the cell by an ABC transporter that includes a periplasmic solute-binding protein (FepB) and the integral membrane proteins FepC, G, and D.
FepG
FepD
Cell membrane
FepC
FepC
ATPase
ATP
ATP
ADP + P
ADP + P
3. Inside the cell, the iron is released and reduced to Fe2+.
Fe3+
+
Fe2+

FIGURE 4.9 ■ Siderophores and iron transport.

without really moving a substance uphill. Group translocation alters the substrate during transport by attaching a new group (for example, phosphate) to it. Because the modified nutrient inside the cell is chemically different from the related compound outside, the parent solute entering the cell is always moving down its concentration gradient, regardless of how much solute has already been transported. Note that this process uses energy to chemically alter the solute. ABC transporters and group translocation systems both involve active transport, but group translocation systems are not ABC transporters.

The **phosphotransferase system (PTS)** is a well-characterized group translocation system present in many bacteria. It uses energy from phosphoenolpyruvate (PEP), an intermediate in glycolysis, to attach a phosphate to specific sugars during their transport into the cell. Glucose, for example, is converted during transport to glucose 6-phosphate. The system has a modular design that accommodates different substrates. Some protein elements are used by all sugars transported by the PTS, whereas other elements are unique to a given carbohydrate (**Fig. 4.10**). More details about the PTS, including new insights, are given in eTopic 4.1. In Chapter 9 we will see how this physiological system impacts the genetic control of many other systems.

Like prokaryotic cells, eukaryotic cells possess antiporters and symporters, and use ABC transport systems as multidrug efflux pumps, but they also employ another process, called endocytosis, which often precedes nutrient transport across membranes. Endocytosis is discussed in eTopic 4.2 and Appendix 2, Section A2.1.

To Summarize

- **Transport systems** move nutrients across phospholipid bilayer membranes.
- **Facilitated diffusion** helps solutes move across a membrane from a region of high concentration to one of lower concentration.
- **Antiporters** and **symporters** are coupled transport systems in which energy released by moving a driving ion (H^+ or Na^+) from a region of high concentration to one of low concentration is harnessed and used to move a solute against its concentration gradient.
- **ABC transporters** use the energy from ATP hydrolysis to move solutes "uphill," against their concentration gradients.
- **Siderophores** are secreted to bind ferric iron (Fe^{3+}) and transport it into the cell, where it is reduced to the more useful ferrous form (Fe^{2+}). Siderophore-iron complexes enter cells with the help of ABC transporters.
- **Group translocation systems** chemically modify the solute during transport.

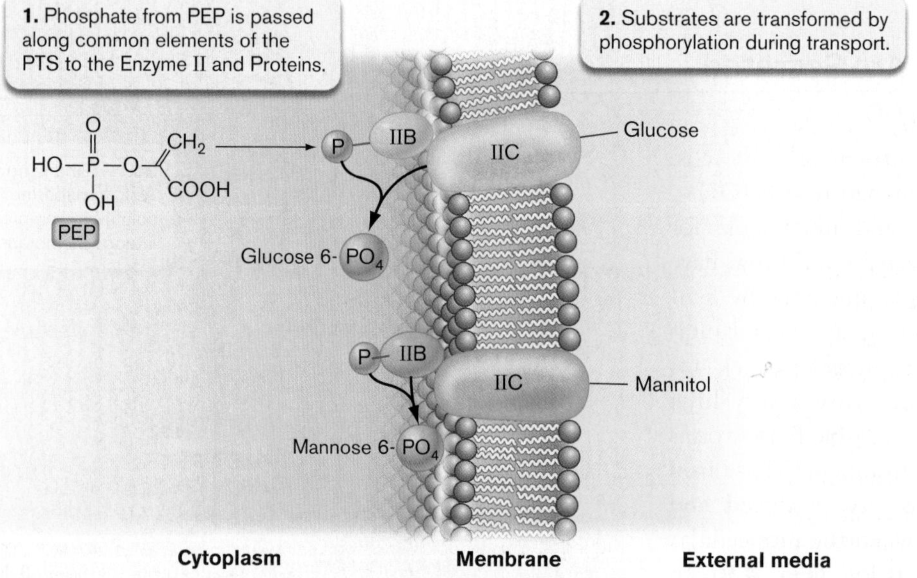
1. Phosphate from PEP is passed along common elements of the PTS to the Enzyme II and Proteins.
2. Substrates are transformed by phosphorylation during transport.
O
HO—P—O
OH
CH2
COOH
PEP
P
IIB
IIC
Glucose
Glucose 6- PO4
P
IIB
IIC
Mannitol
Mannose 6- PO4
Cytoplasm
Membrane
External media

FIGURE 4.10 ■ Group translocation: the phosphotransferase system (PTS) of *E. coli*. The phosphate group from phosphoenolpyruvate (PEP) is ultimately passed along proteins common to all PTS sugar transport systems and to Enzyme II components of the PTS that are specific to individual substrates (such as glucose or mannitol). The substrate is then phosphorylated during transport, making it different from the external sugar. As a result, the sugar is always traveling down its concentration gradient into the cell. A more detailed look at the cellular location of different PTS proteins is found in eTopic 4.1.

4.3

Culturing Bacteria

How are microbes captured and studied? Microbes in nature usually exist in complex, multispecies communities. For detailed studies, though, they must be grown separately in pure culture. It is nothing short of amazing—and humbling—that after 120 years of trying to grow microbes in the laboratory, we have succeeded in culturing only 0.1% of the microorganisms around us. The vast majority of the microbial world has yet to be tamed. Since the time of Koch in the late nineteenth century, microbiologists have used the same fundamental techniques to culture bacteria in the laboratory, although innovative methods now enable us to culture microbes under conditions resembling their native habitat (discussed in Chapter 21).

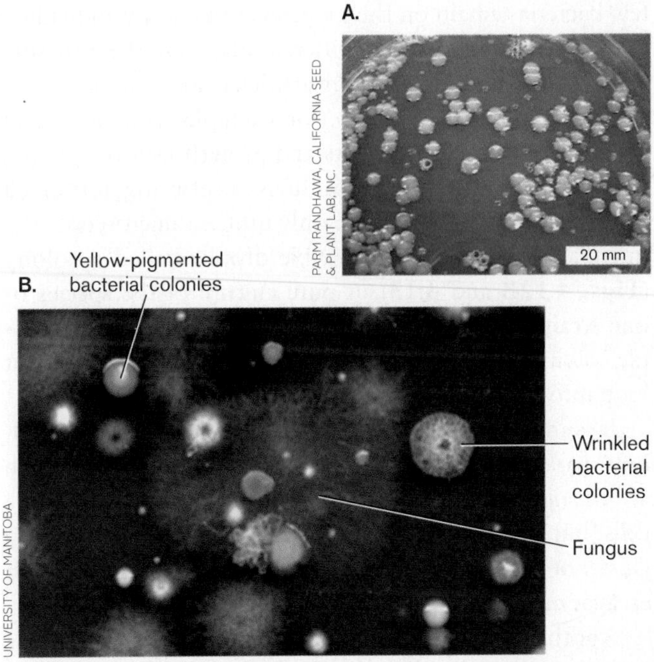
A.
PARM RANDHAWA, CALIFORNIA SEED & PLANT LAB, INC.
20 mm
B.
Yellow-pigmented bacterial colonies
Wrinkled bacterial colonies
Fungus
UNIVERSITY OF MANITOBA

FIGURE 4.11 ■ Separation and growth of microbes on an agar surface. A. Colonies (diameter 1–5 mm) of *Acidovorax citrulli* separated on an agar plate. *A. citrulli* is a plant pathogen that causes watermelon fruit blotch. **B.** A mixture of yellow-pigmented bacterial colonies, wrinkled bacterial colonies, and fungus separated by dilution on an agar plate. Over time the fungal colony overgrew adjacent bacterial colonies.

Bacteria Are Grown in Culture Media

For those organisms that can be cultured, we have access to a variety of culturing techniques that can be used for different purposes. Bacterial culture media may be either liquid or solid. A liquid, or broth, medium, in which organisms can move about freely, is useful for studying the growth characteristics of a single strain of a single species (that is, a **pure culture**). Liquid media are also convenient for examining growth kinetics and microbial biochemistry at different phases of growth. Solid media, usually gelled with agar, are useful for trying to separate mixtures of different organisms as they are found in the natural environment or in clinical specimens.

Dilution Streaking and Spread Plates

Solid media are basically liquid media to which a solidifying agent has been added. The most versatile and widely used solidifying agent is agar (for the development of agar medium, see Section 1.3). Derived from seaweed, agar forms an unusual gel that liquefies at 100°C but does not solidify again until cooled to about 40°C. Liquefied agar medium poured into shallow, covered petri dishes cools and hardens to provide a large, flat surface on which a mixture of microorganisms can be streaked to separate individual cells. Each cell will divide and grow to form a distinct, visible colony of cells (**Fig. 4.11**). As shown in **Figure 4.12** ▶, a drop of liquid culture is collected with an inoculating loop and streaked across the agar plate surface in a pattern called **dilution streaking**. Organisms fall off the loop as it moves along the agar surface. Toward the end of the streak

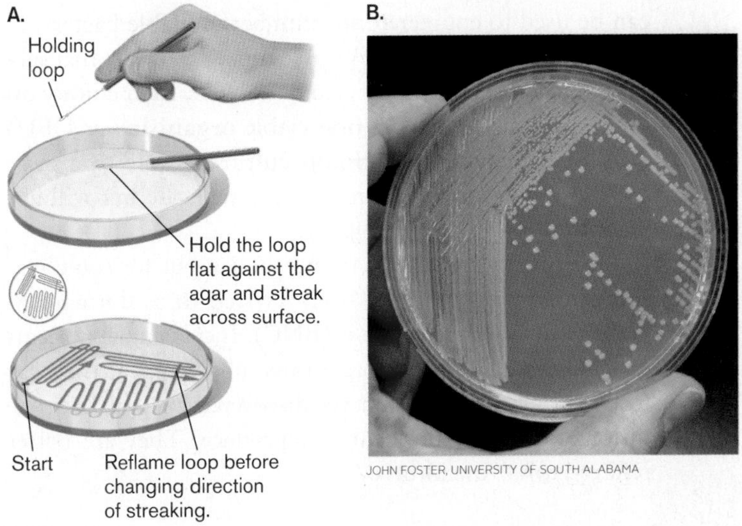
A.
Holding loop
Hold the loop flat against the agar and streak across surface.
Start
Reflame loop before changing direction of streaking.
B.
JOHN FOSTER, UNIVERSITY OF SOUTH ALABAMA

FIGURE 4.12 ■ Dilution streaking technique. A. A liquid culture is sampled with a sterile inoculating loop and streaked across the plate in three to four areas, with the loop flamed between areas to kill bacteria still clinging to the loop. Dragging the loop across the agar diminishes the number of organisms clinging to the loop until only single cells are deposited at a given location. **B.** *Salmonella enterica* culture obtained by dilution streaking. ▶

few bacteria remain on the loop, so at that point individual cells will land and stick to different places on the agar surface. If the medium, whether artificial (for example, laboratory medium) or natural (for example, crab carapace) contains the proper nutrients and growth factors, a single cell will multiply into many millions of offspring, forming a **microcolony**. At first visible only under a microscope, the microcolony grows into a visible droplet called a **colony** (**Figs. 4.12B** and **4.13**). A pure culture of the species or one strain of a species can be obtained by touching a single colony with a sterile inoculating loop and inserting that loop into fresh liquid medium.

It is important to note that the "one cell equals one colony" paradigm does not hold for all bacteria. Organisms such as *Streptococcus* or *Staphylococcus* usually do not exist as single cells, but occur as chains or clusters of several cells. Thus, a cluster of ten *Staphylococcus* cells will form only one colony on an agar medium and is called a colony-forming unit (CFU).

Another way to isolate pure colonies is the **spread plate** technique. Starting from a liquid culture of bacteria, a series of tenfold dilutions is made, and a small amount of each dilution is placed directly on the surface of separate agar plates (**Fig. 4.14**). The sample is spread over the surface of the plate with a heat-sterilized, bent glass rod. The early dilutions, those containing the most bacteria, will produce **confluent** growth that covers the entire agar surface. Later dilutions, containing fewer and fewer organisms, yield individual colonies. As we will see later, spread plates not only enable us to isolate pure cultures but also can be used to enumerate the number of **viable** bacteria in the original growth tube. A viable organism is one that successfully replicates to form a colony. **Thus, each colony on an agar plate represents one viable organism (or CFU) present in the original liquid culture.** (Note that there are many bacteria in nature that can replicate but will not form colonies on an agar plate.)

Organisms that appear to metabolize but for some reason cannot replicate have been referred to as dormant or **viable but nonculturable (VBNC)**. Increasingly, it seems probable that all VBNC organisms simply appear that way to us because we have not yet discovered the culture conditions necessary for them to reproduce. They are better referred to as "uncultured."

Complex versus Synthetic Media

Bacteria can be grown in nutrient-rich but poorly defined **complex media** (or "rich media") or in precisely defined **synthetic media**.

Recipes for complex media usually contain several poorly defined ingredients, such as yeast extract or beef extract, whose exact composition is unknown. These additives

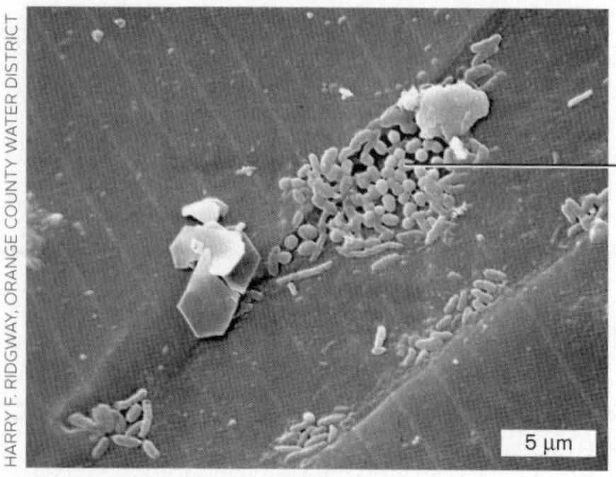
HARRY F. RIDGWAY, ORANGE COUNTY WATER DISTRICT
Microcolony
5 μm

FIGURE 4.13 ■ Bacterial colonies. A 3-day-old microcolony (biofilm) on the surface of a membrane used to treat municipal wastewater in Fountain Valley, California (SEM).

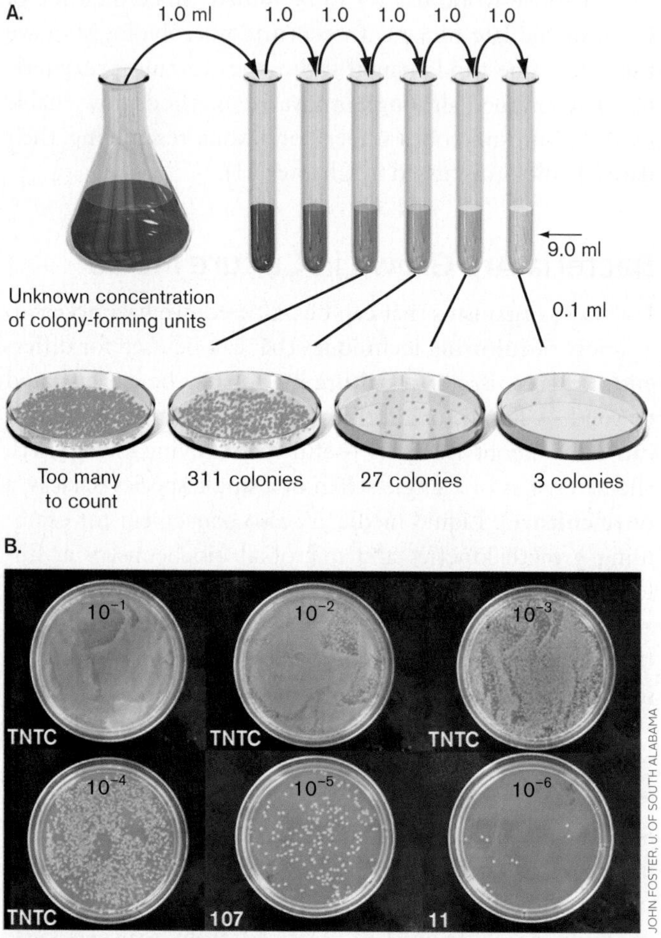
A.
1.0 ml
1.0
1.0
1.0
1.0
1.0
9.0 ml
0.1 ml
Unknown concentration of colony-forming units
Too many to count
311 colonies
27 colonies
3 colonies
B.
10^{-1}
10^{-2}
10^{-3}
TNTC
TNTC
TNTC
10^{-4}
10^{-5}
10^{-6}
TNTC
107
11
JOHN FOSTER, U. OF SOUTH ALABAMA

FIGURE 4.14 ■ Tenfold dilutions, plating, and viable counts. **A.** A culture containing an unknown concentration of cells is serially diluted. One milliliter (ml) of culture is added to 9.0 ml of diluent broth and mixed, and then 1 ml of this $\frac{1}{10}$ dilution is added to another 9.0 ml of diluent (10^{-2} dilution). These steps are repeated for further dilution, each of which lowers the cell number tenfold. After dilution, 0.1 ml of each dilution is spread onto an agar plate. **B.** Plates prepared as in (A) are incubated at 37°C to yield colonies. By multiplying the number of countable colonies (11 colonies on the 10^{-6} plate) by 10, you get the number of cells in 1.0 ml of the 10^{-6} dilution. Multiplying that number by the reciprocal of the dilution factor, you can calculate the number of cells (colony-forming units, or CFUs) per milliliter in the original broth tube ($11 \times 10^1 \times 10^6 = 1.1 \times 10^8$ CFUs/ml). TNTC = too numerous to count.

include a rich variety of amino acids, peptides, nucleosides, vitamins, and some sugars. Some organisms are particularly fastidious, requiring that components of blood be added to a basic complex medium. With these additions, the complex medium is called an **enriched medium**.

Complex media provide many of the chemical building blocks that a cell would otherwise have to synthesize on its own. For example, instead of making proteins that synthesize tryptophan, all the cell needs is a membrane transport system to harvest prefabricated tryptophan from the medium. Likewise, fastidious organisms that require blood in their media may reclaim the heme released from red blood cells as their own, using it as an "enzyme prosthetic group," a group critical to enzyme function (for example, the heme group in cytochromes). All of this saves the scavenging cell a tremendous amount of energy, and as a result, bacteria tend to grow fastest in complex media.

An argument can be made that complex media mimic the rich environments that pathogens encounter in animal hosts. However, the metabolism of a microbe growing in a complex medium is hard to characterize. How would you know whether *E. coli* possesses the ability to make tryptophan if the bacterium grew only in complex media? Questions like this can be investigated by studies of organisms that can be grown in fully defined synthetic media. In preparing a synthetic medium, we start with water and then add various salts, carbon, nitrogen, and energy sources in precise amounts. For self-reliant organisms like *E. coli* or *Bacillus subtilis,* that is all that's needed. Other organisms, such as *Shigella* species or mutant strains of *E. coli* or *B. subtilis,* require additional ingredients to satisfy requirements imposed by the absence of specific metabolic pathways.

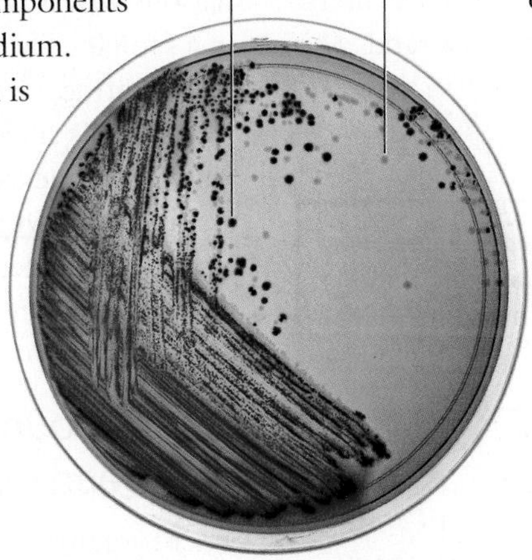
Lactose-fermenting (Lac⁺)
Nonfermenting (Lac⁻)
JOHN FOSTER, UNIVERSITY OF SOUTH ALABAMA

FIGURE 4.15 ■ MacConkey medium, a culture medium both selective and differential. Dilution streak of a mixture of Lac^+ and Lac^- bacteria. Only Gram-negative bacteria grow on lactose MacConkey (selective). Only a species capable of fermenting lactose produces pink colonies (differential), because only fermenters can take up the neutral red and peptones that are also in the medium. Gram-negative nonfermenters appear as uncolored colonies.

Thought Question

4.4 What would be the phenotype (growth characteristic) of a cell that lacks the *trp* genes (genes required for the synthesis of tryptophan)? What would be the phenotype of a cell missing the *lac* genes (genes whose products catabolize the carbohydrate lactose)?

Selective and Differential Media

Microorganisms are remarkably diverse with respect to their metabolic capabilities and resistance to certain toxic agents. These differences are exploited in **selective media**, which favor the growth of one organism over another, and in **differential media**, which expose biochemical differences between two species that grow equally well. For example, Gram-negative bacteria, with their outer membrane, are much more resistant than Gram-positive bacteria to detergents like bile salts and certain dyes, such as crystal violet. A solid medium containing bile salts and crystal violet is considered selective because it favors the growth of Gram-negative over Gram-positive bacteria.

On the other hand, a differential medium is needed to distinguish between organisms that differ not in the ability to grow, but in a particular biochemical aspect. For example, *E. coli* and *Salmonella enterica* are both Gram-negative, but only *E. coli* can ferment lactose. Both can be grown on solid media containing lactose, the dye called "neutral red," and a nonfermentable carbon source like peptone. *E. coli,* however, ferments the lactose and produces acidic end products. The lower pH surrounding an *E. coli* colony causes the dye to enter the cells, turning the colony red. In stark contrast, *S. enterica,* a major cause of diarrhea, will grow on nonfermentable peptides but cannot ferment lactose. Consequently, acidic end products are not produced by *S. enterica,* so the colonies remain white, their natural color. In this example, growth in differential media easily distinguishes colonies of lactose fermenters from nonfermenters.

Several media used in clinical microbiology are both selective and differential. **MacConkey medium**, for example, selects for growth of Gram-negative bacteria because it contains bile salts and crystal violet, which prevent the growth of Gram-positives. The medium also includes lactose, neutral red, and peptones to differentiate lactose fermenters from nonfermenters. Thus, a culture grown in MacConkey medium will consist of Gram-negative organisms that can be identified as lactose fermenters (red colonies) or nonfermenters (uncolored colonies) (**Fig. 4.15**).

Thought Question

4.5 If lactose was left out of MacConkey medium (**Fig. 4.15**), would lactose-fermenting *E. coli* bacteria grow, and if so, what color would their colonies be?

This medium is of particular benefit in diagnosing the etiology (cause) of diarrheal disease because most normal flora that grow on this medium are lactose fermenters, whereas two important pathogens, *Salmonella* and *Shigella,* are lactose nonfermenters.

Thought Question

4.6 The addition of sheep's blood to agar produces a very rich medium called blood agar. Do you think blood agar is a selective medium? A differential medium? *Hint:* Some bacteria can lyse red blood cells.

To Summarize

- **Microbes in nature** usually exist in complex, multispecies communities, but for detailed studies they must be grown separately in pure culture.
- **Bacteria can be cultured** on solid or liquid media.
- **Minimal defined media** contain only those nutrients essential for growth.
- **Complex**, or **rich, media** contain many nutrients. Other media exploit specific differences between organisms and can be defined as **selective**, **differential**, or both (for example, **MacConkey medium**).

4.4 Counting Bacteria

How do we determine if a lake is contaminated with fecal bacteria or whether our peanut butter contains *Salmonella*? We have to identify the organism present, of course, but then we have to count how many there are to understand the extent of contamination. In fact, there are many other reasons why it is important to know the number of bacteria in a sample. These will become obvious as you read further.

Counting or quantifying organisms invisible to the naked eye is surprisingly difficult, because each of the available techniques measures a different physical or biochemical aspect of growth. Thus, a cell density value (given as cells per milliliter) derived from one technique will not necessarily agree with the value obtained by a different method.

Direct Counting of Living and Dead Cells

Microorganisms can be counted directly using a microscope. A dilution of a bacterial culture is placed on a special microscope slide called a "hemocytometer" (or, more specifically for bacteria, a "Petroff-Hausser counting chamber") (**Fig. 4.16**). Etched on the surface of the slide is a grid of precise dimensions, and placing a coverslip over the grid forms a space of precise volume. The number of organisms counted within that volume is used to calculate the concentration of cells in the original culture.

Thought Question

4.7 Use the information in **Figure 4.16** to determine the concentration (in cells per milliliter) of bacteria shown.

However, "seeing" an organism under the microscope does not mean that the organism is alive, because living and dead cells are indistinguishable by this basic approach. Living cells may be distinguished from dead cells by fluorescence microscopy using fluorescent chemical dyes, as discussed in Chapter 2. For example, propidium iodide, a red

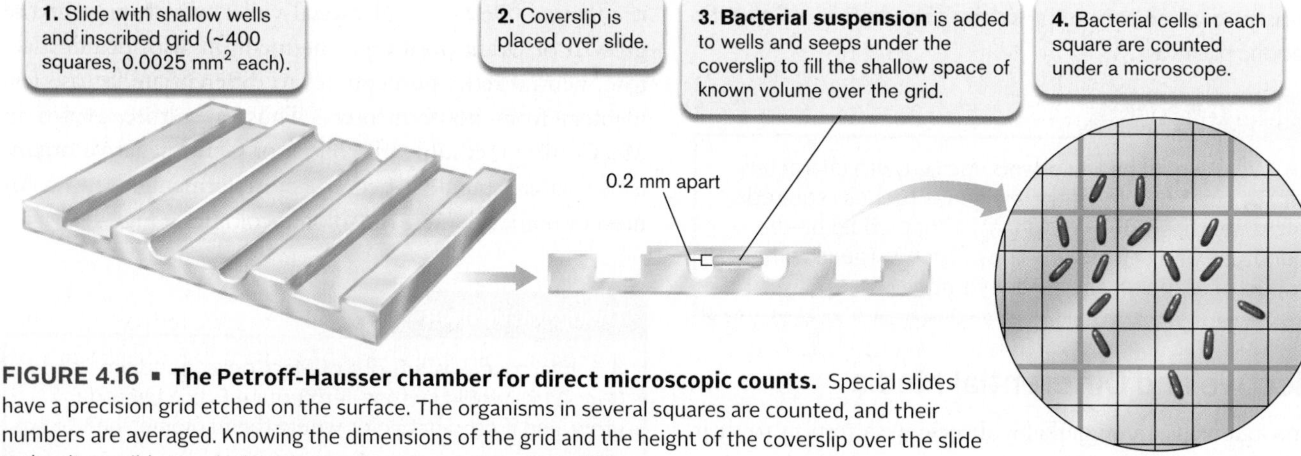
1. Slide with shallow wells and inscribed grid (~400 squares, 0.0025 mm² each).
2. Coverslip is placed over slide.
3. Bacterial suspension is added to wells and seeps under the coverslip to fill the shallow space of known volume over the grid.
4. Bacterial cells in each square are counted under a microscope.
0.2 mm apart

FIGURE 4.16 ■ The Petroff-Hausser chamber for direct microscopic counts. Special slides have a precision grid etched on the surface. The organisms in several squares are counted, and their numbers are averaged. Knowing the dimensions of the grid and the height of the coverslip over the slide makes it possible to calculate the number of organisms in a milliliter.

dye, intercalates between DNA bases but cannot freely penetrate the energized membranes of living cells. Thus, only dead cells stain red under a fluorescence scope. Another dye, Syto-9, enters both living and dead cells, staining them both green. By combining Syto-9 with propidium iodide, living and dead cells can be distinguished: Living cells stain green, whereas dead cells appear orange or yellow because both dyes enter and Syto-9 (green) plus propidium (red) appears yellow (**Fig. 4.17**).

Direct counting without microscopy can be achieved using an electronic technique that not only counts but also separates populations of bacterial cells based on their distinguishing properties. The instrument is called a **fluorescence-activated cell sorter (FACS)**. In the FACS technique, bacterial cells that synthesize a fluorescent protein (such as cyan fluorescent protein; see Section 3.5), or that have been labeled with a fluorescent antibody or chemical, are passed through a small orifice, as in the Coulter counter, and then past a laser (**Fig. 4.18A**). Detectors measure light scatter in the forward direction (an indicator of particle size), and to the side (which indicates shape or granularity). In addition, the laser activates the fluorophore in the fluorescent antibody, and a detector measures fluorescence intensity.

Within a single culture, the FACS technique enables us to use cell size and the level of fluorescence (since one subpopulation may fluoresce more than or less than another) to identify and count different populations of cells. For example, by placing the green fluorescent protein gene (*gfp*) under the control of the regulatory DNA sequences of a bacterial gene (making a gene fusion), researchers can count the cells expressing that gene by using FACS analysis; this analysis, in turn, makes it possible to determine what conditions allow expression of the gene and whether all cells in the population express that gene at the same time and to the same extent (**Fig. 4.18B**).

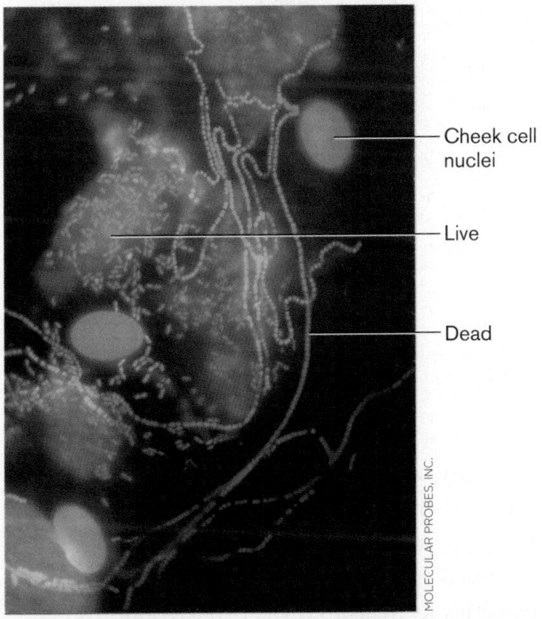
Cheek cell nuclei
Live
Dead
MOLECULAR PROBES, INC.

FIGURE 4.17 ▪ Live/dead stain. Live and dead bacteria visualized on freshly isolated human cheek epithelial cells using the LIVE/DEAD *Bac*Light Bacterial Viability Kit. Dead bacterial cells fluoresce orange or yellow because propidium (red) can enter the cells and intercalate the base pairs of DNA. Live cells fluoresce green because Syto-9 (green) enters the cell. The faint green smears are the outlines of cheek cells.

Viable Counts

Viable cells, as noted previously, are those that can replicate and form colonies on a plate. To obtain a viable cell count, dilutions of a liquid culture can be plated directly on an agar surface (as in **Fig. 4.14**) or added to liquid agar cooled to about 42°C–45°C. The agar is subsequently poured into an empty petri plate (this is called the **pour plate technique**), where the agar cools further and solidifies. Because many bacteria resist short exposures to that temperature, individual cells retain the ability to form colonies on, and in, the pour plate. After colonies form, they are counted, and the original cell number is calculated. For example, if 100 colonies are observed in a pour plate made with 100 microliters (μl) of a 10^{-3} dilution of a culture, then there were 10^6 organisms per milliliter in the original culture.

Although viable counts are widely used in research, there are problems using this method to measure cell number and determine cultural characteristics. One issue is that colony counting does not reflect cell size or growth stage. Even more problematic, colony counts usually underestimate the number of living cells in a culture. As noted earlier, metabolically active cells that do not form colonies on agar plates will typically not be counted as alive. Cells damaged for one reason or another, while still alive, may be too compromised to divide. Comparing a viable count with a direct count obtained from a live/dead stain can expose the presence of damaged cells. Organisms that grow in chains, such as *Streptococcus*, pose another problem because each colony originates from a group of cells, not a single cell. Consequently, actual cell numbers will be underestimated and are reported as colony-forming units (CFUs) rather than cells.

Biochemical Assays

In contrast to methods that visualize individual cells, assays of cell biomass, protein content, or metabolic rate measure the overall size of a population of cells. The most straightforward but time-consuming biochemical approach to monitoring population growth is to measure the dry

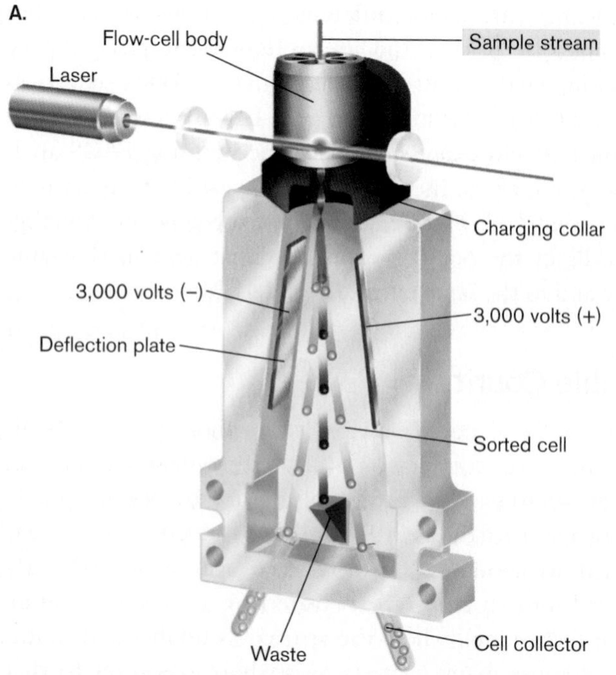
A.
Flow-cell body
Sample stream
Laser
Charging collar
3,000 volts (–)
3,000 volts (+)
Deflection plate
Sorted cell
Waste
Cell collector

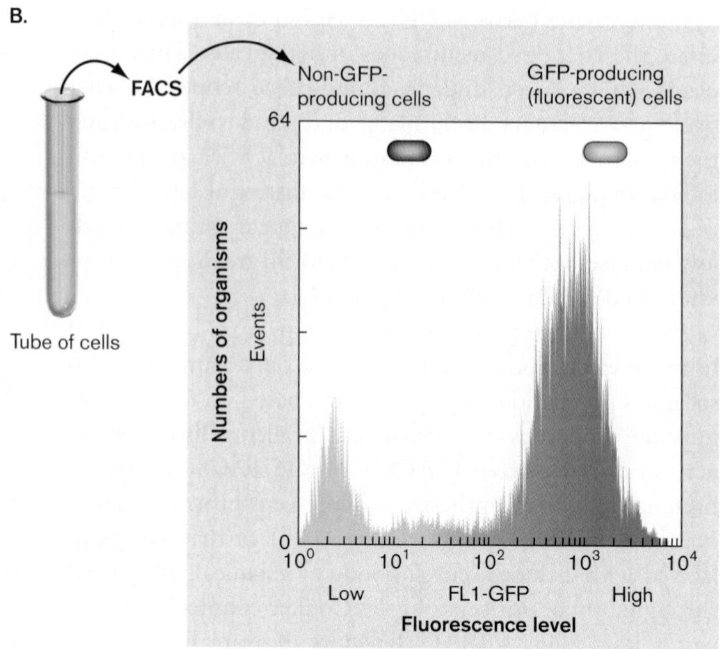
B.
FACS
Tube of cells
Non-GFP-producing cells
GFP-producing (fluorescent) cells
64
0
Numbers of organisms
Events
10^0
10^1
10^2
10^3
10^4
Low
FL1-GFP
High
Fluorescence level

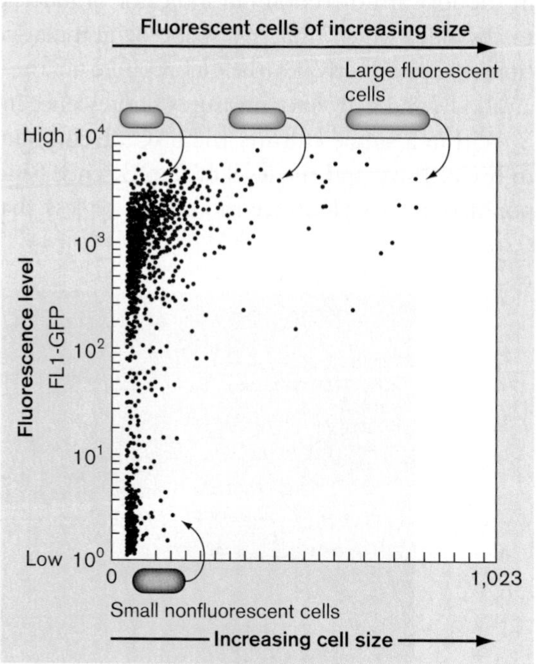
Fluorescent cells of increasing size
Large fluorescent cells
High 10^4
10^3
10^2
10^1
Low 10^0
Fluorescence level
FL1-GFP
0
1,023
Small nonfluorescent cells
Increasing cell size

FIGURE 4.18 ▪ Fluorescence-activated cell sorter (FACS). **A.** Schematic diagram of a FACS apparatus (bidirectional sorting). **B.** Separation of GFP-producing *E. coli* from non-GFP-producing *E. coli*. The low-level fluorescence in the cells on the left is baseline fluorescence (autofluorescence). The scatterplot displays the same FACS data, showing the size distribution of cells (*x*-axis) with respect to the level of fluorescence (*y*-axis). The larger cells may be cells that are about to divide.

weight of a culture. Cells are collected by centrifugation, washed, dried in an oven, and weighed. Because bacterial cells weigh very little, a large volume of culture must be harvested to obtain measurements, making this technique quite insensitive. A more accurate alternative is to measure increases in protein levels, which correlate with increases in cell number. Protein levels are more easily measured with relatively sensitive assays.

Optical Density: Growth in Real Time

The phenomenon of light scattering was introduced in Chapter 2. Recall that individual bacteria could not be resolved from the detection of scattered light by the unaided human eye, but the presence of numerous bacteria in a tube of medium could be detected by its cloudy appearance. The decrease in intensity of a light beam due to the scattering of light by a suspension of particles is measured as **optical density**.

The optical density of light scattered by bacteria is a very useful tool for estimating population size. The method is quick and easy, but because light scattering is a complex function of cell number and cell volume, optical density provides only an approximate result that must be corrected using a standard curve. Typically, a standard curve is obtained that plots viable counts for cultures versus optical densities as measured by a spectrophotometer. Thereafter, the optical density of a growing culture is measured, and the standard curve is used to estimate cell number at any given point during growth.

One problem with using optical density to estimate cell numbers is that a cell's volume can vary depending on its growth stage, altering its light-scattering properties. Thus, the cell number estimated by the standard curve may deviate from the true number. Another problem is that dead cells also scatter light. Clearly, using optical density to estimate viable count can be misleading, especially when measuring populations in stationary phase.

To Summarize

- **Microorganisms in culture may be counted directly** under a microscope, with or without staining, or by use of a fluorescence-activated cell sorter (FACS).
- **Microorganisms can be counted indirectly**, as in viable counts and measurement of dry weight, protein levels, or optical density.
- **A viable bacterial organism** is defined as being capable of replicating. Viable counts can be determined for bacteria able to form colonies on a solid-medium surface.

4.5

The Growth Cycle

How do microbes grow? What determines their rate of growth? And when does growth restart in a nongrowing population? In nature, the answers to these questions are extremely complex, because most microbes exist in complex, mixed communities fixed to solid surfaces. Yet these same fixed-location multicellular communities can send off **planktonic cells**, free-living organisms that grow and multiply on their own.

Nevertheless, all species at one time or another exhibit both rapid growth and nongrowth, as well as many phases in between. For clarity, we present here the principles of rapid growth, while bearing in mind the actual diversity of growth situations in nature. We care about growth because how rapidly a microbe grows influences how fast a pathogen causes disease, or how quickly contaminated food spoils. Similarly, a bacterial species that consumes oil is useful for cleaning up oil spills only if the organism can grow rapidly.

Survival of any species ultimately depends on its ability to make more of its own kind. A typical bacterium that can be cultured in the laboratory grows by increasing in length and mass, which facilitates expansion of its nucleoid as its DNA replicates (see Section 3.5). As DNA replication nears completion, the cell, in response to complex genetic signals, begins to synthesize an equatorial septum that ultimately separates the two daughter cells. In this overall process, called **binary fission**, one parental cell splits into two equal daughter cells (**Fig. 4.19A**).

Although a majority of culturable bacteria divide symmetrically in two equal halves, some species divide asymmetrically. For example, the bacterium *Caulobacter* forms a stalked cell that remains fixed to a solid surface but reproduces by budding from one end to produce small, unstalked motile cells. The marine organism *Hyphomicrobium* also replicates asymmetrically by budding, releasing a smaller cell from a stalked parent (**Fig. 4.19B**).

Eukaryotic microbes divide by a special form of cell fission involving mitosis, the segregation of pairs of chromosomes within the nucleus (see Section 20.2 and Appendix 2). Some eukaryotes also undergo more complex life cycles involving budding and diverse morphological forms. Nevertheless, many eukaryotic microbes and multicellular organisms—we ourselves among them—exhibit the same growth patterns seen in bacteria.

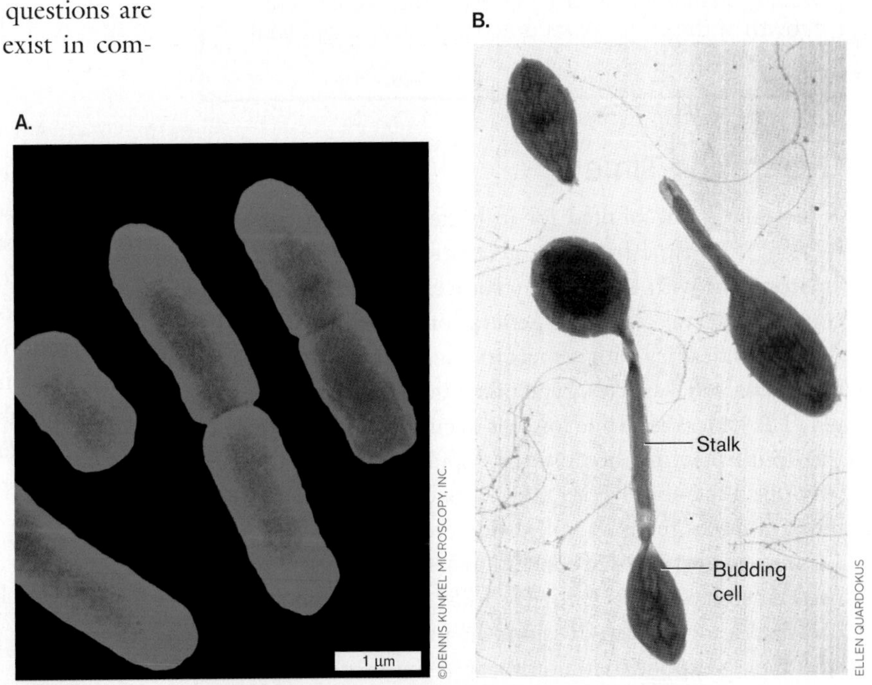
A.
B.
1 µm
Stalk
Budding cell
©DENNIS KUNKEL MICROSCOPY, INC.
ELLEN QUARDOKUS

FIGURE 4.19 ■ Symmetrical and asymmetrical cell division. A. Symmetrical cell division, or binary fission, in *Lactobacillus* sp. (SEM). **B.** Asymmetrical cell division via budding in the marine bacterium *Hyphomicrobium* (approx. 4 µm long).

Exponential Growth

The process of reproduction has implications for growth not only of the individual, but also of populations. If we assume that growth is unbounded, what happens to the population? The unlimited growth of any population obeys a simple law: The **growth rate**, or rate of increase in cell numbers or biomass, is proportional to the population size at a given time. Such a growth rate is called "exponential" because it generates an exponential curve, a curve whose slope increases continually.

How does binary fission of cells generate an exponential curve? If each cell produces two cells per generation, then the population size at any given time is proportional to 2^n, where the exponent n represents the number of generations (that is, cell divisions in which offspring replace parents) that have taken place between two time points. Thus, cell number rises exponentially. Many microbes, however, have replication cycles based on numbers other than 2. For example, some cyanobacteria form cell aggregates that divide by multiple fission, releasing dozens of daughter cells. The cyanobacterium enlarges without dividing, and then suddenly divides many times without separating. The cell mass breaks open to release hundreds of progeny cells.

Thought Question

4.8 A virus such as influenza virus might produce 800 progeny virus particles from one infected host cell. How would you mathematically represent the exponential growth of the virus? What practical factors might limit such growth?

Generation Time

A variable not accounted for in bacterial growth curves is the length of time from one generation to the next. In an environment with unlimited resources, bacteria divide at a constant interval called the **generation time**. The length of that interval varies with respect to many parameters, including the bacterial species, type of medium, temperature, and pH. The generation time for cells in culture is also known as the **doubling time**, because the population of cells doubles over one generation. For example, one cell of *E. coli* placed into a complex medium will divide every 20 minutes. After 1 hour of growth (three generations), that one cell will have become eight (1 to 2, 2 to 4, 4 to 8). Because cell number (N) doubles with each division, the increase in cell number over time is exponential, not linear. A linear increase would occur if cell number rose by a fixed amount after every generation (for example, 1 to 2, 2 to 3, 3 to 4).

Why do we care about generation time? Generation time is important when we're trying to understand how rapidly a pathogen can cause disease symptoms. A fast-growing species will cause disease rapidly, within days, while a slower-growing organism may take weeks or months. In biotechnology, how fast a producing organism grows will affect how quickly a commercially useful by-product can be made. Let's review how we calculate generation time.

Starting with any number of organisms (N_0), the number of organisms after n generations will be $N_0 \times 2^n$. For example, a single cell after three generations ($n = 3$) will produce

$$1 \text{ cell} \times 2^3 = 8 \text{ cells}$$

The number of generations that an exponential culture undergoes in a given time period can be calculated if the number of cells at the start of the period (N_0) and the number of cells at the end of the period (N_t) are known. Methods such as viable counts are used to make those determinations. To simplify calculations, instead of using base 2 logarithms, the $\log_2$ expressions are converted to base 10 through division by a factor of $\log_{10} 2$, which is approximately 0.301. Thus,

$$N_t = \mathrm{N_0} \times 2^{\mathrm{n}}$$

can be expressed as

$$\log N_t = \log N_0 + n \log 2$$

and when converted to log 10 becomes

$$\log_{10} N_t = \log_{10} N_0 + n \log_{10} 2$$

$$n \log_{10} 2 = \log_{10} N_t - \log_{10} N_0$$

$$n = (\log_{10} N_t - \log_{10} N_0) \div \log_{10} 2$$

$$n = \log_{10} (N_t/N_0) \div \log_{10} 2$$

$$n = \log_{10} (N_t/N_0) \div 0.301$$

In practice, exponential growth occurs only for a short period when all nutrients are in full supply and the concentration of waste products has not become a limiting factor.

The rate of exponential growth can be expressed as the mean **growth rate constant** (k), which is the number of generations (n) per unit time (usually generations per hour). Knowing the number of organisms at time zero (N_0) and the number of organisms after incubation time t (N_t), we can calculate k as follows:

$$k = n/t = (\log 2\ N_t - \log 2\ N_0)/t$$

$$k = \log 2\ (N_t / N_0)/t$$

or when simplified using log10 conversion:

$$k = n/t = (\log_{10} N_t - \log_{10} N_0) \div 0.301t$$

$$k = \log_{10} (N_t/N_0)/0.301t$$

Once k is known, calculating generation time is simple. For instance, a culture with a mean growth rate constant (k) of 3 generations/hour has a mean generation time of

20 minutes (60 minutes/3 generations). If $k = 0.5$ generation per hour, on the other hand, then the mean generation time is 2 hours (60 minutes/0.5 generation). In other words, the **mean generation time** (g) in hours is the reciprocal of the mean growth rate constant (k):

$$g = 1/k$$

The growth rate constant can also be calculated from the slope of $\log_{10} N$ over time, where N is a relative measure of culture density, such as the optical density measured in a spectrophotometer (**Fig. 4.20A**). The units of N do not matter, because we are always looking at ratios of cell numbers relative to an earlier level (N_1/N_0). For example, we can use the following series of optical density (OD) measurements to determine N:

Time (min)	OD_{600}	$\log_{10} OD_{600}$
0	0.05	–1.30
15	0.08	–1.10
30	0.13	–0.89
45	0.20	–0.69
60	0.33	–0.48

If we plot $\log_{10} OD_{600}$ versus time, we obtain a line with a slope of 0.0136 per minute. The growth rate constant, in doublings per hour, becomes

$$k = (0.0136/\text{min})(60\ \text{min/h}) \div 0.301$$
$$= 2.7 \text{ generations per hour}$$

The steeper the slope, the faster the organisms are dividing.

Thought Questions

4.9 Suppose you ingest 20 cells of *Salmonella enterica* in a peanut butter cookie. They all survive the stomach and enter the intestine. Suppose further that the sum total of subsequent bacterial replication and death (caused by the host) produces an average generation time of 2 hours and you will feel sick when there are 1,000,000 bacteria. How much time will elapse before you feel sick?

4.10 Suppose one cell of the nitrogen fixer *Sinorhizobium meliloti* colonizes a plant root. After 5 days (120 hours), there are 10,000 bacteria fixing N_2 within the plant cells. What is the bacterial doubling time?

The mathematics of exponential growth is relatively straightforward, but remember that microbes grow differently in pure culture (very rare in nature) than they do in mixed communities, where neighboring cells produce all kinds of substances that may feed or poison other microbes. In mixed communities, the microbes may grow planktonically (floating in liquid), as in the open ocean, or as a biofilm on solid matter suspended in that ocean. In each instance, the mathematics of exponential growth applies, at least until the community reaches a density at which different species begin to compete.

Thought Question

4.11 It takes 40 minutes for a typical *E. coli* cell to completely replicate its chromosome and about 20 minutes to prepare for another round of replication. Yet the organism enjoys a 20-minute generation time growing at 37°C in complex medium. How is this possible? *Hint:* How might the cell overlap the two processes?

Stages of Growth in Batch Culture

Exponential growth never lasts indefinitely, because nutrient consumption and toxic by-products eventually slow the growth rate until it halts altogether. The simplest way to model the effects of changing conditions is to culture bacteria in liquid medium within a closed system, such as a flask or test tube. This is called **batch culture**. In batch culture, no fresh medium is added during incubation; thus, nutrient concentrations decline and waste products accumulate during growth.

The changing conditions of a batch culture profoundly affect bacterial physiology and growth, and illustrate the remarkable ability of bacteria to adapt to their environment. As medium conditions deteriorate, alterations occur in membrane composition, cell size, and metabolic pathways, all of which impact generation time. Microbes possess intricate, self-preserving genetic and metabolic mechanisms that slow growth before their cells lose viability. Because many bacteria replicate by binary fission, plotting culture growth (as represented by the logarithm of the cell number) versus incubation time makes it possible to see the effect of changing conditions on generation time and reveals the stages of growth shown in **Figure 4.20B**—namely, lag phase, log phase, stationary phase, and death phase.

Lag phase. Cells transferred from an old culture to fresh growth media need time to detect their environment, express specific genes, and synthesize components needed to institute rapid growth. As a result, bacteria inoculated into fresh media typically experience a lag period, or **lag phase**, during which cells do not divide. Several factors influence lag phases. Cells taken from an aged culture may be damaged and require time for repair. Carbon, nitrogen, or energy sources different from those originally used by the seed culture must be sensed, and the appropriate enzyme systems must be synthesized. The length of lag phase varies, depending on the age of the culture, changes in temperature, and the differences between the new and old media (such as changes in nutrient levels, pH, and salt concentrations). For

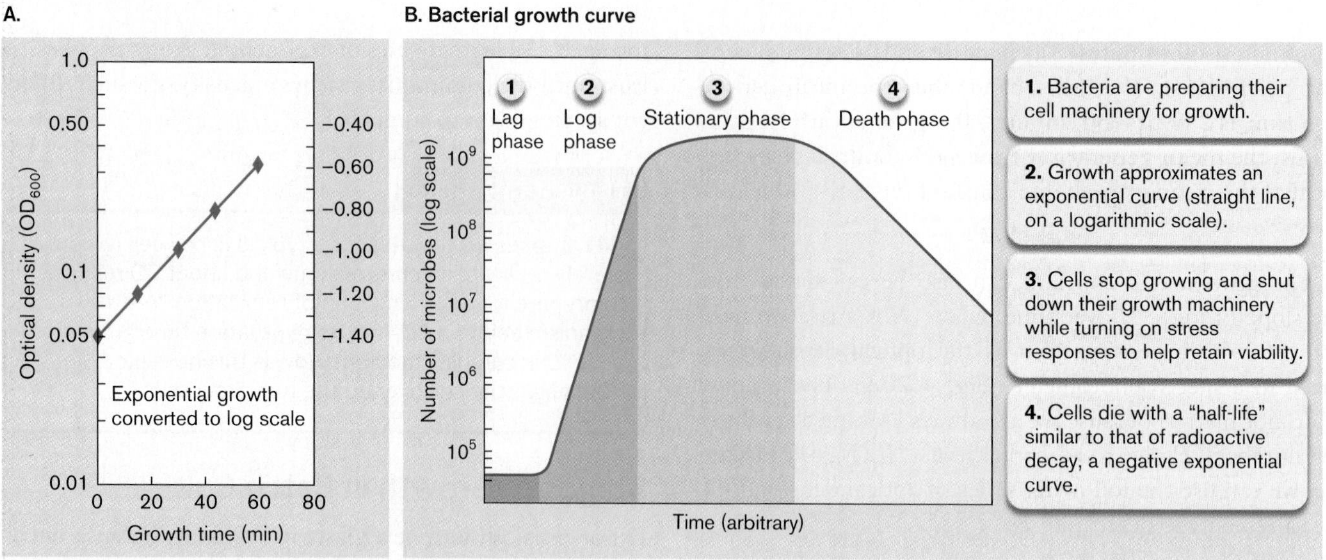
A.
B. Bacterial growth curve
1.0
0.50
0.1
0.05
0.01
-0.40
-0.60
-0.80
-1.00
-1.20
-1.40
Optical density (OD600)
Exponential growth
converted to log scale
0
20
40
60
80
Growth time (min)
1
Lag
phase
2
Log
phase
3
Stationary phase
4
Death phase
10⁹
10⁸
10⁷
10⁶
10⁵
Number of microbes (log scale)
Time (arbitrary)
1. Bacteria are preparing their cell machinery for growth.
2. Growth approximates an exponential curve (straight line, on a logarithmic scale).
3. Cells stop growing and shut down their growth machinery while turning on stress responses to help retain viability.
4. Cells die with a "half-life" similar to that of radioactive decay, a negative exponential curve.

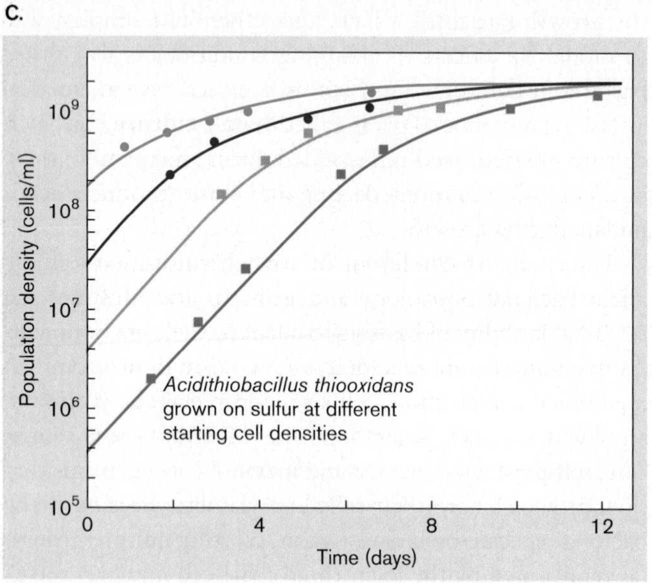
C.
10⁹
10⁸
10⁷
10⁶
10⁵
Population density (cells/ml)
Acidithiobacillus thiooxidans
grown on sulfur at different
starting cell densities
0
4
8
12
Time (days)

FIGURE 4.20 ■ Bacterial growth curves. A. Theoretical growth curve of a bacterial suspension measured by optical density (OD) at a wavelength of 600 nm. **B.** Phases of bacterial growth in a typical batch culture. **C.** Published growth curves of *Acidithiobacillus thiooxidans,* an acidophile that oxidizes sulfur to sulfuric acid. Whatever the starting cell density, the culture grows exponentially until it runs out of sulfur, at which point it enters the stationary phase. *Source:* C. Yasuhiro Konishi et al. 1995. *Appl. Environ. Biol.* **61**:3617.

example, transferring cells from one complex medium to a fresh complex medium results in a very short lag phase, whereas cells grown in a complex medium and then plunged into a minimal defined medium experience a protracted lag phase, during which time they readjust to synthesize all the amino acids, nucleotides, and other metabolites originally supplied by the complex medium.

Thought Question

4.12 **Figure 4.20C** shows growth curves for different population densities of *Acidithiobacillus thiooxidans* when the concentration of sulfur in the medium is constant. Draw the growth curves you would expect to see if the initial population density was constant but the concentration of sulfur varied.

Early log, or exponential, phase. Once cells have retooled their physiology to accommodate the new environment, they begin to grow exponentially and enter what is called **exponential**, or **logarithmic (log), phase**. Exponential growth is balanced growth, in which all cell components are synthesized at constant rates relative to each other. At this stage, represented by the linear part of the growth curve, cells are growing and dividing at the maximum rate possible based on the medium and growth conditions provided (such as temperature, pH, and osmolarity). Cells are largest at this stage of growth. If cell division were synchronized and all cells divided at the same time, the growth curve during this period would appear as a series of steps with cell numbers doubling instantly after every generation time. But batch cultures are not synchronous. Every cell has an equal generation time, but each cell divides at a slightly different moment, making the cell number rise smoothly.

Cells enjoying balanced, exponential growth are temporarily thrown into metabolic chaos (unbalanced growth) when their medium is abruptly changed. Nutritional downshift (moving cells from a good carbon source such as glucose to a poorer carbon source such as succinate) or nutritional upshift (moving cells to a better carbon source) casts cells into unbalanced growth. Downshifting to a carbon source with a lower energy yield means not only that a different set of enzymes must be made and employed to use the carbon source, but also that the previous high rate of macromolecular synthesis (such as ribosome synthesis) used to support a fast generation time is now too rapid relative to the lower energy yield. Failure to adjust will lead to increased mistakes in RNA, protein, and DNA synthesis, depletion of key energy stores, and ultimately death.

How do microbes adjust to nutrient shifts? During downshifts, energy levels (ATP) drop faster than macromolecular synthesis systems can stop making macromolecules. Molecular fail-safe systems quickly slow rates of macromolecular synthesis until DNA, RNA, ribosome, and cell wall synthesis come into balance with the rest of metabolism. In contrast, nutritional upshift means that the cell will start making more energy than it needs at its current growth rate, in which case cell metabolism will again be out of balance with cell division. The same fail-safe mechanisms used during downshift will, during upshift, kick macromolecular synthesis into a higher gear, once again establishing balanced growth. Nutritional upshift causes cells to reenter log phase, but with a shorter generation time.

Late log phase. As cell density (number of cells per milliliter) rises during log phase, the rate of doubling eventually slows, and a new set of growth phase–dependent genes is expressed. At this point, some species can also begin to detect the presence of others by sending and receiving chemical signals in a process known as quorum sensing (discussed in Chapter 10).

Stationary phase. Eventually, cell numbers stop rising, owing to the lack of a key nutrient or the buildup of waste products. The rate of cell division then equals the rate of cell death. Stationary phase conditions occur for bacteria grown in a complex medium when cell density rises above 10^9 cells per milliliter, but they can occur at lower cell densities if nutrients are limiting. At this point, the growth curve levels off and the culture enters what is called **stationary phase**. In contrast to bacteria, eukaryotic microorganisms, such as protozoa, enter stationary phase at much lower cell numbers, usually around 10^6 organisms per milliliter. Eukaryotic microbes reach stationary phase sooner than bacteria simply because they are bigger. Bigger cells use more nutrients and run out of them sooner.

If they did not change their physiology, microbes would be very vulnerable once entering stationary phase. Cells in stationary phase are not as metabolically nimble as cells in exponential phase, so damage from oxygen radicals and the toxic by-products of metabolism readily kill them. As an avoidance strategy, some bacteria differentiate into very resistant spores in response to nutrient depletion (see Section 4.7), while other bacteria undergo less dramatic but very effective molecular reprogramming. The microbial model organism *E. coli,* for example, adjusts to stationary phase by decreasing its size, thus minimizing the volume of its cytoplasm compared to the volume of its nucleoid. Fewer nutrients are then required to sustain the smaller cell. New stress resistance enzymes are also synthesized to handle oxygen radicals, protect DNA and proteins, and increase cell wall strength through increased peptidoglycan cross-linking. As a result, *E. coli* cells in stationary phase become more resistant to heat, osmotic pressure, pH changes, and other stresses that they might encounter while waiting for a new supply of nutrients.

Death phase. Without reprieve in the form of new nutrients, cells in stationary phase will eventually succumb to toxic chemicals present in the environment. Like the growth rate, the **death rate**—the rate at which cells die—is logarithmic. The death rate, however, is a negative exponential function. Recall that the increase in cell number during exponential (log) phase is a positive exponential function of time. In **death phase**, the number of cells that die in a given time period is proportional to the number that existed at the beginning of the time period. Determining microbial death rates is critical to the study of food preservation and to antibiotic development (further discussed in Chapters 5, 16, and 27). Although death curves are basically logarithmic, exact death rates are difficult to define because mutations arise that promote survival, and some cells grow by cannibalizing others. Consequently, the death phase is extremely prolonged. In fact, a portion of the cells will often survive for months, years, or even decades.

Examples of growth curves. In **Figure 4.20C** we see an actual bacterial growth curve from a study of *Acidithiobacillus thiooxidans,* an organism that oxidizes sulfur to sulfuric acid and grows below pH 1. Growth curves are shown for several different starting concentrations of bacteria. In each case, the bacteria grow exponentially for several days, until they have exhausted the sulfur in their growth medium. Their growth then slows until they enter stationary phase.

Note that the nice, smooth exponential phase shown in **Figure 4.20B** does not always hold for microbes growing in natural environments or in complex laboratory media containing multiple carbon and energy sources. Some bacterial species produce odd-looking growth curves in complex media as the population depletes one carbon source and must switch physiology to use another. Even *E. coli* doesn't really experience a uniform log-phase metabolism growing in complex medium; instead, it smoothly transitions through a series of metabolic states.

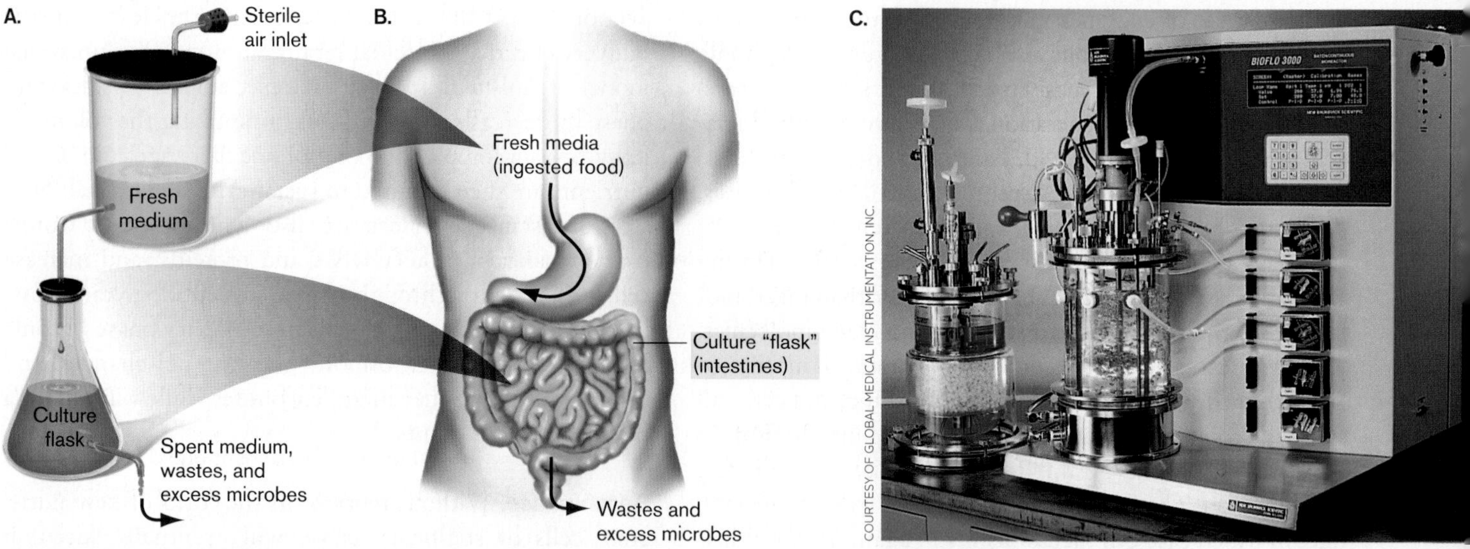
A.
Sterile
air inlet
Fresh
medium
Culture
flask
Spent medium,
wastes, and
excess microbes
B.
Fresh media
(ingested food)
Culture "flask"
(intestines)
Wastes and
excess microbes
C.
COURTESY OF GLOBAL MEDICAL INSTRUMENTATION, INC.

FIGURE 4.21 ■ Chemostats and continuous culture. **A.** The basic chemostat ensures logarithmic growth by constantly adding and removing equal amounts of culture media. **B.** The human gastrointestinal tract is engineered much like a chemostat, in that new nutrients are always arriving from the throat while equal amounts of bacterial culture exit in fecal waste. **C.** A modern chemostat.

Thought Questions

4.13 What can happen to the growth curve when a culture medium contains two carbon sources, if one is a preferred carbon source of growth-limiting concentration and the second is a nonpreferred source?

4.14 How would you modify the equations describing microbial growth rate to describe the rate of death?

4.15 Why are cells in log phase larger than cells in stationary phase?

Continuous Culture

In the classic growth curve that develops in closed systems, the exponential phase spans only a few generations. In open systems, however, where fresh medium is continuously added to a culture and an equal amount of culture is constantly siphoned off, bacterial populations can be maintained in exponential phase at a constant cell mass for extended periods of time. In this type of growth pattern, known as **continuous culture**, all cells in a population achieve a steady state, which permits a detailed analysis of microbial physiology at different growth rates. The **chemostat** is a continuous culture system in which the diluting medium contains a limiting amount of an essential nutrient (**Fig. 4.21**). Increasing the flow rate increases the amount of nutrient available to the microbe. The more nutrient available, the faster a cell's mass will increase. Because cell division is triggered at a defined cell mass, it follows that the growth rate in a chemostat is directly related to the dilution rate, or flow rate (milliliters per hour divided by the vessel volume). The more nutrient a culture receives as a result of increasing flow rate, the faster those cells can replicate (that is, the shorter the generation time).

The complex relationships among dilution rate, cell mass, and generation time in a chemostat are illustrated in **Figure 4.22**. The curves in this figure represent a typical experimental result, which can vary with organism,

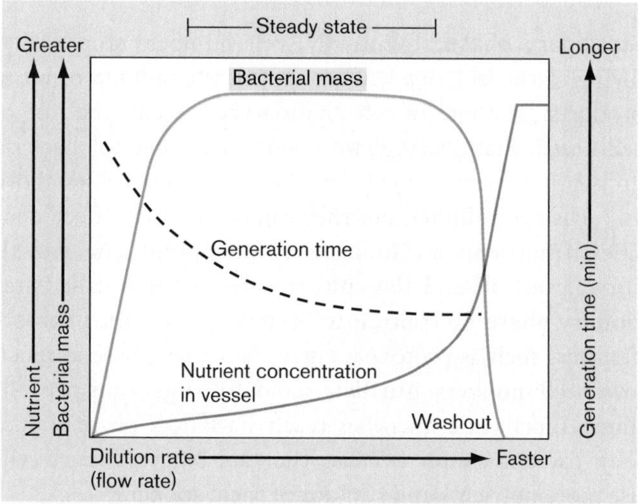
Steady state
Greater
Longer
Bacterial mass
Generation time
Nutrient concentration
in vessel
Washout
Nutrient
Bacterial mass
Generation time (min)
Dilution rate
(flow rate)
Faster

FIGURE 4.22 ■ Relationships between dilution rate, cell mass, and generation time. As the dilution rate increases in a chemostat (meaning, as more nutrient is fed to the culture), the generation time decreases (the cells divide more quickly) and the cell mass of the culture increases. This pattern continues until the rate of dilution exceeds the division rate, at which point cells are washed from the vessel faster than they can be replaced by division and the cell mass decreases. The *y*-axis varies depending on the curve, as labeled.

limiting nutrient, temperature, and so on. Depending on the experimental conditions, the shapes of the curves may change, but the general relationships will remain similar. Note that at moderate dilution rates (defined differently for each species), an increased flow rate of medium through the system increases division rate (thus, generation time decreases as the cells divide faster). A constant cell mass, or density, is maintained over a range of flow rates because the amount of culture (and cells) removed from the vessel exactly compensates for the increased rate of cell division.

At faster and faster flow rates, cells are eventually removed more quickly than they can be replenished by division, so cell density (cell mass) decreases in the vessel—a phenomenon called "washout." Notice, in contrast, that at very low dilution (flow) rates, an increase in rate will actually increase cell density. The reason is that the nutrient is so limiting at extremely low dilution rates that cell mass cannot increase to the point necessary for division. If more nutrient becomes available, the system can maintain a higher cell mass and cell density because the cells are able to divide faster than they are removed. With slow flow rates, the removal of fluid is not a significant factor in determining cell density.

Continuous cultures are used to study large numbers of cells at a constant growth rate and cell mass for both research and industrial applications. Most bacteria in nature grow at very slow rates because of low nutrient levels—a situation that can be mimicked in a chemostat.

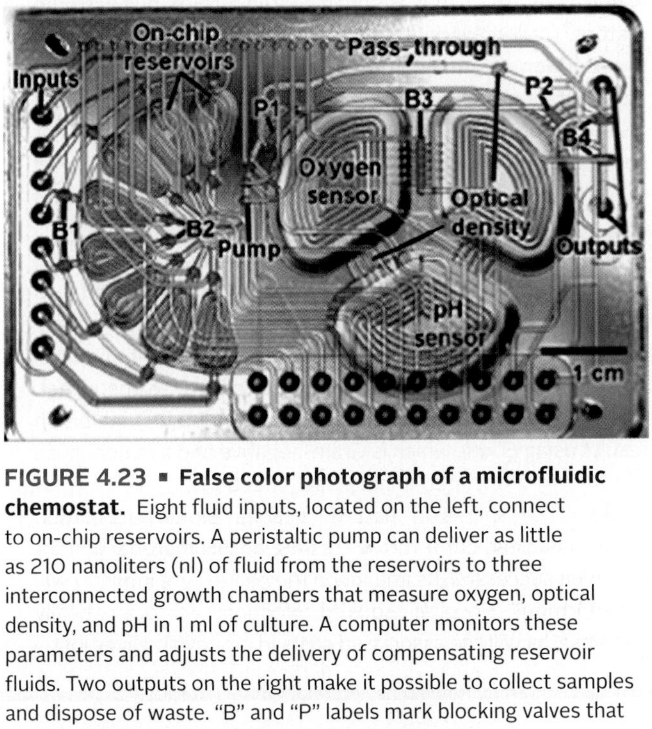
On-chip reservoirs
Pass-through
Inputs
P1
B3
P2
B4
Oxygen sensor
Optical density
B1
B2
Pump
Outputs
pH sensor
1 cm

KEVIN LEE ET AL. 2011. *LAB CHIP* **11**: 1730–1739.

FIGURE 4.23 ▪ False color photograph of a microfluidic chemostat. Eight fluid inputs, located on the left, connect to on-chip reservoirs. A peristaltic pump can deliver as little as 210 nanoliters (nl) of fluid from the reservoirs to three interconnected growth chambers that measure oxygen, optical density, and pH in 1 ml of culture. A computer monitors these parameters and adjusts the delivery of compensating reservoir fluids. Two outputs on the right make it possible to collect samples and dispose of waste. "B" and "P" labels mark blocking valves that control fluid movements through the chambers.

The physiology of these cells is quite different from what is typically observed with batch culture.

One disadvantage of the continuous culture apparatus shown in **Figure 4.21C** is the large volume of media required to maintain a 500-ml culture chamber over a period of days or months. To solve this problem, microfluidic principles have been used to miniaturize the process onto a 5-cm computer-controlled lab chip (**Fig. 4.23**). One milliliter of culture moves along three chambers that monitor oxygen, optical density, and pH. Culture conditions are maintained or changed by a computer that controls delivery of media and chemicals from on-chip reservoirs. The heightened ability to control culture parameters in small volumes can be used to characterize genetic switches, environmental shock responses, directed evolution, co-metabolism, and the dynamics of mixed cultures.

To Summarize

- The **growth cycle** of organisms grown in liquid batch culture consists of lag phase, log phase, stationary phase, and death phase.
- **The physiology of a bacterial population** changes with growth phase.
- **Continuous culture** can be used to sustain a population of bacteria at a specified growth rate and cell density.

4.6

Biofilms

Can bacteria collaborate? Bacteria are typically thought of as unicellular, but many, if not most, bacteria in nature form specialized, surface-attached, collaborative communities called **biofilms**. Indeed, within aquatic environments bacteria are found mainly associated with surfaces—a fact that underscores the importance of biofilms in nature. Biofilms also play critical roles in microbial pathogenesis and environmental quality, and they cost the nation billions of dollars each year in equipment damage, product contamination, and medical infections. For example, pseudomonad or staphylococcal biofilms can damage ventilators used to assist respiration. Biofilms can also form inside indwelling catheter tubes that deliver fluids and medications into patients. In both of these examples, the biofilms serve as direct sources of infection, so developing ingenious ways to prevent biofilm formation on medical instrumentation is a major goal of biomedical research (**Special Topic 4.1**).

Special Topic 4.1: Sharks and Biofilms Don't Mix

What can a microbiologist learn from a shark? Quite a lot, it turns out. The lesson begins with knowing that hospital patients sometimes have thin, sterile, plastic tubes called catheters inserted into their veins as a way to deliver fluids and drugs. If the catheter is not inserted using proper aseptic technique, however, contaminating skin microbes (for example, *Staphylococcus epidermidis;* **Fig. 1**) can build biofilms on the inner walls of the tube. Fluid passing through the contaminated tube will carry dispersal cells from that biofilm into the patient's bloodstream; the result is a bloodstream infection (septicemia). The organism might also lodge in a heart valve and form a more dangerous biofilm. The vegetation that forms around that biofilm causes endocarditis (**Fig. 2**). Because septicemia and endocarditis are very dangerous infections, clinicians are keen to find ways to limit biofilm formation in catheters.

Some scientists are now testing a 400-million-year-old material to see if it can prevent catheter-related infections in hospitals. That 400-million-year-old material is sharkskin. If you examine the skin of most slow-moving marine animals you will often see other life-forms attached to it. Whales, for instance, have barnacles, as do ships or most anything that spends its life in the sea. Sharks, however, do not. In fact, sharks are remarkably free from any kind of surface growth. Anthony Brennan from the University of Florida noted this and wondered what it was about sharkskin that was so uninviting to other creatures.

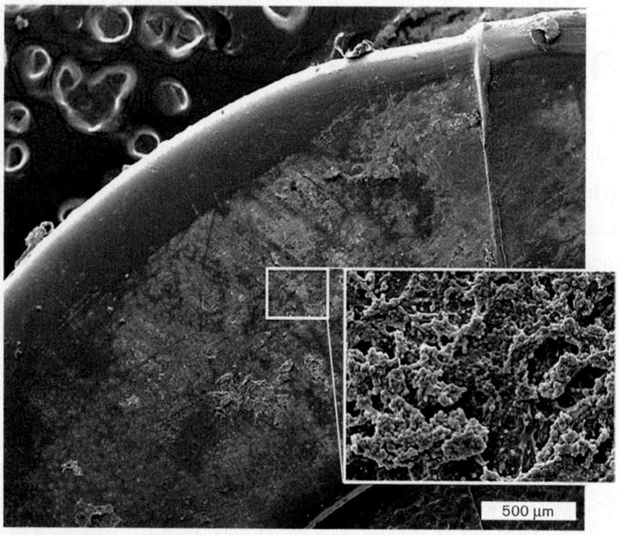
500 µm

FIGURE 1 ■ Biofilm on inner flow connector used for catheters.

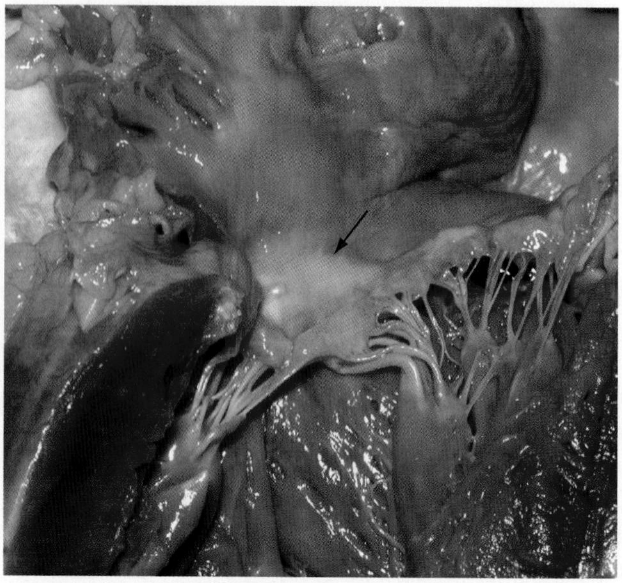

FIGURE 2 ■ Endocarditis of a heart valve. The body's response to a heart valve biofilm is to form a vegetation around it, which limits valve function. A vegetation is seen on the right heart valve in this image.

When examined by scanning EM, sharkskin had an interesting ribbed, diamond-shaped pattern of denticles, which Brennan and his colleagues reproduced on silicone sheets and wafers (**Fig. 3A**). When Brennan coated a ship's hull with the material (called Sharklet), the barnacles would not attach.

Attention eventually turned to what effect, if any, Sharklet might have on biofilm formation. To test this, Brennan and his colleagues (**Fig. 3B**) submerged patterned wafers and flat surface (control) wafers in cultures containing 10^7 CFUs/ml of *Staphylococcus aureus* (which is Gram-positive) for 21 days (**Fig. 4**). Fourteen days is the typical time frame for biofilms to form on indwelling catheters. The results were striking. After 14 days, biofilms covered 54% of the flat surface (control) wafers, but only 7% of the patterned Sharklet wafers. Curiously, other types of patterns did not inhibit biofilms, and a 2-µm distance between ribs was critical to the effect. The group has successfully repeated these biofilm inhibition results using *E. coli,* which is Gram-negative and a major cause of catheter-associated urinary tract infections.

How the sharkskin pattern inhibits biofilm formation remains unclear, but if further testing of this material continues to be successful, its impact on the health care industry will be enormous. Hospital-acquired infections will dramatically decrease, as will the associated costs in money and lives lost.

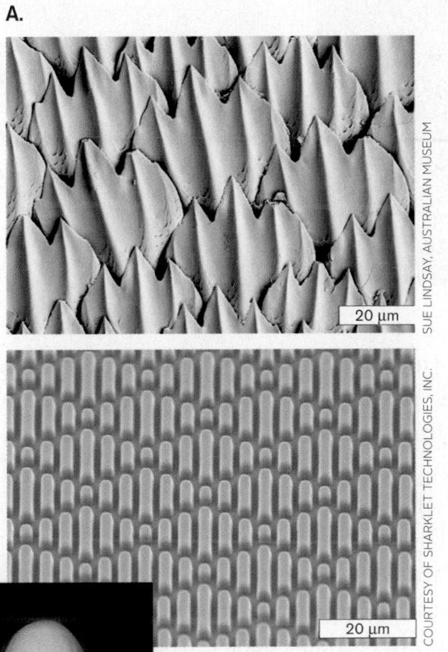
A.
20 µm
SUE LINDSAY, AUSTRALIAN MUSEUM
20 µm
COURTESY OF SHARKLET TECHNOLOGIES, INC.

B.
COURTESY OF ANTHONY BRENNAN

FIGURE 3 ■ Antimicrobial surface pattern. **A.** Comparison of sharkskin (top) to artificial Sharklet (bottom). **B.** Anthony Brennan, who conceived Sharklet.

Weblink Sharklet on *Nova* (*see ebook*)

RESEARCH QUESTION

Where do organisms that form biofilms on urinary catheters originate? You suspect that infections resulting from catheter biofilms have something to do with the bacteria migrating along a catheter surface. Design an experiment to prove that Sharklet inhibits migration of *E. coli* using agar, a strip of Sharklet, and a smooth strip of similar material.

Reddy, Shravanthi T., Kenneth K. Chung, Clinton J. McDaniel, Rabih O. Darouiche, Jaime Landman, et al. 2011. Micropatterned surfaces for reducing the risk of catheter-associated urinary tract infection: An *in vitro* study on the effect of Sharklet micropatterned surfaces to inhibit bacterial colonization and migration of uropathogenic *Escherichia coli*. *Journal of Endourology* **25**:1547–1552.

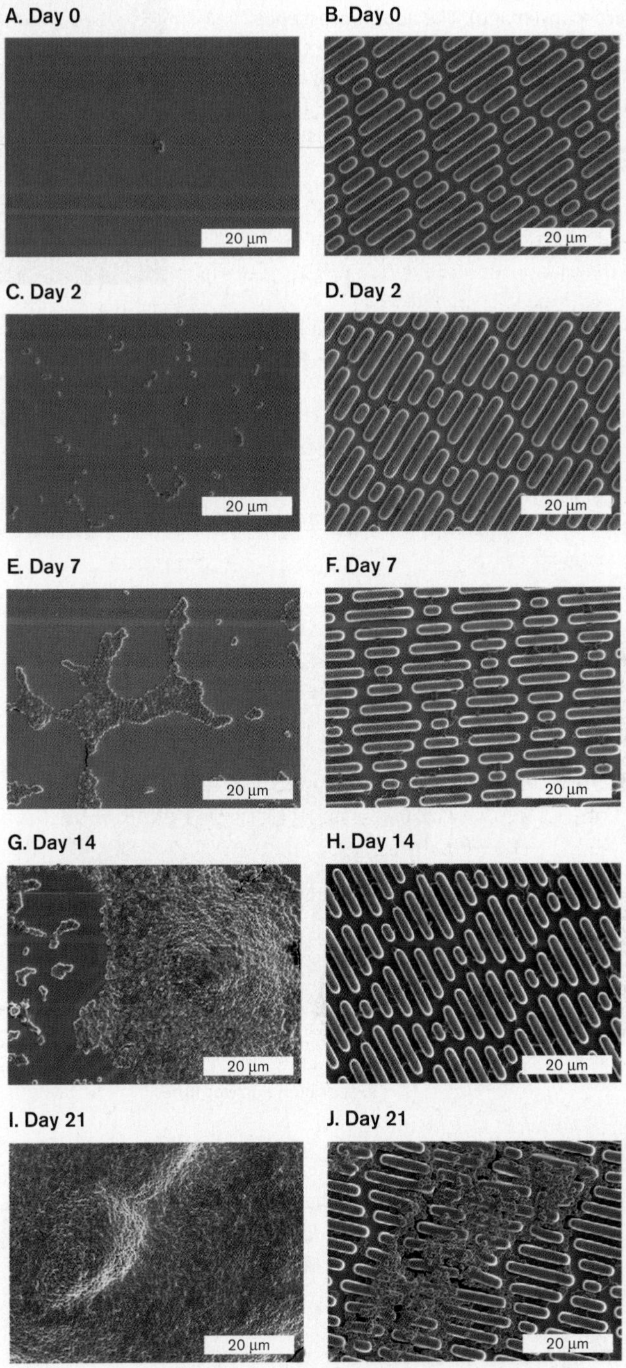
A. Day 0
B. Day 0
C. Day 2
D. Day 2
E. Day 7
F. Day 7
G. Day 14
H. Day 14
I. Day 21
J. Day 21
20 µm

FIGURE 4 ■ The effect of the sharkskin micropattern on biofilm formation. Flat (left column) and patterned (right) wafers were submerged in media containing 10^7 CFUs/ml of *Staphylococcus aureus* for the time indicated. SEMs of the wafers show the amount of biofilm (red) formed on each wafer.

Source: Chung et al. 2007. *Biointerphases* **2**:89–94.

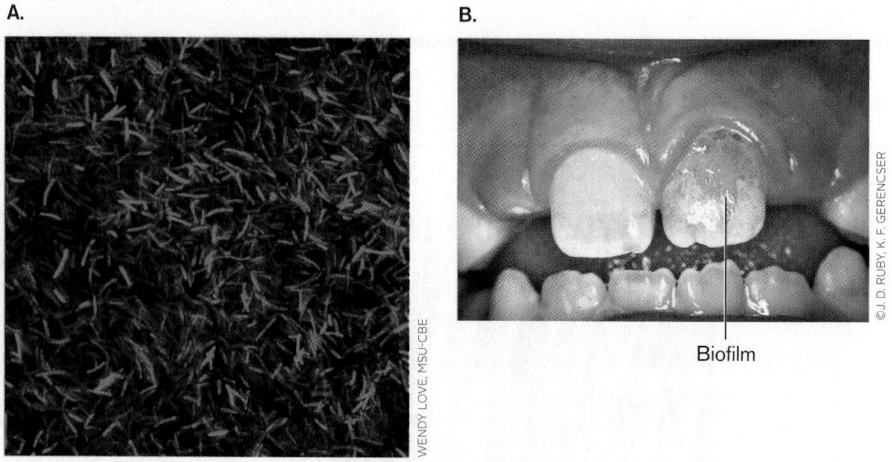
A.
B.
WENDY LOVE, MSU-CBE
©J. D. RUBY, K. F. GERENCSER
Biofilm

FIGURE 4.24 ▪ Biofilms. **A.** A thermophilic microbial mat found attached to a rock in Yellowstone National Park (confocal laser scanning microscopy image). The autofluorescence of these cyanobacterial cells of the genus *Synechococcus* was generated with a 568-nm krypton laser and a 590LP filter. **B.** The biofilm that forms on teeth is called plaque.

Biofilms: Multicellular Microbes?

Biofilms, which can be constructed by a single species or by multiple, collaborating species, can form on a range of organic and inorganic surfaces (**Fig. 4.24**). The Gram-negative bacterium *Pseudomonas aeruginosa,* for example, can form a single-species biofilm on the lungs of cystic fibrosis patients or on medical implants. Distinct stages in biofilm development (the biofilm "life cycle") include initiation, maturation, maintenance, and dissolution (or dispersal). Bacterial biofilms form when nutrients are plentiful. The goal of biofilms in nature is to stay where food is abundant. Why should a microbe travel off to hunt for food when it is already available? Once nutrients become scarce, however, individuals detach from the community to forage for new sources of nutrients.

Biofilms in nature can take many different forms and serve different functions for different species. Confocal images of biofilms are shown in **Figure 4.25** (inset) and **Figure 2.30**.

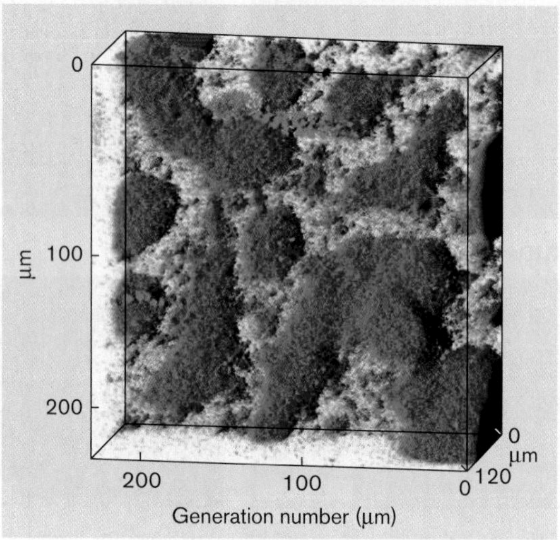
0
100
200
μm
200
100
0
0
μm
120
Generation number (μm)

FIGURE 4.25 ▪ Biofilm development. The stages of biofilm development in *Pseudomonas,* which generally apply to the formation of many kinds of biofilms. **Inset:** A mucoid environmental strain of *P. aeruginosa* produces uneven, lumpy biofilms. Cells in the biofilm were stained green with the fluorescent DNA-binding dye Syto-9 (3D confocal laser scanning microscopy). *Source:* H. C. Flemming and J. Wingender. 2010. *Nat. Rev. Microbiol.* **8**:623–633.

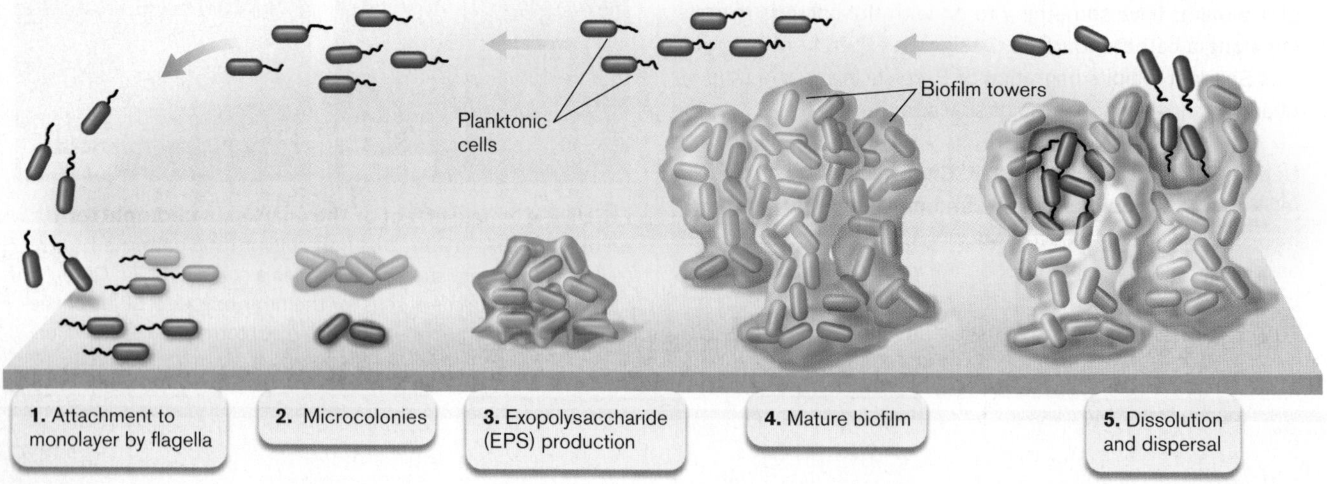
Planktonic cells
Biofilm towers
1. Attachment to monolayer by flagella
2. Microcolonies
3. Exopolysaccharide (EPS) production
4. Mature biofilm
5. Dissolution and dispersal

The formation of biofilms can be cued by different environmental signals in different species. These signals include pH, iron concentration, temperature, oxygen availability, and the presence of certain amino acids. Nevertheless, a common pattern emerges in the formation of many kinds of biofilms (**Fig. 4.25** ▶).

First, the specific environmental signal induces a genetic program in planktonic cells. The planktonic cells then start to attach to nearby inanimate surfaces by means of flagella, pili, lipopolysaccharides, or other cell surface appendages, and they begin to coat that surface with an organic monolayer of polysaccharides or glycoproteins to which more planktonic cells can attach. At this point, cells may move along surfaces using a **twitching motility** that involves the extension and retraction of a specific type of pilus. Ultimately, they stop moving and firmly attach to the surface. As more and more cells bind to the surface, they can begin to communicate with each other by sending and receiving chemical signals in a process called **quorum sensing**. These chemical signal molecules are continually made and secreted by individual cells. Once the population reaches a certain number (analogous to an organizational "quorum"), the chemical signal reaches a specific concentration that the cells can sense. This concentration triggers genetically regulated changes that cause cells to bind tenaciously to the substrate and to each other. Quorum sensing serves the biofilm in many ways. Among other functions, quorum sensing has recently been implicated in the increased resistance of biofilms to antibiotics and to phagocytosis by white blood cells—subjects we explore in eTopic 4.3 and Special Topic 27.1.

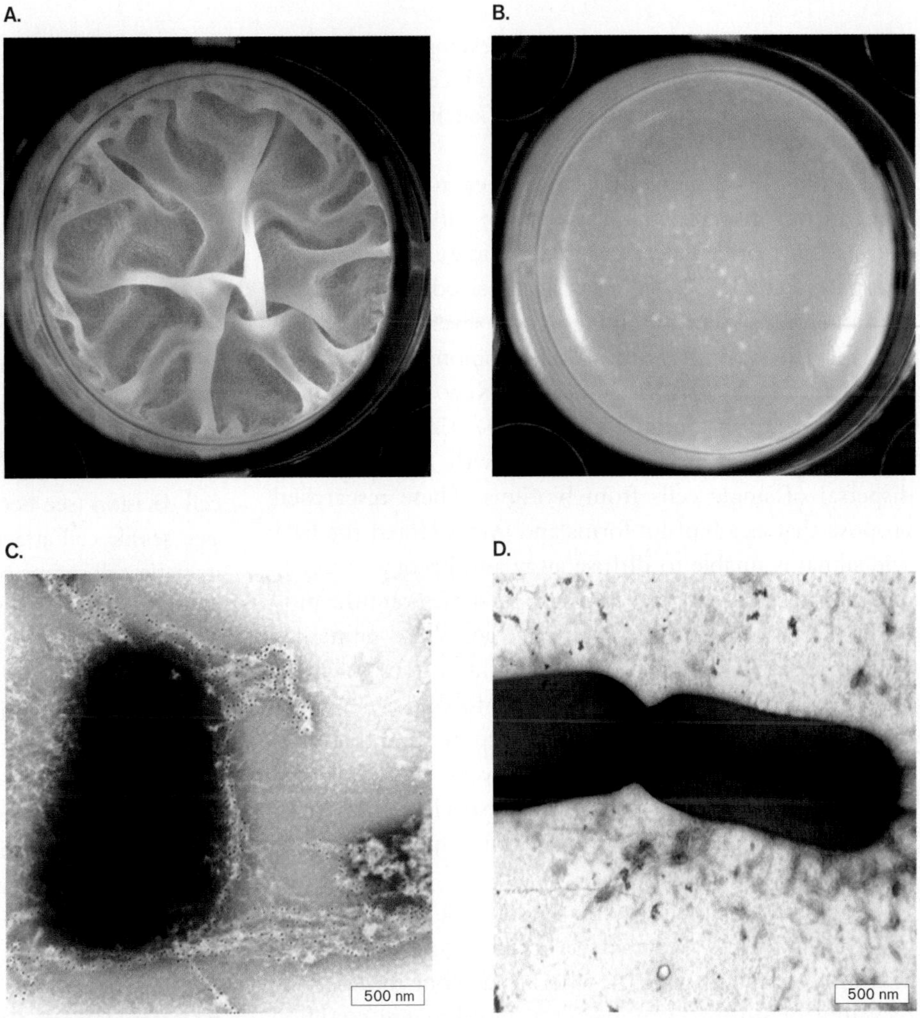
A.
B.
C.
500 nm
D.
500 nm

FIGURE 4.26 ■ Floating biofilm (pellicle) formation of *Bacillus subtilis*. Cells were grown in a broth for 48 hours without agitation at 30°C. The pellicles formed by wild-type and *tasA* mutant *B. subtilis* are strikingly different. Wild-type pellicles are extremely wrinkly **(A)**, whereas *tasA* mutant pellicles are flat and fragile **(B)**. **C, D.** Electron micrographs of TasA fibers between wild-type **(C)** and *tasA* mutant **(D)** cells. *Source:* Diego Romero et al. 2010. *PNAS* **107**: 2230–2234

Once a microcolony has been established (see **Fig. 4.25**), the cells form a thick extracellular matrix of polysaccharide polymers and entrapped organic and inorganic materials. These **exopolysaccharides (EPSs)**, such as alginate produced by *P. aeruginosa* and colanic acid produced by *E. coli*, increase the antibiotic resistance of residents within the biofilm. As the biofilm matures, the amalgam of adherent bacteria and matrix takes on complex 3D forms such as columns and streamers, forming channels through which nutrients flow. Sessile (nonmoving) cells in a biofilm chemically "talk" to each other in order to build microcolonies and keep water channels open.

Bacillus subtilis also spins out a fibril-like amyloid protein called TasA, which tethers cells and strengthens the biofilm. The effect of TasA on a floating biofilm formed on liquid medium is evident in **Figure 4.26**.

When a sessile biofilm (or a part of it) begins to starve or experiences oxygen depletion, some cells begin to make enzymes that dissolve the matrix. *P. aeruginosa*, for instance, produces an alginate lyase that can strip away the EPS. Cells of *B. subtilis* sever their links to TasA fibers by incorporating D-amino acids (for example, D-tyrosine, D-methionine, and D-leucine) into peptidoglycan. The sessile biofilm then sends out "scouts" called dispersal cells to initiate new biofilms. Dispersal cell development includes turning off genes whose products produce EPS and, for bacteria capable of motility, activating flagella genes needed for quick dispersal.

An intracellular signal molecule critical to biofilm development and dispersal in many (if not all) bacteria is an unusual nucleotide called cyclic dimeric guanosine monophosphate (cdiGMP), which is synthesized and degraded by numerous enzymes. High cdiGMP levels promote biofilm formation, whereas low levels promote dispersal. Scientists from the State University of New York have also found that an unsaturated fatty acid, *cis*-2-decenoic acid, produced by *P. aeruginosa* during growth, can trigger the dispersal of single cells from biofilms. These researchers propose that as a biofilm forms and grows larger, the fatty acid signal is unable to diffuse away and becomes trapped in the biofilm's nooks and crannies. At some point the molecule reaches a concentration that causes the enmeshed cells to release enzymes that degrade the matrix polymers. Dissolution of matrix frees some cells to find a new location with more plentiful nutrients. Keep in mind that most biofilms in nature are consortia of several species. Multispecies biofilms certainly demand interspecies communication, and individual species may perform specialized tasks in the community.

Organisms adapted to life in extreme environments also form biofilms. Members of Archaea form biofilms in acid mine drainage (pH 0), where they contribute to the recycling of sulfur; cyanobacterial biofilms are common in thermal springs. Suspended particles called "marine snow" are found in oceans and appear to be floating biofilms comprising many organisms that have not yet been identified. The particles appear capable of methanogenesis, nitrogen fixation, and sulfide production, indicating that the architecture of biofilms makes it possible for anaerobic metabolism to occur in an otherwise aerobic environment.

Weblinks Biofilms and twitching motility (*see ebook*)

To Summarize

- **Biofilms** are complex, multicellular, surface-attached microbial communities.
- **Chemical signals** enable bacteria to communicate (**quorum sensing**) and in some cases to form biofilms.
- **Biofilm development** involves the adherence of cells to a substrate, the formation of microcolonies, and, ultimately, the formation of complex channeled communities that generate new planktonic cells.

4.7

Cell Differentiation

Can bacteria change? Many bacteria faced with environmental stress undergo complex molecular reprogramming that includes changes in cell structure. Some species, like *E. coli,* experience relatively simple changes in cell structure, such as the formation of smaller cells or thicker cell surfaces. However, select species, such as *Caulobacter crescentus,* undergo elaborate cell differentiation processes. *Caulobacter* cells convert from the swimming form to the holdfast form before cell division (see Section 3.6). Each cell cycle then produces one sessile cell attached to its substrate by a holdfast, while its sister cell swims off in search of another habitat.

Eukaryotic microbes also undergo highly complex life cycles. For example, *Dictyostelium discoideum* is a seemingly unremarkable ameba that grows as separate, independent cells. When challenged by adverse conditions such as starvation, however, *D. discoideum* secretes chemical signal molecules that choreograph a massive interaction of individuals to form complex multicellular structures. The developmental cycle of this organism may be compared to that of other eukaryotic microbes that are human parasites (discussed in Chapter 20).

Endospores Are Bacteria in Suspended Animation

Certain Gram-positive genera, including important pathogens such as *Clostridium tetani* (tetanus), *Clostridium botulinum* (botulism), and *Bacillus anthracis* (anthrax), have the remarkable ability to develop dormant spores that are heat and desiccation resistant. Spores are particularly hearty because they do not grow and do not need nutrients until they germinate. Resistance to heat and desiccation (and its lethal toxin) makes *B. anthracis* spores a potential bioweapon.

Most of what we know about bacterial sporulation comes from the Gram-positive soil bacterium *Bacillus subtilis.* When growing in rich media, this microbe undergoes normal vegetative growth and can replicate every 30–60 minutes. However, starvation initiates an elaborate 8-hour genetic program that directs an asymmetrical cell division process and ultimately yields a spore (see Section 10.4).

As shown in **Figure 4.27** ▶, sporulation can be divided into seven discrete stages based primarily on cell morphology. Stage 0 (not shown) represents the point at which the vegetative cell "decides" to use one of two potential polar division sites to begin septum formation instead of the central division site used for vegetative growth. In stage I,

A.

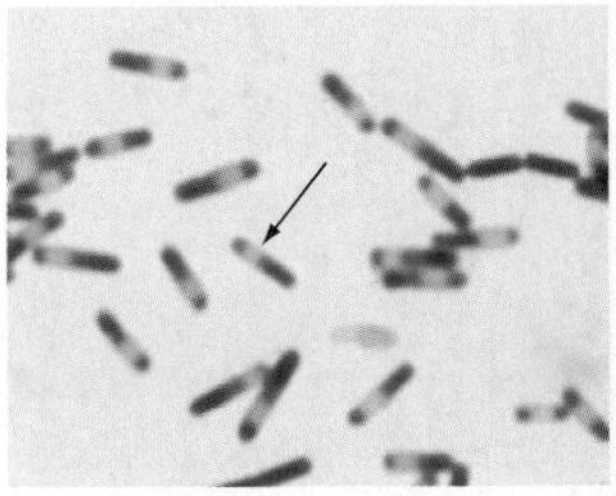

CDC HTTP://PHIL.CDC.GOV/PHIL/HOME.ASP

B.

COURTESY OF PETER SETLOW

C.

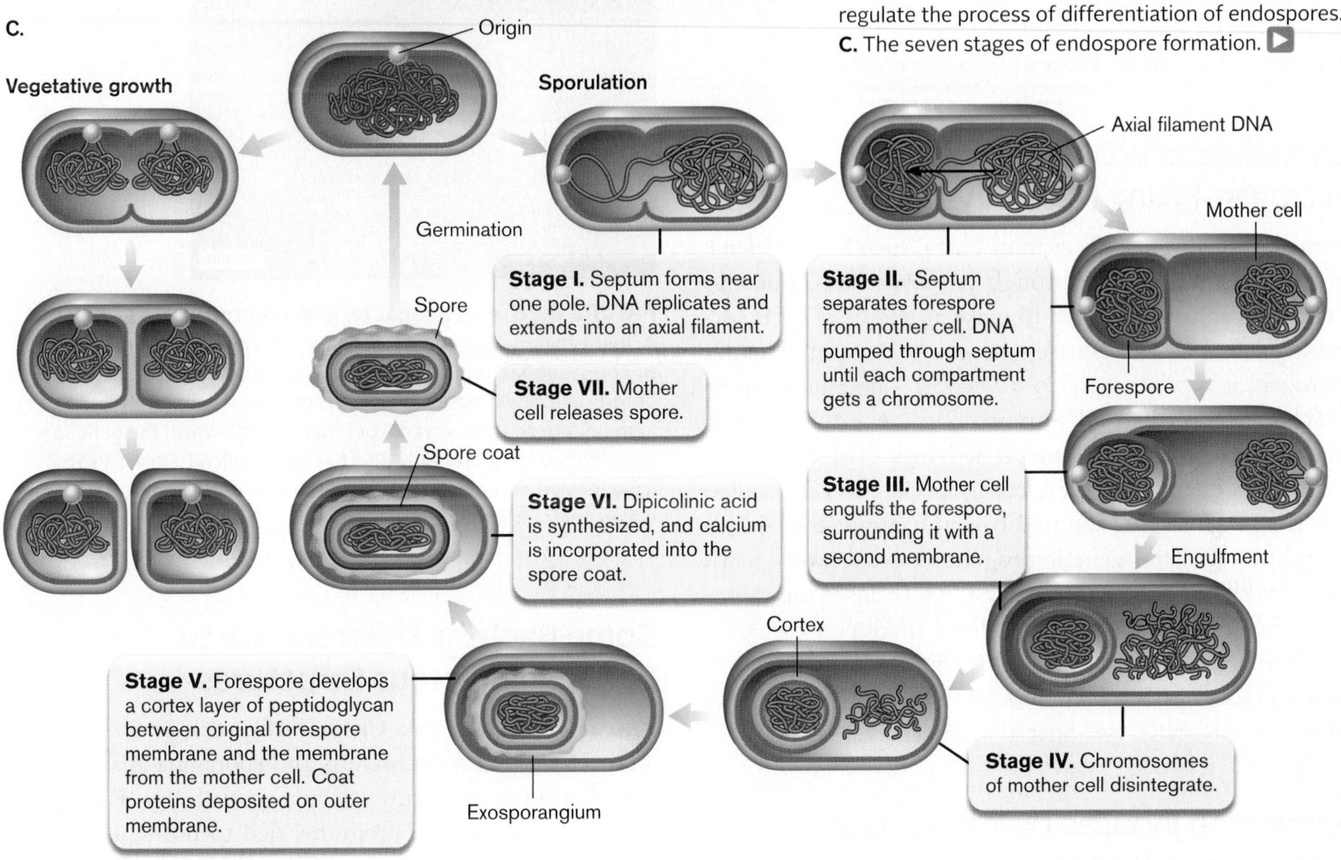
Origin
Vegetative growth
Sporulation
Axial filament DNA
Germination
Mother cell
Spore
Stage I. Septum forms near one pole. DNA replicates and extends into an axial filament.
Stage II. Septum separates forespore from mother cell. DNA pumped through septum until each compartment gets a chromosome.
Stage VII. Mother cell releases spore.
Forespore
Spore coat
Stage VI. Dipicolinic acid is synthesized, and calcium is incorporated into the spore coat.
Stage III. Mother cell engulfs the forespore, surrounding it with a second membrane.
Engulfment
Cortex
Stage V. Forespore develops a cortex layer of peptidoglycan between original forespore membrane and the membrane from the mother cell. Coat proteins deposited on outer membrane.
Exosporangium
Stage IV. Chromosomes of mother cell disintegrate.

FIGURE 4.27 ■ Endospore formation. **A.** Photomicrograph of *Clostridium difficile* endospores. The cells (approx. 3 μm long) are stained with crystal violet. The cells are blue, whereas the spores are colorless and located inside the cells. A spore stain in which spores appear blue-green is shown in Fig. 2.24B. **B.** Peter Setlow (right) of the University of Connecticut figured out how proteins regulate the process of differentiation of endospores. **C.** The seven stages of endospore formation.

the DNA is replicated and stretched into a long axial filament that spans the length of the cell. There are two chromosome copies at this point. Ultimately, one of the polar division sites wins out, and in stage II septation occurs, dividing the cell into two unequal compartments: the smaller **forespore**, which will ultimately become the spore, and the larger **mother cell**, from which the forespore is derived. Each compartment eventually contains one of the replicated chromosomes. (Most of the chromosome in the forespore has to be pumped in after septum formation.)

In stage III of sporulation, the mother cell membrane engulfs the forespore. Next, the mother cell chromosome is destroyed (stage IV)and a thick peptidoglycan layer (cortex) is placed between the two membranes surrounding the forespore protoplast (stage V). Layers of coat proteins are then deposited on the outer membrane in stage V. Stage VI completes the development of spore resistance to heat and chemical insults. This last process includes the synthesis of dipicolinic acid (which stabilizes and protects spore DNA) and the uptake of calcium into the coat of the spore. Finally, the mother cell, now called a sporangium, releases the mature spore (stage VII).

Spores are resistant to many environmental stresses that would kill vegetative cells. Spores owe this resistance, in part, to their desiccation (they have only 10%–30% of a vegetative cell's water content). But, as discovered by Peter Setlow and colleagues, spores are also packed with small acid-soluble proteins (SASPs) that bind to and protect DNA. The SASP coat protects the spore's DNA from damage by ultraviolet light and various toxic chemicals.

A fully mature spore can exist in soil for at least 50–100 years, and spores have been known to last thousands of years. Once proper nutrient conditions arise, another genetic program, called **germination**, is triggered to wake the dormant cell, dissolve the spore coat, and release a viable vegetative cell.

While sporulation is an effective survival strategy, bacteria that sporulate actually go out of their way not to sporulate—even to the point of cannibalism (see Current Research Highlight). During nutrient limitation these bacteria will secrete proteins that can kill their siblings and then use the released nutrients to prevent starvation and, thus, sporulation. When (and only when) that strategy fails, they sporulate.

Weblinks Sporulation, Richard Losick (*see ebook*)

Cyanobacteria Differentiate into Nitrogen-Fixing Heterocysts

Some of the autotrophic cyanobacteria, such as *Anabaena*, not only make oxygen through photosynthesis, but also "fix" atmospheric nitrogen to make ammonia. This is surprising because nitrogenase, the enzyme required to fix nitrogen, is very sensitive to oxygen, so one might expect that photosynthesis and nitrogen fixation would be two mutually exclusive physiological activities. *Anabaena* have solved this dilemma by developing specialized cells, called heterocysts, that function in nitrogen fixation (**Fig. 4.28**). A tightly regulated genetic program converts every tenth photosynthetic cell to a heterocyst, which loses the capacity to fix CO_2 and forms a specialized envelope to limit O_2 access and allow N_2 fixation. How this organism produces such a precise spacing of heterocysts is currently the subject of intensive research (see Special Topic 2.1).

Starvation Induces Differentiation into Fruiting Bodies

Certain species of bacteria, in the microbial equivalent of a barn raising, produce architectural marvels called fruiting bodies. The Gram-negative species *Myxococcus xanthus* uses a **gliding motility** (involving a type of pilus, not a flagellum) to travel on surfaces as individuals or to move together as a mob (**Fig. 4.29**). Starvation triggers a developmental cycle in which 100,000 or more individuals aggregate, rising into a mound called a fruiting body. At this point, the system resembles a stage in *Pseudomonas aeruginosa* biofilm formation. However, myxococci within the interior of the fruiting body differentiate into thick-walled, spherical spores that are released into the surroundings. The random dispersal of spores is an attempt to find new sources of nutrients. The changes involved in this differentiation process require many cell-cell interactions and a complex genetic program that we do not fully understand just yet.

Weblinks Myxococcus fruiting body (*see ebook*)

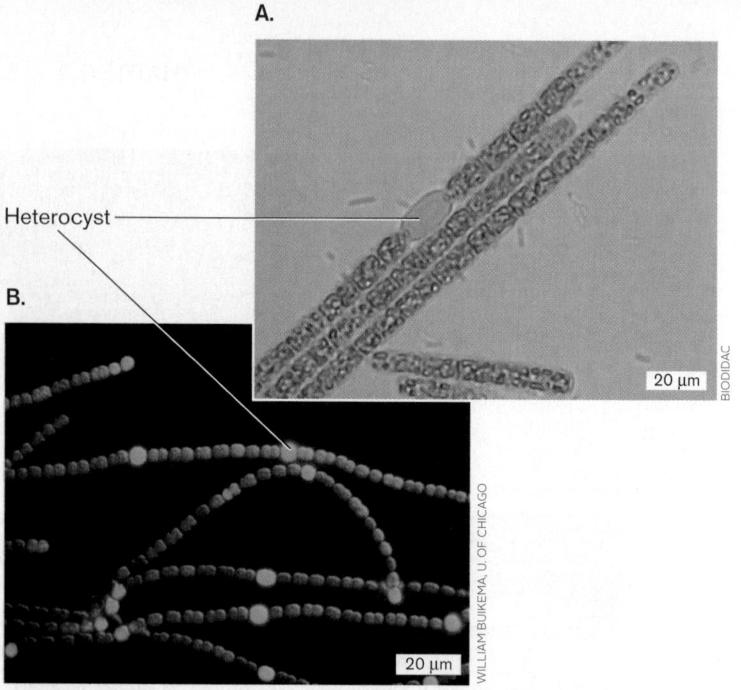
A.
Heterocyst
B.
20 μm
BIODIDAC
20 μm
WILLIAM BUIKEMA, U. OF CHICAGO

FIGURE 4.28 ■ Cyanobacteria and heterocyst formation. **A.** Light-microscope image of cyanobacterium *Phanizomenon.* **B.** The cyanobacterium *Anabaena.* The expression of genes in heterocysts is different from their expression in other cells. All cells in the figure contain a cyanobacterial gene to which the gene for green fluorescent protein (GFP) has been spliced. Only cells that have formed heterocysts are expressing the fused gene, which makes the cell fluoresce bright green.

Some Bacteria Differentiate to Form Filamentous Structures

The actinomycetes (see Chapter 18), such as *Streptomyces*, are bacteria that form mycelia and sporangia analogous to the filamentous structures of eukaryotic fungi (**Fig. 4.30**). Several developmental programs tied to nutrient availability are at work in this process (**Fig. 4.31**). Under favorable nutrient conditions, a germ tube emerges from a germinating spore, grows from its tip (tip extension), and forms branches that grow along the surface of its food source. This type of growth produces an intertwined network of long multinucleate filaments (**hyphae**; singular, **hypha**) collectively called substrate **mycelia** (singular, **mycelium**). After a few days, new genes are activated that cause the hyphae to grow upward, rising above the surface to form aerial mycelia. Compartments at the tips of these aerial hyphae contain 20–30 copies of the genome. Aerial hyphae stop growing as nutrients decline, triggering a developmental program that synthesizes antibiotics. Meanwhile, the older ends of the filaments senesce, and their decomposing cytoplasm attracts scavenger microbes—which are killed by the antibiotics. The younger streptomycete cells then feast on the dead scavengers.

The aerial hyphae then produce spores (arthrospores) that are fundamentally different from endospores. This program lays down multiple septa that subdivide the

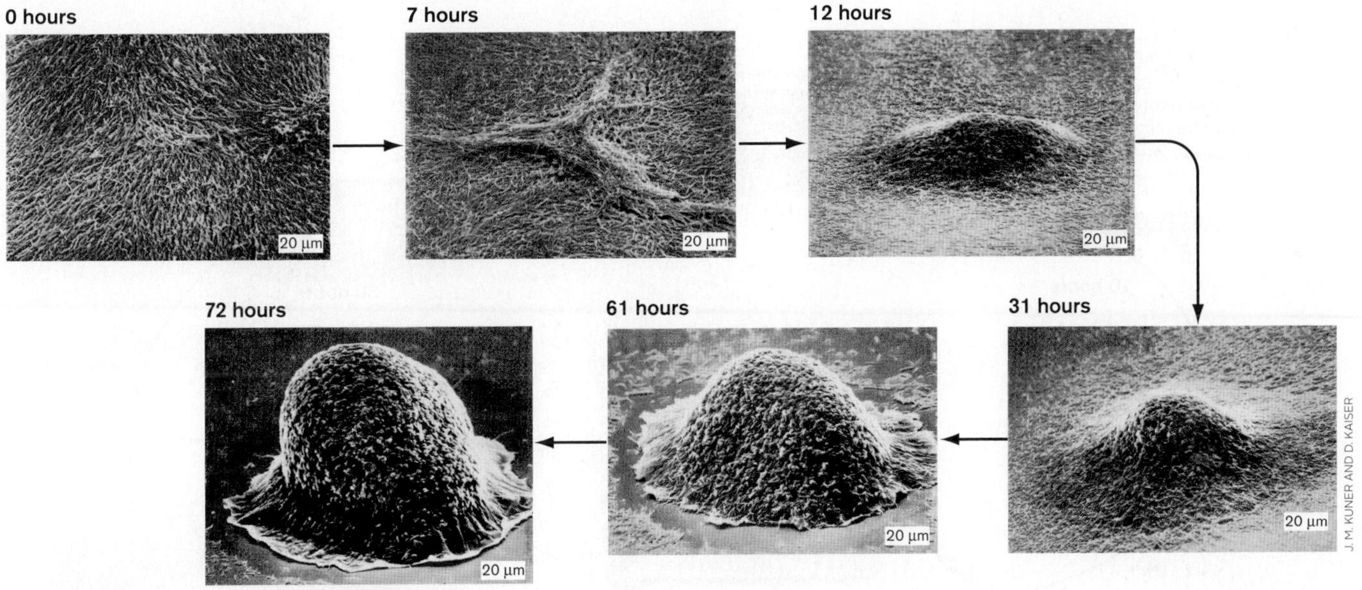
0 hours
7 hours
12 hours
31 hours
61 hours
72 hours
20 µm
J. M. KUNER AND D. KAISER

FIGURE 4.29 ▪ *Myxococcus* swarm, erecting a fruiting body. Approximately 100,000 cells begin to aggregate, and over the course of 72 hours they erect a fruiting body.

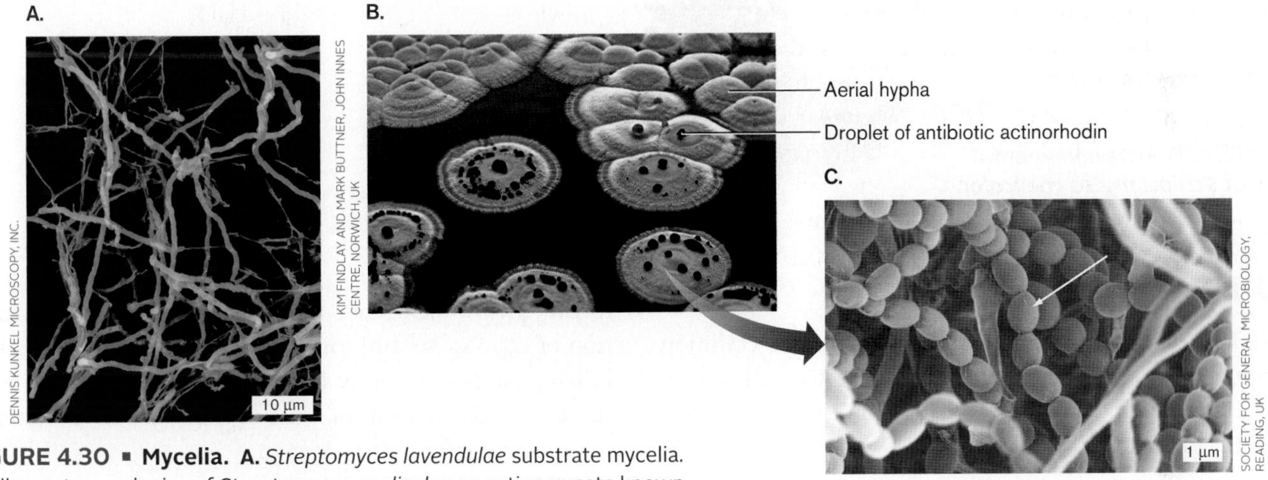
A.
B.
C.
Aerial hypha
Droplet of antibiotic actinorhodin
10 µm
1 µm
DENNIS KUNKEL MICROSCOPY, INC.
KIM FINDLAY AND MARK BUTTNER, JOHN INNES CENTRE, NORWICH, UK
SOCIETY FOR GENERAL MICROBIOLOGY, READING, UK

FIGURE 4.30 ▪ Mycelia. A. *Streptomyces lavendulae* substrate mycelia. **B.** Filamentous colonies of *Streptomyces coelicolor,* an actinomycete known for producing antibiotics (blue pigment in water droplets). **C.** *Streptomyces* aerial hyphae. The arrow points to a hyphal spore (approx. 1 µm each).

compartment into single-genome prespores. The shape of the prespore then changes, its cell wall thickens, and deposits are made in the spore that increase resistance to desiccation. These organisms are of tremendous interest, both for their ability to make antibiotics and for their fascinating developmental programs.

To Summarize

- **Microbial development** involves complex changes in cell forms.
- **Endospore development** in *Bacillus* and *Clostridium* involves the production of dormant, stress-resistant endospores.
- **Heterocyst development** enables cyanobacteria to fix nitrogen anaerobically while maintaining oxygenic photosynthesis.
- **Multicellular fruiting bodies** in *Myxococcus* and *Dictyostelium* and **mycelia** in actinomycetes develop in response to starvation, dispersing dormant cells to new environments.

Thought Question

4.16 How might *Streptomyces* and *Actinomyces* species avoid "committing suicide" when they make their antibiotics?

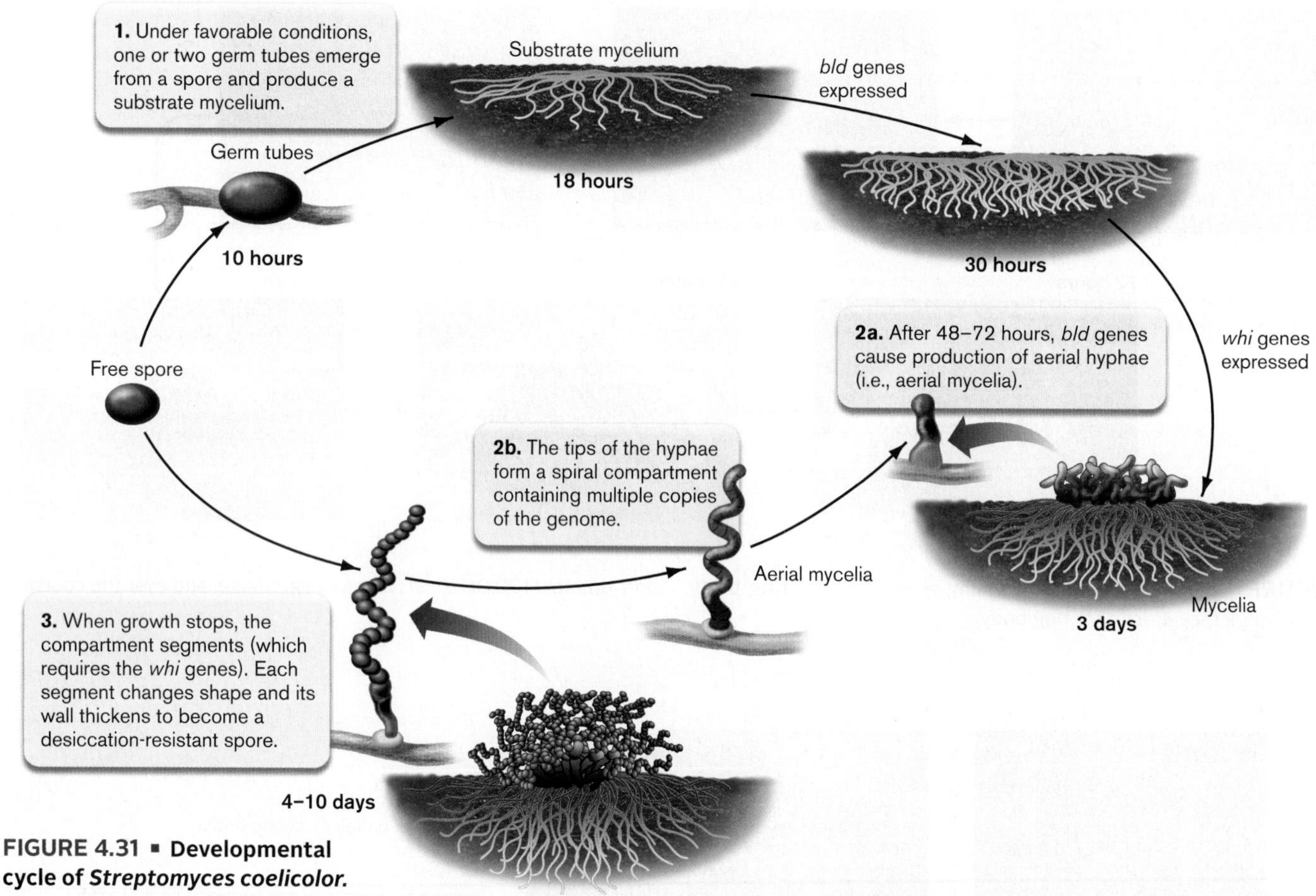

1. Under favorable conditions, one or two germ tubes emerge from a spore and produce a substrate mycelium.
Substrate mycelium
bld genes expressed
Germ tubes
18 hours
10 hours
30 hours
2a. After 48–72 hours, *bld* genes cause production of aerial hyphae (i.e., aerial mycelia).
whi genes expressed
Free spore
2b. The tips of the hyphae form a spiral compartment containing multiple copies of the genome.
Aerial mycelia
Mycelia
3 days
3. When growth stops, the compartment segments (which requires the *whi* genes). Each segment changes shape and its wall thickens to become a desiccation-resistant spore.
4–10 days

FIGURE 4.31 ■ Developmental cycle of *Streptomyces coelicolor.*

Concluding Thoughts

All systems of microbial development share a common theme. They are all triggered in response to a change in their environment, such as the depletion of resources, desiccation, or changes in temperature. Microbial responses to the environment have major implications for the function of ecosystems and for the microbial communities that inhabit the human body for good or ill. In Chapter 5 we survey the mechanisms of some of the major environmental responses of microbes.

CHAPTER REVIEW

Review Questions

1. What nutrients do microbes need to grow?
2. Explain the differences between autotrophy, heterotrophy, phototrophy, and chemotrophy.
3. Explain the basics of the carbon and nitrogen cycles.
4. Describe the various mechanisms of transporting nutrients in prokaryotes and eukaryotes. What are facilitated diffusion, coupled transport, ABC transporters, group translocation, and endocytosis?
5. Why is it important to grow bacteria in pure culture?
6. Under what circumstances would you use a selective medium? A differential medium?
7. What factors define the growth phases of bacteria grown in batch culture?
8. Describe the important features of biofilms.
9. Name three kinds of bacteria that differentiate, and give highlights of the differentiation processes.

Thought Questions

1. Bile salts are used in certain selective media. What are bile salts, and why might they be more harmful to Gram-positive organisms than to Gram-negatives?
2. Why is *Rickettsia prowazekii*, which can grow only in the cytoplasm of a eukaryotic cell, considered a living organism but viruses are not?
3. Suppose 1,000 bacteria are inoculated in a tube containing a minimal salts medium, where they double once an hour, and 10 bacteria are inoculated into rich medium, where they double in 20 minutes. Which tube will have more bacteria after 2 hours? After 4 hours?
4. An exponentially growing culture has an optical density at 600 nm (OD_{600}) of 0.2 after 30 minutes and an OD_{600} of 0.8 after 80 minutes. What is the doubling time?
5. What are the generation times for (a) *Clostridium perfringens*, the cause of gas gangrene; (b) *Mycobacterium leprae*, the cause of leprosy; (c) *Thermus aquaticus*, a hot-springs bacterium; and (d) *Psychromonas antarcticus*, a cold-loving organism that grows at high pressure?
6. Mercuric ions (Hg^{2+}) and methylmercury [$(CH_3Hg)^+$] are major, human-generated, toxic contaminants of water and soil. Some species of bacteria can bioremediate these compounds by transporting them into the cell and reducing them to elemental mercury (Hg^0). One of the transport proteins is called MerC. Use an Internet search engine to determine how many organisms have a MerC homolog.
7. Environmental bacteria were isolated from the water reservoirs of insect-digesting pitcher plants. The isolated strains were cultured in tryptone–yeast extract broth, which contains many different peptide and carbohydrate nutrients. The different growth curves obtained are shown in the graph. Explain how these curves differ from the "standard" growth curve and propose hypotheses as to why they differ.

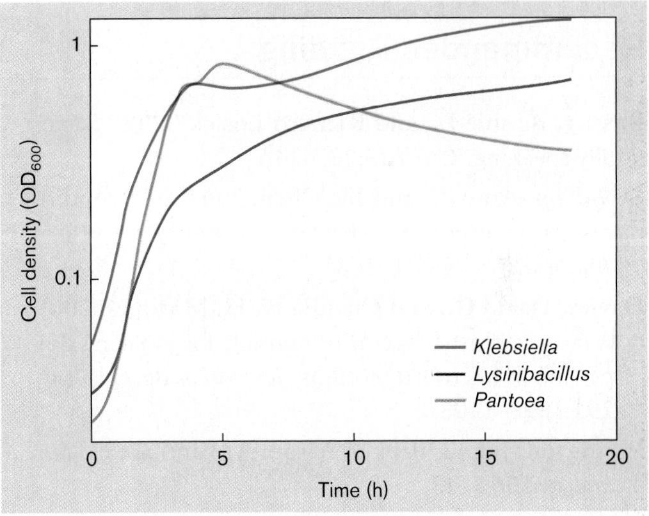
Cell density (OD_{600})
1
0.1
0
5
10
15
20
Time (h)
Klebsiella
Lysinibacillus
Pantoea

Key Terms

ABC transporter (130)
antiport (128)
ATP synthase (125)
autotroph (123, 124)
axenic growth (121)
batch culture (141)
binary fission (139)
biofilm (145)
chemoautotroph, chemoautotrophy (123, 124)
chemoheterotrophy (124)
chemolithotroph (123)
chemostat (144)
chemotrophy (124)
cofactor (120)
colony (134)
complex medium (134)
confluent (134)
continuous culture (144)
coupled transport (128)
death phase (143)
death rate (143)
defined minimal medium (121)
denitrification (126)
differential medium (135)
dilution streaking (133)
doubling time (140)
electrochemical potential (125)
electrogenic (128)
electroneutral (128)
enriched medium (135)
essential nutrient (120)
exopolysaccharide (EPS) (149)
exponential phase (142)
facilitated diffusion (127)
fluorescence-activated cell sorter (FACS) (137)
forespore (151)
generation time (140)
germination (151)
gliding motility (152)
group translocation (131)
growth factor (121)
growth rate (140)
growth rate constant (140)
heterotroph (123, 124)
hypha (152)
lag phase (141)

lithotroph (123, 124)
logarithmic (log) phase (142)
MacConkey medium (135)
macronutrient (120)
mean generation time (141)
membrane potential (125)
microcolony (134)
micronutrient (120)
mixotrophic (127)
mother cell (151)
mycelium (152)
nitrification (126)
nitrogen-fixing bacterium (126)
optical density (138)
organotroph (124)
permease (127)
phosphotransferase system (PTS) (132)
photoautotroph, photoautotrophy (123, 124)
photoheterotrophy (124)
phototrophy (124)
planktonic cell (139)
pour plate technique (137)
proton potential (proton motive force) (125)
pure culture (133)
quorum sensing (149)
selective medium (135)
siderophore (131)
spread plate (134)
stationary phase (143)
substrate-binding protein (130)
symbiont (126)
symport (128)
synthetic medium (134)
twitching motility (149)
viable (134)
viable but nonculturable (VBNC) (134)

Recommended Reading

Bassler, Bonnie L., and Richard Losick. 2006. Bacterially speaking. *Cell* **125**:237–246.

Davidson, Amy L., and Jue Chen. 2004. ATP-binding cassette transporters in bacteria. *Annual Reviews in Biochemistry* **73**:241–268.

Davies, David G., and Claudia N. H. Marques. 2009. A fatty acid messenger is responsible for inducing dispersion in microbial biofilms. *Journal of Bacteriology* **191**:1393–1403.

Errington, John. 2010. From spores to antibiotics. *Microbiology* **156**:1–13.

Gonzalez-Pastor, Jose E. 2011. Cannibalism: A social behavior in sporulating *Bacillus subtilis. FEMS Microbiology Review* **35**:415–424.

Lee, Kevin S., Paolo Boccazzi, Anthony J. Sinskey, and Rajeev J. Ram. 2011. Microfluidic chemostat and turbidostat with flow rate, oxygen, and temperature control for dynamic continuous culture. *Lab on a Chip* **11**:1730–1739.

Lopian, Livnat, Yair Elisha, Anat Nussbaum-Shochat, and Orna Amster-Choder. 2010. Spatial and temporal organization of the *E. coli* PTS components. *EMBO Journal* **29**:3630–3645.

McDougald, Diane, Scott A. Rice, Nicolas Barraud, Peter D. Steinberg, and Staffan Kjelleberg. 2012. Should we stay or should we go: Mechanisms and ecological consequences for biofilm dispersal. *Nature Reviews: Microbiology* **10**:39–50.

Nystrom, Thomas. 2004. Stationary-phase physiology. *Annual Review of Microbiology* **58**:161–181.

Omsland, Anders, Diane C. Cockrell, Dale Howe, Elizabeth R. Fischer, Kimmo Virtaneva, et al. 2009. Host cell-free growth of the Q fever bacterium *Coxiella burnetii. Proceedings of the National Academy of Sciences USA* **106**:4430–4434.

Piggot, Patrick J., and David W. Hilbert. 2004. Sporulation of *Bacillus subtilis. Current Opinion in Microbiology* **7**:579–586.

Romero, Diego, Hera Vlamakis, Richard Losick, and Roberto Kolter. 2011. An accessory protein required for anchoring and assembly of amyloid fibres in *B. subtilis* biofilms. *Molecular Microbiology* **80**:1155–1168.

Saier, Milton H., Jr., Bin Wang, Wei Hao Zheng, Eric I. Sun, and Ming Ren Yen. 2010. Mosaic energy-coupled transporters. *Microbe* **5**:105–109.

Skaar, Eric P. 2010. The battle for iron between bacterial pathogens and their vertebrate hosts. *PLoS Pathogens* **6**:e1000949.

Skerker, Jeffrey M., and Michael T. Laub. 2004. Cell-cycle progression and the generation of asymmetry in *Caulobacter crescentus. Nature Reviews. Microbiology* **2**:325–337.

CHAPTER 5

Environmental Influences and Control of Microbial Growth

5.1 Environmental Limits on Growth

5.2 Adaptation to Temperature

5.3 Adaptation to Pressure

5.4 Water Activity and Salt

5.5 Adaptation to pH

5.6 Oxygen and Other Electron Acceptors

5.7 Nutrient Deprivation and Starvation

5.8 Physical, Chemical, and Biological Control of Microbes

Microbes have both the fastest and the slowest growth rates of any known organism. Some hot-spring bacteria can double in as little as 10 minutes, whereas deep-sea sediment microbes may take as long as 100 years. What determines these differences in growth rate? Nutrition is one factor, but niche-specific physical parameters like temperature, pH, and osmolarity are equally important.

A microbe's physiology is geared to work only within a narrow range of physical parameters. But in nature, the environment can quickly change. Many marine microbes, for instance, can move from deep-sea cold to the searing heat of a thermal vent. How do these organisms survive? Stopgap measures called stress survival responses help, but some species evolve to thrive, not just survive, in extreme environments. How do these so-called extremophiles grow under conditions that would kill most living things? In Chapter 5 we explore the limits of microbial growth and show how this knowledge helps us control the microbial world.

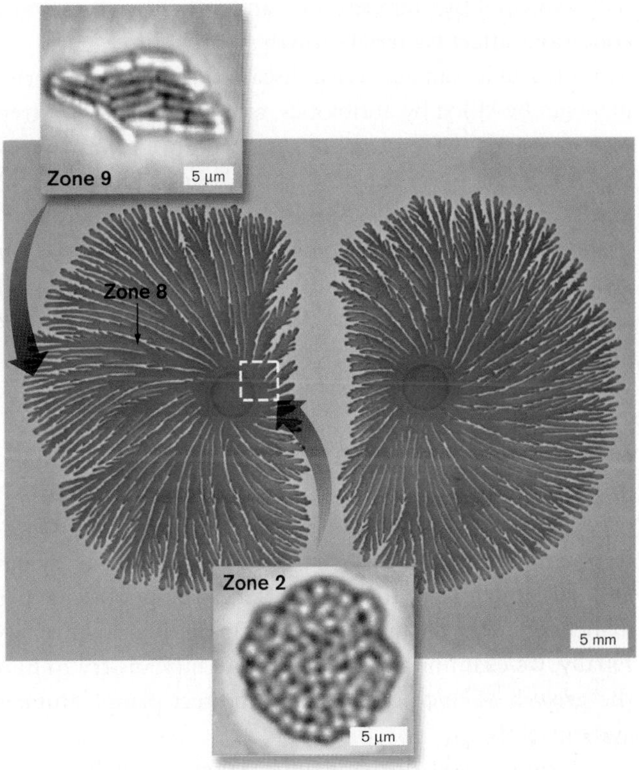

Zone 9
5 µm
Zone 8
Zone 2
5 mm
5 µm

CURRENT RESEARCH **highlight**

Microbial family feud. A curious form of sibling rivalry has been discovered in the microbial world. *Paenibacillus dendritiformis* is a social Gram-positive species whose colonies form elaborate branching patterns on agar. The two colonies of this species shown here were started near each other. As "fingers" from the two colonies approached, a secreted protein stopped their advance by turning normally motile bacilli (zone 9, top inset) into nonmotile cocci (zone 2, bottom inset). The two cell forms have dramatically different metabolic and antibiotic resistance profiles. This intriguing shape-shift mechanism is reversible and may be widespread among bacterial species. *Source:* A. Be'er et al. 2011. *MBio* **2**:e00069-11

In 1998, apple and pear growers in Washington and Oregon lost crops worth an estimated $68 million owing to outbreaks of fire blight, a devastating bacterial disease. The causative agent, *Erwinia amylovora,* destroys apple and pear trees, making them appear as if torched by fire. How could growth of these bacteria be stopped? You might think cold would do it because fire blight progresses only when temperatures rise above 18°C (65°F). Cold will stop growth but, unfortunately, will not kill these bacteria. We will explain how temperature and other environmental conditions affect bacterial growth shortly.

Fortunately, during active disease the fire blight pathogens can be killed by antibiotics, although only two, streptomycin and oxytetracycline, are approved for use on plants. Antibiotics were part of the solution for the apple and pear growers in 1998. But resistance to streptomycin has already developed in some strains of *Erwinia amylovora,* underscoring why we must study all aspects of bacterial growth in order to control pathogens of plants, animals, and humans. How antibiotics are used to control bacterial growth is also discussed in this chapter.

We begin this chapter by describing how physical and chemical changes in the environment modify the growth of different groups of microbes. We also explore how microorganisms adapt to different environments in ways both transient (involving temporary expression of inactive genes) and permanent (modifications of the gene pool). The permanent genetic changes have led to biological diversity. Finally, we examine the different ways humans try to limit the growth of microorganisms to protect plants, animals, and ourselves.

As you proceed through this chapter, you will encounter two recurring themes: that different groups of microbes live in vastly different environments, and that microbes can respond in diverse ways when confronted by conditions outside their niche, or comfort zone. Chapter 21 builds on this information to discuss how communities of microbes interact with their environment, and presents current research methods used to study microbial ecology.

5.1

Environmental Limits on Growth

What is normal? With our human frame of reference, we tend to think that "normal" growth conditions are those found at sea level with a temperature between 20°C and 40°C, a near-neutral pH, a salt concentration of 0.9%, and ample nutrients. Any ecological niche outside this window is labeled "extreme," and the organisms inhabiting them are called **extremophiles** (R. MacElroy first used the term "extremophile" in 1974). Extremophiles are microbes (bacteria, archaea, and some eukaryotes) that are able to grow in extreme environments. For example, one group of organisms can grow at temperatures above the boiling point of water (100°C), while another group requires a strongly acidic (pH 2) environment to grow. According to our definition of what is "normal," conditions on Earth when life began were certainly extreme. The earliest microbes, then, likely grew in these extreme environments. Organisms that grow under conditions that seem normal to humans likely evolved from an ancient extremophile that gradually adapted as the environment evolved to that of our present-day Earth.

A single environment can simultaneously encompass multiple extremes. In Yellowstone National Park, for instance, an extreme acid pool can be found next to an extreme alkali pool, both at extremely high temperatures. Thus, extremophiles typically evolve to survive multiple extreme environments.

Extremophiles may provide insight into the workings of extraterrestrial microbes we may one day encounter, since outer space certainly qualifies as an extreme environment. Our experiences with extremophiles should alert us to the dangers of underestimating the precautions necessary in handling extraterrestrial samples. For example, we should assume that irradiation will not sterilize samples from future planetary or interstellar missions. Such treatments do not even kill *Deinococcus radiodurans,* an extremophile found on Earth.

Weblinks Extremophiles (*see ebook*)

How do we even begin to study organisms that grow in boiling water or sulfuric acid solutions or organisms that we cannot even culture in the laboratory? How do we dissect the molecular response of organisms to changes in an environment such as acid rain or to changes in the human body? Genome sequences present new opportunities for investigating these questions.

Bioinformatic analysis, which uses the DNA sequence of a gene to predict the function of its protein product, allows us to study the biology of organisms that we cannot culture. First, genome sequences from novel organisms are amplified by the polymerase chain reaction (PCR, a technique that multiplies a small sample of DNA; see Section 7.6). PCR amplification can be done directly from the natural environment in which the organisms are found and the sequences compared with those of model systems, such as *E. coli,* whose biochemistry is well known. Genomic comparison quickly reveals whether an organism under study may possess

PHOTO PROVIDED BY ROBERT KELLY

FIGURE 5.1 ■ Bioreactor used to grow thermophilic microorganisms. Dr. Robert Kelly and students stand next to a 20-liter bioreactor that they use for the engineering analysis of biofuel-producing microbes. From left to right: Aaron Hawkins, Andrew Loder, Hong Lian, Dr. Robert Kelly, and Yejun Han.

specific metabolic pathways and regulatory responses. The ability to "see" biochemical pathways by peeking at an extremophile's genome has led Robert Kelly and colleagues at North Carolina State University to mix genes and enzymes from different extremophiles in an attempt to engineer a new organism able to convert abundantly available CO_2 and H_2 directly into high-energy liquid fuels (**Fig. 5.1**).

Techniques that examine the expression of all the genes in a genome or all the proteins in the cell allow us to study the response of an organism to changing environments, such as fluctuations in temperature or pH (Chapter 3). Knowing which genes and proteins are expressed in a given situation reveals how microbes grow under different conditions and defend themselves against environmental stresses. These techniques of molecular analysis are discussed further in Chapter 10.

We have already mentioned the fundamental physical conditions (temperature, pH, osmolarity) that define an environment and favor (select for) the growth of specific groups of organisms. Within a given microbial community, each species is further localized to a specific niche defined by a narrower range of environmental factors. Species find their niche because every protein and macromolecular structure within a cell is affected by changes in environmental conditions. For example, a single enzyme works best under a unique set of temperature, pH, and salt conditions because those conditions allow it to fold into its optimum shape, or conformation. Deviations from these optimal conditions cause the protein to fold a little differently and become less active. While not all enzymes within a cell boast the same physical optima, these optima must at least be similar and matched to the organism's environment for the organism to function effectively.

As may be surmised from the preceding discussion, microbes are commonly classified by their environmental niche. **Table 5.1** summarizes these environmental classes.

TABLE 5.1

Basic environmental classification of microorganisms.

Environmental parameter	Classification			
Temperature	Hyperthermophile* (growth above 80°C)	Thermophile* (growth between 50°C and 80°C)	Mesophile (growth between 15°C and 45°C)	Psychrophile* (growth below 15°C)
pH	Alkaliphile* (growth above pH 9)	Neutralophile (growth between pH 5 and pH 8)	Acidophile* (growth below pH 3)	
Osmolarity	Halophile* (growth in high salt, >2-M NaCl)			
Oxygen	Aerobe (growth only in O_2)	Facultative (growth with or without O_2)	Microaerophile (growth only in small amounts of O_2)	Anaerobe (growth only without O_2)
Pressure	Barophile* (growth at high pressure, greater than 380 atm)		Barotolerant (growth between 10 and 495 atm)	

*Considered extremophiles.

To Summarize

- **Extremophiles** inhabit fringe environments with conditions that do not support human life.
- **The environmental habitat** (such as high salt or acidic pH) inhabited by a particular species is defined by the tolerance of that organism's proteins and other macromolecular structures to the physical conditions within that niche.
- **Global approaches** used to study gene expression allow us to view how organisms respond to changes in their environment.

5.2

Adaptation to Temperature

How do microbes react to hot and cold? Unlike humans (and mammals in general), microbes cannot control their temperature; thus, bacterial cell temperature matches that of the immediate environment. Because temperature affects the average rate of molecular motion, changes in temperature impact every aspect of microbial physiology, including membrane fluidity, nutrient transport, DNA stability, RNA stability, and enzyme structure and function. Every organism has an "optimum" temperature at which it grows most quickly, as well as minimum and maximum temperatures that define the limits of growth. These limits are imposed, in part, by the thousands of proteins in a cell, all of which must function within the same temperature range. The fastest growth rate for a species occurs at temperatures where all of the cell's proteins work most efficiently as a group to produce energy and synthesize cell components. Growth stops when rising temperatures cause critical enzymes or cell structures (such as the cell membrane) to fail. At cold temperatures, growth ceases because enzymatic processes become too sluggish and the cell membrane becomes less fluid. The membrane needs to remain fluid so that it can expand as cells grow larger and so that proteins needed for solute transport can be inserted into the membrane.

Growth Rate and Temperature

In general, microbes that grow at higher temperatures can achieve higher rates of growth (**Fig. 5.2B**). Remarkably, the relationship between the maximum growth temperature and the growth rate constant *k* (the number of generations per hour; see Section 4.5) obeys the Arrhenius equation for simple chemical reactions (**eTopic 5.1**). The general result of the Arrhenius equation is that growth rate roughly doubles for every 10°C rise in temperature (**Fig. 5.2A**). This is the same relationship observed for most chemical reactions.

At the upper and lower limits of the growth range, however, the Arrhenius effect breaks down. Critical proteins denature at high temperatures, whereas lower temperatures decrease membrane fluidity and limit the conformational

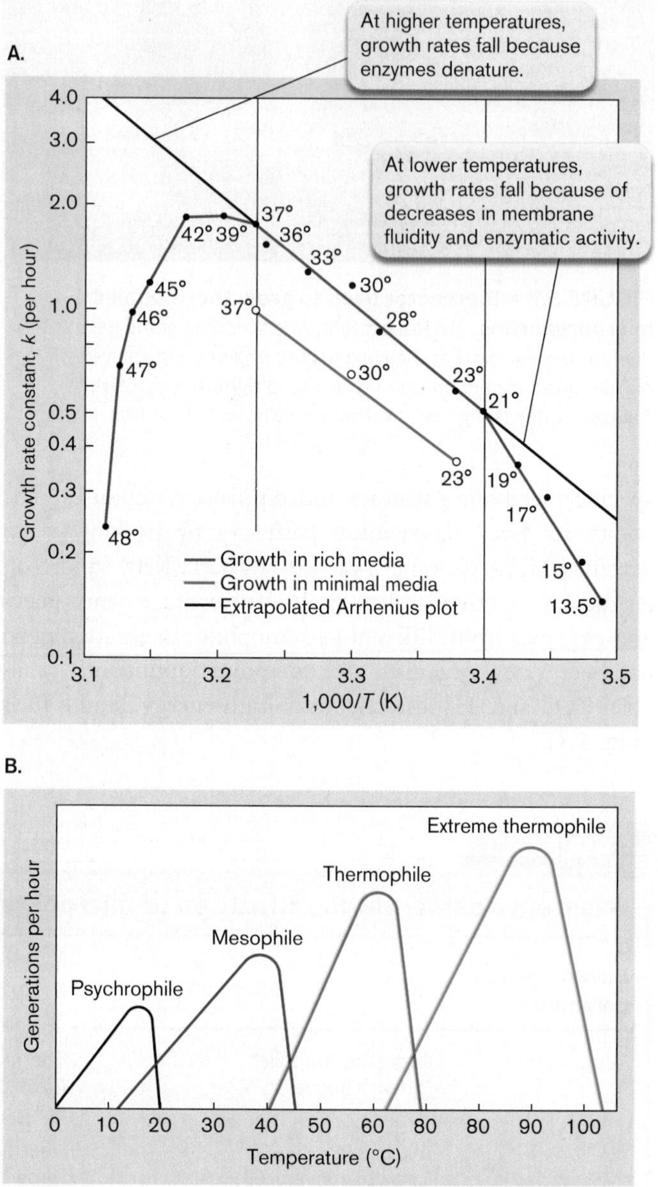
A.
At higher temperatures, growth rates fall because enzymes denature.
At lower temperatures, growth rates fall because of decreases in membrane fluidity and enzymatic activity.
Growth rate constant, k (per hour)
4.0
3.0
2.0
1.0
0.5
0.4
0.3
0.2
0.1
48°
47°
46°
45°
42°
39°
37°
36°
33°
30°
28°
23°
21°
19°
17°
15°
13.5°
37°
30°
23°
Growth in rich media
Growth in minimal media
Extrapolated Arrhenius plot
3.1
3.2
3.3
3.4
3.5
1,000/T (K)
B.
Extreme thermophile
Thermophile
Mesophile
Psychrophile
Generations per hour
0
10
20
30
40
50
60
80
80
90
100
Temperature (°C)

FIGURE 5.2 ■ Relationship between temperature and growth rate. A. Growth rate constant (*k*) of the enteric organism *Escherichia coli* is plotted against the inverse of the growth temperature on the Kelvin scale (1,000/*T* is used to give a convenient scale on the *x*-axis). This is a more detailed view of a mesophilic growth temperature curve. As temperature rises above or falls below the optimum range, growth rate decreases faster than is predicted by the Arrhenius equation. **B.** The relationship between temperature and growth rates of different groups of microbes. Note that the peak growth rate increases linearly with temperature and obeys the Arrhenius equation. *Source:* Part A from Sherrie L. Herendeen et al. 1979. *J. Bacteriol.* **139**:185

mobility of enzymes, thereby lowering their activities. As a result, growth fails to occur at temperature extremes. The typical temperature growth range for most bacteria spans the organism's optimal growth temperature by 30–40 degrees, but some organisms have a much narrower tolerance. Even within a species, we can find mutants that are more sensitive to one extreme or the other (heat sensitive or cold sensitive). These mutations often define key molecular components of stress responses, such as the heat-shock proteins (discussed in Chapter 10).

Thermodynamic principles limit a cell's growth to a narrow temperature range. For example, heat increases molecular movement within proteins. Too much or too little movement will interfere with enzymatic reactions. A great diversity exists among microbes because different groups have evolved to grow within very different thermal ranges. A species grows within a specific thermal range because its proteins have evolved to tolerate that range. Outside that range, proteins will denature or function too slowly for growth. The upper limit for protists is around 50°C, while some fungi can grow at temperatures as high as 60°C. Prokaryotes, however, have been found to grow at temperatures ranging from below 0°C to above 100°C. Temperatures over 100°C are usually found near thermal vents deep in the ocean. Vent water temperature can rise to 350°C, but the pressure is sufficiently high to keep water in the liquid state.

Thought Question

5.1 Why haven't cells evolved so that all their enzymes have the same temperature optimum? If they did, wouldn't they grow even more rapidly?

Microorganisms Are Classified by Growth Temperature

Using range of growth temperature, microorganisms can be classified as mesophiles, psychrophiles, or thermophiles (**Fig. 5.2B**).

Mesophiles include the typical "lab rat" microbes, such as *Escherichia coli* and *Bacillus subtilis*. Their growth optima range between 20°C and 40°C, with a minimum of 15°C and a maximum of 45°C. Because they are easy to grow and because all human pathogens are mesophiles, much of what we know about protein, membrane, and DNA structure came from studying this group of organisms. However, detailed 3D views of protein structures are frequently based on studies of two other classes of organisms whose optimum growth temperature ranges flank that of the mesophiles—namely, **psychrophiles** (on the low-temperature side) and **thermophiles** (on the high-temperature side). Because of their more stable folding structures, proteins from the thermophilic extremophiles are generally easier to crystallize than those of mesophiles or psychrophiles, so it is possible to determine their structures by X-ray crystallography (see Section 2.7).

Psychrophiles are microbes that grow at temperatures as low as –10°C, but their optimum growth temperature is usually around 15°C. Psychrophiles are prominent

A.

5 µm

ASIM BEJ, UNIVERSITY OF ALABAMA AT BIRMINGHAM

B.

ASIM BEJ, UNIVERSITY OF ALABAMA AT BIRMINGHAM

FIGURE 5.3 ▪ Psychrophilic environments and microbes. **A.** The continent of Antarctica is an extreme environment populated by many species of psychrophilic microorganisms, most of them unknown. In addition to being brutally cold, this extreme environment is nutrient poor and subject to high levels of solar UV irradiation. **Inset:** Asim Bej, University of Alabama at Birmingham, collects samples from the ecosystem at Schirmacher Oasis (location as marked). Genome sequences from the captured microbes, like those shown in (B), reveal the composition and metabolic capabilities of the South Pole microbiome. **B.** Psychrotolerant *Flavobacterium* (grows between 0°C and 22°C) from a South Pole lake made from glacial meltwater in summer (high temperature = 0.9°C; SEM). Novel compounds made by members of the polar microbiome are screened for anticancer and antimicrobial potential.

members of microbial communities beneath icebergs in the Arctic and Antarctic (**Fig. 5.3**). In addition to true psychrophiles, there are cold-resistant mesophiles (or psychrotolerant bacteria) that grow between 0°C and 35°C. Psychrotolerant bacteria cause milk to spoil in the refrigerator. Even some pathogens, such as *Listeria monocytogenes* (one cause of food poisoning and septic abortions), can grow at refrigeration temperatures.

Why do these organisms grow so well in the cold? One reason is that the proteins of psychrophiles are more flexible than those of mesophiles and require less energy (heat) to function. Of course, the downside to the increased flexibility of psychrophilic proteins is that they denature at lower temperatures than their mesophilic counterparts. As a result, psychrophiles grow poorly, if at all, when temperatures rise above 20°C. Another reason psychrophiles favor cold is that their membranes are more fluid at low temperatures (because they contain a high proportion of unsaturated fatty acids); at higher temperatures their membranes are too flexible and fail to maintain cell integrity. Finally, bacteria and archaea that grow at 0°C in glaciers also contain antifreeze proteins and other cryoprotectants (such as trehalose) that can depress the freezing point by 2°C. So, although these organisms can grow in ice, they will not freeze. Interestingly, some psychrophilic and psychrotolerant bacteria actually stimulate ice formation in their surrounding environment (**Special Topic 5.1**).

Weblinks Video of ice nucleation by *Pseudomonas syringae* (*see ebook*)

Psychrophilic enzymes are of commercial interest because their ability to carry out reactions at low temperature is useful in food processing and bioremediation. Enzymes help brew beer more quickly, break down lactose in milk, and can remove cholesterol from various foods. The production of foods at lower temperatures is beneficial, too, because the lower processing temperatures minimize the growth of typical mesophiles that degrade and spoil food. Genetically engineered psychrophilic organisms can safely degrade toxic organic contaminants (for example, petroleum) in the cold. Arctic environments are particularly sensitive to pollution because contaminants are slow to degrade in the freezing temperatures. Consequently, the ability to seed arctic oil spills with psychrophilic organisms armed with petroleum-degrading enzymes could rapidly restore contaminated environments.

Thermophiles (**Fig. 5.4**) are species adapted to growth at high temperatures (typically 55°C and higher). **Hyperthermophiles** grow at temperatures as high as 121°C, which occur under extreme pressure (for example, at the ocean floor). These organisms flourish in hot environments such as composts or near thermal vents that penetrate Earth's crust on the ocean floor and on land (for example, hot springs). The thermophile *Thermus aquaticus* was the first source of a high-temperature DNA polymerase used for PCR amplification of DNA. *T. aquaticus* was discovered in a hot spring at Yellowstone National Park by microbiologist Thomas Brock, a pioneer in the study of thermophilic organisms. Its application to the polymerase chain reaction has revolutionized molecular biology (discussed in Chapter 7).

Weblinks Strain 21 (*see ebook*)

Extreme thermophiles often have specially adapted membranes and protein sequences. The thermal limits of these structures determine the specific high-temperature ranges in which various species can grow. Because enzymes in thermophiles (thermozymes) do not unfold as easily as mesophilic enzymes, they more easily hold their shape at higher temperatures. Thermophilic enzymes are stable, in

A.

©MOMATIUK-EASTCOTT/CORBIS

B.

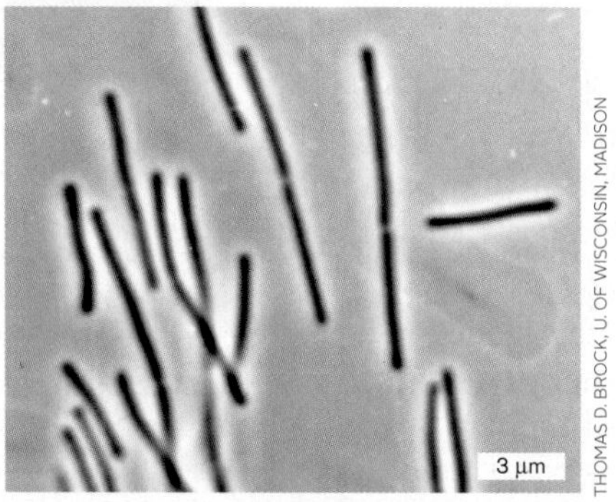
3 µm
THOMAS D. BROCK, U. OF WISCONSIN, MADISON

C.

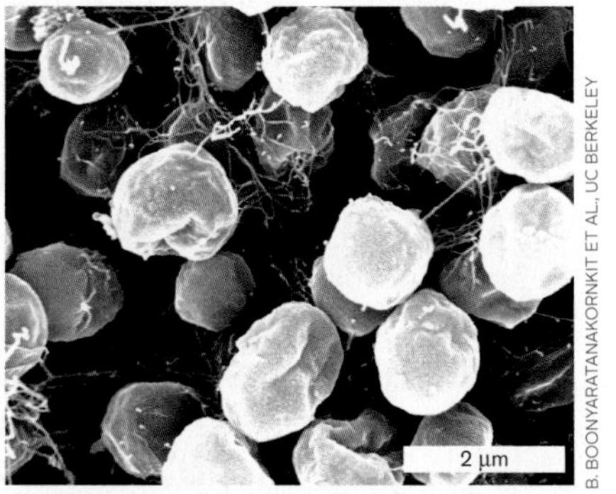
2 µm
B. BOONYARATANAKORNKIT ET AL., UC BERKELEY

FIGURE 5.4 ▪ Thermophilic environments and thermophiles. A. Yellowstone National Park hot spring. **B.** *Thermus aquaticus*, a hyperthermophile first isolated at Yellowstone by Thomas Brock. Cell length varies from 3 to 10 µm. **C.** Thermophile *Methanocaldococcus jannaschii*, grown at 78°C and 30 psi.

Special Topic 5.1: It's Raining Bacteria

How clouds, rain, and snow form has intrigued children and scientists for millennia. Little did anyone know that bacteria are major players in these processes. David Sands at Montana State University first hypothesized a link between rainfall and psychrotolerant bacteria in 1982. Though the scientific community initially scoffed at Sands' idea, he has since been proven right. Without bacteria (or some other tiny particles), clouds would never form, because water vapor droplets are too small—about 250 nm (approximately one ten-thousandth of an inch) or less. A powerful surface tension forms the small curved surfaces on each droplet. Water vapor that goes into the liquid state must overcome this surface tension to form the larger droplets that make up clouds. Left alone, water vapor would never form clouds and it would never rain.

It would never snow, either. Ice formation in clouds is required for snow and even most rainfall. At temperatures above –40°C, however, ice formation is not spontaneous. Tiny catalysts, known as ice nucleators, are required. Scientists knew for years that a protein in the outer membrane of certain bacterial plant pathogens, such as the psychrotolerant *Pseudomonas syringae* and *Erwinia* species, have the capacity to freeze pure water at temperatures as warm as –1°C. The protein binds water molecules in an ordered arrangement, providing a nucleating template that enhances ice crystal formation (**Fig. 1**). The ice crystals that form at this relatively warm temperature break the cell walls of the plants and release nutrients that the bacteria use as food. Ice-nucleating bacteria have been found at altitudes of several kilometers and have been documented in rain and snowfall.

A team led by Brent Christner at Louisiana State University in Baton Rouge has shown the ubiquity of these rainmaking microbes by looking at fresh snow collected at various mid- and high-latitude locations in North America, Europe, and Antarctica. They filtered the snow samples to remove particles, put those particles into containers of pure water, and slowly lowered the temperature, watching closely to see when the water froze. The higher the freezing temperature of any given sample, the greater the number of nuclei and the more likely they were to be biological in nature. To tease apart these two effects, the team treated the water samples with heat or chemicals to kill any bacteria inside, and again checked the freezing temperatures of the samples. In this way they found between 4 and 120 ice nucleators per liter of melted snow. Some 69%–100% of these particles appeared to be biological. In another development, in 2011 Alex Michaud (**Fig. 1C**), a graduate student at Montana State University at the time, discovered living bacteria buried at the core of hailstones that bombarded MSU (one cracked the windshield of his car).

Why is this important? The ability to initiate freezing means that rainmaking bacteria can spur showers as a way of dispersing themselves worldwide. The bacteria facilitate cloud formation, clouds can move large distances in wind currents, and the resulting rain or snow will deposit the bacteria far afield from where they started. So, while climate can affect microbes, microbes can, in turn, affect the weather.

RESEARCH QUESTION

If bacteria can serve as ice nucleation particles, why don't the bacteria freeze and damage themselves?

Achberger, A. M., T. I. Brox, M. L. Skidmore, and B. C. Christner. 2011. Expression and partial characterization of an ice-binding protein from a bacterium isolated at a depth of 3,519 m in the Vostok ice core, Antarctica. *Frontiers in Microbiology* **2**:255.

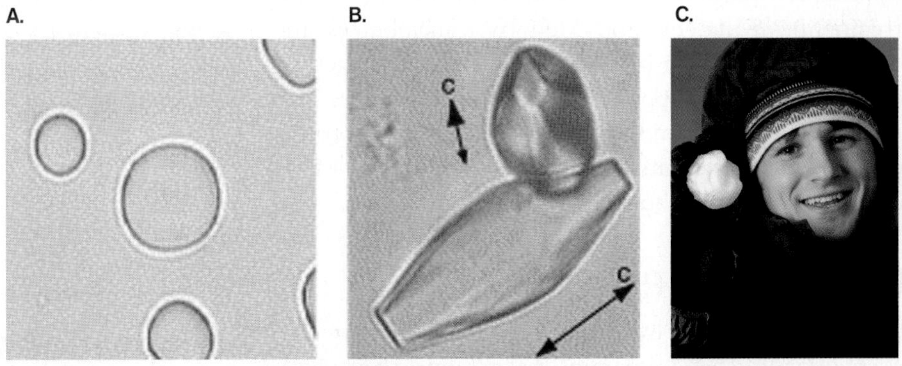
A.
B.
C.

FIGURE 1 ▪ Ice crystallization by a bacterial ice nucleation protein. Photomicrographs of the ice crystal observed for **(A)** solvent buffer alone ([Tris-HCl] = 20 mM (pH 8.0), [EDTA] = 1 mM, [NaCl] = 0.5 M), and **(B)** approximately 400 μM of the ice nucleation protein INP96 from *Pseudomonas syringae* in the buffer. **C.** Alex Michaud, a graduate student at Montana State University, holding one of the hailstones in which he found living bacteria.
Source: Parts A and B from Yoshihiro Kobashigawa et al. 2005. *FEBS Lett.* **579**:1493–1497. © 2005 Federation of European Biochemical Societies; part C from Evelyn Boswell. May 27, 2011. *MSU News Service* (http://montana.edu/cpa/news/nwview.php?article=9907).

part, because they contain relatively low amounts of glycine, a small amino acid that contributes to an enzyme's flexibility (glycines do not contain side chains, so they cannot form stabilizing intramolecular bonds). In addition, the amino termini of proteins in these organisms often are "tied down" by hydrogen bonding to other parts of the protein, making them harder to denature.

Like all microbes, thermophiles have chaperone proteins that help refold other proteins as they undergo thermal denaturation. Thermophile genomes are packed with numerous DNA-binding proteins that stabilize DNA. In addition, these organisms possess special enzymes that function to tightly coil DNA in a way that makes it more thermostable and less likely to denature (think of a coiled phone cord that has twisted and bunched up on itself).

Special membranes also help give cells additional stability at high temperatures. Unlike the typical lipid bilayers of mesophiles, the membranes of thermophiles manage to "glue" together parts of the two hydrocarbon layers that point toward each other, making them more stable. They do this by incorporating more saturated linear lipids into their membranes. Saturated lipids form straight hydrocarbon tails that align well with neighboring lipids and form a highly organized structure stable to heat. The membranes of mesophiles are composed mostly of unsaturated lipids that bend against each other and align poorly. Consequently, the membranes of mesophiles are more fluid at lower temperatures.

The membranes of hyperthermophilic Archaea impart an amazing level of heat resilience by being lipid *monolayers*, not bilayers (Figure A2.3B, Fig 3.11). Lipid bilayers peel apart under withering heat. Monolayers, built for extremophile living, do not. Monolayer membranes are heat stable because long hydrocarbon chains *directly* tether glycerophosphates on opposite sides of the membrane. The chains (C40 length) do not contain fatty acids, but isoprene units bonded by ether linkages to glycerol phosphate. More on thermophiles can be found in chapter 19.

The Heat-Shock Response

As insurance against extinction, most microorganisms possess elegant genetic programs that remodel their physiology to one that can temporarily survive inhospitable conditions. Rapid temperature changes experienced during growth activate batches of stress response genes, resulting in the **heat-shock response** (discussed in Chapter 10). The protein products of these heat-activated genes include chaperones that maintain protein shape and enzymes that change membrane lipid composition. The heat-shock response, first identified in *E. coli* by Tetsuo Yamamori and Takashi Yura in 1982, has since been documented in all living organisms examined thus far.

Thought Question

5.2 If microbes lack a nervous system, how can they sense a temperature change?

The emergence of different branches of life reflects a narrowing of tolerance to heat. Different archaeal species, for example, can grow in extremely hot or extremely cold temperatures, and some can grow in the middle range. Bacteria, for the most part, tolerate temperatures between the archaeal extremes. Eukaryotes are even less temperature tolerant than bacteria, with individual species capable of growth between 10°C and 65°C. As we will see, the evolutionary relationships corresponding to temperature also hold for other environmental conditions.

To Summarize

- **Different species** exhibit different optimal growth values of temperature, pH, and osmolarity.
- **The Arrhenius equation** applies to the growth of microorganisms: Within a specific growth temperature range, the growth rate doubles for every 10°C rise in temperature.
- **Membrane fluidity** varies with the composition of lipids in a membrane, which in turn dictates the temperature at which an organism can grow.
- **Mesophiles, psychrophiles, and thermophiles** are groups of organisms that grow at moderate, low, and high temperatures, respectively.
- **The heat-shock response** produces a series of protective proteins in organisms exposed to temperatures near the upper edge of their growth range.

5.3 Adaptation to Pressure

Living creatures at Earth's surface (sea level) are subjected to a pressure of 1 atmosphere (atm), which is equal to 0.101 megapascal (MPa) or 14 pounds per square inch (psi). At the bottom of the ocean, however, thousands of meters deep, hydrostatic pressure averages a crushing 400 atm and can reach as high as 1,000 atm (110 MPa, or 15,000 psi) in ocean trenches (**Fig. 5.5**). Organisms adapted to grow at these overwhelmingly high pressures are called **barophiles** or **piezophiles**. From the curves in **Figure 5.6**, notice that barophiles actually require elevated pressure to

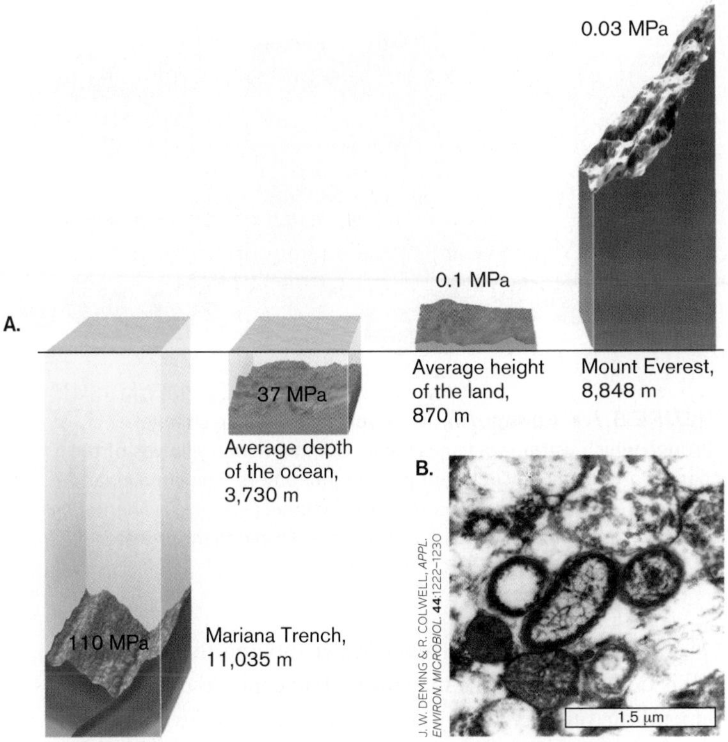
A.
0.03 MPa
0.1 MPa
37 MPa
110 MPa
Average height of the land, 870 m
Mount Everest, 8,848 m
Average depth of the ocean, 3,730 m
Mariana Trench, 11,035 m
B.
J. W. DEMING & R. COLWELL, APPL. ENVIRON. MICROBIOL. 44:1222–1230
1.5 μm

FIGURE 5.5 ■ Barophilic environments and piezophiles. **A.** Ocean depths. The deepest part of the ocean is at the bottom of the Mariana Trench, a depression in the floor of the western Pacific Ocean, just east of the Mariana Islands. The Mariana Trench is 2,500 km (1,554 miles) long and 70 km (44 miles) wide. Near its southwestern extremity, about 340 km (210 miles) southwest of Guam, lies the deepest point on Earth. This point, referred to as the "Challenger Deep," plunges to a depth of 11,035 meters (nearly 7 miles). The pressure there (110 MPa) is over 1,000 times higher than what we experience on land (0.1 MPa). **B.** Barophile *Shewanella violacea.*

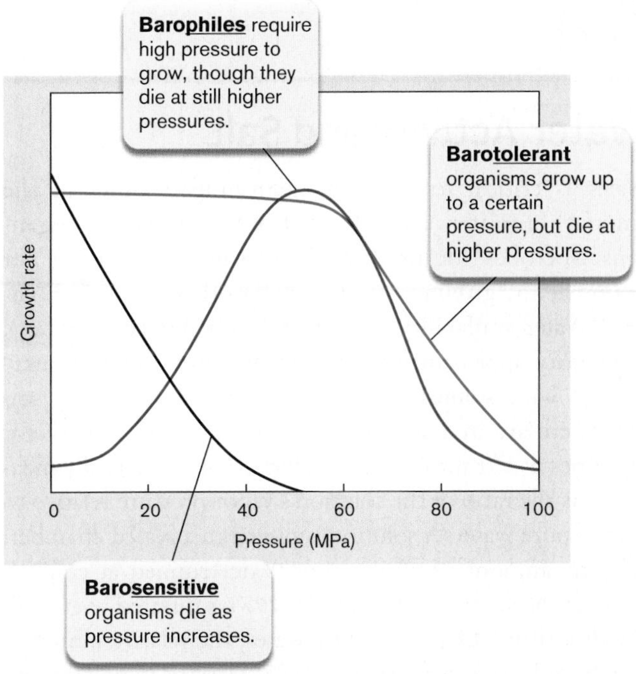
Barophiles require high pressure to grow, though they die at still higher pressures.
Barotolerant organisms grow up to a certain pressure, but die at higher pressures.
Growth rate
0
20
40
60
80
100
Pressure (MPa)
Barosensitive organisms die as pressure increases.

FIGURE 5.6 ■ Relationship between growth rate and pressure.

grow, while barotolerant organisms grow well in the range of 1–50 MPa, but their growth falls off thereafter.

Many barophiles are also psychrophilic, because the average temperature at the ocean's floor is 2°C. However, barophilic hyperthermophiles form the basis of thermal vent communities that support symbiotic worms and giant clams (see Chapter 21).

How bacteria can survive pressures of 80–100 MPa (12,000 psi) is still a mystery. It is known, though, that increased hydrostatic pressure and cold temperatures reduce membrane fluidity. Because fluidity of the cell membrane is critical to survival, the phospholipids of deep-sea bacteria commonly have high levels of polyunsaturated fatty acids to increase membrane fluidity. It is thought that in addition to these membrane changes, internal structures must also be pressure adapted. For example, ribosomes in the barosensitive organism *E. coli* dissociate at pressures above 60 MPa, so barophiles must contain uniquely designed ribosome structures that can withstand pressures even higher than that.

Specific applications for barotolerant proteins and pressure-regulated genes have not yet been developed. However, many food processes are carried out at high pressure to minimize bacterial contamination (destructive bacteria will not tolerate the pressure), so it is expected that barophiles will ultimately offer useful biotechnology products, such as enzymes that will carry out food processing at high pressure. High-pressure processing has been used on cheeses, yogurt, lunch meats, and oysters to kill contaminating bacteria without destroying the flavor or texture of the food.

To Summarize

- **Barophiles (piezophiles)** can grow at pressures up to 1,000 atm but fail to grow at low pressures.
- **Membrane fluidity** can be compromised at high pressures and cold temperatures. Specially designed membranes and protein structures are thought to enable the growth of barophiles.

Thought Question

5.3 What could be a relatively simple way to grow barophiles in the laboratory?

5.4

Water Activity and Salt

Water is critical to life, but environments differ in the amount of water actually available to growing organisms. Microbes, for instance, can use only water that is not bound at any given instant to ions or other solutes in solution. Water availability is measured as **water activity** (a_w), a quantity approximated by concentration. Because interactions with solutes lower water activity, the more solutes there are in a solution, the less water is available for microbes to use for growth. Water activity is typically measured as the ratio of the solution's vapor pressure relative to that of pure water. A solution is placed in a sealed chamber, and the amount of water vapor is determined at equilibrium. If the air above the sample is 97% saturated relative to the moisture present over pure water, the relative humidity is 97% and the water activity is 0.97. Most bacteria growing on land or in freshwater habitats require water activity to be greater than 0.91 (the water activity of seawater). Fungi can tolerate water activity levels as low as 0.86.

Osmotic Stress

Osmolarity is a measure of the number of solute molecules in a solution and is inversely related to a_w. The more particles there are in a solution, the greater the osmolarity and the lower the water activity (see Appendix 2). Osmolarity is important for the cell because it is related to water activity and also because a semipermeable membrane surrounds microbial cells, so osmolarity inside the cell can be, and often is, different from osmolarity outside. The principles of physical chemistry dictate that solute concentrations in two chambers separated by a semipermeable membrane will tend to equilibrate. Equilibrating osmolarity across a semipermeable cell membrane, which does not allow the movement of solutes, requires the movement of water. In hypertonic medium, where the external osmolarity is higher than the internal, water will leave the cell in an attempt to equalize osmolarity across the membrane. In contrast, suspending a cell in a hypotonic medium (one of lower osmolarity than the cell) will cause an influx of water (see Appendix 2, Fig. A2.5).

The movement of water across cell membranes does not occur primarily by simple diffusion. Special membrane water channels formed by proteins called aquaporins enable water to traverse the membrane much faster than by unmediated diffusion and help protect cells against osmotic stress (**Fig. 5.7**). However, too much water moving in or out of a cell is detrimental. Cells may ultimately explode or implode, depending on the direction the water moves. Even bacteria with a rigid cell wall suffer. They may not explode like a human cell, but the forces placed on the cell wall are great.

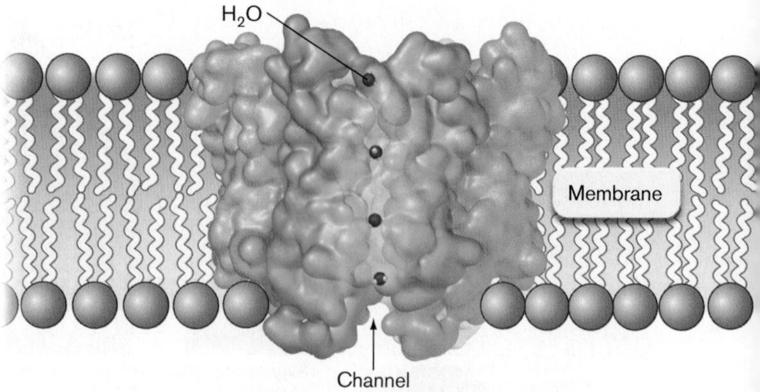

H_2O
Membrane
Channel

FIGURE 5.7 ■ Aquaporin. Transverse view of the channel through which water molecules move. Complementary halves of the channel are formed by adjacent protein monomers. The curvilinear, size-selective (~4 ± 0.5 nm) core of the channel (~18 nm) is lined primarily by hydrophobic residues that speed movement or water through the open pore. (PDB code: 1J4N)

Cells Minimize Osmotic Stress

In addition to moving water, microbes have at least two other mechanisms to minimize osmotic stress across membranes. When stranded in a hypertonic medium (higher osmolarity than the cell), bacteria try to protect their internal water from leaving the cell by synthesizing or importing compatible solutes that increase intracellular osmolarity. Compatible solutes are small molecules that do not disrupt normal cell metabolism (as high levels of Na^+ would do), even when present at high intracellular concentrations. Increasing intracellular levels of these compounds (such as proline, glutamic acid, potassium, or betaine) elevates cytoplasmic osmolarity without any detrimental effects, making it unnecessary for water to leave the cell.

Cells also contain pressure-sensitive (mechanosensitive) channels that can be used to leak solutes out of the cell. It is believed that these channels are activated by rising internal pressures in cells immersed in a hypotonic medium (lower osmolarity than the cell). When activated, the channels allow solutes to escape, which lowers internal osmolarity and prevents too much water from entering the cell.

Outside their osmotic comfort range—that is, where the aforementioned housekeeping strategies become ineffective at controlling internal osmolarity—microbes launch a global response in which cellular physiology is transformed to tolerate brief encounters with potentially lethal salt (or other solute) concentrations. Some changes are similar to those provoked by heat shock, such as the increased synthesis of chaperones that protect critical cell proteins from denaturation. Other changes include alterations in outer membrane pore composition (for Gram-negative organisms).

Thought Question

5.4 How might the concept of water availability be used by the food industry to control spoilage?

Halophiles Require High Salt

Some species of archaea have evolved to require high salt (NaCl) concentration. These are called **halophiles** (**Fig. 5.8**). In striking contrast to most bacteria, which prefer salt concentrations from 0.1 to 1 M (0.2%–5% NaCl), the extremely halophilic archaea can grow at an a_w of 0.75 and actually require 2- to 4-M NaCl (10%–20% NaCl) to grow. For comparison, seawater is about 3.5% NaCl. All cells, even halophiles, prefer to keep a relatively low intracellular Na^+ concentration because some solutes are moved into the cell by symport with Na^+. To achieve a low internal Na^+ concentration, halophilic microbes use special ion pumps to excrete sodium and replace it with other cations, such as potassium, which is a compatible solute. In fact, the proteins and cell components (for example, ribosomes) of halophiles require remarkably high intracellular potassium levels to maintain their structure.

To Summarize

- **Water activity** (a_w) is a measure of how much water in a solution is available for a microbe to use.
- **Osmolarity** is a measure of the number of solute molecules in a solution and is inversely related to a_w.
- **Aquaporins** are membrane channel proteins that allow water to move quickly across membranes to equalize internal and external pressures.
- **Compatible solutes** are used to minimize pressure differences across the cell membrane.
- **Mechanosensitive channels** can leak solutes out of the cell when internal pressure rises.
- **Halophilic organisms** grow best at high salt concentration.

5.5

Adaptation to pH

As with salt and temperature, the concentration of hydrogen ions (H^+)—actually, hydronium ions (H_3O^+)—also has a direct effect on the cell's macromolecular structures. Extreme concentrations of either hydronium or hydroxide ions (OH^-) in a solution will limit growth. In other words, too much acid or base is harmful to cells. Despite this sensitivity to pH extremes, living cells tolerate a greater range

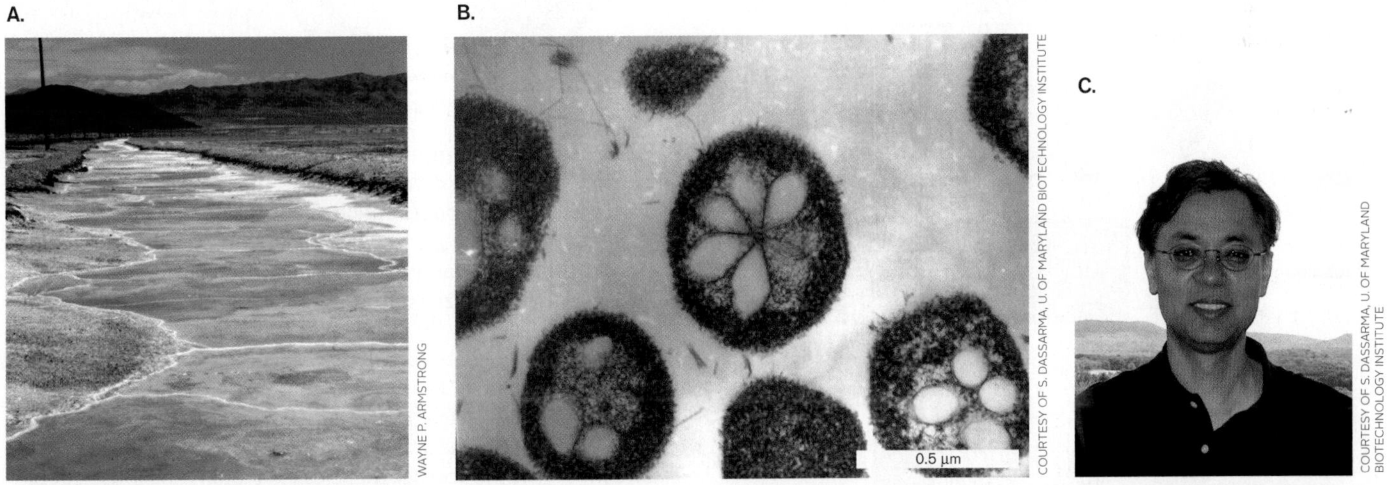
A.
WAYNE P. ARMSTRONG
B.
0.5 μm
COURTESY OF S. DASSARMA, U. OF MARYLAND BIOTECHNOLOGY INSTITUTE
C.
COURTESY OF S. DASSARMA, U. OF MARYLAND BIOTECHNOLOGY INSTITUTE

FIGURE 5.8 ▪ Halophilic salt flats and halophilic bacteria. A. The halophilic salt flats along Highway 50 east of Fallon, Nevada, are colored pinkish red by astronomical numbers of halophilic bacteria. **B.** The archaeon *Halobacterium* sp. (TEM). Cross section; cell width, 0.5–0.8 μm. **C.** Shiladitya DasSarma and colleagues at the University of Maryland completed the genome sequence of *Halobacterium* species NRC-1. They demonstrated novel features of archaeal genetics, including intriguing similarities with molecular regulatory structures in eukaryotes.

in environmental concentration of H^+ than of virtually any other chemical substance. *E. coli*, for example, tolerates a pH range of 2–10, a 10-million-fold difference (but grows only between pH 4.5 to 9). For a brief review of pH, refer to Appendix 1, Section A1.7.

pH Optima, Minima, and Maxima

The charges on various amino or carboxyl groups within a protein help forge the intramolecular bonds that dictate protein shape and thus protein activity. Because H^+ concentration, $[H^+]$, affects the protonation of these ionizable groups, changing the pH can alter the charges on these groups, which in turn changes protein structure and activity. The result is that all enzyme activities exhibit optima, minima, and maxima with regard to pH, much as they do for temperature. As we saw with temperature, groups of microbes have evolved to inhabit diverse niches, for which pH values can range from 0 to 11.5 (**Fig. 5.9**). However, species differences in optimum growth pH are not dictated by the pH limits at which critical cell proteins function.

Generally speaking, the majority of enzymes, regardless of the pH at which their source organism thrives, tend to operate best between pH 5 and 8.5 (which, if you think about it, is still a 3,000-fold range in hydrogen ion concentration). Yet many microbes grow in even more acidic or alkaline environments.

Unlike its temperature, the intracellular pH of a microbe, as well as its osmolarity, is not necessarily the same as that of its environment. Biological membranes are relatively impermeable to protons—a fact that allows the cell to maintain an internal pH compatible with protein function when growing in extremely acidic or alkaline environments. When the difference between the intracellular and extracellular pH (ΔpH) is very high, protons can leak through either directly or via proteins that thread the membrane. Excessive influx or efflux of protons can cause problems by altering internal pH.

Membrane-permeant organic acids, also called weak acids (discussed in Chapter 3), can accelerate the leakage of H^+. Unlike H^+, the uncharged form of an organic acid (HA) can freely permeate cell membranes and dissociate intracellularly, releasing a proton that then acidifies internal pH (**Fig. 5.10**). The extent of the pH drop depends on the buffering capacity of the cell's proteins. This shuttling of protons can turn a relatively mild external pH level (say, pH 6) into a deadly acid stress. A naturally occurring example of organic acid stress is the lactic acid produced by lactobacilli during the formation of yogurt. The buildup of lactic acid limits the bacterial growth, leaving yogurt with plenty of food value. The food industry has taken advantage

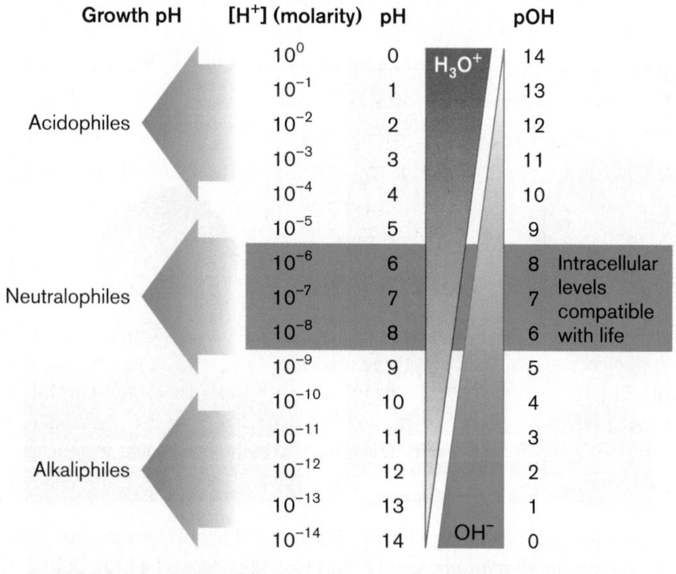

Growth pH
[H+] (molarity)
pH
pOH
Acidophiles
Neutralophiles
Alkaliphiles
H3O+
OH−
Intracellular levels compatible with life

FIGURE 5.9 ■ Classification of organisms according to their optimum growth pH. pOH is the $\log_{10}$ of the reciprocal of the hydroxide ion (OH^-) ion concentration; that is, $pOH = -\log[OH^-]$.

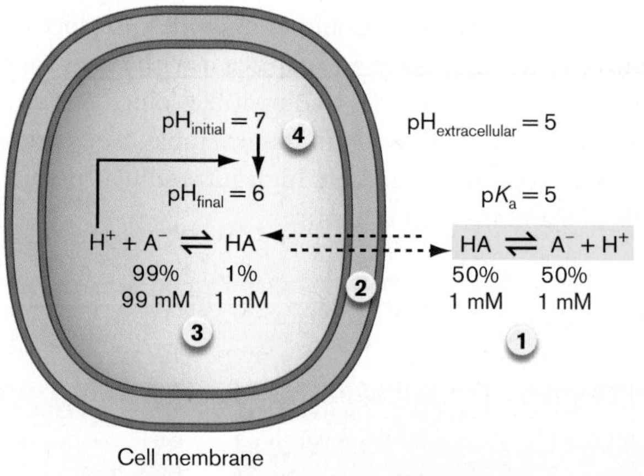

pH initial = 7
pH final = 6
pH extracellular = 5
pKa = 5
H+ + A− ⇌ HA
99% 1%
99 mM 1 mM
HA ⇌ A− + H+
50% 50%
1 mM 1 mM
Cell membrane

FIGURE 5.10 ■ Membrane-permeant organic acids depress internal pH. In this contrived example, the organic acid (HA) has an extracellular concentration of 2 mM and has a dissociation constant of pK_a 5 (1). Because the medium is pH 5, half of the acid is protonated (undissociated, or un-ionized) and half is dissociated (ionized). The un-ionized form, because it is uncharged, diffuses across the membrane (2) to establish equilibrium between the inside and outside of the cell. However, because the inside of the cell is pH 7 (two units above the external pH), 99% of the acid will dissociate, which lowers the internal HA concentration, so more HA enters the cell in search of equilibrium (3). Because neither the ionized form of the acid nor the released proton can diffuse out of the cell, both accumulate. At equilibrium, the concentrations of HA inside and outside the cell are equal, but for every HA that enters the cell, ionization of HA has yielded 99 A^- molecules and an equal number of protons. The protons lower the internal pH to an extent that depends on the buffering capacity of cellular proteins (4).

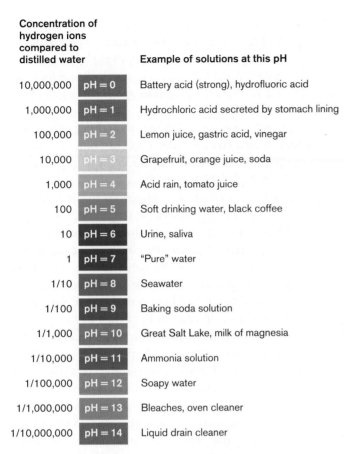

Concentration of hydrogen ions compared to distilled water		Example of solutions at this pH
10,000,000	pH = 0	Battery acid (strong), hydrofluoric acid
1,000,000	pH = 1	Hydrochloric acid secreted by stomach lining
100,000	pH = 2	Lemon juice, gastric acid, vinegar
10,000	pH = 3	Grapefruit, orange juice, soda
1,000	pH = 4	Acid rain, tomato juice
100	pH = 5	Soft drinking water, black coffee
10	pH = 6	Urine, saliva
1	pH = 7	"Pure" water
1/10	pH = 8	Seawater
1/100	pH = 9	Baking soda solution
1/1,000	pH = 10	Great Salt Lake, milk of magnesia
1/10,000	pH = 11	Ammonia solution
1/100,000	pH = 12	Soapy water
1/1,000,000	pH = 13	Bleaches, oven cleaner
1/10,000,000	pH = 14	Liquid drain cleaner

FIGURE 5.11 ▪ pH values of common substances.

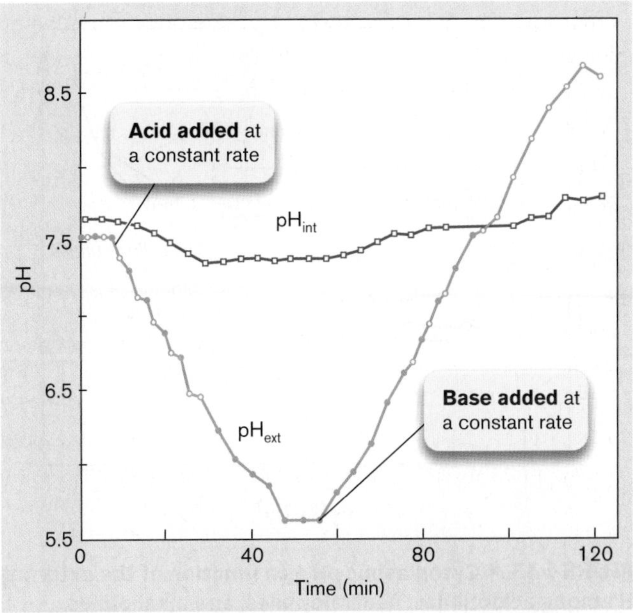

Acid added at a constant rate
Base added at a constant rate
pH$_{int}$
pH$_{ext}$
8.5
7.5
6.5
5.5
pH
0
40
80
120
Time (min)

FIGURE 5.12 ▪ Maintaining internal pH (pH homeostasis) over a wide range of external pH. Internal pH of the neutralophile *Escherichia coli* measured following the addition of acid (HCl, at t = 10 min) to change external pH (circles) and subsequent addition of base (NaOH, at t = 52 min). In this experiment, internal pH (pH$_{int}$) was determined using nuclear magnetic resonance to measure changes in methyl phosphate (open circles/squares). The two phosphate species titrate over different pH ranges. *Source:* Joan L. Slonczewski et al. 1981. *PNAS* **78**:6271

of this phenomenon by preemptively adding citric acid or sorbic acid to certain foods. This practice allows manufacturers to control microbial growth under pH conditions that do not destroy the flavor or quality of the food. **Figure 5.11** provides the pH values of various everyday items. Food microbiology is discussed further in Chapter 16.

Neutralophiles, Acidophiles, and Alkaliphiles Grow in Different pH Ranges

Cells have evolved to live under different pH conditions not by drastically changing the pH optima of their enzymes but by using novel pH homeostasis strategies that maintain intracellular pH above pH 5 and below pH 8, even when the cell is immersed in pH environments well above or below that range.

Three classes of organisms are differentiated by the pH of their growth range: neutralophiles, acidophiles, and alkaliphiles.

Neutralophiles generally grow between pH 5 and pH 8, and include most human pathogens. Many neutralophiles, including *E. coli* and *Salmonella enterica,* adjust their metabolism to maintain an internal pH slightly above neutrality, which is where their enzymes work best. They maintain this pH even in the presence of moderately acidic or basic external environments (**Fig. 5.12**). Other neutralophiles allow their internal pH to fluctuate with external pH but usually maintain a pH difference (ΔpH) of about 0.5 pH unit across the membrane at the upper and lower limits of growth pH. The ΔpH value is an important component of the transmembrane proton potential, a source of energy for the cell (see Chapter 14).

Note: The older term "neutrophile" used for this group of organisms is similar to the descriptor for a specific type of white blood cell ("neutrophil"). To avoid confusion, the term "neutrophil" should be reserved for the white blood cell and the term "neutralophile" used to designate microbes with growth optima near neutral pH (pH 7).

Acidophiles are bacteria and archaea that live in acidic environments. They are often chemoautotrophs (lithotrophs) that oxidize reduced metals and generate strong

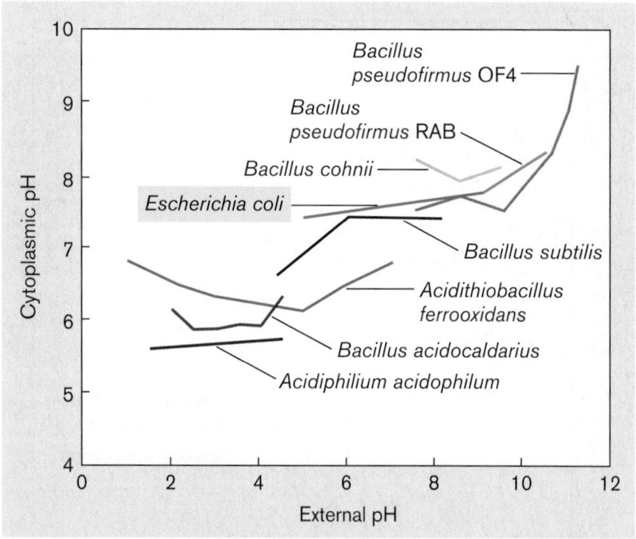
Cytoplasmic pH
External pH
10
9
8
7
6
5
4
0
2
4
6
8
10
12
Bacillus pseudofirmus OF4
Bacillus pseudofirmus RAB
Bacillus cohnii
Escherichia coli
Bacillus subtilis
Acidithiobacillus ferrooxidans
Bacillus acidocaldarius
Acidiphilium acidophilum

FIGURE 5.13 ■ Cytoplasmic pH as a function of the external pH among acidophiles, neutralophiles, and alkaliphiles.
Source: Joan L. Slonczewski et al. 2009. *Adv. Microb. Physiol.* **55**:1–79

acids such as sulfuric acid. Consequently, they grow between pH 0 and pH 5. Acidophiles generally maintain an internal pH that is considerably more acidic than that of neutralophiles but still less acidic than their growth environment (**Fig. 5.13**). The ability to grow at this pH is due partly to altered membrane lipid profiles (high levels of tetraether lipids) that decrease proton permeability, as well as to ill-defined proton extrusion mechanisms. Often, an organism that is an extremophile with respect to one environmental factor is an extremophile with respect to others. *Sulfolobus acidocaldarius,* for example, is a thermophile and an acidophile (**Fig. 5.14**). It uses sulfur as an energy source and grows in acidic hot springs rich in sulfur.

Alkaliphiles occupy the opposite end of the pH spectrum, growing best at values ranging from pH 9 to pH 11. They are commonly found in saline soda lakes, which have high salt concentrations and pH values as high as pH 11.

Soda lakes, like Lake Magadi in Kenya's Great Rift Valley (**Fig. 5.15A**), are steeped in carbonates, which explains their extraordinarily alkaline pH. An alkaliphilic organism first identified in Lake Magadi is *Halobacterium salinarum* (also known as *Natronobacterium gregoryi*), a halophilic archaeon (**Fig. 5.15B**).

The cyanobacterium *Spirulina* is another alkaliphile that grows in soda lakes. Its high concentration of carotene gives the organism a distinctive pink color (note the color of the lake in **Fig. 5.15A**). *Spirulina* is also a major food for the famous pink flamingos indigenous to these African lakes and is, in fact, the reason pink flamingos are pink. After the birds ingest these organisms, digestive processes release the carotene pigment to the circulation, which then deposits it in the birds' feathers, turning them pink (**Fig. 5.15C**). Humans also consume *Spirulina,* as a health food supplement, but do not turn pink, because the cyanobacteria are only a small component of their diet.

The internal enzymes of alkaliphiles, like those of acidophiles, exhibit rather ordinary pH optima (around pH 8). The key to the survival of alkaliphiles is the cell surface barrier that sequesters fragile cytoplasmic enzymes away from harsh extracellular pH. Key structural features of the cell wall, such as the presence of acidic polymers and an excess of hexosamines in the peptidoglycan, appear to be essential. The reason is unclear. At the membrane, some alkaliphiles also possess a high level of diether lipids (more stable than ester-linked phospholipids), which prevent protons from leaking out of the cell (see Section 3.3).

Because external protons are in such short supply at alkaline pH, most alkaliphiles use a sodium motive force in addition to a proton motive force to do much of the work of the cell (see Section 14.2). They also rely heavily on Na^+/H^+ antiporters (see Section 4.2) to bring protons into the cell. This H^+ influx keeps the internal pH well below the extremely alkaline external pH. This is partly why many alkaliphiles are resistant to high salt (NaCl) concentrations: Sodium ions are expelled while protons are sucked in.

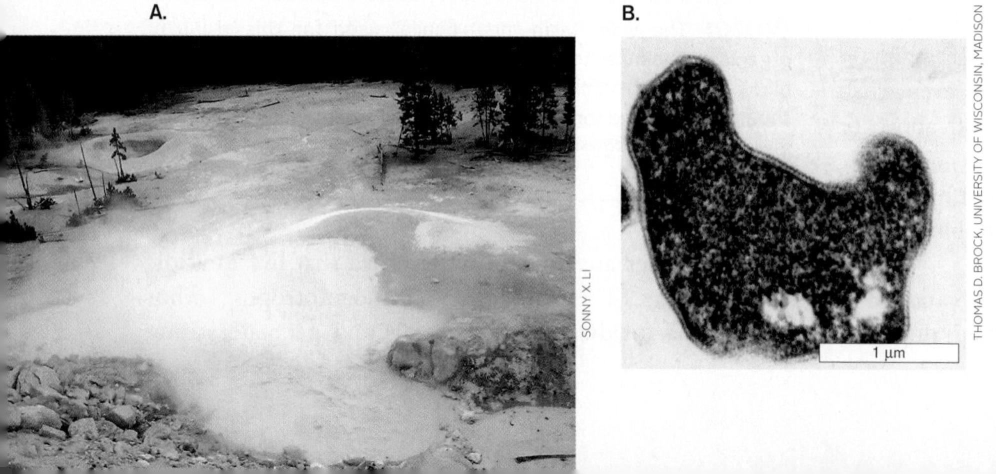
A.
B.
SONNY X. LI
THOMAS D. BROCK, UNIVERSITY OF WISCONSIN, MADISON
1 μm

FIGURE 5.14 ■ Sulfur Caldron acid spring and *Sulfolobus acidocaldarius.* **A.** Sulfur Caldron, in the Mud Volcano area of Yellowstone National Park, is one of the most acidic springs in the park. It is rich in sulfur and in *Sulfolobus,* a bacterium that thrives in hot, acidic waters with temperatures from 60°C to 95°C and a pH of 1–5. **B.** Thin-section electron micrograph of *S. acidocaldarius.* Under the electron microscope, the organisms appear as irregular spheres that are often lobed.

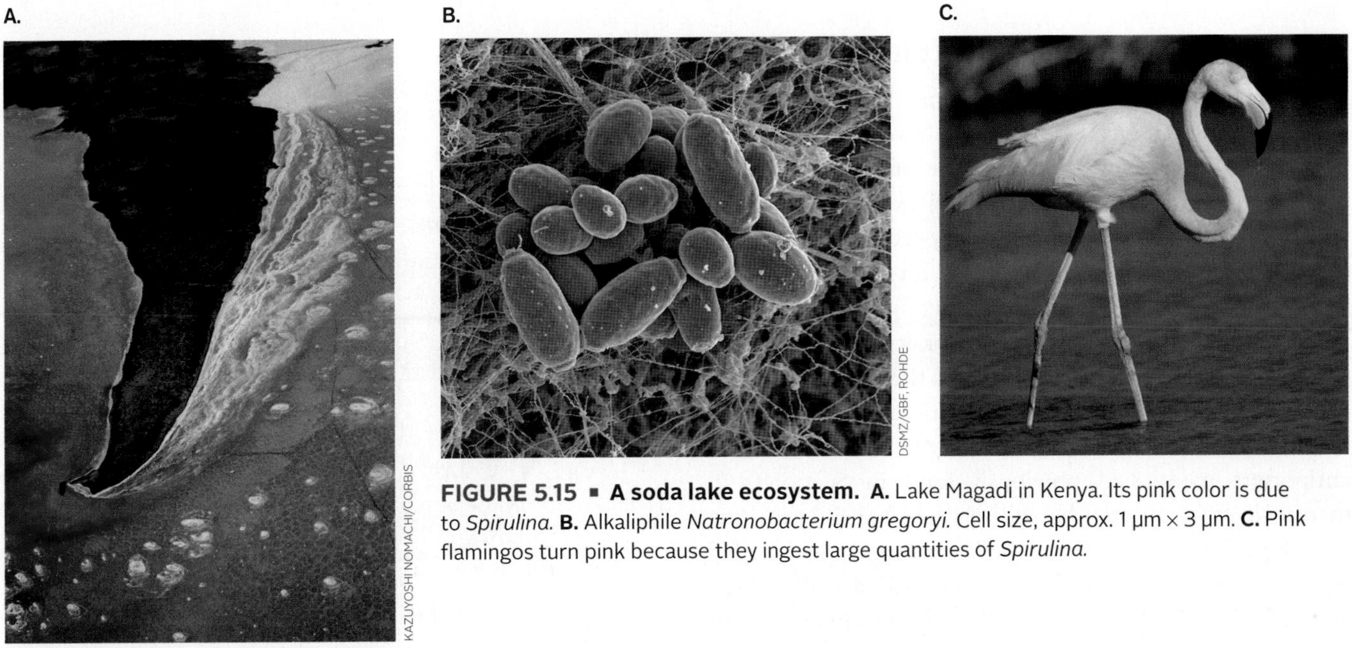
A.
B.
C.
KAZUYOSHI NOMACHI/CORBIS
DSMZ/GBF, ROHDE
WOLFGANG KAEHLER/CORBIS

FIGURE 5.15 ■ A soda lake ecosystem. **A.** Lake Magadi in Kenya. Its pink color is due to *Spirulina*. **B.** Alkaliphile *Natronobacterium gregoryi*. Cell size, approx. 1 μm × 3 μm. **C.** Pink flamingos turn pink because they ingest large quantities of *Spirulina*.

Important aspects of sodium circulation in alkaliphiles are depicted in **Figure 5.16**.

In contrast to proteins within the cytoplasm, enzymes secreted from alkaliphiles are able to work in very alkaline environments. The inclusion of base-resistant enzymes like proteases, lipases, and cellulases in laundry detergents helps get our "whites whiter and our brights brighter." Other commercially useful alkaliphilic enzymes include cyclodextrin glucanotransferase, which produces cyclodextrins from starch (discussed in eTopic 5.2).

Thought Question

5.5 Recall from Section 4.2 that an antiporter couples movement of one ion down its concentration gradient with movement of another molecule uphill, against its gradient. If this is true, how could a Na^+/H^+ antiporter work to bring protons into a haloalkaliphile growing in high salt at pH 10? Since the Na^+ concentration is lower inside the cell than outside and the H^+ concentration is higher inside than outside (see **Fig. 5.16**), both ions are moving against their gradients.

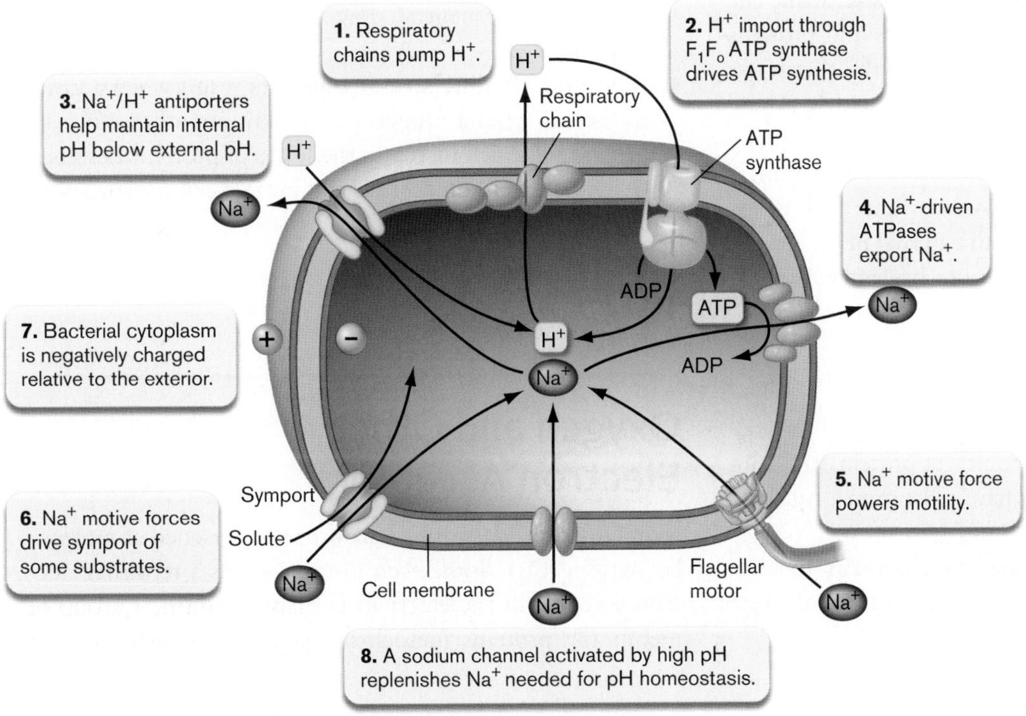
1. Respiratory chains pump H+.
2. H+ import through F1Fo ATP synthase drives ATP synthesis.
3. Na+/H+ antiporters help maintain internal pH below external pH.
4. Na+-driven ATPases export Na+.
5. Na+ motive force powers motility.
6. Na+ motive forces drive symport of some substrates.
7. Bacterial cytoplasm is negatively charged relative to the exterior.
8. A sodium channel activated by high pH replenishes Na+ needed for pH homeostasis.
H+
Na+
Respiratory chain
ATP synthase
ADP
ATP
Symport
Solute
Cell membrane
Flagellar motor
+
–

FIGURE 5.16 ■ Na^+ circulation in alkaliphiles. Cells of alkaliphiles are designed to use Na^+ in place of H^+ to do some of the work of the cell. They require an inwardly directed sodium gradient that can be used to rotate flagella and transport nutrient solutes. However, a proton motive force is still required to generate energy (ATP). The Na^+/H^+ antiporter is used to keep internal pH lower than external pH. Sodium homeostasis inside the cell is maintained by the net effects of efflux and influx mechanisms.

pH Homeostasis and Acid Tolerance

When cells are placed in pH conditions below their optimum, protons can enter the cell and lower internal pH to lethal levels. Microbes can prevent the unwanted influx of protons in a variety of ways. *E. coli*, for example, can counter proton influx by transporting a variety of cations such as K^+ or Na^+. How cation transport accomplishes H^+ efflux is unclear. Some evidence suggests a link to the role of K^+ in osmoprotection. At the other extreme, under extremely alkaline conditions, the cells can use the Na^+/H^+ antiporters mentioned previously (and in Section 4.2) to recruit protons into the cell in exchange for expelling Na^+ (**Fig. 5.17**). Some organisms can also change the pH of the medium using various amino acid decarboxylases and deaminases. For instance, *E. coli* consumes organic acids when growing at low pH, but produces these acids while trying to grow under alkaline conditions. *Helicobacter pylori*, the causative agent of gastric ulcers, employs an exquisitely potent urease to generate massive amounts of ammonia, which neutralizes the acid pH environment. These acid stress and alkaline stress protection systems are usually not made or at least do not become active until the cell encounters an extreme pH.

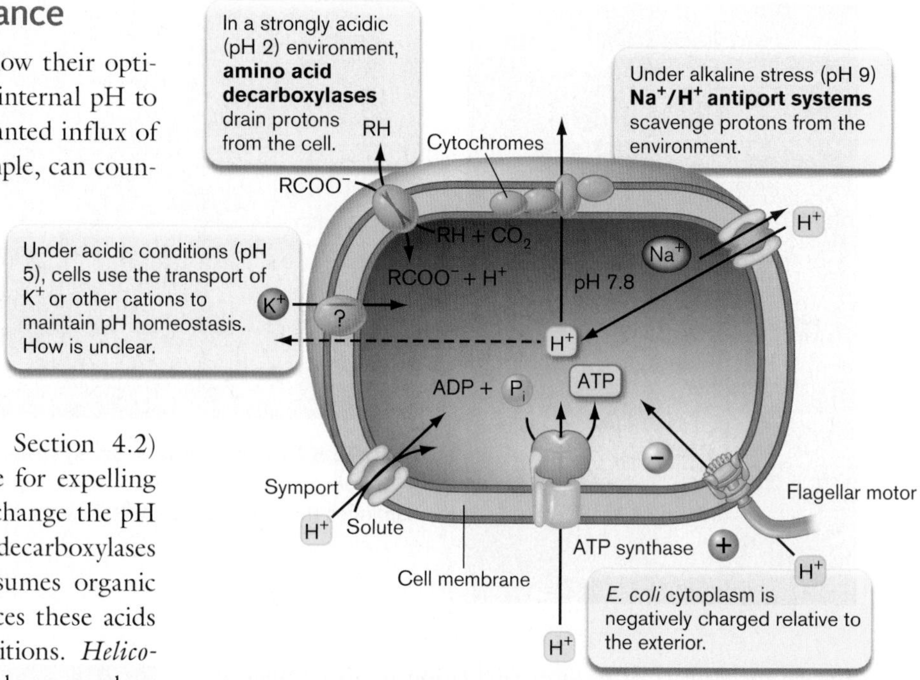
In a strongly acidic (pH 2) environment, **amino acid decarboxylases** drain protons from the cell.
RH
Cytochromes
Under alkaline stress (pH 9) **Na⁺/H⁺ antiport systems** scavenge protons from the environment.
RCOO⁻
RH + CO2
H⁺
Na⁺
RCOO⁻ + H⁺
pH 7.8
Under acidic conditions (pH 5), cells use the transport of K⁺ or other cations to maintain pH homeostasis. How is unclear.
K⁺
?
H⁺
ADP + Pi
ATP
Symport
Flagellar motor
H⁺
Solute
ATP synthase
Cell membrane
H⁺
E. coli cytoplasm is negatively charged relative to the exterior.
H⁺

FIGURE 5.17 ■ Proton circulation and pH homeostasis. A typical *E. coli* cell uses various proton transport strategies to maintain an internal pH near pH 7.8 in the face of different external pH stresses. Proton pumping through cytochromes also establishes a proton gradient, which drives flagellar rotation and solute transport.

Many, if not all, microbes also possess an emergency global response system referred to as acid tolerance or acid resistance. In a process analogous to the heat-shock response, bacterial physiology undergoes a major molecular reprogramming in response to hydrogen ion stress. The levels of a large number of proteins increase, while the levels of others decrease. Many of the genes and proteins involved in the acid stress response overlap with other stress response systems, including the heat-shock response. These physiological responses include modifications in membrane lipid composition, enhanced pH homeostasis, and numerous other changes with unclear purpose. Some pathogens, such as *Salmonella,* sense a change in external pH as part of the signal indicating that the bacterium has entered a host cell environment (see **eTopic 5.3**).

To Summarize

- **Hydrogen ion concentration** affects protein structure and function. Thus, enzymes have pH optima, minima, and maxima.
- **Microbes use pH homeostasis mechanisms** to keep their internal pH near neutral when in acidic or alkaline media.
- **Adding weak acids** to certain foods undermines bacterial pH homeostasis mechanisms, thereby preventing food spoilage and killing potential pathogens.
- **Neutralophiles, acidophiles, and alkaliphiles** prefer growth under neutral, low, and high pH conditions, respectively.
- **Acid and alkaline stress responses** result when a given species is placed under pH conditions that slow its growth. The cell increases the levels of proteins designed to mediate pH homeostasis and protect cell constituents.

5.6

Oxygen and Other Electron Acceptors

Many microorganisms can grow in the presence of molecular oxygen (O_2). Some even use oxygen as a terminal electron acceptor in the **electron transport chain**, a group of membrane proteins (cytochromes) that can help convert

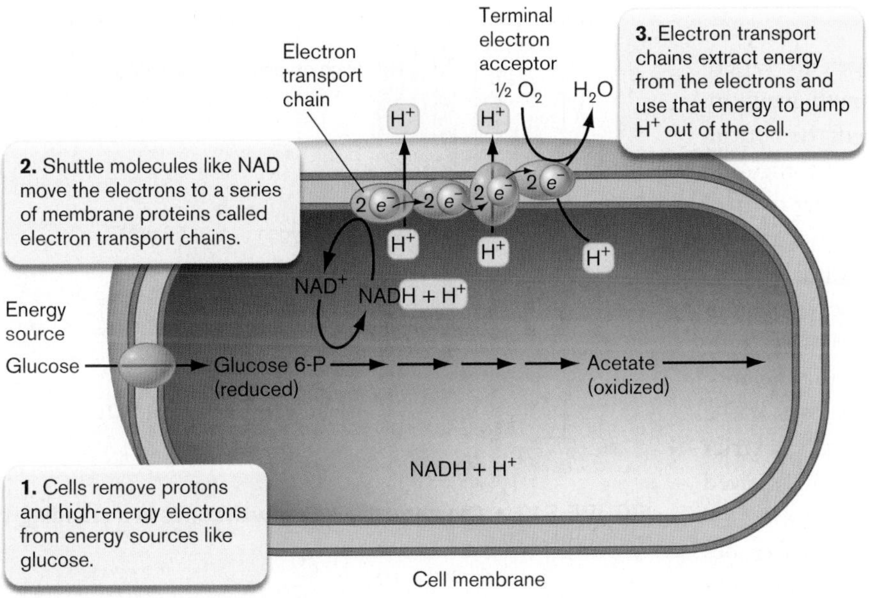
Electron transport chain
Terminal electron acceptor
½ O_2
H_2O
3. Electron transport chains extract energy from the electrons and use that energy to pump H^+ out of the cell.
H^+
2. Shuttle molecules like NAD move the electrons to a series of membrane proteins called electron transport chains.
2 e^-
NAD^+
NADH + H^+
Energy source
Glucose
Glucose 6-P (reduced)
Acetate (oxidized)
NADH + H^+
1. Cells remove protons and high-energy electrons from energy sources like glucose.
Cell membrane

FIGURE 5.18 ■ The role of oxygen as a terminal electron acceptor in respiration. Pumping protons out of the cell by electron transport chains produces more positive charges outside the cell than inside, resulting in an electrochemical gradient (also called proton motive force). An electron from a high-energy source is passed from one member of the chain to the next. With each transfer of an electron to the next member of the chain, more energy is extracted and converted to proton motive force. At the end of the chain, the electron must be passed to a final (terminal) electron acceptor (for example, O_2), thus clearing the path for the next electron. This net process is called respiration.

energy trapped in nutrients to a biologically useful form. The use of O_2 as the terminal electron acceptor is called **aerobic respiration** (see Chapter 14).

Oxygen Has Benefits and Risks

Electrons pulled from various energy sources (for example, glucose) possess intrinsic energy that cytochromes incrementally extract and then use to move protons out of the cell. This unequal distribution of H^+ across the membrane produces a transmembrane electrochemical gradient, a sort of "biobattery" called the proton motive force (details are discussed in Section 14.2). Once the cell has drained as much energy as possible from an electron, that electron must be passed to a final (terminal) electron acceptor molecule that diffuses away in the medium. This clears the way for another electron to be passed down the electron transport chain (**Fig. 5.18**).

One such terminal electron acceptor is dissolved oxygen. However, oxygen and its breakdown products are dangerously reactive—a serious problem for all cells. As a result, different species have evolved to either tolerate or avoid oxygen altogether. **Table 5.2** gives examples of microbes that grow at different levels of oxygen.

TABLE 5.2

Examples of aerobes and anaerobes.

Aerobic	Facultative	Microaerophilic	Anaerobic
Neisseria spp. Causative organisms of meningitis, gonorrhea	*Escherichia coli* Normal gut biota; additional pathogenic strains	*Helicobacter pylori* Causative organism of gastric ulcers	*Azoarcus tolulyticus* Degrades toluene
Pseudomonas fluorescens Found in soil; degrade TNT and aromatic hydrocarbons	*Saccharomyces cerevisiae* Yeast; used in baking	*Lactobacillus* spp. Ferments milk to form yogurt	*Bacteroides* spp. Normal gut biota
Azotobacter spp. Soil microorganisms; fix atmospheric nitrogen	*Bacillus anthracis* Cause of anthrax	*Campylobacter* spp. One cause of gastroenteritis	*Clostridium* spp. Soil microorganisms; causative agents of tetanus and botulism
Rhizobium spp. Soil microorganisms; plant symbionts	*Vibrio cholerae* Cause of cholera	*Treponema pallidum* Cause of syphilis	*Actinomyces* spp. Soil microorganisms; synthesize antibiotics
	Staphylococcus spp. Found on skin; causes boils		*Desulfovibrio* spp. Reduce sulfate

Aerobes vs. Anaerobes

The relationships between microbes and oxygen are varied. **Figure 5.19** illustrates a test tube with growth medium. The top of the tube, closest to air, is oxygenated; the bottom of the tube has much lower levels of oxygen. Some microbes grow only at the top of the tube, while others prefer to grow at the bottom; the distributions are based on each organism's relationship with oxygen. A **strict aerobe** is an organism that not only exists in oxygen but also uses oxygen as a terminal electron acceptor. The strict aerobe grows only with oxygen present and consumes oxygen during metabolism (aerobic respiration). An aerobe will grow only at the top of the tube shown in **Figure 5.19**. In contrast, a **strict anaerobe** dies in the least bit of oxygen (>5 μM dissolved O_2). Strict anaerobes do not use oxygen as an electron acceptor, but this is not why they die in air. Some anaerobes die because they are vulnerable to reactive oxygen molecules (also called reactive oxygen species, or ROS) produced by their own metabolism. Other anaerobes have enzymes that can protect them from ROS, but the dissolved oxygen raises the redox potential to a point that interferes with the use of alternative (non-oxygen) electron acceptors the organism needs to make energy. Anaerobes will grow at the bottom of the tube shown in **Figure 5.19**. As we will discuss later, some bacteria previously considered strict anaerobes are not so strict.

What makes reactive oxygen species? Any organism that possesses NADH dehydrogenase 2—aerobe or anaerobe—will, in the presence of oxygen, inadvertently autooxidize the FAD (flavin adenine dinucleotide) cofactor within the enzyme and produce dangerous amounts of superoxide radicals ($^{\bullet}O_2^-$; **Fig. 5.20**). Superoxide will degrade to hydrogen peroxide (H_2O_2), another reactive molecule. Iron, present as a cofactor in several enzymes, can then catalyze a reaction with hydrogen peroxide to produce the highly toxic hydroxyl radical ($^{\bullet}OH$). All of these molecules seriously damage DNA, RNA, proteins, and lipids. Consequently, oxygen is actually an extreme environment in which survival requires special talents. Aerobes destroy reactive oxygen species with an ample supply of enzymes such as superoxide dismutase (to remove superoxide) and peroxidase and catalase (to remove hydrogen peroxide). Aerobes also have resourceful enzyme systems that detect and repair macromolecules damaged by oxidation.

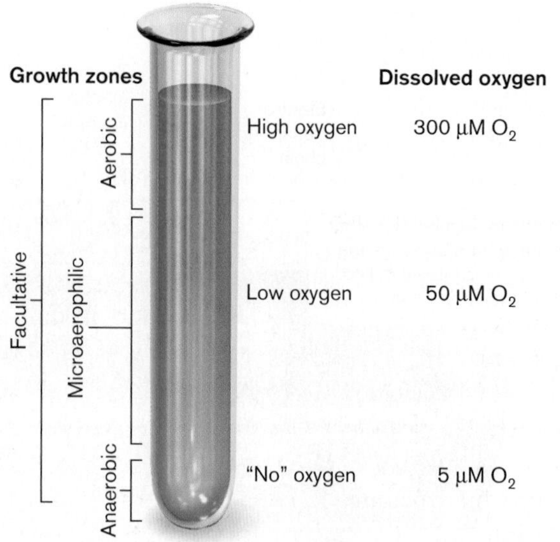
Growth zones
Dissolved oxygen
Aerobic
High oxygen
300 μM O_2
Facultative
Microaerophilic
Low oxygen
50 μM O_2
Anaerobic
"No" oxygen
5 μM O_2

FIGURE 5.19 ■ Oxygen-related growth zones in a standing test tube.

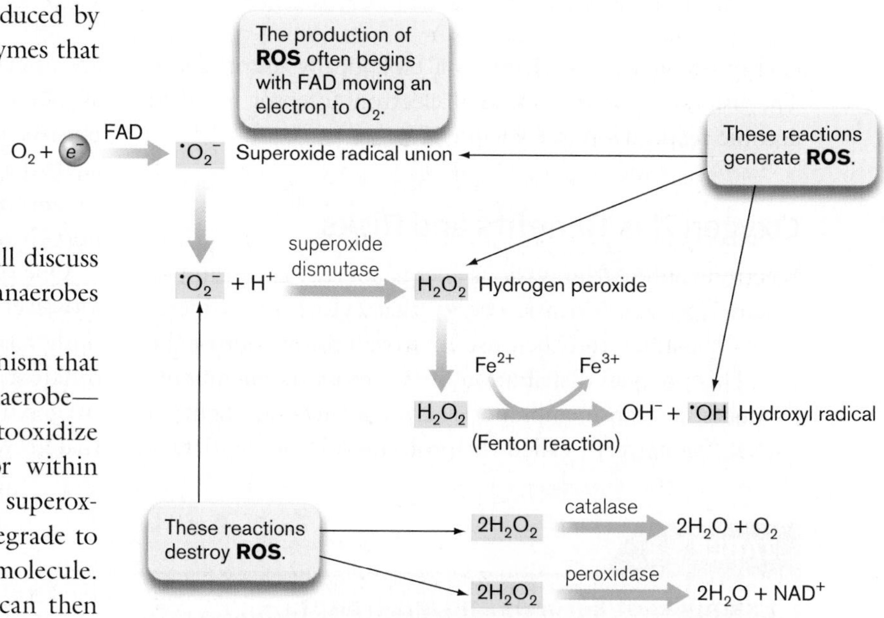
The production of **ROS** often begins with FAD moving an electron to O_2.
O_2 + e^-
FAD
$^{\bullet}O_2^-$ Superoxide radical union
These reactions generate **ROS**.
superoxide dismutase
$^{\bullet}O_2^-$ + H^+
H_2O_2 Hydrogen peroxide
Fe^{2+}
Fe^{3+}
H_2O_2
OH^- + $^{\bullet}OH$ Hydroxyl radical
(Fenton reaction)
These reactions destroy **ROS**.
$2H_2O_2$
catalase
$2H_2O + O_2$
$2H_2O_2$
peroxidase
$2H_2O + NAD^+$

FIGURE 5.20 ■ Generation and destruction of reactive oxygen species (ROS). The autooxidation of flavin adenine dinucleotide (FAD) and the Fenton reaction occur spontaneously to produce superoxide and hydroxyl radicals, respectively. The other reactions require enzymes. FAD is a cofactor for a number of enzymes (for example, NADH dehydrogenase 2). Catalase and peroxidase do not produce ROS but detoxify hydrogen peroxide.

Anaerobes vs. Facultative Microbes

Anaerobic microbes fall into several categories. Some anaerobes actually do respire using electron transport systems, but instead of using oxygen, they rely on alternative terminal electron acceptors like nitrate (NO_3^-) to conduct **anaerobic respiration** and produce energy. Anaerobes of another ilk do not possess cytochromes, cannot respire, and so must rely on carbohydrate **fermentation** for energy (that is, they conduct **fermentative metabolism**). In fermentation, ATP energy is produced through substrate-level phosphorylation in a process that does not involve cytochromes. In either case, tolerance for ROS is low.

Facultative organisms are microbes that can live with or without oxygen and grow throughout the tube shown in **Figure 5.19**. **Facultative anaerobes** (such as *E. coli*) possess enzymes that destroy toxic oxygen by-products, but they have both fermentative and respiratory potential. Whether a member of this group uses aerobic respiration, anaerobic respiration, or fermentation depends on the availability of oxygen and the amount of carbohydrate present. **Aerotolerant anaerobes** use only fermentation to provide energy but contain superoxide dismutase and catalase (or peroxidase) to protect them from ROS. These enzymes allow aerotolerant anaerobes to grow in air (containing oxygen) while retaining a fermentation-based (anaerobic) metabolism. Microorganisms that possess decreased levels of superoxide dismutase and/or catalase will be **microaerophilic**, meaning they will grow only at low oxygen concentrations.

Some organisms previously considered anaerobes (for example, *Bacteroides fragilis*) are really transiently aerotolerant because they possess low levels of ROS protective enzymes and can even use very low levels of oxygen as terminal electron acceptors. *B. fragilis*, part of the normal gastrointestinal microbiota, may even help lower O_2 levels in the intestine.

The fundamental composition of all cells reflects their evolutionary origin as anaerobes. Lipids, nucleic acids, and amino acids are all highly reduced—which is why our bodies are combustible. We never would have evolved that way if molecular oxygen had been present from the beginning. Even today, the majority of all microbes are anaerobic, growing buried in the soil, within our anaerobic digestive tract, or within biofilms on our teeth.

Thought Questions

5.6 If anaerobes cannot live in oxygen, how do they incorporate oxygen into their cellular components?

5.7 How can anaerobes grow in the human mouth when there is so much oxygen there?

Culturing Anaerobes in the Laboratory

Many anaerobic bacteria cause horrific human diseases, such as tetanus, botulism, and gangrene. Some of these organisms or their secreted toxins are even potential weapons of terror (for example, *Clostridium botulinum*). Because of their ability to wreak havoc on humans, culturing these microorganisms was an early goal of microbiologists. Despite the difficulties involved, conditions were eventually contrived in which all, or at least most, of the oxygen could be removed from a culture environment.

Three oxygen-removing techniques are used today. Special reducing agents (for example, thioglycolate) or enzyme systems (such as Oxyrase) that eliminate dissolved oxygen can be added to ordinary liquid media. Anaerobes can then grow beneath the culture surface. A second, very popular way to culture anaerobes, especially on agar plates, is to use an anaerobe jar (**Fig. 5.21A**). Agar plates streaked with the organism are placed into a sealed jar with a foil packet that releases H_2 and CO_2 gases. A palladium packet hanging from the jar lid catalyzes a reaction between the H_2 and O_2 in the jar to form H_2O and effectively removes O_2 from the chamber. The CO_2 released is required by some reactions to produce key metabolic intermediates. Some microaerophilic microbes, like the pathogens *Helicobacter pylori* (the major

A.

Catalyst in lid mediates reaction. $H_2 + ½O_2 \rightarrow H_2O$
GasPak envelope generates H_2 and CO_2.
GasPak
JACK BOSTRACK/VISUALS UNLIMITED

B.

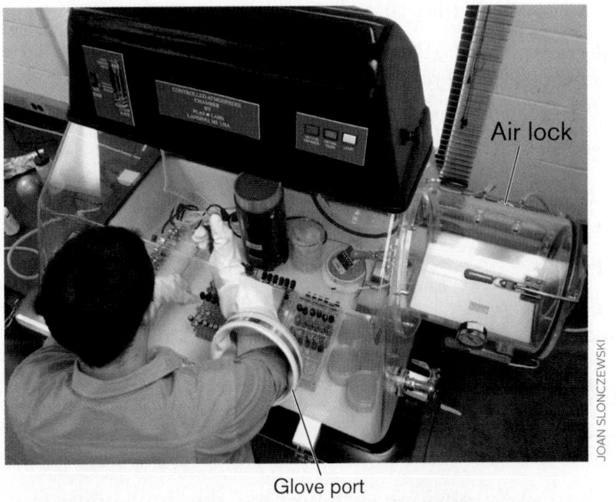
Air lock
Glove port
JOAN SLONCZEWSKI

FIGURE 5.21 ■ Anaerobic growth technology. A. An anaerobe jar. **B.** Student researcher using an anaerobic chamber with glove ports.

cause of stomach ulcers) and *Campylobacter jejuni* (a major cause of diarrhea), require low levels of O_2 but elevated amounts of CO_2. These conditions are obtained by using similar gas-generating packets.

For strict anaerobes exquisitely sensitive to oxygen, even more heroic efforts are required to establish an oxygen-free environment. A special anaerobic glove box must be used in which the atmosphere is removed by vacuum and replaced with a precise mixture of N_2 and CO_2 gases (**Fig. 5.21B**).

Thought Question

5.8 What evidence led people to think about looking for anaerobes? *Hint:* Look up "Spallanzani," "Pasteur," and "spontaneous generation" on the Internet.

To Summarize

- **Oxygen is a benefit to aerobes**, organisms that can use it as a terminal electron acceptor to extract energy from nutrients.
- **Oxygen is toxic** to all cells that do not have enzymes capable of efficiently destroying the reactive oxygen species (ROS)—for example, anaerobes.
- **Anaerobic metabolism** can be either **fermentative** or **respiratory**. Anaerobic respiration requires the organism to possess cytochromes that can use compounds other than oxygen as terminal electron acceptors.
- **Aerotolerant anaerobes** grow in either the presence or the absence of oxygen, but use fermentation as their primary, if not only, means of gathering energy. These microbes also have enzymes that destroy reactive oxygen species, allowing them to grow in oxygen.
- **Facultative anaerobes** grow with or without oxygen and have enzymes that destroy reactive oxygen species. In addition, they possess both the ability for fermentative metabolism and respiration (anaerobic and aerobic). They can use oxygen as a terminal electron acceptor.

5.7

Nutrient Deprivation and Starvation

It is intuitively obvious that limiting the availability of a carbon source or other essential nutrient will limit growth. Not so obvious are the dramatic molecular events that cascade through a starving cell. Optimizing growth rate at suboptimal nutrient levels is an important aim of free-living bacteria, given that intestinal, soil, and marine environments rarely offer excess nutrients.

Starvation Activates Survival Genes

Numerous gene systems are affected when nutrients decline (see Section 10.3). Growth rate slows, and daughter cells become smaller and begin to experience what is called a "starvation" response, in which the microbe senses a dire situation developing but still strives to find new nourishment. The resulting metabolic slowdown generates increased concentrations of critically important small signal molecules, such as cyclic adenosine monophosphate (cyclic AMP or cAMP) and guanosine tetraphosphate (ppGpp), which globally transform gene expression. The highly soluble nature of these small molecules means they can quickly diffuse throughout the cell, promoting a fast response. During this metabolic retooling, transport systems for potential nutrients are produced even if the matching substrates are unavailable. Cells begin to make and store glycogen, presumably as an internal emergency store in case no other nutrient is found. Some organisms growing on nutrient-limited agar plates can even form colonies with intricate geometrical shapes that help the population cope, in some unknown way, with nutrient stress (**Fig. 5.22**).

As a cultural environment progressively worsens, the organism prepares for famine by activating many different stress survival genes. The products of these genes afford protection against stressors such as reactive oxygen radicals or temperature and pH extremes. No cell can predict the precise stresses it might encounter while incapacitated, so it is advantageous to be prepared for as many as possible. As described in Section 4.7, some species undergo elaborate developmental processes that ultimately produce dormant spores.

Microbes Encounter Multiple Stresses in Real Life

Bacterial stress responses have traditionally been studied in terms of individual stresses. *Escherichia coli,* for example, synthesizes a specific set of proteins when exposed to high temperature and a different set of proteins when exposed to high salt. Some proteins, however, may be highly expressed under both conditions, but each stress response also includes proteins unique to that stress.

In the world outside of the laboratory, by contrast, environmental situations can be quite complex, involving multiple, not just single, stresses. An organism could

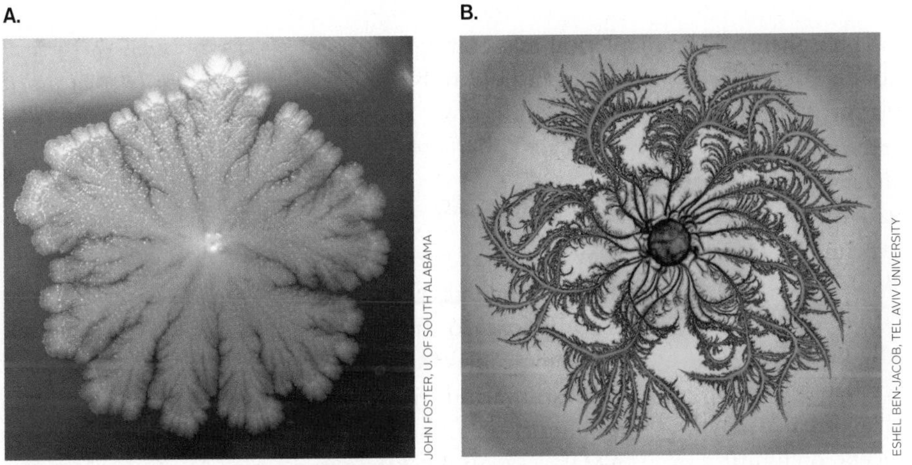
A.
B.
JOHN FOSTER, U. OF SOUTH ALABAMA
ESHEL BEN-JACOB, TEL AVIV UNIVERSITY

FIGURE 5.22 ■ Effects of starvation on colony morphology. A. Starving *E. coli* colony (6 cm diameter). **B.** *Paenibacillus dendritiformis* C morphotype grown on hard agar (1.75%) under starvation conditions. The colony consists of branches with chiral twists (colored green), all with the same handedness. The added coloration indicates time of growth (yellow = oldest; red = most recent).

simultaneously undergo carbon starvation in a high-salt, low-pH environment. A classic study by Kelly Abshire and Fred Neidhardt examined this situation using the pathogen *Salmonella enterica*, a cause of diarrhea. *S. enterica* invades human macrophage cells and survives in phagocytic vacuoles, where numerous stresses, such as low pH, oxidative stress, and nutrient limitations, are simultaneously imposed on the bacteria. Comparing the proteins synthesized by *Salmonella* growing in this compartment with the proteins synthesized under single stresses in the laboratory revealed an unexpected response pattern. Although many stress-related proteins were induced in the intracellular environment, no one set of stress-induced proteins was induced in its entirety. Furthermore, several bacterial proteins were induced by growth only within the macrophage phagolysosome, suggesting the presence of unknown intracellular stresses. Thus, caution is advised when trying to predict cell responses to real-world situations based solely on controlled laboratory studies that alter only single parameters.

Humans Influence Microbial Ecosystems

Natural ecosystems are typically low in nutrients (oligotrophic; eTopic 5.4) but teem with diversity, so that numerous species compete for the same limiting nutrients. Maximum diversity in a given ecosystem is maintained, in part, by the different nutrient-gathering profiles of competing microbes. Imagine a scenario in which microbe A is better than microbe B at gathering phosphate when phosphate levels are low, but microbe B is superior to A at culling limited quantities of nitrogen (**Fig. 5.23**).

Thought Question

5.9 What happens if excess nitrogen is added to the culture illustrated in **Figure 5.23**? What about excess phosphate and excess nitrogen?

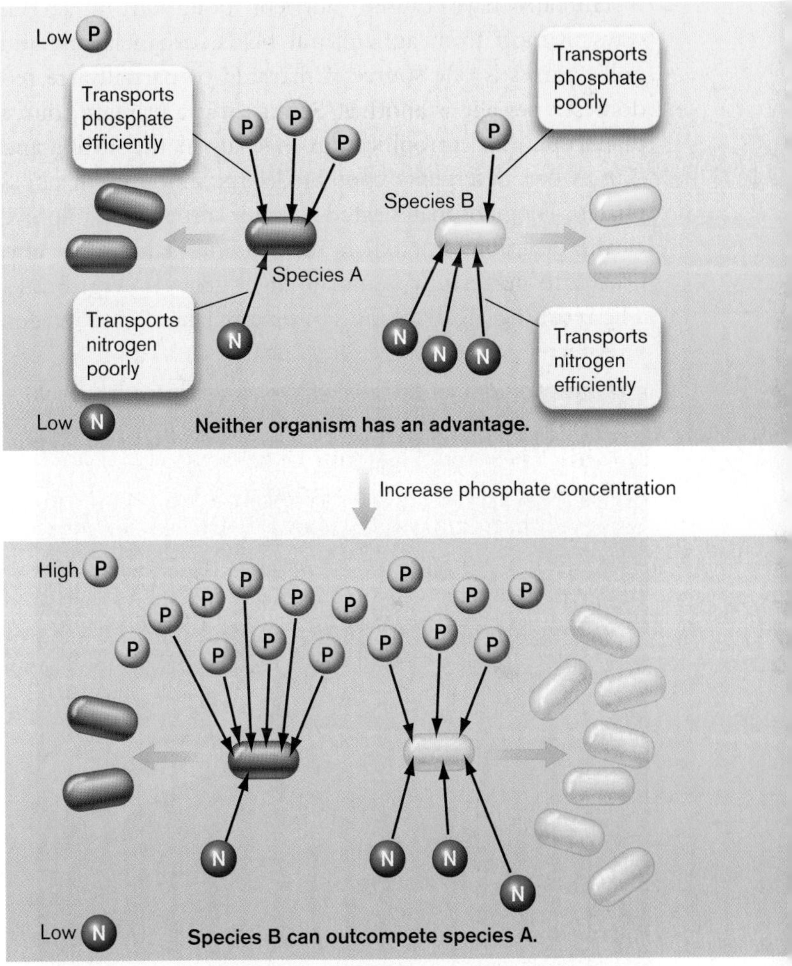
Low P
Transports phosphate efficiently
Transports phosphate poorly
Species B
Species A
Transports nitrogen poorly
Transports nitrogen efficiently
Low N
Neither organism has an advantage.
Increase phosphate concentration
High P
Low N
Species B can outcompete species A.

FIGURE 5.23 ■ Maintaining microbial diversity through nutrient limitation. Species A and species B require both nitrogen and phosphate to grow. Species A transports phosphate efficiently, but nitrogen poorly, whereas species B transports phosphate poorly, but nitrogen efficiently. **Top:** Neither species dominates, because each has a limiting nutrient (N = nitrogen; P = phosphate). **Bottom:** Following phosphate eutrophication, species B will outgrow species A because both species have enough phosphate, but B can more efficiently assimilate the limited quantity of nitrogen.

Neither organism dominates when both phosphate and nitrogen are in short supply. So, even though B can harvest more nitrogen than A, it cannot outgrow A, because the phosphate concentration is low. However, if phosphate is suddenly increased so that it is no longer limiting for either organism, the species better adapted to low nitrogen (microbe B) will outgrow and overwhelm the other. The sudden infusion of large quantities of a formerly limiting nutrient, a process called **eutrophication**, can lead to a "bloom" of microbes. Organisms initially held in check by the limiting nutrient now exhibit unrestricted growth, consuming other nutrients to a degree that threatens the existence of competing species. The concept of limiting nutrients is covered in Chapter 21.

Humans have caused nutrient pollution in several ways. Runoff from agricultural fields, urban lawns, and golf courses is one source. Untreated or partially treated domestic sewage is another. Sewage was a primary source of phosphorus eutrophication of lakes in the 1960s and '70s, when detergents contained large amounts of phosphates. The phosphates acted as water softeners to improve cleaning action, but when washed into lakes they also proved to powerfully stimulate algal growth (**Fig. 5.24**). The resulting algal "blooms" in many lakes led to oxygen depletion and consequent fish kills. Many native fish species disappeared, to be replaced by species more tolerant of the new conditions.

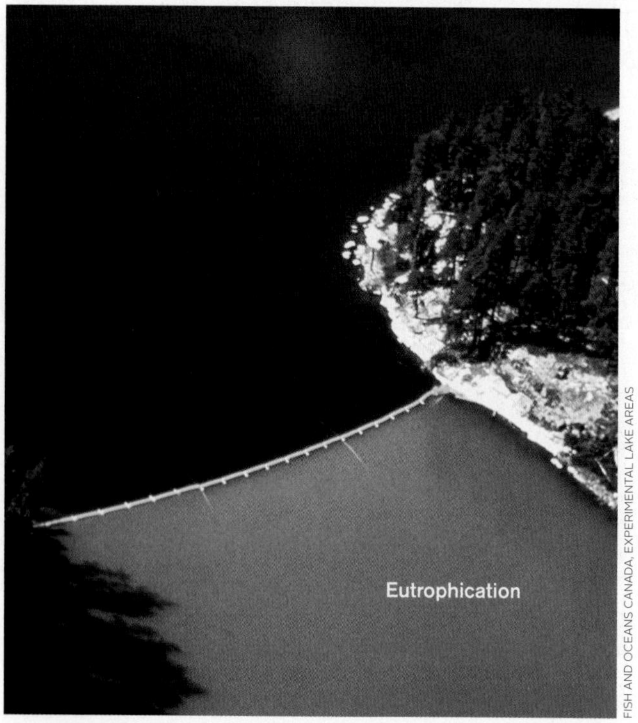
Eutrophication
FISH AND OCEANS CANADA, EXPERIMENTAL LAKE AREAS

FIGURE 5.24 ■ Eutrophication. Algal bloom in an experimental lake resulting from phosphate eutrophication. A divider curtain separates the lake and prevents the water from mixing. The bright green color results from cyanobacteria growing on phosphorus added to the near side of the curtain.

To Summarize

- **Starvation** is a stress that can elicit a molecular response in many microbes. Enzymes are produced to increase the efficiency of nutrient gathering and to protect cell macromolecules from damage.
- **The starvation response** is usually triggered by the accumulation of small signal molecules such as cyclic AMP or guanosine tetraphosphate.
- **Human activities can cause eutrophication**, which damages delicately balanced ecosystems by introducing nutrients that can allow one member of the ecosystem to flourish at the expense of other species.

5.8

Physical, Chemical, and Biological Control of Microbes

We have seen how microbes live; now, how do they die? A primary goal of our health care system is to control or kill microbes that can potentially harm us. Within the recent past, infectious disease was an imminent and constant threat to most of the human population. The average family in the United States prior to 1900 had four or five children, but parents could expect half of them to succumb to deadly infectious diseases. What today would be a simple infected cut in years past held a serious risk of death, and a trip to the surgeon was tantamount to playing Russian roulette with an unsterilized scalpel. Improvements in sanitation procedures and antiseptics and the advent of antibiotics have, to a large degree, curtailed the incidence and lethal effects of many infectious diseases. Success in this endeavor has played a major role in extending life expectancy and in contributing to the population explosion.

A variety of terms are used to describe antimicrobial control measures. The terms convey subtle, yet vitally important, differences in various control strategies and outcomes.

- **Sterilization** is the process by which all living cells, spores, and viruses are destroyed on an object.
- **Disinfection** is the killing, or removal, of disease-producing organisms from inanimate surfaces; it does

not necessarily result in sterilization. Pathogens are killed, but other microbes may survive.

- **Antisepsis** is similar to disinfection, but it applies to removing pathogens from the surface of living tissues, like the skin. Antiseptic chemicals are usually not as toxic as disinfectants, which frequently damage living tissues.
- **Sanitation**, closely related to disinfection, consists of reducing the microbial population to safe levels and usually involves both cleaning and disinfecting an object.

Antimicrobials can also be classified on the basis of the specific groups of microbes destroyed, leading to the terms "microbicide," "bactericide," "algicide," "fungicide," and "virucide." Furthermore, these agents can be classified as either "-static" (inhibiting growth) or "-cidal" (killing cells). For example, antibacterial agents may be **bacteriostatic** or **bactericidal**. Chemical substances are **germicidal** if they kill pathogens (and many nonpathogens), but germicidal agents do not necessarily kill spores.

Although these descriptions emphasize the killing of pathogens, it is important to note that antimicrobial agents can also kill or prevent the growth of nonpathogens. Many public health standards are based on total numbers of microorganisms on an object, regardless of pathogenic potential. For example, to gain public health certification, the restaurants we frequent must demonstrate low numbers of bacteria (pathogenic or not) in their food preparation areas.

Thought Question

5.10 Why would a bacteriostatic antibiotic that only inhibits growth of a pathogenic bacterium be useful for treating an infection? *Hint:* Is the human body a quiet bystander during an infection?

Cells Treated with Antimicrobials Die at a Logarithmic Rate

Exposing microbes to lethal chemicals or conditions does not instantly kill all microorganisms. Microbes die according to a negative exponential curve, where cell numbers are reduced in equal fractions at constant intervals. The efficacy of a given lethal agent or condition is measured as **decimal reduction time (D-value)**, which is the length of time it takes that agent (or condition) to kill 90% of the population (a drop of one log unit, or a drop to 10% of the original value). **Figure 5.25** illustrates the exponential death profile of a bacterial culture heated to 100°C. The D-value is a little over 1 minute. The food industry uses several other parameters (such as z-factor) to evaluate the efficiency of killing (Section 16.5).

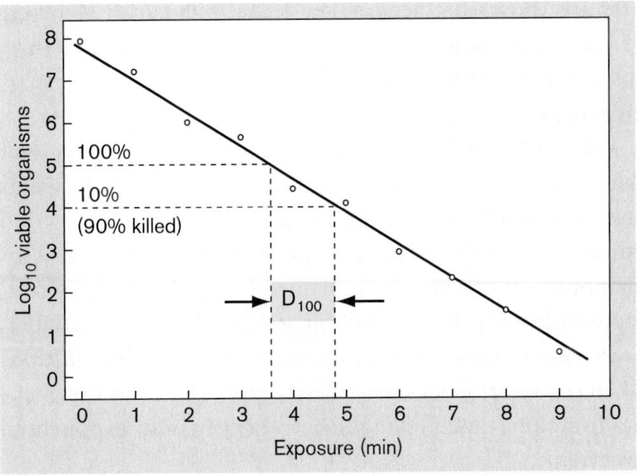
Log10 viable organisms
100%
10%
(90% killed)
D100
0 1 2 3 4 5 6 7 8 9 10
Exposure (min)

FIGURE 5.25 ■ The death curve and the determination of D-values. Bacteria were exposed to a temperature of 100°C, and survivors were measured by viable count. The D-value is the time required to kill 90% of cells (that is, the time it takes for the viable cell count to drop by one $\log_{10}$ unit). In this example, the D-value at 100°C (D_{100}) is approximately 1 minute.

Thought Question

5.11 If a disinfectant is added to a culture containing 1×10^6 CFUs per milliliter and the D-value of the disinfectant is 2 minutes, how many viable cells are left after 4 minutes of exposure?

Several factors influence the ability of an antimicrobial agent to kill microbes. These include the initial population size (the larger the population, the longer it takes to decrease it to a specific number), the population composition (are spores involved?), the concentration of the antimicrobial agent, and the duration of exposure. Although the effect of concentration seems intuitively obvious, only over a narrow range is an increase in concentration matched by an increase in death rate. Increases above a certain level might not accelerate killing at all. For example, 70% ethanol is actually better than pure ethanol at killing organisms, because some water is needed to help ethanol penetrate cells. The ethanol then dehydrates cell proteins.

Why, then, is death a logarithmic function? Why don't all cells in a population die instantly when treated with lethal heat or chemicals? The reason is based, in part, on the random probability that an agent will cause a lethal "hit" in a given cell. Cells contain thousands of different proteins and thousands of molecules of each. Not all proteins and not all genes in a chromosome are damaged by an agent at the same time. Damage accumulates. Only when enough molecules of an essential protein or a gene encoding that protein are damaged will the cell die. Cells that die first are those that accumulate lethal hits early. Members of the population that die later have, by random chance, absorbed more hits on nonessential proteins or genes, sparing the essential ones.

Why, if 90% of a population is killed in 1 minute, aren't the remaining 10% killed in the next minute? It seems logical that all should have perished. Yet after the second minute, 1% of the original population remains alive. This phenomenon can also be explained by the random hit concept. Although there are fewer viable cells after 1 minute, each has the same random chance of having a lethal hit as when the treatment began. Thus, death rate is an exponential function, much like radioactive decay is an exponential function.

A final consideration is the overall fitness of individual cells. It is a mistake to assume that all cells in a population are identical. At any given time, for instance, one cell may express a protein that another cell has just stopped expressing (for example, superoxide dismutase). In that instant, the first cell might contain a bit more of that protein. If the protein is essential or confers a level of stress protection (such as against superoxide), the cell can absorb more punishment before it is dispatched. The presence of lucky individuals expressing the right repertoire of proteins might also explain why death curves commonly level off after a certain point.

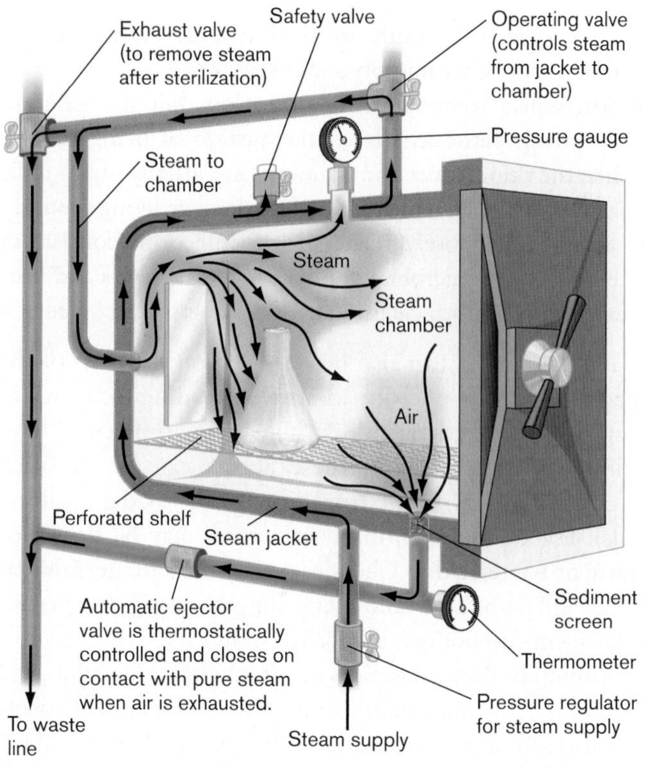
Exhaust valve (to remove steam after sterilization)
Safety valve
Operating valve (controls steam from jacket to chamber)
Pressure gauge
Steam to chamber
Steam
Steam chamber
Air
Perforated shelf
Steam jacket
Automatic ejector valve is thermostatically controlled and closes on contact with pure steam when air is exhausted.
Sediment screen
Thermometer
Pressure regulator for steam supply
To waste line
Steam supply

FIGURE 5.26 ■ Steam autoclave.

Physical Agents That Kill Microbes

Physical agents are often used to kill microbes or control their growth. Commonly used physical control measures include temperature extremes, pressure (usually combined with temperature), filtration, and irradiation.

High temperature and pressure. Even though microbes were discovered less than 400 years ago, thermal treatment of food products to render them safe has been practiced for over 5,000 years. Moist heat is a much more effective killer than dry heat, thanks to the ability of water to penetrate cells. Many bacteria, for instance, easily withstand 100°C dry heat but not 100°C boiling water. We humans are not so different, finding it easier to endure a temperature of 32°C (90°F) in dry Arizona than in humid Louisiana.

While boiling water (100°C) can kill most vegetative (actively growing) organisms, spores are built to withstand this abuse, and thermophiles prefer it. Killing spores and thermophiles usually requires combining high pressure and temperature. At high pressure, the boiling point of water rises to a temperature rarely experienced by microbes living at sea level. Even endospores quickly die under these conditions. This combination of pressure and temperature is the principle behind sterilization using the steam autoclave (**Fig. 5.26**). Standard conditions for steam sterilization are 121°C (250°F) at 15 psi for 20 minutes—a set of conditions that experience has taught us will kill all spores (except those of some thermophiles, which do not affect food or human health). These are also the conditions produced in pressure cookers used for home canning of vegetables.

Failure to adhere to these heat and pressure parameters can have deadly consequences, even in your own home. For instance, *Clostridium botulinum* is a spore-forming soil microbe that commonly contaminates fruits and vegetables used in home canning. The improper use of a pressure cooker while canning these goods will allow spores of this pathogen to survive. Once the can or jar is cool, the spores will germinate and begin producing their deadly toxin. All of this happens while the canned goods sit on a shelf waiting to be opened and consumed. Once ingested, the toxin makes its way to the nervous system and paralyzes the victim. Several incidents of this disease, called botulism, occur each year in the United States. (For more on food poisoning, see Chapter 16.)

Thought Question

5.12 How would you test the killing efficacy of an autoclave?

Pasteurization. Originally devised by Louis Pasteur to save products of the French wine industry from devastating bacterial spoilage, **pasteurization** today involves heating a particular food (such as milk) to a specific temperature long enough to kill *Coxiella burnetii,* the causative agent of Q fever, the most heat-resistant non-spore-forming pathogen known. In the process, pasteurization will also kill other disease-causing microbes. Many different time and temperature combinations can be used for pasteurization. The LTLT (low-temperature/long-time) process involves bringing milk to a temperature of 63°C (145°F) for 30 minutes. In contrast, the HTST (high-temperature/short-time) method (also called "flash pasteurization") brings the milk to a temperature of 72°C (161°F) for only 15 seconds. Both processes accomplish the same thing: the destruction of *C. burnetii* and other bacteria.

Cold. Low temperatures have two basic purposes in microbiology: to temper growth and to preserve strains. Bacteria not only grow more slowly in cold, but also die more slowly. Refrigeration temperatures (4–8°C, or 39–43°F) are used for food preservation because most pathogens are mesophilic and grow poorly, if at all, at those temperatures. One exception is the Gram-positive bacillus *Listeria monocytogenes,* which can grow reasonably well in the cold and causes disease when ingested.

Long-term storage of bacteria usually requires placing solutions in glycerol at very low temperatures (–70°C). Glycerol prevents the production of razor-sharp ice crystals that can pierce cells from without or within. This deep-freezing suspends growth altogether and keeps cells from dying. Another technique, called **lyophilization**, freeze-dries microbial cultures for long-term storage. In this technique, cultures are quickly frozen at very low temperatures (quick-freezing also limits ice crystal formation) and placed under vacuum, where the resulting sublimation process removes all water from the media and cells, leaving just the cells in the form of a powder. These freeze-dried organisms remain viable for years. Finally, viruses and mammalian cells must be kept at extremely low temperatures (–196°C), submerged in liquid nitrogen. Liquid nitrogen freezes cells so quickly that ice crystals do not have time to form.

Filtration. Filtration through micropore filters with pore sizes of 0.2 μm can remove microbial cells, but not viruses, from solutions. Samples from 1 ml to several liters can be drawn through a membrane filter by vacuum or can be forced through it using a syringe (**Fig. 5.27**). Filter sterilization has the advantage of avoiding heat, which can damage the material sterilized. Strictly speaking, though, the solutions are not really sterile, because these filters do not trap viruses.

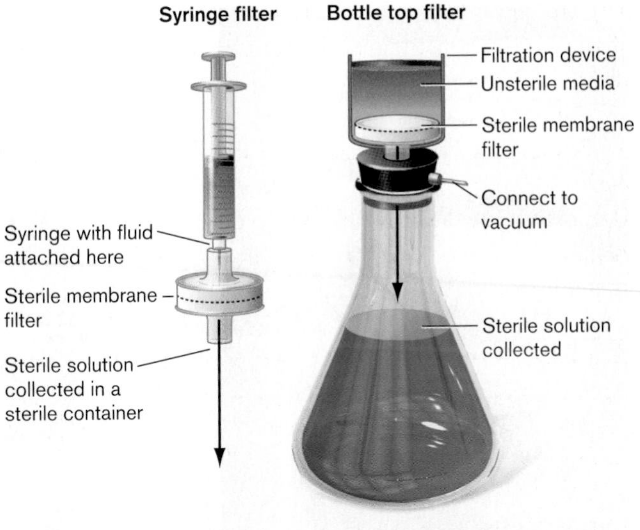
Syringe filter
Bottle top filter
Filtration device
Unsterile media
Sterile membrane filter
Connect to vacuum
Syringe with fluid attached here
Sterile membrane filter
Sterile solution collected
Sterile solution collected in a sterile container

FIGURE 5.27 ▪ Membrane filtration apparatus.

Air can also be sterilized by filtration. This process forms the basis of several personal protective devices. A surgical mask is a crude example, while **laminar flow biological safety cabinets** are more elaborate (and more effective). These cabinets force air through high-efficiency particulate air (HEPA) filters and remove over 99.9% of airborne particulate material 0.3 μm in size or larger. Biosafety cabinets are critical to protect individuals working with highly pathogenic material (**Fig. 5.28A** and **B**). Newer technologies have been developed that embed antimicrobial agents or enzymes directly into the fibers of the filter (**Fig. 5.28C**). Organisms entangled in these fibers are not just trapped; they are attacked by the antimicrobials and lyse.

Irradiation. Public health authorities worldwide are increasingly concerned about food contaminated with pathogenic microorganisms such as *Salmonella* species, *E. coli* O157:H7, *Listeria monocytogenes,* and *Yersinia enterocolitica*. Irradiation, the bombardment of foods with high-energy electromagnetic radiation, has long been a potent, if politically sensitive, strategy for sterilizing food after harvesting. The food consumed by NASA astronauts, for example, has for some time been sterilized by irradiation as a safeguard against food-borne illness in space, but public pressure has severely limited the mainstream use of this technology over concerns about product safety and exposure of workers. Resistance to irradiation largely disappeared, however, after 2001, when anthrax spores were mailed across the eastern United States and irradiation was used to sterilize our mail. Foods do not become radioactive when irradiated, but concerns remain that reactive molecules potentially dangerous to humans may be produced

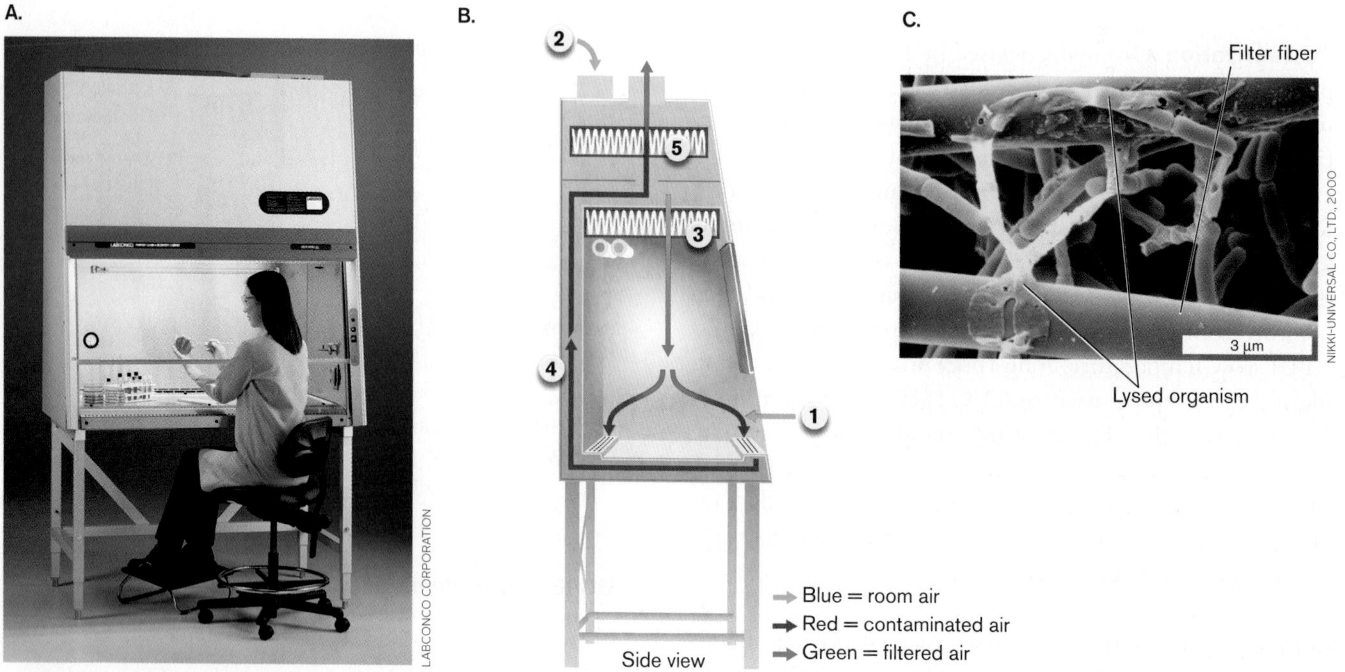
A.
B.
C.
Filter fiber
Lysed organism
3 μm
NIKKI-UNIVERSAL CO., LTD, 2000
LABCONCO CORPORATION
Side view
Blue = room air
Red = contaminated air
Green = filtered air

FIGURE 5.28 ■ Biological safety cabinet. **A.** A scientist examines a sample under the hood. **B.** Schematic of the safety cabinet. Air from the room enters the cabinet through the cabinet opening (1), or is pumped in (2) through a HEPA filter (3). It then passes behind the negative-pressure exhaust plenum (4) and is passed from the cabinet through another HEPA filter (5). **C.** Immobilized enzyme filter. The primary function of this enzyme filter is to kill airborne microorganisms caught on the surface of the filter, thus protecting against secondary contamination from microorganisms in air filtration systems. The photo shows lysed bacteria (*Bacillus subtilis*). The cell walls have been hydrolyzed by enzymatic action, and cell membranes are broken as a result of osmotic pressure pushing outward against the membrane.

when high-energy particles are absorbed by the natural chemicals in the food.

Aside from ultraviolet light, which, owing to its poor penetrating ability, is useful only for surface sterilization, there are three other sources of irradiation: gamma rays, electron beams, and X-rays. Radiation dosage is usually measured in a unit called the gray (Gy), which is the amount of energy transferred to the food, microbe, or other substance being irradiated. A single chest X-ray delivers roughly half a milligray (1 mGy = 0.001 Gy). To kill *Salmonella*, freshly slaughtered chicken can be irradiated at up to 4.5 kilograys (kGy)—about 7 million times the energy of a single chest X-ray. Recently the Food and Drug Administration also approved the use of irradiation (4 kGy) on lettuce and spinach, frequent sources of diarrhea-producing *Escherichia coli*. Water runoff from cattle pastures can carry fecal bacteria into adjacent fields. Postharvest, irradiation will kill microbes that can cause disease or shorten the shelf life of produce.

When microbes present in food are irradiated, water and other intracellular molecules absorb the energy and form transient reactive chemicals that damage DNA. Unless the organism repairs this damage, it will die while trying to replicate. Microbes differ greatly in their sensitivity to irradiation, depending on the size of their genome, the rate at which they can repair damaged DNA, and other factors. It also matters if the irradiated food is frozen or fresh, as it takes a higher dose of radiation to kill microbes in frozen foods.

The size of the DNA "target" is a major factor in radiation efficacy. Parasites and insect pests, which have large amounts of DNA, are rapidly killed by extremely low doses of radiation, typically with D-values of less than 0.1 kGy (in this instance, the D-value is the dose of radiation needed to kill 90% of the organisms). It takes more radiation to kill bacteria (D-values in the range of 0.3–0.7 kGy) because they have less DNA per cell unit (less target per cell). It takes even more radiation to kill a bacterial spore (D-values on the order of 2.8 kGy) because they contain little water, the source of most ionizing damage to DNA. Viral pathogens have the smallest amount of nucleic acid, making them resistant to irradiation doses approved for foods (viruses have D-values of 10 kGy or higher). Infectious agents that do not contain nucleic acids are an even bigger problem. Prions, for example, are misfolded brain proteins that "self-replicate" and cause neurodegenerative diseases (see Section 26.6). Because prions do not contain nucleic acids, the agent can be inactivated by irradiation only at extremely high doses. Thus, irradiation of food is effective in eliminating parasites and bacteria, but is woefully inadequate for eliminating viruses or prions.

Note Electromagnetic radiation emitted by microwave ovens does not directly kill bacteria. However, the heat generated when electromagnetic radiation excites water molecules in an organism will kill the organism if the temperature attained is high enough.

Some Bacteria Are Highly Resistant to Physical Control Measures

Deinococcus radiodurans could be nicknamed "Conan the bacterium" and designated as a poster microbe for extremophiles (**Fig. 5.29**). It was discovered in 1956 in a can of meat that had spoiled despite having been sterilized by radiation. The microbe has the greatest ability to survive radiation of any known organism. The amount of radiation it can handle suggests that *D. radiodurans* could even survive an atomic blast. The bacterium's ability to withstand radiation may have evolved as a side effect of developing resistance to extreme drought, since dehydration and radiation produce similar types of DNA damage.

Weblinks *Deinococcus (see ebook)*

Research by microbiologist John Battista has shown that *D. radiodurans* possesses an unusual capacity for repairing its damaged DNA, although the precise mechanisms remain a mystery. One possible mechanism is that a damaged chromosome in one member of the quartet shown in **Figure 5.29A** can restore itself using DNA from another member of the quartet. Michael Daly, a pathologist at the Uniformed Services University of the Health Sciences in Maryland, has shown that *Deinococcus* also protects its proteins, which are even more susceptible targets of radiation than is DNA, by accumulating large amounts of manganese. The high intracellular concentration of manganese removes the highly damaging free radicals generated by radiation.

On the basis of this research, *D. radiodurans* was genetically engineered to treat radioactive mercury-contaminated waste from nuclear reactors—a process called **bioremediation** (discussed in Chapter 22). The genes for mercury conversion were spliced from a strain of *E. coli* resistant to particularly toxic forms of mercury and inserted into *D. radiodurans.* The genetically altered superbug was able to withstand the ionizing radiation and transform toxic waste into forms that could be removed safely. Fortunately, there is little need to worry about its becoming a superpathogen, because the organism does not cause disease and is susceptible to antibiotics.

Chemical Agents

Disinfection by physical agents is very effective, but in numerous situations their use is impractical (kitchen countertops) or plainly impossible (skin). In these instances, chemical agents are the best approach. A number of factors influence the efficacy of a given chemical agent. These include:

- **The presence of organic matter.** A chemical placed on a dirty surface will bind to the inert organic material present, lowering the agent's effectiveness against microbes. It is not always possible to clean a surface prior to disinfection (as in a blood spill), but the presence of organic material must be factored into estimates of how long to disinfect a surface or object.
- **The kinds of organisms present.** Ideally, the agent should be effective against a broad range of pathogens.
- **Corrosiveness.** The disinfectant should not corrode the surface or, in the case of an antiseptic, damage skin.
- **Stability, odor, and surface tension.** The chemical should be stable during storage, possess a neutral or pleasant odor, and have a low surface tension so that it can penetrate cracks and crevices.

The phenol coefficient. Phenol, first introduced by Joseph Lister in 1867 to reduce the incidence of surgical infections, is no longer used as a disinfectant, because of its toxicity, but its derivatives, such as cresols and orthophenylphenol, are still in use. The household product Lysol is a mixture of phenolics. Phenolics are useful disinfectants because they denature proteins, are effective in the presence of organic material, and remain active on surfaces long after application.

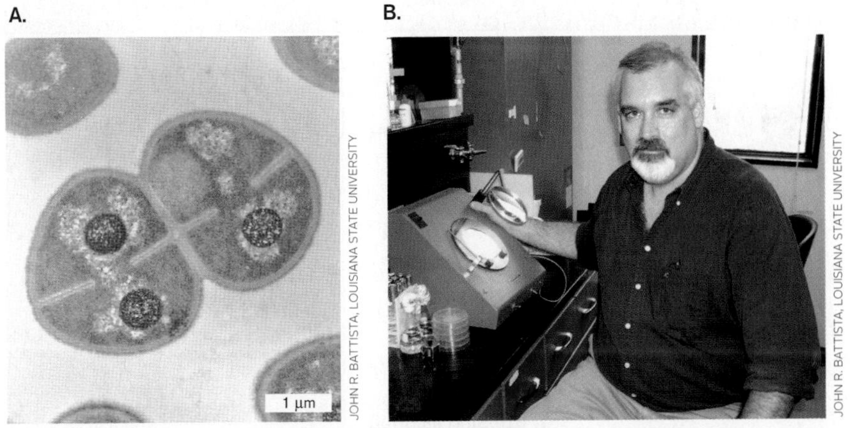
A.
B.
1 μm
JOHN R. BATTISTA, LOUISIANA STATE UNIVERSITY
JOHN R. BATTISTA, LOUISIANA STATE UNIVERSITY

FIGURE 5.29 ■ ***Deinococcus radiodurans.*** **A.** The amount of radiation this organism can survive is equivalent to that of an atomic blast. The nature of the dark inclusion bodies in three of the four cells in the quartet is currently not known. **B.** John Battista of Louisiana State University showed that *D. radiodurans* has exceptional capabilities for repairing radiation-damaged DNA.

Although no longer used as a disinfectant, the potency of phenol makes it the benchmark against which other disinfectants are measured. The **phenol coefficient test** consists of inoculating a fixed number of bacteria—for example, *Salmonella enterica* or *Staphylococcus aureus*—into dilutions of the test agent. At timed intervals, samples are withdrawn from each dilution and inoculated into fresh broth (which contains no disinfectant). The phenol coefficient is based on the highest dilution (lowest concentration) of a disinfectant that will kill all the bacteria in a test after 10 minutes of exposure but leaves survivors after only 5 minutes of exposure. This concentration is known as the maximum effective dilution. Dividing the reciprocal of the maximum effective dilution for the test agent (for example, ethyl alcohol) by the reciprocal of the maximum effective dilution for phenol gives the phenol coefficient (**Table 5.3**). For example, if the maximum effective dilution for agent X is $^1/_{900}$ and that of phenol is $^1/_{90}$, then the phenol coefficient of X is 900/90 = 10; the higher the coefficient, the higher the efficacy of the disinfectant.

TABLE 5.3

Phenol coefficients for various disinfectants.

Chemical agent	*Staphylococcus aureus*	*Salmonella enterica*
Phenol	1.0	1.0
Chloramine	133.0	100.0
Cresols	2.3	2.3
Ethyl alcohol	6.3	6.3
Formalin	0.3	0.7
Hydrogen peroxide	—	0.001
Lysol	5.0	3.2
Mercury chloride	100.0	143.0
Tincture of iodine	6.3	5.8

Commercial Disinfectants

Ethanol, iodine, chlorine, and surfactants (for example, detergents) are all used to reduce or eliminate microbial content from commercial products (**Fig. 5.30**). The first three are compounds that damage proteins, lipids, and DNA. Highly reactive iodine complexed with an organic carrier forms an iodophor, a compound that is water-soluble, stable, nonstaining, and capable of releasing iodine slowly to avoid skin irritation. Wescodyne and Betadine (trade names) are iodophors used, respectively, for the surgical preparation of skin and for wounds. Chlorine is another highly reactive disinfectant with universal application. It is recommended for general laboratory and hospital disinfection and kills the HIV virus.

Detergents can also be antimicrobial agents. The hydrophobic and hydrophilic ends of detergent molecules (the coexistence of which makes the molecules amphipathic) will emulsify fat into water. Cationic (positively charged) but not anionic (negatively charged) detergents are useful as disinfectants because the positive charges can gain access to the negatively charged bacterial cell and disrupt

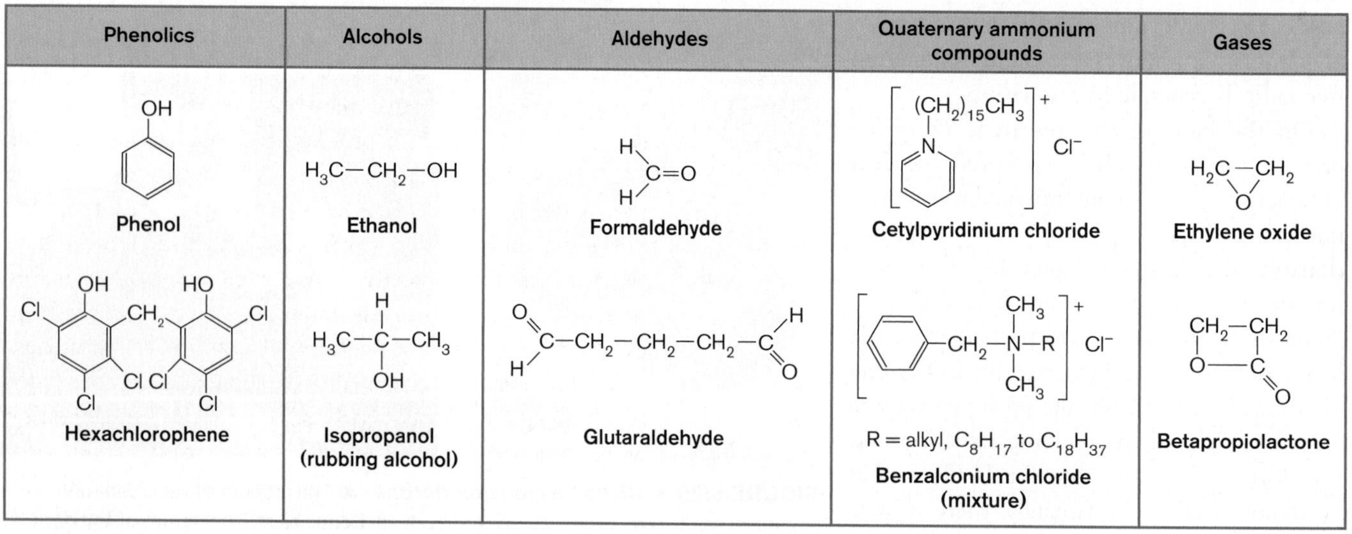
Phenolics
Alcohols
Aldehydes
Quaternary ammonium compounds
Gases
Phenol
Ethanol
Formaldehyde
Cetylpyridinium chloride
Ethylene oxide
Hexachlorophene
Isopropanol (rubbing alcohol)
Glutaraldehyde
R = alkyl, C8H17 to C18H37
Benzalconium chloride (mixture)
Betapropiolactone

FIGURE 5.30 ■ Structures of some common disinfectants and antiseptics.

membranes. Anionic detergents are not antimicrobial but do help in the mechanical removal of bacteria from surfaces.

Low-molecular-weight aldehydes such as formaldehyde (HCHO) are highly reactive, combining with and inactivating proteins and nucleic acids. This characteristic makes them useful disinfectants.

Disposable plasticware like petri dishes, syringes, sutures, and catheters are not amenable to heat sterilization or liquid disinfection. These materials are best sterilized using antimicrobial gases. Ethylene oxide gas (EtO) is a very effective sterilizing agent; it destroys cell proteins, is microbicidal and sporicidal, and rapidly penetrates packing materials, including plastic wraps. Using an instrument resembling an autoclave, EtO at 700 milligrams per liter (mg/l) will sterilize an object after 8 hours at 38°C or 4 hours at 54°C if the relative humidity is kept at 50%. Unfortunately, EtO is explosive. A less hazardous gas sterilant is betapropiolactone. It does not penetrate as well as EtO, but it decomposes after a few hours, which makes it easier to dispose of than EtO.

A new procedure, known as gas discharge plasma sterilization, may replace EtO because it is less harmful to operators. Gas discharge plasma is made by passing certain gases through a radio frequency electrical field to produce highly reactive chemical species that can damage membranes, DNA, and protein. It is not yet widely used.

Some attempts have been made to use more sophisticated microbicides as topical agents to prevent the spread of sexually transmitted infections. Several chemical microbicides have proved effective against the agents that cause gonorrhea, herpes, and chlamydia. Some are currently undergoing clinical trials as vaginal gels to determine whether the compounds can quell the spread of human immunodeficiency virus (HIV). Initial results are encouraging.

Antimicrobial touch surfaces. Despite widespread use of disinfectants and antibiotics by medical personnel, hospital-acquired infections remain a major concern. A promising, new antimicrobial technology that may help reduce infections in hospitals and elsewhere involves embedding antimicrobial compounds such as copper (Cu) in the surfaces people touch. Upon contact with bacteria, metallic copper releases toxic Cu^+ ions that trigger the lysis of bacterial membranes within minutes, although the mechanism involved is unclear (neither DNA damage nor reactive oxygen species are involved). It is thought that incorporating metallic copper into objects such as handrails, door releases, hospital bed rails, and the arms of visitor's chairs would kill microbes deposited by one person before those microbes could be transmitted to someone else.

Bacteria Can Develop Resistance to Disinfectants

It is widely known that bacteria can develop resistance to antibiotics used to treat infections. This is a serious concern in the medical community. So, one might wonder whether bacteria can also develop resistance against chemical disinfectants used to <u>prevent</u> infections. The answer is yes—and no. It is difficult for a bacterium to develop resistance to chemical agents that have multiple targets and can easily diffuse into a cell. Iodine, for example, has both of these characteristics. However, disinfectants that have multiple targets at <u>high</u> concentrations may have only a single target at lower concentrations—a situation that can foster the development of resistance. For instance, triclosan (a halogenated bisphenol compound used in many soaps and deodorants) targets several cell constituents, making it nicely bactericidal at high concentrations. However, at low concentrations triclosan only inhibits fatty acid synthesis and is merely bacteriostatic. Organisms have developed resistance to triclosan at low concentrations by altering the fatty acid synthesis protein normally targeted by triclosan.

Low-level resistance also can be achieved through membrane-spanning, multidrug efflux pumps (described in Section 4.2). For instance, the MexCD-OprJ efflux system of *Pseudomonas aeruginosa*, a Gram-negative bacterium that causes infections in burn and cystic fibrosis patients, can pump several different biocides, detergents, and organic solvents out of the cell, thereby reducing their efficacy. This finding and other reports of *Pseudomonas* gaining resistance to disinfectants have led many clinicians to advocate caution in the widespread use of certain chemical disinfectants.

Biofilm formation is another ingenious way bacteria survive exposures to disinfectants. Biofilms are 3D communities of bacterial cells attached to a solid surface (Section 4.6). Biofilms protect cells in several ways at once. For example, the extracellular matrix proteins and polysaccharides that hold biofilms together also bind disinfectants, slowing their penetration into the deeper recesses of the structure. Slower penetration means that cells deep in the biofilm have time to activate protective stress response systems before destructive levels of disinfectant reach them. Biofilms also exhibit stratified growth patterns based in part on nutrient access. Cells at the periphery have ample access to oxygen and nutrients while cells within do not. Nutrient starvation worsens the farther a cell is from the surface. Because nutrient starvation activates stress response systems, each biofilm has stratified layers of increasingly more stress-resistant cells that can better tolerate chemical insults.

20 μm
A. BRIDIER ET AL. 2011. BIOFOULING 27: 1017–1032

FIGURE 5.31 ▪ Mixed biofilm. 3D projection of a mixed 24-hour biofilm of *E. coli* expressing mCherry fluorescent protein (red) and *Pseudomonas aeruginosa* expressing green fluorescent protein (GFP; green). Mixed-population biofilms can have properties distinct from monospecies biofilms.

Finally, biofilms, which contain multiple species in nature, are opportunities for protective, interspecies collaborations (**Fig. 5.31**). Protective enzymes from one species could protect a nonproducing species from a chemical insult, much as a big brother protects a little brother from a bully. Multispecies biofilms can also be more massive than monospecies biofilms. The food pathogen *Escherichia coli* O157:H7 forms a biofilm with 400 times more volume when grown with *Acinetobacter calcoaceticus,* an organism found in meatpacking plants. Increased volume alone will slow the penetration of the disinfectant and protect the collective.

Antibiotics Selectively Control Bacterial Growth

Antibiotics as made in nature are chemical compounds synthesized by one microbe that selectively kill other microbial species. Naturally occurring antibiotics act like tiny molecular land mines. As a defense against competitors, some organisms secrete antibiotic compounds into their surrounding environment, where they remain until encountered by an intruder. If the interloper is susceptible to the antibiotic, the compound will target specific structures or proteins, disrupting their function. The target cell is either rendered helpless (by bacteriostatic antibiotics) or made nonviable (by bactericidal compounds). (See Chapter 27 for detailed modes of action.) When purified and administered to patients suffering from an infectious disease, these antibiotics can produce seemingly miraculous recoveries.

As we saw in Chapter 1, penicillin (**Fig. 5.32**), produced by *Penicillium notatum,* was discovered serendipitously in 1929 by Alexander Fleming. **Figure 5.32A** shows this molecule, which mimics a part of the cell wall. Because of this mimicry, penicillin binds to biosynthetic proteins involved in peptidoglycan synthesis and prevents cell wall formation. The drug is bactericidal because actively growing cells lyse without the support of the cell wall (**Fig. 5.33**). Other antibiotics target protein synthesis, DNA replication, cell membranes, and various enzyme reactions. These interactions are described throughout Parts 1–3 of this text.

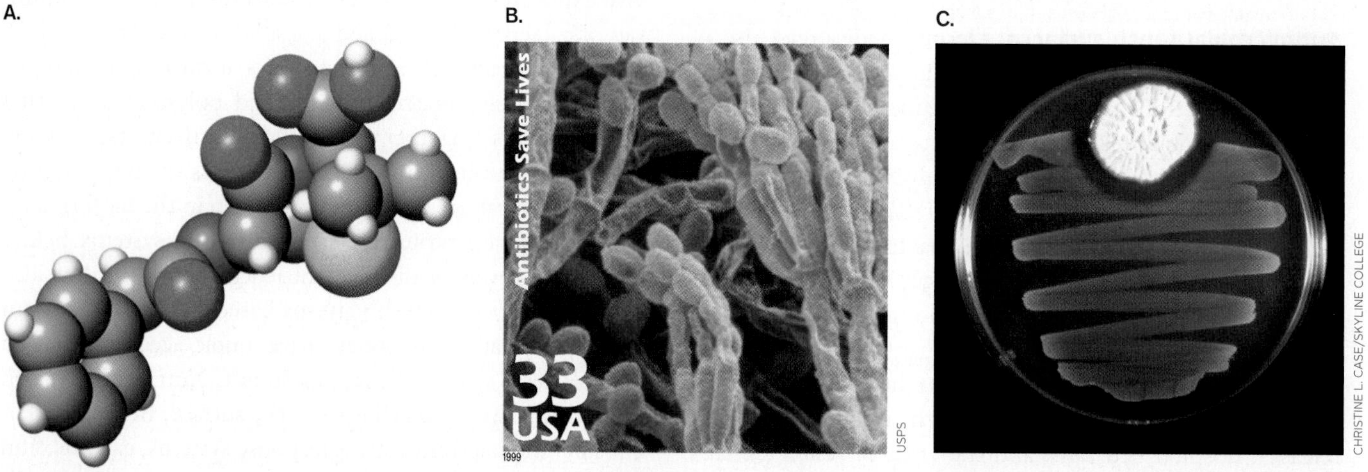
A.
B.
C.
Antibiotics Save Lives
33
USA
1999
USPS
CHRISTINE L. CASE/SKYLINE COLLEGE

FIGURE 5.32 ▪ Penicillin. A. Space-filling model of the penicillin G molecule produced by the *Penicillium* mold. **B.** U.S. postage stamp from 1999 showing *Penicillium notatum.* **C.** *P. notatum* culture. The mold (the white powdery circle at the top) excretes penicillin, which inhibits the growth of *Staphylococcus aureus.*

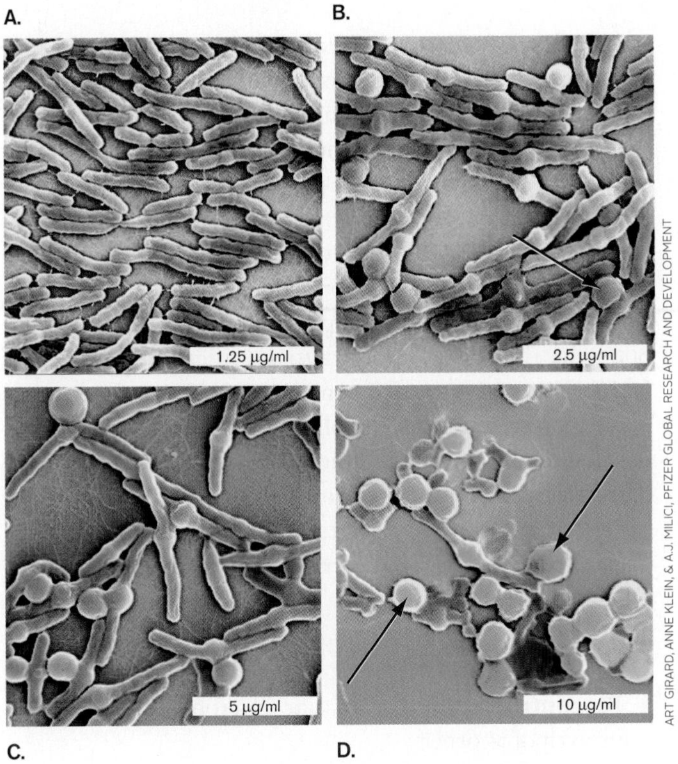
A.
B.
1.25 µg/ml
2.5 µg/ml
5 µg/ml
10 µg/ml
C.
D.
ART GIRARD, ANNE KLEIN, & A.J. MILICI, PFIZER GLOBAL RESEARCH AND DEVELOPMENT

FIGURE 5.33 ■ Effect of ampicillin (a penicillin derivative) on *E. coli*. Cells were incubated for 1 hour at the antibiotic concentrations shown. Swollen areas of cells in panels (B)–(D) reflect weakening cell walls (arrows). Cells shown are approx. 2 µm long.

So how do antibiotic-producing microbes avoid suicide? In some instances, the producing organism lacks the target molecule. *Penicillium* mold, for instance, lacks peptidoglycan and is immune to penicillin by default. Some bacteria produce antimicrobial compounds that target other members of the same species. In this case, the producing organism can modify its own receptors to no longer recognize the compound (as with some bacterial colicins).

Another strategy is to modify the antibiotic if it reenters the cell. This is the case with streptomycin produced by *Streptomyces griseus*. Streptomycin inhibits bacterial protein synthesis and does not discriminate between the protein synthesis machinery of *Streptomyces* and that of others. However, while making enzymes that synthesize and secrete streptomycin, *S. griseus* simultaneously makes the enzyme streptomycin 6-kinase, which remains locked in the cell. If any secreted streptomycin reenters the cell, this enzyme renders the drug inactive by attaching a phosphate group to it.

Because many microorganisms have become resistant to commonly used antibiotics, pharmaceutical companies are continually using a variety of drug discovery approaches to search for new antibiotics. Traditional procedures include scouring soil and ocean samples collected from all over the world for new antibiotic-producing organisms and chemically redesigning existing antibiotics so that they can bypass microbial resistance strategies. These procedures are now supplemented by "mining the genomes" of microbes for potential drug targets. The frontiers of chemistry include rational drug design, which relies on computer-based methods for predicting the structure and function of potential new antibiotics.

To Summarize

- **Physical and chemical agents** kill microbes by denaturing proteins or DNA, or by disrupting lipid bilayers.
- **Sterilization** kills all living organisms.
- **Antisepsis** is the removal of potential pathogens from the surfaces of living tissues.
- **Antimicrobial** compounds can be **bacteriostatic** or **bactericidal**.
- **The D-value** is the time (or dose, for irradiation) it takes an antimicrobial treatment to reduce the numbers of organisms to 10% of the original value.
- **An autoclave** uses high pressure to achieve temperatures that will sterilize objects.
- **Food can be preserved** by pasteurization, refrigeration, filtration, and irradiation.
- **Chemical disinfectants** are compared to one another on the basis of the phenol coefficient.
- **Bacterial resistance to disinfectants** is influenced by the concentration of the disinfectant, the number of targets, the presence of multidrug efflux pumps, and biofilm formation.
- **Antibiotics** are compounds produced by one living microorganism that kill other microorganisms.

Biological Control of Microbes

Pitting microbe against microbe is an effective way to prevent disease in humans and animals. One of the hallmarks of a healthy ecosystem is the presence of a diversity of organisms. This is true not only for tropical rain forests and coral reefs, but also for the complex ecosystems of human skin and the intestinal tract. In these environments, the presence of harmless microbial communities can retard the growth of undesired pathogens. The pathogenic fungus *Phytophthora cinnamomi*, for example, causes root rot in plants but is biologically controlled by *Myrothecium* fungi. Naturally occurring *Staphylococcus* species on human skin produce short-chain fatty acids that retard the growth of pathogenic strains. Another illustration is the human intestine, which is populated by as many as 500 microbial species. Most of these species are nonpathogenic organisms that exist in symbiosis with their human host. Vigorous competition with members of the normal intestinal microbiota and the production of permeant weak acids by fermentation helps control the growth of numerous pathogens. A prominent example of this phenomenon is the pathogen *Clostridium difficile*, whose growth is normally kept in check by gut microbiota.

Microbial competition has been widely exploited for agricultural purposes and to improve human health through the intake of **probiotics**. In general, a probiotic is a food or supplement that contains live microorganisms and improves intestinal microbial balance. Newborn baby chicks, for instance, are fed a microbial cocktail of normal gut microbes designed to quickly colonize the intestinal tract and prevent colonization by *Salmonella,* a frequent contaminant of factory-farmed chicken. In another example, *Lactobacillus* and *Bifidobacterium* have been used to prevent and treat diarrhea in children.

Russian biologist Ilya Mechnikov, winner of the 1908 Nobel Prize in Physiology or Medicine, was the first to suggest that a high concentration of lactobacilli in the gut microbiome is important for health and longevity in humans. Yogurt is a probiotic that contains *Lactobacillus acidophilus* and a number of other lactobacilli. It is often recommended as a way to restore a normal balance to gut microbiota (for example, after the microbiota has been disturbed by antibiotic treatment), and it appears useful in the treatment of inflammatory bowel disease.

Phage therapy, another biocontrol method, was first described in 1907 by Félix d'Herelle at France's Pasteur Institute, long before antibiotics were discovered. Bacteriophages are viruses that prey on bacteria (discussed in Section 6.1). Each bacterial species is susceptible to a limited number of specific phages. Because a phage infection often causes bacterial lysis, it was considered feasible to treat infectious diseases with a phage targeted to the pathogen. At one time, doctors used phages as medical treatment for illnesses ranging from cholera to typhoid fever. In some cases, a liquid containing the phage was poured into an open wound. In other cases, phages were given orally, introduced via aerosol, or injected. Sometimes the treatments worked; sometimes they did not. When antibiotics came into the mainstream, phage therapy largely faded. Now that strains of bacteria resistant to standard antibiotics are on the rise, the idea of phage therapy has enjoyed renewed interest from the worldwide medical community.

The animal health company Elanco recently developed the first phage product to specifically target *E. coli* O157:H7, a pathogen that can contaminate hamburger and cause bloody diarrhea. The phage product, called Finalyse, is sprayed onto cattle 1–4 hours before slaughter. *E. coli* that survive steam pasteurization and the acid washing of cattle carcasses will themselves be infected and killed by the phage, reducing the consumer's risk of disease.

To Summarize

- **Biocontrol** is the use of one microbe to control the growth of another.
- **Probiotics** contain certain microbes that, when ingested, aim to restore balance to the intestinal microbiome.
- **Phage therapy** offers a possible alternative to antibiotics in the face of rising antibiotic resistance.
- **Disinfection** kills pathogens on inanimate objects.

Concluding Thoughts

Microbiology as a science was founded on the need to understand and control microbial growth. The initial impetus was to control the diseases of humans, as well as the diseases of plants and animals. But, as we will see in later chapters, microbiology has developed into a science that has helped us understand the molecular processes of life. Concepts such as biological diversity, food microbiology, microbial disease, and antibiotics will be revisited in later chapters.

CHAPTER REVIEW

Review Questions

1. Explain the nature of extremophiles and discuss why these organisms are important.
2. What parameters define any growth environment?
3. List and define the classifications used to describe microbes that grow in different physical growth conditions.
4. What do thermophiles have to do with PCR technology?
5. Why is water activity important to microbial growth? What changes water activity?
6. How do cells protect themselves from osmotic stress?
7. Why do changes in H^+ concentration affect cell growth?
8. How do acidophiles and alkaliphiles manage to grow at the extremes of pH?
9. If an organism can live in an oxygenated environment, does that mean the organism uses oxygen to grow? If an organism can live in an anaerobic environment, does that mean it cannot use oxygen as an electron acceptor? Why or why not?
10. What happens when a cell exhausts its available nutrients?
11. List and briefly explain the various means by which humans control microbial growth. What is a D-value? What is a phenol coefficient?
12. How do microbes prevent the growth of other microbes?

Thought Questions

1. Given a natural lake environment with 100 species of bacteria, why does the species with the fastest generation time not overwhelm the others? Or does it?
2. *Escherichia coli* is a facultative species, able to grow with or without oxygen. What would it take to make this organism an anaerobe?
3. Two spore formers, *Bacillus stearothermophilus* and *Bacillus coagulans*, have D-values at 121°C of 5 minutes and 0.07 minute, respectively. How could the spores from these organisms have such different D-values? *Hint:* Find the optimum growth temperatures for these organisms.
4. Phage therapy is touted by some as a solution to antibiotic resistance. Explain why phage therapy may not be able to solve this problem.
5. With respect to bacteria, is the physiological effect of HCl at pH 4 the same as that of an organic acid at pH 4?

Key Terms

acidophile (169)
aerobic respiration (173)
aerotolerant anaerobe (175)
alkaliphile (170)
anaerobic respiration (174)
antibiotic (186)
antisepsis (179)
bactericidal (179)
bacteriostatic (179)
barophile (164)
bioremediation (183)
decimal reduction time (D-value) (179)
disinfection (178)
electron transport chain (172)
eutrophication (178)
extremophile (158)
facultative (175)
facultative anaerobe (175)
fermentation (fermentative metabolism) (174)
germicidal (179)
halophile (167)
heat-shock response (164)
hyperthermophile (162)
laminar flow biological safety cabinet (181)
lyophilization (181)
membrane-permeant organic acid (168)
mesophile (161)
microaerophilic (175)
neutralophile (169)
osmolarity (166)
pasteurization (181)
phenol coefficient test (184)
piezophile (164)
probiotic (188)
psychrophile (161)
sanitation (179)
sterilization (178)
strict aerobe (174)
strict anaerobe (174)
thermophile (161)
water activity (166)

Recommended Reading

Atomi, Haruyuki. 2005. Recent progress towards the application of hyperthermophiles and their enzymes. *Current Opinion in Chemical Biology* **9**:163–173.

Blasius, Melanie, Ulrich Hubscher, and Suzanne Sommer. 2008. *Deinococcus radiodurans*: What belongs to the survival kit? *Critical Reviews in Biochemistry and Molecular Biology* **43**:221–238.

Christner, Brent C. 2012. Cloudy with a chance of microbes. *Microbes* 7:70–75.

Daly, Michael, Elena Gaidamakova, Vera Matrosova, Alexander Vasilenko, Min Zhai, et al. 2007. Protein oxidation implicated as the primary determinant of bacterial radioresistance. *PLoS Biology* **5**:769–779.

D'Amico, Salvino, Tony Collins, Jean-Claude Marx, Georges Feller, and Charles Gerday. 2006. Psychrophilic microorganisms: Challenges for life. *EMBO Reports* 7:385–389.

Edgar, Rotem, Nir Friedman, Shahar Molshanski-Mor, and Udi Qimron. 2012. Reversing bacterial resistance to antibiotics by phage-mediated delivery of dominant sensitive genes. *Applied and Environmental Microbiology* **78**:744–751.

Horikoshi, Koki. 1998. Barophiles: Deep-sea microorganisms adapted to an extreme environment. *Current Opinion in Microbiology* **1**:291–295.

Horikoshi, Koki. 1999. Alkaliphiles: Some applications of their products for biotechnology. *Microbiology and Molecular Biology Review* **63**:735–750.

Kota, Swathi, and Hari S. Misra. 2008. Identification of a DNA processing complex from *Deinococcus radiodurans. Biochemistry and Cell Biology* **86**:448–458.

Krulwich, Terry. A., George Sachs, and Etana Padan. 2011. Molecular aspects of bacterial pH sensing and homeostasis. *Nature Reviews. Microbiology* **9**:330–343.

Romero, Diego, Claudio Aguilar, Richard Losick, and Roberto Kolter. 2010. Amyloid fibers provide structural integrity to *Bacillus subtilis* biofilms. *Proceedings of the National Academy of Sciences USA* **107**:2230–2234.

Slonczewski, Joan, James A. Coker, and Shiladitya DasSarma. 2010. Microbial growth with multiple stressors. *Microbe* **5**:110–116.

Yacoby, Iftach, and Itai Behar. 2008. Targeted filamentous bacteriophages as therapeutic agents. *Expert Opinion on Drug Delivery* **5**:321–329.

CHAPTER 6
Viruses

6.1 What Is a Virus?

6.2 Virus Structure

6.3 Viral Genomes and Classification

6.4 Bacteriophage Replication

6.5 Animal and Plant Virus Replication

6.6 Culturing Viruses

6.7 Viral Ecology

All kinds of cells, including bacteria, eukaryotes, and archaea, can be infected by particles called viruses. Viruses are the smallest known units of reproduction, some with genomes of fewer than ten genes. A virus must infect a host cell and subvert the cell's machinery to reproduce more virus particles, usually killing the host cell. But some viral genomes are copied into the host genome, where they replicate silently with the host. DNA from ancient viral insertions evolved into many portions of the human genome. Most animals, plants, and fungi have evolved symbiotic relationships with viruses that may benefit their hosts. And researchers now engineer viruses to attack tumors, and to conduct gene therapy.

Virus structure varies in complexity from single RNA molecules called viroids to spore-like packages containing multiple layers of protein and membranes. Some viruses resemble degenerate cells. How can such particles, barely large enough to be detected by an electron microscope, destroy entire cells and multicellular organisms? How can some viruses integrate their genomes into the genome of their host, replicating with their host for future generations? And in ecosystems, how do viruses cycle nutrients and sustain host diversity?

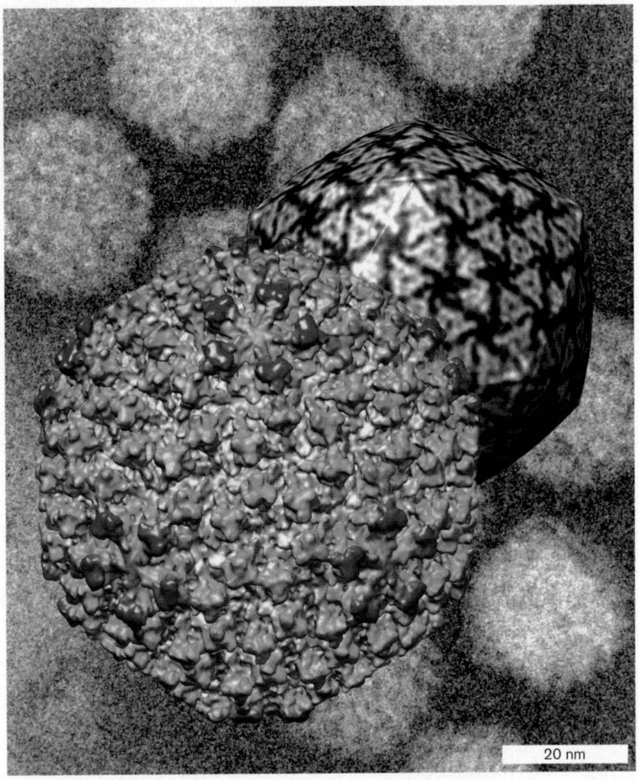

20 nm

CURRENT RESEARCH highlight

A new virus carried by insects. Espirito Santo virus was discovered in 2012, replicating together with dengue virus in insect tissue culture. A new type of birnavirus, Espirito Santo virus was found in a dengue virus sample obtained from a patient in the state of Espirito Santo, Brazil. Dengue virus is transmitted to humans by mosquitoes, causing an illness known as dengue fever or "breakbone fever." Common in tropical countries, dengue fever has now reached the southern United States. The pathogenicity of Espirito Santo virus is yet unknown; so far, the newly characterized birnavirus has been cultured only in insect cells. Ricardo Vancini and colleagues observed the icosahedral capsid of Espirito Santo virus by cryo-EM (background of particles) and computed a 3D model. The surface rendering is colored radially, yellow (radius 30 nm) to blue (radius 37 nm), along the threefold axis, showing 260 trimeric protrusions that extend approximately 5 nm from the viral surface. *Source:* Ricardo Vancini et al. 2012. *J. Virol.* **86**:2390.

In Key West, Florida, a vacationer presents with mild fever and headache. The fever may subside—or it may intensify, with muscle aches, joint pain, and a spreading rash (**Fig. 6.1A**). The disease, dengue fever, is also known as "breakbone fever." In 1% of cases, dengue fever leads to massive hemorrhages, bleeding, and death. The cause of the disease is dengue virus—a small flavivirus, only 50 nm in diameter, with a genome of just 11,000 bases of single-stranded RNA (**Fig. 6.1B**). Where did this tiny, deadly pathogen come from? Spread by an *Aedes* mosquito (**Fig. 6.1C**), dengue fever infects 100 million people annually in tropical regions of South America, Africa, and Asia. In 2009, the first case in over 60 years was diagnosed in Florida. Since then, periodic outbreaks have affected Florida and coastal Texas. The expanded range of dengue may be associated with the northward spread of the mosquitoes *A. aegyptii* and *A. albopictus.*

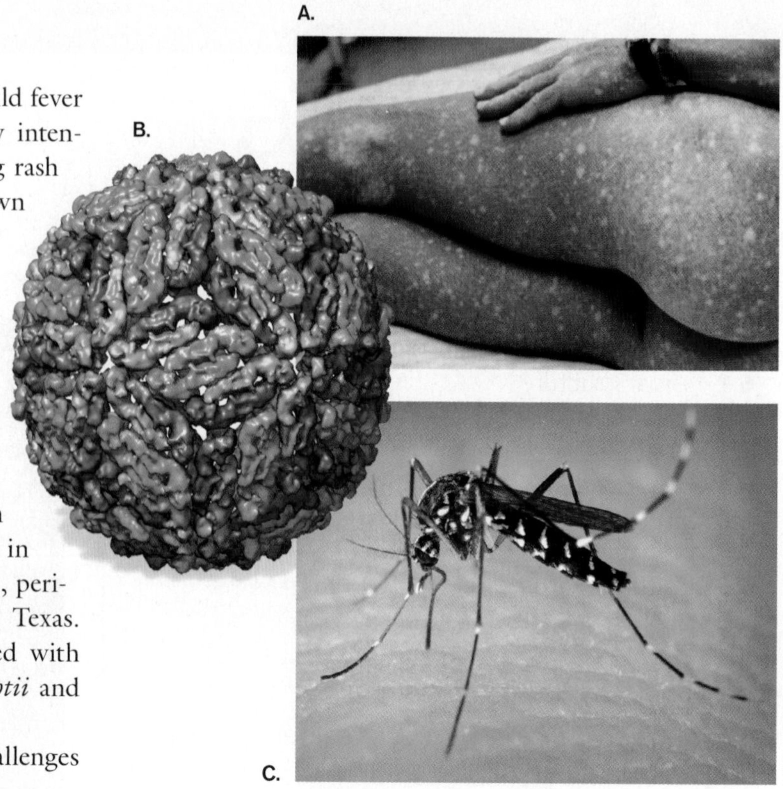
A.
B.
C.
HTTP://FARM3.STATICFLICKR.COM/2127/2344898589_45FBAA185C.JPG
HTTP://VACCINENEWSDAILY.COM/NEWS/214483-FLORIDA-REPORTS-INCREASE-IN-DENGUE-FEVER-INFECTIOINS/

FIGURE 6.1 ■ Dengue fever. A. Rash typical of dengue fever. **B.** Dengue virus. Crystallographic model (PDB code: 1K4R). Virion diameter, 50 nm. **C.** *Aedes* sp. mosquitoes transmit the dengue virus.

The case of dengue virus illustrates several challenges that scientists face in the study of viruses.

- **Viral infections.** Viruses cause many human diseases, including frightening outbreaks and epidemics, from dengue to influenza and AIDS. Understanding viral form and function helps us devise treatments and vaccines.
- **Virus culture in the laboratory.** To study viruses requires coculture with a host cell—in the case of dengue virus, insect cell culture. Host cell culture poses special challenges, such as avoiding viral contaminants.
- **Viral ecology.** Viruses interact with alternate hosts such as mosquito vectors; and they transfer genes among host cells. Viruses fill essential niches in the environment, particularly in marine ecosystems, where they cycle carbon and select for host species diversity.

In research, viruses have provided both tools and model systems for our discovery of the fundamental principles of molecular biology. The first genes mapped, the first regulatory switches defined, and the first genomes to be sequenced were all those of viruses. Vectors for gene cloning and gene therapy today continue to be derived from viruses. Our understanding of viruses, particularly bacterial viruses, called bacteriophages, provides a background for the molecular biology we will encounter in Part 2 of this book (Chapters 7–12). And remarkably, human viruses are now being engineered to destroy tumors and to deliver gene therapy (discussed in Chapter 11).

In Chapter 6 we introduce the major themes of virus structure and function, and the fundamental challenges that all viruses face: genome packaging, cell attachment and entry, and the strategies that viruses employ to divert the metabolism of their host cell for their own reproduction. The molecular biology of viral infection and replication is explored further in Chapter 11, and viral disease pathology and epidemiology are discussed in Chapters 25–27.

6.1

What Is a Virus?

A **virus** is a noncellular particle that infects a host cell, where it reproduces. The virus particle, or **virion**, consists of an infective nucleic acid (either DNA or RNA) contained within a protective shell made of protein, called the **capsid**. For some viruses, the capsid is further enclosed by an envelope composed of membrane and embedded proteins. The virion may have a molecular delivery device to transfer its genome into the host cell.

The preceding definition and description may sound simple, but in fact they pose thorny problems, as we will see.

Our concept of a virus has developed over the past century. The Russian botanist Dmitri Ivanovsky (1864–1920) and the Dutch microbiologist Martinus Beijerinck (1851–1931) first proposed the existence of viruses as soluble infectious agents that passed through a filter too small for cells to pass. A virus that infects bacterial cells (bacteriophage) was first observed in 1915 by English bacteriologist Frederick William Twort (1877–1950). The first bacteriophage replication cycle was characterized by French microbiologist Félix d'Herelle (1873–1949). In animals and plants, some viruses cause tumors, as discovered by American cancer researcher Peyton Rous (1879–1970), who won the 1966 Nobel Prize in Physiology or Medicine. The first tumor-causing virus that was discovered turned out to be an RNA retrovirus of birds called Rous sarcoma virus (RSV).

Viral diseases led to development of some of the first vaccines (discussed in Chapter 24). Poliovirus is famous for causing the outbreaks of poliomyelitis that swept the United States during the first half of the twentieth century. The virus enters humans orally from fecal sources such as contaminated water. In most cases, poliovirus infection causes a mild gastrointestinal disease, but the virus can invade motor neurons of the spinal cord and lower brain, resulting in paralysis. Tens of thousands of Americans were infected, including many young children, and in some cases desperate quarantine measures led to banning all children from entering a town (**Fig. 6.2A**). President Franklin Roosevelt (**Fig. 6.2B**), himself a victim of the disease, established a national foundation called the March of Dimes to develop a vaccine. Support from the March of Dimes led Jonas Salk and colleagues at the University of Pittsburgh to develop the first vaccine against polio in 1952. The spectacular success of the March of Dimes set the pattern for future public support for research on more challenging diseases such as cancer and AIDS. Today, the poliovirus approaches extinction, thanks to a global campaign of vaccination.

A. Quarantine notice **B. President Franklin D. Roosevelt**

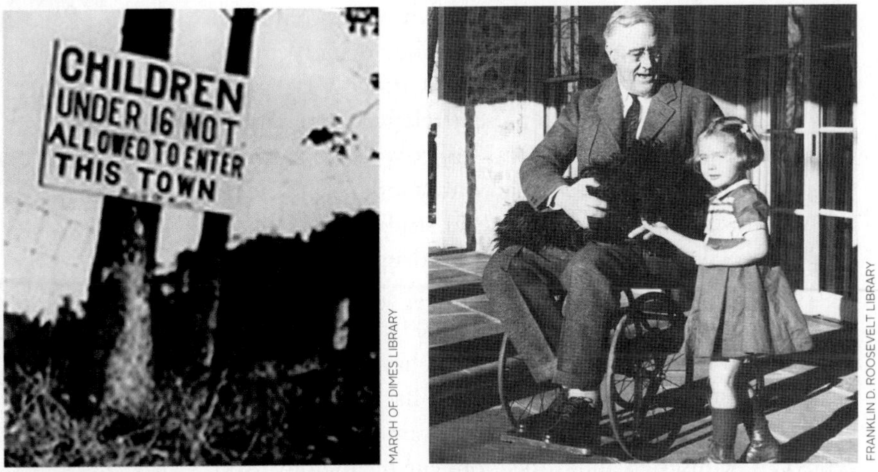
CHILDREN
UNDER 16 NOT
ALLOWED TO ENTER
THIS TOWN
MARCH OF DIMES LIBRARY
FRANKLIN D. ROOSEVELT LIBRARY

FIGURE 6.2 ■ Polio in the United States. A. Board of Health quarantine notice, around 1900, aiming to exclude polio carriers. **B.** The most famous polio victim, President Franklin Delano Roosevelt, with his granddaughter Ruthie Bie and his dog Fala. Roosevelt's legs were paralyzed permanently, and he used a wheelchair throughout his presidency.

What kinds of genomes do viruses have? In 1952, Alfred Hershey (1908–1997) and Martha Chase (1927–2003) tested the incorporation of ^{32}P-radiolabeled phosphate versus ^{35}S-radiolabeled amino acids into the genetic material of a bacteriophage, T2. When the phage T2 infected *E. coli*, by injecting its ^{32}P-labeled genome, only ^{32}P label was found in the infected cells—no ^{35}S (which labeled protein). This result confirmed that the genetic material was DNA, not proteins. Other viruses have genomes of RNA—for example, tobacco mosaic virus (TMV). The structure of RNA from TMV was first solved by British crystallographer Rosalind Franklin (1920–1958). The first viral genome, that of bacteriophage phiX174, was sequenced by Fred Sanger, for which he earned the 1980 Nobel Prize in Chemistry. The sequencing of viral genomes and the discovery of viruses as large as cells yield surprising clues as to how viruses originated, and how they evolve (discussed in Section 6.3).

Viruses Infect All Forms of Life

Where do we find viruses? Viruses are ubiquitous in all environments. Different viruses infect every taxonomic group of organisms, including bacteria, eukaryotes, and archaea. In marine ecosystems, viruses act as major predators and sequester large amounts of nutrients. For humans, viruses cause many forms of illness, whose influence on our history and culture would be hard to overstate. More people died of influenza in the global epidemic of 1918 than in the battles of World War I. In the past 30 years, the AIDS pandemic caused by HIV has killed nearly 30 million people worldwide. The blood-borne liver virus hepatitis C infects 200 million people worldwide, and causes more deaths in the United States than AIDS.

Viruses are part of our daily lives. The most frequent infections of college students are due to respiratory pathogens such as rhinovirus (the common cold) and Epstein-Barr virus (infectious mononucleosis), as well as sexually transmitted viruses such as herpes simplex virus (HSV) and papillomavirus (genital warts). Viruses also impact human industry; for example, bacteriophages (literally, "bacteria-eaters") infect cultures of *Lactococcus* during the production of yogurt and cheese. Plant pathogens such as cauliflower mosaic virus and rice dwarf virus cause substantial losses in agriculture.

Many viruses insert their own genomes into that of their host cell. Viral gene transfer is one of the most important processes that mediates evolution of cell genomes. Indeed, viral genomes are the ancestral source of about a tenth of the human genome.

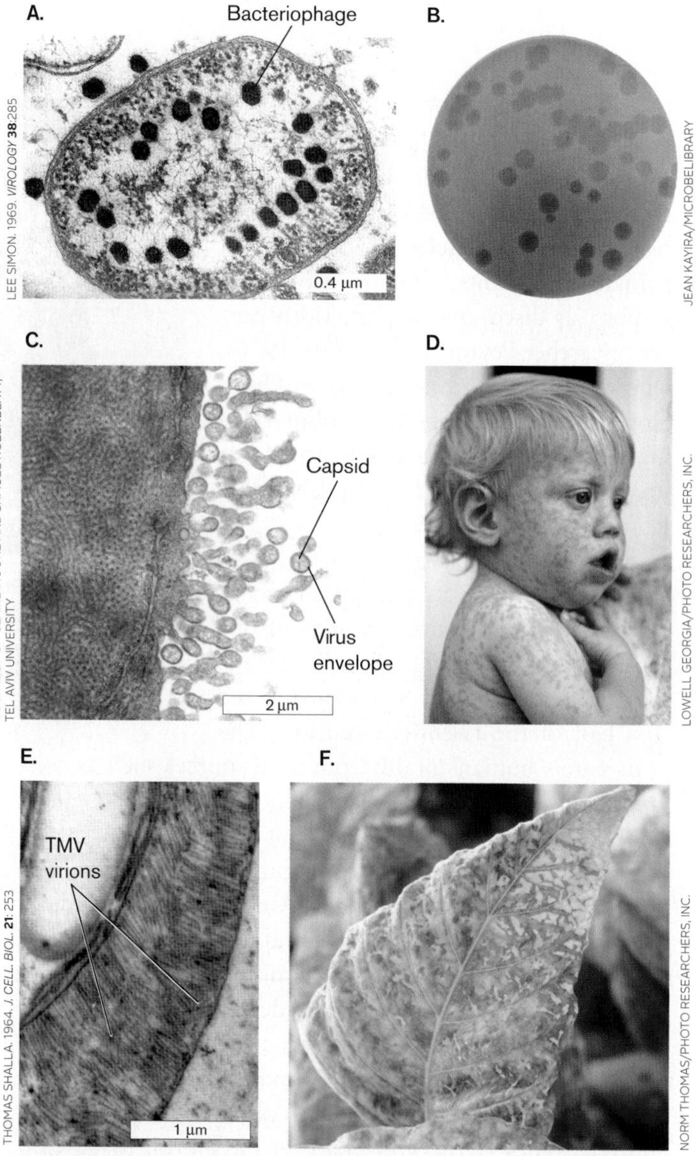
A.
Bacteriophage
0.4 μm
LEE SIMON. 1969. VIROLOGY 38:285
B.
JEAN KAYIRA/MICROBELIBRARY
C.
Capsid
Virus envelope
2 μm
COURTESY OF CLAIRE MOORE AND SHMUEL ROZENBLATT, TEL AVIV UNIVERSITY
D.
LOWELL GEORGIA/PHOTO RESEARCHERS, INC.
E.
TMV virions
1 μm
THOMAS SHALLA. 1964. J. CELL BIOL. 21:253
F.
NORM THOMAS/PHOTO RESEARCHERS, INC.

FIGURE 6.3 ■ Virus infections. **A.** Bacteriophage T2 particles pack in a regular array within an *E. coli* cell (TEM). **B.** Bacteriophage infection forms plaques of lysed cells on a lawn of bacteria. **C.** Measles virions bud out of human cells in tissue culture (TEM). **D.** Child infected with measles shows a rash of red spots. **E.** Tobacco leaf section is packed with tobacco mosaic virus particles. **F.** Tomato leaf infected by tobacco mosaic virus shows mottled appearance.

In contrast to our vast arsenal of antibiotics (effective against bacteria), the number of antiviral drugs remains small. Because the machinery of viral growth is largely that of the host cell, viruses present relatively few targets that can be attacked by antiviral drugs without harming the host. An exception is human immunodeficiency virus (HIV), the cause of AIDS. Molecular studies of HIV have yielded several major classes of antiviral drugs, such as AZT and protease inhibitors. The molecular biology behind anti-HIV drugs is discussed in Chapter 11.

Despite their lethal potential, viruses have made surprising contributions to medical research. Bacteriophages are used as **cloning vectors**, small genomes into which foreign genes can be inserted and cloned for gene technology. And human viruses such as HIV are being modified to serve as vectors for human gene therapy (Section 11.6).

How do viruses infect different kinds of organisms? Viruses that infect bacteria are known as **bacteriophages** or **phages**. An example is bacteriophage T2, which infects *Escherichia coli* (**Fig. 6.3A**). The T2 and T4 phages have a capsid with a tail that inserts the viral genome into the host cell, where it directs the reproduction of progeny virions. Virions are released when the host cell lyses. As cells lyse, their disappearance can be observed as a **plaque**, a clear spot against a lawn of bacterial cells (**Fig. 6.3B**). Each plaque arises from a single virion or phage particle that lyses a host cell and spreads progeny to infect adjacent cells. Plaques can be counted to represent individual infective virions from a phage suspension.

A virus that infects humans is the measles virus (**Fig. 6.3C**). The measles virus has an envelope that is derived from the host cell plasma membrane as the virus exits the host cell. As the virus infects a new host cell, the envelope fuses with the host cell plasma membrane, releasing the viral contents into the cytoplasm of the cell. After replicating within the infected cell, newly formed measles virions become enveloped by host cell membrane as they bud out of the host cell. The spreading virus generates a rash of red spots on the skin of infected patients (**Fig. 6.3D**) and can be fatal (one in 500 cases).

Plants are infected by viruses such as tobacco mosaic virus (TMV). Within the plant cell, virions accumulate to high numbers (**Fig. 6.3E**) and travel through interconnections to neighboring cells. Infection by tobacco mosaic virus results in mottled leaves and stunted growth (**Fig. 6.3F**). Plant viruses cause major economic losses in agriculture worldwide.

Each species of virus infects a particular group of host species, known as the **host range**. Some viruses can infect only a single species; for example, HIV infects only humans. Close relatives of humans, such as the chimpanzee, are not infected by HIV, although they are susceptible to a highly related virus, simian immunodeficiency virus (SIV). By contrast, West Nile virus, transmitted by mosquitoes, infects many species of birds and mammals. West Nile virus has a much broader host range than HIV and SIV.

Thought Question

6.1 Which viruses do you know that have a narrow host range, and which have a broad host range?

Weblinks International Committee on Taxonomy of Viruses *ictvonline.org*

Viral Genomes

What does a viral infection accomplish? Compared to living cells, a virus is a stripped-down entity that propagates its genome with minimal extragenomic biomass. Viral replication exemplifies the central role of information in biological reproduction. The propagation of viruses is mimicked by the spread of "computer viruses," whose information "infects" computer memory (**Fig. 6.4**). When a biological virus infects a host cell, the information in its genome subverts the host cell machinery to produce multiple copies of the virus; the multiple copies then escape to infect more host cells. Similarly, when a computer virus infects a host computer, its program code subverts the host to produce multiple copies of the virus, which then escape to infect more host computers. Computer viruses generate epidemics analogous to those of biological viruses. The virus's code can even be designed to "mutate" in order to foil the "immune system" of antivirus software. And some computer viruses bury their code within the core operating system. Fundamentally, both biological viruses and computer viruses propagate sequences of information.

Small viruses commonly have a small genome, encoding fewer than ten genes. For example, cauliflower mosaic virus (diameter 50 nm) has a genome encoding only seven genes (**Fig. 6.5A**), which actually overlap each other in sequence. This overlap in sequence is made possible by the use of different **reading frames**, start positions to define the first base of the codon for translation to amino acid sequence. Many viral genomes are encoded by RNA. The RNA genome of avian leukosis virus (**Fig. 6.5B**) has protein-encoding genes grouped by functional categories of core capsid, replicative enzymes, and envelope proteins (proteins embedded in the envelope phospholipid bilayer).

What kind of genomes are found in large viruses? Large viruses such as bacteriophage T4 (200 nm × 80 nm) and herpes viruses (diameter 100–200 nm) may have more than a hundred genes dispersed around the chromosome, similar to the genomes of bacteria. A giant virus called "mimivirus" (diameter 300 nm), which infects amebas and may cause human pneumonia, is itself as large as some bacteria (**Fig. 6.6**) and has a bacterium-sized genome of 1.2 million

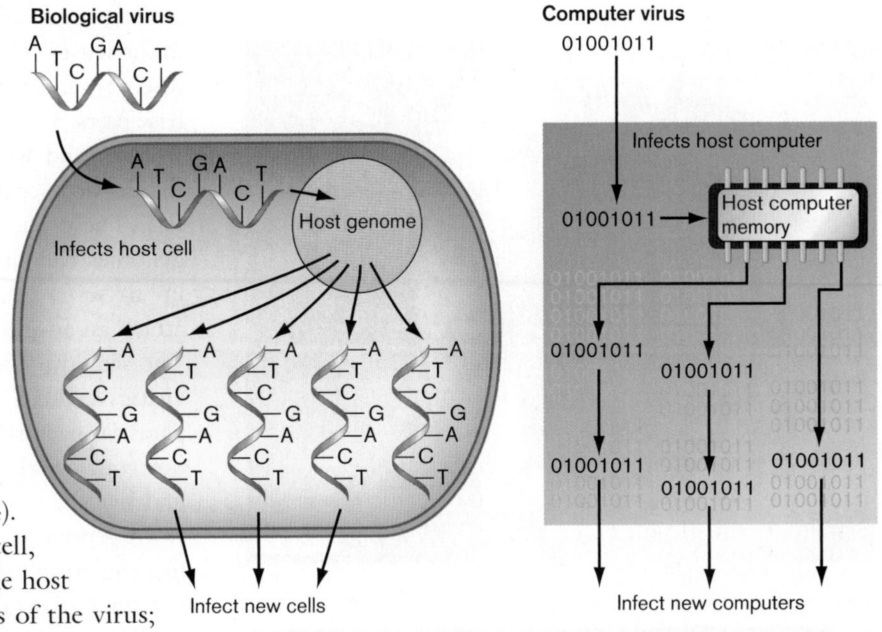
Biological virus
Infects host cell
Host genome
Infect new cells
Computer virus
01001011
Infects host computer
Host computer memory
Infect new computers

FIGURE 6.4 ■ Biological viruses and computer viruses. From the standpoint of information, the behavior of biological viruses and computer viruses is analogous.

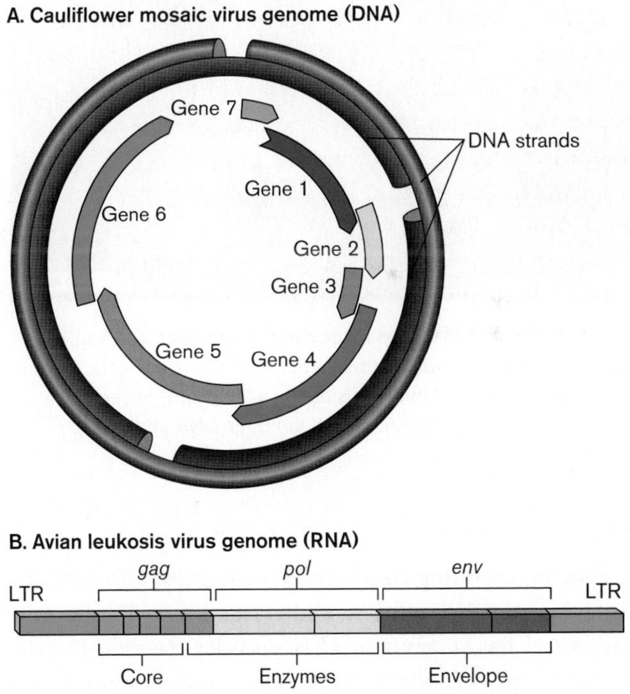
A. Cauliflower mosaic virus genome (DNA)
Gene 7
Gene 1
Gene 2
Gene 3
Gene 4
Gene 5
Gene 6
DNA strands
B. Avian leukosis virus genome (RNA)
LTR
gag
pol
env
LTR
Core
Enzymes
Envelope

FIGURE 6.5 ■ Simple viral genomes. A. Cauliflower mosaic virus has a circular genome of double-stranded DNA, whose strands are interrupted by nicks. The genome encodes seven overlapping genes. **B.** Avian leukosis virus, a single-stranded RNA retrovirus resembling eukaryotic mRNA. Three genes (*gag, pol,* and *env*) encode polypeptides that are eventually cleaved to form a total of nine functional products. LTR = long terminal repeat.

A.

B.

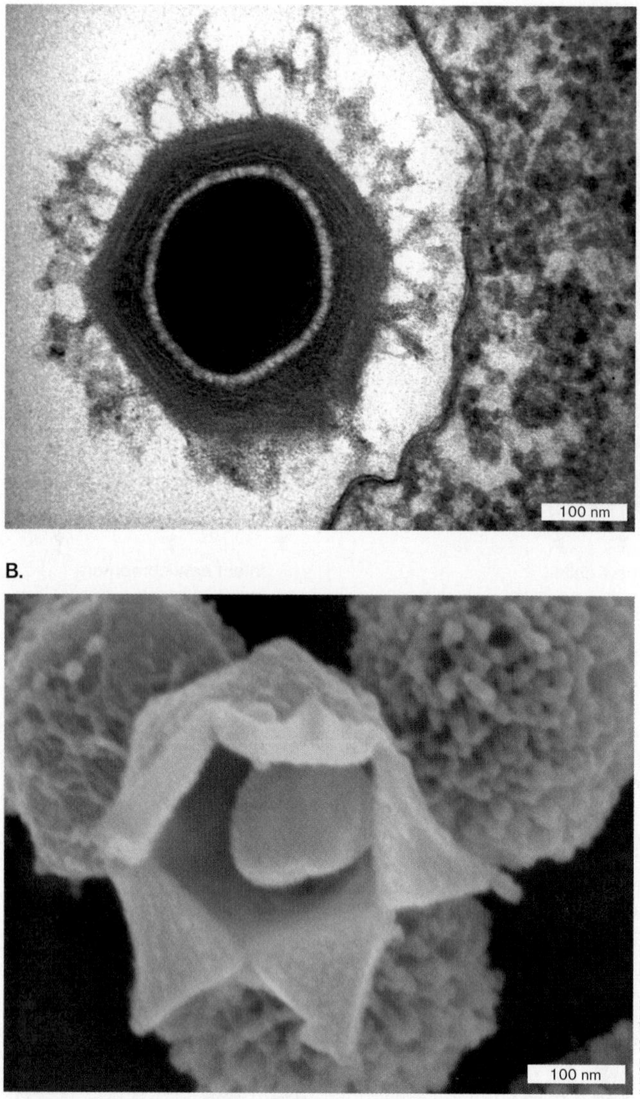
100 nm
100 nm

DIDIER RAOULT, MEDITERRANEAN UNIVERSITY, MARSEILLE

IMAGE COURTESY OF ABRAHAM MINSKY, WEIZMANN INSTITUTE OF SCIENCE. IN: JAMES VAN ETTEN. 2011. *AM. SCI.* **99**:304

FIGURE 6.6 ■ Mimivirus infecting an ameba. A. About 300 nm across, the mimivirus is larger than some bacteria (TEM). **B.** An engulfed mimivirus opens a "stargate" to release its membrane-enclosed package of genomic DNA and viral enzymes (SEM).

base pairs, encoding over a thousand genes. Discovered in 2003 by French virologist Didier Raoult and colleagues, mimivirus has a genome that encodes numerous cellular functions, including DNA repair and protein folding by chaperones. The virus gains entry to its ameba host by phagocytosis, because its large particle size makes the virus particle resemble a bacterium that the ameba could engulf for food. Once taken up, the mimivirus opens a five-sided "stargate" that releases a sac of viral enzymes and DNA. Within the ameba's cytoplasm, the viral enzymes generate a novel organelle, a "virus factory" that produces progeny mimiviruses. In 2013, an even larger virus of amebas was discovered: Pandoravirus, with a genome of 2.5 million base pairs.

How and where did viruses originate? Did they form within host cells, or did they arise as independent entities? These questions are highly controversial. Giant viral genomes encode numerous metabolic enzymes, as well as up to seven aminoacyl-tRNA synthetases (enzymes that attach an amino acid to the transfer RNA for translation). The size and complexity of giant viruses argue that these viruses evolved by reductive evolution of an intracellular parasitic bacterium. Similar arguments are made (though not conclusively) for other large viruses, such as smallpox and herpes. On the other hand, small viruses with small RNA genomes, such as influenza virus and HIV, look more like something that arose from parts of a cell. For example, the key retroviral enzyme reverse transcriptase (which copies RNA to DNA) shows homology to a cellular enzyme, telomerase, which maintains chromosome ends. Viral origins are discussed further in **eTopic 6.1**.

Viroids and Prions

Are there any infectious agents even simpler than small viruses? For some infectious agents, the nucleic acid genome is itself the entire infectious particle; there is no protective capsid. Such agents are called **viroids**. Most viroids are RNA molecules that infect plants, including many kinds of fruits and vegetables. For example, citrus viroids have caused economic losses in the citrus industry of the United States, Australia, and Israel.

A well-known example of a viroid is the potato spindle tuber viroid (**Fig. 6.7**). This viroid consists of a circular, single-stranded molecule of RNA that doubles back on itself to form base pairs interrupted by short, unpaired loops. This unusual circularized form avoids breakdown by host RNase enzymes. The RNA folds up into a globular structure that interacts with host cell proteins. The RNA is replicated by host RNA polymerase. The host RNA polymerase normally requires a DNA template, but the replication process is modified during viroid infection. During viroid infection, the RNA polymerase thus replicates progeny copies of the viroid, which encodes no products other than itself. Viroids can cause as much host destruction as "true viruses," and some authors, particularly in plant pathology, classify them as "viruses without capsids."

Some viroids have catalytic ability, comparable to enzymes made of protein. An RNA molecule capable of catalyzing a

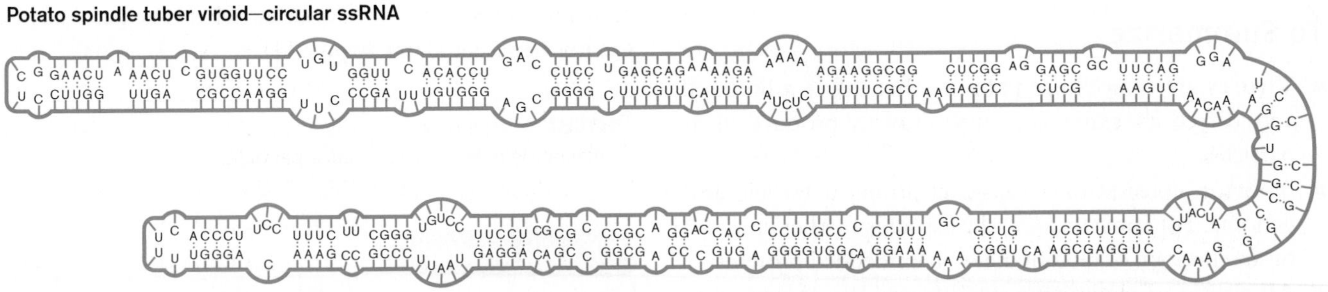
Potato spindle tuber viroid—circular ssRNA

FIGURE 6.7 ▪ Viroids: infective RNA. Potato spindle tuber viroid consists of a circular single-stranded RNA (ssRNA) that hybridizes internally. A viroid encodes no genes, but it hijacks the plant cell's RNA-dependent RNA polymerase to replicate itself.

reaction is called a ribozyme (discussed in Chapter 8). Viral ribozymes may be able to cleave themselves or other specific RNA molecules. Their ability to cleave very specific RNA sequences has applications in medical research, such as cleaving human mRNA involved in cancer.

Can an infectious agent propagate without its own genome? A remarkable class of infectious agents is believed to consist solely of protein. These agents, known as **prions**, are thought to be aberrant proteins arising from the host cell. Prions gained notoriety when they were implicated in brain infections such as Creutzfeldt-Jakob disease, a variant form of which is known as "mad cow" disease because it may be transmitted through defective proteins in beef from diseased cattle. Other diseases believed to be caused by prion transmission include scrapie, a disease of sheep; and kuru, a degenerative brain disease found in a tribe of people who customarily consumed the brains of deceased relatives.

In prion-associated diseases, the infective agent is unaffected by treatments that destroy RNA or DNA, such as nucleases or UV irradiation. A prion is an aberrant form of a normally occurring cell protein that assumes an abnormal conformation or tertiary structure (**Fig. 6.8A**). The prion form of the protein acts by binding to normally folded proteins of the same class and altering their conformation to that of the prion. The multiplying prion then alters the conformation of other normal subunits, forming harmful aggregates in the cell and ultimately leading to cell death. In the brain, prion-induced cell death leads to tissue deterioration and dementia (**Fig. 6.8B**).

A prion disease can be initiated by infection with an aberrant protein. More rarely, the cascade of protein misfolding can start with the spontaneous misfolding of an endogenous host protein. The chance of spontaneous misfolding is greatly increased in individuals who inherit certain alleles encoding the protein; thus, spontaneous prion diseases can be inherited genetically. An intriguing candidate for a prion is the defective tau particle of Alzheimer's disease. While the mechanism remains unclear, research published in 2012 shows that a tau protein expressed in mice propagates through the brain in a pattern similar to that of a brain infection. Prion diseases are unique in that they can be transmitted by an infective protein instead of by DNA or RNA; and they propagate the conformational change of existing molecules without synthesizing entirely new infective molecules.

A.

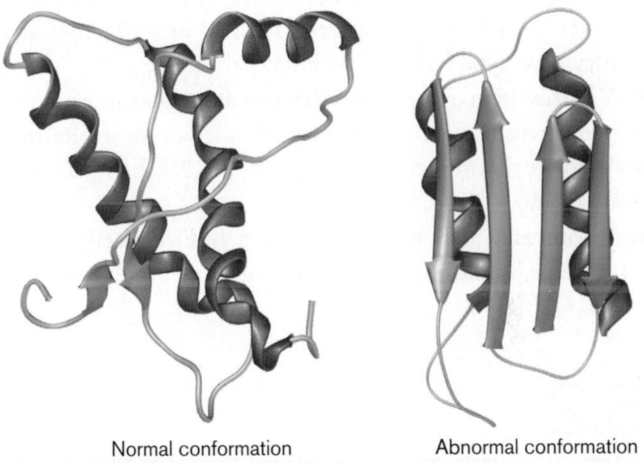
Normal conformation
Abnormal conformation

B.

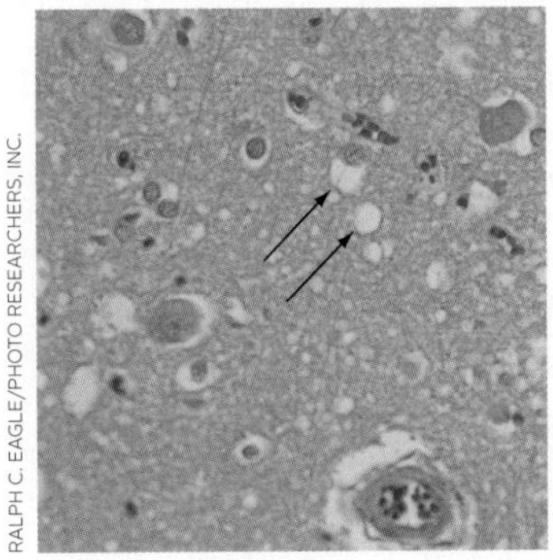
RALPH C. EAGLE/PHOTO RESEARCHERS, INC.

FIGURE 6.8 ▪ Prion disease. **A.** The normal conformation of a prion, compared to the abnormal conformation. The abnormal form "recruits" normally folded proteins and changes their conformation into the abnormal form. (PDB code: 1AG2) **B.** Section of a human brain showing "spongiform" holes typical of Creutzfeldt-Jakob disease.

To Summarize

- **Viruses** are noncellular particles that infect a host cell and direct its expression apparatus to produce virus particles.
- A **virion** consists of a capsid enclosing a nucleic acid genome. For some viruses, the capsid is further enclosed by an envelope of membrane with embedded proteins.
- **All classes of organisms are infected by viruses.** Usually the hosts are limited to a particular host range of closely related strains or species.
- **Viruses contain infective genomes** that take over a cell, reprogramming its cell machinery to make progeny virus particles (virions). Some viral genomes consist of fewer than 10 genes; others have 100 or 200 genes and may have evolved from a cellular parasite.
- **Viruses transfer genes between host genomes.** Viral gene transfer is a major source of cell genome evolution.
- **Viroids** that infect plants consist of RNA hairpins with no capsid.
- **Prions** are infectious proteins that induce a cell's native proteins to fold incorrectly and impair cell function.

6.2

Virus Structure

Understanding virus structure is crucial for devising vaccines and drug therapies. For example, the Gardasil vaccine for human papillomavirus, HPV (recommended for all children before adolescence), is composed of capsid proteins from four different HPV strains. The structure of a virion or virus particle achieves two goals: It keeps the viral genome intact, and it enables infection of the appropriate host cell. First, the stable capsid protects the viral genome from degradation and enables it to be transmitted outside the host. Second, in order for the viral genome to reproduce, the virion must either insert its genome into the host cell or disassemble within the host. In the process, the original particle loses its stable structure and its own identity as such. But if viral reproduction succeeds, then it yields numerous progeny virions.

The form of the virion depends on the species of virus. The virion may be symmetrical or asymmetrical, or combine aspects of both. The papillomavirus virion, for example, consists solely of a capsid containing a genome. In other virions, such as the herpes viruses, the capsid is enclosed by an envelope membrane. In some species, the capsid includes a "motor" device to deliver the genome into the host (presented in Chapter 11).

Note: In an enveloped virus, the protein capsid containing the viral genome may be called a **core particle**.

Thought Question

6.2 What will happen if a virus particle remains intact within a host cell and fails to release its genome?

Symmetrical Virus Particles

In a symmetrical virus, the capsid may be one of two types: icosahedral or filamentous (helical). Some have an intermediate form, such as the HIV core. The advantage of geometrical symmetry is that it provides a way to form a package out of repeating protein units generated by a small number of genes and encoded by a short chromosomal sequence. The smaller the viral genome, the more genome copies can be synthesized from the host cell's limited supply of nucleotides. Nevertheless, some symmetrical viruses, such as herpes virus and mimivirus, have much larger genomes. Large genomes offer a greater range of functions for viral components.

Icosahedral viruses. Many viruses package their genome in an **icosahedral** (20-sided) **capsid**. An icosahedral capsid takes the form of a polyhedron with 20 identical triangular faces. In the capsid, each triangle can be composed of three identical but asymmetrical protein units. An icosahedral capsid is found in the herpes simplex virus (**Fig. 6.9A**). Each triangular face of the capsid is determined by the same genes encoding the same protein subunits. The actual form of viral subunits can vary greatly, generating very different complex shapes for different viral species. But no matter what the pattern of subunits in the triangular unit, the structure overall shows the rotational symmetry of an icosahedron (**Fig. 6.9B**): threefold symmetry around the axis through two opposed triangular faces, fivefold symmetry around an axis through opposite points, and twofold symmetry around an axis through opposite edges.

Thought Question

6.3 For a viral capsid, what is the advantage of an icosahedron, as shown in **Figure 6.9** instead of some other polyhedron?

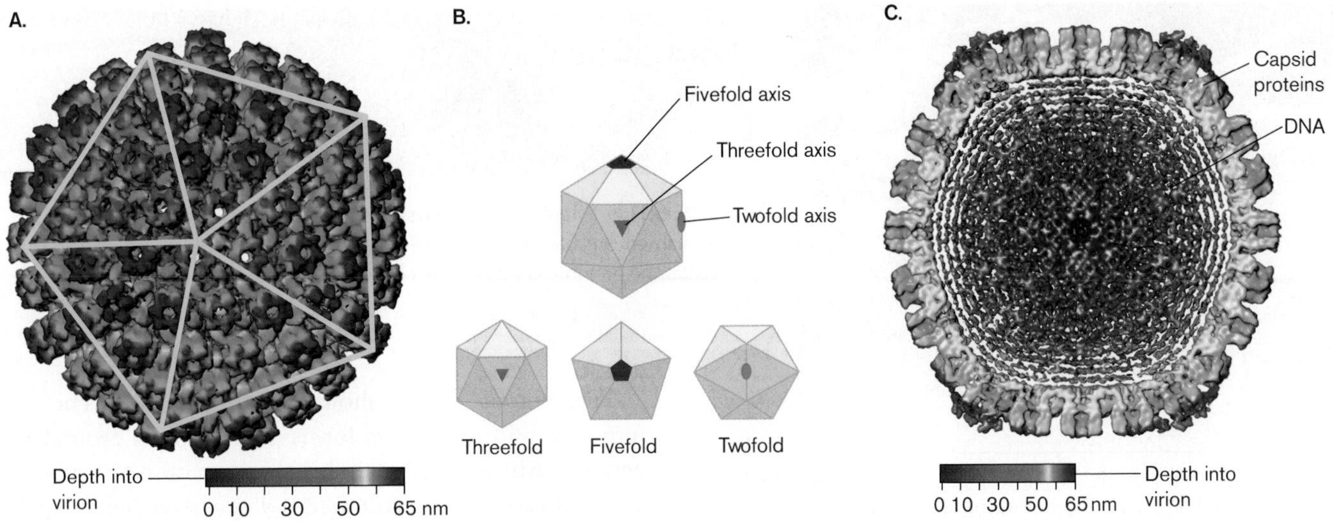
A.
B.
C.
Fivefold axis
Threefold axis
Twofold axis
Threefold
Fivefold
Twofold
Capsid proteins
DNA
Depth into virion
0 10 30 50 65 nm
Depth into virion
0 10 30 50 65 nm

FIGURE 6.9 ■ Herpes: icosahedral capsid symmetry. A. Icosahedral capsid of herpes simplex 1 (HSV-1), with envelope removed. Imaging of the capsid structure is based on computational analysis of cryo-TEM. Images of 146 virus particles were combined digitally to obtain this model of the capsid at 2-nm resolution. **B.** Icosahedral symmetry includes fivefold, threefold, and twofold axes of rotation. **C.** The icosahedral capsid contains spooled DNA. *Source:* A, C. Z. Hong Zhou et al. 1999. *J. Virol.* **73**:3210.

Virus particles can be observed by standard transmission electron microscopy (TEM), but the details of capsid structure as in Figure 6.9A require digital reconstruction of cryo-EM images (discussed in Section 2.6). Recall from Chapter 2 that in cryo-EM, the viral samples for TEM are prepared flash-frozen, preventing the formation of ice crystals. Flash-freezing enables observation without stain. The electron beams penetrate the object; thus, images of individual capsids actually provide a glimpse of the virus's internal contents. By digitally combining and processing cryo-TEM images from a number of capsids, a 3D reconstruction is built for the entire virus particle. Within the icosahedral capsid, the herpes genome is spooled and tightly packed. **Figure 6.9C** shows that the double-stranded DNA of the herpes genome is packed under high pressure. A molecular motor powered by ATP drives viral DNA into the host nucleus (discussed in Chapter 11).

In some icosahedral viruses, such as HIV, the capsid (called a core particle) is enclosed in an **envelope** composed of plasma membrane from the host cell in which the virion formed. Other viruses, such as herpes virus, derive their envelope from intracellular membranes, such as the nuclear membrane or endoplasmic reticulum. The envelope and capsid contents of herpes virus are shown in **Figure 6.10**.

The space between the capsid and the envelope contains "tegument" proteins, such as enzymes that interact with the host cell to depress its defense responses. Tegument proteins are expressed during infection of a host cell, and then get packaged in the virion. Both viral and host proteins may be packaged as tegument.

The mature envelope bristles with glycoprotein **spike proteins** that attach it to the capsid. The spike proteins enable the virus to attach and infect the next host cell. Further details of herpes structure and function are presented in Section 11.5.

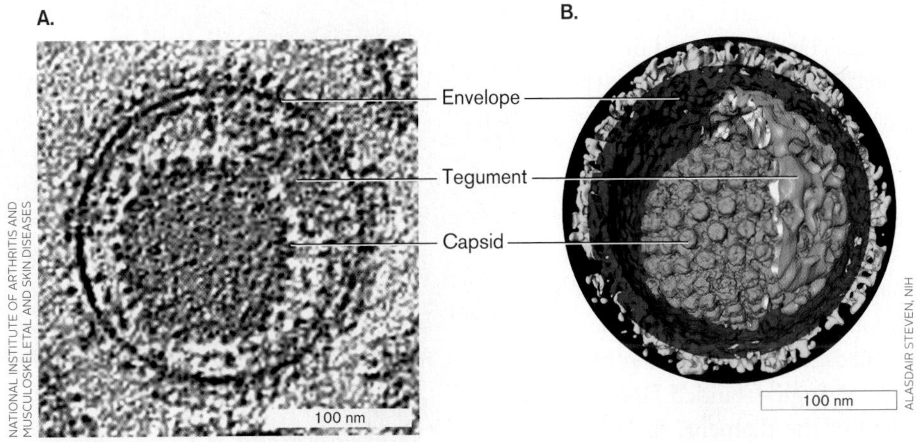
A.
B.
Envelope
Tegument
Capsid
100 nm
100 nm
NATIONAL INSTITUTE OF ARTHRITIS AND MUSCULOSKELETAL AND SKIN DISEASES
ALASDAIR STEVEN, NIH

FIGURE 6.10 ■ Envelope and tegument surround the herpes capsid. A. Section showing envelope and tegument proteins surrounding capsid (cryo-TEM). **B.** Cutaway reconstruction: spike envelope glycoproteins (yellow), envelope membrane (dark blue), tegument proteins (orange), capsid (light blue).

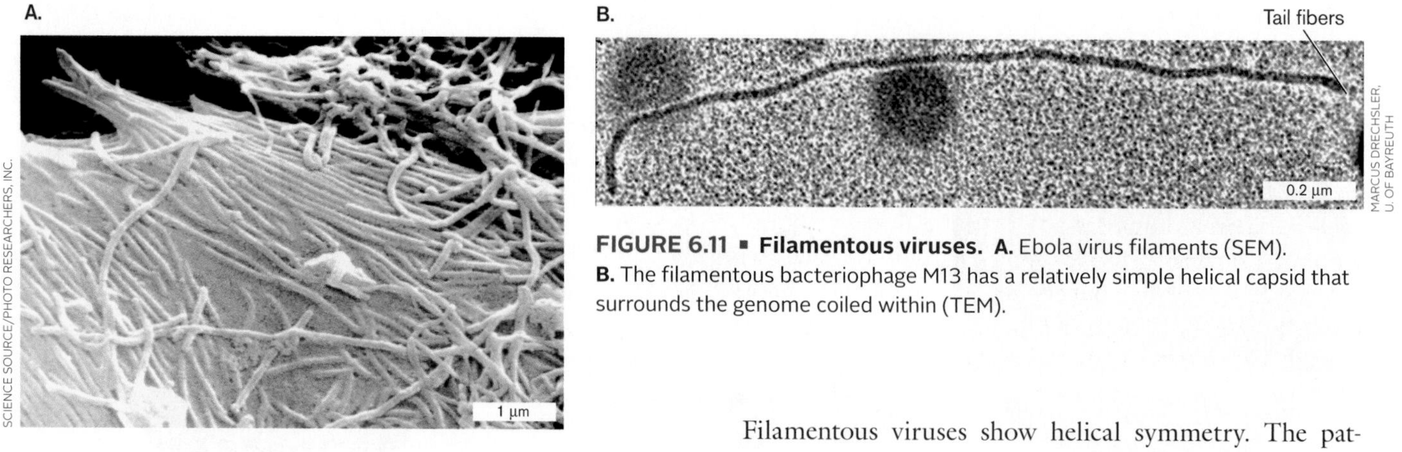
A.
1 µm
SCIENCE SOURCE/PHOTO RESEARCHERS, INC.
B.
Tail fibers
0.2 µm
MARCUS DRECHSLER, U. OF BAYREUTH

FIGURE 6.11 ■ Filamentous viruses. A. Ebola virus filaments (SEM). **B.** The filamentous bacteriophage M13 has a relatively simple helical capsid that surrounds the genome coiled within (TEM).

Note: Distinguish the viral envelope (phospholipid membrane derived from a host cell membrane) from the bacterial cell envelope (protective layers outside the bacterial cell membrane). The bacterial envelope is discussed in Chapter 3.

Filamentous viruses. A second major category of virus structure is that of **filamentous viruses**. A famous example is Ebola virus, which causes a swiftly fatal disease of humans and related primates (**Fig. 6.11A**). A filamentous bacteriophage that infects *E. coli* is phage M13 (**Fig. 6.11B**). Filamentous bacteriophages cause problems for human medicine and industry. One infects the Gram-positive species *Propionibacterium freudenreichii,* a key fermenting agent for Swiss cheese. Another filamentous phage, CTXϕ, integrates its sequence into the genome of *Vibrio cholerae,* where it carries the deadly toxin genes required for cholera. On the other hand, filamentous phages have been used in nanotechnology to nucleate the growth of crystalline "nanowires" for electronic devices.

The filamentous bacteriophage M13 (**Fig. 6.11B**) consists of a relatively simple capsid of protein monomers. The monomers are stacked around a coiled genome consisting of a circle of single-stranded DNA. At one end of the filament, short tail fibers mediate specific attachment to the host. The filament attaches to the F pilus of an *E. coli* bacterium containing an F plasmid that encodes pili.

Filamentous viruses show helical symmetry. The pattern of capsid monomers forms a helical tube around the genome, which usually winds helically within the tube. In a helical capsid, the genome is a single-stranded DNA (as in phage M13) or RNA (as in tobacco mosaic virus). **Figure 6.12** shows how the RNA strand of tobacco mosaic virus winds in a spiral within a tube of capsid monomers laid down in a spiral array. Such a tube can be imagined as a planar array of subunits that coils around such that each row connects to the row above, generating a spiral. The length of the helical capsid may extend up to 50 times its width, generating a flexible filament.

Unlike an icosahedral capsid, which usually has a fixed size, a helical capsid can vary in length to accommodate different lengths of nucleic acid. Furthermore, some viruses

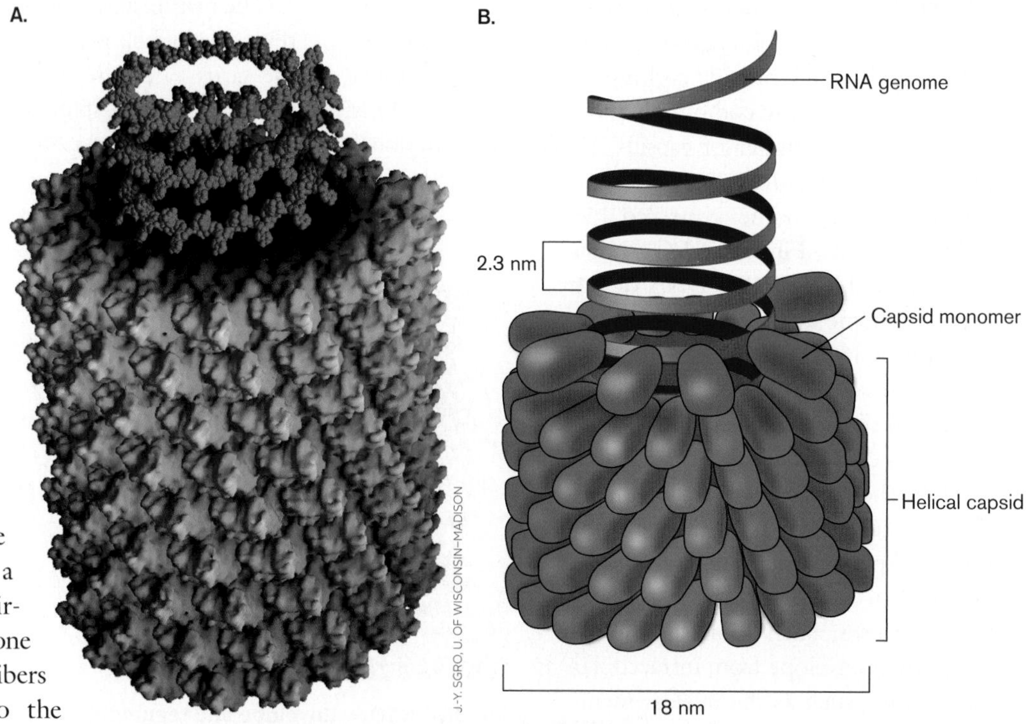
A.
J.-Y. SGRO, U. OF WISCONSIN–MADISON
B.
RNA genome
2.3 nm
Capsid monomer
Helical capsid
18 nm

FIGURE 6.12 ■ Tobacco mosaic virus: helical symmetry. A. The helical filament of tobacco mosaic virus (TMV) contains a single-stranded RNA genome coiled inside. **B.** Components of TMV virion.

package several genome segments into separate helical capsids. For example, influenza virus packages several different genome segments into separate helical packages of different sizes, contained together within a membrane envelope. Separate chromosome packaging enables influenza virus to pack different numbers of RNA segments into different virions. A surprising proportion of influenza progeny virions are thus defective, but the process enables rapid evolution of new strains (discussed further in Chapter 11).

Complex and Asymmetrical Virus Particles

Some bacteriophages have complex multipart structures. Bacteriophages often supplement the icosahedral capsid or head coat with an elaborate delivery device. For example, bacteriophage T4 (**Fig. 6.13A**) has an icosahedral "head" containing the pressure-packed DNA, attached to a helical "neck" that channels the nucleic acid into the host cell.

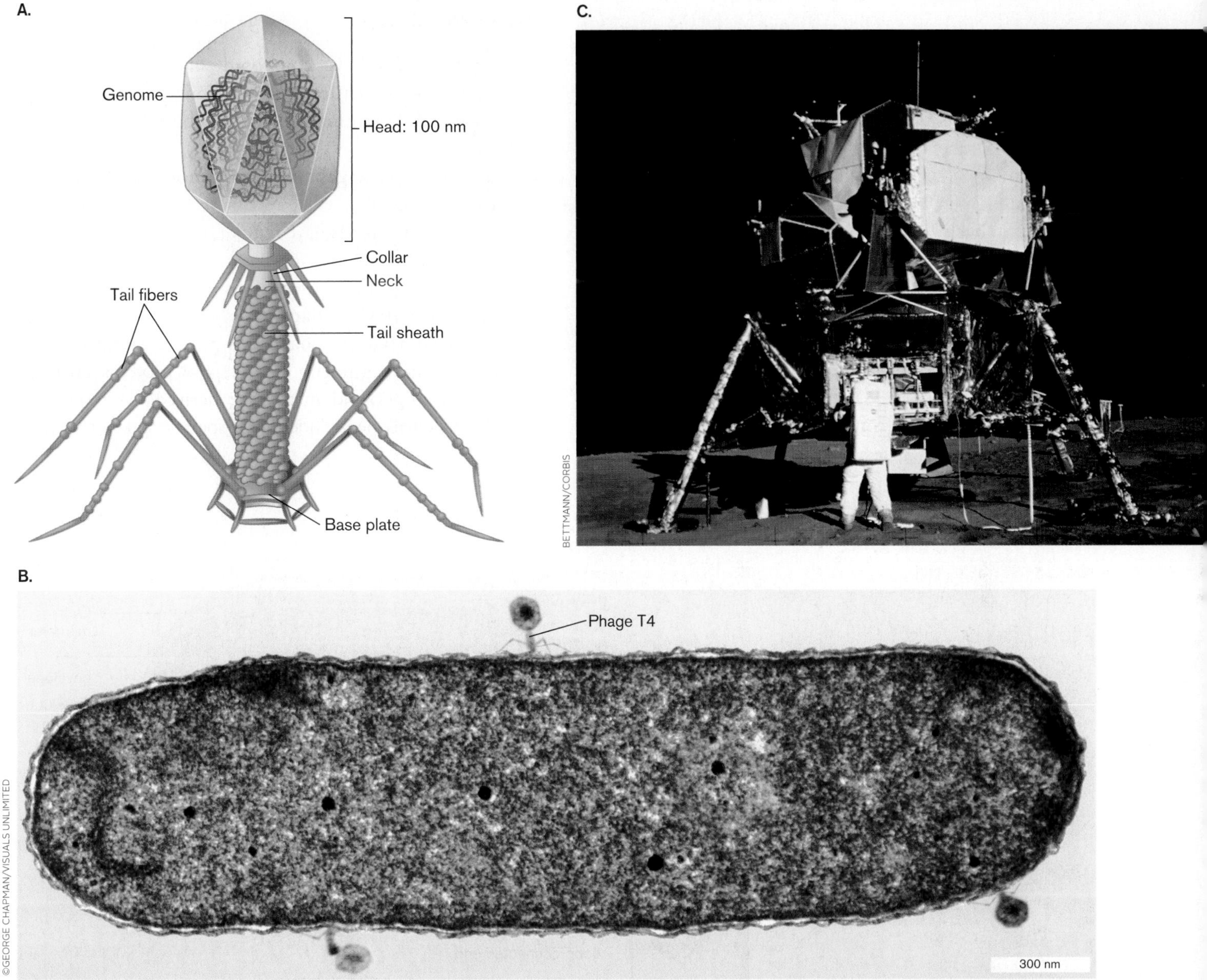
A.
Genome
Head: 100 nm
Collar
Neck
Tail fibers
Tail sheath
Base plate
B.
Phage T4
300 nm
©GEORGE CHAPMAN/VISUALS UNLIMITED
C.
BETTMANN/CORBIS

FIGURE 6.13 ■ Bacteriophage T4 capsid. A. Phage T4 particle with protein capsid containing packaged double-stranded DNA genome. The capsid is attached to a sheath with tail fibers that facilitate attachment to the surface of the host cell. After attachment, the sheath contracts and the core penetrates the cell surface, injecting the phage genome. **B.** *E. coli* infected by phage T4 (colorized blue, TEM). **C.** The structure of phage T4 resembles an *Apollo* lunar module that landed on the moon in 1969.

The tail baseplate has six jointed tail fibers that stabilize the structure on the host cell surface (**Fig. 6.13A**). The structure of phage T4 was first observed by microbiologists during the rise of the NASA space program, and its form was compared to that of the "lunar module" that landed on the moon (**Fig. 6.13C**). While the lunar module explored the moon, microbiologists used phages to "explore" the molecular biology of their host bacteria (discussed in Chapters 7 and 9).

Weblinks T4 bacteriophage cell-puncturing device, a 3D interactive tutorial (*see ebook*)

Other viruses have a capsid that is asymmetrical. In poxviruses, such as vaccinia and smallpox (**Fig. 6.14A**), the double-stranded DNA genome is stabilized by covalent connection of its two strands at each end (**Fig. 6.14B**). Instead of a rigid capsid, the DNA is enclosed loosely by a core envelope studded with spike proteins, surrounded by an outer membrane. The core envelope also encloses a large number of accessory proteins needed early in viral infection, such as initiation proteins for the transcription of viral genes and RNA-processing enzymes that modify viral mRNA molecules.

Asymmetrical DNA viruses usually contain a large number of accessory proteins. By contrast, the first viruses to be studied, including TMV and poliovirus, had extremely simple structures that consisted only of nucleic acid and packaging proteins. These models led to the conclusion that a virion consists solely of a packaged genome. Further research, however, revealed that other kinds of viruses contain enzymes and regulatory proteins—encoded by the virus, its host, or both. The proteins may be found either inside the capsid or in the tegument between the capsid and the envelope. Examples include the reverse transcriptase and protease enzymes of HIV and the dozen different enzymes contained by vaccinia poxvirus. Large asymmetrical viruses contain so many enzymes that they appear to have evolved from degenerate cells.

To Summarize

- **A viral capsid** is composed of repeated protein subunits—a structure that maximizes the structural capacity while minimizing the number of genes needed for construction.
- **The capsid packages the viral genome** and delivers it into the host cell.
- **Icosahedral capsids** have regular, icosahedral symmetry.
- **Filamentous (helical) capsids** have uniform width, generating a flexible filamentous virion.
- **Enveloped viruses** consist of a protein capsid (or core particle) and tegument or accessory proteins enclosed within phospholipid membrane derived from the host cell. The envelope includes virus-specific spike proteins.

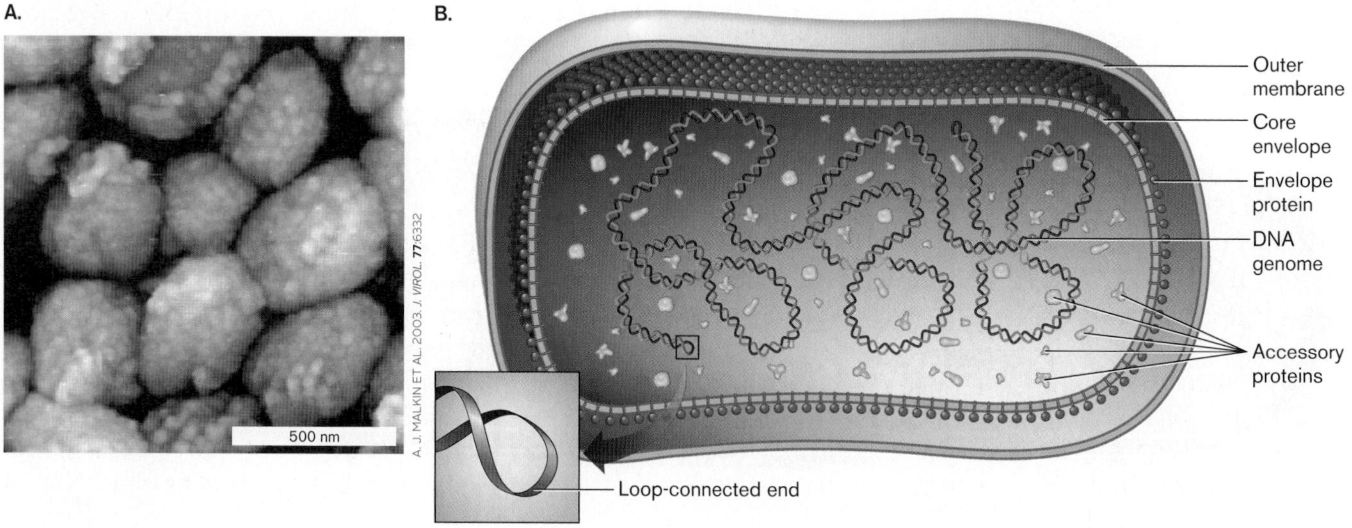
A.
B.
500 nm
A. J. MALKIN ET AL. 2003. J. VIROL. 77:6332
Outer membrane
Core envelope
Envelope protein
DNA genome
Accessory proteins
Loop-connected end

FIGURE 6.14 ■ Vaccinia poxvirus. A. Vaccinia virions observed in aqueous medium by atomic force microscopy (AFM). **B.** A pox virion includes an outer membrane and a core envelope membrane containing envelope proteins enclosing the double-stranded DNA genome and accessory proteins. The DNA is stabilized by a hairpin loop at each end.

6.3

Viral Genomes and Classification

Living organisms today are classified by relatedness of their gene sequences. Genetic relatedness can also be used to compare closely related viruses, such as herpes viruses. The definition of a virus species, however, is problematic, given the small size and high mutability of viral genomes and the ability of different viruses to recombine their genome segments within an infected host cell. Furthermore, not all viruses are monophyletic—that is, descended from a common ancestor. In fact, different classes of viruses appear to have evolved from different sources—for example, from parasitic cells or from host cell components such as DNA replication enzymes. Viruses are classified by genome composition, virion structure, and host range.

The International Committee on Taxonomy of Viruses

For purposes of study and communication, a working classification system has been devised by the International Committee on Taxonomy of Viruses (ICTV). The ICTV classification system is based on several criteria:

- **Genome composition.** The nucleic acid of the viral genome can vary remarkably with respect to physical structure: It may consist of DNA or RNA, it may be single- or double-stranded, it may be linear or circular, and it may be whole or **segmented** (that is, divided into separate "chromosomes"). Genomes are classified by the Baltimore method (discussed next).
- **Capsid symmetry.** The protein capsid or core particle may be helical or icosahedral, with various levels of symmetry.
- **Envelope.** The presence of a host-derived envelope, and the envelope structure if present, are characteristic of related viruses.
- **Size of the virus particle.** Related viruses generally share the same size range; for example, enteroviruses such as poliovirus are only 30 nm across (about $\frac{1}{30}$ the size of bacteria such as *E. coli*), whereas poxviruses are 200–400 nm, as large as a small bacterium.
- **Host range.** Closely related viruses usually infect the same or related hosts. However, viruses with extremely different hosts can show surprising similarities in genetics and structure. For example, both rabies virus and potato yellow dwarf virus are enveloped, bullet-shaped viruses of the rhabdovirus family.

Note: In nomenclature, families of viruses are designated by Latin names with the suffix "-viridae": for example, *Papillomaviridae.* Nevertheless, the common forms of such family names are also used, for example, the "papillomaviruses." Within a family, a virus species is simply capitalized, as in "Papillomavirus."

The Baltimore Virus Classification

Of the classification criteria just described, many virologists consider genome composition the most fundamental; that is, viruses of the same genome class (such as double-stranded DNA) are more likely to share ancestry with each other than with viruses of a different class of genome (such as RNA). In 1971, David Baltimore proposed that the primary distinctions among classes of viruses be the genome composition (RNA or DNA) and the route used to express messenger RNA (mRNA). Baltimore, together with Renato Dulbecco and Howard Temin, was awarded the 1975 Nobel Prize in Physiology or Medicine for discovering how tumor viruses cause cancer.

All cells and viruses need to make messenger RNA to produce their fundamental protein components. The production of mRNA from the viral genome is central to a virus's ability to propagate its kind. Cellular genomes always make mRNA by copying double-stranded DNA. For viruses, however, different kinds of genomes require fundamentally different mechanisms to produce mRNA. The different means of mRNA production generate distinct groups of viruses with shared ancestry.

So far, the genome composition and mechanisms of replication and mRNA expression define seven fundamental groups of viral species (**Fig. 6.15**). Examples of these seven fundamental groups are shown in **Table 6.1**. A more extensive taxonomic survey of viral species is presented in **Appendix 4**.

Group I. Double-stranded DNA viruses such as the herpes and smallpox viruses make their own DNA polymerase or use that of the host for genome replication. Their genes can be transcribed directly by a standard RNA polymerase, in the same way that a cellular chromosome would be transcribed. The RNA polymerase used can be that of the host cell, or it can be encoded by the viral genome.

Group II. Single-stranded DNA viruses such as canine parvovirus require the host DNA polymerase to generate the complementary DNA strand. The double-stranded DNA can then be transcribed by host RNA polymerase.

Group III. Double-stranded RNA viruses require a viral **RNA-dependent RNA polymerase** to generate messenger RNA by transcribing directly from the

A.

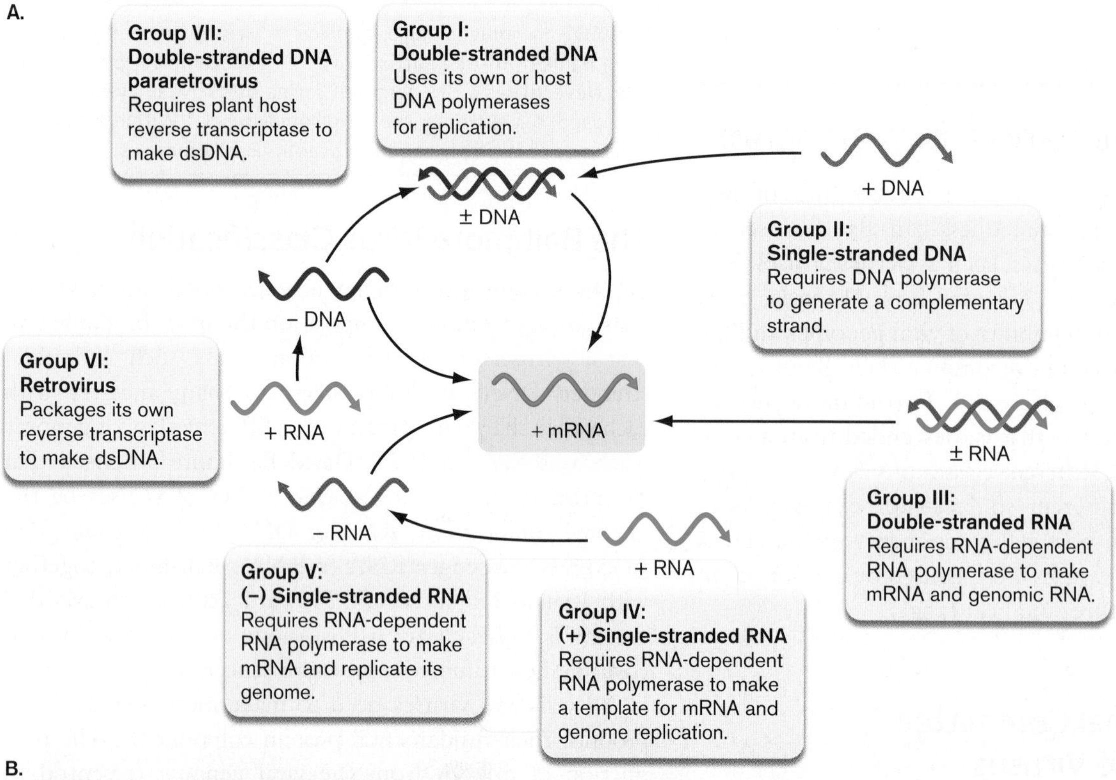
Group VII:
Double-stranded DNA pararetrovirus
Requires plant host reverse transcriptase to make dsDNA.
Group I:
Double-stranded DNA
Uses its own or host DNA polymerases for replication.
± DNA
+ DNA
Group II:
Single-stranded DNA
Requires DNA polymerase to generate a complementary strand.
– DNA
Group VI:
Retrovirus
Packages its own reverse transcriptase to make dsDNA.
+ RNA
+ mRNA
± RNA
Group III:
Double-stranded RNA
Requires RNA-dependent RNA polymerase to make mRNA and genomic RNA.
– RNA
+ RNA
Group V:
(–) Single-stranded RNA
Requires RNA-dependent RNA polymerase to make mRNA and replicate its genome.
Group IV:
(+) Single-stranded RNA
Requires RNA-dependent RNA polymerase to make a template for mRNA and genome replication.

B.

CALIFORNIA INSTITUTE OF TECHNOLOGY

FIGURE 6.15 ■ Baltimore classification of viral genomes. **A.** Seven categories of viral genome composition and replication mechanism:

I. Double-stranded DNA is transcribed to mRNA.

II. Single-stranded DNA generates a double-stranded form within the host cell, which is transcribed to mRNA.

III. Double-stranded RNA makes mRNA using RNA-dependent RNA polymerase.

IV. Single-stranded RNA (+) makes a complementary (–) strand, which is transcribed to mRNA.

V. Single-stranded RNA (–) is transcribed to mRNA.

VI. Retrovirus: Single-stranded RNA (+) is reverse-transcribed to DNA, which is transcribed to mRNA.

VII. Pararetrovirus: Double-stranded DNA (dsDNA) is transcribed to mRNA, which is reverse-transcribed to regenerate viral genomes for packaging into virions.

B. David Baltimore (left), with a graduate student at the California Institute of Technology. Baltimore won the 1975 Nobel Prize in Physiology or Medicine for his work on retroviruses; co-winners were Renato Dulbecco and Howard Temin.

RNA genome. Since the RNA polymerase is required immediately upon infection, such viruses package a viral RNA polymerase with their genome before exiting the host cell. A major class of double-stranded RNA viruses is the reoviruses, including rotavirus, a cause of diarrhea in children. Another reovirus has been engineered to destroy human tumors without infecting normal cells. This "oncolytic" reovirus, called Reolysin, is now in clinical trials for cancer therapy.

Group IV. (+) sense single-stranded RNA viruses consist of a positive-sense (+) strand (the coding strand) that can serve directly as mRNA to be translated to viral proteins. Replication of the RNA genome, however, requires synthesis of the template (–) strand [complementary to the (+) strand] by a viral RNA-dependent RNA polymerase to form a double-stranded RNA intermediate. Positive-sense (+) RNA viruses include enteroviruses such as poliovirus, the

TABLE 6.1

Groups of viruses—Baltimore classification. (Expanded table appears in Appendix 4.)

Virus example	Taxonomic group with examples
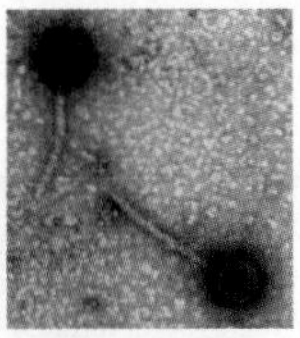 Phage lambda	**Group I. Double-stranded DNA viruses** ± DNA **Bacteriophage lambda** infects *Escherichia coli.* **Herpes viruses** cause chickenpox, genital infections, and birth defects. **Human papillomavirus** strains cause warts and tumors. **Mimivirus,** one of the largest known viruses, infects amebas.
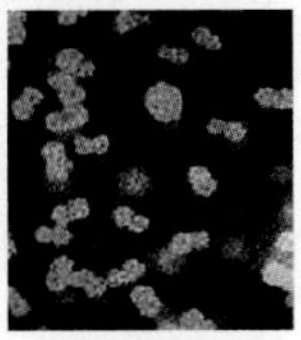 Geminivirus	**Group II. Single-stranded DNA viruses** + DNA **Bacteriophage M13** infects *E. coli.* **Parvoviruses** cause disease in cats, dogs, and other animals. **Geminiviruses** infect tomatoes and other plants.
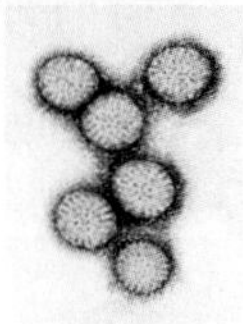 Rotavirus	**Group III. Double-stranded RNA viruses** ± RNA **Birnaviruses** infect fish. **Reoviruses** such as rotavirus cause severe diarrhea in infants. Other reoviruses are in clinical trials to fight tumors (oncolysis). **Varicosaviruses** infect plants.
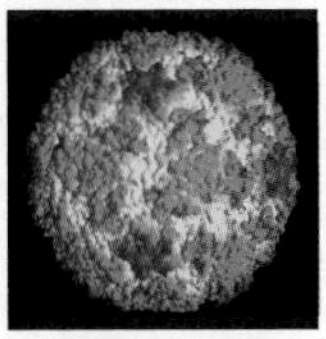 Rhinovirus	**Group IV. (+) sense single-stranded RNA viruses** + RNA **Tobacco mosaic virus** infects plants. **Coronaviruses** such as SARS cause severe respiratory disease. **Flaviviruses** cause hepatitis C, West Nile disease, yellow fever, and dengue fever. **Poliovirus** infects human intestinal epithelium and nerves.
 Rabies virus	**Group V. (–) sense single-stranded RNA viruses** – RNA **Orthomyxoviruses** cause influenza. **Filoviruses** such as Ebola virus cause severe hemorrhagic disease. **Paramyxoviruses** cause measles and mumps. **Rhabdovirus** causes rabies.
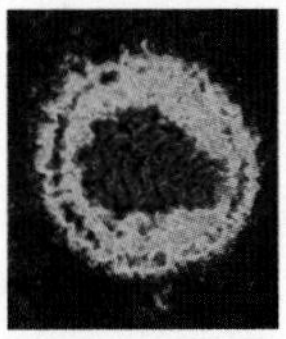 Human immuno-deficiency virus	**Group VI. Retroviruses (RNA reverse-transcribing viruses)** + RNA → DNA **Feline leukemia virus (FeLV), Rous sarcoma virus (RSV),** and **avian leukosis virus (ALV)** cause cancer. **Lentiviruses** include **human immunodeficiency virus (HIV),** the cause of AIDS; and **simian immunodeficiency virus (SIV),** the cause of simian AIDS.
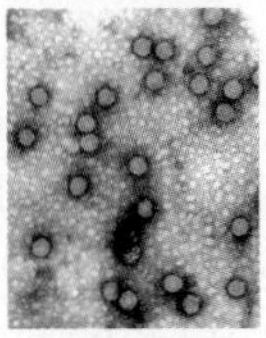 Caulimovirus	**Group VII. Pararetroviruses (DNA reverse-transcribing viruses)** DNA ⇄ + RNA **Caulimoviruses** (such as cauliflower mosaic virus, or CaMV) infect many kinds of vegetables. CaMV provides the best vector tools for plant biotechnology. **Hepadnaviruses** such as hepatitis B virus infect the human liver.

coronaviruses, and the flaviviruses that cause West Nile encephalitis and hepatitis C.

Group V. (–) sense single-stranded RNA viruses such as influenza virus have genomes that consist of template, or "negative-sense," RNA. Thus, they need to package a viral RNA-dependent RNA polymerase for transcribing (–) RNA to (+) mRNA. The (–) strand RNA viral genomes are often segmented; that is, they consist of multiple separate linear chromosomes—a key factor in the evolution of killer strains of influenza (see Section 11.3).

Group VI. Retroviruses, or **RNA reverse-transcribing viruses**, such as HIV and feline leukemia virus, have genomes that consist of (+) strand RNA. Instead of RNA polymerase, they package a **reverse transcriptase**, which transcribes the RNA into a double-stranded DNA (for details, see Section 11.4). The double-stranded DNA is then integrated into the host genome, where it directs the expression of the viral genes.

Group VII. Pararetroviruses, or **DNA reverse-transcribing viruses**, have a replication cycle that requires reverse transcriptase. For example, hepatitis B virus (a hepadnavirus) first copies its double-stranded DNA genomes into RNA, and then reverse-transcribes the RNA to progeny DNA using a reverse transcriptase packaged in the original virion. In contrast, plant pararetroviruses, such as cauliflower mosaic virus (CaMV, a caulimovirus), generate an RNA intermediate that replicates using a reverse transcriptase made by the host cell. Many plant genomes include a gene for reverse transcriptase. Cauliflower mosaic virus is of enormous agricultural significance for its use as a vector to construct pesticide-resistant food crops.

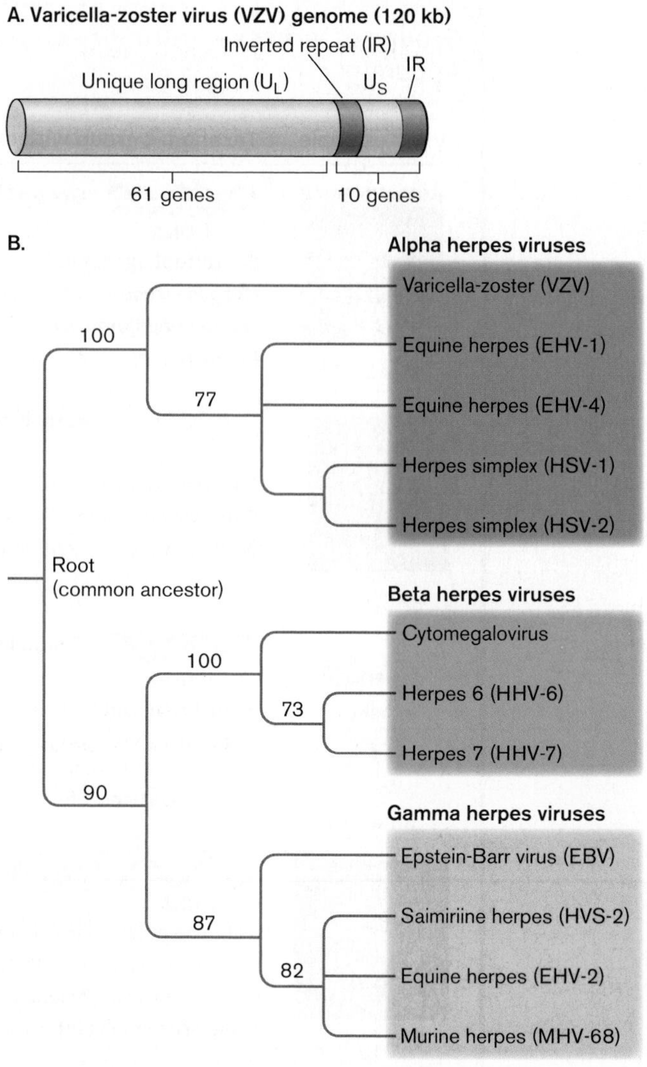

A. Varicella-zoster virus (VZV) genome (120 kb)
Inverted repeat (IR)
Unique long region (U_L)
U_S
IR
61 genes
10 genes
B.
Alpha herpes viruses
Varicella-zoster (VZV)
Equine herpes (EHV-1)
Equine herpes (EHV-4)
Herpes simplex (HSV-1)
Herpes simplex (HSV-2)
100
77
Root
(common ancestor)
Beta herpes viruses
Cytomegalovirus
Herpes 6 (HHV-6)
Herpes 7 (HHV-7)
100
73
90
Gamma herpes viruses
Epstein-Barr virus (EBV)
Saimiriine herpes (HVS-2)
Equine herpes (EHV-2)
Murine herpes (MHV-68)
87
82

FIGURE 6.16 ■ Phylogeny of herpes viral genomes. **A.** Genome structure of human varicella-zoster virus (VZV), the causative agent of chickenpox. **B.** Phylogeny of human and animal herpes viruses, based on whole-genome sequence analysis comparing clusters of orthologous groups of genes. Numbers measure percentage of DNA sequence shared by two divergent genomes.

Molecular Evolution of Viruses

The phylogeny, or genetic relatedness, of viruses can be determined within families. For example, the herpes family includes double-stranded DNA viruses that cause several human and animal diseases, such as chickenpox, oral and genital herpes infection, and respiratory and genital infections in horses. Herpes genomes consist of double-stranded DNA, 120–220 kilobases (kb) encoding about 70–200 genes; an example is that of varicella-zoster virus, the causative agent of chickenpox (**Fig. 6.16A**). The genome includes two "unique" segments of genes, one long and one short (U_L and U_S), joined by two inverted repeats (IRs). Other herpes genomes share similar structure, though they differ in gene order and IR position.

The relatedness of different herpes viruses that evolved from a common ancestor can be measured by comparing their genome sequences. Comparison is based on **orthologous genes**, or **orthologs**. Orthologs are genes of common ancestry in two genomes that share the same function (a topic discussed in Chapter 8). For cellular organisms, we often use the ribosomal RNA gene sequences to measure relatedness. Viruses have no ribosomal RNA, but closely related viruses share other orthologous genes. In pairs of orthologs, the amount of difference in sequence correlates approximately with the time following divergence from a common ancestor (a topic discussed in Chapter 17). A tree of genomic divergence (or phylogeny) was devised for herpes viruses (**Fig. 6.16**). The comparison of all gene pairs places herpes strains into three related classes designated alpha, beta, and gamma (**Fig. 6.16B**). The alpha class includes human varicella-zoster virus and the oral and genital herpes viruses (HSV-1 and HSV-2), as well as equine herpes virus. The beta class includes cytomegalovirus, a common cause of congenital infections (present at birth), as well as two lesser-known viruses. The gamma class includes Epstein-Barr virus (the cause of infectious mononucleosis), as well as several viruses of animals. Divergence of phylogenetic trees is discussed further in Chapter 17.

Gene comparison generates a tree for closely related viruses such as herpes. But how can we assess phylogeny of more distantly related viruses that share no genes—and even have genomes of different kinds of nucleic acids? Unlike cells, viruses do not possess genes common to all species. Furthermore, viral genomes are highly mosaic; that is, they evolved from multiple sources. Mosaic genomes result from recombination or reassortment of chromosomes from different viruses coinfecting a host.

In some cases, phylogeny is inconsistent with the fundamental chemical composition of the genome. For example, some DNA bacteriophages actually share closer ancestry with RNA bacteriophages than they do with DNA animal viruses. This is because two or more viruses can coinfect a cell and exchange genetic components. Thus, the genomic content of a virus can be influenced by its host range.

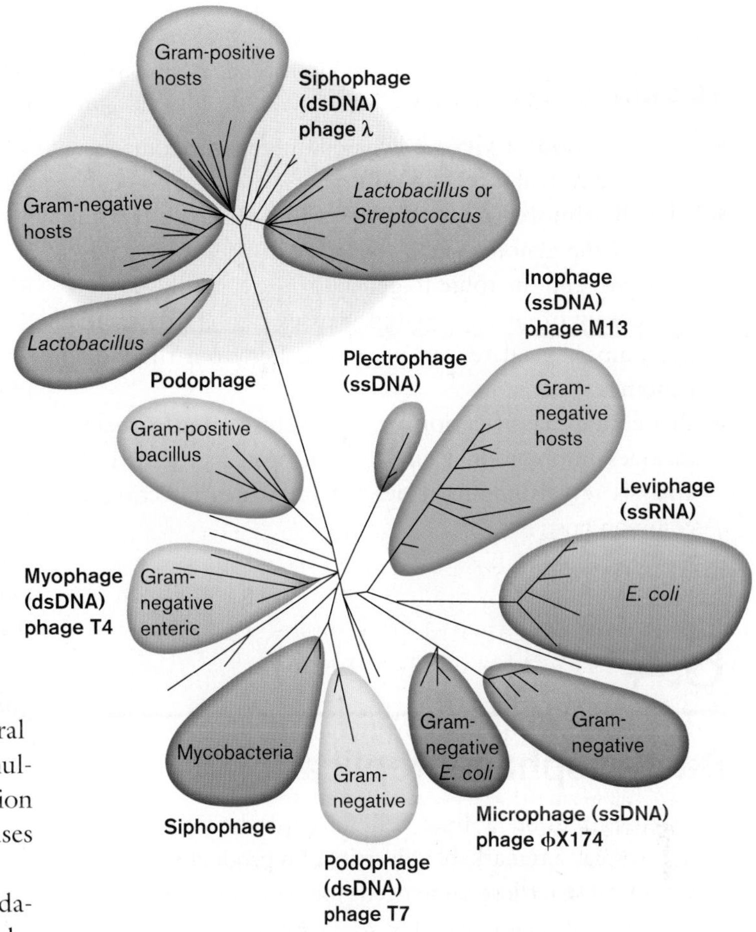

Gram-positive hosts
Siphophage (dsDNA) phage λ
Gram-negative hosts
Lactobacillus or Streptococcus
Inophage (ssDNA) phage M13
Lactobacillus
Plectrophage (ssDNA)
Podophage
Gram-negative hosts
Gram-positive bacillus
Leviphage (ssRNA)
Myophage (dsDNA) phage T4
Gram-negative enteric
E. coli
Mycobacteria
Gram-negative E. coli
Gram-negative
Gram-negative
Siphophage
Microphage (ssDNA) phage ϕX174
Podophage (dsDNA) phage T7

FIGURE 6.17 ▪ The bacteriophage proteomic tree. Comparison of all proteins encoded by each genome predicts distinct groups of bacteriophages. Within each group, there are subgroups of phages with shared hosts, since sharing of hosts facilitates genetic recombination and horizontal transfer of genes between different phages. *Source:* Based on Forest Rohwer and Rob A. Edwards. 2002. *J. Bacteriol.* **184**:4529.

Thought Question

6.4 How could viruses with different kinds of genomes (RNA versus DNA) combine and share genetic content in their progeny?

As more viral genomes are sequenced, classification methods have been devised to take advantage of sequence information without requiring gene products common to all species. One promising approach is that of **proteomics**, analysis of the **proteome**, the proteins encoded by genomes (discussed in Chapter 8). Proteins are identified through biochemical analysis of virus particles and through bioinformatic analysis of protein sequences encoded in the genomes. Proteomic analysis is useful for distantly related viruses because protein sequences often show relatedness that is obscured in the nucleic acid sequence by silent mutations (discussed in Chapter 16). Even if only a few proteins are shared by any two viruses, statistical comparison of all proteins in a set of viral species reveals underlying degrees of relatedness.

An example of viral classification based on proteomic analysis is that of the "proteomic tree" of bacteriophages proposed by Forest Rohwer and Rob Edwards (**Fig. 6.17**). Unlike earlier trees based on a single common gene sequence, the proteomic tree is based on the statistical comparison of phage protein sequences predicted by the genomic DNA of many different species of phages. The proteomic analysis predicts major evolutionary categories of phage species that share common host bacteria. For example, phages that infect Gram-negative hosts will show more genetic commonality with each other than with phages that infect Gram-positive hosts. Shared hosts have a significant impact on phage evolution because coinfecting a host enables phages to exchange genes.

To Summarize

- **Classification of viruses** is based on genome composition, virion structure, and host range.
- **The Baltimore virus classification** emphasizes the form of the genome (DNA or RNA, single- or double-stranded) and the route to generate messenger RNA.
- **Phylogeny of closely related viruses** can be calculated by comparing all related pairs of genes from the viral genomes.
- **Proteomic classification of distantly related viruses** includes information from all viral proteins. Statistical analysis reveals common descent of viruses infecting a common host.

6.4

Bacteriophage Replication

All viruses require a host cell for reproduction. While viruses display a remarkable diversity of reproductive strategies, they all face these same needs for host infection:

- **Host recognition and attachment.** Viruses must contact and adhere to a host cell that can support their particular reproductive strategy.
- **Genome entry.** The viral genome must enter the host cell and gain access to the cell's machinery for gene expression.
- **Assembly of virions.** Viral components must be expressed and assembled. Components usually "self-assemble"; that is, the joining of their parts is favored thermodynamically.
- **Exit and transmission.** Progeny virions must exit the host cell, and then reach new host cells to infect. In the case of multicellular organisms, the virus must eventually reach other multicellular hosts (as discussed in Chapter 26).

Bacteriophages Attach to Host Cells

To commence an infectious cycle, bacteriophages need to contact and attach to the surface of an appropriate host cell. Contact and attachment are mediated by **cell-surface receptors**, proteins on the host cell surface that are specific to the host species and that bind to a specific viral component.

Receptor proteins. The cell-surface receptor for a virus is actually a protein with an important function for the host cell, but the virus has evolved to take advantage of the protein. An extensively studied model system of virus-receptor binding is that of bacteriophage lambda infecting *E. coli* (see **Table 6.1**, top row). The phage lambda virion attaches specifically to the maltose porin, an outer-membrane porin that transports maltose into the cell (**Fig. 6.18**). Although the protein is often called "lambda receptor protein," it actually evolved in the host as a way to obtain the sugar maltose to metabolize. Thus, natural selection maintains the maltose porin in *E. coli,* despite the danger of phage infection.

The precise domain of the maltose porin that binds to phage lambda was defined experimentally by mutations in *E. coli* that cause amino acid substitution in the protein. Some of the mutant *E. coli* strains were resistant to phage lambda infection. The mutations that conferred host resistance mapped to the domain of maltose porin that binds the phage capsid.

Phage Reproduction within Host Cells

Historically, the replication cycles of bacteriophages have provided some of the most fundamental insights in molecular biology. In 1952, Alfred Hershey and Martha Chase showed that the transmission of DNA by a bacteriophage to a host cell led to the production of progeny bacteriophages, thus confirming that DNA is the hereditary material. In

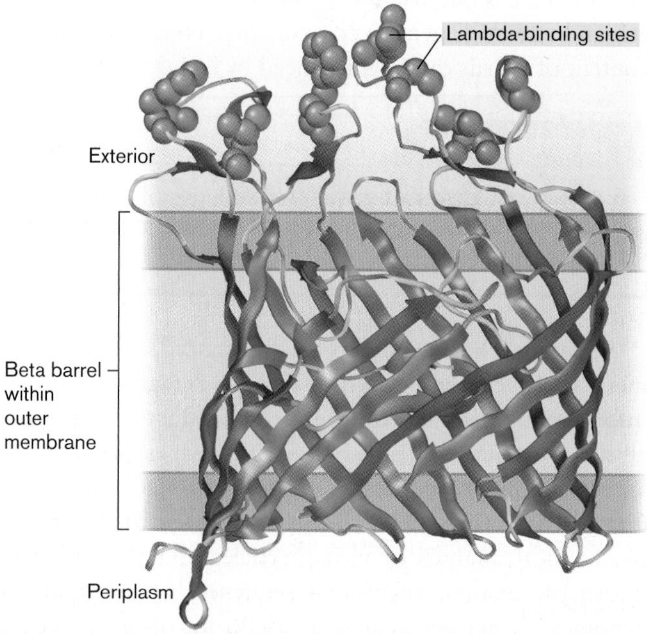
Lambda-binding sites
Exterior
Beta barrel within outer membrane
Periplasm

FIGURE 6.18 ■ Maltose porin is *E. coli* host receptor for bacteriophage lambda. The "beta barrel" of maltose porin (blue) is buried in the outer membrane. The phage-binding sites (green) were identified by amino acid substitution mutations that prevent phage binding and confer host resistance to lambda. (PDB code: 1MAL)

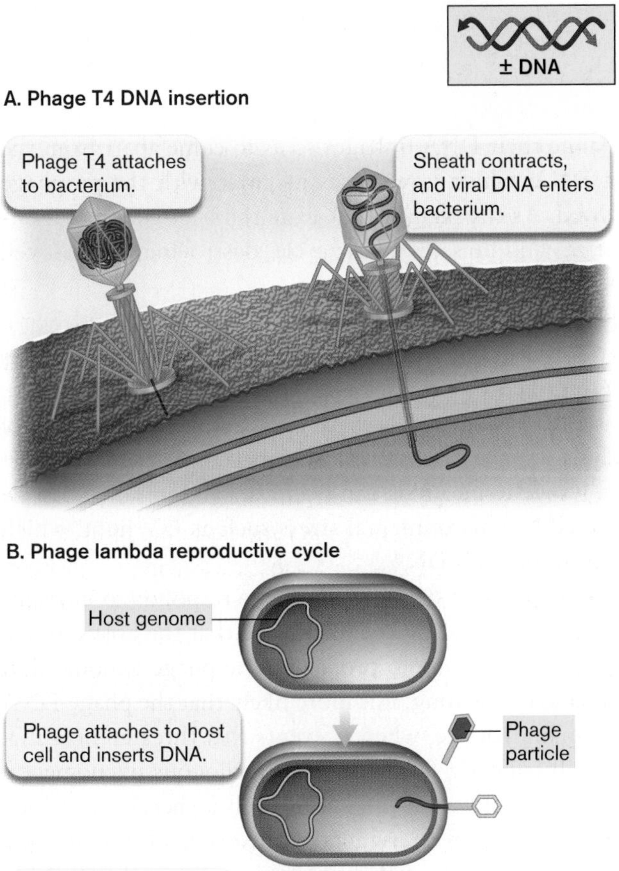
± DNA
A. Phage T4 DNA insertion
Phage T4 attaches to bacterium.
Sheath contracts, and viral DNA enters bacterium.
B. Phage lambda reproductive cycle
Host genome
Phage attaches to host cell and inserts DNA.
Phage particle

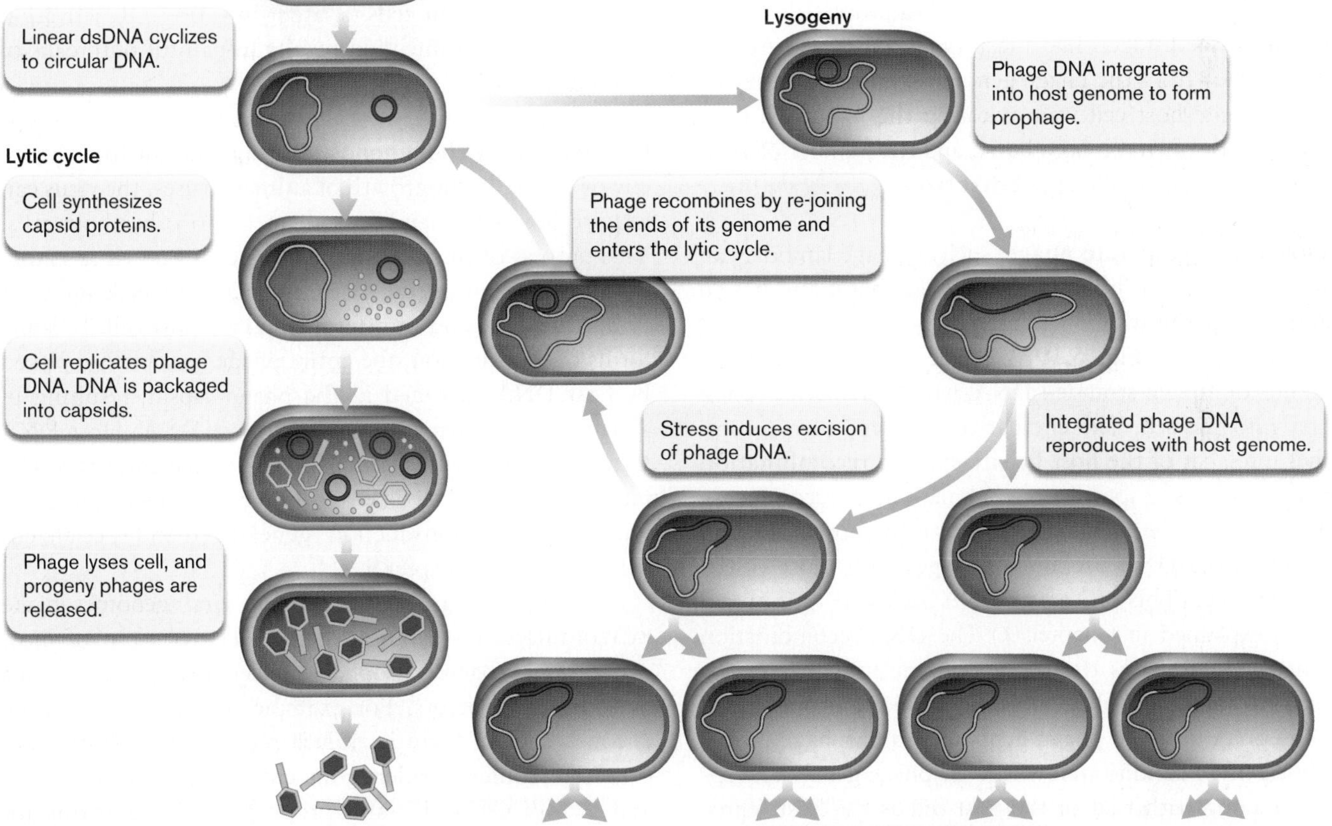
Linear dsDNA cyclizes to circular DNA.
Lysogeny
Phage DNA integrates into host genome to form prophage.
Lytic cycle
Cell synthesizes capsid proteins.
Phage recombines by re-joining the ends of its genome and enters the lytic cycle.
Cell replicates phage DNA. DNA is packaged into capsids.
Stress induces excision of phage DNA.
Integrated phage DNA reproduces with host genome.
Phage lyses cell, and progeny phages are released.
Integrated phage DNA replicates with host genome.

FIGURE 6.19 ▪ Bacteriophage reproduction: lysis and lysogeny. **A.** Phage T4 attaches to the cell surface by its tail fibers and then contracts to inject its DNA. The empty capsid remains outside as a "ghost." **B.** Lysis occurs when the phage genome reproduces progeny phage particles, as many as possible, and then lyses the cell to release them. In phage lambda, lysogeny can occur when the phage genome integrates itself into that of the host. The phage genome is replicated along with that of the host cell. The phage DNA, however, can direct its own excision by expressing a site-specific DNA recombinase. This excised phage chromosome then initiates a lytic cycle.

1950, André Lwoff and Antoinette Gutman showed that a phage genome could integrate itself within a bacterial genome—the first recognition that genes could enter and leave a cell's genome. Other fundamental concepts of the genetic unit and the basis of gene transcription came from experiments on bacteriophages, as discussed in Chapters 7–9.

Most bacteriophages (phages) inject only their genome into a cell through the cell envelope, thus avoiding the need for the capsid to penetrate the molecular barrier of the cell wall. For example, the phage T4 virion has a sheath that contracts, bringing the head near the cell surface to inject its DNA (**Fig. 6.19A**). The pressure of the spooled DNA—as high as 50 atmospheres (atm)—is released, expelling the DNA into the cell. After the genome has been inserted, the phage capsid remains outside, attached to the cell surface. The empty capsid is termed a "ghost" because of its pale appearance in an electron micrograph.

The lytic cycle. In a **lytic cycle**, when a phage particle injects its genome into a cell, it immediately reproduces as many progeny phage particles as possible. The process of reproduction involves replicating the phage genome, as well as expressing phage mRNA to make enzymes and capsid proteins. Some phages, such as T4, digest the host DNA to increase the efficiency of phage production. Phage particles assemble, and the host cell lyses, releasing progeny phages.

Phage T4 reproduces entirely by the lytic cycle and thus is called a "virulent" phage. Other phages, such as phage lambda, have the options of reproducing by lysis or by lysogeny (**Fig. 6.19B** ▶).

Lysis. First, the phage lambda adsorbs to its host receptor (maltose porin) and inserts its double-stranded DNA into the host cytoplasm. The phage genes are then expressed by the host cell RNA polymerase and ribosomes. "Early genes" are expressed early in the lytic cycle. Other phage-expressed proteins then work together with the cellular enzymes and ribosomes to replicate the phage genome and produce phage capsid proteins. The capsid proteins self-assemble into capsids and package the phage genomes—a process that takes place in defined stages, like a factory assembly line. At last, a "late gene" from the phage genome expresses an enzyme that lyses the host cell wall, releasing the mature virions. **Lysis** is also referred to as a burst, and the number of virus particles released is called the **burst size**.

Lysogeny. A **temperate phage**, such as phage lambda, can infect and lyse cells like a virulent phage, but it also has an alternative pathway: to integrate its genome into that of the host cell (see **Fig. 6.19B**). Phage lambda has a linear genome of double-stranded DNA, which circularizes upon entry into the cell. The circularized genome then recombines into that of the host by **site-specific recombination** of DNA. In site-specific recombination, a recombinase enzyme aligns the phage genome with the host DNA and exchanges the DNA backbone linkages with those of the host genome. (This process of DNA backbone recombination is explained in Chapter 9.) The DNA recombination event thus integrates the phage genome into that of the host. The integrated phage genome is called a **prophage**.

Integration of the phage genome as a prophage results in **lysogeny**, a condition in which the phage genome is replicated along with that of the host cell as the host reproduces (**Fig. 6.19B**). Implicit in the term "lysogeny," however, is the ability of such a strain to spontaneously generate a lytic burst of phage. For lysis to occur, the prophage (integrated phage genome) directs its own excision from the host genome by an intramolecular process of site-specific recombination. The two ends of the phage genome exchange their DNA linkages so as to come apart from the host DNA, which now closes its circle with the prophage removed. As the phage DNA exits the host genome, it circularizes and initiates a lytic cycle, destroying the host cell and releasing phage particles.

How does a lysogen "decide" to reactivate and begin a lytic cycle? The "decision" between lysogeny and lysis is determined by proteins that bind DNA and repress the transcription of genes for virus replication. (The genetics of the lysis-lysogeny decision is described in **eTopic 10.5**.) Exit from lysogeny into lysis can occur at random, or it can be triggered by environmental stress such as UV light, which damages the cell's DNA.

The regulatory switch of lysogeny responds to environmental cues indicating the likelihood that the host cell will survive and continue to propagate the phage genome. If a cell's growth is strong, it is more likely that the phage DNA will remain inactive, whereas events that threaten host survival will trigger a lytic burst. An analogous phenomenon occurs in animal viral infections such as herpes, in which environmental stress triggers reactivation of a virus that was dormant within cells (a latent infection). Reactivation of a latent herpes infection results in painful outbreaks of skin lesions.

Viruses transfer host genes. During the exit from lysogeny or during latent growth of animal viruses, the virus can acquire host genes and pass them on to other host cells. This process of transferring host genes is known as **transduction**. A transducing bacteriophage can pick up a bit of host genome and transfer it to a new host cell. In some forms of transduction, the entire phage genome is replaced by host DNA packaged in the phage capsid, resulting in a virus particle that transfers only host DNA. Host DNA transferred by viruses can become permanently incorporated into the infected host genome. The mechanisms of phage-mediated transduction (generalized and specialized) are discussed in Chapter 9.

The integration and excision of viral genomes make extraordinary contributions to the evolution of host genomes. Lysogenic prophages often express key toxins and virulence factors. For example, the CTXϕ prophage encodes cholera toxin in *Vibrio cholerae*; the Shiga toxin prophage confers virulence on enteric pathogens *Shigella* and *E. coli* O157:H7; and prophages encode toxins for *Corynebacterium diphtheriae* (diphtheria) and *Clostridium botulinum* (botulism). In natural environments, phage transduction mediates much of the recombination of bacterial genomes. In the laboratory, the ability of phages to transfer genes provided the first vectors for recombinant DNA technology (Chapter 12).

When viral transfer of genes was first discovered in the 1960s, multicellular eukaryotes were thought to have a much lower tolerance for genome change. We have since learned, however, that much of the human genome also shows evidence of gene transfer mediated by ancient viruses, including retroviruses similar to HIV (discussed in Chapter 11).

Slow release. The slow-release replication cycle differs from lysis and lysogeny in that phage particles reproduce without destroying the host cell (**Fig. 6.20**). Slow release is performed by filamentous phages such as phage M13. In slow-release replication, the single-stranded circular DNA of M13 serves as a template to synthesize a double-stranded intermediate. The double-stranded intermediate slowly generates single-stranded progeny genomes, which are packaged by supercoiling and coating with capsid proteins. The phage particles then extrude through the cell envelope without lysing the cell. The host cell continues to reproduce, though more slowly than uninfected cells do because much of its resources are diverted to virus production.

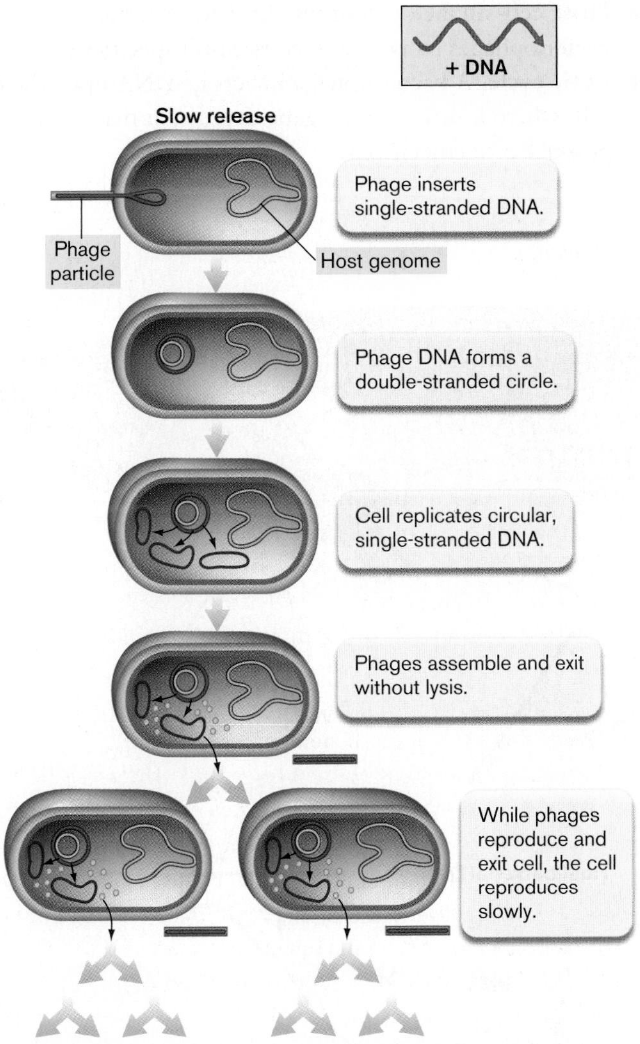
+ DNA
Slow release
Phage inserts single-stranded DNA.
Phage particle
Host genome
Phage DNA forms a double-stranded circle.
Cell replicates circular, single-stranded DNA.
Phages assemble and exit without lysis.
While phages reproduce and exit cell, the cell reproduces slowly.

FIGURE 6.20 ■ Bacteriophage replication cycle: slow release. In the slow-release replication cycle, a filamentous phage produces phage particles without lysing the cell. The host continues to reproduce itself, but more slowly than uninfected cells do because much of its resources are being used to make phages.

Some phage genomes consist of RNA. RNA phages include the leviviruses, such as phage MS2 and Qβ. The RNA phages replicate their genomes through a double-stranded RNA intermediate, analogous to the double-stranded DNA intermediate of the filamentous phage M13, but RNA phages lyse their hosts. RNA phages have very simple genomes composed of only three or four genes. These genes encode a replicase, a coat protein, and a maturation factor. The RNA genomes of these phages are of historical interest as model systems to study ribosomal translation.

Thought Question

6.5 What are the relative advantages and disadvantages (to a virus) of the slow-release strategy, compared with the strategy of a temperate phage, which alternates between lysis and lysogeny?

Bacterial Host Defenses

In natural environments, viruses commonly outnumber cellular microbes by tenfold or more. So how have host cells evolved to defend themselves? Several remarkable resistance mechanisms have evolved. Their molecular basis is explained in greater detail in Chapters 7 and 9.

Genetic resistance. All bacteria acquire random mutations in their genomes as they reproduce (Chapter 7). Under attack by bacteriophages, bacterial populations undergo natural selection; mutants that happen to be harder to infect will survive. Resistance can occur through a gene that encodes an altered host receptor protein, which fails to bind the viral coat protein. Alternatively, a different cellular protein evolves to block phage binding to the receptor. An evolutionary "arms race" occurs, in which phages may evolve enzymes that cleave the host defense molecules.

Restriction endonucleases. Bacteria modify their DNA by adding methyl groups to bases within certain sequences. The bacteria then express "restriction endonucleases," enzymes that cleave DNA lacking the methylated patterns—which includes potential virus DNA (Chapter 7). However, phage genomes composed of RNA or of modified DNA (such as phage T4, discussed in Chapter 11) escape cleavage by these enzymes.

CRISPR: A bacterial immune system. Amazingly, bacteria possess an adaptive defense against viruses that is analogous to an immune system. (Adaptive immunity of humans is presented in Chapter 24.) The bacterial adaptive defense involves short DNA sequences homologous to DNA of phages that could infect the cell. The sequences are called clustered regularly interspaced short palindromic repeats (**CRISPR**). When a phage attacks a bacterium, if bacterial enzymes succeed in destroying the phage DNA, they may copy a tiny piece of it as a CRISPR segment, inserted as a "spacer" at the head of a long line of about 30 CRISPR sequences (**Fig. 6.21**).

Thought Question

6.6 How might a phage evolve resistance to the CRISPR host defense (outlined in **Fig. 6.21**)?

Now the adapted host cell "remembers" infection by the specific phage—along with many other previous phages from previous infections that had inserted other spacers. The next time the adapted host cell is attacked by the same phage, all of its genomic CRISPR sequences are expressed as RNA. The CRISPR RNA is cleaved into small sequences (crRNA) containing one spacer from the original bacterial DNA. The crRNA binds to the "Cascade" protein complex (or Cas complex), which now detects phage DNA homologous to its virus-derived crRNA. The Cas-crRNA complex proceeds to cleave the phage DNA, preventing phage replication. For more details of this exciting molecular defense system, see Chapter 9.

To Summarize

- **Host cell-surface receptors** mediate the attachment of bacteriophages to a cell and confer host specificity.
- **Lytic cycle.** A bacteriophage injects its DNA into a host cell, where it utilizes host gene expression machinery to produce progeny virions.

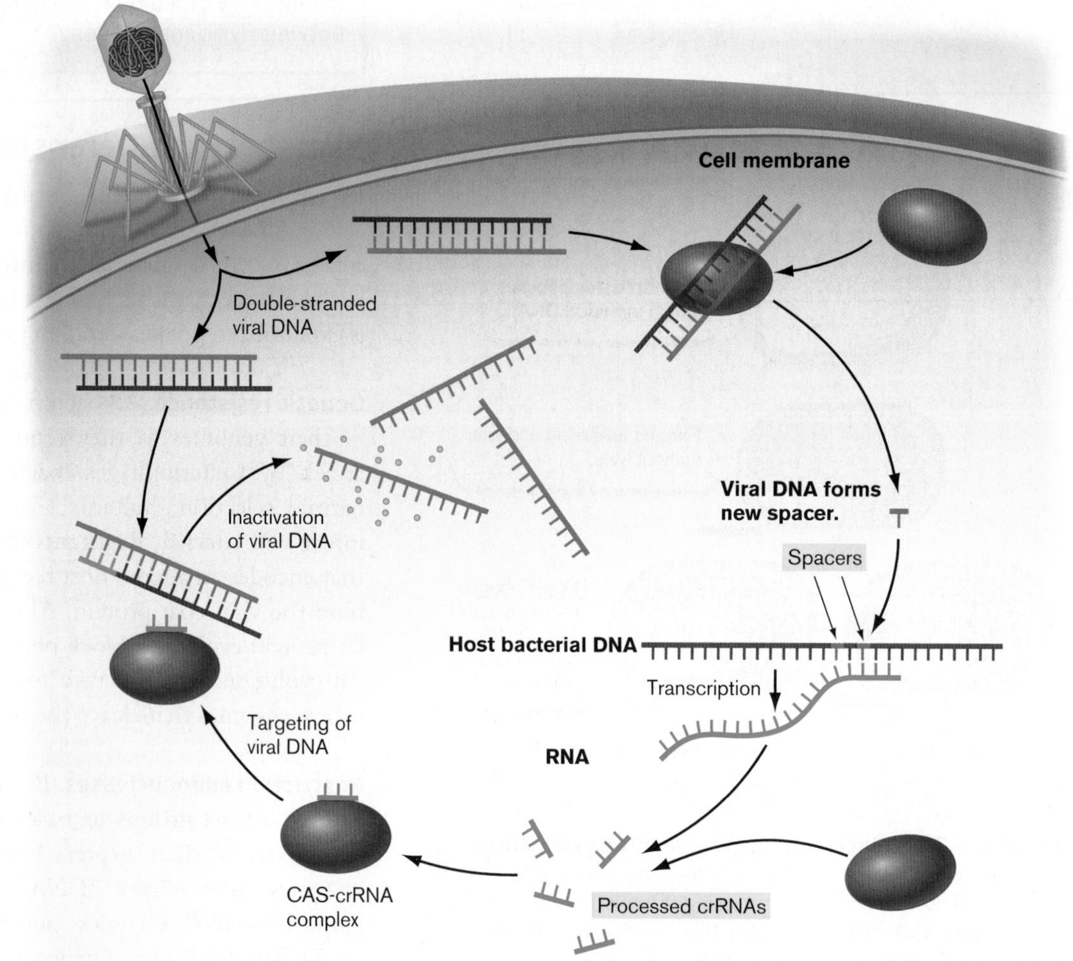
Cell membrane
Double-stranded viral DNA
Inactivation of viral DNA
Viral DNA forms new spacer.
Spacers
Host bacterial DNA
Transcription
RNA
Targeting of viral DNA
CAS-crRNA complex
Processed crRNAs

FIGURE 6.21 ■ CRISPR defense of bacterial cell. A piece of phage DNA gets copied as a "spacer" into the host genome. If the bacterium survives infection, later reinfection by the same kind of phage causes transcription of the spacers into CRISPR RNA. A processed spacer (crRNA) joins the Cas complex to recognize and cleave the phage DNA.

- **Lysogeny.** Some bacteriophages can insert their genome into that of the host cell, which then replicates the phage genome along with its own. A lysogenic bacterium can initiate a lytic cycle.
- **Cells evolve through viral gene transfer.** Genes are transferred by phage processes of transduction and lysogeny.
- **Slow release.** Some bacteriophages use the host machinery to make progeny that bud from the cell slowly, slowing growth of the host without lysis.
- **Bacterial host defense.** Bacteria have evolved several forms of defense against bacteriophage infection, such as altered receptor proteins, restriction endonucleases, and CRISPR integration of phage DNA sequences.

6.5

Animal and Plant Virus Replication

Animal and plant viruses solve problems similar to those faced by bacteriophages: host attachment, genome entry and gene expression, virion assembly, and virion release. The more complex structure of eukaryotic cells, however, requires greater complexity and diversity of viral replication cycles than we see in bacteriophages. Viral reproduction may involve intracellular compartments such as the nucleus or secretory system and may depend on tissue and organ development in multicellular organisms. Study of animal virus replication reveals potential targets for antiviral drugs, such as protease inhibitors for HIV (discussed in Chapter 11).

Note: The virus replication cycles in this chapter are simplified. For greater molecular detail, see Chapter 11.

Animal Viruses Show Tissue Tropism

Like bacteriophages, animal viruses evolve by the fitness advantage of binding specific receptor proteins on their host cell. An example of a human virus-receptor interaction is that of rhinovirus (see **Table 6.1**, Group IV), which causes the common cold. Rhinovirus attaches to ICAM-1, a human glycoprotein (protein with sugar chains) needed for intercellular adhesion (**Fig. 6.22A**). The rhinovirus binds to a domain of ICAM-1 essential for ICAM-1 to bind a lymphocyte protein called integrin.

The host receptors play a key role in determining the host range, the group of host species permitting infection. Within a host, receptor molecules can also determine the viral **tropism**, or ability to infect a particular tissue type. Some viruses, such as Ebola virus, exhibit broad tropism, infecting many kinds of host tissues, whereas others, such as papillomavirus, show tropism for only one type (in the case of papillomavirus, the epithelial tissues). Tropism may depend on the virus's ability to interact with the host cytoplasm, or it may require the presence of an appropriate host cell receptor protein that can bind the viral surface attachment protein. For example, poliovirus infects only a specific class of human cells that display the immunoglobulin-like receptor protein PVR. Mice lack the PVR protein on their cell surfaces, and they are not normally infected by polio, but when transgenic mice were engineered to express PVR on their cells, the mice could then be infected.

Tissue specific receptors determine the host tropism of avian influenza strain H5N1. The H5N1 strain infects birds by binding to a glycoprotein receptor on cell surfaces of the avian respiratory tract. The H5N1 strain requires a receptor protein with a sialic acid sugar chain terminating in galactose linked at the C-3 position (α2,3), common in the avian respiratory tract. In humans, however, most nasal upper respiratory cells have receptors with galactose linked at the C-6 position (α2,6). Human cells with the C-3 linkage are more common in the lower respiratory tract. That is why avian influenza H5N1 infection of humans has been relatively rare. However, only a small mutation in the H5N1 envelope protein could enable it to bind to the (α2,6) receptor more effectively, allowing rapid transmission between humans. The molecular basis of influenza infection is discussed further in Chapter 11.

Thought Question

6.7 How could humans undergo natural selection for resistance to rhinovirus infection? Is such evolution likely? Why or why not?

Genome entry and uncoating. Most animal viruses, unlike bacteriophages, enter the host cell as virions. The contents of the virion (genome and matrix proteins) may then interact with the cell in several different ways. For example, measles virus, a paramyxovirus, enters the cell by binding host receptor proteins, which causes the viral envelope to fuse with the host cell membrane (**Fig. 6.22B**). The measles RNA genome coated by nucleocapsid proteins is released directly into the

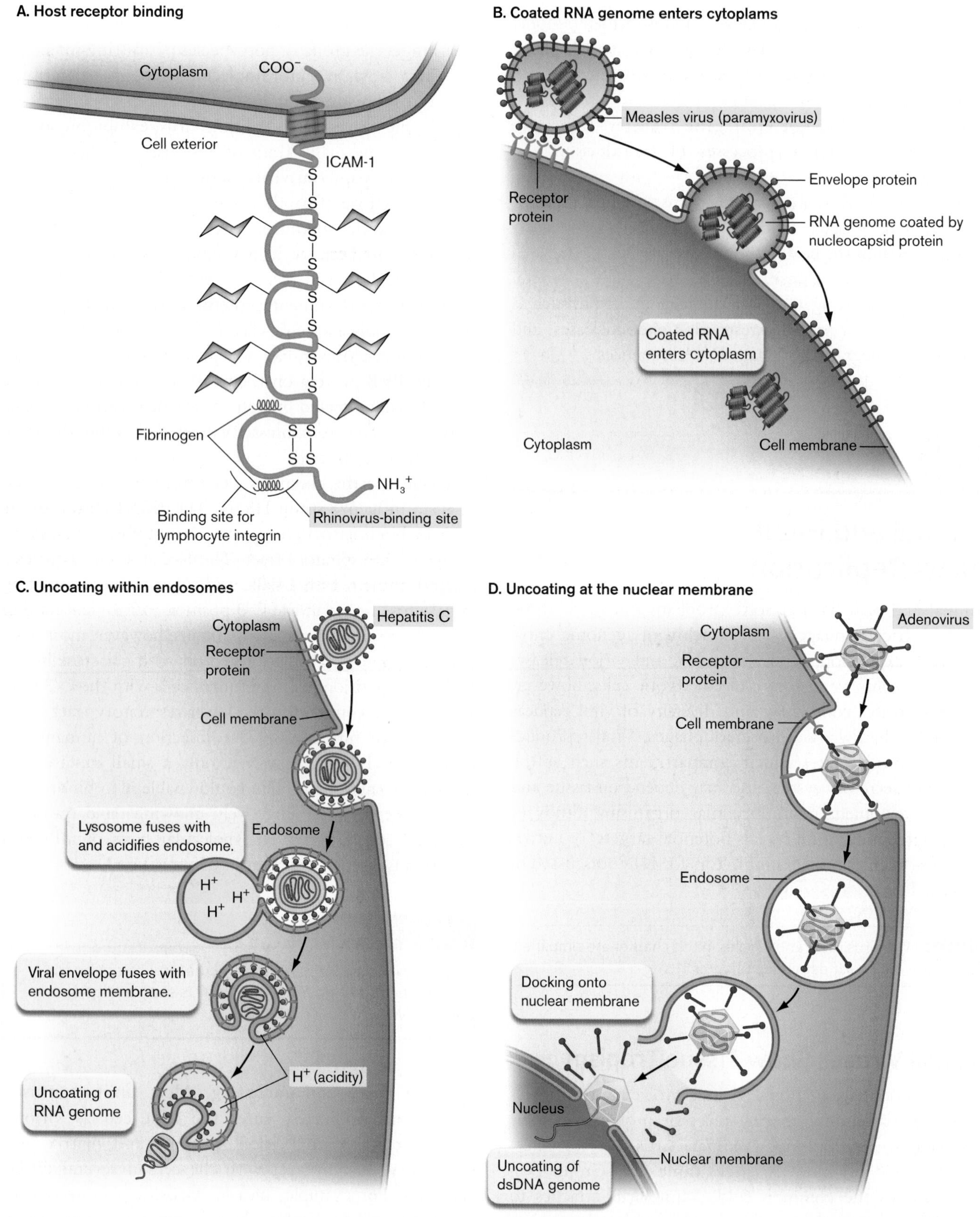
A. Host receptor binding
Cytoplasm
COO⁻
Cell exterior
ICAM-1
Fibrinogen
NH₃⁺
Binding site for lymphocyte integrin
Rhinovirus-binding site
B. Coated RNA genome enters cytoplams
Measles virus (paramyxovirus)
Envelope protein
Receptor protein
RNA genome coated by nucleocapsid protein
Coated RNA enters cytoplasm
Cytoplasm
Cell membrane
C. Uncoating within endosomes
Cytoplasm
Hepatitis C
Receptor protein
Cell membrane
Lysosome fuses with and acidifies endosome.
Endosome
H⁺
Viral envelope fuses with endosome membrane.
H⁺ (acidity)
Uncoating of RNA genome
D. Uncoating at the nuclear membrane
Cytoplasm
Adenovirus
Receptor protein
Cell membrane
Endosome
Docking onto nuclear membrane
Nucleus
Nuclear membrane
Uncoating of dsDNA genome

FIGURE 6.22 ■ Receptor binding and genome uncoating. A. Rhinovirus attaches to the intercellular adhesion molecule (ICAM-1), a glycoprotein required by the host cells to bind a lymphocyte integrin, a cell surface matrix protein required for cell-cell adhesion. After binding a specific receptor on the host cell membrane, an animal virus enters the cell, where its genome is uncoated. **B.** Coated RNA genome enters cytoplasm. **C.** Uncoating within endosomes. **D.** Uncoating at the nuclear membrane. *Source:* Part A based on J. Bella et al. 1998. *PNAS* **95**:4140.

cytoplasm. For other types of viruses, the entire virion is internalized within an endosome. The internalized capsid undergoes **uncoating**, a process in which the capsid comes apart, releasing the viral genome into the cytoplasm. Uncoating occurs for a flavivirus such as hepatitis C virus (**Fig. 6.22C**). The hepatitis C virion is first taken up by **endocytosis**. In endocytosis, the cell membrane forms a vesicle around the virion and engulfs it, forming an endocytic vesicle. The endocytic vesicle fuses with a lysosome, whose acidity activates entry of the capsid into the cytoplasm. The capsid then comes apart, and the viral genome is uncoated.

Yet other kinds of viruses, such as adenovirus, enter the cell by endocytosis but require transport to the nucleus (**Fig. 6.22D**). Following endocytosis and lysosome fusion, the adenoviral genome loses some of its capsid proteins (partial uncoating). A capsid protein then disrupts the endocytic membrane, allowing the remaining capsid to exit. The capsid then docks at a nuclear pore and injects its DNA genome into the nucleus. The adenoviral DNA is replicated by its own adenoviral DNA polymerase (carried in the virion), but it uses host nuclear histones for DNA packing, as well as host transcription factors available in the nucleus.

Animal Virus Replication Cycles

How does a virus replicate within an animal cell? The replication cycle of a given virus depends on the form of the viral genome. A DNA genome can use some or all of the host replication enzymes. An RNA genome, however, must encode either an RNA-dependent RNA polymerase to generate an RNA template or a reverse transcriptase to generate a DNA template (in the case of retroviruses).

DNA virus replication. An example of a double-stranded DNA virus is human papillomavirus (HPV) (see **Table 6.1**, Group I), the cause of genital warts (**Fig. 6.23A**). HPV is the most common sexually transmitted disease in the United States and one of the most common worldwide. Certain strains infect the skin, whereas others infect the mucous membranes through genital or anal contact (sexual transmission). Like phage lambda, papillomavirus has an active reproduction cycle and a dormant cycle in which the viral genome integrates into that of the host.

HPV initially infects basal epithelial cells (**Fig. 6.23B**). In the basal cells, papilloma virions enter the cytoplasm by receptor binding and membrane fusion. The virion then undergoes uncoating via disintegration of the protein capsid or coat, releasing its circular, double-stranded genome

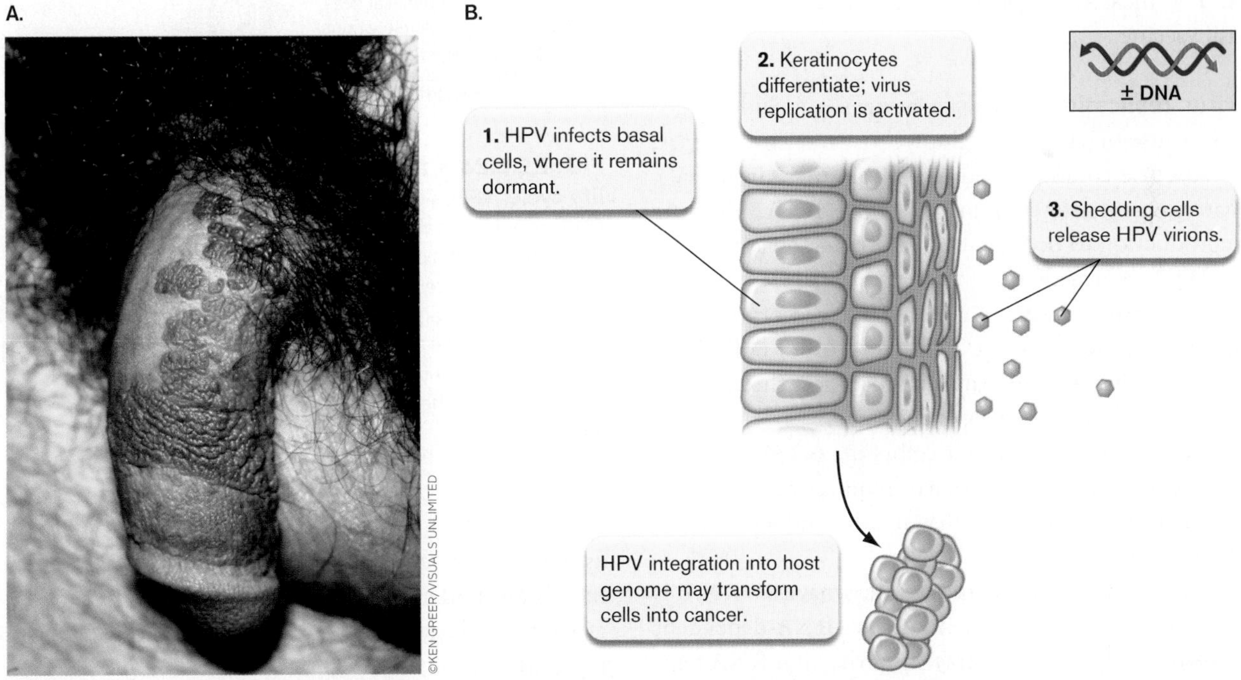
A.
©KEN GREER/VISUALS UNLIMITED
B.
2. Keratinocytes differentiate; virus replication is activated.
± DNA
1. HPV infects basal cells, where it remains dormant.
3. Shedding cells release HPV virions.
HPV integration into host genome may transform cells into cancer.

FIGURE 6.23 ▪ Human papillomavirus. A. Certain strains of human papillomavirus (HPV) cause warts on the genitals or anus. **B.** HPV infects basal epithelial cells, where the DNA uncoats but remains dormant. As cells differentiate, new virions are synthesized and released by shedding cells. Some HPV proteins transform host cells into cancer cells.

(**Fig. 6.24**). The uncoated viral genome enters the nucleus, where it gains access to host DNA and RNA polymerases.

The process of HPV reproduction is complicated by the developmental progression of basal cells into keratinocytes (mature epithelial cells), and ultimately cells to be shed or sloughed off from the surface (see **Fig. 6.23B**). Viral replication is largely inhibited until the basal cells start to differentiate into keratinocytes. Host cell differentiation induces the viral DNA to replicate and undergo transcription by host polymerases. The mRNA transcripts then exit the nuclear pores, as do host mRNAs, for translation in the cytoplasm. The translated capsid proteins, however, return to the nucleus for assembly of the virion. Nuclear virion assembly is typical of DNA viruses (with the exception of poxviruses, which replicate entirely in the cytoplasm).

How do HPV virions disseminate; and how may they cause cancer? As the keratinocytes containing HPV complete differentiation, the cells start to come apart and are shed from the epithelial surface. Cells release HPV virions during this shedding process. But in the basal cells HPV has an alternative pathway of integrating its genome into that of the host (analogous to phage lysogeny). The integrated genome can transform host cells into cancer cells. Transformation occurs through increased expression of viral oncogenes (cancer-causing genes). The oncogenes inhibit the expression of host tumor suppressor genes. Certain HPV strains are more likely to cause cancer than others. The most common strains are preventable by the Gardasil vaccine.

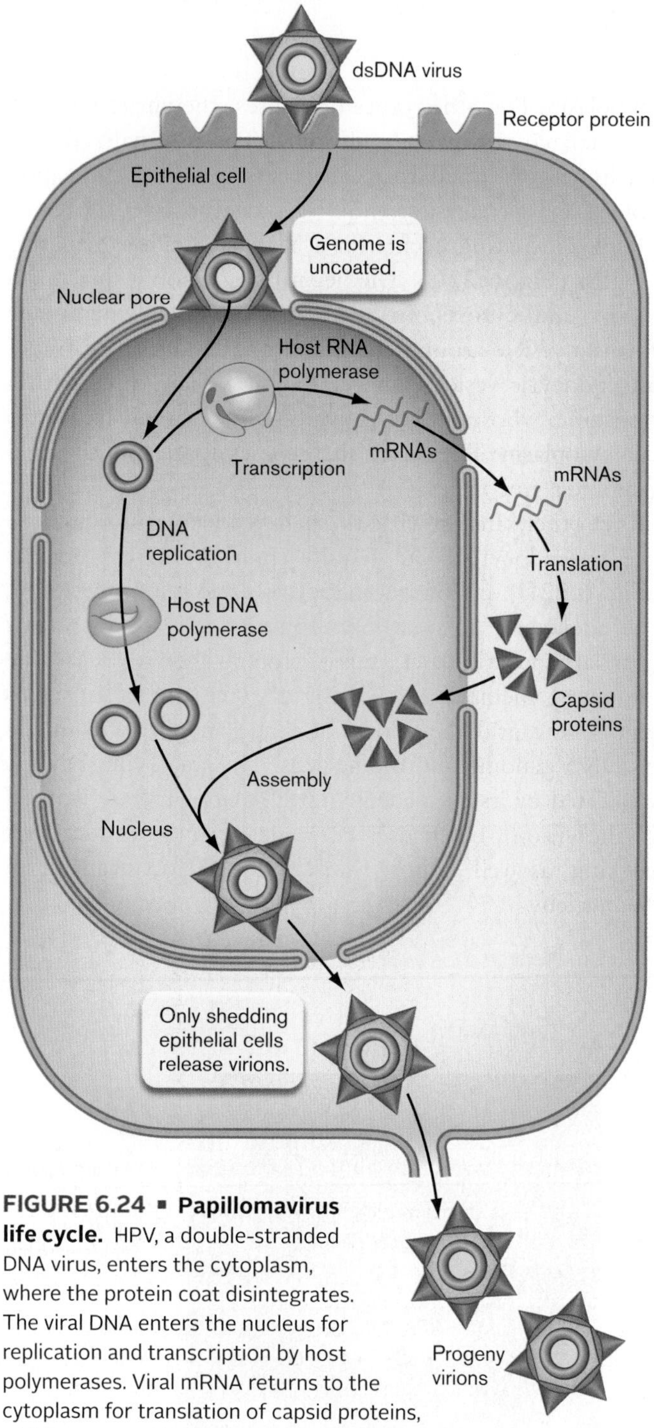
dsDNA virus
Receptor protein
Epithelial cell
Genome is uncoated.
Nuclear pore
Host RNA polymerase
mRNAs
Transcription
mRNAs
DNA replication
Translation
Host DNA polymerase
Capsid proteins
Assembly
Nucleus
Only shedding epithelial cells release virions.
Progeny virions

FIGURE 6.24 ■ Papillomavirus life cycle. HPV, a double-stranded DNA virus, enters the cytoplasm, where the protein coat disintegrates. The viral DNA enters the nucleus for replication and transcription by host polymerases. Viral mRNA returns to the cytoplasm for translation of capsid proteins, which return to the nucleus for assembly of virions.

RNA virus replication. The picornaviruses include poliovirus, the cause of paralytic poliomyelitis; and rhinoviruses, which cause the common cold (see **Table 6.1**, Group IV). Picornavirus genomes contain (+) strand RNA, allowing viral reproduction to occur entirely in the cytoplasm, without any DNA intermediates.

A picornavirus binds to a surface receptor, such as ICAM-1 for rhinovirus or the PVR receptor for poliovirus. The (+) strand RNA is uncoated by insertion through the cell membrane into the cytoplasm—much as a bacteriophage inserts its genome into a cell (**Fig. 6.25**). The role of endocytosis is debated; poliovirus requires no endocytosis, but the rhinovirus genome may require endocytosis and low-pH induction.

After uncoating, a gene in the picornavirus RNA is translated by host ribosomes to make RNA-dependent RNA polymerase. The polymerase uses the viral RNA template to make (–) strand RNA. The (–) strand RNA then serves as a template for other viral mRNAs, as well as progeny genomic RNA, which is replicated in virus-induced vesicles from the endoplasmic reticulum. Capsid proteins are synthesized by host ribosomes, and the capsids self-assemble in the cytoplasm. Virions then assemble at the cell membrane and are released by subverting lysosomes, which attempt to digest them. For greater detail of the poliovirus replication cycle, see **eTopic 11.1**.

Note that other kinds of RNA viruses, such as influenza virus, encapsidate a (–) strand genome. In this case, the (–) strand must serve as the template to generate mRNA, as well as a (+) secondary template for (–) strand progeny

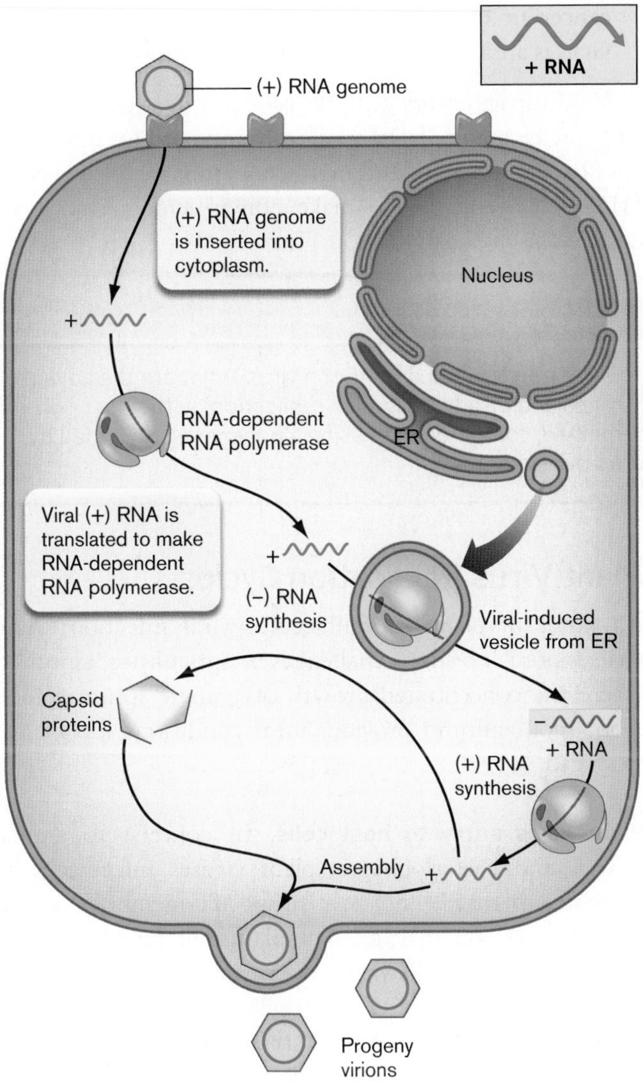
+ RNA
(+) RNA genome
(+) RNA genome is inserted into cytoplasm.
Nucleus
RNA-dependent RNA polymerase
ER
Viral (+) RNA is translated to make RNA-dependent RNA polymerase.
(–) RNA synthesis
Viral-induced vesicle from ER
Capsid proteins
+ RNA
(+) RNA synthesis
Assembly
Progeny virions

FIGURE 6.25 ■ **Picornavirus life cycle.** A picornavirus inserts its (+) strand RNA into the cell. Reproduction occurs entirely in the cytoplasm. A key step is the early translation of a viral gene to make RNA-dependent RNA polymerase. The polymerase uses the picornavirus RNA template to make (–) strand RNA, which then serves as a template for other viral mRNAs, as well as progeny genomic RNA, which is replicated in virus-induced vesicles from the endoplasmic reticulum.

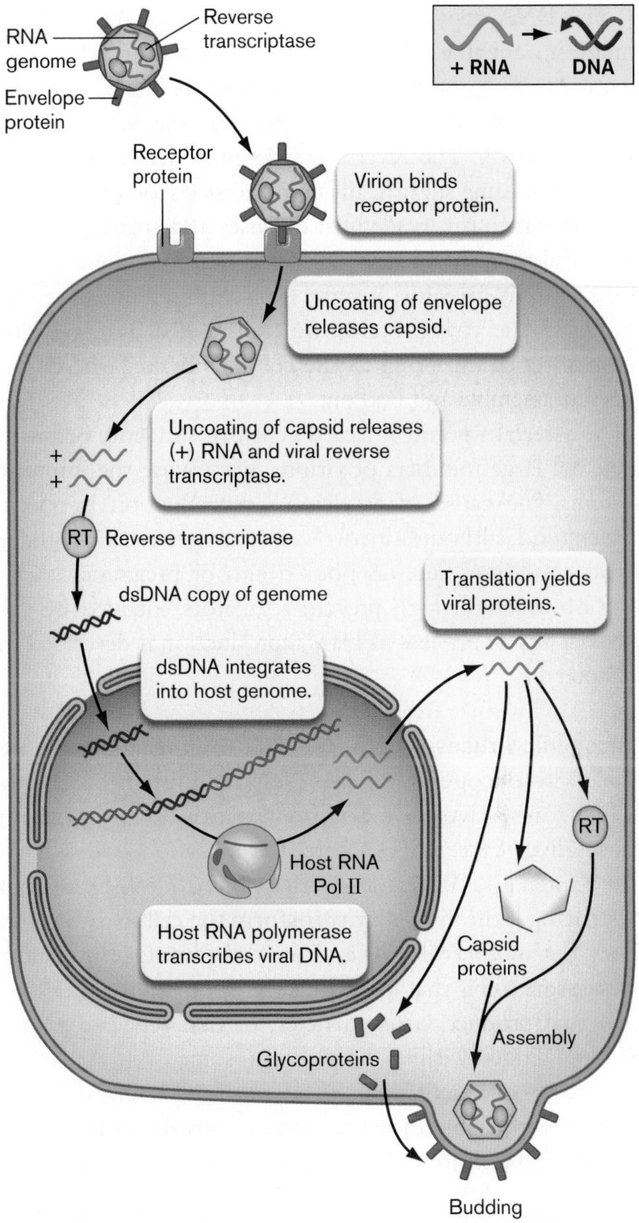
RNA genome
Reverse transcriptase
Envelope protein
+ RNA
DNA
Receptor protein
Virion binds receptor protein.
Uncoating of envelope releases capsid.
Uncoating of capsid releases (+) RNA and viral reverse transcriptase.
RT Reverse transcriptase
dsDNA copy of genome
Translation yields viral proteins.
dsDNA integrates into host genome.
Host RNA Pol II
Host RNA polymerase transcribes viral DNA.
RT
Capsid proteins
Assembly
Glycoproteins
Budding

FIGURE 6.26 ■ **Retrovirus life cycle.** A retrovirus such as human immunodeficiency virus (HIV) uses reverse transcriptase to copy its RNA into double-stranded DNA. The DNA then enters the nucleus to recombine in the host genome, where a host RNA polymerase generates viral mRNA and viral genomic RNA. The viral mRNA enters the cytoplasm for translation. Viral coat proteins are transported by the endoplasmic reticulum to the cell membrane, where virions assemble and bud out.

genomes. Influenza virus replication includes other interesting molecular complications, discussed in Chapter 11.

RNA retroviruses. Retroviruses include human immunodeficiency virus (HIV), the causative agent of AIDS; and feline leukemia virus (FeLV), the cause of feline leukemia, a disease that commonly afflicts domestic cats (see **Table 6.1**, Group VI). A retrovirus such as HIV uses a reverse transcriptase to make a DNA copy of its RNA genome (**Fig. 6.26**). Instead of being translated from an early gene, the reverse transcriptase is actually carried within the virion, bound to the RNA genome with a primer in place. The virion contains two (+) strand RNA copies of the HIV genome, each carrying its own reverse transcriptase. Virions infect helper T cells and other lymphocytes of the active immune system (discussed in Chapter 24). After uncoating in the cytoplasm, the viral RNA is copied into double-stranded DNA.

The DNA copy then enters the nucleus, where a viral integrase enzyme integrates the DNA into the host genome. To generate virions, a host RNA polymerase transcribes the viral genome into viral mRNA and viral genomic RNA. The reverse transcriptase has a high error rate, generating slightly different versions of the virus—some of which can evade host defenses and antiviral drugs. The viral mRNA reenters the cytoplasm for translation to produce coat proteins, reverse transcriptase, and envelope proteins. The coat proteins are transported by the endoplasmic reticulum (ER) to the cell membrane, where virions self-assemble and bud out.

At a certain point, infected cells can suddenly begin to generate large numbers of virions, destroying the immune system. The cause of accelerated reproduction is poorly understood, although it involves cytokines released under stress conditions such as poor health or pregnancy. Activation occurs through protein regulators encoded by the virus. The full process of HIV reproduction is described in Chapter 11.

Oncogenic viruses. As many as 20% of human cancers are caused by **oncogenic viruses**, such as Epstein-Barr virus (which causes lymphomas) and hepatitis C virus (which causes liver cancer). (Hepatitis C replication is discussed in Chapter 11.) When these viruses infect a cell, instead of destroying it the virus may **transform** the cell to divide and grow out of control. For a virus, the advantage of cancer transformation is that it expands the population of infected cells that produce virus particles, or that replicate a viral genome hidden within a host chromosome.

How do oncogenic viruses cause cancer? Several different mechanisms enable different types of viruses to transform normal host cells to proliferate abnormally and form tumors.

- **Oncogenes.** A retrovirus such as feline leukemia virus (FeLV) may carry an **oncogene**, which can transform the host cells. Usually an oncogene encodes an abnormal form of a host protein that controls cell proliferation.
- **Genome integration.** Certain DNA viruses, such as human papillomavirus, can integrate their genome into a host chromosome. The integrated viral genome expresses proteins that stimulate host cell division and may ultimately lead to growth of tumors.
- **Cell cycle control.** Oncogenic viruses such as papillomaviruses express oncogenes for proteins that interact with cell cycle controls and can stimulate uncontrolled growth. Even without integrating into the host chromosome, other viruses cause cancer via oncogenes. For example, the polyomaviruses are DNA viruses that cause cancer in immunosuppressed people, and hepatitis C virus is an RNA virus that causes liver cancer.

Viral capacities for gene transfer and host genome control may be manipulated artificially and used for gene therapy. In fact, some of the most dangerous viruses, such as HIV, are engineered to make nonvirulent gene delivery devices (discussed in Section 11.6).

Thought Question

6.8 From the standpoint of a virus, what are the advantages and disadvantages of replication by the host polymerase, compared with using a polymerase encoded by its own genome?

Plant Virus Replication Cycles

All kinds of plants are subject to viral infection. Plant viruses pose enormous challenges to agriculture, especially where the concentrated growth of a single strain of food crop (monoculture) provides ideal conditions for a virus to spread.

Plant virus entry to host cells. In contrast to animal viruses and bacteriophages, plant viruses infect cells by mechanisms that do not involve specific membrane receptors. The reason may be that plant cell membranes are covered by thick cell walls impenetrable to virion uptake or genome insertion. Thus, the entry of plant viruses usually requires **mechanical transmission**, nonspecific access through physical damage to tissues, such as abrasions of the leaf surface caused by a feeding insect. Mechanical transmission of plant viruses is limited by the cell wall. Most plant viruses gain entry to cells by one of three routes:

- **Contact with damaged tissues.** Viruses such as tobacco mosaic virus appear to require nonspecific entry into broken cells.
- **Transmission by an animal vector.** Insects and nematodes transmit many kinds of plant viruses. For example, the geminiviruses are inoculated into cells by plant-eating insects such as aphids, beetles, and grasshoppers.
- **Transmission through seed.** Some plant viruses enter the seed and infect the next generation.

An economically important plant virus is the potyvirus called plum pox virus (**Table 6.1**, Group IV), a major pathogen of plums, peaches, and other stone fruits. Plum pox virus, a (+) strand RNA virus, is transmitted by

A.
B.
C.
D.
FREDERICK E. GILDOW, PENNSYLVANIA STATE UNIVERSITY
SCOTT BAUER, AGRICULTURAL RESEARCH SERVICE, USDA
FREDERICK E. GILDOW, PENNSYLVANIA STATE UNIVERSITY
MARINA BARBA, ISTITUTO SPERIMENTALE PER LA PATOLOGIA VEGETALE

FIGURE 6.27 ■ Plum pox is caused by potyvirus. A. Potyvirus, a filamentous (+) strand RNA virus, approximately 800 nm in length (TEM). **B.** Potyvirus is transmitted by aphids, which suck the plant sap and release the virus into the damaged tissues. **C.** Streaking of flowers caused by potyvirus infection. **D.** Ring-shaped pockmarks appear on the infected fruit and the stone inside.

aphids (**Fig. 6.27**). Following infection, the spread of the virus generates streaked leaves and flowers, as well as ring-shaped pockmarks on the surfaces of the fruit and of the stone within.

Plant virus transmission through plasmodesmata. Within a plant, the thick cell walls prevent a lytic burst or budding out of virions. Instead, plant virions spread to uninfected cells by traveling through **plasmodesmata** (singular, **plasmodesma**). Plasmodesmata are membrane channels that connect adjacent plant cells (**Fig. 6.28**). The outer channel connects the cell membranes of the two cells; the inner channel connects the endoplasmic reticulum.

Passage through the plasmodesmata requires action by movement proteins whose expression is directed by the viral genome. In some cases, the movement proteins transmit the entire plant virion; in other cases, only the nucleic acid itself is small enough to pass through. The infection strategies of plant viral genomes may have features in common with those of viroids, which lack capsids altogether.

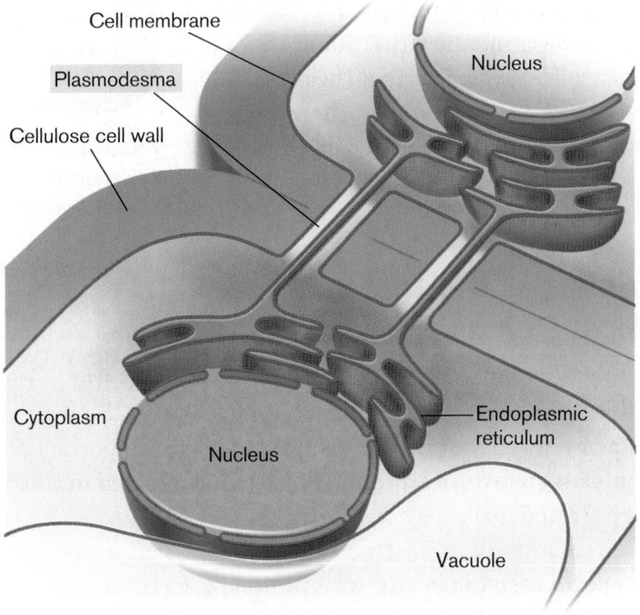
Cell membrane
Plasmodesma
Nucleus
Cellulose cell wall
Cytoplasm
Nucleus
Endoplasmic reticulum
Vacuole

FIGURE 6.28 ■ Plant cells connected by plasmodesmata. Plasmodesmata offer a route for plant viruses to reach uninfected cells.

DNA pararetroviruses. Pararetroviruses possess a DNA genome that requires transcription to RNA in the cytoplasm, followed by reverse transcription to form DNA genomes for progeny virions. Some pararetroviruses, such as hepadnavirus, infect humans, but the best-known pararetrovirus is cauliflower mosaic virus (CaMV), a caulimovirus (see **Table 6.1**, Group VII). CaMV is an important tool for biotechnology because it has a highly efficient promoter for gene transcription, allowing high-level expression of cloned genes. Vectors derived from caulimoviruses are used to construct transgenic plants.

CaMV is transmitted by secretions from an insect whose bite damages plant tissues, providing access to the cytoplasm

(**Fig. 6.29**). The CaMV genome moves from the cytoplasm to the cell nucleus through a nuclear pore. Within the nucleus, two promoters on its DNA genome direct transcription to RNA. The two RNA transcripts exit the nucleus for translation by host ribosomes to make viral proteins. A host reverse transcriptase, present in plant cells, copies the RNA into DNA viral genomes. After virions are assembled in the cytoplasm, movement proteins help transfer them through plasmodesmata into an adjacent cell.

A CaMV promoter sequence is commonly used in gene transfer vectors for plant biotechnology because transcription of the gene (such as one that confers pesticide resistance on the host) linked to the viral promoter is very efficient. In the field, 10% of cruciferous vegetables are typically infected with cauliflower mosaic virus. Some critics of gene technology fear that the prevalence of the CaMV promoter in transgenic crops may lead to the evolution of new pararetroviruses.

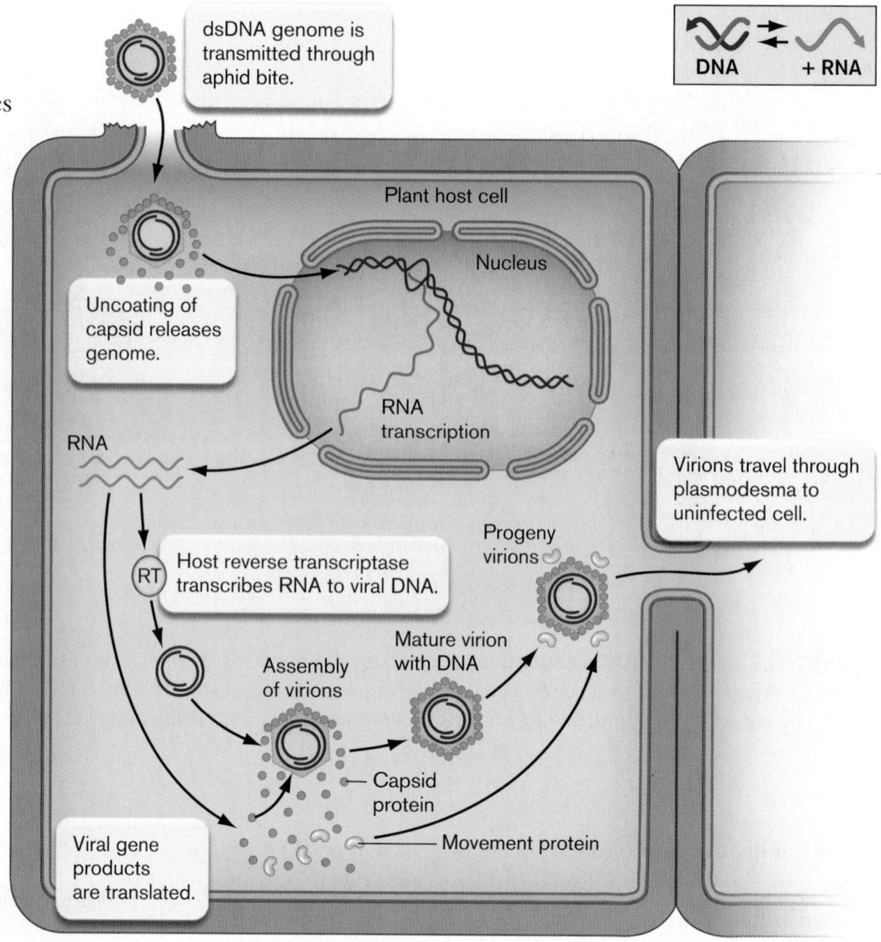
DNA
+ RNA
dsDNA genome is transmitted through aphid bite.
Plant host cell
Nucleus
Uncoating of capsid releases genome.
RNA
RNA transcription
Virions travel through plasmodesma to uninfected cell.
Progeny virions
RT
Host reverse transcriptase transcribes RNA to viral DNA.
Mature virion with DNA
Assembly of virions
Capsid protein
Viral gene products are translated.
Movement protein

FIGURE 6.29 ▪ Caulimovirus life cycle. The cauliflower mosaic virus (CaMV), a caulimovirus, uses host RNA polymerase to copy its DNA into RNA and uses host reverse transcriptase (RT) to make DNA copies. The DNA genomes are packaged into progeny virions that use virus-encoded movement proteins to travel through plasmodesmata to an adjacent uninfected cell.

Animal and Plant Host Defenses

How do animals and plants defend themselves from virus infection? Since viruses are ubiquitous, a wide range of defense mechanisms have evolved. Defenses important for humans are part of our immune system, presented in Chapters 23 and 24.

Genetic resistance. As we saw for bacteria, animal and plant hosts continually experience mutations, some of which lead to strains that resist viral infection by halting adsorption or some other key step of the virus's replication cycle. When a virus becomes widespread, natural selection favors resistant strains. But commercial livestock and crops typically consist of "monoculture" in which no resistant variants are available. Thus, when an outbreak occurs, an entire crop may be destroyed. To save a crop, it may be interbred with a wild strain that possesses genes conferring resistance.

An example of genetic resistance in humans is resistance to HIV/AIDS. In populations where HIV prevalence is high, rare individuals emerge whose cells are not infected, or are infected at such low rates that the virus is eventually cleared. The basis of this resistance is a defective allele encoding a T-lymphocyte cell-surface protein that acts as a coreceptor, which is required for binding of virus HIV-1. The role of coreceptors in HIV infection and resistance is discussed further in Chapter 11.

Immune system. The immune system of humans and other animals possesses extensive cellular machinery to thwart viral infection. An example of "innate immunity" is the class of proteins called interferons, which recognize general signs of viral infections such as the presence of double-stranded RNA (Chapter 23). For "adaptive immunity," viral proteins expressed in the cell membrane of an infected cell are recognized by specific antibodies that stimulate immune cells to destroy the infected cell and halt its viral production. The antibodies recognize a specific virus strain, such as a strain of influenza virus to which the individual has been exposed previously. For example, during the 2009 epidemic of flu, many individuals over the age of 50 had some protection arising from exposure to a similar strain in an earlier epidemic (Chapter 24).

RNA interference. RNA interference, or RNAi, is a mechanism by which mRNA molecules expressed by a viral genome are recognized by a host protein-RNA complex that shuts down further expression. RNA interference was first discovered in plants, where the system is most extensive; but it is now known to be widespread among all eukaryotes and archaea (discussed in Chapter 9). The mechanisms of RNA interference are now being engineered for use in gene therapy to halt gene expression in cancer and in inherited diseases.

To Summarize

- **Host cell-surface receptors** mediate animal virus attachment to a cell and confer host specificity and tropism.
- **Animal DNA viruses** either inject their genome or enter the host cell by endocytosis. The viral genome requires uncoating for gene expression.
- **RNA viruses** use an RNA-dependent RNA polymerase to transcribe their messenger RNA.
- **Retroviruses** use a reverse transcriptase to copy their genomic sequence into DNA for insertion in the host chromosome.
- **Oncogenic viruses** transform the host cell to become cancerous. Mechanisms of oncogenesis by different types of viruses include insertion of an oncogene into the host genome; integration of the entire viral genome; and expression of viral proteins that interfere with host cell cycle regulation.
- **Plant viruses** enter host cells by transmission through a wounded cell surface or an animal vector. Plant viruses travel to adjacent cells through plasmodesmata.
- **Pararetroviruses** contain DNA genomes but generate an RNA intermediate that requires reverse transcription to DNA for progeny virions.

6.6 Culturing Viruses

To learn how microbes grow, we culture them in the laboratory. So how do we culture a virus? A complication of virus culture is the need to grow the virus within a host cell. Therefore, any virus culture system must be a double culture of host cells plus viruses. Culturing viruses of multicellular animals and plants involves additional complications, as viruses show tropism for particular tissues or organs. Viruses may replicate in tissue culture; but the tissue culture does not show all the properties of an organ within a living organism. Therefore, a viral infection in tissue culture will miss some of the virulence factors needed to infect an animal.

Batch Culture

Batch culture, or culture in an enclosed vessel of liquid medium, enables growth of a large population of viruses for study. Bacteriophages can be inoculated into a growing culture of bacteria, usually in a culture tube or a flask. The culture fluid is then sampled over time and assayed for phage particles. The growth pattern usually takes the form of a step curve (**Fig. 6.30**).

To observe one cycle of phage reproduction, phages are added to host cells at a **multiplicity of infection** (**MOI**, ratio of phage to cells) such that every host cell is infected. The phage particles immediately adsorb to surface receptors of host cells and inject their DNA. As a result, virions are virtually undetectable in the growth medium for a short period after infection; this is called the **eclipse period**. For some species, it is possible to distinguish between the eclipse period and a **latent period**, which includes the eclipse period plus the time during which progeny viruses have been formed but remain trapped within the cell. In

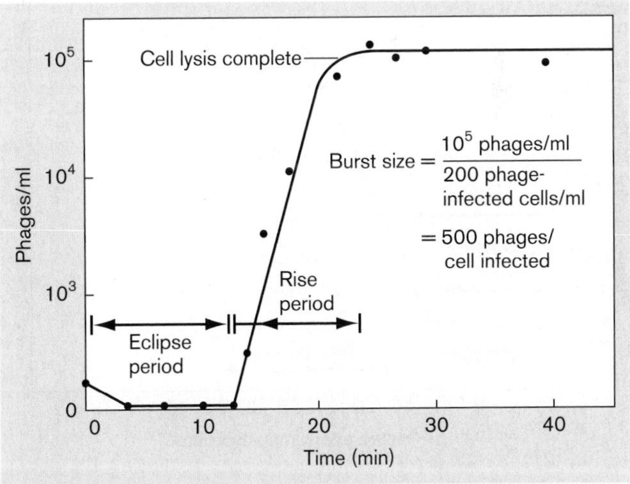
Cell lysis complete
Burst size = 10^5 phages/ml / 200 phage-infected cells/ml
= 500 phages/cell infected
Rise period
Eclipse period
Phages/ml
10^5
10^4
10^3
0
0
10
20
30
40
Time (min)

FIGURE 6.30 ■ One-step growth curve for a bacteriophage. After initial infection of a liquid culture of host cells, the titer of virus drops near zero as all virions attach to the host. During the eclipse period, progeny phages are being assembled within the cell. As cells lyse (the rise period), virions are released until they reach the final plateau. The infectious cycle is typically complete within less than an hour.

animal viruses, the latent period is particularly significant because large numbers of virions usually generate progeny through budding out of the host cell (**Fig. 6.31**).

Note: The "latent period" of a lytic virus is the period between initial phage-host contact and the first appearance of progeny phage. This must be distinguished from the "latent infection" of a virus that maintains its genome within a host cell without reproducing virions.

As cells begin to lyse and liberate progeny viruses, the culture enters the **rise period**, during which virus particles appear in the growth medium. The rise period ends when all the progeny viruses have been liberated from their host cells. If the number of viruses that go on to inoculate additional host cells is small, then the virus concentration at the end point divided by the original concentration of inoculated phage approximates the **burst size**—that is, the number of viruses produced per infected host cell. We can estimate the burst size by dividing the concentration of progeny virions by the concentration of inoculated virions, assuming that all the original virions infect a cell.

The burst size, together with the cell density prior to lysis, determines the concentration of the resultant suspension of virus particles, called a **lysate**. In the case of bacteriophages, a lysate of phage particles can be extremely stable, remaining infective at room temperature for many years. Eukaryotic viruses, however, tend to be less stable and need to be maintained in culture or deep freeze.

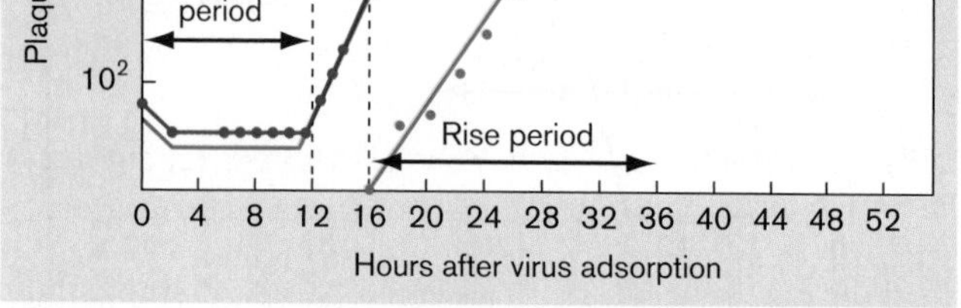
10^{10}
10^8
10^6
10^4
10^2
Plaque-forming units/ml
Intracellular virions
Extracellular virions
Latent period
Eclipse period
Rise period
0 4 8 12 16 20 24 28 32 36 40 44 48 52
Hours after virus adsorption

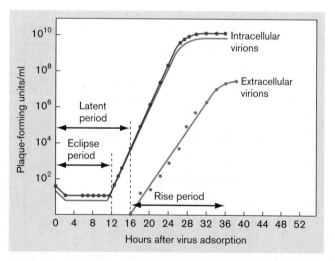

FIGURE 6.31 ■ One-step growth curve for a virus. The titer of extracellular virus drops to near zero during the latent period, as all virions adsorb to the host. Then progeny virions begin to emerge by budding out from the infected cell. For an animal virus, the growth curve may take hours to level off; the "burst" event is not defined as clearly as for phages.

Thought Question

6.9 Why does bacteriophage reproduction give a step curve, whereas cellular reproduction generates an exponential growth curve? Could you design an experiment in which viruses generate an exponential curve? Under what conditions does the growth of cellular microbes give rise to a step curve?

Tissue Culture of Animal Viruses

In the case of animal and plant viruses, the multicellular nature of the host is an important factor in the pathology and transmission of the pathogen (discussed in Chapters 25 and 26). Animal viruses can be cultured within whole animals by serial inoculation, where virus is transferred from an infected animal to an uninfected one. Culture within animals ensures that the virus strain maintains its original **virulence** (ability to cause disease). But the process is expensive and laborious, involving the large-scale use of animals.

A historic event in 1949 was the first successful growth of a virus in tissue culture. Poliovirus, the causative agent of the devastating childhood disease poliomyelitis, was grown in human cell tissue culture (**Fig. 6.32**) by John F. Enders, Thomas J. Weller, and Frederick Robbins at Children's Hospital in Boston. As heralded that year in *Scientific American*, "It means the end of the 'monkey era' in poliomyelitis research. . . . Tissue-culture methods have provided virologists with a simple in vitro method for testing a multitude of chemical and antibiotic agents." Since then, tissue culture has remained the most effective way to study the molecular biology of animal and plant viruses and to develop vaccines and antiviral agents.

Some viruses can be grown in a tissue culture of cells growing confluently on a surface. The cells must be "immortalized"—that is, genetically altered—to continue cell division indefinitely. The fluid bathing the tissue layer is sampled for virus concentration. As in the case of bacteriophage batch culture, we can define an eclipse period, a latent period before appearance of the first progeny virions in the culture fluid, and a rise period. In tissue culture, the time course of animal virus replication is usually much longer (hours or days) than that of bacteriophages (typically less than an hour under optimal conditions). The burst size of animal viruses, however, is typically several orders of magnitude larger than that of phages. The reason for the larger burst size is probably that the volume of a host cell is much larger than that of a bacterial host, thus providing a larger supply of materials to build virions.

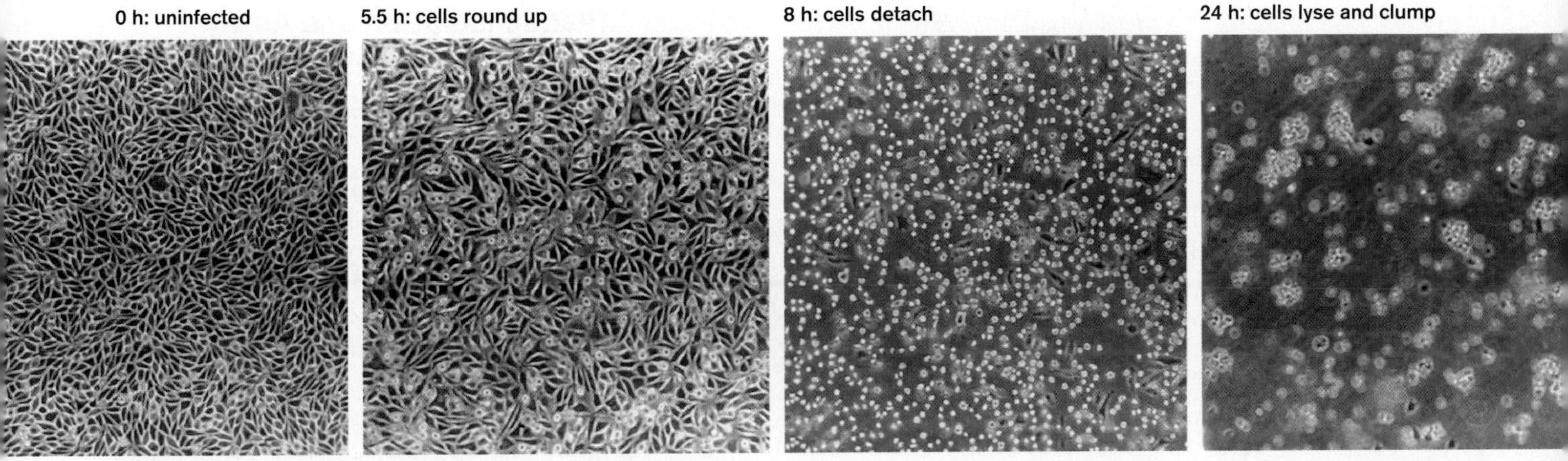
0 h: uninfected
5.5 h: cells round up
8 h: cells detach
24 h: cells lyse and clump

FIGURE 6.32 ▪ Poliovirus replication in human tissue culture. Before infection (0 hours), the cultured cells grow in a smooth layer. At 5.5 hours after infection, cells are starting to retract and round up. At 8 hours, infected cells detach from the culture dish. By 24 hours, cells have lysed or in some cases clumped with other cells. (cell length, approx. 50 μm). *Source:* R. Compans, Emory University, School of Medicine

Not all animal viruses exhibit growth that can be represented by a step curve. Some species, called slow viruses, bud off virions relatively slowly, without immediately lysing the host cell. They may even transmit from cell to cell through cytoplasmic nanotubes (introduced in Chapter 3), which avoid viral destruction by the immune system. A well-known family of slow viruses is the **lentiviruses** (literally, "slow viruses"), a group of retroviruses that includes HIV. HIV and related retroviruses are known for their long incubation periods, in some cases many years, during which time extremely low concentrations of virions are found in the blood.

Plaque Isolation and Assay of Bacteriophages

For the investigation of cellular microbes, an important tool is the culturing of individual colonies on a solid substrate that prevents dispersal throughout the medium, as described in Chapters 1 and 4. Plate culture of colonies enables us to isolate a population of microbes descended from a common progenitor. But viruses cannot be isolated as "colonies." The reason is that although viruses can be obtained at incredibly high concentrations, they disperse in suspension. Even on a solid medium, viruses never form a solid visible mass comparable to the mass of cells that constitutes a cellular colony.

In viral plate culture, viruses from a single progenitor lyse their surrounding host cells, forming a clear area called a **plaque**. The **plaque assay** for lytic bacteriophages was invented early in the twentieth century by the French microbiologist Félix d'Herelle.

To perform a plaque assay of bacteriophages, a diluted suspension of bacteriophages is mixed with bacterial cells in soft agar, and the mixture is then poured over a nutrient agar plate (**Fig. 6.33**). Where no bacteriophages are present, the bacteria grow homogeneously as a "lawn," an opaque sheet over the surface (confluent growth). Where there is a bacteriophage, it infects a cell, replicates, and spreads progeny phages to adjacent cells, killing them as well (**Fig. 6.34**). The loss of cells results in a round, clear area seemingly cut out of the bacterial lawn. Plaques can be counted and used to calculate the concentration of phage particles, or **plaque-forming units (PFUs)**, in a given suspension of liquid culture. The liquid culture can be analyzed by serial dilution in the same way one would analyze a suspension of bacteria.

Can a plaque assay be used to titer bacteriophages that undergo lysogeny? A temperate or lysogenic phage suspension will make plaques because only a small minority of the bacterial cells become lysogens; most undergo lytic burst. But the lysogen's plaques are cloudy because they contain a few lysogenized cells. The prophage in the host genome protects the cells from subsequent infection and lysis by another phage.

Plaques offer a convenient way to isolate a recombinant DNA molecule contained in a bacteriophage vector. In **Figure 6.34B**, the blue plaques result from a phage vector carrying the gene encoding the enzyme beta-galactosidase. This enzyme converts a colorless compound into a blue dye. When the indicator gene is interrupted by an inserted recombinant gene, the phage produces white plaques, which indicate the successful production of recombinant DNA phages.

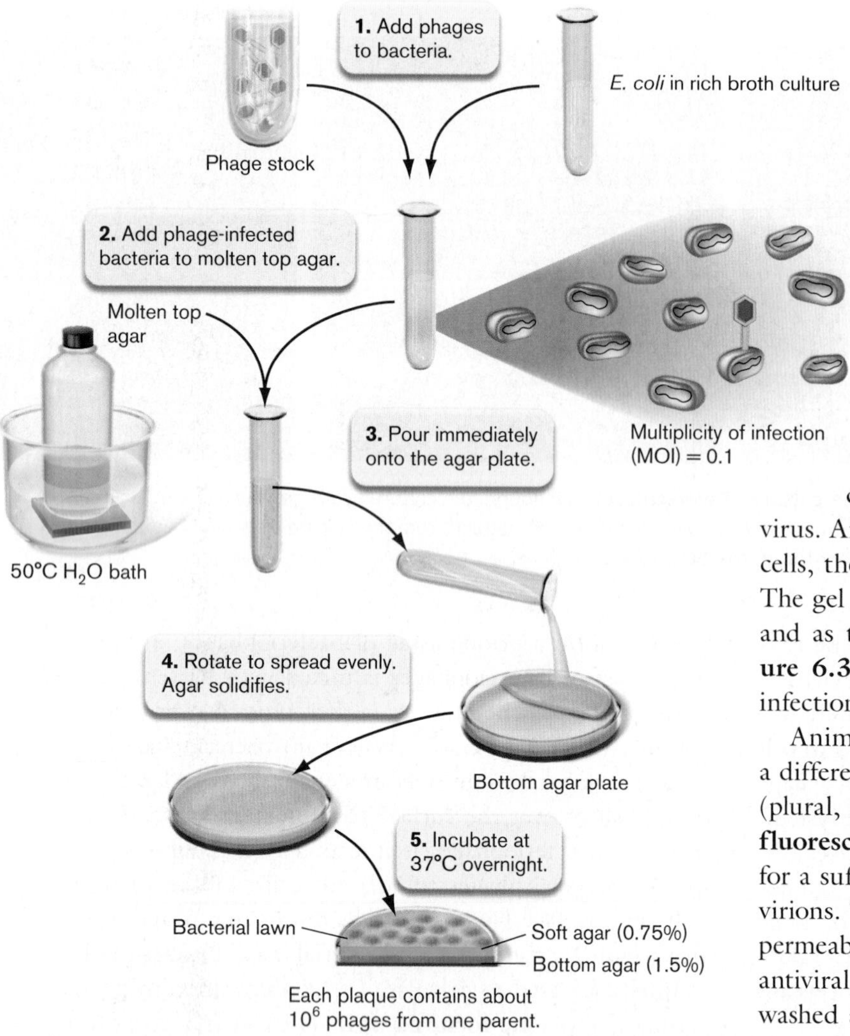

1. Add phages to bacteria.
Phage stock
E. coli in rich broth culture
2. Add phage-infected bacteria to molten top agar.
Molten top agar
Multiplicity of infection (MOI) = 0.1
3. Pour immediately onto the agar plate.
50°C H2O bath
4. Rotate to spread evenly. Agar solidifies.
Bottom agar plate
5. Incubate at 37°C overnight.
Bacterial lawn
Soft agar (0.75%)
Bottom agar (1.5%)
Each plaque contains about 10^6 phages from one parent.

FIGURE 6.33 ▪ Plating a phage suspension to count isolated plaques. A suspension of bacteria in rich broth culture is inoculated with a low proportion of phage particles (multiplicity of infection is approximately 0.1). This means that only a few of the bacteria become infected immediately, while the rest continue to grow. Each plaque arises from a single infected bacterium that bursts, its phage particles diffusing to infect neighboring cells.

Plaque Isolation and Assay of Animal Viruses

For animal viruses, the plaque assay has to be modified because it requires infection of cells in tissue culture. Tissue culture usually involves growth of cells in a monolayer on the surface of a dish containing fluid medium, which would quickly disperse any viruses released by lysed cells. To solve this problem, in 1952 Renato Dulbecco, at the California Institute of Technology, modified the tissue culture procedure for plaque assays (**Fig. 6.35A**). In Dulbecco's method, the tissue culture with liquid medium is first inoculated with virus. After sufficient time to allow for viral attachment to cells, the fluid is removed and replaced by a gel medium. The gel retards the dispersal of viruses from infected cells, and as the host cells die, plaques can be observed. **Figure 6.35B** shows a plate culture of human coronavirus infection of colon carcinoma cells.

Animal viruses that do not kill their host cells require a different kind of assay based on identification of a **focus** (plural, **foci**), a group of cells infected by the virus. In a **fluorescent-focus assay**, the infected cells are incubated for a sufficient period to allow the production of progeny virions. The plasma membranes of the cells are then made permeable by treatment with an organic solvent, and an antiviral antibody is added. Unattached antibodies are then washed away, and a second antibody is added that recognizes the first antibody molecule. The second antibody is conjugated to a fluorophore whose fluorescence reveals the foci of cells, each of which has arisen from a single virion in the original inoculum.

Another kind of fluorescent-focus assay uses expression of a fluorescent reporter protein such as green fluorescent protein (GFP). The viruses are genetically engineered to

FIGURE 6.34 ▪ Phage plaques on a lawn of bacteria. A. Phage lambda plaques on a lawn of *Escherichia coli* K-12. **B.** Plaques of recombinant phage M13 on *E. coli*. The original phage expresses beta-galactosidase, an enzyme that makes a blue product (blue plaques). White plaques are produced by phage particles whose genome is recombinant (contains a cloned gene interrupting the gene for beta-galactosidase).

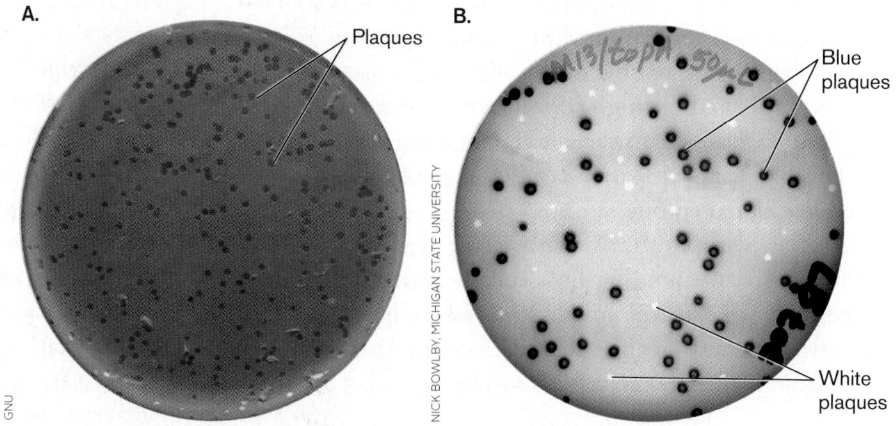

A.
Plaques
GNU
B.
Blue plaques
White plaques
NICK BOWLBY, MICHIGAN STATE UNIVERSITY

A.

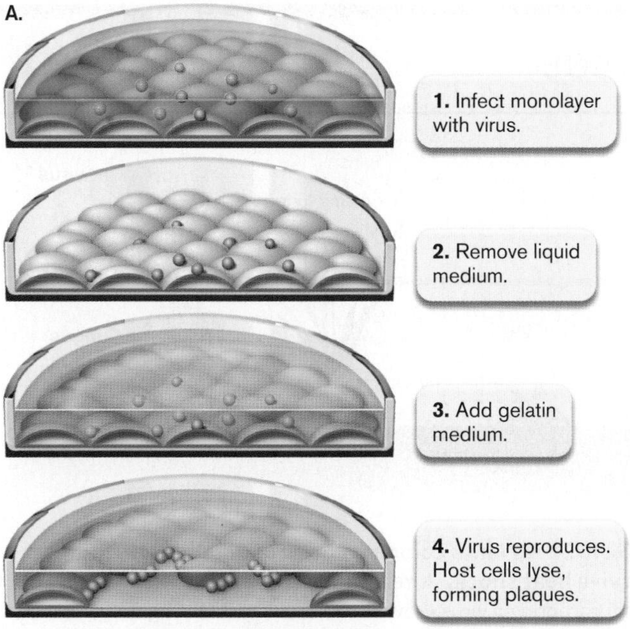
1. Infect monolayer with virus.
2. Remove liquid medium.
3. Add gelatin medium.
4. Virus reproduces. Host cells lyse, forming plaques.

B.

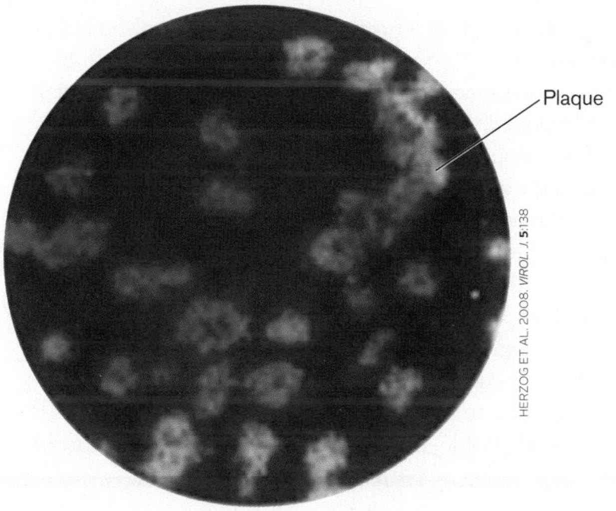
Plaque
HERZOG ET AL. 2008. VIROL. J. 5:138

FIGURE 6.35 ■ Plate culture of animal viruses. A. Modified plaque assay for animal viruses. The gelled medium retards the dispersal of progeny virions from infected cells, restricting new infections to neighboring cells. The result is a visible clearing of cells (a plaque) in the monolayer. **B.** Plaque assay in which human coronavirus suspension was plated on a monolayer of colon carcinoma cells in tissue culture.

contain a GFP gene fusion—a labeling technique discussed in Chapter 3. The GFP fluorescence reveals foci of virus-infected cells (**Fig. 6.36A**).

A focus assay can also be used to isolate oncogenic viruses, which transform their host cells into cancer cells. Cancer cells lose contact inhibition; they grow up in a pile instead of remaining in the normal monolayer. These piles of transformed cells, or transformed foci, can easily

A.

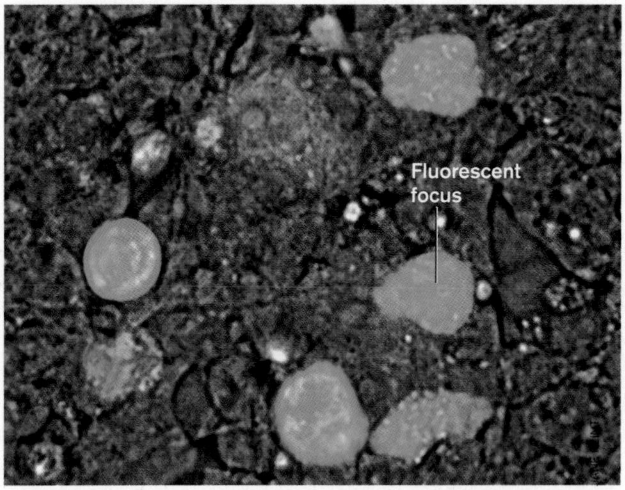
Fluorescent focus

B.

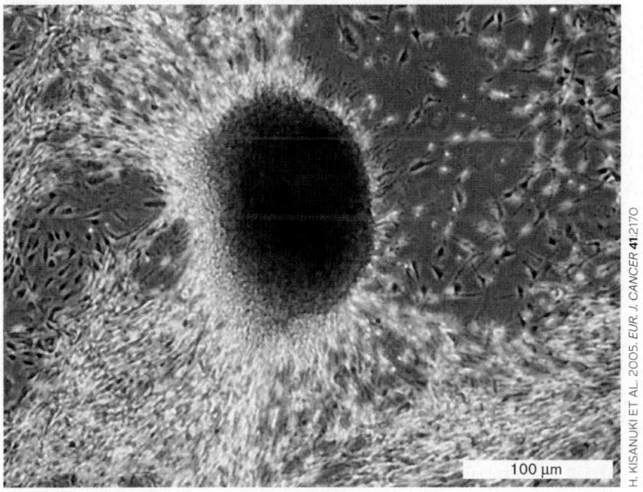
100 μm
H. KISANUKI ET AL. 2005. EUR. J. CANCER 41:2170

FIGURE 6.36 ■ Focus assays of animal viruses. A. GFP fluorescent-focus assay. **B.** Transformed-focus (cancer-forming) assay of an oncogenic virus. Transformed cells grow in an uncontrolled manner, which can produce a tumor.

be visualized and counted. This procedure is known as the **transformed-focus assay** (**Fig. 6.36B**).

To Summarize

- **Culturing viruses** requires growth in host cells. Bacteriophages may be cultured either in batch culture or as isolated plaques on a bacterial lawn.
- **Batch culture** of viruses generates a step curve.
- **Animal viruses** are cultured within animals or as plaques in a tissue culture.
- **Fluorescent-focus assays** reveal foci of virus-infected cells.
- **Oncogenic viruses** are cultured as foci of cancer-transformed host cells.

Special Topic 6.1: The Good Viruses

Do most viruses harm their host? This view would be questioned by Marilyn Roossinck, plant virologist at the Samuel Roberts Noble Foundation in Oklahoma. The vast majority of "viruses" in nature may actually be reproducing at low levels within their hosts, without harming the host overall. But research emphasizes those viruses that cause disease. For example, when scientists first isolated viruses from monkeys, they characterized 50 viruses, SV-1 through SV-50. But only one virus, SV-40, caused tumors—and that was the only one of the 50 viruses studied further. Roossinck estimates that human bodies may harbor perhaps 20,000 different kinds of viruses, of which only a few hundred may cause disease.

Roossinck focuses her research on viruses of plants and of fungi. She takes particular interest in viruses that support their hosts in a mutualistic relationship. A particularly interesting case is that of a mysterious virus she identified in a fungus, *Curvularia protuberata,* that grows associated with a panic grass, *Dichanthelium lanuginosum,* in Yellowstone National Park. The fungus, *C. protuberata,* grows as an "endophyte" within the vasculature of the grass—a highly common form of symbiosis (discussed in Chapter 21). What is remarkable about this fungus-plant pair is that each partner confers thermal tolerance (heat resistance) on the other (**Fig. 1**). As shown by colleagues Regina S. Redman and Russell Rodriguez, together the plant and fungus grow at temperatures as high as 65°C—which commonly occur in the vicinity of Yellowstone's thermal springs. Separately, the fungus and plant each grow at temperatures no higher than 38°C, typical of their species.

Roossinck knew that fungi harbor many diverse viruses, and that viruses often mediate fungus-plant interactions.

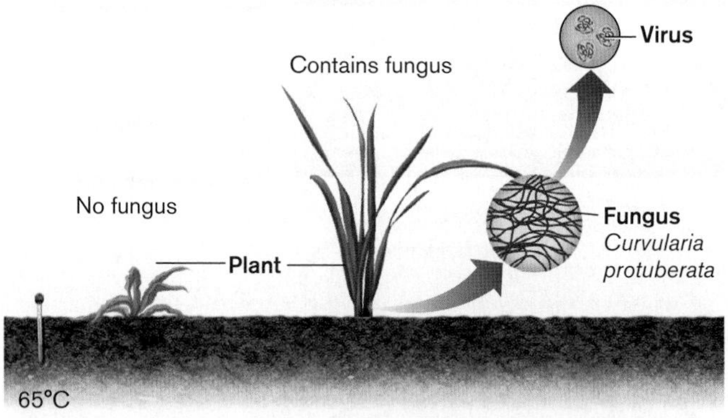
Virus
Contains fungus
No fungus
Plant
Fungus
Curvularia protuberata
65°C

FIGURE 1 ▪ Plants colonized by virus-infected fungus survive heat shock. A Yellowstone grass colonized by fungi harboring a virus grows taller than grass lacking fungi.
Source: Modified from Marilyn Roossinck. 2011. *Nat. Rev. Microbiol.* **9**:99.

She hypothesized that a fungal virus might be involved in the thermotolerance. To hunt for a virus, she used northern blot (described in Appendix 3) to analyze the fungal RNA for small double-stranded segments (dsRNA) typical of fungal viruses (Group III genome, double-stranded RNA). The northern blot revealed two viral dsRNAs that form the genome of a previously unknown virus. She ultimately named the virus Curvularia thermal tolerance virus (CThTV).

But does thermal tolerance require the virus? To answer this question, Roossinck had to isolate CThTV virions from the virus's fungal host (**Fig. 2A**). She then conducted experiments comparing plants colonized by fungi with and without the virus. She found that only the fungi infected with CThTV could enable the plant *Dichanthelium lanuginosum*

6.7

Viral Ecology

What roles do viruses play in ecosystems? The most common aspect of viral ecology that we hear about is the persistence of human viruses in natural ecosystems. Persistence in "wild" populations provides a reservoir of infection that hinders eradication, especially when associated with an insect vector (carrier organism) such as a tick or a mosquito. New viruses evolving in "wild" environments impact human health, and that of agricultural plants and animals.

From a broader perspective, viruses exist throughout all ecosystems and make surprising positive contributions. In some ecosystems, such as marine water or terrestrial soil, virus particles may outnumber all host cells by tenfold or more. Viral ecologists consider viruses as living entities, playing roles comparable to those of cellular organisms, such as "predator" or "parasite." Viruses mediate their host population size in ways that increase species diversity. And many viruses interact positively with hosts, by expressing host genes and by protecting hosts from other organisms. For example, in the human body an endogenous (permanently integrated) retrovirus expresses a protein needed for placental development. Research on virus-host mutualism is presented in **Special Topic 6.1**. Mutualism and other topics of ecology are presented in detail in Chapters 21 and 22.

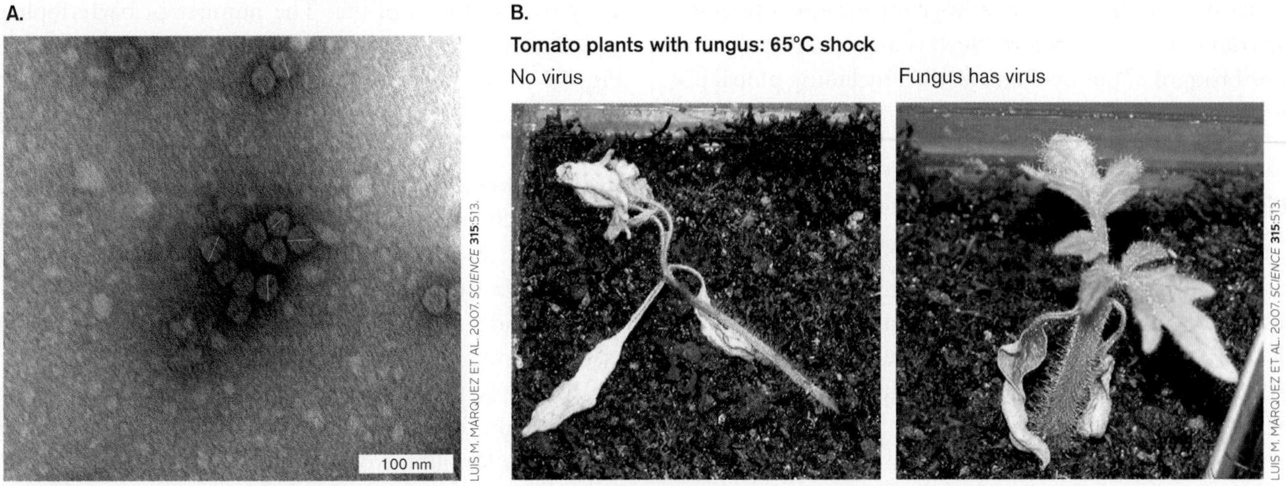
A.
100 nm
LUIS M. MÁRQUEZ ET AL. 2007. SCIENCE 315:513.
B.
Tomato plants with fungus: 65°C shock
No virus
Fungus has virus
LUIS M. MÁRQUEZ ET AL. 2007. SCIENCE 315:513.

FIGURE 2 ▪ Virus confers thermotolerance on tomato plants. A. Virus (CThTV) isolated from a fungus that confers thermotolerance. **B.** Tomato plants with fungus lacking virus die after exposure at 65°C; with virus present, plants survive.

to grow at high temperature. Remarkably, she found that the virus-infected fungus could also confer thermal tolerance on tomato plants (**Fig. 2B**). Thus, the viral-fungus mutualism may confer a specific thermotolerance property on various plant species—including plants important for agriculture.

Roossinck is now extending her work to identify the mechanism of thermotolerance in the three-way mutualism, which may have practical applications as farmlands heat up through climate change. She also seeks evidence of additional kinds of mutualism—for example, viruses that mediate salt tolerance. Above all, she seeks to promote our awareness of the potential contributions of "good" viruses.

RESEARCH QUESTION

Researchers hypothesize that the mechanism by which the CThTV-infected fungus protects the plant is to inactivate the reactive oxygen ions produced by the plant's stress response to high temperature. Can you propose an experiment to test this hypothesis?

Luis M. Márquez, Regina S. Redman, Russell J. Rodriguez, and Marilyn J. Roossinck. 2007. A virus in a fungus in a plant: Three-way symbiosis required for thermal tolerance. *Science* **315**:513–515.

Emergence of Viral Pathogens

Certain human-infecting viruses are well known to persist in the wild, such as rabies virus and West Nile virus. Their persistence requires broad host range; rabies infects many different mammals, whereas West Nile virus infects birds as well as humans and horses. Understanding the epidemiology of rabies or of West Nile encephalitis requires understanding the behavior and seasonal migration patterns of wild organisms. But how does a seemingly "new" virus emerge to sicken humans, such as SARS coronavirus in 2002? In the case of SARS, the coronavirus infecting humans was eventually traced genetically to viruses found in bats and in civet cats marketed for food, in the area of Guangdong, China, where the human outbreak began. Human consumption of wildlife is also suspected, though not proven, to have played a role in the "jump" of HIV from an ape to humans.

Some emerging viruses arise as variants of endemic milder pathogens. Viruses long associated with a host, such as the common-cold viruses (rhinoviruses), tend to have evolved a moderate disease state that provides ample opportunities for host transmission. A virus that "jumps" from an animal host, however, may cause a more acute syndrome with higher mortality. The best-known cases are the exceptionally virulent emerging strains of influenza, which generally result from intracellular recombination of human strains with strains from pigs or ducks (discussed in Chapter 11). For example, in 2013 the avian influenza strain H7N9 emerged from poultry in China, where it killed several people before it was contained. Changing distribution patterns of insect vectors and animal hosts can generate new epidemics of a pathogen in regions where the virus could not spread before. Such changes in distribution can be brought about by many factors, including global climate change (**eTopic 6.2**).

Viral Roles in Ecosystems

Viruses fill important niches in all ecosystems. "Acute" viruses (which rapidly kill their hosts) act as predators or parasites to limit host populations, and they recycle nutrients from their host bodies. "Persistent" viruses (which reproduce or persist without killing the host) can have more subtle and complex effects.

Limiting host population density. An increase in host population density increases the rate of transmission of viral pathogens. As the host population declines, viruses are less likely to find a new host before they lose infectivity; and the few remaining hosts have undergone selection for resistance. Thus, viruses can limit host density without extinction of the host. Viral limitation is best documented in the case of marine phytoplankton (algae and protists). In animals, viruses participate in population decline and resurgence. Cowpox virus circulating among woodland rodents can cause a decline in the population of voles and mice. Another case is the myxovirus introduced into Australia to curb the population of rabbits (which had been previously introduced by British colonists). The myxovirus did cause the rabbit population to decline, although the rabbits evolved resistance and their population later rebounded.

Selecting for host diversity. Each viral species has a limited host range and requires a critical population density to sustain the chain of infection. In marine phytoplankton, a virus limits its host species to a population density far lower than what is sustainable by the available resources, such as light for photosynthesis. The resources then support other species resistant to the given virus (but susceptible to others). Thus, overall, marine viruses prevent the dominance of any one species and foster the evolution of many distinct host species. This explains the exceptional diversity of phytoplankton ranging from silica-shelled diatoms to toxin-producing dinoflagellates.

In the oceans, viruses are the most numerous and genetically diverse forms of life. The number of bacteriophages and algal viruses can reach 10^7 (10 million) per milliliter. **Figure 6.37** shows examples of marine viruses infecting phytoplankton such as the algae *Emiliana huxleyi* and *Pavlova virescens.* When marine algae overgrow, generating an algal bloom, viruses play a decisive role in controlling the bloom, such as the overgrowth of *Emiliana huxleyi* shown in **Figure 6.38**. Consumer organisms apparently cannot grow fast enough to control such blooms, but viruses spread rapidly through the population. By lysing the algae as they grow, viruses return algal carbon and minerals to the surface water before the algae starve to death and their bodies sink. When biomass sinks, its minerals become unavailable for phototrophs and their consumers.

Thought Question

6.10 Suppose a certain virus depletes the population of an algal bloom. If some of the algae are selected for resistance, will they grow up again and dominate the producer community?

Persistent viruses may help their hosts. As we survey natural ecosystems—including that of the human body—we find that acute viruses, which kill their hosts, are outnumbered by persistent viruses, which are "along for the ride." Persistent viruses are poorly understood, compared to acute viruses. In some cases the persistent virus causes initial illness followed by dormancy, while in other cases no

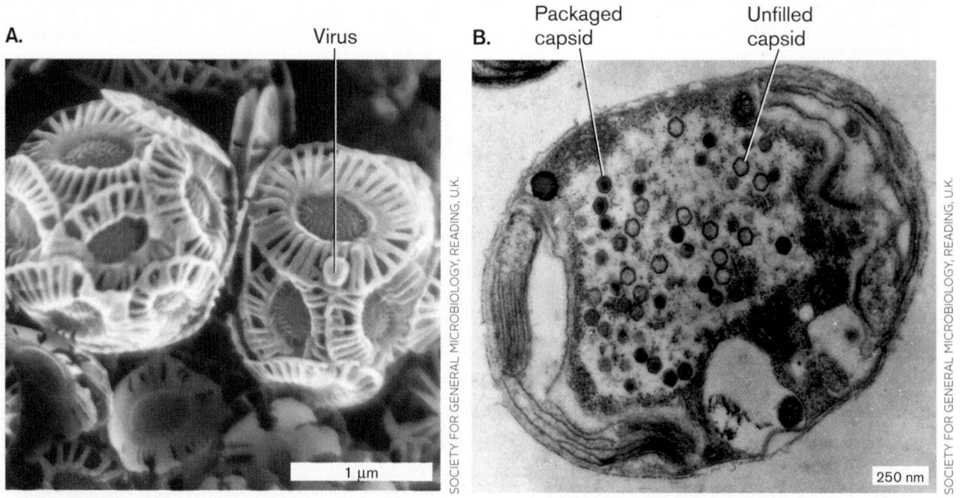
A.
Virus
1 μm
SOCIETY FOR GENERAL MICROBIOLOGY, READING, U.K.
B.
Packaged capsid
Unfilled capsid
250 nm
SOCIETY FOR GENERAL MICROBIOLOGY, READING, U.K.

FIGURE 6.37 ■ Viruses infect algae. A. A virus attaches to the surface of a marine phytoplankton, *Emiliana huxleyi* (SEM). **B.** Progeny virions assemble within the phytoplankton *Pavlova virescens.*

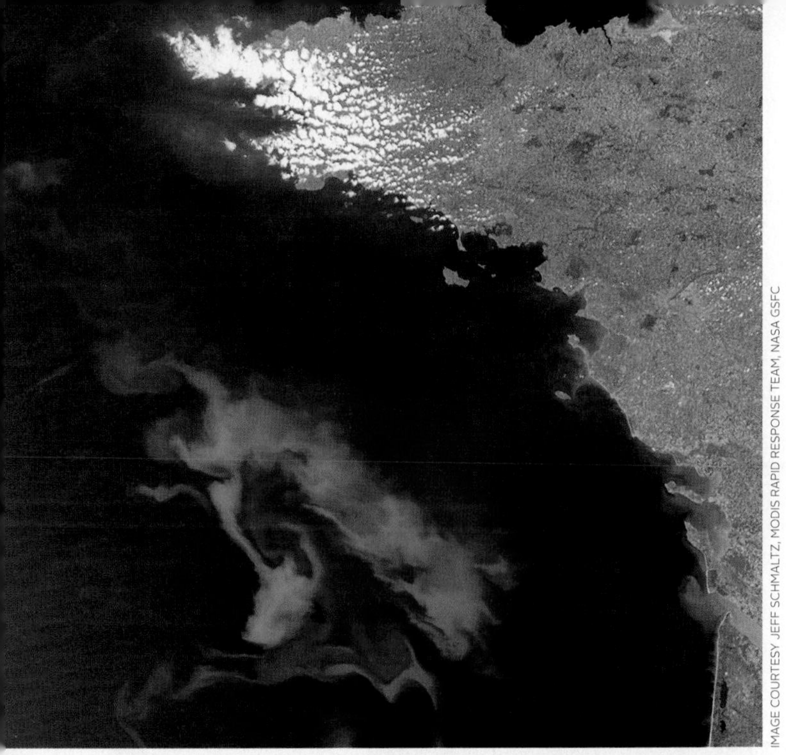
IMAGE COURTESY JEFF SCHMALTZ, MODIS RAPID RESPONSE TEAM, NASA GSFC

FIGURE 6.38 ▪ Marine algal bloom controlled by viruses. Bloom of the alga *Emiliana huxleyi* off Plymouth, England, detected by satellite remote sensing. The pale clouds in the water are the reflected light from billions of calcite plates, or "coccoliths," that coat each algal cell.

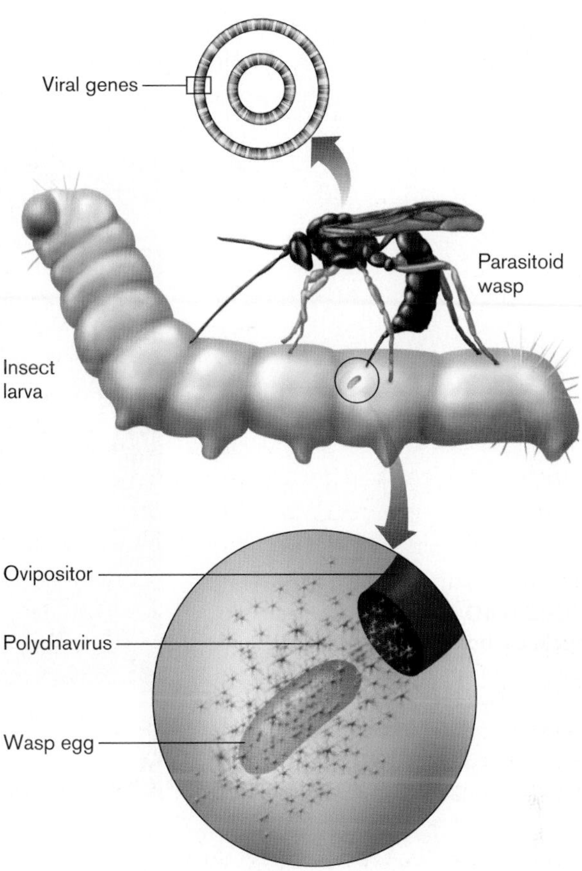
Viral genes
Parasitoid wasp
Insect larva
Ovipositor
Polydnavirus
Wasp egg

FIGURE 6.39 ▪ The relationship of polydnaviruses, wasps, and caterpillars. Parasitoid wasps lay their eggs inside a living insect caterpillar. When a female wasp deposits her eggs inside the caterpillar, she also deposits her symbiogenic polydnavirus virions. The virions express wasp genes in the caterpillar, where they prevent the encapsulation process that would otherwise wall off the wasp egg and kill it.

illness may be apparent. For example, humans harbor many obscure herpes and cold-type viruses whose effects are unknown. Similarly, apes related to humans harbor many retroviruses with few effects, despite the close relatedness of these viruses to HIV, which causes AIDS in humans.

An intriguing feature of persistent viruses is their evolution of properties that contribute to the fitness of the host. For example, some bacteria harbor a bacteriophage whose genome expresses a toxin that kills off competing bacteria. Similarly, animals such as monkeys carrying persistent viruses may disseminate the virus to a competitor species in which the virus is lethal. In other cases, a specific virus evolves a mutualistic relationship with a host that enables the host to parasitize a third organism (**Fig. 6.39**). One example is the relationship of polydnaviruses, wasps, and caterpillars. Parasitoid wasps lay their eggs inside a living insect caterpillar, where the larvae must hatch and feed without stimulating the caterpillar's defense. When a wasp deposits her eggs inside the caterpillar, she also deposits her mutualistic polydnavirus virions. The polydnavirus virions express wasp genes in the tissues of the caterpillar. The viral-expressed genes encode proteins that prevent the caterpillar cells from "encapsulation," a process that would otherwise wall off the wasp egg and kill it.

An exciting possibility for agriculture is that viruses can help an organism survive in an extreme environment. For research on such virus-host mutualism, see **Special Topic 6.1**.

Quantitative ecology. Is it possible to measure and quantify the impact of viruses on ecosystems? This daunting task can at least be attempted through sequencing the genomes of virus populations in natural environments (metagenomic sequencing, discussed in Chapter 7). Viruses, particularly bacteriophages, are equally abundant in coastal and aquatic habitats, and they reach numbers tenfold higher in soil. In coastal waters, the populations of DNA viruses turn over daily, and they lyse 20%–100% of the bacteria, thus serving as major consumers. It was thought that biofilms on marine snow (suspended organic particles) might be protected from virus infection by their polysaccharide matrix, but biofilm structures also provide receptors for viral recognition and attachment. In fact, viral infection rates for biofilm bacteria equal the infection rates of planktonic (floating) cells. Various species of viruses infect host organisms at all levels of the food chain; in a wetland, for example, besides bacteriophages infecting bacteria, viruses infect diatoms, insects,

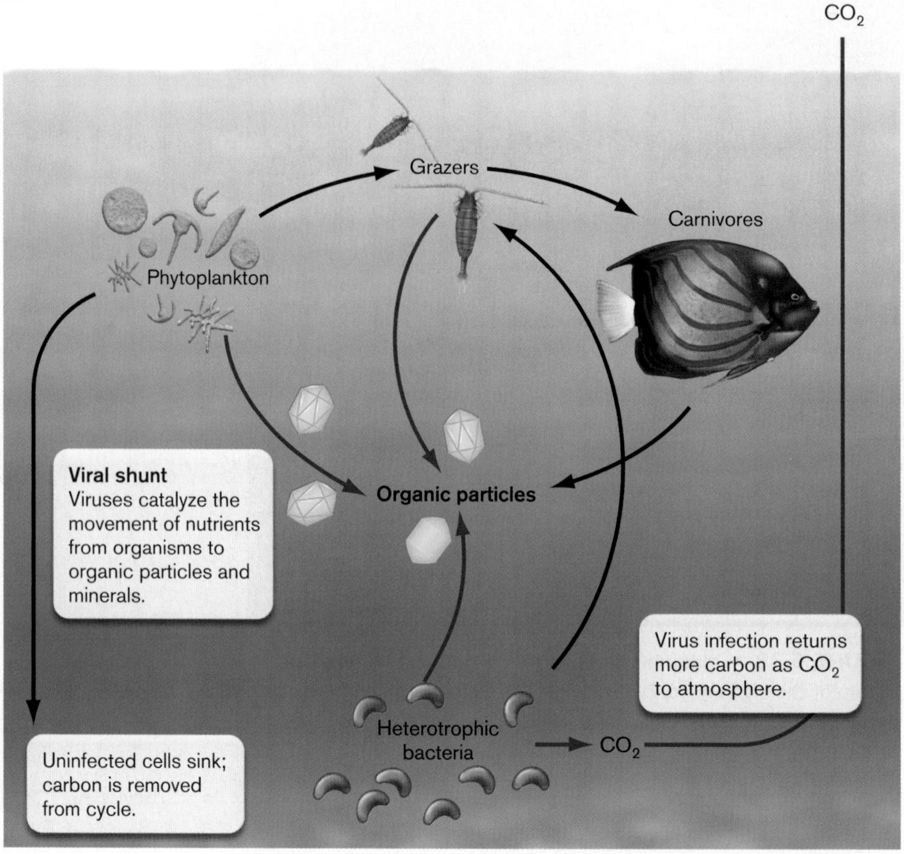

CO_2
Grazers
Carnivores
Phytoplankton
Viral shunt
Viruses catalyze the movement of nutrients from organisms to organic particles and minerals.
Organic particles
Virus infection returns more carbon as CO_2 to atmosphere.
Heterotrophic bacteria
CO_2
Uninfected cells sink; carbon is removed from cycle.

FIGURE 6.40 ■ Marine viruses recycle carbon. Viruses divert the flow of carbon from marine phytoplankton and bacteria away from consumers and into organic particles (detritus). The organic particles are consumed by bacteria that respire, releasing CO_2.

vertebrates (including fish and amphibians), and plants such as reeds and salt marsh grasses. Wetland plants may serve as reservoirs or as overwintering hosts for viruses harmful to agriculture.

A common measure of the influence of viruses in a given habitat is the virus-to-bacteria ratio. In marine water, the ratio has been estimated at values of 1–100 viruses per bacterium. Similar ratios are found in forested soil and in wetland soil. Presumably, these viruses await contact with a susceptible host, since virus particles lack any motility of their own.

In agricultural soil, the ratio can be as high as 1,000 viruses per bacterium. The reasons for this viral density in agricultural soil are not known, but a statistical correlation shows that viral numbers in soil increase with dryness and low content of organic matter. In experimental tests in which carbon- and nitrogen-containing nutrients are added to a model ecosystem, bacterial numbers increase rapidly, while viral increase lags. A possible explanation for the lag in virus production is that ample water and nutrients, which favor bacterial growth, also favor bacteriophage lysogeny, in preference to lytic burst and release of phages.

On a global scale, viruses play a significant role in the carbon balance. Models of ocean carbon flux typically emphasize the consumption of phytoplankton by grazers and the consumption of grazers by carnivores (**Fig. 6.40**). At each level, some of the carbon fixed as biomass sinks to the ocean floor, removed from the carbon cycle. But at each level, viral infection and lysis converts the bodies of phytoplankton, grazers, and carnivores into detritus consisting of small organic particles. The organic material is consumed by bacteria, whose rapid respiration releases much CO_2. Thus, carbon is diverted from the ocean sink into CO_2 returning to the atmosphere—a factor that must be considered in models of global warming.

To Summarize

- **Environmental change** results in new emerging viral pathogens.
- **Acute virus infection limits host population density.** Virus-associated mortality may lead to increased genetic diversity of host species.
- **Persistent viruses remain in hosts**, where they may evolve traits that confer positive benefits in a virus-host mutualism.
- **Virus-to-bacteria ratios** range from 1–100 in marine and aquatic environments, to 1–1,000 in soils.
- **Marine viruses** infect most phytoplankton, releasing their minerals in the upper water, where they are available for the next phototrophs. Viral activity substantially impacts the global carbon balance.

Concluding Thoughts

In this chapter we have covered broadly the structure and function of all kinds of viruses. As devastating as viruses can be to their hosts, they also provide engines of genomic change (see Chapter 9) and ecological balance, including mutualistic relationships with a host (see Chapter 21). Understanding virus function prepares us to discuss microbial genetics in Chapters 7–10. Chapter 11 then explores in depth the molecular biology of viruses that cause human disease. Viral molecular biology is a field of growing importance for medical and agricultural research, for genetic engineering and gene therapy, and for understanding natural ecosystems.

CHAPTER REVIEW

Review Questions

1. Compare and contrast the form of icosahedral and filamentous (helical) viruses, citing specific examples.
2. How do viral genomes gain entry into cells in bacteria, plants, and animals?
3. Explain the key structural features that define the seven Baltimore groups of viral genomes. Explain the consequences of each structure for viral replication.
4. How do viral genomes interact with host genomes, and what are the consequences for host evolution?
5. Compare and contrast the lytic, lysogenic, and slow-release replication cycles of bacteriophages. What are the strengths and limitations of each?
6. Compare and contrast the replication cycles of RNA viruses and DNA viruses in animal hosts. What are the strengths and limitations of each?
7. Explain the plate titer procedure for enumerating viable bacteriophages. How must this procedure be modified to titer animal viruses? Oncogenic viruses?
8. Explain how a pure isolate of a virus can be obtained. How do the procedures differ from those used for isolating bacteria?
9. Explain the generation of the step curve of virus proliferation. Why is virus proliferation generally observed as a single step, or generation, in contrast to the life cycles of cellular microbes, outlined in Chapter 4?
10. Explain the key contributions of viruses to natural ecosystems. What may happen in an ecosystem where viruses are absent or fail to cause significant infection?

Thought Questions

1. Discuss the functions of different structural proteins of a virion, such as capsid, nucleocapsid, tegument, and envelope proteins. How do these functions compare and contrast with functions of cellular proteins?
2. What are the relative advantages of the virulent phage replication cycle of phage T4, the lysis-lysogeny options of phage lambda, and the slow-release replication of phage M13? Under what conditions might each strategy be favored over the others?
3. Given the basis of viral tropism, how might an animal evolve traits that confer resistance to a virus infection?
4. If viruses return a substantial fraction of marine CO_2 to the atmosphere, can you imagine any ways to modulate virus proliferation so as to divert the carbon into sedimenting biomass?

Key Terms

bacteriophage (194)
batch culture (221)
burst size (210, 222)
capsid (192)
cell-surface receptor (208)
cloning vector (194)
core particle (198)
CRISPR (212)
DNA reverse-transcribing virus (206)
eclipse period (221)
endocytosis (215)
envelope (199)
filamentous virus (200)
fluorescent-focus assay (224)
focus (224)
host range (194)
icosahedral capsid (198)
latent period (221)
lentivirus (223)
lysate (222)
lysis (210)
lysogeny (210)
lytic cycle (210)
mechanical transmission (218)
multiplicity of infection (MOI) (221)
oncogene (218)
oncogenic virus (218)
orthologous gene (ortholog) (206)
pararetrovirus (206, 219)
phage (194)
plaque (194, 223)
plaque assay (223)
plaque-forming unit (PFU) (223)
plasmodesma (219)
prion (197)
prophage (210)
proteome (207)
proteomics (207)
reading frame (195)
retrovirus (206, 217)
reverse transcriptase (206)
rise period (222)
RNA-dependent RNA polymerase (203)
RNA reverse-transcribing virus (206)
segmented genome (203)
site-specific recombination (210)
spike protein (199)
temperate phage (210)
transduction (210)
transform (218)
transformed-focus assay (225)
tropism (213)
uncoating (215)
virion (192)
viroid (196)
virulence (222)
virus (192)

Recommended Reading

Edwards, Robert A., and Forest Rohwer. 2005. Viral metagenomics. *Nature Reviews. Microbiology* **3**:504–510.

Grünewald, Kay, Prashant Desai, Dennis C. Winkler, J. Bernard Heymann, David M. Belnap, et al. 2003. Three-dimensional structure of herpes simplex virus from cryo-electron tomography. *Science* **302**:1396–1398.

Harris, Audray, Giovanni Cardone, Dennis C. Winkler, J. Bernard Heymann, Matthew Brecher, et al. 2006. Influenza virus pleiomorphy characterized by cryoelectron tomography. *Proceedings of the National Academy of Sciences USA* **103**:19123–19127.

Horvath, Phillippe, and Randolphe Barrangou. 2010. CRISPR/Cas, the immune system of bacteria and archaea. *Science* **327**:167–170.

Labrie, Simon J., Julie E. Sampson, and Sylvain Moineau. Bacteriophage resistance mechanisms. 2010. *Nature Reviews. Microbiology* **8**:317–327.

Moineau, S., D. Tremblay, and S. Labrie. 2002. Phages of lactic acid bacteria: From genomics to industrial applications. *ASM News* **68**:388–391.

Ptashne, Mark. 2004. *Genetic Switch: Phage Lambda Revisited.* Cold Spring Harbor Laboratory Press, Cold Spring Harbor, NY.

Raoult, Didier, Stéphane Audic, Catherine Robert, Chantel Abergel, and Patricia Renesto. 2004. The 1.2-megabase genome sequence of Mimivirus. *Science* **306**:1344–1350.

Roossinck, Marilyn J. 2011. The good viruses: Viral mutualistic symbioses. *Nature Reviews. Microbiology* **9**:99.

Srinivasiah, Sharath, Jaysheel Bhavsar, Kanika Thapar, Mark Liles, Tom Schoenfeld, et al. 2008. Phages across the biosphere: Contrasts of viruses in soil and aquatic environments. *Research in Microbiology* **159**:349–357.

Suttle, Curtis A. 2007. Marine viruses—major players in the global ecosystem. *Nature Reviews. Microbiology* **5**:801–812.

Trask, Shane D., Sarah M. McDonald, and John T. Patton. 2012. Structural insights into the coupling of virion assembly and rotavirus replication. *Nature Reviews. Microbiology* **10**:165–177.

Wommack, K. Eric. 2010. Viral ecology: Old questions, new challenges. *Microbiology Today,* 96.

Xiao, Chuan, Yurii G. Kuznetsov, Siyang Sun, Susan L. Hafenstein, Victor A. Kostyuchenko, et al. 2009. Structural studies of the giant Mimivirus. *PLoS Biology* 7:e1000092.

COURTESY OF CHRISTINE JACOBS-WAGNER

Christine Jacobs-Wagner, Professor of Molecular, Cellular and Developmental Biology, Yale University; Investigator of the Howard Hughes Medical Institute.

AN INTERVIEW WITH:

Christine Jacobs-Wagner

The Thrill of Discovery in Molecular Microbiology

Christine Jacobs-Wagner has taught for 12 years at Yale University. With her students and postdocs, she has made striking discoveries by studying the bacterium *Caulobacter crescentus,* which generates two very different cell forms during each round of division. As a graduate student at the University of Liège, Belgium, she won the General Electric & *Science* Prize for her work on beta-lactam antibiotic resistance and cell wall sensing in Gram-negative bacteria. At Yale, she was designated an investigator of the Howard Hughes Medical Institute, one of the nation's largest medical philanthropic organizations.

Why did you decide to study microbiology?

Because microorganisms are fascinating creatures! The more I learn about them, the more I marvel. It often seems as if there isn't a thing microorganisms cannot do. They are extremely sensitive and responsive to their physical and chemical environment. They can multiply at a blazing speed, yet they replicate and segregate their genetic material with an exquisite temporal and spatial accuracy. Some of them can thrive under ridiculously harsh conditions, in Antarctic ice or boiling-hot deep-sea vents. Bacteria like *Deinococcus radiodurans* can withstand 500 times more gamma radiation than what could kill a human!

Microbes have been studied for a very long time, and we have learned a great deal; yet they are still full of surprises. For example, bacteria are more than tiny mixed bags of molecules. They exhibit a remarkably elaborate internal organization, which has revolutionized the way we view and study bacteria. For instance, we have shown that *Caulobacter* localizes key signaling proteins at opposite ends of the cell to monitor the progression of cell division. This can be important for cell cycle regulation and the initiation of different developmental programs in daughter cells. Protein localization at the cell ends can also be critical for many important bacterial processes, including chromosome segregation, cell division, motility, and pathogenesis, just to name a few.

How did the General Electric & *Science* Prize influence the development of your career?

This is a prize that recognizes the contribution of graduate students, and it is wonderful that GE & *Science* and other foundations acknowledge their work in the form of a prize. More than anything, the GE & *Science* prize helped me realize I was doing work that others found important and interesting. It gave me confidence. My graduate career was unusual in that I spent time in five different labs in four different countries, each time learning the unique expertise that each lab offered—for example, biochemical characterization in Jean-Marie Frère's lab in Belgium, and transcriptional regulation in Staffan Normark's lab in Sweden.

How did you choose to study *Caulobacter crescentus*?

I was drawn to *Caulobacter* because of its tractability to study the bacterial cell cycle, given the ease of obtaining cell populations that are synchronized with respect to the cell cycle. The ability to synchronize cell populations is very important; it is a key reason why the budding yeast has contributed so much to our understanding of the eukaryotic cell cycle. Furthermore, *Caulobacter* is an excellent model to study both cellular polarization and development because its cell cycle is coordinated with a highly polarized developmental program that culminates in an asymmetrical division and the generation of daughter cells with different cell fates. Applied and environmental microbiologists are also developing *Caulobacter* as a living detector of various toxic compounds and as a multipurpose bioremediation agent.

Why do bacteria have a cytoskeleton? How does it relate to the cytoskeleton of animal cells?

I think that bacteria have a cytoskeleton for many of the same reasons that our own human cells do: to organize their cytoplasm, to govern their cell shape and size, and to generate force (for example, for DNA segregation). These activities are important for cellular life regardless of its origin, and a cytoskeleton seems like a good tool to perform these important functions.

Some elements of the bacterial cytoskeleton, such as actin and tubulin homologs, are evolutionarily linked to the animal cytoskeleton.

What current research question are you pursuing, and how do students participate?

We study how cellular organization is achieved and regulated in bacteria. In particular, we are interested in the spatiotemporal mechanisms involved in cell division, chromosome segregation, cell cycle regulation, and cell morphogenesis. Undergraduate and graduate students are an integral part of our research team. They each have an independent research project, which allows them to explore their ideas and use their creativity. Yet each student's project is sufficiently linked to others in the lab to foster interaction and intellectual exchange among lab members. One of the beauties of working with a highly tractable bacterial model is that students can learn many techniques, test numerous ideas, and explore different directions in a reasonable amount of time.

Last year, my postdoc Sebastian Poggio and my undergraduate student Constantin (Nick) Takacs took advantage of Nick's school break to spend a couple of weeks in Professor Waldemar Vollmer's lab at Newcastle University in England. Within that short amount of time, Nick and Sebastian, with the help of Professor Vollmer and one of his students, fully characterized the complex composition and structure of the *Caulobacter* peptidoglycan cell wall. They brought back their newly acquired expertise in high-performance liquid chromatography and mass spectrometry here to Yale so that now our lab can do this type of technically demanding work in-house. Everybody wins!

What genetic and molecular tools do you use? How does your research combine genetics and advanced microscopy?

We often cannot anticipate which approach will be most successful, so we try to use every tool at our disposal, at times simultaneously, to achieve our goal. For example, Nora Ausmees, as a postdoctoral fellow in my lab, led the discovery of bacterial intermediate filaments by visually screening *Caulobacter* mutants with cell shape defects. Intermediate filaments are major cytoskeletal components of human cells, and over 30 human diseases

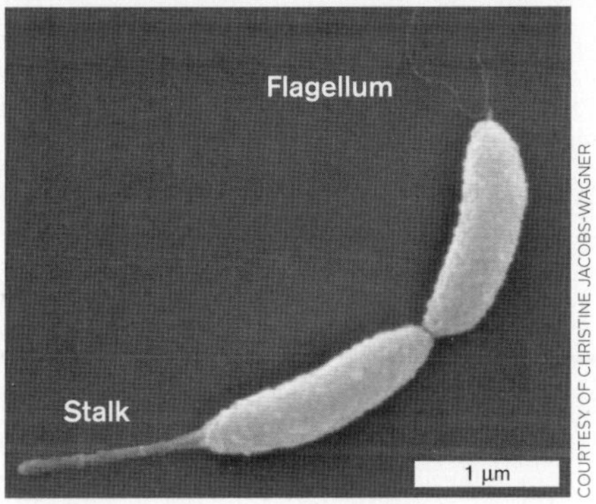
Flagellum
Stalk
1 μm
COURTESY OF CHRISTINE JACOBS-WAGNER

Caulobacter crescentus (SEM) undergoing cell division to form stalked cell and flagellated cell.

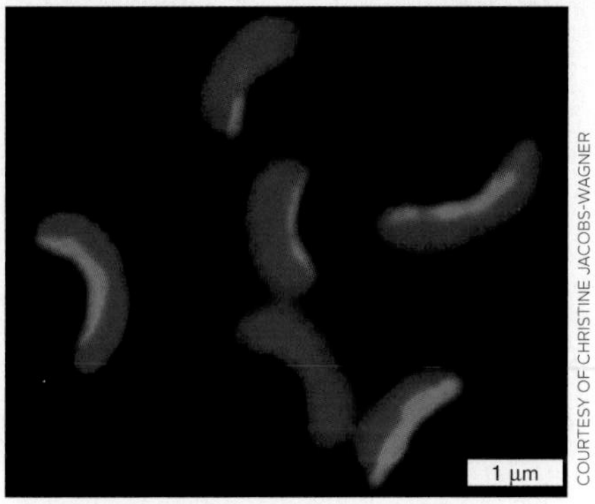
1 μm
COURTESY OF CHRISTINE JACOBS-WAGNER

C. crescentus cells; intermediate filaments are labeled by pink fluorescence.

(such as progeria and some cardiomyopathies and lipodystrophies) have been associated with their malfunction. Yet, little is known about their basic properties, such as their molecular function, assembly, and regulation within cells. *Caulobacter* now offers a highly tractable experimental model system to study intermediate-filament biology.

Does your work have implications for the developmental cycle of human pathogens?

Many of the developmental and cell cycle pathways that we and others study are essential for viability and are present in close relatives of *Caulobacter,* including dangerous pathogens (for example, *Brucella* and *Rickettsia* species, which cause a variety of diseases, including brucellosis and typhus). These pathogens are far less tractable, and *Caulobacter* research has stimulated exciting, ongoing research on how the pathways we study affect the cell cycle and development of these pathogens in their hosts.

What advice do you have for today's students?

I would strongly recommend that students spend some time in a microbiology lab. Whatever they aspire to do in the future, this experience will give them a better understanding of the world they live in. Microbes are virtually omnipresent, and they are indispensable to our well-being and prosperity, even while some pose threats to our health and economy. Doing some microbiological research is also one of the best ways to learn hypothesis-driven reasoning and to develop skills in critical thinking. The speed at which microbes grow and can be manipulated allows students to design experiments and test hypotheses even when time is limited.

For students who already work in a lab, I suggest that they explore the possibility of visiting another lab with distinct expertise to learn a new approach that can directly help their research project. I did this when I was an undergraduate student, a graduate student, and a postdoctoral fellow. Each experience generated a huge boost in my research; enhanced my skill set; created new, long-lasting friendships; and generated connections to important people who have since supported my career.

How does your family relate to your work? Do you have interesting pursuits outside of science?

My mother has been a constant source of support throughout my life, which has helped a lot. The daily support comes from my husband, Matt, who is incredibly enthusiastic about my career. He is not a scientist (he is an e-learning developer), yet he plays an active role in my work that ranges from building and maintaining our lab website to being my top adviser on nonscientific issues, such as running a lab or raising money. Having someone close who listens and gives good input has been extremely valuable.

CHAPTER 7
Genomes and Chromosomes

7.1 DNA: The Genetic Material

7.2 Genome Organization

7.3 DNA Replication

7.4 Plasmids

7.5 Eukaryotic Chromosomes

7.6 DNA Sequence Analysis

A genome is all the genetic information that defines an organism. For over a hundred years, we have known that chromosomes consist of DNA, yet we have only recently been able to sequence complete genomes. We are now in the postgenomic era, in which biological research is driven, in large part, by the knowledge of complete genome sequences.

Microbial genomes consist of one or more DNA chromosomes. Many bacteria have a single, circular chromosome, but some species have multiple chromosomes or even a mixture of linear and circular chromosomes. For example, the genome of *Sinorhizobium meliloti,* a major contributor to nitrogen fixation, consists of three circular chromosomes.

Chapter 7 addresses fundamental concepts as well as new and exciting ideas about DNA and genomes. What is DNA? How is it packaged in the cell? Because DNA is essential to life, nature must have devised a remarkable machine to replicate it. But how does this machine work? And do different species use different machines? Microbiology has played a leading role in answering these difficult questions, thus opening the door to postgenomic studies of all living things.

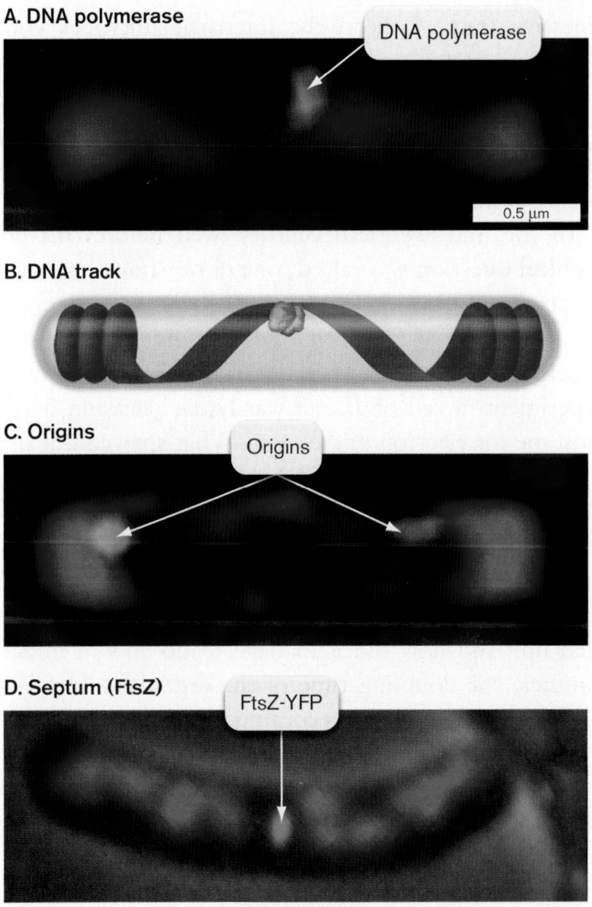

A. DNA polymerase
DNA polymerase
0.5 μm
B. DNA track
C. Origins
Origins
D. Septum (FtsZ)
FtsZ-YFP

CURRENT RESEARCH highlight

Tracking a replicating chromosome. A DNA polymerase complex (panel A; green, GFP-labeled) in *Bacillus subtilis* is anchored at midcell (the equator of the cell), where it synthesizes two daughter chromosomes (red). The newly synthesized DNA is red because fluorescently labeled nucleotides were incorporated during replication. Daughter chromosomes emerge from midcell and form helical structures that extend toward each pole. Panel B is a conceptual illustration of panel A. Panel C shows that the replication origins for each new chromosome (green; tagged with origin-specific GFP) also move toward opposite poles. In panel D, replicated chromosomes (red) clear midcell, and the division septum forms (green, FtsZ-YFP). Mechanisms that choreograph nucleoid movement are elegant but still not fully understood. *Source:* Parts A–C from Ida Berlatzky et al. 2008. *PNAS* **105**:14136; part D from Christopher D. Rodrigues and Elizabeth J. Harry. 2012. *PLoS Genet.* **8**:e1002561.

What makes each species unique is its genome, the sum total of its genes. But, what comprises that "sum" may be more than you think. For instance, the human genome is supplemented by the genomes of all the microbes in the digestive tract. Collectively, intestinal microbes contain 100 times more genes than the human genome and provide key metabolic capabilities that humans lack, such as the ability to digest complex plant materials. Some scientists even argue that these microbial genomes should count as part of the human genome.

In the mid-twentieth century, well before this philosophical question was raised, one of our intestinal bacteria became the focus of efforts to understand genes and genetics. The choice of *Escherichia coli* was fortunate, because its genetics proved to be especially pliable. In one remarkable experiment, a cell of *E. coli* was lysed, releasing its chromosome for electron microscopy. What spewed out of this single cell was a strand of DNA 1,500 times longer than *E. coli* itself (**Fig. 7.1**). Scientists wondered how this enormous molecule could fit into a single cell, and how an enzyme could duplicate that much DNA (over 4.6 million base pairs) without the whole DNA molecule getting tangled up. And how did it manage to do this in under 20 minutes, the doubling time of the organism? Half a century later, those questions continue to intrigue us.

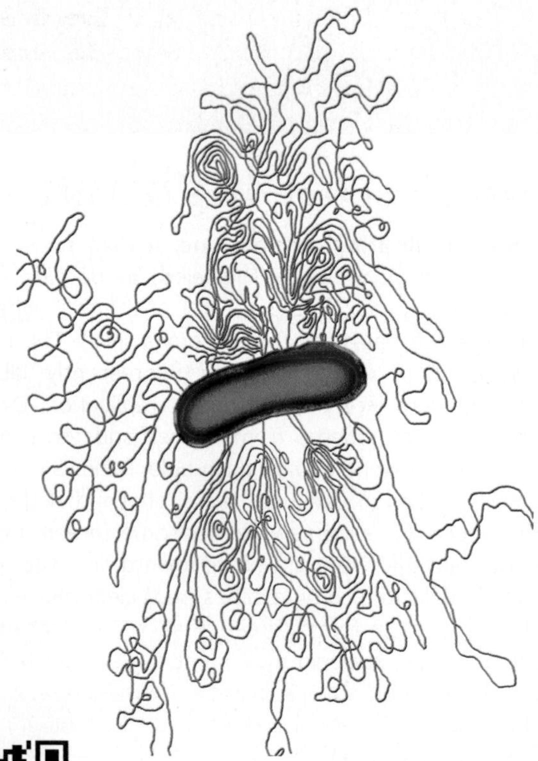

GOPAL MURTI/VISUALS UNLIMITED

FIGURE 7.1 ■ Osmotically disrupted bacterial cell with its DNA released. Length of bacterium is approximately 2 micrometers.

An important concept in molecular biology is **information flow**—namely, how information stored in DNA gets passed to new generations and becomes protein. Three processes are required: **DNA replication** copies the information held in a DNA sequence and makes a new DNA molecule that can be passed to offspring. **Transcription** converts the same DNA information into a different, "readable" form called RNA. **Translation** then "reads" the information in RNA to make protein.

In this chapter we explore the nature of DNA, the structure of prokaryotic genomes, and the mechanisms used for their replication. We also examine how microbial enzymes that manipulate DNA were used to develop the fundamental techniques of present-day biotechnology, such as DNA restriction analysis, DNA sequencing, and the polymerase chain reaction (PCR). The expression of genes to make RNA and protein is discussed in Chapter 8. The remaining chapters of Part 2 discuss gene recombination and transmission (Chapter 9), the regulation of gene expression (Chapter 10), the specialized genetic mechanisms of viruses (Chapter 11), and the use of microbial genetic tools for research and biotechnology (Chapter 12). From the advances of the last 50 years, a new molecular and genomic perspective of biology has emerged.

7.1

DNA: The Genetic Material

From where do genetic traits come and how are they passed? The ability of plants and animals to transfer genetic traits was known well before we knew that DNA was the carrier of genetic information. Early researchers understood the Mendelian model of gene inheritance and how it dictated the passage of traits from parent to offspring in eukaryotes. This transfer of genes from parent to offspring is known as **vertical transmission**. Bacteria, however, appeared to operate differently, because in addition to vertical transmission, they are able to conduct **horizontal transmission**—the transfer of small pieces of genetic information from one cell into another. The study of horizontal gene transfer in bacteria ultimately led to the discovery of horizontal transfer throughout animals and plants, albeit on a slower timescale because of their longer generation times. Mechanisms of horizontal transmission are discussed in Chapter 9.

Our path to understanding DNA began in 1928 when Frederick Griffith (1879–1941) discovered that he could kill mice with live but seemingly harmless (avirulent) *Streptococcus pneumoniae* if he coinjected the mice with dead cells from a virulent strain of the bacteria. Something from the dead bacteria transformed the innocuous live bacteria into killers. This form of horizontal gene transfer, called **transformation**, led to the discovery in 1944 that genetic information is embedded in the base sequence of deoxyribonucleic acid (DNA) (see Section 1.6). Another example of transformation is the toxic shock gene that can be moved horizontally among strains of *Staphylococcus aureus*. As a result of these and many other experiments, we now appreciate that chromosomes are made of contiguous packets of information, called **genes**.

Genes are units of information composed of a sequence of DNA nucleotides of four different types: adenine (A), guanine (G), thymine (T), and cytosine (C). (For review at the level of introductory biology, see Appendix 1.) A **structural gene** is a string of nucleotides that can be decoded by an enzyme to produce a functional RNA molecule. A structural gene usually produces an RNA molecule that in turn encodes a protein. A **DNA control sequence**, on the other hand, regulates the expression of a structural gene. DNA control sequences do not encode RNA or protein (so they are not really genes), but they do regulate the RNA production from an adjacent structural gene. Control sequences include promoters that launch RNA synthesis from a structural gene, and binding sites for regulatory proteins that can activate or inactivate that promoter.

As noted previously, the entire genetic complement of DNA in a cell that defines it as an organism is called its **genome**. Our primary goal here in Chapter 7 is to convey what is known about how genomes are maintained and replicated. Mining the informational content of DNA sequences and genes is described more fully in Chapters 8 and 10.

Bacterial Genomes

In the early twentieth century, the chromosomes of bacteria, unlike those of eukaryotes, could not be observed by light microscopy. The reason is that bacteria, unlike eukaryotes, do not undergo mitosis, a process in which chromosomes are condensed and thickened about a thousandfold, making them visible. Important clues to bacterial chromosome structure were gleaned, however, from painstaking genetic studies. In the 1950s it was discovered that **conjugation**, a horizontal gene transfer mechanism requiring cell-to-cell contact, could transfer large segments of some bacterial chromosomes—not all at once, as in the established Mendelian model of plants and animals, but sequentially over a period of time (it takes 100 minutes to move the entire *E. coli* chromosome from one cell to another).

Thus, even though the bacterial chromosome could not be seen, conjugation allowed genes to be mapped relative to each other according to time of transfer. For example, a donor strain that can synthesize the amino acids alanine and proline can directly transfer the encoding genes to a recipient cell defective in those genes. The transfer process is nonspecific, so any gene can be transferred in this way. Completion of transfer also requires **recombination**, in which the donor DNA fragment replaces the recipient DNA fragment. Successful transfer of the genes for amino acid synthesis enables the formerly defective recipient to form colonies on minimal media lacking either amino acid. But because transfer occurs over time from a fixed starting point (that is, not all genes are transferred at the same time), it might take one gene 10 minutes of cell contact to be transferred while the second gene, farther away from the starting point of DNA transfer, takes an additional 20 minutes.

Because eukaryotic chromosomes are linear, scientists initially expected that bacterial chromosomes would be linear too. However, the early genetic maps drawn from conjugation experiments just would not fit together in a manner consistent with a linear model—for the simple reason that the bacterial chromosome in *E. coli*, the organism under study at the time, is circular. We now know that many bacteria and archaea have circular chromosomes, although some, like the Lyme disease agent *Borrelia burgdorferi*, have linear chromosomes and others, like *Agrobacterium tumefaciens*, have a mixture of circular and linear chromosomes (**Table 7.1**).

The size range of genomes across the phylogenetic tree, from viruses to humans, is enormous. Generally speaking, the simpler the organism, the smaller its genome. At some point, as DNA content is trimmed, the organism loses independence and must parasitize another organism, thus begging the question "What constitutes a 'minimal' genome?" or "How 'lean' can a chromosome become and still support independent growth?"

To Summarize

- **A genome** is all of the genetic information that defines an organism.
- **Genomes** of bacteria and archaea are made up of chromosomes and plasmids consisting of DNA.
- **Chromosomes** of bacteria and archaea can be circular or linear, as can plasmids.
- **Functional units** of DNA sequences include structural genes and regulatory sequences.

TABLE 7.1

Genomes of representative bacteria and archaea.

Species (strain)	Chromosome(s)[a] (kilobase pairs, kb)	Plasmid(s)[a] (kb)	Total (kb)
Bacteria	**Circular and linear**	**Circular**	
Mycobacterium tuberculosis Tuberculosis	4,400		4,400
Mycoplasma genitalium Normal flora, human skin	580		580
Burkholderia cepacia Respiratory infections in immune compromised patients	3,870 + 3,217 + 876	93	8,056
Escherichia coli K-12 (W3110) Model strain for *E. coli* proteomics	4,600		4,600
Anabaena species (PCC 7120) Cyanobacteria: major photosynthetic producer of carbon source for aquatic ecosystems	6,370	110 + 190 + 410	7,080
Borrelia burgdorferi Lyme disease	911	21 plasmids with sizes between 9 and 58	>1,250
Agrobacterium tumefaciens Tumors in plants; genetic engineering vector	2,840 + 2,070	214 + 542	5,666
Archaea			
Methanocaldococcus jannaschii Methanogen from thermal vent	1,660	16 + 58	1,734
Haloarcula marismortui Halophile from volcanic vent	3,130 + 288	33 + 33 + 39 + 50 + 155 + 132 + 410	4,270
	1,000 kb	500 kb	

[a]Purple circles and lines indicate relative sizes of genomic elements and whether these are circular or linear. Size bars are provided under each column.

7.2

Genome Organization

Do microbial genomes differ in structure or functional organization? New techniques for constructing physical maps of genomes and determining the sequences of whole genomes have revealed tremendous diversity in the size and organization of prokaryotic genomes.

Genomes Vary in Size

Bacterial and archaeal chromosomes range in size from 490 to 9,400 kilobase pairs (kb). For comparison, eukaryotic chromosomes range from 2,900 kb (Microsporidia) to over

100,000,000 kb (flowering plants). The human genome is over 3,000,000 kb.

Note: The designation "kb" can refer to the length of a double-stranded or single-stranded DNA molecule. A bacterial genome is, by definition, double-stranded. Some viral genomes can be single-stranded.

One of the smallest cellular genomes sequenced thus far is that of *Mycoplasma genitalium*. These pathogens rely on their host environment for many products but can still grow outside a host cell. The complete genome of *M. genitalium* consists of only 580 kb and encodes 480 proteins. It lacks the genes required for many biosynthetic functions, including the synthesis of amino acids, the construction of cell walls, and a functional tricarboxylic acid (TCA) cycle. In contrast, free-living bacteria that can grow in soil have larger genomes and dedicate many genes to the synthesis or acquisition of amino acids or TCA cycle intermediates. Even different strains of one species, such as *Salmonella enterica*, may vary considerably in gene distribution (**Fig. 7.2**).

Note: Current *Salmonella* nomenclature lumps many former species into one—*Salmonella enterica*—and downgrades former species names to serovar (serological variant) status. Serovar designations are based on using antibodies to detect strain differences in a surface structure, such as lipopolysaccharide O antigen for *Salmonella*. The serovar names are not italicized, and their first letters are capitalized (for example, *S. enterica* serovar Typhi, sometimes written *S.* Typhi).

Another feature that distinguishes bacterial and archaeal genomes from those of eukaryotes is the amount of so-called noncoding DNA. Many, but not all, eukaryotes contain huge amounts of noncoding DNA scattered between genes. In some species (such as humans), over 90% of the total DNA is noncoding. Some noncoding regions include **enhancer** sequences needed to drive transcription of eukaryotic promoters and DNA expanses that separate enhancers. Enhancer sequences can function at large distances from the gene they regulate. A **promoter** is the DNA sequence immediately in front of, and sometimes within, a gene needed to activate the gene's expression. Most of the noncoding spacers appear to be remnants of genes lost over the course of evolution and pieces of defunct viral genomes. Noncoding regions, however, may provide raw material for future evolution.

In contrast to many eukaryotes, Bacteria and Archaea tend to have very little noncoding DNA (typically less than 15% of the genome). Archaeal genomes do, however, contain a few genes with internal noncoding DNA sequences that resemble the introns of eukaryotes.

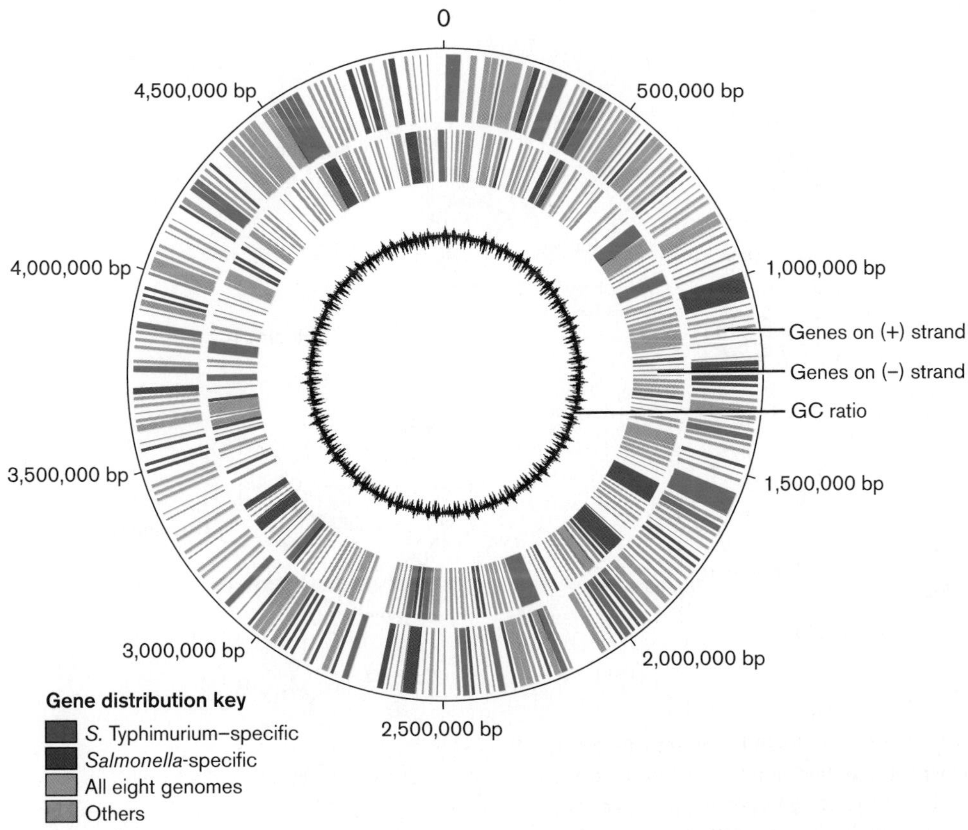
0
500,000 bp
1,000,000 bp
1,500,000 bp
2,000,000 bp
2,500,000 bp
3,000,000 bp
3,500,000 bp
4,000,000 bp
4,500,000 bp
Genes on (+) strand
Genes on (–) strand
GC ratio
Gene distribution key
S. Typhimurium–specific
Salmonella-specific
All eight genomes
Others

FIGURE 7.2 ■ The genome of *Salmonella enterica* serovar Typhimurium LT2. The circular chromosome. Base pairs (bp) are indicated around the perimeter. The two outer, multicolored circles indicate genes encoded on the separate strands, which means they are transcribed in opposite directions. Color codings represent homologies of eight compared species. Blue indicates genes present only in *Salmonella* Typhimurium LT2. Orange indicates that the gene has a close homolog in all eight genomes compared. Green indicates genes with a close homolog in at least one other *Salmonella* (*S.* Typhi, *S.* Paratyphi A, *S.* Paratyphi B, *S. arizonae,* or *S. bongori*) but not in *E. coli* K-12, *E. coli* O157:H7, or *Klebsiella pneumoniae.* Gray indicates other combinations. The black, inner circle is the GC content of the LT2 DNA (peaks pointing outward and inward indicate GC-rich and AT-rich areas, respectively).

Weblinks Genome sizes (*see ebook*)

Functional Units of Genes

In the simplest case, a gene can operate independently of other genes. The RNA produced from a single gene is said to be "monocistronic," which means it codes for one protein. Alternatively, a gene may exist in tandem with other genes in a unit called an **operon** (**Fig. 7.3A**). All genes in an operon are situated head to tail on the chromosome and are controlled by a single regulatory sequence located in front of the first gene. The single RNA molecule produced from the operon contains all the information from all the genes in that operon and is called "polycistronic."

On a functional level, a collection of genes and operons at different positions on the chromosome form a **regulon** when they have a unified biochemical purpose (such as amino acid biosynthesis) and are regulated by the same regulatory protein (**Fig. 7.3B**). The various mechanisms regulating expression of these functional genetic units are discussed in Chapter 10.

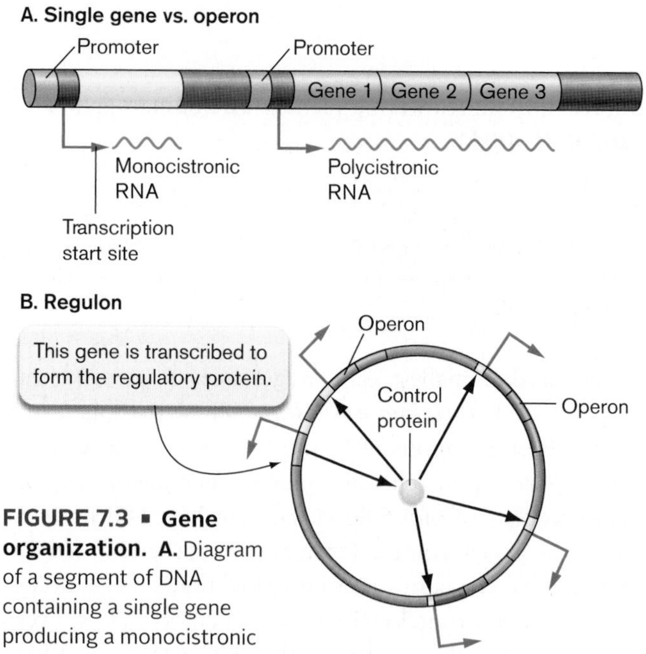
A. Single gene vs. operon
Promoter
Promoter
Gene 1
Gene 2
Gene 3
Monocistronic RNA
Polycistronic RNA
Transcription start site
B. Regulon
This gene is transcribed to form the regulatory protein.
Operon
Control protein
Operon

FIGURE 7.3 ■ Gene organization. A. Diagram of a segment of DNA containing a single gene producing a monocistronic message (which codes for one protein) and an operon of three genes producing a polycistronic message (from which three different proteins are made, one corresponding to each gene). **B.** Diagram of a circular bacterial genome, containing genes and operons coordinately controlled by a single regulatory protein.

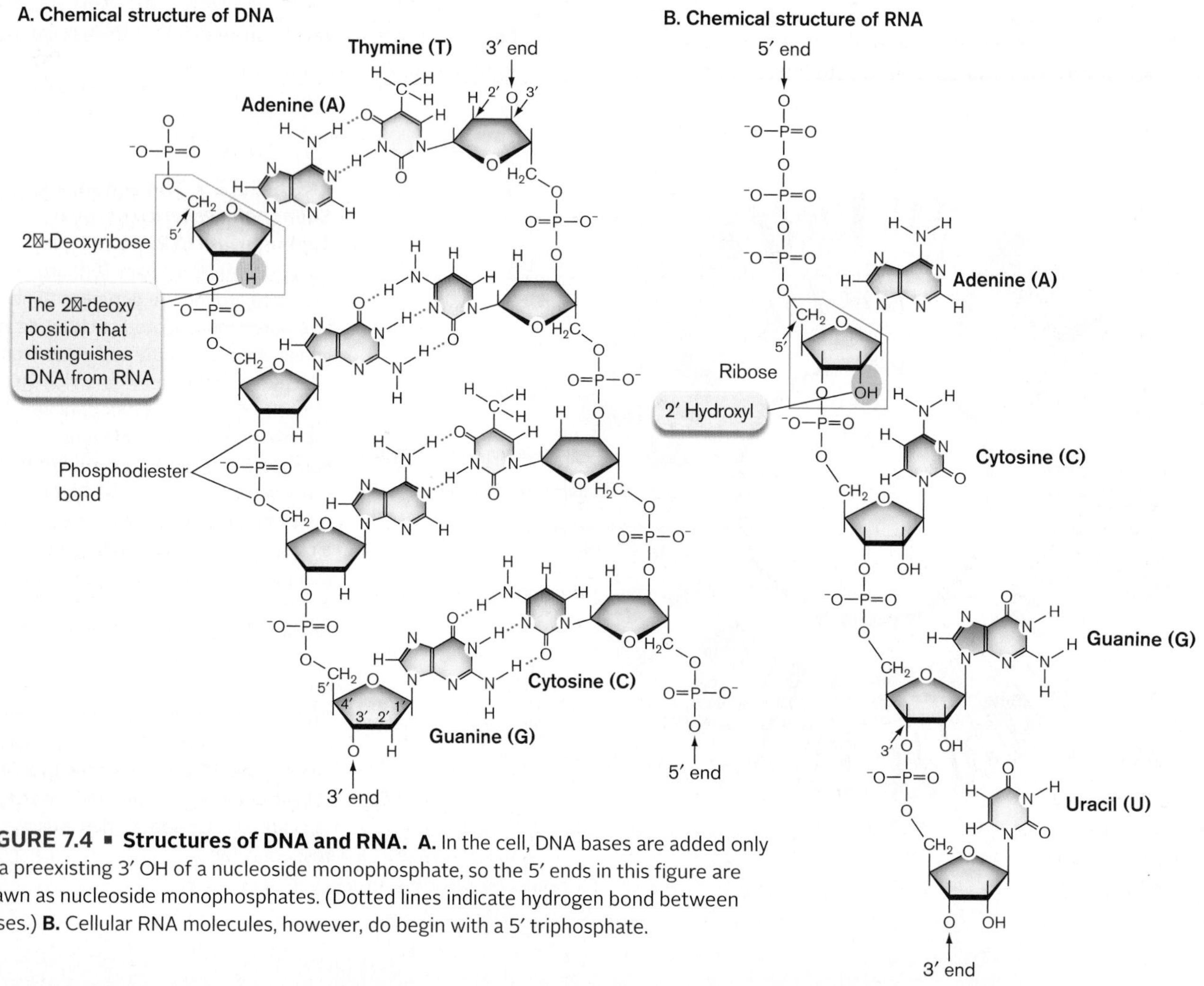
A. Chemical structure of DNA
Thymine (T)
3′ end
Adenine (A)
2′-Deoxyribose
The 2′-deoxy position that distinguishes DNA from RNA
Phosphodiester bond
Cytosine (C)
Guanine (G)
3′ end
5′ end
B. Chemical structure of RNA
5′ end
Adenine (A)
Ribose
2′ Hydroxyl
Cytosine (C)
Guanine (G)
Uracil (U)
3′ end

FIGURE 7.4 ■ Structures of DNA and RNA. A. In the cell, DNA bases are added only to a preexisting 3′ OH of a nucleoside monophosphate, so the 5′ ends in this figure are drawn as nucleoside monophosphates. (Dotted lines indicate hydrogen bond between bases.) **B.** Cellular RNA molecules, however, do begin with a 5′ triphosphate.

Note: In bacteria, the names of genes are given as a three-letter abbreviation of the encoded enzyme's name (for example, the gene *dam* encodes deoxyadenosine methylase) or the function of related genes (for example, genes designated *proA, proB,* and *proC* encode enzymes involved in proline biosynthesis). Bacterial gene names are written in italics with lowercase letters. For example, the genes involved in catabolizing lactose are the *lac* genes. If several genes are involved in the pathway, a fourth letter, capitalized, is used. Thus, the three genes *lacZ, lacY,* and *lacA* are all associated with lactose catabolism. When speaking of a gene product, a nonitalic (roman) font is used, and the first letter is capitalized. Thus, *lacZ* is the gene and LacZ is the protein product of that gene.

Weblinks Microbial Genome Database (*see ebook*)

DNA Function Depends on Chemical Structure

DNA is composed of four different nucleotides linked by a phosphodiester backbone (**Fig. 7.4A**). Each nucleotide consists of a nitrogenous base (also called a nucleobase) attached through a ring nitrogen to carbon 1 of 2-deoxyribose in the phosphodiester backbone. The 2-deoxy position that distinguishes DNA from RNA is highlighted in the figure. A **phosphodiester bond** (also marked in **Fig. 7.4A**) joins adjacent deoxyribose molecules in DNA to form the phosphodiester backbone. Phosphodiester bonds link the 3′ carbon of one ribose to the 5′ carbon of the next ribose. The two backbones are **antiparallel** so that at either end of the DNA molecule, one DNA strand ends with a 3′ hydroxyl group and the complementary strand ends with a 5′ phosphate. This antiparallel arrangement is necessary so that complementary bases protruding from the two strands can pair properly via hydrogen bonding. Base pairing is not possible if DNA strands are modeled in a parallel arrangement.

The nucleobases in DNA are planar heteroaromatic structures stacked perpendicular to the phosphodiester backbone and parallel to each other (see **Fig. 7.4A**). **Purines** (bicyclic nucleobases; adenine or guanine) pair with **pyrimidines** (monocyclic nucleobases; thymine or cytosine). Under physiological conditions of salt (about 0.85%) and pH (pH 7.8), the hydrogen bonding of the bases permits adenine to pair only with thymine (via two hydrogen bonds) and likewise guanine with cytosine (via three hydrogen bonds). These complementary base interactions enable the two phosphodiester backbones to wrap around each other to form the classic double helix, or duplex.

The thousands of H-bonds that form between purines and pyrimidines along the interior of a DNA duplex (**Fig. 7.5**) make the bonding of the two complementary strands of DNA highly specific, so that a duplex is formed only between complementary strands. Although H-bonds govern the specificity of strand pairing, the thermal stability of the helix is due predominantly to the stacking of the hydrophobic base pairs. Stacking enables interactions between the aromatic rings of these base pairs so that water and ions can interact with the negatively charged, hydrophilic phosphate backbone and avoid the hydrophobic interior of the double helix.

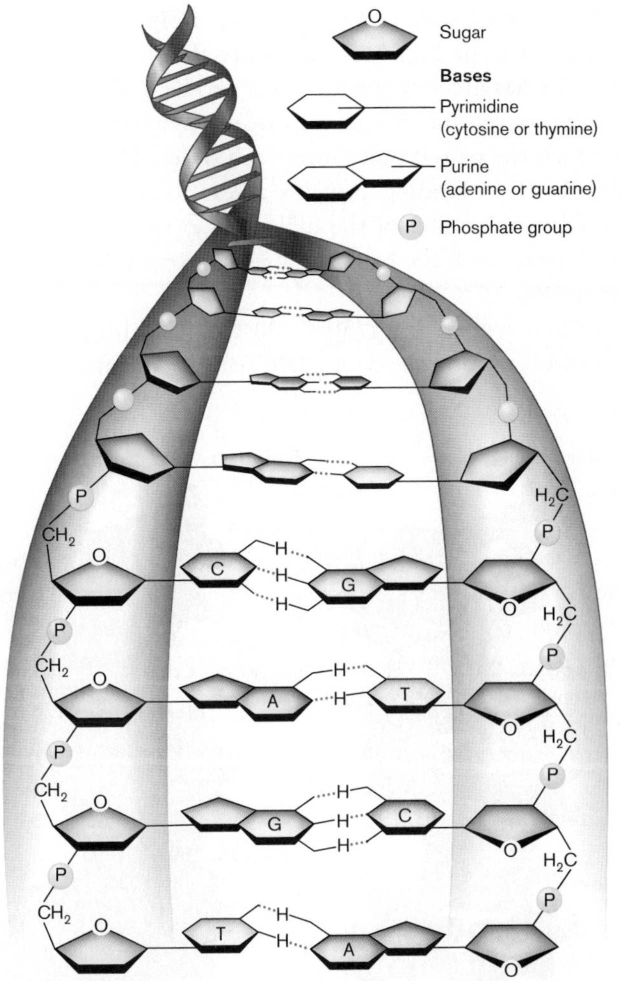
Sugar
Bases
Pyrimidine (cytosine or thymine)
Purine (adenine or guanine)
P Phosphate group
C
G
A
T
G
C
T
A

FIGURE 7.5 ■ Progressive magnification of the DNA helix. The top half of the molecule illustrates the typical double-helix structure that becomes magnified in the lower half to show individual bases and the hydrogen bonding that holds the two strands together.

Thought Question

7.1 What do you think happens to two single-stranded DNA molecules isolated from different genes when they are mixed together at very high concentrations of salt? *Hint:* High salt concentrations favor bonding between hydrophobic groups.

In the space-filling model of DNA in **Fig. 7.6A** and the contour map in **Figure 7.6B**, notice that the DNA double helix has grooves: a wide major groove and a narrower minor groove. The two grooves are generated by the angles at which the paired bases meet each other. These grooves provide DNA-binding proteins access to base sequences buried in the center of the molecule, so that the proteins can interact with the bases without the strands being separated (**Fig. 7.7**).

At high temperatures (50°C–90°C), the hydrogen bonds in DNA break and the duplex falls apart, or **denatures**, into two single strands. The temperature required to denature a DNA molecule depends on the GC/AT ratio of a sequence. More energy is required to break the three H-bonds of a GC base pair than the two H-bonds of an AT base pair. Thus, DNA with a high GC content requires a higher denaturing temperature than similar-sized DNA with a lower GC content. The black center ring in **Figure 7.2** illustrates how the GC content of bases can change around a single chromosome.

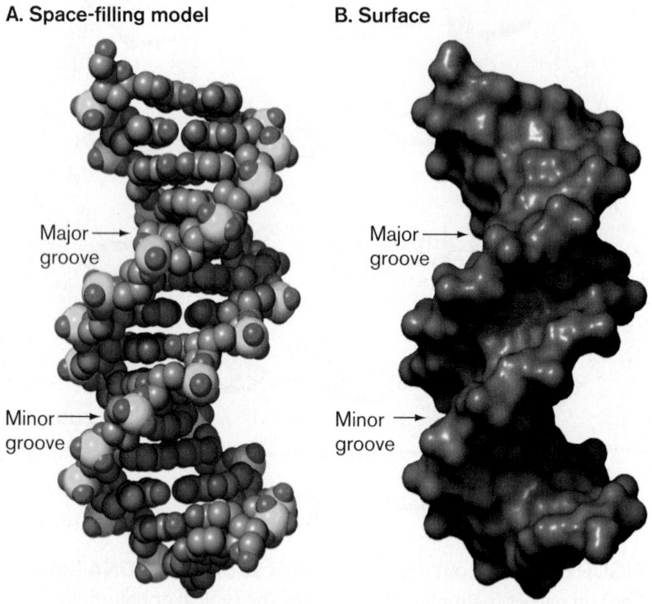
A. Space-filling model
B. Surface
Major groove
Minor groove
Major groove
Minor groove

FIGURE 7.6 ■ Models of DNA. A. Space-filling model of DNA. **B.** DNA surface modeled using nuclear magnetic resonance. (PDB code: 1K8J)

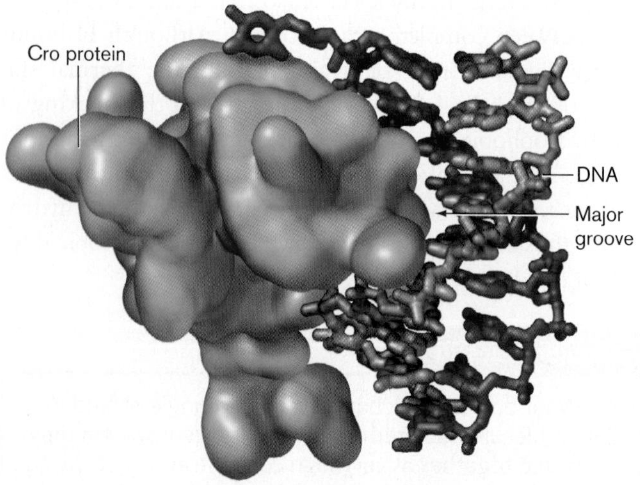
Cro protein
DNA
Major groove

FIGURE 7.7 ■ Protein that recognizes DNA. Cro repressor protein binds DNA within the major groove. (PDB code: 6CRO)

When DNA has been heated to the point of strand separation, lowering the temperature permits the two single strands to find each other and reanneal into a stable double helix. The kinetics of DNA renaturation is much slower than that of denaturation, because renaturation is a random, hit-or-miss process of complementary sequences finding each other. This melting/reannealing property of DNA is exploited in a number of molecular techniques (see the discussion of the polymerase chain reaction in Section 7.6). Note, however, that bacteria and archaea growing at extreme pH or temperature protect their DNA from denaturation through the use of remarkable DNA-binding proteins such as the archaeal histone proteins Hmf and Htz. The role of DNA-binding proteins in microbial survival is the subject of considerable research.

Thought Question

7.2 How do the kinetics of denaturation and renaturation depend on DNA concentration?

RNA Differs Slightly from DNA

Why do cells make both RNA and DNA? As we learned in Chapter 3, the growing cell makes temporary copies of its genes in the form of RNA (ribonucleic acid) molecules that direct the synthesis of proteins. DNA in a cell is usually double-stranded, whereas RNA is usually single-stranded. DNA and RNA are chemically similar, except that in RNA the sugar ribose replaces deoxyribose and the pyrimidine base uracil replaces thymine (**Fig. 7.4B**). Functionally, these two differences prevent enzymes meant to work on DNA, such as DNA polymerases, from acting on RNA. They also prevent RNA nucleases (RNases) from degrading DNA. However, uracil can still base-pair with adenine, which means that hybrid RNA-DNA double-stranded molecules can form (hybridize) when base sequences are complementary. In fact, this **hybridization** is a necessary step in the decoding of genes to make proteins.

Although RNA molecules are commonly thought of as single-stranded, all RNA molecules in nature have RNA-RNA double-stranded regions called "hairpins." Hairpin

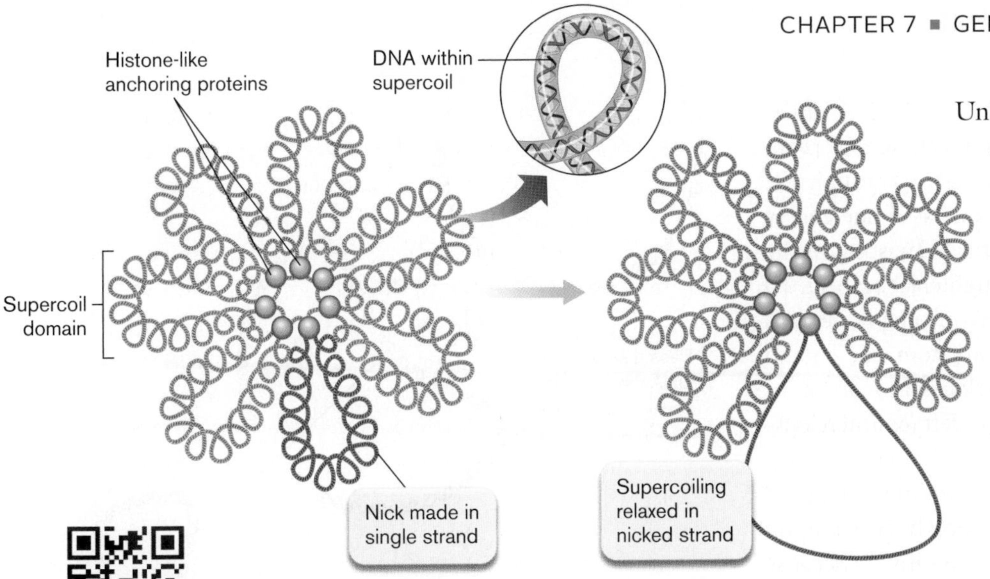
Histone-like anchoring proteins
DNA within supercoil
Supercoil domain
Nick made in single strand
Supercoiling relaxed in nicked strand

FIGURE 7.8 ▪ Bacterial nucleoid. A nucleoid showing domain loops after gentle release from cells. The single-strand nick unwinds (relaxes) only one loop.

structures form when complementary nucleotide sequences within the primary RNA sequence bend back and hybridize. These double-stranded hairpins have a variety of biological functions.

Bacterial Chromosomes Are Compacted into a Nucleoid

The chromosome of *E. coli* has over 4.6 million bases in one strand, or over 9 million, counting both strands. This is a huge molecule. At the normal pH of the cell (7.8), all the phosphates in the backbone (all 9 million of them) are unprotonated and negatively charged, so this one molecule contributes greatly to the overall negative charge of the cytoplasm.

Figure 7.1 shows DNA spewing out of a damaged bacterial cell. Laid out, the chromosome is 1,500 times longer than the cell. It is obvious from this photomicrograph that an intact, healthy cell must compact a huge bundle of DNA into a very small volume. DNA is the second-largest molecule in the cell (only peptidoglycan is larger) and comprises a large portion of a bacterial cell's dry mass—about 3%–4%. While packaging 3% of a cell's dry weight may not seem like a challenge, realize that DNA is further confined only to ribosome-free areas of the cell, so the chromosome-packing density reaches about 15 mg/ml. In a test tube, DNA at 15 mg/ml is almost a gel, so how can anything move inside a cell? And how does all this DNA keep from getting hopelessly tangled?

As introduced in Chapter 3, cells pack their DNA into a manageable form that still allows ready access to DNA-binding proteins. Although bacteria lack a nuclear membrane, they pack their DNA into a series of protein-bound domains collectively called the **nucleoid** (see Section 3.4).

Unlike the compact nucleus of eukaryotes, the bacterial nucleoid is distributed throughout the cytoplasm.

DNA Supercoiling Compacts the Chromosome

A nucleoid gently released from *E. coli* appears as 30–100 tightly wound loops (**Fig. 7.8**). The boundaries of each loop are defined by anchoring proteins called histone-like proteins for their similarity to histones, the DNA-binding proteins of eukaryotes. The double helix within each domain is itself helical, or supercoiled. The easiest way to envision supercoiling is to picture a coiled telephone cord. After much use, a phone cord twists, or supercoils, upon itself. Supercoiled phone cords are quite compact, taking up less space than a relaxed cord. Circular DNA works the same way—a property used by the cell to pack its chromosome.

Note that DNA cannot form supercoils unless its ends are tethered. In a circular chromosome, the DNA ends are tethered to each other. Introducing an extra twist by breaking one or both strands, twisting one end, and then resealing the strands means that the increased, or decreased, torsional (twisting) stress is trapped in the final circular molecule. It cannot then spontaneously unwind. Details are discussed later in the chapter.

Remarkably, the nucleoid with its 30–100 loops (or "domains") can maintain different loops at different superhelical densities. The independence of supercoiled domains was demonstrated by introducing one single-strand nick in the phosphodiester backbone of one domain (see **Fig. 7.8**). You can do this by adding very small amounts of a nuclease (an enzyme that cleaves a nucleic acid). The ends of the broken strand, driven by the energy inherent in the supercoil, rotate about the unbroken complementary strand of the duplex and relax the supercoil. A single nick in a genome, however, only removes supercoils from one domain. How is this possible if the chromosome is one circular molecule? The unaffected chromosomal domains remained supercoiled because they were constrained at their bases by anchoring proteins, such as HU and HN-S (histone-like proteins), that prevent rotation (for nucleoid organization, see Fig. 3.26).

How does DNA achieve the supercoiled state? The bacterial cell produces enzymes that can twist DNA into supercoils and other enzymes that relieve supercoils. A single twist introduced into a small (300-bp) circular DNA molecule forms a single supercoil as shown in **Figure 7.9**. A supercoiling enzyme makes a double-strand break at one point in the circle, passes another part of the DNA through the break, and reseals it. The result is the same as if one end of the broken circle were twisted one full turn. Twisting in the opposite direction of the helical turn tightens the helix by adding more turns (overwinding). Think of the phone cord again. Look down the length of the cord from one end. If the cord turns left to right (that is, clockwise) down its length, then twist it from one end right to left (counterclockwise). This increases the number of twists.

The resulting torsional stress of overwinding is relieved when the DNA (or phone cord) subsequently twists upon itself, introducing positive supercoils ("positive" because the DNA is overwound). In contrast, negative supercoils are formed if one end of a DNA molecule is turned in the same direction as the helix (thereby underwinding the DNA). In terms of the phone cord, if the cord naturally turns clockwise down its length, then turn one end clockwise several more times. Torsional stress results in this case as well, because the maneuver tries to decrease the number of turns in the helix. To maintain the same number of turns, the molecule must supercoil in the opposite direction and form negative supercoils, which reduce the torsional strain.

To put this in context, most DNA in nature is right-handed. Right-handed helical DNA turns clockwise when you look down the length of the double strand. Rotating one end clockwise will underwind the DNA. The underwound double-stranded DNA (dsDNA) will twist counterclockwise in a negative supercoil to relieve the stress. In contrast, rotating one end counterclockwise will overwind

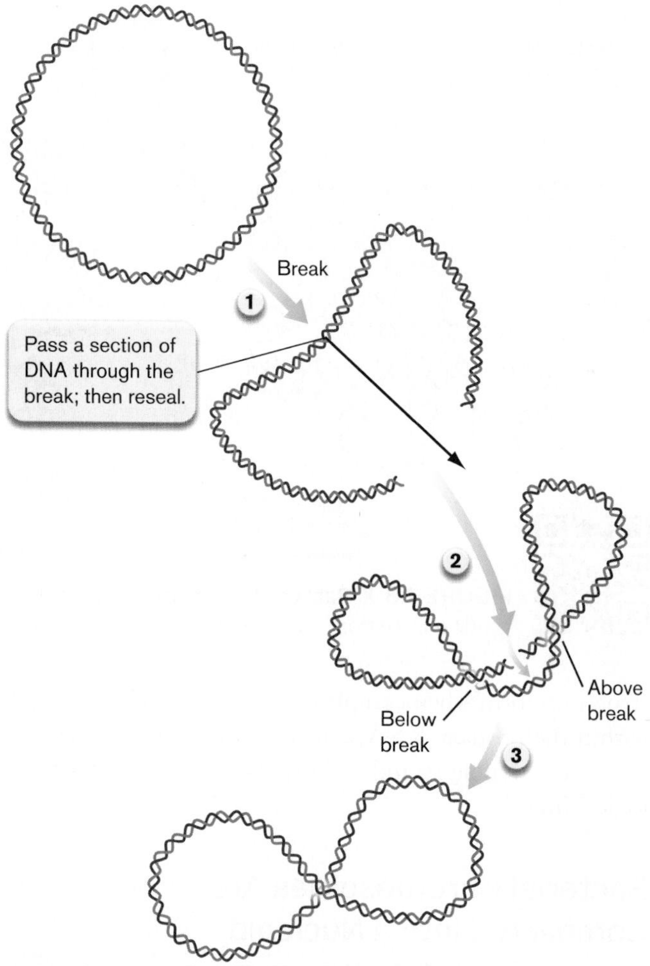
Break
1
Pass a section of DNA through the break; then reseal.
2
Above break
Below break
3

FIGURE 7.9 ▪ Supercoiling of 300-bp circular DNA. A supercoil can be introduced into a double-stranded, circular DNA molecule by (1) cleaving both strands at one site in the molecule, (2) passing an intact part of the molecule between ends of the cut site, and (3) reconnecting the free ends.

FIGURE 7.10 ▪ Mechanism of action for type I topoisomerases (topo I of *E. coli*). Topoisomerase I relaxes a negatively supercoiled DNA molecule. From left to right: Circular, negatively supercoiled, double-stranded DNA (underwound) is nicked in a single strand by topoisomerase I. Release of intrinsic energy reintroduces a helical turn, and the molecule is released with one less negative supercoil.

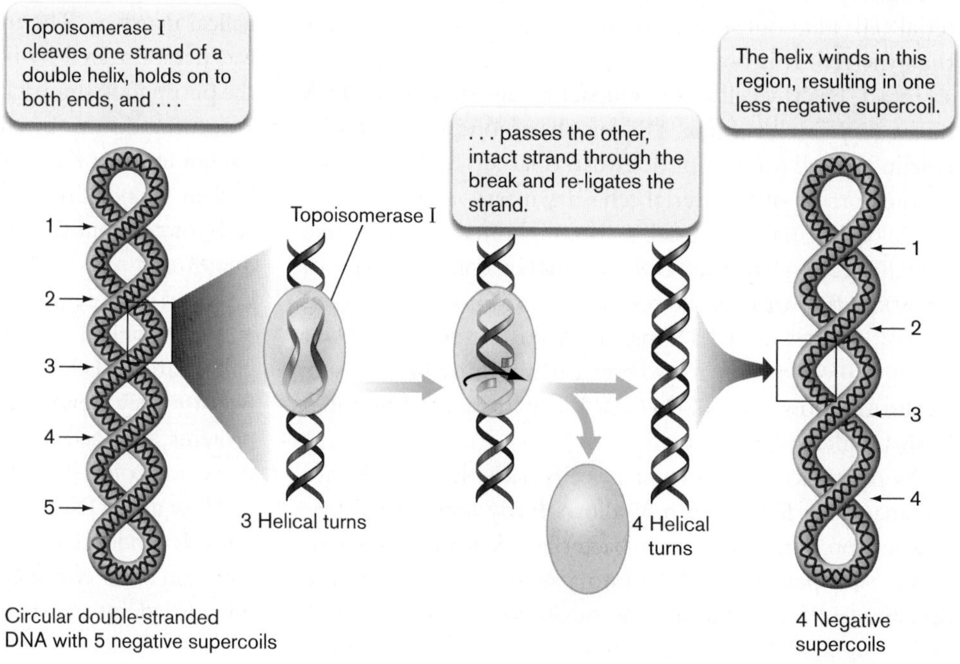
Topoisomerase I cleaves one strand of a double helix, holds on to both ends, and . . .
. . . passes the other, intact strand through the break and re-ligates the strand.
The helix winds in this region, resulting in one less negative supercoil.
Topoisomerase I
1
2
3
4
5
3 Helical turns
4 Helical turns
Circular double-stranded DNA with 5 negative supercoils
4 Negative supercoils

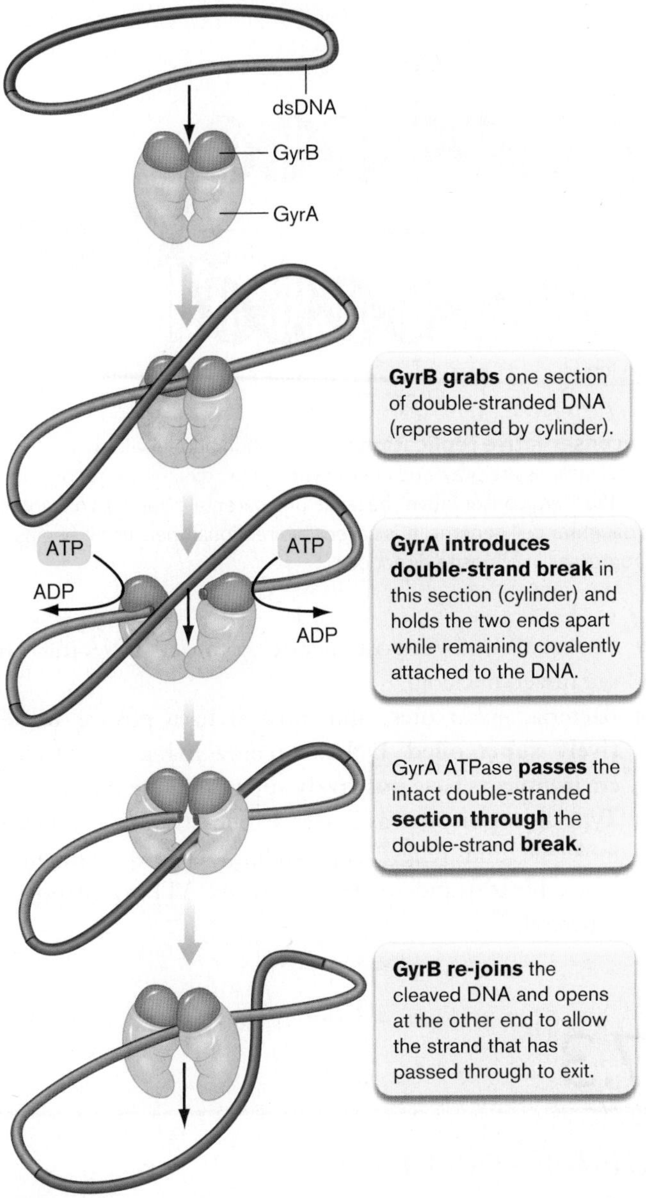
dsDNA
GyrB
GyrA
GyrB grabs one section of double-stranded DNA (represented by cylinder).
ATP
ATP
ADP
ADP
GyrA introduces double-strand break in this section (cylinder) and holds the two ends apart while remaining covalently attached to the DNA.
GyrA ATPase passes the intact double-stranded section through the double-strand break.
GyrB re-joins the cleaved DNA and opens at the other end to allow the strand that has passed through to exit.

FIGURE 7.11 ▪ Mechanism of action for type II topoisomerases (DNA gyrase of *E. coli*). The gyrase enzyme grabs DNA and, in an ATP-dependent process, introduces a double-strand break, passes another part of the double helix through the break, and then reseals the break. The end result is the introduction of a negative supercoil.

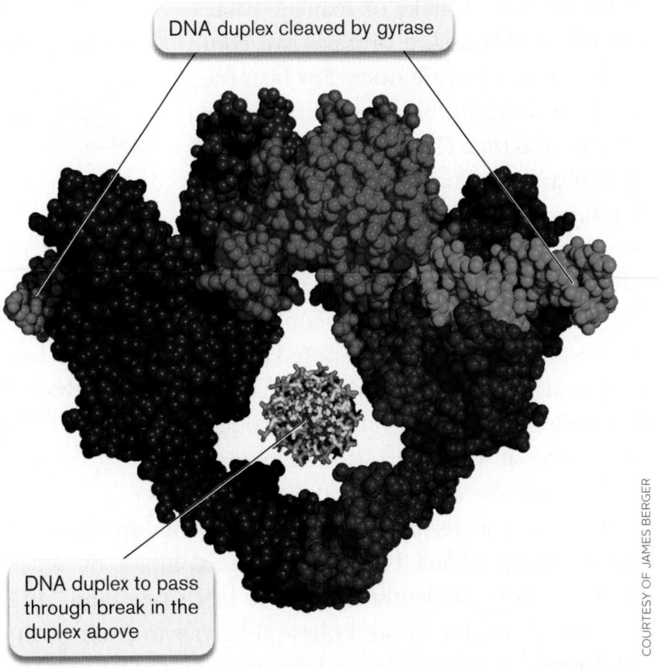
DNA duplex cleaved by gyrase
DNA duplex to pass through break in the duplex above
COURTESY OF JAMES BERGER

FIGURE 7.12 ▪ Three-dimensional representation of DNA gyrase. The 3D model of gyrase was determined using data from X-ray-crystallographic studies (blue and red regions). The gyrase complex is modeled in the process of gripping a broken DNA duplex (shown in green) and transporting a second duplex (the multicolored rosette).

the DNA. The overwound dsDNA will twist clockwise in a positive supercoil to relieve stress.

The nucleoids of bacteria and most archaea, as well as the nuclear DNA of eukaryotes, are kept negatively supercoiled. Because the DNA is underwound, the two strands of negatively supercoiled DNA are easier to separate than those of positively supercoiled DNA. This is important for transcription enzymes, like RNA polymerase, that must separate strands of DNA to make RNA.

Note, however, that some archaeal species living in acid at high temperature have nucleoids that are positively supercoiled to keep DNA double-stranded in these inhospitable environments (discussed next). Positively supercoiled DNA is harder to denature, because it takes excess energy to separate overwound DNA.

Topoisomerases Supercoil DNA

Supercoiling changes the topology of DNA. Topology is a description of how spatial features of an object are connected to each other. Thus, enzymes that change DNA supercoiling are called **topoisomerases**. To maintain proper DNA supercoiling levels, a cell must delicately balance the activities of two types of topoisomerases. Type I topoisomerases cleave only one strand of a double helix, while type II enzymes cleave both strands. Type I enzymes are generally used to relieve or unwind supercoils, while type II enzymes use energy to introduce them. **Figures 7.10** and **7.11** illustrate the mechanisms used by type I and type II topoisomerases, respectively. Type I enzymes are usually single proteins, while type II enzymes have multiple subunits. An example of a type II topoisomerase is DNA gyrase, whose function is to introduce negative supercoils in DNA (see **Fig. 7.11**). The active gyrase complex is a tetramer composed of two GyrA and two GyrB proteins. **Figure 7.12** shows a 3D representation of DNA gyrase in the midst of generating a supercoil.

Enzymes that make or manage bacterial DNA, RNA, and proteins are common targets for antibiotics. For instance, the **quinolone** antibiotics specifically target bacterial type II topoisomerases. They do not affect eukaryotic topoisomerases. A modern quinolone, ciprofloxacin, was the treatment of choice for anthrax pneumonia during the 2001 anthrax attacks. The progenitors of this drug family, nalidixic and oxolinic acids, were used to map *gyrA* and *gyrB*, the first drug resistance genes identified in *E. coli*. The modern successors of these drugs, the fluoroquinolones, are among the most widely used antimicrobials in the world. These drugs do not block topoisomerase action but stabilize the complex in which DNA gyrase is covalently attached to DNA (see **Fig. 7.11**). The stuck complex forms a physical barrier in front of the DNA replication complex, and the bacterial cell dies.

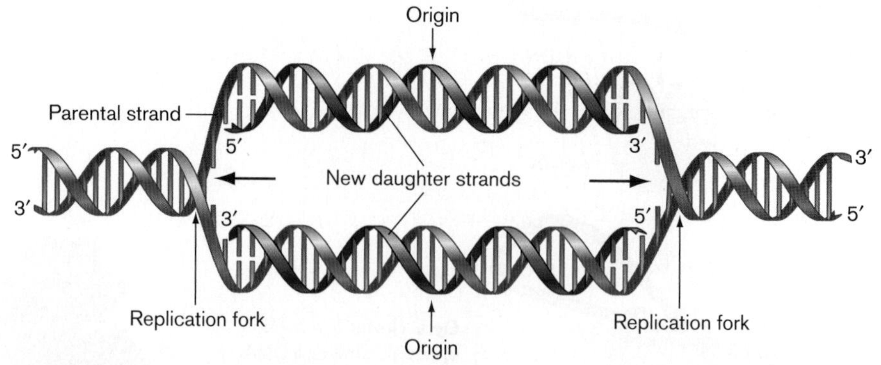
Origin
Parental strand
5′
3′
5′
3′
New daughter strands
3′
5′
3′
5′
Replication fork
Replication fork
Origin

FIGURE 7.13 ■ Semiconservative replication. A replication bubble with two replication forks. Parental strands are gray, and newly synthesized daughter strands are purple. Replication is called "semiconservative" because one parental strand is conserved and inherited by each daughter cell genome. It is called "bidirectional" because it begins at a fixed origin and progresses in opposite directions.

Extreme thermophiles (hyperthermophilic archaea) possess an unusual gyrase called reverse DNA gyrase. In contrast to the DNA gyrase from mesophiles, reverse gyrase introduces positive supercoils into the chromosome. It is proposed that tightening the coil helps protect the chromosome against thermal denaturation. Because the DNA has extra turns, it takes more energy (heat) to separate the strands.

Thought Questions

7.3 DNA gyrase is essential to cell viability. Why, then, are nalidixic acid–resistant cells that contain mutations in *gyrA* still viable?

7.4 Bacterial cells contain many enzymes that can degrade linear DNA. How, then, do linear chromosomes in organisms like *Borrelia burgdorferi* (the causative agent in Lyme disease) avoid degradation?

To Summarize

- **Noncoding DNA** can constitute a large amount of a eukaryotic genome, while prokaryotes have very little noncoding DNA.
- **DNA is composed of two antiparallel chains** of purine and pyrimidine nucleotides in which phosphate links the 5′ carbon of one nucleotide with the 3′ carbon of the next in the chain. The result is a double helix containing a deep major groove and a more shallow minor groove.
- **Hydrogen bonding and interactions between the stacked bases** hold together complementary strands of DNA.
- **Supercoiling** by topoisomerases compacts DNA into an organized nucleoid.
- Bacteria, eukaryotes, and most archaea possess **negatively supercoiled DNA**. Archaea living in extreme environments have **positively supercoiled genomes**.
- **Type I topoisomerases** cleave one strand of a DNA molecule and relieve supercoiling; **type II enzymes** cleave both strands of DNA and use ATP to introduce supercoils.

7.3 DNA Replication

How do bacteria replicate their DNA quickly and minimize errors? Microbial DNA needs to replicate itself as accurately and as quickly as possible so that the organism can grow and compete with other species. Replication efficiency is one reason why bacterial pathogens such as *Salmonella* can cause disease so quickly after ingestion. In this respect, bacteria differ from multicellular organisms, which need to regulate cell division carefully within their tissues because unregulated growth within tissues leads to cancer. The process of bacterial replication involves an amazing number of proteins and genes coming together in a complex machine. A list of 20 DNA replication components can be found in eTopic 7.1. Operation of the replication complex is all the more remarkable considering that some bacteria, such as thermophilic *Bacillus* species that live in hot springs, can double their population in less than 15 minutes.

The molecular details of bacterial DNA replication are important because they provide targets for new antibiotics and tools for biotechnology, such as the polymerase chain reaction (PCR) (Section 7.6). In addition, the proteins of

DNA repair have homologs in the human genome, defects which produce inherited human diseases, such as xeroderma pigmentosum, that predispose the carrier to certain cancers.

Overview of Bacterial DNA Replication

To replicate a molecule containing millions of base pairs poses formidable challenges. How does replication begin and end? How can two identical copies be generated? How is accuracy checked and maintained?

Semiconservative replication. Replication of cellular DNA in most cases is **semiconservative**, meaning that each daughter cell receives one parental strand and one newly synthesized strand (**Fig. 7.13**). At the **replication fork**, the advancing DNA synthesis machine separates the parental strands while extending the new, growing strands. The semiconservative mechanism provides a means for each daughter duplex to be checked for accuracy against its parental strand.

Enzymes that synthesize DNA or RNA can connect nucleotides together only in a 5′-to-3′ direction. That is, every newly made strand begins with a 5′ triphosphate and ends with a 3′ hydroxyl group (see **Fig. 7.4**). A polymerase (a chain-lengthening enzyme complex) fastens the 5′ alpha-phosphate of an incoming nucleoside triphosphate to the 3′ hydroxyl end of the growing chain, thus forming a phosphodiester bond. (The alpha-phosphate is the phosphate closest to the sugar.) This 5′-to-3′ enzymatic constraint produces an interesting mechanistic puzzle: If polymerases can synthesize DNA only in a 5′-to-3′ direction and the two phosphodiester backbones of the double helix are antiparallel, then how are both strands of a moving replication fork synthesized simultaneously? One strand presents no problem, because it is synthesized in a 5′-to-3′ direction toward the fork, but synthesizing the other new strand in a 5′-to-3′ direction would seem to dictate that it move away from the fork (**Fig. 7.14**). How is this possible?

Thought Question

7.5 Suppose you can label DNA in a bacterium by growing cells in medium containing either ^{14}N nitrogen or the heavier isotope, ^{15}N; you can isolate pure DNA from the organism; and you can subject DNA to centrifugation in a cesium chloride solution, a solution that forms a density gradient when subjected to centrifugal force, thereby separating the light (^{14}N) and heavy (^{15}N) forms of DNA to different locations in the test tube. Given these capabilities, how might you prove that DNA replication is semiconservative?

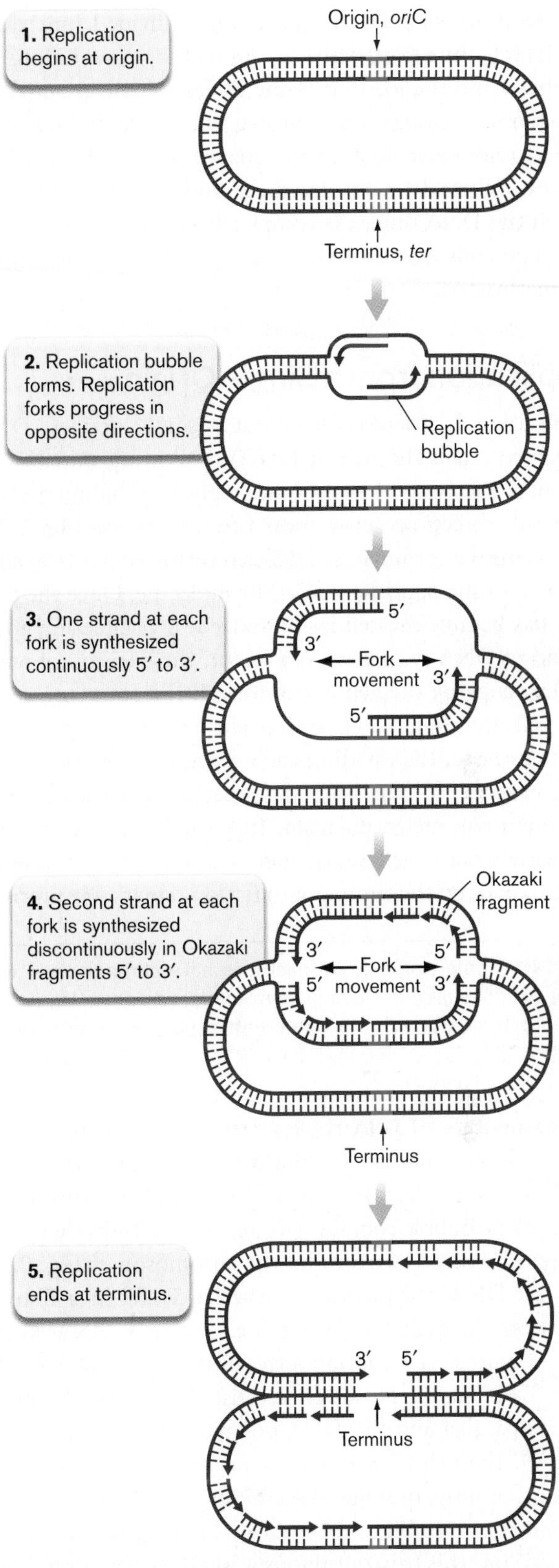
1. Replication begins at origin.
Origin, oriC
Terminus, ter
2. Replication bubble forms. Replication forks progress in opposite directions.
Replication bubble
3. One strand at each fork is synthesized continuously 5′ to 3′.
5′
3′
Fork movement
3′
5′
4. Second strand at each fork is synthesized discontinuously in Okazaki fragments 5′ to 3′.
Okazaki fragment
3′
5′
Fork movement
5′
3′
Terminus
5. Replication ends at terminus.
3′
5′
Terminus

FIGURE 7.14 ■ Comparing direction of fork movement with direction of DNA synthesis.

The process of DNA replication is divided into three phases: (1) initiation, which is the melting (unwinding) of the helix and the loading of the DNA polymerase enzyme complex; (2) elongation, which is the sequential addition of deoxyribonucleotides from deoxyribonucleoside triphosphates, followed by proofreading; and (3) termination, in which the DNA duplex is completely duplicated, the negative supercoils are restored, and key sequences of new DNA are methylated.

Replication from a Single Origin

Replication in bacteria begins at a single defined DNA sequence called the **origin (*oriC*)**. Following initiation, a circular bacterial chromosome replicates bidirectionally (in both directions away from the origin; see **Fig. 7.14**, step 1) until it terminates at defined **termination (*ter*) sites** located on the opposite side of the molecule. Once the process has begun, the cell is committed to completing a full round of DNA synthesis. As a result, the decision of when to start copying the genome is critical. If it starts too soon, the cell accumulates unneeded chromosome copies; if it starts too late, the dividing cell's septum "guillotines" the chromosome, killing both daughter cells. Consequently, elaborate fail-safe mechanisms link the initiation of DNA replication with cell mass, generation time, and cellular health, making the timing of initiation remarkably precise.

Note: Some single-celled microbes have chromosomes with more than one origin. The archaeon *Sulfolobus acidocaldarius*, for instance, has a single chromosome with three replication origins.

Fundamentals of DNA replication. The basic process of chromosome replication is outlined in **Figure 7.14**. After initiation of replication, a replication bubble forms at the origin. The bubble contains two replication forks that move in opposite directions around the chromosome (**Fig. 7.14**, step 2). DNA polymerases synthesize DNA in a 5′-to-3′ direction. At each fork, therefore, one new DNA strand can be synthesized continuously until the terminus region (**Fig. 7.14**, step 3). However, because the two DNA strands are antiparallel and the DNA polymerases synthesize only 5′ to 3′, the other daughter strand has to be synthesized discontinuously, in stages—seemingly backward relative to the moving fork (**Fig. 7.14**, step 4). The fragments of DNA formed on this discontinuously synthesized strand are called **Okazaki fragments**, after the Japanese scientists, Reiji and Tsuneko Okazaki, who discovered them. As we will discuss later, the Okazaki fragments are progressively stitched together to make a continuous, unbroken strand. Ultimately, the two replicating forks meet at the terminal sequence (**Fig. 7.14**, step 5), and the two daughter chromosomes separate.

Overall, copying the entire chromosome in *E. coli* takes about 40 minutes. Chromosome-partitioning processes then move each chromosome to different ends of the cell so that a cell wall can form at midcell (the cell "equator"). Once the cell wall is complete (the amount of time varies but is generally about 20 minutes), the two daughter cells, with their new chromosomes, can separate. So you might imagine the whole process from the start of replication to cell separation would take about 60 minutes for *E. coli*.

Under optimal growth conditions, however, *E. coli* cells divide in 20 minutes. How is this possible if replication takes at least 40 minutes? The answer is that a partially replicated chromosome can start new rounds of replication at the two daughter origins even before the first round is complete. By overlapping generations in this way, the cell can accommodate the 40-minute DNA replication time within the 20-minute generation time that was discussed in Section 3.4. During these rapid cell divisions, the DNA-partitioning mechanisms ensure that the two actively replicating chromosomes move to different ends of the cell, which keeps them from being severed when the cell septum forms.

Now let's examine each step in molecular detail to answer some important questions about mechanism.

Initiating Replication

What determines when replication begins? Initiation is controlled by DNA methylation, and by the binding of a specific initiator protein to the origin sequence. Further molecular events load the elaborate DNA polymerase complex and generate the first RNA primer for the new DNA strand. **Figure 7.15** presents an overview of the initiation process.

DNA methylation controls timing. The chromosomal origin of *E. coli* is a sequence of 245 base pairs designated *oriC*. It is subject to a critical molecular control mechanism that dictates the precise timing of replication initiation. Initiation of replication at *oriC* is activated by one protein, DnaA, and inhibited by another, SeqA. Immediately after a cell has divided, the level of active DnaA (DnaA bound to ATP) is low, and the inhibitor SeqA will bind to *oriC* and prevent ill-timed initiations (before the cell has grown enough to divide again).

How does SeqA know to bind just after the origin has replicated? The key is DNA methylation. *E. coli* uses the enzyme deoxyadenosine methylase (Dam) to attach a methyl group to the adenine residue at position N-6

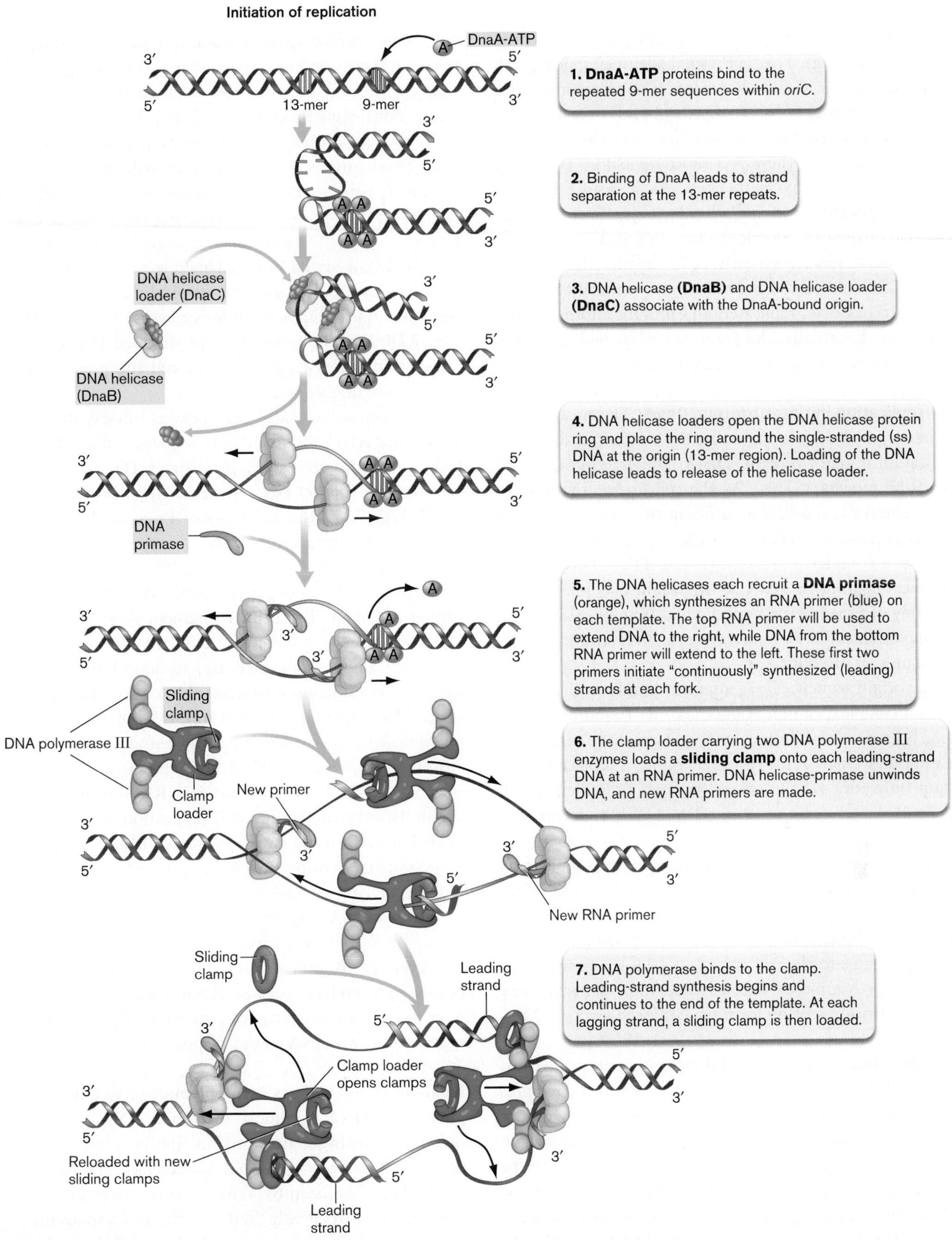
Initiation of replication
DnaA-ATP
A
3′
5′
5′
3′
13-mer
9-mer
1. DnaA-ATP proteins bind to the repeated 9-mer sequences within oriC.
2. Binding of DnaA leads to strand separation at the 13-mer repeats.
DNA helicase loader (DnaC)
DNA helicase (DnaB)
3. DNA helicase (DnaB) and DNA helicase loader (DnaC) associate with the DnaA-bound origin.
4. DNA helicase loaders open the DNA helicase protein ring and place the ring around the single-stranded (ss) DNA at the origin (13-mer region). Loading of the DNA helicase leads to release of the helicase loader.
DNA primase
5. The DNA helicases each recruit a DNA primase (orange), which synthesizes an RNA primer (blue) on each template. The top RNA primer will be used to extend DNA to the right, while DNA from the bottom RNA primer will extend to the left. These first two primers initiate "continuously" synthesized (leading) strands at each fork.
Sliding clamp
DNA polymerase III
Clamp loader
New primer
6. The clamp loader carrying two DNA polymerase III enzymes loads a sliding clamp onto each leading-strand DNA at an RNA primer. DNA helicase-primase unwinds DNA, and new RNA primers are made.
New RNA primer
Sliding clamp
Leading strand
7. DNA polymerase binds to the clamp. Leading-strand synthesis begins and continues to the end of the template. At each lagging strand, a sliding clamp is then loaded.
Clamp loader opens clamps
Reloaded with new sliding clamps
Leading strand

FIGURE 7.15 ■ Initiation of DNA replication. The start of DNA replication is precisely timed and linked to the ratio of DNA to cell mass. In *E. coli*, the initiator protein DnaA accumulates during growth and then triggers the initiation of replication. It begins with DnaA-ATP complexes binding to 9-mer (9-bp) repeats upstream of the origin (13-mer region). The binding of DnaA-ATP (along with other proteins not shown) first causes the DNA to loop in preparation for being melted open by the helicase (DnaB).

(**Fig. 7.4A**; see the N of the NH_2 group attached to the six-membered ring). The methylated adenines appear in all GATC sequences. GATC sequences (the recognition sites of Dam methylase) are scattered throughout the chromosome and occur on both strands. Just after the origin has replicated, however, there is a short lag before the newly synthesized strand is methylated. As a result, the origin is temporarily hemimethylated—a situation in which only one of the two complementary strands is methylated. Because SeqA has a high affinity for hemimethylated origins, this inhibitor will bind most tightly immediately after the origin has been replicated and prevent another initiation event. Eventually, the Dam methylase will methylate the new strand and decrease SeqA binding.

The replication initiator protein, DnaA. The onset of the initiation phase is determined by the concentration of the replication initiator protein, DnaA. DnaA recognizes specific 9-bp repeats at *oriC*. As the cell grows, the level of active DnaA rises until it is sufficient to bind to a series of 9-bp repeats at *oriC* (**Fig. 7.15** ▶, step 1). DnaA actually binds as a complex with ATP (DnaA-ATP). This binding facilitates melting of DNA at the origin and initiates the assembly of a membrane-bound replication hyperstructure, a complex assembly of numerous proteins forming a functional unit, at midcell.

The origin sequence (*ori*), after moving through the replication complex, cannot trigger another round of replication, because of inhibition by SeqA and decreasing levels of unbound DnaA-ATP. Another round of replication can begin only after (1) the origin becomes fully methylated, (2) SeqA dissociates, and (3) the DnaA-ATP concentration rises.

What happens to new replication origins after replication begins? **Figure 7.16** reveals that, even though the origin starts in the center of the cell, the newly replicated origins (green fluorescence) move toward opposite cell poles. Moving origins to the cell poles is part of the partitioning mechanism that moves chromosomes out of harm's way from the division septum forming at midcell (Section 3.6).

Initiation requires RNA polymerases. An unexpected feature of DNA replication is that its initiation actually requires RNA polymerases. The housekeeping RNA polymerase helps separate the strands of the DNA helix at the origin. A second RNA polymerase, DNA primase, produces the short primer, or starter, sequences needed to synthesize new DNA. The housekeeping RNA polymerase used to make most of the RNA in the cell (discussed in Chapter 8) transcribes DNA at *oriC* to produce RNA that helps separate the two DNA strands (**Fig. 7.15**, step 2; this RNA polymerase is not shown). Strand separation allows a special DNA helicase (protein DnaB), in association with a DNA helicase loader (DnaC), to bind the two replication forks formed during bidirectional replication (**Fig. 7.15**, step 3).

As the lead protein of the replication machine, the helicase (DnaB) uses energy from ATP hydrolysis to unwind the DNA helix as the DNA moves into the DNA polymerase replicating complex. The ringlike DnaB is assembled around one DNA strand at each replication fork. After loading DnaB at the origin, DnaC is released (**Fig. 7.15**, step 4). Coincident with the unwinding of DNA, small single-stranded DNA-binding proteins (SSBs, seen in **Fig. 7.18**) coat the exposed single-stranded DNA, protecting it from nuclease patrolling the cell and preventing reformation of double-stranded DNA.

DNA-dependent DNA polymerases possess the unique ability to "read" the nucleotide sequence of a DNA template and synthesize a complementary DNA strand. The discovery of this activity earned Arthur Kornberg (1918–2007; **Fig. 7.17**) the 1959 Nobel Prize in Physiology or Medicine. As remarkable as these enzymes are, no DNA polymerase can start synthesizing DNA unless there is a preexisting DNA or RNA fragment to extend—that is, a primer fragment. The primer fragment possesses a 3′ OH end that can receive incoming deoxyribonucleotides. Consequently, once DnaB (helicase) is bound to DNA, the next step in initiation is to make RNA primers at each fork (**Fig. 7.15**, step 5). In contrast to DNA polymerases, RNA polymerases can synthesize RNA without a primer. A specific RNA polymerase called DNA **primase** (DnaG) synthesizes short (10–12 nucleotides) RNA primers at the origin, thereby launching DNA replication. One primase is loaded at each of the two replication forks. Note that primase is different from the RNA polymerase involved in initially separating the DNA strands at the origin.

Why do DNA polymerases require RNA primers? RNA primers probably remain from when all life was RNA based. However, DNA polymerases did not evolve to become primer independent, because RNA primers help limit the start of replication to true origins rather than to wherever segments of two DNA strands happen to melt.

A sliding clamp tethers DNA polymerase to DNA. At this point, the DNA is almost ready for DNA polymerase. But first a **sliding clamp** protein (the beta subunit) must be loaded to keep the DNA polymerase affixed to the DNA (**Fig. 7.15**, step 6). Without this clamp, DNA polymerase would frequently "fall off" the DNA molecule (see eTopic 7.2). A multisubunit complex (called the clamp-loading complex) places the beta clamp, along with an attached pair of DNA polymerase molecules, onto DNA.

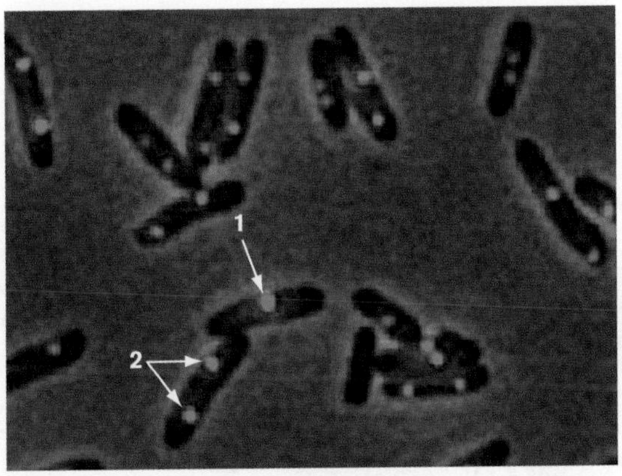
1
2

KRUSE, ET AL. 2006. *GENES AND DEVELOPMENT* **20**:113. PHOTO BY KENN GERDES.

FIGURE 7.16 ■ Movement of newly replicated origins. This photomicrograph uses fluorescent labeling to monitor the location of *E. coli* origins. The *oriC* loci on chromosomes were visualized through a GFP protein that binds only at *oriC*. Cells that have not started to replicate have a single *oriC* located midcell (1). Cells in the process of replicating their DNA show two *oriC* loci moving toward opposite cell poles (2).

COURTESY OF STANFORD UNIVERSITY

FIGURE 7.17 ■ Arthur Kornberg and his wife Sylvy Kornberg, circa 1960. A biochemist in her own right, Sylvy contributed in the effort to purify DNA polymerase I.

DNA polymerase (specifically DNA Pol III; discussed later) can then bind to the 3′ OH terminus of the primer RNA molecule and begin to synthesize new DNA (**Fig. 7.15**, step 7). The molecular structure of the beta clamp loaded onto DNA is shown in **eTopic 7.2**.

Elongation of Replicating DNA

Escherichia coli contains five different DNA polymerase proteins, designated Pol I through Pol V. All polymerases catalyze the synthesis of DNA in the 5′-to-3′ direction. However, only Pol III and Pol I participate directly in chromosome replication. The other polymerases conduct operations to rescue stalled replication forks and repair DNA damage.

DNA polymerase III. The main replication polymerase, Pol III, is a complex, multicomponent enzyme. Pol III was discovered by Arthur Kornberg, whose 1959 Nobel Prize had been awarded for his initial discovery, of Pol I (**Fig. 7.17**). Because of its complex nature, Pol III is often referred to as a "molecular machine." The DNA synthesis activity of Pol III is held in the alpha subunit of the complex, while other subunits are used for improving fidelity (accuracy of replication) and processivity (a measure of how long the polymerase remains attached to, and replicates, a template). The Pol III epsilon subunit (DnaQ), for example, contains a **proofreading** activity that corrects mistakes and improves fidelity.

Proofreading activities within DNA polymerases scan for mispaired bases that have been inappropriately added to a growing chain. A mispaired base mistakenly linked by a phosphodiester bond to a growing DNA chain is more mobile than the correct base because it does not properly hydrogen-bond to the template base. This motion halts DNA elongation by Pol III because the base is not properly positioned at the enzyme's active site. This stalling of Pol III activity triggers an intrinsic 3′-to-5′ **exonuclease** activity in the epsilon subunit. Exonucleases degrade DNA starting from either the 5′ end or the 3′ end. The exonuclease activity of Pol III cleaves the phosphodiester bond, releasing the improperly paired base from the growing chain. Once the wayward mispaired base has been excised, Pol III can resume elongation.

Both DNA strands are elongated simultaneously. After initiation, each replication fork contains one elongating 5′-to-3′ strand called the "leading strand" (look back at **Fig. 7.15**, step 7). But how is the opposite strand at each fork replicated? There are no known DNA polymerases capable of synthesizing DNA in the 3′-to-5′ direction, which would seem to be needed if both strands are to be synthesized simultaneously. DNA synthesis of one strand continuing all the way back to the origin is not a solution, because it would leave the unreplicated strand at each fork exposed to possible degradation for too long and would double the time needed to complete DNA replication. The cell has solved this dilemma by coordinating the activity of <u>two</u> DNA Pol III enzymes in one complex—one for each strand. The two associated Pol III complexes, together with DNA primase and helicase, form the **replisome**. As the dsDNA unwinds at the fork, the problem strand loops out and primase (DnaG) synthesizes a primer.

Elongation of DNA synthesis

1. The **leading-strand DNA Pol III** enzyme replicates the leading strand. **SSBs** cover and protect the unreplicated single strand. The DNA helicase remains on the lagging strand, unwinding the dsDNA moving into the replisome complex.

2. Lagging-strand DNA polymerase synthesizes the lagging strand, which loops out after passing through the polymerase.

3. After DNA helicase has moved approximately 1,000 bases, another **RNA primer** is synthesized on each lagging strand.

4. When the lagging-strand polymerase bumps into the 5′ end of a previously synthesized fragment, the **DNA polymerase is released** and the clamp is disengaged.

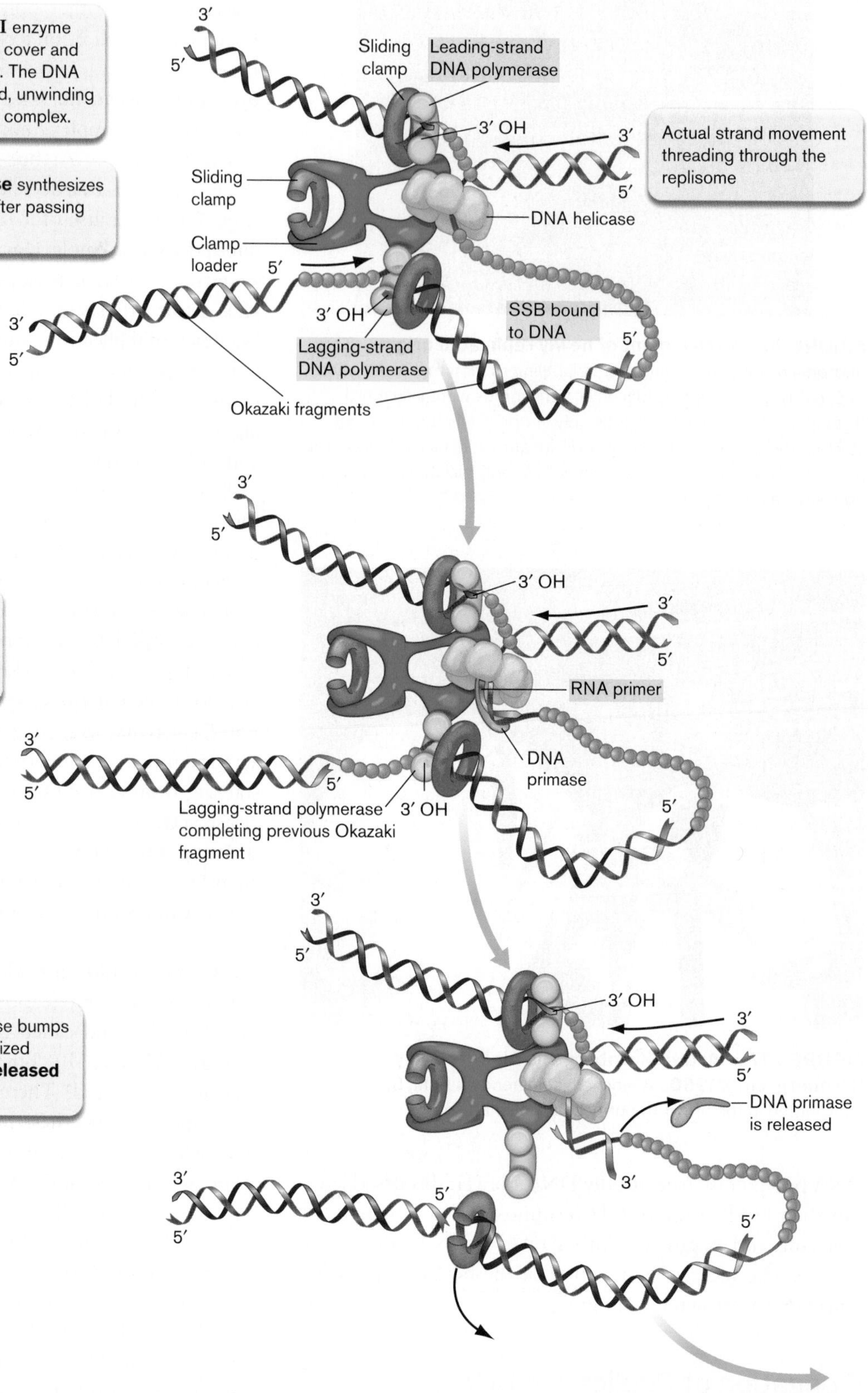
3′
5′
Sliding clamp
Leading-strand DNA polymerase
3′ OH
3′
Actual strand movement threading through the replisome
Sliding clamp
5′
DNA helicase
Clamp loader
5′
3′
5′
3′ OH
SSB bound to DNA
5′
Lagging-strand DNA polymerase
Okazaki fragments
3′
5′
3′ OH
3′
5′
RNA primer
3′
5′
5′
DNA primase
Lagging-strand polymerase completing previous Okazaki fragment
3′ OH
5′
3′
5′
3′ OH
3′
5′
DNA primase is released
3′
3′
5′
5′
5′

FIGURE 7.18 ■ The DNA polymerase dimer acting at a replication fork. The leading and lagging strands are synthesized simultaneously in the 5′-to-3′ direction. For clarity, the beta clamp on the lagging strand is shown on the opposite side of Pol III as compared to its position on the leading strand.

The second Pol III enzyme binds to the primed section of the loop and synthesizes DNA in the 5′-to-3′ direction (imagine the lower template strand in **Fig. 7.18** threading from left to right, through the polymerase ring). All the while, the second polymerase is moving in parallel with the first polymerase (on the leading strand) relative to the fork (**Fig. 7.18**, step 1). Realize that, in actuality, the replisome remains at a fixed, midcell location in the cell, attached to the membrane, and the template DNA threads through it as depicted in the chapter opener figure.

Note that simultaneous extension of the two strands requires that synthesis of the looped strand must lag behind the leading strand and that new RNA primers must be synthesized by primase (DnaG) every 1,000 bases or so. Thus, the lagging strand is synthesized discontinuously, in pieces called Okazaki fragments, while the leading strand can be synthesized continuously. As the leading strand moves forward, advancing the fork, there remains a long stretch of lagging strand complementary to the already replicated leading strand. This lagging strand remains single-stranded but protected by single-stranded DNA-binding proteins (SSBs) (**Fig. 7.18**, step 2). After about 1,000 bases, DNA primase reenters and synthesizes a new RNA primer in anticipation of lagging-strand DNA synthesis (**Fig. 7.18**, step 3). At some point, the lagging-strand polymerase bumps into the 5′ end of the previously synthesized fragment. This interaction causes DNA polymerase to disengage from that strand (**Fig. 7.18**, step 4), and the clamp loader loads a new clamp near the new RNA primer (**Fig. 7.18**, step 5). The DNA polymerase binds to that clamp and begins synthesizing another Okazaki fragment (**Fig. 7.18**, step 6). This process repeats every 1,000 bases or so around the chromosome.

The model just presented assumes that the replisome contains two DNA polymerase III molecules. Recent findings show, however, that the replisome actually consists of three polymerases—one on the leading strand and two on the lagging strand. The second polymerase on the lagging strand comes into play only when a large gap of unreplicated DNA remains on the lagging strand. For simplicity, the third polymerase was not included in the model.

DNA polymerase I. Discontinuous DNA synthesis results in a daughter strand containing long stretches of DNA punctuated by tiny patches of RNA primers. This RNA must be replaced with DNA to maintain chromosome integrity. To remove the RNA, cells typically use the 5′ to 3′ exonuclease activity of Pol I, or an RNase enzyme specific for RNA-DNA hybrid molecules (called RNase H). A DNA Pol I enzyme then synthesizes a DNA patch using the 3′ OH end of the preexisting DNA fragment as a priming site (**Fig. 7.19**). When DNA Pol I reaches the next fragment, the enzyme removes the 5′ nucleotide and resynthesizes it. This process of replicating DNA increases accuracy and decreases mutations.

Once DNA Pol I stops synthesizing, it cannot join the 3′ OH of the last added nucleotide with the 5′ phosphate of the abutting fragment. The resulting nick in the phosphodiester backbone is repaired by **DNA ligase**, which in *E. coli* and many other bacteria uses energy gained by cleaving nicotinamide adenine dinucleotide (NAD) to form the phosphodiester bond (see **Fig. 7.19**). NAD is not used in its usual way, as a reductant that oxidizes substrates. Energy inherent in the diphosphate bond of NAD is captured upon cleavage by DNA ligase and used to re-join the 3′ OH and 5′ phosphate ends present at the nick. DNA ligase from

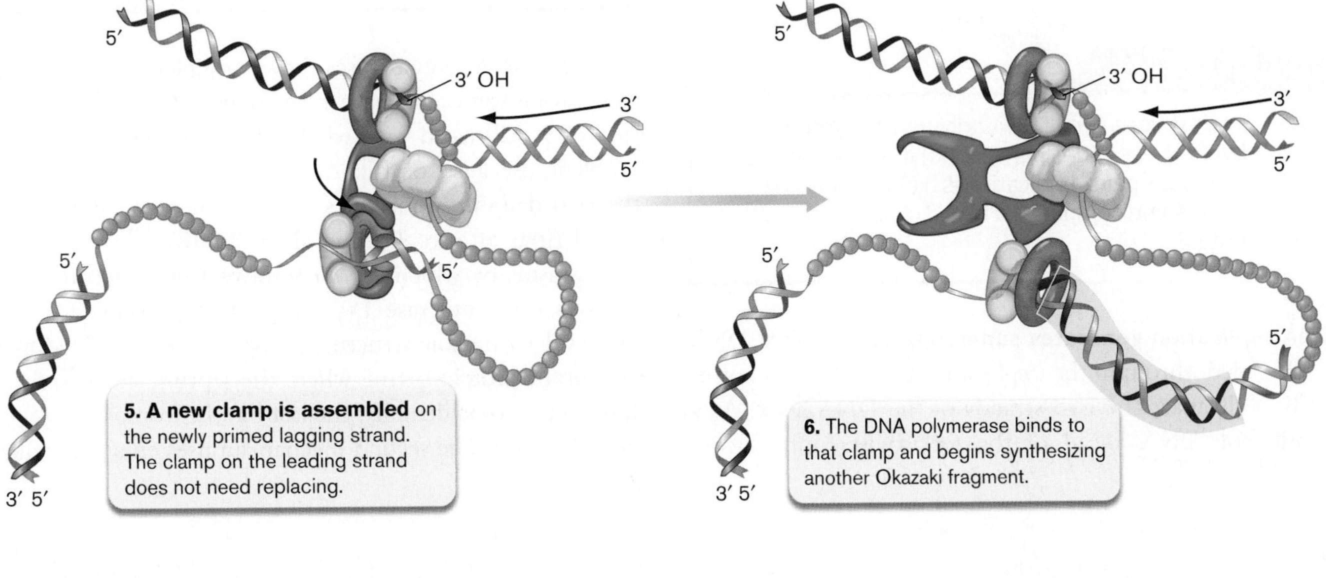
3′
5′
3′ OH
3′
5′
5′
5′
3′ 5′
5. A new clamp is assembled on the newly primed lagging strand. The clamp on the leading strand does not need replacing.
3′
5′
3′ OH
3′
5′
5′
5′
3′ 5′
6. The DNA polymerase binds to that clamp and begins synthesizing another Okazaki fragment.

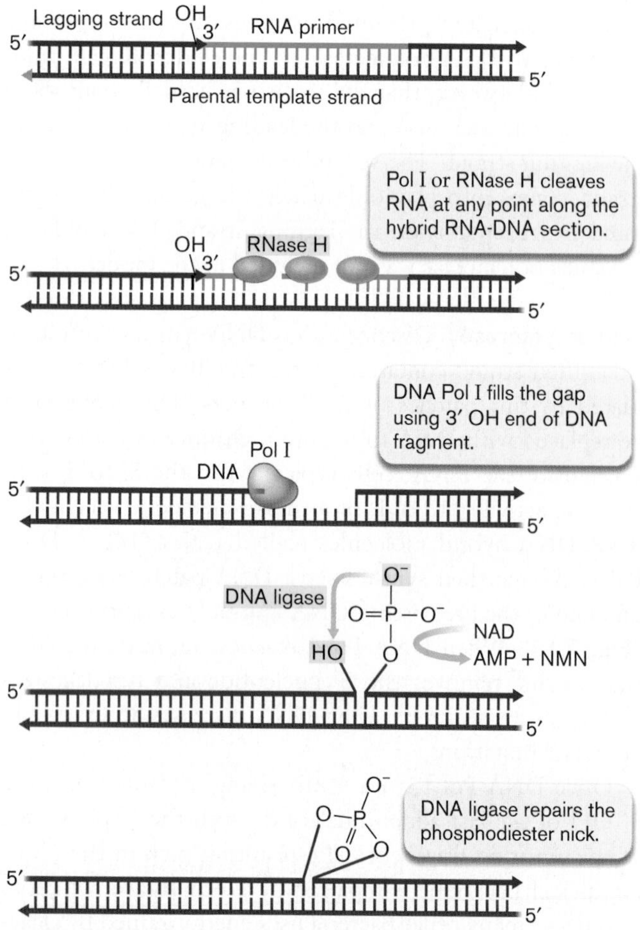

Lagging strand
OH
3′
RNA primer
5′
5′
Parental template strand
Pol I or RNase H cleaves RNA at any point along the hybrid RNA-DNA section.
OH
RNase H
3′
5′
5′
DNA Pol I fills the gap using 3′ OH end of DNA fragment.
Pol I
DNA
5′
5′
DNA ligase
HO
NAD
AMP + NMN
5′
5′
DNA ligase repairs the phosphodiester nick.
5′
5′

FIGURE 7.19 ■ Removing the RNA primer. The 3′-to-5′ exonuclease activity of Pol I or RNase H cleaves the RNA primer (blue). In either case, DNA polymerase I uses the preexisting 3′ OH end of the DNA fragment to fill the gap. Finally, DNA ligase repairs the phosphodiester nick using energy derived from the cleavage of NAD. AMP = adenosine monophosphate; NMN = nicotinamide monophosphate.

eukaryotes and some other microbes uses ATP rather than NAD in this capacity.

Thought Question

7.6 How fast does *E. coli* DNA polymerase synthesize DNA (in nucleotides per second) if the genome is 4.6 million base pairs, replication is bidirectional, and the chromosome completes a round of replication in 40 minutes?

DNA replication generates supercoils. As template DNA is threaded through the replisome, the helicase continually pulls apart the two strands of the DNA helix. As a result, the DNA ahead of the fork twists, introducing positive supercoils. (Try this yourself. Twist two pieces of string together, staple one end of the pair to a piece of cardboard, and then pull the two strands apart from the free end. Notice the supercoiling that takes place beyond, or downstream of, the moving fork.) The increasing torsional stress in the chromosome could stop replication by making strand separation more and more difficult. What prevents the buildup of torsional stress is the DNA gyrase (**Fig. 7.11**) that is located ahead of the fork, removing the positive supercoils as they form.

Terminating Replication

Bidirectional replication of a circular bacterial chromosome results in the two membrane-associated replication machines attempting to replicate through the same DNA sequences 180° from the origin—that is, halfway around the chromosome. What tells the polymerases to stop? The *E. coli* chromosome has as many as 10 terminator sequences (*ter*) that polymerases enter but rarely, if ever, leave (**Fig. 7.20A**). One set of terminators deals solely with the clockwise-replicating polymerase, while the other set halts DNA polymerases replicating counterclockwise relative to the origin. A protein called Tus (terminus utilization substance) binds to these sequences. Tus stops DnaB helicase activity and, thus, polymerization. Multiple terminator sites ensure that the polymerase complex does not escape and continue replicating DNA. Which terminator site is used depends in part on whether replication of one fork has lagged behind replication of the other.

Thought Question

7.7 Reexamine **Figure 7.16.** If you GFP-tagged a protein bound to the ter region, where would you expect fluorescence to appear in the cell with two origins?

Once DNA polymerases have completed duplicating the chromosome and have been removed at the *ter* sites, the cell is still faced with what could be called a "knotty" problem. Because of the topology of the chromosome, the two daughter molecules will appear as a **catenane** of linked rings after replication (**Fig. 7.20B**). There are two mechanisms by which *E. coli* resolves this structure. One involves topoisomerase IV, a type II topoisomerase that can resolve catenane structures. Topo IV is stimulated by a membrane protein sitting where the septum starts to form and where, coincidentally, the last bit of chromosome replication occurs. The second mechanism uses enzymes called

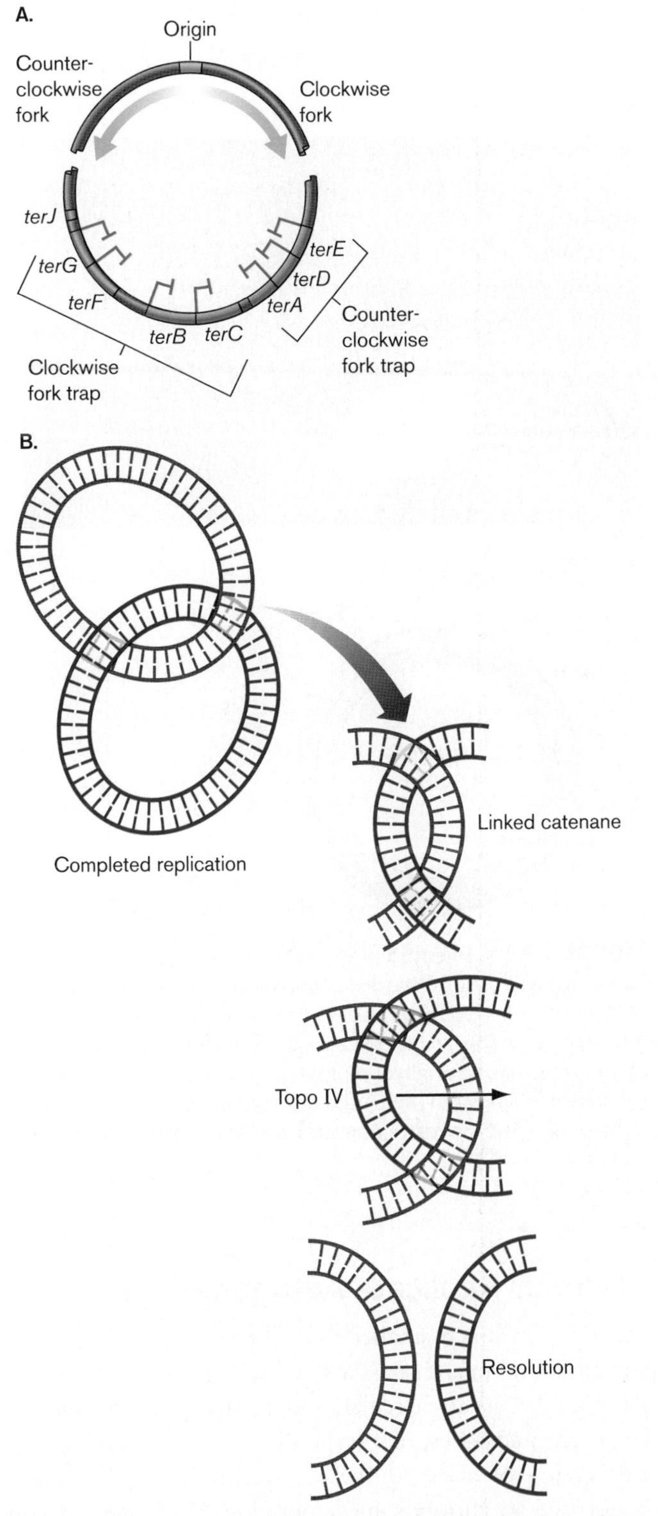
A.
Origin
Counter-
clockwise
fork
Clockwise
fork
terJ
terG
terF
terB
terC
terA
terD
terE
Counter-
clockwise
fork trap
Clockwise
fork trap
B.
Completed replication
Linked catenane
Topo IV
Resolution

FIGURE 7.20 ▪ Terminating replication of the chromosome. **A.** Terminator regions for DNA replication on the *E. coli* chromosome. Replication forks moving clockwise are trapped by *terJ, terG, terF, terB,* and *terC*. Counterclockwise-moving forks are trapped by *terA, terD,* and *terE*. Bent T's indicate which advancing polymerase is blocked. **B.** Resolution of DNA replication catenanes. On the left is a linked-chromosome catenane. The highlighted area is enlarged on the right. Topoisomerase IV catalyzes a breaking and re-joining event (like DNA gyrase) that passes the chromosomes through each other, resolving the link.

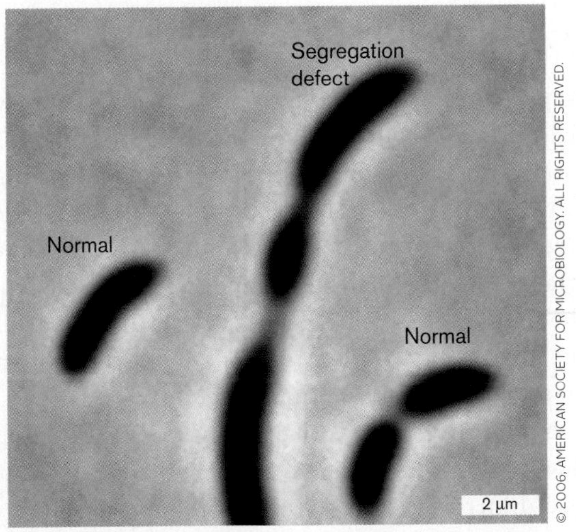
Segregation
defect
Normal
Normal
2 µm

FIGURE 7.21 ▪ Effect of a *xerD* mutation on *Caulobacter* cell division. *Caulobacter* replicates by producing a motile cell at one end of a nonmotile, fixed, stalked cell. The photomicrograph shows cells of a *xerD* mutant. Cells to the right are dividing normally using Topo IV to resolve catenanes. The middle cell is elongated with multiple constrictions because of the *xerD* defect in chromosomal segregation. Even though TopoIV is present in the *xerD mutant*, both mechanisms are needed to ensure that linked replicating chromosomes always clear the division plane. *Source:* R. B. Jensen. 2006. *J. Bacteriol.* **188**:6016–6019.

XerC and XerD, which recognize a specific site (called *dif*) on both DNA molecules and catalyze a series of cutting and re-joining steps that essentially pass one molecule through the other. Once the two daughter chromosomes have resolved and move toward the poles, the cell can begin to divide, forming the cell septum. **Figure 7.21** illustrates what happens with a *Caulobacter* mutant lacking XerD.

Thought Questions

7.8 Individual cells in a population of *E. coli* typically initiate replication at different times (asynchronous replication). However, depriving the population of a required amino acid can synchronize reproduction of the population. Ongoing rounds of DNA synthesis finish, but new rounds do not begin. Replication stops until the amino acid is once again added to the medium—an action that triggers simultaneous initiation in all cells. Why does this synchronization of reproduction happen?

7.9 The antibiotic rifampin inhibits transcription by RNA polymerase, but not by primase (DnaG). What happens to DNA synthesis if rifampin is added to a synchronous culture?

Weblinks DNA replication in archaea (*see ebook*)

To Summarize

- **Replication is semiconservative**, with newly synthesized strands lengthening in a 5′-to-3′ direction. Replication consists of initiation, elongation, and termination steps.
- **Replication is initiated** from a fixed DNA origin attached to the cell membrane. Initiation depends on the mass and size of the growing cell. It is controlled by the accumulation of initiator and repressor proteins and by methylation at the origin.
- **Elongation** requires that primase (DnaG) must lay down an RNA primer, DNA polymerase III must act as a dimer at each replication fork, and a sliding clamp must keep DNA Pol III attached to the template DNA molecule.
- **The 3′-to-5′ proofreading** activity of Pol III corrects accidental errors during polymerization.
- **DNA ligase** joins Okazaki fragments in the lagging strand.
- **Two replisomes, each containing a pair of DNA Pol III complexes**, are fixed at the membrane, and DNA feeds through them.
- **Termination** involves stopping replication forks halfway around the chromosome at *ter* sites that inhibit helicase (DnaB) activity.
- **Ringed catenanes** formed at the completion of replication are separated by topoisomerase IV and the proteins XerC and XerD.

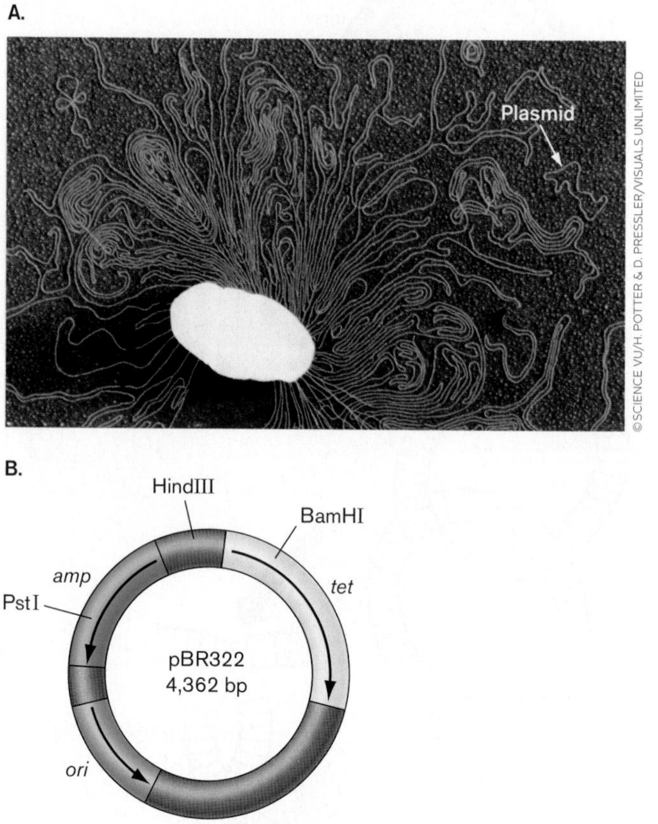
A.
Plasmid
©SCIENCE VU/H. POTTER & D. PRESSLER/VISUALS UNLIMITED
B.
HindIII
BamHI
amp
tet
PstI
pBR322
4,362 bp
ori

FIGURE 7.22 ■ Plasmid map. A. Note the huge difference in size between a circular plasmid DNA molecule and chromosomal DNA after both are gently released from a cell (approx. 1 μm). The arrow points to the circular plasmid. **B.** Map of plasmid pBR322. This plasmid contains an origin of replication (*ori*) and genes encoding resistance to ampicillin (*amp*) and tetracycline (*tet*). The locations of three unique 6-bp restriction sites are indicated (HindIII, BamHI, and PstI).

7.4

Plasmids

Two kinds of extragenomic DNA molecules can interact with bacterial genomes: horizontally transferred **plasmids** and the genomes of bacteriophages (viruses that infect bacterial cells). Plasmid-encoded functions can contribute to the physiology of the cell (for example, antibiotic resistance). In some cases the plasmid or phage DNA itself will integrate into the bacterial genome (see Section 9.2). Eukaryotic genomes and the genomes of eukaryote-specific plasmids and viruses are similarly subject to sharing proteins and chromosomal interactions, such as insertions. Viruses of eukaryotes and their applications in genetic engineering are discussed in greater detail in Chapters 11 and 12. This section discusses aspects of plasmid replication. Discussion of bacteriophage replication can be found in Chapter 6 and **eTopic 7.3**.

Plasmids Replicate Autonomously

Plasmids are much smaller than chromosomes (**Fig. 7.22**) and are found in archaea, bacteria, and eukaryotic microbes. Plasmids are usually circular, and circular plasmids, like circular chromosomes, are typically negatively supercoiled. Replication of many of these extrachromosomal elements is not tied to chromosome replication. Each plasmid contains its own origin sequence for DNA replication, but only a few of the genes needed for replication. Thus, even when the timing of plasmid replication is not linked to that of the chromosome, many of the proteins used for plasmid replication are actually host enzymes. Which host proteins are used depends on the plasmid.

Plasmids can replicate in two different ways. Bidirectional replication starts at a single origin and occurs in two directions simultaneously, while rolling-circle replication (**Fig. 7.23**) is unidirectional. In rolling-circle

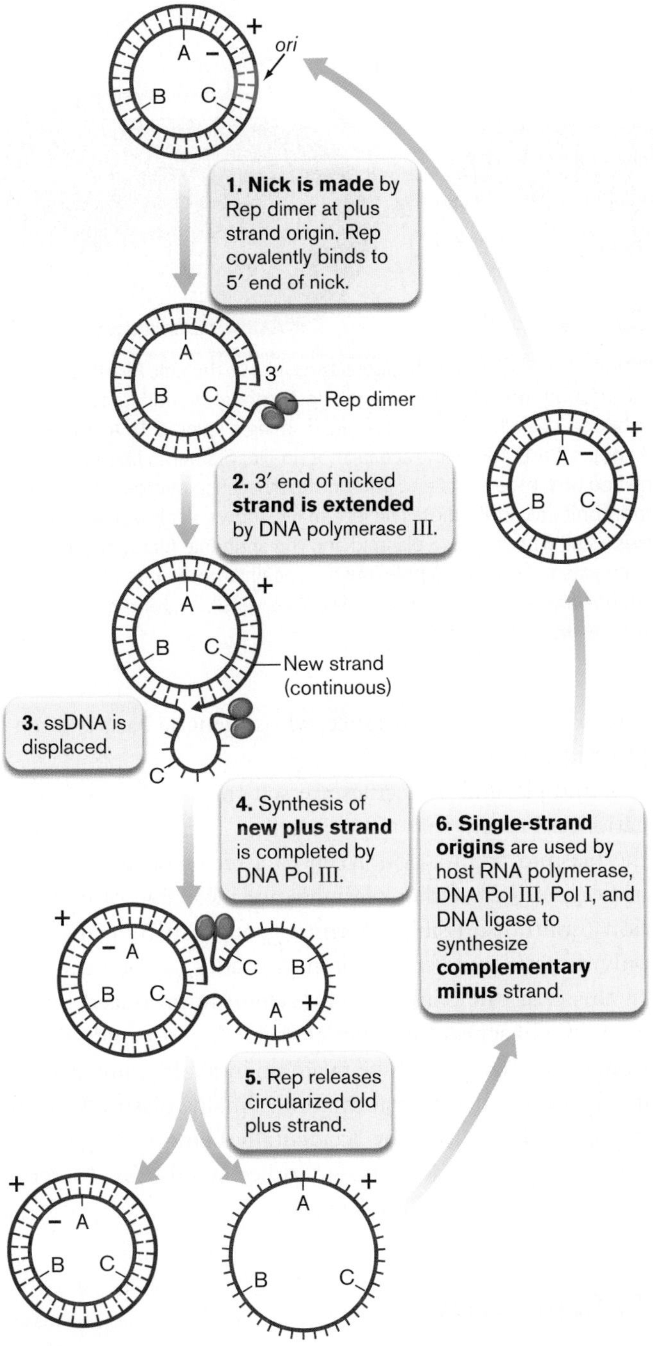
ori
A
B
C
+
1. Nick is made by Rep dimer at plus strand origin. Rep covalently binds to 5′ end of nick.
3′
Rep dimer
2. 3′ end of nicked strand is extended by DNA polymerase III.
New strand (continuous)
3. ssDNA is displaced.
4. Synthesis of new plus strand is completed by DNA Pol III.
6. Single-strand origins are used by host RNA polymerase, DNA Pol III, Pol I, and DNA ligase to synthesize complementary minus strand.
5. Rep releases circularized old plus strand.

FIGURE 7.23 ▪ Plasmid replication: rolling-circle model. Replication begins at a fixed origin but moves in only one direction. As the polymerase moves, it displaces one strand as it replicates the other. At the end of a round, the polymerase releases the displaced strand as a single-stranded circle that is then replicated.

replication, a replication initiator (RepA, encoded by a plasmid gene) binds to the origin of replication, nicks one strand, and holds on to one end (5′ PO_4) of it while the other end (3′ OH) serves as a primer for host DNA polymerase to replicate the intact, complementary strand. The RepA initiator protein recruits a helicase that unwinds DNA, which becomes coated by single-stranded DNA-binding proteins. As replication proceeds, the nicked strand progressively peels off without replicating until the strand is completely displaced. Then the two ends of the nicked single strand are re-joined by the Rep protein and released. The single-stranded circular molecule is protected by host single-stranded DNA-binding proteins until host enzymes replicate a complementary strand and regenerate a double-stranded molecule. Although most known plasmids use only one of the two replication strategies (bidirectional or unidirectional), a few can use either one, depending on the circumstance of the cell.

Plasmids Have Tricks to Ensure Their Inheritance

If plasmids replicate autonomously, can cells easily lose their plasmids? When a plasmid-containing host cell divides, is it just chance that determines whether both daughter cells inherit the plasmid? In some instances, the answer is yes. Some plasmids make many copies of themselves before the bacterial cell divides. These are called high-copy-number plasmids and can attain 50–700 copies per cell, depending on the plasmid. As a result, there is a high probability that both daughter cells will, by random chance, contain at least one plasmid.

Not all plasmids are high-copy-number, however. Low-copy-number plasmids limit how many copies they make to avoid draining the cell of energy. Cells forced to waste energy making many plasmid copies could be at a growth disadvantage when competing with plasmid-less cells in a natural environment. Low-copy-number plasmids have evolved clever partitioning systems that ensure both daughter cells will contain copies of the plasmid. **Figure 7.24** illustrates a model for how one type of partitioning works for plasmid R1, a *Salmonella* plasmid that imparts multidrug antibiotic resistance to host bacteria. The process employs three known plasmid-encoded genes: *parC, parM,* and *parR*. The *parC* DNA sequence is analogous to the centromere of eukaryotic chromosomes. The DNA-binding protein ParR binds to the *parC* sequence and forms a ParR-*parC* plasmid complex. ParM protein is an actin-like molecule that forms long filaments as it hydrolyzes ATP. The ParM filaments are dynamically unstable, constantly elongating and shortening. However, each end of a ParM actin-like filament can bind to a ParR-*parC* plasmid complex. When both ends of a ParM filament contact ParR-*parC* plasmid complexes, the filament stabilizes and elongates until the two plasmids hit the opposite poles of the cell. The plasmid is thus dislodged from the filament, and the filament quickly dissociates.

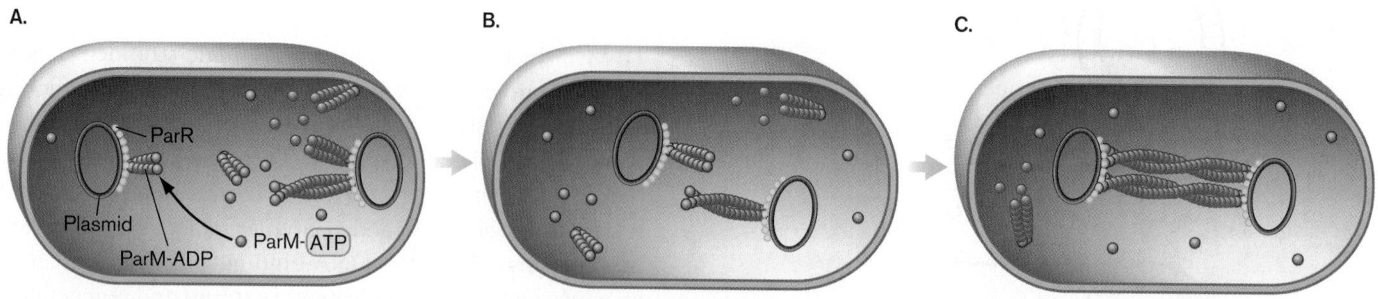
A.
B.
C.
ParR
Plasmid
ParM-ADP
ParM-ATP

FIGURE 7.24 ■ Molecular model of plasmid segregation. **A.** Nucleation of new filaments happens throughout the cell. Filaments attached to one plasmid will search for a second plasmid. **B.** Plasmids diffuse around the cell until they get close enough to encounter each other. **C.** When two plasmids come within close proximity, filaments will be bound at each end by a plasmid, forming a spindle. **D.** As these stabilized filaments polymerize, the two plasmids will be forced to opposite poles. ParM forms intracellular actin-like filaments like those shown in the two images at far right. ParM localization was seen with combined phase-contrast and immunofluorescence microscopy using rabbit anti-ParM antibodies and fluorescent Alexa 488–conjugated goat antirabbit IgG antibodies. The left image shows a cell with a *parM* mutant plasmid unable to make an actin-like filament. The right image shows a cell with a *parM+* plasmid and the actin-like filament (green) produced to partition daughter plasmids (equivalent to panel D). **E.** After each plasmid reaches a pole, pushing against both ends of the cell causes the filament to dissociate from the plasmid at one end and quickly depolymerize. *Source:* Adapted from Christopher S. Campbell and R. Dyche Mullins. 2007. *J. Cell Biol.* **179**:1059–1066, Fig. 4. Micrographs from J. Møller-Jensen et al. 2002. *EMBO J.* **21**:3119–3127.

Some plasmids ensure their inheritance by carrying genes whose functions benefit the host bacterium under certain conditions. For instance, bacterial plasmids can carry genes that confer antibiotic resistance (discussed in Chapter 28). As long as the antibiotic is present in the environment, any cell that loses the plasmid will be killed or stop growing. Antibiotic resistance plasmids benefit bacteria, but they are a major problem for modern hospitals, where plasmids carrying multiple drug resistance genes are transmitted from harmless bacteria into pathogens. On the other hand, plasmids (such as pBR322) containing drug resistance genes are the workhorses of genetic technology and have benefited society tremendously.

Other kinds of host survival genes carried by plasmids include genes providing resistance to toxic metals, genes encoding toxins that aid pathogenesis, and genes encoding proteins that enable symbiosis. These are discussed in later chapters. Most of the genes involved in the nitrogen-fixing symbiosis of *Rhizobium,* for example, are plasmid-borne. Far from being freeloaders, plasmids often contribute significantly to the physiology of an organism.

Some plasmids, however, use what could be called "strong-arm measures" to assure their maintenance. These plasmids come equipped with self-preservation genes, called "addiction modules," that force the host cell to keep the plasmid or die (**eTopic 7.4**).

Plasmids Are Transmitted between Cells

Plasmids have played a pivotal role in evolution because of how easily they can be passed back and forth between bacterial species. This is particularly evident today in the spread of antibiotic resistance, whose genetic basis is often a plasmid-encoded gene.

Some plasmid molecules are self-transferable via conjugation, a process that requires cell-to-cell contact to move the plasmid from a donor cell to a recipient (as discussed in Section 9.2). Other plasmids are incapable of conjugation (nontransmissible). A third group can be transferred only if a self-transferable plasmid resides in the same cell. In this case, the conjugation mechanism produced by one plasmid will act on the other plasmid. Any plasmid released from dead cells can also be taken up intact by some bacteria in a process called transformation. Finally, plasmids can be transmitted in nature by accidentally being packaged into bacteriophage head coats—in other words, by bacteriophage transduction (Section 9.2).

To Summarize

- **Plasmids are autonomously replicating** circular or linear DNA molecules that are part of a cell's genome.
- **Plasmids replicate** by rolling-circle and/or bidirectional mechanisms.
- **Plasmids can be transferred** between cells.

7.5

Eukaryotic Chromosomes

The chromosomes of eukaryotic microbes such as protists and algae have much in common with those of prokaryotes.

D. E.

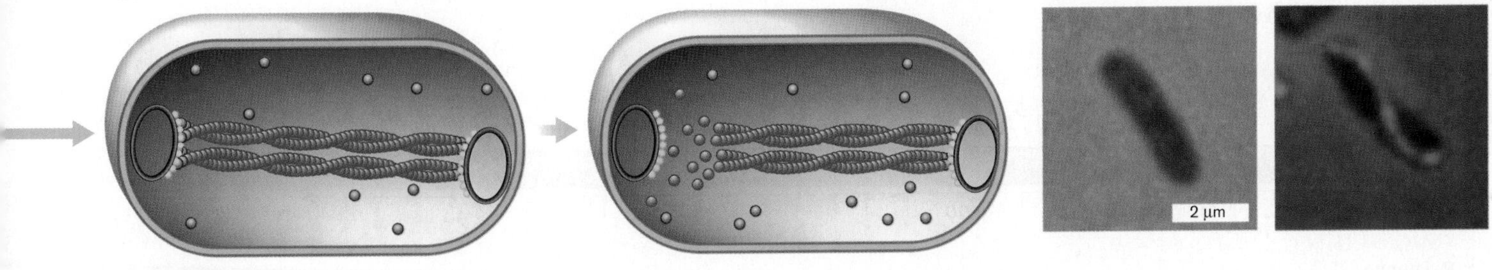
2 μm

Both consist of double-stranded DNA, for example, and both usually replicate bidirectionally. Nevertheless, important differences exist as well, particularly in genome structure. Eukaryotic chromosomes are linear and contained within a nucleus, and their replication involves mitosis (reviewed in Appendix 2).

Eukaryotic Genomes Are Large and Linear

Overall, the genomes of eukaryotic nuclei are larger than those of bacteria, sometimes by several orders of magnitude. For example, *Saccharomyces cerevisiae* and *Mycoplasma pneumoniae* genomes are 12,052 kb and 800 kb, respectively. In addition, most eukaryotic chromosomes are linear, whereas most (though not all) bacterial chromosomes are circular. Eukaryotic chromosomes require segregation by mitosis, in order to ensure that each daughter cell receives the correct combination of daughter chromosomes.

Because their genomes are linear, eukaryotes require a special process to duplicate the chromosome ends, providing a primer for the lagging strand. The primer is provided by a special enzyme called telomerase. At each end of the chromosome (called a telomere), there often is insufficient room on the lagging strand to add an appropriate RNA primer. So after each round of replication, the chromosome would be a little shorter and information would eventually be lost. Telomerase is actually a reverse transcriptase, an enzyme that reads RNA as a template to synthesize a complementary DNA molecule. Telomerase uses an intrinsic RNA (an RNA that is part of the enzyme) as a template to add DNA repeat sequences to the end of a lagging strand forming the telomere. The telomere allows for an RNA primer to be synthesized so that the end of the chromosome can be replicated. In this way, genetic information is not lost during division.

Telomerase may have evolved from the ancient progenitor cells that contained RNA rather than DNA genomes (see Section 17.2). For cells with RNA genomes to evolve modern chromosomes made of DNA, a reverse transcriptase must have been necessary. The reverse transcriptase may also be the evolutionary source of retroviruses (see eTopic 6.1).

Eukaryotic cells pack their DNA within the confines of a nucleus, using a series of proteins called **histones**. Histones are rich in arginine and lysine, so they are positively charged, basic proteins that easily bind to the negatively charged DNA. The DNA becomes wrapped around the histones to form units called nucleosomes. For protection and condensation, all eukaryotic chromosomes are packaged by histones, which play a regulatory role through methylation and acetylation. Bacteria, too, have DNA-packaging proteins, but they are more diverse and less essential for function (see Sections 3.4 and 7.2).

The detailed structures of the genomes of eukaryotes and prokaryotes reveal surprising differences (**Fig. 7.25**). The genome of the bacterium *E. coli* is packed with genes (green in the figure) encoding proteins or RNA molecules. These genes are separated by very little unused or noncoding DNA (purple), and by only an occasional mobile element, such as an insertion sequence, that can move from one DNA molecule to another. As we learned in Section 7.2, most prokaryotic genes are organized in coordinately regulated operons; one example is *thrABC*, which encodes three enzymes needed to synthesize the amino acid threonine. In contrast, 95% of the human genome consists of noncoding sequences composed largely of regulatory sequences and the fossil genomes of ancient viruses. The coding genes are separated by large stretches of noncoding sequences and usually are not situated together in operons. Moreover, coding genes in the human genome are interrupted by **introns** (DNA within a gene that is not part of the coding sequence for a protein, shown as yellow in **Fig. 7.25**) and ancient gene duplications that have decayed into nonfunctional, vestigial **pseudogenes** (orange). Bacteria also have pseudogenes, but they account for much less of the genome than in eukaryotes. The more rapid replication of bacteria causes pseudogenes to be lost more quickly.

Note: Pseudogenes differ from noncoding DNA in that at least part of a pseudogene's sequence is similar to that of a gene with a known function. Also note that some pseudogenes express mRNA, but the proteins produced have no obvious function.

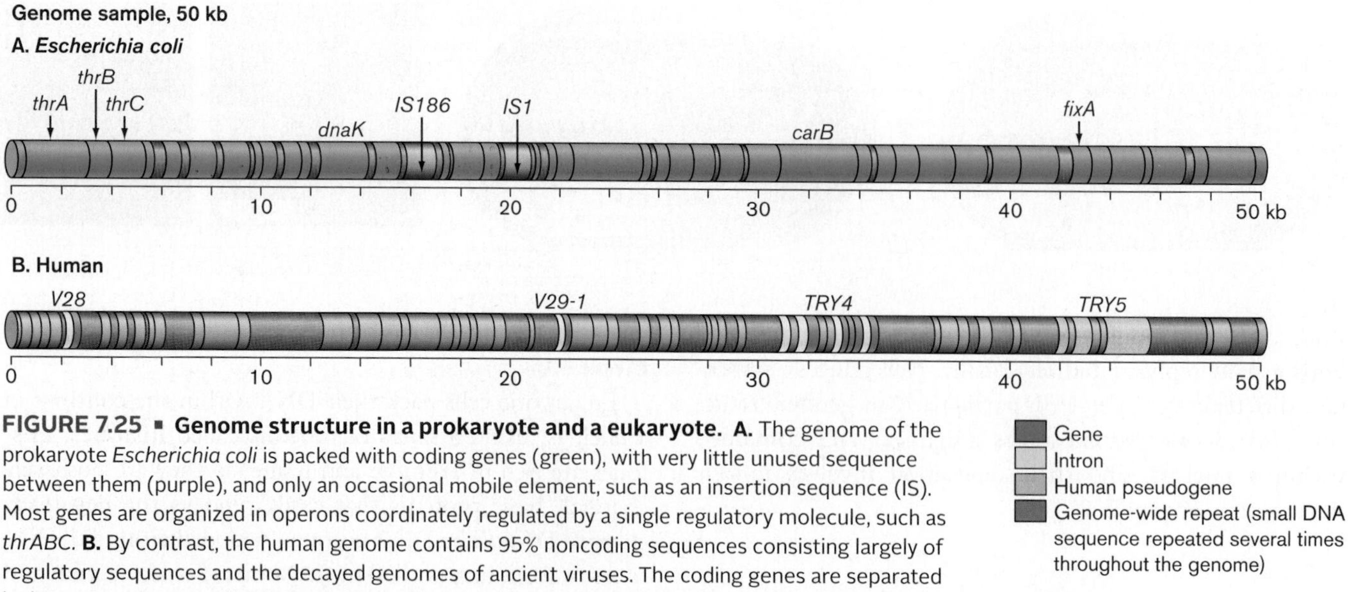
Genome sample, 50 kb
A. *Escherichia coli*
thrB
thrA
thrC
IS186
IS1
dnaK
carB
fixA
0
10
20
30
40
50 kb
B. Human
V28
V29-1
TRY4
TRY5
0
10
20
30
40
50 kb
Gene
Intron
Human pseudogene
Genome-wide repeat (small DNA sequence repeated several times throughout the genome)

FIGURE 7.25 ■ Genome structure in a prokaryote and a eukaryote. A. The genome of the prokaryote *Escherichia coli* is packed with coding genes (green), with very little unused sequence between them (purple), and only an occasional mobile element, such as an insertion sequence (IS). Most genes are organized in operons coordinately regulated by a single regulatory molecule, such as *thrABC*. **B.** By contrast, the human genome contains 95% noncoding sequences consisting largely of regulatory sequences and the decayed genomes of ancient viruses. The coding genes are separated by large stretches, usually are not organized in operons, and tend to be interrupted by introns (yellow). Ancient gene duplications have decayed into pseudogenes (orange).

The amount of non protein-coding DNA present in a eukaryote varies with the complexity of the species (for example, yeasts contain a much lower percentage of non protein-coding DNA than humans have). Eukaryotes may have retained non protein-coding DNA in their genomes because they rely heavily on it for survival, probably for untranslated regulatory RNAs (discussed in Chapter 10).

By themselves, promoters of eukaryotic genes have very low activity and require enhancer DNA sequences to drive transcriptional activity. Enhancer sequences act at long distances from promoters, and scientists suspect that once enhancers became important, it was necessary to place enough DNA between them to reduce the activation of other, unrelated but adjacent promoters.

Archaeal Genomes Combine Features of Bacteria and Eukaryotes

Like bacteria, archaea have polygenic operons, and their reproduction is predominantly asexual. Archaea are true prokaryotes because their cells lack a nuclear membrane and they generally have a single circular chromosome. In most species of archaea, however, the structures of the DNA-packing proteins, RNA polymerase, and ribosomal components more closely resemble those of eukaryotes. Even the DNA polymerase and the origin recognition sequence show greater similarity to those of eukaryotes. Finally, it should be noted that archaeal genomes encode certain unique components, such as the metabolic pathway of methanogenesis. (For more on archaea, see Chapter 19.)

What experimental data allow us to make such comparisons? Overwhelmingly, we rely on new data from the growing number of microbial genomes sequenced. Comparison of genomes has revolutionized our understanding of evolutionary relationships among microbes.

We will now examine the tools of DNA sequence analysis that have made these studies possible. The most important of these tools—restriction mapping, DNA sequencing, and amplification by the polymerase chain reaction—actually harness the molecular apparatus used by bacteria to replicate or protect their own chromosomes.

To Summarize

- **Eukaryotic chromosomes** are always linear, double-stranded DNA molecules that replicate by mitosis.
- A **reverse transcriptase** called telomerase is needed to finish replicating the ends of a eukaryotic chromosome.
- **Histones** (eukaryotic DNA-packing proteins) play a critical role in forming chromosomes.
- **Introns and pseudogenes** are noncoding DNA sequences that make up a large portion of eukaryotic chromosomes.
- **Archaeal chromosomes** resemble those of bacteria in size and shape, but archaeal DNA polymerases are more closely related to eukaryotic enzymes.

7.6

DNA Sequence Analysis

How do we study a genome? We have just described the core concepts of DNA structure, packaging, and replication. This knowledge is crucial to understanding genomics and the fundamentals of genomic analysis. But what about the basic techniques used to manipulate DNA? These include isolating genomic DNA from cells, snipping out DNA fragments with surgical precision, splicing them into plasmid vehicles, and reading their nucleotide sequences. These are the techniques that drove the genomic revolution.

DNA Isolation and Purification

The chemical uniformity of DNA means that simple and reliable purification methods can be used to isolate it. A variety of techniques are used to extract DNA from bacterial cells. The cells may be lysed using lysozyme to degrade the peptidoglycan of the cell wall, followed by treatment with detergents to dissolve the cell membranes. Next, most of the proteins are removed by precipitating them in a high-salt solution. The precipitated proteins are then removed by centrifugation, and the cleared lysate containing DNA is passed through a column containing a silica resin that specifically binds DNA. The remaining proteins are washed out of the column because they do not stick, and the DNA is eluted with water. The DNA is then concentrated via alcohol precipitation (ethanol or isopropanol). DNA may be precipitated from aqueous solution by adding ethanol and salt because the charge density of the phosphates makes the molecule particularly insoluble in nonaqueous solvents. The precipitated DNA is then redissolved in water or a weak buffer. At this point, the extracted DNA can be examined with a variety of analytical tools. There are other methods for isolating plasmid DNA. (See **eTopic 7.5** for a discussion of equilibrium density gradient centrifugation and its use in the classic Meselson-Stahl experiment.)

Weblinks DNA purification protocols (*see ebook*)

Restriction Endonuclease Digestion

Although the method has largely been supplanted in recent years by PCR techniques (discussed later in this section), cloning genes for study typically required cutting the chromosome into distinct pieces, or fragments, that could be selectively "plucked" from the sea of other genes composing the chromosome. Fortunately, bacteria themselves provided a way to produce these small fragments (discussed in Section 9.2). The bacterial proteins involved are called **restriction endonucleases**, which naturally function to clip the "foreign DNA" of invading plasmids and phages at specific points. Different species of bacteria use restriction endonucleases that recognize different DNA sequences.

The most useful restriction endonucleases for molecular biology recognize specific base sequences (usually four to six bases in length) in a DNA molecule and cleave both phosphodiester backbones at locations either near or within the center of the site. The cut may produce blunt ends or staggered ends; in staggered ends, the top strand is cut at one end of the site and the bottom strand cut at the other (**Fig. 7.26A**). Staggered ends are also called "cohesive ends" because the protruding strand of one of those ends can base-pair with complementary protruding strands from any DNA fragment cut with the same restriction endonuclease, regardless of the source organism. The ability of cohesive ends from different organisms to base-pair makes recombinant DNA technology possible.

Notice in **Figure 7.26A** that each recognition sequence, also called a restriction site, is a **palindrome** in which the top and bottom strands read the same in the 5′-to-3′ direction. This arrangement is typical of the sequences recognized by one of the three general classes of restriction endonucleases. Palindromic sequences allow an enzyme to attack each strand of a duplex using the same substrate recognition site. But how do bacteria making these scissor-like DNA enzymes cut foreign DNA and not destroy their own? They protect their chromosome with a methylating enzyme (a modification enzyme) that places methyl groups on both strands of the same DNA sequence recognized by the restriction endonuclease. A restriction endonuclease cannot cut a site if either one or both of the strands are methylated. Thus, newly replicated double-stranded molecules are protected, since one strand, the template, remains methylated at all times. Foreign DNA, such as phage DNA, that originates from a cell with one type of restriction modification system will <u>not</u> be protected upon entering another cell with a different restriction modification system, because the foreign DNA will <u>not</u> be methylated in the right places. As a result, foreign DNA is more likely to be destroyed than to undergo protective methylation.

Restriction endonucleases from hundreds of bacteria are now commercially available for analyzing DNA. A few examples of these enzymes are shown in **Figure 7.26A** (for method of agarose gel electrophoresis, see Appendix 3). Agarose gels can be used to analyze the DNA fragments obtained by treatment with these enzymes. In the gel shown in **Figure 7.26B**, each lane represents a different,

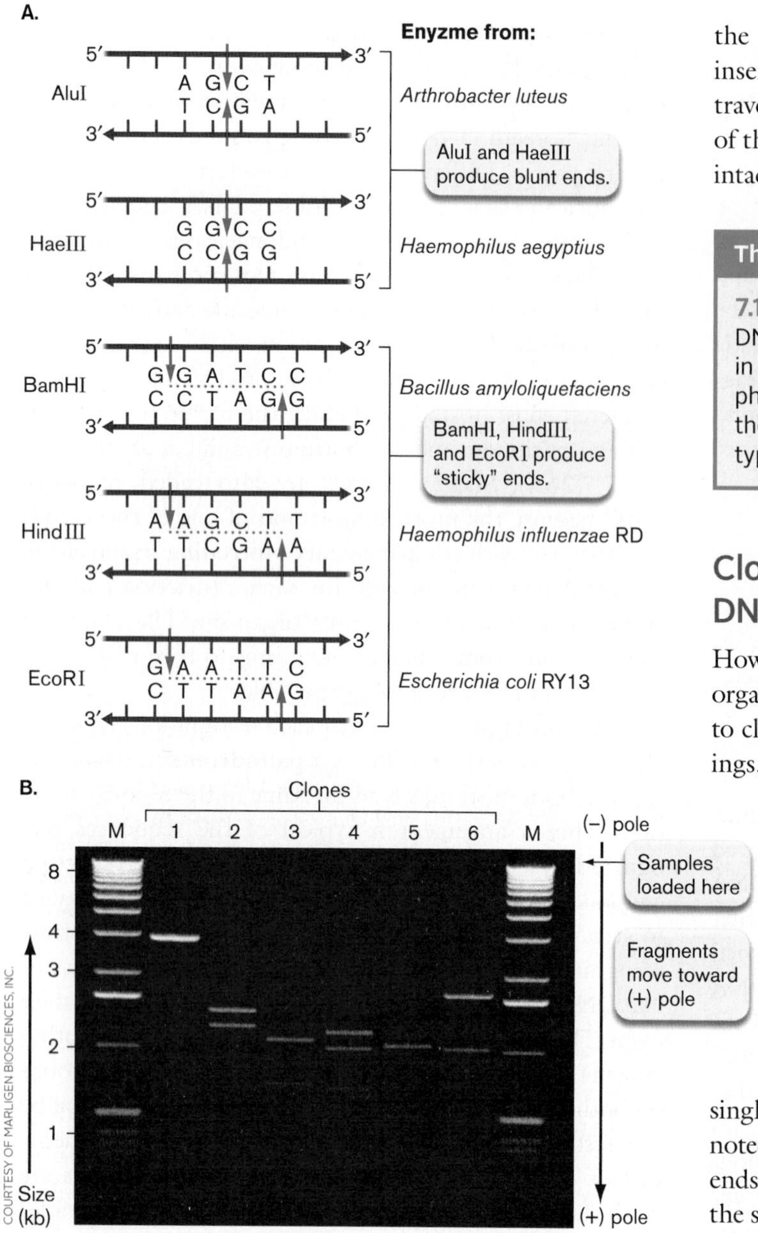
A.
Enzyme from:
AluI
A G C T
T C G A
Arthrobacter luteus
AluI and HaeIII produce blunt ends.
HaeIII
G G C C
C C G G
Haemophilus aegyptius
BamHI
G G A T C C
C C T A G G
Bacillus amyloliquefaciens
BamHI, HindIII, and EcoRI produce "sticky" ends.
HindIII
A A G C T T
T T C G A A
Haemophilus influenzae RD
EcoRI
G A A T T C
C T T A A G
Escherichia coli RY13
5′
3′
B.
Clones
M 1 2 3 4 5 6 M
(–) pole
Samples loaded here
Fragments move toward (+) pole
(+) pole
8
4
3
2
1
Size (kb)
COURTESY OF MARLIGEN BIOSCIENCES, INC.

FIGURE 7.26 ■ Bacterial DNA restriction endonucleases. **A.** Target sequences of sample enzymes. The names of these enzymes reflect the genus and species of the source organism. For example, EcoRI comes from *Escherichia coli.* **B.** Agarose-gel size analysis of EcoRI-restricted DNA fragments. "M" refers to marker fragments of known size. Smaller fragments move toward the bottom faster than large fragments. DNA bands are viewed using chemical stains that bind DNA and fluoresce under UV light.

homogeneous population of DNA molecules cut with the same restriction endonuclease. Because DNA is negatively charged, all DNA molecules travel to the positive pole during electrophoresis. Pore sizes in the agarose are such that they allow small molecules to speed through the gel, while larger molecules take longer to move. Thus, DNA fragments in this gel separate on the basis of size—the smaller the fragment, the farther it travels down the lane. The inserted DNA fragments were of different sizes, so they travel different distances in the gel. The sum of the sizes of the fragments in each lane yields the size of the original, intact, uncut molecule.

Thought Question

7.10 **Figure 7.26B** described how agarose gels separate DNA molecules. If you ran a highly supercoiled plasmid in one lane of the gel and the same plasmid with one phosphodiester bond cut in another lane, where would the two molecules end up relative to each other in this type of gel?

Cloning: The Birth of Recombinant DNA Technology

How did scientists ever conceive of cloning a gene from one organism into the DNA of an unrelated organism? The idea to clone genes arose from several seemingly unrelated findings. The discovery of restriction endonucleases in the late 1960s was initially considered interesting but esoteric, and its significance was grossly underestimated at the time. Only later would the full impact of these enzymes on biological research be realized. The beginning of the recombinant DNA revolution can be traced to 1972. It was known at that time that plasmids were small circular DNA molecules capable of independent replication in bacterial cells. Some of the known plasmids contained a single site for certain restriction endonucleases. As already noted, many restriction endonucleases generate cohesive ends that can base-pair with the ends of any DNA cut with the same enzyme.

The significance of these facts went unrecognized until Stanley Cohen (**Fig. 7.27A**), Herb Boyer (**Fig. 7.27B**), and Stanley Falkow were relaxing with colleagues in a Waikiki deli after a scientific meeting in 1972. In something of a "eureka" moment, the scientists suddenly realized that it might be possible to use a restriction endonuclease (such as EcoRI) to cut a piece of DNA from one organism's chromosome and graft it in vitro into a plasmid cut in a single place with the same enzyme (**Fig. 7.27C**). They also knew that DNA ligase (see **Fig. 7.19**) would seal the fragment to the plasmid and form a new artificial DNA molecule. This gene, as part of the recombinant molecule, could then be introduced into *E. coli* by transformation (discussed in Section 9.2) and be produced in large numbers through plasmid replication. Three months later, the strategy worked. This seminal work was published by Boyer and Cohen

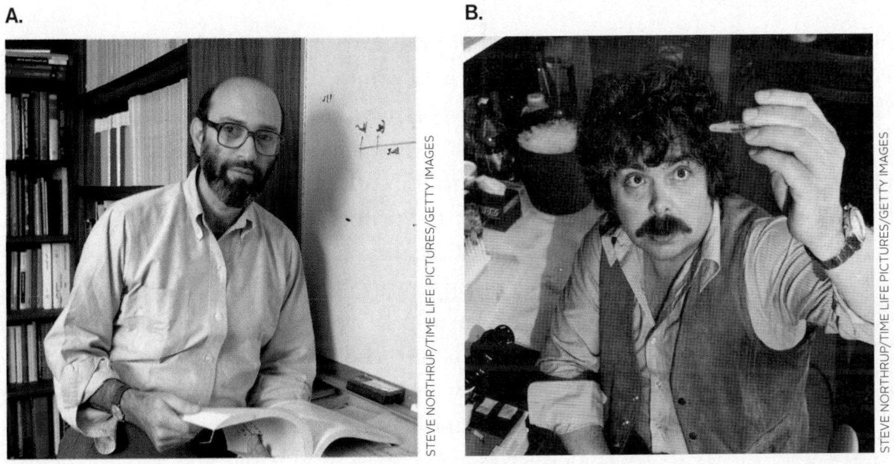
A.
B.
STEVE NORTHRUP/TIME LIFE PICTURES/GETTY IMAGES
STEVE NORTHRUP/TIME LIFE PICTURES/GETTY IMAGES

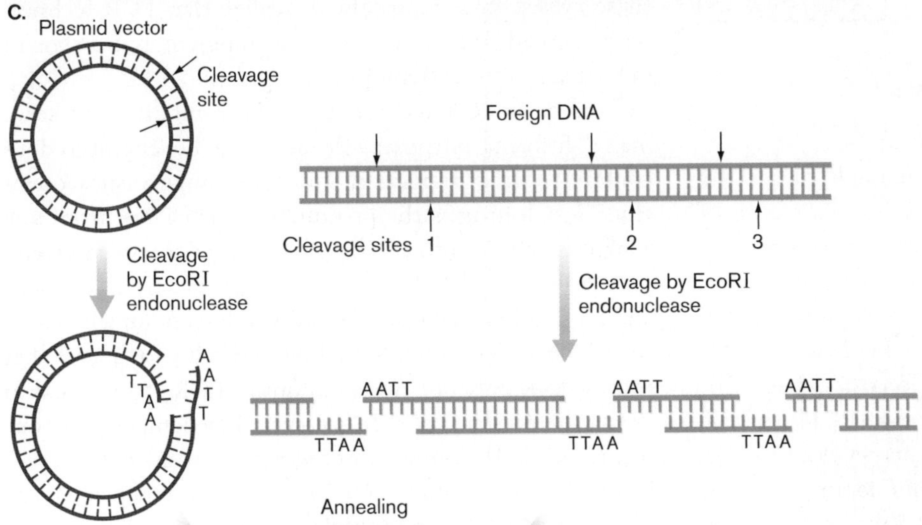
C.
Plasmid vector
Cleavage site
Foreign DNA
Cleavage sites 1
2
3
Cleavage by EcoRI endonuclease
Cleavage by EcoRI endonuclease
AATT
TTAA

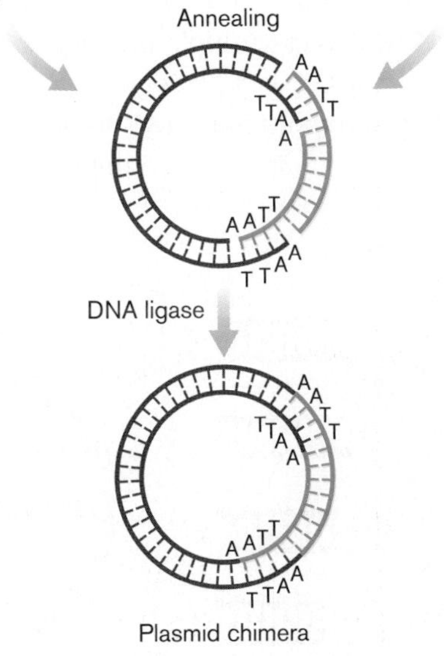
Annealing
DNA ligase
Plasmid chimera

FIGURE 7.27 ▪ Formation of recombinant DNA molecules. Stanley Cohen **(A)** of Stanford University and Herbert Boyer **(B)** of UC San Francisco, two of the founders of recombinant DNA technology. **C.** A plasmid vector is cut with a restriction endonuclease (EcoRI) that produces cohesive ends. The same enzyme is used to digest foreign DNA, producing a series of fragments, all with identical cohesive ends. The cut vector and foreign DNA fragments are mixed. Cohesive ends of the foreign DNA and vector anneal to form a chimeric molecule containing two nicks. DNA ligase is used to seal the nicks and connect the phosphodiester backbones. The resulting population of plasmids represents the entire genome of the donor species (the genome library).

along with students Annie Chang and Robert Helling. Over the next decade, the technique became known as "gene cloning," an immensely powerful technology that thrust biology into the genomic era.

Genome libraries (also called clone libraries or clone pools) containing all the genes in an organism's chromosome are routinely made today. To make a genome library, an entire chromosome is digested with a single restriction endonuclease, and the resulting fragments are randomly ligated into a plasmid cloning vector. The resulting population of plasmids (the genome library) is composed of multiple copies of recombinant plasmids containing different gene inserts.

Weblinks Cloning strategies (*see ebook*)

One problem with early cloning techniques was that *E. coli* plasmid origins do not initiate replication in most other bacterial species or in eukaryotes or archaea. This means that cloned genes from these groups of organisms could be studied only in *E. coli*. The problem with studying these foreign genes in *E. coli* is that the RNA processing and protein modifications naturally carried out in a eukaryote, for instance, are not possible in *E. coli* because *E. coli* lacks those systems. Nor can the function of these eukaryotic proteins be tested in the prokaryotic cell. The solution to these problems was to use a **shuttle vector**. A shuttle vector is a plasmid that contains a replication origin compatible with *E. coli* and a second origin that will allow the plasmid to replicate in an unrelated bacterial species, or in a eukaryote or archaeon. In this way, the cloned gene can be genetically manipulated in *E. coli* using well-characterized molecular protocols (that is, introducing specific mutations) and then can be reintroduced into the source organism to study function.

Thought Question

7.11 Given that *E. coli* possesses restriction endonucleases that cleave DNA from other species, how is it possible to clone a gene from one organism to another without the cloned gene sequence being degraded?

PCR Amplifies Specific Genes from Complex Genomes

Imagine what you could do if you could amplify a specific gene from a genome into millions of copies in a matter of one or two hours. This is now possible and even routine in many laboratories around the world. The technique, which relies on thermostable DNA polymerases isolated from thermophiles, has revolutionized many fields, including biological research, medicine, and forensics. The technique is the **polymerase chain reaction (PCR)**, briefly described in Chapter 1. The basic PCR process, outlined in **Figure 7.28**, can produce over a millionfold amplification of target DNA. Using PCR, large quantities of a specific DNA sequence can be produced within a few hours.

In this technique, specific oligonucleotide primers (usually between 20 and 30 bp) are annealed to known DNA sequences flanking the target gene. A thermostable DNA polymerase uses these primers to replicate the target DNA. Heat-stable DNA polymerases must be used in this process because the reaction mixture must be subjected to repeated cycles of heating to 95°C (to separate DNA strands so that they are available for primer annealing), cooling to 55°C (to allow primer annealing), and heating to 72°C (the optimum reaction temperature for the polymerase). The polymerase Taq, from the thermophile *Thermus aquaticus*, is often used for this purpose (polymerases from other thermophiles are also used). Because PCR reactions require 25–30 heating and cooling cycles, a machine called a thermocycler is used to reproducibly and rapidly deliver these cycles. It is important to realize that PCR technology emerged from studies of a seemingly obscure class of microorganisms: extremophiles.

This basic PCR technique has been modified to serve many different purposes. Primers can be engineered to contain specific restriction sites that simplify subsequent cloning. **Cloning** is the process by which DNA from one source is spliced into DNA from another source. **Restriction sites** are small sequences (4–8 bp) that are recognized and cleaved by enzymes called restriction endonucleases. If the primers used for PCR are highly specific for a gene that is present in only one microorganism, PCR can be used to detect the presence of that organism in a complex environment, such as the presence of the pathogen *E. coli* O157:H7 in hamburger. Multiplex PCR involving several primer sets can be used to detect multiple pathogens in a single reaction (Appendix 3).

PCR has profoundly impacted human society. By making it possible to amplify the tiniest amounts of DNA

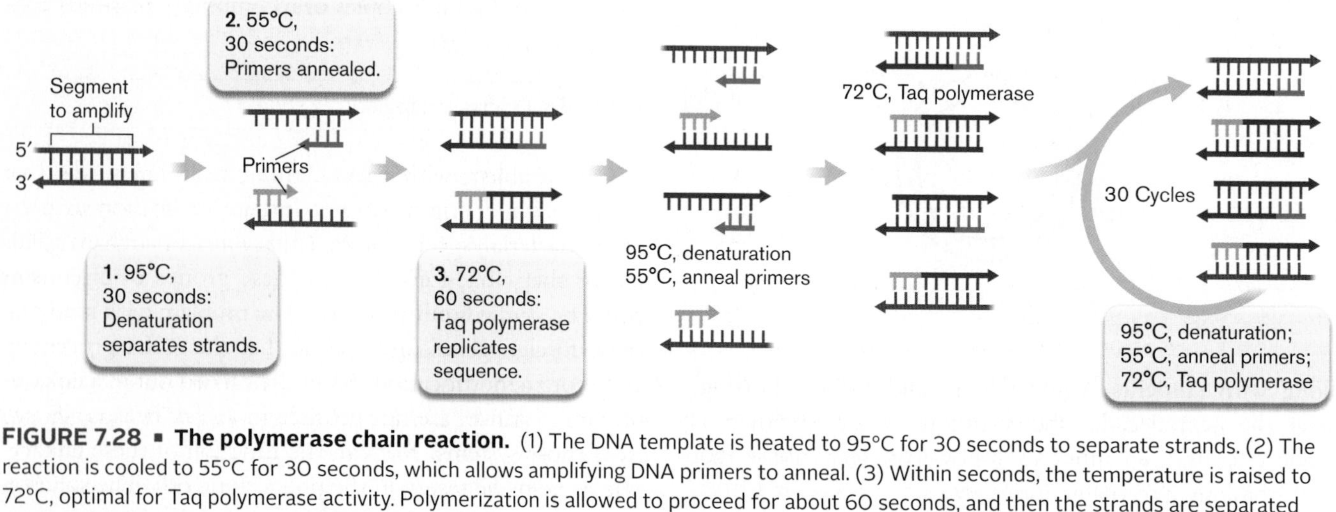
2. 55°C,
30 seconds:
Primers annealed.
Segment to amplify
5′
3′
Primers
1. 95°C,
30 seconds:
Denaturation separates strands.
3. 72°C,
60 seconds:
Taq polymerase replicates sequence.
95°C, denaturation
55°C, anneal primers
72°C, Taq polymerase
30 Cycles
95°C, denaturation;
55°C, anneal primers;
72°C, Taq polymerase

FIGURE 7.28 ■ The polymerase chain reaction. (1) The DNA template is heated to 95°C for 30 seconds to separate strands. (2) The reaction is cooled to 55°C for 30 seconds, which allows amplifying DNA primers to anneal. (3) Within seconds, the temperature is raised to 72°C, optimal for Taq polymerase activity. Polymerization is allowed to proceed for about 60 seconds, and then the strands are separated again at 95°C to prepare for another cycle. Each polymerization step increases (amplifies) the target sequence until, ultimately, only the fragment bounded by the primers is amplified. Thus, from a single DNA molecule, potentially 10^{30} copies of a fragment can be made.

contaminating a crime scene, PCR has changed our judicial system by providing conclusive evidence in court cases where no evidence would have existed previously. Increasingly, individual human genomes are being sequenced as a standard medical test—with profound ethical and societal implications. The hopes and fears raised by the invention of PCR-based genomic sequencing inspired the science fiction film *Gattaca*, directed by Andrew Niccol, depicting an imaginary future in which everyone's destiny is determined by his or her DNA sequence. This knowledge could impact which jobs are available to someone, whether insurance coverage can be withheld, and even whether two individuals can marry.

Thought Question

7.12 PCR is a powerful technique, but a sample can be easily contaminated and the wrong DNA amplified—perhaps sending an innocent person to jail. What might you do to minimize this possibility?

DNA Sequencing

Even with the development of gene cloning, our transition into the genomic age would not have been possible without the ability to rapidly sequence entire genomes. Previous DNA sequencing methods used an ingeniously simple strategy but were not rapid. The dideoxy chain termination method developed by Fred Sanger in the 1970s relied on the fact that the 3′ hydroxyl group on 2′-deoxyribonucleotides (shown in **Fig. 7.4A**) is absolutely required for a DNA chain to grow. Thus, incorporation of a 2′,3′-dideoxynucleotide (**Fig. 7.29**) into a growing chain prevents further elongation, because without the 3′ OH, extension of the phosphodiester backbone is impossible.

The dideoxy sequencing method begins with cloning a gene to be sequenced into a plasmid vector. Next, a short (20- to 30-bp) oligonucleotide DNA primer molecule is designed to anneal at a site immediately adjacent to the cloned insert to be sequenced. The mixture is first heated to 95°C to separate the DNA strands and then cooled, which allows the primer to bind. DNA polymerase can then build on this primer and synthesize DNA using the cloned insert as a template. A small amount of dideoxyadenosine triphosphate (dideoxy ATP) is included in the DNA synthesis reaction, which already contains all four normal deoxyribonucleotides. Because normal 2′-deoxyadenosine is present at high concentration, chain elongation usually proceeds normally. However, the sequencing reaction contains just enough dideoxy ATP to make DNA polymerase occasionally substitute the dead-end base for

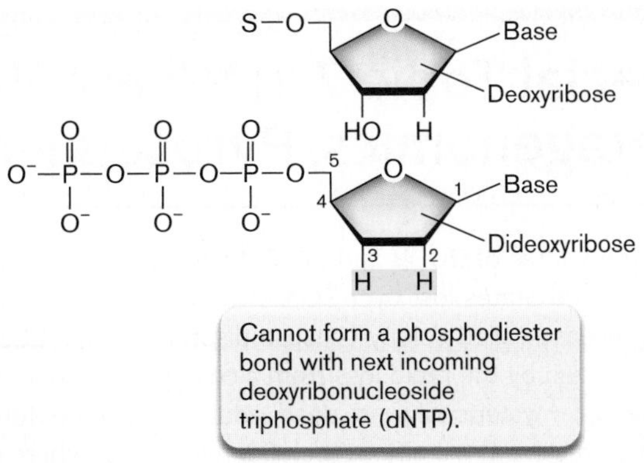
S—O
Base
Deoxyribose
HO H
Base
Dideoxyribose
H H
Cannot form a phosphodiester bond with next incoming deoxyribonucleoside triphosphate (dNTP).

FIGURE 7.29 ▪ Dideoxynucleotide. DNA nucleotides (top) lack the 2′ OH group on ribose but retain the 3′ OH needed for DNA synthesis. Dideoxynucleotides (bottom) lack both the 2′ OH and the 3′ OH. Thus, the 5′ phosphate of a dideoxynucleotide can still form a phosphodiester bond to a growing chain but lacks the 3′ OH acceptor for a new incoming nucleotide. Review Figure 7.4 to see how missing the 3′ OH group would terminate DNA synthesis.

the natural one, at which point the chain stops growing. The result is a population of DNA strands of varying size, each one truncated at a different adenine position. The Sanger method of DNA sequencing used fluorescently tagged dideoxy nucleotides and is described in Appendix 3. This method has been replaced by even more rapid techniques, all of which include an element of the Sanger chain termination approach.

Sequencing an Entire Genome

The old way of sequencing genomes was to use restriction endonucleases to fragment a genome and then clone those fragments into vectors such as plasmids that can hold 40–50 kb of insert DNA, or bacterial artificial chromosomes that harbor inserts of 150–1,000 kb. Each plasmid then had to be sequenced separately, so it took months to sequence an entire genome. New sequencing technologies, called "next-generation sequencing," have combined the power of robotics, computers, and fluidics such that an entire bacterial genome can be sequenced within a few days.

One of these techniques, pyrosequencing, is described in **Special Topic 7.1**. Another popular technique is called **sequencing by synthesis** (a technology developed by Solexa, a company now part of Illumina, Inc.); the process is outlined in **Figure 7.30**. Basically, the genome to be sequenced is sonically fragmented into 100- to 200-bp segments and different linker oligonucleotides are ligated to each end. Small fragments are used because the technique can sequence only 100–200 bp from any one fragment. Strands of each fragment are then separated and the mixture added to an optical flow cell. The fragments are

Special Topic 7.1: Where Have All the Bees Gone? Metagenomics, Pyrosequencing, and Nature

Over the winter of 2009 and 2010, beekeeping operations in the United States lost over 30% of their hives to a devastating disease called colony collapse disorder (CCD). CCD is characterized by the rapid loss from a colony of its adult bee population. Mysteriously, no dead adult bees were found inside or in close proximity to the affected colony. Where did they go? Since more than 90 cultivated crops depend on the honeybee (*Apis mellifera*) for pollination, CCD could lead to dramatic food shortages. The cause of CCD remains a mystery but probably is the result of a combination of factors, one of which is an infection. When combs from affected bee colonies are sterilized by irradiation, they can be repopulated with new bees. A popular hypothesis is that infected bees leaving the hive become disoriented because of the infection and cannot find their way back to the colony.

A major step toward identifying the pathogen(s) has come from an elegant metagenomic strategy. The scientific team conducting the study randomly sequenced nucleic acid obtained from CCD colonies and compared the sequences they found with those obtained from normal bee colonies and from royal jelly (a honeybee secretion used to feed larvae). The idea was that DNA sequences unique to CCD colonies would most likely come from the causative infectious agent. Unique sequences were indeed found. Comparing these sequences with massive databases of known, sequenced genomes showed a strong correlation with CCD for one organism: the Israeli acute paralysis virus of bees.

Carrying out this herculean task by conventional DNA sequencing techniques would have required a huge number of DNA sequencing reactions and would likely have taken many years to complete. The scientists, however, were able to accomplish the task in a matter of months using a new sequencing technique called **pyrosequencing**. The basic process is outlined in **Figure 1**.

Genomic DNA is isolated from a source, fragmented to 100 bp, ligated to adapter oligonucleotides containing a known sequence, and separated into single strands (**Fig. 1A**, step 1). The adapter oligonucleotides contain known sequences that are used to prime subsequent PCR and DNA sequencing steps. Single-stranded fragments are then bound to microscopic beads under conditions that favor binding of only one fragment per bead. Each bead is then captured in an individual droplet of an emulsion consisting of PCR reaction mixture in oil. After PCR amplification within each droplet, each bead carries 10 million copies of a unique dsDNA template (**Fig. 1A**, step 2). The emulsion is broken, the DNA strands are denatured, and beads carrying single-stranded DNA clones are deposited into microscopic wells of a fiber-optic slide (step 3). Smaller beads carrying immobilized enzymes required for pyrophosphate sequencing are deposited into each well (step 4). Thousands of sequencing reactions can be carried out simultaneously, and no gels are required.

The actual pyrosequencing process relies on the fact that pyrophosphate (PP_i) is released from any deoxyribonucleoside triphosphate when DNA polymerase adds it to a growing DNA chain. ATP sulfurylase, another enzyme in the reaction mix, combines PP_i with adenosine phosphosulfate to make ATP. The ATP is used by luciferase, also in the mix, to generate light, which is picked up by a detector. Light generation indicates that a nucleotide has been added to the growing DNA chain. Repeatedly flooding the fiber-optic slide with one deoxynucleotide at a time (dTTP, dCTP, dGTP, or dATPαS, which is not a luciferase substrate but is a polymerase substrate) causes light to be emitted in any well where the complementary nucleotide is present at the 3′ end of the template. The nucleotide is added and pyrophosphate is liberated. In a typical run, pyrosequencing can generate 25 million bases of DNA sequence and costs about 3 cents per 1,000 bases. The comparative cost for Sanger sequencing is about $1 per 1,000 bases.

Though candidate pathogens have now been associated with CCD (fungi of the group Microsporidia have also been implicated), much work remains to prove the connection. Meanwhile, the bees are still disappearing.

RESEARCH QUESTION

Recent evidence suggests that the global spread of the ectoparasitic mite *Varroa destructor* is a major contributing factor to CCD. How might mites and viruses collaborate in the spread of this disease?

Martin, Stephen J., Andrea C. Highfield, Laura Brettell, Ethel M. Villalobos, Giles E. Budge, et al. 2012. Global honey bee viral landscape altered by a parasitic mite. *Science* **336**:1304–1306.

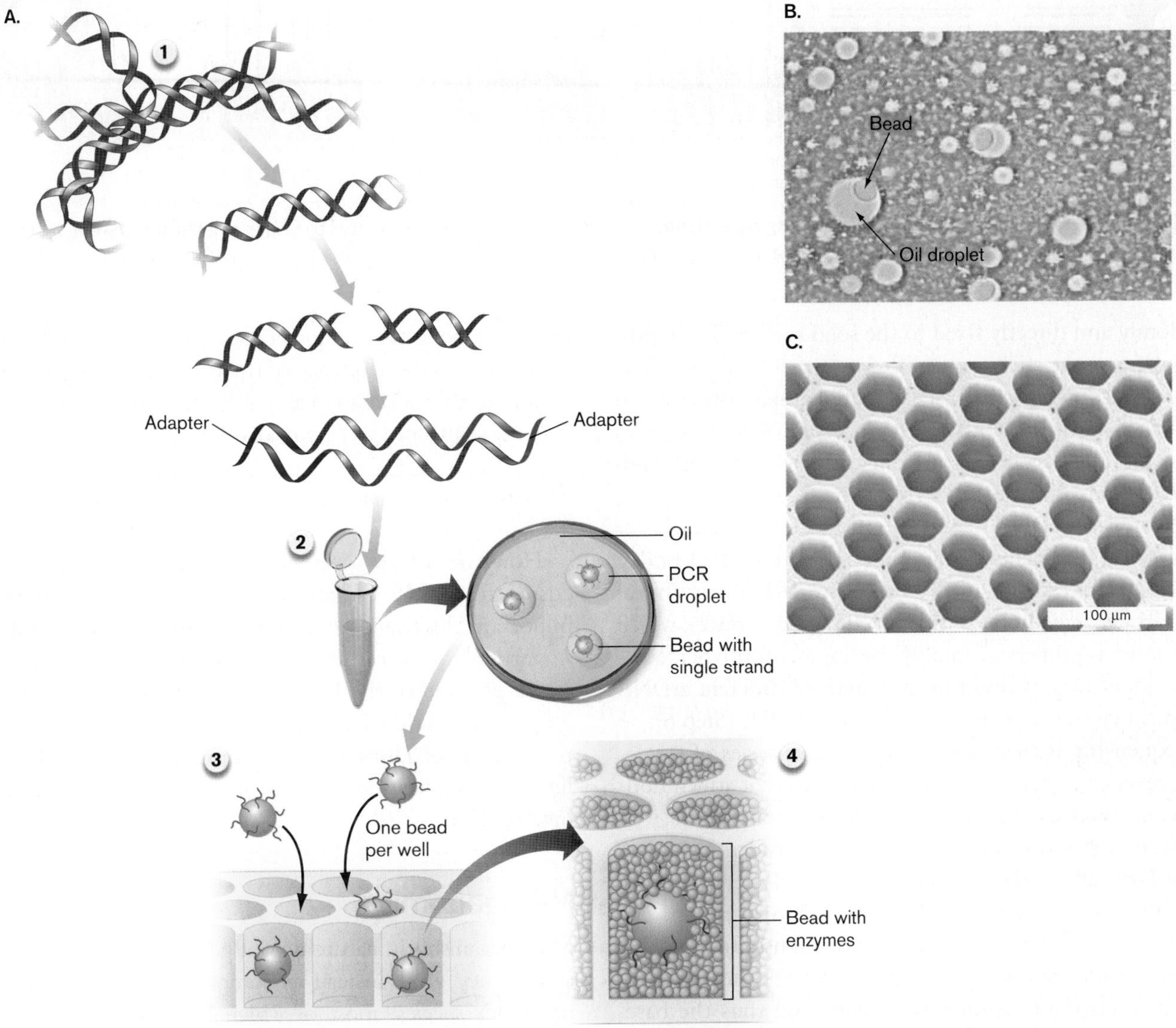
A.
1
Adapter
Adapter
2
Oil
PCR
droplet
Bead with
single strand
3
One bead
per well
4
Bead with
enzymes
B.
Bead
Oil droplet
C.
100 μm

FIGURE 1 ■ Sample preparation for pyrosequencing. A. (1) Genomic DNA is isolated, fragmented, ligated to adapters, and separated into single strands. (2) Fragments are bound to beads under conditions that favor one fragment per bead, the beads are captured in the droplets of a PCR reaction mixture in oil emulsion, and PCR amplification occurs within each droplet. Amplification results in beads that each carry 10 million copies of a unique DNA template. (3) Next, the emulsion is broken, the DNA strands are denatured, and the beads, now carrying single-stranded DNA clones, are deposited into wells of a fiber-optic slide. (4) Smaller beads carrying immobilized enzymes required for pyrophosphate sequencing are deposited into each well. **B.** Micrographs of emulsion (step 2 in part A) showing droplets containing a bead and empty droplets. The thin arrow points to a 28-μm bead; the thick arrow points to an approximately 100-μm droplet. **C.** Scanning electron micrograph of a portion of a fiber-optic slide, showing fiber-optic cladding (jacket) and wells before bead deposition (step 3 in part A). *Source:* Adapted from Marcel Margulies et al. 2005. *Nature* **437**:376–380.

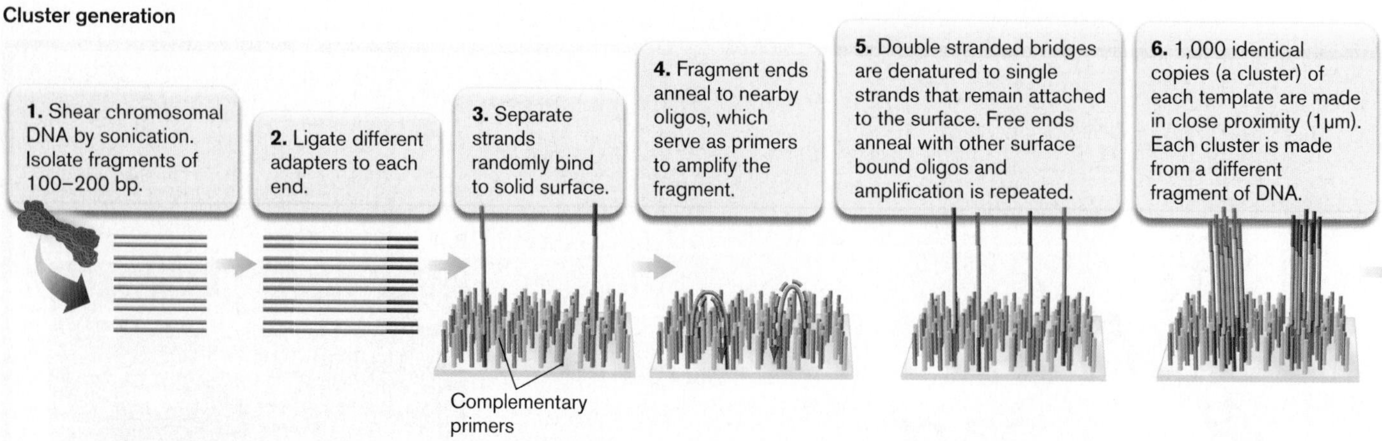

Cluster generation
1. Shear chromosomal DNA by sonication. Isolate fragments of 100–200 bp.
2. Ligate different adapters to each end.
3. Separate strands randomly bind to solid surface.
4. Fragment ends anneal to nearby oligos, which serve as primers to amplify the fragment.
5. Double stranded bridges are denatured to single strands that remain attached to the surface. Free ends anneal with other surface bound oligos and amplification is repeated.
6. 1,000 identical copies (a cluster) of each template are made in close proximity (1 μm). Each cluster is made from a different fragment of DNA.
Complementary primers

FIGURE 7.30 ▪ DNA sequencing; sequencing by synthesis. Steps 1–6: Generation of clusters by bridge amplification. **Steps 7–10:** Sequencing by synthesis using reversible fluorescent termination.

randomly and directly fixed to the solid surface. Each individual fragment sticks to a different area of the cell. The glass surface also has a dense lawn of oligonucleotides fixed at their 5′ ends to the glass surface. These oligonucleotides are complimentary to the fragment linker ends, although interaction does not occur at this point.

Next, each ssDNA fragment is turned into a tight cluster of identical fragments through a series of so-called bridge amplifications (Steps 4 and 5). Ends of each fragment anneal to nearby matched oligos (fixed to the glass surface), which then serve as primers to amplify the fragments further. Multiple amplifications result in thousands of different ssDNA fragment clusters dotting each lane of each slide (Step 6).

Sequencing is then done in a repetitive series of single-step, reversible chain termination reactions that attach fluorescently labeled nucleotides one at a time to the clusters. Because each individual cluster is a tuft of identical DNA molecules, all of the molecules in that cluster will have the same tagged base added and will fluoresce the same color. Note that the sequence of either stand can be determined by adding one or the other primer. After a base is added, a snapshot captures the colors, and thus the base, added to each cluster. Next, the fluorescent marker (fluor) is removed, which also reverses chain termination, and the slide is again flooded with tagged bases. Snapshots taken after each sequencing round sequentially capture the fluorescent colors of each cluster as each base is added.

The Sanger dideoxy method could sequence 1,000 bp per template, but it required considerable time, effort, and space to do so. In contrast, each flow-cell lane in the sequencing-by-synthesis technique produces 10–120 million DNA fragments, thus generating up to 600 gigabases (Gb) of DNA sequence per run (1 Gb = 1 billion bases). Each read is short (about 100 bases per template), but because millions of templates are read, the massively parallel sequencing yields hundreds of millions of bases. Computer programs then find the overlaps among the millions of sequence results and assemble them into the sequence of the entire chromosome. Estimates are that soon anyone will be able to have a personal genome sequenced for about $1,000.

Weblinks Sequencing by synthesis (*see ebook*)

How the information from sequences is mined is discussed further in Chapter 8. Suffice it to say that many important discoveries have resulted from this technology. A particularly intriguing discovery was finding that the entire genome of the bacterium *Wolbachia* (which infects 20% of the world's insect population) is embedded within the genome of *Drosophila,* the common fruit fly. The evolutionary implications of this discovery are under intense investigation.

Metagenomics

Microorganisms in nature do not generally exist as pure cultures. They grow instead as complex consortia containing numerous species, most of which have never been grown in a laboratory. With newer and ever-faster DNA sequencing methodologies and PCR techniques, we can now "peer" deeper into the natural world. Like the explorers of old, scientists are finding new species never before seen and are beginning to understand the intricate ways bacterial species interact to form what could be called biological support groups.

Metagenomics is the use of modern genomic techniques to study microbial communities directly in their natural environments, bypassing the need for isolating and cultivating individual species in the laboratory. The field began with the culture-independent retrieval of 16S rRNA genes from the environment by amplifying DNA directly from environmental samples using PCR protocols. Since genes encoding 16S rRNA from small ribosomal subunits

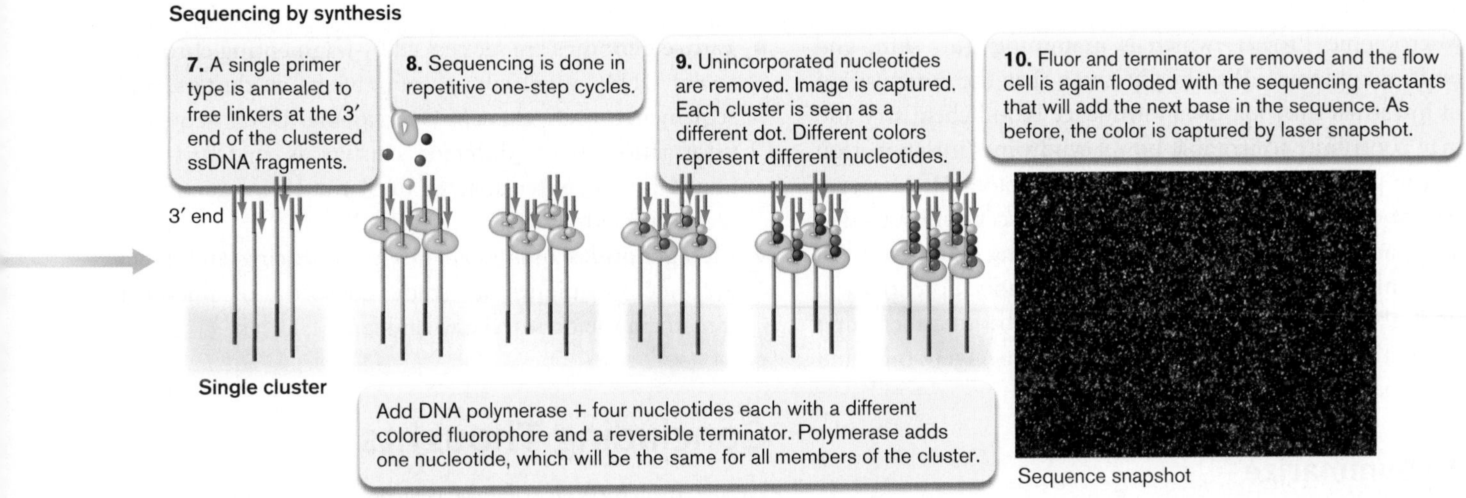
Sequencing by synthesis
7. A single primer type is annealed to free linkers at the 3′ end of the clustered ssDNA fragments.
8. Sequencing is done in repetitive one-step cycles.
9. Unincorporated nucleotides are removed. Image is captured. Each cluster is seen as a different dot. Different colors represent different nucleotides.
10. Fluor and terminator are removed and the flow cell is again flooded with the sequencing reactants that will add the next base in the sequence. As before, the color is captured by laser snapshot.
3′ end
Single cluster
Add DNA polymerase + four nucleotides each with a different colored fluorophore and a reversible terminator. Polymerase adds one nucleotide, which will be the same for all members of the cluster.
Sequence snapshot

(also called small-subunit rRNA) are highly conserved, it was possible to determine whether a given sequence of 16S rRNA represented a new species and to which known species it was most closely related. Since then, metagenomics has revolutionized microbiology by shifting the focus even further away from clonal isolates capable of growth in the lab toward the estimated 99% of microbial species that cannot currently be cultivated. Since 2003, hundreds of metagenomic projects have been carried out or are ongoing. Examples of the metagenomic analysis of ecosystems are presented in Chapter 21.

Originally, a metagenomic project began with construction of a genome library from DNA sequences retrieved from an environmental sample (soil, for example). Many sequence reads derived from a library remained as unassembled one-pass sequences because of the variety of sizes of environmental genomes and their abundance. Indeed, half of the reads for the Sargasso Sea data set and all of the reads for the Minnesota soil data set were unassembled sequences of roughly 700 bp.

Current metagenomic protocols generally use 16S rRNA sequence hits to deduce the various species of organisms present in a sample. The 16S information can also help determine how the many unknown organisms present in that sample relate to known species. But random sequencing will also identify most, if not all, of the other genes present in all the organisms that the consortium comprises. All of these genes make up what is known as the **metagenome**—that is, the combined yet unsorted genomic potential of all the organisms in the sample. Thus, metagenomics not only identifies the principal partners of a microbial consortium but also reveals many of their metabolic capabilities. Communities of microorganisms can actually form new metabolic pathways in which a useless by-product of one member can be used by another member in a reaction whose product can benefit a third member.

The science has even progressed to the point that if the sample contains a low number of species, then whole-genome shotgun (WGS) sequencing methods, such as pyrosequencing (see **Special Topic 7.1**) or sequencing by synthesis, can reconstruct complete or nearly complete genomes for those species, whether bacteria, archaea, or viruses. The key is assembling hundreds of thousands of sequences generated by random sequencing into larger, overlapping sequences called **contigs**. Powerful new genome-assembling computer algorithms make it possible to simultaneously build complete or nearly complete genomes from a mixture of microbes taken from a single sample.

A major strength of metagenomics is its potential for serendipitous discovery. One example where metagenomic data led to an unexpected finding came from studies of archaeal populations in acid mine drainage. DNA sequence information gained from these studies was used to show that genetic recombination occurs at a much higher frequency than previously predicted and is the primary evolutionary force shaping these populations. In another example, data from the Pacific and Atlantic oceans revealed that the greatest variation within populations of *Prochlorococcus* (the most abundant photosynthetic organism in the ocean) occurs in genomic "islands." These islands are discrete regions in the genome that are believed to be hot spots for genomic innovation. They are derived in part from genes horizontally transferred from other organisms by viruses.

The power of metagenomics has also been directed toward solving the human microbiome. There are 100 trillion microbes (10 microbial cells for every one human cell) in the human body, but we do not know the identity of most of them and are ignorant as to how to grow them in the laboratory. How we interact with these organisms and they with us is a major question in modern microbiology—one that has led to the formation of the Human

Microbiome Project, which is examining gut, skin, and oral microbiomes. We already know that the composition of intestinal microbiota can influence blood chemistry and may contribute to irritable bowel syndrome (inflammation of the gastrointestinal tract), but microbiota are also suspected of influencing an individual's susceptibility to diabetes and obesity. Consequently, knowing the identity of these microbes and their collective metabolic potential will undoubtedly lead to new advances in the treatment and prevention of human diseases. More on metagenomics and microbial ecology is found in section 21.1.

To Summarize

- **DNA restriction endonucleases** used for DNA analysis cleave DNA at specific recognition sequences, which are usually 4–6 bp in length and produce either a blunt or a staggered cut.
- **Agarose-gel electrophoresis** will separate DNA molecules on the basis of their size.
- **Restriction endonuclease–digested DNA** molecules were first cloned into plasmids in the early 1970s.
- **The polymerase chain reaction (PCR)** will amplify one or two copies of a specific gene a millionfold.
- **The most commonly used DNA sequencing** technique (the Sanger technique) uses dideoxyribonucleotides—nucleotides that lack both the 2′ OH and 3′ OH groups—to terminate chain elongation.
- **Entire genomes** are sequenced by fragmenting chromosomal DNA, amplifying the fragments in clusters, and sequencing each cluster. Overlapping DNA sequence information from different clusters is matched and joined by computer analysis until the entire genome is reconstructed.
- **Metagenomics** uses rapid DNA sequencing and other genomic techniques to study consortia of microbes directly in their natural environment.

Concluding Thoughts

We used to think of bacteria as simple, single-celled organisms. In fact, they are far from simple. Complexity is evident in the elegant mechanisms they use to organize and replicate their genomes. The next three chapters will expand on this idea and discuss the way bacteria make proteins, regulate genes, exchange DNA, and, in the process, evolve. Knowing how a bacterial cell replicates and genetically controls the repertoire of available proteins and enzymes provides a unique perspective from which to study the physiology of microbial growth, explored in Part 3 of this textbook. Furthermore, understanding the bacterial genome allows us to investigate the physiology of pathogenic, as well as nonculturable, organisms and gain greater insight into the evolutionary diversity of species that we will discuss in Part 4.

CHAPTER REVIEW

Review Questions

1. What is the difference between vertical and horizontal gene transmission?
2. Explain the structural types of bacterial genomes. What is a structural gene?
3. Describe the three functional levels of gene organization.
4. What are the differences between DNA and RNA?
5. Explain DNA supercoiling. Why is it important to microbial genomes?
6. Discuss the mechanisms of topoisomerases. What drug targets a topoisomerase?
7. What are the basic mechanisms involved in DNA replication?
8. How does the bacterial cell regulate the initiation of chromosome replication?
9. What is the clamp loader? Primase? DNA helicase (DnaB)? Helicase loader (DnaC)? DNA proofreading?
10. How is the problem of replicating both strands at a replication fork solved?
11. What is a catenane? What does it have to do with DNA replication?
12. Explain the polymerase chain reaction (PCR).
13. What is rolling-circle replication?
14. What is DNA restriction and modification? Why is it important to bacteria? Why is it important to forensic scientists?
15. Explain next generation DNA sequencing.

Thought Questions

1. When you sequence a genome, the DNA used is in fragments. How do you know where the base pairs in the genome are located?
2. During rapid growth, why would a bacterial cell die if an antibiotic drug formed, as the text says, "a physical barrier in front of the DNA replication complex"?
3. Why would bacteria, which replicate more quickly than eukaryotic organisms, lose pseudogenes more rapidly than eukaryotes do?

Key Terms

antiparallel (243)
catenane (256)
cloning (266)
conjugation (239)
contig (271)
denature (244)
DNA control sequence (239)
DNA ligase (255)
DNA replication (238)
enhancer (241)
exonuclease (253)
gene (239)
genome (239)
genome library (265)
histone (261)
horizontal transmission (238)
hybridization (244)
information flow (238)
intron (261)
metagenome (271)
metagenomics (270)
nucleoid (245)
Okazaki fragments (250)
operon (242)
origin (*oriC*) (250)
palindrome (263)
phosphodiester bond (243)
plasmid (258)
polymerase chain reaction (PCR) (266)
primase (252)
promoter (241)
proofreading (253)
pseudogene (261)
purine (243)
pyrimidine (243)
pyrosequencing (268)
quinolone (248)
recombination (239)
regulon (242)
replication fork (249)
replisome (253)
restriction endonuclease (263)
restriction site (266)
semiconservative (249)
sequencing by synthesis (267)
shuttle vector (265)
sliding clamp (252)
structural gene (239)
termination (*ter*) site (250)
topoisomerase (247)
transformation (239)
transcription (238)
translation (238)
vertical transmission (238)

Recommended Reading

Boeneman, Kelly, and Elliott Crooke. 2005. Chromosomal replication and the cell membrane. *Current Opinion in Microbiology* **8**:143–148.

Campbell, Christopher S., and R. Dyche Mullins. 2007. In vivo visualization of type II plasmid segregation: Bacterial actin filaments pushing plasmids. *Journal of Cell Biology* **179**:1059–1066.

Cox-Foster, Diana L., Sean Conlan, Edward C. Holmes, Gustavo Palacios, Jay D. Evans, et al. 2007. A metagenomic survey of microbes in honey bee colony collapse disorder. *Science* **318**:283–287.

Dame, Remus T., Olga J. Kalmykowa, and David C. Grainger. 2011. Chromosomal macrodomains and associated proteins: Implications for DNA organization and replication in Gram-negative bacteria. *PLoS Genetics* **7**:e1002123.

Dillon, Shane, and Charles J. Dorman. 2010. Bacterial nucleoid-associated proteins, nucleoid structure, and gene expression. *Nature Reviews. Microbiology* **8**:185–195.

Duggin, Ian G., R. Gerry Wake, Stephen D. Bell, and Thomas M. Hill. 2008. The replication fork trap and termination of chromosome replication. *Molecular Microbiology* **70**:1323–1333.

Falkow, Stanley. 2001. I'll have chopped liver please, or how I learned to love the clone. *ASM News* **67**:555.

Georgescu, Roxana E., Isabel Kurth, and Mike E. O'Donnell. 2011. Single-molecule studies reveal the function of a third polymerase in the replisome. *Nature Structural and Molecular Biology* **19**:113–116.

Katayma, Tsutomu, Shogo Ozaki, Kenji Keyamuaa, and Kazuyuki Fujimitsu. 2010. Regulation of the replication cycle: Conserved and diverse regulatory systems for DnaA and oriC. *Nature Reviews. Microbiology* **8**:163–170.

Kirkpatrick, Clare L., and Patrick H. Viollier. 2010. A polarity factor takes the lead in chromosome segregation. *EMBO Journal* **29**:3035–3036.

Matsunaga, Fujihiko, Kie Takemura, Masaki Akita, Akinori Adachi, Takeshi Yamagami, et al. 2010. Localized melting of duplex DNA by Cdc6/Orc1 at the DNA replication origin in the hyperthermophilic archaeon *Pyrococcus furiosus. Extremophiles* **14**:21–31.

McHenry, Charles S. 2011. Bacterial replicases and related polymerases. *Current Opinion in Chemical Biology* **15**:587–594.

Mott, Melissa L., and James M. Berger. 2007. DNA replication initiation: Mechanisms and regulation in bacteria. *Nature Reviews. Microbiology* **5**:343–354.

Possoz, Christophe, Ivan Junier, and Oliver Espeli. 2012. Bacterial chromosome segregation. *Frontiers in Bioscience* **17**:1020–1034.

Rimsky, Sylvie, and Andrew Travers. 2011. Pervasive regulation of nucleoid structure and function by nucleoid-associated proteins. *Current Opinion in Microbiology* **14**:136–141.

Robinson, Andrew, and Antoine M. van Oijen. 2013. Bacterial replication, transcription and translation: mechanistic insights from single-molecule biochemical studies. *Nature Reviews Microbiology* **11:** 303–315.

Vologodskii, Alexander. 2010. DNA supercoiling helps unlink sister duplexes after replication. *Bioessays* **32**:9–12.

CHAPTER 8
Transcription, Translation, and Bioinformatics

8.1 RNA Polymerases and Sigma Factors
8.2 Transcription of DNA to RNA
8.3 Translation of RNA to Protein
8.4 Protein Modification and Folding
8.5 Protein Degradation: Cleaning House
8.6 Secretion: Protein Traffic Control
8.7 Bioinformatics: Mining the Genomes

The cell accesses the vast store of data in its genome by reading a DNA template to make an RNA copy (transcription), and by decoding the RNA to assemble protein (translation). The molecular machines that carry out these processes have been studied relentlessly, yet they still hold secrets.

After translation, each polypeptide must fold into its proper shape and find its correct cellular or extracellular location. How does the cell do this? And what does the cell do with proteins it no longer needs? As we will discuss, elegant chaperone pathways help fold proteins, complex transport systems move proteins out of the cytoplasm, and regulated proteolytic systems properly dispose of unneeded or damaged proteins. What emerges is a picture of remarkable biomolecular integration, controlled to maintain balanced growth and ensure survival.

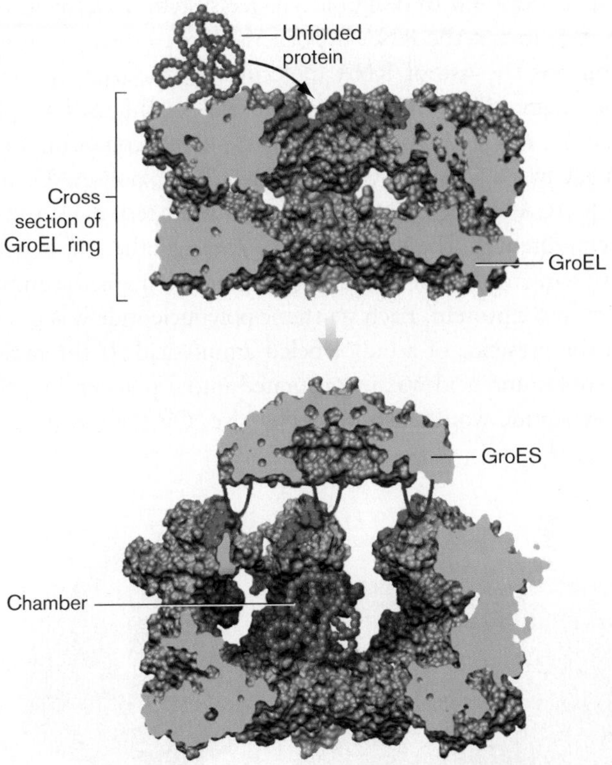

Unfolded protein
Cross section of GroEL ring
GroEL
GroES
Chamber

CURRENT RESEARCH highlight

The shifty chaperone. Proteins newly made or accidentally denatured in the cell need other proteins, called chaperones, to help them fold properly. Folding new proteins and rescuing (or destroying) damaged proteins are critical for cell growth and survival. But how do chaperones carry out their jobs? Daniel Clare and colleagues shed new light on this question. This graphic, derived from cryo-electron microscopy, shows the stunning shape-shifting changes made by the *E. coli* GroEL chaperone after capturing an unfolded substrate protein. The substrate enters one end of the GroEL barrel formed by two rings (top panel, cross section side view) and sticks to hydrophobic sites at the top of the rings. Binding of ATP to GroEL causes ring proteins (shown in red and orange) to twist and elevate (bottom panel). The shift allows the GroES cap protein to dock with GroEL (bottom in panel B). Docking seals the chamber and dislodges the substrate protein into the chamber for refolding. *Source:* Modified from Daniel Clare et al. 2012. *Cell* **149**:113–123.

Although it was clear by 1960 that the genetic code was embedded in DNA, the code itself and the mechanism by which the code produces protein remained mysteries. RNA was a suspected, but unproven, intermediate. The groundbreaking biochemist Marshall Nirenberg and Heinrich Matthaei, a postdoctoral student of Nirenberg's at the National Institutes of Health (**Fig. 8.1A**), set out in 1959–60 to design a cell-free system (a cell lysate of *E. coli*) to test the RNA hypothesis. Key to their investigation was the use of RNA molecules synthesized by Maxine Singer (**Fig. 8.1B**). These molecules contained simple, known, repeated sequences, such as poly-A (consisting only of adenylic acid), poly-U (polyuridylic acid), poly-AAU, and poly-ACAC. These RNA molecules were tested in a cell-free protein-synthesizing system to see whether the repetitive sequence might direct incorporation of a specific amino acid into a protein. Each synthetic polynucleotide was tested in the presence of a radiolabeled amino acid. If the radioactive amino acid was incorporated into a polypeptide, the polypeptide would also be radioactive. On the morning of May 27, 1960, the results of experiment 27Q showed that the poly-U RNA specified the assembly of radioactive polyphenylalanine. It was the first break in the genetic code.

B.

A.

COURTESY OF MAXINE SINGER

COURTESY OF NATIONAL LIBRARY OF MEDICINE/ NATIONAL INSTITUTES OF HEALTH

FIGURE 8.1 ■ Many scientists have contributed to our understanding of the genetic code. A. Heinrich Matthaei (left) and Marshall Nirenberg (right) were the first to crack the genetic code. An early hand-drawn model of what would eventually be known as translation is behind them. **B.** Maxine Singer, a key contributor to the genetic code experiments, also helped develop guidelines for recombinant DNA research.

News that Nirenberg's poly-U experiment had determined the first "word" of the genetic code was an international media event. In January 1962, the *Chicago Sun-Times* announced: "No stronger proof of the universality of all life has been developed since Charles Darwin's *The Origin of Species*." For his work, Nirenberg, along with Har Gobind Khorana (who developed methods for making synthetic nucleic acids) and Robert Holley (who solved the structure of yeast transfer RNA), won the 1968 Nobel Prize in Physiology or Medicine. For more on the discovery of the genetic code, see Special Topic 1.1 and **eTopic 1.2**.

Since Nirenberg's breakthrough, we have learned a tremendous amount about genes and proteins, and in this electronic age we can use the information in ways never before possible. A new discipline called bioinformatics uses powerful computer technologies to store and analyze gene and protein sequences. Bioinformatic programs can compare a new gene sequence with hundreds of thousands of known genes. These comparisons help predict for any given microbe what genes it has, what proteins it makes, and even what food it consumes. Patterns in DNA sequences across species also allow us to pose new questions about microbial life, disease, and evolution.

In Chapter 8 we explore the way microbes, primarily bacteria, interpret the nucleotide sequence of DNA and convert it to proteins. From there we look at what the cell does with those proteins once they are made. Specific proteins are targeted for movement into the periplasm, while other proteins must be inserted into membranes. We also show how damaged proteins are selectively degraded. We conclude with bioinformatics, now an essential tool of the new microbiology. Chapters 9 and 10 then discuss how microbes respond to their environment by controlling the expression of their genes.

8.1

RNA Polymerases and Sigma Factors

To survive and reproduce, every cell needs to access information encoded within DNA. How does this happen? Chromosomal DNA is large and cumbersome, so the first step in the process is to make multiple copies of the information in snippets of RNA that can move around the cell and, like disposable photocopies, be destroyed once the encoded protein is no longer needed. This copying of DNA to RNA is called **transcription**.

RNA Polymerase Transcribes DNA to RNA

An enzyme complex called RNA polymerase, also known as DNA-dependent RNA polymerase, carries out transcription, making RNA copies (called **transcripts**) of a DNA template. The DNA **template strand** specifies the base sequence of the new complementary strand of RNA. Like DNA replication, transcription to RNA occurs in three stages:

1. **Initiation**, which is the binding of RNA polymerase to the beginning of the gene, followed by melting of the helix and synthesis of the first nucleotide of the RNA
2. **Elongation**, the sequential addition of ribonucleotides from nucleoside triphosphates to the 3′ OH end of a growing chain
3. **Termination**, in which sequences at the end of the gene trigger release of the polymerase and the completed RNA molecule

RNA polymerase in bacteria consists of a "core polymerase" and a sigma factor. Core polymerase contains the proteins required to elongate an RNA chain. **Sigma factor** is a protein needed only for initiation of RNA synthesis, not for its elongation. Together, core polymerase plus sigma factor are called "holoenzyme."

Core RNA polymerase is a complex of four different subunits: two alpha (α) subunits, one beta (β) subunit, and one beta-prime (β′) subunit for every core enzyme (**Fig. 8.2**). The beta-prime subunit houses the Mg^{2+}-containing catalytic site for RNA synthesis, as well as sites for the ribonucleotide substrates, the DNA substrates, and the RNA products. The 3D structure of RNA polymerase resembles a hand, whose "fingers" consist of the beta and beta-prime subunits. DNA fits into a cleft formed by the beta and beta-prime subunits (**Fig. 8.2**). The alpha subunit assembles the other two subunits into a functional complex and communicates through physical "touch" with various regulatory proteins that can bind DNA. The resulting protein-protein communications tell RNA polymerase what to do after binding DNA.

Promoter binding. Where does a gene begin and end? The helical structure of a DNA sequence appears largely homogeneous compared to proteins, which bend and fold in complex ways. In fact, without sigma factor, the core RNA polymerase binds and releases DNA at random. Yet there must be a mechanism that tells RNA polymerase where a gene starts, because random transcription is extremely wasteful.

The proteins that guide RNA polymerase to genes are the sigma (σ) factors (**Fig. 8.2**). A sigma factor binds RNA polymerase through the beta and beta-prime subunits and then helps the core enzyme detect a specific DNA sequence, called a **promoter**, which signals the beginning of a gene. Later in this section we will discuss how sigma recognizes a promoter.

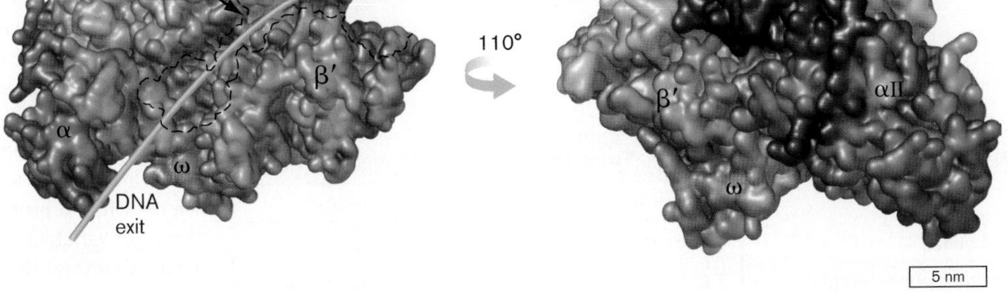
σ3
σ2
σ1
σ4
β
DNA channel
β′
α
ω
DNA exit
110°
β
αI
β′
αII
ω
5 nm

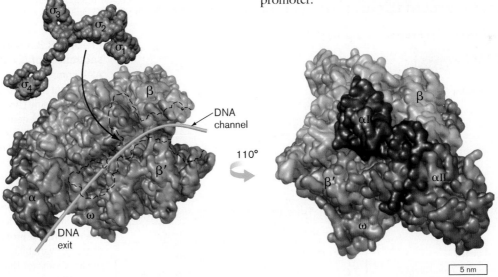

FIGURE 8.2 ■ Subunit structure of RNA polymerase. Two views of RNA polymerase. The channel for the DNA template is shown by the yellow line. Subunits (αI, αII, β, β′, and ω) are color-coded dark green, medium green, light green, cyan, and gold, respectively. The function of the omega (ω) subunit is currently unclear. Recent evidence suggests it may have a role in sigma factor competition for core polymerase. Sigma factor (red), which recognizes promoters on DNA, is shown separate from core polymerase in panel A. Different functional areas of sigma are labeled sigma 1 through sigma 4. Sigma factor interacts with the alpha, beta and beta′ subunits. The molecule in (A) is rotated 110° to give the image in (B). To view stereo images of RNA polymerase, locate code 1L9Z in the RCSB Protein Data Bank on the Internet.
Source: Robert D. Finn et al. 2000. *EMBO J.* **19**:6833–6844.

A single bacterial species can make several different sigma factors, with each sigma factor helping core RNA polymerase find the start of a different subset of genes. However, a single core polymerase complex can bind only one sigma factor at a time. By acting as a sort of "seeing-eye" protein, sigma factors help RNA polymerase recognize different classes of promoter sequences. The specific sigma factor used to initiate transcription of a given gene will vary, depending on the gene and the environmental signals needed to initiate transcription of that gene. But regardless of which sigma factor is used, core RNA polymerase is required for all gene transcription.

Sigma factors regulate major physiological responses. Individual sigma factors coordinately control genes involved in nitrogen metabolism, flagellar synthesis, heat stress, starvation, sporulation, and many other physiological responses. So far, over 100 sigma factor genes have been sequenced from numerous species. Although they all have different amino acid sequences, they also show sequence similarities that make them recognizable as sigma factors. We can determine the DNA sequence recognized by a given sigma factor by comparing known promoter sequences of different genes whose expression requires the same sigma. Similarities among these different promoter DNA sequences define a **consensus sequence** likely recognized by the sigma factor (**Fig. 8.3A** and **B**). A consensus sequence consists of the most likely base (or bases) at each position of the predicted promoter. Although the promoter is a double-stranded DNA (dsDNA) sequence, convention is to present the promoter as the single-stranded DNA (ssDNA) sequence of the sense strand, which has the same sequence as the RNA product.

Some positions in a consensus sequence are highly conserved, meaning the same base is found in that position in every promoter. Other, less conserved positions can be occupied by different bases. Few promoters will actually have the most common base at every position. Even highly efficient promoters usually differ from the consensus at one or two positions.

A. Strong *E. coli* promoters

```
                                -35                        -10              +1
tyr tRNA     TCTCAACGTAACACTTTACAGCGGCG••CGTCATTTGATATGATGC•GCCCCGCTTCCCGATAAGGG
rrn D1       GATCAAAAAAATACTTGTGCAAAAAA••TTGGGATCCCTATAATGCGCCTCCGTTGAGACGACAACG
rrn X1       ATGCATTTTTCCGCTTGTCTTCCTGA••GCCGACTCCCTATAATGCGCCTCCATCGACACGGCGGAT
rrn (DXE)2   CCTGAAATTCAGGGTTGACTCTGAAA••GAGGAAAGCGTAATATAC•GCCACCTCGCGACAGTGAGC
rrn E1       CTGCAATTTTTCTATTGCGGCCTGCG••GAGAACTCCCTATAATGCGCCTCCATCGACACGGCGGAT
rrn A1       TTTTAAATTTCCTCTTGTCAGGCCGG••AATAACTCCCTATAATGCGCCACCACTGACACGGAACAA
rrn A2       GCAAAAATAAATGCTTGACTCTGTAG••CGGGAAGGCGTATTATGC•ACACCCCGCGCCGCTGAGAA
λ PR         TAACACCGTGCGTGTTGACTATTTTA•CCTCTGGCGGTGATAATGG••TTGCATGTACTAAGGAGGT
λ PL         TATCTCTGGCGGTGTTGACATAAATA•CCACTGGCGGTGATACTGA••GCACATCAGCAGGACGCAC
T7 A3        GTGAAACAAAACGGTTGACAACATGA•AGTAAACACGGTACGATGT•ACCACATGAAACGACAGTGA
T7 A1        TATCAAAAAGAGTATTGACTTAAAGT•CTAACCTATAGGATACTTA•CAGCCATCGAGAGGGACACG
T7 A2        ACGAAAAACAGGTATTGACAACATGAAGTAACATGCAGTAAGATAC•AAATCGCTAGGTAACACTAG
fd VIII      GATACAAATCTCCGTTGTACTTTGTT••TCGCGCTTGGTATAATCG•CTGGGGGTCAAAGATGAGTG
```

B. Consensus sequences of σ^{70} promoters

```
-35 region             -10 region
TTGACAT ——17 ± 1 bp—— TATAAT
```

C. Sequence of *lac* promoter

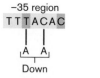

```
-35 region     -10 region
TTTACAC        TATGTT
  |  |            ||
  A  A            AA
  Down            Up
```

D.

Sigma factor	Promoter recognized	Promoter consensus sequence	
		−35 region	−10 region
RpoD σ^{70}	Most genes	TTGACAT	TATAAT
RpoH σ^{32}	Heat shock–induced genes	TCTCNCCCTTGAA	CCCCATNTA
RpoF σ^{28}	Genes for motility and chemotaxis	CTAAA	CCGATAT
RpoS σ^{38}	Stationary-phase and stress response genes	TTGACA	TCTATACTT
		−24 region	−12 region
RpoN σ^{54}	Genes for nitrogen metabolism and other functions	CTGGNA	TTGCA

FIGURE 8.3 ▪ –10 and –35 sequences of *E. coli* promoters. **A.** Alignment of sigma-70 (σ^{70})-dependent promoters from different genes [tyrosine (tyr) tRNA; several ribosomal RNA genes (rrn); and genes from lambda, T7, and fd phages]. Yellow indicates conserved nucleotides; brown denotes a transcript start site (+1). **B.** The alignment in (A) was used to generate consensus sequences of σ^{70}-dependent promoters (red-screened letters indicate nucleotide positions where different promoters show a high degree of variability). **C.** Mutations in the *lac* promoter that affect promoter strength (*lac* genes encode proteins that are used to metabolize the carbohydrate lactose). Some mutations can cause decreased transcription (called "down mutations"), while others cause increased transcription ("up mutations"). **D.** Some *E. coli* promoter sequences recognized by different sigma factors.

How does a sigma factor control gene expression? Microbes are guarded by an array of protein sensors that continually sample the internal and external environments. These sensors "look" for chemical deficiencies or dangers and, when triggered, direct the cell to increase synthesis or decrease destruction of the appropriate sigma factor. Accumulation of a specialty sigma factor dislodges other sigma factors from core polymerase and redirects RNA polymerase to the promoters of genes whose products can best solve the problem sensed. Regulation of multiple genes by a single sigma factor enables the cell to coordinately control those genes simply by regulating when a given sigma factor accumulates (Section 10.4). This control is critical to survival in constantly changing environments.

Thought Question

8.1 If each sigma factor recognizes a different promoter, how does the cell manage to transcribe genes that respond to multiple stresses, each involving a different sigma factor?

Every cell has a "housekeeping" sigma factor that keeps essential genes and pathways operating. In the case of *E. coli* and other rod-shaped, Gram-negative bacteria, that factor is sigma-70, so named because it is a 70-kilodalton (kDa) protein (its gene designation is *rpoD*). Genes recognized by sigma-70 all contain similar promoter sequences that consist of two parts. The DNA base corresponding to the start of the RNA transcript is called nucleotide +1 (+1 nt). Relative to this landmark, promoter sequences are characteristically centered at –10 and –35 nt before the start of transcription (see **Fig. 8.3**). Other sigma factors typically recognize different consensus sequences at one or both of these positions (or at nearby locations in some cases), or they shape the overall polymerase complex to bind the promoter.

How do sigma factors, or any other DNA-binding proteins for that matter, recognize specific DNA sequences when the DNA is a double helix? The phosphodiester backbone is quite uniform, and the interior of paired bases appears inaccessible. Recognition of DNA sequences is possible, however, because the alpha-helical portions of DNA-binding proteins recognize (via noncovalent bonding) base side groups that protrude from the base into the major and minor grooves of DNA. Portions of sigma-70 from *E. coli* wrap around DNA, allowing certain parts of the protein to fit into DNA grooves.

A key to understanding how sigma factors recognize different promoters is their structure. **Figure 8.4** shows the holoenzyme RNA polymerase complex positioned at a promoter and the points where sigma factor contacts DNA.

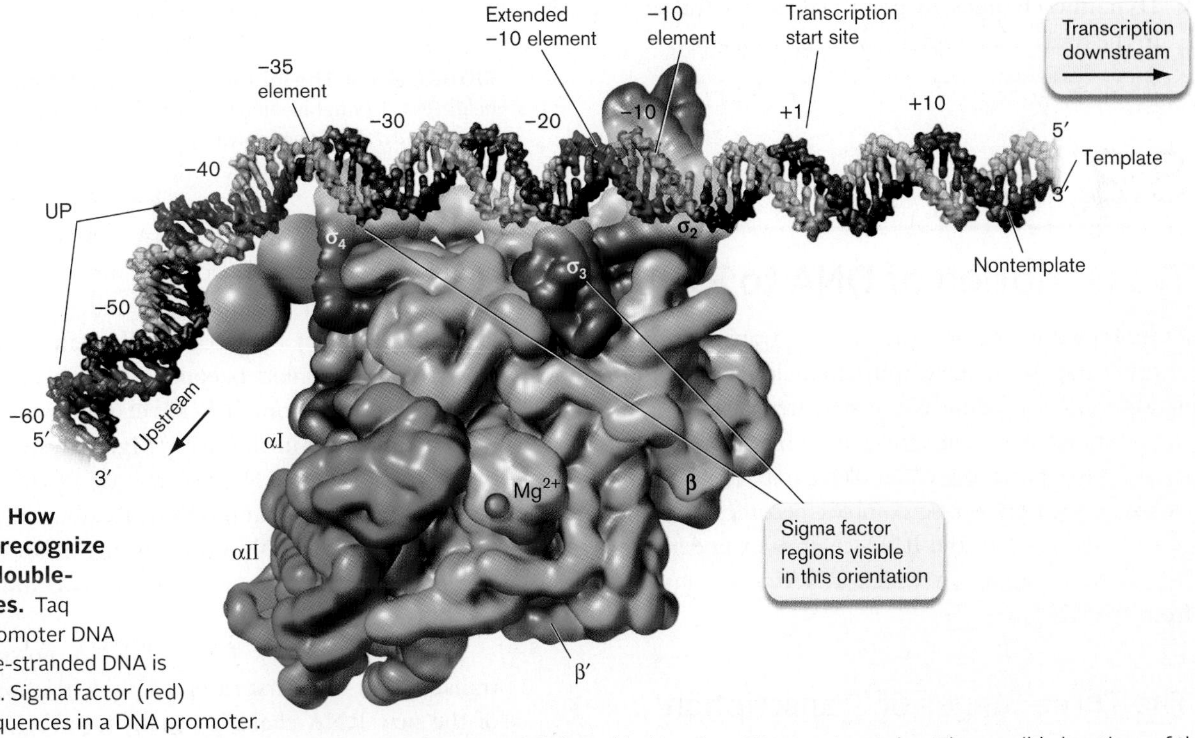
Extended
–10 element
–10
element
Transcription
start site
Transcription
downstream
–35
element
–30
–20
–10
+1
+10
5′
Template
3′
–40
UP
σ_4
σ_2
σ_3
Nontemplate
–50
–60
5′
3′
Upstream
αI
αII
Mg^{2+}
β
β′
Sigma factor
regions visible
in this orientation

FIGURE 8.4 ■ How sigma factors recognize specific DNA double-helix sequences. Taq holoenzyme–promoter DNA complex. Double-stranded DNA is shown as atoms. Sigma factor (red) binds to two sequences in a DNA promoter. Region 2 of sigma binds to the –10 region of promoters, while region 4 of sigma binds to the –35 promoter region. The possible locations of the two alpha subunit C-terminal domains (drawn as green spheres labeled I and II) on the DNA UP elements are illustrated. The UP element is a DNA sequence that increases transcription. The location of magnesium ion in the polymerase active site is shown. (PDB codes: 1L9Z, 1K8J) *Source:* Parts A and C from Seth A. Darst. 2001. *Curr. Opin. Struct. Biol.* **11**:155–162.

Amino acid sequence comparisons indicate that sigma factors generally contain four highly conserved regions (**Fig. 8.2** and **8.4**). Part of region 2 of the sigma-70 family recognizes –10 sequences, whereas region 4 recognizes the –35 sites. A part of region 1 helps separate the DNA strands in preparation for RNA synthesis.

Thought Questions

8.2 Imagine two different sigma factors with different promoter recognition sequences. What would happen to the overall gene expression profile in the cell if one sigma factor were artificially overexpressed? Could there be a detrimental effect on growth?

8.3 Why might some genes contain multiple promoters, each one specific for a different sigma factor?

To Summarize

- **RNA polymerase holoenzyme**, consisting of core RNA polymerase and a sigma factor, executes transcription of a DNA template strand.
- **Sigma factors** help core RNA polymerase locate consensus promoter sequences near the beginning of a gene. The sequences identified and bound by *E. coli* sigma-70 are located –10 and –35 bp upstream of the transcription start site.
- **Dynamic changes in gene expression follow** changes in the relative levels of different sigma factors.

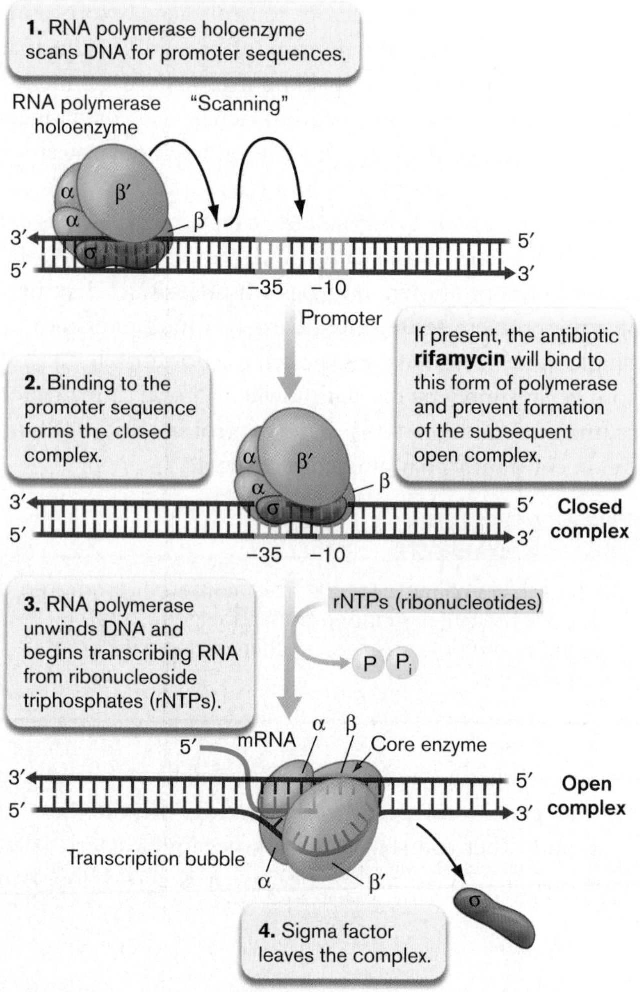

1. RNA polymerase holoenzyme scans DNA for promoter sequences.
RNA polymerase holoenzyme
"Scanning"
α
β′
α
β
σ
3′
5′
5′
3′
–35
–10
Promoter
2. Binding to the promoter sequence forms the closed complex.
If present, the antibiotic rifamycin will bind to this form of polymerase and prevent formation of the subsequent open complex.
Closed complex
3. RNA polymerase unwinds DNA and begins transcribing RNA from ribonucleoside triphosphates (rNTPs).
rNTPs (ribonucleotides)
P
P_i
mRNA
Core enzyme
Open complex
Transcription bubble
4. Sigma factor leaves the complex.

FIGURE 8.5 ■ The initiation of transcription. Sigma factor helps RNA polymerase find promoters, but is discarded after the first few RNA bases are polymerized. (Omega is not shown.)

8.2

Transcription of DNA to RNA

As described in the previous section, transcription consists of three stages—namely, initiation, elongation, and termination. Sigma binding is the first step of initiation. Once the promoter has been activated by binding with its sigma factor, RNA polymerase completes initiation, in which the closed RNA polymerase complex becomes an open complex. In elongation, the RNA chain is extended. The final step is termination, in which RNA polymerase detaches from the DNA.

The Three Stages of Transcription

Initiation of transcription. RNA polymerase constantly scans DNA for promoter sequences (**Fig. 8.5**, step 1). It binds DNA loosely and comes off repeatedly. Once bound to the promoter, RNA polymerase holoenzyme forms a loosely bound, closed complex with DNA, which remains annealed and double-stranded (that is, unmelted) (**Fig. 8.5**, step 2). To successfully transcribe a gene, this closed complex must become an open complex through the unwinding of one helical turn, which causes DNA to become unpaired in this area (**Fig. 8.5**, step 3). After promoter unwinding, RNA polymerase in the open complex becomes tightly bound to DNA. Region 1 of the sigma factor is important for DNA unwinding, as is a "rudder" complex in beta-prime that separates the two strands (shown in **Fig. 8.6A**).

The open-complex form of RNA polymerase begins transcription. The first ribonucleoside triphosphate (rNTP) of the new RNA chain is usually a purine (A or G). The purine base-pairs to the position designated +1 on the DNA template, which marks the start of the gene. As the enzyme complex moves along the template, subsequent

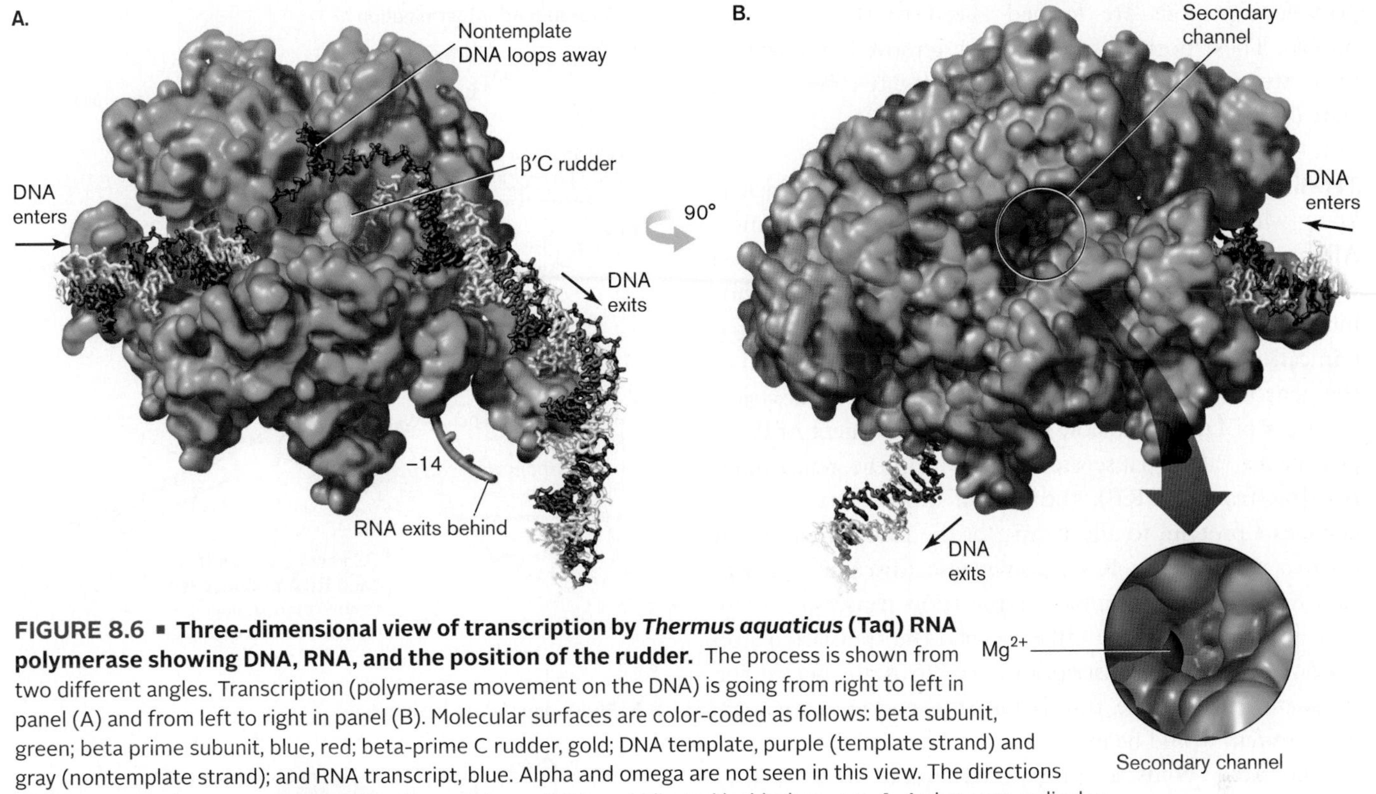
A.
Nontemplate
DNA loops away
β′C rudder
DNA
enters
DNA
exits
−14
RNA exits behind
90°
B.
Secondary
channel
DNA
enters
DNA
exits
Mg2+
Secondary channel

FIGURE 8.6 ▪ Three-dimensional view of transcription by *Thermus aquaticus* (Taq) RNA polymerase showing DNA, RNA, and the position of the rudder. The process is shown from two different angles. Transcription (polymerase movement on the DNA) is going from right to left in panel (A) and from left to right in panel (B). Molecular surfaces are color-coded as follows: beta subunit, green; beta prime subunit, blue, red; beta-prime C rudder, gold; DNA template, purple (template strand) and gray (nontemplate strand); and RNA transcript, blue. Alpha and omega are not seen in this view. The directions of the entering downstream DNA and exiting upstream DNA are indicated by black arrows. **A.** A view perpendicular to the main active-site channel, which runs roughly horizontal. Parts of the protein structure are colored. **B.** In this view, the structure is rotated 90° relative to (A). With the entering DNA removed for clarity, Mg^{2+} is seen in the catalytic site. The blowup shows the secondary channel where nucleotides (orange) enter, as well as a part of the RNA strand (blue). (PDB codes: 1L9Z and 1K8J) *Source:* Seth A. Darst. 2001. *Curr. Opin. Struct. Biol.* **11**:155–162.

rNTPs diffuse through a channel in the polymerase and into position at the DNA template (**Fig. 8.6B**). After the first base is in place, each subsequent rNTP transfers a ribonucleoside monophosphate (rNMP) to the growing chain while releasing a pyrophosphate (PP_i):

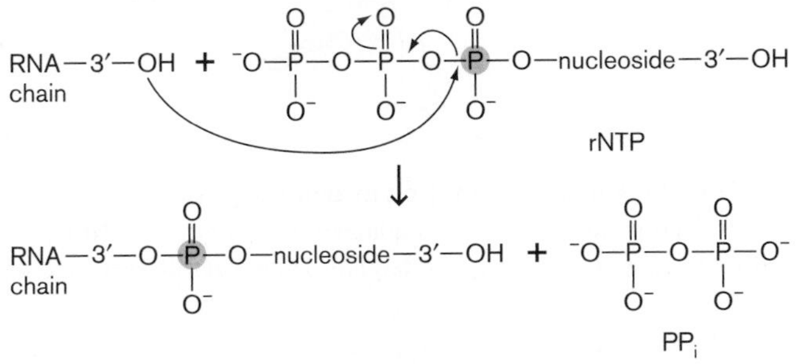
RNA—3′—OH
chain
+
⁻O—P—O—P—O—P—O—nucleoside—3′—OH
rNTP
RNA—3′—O—P—O—nucleoside—3′—OH
chain
+
⁻O—P—O—P—O⁻
PPi

The "energy released" (actually the free energy change) following cleavage of the rNTP triphosphate groups is used to form the phosphodiester link to the growing polynucleotide chain. (Discussion of free energy change, ΔG, can be found in A1.7 and Chapter 13.)

As seen in **Figure 8.6**, the beta-prime subunit forms part of the pocket, or cleft, in which the DNA template is read. Another part of the beta-prime cleft is involved in hydrolyzing rNTPs. Deep at the base of the cleft is the active site of polymerization, defined by three evolutionarily conserved aspartate residues that chelate (bond with) two magnesium ions (Mg^{2+}). As we saw for DNA polymerases, these metal ions play a key role in catalyzing the polymerization reaction. (Mg^{2+} links the 3′ OH group with an aspartate at the active site, and lowers the pKa of the 3′ OH. The lower pKa facilitates dissociation of the proton and formation of the phosphodiester bond.)

Weblinks 3D view of RNA polymerase interaction with DNA (*see ebook*)

Elongation of RNA transcripts. A transcribing complex retains the sigma factor until about nine bases have been joined, and then it dissociates (**Fig. 8.5**, step 4). The newly liberated sigma factor can then reassociate with an unbound core RNA polymerase to direct another round of promoter binding. Meanwhile, the original RNA polymerase continues to move along the template, synthesizing RNA at approximately 45 bases per second. A 17-bp transcription bubble forms as DNA unwinds ahead of the moving polymerase complex. Because of this unwinding, positive supercoils are formed ahead of the advancing bubble. These positive supercoils are removed by enzymes that return the DNA to its normal negative supercoiled state (see Section 7.2).

Termination of transcription. How does RNA polymerase know when to stop? Again, the secret is in the sequence. All bacterial genes use one of two known transcription termination signals, called Rho-dependent and Rho-independent. **Rho-dependent** termination relies on a protein called Rho and an ill-defined sequence at the 3′ end of the gene that appears to be a strong pause site. Pause sites are areas of DNA that slow or stall the movement of RNA polymerase. Some pause sites occur within the gene's **open reading frame (ORF)**, the part of a gene that actually encodes a protein, to allow time for the ribosome to catch up to a transcribing RNA polymerase (discussed in Section 8.3). The transcription termination pause site, however, is located after the ORF, beyond the translation stop codon, because if transcription were to cease before the ribosome reached the translation stop codon, an incomplete protein would be made.

Rho factor binds to an exposed region of RNA after the ORF at GC-rich sequences that lack obvious secondary structure. This is the transcription terminator pause site. Rho monomers assemble as a hexamer around the RNA (**Fig. 8.7A**). Then, like a person pulling a raft to shore by a rope tied to a tree, Rho pulls itself to the paused RNA polymerase by wrapping downstream RNA around itself via intrinsic ATPase activity. Once Rho touches the polymerase, an RNA-DNA helicase activity built into Rho appears to unwind the RNA-DNA heteroduplex, which releases the completed RNA molecule and frees the RNA polymerase.

The second type of termination event, called **Rho-independent**, occurs in the absence of Rho. Rho-independent termination requires a GC-rich region of RNA roughly 20 bp upstream from the 3′ terminus, as well as four to eight consecutive uridine residues at the terminus. The GC-rich sequence contains complementary bases and forms an RNA stem, or stem loop, a structure that "grabs" RNA polymerase, causing it to pause (**Fig. 8.7B**). A protein called NusA stimulates a transcription pause at these sites. While the polymerase is paused, the DNA-RNA duplex is weakened because the poly-U/poly-A base pairs at the 3′ terminus contain only two hydrogen bonds per pair, so they are easier to melt. Melting the hybrid molecule releases the transcript and halts transcription. The pause in polymerase movement, and thus transcription, is important to prevent the formation of tighter base-pairing downstream of the UA region.

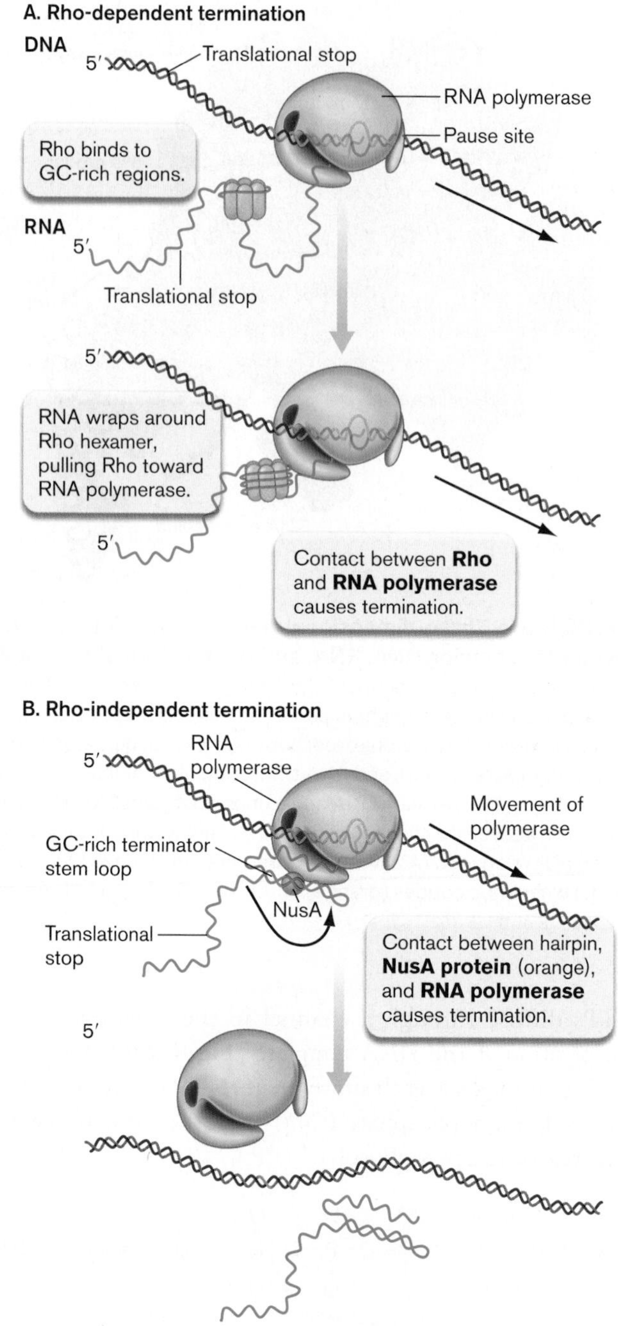
A. Rho-dependent termination
DNA
5′
Translational stop
RNA polymerase
Pause site
Rho binds to GC-rich regions.
RNA
5′
Translational stop
5′
RNA wraps around Rho hexamer, pulling Rho toward RNA polymerase.
5′
Contact between Rho and RNA polymerase causes termination.
B. Rho-independent termination
5′
RNA polymerase
Movement of polymerase
GC-rich terminator stem loop
NusA
Translational stop
Contact between hairpin, NusA protein (orange), and RNA polymerase causes termination.
5′

FIGURE 8.7 ■ The termination of transcription.
A. Rho-dependent termination requires Rho factor but not NusA.
B. Rho-independent termination requires NusA but not Rho.

Antibiotics That Affect Transcription

To be useful in medicine, all antibiotics must meet two fundamental criteria: They must kill or retard the growth of a pathogen, and they must not harm the host. Antibiotics used against bacteria, for instance, must selectively attack features of bacterial targets not shared with eukaryotes.

Many antibiotics possess highly specific modes of action, binding to and altering the activity of one specific "machine" of one type of cell.

One example is the antibiotic rifamycin B (**Fig. 8.8A**), produced by the actinomycete *Amycolatopsis mediterranei* (**Fig. 8.8B**). Rifamycin B selectively targets bacterial RNA polymerase and binds to the bacterial RNA polymerase beta subunit near the Mg^{2+} active site, thus blocking the exit channel (**Fig. 8.8C**). RNA polymerase can carry out two or three polymerization steps, but then it stops because the nascent RNA cannot exit. Rifamycin does not prevent RNA polymerase from binding to promoters, and because an mRNA molecule already passing through the exit channel masks the rifamycin-binding site, the drug cannot bind to RNA polymerases already transcribing DNA. The semisynthetic derivative of this drug, rifampin, is used for treating tuberculosis and leprosy and is given to contacts (family members or roommates) of individuals with bacterial meningitis caused by *Neisseria meningitidis.*

Actinomycin D (**Fig. 8.9A**) is another antibiotic produced by an actinomycete. Its phenoxazone ring is a planar structure that "squeezes," or intercalates, between GC base pairs within DNA. The side chains then extend in opposite directions along the minor groove (**Fig. 8.9B**). Because it mimics a DNA base, actinomycin D blocks the elongation phase of transcription; but because it binds any DNA, it is not selective for bacteria. It can be used to treat human cancers because cancer cells replicate rapidly, but it has severe side effects because it inhibits DNA synthesis in normal cells too. (Further discussion of antibiotics can be found in Chapter 27.)

Antibiotic Resistance Helped Reveal Components of RNA Synthesis Machines

How did scientists determine the details of RNA and protein synthesis? The processes were observed in cell-free systems. For example, transcription and translation were observed in the test tube using purified RNA polymerase and ribosomes. In addition, the various protein subunits of RNA polymerase were separated from each other and mixed together again to make an active complex.

A.

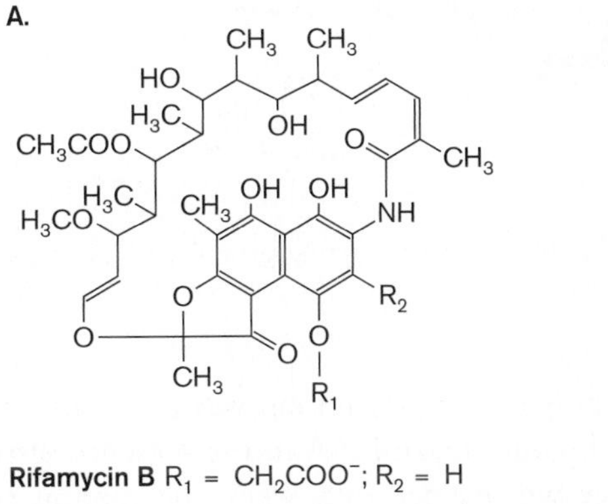
Rifamycin B R1 = CH2COO−; R2 = H
Rifampin R1 = H; R2 = CH=N+ N−CH3

B.

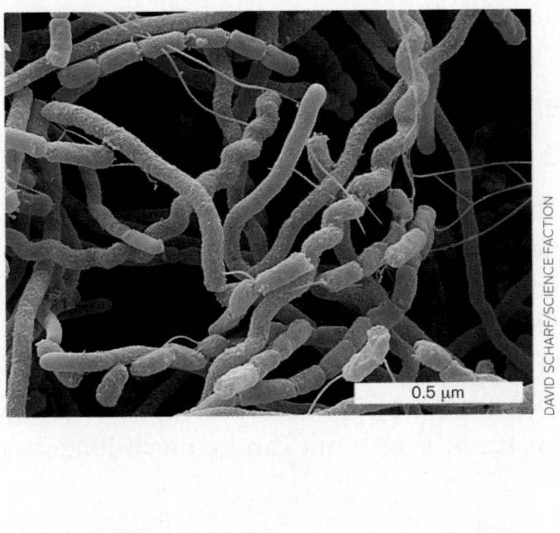
0.5 µm
DAVID SCHARF/SCIENCE FACTION

C.

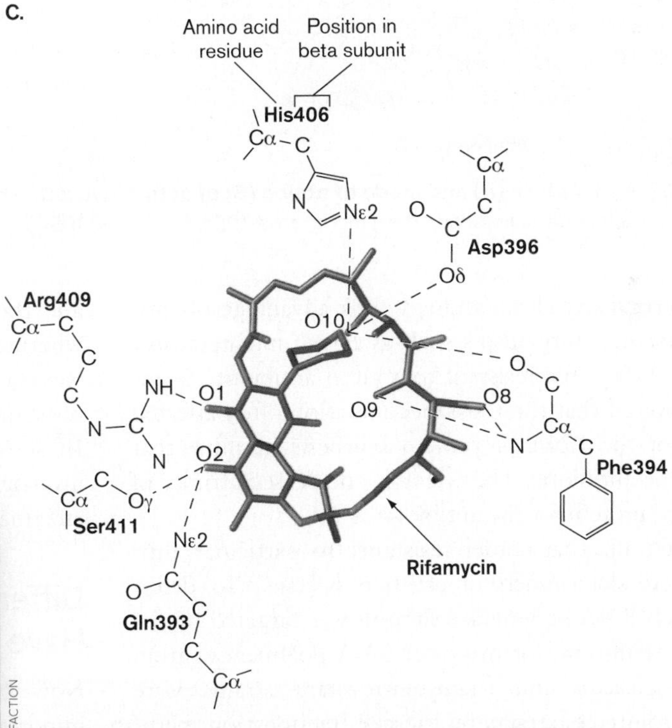
Amino acid residue
Position in beta subunit
His406
Asp396
Arg409
Ser411
Gln393
Phe394
Rifamycin

FIGURE 8.8 ■ Structure and mode of action of rifamycin. **A.** Structure of rifamycin. The R groups indicated are added to alter the structure and pharmacology of the basic structure. **B.** Electron micrograph of *Amycolatopsis.* **C.** Contact points between rifamycin and residues in the beta subunit of RNA polymerase.

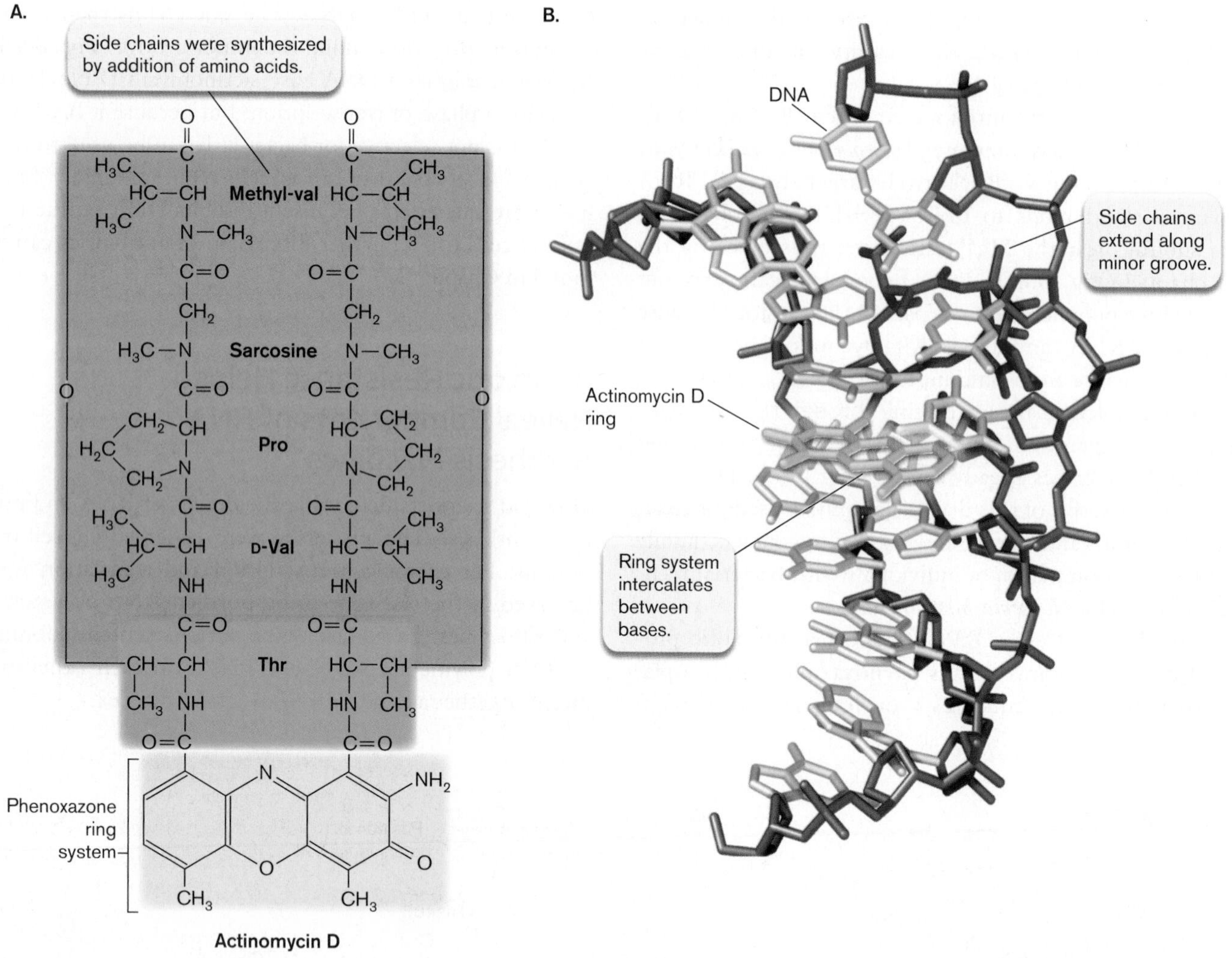
A.
Side chains were synthesized by addition of amino acids.
Methyl-val
Sarcosine
Pro
D-Val
Thr
Phenoxazone ring system
Actinomycin D
B.
DNA
Side chains extend along minor groove.
Actinomycin D ring
Ring system intercalates between bases.

FIGURE 8.9 ■ Structure (A) and mode of action (B) of actinomycin D. This antibiotic inserts its ring structure between parallel DNA bases and wraps its side chains along the minor groove. (PDB code for B: 1DSC)

One particularly clever strategy took advantage of antibiotic resistance. Organisms such as *E. coli* mutate at low frequency to become resistant to a given antibiotic. Scientists discovered that the resistance mutation often affected the part of the RNA or protein synthesis machine that binds to the antibiotic. The effect is a decreased affinity of that altered protein for the antibiotic.

The subunits that confer resistance to particular antibiotics were determined in cell-free systems. To determine which RNA polymerase subunit was targeted by the antibiotic rifamycin, for instance, RNA polymerases from rifamycin-sensitive and rifamycin-resistant strains were purified from cell extracts by the cell fractionation methods introduced in Chapter 3. Then component parts of the polymerase were separated, and a chimeric RNA polymerase was reassembled (like a 3D jigsaw puzzle) using all but one of the protein subunits from the sensitive strain. The missing subunit was supplied by the resistant strain, and the polymerase preparation was tested in vitro to see whether it could make RNA in the presence of rifamycin. Resistance will happen only when the subunit conveying resistance has been added—in this case, the beta subunit. In subsequent in vitro assays and X-ray crystallography, investigators used this information to learn more about the enzymatic mechanism of the beta subunit.

Different Classes of RNA Have Different Functions

Now that we have shown how RNA is synthesized, it is important to note that not all RNA molecules are translated to protein. There are several classes of RNA, each designed for a different purpose (**Table 8.1**). The class of RNA molecules that encode proteins is called **messenger RNA (mRNA)**. Molecules of mRNA average 1,000–1,500 bases in length but can be much longer or shorter,

TABLE 8.1

Classes of RNA in *E. coli*.[a]

RNA class	Function	Number of types	Average size	Approximate half-life	Unusual bases
mRNA (messenger RNA)	Encodes protein	Thousands	1,500 nt	3–5 minutes	No
rRNA (ribosomal RNA)	Synthesizes protein as part of ribosome	3	5S, 120 nt; 16S, 1,542 nt; 23S, 2,905 nt	Hours	Yes
tRNA (transfer RNA)	Shuttles amino acids	41 (86 genes)	80 nt	Hours	Yes
sRNA[b] (small RNA)	Controls transcription, translation, or RNA stability	20–30	<100 nt	Variable	No
tmRNA (properties of transfer and messenger RNA)	Frees ribosomes stuck on damaged mRNA	Roughly one per species	300–400 nt	3–5 minutes	No
Catalytic RNA	Carries out enzymatic reactions (e.g., RNase P)	?	Varies	3–5 minutes	No

[a]Six major classes of RNA exist in all bacteria. They differ in their function, quantity, average sizes, and half-lives, and whether they contain modified bases.

[b]Small RNAs include antisense RNA and micro-RNA.

depending on the size of the protein they encode. Another class of RNA, called **ribosomal RNA (rRNA)**, forms the scaffolding on which ribosomes are built. As will be discussed later, rRNA also forms the catalytic center of the ribosome. **Transfer RNA (tRNA)** molecules ferry amino acids to the ribosome. A unique property of rRNA and tRNA is the presence of unusual modified bases not found in other types of RNA.

The fourth important class of RNA is called **small RNA (sRNA)**. Molecules of sRNA do not encode proteins but are used to regulate the stability or translation of specific mRNAs into proteins (discussed in Chapter 10). As will be discussed later in more detail, all mRNA molecules contain untranslated leader sequences that precede the actual coding region. Complementary sequences within some of these RNA leader regions snap back on themselves to form double-stranded stems and stem loop structures that can obstruct ribosome access and limit translation. Regulatory sRNAs can base-pair with these regions in mRNA and either disrupt or stabilize intrastrand stem structures.

The final two classes of RNA are **tmRNA**, which has properties of both tRNA and mRNA (discussed later), and **catalytic RNA**. Catalytic RNA is usually found associated with proteins in which the enzymatic (catalytic) activity actually resides in the RNA portion of the complex rather than in the protein. All of these functional RNAs may represent remnants of the ancient "RNA world," where the earliest ancestral cells were built of RNA parts whose functions were later assumed by proteins.

RNA Stability

Once released from RNA polymerase, most prokaryotic mRNA transcripts are doomed to a short existence owing to their degradation by intracellular RNases. Why would a cell tolerate this seemingly wasteful practice? Remember, bacteria face extremely rapid changes in their environment, including changes in temperature, salinity, or nutrients, to name a few. To survive, cells must be prepared to react quickly and halt synthesis of superfluous or even detrimental proteins. An effective way to do this is to rapidly destroy the mRNAs that encode those proteins. Unfortunately, rapid mRNA degradation makes it necessary to continually transcribe those genes as long as their products are needed.

RNA stability is measured in terms of half-life, which is the length of time the cell needs to degrade half the molecules of a given mRNA species. The average half-life for mRNA is 1–3 minutes but can be as little as 15 seconds. Other RNAs can be fairly stable, with half-lives of 10–20 minutes or longer. However, the stabilities of the different kinds of RNAs differ drastically. The mRNAs are usually unstable compared to rRNAs and tRNAs, which contain

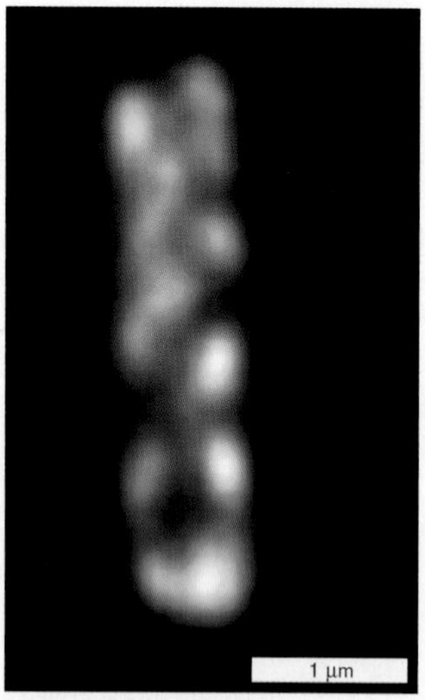
1 μm

FIGURE 8.10 ■ RNA degradosomes are arranged as a helix in the cell. Fluorescent micrograph of an RNase E–YFP fusion protein in *Escherichia coli.* RNase E, part of the degradosome, is organized as helical cytoskeletal structures along the cytoplasmic membrane. Source: From Aziz Taghbalout and Lawrence Rothfield. 2007. RNaseE and the other constituents of the RNA degradosome are components of the bacterial cytoskeleton. *Proceedings of the National Academy of Sciences USA* **104**: 1667–1672.

modified bases less susceptible to RNase digestion. These more stable RNA molecules have half-lives of hours.

Recent discoveries show that components of the major RNA-degrading machine of *E. coli*—a four-protein complex including an RNase, an RNA helicase, and two metabolic enzymes—are organized in a helical cytoskeletal structure associated with the cell membrane (**Fig. 8.10**). These findings suggest that the RNA degradosome is compartmentalized within the cell. By preventing free access of the RNase to cytoplasmic RNA, compartmentalization may enable regulation of RNA degradation.

To Summarize

- **A closed complex** occurs on initial binding of RNA polymerase to promoter DNA.
- **Opening (melting) of the DNA** strands produces a "bubble" of DNA around the polymerase and forms the open complex.
- **Rifamycin B** is an antibiotic that selectively inhibits transcription initiation by bacterial RNA polymerase.
- **Actinomycin D** is an antibiotic that nonselectively inhibits transcription elongation.
- **Rho-dependent and Rho-independent** mechanisms mediate transcription termination.
- **Messenger RNA** (mRNA) molecules encode proteins.
- **Ribosomal RNAs** (rRNAs) are stable molecules found in ribosomes and contain unusual modified bases.
- **Transfer RNAs** (tRNAs) are stable molecules that bind amino acids and contain modified bases.
- **Small RNA** molecules can regulate gene expression (sRNA), have catalytic activity (catalytic RNA), or function as a combination of tRNA and mRNA (tmRNA).
- **Half-lives of RNA** species vary within the cell.

8.3

Translation of RNA to Protein

Once a gene has been copied into mRNA, the next stage is **translation**, the decoding of the RNA message to synthesize protein. An mRNA molecule can be thought of as a sentence in which triplets of nucleotides, called **codons**, represent individual words, or amino acids (**Fig. 8.11**). Ribosomes are the machines that read the language of mRNA and convert, or translate, it into protein. Translation involves numerous steps, the first of which is the search by ribosomes for the beginning of an mRNA protein-coding region. Because the code consists of triplet codons, a ribosome must start translating at precisely the right base (in the right frame) or the product will be gibberish. Before we discuss how the ribosome finds the right reading frame, let's review the code itself and the major players in translation.

The Genetic Code and tRNA Molecules

When learning a new language, we need a dictionary that converts words from one language into the other. In the case of gene expression, how do we convert from the four-base language of RNA to the protein language of 20 amino acids? Through painstaking effort, Marshall Nirenberg, Har Gobind Khorana, and colleagues cracked the molecular code and found that each codon (a triplet of nucleotides) represents an individual amino acid. Remarkably, and with few exceptions, the code operates universally across species. (Exceptions include UGA, which is normally read as a stop codon but encodes tryptophan in vertebrate mitochondria, *Saccharomyces,* and mycoplasmas; and CUG, which encodes serine instead of leucine in *Candida albicans.*) In the code presented in **Figure 8.11**, we can see that most amino acids have multiple codon synonyms. Glycine (Gly), for example, is translated from four different codons, and leucine (Leu) from six, but methionine (Met) and tryptophan (Trp) each

1st Base	2nd Base U: RNA codon	U: Amino acid	2nd Base C: RNA codon	C: Amino acid	2nd Base A: RNA codon	A: Amino acid	2nd Base G: RNA codon	G: Amino acid	3rd Base
U	U U U	Phe	U C U	Ser	U A U	Tyr	U G U	Cys	
U	U U C	Phe	U C C	Ser	U A C	Tyr	U G C	Cys	
U	U U A	Leu	U C A	Ser	U A A	Stop	U G A	Stop	
U	U U G	Leu	U C G	Ser	U A G	Stop	U G G	Trp	
C	C U U	Leu	C C U	Pro	C A U	His	C G U	Arg	
C	C U C	Leu	C C C	Pro	C A C	His	C G C	Arg	
C	C U A	Leu	C C A	Pro	C A A	Gln	C G A	Arg	
C	C U G	Leu	C C G	Pro	C A G	Gln	C G G	Arg	
A	A U U	Ile	A C U	Thr	A A U	Asn	A G U	Ser	
A	A U C	Ile	A C C	Thr	A A C	Asn	A G C	Ser	
A	A U A	Ile	A C A	Thr	A A A	Lys	A G A	Arg	
A	A U G	Met	A C G	Thr	A A G	Lys	A G G	Arg	
G	G U U	Val	G C U	Ala	G A U	Asp	G G U	Gly	
G	G U C	Val	G C C	Ala	G A C	Asp	G G C	Gly	
G	G U A	Val	G C A	Ala	G A A	Glu	G G A	Gly	
G	G U G	Val	G C G	Ala	G A G	Glu	G G G	Gly	

FIGURE 8.11 ■ The genetic code. Codons within a single box encode the same amino acid. Color-highlighted amino acids are encoded by codons in two boxes. Stop codons are highlighted pink. Often, single-letter abbreviations for amino acids are used to convey protein sequences: Ala (A); Arg (R); Asn (N); Asp (D); Cys (C); Gln (Q); Glu (E); Gly (G); His (H); Ile (I); Leu (L); Lys (K); Met (M); Phe (F); Pro (P); Ser (S); Thr (T); Trp (W); Tyr (Y); Val (V).

have only one codon. Because more than one codon can encode the same amino acid, the code is said to be degenerate or redundant. Notice that in almost every case, synonymous codons differ only in the last base. How the cell handles this degeneracy (redundancy) in the code will be explained later.

Only 61 out of a possible 64 codons specify amino acids. Four of these act as codons that can mark the beginning of the protein "sentence." In order of preference, they are AUG (90%), GUG (8.9%), UUG (1%), and CUG (0.1%). Because of its inefficiency in starting translation, the use of a rare start codon, such as CUG, limits the translation of any ORF in which it is found. Although these **start codons** are used to begin all proteins, they are not restricted to that role; they also encode amino acids found in the middle of coding sequences. Consequently, there must be something else about mRNA that signifies whether an AUG codon, for example, marks the start of a protein or resides internally, where it codes for an amino acid (as will be discussed shortly).

The remaining three triplets (UAA, UAG, and UGA) are equally important, for they tell the ribosome when to stop reading a gene sentence. Called **stop codons**, they trigger a series of events (to be described later) that dismantle ribosomes from mRNA and release a completed protein.

Thought Question

8.4 How might the redundancy of the genetic code be used to establish evolutionary relationships between different species? *Hints:* 1. Genomes of different species have different overall GC content. 2. Within a given genome one can find segments of DNA sequence with a GC content distinctly different from the rest of the genome.

The decoder (or adapter) molecules that convert the language of RNA (codons) to the language of proteins (amino acids) are the tRNAs. Their job is to travel through the cell and return to the ribosome with amino acids in tow.

A.
5′ GCGGAUUUAGCUCAGDDGGGAGAGCMCCAGACUGAAYAUCUGGAGMUCCUGUGTΨCGAUCCACAAUUCGCACCA 3′

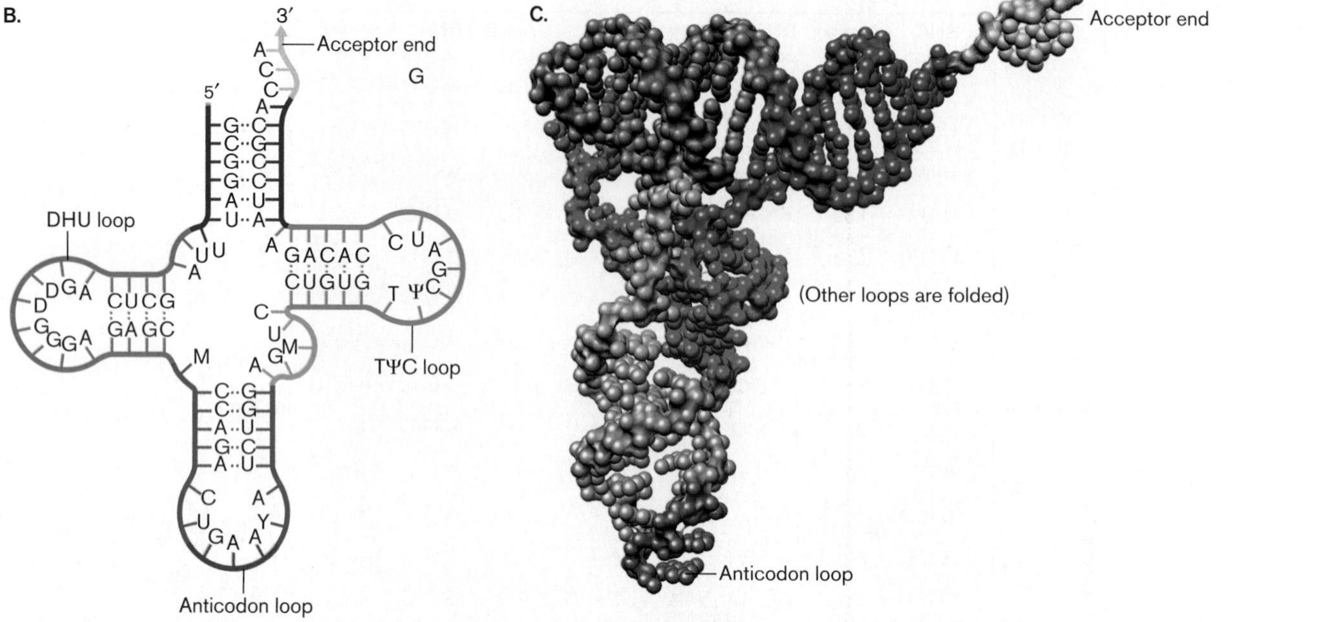
B.
3′
Acceptor end
5′
DHU loop
TΨC loop
Anticodon loop
C.
Acceptor end
(Other loops are folded)
Anticodon loop

FIGURE 8.12 ■ Transfer RNA. **A.** Primary sequence. The letters D, M, Y, T, and Ψ stand for modified bases found in tRNA. **B.** Cloverleaf structure. DHU (or D) is dihydrouracil, which occurs only in this loop; TΨC consists of thymine, pseudouracil, and cytosine bases that occur as a triplet in this loop. The DHU and TΨC loops are named for the modified nucleotides that are characteristically found there. **C.** Three-dimensional structures. The anticodon loop binds to the codon, while the acceptor end binds to the amino acid. (PDB code: 1GIX)

Approximately 80 bases in length, a tRNA laid flat looks like a clover leaf with three loops (**Fig. 8.12A** and **B**). In nature, however, the molecule folds into the "boomerang-like" 3D structure shown in **Figure 8.12C**. The very bottom or middle loop of all tRNA molecules (as depicted in **Fig. 8.12B** and **C**) harbors an **anticodon** triplet that base-pairs with codons in mRNA (**Fig. 8.13**). As a result, this loop is called the anticodon loop. Notice that codon and anticodon pairings are aligned in an antiparallel manner. Most tRNA molecules begin with a 5′ G, and all end with a 3′ CCA, to which amino acids attach (see **Figs. 8.12** and **8.13**). Because the 3′ end of the tRNA accepts the amino acid, it is called the acceptor end.

Transfer RNA molecules contain a large number of unusual, modified bases, which accounts for the strange letter codes seen in **Figure 8.12A** (D, M, Y, T, and Ψ). Structures of some of the odd bases are shown in **Figure 8.14**. Wybutosine (yW), for example, has three rings instead of the two found in a normal purine. How do these unusual bases end up in tRNA? During transcription of the tRNA genes, normal, unmodified bases are incorporated into the transcript. Some of these are modified later by specific enzymes to make inosine and other odd bases. The remarkable stability of tRNA molecules is explained, in part, by these unusual bases, because they are poor substrates for RNases.

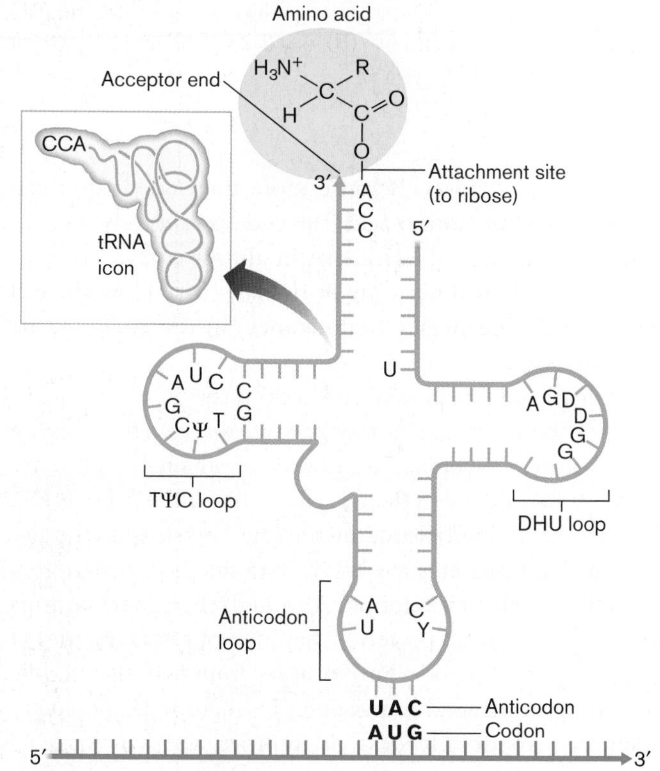
Amino acid
Acceptor end
CCA
tRNA icon
Attachment site (to ribose)
3′
5′
TΨC loop
DHU loop
Anticodon loop
UAC
Anticodon
AUG
Codon
5′
3′

FIGURE 8.13 ■ Codon-anticodon pairing. The tRNA anticodon consists of three nucleotides at the base of the anticodon loop. The anticodon hydrogen-bonds with the mRNA codon in an antiparallel fashion. This tRNA is "charged" with an amino acid covalently attached to the 3′ end.

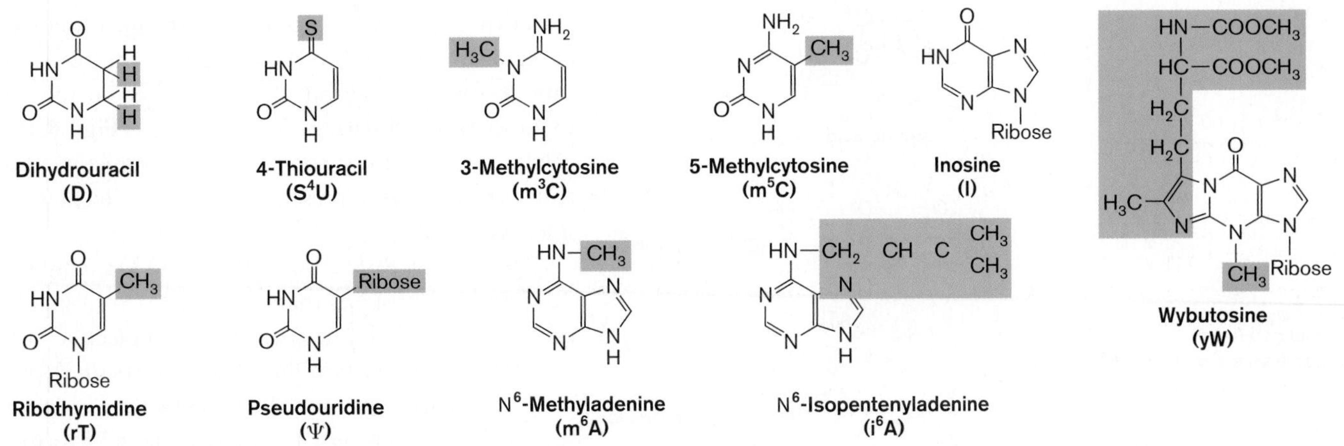
Dihydrouracil (D)
4-Thiouracil (S^4U)
3-Methylcytosine (m^3C)
5-Methylcytosine (m^5C)
Inosine (I)
Wybutosine (yW)
Ribothymidine (rT)
Pseudouridine (Ψ)
N^6-Methyladenine (m^6A)
N^6-Isopentenyladenine (i^6A)
Ribose

FIGURE 8.14 ■ Modified bases present in tRNA and rRNA molecules. Modifications are highlighted in green.

In the cloverleaf structure of tRNA, two of the loops are named after modifications that are invariantly present in those loops. One is called the TΨC loop because this loop in every tRNA has the nucleotide triplet thymidine, pseudouridine (Ψ), and cytidine. The other loop always contains dihydrouridine and is called the DHU loop (see **Figs. 8.12** and **8.13**). These loops are recognized by the enzymes that match tRNA to the proper amino acid.

As noted previously, the genetic code is redundant in that many amino acids have codon synonyms. Redundancy occurs primarily in the third position of the codon (for example, UUU and UUC both encode phenylalanine). A single tRNA can recognize both codons because of "wobble" in the first position of the anticodon, which corresponds to the third position of the codon (remember, as with DNA, base-pairing between RNA strands is antiparallel). The wobble is due in part to the curvature of the anticodon loop and to the use of an unusual base (inosine) at this position in some tRNA molecules. The wobble structure allows one anticodon to pair with several codons differing only in the third position.

Aminoacyl-tRNA Synthetases Attach Amino Acids to tRNA

When a tRNA GCG anticodon, for instance, pairs with a CGC codon, the ribosome has no way of checking that the tRNA is attached (that is, charged) to the "correct" amino acid (arginine in this case). Consequently, each tRNA must be charged with the proper amino acid before it encounters the ribosome. How are amino acids correctly matched to tRNA molecules and affixed to their 3′ ends? The charging of tRNAs is carried out by a set of enzymes called **aminoacyl-tRNA synthetases**. Each cell has generally 20 of these "match and attach" proteins, one for each amino acid. Some bacteria, however, have only 18, choosing instead to modify already charged glutamyl-tRNA and aspartyl-tRNA by adding an amine to make glutamine and asparagine derivatives. Each aminoacyl tRNA synthetase recognizes all the tRNA molecules that recognize the same amino acid.

Every aminoacyl-tRNA synthetase enzyme has a specific binding site for its cognate (matched) amino acid. Each enzyme also has a site that recognizes the target tRNA and an active site that joins the carboxyl group of the amino acid to the 3′ OH (class II synthetases) or 2′ OH (class I synthetases) of the tRNA by forming an ester (**Fig. 8.15**). The tRNA synthetase disengages after the amino acid has been attached to the tRNA. Amino acids initially attached to the 2′ OH are then moved to the 3′ OH position.

Each aminoacyl-tRNA synthetase must recognize its own tRNA but not bind to any other tRNA. Specificity is based on recognizing unique features of the tRNA anticodon loop, the TΨC loop, and the DHU loop. Thus, each tRNA has its own set of interaction sites that match only the proper aminoacyl-tRNA synthetase.

The Ribosome, a Translation Machine

The decoding process of the ribosome may be compared to a language translation machine that converts one language into another. The ribosome can be viewed as translating the language of the mRNA code into functional protein sequences that conduct the activities of the cell. Ribosomes are composed of two complex subunits, each of which includes rRNA and protein components. In prokaryotes, the subunits are named 30S and 50S for their "size" in Svedberg units. A Svedberg unit reflects the rate at which a molecule sediments

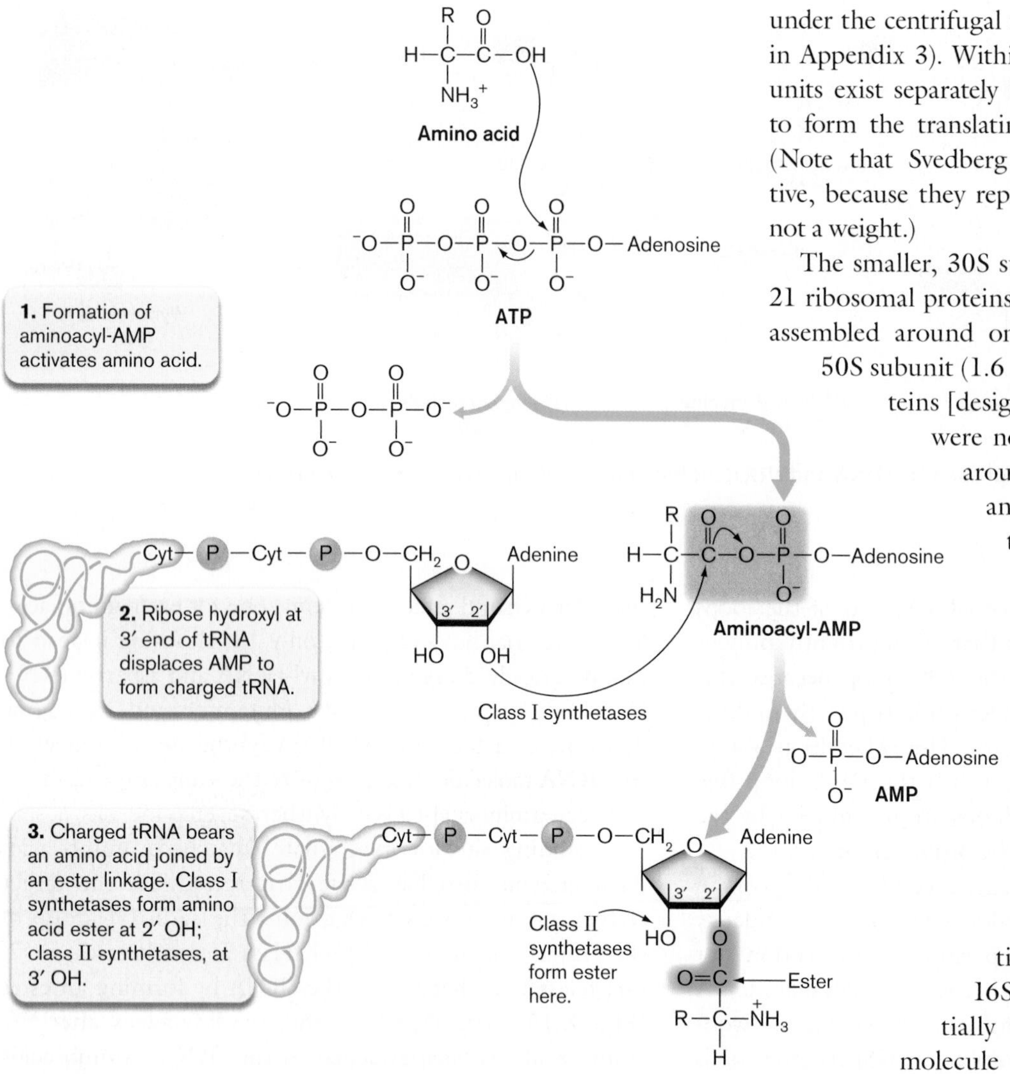
R O
H—C—C—OH
NH_3^+
Amino acid
$^-$O—P—O—P—O—P—O—Adenosine
ATP
1. Formation of aminoacyl-AMP activates amino acid.
Cyt—P—Cyt—P—O—CH_2
Adenine
2. Ribose hydroxyl at 3′ end of tRNA displaces AMP to form charged tRNA.
HO OH
Class I synthetases
Aminoacyl-AMP
AMP
3. Charged tRNA bears an amino acid joined by an ester linkage. Class I synthetases form amino acid ester at 2′ OH; class II synthetases, at 3′ OH.
Class II synthetases form ester here.
Ester
R—C—NH_3

FIGURE 8.15 ■ Charging of tRNA molecules by aminoacyl-tRNA synthetases. At the end of this process each amino acid is attached to the 3′ end of CCA on a specific tRNA molecule.

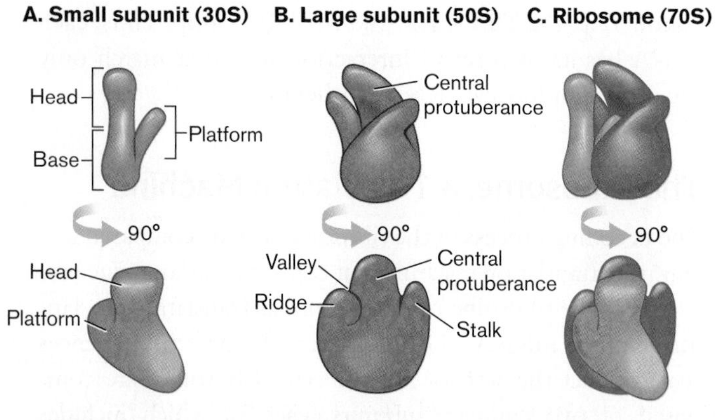
A. Small subunit (30S)
B. Large subunit (50S)
C. Ribosome (70S)
Head
Base
Platform
Central protuberance
90°
90°
90°
Head
Platform
Valley
Ridge
Central protuberance
Stalk

FIGURE 8.16 ■ Schematic ribosome structure. Rotated views of the 30S subunit **(A)**, the 50S subunit **(B)**, and the 70S ribosome complex **(C)**. Note that the platform of the 30S subunit fits into the valley of the 50S subunit when forming the 70S ribosome.

under the centrifugal force of a centrifuge (discussed in Appendix 3). Within the living cell, the two subunits exist separately but come together on mRNA to form the translating 70S ribosome (**Fig. 8.16**). (Note that Svedberg units are not directly additive, because they represent a rate of sedimentation, not a weight.)

The smaller, 30S subunit (900,000 Da) contains 21 ribosomal proteins (named S1–S21; S = "small") assembled around one 16S rRNA molecule. The 50S subunit (1.6 million Da) consists of 31 proteins [designated L1–L34 (three numbers were not used); L = "large"] formed around two rRNA molecules (5S and 23S). **Figure 8.17** presents the 3D spatial arrangement of rRNA and protein in the 50S subunit. Note that the majority of the ribosome is RNA.

How does such a complex molecular machine get built? As discussed in greater detail in **eTopic 8.1**, assembly begins with the transcription of ribosomal RNA genes (DNA), sometimes referred to as rDNA. The 16S, 23S, and 5S rRNAs are initially transcribed as a single RNA molecule from a common DNA promoter, and they are eventually processed into separate rRNA molecules. Simultaneous with rRNA transcription and processing, the ribosomal proteins begin to assemble, finding their places nestled in the secondary rRNA structures (see **Fig. 8.18**). Ribosomal proteins bind sequentially in waves, the first wave providing new binding sites for a second wave of different proteins. Thus, the ribosome is built by precise, timed molecular interactions that occur between RNA sequences, between RNA and protein, and between protein and protein.

Thought Question

8.5 The synthesis of ribosomal RNAs and ribosomal proteins represents a major energy drain on the cell. How might the cell regulate the synthesis of these molecules when the growth rate slows because amino acids are limiting? *Hint:* Predict what happens to any translating ribosome when an amino acid is limiting. Look up RelA protein on the Internet.

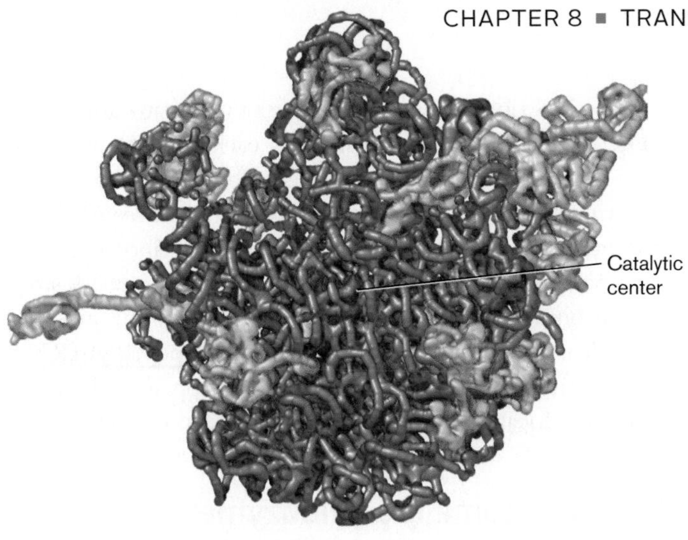
Catalytic
center

FIGURE 8.17 ■ RNA-protein interfaces in the large (50S) ribosome subunit. rRNA is blue; proteins are gold. (PDB code: 1GIY)

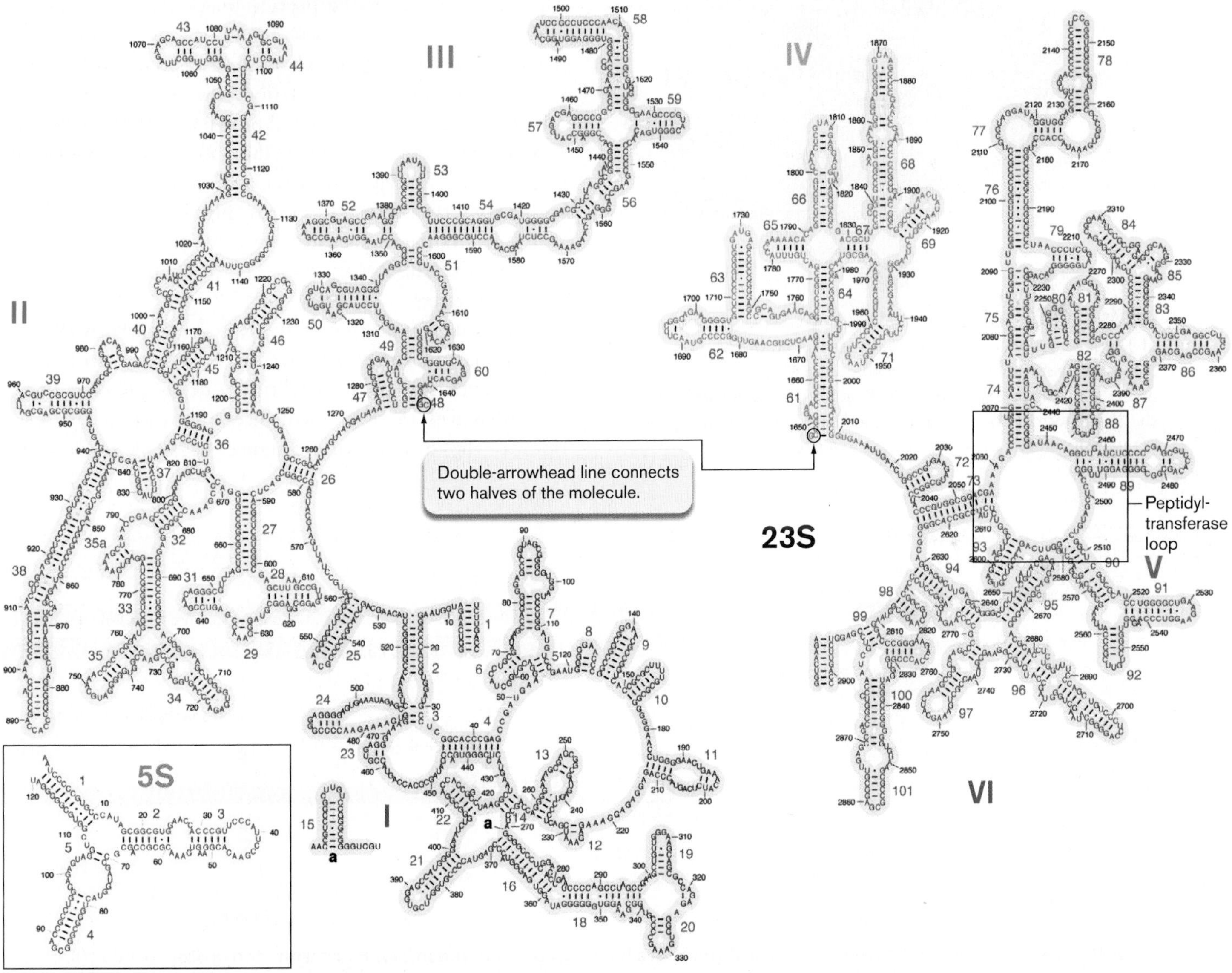
I
II
III
IV
V
VI
23S
5S
Double-arrowhead line connects two halves of the molecule.
Peptidyl-transferase loop

FIGURE 8.18 ■ Secondary structure of the 5S and 23S rRNA. Note the many double-stranded hairpin structures that fold into domains (Roman numerals). Nucleotides are numbered in black; stem structures are numbered in blue. During the building of the ribosome, proteins bind to specific sites on the RNA molecule. The proteins bind to some of the secondary and tertiary rRNA structures and help form the key tertiary 23S structure. Domains where various ribosomal proteins bind are indicated by different colors. The 23S RNA is split where indicated to provide space for all secondary structures.

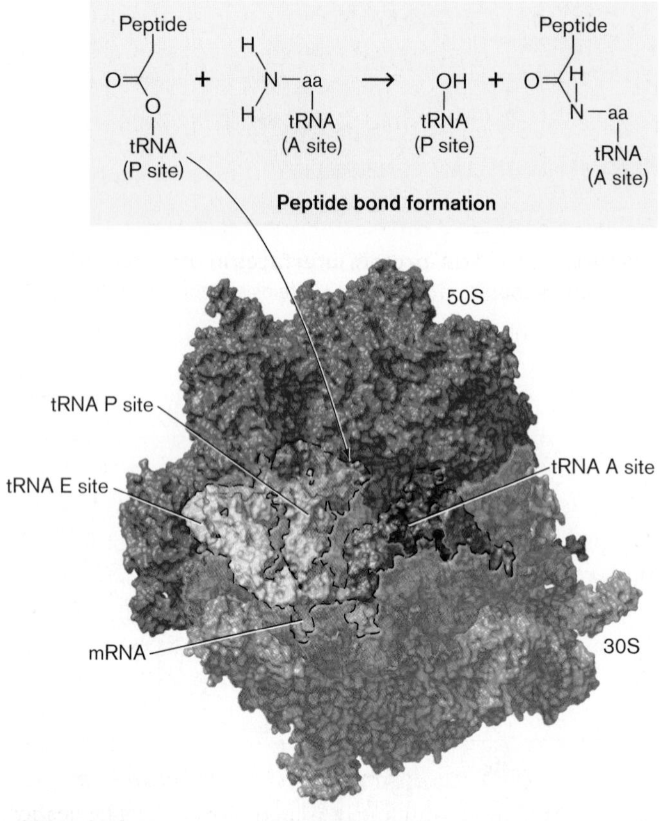

Peptide
H
N—aa
H
tRNA
(A site)
OH
tRNA
(P site)
Peptide
H
N—aa
tRNA
(A site)
O
tRNA
(P site)
Peptide bond formation
50S
tRNA P site
tRNA E site
tRNA A site
mRNA
30S

FIGURE 8.19 ■ Binding of tRNA. X-ray-crystallographic model of *Thermus aquaticus* ribosome with associated tRNAs. 50S is red, 30S is magenta, and tRNAs in the A, P, and E sites are blue, purple, green, and yellow, respectively. (PDB codes: 1GIX and 1GIY) Inset shows the formation of a peptide bond between the peptidyl-tRNA in the P site and aminoacyl-tRNA in the A site.

The 70S ribosome harbors three binding sites for tRNA (**Fig. 8.19**). During translation, each tRNA molecule moves through the sites in an assembly-line fashion, first entering at one site, and then moving progressively through the others before being jettisoned from the ribosome. The first position is the aminoacyl-tRNA **acceptor site (A site)**, which binds an incoming aminoacyl-tRNA. The growing peptide is attached to tRNA berthed in the **peptidyl-tRNA site (P site)**. Finally, a tRNA recently stripped of the polypeptide is located in the **exit site (E site)**.

The Ribosome Is a "Ribozyme"

The ribosome makes the peptide bonds that stitch amino acids together, and it does so by using a remarkable enzymatic activity called **peptidyltransferase**, present in the 50S subunit. Contrary to expectation, the peptidyltransferase activity resides not in ribosomal proteins, but in the rRNA. Peptidyltransferase is actually a **ribozyme** (an RNA molecule that carries out catalytic activity), and it is part of 23S rRNA highly conserved across all species (**Fig. 8.18**, part V). Proteins surrounding this active center offer structural assistance to ensure that the RNA is folded properly and interact with tRNA substrates.

The sequences of ribosomal RNAs from all microbes are very similar, in large part because rRNA plays an important catalytic role in the activities of ribosomes. But there are differences in rRNA sequences that increase in relation to the evolutionary distance between species. As

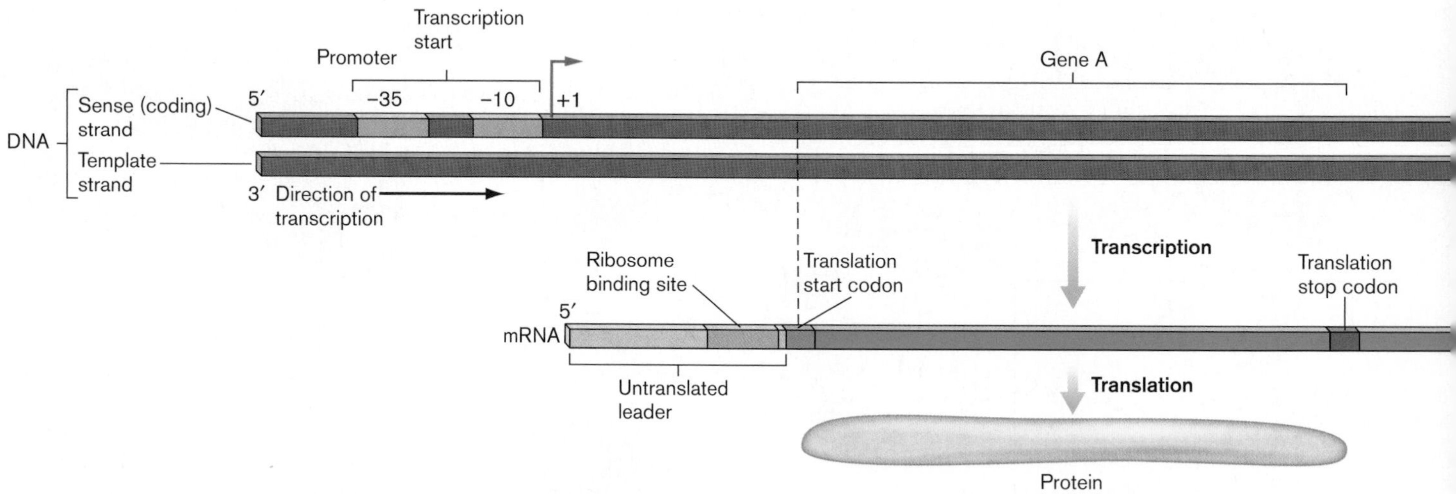

Transcription
start
Promoter
Gene A
−35
−10
+1
5′
Sense (coding)
strand
DNA
Template
strand
3′ Direction of
transcription
Transcription
Ribosome
binding site
Translation
start codon
Translation
stop codon
5′
mRNA
Untranslated
leader
Translation
Protein

FIGURE 8.20 ■ Alignment between structural genes in a bacterial operon, the mRNA transcript, and protein products. RNA polymerase binds to the −10 and −35 regions of DNA and, reading the template strand, begins RNA transcription at nucleotide +1. Messenger RNA has an untranslated region of variable length. Each part of the transcript encoding a protein has a ribosome-binding site (RBS). Ribosomes begin translating a few nucleotides downstream from each RBS at a codon corresponding to the amino terminus of each translated protein. In this figure, the term "gene" refers to the region of DNA corresponding to the entire mRNA transcript, including upstream and downstream untranslated areas.

a result, rRNA serves as a molecular clock that measures the approximate time since two species diverged. More information on molecular clocks can be found in Section 17.3.

How Do Ribosomes Find the Right Reading Frame?

Every mRNA has three potential reading frames, depending on which base the ribosome happens to start with. Having a one in three chance of randomly picking the right frame on the template, a ribosome that chooses the wrong frame would produce proteins with totally different amino acid sequences. In addition, a shifted reading frame often generates an inadvertent stop codon that prematurely terminates the peptide; thus, the translated protein would not even have the right length. So how does the ribosome find the right frame?

The key is that each mRNA contains additional sequences upstream of the segment that encodes protein. Upstream, untranslated leader RNA plays a critical role in directing the bacterial ribosome to the right place. The upstream leader RNA contains a purine-rich sequence with the consensus 5′-AGGAGGU-3′ located four to eight bases upstream of the start codon in *E. coli*. This upstream sequence is called the **ribosome-binding site** or **Shine-Dalgarno sequence**, after Lynn Dalgarno and her student John Shine at Australian National University, who discovered it in 1974. The Shine-Dalgarno site is complementary to a sequence at the 3′ end of 16S rRNA (5′-ACCUCCU-3′), found in the 30S ribosomal subunit. Binding of mRNA to this site positions the start codon (such as AUG or GUG) precisely in the ribosome P site, ready to pair with an *N*-formylmethionyl-tRNA.

Defining a gene. We have described the process of transcription and are about to translate a resulting mRNA. It will help to illustrate the alignment between the DNA region comprising a structural gene (a gene encoding a protein), and the resulting mRNA transcript containing translation signals and the protein-coding region. **Figure 8.20** shows the "sense" and "template" DNA strands of a two-gene operon. The sequence of the sense strand matches that of the mRNA transcript but with T's substituting for U's. The template strand is the strand actually "read" by RNA polymerase. Note that +1 marks the DNA base where the mRNA transcript starts and that a single transcript includes both operon genes (polycistronic). In the mRNA transcript, an untranslated "leader" sequence precedes the gene A protein-coding region, located between translation start and stop signals (discussed later). Downstream of the translation stop signal for gene B lies an untranslated "trailer." The leader and trailer sequences help regulate gene expression. As you proceed in this chapter, refer back to **Figure 8.20** to clarify important connections between a DNA operon, its composite genes, the polycistronic transcript, and the resulting proteins.

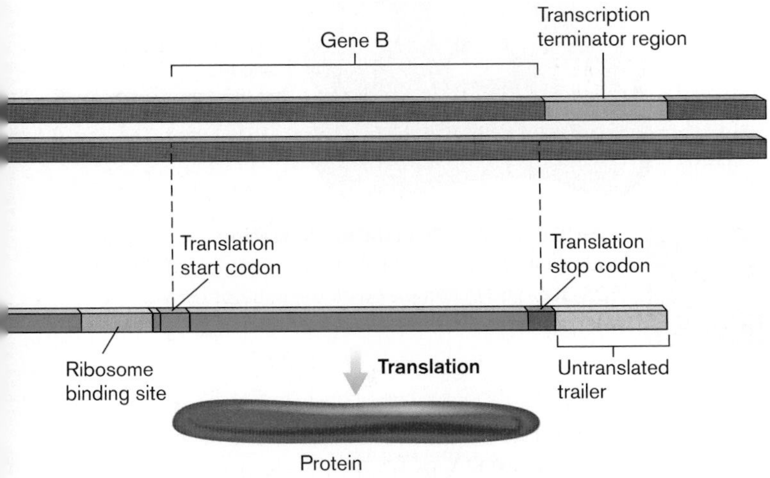

Gene B
Transcription terminator region
Translation start codon
Translation stop codon
Ribosome binding site
Translation
Untranslated trailer
Protein

Thought Question

8.6 **Fig. 8.20** illustrates an operon and its relationship to transcripts and protein products. Imagine a mutation that stops translation (TAA, for example) was substituted for a normal amino acid about midway through the DNA sequence encoding gene A. What happens to the expression of the gene A and gene B proteins?

Note: Eukaryotic ribosomes have a different mechanism for finding the start codon. They generally start translating at the first AUG after the 5′ cap. The 5′ cap is a 7-methylguanosine added post-transcriptionally to the 5′ end of eukaryotic mRNA molecules. The normal role of the cap is thought to be as a signal to transport RNA out of the nucleus and/or to stabilize mRNA.

Although Dalgarno and Shine predicted the importance of the ribosome-binding site in 1974, experimental evidence that this mRNA region truly bound 16S rRNA

COURTESY OF JOAN STEITZ

FIGURE 8.21 ■ Joan Steitz (second from left) at Yale University with some of her students. Steitz and her colleagues provided key evidence for the existence of the ribosome-binding site on mRNA.

was lacking until Joan Steitz (**Fig. 8.21**) supplied it in 1975 with a simple yet elegant experiment. The experiment demonstrated that a radiolabeled RNA fragment taken from the 5′ end of mRNA binds to the 3′ end of 16s rRNA (see eTopic 8.2).

The Three Stages of Protein Synthesis

Once the ribosome has been properly positioned on the message, peptide synthesis can begin. Like transcription, peptide synthesis has three stages: initiation, which brings the two ribosomal subunits together, placing the first amino acid in position; elongation, which sequentially adds amino acids as directed by the mRNA transcript; and termination. Several steps in this process can be inhibited by certain antibiotics (described later in this section).

Initiation of translation. In bacteria such as *E. coli*, initiation of protein synthesis requires three small proteins called initiation factors (IF1, IF2, and IF3). In archaea (for example, *Methanocaldococcus jannaschii*), initiation is much more complex, involving six different IF proteins. For simplicity, we will limit discussion to the process in bacteria.

IF3 first brings mRNA and the 30S ribosome subunit together, allowing the ribosome-binding site to find its complementary site on 16S rRNA (**Fig. 8.22**). Next, IF1 binds and blocks the A site. IF2 bound to GTP then escorts the initiator *N*-formylmethionyl-tRNA (fMet-tRNA) to the start codon located at what will be the P site.

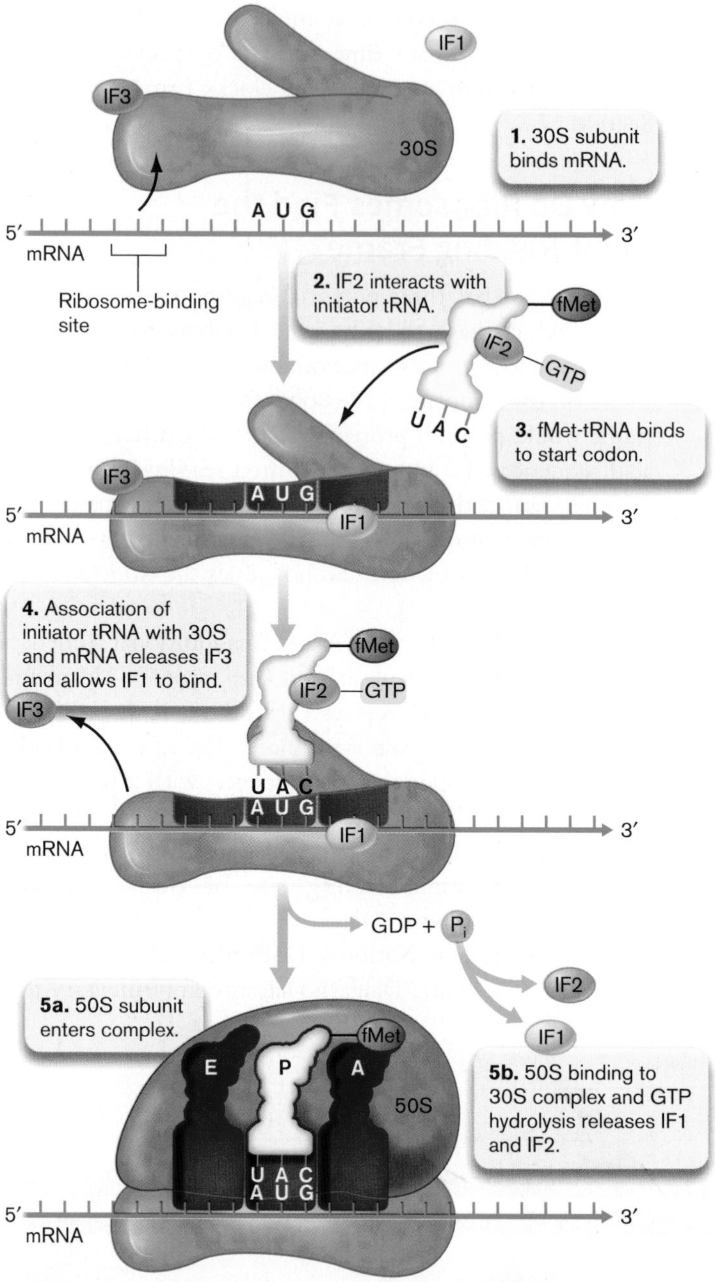
IF1
IF3
30S
1. 30S subunit binds mRNA.
AUG
5′
3′
mRNA
Ribosome-binding site
2. IF2 interacts with initiator tRNA.
fMet
IF2
GTP
UAC
3. fMet-tRNA binds to start codon.
4. Association of initiator tRNA with 30S and mRNA releases IF3 and allows IF1 to bind.
GDP + Pi
5a. 50S subunit enters complex.
E
P
A
50S
5b. 50S binding to 30S complex and GTP hydrolysis releases IF1 and IF2.

FIGURE 8.22 ■ Translation initiation. The end result of translation initiation is assembly of the 50S-30S-mRNA complex with the initiator tRNA-fMet set in the P site. See text for details.

N-formylmethionyl-tRNA binds to all start codons and is the only aminoacyl-tRNA to bind directly to the P site. Once the initiator tRNA is in place, IF3 is released. The 50S subunit then docks to the 30S subunit, GTP is hydrolyzed, and IF1 and IF2 are released. The ribosome is now "locked and loaded."

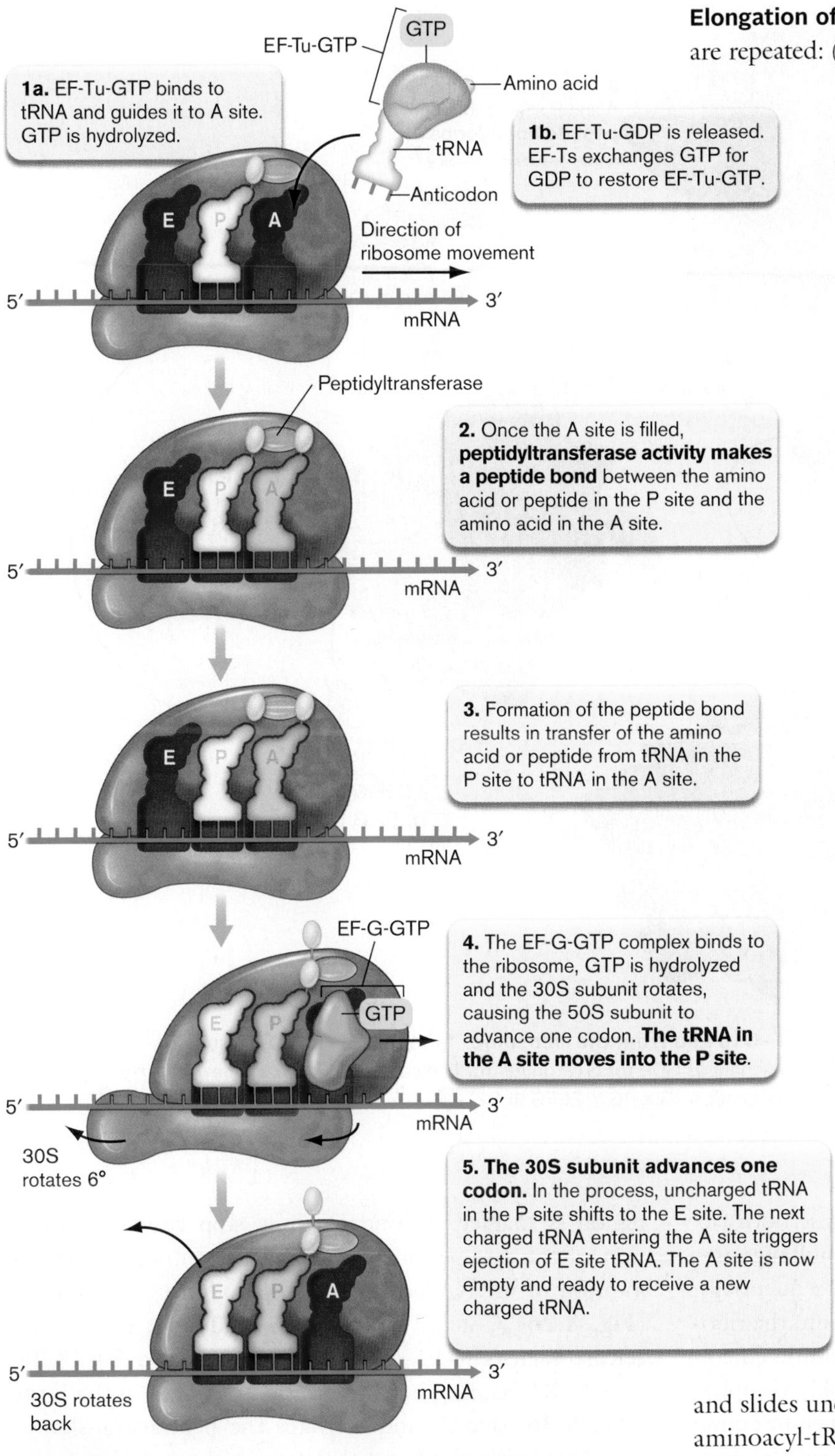
EF-Tu-GTP
GTP
Amino acid
1a. EF-Tu-GTP binds to tRNA and guides it to A site. GTP is hydrolyzed.
tRNA
Anticodon
1b. EF-Tu-GDP is released. EF-Ts exchanges GTP for GDP to restore EF-Tu-GTP.
E
P
A
Direction of ribosome movement
5′
3′
mRNA
Peptidyltransferase
2. Once the A site is filled, peptidyltransferase activity makes a peptide bond between the amino acid or peptide in the P site and the amino acid in the A site.
3. Formation of the peptide bond results in transfer of the amino acid or peptide from tRNA in the P site to tRNA in the A site.
EF-G-GTP
GTP
4. The EF-G-GTP complex binds to the ribosome, GTP is hydrolyzed and the 30S subunit rotates, causing the 50S subunit to advance one codon. The tRNA in the A site moves into the P site.
30S rotates 6°
5. The 30S subunit advances one codon. In the process, uncharged tRNA in the P site shifts to the E site. The next charged tRNA entering the A site triggers ejection of E site tRNA. The A site is now empty and ready to receive a new charged tRNA.
30S rotates back

FIGURE 8.23 ■ Elongation of the peptide. At the end of each elongation cycle the ribosome has moved forward by one codon, which positions the tRNA with nascent peptide in the P site and the tRNA from which the peptide was passed in the E site, ready for ejection.

Elongation of the peptide. In elongation, three basic steps are repeated: (1) An aminoacyl-tRNA binds to the A (acceptor) site; (2) a peptide bond forms between the new amino acid and the growing peptide pre-positioned in the P site; and (3) the message must move by one codon (**Fig. 8.23**). First an elongation factor (EF-Tu) associates with GTP to form a complex (EF-Tu-GTP) that binds to charged aminoacyl-tRNAs (except for the initiator tRNA). Sequences within the 50S rRNA recognize features of the EF-Tu-GTP–aminoacyl-tRNA complex, guiding it into the A site (**Fig. 8.23**, step 1). Correct selection of a tRNA is guided in large part by codon-anticodon pairing, but conformational changes sensed by various ribosomal proteins customize the fit. If the fit is not perfect, the tRNA is rejected.

Once an aminoacyl-tRNA is in the A site, a peptide bond is formed between the amino acid moored there and the terminal amino acid linked to tRNA in the P site (**Fig. 8.23**, step 2). Simultaneously, a GTP is hydrolyzed and the resulting EF-Tu-GDP is expelled. Peptide bond formation effectively transfers the peptide from the tRNA in the P site to tRNA in the A site (**Fig. 8.23**, step 3). Finally, for protein synthesis to continue, the ribosome must advance by one codon, which moves the peptidyl-tRNA from the A site into the P site, leaving the A site vacant. The process, called **translocation**, involves another elongation factor, EF-G, associated with GTP. EF-G-GTP binds to the ribosome, GTP is hydrolyzed, and the 50S subunit ratchets ahead on the message by one codon (**Fig. 8.23**, step 4). The 30S subunit then follows (**Fig. 8.23**, step 5). This maneuver opens up the A site, moves the peptidyl-tRNA into the P site, and slides uncharged tRNA into the E (exit) site. The next aminoacyl-tRNA that enters the A site stimulates a conformational change in the ribosome that telegraphs through to the E site and ejects the uncharged tRNA.

Figure 8.24 shows, in 3D, how EF-Tu and EF-G cycle on and off the ribosome. Note that the two factors bind to the same site. This reflects a structural similarity (or **molecular mimicry**) between EF-G-GTP and the

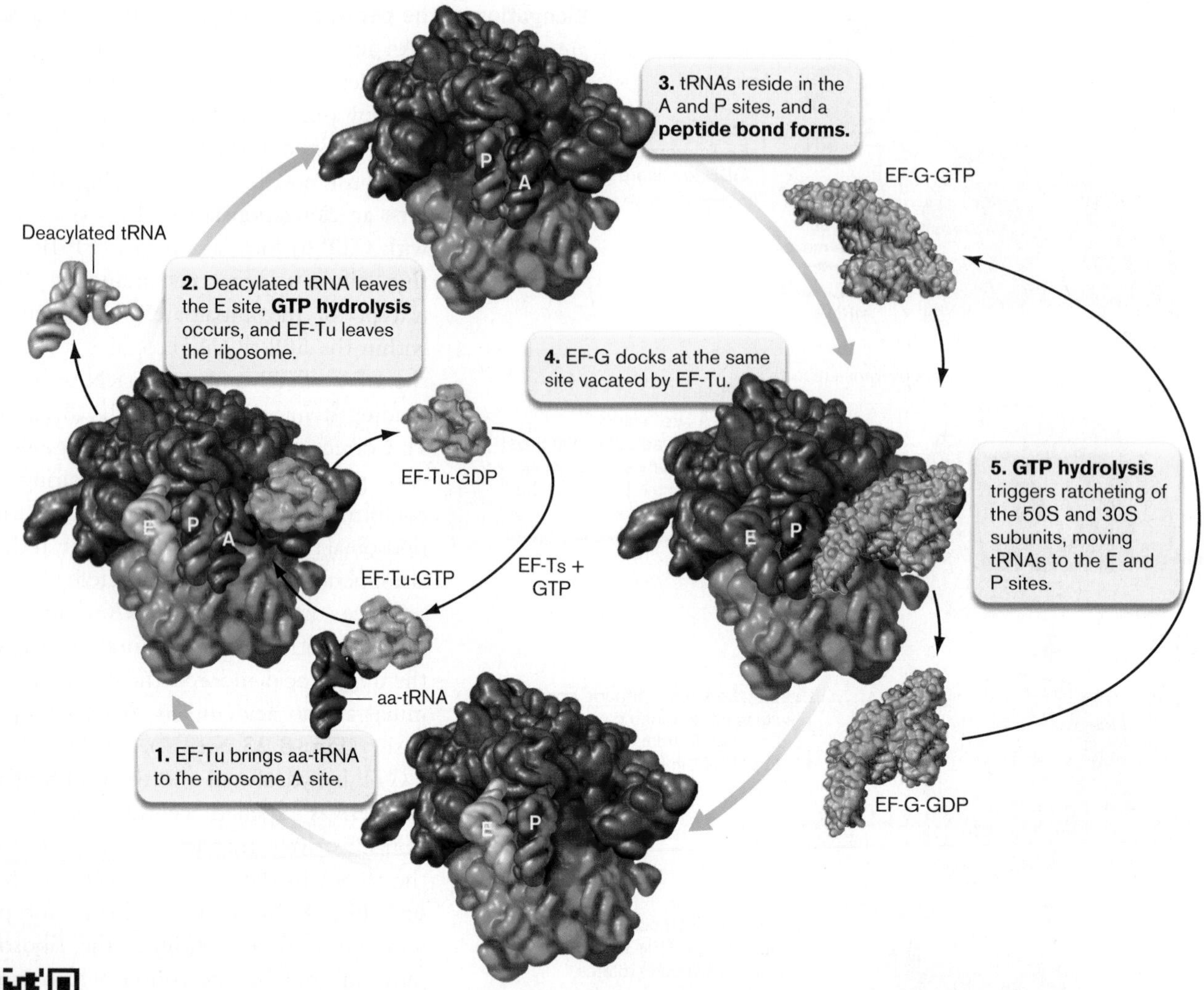
3. tRNAs reside in the A and P sites, and a peptide bond forms.
EF-G-GTP
Deacylated tRNA
2. Deacylated tRNA leaves the E site, GTP hydrolysis occurs, and EF-Tu leaves the ribosome.
4. EF-G docks at the same site vacated by EF-Tu.
EF-Tu-GDP
5. GTP hydrolysis triggers ratcheting of the 50S and 30S subunits, moving tRNAs to the E and P sites.
EF-Ts + GTP
EF-Tu-GTP
aa-tRNA
1. EF-Tu brings aa-tRNA to the ribosome A site.
EF-G-GDP
P
A
E
P
A
E
P
E
P

FIGURE 8.24 ■ Cycling of EF-Tu and EF-G to and from the ribosome. Note that EF-Tu-GDP in step 2 is recycled to EF-Tu-GTP by interacting with EF-Ts. EF-Ts exchanges GDP for GTP on EF-Tu. The colors of tRNA change when moving from step 5 back to step 1 to start a new cycle. (PDB codes: 1GIX, 1GIY, 2EFG, 1LS2)

EF-Tu-GTP–aminoacyl-tRNA complex. Because both factors bind to the same site, they cannot do so simultaneously, so they must cycle on and off the ribosome sequentially. **Figure 8.25** illustrates the path of mRNA into the ribosome and the nascent (that is, incomplete) peptide emerging from it. While this process seems incredibly complex, the ribosomes of *E. coli* manage to link together 16 amino acids per second. Scientists have now visualized the staggered movements of a single ribosome as it translates an mRNA molecule (**Special Topic 8.1**).

Termination of translation. Eventually, the ribosome arrives at the end of the coding region, but not the end of the RNA. As noted previously, the end of the coding region is marked by one of three stop codons. Formation of the last peptide bond and the subsequent translocation of mRNA brings the stop codon into the A site (**Fig. 8.26**, step 1). No tRNA binds, but one of two **release factors** (RF1 or RF2) will enter (also **Fig. 8.26**, step 2). RF binding leads to ejection of tRNA in the E site (**Fig. 8.26**, step 2) and activates the peptidyltransferase, thereby cutting the bond that tethers the completed peptide to tRNA in the P site (**Fig. 8.26**, step 3).

With the protein released, the ribosome must disassemble. RF3 causes RF1 or RF2 to depart the ribosome (**Fig. 8.26**, step 4). Then, ribosome recycling factor (RRF), along with EF-G, binds at the A site, and an accompanying GTP hydrolysis undocks the two ribosomal

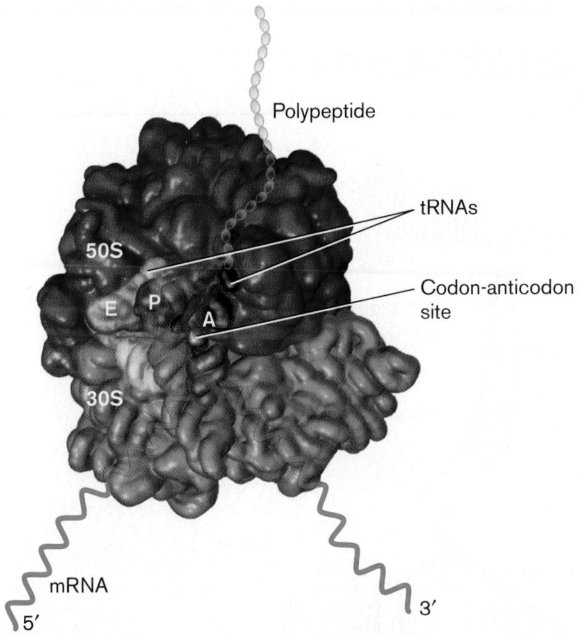
Polypeptide
tRNAs
50S
E
P
A
Codon-anticodon site
30S
mRNA
5′
3′

FIGURE 8.25 ■ Orientation of tRNA molecules within the ribosome and tracks of mRNA and the nascent polypeptide. The mRNA travels along the 30S subunit, and the growing peptide exits from a channel formed in the 50S subunit. (PDB codes: 1GIX and 1GIY)

subunits (**Fig. 8.26**, step 5). IF3 then reenters the 30S subunit to replace the remaining uncharged tRNA and mRNA (**Fig. 8.26**, step 6), thereby preventing the 30S and 50S subunits from redocking. The liberated ribosomal subunits are now free to diffuse through the cell, ready to bind yet another mRNA and begin the translation sequence anew.

Thought Question

8.7 While working as a member of a pharmaceutical company's drug discovery team, you find that a soil microbe snatched from the jungles of South America produces an antibiotic that will kill even the most deadly, drug-resistant form of *Enterococcus faecalis,* which is an important cause of heart valve vegetations in bacterial endocarditis. Your experiments indicate that the compound stops protein synthesis. How could you more precisely determine the antibiotic's mode of action? *Hint:* Can you use mutants resistant to the antibiotic?

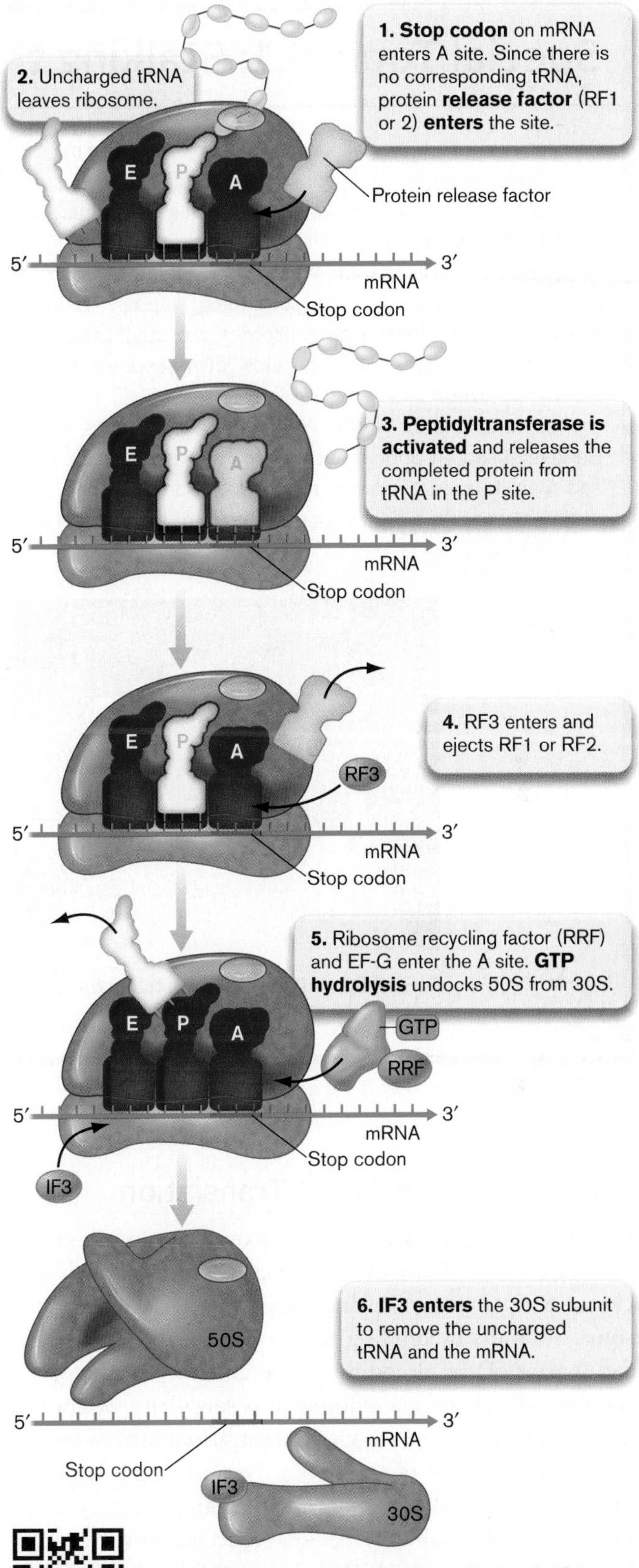
2. Uncharged tRNA leaves ribosome.
1. Stop codon on mRNA enters A site. Since there is no corresponding tRNA, protein release factor (RF1 or 2) enters the site.
Protein release factor
E
P
A
5′
3′
mRNA
Stop codon
3. Peptidyltransferase is activated and releases the completed protein from tRNA in the P site.
4. RF3 enters and ejects RF1 or RF2.
RF3
5. Ribosome recycling factor (RRF) and EF-G enter the A site. GTP hydrolysis undocks 50S from 30S.
GTP
RRF
IF3
6. IF3 enters the 30S subunit to remove the uncharged tRNA and the mRNA.
50S
30S

FIGURE 8.26 ■ Termination of translation. The completed protein is released, and the ribosome subunits are recycled.

Special Topic 8.1: Stalking the Lone Ribosome

Recent advances in crystallography and cryo-electron microscopy have strengthened our understanding of ribosome structure and function. However, it has been extremely difficult to follow the steps of ribosomes during translational elongation in the test tube, where thousands of ribosomes are usually needed to see something happen. Because the dynamics of individual ribosomes are stochastic (random), the functions of thousands of ribosomes cannot be synchronized.

Recently, Ignatio Tinoco (**Fig. 1**) and colleagues at UC Berkeley used an optical tweezer to trap a single ribosome and map its progress along an mRNA, codon by codon. An optical tweezer is a tightly focused beam of light (laser) able

IGNATIO TINOCO

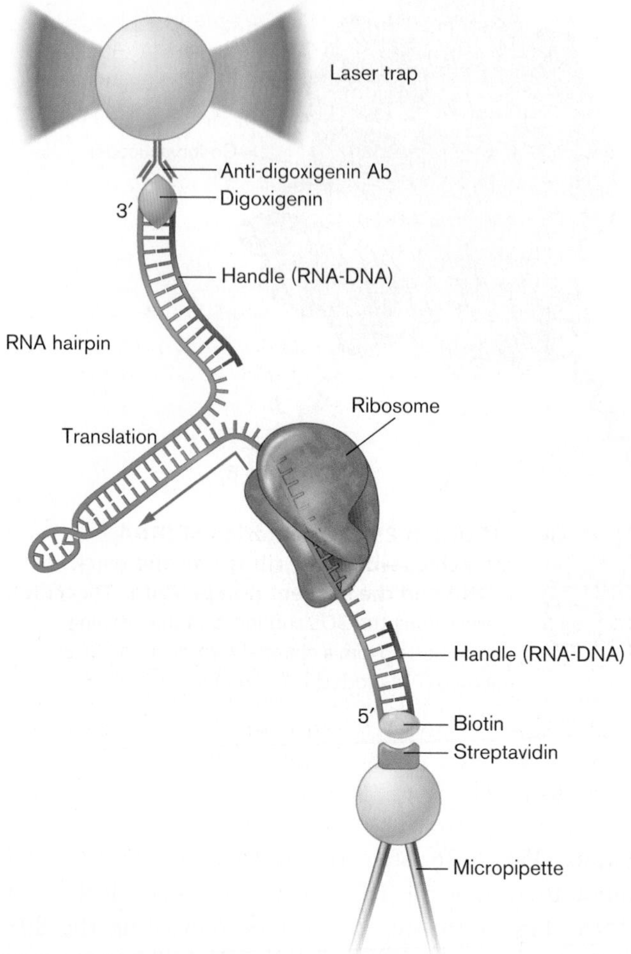
Laser trap
Anti-digoxigenin Ab
Digoxigenin
3′
Handle (RNA-DNA)
RNA hairpin
Ribosome
Translation
Handle (RNA-DNA)
5′
Biotin
Streptavidin
Micropipette

FIGURE 1 ■ The ribosome trap concept.
Inset: Ignatio Tinoco at UC Berkeley.

Antibiotics That Affect Translation

Streptomycin (**Fig. 8.27A**), a well-known member of the aminoglycoside family of antibiotics, is produced by a species of *Streptomyces*. The drug targets bacterial small ribosomal subunits by binding to a region of 16S rRNA that forms part of the decoding A site and to protein S12, a protein critical for maintaining the specificity of codon-anticodon binding. At high concentrations, streptomycin stops the initiation of translation because it binds to the 30S-mRNA-tRNA initiation complex and prevents binding of the 50S subunit. At low concentrations, translation can go on but the A site becomes "sloppy," permitting illicit codon-anticodon matchups that result in a mistranslated protein sequence.

Bacteria usually become resistant to streptomycin by spontaneously mutating the S12 gene (*rpsL*), although mutations in 16S rRNA can also prevent binding. The altered S12 protein maintains function but will not bind streptomycin. Some bacteria gain resistance by acquiring an aminoglycoside phosphotransferase that modifies streptomycin so that it cannot bind to its target. Additional mechanisms are discussed in Chapter 27. Other therapeutically important aminoglycosides include gentamicin, kanamycin, and amikacin.

Tetracycline (**Fig. 8.27B**) also targets the 30S ribosomal subunit, but instead of preventing formation of the 70S complex (as with streptomycin), it outright prevents binding of aminoacyl-tRNA to the A site and stops to hold microscopic particles stable in three dimensions. **Figure 1** illustrates the trap and how it works. An RNA strand was made with a digoxigenin molecule affixed at the 3′ end and a biotin molecule at the 5′ end. Tiny polystyrene beads coated with streptavidin (which binds to biotin) or with antibodies to digoxigenin were used to trap separate ends of the RNA molecule. The bead attached to the 3′ end was placed in a laser trap, and the bead attached to the 5′ end was held in place by a micropipette tip. A single ribosome was stalled at the 5′ side of the mRNA hairpin construct by leaving out components needed for translation. Once a translating mixture of appropriately charged tRNAs, translocation factors, and GTP was added, the progress of the unstalled, translating ribosome was monitored by measuring the change in size of the hairpin, which shortens as the ribosome moves through it.

Results are shown in **Figure 2**, which measures change in extension versus time. Strikingly, the extension shows a repeated step-pause pattern. Every arrow indicates a discrete step where the ribosome has progressed by one codon (three nucleotides), which, in turn, decreases the hairpin by six nucleotides as the hairpin unwinds. Each step took place in 0.078 second, with punctuating pauses lasting 2.8 seconds. The pauses are thought to represent the collective time it takes to introduce a new charged tRNA into the A site, catalyze peptide bond formation, and begin translocation. These studies have revealed for the first time that translation occurs not in a continuous manner, but as a series of translocation-pause events.

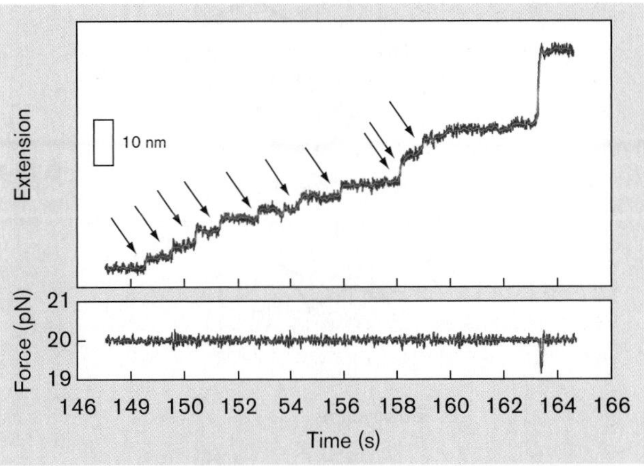
Extension
10 nm
Force (pN)
21
20
19
146 149 150 152 54 156 158 160 162 164 166
Time (s)

FIGURE 2 ■ Results from a ribosome trap experiment. Arrowheads indicate where the ribosome has moved by one codon, shortening the hairpin but lengthening the distance between the two beads. pN = piconewton = 10^{-12} newton. *Source:* Jin-Der Went et al. 2008. *Nature* **452**:598–603.

Research Question

Secondary hairpin structures are common in mRNA undergoing translation. What effect would GC content have on translation? Design an experiment that would test your idea.

Qu, Xiaohui, Jin-Der Wen, Laura Lancaster, Harry F. Noller, Carlos Bustamante, et al. 2011. The ribosome uses two active mechanisms to unwind messenger RNA during translation. *Nature* **475:**118-121

protein synthesis. It binds to 16S rRNA, right above the A site for incoming tRNA. Resistance to tetracycline can be conferred by an efflux transport system that effectively removes the antibiotic from the bacterial cell. Genes encoding this drug resistance transport system are usually carried on mobile genetic elements like plasmids that can be passed from cell to cell (see Section 9.2). Other resistance mechanisms are described in Chapter 27.

Another species of *Streptomyces* produces chloramphenicol (**Fig. 8.27C**), which attacks the 50S subunit. It binds to 23S rRNA at the peptidyltransferase active site and inhibits peptide bond formation. Resistance to this drug comes from an ability to synthesize an enzyme, chloramphenicol acetyltransferase, that modifies chloramphenicol in a way that destroys its activity.

Erythromycin, made by *Streptomyces erythraeus,* is one of a large group of related antibiotics called macrolides, whose hallmark is a large lactone ring of 12–22 carbon atoms (**Fig. 8.27D**). They all affix themselves to the 50S subunit by binding to protein L15 and 23S rRNA in the peptidyltransferase cavity. Binding triggers an abortive translocation step that evicts peptidyl-tRNA from the P site while preventing peptide bond formation. In some organisms, mutational alterations in L15 decrease macrolide binding and confer resistance to these drugs. Other microbes reduce erythromycin binding by methylating the relevant area of 23S rRNA using an enzyme produced from another mobile genetic element.

Other translation-targeting antibiotics interfere with mRNA binding to the ribosome (kasugamycin), prevent

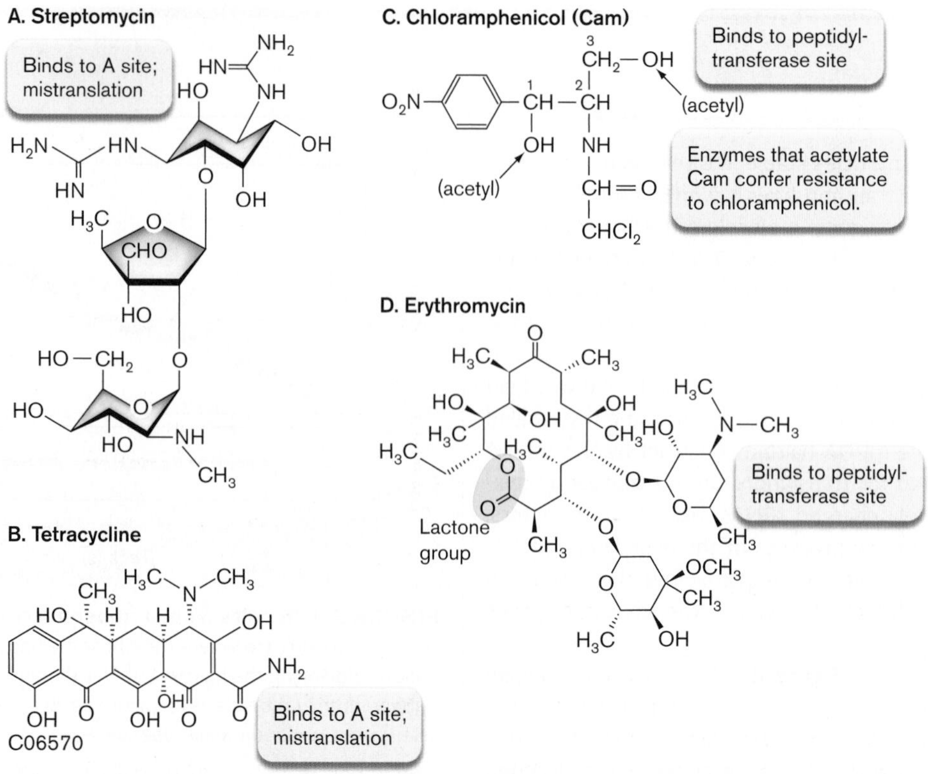
A. Streptomycin
Binds to A site; mistranslation
B. Tetracycline
Binds to A site; mistranslation
C06570
C. Chloramphenicol (Cam)
Binds to peptidyl-transferase site
(acetyl)
(acetyl)
Enzymes that acetylate Cam confer resistance to chloramphenicol.
D. Erythromycin
Binds to peptidyl-transferase site
Lactone group

FIGURE 8.27 ▪ Antibiotics that inhibit protein synthesis in bacteria. Streptomycin **(A)** and tetracycline **(B)** bind to the A site, causing mistranslation, whereas chloramphenicol **(C)** and erythromycin **(D)** bind to the peptidyltransferase site, thus inhibiting peptide bond formation.

translocation by targeting EF-G (fusidic acid), or use structural similarity to tRNA (molecular mimicry) to trick peptidyltransferase into action without having a bona fide tRNA in the A site (puromycin). We chronicle the discovery and use of antibiotics more completely in Chapter 27.

Bacterial Transcription and Translation Are Coupled

Many genes in bacterial chromosomes are arranged and transcribed in tandem and produce a polycistronic mRNA (see Section 7.2). Remember that the beginning of each gene in a polycistronic mRNA carries its own ribosome-binding site. This means that different ribosomes can bind simultaneously to the start of each cistron within a polycistronic mRNA.

Before RNA polymerase has even finished making an mRNA molecule, ribosomes will bind to the 5′ end of the mRNA and begin translating protein. This is called coupled transcription and translation (**Fig. 8.28**). Transcriptional-translational coupling in bacteria makes it possible to use translation as a means of regulating transcription, a process that is explained further in Chapter 10. Note, however, that most translation, at least in *E. coli,* occurs at the cell ends, which are ribosome-rich and have little to no DNA. Thus, although transcriptional-translational coupling happens when transcripts are first made, most translation occurs on completed transcripts.

Eukaryotic microbes, on the other hand, use separate cellular compartments to carry out most of their transcription and translation. They transcribe genes in the nucleus and, after splicing (in which internal, noncoding parts of the mRNA are removed) and other forms of processing, transport mRNA to the cytoplasm, where the majority of translation occurs. However, a small amount of translation is carried out in the eukaryotic nucleus, where it is also coupled to transcription.

In prokaryotes, the coupling of transcription and translation presents a potential problem. Ribosomes generally travel along mRNA more slowly than mRNA is generated by RNA polymerase, so RNA polymerase can potentially scoot ahead of the ribosome and leave large tracts of RNA unprotected and susceptible to nucleases. Unintentional cleavage of the RNA between the ribosome and polymerase would separate the two macromolecular machines and destroy the message. The cell handles this problem by modulating transcriptional speed. The rate of RNA synthesis averages about 45 nt per second, which roughly equals

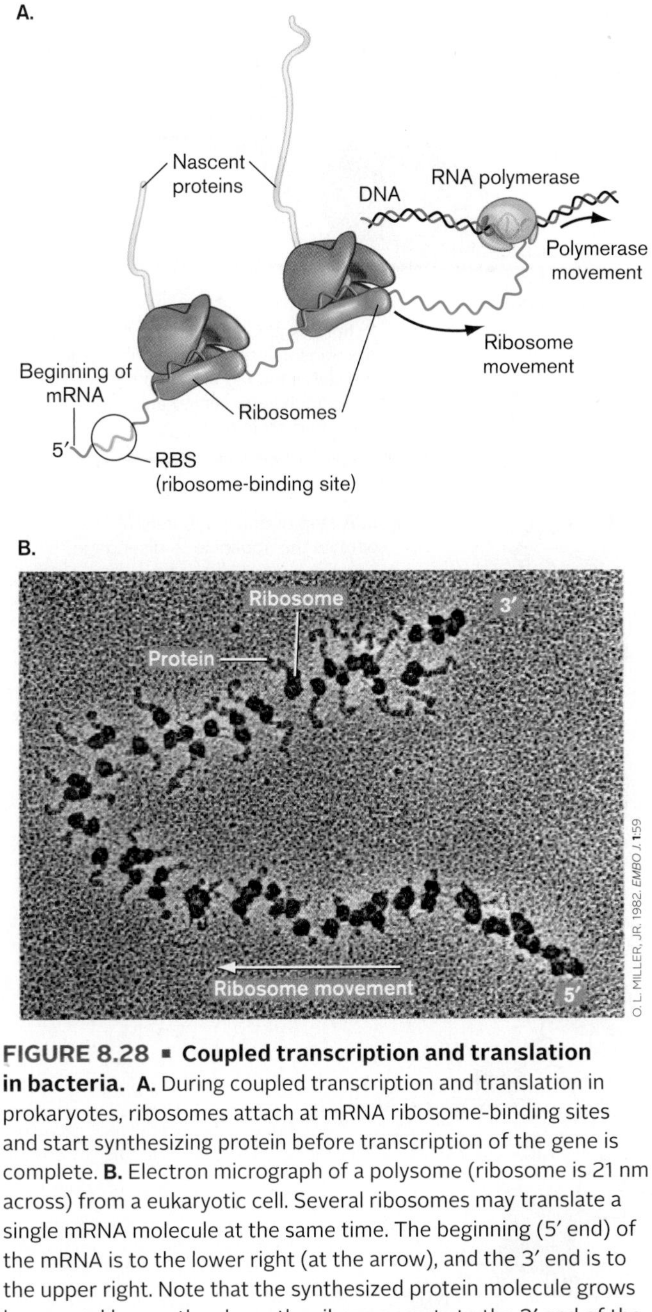
A.
Nascent proteins
DNA
RNA polymerase
Polymerase movement
Ribosome movement
Beginning of mRNA
Ribosomes
5′
RBS (ribosome-binding site)
B.
Ribosome
3′
Protein
Ribosome movement
5′
O. L. MILLER, JR. 1982. EMBO J. 1:59

FIGURE 8.28 ■ Coupled transcription and translation in bacteria. **A.** During coupled transcription and translation in prokaryotes, ribosomes attach at mRNA ribosome-binding sites and start synthesizing protein before transcription of the gene is complete. **B.** Electron micrograph of a polysome (ribosome is 21 nm across) from a eukaryotic cell. Several ribosomes may translate a single mRNA molecule at the same time. The beginning (5′ end) of the mRNA is to the lower right (at the arrow), and the 3′ end is to the upper right. Note that the synthesized protein molecule grows longer and longer the closer the ribosome gets to the 3′ end of the mRNA where the protein molecule is most clearly seen. Polysomes also occur in prokaryotes.

the average rate of translation (16 amino acids per second, which represents 48 nt/s). However, these speeds vary. For example, RNA polymerase pauses momentarily at sites rich in GC content (GC base pairs, with three hydrogen bonds, are harder to melt than AT pairs, which have only two hydrogen bonds). Once sigma factor exits the transcription complex, proteins called NusA and NusG enter the complex and can actually lengthen the pause in RNA polymerase activity to allow time for the trailing protein-synthesizing ribosome to catch up to the polymerase. Pausing allows the ribosome to follow RNA polymerase closely and protect the RNA.

Thought Questions

8.8 How might one gene code for two proteins with different amino acid sequences?

8.9 Why involve RNA in protein synthesis? Why not translate directly from DNA?

8.10 Codon 45 of a 90-codon gene was changed into a translation stop codon, producing a shortened (truncated) protein. What kind of mutant cell could produce a full-length protein from the gene without removing the stop codon? *Hint:* What molecule recognizes a codon?

Unsticking Stuck Ribosomes: tmRNA and Protein Tagging

Sometimes an RNase will shear off the 3′ end of a message, removing the stop codon before the translation is complete. What happens after a ribosome has finished translating this damaged mRNA molecule? Without a stop codon, there is nothing to trigger ribosome release when the ribosome reaches the end of the message. So the ribosome is stuck at the end of the mRNA with a peptidyl-tRNA in the ribosome P site.

The answer to this problem is a translation rescue molecule, the previously mentioned tmRNA, which has the properties of both tRNA and mRNA. One end of the tmRNA molecule has an attached amino acid and a folded structure that resembles a tRNA. This aminoacyl end acts like tRNA, entering the unoccupied ribosome A site, where a peptide bond then forms between the stalled polypeptide (in the P site) and the emergency amino acid on the tmRNA. Another section of the tmRNA is then read as mRNA, which adds a tag of 12 or so amino acids to the now dislodged peptide (SsrA) (**Fig. 8.29A**). These amino acids, called a proteolysis tag, predestine the protein for destruction. A stop codon present in tmRNA triggers peptide release and ribosome disassembly. A helper protein (SspB) recognizes the proteolysis tag and brings the useless aberrant polypeptide to the ClpXP protease (discussed in Section 8.6) for degradation. **Figure 8.29B** shows this process for tmRNA from *E. coli*.

Thought Question

8.11 What would happen if the tyrosine residue UAC in the mRNA-like domain of tmRNA was altered by mutation to UAA?

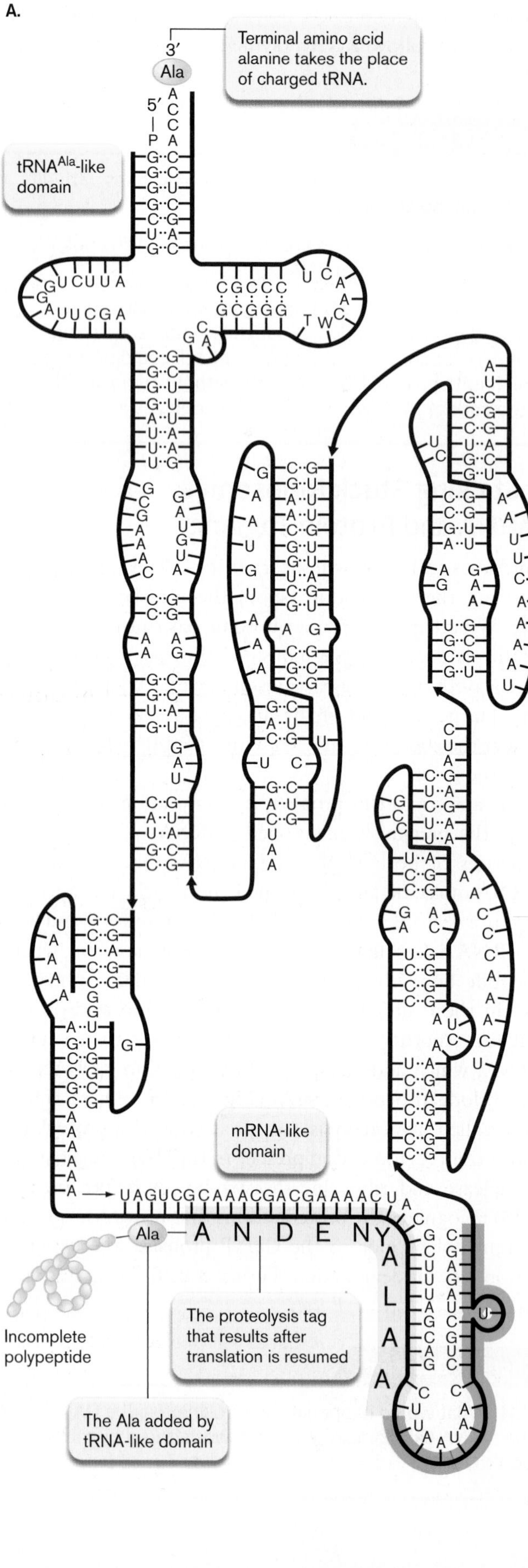
A.
3′
Ala
5′
P
Terminal amino acid alanine takes the place of charged tRNA.
tRNA^{Ala}-like domain
mRNA-like domain
UAGUCGCAAACGACGAAAACU
Ala — A N D E N Y A L A A
Incomplete polypeptide
The proteolysis tag that results after translation is resumed
The Ala added by tRNA-like domain

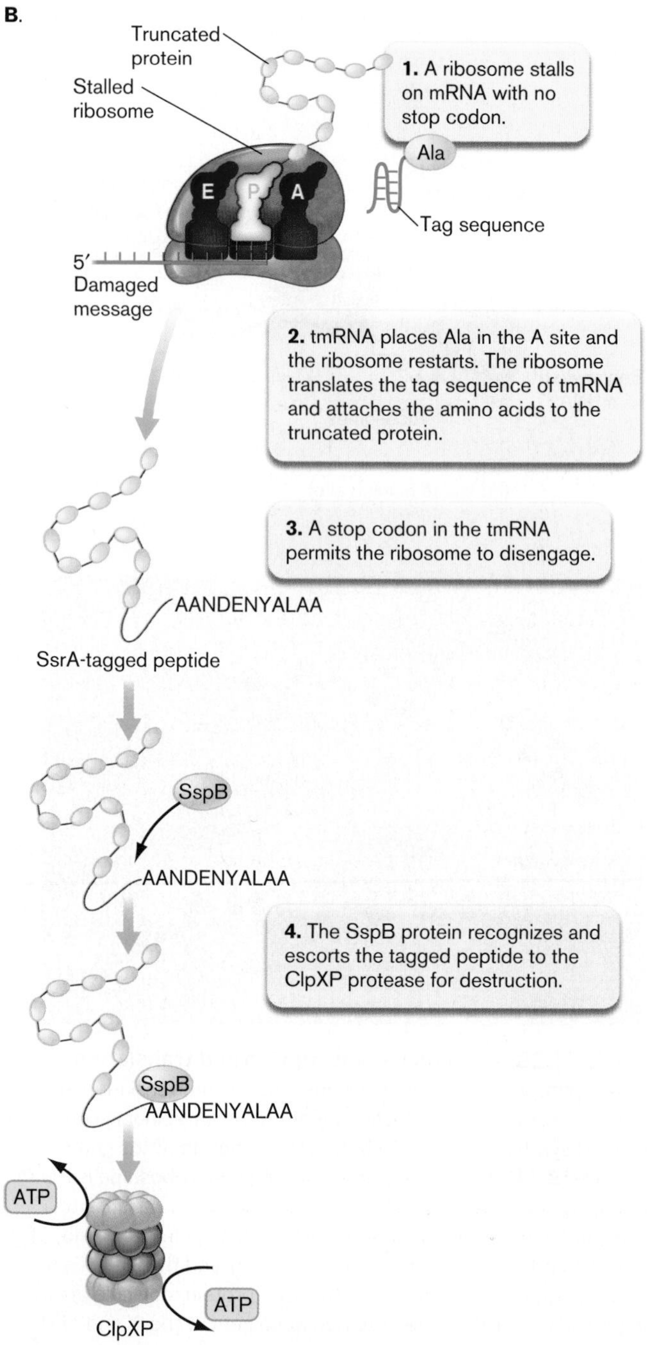
B.
Truncated protein
Stalled ribosome
1. A ribosome stalls on mRNA with no stop codon.
Ala
E
P
A
Tag sequence
5′
Damaged message
2. tmRNA places Ala in the A site and the ribosome restarts. The ribosome translates the tag sequence of tmRNA and attaches the amino acids to the truncated protein.
3. A stop codon in the tmRNA permits the ribosome to disengage.
AANDENYALAA
SsrA-tagged peptide
SspB
AANDENYALAA
4. The SspB protein recognizes and escorts the tagged peptide to the ClpXP protease for destruction.
SspB
AANDENYALAA
ATP
ATP
ClpXP

FIGURE 8.29 ■ **tmRNA and protein tagging. A.** Structure of SsrA, a type of tmRNA in *E. coli.* The tRNA-like domain is at the top left, and the mRNA-like portion encoding the proteolysis tag is at the bottom. The alanine (Ala) connecting the incomplete peptide to the tag was added from the tRNA end. The stop codon is highlighted orange. **B.** Mechanism of tmRNA tagging and degradation in *E. coli.*

To Summarize

- **Triplet nucleotide codons** in mRNA encode specific amino acids. **Transfer RNA molecules** interpret the genetic code and bring specific amino acids to the A (acceptor) site in the ribosome.
- **Specific codons** mark the beginning and end of a gene. The Shine-Dalgarno sequence in mRNA located before the start codon helps the ribosome find the correct reading frame in the mRNA.
- **Initiation of protein synthesis** in bacteria requires three initiation factors that bring the ribosomal subunits together on an mRNA molecule.
- **Peptidyltransferase** activity of the ribosome is carried out by ribosomal RNA, not protein. The peptide elongates when the ribosome ratchets one codon length along the mRNA.
- **Translation terminates** upon reaching a stop codon. The ribosome pauses because it cannot find an appropriate tRNA. A release factor enters the A site and triggers peptidyltransferase activity, thus freeing the completed protein from tRNA in the P site.
- **Ribosome release factor** and EF-G bind to the A site to dissociate the two ribosomal subunits from the mRNA.
- **Antibiotics that affect translation** can prevent 70S ribosome formation (streptomycin), inhibit aminoacyl-tRNA binding to the A site (tetracycline), interfere with peptidyltransferase (chloramphenicol), trigger peptidyltransferase prematurely (puromycin), cause translocation to abort (erythromycin), or prevent translocation (fusidic acid).
- **Transcription and translation** in prokaryotes are coupled.
- **RNA polymerase pauses** during transcription to allow the slower-translating ribosomes to stay close. This pause minimizes exposure of mRNA to degradative cellular enzymes.
- **tmRNA** rescues ribosomes stuck on damaged mRNA that lacks a stop codon.

8.4

Protein Modification and Folding

Once a protein is made, is it functional? For many proteins, translation is not the last step in producing a functional molecule. Cell function often requires that a protein be modified after translation to achieve an appropriate 3D structure or to regulate the activity of the protein. Primary, secondary, and tertiary structures of proteins can be modified after the primary protein sequence has been assembled by the ribosomes.

Protein Structure May Be Modified after Translation

Completed proteins released from the ribosome contain *N*-formylmethionine (fMet) at the N terminus (as previously described). With some proteins, the *N*-formyl group is "surgically" removed from the N terminus by methionine deformylase, leaving methionine. Alternatively, methionyl aminopeptidase will remove the entire amino acid. *N*-formylmethionine (fMet) peptides are produced only by bacteria and mitochondria, not by archaea or by the cytoplasmic ribosomes of eukaryotes. This is important to humans because our white blood cells can detect fMet peptides as a sign of invading bacteria or of necrotic (dying) host cells releasing mitochondria. The fMet peptides are detected at incredibly low concentrations (around 10^{-12} M).

Some proteins undergo other types of processing in which acetyl groups or AMP can be attached to change their functions, and proteolytic cleavages may either activate or inactivate a protein. Examples of modified bacterial enzymes include glutamine synthetase (adenylylation), isocitrate dehydrogenase (phosphorylation), and a variety of ribosomal proteins (acetylation). These groups directly regulate enzyme activity (glutamine synthetase and isocitrate dehydrogenase) or alter tertiary structure (ribosomal proteins).

Proteins Must Be Correctly Folded

As a new protein emerges from the ribosome, how does it fold into exactly the correct shape to do its job? Christian Anfinsen (1916–1995) won the 1972 Nobel Prize in Chemistry for demonstrating that, for some proteins, folding is governed solely by the protein itself. In other words, the optimal 3D structure of a protein is determined solely by the linear sequence of amino acid residues. But three decades later, other scientists discovered that the folding of many proteins requires assistance from other proteins. These helper proteins are called **chaperones** (or chaperonins). Chaperones associate with target proteins during some phase of the folding process and then dissociate, usually after folding of the target protein is completed. Although there is some specificity, a given chaperone can help fold many different types of proteins.

The major chaperone family in most species includes GroEL (60 kDa), GroES (10 kDa), DnaK (70 kDa), DnaJ (40 kDa), and trigger factor (48 kDa). Because their levels in *E. coli* increase in response to high-temperature stress, these chaperones were originally named **heat-shock proteins (HSPs)** and are, in fact, more resistant to heat denaturation than is the average protein. Representatives of these chaperones are found in all species. Because their molecular weights are similar, DnaK examples throughout nature are called HSP70s, while homologs of GroEL and DnaJ are called, respectively, HSP60s and HSP40s. Chaperones increase in response to heat stress because they are needed to help refold heat-damaged proteins.

The GroEL and GroES chaperones form a stacked ring with a hollow center like a barrel (**Fig. 8.30**). The chaperoned protein fits inside. The small capping protein GroES controls entrance to the chamber as shown in the chapter opener figure and **Figure 8.30A**. Cycles of ATP binding and hydrolysis cause conformational changes within the chamber that can reconfigure target proteins. DnaK (HSP70) chaperones have a very different structure (**Fig. 8.30B**). They do not form rings like the GroEL and GroES chaperones, but can clamp down on a peptide to assist folding. Proteins emerging from a bacterial ribosome enter a folding pathway that involves a hierarchy of these chaperones.

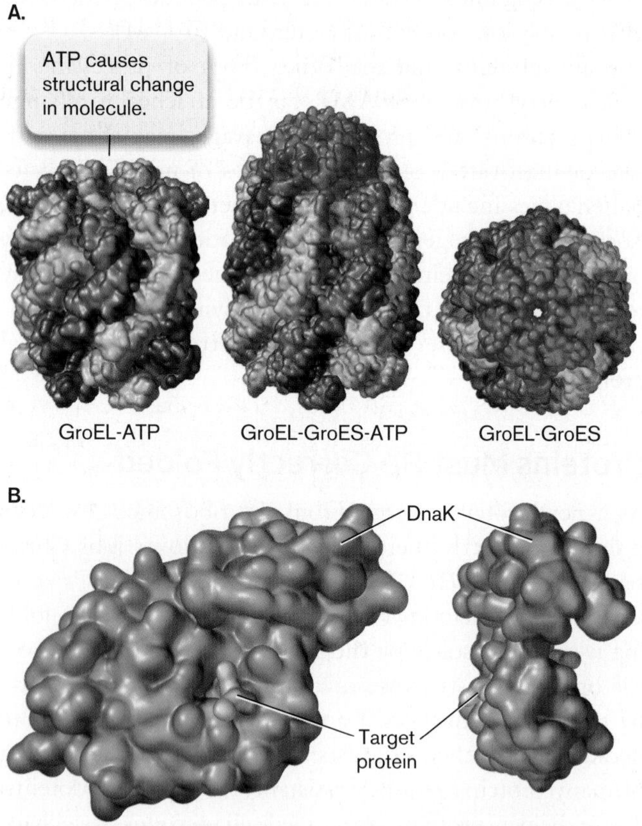
A.
ATP causes structural change in molecule.
GroEL-ATP
GroEL-GroES-ATP
GroEL-GroES
B.
DnaK
Target protein

FIGURE 8.30 ■ *E. coli* GroEL-GroES and DnaK (HSP70) structures. A. Three-dimensional reconstructions of GroEL-ATP, GroEL-GroES-ATP, and GroEL-GroES from cryo-EM. The first two panels are side views; the third panel is a top view. GroES is red. The chaperoned protein fits inside the GroEL chamber. (PDB codes: 2C7E, 1PCQ) **B.** DnaK clamping down on a peptide (yellow). (PDB code: 1DKX)

To Summarize

- **Protein modifications** are made after translation.
- **The N-terminal amino acid** (*N*-formylmethionine, fMet) can be removed by methionyl aminopeptidase, or the formyl group only can be removed by methionine deformylase.
- **An inactive precursor protein** can be cleaved into a smaller active protein, or other groups can be added to the protein (for example, phosphate or AMP).
- **Chaperone proteins** help translated proteins fold properly.

8.5

Protein Degradation: Cleaning House

What happens when a cell no longer needs a specific protein or when a cell synthesizes a protein with incorrect amino acids? These useless proteins must be degraded to maintain cellular health. Because cellular needs are constantly changing, proteins no longer needed can adversely affect the cell. This is particularly true of regulatory proteins, whose concentrations must change with time or in response to alterations in the cellular condition.

Many normal proteins contain degradation signals called **degrons** that dictate the stability of a protein. The **N-terminal rule** describes one type of degron. The rule states that the N-terminal amino acid of a protein directly correlates with its stability. For example, proteins beginning with arginine, lysine, or phenylalanine experience a short half-life (2 minutes or less), whereas proteins with aspartic acid, glutamic acid, or cysteine in the lead position enjoy a much longer half-life (more than 10 hours). Why this correlation exists is not clear.

Abnormally folded proteins are recognized by proteases in part because hydrophobic regions that are normally buried within the protein's 3D structure become exposed.

The protein is progressively degraded into smaller and smaller pieces by a series of these proteases. Initial cuts, usually involving ATP-dependent endoproteases like Lon protein or ClpP, are followed by digestion with tripeptidases and dipeptidases. Endopeptidases cleave proteins somewhere within the sequence but not from the ends of the sequence. Many peptidases use ATP hydrolysis to help unfold the target protein prior to digestion. Unfolding is necessary for the protein to slide into a barrel-shaped protease such as ClpAP, ClpXP, or the proteasomes of archaea (20S) and eukaryotic microbes (26S) (**Figs. 8.31** and **8.32**). Proteasomes (different from proteases) are complex

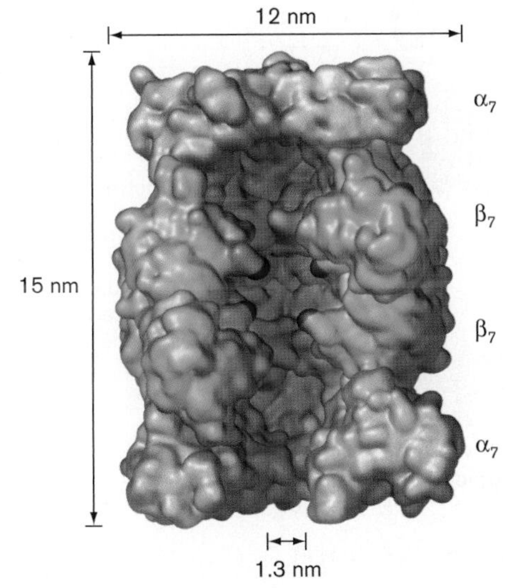
12 nm
15 nm
1.3 nm
α_7
β_7
β_7
α_7

FIGURE 8.31 ■ Protein degradation machines. Predicted structure of the 20S proteasome from the methanoarchaeon *Methanosarcina thermophila.* The active sites involved in peptide bond cleavage are indicated in red. (PDB code: 1GOU)

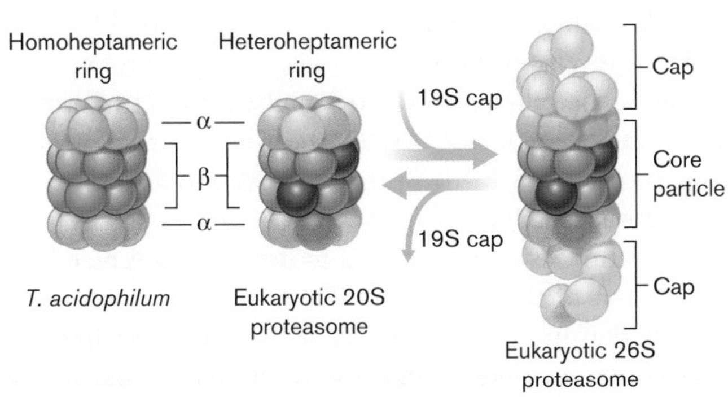
Homoheptameric ring
Heteroheptameric ring
α
β
α
19S cap
19S cap
Cap
Core particle
Cap
T. acidophilum
Eukaryotic 20S proteasome
Eukaryotic 26S proteasome

FIGURE 8.32 ■ Proteolysis machines from archaea and eukarya. Structural comparison of the 20S proteasome from an archaeon (*Thermoplasma acidophilum*), the 20S equivalent in a eukaryote, and the 26S eukaryotic proteasome. The 19S cap is a multisubunit regulatory structure.

protein-degrading organelles found primarily in eukaryotes and archaea, although a few bacteria, such as the lung pathogen *Mycobacterium tuberculosis,* have them. The protein-degrading enzymes are classified as serine, cysteine, or threonine proteases, depending on the key residue in their active sites.

There is a striking resemblance between archaeal and eukaryotic ATP-dependent proteosomes (**Fig. 8.32**). The core particle of the eukaryotic proteasome contains two copies of 14 different proteins assembled in four stacked rings, each containing seven proteins (heteroheptameric). The core particle is capped at the top and bottom by the regulatory particles, also made from 14 proteins that are different from those in the core. Six of those proteins are ATPases, while others recognize a ubiquitin peptide tag (a 76-amino-acid peptide) placed on doomed proteins (see **eTopic 8.3**). The purpose of this cap is to unfold the target protein and inject it into the 20S proteasome barrel. Archaea also contain a proteasome, but it comprises 14 copies of two different proteins assembled in four stacked rings, each containing seven proteins (homoheptameric). Unlike eukaryotic proteasomes, the archaeal proteasome has no regulatory particle, and substrate proteins are not ubiquitinated. Bacterial proteins are not ubiquitinated either, as a rule, although *Mycobacterium tuberculosis* provides an exception. These organisms place a 26-amino-acid peptide (PUP) on proteins targeted for degradation. **Figure 8.32** shows a side-by-side comparison of the archaeal proteasome and the eukaryotic 20S and 26S proteasomes.

Bacteria contain Clp proteases, which have a proteolytic core made of two homoheptameric rings of the protein ClpP. These proteases also have interchangeable homohexameric ATPase caps made of ClpX, ClpA, ClpB, or ClpC, each of which recognizes different substrates. The accessory proteins recognize and present different substrate proteins to the ClpP protease, thereby regulating which proteins are degraded.

What happens to proteins damaged by stress? Are they always degraded? Microbes are constantly exposed to environmental insults such as high temperature or pH extremes, which damage proteins and cause them to misfold. As an energy-saving device and to prevent interruption of protein function, injured proteins go through a kind of triage process that evaluates whether they are salvageable or must be destroyed before they can endanger the cell. Chaperones constantly hunt for misfolded (or otherwise damaged) proteins and attempt to refold them. But if the protein is released from a chaperone and remains misfolded, it can, by chance, either reengage the chaperone or bind a protease that destroys it (**Fig. 8.33**). This fold-or-destroy triage

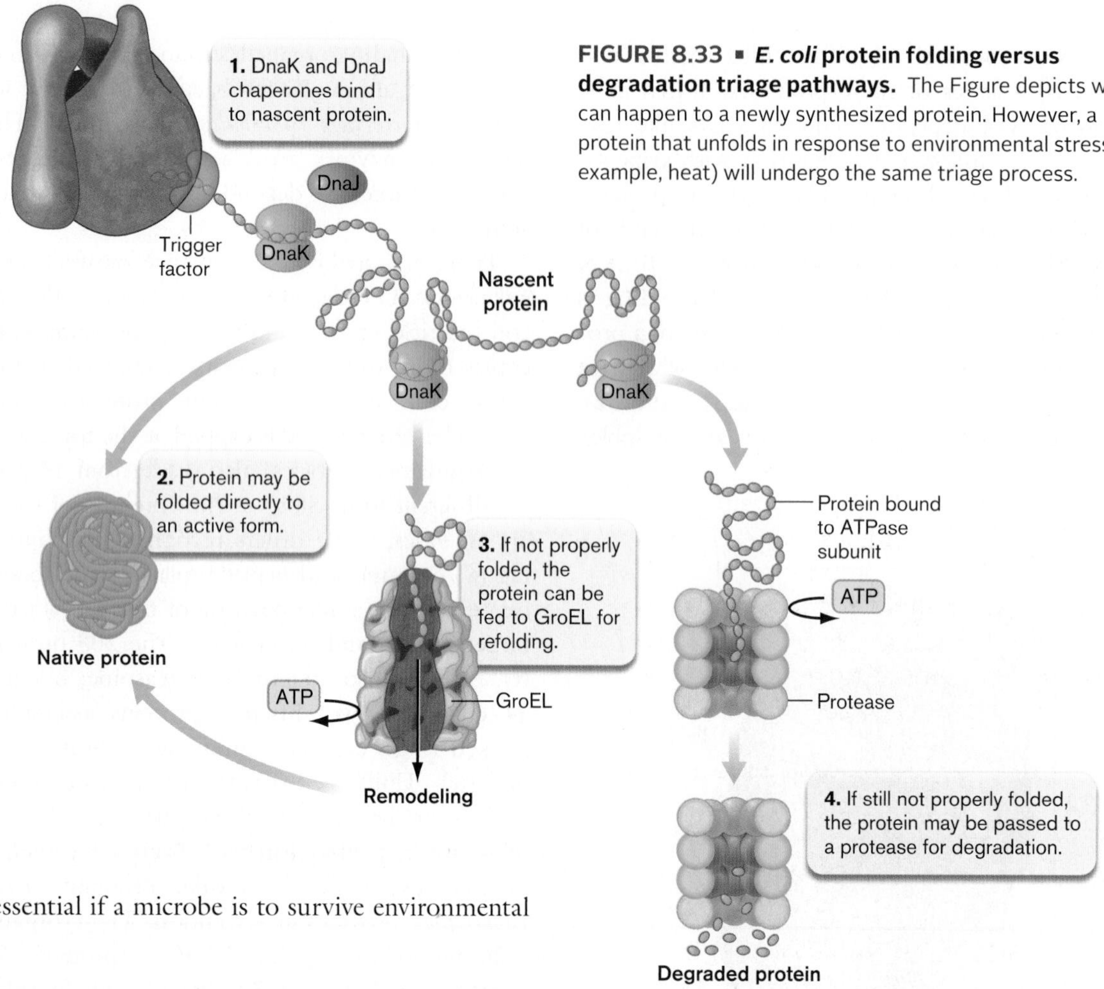
1. DnaK and DnaJ chaperones bind to nascent protein.
DnaJ
Trigger factor
DnaK
Nascent protein
DnaK
DnaK
2. Protein may be folded directly to an active form.
Native protein
3. If not properly folded, the protein can be fed to GroEL for refolding.
ATP
GroEL
Remodeling
Protein bound to ATPase subunit
ATP
Protease
4. If still not properly folded, the protein may be passed to a protease for degradation.
Degraded protein

FIGURE 8.33 ■ *E. coli* protein folding versus degradation triage pathways. The Figure depicts what can happen to a newly synthesized protein. However, a protein that unfolds in response to environmental stress (for example, heat) will undergo the same triage process.

system is essential if a microbe is to survive environmental stress.

To Summarize

- **All proteins in all cells are eventually degraded** by specific devices such as proteases or proteasomes.
- **The N-terminal rule** describes one type of degradation signal (degron) that marks the half-life of a protein (that is, how long it takes 50% of the protein to degrade).
- **ATP-dependent proteases** such as Lon or ClpP usually initiate the degradation of a large protein.
- **Damaged proteins** randomly enter chaperone-based refolding pathways or degradation pathways until the protein is repaired or destroyed.

8.6

Secretion: Protein Traffic Control

Microorganisms, especially Gram-negative bacteria, face a challenge in delivering proteins to different target locations in the cell. Recall that Gram-negative microbes are surrounded by two layers of membrane (the inner membrane, or cell membrane, and an outer membrane), between which lies a periplasmic space (see Section 3.4). Many proteins are specifically destined for one or another of these cell compartments. Other proteins are secreted completely out of the cell into the surrounding environment (for example, hemolysins that lyse red blood cells). But how do these diverse proteins know where to go? Protein traffic out of the cell is directed by an elaborate set of protein secretion systems. Each system selectively delivers a set of proteins originally made in the cytoplasm to various extracytoplasmic locations.

The term "secretion" is used to describe movement of a protein out of the cytoplasm. There are protein secretion systems that move proteins out of the cytoplasm into the cytoplasmic membrane and across the membrane to the periplasm, others that move proteins to the outer membrane, and still others that deliver proteins across both of the membranes and into the surrounding environment. An added complication of protein export is that periplasmic proteins are usually delivered unfolded into the periplasm and require another set of chaperones to fold properly in this cell compartment.

Protein Export Out of the Cytoplasm

Proteins destined for the bacterial cell membrane (such as membrane transport proteins) or envelope regions—periplasm (binding proteins), outer membrane (porins), or extracellular spaces (proteases)—require special export systems. These systems manage to move hydrophilic proteins through one or more hydrophobic membrane barriers. Proteins meant for the inner membrane (for example, cytochromes) are tagged, as part of the open reading frame, with very hydrophobic N-terminal **signal sequences** of 15–30 amino acids. Signal sequences tether nascent proteins to the membrane and confer conformations that allow the proteins to melt into the fabric of the membrane. Inner membrane proteins also contain hydrophobic transmembrane-spanning regions (20–25 amino acids) that aid in this insertion process. These hydrophobic regions are important because they are compatible with the hydrophobicity of the membrane itself. A nutrient transport protein often has 12 such membrane-spanning regions, which weave back and forth across the membrane.

One special export system begins with a complex called the **signal recognition particle (SRP)**, which targets proteins for inner membrane insertion. A second export mechanism uses a protein called trigger factor that assists proteins destined for the periplasm. (Trigger factor was mentioned earlier as a chaperone.) These two protein traffic pathways converge on a general secretion complex composed of three proteins, collectively called the SecYEG translocon, embedded in the cell (inner) membrane. Depending on the exported protein, SecYEG will assist export to the periplasm or insertion into the membrane.

Protein Export to the Cell Membrane

The pathway leading proteins to the inner (cell) membrane begins with SRP. In *E. coli*, SRP consists of a 54-kDa protein (Ffh) complexed with a small RNA molecule (*ffs*). SRP binds to the signal sequences of integral cell membrane proteins as they are being translated (**Fig. 8.34**) and halts further translation in the cytoplasm. The nascent protein with its paralyzed nontranslating ribosome is delivered to the membrane-embedded protein FtsY, where translation resumes. The partially translated protein is now subject to one of two fates: It may be cotranslationally inserted directly into the cell membrane, meaning that the protein is inserted even as it is still being translated, or it may be completely synthesized, after which it is delivered to the SecYEG translocon for insertion. The route to membrane insertion depends on the protein.

Surprising recent evidence indicates that cotranslational insertion of proteins into membranes can also affect nucleoid structure. If a transcript encoding a membrane protein is simultaneously being transcribed and translated, and the nascent protein is being inserted into the membrane (called transertion), then the gene and associated genome can be pulled toward the membrane. This movement may loosen nucleoid structure and ensure that regulatory proteins, ribosomes, and RNA polymerase have ready access to the genome.

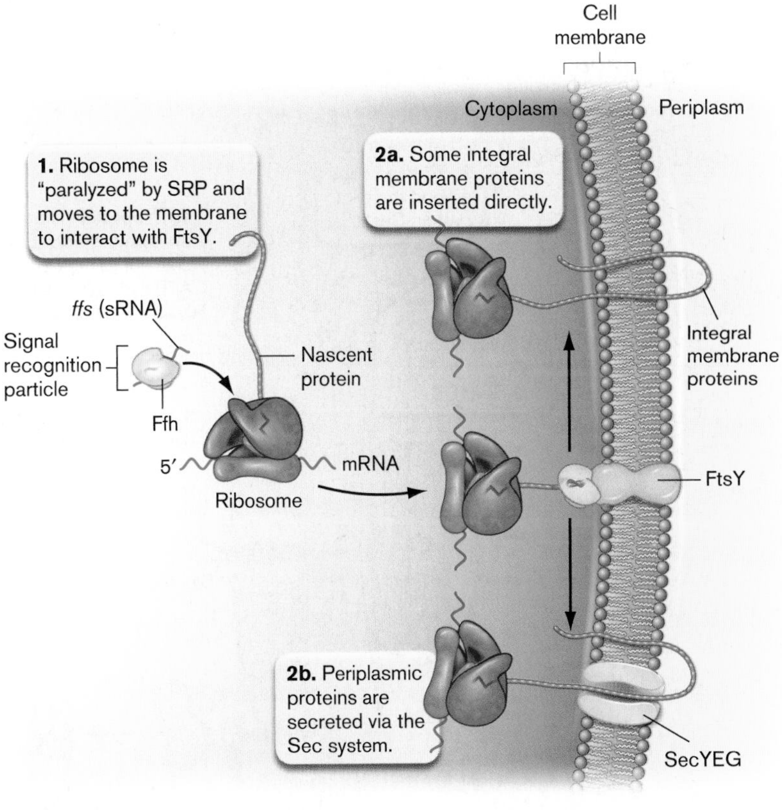
Cell membrane
Cytoplasm
Periplasm
1. Ribosome is "paralyzed" by SRP and moves to the membrane to interact with FtsY.
2a. Some integral membrane proteins are inserted directly.
ffs (sRNA)
Signal recognition particle
Ffh
Nascent protein
5′
mRNA
Ribosome
Integral membrane proteins
FtsY
2b. Periplasmic proteins are secreted via the Sec system.
SecYEG

FIGURE 8.34 ▪ SRP and cotranslational export in *E. coli*. A ribosome "paralyzed" by SRP does not resume translating protein until encountering FtsY in the membrane. Translation can then recommence. Some proteins designated for integral membrane location are inserted directly (top). Other integral membrane proteins and proteins destined for the periplasm are inserted or secreted via the Sec system (bottom).

Protein Export to the Periplasm: The Sec-Dependent General Secretion Pathway

The periplasm contains important proteins that bind nutrients for transport into the cell and other proteins that carry out enzymatic reactions. For example, one form of superoxide dismutase (SOD), an enzyme that degrades superoxide, is a periplasmic protein in *Salmonella enterica* and other Gram-negative bacteria. Many periplasmic proteins, such as SOD and maltose-binding protein (which imports the sugar maltose), are delivered to the periplasm by a common pathway called the general secretion pathway.

There are several steps in the general secretion pathway. First, the peptide is completely translated in the cytoplasm (**Fig. 8.35**, step 1). Trigger factor interacts with newly synthesized protein as it exits the ribosome and keeps pre-secreted proteins in a loosely folded conformation, awaiting interaction with the next component of the secretion machinery. The completed pre-secretion protein is then captured by a piloting protein called SecB (**Fig. 8.35**, step 2), which unfolds the pre-secretion protein and delivers it to SecA, a protein peripherally associated with the membrane-spanning SecYEG translocon (**Fig. 8.35**, step 3). Keeping a pre-secretion protein unfolded in the cytoplasm assists the secretion process because sliding an unfolded protein through a membrane is far easier than trying to deliver a folded one.

The SecA ATPase appears to act like a plunger (**Fig. 8.35**, step 4). It inserts deep into the SecYEG channel, shoving about 20 amino acids of the target export protein into the channel. ATP hydrolysis causes SecA to release the protein and withdraw (**Fig. 8.35**, step 5). At this point, SecA can bind fresh ATP, rebind the target protein, and reinsert, pushing another 20 amino acids through. Proteins needed in the periplasm have cleavable signal sequences at their amino-terminal ends. Immediately following translocation of the amino-terminal sequence into the periplasm, the sequence is snipped off by periplasmic signal peptidases (LepB is one of several examples in *E. coli*). This cleaving completes translocation and releases the mature protein into the periplasm (**Fig. 8.35**, step 6). Signal peptidases, however, will not cleave signals from proteins destined to stay embedded within the membrane (integral membrane proteins).

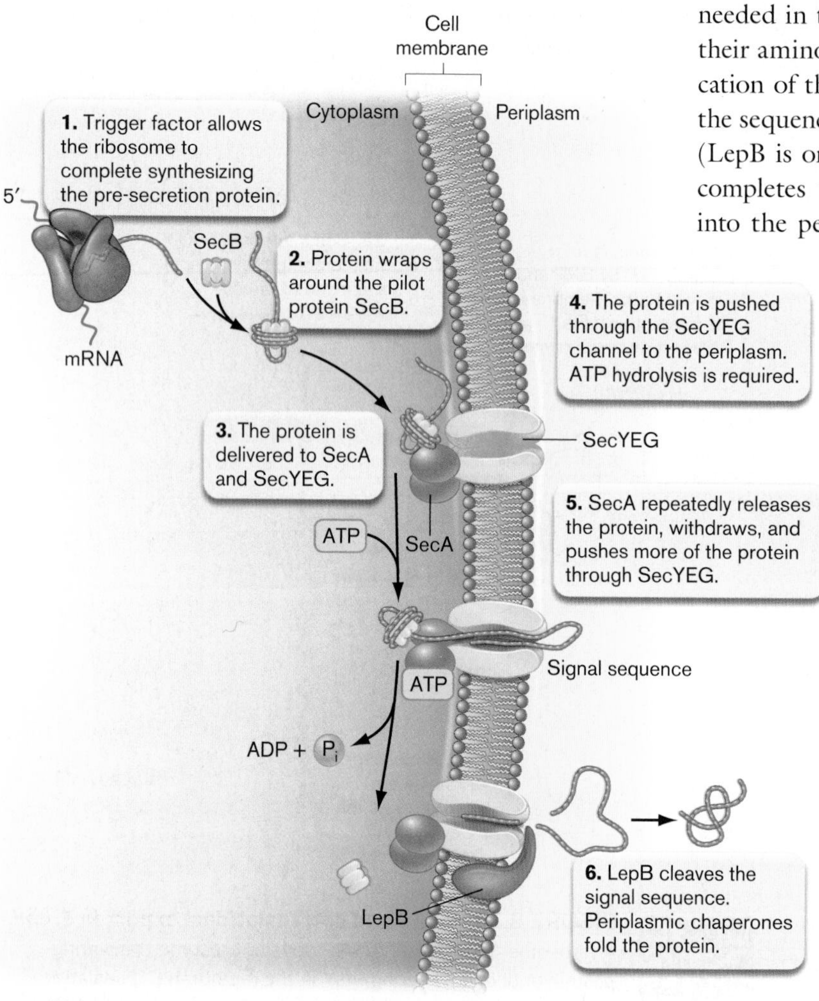

Cell membrane
Cytoplasm
Periplasm
1. Trigger factor allows the ribosome to complete synthesizing the pre-secretion protein.
5′
SecB
2. Protein wraps around the pilot protein SecB.
mRNA
4. The protein is pushed through the SecYEG channel to the periplasm. ATP hydrolysis is required.
3. The protein is delivered to SecA and SecYEG.
SecYEG
5. SecA repeatedly releases the protein, withdraws, and pushes more of the protein through SecYEG.
ATP
SecA
Signal sequence
ATP
ADP + P
6. LepB cleaves the signal sequence. Periplasmic chaperones fold the protein.
LepB

FIGURE 8.35 ▪ The SecA-dependent general secretion pathway. This pathway exports many proteins across the cell membranes of Gram-negative and Gram-positive bacteria.

Note: "Translocation" can refer to the movement of a ribosome along mRNA, or it can describe the movement of a protein from one cell compartment (cytoplasm) to another (periplasm).

Periplasmic proteins delivered by the Sec system arrive unfolded and inactive. Because the folding chaperones mentioned earlier are cytoplasmic, periplasmic proteins need another set of dedicated periplasmic chaperones to guide their tertiary folding. Another problem with periplasmic proteins is that the oxidizing environment of aerobic cells can oxidize cysteines within a protein and produce inappropriate cysteine disulfide bonds that destroy enzyme function. Special periplasmic disulfide reductases are required to reduce these S–S bonds back to two SH groups. Many periplasmic proteins, however, need certain disulfide bonds to be active, so the periplasm also contains a disulfide bond catalyst (DsbA) to make those bonds.

Eukaryotic microbes such as the yeast *Saccharomyces cerevisiae* also possess secretion systems that move proteins to the membrane and beyond. However, the eukaryotic Sec systems are more complex and are evolutionarily distinct from bacterial Sec systems. Archaeal secretion systems are actually more similar to those of eukaryotes than to those of bacteria.

Export of Prefolded Proteins to the Periplasm

In a dramatic departure from Sec-dependent transport systems, some proteins, like TorA, a component of an anaerobic respiratory chain, can be transported fully folded across the membrane to their periplasmic destination. These proteins contain the amino acid motif RRXFXK within their N-terminal signal sequence (where R = arginine, F = phenylalanine, K = lysine, and X = any amino acid). This sequence, called the "twin arginine motif" because it begins with two arginines, targets the protein to the membrane-embedded twin arginine translocase (TAT), a transport protein that ships fully folded proteins across the cell membrane to the periplasm (**Fig. 8.36**). Whereas Sec-dependent transport is ATP driven, the TAT system is powered by the proton motive force (see Chapters 4 and 14).

Journeys to the Outer Membrane

Outer membrane proteins (OMPs) are made in the cytoplasm and exported to the periplasm by the SecA-dependent secretion system. Some OMPs have hydrophobic C-terminal signal sequences that facilitate insertion into the outer membrane, but all have a beta-barrel structure that ultimately suits them for outer membrane placement (for instance, see TolC, **Fig. 8.37**). Periplasmic chaperones prevent aggregation of OMPs as they traverse the periplasm and deliver the proteins to a multisubunit, outer membrane machine called the BAM (beta barrel assembly machine) complex that facilitates OMP assembly in the outer membrane. Although the players seem to be known, the mechanism by which beta-barrels are folded and inserted into the OM bilayer remains unclear.

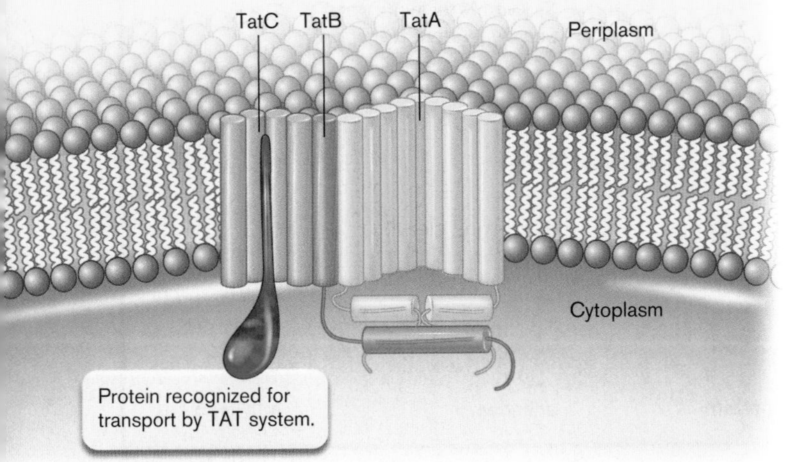
TatC
TatB
TatA
Periplasm
Cytoplasm
Protein recognized for transport by TAT system.

FIGURE 8.36 ▪ The twin arginine translocase (TAT). A commonly accepted model for the Tat protein translocase, which includes proteins TatA, TatB, and TatC.

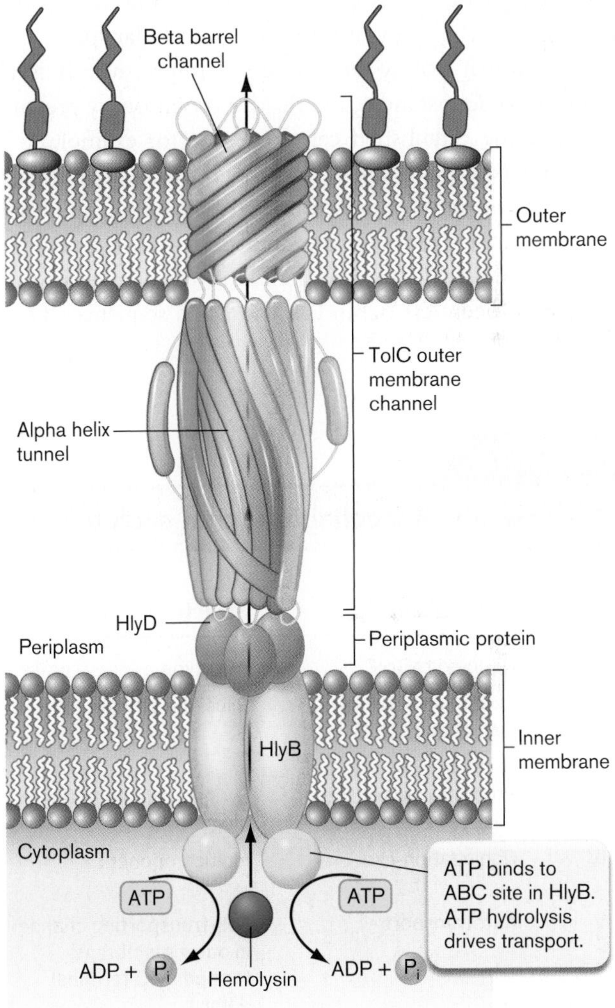
Beta barrel channel
Outer membrane
TolC outer membrane channel
Alpha helix tunnel
HlyD
Periplasm
Periplasmic protein
HlyB
Inner membrane
Cytoplasm
ATP
ATP
ATP binds to ABC site in HlyB. ATP hydrolysis drives transport.
ADP + P_i
Hemolysin
ADP + P_i

FIGURE 8.37 ▪ Type I secretion: the HlyABC transporter. Hemolysin (HlyA) is transported directly from the cytoplasm into the extracellular medium through a multicomponent ABC transport system. The HlyB and D proteins are dedicated to HlyA transport. TolC is shared with other transport systems. Not drawn to scale. *Source:* Modified from Moat et al. 2002. *Microbial Physiology*, 4th ed. Wiley-Liss.

Journeys through the Outer Membrane

For many reasons, Gram-negative bacteria need to export proteins completely out of the cell and into their surrounding environment. Some exported proteins digest extracellular peptides for carbon and nitrogen sources; others act as free-floating toxins that bind and kill host cells. Still others are injected directly into eukaryotic cells by pathogenic or symbiotic microbes to commandeer host metabolic processes. Seven elegant secretion systems, identified as type I–type VII, have evolved to ship these proteins out of the cell (**Table 8.2**). A few start with the Sec system just to get the protein into the periplasm, where dedicated outer membrane systems take over and complete export. Other systems provide nonstop service, delivering the protein directly from the cytoplasm to the extracellular space.

The diversity of system design is impressive. It is the result, in some instances, of selective evolutionary pressures appropriating established cell processes (for example, pilus assembly). New systems evolve through the accidental duplication of one set of genes followed by random mutations that innovatively redesign the duplicated set for a new purpose. We know this because the footprints of genetic divergence have been left behind in the DNA sequence. Type I secretion is described here. Other systems will be covered during the discussion of pathogenesis in Chapter 25.

Type I Protein Secretion

Chapter 4 describes the family of ATP-binding cassette (ABC) influx transporters, whose signature is an amino acid motif that binds ATP. (The term "cassette" refers to a sequence of amino acids that is conserved in many proteins with similar functions.) In addition to ABC influx transporters, similar ABC transporters function in the opposite direction to export various toxins, proteases, and lipases, as well as antimicrobial drugs (multidrug efflux transporters). These ABC transporters are the simplest of the protein secretion systems and make up what is called type I protein secretion (for example, the Hly system which secretes hemolysin; **Fig. 8.37**▶). Type I systems all have three protein components, one of which contains an ATP-binding cassette. One component is an outer membrane channel, the second is an ABC protein at the inner membrane, and the third is a periplasmic protein lashed to the inner membrane.

Proteins secreted through type I systems never contact the periplasm, because they pass through a continuous channel that extends from the cytoplasm to the outer

TABLE 8.2

Comparison of mechanisms that secrete proteins across the outer membrane.[a]

Type	Mechanism	Structure	Location of protein substrate	Location of secretory signals	Number of components
I	Coupled to TolC	ABC type	Cytoplasm	N terminus, not cleaved	3
II	Extending and contracting	Pilus-like structure	Periplasm	Cleaved N terminus for Sec-dependent transport	12–16
III	Proteins injected into host cell cytoplasm	Syringe, related to flagella biogenesis	Cytoplasm	None	20
IV	Conjugation-like	Multicomponent	Some are cytoplasmic, others are periplasmic	None	8 or 9
V	Autotransport	Self-transporting channel in outer membrane formed by C-terminal domain	Periplasm	N terminus, cleaved	1
VI	Proteins injected directly into host cytoplasm	Contractile piston, phage tail origin	Cytoplasm	None	12–20
VII	Unknown	Unknown	Cytoplasm	Unknown	At least 4

[a]The seven classes are grouped on the basis of their structure. Types II, III, and IV are evolutionarily related to mechanisms that assemble pili (type II) or flagella (type III) or that carry out conjugation (type IV). Type VI systems are related to phage tail proteins. Type VII is unrelated to the rest. Some systems pick up protein substrates from the cytoplasm and transport them across both membranes. Substrates for other systems are collected in the periplasm. Substrates also differ as to the presence of N-terminal signal sequences.

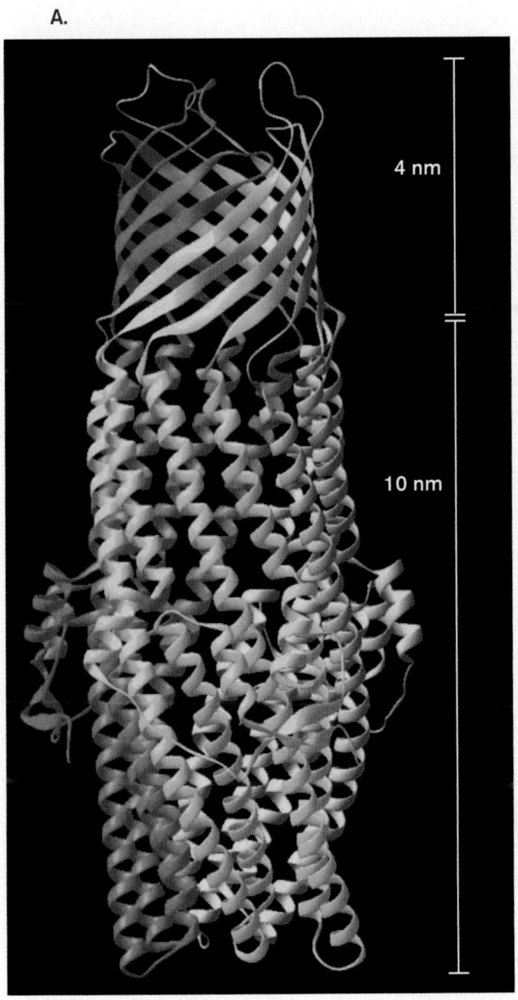
A.
4 nm
10 nm

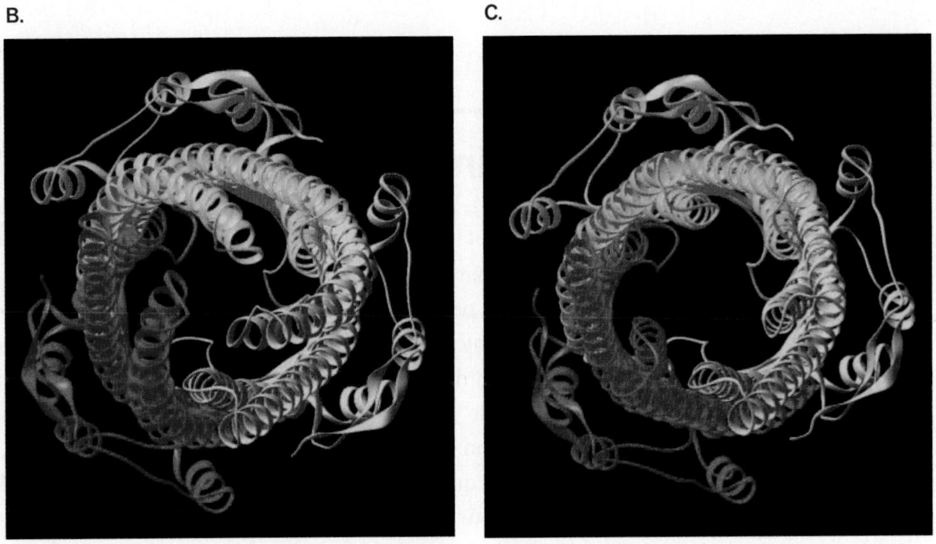
B.
C.

Vassilis Koronakis et al. 2000. *Nature* **405**:914–919

FIGURE 8.38 ■ TolC protein structure. **A.** The beta barrel channel spans the outer membrane (not shown), and the alpha helix tunnel extends into the periplasm. The inner membrane would be located at the bottom of this figure. Three monomers make up the channel. **B.** A view of the crystal structure from the periplasmic end of the molecule, looking down the threefold symmetry axis. The alpha helix tunnel is closed. **C.** A hypothetical model of how the tunnel might be opened, with the same orientation as in (B).

membrane. The inner membrane and periplasmic subunits are generally substrate specific, but numerous ABC export systems share the channel protein TolC. TolC is an intriguing protein composed of a beta barrel channel embedded in the outer membrane and an alpha helix tunnel spanning the periplasm. **Figure 8.38** illustrates how the tunnel might open and close. The type I transport shown in **Figure 8.37** exports a hemolysin (HlyA) from *E. coli* that lyses red blood cell membranes. HlyB and HlyD are the ABC and periplasmic components, respectively.

Six other protein secretion systems are briefly summarized in **Table 8.2**. Some move proteins directly from the cytoplasm to the outside, similar to the type I system, whereas others pick up proteins deposited in the periplasm by the Sec system. They all play important roles in microbial pathogenesis and are more fully discussed in Chapter 25.

Weblinks Chaperone-assisted protein folding (*see ebook*)

To Summarize

- **Special protein export mechanisms** are used to move proteins to the inner membrane, the periplasm, the outer membrane, and the extracellular surroundings.
- **N-terminal amino acid signal sequences** help target membrane proteins to the membrane.
- **The general secretory system** involving the SecYEG translocon can move unfolded proteins to the inner membrane or periplasm.
- **The signal recognition particle (SRP)** pauses the translation of a subset of proteins that will be placed into the membrane.
- **SecB protein** binds to certain unfolded proteins that will eventually end up in the periplasm, and pilots them to the SecYEG translocon.
- **The twin arginine translocase (TAT)** can move a subset of already folded proteins across the inner membrane and into the periplasm.
- **Type I secretion systems** are ATP-binding cassette (ABC) mechanisms that move certain secreted proteins directly from the cytoplasm to the extracellular environment.

8.7

Bioinformatics: Mining the Genomes

We have described how the information within a genome is deciphered by the cell to produce proteins and how a subset of those proteins find their way to the membrane and beyond. But the experiments that defined these processes cannot be performed in the vast majority of microbes, which are unculturable. In Section 7.6 we discussed how we can quickly and efficiently sequence entire genomes. This knowledge, combined with our understanding of protein structure (amino acid sequences) and the relationships between protein structure and function, has brought us to the postgenomic era. We can now call on the vast store of information gathered over the last century to make predictions about the genetics and physiology of microbes even when we cannot grow them in the laboratory. The following sections reveal how these predictions are made.

Annotating the Genome Sequence

Chapter 7 explained how the DNA sequence of a genome is obtained. Knowing the sequence is just a first step; meaningful information comes from **annotation** of the DNA sequence—that is, understanding what the sequence means. Annotation is analogous to identifying separate sentences and words in an unknown language. For most languages, you would look for punctuation marks like periods or exclamation points to identify sentences and then scan for blank spaces between letters to signify the beginning and end of a word. You may not initially know what the sentence says, but defining what makes a sentence is the first step.

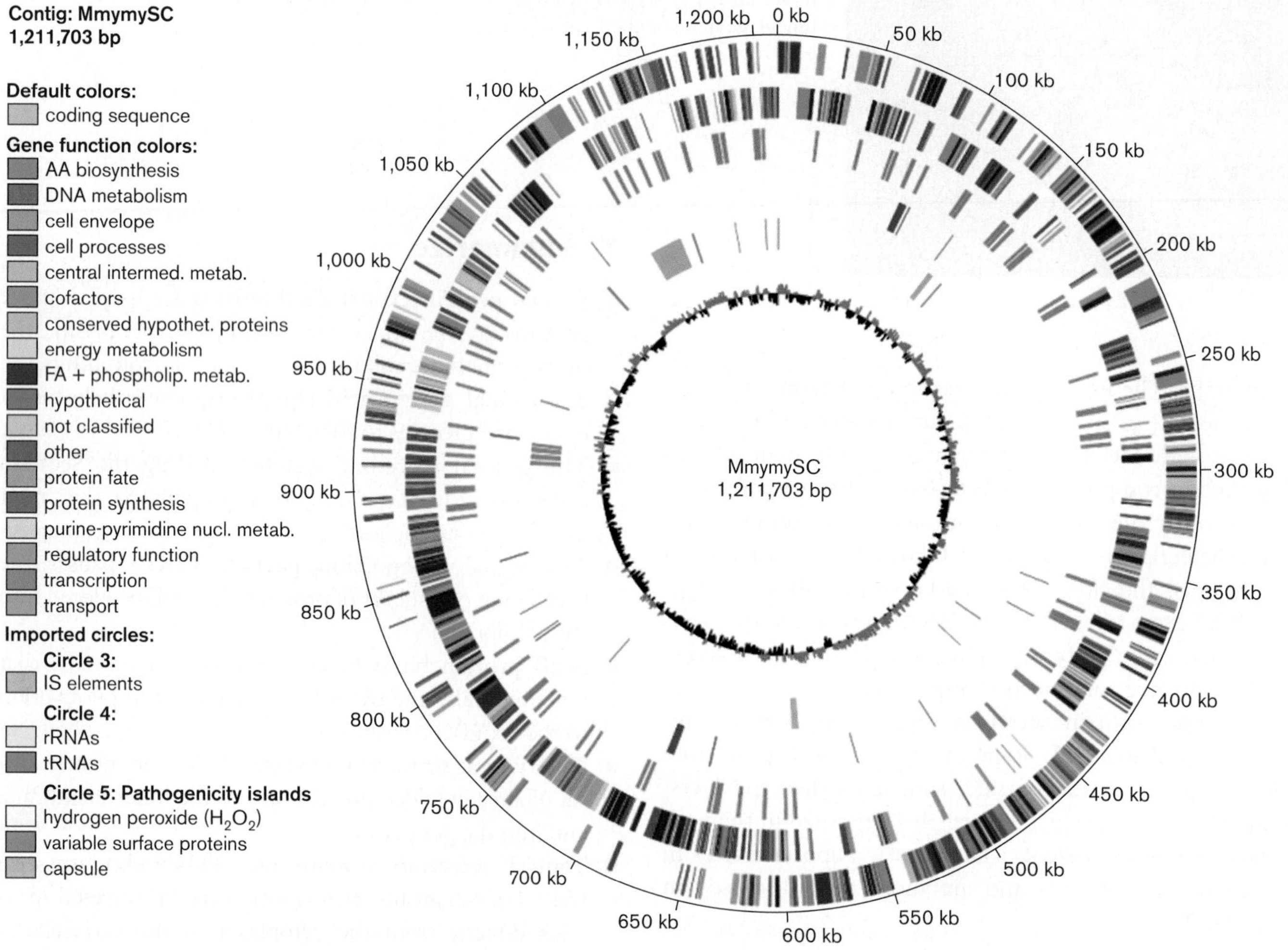
Contig: MmymySC
1,211,703 bp
Default colors:
coding sequence
Gene function colors:
AA biosynthesis
DNA metabolism
cell envelope
cell processes
central intermed. metab.
cofactors
conserved hypothet. proteins
energy metabolism
FA + phospholip. metab.
hypothetical
not classified
other
protein fate
protein synthesis
purine-pyrimidine nucl. metab.
regulatory function
transcription
transport
Imported circles:
Circle 3:
IS elements
Circle 4:
rRNAs
tRNAs
Circle 5: Pathogenicity islands
hydrogen peroxide (H_2O_2)
variable surface proteins
capsule
1,200 kb
0 kb
50 kb
100 kb
150 kb
200 kb
250 kb
300 kb
350 kb
400 kb
450 kb
500 kb
550 kb
600 kb
650 kb
700 kb
750 kb
800 kb
850 kb
900 kb
950 kb
1,000 kb
1,050 kb
1,100 kb
1,150 kb
MmymySC
1,211,703 bp

FIGURE 8.39 ■ Circular display map of the *Mycoplasma mycoides* genome (1,211,703 bp). The color code indicates gene clustering by function. The two outer rings represent genes transcribed from different DNA strands. The innermost circle uses peaks to show GC content levels above 50% (red) or below 50% (black).

Annotation of a DNA sequence includes finding the start and stop sites of the genes, as well as predicting the function of the gene product. Computers use established rules to mark potential genes—open reading frames (ORFs). Then, similarities are sought between the deduced amino acid sequences of those ORFs and the sequences of proteins with known functions. Similarities are used to infer the function of the unknown ORF. Genes encoding transfer RNAs and ribosomal RNAs do not encode proteins but can be identified because of the conservation of these sequences across vast phylogenetic distances. The results of annotation can be represented in many different forms. **Figure 8.39** shows a commonly used method of presenting the results—namely, a circular display map (in this case of the *Mycoplasma mycoides* genome) with genes color-coded by their predicted function.

Since 1998, the complete genomes of over 225 microbial species have been published, and many others have been partially sequenced. This wealth of information has spawned a new discipline, called **bioinformatics**, dedicated to comparing genes of different species. Data from bioinformatics enable scientists to make predictions about an organism's physiology and evolutionary development, even if the organism cannot be grown in the laboratory. The magnitude of this achievement, and of the job ahead, can be appreciated if we recall that fewer than 1% of microorganisms can be cultured. Bioinformatics has forever changed how the science of microbiology is conducted.

Weblinks Sequenced genomes (*see ebook*)

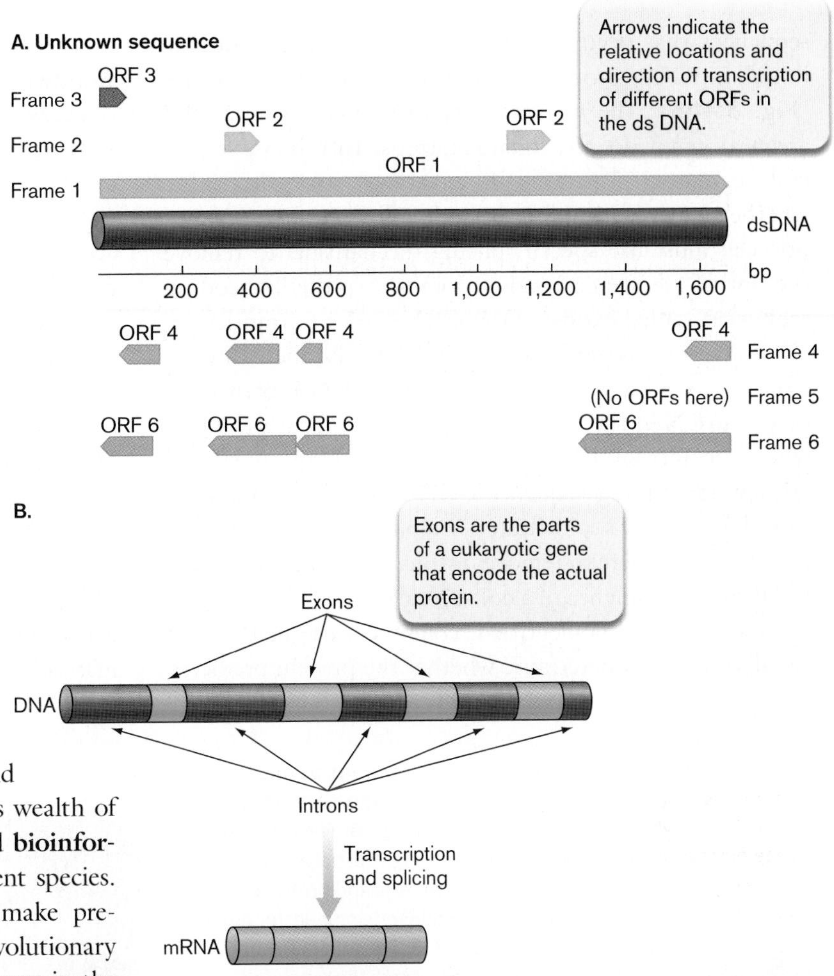
A. Unknown sequence
Arrows indicate the relative locations and direction of transcription of different ORFs in the ds DNA.
ORF 3
Frame 3
ORF 2
ORF 2
Frame 2
ORF 1
Frame 1
dsDNA
bp
200
400
600
800
1,000
1,200
1,400
1,600
ORF 4
ORF 4
ORF 4
ORF 4
Frame 4
(No ORFs here)
Frame 5
ORF 6
ORF 6
ORF 6
ORF 6
Frame 6
B.
Exons are the parts of a eukaryotic gene that encode the actual protein.
Exons
DNA
Introns
Transcription and splicing
mRNA

FIGURE 8.40 ■ Predicting open reading frames (ORFs) in a DNA sequence. A. Each predicted ORF in this 1,600-bp bacterial sequence begins with AUG or GUG (potential translation start codons in mRNA) and ends with a translation terminator codon. The computer can translate in all six reading frames (three on the top strand, three on the bottom strand). Determining which is the real ORF requires additional information, such as potential ribosome-binding sites and a transcriptional start site. **B.** In eukaryotes, finding ORFs is complicated by the presence of noncoding DNA sequences called introns. The actual ORFs (exons) are interrupted by introns, which must be spliced out of the primary RNA message before the final mRNA leaves the nucleus.

Computer Analysis and Web Science

Lengthy DNA sequences are analyzed by computer programs such as ORF Finder. The program uses the universal genetic code to deduce all possible protein sequences that could be formed in all reading frames on RNA molecules transcribed from either direction on the chromosome (**Fig. 8.40A**). An open reading frame (ORF), the equivalent of a sentence in our analogy, is defined as a DNA sequence that can potentially encode a string of amino acids of minimum length—say, 50 residues. The 50 residues of the ORF could encode a 5,500-Da (5.5-kDa) protein, since the average weight of an amino acid is 110 Da. Each ORF begins with a translation start codon (usually ATG, or more rarely, GTG or TTG). A translation start codon marks where ribosomes start to read a messenger RNA molecule. An ORF ends with a termination codon (in the DNA, these are TAA, TAG, or TGA), which, as UAA, UAG, or UGA in RNA, signals the ribosome to stop translation. In addition, the computer can identify an ORF by looking for potential ribosome-binding sites upstream of the start codon, but these ribosome-binding sites may differ between species, so finding one is not essential to declaring the presence of an ORF.

While the preceding methods work for prokaryotes, identifying ORFs in eukaryotes is more difficult. In eukaryotes, most genes contain **introns** (long, noncoding sequences that occur in the middle of genes), as do the initial mRNA products produced from those genes (**Fig. 8.40B**). The vast majority of known bacterial and archaeal genes do not contain introns. Introns serve several regulatory functions in eukaryotes that determine whether a protein product is made. However, the eukaryotic cell must use special splicing mechanisms to remove the intron sequences and re-join the protein-encoding sequences, called **exons**, of the mRNA prior to translation. (In eukaryotic organisms, the RNA transcribed directly from DNA is called the **preliminary mRNA transcript** or **pre-mRNA**; the "final" mRNA transcript is produced when the introns have been spliced out.) Because there are few identifying sequence characteristics that mark an intron, extremely sophisticated computer analysis is needed to determine where to remove introns in order to derive the ORF coding sequence of a eukaryotic gene.

Once an ORF is identified, computers use mathematical algorithms to determine whether the protein predicted by the ORF resembles any other protein deposited in the worldwide databases or, even better, resembles any protein of known function. By "resemble," we mean that the query protein possesses amino acid sequences that are identical or functionally similar to those found in other proteins. In a functionally similar sequence, certain amino acids are replaced by similar amino acids—for example, isoleucine for leucine. We know from considerable experimental precedent that proteins with the same function, regardless of species, usually have critical amino acid sequences in common because they evolved from the same ancestral sequence. Sequences common to two different protein or DNA molecules are called homologous sequences.

Figure 8.41 presents a sequence alignment whose goal was to identify cardiolipin synthase in the protozoan parasite *Trypanosoma brucei* (**Figure 8.41A**), the causative agent of human African sleeping sickness. This parasite is carried between human hosts by the tsetse fly and dramatically changes physiology in the process. In the human

FIGURE 8.41 ■ Multiple-sequence alignment of bacterial-like cardiolipin synthases (Cls) in protozoan parasites. **A.** *Trypanosoma brucei* next to a red blood cell. **B.** The program ClustalW2 (available online) was used to align partial amino acid sequences of putative Cls genes from bacteria and protozoa containing the conserved motifs. The code for single-letter amino acid designations is provided in Figure 8.12. The minimal motif typical for bacterial-type Cls is indicated in red; the extended conserved sequence is highlighted in blue. These proteins are called orthologs because they have similar sequences and carry out the same biochemical reaction. Asterisks indicate amino acid identity, and colons indicate amino acid similarity. Deletions (sequences missing in one or more orthologs) are indicated with dashes.
Source: Part B modified from M. Serricchio and P. Bütikofer. 2012. *PNAS* **109**:E954–E961.

A.

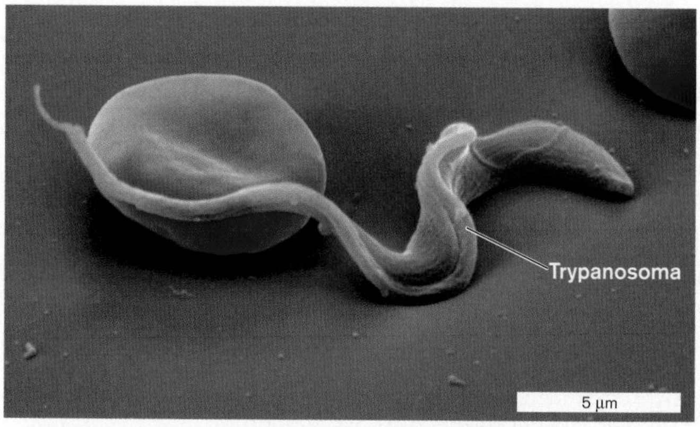
Trypanosoma
5 µm

EYE OF SCIENCE/PHOTO RESEARCHERS, INC.

B.

```
Escherichia coli           LRQHRKMIMIDNYIAYTGSMNMVDPRYFKQDAGVGQWIDLM-ARMEGPIATAMGII  278
Streptococcus pneumoniae   -RDHRKILIVDGQIAYTGGVNLADEYINHVER-FGYWKDSG-IRLDGLAVKALTRL  303
Trypanosoma brucei         IRNHRKILIVDNKLGFCGGLNIGDEYCGTSEGGNGRFRDTH-CVVMGPAVMHMREV  259
Trypanosoma congolense     IRNHRKILIVDNKLGFCGGLNIGNEYCGKEKGGSGRFRDTH-CAVMGPAVAHLREV  259
Trypanosoma cruzi          IRNHRKILIVDDKIGFCGGLNIGNEYCGRSQGGTGRFRDTH-CSVVGPAVAHLREA  262
Leishmania major           LRNHRKILLVDASQGFCGGLNIGNEYCGKEAGGTGKFRDTH-CSVIGPAAAHLAEV  302
Plasmodium falciparum      FRDHRKILIVDN-TAYCGSMNVAENVFPSEIFHEYEEGDEREEDTEYSEEIGKKKK  243
Toxoplasma gondii          FRDHRKNLIVDGKVAFVGSLNVSEDAVGEKFGGRNHFYDLH-VRLRGPAVKALESL  331
                            *:*** :::*   .: *.:* :              *        .

Escherichia coli           GVKIYQFEGG-LLHTKSVLVDGELSLVGTVNLDMRSLVLNFEITLAIDDKGFGADL  446
Streptococcus pneumoniae   GVRIYEYSPG-FIHSKQMLVDEDFAVVGTINLDYRSLVHHYENAVLLYKTPSIREI  471
Trypanosoma brucei         GVRVYEYKGGQVMHAKTVVADSIWTSIGSYNWDMMSN-KNMEICVCHLDHNLALTM  581
Trypanosoma congolense     GVRVYEYKGEQVMHAKTVVVDSIWSSIGSYNWDMMSN-KNMEVCLCHLDYDLARVM  580
Trypanosoma cruzi          GVRIYEYKGDQIMHAKTVVVDSIWCSVGSYNWDMMSN-KNMEVCLCHLGYEMAREM  578
Leishmania major           GVKIYEFQGQQIMHAKTVVVDSVWCSIGSYNWDLMSN-RNLEVCLCHLDLEVAHSM  690
Plasmodium falciparum      GSMNFYFFQKKHCHAKNLVVDNLWCSIGSYNWDRFSSRRNLEVMISIFDKKICDKF  520
Toxoplasma gondii          GDASVYFLTSRHCHAKNIVVDHLWSTTGSFNFDRFSSRRNMEVLVAFLDPGIALKF  744
                           *     :      *:* ::.*      *: * *  *   : *  :          :
```

bloodstream, *T. brucei* mitochondria are inactive, but in the tsetse fly the parasite's mitochondrial membranes suddenly incorporate the lipid cardiolipin and become fully active. This developmental process led Mauro Serricchio and Peter Bütikofer from the University of Bern in Switzerland to search the *T. brucei* genome for homologs of eukaryotic cardiolipin synthases. That search failed. However, a subsequent search of bacterial homologs, shown in **Figure 8.41B**, found genes in *T. brucei* and other protozoan parasites that looked similar to bacterial cardiolipin synthases (*E. coli* and *Streptococcus pneumoniae*). Thus, a bacterial-like enzyme synthesizes cardiolipin in these parasitic eukaryotes.

Note that the highlighted areas in **Figure 8.41** represent regions of identity or similarity at similar positions. "Identity" means that the same amino acid is present in each protein, whereas "similarity" means that chemically related amino acids are substituted; for example, asparagine (N) is substituted by the similar aspartate (D), or isoleucine (I) is substituted by leucine (L).

On the basis of this homology, the predicted cardiolipin gene was deleted from the *T. brucei* genome. Cardiolipin synthesis stopped, mitochondrial morphology changed, and cytochrome activity decreased. As we will discuss in Chapter 27, the presence of a bacterial-like cardiolipin synthase in *T. brucei* makes this enzyme a potential selective drug target. A drug with selective toxicity will inhibit the bacterial enzyme, and thus the parasite, but will not affect host eukaryotic enzyme.

Alignments can be done either with base sequences in DNA or with protein sequences deduced from the DNA sequence. Because of the degeneracy of the code, however, it is usually easier to identify evolutionary relationships between genes from distantly related species by comparing their protein products rather than the DNA sequence. Because one amino acid can be encoded by more than one DNA codon, a DNA strand (nontemplate) encoding the peptide Ala-Leu-Ser could be 5′ GCT CCT TCC or 5′ GCA CCG TCA. It is easier to see the homology using the deduced peptide. On the other hand, not all genes encode proteins. Prime examples are the genes encoding stable ribosomal RNA molecules. The DNA sequences for these rRNAs are highly conserved across species, and their similarities reveal homology even among the three ancestral domains (Bacteria, Archaea, and Eukarya).

Note: As first mentioned in Section 8.3, do not confuse template and nontemplate strands of DNA for a given gene. The template strand, read 3′ to 5′ by RNA polymerase, is transcribed to make the 5′-to-3′ RNA transcript. The nontemplate DNA strand, also called the sense strand, has the same sequence as the RNA transcript, with the substitution of T's for U's.

Homologs, Orthologs, and Paralogs

Homologies found between genes or proteins suggest an evolutionary relatedness. Genes or proteins that are **homologous** probably evolved from a common ancestral gene. Homologous genes or proteins can be classified as either orthologous or paralogous. **Orthologous** genes have functions essentially the same but occur in two or more different species. For example, the gene for glutamine synthase (*gltB*) in *E. coli* is orthologous to *gltB* in *Vibrio cholerae* and to *gltB* in the Gram-positive bacterium *Bacillus subtilis.* **Paralogous** genes arise by duplication within the same species (or progenitor) but evolve to carry out different functions (**Fig. 8.42**). For example, genes encoding type III

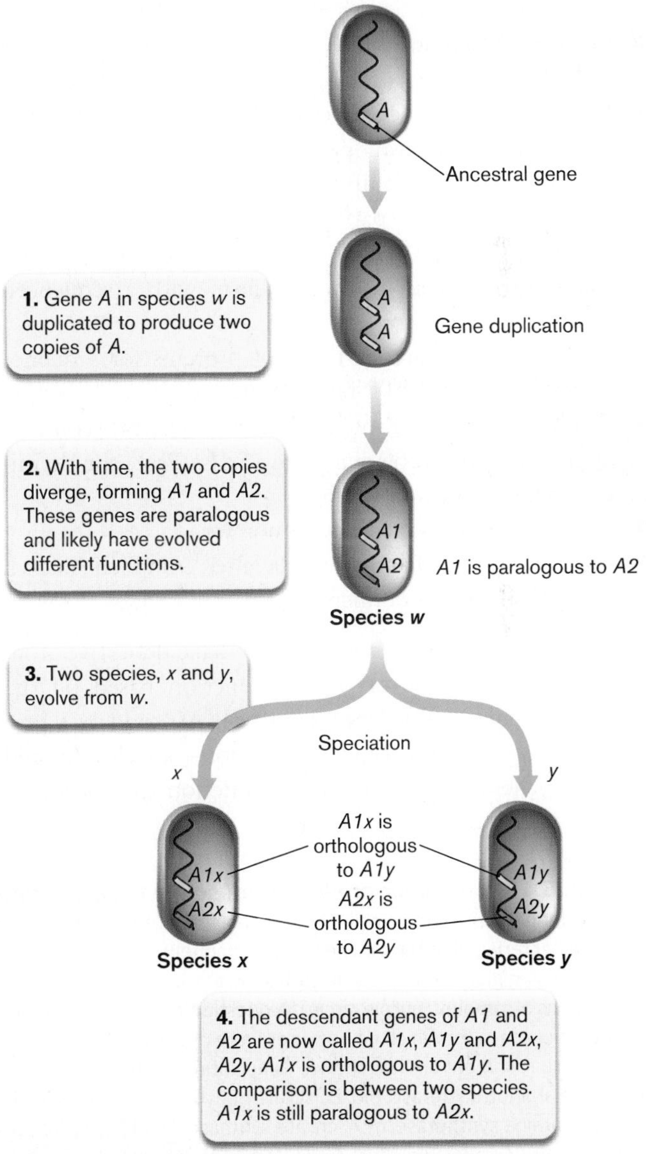
A
Ancestral gene
1. Gene A in species w is duplicated to produce two copies of A.
A
A
Gene duplication
2. With time, the two copies diverge, forming A1 and A2. These genes are paralogous and likely have evolved different functions.
A1
A2
A1 is paralogous to A2
Species w
3. Two species, x and y, evolve from w.
Speciation
x
y
A1x
A2x
A1x is orthologous to A1y
A2x is orthologous to A2y
A1y
A2y
Species x
Species y
4. The descendant genes of A1 and A2 are now called A1x, A1y and A2x, A2y. A1x is orthologous to A1y. The comparison is between two species. A1x is still paralogous to A2x.

FIGURE 8.42 ■ Paralogous versus orthologous genes. An ancestral gene can undergo a duplication to evolve an orthologous or paralogous gene.

protein secretion systems, which export virulence proteins, have evolved from paralogous genes encoding flagellar biogenesis. The two sets of genes share a common ancestor but have since evolved completely different functions (in this case, secretion and motility). Paralogous genes are maintained in a microbial genome because their distinct functions contribute to the organism's adaptive potential. Duplicate genes of identical function usually result in loss of one copy or the other by degenerative evolution (see Chapter 9).

Many computer programs and resources used to analyze DNA and protein sequences are freely available on the Web. Here are some useful programs and websites:

- **BLAST NCBI** (Basic Local Alignment Search Tool) compares a sequence of interest with all other DNA or protein sequences deposited in sequence databases.
- **Multiple Sequence Alignment EBI** aligns sequences of genes identified as homologous by BLAST analysis.
- **KEGG** (Kyoto Encyclopedia of Genes and Genomes) outlines biochemical pathways in many sequenced organisms. Areas within this site graphically display reference biochemical pathways (for example, glycolysis) and then indicate which proteins in that pathway are predicted to occur in any organism with a sequenced genome.
- **Motif Search** searches DNA or proteins for sequence signatures such as ATP-binding sites.
- **ExPASy** (Expert Protein Analysis System) contains many molecular tools, including Swiss-Prot, the definitive index of known proteins.
- ***coli*BASE** focuses on *E. coli* genomics.
- **Joint Genome Institute** has also compiled known genome sequences of microbes and eukaryotes.

A word of caution: It is enticing to make definitive proclamations about gene or protein function based on the computer analysis of a DNA sequence. As good as a prediction may seem, it is only a prediction—a well-educated guess. Biochemical confirmation of function must be made, if possible. Nevertheless, the predictions we can make are powerful. In one recent example, a metabolic model was constructed by genome annotation for *Helicobacter pylori*, the causative agent of gastritis (ulcers). The *H. pylori* model (**Fig. 8.43**) was used to predict which amino acids the organism would require from its host (because the predicted metabolic pathway appears to be missing from the bacterium) and which genes might be essential (because the host cannot provide their function or products). This is an example of what has been called **functional genomics**. Functional genomics is an integrative process in which bioinformatic approaches enable scientists to make predictions about function for a set of genes and then test those predictions experimentally. In another application, some investigators are trying to predict the smallest number of genes required for a living cell (see **eTopic 8.4**).

Thought Question

8.12 An ORF 1,200 bp in length could encode a protein of what size and molecular weight? *Hint:* Find the molecular weight of an average amino acid.

The information age has spawned what could be called Web biology, an *in silico* science that has provided new insight into evolution, physiology, and pathogenesis. Bioinformatics has not only confirmed what we know in greater detail, but also revealed new information, such as the existence of ORFs with no known function. The sequencing of entire genomes has enabled a greater understanding of sequence similarity or homology between genes or proteins, and has provided a foundation for understanding evolutionary relationships. For example, the sequencing of the alphaproteobacterium *Rickettsia prowazekii* revealed many similarities to genes within eukaryotic mitochondrial DNA. This finding led to the conclusion that eukaryotic mitochondria were derived from a rickettsial predecessor that, instead of becoming food for its eukaryotic host, became an endosymbiont (an intracellular symbiotic organism).

FIGURE 8.43 ■ Genomic predictions of solute transport and metabolic pathways of *Helicobacter pylori* strain 26695. The large rectangle represents a cell. The transporters around the periphery were predicted by sequence homology to transporters in Gram-negative bacteria (Chapter 4). The main components of *H. pylori* central metabolism are presented within the rectangle. This scheme is based on using glucose as the sole carbohydrate source. Urease, a multisubunit enzyme crucial for the survival of *H. pylori* at acidic pH (<7), is indicated as a complex (purple circle). A question mark is attached to pathways that could not be completely elucidated. Red arrows represent pathways or steps for which no enzymes were predicted from the sequence. Pathways for macromolecular biosynthesis (RNA, DNA, and fatty acids) were found but are not shown. Gene abbreviations: *aceEF*, pyruvate dehydrogenase; *ackA*, acetate kinase; *acnB*, aconitase B; *aspC*, aspartate aminotransferase; *dld*, D-lactate dehydrogenase; *frdABC*, fumarate reductase; *fumA*, fumarase; *gdhA*, glutamate dehydrogenase; *glnA*, glutamine synthetase; *gltA*, citrate synthase; *hydABC*, hydrogenase complex; *icd*, isocitrate dehydrogenase; *pfl*, pyruvate formate lyase; *por*, pyruvate ferredoxin oxidoreductase; *ppc*, phosphoenolpyruvate carboxylase; *pps*, phosphoenolpyruvate synthase; and *pta*, phosphate acetyltransferase.

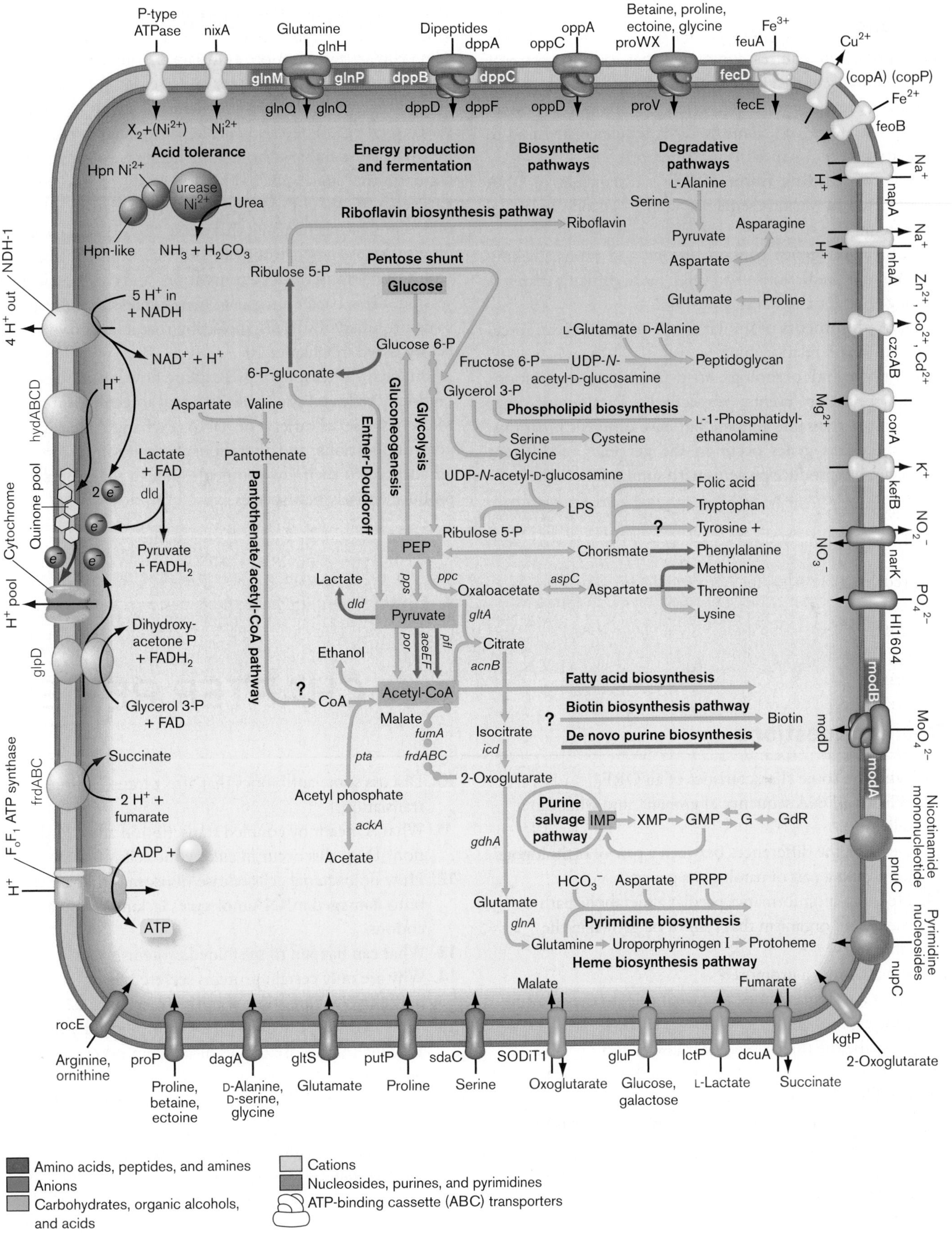
P-type ATPase
nixA
Glutamine
glnH
Dipeptides
dppA
oppA
oppC
Betaine, proline, ectoine, glycine
proWX
Fe3+
feuA
Cu2+
glnM
glnP
dppB
dppC
fecD
(copA) (copP)
Fe2+
glnQ
glnQ
dppD
dppF
oppD
proV
fecE
feoB
X2+(Ni2+)
Ni2+
Acid tolerance
Energy production and fermentation
Biosynthetic pathways
Degradative pathways
Hpn Ni2+
urease Ni2+
Urea
Hpn-like
NH3 + H2CO3
Riboflavin biosynthesis pathway
Riboflavin
L-Alanine
Serine
Pyruvate
Asparagine
Aspartate
Glutamate
Proline
Pentose shunt
Ribulose 5-P
Glucose
Glucose 6-P
6-P-gluconate
Fructose 6-P
UDP-N-acetyl-D-glucosamine
L-Glutamate D-Alanine
Peptidoglycan
Glycerol 3-P
Phospholipid biosynthesis
L-1-Phosphatidyl-ethanolamine
Serine
Cysteine
Glycine
Aspartate
Valine
Pantothenate
Gluconeogenesis
Glycolysis
Entner-Doudoroff
Pantothenate/acetyl-CoA pathway
UDP-N-acetyl-D-glucosamine
LPS
Folic acid
Tryptophan
Tyrosine +
Phenylalanine
Ribulose 5-P
PEP
Chorismate
Methionine
Threonine
Lysine
Lactate
pps
ppc
aspC
Oxaloacetate
Aspartate
dld
Pyruvate
gltA
por
aceEF
pfl
Citrate
Ethanol
acnB
Acetyl-CoA
Fatty acid biosynthesis
CoA
Biotin biosynthesis pathway
Biotin
Malate
fumA
Isocitrate
De novo purine biosynthesis
icd
pta
frdABC
2-Oxoglutarate
Acetyl phosphate
Purine salvage pathway
IMP
XMP
GMP
G
GdR
ackA
gdhA
Acetate
HCO3−
Aspartate
PRPP
Glutamate
glnA
Pyrimidine biosynthesis
Glutamine
Uroporphyrinogen I
Protoheme
Heme biosynthesis pathway
Malate
Fumarate
NDH-1
4 H+ out
5 H+ in + NADH
NAD+ + H+
H+
hydABCD
Lactate + FAD
dld
Pyruvate + FADH2
Quinone pool
Cytochrome
2 e−
e−
H+ pool
Dihydroxy-acetone P + FADH2
glpD
Glycerol 3-P + FAD
Succinate
frdABC
2 H+ + fumarate
FoF1 ATP synthase
ADP +
ATP
H+
Na+
H+
napA
Na+
H+
nhaA
Zn2+, Co2+, Cd2+
czcAB
Mg2+
corA
K+
kefB
NO2−
NO3−
narK
PO42−
HI1604
modB
modD
MoO42−
modA
Nicotinamide mononucleotide
pnuC
Pyrimidine nucleosides
nupC
rocE
Arginine, ornithine
proP
Proline, betaine, ectoine
dagA
D-Alanine, D-serine, glycine
gltS
Glutamate
putP
Proline
sdaC
Serine
SODiT1
Oxoglutarate
gluP
Glucose, galactose
lctP
L-Lactate
dcuA
Succinate
kgtP
2-Oxoglutarate
Amino acids, peptides, and amines
Anions
Carbohydrates, organic alcohols, and acids
Cations
Nucleosides, purines, and pyrimidines
ATP-binding cassette (ABC) transporters

To Summarize

- **Annotation** requires computers that look for patterns in DNA sequences. Annotation predicts regulators, ORFs, rDNA, and tRNA genes. Similarities in protein sequence (deduced from the DNA sequence) are used to predict protein structure and function.
- An **open reading frame (ORF)** is a sequence of DNA predicted by various sequence cues to encode an actual protein.
- **Eukaryotic genes** contain introns and exons, making computer predictions of an ORF more difficult than it is for bacterial or archaeal genes.
- **DNA alignments** of similar genes or proteins can reveal evolutionary relationships.
- **Paralogs and orthologs** arise from gene duplications and speciation events, respectively. Paralogous genes coexist in the same genome but have different functions. Orthologous genes occur in the genomes of different species but produce proteins with similar functions.

Concluding Thoughts

The cellular transcription, translation, and secretion pathways described in this Chapter are essential for life because they efficiently assemble biochemical pathways without wasting energy. They also enable pathogens to deliver toxins that can subdue a host, and help microbes in mixed communities make antibiotics to eliminate competitors. Efficiency in these processes is maintained through elaborate control mechanisms that sense the organism's physiological state and environment and then trigger changes in replication, transcription, translation, and/or protein processing. How bacteria regulate gene expression in response to environmental stimuli, including threats to survival, will be discussed in Chapter 10.

But first, in Chapter 9, we explore how natural selection randomly redesigns genomes to adapt to ecological niches. Microbes use a variety of DNA exchange mechanisms, gene duplications, and alterations to evolve into forms better adapted to their environments and, in the process, may produce entirely new species.

CHAPTER REVIEW

Review Questions

1. What are some characteristics of an ORF?
2. What is a DNA sequence alignment, and what can it tell you?
3. Describe the differences between a pair of orthologous genes and a pair of paralogous genes.
4. How can bioinformatics predict a metabolic pathway for an organism that cannot be grown in the laboratory?
5. What defines a promoter?
6. What are sigma factors, and what role do they play in gene expression?
7. Describe the three stages of transcription.
8. Explain the degeneracy of the genetic code. What is the wobble in codon-anticodon recognition?
9. Describe the stages of protein synthesis. Why is the ribosome called a ribozyme?
10. Discuss some antibiotics that affect transcription or translation.
11. What is meant by coupled transcription and translation? Does this occur in eukaryotic cells?
12. How do bacterial cells release ribosomes that are stuck onto damaged mRNA molecules lacking termination codons?
13. What can happen to misfolded proteins?
14. Why are only certain proteins secreted from the bacterial cell? What are some secretion mechanisms?
15. In what major way do proteins transported by the twin arginine translocase (TAT) differ from other exported proteins?
16. Compare protein degradation in eukaryotes and bacteria.
17. What is annotation? How does it apply to bioinformatics?

Thought Questions

1. The process of transcription generates positive supercoils in front of the polymerase as it moves along a DNA template. Why doesn't the DNA in front of the polymerase become so knotted that the polymerase can no longer separate the DNA strands?
2. Why do cells secrete some proteins into their environments?
3. Type I protein secretion systems transport certain proteins from the cytoplasm of Gram-negative bacteria directly to the outside of the cell, across two membranes. How might the system "know" which proteins to transport?

Key Terms

acceptor site (A site) (292)
aminoacyl-tRNA synthetase (289)
annotation (312)
anticodon (288)
bioinformatics (313)
catalytic RNA (285)
chaperone (303)
codon (286)
consensus sequence (278)
degron (304)
exit site (E site) (292)
exon (314)
functional genomics (316)
heat-shock protein (HSP) (304)
homologous (315)
intron (313)
messenger RNA (mRNA) (284)
molecular mimicry (295)
N-terminal rule (304)
open reading frame (ORF) (282)
orthologous (315)
paralogous (315)
peptidyl-tRNA site (P site) (292)
peptidyltransferase (292)
preliminary mRNA transcript (pre-mRNA) (314)
promoter (277)
release factor (296)
Rho-dependent (282)
Rho-independent (282)
ribosomal RNA (rRNA) (285)
ribosome-binding site (293)
ribozyme (292)
Shine-Dalgarno sequence (293)
sigma factor (277)
signal recognition particle (SRP) (307)
signal sequence (307)
small RNA (sRNA) (285)
start codon (287)
stop codon (287)
template strand (277)
tmRNA (285)
transcript (277)
transcription (277)
transfer RNA (tRNA) (285)
translation (286)
translocation (295)

Recommended Reading

Bakshi, S., A. Siryaporn, M. Goulian, and J. C. Weisshaar. 2012. Superresolution imaging of ribosomes and RNA polymerase in live *Escherichia coli* cells. *Molecular Microbiology* **85**:21–38.

Basler, Michael, Martin Pilhofer, Gregory P. Henderson, Grant J. Jensen, and John J. Mekalanos. 2012. Type VI secretion requires a dynamic contractile phage tail-like structure. *Nature* **483**:182–186.

Burger, Adelle, Chris Whiteley, and Aileen Boshoff. 2011. Current perspectives of the *Escherichia coli* RNA degradosome. *Biotechnology Letters* **33**:2337–2350.

Dalbey, Ross E., and Andreas Kuhn. 2012. Protein traffic in Gram-negative bacteria—how exported and secreted proteins find their way. *FEMS Microbiology Reviews* **36**:1033–1045.

Gowrishankar, Jayaraman, and Rajendran Harinarayanan. 2004. Why is transcription coupled to translation in bacteria? *Molecular Microbiology* **54**:598–603.

Haugen, Shanil P., Wilma Ross, and Richard L. Ghorse. 2008. Advances in bacterial promoter recognition and its control by factors that do not bind DNA. *Nature Reviews. Microbiology* **6**:507–519.

Janssen, Brian D., and Christopher S. Hayes. 2012. The tmRNA ribosome-rescue system. *Advances in Protein Chemistry and Structural Biology* **86**:151–191.

Kaczanowska, Magdalena, and Monica Rydén-Aulin. 2007. Ribosome biogenesis and the translation process in *Escherichia coli*. *Microbiology and Molecular Biology Reviews* **71**:477–494.

Marlovits, Thomas C., and C. Erec Stebbins. 2010. Type III secretion systems shape up as they ship out. *Current Opinions in Microbiology* **13**:47–52.

Murakami, Katsuhiko S., Shoko Masuda, Elizabeth A. Campbell, Oriana Muzzin, and Seth Darst. 2002. Structural basis of transcription initiation: RNA polymerase holoenzyme-DNA complex. *Science* **296**:1285–1290.

Preissler, Steffen, and Elke Deuerling. 2012. Ribosome-associated chaperones as key players in proteostasis. *Trends in Biochemical Sciences* **37**:274–283.

Proshkin, Sergey, A. Rachid Rahmouni, Alexander Mironov, and Evgeny Nudler. 2010. Cooperation between translating ribosomes and RNA polymerase in transcription elongation. *Science* **328**:504–508.

Schmeing, T. Martin, and Venkatraman Romakrishnan. 2009. What recent ribosome structures have revealed about the mechanism of translation. *Nature* **461**:1234–1242.

Wen, Jin-Der, Laura Lancaster, Courtney Hodges, Ana-Carolina Zeri, Shige H. Yoshimura, et al. 2008. Following translation by single ribosomes one codon at a time. *Nature* **452**:598–603.

Wojtas, Magdelena, Bibiana Peralta, Marina Ondiviela, Maria Mogni, Stephen D. Bell, et al. 2011. Archaeal RNA polymerase: The influence of the protruding stalk in crystal packing and preliminary biophysical analysis of the Rpo13 subunit. *Biochemical Society Transactions* **39**:25–30.

CHAPTER 9

Gene Transfer, Mutations, and Genome Evolution

9.1 The Mosaic Nature of Genomes

9.2 Gene Transfer

9.3 Recombination

9.4 Mutations

9.5 DNA Repair

9.6 Mobile Genetic Elements

9.7 Genome Evolution

DNA is not a static molecule. DNA sequences change over generations through various mutations, DNA rearrangements, and gene transfers between species. Bacterial and archaeal genomes sometimes shuttle large clusters of genes between members of different taxonomic domains. All this interspecies and intraspecies DNA traffic has led ecologists to expand the definition of the "microbial genome" to include DNA in a cell plus DNA from other organisms to which the microbe has potential access.

Important questions come to mind when we contemplate the consequences of DNA plasticity. For example, if adding, subtracting, and mutating genes all come at some cost, can the cell protect itself from undergoing too many drastic changes? Are some mutations good and others bad? Why would nature allow this much genetic uncertainty? And do we find bacterial DNA embedded in our own DNA? This chapter explores these long-standing evolutionary questions and shows how microbial genomes continually change, and change us.

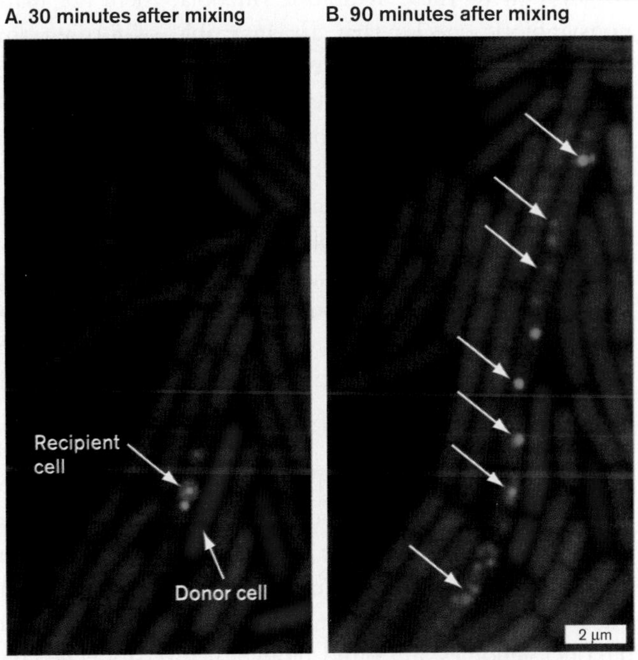

A. 30 minutes after mixing
B. 90 minutes after mixing
Recipient cell
Donor cell
2 µm

CURRENT RESEARCH highlight

Transposons with benefits. Two important contributors to evolution are conjugation, in which cell-to-cell contact is required to transfer DNA from one cell to another, and transposons, which are genetic elements that can "jump" from one part of a genome to another. Conjugative transposons can both "jump" and transfer between cells. In 2011, Alan Grossman's laboratory at MIT discovered that the transfer of conjugative transposons occurs most quickly between cells in the same chain (members of each chain arose via binary fission). Shown in this Figure is the rapid transfer of a conjugative transposon through a chain of *Bacillus subtilis* cells. The original transposon donor is red (mCherry fluorescent protein). In panel A, the conjugative transfer of a transposon between two parallel cells produces a focal green spot in the recipient. One hour later (panel B), the transposon has rapidly moved between cells in the same chain but not to any adjacent cells. Intrachain transfer likely accelerates the spread of conjugative elements within microbial communities. *Source:* Ana Babic et al. 2011. *mBio* **2**: e00027-11.

We think of evolution as taking place over thousands if not millions of years. In New Zealand, however, John Sullivan and Clive Ronson witnessed evolution taking place within a mere decade. These scientists were studying the microbe *Mesorhizobium loti*, a bacterial species that establishes symbiotic relationships with plants by forming nitrogen-fixing nodules on the plants' roots. The genes encoding symbiosis are located as a group on the mesorhizobial chromosome. In addition to these symbiotic rhizobia, there are many nonsymbiotic species that are incapable of forging this relationship. In a remarkable experiment started in 1986, Sullivan and Ronson inoculated a single strain of *M. loti* into an area of land devoid of natural nodulating rhizobia. Seven years later, they discovered that the area contained many genetically diverse symbiotic mesorhizobia now able to nodulate the flowering plant *Lotus corniculatus*. These microbes did not exist before the experiment. The scientists found that the 500-kb genome segment encoding symbiosis had somehow made its way from *M. loti* into these other bacteria—essentially generating new species. How did this happen so quickly?

In Chapter 9 we address the mechanisms that mediate transient as well as heritable DNA movements between species. Next, we describe the competing processes of mutagenesis and mutation repair required for self-preservation and evolutionary change. We close the chapter by explaining how all these processes collaborate over millennia to remodel genomes and build biological diversity. The consequences of gene flow during evolution will be discussed in Chapter 17.

9.1

The Mosaic Nature of Genomes

Genomic analysis has revealed that over the millennia, microbes have undergone extensive gene loss and gain. Archaea, for example, arose from a common eukaryotic-archaeal phylogenetic branch and, as a result, possess many traits in common with eukaryotes, such as the structure and function of their DNA and RNA polymerases. However, many archaeal genes whose products are involved with intermediary metabolism look purely bacterial. In fact, 37% of the proteins found in the archaeon *Methanocaldococcus jannaschii* are found in all three domains—Archaea, Eukarya, and Bacteria. Another 26% are otherwise found only among bacteria, while a mere 5% are confined to archaea and eukaryotes. Archaea, then, enjoy a mixed heritage.

Another surprise arising from bioinformatic studies is the mosaic nature of the *E. coli* genome and, indeed, of all microbial genomes. Though we have intensively studied *E. coli* for over 100 years, we now find that the organism's DNA is rife with pathogenicity islands, fitness islands, inversions, deletions (when compared to similar species), paralogous genes, and orthologous genes. How did all this genomic blending happen? The answer appears to involve heavy gene traffic between species (horizontal gene transfer), recombination events occurring within a species, and a variety of mutagenic and DNA repair strategies. All of these processes accelerated natural selection, where trial and error shaped genome content.

9.2

Gene Transfer

In 1928, a perceptive English medical officer, Frederick Griffith (1879–1941), found that he could kill mice by injecting them with <u>dead</u> cells of a virulent pneumococcus (*Streptococcus pneumoniae*, a cause of pneumonia), together with <u>live</u> cells of a nonvirulent mutant. Even more extraordinary was that he recovered live, <u>virulent</u> bacteria from the dead mice. Were the dead bacteria brought back to life? Unfortunately, Griffith was killed by a German bomb during an air raid on London in 1941 and never learned the answer.

In a landmark series of experiments published in 1944, Oswald Avery (1877–1955), Colin MacLeod (1909–1972), and Maclyn McCarty (1911–2005) proved that Griffith's experiment was not a case of resurrecting dead cells, but involved the transfer of DNA released from the virulent, dead strain <u>into</u> the harmless living strain of *S. pneumoniae*—an event that transformed the live strain into a killer. The process of importing free DNA into bacterial cells is now known as **transformation**.

Transformation provided the first clue that gene exchange can occur in microorganisms. But why bacteria carry out this and other forms of gene transfer was not fully appreciated until much later. Genome comparisons now show that the fundamental purpose of bacterial gene transfer is to acquire genes that might be useful as the environment changes.

Before we can discuss how bacterial genomes evolve, we need to understand the various gene exchange mechanisms available to them. Following the explanations of gene exchange, we will discuss the mechanisms that incorporate newly acquired DNA into genomes (for example, recombination).

Transformation of Naked DNA

How do live bacteria import DNA from dead bacteria? Many bacteria can import DNA fragments and plasmids released from nearby dead cells via the process of transformation. Natural transformation is conferred by specific protein complexes called **transformasomes**. Organisms in which transformation is a natural part of the growth cycle include Gram-positive bacteria like *Streptococcus* and *Bacillus,* as well as Gram-negative species such as *Haemophilus* and *Neisseria*.

Other bacteria, however, such as *E. coli* and *Salmonella,* do not possess the equipment needed to import DNA naturally. They require artificial manipulations to drive DNA into their cells. These laboratory techniques include perturbing the membrane by chemical ($CaCl_2$) and electrical (**electroporation**) methods. $CaCl_2$ alters the membrane, making these cells chemically competent, which allows DNA to pass. Electroporation, on the other hand, uses a brief electrical pulse to "shoot" DNA across the membrane.

Why do species such as *Neisseria* undergo natural transformation? First, species that indiscriminately import DNA may use the transformed DNA as food. Second, the more finicky species that transform only compatible DNA sequences may use DNA released from dead compatriots to repair their own damaged genomes. Note that this process is random. A cell does not choose which genes to import or keep. Finally, transformation can influence evolution, enabling species to adjust to new environments by acquiring new genes from other species in a process called **horizontal gene transfer**. It is called horizontal gene transfer to distinguish it from vertical transfer, the generational passing of genes from parent to offspring (as in cell division). If horizontal acquisition of a gene system improves the competitiveness of the cell, the new genes will be retained and the descendants will have evolved into a new kind of organism. Transformation may have enabled pathogens such as *Neisseria gonorrhoeae,* the cause of gonorrhea, to acquire genes whose protein products now help the organism evade the host immune system.

Gram-positive organisms transform DNA using a transformasome complex. Natural transformation in Gram-positive organisms typically involves the growth phase–dependent assembly of a transformasome complex across the cell membrane (**Fig. 9.1**). The transformasome is composed of a binding protein that captures extracellular DNA floating in the environment, plus proteins that form a transmembrane pore. The complex also includes a nuclease that degrades one strand of a double-stranded DNA molecule while pulling the other strand intact through the pore and into the cell. Once inside, the strand can be incorporated into the chromosome by recombination—a process we will discuss in Section 9.3.

Once the transformasome is assembled, the cell is **competent**, meaning that it can import free DNA fragments and incorporate them into its genome. What triggers growth phase–dependent competence? For some Gram-positive bacteria, competence for transformation is generated by a chemical conversation (quorum sensing; Sections 4.6 and 10.8) that takes place between members of the culture. Every individual in a growing population produces and secretes a small, 15- to 20-amino-acid peptide, generically called **competence factor (CF)**, that progressively accumulates in the medium until it induces a genetic program that makes the population competent (**Fig. 9.1**, step 1). The sequence of the competence factor peptide is unique to each species, as are the specifics of the induction process. For *Streptococcus pneumoniae,* the level of competence factor in the medium increases (**Fig. 9.1**, step 2) as the population increases—that is, as the cell density increases.

Above a certain concentration threshold, competence factor is able to bind to a sensory protein built into the cell membrane (ComD for *S. pneumoniae*). This binding begins what is called a phosphorylation cascade (the passing of a phosphate group from one protein to another; eTopic 4.1). In the phosphorylation cascade, the sensory protein uses ATP to phosphorylate itself and then passes the phosphate to a cytoplasmic regulatory protein, ComE, which stimulates expression of a novel sigma factor, SigH (**Fig. 9.1**, step 3). The resulting sigma factor is specifically used to transcribe genes encoding the transformasome (**Fig. 9.1**, step 4). The protein products of these genes are assembled at the membrane, and the cell becomes competent (**Fig. 9.1**, step 5).

Why would organisms regulate transformation competence by quorum sensing? One hypothesis holds that cells are unlikely to encounter stray DNA when growing in dilute natural environments such as ponds, where other bacteria are scarce. So, in this situation, why waste energy making the transformasome? However, when these same cells are growing at high density, as in a biofilm, they are more likely to encounter DNA released from dying neighbors. This is DNA they could use to repair their own damaged genomes, to consume as food, or to sample for a new survival mechanism, should the DNA come from a different species present in a biofilm consortium. Regulation by quorum sensing would ensure that the transformasome

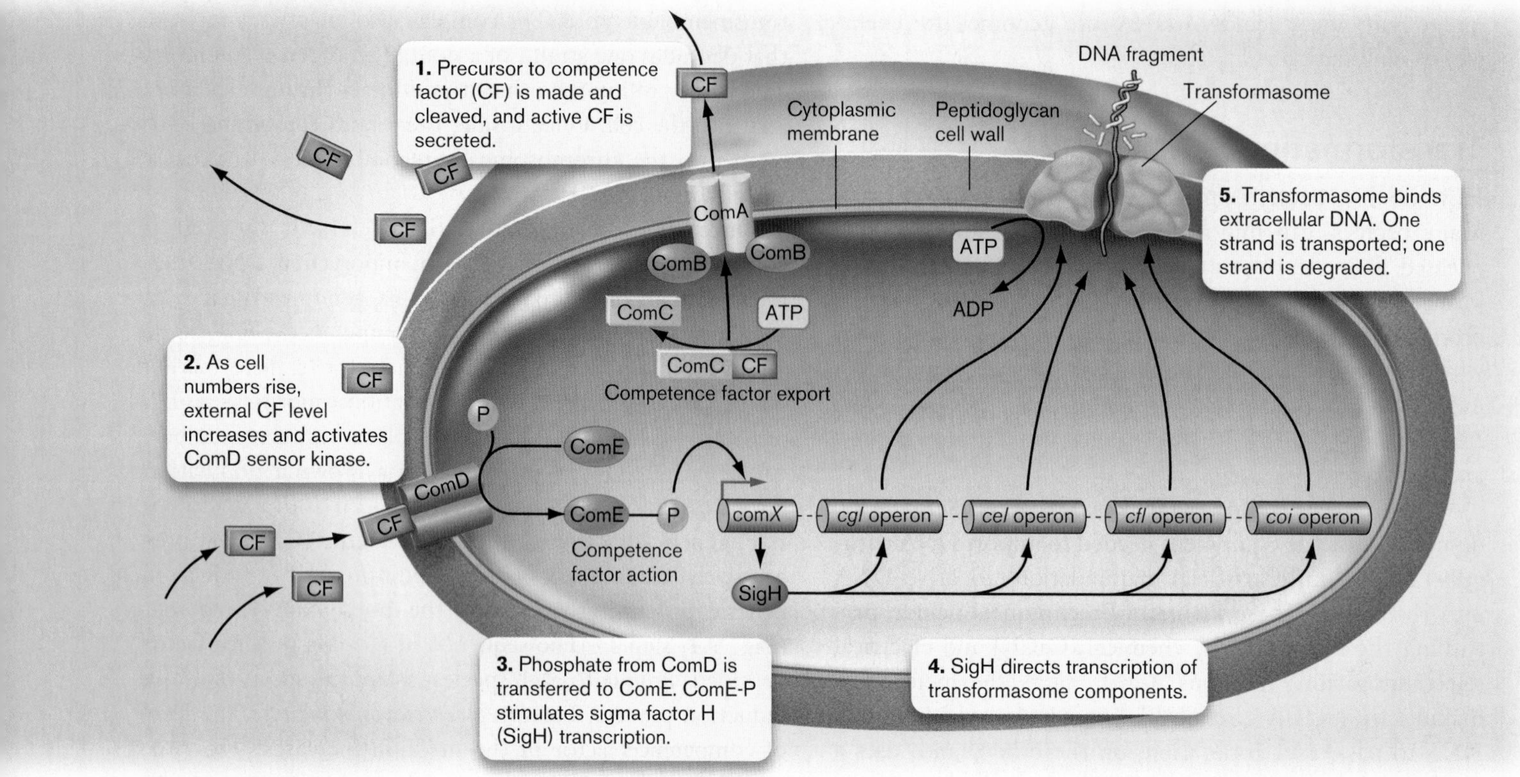
1. Precursor to competence factor (CF) is made and cleaved, and active CF is secreted.
CF
Cytoplasmic membrane
Peptidoglycan cell wall
DNA fragment
Transformasome
5. Transformasome binds extracellular DNA. One strand is transported; one strand is degraded.
ComA
ComB
ComB
ComC
ATP
ATP
ADP
ComC CF
Competence factor export
2. As cell numbers rise, external CF level increases and activates ComD sensor kinase.
P
ComE
ComD
ComE P
Competence factor action
comX
cgl operon
cel operon
cfl operon
coi operon
SigH
3. Phosphate from ComD is transferred to ComE. ComE-P stimulates sigma factor H (SigH) transcription.
4. SigH directs transcription of transformasome components.

FIGURE 9.1 ▪ Transformation in *Streptococcus.* The process of transformation in this organism begins with the synthesis of a signal molecule (competence factor, CF) and concludes with the import of a single-stranded DNA strand through a transformasome complex.

would not form until there was a good chance that free foreign DNA was available.

Thought Question

9.1 Figure 9.1 illustrates the process of transformation in *Streptococcus pneumoniae.* Will a mutant of *Streptococcus* lacking comD be able to transform DNA?

Gram-negative species transform DNA without competence factors. Gram-negative species capable of natural transformation do not appear to make competence factors. Either they are always competent, like *Neisseria,* or they become competent when starved (for example, *Haemophilus*). Competence development does not depend on cell-cell communication. Gram-negative bacteria also must overcome a barrier to DNA importation not faced by Gram-positive bacteria—namely, the outer membrane. Thus, Gram-negative organisms cannot use the cytoplasmic membrane transformasome that is employed by Gram-positive microbes. *Neisseria* species, for example, appear to import DNA through a system paralogous to type IV pilus assembly (discussed in Section 8.6). Type IV pili reversibly assemble and disassemble at the cell surface, expanding and contracting as a result. This happens constantly during the growth of a culture and can help cells move across solid surfaces. During transformation in species of *Neisseria,* disassembly of a pilus is thought to drag transforming DNA into the cell and across the two membranes.

In another departure from the Gram-positive example, transformation in species of *Haemophilus* and *Neisseria* is species and sequence specific, thereby limiting gene exchange between different genera. Specificity is due to particular sequences in DNA that are recognized by part of the uptake apparatus. However, not all Gram-negative competence systems display such specificity; *Acinetobacter calcoaceticus,* a lung pathogen, is able to take up DNA from any source at very high frequency. This is possible because the DNA-binding part of its transformation system does not need to recognize a specific nucleotide sequence.

Gene Transfer by Conjugation

Conjugation, or "bacterial sex," requires cell-cell contact typically initiated by a special pilus protruding from a donor cell (**Fig. 9.2**). The pilus can also dramatically contribute to biofilm formation (**eTopic 9.1**). Conjugation occurs in many species of bacteria and archaea, even

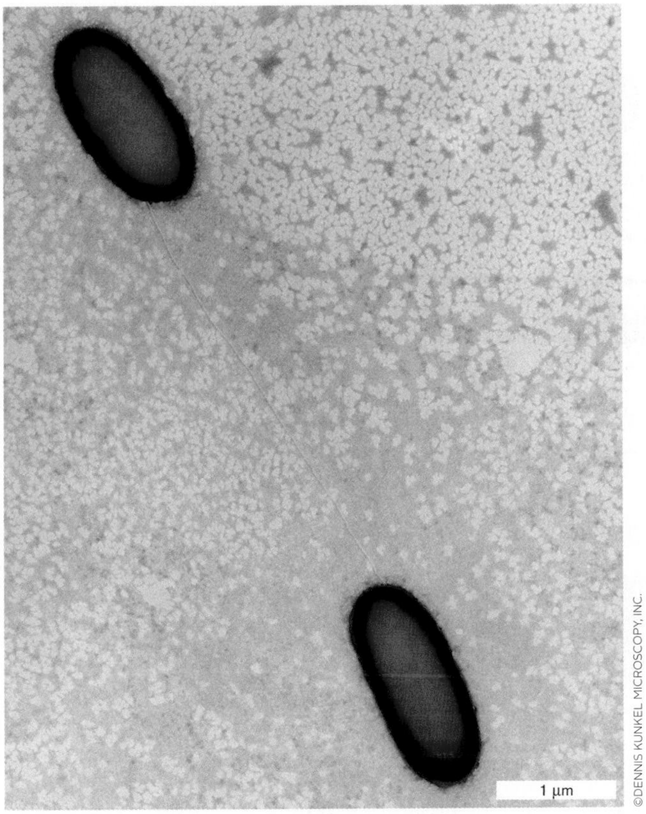
1 μm
©DENNIS KUNKEL MICROSCOPY, INC.

FIGURE 9.2 ■ Sex pilus connecting two *E. coli* cells. Pseudocolor added.

in hyperthermophiles such as *Sulfolobus* species. In nature, mixed-species biofilms are thought to enable DNA transfer between different species, albeit at a low frequency. As for the mechanism of DNA transfer by conjugation, most species use type IV–like protein secretion systems because a protein is transferred along with DNA (as described later and in Section 8.6).

In the case of the Gram-negative *E. coli,* the tip of the specialized plasmid-encoded sex pilus attaches to a receptor on the recipient cell and then contracts, drawing the two cells closer. The two cell envelopes fuse and generate a conjugation complex, similar to the transformation complex, through which single-stranded DNA passes from donor to recipient. The DNA passes across the membranes through a conjugation complex.

Bacterial conjugation requires the presence of special transferable plasmids that usually contain all the genes needed for pilus formation and DNA export. In some cases, plasmids can also contain genes encoding antibiotic resistance. A well-studied, transferable plasmid in *E. coli* is called **fertility factor**, or **F factor** (**Fig. 9.3**). F factor can transfer itself to another recipient cell, but if it integrates into the chromosome of its host, it can also transfer host genes. To carry out plasmid replication and DNA transfer, F factor uses two replication origins, *oriV* and *oriT,* located at different positions on the plasmid. The origin *oriV* is used to replicate and maintain the plasmid in nonconjugating cells, whereas *oriT* is used only to replicate DNA during DNA transfer. F factor also contains several *tra* genes, whose protein products carry out the DNA transfer process.

Conjugation begins with cell-cell contact between a donor **F^+ cell** that carries the plasmid and a recipient **F^- cell** (Fig. 9.3, step 1). Formation of a fused membrane conjugation conduit (Fig. 9.3, step 2) triggers synthesis of an F factor–encoded helicase/endonuclease (also called TraI or relaxase) that nicks the phosphodiester backbone at oriT (Fig. 9.3, step 3). TraI forms a relaxosome complex with other F factor–encoded proteins. The relaxosome unwinds donor double-stranded DNA and begins the transfer process. The blowup at step 3 in Figure 9.3 shows a model of the multifactor relaxosome complex at a membrane bridge. The translocated TraI relaxase remains bound to the 5′ end of the nicked strand, and the DNA-protein complex is transferred through a pore protein in the recipient. The translocated relaxase and the 5′ end of the strand remain in the membrane, while the rest of the strand is transferred through the pore (Fig. 9.3, step 4). The intact circular strand remains in the donor. DNA polymerase III (Pol III) is then recruited to oriT in the donor, where replication begins (Fig. 9.3, step 4). In contrast to bidirectional replication used to duplicate the chromosome, replication for conjugative transfer occurs unidirectionally by rolling-circle replication (Fig. 9.3, step 5) (see Section 7.4).

DNA polymerase III synthesis of the replacement strand in the donor cell drives movement of the transferred strand through the pore. The polymerase uses the untransferred intact circular strand as a template (see blowup at step 5 in Fig. 9.3). After secretion into the recipient cell, the transferred strand also replicates and becomes double-stranded. Because the 5′ end of the transferred strand enters first, replication of the complementary strand (5′ to 3′) is, by necessity, discontinuous, like a lagging strand.

Once the transfer is complete, the relaxase re-ligates the 5′ end it was holding to the 3′ tail of the transferred strand. Thus, the last portion of F factor moved to the recipient is *oriT.* Recircularization of the replicated F factor converts the F^- recipient cell to a new F^+ donor cell (Fig. 9.3, step 6). Ultimately, the conjugation complex spontaneously comes apart, and the membranes seal. This transfer process is very quick, taking less than five minutes to transfer the entire 110-kb F factor.

Is it possible for two donors to exchange DNA with each other? The answer is generally no, donors have mechanisms in place to prevent this. For example, a membrane protein

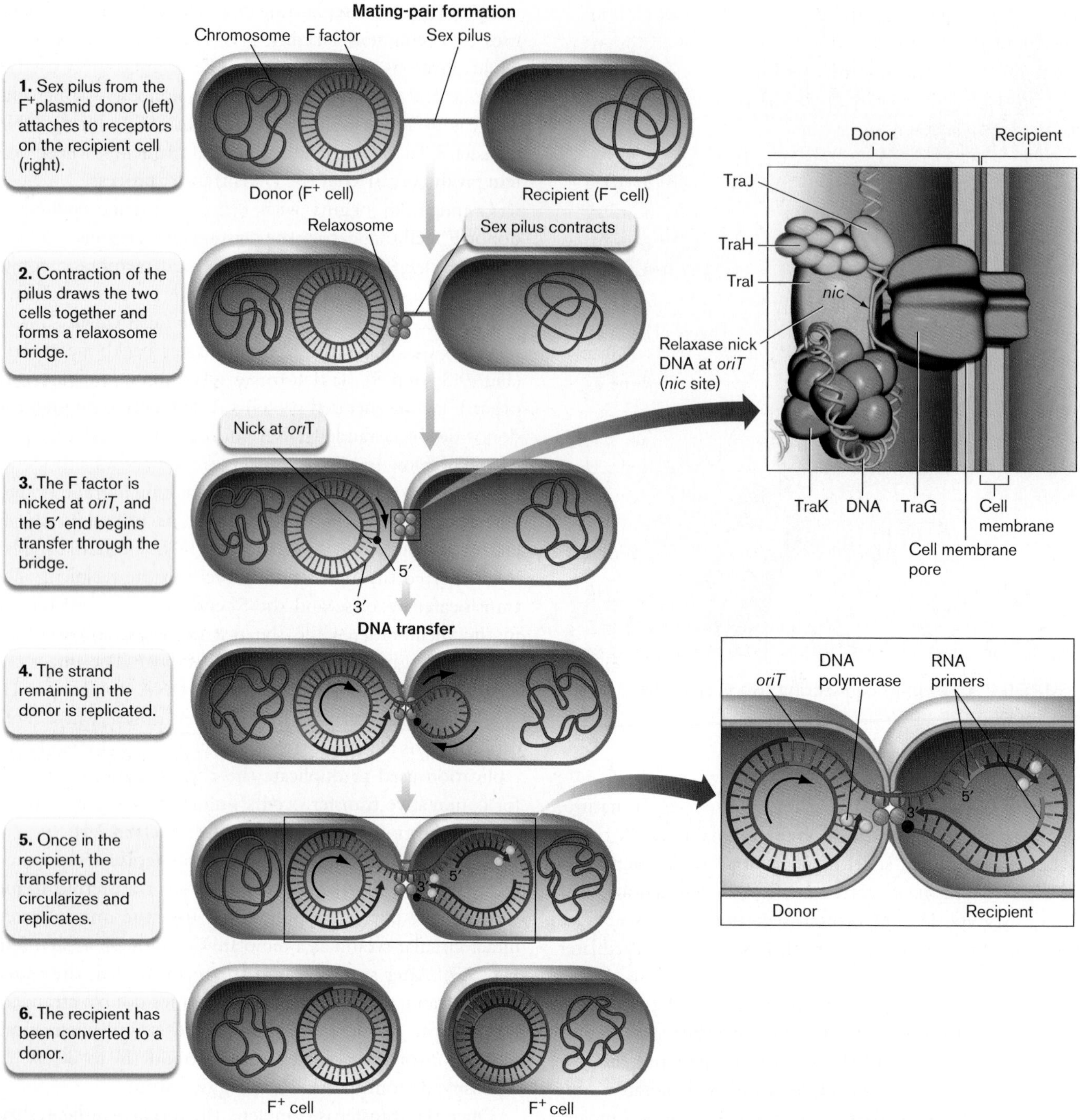
Mating-pair formation
Chromosome
F factor
Sex pilus
1. Sex pilus from the F+plasmid donor (left) attaches to receptors on the recipient cell (right).
Donor (F+ cell)
Recipient (F− cell)
Relaxosome
Sex pilus contracts
2. Contraction of the pilus draws the two cells together and forms a relaxosome bridge.
Nick at oriT
3. The F factor is nicked at oriT, and the 5′ end begins transfer through the bridge.
5′
3′
DNA transfer
4. The strand remaining in the donor is replicated.
5. Once in the recipient, the transferred strand circularizes and replicates.
6. The recipient has been converted to a donor.
F+ cell
F+ cell
Donor
Recipient
TraJ
TraH
TraI
nic
Relaxase nick DNA at oriT (nic site)
TraK
DNA
TraG
Cell membrane
Cell membrane pore
oriT
DNA polymerase
RNA primers
5′
3′
Donor
Recipient

FIGURE 9.3 ■ The conjugation process. Some plasmids, such as F factor in *E. coli,* can mediate cell-to-cell transfer of DNA. The plasmid is nicked at the *nic* site (step 3), and the 5′ end is transferred to the other cell. Purple arrowheads mark the 3′ end of replicating DNA. The inset at step 3 shows a current model of the DNA secretion apparatus. The relaxosome complex is composed of TraH, TraI (the helicase/endonuclease), TraJ, and TraK. The blowup illustrates the step in which the TraI endonuclease is about to nick the donor plasmid. The 5′ end of the nick will move through the pore and remain attached to the membrane while the rest of the single-stranded DNA passes into the recipient. *Source:* Step 3 blowup adapted from G. Schröder et al. 2002. *J. Bacteriol.* **184**:2767–2779.

encoded by the F factor will inhibit formation of a conjugation complex with another donor that possesses the same protein. This mechanism prevents the pointless transfer of a plasmid to a cell that already has that plasmid. There are exceptions, however.

The F-factor plasmid can integrate into the chromosome. Once inside the recipient cell, F factor can remain an independent plasmid or can be integrated into the chromosome of the recipient cell (**Fig. 9.4**). Integration of the plasmid into the genome involves recombination between two circular DNA duplexes, a process explained in Section 9.3. Because the plasmid can exist in extrachromosomal and integrated forms, it is sometimes called an **episome** (*epi*, Greek for "over"; and *some* from "chromosome").

When the chromosome has integrated an F factor, the cell is designated an **Hfr**, or high-frequency recombination strain. The cells are called "high frequency" because there are more of them capable of transferring chromosomal DNA in an Hfr population than in an F^+ culture, where very few Hfr cells are present. It does not refer to the speed with which individual DNA molecules are transferred. Because the F factor is inserted into the chromosome, an Hfr cell is capable of transferring all or part of the chromosome into a recipient cell. Essentially, the entire chromosome becomes the F factor. However, the integrated F factor is actually the last bit of DNA transferred during Hfr conjugation. Because it takes nearly 100 minutes to transfer the entire *E. coli* chromosome (as opposed to only 5 minutes for the free F plasmid), the integrated F factor is rarely transferred before the conjugation bridge breaks, and the recipient almost never becomes an F^+ or Hfr cell. How Hfr conjugation was used in the past to map *E. coli* genes is described in eTopic 9.2. Today, rapid DNA sequencing is used to map genes in any organism. Nevertheless, conjugation remains important for evolution and as a tool for scientists to move genes between species.

An integrated F factor can excise from the chromosome. Another type of gene shuffling can occur when an integrated

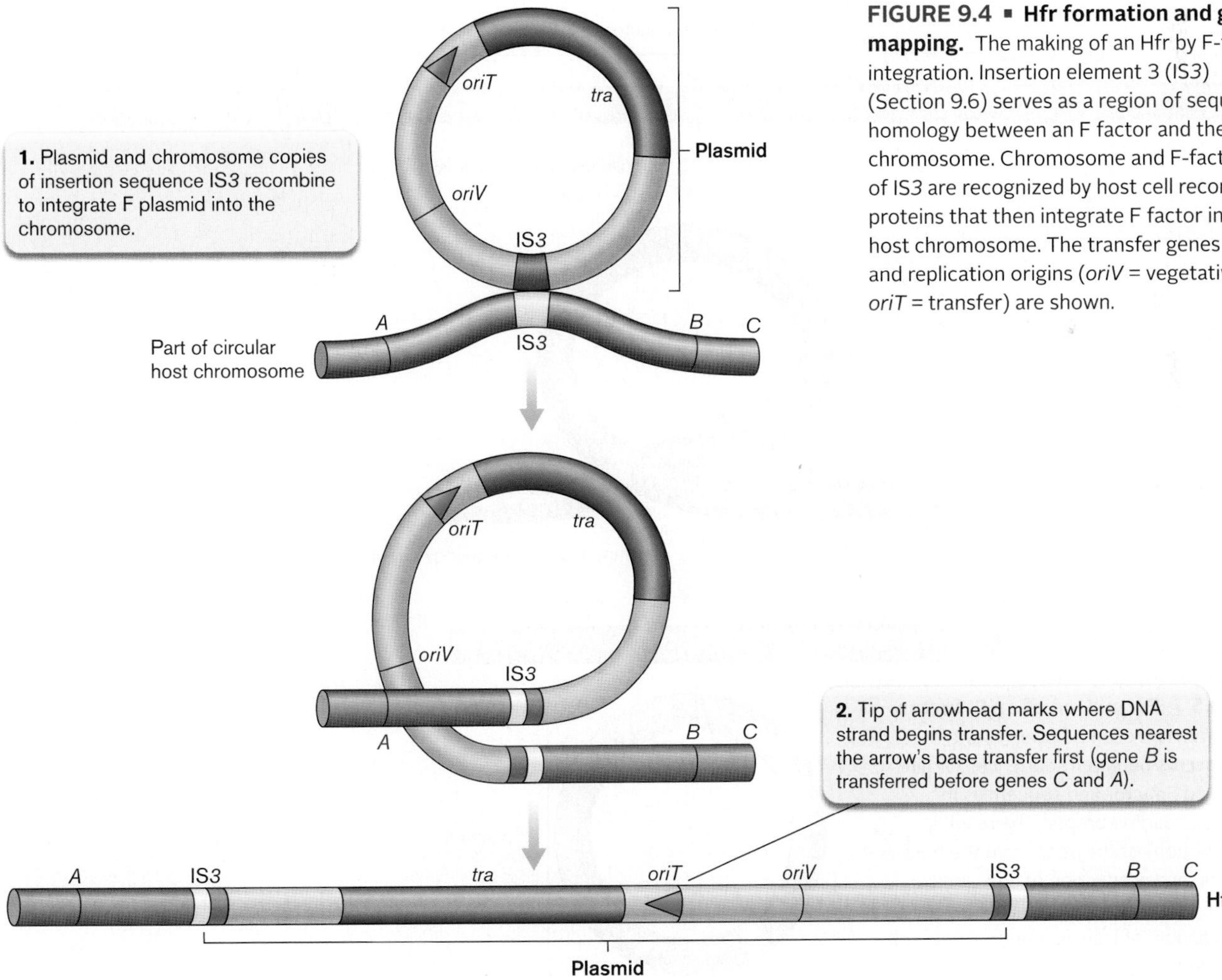
1. Plasmid and chromosome copies of insertion sequence IS3 recombine to integrate F plasmid into the chromosome.
oriT
tra
oriV
IS3
Plasmid
A
B
C
IS3
Part of circular host chromosome
oriT
tra
oriV
IS3
A
B
C
2. Tip of arrowhead marks where DNA strand begins transfer. Sequences nearest the arrow's base transfer first (gene B is transferred before genes C and A).
A
IS3
tra
oriT
oriV
IS3
B
C
Hfr
Plasmid

FIGURE 9.4 ■ Hfr formation and gene mapping. The making of an Hfr by F-factor integration. Insertion element 3 (IS3) (Section 9.6) serves as a region of sequence homology between an F factor and the host chromosome. Chromosome and F-factor copies of IS3 are recognized by host cell recombination proteins that then integrate F factor into the host chromosome. The transfer genes (*tra*) and replication origins (*oriV* = vegetative; *oriT* = transfer) are shown.

F factor subsequently excises from the chromosome via host recombination mechanisms. The F factor is usually excised completely and restored to its original form. Occasionally, however, it is excised along with some neighboring DNA from the host chromosome, yielding a product that remains an F-factor plasmid but also contains some chromosomal DNA. The derivative F plasmid containing host DNA is called an **F-prime (F′) plasmid** or **F-prime (F′) factor** (**Fig. 9.5**).

In contrast to chromosomal genes, whose transfer is directed by Hfr, genes hitchhiking on an F′ plasmid do not have to recombine into the recipient chromosome to be maintained. The extra genes can be expressed as part of the F′ plasmid. The result is called a "partial diploid situation," in which the conjugal recipient of the F′ factor contains two copies of those few genes—one set on the chromosome, the other on the F′ factor. Partial diploids are useful to the cell because the second copy of the gene can be the raw material that evolves into a new gene. And they are useful to geneticists because they can reveal whether a mutant allele of a gene is dominant over a normal (wild-type) allele. Partial diploids can also help geneticists understand gene organization.

There are many different types of plasmids in the microbial world. Most are not transferable by themselves. However, one group of plasmids that cannot transfer themselves can be mobilized if a transferable plasmid is also present in the same cell. Mobilizable plasmids usually contain an *oriT*-like DNA replication origin recognized by the conjugation apparatus of the transferable plasmid. As a result, when the transferable plasmid begins conjugating, so does the mobilizable plasmid. This is one way in which antibiotic resistance genes on plasmids called R factors can be spread throughout a microbial population.

Thought Question

9.2 Transfer of an F factor from an F^+ cell to an F^- cell converts the recipient to F^+. Why doesn't transfer of an Hfr do the same?

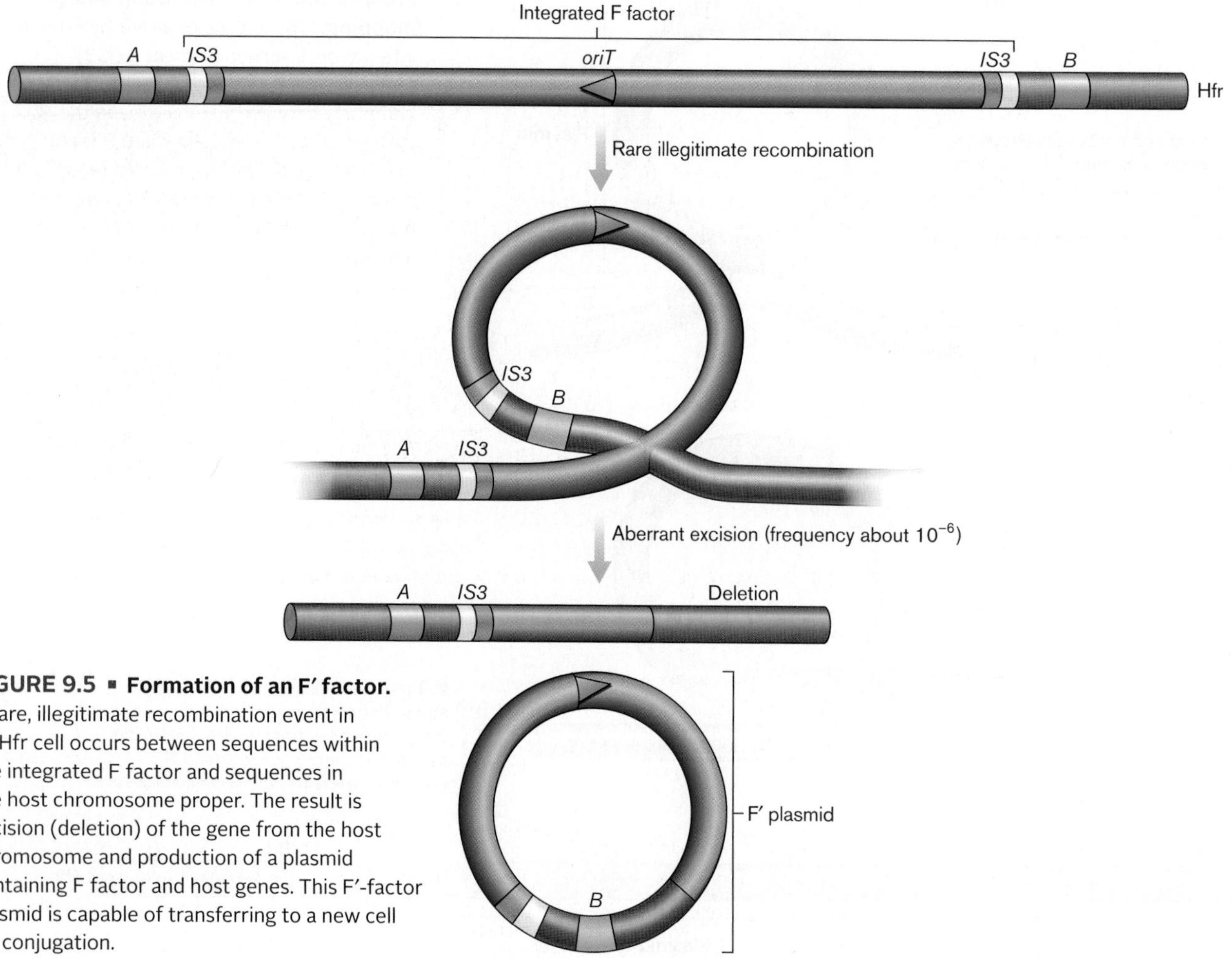
Integrated F factor
A
IS3
oriT
IS3
B
Hfr
Rare illegitimate recombination
IS3
B
A
IS3
Aberrant excision (frequency about 10^{-6})
A
IS3
Deletion
F′ plasmid
B

FIGURE 9.5 ▪ Formation of an F′ factor. A rare, illegitimate recombination event in an Hfr cell occurs between sequences within the integrated F factor and sequences in the host chromosome proper. The result is excision (deletion) of the gene from the host chromosome and production of a plasmid containing F factor and host genes. This F′-factor plasmid is capable of transferring to a new cell via conjugation.

DNA transfers between Bacteria and Eukarya. In eukaryotes, the sexual exchange of genes usually occurs only within a single species. Microbes, on the other hand, are more promiscuous. Mating among different microbial genera is common and perhaps even desirable from an evolutionary viewpoint. For example, *Salmonella* can conduct interspecies mating with *E. coli*. Sharing genes between species allows each one to sample genes from the other and keep genes (through natural selection) that increase fitness.

What may be surprising is that some bacteria can actually transfer genes across biological domains. One striking example is *Agrobacterium tumefaciens*, which causes crown gall disease in plants. This Gram-negative plant pathogen, common to the rhizosphere (the area around root surfaces), contains a tumor-inducing (Ti) plasmid that can be transferred via conjugation to plant cells. These bacteria detect and swim toward phenolic wound compounds released by damaged plant cells. The phenolic compounds signal where the microbe can gain access to plant cells. Subsequent transfer, integration, and expression of the Ti plasmid in the plant cell genome trigger the release of plant hormones that stimulate tumorous growth of the plant (**Fig. 9.6**). Plant cells within the tumor release amino acid derivatives, called "opines," that the microbe can then use as a source of carbon and nitrogen.

The unique mode of action of *A. tumefaciens* has made this bacterium an indispensable tool for plant breeding. Any desired genes, such as insecticidal toxin genes (like those produced by *Bacillus thuringiensis*) or herbicide resistance genes, can be engineered into the bacterial plasmid DNA and thereby inserted into the plant genome (discussed in Chapter 16). The use of *Agrobacterium* not only shortens the conventional plant-breeding process, but also allows entirely new (nonplant) genes to be engineered into crops.

Scientists recently discovered that interdomain transfer of DNA may be more common than was previously realized. There is evidence of massive DNA transfers from some bacteria to their eukaryotic hosts (**Special Topic 9.1**). Mark Anderson and Han Seifert from Northwestern University suggest that transfer also takes place in the opposite direction, from humans to bacteria. They found that up to 11% of *Neisseria gonorrhoeae* isolates (the cause of gonorrhea) contain human-derived DNA sequences known as L1, or long interspersed nuclear elements (LINEs). LINE elements are human DNA sequences that copy and insert themselves from one gene into another. Since the L1 sequence in *N. gonorrhoeae* was not found in the closely related species *N. meningitidis*, the transfer from L1 to *N. gonorrhoeae* probably occurred relatively recently in evolutionary history (after *N. gonorrhoeae* and *N. meningitidis* split from their common ancestor). The function of *Neisseria* L1 is currently unknown.

A.

ROBERT CALENTINE/VISUALS UNLIMITED

B.

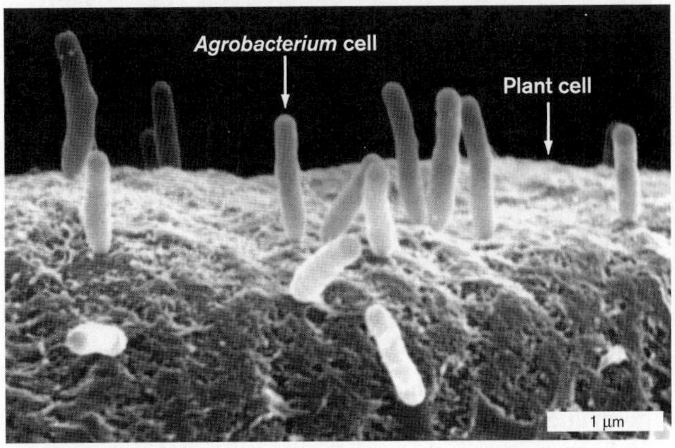
Agrobacterium cell
Plant cell
1 µm
MARTHA HAWES, UNIVERSITY OF ARIZONA

FIGURE 9.6 ■ An example of gene transfer between bacteria and plants. A. Crown gall disease tumor caused by the bacterium *Agrobacterium tumefaciens*. **B.** Electron micrograph of *A. tumefaciens* attached to plant cells.

Note: Conjugation in bacteria and archaea is mechanistically different from conjugation among eukaryotic microbes. Some eukaryotic microbes exchange nuclei through a structure, called a conjugation bridge, very different from that of bacteria (see Chapter 20).

Special Topic 9.1: There's a Bacterial Genome Hidden in My Fruit Fly

In the 1986 movie *The Fly,* the DNA of a fly becomes mixed with the DNA of a man in a teleportation experiment gone horribly wrong. Predictably, the resulting "flyman" and his girlfriend have trouble coping with his horrible transmogrification. Originally thought to be solely the stuff of science fiction, we now know that interdomain genome transfers do occur in nature. Bacterial DNA can actually become part of a eukaryotic genome. The transfer of Ti plasmid from *Agrobacterium tumefaciens* to plant cells is one well-documented example. However, there are other intriguing biological situations that provide opportunities for more frequent interdomain transfers.

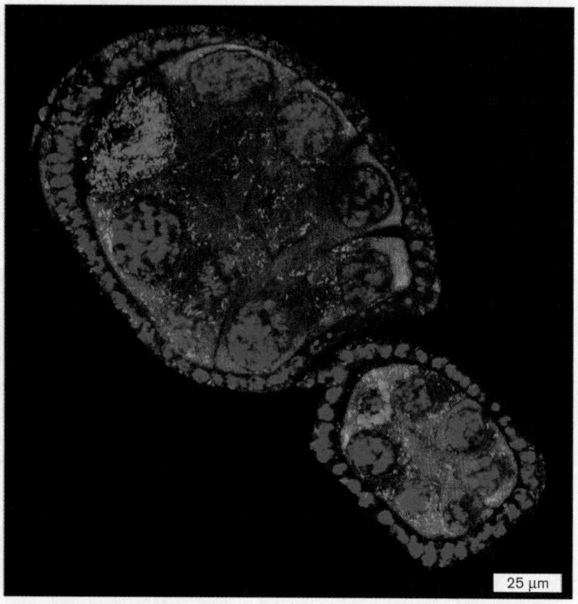
25 µm

FIGURE 1 ■ *Wolbachia* within the *Drosophila* female germ line. The intracellular alphaproteobacterium *Wolbachia* (red) accumulates in the germ-line cell that will form the fruit fly egg and embryo, and is thereby transmitted to the next host generation. Green indicates germ-line cytoplasm; blue indicates DNA in this fluorescent microscopy image.

One of these situations involves bacterial endosymbionts that have evolved to live within eukaryotic host cells. *Wolbachia pipientis* is a maternally inherited bacterial endosymbiont that infects a wide range of arthropods, including at least 20% of insect species, as well as filarial nematodes (discussed in Section 17.6). *Wolbachia* is passed from one generation of insect to the next because it lives inside the eggs of the insect (**Fig. 1**). Scientists recognized that the presence of this bacterium in developing gametes provided a unique opportunity for horizontal transfer from the bacterium to the eukaryotic chromosome.

Startling work has recently shown that *Wolbachia*-to-host genome transfer has actually occurred. *Wolbachia* DNA inserts have been discovered deep within the genomes of diverse invertebrate species, including wasps, fruit flies, and nematodes. **Figure 2** shows the presence of a particular *Wolbachia* gene embedded in a *Drosophila* (fruit fly) chromosome extracted from a fly cured of the bacterium by antibiotics. In fact, the entire *Wolbachia* genome was transferred, and many of the genes transferred were transcribed, though it is unclear whether they have any effect on fly metabolism. This finding is extremely important from an evolutionary viewpoint. Researchers first thought the *Wolbachia* DNA was a contaminant because eukaryotic genome projects typically include bacterial cloning vector sequences that must be excluded. But the actual presence of a whole bacterial genome within *Drosophila* forced a reexamination of the excluded sequences.

Research Question

Aside from potential effects on fly physiology, what other benefit might chromosome-embedded *Wolbachia* DNA provide Drosophila?

Hussain, Mazhar, Guangjin Lu, Shessy Torres, Judith H. Edmonds, Brian H. Kay, et al. 2013. Effect of Wolbachia on replication of West Nile virus in a mosquito cell line and adult mosquitoes. *J Virol* **87:** 851–858.

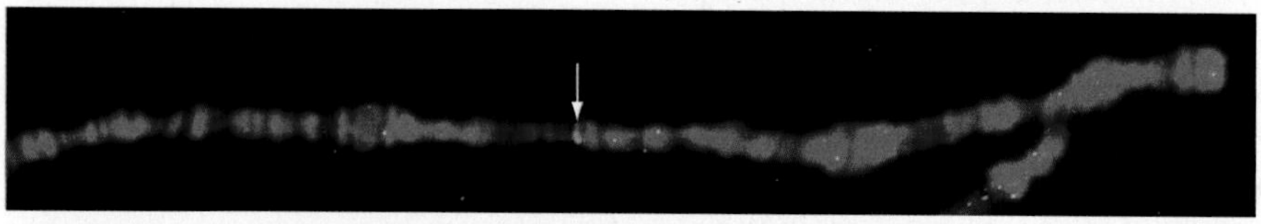

FIGURE 2 ■ *Wolbachia* DNA in a *Drosophila* chromosome. A fluorescein-stained probe for the *Wolbachia* gene WD0484 is seen bound to a unique location (green, arrow) on chromosome 2L of *Drosophila ananassae* stained with propidium iodide (red). Chromosome 2L is 2 megabase pairs (Mbp). *Source:* Hotopp et al. 2007. *Science* **317**:1753–1756.

Gene Transfer by Phage Transduction

Bacteriophages harbor their own genomes, distinct from those of their host cells, but could phages also pluck host genes from one cell and move them to another? Norton Zinder (1928–2012) first suspected that bacteriophages could taxi chromosomal DNA between cells. To test this idea, he grew *Salmonella* in two tubes separated by a fine filter through which viruses could pass, but not bacteria. The experiments, reported in 1966, showed that a filterable agent, or virus, could carry genetic material between bacterial strains; direct contact between bacteria was unnecessary. Thus, bacteriophages can accidentally move bacterial genes between cells as an offshoot of the phage life cycle (see Section 6.4 and eTopic 7.3). The process in which bacteriophages carry payloads of host DNA from one cell to another is known as **transduction**. There are two basic types of transduction: generalized and specialized. **Generalized transduction** can take any gene from a donor cell and transfer it to a recipient cell, whereas **specialized transduction** (also known as restricted transduction) can transfer only a few closely linked genes between cells.

How does transduction happen? Bacteriophages capable of generalized transduction have trouble distinguishing their own DNA from that of the host when attempting to package DNA into their capsids, so pieces of bacterial host DNA accidentally become packaged in the phage capsid instead of phage DNA. In the case of *Salmonella* P22 phage, the packaging system recognizes a certain DNA sequence on P22 DNA called a *pac* site. During rolling-circle replication, P22 DNA forms long concatemers containing many P22 genomes arranged in tandem. The *pac* site defines the ends of the phage genome, marking where the packaging system cuts the P22 DNA and starts packaging DNA into the next empty phage head. However, certain DNA sequences on the *Salmonella* chromosome also "look" like *pac* sites. As a result, the packaging system sometimes mistakenly packages host DNA instead of phage DNA.

The result of this mistake is that 1% of the phage particles in any population of P22 do not contain any phage DNA but carry host DNA plucked from around the chromosome (**Fig. 9.7**). The phages that carry host DNA are called transducing particles. Any single transducing particle will contain only one segment of host DNA, but different particles in the phage population will contain different segments of host DNA. When a transducing particle injects its DNA into a cell, no new phages are made, but the hijacked host DNA can recombine, or exchange, with sequences in the host chromosome of the newly infected cell, thus changing the genetic makeup of the recipient.

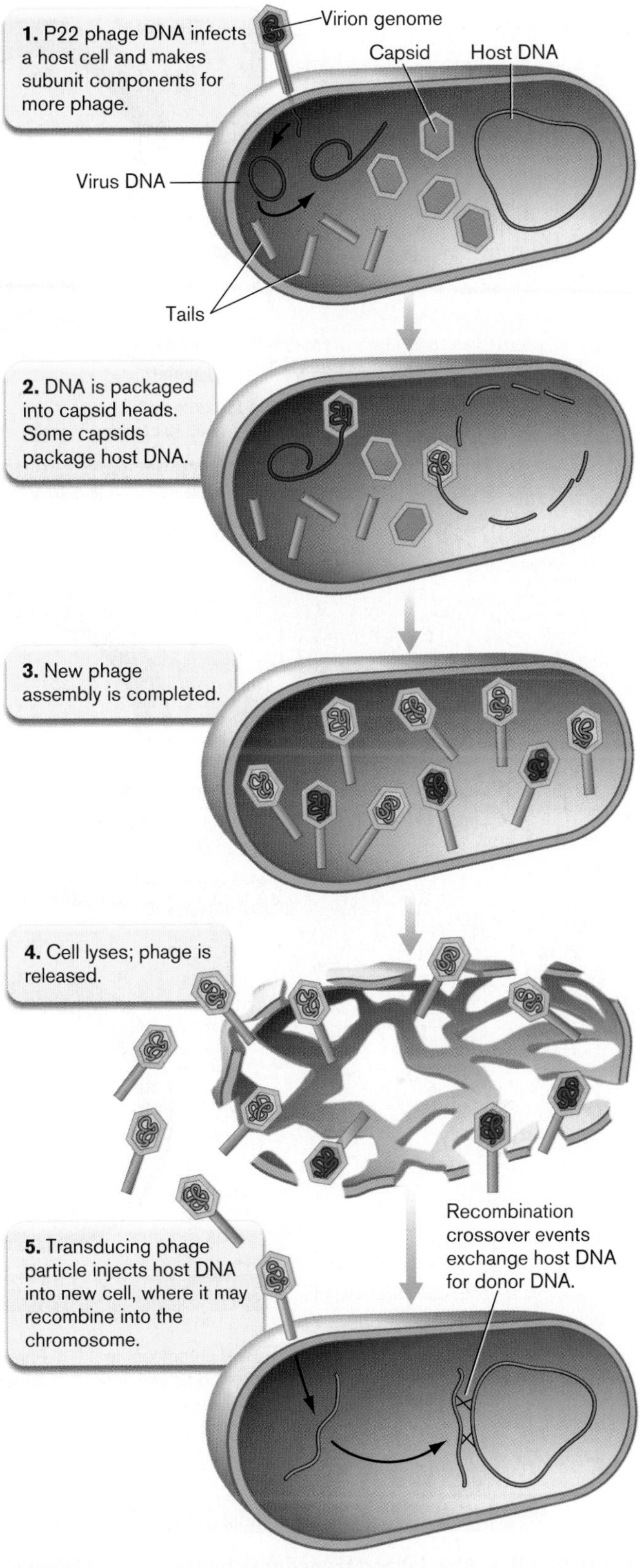
1. P22 phage DNA infects a host cell and makes subunit components for more phage.
Virion genome
Capsid
Host DNA
Virus DNA
Tails
2. DNA is packaged into capsid heads. Some capsids package host DNA.
3. New phage assembly is completed.
4. Cell lyses; phage is released.
5. Transducing phage particle injects host DNA into new cell, where it may recombine into the chromosome.
Recombination crossover events exchange host DNA for donor DNA.

FIGURE 9.7 ■ Generalized transduction. Generalized transduction by phage vectors can move any segment of donor chromosome to a recipient cell. The number of genes transferred in any one phage capsid is limited, however, to what can fit in the phage head.

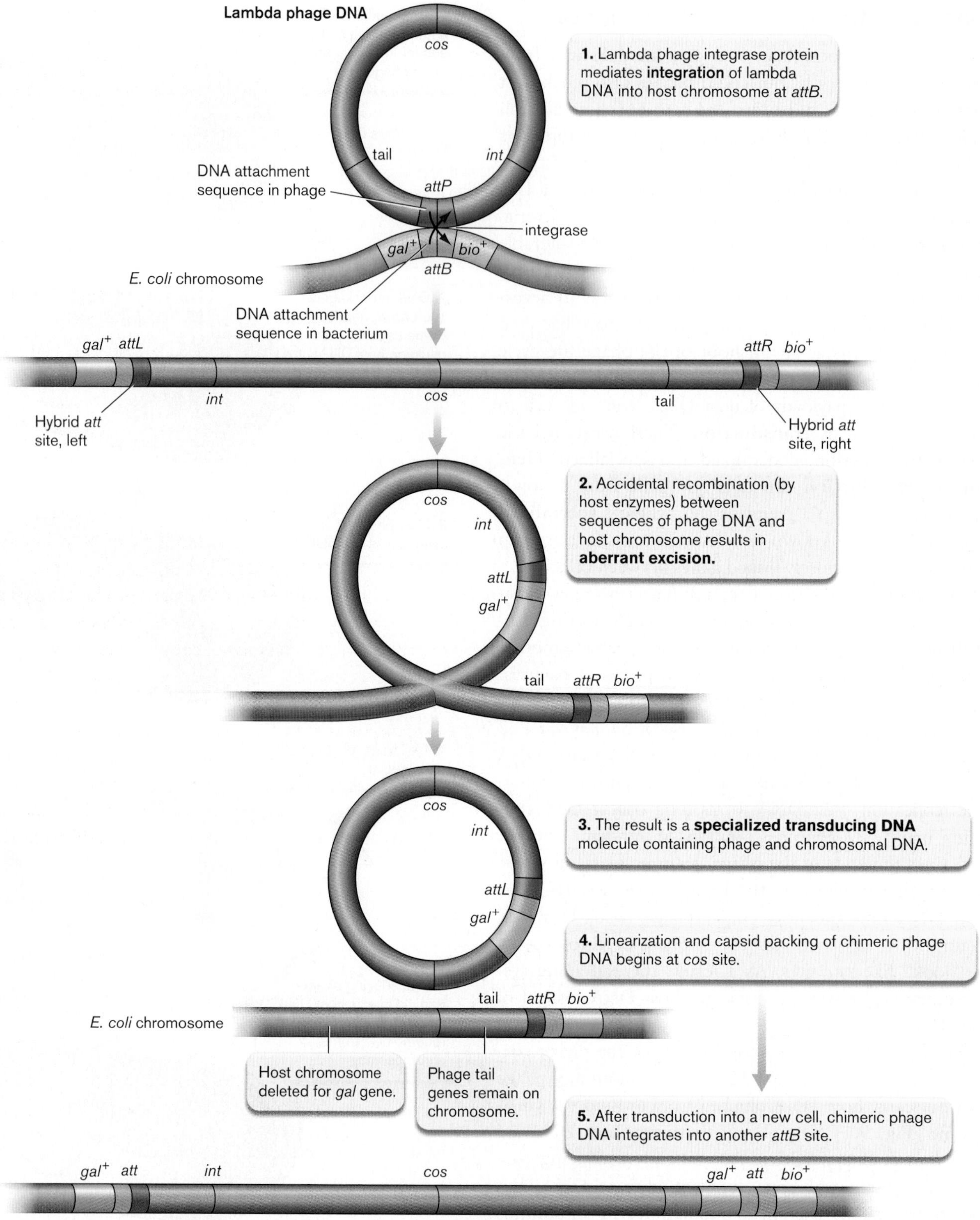

Lambda phage DNA
cos
1. Lambda phage integrase protein mediates integration of lambda DNA into host chromosome at attB.
tail
int
DNA attachment sequence in phage
attP
integrase
gal+
bio+
attB
E. coli chromosome
DNA attachment sequence in bacterium
gal+ attL
int
cos
tail
attR bio+
Hybrid att site, left
Hybrid att site, right
2. Accidental recombination (by host enzymes) between sequences of phage DNA and host chromosome results in aberrant excision.
cos
int
attL
gal+
tail attR bio+
3. The result is a specialized transducing DNA molecule containing phage and chromosomal DNA.
cos
int
attL
gal+
4. Linearization and capsid packing of chimeric phage DNA begins at cos site.
E. coli chromosome
tail attR bio+
Host chromosome deleted for gal gene.
Phage tail genes remain on chromosome.
5. After transduction into a new cell, chimeric phage DNA integrates into another attB site.
gal+ att
int
cos
gal+ att bio+

FIGURE 9.8 ▪ Specialized transduction is restricted to moving host genes flanking the phage attachment site. The number of genes transferred is limited by the size of the phage head. The resulting transductant chromosome becomes partially diploid for the transferred gene (in this case, *gal*).

Thought Question

9.3 A prophage genome that is embedded in the chromosome of a lysogenic cell will prevent the replication of any new phage DNA (from the same phage group) that has infected the cell. By preventing the replication of superinfecting phage DNA, the prophage genome protects itself from destruction by lysis. Is a lysogen also protected from generalized transduction carried out by a similar phage? For example, can a P22 lysogen of *Salmonella* be transduced by P22 grown on a different strain of *Salmonella*? Explain why or why not.

Specialized (restricted) transduction is a phage-mediated gene transfer mechanism that resembles the formation of the F′ factors previously described. In contrast to generalized transduction, specialized transduction can move only a limited number of host genes. *E. coli* phage lambda (discussed in Chapter 6) provides the classic example of specialized transduction. Lambda phage DNA is linear when it first enters the cell; then it circularizes at cohesive ends called *cos* sites. Next, a small DNA sequence in lambda called *attP* can recombine with a similar host DNA sequence (called *attB*) located between the *gal* (galactose catabolism) and *bio* (biotin synthesis) genes on the *E. coli* chromosome (**Fig. 9.8**, step 1). This process, carried out by the phage integrase protein, produces a chromosome with an integrated phage genome, referred to as a prophage, flanked by chimeric *att* sites called *attL* and *attR*. These sites are called chimeric because each one is made half from the bacterial and half from the phage *att* sites. Prophage DNA remains latent until something happens to the cell to activate it.

Specialized transduction begins with the reactivation of this prophage DNA, usually by DNA damage. Most of the time, recombination mechanisms involving phage enzymes excise the lambda DNA precisely, so that the chromosome and viral DNAs are restored to their native states. The viral DNA will then replicate and make more phage particles containing normal phage DNA. On rare occasions, however, improper excision (mediated by host recombination enzymes) can take place between host DNA sequences that lie adjacent to the phage insertion site (*attB* in **Fig. 9.8**, step 2) and similar DNA sequences within the prophage. Improper excision yields a virus that will lack a few viral genes (tail genes missing in **Fig. 9.8**, step 3), but will include host genes lying adjacent to the phage attachment site (the galactose utilization gene *gal* in **Fig. 9.8**, step 3). With some phages, the specialized transducing particles can replicate unaided. However, in **Figure 9.8** the result is a defective, specialized transducing phage DNA (lambda d*gal*, or λd*gal*). Specialized transducing phage particles that are defective cannot replicate by themselves, but require the presence of a helper phage to supply missing gene products.

Once formed, specialized transducing phages can deliver the hybrid DNA molecule to a new recipient cell. This is the transduction process. Once in that cell, the phage DNA can integrate into the host *attB* site, carrying the donor host gene(s) with it (**Fig. 9.8**, steps 4 and 5). The result will be another partial diploid situation in which the new recipient contains two copies of a host gene: one originally present on its chromosome, and one brought in by the transducing DNA.

DNA Restriction and Modification

There are dangers to the cell associated with the indiscriminate transfer of DNA between bacteria. The most obvious risk involves bacteriophages whose goal is to replicate at the expense of target cells. A bacterium that can digest invading phage DNA while protecting its own chromosome has a far better chance of surviving in nature than do cells unable to make this distinction. Beyond the threat of overt destruction by phages, however, the unrestricted incorporation of foreign DNA can be an energetic drain on the cell. Genes encoding competing products, useless products, or products that lack regulatory restraints would squander resources and lower the overall fitness of the cell. As a result, bacteria have developed a kind of "Halt! Who goes there?" approach to gene exchange. It is an imperfect approach, however, that still leaves room for beneficial genetic exchanges.

This protection system, called "restriction and modification," involves the enzymatic cleavage (restriction) of alien DNA and the protective methylation (modification) of self DNA (**Fig. 9.9**). Most bacteria produce DNA **restriction**

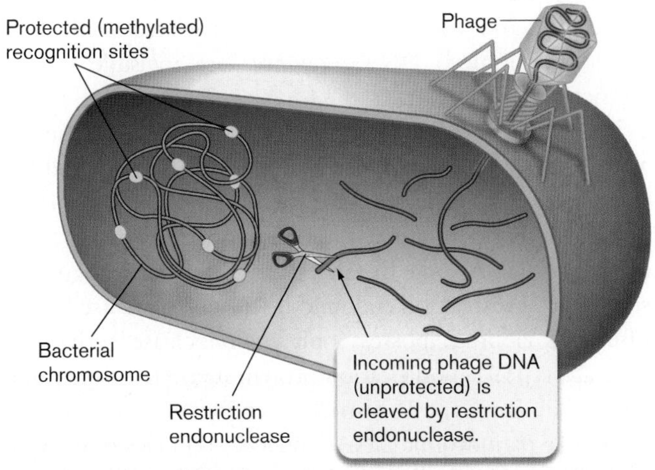
Protected (methylated) recognition sites
Phage
Bacterial chromosome
Restriction endonuclease
Incoming phage DNA (unprotected) is cleaved by restriction endonuclease.

FIGURE 9.9 ■ Restriction of invading phage DNA. Phage DNA is injected into a host, where restriction endonucleases can digest it. Host DNA is protected because specific methylations of its own DNA prevent the enzymes from cutting it.

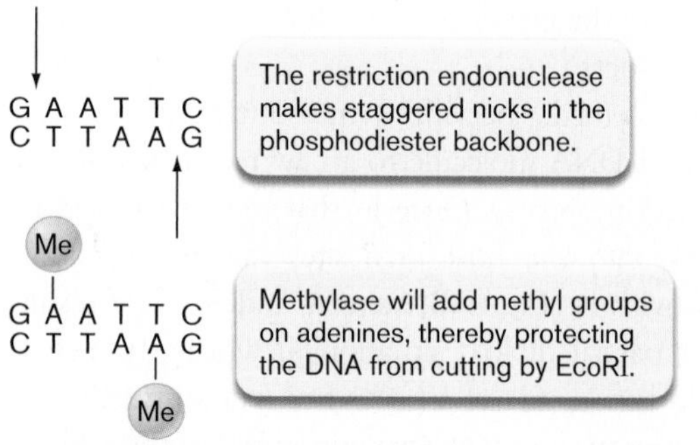
G A A T T C
C T T A A G
The restriction endonuclease makes staggered nicks in the phosphodiester backbone.
Me
G A A T T C
C T T A A G
Me
Methylase will add methyl groups on adenines, thereby protecting the DNA from cutting by EcoRI.

FIGURE 9.10 ■ EcoRI restriction site. EcoRI is an example of a class II restriction-modification system. Shown here are cleavage (top) and methyl modifications (bottom). The DNA sequence shown is specifically recognized by the endonuclease and methylase enzymes.

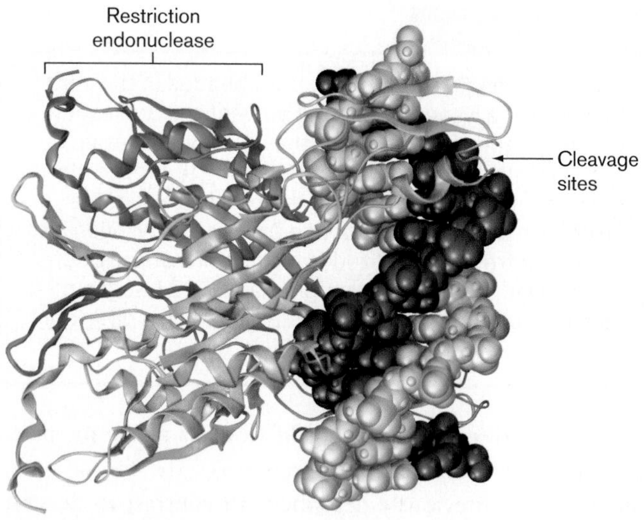
Restriction endonuclease
Cleavage sites

FIGURE 9.11 ■ DNA restriction endonuclease attached to a DNA restriction sequence. The part of the protein shown in gold in this side view is about to cut the dark gray DNA backbone at the top. The light blue part of the protein will cut the light gray DNA at the bottom. (PDB code: 1QRI)

endonucleases (also known as restriction enzymes), enzymes that recognize specific short DNA sequences (known as recognition sites) and cleave DNA at or near those sequences (**Fig. 9.10**). A structural model of the restriction endonuclease EcoRI cleaving its target DNA sequence is shown in **Figure 9.11**.

Given that the producing organism's own chromosomes contain these same restriction sequences, how do bacteria avoid "committing suicide"? They protect themselves by producing matching modification enzymes that use *S*-adenosylmethionine to attach methyl groups to those same sequences. A 3D model of a methylation enzyme interacting with a DNA recognition site is shown in **Figure 9.12**. This modification makes the sequence invisible to the cognate (matched) restriction endonuclease (the "A" no longer looks like adenine to the restriction endonuclease). Only one strand of the sequence needs to be methylated to protect the duplex from cleavage; thus, even newly replicated, and consequently hemimethylated (only one strand methylated) DNA sequences are invisible to the restriction endonuclease.

There are three main types of restriction endonucleases (**Table 9.1**), called types I, II, and III. Types I and III restriction endonucleases have their restriction and modification activities combined in one multifunctional protein and cleave DNA some distance away from the recognition site. Type II restriction endonucleases (which are used most often for cloning) possess only endonuclease activity; a separate type II modification protein methylates the same restriction site. Type II restriction endonucleases generally recognize palindromic DNA sequences and cleave at those sites. **Figure 9.11** illustrates how the class II restriction endonuclease EcoRI holds the DNA as it makes a staggered cut at the cleavage (restriction) site.

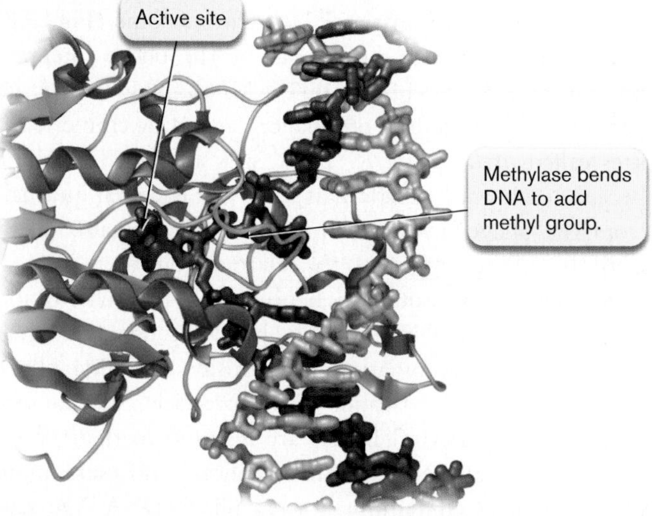
Active site
Methylase bends DNA to add methyl group.

FIGURE 9.12 ■ DNA methylase used for DNA modification. DNA methylase adds a methyl group to a restriction site, changing the site's conformation so that it is no longer recognized by the restriction endonuclease. The methylase bends the DNA helix to bring the target base into the active site (left side). (PDB code: 1DCT)

Note: Restriction endonucleases are named according to the species from which the enzyme was isolated. Thus, EcoRI is an enzyme from *E. coli*. Previously written as *Eco*RI, these enzymes no longer have the first three letters italicized.

TABLE 9.1

Main types of restriction-modification systems.

Restriction-modification system	Restriction and modification activities	Number of subunits	Recognition site characteristics	Cleavage and modification sites	Examples
Type I	Present in one multifunctional protein	Three different subunits	5–7 bp, asymmetrical	Located 100 bp or more from recognition site	EcoK in *E. coli*; StyLTIII in *Salmonella enterica*
Type II	Separate methylase and restriction enzymes	One or two identical subunits per activity	4–6 bp, palindromic	At or near recognition site	EcoRI in *E. coli*; HindIII in *Haemophilus influenzae*
Type III	Present in one multifunctional protein	Two different subunits	5–7 bp, asymmetrical	Located 24–26 bp from recognition site	Eco571 in *E. coli*; BceSI in *Bacillus cereus*

Thought Questions

9.4 How do you think phage DNA containing restriction sites evades the restriction-modification screening systems of its host?

9.5 Each type I restriction-modification system includes separate host specificity determinant (Hsd) proteins for restriction (HsdR), modification (HsdM), and sequence recognition (HsdS). Can a plasmid grown in a wild-type strain be used to transform a strain defective in *hsdR*? What about an *hsdS* mutant? Can a plasmid grown in an *hsdM* mutant strain be transformed into an *hsdR* defective strain?

CRISPR Cassettes: Small RNA-Guided Defense Systems

One of the most formidable wars on Earth is the one waged by phages and viruses against Bacteria and Archaea, respectively. Phage predators outnumber prokaryotic prey by 10:1 and destroy between 4% and 50% of prokaryotic life-forms. In an attempt to save themselves, Bacteria and Archaea have developed intriguing defense systems. One system masks receptors for phage attachment. A second uses enzymes to digest invading phage DNA while strategically protecting their own (see the preceding discussion). A third, newly recognized adaptive system, first introduced in Chapter 6, is called **CRISPR** (clustered regularly interspaced short palindromic repeats), in which an organism that manages to survive a phage attack captures a piece of the invader's genome and wields it as defense against future attack. Many species of Bacteria and Archaea possess one or more of these CRISPR loci.

CRISPR anatomy. A CRISPR locus on a bacterial chromosome (**Fig. 9.13**) is composed of short direct-repeat sequences (averaging 32 bp) separated by spacers of uniform length (20–72 bp, depending on the species). Although the sequences of direct repeats are, by definition, nearly identical, the sequences of the spacers are different from one another. Note that repeats and spacers do not encode proteins. Near these sequence clusters lie CRISPR-associated gene families (*cas1–cas4*) that do encode proteins. A single species can have one or more of these *cas* genes, as well as *cas* subtype genes (called *cse* for *E. coli*). Some CRISPR loci also contain a set of genes called the RAMP (repeat-associated mysterious proteins) module, which includes the gene encoding a putative polymerase.

CRISPR function. Clues about the function of the variable spacer regions were uncovered using bioinformatics, which revealed that some spacers bear sequence homology to bacteriophage or plasmid genes. It turned out that cells harboring these spacers were immune to the corresponding invaders, but related species lacking these spacers were susceptible. Thus, CRISPR is perceived as a primitive microbial immune system.

How does a CRISPR locus work? Details are still being defined, but in the current model, bacteria with a CRISPR locus acquire new spacers by incorporating a piece of an

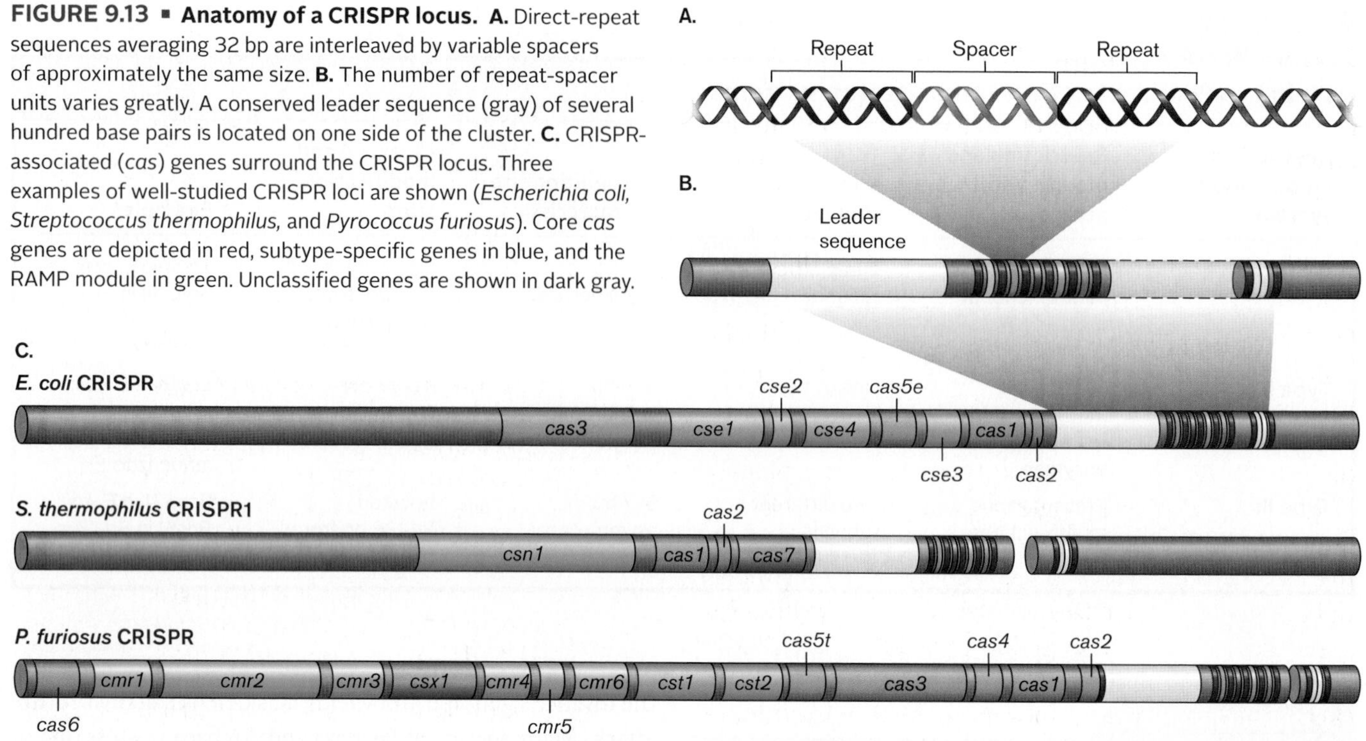
A.
Repeat
Spacer
Repeat
B.
Leader
sequence
C.
E. coli CRISPR
cas3
cse1
cse2
cse4
cas5e
cas1
cse3
cas2
S. thermophilus CRISPR1
csn1
cas1
cas2
cas7
P. furiosus CRISPR
cmr1
cmr2
cmr3
csx1
cmr4
cmr6
cst1
cst2
cas5t
cas3
cas4
cas1
cas2
cas6
cmr5

FIGURE 9.13 ■ Anatomy of a CRISPR locus. **A.** Direct-repeat sequences averaging 32 bp are interleaved by variable spacers of approximately the same size. **B.** The number of repeat-spacer units varies greatly. A conserved leader sequence (gray) of several hundred base pairs is located on one side of the cluster. **C.** CRISPR-associated (*cas*) genes surround the CRISPR locus. Three examples of well-studied CRISPR loci are shown (*Escherichia coli, Streptococcus thermophilus,* and *Pyrococcus furiosus*). Core *cas* genes are depicted in red, subtype-specific genes in blue, and the RAMP module in green. Unclassified genes are shown in dark gray.

invader's DNA (**Fig. 9.14A**; spacer acquisition). One or more Cas proteins may cleave part of an invading phage genome and integrate it as the lead spacer in the CRISPR region. It is then thought that the CRISPR locus is transcribed starting from an upstream leader sequence. The RNA transcript is then cleaved and trimmed (processed) by some of the Cas products into small RNAs composed of a single spacer sequence (crRNA; **Fig. 9.14B**; crRNA processing stage). Each crRNA is then thought to bind to a homologous sequence from an infecting phage or plasmid

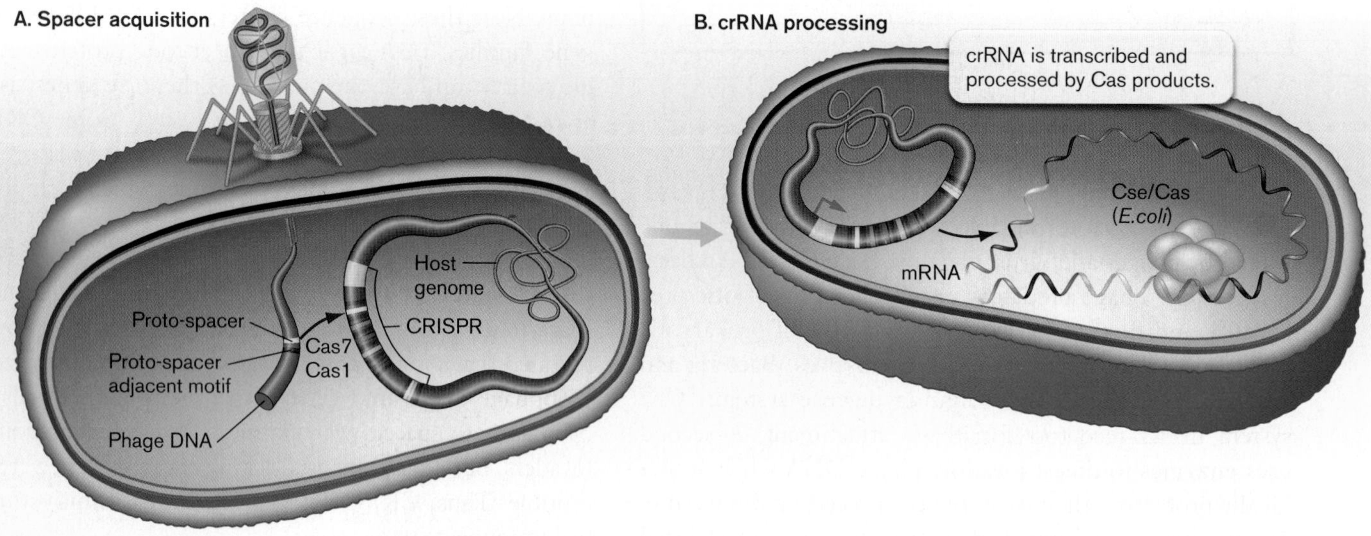
A. Spacer acquisition
Host
genome
CRISPR
Proto-spacer
Proto-spacer
adjacent motif
Cas7
Cas1
Phage DNA
B. crRNA processing
crRNA is transcribed and
processed by Cas products.
Cse/Cas
(E.coli)
mRNA

FIGURE 9.14 ■ An overall model of CRISPR/Cas activity. During spacer acquisition **(A)**, sequence elements from invading nucleic acids (yellow) become incorporated at the leader-proximal end of the CRISPR locus. In the processing stage **(B)**, the locus is transcribed and processed into mature crRNA containing an 8-nt repeat tag and a single spacer unit. During the effector stage **(C)**, the mature crRNAs in complex with Cas proteins lead to the degradation of complementary invading nucleic acids. *Source:* Adapted from **Karginov, Fedor V., and Gregory J. Hannon.** 2010. The CRISPR system: Small RNA-guided defense in bacteria and archaea. *Mol. Cell* **37**:7–19 (Fig. 4).

and directs degradation of the foreign DNA by another Cas protein. The process may be aided by RAMP module proteins (**Fig. 9.14C**; effector stage). As a result, the infected cell is spared destruction. One consequence of limiting phage infections is that CRISPR may also limit the transfer of genetic material between cells by transduction.

Biofilms and CRISPR. Another function of CRISPR is to banish a lysogenic cell from a biofilm, so that phage produced by the lysogen cannot kill the rest of the biofilm community. **Figure 9.15** demonstrates this phenomenon with *Pseudomonas aeruginosa*.

The first panel in **Figure 9.15** shows that a lysogenized (infected) cell of *P. aeruginosa* does not form a biofilm. If this lysogen were a single cell among many biofilm-producing nonlysogens, then it would not associate with the biofilm. This self-exile saves the biofilm from infection, should the phage become active. The CRISPR locus is essential for this loss, since disruption of the CRISPR or several *cas* genes restores a lysogen's ability to form biofilms (**Fig. 9.15**, second panel). Introducing a wild-type copy of the gene (**Fig. 9.15**, third panel) into the mutant cell once again prevents biofilm formation. Thus, a mechanism in which an infected cell would impose self-exile would save the population. Why this particular CRISPR does not prevent lysogeny is currently unclear.

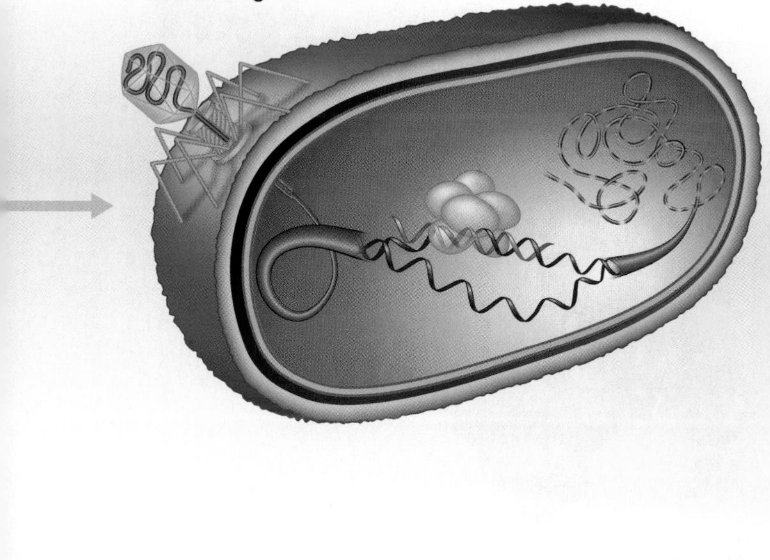
C. Effector stage

Wtz Lys.
Δ*csy4* Lys.
Δ*csy4* + *csy4* Lys.
JANE SANDERS MILLER

FIGURE 9.15 ■ The effect of CRISPR on *Pseudomonas aeruginosa* biofilm formation. Shown are upside-down tubes in which *P. aeruginosa* lysogenized with DMS3 phage were grown. Cells that form a biofilm stick to the side of the tube and are not dislodged by washing. The biofilm is revealed after staining with crystal violet. The left-hand panel shows that the lysogenized wild-type (Wt) *P. aeruginosa* does not form a substantial biofilm. The middle panel shows that deleting a Cas-encoding gene (called *csy4*) allows biofilm formation (a nonlysogen of *P. aeruginosa* would also form this biofilm). The right-hand panel shows that introducing a plasmid that expresses *csy4* into the mutant restored the antibiofilm phenotype. A "Lys" following the genotype designation indicates a strain lysogenized with DMS3.

To Summarize

- **Transformation** is the uptake by living cells of free-floating DNA from dead, lysed cells.
- **Competence** for transformation in some organisms is triggered by a genetically programmed physiological change.
- **Conjugation** is a DNA transfer process mediated by a transferable plasmid that requires cell-cell contact and formation of a protein complex between mating cells.
- **Some bacteria can transfer DNA across phylogenetic domains.** For example, *Agrobacterium tumefaciens* conjugates with plant cells.
- **Transduction** is the process whereby bacteriophages transfer fragments of bacterial DNA from one bacterium to another. In **generalized transduction**, a phage preparation can move any gene in a bacterial genome to another bacterium. In **specialized transduction**, a phage can move only a limited number of bacterial genes.
- **Restriction endonucleases** protect bacteria from invasion by foreign DNA. Restriction-modification enzymes methylate restriction target sites in the host DNA to prevent self-digestion.
- **The CRISPR cassette system** is a small interfering RNA system in which an organism captures a piece of an invader's DNA and uses it to fend off future attacks.

9.3

Recombination

Once a new piece of DNA has entered a cell through one of the gene exchange systems, what happens to it? The answer depends on the nature of the acquired DNA. If the DNA is a plasmid, capable of autonomous replication, it can coexist in the cell separate from the host chromosome. If the DNA is incapable of autonomous replication, however, then it may be incorporated into the chromosome through recombination or, as is usually the case, it may be degraded by nucleases.

Two DNA molecules in a single cell can recombine by one of two main mechanisms: generalized or site-specific recombination. **Generalized recombination** requires that the two recombining molecules have a considerable stretch of homologous DNA sequence. **Site-specific recombination**, on the other hand, requires very little sequence homology between the recombining DNA molecules, but does require a short (10–20 bp) sequence recognized by the recombination enzyme.

Generalized Recombination and RecA

Enzymes participating in generalized recombination are able to find and align homologous stretches of DNA in two different DNA molecules and catalyze an exchange of strands. The result can be a **cointegrate** molecule in which a single recombination event (that is, a single crossover) joins the two participating DNA molecules. For instance, when two circular DNA molecules are joined by a single recombination event, the two molecules are added together and form one large circular molecule (for an example, see **Fig. 9.4**). A second crossover separated by some distance from the first, however, will lead to an equal exchange of DNA in which neither molecule increases in length.

Why is recombination advantageous? There are three probable functions for generalized recombination in the microbial cell:

- Recombination probably first evolved as an internal method of DNA repair, useful to fix mutations or restart stalled replication forks. This role does not involve foreign DNA.
- Cells with damaged chromosomes use DNA donated by others of the same species to repair their damaged genes.
- Recombination is part of a "self-improvement" program that samples genes from other organisms for an ability to enhance the competitive fitness of the cell.

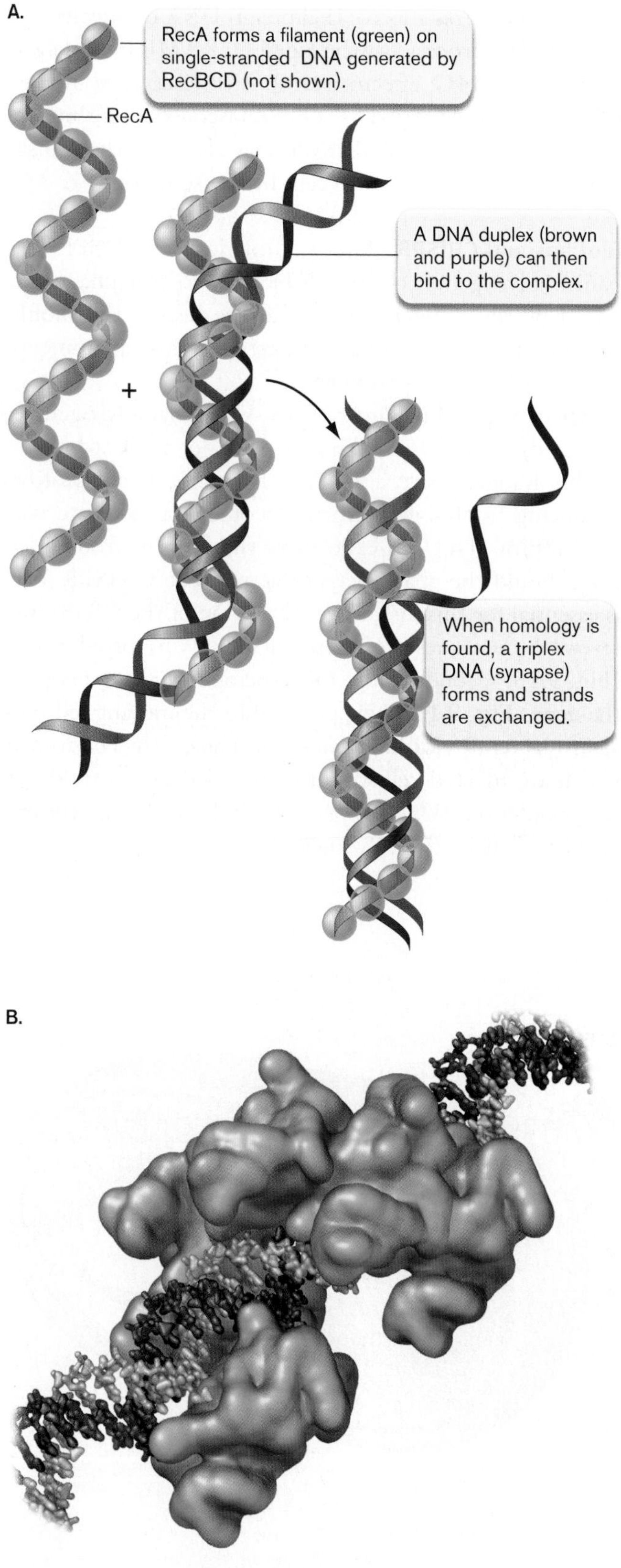
A.
RecA forms a filament (green) on single-stranded DNA generated by RecBCD (not shown).
RecA
+
A DNA duplex (brown and purple) can then bind to the complex.
When homology is found, a triplex DNA (synapse) forms and strands are exchanged.
B.

FIGURE 9.16 ■ RecA protein catalyzing recombination. **A.** The standard mechanism of recombination. **B.** A molecular model of RecA protein filament catalyzing recombination. The DNA is light gray and dark gray. (PDB codes: 1NO3 and 1K8J)

The central, but by no means only, player in generalized recombination is a protein called RecA. RecA molecules are also called synaptases because they are able to scan DNA molecules for homology and align the homologous regions, forming a triplex DNA molecule, or synapse (**Fig. 9.16**). Homologs of RecA are found in many other species.

The clearest model of RecA-mediated recombination comes from *E. coli* (**Figs. 9.16** and **9.17**). Before RecA

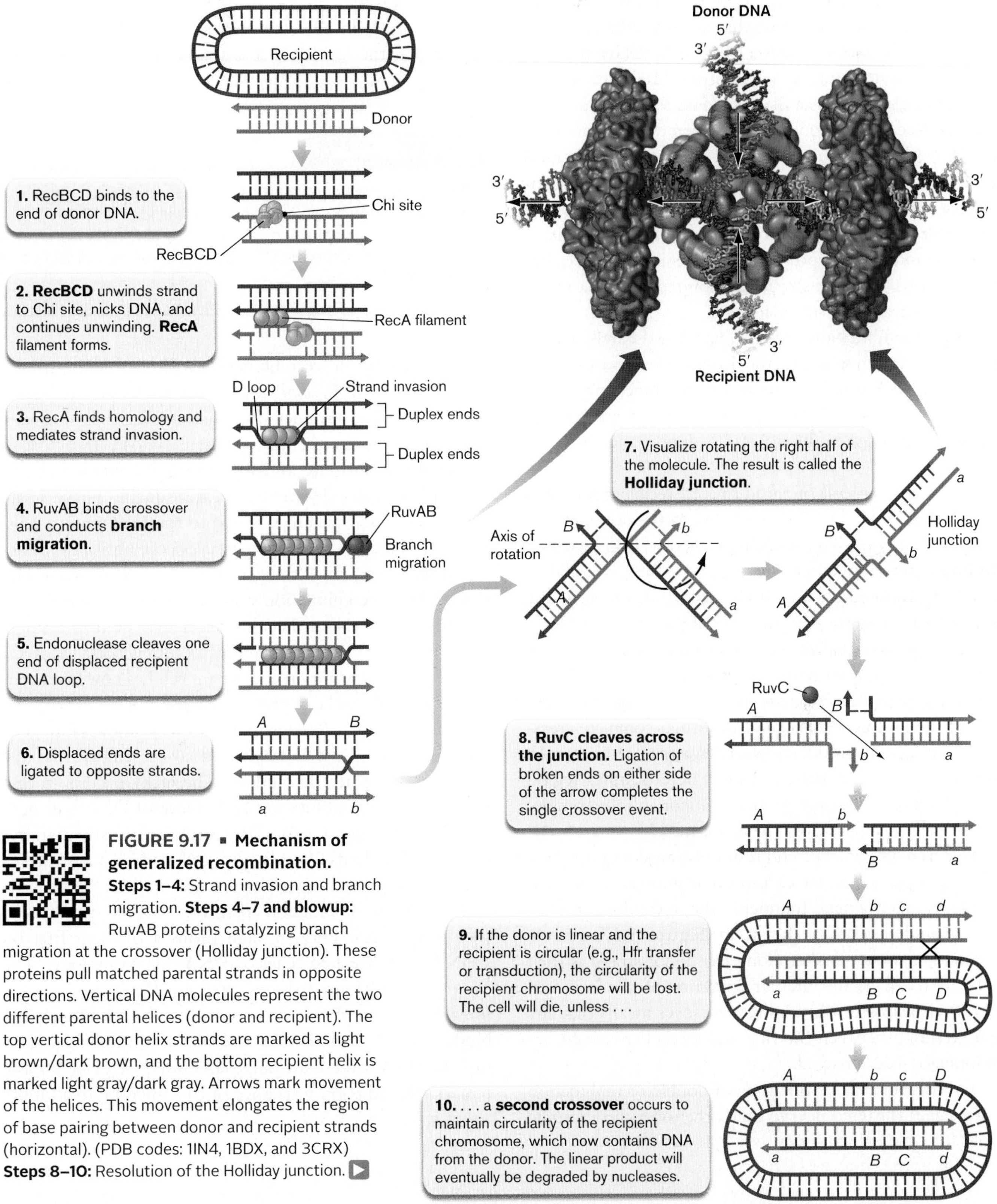

Recipient
Donor
1. RecBCD binds to the end of donor DNA.
Chi site
RecBCD
2. RecBCD unwinds strand to Chi site, nicks DNA, and continues unwinding. RecA filament forms.
RecA filament
D loop
Strand invasion
3. RecA finds homology and mediates strand invasion.
Duplex ends
Duplex ends
4. RuvAB binds crossover and conducts branch migration.
RuvAB
Branch migration
5. Endonuclease cleaves one end of displaced recipient DNA loop.
6. Displaced ends are ligated to opposite strands.
A
B
a
b
Donor DNA
5′
3′
3′
5′
3′
5′
5′
3′
Recipient DNA
7. Visualize rotating the right half of the molecule. The result is called the Holliday junction.
Axis of rotation
Holliday junction
RuvC
8. RuvC cleaves across the junction. Ligation of broken ends on either side of the arrow completes the single crossover event.
9. If the donor is linear and the recipient is circular (e.g., Hfr transfer or transduction), the circularity of the recipient chromosome will be lost. The cell will die, unless . . .
A b c d
a B C D
10. . . . a second crossover occurs to maintain circularity of the recipient chromosome, which now contains DNA from the donor. The linear product will eventually be degraded by nucleases.
A b c D
a B C d

FIGURE 9.17 ■ Mechanism of generalized recombination. **Steps 1–4:** Strand invasion and branch migration. **Steps 4–7 and blowup:** RuvAB proteins catalyzing branch migration at the crossover (Holliday junction). These proteins pull matched parental strands in opposite directions. Vertical DNA molecules represent the two different parental helices (donor and recipient). The top vertical donor helix strands are marked as light brown/dark brown, and the bottom recipient helix is marked light gray/dark gray. Arrows mark movement of the helices. This movement elongates the region of base pairing between donor and recipient strands (horizontal). (PDB codes: 1IN4, 1BDX, and 3CRX) **Steps 8–10:** Resolution of the Holliday junction.

can find homology between two DNA molecules, the donor double-stranded DNA molecule is converted to a single strand by the RecBCD enzyme. The RecBCD complex enters at the end of a DNA fragment and begins to unwind it (**Fig. 9.17**, steps 1 and 2). The complex changes activity when it encounters 8-bp sequences called Chi (crossover hot-spot instigator) that are scattered throughout most DNA molecules. RecBCD nicks DNA at a Chi site and continues unwinding the strand. RecA then loads onto the single strand as a filament (**Fig. 9.17**, step 2).

When RecA finds homology between donor and recipient DNA (the minimum required is about 50 bp), strand invasion occurs (**Fig. 9.17**, step 3), in which the donor single-stranded DNA invades the homologous region in the double-stranded DNA recipient molecule and displaces its like strand. At this point, a single-strand crossover has been made. The single-strand crossover produces a molecule with four double-stranded ends.

Other recombination proteins, RuvA and RuvB, assemble at the crossover point and extend the invasion in a process called "strand assimilation" or branch migration (**Fig. 9.17**, step 4 and blowup). This process extends the base-pairing between homologous donor and recipient strands.

Ultimately, the end of the displaced recipient strand is cleaved (**Fig. 9.17**, step 5) and ligated to the donor strand (**Fig. 9.17**, step 6). The resulting structure is known as a **Holliday junction**, after the scientist (Robin Holliday) who first proposed it (**Fig. 9.17**, step 7; also seen in the blowup). The final step is resolution of the crossover Holliday junction. To visualize this, we mentally rotate the right half of the Holliday structure as shown in **Figure 9.17**, step 7. RuvC protein cleaves across the junction (**Fig. 9.17**, step 8), and the products are ligated to form a complete crossover. The ligated product represents a true crossover. In each case, notice that there is a small **heteroduplex** region in which one strand comes from the donor and the complementary region comes from the recipient.

If we were viewing the end result of transduction, where the recipient is a circular chromosome and the donor is a double-stranded linear fragment, the recombinant molecule would appear as shown in **Figure 9.17**, step 9. A second crossover farther along the donor strand is then required to maintain circular integrity of the recipient (**Fig. 9.17**, step 10). This double crossover means that any genes residing between the first and second crossovers are simultaneously exchanged.

One consequence of this type of double recombination is that genes clustered next to each other on a chromosome can be recombined as a group into a recipient's genome. In the case of transduction, a piece of DNA from the donor bacterium replaces a segment of DNA in the recipient host. One gene in the segment is said to be cotransduced with another. The closer the two genes are physically on the chromosome, the higher the probability is that they will be cotransduced, because the closer two genes are to each other, the less likely it is that a crossover will occur between them.

Thought Question

9.6 In a transductional cross between an $A^+B^+C^+$ genotype donor and an $A^-B^-C^-$ genotype recipient, 100 A^+ recombinants were selected. Of those 100, 15% were also B^+, while 75% were C^+. Is gene B or gene C closer to gene A?

The benefit of recombination can be demonstrated using a culture of *E. coli* that contains a defective *lacZ* gene. The *lacZ*$^+$ gene product (beta-galactosidase) is normally used to catabolize the carbohydrate lactose. If lactose is the only carbohydrate available, the *lacZ* mutant cell culture fails to grow. However, if a transducing phage lysate grown on a *lacZ*$^+$ strain is added to the *lacZ* mutant cells, the subpopulation of phage particles containing the functional *lacZ*$^+$ can inject the gene into the mutant (transduction), where recombination systems can exchange it for the defective gene. (Note that the defect may be only a single base change, so there is plenty of homology left for RecA to recognize.) The new recombinant cell has now been converted to *lacZ*$^+$ and can grow on lactose.

When trying to understand recombination (and gene exchange overall), it is important to remember that the processes are generally random. The lucky bacterium that receives and recombines the right piece of DNA will benefit. This is why a genetic experiment usually requires screening hundreds of millions of cells to find a few recombinants that grow into visible colonies.

Not all recombination pathways follow the model shown, in which RecA protein forms a filament first on ssDNA and then searches dsDNA for homology. The extremely radiation-resistant species *Deinococcus radiodurans*, for example, does the exact opposite. Its RecA first binds dsDNA and searches ssDNA for homology. This inverse DNA strand exchange accounts, in part, for the remarkable radiation resistance of this microbe (read more about it in eTopic 9.3).

Site-Specific Recombination Is RecA Independent

In contrast to generalized recombination mechanisms that require RecA protein and can recombine any region of the chromosome, site-specific recombination does not utilize RecA and moves only a limited number of genes. This form of recombination involves very short regions of homology between donor and target DNA molecules. Dedicated enzyme systems specifically recognize those sequences and catalyze a crossover between them to produce a cointegrate molecule. The integration of phage lambda, described in Section 9.2, is one example of site-specific recombination involving a 15-bp *att* site and the integrase enzyme (see **Fig. 9.8**).

Other examples of site-specific recombination are flagellar **phase variation** in the pathogen *Salmonella enterica* and phase variation in the expression of type I pili in *E. coli*. In these systems, the sequence of a segment of DNA (called a cassette) is inverted (flipped), resulting in the on-or-off regulation of adjacent gene expression. Phase variation is frequently employed by pathogens to evade the host immune system by changing the expression of cell-surface proteins. For more on flagellar phase variation, see Section 10.6.

Note: A DNA cassette can be moved from one place to another, or inverted within the genome. In the case of a kanamycin resistance cassette, even the inverted form maintains function. Both in vivo and in vitro mechanisms can be used to move gene cassettes.

To Summarize

- **Recombination** is the process by which DNA sequences can be exchanged between DNA molecules.
- **General recombination** involves large regions of sequence homology between recombining DNA molecules.
- **RecA synaptase** mediates generalized recombination.
- **The extremely radiation-resistant *Deinococcus radiodurans*** is thought to use RecA protein to patch together homologous ends of fragmented DNA in a way that reconstructs the chromosome after extreme radiation damage.
- **Site-specific recombination** requires little homology between donor and recipient DNA molecules. Site-specific recombination enables phage DNA to integrate into bacterial chromosomes and is a process that can turn on or turn off certain genes, as in flagellar phase variation in *Salmonella*.

9.4

Mutations

What are mutations, and how do they affect evolution? Any permanent, heritable alteration in a DNA sequence, whether harmful, beneficial, or neutral, is called a **mutation**. There are two basic requirements to produce a heritable mutation: Namely, there must be a change in the base sequence, and the cell must fail to repair the change before the next round of replication. Repair mechanisms will be discussed in Section 9.5. Although it is tempting to think that all mutations destroy a protein's activity, this is not the case. A given mutation may or may not affect the informational content or phenotype of the organism.

Mutations come in several different physical and structural forms:

- A **point mutation** is a change in a single nucleotide (**Fig. 9.18A** and **B**). Among point mutations, changing a purine to a different purine or a pyrimidine to a different pyrimidine is called a **transition**, while swapping a purine for a pyrimidine (or vice versa) is a **transversion**.
- **Insertions** and **deletions** involve, respectively, the addition or subtraction of one or more nucleotides (**Fig. 9.18C** and **D**), making the sequence either longer or shorter than it was originally.
- An **inversion** occurs when a fragment of DNA is flipped in orientation relative to DNA on either side (**Fig. 9.18E**).
- A **reversion** occurs when a sequence altered by mutation returns to its original sequence.

Mutations can also be categorized into several informational classes (refer to Figure 8.11 for the genetic code). Mutations that do not change the amino acid sequence of a translated open reading frame (ORF) are called **silent mutations**. For example, a point mutation changing TTT to TTC in the sense DNA strand (corresponding to a UUU-to-UUC codon change in mRNA) still codes for phenylalanine. Thus, even though the DNA sequence has changed, the protein sequence remains the same (a synonymous substitution). However, if the UUU codon were changed to UUA (a U-to-A transversion), then the protein would have a leucine where a phenylalanine had been (see **Fig. 9.18A**). This type of mutation is a **missense mutation** because the amino acid sequence of the protein has changed. (Remember: RNA polymerase makes mRNA

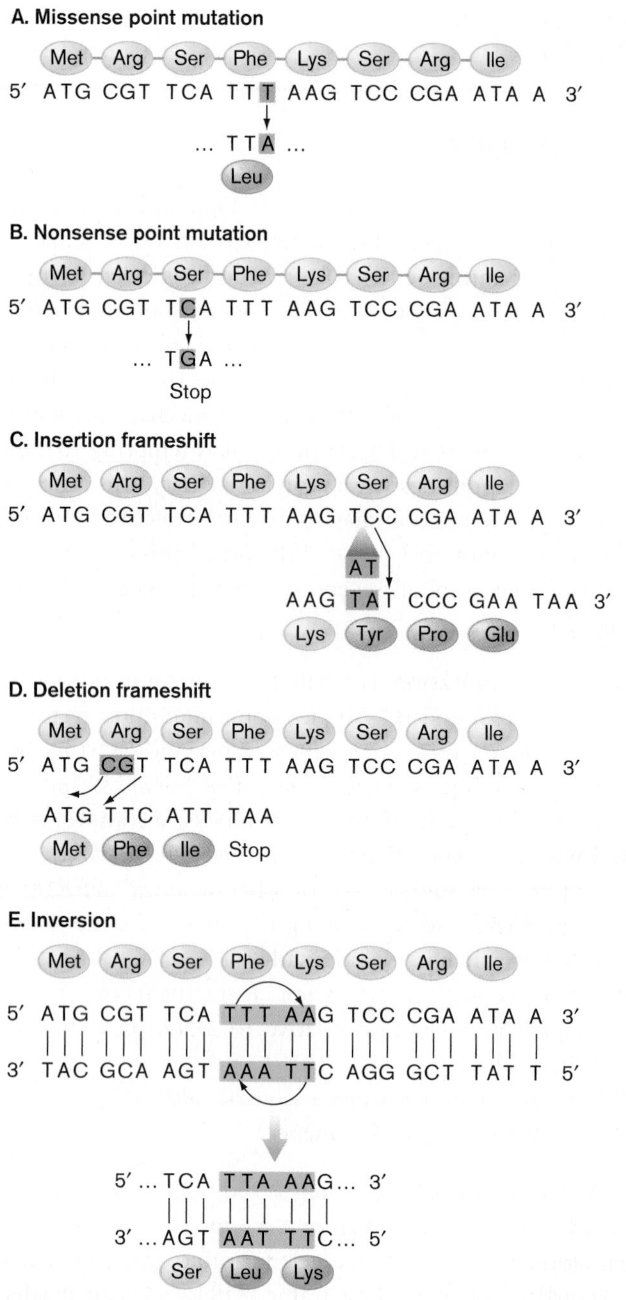
A. Missense point mutation
Met Arg Ser Phe Lys Ser Arg Ile
5′ ATG CGT TCA TTT AAG TCC CGA ATA A 3′
... TTA ...
Leu
B. Nonsense point mutation
Met Arg Ser Phe Lys Ser Arg Ile
5′ ATG CGT TCA TTT AAG TCC CGA ATA A 3′
... TGA ...
Stop
C. Insertion frameshift
Met Arg Ser Phe Lys Ser Arg Ile
5′ ATG CGT TCA TTT AAG TCC CGA ATA A 3′
AT
AAG TAT CCC GAA TAA 3′
Lys Tyr Pro Glu
D. Deletion frameshift
Met Arg Ser Phe Lys Ser Arg Ile
5′ ATG CGT TCA TTT AAG TCC CGA ATA A 3′
ATG TTC ATT TAA
Met Phe Ile Stop
E. Inversion
Met Arg Ser Phe Lys Ser Arg Ile
5′ ATG CGT TCA TTT AAG TCC CGA ATA A 3′
3′ TAC GCA AGT AAA TTC AGG GCT TAT T 5′
5′ ... TCA TTA AAG... 3′
3′ ... AGT AAT TTC... 5′
Ser Leu Lys

FIGURE 9.18 ▪ Changes in a DNA sequence that result in different classes of mutations. A. Missense mutation. The T-to-A transversion is a point mutation that converts the phenylalanine codon TTT to the leucine codon TTA. **B.** Nonsense point mutation. The C-to-G transversion converts the TCA codon encoding serine, to the TGA stop codon. **C.** Insertion frameshift. The addition of AT into the middle of the TCC serine codon causes a shift in the reading frame that changes all the downstream amino acids. **D.** Deletion frameshift. Removing two nucleotides from the arginine codon also causes a frameshift. **E.** Inversion. Rotating a DNA sequence 180° relative to adjacent sequences changes the amino acids produced.

by reading the DNA template strand, while the complementary, or sense, DNA strand has the same sequence as the mRNA.)

The amino acid substitution resulting from a missense mutation may or may not alter protein function. The outcome depends on the structural importance of the original amino acid and how close in structure the replacement amino acid is to the original. Missense mutations result in either conservative amino acid replacements, in which the new amino acid is structurally similar to the original (for example, leucine is substituted for isoleucine), or nonconservative replacements, in which a very different amino acid is substituted (for example, tyrosine for alanine). A missense change may decrease or eliminate the activity of the protein (a **loss-of-function mutation**), or it may make the protein more active. It could even gain a new activity, such as an expanded substrate specificity or a completely different substrate specificity (these are called **gain-of-function mutations**). A mutation that eliminates function is known as a **knockout mutation** (see **Fig. 9.18B**). Knockout mutations can include multiple-base insertions and deletions, as well as nonsense mutations.

A **nonsense mutation** is a point mutation that changes an amino acid codon into a translation termination codon—for example, UCA (serine) to UAA. The result will be a truncated protein most likely lacking any function—another example of a knockout mutation. Typically, these defective, truncated proteins are degraded by cellular proteases (Section 8.5).

Insertions and deletions can alter the reading frame of the DNA sequence (see **Fig. 9.18C** and **D**). Remarkable as it is, the ribosome simply reads RNA sequences one codon at a time, stringing amino acids together in the process. It translates each codon "word" but cannot understand the overall protein "sentence." It does not recognize when bases have been added or removed through mutation; instead, it keeps reading the sequence in triplets. If the number of bases inserted or deleted is not a multiple of three, the ribosome will read the wrong triplets. The result is a **frameshift mutation**, which produces a garbled protein "sentence." This often causes the ribosome to encounter a premature stop codon, originally in a different reading frame. If the insertion or deletion involves multiples of three bases, the reading frame is not changed, but one or more amino acids are added or removed.

An inversion mutation flips a DNA sequence (see **Fig. 9.18E**). Imagine the highlighted sequence rotating 180° while the adjacent sequences (in black) remain right where they are. The rotation would retain the 5′-to-3′ polarity in the new molecule. But if the inversion occurred within a gene, it would likely change the codons in the area and alter the resulting protein. What is shown in **Figure 9.18E** is a small inversion; however, inversions often involve large tracts of DNA encompassing several genes. If an entire gene with its promoter inverts, the gene will likely remain functional, and its encoded protein may very well still be made. Inversions occur within a genome as a result of recombination events between similar DNA sequences or as a consequence of mobile genetic elements jumping between different areas of a genome.

Mutations can affect both the genotype and the phenotype of an organism. The genotype of an organism reflects its genomic sequences. Regardless of whether a mutation causes a change in a biochemical trait (phenotype), every mutation causes a change in the genotype. In contrast to genotype, phenotype comprises only observable characteristics, such as biochemical, morphological, or growth traits. For instance, consider the arginine biosynthesis gene *argA*. The product of this gene allows an organism to make enough arginine to support growth in media devoid of arginine. Any mutation in *argA* is a genotypic change, but if the result is a synonymous amino acid substitution, then there is no phenotypic change—the organism can still make its own arginine. However, if the mutation destroys the activity of the *argA* product, then the microbe can no longer make arginine, and the amino acid must be supplied in the medium. This is a phenotypic change too. A mutant that has lost the ability to synthesize a substance required for growth is called an auxotroph.

It is also important to realize that small mutations (such as a single base substitution) can have large effects on phenotype, whereas large mutations (such as the insertion of a 5-kb transposon between two genes) may have little or no effect. To illustrate, consider that a single point mutation in the *hpr* gene of the phosphotransferase sugar transport system (discussed in Section 4.2 and **eTopic 4.1**) will render a bacterium incapable of growing on many sugars, but inserting 5 kb of DNA just past the *hpr* stop codon will have no effect on cell growth. For mutations, it's often not the size that counts; it's the location.

Mutations Arise by Diverse Mechanisms

Although DNA is quite stable, especially compared to mRNA, which is rapidly degraded, DNA is susceptible to damage inflicted by a variety of physical and chemical agents. For example, irradiation by X-rays can cause a massive number of DNA strand breaks. When this happens, the integrity of the chromosome is lost and the cell dies. Other agents can directly modify the bases in DNA while leaving its overall structure intact. In this case, the modified bases have altered hydrogen bond base-pairing properties that result in the incorporation of an inappropriate base during replication. When that happens, a mutation occurs.

Mutations can be caused by **mutagens**, chemical agents that can damage DNA (**Table 9.2**). Even in the absence of a mutagen, though, mutations arise spontaneously. Because

TABLE 9.2

Mutagenic agents and their effects.

Mutagenic agent	Effects
Chemical agent	
Base analog *Examples*: caffeine, 5-bromouracil	Substitutes "look-alike" molecule for the normal nitrogenous base during DNA replication: point mutation
Alkylating agent *Example*: nitrosoguanidine	Adds an alkyl group, such as methyl group ($-CH_3$), to nitrogenous base, resulting in incorrect pairing: point mutation
Deaminating agent *Examples*: nitrous acid, nitrates, nitrites	Removes an amino group ($-NH_2$) from a nitrogenous base: point mutation
Acridine derivative *Examples*: acridine dyes, quinacrine	Inserts (intercalates) into DNA ladder between backbones to form a new rung, distorting the helix: can cause frameshift mutations
Electromagnetic radiation	
Ultraviolet rays	Link adjacent pyrimidines to each other, as in thymine dimer formation, thereby impairing replication; lethal if not repaired
X-rays and gamma rays	Ionize and break molecules in cells to form free radicals, which in turn break DNA; lethal if not repaired

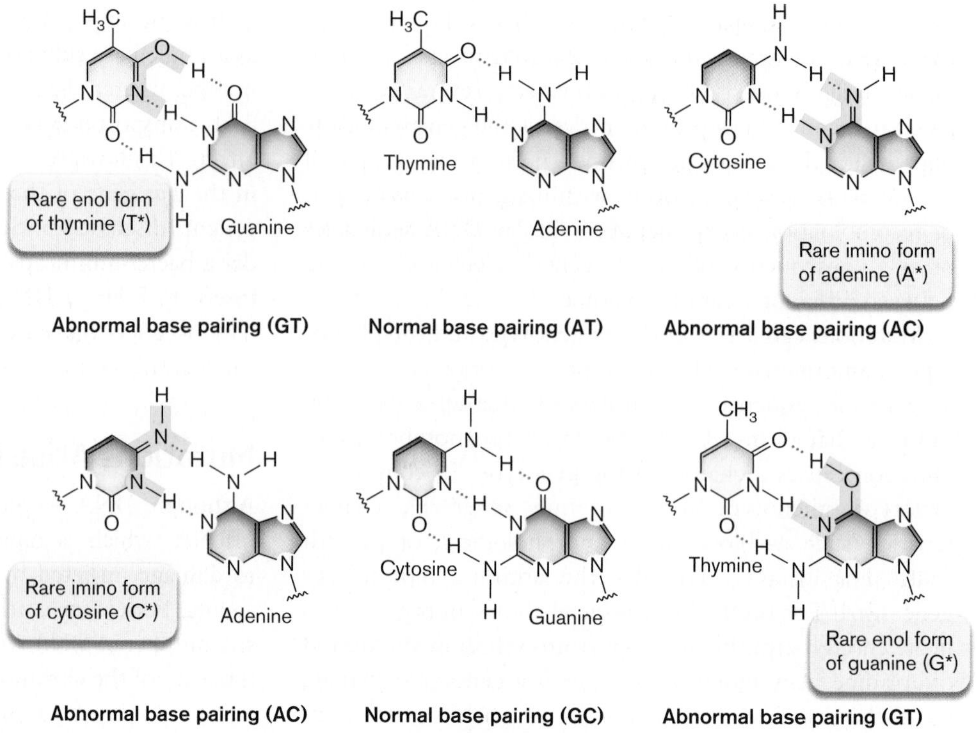
Rare enol form of thymine (T*)
Guanine
Abnormal base pairing (GT)
Thymine
Adenine
Normal base pairing (AT)
Cytosine
Rare imino form of adenine (A*)
Abnormal base pairing (AC)
Rare imino form of cytosine (C*)
Adenine
Abnormal base pairing (AC)
Cytosine
Guanine
Normal base pairing (GC)
Thymine
Rare enol form of guanine (G*)
Abnormal base pairing (GT)

FIGURE 9.19 ■ Rare tautomeric forms of bases have altered base-pairing properties. Cytosine, which normally base-pairs with guanine, will pair, in its imino form, with adenine. The imino form of adenine will also base-pair with cytosine, rather than with thymine, its normal complementary base. Likewise, the enol form of thymine base-pairs with guanine, while the enol form of guanine binds thymine. These tautomeric transitions can lead to permanent mutations.

DNA proofreading and repair pathways are so efficient, spontaneous mutations are rare, having a frequency of occurrence ranging from 10^{-6} to 10^{-8} per cell division in a given gene (*E. coli*).

Spontaneous mutations in a genome arise for many reasons—for example, tautomeric shifts in the chemical structure of the bases (**Fig. 9.19**). Tautomeric shifts involve a change in the bonding properties of amino ($-NH_2$) and keto (C═O) groups. Normally, the amino and keto forms predominate (over 85%), but when an amino group shifts to an imino (═NH) group, for example, then base-pairing changes. A cytosine that normally base-pairs with guanine will, in its rare imino form, base-pair with adenine. Tautomeric shifts that occur during DNA replication will increase the number of mutational events. Even though the replication apparatus is very accurate with various proofreading and repair functions, such as the 3′-to-5′ exonuclease activity of DNA polymerase III (see the discussion of DnaQ in Section 7.3), mistakes do occur, albeit at a very low rate.

Besides misincorporation mistakes, naturally occurring intracellular chemical reactions with water can damage DNA. These endogenous reactions are an important source of spontaneous mutations. For example, cytosine spontaneously deaminates to yield uracil (**Fig. 9.20**). The result is a GC-to-AT transition. In addition, purines are particularly susceptible to spontaneous loss from DNA via breakage of the glycosidic bond connecting the base to the sugar backbone (**Fig. 9.21**). The result of this loss is the formation of an **apurinic site** (one missing a purine base) in the DNA. Lack of a purine would obviously hinder transcription and replication.

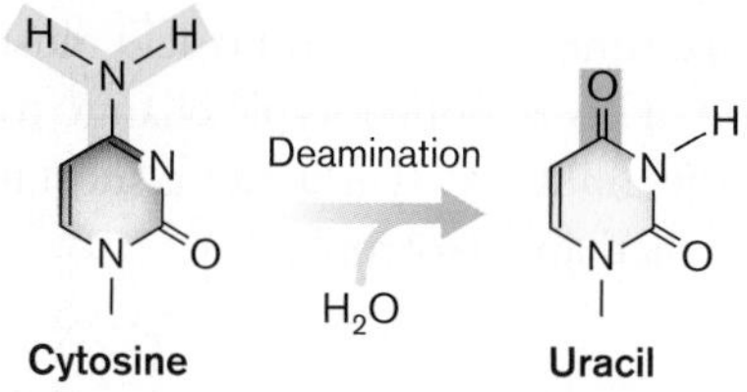
Deamination
H2O
Cytosine
Uracil

FIGURE 9.20 ■ Spontaneous deamination of cytosine. Oxidative deamination changes cytosine to uracil, which will base-pair with adenine. The result is a transition mutation.

DNA can be damaged by metabolic activities of the cell that produce reactive oxygen species, such as hydrogen peroxide (H_2O_2), superoxide radicals ($^{\bullet}O_2^{-}$), and hydroxyl radicals ($^{\bullet}OH$). Even though bacteria have biochemical mechanisms to detoxify reactive oxygen species, the systems can be overwhelmed sometimes. Oxidative damage causes the production of thymidine glycol or 8-oxo-7-hydrodeoxyguanosine in DNA (**Fig. 9.22**).

Naturally occurring intracellular methylation agents (for example, *S*-adenosylmethionine) can spontaneously methylate DNA to produce a variety of altered bases. The spontaneous methylation of the N-7 position of guanine, for example, weakens the glycosidic bond and spontaneously releases the base (forming an apurinic site) or opens the imidazole ring (forming a methylformamide pyrimidine). In addition to mispairing, some of these spontaneous events can lead to major chromosomal rearrangements, such as duplications, inversions, and deletions. Mutagens tend to increase the mutation rate by increasing the number of mistakes in a DNA molecule, as well as by inducing repair pathways that themselves introduce mutations (discussed in Section 9.5).

Ultraviolet (UV) light will produce striking structural alterations in DNA molecules. Pyrimidines (more than purines) are highly susceptible to UV radiation. The energy absorbed by a pyrimidine hit with UV light boosts the energy of its electrons to the point where the molecule is unstable. If two pyrimidines are neighbors on a single DNA strand, their energized electrons can react to form a four-membered cyclobutane ring. The result is a pyrimidine dimer that will block replication and transcription (**Fig. 9.23**).

Although there are numerous ways DNA can be damaged, the cell can repair that damage before it becomes fixed as a mutation. But the repair mechanisms are not perfect. Repair errors contribute heavily to the formation of heritable mutations and, thus, to evolution. Although most mutations decrease the fitness of a species, some can improve fitness by, among other things, enabling an organism to more efficiently use or compete for available food sources.

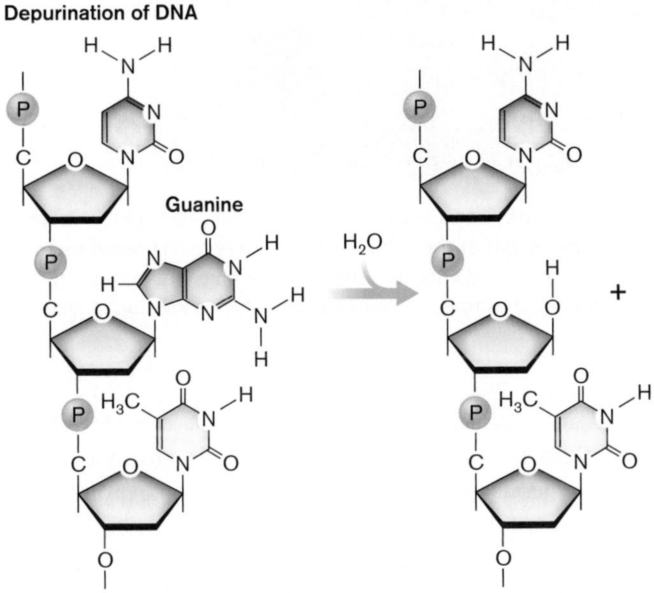
Depurination of DNA
Guanine
H_2O
+

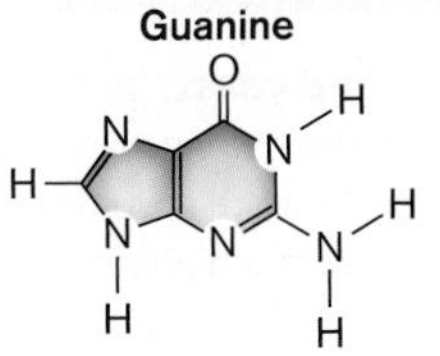
Guanine

FIGURE 9.21 ■ Spontaneous formation of an apurinic site.

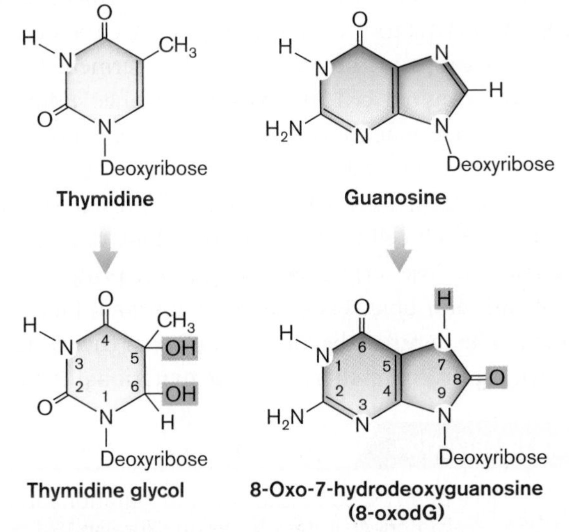
Deoxyribose
Thymidine
Deoxyribose
Guanosine
Deoxyribose
Thymidine glycol
Deoxyribose
8-Oxo-7-hydrodeoxyguanosine
(8-oxodG)

FIGURE 9.22 ■ Examples of damage caused by reactive oxygen species. Growth in the presence of oxygen leads to the production of reactive oxygen species that can modify nucleotide residues. The modifications can interfere with polymerase function and stop replication or interfere with the transcription of affected genes. The blue-highlighted groups are modifications to thymidine and guanosine residues.

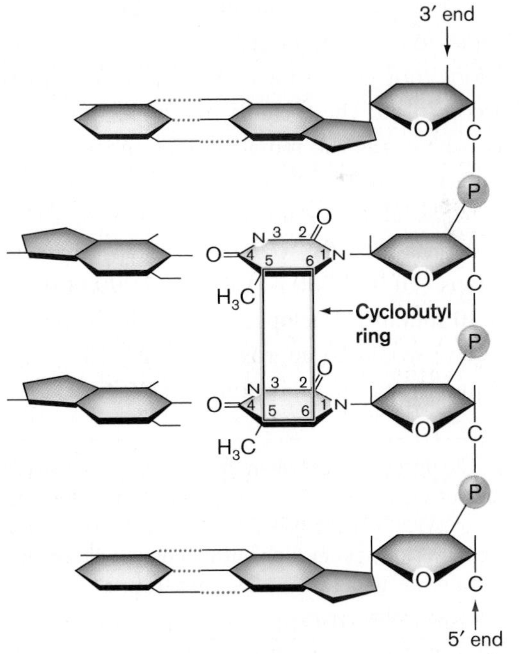
3′ end
Cyclobutyl
ring
5′ end

FIGURE 9.23 ■ Production of a pyrimidine dimer. The energy from UV irradiation can be absorbed by pyrimidine molecules. The excited electrons of carbons 5 and 6 on adjacent pyrimidines can then be shared to form a four-membered cyclobutane ring between adjacent pyrimidines. The pyrimidine dimer blocks replication and transcription.

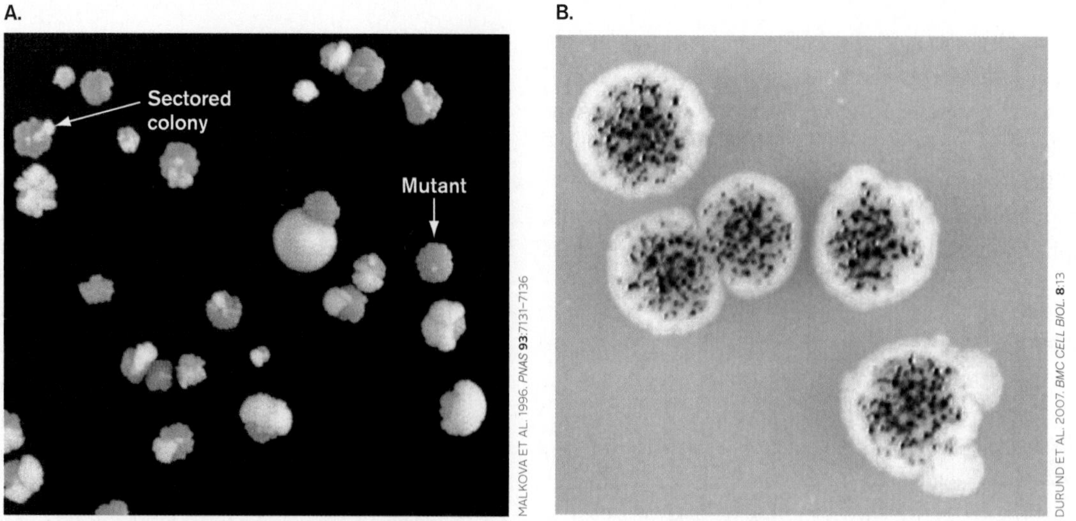
A.
Sectored colony
Mutant
MALKOVA ET AL. 1996. PNAS 93:7131–7136
B.
DURUND ET AL. 2007. BMC CELL BIOL. 8:13

FIGURE 9.24 ■ Sectored colonies of bacteria and yeast. A. Sectored colonies of yeast. On this medium, normal yeast cells form white colonies. The occasional mutant cell that develops as the colony grows will replicate as part of the colony but forms a reddish orange sector. **B.** *E. coli* mutator strain. Cells are lactose-negative mutants that do not ferment lactose and do not turn the indicator blue. But because the strain has a mutation in one of the mutator genes, there is a high rate of reversion to the lactose fermenter strain, which shows up as blue papillae on the white colony.

Mutation Rates and Frequencies

The **mutation rate** for a given gene is often defined as the number of mutations formed per cell division. Since the number of cells in a typical culture starts out very low and grows to a very high density, the number of cell divisions is approximately equal to the number of cells in the culture. For example, to go from one cell to eight cells requires seven cell divisions (not generations). Mutation rate can be estimated according to the formula $a = m$ per number of cell divisions, where a is the mutation rate and m is the number of mutations that occur as the number of cells increases from N_0 to N. Thus, a culture growing from 10^3 (1,000) cells to 10^7 (10,000,000) cells requires approximately 10^7 cell divisions (10,000,000 – 1,000 = 9,999,000 cell divisions). If 10 mutants developed in a specific gene, then the mutation rate would be approximately $10/10^7 = 1 \times 10^{-6}$. More on mutation rate can be found in **eTopic 9.4**.

Note: The difference between the number of cell divisions and the number of generations can be appreciated by looking at the replication of 1,000 cells to give 2,000 cells. There are 1,000 cell divisions, but only one generation. Although you will see the phrase "mutation rate per generation," "per generation" in that context really means per cell division.

A less complicated calculation is **mutation frequency**. Mutation frequency tells us how many mutant cells are present in a population. It is calculated simply as the ratio of mutants per total cells in the population.

The colonies in **Figure 9.24** graphically illustrate the result of a high mutation rate. Single yeast cells placed on an agar surface multiplied to form the colonies shown (**Fig. 9.24A**). As the yeast cell replicates to form a colony, a spontaneous mutant may develop at some point. The mutant yeast is unable to metabolize a particular compound, which then accumulates and turns the cell red. As the mutant cell increases in number alongside normal cells, a sectored colony forms.

In the case of the bacteria, the medium contains an indicator that turns blue if the organism can ferment lactose (**Fig. 9.24B**). This tool helps scientists follow the development of mutations in the gene needed to ferment lactose. Because the original cell in this case was Lac^- (due to a mutation in that gene), most of the colony is white. However, the strain is a mutator strain defective in a DNA repair gene (Section 9.5). This defect causes a high rate of reversion to Lac^+. Each cell in the growing colony that reverted to become a lactose fermenter produced offspring that can turn the indicator blue. These mutants appear as blue papillae (nipple-like outgrowths) on the surface of the white colony. Each papilla represents a separate mutational event.

Thought Question

9.7 You have been asked to calculate the mutation rate for a gene in *Salmonella enterica*, so you dilute an 18-hour broth culture to 1×10^4 cells/ml and let it grow to 1×10^9 cells/ml. At that point you dilute and plate cells on the appropriate agar medium and estimate that there were 10,000,000 mutants in the culture. What you don't know is that the original dilution of 1×10^4 cells/ml already had 10 mutants defective in that gene. Does that affect your estimate of the mutation rate? If so, then how?

Mutagens Can Be Identified Using Bacterial "Guinea Pigs"

In a world where we are continually exposed to new chemicals, it is important to determine which chemicals are potential mutagens. Bruce Ames and his colleagues invented a simple alternative that uses bacteria as a rapid initial screen, for which Ames received the National Medal of Science in 1998. The method, called the Ames test, relies on a mutant of *Salmonella enterica* that is defective in the *hisG* gene, whose product is involved in histidine biosynthesis (**Fig. 9.25**). The *hisG* mutant cannot grow on a minimal defined medium lacking histidine. However, if a reversion mutation occurs in the *hisG* gene and restores the gene to its original functional state (phenotypic reversion), the new mutant cell will form a colony even in the absence of histidine. This method is called a reversion test.

Ames used this *his* reversion test to screen compounds for potential mutagenicity. If a chemical can mutate one gene (*hisG*), it can potentially affect any gene. As shown in **Figure 9.25**, a mutagen-containing disk is placed in the middle of an agar plate spread with the original *his* mutant. As the mutagen diffuses into the medium and causes reversion mutations, colonies start to appear, forming a ring around the disk. The longer the plate is incubated, the more revertants are produced.

The plate reversion test goes only so far, however. The technique will not expose "cloaked" mutagens that require processing by mammalian enzymes to become mutagenic. In humans, many nonmutagenic chemicals can be transformed into potent mutagens by the liver. The liver is the chief organ for detoxifying the body—a task that liver enzymes accomplish by chemically modifying foreign substances. To identify "cloaked" mutagens, the basic Ames technique was modified by treating a potential mutagen with a rat liver extract before applying the mixture to bacteria for the *his* reversion test (**Fig. 9.26**). If the enzymes of the liver convert the compound to a mutagen (they do this accidentally, not on purpose), then His^+ revertant colonies will be seen. This method accelerates the process of drug discovery by providing an inexpensive preliminary screen

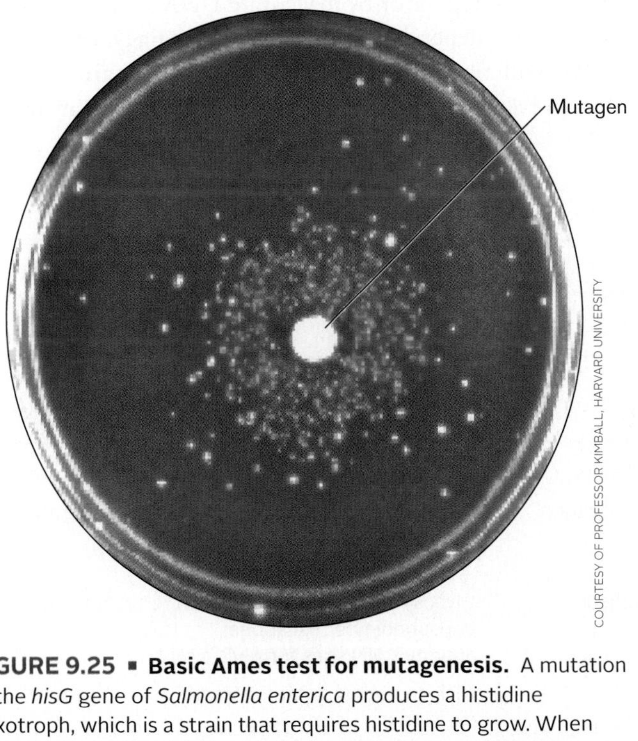
Mutagen
COURTESY OF PROFESSOR KIMBALL, HARVARD UNIVERSITY

FIGURE 9.25 ■ Basic Ames test for mutagenesis. A mutation in the *hisG* gene of *Salmonella enterica* produces a histidine auxotroph, which is a strain that requires histidine to grow. When plated onto medium lacking histidine, only a few spontaneous mutations (reversions) occur that reverse the mutation so that the cell can make histidine again. These cells form colonies on medium lacking histidine. A mutagen in a paper disk placed in the center of the plate will diffuse and, over time, increase the number of prototrophic revertants seen orbiting the disk.

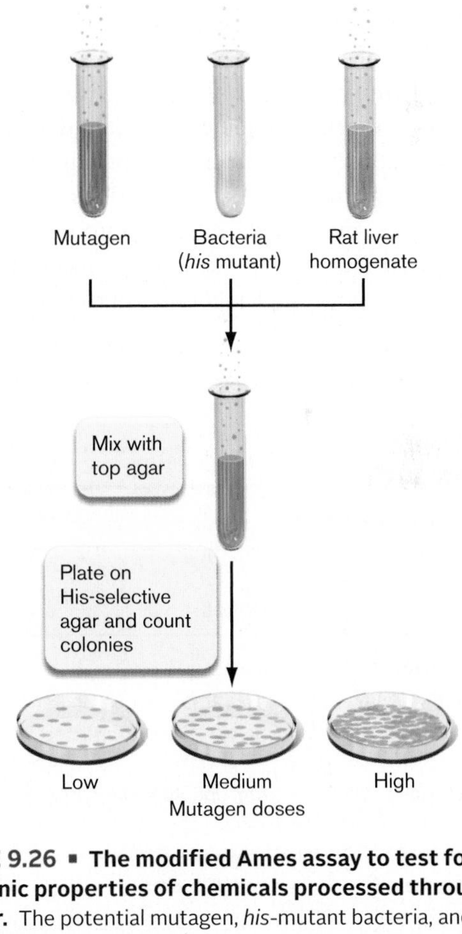
Mutagen
Bacteria (*his* mutant)
Rat liver homogenate
Mix with top agar
Plate on His-selective agar and count colonies
Low
Medium
High
Mutagen doses

FIGURE 9.26 ■ The modified Ames assay to test for the mutagenic properties of chemicals processed through the liver. The potential mutagen, *his*-mutant bacteria, and liver homogenate are combined and mixed with agar. The combination is poured into a petri plate. If the liver extract enzymes act on the test compound and the metabolites produced are mutagenic, then increasing numbers of His^+ revertants will be observed with increasing doses of mutagen. If the compound is not mutagenic, few relevant colonies will be seen on any plate.

for weeding out mutagenic chemicals before more expensive animal testing is undertaken.

More recent assays for mutagenicity involve transgenic mice. Mice are engineered to have the bacterial *lacZ* gene, for instance, inserted into their chromosomes. The *lacZ* gene encodes an enzyme that is easily assayed, so it is useful as a reporter for mutagenic activity. This reporter gene will be distributed throughout the mouse in all organs. A potential mutagen is administered to the mouse, and after some time the mouse is sacrificed, its organs are removed, and the DNA is extracted and examined for mutations arising in the *lacZ* gene. This approach allows scientists to track where the mutagen is distributed in the mouse and to determine whether certain organs convert harmless precursor chemicals into dangerous mutagenic compounds.

To Summarize

- **A mutation** is any heritable change in DNA sequence, regardless of whether there is a change in gene function.
- **Genotype** reflects the genetic makeup of an organism, whereas **phenotype** reflects its physical traits.
- **Classes of mutations** include silent mutations, missense mutations, nonsense mutations, point mutations, insertions, deletions, and frameshift mutations.
- **Spontaneous mutations** reflect tautomeric shifts in DNA nucleotides during replication, accidental incorporation of noncomplementary nucleotides during replication, or "natural" levels of chemical or physical (irradiation) mutagens in the environment.
- **Chemical mutagens** can alter purine and pyrimidine structure and change base-pairing properties.
- **The mutagenicity** of a chemical can be assessed by its effect on bacterial cultures.

9.5

DNA Repair

Microorganisms are equipped with a variety of molecular tools that repair DNA damage before the damage becomes a heritable mutation (see **Table 9.3**). The type of repair mechanism used (and when it is used) depends on two things: the type of mutation needing repair and the extent of damage involved. Some mechanisms excise whole fragments of DNA that contain damaged bases; others precisely excise the damaged bases or directly reverse the damage. After extensive damage, special "emergency" DNA polymerases are expressed that sacrifice replication accuracy to rescue the damaged genome. Whether damage is introduced by mutagens or by inaccurate DNA synthesis, microbial survival depends on the ability to repair DNA.

We will first discuss the **error-proof repair** pathways that prevent mutations. These include methyl mismatch

TABLE 9.3

Types of DNA repair.

System	Genes	Mutations recognized	Repair mechanism	Result
Photoreactivation	*phrB*	Pyrimidine dimers	Cyclobutane ring cleaved	Accurate
Nucleotide excision	*uvrABCD*	Helical destabilization (e.g., pyrimidine dimers)	Patch of nucleotides excised	Accurate
Base excision	*fpg, ung, tag, mutY, nfo*	Various modified bases	Glycosylases remove base from phosphodiester backbone; apurinic (AP) sites formed	Accurate
Methyl mismatch	*mutHSL, dam*	Transitions, transversions	Nick on nonmethylated strand, excision of nucleotides	Accurate
Recombination	*recA*	Single-strand gaps	Recombination	Accurate
Translesion bypass synthesis	*umuDC*	Gaps	Part of SOS system	Error-prone (generates mutations)

repair, photoreactivation, nucleotide excision repair, base excision repair, and recombinational repair. We will then discuss **error-prone repair** pathways. These pathways risk introducing mutations and operate only when damage is so severe that the cell has no other choice but to die.

Error-Proof Repair Pathways

Methyl mismatch repair. What happens if DNA polymerase simply makes a mistake and incorporates a normal but incorrect base? Although proofreading functions in DNA polymerases are evolutionarily designed to prevent the misincorporation of bases during replication, misincorporation does occur. Curiously, far fewer mutations arise than one would predict from the inherent error rate of DNA polymerase III (approximately one mistake per 10^8 bases synthesized, after proofreading). However, the mutation rate in a live cell is actually only 10^{-10} per base pair replicated. How can a cell repair a mutation after it has been introduced during replication? One method is **methyl mismatch repair**, which is based on recognizing the methylation pattern in DNA bases. As discussed in Section 7.3, many bacteria tag their parental DNA by methylating it at specific sites. In *E. coli*, for example, deoxyadenosine methylase (Dam) methylates the palindromic sequence GATC to produce $GA^{ME}TC$. The Dam methylase does this soon, but not immediately, after replication of a DNA sequence.

When DNA polymerase misincorporates a base during replication, a mismatch forms between the incorrect base in the newly synthesized but unmethylated strand and the correct base residing in the parental, methylated strand. Methyl-directed mismatch repair enzymes (MutS, MutL, and MutH) bind to the mismatch. MutS first binds to the mismatch and recruits MutL and MutH. MutL recognizes the methylated strand ($GA^{ME}TC$) and brings it in a loop to meet MutS and MutH; and then MutH cleaves the unmethylated strand containing the mutation, near the GATC sequence (**Fig. 9.27**). A DNA helicase called UvrD then unwinds the cleaved strand, exposing it to a variety of exonucleases. The result is a gap that is filled in by DNA polymerase I (Pol I) and sealed by DNA ligase.

The methyl-directed mismatch repair proteins (and genes) are called Mut (and *mut*) because a high mutation rate results in strains that are defective in one of these proteins. A bacterial strain with a high mutation rate is called a **mutator strain**.

Note: Not all DNA methylations signal methyl mismatch repair. DNA methylation by restriction-modification systems, for example, prevents cleavage by DNA restriction endonucleases, but is not used to designate parental DNA after replication.

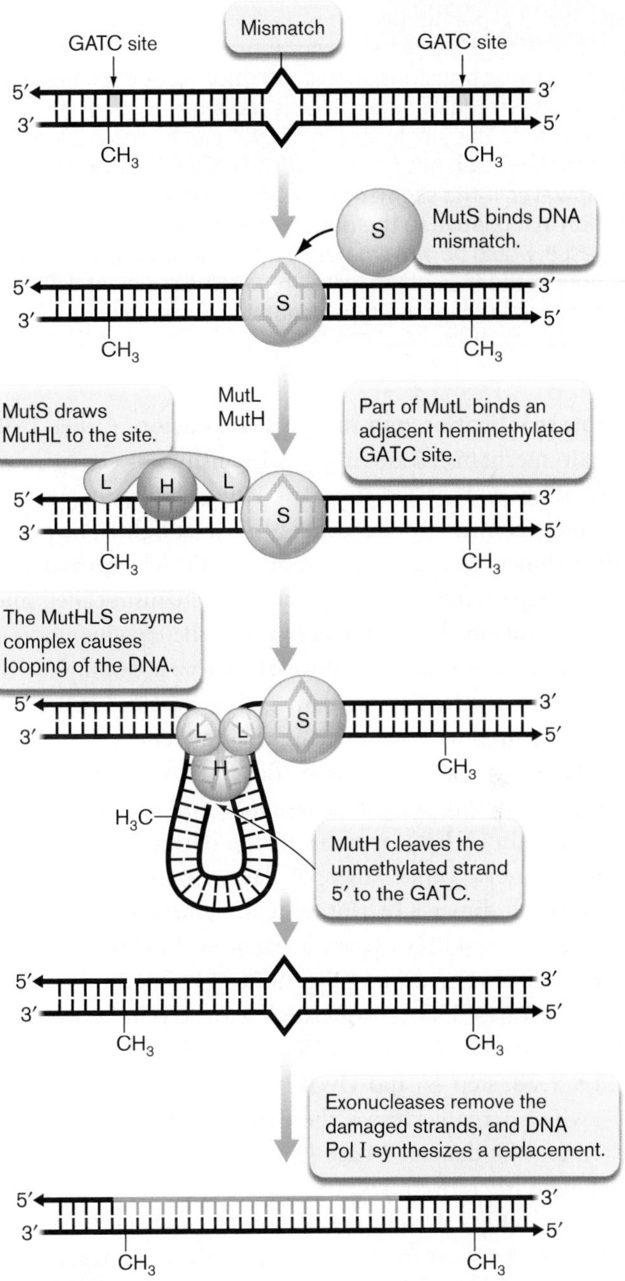
Mismatch
GATC site
GATC site
5′
3′
3′
5′
CH3
CH3
MutS binds DNA mismatch.
S
MutL
MutH
MutS draws MutHL to the site.
Part of MutL binds an adjacent hemimethylated GATC site.
L
H
L
The MutHLS enzyme complex causes looping of the DNA.
H3C
MutH cleaves the unmethylated strand 5′ to the GATC.
Exonucleases remove the damaged strands, and DNA Pol I synthesizes a replacement.

FIGURE 9.27 ▪ Methyl mismatch repair. The bacterial cell, in this case *E. coli,* can use specific methylations on DNA to recognize parental DNA strands for preferential DNA repair. Newly replicated strands are not immediately methylated. So, when a mismatch is found, the mismatch repair system views the newly synthesized strand as suspect and replaces the section of unmethylated DNA encompassing the mismatch. The steps following cleavage are similar to what is shown in Figure 9.28 (steps 7 and 8).

Thought Question

9.8 During transformation in *Streptococcus pneumoniae*, a single strand of donor DNA is transferred into the recipient. This single strand of DNA can be recombined into the recipient's genome at the homologous area. However, if the segment of recipient DNA had a mutation in it relative to the donor DNA strand, then the result would be a mismatch after recombination. Explain how, in the face of mismatch repair, the recipient cell ever retains the wild-type sequence from the donor.

Photoreactivation and nucleotide excision. Other DNA repair mechanisms do not distinguish between parental and newly synthesized strands. Many microbes in the environment commonly encounter ultraviolet light. The pyrimidine dimers that form as a result of ultraviolet irradiation can be repaired by a light-activated mechanism called **photoreactivation**. In photoreactivation, the enzyme photolyase binds to the dimer and cleaves the cyclobutane ring linking the two adjacent, damaged nucleotides. The damage is repaired without any bases being excised.

While photolyase is specific for pyrimidine dimer repair and requires light, a system called **nucleotide excision repair (NER)** operates in the dark (or light) and is used to excise other kinds of damaged DNA, as well as pyrimidine dimers. In this system, a three-subunit endonuclease (UvrABC) excises a patch of 12–13 nucleotides that includes the dimer (**Fig. 9.28**). The basic mechanism involves UvrA, associated with UvrB, recognizing the damaged base (**Fig. 9.28**, steps 1–3). UvrA is ejected (**Fig. 9.28**, step 4), and UvrB binds UvrC (**Fig. 9.28**, step 5), which actually cleaves the damaged strand at two sites flanking the damaged base (**Fig. 9.28**, step 6). The small fragment is removed, and the gap is repaired by DNA polymerase I (**Fig. 9.28**, steps 7 and 8). Nucleotide excision repair is also used to excise other kinds of damaged DNA besides pyrimidine dimers.

Nucleotide excision repair can be enhanced by ongoing RNA transcription. Bacteria and eukaryotic cells preferentially repair genes that are being transcribed. The preferential repair of transcriptionally active genes makes sense, since a cell unable to transcribe such a gene because of recent DNA damage would be at a survival disadvantage. An RNA polymerase that encounters a UV dimer or other unrecognizable base on a template DNA strand will stall during transcription. The stalled RNA polymerase is then recognized by a protein that mediates a process called transcription-coupled repair. The transcription-coupled repair protein then snares a nearby UvrAB to begin nucleotide excision repair.

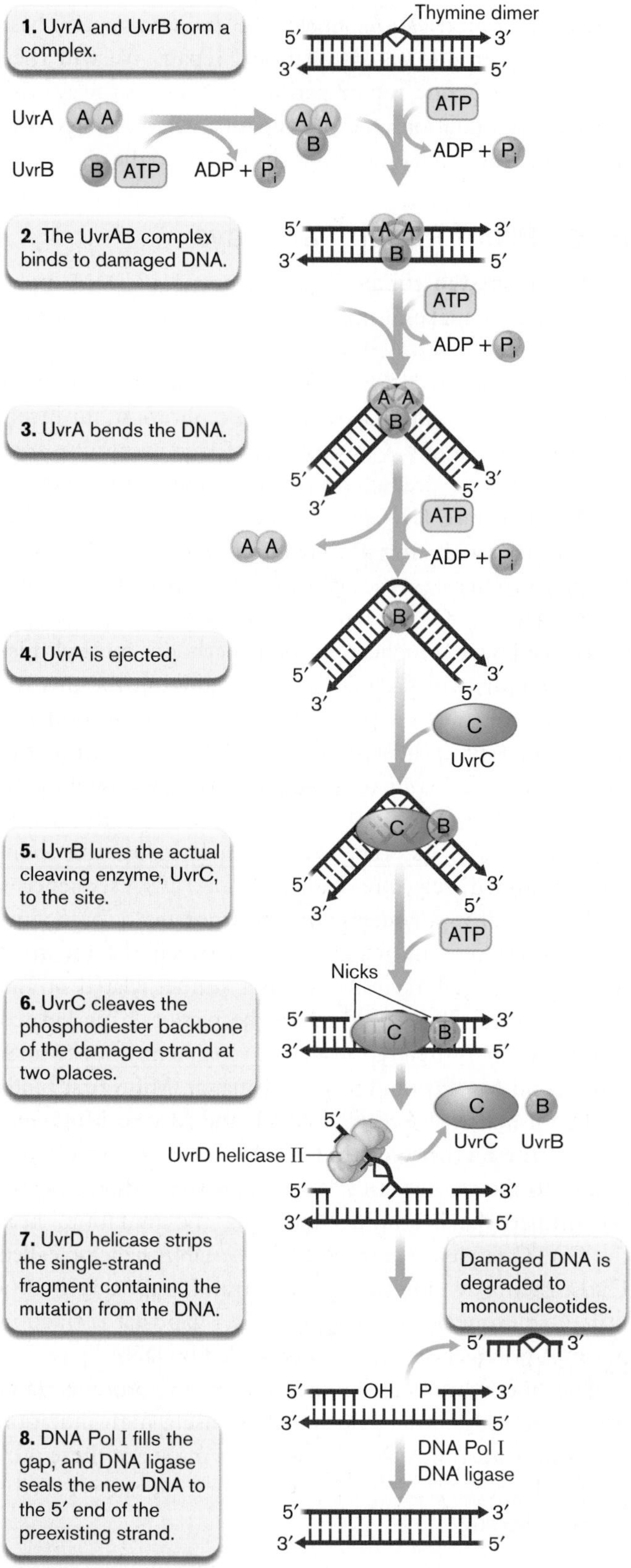
1. UvrA and UvrB form a complex.
Thymine dimer
UvrA
UvrB
ATP
ADP + Pi
2. The UvrAB complex binds to damaged DNA.
3. UvrA bends the DNA.
4. UvrA is ejected.
UvrC
5. UvrB lures the actual cleaving enzyme, UvrC, to the site.
Nicks
6. UvrC cleaves the phosphodiester backbone of the damaged strand at two places.
UvrD helicase II
7. UvrD helicase strips the single-strand fragment containing the mutation from the DNA.
Damaged DNA is degraded to mononucleotides.
OH
P
8. DNA Pol I fills the gap, and DNA ligase seals the new DNA to the 5′ end of the preexisting strand.
DNA Pol I
DNA ligase

FIGURE 9.28 ■ Nucleotide excision repair. This repair mechanism cuts out a single-stranded segment of DNA containing the mutation and uses DNA replication to repair the gap.

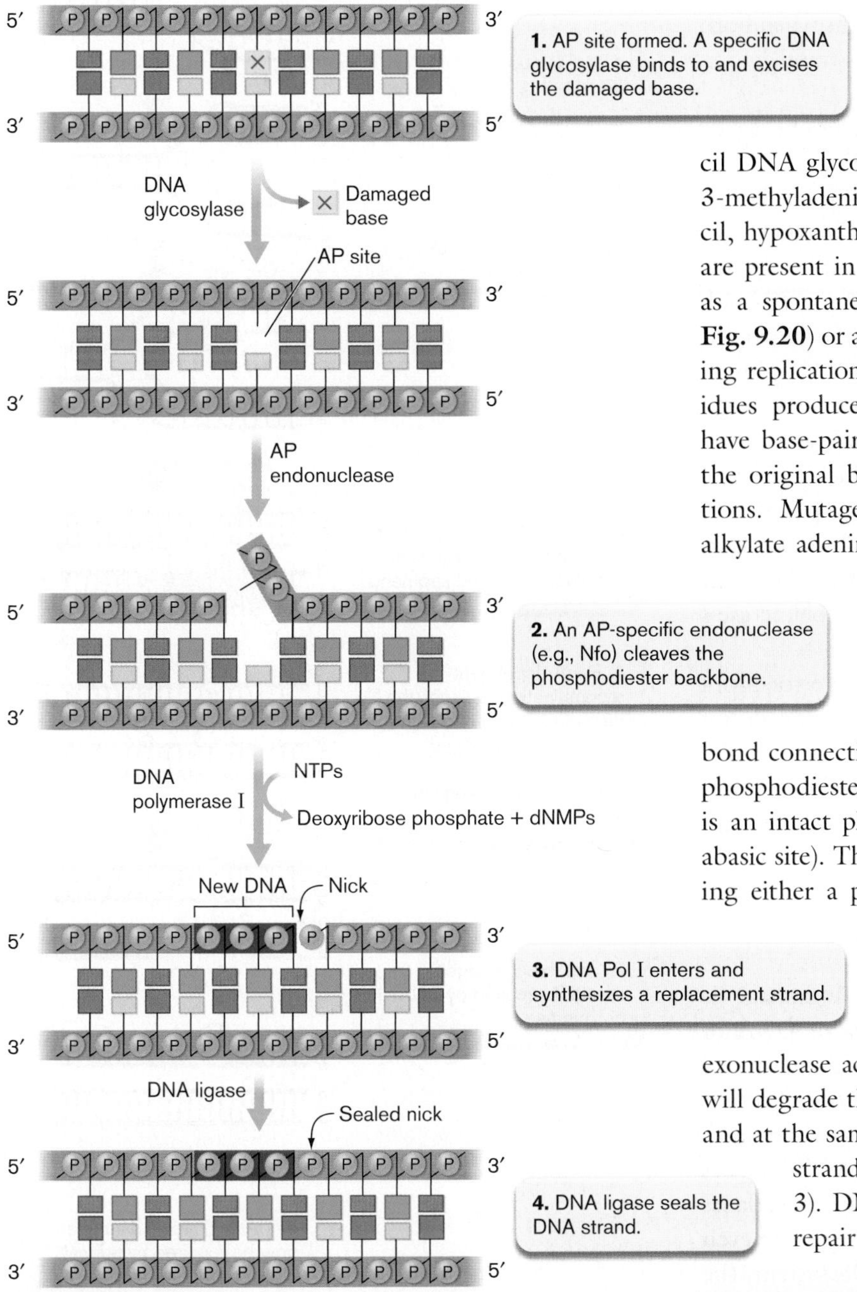
1. AP site formed. A specific DNA glycosylase binds to and excises the damaged base.
DNA glycosylase
Damaged base
AP site
AP endonuclease
2. An AP-specific endonuclease (e.g., Nfo) cleaves the phosphodiester backbone.
DNA polymerase I
NTPs
Deoxyribose phosphate + dNMPs
New DNA
Nick
3. DNA Pol I enters and synthesizes a replacement strand.
DNA ligase
Sealed nick
4. DNA ligase seals the DNA strand.
5′
3′
P

FIGURE 9.29 ■ Base excision repair. Specialized enzymes can recognize specific damaged bases and remove them from DNA without breaking the phosphodiester backbone. The result is an abasic site (AP = apurinic or apyrimidinic) that can be recognized and cleaved by a specific AP endonuclease. This AP site allows DNA polymerase I to synthesize a replacement strand containing the proper base. dNMPs = deoxyribonucleoside monophosphates; NTPs = nucleoside triphosphates.

Base excision repair snips damaged bases from DNA. Although nucleotide excision repair can recognize and repair several types of damage, it does not work in all instances. Another error-proof process, known as **base excision repair (BER)**, employs a battery of glycosylase enzymes that can recognize and clip certain damaged bases from the phosphodiester backbone. Uracil DNA glycosylase, hypoxanthine DNA glycosylase, and 3-methyladenine glycosylase recognize, respectively, uracil, hypoxanthine, and 3-methyladenine when these bases are present in DNA. Uracil can be found in DNA either as a spontaneous deamination product of cytosine (see **Fig. 9.20**) or as a result of inappropriate incorporation during replication. Spontaneous deamination of adenine residues produces hypoxanthine. Uracil and hypoxanthine have base-pairing properties very different from those of the original bases, so their formation can lead to mutations. Mutagens such as methyl methanesulfonate can alkylate adenine to produce the adduct 3-methyladenine. A 3-methyladenine residue will totally block DNA replication, making this a lethal form of damage. Consequently, repairing these mutations is vital for survival.

The glycosylases just noted cleave the bond connecting the damaged base to deoxyribose in the phosphodiester backbone (**Fig. 9.29**, step 1). The result is an intact phosphodiester backbone missing a base (an abasic site). The site is called an **AP site** because it is missing either a purine (apurinic) or a pyrimidine (apyrimidinic). The next step in base excision repair involves **AP endonucleases** that specifically cleave the phosphodiester backbone at AP sites (**Fig. 9.29**, step 2). The 5′-to-3′ exonuclease activity of the gap-filling DNA polymerase I will degrade the cleaved strand downstream of the AP site and at the same time synthesize in its stead a replacement strand containing the proper base (**Fig. 9.29**, step 3). DNA ligase seals the remaining nick, and the repair process is complete (**Fig. 9.29**, step 4).

Thought Question

9.9 It has been reported that hypermutable bacterial strains are overrepresented in clinical isolates. Out of 500 isolates of *Haemophilus influenzae*, for example, 2%–3% were mutator strains having mutation rates 100–1,000 times higher than lab reference strains. Why might mutator strains be beneficial to pathogens?

All of the repair processes described thus far (mismatch repair, photoreactivation, nucleotide excision repair, and base excision repair) are error-proof pathways that rely on the presence of a good template strand opposite the damaged strand. Replication of the template strand is used to replace the damaged bases accurately. But what happens when both strands are damaged or when there is only one strand?

Recombinational repair. **Figure 9.30** illustrates one repair mechanism that engages when DNA replication takes place before a UV-induced dimer can be excised by nucleotide excision repair. Because DNA polymerase III (the main polymerase for chromosome replication) cannot decode a dimer, it skips over the damaged area and restarts with a new RNA primer, leaving a gap opposite the dimer. At this point, the Uvr complex cannot excise the UV dimer, because chromosome integrity would be lost. However, the RecA protein involved with recombination can bind to the gap and initiate a genetic exchange in which a piece of the undamaged, properly replicated strand is spliced into the gap. This is called **recombinational repair** and occurs mostly near replication forks. Once the gap is filled, the Uvr complex can remove the dimer and DNA polymerase I will fill the gap in the other strand.

The recombinational repair system is also an error-proof repair pathway. Note that the gap formed in the donor strand is easily replaced by the gap-filling DNA polymerase I. Recombinational repair is not limited to pyrimidine dimers. It will work on any damage that causes gaps during replication.

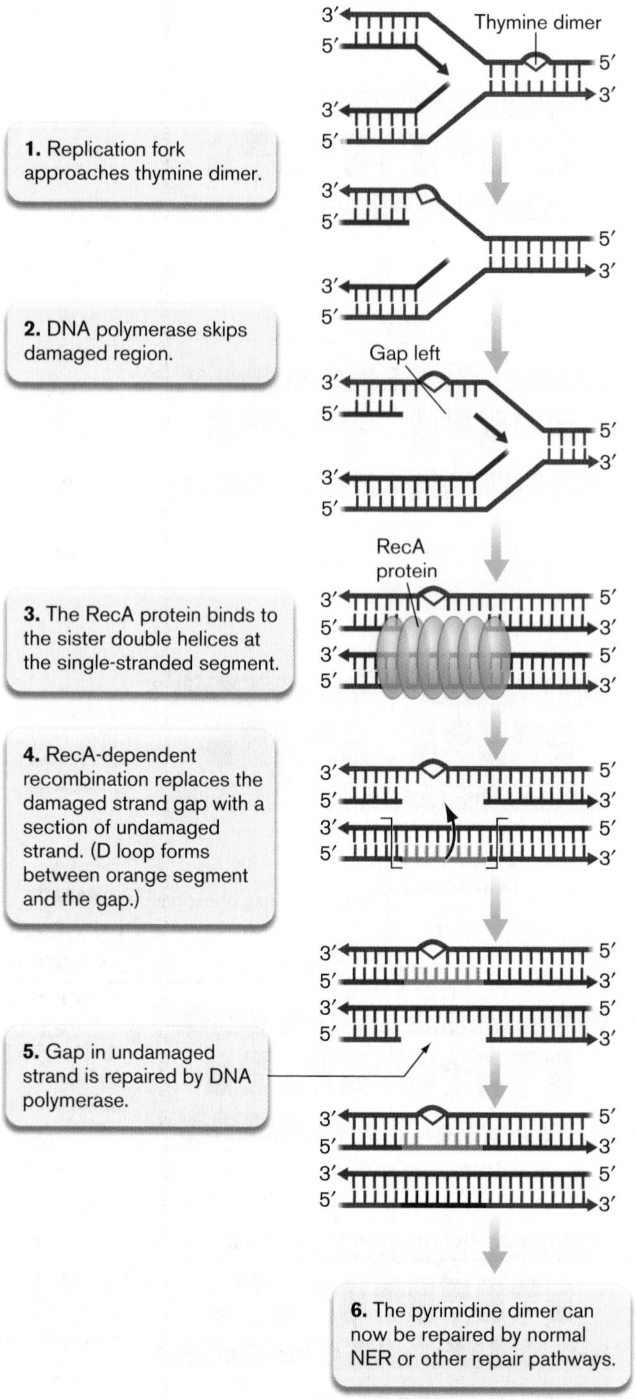
Thymine dimer
1. Replication fork approaches thymine dimer.
2. DNA polymerase skips damaged region.
Gap left
RecA protein
3. The RecA protein binds to the sister double helices at the single-stranded segment.
4. RecA-dependent recombination replaces the damaged strand gap with a section of undamaged strand. (D loop forms between orange segment and the gap.)
5. Gap in undamaged strand is repaired by DNA polymerase.
6. The pyrimidine dimer can now be repaired by normal NER or other repair pathways.

FIGURE 9.30 ▪ Recombinational repair. Recombination can be used to repair DNA damage of replicating DNA when one daughter strand is undamaged. A single-stranded segment of the undamaged daughter strand can be used to replace a gap in the damaged daughter strand.

Error-Prone DNA Repair

SOS ("Save Our Ship") repair. When DNA damage is extensive, repair strategies that excise pieces of DNA or that require recombination will destroy the circularity of the chromosome and kill the cell. To save the chromosome, and itself, the cell must take more drastic measures and induce the **SOS response**, a system that introduces mutations into severely damaged DNA. The cell relaxes replication fidelity to maintain a circular chromosome even if incorrect bases are introduced. In the SOS system, the RecA protein senses the extent of DNA damage by monitoring the level of single-stranded DNA produced. For example, excessive ultraviolet irradiation produces numerous ssDNA gaps because DNA polymerase cannot replicate through pyrimidine dimers.

Formation of RecA filaments (see **Fig. 9.16**) on ssDNA activates a second function of RecA, called coprotease activity, that stimulates autodigestion of the LexA repressor, a protein that normally prevents SOS activation. LexA protein binds to the promoters of DNA repair genes and prevents their transcription (**Fig. 9.31A** and **B**). Cleavage of LexA then unleashes the production of DNA repair enzymes. Among these enzymes are two "sloppy" DNA polymerases that lack proofreading activity: UmuDC (also called DNA polymerase V, or Pol V) and DinB (also called DNA polymerase IV, or Pol IV) (**Fig. 9.31C**). The UmuDC enzyme is perfect for replicating through damaged bases (a process called translesion replication) because it sacrifices accuracy for continuity. When the enzyme encounters an undecipherable damaged base, it will insert whatever nucleotide is available. The result, of course, will be numerous

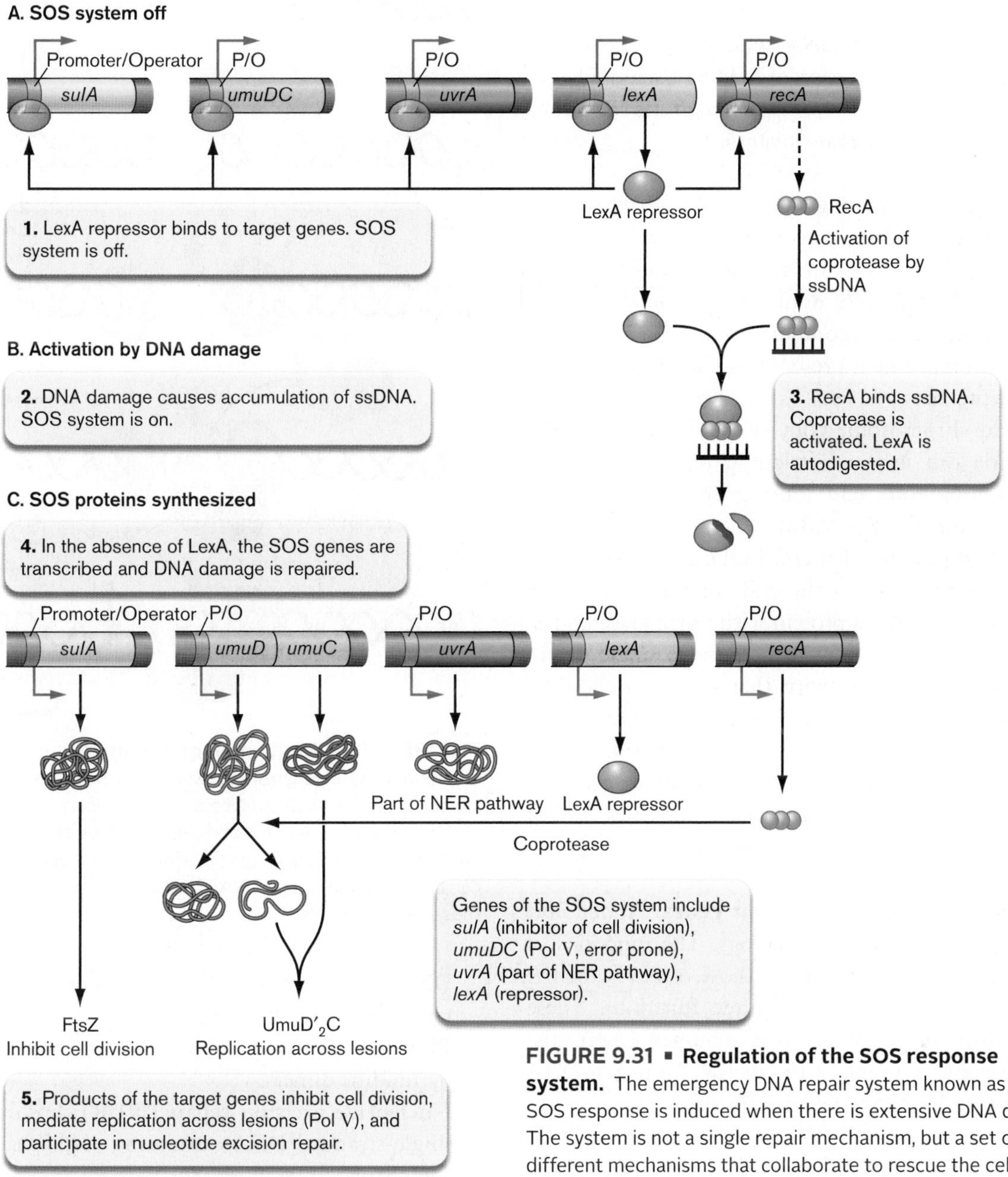
A. SOS system off
Promoter/Operator
P/O
sulA
umuDC
uvrA
lexA
recA
LexA repressor
RecA
1. LexA repressor binds to target genes. SOS system is off.
Activation of coprotease by ssDNA
B. Activation by DNA damage
2. DNA damage causes accumulation of ssDNA. SOS system is on.
3. RecA binds ssDNA. Coprotease is activated. LexA is autodigested.
C. SOS proteins synthesized
4. In the absence of LexA, the SOS genes are transcribed and DNA damage is repaired.
umuD
umuC
Part of NER pathway
LexA repressor
Coprotease
Genes of the SOS system include sulA (inhibitor of cell division), umuDC (Pol V, error prone), uvrA (part of NER pathway), lexA (repressor).
FtsZ
Inhibit cell division
UmuD′2C
Replication across lesions
5. Products of the target genes inhibit cell division, mediate replication across lesions (Pol V), and participate in nucleotide excision repair.

FIGURE 9.31 ■ Regulation of the SOS response system. The emergency DNA repair system known as the SOS response is induced when there is extensive DNA damage. The system is not a single repair mechanism, but a set of different mechanisms that collaborate to rescue the cell.

permanent mutations, but the benefit is that the cell has a chance to live if it can tolerate the mutations. This high mutation frequency is a calculated risk. The cell really has no other option, however, because housekeeping DNA polymerases like Pol III cannot move through damaged areas of chromosomal DNA. The choice really is "mutate or die." (Note that the term "housekeeping" is applied to proteins or enzymes that keep the cell running at all times.)

When the cell is severely compromised by mutation, it temporarily stops dividing. Like a pit stop during an auto race, this pause in cell division allows time for repair enzymes to fix the damage. The product of another SOS-regulated gene, called *sulA,* causes this pause by binding to the FtsZ cell division protein, keeping it from initiating cell division (see Section 3.5). Once the damage has been repaired, RecA coprotease is inactivated and LexA repressor accumulates, turning off all the SOS genes, including *sulA*. The lingering SulA protein is then degraded by a protease called Lon. The protein is called Lon because when it is missing, SulA inappropriately accumulates, inhibits cell division, and produces long filamentous cells.

The coprotease activity of RecA can also activate prophages by stimulating autocleavage of proteins that prevent phage replication. This phenomenon is used by some organisms to displace competitors in an environmental niche. For instance, *Streptococcus pneumoniae,* a cause of pneumonia, displaces *Staphylococcus aureus* in the human nasopharynx by producing hydrogen peroxide. Hydrogen peroxide damages *S. aureus* DNA, which activates the SOS response and triggers replication of resident bacteriophages that lyse the cell. The loss of *Staphylococcus aureus* means more room for *Streptococcus pneumoniae*.

Nonhomologous end joining. Double-strand breaks in DNA are particularly dangerous to a cell because the loss of chromosome integrity is immediate. In rapidly growing bacteria such as *E. coli*, double-strand breaks can be repaired by homologous recombination as long as another copy of the chromosome is present to guide the repair. But in slow-growing bacteria such as *Mycobacterium tuberculosis*, a second chromosome copy is usually not available. These bacteria can use an intriguing repair mechanism called **nonhomologous end joining (NHEJ),** which is also used by mammals (**Fig. 9.32**).

Two bacterial proteins, Ku and LigD, carry out NHEJ repair. Ku protein binds to the ends of a double-strand break and recruits LigD, a protein with polymerase and 3′ exonuclease activities that fill in or remove single-strand overhangs, and a ligase activity that joins two double-strand breaks. Because the system does not require homology, the system can be error prone, causing the loss or addition of a few nucleotides at the break site or even the joining of two previously unlinked DNA molecules. NHEJ has been reported in *Mycobacterium* and *Bacillus* species.

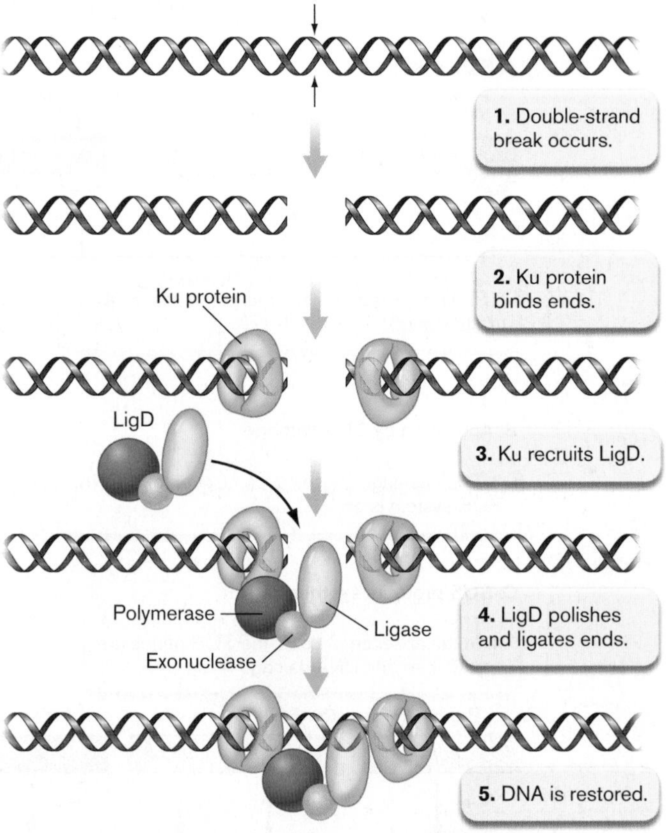
1. Double-strand break occurs.
2. Ku protein binds ends.
Ku protein
LigD
3. Ku recruits LigD.
Polymerase
Ligase
Exonuclease
4. LigD polishes and ligates ends.
5. DNA is restored.

FIGURE 9.32 ▪ Nonhomologous end joining. NHEJ, a common repair mechanism in eukaryotes, has been documented in some bacteria, such as *Mycobacterium tuberculosis.* These pathogens survive within macrophages where they are exposed to antibacterial compounds (nitric oxide and hydrogen peroxide) that can cause double-strand DNA breaks. NHEJ mechanisms can repair those dsDNA breaks.

Human homologs of bacterial repair genes. About 30% of *E. coli* genes have human homologs. The functions of the human genes may be similar to those of *E. coli,* or the human genes may have newly acquired functions. These genes often turn out to play key roles in human genetic diseases. An example is *mutS*, of the methyl mismatch DNA repair system. In humans, the ancestor of *mutS* evolved to have multiple repair functions, as well as to participate in postmeiotic segregation. Defects in this repair mechanism have been linked to increased risk of certain colon cancers. Numerous other human genetic diseases are caused by mutations in homologs of bacterial repair genes. For example, defective excision repair genes in humans cause xeroderma pigmentosum, a disease that causes blindness and skin cancers. A deficiency in transcription-coupled repair causes Cockayne syndrome in humans, a devastating neurodegenerative disease.

To Summarize

- **DNA repair pathways** in microorganisms include error-proof and error-prone mechanisms.
- **Methyl mismatch repair** uses methylation of the parental DNA strand to distinguish it from newly replicated DNA. The premise is that the parental strand will contain the proper DNA sequence.
- **Photoreactivation** cleaves the cyclobutane rings of pyrimidine dimers.
- **Nucleotide excision repair (NER)** clips out a patch of single-stranded DNA containing certain types of damaged bases.
- **Base excision repair (BER)** excises structurally altered bases without cleaving the phosphodiester backbone. The resulting AP (apurinic or apyrimidinic) site is targeted by AP nucleases.
- **Recombinational repair** takes place at a replication fork. A "good" strand of DNA is used to replace a homologous damaged strand.
- **Extensive DNA damage leads to induction of the SOS response**, producing increased levels of the error-proof repair systems, as well as error-prone translesion bypass DNA polymerases that introduce mutations.
- **Nonhomologous end joining (NHEJ) mechanisms** repair double-strand DNA breaks in some, but not all, bacteria.

9.6

Mobile Genetic Elements

In 1948, Barbara McClintock (1902–1992) noticed that certain genetic traits of corn defied the laws of Mendelian inheritance. The genes encoding these traits, sometimes called "jumping genes," seemed to hop from one chromosome to another. Although McClintock's theories were provocative at the time, we now know that these types of genes, referred to as **transposable elements**, exist in virtually all life-forms and can move both within and between chromosomes.

Nancy Kleckner, Russel Chan, Bik-Kwoon Tye, and David Botstein described the presence of transposable elements in bacteria in 1975 (**Fig. 9.33**). In this classic study, a transposable element carrying a tetracycline (Tc) resistance gene was found inserted into the P22 phage genome. P22 is a lysogenic phage that infects *Salmonella*. Recall that a lysogenic phage can integrate into the host chromosome and remain dormant for long periods of time.

The experiment used a defective P22 phage (carrying the Tc resistance gene) that could replicate in only certain strains of *Salmonella*. When this phage was used to infect nonpermissive strains of *Salmonella* (strains that would not allow the phage to grow or lysogenize), the *Salmonella* still became tetracycline resistant at a high frequency. The simple explanation was that the tetracycline resistance (Tc^R) element translocated ("jumped") out of the P22 DNA and into one of a number of locations on the *Salmonella* chromosome, sometimes inserting within a structural gene, thereby generating a mutation and phenotypic change. Today we know that these mobile DNA elements have contributed to remodeling genomes during the evolution of all species.

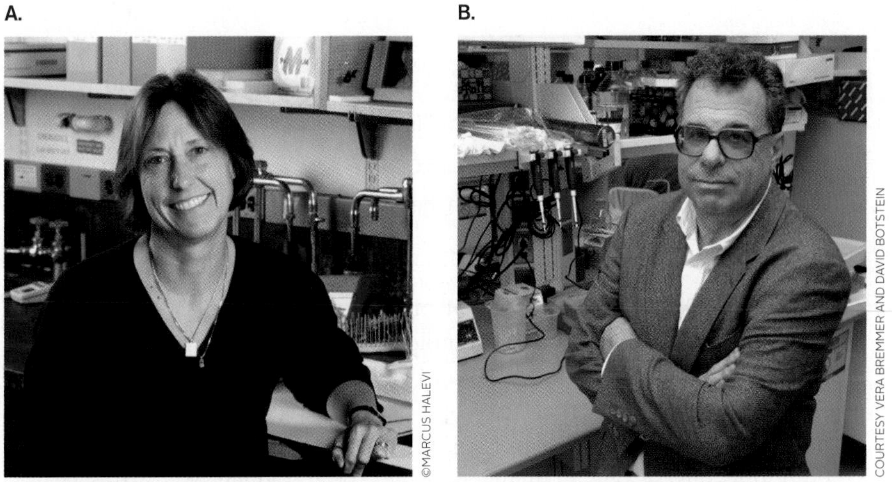
A.
B.
©MARCUS HALEVI
COURTESY VERA BREMMER AND DAVID BOTSTEIN

FIGURE 9.33 ▪ Two of the discoverers of bacterial transposition. A. Nancy Kleckner, Harvard University. **B.** David Botstein, Princeton University.

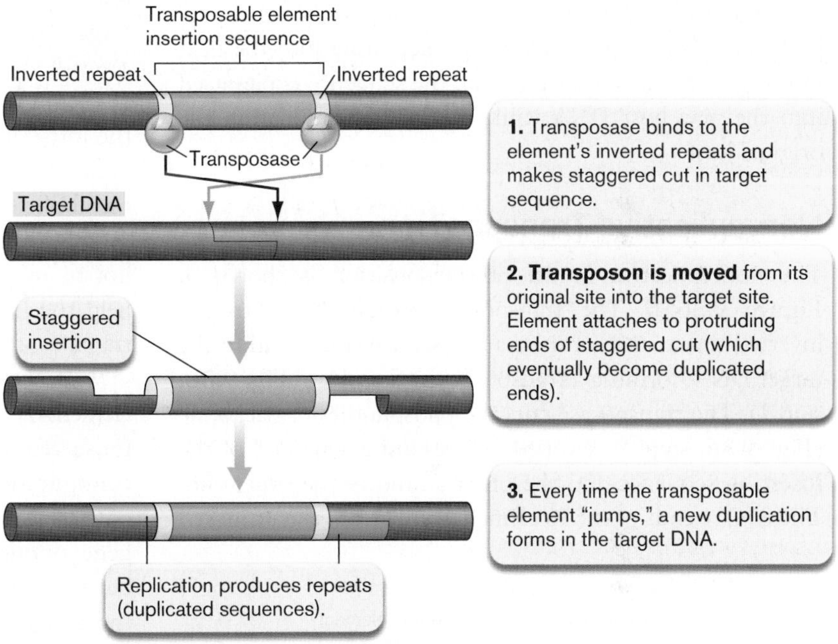
Transposable element insertion sequence
Inverted repeat
Inverted repeat
Transposase
Target DNA
1. Transposase binds to the element's inverted repeats and makes staggered cut in target sequence.
Staggered insertion
2. Transposon is moved from its original site into the target site. Element attaches to protruding ends of staggered cut (which eventually become duplicated ends).
3. Every time the transposable element "jumps," a new duplication forms in the target DNA.
Replication produces repeats (duplicated sequences).

FIGURE 9.34 ▪ Basic transposition and the origin of target site duplication. Enzymes that catalyze transposition generate duplications in the target site by ligating the ends of the insertion element to the long ends of a staggered cut at the target DNA site.

Note: The symbol :: is a convention representing insertion. The element appearing after the two colons has been inserted into the element noted in front of the two colons. For example, *proA*::Tn*10* means that Tn*10* (a transposable element) has been inserted into the *proA* gene involved in proline synthesis.

Transposable elements are not autonomous; unlike plasmids, they are incapable of existing outside of a larger DNA molecule. They exist only as hitchhikers integrated into some other DNA molecule (**Fig. 9.34**). All transposable elements include a gene encoding a **transposase**, an enzyme that catalyzes the transfer or copying of the element from one DNA molecule into another. A simple transposable element (700–1,500 bp) called an **insertion sequence (IS)** consists of a transposase gene flanked by short inverted-repeat sequences that are targets of the transposase (see **Fig. 9.34**).

Note: An inverted repeat is a DNA sequence identical to another downstream sequence. The repeats have reversed sequences and are separated from each other by intervening sequences:

5′-AATCGAT . . . . ATCGATT-3′

When no nucleotides intervene between the inverted sequences, the structure is called a palindrome.

Transposition is the process of moving a transposable element within or between DNA molecules. During transposition, a short DNA sequence on the target DNA molecule is duplicated so that one copy of the sequence will flank each end of the element (**Fig. 9.34**). The transposase randomly selects one of many possible target sequences where it will move the insertion element.

IS elements transfer by one of two mechanisms—namely, replicative or nonreplicative transposition (**Fig. 9.35**). In nonreplicative transposition, the insertion sequence excises itself out of one host DNA while integrating into the next DNA. In replicative transposition, the sequence copies itself into the new host DNA while a copy remains within the original host.

Nonreplicative Transposition

The nonreplicative model of transposition is shown in **Figure 9.36**. The transposase protein binds to the inverted-repeat ends of the transposable element and to the target DNA, forming a transpososome complex (**Fig. 9.36**, step 1). The transposase cuts the phosphodiester backbone (**Fig. 9.36**, step 2), severing one strand at one end of the insertion sequence and the other strand at the other end. The 3′ OH ends of the IS element then attack the unnicked strands in a transesterification reaction that produces hairpin structures (**Fig. 9.36**, step 3). In this instance, joining the ends of the double-stranded molecule produces a single strand with a hairpin.

After the host carrier DNA is ejected from the transpososome, the hairpin ends are renicked and the 3′ OH ends attack the target DNA molecule in a staggered manner (**Fig. 9.36**, step 4). The element has successfully "jumped" from one molecule to another without replicating. Note that the target DNA segment replicates as a result of the insertion process, thereby producing a copy of the target sequence at each end of the IS element (**Fig. 9.36**, step 5).

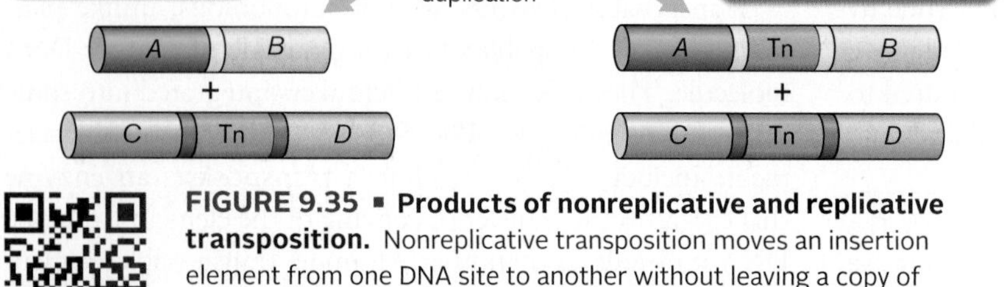

Duplication from previous jump
Transposable element
A
Tn
B
Nonreplicative tranposition: Transposable element jumps from one site to another.
Target DNA
C
D
Target sequence for insertion and duplication
Replicative transposition: Transposable element is copied. One copy remains in original site.
A
B
+
C
Tn
D
A
Tn
B
+
C
Tn
D

FIGURE 9.35 ■ Products of nonreplicative and replicative transposition. Nonreplicative transposition moves an insertion element from one DNA site to another without leaving a copy of the element at the original site. Replicative transposition leaves the element at the original site and moves a replicated copy to the new site.

Replicative Transposition

Figure 9.37 shows how a transposable element in a plasmid can bring about cointegration of the plasmid into a target DNA molecule during replicative transposition. As a result of replication of the IS element, the cointegrate molecule will contain two copies of the element flanking the integrated plasmid. Thus, transposition can provide an organism with a whole set of genes that might extend its metabolic capacity, provide drug resistance, or serve as the starting point for gene divergence through mutation. Do not be misled—transposition is not a process designed to help the host; rather, it evolved to ensure survival of the transposon. The benefit to the host is accidental.

Transposons are more complex than simple insertion elements, because they carry other genes in addition to those required for transposition. They come in different configurations. A **composite transposon** typically consists of two insertion sequences that flank an antibiotic resistance gene, or one or more catabolic genes (for example, genes for benzene catabolism). Often the interior inverted repeats of the IS elements have degenerated, so the transposase acts on primarily the two outermost inverted repeats, causing the whole transposon to move as one unit. Tn*10* is one such example (**Fig. 9.38A**). **Complex transposons** have, as the name suggests, a more complex organization. Tn*3*, for instance, not only possesses transposase and antibiotic resistance genes, but also includes a gene whose product, called resolvase, specifically unlinks (resolves) the cointegrate produced by replicative transposition (**Fig. 9.38B**). The transposase acts on the inverted-repeat ends of the element, causing simultaneous insertion and replication. Resolvase is a site-specific recombinase that mediates recombination at the internal Tn*3* *res* sites. The result of this

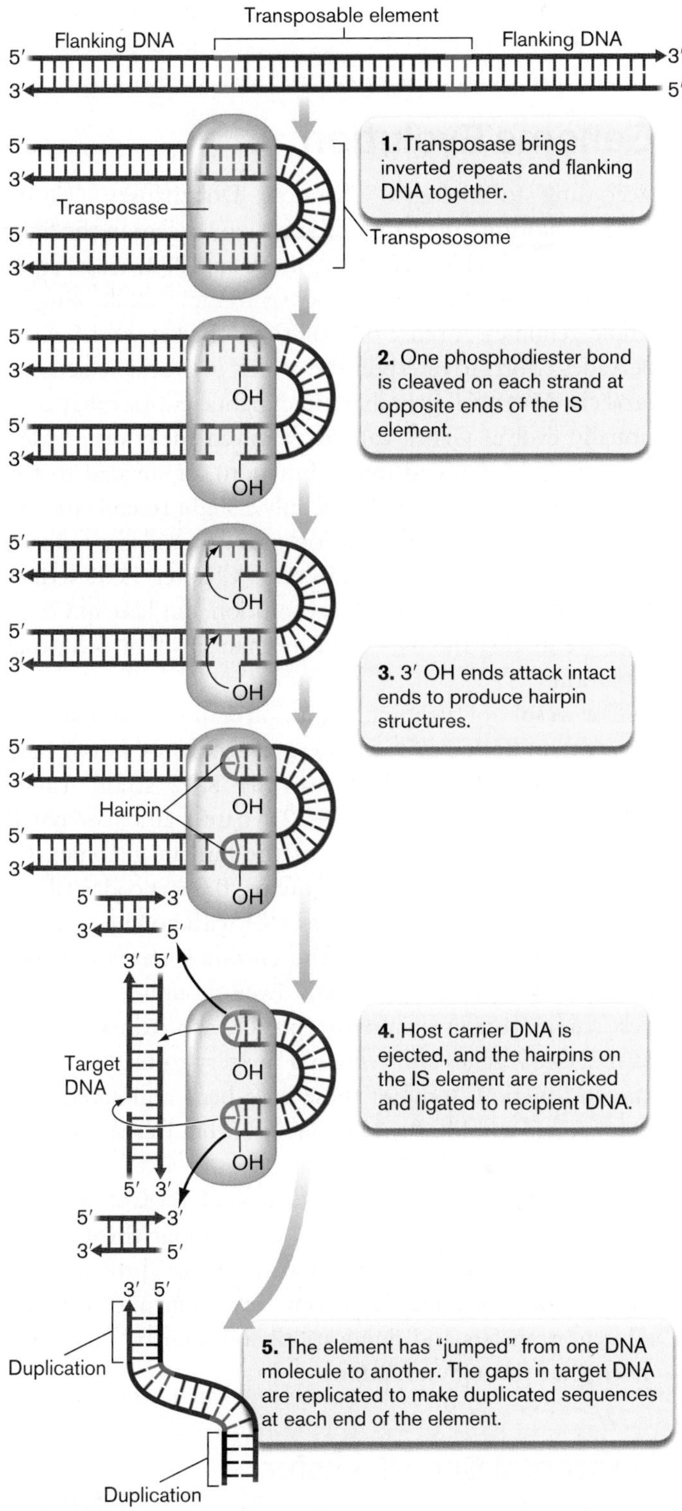
Transposable element
Flanking DNA
Flanking DNA
5′
3′
3′
5′
Transposase
Transpososome
1. Transposase brings inverted repeats and flanking DNA together.
OH
2. One phosphodiester bond is cleaved on each strand at opposite ends of the IS element.
3. 3′ OH ends attack intact ends to produce hairpin structures.
Hairpin
Target DNA
4. Host carrier DNA is ejected, and the hairpins on the IS element are renicked and ligated to recipient DNA.
Duplication
5. The element has "jumped" from one DNA molecule to another. The gaps in target DNA are replicated to make duplicated sequences at each end of the element.
Duplication

FIGURE 9.36 ▪ Nonreplicative transposition. The transpososome complex includes the transposase binding to the ends of the transposable element and the target DNA. The black DNA segments marked "Duplication" in the last panel correspond to the duplications represented in Figure 9.34.

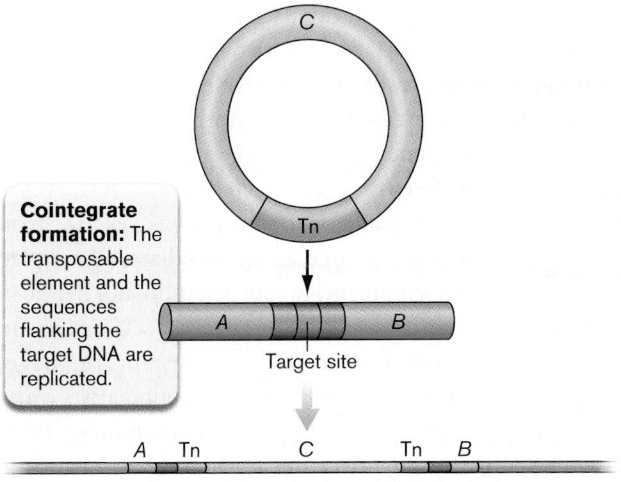
C
Tn
Cointegrate formation: The transposable element and the sequences flanking the target DNA are replicated.
A
B
Target site
A Tn C Tn B

FIGURE 9.37 ▪ Transposition-mediated cointegrate formation between two DNA elements. During replicative transposition, the insertion element is copied and simultaneously ligated to the target site. The result is a transient cointegrate molecule in which donor DNA and target DNA become one molecule.

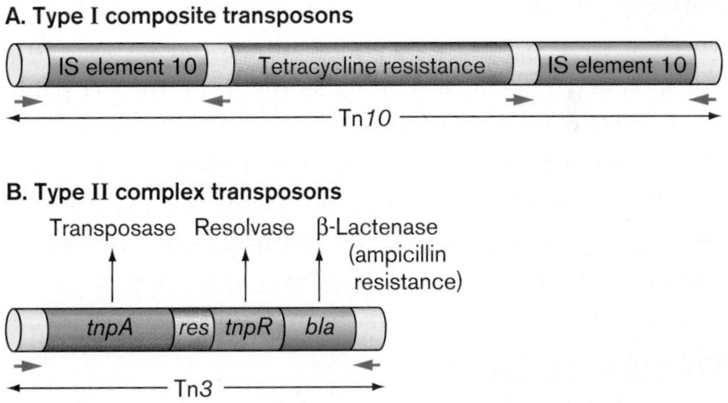
A. Type I composite transposons
IS element 10
Tetracycline resistance
IS element 10
Tn*10*
B. Type II complex transposons
Transposase
Resolvase
β-Lactenase (ampicillin resistance)
tnpA
res
tnpR
bla
Tn*3*

FIGURE 9.38 ▪ Examples of composite and complex transposons. In these depictions the yellow fragments represent inverted repeats and the red arrows indicate their orientation. **A.** In Tn*10* the internal repeats contain mutations that prevent the individual IS elements from jumping off on their own. **B.** TnpR is a repressor of the Tn*3 tnpA* transposase gene and is the enzyme that recognizes the *res* site required to resolve the cointegrates formed during transposition.

recombination is restoration of the two original circular replicons, each one now containing a copy of the transposon.

Thought Question

9.10 Diagram how a composite transposon like Tn*10* can generate inversions or deletions in target DNA during transposition. *Hint:* This happens when transposition within the chromosome occurs using the inverted repeats closest to the tetracycline resistance gene (see **Fig. 9.38A**). If you draw it right, you will see, in the end, that the tetracycline resistance gene is lost.

All of the transposition events mentioned so far involve transfer between DNA molecules within the same cell. Some transposons, called **conjugative transposons**, are able to transfer from one cell to another by conjugation (see the chapter opener figure). One example is the transposon Tn*916* of *Enterococcus,* which encodes tetracycline resistance. These conjugative transposons excise from donor DNA and form a circular intermediate, like a plasmid, just prior to conjugation, but they do not replicate autonomously. They must be part of a larger self-replicating DNA entity (for example, plasmid, chromosome, or phage). Once transferred into a new host, the conjugative transposon integrates into the recipient chromosomes. Tn*916* and some other conjugative transposons evolved to transfer when a population is facing danger. For example, in the absence of tetracycline, only a few cells in a population may have Tn*916*. However, the presence of just a small amount of tetracycline triggers conjugative transfer of the element, spreading it throughout the population.

The study of transposons may provide insight into the workings of the human immunodeficiency virus (HIV), which causes AIDS. The structure of transposase from transposon Tn*5* bears a striking similarity to the HIV integrase needed to embed the reverse-transcribed HIV viral DNA into the human genome (see **eTopic 9.5** and Section 11.4). Another type of mobile element, called an integron, has evolved in bacteria to capture DNA cassettes that encode antibiotic resistance genes (see **eTopic 9.6**).

To Summarize

- **Transposable elements and insertion sequences** ("jumping genes") move from one DNA molecule to another, usually without replicating separately (that is, they are not plasmids).
- **Transposase** is an enzyme that forms a transpososome complex with the transposable element and target DNAs.
- **Insertion sequences** are simple DNA transposable elements containing a transposase gene flanked by short inverted-repeat sequences.
- **Transposable elements move** by nonreplicative or replicative mechanisms.
- **Transposons** are complex transposable elements carrying additional genes (encoding, for example, drug resistance).
- **Composite transposons** have two duplicate insertion sequence elements that flank additional genes. Often the interior inverted repeats of these elements contain mutations.
- **Transposons can carry a variety of genes**, including antibiotic resistance genes.

9.7

Genome Evolution

According to biologist Theodore Dobzhansky (1900–1975), "Nothing makes sense in biology except in the light of evolution." This statement, made in 1972, rings especially true now that we have access to rapid sequencing of whole genomes. A sense of evolution, of what went before, pervades efforts to interpret the results of the many genome projects. Microbial genomes are dynamic entities that continually evolve. Genes can be removed (deletion), added (insertion), rearranged (recombination), or divided to the point where the genome bears only a slight resemblance to what it once was. The evolution of a genome is a random process driven by natural selection. Chapter 17 describes the many features of microbial evolution, but here in Chapter 9 we discuss the molecular mechanisms that drive that evolution.

The result of genome evolution can be seen in the genomes of *Escherichia coli*. Several strains of *E. coli* have been sequenced. The first was the K-12 strain (called MG1655). It contains 4,639,221 base pairs (compared with the 3 billion base pairs of the human genome). Of this genome, 87.8% encodes proteins, 0.8% encodes tRNA and rRNA, and another 0.7% is DNA with no known function. Approximately 11% of the chromosome is involved with various forms of regulation. Even though *E. coli* is the best-studied organism on the planet, approximately 24% of the 4,489 genes still remain a mystery to us, having no known function (some of these have been annotated with vague functions, like "oxidoreductase," but lack supporting experimental evidence).

Close examination of the sequences reveals how the current organism we call *E. coli* came about through the evolution of an ancestral genome. Two basic processes are thought to contribute to genome restructuring: horizontal gene transfers and duplications followed by functional divergence through mutation.

Horizontal Gene Transfer

Microbial genomes evolve by randomly assembling an eclectic array of genes from many sources. For example, *E. coli* strain O157:H7, the culprit in several fatal outbreaks of food-borne disease in the United States and Europe, contains 1,387 genes lacking in strain K-12. These additional genes represent about 25% of the O157:H7 genome and encode virulence factors, metabolic pathways, and prophages (phage genes integrated into host chromosomes), all of which were acquired from other species. Until 1990, it was thought that bacterial genomes were rather static; one strain of *E. coli* was thought to be very similar to any other strain. But because of genomic comparisons like this, we now know that enterobacteria are subject to a great deal more interspecies recombination than was previously thought.

Evidence for horizontal transfer. More evidence of gene shuffling comes from comparing the proportion of GC base pairs along the chromosome (also known as GC content). On average, the *E. coli* genome consists of 50.8% GC pairs. But 15% of the K-12 genome and 26% of the O157:H7 genome show GC proportions that differ significantly from the rest of the genome and show a different codon usage, suggesting that these genes came from other bacterial lines or species and were acquired by *E. coli* more recently. These horizontal gene transfers, occurring via conjugation, transduction, or transformation, are estimated to happen at the rate of about 16 kb, or 0.4% of the *E. coli* genome, every million years. A sudden change in the GC content within a chromosome is the equivalent of an evolutionary "footprint" marking a horizontal gene transfer.

Horizontal gene transfer appears to be a primary force in microbial evolution. Some estimates place the level of recombination between species at about 100 times greater than the rate of mutation. In *E. coli,* it has been estimated that nearly 20% of its genome may have originated in other microbes. Gene acquisitions performed by *E. coli* have played an important role in the bacterium's ecological plasticity and contributed to its success as a pathogen.

Genome rearrangement is not restricted to *E. coli,* of course. All life-forms enjoy this form of genetic "gambling." Every species of microbe is thought to have a core gene pool that includes the minimal set of genes needed for independent growth and replication. Every genome is also thought to possess a flexible gene pool that is not common to all strains within the species (**Table 9.4**). Therefore, differences in genome size between strains are usually thought to reflect variations in the flexible gene pool brought about by loss (deletion) or gain (insertion) of chromosomal DNA.

Genomic islands are large, transferred genetic elements containing numerous genes that serve a common function. For example, a **pathogenicity island** is a genomic island that encodes virulence factors. They are found in bacteria pathogenic for plants and animals, but not in related nonpathogenic strains. *Salmonella enterica* serovar Typhimurium, for example, contains five pathogenicity islands, each with a role specific to a different stage of disease. Pathogenicity islands are often flanked by boundary regions such as direct repeats or insertion elements, usually found near tRNA genes (**Fig. 9.39**). The association of pathogenicity islands with tRNA genes may reflect the fact that some tRNA genes serve as attachment sites for the integration of some phage DNAs or transposons. In addition, some tRNA genes are dispensable, since there are often multiple copies of a given tRNA gene within a genome. The pathogenicity island as a whole is often genetically unstable, subject to further rearrangements or deletions. In addition to virulence genes, pathogenicity islands contain mobility genes encoding transposases, integrases, or other enzymes associated with recombination.

TABLE 9.4

Composition of bacterial genomes.

Core gene pool		Flexible gene pool	
DNA elements	**Encoded features**	**DNA elements**	**Encoded features**
Chromosomes and some plasmids	Ribosomes	Genomic islets	Pathogenicity
	Cell envelope	Bacteriophages	Antibiotic resistance
	Key metabolic pathways	Plasmids	Secretion
	DNA replication	Integrons	Symbiosis
	Nucleotide turnover	Transposons	Degradation
	Genomic islands		Secondary metabolism; restriction-modification; transposases/integrases

Source: U. Dobrindt et al. 2002. *Curr. Top. Microbiol. Immunol.* **264**:157–175.

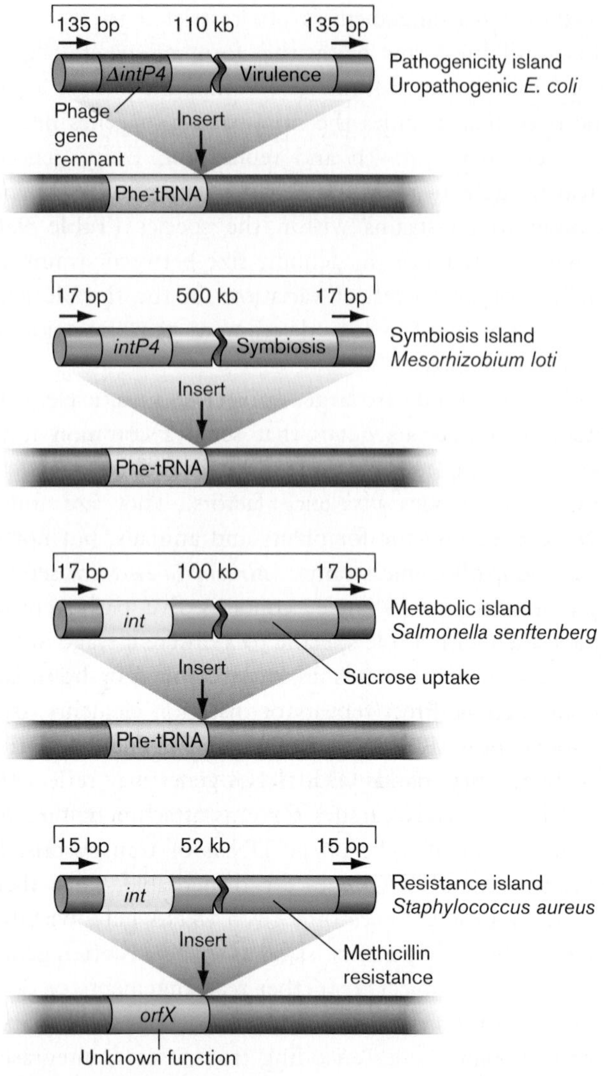
135 bp
110 kb
135 bp
ΔintP4
Virulence
Pathogenicity island
Uropathogenic E. coli
Phage gene remnant
Insert
Phe-tRNA
17 bp
500 kb
17 bp
intP4
Symbiosis
Symbiosis island
Mesorhizobium loti
Insert
Phe-tRNA
17 bp
100 kb
17 bp
int
Metabolic island
Salmonella senftenberg
Insert
Sucrose uptake
Phe-tRNA
15 bp
52 kb
15 bp
int
Resistance island
Staphylococcus aureus
Insert
Methicillin resistance
orfX
Unknown function

FIGURE 9.39 ■ Genomic islands from different microorganisms. Pathogenicity islands are often found inserted adjacent to tRNA genes and have direct repeats at their ends (arrows). Functions for each island are identified.

Another kind of genomic island, called a "mobile symbiosis island," helps a microbe enter into symbiotic relationships. For example, *Mesorhizobium loti* is a symbiotic bacterium that lives in nodules on the roots of legumes. This root nodule bacterium derives nourishment from the plant in exchange for fixing nitrogen from the atmosphere that the plant can use. Several species of rhizobia carry the genes for nodulation and nitrogen fixation on large plasmids. Others, like *Bradyrhizobium* species, have symbiosis genes as part of their genomes, but clustered in a way that suggests they were once derived from horizontal transfer. These gene clusters appear to have lost whatever mobility they once had. The symbiosis genes of *M. loti,* however, are contained on a mobile symbiosis island. These were the genes horizontally transferred in the Sullivan and Ronson field experiment described at the start of this chapter.

Pathogenicity islands and symbiosis islands alike encode remarkable protein secretion systems (called type III and type IV secretion systems) capable of injecting effector proteins directly from the bacterium straight into a eukaryotic cell. Once inside the target cell, these effector proteins alter the function of eukaryotic cell proteins, making the plant, in the case of symbiosis, more accepting of the symbiont. Types III and IV secretion systems are discussed further in Chapter 25.

Other genomic islands are called "fitness islands." Fitness islands encode functions that are important to survival in the environment. *E. coli,* for example, possesses an acid fitness island that enables the organism to survive at pH 2, a condition found in the human stomach (see Section 12.1).

Genome Reduction

It is important to note that evolution toward pathogenicity involves gene loss as well as gene acquisition. The large-scale loss of genes through evolution is known as genome reduction. For example, pathogenic shigellae (the cause of bacillary dysentery) exhibit chromosomal "black holes," regions lacking genes that occur in the closely related *E. coli*. Absence of these genes is required for a fully virulent phenotype (why is not clear). Similarly, the genome of the organism that causes whooping cough, *Bordetella pertussis,* is 100 kb smaller than the genome of the less pathogenic *Bordetella bronchiseptica*—another example of evolution by genome reduction. Genome reduction is discussed in Chapter 17.

Sometimes evidence of genome reduction in the making can be observed. Many genomes contain pseudogenes, genes that by homology appear to encode an enzyme but are nonfunctional because a portion is missing through deletion. These appear to be the remnants of genes that were useful to an ancestral species but were made superfluous when one of its descendants adapted to a new environmental niche. When evolutionary pressure to retain a functional form of the gene no longer exists, the imperative to repair mutations within that gene also evaporates. The pseudogene, therefore, is in the process of being eliminated. One pathogen caught in the act of losing its genes due to its lifestyle is *Mycobacterium leprae,* the cause of Hanson's disease (leprosy). Over half of the *M. leprae* genome is made of pseudogenes, presumably formed from genes it no longer needs after becoming an obligate intracellular pathogen.

Duplications and Divergence

Gene duplication is the most important mechanism for generating new genes and biochemical processes. Duplication frees a gene from its previous functional constraints and allows divergent evolution through mutation. Duplications can arise in several ways, including the transposition of a transposable element and recombination after replication between multiple direct or inverted-repeat sequences located in the chromosome. **Figure 9.40** shows a duplication formed by recombination between direct repeats of chromosomal sequences. Superfamilies of proteins arising from divergent evolution share structural and functional features but may catalyze different reactions. One example is the ABC superfamily of ABC transporters. Homology between members of the ABC family may be as little as 10%, suggesting that once an enzyme adopts a new function, the sequence diverges rapidly. Thus, gene duplications and mutation-driven divergence are processes that directly generate orthologous and paralogous genes, discussed in Chapter 8.

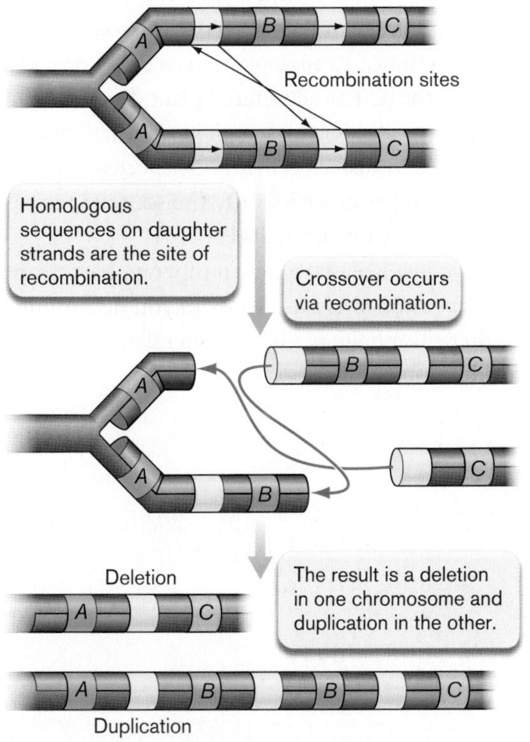
A
B
C
Recombination sites
Homologous sequences on daughter strands are the site of recombination.
Crossover occurs via recombination.
Deletion
The result is a deletion in one chromosome and duplication in the other.
Duplication

FIGURE 9.40 ■ Duplication during replication. Replicating DNA with three genes: *A*, *B*, and *C*. Yellow segments denote areas with sequence homology. Black arrows represent orientation. Crossover lines indicate a recombination event between homologous areas in different places on daughter strands. Red arrows indicate where the strands reconnect after recombination.

Thought Question

9.11 Gene homologs of *dnaK* encoding the heat-shock chaperone HSP70 exist in all three domains of life. All bacteria contain HSP70, but only some species of archaea encode a *dnaK* homolog. The archaeal homologs are closely related to those of bacteria. Knowing this information, how do you suppose *dnaK* genes arose in archaea?

To Summarize

- **Horizontal gene transfers** occur between species by conjugation, transformation, and transduction.
- A DNA sequence with a **GC content** different from that of flanking chromosomal DNA is one sign of horizontal gene transfer.
- **Pathogenicity islands** are the result of horizontal gene transfers that improve the pathogenicity of a recipient. Fitness islands are similar to pathogenicity islands but improve the environmental survival characteristics of the recipient.
- **Superfamilies of functional proteins** result from gene duplications and mutations that cause divergent evolution of function.

Concluding Thoughts

Fortuitous movements of genes between species through conjugation (cell-cell contact), transformation (DNA uptake), transduction (phage vector), and transposition (jumping genes) have contributed in a fundamental way to evolution. These mechanisms caused evolutionary leaps and enabled ancestral bacteria to develop (and trade) the symbiotic and pathogenic mechanisms needed to inhabit new ecological niches. DNA repair mechanisms that prevent the establishment of deleterious mutations nevertheless allow some mutations to occur. An imperfect DNA repair mechanism offers the cell the opportunity to throw the genetic dice. The vast majority of mutations will be detrimental, but a key few may help the variant strain outcompete its neighbors in its ecosystem. Given our new awareness of genetic mobility, how do we now define a species? For instance, numerous strains of *Helicobacter pylori* show large differences in gene sequence and organization. Should each be called a different species? Currently, we use an arbitrary setting of ribosomal RNA gene divergence to assign species, but some wonder whether this is still a valid landmark. For further discussion of microbial evolution, see Chapter 17.

CHAPTER REVIEW

Review Questions

1. What are the basic ways microorganisms exchange DNA?
2. Discuss horizontal versus vertical gene transfer.
3. Describe competence and how it comes about in a population.
4. What is an F factor, and how does it (and other factors like it) contribute to gene exchange?
5. What is a partial diploid? Explain how a cell can become a partial diploid.
6. What does microbial gene exchange have to do with the plant disease called crown gall disease?
7. Compare specialized versus generalized transduction.
8. Discuss how bacteria protect themselves from invading bacteriophages.
9. What is the value of recombination to a species?
10. List the major proteins that contribute to recombination, and specify their roles in the process.
11. Define "Holliday junction," "strand assimilation," "cointegrate," and "cotransduction."
12. List and explain the different types of mutations.
13. How are genotype and phenotype different?
14. Describe six different DNA repair mechanisms. Which ones contribute to mutations?
15. Explain the basic process of transposition. Why are insertion sequences always flanked by direct repeats of host DNA? How are transposons different from plasmids?
16. What are pathogenicity islands and fitness islands? What are their characteristics?

Thought Questions

1. Calculate the mutation rate for a gene in which the number of mutations increases from 0 to 500 as the cell number increases from 10^1 to 10^8.
2. Why does the competence factor for initiating synthesis and assembly of the transformasome have to be exported out of the cell to bind ComD? Why doesn't the molecule bind internally?
3. Rapidly growing bacteria can initiate a second round of replication before a previous round is completed. Would such rapid growth introduce more mutations (due to replication errors) than slower growth by the same species?
4. *Agrobacterium tumefaciens* Ti plasmid has been used as a tool to genetically modify plants. Why would a plant biologist use a tumor-causing plasmid to breed new plants? Wouldn't the genetically altered plant develop a tumor?
5. Though we can identify sets of genes that have been horizontally transferred from one species of microbe to another, rarely can we identify the source species. What might account for this failure?
6. You have just isolated a new temperate bacteriophage for *Salmonella enterica*. How can you determine whether this phage mediates generalized or specialized transduction?

Key Terms

AP endonuclease (351)
AP site (351)
apurinic site (344)
base excision repair (BER) (351)
cointegrate (338)
competence factor (CF) (323)
competent (323)
complex transposon (356)
composite transposon (356)
conjugation (324)
conjugative transposon (358)
CRISPR (335)
deletion (341)
electroporation (323)
episome (327)
error-prone repair (349)
error-proof repair (348)
F^- cell (325)
F^+ cell (325)
F-prime (F′) factor (328)
F-prime (F′) plasmid (328)
fertility factor (F factor) (325)
frameshift mutation (342)
gain-of-function mutation (342)
generalized recombination (338)
generalized transduction (331)
genomic island (359)
heteroduplex (340)
Hfr (327)
Holliday junction (340)
horizontal gene transfer (323)
insertion (341)
insertion sequence (IS) (355)
inversion (341)
knockout mutation (342)
loss-of-function mutation (342)
methyl mismatch repair (349)
missense mutation (341)
mutagen (343)
mutation (341)
mutation frequency (346)
mutation rate (346)
mutator strain (349)
nonhomologous end joining (NHEJ) (354)
nonsense mutation (342)
nucleotide excision repair (NER) (350)
pathogenicity island (359)
phase variation (341)
photoreactivation (350)
point mutation (341)
recombinational repair (352)
restriction endonuclease (333)
reversion (341)
silent mutation (341)
site-specific recombination (338)
SOS response (352)
specialized transduction (331)
transduction (331)
transformasome (323)
transformation (322)
transition (341)
transposable element (355)
transposase (355)
transposition (356)
transposon (356)
transversion (341)

Recommended Reading

Babic, Ana, Ariel B. Lindner, Marin Vulic, Eric J. Stewart, and Miroslav Radman. 2008. Direct visualization of horizontal gene transfer. *Science* **319**:1533–1536.

Bridges, Bryn A. 2005. Error-prone DNA repair and translesion synthesis: Focus on the replication fork. *DNA Repair* **4**:618–619.

Chen, Inês, Peter J. Christie, and David Dubnau. 2005. The ins and outs of DNA transfer in bacteria. *Science* **310**:1456–1460.

Dini-Andreote, Francisco, Fernando D. Andreote, Welington L. Araujo, Jack T. Trevors, and Jan D. van Elsas. 2012. Bacterial genomes: Habitat specificity and uncharted organisms. *Microbial Ecology* **64**:1–7. doi:10.1007/s00248-012-0017-y.

Doyle, Marie, Maria Fookes, Al Ivens, Michael W. Mangan, John Wain, et al. 2007. An H-NS-like stealth protein aids horizontal DNA transmission in bacteria. *Science* **5**:654–666.

Dunning Hotopp, Julie C., Michael E. Clark, Deodoro C. Oliveira, Jeremy M. Foster, Peter Fischer, et al. 2007. Widespread lateral gene transfer from intracellular bacteria to multicellular eukaryotes. *Science* **317**:1753–1756.

Fraser, Christophe, Eric J. Alm, Martin F. Polz, Brian G. Spratt, and William P. Hanage. 2009. The bacterial species challenge: Making sense of genetic and ecological diversity. *Science* **323**:741–746.

Frost, Laura S., Raphael Leplae, Anne O. Summers, and Ariane Toussaint. 2005. Mobile genetic elements: The agents of open source evolution. *Nature Reviews. Microbiology* **3**:722–732.

Krüger, Nora-Johanna, and Kerstin Stingl. 2011. Two steps away from novelty—principles of bacterial DNA uptake. *Molecular Microbiology* **80**:860–867.

Lim, Yin Mei, Ad J. Groof, Mrinal K. Bhattacharjee, David H. Figurski, and Eric A. Schon. 2008.

Bacterial conjugation in the cytoplasm of mouse cells. *Infection and Immunity* **76**:5110–5119.

Medini, Duccio, Claudio Donati, Hervé Tettelin, Vega Masignani, and Rino Rappuoli. 2005. The microbial pan-genome. *Current Opinion in Genetics and Development* **15**:589–594.

Selva, Laura, David Viana, Gill Regev-Yochay, Krzysztof Trzcinski, Juan M. Corpa, et al. 2009. Killing niche competitors by remote-control bacteriophage induction. *Proceedings of the National Academy of Sciences USA* **106**:1234–1238.

Thomas, Christopher, and Kaare M. Neilsen. 2005. Mechanisms of, and barriers to, horizontal gene transfer between bacteria. *Nature Reviews. Microbiology* **3**:711–721.

Tippin, Brigette, Phuong Pham, and Myron F. Goodman. 2004. Error-prone replication for better or worse. *Trends in Microbiology* **12**:288–295.

Truglio, James J., Deborah L. Croteau, Bennett Van Houten, and Caroline Kisker. 2006. Prokaryotic nucleotide excision repair: The UvrABC system. *Chemical Reviews* **106**:233–252.

CHAPTER 10
Molecular Regulation

10.1 Regulating Gene Expression

10.2 Paradigm of the Lactose Operon

10.3 Other Systems of Operon Control

10.4 Sigma Factor Regulation

10.5 Regulatory RNAs

10.6 DNA Rearrangements: Phase Variation by Shifty Pathogens

10.7 Integrated Control Circuits

10.8 Quorum Sensing: Chemical Conversations

10.9 Transcriptomics and Proteomics

A bacterial genome encodes thousands of different proteins needed to handle many different environmental contingencies. Some proteins, such as RNA polymerase, are always required for growth. Most proteins, however, are needed only under a limited set of conditions. Proteins that degrade the sugar lactose, for example, are useful only when lactose is present. Likewise, *Vibrio cholerae* needs to make cholera toxin only when it's growing inside the human body, not when it's living in the ocean. To compete successfully with others, the microbe must not waste energy making unneeded proteins. Cells achieve molecular efficiency by using elegant control systems that selectively increase or decrease gene transcription, mRNA translation, or mRNA degradation, as well as by degrading or sequestering regulatory proteins. But how does a cell "know" when to alter expression?

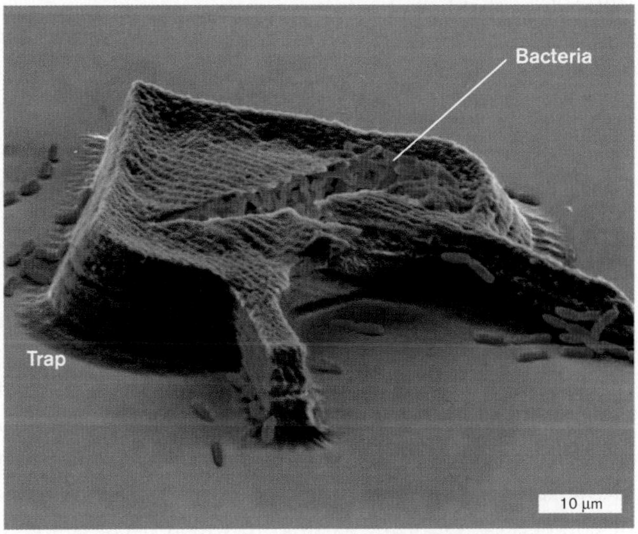

Bacteria
Trap
10 µm

CURRENT RESEARCH **highlight**

Bacterial lobster traps. Bacteria chemically "talk" to each other by secreting signal molecules (quorum sensing, QS). QS allows bacteria to engage in a variety of social behaviors, like forming biofilms. Although QS is widely studied, it is unclear whether population size, cell density, and/or medium flow rate determine the extracellular signal concentration required for QS. To assess these factors, researchers devised bacterial "lobster traps" made of photo-cross-linked protein. A single bacterium enters a 2- or 6-picoliter chamber. The chamber is then subjected to different medium flow rates while the organism multiplies to a density and population dictated by trap size. When QS is triggered, the organisms fluoresce green (look through the tear in the trap pictured here). Data from these traps proved that an increase in population size at constant flow rate, or a decrease in flow rate at constant population size will stimulate QS-dependent expression of green fluorescent protein (GFP). Bacteria, then, are "aware" of cell density, population size, and flow rate, and respond accordingly.

Source: From J. Connell et al. 2012. *MBio* **1**:e00201–00210.

During the day, the Hawaiian bobtailed squid, *Euprymna scolopes,* remains buried in the sand of shallow reef flats around Hawaii. After sunset, the animal emerges from its hiding place and begins its search for food. As it swims in the moonlit night, its light organ projects light downward in an apparent attempt to camouflage the squid from predatory fish swimming below. Looking up, the fish see only light (called counterillumination), not a squid's shadow moving against the surface-filtered light of the moon. What does this have to do with microbiology? Inside the light organ are luminescent bacteria called *Aliivibrio fischeri* (formerly *Vibrio fischeri*). The bacteria, not the squid, produce the light. However, these microbes do not constantly glow. They light up only when their cell density rises above a certain level, at which time the genes needed to make light are "turned on." This intricate level of genetic control by such a small, seemingly simple bacterium is truly amazing.

Microbes use numerous mechanisms to sense their internal and external environments and then translate that information into action. The cell's surface, for example, contains an array of sensing proteins that monitor osmolarity, pH, temperature, and the chemical content of the surroundings, or that detect the presence of a host or competitor. Quorum sensing, in which bacteria secrete and sense chemical signal molecules, makes it possible for members of microbial communities to communicate and cooperate.

All these sensing mechanisms orchestrate the synthesis of waves of proteins whose concentrations change with changing environments. Even virulence genes have tapped into these sensing systems, using them to determine whether the microbe has entered a susceptible host. Chapter 10 explores the fundamental principles of gene regulation in microbes and discusses how individual systems are woven into a global regulatory network that interconnects many processes throughout the cell.

10.1

Regulating Gene Expression

A cell must monitor two compartments in order to know when to alter gene expression and adjust its physiology: the cytoplasm within the cell and the environment outside. Intracellularly, the concentrations of vitamins, amino acids, and nucleotides must be sufficient and balanced to supply the biosynthetic and energetic needs of the cell. To achieve balance, the cell must control de novo (new) synthesis of these compounds and "know" what carbon and energy sources are present in order to assemble the proper catabolic pathway.

The microbe also needs to sense hazardous conditions outside the cell and to distinguish whether it is floating in a pond of water, a gastrointestinal tract, or a mammalian host cell. Once the environment is sensed, the cell can change the repertoire of genes it expresses to meet its needs or to protect itself.

Cells use different mechanisms to sense and respond to conditions within the cell and outside the cell membrane. Sensing conditions within the cell is relatively straightforward. **Regulatory proteins** bind specific small-molecular-weight compounds called ligands (**Fig. 10.1A**). Different regulatory proteins bind different ligands. For example, one regulator will bind a carbohydrate ligand and alert the cell that a new potential carbon source is available, while a different regulator will sense whether there is enough of the amino acid tryptophan present in the cytoplasm to carry out protein synthesis. The ligand, once bound, then alters the ability of the regulatory protein to latch onto specific DNA regulatory sequences located near the promoters of target genes.

General Concepts of Transcriptional Control

Genes encoding regulatory proteins are usually, but not always, transcribed separately from the target gene (**Fig. 10.1A**). Regulatory proteins come in two forms: **repressors** and **activators** (not to be confused with activator sequences). Repressor proteins bind to regulator sequences and prevent the transcription of target genes—an event known as **repression**. Repression happens in one of two ways (scenarios 1 and 2 in **Fig. 10.1B**), depending on the repressor. In scenario 1, the repressor binds DNA sequences by itself and prevents transcription. Relief from repression requires that a specific ligand, called an **inducer**, bind to the repressor protein, causing it to release from the DNA sequence. Because a small inducer molecule is required, the increased expression of the target gene is called **induction**. The lactose operon, discussed later, is one example of an inducible system.

Note: DNA regulatory sequences are sometimes called "operator sequences" if binding decreases expression of the target genes, or "activator sequences" if binding increases expression.

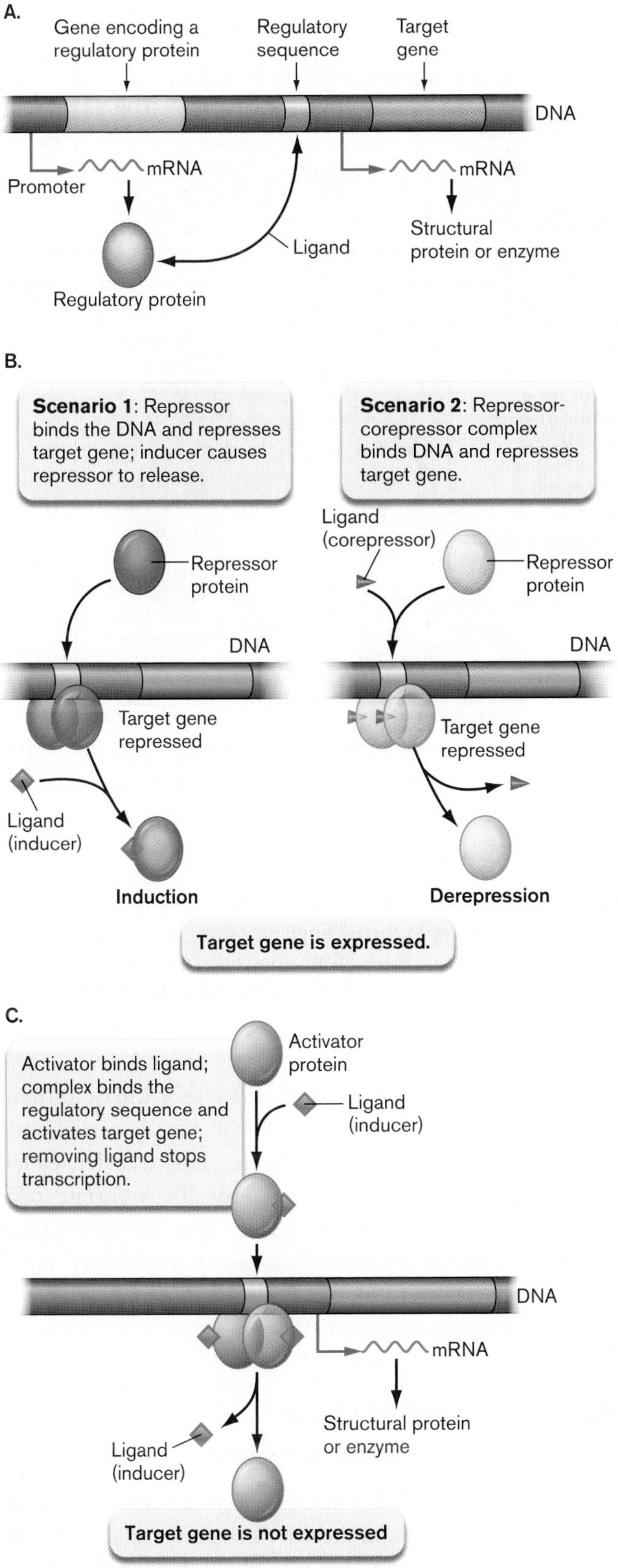
A.
Gene encoding a regulatory protein
Regulatory sequence
Target gene
DNA
Promoter
mRNA
mRNA
Ligand
Structural protein or enzyme
Regulatory protein
B.
Scenario 1: Repressor binds the DNA and represses target gene; inducer causes repressor to release.
Scenario 2: Repressor-corepressor complex binds DNA and represses target gene.
Repressor protein
Ligand (corepressor)
Repressor protein
DNA
DNA
Target gene repressed
Target gene repressed
Ligand (inducer)
Induction
Derepression
Target gene is expressed.
C.
Activator binds ligand; complex binds the regulatory sequence and activates target gene; removing ligand stops transcription.
Activator protein
Ligand (inducer)
DNA
mRNA
Structural protein or enzyme
Ligand (inducer)
Target gene is not expressed

FIGURE 10.1 ■ General aspects of transcriptional regulation by repressor and activator proteins. **A.** Schematic drawing of a regulatory system. The regulatory gene may be located near or quite far away from the target gene on the chromosome. When located nearby, it is often, but not always, transcribed separately from the target gene. The product of the regulatory gene (the regulatory protein) binds to DNA sequences near the promoter of the target gene and controls whether transcription occurs. Specific ligands influence binding. **B.** Repressor proteins bind to DNA sequences and prevent transcription. One type of repressor (scenario 1) releases from the target DNA sequence after binding a chemical ligand (called an inducer) that is specific for that repressor. The target gene is thereby induced. The second type of repressor (scenario 2) must bind a chemical ligand, called a corepressor, before it binds to DNA. **C.** Activator proteins generally bind to specific chemical ligands in the cytoplasm before the protein can bind to DNA sequences near target genes.

Other repressor proteins (scenario 2) do not bind well to DNA regulatory sequences unless they first bind a small ligand called a **corepressor**. As the ligand disappears from the cell, it is no longer available to bind to the repressor protein. When this happens, the repressor releases from the DNA and the target gene is expressed. This process is called **derepression** rather than induction. The tryptophan operon, also discussed later, is an example of a repression/derepression system.

Activator proteins also bind DNA but stimulate transcription by touching an RNA polymerase stuck at a nearby promoter, spurring it to initiate transcription (**Fig. 10.1C**). Most activator proteins bind poorly to DNA sequences unless an inducer is present. When the intracellular concentration of inducer falls, the activator protein (without inducer) either leaves the DNA or moves to a nearby site from which it can no longer contact RNA polymerase. As a result, transcription of that target gene stops.

Figure 10.2 illustrates how one repressor protein, CI from lambda phage, binds to DNA. The protein forms a dimer, and one part of each molecule, called the DNA-binding domain, interacts with DNA in the major groove.

It is important to realize that all DNA-binding proteins can actually bind at a low level to any DNA sequence. However, each individual binding protein will bind more tightly to a specific sequence near a particular promoter. Thus, DNA binding proteins exhibit a relative specificity for DNA target sequences, not absolute specificity. These proteins essentially "scan" DNA for high affinity binding sites. Relative specificity helps explain how a promoter, through evolution, "recruits" the service of a regulatory protein when a nearby weak binding site sequence changes by mutation into a tighter binding site.

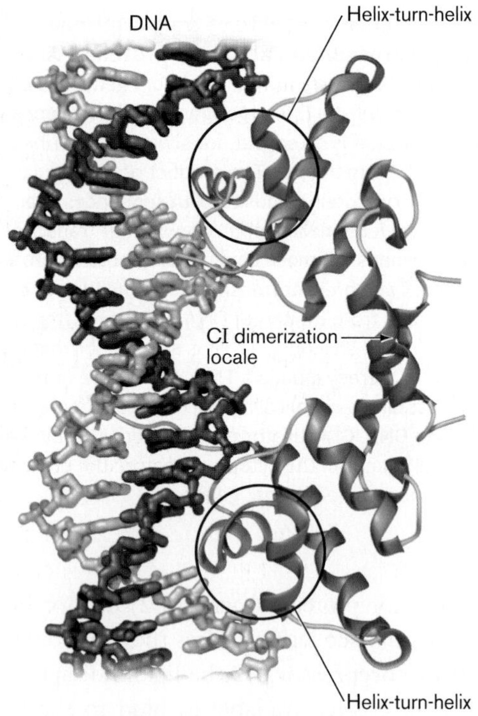
DNA
Helix-turn-helix
CI dimerization locale
Helix-turn-helix

FIGURE 10.2 ▪ Binding of a repressor protein to DNA. The dimer of lambda CI repressor binding to DNA. Note the helix-turn-helix motif located in two successive major grooves. Helices from these motifs are brown and purple, whereas the turns are green. (PDB code: 1LMB)

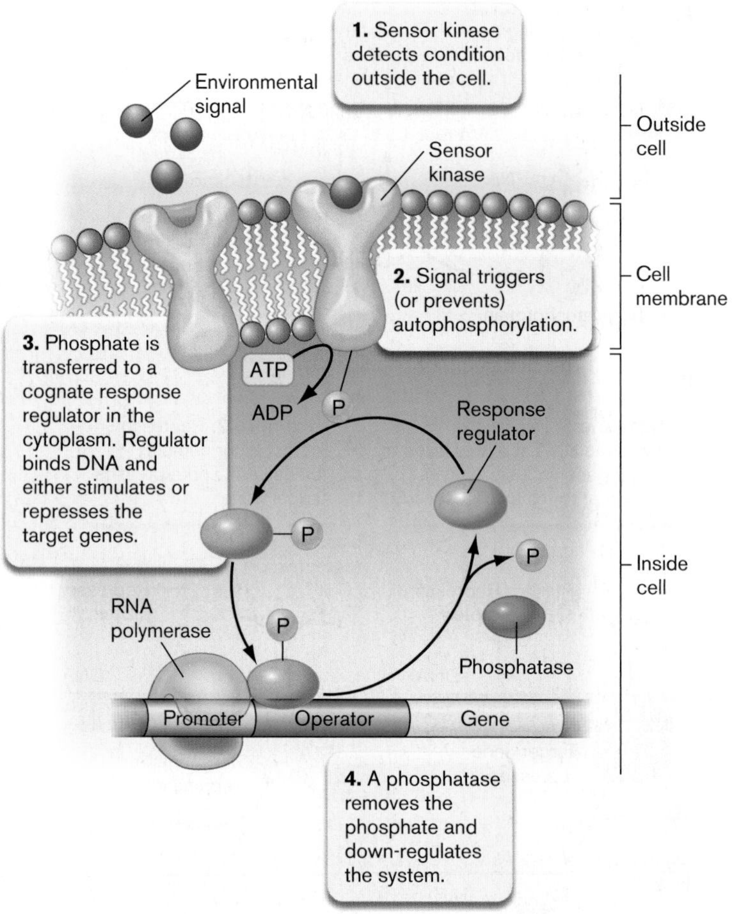
1. Sensor kinase detects condition outside the cell.
Environmental signal
Sensor kinase
Outside cell
Cell membrane
2. Signal triggers (or prevents) autophosphorylation.
3. Phosphate is transferred to a cognate response regulator in the cytoplasm. Regulator binds DNA and either stimulates or represses the target genes.
ATP
ADP
P
Response regulator
P
P
Inside cell
RNA polymerase
P
Phosphatase
Promoter
Operator
Gene
4. A phosphatase removes the phosphate and down-regulates the system.

FIGURE 10.3 ▪ Two-component signal transduction systems sense the external environment. A transmembrane sensor kinase protein senses an environmental condition outside the cell (as in Gram-positive bacteria) or in the periplasm (as in Gram-negative bacteria).

Sensing the Extracellular Environment

Sensing what goes on outside the cell is more challenging than sensing intracellular conditions because intracellular regulatory proteins cannot reach through the membrane and touch what is outside. A common mechanism used by Gram-positive and Gram-negative organisms to transmit information from outside to inside the cell relies on a series of two-member protein phosphorylation relay systems called **two-component signal transduction systems**. Each protein pair will activate different sets of genes. The first protein in each relay, the **sensor kinase**, spans the membrane (**Fig. 10.3**). Kinases transfer phosphate from ATP to a protein. The sensory domain of most sensor kinase proteins contacts the outside environment (or periplasm), while the other end (the kinase domain) protrudes into the cytoplasm.

Each sensor protein recognizes a different molecule or condition (for example, PhoQ in *Salmonella* senses magnesium). Once activated, the external sensory domain triggers a conformational change in the kinase domain that activates a self-phosphorylation reaction. Phosphate from ATP is attached to a specific histidine residue located in a different part of the protein. Then, like two relay runners passing a baton, the phosphorylated sensor kinase protein passes the phosphate to a cognate (matched) cytoplasmic protein called a **response regulator**. This transfer, called transphosphorylation, occurs at a specific aspartate residue within the response regulator. The phosphorylated response regulator commonly binds to regulatory DNA sequences in front of one or more specific genes and activates or represses expression.

Microbes Control Gene Expression at Several Levels

Hundreds of intracellular and extracellular sensing mechanisms monitor the overall health of the cell and its environment. But how do these systems control gene expression?

It is important to note that expression of genes and operons and their products can be controlled at various levels. Different levels of control offer different advantages to the cell. The major levels of control can be categorized as DNA sequence controls, transcriptional controls, translational controls, and posttranslational modifications. In general, DNA sequence–level control is the most drastic and the least reversible, whereas control at the protein (translational/posttranslational) level is the most rapid and most reversible. Examples of these control levels are summarized here and then discussed in detail later in the chapter.

- **Alterations of DNA sequence.** Some microbes use random or programmed changes of DNA sequence to activate or disable a particular gene. One example is phase variation, in which DNA rearrangement involving the reversible flipping of a DNA segment enables a pathogen to turn on or turn off expression of cell-surface proteins to evade the host immune system.
- **Control of transcription.** Many types of gene regulation occur at the level of transcription. The most common mechanisms of transcriptional control in prokaryotes involve operons. Recall that an operon is a string of two or more genes in a chromosome that are expressed from a common promoter in front of the first gene in the operon. Genes in an operon are coordinately regulated by protein repressors, activators, and sigma factors, as well as by small RNAs (**sRNAs**). Coordinate regulation means that the expression levels of all genes in the operon increase or decrease simultaneously.
- **Control of mRNA stability.** Levels of specific mRNA molecules are also regulated by RNase activity, which degrades some mRNA molecules as fast as they are transcribed. In some, but not all, instances, sRNA molecules bind RNA transcripts and help or hinder degradation.
- **Translational control.** Translation by ribosomes can be regulated by translation initiation sequences in the mRNA, which recognize specific translational repressor proteins. Translational control mechanisms are often coupled to transcriptional mechanisms, offering "fine-tuning" of operon control. Attenuation, for instance, is a mechanism that uses translation to sense the level of an amino acid in the cell and then increase or decrease transcription of the genes encoding the enzymes that synthesize the amino acid.
- **Posttranslational control.** Once proteins are made, their activity can be controlled by modifying their structure—for example, by protein cleavage, phosphorylation, methylation, or acetylation. These modifications can activate, deactivate, or even lead to the destruction of the protein.

In the past, scientists focused on regulator proteins as the primary mediator of expression control. Recent years, however, have shown that a variety of regulatory RNA molecules also play critical roles. The latest addition are the antisense RNA molecules that control gene expression by affecting translation, influencing mRNA degradation, or even by stopping transcription through the simple act of being made (for instance, colliding polymerases).

All of the previously described single control measures can also be used in combination by the cell to build integrated control circuits to coordinately regulate multiple systems throughout the cell—a key to homeostasis and balanced growth. We will discuss examples of integrated control circuits, such as the metabolic/genetic system of nitrogen regulation (see Section 10.7). But first we will explore model examples of control, most of which function at the transcriptional level, the level best understood.

To Summarize

- **Regulatory proteins** help a cell sense changes in its internal environment and alter gene expression to match.
- **Repressor and activator proteins** bind to operator and activator DNA sequences, respectively, in front of target genes. Repressors prevent transcription, whereas activator proteins stimulate transcription.
- **Two-component signal transduction systems** also help the cell sense and respond to its environment, both inside and outside.
- **Gene expression** can be controlled at several levels: DNA sequence, transcription, translation, mRNA stability or posttranslation (by modifying the protein).
- **Antisense RNAs** also play key roles in altering cell physiology by affecting the transcription, mRNA turnover, and translation of specific genes or gene products.
- **Integrated control circuits** connect many individual regulatory systems to coordinate regulation throughout the cell.

10.2

Paradigm of the Lactose Operon

In 1961, French scientists Jacques Monod (1910–1976) and François Jacob proposed the revolutionary idea that

A.
B.
C.
COURTESY OF THE ARCHIVES, CALIFORNIA INSTITUTE OF TECHNOLOGY
©BETTMANN/CORBIS
COURTESY OF ANDRÉ LWOFF

FIGURE 10.4 ■ Discoverers of gene regulation. Jacques Monod **(A)**, François Jacob **(B)**, and André Lwoff **(C)** all worked at the Pasteur Institute and won the 1965 Nobel Prize in Physiology or Medicine for their groundbreaking work on induction and gene regulation.

genes could be regulated (**Fig. 10.4A** and **B**). They, among others, noticed that the enzyme used by *Escherichia coli* to consume the carbohydrate lactose was produced only when lactose was added to growth media. The term "induction" was coined to describe this phenomenon. The enzymes required to metabolize glucose, however, were different. They were always present (that is, constitutive). The groundbreaking work of Jacob and Monod, and of André Lwoff (1902–1994) (**Fig. 10.4C**) for his study of phage lysogeny, won them a Nobel Prize and launched the field of gene regulation, a scientific realm where discoveries continue to surprise us. Still, it took many years after Monod and Jacob's discovery to learn exactly how the lactose-degrading enzyme beta-galactosidase is induced. Many of the concepts presented here apply to numerous other bacterial gene systems, including some required for *Erwinia* infection of plants and *Streptococcus pneumoniae* infection of humans.

Lactose Induces Expression of the *lacZYA* Operon

Figure 10.5A shows the genes in *E. coli* that encode the simple regulatory circuit for lactose catabolism. "Catabolism" is a general term to describe the degradation of an organic food source. The genes *lacZ, lacY,* and *lacA* form an **operon**, which is a group of genes cotranscribed from a common promoter. The *lacY* gene encodes lactose permease (LacY), an integral membrane protein that imports (transports) lactose from the extracellular environment. The product of *lacZ,* beta-galactosidase (LacZ), intracellularly cleaves the disaccharide lactose into its component parts: glucose and galactose (**Fig. 10.6**). Both of these sugars are subsequently degraded by the enzymes of glycolysis (see Section 13.5). To break down lactose in this way requires synthesis of both beta-galactosidase and lactose permease; without these two gene products, the catabolic energy of lactose is unavailable to the cell.

The role of *lacA,* which encodes thiogalactoside transacetylase (LacA), is unclear. It is not needed to ferment lactose but may detoxify a harmful by-product of lactose metabolism. As you can see, even well-studied systems such as the *lac* operon still pose unanswered questions.

Note: "Lactose operon," "*lac* operon," and "*lacZYA* operon" all refer to the same system.

Thought Question

10.1 If the gene *lacZ* has a nonsense mutation in its open reading frame, will *lacY* still be translated?

In the absence of lactose, the *lac* operon is transcribed at extremely low levels (fewer than ten molecules of LacZ per cell). The reason is that transcription of *lacZYA* is repressed by the protein product of the regulator gene, *lacI.* The *lacI* gene is situated immediately upstream of *lacZYA* and is transcribed from a different promoter. A tetramer of LacI repressor protein forms in the cell and binds to two operator regions of DNA. One operator sequence is called *lacO,* which partially overlaps the *lacZYA* promoter

(P_{lacZYA} or *lacP*). The second operator site occurs within *lacI* and is called *lacO*$_I$ (**Fig. 10.5A** ▶). The lac operators control whether the three structural genes are transcribed. The LacI tetramer simultaneously binds the lacO and lacO$_I$ operators (a dimer at each site). As a result, the intervening DNA loops out (**Figs. 10.5B** and **10.7A**) and prevents RNA polymerase from continuing transcription into the structural lacZYA genes. (A third binding site for LacI exists but is not important to our discussion here.)

Once lactose is added to the medium, the lac operon is expressed at 100-fold higher levels. How does lactose get into the cell to induce the operon if the operon encoding its transport protein is repressed? As noted already, even when repressed, the *lacZYA* operon is transcribed at a low level. (Most genes are expressed at low constitutive levels even when uninduced.) This means that a small amount of lactose can be transported into the cell because a few molecules of the lactose transporter LacY are made. The tiny amount of beta-galactosidase (LacZ) expressed in uninduced cells is also important. At this low level, the enzyme does not completely cleave the glycosidic bond of lactose, but rearranges it to form allolactose, the form of lactose that activates the operon (whose structure is shown in **Fig. 10.6**). Allolactose binds the repressor and "unlocks" the protein (by altering the conformation of LacI), so it is released from the operator (**Fig. 10.5C**). Once this happens, RNA polymerase guided by sigma factor (not shown in **Fig. 10.5C**) can find the *lac* promoter sequences and initiate transcription of the *lacZYA* structural genes.

Activation of Transcription by cAMP-CRP

Another important mechanism that governs the level of *lacZYA* transcription in *E. coli* involves a small molecule called cyclic AMP (cAMP), which accumulates when a cell is starved for carbon. Cyclic AMP is a derivative of AMP in which the 5′ phosphate is linked to the 3′ OH group of the ribose, making a cyclic structure. Made by the enzyme adenylyl cyclase, cyclic AMP controls the expression of many genes by combining with a dimeric regulatory protein

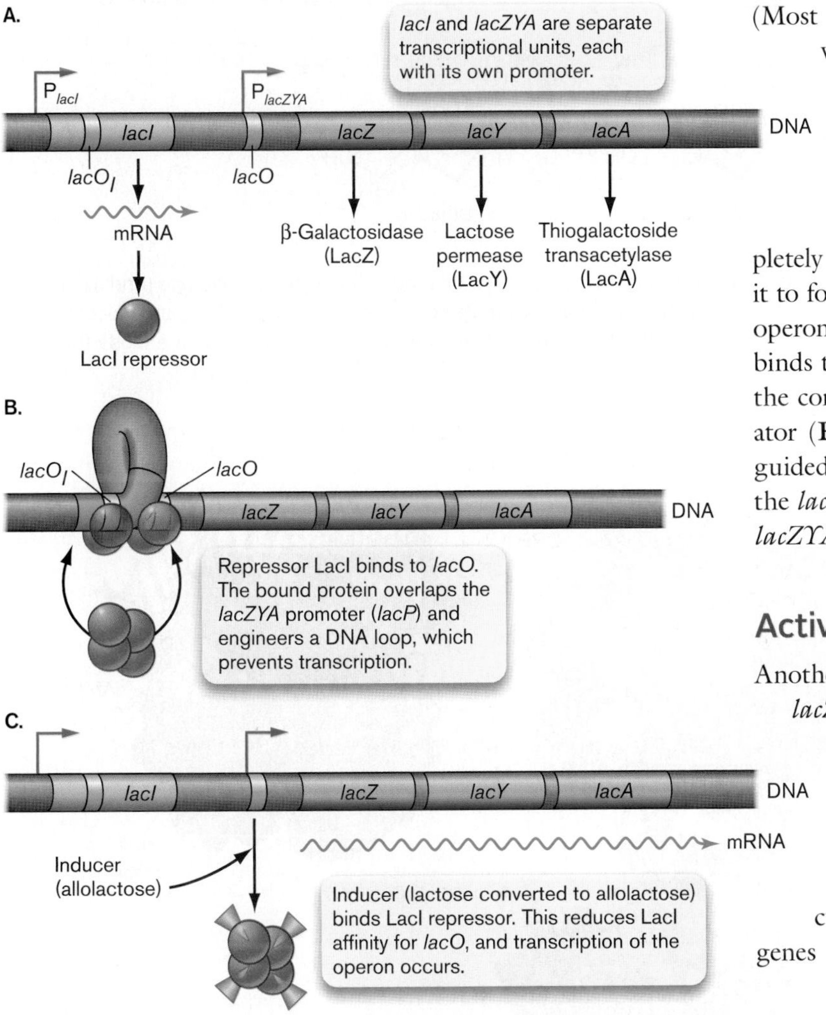
A.
lacI and lacZYA are separate transcriptional units, each with its own promoter.
P lacI
P lacZYA
lacI
lacZ
lacY
lacA
DNA
lacO I
lacO
mRNA
β-Galactosidase (LacZ)
Lactose permease (LacY)
Thiogalactoside transacetylase (LacA)
LacI repressor
B.
lacO I
lacO
lacZ
lacY
lacA
DNA
Repressor LacI binds to lacO. The bound protein overlaps the lacZYA promoter (lacP) and engineers a DNA loop, which prevents transcription.
C.
lacI
lacZ
lacY
lacA
DNA
mRNA
Inducer (allolactose)
Inducer (lactose converted to allolactose) binds LacI repressor. This reduces LacI affinity for lacO, and transcription of the operon occurs.

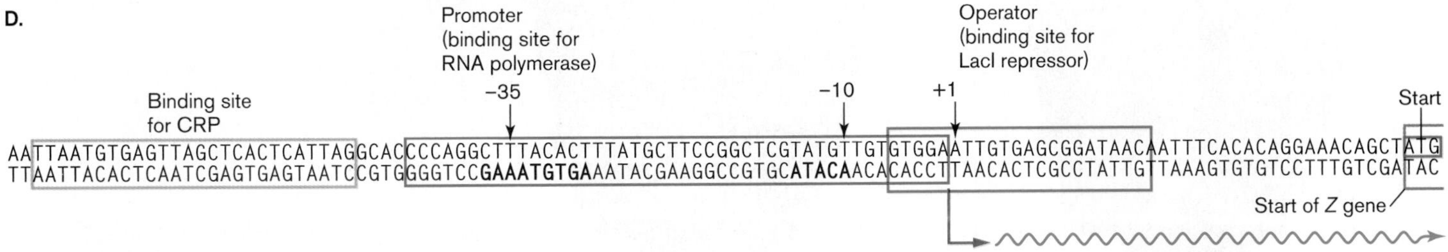
D.
Promoter (binding site for RNA polymerase)
Operator (binding site for LacI repressor)
Binding site for CRP
−35
−10
+1
Start
AATTAATGTGAGTTAGCTCACTCATTAGGCACCCCAGGCTTTACACTTTATGCTTCCGGCTCGTATGTTGTGTGGAATTGTGAGCGGATAACAATTTCACACAGGAAACAGCTATG
TTAATTACACTCAATCGAGTGAGTAATCCGTGGGGTCCGAAATGTGAAATACGAAGGCCGTGCATACAACACACCTTAACACTCGCCTATTGTTAAAGTGTGTCCTTTGTCGATAC
Start of Z gene

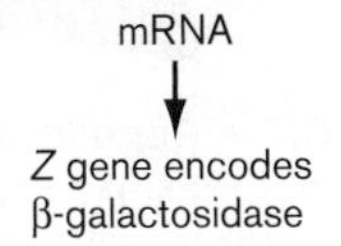
mRNA
Z gene encodes β-galactosidase

FIGURE 10.5 ■ **Transcriptional induction of the lactose operon. A.** Organization of the operon. Bent arrows mark promoters. Green indicates LacI protein-binding sites on DNA. **B.** The LacI tetrameric repressor binds to specific DNA sites (the operator: *lacO*). **C.** The inducer allolactose (an altered form of lactose made by low levels of beta-galactosidase) removes the repressor LacI and allows expression of *lacZYA*. **D.** DNA sequence of *lac* control region. Boxes enclose sequences important to binding of CRP (defined later), RNA polymerase, and LacI. ▶

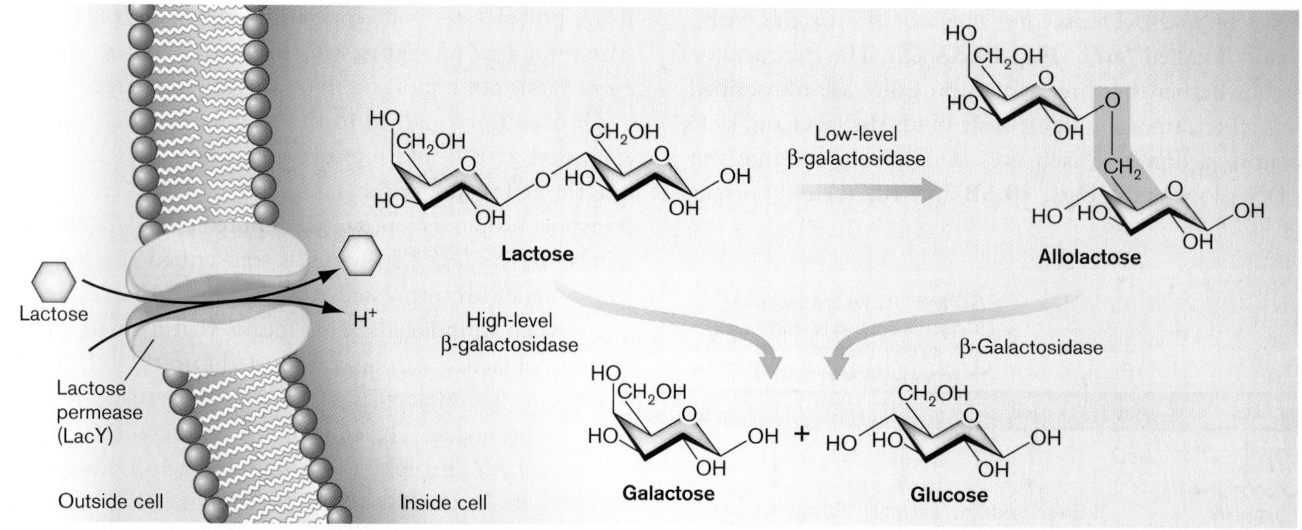
Lactose
Lactose permease (LacY)
H+
Outside cell
Inside cell
Lactose
Low-level β-galactosidase
Allolactose
High-level β-galactosidase
β-Galactosidase
Galactose
Glucose

FIGURE 10.6 ■ Lactose transport and catabolism. A dedicated lactose permease uses proton motive force to move lactose (and a proton) into the cell. Once there, the enzyme beta-galactosidase can cleave the disaccharide into its component parts (galactose and glucose) or alter the linkage between the monosaccharides to produce allolactose, an important chemical needed to induce the genes that encode this pathway.

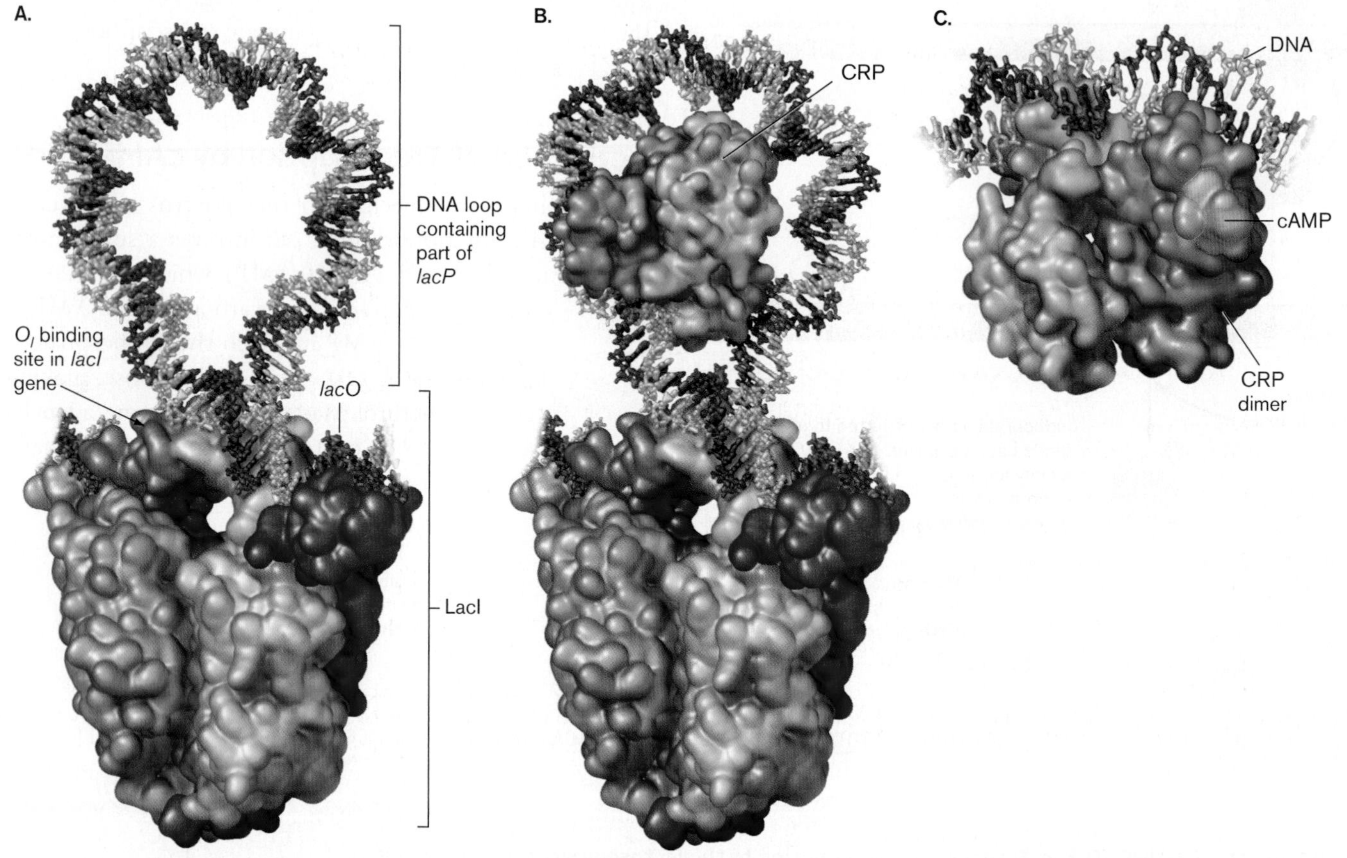
A.
B.
C.
DNA loop containing part of lacP
O1 binding site in lacI gene
lacO
LacI
CRP
DNA
cAMP
CRP dimer

FIGURE 10.7 ■ Regulatory protein interactions with DNA at the *lacZYA* control region. A. LacI repressor binds to the *lacO* region at two points, so that the DNA forms a loop. (PDB codes: 1Z04, 1K8J) **B.** If no lactose is present, LacI will prevent cAMP-CRP from activating transcription. (PDB codes: 1Z04, 1K8J, 2CGP) **C.** If only lactose is present (no glucose), cAMP-CRP binds to the binding site near the *lac* promoter, bends DNA, and interacts with RNA polymerase. (PDB code: 2CGP)

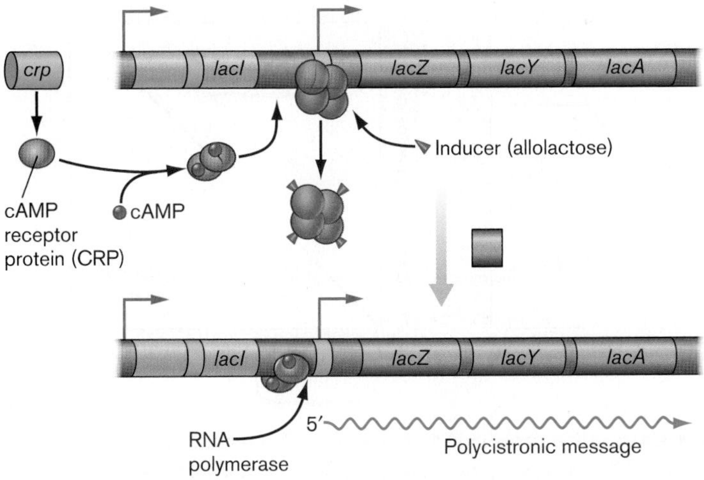

crp
lacI
lacZ
lacY
lacA
Inducer (allolactose)
cAMP receptor protein (CRP)
cAMP
lacI
lacZ
lacY
lacA
RNA polymerase
5′
Polycistronic message

FIGURE 10.8 ■ Activation of the *lacZYA* operon by cAMP and CRP. Although the inducer allolactose removes the repressor LacI and allows expression of *lacZYA,* maximum expression requires the presence of cAMP and cAMP receptor protein (CRP), which bind to a separate site at the *lacZYA* promoter. Bound CRP can interact with RNA polymerase and increase the rate of transcription initiation.

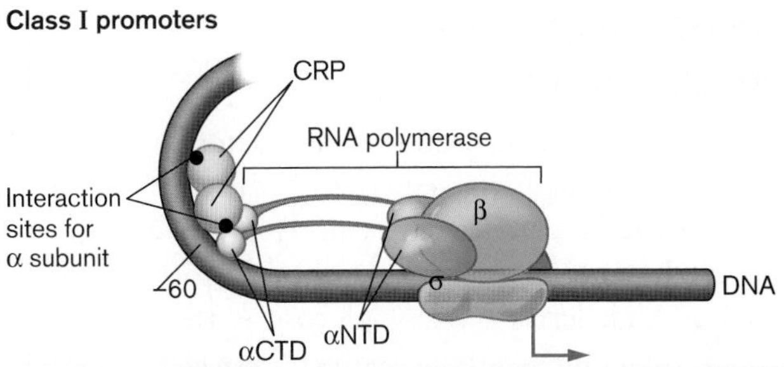

Class I promoters
CRP
RNA polymerase
Interaction sites for α subunit
β
–60
σ
DNA
αCTD
αNTD

Abbreviations key

CRP: cAMP receptor protein
αCTD: C-terminal domain of RNA polymerase alpha subunit
αNTD: N-terminal domain of RNA polymerase alpha subunit

FIGURE 10.9 ■ CRP interactions with RNA polymerase. Promoters like the one at the *lacZYA* operon possess a CRP-binding site positioned about –60 bp from the transcriptional start. CRP on these promoters interacts with the alpha subunit C-terminal domain (αCTD) of RNA polymerase. *Source:* Adapted from Moat et al. 2002. *Microbial Physiology,* 4th ed. Wiley-Liss.

called cAMP receptor protein (CRP). The cAMP-CRP complex binds to specific DNA sequences located near many bacterial genes and modifies their transcription, usually acting as an activator. This is the case for the *lacZYA* operon (**Figs. 10.8** and **10.5C**).

Note: The original name for CRP was "catabolite activator protein" (CAP), given to reflect its role in catabolite repression. Though some investigators still use the "CAP" designation, the term "CRP" is generally preferred today. Thus, the gene encoding the protein is called *crp*.

How does the cAMP-CRP complex ultimately activate the expression of *lacZYA*? RNA polymerase cannot easily form an open complex (see Section 8.2) and is essentially stuck at *lacP* and other CRP-dependent promoters, even in the absence of a LacI repressor. This problem is overcome when the cAMP-CRP complex binds to a 22-bp DNA sequence located –60 bp from the start of transcription (just upstream of the *lacZYA* operator), causing the DNA to bend (see **Figs. 10.5D** and **10.7C**). CRP can then directly interact with the alpha subunit of RNA polymerase bound at the *lac* promoter and activate transcription (**Fig. 10.9** and **eTopic 10.1**).

The C terminus of the RNA polymerase alpha subunit can bind DNA sequences near certain promoters and inhibit RNA polymerase movement. When cAMP-CRP touches regions in the C terminus of the alpha subunit, this inhibition is relieved. Contact between CRP and the alpha subunit is enough to stimulate class I promoters such as *lacP.* Notice in **Figure 10.7B** that when neither glucose nor lactose is present, cAMP-CRP cannot by itself activate *lacZYA* transcription, because it binds within the DNA loop formed by the LacI (repressor) tetramer, so there is no room for RNA polymerase to bind to DNA.

Note: Recall that a gene name written in roman (not italic) type and with a capital first letter is really the name of the protein product of that gene. Thus, *lacZ* is the gene, but LacZ is the protein. In presenting a genotype, gene names written with a superscript + are considered wild-type genes. Gene names written without a superscript + are considered mutant genes.

Thought Questions

10.2 Null mutations completely eliminate the function of a mutated gene. Predict the effect of the following null mutations on the induction of beta-galactosidase by lactose, and predict whether the *lacZ* gene is expressed at high or low levels in each case: *lacI, lacO, lacP, crp,* and *cya* (the gene encoding adenylyl cyclase). What effect will those mutations have on catabolite repression?

10.3 Predict what will happen to the expression of *lacZ* under the following partial diploid conditions (see Section 9.2). The genotypes of these strains are presented as chromosomal genes/plasmid gene. (a) $lacI^-$ $lacO^+P^+Z^+Y^+A^+$/plasmid $lacI^+$; (b) $lacO^-$ $lacI^+P^+Z^+Y^+A^+$/ plasmid $lacO^+$; (c) crp^- $lacI^+O^+P^+Z^+Y^+A^+$/plasmid crp^+.

Glucose Represses the *lac* Operon

What happens if, in addition to lactose, the medium contains an alternative carbon source, such as glucose? Enzymes for glucose catabolism (glycolysis) are always produced at high levels because glucose is the favored catabolite (carbon source), providing the quickest source of energy. Many carbohydrates, including lactose, must first be converted to glucose to be catabolized. So, should *E. coli* forestall induction of the *lac* operon while glucose is present, in the interest of greater efficiency? In fact, this is precisely what happens. In what is known as **catabolite repression**, an operon enabling catabolism of one nutrient is repressed by the presence of a more favorable catabolite (commonly glucose).

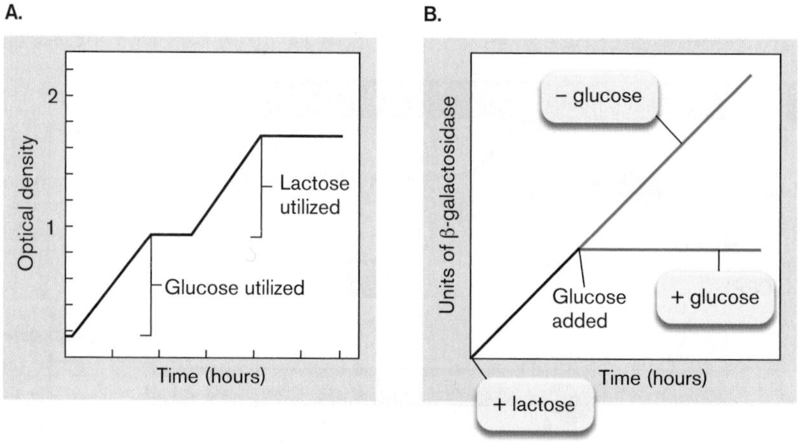
A.
Optical density
2
1
Lactose utilized
Glucose utilized
Time (hours)
B.
Units of β-galactosidase
– glucose
Glucose added
+ glucose
+ lactose
Time (hours)

FIGURE 10.10 ■ Catabolite repression of the *lacZYA* operon. A. The diauxic growth curve of *E. coli* growing on a mixture of glucose and lactose. **B.** Glucose repression of LacZ (beta-galactosidase) production. Lactose was added at the beginning of the experiment to parallel cultures, and beta-galactosidase activity was measured. At the point indicated, glucose was added to one culture. The synthesis of beta-galactosidase continues to increase in the culture without glucose but stops in the culture with glucose.

When glucose and lactose are both present in the medium, cells grow by breaking down glucose until the glucose is depleted; then growth stops. At this point the cells sense a need for carbon and induce the *lacZYA* operon (because lactose is present). This allows them to begin consuming lactose and to reinitiate growth. The biphasic curve of a culture growing on two carbon sources is often called **diauxic growth** (**Fig. 10.10A**). But if lactose was present from the start, why was *lacZYA* turned off? It was turned off because glucose indirectly prevents the induction of *lacZYA*. **Figure 10.10B** illustrates that even when *lacZYA* is already induced, adding glucose stops (or represses) induction.

Glucose Causes Catabolite Repression via Inducer Exclusion

Failure of lactose to induce *lacZYA* during growth on glucose (**Fig. 10.10B**; red horizontal line) is due mainly to the fact that growth on glucose keeps lactose out of the cell. This phenomenon is known as **inducer exclusion**. If lactose cannot enter the cell, the *lacZYA* operon cannot be induced. The key to inducer exclusion is that a component of the glucose transport system (phosphotransferase system, or PTS; eTopic 4.1), while transporting glucose, will bind to and inhibit LacY permease (**Figure 10.11**).

The PTS system transfers a phosphate from phosphoenolpyruvate (PEP) along a series of proteins to glucose during transport. When glucose is not present the proteins remain phosphorylated and LacY remains active (**Figure 10.11A**). When glucose *is* present, however, the glucose transport proteins continually transfer phosphate to glucose and, so, are usually left without phosphate (**Figure 10.11B**). Unphosphorylated enzyme IIA interacts with LacY in the membrane and inhibits LacY activity. So, lactose cannot enter the cell and the *lac* operon remains uninduced.

Transport of some sugars through the PTS also affects adenylyl cyclase activity, but this mechanism is not a major cause of glucose/lactose diauxic growth (see eTopic 10.2).

Thought Question

10.4 Researchers often use isopropyl-β-D-thiogalactopyranoside (IPTG) rather than lactose to induce the *lacZYA* operon. IPTG resembles lactose, which is why it can interact with the LacI repressor, but it is not degraded by beta-galactosidase. Why do you think the use of IPTG is preferred in these studies?

How Do We Study Protein-DNA Binding?

The search for DNA-binding sites often begins with bioinformatics. DNA sites that bind regulatory proteins typically exhibit a sequence symmetry that allows unambiguous recognition. Dimers of a regulatory protein bind to the DNA, with each member of the duo binding half of the symmetrical DNA sequence. Symmetry usually involves an inverted repeat (**Fig. 10.12**). Once a potential regulatory protein and a presumed target gene have been identified, the DNA sequence extending 200–300 bp upstream of the target gene's transcriptional start site is examined using computer

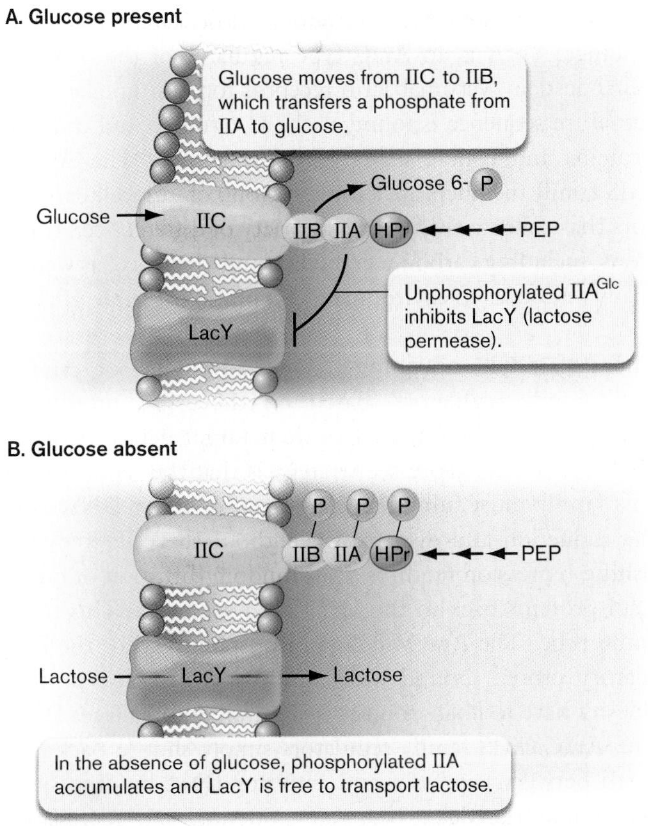
A. Glucose present
Glucose moves from IIC to IIB, which transfers a phosphate from IIA to glucose.
Glucose 6- P
Glucose
IIC
IIB IIA HPr
PEP
Unphosphorylated IIA^Glc inhibits LacY (lactose permease).
LacY
B. Glucose absent
P P P
IIC
IIB IIA HPr
PEP
Lactose
LacY
Lactose
In the absence of glucose, phosphorylated IIA accumulates and LacY is free to transport lactose.

FIGURE 10.11 ▪ Glucose transport via the phosphotransferase system inhibits LacY (lactose permease). **A.** Phosphoenolpyruvate (PEP) contributes a phosphate to components of the PTS system. The phosphate is eventually passed to glucose during transport. When glucose is present, the level of phosphorylated IIA^{Glc} is low because glucose siphons off the phosphate. IIA^{Glc} then interacts with and inhibits LacY (lactose permease) activity, thereby excluding transport of inducer (inducer exclusion). **B.** In the absence of glucose, the phosphorylated forms of glucose-specific IIA^{Glc} and $IIBC^{Glc}$ accumulate because they cannot pass the phosphate to substrate (there is no glucose). LacY functions in this situation to transport lactose.

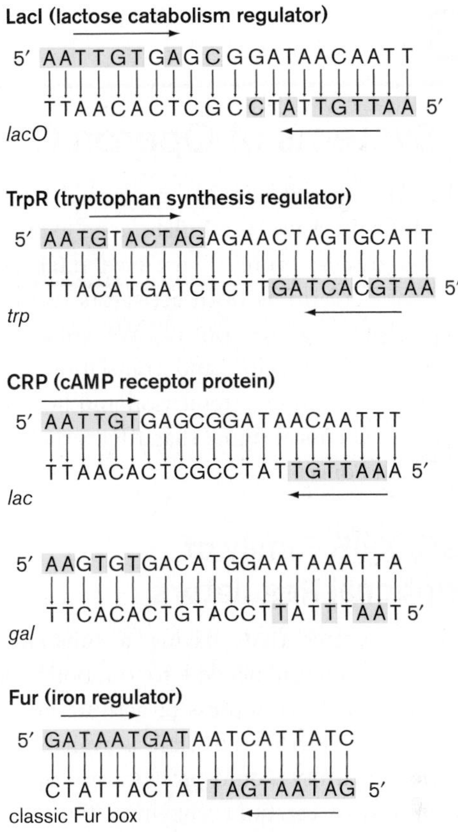
LacI (lactose catabolism regulator)
5′ AATTGTGAGCGGATAACAATT
TTAACACTCGCCTATTGTTAA 5′
lacO
TrpR (tryptophan synthesis regulator)
5′ AATGTACTAGAGAACTAGTGCATT
TTACATGATCTCTTGATCACGTAA 5′
trp
CRP (cAMP receptor protein)
5′ AATTGTGAGCGGATAACAATTT
TTAACACTCGCCTATTGTTAAA 5′
lac
5′ AAGTGTGACATGGAATAAATTA
TTCACACTGTACCTTATTTAAT 5′
gal
Fur (iron regulator)
5′ GATAATGATAATCATTATC
CTATTACTATTAGTAATAG 5′
classic Fur box

FIGURE 10.12 ▪ Examples of DNA regulatory sequences. The sequences shown are located upstream of the genes noted in italics. Inverted repeats are shown in yellow. Note that in some inverted repeats, occasionally a base is not repeated (only those bases that repeat are highlighted in the diagram). Arrows indicate the direction of symmetry.

programs designed to reveal possible binding sites. Finding an inverted repeat within 100–200 bases of a promoter suggests the presence of a regulatory binding site. Finding very similar sequences in front of two or more different promoters controlled by the same regulatory protein is stronger evidence that, in fact, the sequences represent regulatory binding sites. Proof ultimately requires showing that the protein will bind a piece of DNA containing this sequence. Protein-DNA binding can be demonstrated in several ways, as described in Section 12.3.

To Summarize

- **The lactose utilization *lacZYA* operon** of *E. coli* was the first gene regulatory system described.
- **LacI binds as a tetramer** to the operator region and represses the *lac* operon by preventing open complex formation by RNA polymerase.
- **Beta-galactosidase (LacZ)**, when at low concentration, cleaves and rearranges lactose to make the inducer allolactose.
- **Allolactose complexes to LacI**, reducing repressor affinity for the operator and allowing induction of the operon.
- **The cAMP-CRP complex** binds near the *lac* promoter region and stimulates RNA polymerase through CRP interaction with the C-terminal domain of the RNA polymerase alpha subunit. cAMP-CRP regulates many types of operons.
- **In catabolite repression**, a preferred carbon source (for example, glucose) prevents the induction of an operon (for example, *lac*) that enables catabolism of a different carbon source (lactose). Glucose transport through the PTS system causes catabolite repression by inhibiting LacY permease activity (inducer exclusion) and lowers cAMP levels.

10.3

Other Systems of Operon Control

Besides the *lac* operon model, bacteria have surprisingly many different ways to regulate gene expression. In this section we discuss proteins that have dual regulatory functions—that perform as both activators (positive control) and repressors (negative control). We will also see how the coupling of transcription and translation is used to regulate gene expression by attenuation and how an idling ribosome sends a chemical signal that affects the expression of many genes and operons.

The AraC/XylS Family of Transcriptional Regulators

What could be better than having a substrate induce expression of the operon needed to catabolize it? How about a regulator that can repress or activate gene expression, depending on whether substrate is available? A regulator with this versatility would provide very tight control over the synthesis of catabolic enzymes. One such regulator, called AraC, regulates the genes encoding arabinose catabolism. When arabinose is absent, AraC represses expression of the genes that break down arabinose; when arabinose is present, AraC activates these same genes. The products of these genes ultimately convert the five-carbon sugar L-arabinose to D-xylulose 5-phosphate, an intermediate in the pentose phosphate shunt, a pathway that provides reducing energy for biosynthesis (see Section 13.5).

Computer analysis of microbial genomes reveals a family of over 10,000 regulators with homology to AraC and the closely related XylS activator (xylose catabolism). For instance, YbtA is an AraC-type regulator of the *Yersinia pestis* pesticin/yersiniabactin receptor for iron uptake. This signature sequence is found at the C-terminal end of these proteins and contains DNA-binding motifs. The AraC/XylS family members are a diverse group of intracellular sensors that collectively regulate a variety of different cell functions, including carbon metabolism and virulence, as well as many responses to environmental conditions (**Table 10.1**).

The AraC/XylS advantage. One advantage of AraC/XylS family regulators, based on the AraC prototype, is that they remain affixed to their target genes. A drawback of the LacI repressor strategy is that the repressors in this family must fully dissociate from operator DNA during induction and disperse throughout the cell. Reestablishing repression requires slow, random diffusion of bulky LacI proteins back to the *lacO* DNA sequence. This takes some time. The AraC/XylS family strategy keeps the regulatory protein bound near target operons. The protein doesn't have to float around "searching" for its target DNA site. AraC/XylS family regulators simply shuttle back and forth between repressor and activator DNA-binding sites, shortening the delay between induction and repression. The known small chemical inducer molecules for these regulators quickly diffuse through the cytoplasm to find their cognate regulatory proteins already camped at target promoters.

The AraC strategy. The model for regulation by AraC/XylS regulators is based primarily on AraC itself, the most intensely studied family member. AraC is a 33-kDa protein with a DNA-binding C-terminal domain attached by a flexible linker peptide to the N-terminal dimerization domain

TABLE 10.1

Examples of the AraC/XylS family of transcriptional regulators.

Regulator	Organism	Function
ExsA	*Pseudomonas aeruginosa* (human pathogen)	Controls type III secretion system in response to Ca^{2+}
ToxT	*Vibrio cholerae* (human pathogen)	Controls several virulence genes, including one encoding cholera toxin
TxtR	*Streptomyces scabies* (taproot pathogen)	Controls synthesis of thaxtomin, a plant toxin
RipA	*Corynebacterium glutamicum* (soil bacterium)	Represses aconitase and other iron-containing proteins
YbtA	*Yersinia pestis* (human pathogen)	Controls iron uptake transporters
NitR	*Rhodococcus rhodochrous* (plant pathogen)	Regulates synthesis of indole-3-acetic acid

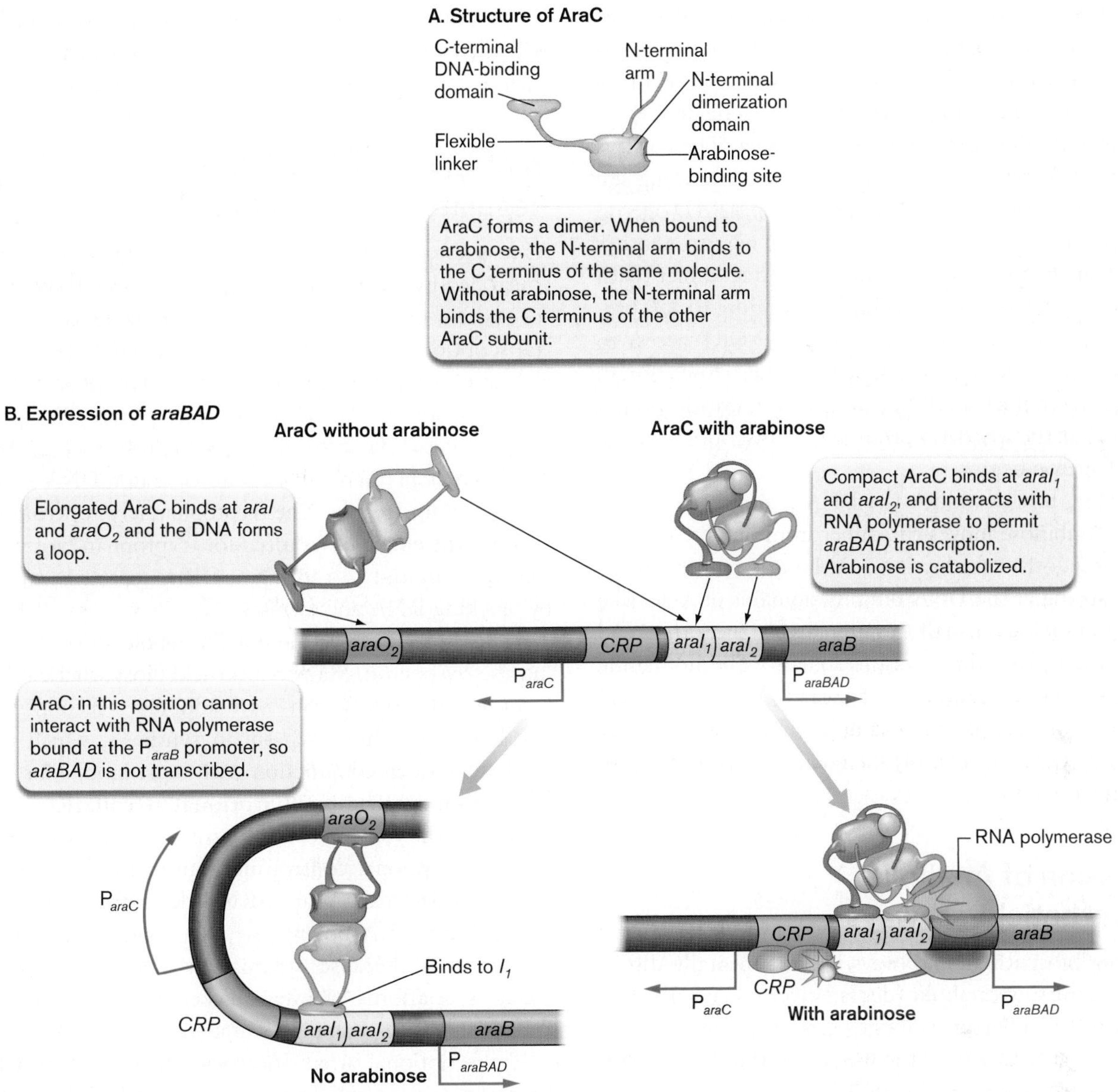
A. Structure of AraC
C-terminal DNA-binding domain
N-terminal arm
N-terminal dimerization domain
Flexible linker
Arabinose-binding site
AraC forms a dimer. When bound to arabinose, the N-terminal arm binds to the C terminus of the same molecule. Without arabinose, the N-terminal arm binds the C terminus of the other AraC subunit.
B. Expression of araBAD
AraC without arabinose
AraC with arabinose
Elongated AraC binds at araI and araO2 and the DNA forms a loop.
Compact AraC binds at araI1 and araI2, and interacts with RNA polymerase to permit araBAD transcription. Arabinose is catabolized.
araO2
CRP
araI1
araI2
araB
ParaC
ParaBAD
AraC in this position cannot interact with RNA polymerase bound at the ParaB promoter, so araBAD is not transcribed.
Binds to I1
No arabinose
RNA polymerase
With arabinose

FIGURE 10.13 ■ Regulation of the *araBAD* operon by the AraC regulator. A. View of AraC showing dimerization and DNA-binding domains. **B.** Alternative conformations of AraC dimers. The different conformations change the location of where the dimer can bind DNA. The region shown is about 400 bp. Note that O_2 is an operator, and I_1 and I_2 are other DNA-binding sites that, when occupied by AraC, induce expression.

(**Fig. 10.13A**). AraC forms a dimer in vivo that can assume one of two conformations depending on whether arabinose is available. When arabinose is absent, the dimer exists in a rigid, elongated form because the N-terminal arm of each monomer binds to its own C-terminal domain. This form represses expression of the *araBAD* operon for arabinose degradation. When arabinose is present, however, the dimer assumes a more compact form because the N-terminal arm of one monomer binds to the C-terminal domain of the other monomer. This compact form activates expression of *araBAD* (**Fig. 10.13B**).

Thought Question

10.5 When the *lacI* gene of *E. coli* is missing because of mutation, the *lacZYA* operon is highly expressed regardless of whether lactose is present in the medium. Based on **Figure 10.13** and its illustration of arabinose operon expression, what would happen to *araBAD* expression if AraC were missing? Why?

How does AraC act as both repressor and activator? There are three important DNA-binding sites to which

AraC can bind. Two of them, O_2 and I_1, are widely separated and flank a binding site for cAMP-CRP that, as discussed in Section 10.2, is a global activator of many bacterial genes, including *araBAD*. The elongated AraC dimer (no arabinose) can bind these two sites. The result is a looped DNA structure that blocks the CRP-binding site and limits activation by cAMP-CRP. The *araBAD* operon is not expressed. However, the I_1 site sits next to another site, I_2, near the promoter. The compact AraC dimer (with arabinose) can bind only to these sites. Binding to I_1I_2 prevents DNA loop formation, allows cAMP-CRP access to the CRP DNA-binding site, and brings AraC close enough to physically contact an RNA polymerase already bound (but stuck) at the *araBAD* promoter. Transcription of the *araBAD* operon begins.

According to this model, the DNA region is rarely free of AraC. Arabinose moves the dimer from one site (repression) to the other (induction). Although family resemblance is strong at the DNA-binding domains of AraC-like proteins, homology usually evaporates at the other end of the protein, the dimerization domain. The nonfamily domain is where these proteins appear to bind or respond to a particular ligand (for example, arabinose). But for most of the AraC/XylS family members, the identity of the ligand remains a mystery.

Repression of Anabolic (Biosynthetic) Pathways

Repressing biosynthetic pathways is fundamentally different from repressing catabolic (degradative) systems. Repressor proteins that control catabolic pathways, such as lactose degradation, typically bind the initial substrate or a closely related product (for example, allolactose in the case of the *lac* operon). Binding the substrate decreases repressor protein affinity for operator DNA. Thus, increased concentration of the substrate or inducer actually removes the repressor from the operator and derepresses expression of the operon. This derepression makes sense because the cell "wants" to make the enzymes that use the substrate as a carbon and energy source.

In contrast, genes encoding biosynthetic enzymes are regulated by repressors (called inactive aporepressors) that must bind the end product of the pathway (for example, tryptophan for the *trp* operon) to become active repressors. The pathway product that binds the aporepressor is called a corepressor. Binding of the corepressor (end product) to the repressor increases the repressor's affinity for the operator sequence upstream of the target gene or operon. As the concentration of end product, such as an amino acid or nucleotide, increases in the cell beyond what is needed to support growth, the cell will shut the biosynthetic pathway down and not waste energy making a superfluous pathway or compound.

In Attenuation, Translation Regulates Transcription

As just noted, many amino acid biosynthetic pathways are controlled by transcriptional repression, in which a repressor protein binds to a DNA operator sequence to prevent transcription. For instance, when internal tryptophan levels exceed cellular needs, the excess tryptophan (acting as a corepressor) will bind to an inactive repressor protein, TrpR, converting it to an active repressor (**Fig. 10.14**). TrpR repressor then binds to an operator DNA sequence positioned upstream of the tryptophan (*trp*) operon, which encodes the enzymes required for tryptophan biosynthesis. Repressor bound to the *trp* operator represses expression by blocking RNA polymerase.

Repression, however, is not the whole story in regulating the *trp* operon. Many amino acid biosynthetic operons, including the *trp* operon, have adopted a second strategy for down-regulating tryptophan synthesis, which can be used alone or in conjunction with repression. This second mechanism is called **transcriptional attenuation**. Attenuation uses the ribosome as a sensor of amino acid levels. It is a transcriptional control mechanism in which the ability to translate part of an mRNA determines whether the transcribing RNA polymerase can continue downstream transcription. Because attenuation halts transcription in progress, it affords an even quicker response to changing amino acid levels than does simple repression.

Transcriptional attenuation was discovered by Charles Yanofsky and his colleagues at Stanford University. While examining the beginning of the *trp* operon in *E. coli*, they discovered an odd DNA region located between the *trp* operator and *trpE*, the first structural gene. This region, called the **leader sequence**, does not directly control the biosynthesis of tryptophan, but instead determines whether RNA polymerase, already authorized to begin transcription, can proceed into the *trp* structural genes. The leader sequence encodes a peptide, but the peptide has no enzymatic function. The importance of the leader sequence lies in a pair of tryptophan codons (UGGUGG) embedded within it. The act of translating those codons couples the intracellular level of tryptophan (measured as charged tryptophanyl-tRNA) to transcription. If the level of tryptophan is sufficient to maintain a level of charged tryptophanyl-tRNA adequate to support growth at a rapid rate, then the mRNA of the leader region jettisons RNA

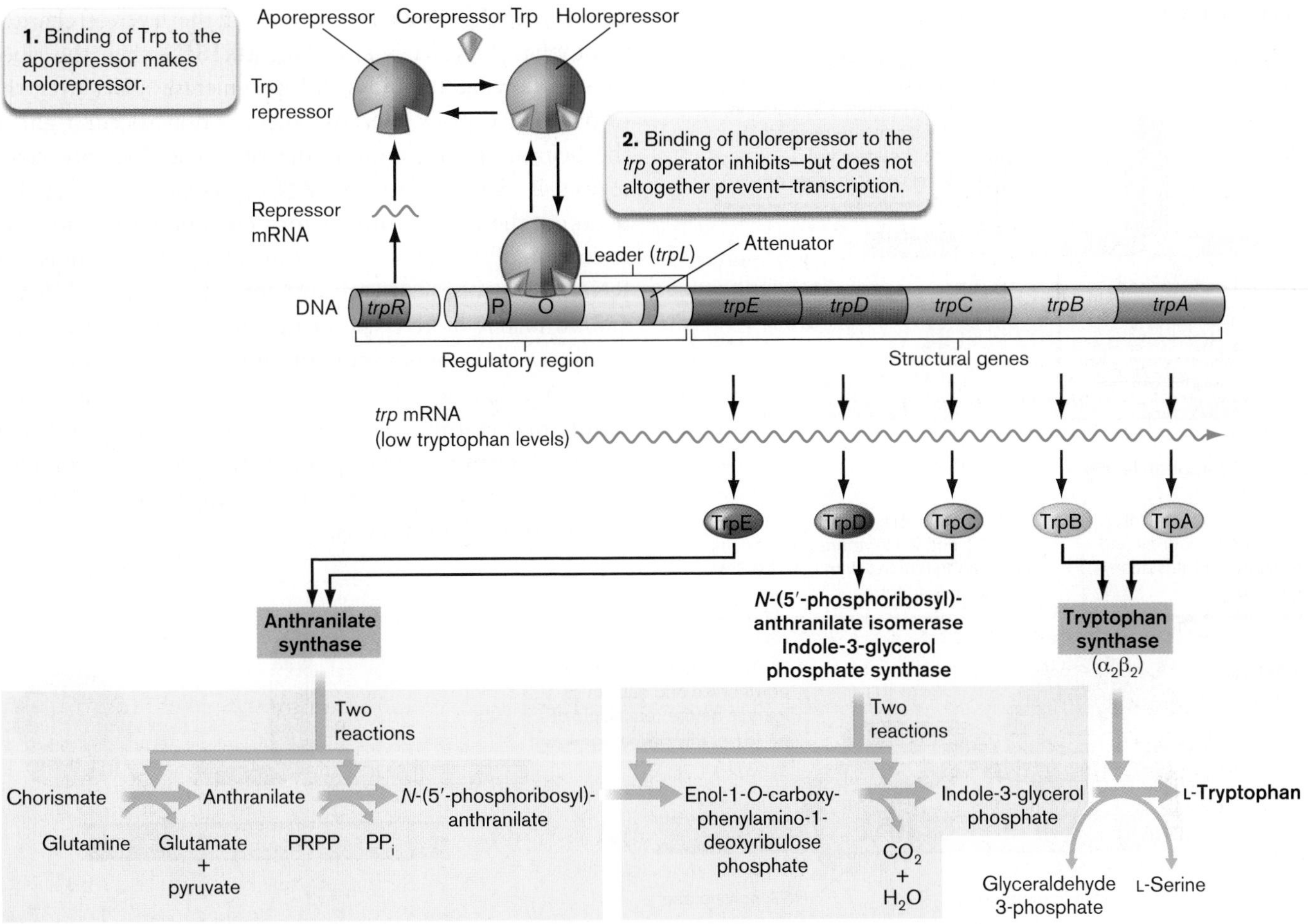
1. Binding of Trp to the aporepressor makes holorepressor.
Aporepressor
Corepressor Trp
Holorepressor
Trp repressor
2. Binding of holorepressor to the trp operator inhibits—but does not altogether prevent—transcription.
Repressor mRNA
Leader (trpL)
Attenuator
DNA
trpR
P
O
trpE
trpD
trpC
trpB
trpA
Regulatory region
Structural genes
trp mRNA
(low tryptophan levels)
TrpE
TrpD
TrpC
TrpB
TrpA
Anthranilate synthase
N-(5′-phosphoribosyl)-anthranilate isomerase
Indole-3-glycerol phosphate synthase
Tryptophan synthase
(α2β2)
Two reactions
Two reactions
Chorismate
Anthranilate
N-(5′-phosphoribosyl)-anthranilate
Enol-1-O-carboxy-phenylamino-1-deoxyribulose phosphate
Indole-3-glycerol phosphate
L-Tryptophan
Glutamine
Glutamate + pyruvate
PRPP
PPi
CO2 + H2O
Glyceraldehyde 3-phosphate
L-Serine

FIGURE 10.14 ■ The tryptophan biosynthetic pathway in *E. coli* and repression of the tryptophan operon. The tryptophan biosynthetic enzymes and their encoding genes. TrpR aporepressor (inactive repressor) binds excess tryptophan when intracellular concentration exceeds need. The holorepressor (active repressor) then binds to the *trp* operator and prevents transcription. Note the long polycistronic message in blue. Repression lowers expression about 100-fold. PRPP = phosphoribosylpyrophosphate.

polymerase before it reaches *trpE,* the first structural gene involved in the biosynthesis of tryptophan.

The attenuation mechanism hinges on four complementary nucleotide stretches within the leader mRNA. These regions, numbered 1–4, can base-pair to form competing stem loop structures (**Fig. 10.15A**). Two of the stem loop structures are critical to the mechanism. These are the **anti-attenuator stem loop** formed by regions 2 and 3 and the **attenuator stem loop** (or terminator stem loop) formed by regions 3 and 4. If the 3:4 attenuator stem loop forms, then the RNA polymerase is ejected and transcription stops (**Fig. 10.15B**). Formation of the 2:3 anti-attenuator stem loop, however, prevents formation of the 3:4 stem loop because the 2:3 stem is longer than the 3:4 stem and, thus, more thermodynamically stable. The anti-attenuator stem allows RNA polymerase to transcribe into *trpE* (**Fig. 10.15C**). What controls which stem loop forms?

High tryptophan levels. Because the ribosome is very large, it can barrel through RNA stem loop structures. When the cell is replete with charged tryptophanyl-tRNA and needs no more (**Fig. 10.15B**), the ribosome quickly translates through the key tryptophan codons but runs into a translation stop codon between regions 1 and 2. The ribosome stalls in this position, enveloping region 2 and preventing formation of the 2:3 stem. As a result, once RNA polymerase transcribes through region 4, the 3:4 attenuator stem snaps together. This structure interacts with the RNA polymerase ahead of it and halts transcription. As you would expect, the ribosome dissociates after reaching the region 2 stop codon, but because the 3:4 stem loop is already in place, the anti-attenuator 2:3 stem loop does not form. Subsequent ribosome release leads to formation of a 1:2 stem structure, precluding all possibility of regions 2 and 3 annealing.

Low tryptophan levels. However, if the level of charged tryptophanyl-tRNA is low (**Fig. 10.15C**), then the ribosome following behind RNA polymerase stalls over the tryptophan codons. Because these codons occur right at the beginning of region 1, the ribosome does not cover region 2. So, as soon as RNA polymerase transcribes region 3, the 2:3 anti-attenuator stem loop forms and stops formation of the 3:4 attenuator stem loop. The result is that RNA polymerase can continue into the structural genes, and ultimately more tryptophan is made. (A new ribosome binds to a ribosome-binding site at the *trpE* message.)

Realize that for the *trp* operon, attenuation is a fine-tuning mechanism. The repressor provides the majority of control. However, transcriptional attenuation is a common regulatory strategy used to control many operons that code for amino acid biosynthesis.

A. Stem loop structures in attenuator region

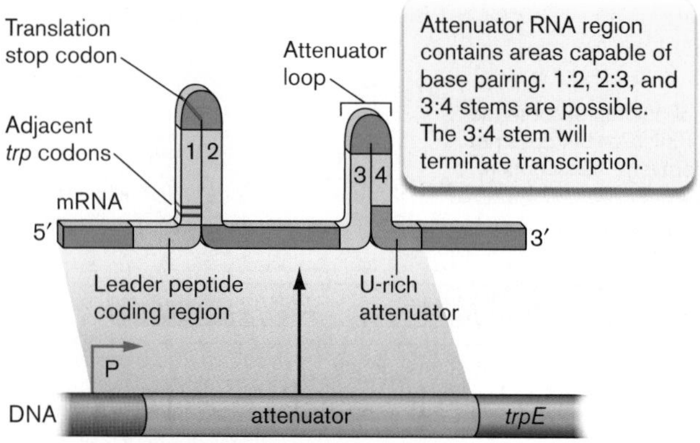
Translation stop codon
Attenuator loop
Attenuator RNA region contains areas capable of base pairing. 1:2, 2:3, and 3:4 stems are possible. The 3:4 stem will terminate transcription.
Adjacent trp codons
1 2
3 4
mRNA
5′
3′
Leader peptide coding region
U-rich attenuator
P
DNA
attenuator
trpE

B. High tryptophan levels

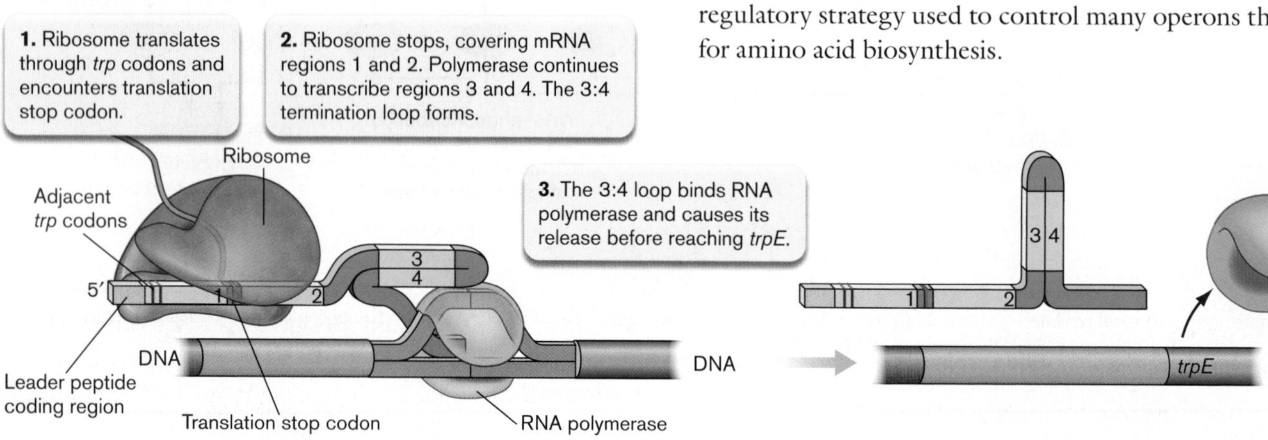
1. Ribosome translates through trp codons and encounters translation stop codon.
2. Ribosome stops, covering mRNA regions 1 and 2. Polymerase continues to transcribe regions 3 and 4. The 3:4 termination loop forms.
Ribosome
Adjacent trp codons
3. The 3:4 loop binds RNA polymerase and causes its release before reaching trpE.
5′
DNA
Leader peptide coding region
Translation stop codon
RNA polymerase
DNA
trpE

C. Low tryptophan levels

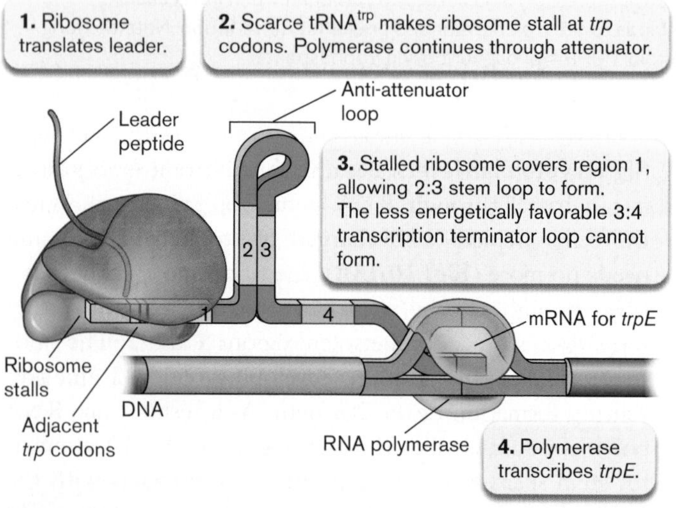
1. Ribosome translates leader.
2. Scarce tRNAtrp makes ribosome stall at trp codons. Polymerase continues through attenuator.
Anti-attenuator loop
Leader peptide
3. Stalled ribosome covers region 1, allowing 2:3 stem loop to form. The less energetically favorable 3:4 transcription terminator loop cannot form.
2 3
4
mRNA for trpE
Ribosome stalls
DNA
Adjacent trp codons
RNA polymerase
4. Polymerase transcribes trpE.

D.

AP PHOTO/J. SCOTT APPLEWHITE

FIGURE 10.15 ■ The transcriptional attenuation mechanism at the *trp* operon. **A.** Relationship between the mRNA attenuator region and encoding DNA. **B.** Attenuation when *E. coli* is growing in high tryptophan concentrations. **C.** Transcriptional read-through when *E. coli* is growing in low tryptophan concentrations. $tRNA^{trp}$ = tryptophanyl-tRNA. **D.** Charles Yanofsky (Stanford University) was instrumental in discovering attenuation and several other gene regulatory mechanisms while studying tryptophan metabolism. Here he is seen receiving the National Medal of Science from President George W. Bush in 2003.

Note: Even though translation is part of the attenuation control mechanism, attenuation is not considered translational control. The reason is that RNA polymerase, rather than the ribosome, is the target of the control. Translational control of gene expression is discussed later.

When Energy Sources Dwindle, Ribosome Synthesis Slows

During transitions from nutrient-rich to nutrient-poor conditions, microbes must contend with dramatic fluctuations in growth rate. This presents a problem. When a cell is growing rapidly, its molecular machinery is geared for peak performance, and the synthesis of new ribosomes is frenetic, trying to keep pace with rapid cell division. The more ribosomes a cell contains, the faster that cell can make new proteins and the faster it can grow. But what happens when the party's over—when poor carbon and energy sources cannot supply enough energy to maintain rapid cell division? Without a way of curbing ribosome construction, cells would soon fill with idle ribosomes. Under these conditions, bacteria undergo a process called the **stringent response**. The stringent response causes a decrease in the number of rRNA transcripts made for ribosome assembly and alters the expression of numerous other genes.

In the stringent-response strategy, idling ribosomes produce a signal molecule called guanosine tetraphosphate (ppGpp), which interacts with RNA polymerase and lowers its ability to transcribe genes encoding ribosomal RNA (**Fig. 10.16**). How is ppGpp made? When an uncharged tRNA binds at the ribosome A site, which can happen during amino acid starvation, a ribosome-associated protein called RelA transfers phosphate from ATP to GTP to form ppGpp. This signal nucleotide interacts with the beta subunit of RNA polymerase and diminishes its recognition of promoters for operons producing rRNA and tRNA. The result is the down-regulation of rRNA and tRNA synthesis. The less rRNA scaffolding there is available for building ribosomes, the fewer ribosomes will be produced.

This raises another question. Even though the synthesis of ribosomal RNA has been curtailed, won't the cell continue to waste resources on the synthesis of ribosomal proteins? It turns out that some ribosomal proteins can bind to the mRNA that encodes them and inhibit translation. So, when there is less rRNA in the cell, free ribosomal proteins accumulate in the cytoplasm, unassociated with ribosomes. These excess ribosomal proteins begin to bind to their own mRNA molecules and inhibit the translation of their own coding regions, as well as the coding regions of other ribosomal proteins residing on the same polycistronic mRNA. This process is called **translational control**

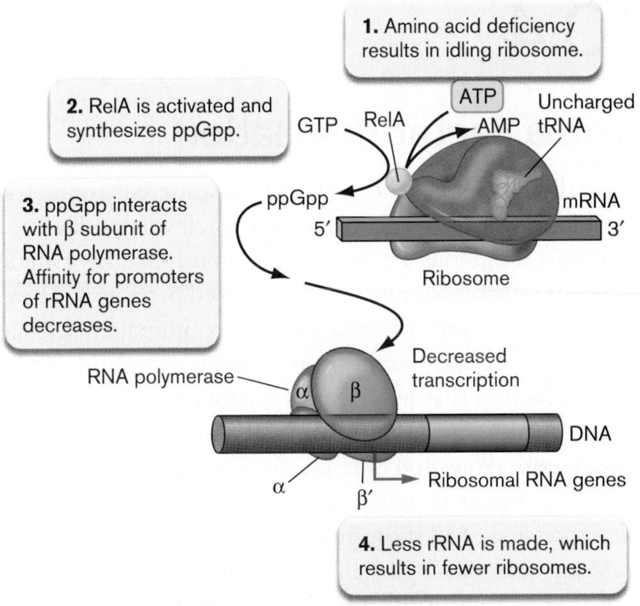
1. Amino acid deficiency results in idling ribosome.
2. RelA is activated and synthesizes ppGpp.
GTP
RelA
ATP
AMP
Uncharged tRNA
3. ppGpp interacts with β subunit of RNA polymerase. Affinity for promoters of rRNA genes decreases.
ppGpp
5′
mRNA
3′
Ribosome
RNA polymerase
α
β
Decreased transcription
DNA
α
β′
Ribosomal RNA genes
4. Less rRNA is made, which results in fewer ribosomes.

FIGURE 10.16 ▪ Ribosome-dependent synthesis of guanosine tetraphosphate and the stringent response.

because regulation affects the translation of an mRNA by ribosomes rather than transcription by RNA polymerase.

In this section we have described several ways bacteria control operon expression. There are many more examples, such as the riboswitches discussed in **eTopic 15.4**.

To Summarize

- **DNA-binding proteins** often recognize symmetrical DNA sequences.
- **Many anabolic pathway genes** (for example, for tryptophan biosynthesis) are repressed by the end product of the pathway (for example, the amino acid), which binds to a corepressor that inhibits transcription.
- **AraC-like proteins** are a large family of regulators, present in many bacterial species, that can activate or repress operons by assuming different conformations.
- **Attenuation** is a transcriptional regulatory mechanism in which translation of a leader peptide affects transcription of a downstream structural gene.
- **Idling ribosomes** synthesize the signal molecule ppGpp, which interacts with the beta subunit of RNA polymerase and decreases the affinity of RNA polymerase for rRNA gene promoters and a variety of other gene promoters needed for rapid growth. The result, called the **stringent response**, is lower transcription of these genes, reduced levels of rRNA, and a decrease in the number of ribosomes.

10.4

Sigma Factor Regulation

A **regulon** is a set of genes and operons scattered around the chromosome with related functions (such as tyrosine biosynthesis). Regulons are coordinately controlled by a single repressor or activator protein, as described previously. However, in many cases bacteria need to coordinately activate large sets of genes, operons, and regulons of seemingly disparate function that nevertheless collaborate to aid survival in particularly hostile environments. One way to indirectly regulate the expression of large sets of genes is to first regulate the synthesis or activity of the sigma factor that directs the expression of all those genes.

Many bacteria employ alternative sigma factors to direct transcription of distinct sets of genes (see Section 8.2). For example, many Gram-negative bacteria, such as *E. coli* or *Salmonella enterica,* use the sigma factor called sigma S (also called sigma-38, RpoS, σ^S, or σ^{38}) to initiate the transcription of a variety of genes associated with survival in stationary phase. The level of sigma S is low in exponentially growing cells but rises dramatically as cells enter stationary phase.

Sigma Factor Activity Can Be Regulated

The cell can regulate an enzymatic reaction by changing the amount of the enzyme in a cell or by changing the activity of an enzyme already present. The same is true for sigma factors. Some microbes use **anti-sigma factor** proteins to inhibit sigma factor activity. Anti-sigma factor proteins target specific sigma factors, blocking sigma access to core RNA polymerase). This strategy prevents expression of that sigma factor's target genes.

Salmonella uses an anti-sigma factor (FlgM) to time events involved in constructing flagella. The transcription factor sigma F (also called sigma-28, RpoF, σ^F, or σ^{28}) is required to synthesize proteins used in the last stages of flagellar biosynthesis. The anti-sigma factor FlgM, however, keeps sigma F function at bay until membrane assembly of the flagellar basal bodies is complete. Once completed, the basal body selectively secretes the anti-sigma factor from the cell. This frees intracellular sigma F to direct transcription of the final set of flagellar assembly genes.

But what happens when the blocked sigma factor is suddenly needed? How do cells counter anti-sigma factors? Some systems counter anti-sigma factors with **anti-anti-sigma factors**, which bind the blocking protein, like a decoy, and free the sigma factor to join core RNA polymerase. Other systems link the liberation of a bound sigma factor with a cell cycle event or with the assembly of a structure. *Bacillus* species, for instance, produce spores. During the process of making a spore, the cell couples formation of a forespore cross-wall (see Section 4.7) to the activation of an anti-anti-sigma factor, which triggers the release of a whole cascade of new sigma factors needed only to complete spore formation (discussed later in this section).

Sigma Factor Translation and Degradation Are Regulated

The synthesis or accumulation of sigma factors can also be regulated. Regulation of the heat-shock sigma factor sigma H (also called sigma-32, RpoH, σ^H, or σ^{32}) in *E. coli* illustrates how the synthesis of a sigma factor can be controlled at the level of translation and degradation (**Fig. 10.17**). Excessive heat, above 42°C for *E. coli,* causes proteins to denature and membrane structure to deteriorate. All cells subjected to heat above their comfort zone (optimal growth range) will express a set of proteins called heat-shock proteins. These proteins include chaperones that refold damaged proteins, and a variety of other proteins that affect DNA and membrane integrity. The transcription of many *E. coli* heat-shock genes requires sigma H. So, one of the first consequences of growth at elevated temperature is an increase in the amount of sigma H present in the cell.

The level of sigma H is regulated by two temperature-dependent mechanisms; one controls the speed (rate) of sigma H synthesis, while the other determines its rate of degradation. The gene encoding sigma H is *rpoH.* At 30°C, *rpoH* mRNA adopts a secondary structure at the 5′ end that buries a ribosome-binding site, so *rpoH* mRNA is poorly translated. A sudden rise in temperature melts this secondary structure and exposes the ribosome-binding site, allowing translation to occur more readily. Thus, heat shock increases sigma H synthesis, which in turn increases transcription of the heat-shock genes whose products include chaperones and proteases.

Proteolysis also controls sigma factor accumulation. At 30°C, the *rpoH* message is poorly translated, owing to secondary structure, as previously described, but some sigma H protein is made. Inappropriate expression of heat-shock genes at 30°C is prevented by the DnaK-DnaJ-GrpE chaperone system, which interacts with sigma H and shuttles it to various proteases for digestion (see **Fig. 10.17**). At 42°C, however, proteolysis of sigma H decreases, and sigma H is allowed to accumulate. Degradation decreases because at the higher temperature, the chaperones are siphoned away

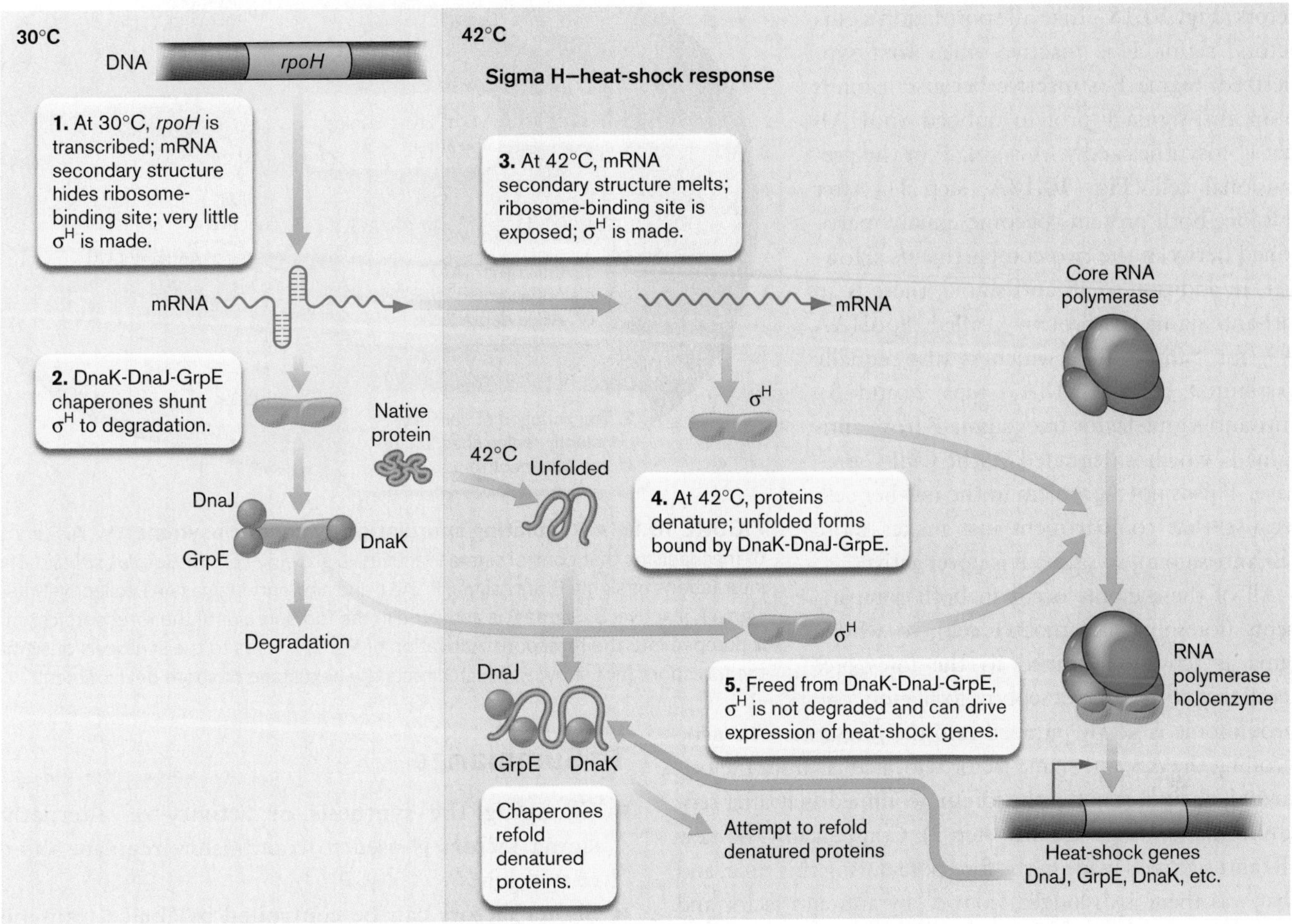
30°C
DNA
rpoH
1. At 30°C, rpoH is transcribed; mRNA secondary structure hides ribosome-binding site; very little σH is made.
mRNA
2. DnaK-DnaJ-GrpE chaperones shunt σH to degradation.
Native protein
DnaJ
GrpE
DnaK
Degradation
42°C
Sigma H—heat-shock response
3. At 42°C, mRNA secondary structure melts; ribosome-binding site is exposed; σH is made.
mRNA
σH
42°C
Unfolded
4. At 42°C, proteins denature; unfolded forms bound by DnaK-DnaJ-GrpE.
σH
DnaJ
5. Freed from DnaK-DnaJ-GrpE, σH is not degraded and can drive expression of heat-shock genes.
GrpE
DnaK
Chaperones refold denatured proteins.
Attempt to refold denatured proteins
Core RNA polymerase
RNA polymerase holoenzyme
Heat-shock genes
DnaJ, GrpE, DnaK, etc.

FIGURE 10.17 ▪ The heat-shock response of *E. coli.* Two mechanisms control sigma H levels. The small amount of sigma H that can be made at 30°C is met by the DnaK-DnaJ-GrpE chaperone system and shuttled toward degradation. At 42°C, however, misfolded cytoplasmic proteins siphon off the chaperone trio and release sigma H to direct transcription of the heat-shock genes.

from sigma H by the many other heat-denatured proteins formed. The new goal of the chaperones is to refold and rescue those damaged proteins. Chaperone redeployment frees sigma H to transcribe the heat-shock genes, which include the chaperone genes *dnaK, dnaJ,* and *grpE.* These genes have other promoters that drive basal expression. Thus, as the temperature rises, the amount of sigma H is increased by two temperature-dependent mechanisms: One increases translation by exposing the ribosome-binding site, while the second redeploys chaperones that direct its proteolysis.

In Sporulation, Different Sigma Factors Are Activated in the Mother Cell and Forespore

Bacillus and *Clostridium* species such as *B. anthracis* (anthrax) and *C. tetani* (tetanus) are soil organisms that produce spores when nutrients become scarce. Sporulation requires an asymmetrical cell division in which one of the compartments becomes the spore (see Section 4.7). The regulation of this system is quite complex, involving programs of transcription in the mother cell that are separate from those in the forespore compartment. How is this possible? Part of the answer is that at the time of septum formation, when the forespore is first produced, only 30% of the chromosome (the part near the replication origin) is actually inside the forespore. The remainder of the chromosome is slowly pumped in over a 15-minute period. The sporulating cell takes advantage of this delay to selectively express different genes in different compartments.

For example, sigma F is present in both forespore and mother cell compartments, but it directs the transcription of select genes only in the forespore. Compartment-dependent activation starts with an intricate timing mechanism that involves anti-sigma and anti-anti-sigma

A. Vegetative cell—σ^F is inactive

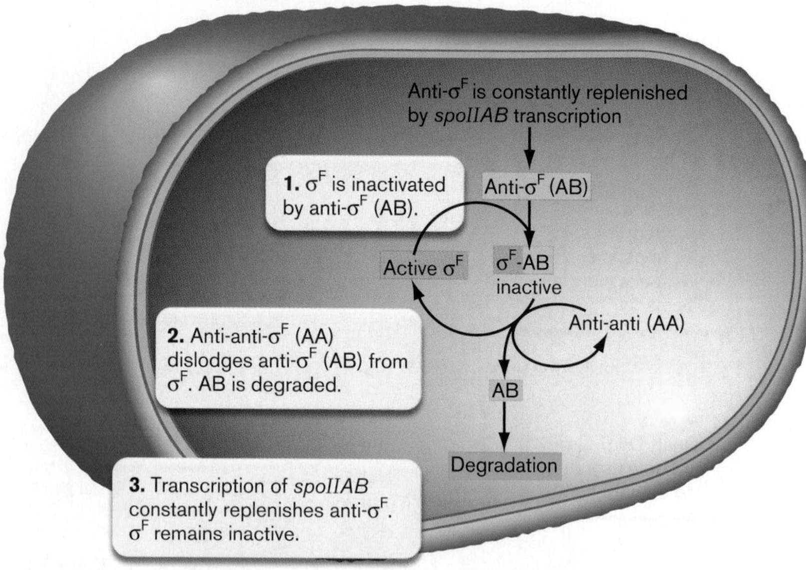

Anti-σF is constantly replenished by spoIIAB transcription
1. σF is inactivated by anti-σF (AB).
Anti-σF (AB)
Active σF
σF-AB inactive
Anti-anti (AA)
2. Anti-anti-σF (AA) dislodges anti-σF (AB) from σF. AB is degraded.
AB
Degradation
3. Transcription of spoIIAB constantly replenishes anti-σF. σF remains inactive.

FIGURE 10.18 ■ Regulating sporulation by genetic asymmetry. A. Summary of mechanisms that control sigma F activity in growing cells of *Bacillus subtilis*. The interactions of sigma F, anti-sigma F (AB), and anti-anti-sigma (AA) collectively keep sigma F inactive. **B.** Sigma F is activated in the forespore until the forespore chromosome is pumped into the forespore. Activation of sigma F leads to the synthesis of sigma G in the forespore (not shown), which directs the next stage of spore development.

factors (**Fig. 10.18**). Like all sporulation sigma factors, sigma F is inactive when first synthesized. Sigma F is inactive because it binds to an anti-sigma F protein dubbed SpoIIAB that is cosynthesized with sigma F in the predivisional cell (**Fig. 10.18A**, step 1). After division, both proteins become equally partitioned between the two compartments. However, in addition to an anti-sigma, there is an anti-anti-sigma F protein, called SpoIIAA (AA for "anti-anti"), which is also equally distributed (**Fig. 10.18A**, steps 2 and 3). Anti-anti sigma factor frees sigma F from anti-sigma F, which is degraded by the ClpPC protease. This is not a problem in the mother cell, because that compartment just makes more AB (anti-sigma), so sigma F is never active.

All of these events occur in both compartments (forespore and mother cell), so why is sigma F activation limited to the forespore? Recall that at the time of septum formation, one chromosome is slowly pumped into the forespore. The gene encoding the AB anti-sigma factor is located far from the origin and, as such, is not in the forespore immediately after septum formation (**Fig. 10.18B**, step 2). Consequently, no new AB (anti-sigma) is made in the forespore during this time, and what was there is dislodged from σ^F by anti-anti factor and degraded by ClpPC (**Fig. 10.18B**, steps 3 and 4). Depletion of anti-sigma in the forespore frees sigma F to act in the forespore compartment. Liberated sigma F transcribes the genes needed for subsequent sporulation steps; these genes reside in the part of the genome trapped early on in the forespore. Eventually, the *spoIIAB* gene enters the forespore and replenishes anti-σ^F.

Elegant proof for this model was obtained in the laboratory of Richard Losick, a leading scientist in the field of microbial development. In Losick's experiment, the gene for SpoIIAB (anti-sigma) was moved from its normal location far from the origin to a position near the origin. As a result, the anti-sigma F gene entered the forespore early, continually replenished anti-sigma F lost by degradation, and stopped sporulation.

The simple, unequal distribution of the chromosome during septal formation is heavily exploited during sporulation to trigger differential gene expression in two different cell compartments.

Thought Question

10.6 Predict the phenotype of a *spoIIAA* mutant that completely lacks SpoIIA anti-anti-sigma factor (see **Fig. 10.18**).

To Summarize

- **Changing the synthesis or activity of alternative sigma factors** is used to coordinately regulate sets of related genes.
- **Sigma factors can be controlled** by altered transcription, translation, proteolysis, and anti-sigma factors.
- **Secondary structures** at the 5′ end of mRNA can obscure access to ribosome-binding sites. Conditions that melt the secondary structures will increase translation of the sigma factor.
- **Chaperones** can direct certain sigma factors toward degradation by proteases. Conditions that draw those chaperones away from the sigma factor will allow the sigma factor to accumulate.
- **Sporulation** relies on the hierarchical activation of a series of different sigma factors.
- **Temporary asymmetrical distribution** of the chromosome between the forespore and the mother cell results in differential activation of compartment-specific alternative sigma factors.
- **Sigma F is a forespore-specific sigma factor** regulated by anti-sigma and anti-anti-sigma factors. Sigma F is preferentially activated by temporary asymmetrical distribution of the chromosome during the early stages of sporulation.

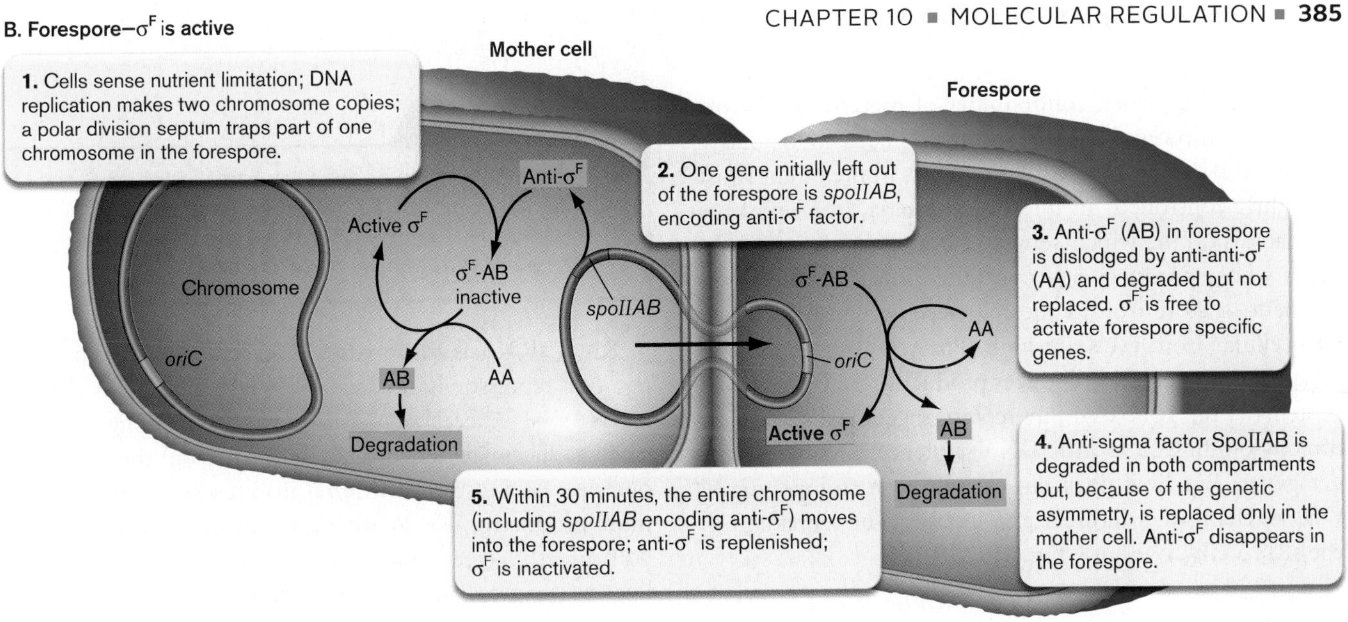
B. Forespore–σF is active
Mother cell
Forespore
1. Cells sense nutrient limitation; DNA replication makes two chromosome copies; a polar division septum traps part of one chromosome in the forespore.
2. One gene initially left out of the forespore is spoIIAB, encoding anti-σF factor.
3. Anti-σF (AB) in forespore is dislodged by anti-anti-σF (AA) and degraded but not replaced. σF is free to activate forespore specific genes.
4. Anti-sigma factor SpoIIAB is degraded in both compartments but, because of the genetic asymmetry, is replaced only in the mother cell. Anti-σF disappears in the forespore.
5. Within 30 minutes, the entire chromosome (including spoIIAB encoding anti-σF) moves into the forespore; anti-σF is replenished; σF is inactivated.
Anti-σF
Active σF
σF-AB
inactive
Chromosome
oriC
AB
AA
Degradation
spoIIAB
σF-AB
AA
oriC
Active σF
AB
Degradation

10.5

Regulatory RNAs

Are regions of the genome that are not traditional genes still important? What about the non-template strands of genes, do they have a function other than having complementarity to the template strand? A surprising fraction of a bacterial chromosome does not encode mRNA, rRNA, or tRNA, but instead makes untranslated RNA with regulatory functions. Regions between genes (intergenic) can encode small RNA (**sRNA**) molecules (100–200 nt) that affect the expression of other genes. In addition to intergenic sRNAs, there are so-called ***cis*-antisense RNA** molecules. A *cis*-antisense RNA is transcribed from the DNA strand opposite the mRNA-encoding template strand (see Fig. 8.20). *Cis*-antisense regulatory transcripts can base-pair with cognate sense mRNAs and control their expression. Regulatory RNA molecules typically affect gene expression posttranscriptionally, either by interacting with proteins or by binding to complementary sequences of target transcripts in ways that stimulate or prevent translation. Regulatory RNAs help control a variety of processes, such as plasmid replication, transposition, phage development, viral replication, bacterial virulence, environmental stress responses, and developmental control in lower eukaryotes.

sRNA Molecules Expand the Reach of Protein Regulators

Large regulatory proteins (for example, LacI or TrpR) typically control only a few genes, but their synthesis is energetically expensive. Small RNAs expand a protein's regulatory reach and, thus, maximize return on investment. The benefits of sRNAs are that they do not require protein synthesis to control gene expression, they diffuse rapidly, and they typically act on preexisting messages. Small RNAs represent one of the most economical ways to inhibit gene expression. The locations of some sRNA genes identified in *E. coli* are shown in **Figure 10.19**.

One example of an sRNA molecule in *E. coli* is RhyB, which regulates iron storage by expanding the reach of Fur, the ferric uptake regulator. Iron is hugely important for pathogens growing in the human body and for soil quality

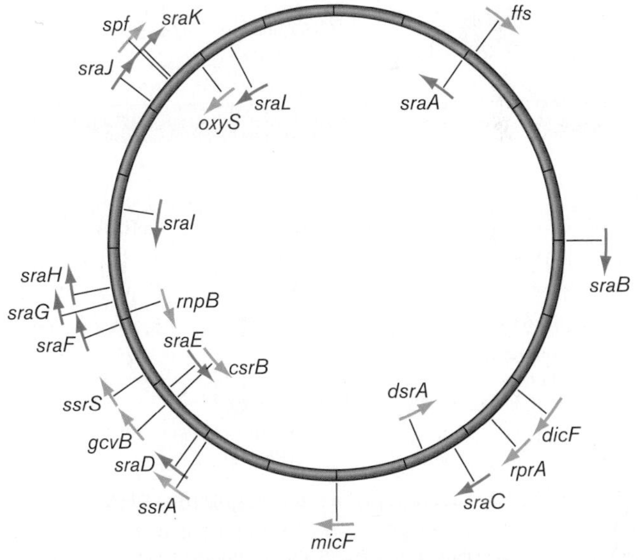
spf
sraK
ffs
sraJ
sraL
oxyS
sraA
sraI
sraH
sraB
sraG
rnpB
sraF
sraE
csrB
ssrS
dsrA
dicF
gcvB
sraD
rprA
ssrA
sraC
micF

FIGURE 10.19 ■ Predicted sRNA genes in *E. coli*. Locations of various sRNA genes identified in the *E. coli* genome. Orange arrows depict sRNAs of proven function; red arrows indicate predicted sRNAs of unknown function. Arrowheads indicate the direction of transcription.

by determining which communities of bacteria and plants can grow. However, too much iron increases oxidative stress and damages the cell. As a result, iron content must be tightly controlled. In intestinal bacteria, iron uptake is regulated by Fur, which senses iron and regulates several genes whose products either scavenge iron from the environment or store iron in the cell (see Section 4.2). When intracellular iron levels are high, Fur repress expression of scavenging genes but induces production of iron storage proteins (**Fig. 10.20A**). Fur represses gene expression by directly binding to specific DNA sequences in front of target genes. But how does Fur activate iron storage genes? One mechanism is indirect and involves Fur repression of the sRNA RhyB (90 nt) (**Fig. 10.20B**).

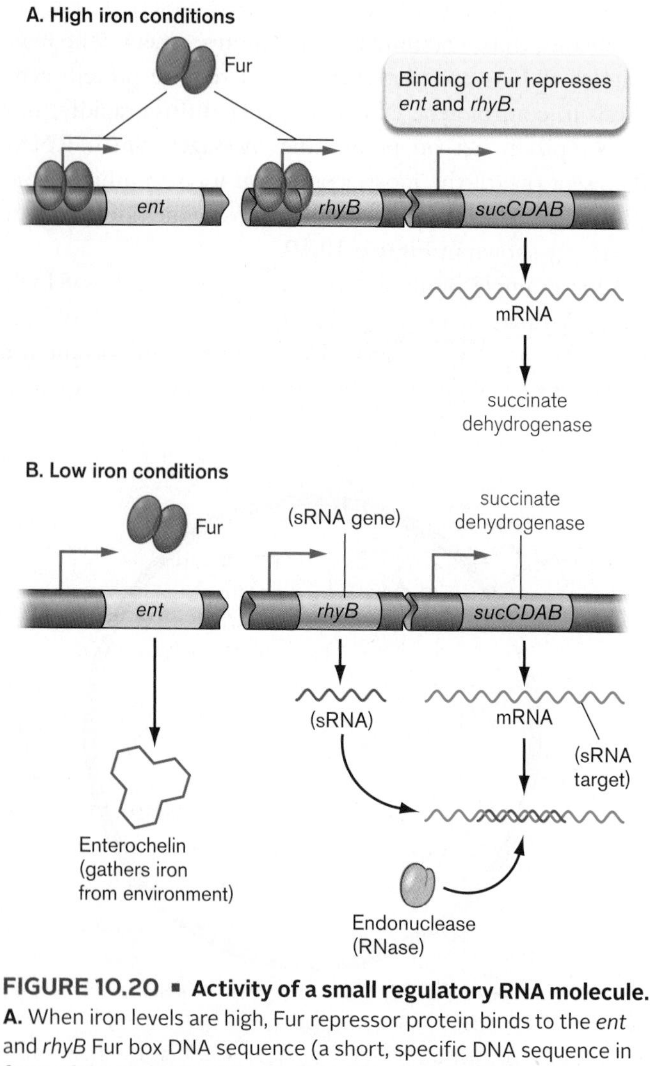
A. High iron conditions
Fur
Binding of Fur represses ent and rhyB.
ent
rhyB
sucCDAB
mRNA
succinate dehydrogenase
B. Low iron conditions
Fur
(sRNA gene)
succinate dehydrogenase
ent
rhyB
sucCDAB
(sRNA)
mRNA
(sRNA target)
Enterochelin (gathers iron from environment)
Endonuclease (RNase)

FIGURE 10.20 ▪ Activity of a small regulatory RNA molecule. **A.** When iron levels are high, Fur repressor protein binds to the *ent* and *rhyB* Fur box DNA sequence (a short, specific DNA sequence in front of the genes regulated by Fur) and represses their expression. Enterochelin is no longer made, but the *sucCDAB* message encoding succinate dehydrogenase can be translated. **B.** Under low iron conditions, RhyB sRNA is expressed. RhyB sRNA binds to the *sucCDAB* message and renders it susceptible to an RNase.

Thought Question

10.7 The relationship between the small RNA rhyB, the iron regulatory protein Fur, and succinate dehydrogenase is shown in Figure 10.20. Based on this regulatory circuit, will a *fur* mutant grow on succinate?

RhyB sRNA is made when iron levels are low because the Fur repressor is not active. RhyB then helps destroy mRNAs of several iron-storing and iron-using proteins (such as succinate dehydrogenase made from the *sucCDAB* operon). As a result, dwindling iron reserves can be put to more productive use. When iron is plentiful, however, Fur will directly repress *rhyB*. The lack of RhyB sRNA stabilizes the expression of the iron storage genes. Thus, succinate dehydrogenase is made, binds intracellular iron, and enables the use of succinate as a carbon and energy source.

Figure 10.21 illustrates the mechanism of another sRNA. The Gram-positive pathogen *Staphylococcus aureus* produces an sRNA, called RNAIII (514 nt long), that controls a large number of virulence genes. The 3′ end of RNAIII forms hairpin loops that can interact with hairpin loops at the 5′ ends of target mRNA molecules, one of which contains a ribosome-binding site. Sections of the loops hybridize, block the mRNA ribosome-binding site, and inhibit translation. The RNA-RNA hybrid is also an alluring target for a ribonuclease that will degrade the message. RNAIII is expressed after the bacterial population reaches a certain density during infection (see the discussion of quorum sensing in Section 10.8). It then inhibits the translation of, and helps degrade, mRNAs required earlier in the infection (for example, those for some exotoxins and surface attachment proteins).

How can you find sRNA genes in a genome? Genes encoding sRNA usually have a recognizable promoter but lack ribosome-binding sites and translational start codons that mark ORFs (see Sections 8.1 and 8.2). The laboratories of Susan Gottesman and Gisela Storz at the National Institutes of Health, or NIH (**Fig. 10.22**), discovered that genes encoding known sRNA molecules always occur in intergenic regions between genes encoding ORFs and exhibit high degrees of homology between related species (greater than 80%). Today, deep sequencing of RNA molecules in an organism will quickly reveal the presence of intergenic genes encoding sRNAs (see Sections 7.6 and 12.2). Characterizing their regulatory role is more problematic. Examples of some sRNAs are listed in **Table 10.2**.

Compared with prokaryotes, eukaryotic microbes possess an even broader array of small functional RNA molecules, including those of the spliceosome (similar to the prokaryotic CRISPR described in Section 9.2) and of various cytoplasmic

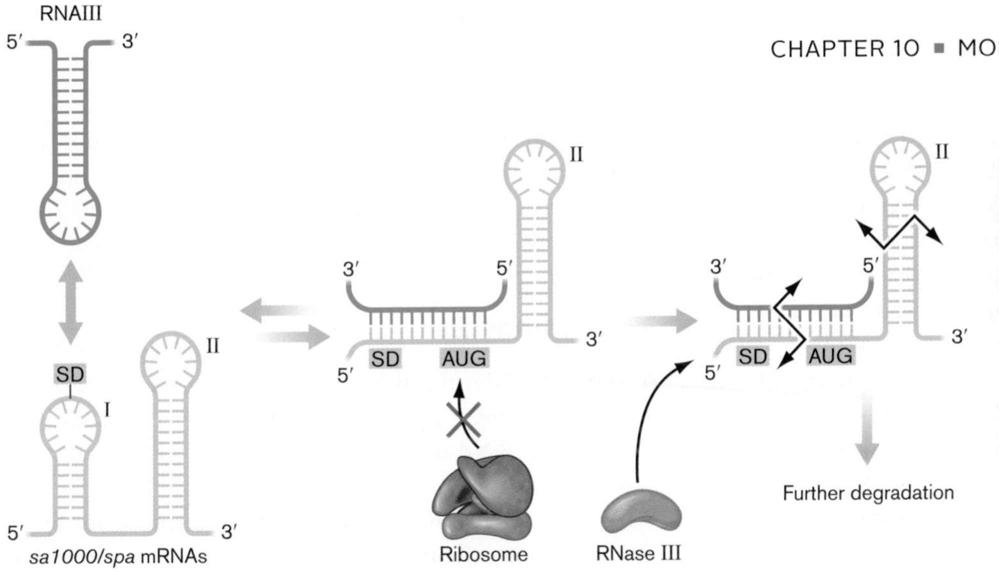
RNAIII
5′
3′
II
3′
5′
3′
SD
AUG
5′
II
3′
5′
3′
SD
AUG
5′
SD
I
II
5′
3′
sa1000/spa mRNAs
Ribosome
RNase III
Further degradation

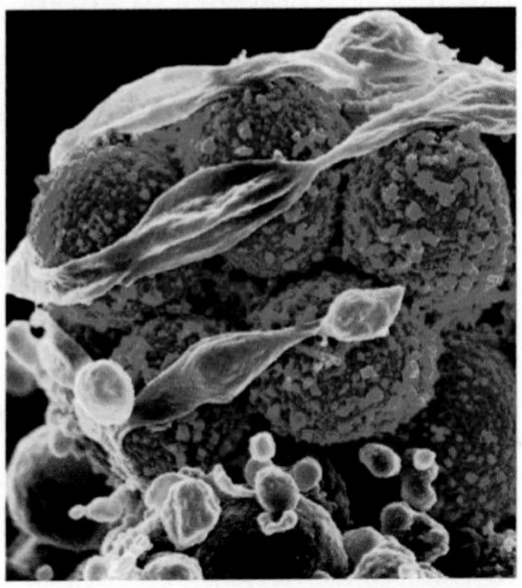

FIGURE 10.21 ■ Model for small RNA function. RNAIII in *Staphylococcus aureus* binds to its target mRNAs (*sa1000/spa*) via loop-loop interactions. The initial base pairings expand, leading to the formation of an extended duplex. Binding of RNAIII hinders ribosome binding and promotes the access to RNase III. SD (the Shine-Dalgarno ribosome-binding site) and AUG are marked green, RNAIII is in blue, and the mRNA target is in light blue. Bent arrows indicate degradation. **Inset:** SEM showing methicillin-resistant *S. aureus,* or MRSA (red), killing a human neutrophil. *Source:* Modified from S. Boisset et al. 2007. *Genes Dev.* **21**:1353–1366; inset photo from Adam D. Kennedy et al. 2008. *PNAS* cover image **105**(4). (c) 2008 by The National Academy of Sciences of the U.S.

A.

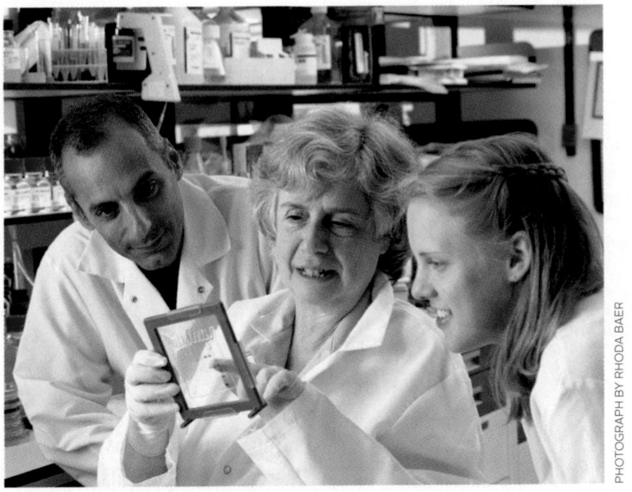
PHOTOGRAPH BY RHODA BAER

B.

COURTESY OF GISELA STORZ

FIGURE 10.22 ■ Susan Gottesman (A, center) and Gisela Storz (B), both from the National Institutes of Health, played instrumental roles in establishing the importance of sRNA molecules in bacteria.

TABLE 10.2

Examples of sRNA molecules.

sRNA	Function
CsrB	Regulates carbon storage
CyaR	Represses porin OmpX
IsrA	Regulates SOS response
GadY	Regulates acid resistance
OxyS	Regulates oxidative stress gene expression
RhyB	Expands Fur repressor
SprD	Regulates sbi (*Staphylococcus aureus* binder of IgG) immune evasion molecule
RNAIII	Regulates global regulator of *agr*-controlled virulence genes; also encodes a delta hemolysin
Qrr	Regulates quorum sensing

protein-RNA complexes. The genomes of eukaryotes such as yeast reveal thousands of potential sRNA sequences.

Effects of *cis*-Antisense RNA Are Local

We are increasingly finding that the sense DNA strands from many protein-encoding genes are transcribed to make so-called *cis*-antisense RNAs (**asRNAs**) and that these RNAs have regulatory functions. Typically, asRNAs are 700–3,000 nt long, originate within a protein-coding gene, and affect <u>only</u> that gene. These features distinguish asRNA from sRNA. Deep sequencing of the *E. coli* transcriptome revealed the presence of about 1,000 different *cis*-antisense RNA genes out of 4,290 protein-encoding genes. In fact, every microbe examined produced asRNA molecules.

Different asRNAs have different effects on their target gene. When bound to their sense mRNA counterparts, **asRNAs** can engineer attenuator loops that stop transcription, prevent mRNA translation, or trigger mRNA degradation. Even the simple act of asRNA transcription can produce collisions between converging RNA polymerase complexes that prematurely terminate transcription. An example of the collision mechanism was found for *Clostridium acetobutylicum*. The *ubiG-mccBA* operon encodes proteins needed to convert methionine to cysteine (**Fig. 10.23**) but is expressed only when methionine levels are high. When cytoplasmic methionine concentration is low, the RNA polymerase that transcribes the antisense *mccA* gene will collide with, and stop, the RNA polymerase transcribing the operon. The sense MccA transcript is not made, and the premature transcript is degraded. However, when methionine concentration is high, S-adenosylmethionine (made from methionine) binds to the asRNA at the S-box (a riboswitch; **eTopic 15.4**) and stabilizes a transcription termination loop. The *mccA* asRNA is not produced, so the operon mRNA is completed, MccA is made, and cysteine is synthesized. New research suggests that asRNAs in bacteria have a broad impact on microbial processes.

To Summarize

- **Small regulatory RNAs** found within bacterial intergenic regions regulate the transcription or stability of specific mRNA molecules and broaden the reach of protein regulators.
- **A *cis*-antisense RNA** is produced from the sense DNA strand of a protein-encoding gene and affects expression of only that gene. **The antisense nature** of regulatory RNAs allows these molecules to bind target mRNA. Binding can either stabilize the target mRNA or make it susceptible to degradation.
- **Identifying sRNAs** is more complex than identifying ORFs.
- **Eukaryotic microbes** use small interfering RNA to silence genes.

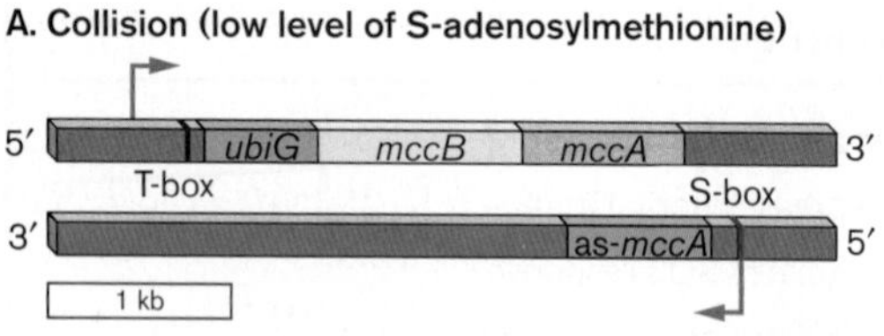
A. Collision (low level of S-adenosylmethionine)
5′
ubiG
mccB
mccA
3′
T-box
S-box
3′
as-mccA
5′
1 kb

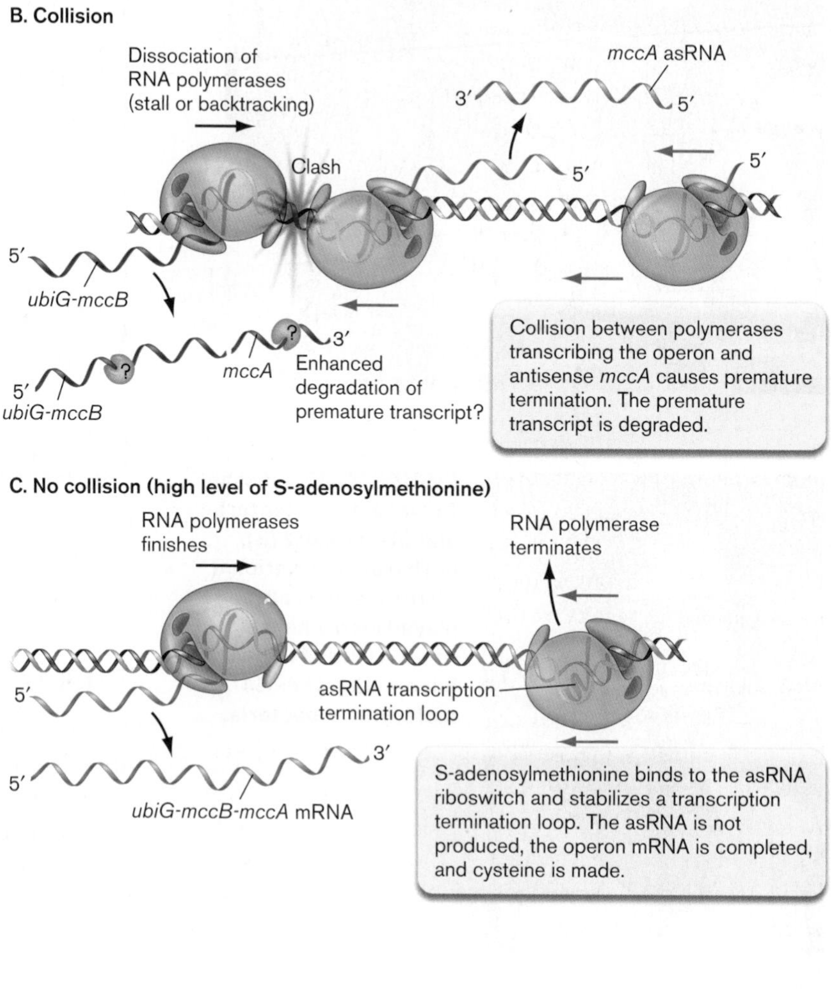
B. Collision
Dissociation of RNA polymerases (stall or backtracking)
mccA asRNA
3′
5′
Clash
5′
5′
5′
ubiG-mccB
?
?
3′
mccA
Enhanced degradation of premature transcript?
5′
ubiG-mccB
Collision between polymerases transcribing the operon and antisense mccA causes premature termination. The premature transcript is degraded.
C. No collision (high level of S-adenosylmethionine)
RNA polymerases finishes
RNA polymerase terminates
5′
asRNA transcription termination loop
3′
5′
ubiG-mccB-mccA mRNA
S-adenosylmethionine binds to the asRNA riboswitch and stabilizes a transcription termination loop. The asRNA is not produced, the operon mRNA is completed, and cysteine is made.

FIGURE 10.23 ▪ A *cis*-antisense RNA gene produces colliding RNA polymerases. The MccA protein from the *Clostridium acetobutylicum ubiG-mccB-mccAB* operon helps convert methionine to cysteine when cellular methionine levels are high. **A.** Gene arrangement for *ubiG-mccB-mccA* operon. The S box (specific for S-adenosylmethionine) and T box (specific for cysteine) sequences are riboswitches. Red arrows indicate promoters and transcription starts for the *ubiG* operon and the antisense gene. **B.** When methionine levels are low, the RNA polymerase (right) transcribing the antisense *mccA* gene collides with the RNA polymerase (left) transcribing the operon and causes premature termination. (Black and red horizontal arrows indicate direction of polymerase movement.) The MccA transcript is not made and the premature transcript is degraded. **C.** When methionine concentration is high, S-adenosylmethionine binds to a region of the asRNA called a riboswitch (S-box). This binding stabilizes a termination loop that stops antisense MccA transcription. Transcription of the operon is completed and cysteine can be made.

10.6

DNA Rearrangements: Phase Variation by Shifty Pathogens

All of the operon mechanisms previously described involve binding reactions between DNA, RNA, and proteins. More drastic means of control involve altering the DNA sequence itself.

Like a chameleon changing its color, some microbes use gene regulation to periodically change their appearance in a process called **phase variation**. Phase variation involves changing the amino acid composition of a particular protein on the bacterial surface. Bacteria that infect mammals, for example, must contend with immune systems whose mission is to destroy any and all invaders (discussed in Chapter 24). A central feature of immunity is the production of antibodies that recognize and bind foreign structures like microbial proteins and lipopolysaccharides (LPS). An infection will trigger the production of antibodies specific to the invader's component parts, such as pili, flagella, and LPS. Antibodies that bind to these microbial surface structures are especially useful for clearing an infection. Some pathogens, however, confound the immune system by changing the composition of these structures while the infection is in progress. This "shape-shifting" by the microbe, called **immune avoidance**, renders useless those antibodies specific for the old structure. The embattled immune system must start all over again making new antibodies, thus prolonging the course of infection.

Gene Inversion: An On-Off Switch

Flagellar phase variation in the Gram-negative bacterium *Salmonella enterica* is a classic example of the regulatory genre known as gene inversion. Gene inversion is a recombinational event that flips the orientation of a gene or DNA segment in the chromosome, thereby turning that gene, or an adjacent gene, on or off. A major cause of diarrhea, *S. enterica* periodically changes the type of protein, called flagellin, used to make its flagella. Each organism has two genes, widely separated on the chromosome, that encode different forms of flagellin. The mechanism of the switch is a site-specific DNA recombination that turns off one gene while turning on the other. The target of the switch is a 993-bp DNA fragment (or cassette), called the H region, that contains an outwardly directed promoter and a gene called *hin*, whose product, Hin recombinase (also called Hin invertase), mediates the recombination (**Fig. 10.24**). In antigenic parlance, the term "H antigen" refers to flagella, so the acronym Hin stands for H inversion. The DNA cassette is flanked by short (26-bp) inverted repeats called *hixL* (left) and *hixR* (right).

Note: An **inverted repeat** is a sequence found in identical (but inverted) forms at two sites on the same double helix (for example, 5′-ATCGATCGnnnnnnnnnnnnnnCGATCGAT-3′). A **direct repeat** is a sequence found in identical form at two sites on the same double helix (for example, 5′-ATCGATCGnnnnnnnnnnnnnnATCGATCG-3′). A **tandem repeat** is a direct repeat without any intervening DNA sequence (for example, ATCGATCGATCGATCGATCGATCG).

The Hin recombinase collaborates with other less specific DNA remodeling proteins, such as Fis, to engineer an association between the 26-bp left (*hixL*) and right (*hixR*) ends of the invertible DNA element. The two ends, each bound to a Hin monomer, are brought together by Hin-Hin protein interactions. DNA within the cassette then forms a loop. Hin cuts within the center of each *hix* site, producing staggered ends. An exchange of Hin subunits is thought to lead to strand inversion, so that the orientation of the DNA cassette is reversed relative to the flanking DNA on either side.

In one orientation, the outwardly directed promoter of the H region directs expression of H2 flagellin (encoded by *fljB*) and a repressor (FljA) that prevents transcription of the other flagellin gene, *fliC* (see **Fig. 10.24A**). After the switch, however, the promoter points in the wrong direction, so there is no production of H2 flagellin or FljA, the repressor of *fliC* (see **Fig. 10.24D**). Lacking this repressor, the *fliC* flagellin gene is expressed. Thus, H1 flagellin (present in phase 1 cells) is synthesized instead of H2 flagellin (present in phase 2 cells). The amino acid sequences, and thus the antigenicity, of the two flagellar proteins are different, so this mechanism allows *Salmonella* to change its appearance to a host immune system. In each generation, the rate of the reversible switch varies from about one cell in 10^3 to one in 10^5. Note, however, that this switch would not accomplish much if the infecting population of bacteria started out mixed—that is, producing both types of flagella. The initial infection must be of one phenotype.

Slipped-Strand Mispairing

A different type of phase variation relies on multiple, short sequence repeats in a gene. The repeats "confuse" DNA polymerase as it replicates, causing it to slip occasionally

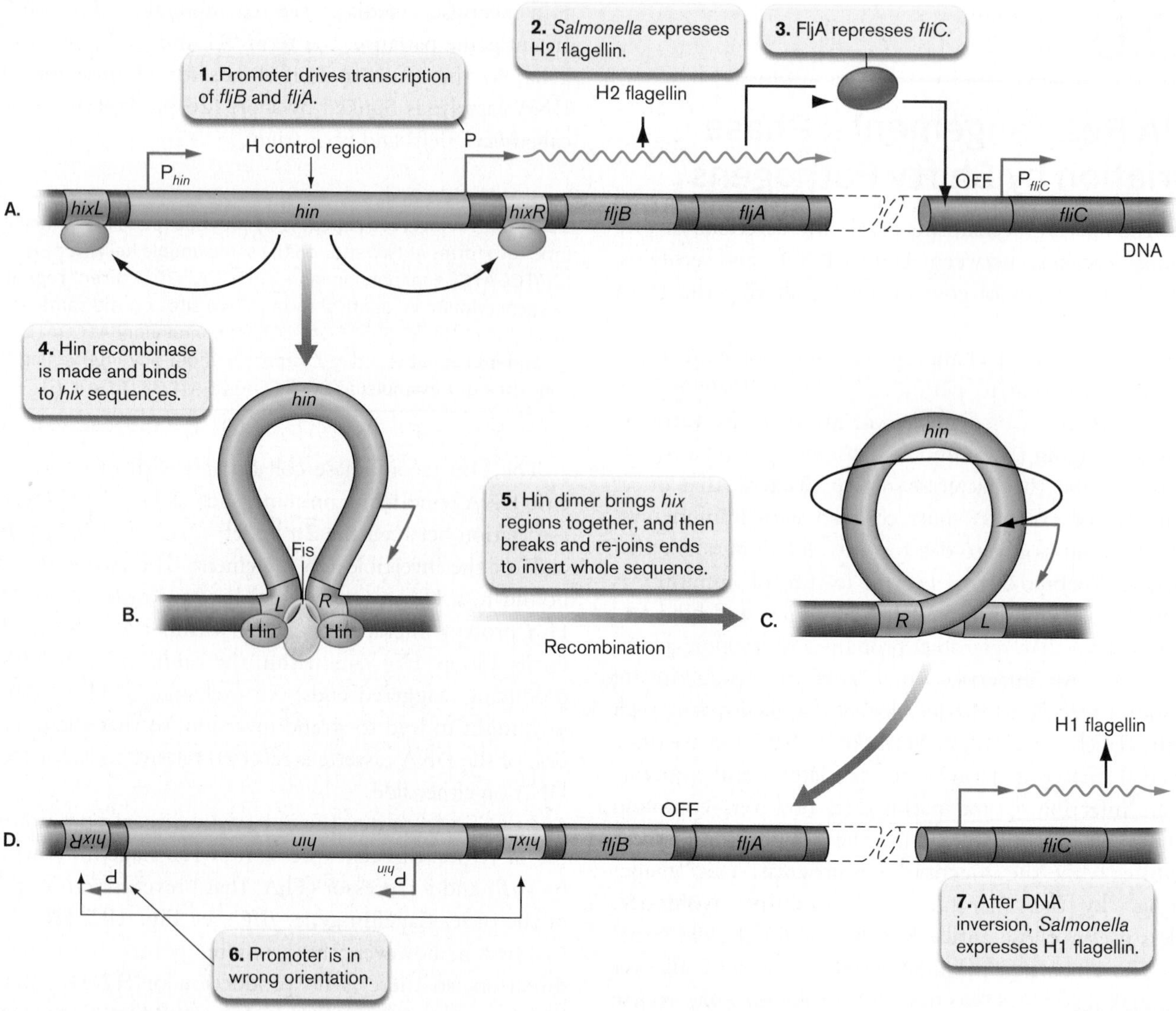

1. Promoter drives transcription of fljB and fljA.
2. Salmonella expresses H2 flagellin.
3. FljA represses fliC.
H2 flagellin
H control region
P
P_hin
OFF
P_fliC
A.
hixL
hin
hixR
fljB
fljA
fliC
DNA
4. Hin recombinase is made and binds to hix sequences.
hin
Fis
L
R
Hin
Hin
B.
5. Hin dimer brings hix regions together, and then breaks and re-joins ends to invert whole sequence.
Recombination
C.
hin
R
L
H1 flagellin
OFF
D.
hixR
hin
hixL
fljB
fljA
fliC
P
P_hin
6. Promoter is in wrong orientation.
7. After DNA inversion, Salmonella expresses H1 flagellin.

FIGURE 10.24 ■ Phase variation of flagellar proteins in *Salmonella enterica*. An invertible region containing a promoter controls the expression of two unlinked flagellar protein genes. In one orientation **(A)**, the promoter drives synthesis of H2 flagellin (*fljB*) and a repressor (FljA) of the H1 flagellin gene (*fliC*). Action by the Hin recombinase causes the segment to invert **(B, C)**, thereby removing the promoter. Because the repressor FljA is not formed, the gene for H1 can be expressed **(D)**.

during replication. Slippage either adds a repeat to, or deletes a repeat from, the gene and alters its translational reading frame. If the mRNA produced during transcription is out of frame, the protein is not made. This random process can alternately turn a gene off and then back on again in subsequent generations. A detailed example is discussed in **eTopic 10.3** using the *opa* genes of *Neisseria gonorrhoeae,* the causative agent of gonorrhea. Eukaryotic microbes, especially pathogenic sporozoa, possess elaborate mechanisms of phase variation. The trypanosome that causes "sleeping sickness" undergoes extensive genetic shuffling and mutation of its coat proteins over successive generations, essentially overwhelming the host immune system by presenting every possible form of antigen.

Thought Question

10.8 While viewing **Figure 10.24**, imagine the phenotype of a cell in which *fljB* has been deleted so that *fljA* is still expressed. Would cells be motile? What type of flagella would be produced? Would the cells undergo phase variation? What would happen if *fliC* was deleted?

To Summarize

- **In phase variation**, the structure of a bacterial protein reversibly changes from one form to another via a genetic mechanism.

- **An invertible promoter switch** regulates two genes of *Salmonella enterica* encoding two alternative structural types of flagellin.
- **Slippage while replicating** through repetitive DNA sequences, a process called **slipped-strand mispairing**, can reversibly activate or inactivate a series of outer membrane proteins in *Neisseria gonorrhoeae.*

10.7

Integrated Control Circuits

The regulatory circuits outlined to this point have involved mostly single-switch mechanisms. But microbes build on these mechanisms by combining multiple switches with redundant feedback controls that can form a kind of integrated gene circuit. Integrated circuits can send a virus down alternate lifestyle paths (lysis or lysogeny), couple the genetic and biochemical control of a metabolic pathway (ammonia assimilation), or time the events of a developmental cycle such as sporulation.

The deeper we delve into the regulatory workings that govern the expression of a cell's genome, the more clearly we see the hierarchy of control. Groups of simple control circuits are collectively regulated by other, more complex circuits, which are governed, in turn, by an even higher regulatory authority. One illustration of this hierarchy was already presented—namely, the role of cAMP and CRP in regulating lactose and arabinose catabolism. The *lacZYA* and *araBAD* operons are each controlled separately by dedicated regulatory switches—LacI and AraC, respectively. At a higher level, however, both operons are coregulated by cAMP and CRP in response to energy status (whether or not glucose is present and utilized). The systems we are about to discuss represent increasingly complex examples of nature's inventiveness and need for control.

The first example we will discuss here actually has nothing to do with transcriptional/translational or posttranslational control of protein production. It does illustrate the concept of two-component signal transduction, whereby one protein (the sensor kinase) senses a change in the environment and transmits a signal (via phosphorylation) to the second component in the system (the response regulator). Two-component signal transduction systems can be used to control transcription, but they can also be used to affect the physiology and behavior of a cell without changing transcription. The example we will use is chemotaxis, the ability of a bacterium to move toward environments favorable to growth and away from harsh environments.

Chemotaxis: How Bacteria Know Where They Are Going

Bacteria that use flagella to move do not simply dash about "hoping" they will swim in a favorable direction. They have a sense about where they want to go and where not to go. Bacteria such as *E. coli* purposefully move toward certain amino acids and carbohydrates present in media. They swim toward these chemicals by sensing their gradients. **Chemotaxis** is the ability of an organism to sense chemical gradients and modify its motility in response.

Before we discuss how these chemical gradients are sensed, let's first talk about how bacteria can change direction. Recall that the flagellar motor can rotate clockwise or counterclockwise (Section 3.7, Fig. 3.45) and that, on any one bacterium, all motors coordinate their rotations. Thus, all flagella on a single cell will rotate clockwise (or counterclockwise), and then suddenly switch to the opposite rotation. In the counterclockwise mode, all flagella sweep behind the cell, forming a rotating bundle that propels the organism forward in what is called smooth swimming or a "run." If enough flagellar rotors suddenly switch to a clockwise rotation, the bundle is disrupted and the bacterium "tumbles" in a random fashion. Once the flagellar rotors switch back to counterclockwise rotation, the bacterium will have reoriented itself in a different position, changing its swimming direction.

The key to chemotaxis is a mechanism that suppresses the number of tumbles an organism makes when it moves from a lower to higher concentration of an attractant chemical. For instance, an organism moving toward an attractant may tumble only twice in five seconds. In contrast, an organism moving in the wrong direction (that is, toward a lower concentration of attractant) may tumble eight times in five seconds. If, all of a sudden, the organism finds itself going in the right direction, the sensory transduction system will suppress the number of tumbles that occur, and the cell will continue moving in the right direction.

This is the same strategy you use to find a hamburger that you can smell cooking on a grill in your neighborhood but cannot see. You randomly walk around until the smell gets stronger, and then you keep moving in that direction until you happen upon it.

Weblinks Video showing transition from bundled to dispersed flagella (*see ebook*)

How does the bacterium suppress tumble frequency? Let's start at the end of the system and work backward. The natural bias of the flagellar rotor is counterclockwise rotation (in other words, smooth swimming). In order to tumble, the cell phosphorylates a protein called CheY. CheY-P interacts with a rotor protein and causes the rotor to reverse direction to turn clockwise (**Fig. 10.25A**, step 1). CheA protein is the kinase that phosphorylates CheY and is the "key" to chemotaxis. The more CheA kinase activity there is, the more CheY-P is made, resulting in a higher frequency of tumbling (reorienting) events. Another protein, CheZ, continually dephosphorylates CheY. When *E. coli* senses an attractant chemical like an amino acid at the cell surface, the result is decreased CheA kinase activity, which leads to lower CheY-P levels. The presence of fewer CheY-P molecules allows longer periods of counterclockwise rotation and, consequently, extended smooth swimming toward the attractant.

This makes sense, but how does a cell "know" that it has moved into an area with attractant? The answer begins with clusters of special membrane-spanning proteins called **methyl-accepting chemotaxis proteins** (**MCPs**, or chemoreceptors) located at the poles of *E. coli* and other bacteria (**Fig. 10.25A**, step 1). The periplasmic domains of different MCPs bind to different attractant molecules that flow into the periplasm as the cell moves. The cytoplasmic

A.

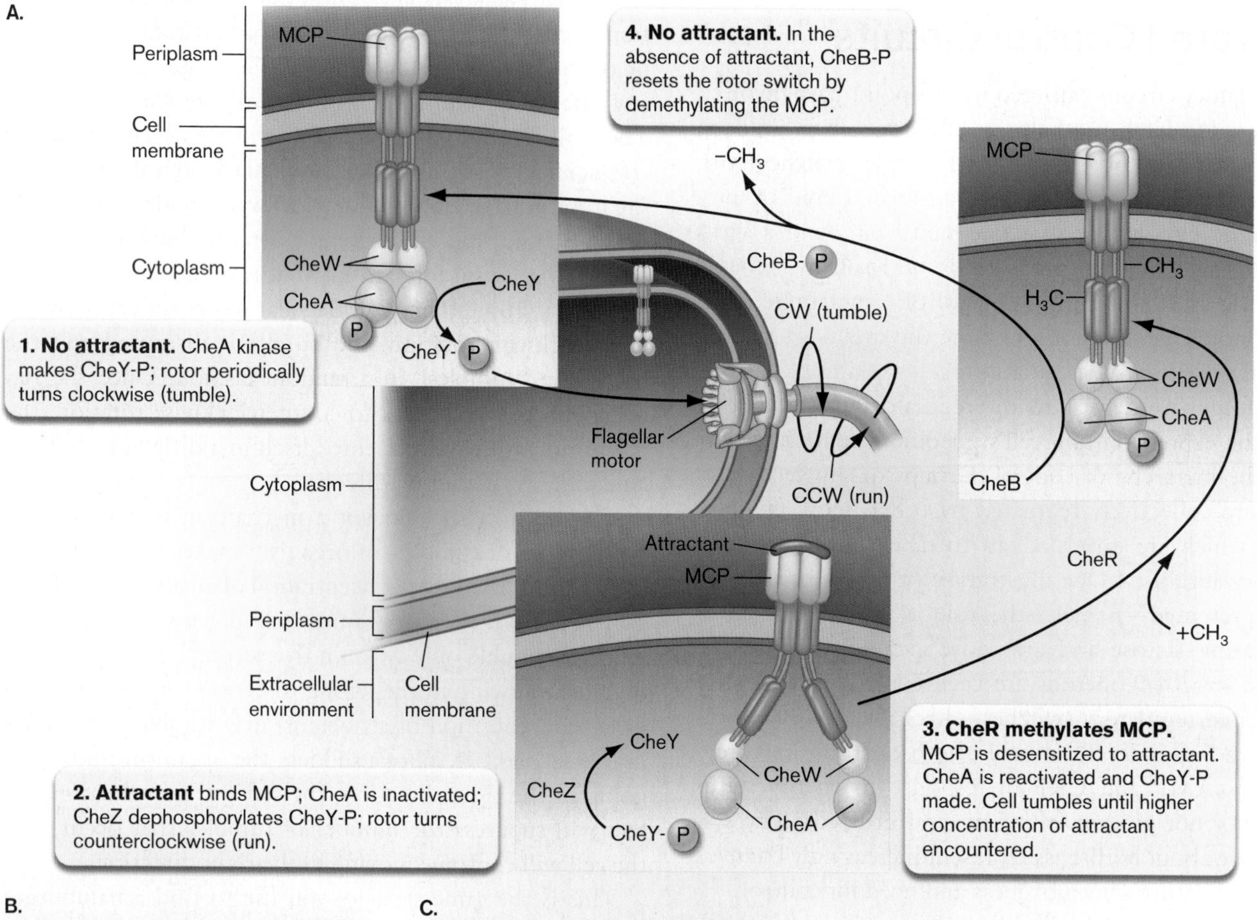
Periplasm
Cell membrane
Cytoplasm
MCP
CheW
CheA
P
CheY
CheY-P
1. No attractant. CheA kinase makes CheY-P; rotor periodically turns clockwise (tumble).
4. No attractant. In the absence of attractant, CheB-P resets the rotor switch by demethylating the MCP.
$-CH_3$
CheB-P
CW (tumble)
Flagellar motor
CCW (run)
MCP
$-CH_3$
H_3C-
CheW
CheA
P
CheB
CheR
$+CH_3$
Cytoplasm
Periplasm
Extracellular environment
Cell membrane
Attractant
MCP
CheY
CheZ
CheY-P
CheW
CheA
2. Attractant binds MCP; CheA is inactivated; CheZ dephosphorylates CheY-P; rotor turns counterclockwise (run).
3. CheR methylates MCP. MCP is desensitized to attractant. CheA is reactivated and CheY-P made. Cell tumbles until higher concentration of attractant encountered.

B.

COURTESY OF JOHN EMERSON

C.

COURTESY OF HOWARD C. BERG

FIGURE 10.25 ▪ Chemotaxis. **A.** Chemotaxis signaling pathway in *E. coli.* Ann Stock, Robert Wood Johnson Medical School **(B)**, and Howard C. Berg, Harvard University **(C)**, are major contributors to the fields of signal transduction, motility, and chemotaxis.

A. B.

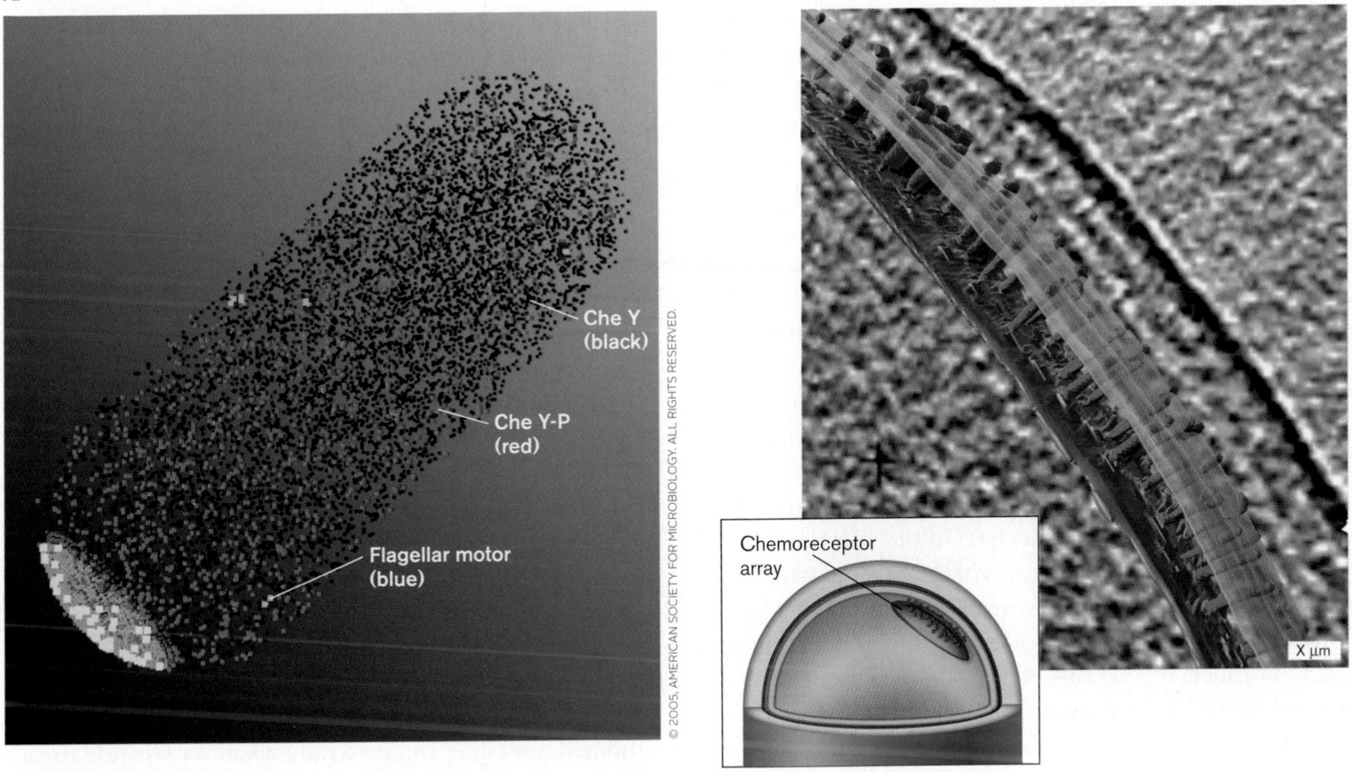
Che Y
(black)
Che Y-P
(red)
Flagellar motor
(blue)
Chemoreceptor
array
X µm

FIGURE 10.26 ■ Location of chemotaxis proteins. A. In this computer-generated view of *E. coli* chemotaxis proteins, notice the MCP cluster of receptors at one end of the cell (orange and yellow). In the absence of attractant chemicals, the complex generates a signal that changes the rotation of flagellar motors in the cell wall (light blue and purple). The signal is carried by the diffusion of a small protein, CheY (black), in its phosphorylated form, CheY-P (red) (note the gradient of CheY-P). **B.** Chemoreceptor array structure at a pole of *E. coli.* The tight hexagonal array of Tsr serine-detecting chemoreceptors (red) is held together at the cytoplasmic face by signal transduction proteins CheA and CheW (blue). Cryo-EM tomography. *Source:* Part A from Karen Lipkow et al. 2005. *J. Bacteriol.* **187**:45–53 (cover photo); part B from Peijun Zhang et al. 2007. *PNAS* **104**:3777–3781.

domain of each MCP binds to the CheA kinase noted earlier (via an intermediary protein, CheW) and controls CheA activity. When a chemoattractant chemical, such as serine, binds to the periplasmic side of the MCP, the conformation of the cytoplasmic domain changes and inhibits CheA kinase activity (**Fig. 10.25A**, step 2). The CheY-P level then decreases (because CheZ continues to dephosphorylate CheY-P), and smooth swimming ensues.

This, too, makes sense when an organism first encounters a chemoattractant, but how does the cell know to keep moving into even higher concentrations? As it happens, the conformational change in the MCP that inactivates CheA kinase activity also subjects the cytoplasmic side of the MCP to methylation (via *S*-adenosylmethionine) by CheR methylase (**Fig. 10.25A**, step 3). Methylation of glutamate residues in the cytoplasmic struts of the MCP reactivates the CheA kinase but desensitizes the MCP so that an even higher concentration of attractant must be present in order to bind the MCP and inhibit CheA kinase. So, if the organism is moving into a higher concentration of attractant, CheA is again inactivated, tumbling is suppressed, and the cell prolongs the run. If it is not moving into higher concentrations of attractant, CheY-P is made again and the organism will tumble. If the bacterium moves to a lower concentration of attractant, the cell will keep tumbling at a high frequency until it moves back into a higher concentration, at which point tumbling is suppressed to allow a smooth run.

When the cell moves away from attractant, the methylation switch is reset by another protein, CheB-P, which removes methyl groups so that the system is resensitized to attractant (**Fig. 10.25A**, step 4). The system is elegantly fine-tuned in that CheA kinase is also the protein that phosphorylates CheB. So, as the cell moves away from attractant, CheA kinase phosphorylates CheY to produce tumble and phosphorylates CheB to reset the sensitization switch.

MCP sensory proteins are not uniformly distributed around the cell surface, but assemble as clusters at both poles of *E. coli.* This clustering seems logical because the bacterium points with its pole in the direction it swims. **Figure 10.26** depicts one such MCP cluster (yellow) at one pole of the bacterium and illustrates the diffusion of CheY-P toward the opposite pole. CheY-P will interact with flagellar rotors all along the cell surface, causing tumble. The physical structure of a chemoreceptor array is shown in **Figure 10.26B**.

It is interesting to note that while the basic chemotactic mechanism is evolutionarily conserved in bacteria, different genera use it differently. In the Gram-positive *Bacillus subtilis,* for example, ligand binding to an MCP stimulates CheA kinase, and CheY-P stimulates counterclockwise flagellar rotation and, thus, extended runs. This mechanism is the exact opposite of that used in *E. coli*.

Chemotaxis is important for some obvious and perhaps not so obvious reasons. It clearly provides a useful survival strategy in nature, keeping bacteria moving toward nutrient and away from trouble (for example, toxic compounds). For natural commensal bacteria or pathogens, however, chemotaxis can also be used to move the organism toward a cell surface to which it can attach. For example, the intestinal lining can exude chemical attractants that, like a beacon, will lead the bacteria in the intestine toward the cell surface where they can attach. In sum, chemotactic sensory perception plays a major role in structuring microbial communities, in affecting microbial activities, and in influencing various microbial interactions with the surroundings. Other important regulatory mechanisms operate without affecting gene expression. One intriguing system involves toxin-antitoxin modules described in **eTopic 10.4**.

Thought Question

10.9 Using antibody to flagella, we can tether a cell of *E. coli* to a glass slide via a single flagellum. Looking through a microscope, you will then see the bacterium rotate in opposite directions as the flagellar rotor switches from clockwise to counterclockwise rotation and back again. (See the "tethered *E. coli*" video at the following Internet link.) Which way will the bacillus rotate when an attractant is added? What would the interval between clockwise and counterclockwise switching be if you tethered the mutants *cheY, cheA, cheZ,* and *cheR* to the slide and then added attractant?

Weblinks Rowland Institute at Harvard: video of tethered *E. coli* (*see ebook*)

Coupling Metabolic and Genetic Control: Nitrogen Regulation

Another mechanism by which levels of key compounds in the cell are regulated is to link biochemical control of a metabolic pathway (that is, whether an enzyme is active or inactive) to the genetic regulation of that pathway (which determines whether the enzyme is made). The assimilation of nitrogen is a classic example of coupled regulation (see Section 15.5).

All cells require nitrogen to grow. Thus, when nitrogen becomes limiting, bacteria activate pathways to gather nitrogen from the environment. The indicator of nitrogen abundance turns out to be the cellular levels of two amino acids: glutamate and glutamine. Actually, it is the intracellular ratio between 2-oxoglutarate (the precursor of glutamic acid) and glutamine levels, but to make the discussion easier we will use glutamate/glutamine levels. Put simply, when glutamine is in excess, the cell has plenty of nitrogen; when glutamate (or 2-oxoglutarate) is in abundance, the organism is nitrogen starved.

During times of nitrogen abundance, cells accumulate a store of nitrogen via glutamine synthetase (GlnA). Glutamine synthetase uses the energy released from ATP hydrolysis to assimilate NH_4^+ into glutamic acid and produce glutamine (**Fig. 10.27A**, step 1). However, if the cell makes too much active glutamine synthetase, then all of the cell's glutamate will be converted to glutamine, and not enough glutamate will remain for protein synthesis. To prevent a glutamate shortage, glutamine, when present at high levels relative to glutamic acid, signals the cell to stop making GlnA and inactivate whatever GlnA is already present. Thus, the glutamate-to-glutamine ratio is conserved. *E. coli* uses a single coupling protein called GlnB to regulate both the transcription of *glnA* and the biochemical activity of its product, glutamine synthetase. GlnB allows cells to balance the activation/inactivation of glutamine synthetase activity with its rate of synthesis.

We begin by explaining transcriptional control of the system and then show how GlnB links transcriptional and biochemical controls. Genetic control of *glnA* transcription is managed by the products of the nitrogen regulator genes *ntrB* and *ntrC* and a dedicated 54-kDa sigma factor, sigma N (also called sigma-54 or RpoN), encoded by the gene *rpoN*. NtrB and NtrC form a two-component signal transduction system. NtrB is a sensor kinase that phosphorylates the response regulator NtrC when glutamine levels are low (**Fig. 10.27A**, step 2). The phosphorylated form of NtrC (NtrC-P) induces expression of *glnA* and increases the level of glutamine synthetase in the cell (**Fig. 10.27A**, step 3). Note that contrary to the depiction in **Figure 10.27**, the NtrC-P regulator binds not to an operator region but to an enhancer sequence. Unlike operators, which must lie close to the promoter they control, enhancers can influence target gene expression from great distances (1 or 2 kb). Bound to its enhancer sequence, NtrC-P activates transcription of *glnA* through direct interactions with sigma N RNA polymerase bound at the *glnA* promoter, resulting in a DNA loop (**Fig. 10.28**). As a result, glutamine synthetase is made, nitrogen (in the form of ammonia) is assimilated, and glutamine is produced.

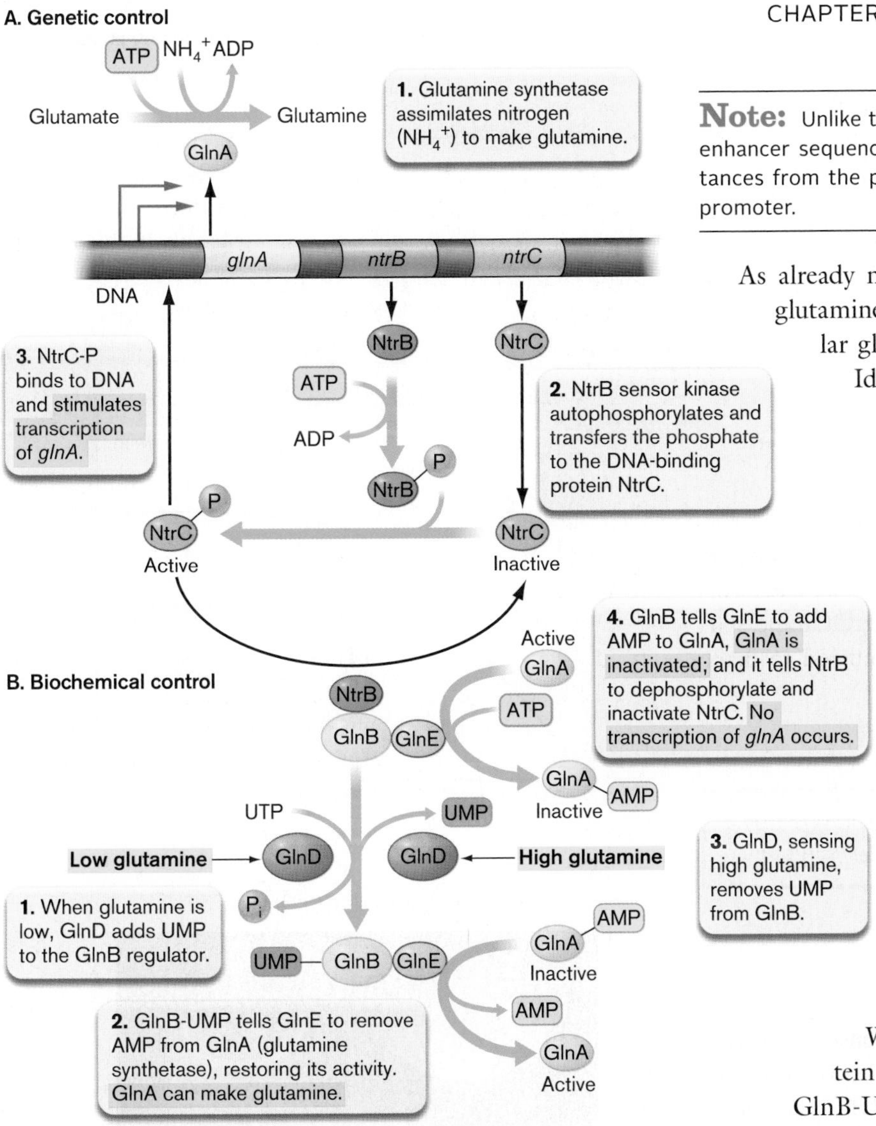

A. Genetic control
ATP
NH_4^+ ADP
Glutamate
Glutamine
GlnA
1. Glutamine synthetase assimilates nitrogen (NH_4^+) to make glutamine.
glnA
ntrB
ntrC
DNA
NtrB
NtrC
3. NtrC-P binds to DNA and stimulates transcription of glnA.
ATP
ADP
P
NtrB
2. NtrB sensor kinase autophosphorylates and transfers the phosphate to the DNA-binding protein NtrC.
P
NtrC
Active
NtrC
Inactive
B. Biochemical control
Active
GlnA
4. GlnB tells GlnE to add AMP to GlnA, GlnA is inactivated; and it tells NtrB to dephosphorylate and inactivate NtrC. No transcription of glnA occurs.
NtrB
ATP
GlnB
GlnE
GlnA
AMP
Inactive
UTP
UMP
Low glutamine
GlnD
GlnD
High glutamine
3. GlnD, sensing high glutamine, removes UMP from GlnB.
1. When glutamine is low, GlnD adds UMP to the GlnB regulator.
P_i
UMP
GlnB
GlnE
AMP
GlnA
Inactive
AMP
2. GlnB-UMP tells GlnE to remove AMP from GlnA (glutamine synthetase), restoring its activity. GlnA can make glutamine.
GlnA
Active

FIGURE 10.27 ■ Regulation of nitrogen metabolism. **A.** Genetic control. NtrB and NtrC form a cytoplasmic two-component signal transduction system that controls *glnA* transcription. **B.** Biochemical control. GlnB coordinates the function of NtrB (transcriptional control) and GlnE (metabolic control).

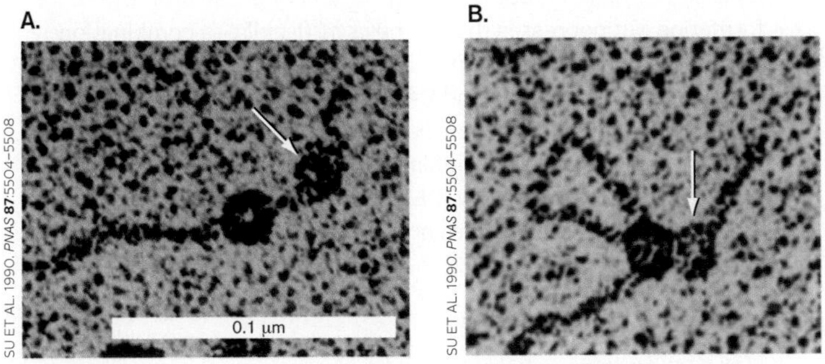

A.
B.
0.1 μm
SU ET AL. 1990. PNAS 87:5504–5508
SU ET AL. 1990. PNAS 87:5504–5508

FIGURE 10.28 ■ Visualizing NtrC function. NtrC binds to an enhancer sequence a great distance from the promoter of *glnA*. **A.** Electron micrograph of the NtrC protein (gray circle marked by arrow) and sigma-54 RNA polymerase (black circle) bound to their respective sites on the DNA but not in contact. **B.** Looping of the DNA brings the proteins in contact. In this configuration, NtrC will activate transcription.

Note: Unlike the DNA loop formed by AraC (see **Fig. 10.13B**), enhancer sequences can be moved experimentally to great distances from the promoter and still affect the expression of the promoter.

As already noted, a dire consequence of unrestricted glutamine production is the depletion of intracellular glutamate levels relative to glutamine levels. Ideally, when glutamine levels are in excess, the cell stops making GlnA and inactivates whatever GlnA has already been made. That way the cell can preserve the glutamate/glutamine balance. Conversely, when glutamine levels are limiting, the cell resumes making GlnA and makes sure it is active. The GlnB regulator coordinates these genetic and biochemical controls by modulating the phosphorylation of NtrC (thereby influencing transcription of *glnA*) and by altering GlnA activity.

The direction of control centers on whether a UMP group (UMP stands for "uridine monophosphate") is attached to GlnB (uridylylation). This is a form of posttranslational control. When glutamine levels are low, GlnD protein uridylylates GlnB (**Fig. 10.27B**, step 1). GlnB-UMP allows *glnA* transcription and activates preexisting GlnA enzyme (**Fig. 10.27B**, step 2). Conversely, when internal stores of glutamine are high relative to glutamate, GlnD removes UMP from GlnB (**Fig. 10.27B**, step 3). GlnB then halts *glnA* transcription and inactivates preexisting GlnA enzyme.

To carry out these functions, GlnB interacts with two proteins: the NtrB sensor kinase and another protein, called GlnE (**Fig. 10.27B**, step 4). GlnB (the form when glutamine levels are high) causes NtrB to dephosphorylate NtrC, which effectively halts *glnA* transcription. GlnB also compels GlnE to covalently add an AMP (adenosine monophosphate) to GlnA (another posttranslational control process called adenylylation). This modification inactivates whatever GlnA remains in the cell. The result is that ammonia is no longer assimilated and the glutamate/glutamine ratio is protected.

When glutamine levels are low relative to glutamate, GlnB-UMP allows the phosphorylation of NtrC by NtrB. So, NtrC-P is made and stimulates the synthesis of glutamine synthetase. In addition,

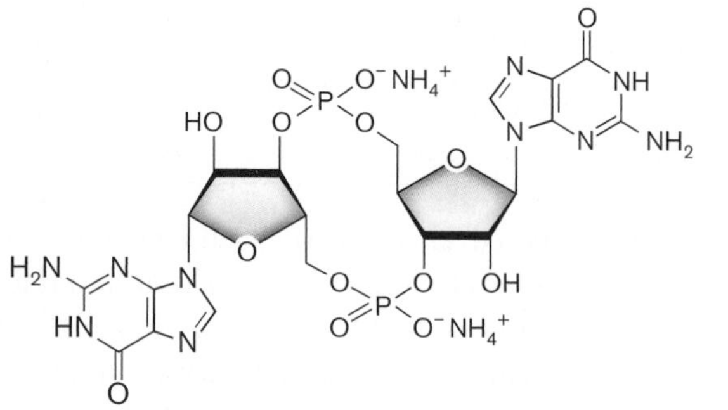

FIGURE 10.29 ■ Cyclic di-GMP [bis-(3′-5′)-cyclic dimeric guanosine monophosphate (cdiGMP)].

GlnB-UMP causes GlnE protein to remove AMP from preexisting glutamine synthetase, thus activating it. The result is that transcription of *glnA* is enhanced, glutamine synthetase is activated, and ammonia is assimilated to make glutamine once again.

Thought Question

10.10 Predict the phenotype of a *glnB* mutant. Will it be a glutamine auxotroph? What about a *glnD* or *glnE* mutant?

Biofilms and Second Messengers: Cyclic Di-GMP

To change cell physiology, some stress-sensing proteins produce **second messenger** molecules, which bind to regulatory (or effector) proteins and change their activities. We have already discussed two second messenger molecules: cAMP (involved in carbon/energy metabolism; Section 10.2) and ppGpp (ribosome synthesis; Section 10.3). These molecules freely diffuse through the bacterial cell in search of effector proteins. Another important second messenger is cyclic di-GMP (cdiGMP) (**Fig. 10.29**), which is one key to timing biofilm formation. *E. coli* cells transition between a motile, single-cell state (planktonic) and an adhesive multicellular biofilm. The transition can be seen clearly in batch cultures. The highly motile state appears during post-exponential growth when nutrient limitation forces *E. coli* to "forage" for food. When resources diminish further (stationary phase), the organism changes strategy: growth slows, motility decreases (a huge energy savings), and the synthesis of adhesins such as curli pili increases. Once the cell adheres to a surface, the cells start synthesizing exopolysaccharide matrix (Chapter 4). The molecule cdiGMP coordinates the transition by repressing flagellar synthesis genes and activating biofilm-promoting genes such as those encoding curli pili.

How is coordination achieved? Synthesis of cdiGMP from GTP is carried out by many diguanylate cyclases

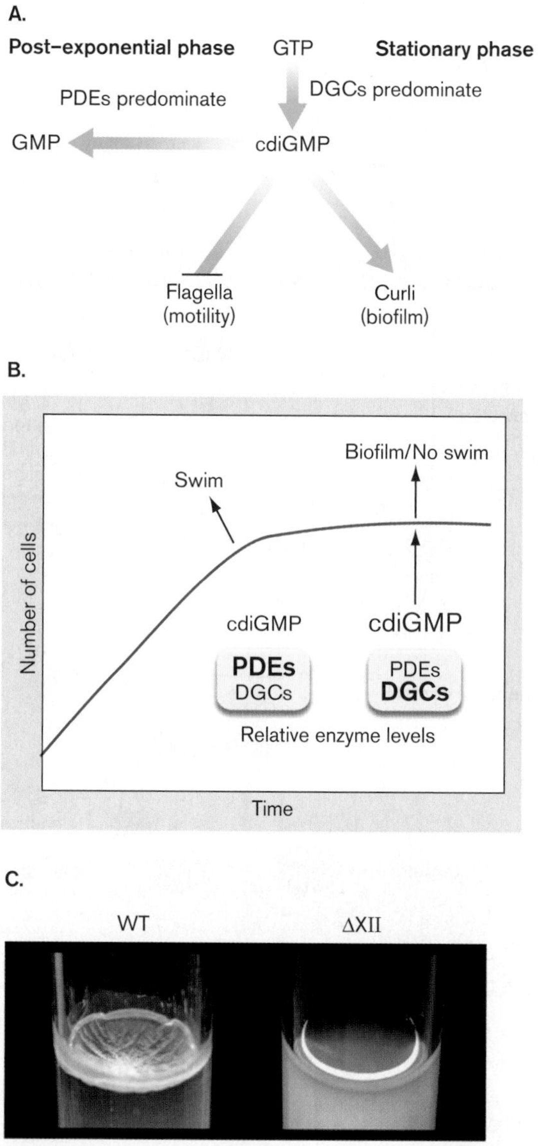
A.
Post-exponential phase
GTP
Stationary phase
PDEs predominate
DGCs predominate
GMP
cdiGMP
Flagella (motility)
Curli (biofilm)
B.
Biofilm/No swim
Swim
Number of cells
cdiGMP
cdiGMP
PDEs
DGCs
PDEs
DGCs
Relative enzyme levels
Time
C.
WT
ΔXII

FIGURE 10.30 ■ cdiGMP coordinates the switch from planktonic growth to biofilm formation. A. Cyclic di-GMP (cdiGMP) is synthesized by diguanylate cyclases (DGCs), which predominate in stationary phase, and is degraded by phosphodiesterases (PDEs), which predominate in late exponential phase. **B.** The relative levels of PDEs and DGCs at different growth phases. The higher cdiGMP level in stationary phase promotes curli formation but depresses the synthesis of flagella—a combination that favors biofilm formation. In late exponential phase, the low level of cdiGMP favors motility and the search for more food. **C.** Surface biofilm production by *Salmonella*. Wild-type (WT) organisms produce a distinct surface biofilm on the liquid medium, whereas a mutant (ΔXII) lacking all GGDEF domain proteins (and therefore unable to make cdiGMP) does not form a biofilm. *Source:* C. Solano et al. 2009. *PNAS* **106**:7997–8002.

(DGCs) in the cell, all of which contain the amino acid motif GGDEF (see Figure 8.12 for explanation of the single-letter amino acid abbreviations). Dedicated phosphodiesterases (PDEs) with an EAL motif degrade cdiGMP to GMP (**Fig. 10.30A**). *Salmonella* and *Pseudomonas* contain 19 and 41 different GGDEF/EAL proteins, respectively. Each DGC and PDE protein becomes activated by a different signal, thus increasing or decreasing cdiGMP levels. As shown in **Figure 10.30B**, PDEs predominate in post-exponential phase (low cdiGMP), while DGCs predominate in stationary phase (high cdiGMP). Thus, highly motile cells are made in post-exponential phase while sessile, adherent cells are produced in stationary phase. **Figure 10.30C** shows how preventing cdiGMP synthesis by deleting all known GGDEF motif proteins halts biofilm formation in *Salmonella*.

Cyclic di-GMP is a ubiquitous second messenger in bacteria and is not found in eukaryotes or archaea. But why does one cell possess so many GGDEF and EAL proteins? One hypothesis is that cognate pairs of these proteins may operate at separate locations within a single cell and generate localized changes in cdiGMP concentrations. Multiple, focused cdiGMP gradients could produce different outputs in different parts of a single cell.

There are numerous other examples of integrated control circuits. A classic one involves the "lifestyle" decision made by bacteriophage lambda, which is described in eTopic 10.5.

To Summarize

- **Bacterial genes are regulated by a hierarchy of regulators** that form integrated gene circuits.
- **Gene switches** can regulate "decision-making" processes in microbes.
- **Chemotaxis** is a behavior in which motile microbes swim toward favorable environments (chemoattractants) or away from unfavorable environments (chemorepellents).
- **The direction of flagellar motor rotation** determines the type of movement. Counterclockwise rotation results in smooth swimming; clockwise rotation results in tumbling.
- **Random movement toward an attractant** causes a drop in CheY-P levels, which allows counterclockwise rotation and smooth swimming.
- **Methyl-accepting chemotaxis proteins (MCPs)** clustered at cell poles bind chemoattractants and initiate a series of events lowering CheY-P levels. Reversible methylation or demethylation of MCPs desensitizes or sensitizes MCPs, respectively.
- **The NtrB-NtrC two-component signal transduction system** regulates nitrogen assimilation by altering transcription of the gene encoding glutamine synthetase (*glnA*).
- **The protein GlnB** links the biochemical control and genetic regulation of nitrogen assimilation.
- **The second messenger cyclic di-GMP (cdiGMP)** is made by numerous proteins containing a GGDEF amino acid motif. Many cellular functions are influenced by cdiGMP, including biofilm formation and motility.

10.8

Quorum Sensing: Chemical Conversations

A discovery that fundamentally changed the way we think about microbes was made while studying *Aliivibrio fischeri*, a peculiar marine microorganism that colonizes the light organ of the Hawaiian squid (*Euprymna scolopes*) (**Fig. 10.31**). As discussed at the beginning of this chapter, *A. fischeri* is bioluminescent but glows only at high cell densities—a situation achieved naturally in the squid's

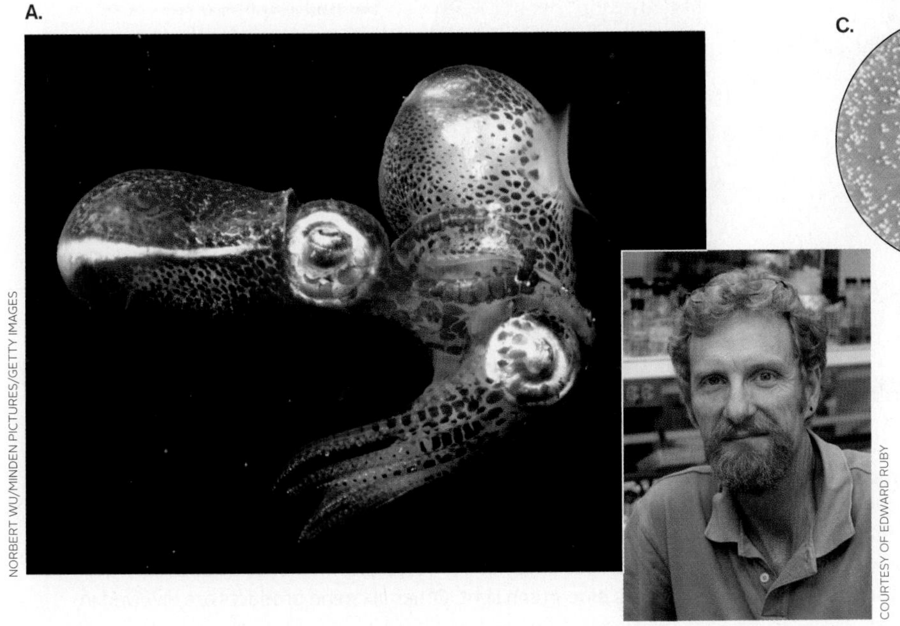
A.
NORBERT WU/MINDEN PICTURES/GETTY IMAGES
COURTESY OF EDWARD RUBY

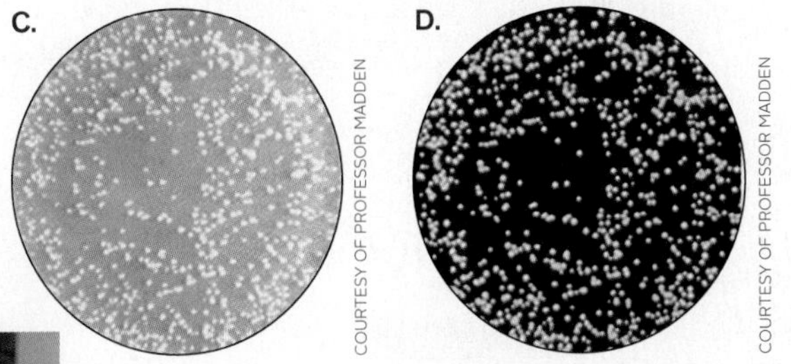
C.
D.
COURTESY OF PROFESSOR MADDEN
COURTESY OF PROFESSOR MADDEN

FIGURE 10.31 ▪ Visual demonstration of quorum sensing. A. The luminescent bacterium *Aliivibrio fischeri* (formerly *Vibrio fischeri*) colonizes the light organ of the Hawaiian squid (*Euprymna scolopes*). Shown is an *E. scolopes* pair mating at night. The light organ is deep inside each squid and therefore cannot be seen. **B.** Edward Ruby (University of Hawaii) has studied various aspects of the symbiotic relationship between *A. fischeri* and its squid host. **C.** When colonies of *A. fischeri* are observed in a well-lit place, the light emitted by the bacteria is not visible. **D.** If the same colonies are viewed in darkness, the intensity of luminescence is remarkable.

colonized light organ or artificially in a test tube (**Fig. 10.31C** and **D**). What accounts for the dependence of gene expression on cell density? How do cells know they are crowded?

The phenomenon was originally dubbed **quorum sensing** because it seemed akin to parliamentary rules of order that require a minimum number of members (a quorum) to be present at a meeting in order to conduct business. In the microbial world, however, gene regulation is only loosely associated with actual cell numbers. Induction of a quorum-sensing gene system really requires the accumulation of a secreted small molecule called an **autoinducer** (usually a homoserine lactone, although Gram-positive organisms are known to use short peptides). The more cells there are in a discrete space, the faster the critical level of autoinducer is reached (see the chapter opener).

At a certain extracellular concentration, the secreted autoinducer reenters cells and binds to a regulatory molecule. In the case of *A. fischeri*, the regulatory molecule is LuxR (**Fig. 10.32** ▶). The LuxR-autoinducer complex activates transcription of the luciferase target genes that confer bioluminescence. Luciferase is the enzyme that produces light in *A. fischeri*. The apparent cell density requirement can be bypassed by simply adding purified autoinducer to a low-density cell culture. Many other microbes use these chemical languages to coordinate behavior of the population (**Table 10.3**). For instance, quorum sensing is used by pathogens to time the production of virulence factors for optimum effect on the host.

Quorum Sensing and Pathogenesis

Pseudomonas aeruginosa is a human pathogen that commonly infects patients with cystic fibrosis, a genetic disease of the lung. The organism forms a biofilm in the lung and secretes virulence factors (such as proteases and other degradative enzymes) that destroy lung tissues (and thereby severely compromise lung function). These virulence proteins, however, are not made until cell density is fairly high—that is, at a point where the organism might have a chance of overwhelming its host. Made too early, virulence proteins would alert the host to launch an immune response. The induction mechanism involves two interconnected quorum-sensing systems, called Las and Rhl, both composed of regulatory proteins homologous to LuxR and LuxI of *Aliivibrio fischeri*. Many pathogens besides *Pseudomonas* appear to use chemical signaling to control virulence genes. These include *Salmonella, Escherichia coli, Vibrio cholerae,* the plant symbiont *Rhizobium*, and many others.

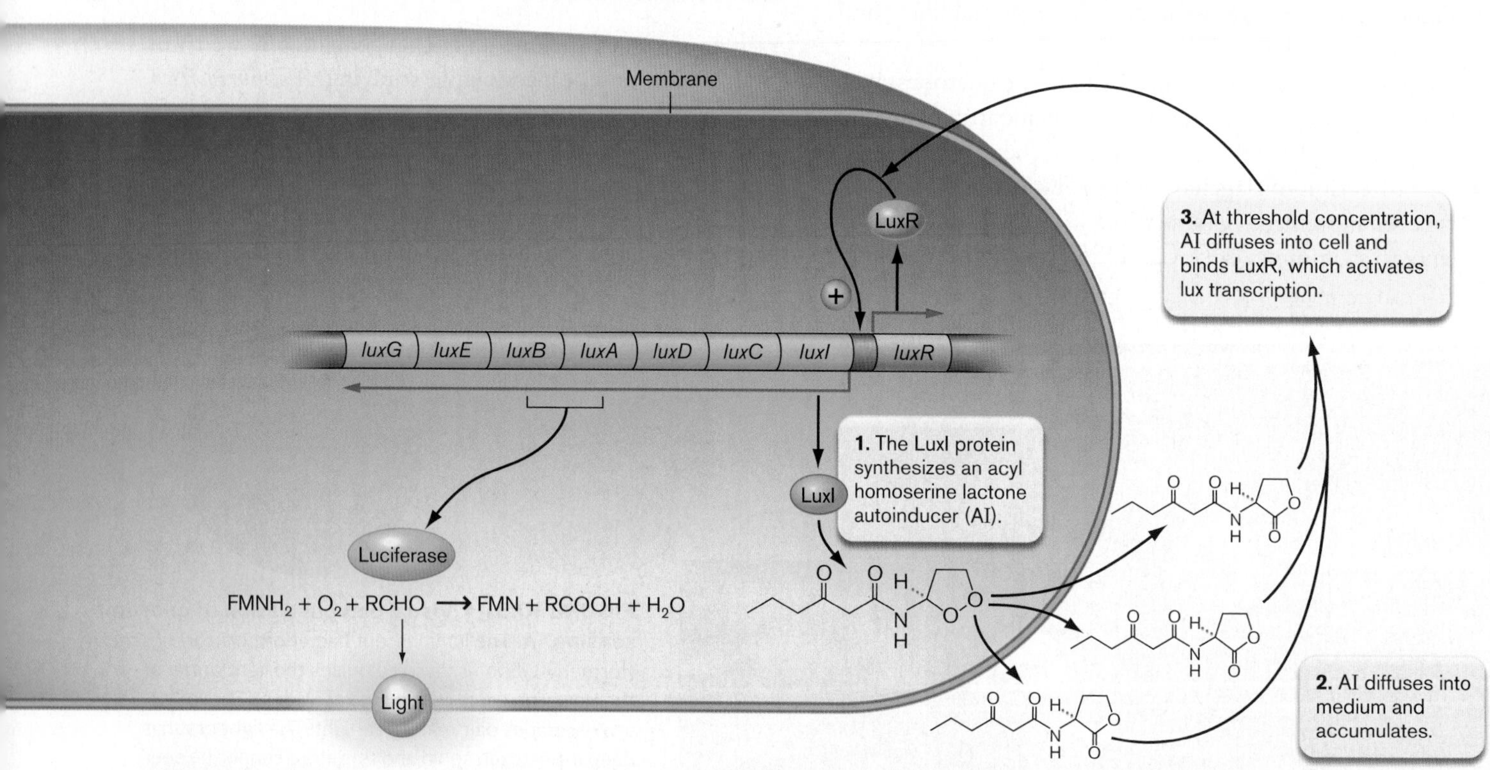

Membrane
LuxR
+
luxG
luxE
luxB
luxA
luxD
luxC
luxI
luxR
3. At threshold concentration, AI diffuses into cell and binds LuxR, which activates lux transcription.
1. The LuxI protein synthesizes an acyl homoserine lactone autoinducer (AI).
LuxI
Luciferase
FMNH2 + O2 + RCHO ⟶ FMN + RCOOH + H2O
Light
2. AI diffuses into medium and accumulates.

FIGURE 10.32 ▪ Microbial communication through quorum sensing. The *lux* system of *Aliivibrio fischeri* mediates that organism's bioluminescence. Synthesis and accumulation of an autoinducer (AI) triggers expression of the *lux* operon. The greater the cell number and the smaller the container, the faster AI will accumulate. The resulting luciferase enzymes catalyze bioluminescence. The luciferase reaction, catalyzed by LuxA and LuxB, uses oxygen and reduced flavin mononucleotide (FMN) to oxidize a long-chain aldehyde (RCHO) and, in the process, produces blue-green light. Other *lux* gene products are involved in synthesis of the aldehyde. ▶

TABLE 10.3

Examples of microbial quorum-sensing systems.

System	Organism	Autoinducer family	Function
LuxR/LuxI	*Aliivibrio fischeri*	Homoserine lactone	Bioluminescence
LuxQ/LuxS	*Vibrio harveyi*	Furanosyl borate diester	Bioluminescence
LuxN/LuxLM	*Vibrio harveyi*	Homoserine lactone	Bioluminescence
LasR/LasI	*Pseudomonas aeruginosa*	Homoserine lactone	Exoenzyme production
Rhl	*Pseudomonas aeruginosa*	Homoserine lactone	Exoenzyme production
Agr	*Staphylococcus*	Peptide	Exotoxin production
StrR	*Streptomyces griseus*	γ-Butyrolactone	Aerial hyphae; antibiotic production
TraR/TraI	*Agrobacterium tumefaciens*	Homoserine lactone	Conjugation
YpeR/YpeI	*Yersinia pestis*	Unknown	Unknown
SdiA	*Escherichia coli*	Unknown	Cell division

Structures of different autoinducer families:

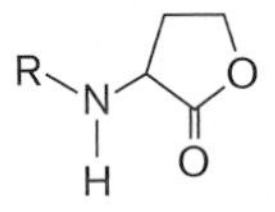
R
N
H
O
O

Homoserine lactone family

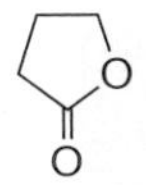
O
O

γ-Butyrolactone

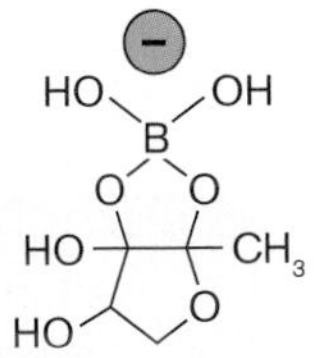
HO
OH
B
O
O
HO
CH3
HO
O

Furanosyl borate diester

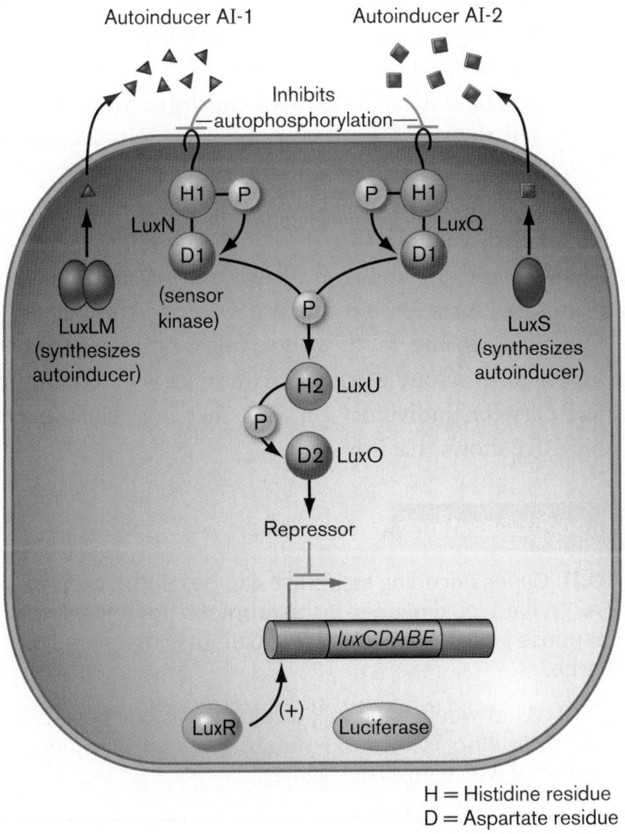
Autoinducer AI-1
Autoinducer AI-2
Inhibits
autophosphorylation
H1
P
P
H1
LuxN
LuxQ
D1
D1
(sensor
kinase)
P
LuxLM
(synthesizes
autoinducer)
LuxS
(synthesizes
autoinducer)
H2
LuxU
P
D2
LuxO
Repressor
luxCDABE
(+)
LuxR
Luciferase
H = Histidine residue
D = Aspartate residue

FIGURE 10.33 ▪ The two quorum-sensing systems of *Vibrio harveyi*. In the absence of autoinducers (AI-1 and AI-2), both sensor kinases trigger converging phosphorylation cascades that end with the phosphorylation of LuxO. Phosphorylated LuxO (LuxO-P) activates a repressor that inhibits expression of the luciferase genes. As autoinducer concentrations increase, they inhibit autophosphorylation of the sensor kinases and the phosphorylation cascade. As a result, repressor levels decrease, allowing the LuxR protein to activate the *lux* operon.

Interspecies Communication

Some microbial species not only chemically talk among themselves but can communicate with other species. *Vibrio harveyi,* for example, uses two different, but converging, quorum-sensing systems to coordinate control of its luciferase. Both sensing pathways are very different from the *A. fischeri* system. One utilizes an acyl homoserine lactone (AHL) as an autoinducer (AI-1) to communicate with other *V. harveyi* cells. The second system produces a different autoinducer (AI-2) that contains borate. Because many species appear to produce this second signal molecule, it is thought that mixed populations of microbes use it to "talk" to each other.

In the case of *V. harveyi,* specific membrane sensor kinase proteins are used to sense each autoinducer (**Fig. 10.33**). At low cell densities (no autoinducer), both sensor kinases

©DENISE APPLEWHITE

FIGURE 10.34 ■ Bonnie Bassler of Princeton University. Bonnie Bassler (left) was instrumental in characterizing interspecies communication among bacteria.

COURTESY OF UNIVERSITY OF IOWA MEDICAL PHOTOGRAPHY UNIT

FIGURE 10.35 ■ Peter Greenberg, one of the pioneers of cell-cell communication research. Peter Greenberg, first at the University of Iowa and now at the University of Washington, has studied quorum sensing in *Vibrio* species and various other pathogenic bacteria, such as *Pseudomonas*.

initiate phosphorylation cascades that converge on a shared response regulator, LuxO, to produce phosphorylated LuxO. LuxO-P activates a repressor of the *lux* genes. Thus, at low cell densities the culture does not display bioluminescence. At high cell density, the autoinducers prevent signal transmission by inhibiting phosphorylation. The cell stops making repressor, thereby allowing another protein, LuxR (not a homolog of the *A. fischeri* LuxR), to activate the *lux* operon. The "lights" are turned on. Bonnie Bassler (**Fig. 10.34**) and Peter Greenberg (**Fig. 10.35**) are two of the leading scientists whose studies revealed the complex elegance of quorum sensing in *Vibrio* and *Pseudomonas* species. Other organisms, such as *Salmonella*, have been shown to activate the AI-2 pathway of *V. harveyi*, dramatically supporting the concept of cross-species communication.

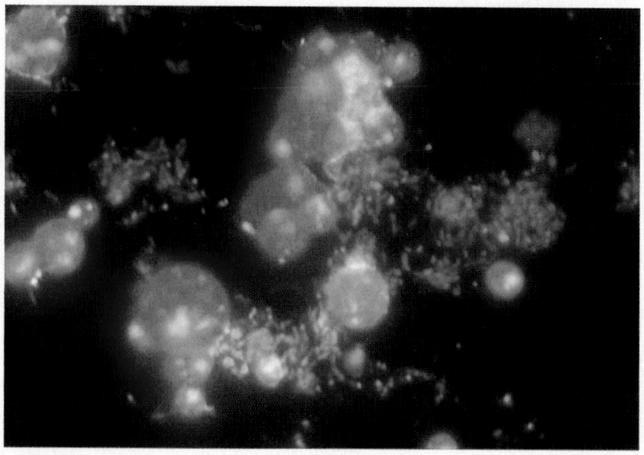

REPRINTED WITH PERMISSION FROM AAAS

FIGURE 10.36 ■ *Enteromorpha* zoospores. These algae (red) attach to biofilm-producing bacteria (blue) in response to lactones produced by the bacteria.

A report by Ian Joint and his colleagues showed that bacteria can even communicate across the prokaryotic-eukaryotic boundary. The green seaweed *Enteromorpha* (a eukaryote) produces motile zoospores that explore and attach to *Vibrio anguillarum* bacterial cells in biofilms (**Fig. 10.36**). They attach and remain there because the bacterial cells produce AHL molecules that the zoospores sense. Part of the evidence for this interdomain communication involved showing that the zoospores would even attach to biofilms of *E. coli* carrying the *Vibrio* genes for the synthesis of AHL. The implications of possible interdomain conversations are staggering. Do our normal flora "speak" to us? Do we "speak" back? For further discussion of molecular communication between prokaryotes and eukaryotes, see Chapter 21.

This knowledge begs the question, Why would a squid want to harbor this bioluminescent microbe? Buried in the sand by day, the squid emerges at night from its safe hiding place to hunt for food. In moonlight, the squid would appear as a dark silhouette from below, marking it as easy prey for predators. It is thought that the squid camouflages itself by projecting light downward from its light organ, giving the appearance of surface-filtered light to the predators below.

Quorum sensing is a wonderful way to coordinate group behaviors, but what about more private communication between individual cells. Is that possible? **Special Topic 10.1** shows that it is.

Thought Questions

10.11 Genes encoding luciferase can be used as "reporters." What would happen if the promoter for an SOS response gene was fused to the luciferase open reading frame?

10.12 What would happen if a culture was coinoculated with a *Aliivibrio fischeri luxI* mutant and a *luxA* mutant, neither of which produces light?

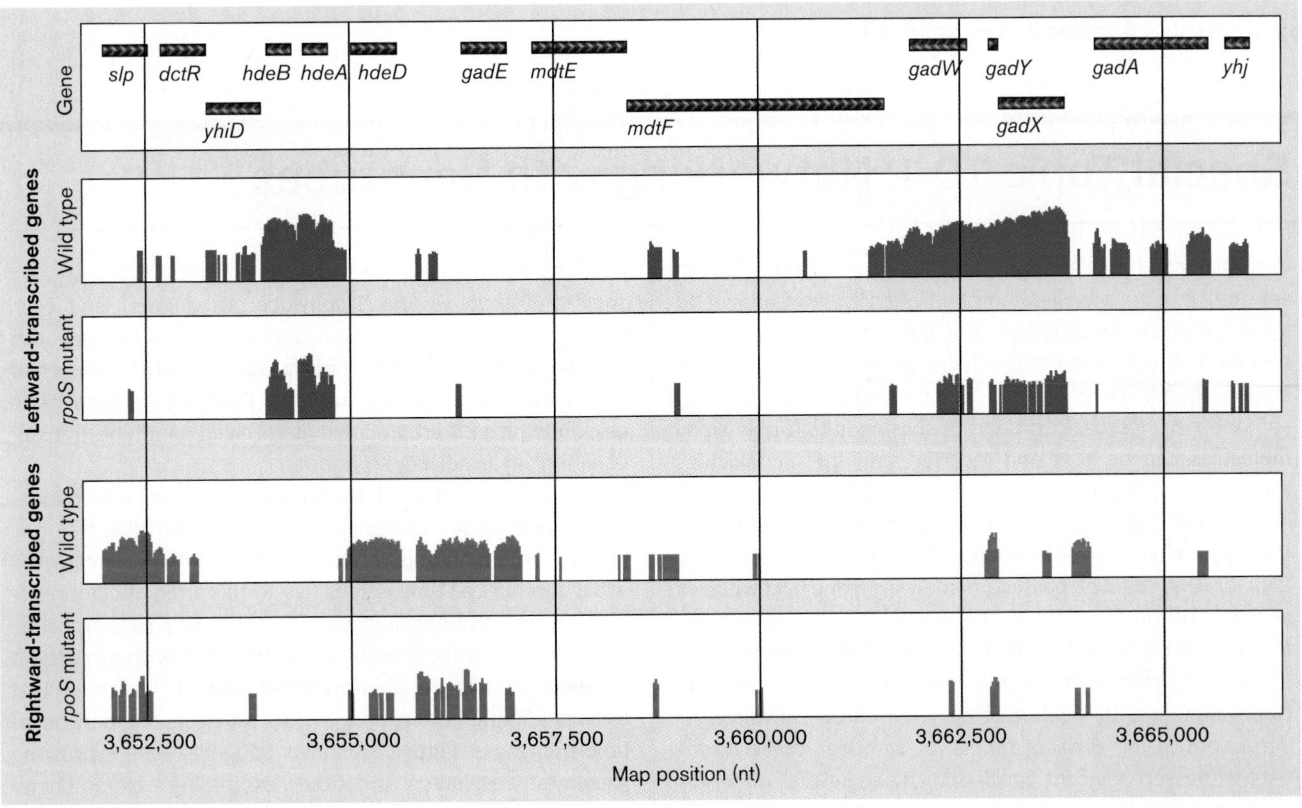
Gene
slp
dctR
hdeB
hdeA
hdeD
gadE
mdtE
yhiD
mdtF
gadW
gadY
gadA
yhj
gadX
Leftward-transcribed genes
Wild type
rpoS mutant
Rightward-transcribed genes
Wild type
rpoS mutant
3,652,500
3,655,000
3,657,500
3,660,000
3,662,500
3,665,000
Map position (nt)

FIGURE 10.37 ▪ Transcriptomics of *Escherichia coli* acid resistance genes. *E. coli* can survive pH 2 exposures for long periods of time—a property known as acid resistance. This capability requires numerous genes, several of which are shown here. The X axis indicates the nucleotide map position on the *E. coli* genome. The top panel indicates the genes expressed in this area. Red-arrow genes are transcribed from left to right; blue-arrow genes are transcribed in the opposite direction. Deep sequencing was used to quantify the relative numbers of transcripts made from each gene. Each vertical bar in the bottom four panels represents a portion of a transcript. The data compare the expression of the same genes in wild-type (panels 2 and 4) and RpoS-deficient strains (panels 3 and 5). RpoS encodes a sigma factor required for optimal expression of these genes. Note that the leftward and rightward transcripts of all of these genes are decreased in the *rpoS* mutant. The experiments actually capture data from all genes expressed from the genome, but only a portion is shown. *Source:* Modified from University of Oklahoma Gene Expression Database.

10.9

Transcriptomics and Proteomics

The original studies exposing many regulatory systems discussed in this chapter were accomplished by examining only a small number of genes at any one time. Now, however, new technologies enable scientists to examine thousands of genes or proteins in a single experiment.

Transcriptomics

A pebble tossed into a pond makes a wave that spreads and ultimately disturbs the entire pond. Similarly, altering one biochemical pathway in a cell produces a physiological "wave" that impacts, to a greater or lesser degree, all other pathways. What typically initiate and propagate a physiological "wave" are transcriptional changes in gene expression. Scientists now have the ability to see these events unfold by monitoring the expression of every gene in a cell—simultaneously. The result is known as the **transcriptome**.

This global view of gene expression is possible using next-generation, high-throughput DNA sequencing technology (Section 7.6). Basically, all RNA is extracted from a culture of cells, and then stable RNAs (tRNA and rRNA) comprising the bulk of RNA are removed by hybridization. The remaining RNA molecules are fragmented by sonic disruption and converted to complementary DNA (cDNA) with reverse transcriptase. DNA adapters are ligated to each end of the cDNA molecules. Each molecule, with or without amplification, is then sequenced in a high-throughput manner (for example, pyrosequencing or Illumina's sequencing by synthesis) to obtain short sequences from one end or both ends of each fragment (corresponds to transcript copy). The reads, millions of them, are typically 30–400 bp. A computer program aligns each sequence with its genome position (gene) and then tabulates the number of fragments sequenced at any one location. The number of RNA sequences (RNA-seq data) produced for a particular gene is proportional to how much that gene was expressed.

Comparing results between cells grown under different conditions, or between a mutant and a wild-type strain, will expose how the expression of every gene is influenced by the change (**Fig. 10.37**). Derek Bickhart and David Benson

Special Topic 10.1: Networking with Nanotubes

Microbial communication by secreted molecules (quorum sensing) is now a well-established phenomenon known to synchronize group behaviors. But quorum sensing is a system more like a loudspeaker than a phone. Everyone in the neighborhood, friend or foe, can potentially "hear" the signal. Another drawback to quorum sensing is that only small molecules can be sent and received, and these molecules are subject to degradation by extracellular factors. But what if microbes had a one-on-one communication system (like texting) in which not only signal molecules, but also enzymes, DNA, or RNA could be passed privately from one bacterium to another without ever facing the extracellular environment—temporarily changing the phenotype of the receiving cell?

Such a system was recently discovered by Gyanendra P. Dubey and Sigal Ben-Yehuda from the Hebrew University of Jerusalem. Using *Bacillus subtilis* as a model, these scientists witnessed adjacent bacilli exchanging cellular materials such as green fluorescent protein (GFP), while cells separated by a very small distance could not. Electron microscopy revealed how this was possible. Unexpectedly, the adjacent cells were connected by small tubes called nanotubes (**Fig. 1**). These tubes were much larger (about 100 nm) than the bore of an average pilus (about 5 nm). Not only were proteins transferred, but nanotubes also served as portals through which DNA, RNA, and other molecules could be transferred. Unlike quorum sensing, nanotube communications were private, between one cell and another, although small, multicellular networks connected by nanotubes were also observed.

Dubey and Ben-Yehuda found that nanotubes could also form between different species of bacteria. Nanotube connections developed between two different Gram-positive species (*B. subtilis* and *Staphylococcus aureus*), and even between Gram-positive (*B. subtilis*) and Gram-negative (*E. coli*) species (**Fig. 2A**). **Figure 2B** shows the transfer of GFP from a *gfp*$^+$ *B. subtilis* to *E. coli*. The faint green speckled cells in **Figure 2B** are *E. coli* that received nanotube transfer of GFP from an adjacent *B. subtilis*.

Nanotube transfer of proteins could prove very beneficial to members of mixed-species biofilms. Imagine that one cell type in a biofilm is resistant to the antibiotic lincomycin, while another cell type is sensitive to this antibiotic but resistant to chloramphenicol. Could nanotubes help cells exchange resistance proteins, thereby providing each with a transient antibiotic resistance phenotype? **Figure 3** illustrates that transient nonhereditary phenotypes can be acquired by nanotube exchanges. **Figure 3A** shows two cells—one expressing lincomycin resistance, the other expressing chloramphenicol resistance—illustrating how the resistance proteins (and possibly mRNA) can exchange. **Figure 3B** shows the result of an actual experiment. *B. subtilis* Cm^R cells (P1) and Lin^R cells (P2) were grown on LB complex agar as individual or mixed patches (column 1). Patches were transferred to fresh media containing the drugs indicated. Notice that only the P1 + P2 mixed cells grew on the medium containing both Cm and Lin. To see whether the resistance phenotype was heritable, material from the patch was streaked for single colony isolation on LB (no drug), and individual colonies (arising from single cells) were tested for drug resistance. Amazingly, each colony exhibited resistance to one or the other drug only, indicating that double drug resistance was a transient, nonheritable phenotype (not the result of DNA transfer).

A.

B.

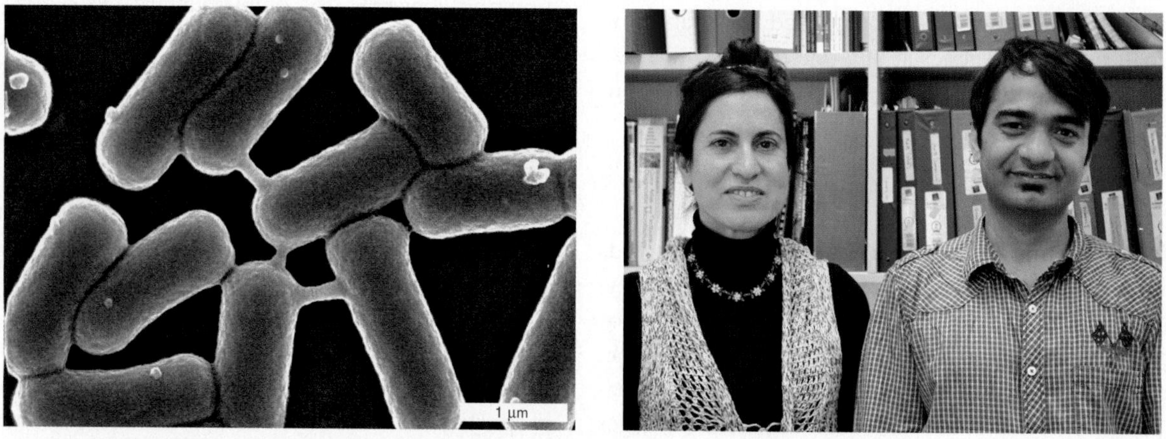
1 μm

FIGURE 1 ■ Nanotubes form between neighboring *Bacillus subtilis* cells. A. Cells were grown on agar for 6 hours and visualized by high-resolution SEM. Notice that several cells are connected. **B.** The research team of Professor Sigal Ben-Yehuda (left) and Gyanendra P. Dubey discovered nanotubes.

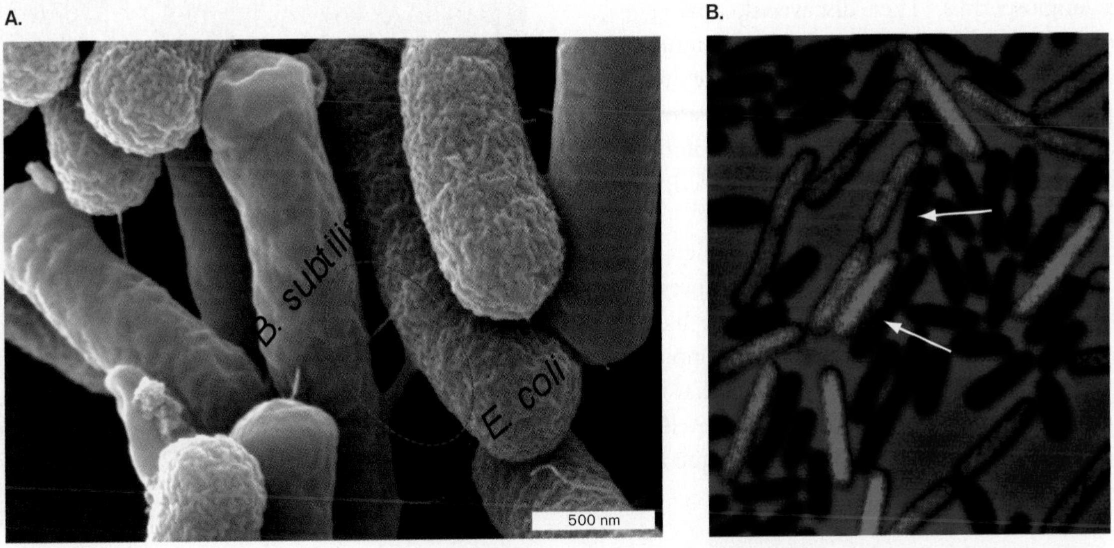
A.
B.
B. subtilis
E. coli
500 nm

FIGURE 2 ■ Interspecies nanotubes. A. SEM showing nanotubes forming between *Bacillus subtilis* and *Escherichia coli.* **B.** *B. subtilis* containing GFP and *E. coli* without GFP were plated together on agarose. The fluorescent image was taken 30 minutes after plating. The bright green organisms are *B. subtilis,* but notice that *E. coli* cells adjacent to *B. subtilis* have started to turn green (arrows) because of the transfer of GFP protein.

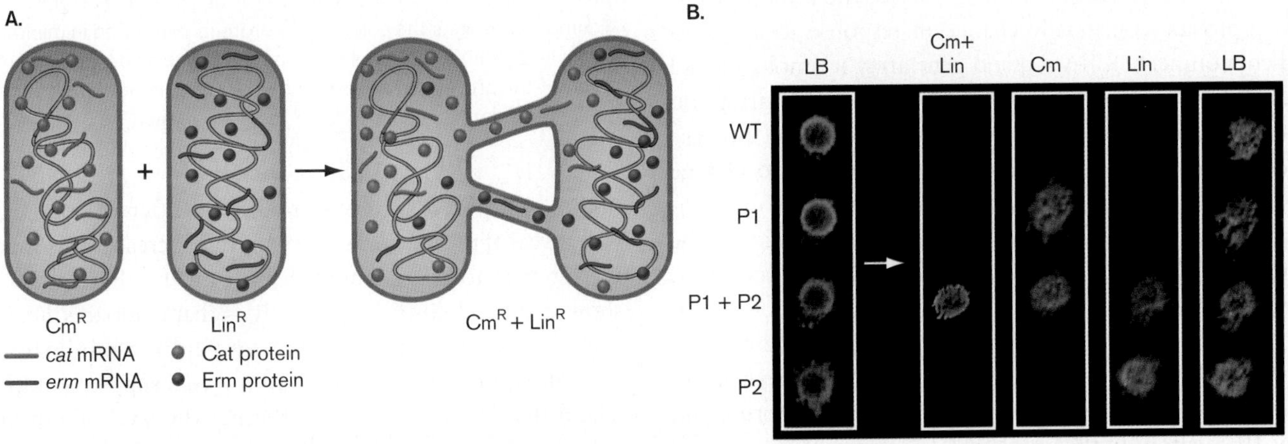
A.
+
Cm^R
Lin^R
Cm^R + Lin^R
cat mRNA
erm mRNA
Cat protein
Erm protein
B.
LB
Cm+ Lin
Cm
Lin
LB
WT
P1
P1 + P2
P2

FIGURE 3 ■ Transient nonhereditary phenotypes acquired by nanotubes. A. Chloramphenicol-resistant (Cm^R) and lincomycin-resistant (Lin^R) cells forming nanotube connections that exchange resistance proteins and possibly mRNA. **B.** *Bacillus subtilis* wild-type cells (WT), Cm^R cells (P1), and Lin^R cells (P2) were grown for 4 hours on LB complex agar individually or mixed (column 1). Patches were transferred to fresh media without or with the drugs indicated (columns 2–5).

This mechanism of directly sharing resistance proteins could be extremely valuable to a biofilm under assault by multiple agents. Communicating by nanotubes enables bacteria a straightforward immediate transfer of information across inherent species barriers. Moreover, molecules channeled by nanotubes are protected from degrading enzymes and harsh environmental conditions. Though untested, nanotubes might also be an efficient defensive strategy used to kill competitors by directly delivering toxic molecules.

Research Question

Knowing that nanotubes can temporarily transfer antibiotic resistance from one cell to a second cell, how might you select for a mutant that is defective in nanotube formation? Discuss one potential pitfall with your approach.

Gyanendra P. Dubey, Sigal Ben-Yehuda. 2011. Intercellular Nanotubes Mediate Bacterial Communication. Cell 144:590-600.

from the University of Connecticut recently used RNA deep sequencing to examine how age affects gene expression changes in *Frankia,* an actinobacterium that forms nitrogen-fixing nodules in certain plants (angiosperms). They discovered, among other things, that transposase genes were more highly expressed in nitrogen-fixing *Frankia* (relative to ammonia grown cells) and in aging cultures (relative to young). Thus, N_2 fixation and aging, two conditions faced in a root nodule, stimulate major transcriptome modifications.

Another powerful technique, called DNA microarray, can also simultaneously examine the expression of every gene in a cell. For an organism like *Salmonella* Typhi, the causative agent of typhoid fever, this requires seeing over 4,600 genes. What is amazing is that all this can be done on a slide the size of a postage stamp. The microarray technique is described in Appendix 3.

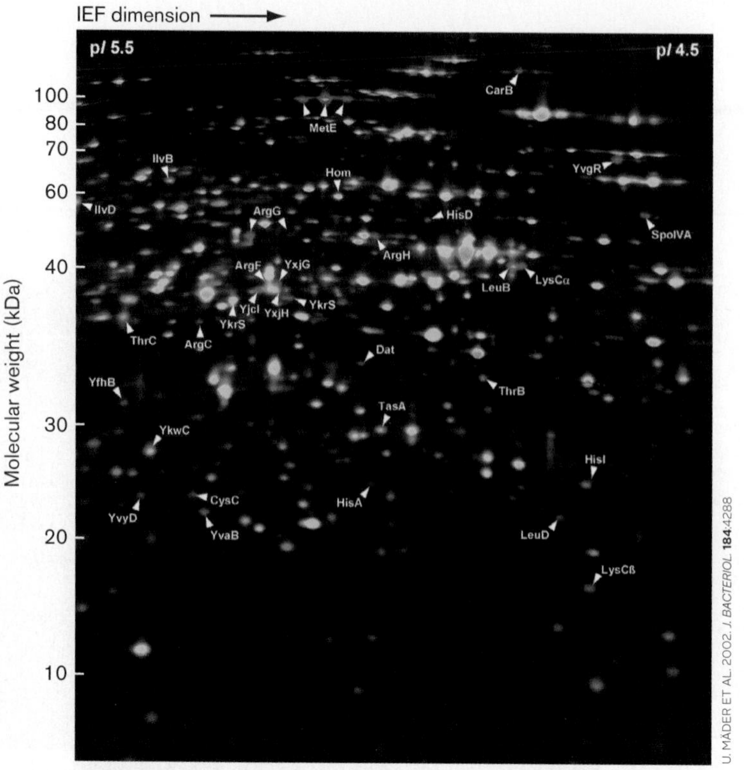
IEF dimension
pI 5.5
pI 4.5
Molecular weight (kDa)
100
80
70
60
40
30
20
10
CarB
MetE
IlvB
Hom
YvgR
IlvD
ArgG
HisD
SpoIVA
ArgH
ArgF
YxjG
LeuB
LysCα
YkrS
ThrC
ArgC
Dat
YfhB
ThrB
TasA
YkwC
HisI
CysC
HisA
YvyD
YvaB
LeuD
LysCß
U. MADER ET AL. 2002. J. BACTERIOL. 184:4288

FIGURE 10.38 ▪ Proteomic profile of *Bacillus subtilis* cells grown in minimal media with and without mixed amino acid supplementation (casamino acids). The isoelectric focusing (IEF) gradient used in the first dimension was pH 4.5–5.5; this represents only a part of the entire proteome. The figure is the result of dual-channel image analysis of silver-stained gels. A computer assigns the color red to proteins expressed in minimal media and green to proteins expressed in amino acid–supplemented media. If the proteins are expressed under both conditions, the red and green colors combine to form yellow or orange.

Proteomics

If all the expressed mRNAs in a microbe make up its transcriptome, then all the expressed proteins form its **proteome**. Unlike the DNA genome, which is fixed, the transcriptome and proteome continually change in response to a changing environment. RNA-seq and gene array technologies can capture what happens transcriptionally, but gene arrays do not necessarily reveal which proteins are expressed, because stress often causes the translation of a message to change without changing its transcription. Thus, the level of a protein can increase significantly without any change in the amount of mRNA. In addition, posttranslational modifications of proteins (for example, acetylation, phosphorylation) can take place in response to different environments. The powerful techniques of 2D gel electrophoresis (introduced in Section 3.1) and mass spectrometry are used to capture and view fluctuations in the proteome.

Two-dimensional gel electrophoresis. The science of 2D gels is described in Appendix 3. Briefly, the technique uses polyacrylamide gel electrophoresis to separate proteins from cell extracts in one dimension by their charge and in a second dimension by their molecular weights. When combined, isoelectric focusing and SDS-PAGE display a majority of the cell's proteins in a 2D array, as shown in **Figure 10.38** (some proteins do not appear because their quantities or solubilities are too low). The proteins are visualized by staining with fluorescent dyes after separation. Subsequent computer analysis of the proteome patterns obtained from cells grown under two different conditions will reveal those proteins whose levels increase or decrease in response to the changing environment. **Figure 10.38** shows a dual-channel image analysis that compares *Bacillus subtilis* proteomes from cultures grown in a minimal glucose medium with or without amino acid supplementation. The different colors indicate whether the level of a protein is higher, lower, or the same in the two cultures. Proteins of interest can be plucked from the gel and their identities determined though mass spectrometry techniques, discussed next (see also Appendix 3).

Mass spectrometry. Knowing the sequence of an organism's genome gives us a wealth of useful information and new ways to probe its physiology. One of those ways involves mass spectrometry, which experimentally determines the exact mass of an unknown protein or peptide fragment and uses that mass to identify the protein. The process is easy for a single protein picked from a gel, but new techniques make it possible to identify and quantify

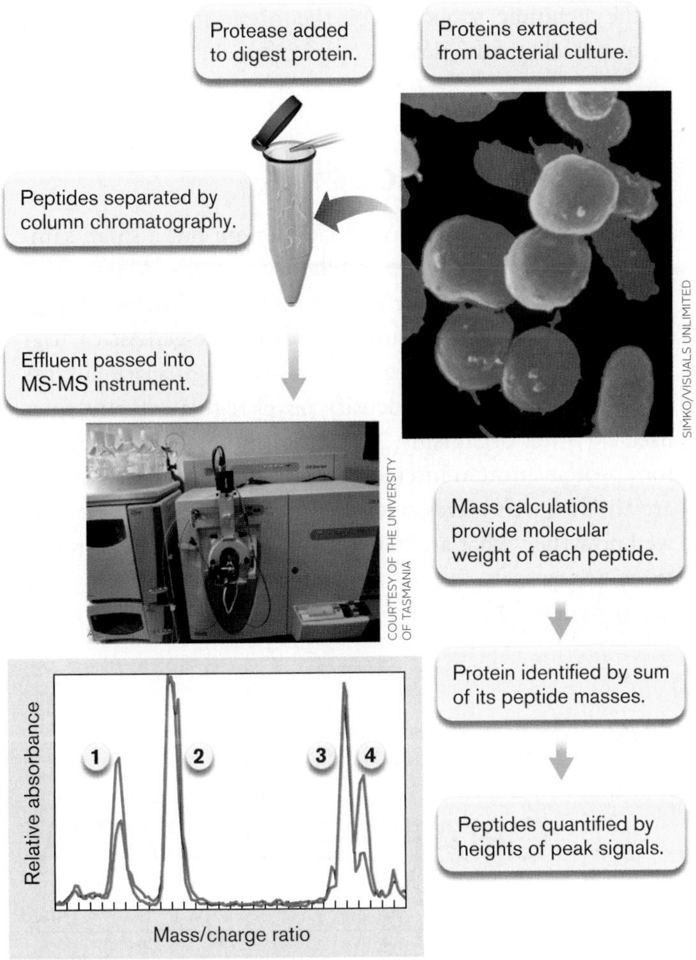
Protease added
to digest protein.
Proteins extracted
from bacterial culture.
Peptides separated by
column chromatography.
Effluent passed into
MS-MS instrument.
SIMKO/VISUALS UNLIMITED
COURTESY OF THE UNIVERSITY
OF TASMANIA
Mass calculations
provide molecular
weight of each peptide.
Protein identified by sum
of its peptide masses.
Peptides quantified by
heights of peak signals.
Relative absorbance
Mass/charge ratio
1
2
3
4

FIGURE 10.39 ■ Identifying proteins directly from whole cell extracts by mass spectrometry. Proteins are extracted from a bacterial culture and digested into peptides with trypsin. The resulting peptides are separated by column chromatography and analyzed by mass spectrometry. In a process called tandem mass spectrometry (MS-MS), the mass of each peptide is determined, and then selected peptides are subjected to additional fragmentation by ion spray. Each resulting peptide fragment will differ in size by one or more amino acids. Knowing the mass of each amino acid and the masses of the different peptide fragments allows one to extrapolate the sequence of the original peptide. Then the MS-MS sequences obtained for all of the tryptic peptides are compared by computer to all the predicted ORFs in a microbial genome. If one ORF contains all the peptides, a match is declared and the protein is identified. Quantitation can be achieved under the proper circumstances by comparing peak heights of MS results. In the example, two samples (red and green) are overlaid. Peaks for peptides 2 and 3 are equivalent, whereas red peaks for 1 and 4 are up and down relative to the respective green peaks (*x*-axis measures the ratio of a peptide's mass to its charge).

proteins from whole-cell extracts. To begin the analysis, proteins present in cell extracts are digested into distinct peptide fragments with a site-specific protease (trypsin) (**Fig. 10.39**). The fragment mixture is passed through an analytical column that separates peptides according to differences in hydrophobicity or some other parameter. The column effluent is then directly fed through tandem mass spectrometry (MS-MS) instrumentation. In MS-MS, each proteolytic fragment is subfragmented by ionization to produce progressively smaller secondary fragments missing one or more amino acids. Because the weight of each amino acid is distinct, MS-MS analysis can determine the amino acid sequence of the initial proteolytic fragment. Sophisticated computer programs identify the proteins by comparing the amino acid sequences of each protein fragment with the predicted sequences of proteolytic fragments from all ORFs in a genome.

Quantifying changes in protein levels with MS-MS is complex. Put simply, the computer can quantify relative protein levels in two samples either by comparing the signal peak heights of a peptide or by tabulating the number of times a fragment was found and sequenced (**Fig. 10.39**). Dörte Becher's laboratory in Germany recently used this technique to characterize the stress response of *Staphylococcus aureus*, a common human pathogen, to nitric oxide (NO), a toxic reactive oxygen species produced by the immune response. Becher's efforts uncovered over 1,400 proteins, 57 of which responded to the presence of NO.

Thought Question

10.13 Can tandem mass spectrometry (MS-MS) identify where modifications such as phosphorylation or acetylation happen on proteins?

Transcriptomic and proteomic technologies have been used for many microbial applications. One fascinating example centers on the transcriptional events that occur during the developmental cycle of *Caulobacter crescentus*. This organism progresses from a nonreplicating swimming form (swarmer cell) to a sessile stalked form that can replicate to make more swarming offspring. DNA microarray and proteomic analysis have revealed several potential regulators of this complex developmental process. Bioinformatic approaches have also been applied to human body odor. *Corynebacterium jeikeium* strain K411 degrades skin lipids to make volatile odorous products. Helena Barzantny and colleagues reconstructed a putative transcriptional regulatory network from the complete DNA sequence analysis of this organism. The result was a network of 48 regulators and 674 gene regulatory interactions that provide an early step in understanding how body odor develops.

To Summarize

- **The transcriptome and proteome** constitute all of a cell's mRNA molecules and proteins, respectively. Transcriptomes and proteomes change as environmental conditions change.
- **DNA microarrays** and **RNA-seq next-generation sequencing** is used to monitor the levels of thousands of individual mRNA molecules made during growth.
- **Two-dimensional gels** separate proteins by isoelectric point and molecular weight. They offer a snapshot of the proteome at any given point of growth or under any given growth condition.
- **Mass spectrometry** can determine the exact molecular weight of a protein (or the sequence of a component peptide) taken from an otherwise anonymous spot on a 2D gel. The exact molecular weight or peptide sequence is compared against the database of ORFs deduced from the genomic sequence of the organism. A successful match, therefore, determines the protein's identity.

Concluding Thoughts

The systems discussed in this chapter are but a small sampling of the known gene regulatory systems. Many more await discovery. How the cell coordinates all these systems is still a matter of conjecture. Scientists, like genetic cartographers, are now trying to map all the regulatory circuits of *E. coli,* as well as to identify novel regulatory strategies used by other organisms. The structural biology of regulatory protein-protein interactions and protein-DNA binding are the research waves of the future because, even though we know the "what and where" of many molecular interactions, we remain, for the most part, ignorant of how they occur structurally.

CHAPTER REVIEW

Review Questions

1. List regulatory mechanisms discussed in this chapter that work at the DNA level, transcriptional level, translational level, and posttranslational level.
2. How does lactose induce the *lacZYA* operon?
3. If *lacY* is induced only when lactose is present, how does external lactose induce the system?
4. How does tryptophan repress the tryptophan operon?
5. Describe a two-component signal transduction system.
6. Discuss how glucose impacts the utilization of lactose as a carbon source.
7. Name a regulatory protein that can activate and repress an operon's expression. How does it do that?
8. What is the regulatory mechanism that uses translation to control transcription? How does it work?
9. When growth of *E. coli* slows, how does the cell slow its production of ribosomes?
10. Describe four ways that sigma factor production/activity can be regulated.
11. Discuss the different forms of small regulatory RNA molecules.
12. How do bacterial cells couple genetic control and biochemical control in terms of nitrogen assimilation?
13. Describe how *E. coli* senses a chemical gradient and changes its behavior.
14. What is quorum sensing?
15. Describe transcriptomics and proteomics. Can you think of a situation in which a protein could be shown to increase in response to a stress but that same stress would not increase synthesis of the mRNA encoding the protein?

Thought Questions

1. What will happen to the expression of the tryptophan operon if you replace the key tryptophan codons in the attenuator region with tyrosine codons?
2. Adding tryptophan to *E. coli* will cause repression of the *trp* operon genes. Mutations in the *trpR* repressor gene and the *trp* operator will have the same phenotype; that is, adding tryptophan will no longer repress expression of the *trp* genes. What will happen to the phenotype if you transform each mutant with a plasmid carrying the wild-type *trpR* gene or the wild-type *trp* operator region?
3. The mosquito that transmits yellow fever, *Aedes aegypti*, requires water-filled human-made containers for egg laying. Normally, gravid females deposit their eggs in multiple containers but are particularly stimulated to deposit on fermenting leaves in water. Scientists tested the relative egg deposits made by gravid mosquitoes given a choice between a container with sterile water and a second container inoculated with 14 bacterial species isolated from a bamboo infusion (which was previously shown to be a favorite deposit medium). Ninety percent of the eggs were deposited on the surface of the bacterial infusion, and only 10% were deposited on the sterile water. How did the presence of the bacteria influence the choice?
4. This chapter has outlined the many ways bacteria sense their environment, their neighbors, and their location (rock, intestine, ocean). Can we then say that bacteria are conscious?

Key Terms

activator (366)
anti-anti-sigma factor (382)
anti-attenuator stem loop (379)
anti-sigma factor (382)
attenuator stem loop (379)
autoinducer (398)
catabolite repression (374)
chemotaxis (391)
cis-antisense RNA (385)
corepressor (367)
derepression (367)
diauxic growth (374)
direct repeat (389)
immune avoidance (389)
inducer (366)
inducer exclusion (374)
induction (366)
inverted repeat (389)
leader sequence (378)
methyl-accepting chemotaxis protein (MCP) (392)
operon (370)
phase variation (389)
proteome (404)
quorum sensing (398)
regulatory protein (366)
regulon (382)
repression (366)
repressor (366)
response regulator (368)
second messenger (396)
sensor kinase (368)
stringent response (381)
tandem repeat (389)
transcriptional attenuation (378)
transcriptome (401)
translational control (381)
two-component signal transduction system (368)

Recommended Reading

Bandara, H. M., O. L. Lam, L. J. Jin, and L. Samaranayake. 2012. Microbial chemical signaling: A current perspective. *Critical Reviews in Microbiology* **38**:217–249.

Battesti, Aurelia, Nadim Majdalani, and Susan Gottesman. 2011. The RpoS-mediated general stress response in *Escherichia coli. Annual Review of Microbiology* **65**:189–213.

Boisset, Sandrine, Thomas Geissmann, Eric Huntzinger, Pierre Fechter, Nadia Bendridi, et al. 2007. *Staphylococcus aureus* RNAIII coordinately represses the synthesis of virulence factors and the transcription regulator Rot by an antisense mechanism. *Genes and Development* **21**:1353–1366.

Dalebroux, Zachary D., and Michelle S. Swanson. 2012. ppGpp: Magic beyond RNA polymerase. *Nature Reviews. Microbiology* **10**:203–212.

Fozo, Elizabeth M., Matthew R. Hemm, and Gisela Storz. 2008. Small toxic proteins and the antisense RNAs that repress them. *Microbiology and Molecular Biology Reviews* **72**:579–589.

Henke, Jennifer M., and Bonnie L. Bassler. 2004. Bacterial social engagements. *Trends in Cell Biology* **14**:648–656.

Higgins, Douglas, and Jonathan Dworkin. 2012. Recent progress in *Bacillus subtilis* sporulation. *FEMS Microbiology Reviews* **36**:131–148.

Jørgensen, Mikkel G., Deo P. Pandey, Milena Jaskolska, and Kenn Gerdes. 2009. HicA of *Escherichia coli* defines a novel family of translation-independent mRNA interferases in bacteria and archaea. *Journal of Bacteriology* **191**:1191–1199.

Kamp, Heather D., and Daniel E. Higgins. 2011. A protein thermometer controls temperature-dependent transcription of flagellar motility genes in *Listeria monocytogenes. PLoS Pathogens* 7:e1002153.

Lewis, Mitchell. 2005. The *lac* repressor. *Critical Reviews in Biology* **328**:521–548.

Merino, Enrique, and Charles Yanofsky. 2005. Transcription attenuation: A highly conserved regulatory strategy used by bacteria. *Trends in Genetics* **21**:260–264.

Miller, Lance D., Matthew H. Russell, and Gladys Alexandre. 2009. Diversity in bacterial chemotactic responses and niche adaptation. *Advances in Applied Microbiology* **66**:53–75.

Parker, Christopher T., and Vanessa Sperandio. 2009. Cell-to-cell signalling during pathogenesis. *Cellular Microbiology* **11**:363–369.

Ponnusamy, Longanathan, Ning Xu, Satoshi Nojima, Dawn M. Wesson, Coby Schal, et al. 2008. Identification of bacteria and bacteria-associated chemical cues that mediate oviposition site preferences by *Aedes aegypti. Proceedings of the National Academy of Sciences USA* **105**:9262–9267.

Povolotsky, Tatyana L., and Regine Hengge. 2011. "Life-style" control networks in *Escherichia coli*: Signaling by the second messenger c-di-GMP. *Journal of Biotechnology* **160**(1–2):10–16. doi:10.1016/j.jbiotec.2011.12.024.

Ryan, Robert P., Tim Tolker-Nielsen, and J. Maxwell Dow. 2012. When the PilZ don't work: Effectors for cyclic di-GMP action in bacteria. *Trends in Microbiology* **20**:235–242.

Schleif, Robert. 2010. AraC protein, regulation of the L-arabinose operon in *Escherichia coli*, and the light switch mechanism of AraC action. *FEMS Microbiology Reviews* **34**:779–796.

Sesto, Nina, Omri Wurtzel, Cristel Archambaud, Rotem Sorek, and Pascale Cossart. 2013. The excludon: a new concept in bacterial antisense RNA-mediated gene regulation. *Nature Reviews Microbiology* **11**: 75–82.

Staron, Anna, and Thorsten Mascher. 2010. Extracytoplasmic function sigma factors come of age. *Microbe* **5**:164–170.

Sterberg, Sofia, Teresa del Peso-Santos, and Victoria Shingler. 2011. Regulation of alternative sigma factor use. *Annual Review of Microbiology* **65**:37–55.

Vink, Cornelis, Gloria Rudenko, and H. Steven Seifert. 2012. Microbial antigenic variation mediated by homologous DNA recombination. *FEMS Microbiology Reviews* **36**:917–948.

Williams, Paul. 2007. Quorum sensing, communication and cross-kingdom signalling in the bacterial world. *Microbiology* **153**:3923–3938.

Yang, Ji, Marija Tauschek, and Roy M. Robins-Browne. 2011. Control of bacterial virulence by AraC-like regulators that respond to chemical signals. *Trends in Microbiology* **19**:128–135.

CHAPTER 11
Viral Molecular Biology

11.1 Phage T4: The Classic Molecular Model

11.2 Hepatitis C: (+) Strand RNA Virus

11.3 Influenza Virus: (–) Strand RNA Virus

11.4 Human Immunodeficiency Virus (HIV): Retrovirus

11.5 Herpes Simplex Virus: DNA Virus

11.6 Gene Therapy with Viruses

Viruses such as influenza virus, hepatitis virus, and HIV sicken hundreds of millions of people worldwide. To prevent or treat these infections, we face the challenge of viral diversity. Viral genomes range from a single nucleic acid coiled inside a protein tube to a cell-like package composed of multiple layers of protein and membrane. Viral genomes may include fewer than 10 genes or more than 200. Herpes and other latent viruses sequester their genomes within that of the host, causing devastating disease. At the same time, latent viral genomes contribute in surprising ways to human evolution. And the viruses most deadly to humans can be converted to vectors that cure a human disease.

Molecular biology helps us address urgent questions: How do viruses trick host cell defenses into permitting their entry and replication? How can we prevent viruses from gaining access to cells, or halt their replication once they do? The molecular mechanisms of viral infection reveal targets for antiviral drugs—and help us design life-saving gene transfer vectors.

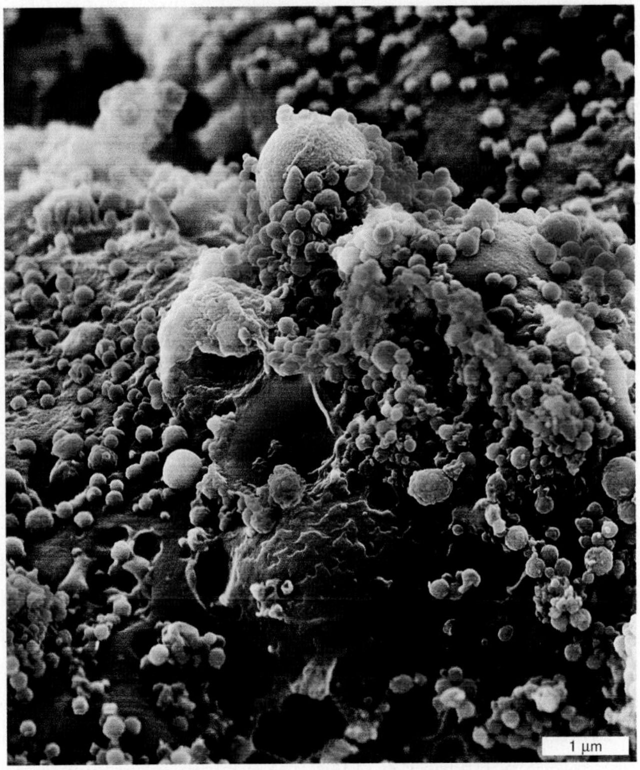

1 µm

LENNART NILSSON/SCANPIX

CURRENT RESEARCH highlight

Avian influenza virions. Humans infected with the avian influenza strain H5N1 show a death rate of greater than 50%. The tiny influenza virions (colorized blue in the photo) infect the much larger human host cell (colorized red; SEM). The human transmission rate for H5N1 influenza is poor because the H5 version of the viral envelope protein hemagglutin is adapted to cell-surface receptor proteins of birds, which differ slightly from those of humans. In 2011, however, virologist Ron Fouchier, of Erasmus MC, Rotterdam, reported generating a strain of H5N1 showing airborne transmission in ferrets, which are considered a good model system for human flu transmission. The mutant strain was obtained by "laboratory evolution" in which an H5N1 strain was passaged from ferret to ferret in ten transmission cycles. Surprisingly, the genome of the highly transmissible strain showed only two amino acid differences from the original strain. When the authors tried to publish their result in the journal *Science*, a U.S. government agency advised the journal to redact (remove) key information that terrorists might use to make a bioweapon. This startling incident marks the first time that the U.S. government recommended censorship of a biological manuscript. *Source:* Sander Herfst et al. 2012. *Science* **336**: 1534.

In 2009, around the globe people fell ill from a mysterious strain of influenza virus. The strain was called H1N1 for the versions of its envelope protein genes hemagglutinin and neuraminidase (**Fig. 11.1**). The virus was also called "swine flu" because some of its genes originated in influenza viruses infecting pigs. But travelers soon brought the strain to other countries, where healthy people succumbed. What could explain the new strain's virulence? The new influenza virion had a combination of envelope proteins never experienced before, to which people lacked immunity. Fortunately, prompt public health measures lessened the impact of the 2009 strain. But even so, seasonal strains of influenza kill half a million people each year. Meanwhile, other deadly viruses, such as HIV and hepatitis C virus, infect tens of millions of people annually. The envelope protein of HIV mutates so fast that new strains arise within a single infected patient, overcoming the body's immune defenses against them.

The envelope proteins of influenza virus and of HIV offer just two examples of the importance of viral molecular biology in medicine. In Chapter 10 we discussed how molecular mechanisms such as repressors and DNA inversions regulate the function of microbial cells. Chapter 11 presents the molecular mechanisms of viral infection of cells. We assume a fundamental understanding of the nature of viruses, as presented in Chapter 6. Here we explore in greater depth the molecular basis of infection. First we consider the classic bacteriophage T4, whose virion assembly provides a key model for viral development at the molecular level. We then examine in detail four viruses that infect humans: hepatitis C, influenza, HIV, and herpes viruses. This selection of viruses includes, respectively, a (+) strand RNA virus, a (–) strand RNA virus, a retrovirus, and a large DNA virus. We will see that all viral infection processes share common themes but also use molecular mechanisms unique to each virus. Finally, in Section 11.6 we look at how the mechanisms of lethal viruses may actually be harnessed for lifesaving gene therapy.

This chapter focuses on the fundamental biology of viral infection and propagation. The consequences of viral disease for human patients are discussed in Chapters 25 and 26.

Note: Poliovirus molecular biology is presented in detail in **eTopic 11.1**.

11.1

Phage T4: The Classic Molecular Model

Where do we find bacteriophages? Phage T4 (**Fig. 11.2**) is one of hundreds of tailed phages that have evolved to make a home in the human gut, where they infect Gram-negative enteric bacteria. More distantly related tailed phages permeate the soil and swarm in marine waters; for example, sev-

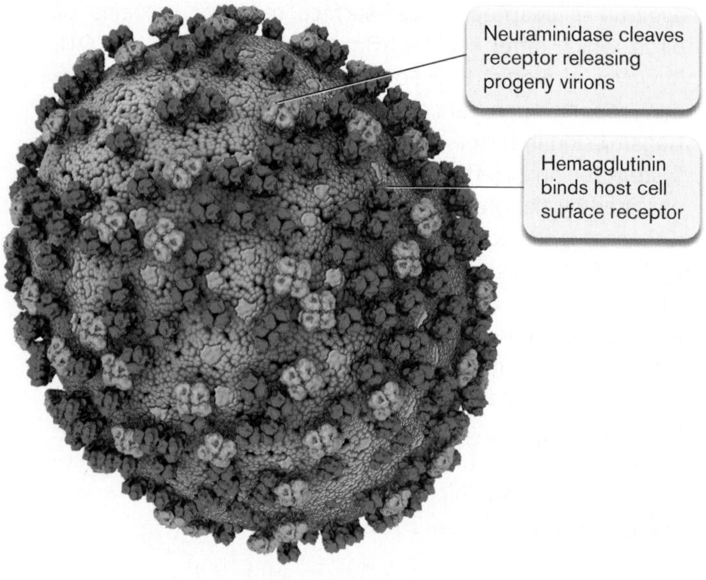
Neuraminidase cleaves receptor releasing progeny virions
Hemagglutinin binds host cell surface receptor

FIGURE 11.1 ■ Influenza virion. Surface receptors hemagglutinin (HA) and neuraminidase (NA) acquire mutations that lead to virulent new strains of influenza.

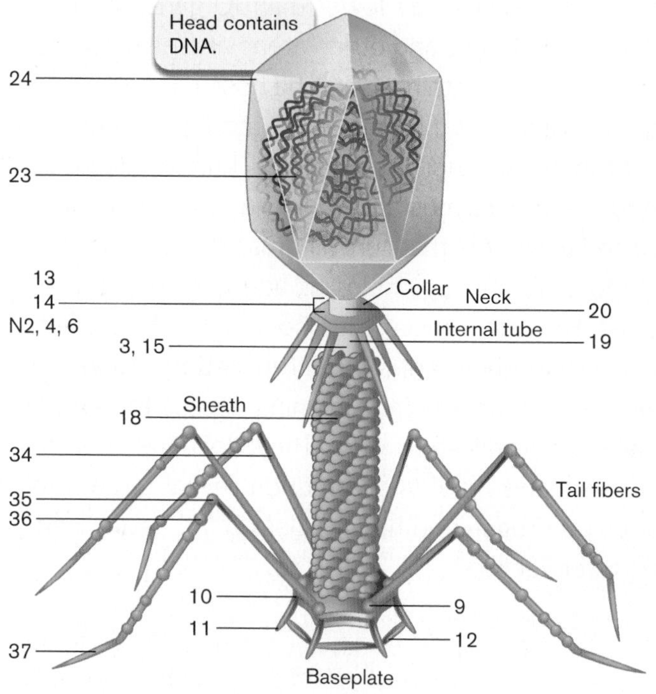
Head contains DNA.
24
23
13
14
N2, 4, 6
3, 15
Collar
Neck
20
Internal tube
19
Sheath
18
34
35
36
Tail fibers
10
11
9
12
37
Baseplate

FIGURE 11.2 ■ Structure of a phage T4 capsid (simplified). Proteins are assigned to specific genes (numbered) that are mapped on the chromosome in Figure 11.3.

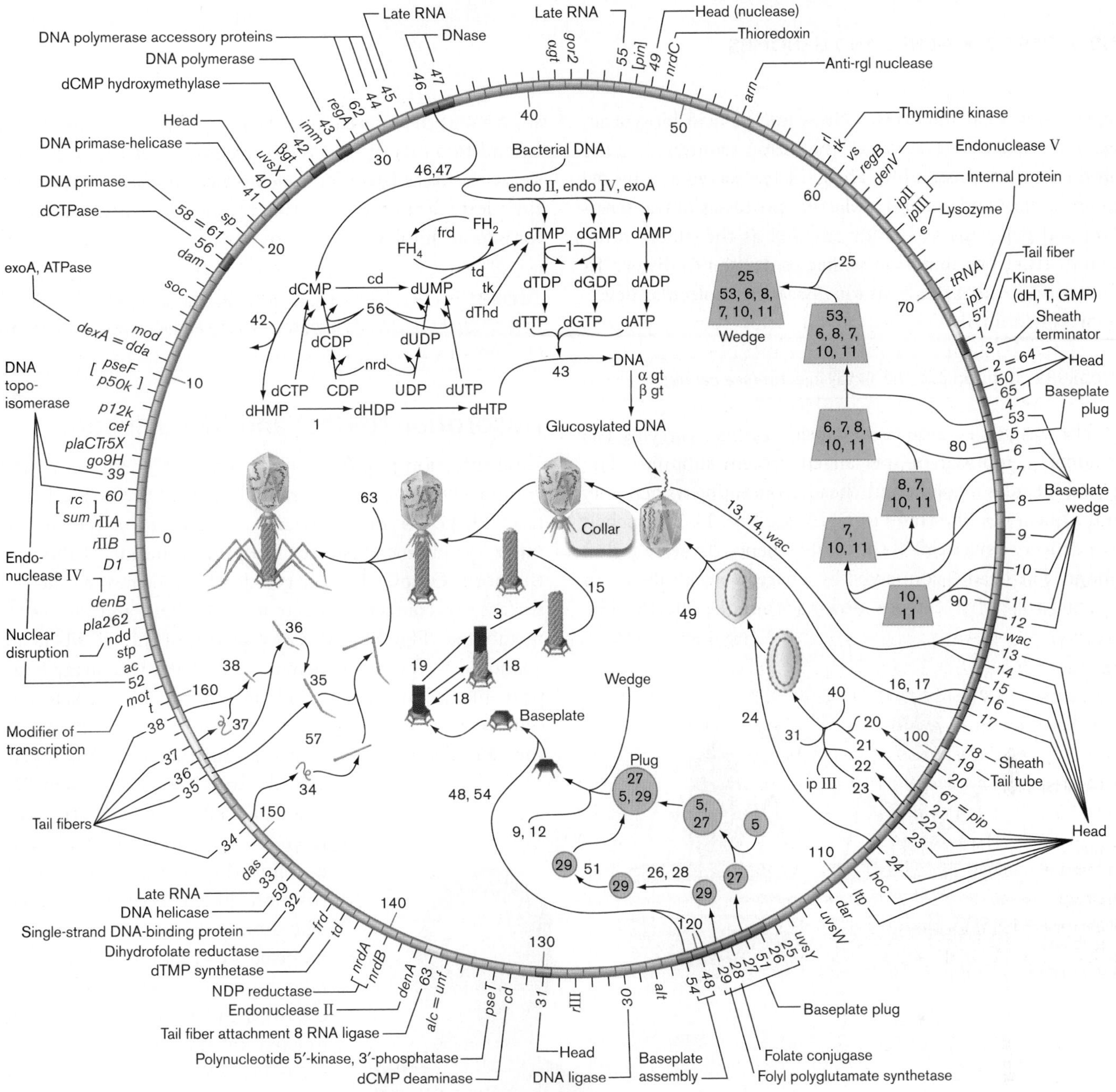
Late RNA
DNase
Late RNA
Head (nuclease)
Thioredoxin
Anti-rgl nuclease
DNA polymerase accessory proteins
DNA polymerase
dCMP hydroxymethylase
Head
DNA primase-helicase
DNA primase
dCTPase
exoA, ATPase
DNA topo-isomerase
Endo-nuclease IV
Nuclear disruption
Modifier of transcription
Tail fibers
Thymidine kinase
Endonuclease V
Internal protein
Lysozyme
Tail fiber
Kinase (dH, T, GMP)
Sheath terminator
Head
Baseplate plug
Baseplate wedge
Sheath
Tail tube
Head
Late RNA
DNA helicase
Single-strand DNA-binding protein
Dihydrofolate reductase
dTMP synthetase
NDP reductase
Endonuclease II
Tail fiber attachment 8 RNA ligase
Polynucleotide 5′-kinase, 3′-phosphatase
dCMP deaminase
Head
DNA ligase
Baseplate assembly
Baseplate plug
Folate conjugase
Folyl polyglutamate synthetase
Bacterial DNA
endo II, endo IV, exoA
dTMP dGMP dAMP
dTDP dGDP dADP
dTTP dGTP dATP
dCMP
dUMP
dCDP
dUDP
dCTP CDP UDP dUTP
dHMP dHDP dHTP
DNA
α gt
β gt
Glucosylated DNA
Collar
Wedge
Baseplate
Plug
13, 14, wac

FIGURE 11.3 ▪ Phage T4 genome, showing functions of mapped genes. The gene numbers correspond to the steps of assembly in Figure 11.2. *Source:* Based on Matthews et al. 1983. *Bacteriophage T4*, ASM Press, Washington, DC.

eral thousand new T4-like capsid genes were identified by a global ocean sampling expedition (discussed in Section 21.4). Wherever you find bacteria you will find bacteriophages.

Historically, phages provided the first living systems simple enough to dissect at the molecular level. Phages yielded fundamental discoveries in gene regulation, including the classic case of lambda lysogeny (discussed in eTopic 10.5). The first model for the genetic analysis of animal development was a phage system, the assembly of the "tailed phage" T4 (*Myoviridae*). Today, cultured in bacteria, phages remain easier and less expensive to study than animal viruses. As recently as 2012, the ATP-driven motor for DNA packaging of phage T4 was reported as a model for DNA packaging of herpes virus. Phages provide new tools for antibacterial agents and new ways to make vaccines. They are even used in nanotechnology, to assemble wires in microscopic devices.

Phage T4 Structure

Phage T4 may be the most complicated noncellular reproductive unit that has been characterized. The T4 genome of 169 kilobase pairs (kb) specifies approximately 300 genes (**Fig. 11.3**), many of which encode enzymes that

Thought Question

11.1 The phage T4 chromosome is a linear piece of DNA, yet the genomic map of T4 (as shown in **Fig. 11.3**) is circular. Why?

replace host cell functions such as nucleic acid biosynthesis. The intricate structure of the virion requires 20 gene products just for assembly. Thus, T4 has served as a model to probe the fundamental molecular processes of transcription and gene recombination, as well as the global function of cells. The mutation strategies devised to dissect T4 function provided the basis for dissecting molecular development in animals.

Weblinks 3D tutorial on the T4 tail injector (*see ebook*)

The phage T4 virion has a **capsid**, a device carrying the genome, composed of specialized protein subunits. The capsid includes a polyhedral "head" containing its genome, and a linear double strand of DNA (see **Fig. 11.2**) attached to a delivery device called the "tail"—hence the term **tailed phage**. The head is connected by a narrow neck tube to an internal tube that extends down within the sheath. The sheath plus the internal tube make up the tail. From the tail extend two sets of fibers—six connected at the neck, and six at the baseplate. The elegant mechanical structure and function of this "molecular motor" was an early inspiration for nanotechnology.

Note: A very different type of phage, the slow-release filamentous phage M13, with its nanotechnology applications, is discussed in **eTopic 11.2**.

Adsorption to Host and DNA Injection

Viral infection requires a molecular "fit" between a viral component and one or more molecules of the host cell surface. Phage T4 infects strains of *Escherichia coli*, whose outer membrane possesses lipopolysaccharide (LPS) and the porin OmpC. The phage adsorbs (attaches) to the cell surface by contact between its tail fibers and the outer membrane (**Fig. 11.4A**). Once several of the tail fibers are anchored to the outer membrane by LPS, the phage baseplate makes contact. The contact between baseplate and cell outer membrane generates a conformational change in the tail. The outer tube of the tail (the sheath) is a helical tube of subunits of protein P18 (**Fig. 11.4B**). The sheath contracts by shortening and widening, enabling the internal tube to push through, like a molecular syringe. The dimensions of the sheath were modeled by combining X-ray crystallography of protein P18 with cryo-EM images of tails that were isolated in the extended or contracted position. This impressive model was built by Michael Rossmann (**Fig. 11.4A**, inset) and colleagues at Purdue University. Rossmann's lab developed techniques to visualize many challenging viral structures, including whole virions such as rhinovirus (cold virus), dengue fever flavivirus, and canine parvovirus. These viral structures may now be used to test new antiviral drugs.

The internal tube of the tail is capped by a needlelike "injector," composed of proteins P27 and P5. As the sheath surrounding the internal tube contracts, the injector pokes through the host outer membrane and penetrates

A.

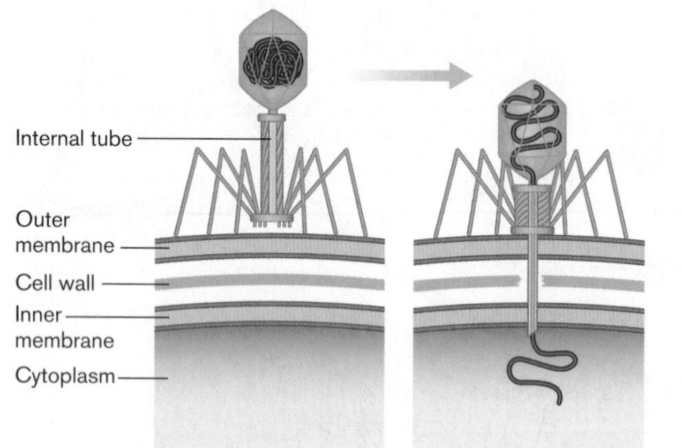
Internal tube
Outer membrane
Cell wall
Inner membrane
Cytoplasm

Michael Rossman
DAVE UMBERGER

B.

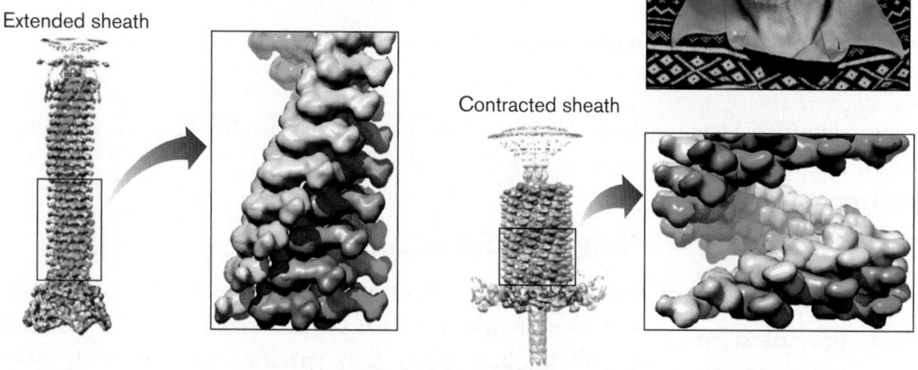
Extended sheath
Contracted sheath

FIGURE 11.4 ▪ Phage T4 adsorption and DNA injection. A. First the tail fibers contact the outer membrane; then the baseplate completes the connection. The sheath (composed of protein P18) contracts, the internal tube (composed of protein P19) penetrates the outer membrane, and then the peptidoglycan is digested. The tube extends to the inner membrane, and the head releases DNA under packing pressure, forcing it into the cytoplasm. **B.** Sheath contraction, based on crystallography of P18 and on cryo-EM of extended and contracted tails. Model devised by Michael Rossmann (A, inset) and colleagues at Purdue University. *Source:* Part B modified from Anastasia Aksyuk et al. 2009. *EMBO J.* **28**:821.

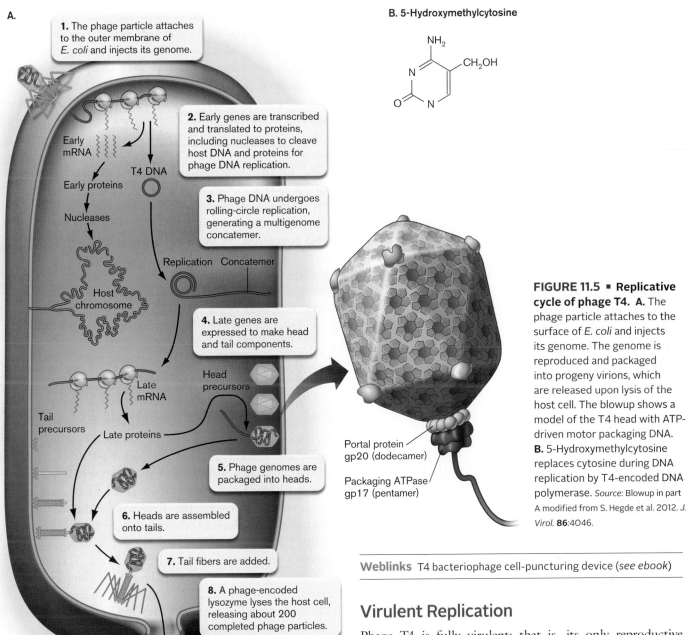

A.

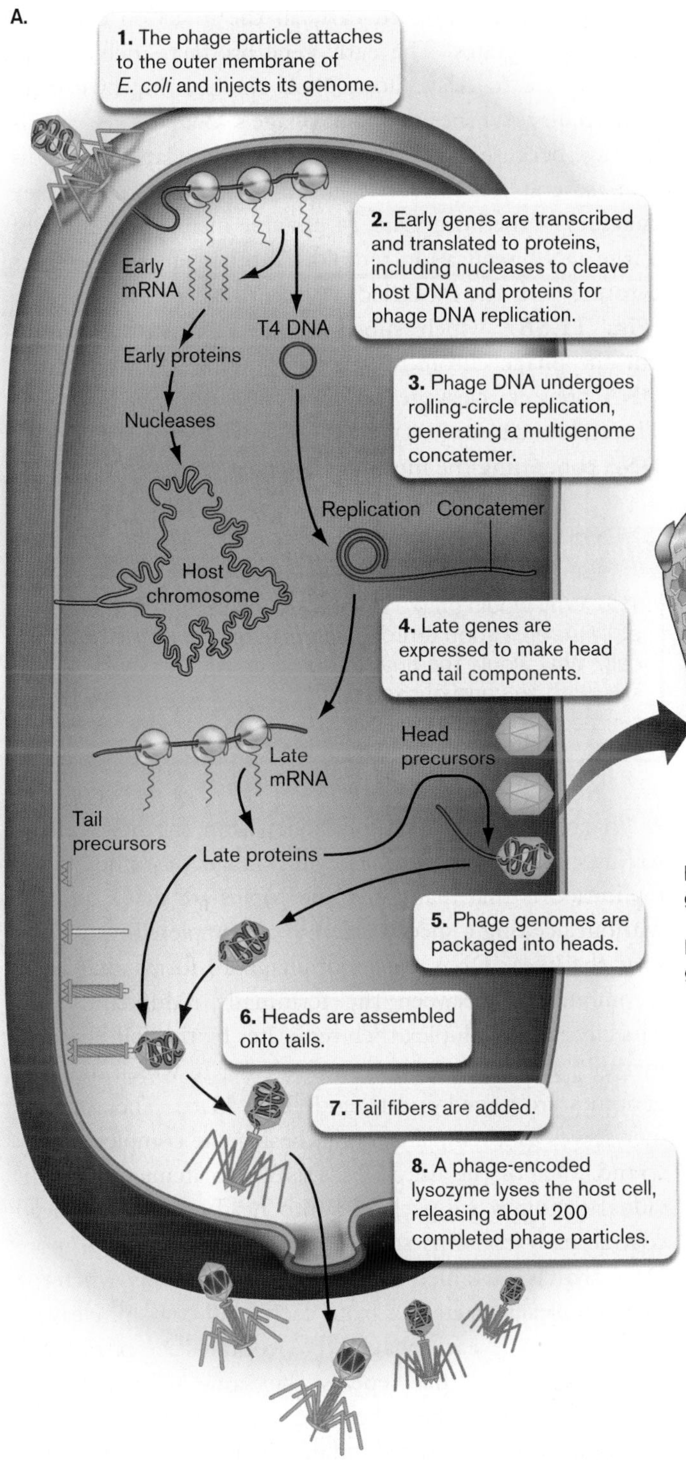
1. The phage particle attaches to the outer membrane of E. coli and injects its genome.
Early mRNA
T4 DNA
Early proteins
Nucleases
2. Early genes are transcribed and translated to proteins, including nucleases to cleave host DNA and proteins for phage DNA replication.
3. Phage DNA undergoes rolling-circle replication, generating a multigenome concatemer.
Replication
Concatemer
Host chromosome
4. Late genes are expressed to make head and tail components.
Late mRNA
Head precursors
Tail precursors
Late proteins
5. Phage genomes are packaged into heads.
6. Heads are assembled onto tails.
7. Tail fibers are added.
8. A phage-encoded lysozyme lyses the host cell, releasing about 200 completed phage particles.

B. 5-Hydroxymethylcytosine

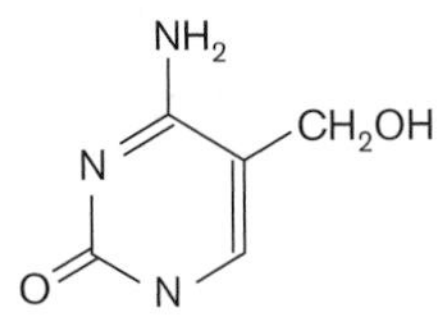
NH2
CH2OH
N
O
N

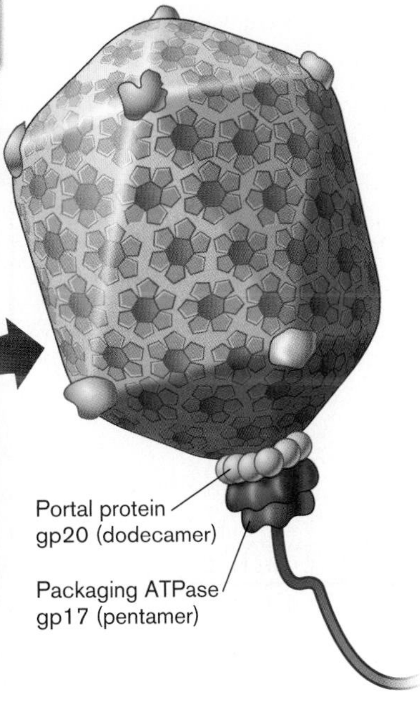
Portal protein gp20 (dodecamer)
Packaging ATPase gp17 (pentamer)

FIGURE 11.5 ▪ Replicative cycle of phage T4. A. The phage particle attaches to the surface of *E. coli* and injects its genome. The genome is reproduced and packaged into progeny virions, which are released upon lysis of the host cell. The blowup shows a model of the T4 head with ATP-driven motor packaging DNA. **B.** 5-Hydroxymethylcytosine replaces cytosine during DNA replication by T4-encoded DNA polymerase. *Source:* Blowup in part A modified from S. Hegde et al. 2012. *J. Virol.* **86**:4046.

the cell wall. The peptidoglycan surrounding the injector is then digested by a lysozyme protein within the needle tip. The hole in the cell wall enables the entire tail to penetrate the cell wall and inner membrane. When the tip of the tail tube penetrates the inner membrane, the phage DNA is released and forced out of the head under pressure. Double-stranded DNA viruses are packed at a pressure of up to 50 atmospheres (atm), some of the highest pressures found in living organisms.

Weblinks T4 bacteriophage cell-puncturing device (*see ebook*)

Virulent Replication

Phage T4 is fully virulent; that is, its only reproductive option is to assemble progeny virions while destroying a host cell (**Fig. 11.5A**). By contrast, other tailed phages, such as phage lambda, can undergo lysogeny, the integration of their genome into that of their host cell. The integrated phage genome then replicates passively with the host. Lysogeny is described in Chapter 6, and the genetic regulation of phage lambda is presented in **eTopic 10.5**.

When phage T4 DNA enters the host cytoplasm, a set of phage genes is activated for transcription. Phage gene activation involves DNA-binding regulators like those of bacterial gene regulation (discussed in Chapters 8 and 10). The phage genes are transcribed by the host cell RNA polymerase and translated by the host ribosomes. These host components, however, are supplemented by phage-encoded components such as tRNAs.

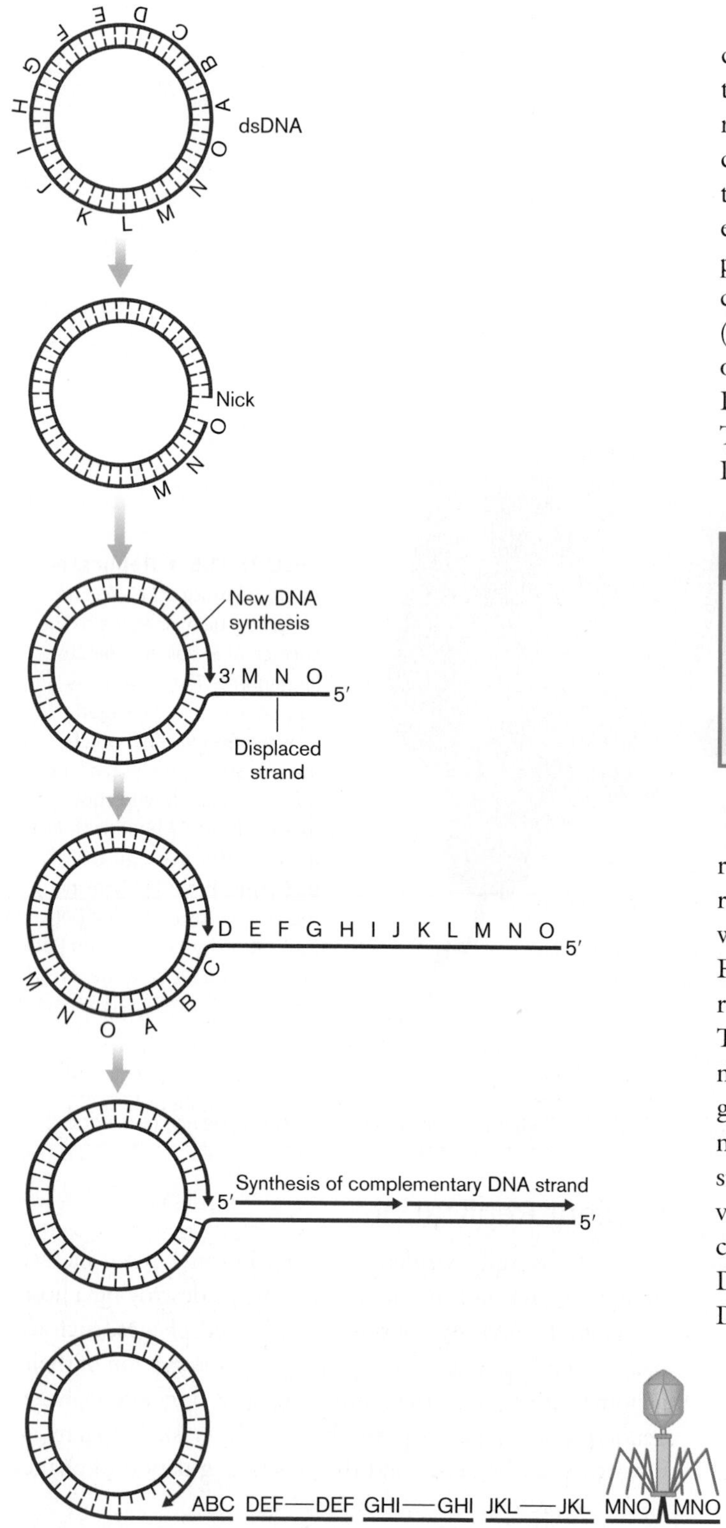

dsDNA
Nick
New DNA synthesis
3′ M N O
5′
Displaced strand
D E F G H I J K L M N O
5′
Synthesis of complementary DNA strand
5′
5′
ABC DEF—DEF GHI—GHI JKL—JKL MNO MNO

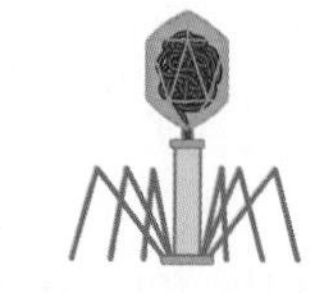

FIGURE 11.6 ■ Rolling-circle replication of phage T4. Initially, the linear genome circularizes; then it replicates as a concatemer. Out of the concatemer, the individual genomes are packaged into head coats and then cleaved. Each encapsidated DNA contains an end-duplicated 3% of its genome.

Phage genes expressed early in the infection cycle are called **early genes**. The early-gene products include proteins needed to cleave host DNA, thus halting host macromolecular synthesis. (The phage's own DNA avoids cleavage because its cytosine bases are methylated.) In addition, several phage DNA synthesis enzymes replace key enzymes of the host. One T4 enzyme increases the rate of phage DNA replication tenfold. Another enzyme replaces cytosine with the modified base 5-hydroxymethylcytosine (**Fig. 11.5B**), which substitutes for cytosine throughout the phage chromosome. This modification of phage DNA prevents cleavage by viral and host endonucleases. The endonucleases (enzymes that cleave DNA) fail to "fit" DNA containing the modified cytosine.

Thought Question

11.2 Why would phage T4 production require a rate of DNA replication tenfold higher than that of the host cell? Why would the phage substitute all of its cytosine with an unusual base that requires more energy to synthesize?

Phage T4 DNA is synthesized within the host cell by rolling-circle replication. The advantage of rolling-circle replication is that many genome copies are made quickly, without needing a special enzyme to complete linear ends. First the linear DNA duplex of phage T4 forms a circle by recombination between the terminally redundant ends. The circularized duplex then replicates by the rolling-circle method, generating a linear concatemer in which multiple genomes are joined end to end (**Fig. 11.6**). The concatemer serves as a template to synthesize the complementary strand. Out of the linear T4 DNA concatemer, the individual genomes are packaged into head coats. Each head coat actually contains sufficient volume to pack 3% more DNA than is contained in a phage genome. Thus, when the DNA duplexes extending from each head coat are cleaved, each phage DNA contains 3% extra DNA that repeats the same sequence from the other end. (This is called "terminal redundancy.") Each cleavage of the concatemer must then occur another 3% farther along the genome and duplicate this next piece instead. Thus, every T4 DNA molecule ends up with a different 3% terminal repeat at some point throughout the genome.

Ultimately, the **late genes** are induced to produce the capsid and tail proteins that assemble at the membrane to make mature phage. During **assembly**, a key problem is how to get the entire DNA genome stuffed into the phage head. The process of DNA packaging actually requires energy spent by an ATP-driven "nanomotor" (**Fig. 11.5A**, blowup). We can actually "watch" the DNA being pulled into the phage head, by using the technique of optical tweezers, a form of microscopy in which an object is trapped within a laser beam. The phage head and the distal end of the entering DNA are each attached to a microsphere; then the two microspheres are observed to approach each other, as the DNA between them is pulled into the head.

The filled heads are attached to tails, and finally the various tail, collar, and capsid substructures are assembled. One phage infection yields about 200 complete phage particles per cell. At last, an unknown signal triggers **lysis**. A late gene encodes a lysozyme that digests the cell wall, releasing the phages into their surroundings. If the lysozyme gene is defective, all the progeny phage particles remain trapped within the host cell.

The temporal separation of early genes and late genes is common to the replication cycles of all kinds of phages, as well as viruses of eukaryotes. Viruses differ, however, in the size of their genomes; some, such as human immunodeficiency virus, or HIV (discussed in Section 11.4), contain only a small number of genes, whereas others, such as phage T4, encode products that duplicate host functions of DNA synthesis, ribosomal proteins, tRNA, and biosynthetic enzymes such as dihydrofolate reductase. Some phage-encoded enzymes, such as the T4 DNA polymerase, modify a cellular function to favor phage reproduction.

All viral infections need to coordinate the actions of the viral components with a much greater number of host components, such as those of the transcription and translation apparatus. In some cases, a virus actually evolves to "pick up" a host gene encoding a useful product—and then finds a novel use for that product. The phage T4 genome has acquired many host genes through recombination with scraps of host DNA. For example, the phage-expressed enzyme dihydrofolate reductase evolved from a host enzyme used to reduce the cofactor folic acid for use in biosynthesis. But the phage-encoded enzyme has a completely different use—assembly with folate as structural parts of the T4 injector baseplate.

Phage Particles Self-Assemble

The assembly of phage T4 particles within the host cytoplasm offers exceptional opportunities to visualize a molecular pathway. Each phage particle is assembled by convergence of three pathways involving the head coat, the tail, and the tail fibers (**Fig. 11.7**). All of these stages can be isolated

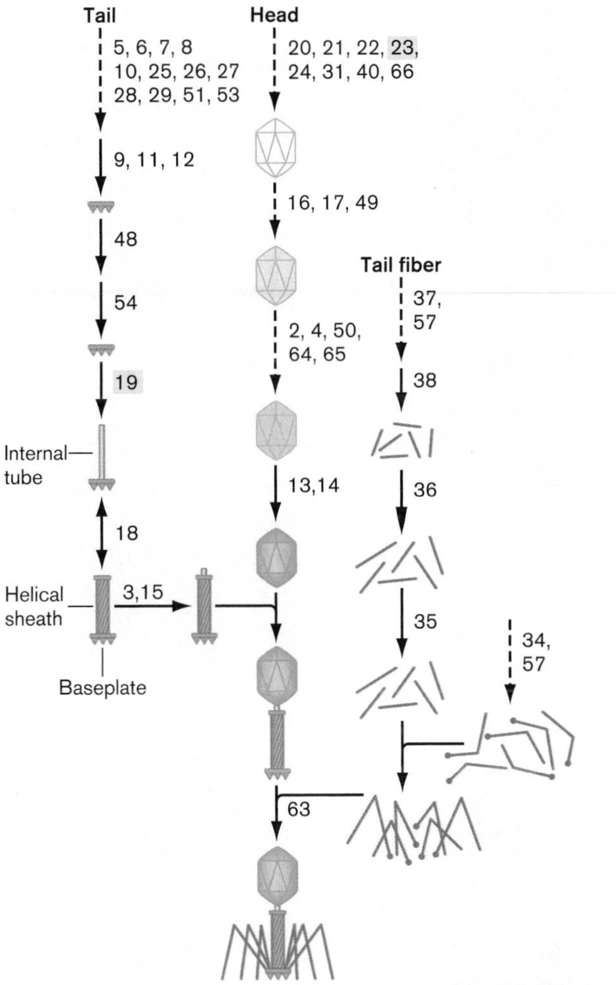
Tail
5, 6, 7, 8
10, 25, 26, 27
28, 29, 51, 53
9, 11, 12
48
54
19
Internal tube
18
Helical sheath
3,15
Baseplate
Head
20, 21, 22, 23,
24, 31, 40, 66
16, 17, 49
2, 4, 50,
64, 65
13,14
63
Tail fiber
37,
57
38
36
35
34,
57

FIGURE 11.7 ■ Phage T4 assembly. The phage T4 capsid assembles automatically in a predetermined order from parts encoded by phage genes (numbers correspond to the gene numbers in Fig. 11.3). Genes 19 and 23 (highlighted) are mutated in Fig. 11.8.

and observed within infected cells by electron microscopy (**Fig. 11.8**). But with all these many steps, how did we determine the order of assembly and identify the coding genes?

The stages of phage assembly were discovered using strains that carry mutations in various genes that encode proteins essential for development (see **Fig. 11.7**). In some cases, defective assembly leads to a bizarrely altered shape of the particle, such as a phage with a "giant head" (**Fig. 11.8A**). In other cases, a defective gene simply prevents progression in the pathway, halting assembly at an unfinished stage. For example, **Figure 11.8B** shows the phage particles found in a cell infected by a phage defective for gene 23. The unfinished particles consist of tail structures only. This defect occurs because gene 23 is required for assembly of the head, which must be completed before attachment to the tail. In **Figure 11.8C**, which shows

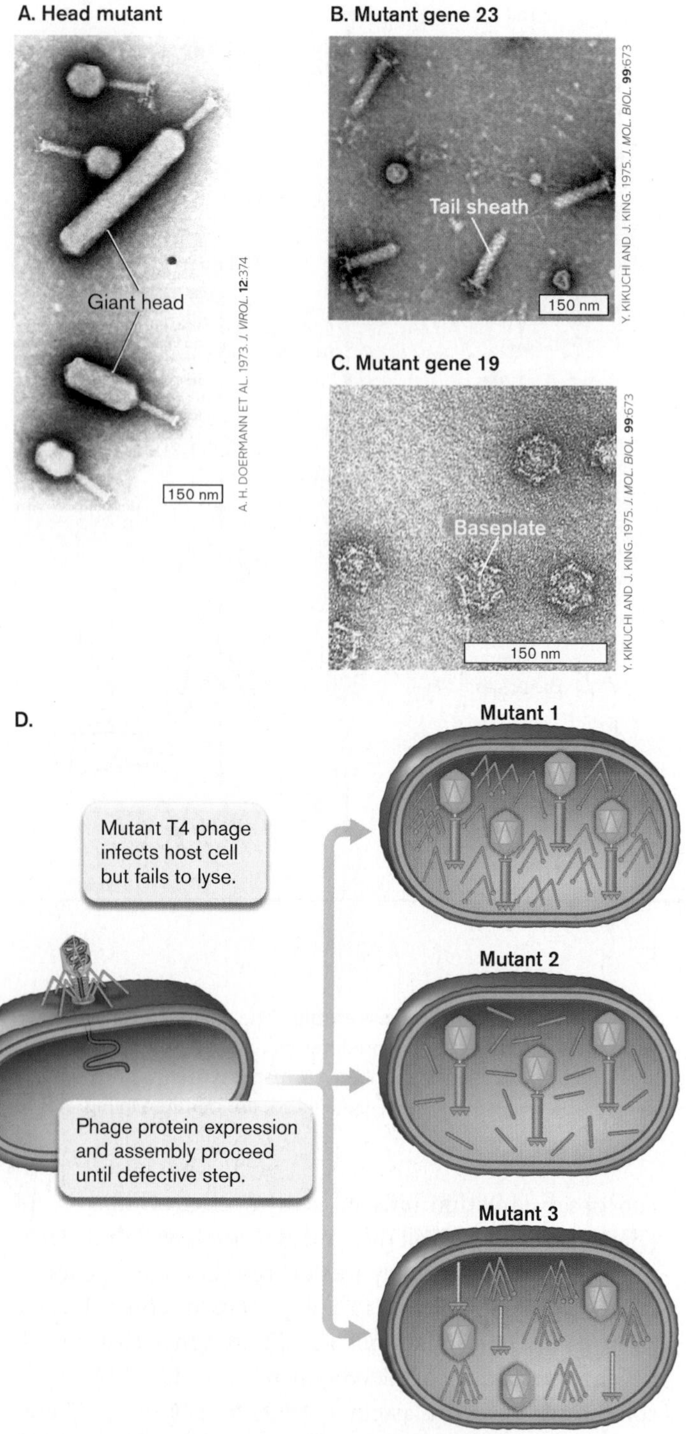
A. Head mutant
Giant head
150 nm
A. H. DOERMANN ET AL. 1973. J. VIROL. 12:374
B. Mutant gene 23
Tail sheath
150 nm
Y. KIKUCHI AND J. KING. 1975. J. MOL. BIOL. 99:673
C. Mutant gene 19
Baseplate
150 nm
Y. KIKUCHI AND J. KING. 1975. J. MOL. BIOL. 99:673
D.
Mutant T4 phage infects host cell but fails to lyse.
Phage protein expression and assembly proceed until defective step.
Mutant 1
Mutant 2
Mutant 3

FIGURE 11.8 ■ Developmental mutants of phage T4. Analyzing such mutants made it possible to decipher the genetic pathway governing assembly of the structure. **A.** Mutations in developmental genes lead to variant capsid morphology such as "giant head." **B.** Gene 23 defect results in tails without heads. **C.** Gene 19 defect results in baseplates without tails. **D.** Design of experiment using temperature-dependent mutations to identify steps of T4 assembly.

phages defective for gene 19, we see only baseplates because the product of gene 19 is needed to build the tail on the baseplate. In this case, assembled heads are found near the cell membrane.

Thought Question

11.3 Which numbered genes from **Figure 11.7** might be mutated in each of the defective phage populations presented in **Figure 11.8D**?

An interesting problem arises in designing an experimental strategy: How do we grow populations of phages whose particles fail to complete assembly? The answer lies in using **conditional lethal mutations**—that is, mutations that cause a lethal defect under one growth condition but permit growth under a second condition. Two classes of conditional lethal mutant strains of phage T4 were obtained, through mutagenesis, by Robert Edgar and Richard Epstein in the early 1960s. These two classes are:

- **Temperature-sensitive mutants.** Mutants that are temperature sensitive can grow at one temperature but fail to grow at another. For example, a point mutation may result in a protein product that is stable at 25°C but denatures at 42°C, so the phage can reproduce at 25°C but not at 42°C.
- **Nonsense mutations countered by suppressor tRNA.** A mutant that contains a nonsense codon (stop codon) in the middle of a gene prematurely terminates translation of the protein. Such a mutant, however, can replicate in a strain of *E. coli* that makes a certain type of mutant tRNA. The mutant tRNA contains an anticodon that matches the stop codon sequence, which allows the tRNA to place an amino acid into the growing protein. This kind of tRNA is called a **nonsense suppressor** because the tRNA anticodon mutation suppresses the effect of the nonsense mutation in the phage. (Nonsense mutations are discussed further in Chapter 9.)

Through many years of analysis of phage mutants, Edgar and Epstein and their colleagues showed that conditional lethal mutations could be used to dissect a developmental pathway as complex as that of phage T4. Their work laid the foundation for the molecular analysis of development in animals and plants and the discovery of the molecular basis of inherited diseases. With whole-genome sequences in hand, we use conditional lethal mutations and suppressor mutations to decipher how all the gene products work together. For example, genetic analysis of development in the fruit fly *Drosophila* uses temperature-sensitive mutations to dissect mechanisms of gene regulation.

To Summarize

- **The phage T4 virion** consists of a head containing its DNA genome and accessory proteins, a tail composed of an internal tube with a sheath, and tail fibers.
- **T4 is adsorbed** by binding of the tail fibers to bacterial outer membrane receptors, followed by binding of the tail baseplate.
- **Injection of T4 DNA** through the internal tube leads to lytic infection. Early genes encode a DNA polymerase and a DNase to cleave host DNA.
- **Rolling-circle replication** generates progeny genomes linked in a concatemer. The concatemer is cut with an offset, so that each linear genome has a slight overlap, cut at a different position in the sequence.
- **Virions are self-assembled**, including packaging of DNA into the head, in arrays beneath the inner membrane.
- **Cell lysis** is caused by a lysozyme expressed by late genes in the T4 genome.
- **The phage T4 assembly process** was dissected in studies of temperature-sensitive mutants and nonsense mutants with suppressor tRNA. This strategy of genetic dissection was used as a model for experiments on animal development.

11.2

Hepatitis C: (+) Strand RNA Virus

The (+) strand RNA viruses are the largest group of viruses known to infect humans and other animals (**Table 11.1**). Examples include the picornaviruses, such as the famous poliovirus and related enteroviruses; the Noroviruses that are the bane of cruise ships; and the insect-borne West Nile and dengue viruses. West Nile and dengue are flaviviruses, a group that includes hepatitis C virus (HCV).

HCV is a blood-borne virus of the liver that infects 200 million people worldwide and now kills more people in the United States than AIDS does (up to 15,000 deaths per

TABLE 11.1

(+) strand RNA virus species (examples).

Family/genus	Virus	Disease	Host
Caliciviridae			
Norovirus	Norwalk viruses	Gastroenteritis	Humans
Coronaviridae	SARS and MERS coronaviruses	Severe acute respiratory syndrome	Humans
Flaviviridae			
Flavivirus	Hepatitis C virus	Hepatitis, liver cancer	Humans
	Dengue virus	Dengue fever ("breakbone" fever)	Humans
	West Nile virus	Encephalitis	Humans, horses
Picornaviridae			
Enterovirus	Coxsackievirus	Common cold, myocarditis	Humans
	Echovirus	Meningitis, paralysis, encephalitis	Humans
	Poliovirus (PV)	Meningitis, paralysis	Humans
Rhinovirus	Human rhinovirus (HRV)	Common cold	Humans
Aphthovirus	Foot-and-mouth disease virus (FMDV)	Foot-and-mouth disease	Cattle, swine
Hepatovirus	Hepatitis A virus (HAV)	Hepatitis	Humans
Togaviridae	Rubella virus	German measles	Humans

year). In some countries, unscreened blood transfusions and reuse of medical supplies lead to high prevalence (proportion of individuals infected) of HCV—as high as 14% in Egypt. In the United States, an estimated 1% of the population carries HCV, acquired primarily through blood contact such as transfusions and injection drug use, as well as by sexual transmission. For unknown reasons, the prevalence is three times higher for people born between 1945 and 1965. Many have no recollection of risk factors and are unaware of their infection—which they could pass on to others. In 2012, the Centers for Disease Control and Prevention (CDC) recommended HCV testing for all members of this age group.

Surprisingly, little is known of the virus, in part because when HCV infections first emerged, their symptoms were hard to distinguish from those of two other viruses: hepatitis A and B. The potential consequences of HCV are particularly serious because they include cancer of the liver, now a leading cause of liver transplants. In the United States, HCV is particularly prevalent among veterans of the armed forces, where rates of liver cancer may now outnumber colon cancer.

About 15%–20% of infected individuals actually clear the virus, but in other patients HCV can persist for many decades without symptoms. We don't know why 10%–15% of these patients go on to exhibit symptoms such as cirrhosis, the scarring of the liver leading to liver failure, and liver cancer. For those who develop symptoms, treatment includes the antiviral agent ribavirin (a nucleoside analog that inhibits genome replication) combined with a weekly injection of interferon-alpha (an antiviral protein of the innate immune system, discussed in Chapter 24). The treatment has severe side effects and works for only about half of those treated (**Fig. 11.9A**). New drugs are being developed, but the virus mutates quickly into new resistant strains. Clearly, we need to know more about HCV to develop effective therapies and a vaccine.

How do we discover new antiviral agents for HCV? The virus infects only humans and chimpanzees, so testing antiviral agents has been a challenge. While chimpanzees have yielded useful data, their study increasingly draws criticism from advocates for special treatment of these apes. Recently, mice have been engineered as a model system for HCV (discussed below).

A major challenge has been to grow the virus in cultured human cells. Between 2006 and 2008, several laboratories at last developed culture systems for HCV in hepatocarcinoma lines, as well as in normal hepatocyte tissue culture. **Figure 11.9B** shows a TEM section through a cultured hepatocyte infected by HCV. Each virion contains an RNA genome packaged in an icosahedral capsid, surrounded by an envelope membrane containing glycoproteins. After binding to host cell surface proteins, the virions become internalized by the cytoplasm as discussed below. Tissue culture enables us to characterize molecular targets for drugs—but culture does not replicate all aspects of infection of organs within an organism.

A.

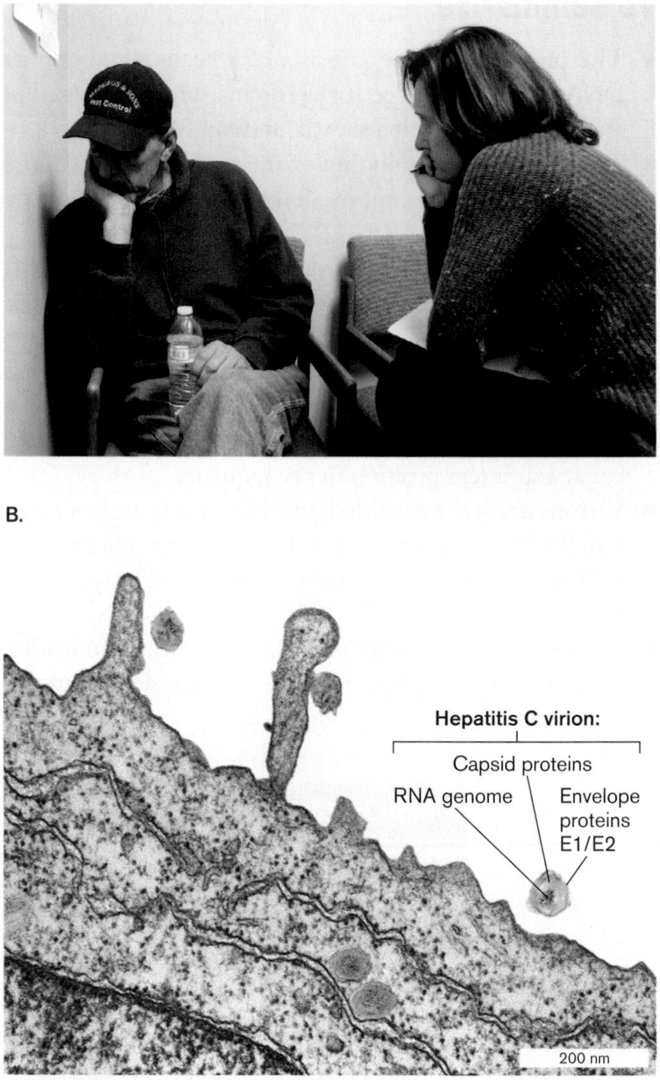
Hepatitis C virion:
Capsid proteins
RNA genome
Envelope proteins E1/E2
200 nm

B.

BILL HACKWELL

THOMAS DEERINCK, NCMIR/SCIENCE SOURCE/PHOTO RESEARCHERS, INC.

FIGURE 11.9 ■ Hepatitis C virus infection. A. An individual infected with hepatitis C learns from his physician (at right) that the virus has reemerged 6 months after treatment. The physician, Dr. Diana Sylvestre, at UC San Francisco, emphasizes community-based factors in treatment success. **B.** Hepatitis C viruses infecting cultured liver cells (purple = cytoplasm; blue = ER membranes). Virions consist of a core of RNA (ribonucleic acid, brown) enclosed in a capsid (green), and surrounded by a glycoprotein envelope (yellow). Colorized TEM.

HCV Virion Structure and Genome

HCV virions pose several challenges for structural studies. First, the virions are elusive; while an infected individual may produce 10^{12} virions per day, the virions rapidly decay and become incapable of infection; their half-life in the blood is only a few hours. Virions have been visualized by TEM and SEM, but no detailed model of a virion has yet been obtained. (See, for example, the cryo-EM models of poliovirus and herpes virus in Chapter 6.) One reason for the difficulties in visualizing HCV may be that within blood serum, the virions pick up a coating of low-density lipoproteins, which interferes with sample preparation.

The virion is small, appearing in electron micrographs to be 40–70 nm in diameter. It consists of an icosahedral core particle surrounded by an envelope with spike proteins. Overall, the virion is composed of just three types of structural proteins: the protein C monomer, forming the icosahedral core that packages the RNA genomes; and envelope proteins E1 and E2, which form club-shaped dimers (**Fig. 11.10**). The genome encodes seven additional nonstructural proteins (designated NS) that function in the host cytoplasm during viral replication. The most crucial of these is NS5B, the RNA-dependent RNA polymerase, which synthesizes (–) strand template and (+) strand progeny genomes. Other nonstructural proteins include proteases (NS2, NS3, NS4A), RNA helicase (a second activity of protein NS3), interferon resistance (NS5A), and a function in the replication factories where virions assemble (NS4B).

But how is HCV viral RNA translated? The viral genomic RNA looks nothing like a host mRNA; it lacks the

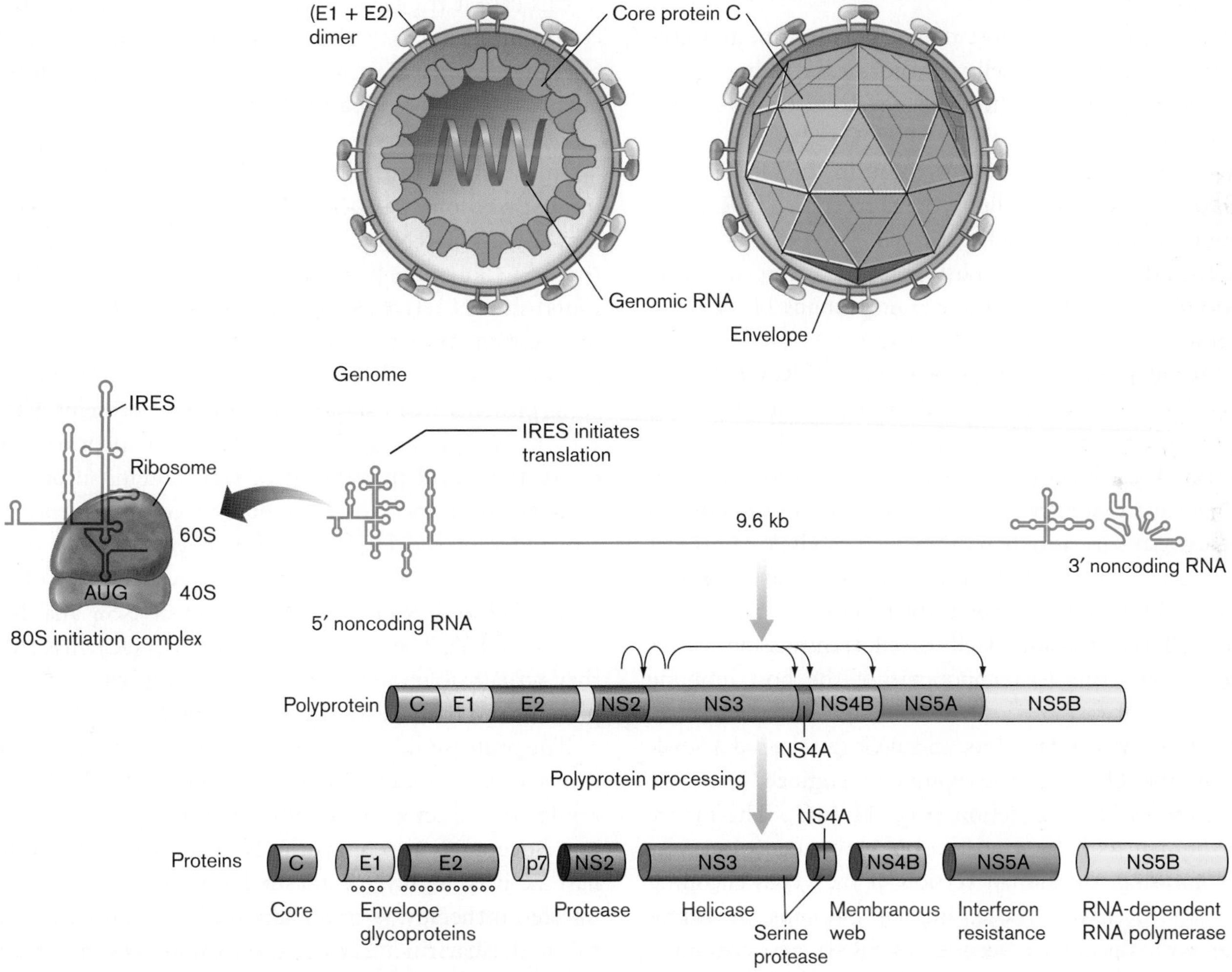
(E1 + E2) dimer
Core protein C
Genomic RNA
Envelope
Genome
IRES
Ribosome
60S
AUG
40S
80S initiation complex
IRES initiates translation
9.6 kb
3′ noncoding RNA
5′ noncoding RNA
Polyprotein
C
E1
E2
NS2
NS3
NS4B
NS5A
NS5B
NS4A
Polyprotein processing
NS4A
Proteins
C
E1
E2
p7
NS2
NS3
NS4B
NS5A
NS5B
Core
Envelope glycoproteins
Protease
Helicase
Serine protease
Membranous web
Interferon resistance
RNA-dependent RNA polymerase

FIGURE 11.10 ■ Hepatitis C virus structure and genome. The (+) strand RNA genome is wound inside an icosahedral core composed of a single monomer type, protein C. The envelope, derived from host membrane lipids, contains paired glycoproteins E1 and E2. Translation is initiated by the internal ribosome entry site (IRES; blowup), to synthesize one polyprotein. The polyprotein is processed (cleaved by proteases) to generate three structural proteins (C, E1, E2), plus seven nonstructural (NS) proteins that fill various functions during infection.

5′ "cap" and 3′ poly-A "tail" signals that bind to a eukaryotic ribosome. This problem is solved differently by different kinds of viruses. Some RNA viruses, such as influenza virus, cleave the cap from a host mRNA to attach to their own (discussed in Section 11.3). But HCV uses a different mechanism. The HCV genome starts with a 5′ **internal ribosome entry site (IRES)**. The IRES contains numerous stem loop structures in which the RNA doubles back and forms short regions of A-form duplex. These stems and loops twist around each other, forming a globular structure that has evolved to "fit" the ribosome in place of the "cap" used by standard host mRNA (see the blowup in **Fig. 11.10**). In the case of HCV, the IRES replaces most of the eukaryotic initiation factors as well; thus it initiates translation with high efficiency.

HCV Attachment and Host Cell Entry

What determines HCV host range and tropism, and how does the virion infect a cell? A major factor in host range and tissue tropism is the host cell surface receptors needed to bind the HCV virion (**Fig. 11.11A**). The serum lipoproteins surrounding the HCV virion in the bloodstream bind to the low-density lipoprotein receptor (LDLR) on the luminal surface of the hepatic cell. Two other host proteins (CD81 and SR-B1) must bind E2 before infection can occur. Antibodies to these host proteins block HCV infection.

After initial binding to the proteins LDLR, CD81, and SR-B1, HCV then binds to additional proteins that form part of the "tight junction" between cells: claudin-1 and occludin. Tight junctions are structures that join adjacent cells within a tissue, preventing penetration of the luminal contents. But surprisingly, viruses such as HCV can target the tight junction for host entry. It has also been proposed that progeny virions use the tight junction as a route for cell-to-cell transmission. Cell-to-cell transmission avoids exposing the virus to components of the host immune system.

In 2011, Alexander Ploss and colleagues used knowledge of the HCV host receptors to engineer a mouse that supports HCV infection (**Fig. 11.11C**). The mouse was "humanized" by transfection with adenoviral vectors expressing the human version of the genes encoding CD81 and occludin. These "human" proteins, in combination with the mouse versions of SR-B1 and claudin-1, enabled successful liver infection by HCV. The infection was visualized in the animal using a special HCV strain (HCV-CRE) modified to induce expression of the luciferase reporter, which generates bioluminescence. A drawback of the mouse system was that it supports only HCV uptake, but not HCV replication and virion release; but the achievement represents the first breakthrough in devising an immunocompetent model animal to test new anti-HCV drug therapies.

Replication Cycle of HCV

How can a virus expressing as few as ten genes commandeer the entire machinery of a host cell? As HCV infects its host, its replication cycle includes many processes in which viral components trick the host into cooperating (**Fig. 11.12**). This viral trickery is a recurring theme of all virus infections.

As the HCV virion binds host receptors at or near a tight junction (**Fig. 11.12**, step 1), the virion becomes endocytosed by the host cell membrane. The endocytic vesicle fuses with a lysosome (step 2). The acidity of the lysosome generally functions to counteract pathogenic microbes, but in this case it triggers the virion envelope to fuse with the endosome membrane. The membrane fusion generates an opening that releases the core particle into the cytoplasm (step 3). The core disassembles, releasing the (+) strand RNA genome (step 4).

In the cytoplasm, the IRES of the RNA genome binds to host ribosomes associated with the endoplasmic reticulum (ER). The RNA genome is translated to a **polyprotein** (step 5). Polyprotein synthesis is a common mechanism found in RNA viruses. The polyprotein must be cleaved into separate proteins (step 6). First, a host enzyme, endoplasmic reticulum (ER) signal peptidase, cleaves the structural proteins (core subunit E, envelope proteins E1 and E2). (The normal function of the signal peptidase is to cleave the signal peptide off of cell proteins after signal-directed transport to a membrane compartment.) The remainder of the HCV polyprotein gets cleaved by two virus-encoded proteases, composed respectively of subunits NS2 and NS3 (NS2/NS3) and of NS3 and NS4A (NS3/NS4A). The viral proteases are self-cleaving; that is, they actually fold into their active domains within the polyprotein, and then cleave themselves out.

The protease subunit NS3 is the target of two important drugs: telaprevir and boceprevir. Telaprevir binds the serine residue of the active site of NS3, preventing substrate binding. Boceprevir also binds the NS3 active site. Both drugs have activity highly specific to HCV, unlike the ribavirin interferon therapy. A combination of both protease inhibitors with ribavirin interferon greatly improves the cure rate, and may decrease the duration of severe side effects.

Most of the virus-encoded proteins sit within the membranes of vesicles budding out of the ER. The formation of these viral vesicles is poorly understood, but it is thought that the vesicles detach from the ER and move into the cytoplasm, forming a "membranous web" containing replication

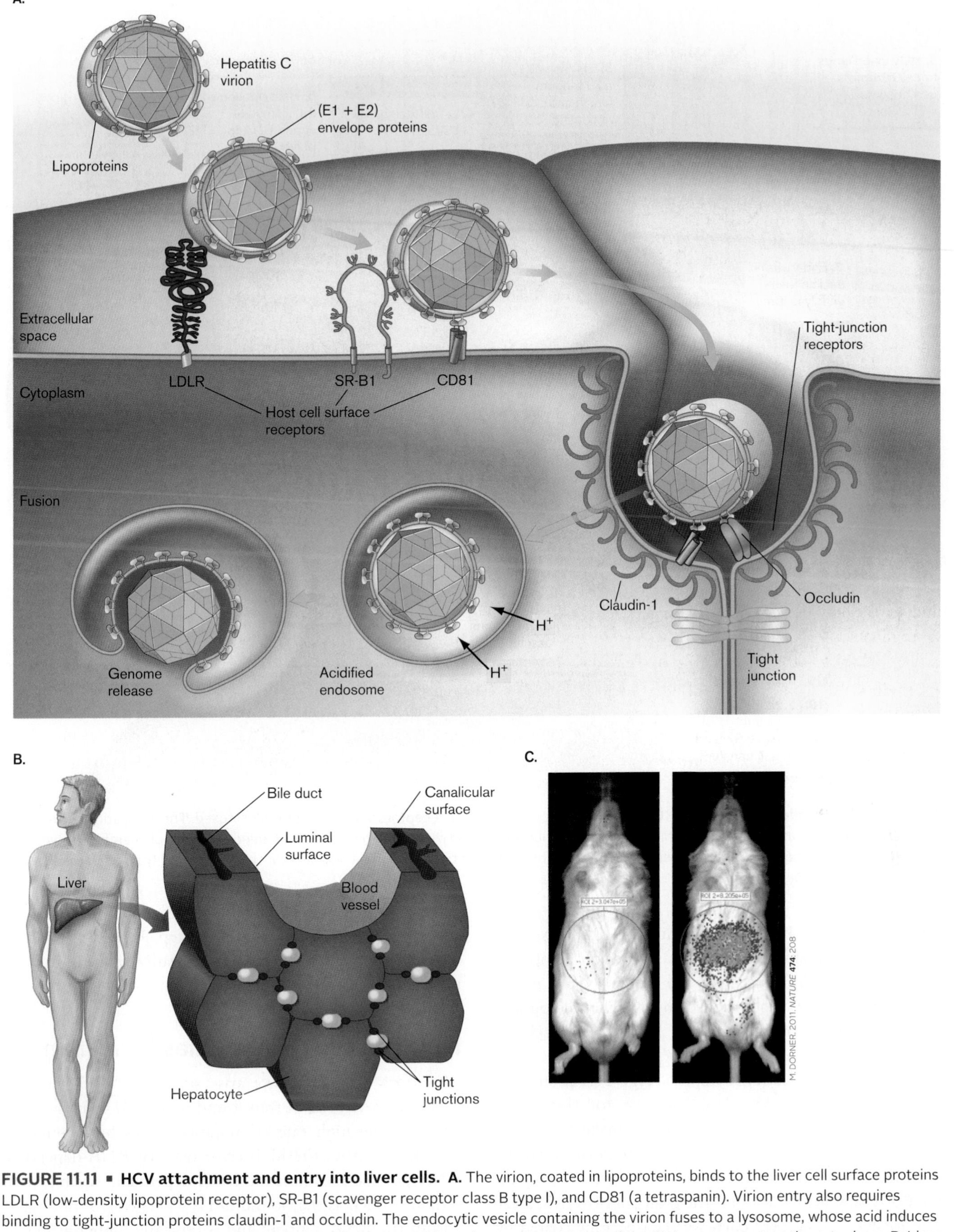
A.
Hepatitis C virion
(E1 + E2) envelope proteins
Lipoproteins
Extracellular space
Cytoplasm
LDLR
SR-B1
CD81
Host cell surface receptors
Tight-junction receptors
Fusion
Claudin-1
Occludin
H+
H+
Tight junction
Genome release
Acidified endosome
B.
Bile duct
Canalicular surface
Luminal surface
Liver
Blood vessel
Hepatocyte
Tight junctions
C.
M. DORNER. 2011. NATURE 474:208

FIGURE 11.11 ■ HCV attachment and entry into liver cells. A. The virion, coated in lipoproteins, binds to the liver cell surface proteins LDLR (low-density lipoprotein receptor), SR-B1 (scavenger receptor class B type I), and CD81 (a tetraspanin). Virion entry also requires binding to tight-junction proteins claudin-1 and occludin. The endocytic vesicle containing the virion fuses to a lysosome, whose acid induces viral envelope fusion with the vesicle membrane. The core particle undergoes uncoating, and the RNA genome enters the cytoplasm. **B.** Liver cells are connected by tight junctions, which is where HCV binds to infect. **C.** A mouse made to express the human version of two receptor proteins, CD81 and occludin, is successfully infected by HCV. Color indicates intensity of bioluminescence from a luciferase reporter gene activated by HCV. *Source:* Part A modified from D. Moradpour. 2007. *Nat. Rev. Microbiol.* **5**:453; part B, from T. Pietschmann. 2009. *Nature* **4578**:797.

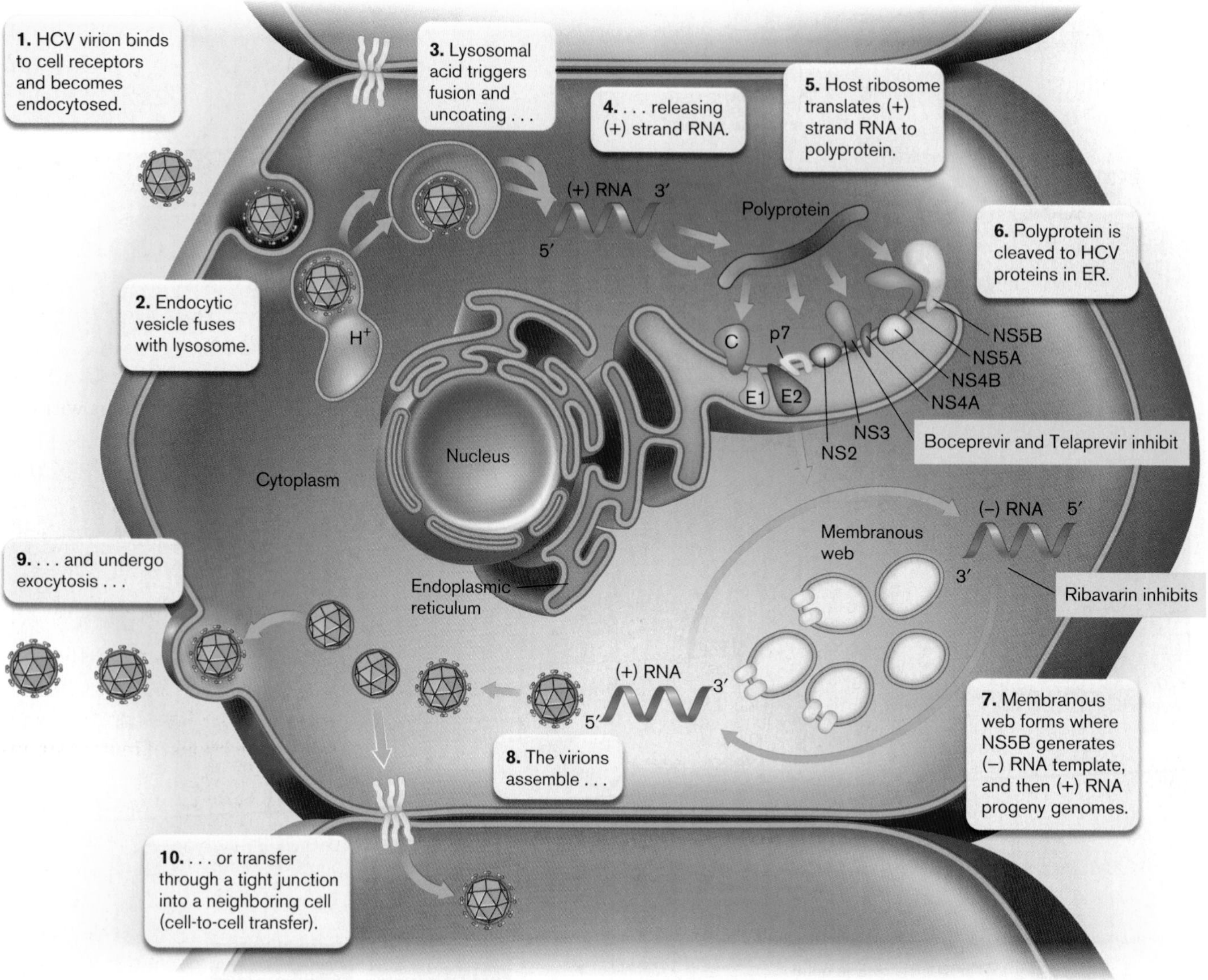
1. HCV virion binds to cell receptors and becomes endocytosed.
2. Endocytic vesicle fuses with lysosome.
3. Lysosomal acid triggers fusion and uncoating . . .
4. . . . releasing (+) strand RNA.
5. Host ribosome translates (+) strand RNA to polyprotein.
6. Polyprotein is cleaved to HCV proteins in ER.
7. Membranous web forms where NS5B generates (–) RNA template, and then (+) RNA progeny genomes.
8. The virions assemble . . .
9. . . . and undergo exocytosis . . .
10. . . . or transfer through a tight junction into a neighboring cell (cell-to-cell transfer).
H+
(+) RNA 3′
5′
Polyprotein
C
p7
E1
E2
NS2
NS3
NS4A
NS4B
NS5A
NS5B
Boceprevir and Telaprevir inhibit
Nucleus
Cytoplasm
Endoplasmic reticulum
Membranous web
(–) RNA 5′
3′
Ribavarin inhibits
(+) RNA
3′
5′

FIGURE 11.12 ▪ Replication cycle of HCV. The HCV virion binds to cell receptors and becomes endocytosed. The replication cycle includes lysosome fusion and uncoating, translation by host ribosomes, and replication by viral RNA-dependent RNA polymerase. The virions assemble and exit the host cell by exocytosis, or by transfer through a tight junction into a neighboring cell (cell-to-cell transfer). *Source:* Modified from Darius Moradpour, 2007. *Nature Reviews Microbiology* **5**:453.

complexes (**Fig. 11.12**, step 7). **Figure 11.13** shows a fluorescence micrograph in which the replication complexes are visualized by GFP fused to viral protein NS5A, which becomes part of the replication complex. The micrograph has been overlaid by white tracks marking intracellular movement of individual replication complexes.

Within each replication complex (**Fig. 11.12**, step 7), the (+) strand RNA serves as template for the RNA-dependent RNA polymerase (NS5B) to make (–) strand RNA. The (–) strand RNA serves as a template to make (+) strand RNA to package into new core particles (step 8). The RNA polymerase NS5B is the target of ribavarin, a key drug for HCV. The core particles then become coated by vesicle membranes containing E1 and E2. Vesicles carry the progeny virions to the cell membrane, where they fuse and the progeny virions are released (step 9). Alternatively, it has been proposed that some virions subvert the tight junctions for cell-to-cell transfer (step 10).

Mutation and Quasispecies Formation

Why does drug resistance arise so fast for HCV and for other RNA viruses? A trait found in all RNA viruses is an exceptionally high rate of mutation. This high mutation rate derives from the high error rates of RNA-dependent RNA polymerases (and of DNA-dependent reverse transcriptases such as that found in HIV). The error rate varies within the RNA genome and is typically higher at hypervariable (HR) regions within the envelope proteins exposed to the immune system. The high rate of mutation leads to

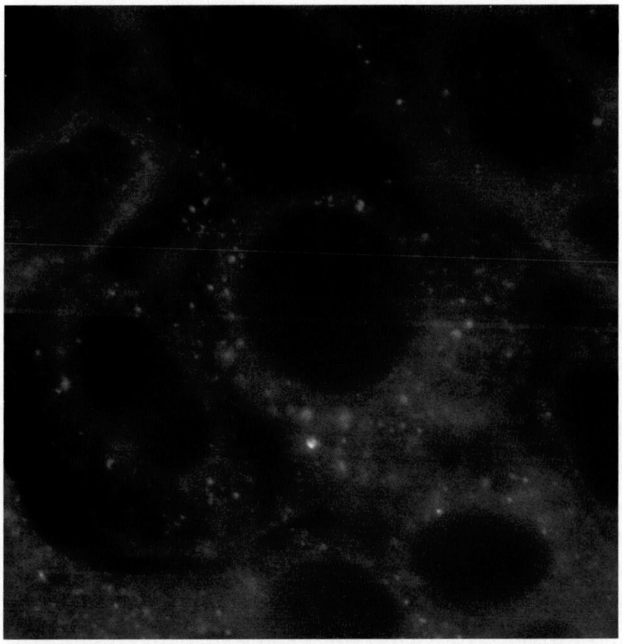
SOURCE: BENNO WOLK ET AL. 2008. *J. VIROL.* **82**:10519

FIGURE 11.13 ■ HCV replication complexes. Membranous replication complexes surround the nucleus of a transformed liver cell in tissue culture. The cultured cells have been engineered to express viral protein NS5A fused to green fluorescent protein (GFP). When the cells are infected with HCV, the NS5A-GFP proteins are expressed and incorporated into the membranous HCV replication complexes. White lines indicate the tracked motion of the complexes.

production of exceptionally diverse progeny—so diverse that many progeny are actually defective in one or more aspects of viral propagation. For this reason, HCV propagated in tissue culture leads to virus that produces exceptionally high numbers of virions, but the virions show reduced infection rates within an organism.

Another source of genetic change in HCV is recombination. Over 10% of patients with HCV show evidence of genome recombination between different HCV strains. For recombination to occur, different virions must coinfect a common cell. Recombination then occurs by "template switching." Template switching occurs when the RNA-dependent RNA polymerase complex detaches from the viral RNA template, still attached to its nascent RNA product, and then attaches to a new template at the same position and continues RNA synthesis. This molecular process differs from the recombination of duplex DNA chromosomes because no Holliday junction can occur, as shown in Chapter 9. But the genetic result is similar: A recombinant progeny chromosome is formed, containing nucleic acid from two different parental virions.

Within a single infected patient, the high mutation rates generate multiple virus strains with differing properties of replication, tissue tropism, and resistance to antibiotics. This dynamic population of diverse mutant strains is called a **quasispecies**. The term "quasispecies" acknowledges the limitations of characterizing consistent traits of all the individuals, since within the population individuals differ widely, and their differences change with time. Furthermore, some researchers argue that virions within a quasispecies interact cooperatively on a functional level, by serving complementary roles in the disease state, and thus collectively define the traits of the viral population. Clearer evidence for functional cooperation of a viral quasispecies is shown in HIV (discussed in Section 11.4).

For hepatitis C, an example of a quasispecies evolving within a patient is illustrated in **Figure 11.14**. As HCV proliferates in the patient, mutations accumulate within the hypervariable region 1 (HVR1) of envelope protein E2. Within 2 weeks of starting interferon treatment, the quasispecies shows hugely different proportions of mutant strains (indicated by different colored bars). Inevitably, some of the new strains turn out to be more resistant to interferon than the original parental strain was. Thus, after interferon treatment stops, the quasispecies yet again shows vastly different strains—some of which lead to rapid resurgence of the virus. For patients enduring weeks of treatment with severe side effects, this rapid return of the virus is discouraging news. Nevertheless, it is hoped that, as we will see for HIV (Section 11.4), new therapies will improve treatment, as well as help prevent the spread of HCV.

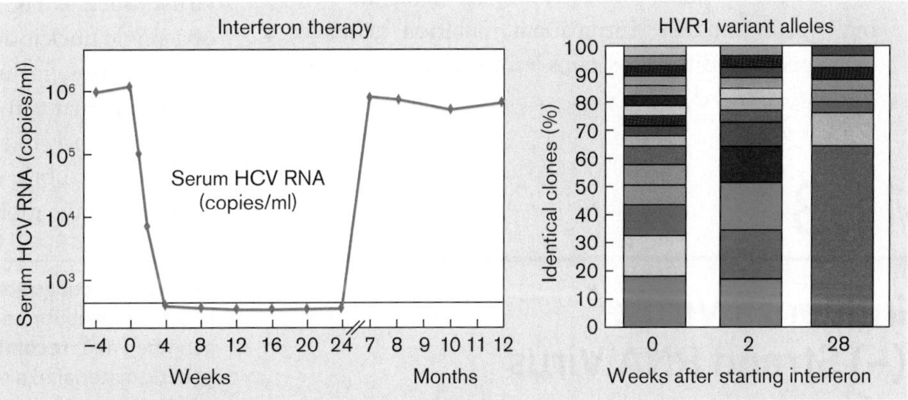
Interferon therapy
Serum HCV RNA (copies/ml)
10⁶
10⁵
10⁴
10³
Serum HCV RNA
(copies/ml)
−4 0 4 8 12 16 20 24 7 8 9 10 11 12
Weeks
Months
HVR1 variant alleles
Identical clones (%)
100
90
80
70
60
50
40
30
20
10
0
0
2
28
Weeks after starting interferon

FIGURE 11.14 ■ Quasispecies formation in HCV. As HCV proliferates in the patient, the E2 hypervariable region 1 (HVR1) accumulates mutations leading to diverse alleles with different sensitivities to antiviral therapy. Colored bars represent variant alleles with mutant RNA sequences. *Source:* Patrizia Farci. 2002. *PNAS* **99**:3081.

Thought Question

11.4 If HCV virions develop a quasispecies under strong selection for propagation within a single host, how do isolated progeny virions retain the ability to initiate infection of a new host?

To Summarize

- **Hepatitis C virus (HCV) has a (+) strand RNA genome** contained in an icosahedral core surrounded by an envelope with dimers of envelope proteins E1 and E2.
- **HCV virions bind several hepatic cell surface proteins**, including two associated with tight junctions. Virions are endocytosed and acidified by lysosomes. The core particles enter the cytoplasm and undergo uncoating.
- **The HCV genome** is translated from the 5′ IRES, generating a polyprotein. Within the polyprotein, the structural proteins (C, E1, E2) are cleaved by ER signal peptidase, and then the nonstructural proteins are cleaved by self-cleaving viral proteases NS2/NS3 and NS3/NS4A.
- **HCV structural and nonstructural proteins collect in vesicular membranes.** The membranes form a "membranous web" in the cytoplasm, where replication factories replicate viral RNA using RNA-dependent RNA polymerase NS5B, and assemble proteins into virions.
- **Progeny HCV virions** can be released by exocytosis, or they may undergo cell-to-cell transmission. Cell-to-cell transmission evades the host immune system.
- **HCV mutates rapidly and forms a quasispecies.** Mutation occurs by point mutation and by "template switching" recombination. Intrapatient mutation generates a quasispecies of different strains leading to drug resistance.

11.3

Influenza Virus: (–) Strand RNA Virus

Some viruses package a (–) strand RNA genome, whose replication cycle is more complex than that of (+) strand RNA viruses. Important (–) strand RNA viruses include Ebola virus and measles virus. A major global health concern for humans and livestock is the orthomyxovirus influenza. Influenza A virus is one of the most common life-threatening viruses in the United States. Each year, influenza A infects approximately 10% of the U.S. population, causing about 36,000 deaths annually. The elderly are most susceptible, but in an epidemic year, mortality rises among young people.

Influenza shows a cyclic appearance of extremely virulent strains that cause pandemic mortality, such as the famous pandemic of 1918, which infected 20% of the world's population and killed more people than World War I did. The 1918 strain arose as a mutant form of an influenza strain infecting birds. In 2006, a similar avian influenza strain emerged as a potential source of the world's next influenza pandemic. As of this writing, however, the avian strain has not yet mutated to a form readily transmitted between humans. In 2009, a highly transmissible strain related to swine influenzas ("swine flu") spread rapidly around the world but caused relatively mild illness. A future strain might emerge combining the high transmission seen in swine flu with the high human mortality seen in the avian strain.

Highly virulent strains of influenza usually result from **reassortment** between distantly related strains. The reassortment process is enhanced by a particular feature of their molecular biology—namely, the **segmented genome**. A segmented viral genome consists of multiple separate nucleic acids, like the multiple chromosomes of a eukaryotic cell. The influenza genome consists of eight separate linear (–) strands of RNA (**Figs. 11.15** and **11.16**). These eight strands, encoding eight different products, can reassort from different strains to generate a novel hybrid strain. Because influenza genomes are capable of reassortment, they generate drastically new strains more quickly than do viruses such as HCV or poliovirus, whose genome consists of a single nucleic acid molecule. At the same time, like the other viruses, influenza viruses continually acquire small mutations that can lead to new phenotypes with respect to drug resistance and host range. The double threat of drastic change and subtle smaller changes explains why influenza presents a huge global challenge for public health.

Note: Distinguish between **reassortment** (two different viruses contribute separate genome segments to a reassortant genome) and **recombination** (two different viruses contribute genetic material to a recombinant molecule.)

Weblinks CDC page on influenza virus (*see ebook*)

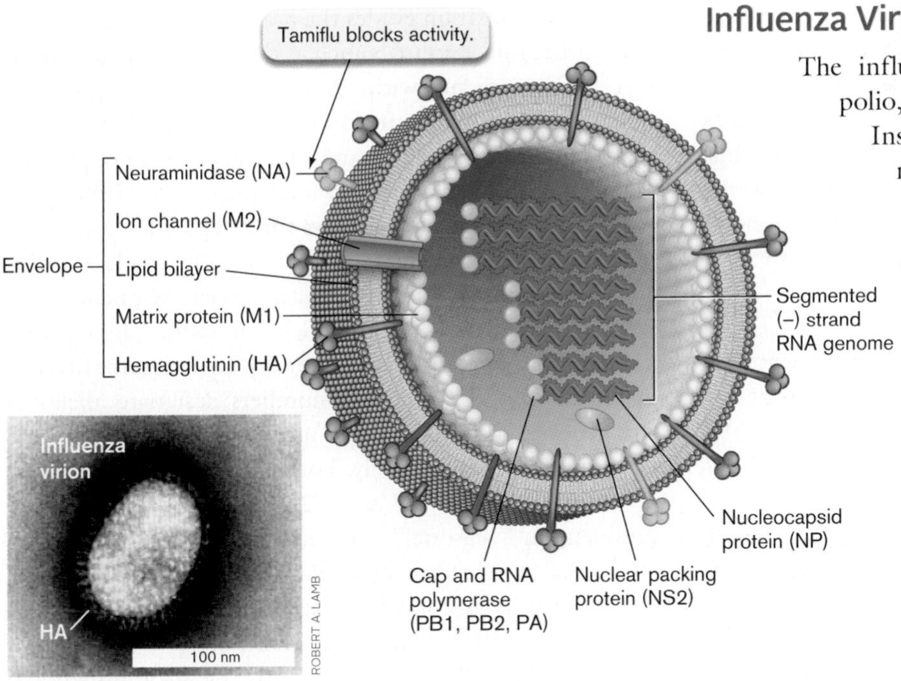
Tamiflu blocks activity.
Envelope
Neuraminidase (NA)
Ion channel (M2)
Lipid bilayer
Matrix protein (M1)
Hemagglutinin (HA)
Segmented (–) strand RNA genome
Nucleocapsid protein (NP)
Cap and RNA polymerase (PB1, PB2, PA)
Nuclear packing protein (NS2)
Influenza virion
HA
100 nm
ROBERT A. LAMB

FIGURE 11.15 ▪ **Structure of influenza A.** Diagram of virion structure showing envelope (colored tan), envelope proteins, matrix protein (yellow), RNA segments (blue) with attached polymerase, and the nuclear packing protein NS2. **Inset:** Influenza A virions (TEM). The brush-like border coating the envelope consists of glycoproteins, hemagglutinin (HA), and neuraminidase (NA).

Influenza Virion Structure and Genome

The influenza virion, unlike that of HCV or polio, has no geometrical capsid (**Fig. 11.15**). Instead, its (–) strand RNA genome segments are individually coated by **nucleocapsid proteins (NPs)**. The term "nucleocapsid" refers generally to proteins coating a viral genome and packaged within or as part of the viral capsid. Each NP-coated RNA segment also possesses a bound RNA-dependent RNA polymerase complex. Thus, unlike poliovirus and most bacteriophages, the influenza virion contains more than its genome; it also contains its own RNA polymerase enzyme, bound to each RNA segment and poised for synthesis. During infection, each (–) strand RNA segment must be transcribed to a (+) strand mRNA for translation by eukaryotic enzymes.

The NP-coated RNA segments are loosely contained by a shell of **matrix proteins** (M1). The matrix layer is further enclosed by the envelope. The envelope derives from the phospholipid membrane of the host cell, which incorporates viral

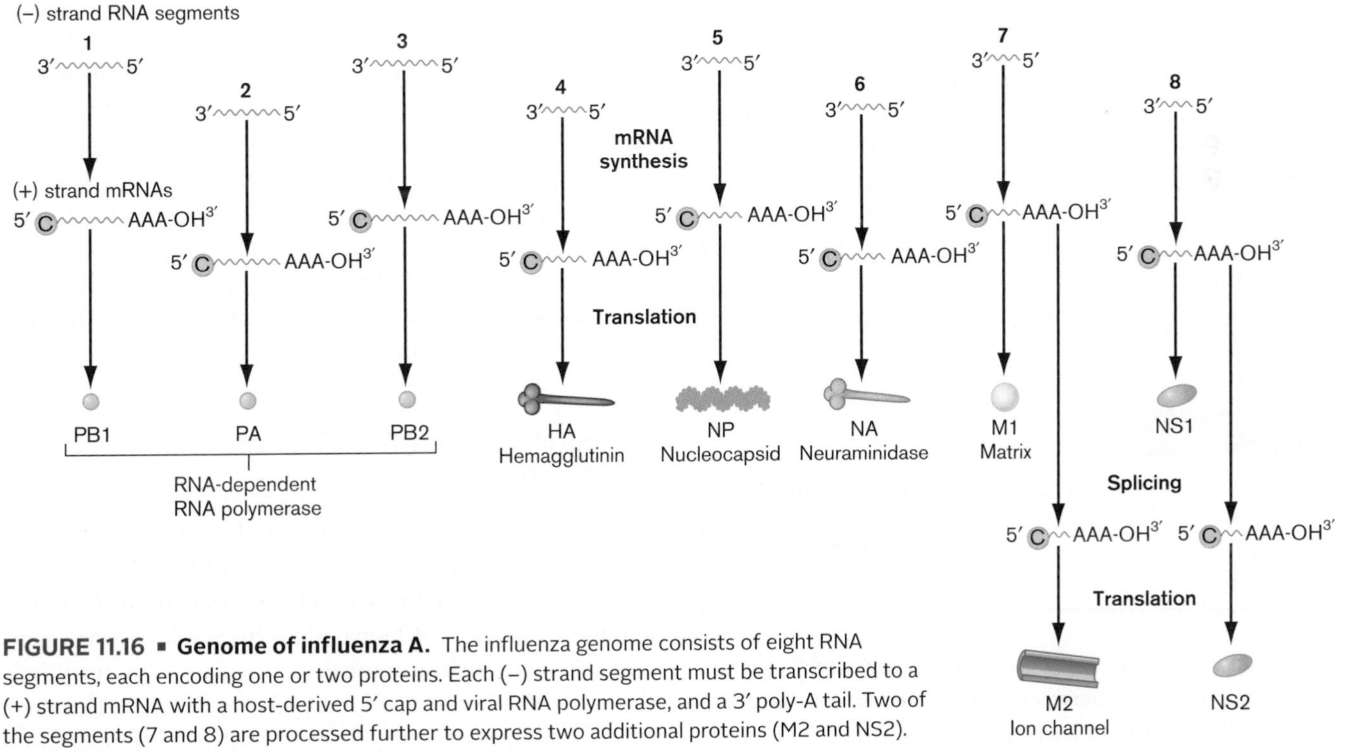
(–) strand RNA segments
1 2 3 4 5 6 7 8
3′ 5′
mRNA synthesis
(+) strand mRNAs
5′ C AAA-OH3′
Translation
PB1
PA
PB2
RNA-dependent RNA polymerase
HA Hemagglutinin
NP Nucleocapsid
NA Neuraminidase
M1 Matrix
NS1
Splicing
Translation
M2 Ion channel
NS2

FIGURE 11.16 ▪ **Genome of influenza A.** The influenza genome consists of eight RNA segments, each encoding one or two proteins. Each (–) strand segment must be transcribed to a (+) strand mRNA with a host-derived 5′ cap and viral RNA polymerase, and a 3′ poly-A tail. Two of the segments (7 and 8) are processed further to express two additional proteins (M2 and NS2).

proteins such as hemagglutinin (HA) and neuraminidase (NA). These viral envelope proteins join the membrane to the enclosed matrix, maintaining its structure. When a newly formed virion exits the host cell, neuraminidase acts as an enzyme to cleave a cell membrane glycoprotein, thus releasing the virion outside the cell. Neuraminidase can be blocked by the antiviral agent Tamiflu (oseltamivir), one of the main drugs available to treat influenza.

Thought Question

11.5 Why does influenza virus have to provide its polymerase ready-made, whereas hepatitis C does not?

Note: In figures of this chapter, AAA-OH$^{3'}$ represents a poly-A tail consisting of a variable number of adenine nucleotides with a 3′ hydroxyl end.

Given that viral infection requires all eight segments, how does the assembly mechanism package exactly eight segments, one of each? In fact, the segments are packaged at random, resulting in a vast majority of defective particles. The fraction of influenza particles observed to be capable of infection is consistent with random packaging of any eight segments into a capsid. The theoretical fraction of infectious virions would be given by the following equation:

$$\frac{8!}{8^8} = \frac{8\times7\times6\times5\times4\times3\times2\times1}{16,777,216} = \frac{1}{416}$$

One in 400 may sound tremendously inefficient, but if 10,000 virions are produced, there will be more than enough infective particles to propagate the virus. Furthermore, animal viruses require a relatively small number of progeny to ensure infecting the neighboring cells of a multicellular tissue, as compared to phages or viruses of single-celled organisms. Thus, influenza virus may avoid the energetic expense of an accurate packaging mechanism for its segmented genome.

Thought Question

11.6 Explain why the expression $8!/8^8$ approximates the proportion of infective particles of influenza virus. What assumption might be changed to make the proportion greater or less?

The key advantage of a segmented genome is that it enables reassortment between two strains coinfecting the same cell. Reassortment leads to "antigenic shift," in which a very different strain evades the host immune system. Even strains that infect animals such as ducks or swine may reassort their segments with those of a coinfecting human virus. Influx of genes from a distantly related strain can sharply increase virulence and mortality. For example, the 1968 Hong Kong flu strain, which killed over 33,000 people in the United States, derived three segments from avian strains. Major epidemics of exceptionally virulent influenza arise from reassortment with genome segments from strains that evolved within ducks or swine, agricultural animals that live in close proximity to humans. In each genome, the "H" and "N" numbers designate alleles of the genes encoding envelope proteins hemagglutinin and neuraminidase, respectively. For example, the Hong Kong flu strain had alleles H3 and N2 (which have since become prevalent in "seasonal" human flu strains). The "avian" strain that emerged in 2006 had alleles H5 and N1. The "swine" strain in 2008 had alleles H1 and N1, similar to the pandemic 1918 strain (**Fig. 11.17A**).

Figure 11.17B shows how reassortment of human and avian genome segments evolved into the H1N1 swine flu strain that caused a global pandemic in 2008. In 1979, in Europe, an avian flu strain was found to have "jumped" into swine (the "avian-like" swine strain). In 1992, a triple-reassortant strain was identified that included segments PB2 and PA (encoding RNA-dependent RNA polymerase) from an avian virus; PB1 (polymerase subunit), NP (nucleocapsid), and M (matrix protein) from a swine virus; and PB, H, and N from a human seasonal strain of influenza A. Since the H and N envelope proteins came from a human strain, the triple reassortant could be transmitted readily between humans. In subsequent years, the strain evolved further, with introduction of N and M genes from the avian-like swine strain. This new reassortant strain was a major antigenic shift, unrecognizable by most human immune systems, and spread widely. Yet another reassortant avian strain, H7N9, emerged in China in 2013. So far, these avian strains have been contained—but the next time, we may be less fortunate.

Influenza Attachment and Host Cell Entry

What determines which animals can be infected by a given flu strain? One factor is the requirement for a host cellular protease to cleave the hemagglutinin protein on the virion envelope. Cleavage of hemagglutinin enables a small peptide called the **fusion peptide** to mediate viral entry into the host cell (discussed shortly; see **Fig. 11.19**). The presence of the protease is one **host factor** that determines

A.

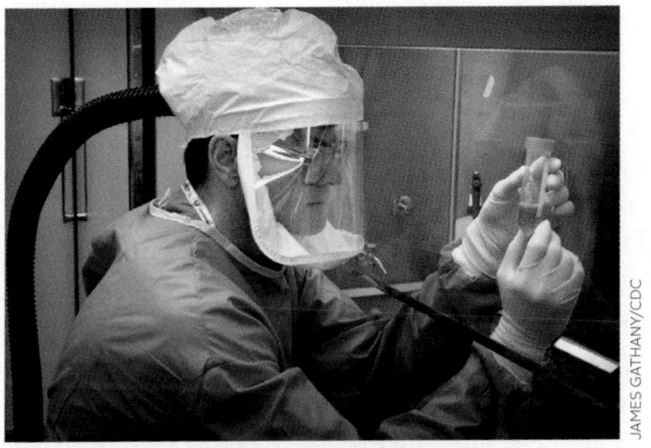
JAMES GATHANY/CDC

FIGURE 11.17 ■ Reassortment between human, avian, and swine strains generates exceptionally virulent influenza strains. A. The reconstructed strain of the 1918 pandemic influenza virus is studied by Terrence Tumpey, a microbiologist at the Centers for Disease Control and Prevention, Atlanta. The 1918 strain may hold clues to combating the "avian influenza" strain today. **B.** The 2009 strain of H1N1 influenza arose from a series of reassortments of the eight RNA segments from avian, swine, and human influenza strains. The numbers following "H" and "N" refer to different alleles of the genes encoding hemagglutinin and neuraminidase, respectively. Each round dot represents one of the eight gene segments. *Source:* Part B modified from B. Gavin Smith et al. 2009. *Nature* **459**:1122.

B.

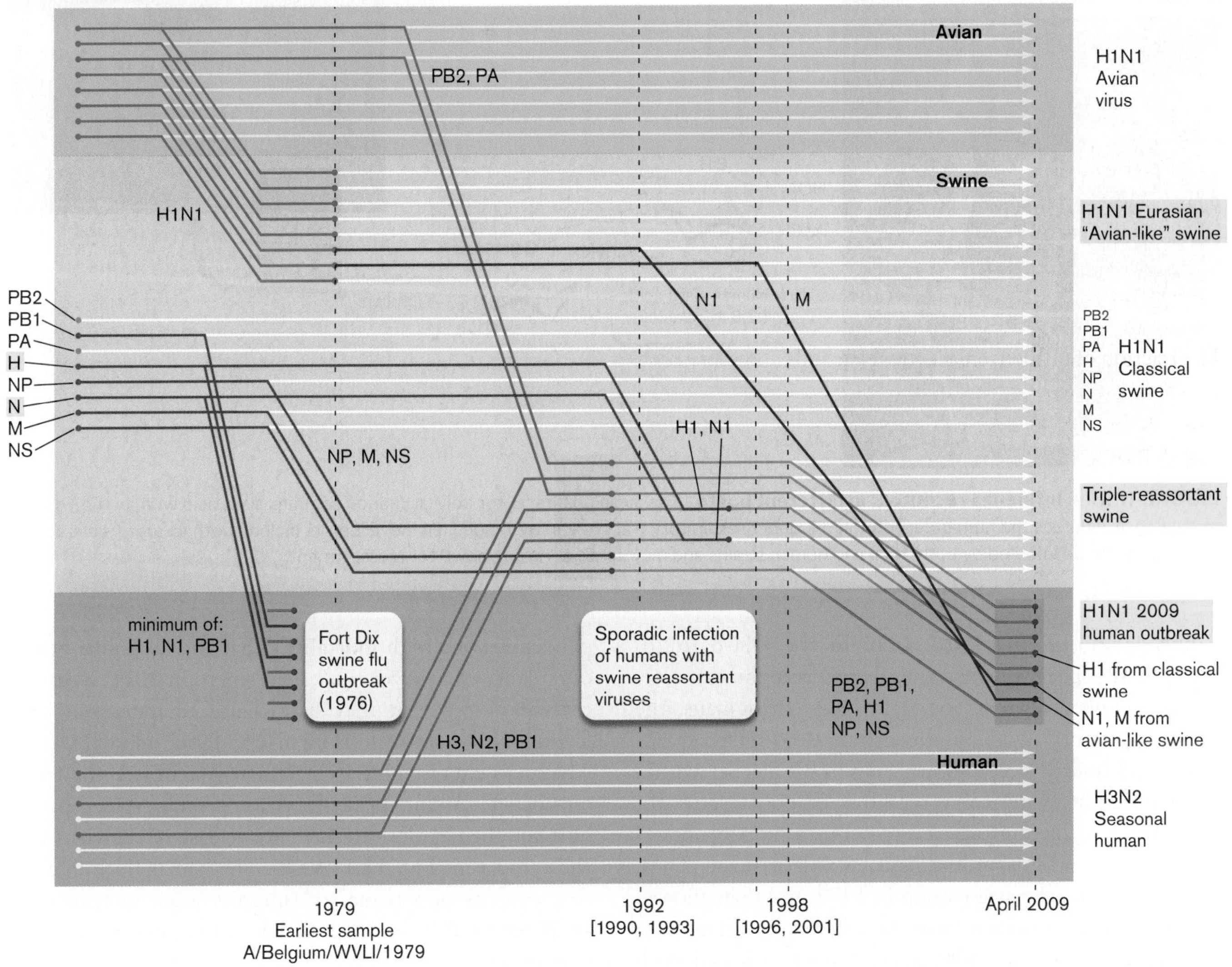
Avian
H1N1
Avian
virus
PB2, PA
Swine
H1N1 Eurasian
"Avian-like" swine
H1N1
PB2
PB1
PA
H
NP
N
M
NS
N1
M
PB2
PB1
PA
H
NP
N
M
NS
H1N1
Classical
swine
H1, N1
NP, M, NS
Triple-reassortant
swine
minimum of:
H1, N1, PB1
Fort Dix
swine flu
outbreak
(1976)
Sporadic infection
of humans with
swine reassortant
viruses
PB2, PB1,
PA, H1
NP, NS
H1N1 2009
human outbreak
H1 from classical
swine
N1, M from
avian-like swine
H3, N2, PB1
Human
H3N2
Seasonal
human
1979
Earliest sample
A/Belgium/WVLI/1979
1992
[1990, 1993]
1998
[1996, 2001]
April 2009

which kind of host may be infected, and which tissues within the host support viral replication. Host factors of many kinds mediate viral infections.

Another important host factor for influenza is cell-surface glycoproteins that contain a terminal sialic acid (**Fig. 11.18**). The sialic acid polysaccharide of the glycoprotein binds hemagglutinin, attaching the virion and enabling endocytosis. The precise structure of the sialic acid host receptor may determine whether a strain such as avian influenza H5N1 will spread directly between humans. For example, the sialic acid connection in the receptor polysaccharide can involve different OH groups of the sugar galactose:

Avian sialic acid:
α2,3 to galactose
Sialic acid
Galactose
Receptor protein
"Mixing bowl"
Swine sialic acid:
α2,3 or α2,6 to galactose
Sialic acid
Galactose
Human sialic acid:
α2,6 to galactose
Receptor protein
©ZERO CREATIVES/CULTURA/CORBIS
©CHRISTOPHER J. MORRIS/CORBIS
NATURE 483: S35

FIGURE 11.18 ■ Influenza receptors in different hosts. The avian host receptor polysaccharide contains sialic acid with an α2,3 bond to galactose, whereas the human receptor in the upper respiratory tract has an α2,6 bond. Swine receptors include both forms of sialic acid; thus, swine can be infected by both avian and human strains and may act as a "mixing bowl" for reassortment.

a linkage to the OH-3 (α2,3) or to the OH-6 (α2,6) (**Fig. 11.18**). The influenza strain H5N1 recognizes mainly the α2,3-linked protein, found in birds. In humans, the upper respiratory tract contains mainly α2,6-linked receptors; α2,3-linked receptors are found only deeper within the lungs. But swine carry receptors of both types. For this reason, swine are believed to act as a "mixing bowl" for strains from birds and humans, as well as swine. Thus, swine incubated the avian strain in 1979, and then allowed later reassortment with human-flu genome segments, leading eventually to the 2009 strain—which sickened both pigs and humans. Today, large swine facilities are monitored for appearance of novel reassortant strains.

The avian influenza strain H5N1 causes exceptionally high mortality in humans, but it is rarely transmitted from one person to another. More rapid transmission could arise from antigenic drift, a small mutation in the gene encoding avian hemagglutinin. Rapid person-to-person transmission of H5N1 might cause an influenza pandemic with high mortality. That is why public health organizations were so concerned when, in 2011, researchers announced that they had identified mutations conferring H5N1 transmission in the ferret model system, which closely resembles the human system (see the chapter opener). As the influenza virion is endocytosed by the host cell, hemagglutinin mediates release of the virus into the cytoplasm (**Fig. 11.19A** ▶). The hemagglutinin complex consists of a trimer of subunits, each of which has an N-terminal fusion peptide. A fusion peptide is a portion of an envelope protein (cleaved from hemagglutinin in the case of influenza virus) that changes conformation so as to facilitate envelope fusion with the host cell membrane. Before membrane contact, the fusion peptides are buried within the core of the hemagglutinin (HA) trimer (**Fig. 11.19B**). Each HA C-terminal domain binds a sialic acid receptor in the host membrane.

When the virion is taken up by endocytosis, the endocytic vesicle fuses with a lysosome and its interior

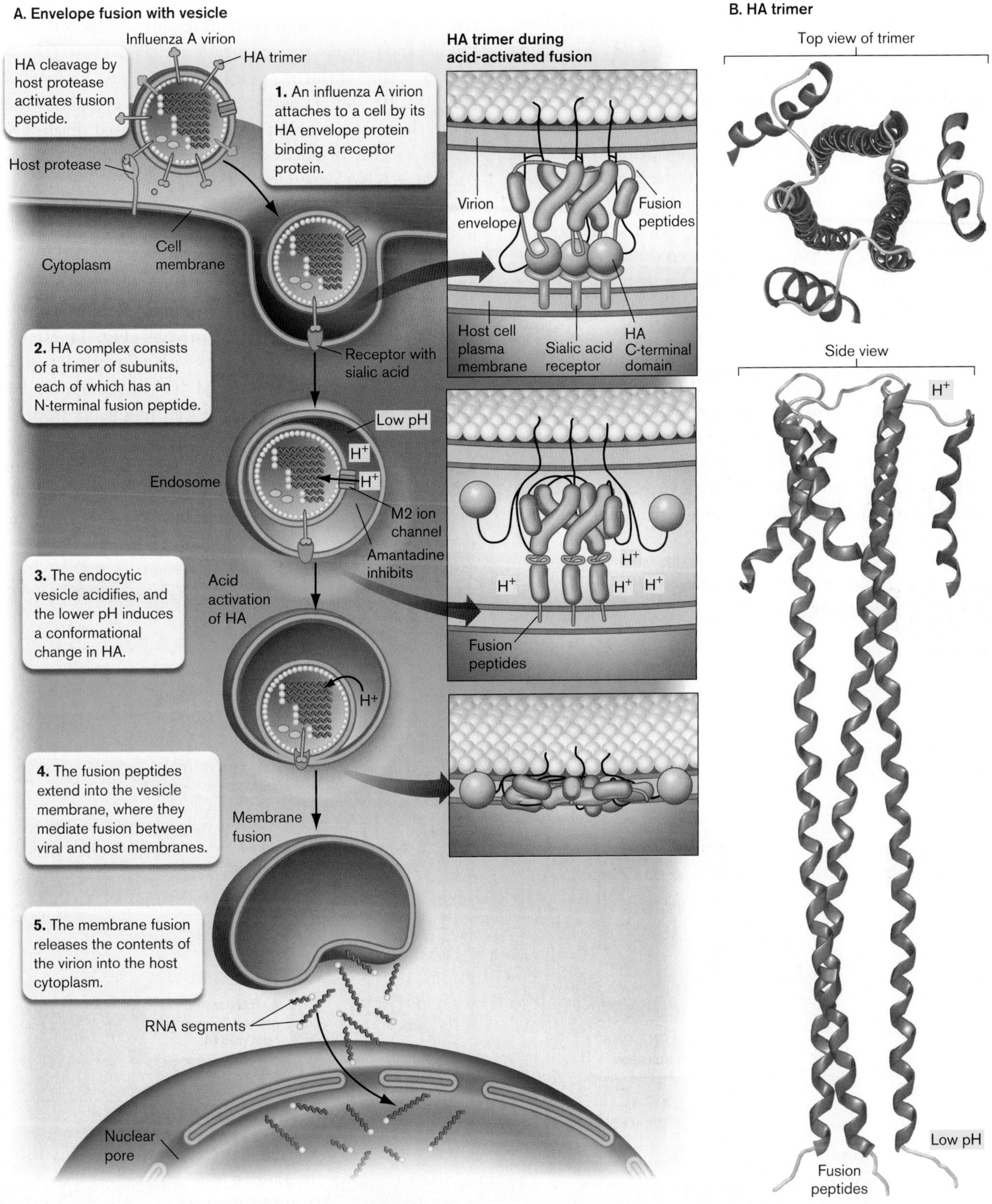
A. Envelope fusion with vesicle
Influenza A virion
HA trimer
HA cleavage by host protease activates fusion peptide.
Host protease
1. An influenza A virion attaches to a cell by its HA envelope protein binding a receptor protein.
Cytoplasm
Cell membrane
2. HA complex consists of a trimer of subunits, each of which has an N-terminal fusion peptide.
Receptor with sialic acid
Low pH
H+
Endosome
M2 ion channel
Amantadine inhibits
3. The endocytic vesicle acidifies, and the lower pH induces a conformational change in HA.
Acid activation of HA
4. The fusion peptides extend into the vesicle membrane, where they mediate fusion between viral and host membranes.
Membrane fusion
5. The membrane fusion releases the contents of the virion into the host cytoplasm.
RNA segments
Nuclear pore
HA trimer during acid-activated fusion
Virion envelope
Fusion peptides
Host cell plasma membrane
Sialic acid receptor
HA C-terminal domain
Fusion peptides
B. HA trimer
Top view of trimer
Side view
Low pH
Fusion peptides

FIGURE 11.19 ■ **Entry and acid activation of influenza A virion. A.** The hemagglutinin (HA) envelope protein of an influenza A virion attaches to a host cell (1) binding a sialic acid receptor protein. HA complex consists of a trimer of subunits, each of which has an N-terminal fusion peptide. When the virion is taken up by endocytosis, the interior of the endocytic vesicle acidifies, and the lower pH induces a conformational change in HA (2, 3). The receptor-binding domains fold back, and the fusion peptides extend into the vesicle membrane, where they mediate fusion between viral and host membranes (4). The membrane fusion releases the contents of the virion into the host cytoplasm (5). **B.** Model of the core portion of the HA trimer (top and side views). (PDB code: 1HTM) acidifies. The lowered pH (increased H^+ concentration) drives H^+ ions into the virion through the M2 ion channel, causing the matrix proteins to dissociate from the NP-coated RNA. M2 is the target of amantadine, one of the first anti-influenza drugs; unfortunately, most strains today have evolved resistance to the drug. Low pH also induces a conformational change shifting the C-terminal ends back and the N-terminal fusion peptides outward to face the vesicle membrane. The peptides extend into the membrane, where they mediate fusion between viral and host membranes. The fusion process expels the contents of the virion into the host cytoplasm.

Replication Cycle of Influenza A

The replication of influenza virus is more complex than that of hepatitis C virus because the viral components travel in and out of the nucleus. Unlike HCV, which injects only its chromosome for translation, influenza virus releases several enzymes and structural proteins along with its genetic material. In addition, envelope proteins require transport through the ER and Golgi to the cell membrane (**Fig. 11.20** ▶).

After the influenza virion undergoes endocytosis (**Fig. 11.20**, step 1) and release into the cytoplasm (step 2), all the viral (–) RNA segments are released. Each RNA retains its coat of nucleocapsid proteins, as well as a prepackaged RNA-dependent RNA polymerase. The NP-coated RNA segments individually pass through a nuclear pore into the nucleus (step 3).

Synthesis of (+) strand mRNA. Within the nucleus, each genomic (–) RNA segment with its prepackaged polymerase synthesizes (+) strand RNA for mRNA (**Fig. 11.20**, step 4). Each mRNA synthesis initiates with a 7-methylguanine-"capped" RNA fragment (labeled "C" in the figure). The influenza polymerase obtains the cap fragments from the host by cleaving them from host nuclear pre-mRNA—a process quaintly known as "cap snatching." The (+) strand mRNA molecules return to the cytoplasm for translation (step 5), using the snatched cap to bind the host ribosome.

FIGURE 11.20 ■ Replication of influenza virus. ▶

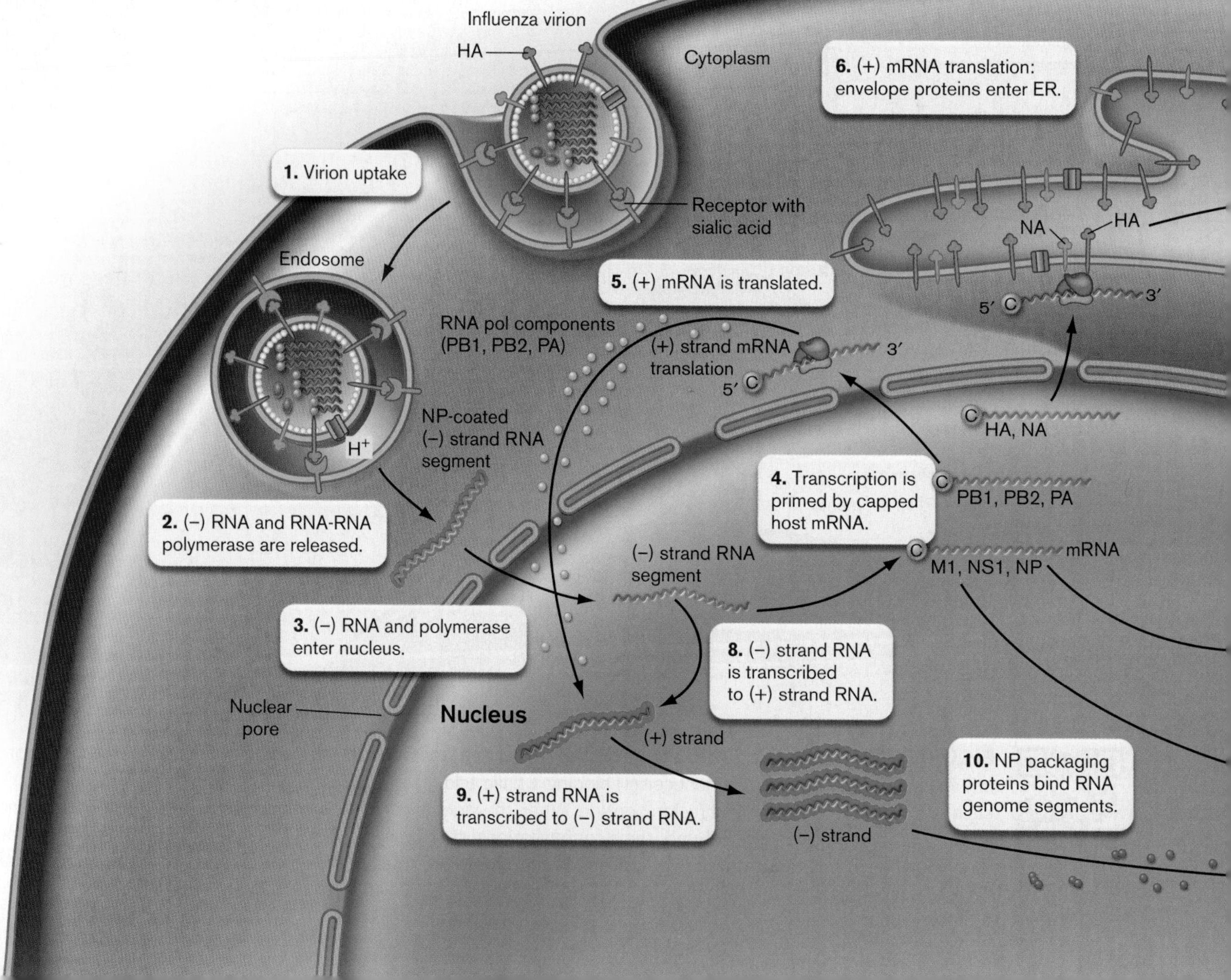
Influenza virion
HA
Cytoplasm
6. (+) mRNA translation: envelope proteins enter ER.
1. Virion uptake
Receptor with sialic acid
NA
HA
Endosome
5. (+) mRNA is translated.
5′
3′
RNA pol components (PB1, PB2, PA)
(+) strand mRNA translation
3′
5′
NP-coated (–) strand RNA segment
HA, NA
H^+
4. Transcription is primed by capped host mRNA.
PB1, PB2, PA
2. (–) RNA and RNA-RNA polymerase are released.
(–) strand RNA segment
mRNA
M1, NS1, NP
3. (–) RNA and polymerase enter nucleus.
8. (–) strand RNA is transcribed to (+) strand RNA.
Nuclear pore
Nucleus
(+) strand
10. NP packaging proteins bind RNA genome segments.
9. (+) strand RNA is transcribed to (–) strand RNA.
(–) strand

(Recall that HCV, lacking the cap, uses an IRES to bind the ribosome.) The RNA segments encoding envelope proteins attach to the ER for protein synthesis and transport to the host cell membrane (step 6). The newly synthesized nucleocapsid proteins (NPs), as well as RNA-dependent RNA polymerase components, subsequently return to the nucleus (step 7). Other genome-packaging proteins (M1 and NS2) also return to the nucleus.

Synthesis of (+) strand and (–) strand genomic RNA. Back in the nucleus, the original (–) strand RNA segments also serve as templates for RNA synthesis without cap snatching (**Fig. 11.20**, step 8). The uncapped (+) strand RNA then becomes coated with the newly made NP subunits imported from the cytoplasm. The NP-coated (+) strand serves as template to synthesize (–) strand RNA (step 9), which also becomes coated with NP (step 10).

The NP-coated (–) RNA associates with a newly made polymerase for a future cycle of viral replication. The RNA is then complexed with matrix protein (M1) and nuclear packaging protein (N2), proteins that were imported from the cytoplasm earlier. At last, the fully packaged (–) RNA segments exit the nucleus to the cytoplasm (step 11), where they approach the cell membrane for packaging into progeny virions (step 12).

Envelope synthesis and assembly. The envelope proteins synthesized at the ER include hemagglutinin (HA) and neuraminidase (NA). Within the ER lumen, these proteins are glycosylated by host enzymes and then transferred to the Golgi (**Fig. 11.20**, step 13) for export to the cell membrane (step 14). Within the cell membrane, the envelope proteins assemble around a group of (–) RNA segments complexed with their matrix and packaging proteins, completing the virion particle (step 15).

To exit the cell, the virion buds out (step 16). Viral exit requires a final step of host release: Neuraminidase (the envelope protein N) cuts the sialic acid link of the host glycoproteins (step 17), releasing the virion out into the bloodstream. This host release activity of neuraminidase is inhibited by oseltamivir (Tamiflu), the main antiviral agent currently useful against influenza. Tamiflu remains useful today—but resistant strains of the 2009 H1N1 virus have emerged, so we urgently need new antivirals ahead of the next influenza pandemic.

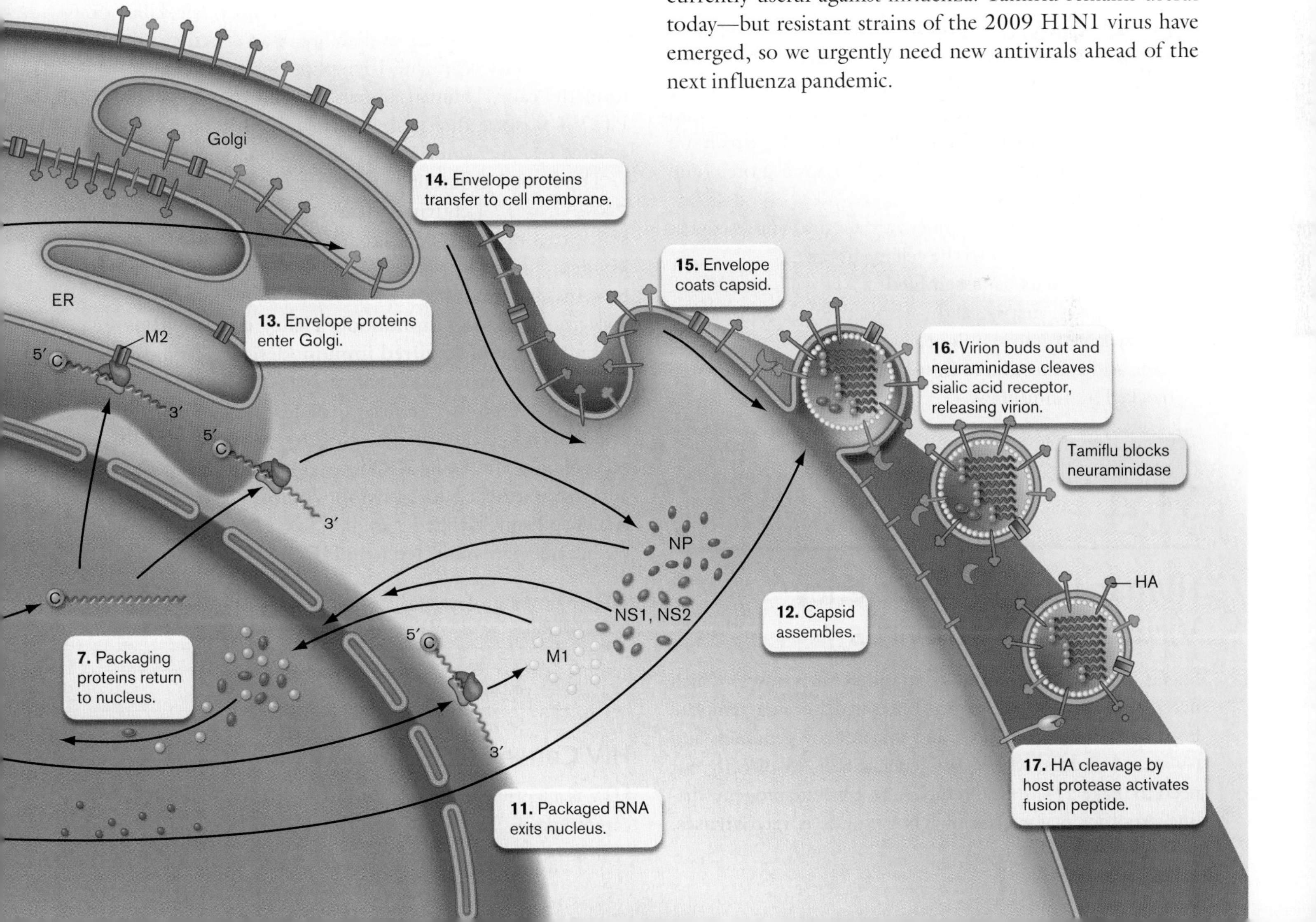
Golgi
14. Envelope proteins transfer to cell membrane.
15. Envelope coats capsid.
ER
M2
13. Envelope proteins enter Golgi.
5′ C
3′
5′ C
3′
16. Virion buds out and neuraminidase cleaves sialic acid receptor, releasing virion.
Tamiflu blocks neuraminidase
NP
HA
C
NS1, NS2
12. Capsid assembles.
7. Packaging proteins return to nucleus.
5′ C
M1
3′
17. HA cleavage by host protease activates fusion peptide.
11. Packaged RNA exits nucleus.

To Summarize

- **Influenza virus** causes periodic pandemics of respiratory disease. New virulent strains arise through reassortment of human and avian strains.
- **The influenza virus consists of segmented (–) strand RNA.** Each segment is packaged with nucleocapsid proteins. Segments from different strains reassort through coinfection.
- **Nucleocapsid and matrix proteins** enclose the RNA segments of the influenza virus. The matrix is enclosed by an envelope containing spike proteins.
- **Envelope HA proteins mediate virion attachment.** The HA protein includes a fusion peptide that undergoes conformational change to cause fusion between the viral envelope and host cell membrane. For influenza, the virion is internalized by endocytosis.
- **Lysosome fusion with endosomes** triggers viral envelope fusion with the endosome membrane. The viral genome and proteins are then released into the cytoplasm. Viral (–) strand RNA segments attached to RNA–RNA polymerase enter the nucleus.
- **Influenza mRNA synthesis initiates with a capped RNA fragment**, cleaved from host mRNA. The capped viral mRNAs return to the cytoplasm for translation.
- **Genomic RNA synthesis generates a (+) strand RNA as a template for (–) RNA strands**, which are then packaged in newly made nucleocapsid protein and exported to the cytoplasm.
- **Envelope proteins** of influenza virus are synthesized at the ER for transport to the cell membrane.
- **Influenza virus is assembled** at the cell membrane, where capsid, matrix, and (–) strand RNA components are packaged into envelope. Neuraminidase cleaves the sialic acid connection to host glycoproteins—the key step blocked by Tamiflu.

11.4

Human Immunodeficiency Virus (HIV): Retrovirus

So far we have discussed (+) strand RNA viruses, which need to generate a (–) strand intermediate as a template both for (+) strand mRNA and for progeny genomes; and (–) strand viruses, which can generate mRNA directly but need to make a (+) strand template to generate progeny virions. Another major class of RNA viruses is **retroviruses**. Retroviruses are so called because they reverse the normal order of synthesis to copy their RNA into a double-stranded DNA, which is then integrated into the host genome. As introduced in Chapter 6, the RNA-to-DNA synthesis requires a key enzyme, reverse transcriptase (RT), the target of the major anti-AIDS drug azidothymidine (AZT).

Weblinks: UNAIDS (*see ebook*)

Retroviruses form a large family of viruses that infect animals; for examples, see **Table 11.2**. Retroviruses are divided into two major groups: the "simple retroviruses," whose genomes include just four genes; and the lentiviruses, with more complex genomes (discussed below). Simple retroviruses cause tumors and leukemias; for example, **Rous sarcoma virus (RSV)** was the first virus demonstrated to cause cancer.

The DNA insertion mechanism of retroviruses is ideally suited to alter host gene regulation, either by decoupling a host gene from its regulatory sites or by inserting an oncogene (an altered host gene that causes cancer). Another retrovirus, **feline leukemia virus (FeLV)**, is a major veterinary pathogen. FeLV remains America's number one killer of outdoor cats, despite the availability of an effective vaccine. Related to FeLV is primate T-lymphotrophic virus (PTLV-1), formerly called human T-cell leukemia virus (HTLV). PTLV-1 was the first retrovirus identified in humans, discovered by American virologist Robert Gallo in 1980.

The second group of retroviruses is the **lentiviruses**, or "slow viruses." Lentiviruses cause infections that progress slowly over many years. Lentiviral genomes include the four key genes of retroviruses, plus genes for several regulatory proteins that modulate host interactions. The most famous lentivirus is **human immunodeficiency virus (HIV)**, the causative agent of **acquired immunodeficiency syndrome (AIDS)**. The virus HIV and its causative role in AIDS were discovered by French virologist Luc Montagnier, building on Gallo's studies of retroviruses. In 2010, according to the United Nations, 34 million people globally were estimated to be living with HIV. That year, the AIDS epidemic claimed 2 million lives, and 3 million people became newly infected with HIV.

Weblinks: Feline leukemia virus, Cornell Feline Health Center (*see ebook*)

Weblinks: HIV InSite, UC San Francisco (*see ebook*)

HIV Causes AIDS

HIV is a lentivirus that evolved from viruses infecting African monkeys. Two major types are recognized: HIV-1, the

TABLE 11.2

Retroviruses of animals (examples).

Genus	Virus	Disease	Host
Simple retroviruses			
Alpharetrovirus	Avian leukosis virus (ALV)	Leukemia	Birds
	Rous sarcoma virus (RSV)	Sarcoma (tumor)	Birds
Betaretrovirus	Mouse mammary tumor virus (MMTV)	Mammary tumor	Mice
Gammaretrovirus	Feline leukemia virus (FeLV)	Lymphoma, immune deficiency	Cats
	Moloney murine leukemia virus (MMLV)	Leukemia	Mice
Deltaretrovirus	Bovine leukemia virus (BLV)	Leukemia	Cattle
	Primate T-lymphotrophic virus (PTLV-1) [formerly human T-cell leukemia virus (HTLV)]	Leukemia	Humans
Epsilonretrovirus	Walleye dermal sarcoma virus (WDSV)	Sarcoma	Fish
Lentiviruses	Human immunodeficiency virus (HIV-1, HIV-2)	AIDS	Humans
	Simian immunodeficiency virus (SIV)	Simian AIDS	Monkeys
	Equine infectious anemia virus (EIAV)	Anemia	Horses
	Maedi-Visna virus (MV)	Neurological disease	Sheep

cause of most infections at present; and HIV-2, another type that appears to have evolved independently from another strain infecting monkeys. The virus is transmitted through blood and through genital or oral-genital contact. HIV can hide in the host cell for many years, with only gradual buildup of viruses, most of which are eliminated by the host. Eventually, however, the virus destroys the body's T lymphocytes, leaving the host defenseless against many organisms that normally would be harmless.

The first cases of AIDS in the United States were reported in 1981. Thirty years later, HIV infects over a million Americans, and more than half a million have died of AIDS. Worldwide, HIV infects one in every 100 adults, equally among women and men. In developed countries, treatment effectively prolongs the life of people with AIDS. Treatment involves mixtures of antiretroviral drugs—all discovered based on the molecular mechanisms of HIV infection, described in this chapter. HIV treatment with multiple drugs is called "highly active antiretroviral therapy," or HAART. But the HAART drugs have side effects and remain too expensive for the majority of those infected worldwide. In some countries, such as South Africa, approximately 20% of adults test positive for HIV.

A surprising benefit of HAART was discovered through humanitarian treatment programs in Africa, such as the President's Emergency Plan for AIDS Relief (PEPFAR), initiated in 2003 by President George W. Bush and continued under President Barack Obama. The PEPFAR program showed that, in communities where HAART is delivered to all members, including those of early infection status, the virus production decreases so far that transmission declines—and HIV could ultimately be eliminated. Thus, testing an entire population, and treating those infected, might actually wipe out AIDS. In the United States, since 2012 the CDC recommends routine HIV testing for all people 13 years of age and older.

History of AIDS

A historical view of AIDS in the United States emerges in the book *And the Band Played On* by Randy Shilts, adapted as an award-winning film in 1993. The book and film show how American society failed for many years to grasp the significance of AIDS because the syndrome first appeared in societal groups considered marginal (homosexual men and certain ethnic immigrants), although it spread to all social classes. In addition, the virus proved extremely difficult to detect and grow in culture. The discovery of HIV sparked controversy in the scientific community because Gallo failed to acknowledge his use of a virus-producing cell line from Montagnier, the first to isolate HIV-1. Since that time, the two scientists and many others have collaborated to develop a test for HIV-1 infection and to search

A. Luc Montagnier and Robert Gallo

B. Françoise Barré-Sinoussi

JOHN MOTTEM/AFP/GETTY IMAGES
PHOTOS12.COM/COLLECTION CINEMA

C. HIV virions

D. HIV core particle

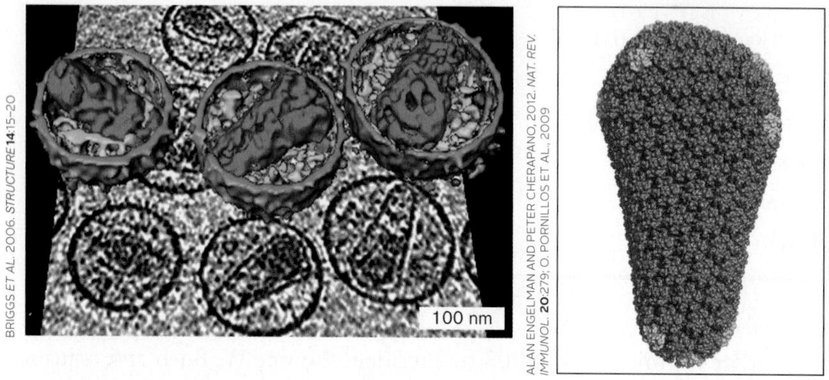
100 nm
BRIGGS ET AL. 2006. STRUCTURE 14:15–20
ALAN ENGELMAN AND PETER CHERAPANO, 2012. NAT. REV. IMMUNOL. 20:279; O. PORNILLOS ET AL., 2009

FIGURE 11.21 ■ HIV discovery. A. Luc Montagnier and Robert Gallo agree to collaborate on development of an AIDS vaccine, 2002. **B.** Françoise Barré-Sinoussi, at the Pasteur Institute, who worked with Montagnier to discover the virus that causes AIDS. **C.** HIV virions (cryo-EM tomography) show the cone-shaped capsids (colorized red), proteases and host-derived proteins (yellow), and envelopes (blue). **D.** Core particle, based on cryo-EM and X-ray crystallography.

for a vaccine (**Fig. 11.21A**). In 2008, the Nobel Prize in Physiology or Medicine was awarded to Montagnier and his associate Françoise Barré-Sinoussi (**Fig. 11.21B**) for the discovery of HIV and its role in AIDS.

Why can we find no vaccine for HIV when safe and effective vaccines were obtained long ago for major killers such as smallpox and polio and even for the feline retrovirus FeLV? The answers to this question are complex.

- **High mutation rate.** The mutation rate of HIV is among the highest known for any virus. Within one patient, the virus evolves into a quasispecies whose different strains attack different organs and predominate at different stages of the disease.
- **Complex regulation of replication.** The lentiviruses express a greater number of regulator proteins than do the "simple" retroviruses. These complex regulatory options of a lentivirus enable HIV to hide itself within host cells.

While a cure remains elusive, studying the molecular biology of HIV has led to successful treatments that extend life span and improve quality of life while decreasing the rate of transmission. And the very traits that make HIV so insidious a pathogen also make it an effective vector for gene therapy (see Section 11.6).

HIV Structure and Genome

The structure of HIV as visualized by TEM consists of an electron-dense **core particle** (or capsid) surrounded by a phospholipid envelope (**Fig. 11.21C**). The conical core is composed of capsid (CA) subunits whose arrangement is partly icosahedral (**Fig. 11.21D**). The membrane around the core contains **spike proteins**, which join the membrane to the matrix, as in influenza virus. The envelope forms around the core from host cell membrane, when a progeny virion is budding out. In HIV, budding out does not rapidly lyse the cell but has other devastating consequences for cell function (discussed below).

The HIV core. The core contains two distinct single-stranded copies of the RNA genome (**Fig. 11.22A**). Unlike influenza, each RNA contains a complete "map" of HIV genes. The two RNA genomes can have different sequences arising from distinct replication events; thus the HIV virion can be considered "diploid."

Each RNA genome is coated with nucleocapsid proteins (NC) similar in function to the NP proteins of influenza virus. Unlike influenza virus, each RNA of HIV requires a primer for DNA synthesis: a tRNA derived from the previously infected host cell. The host-derived tRNA is packaged in place on the RNA template, ready to go. The primed and packaged RNA is contained within the core composed of CA subunits. The core also contains about 50 copies of reverse transcriptase (RT) and protease (PR), as well as a DNA integration factor (IN). Unique to type HIV-1, subunits of a host chaperone named cyclophilin A are incorporated into the structure—about one for every ten core subunits. An HIV-1 mutant that fails to incorporate cyclophilin A can attach to a host cell and insert its capsid, but the core fails to come apart, and infection is halted.

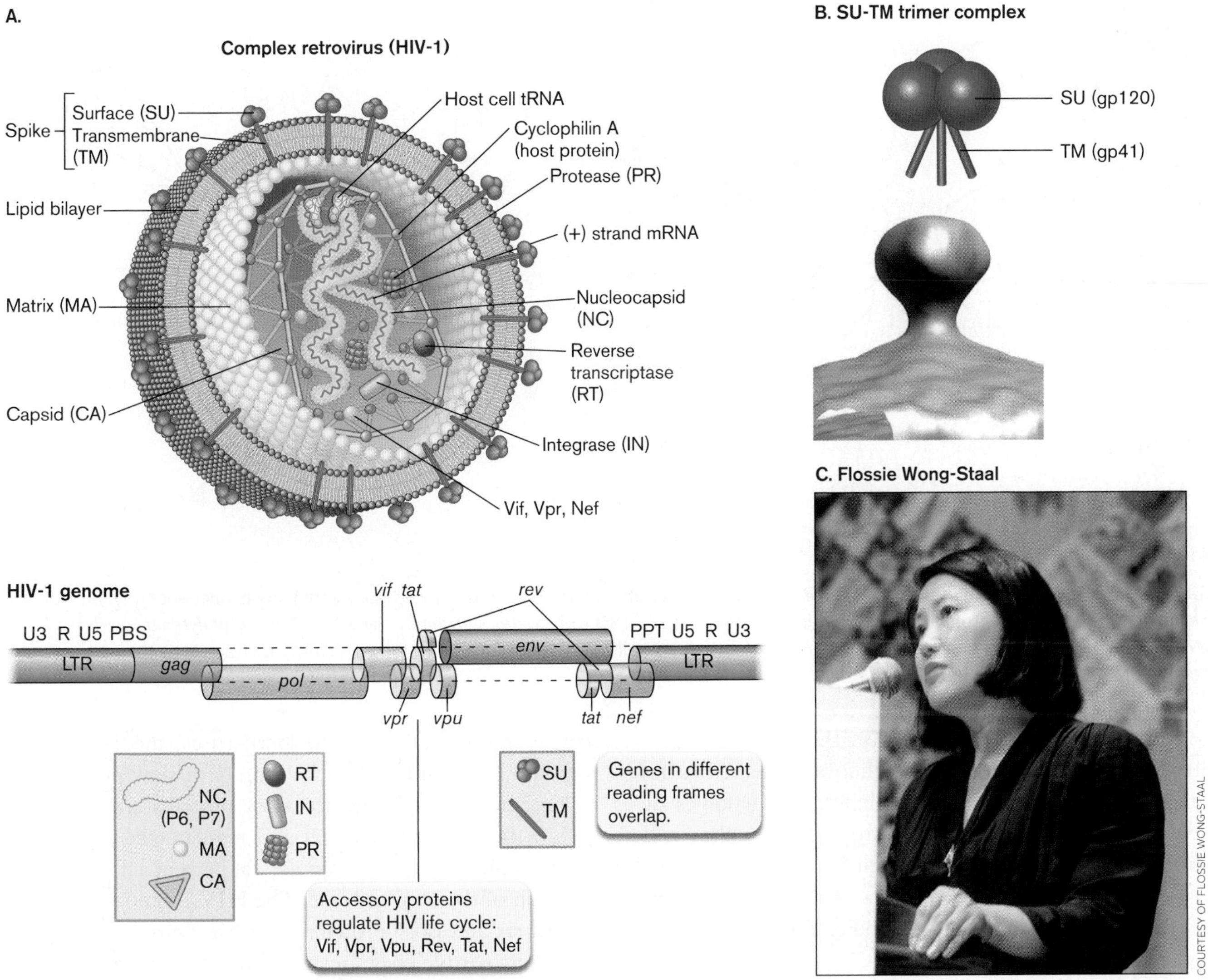
A.
Complex retrovirus (HIV-1)
Spike
Surface (SU)
Transmembrane (TM)
Lipid bilayer
Matrix (MA)
Capsid (CA)
Host cell tRNA
Cyclophilin A (host protein)
Protease (PR)
(+) strand mRNA
Nucleocapsid (NC)
Reverse transcriptase (RT)
Integrase (IN)
Vif, Vpr, Nef
HIV-1 genome
vif tat
rev
U3 R U5 PBS
PPT U5 R U3
LTR
gag
pol
env
LTR
vpr
vpu
tat
nef
NC (P6, P7)
MA
CA
RT
IN
PR
SU
TM
Genes in different reading frames overlap.
Accessory proteins regulate HIV life cycle: Vif, Vpr, Vpu, Rev, Tat, Nef
B. SU-TM trimer complex
SU (gp120)
TM (gp41)
C. Flossie Wong-Staal
COURTESY OF FLOSSIE WONG-STAAL

FIGURE 11.22 ■ HIV-1 structure and genome. A. Internal structure of virion, color-coded to match the genome. In the genome sequence, the staggered levels indicate three different reading frames. Each virion contains two copies of the RNA genome plus multiple copies of reverse transcriptase (RT) and protease (PR) enclosed within a conical capsid (CA subunits plus subunits of a host protein, cyclophilin A). The capsid is surrounded by a matrix (MA subunits), which reinforces the host-derived phospholipid membrane, pegged by spike proteins (SU, TM). The genome also encodes six accessory proteins that are expressed within the infected host cell and regulate the replicative cycle. **B.** Envelope spike complex is a trimer of SU and TM subunits (drawing based on cryo-EM tomography). **C.** Flossie Wong-Staal, pioneering AIDS researcher at UC San Diego, was the first to clone the HIV genome. Wong-Staal now pursues gene therapy approaches to AIDS prevention and develops lentiviral gene vectors. *Source:* Part A modified from Briggs et al. 2006. *Structure* **14**:15–20; part B modified from P. Zhu. 2008. *PLoS Pathog.* **11**:e1000203.

The core is surrounded by a matrix (MA subunits), which reinforces the host-derived phospholipid membrane. The membrane is pegged by spike proteins composed of the envelope subunits TM and SU (**Fig. 11.22B**). As in influenza virus, the spike proteins play crucial roles in host attachment and entry.

The origin and evolution of HIV. Where did HIV come from? The origin of HIV has been traced back to the early twentieth century, based on comparison with sequences of related viruses infecting other primates, called simian immunodeficiency viruses (SIV) (**Fig. 11.23A**). Sequence comparison of different strains of HIV and SIV reveals that an immunodeficiency virus actually entered the human population more than once, from SIV strains derived from related primates in Africa. It is thought that human consumption of primates for meat may have introduced SIV strains that then adapted to human infection. Today, the vast majority of HIV-infected patients show the HIV-1 strain M; but some people have been infected by HIV-2, which derived independently from another SIV strain.

HIV is the most rapidly evolving pathogen known. Within an infected patient, viral mutation generates about one mutation per progeny virion. Such rapid mutation generates a quasispecies of virus types, even more complex than the quasispecies generated by hepatitis C. In advanced

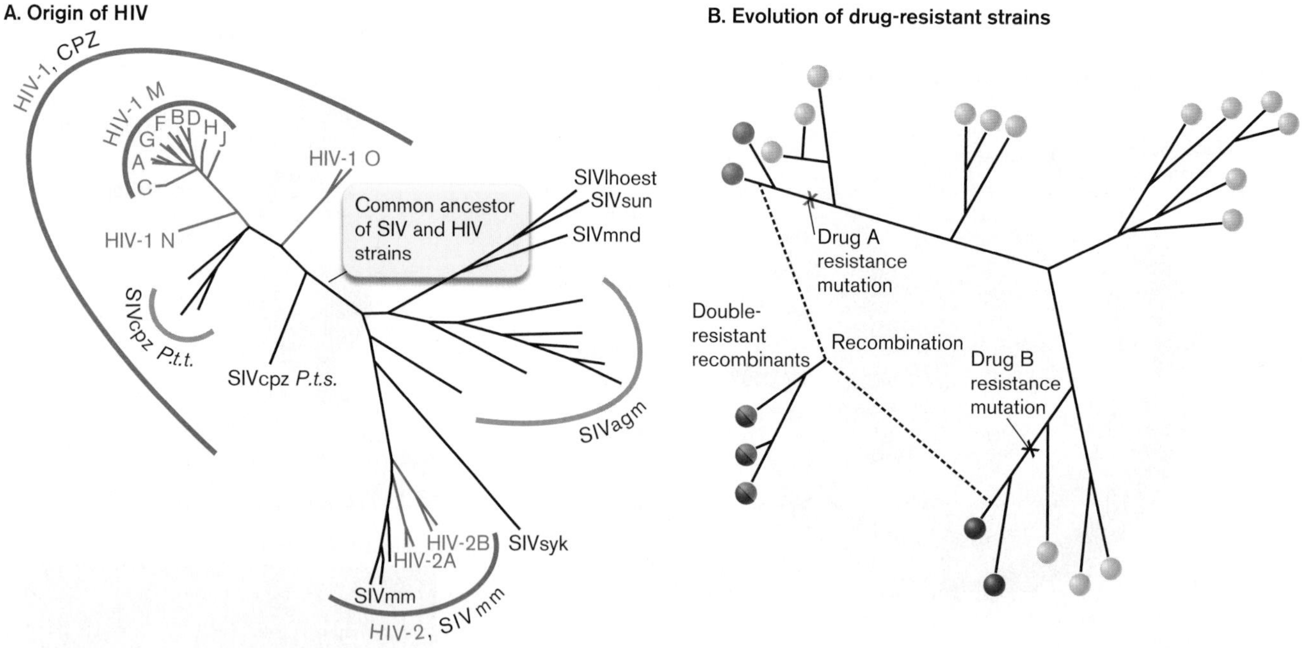
A. Origin of HIV
HIV-1, CPZ
HIV-1 M
F B D H
G
A
C
J
HIV-1 O
HIV-1 N
Common ancestor of SIV and HIV strains
SIVlhoest
SIVsun
SIVmnd
SIVcpz P.t.t.
SIVcpz P.t.s.
SIVagm
HIV-2B
HIV-2A
SIVsyk
SIVmm
HIV-2, SIVmm
B. Evolution of drug-resistant strains
Drug A resistance mutation
Double-resistant recombinants
Recombination
Drug B resistance mutation

FIGURE 11.23 ▪ Origin and evolution of HIV. A. Strains of HIV and SIV (simian immunodeficiency virus) arose independently multiple times over several decades, from a common origin in monkeys. *P.t.s.* = *Pan troglodytes schweinfurthii*; *P.t.t.* = *Pan troglodytes troglodytes*. **B.** When HIV infects a patient, different drugs may select strains with different resistance mutations. The mutant strains may then recombine to generate a double-resistant strain.

cases of AIDS, the different viral types then infect different cell types all over the body, including brain cells. Besides mutation, HIV virions can recombine their genomes by disengaging reverse transcriptase from the template and resuming DNA synthesis upon another template. The recombination of different mutants can generate strains resistant to multiple antiviral agents (**Fig. 11.23B**).

The HIV genome. The genome of HIV (**Fig. 11.22A**) was first cloned for molecular study by Flossie Wong-Staal, now at UC San Diego (**Fig. 11.22C**). Born in China in 1947, Wong-Staal immigrated to the United States and then worked with Gallo on the early discoveries of HIV. Wong-Staal now pursues gene therapy approaches to combat HIV infection and is developing retroviral vectors for human gene therapy (discussed in Section 11.6).

The HIV genome includes three main open reading frames that are found in all retroviruses: *gag*, *pol*, and *env*. The *gag* sequence encodes capsid, nucleocapsid, and matrix proteins; *pol* encodes reverse transcriptase (RT), integrase, and protease; and *env* encodes envelope proteins. The *gag* and *pol* sequences overlap, but because they are translated in different reading frames (different ways to align the triplet code), the ribosome expresses each independently of the other. During infection, each reading frame is transcribed and translated as a polyprotein; then, at subsequent stages, each polyprotein is cleaved by proteases to form the mature products.

In HIV-1, the *gag* and *pol* sequences overlap, and *env* overlaps with genes encoding **accessory proteins**, proteins that modify and regulate retroviral infection. Accessory proteins are unique to lentiviruses, mediating host response and maintaining the long-term slow progression of lentiviral disease. The accessory proteins are expressed within the infected host cell and regulate the replication cycle (**Table 11.3**). For example, Tat protein activates transcription of the viral genome. The HIV-1 genome encodes at least six accessory proteins—a greater number than any other retrovirus. They are major targets for research and drug discovery aimed at preventing HIV proliferation.

HIV Attachment and Host Cell Entry

Like other viruses, HIV needs to recognize specific receptor molecules on the surface of its target cells. The primary receptor for HIV is the CD4 surface protein on CD4 T lymphocytes (T cells). The normal function of CD4 surface proteins is to connect the T cell with an antigen-presenting cell, which activates the T cell to turn on B-cell production of antibodies (discussed in Section 24.5). Disruption of this antibody production is the main cause of the AIDS-related susceptibility to opportunistic infections. Note, however, that CD4 proteins appear on many other cell types, such as microglia (macrophage-like cells in the central nervous system) and Langerhans cells (immune cells of the epidermis). Their presence may make other cells susceptible to infection by HIV.

Spike proteins mediate membrane fusion. The binding of HIV to CD4 receptors involves the envelope spike protein SU (**Fig. 11.24**). Spike proteins are the main external proteins accessible to the host immune system.

TABLE 11.3

Accessory proteins of HIV-1.

Protein	Function	Effect of mutation
Vif	Virion component: • Protects reverse transcriptase from error-inducing host protein APOBEC3G.	Virions produced are noninfective
Vpr	Virion component: • Transcription factor, activates HIV transcription during G_2 phase of cell cycle; arrests T-cell growth. • Imports DNA across nuclear membrane; avoids need to infect rapidly dividing cells in which mitosis dissolves the nucleus.	Lower production of virions
Nef	Virion component: • Internalizes and degrades CD4 receptors, to avoid superinfection by more HIV virions, and to lessen immune response to the infected cell. • Decreases expression of major histocompatibility complex (MHC) proteins that stimulate cytotoxic T cells.	Slower progression to AIDS
Vpu	Membrane protein: • Degrades CD4, releasing bound spike proteins. • Promotes virion assembly and release from cell-surface tetherins.	Early death of host cell; lower production of virions
Rev	Nuclear phosphoprotein, combines with host cell proteins: • Stabilizes certain mRNAs in nucleus. • Exports mRNA out of nucleus into the cytoplasm, inducing shift from latent phase to virion-producing phase.	Failure of infection
Tat	Transcription factor: • Binds TAR site on nascent RNA to activate transcription. • Associates with histone acetylases and kinases to activate transcription of integrated viral DNA.	Failure of chromosome replication

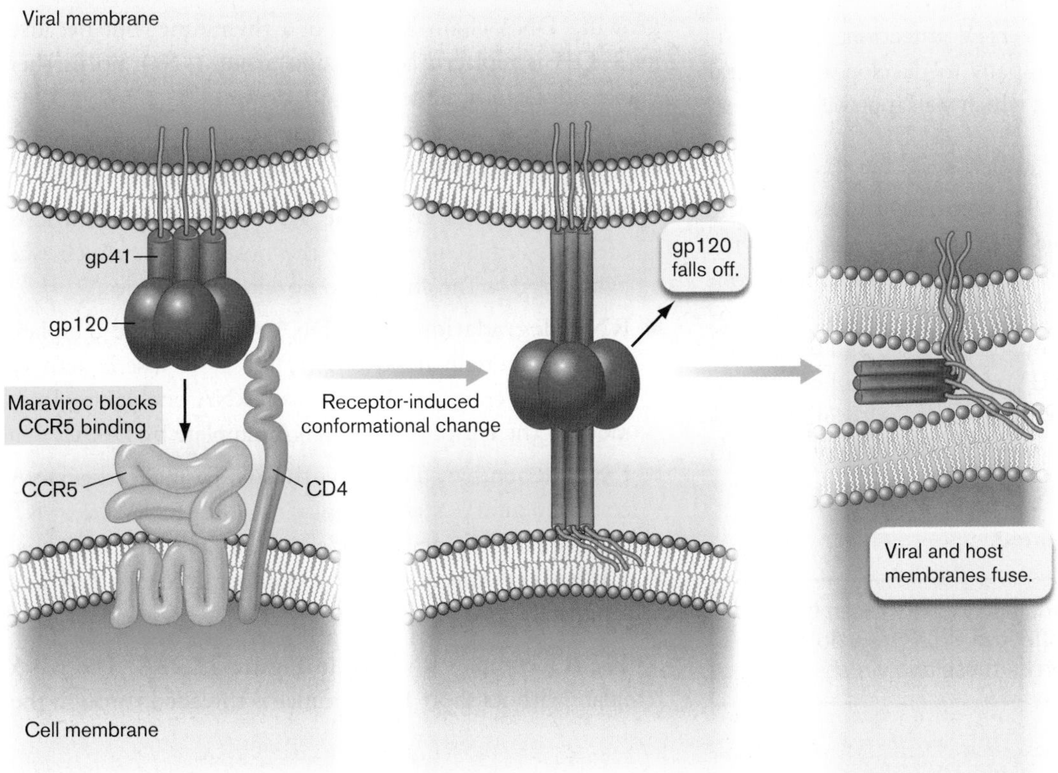
Viral membrane
gp41
gp120
Maraviroc blocks CCR5 binding
CCR5
CD4
Receptor-induced conformational change
gp120 falls off.
Viral and host membranes fuse.
Cell membrane

FIGURE 11.24 ■ HIV attachment to host cell. The SU (gp120) subunits of the spike protein complex attach to the receptor (CD4 cell-surface protein) and to CCR5. The SU subunits then fall off, and the TM (gp41) fusion peptides extend into the host cell membrane. The fusion peptides contract, pulling the membranes together.

HIV attachment to the cell membrane requires a fusion peptide rearrangement similar to that of influenza virus, except that it occurs at the cell surface (**Fig. 11.24**). When SU (gp120) binds to CD4, the spike transmembrane component TM (gp41) unfolds and extends its fusion peptide into the host cell membrane. In addition, SU binds to secondary receptors in the membrane called **chemokine receptors (CCRs)**, such as the macrophage receptor CCR5. Chemokines are signal molecules for the immune system, but their receptor proteins can bind viruses that evolve to take advantage of them. After the spike protein SU binds receptors and the TM fusion peptide inserts, the HIV-1 envelope fuses with the plasma membrane. A CCR such as CCR5 is also called a **coreceptor**, a protein acting with CD4 to bind the HIV spike proteins.

The requirement for CCR attachment varies among different types of HIV. Some chemokine receptors are found on neurons, and their involvement in HIV infection may mediate the neurological disorders seen in AIDS. Furthermore, the predominance of different viral envelope types with different receptor preferences varies over the course of infection. As HIV evolves a quasispecies (as we saw for hepatitis C), the virions target the CCR5 receptor early in infection, whereas later-evolved virions target a different host surface protein, called CXCR4. These "X4" virions can infect early-stage T cells that do not yet carry CCR5; thus the AIDS disease accelerates. The X4 virions are less infective when transmitted; thus, most new infections start with a CCR5 strain.

An exciting discovery was that individuals who lack the CCR5 protein because of a genetic defect show a high degree of resistance to HIV infection. This finding prompted the Pfizer company to develop an antiviral blocker of CCR5, maraviroc, which was approved for therapy in 2007 (see eTopic 11.3).

After HIV binds to membrane receptors, how does its genome enter the cell? The HIV envelope fuses with the cell membrane, enabling the HIV core to enter the cytoplasm directly. This HIV entry mechanism differs from that of influenza virus, in which endocytosis and lysosome fusion are required to open the capsid and release the genome into the cell. The HIV core (composed of CA and the host-derived cyclophilin A) dissolves, releasing the two RNA genomes, along with associated viral enzymes, into the cytoplasm.

Thought Question

11.7 How do attachment and entry of HIV resemble attachment and entry of influenza virus? How do attachment and entry differ between these two viruses?

The two RNA genomes each possess a 5′ "cap" and a 3′ poly-A "tail" that enable them to mimic host nuclear mRNA. Each RNA is hybridized to a tRNA primer for DNA synthesis. The primer is a lysine-specific tRNA from the previous infected cell. The primer might be expected to hybridize at the 3′ end of the template, where its 3′ OH "points" toward the opposite end, positioned to synthesize all the way down. Surprisingly, however, the 3′ OH end of the tRNA actually binds near the 5′ end, where initially it can generate only a brief sequence. These early sequences bind key regulatory factors for transcription and for DNA insertion into the host genome (discussed next).

Reverse Transcriptase Copies RNA to DNA

A retrovirus, unlike other RNA viruses, must integrate its entire genome into the host genome in order to replicate viral genomes and produce new progeny virions. Thus, the RNA genome needs to serve as a template to synthesize a DNA complement; but then the original RNA template must be degraded and replaced by a DNA strand, for host integration. All these processes are accomplished by **reverse transcriptase (RT)**, the defining enzyme of a retrovirus. Reverse transcriptase is the source of the high error rate of retroviral replication—on average, one or two errors per copy of HIV. This high error rate generates the quasispecies of different strains within an HIV-infected person.

Reverse transcriptase is the target of the first clinically useful drug to treat HIV infection, the nucleotide analog **azidothymidine (AZT)**. AZT is incorporated into the growing DNA chain in place of a thymidine, but because its 3′ OH is replaced by an azido group ($-N_3$), no further nucleotides can be added.

The reverse transcriptase complex actually possesses three different activities:

- **DNA synthesis from the RNA template.** Synthesis of DNA is first primed by the host tRNA, which was hybridized to the chromosome within the virion.
- **RNA degradation.** After DNA synthesis, the template RNA is gradually removed through an RNase H activity of the RT complex. Removal of RNA enables replacement of the entire original RNA template by DNA.
- **DNA-dependent DNA synthesis.** To make the DNA complementary strand replacing the RNA, the RT needs to use the newly made DNA as its template. Thus, RT has the rare ability to use either DNA or RNA as template.

The RT complex is shown in **Figure 11.25A**. The RNA template with its short RNA primer is threaded through the

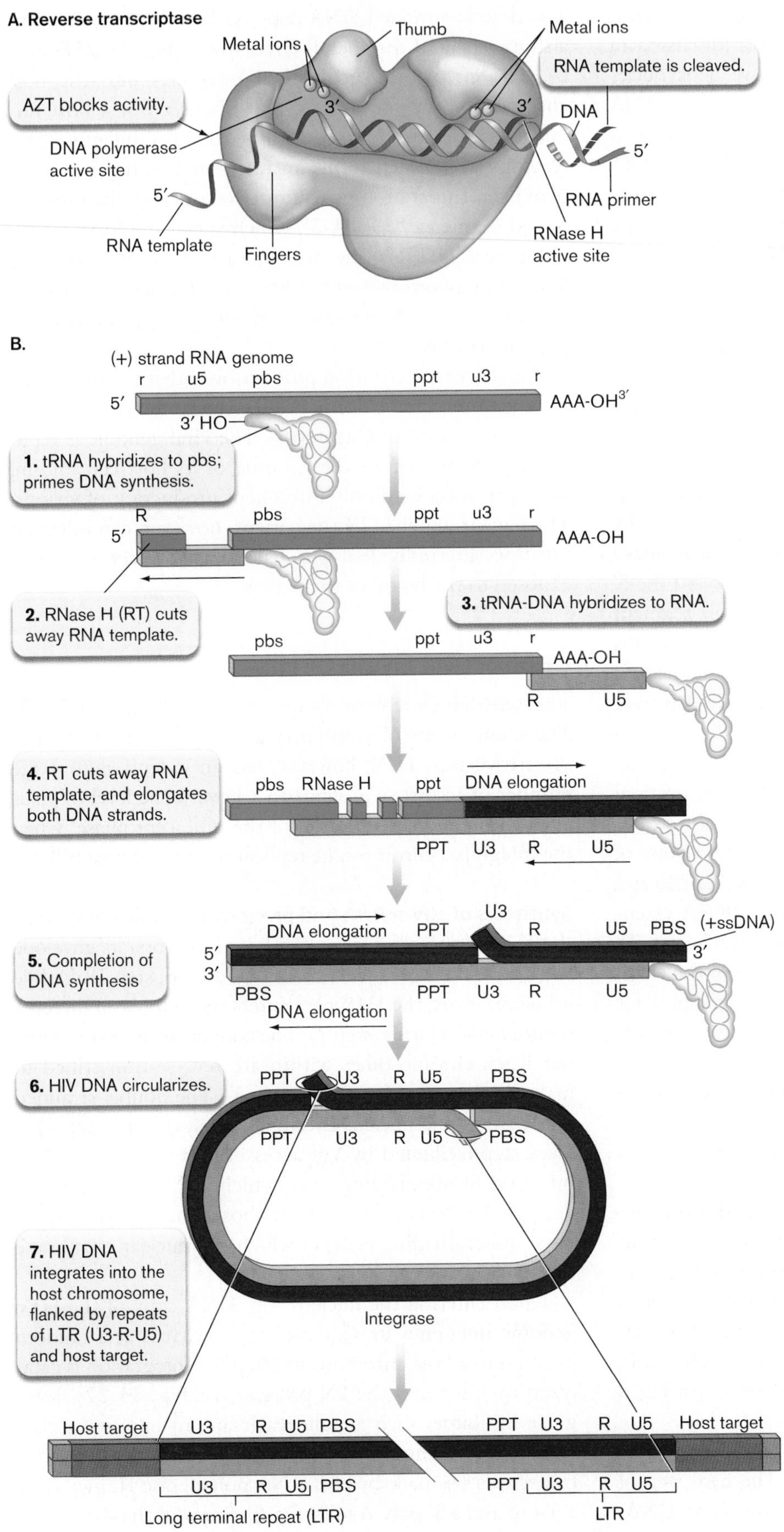
A. Reverse transcriptase
Metal ions
Thumb
Metal ions
RNA template is cleaved.
AZT blocks activity.
3′
3′
DNA
DNA polymerase active site
5′
5′
RNA primer
RNA template
Fingers
RNase H active site
B.
(+) strand RNA genome
r u5 pbs ppt u3 r
5′
AAA-OH3′
3′ HO
1. tRNA hybridizes to pbs; primes DNA synthesis.
R pbs ppt u3 r
5′
AAA-OH
2. RNase H (RT) cuts away RNA template.
3. tRNA-DNA hybridizes to RNA.
pbs ppt u3 r
AAA-OH
R U5
4. RT cuts away RNA template, and elongates both DNA strands.
pbs RNase H ppt DNA elongation
PPT U3 R U5
U3
DNA elongation PPT R U5 PBS (+ssDNA)
5. Completion of DNA synthesis
5′
3′
3′
PBS PPT U3 R U5
DNA elongation
6. HIV DNA circularizes.
PPT U3 R U5 PBS
PPT U3 R U5 PBS
7. HIV DNA integrates into the host chromosome, flanked by repeats of LTR (U3-R-U5) and host target.
Integrase
Host target U3 R U5 PBS PPT U3 R U5 Host target
U3 R U5 PBS PPT U3 R U5
Long terminal repeat (LTR)
LTR

FIGURE 11.25 ■ Reverse transcription of the HIV genome and integration into host DNA. A. The RNA template with its short RNA primer is threaded through the RT complex between the "thumb" and "fingers"—a configuration typical of other RNA polymerases. The RT complex adds successive deoxynucleotides from dNTPs (deoxyribonucleoside triphosphates), starting at the 3′ OH end of the RNA primer (as in regular DNA synthesis). As DNA elongates, the RNA template is cleaved from behind by RNase H activity. **B.** The tRNA primer (1) initiates a short sequence of DNA complementary to the 5′ end of the HIV chromosome (2). The corresponding template is then degraded. The original RNA template has repeated ends (r), so the exposed DNA end (extended tRNA primer of the second copy) then hybridizes to the 3′ end (3). DNA now elongates along the rest of the chromosome (4), up to the primer-binding site (pbs) for mRNA transcription. The remaining RNA template is degraded by RT and replaced by complementary DNA up to the PPT-U3 junction (5). The completed double-stranded DNA copy of the HIV genome, plus a short repeat of the 5′ end (U3-R-U5), circularizes (6) and integrates into the host chromosome (7), mediated by integrase (IN). The LTR (U3-R-U5) and "host target" sequences generate repeats, flanking the integrated HIV genome (provirus).

RT complex between the "thumb" and "fingers"—a configuration typical of other RNA polymerases (discussed in Chapter 8). The RT complex adds successive deoxynucleotides from dNTPs starting at the 3′ OH end of the RNA primer. As DNA elongates, however, the RNA template is cleaved from behind by the RT complex. Thus, the new DNA actually replaces preexisting RNA sequence. This "destructive replication" is unique to retroviruses. The details are important, as they suggest possible targets for new antiviral drugs.

Weblinks Reverse transcriptase tutorial (Biomolecules at Kenyon) (*see ebook*)

Initiation of reverse transcription. **Figure 11.25B** outlines the entire process of reverse transcription in HIV-1, from primed RNA template to integration of duplex DNA. First, the host-derived tRNA primer initiates synthesis (by reverse transcriptase, RT) of a DNA strand complementary to the RNA chromosome. DNA is elongated toward the 5′ end of the HIV chromosome, generating a short segment (**Fig. 11.25B**, step 1). The original RNA template for this short segment is then degraded by the RNase H activity of RT, leaving only the DNA extension off the tRNA primer (step 2).

The new DNA primes the second template. The original RNA template had repeated ends (labeled "r," lowercase, for RNA in **Fig. 11.25B**), and the exposed DNA copy of the 5′ end has a complementary sequence ("R," uppercase, for DNA). The "R" DNA from the second tRNA extension hybridizes to the 3′ end of the original RNA (step 3). The hybridized DNA elongates along the rest of the chromosome (step 4), up to the primer-binding site (pbs) for mRNA transcription. DNA completion is followed by degradation of the remaining RNA template, except for occasional short fragments to serve as primers, such as the polypurine tract (ppt). A complementary DNA strand is then synthesized through the PPT primer, leaving a nick at U3 (step 5).

Interestingly, human host cells have evolved a protein, APOBEC3G, that deaminates viral cytosines and thus increases the error rate of reverse transcriptase. APOBEC3G can be packaged into progeny virions, decreasing production of infective virions in the next host. However, HIV has evolved an accessory protein, Vif (see **Table 11.3**), that binds APOBEC3G to prevent it from being packaged into virions.

DNA integration into the host genome. The final step of genome processing requires integration into host DNA. The double-stranded DNA copy of the HIV genome (plus a short repeat of the 5′ end) circularizes (**Fig. 11.25B**, step 6). The circular molecule then integrates into the host chromosome by site-specific recombination at a host target sequence (step 7), mediated by the HIV integrase protein (IN). This step forms an integrated viral genome, or **provirus**. Integration generates two copies of the provirus 5′ end (sequence U3-R-U5), which is called a **long terminal repeat (LTR)**. The provirus and LTR ends are also flanked by two copies of the host target sequence. Proviral sequences can now be expressed, directing production of progeny virions.

An alternative to virion production is that the integrated HIV genome can lie dormant, like the lambda prophage in *E. coli* (discussed in Chapter 6). The viral genome is replicated passively within the genome of its host cell, hiding for many years with only infrequent production of virions. The few virions shed by the patient, however, can infect an unsuspecting individual who has sexual contact with or is exposed to the blood of the patient.

Replication Cycle of HIV

The steps of HIV replication are outlined in **Figure 11.26**. The main points of viral entry and replication are typical of retroviruses. HIV, however, has an exceptionally large number of accessory proteins that govern the level of virus production and the duration of the quiescent phase, when the integrated chromosome replicates with the host cell.

Synthesis of HIV mRNA and progeny genomic RNA. After the HIV virion attaches to the host receptors, its envelope fuses with the host membrane (**Fig. 11.26**, step 1). Unlike influenza virus, the HIV core enters the cytoplasm directly, without endocytosis (step 2). The core partly uncoats, while the RNA chromosomes within are reverse-transcribed to make double-stranded DNA (step 3). The double-stranded DNA enters the nucleus through a nuclear pore (step 4)—a key step facilitated by Vpr accessory protein. Vpr enables infection of nondividing cells, which only lentiviruses can do; other retroviruses, such as those causing lymphoma, must infect dividing cells, in which the nuclear membrane dissolves during mitosis.

Upon entering the nucleus, the DNA copy of the HIV genome integrates its sequence as a provirus at a random position in a host chromosome (step 5). Integration is catalyzed by integrase (the IN protein; see **Fig. 11.22**). Integrase inhibitors such as raltegravir are an important class of anti-HIV drugs. Within the nucleus, full-length RNA transcripts are made by host RNA polymerase II, including a 5′ cap and a 3′ poly-A tail (step 6). Some of the RNAs exit

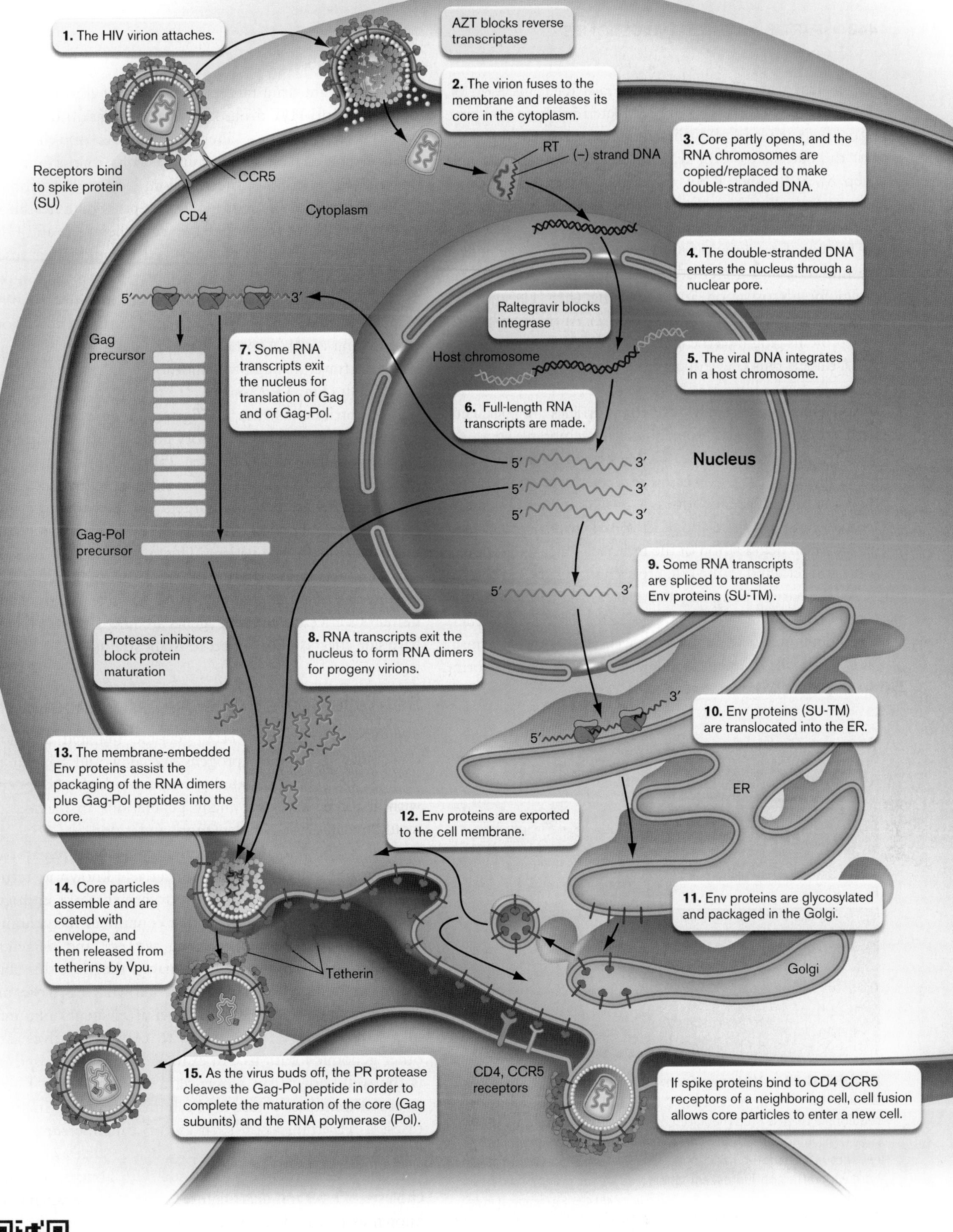
1. The HIV virion attaches.
AZT blocks reverse transcriptase
2. The virion fuses to the membrane and releases its core in the cytoplasm.
RT
(–) strand DNA
3. Core partly opens, and the RNA chromosomes are copied/replaced to make double-stranded DNA.
Receptors bind to spike protein (SU)
CCR5
CD4
Cytoplasm
4. The double-stranded DNA enters the nucleus through a nuclear pore.
5′
3′
Raltegravir blocks integrase
Gag precursor
7. Some RNA transcripts exit the nucleus for translation of Gag and of Gag-Pol.
Host chromosome
5. The viral DNA integrates in a host chromosome.
6. Full-length RNA transcripts are made.
Nucleus
Gag-Pol precursor
9. Some RNA transcripts are spliced to translate Env proteins (SU-TM).
Protease inhibitors block protein maturation
8. RNA transcripts exit the nucleus to form RNA dimers for progeny virions.
10. Env proteins (SU-TM) are translocated into the ER.
13. The membrane-embedded Env proteins assist the packaging of the RNA dimers plus Gag-Pol peptides into the core.
ER
12. Env proteins are exported to the cell membrane.
14. Core particles assemble and are coated with envelope, and then released from tetherins by Vpu.
11. Env proteins are glycosylated and packaged in the Golgi.
Tetherin
Golgi
15. As the virus buds off, the PR protease cleaves the Gag-Pol peptide in order to complete the maturation of the core (Gag subunits) and the RNA polymerase (Pol).
CD4, CCR5 receptors
If spike proteins bind to CD4 CCR5 receptors of a neighboring cell, cell fusion allows core particles to enter a new cell.

FIGURE 11.26 ▪ HIV replicative cycle. The HIV virion attaches its receptor and fuses with the host cell membrane, releasing its contents in the cytoplasm to undergo a replicative cycle.

the nucleus (step 7) to serve as mRNA for translation of polyproteins. Polyproteins are translated in alternative versions, such as Gag-Pol. Other full-length RNA transcripts exit the nucleus to form RNA dimers for progeny virions (step 8). Still other RNA transcripts within the nucleus are cut and spliced to complete the *env* gene sequence for translation of envelope proteins (step 9).

The Env (envelope) proteins are made within the endoplasmic reticulum (ER) (step 10). They pass through the Golgi for glycosylation and packaging (step 11) and are exported to the cell membrane (step 12). At the membrane, Env proteins plug into the core particle as it forms from the RNA dimers plus Gag-Pol peptides (step 13).

Virion assembly and exit. The core particles are packaged with envelope derived from host cell membrane containing Env spike proteins (**Fig. 11.26**, step 14). To escape the host cell, emerging virions require accessory protein Vpu to bind a "tetherin," a host adhesion protein induced by interferon to cause reuptake and digestion of virions. Vpu causes proteosomal degradation of the tetherin. Emerging virions, some still tethered to the cell, are seen in **Figure 11.27A**.

As the virion buds off, the protease (PR) cleaves the Gag-Pol peptide to complete maturation of the core structure containing Gag subunits as well as RNA polymerase (RT) (**Fig. 11.26**, step 15). The Gag subunits now form the conical core structure. Proteases that cleave Gag-Pol offer important drug targets, which have led to the development of anti-HIV drugs known as **protease inhibitors**.

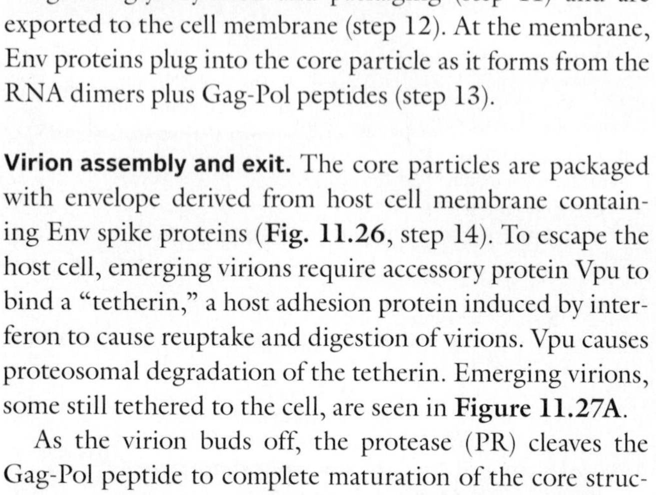
A.
Tetherin
SOURCE: S. NEIL ET AL. 2008. *NATURE* **451**:425
B.
SOURCE: S. SOWINSKI ET AL. 2008. *NAT. CELL BIOL.* **10**:211

FIGURE 11.27 ■ HIV exits from an infected cell. A. As virions emerge, they remain tethered to the cell surface by host tetherins, requiring a release step mediated by accessory protein Vpu (TEM). **B.** A T cell infected with HIV can transfer virions to an uninfected cell through a nanotubular connection. The two T cells are tagged here with different fluorescent labels (red versus green). *Source:* Part A modified from Stuart J. D. Neil et al. 2008. *Nature* **451**:425, Fig. 4e; part B from Stefanie Sowinski et al. 2008. *Nat. Cell Biol.* **10**:211.

However, HIV has alternative means of cell-to-cell transmission that avoid exposing virions to the immune system. One alternative is cell fusion, mediated by binding of Env in the membrane to CD4 receptors on a neighboring cell (see **Fig. 11.26**). The two cells then fuse, and HIV core particles can enter the new cell through their fused cytoplasm. The fusion of many cells can form a giant multinucleate cell called a syncytium. Cell fusion with formation of syncytia enables HIV to infect neighboring cells without ever exiting a cell. Another means of cell-to-cell transmission is to travel through a "nanotube" connection between two T cells (**Fig. 11.27B**).

The intricate scheme in **Figure 11.26** actually omits many functions of HIV accessory proteins that enhance the virulence of HIV infection (see **Table 11.3**). Mutation of genes for accessory proteins often decreases virulence; thus, these proteins are potential targets for chemotherapy. Surprisingly, even a modest decrease of HIV infectivity can have major benefits for the patient. This suggests that HIV is so crippled by its high mutation rate that the slightest interference greatly decreases production of infective virions. Thus, numerous effective antiviral agents are now known—but all have side effects, and all select resistant strains.

Retroelements in the Human Genome

Suppose an integrated HIV genome mutated and lost the ability to produce progeny. What would happen to its genome? The integrated genome would be "trapped" within a cell, and over many generations in its host, it would inevitably accumulate more mutations.

In fact, the human genome is riddled with remains of decaying retroviral genomes, collectively known as retroelements (**Fig. 11.28**). Some retroelements are **endogenous retroviruses**, sequences that contain all the genomic elements of a retrovirus, including *gag, env,* and *pol* genes, yet never generate virions. Presumably, they lost this ability by mutation. Other elements, known as **retrotransposons**, retain only partial retroviral elements but may maintain a reverse transcriptase to copy themselves into other genomic locations. An example of a retrotransposon is the well-known Alu sequence, a short sequence found in about a million copies in the human genome. In some cases, a retrotransposon such as Alu can interrupt a key human gene, leading to a genetic defect such as a defective lipoprotein receptor associated with abnormally high cholesterol level and heart failure. Still other retroelements, known as LINEs (long interspersed nuclear elements) and

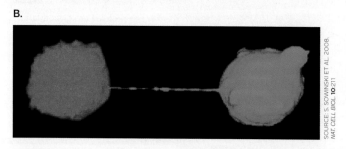

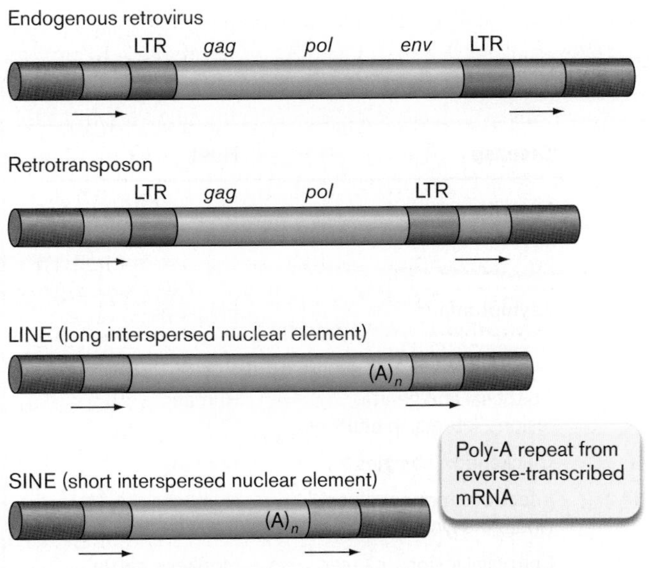
Endogenous retrovirus
LTR
gag
pol
env
LTR
Retrotransposon
LTR
gag
pol
LTR
LINE (long interspersed nuclear element)
(A)n
Poly-A repeat from reverse-transcribed mRNA
SINE (short interspersed nuclear element)
(A)n

FIGURE 11.28 ▪ Retroelements in the human genome. Endogenous retroviruses and other retroelements in the human genome may arise from progressive degeneration of ancestral retroviruses, or they may be progenitors of new retroviruses.

SINEs (short interspersed nuclear elements), show more vestigial remnants of retroviral genomes. Amazingly, about half the sequence of the human genome appears to have originated from viruses and retroelements. The origin of viruses, including their interaction with cellular evolution, is explored in eTopic 6.1.

Could a virus as deadly as HIV be used to improve human health? In fact, the exceptional ability of lentiviruses to deliver their own DNA to human cells makes them our most effective vectors for gene therapy. Lentiviruses are particularly promising for their ability to transfer genes into nonmitotic cells (see Section 11.6).

To Summarize

- **Human immunodeficiency virus (HIV)** causes an ongoing pandemic of acquired immunodeficiency syndrome (AIDS). Molecular biology has led to drugs that control the infection.
- **HIV is a retrovirus** whose RNA genome is reverse-transcribed into double-stranded DNA, which integrates into the DNA of the host cell.
- **The HIV core** contains two different copies of its RNA genome, each bound to a primer (host tRNA) and reverse transcriptase (RT). The core is surrounded by an envelope containing spike protein trimers.
- **HIV binds the CD4 receptor** of T lymphocytes together with the chemokine receptor CCR5. Following virion-receptor binding and envelope-membrane fusion, HIV core particle is released into the cytoplasm, where it partly uncoats.
- **DNA is synthesized from the HIV RNA by reverse transcriptase**, primed by the tRNA. **RNA degradation** allows formation of a double-stranded DNA. Entering the nucleus, the retroviral **DNA integrates** into the host genome.
- **Retroviral mRNAs are exported to the cytoplasm** for translation. Envelope proteins are translated at the ER and exported to the cell membrane.
- **Retroviruses are assembled at the cell membrane**, where virions are released slowly, without lysis. Alternative routes of cell-to-cell transmission involve cell fusion (forming syncytia) or travel through an intercellular nanotube.
- **Accessory proteins regulate virion formation** and the latent phase, in which double-stranded DNA persists without reproduction of progeny virions.
- **Ancient retroviral sequences** persist within animal genomes, including the human genome.

11.5

Herpes Simplex Virus: DNA Virus

Many important viruses of humans and other animals contain genomes of double-stranded DNA (**Table 11.4**). DNA viruses include the causative agents of well-known diseases such as smallpox, chickenpox, and infectious mononucleosis (mono). Most DNA viruses are considerably larger than RNA viruses and encode a wider range of viral enzymes; for example, the vaccinia genome encodes nearly 200 different proteins. The complexity of viruses such as vaccinia and herpes approaches that of small cells.

Herpes viral DNA replicates by mechanisms similar to those used in the replication of prokaryotic and phage genomes, either bidirectionally from an origin of replication (as in bacteria) or by the rolling-circle method (as in phages such as T4). Like cellular genomes, herpes viral DNA replication requires more than a polymerase; enzymes such as helicase, primase, and single-strand binding proteins are also needed. Simian virus 40 (SV40) and Epstein-Barr virus rely entirely on cellular components, whereas poxviruses rely entirely on viral components for DNA replication. Other DNA viruses, such as adenovirus and papillomavirus, combine viral and cellular components.

TABLE 11.4

DNA viruses of animals (examples).

Virus	DNA replication	Disease	Host
Adenoviruses (many strains)	Viral DNA polymerase, single-strand binding protein, and protein primer	Enteritis or respiratory diseases	Humans, other mammals, birds
Papovavirus (simian virus 40, SV40)	Cellular DNA polymerases	Asymptomatic	Monkeys
Herpes viruses			
Herpes simplex virus 1 and 2	All viral components (DNA polymerase, primase, etc.)	Epithelial and genital lesions, latency in neurons	Humans
Varicella-zoster virus	Viral components	Chickenpox, shingles	Humans
Epstein-Barr virus	All cellular components (DNA polymerase, etc.)	Infectious mononucleosis, Hodgkin's lymphoma	Humans
Other strains		Epithelial lesions, cancer	Monkeys, cattle, horses
Papillomaviruses	Viral DNA helicase; cellular polymerase		
Human papillomaviruses (many strains)		Genital warts, cervical and penile cancer, skin warts	Humans
Other papillomaviruses		Warts, cancer	Rabbits, cattle, sheep
Poxviruses	All viral components		
Variola major virus		Smallpox	Humans
Vaccinia virus		Cowpox	Cattle, humans
Other poxviruses		Monkeypox	Monkeys, camels, birds, humans

Herpes Simplex Virus Infects the Oral or Genital Mucosa

An important example of a DNA virus is herpes simplex virus (HSV). Strains HSV-1 and HSV-2 cause one of the most common infections in the United States. Approximately 60% of Americans acquire herpes simplex, usually HSV-1, in epithelial lesions commonly known as cold sores. About 30%–60% acquire genital herpes, usually HSV-2, through sexual contact (oral or vaginal). Genital herpes causes recurrent eruptions of infection in the reproductive tract (**Fig. 11.29**). Many of those infected are unaware of symptoms, but they can still transmit the disease to others.

Herpes simplex virus typically infects cells of the oral or genital mucosa, causing ulcerated sores. The primary infection is epithelial, followed by latent infection within neurons of the ganglia. A common site of infection is the trigeminal ganglion, which processes nerve impulses between the face and eyes and the brain stem.

The latent infection of the ganglia later leads to new outbreaks of virus, often triggered by stress such as menstruation, sunlight exposure, or depression of the immune system. Progeny virions travel back down the dendrites to the epithelia, causing **lytic** infection. In the trigeminal ganglion, herpes reactivation can lead to eye disease or lethal brain infection. In most cases, herpes symptoms can be controlled by antiviral agents such as acyclovir (discussed in Chapter 27). There is no cure or means of preventing future outbreaks. In pregnant women, HSV can be transmitted to the fetus, with serious complications for the child.

Herpes simplex virus is closely related to varicella-zoster virus (VZV), the cause of chickenpox, also an epithelial infection. Varicella, too, can hide in ganglial neurons, emerging decades later to cause painful skin lesions called shingles.

Herpes Simplex Virus Structure

The herpes virion comprises a double-stranded DNA chromosome packed within an icosahedral capsid (**Fig. 11.30A** ▶). The capsid is surrounded by **tegument**, a collection of about 15 different kinds of virus-encoded proteins, as well

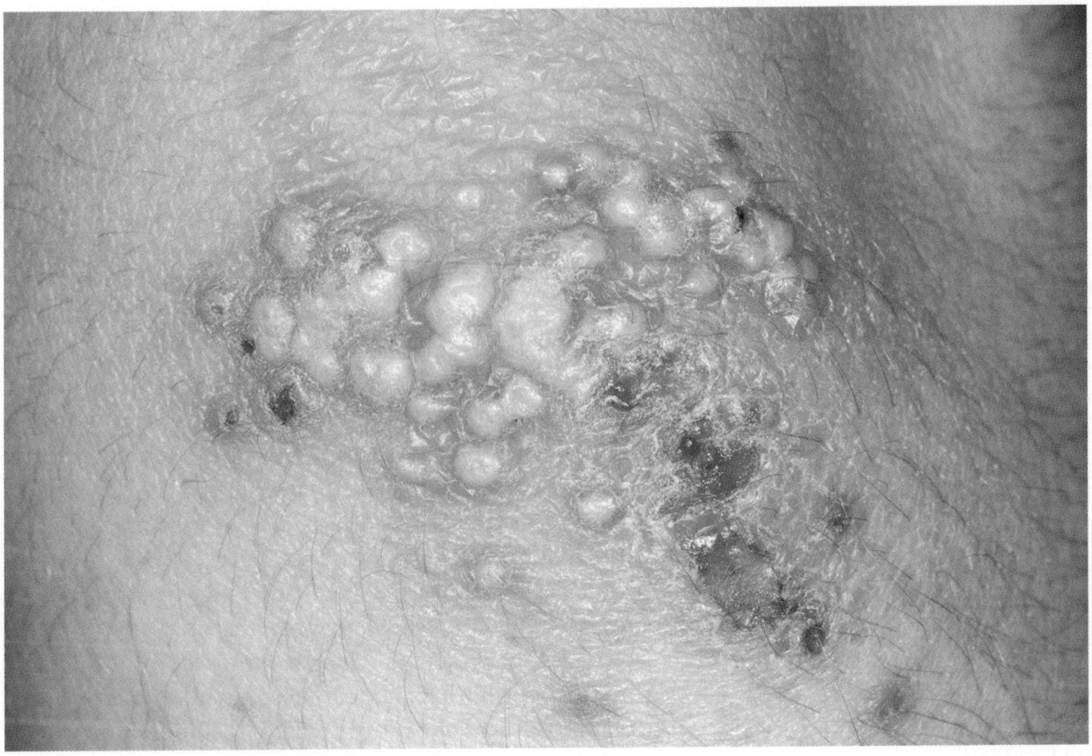

DR. P. MARAZZI/SCIENCE SOURCE

FIGURE 11.29 ■ Genital herpes infection. Lesions on the elbow of an 11-year old, female patient caused by HSV-2 infection.

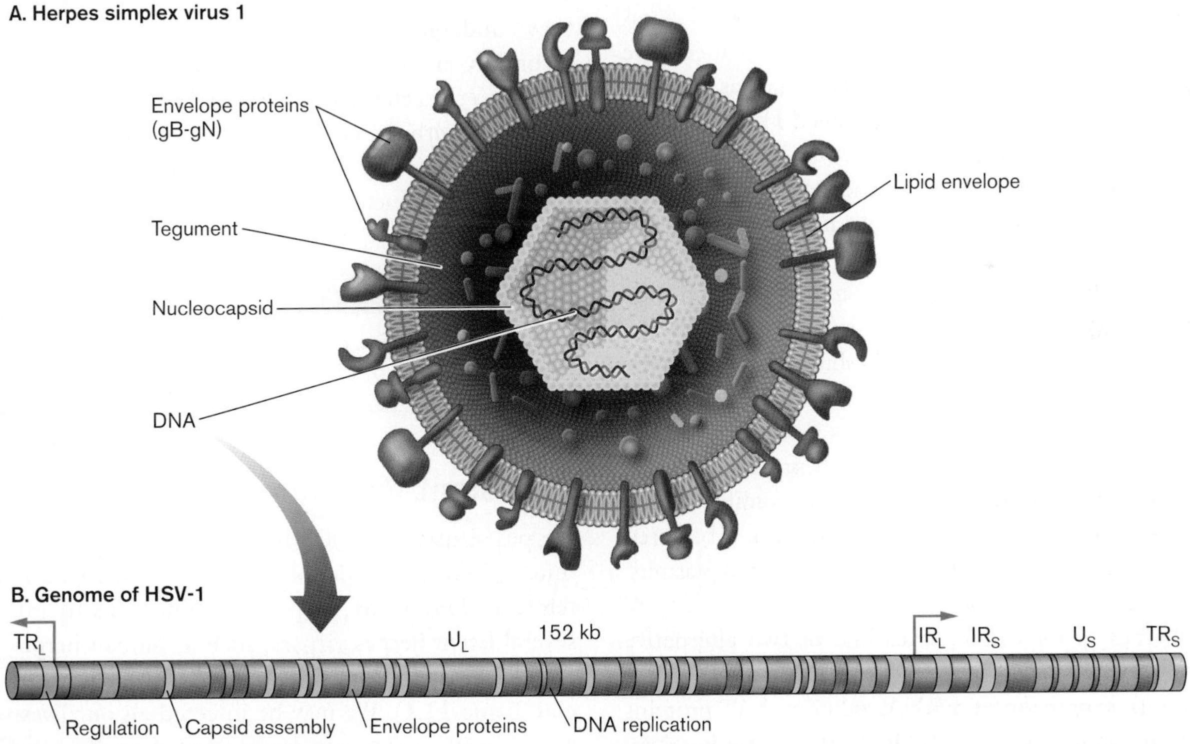
A. Herpes simplex virus 1
Envelope proteins (gB-gN)
Tegument
Nucleocapsid
DNA
Lipid envelope
B. Genome of HSV-1
TR_L
U_L
152 kb
IR_L
IR_S
U_S
TR_S
Regulation
Capsid assembly
Envelope proteins
DNA replication

FIGURE 11.30 ■ Herpes simplex virus 1: virion and genome. A. HSV-1 virion consists of a double-stranded DNA chromosome packaged within an icosahedral capsid. The capsid is surrounded by tegument, a collection of virus-encoded and host-derived proteins. The tegument is contained within a host-derived membrane envelope including several kinds of envelope proteins. **B.** The genome of HSV-1 spans 152,000 base pairs, encoding more than 70 gene products. The HSV sequence consists of two segments: long (U_L) and short (U_S). Each segment contains a unique region (U_L or U_S) flanked by two inverted-repeat regions, terminal (TR_L or TR_S) and internal (IR_L or IR_S), where the two segments meet.

as proteins from the previous host. The tegument is contained within a host-derived membrane envelope with several kinds of spike proteins. The HSV-1 genome spans 152 kb, encoding more than 70 gene products (**Fig. 11.30B**). The sequence includes two unique segments, long (U_L) and short (U_S), each flanked by a terminal repeat (TR_L or TR_S) and an internal repeat (IR_L or IR_S). Within the host, the genome circularizes, so the genetic linkage map appears circular.

Herpes genes and gene regulatory elements resemble those of eukaryotic genes. Each gene has a promoter with eukaryotic control sequences such as the TATA box. Genes are transcribed and translated individually; there is no polyprotein. Genes fall into temporal classes within the viral replication cycle: immediate-early, early, and late expressed genes. Each class has specific regulatory elements. Thus, the virus expresses only the gene products needed for a given phase of infection. An additional class, consisting of the LAT genes, is expressed for latent infection (discussed next).

Herpes Simplex Attachment and Host Cell Entry

Unlike HCV and HIV, the relatively large herpes virion has several envelope proteins that can bind to several alternative receptor molecules on the host cell surface, such as a homolog of tumor necrosis factor receptor called HveA or intercellular adhesion molecules called nectins (**Fig. 11.31** ▶, step 1). As in the case of HIV, the entire herpes capsid enters the cytoplasm (step 2); but unlike HIV, whose core particle partly uncoats, the intact herpes capsid travels down a scaffold of microtubules (step 3) to the nuclear membrane. During this stage, the virion host shutoff factor (Vhs) degrades host mRNA, thus shutting off host protein synthesis. At a nuclear pore complex, the herpes capsid injects its DNA (step 4). The DNA is forced out of the capsid and into the nucleus by the high pressure of double-stranded DNA packing in the capsid, similar to the high-pressure injection of phage T4 DNA into a bacterial cell. The DNA then circularizes (step 5) to form a plasmid-like intermediate.

The herpes genome now takes one of two alternative directions: expression of mRNA for proteins of the infection cycle, or expression of mRNA encoding LAT proteins to maintain latency (step 6). If the latent course is taken, the DNA circle can persist within the cell for decades before switching to lytic infection. Latent infection most commonly occurs in nerve cells, such as those of the trigeminal ganglion.

Replication of Herpes Simplex Virus

In the nucleus, herpes DNA is transcribed to mRNA by host RNA polymerase II (see **Fig. 11.31**). If mRNA for lytic infection is produced, it exits the nucleus to be translated by ribosomes. Many different mRNAs are produced and exported, including those required for "immediate" and "early" stages of infection (**Fig. 11.31**, step 7). The translated proteins return to the nucleus for packaging within capsids.

To generate progeny genomes, the circular DNA is replicated by viral enzymes, including DNA polymerase, single-strand binding protein, and a proofreading endonuclease (step 8). Additional enzymes are provided by the host cell. DNA is replicated by the rolling-circle method, generating a concatemer similar to that of phage T4. Unlike T4, however, the herpes DNA is eventually cut into segments defined by the terminal repeat sequences.

The newly synthesized DNA expresses late-stage mRNA (step 9), which exits the nucleus for translation. Translated envelope proteins are inserted into the ER membrane, through which they migrate to the nuclear membrane (step 10). Other late proteins reenter the nucleus for assembly into capsids containing DNA genomes (step 11). The envelope forms from the outer nuclear membrane (step 12). The virions are then transported through the ER, where they undergo secondary envelopment (step 13). The secondary-enveloped virions move to the Golgi, and ultimately to the cell membrane (step 14). The secondary envelope fuses with the cell membrane, releasing mature virions through exocytosis. Rapid release of virions destroys cells, causing the characteristic sores of herpes infection.

Thought Question

11.8 Compare and contrast the fate of the HSV chromosome with that of the HIV chromosome.

Persistent Viral Infections

Herpes viruses can infect humans and other animals indefinitely, following initiation of latency by the viral LAT proteins (see **Fig. 11.31**, Step 6). Most humans are infected by several latent herpes viruses, such as human herpes viruses 6 and 7, Epstein-Barr virus, and cytomegalovirus (see **Special Topic 11.1**). We may be infected all our lives without knowing it; in fact, our bodies may actually benefit from the presence of these viruses. For example, in 2007, Herbert Virgin and colleagues at the Washington University School of Medicine in St. Louis, showed that mice infected with a gamma herpes virus similar to Epstein-Barr virus

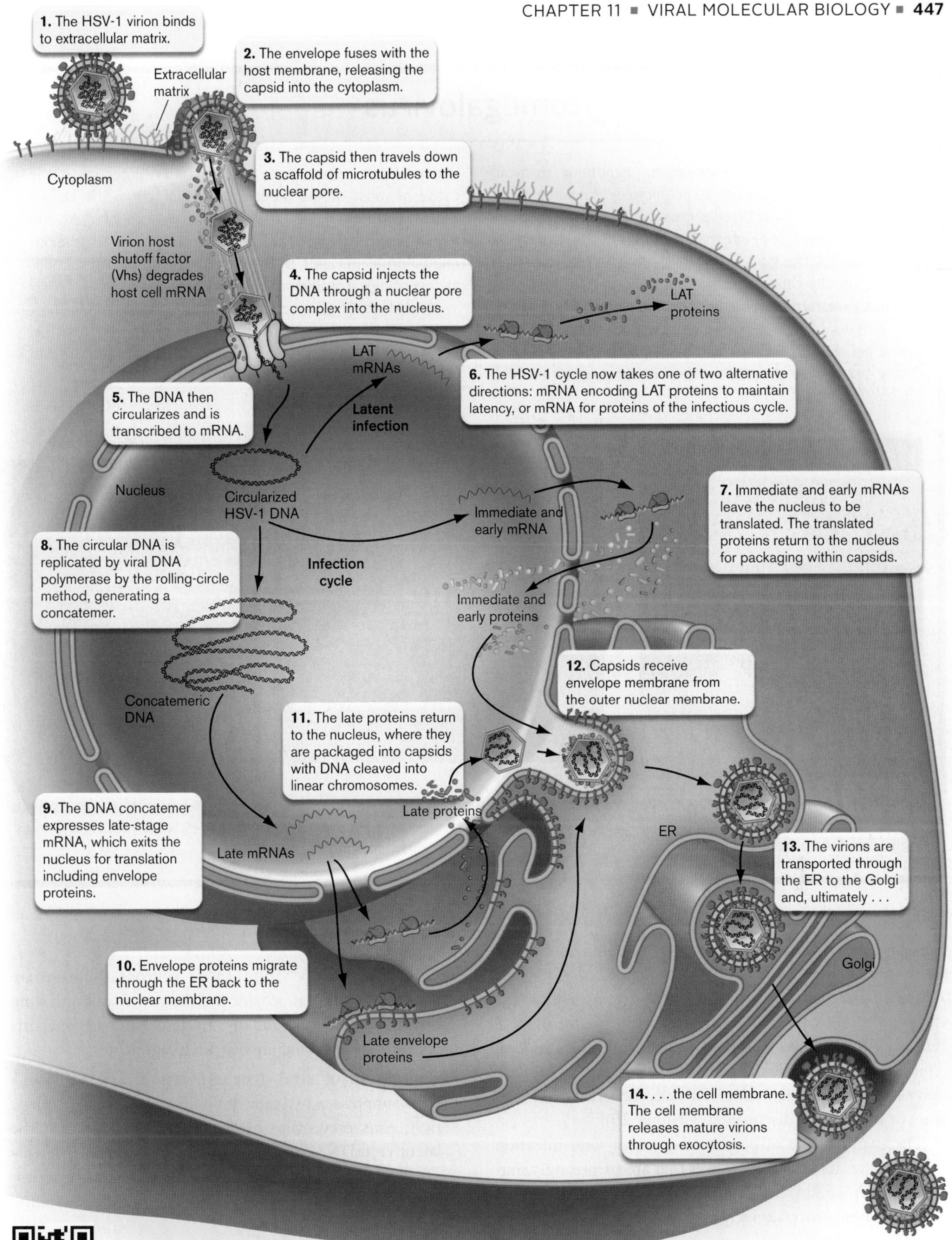
1. The HSV-1 virion binds to extracellular matrix.
Extracellular matrix
2. The envelope fuses with the host membrane, releasing the capsid into the cytoplasm.
Cytoplasm
3. The capsid then travels down a scaffold of microtubules to the nuclear pore.
Virion host shutoff factor (Vhs) degrades host cell mRNA
4. The capsid injects the DNA through a nuclear pore complex into the nucleus.
LAT proteins
LAT mRNAs
6. The HSV-1 cycle now takes one of two alternative directions: mRNA encoding LAT proteins to maintain latency, or mRNA for proteins of the infectious cycle.
5. The DNA then circularizes and is transcribed to mRNA.
Latent infection
Nucleus
Circularized HSV-1 DNA
Immediate and early mRNA
7. Immediate and early mRNAs leave the nucleus to be translated. The translated proteins return to the nucleus for packaging within capsids.
8. The circular DNA is replicated by viral DNA polymerase by the rolling-circle method, generating a concatemer.
Infection cycle
Immediate and early proteins
12. Capsids receive envelope membrane from the outer nuclear membrane.
Concatemeric DNA
11. The late proteins return to the nucleus, where they are packaged into capsids with DNA cleaved into linear chromosomes.
9. The DNA concatemer expresses late-stage mRNA, which exits the nucleus for translation including envelope proteins.
Late proteins
ER
Late mRNAs
13. The virions are transported through the ER to the Golgi and, ultimately . . .
10. Envelope proteins migrate through the ER back to the nuclear membrane.
Golgi
Late envelope proteins
14. . . . the cell membrane. The cell membrane releases mature virions through exocytosis.

FIGURE 11.31 ■ Replicative cycle of HSV-1. The HSV-1 virion binds to receptors on the host cell membrane and releases its capsid in the cytoplasm. The DNA chromosome is transferred into the host nucleus to conduct the replicative cycle.

Special Topic 11.1: Cytomegalovirus

We often think of herpes viruses as something to avoid "catching," like chickenpox or herpes simplex. But some members of this ancient and diverse virus family are hard to avoid. Cytomegalovirus (CMV) infects between 50% and 80% of us before age 40. The virus is transmitted by person-to-person contact, usually causing no symptoms, though exposure is shown by western blotting with antibodies (see Chapter 12). The dangers of CMV become apparent in immunocompromised people—for example, after an organ transplant. And during pregnancy, new CMV exposure or reactivation of a latent infection can expose the fetus. CMV infection is the cause of most of the prenatal birth defects in the United States. About one of every 100 infants is born infected, and a quarter of these will have severe birth defects or developmental problems later. Birth defects may include microcephaly, neurological problems, and loss of hearing or vision.

How does CMV infection cause birth defects? Lee Fortunato at the University of Idaho, Moscow, is studying this question (**Fig. 1A**). Fortunato investigated the effect of CMV infection on human neural progenitor cells (NPCs). The NPCs

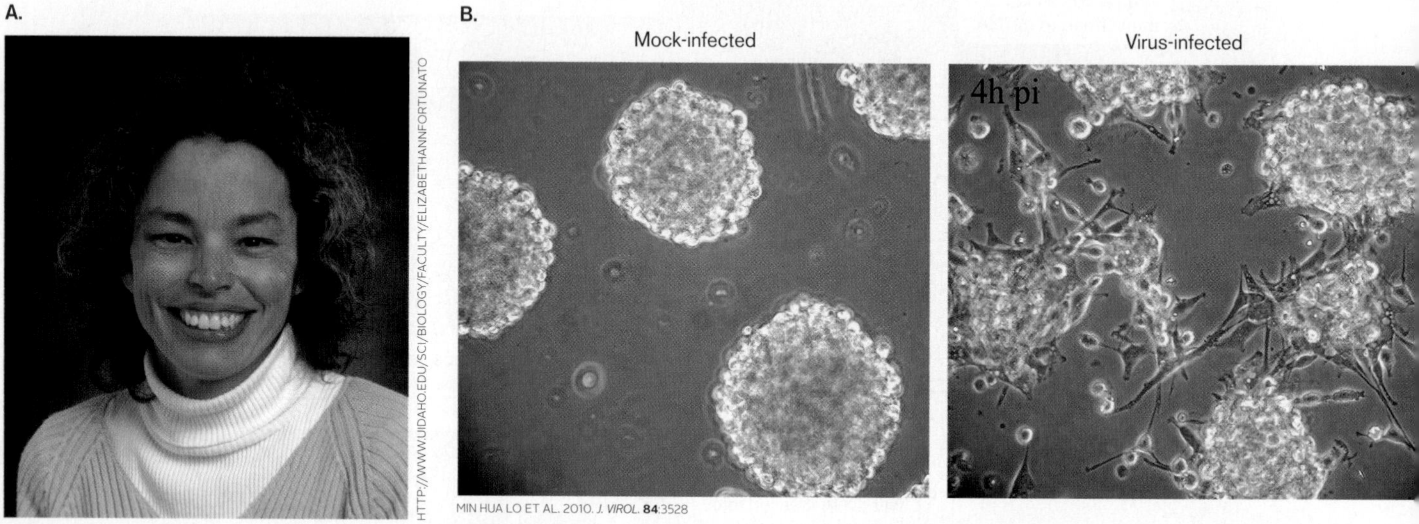
A.
B.
Mock-infected
Virus-infected
4h pi
HTTP://WWW.UIDAHO.EDU/SCI/BIOLOGY/FACULTY/ELIZABETHANNFORTUNATO
MIN HUA LO ET AL. 2010. *J. VIROL.* **84**:3528

FIGURE 1 ▪ Cytomegalovirus (CMV) alters neural development. A. Lee Fortunato studies cytomegalovirus at the University of Idaho, Moscow. **B.** Mock-infected embryonic neural progenitor cells (NPCs, left) grow as clumps of round neurospheres. NPCs infected with virus (right) form long projections and commence migration, a sign of premature differentiation.

resist infection by the bacterial pathogens *Listeria monocytogenes* (the cause of listeriosis) and *Yersinia pestis* (the cause of bubonic plague). The mechanism of antibacterial resistance may involve stimulation of the immune response. Thus, some of our silent herpes viral "partners" may have coevolved with humans in a mutually beneficial relationship, or mutualism (discussed in Chapter 20).

The maintenance of persistent or latent viral infection involves several kinds of processes that are surprisingly similar in DNA viruses such as herpes and in retroviruses such as HIV. These processes include:

- **Infection of cell types suitable for long-term persistence.** After infecting skin cells for rapid viral replication, some HSV virions infect neurons. Neurons are long-lived cells that provide the virus with an everlasting home in the host. Similarly, Epstein-Barr virus persists within long-lived memory B-cell lymphocytes.
- **Regulation of viral gene expression.** The LAT proteins suppress expression of viral genes for lytic replication, thus preventing host cell destruction where the latent viral DNA resides. Suppression occurs by assembly of "heterochromatin," chromosome-associated host proteins that inactivate all viral genes except those encoding LAT proteins.
- **Viral subversion of cellular apoptosis.** To maintain latent infection, a viral DNA must prevent host cell

were obtained from the brain of a premature infant that died of natural causes without CMV infection. When cultured in a dish, the NPCs grow as round cells in a clump, remaining in position. The NPCs can be induced to differentiate (become more advanced neural cells) by being grown in the presence of fibroblasts. But what happens when the cells are infected by CMV? Fortunato found that infected NPCs begin to differentiate prematurely (**Fig. 1B**), by extending their cytoplasm and moving across the dish.

The premature differentiation of NPCs was accompanied by abnormal changes in the expression of key genes for neural development (**Fig. 2**). CMV infection down-regulated expression of these genes: SOX2, a transcription factor needed for eye and brain development; GFAP, a regulator needed for astroglial development; NES, a cytoskeletal regulator needed for differentiation; and DCX, a neuron migration signal needed for brain development. The abnormal down-regulation of these genes was preventable by an antiviral agent, ganciclovir; and was absent in cells infected with UV-irradiated virus, which cannot replicate.

Fortunato's work shows how cytomegalovirus interferes specifically with the differentiation of embryonic cells during neonatal development. It is hoped that improved understanding of this viral process can yield insights that will help prevent future birth defects.

RESEARCH QUESTION

If CMV promotes premature differentiation, would you expect some proteins to be up-regulated, instead of down-regulated? Why? How might you test this hypothesis?

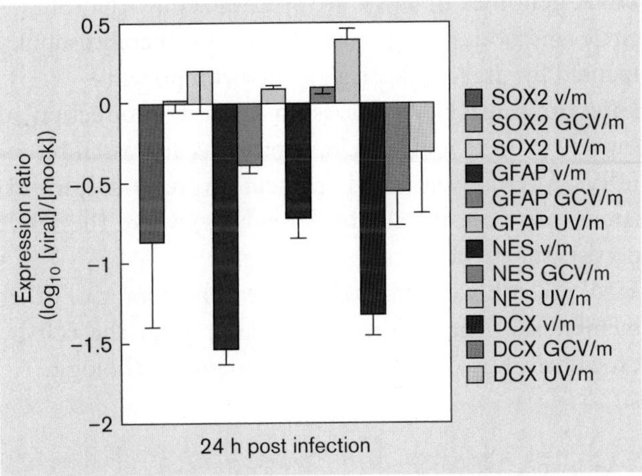
0.5
0
−0.5
−1
−1.5
−2
Expression ratio (log₁₀ [viral]/[mock])
24 h post infection
SOX2 v/m
SOX2 GCV/m
SOX2 UV/m
GFAP v/m
GFAP GCV/m
GFAP UV/m
NES v/m
NES GCV/m
NES UV/m
DCX v/m
DCX GCV/m
DCX UV/m

FIGURE 2 ▪ Viral replication down-regulates normal NPC mRNA expression. The proteins encoded by the down-regulated genes are: sex-determining homeobox 2 (SOX2), glial fibrillary acidic protein (GFAP), nestin (NES), and doublecortin (DCX). Bars represent the fold change in expression due to virus (v/m), to virus inhibited by ganciclovir (GCV/m), and to virus inactivated by UV (UV/m). *Source:* Modified from Min H. Luo et al. 2010. *J. Virol.* **84**:3528.

Luo, Min Hua, Holger Hannemann, Amit S. Kulkarni, Philip H. Schwartz, John M. O'Dowd, et al. 2010. Human cytomegalovirus infection causes premature and abnormal differentiation of human neural progenitor cells. *Journal of Virology* **84**:3528–3541.

apoptosis, a form of programmed self-destruction in response to viral infection. For example, cytomegalovirus prevents apoptosis by mimicking one host apoptosis protein and inhibiting another. Similarly, the HIV retroviral accessory protein Tat prevents apoptosis by inhibiting protein p53, which suppresses tumors through apoptosis.

- **Evasion of immune responses.** Both herpes viruses and retroviruses express proteins that inhibit signaling molecules of the immune system, or that mimic immunosuppressive signals.

The phenomena of viral persistence can actually be manipulated for human benefit by construction of viral vectors for gene transfer into a human genome. Both DNA viruses and retroviruses are adapted for exciting new forms of gene therapy (discussed in Section 11.6).

To Summarize

- **Herpes simplex virus** causes recurring eruptions of sores in the oral or genital mucosa. Initial transmission is by oral or genital contact, followed by eruptions from reactivated virus latent in ganglial neurons.
- **The herpes virion** contains a double-stranded DNA genome, packed in an icosahedral capsid. The capsid is surrounded by numerous matrix proteins and by an envelope.

- **HSV attachment** may involve several alternative receptors. A microtubular scaffold transports the herpes virions to the nucleus, where the DNA genome is inserted. The DNA circularizes for transcription.
- **DNA genomes of HSV** are synthesized by the rolling-circle method, using viral DNA polymerase supplemented by viral and host-generated components.
- **Infectious mRNA expression** leads to production of capsid, matrix, and envelope proteins for assembly of HSV. Alternatively, LAT protein expression leads to latent infection, usually in nerve cells, where the DNA persists silently for months or years.
- **HSV assembly** occurs at the nuclear membrane or other membranes. The virions are released from the cell by exocytosis. Rapid release leads to mucosal pathology.

11.6

Gene Therapy with Viruses

It may seem surprising that the properties that enable certain viruses to cause deadly diseases also make them promising candidates for **gene transfer vectors**. A gene transfer vector is a DNA sequence that can express a recombinant gene within an animal or plant cell, either from a plasmid or from a sequence integrated into the host genome. Gene transfer vectors are now commonly used for experiments probing the function of animal and plant tissues. And exciting new advances enable viral vectors to treat genetic defects.

The viruses used as gene transfer vectors are those whose replicative cycles establish a viral genome within the host nucleus. Some of the first viral vectors were made from double-stranded DNA viruses such as adenoviruses, which cause mild respiratory illness. An adenoviral genome enters the cell nucleus, where it circularizes and replicates separately from the host chromosomes, in a cycle similar to that of herpes viruses. Thus, adenoviral vectors avoid the long-term risks of inserting DNA permanently into the genome of the host cell. But the disadvantage is that the adenoviral genes are eventually lost, and thus the treatment must be repeated. In 2009 an adeno-associated viral vector was used to treat patients suffering vision loss from Leber congenital amaurosis. The vector transferred a gene encoding a light-receptor protein into the retinal cells, restoring vision.

An even more exciting class of gene transfer vectors derives from retroviruses, such as the lentivirus HIV. A **lentiviral vector**, or **lentivector**, integrate genes into a host chromosome, providing longer-lasting therapy (**Fig. 11.32** ▶). The lentiviral vector is engineered to remove viral genes that cause disease, and to express an altered envelope protein from a different virus such as vesicular stomatitis virus (VSV). The VSV envelope protein increases the viral host range and tropism, allowing treatment of a wide range of tissues. The vector contains a human transgene which becomes integrated into the host genome.

Could lentiviral integration be dangerous for the patient? Lentiviral integration into host DNA poses the danger of activating an adjacent proto-oncogene, thus causing cancer; but surprisingly, this problem has been avoided in clinical trials. Furthermore, lentiviral vectors offer a way to integrate DNA into nondividing cells of differentiated tissues such as brain neurons. For example, in 2009 a lentivector derived from HIV halted progression of a fatal brain disease, adrenoleukodystrophy (ALD), in two young boys. The lentivector inserted a gene replacing a defective gene in the boy's blood stem cells. In a different case, in 2012, the first leukemia patients were successfully treated by lentiviral gene therapy. The HIV-derived vector reprogrammed the patients' own T-cells to attack cancerous B-cells. A seven-year-old girl, Emily Whitehead (**Fig. 11.32**) was the first child to be considered fully "cured" of disease by a lentiviral vector.

Safety of Viral Vectors

The use of disease-causing viruses raises important concerns about safety. The viruses might express toxins, induce cancer, or trigger a damaging immune response. These concerns need to be weighed against the risks of the conditions they are used to treat, such as SCID, cystic fibrosis, and cancer. In general, gene therapy is approved only for conditions that are life-threatening and for which alternative therapies are inadequate.

Viruses used for gene therapy are engineered extensively to decrease risks, using techniques of genetic engineering presented in Chapter 12. Safety features of viral vectors include:

- **Deleting virulence genes.** Viral genes that promote disease and virion proliferation, but are not required for establishment of the viral DNA in the nucleus, are deleted from the vector genome. To produce the vector in tissue culture, the viral proliferation genes are provided on helper plasmids.
- **Avoiding genome insertion next to oncogenes.** Vectors engineered from adenoviruses are usually designed to avoid host chromosome integration altogether. The disadvantage of avoiding integration is that the separate

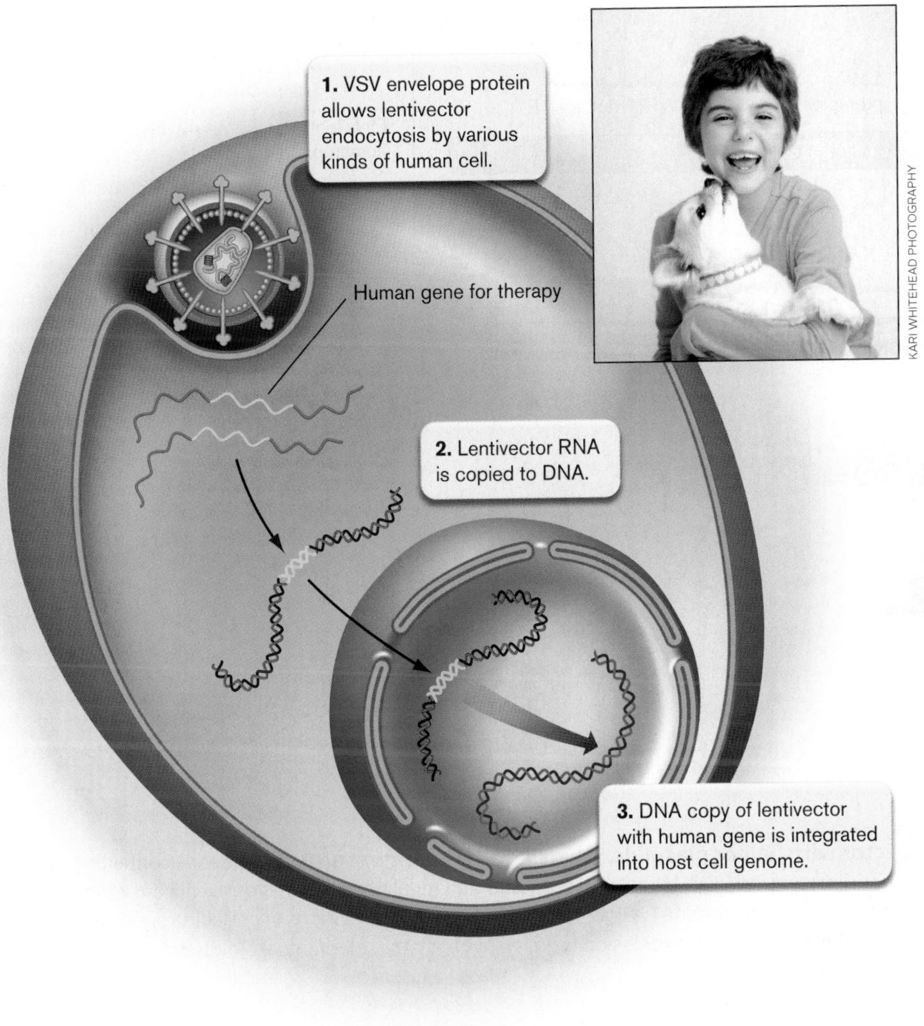
1. VSV envelope protein allows lentivector endocytosis by various kinds of human cell.
Human gene for therapy
2. Lentivector RNA is copied to DNA.
3. DNA copy of lentivector with human gene is integrated into host cell genome.
KARI WHITEHEAD PHOTOGRAPHY

FIGURE 11.32 ■ Lentiviral gene therapy. A lentiviral vector (lentivector) derived from HIV consists of virions coated with a VSV envelope protein that allows uptake by various kinds of cells. The engineered viral genome lacks disease-causing genes, but possesses a transgene needed by the patient. The vector RNA with the transgene becomes copied into DNA and integrated into a host chromosome. Emily Whitehead (inset) was the first child to be considered cured of an illness (B-cell leukemia) by a lentiviral vector.

viral DNA is soon lost from host tissues, and the therapy requires frequent repetition. In lentiviruses, current research aims to integrate the genome far from the site of proto-oncogenes.

- **Altering tissue specificity.** The tropism, or tissue specificity, can be altered by replacing the gene for the viral envelope glycoprotein (spike protein) with the envelope gene from a different virus. For example, a rabies virus glycoprotein can be used to target the vector to brain cells. The alteration of viral tissue specificity by envelope gene replacement is called "pseudotyping." Pseudotyping can be used either to narrow the host range or to broaden it, depending on the needs of the vector.
- **Avoiding germ-line infection.** Current medical standards prohibit alteration of the germ line, the egg and sperm cells that transmit genes to the next generation. Because the long-term risks of gene therapy are unknown, only somatic gene therapy (gene insertion into somatic, or body, cells) is permitted.

How a Lentiviral Vector Works

To construct a safe and effective vector, the HIV genome is modified extensively (**Fig. 11.33**). In the example shown, accessory genes *vpr, vpu, nef,* and *vif,* which encode HIV virulence factors for disease, were removed. Other protein-encoding genes necessary for virion production were put into DNA helper plasmids, to be provided only in tissue culture for vector production. These genes provide the capsid monomer (*gag*), reverse transcriptase (*pol*), envelope glycoprotein (*env*), and a regulator of mRNA export from the nucleus (*rev*). The HIV *env* gene is replaced by an *env* gene from another virus, vesicular stomatitis virus (VSV). The VSV envelope protein has a broad tropism (host range) and thus enables the lentivector to infect a broad range of host cell types. Each helper plasmid drives its gene expression from a well-studied promoter of another virus, such as cytomegalovirus (CMV) or respiratory syncytial virus (RSV). The original HIV vector genome retains only the LTR end sequences (R-U5) required for genome

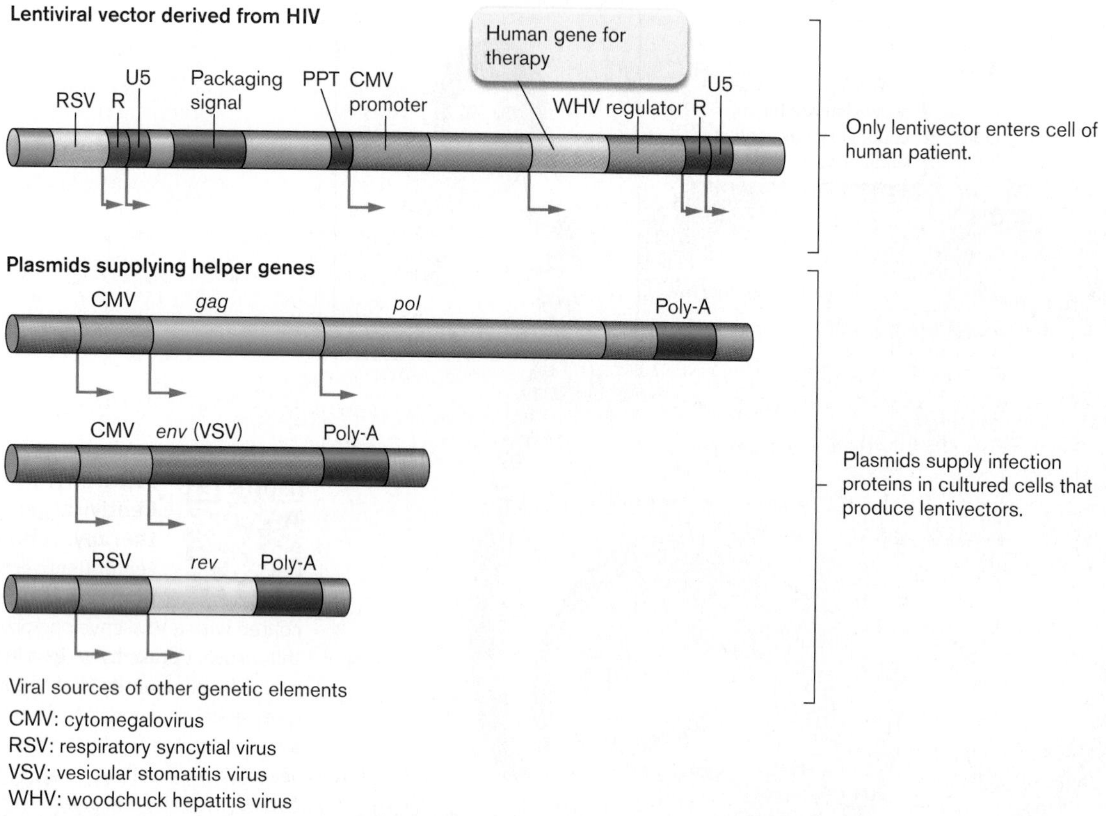
Lentiviral vector derived from HIV
Human gene for therapy
RSV
R
U5
Packaging signal
PPT
CMV promoter
WHV regulator
R
U5
Only lentivector enters cell of human patient.
Plasmids supplying helper genes
CMV
gag
pol
Poly-A
CMV
env (VSV)
Poly-A
RSV
rev
Poly-A
Plasmids supply infection proteins in cultured cells that produce lentivectors.
Viral sources of other genetic elements
CMV: cytomegalovirus
RSV: respiratory syncytial virus
VSV: vesicular stomatitis virus
WHV: woodchuck hepatitis virus

FIGURE 11.33 ■ Lentiviral vector with helper plasmids. The lentivector consists of an RNA sequence containing HIV signal elements required for genomic integration (dark blue), with promoter and regulator elements derived from various other viruses (red) and the therapeutic human transgene (yellow). In order to produce the virions in cell culture, essential HIV genes are provided on DNA helper plasmids. Their expression is driven by regulatory elements from other viruses (red). *Source:* Modified from Blesch. 2003. *Methods* **33**:164.

integration, a packaging signal from the start of *gag*, and the infectivity-enhancing polypurine tract (PPT).

To express the **transgene** (the gene of interest for expression in the host), a CMV promoter was inserted in the HIV-derived RNA vector. To further enhance transgene expression, a genetic enhancer sequence was added from woodchuck hepatitis virus (WHV). Note that a "lentiviral" gene transfer system, in fact, includes genetic elements from a diverse set of human and animal viruses, all found in previous studies to contribute specific properties to infection. Although these genes originate from different viruses, they nevertheless function together like the parts of a machine.

To produce infective virions, the HIV-derived RNA vector plus the three helper plasmids are introduced into a special tissue culture line. The vector and plasmids enter the cells by a process, called **transfection**, in which calcium phosphate treatment promotes the uptake of nucleic acids across the cell membrane. The host tissue culture cells are derived from human embryonic kidney 293T cells containing a gene for a protein from simian virus 40 (SV40) that enables the replication of DNA plasmids containing an SV40 replication site. The 293T cells allow efficient expression of the viral genes on the helper plasmids, as well as a full virion production cycle in which the vector RNA is replicated and packaged into virions.

The integration of the vector into neurons was demonstrated by infecting rats in the hippocampus, a part of the brain important for long-term memory and spatial navigation. In **Figure 11.34** the transgene of the vector expresses green fluorescent protein (GFP). Neurons are identified by a red fluorescent antibody to a neuronal protein, whereas cells expressing GFP from the transgene are labeled green. Neurons expressing both the neuronal protein and the transgene show the combined color, yellow-green.

The successful introduction of the fluorescent transgene led to another experiment, in which the vector introduced nerve growth factor into a rat hippocampus with nerve damage. Nerve growth factor is a protein capable of regenerating damaged or atrophied brain tissue. The nerve growth factor was expressed successfully, and it enhanced the regeneration of nerves. Subsequently, a clinical trial was

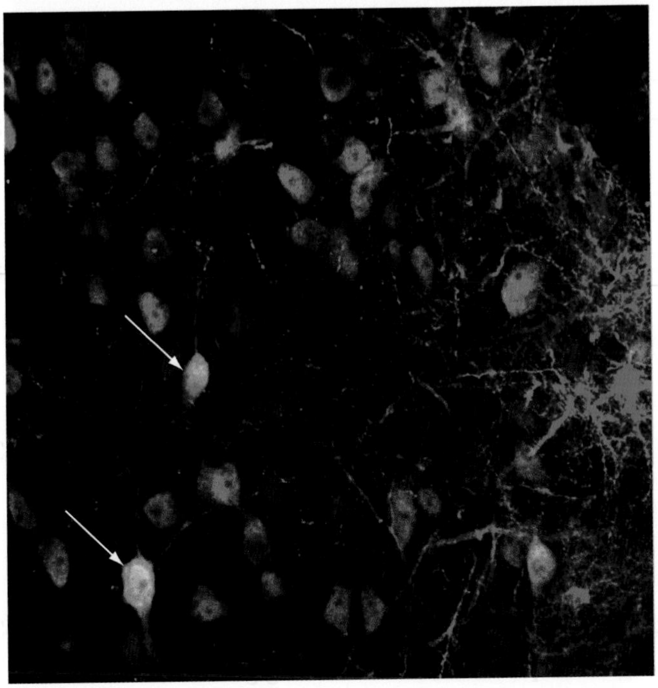

BLESCH ET AL. 2005. *MOL. THER.* **11**:916, FIG. 2B.

FIGURE 11.34 ■ Neurons expressing reporter protein from a lentiviral vector. Rat hippocampal neurons were double-labeled with an antibody binding all neurons (red fluorescence) and an antibody binding the lentiviral reporter protein (green fluorescence). Arrows indicate transformed neurons (yellow-green). Neuron cell bodies are 10–20 μm in length.

begun in humans. The trial involves NGF gene transfer in eight patients with Alzheimer's disease. The aim is to test the safety and effectiveness of lentiviral NGF gene transfer to promote nerve regeneration (nerve degeneration is associated with the disease).

Viruses for Cancer Therapy

A very different potential application of viral gene therapy, still highly experimental, is the use of viruses to destroy tumors. Certain naturally occurring viruses are "oncolytic"; that is, they infect and lyse tumor cells preferentially over normal cells. In this case, viral proliferation is advantageous, as the progeny virions spread to neighboring tumor cells, thus amplifying the effect of treatment. Research has focused on adenoviruses, which have been engineered to target cancers such as glioma (a brain tumor) and breast cancer.

A different virus called reovirus was shown to selectively infect tumor cells. Reovirus is a double-stranded RNA virus that causes mild respiratory infections. The virus replicates to an exceptional level in cancer cells containing mutations in the Ras growth control pathway. In 2011, a form of reovirus called Reolysin, developed by the company Oncolytics Biotech, was introduced to treat head and neck cancers. The treatment entered phase 3 clinical trials. In combination with chemotherapy, Reolysin shows promise as a means of selectively eliminating tumors.

To Summarize

- **Gene transfer** vectors are made from viruses.
- **Virulence genes are deleted**, and efficient promoter sequences from other viruses are inserted.
- **Adenoviral vectors circularize** and replicate separately from the host genome.
- **Lentiviral vectors integrate** within the host genome.
- **Oncolytic viruses selectively infect tumor cells**, providing a way to eliminate cancer.

Concluding Thoughts

In considering the mechanisms of viral infection, a picture emerges of interesting commonalities, as well as intriguing diversity:

- **Viral infection requires host membrane receptors.** Different viruses evolve to take advantage of different cell membrane proteins in order to recognize and gain entry into host cells. Mutation of host receptors confers resistance to the virus.
- **Viruses recruit host cells to replicate their genomes.** Viral genomes are remarkably diverse, including RNA or DNA, single- or double-stranded, linear or circular. Within the host cell, replication proceeds by a variety of methods, with varying dependence on viral and host enzymes.
- **Virus particles (and their genomes) range in size from a few components to assemblages approaching the complexity of cells.** The advantage of simplicity is the minimal requirement for resources. The advantage of complexity is the fine-tuning of infection mechanisms for evading host defenses.

The study of viral molecular biology raises intriguing questions about the nature of viral replication and about viral origins (see **eTopic 6.1**). Research in virology offers hope for new drugs and cures for humanity's worst plagues, as well as devastating diseases of agricultural plants and animals. At the same time, it is sobering to note that despite the enormous volumes we now know about viruses such as HIV and influenza, the AIDS pandemic continues to grow, and we face the likely emergence of a new deadly flu strain. Molecular research can succeed only in partnership with epidemiology and public health (discussed in Chapter 28).

CHAPTER REVIEW

Recommended Reading

Barton, Erik S., Douglas W. White, Jason S. Cathelyn, Kelly A. Brett-McClellan, Michael Engle, et al. 2007. Herpesvirus latency confers symbiotic protection from bacterial infection. *Nature* 447:326–329.

Blesch, Armin, James Conner, Alexander Pfeifer, Mehdi Gasmi, Anthony Ramirez, et al. 2005. Regulated lentiviral NGF gene transfer controls rescue of medial septal cholinergic neurons. *Molecular Therapy* 11:916–[illegible].

Cartier, Nathalie, Salima Hacein-Bey-Abina, Cynthia C. Bartholomae, Gabor Veres, Manfred Schmidt, et al. 2009. Hematopoietic stem cell gene therapy with a lentiviral vector in X-linked adrenoleukodystrophy. *Science* 326:818–823.

Cockrell, Adam S., and Tal Kafri. 2007. Gene delivery by lentivirus vectors. *Molecular Biotechnology* 36:184–204.

Comeau, André M., and Henry M. Krisch. 2008. The capsid of the T4 phage superfamily: The evolution, diversity, and structure of some of the most prevalent proteins in the biosphere. *Molecular Biology and Evolution* 25:1321–1332.

Engelman, Alan, and Peter Cherepanov. 2012. The structural biology of HIV-1: Mechanistic and therapeutic insights. *Nature Reviews Microbiology* 10:279–290.

Frampton, A. R., Jr., W. F. Goins, K. Nakano, E. A. Burton, and J. C. Glorioso. 2005. HSV trafficking and development of gene therapy vectors with applications in the nervous system. *Gene Therapy* 12:891–901.

Heeney, Jonathan L., Angus G. Dalgleish, and Robin A. Weiss. 2006. Origins of HIV and the evolution of resistance to AIDS. *Science* 313:462–466.

Kane, Melissa, and Tatyana Golovkina. 2010. Common threads in persistent viral infections. *Journal of Virology* 84:4116–4123.

Lanford, Robert E., Elisabeth S. Hildebrandt-Eriksen, Andreas Petri, Robert Persson, Morten Lindow, et al. 2010. Therapeutic silencing of microRNA-122 in primates with chronic hepatitis C virus infection. *Science* 327:198–201.

Marcato, Paola, Cheryl A. Dean, Carman A. Giacomantonio, and Patrick W. K. Lee. 2009. Oncolytic reovirus effectively targets breast cancer stem cells. *Molecular Therapy* 17:972–979.

Moradpour, Darius, François Penin, and Charles M. Rice. 2007. Replication of hepatitis C virus. *Nature Reviews Microbiology* 5:453–463.

Raoult, Didier, Stéphane Audic, Catherine Robert, Chantal Abergel, Patricia Renesto, et al. 2004. The 1.2-megabase genome sequence of mimivirus. *Science* 306:1344–1350.

Roan, Nadia R., Jan A. Müller, Haichuan Liu, Simon Chu, Franziska Arnold, et al. 2011. Peptides released by physiological cleavage of semen coagulum proteins form amyloids that enhance HIV infection. *Cell Host & Microbe* 10:541–550.

Rossmann, Michael G., Vadim V. Mesyanzhinov, Fumio Arisaka, and Petr G. Leiman. 2004. The bacteriophage T4 DNA injection machine. *Current Opinion in Structural Biology* 14:171–180.

Smith, Gavin J. D., Dhanasekaran Vijaykrishna, Justin Bahl, Samantha J. Lycett, Michael Worobey, et al. 2009. Origins and evolutionary genomics of the 2009 swine-origin H1N1 influenza A epidemic. *Nature* 459:1122–1126.

Stevens, James, Ola Blixt, Terrence M. Tumpey, Jeffery K. Taubenberger, James C. Paulson, et al. 2006. Structure and receptor specificity of the hemagglutinin from an H5N1 influenza virus. *Science* 312:404–410.

CHAPTER 12

Biotechniques and Synthetic Biology

12.1 Basic Tools of Biotech: A Research Case Study

12.2 Genetic Analyses

12.3 Classic Molecular Techniques

12.4 Viewing the Interactions and Movements of Proteins

12.5 Applied Biotechnology

12.6 Synthetic Biology: Biology by Design

The science of biotechnology uses living organisms or their products to improve human health or to perform specific industrial or manufacturing processes. Achieving these goals, however, requires the constant invention of new technologies. Chapter 12 describes these technologies and explores how they are used.

Although we think of biotechnology as a new field, the earliest biotechnologists actually lived about 10,000 years ago, when our ancestors unwittingly figured out how to use yeast to make alcohol, bread, and dairy products (discussed in Chapter 16). The field changed little over the millennia, until 1928, when the Scotsman Alexander Fleming discovered that microbes produce antibiotics. That watershed event, along with unraveling the structure of DNA and deciphering the genetic code, ushered in the high-tech era of biotechnology. We have since learned a great deal about the molecular biology of the microbes our ancestors used. Now, through genetic engineering and synthetic biology, we can make bacteria produce hormones, engineer vaccines in plants, devise previously unimagined biochemical pathways, and design bacteria that detect explosives and flash a warning.

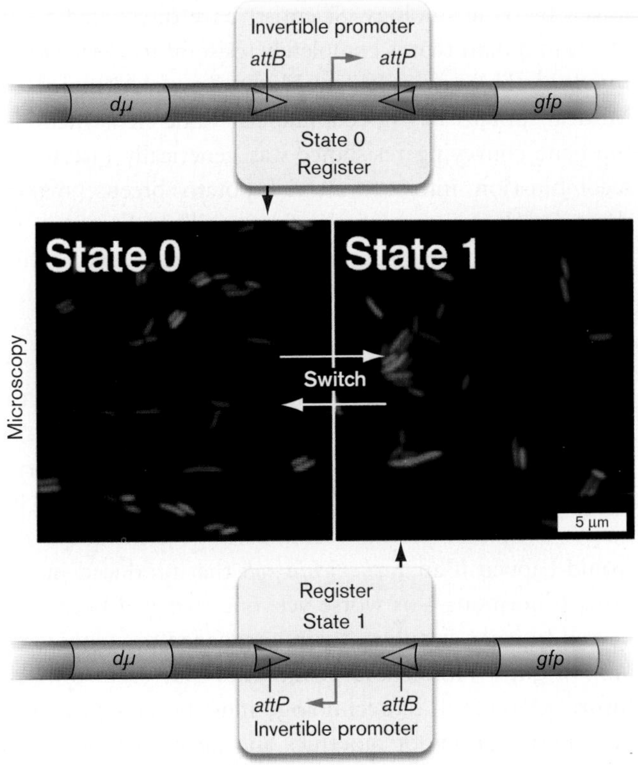

Invertible promoter
attB
attP
dμ
gfp
State 0
Register
State 0
State 1
Switch
Microscopy
5 μm
Register
State 1
dμ
gfp
attP
attB
Invertible promoter

CURRENT RESEARCH **highlight**

Bacterial computers. Synthetic biology is a science that conceives and builds biological systems that do not exist in nature. Scientists link bits and pieces of DNA plucked from the microbial realm to produce novel, chimeric gene circuits or biochemical pathways. Shown is an ***Escherichia coli,*** engineered by Drew Endy of Stanford University, that stores digital information (like a computer) within its chromosome. An invertible promoter was placed between genes for green and red fluorescent proteins and then chemically commanded to switch one way or the other. After the chemical switch signal was removed, the state of the reporter (red or green) remained stable, meaning the information was stored. The fluorescent micrographs show the two distinguishable states for this invertible data register. This bacterial molecular circuit is the first reliable and rewritable DNA inversion-based data storage system. *Source:* From Bonnet et al. 2012. *PNAS* **109**:8884–8889.

Today's potato farmers must spray huge amounts of antifungal agents on their crops to combat potato blight disease, the cause of the Irish potato famine that wiped out 30% of the Irish population in the nineteenth century. Spraying plants as many as 25 times a season with chemicals, the typical U.S. potato farmer spends about $250 per acre to fight the disease—a huge incentive to find less expensive means of prevention. But the problem may be solved by biotechnology. Scientists have discovered a wild Mexican potato that is completely resistant to *Phytophthora infestans,* the cause of potato blight disease. Although the Mexican potato has no commercial value (it is inedible), the gene conveying resistance was genetically inserted by recombination into commercial potato breeds, making them utterly resistant to the fungus. This discovery will eventually improve the health of potato crops (use of these modified potatoes is nearing approval in England), reduce our exposure to currently used antifungal chemicals, and contribute to our understanding of disease resistance in plants. This is but one example of how biotechnology can potentially improve human existence.

Of course, with technological power comes a sobering responsibility, for if nature has taught us anything, it has taught us that tampering with it is risky. For instance, what would happen if an *Escherichia coli* that produced human growth hormone—or worse yet, one that had been engineered to make botulism toxin—managed to colonize the human gut? While these possibilities may seem unlikely, future advances in biotechnology must be guided by serious considerations of bioethics and an eye toward unintended consequences.

We begin our discussion of biotechnology by explaining the techniques of modern molecular biology used to probe the inner workings of microbes. A research case history illustrates how many of these techniques are applied to a single problem. Other applications and the exciting new field of synthetic biology are discussed at the end of the chapter.

12.1

Basic Tools of Biotech: A Research Case Study

Biotechnology encompasses numerous molecular techniques developed over decades. We need to understand these techniques to fully appreciate the potential of biotechnology. Many techniques, such as DNA sequencing and proteomics, are described throughout the book (**Table 12.1**). We will repeatedly refer you to these sections as we describe more cutting-edge techniques in Chapter 12. Here we present many emerging molecular tools being used to answer remaining mysteries of basic biology. However, we begin with some classic molecular techniques not yet covered and show how they were used to expose the mechanisms by which *E. coli* survives in the highly acidic conditions of the human stomach.

Few microbes that grow at neutral pH can survive at pH 2, equivalent to 10-mM HCl, the degree of acidity present in the human stomach. Such a highly acidic environment kills most ingested pathogens, but strains of the intestinal bacterium *E. coli,* such as enterohemorrhagic *E. coli* O157:H7, can survive at pH 2 for hours. This means that a person needs to ingest only a few cells of *E. coli* to become colonized or infected. Thus, the organism is said to have a low infectious dose. What makes *E. coli* so special? The molecular tools that we describe here helped reveal the novel mechanisms of acid resistance used by this microbe and the regulatory networks that control them. Our story begins with the discovery of acid-sensitive *E. coli* mutants that failed to survive at pH 2.

12.2

Genetic Analyses

Transposons Inactivate and Mark Target Genes

As just noted, *E. coli* has a remarkable system of acid resistance. How were genes needed for acid survival of this organism found and their function discovered? An early step taken to gain insight into *E. coli* acid resistance involved isolating mutants defective in that process. The initial goal was to inactivate relevant genes and tag them with an antibiotic resistance marker. To do this, a gene whose product confers resistance to an antibiotic (that is, an antibiotic resistance gene) was randomly inserted into the chromosome. Although we ordinarily think of pathogens as having antibiotic resistance genes (to escape antibiotic therapy), microbiologists have found ways to move these genes into a variety of nonpathogenic bacteria, such as laboratory strains of *E. coli.* When the antibiotic resistance gene inserts into a target gene sequence, the target gene is inactivated and the mutant cell becomes antibiotic resistant.

To generate mutations in genes needed to confer acid resistance, researchers infected a population of *E. coli* with

TABLE 12.1

Techniques described elsewhere in this book.

Molecular technique	Designed use	Location in book
Electrophoresis	To separate DNA, RNA, protein or other molecules based on size or charge	Appendix 3
Fluorescence in situ hybridization (FISH)	To identify specific genes or RNA targets in a permeabilized cell	Chapter 18; Appendix 3
Polymerase chain reaction (PCR)	To amplify specific segments of the genome sequence	Chapter 7
Real-time PCR amplification	To quantify specific nucleic acids (DNA or RNA)	Chapter 12
Restriction digestion	To excise sections of DNA that can be cloned into plasmids or other vectors	Chapter 7
Gene cloning (recombinant DNA)	To make a recombinant DNA molecule (by splicing DNA from one organism into DNA vectors) that can be introduced into a different organism	Chapter 7
DNA sequencing	To read the order of deoxynucleotides in a DNA molecule	Chapter 7
Metagenomics	To use DNA sequencing to discover, identify, and characterize members of a mixed population of microbes in their natural environment	Chapter 7
Transcriptomics	To quantify the environmentally induced expression of RNA molecules from all genes in a genome	Chapter 10
Microarrays	To quantify changes in gene expression using hybridization technology	Chapter 10; Appendix 3
Deep sequencing	To repeatedly and rapidly sequence short segments of DNA or RNA to reconstruct a microbial genome or to quantify transcripts	Chapter 7 (DNA seq) and Chapter 10 (RNA seq)
Proteomics	To quantify the environmentally induced levels of all proteins in a cell	Chapter 10
SDS polyacrylamide gels	To separate proteins according to their molecular size	Chapter 3
Isoelectric focusing	To separate proteins according to their charge	Chapter 10; Appendix 3
Two-dimensional gels	To separate proteins according to charge and size	Chapter 10; Appendix 3
Mass spectrometry	To rapidly determine a protein's mass and (by MS-MS) to determine its amino acid sequence	Chapter 10; Appendix 3
Molecular databases	To search known DNA or protein sequences for similarities to newly sequenced molecules	Chapter 8
Site-directed mutagenesis	To target specific amino acids in a protein for mutagenesis	eTopic 12.7

bacteriophages containing a tetracycline resistance transposon (see Section 9.6). The bacteriophage was a mutant that fails to replicate in most strains of *E. coli,* so for an infected cell to become tetracycline resistant, the transposon had to "hop," or transpose, out of the phage DNA and into the host chromosome. Because target sequences are randomly selected during transposition, each tetracycline-resistant cell contained a transposon randomly inserted into a different gene. As each cell grew into a colony, the colony was picked with a sterile toothpick and transferred into a well of a microtiter plate containing growth medium (**Fig. 12.1A**).

After about 10,000 colonies were transferred to microtiter dishes and grown, a multipronged replicator device (a square with 48 metal prongs arranged in six rows of eight) was used to transfer small amounts of each culture to new microtiter plate wells that contained a pH 2 medium. After several hours of incubation, the multipronged replicator was again used to sterilely transfer a sample of the pH 2 culture from each well to an agar plate. Most of the insertion mutants survived at pH 2 and were able to form colonies on agar. These insertions occurred in genes unimportant for acid resistance. A few mutants, however, were killed. The transposon in these acid-sensitive mutants had inserted into a gene required for acid resistance, rendering it inactive. The acid-sensitive mutants were retrieved from the original growth medium microtiter plates and subjected to further analysis.

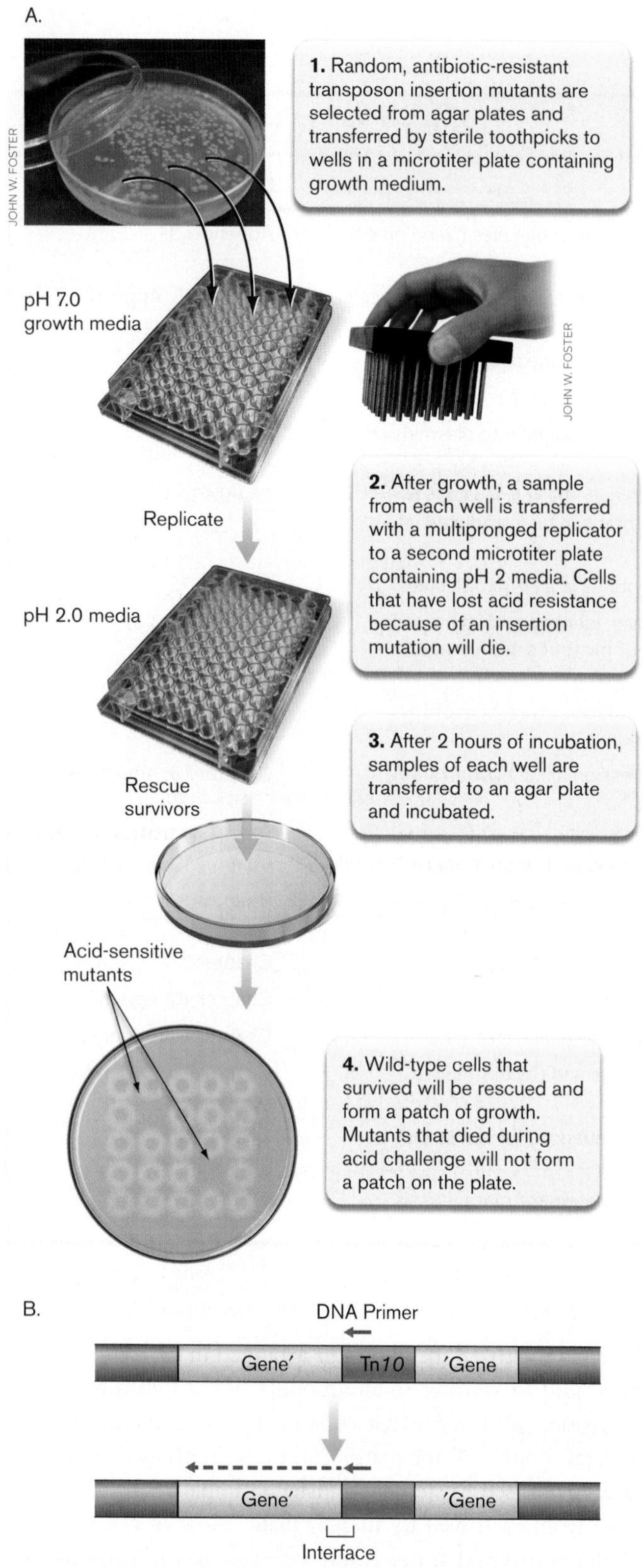
A.
JOHN W. FOSTER
1. Random, antibiotic-resistant transposon insertion mutants are selected from agar plates and transferred by sterile toothpicks to wells in a microtiter plate containing growth medium.
pH 7.0 growth media
JOHN W. FOSTER
Replicate
2. After growth, a sample from each well is transferred with a multipronged replicator to a second microtiter plate containing pH 2 media. Cells that have lost acid resistance because of an insertion mutation will die.
pH 2.0 media
3. After 2 hours of incubation, samples of each well are transferred to an agar plate and incubated.
Rescue survivors
Acid-sensitive mutants
4. Wild-type cells that survived will be rescued and form a patch of growth. Mutants that died during acid challenge will not form a patch on the plate.
B.
DNA Primer
Gene′
Tn10
′Gene
Gene′
′Gene
Interface

FIGURE 12.1 ■ Selection for acid-sensitive mutants. **A.** Mutagenesis with transposons led to the discovery of several *E. coli* genes involved in acid resistance. The technique is applicable to many other biochemical systems. **B.** Strategy to identify the gene target of a transposon insertion. A primer that anneals to a sequence in the transposon is used to amplify the junction between the transposon and the gene into which it has inserted. Subsequent sequencing of the product identifies the gene and the insertion point.

Thought Question

12.1 How would you modify the technique illustrated in **Figure 12.1** to identify mutants defective in the synthesis of an amino acid, such as glutamic acid? *Hint:* You can use a minimal agar medium.

Thought Question

12.2 In this postgenomic era, how might you generally design a strategy to construct a strain of *E. coli* that cannot synthesize histidine?

DNA Sequencing of Insertion Sites and Computer Analysis

Once mutants in a project are identified, it is important to know in which genes the insertions occurred. In our example, an oligonucleotide primer (a short, single-stranded DNA molecule) that anneals to one end of the transposon was used to sequence across the insertion joint and into the adjacent *E. coli* DNA (**Fig. 12.1B**). The DNA sequences obtained were subjected to computer-based homology searches with the sequenced genome of *E. coli*. The matches obtained indicated that insertions had occurred in four genes. Two of these genes encode isoforms of glutamate decarboxylase (GadA and GadB; the isoforms differ by only a few amino acids), which converts glutamic acid to gamma-aminobutyric acid (γ-aminobutyric acid), known as GABA. A third gene produces a glutamate/GABA antiporter (GadC), and a fourth gene encodes a putative DNA-binding protein with no known function, although it was annotated as a possible regulator.

The nature of the genes' functions suggests how the system might protect cells submerged in acid (**Fig. 12.2**). When *E. coli* is placed in a pH 2 environment, the cytoplasmic pH falls to dangerously low levels. Glutamate decarboxylase removes a carboxyl group from glutamate and replaces it with a proton from the cytoplasm, thereby decreasing acidity and elevating cytoplasmic pH. But to continue draining protons from the cytoplasm, *E. coli* must bring in more glutamate. The glutamate/GABA antiporter does this by bringing in glutamate in exchange for expelling GABA, the decarboxylation end product.

Exploring Gene Regulation: Reporter Fusions

Another important question in our research example was whether acid resistance genes are regulated in response to changes in the environment. Are they always expressed,

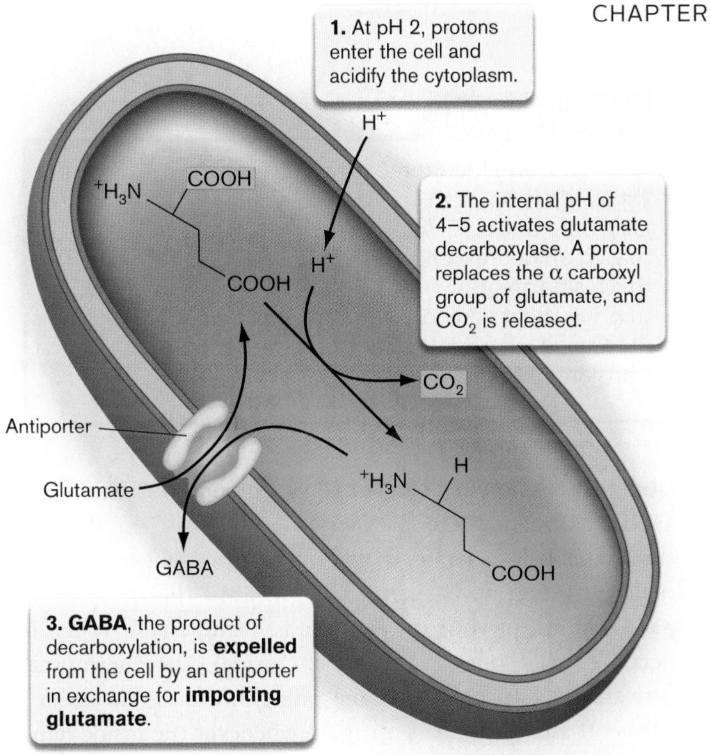

1. At pH 2, protons enter the cell and acidify the cytoplasm.
H^+
^+H_3N
COOH
COOH
H^+
2. The internal pH of 4–5 activates glutamate decarboxylase. A proton replaces the α carboxyl group of glutamate, and CO_2 is released.
CO_2
Antiporter
Glutamate
^+H_3N
H
COOH
GABA
3. GABA, the product of decarboxylation, is expelled from the cell by an antiporter in exchange for importing glutamate.

FIGURE 12.2 ■ Proposed model of acid resistance in *E. coli*. DNA sequence data identified the location of the transposon insertions leading to acid sensitivity. The genes identified included two isoforms of glutamate decarboxylase and a putative antiporter that brings glutamate into the cell in exchange for the decarboxylation product gamma-aminobutyric acid (GABA), which is transported out of the cell. The combined action of the decarboxylase and antiporter removes H^+ from the cell.

and is the cell ever-ready to encounter acid stress? Or does the cell sense when it might encounter life-threatening low pH and only then express acid resistance genes? Several experimental strategies can test whether a given gene is regulated in response to changes in the environment. Recall that the cell can control production of a protein at several levels. The cell can regulate transcription, translation, and the stability of the mRNA or protein (processes described in Chapter 8). How do microbiologists differentiate among these possibilities or determine whether there is any regulation in the first place?

One technique used to examine gene regulation involves linking (or fusing) the promoter of the gene of interest to what is called a **reporter gene**, such as *lacZ* (beta-galactosidase) or *gfp* (GFP, green fluorescent protein; Chapter 2). Reporter genes encode proteins whose activities are easily assayed. The result is called a reporter fusion. Reporter fusions facilitate monitoring the expression of genes whose products are difficult to measure directly. In the case of *lacZ*, beta-galactosidase (β-galactosidase) converts the substrate *o*-nitrophenyl galactoside (ONPG) to a yellow product that is easily measured with a spectrophotometer. Fusions to GFP are detected by a fluorometer, which detects fluorescence.

Operon vs. gene fusions. Can reporter genes distinguish between transcriptional and translational control? There are two general types of reporter fusions that can make this distinction in bacteria—namely, **operon fusions** and **gene fusions**. Operon fusions reveal only transcriptional control of the target gene, whereas gene fusions expose both transcriptional and translational controls. Both types of fusions involve linking a promoter-less reporter gene to the 3′ end of a target gene that still possesses its promoter (**Fig. 12.3**). The result is an artificial operon. Because there is only one promoter, only one mRNA transcript is made encompassing the target and reporter gene sequences. The 5′ end of the newly engineered polycistronic mRNA will contain the target gene transcript followed at the 3′ end by the reporter gene transcript. When the reporter gene is inserted into the middle of the target gene, as shown in **Figure 12.3B**, the downstream part of the target gene is not transcribed, because of a transcriptional terminator located at the end of the reporter.

In both types of fusions, any factor that controls expression of the target gene promoter will also control production of the reporter. The difference between operon fusions (also known as **transcriptional fusions**) and gene fusions (also known as protein or **translational fusions**) is whether the reporter gene carries its own ribosome-binding site (RBS). The reporter gene in an operon fusion contains an RBS, and for this reason the reporter's message is translated independently of the target gene's message. As a result, operon fusions reflect only transcriptional control at the promoter, not control at the level of translation.

In gene fusions, the reporter is missing both its promoter and its ribosome-binding site. So, for the reporter protein to be synthesized, the reporter gene sequence must be joined to the target gene sequence in the proper codon reading frame (**Fig. 12.3C**). Now, not only will the mRNA messages from the target and reporter genes be fused, but the peptides of the truncated target gene and the reporter gene will also be linked. In this case, anything controlling the transcription or translation of the target gene will also control reporter protein levels.

Each type of fusion can be constructed in vitro using plasmids and then transferred into recipient cells in a way that promotes the exchange of the reporter fusion for the resident gene. Once constructed, these fusions allow the researcher to monitor how different growth conditions, or whether suspected regulators, influence the transcription or translation of the gene. These fusions provide convenient phenotypes (for example, lactose fermentation or fluorescence) that can be exploited in mutant hunts to screen for new regulators. Appendix 3 describes how *lacZ* fusions can be used to find regulatory genes.

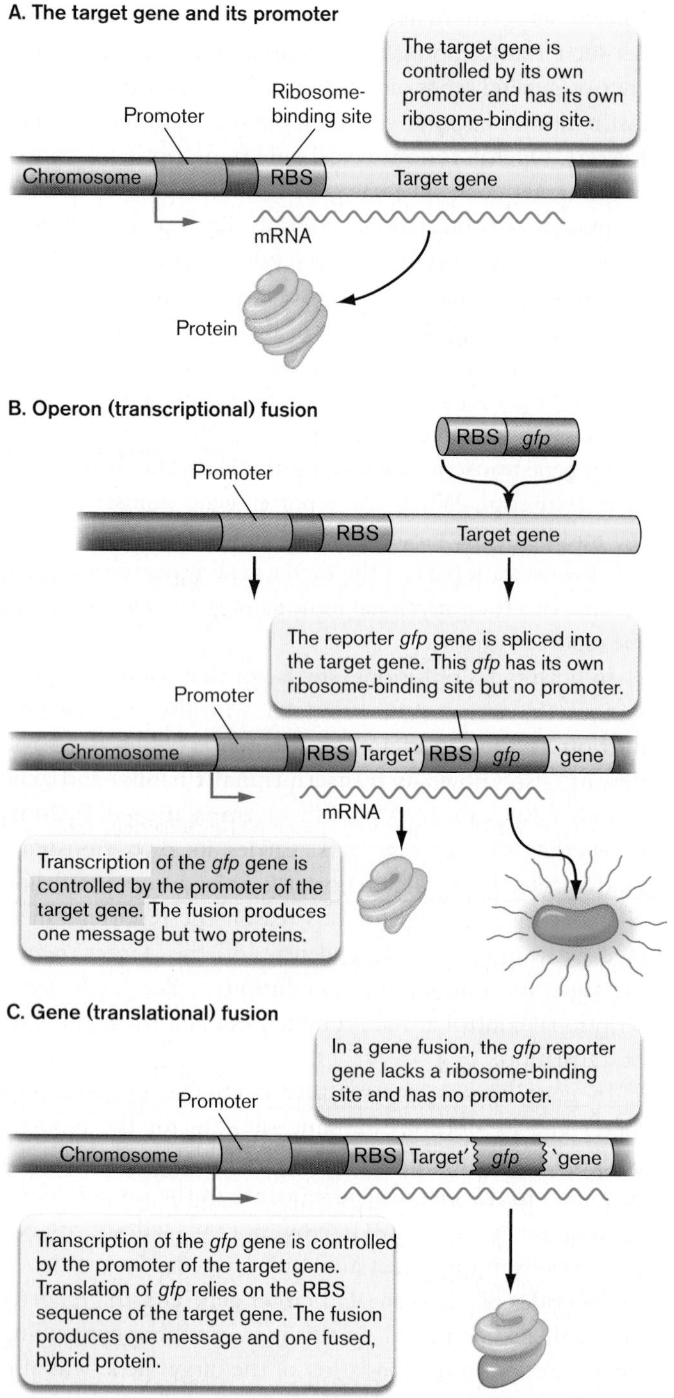
A. The target gene and its promoter
The target gene is controlled by its own promoter and has its own ribosome-binding site.
Promoter
Ribosome-binding site
Chromosome
RBS
Target gene
mRNA
Protein
B. Operon (transcriptional) fusion
RBS gfp
Promoter
RBS
Target gene
The reporter gfp gene is spliced into the target gene. This gfp has its own ribosome-binding site but no promoter.
Promoter
Chromosome
RBS Target′ RBS gfp ′gene
mRNA
Transcription of the gfp gene is controlled by the promoter of the target gene. The fusion produces one message but two proteins.
C. Gene (translational) fusion
In a gene fusion, the gfp reporter gene lacks a ribosome-binding site and has no promoter.
Promoter
Chromosome
RBS Target′ gfp ′gene
Transcription of the gfp gene is controlled by the promoter of the target gene. Translation of gfp relies on the RBS sequence of the target gene. The fusion produces one message and one fused, hybrid protein.

FIGURE 12.3 ▪ Transcriptional (operon) and translational (gene) fusions. Aspects of a gene's regulation can be monitored by fusing a second gene, one that produces an easily assayed protein, to the gene under study. **A.** The gene targeted for construction of a reporter fusion. **B.** Construction of an operon fusion. **C.** Construction of a gene fusion. The *gfp* open reading frame must be fused in-frame to the amino-terminal end of the target gene.

Thought Question

12.3 You have monitored expression of a gene fusion in which *gfp* is fused to *gadA*, the glutamate decarboxylase gene. You find a 50-fold increase in expression of *gadA* (based on fluorescence level) when the cells carrying this fusion are grown in media at pH 5.5 as compared to pH 8. What additional experiments must you do to determine whether the control is transcriptional or translational?

Gene expression in a single cell vs. the population. When a population of bacteria experiences a stress, do all cells simultaneously activate their stress response genes? And once activated, do the genes stay on until the stress dissipates? Think of your own house as a single cell. When it becomes cold outside (and then inside), the thermostat activates your heater. At some point, however, the heater turns off and then turns on again whenever the inside temperature dips. Now imagine looking down at a neighborhood of houses. You see that the heater in one house turns on as another house's heater turns off. The result appears to be a stochastic (random) process of heaters being turned on and off, but the population as a whole stays warm.

Do bacterial cells respond to stress in the same way? What technique could even address this question in single cells? LacZ fusions are inadequate because LacZ enzyme activity is measured as an average of all cells in the population. However, green fluorescent protein (GFP), or derivatives such as yellow fluorescent protein (YFP), fused to a gene do allow scientists to view (in real time) what happens inside single cells within a population. Realizing this, Michael Elowitz and colleagues from the California Institute of Technology used a YFP fusion to elegantly reveal stochastic gene expression in *Bacillus subtilis* subjected to energy stress.

Elowitz and co-workers placed the promoter and ribosome-binding site of the gene encoding sigma B, a major stress sigma factor from *B. subtilis,* in front of *yfp.* Cells in which sigma B was expressed would fluoresce green. A sigma B–YFP cell was then treated with mycophenolic acid to induce energy stress. What happened next was followed using quantitative time-lapse microscopy. Surprisingly, the constant energy stress imposed on the population triggered unsynchronized pulses of sigma B activation in individual cells (**Fig. 12.4**). As the stress increased, the pulses became more frequent. Thus, sigma B controls its target genes through sustained on-off pulsing (like the neighborhood house heaters) rather than by continuous activation.

As shown in this example, GFP fusions are more versatile than LacZ fusions in many ways. GFP fusions can be

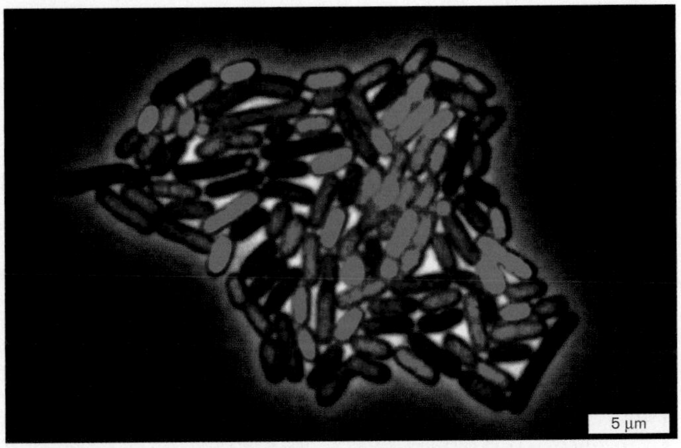
5 µm

JONATHAN W. YOUNG, JAMES C. W. LOCKE, AND MICHAEL B. ELOWITZ

FIGURE 12.4 ■ Flashing bacteria. YFP fusion (shown in green) revealed the stochastic expression of the sigma B gene among single cells of *Bacillus subtilis* subjected to energy stress. Individual cells flash on and off much like a thermostat-controlled house heating system. A movie of the process can be seen at: http://www.sciencemag.org/content/suppl/2011/10/05/science.1208144.DC1/1208144s1.mov

used to track gene expression, identify the location of proteins inside cells, and follow the movement of pathogens within host animals.

Thought Question

12.4 How could you quickly separate cells in the population that express sigma B–YFP from those that do not express this fusion? And what could you learn by separating these subpopulations? *Hint:* A technique presented in Chapter 4 will help.

To Summarize

- **Transposons** can be used to mutate genes and mark the defective gene with antibiotic resistance.
- **The location of a transposon insertion** can be determined by sequencing from the end of the transposon across the insertion junction. The junction sequence is examined by online comparative BLAST analysis to identify the gene in the genome.
- **Annotations of the gene** may provide clues about the function of the protein.
- **Regulation of a gene** can be determined by fusing that gene to an easily assayed reporter gene.
- **Operon fusions** (also called transcriptional fusions) reflect transcriptional control of the subject gene.
- **Gene fusions** (also called translational fusions) reflect transcriptional and translational control.

12.3

Classic Molecular Techniques

There are many instances in which making gene fusions is not possible. For instance, some fusion proteins can aggregate and block membranes. In these situations, deep sequencing and mass spectrometry techniques can be used for whole-cell transcriptomic and proteomic purposes, respectively (Chapter 10). But sometimes examining only a specific gene or protein is desired. Classic molecular techniques can then be used, such as the northern blot (for RNA), the Southern blot (for DNA), and the western blot (for protein). All rely on electrophoresis to separate relevant macromolecules (Section 3.2 and Appendix 3). Details of the northern and Southern blot techniques are found in Appendix 3. Here we briefly review the techniques and show how they helped reveal details of *E. coli* acid resistance.

Northern and Southern Blots

Northern blots are used to analyze the presence, size, and processing of a specific RNA molecule in a cell extract. In the northern blot technique, RNA is extracted from the cell and individual molecules are separated by size using electrophoresis through an agarose-formaldehyde gel (Appendix 3). The separated fragments are transferred by simple capillary action onto a nylon or nitrocellulose membrane, forming the blot. After the RNA has been transferred, the membrane is hybridized (or "probed") with a small, labeled DNA fragment (usually made via PCR) that will anneal to a specific species of mRNA on the blot. A similar probing strategy, called the **Southern blot** technique, is used to detect the presence of specific DNA bands (genes) among electrophoretically separated DNA fragments (Appendix 3).

In our acid resistance example, northern blots were used to ask whether the *E. coli* acid resistance genes are regulated. **Figure 12.5A** shows that the location of the *gadA* gene on the genome differs from that of the *gadB* and *gadC* genes, and that *gadB* and *gadC* form an operon. The northern blot in **Figure 12.5B** reveals that when cells are grown under acidic conditions (low pH), the *gadA* gene and the operon *gadBC* are both induced. The DNA probe used can bind to either the *gadA* or *gadB* mRNA. Because *gadB* and *gadC* form an operon, a longer mRNA molecule appears on the blot. The RNA bands are also more intense in the lane from acid-grown cultures (pH 5.5) than in the lane from cultures grown at high pH (pH 7.7), suggesting that the transcription of these genes is induced by acid. An

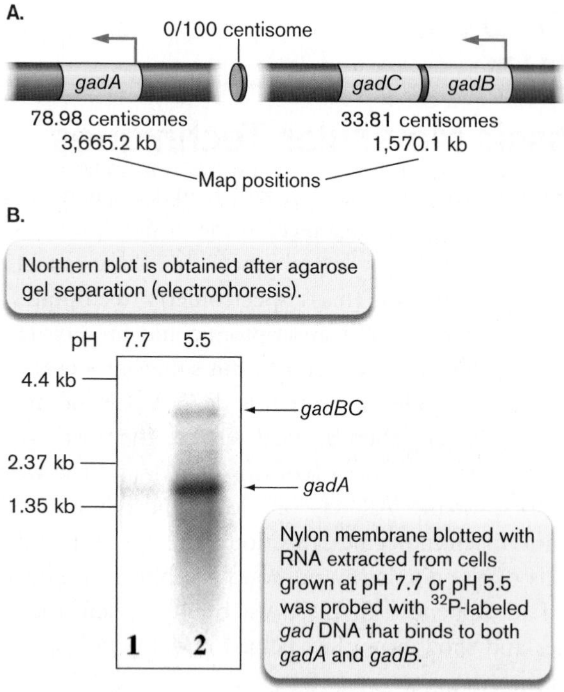

A.
0/100 centisome
gadA
gadC
gadB
78.98 centisomes
3,665.2 kb
33.81 centisomes
1,570.1 kb
Map positions
B.
Northern blot is obtained after agarose gel separation (electrophoresis).
pH 7.7 5.5
4.4 kb
2.37 kb
1.35 kb
gadBC
gadA
1 2
Nylon membrane blotted with RNA extracted from cells grown at pH 7.7 or pH 5.5 was probed with 32P-labeled gad DNA that binds to both gadA and gadB.

FIGURE 12.5 ■ Northern blot to view mRNA levels. Northern blots can be used to monitor the quantity and breakdown of RNA. **A.** Organization of genes encoding acid resistance in *E. coli.* Genes *gadA* and *gadB* encode isoforms of glutamate decarboxylase. Gene *gadC* encodes the putative antiporter. Map positions determined by sequencing are given in centisomes and kilobases (kb). A centisome equals one one-hundredth (1/100) of the chromosome. **B.** Northern blot illustrates that *gadB* and *gadC* are transcribed as an operon (note the size markers on the left). The *gadA* gene is located elsewhere in the genome and is transcribed separately. The data also illustrate that *gadA* and *gadBC* mRNAs accumulate after growth at pH 5.5. The accumulation is due to increased transcription. A second blot using a *gadC* probe was used to show that the larger transcript also contains *gadC* (not shown). *Source:* Modified from Zhuo Ma et al. 2003. *Mol. Microbiol.* **49**:1309–1320.

alternative procedure for quantifying levels (but not sizes) of specific RNA molecules is real-time quantitative PCR, which is discussed later in this section.

Western Blots

Another way to examine gene regulation is to detect the protein products themselves using a technique called the **western blot**. Compared to the northern and Southern blots, the western blot is more frequently used in today's laboratories. To begin with, the protein extract is subjected to SDS polyacrylamide gel electrophoresis (SDS-PAGE) (Section 10.9 and Appendix 3). The protein bands are electrophoretically transferred from the gel to a membrane made of polyvinylidene fluoride (PVDF) (**Fig. 12.6A**). The membrane is then probed with a **primary antibody** directed against the specific protein. Primary antibodies are made by injecting purified protein into animals and then extracting the serum after a few months.

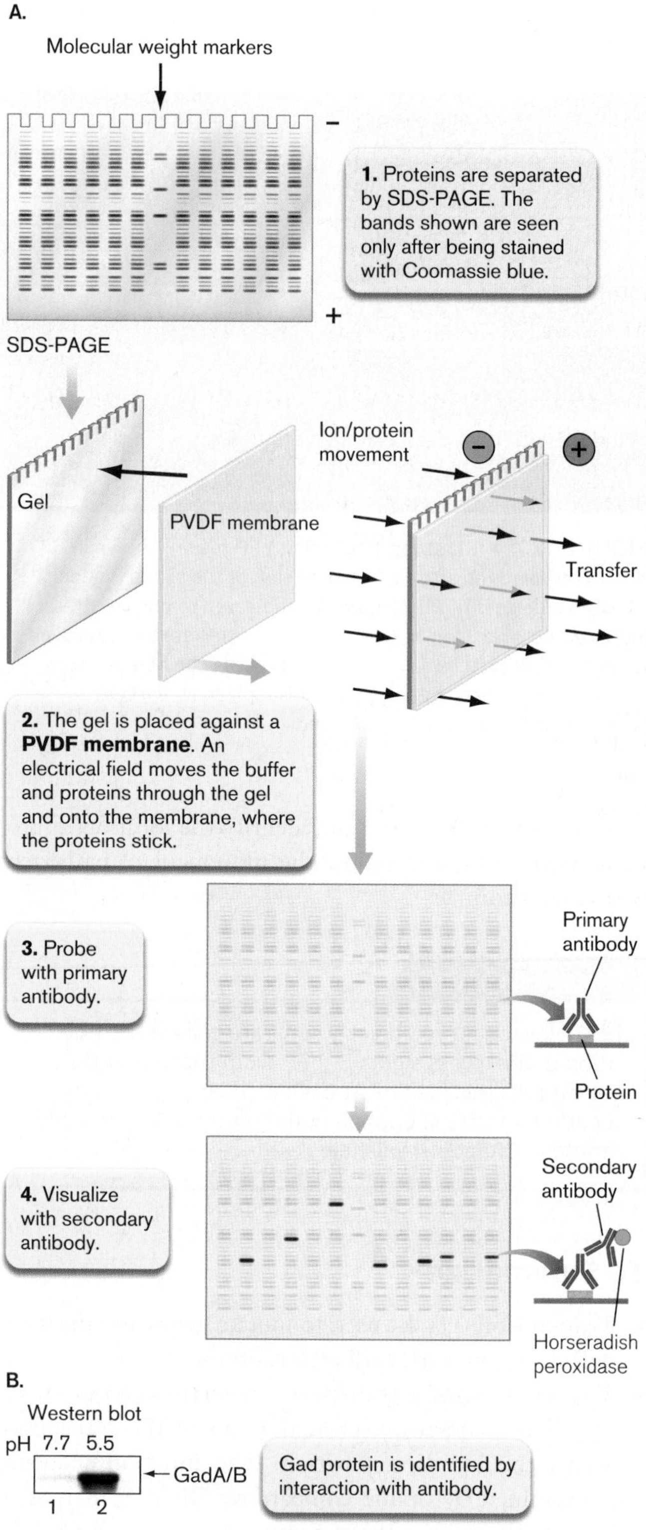

A.
Molecular weight markers
1. Proteins are separated by SDS-PAGE. The bands shown are seen only after being stained with Coomassie blue.
SDS-PAGE
Ion/protein movement
Gel
PVDF membrane
Transfer
2. The gel is placed against a PVDF membrane. An electrical field moves the buffer and proteins through the gel and onto the membrane, where the proteins stick.
3. Probe with primary antibody.
Primary antibody
Protein
4. Visualize with secondary antibody.
Secondary antibody
Horseradish peroxidase
B.
Western blot
pH 7.7 5.5
GadA/B
1 2
Gad protein is identified by interaction with antibody.

FIGURE 12.6 ■ Western blot analysis. Specific proteins within a mixture of many proteins can be identified by western blot. **A.** The western blot procedure. **B.** Western blot analysis of glutamate decarboxylase content of *E. coli* grown at pH 7.7 versus pH 5.5. The production of the protein is induced by growth under acidic conditions. *Source:* Modified from Zhuo Ma et al. 2003. *Mol. Microbiol.* **49**:1309.

Depending on the source of the primary antibody (for example, mouse, rabbit, or other), the membrane is then probed with a **secondary antibody** (for example, anti-mouse IgG antibody) that specifically binds to an amino acid sequence common to all mouse primary antibodies. The secondary antibody may be tagged with horseradish peroxidase that can be used to visualize where the antibody binds to the blot. The result is an antibody "sandwich" in which the primary antibody binds to the target protein and the secondary antibody binds to the primary antibody. Adding hydrogen peroxide (which is reduced by horseradish peroxidase) and luminol (a chemical that emits light when oxidized by horseradish peroxidase) sets off a luminescent reaction wherever an antibody sandwich has assembled. The light emitted is detected by **autoradiography**, which uses X-ray film to capture the light, or phosphorimaging (a phosphorimager is a machine that records the energy emitted as light or radioactivity from gels or membranes). Only protein bands to which the primary antibody has bound will be detected. The technique is used to determine whether, and at what levels, specific proteins are present in a cell extract; it can also reveal whether a protein has been cleaved (processed) from a larger to a smaller form.

In the acid resistance example (**Fig. 12.6B**), western blot analysis was performed to detect glutamate decarboxylase protein. The blot shows that the level of this protein is induced by growth under acidic conditions. Thus, the acid-induced increase in mRNA from transcription (as shown by northern blot) correlates with an increase in translated protein (as shown by western blot).

Thought Question

12.5 What would you conclude if the northern blot of the *gad* genes showed no difference in mRNA levels in cells grown at pH 5.5 and pH 7.7, but the western blot showed more protein at pH 5.5 than at pH 7.7?

DNA Mobility Shifts

Once a gene encoding a putative transcriptional regulatory protein has been identified by sequence annotation, proof of function requires a demonstration that the protein will bind to a DNA sequence present in the target promoter area. One approach is to add purified protein to the putative target DNA fragment and then see whether the protein causes a shift in the electrophoretic mobility of that fragment in a gel. An example of this method, as applied to our acid resistance problem, is illustrated in **Figure 12.7A**. Linear DNA molecules travel in agarose or polyacrylamide gels at an inverse rate relative to their size; that is, larger molecules travel more slowly than smaller ones. If a protein binds to a DNA molecule, the complex is larger than the DNA alone and will travel even more slowly in the gel. This method of analyzing protein-DNA binding is called the **electrophoretic mobility shift assay (EMSA)** (Section 10.2).

In our research example, one of the genes that affected acid resistance (*yhiE*) was annotated in databases as a potential transcriptional regulator. The encoded protein was purified and tested by EMSA for its ability to bind to the promoter regions of the two potential targets, *gadA* and *gadBC*. As shown in **Figure 12.7B**, the purified YhiE protein, renamed GadE because of its role in regulating glutamic acid decarboxylase expression, did bind the *gadA* sequence and retard its mobility. Thus, GadE is a true regulator of the acid resistance genes.

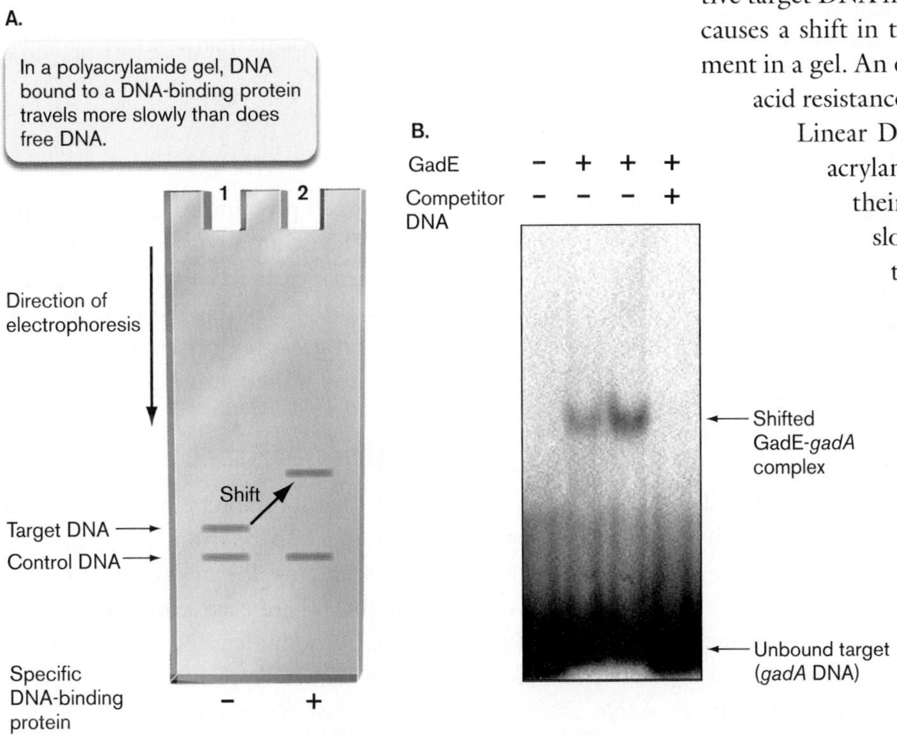
A.
In a polyacrylamide gel, DNA bound to a DNA-binding protein travels more slowly than does free DNA.
1
2
Direction of electrophoresis
Shift
Target DNA
Control DNA
Specific DNA-binding protein
–
+
B.
GadE – + + +
Competitor DNA – – – +
Shifted GadE-*gadA* complex
Unbound target (*gadA* DNA)

FIGURE 12.7 ■ Electrophoretic mobility shift assay (EMSA). Protein-DNA interactions can be monitored by EMSA. **A.** Diagrammatic representation of EMSA showing slowed mobility of a hypothetical DNA-protein complex in a polyacrylamide gel. **B.** Gel shift experiment showing that the GadE regulatory protein binds to a radiolabeled promoter fragment of *gadA*. Lane 1 contains radiolabeled target DNA only. Lanes 2 and 3 contain radiolabeled target DNA and GadE protein. Lane 4 is the same as lane 3 but includes excess unlabeled promoter fragment. The unlabeled fragments effectively outcompete labeled fragments for the protein, so no shift in the radiolabeled fragment is seen. This provides a check on the specificity of the binding. *Source:* Modified from Zhuo Ma et al. 2003. *Mol. Microbiol.* **49**:1309.

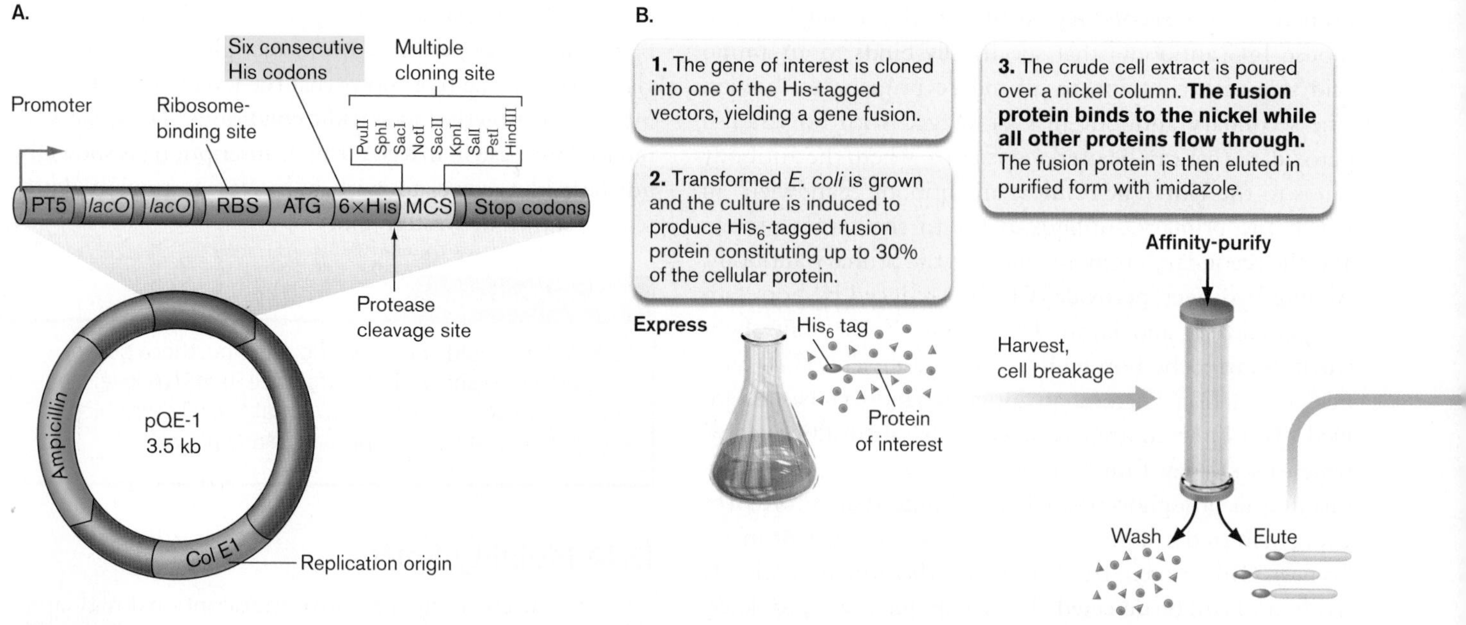
A.
Six consecutive His codons
Multiple cloning site
Promoter
Ribosome-binding site
PvuII
SphI
SacI
NotI
SacII
KpnI
SalI
PstI
HindIII
PT5
lacO
lacO
RBS
ATG
6×His
MCS
Stop codons
Protease cleavage site
Ampicillin
pQE-1
3.5 kb
Col E1
Replication origin
B.
1. The gene of interest is cloned into one of the His-tagged vectors, yielding a gene fusion.
2. Transformed E. coli is grown and the culture is induced to produce His6-tagged fusion protein constituting up to 30% of the cellular protein.
3. The crude cell extract is poured over a nickel column. The fusion protein binds to the nickel while all other proteins flow through. The fusion protein is then eluted in purified form with imidazole.
Affinity-purify
Express
His6 tag
Protein of interest
Harvest, cell breakage
Wash
Elute

FIGURE 12.8 ■ Tagging proteins for easy purification. There are several different strategies for tagging proteins. **A.** His_6 tag vector. The vector contains a promoter from the T5 phage, tandem *lacO* operators [to repress expression of the inserted gene unless isopropyl thiogalactoside (IPTG) is added], a ribosome-binding site (RBS), a start ATG codon, a sequence encoding six histidine residues, and a multicloning site (MCS) to insert the desired open reading frame (ORF). The translational reading frame of the His codons and the codons of the ORF must be the same to yield a protein fusion. **B.** Purification of the fusion protein on a nickel column. **C.** SDS polyacrylamide gel electrophoresis (SDS-PAGE) of fractions from the purification of His_6-GadB. Lane 1: protein molecular weight markers. Lane 2: uninduced cells. Lane 3: induced cells. Lanes 4–9: His_6-GadB eluted from nickel column with imidazole. *Source:* Part C from Zhuo Ma et al. 2003. *Mol. Microbiol.* **49**:1309.

Purifying Proteins by Affinity Tag

How do researchers purify proteins? Today, a protein with a known sequence can be purified in about a week. The trick is to fuse the gene encoding the protein to a DNA sequence encoding a peptide tag that strongly binds a particular small ligand molecule. The target ligand is attached to beads, and the cell extract containing the tagged protein is passed over the beads. The tagged protein binds to the beads, while other proteins do not. This technique, called **affinity chromatography**, allows researchers to essentially "fish" the tagged protein from a complex mixture of proteins present in a cell extract.

There are several commonly used peptide tags. One tag is a series of six histidines, called a His_6 tag, which tightly binds to nickel. A commercially available plasmid used to make His-tagged proteins is shown in **Figure 12.8A**. These plasmids typically include an inducible promoter to control when the fusion protein is produced. To purify the resulting protein, nickel-coated beads are loaded into a column, and then a crude extract containing the tagged protein is added to the top (**Fig. 12.8B**). Proteins that do not bind to the nickel (that is, those without the His_6 tag) are eluted with buffer. The tagged protein molecules that did bind are then eluted from the column with an elution buffer containing imidazole. Imidazole has a stronger affinity for the nickel, so it displaces the His_6-tagged protein from the beads. Analysis of the progressive purification of GadB by the His_6 tag fusion method is shown in **Figure 12.8C**.

Mapping Transcriptional Start Sites

Recall from Section 8.1 that promoters have consensus sequences located at –10 and –35 bp from the transcription start site. To begin searching for upstream promoter sequences, we need to define where each transcript begins. One method to determine transcript length is called **primer extension** (details are provided in Appendix 3). In brief, a single DNA primer is designed to anneal near the 5′ end of an mRNA. The primer is used in a reverse transcription reaction (reverse transcriptase makes DNA from RNA) that extends the primer to the 5′ end (that is, the start) of the message. The result is complementary DNA (cDNA) of a precise length. When electrophoresed next to a DNA sequencing ladder of the region (made using the same primer), the cDNA product aligns with the base correspond-

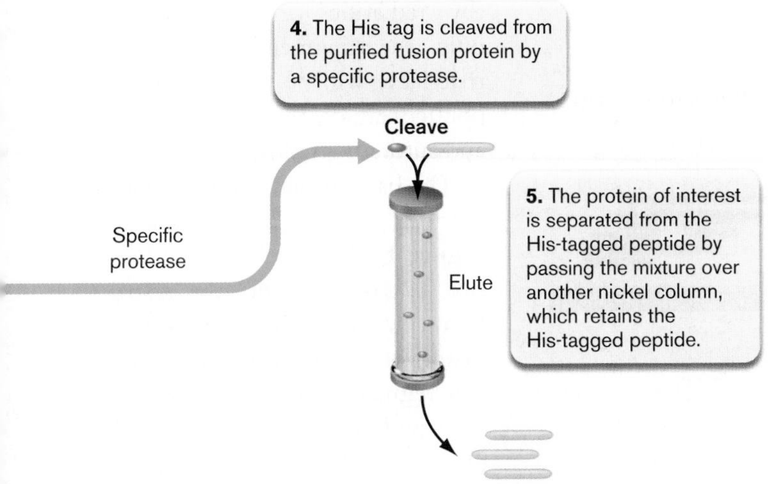
4. The His tag is cleaved from the purified fusion protein by a specific protease.
Cleave
Specific protease
Elute
5. The protein of interest is separated from the His-tagged peptide by passing the mixture over another nickel column, which retains the His-tagged peptide.

C.

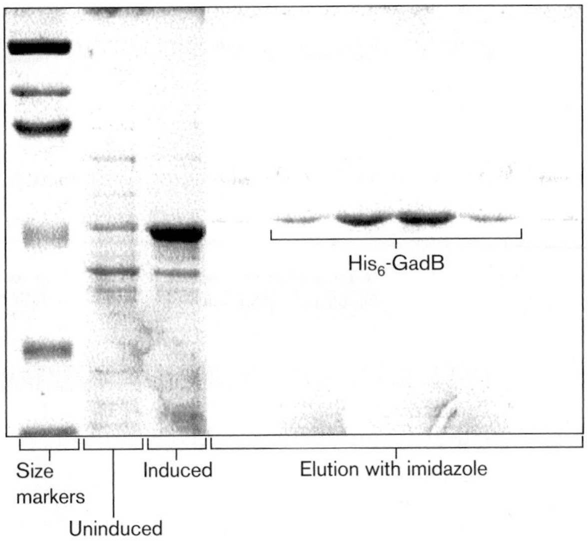
His6-GadB
Size markers
Uninduced
Induced
Elution with imidazole

ing to the transcription start site. Appendix 3 describes how primer extension was used to identify the *gadA* promoter.

Primer extension will identify transcripts for single genes, but how can transcript start sites for an entire genome be determined all at once? This is possible using modern deep sequencing technologies (Section 10.9). Briefly, whole-cell RNA transcripts are extracted from cells and made into small fragments that are converted to cDNA. The cDNA is sequenced using next-generation sequencing platforms. Computer programs then assemble the fragments and align them with the genomic DNA sequence. An example of this strategy is shown in **Figure 12.9A**. The 5′ end of each assembled, nonoverlapping transcript is presumed to be the transcriptional start site. So the segment in the figure contains two transcripts—or does it?

RNA molecules can undergo a type of processing in which a long transcript starting from a bona fide promoter is cleaved into smaller fragments. So, is transcript B in **Figure 12.9A** made from its own promoter, or is it a product made from a larger transcript encompassing A and B?

Recall that the 5′ nucleotide in all primary RNA transcripts is a triphosphate, whereas the first nucleotide of a cleaved (processed) transcript has to be a monophosphate (**Fig. 12.9B**). This distinction led to a novel **differential RNA-sequencing (dRNAseq)** approach that was developed to discover primary transcriptional start sites at a genome-wide scale. It uses the 5′-monophosphate-dependent terminator exonuclease (TEX) that specifically degrades 5′-monophosphorylated RNA species, such as processed RNA (including mature rRNA and tRNA), whereas 5′-triphosphorylated RNA species (primary transcripts) are protected and remain intact. This approach makes it possible to identify primary-transcript start sites by comparing TEX-treated with untreated libraries. The dRNAseq procedure has been used to identify primary transcripts in numerous species, including sRNA in the plant pathogen *Xanthomonas campestris*.

DNA Protection Analysis Identifies DNA-Protein Contacts

Although EMSA identifies DNA fragments bound by a transcriptional regulatory protein, the technique does not reveal which nucleotides are bound. For that level of detail, we need to ask the following: when the DNA fragment binds to a regulatory protein, which nucleotides are protected from nuclease digestion? The protected bases represent what is called the "footprint" of the binding protein. Identifying specific nucleotide-protein interactions between transcriptional regulators and their target sites (for instance, an activator of a pathogenicity operon) provides insight into how these interactions promote or inhibit transcription.

Because a regulatory protein covers DNA at its binding-site sequence, an enzyme that might attack the sequence cannot get close enough to do damage. Deoxyribonuclease I (DNase I), one of many different types of DNases, is one such enzyme. It cleaves phosphodiester bonds between bases. **Figure 12.10** presents the DNase I footprinting technique. In this method, DNase I is added to a DNA fragment at a concentration that can nick each molecule only once during the time of incubation. When the DNA is later denatured by heat, arrays of single-stranded fragments are produced. Each fragment is a different size,

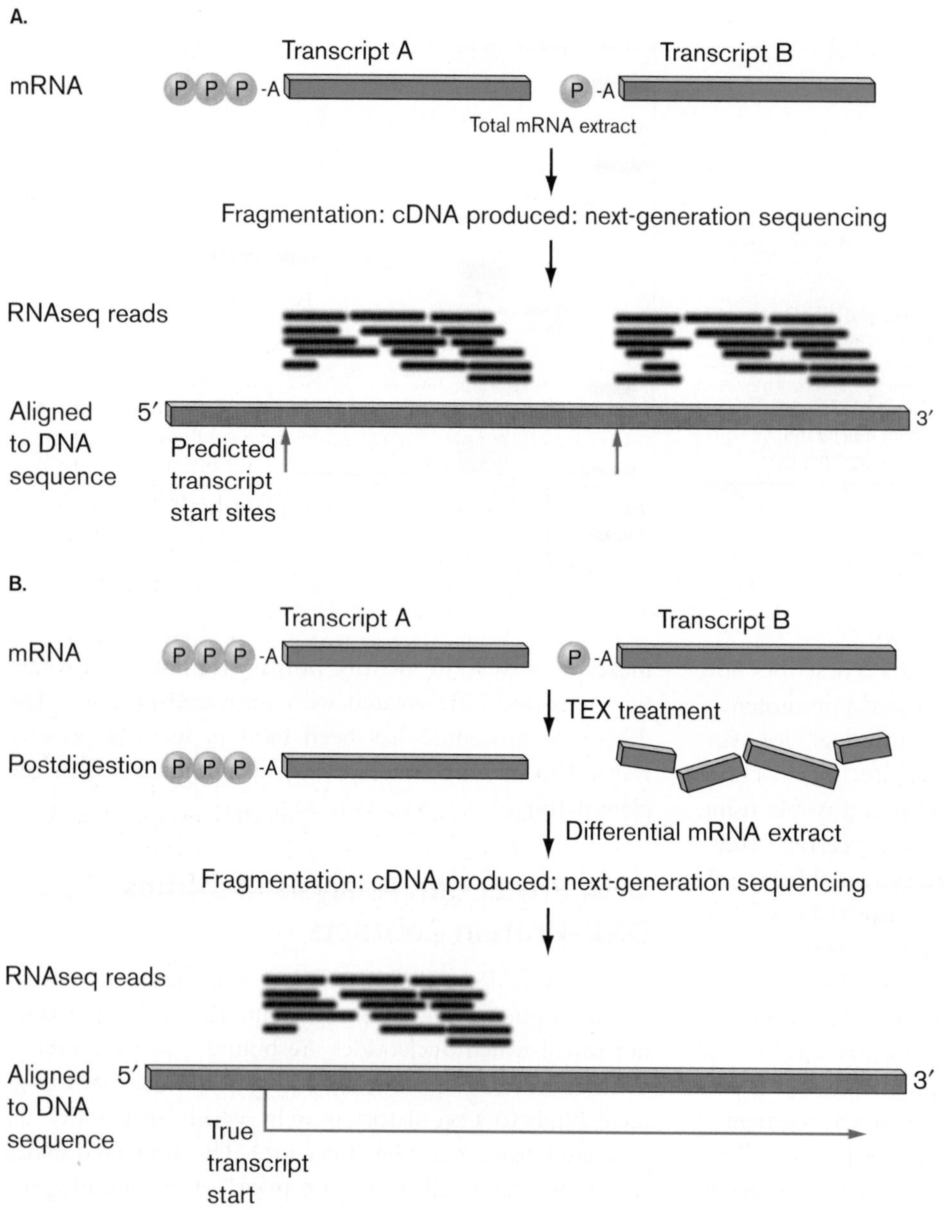
A.
Transcript A
Transcript B
mRNA
P P P -A
P -A
Total mRNA extract
Fragmentation: cDNA produced: next-generation sequencing
RNAseq reads
Aligned to DNA sequence
5′
3′
Predicted transcript start sites
B.
Transcript A
Transcript B
mRNA
P P P -A
P -A
TEX treatment
Postdigestion
P P P -A
Differential mRNA extract
Fragmentation: cDNA produced: next-generation sequencing
RNAseq reads
Aligned to DNA sequence
5′
3′
True transcript start

FIGURE 12.9 ■ RNAseq analysis of transcript start sites. **A.** Primary-transcript mRNA molecules start with a 5′ triphosphate, whereas transcripts resulting from internal cleavage during processing begin with a monophosphate. Next-generation RNAseq (RNA-sequencing) data cannot tell the difference between the two. Black lines above the DNA sense sequence strand represent independent RNAseq reads made from RNA. Red arrows indicate the 5′ ends of transcripts based on the RNAseq data. **B.** RNAseq results after treating mRNA transcripts with the 5′-monophosphate-dependent terminator exonuclease (TEX). TEX destroys processed transcripts, but not fragments containing a 5′ triphosphate end. Comparing the RNAseq results of the whole transcript versus TEX-treated transcripts will reveal true primary-transcript start sites.

depending on where the nick occurred. Because only one end of one strand of the DNA is labeled (by radiation or fluorescence), only the fragments containing that label will be seen following polyacrylamide gel electrophoresis and autoradiography.

The bases protected by a hypothetical protein are shown in **Figure 12.10A** and **B**, while **Figure 12.10C** presents the actual footprint of GadX, another regulator of *E. coli* acid resistance. The regulatory protein GadX covers a region around the promoter of *gadA* and *gadB* (only *gadA* is shown). The region is immediately upstream of the promoter. It is thought that both regulatory proteins GadE (previously described) and GadX communicate with RNA polymerase at the promoter, allowing it to transcribe the acid resistance genes.

DNA protection assays do not definitively show contact points between protein and DNA. Areas of the protein only have to be close enough to the DNA to hinder nuclease access. Protein-DNA interaction sites can be better defined by other, more difficult methods that cause a cross-link between amino acids of the protein and the specific bases they contact in the DNA. Nevertheless, the DNA protection assay remains a useful tool.

Thought Question

12.6 Another regulator of glutamate decarboxylase (Gad) production does not affect the production of *gad* mRNA, but is required to accumulate Gad protein. What two regulatory mechanisms might account for this phenotype?

Real-Time PCR

Another modification of the PCR technique is called **real-time PCR**. Real-time PCR uses fluorescence to monitor the progress of PCR as it occurs (that is, in real time). Data are collected throughout the PCR process rather than just at the end of the reaction. A major advantage of real-time PCR is that it can be used to quantify the level of DNA or RNA in a sample. DNA is quantified by how long it takes to first detect an amplified product while the polymerase chain reaction is still running. The higher the starting copy

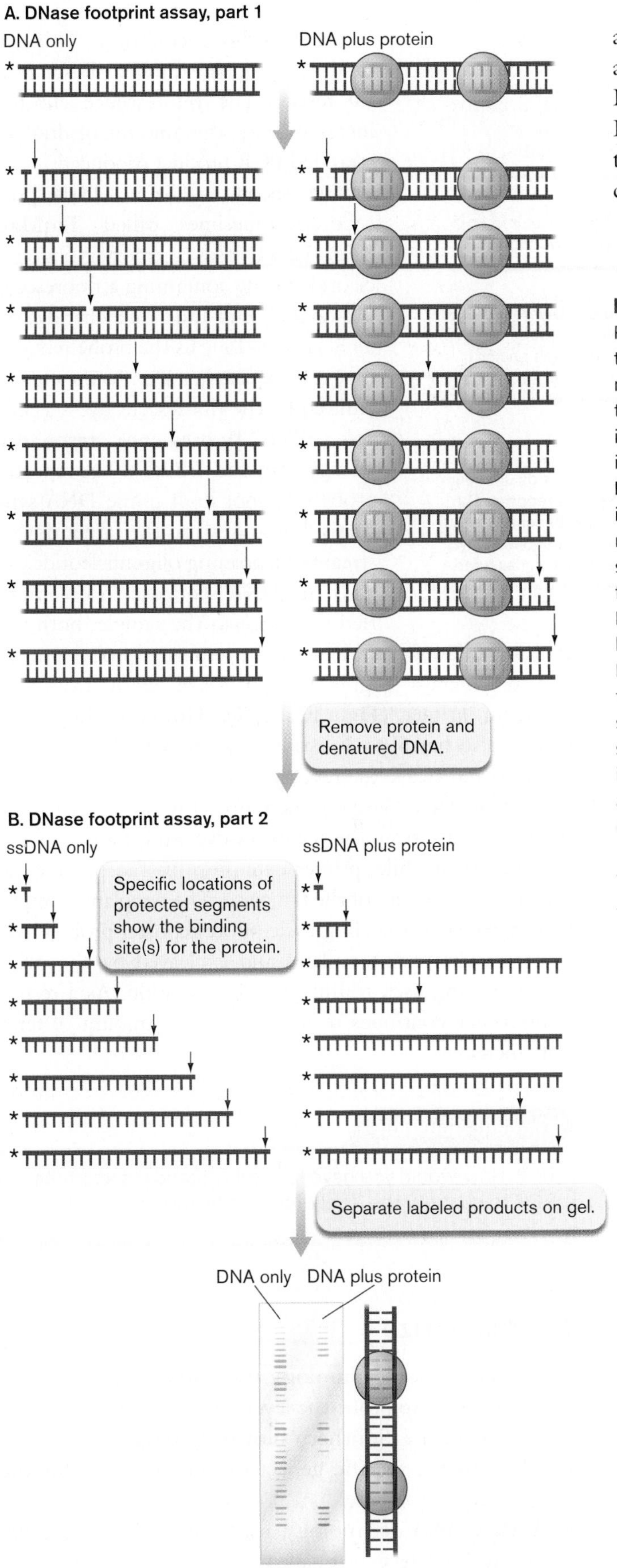
A. DNase footprint assay, part 1
DNA only
DNA plus protein
Remove protein and denatured DNA.
B. DNase footprint assay, part 2
ssDNA only
ssDNA plus protein
Specific locations of protected segments show the binding site(s) for the protein.
Separate labeled products on gel.
DNA only
DNA plus protein

FIGURE 12.10 ■ DNA protection assay using DNase I. Protein-DNA interactions can be revealed by an ability of the protein to protect DNA from nuclease digestion. **A.** DNase I will randomly nick unprotected double-stranded DNA. Conditions are designed so that there will be only one nick per molecule of DNA. Small arrows indicate nicks. To facilitate tracking, only one end of one strand is labeled (asterisk). **Left:** Random cleavage of unprotected DNA. **Right:** DNA-binding proteins block DNase I activity. **B.** The protein is removed, and the strands are separated by heat denaturation and run on polyacrylamide gels. By comparing the two lanes, one can see the "footprint" where the DNA-binding protein protected the fragment from nuclease digestion. **C.** Gel autoradiograph showing DNase I footprint where a regulator of *E. coli* acid resistance, GadX, binds to the *gadA* promoter region. The actual protein used was an MBP (maltose-binding protein)–GadX fusion protein. Numbers to the left indicate the number of bases relative to the transcriptional start (+1). The first two lanes represent the fragment's DNA sequence (G and A lanes only). Numbers above the gel indicate increasing amounts (in picomoles) of GadX regulatory protein added. Orange boxes marked with Roman numerals on the right indicate areas protected by GadX. The pink box marks the sequence bound by GadE, the other regulator. The bent arrow denotes the transcriptional start site. *Source:* Modified from Angela Tramonti et al. 2002. *J. Bacteriol.* **184**:2603.

C.

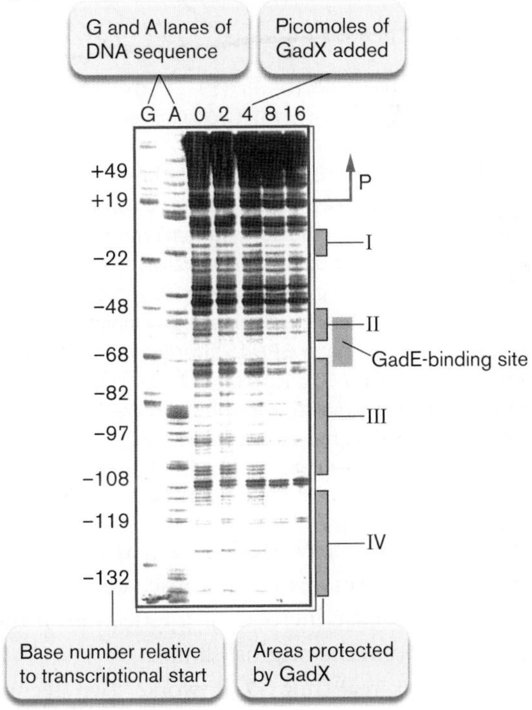
G and A lanes of DNA sequence
Picomoles of GadX added
G A 0 2 4 8 16
+49
+19
−22
−48
−68
−82
−97
−108
−119
−132
P
I
II
GadE-binding site
III
IV
Base number relative to transcriptional start
Areas protected by GadX

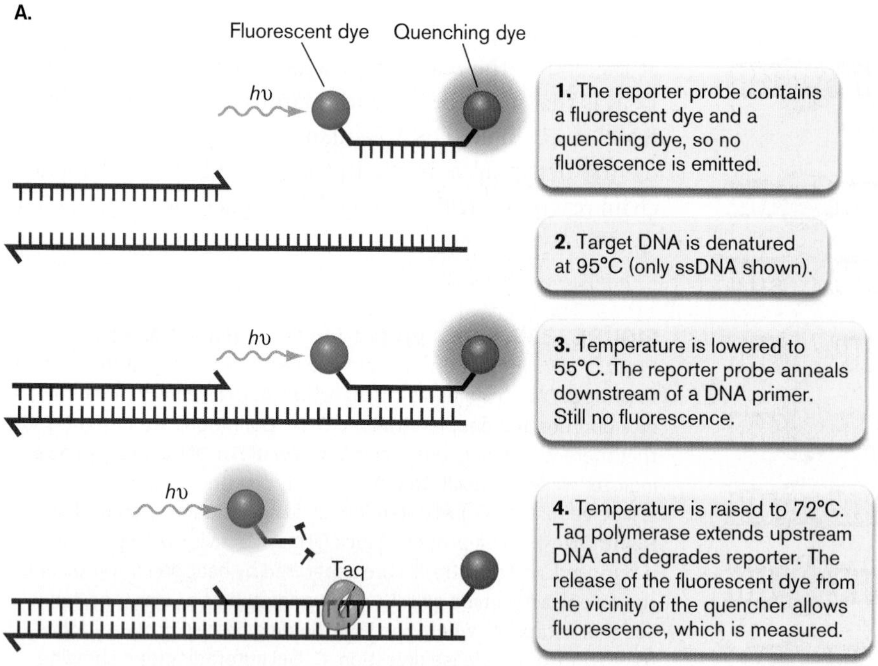
A.
Fluorescent dye
Quenching dye
hυ
1. The reporter probe contains a fluorescent dye and a quenching dye, so no fluorescence is emitted.
2. Target DNA is denatured at 95°C (only ssDNA shown).
hυ
3. Temperature is lowered to 55°C. The reporter probe anneals downstream of a DNA primer. Still no fluorescence.
hυ
Taq
4. Temperature is raised to 72°C. Taq polymerase extends upstream DNA and degrades reporter. The release of the fluorescent dye from the vicinity of the quencher allows fluorescence, which is measured.

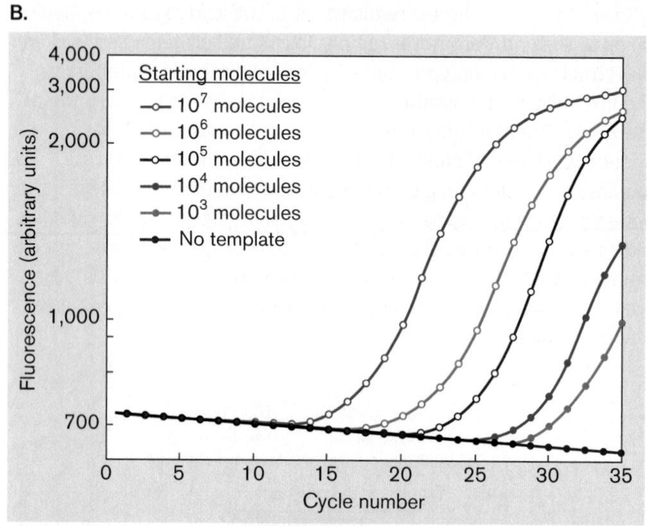
B.
Fluorescence (arbitrary units)
4,000
3,000
2,000
1,000
700
Starting molecules
10^7 molecules
10^6 molecules
10^5 molecules
10^4 molecules
10^3 molecules
No template
0
5
10
15
20
25
30
35
Cycle number

FIGURE 12.11 ■ Real-time PCR. A. The Taq DNA polymerase extends an upstream primer and reaches the downstream reporter probe, where a 5′-to-3′ exonuclease activity degrades the probe, releasing the fluorescent dye from the vicinity of the quencher, so it can now fluoresce. **B.** Amplification plot of *Rhodococcus* with primer BPH4. Different numbers of DNA copies were used as follows: 10^7, 10^6, 10^5, 10^4, 10^3 copies per reaction mixture, and a no-template control. The more DNA copies present in the initial reaction, the sooner (represented in numbers of cycles) fluorescence becomes detectable above background.

number of the nucleic acid target, the sooner a significant increase in fluorescence is observed. Thus, the time at which fluorescence increases is a reflection of the amount of nucleic acid in the original sample.

How does real-time PCR work? Two techniques are commonly used. In one, a compound called SYBR green is added to the reaction mix. This dye binds to double-stranded DNA and fluoresces. The fluorescence emitted increases with the amount of double-stranded PCR product produced.

The second real-time PCR procedure, sometimes called **TaqMan** (**Fig. 12.11**) uses a reporter oligonucleotide probe containing a fluorescent dye on its 5′ end and a quencher dye on its 3′ end. As long as the probe remains intact, the quencher absorbs the energy emitted by the fluorescent dye—a process called **fluorescence resonance energy transfer (FRET)**. The reporter probe does not itself prime DNA synthesis, but anneals to the target downstream of a priming oligonucleotide.

If the DNA sequence to be amplified is present in the sample, both the priming oligonucleotide and the reporter probe anneal. Taq polymerase begins to synthesize DNA from the upstream primer (**Fig. 12.11**). However, Taq polymerase also has 5′-to-3′ exonuclease activity. So, Taq will run into and cleave the reporter probe while continuing to synthesize DNA. Cleaving the reporter probe separates the fluorescent dye from the quencher dye, and fluorescence is emitted. Meanwhile, primer extension by Taq polymerase continues to the end of the template. The template is amplified. After each annealing cycle, more reporter probe binds to the newly made templates and is cleaved by Taq polymerase during each round of polymerization. As a result, fluorescence continues to increase as the amount of template increases.

Thought Question

12.7 How would you have to modify standard real-time PCR to quantify the level of a specific mRNA?

To Summarize

- **Northern blot technology** uses labeled DNA from a specific gene to probe the levels and sizes of RNA made from that gene. **Southern blot technology** uses labeled DNA from a specific gene to probe genomes for the presence of that gene.
- **Western blot technology** uses antibodies to specific proteins to detect the quantity and size of those proteins in cell extracts.

- **Electrophoretic mobility shift assay (EMSA)** examines protein interactions with DNA. **DNase protection assays** reveal the bases in a DNA sequence protected by a DNA-binding protein.
- **Affinity chromatography** purifies fusion proteins for biotechnology purposes.
- **Primer extension analysis** uses labeled DNA probes and reverse transcriptase to identify the 5′ end of a specific transcript (mapping the transcriptional start site). **Differential RNA-sequencing (dRNAseq)** reveals transcript start sites from an entire genome.
- **Real-time PCR** is used to quantify specific DNA or RNA molecules present in cell extracts.

12.4

Viewing the Interactions and Movements of Proteins

When a cell encounters a change in its environment, it is rare that only one gene or protein becomes affected. Rather, the impact on the cell radiates along and between pathways affecting many different genes and proteins. Tracking these global regulatory influences can show how a species adjusts its physiology to better tolerate, or even flourish in, a new environment. Transcriptomic and proteomic techniques described in Chapters 7, 8, and 10 reveal the global changes in mRNA and protein levels that cells undergo when entering new environments. Here we will show how to reveal interaction networks between proteins and genes, and between proteins and proteins. We will also see how to track protein movements through cells.

Whole-Genome DNA-Binding Analysis: ChIP-on-Chip

The techniques discussed in Section 12.3 are great for identifying a protein-DNA binding site where you already suspect that the protein binds. But is there a way to blindly determine all the sites in a genome to which a given protein binds? There is, and the basic process, called **ChIP-on-chip analysis**, is outlined in **Figure 12.12**. In the cell, the DNA-binding protein of interest (a transcription factor, TF) binds to all of its genome target sites. The cells are treated with formaldehyde to cross-link proteins to DNA and keep them together during subsequent purification steps. The DNA, with its bound proteins, is isolated and sheared into small fragments. Next, the transcription factor–DNA complexes are "fished" from the extract using antibodies that bind only that particular TF protein. The antibodies are attached to tiny beads that can be pelleted by centrifugation. Thus, centrifugation will "pull down" a complex of antibody bound to the TF protein, which is bound in turn to its specific DNA-binding sites. DNA fragments that never bound to the TF factor are washed away. This process is called **chromatin immunoprecipitation**, or **ChIP**.

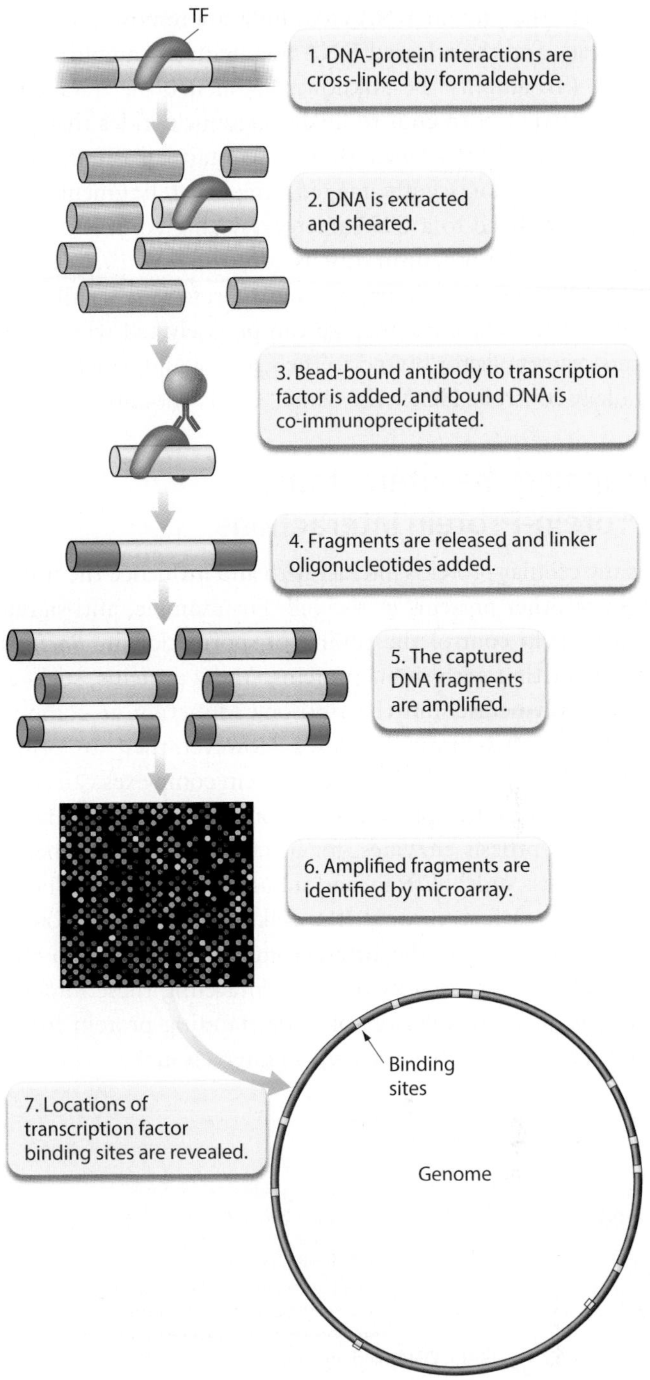
TF
1. DNA-protein interactions are cross-linked by formaldehyde.
2. DNA is extracted and sheared.
3. Bead-bound antibody to transcription factor is added, and bound DNA is co-immunoprecipitated.
4. Fragments are released and linker oligonucleotides added.
5. The captured DNA fragments are amplified.
6. Amplified fragments are identified by microarray.
7. Locations of transcription factor binding sites are revealed.
Binding sites
Genome

FIGURE 12.12 ▪ **ChIP-on-chip technology.**

Next, the protein-DNA cross-links are removed (usually by heat), and the released DNA fragments are amplified by PCR. (To amplify the unknown fragments, the fragments are ligated at both ends to linker oligonucleotides that can be amplified by known primers containing fluorescent dye.) The fluorescently tagged, amplified fragments are then hybridized to a DNA microarray chip containing the genome of the organism (discussed in Section 10.9). Since each spot on the microarray chip represents a small segment of the chromosome, we can precisely tell where the transcription factor bound to the genome and which genes it likely controls. Hence the name "ChIP-on-chip."

Mapping the Interactome: Protein-Protein Interactions

Many cellular proteins interact with and influence the function of other proteins in the cell. For example, anti-sigma factors help control the timing of sporulation in *Bacillus* by interacting with sigma proteins. Many proteins, such as RNA polymerase and the ribosome, function as components of multisubunit complexes. However, there are other intracellular situations in which protein complexes can simplify physiological processes. A complex containing fatty acid biosynthesis enzymes, for instance, allows intermediates to pass quickly from one subunit to another without having to diffuse through the cell. Thus, protein-protein interactions (called the **interactome**) are essential to the normal workings of a living cell. Unraveling these interactions is an invaluable way of understanding protein function. How, then, does one map an interactome?

Yeast two-hybrid analysis. It is important to probe protein-protein interactions in vivo, where the proteins actually work. An ingenious tool to measure in vivo protein-protein interaction is the two-hybrid system. Two-hybrid analysis can be used to mine for unknown "prey" proteins that interact with a known "bait" protein.

While there are many types of two-hybrid techniques, a classic example is the **yeast two-hybrid system** shown in **Figure 12.13**. In yeast two-hybrid analysis, genes encoding two potentially interacting proteins are fused to separated parts of the yeast GAL4 transcription factor, which is normally one protein (**Fig. 12.13A**). One gene is fused to the GAL4 activation domain, while the other gene is fused to the GAL4 DNA-binding domain. If the two proteins of interest interact with each other, then the two parts of GAL4 are brought together. The DNA-binding domain binds to the yeast *GAL1* gene promoter region, and the GAL4 activation domain (towed behind by the interacting proteins) is positioned to bind RNA polymerase and stimulate *GAL1* transcription.

The target reporter gene is typically a chromosomal *GAL1-lacZ* transcriptional fusion, although other reporter genes can be used. When the two hybrid proteins interact, the complex activates transcription of *GAL1-lacZ,* which is visualized on agar plates containing X-Gal. X-Gal is a colorless substrate of beta-galactosidase that, when cleaved, produces a blue product.

The power of this technique is that it can also be used to find unknown proteins that interact with a known protein. A known protein fused to one of the GAL4 domains can be used as "bait" to find the "prey" proteins expressed

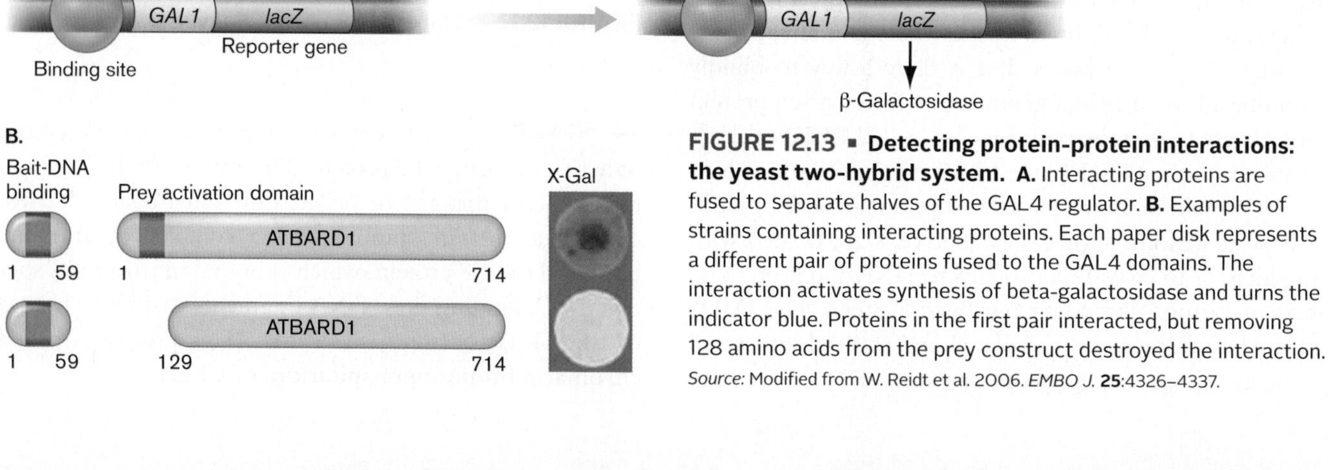
A.
GAL4 activation domain
Interacting proteins
Y
Interacting proteins are fused to separate halves of the GAL4 regulator (green): one is a DNA-binding domain; the other is an activation domain that can recruit yeast RNA polymerase.
X
GAL4 DNA-binding domain
GAL1
lacZ
Reporter gene
Binding site
RNA pol
When the proteins fused to the GAL4 moieties interact, RNA polymerase will be recruited to the promoter of the lacZ reporter gene. The cell will make β-galactosidase.
Transcription
β-Galactosidase
B.
Bait-DNA binding
Prey activation domain
X-Gal
ATBARD1
1
59
1
714
ATBARD1
1
59
129
714

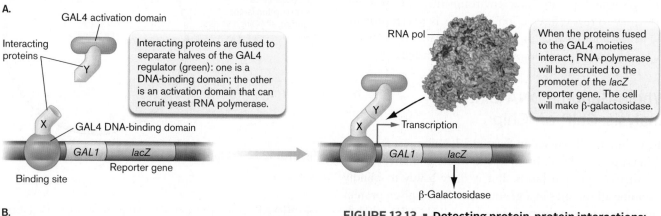

FIGURE 12.13 ■ Detecting protein-protein interactions: the yeast two-hybrid system. A. Interacting proteins are fused to separate halves of the GAL4 regulator. **B.** Examples of strains containing interacting proteins. Each paper disk represents a different pair of proteins fused to the GAL4 domains. The interaction activates synthesis of beta-galactosidase and turns the indicator blue. Proteins in the first pair interacted, but removing 128 amino acids from the prey construct destroyed the interaction. *Source:* Modified from W. Reidt et al. 2006. *EMBO J.* **25**:4326–4337.

by randomly cloned "prey" genes fused to the other GAL4 domain (see Section 7.6). Plasmids containing the bait and prey fusions are transformed into yeast cells, and the transformants are plated onto a medium containing X-Gal. Colonies that express beta-galactosidase are the result of protein-protein interactions occurring between the bait and prey fusion proteins. Sequencing the DNA insertion can then identify the gene encoding the prey protein. An example of two-hybrid analysis is shown in **Figure 12.13B**.

Thought Question

12.8 You suspect that proteins A and C can simultaneously bind to protein B, and that this interaction then allows A to interact with C in the complex. How could you use two-hybrid analysis to determine whether these <u>three</u> proteins can interact?

Affinity chromatography. Another approach used to map interactomes does not involve yeast two-hybrid systems. The alternative strategy uses individual bait proteins to pluck <u>all</u> of that protein's interacting partners out of the cell. This is done by inserting a small-peptide affinity tag at the C-terminal end of the specific "bait" protein. The process is described in **eTopic 12.1**. A protein extract from cells expressing the bait protein is prepared, and the tagged bait protein is purified by affinity chromatography. Any other cell proteins that bind to the tagged protein will be copurified with the bait. Identities of the copurified (interacting) proteins are determined by mass spectrometry. One study used this method to examine 648 bait proteins from *Escherichia coli* and found that 530 of them participated in 5,254 putative protein-protein interactions. A graphical representation of part of the interactome is shown in **Figure 12.14**.

Tracking Cells with Light

The use of green fluorescent protein (GFP) to track the location of DNA polymerase in living cells of *Bacillus subtilis* is described at the start of Chapter 7. GFP technology can also be used to target how microbial proteins move through eukaryotic cells. For instance, the potato virus X

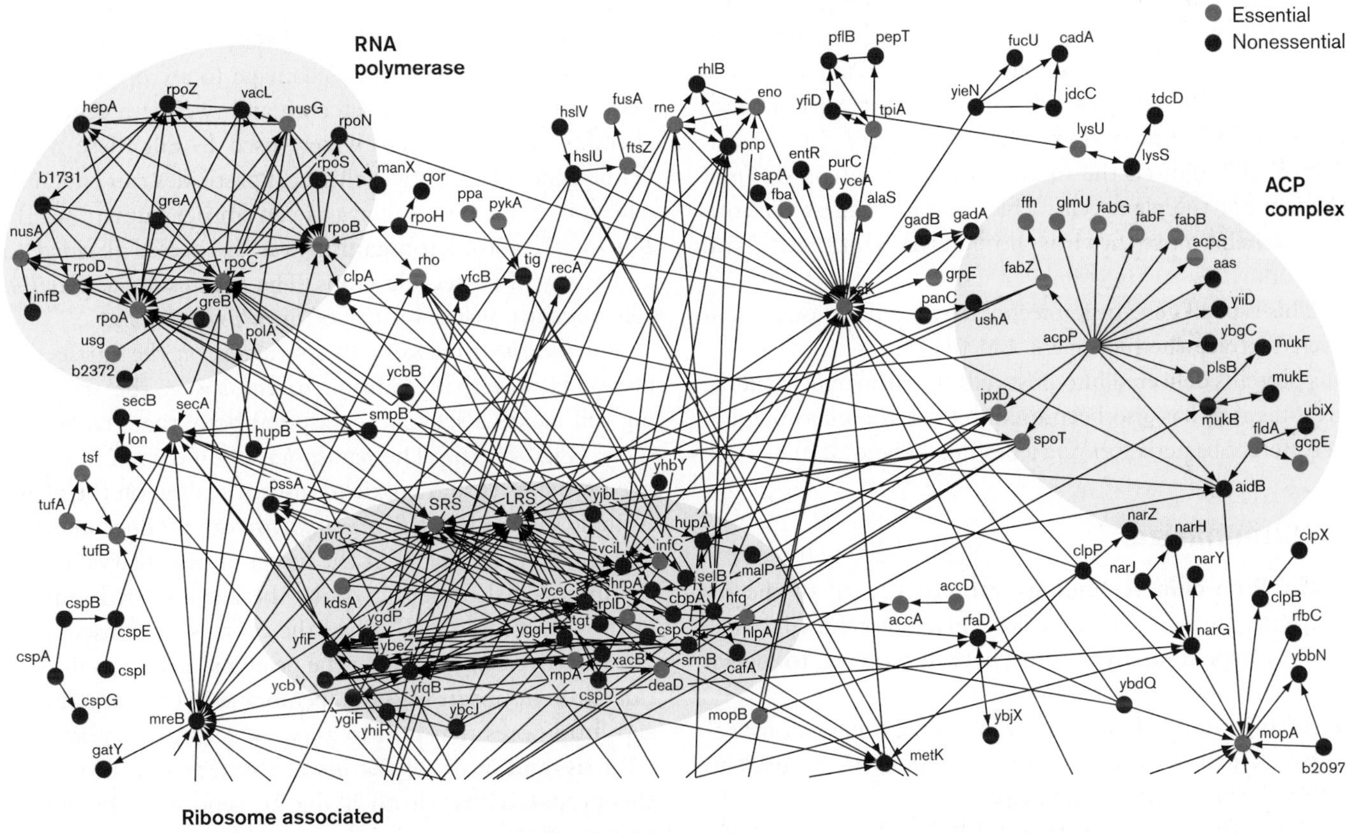
Essential
Nonessential
RNA polymerase
ACP complex
Ribosome associated
hepA
rpoZ
vacL
nusG
rpoN
rpoS
manX
b1731
greA
rpoB
nusA
rpoD
rpoC
infB
rpoA
greB
usg
polA
b2372
qor
rpoH
ppa
pykA
rho
clpA
yfcB
tig
recA
hslV
hslU
fusA
rne
ftsZ
rhlB
eno
pnp
yfiD
pflB
pepT
tpiA
entR
sapA
fba
purC
yceA
alaS
yieN
fucU
cadA
jdcC
tdcD
lysU
lysS
gadB
gadA
grpE
panC
ushA
ffh
glmU
fabG
fabF
fabB
acpS
aas
fabZ
yiiD
ybgC
acpP
mukF
plsB
mukE
mukB
ubiX
fldA
gcpE
ipxD
spoT
aidB
ycbB
smpB
secB
secA
lon
hupB
tsf
tufA
tufB
pssA
SRS
LRS
yjbL
yhbY
hupA
uvrC
vciL
infC
malP
kdsA
yceC
hrpA
selB
rplD
cbpA
hfq
ygdP
tgt
cspC
hlpA
yfiF
ybeZ
yggH
xacB
srmB
cafA
ycbY
yfqB
rnpA
cspD
deaD
ygiF
ybcJ
yhiR
mopB
metK
accD
accA
rfaD
ybjX
narZ
narH
narY
narJ
narG
clpP
clpX
clpB
rfbC
ybbN
ybdQ
mopA
b2097
cspB
cspE
cspA
cspI
cspG
mreB
gatY

FIGURE 12.14 ▪ Protein interaction network of *E. coli*. Partial predicted network. Each small, filled circle represents a gene or protein. Red and blue nodes indicate essential and nonessential proteins, respectively. Interconnecting lines indicate interaction between two proteins. Large, tan circles reflect interactions taking place with RNA polymerase and acyl carrier protein (ACP) (fatty acid biosynthesis). *Source:* From Gareth Butland et al. 2005. *Nature* **433**:531–537.

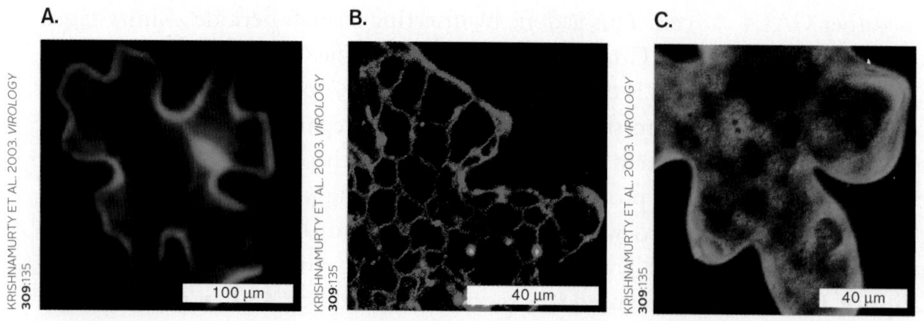
A.
100 µm
B.
40 µm
C.
40 µm
KRISHNAMURTY ET AL. 2003. VIROLOGY 309:135
KRISHNAMURTY ET AL. 2003. VIROLOGY 309:135
KRISHNAMURTY ET AL. 2003. VIROLOGY 309:135

FIGURE 12.15 ■ Using GFP protein to locate the potato virus X protein TBGp3 in the reticulated plant leaf network. Tagging a protein of interest with GFP allows one to determine the location of that protein within a cell or, in this case, a whole organism (a leaf). **A.** Plasmid containing GFP only. Fluorescence is distributed throughout the leaf. **B.** Plasmid containing a GFP-TBGp3 fusion. Fluorescence is restricted to the reticulated network. **C.** Plasmid containing mutant GFP-TBGp3 fusion gene. The mutant protein is not properly targeted to the reticulated network. *Source:* From K. Krishnamurthy et al. 2003. *Virology* **309**:135.

(or potexvirus) is a positive-strand RNA virus (a positive-strand RNA is equivalent to mRNA). One viral protein, called TBGp3, is required for the virus to move between plant cells. Investigators fused GFP to the TBGp3 protein and asked where the protein ended up in the plant leaves. They delivered the genes to plant cells by bombarding leaves with the fusion plasmids. As shown in **Figure 12.15**, GFP alone was diffusely located throughout the leaf. However, the GFP-TBGp3 fusion protein inserted only into the reticulated network of veins. Altering specific residues within the protein via mutation prevented this localization and allowed the fusion protein to remain diffused throughout the plant cells. These results suggest that this protein helps move the virus through the plant's reticulated network.

This type of gene fusion technology has been used extensively to trace the movement and final location of numerous proteins delivered into host cells by pathogenic bacteria. GFP tagging has also been used to track the intermingling of different bacteria forming a biofilm (**eTopic 12.2**).

To Summarize

- **ChIP-on-chip technology** can identify all of the sequences in a genome to which a given protein will bind.
- **Two-hybrid analysis** uses hybrid proteins to detect protein-protein interactions.
- **Affinity-tagged proteins** can expose interactions between many proteins in the cell, and can be used to build protein interaction maps.
- **Green fluorescent protein (GFP)** fused to a cell protein makes it possible to track that protein's location or movement in a cell.

12.5

Applied Biotechnology

It is remarkable how many useful products can be imagined and are being developed using the tools of biotechnology (see Chapter 16). You may be surprised to learn that plants have been engineered to produce microbial vaccines, and that vaccines can be delivered not only by injecting a protein but by simply injecting DNA or RNA that encode the antigenic protein (DNA vaccines are discussed in **eTopic 12.3**). We even have technology that uses a genetically engineered virus to repair a human genetic defect (**eTopic 12.4**). These are all remarkable additions to our scientific toolbox and are already helping to improve human life.

Bacterial Genes Save Crops

Protecting crops such as corn and cotton from hungry insects is an age-old problem. Insects such as moths or beetles can cause widespread damage to a crop in a short period of time. In the nineteenth and twentieth centuries, chemical pesticides were widely used to kill these pests. While generally effective, chemical pesticides can contaminate soil and groundwater and can impact human health. But in 1911, a microorganism was discovered that led to a new way of protecting crops. The organism was *Bacillus thuringiensis,* a close relative of *Bacillus anthracis.*

Bacillus thuringiensis (*Bt*) sporulates on the surfaces of plants. However, in addition to the spore, the sporulating cell makes a separate parasporal body that cradles the spore (**Fig. 12.16**). The parasporal body contains crystallized proteins that are toxic to insects feeding on the plant. A single subspecies of *B. thuringiensis* can produce multiple insecticidal proteins. After the insect or insect larva ingests the insecticidal protein crystals, the alkaline environment in the insect midgut dissolves the crystals, and insect proteases inadvertently activate the proteins. Activated insecticidal proteins insert into the membrane of the midgut cells and form pores that lead to a loss of membrane potential, cell lysis, and death of the insect through starvation. It is also proposed that death is due to septicemia by enteric bacteria that escape the damaged midgut (antibiotic treatment reduces death). Because of its effectiveness, farmers have taken to spreading *B. thuringiensis* directly on their crops, where even dead cells are effective.

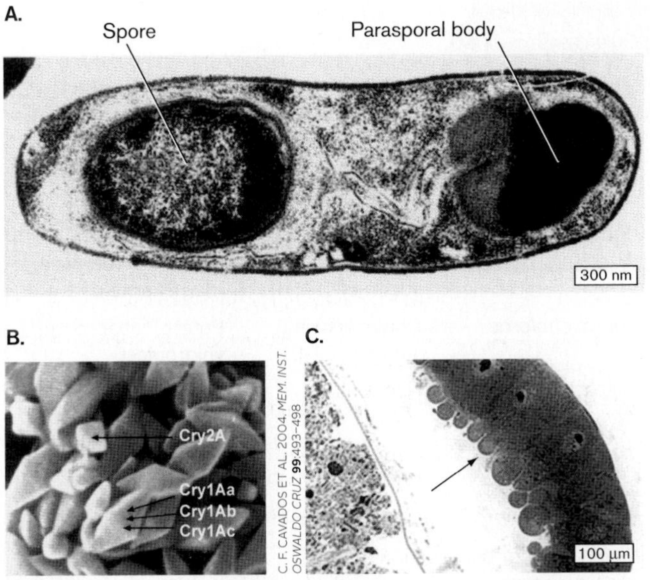
A.
Spore
Parasporal body
300 nm
B.
Cry2A
Cry1Aa
Cry1Ab
Cry1Ac
C.
100 μm
C. F. CAVADOS ET AL. 2004. MEM. INST. OSWALDO CRUZ 99:493–498

FIGURE 12.16 ■ The insecticidal bacterium *Bacillus thuringiensis*. **A.** A cell of *B. thuringiensis* (*Bt*) in the process of sporulation, during which insecticidal proteins are produced and crystallize. **B.** Parasporal crystals from *B. thuringiensis* subsp. *kurstaki*. Four major crystal proteins are formed (Cry1Aa, Cry1Ab, Cry1Ac, and Cry2A). They are used to control caterpillar pests. **C.** Midgut columnar cells of blackfly larvae after a 2-hour treatment with *Bt* toxin. Note the swollen appearance of the cells and the emission of secreted bubbles (arrow). *Source:* Parts A and B courtesy of Brian Federici, American Academy of Microbiology; part C from C. F. Cavados et al. 2004. *Mem. Inst. Oswaldo Cruz* **99**:493–498.

B. thuringiensis is generally considered a highly beneficial pesticide with few downsides. Unlike most insecticides, *B. thuringiensis* insecticides do not have a broad spectrum of activity, so they do not kill animals or beneficial insects—including the natural enemies of harmful insects (predators and parasites) and beneficial pollinators (such as honeybees). Therefore, *B. thuringiensis* integrates well with other natural crop controls. Perhaps the major advantage is that *B. thuringiensis* is essentially nontoxic to people, pets, and wildlife, so it is safe to use on food crops or in other sensitive sites where chemical pesticides can cause adverse effects.

Although spreading *B. thuringiensis* on crops is helpful, an even more effective means of delivering the toxin has been developed. The genes encoding insecticidal proteins have been spliced right into the genomes of certain plants, such as cotton and corn. The gene is placed under the control of a plant promoter, so that the transgenic plant makes its own insecticide. As of 2012, 72% of the cotton and 67% of the corn planted in the United States contain *B. thuringiensis* genes.

Although there have been no reports of adverse effects on humans, these proteins could cause allergic reactions in some people if ingested (much like peanuts, citrus, or shellfish). As a result of this concern, the use of transgenic corn expressing insecticidal proteins has been limited to cattle feed. In the bovine digestive tract, the protein is thoroughly digested to individual amino acids; thus, the protein does not show up in meat products produced from cattle and cannot cause an allergic reaction.

There are other ethical considerations surrounding the development and use of genetically modified plants. For example, are there any consequences if genes from GM crops find their way into other species? What are the risks to birds, insects, and other animals that consume GM plants? These and other ethical questions should be debated and tested.

Thought Question

12.9 Would insect resistance to an insecticidal protein be a concern when developing a transgenic plant? Why or why not? How would you design a transgenic plant to limit the possibility of insects developing resistance?

Vaccine Proteins Produced in Plants

Starting with the war against smallpox in the late 1700s, vaccines have been used to inoculate humans against many terrible diseases (see Chapters 25 and 26). However, the cost of producing them is enormous, and the price, of course, is passed on to those who are vaccinated or to governments that conduct vaccination programs. Because the cost is problematic for impoverished nations, 20% of the world's infants are not vaccinated properly, resulting in over 2 million preventable deaths annually.

One potential solution to this problem was to engineer plants to express vaccine proteins. The gene encoding a vaccine protein is inserted into the genome of a plant so the plant makes the vaccine protein as part of itself. The vaccine gene must be placed under the control of a well-expressed plant promoter (recently, plastid genes have been used). Transgenic plants expressing bacterial or viral virulence proteins can be used in two ways: They can be direct vaccine delivery systems, as through edible fruits and vegetables (such as potatoes, tomatoes, and corn); or, as with transgenic tobacco crops, they can be a cheap way to grow large amounts of the vaccine protein that can be extracted later and used in more conventional ways. Producing vaccine in edible fruits was an attractive idea because the fruit could be grown locally and distributed to residents of poor countries even without refrigeration. But practical and ethical problems have, so far, prevented implementation. The use of engineered plants to mass-produce and purify vaccine proteins, however, especially for use in animals, is happening. Plants used to make vaccines are cited in **Table 12.2**.

Phage Display Technology

Phage display is a technique in which DNA sequences encoding nonphage peptides are cloned into a phage capsid gene. These peptides are synthesized as part of the capsid protein and then "displayed" on the surface of the phage (**Fig. 12.17A**). The DNA sequences can code for random peptides, antibodies cloned from natural sources, or libraries (collections) of mutant enzymes that one wishes to study. Phage display enables scientists to screen for peptide variants with altered properties such as increased affinity for ligand.

The phage vectors used most often are the filamentous bacteriophages fd and M13 (discussed in eTopic 11.2). These phages infect *E. coli* strains containing the F plasmid (the phages attach to the sex pili) and replicate without killing the host cell. The phage particle consists of a sin-

TABLE 12.2

Potential vaccine antigens expressed in plants.

Source of the protein	Vaccine protein or peptide	Plant
Enterotoxigenic *E. coli*	Heat-labile enterotoxin B subunit (LT-B)	Tobacco, potato, tobacco chloroplast, maize kernels
Vibrio cholerae	Cholera toxin B subunit (CT-B)	Potato, tobacco chloroplast
Hepatitis B virus	Hepatitis B surface antigen	Tobacco, potato
Norwalk virus	Capsid protein (NVCP)	Tobacco, potato
Rabies virus	Glycoprotein	Tomato
Foot-and-mouth disease virus	Viral protein I	Alfalfa
Clostridium tetani	TetC	Tobacco chloroplast

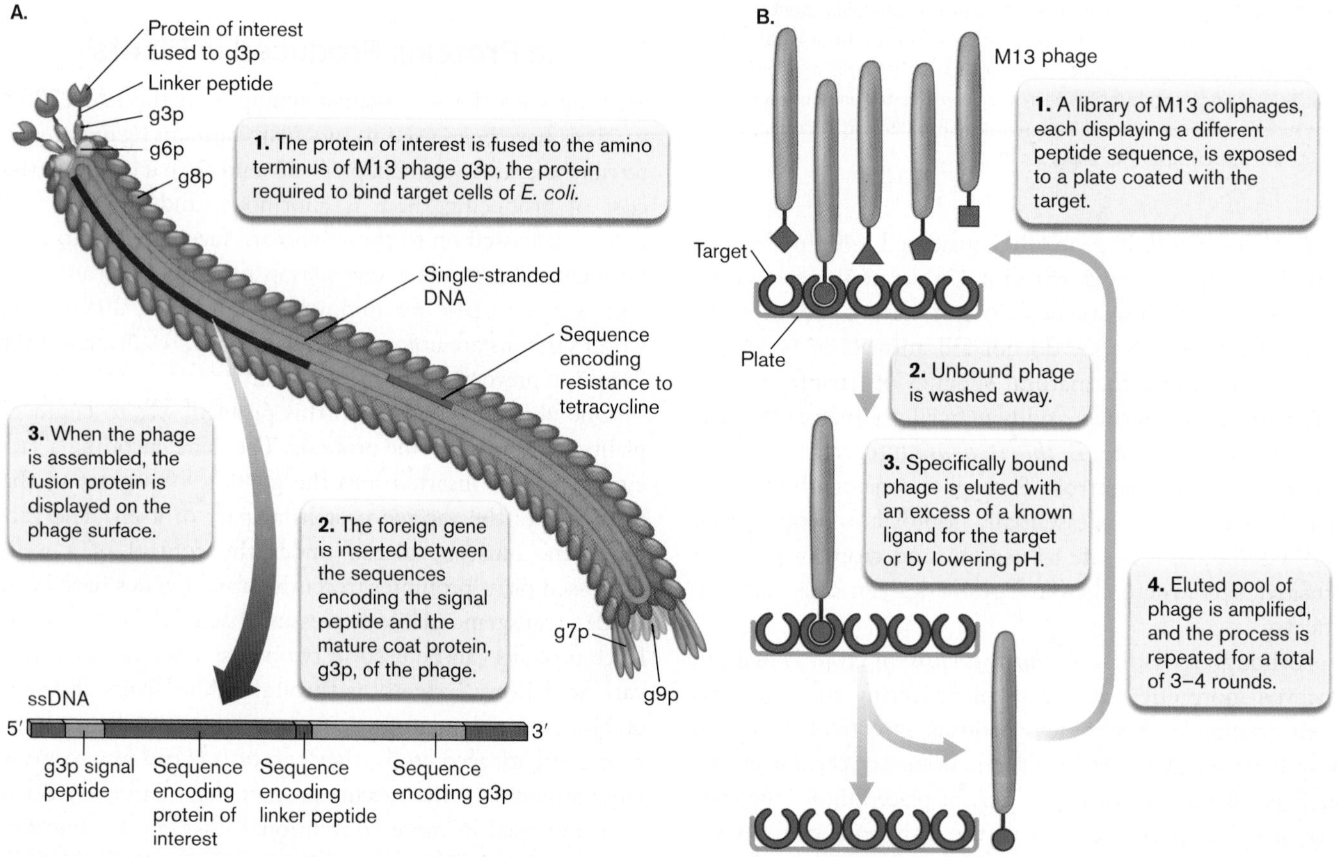
A.
Protein of interest fused to g3p
Linker peptide
g3p
g6p
g8p
1. The protein of interest is fused to the amino terminus of M13 phage g3p, the protein required to bind target cells of E. coli.
Single-stranded DNA
Sequence encoding resistance to tetracycline
3. When the phage is assembled, the fusion protein is displayed on the phage surface.
2. The foreign gene is inserted between the sequences encoding the signal peptide and the mature coat protein, g3p, of the phage.
g7p
g9p
ssDNA
5′
3′
g3p signal peptide
Sequence encoding protein of interest
Sequence encoding linker peptide
Sequence encoding g3p
B.
M13 phage
1. A library of M13 coliphages, each displaying a different peptide sequence, is exposed to a plate coated with the target.
Target
Plate
2. Unbound phage is washed away.
3. Specifically bound phage is eluted with an excess of a known ligand for the target or by lowering pH.
4. Eluted pool of phage is amplified, and the process is repeated for a total of 3–4 rounds.

FIGURE 12.17 ▪ Phage display technology. Phage display is a tool for the directed evolution of peptides. Discovering peptides that bind to specific target proteins is simplified by phage display techniques. **A.** Construction of the phage display protein. The product of gene 8 (g8p) is a small 5.2-kDa protein that forms the cylinder of the capsid. The other coat proteins (g3p, g6p, g7p, and g9p) cap the ends of the cylinder. The g3p protein is an attachment protein. **B.** Biopanning a library of M13 display phage for a peptide that specifically binds to a target molecule.

gle-stranded DNA molecule encapsulated in a long, cylindrical capsid made up of five proteins. The product of gene 3 (g3p)—present in three to five copies—confers phage infectivity. Most phage display libraries have been produced by cloning into gene 3.

Peptides displayed on filamentous phages have been used to select peptides, including antibodies, with high affinities for receptors or antigens (see Section 24.2). Recombinant phages that display high-affinity binding peptides are isolated from libraries in a process called biopanning (**Fig. 12.17B**). In biopanning, the target ligand—for example, a human virus attachment protein—is attached to a plastic plate or other support, and phages displaying random peptides are added. The phages that do not stick to the human virus protein are washed away. Those that do bind are subsequently released by adding free ligand (for example, human virus protein). This process is analogous to the way miners would pan rivers for gold. Several rounds of biopanning are required to obtain high-affinity clones. Following each round, the selected phages are enriched by repropagation in *E. coli*. Phage display was successfully used to identify high-affinity peptides that bind to attachment proteins of some eukaryotic viruses. These virus-binding peptides effectively block viral infection.

Phage display can also be used for directed evolution projects that retool enzymes (directed evolution is discussed in Chapter 17). Natural enzymes, because they were selected by evolution to sustain life, do not necessarily have the stability or catalytic activity that would make them suitable for more harsh biotechnological applications. Phage display, with its ability to screen millions of mutants, can quickly find the variant with the desired trait. An example of how phage display was used to direct the evolution of a more effective beta-lactamase enzyme that cleaves penicillin is described in **eTopic 12.5**. Other in vitro evolution techniques—DNA shuffling and site-directed mutagenesis—are described in **eTopics 12.6** and **12.7**.

Thought Question

12.10 You have just employed phage display technology to select for a protein that tightly binds and blocks the eukaryotic cell receptor targeted by anthrax toxin. When this receptor is blocked, the toxin cannot get into the target cell. You suspect that the protein may be a useful treatment for anthrax. How would you recover the gene following phage display and then express and purify the protein product?

In addition to what we describe here, microbes have been genetically engineered for many other applications that we will discuss in later chapters. These applications include the bioremediation of organic wastes (Chapter 13) and inorganic wastes (Chapter 14); mine leaching (Chapter 14); engineering of producer microbes (Chapter 16); and oil-spill cleanup, mercury removal, and wastewater treatment (**eTopic 22.2**).

To Summarize

- **Genes encoding insecticidal proteins** from *Bacillus thuringiensis* have been engineered into plant genomes to protect crops from insect devastation.
- **Proteins from pathogenic bacteria and viruses** cloned into plants may provide a cost-effective means to vaccinate large populations.
- **Genetically engineered viruses** are being tested as gene therapy vehicles to deliver DNA to correct inherited diseases in humans.
- **Phage display** is a form of in vitro evolution in which DNA sequences encoding random peptides are cloned into phage capsid protein genes. Phage particles displaying peptides with a desired binding property are pulled from the general phage population by biopanning.

12.6

Synthetic Biology: Biology by Design

What if we could, with all our knowledge of cells and genes and gene circuits, make bacteria do things they wouldn't ordinarily do, such as sense explosives in mines and then tell us about it? Or detect pathogens or keep time? Or maybe even remember things—much like a computer (see chapter opening Current Research Highlight)? These are all goals of a new science known as **synthetic biology**. Synthetic biology is the design and construction of new biological parts, devices, and systems for a desired purpose, similar to building electronic circuits.

How do we make these new parts or components? All living organisms already contain an instruction set encoded in DNA that determines what the creature looks like and what it does. Humans have been altering the genetic codes of plants and animals for millennia by selectively breeding individual plants and animals with desirable features. Today, scientists easily take pieces of genetic information from one organism and insert them directly into another, bypassing the need to breed. This is the basis of genetic engineering.

Special Topic 12.1: Bacteria "Learn" to Keep Time and Signal Danger

We know that bacteria "talk" to each other using signal molecules, but with synthetic biology can we also train them to keep time? The answer appears to be, "Yes, we can." Jeff Hasty's laboratory at UC San Diego has constructed a network circuit that relies on quorum sensing to propagate a signal through a population of cells (**Fig. 1**). The network design is shown in **Figure 2**. The organism, *E. coli,* was built to contain the *luxI* gene from *Vibrio fischeri* and the *aiiA* gene from *Bacillus thuringiensis.* LuxI synthase produces an acyl homoserine lactone (AHL) autoinducer (Chapter 10); AiiA is a secreted enzyme that degrades extracellular AHL. Both *luxI* and *aiiA* have a *luxI* promoter that can be activated by LuxR bound to AHL. LuxR was also engineered into this synthetic organism, but it is constitutively expressed. The output signal is a form of GFP, so the cells glow when the circuit is on.

How does it work? **Figure 3** shows what happens when these cells are grown as a streak on an agar plate. The system produces a flash of emission that propagates bidirectionally along the streak and then shuts off. The flash starts at a point in the colony where AHL stochastically accumulates at a low level. The secreted AHL reenters nearby cells, where AHL binds LuxR and activates the expression of all three circuits in the cell cluster. Those cells then glow (see the 100-minute panel in **Fig. 3**). The new burst of AHL diffuses and activates more nearby cells, which also start to glow (the 106- and 118-minute panels in **Fig. 3**). Because the AHL signal diffuses away from the producing cells, the light wave propagates through the streak. However, the AiiA enzyme that degrades AHL is also made and secreted by the glowing cells. AiiA degrades AHL and shuts the system off beginning where the light wave started (the 138-minute panel in **Fig. 3**). The remaining GFP is degraded by cell proteases, and cells stop glowing. But now, without AHL, AiiA is no longer made. Consequently, small amounts of AHL can again accumulate at some point in the streak and the migrating flash begins again (the 170- and 180-minute panels in **Fig. 3**).

FROM JIM HASTY, HASTY@UCSD.EDU

FIGURE 1 ■ Synchronous-clock scientists. Jim Hasty (right) and Arthur Prindle examine a culture used for their synchronous-clock experiments.

The system was then modified to be a sensor for arsenic. To do this, *luxR* was placed under the control of a promoter repressed by ArsR. In the absence of arsenic, flashing was not

Recent technological advances now allow scientists to synthesize and manipulate DNA in ways never before possible. By applying engineering principles to these genetic manipulations, researchers can take components of genes from several different organisms, link them together like Lego blocks, and design new organisms that do new things. For instance, Jay Keasling's laboratory (UC Berkeley) used genes from several species to engineer a new pathway for artemisinin, an important antimalarial drug. By exploiting *E. coli,* Keasling's team eliminated the effort required to chemically synthesize a structurally complex molecule. In another example, Chang Li and colleagues (Harvard Medical School) engineered an *E. coli* that prevented cancer in mice. The new strain contained the invasin gene from *Yersinia pseudotuberculosis* and the hemolysin gene from *Listeria monocytogenes.* The invasin gene allowed the modified *E. coli* to invade mouse cells. The hemolysin enabled delivery of an inhibitory small RNA molecule that reduced expression of a tumor-initiating gene. Amazingly, this novel *E. coli* prevented cancer in this mouse model.

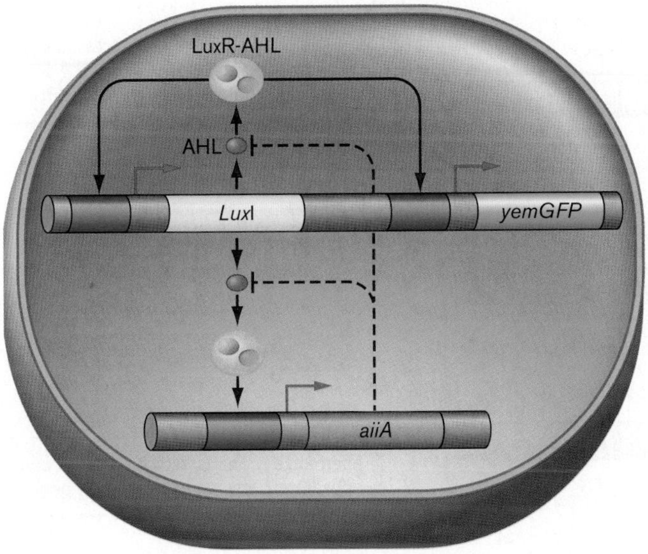
LuxR-AHL
AHL
LuxI
yemGFP
aiiA

FIGURE 2 ■ Network diagram for a synchronized oscillating clock. Arrows indicate activation; "T" lines mark inhibition. Although it was not drawn this way, realize that AiiA protein is secreted and degrades only extracellular AHL. This degradation stops AHL from accumulating extracellularly. The AHL concentration becomes too low to diffuse back into the cell, so *luxI, aiiA,* and *yemGFP* promoters are no longer activated.

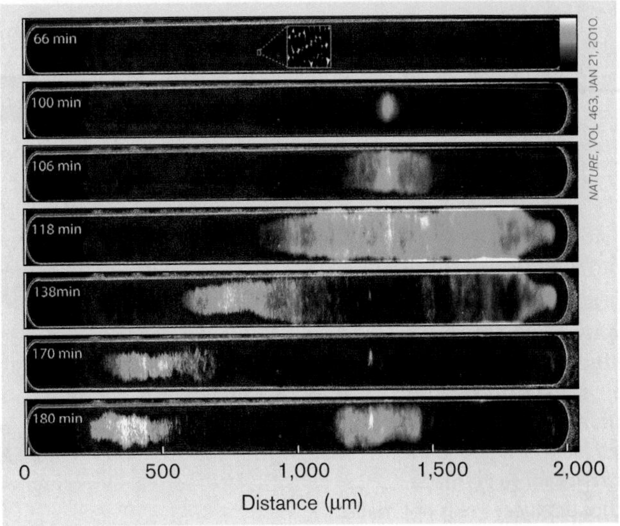
66 min
100 min
106 min
118 min
138min
170 min
180 min
0
500
1,000
1,500
2,000
Distance (μm)
NATURE, VOL 463, JAN 21, 2010

FIGURE 3 ■ Oscillatory flashing from engineered *E. coli.*

seen, because LuxR is needed to activate the *luxP-gfp* fusion. When even small amounts of arsenic were encountered, however, colonies trapped in multiple chambers of a microfluidic device begin to flash. (For a movie of this flashing, go to http://www.nature.com/nature/journal/v481/n7379/extref/nature10722-s3.mov.)

The authors successfully built a liquid crystal display (LCD)–like macroscopic clock that can sense arsenic and alert us to its presence. Given the vast sensing capabilities of bacteria, it now seems possible to build low-cost genetic biosensors able to detect heavy metals and even pathogens in the field.

RESEARCH QUESTION

Can you think of a potential pitfall with using the "arsenic alarm" in real life? How might you overcome this problem? Hint: Look up the effect of arsenic on bacterial cells. What does the ArsR protein regulate?

Prindle, A., P. Samayoa, I. Razinkov, T. Danino, L. S. Tsimring, et al. 2012. A sensing array of radically coupled genetic "biopixels." *Nature* **481**: 39–44.

Principles of Synthetic Biology

The key to synthetic biology is engineering. Among the more visually intriguing ways to apply engineering principles to biological systems involves the use of GFP protein as a visual on-or-off output; that is, cells light up or go dark. If you build the biological circuit correctly, you can make a population of bacteria alert you to the presence of a toxic compound by flashing on and off (**Special Topic 12.1**) or even become a biological computer (see Current Research Highlight).

What engineering principles are employed by synthetic-biology scientists? Many of them are borrowed from the field of electrical engineering and involve "logic gates." An electronic logic gate (as in semiconductors) receives a tiny current as an input and produces voltage as an output. A genetic logic gate is simply a promoter and a gene. An input signal such as a regulatory protein affects the promoter, which drives the output signal (mRNA and protein). The output signal for one gate can be an input signal for another part of the logic circuit. The final output signal of

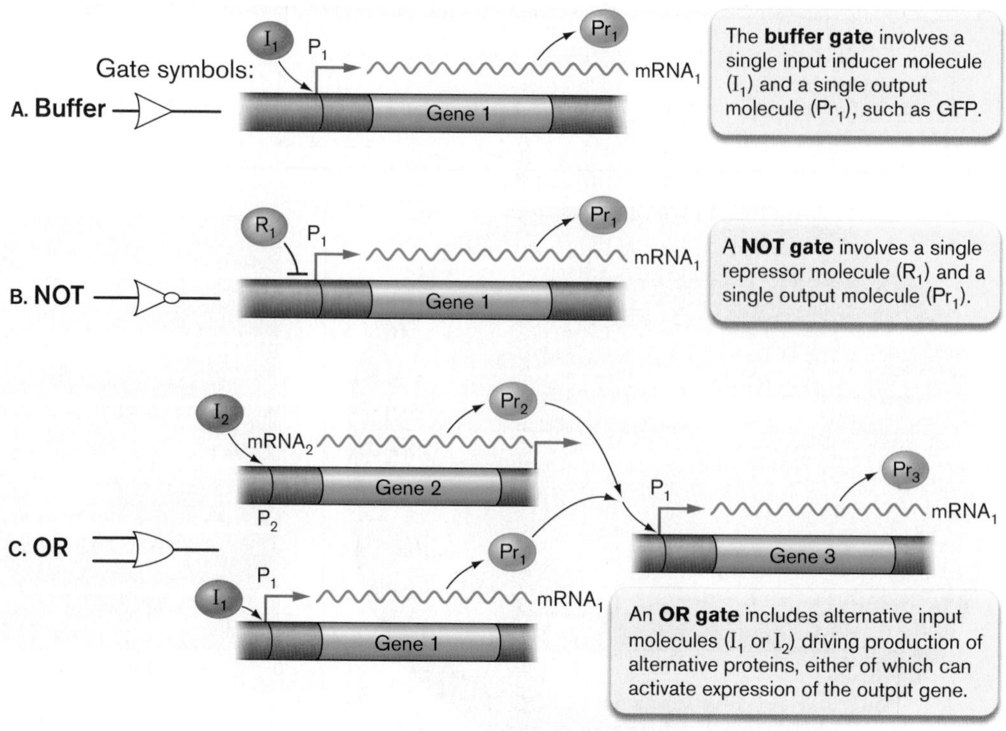
Gate symbols:
A. Buffer
I1
P1
Pr1
mRNA1
Gene 1
The **buffer gate** involves a single input inducer molecule (I1) and a single output molecule (Pr1), such as GFP.
B. NOT
R1
P1
Pr1
mRNA1
Gene 1
A **NOT gate** involves a single repressor molecule (R1) and a single output molecule (Pr1).
C. OR
I2
mRNA2
Pr2
Gene 2
P2
P1
Pr1
I1
mRNA1
Gene 1
P1
Pr3
mRNA1
Gene 3
An **OR gate** includes alternative input molecules (I1 or I2) driving production of alternative proteins, either of which can activate expression of the output gene.

FIGURE 12.18 ▪ Examples of gate symbols used in electronics, and their synthetic-biology equivalents. In A for example, when inducer is added the transcript for Pr1 is made. Without inducer Pr1 is not made.

the circuit does something useful or eye-catching (think GFP fluorescence).

Many of the symbols used in building an electronic circuit are also used when building a complex genetic circuit. **Figure 12.18** shows three common logic gates used in synthetic biology and the symbols that represent them. The first is a "buffer gate" that amplifies signals. For a simple gene, the protein input activates a promoter; a message is then transcribed and a protein is made. With a "NOT gate," a protein input represses a promoter and the output protein is not made. Finally, an "OR gate" involves several genes. For instance, two alternative gene output proteins can activate a third gene. So, in **Figure 12.18C**, the alternative input proteins I_1 and I_2 activate gene 1 or gene 2, respectively, to make product 1 or product 2 (Pr_1 or Pr_2). Either of those output signals can activate gene 3 to make product 3, Pr_3 (such as GFP). Once you understand how these gates work, you can make any kind of circuit.

Toggle Switches

Electrical systems rely heavily on toggle switches to control whether a system is turned on or off. Similarly, synthetic-biology circuits depend on biological toggle switches. **Figure 12.19** illustrates a basic, genetically engineered toggle switch designed to control whether a *gfp* gene is turned on

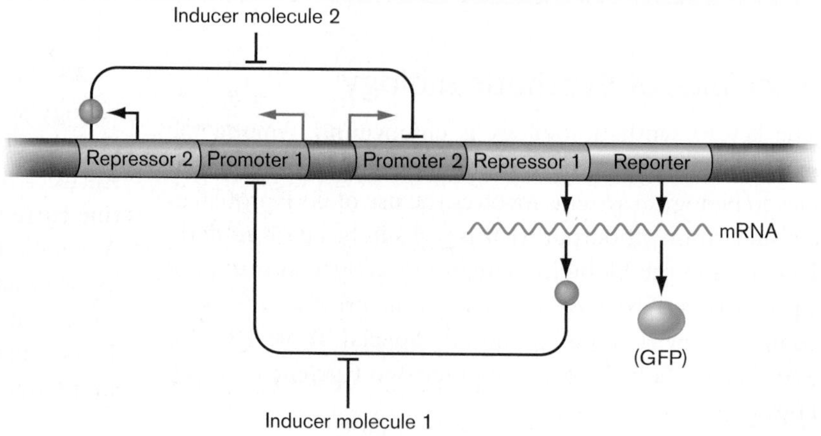
Inducer molecule 2
Repressor 2
Promoter 1
Promoter 2
Repressor 1
Reporter
mRNA
(GFP)
Inducer molecule 1

FIGURE 12.19 ▪ Genetic toggle switch. Notice that repressor 1 and the reporter *gfp* are transcribed colinearly from promoter 2. Repressor 1 inhibits transcription from promoter 1. Conversely, repressor 2 inhibits transcription from promoter 2. Two scenarios are shown. Adding inducer 1 promotes the transcription of promoter 1, which means repressor 2 is made, which shuts down promoter 2 (no fluorescence). Adding inducer 2 activates promoter 2, which means repressor 1 and GFP are made (fluorescence). Repressor 1 halts transcription of promoter 1, and repressor 2 is not made. Adding inducer 1 stops fluorescence, whereas adding inducer 2 triggers fluorescence. What happens if you add them both?

or off. The circuit involves two repressor genes (NOT gates) whose products can repress each other's transcription. The switch depends on whether one of two different inducer signals is present. Inducer 1 inactivates repressor 1, which means repressor 2 is produced. Repressor 2, in turn, stops transcription of the repressor 1 gene and the reporter gene. So, when inducer 1 is added, GFP is not made and the system is stably toggled off. Alternatively, inducer 2 inactivates repressor 2, which means that the genes for repressor 1 and GFP are transcribed and the cell lights up. The system is stably toggled "on" because repressor 1 halts transcription of repressor 2. So, a genetic engineer can control the on-off switch of these cells by adding one or the other inducer. Of course, for this system to toggle, there must be at least a small amount of both repressor proteins made at all times; just enough to bind to inducer molecules and have an effect on gene expression.

Thought Question

12.11 What would happen with a toggle switch if the genetic engineer could set repressor 1 and repressor 2 protein levels to be exactly equal? Imagine that this is done without either inducer present.

A Bacterial Oscillator Switch

An oscillator is another important component of many electronic systems. In an electronic circuit, the oscillator produces a repetitive electronic signal viewed as a wave. Can synthetic-biology scientists also make an oscillating genetic circuit? **Figure 12.20** shows a basic genetic oscillator switch designed by Jim Hasty's laboratory at UC San Diego. The components come from some of the systems discussed in Chapter 10. One component is the gene for the AraC activator protein, which was linked by Hasty's group to a hybrid operator region that included an *ara* activator sequence and a LacI repressor control sequence. The next component was the gene encoding the repressor LacI, which was also spliced to the hybrid *ara* activator/LacI repressor control sequences. The resulting circuit contains negative and positive feedback loops. The *lac* and *ara* operons are discussed in Sections 10.2 and 10.3.

To see how this oscillator works, IPTG (the chemical that inactivates LacI) and L-arabinose (the sugar that activates AraC) are added at the same time. The increase in AraC production drives expression of GFP but also increases LacI repressor, which eventually represses *araC*. However, the IPTG inactivates LacI protein, which allows renewed *araC* and *lacI* expression. As shown in **Figure 12.20B**, the differential activity of the two feedback loops drives oscillation back and forth between fluorescence (on) and no fluorescence (off). Changing the concentrations of arabinose and IPTG will modulate the oscillation frequency, making the circuit tunable. **Special Topic 12.1** presents an example of an oscillatory genetic circuit that keeps time and senses arsenic. Reporter organisms equipped with this circuit could announce the presence of dangerous toxins such as arsenic in water, soil, or a variety of food materials.

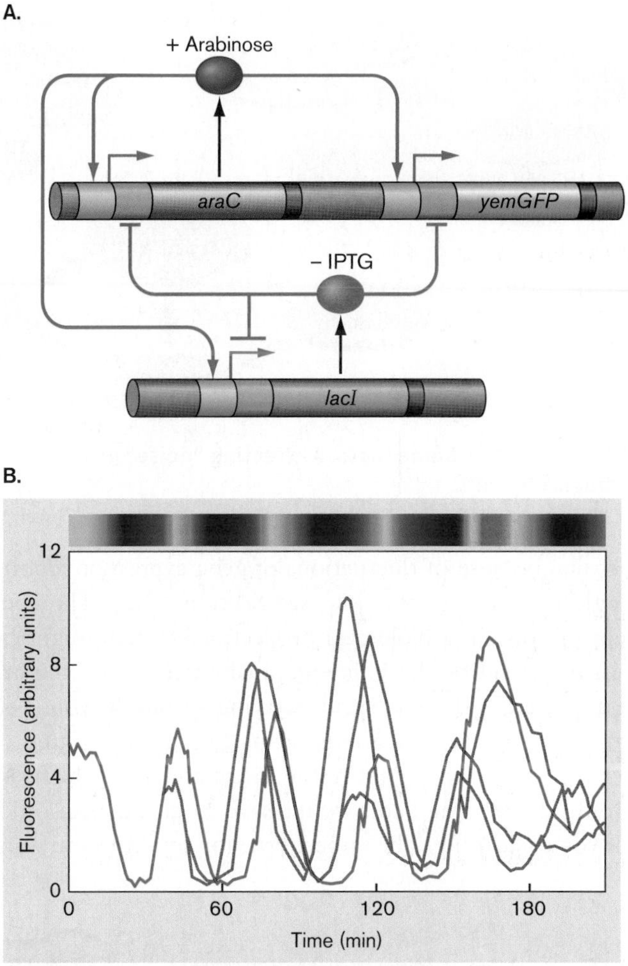
A.
+ Arabinose
araC
yemGFP
– IPTG
lacI
B.
Fluorescence (arbitrary units)
12
8
4
0
0
60
120
180
Time (min)

FIGURE 12.20 ▪ A dual-feedback oscillator constructed in *Escherichia coli*. **A.** Network diagram. A hybrid promoter, $P_{lac/ara}$ (small pink and green boxes), drives the transcription of *araC* and *lacI*, forming positive and negative feedback loops. **B.** Single-cell fluorescence oscillations induced with 0.7% arabinose plus 2-mM IPTG (red line in graph) or 1-mM IPTG (gray lines). The points represent experimental fluorescence values. The top panel represents the intensity of the fluorescence (red = highest; blue = lowest). *Source:* Modified from J. Stricker et al. 2008. *Nature* **456**:516–519.

System Noise

Before we discuss more complex circuits, we should first talk about "noise." **Noise** (variance from a mean) is a problem in any electrical circuit but also arises in biological

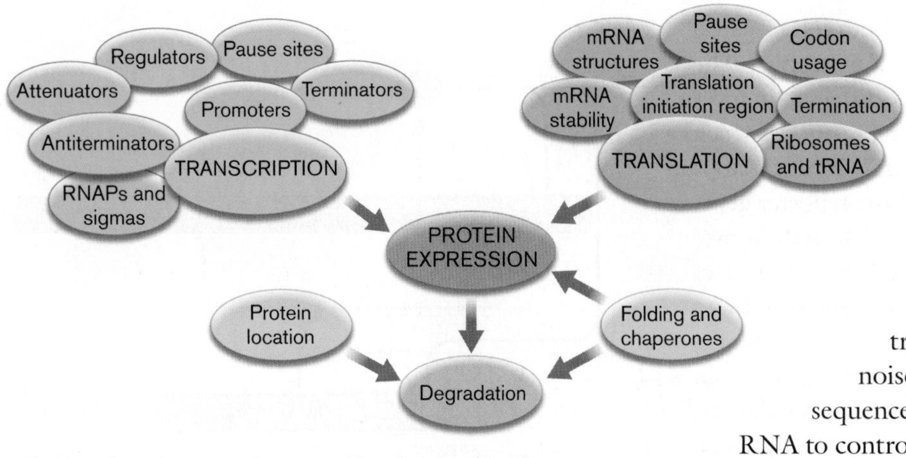
Regulators
Pause sites
Attenuators
Terminators
Promoters
Antiterminators
TRANSCRIPTION
RNAPs and sigmas
mRNA structures
Pause sites
Codon usage
mRNA stability
Translation initiation region
Termination
TRANSLATION
Ribosomes and tRNA
PROTEIN EXPRESSION
Protein location
Folding and chaperones
Degradation

FIGURE 12.21 ■ Some factors affecting "noise" in a biological circuit.

circuits, because of fluctuations of gene expression among single cells in a population (see Section 12.2). The concept of noise in a biological or electrical system might be best explained by the following analogy. Imagine you are at a concert and your friend is trying to talk to you. You can't understand what she is saying because the sound of her voice is not rising above the music and all of the background noise. The band would have to go quiet in order for you to hear her. The same is true for anything you want to measure. The less noise there is the easier it is to measure what rises above it. In bacteria, translation efficiency is a major contributor to noise, although many factors can influence the expression of a gene (**Fig. 12.21**). To be useful, synthetic biological circuits must operate with transcriptional and translational controls balanced in a way that minimizes noise. Choosing the right ribosome-binding sequence and promoter, as well as using a small RNA to control message stability or translation, can all help to decrease noise.

Engineered Riboswitches and Switchboards

Chapter 10 described how the cell can use small RNA molecules to control the stability and translation of mRNA molecules. Synthetic-biology scientists seized upon this concept and have learned to tailor small RNAs, called **riboswitches**, to control the translation of nearly any gene they want. **Figure 12.22** reveals that a basic riboswitch consists of two parts: a *cis*-repressed mRNA (crRNA) and

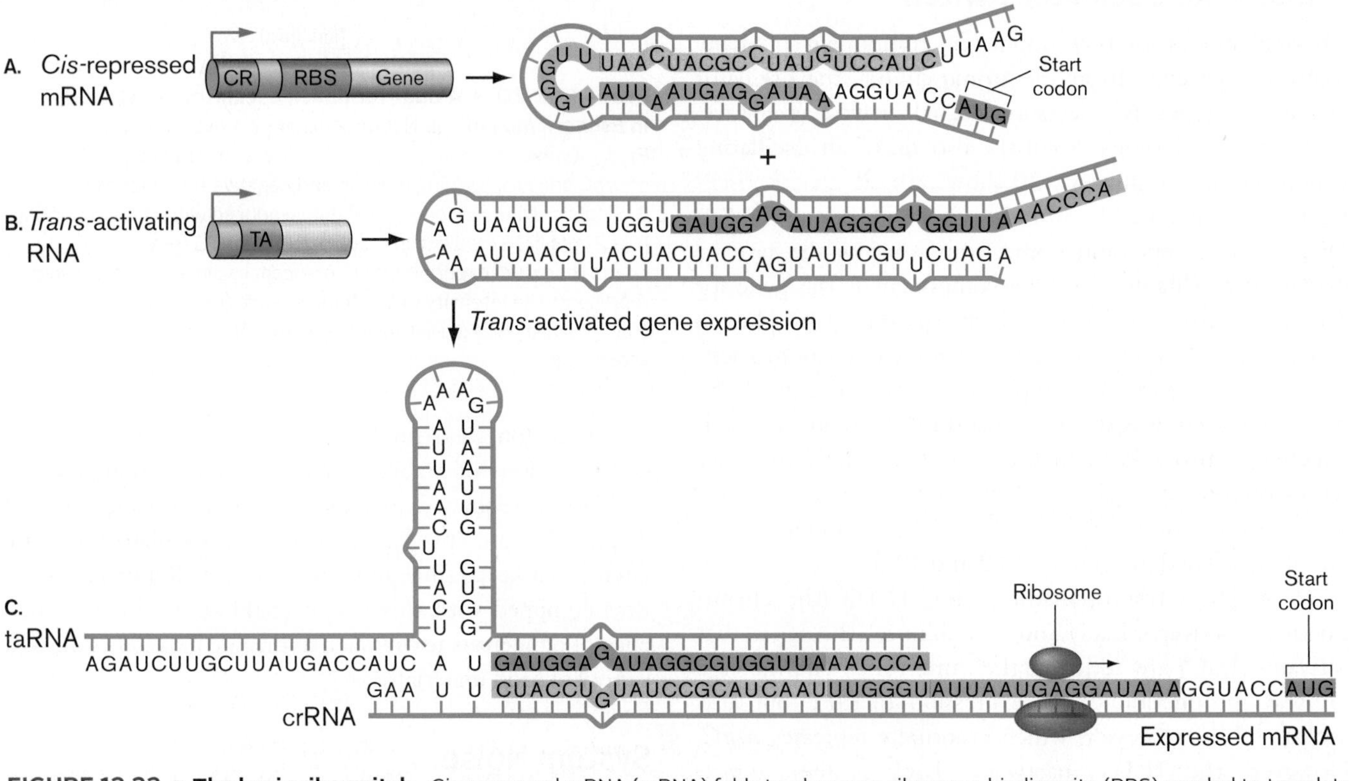
A. Cis-repressed mRNA
CR
RBS
Gene
Start codon
+
B. Trans-activating RNA
TA
Trans-activated gene expression
C.
taRNA
crRNA
Ribosome
Start codon
Expressed mRNA

FIGURE 12.22 ■ The basic riboswitch. *Cis*-repressed mRNA (crRNA) folds to obscure a ribosome-binding site (RBS) needed to translate a downstream output gene (**A**). *Trans*-activating RNA (taRNA), driven by a different promoter (**B**), can base-pair with crRNA (**C**), thus opening up the hairpin to expose the ribosome-binding site. Whatever activates the taRNA will activate translation of the output genes' mRNA.

a *trans*-activating RNA (taRNA). The crRNA sequence is linked to an output gene and, by forming a hairpin, hides a ribosome-binding site to prevent translation. The taRNA, once it is made, will promote translation of the output gene by base-pairing with the crRNA. The base-pairing releases the RBS for translation. Transcription and translation of the output gene can now be manipulated by whatever activates the promoters used to express the crRNA and taRNA genes. This process can reduce noise in a system.

Members of James Collins's laboratory (Harvard University) designed a series of matched crRNA/taRNA riboswitches and linked them to various promoters and output genes. The design and response of these circuits to various environmental signals is shown in **Figure 12.23**. Notice that each riboswitch responded to only one signal.

The Collins group then linked these riboswitches to genes encoding carbon metabolism enzymes and placed them all in the same cell, essentially making a metabolic switchboard that could channel carbon flow

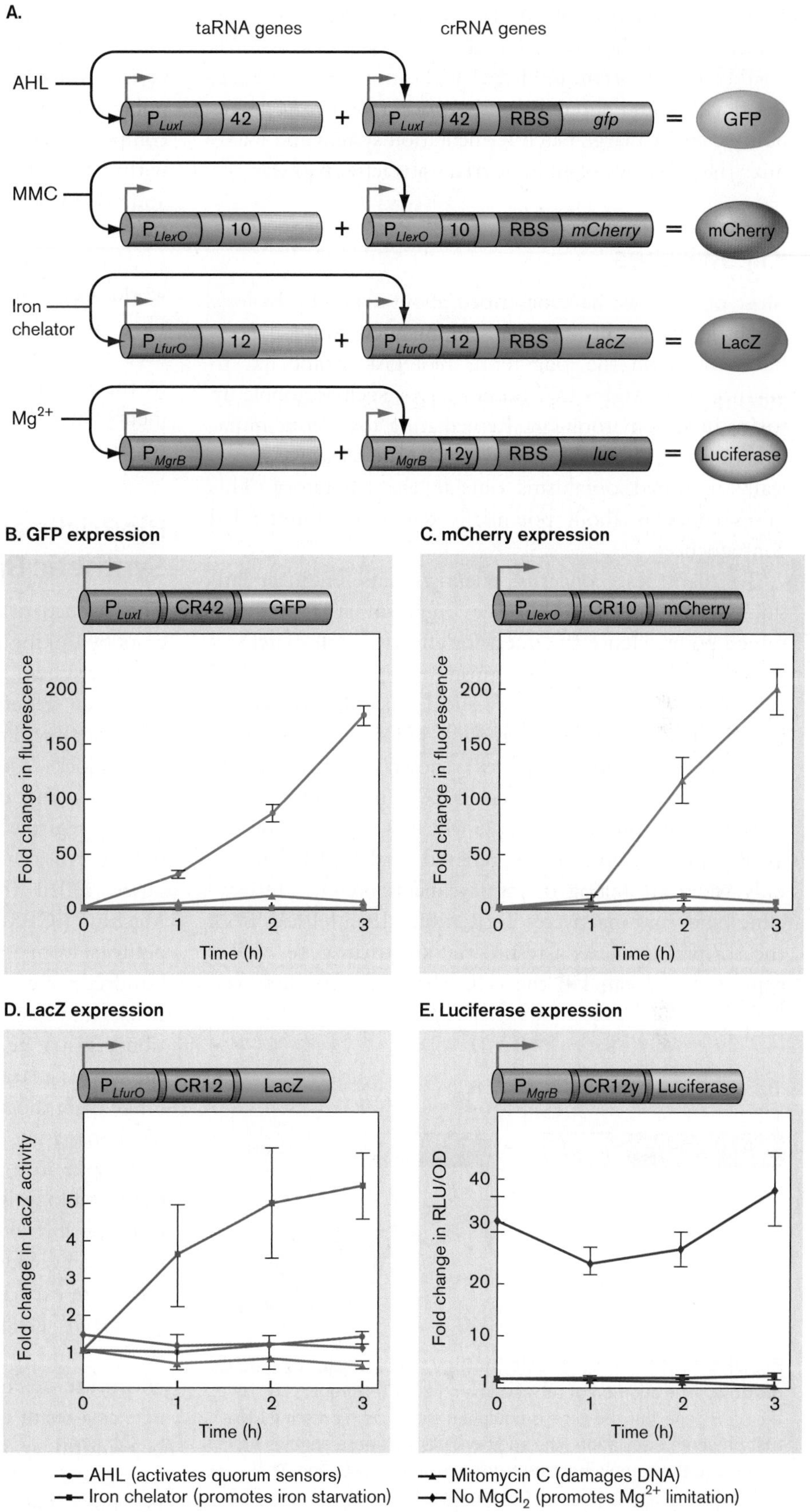
A.
taRNA genes
crRNA genes
AHL
P_{LuxI} 42
P_{LuxI} 42 RBS gfp
GFP
MMC
P_{LlexO} 10
P_{LlexO} 10 RBS mCherry
mCherry
Iron chelator
P_{LfurO} 12
P_{LfurO} 12 RBS LacZ
LacZ
Mg^{2+}
P_{MgrB}
P_{MgrB} 12y RBS luc
Luciferase
B. GFP expression
P_{LuxI} CR42 GFP
Fold change in fluorescence
Time (h)
C. mCherry expression
P_{LlexO} CR10 mCherry
Fold change in fluorescence
Time (h)
D. LacZ expression
P_{LfurO} CR12 LacZ
Fold change in LacZ activity
Time (h)
E. Luciferase expression
P_{MgrB} CR12y Luciferase
Fold change in RLU/OD
Time (h)
AHL (activates quorum sensors)
Iron chelator (promotes iron starvation)
Mitomycin C (damages DNA)
No $MgCl_2$ (promotes Mg^{2+} limitation)

FIGURE 12.23 ■ Riboswitchboard. **A.** Four separate circuits were designed. Each circuit is controlled by a different promoter that senses a different environmental parameter: P_{LuxI} (AHL, acylhomoserine lactone), P_{LlexO} (MMC, mitomycin C), P_{LfurO} (iron), or P_{MgrB} (magnesium). Different riboswitch pairs (labeled 42, 10, 12, and 12y) were used to control the translation of the output reporters for each circuit. **B–E.** These graphs show that each circuit responded to only a single environmental input signal. Inducers for GFP, mCherry, and LacZ (B, C, and D) were added at time 0. The inoculum for the luciferase circuit (E) was grown in low Mg^{2+} and was active even at time 0.

through alternative metabolic pathways depending on which riboswitch was activated. Metabolic switchboards would have important industrial applications. For instance, the device could simultaneously sense a variety of metabolic states in a large batch fermentation system and maximize the efficiency of an industrially attractive pathway.

Kill Switches

Most of what we have described about synthetic biology has little chance of dangerous unintended consequences, but what about the long term? Biologists would like to engineer new strains that do new tasks, such as gobble up toxins in the environment. Remediating toxin contamination in nature, however, would require the release of genetically modified organisms outside the laboratory. This raises concerns about potentially dangerous unintended consequences.

To allay these concerns, scientists must engineer fail-safe mechanisms that kill the organism at a predetermined point. Hence the quest for effective "kill switches." A genetically modified organism equipped with a kill switch can be made to commit suicide once the bacterium's job is done. One proof-of-principle kill switch engineered by synthetic-biology techniques is shown in **Figure 12.24**.

The gene for CcdB, a potent DNA-damaging toxin, was linked to the promoter P_{LtetR}, which is repressed by the TetR protein. Repression is relieved (and CcdB is made) only when an analog of tetracycline is present. Tetracycline binds and inactivates TetR, which then releases from the *ccd* promoter. As a result, the kill toxin gene *ccdB* is expressed and can kill the cell. The engineers, however, added a fail-safe mechanism to prevent the organism from killing itself too early. They made sure that the translation of CcdB mRNA is blocked by a crRNA. A second component of the system, activated by arabinose, encodes the compensatory taRNA that can expose the RBS buried within the crRNA-CcdB message. So, adding tetracycline and arabinose is needed for the cell to effectively produce CcdB and kill itself.

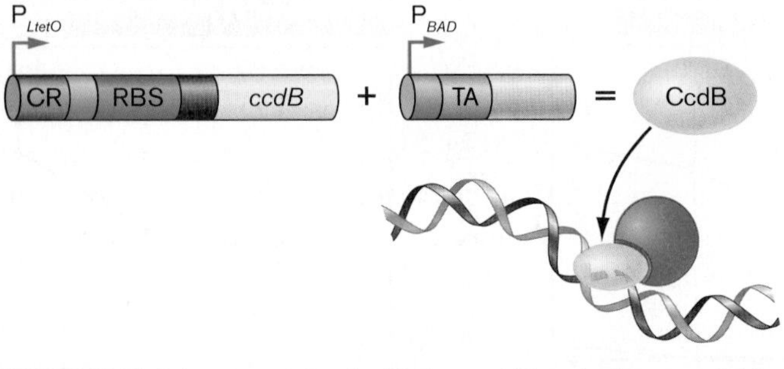

P_{LtetO}
P_{BAD}
CR
RBS
ccdB
+
TA
=
CcdB

FIGURE 12.24 ■ A synthetic-biology "kill switch," or suicide module. The addition of tetracycline relieves TetR repression of the *ccdB* gene, but the gene is equipped with a *cis*-repressing RNA that prevents translation. The addition of L-arabinose enables AraC to activate the complementary *trans*-activating RNA gene. The *ccdB* mRNA is translated, and CcdB causes DNA damage, which kills the cell.

Although this system, as designed, is impractical for real-world use, its success shows that kill switches can be made. How could this kill switch be modified so that it could be used in the real world? What if repression of CcdB was tied to the presence of a toxic product found in the environment? Once the organism destroys the toxin, the kill switch is "thrown," and the bacterium kills itself.

BioBricks and Do-It-Yourself Synthetic Biology

The science of synthetic biology makes new logic circuits by linking promoters from one system to genes from another system. The art of synthetic biology is to combine these new genetic logic circuits in ways that produce new functional systems. But where do all the genetic "Lego-like" blocks come from? Laboratories around the world construct and deposit their "building materials" into a central registry at the Massachusetts Institute of Technology. An annual weekend-long synthetic-biology showdown called the International Genetically Engineered Machine (iGEM) competition is also held at MIT. Competitors from universities over the years have assembled hundreds of connectible pieces of DNA that they call BioBricks and have deposited them in the BioBricks Foundation registry at MIT (the iGEM competition trophy, by the way, is a large aluminum Lego; **Fig. 12.25**). BioBricks range from those that kill cells to one that makes cells smell like bananas.

Despite its potential for good, one concern about synthetic biology is that anyone, theoretically, can do it. In fact, the lure of constructing a new organism, combined with the ever-lowering cost of equipment needed to carry out these experiments, has spawned a community of "do-it-yourself" (DIY) genetic engineers, including high school students and so-called garage scientists. Most of these DIY efforts are well intentioned and could produce useful products as a result of outside-the-box thinking, but we should be mindful, again, of unintended consequences.

Weblinks Websites for BioBrick assembly parts (*see ebook*)

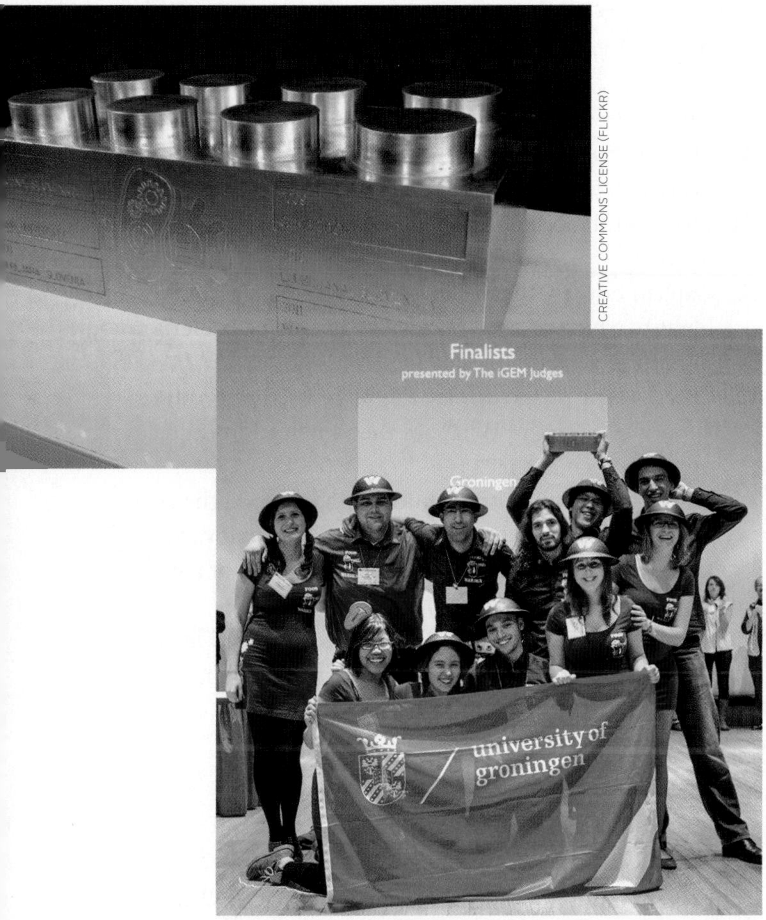
CREATIVE COMMONS LICENSE (FLICKR)
Finalists
presented by The iGEM Judges
university of
groningen
IGEM FOUNDATION AND JUSTIN KNIGHT (CREATIVE COMMONS LICENSE/FLICKR)

FIGURE 12.25 ■ Winners of the 2012 iBEMC competition trophy. The University of Groningen, Netherland, team (bottom) won the 2012 competition by engineering a strain of *Bacillus subtilis* that alerts when meat spoils. The top photo shows the actual Biobrick trophy of winners.

Genome Transplants: The Ultimate in Synthetic Biology

Tinkering with gene circuits is one thing, but carrying out a complete genome transplant is another. We have been able to manipulate small pieces of bacterial genomes for many years via mutation, cloning, and recombination techniques. But manipulating an entire genome has been out of reach—until now. Scientists have taken the genome of one bacterial species, *Mycoplasma mycoides,* moved it to yeast (they added a yeast centromere so it would replicate), modified the genome using yeast genetics, and then transplanted the modified genome into a different species, *Mycoplasma capricolum,* replacing the *M. capricolum* genome. This biotechnological advance is especially exciting because genetic manipulation of *M. mycoides* was previously impossible. Using this technique, we might redesign prokaryotic systems and even engineer new species.

To Summarize

- **Synthetic biology** applies engineering principles to design and constructs new biological parts for a desired purpose.
- **Synthetic-biology circuits use genetic logic gates,** such as buffer gates, NOT gates, and OR gates.
- **A toggle switch**, built from two NOT gates, can turn a gene on or off by sensing two different chemical signals.
- **An oscillating switch** will repeatedly turn on and off in response to a signal.
- **System noise** is the fluctuation of gene expression among single cells of a population. System noise can dull the clarity of a response circuit. **Riboswitches** engineered into a genetic circuit can minimize noise.
- **Genetic kill switches** can eliminate a genetically modified organism when its job is done.
- **Complete genome transplants** will enable the genetic manipulation of organisms in which traditional methods have failed.

Concluding Thoughts

In this chapter we have discussed only some of the available molecular techniques and developing technologies (such as synthetic biology) used to probe biological processes and manipulate them to our benefit. Keep them in mind as we progress deeper into the physiology of microbial growth and the mechanisms of microbial pathogenesis. Many of the approaches described here and in Appendix 3 were used to elucidate the concepts presented in later chapters.

CHAPTER REVIEW

Review Questions

1. What are the important considerations when designing a mutant selection strategy?
2. Once the DNA sequence of a gene is known, what specific methods can be used to gain clues as to the possible function of the gene product?
3. What are the key features of reporter genes used to measure transcriptional control? Translational control?
4. What is the difference between a northern blot and a Southern blot?
5. Explain the principle of a western blot.
6. What techniques are used to determine whether a protein binds to a DNA sequence?
7. Explain the important points for constructing a plasmid that makes His_6-tagged protein.
8. Explain how one can determine the start site of a transcript.
9. How can protein-protein interactions be determined in vitro?
10. Describe two-hybrid analysis. What are bait and prey proteins?
11. How can the cellular location of a protein be tracked in vivo?
12. Explain how PCR can be used in a single reaction to identify which of two species of pathogenic bacteria are present in a food sample.
13. Discuss how phage display might be used to produce a more toxic toxin. What are the ramifications of technologies such as this?
14. Discuss some ways biotechnology has helped the agricultural industry. How has the field impacted vaccine delivery or our approaches to the treatment of metabolic diseases in humans?
15. Define how BioBricks and logic gates are used in synthetic biology.

Thought Questions

1. You have identified two regulatory proteins, RegA and RegB, which affect the expression of gene *yxx*, a gene involved in quorum sensing. Mutants lacking RegA or RegB fail to express *yxx*. Describe two alternative models that would explain this. Next, describe how results from ChIP-on-chip analysis could help distinguish between those possibilities.
2. The Gram-negative bacillus *Shigella dysenteriae*, the agent causing bacillary dysentery, will live and grow inside eukaryotic cells. Despite lacking flagella, the organism moves through the cytoplasm of the host cell by polymerizing host actin at one pole of the cell. This pushes the organism through the cytoplasm with such force that the bacterium will poke into an adjacent cell. You have identified a *Shigella* protein that you think helps mediate actin tail formation and suspect that the protein might localize to one pole of the cell. What biotechnique would you use to test your hypothesis?
3. In Chapter 10 we discussed the *lac* operon and the *Salmonella* phase variation system in which DNA recombination controls a switch between two different flagellin proteins. You would like to understand more about the mechanism of regulation, so you have genetically engineered a strain of *E. coli* to contain the following genes: (1) the *hin* gene controlled by the *lacP* promoter, (2) a transcriptional activator gene flanked by *hix* sites, and (3) a green fluorescent protein gene (*gfp*) controlled by the transcriptional activator. Briefly describe what happens if (a) the strain is grown in lactose, (b) the strain is grown without lactose, (c) you delete the *lacI* gene, and (d) you eliminate the *lac* operator.

Key Terms

affinity chromatography (466)
autoradiography (465)
ChIP-on-chip analysis (471)
chromatin immunoprecipitation (ChIP) (471)
differential RNA-sequencing (dRNAseq) (467)
electrophoretic mobility shift assay (EMSA) (465)
fluorescence resonance energy transfer (FRET) (470)
gene fusion (461)
interactome (472)
noise (481)
northern blot (463)
operon fusion (461)
phage display (476)
primary antibody (464)
primer extension (466)
real-time PCR (468)
reporter gene (461)
riboswitch (482)
secondary antibody (465)
Southern blot (463)
synthetic biology (477)
TaqMan (470)
transcriptional fusion (461)
translational fusion (461)
western blot (464)
yeast two-hybrid system (472)

Recommended Reading

Ahmad, Parvaiz, Muhammad Ashraf, M. Muhammad Younis, Xiangyang Hu, Ashwani Kumar, et al. 2012. Role of transgenic plants in agriculture and biopharming. *Biotechnology Advances* **30**:524–540.

Baumgardner, Jordan, Karen Acker, Oyinade Adefuye, Samuel T. Crowley, Will DeLoache, James O. Dickson, et. al. 2009. Solving a Hamiltonian Path Problem with a bacterial computer. *Journal of Biological Engineering* **3**:11.

Bouveret, Emmanuelle, and Christine Brun. 2012. Bacterial interactomes: From interactions to networks. *Methods in Molecular Biology* **804**:15–33.

Broderick, Nicole A., Courtney J. Robinson, Matthew D. McMahon, Jonathan Holt, Jo Handelsman, et al. 2009. Contributions of gut bacteria to *Bacillus thuringiensis*–induced mortality vary across a range of Lepidoptera. *BMC Biology* **7**:11.

Huang, Johnny X., Sharon L. Bishop-Hurley, and Matthew A. Cooper. 2012. Development of anti-infectives using phage display: Biological agents against bacteria, viruses and parasites. *Antimicrobial Agents and Chemotherapy* **56**:4569–4582.

Locke, James C., Jonathan W. Young, Michelle Fontes, Maria J. Hernandez Jimenez, and Michael B. Elowitz. 2011. Stochastic pulse regulation in bacterial stress response. *Science* **334**:366–369.

Martínez-Garcia, Esteban, and Victor de Lorenzo. 2012. Transposon-based and plasmid-based genetic tools for editing genomes of Gram-negative bacteria. *Methods in Molecular Biology* **813**:267–283.

Pan, Zhiming, Xiaoming Zhang, Shizhong Geng, Ningning Cheng, Lin Sun, et al. 2009. Priming with a DNA vaccine delivered by attenuated *Salmonella typhimurium* and boosting with a killed vaccine confers protection of chickens against infection with the H9 subtype of avian influenza virus. *Vaccine* **27**:1018–1023.

Ruder, Warren C., Ting Lu, and James J. Collins. 2011. Synthetic biology moving into the clinic. *Science* **333**:1248–1252.

Saade, Fadi, and Nikolai Petrovsky. 2012. Technologies for enhanced efficacy of DNA vaccines. *Expert Review of Vaccines* **11**:189–209.

Tomljenovic-Berube, Ana M., David T. Mulder, Matthew D. Whiteside, and Brian K. Coombes. 2010. Identification of the regulatory logic controlling *Salmonella* pathoadaptation by the SsrA-SsrB two-component system. *PLoS Genetics* **6**:e1000875.

Tramonti, Angela, Paolo Visca, Michele De Canio, Maurizio Falconi, and Daniella De Biase. 2002. Functional characterization and regulation of *gadX*, a gene encoding an AraC/XylS-like transcriptional activator of the *Escherichia coli* glutamic acid decarboxylase system. *Journal of Bacteriology* **184**:2603–2613.

Trimble, Cornelia L., Shiwen Peng, Ferdynand Kos, Patti Gravitt, Raphael Viscidi, et al. 2009. A phase I trial of a human papillomavirus DNA vaccine for HPV16+ cervical intraepithelial neoplasia 2/3. *Clinical Cancer Research* **15**:361–367.

COURTESY OF DAN WOZNIAK

Dan Wozniak, College of Medicine, Ohio State University.

AN INTERVIEW WITH:

Dan Wozniak

Polymer Biosynthesis Makes a Pathogenic Biofilm

Dan Wozniak taught for 15 years at the Wake Forest School of Medicine, where he earned the Friend of Students Award for his contribution to the learning environment. He began teaching at the Ohio State University in 2008, where his students investigate how *Pseudomonas aeruginosa* infects the lungs of cystic fibrosis patients. They found that the bacteria synthesize a helical sugar polymer that interconnects cells, forming biofilms that colonize the lung. Wozniak's laboratory is currently working on means to disrupt the biofilm matrix so that these devastating infections might be prevented.

Why did you decide to study microbiology?

While taking a medical microbiology course in college, I became fascinated with how microbes can be beneficial and yet also agents of disease. One of my college professors made learning microbiology fun, in both the laboratory and classroom environment. He was rigorous but fair, and he would only give us credit on our answers if we correctly spelled the name of the organism (both genus and species!). I was also attracted to history and learned how important infectious diseases are in shaping the geography, politics, religion, and economics of many nations, including the United States.

How did you choose to study *Pseudomonas aeruginosa*? What makes this organism interesting?

I began working with *P. aeruginosa* as a graduate student, and the bug captured my attention so much that I have remained working with it since. I've been captivated by this organism because of its extreme versatility. Because of its nutritional adaptability, *P. aeruginosa* is readily isolated from diverse environments such as humans, animals, soil, water, plants, sewage, and hospitals. Some say it is the most abundant organism on Earth! Thus, humans are in continuous contact with the organism, and healthy people normally do not succumb to infections. However, people with a compromised immune system, such as those with cystic fibrosis [CF] and hospitalized patients, are extremely vulnerable to *P. aeruginosa* infections. *P. aeruginosa* is currently the second most common pathogen isolated from hospitalized patients, and it is consistently associated with the highest mortality rate.

How did you use molecular biology to figure out clues about biofilm formation?

When microbes attach to surfaces, they form aggregated communities called biofilms. Biofilm populations show resistance to assault by other cells and chemical agents, which normally kill free individual cells. Our recent work has focused on the matrix or glue that holds biofilm cells together. The long-term goal is to design inhibitors that either prevent synthesis of the matrix or break down an existing biofilm matrix.

We used bioinformatic models to predict which genes in *P. aeruginosa* might be involved in forming the biofilm matrix. Then, using molecular biology and genetics, graduate student Matthew Byrd and research technician Haiping Lu systematically disrupted 15 genes to determine which ones are required for biofilm matrix formation. We also used molecular biology to engineer strains of *P. aeruginosa* that overproduce a polysaccharide component of the matrix called Psl. This allowed us to purify large quantities of the material for structural determination by mass spectrometry and nuclear magnetic resonance. The structure revealed that Psl is a branched pentasaccharide composed of mannose, rhamnose, and glucose (3:1:1 ratio). Dr. Luyan Ma, a research scientist in our group, used specialized staining methods to visualize Psl on individual *P. aeruginosa* cells and within a biofilm. We utilized computer modeling with space-filling data to predict the structure of a single Psl pentasaccharide, as well as four linked repeating units. This modeling revealed clear periodicity that may account for the helical pattern of Psl on individual *P. aeruginosa* cells. We propose a model that the helical distribution of Psl on the surface of *P. aeruginosa* may readily promote interactions with Psl on adjacent bacteria.

We also visualized Psl distribution in biofilms. In a well-defined 3D microcolony structure, the Psl matrix was unevenly distributed. This was visualized in representative horizontal Z-images of a microcolony and a 3D reconstruction of the microcolony. Enhanced Psl staining was observed in the periphery of each microcolony. Direct visualization reveals that Psl is a key scaffolding matrix component and opens up avenues for therapeutics of biofilm-related complications.

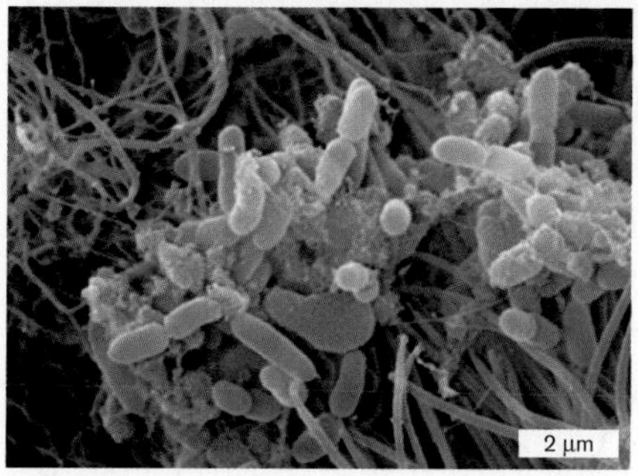
2 µm

COURTESY OF DAN WOZNIAK

Biofilm of *P. aeruginosa* growing on the human trachea.

How might your work help physicians treating cystic fibrosis and other diseases?

Since the matrix provides a critical protective role, as well as a scaffold for the developing biofilm, agents aimed at disrupting the matrix or preventing its synthesis would have therapeutic value. This represents a new target for *P. aeruginosa* and exposes a critical window of opportunity for immunological or chemotherapeutic intervention.

We are also investigating "pathoadaptive" changes that *P. aeruginosa* undergoes during the course of chronic infection in the CF lung. We are interested in identifying host and bacterial determinants that promote persistent infection of *P. aeruginosa* in the CF airway. In addition, we are collaborating with scientists at Wake Forest University Health Sciences to develop a multivalent vaccine against *P. aeruginosa*. The goal is to utilize this vaccine to prevent initial colonization of CF patients and other patient populations at risk for *P. aeruginosa* infections.

Does your work have implications for microbial ecology—for the survival and emergence of pathogens in the environment?

P. aeruginosa is a tremendously versatile organism. The basis for this versatility is a large arsenal of metabolic enzymes, as well as a large genome, which encodes numerous complex regulatory pathways, antimicrobial resistance determinants, and cell surface molecules. To illustrate this diversity, it is almost impossible to purify water to an extent that does not allow for the growth of *P. aeruginosa*! The genus *Pseudomonas* is an essential agent of the rhizosphere, and many pseudomonads are important mediators of plant disease. For many of us, how a normal inhabitant of the soil and water can, in certain situations, become pathogenic is the fundamental question. I suspect most, if not all, of the well-studied *P. aeruginosa* "virulence factors" serve essential roles for this organism in its natural ecological niche.

What advice do you have for today's students?

Become broadly educated in the life, social, and physical sciences. Major breakthroughs in the future will come from multidisciplinary efforts, and students must have a working knowledge of fields outside microbiology. I would also advise students to gain research experience early in their career. I recognize this isn't easy, as research opportunities do not always present themselves, but be persistent and flexible (consider volunteering!). Early exposure to the laboratory environment is critical, as it helps foster development of the scientific method, decision-making, and logic skills, and helps students with career choices.

How does your family relate to your work? Do you have interesting pursuits outside of science?

My parents and grandparents instilled in me a very strong "Midwest work ethic," which has helped me considerably. My father was an engineer who taught me how basic logic and reasoning skills can be applied to seemingly complex problems. My mother and grandmother were avid readers and had an excellent command of the English language. My wife, Brooke, has endured my frequent absences and has been a loyal partner and "sounding board" when I've needed her logical viewpoint. My children, Chris and Jessie, as well as my brother John, keep me honest by always asking me, "Why is this important?" or "Found a cure yet?"

I am also a passionate long-distance runner, having completed seven marathons (Boston twice!). My lab knows that when they can't find me, I'm probably out for a run.

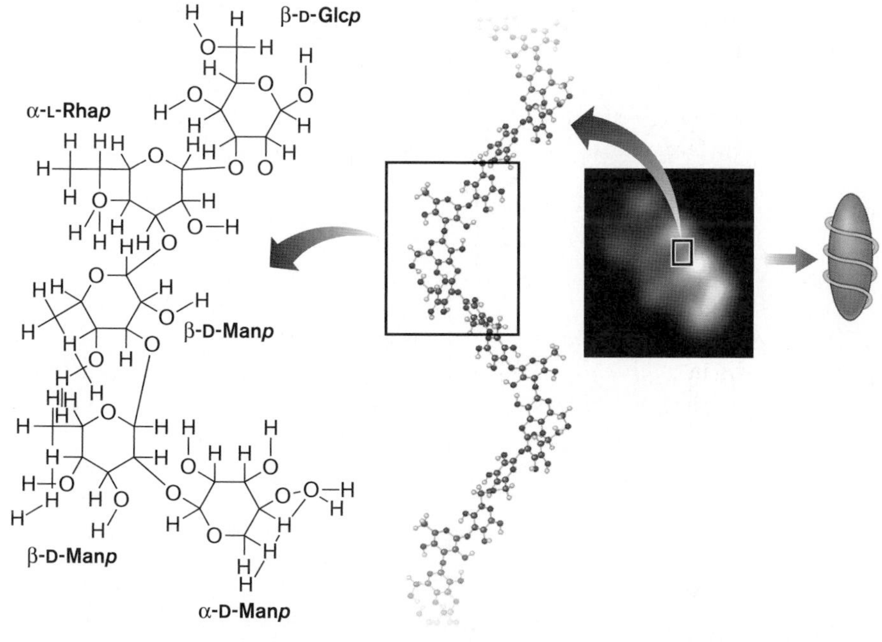
β-D-Glcp
α-L-Rhap
β-D-Manp
β-D-Manp
α-D-Manp

Psl polymer appears to be coiled around the cell surface. Modified from a figure provided by Dan Wozniak.

CHAPTER 13
Energetics and Catabolism

13.1 Energy and Entropy for Life
13.2 Energy in Biochemical Reactions
13.3 Energy Carriers and Electron Transfer
13.4 Catabolism: The Microbial Buffet
13.5 Glucose Breakdown and Fermentation
13.6 The Tricarboxylic Acid (TCA) Cycle
13.7 Aromatic Pollutants

All living cells need energy to move and grow. Growth requires energy to incorporate nonliving substances into new cells. The energy to build cells comes from chemical reactions. For microbes, the variety of such reactions is limitless; virtually any kind of molecule in our biosphere—from hydrogen gas to chlorinated pollutants—can yield energy for some kind of microbe.

To capture and use energy, microbes must regulate their energy-yielding reactions and couple them to biosynthesis, using enzymes. Enzymes direct the transfer of energy onto carriers such as ATP.

Many energy-yielding reactions used by microbes break down complex molecules into smaller ones—a process called catabolism. Microbes catabolize chemicals within our own digestive tracts and all around us, in the soil and water. Collectively, microbes show astonishing potential to catabolize nearly any organic substance, including petroleum, hardwoods, and synthetic polymers. Applications of microbial catabolism range from producing alcohol to generating electricity.

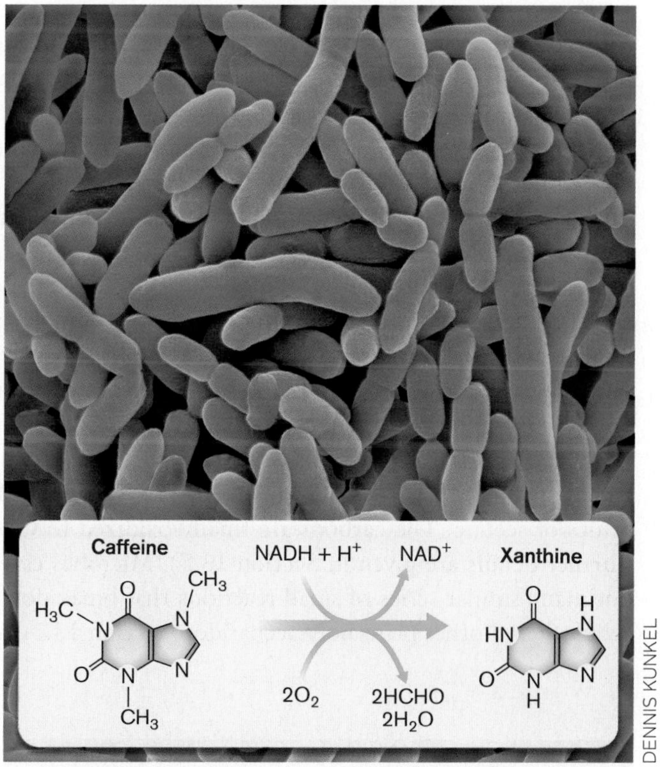

Caffeine
NADH + H⁺
NAD⁺
Xanthine
2O₂
2HCHO
2H₂O
DENNIS KUNKEL

CURRENT RESEARCH **highlight**

Eating caffeine. The soil bacterium *Pseudomonas putida* (colorized TEM) can break down many kinds of complex molecules for energy. Even caffeine—the "energy" beverage provides a literal source of energy. In 2012, Ryan Summers and co-workers reported the catabolic pathway by which bacteria break down caffeine. Converting caffeine to xanthine requires electrons from one molecule of NADH. The xanthine is further broken down to uric acid, and ultimately ammonia and carbon dioxide, in electron-transfer reactions that yield energy for the cell. We can use bacterial catabolism of caffeine to remediate soil contaminated by large-scale coffee production. ***Source:*** Ryan Summers et al. 2012. *J. Bacteriol.* **194**:2041

The U.S. Army manufactures billions of pounds of explosives such as trinitrotoluene (TNT). But waste TNT from the manufacturing process, and from outdated weapons, seeps into the soil. When a munitions plant shuts down, many acres of explosive-contaminated soil need to be remediated. To clean up the site, the army enlists the help of bacteria, by composting (**Fig. 13.1**). Compost bacteria and fungi consume the explosive molecules as food, gaining energy to grow new cells. To prepare compost, we mix nitrogen-rich material (such as explosives or food wastes) with cellulose-rich yard waste or corn husks. The mixed compost materials are aerated to provide oxygen for oxidative breakdown by bacteria and fungi.

So how do bacteria gain energy from TNT or cellulose? These carbon sources break down into a number of smaller products such as carbon dioxide. The products that are more stable (less reactive) and more disordered than the original collection of molecules. Such reactions yield energy. The breakdown of large molecules to a number of smaller ones, yielding energy, is called **catabolism**. Catabolism must release energy in many small steps—not all at once, as in combustion or an explosion. In TNT catabolism (**Fig. 13.1C**), the molecule is first denitrated to toluene and then catabolized (broken down) to two-carbon molecules of acetate. The carbons are finally oxidized to CO_2 (further details are given in Section 13.7). Microbes carry out many similar series of small reactions that break down cellulose and other plant polysaccharides (Section 13.2), as well as fats and amino acids. Microbes catabolize many carbon sources that animals cannot; for example, *Pseudomonas* species of soil bacteria may catabolize benzene, caffeine, or chlorinated pollutants.

Catabolism is one of several major classes of reactions that yield energy for life (**Table 13.1**). In these reactions, energy may derive ultimately from chemical rearrangement of molecules (prefix **chemo**-) or from light absorption (prefix **photo**-). Chemical pathways such as catabolism, in which organic compounds donate electrons to yield energy, are known as **organotrophy** or **chemoorganotrophy**. Organic compounds include the foods we eat—and which our intestinal bacteria help catabolize. Most organotrophs are also **heterotrophs**, organisms that use preformed organic compounds for biosynthesis (a process called **heterotrophy**).

But other bacteria and archaea gain energy from inorganic electron donors such as metals; this chemical pathway is known as **lithotrophy** or **chemolithotrophy** (literally "rock eater"). For example, the archaeon *Pyrodictium occultum* gains energy by oxidizing hydrogen gas with sulfur—an inorganic reaction. Still other microbes, such as marine cyanobacteria, gain energy from light absorption (**phototrophy** or **photosynthesis**). A photoautotroph gains energy solely from light and builds biomass solely from CO_2, whereas a photoheterotroph obtains energy from light but needs organic carbon for growth. You may review these fundamental terms in Section 4.1.

A.

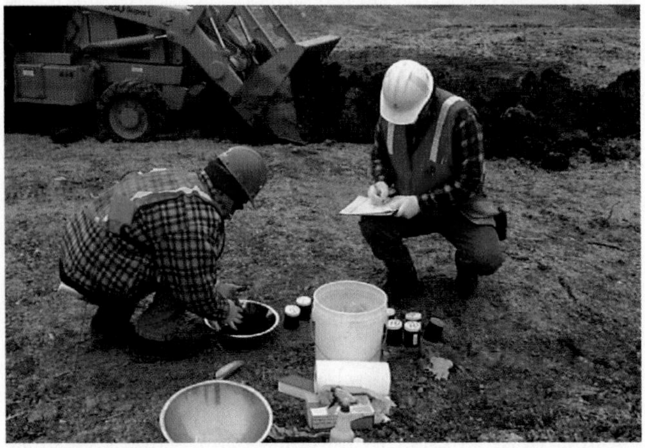

COURTESY OF EPA

B.

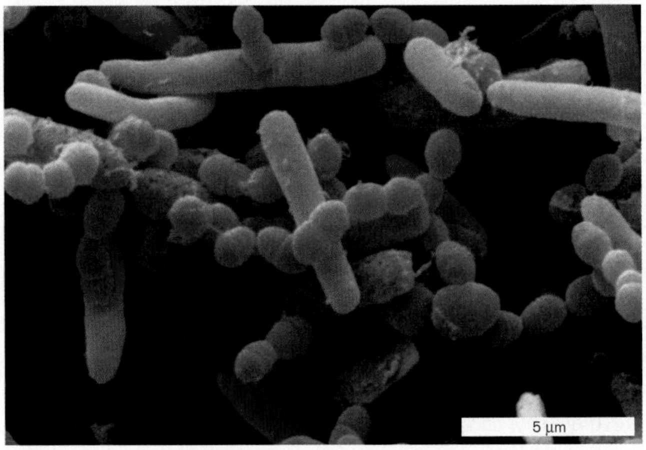
5 µm

SIMKO/VISUALS UNLIMITED, INC.

C.

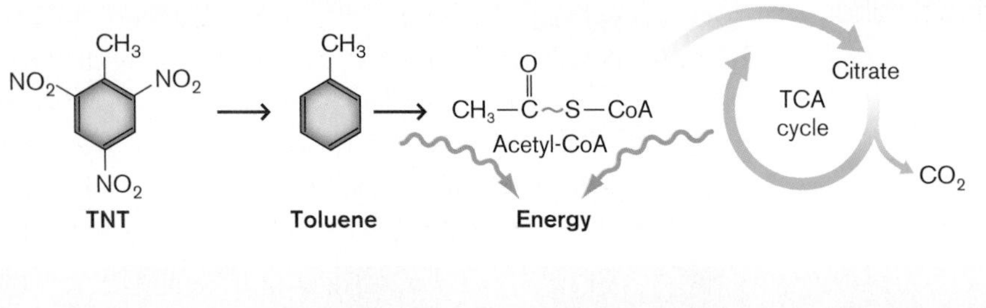
CH3
NO2
NO2
NO2
TNT
CH3
Toluene
O
CH3—C~S—CoA
Acetyl-CoA
Energy
TCA cycle
Citrate
CO2

FIGURE 13.1 ■ Microbial composting. **A.** Composting soil contaminated by explosives at the Joliet Army Ammunition Plant. **B.** Rod-shaped (bacilli) and spherical (cocci) bacteria are examples of microbes found in compost (colorized SEM). **C.** Microbial catabolism of trinitrotoluene (TNT) yields energy.

TABLE 13.1

Energy acquisition in bacteria and archaea.

Energy source	Class of metabolism	Examples of energy-yielding reactions	Electron acceptor	Systems for energy acquisition
		CHEMICAL		
Chemoorganotrophy Organic compounds (at least one C–C bond) donate electrons	**Fermentation** Catabolism	$C_6H_{12}O_6 \rightarrow 2C_3H_6O_3$ (or other small molecules)	Organic	Glycolysis and other catabolism
	Organic respiration Catabolism with inorganic electron acceptor, or with small organic electron acceptor	$C_6H_{12}O_6 + 6H_2O + 6O_2 \rightarrow 6CO_2 + 12H_2O$	O_2	Glycolysis and other catabolism, TCA cycle, and electron transport systems
		$C_6H_{12}O_6 + 6H_2O + 12NO_3^- \rightarrow 6CO_2 + 12H_2O + 12NO_2^-$	NO_3^-, SO_4^{2-}, Fe^{3+}, or other	
Chemolithotrophy Inorganic compounds donate electrons	**Lithotrophy or chemoautotrophy**	Electron donor for respiration is H_2, Fe^{2+}, H_2S, NH_4^+	O_2, NO_3^-, or other	Electron transport system
	Methanogenesis	Electron donor is H_2, $CO_2 + 4H_2 \rightarrow CH_4 + 2H_2O$	CO_2	Methanogenesis
		LIGHT		
Phototrophy Light absorption provides electrons	**Photoautotrophy** Light absorption drives CO_2 fixation	Photolysis of H_2O $6CO_2 + 12H_2O \rightarrow C_6H_{12}O_6 + 6H_2O + 6O_2$		Photosystems I and II
		Photolysis of H_2S, HS^-, or Fe^{2+} $6CO_2 + 12H_2S \rightarrow C_6H_{12}O_6 + 6H_2O + 12S$		Photosystem I or II
	Photoheterotrophy Light absorption without CO_2 fixation	Photolysis of H_2S, HS^-, or cyclic photophosphorylation Light-driven H^+ pump or Na^+ pump	Organic	Photosystem I or II Bacteriorhodopsin or proteorhodopsin

Note: Distinguish the following prefixes for "-trophy" terms.

Carbon source for biomass:

Auto-: CO_2 is fixed and assembled into organic molecules.
Hetero-: Preformed organic molecules are acquired from outside and assembled.

Energy source:

Photo-: Light absorption excites electron to high-energy state.
Chemo-: Chemical electron donors are oxidized.

Electron source:

Litho-: Inorganic molecules donate electrons.
Organo-: Organic molecules donate electrons.

Catabolism and other kinds of energy-yielding reactions enable microbes to "do the work" of growing new cells. The energy released from a reaction is transferred by enzymes to other reactions that build simple molecules into a complex cell. Chapter 13 explains how microbes gain energy from chemical reactions, by transferring electrons between molecules. Chapter 14 explores electron transfer in greater depth, through pathways of organic respiration (oxidation of organic nutrients), lithotrophy (oxidation of inorganic nutrients), and photosynthesis. The energy from all these pathways is used to build cells by pathways of **anabolism** (biosynthesis), the focus of Chapter 15. Finally,

Chapter 16 explores commercial applications of microbial metabolism to produce food and beverages, as well as industrial products and pharmaceuticals.

13.1

Energy and Entropy for Life

Every form of life, from a composting microbe to a human body, uses energy. **Energy** is the ability to do work, such as flagellar propulsion or cell growth. Energy is used to organize proteins and to maintain ion gradients across the cell membrane. Yet the second law of thermodynamics tells us that systems tend to become less ordered and that **entropy**, the disorder, or randomness, of the universe, always increases. So how do cells assemble simpler molecules into complex forms (**Fig. 13.2**)? How does life build order out of disorder?

Microbes Use Energy to Build Order

Since the universe overall becomes more disordered, how can microbes use energy to assemble less-ordered molecules into a complex cell? Organisms build order by spending the energy they gain through catabolism, lithotrophy, or phototrophy (see **Table 13.1**). As order increases, the cell is said to decrease in entropy, or disorder. But the decrease in entropy is local to the cell, and temporary. Ultimately, the cell's energy must be spent as heat, which radiates away, causing entropy to increase. In other words, the local, temporary gain of energy enables a cell to grow. Continued growth requires continual gain of energy and continual radiation of heat. We see this release of heat, for example, in a compost pile, where heat is produced faster than it dissipates. The temperature of compost typically rises to 60°C.

A similar energy economy occurs throughout the biosphere. In Earth's biosphere, the total metabolism of all life must dissipate most energy as heat. Biological heat production is not always obvious, because soil and water provide a tremendous heat sink. But overall, Earth's biosphere behaves as a giant thermal reactor (**Fig. 13.3**). As solar radiation reaches Earth, a small fraction is captured by photosynthetic microbes and plants. The fraction captured is largely in the range of visible light (**Fig. 13.4**), the range of wavelengths in which photon energies are appropriate for the controlled formation and dissociation of molecular bonds. Some bacteria conduct photosynthesis using ultraviolet and near-infrared radiation. At shorter wavelengths (X-rays), chemical bonds are broken indiscriminately; at

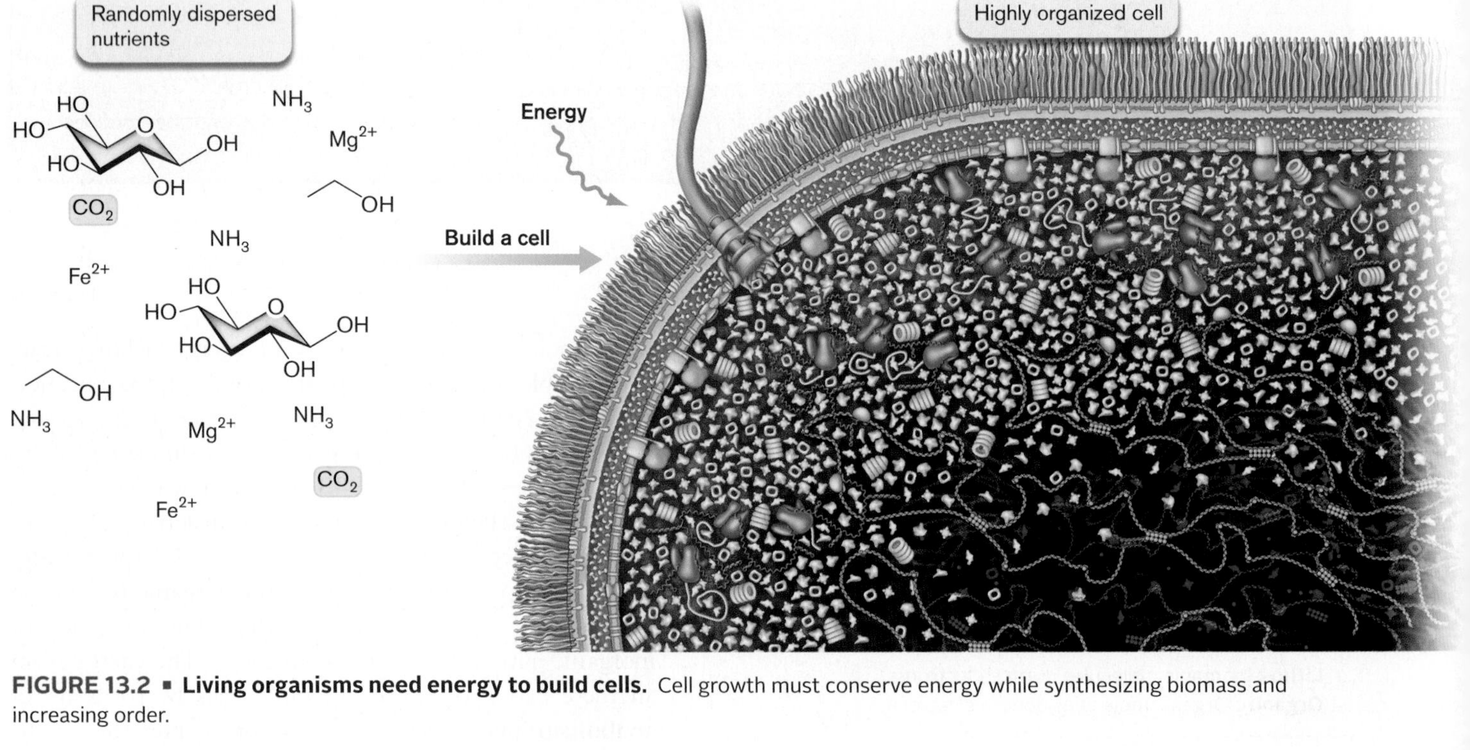
Randomly dispersed nutrients
HO
HO
O
OH
HO
OH
NH3
Mg2+
OH
CO2
NH3
Fe2+
NH3
Mg2+
NH3
CO2
Fe2+
Energy
Build a cell
Highly organized cell

FIGURE 13.2 ■ Living organisms need energy to build cells. Cell growth must conserve energy while synthesizing biomass and increasing order.

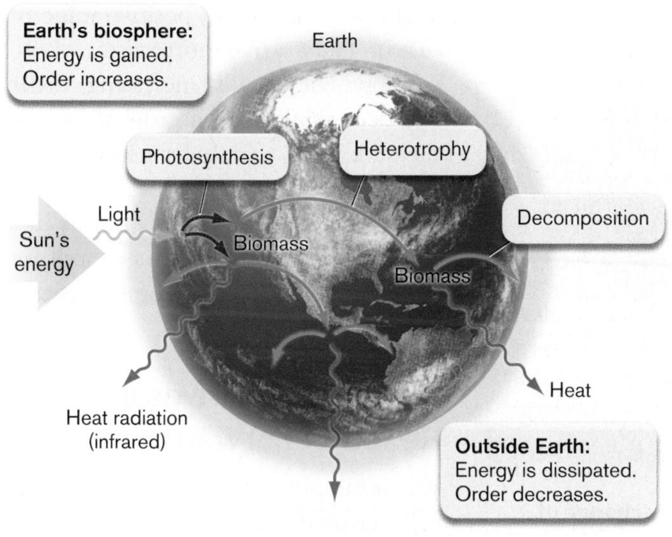
Earth's biosphere:
Energy is gained.
Order increases.
Earth
Photosynthesis
Heterotrophy
Light
Sun's energy
Biomass
Decomposition
Biomass
Heat
Heat radiation (infrared)
Outside Earth:
Energy is dissipated.
Order decreases.

FIGURE 13.3 ■ Solar energy. Solar radiation reaches Earth, where a small fraction is captured by photosynthetic microbes and plants. The microbial and plant biomass enters heterotrophs and decomposers, which convert a small fraction to biomass at each successive level. At each level, the majority of energy is lost, radiated from Earth as heat.

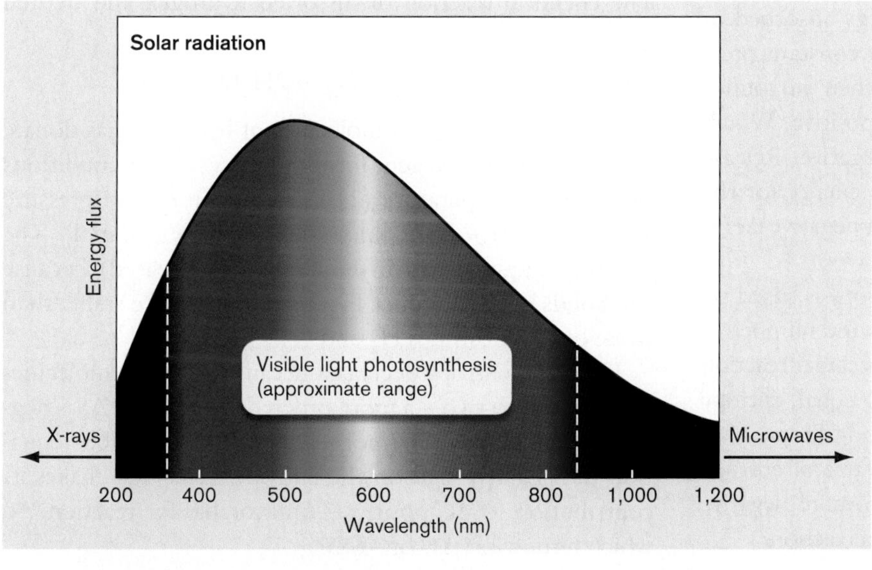
Solar radiation
Energy flux
Visible light photosynthesis (approximate range)
X-rays
Microwaves
200
400
500
600
700
800
1,000
1,200
Wavelength (nm)

FIGURE 13.4 ■ The solar spectrum. The Sun radiates across the spectrum, but the intensity of solar radiation reaching Earth peaks in the range of visible light.

longer wavelengths (microwaves and radio waves), the quantum energy is too low to drive chemical reactions.

Microbial and plant photosynthesis generates biomass, which is consumed by heterotrophs and decomposers. The consumers store a small fraction of their energy in biomass. At each successive level, the majority of energy is lost, radiated from Earth as heat. Despite growth of living organisms on Earth, the universe as a whole becomes more disordered. The complex roles of microbial metabolism in global ecosystems are discussed in Chapters 21 and 22.

Note: The principles of energy change, discussed next, apply to all the reactions of Table 13.1. These reactions are covered in detail in Chapters 13 and 14.

Gibbs Free Energy Change

From **Table 13.1** we see that microbes can harness an enormous variety of chemical reactions for growth. But what determines whether a given reaction can support life?

To provide energy to a cell, a biochemical reaction must go forward from reactants to products. The direction of a reaction can be predicted by a thermodynamic quantity known as the **Gibbs free energy change, ΔG** (also known as free energy change or Gibbs energy change). The concept of ΔG is reviewed in Appendix 1. The ΔG value of a reaction determines how much energy is potentially available to do work, such as to drive rotary flagella, to build a cell wall, or to store accurate information in DNA. The sign of the free energy change, ΔG, determines whether a process may go forward. If ΔG is negative, the process may go forward, whereas positive values mean that the reaction will go in reverse. The sign of ΔG determines which "foods" a microbe can eat or, more precisely, which reactions between available molecules can be harnessed for microbial growth.

Under defined laboratory conditions, calculating the ΔG value of a reaction can predict how much biomass microbes will build (**Fig. 13.5**). The reactions listed in **Figure 13.5B** are various catabolic reactions performed by our intestinal bacteria, as well as bacteria in soil. When oxygen is available, oxidative catabolism of sugars or of short-chain acids yields relatively large amounts of energy. Once oxygen is used up, some bacteria can oxidize food with an alternative electron acceptor such as nitrate (NO_3^-), or else ferment the carbon source with no mineral oxidant. Whatever the reaction, a linear relationship appears between ΔG and the biomass produced. Under natural conditions, however, the relationship becomes more complicated.

A.

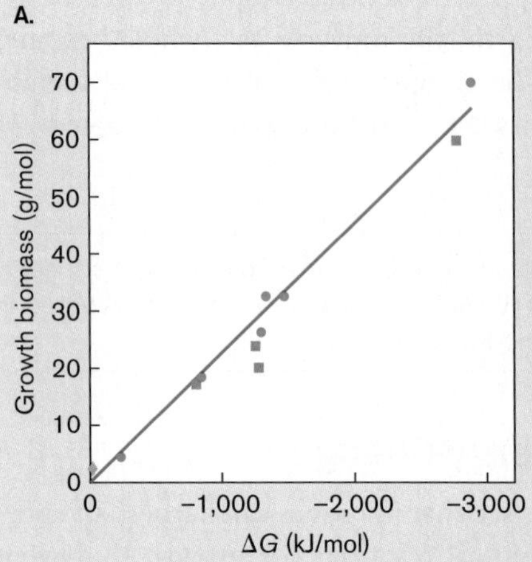

Growth biomass (g/mol)
70
60
50
40
30
20
10
0
0
−1,000
−2,000
−3,000
ΔG (kJ/mol)

B.

Carbon source	Oxidant	ΔG (kJ/mol)	Biomass (g/mol)
Glucose	O_2	−2,883.0	70.54
Propionate	O_2	−1,487.0	32.54
Lactate	O_2	−1,333.0	32.83
Ethanol	O_2	−1,308.0	26.32
Acetate	O_2	−847.0	18.72
Formate	O_2	−234.0	4.99
Glucose	NO_3^-	−2,774.0	60.00
Lactate	NO_3^-	−1,282.0	20.20
Ethanol	NO_3^-	−1,257.0	24.00
Acetate	NO_3^-	−813.0	16.95
Ethanol	None	−14.6	3.50

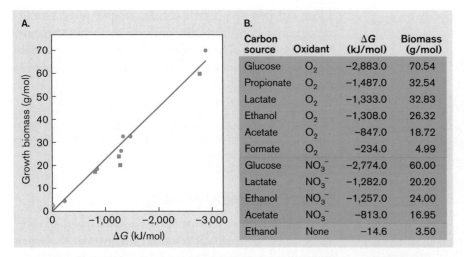

FIGURE 13.5 ■ Bacterial growth biomass depends on free energy change of metabolic reactions. A. Observed biomass of cells grown on given pairs of carbon source and oxidant is plotted as a function of free energy change (ΔG). Product is CO_2, except for ethanol without oxidant (green triangle), whose products are acetate and hydrogen gas. **B.** Energy data used for panel A. Oxidant types (O_2, NO_3^-, none) are color-coded. *Source:* Data from Eric E. Roden and Qusheng Jin. 2011. *Appl. Environ. Microbiol.* **77**:1907.

ΔG includes enthalpy and entropy. The free energy change ΔG has two components:

- **ΔH** = change in **enthalpy**, the heat energy absorbed or released as reactants become products at constant pressure. When reactants absorb heat from their surroundings as they convert to products, ΔH is positive. When, instead, heat energy is released, ΔH is negative. Release of heat (negative value of ΔH) can yield energy for the cell to use. An example of a reaction with negative ΔH is the oxidation of glucose by O_2.
- **ΔS** = change in **entropy**, or disorder. Entropy is based on the number of states of a system, such as the number of possible conformations of a molecule. If a cellular reaction splits one molecule into two, all else being equal, entropy increases; the system is more disordered, and ΔS is positive. Most catabolic reactions have a positive value of entropy change. A positive value of ΔS makes ΔG more negative and increases the potential energy yield of a reaction.

The relationship of the free energy change ΔG with ΔH and ΔS is given by:

$$\Delta G = \Delta H - T\Delta S$$

The overall sign of ΔG depends on its two components: ΔH (the absorption or release of heat energy) and $-T\Delta S$ [the product of entropy change (ΔS) and temperature (T)]. In living organisms, a sufficiently negative ΔH (energy lost as heat) often overrides $-T\Delta S$, the term for increase in order (negative value of ΔS, positive value of $-T\Delta S$). Thus, a living organism, whose development entails increasing order and decreasing ΔS, can grow as long as the sum of its metabolism has a sufficiently negative value of ΔH. The heat loss associated with ΔH is obvious in a compost pile, and it occurs as well for all living organisms and communities.

Negative ΔG Drives a Reaction Forward

An example of a thermodynamically favored reaction is the oxidation of hydrogen gas (H_2) to form water. Hydrogen is oxidized for energy by many kinds of bacteria in soil and water (a form of lithotrophy discussed in Chapter 14). For example, hydrogenotrophic bacteria of the genus *Ralstonia* have been isolated from ultrapure water used for nuclear fuel storage, where radioactive ionization generates H_2. The chemical reaction of dissolved hydrogen and oxygen gases is:

$$2H_2 + O_2 \rightarrow 2H_2O$$

In this reaction, two molecules of hydrogen gas donate four electrons to oxygen, forming water. Under conditions of standard temperature (298 kelvins, or K) and pressure (sea level), ΔH = −572 kilojoules per mole (kJ/mol). The ΔH is strongly negative (much heat is released) because the bonds of the product H_2O are much more stable than those of the substrates H_2 and O_2.

However, entropy decreases because the three molecules are replaced by two—a more ordered state. Thus, ΔS is negative [$\Delta S = -0.327$ kJ/(mol · K)]; and in the Gibbs equation the negative sign on the entropy term $-T\Delta S$ makes its contribution to ΔG positive—unfavorable for reaction. So which term wins: ΔH or $-T\Delta S$?

$$\begin{aligned}\Delta G &= \Delta H - T\Delta S \\ &= -572 \text{ kJ/mol} - (298 \text{ K})\,[-0.327 \text{ kJ/(mol} \cdot \text{K)}] \\ &= -562 \text{ kJ/mol} + 97 \text{ kJ/mol} \\ &= -475 \text{ kJ/mol}\end{aligned}$$

Overall, ΔG is negative, so bacteria with the appropriate enzyme pathways can use the reaction of hydrogen gas with oxygen to provide energy.

Note: The **joule (J)** is the standard SI unit to denote energy. 1 kilojoule (kJ) = 1,000 joules. Another unit commonly used is the kilocalorie (kcal). The conversion factor is: 1 kJ = 0.239 kcal.

Enthalpy and Entropy in Metabolism

How do ΔH and $-T\Delta S$ affect biochemical reactions? Each reaction in a living cell is associated with these two types of energy change, but to differing degrees depending on the reaction. In general, a reaction yields energy for the cell if:

- **Molecular stability increases.** When reactants combine to form products with more stable bonds, the reaction has a negative ΔH. For example, the reaction of a sugar with oxygen has a negative ΔH because the relatively unstable oxygen molecules are reduced to H_2O.
- **Entropy increases.** Reactions in which a complex molecule is broken down to a greater number of smaller molecules increase entropy (positive ΔS, negative $-T\Delta S$). An important class of such product molecules is carboxylic acids, such as lactic acid, in which a H^+ dissociates from a carboxylate ion ($RCOO^-$). Entropy also increases with conversion of a solid reactant to a gas, such as CO_2. For example, glucose may be fermented to ethanol and CO_2, as in the production of alcoholic beverages.

The relative contributions of ΔH and $-T\Delta S$ are important for physiology because they determine the effect of temperature on microbial growth. ΔH-dependent reactions such as glucose oxidation release a lot of heat, causing, for example, the rise in temperature of an aerated compost pile. On the other hand, the entropy component of Gibbs energy, $-T\Delta S$, does not reflect heat loss, but its magnitude grows larger as temperature increases. An extreme case of entropy-driven metabolism is that of acetate conversion to methane and CO_2, conducted by soil archaea such as *Methanosarcina barkeri*. This methanogen's growth is driven solely by the increase in entropy. Because the sign of ΔH is positive, the organism actually absorbs heat and cools its environment. But the larger magnitude of the entropy change $-T\Delta S$ allows the reaction to proceed, yielding net free energy for the cell to grow.

How do we measure ΔG and its components, ΔH and $-T\Delta S$? The amount of energy released by a reaction is measured in an instrument called a **calorimeter**. A calorimeter may be designed to measure thermal energy (heat) released by a reaction of pure chemicals in a test tube, or of metabolizing bacteria—or even a large farm animal (**Fig. 13.6**). The cow's thermal energy includes microbial digestion coupled to its own digestive processes, and to its own cellular biosynthesis; what is measured is residual energy lost as heat. In a calorimeter, the release of heat during a reaction (or collection of reactions) maintained at constant temperature gives a measure of ΔH. Measurement conducted at various temperatures (T) yields the temperature dependence of heat release, from which we calculate $-T\Delta S$. The sum of the two components yields ΔG.

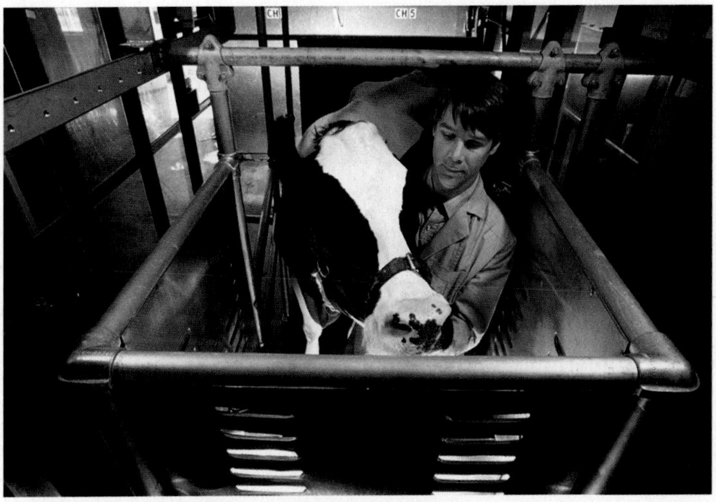

HTTP://WWW.ARS.USDA.GOV/IS/GRAPHICS/PHOTOS/JUL98/K6010-15.HTM

FIGURE 13.6 ■ Calorimeter for animal feed. A cow is prepared for a feeding test within a large animal calorimeter at the U.S. Department of Agriculture (USDA) in Beltsville, Maryland. The test will show how much energy the cow (and its digestive microbes) obtains from feeding on ground corn.

To Summarize

- **Energy** enables cells to build ordered structures out of simple molecules from the environment.
- **Cells release heat continually** because transfer of energy to build biomass is never perfectly efficient.
- **The free energy change (ΔG)** includes enthalpy (ΔH), the heat energy absorbed or released; and $-T\Delta S$, the product of temperature and entropy change (ΔS). Most reactions include both ΔH changes (such as oxidation-reduction) and ΔS changes (such as breakdown to a larger number of products).
- **Negative values of ΔG** show that a reaction can drive the cell's metabolism. The sign of ΔG depends on the relative magnitude of ΔH and $-T\Delta S$.
- **ΔH-driven reactions release heat.**
- **$T\Delta S$-dependent reactions do not release heat,** but they cause a greater change in free energy (ΔG) at higher temperature.

Thought Questions

13.1 Consider glucose catabolism in your blood, where the sugar is completely oxidized by O_2 and converted to CO_2:

$$C_6H_{12}O_6 + 6O_2 \rightarrow 6CO_2 + 6H_2O$$

Do you think this reaction releases greater energy as heat, or by change in entropy? Explain.

13.2 The bacterium *Lactococcus lactis* was voted the official state microbe of Wisconsin, because of its importance for cheese production. During cheese production, *L. lactis* ferments milk sugars to lactic acid:

$$C_6H_{12}O_6 \rightarrow 2C_3H_6O_3 \rightleftharpoons 2C_3H_5O_3^- + 2H^+$$

Large quantities of lactic acid are formed, with relatively small growth of bacterial biomass. Why do you think this occurs? Cheese making usually runs more efficiently at high temperature; why?

13.2

Energy in Biochemical Reactions

For a given reaction in a given environment, many factors determine ΔG. These factors fall into two classes—those intrinsic to the reaction, and those dependent on the environment:

- **Intrinsic properties of a reaction.** The intrinsic properties of a reaction are the changes of ΔH and ΔS contributing to ΔG. We can define standard values of these properties relative to arbitrary standard conditions such as concentration and temperature.
- **Concentrations and environmental factors.** The direction of the reaction depends on the concentrations of reactants and products. An excess of reactants over products makes ΔG more negative (favoring the forward reaction), whereas an excess of product makes ΔG more positive (favoring the reverse reaction). Reaction direction also depends on environmental factors such as temperature, pressure, and ionic strength (salt concentrations).

Standard Reaction Conditions

Scientists commonly present thermodynamic values under standard conditions for temperature, pressure, and concentration. The standard conditions enable scientists to compare the intrinsic properties of reactions. The standard Gibbs free energy change is designated $\Delta G°$. The standard conditions for $\Delta G°$ are as follows:

- The temperature is 298 K (25°C).
- The pressure is 1 atm (standard atmospheric pressure).
- All concentrations of substrates and products are 1 molar (M).

For every chemical compound, a standard $\Delta G°$ of formation from its elements can be determined. **A table of standard $\Delta G°$ values for common biological molecules is presented in Appendix 1.** As shown in Appendix 1, we can use $\Delta G°$ values to obtain the $\Delta G°$ for a molecular reaction under standard conditions, by summing the $\Delta G°$ values of reactants, and subtracting the sum of $\Delta G°$ values of the products.

But the $\Delta G°$ values reported in data tables hold only for isolated reactions under "standard reaction conditions." Standard conditions differ greatly from the actual conditions of living cells. These conditions include temperature, ionic strength, and gas pressure (in the case of gaseous components, such as CO_2), as well as the concentrations of reactants and products. To account for some differences, biochemists add special standard conditions: the hydrogen ion concentration at pH 7, because living cells commonly maintain their cytoplasm within a unit of neutral pH; and the water concentration of 55.5 M for dilute solutions. The free energy change in biochemistry is thus designated $\Delta G°'$.

The additivity of energy change is central to all living metabolism. Additivity makes it possible to "do work" by coupling an energy-yielding reaction to an energy-spending reaction (Section 13.3). **Figure 13.7** shows an

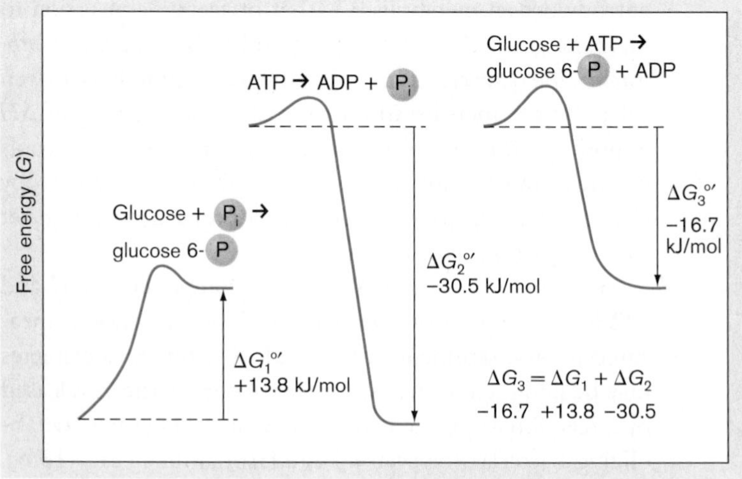

Free energy (G)
Glucose + ATP →
glucose 6-P + ADP
ATP → ADP + P_i
Glucose + P_i →
glucose 6-P
ΔG_1°'
+13.8 kJ/mol
ΔG_2°'
−30.5 kJ/mol
ΔG_3°'
−16.7
kJ/mol
ΔG_3 = ΔG_1 + ΔG_2
−16.7 +13.8 −30.5

FIGURE 13.7 ▪ Calculating the standard free energy of a reaction. The $\Delta G°'$ value for the phosphorylation of glucose by ATP equals the sum of the values for the two coupled reactions: phosphorylation of glucose ($\Delta G_1°'$) and ATP hydrolysis ($\Delta G_2°'$).

example, the initial reaction of glycolysis (Section 13.5), in which ATP phosphorylates glucose to glucose 6-phosphate, catalyzed by the enzyme hexokinase. In this reaction, the loss of a phosphoryl group from ATP yields energy, some of which is captured by the enzyme to transfer the phosphoryl group onto glucose. The reactions and their $\Delta G^{\circ\prime}$ values are summed as follows:

Phosphorylation of glucose

$$C_6H_{12}O_6 + H_2PO_4^- \rightarrow C_6H_{12}O_6 - PO_3^- + H_2O$$

$$\Delta G_1^{\circ\prime} = +13.8 \text{ kJ/mol}$$

ATP hydrolysis

$$\text{ATP} + H_2O \rightarrow \text{ADP} + H_2PO_4^- + H^+$$

$$\Delta G_2^{\circ\prime} = -30.5 \text{ kJ/mol}$$

ATP phosphorylation of glucose

$$C_6H_{12}O_6 + \text{ATP} \rightarrow C_6H_{12}O_6 - PO_3^- + \text{ADP} + H^+$$

$$\Delta G_3^{\circ\prime} = -16.7 \text{ kJ/mol}$$

Note: Distinguish these forms of the Gibbs free energy term:

ΔG = change in Gibbs free energy for a reaction under defined conditions.
ΔG° = ΔG at standard conditions of temperature (298 K) and pressure (1 atm, sea level), with all reactants and products at a concentration of 1 M. (Table of values in Appendix 1.)
$\Delta G^{\circ\prime}$ = ΔG at standard temperature, pressure, and concentrations for ΔG°, plus the biochemically relevant conditions of pH 7 (H^+ concentration of 10^{-7} M) and water (H_2O concentration of 55.5 M; activity = 1). Some biochemists standardize additional factors, such as magnesium ion concentration ($[Mg^{2+}]$ = 1 mM).

Concentrations of Reactants and Products

In living cells, the concentrations of reactants and products usually differ from 1 M; for example, in the *E. coli* cytoplasm, the concentrations of ATP and P_i are about 8 mM. Thus, the actual ΔG of reactions occurring within a cell differs from ΔG° or $\Delta G^{\circ\prime}$.

Consider a reaction in which reactants A and B are reversibly converted to products C and D:

$$A + B \rightleftharpoons C + D$$

Higher concentration of reactants (A or B) drives the reaction forward, whereas higher concentration of products drives it in reverse. So, ΔG includes the ratio of products to reactants:

$$\Delta G = \Delta G^\circ + RT\ln\frac{[C][D]}{[A][B]}$$

$$= \Delta G^\circ + 2.303RT\log\frac{[C][D]}{[A][B]}$$

where R is the gas constant [$R = 8.315 \times 10^{-3}$ kJ/(mol · K)] and T is the absolute temperature in kelvins (298 K, or 25°C). The factor 2.303 converts the logarithm of the ratio of products to reactants from base e to base 10. Note that in $\Delta G^{\circ\prime}$ calculations, the water "activity" (concentration modified by a constant) is set at 1.

The effect of the concentration ratio on ΔG is shown in **Table 13.2**. In reactions at medium temperature (25°C–40°C), a 100-fold increase in the ratio of products to reactants adds about 11 kJ/mol to ΔG. ΔG is then less negative and the reaction less favorable. On the other hand, a 100-fold decrease in the concentration ratio makes ΔG more negative by –11 kJ/mol.

In some environments, a highly negative concentration term can override a positive ΔG°, resulting in a reaction with negative ΔG that microbes can use for energy. For example, in an iron mine the high concentration of reduced iron favors iron-oxidizing microbes. Alternatively, a high temperature may increase the magnitude of the term $2.303RT$ log [products]/[reactants] when it is negative, until it overrides a positive ΔG°. This temperature dependence of ΔG occurs in thermophiles such as *Sulfolobus*, which metabolize sulfur at 90°C by using reactions that could not go forward at lower temperatures.

Another way the direction of reaction may change is for a second metabolic process to remove one of the reaction's products (C or D) as fast as it is produced. The decrease in product concentration could then change the ΔG of the first reaction to a negative value. For example, when our intestinal bacteria break down glucose to pyruvate (see Section 13.5), many of the individual reactions have $\Delta G^{\circ\prime}$ values smaller

TABLE 13.2

Effect of the concentration ratio on ΔG.

Initial ratio of products to reactants: $\frac{[C][D]}{[A][B]}$	Change in ΔG (kJ/mol) at standard temperature (298 K) and atmospheric pressure	Result of change from standard concentrations
10^{-4}	–23	Products increase
10^{-2}	–11	Products increase
1	0	$\Delta G = \Delta G^\circ$
10^{2}	+11	Reactants increase
10^{4}	+23	Reactants increase

than 5 kJ/mol. Their actual direction of reaction in the cell depends on concentrations of products and reactants. Glycolytic reactions with near-zero $\Delta G^{\circ\prime}$ are reversible and can, in fact, participate in the biosynthetic pathway of gluconeogenesis (glucose biosynthesis, discussed in Chapter 15).

A surprising discovery has been that many bacteria and archaea in natural environments grow extremely slowly, using energy-yielding metabolism with values of ΔG approaching zero (that is, near thermodynamic equilibrium). When the actual ΔG (under actual reaction conditions) equals zero, a reaction proceeds equally forward and in reverse, and there is no net change in energy. At equilibrium, the ratio of product and reactant concentrations exactly cancels ΔG° or $\Delta G^{\circ\prime}$:

$$\Delta G = 0 = \Delta G^\circ + 2.303RT\log\frac{[\mathrm{C}][\mathrm{D}]}{[\mathrm{A}][\mathrm{B}]}$$

$$\Delta G^\circ = -2.303RT\log\frac{[\mathrm{C}][\mathrm{D}]}{[\mathrm{A}][\mathrm{B}]}$$

Living cells can never grow exactly at equilibrium ($\Delta G = 0$). But most soil bacteria and archaea gain energy from anaerobic metabolism with near-zero values of ΔG. Such organisms must form biomass very slowly, but if no other metabolism is available, they may outgrow competitors. The discovery of low-ΔG energetics has opened new possibilities for environmental remediation previously thought impossible, such as the anaerobic digestion of complex organic pollutants in contaminated soil (see Section 13.7).

Some of the near-zero anaerobic pathways involve **syntrophy**, an intimate metabolic relationship between two species. Michael McInerney and colleagues at the University of Oklahoma, Norman, show how syntrophy works for microbes that break down pollutants (see **Special Topic 13.1**). In a syntrophy, a bacterium catabolizes reactants with a positive $\Delta G^{\circ\prime}$ value, while H_2-oxidizing organisms such as methanogens complete the reaction with a net negative $\Delta G^{\circ\prime}$. The bacterial catabolism provides H_2 gas, which the partner species consumes, thus keeping H_2 concentration low enough to drive the syntrophic reaction.

Thought Question

13.3 The thermophilic bacteria *Thermus* species grow in deep-sea hydrothermal vents at 80°C. It was proposed that they metabolize formate to bicarbonate ion and hydrogen gas:

$$HCOO^- + H_2O \rightarrow HCO_3^- + H_2$$

But the standard $\Delta G^{\circ\prime}$ is near zero (–2.6 kJ/mol). Under actual conditions, do you think the reaction yields energy? Assume concentrations of 150 mM formate, 20 mM bicarbonate ion, and 10 mM hydrogen gas.

Concentration Gradients

So far, our discussion of energy has assumed an isotropic system, in which concentrations are the same everywhere. But a living cell needs to obtain its molecules, such as sugars, amino acids, and inorganic ions, from outside (discussed in Chapter 4). Moving a scarce substance "uphill" against its concentration gradient—such as iron (Fe^{2+}) needed by a blood pathogen—requires an expenditure of energy, which must be subtracted from the energy yield of subsequent reactions. Alternatively, a concentration gradient across the membrane (such as a H^+ gradient) can be used to store energy for the cell. The hydrogen ion gradient, or proton motive force, is discussed fully in Chapter 14.

A substance dissolved in water diffuses by random movements until its distribution has the same concentration throughout (**Fig. 13.8A**). The random distribution of molecules at uniform concentration represents the state of greatest entropy. Diffusion in the environment ultimately brings nutrients into contact with microbial cells, even cells that lack chemotactic motility to hunt for food. For example,

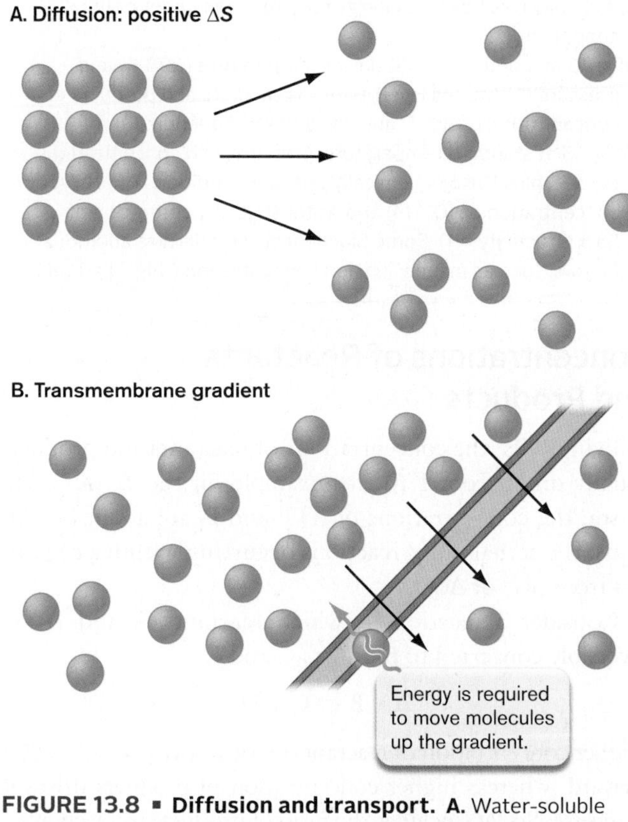
A. Diffusion: positive ΔS
B. Transmembrane gradient
Energy is required to move molecules up the gradient.

FIGURE 13.8 ▪ Diffusion and transport. A. Water-soluble molecules diffuse to uniform concentration throughout the solution: ΔS is positive, $-T\Delta S$ negative; the entropy term favors the process. **B.** If a membrane separating two compartments is permeable, molecules move from a compartment with high concentration to one with low concentration. Energy is required to move molecules up their concentration gradient.

sugars in the food we eat diffuse through our saliva to reach the bacteria growing in biofilms on our teeth.

The cell membrane contains transporters for useful molecules, allowing nutrient molecules to cross. Entropy favors their movement from higher to lower concentration (**Fig. 13.8B**). In most environments, however, the nutrients are at lower concentrations than inside the cell. To obtain these molecules from outside, the bacterial cell must transport them against their gradient—that is, from lower to higher concentration—increasing the concentration difference and thus decreasing entropy. Uptake against a gradient requires an energy source to power transport proteins embedded in the membrane (as shown in Chapter 4). The transporters must spend energy to compensate for the decrease in entropy when a cell transports the nutrient against its concentration gradient.

To Summarize

- **Intrinsic properties of the reaction determine the standard value of $\Delta G°$.** Properties include the molecular stability of reactants and products, and the entropy change associated with product conversion to reactants. By convention, standard conditions are set to define $\Delta G°$ (for biochemists, $\Delta G°'$).
- **Concentrations of reactants and products affect the actual value of ΔG.** The lower the concentration ratio of products to reactants, the more negative the value of ΔG. Environmental factors such as temperature and pressure also affect ΔG.
- **Within living cells, energy-yielding reactions are coupled with energy-spending reactions.** The measurement of energy flow under changing conditions requires calculations more complex than those shown here.
- **A concentration gradient stores energy.** Solutes run down their concentration gradient unless energy is applied to reverse the flow. Gradient energy can be interconverted with energy from chemical reactions.

13.3

Energy Carriers and Electron Transfer

Our ΔG equations show only the total energy of a reaction such as glucose oxidation. If all the energy were released at once, however, it would dissipate as heat without building biomass. In living cells, glucose oxidation never occurs in one step. Instead, the energy yield is divided among a large number of stepwise reactions with smaller energy changes. In this way, the cell can be thought of as "making change" by converting a large energy source to numerous smaller sources that can be "spent" conveniently for cell function and biosynthesis. The "spending" of energy is controlled by enzymes that couple all of the energy-providing reactions to specific energy-spending reactions.

Energy Carriers Gain and Release Energy

Many of the cell's energy transfer reactions involve **energy carriers**. Examples of energy carriers are ATP (adenosine triphosphate) and NADH (the reduced form of nicotinamide adenine dinucleotide). Energy carriers are molecules that gain and release small amounts of energy in reversible reactions. Energy carriers are used to transfer energy in a wide range of biochemical reactions.

Some energy carriers, such as NADH, transfer energy associated with electrons received from a food molecule. A molecule that transfers, or "donates," electrons to another molecule is called an **electron donor** or a reducing agent; a molecule that receives, or "accepts," electrons is called an **electron acceptor**. For example, during glucose catabolism a molecule of glyceraldehyde 3-phosphate transfers a pair of electrons ($2e^-$) with a hydrogen ion (H^+) to NAD^+, forming NADH. NAD^+ is an electron acceptor that receives the electrons; it then becomes the electron donor NADH. Electron donors such as NADH mediate electron transfer from reduced food molecules to a terminal electron acceptor such as oxygen. Energy carriers that transfer electrons are needed for all energy-yielding pathways and for biosynthesis of cell components such as amino acids and lipids.

Note that in living cells, all energy transfer reactions are coupled by enzymes to specific biochemical processes. Without enzyme coupling, energy dissipates and would be lost from the living system.

ATP Carries Energy

Adenosine triphosphate, or **ATP** (**Fig. 13.9A**), is composed of a base (adenine), a sugar (ribose), and three phosphates. Note that adenine-ribose-phosphate (adenosine nucleotide) is equivalent to a nucleotide "link" of RNA. The base adenine is a fundamental molecule of life, one that forms spontaneously from methane and ammonia in experiments simulating the origin of life on early Earth (discussed in Chapters 1 and 17). Like the sugar ribose, ATP is an ancient component of cells, found in all living organisms.

Special Topic 13.1: Microbial Syntrophy Cleans Up Oil

Have you ever wondered how to clean up polluted soil and water? Michael McInerney at the University of Oklahoma, Norman, thinks about cleanup from the point of view of microbial energetics (**Fig. 1**). As the son of a wastewater treatment engineer in Chicago, he learned early on about pollution in the Great Lakes. Later, in Oklahoma, he realized that there was a growing problem of water supplies contaminated by sewage and waste petroleum. He began to study how microbes catabolize sewage components, such as the short-chain acids propionate and butyrate, and aromatic oil compounds such as benzoate.

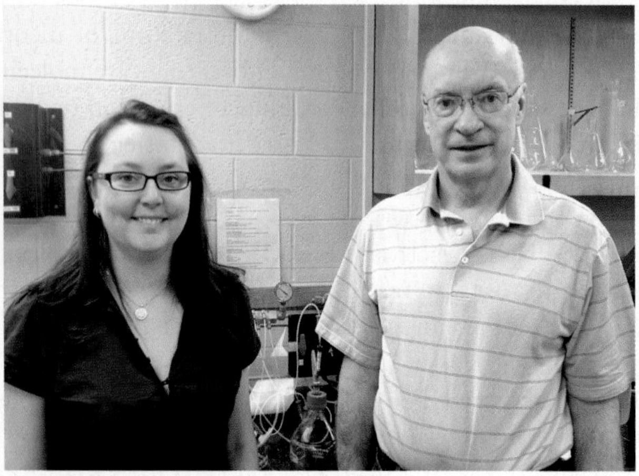

NEIL WOFFORD, UNIVERSITY OF OKLAHOMA

FIGURE 1 ■ Michael McInerney and Jessica Sieber study syntrophy at the University of Oklahoma, Norman.

A fundamental challenge to degrading these pollutants is the lack of oxygen. The molecules seep into the depths of soil or water where any oxygen is quickly used up by microbes capable of respiration. As microbes catabolize the more complex long-chain components of sewage or oil, the remaining short-chain molecules and aromatic rings can be broken down only by transferring leftover electrons onto hydrogen ions, generating H_2. For example, bacteria such as *Syntrophus aciditrophicus* and *Syntrophomonas wolfei* break down substrates such as benzoate into acetate and bicarbonate ions (**Table 1**).

But the hydrogen-generating reactions of *Syntrophus* and *Syntrophomonas* have a positive value of $\Delta G^{\circ\prime}$ (see **Table 1**), and thus they cannot go forward to yield energy. So how can the sign of ΔG be changed for the bacteria to obtain energy? The free energy change can be shifted by rapidly removing H_2 gas, thus lowering the concentration of this key product. Once the equations are balanced, negative values of ΔG are obtained. H_2 can be removed by other microbes that oxidize hydrogen using alternative electron acceptors such as bicarbonate (the archaeal methanogen *Methanospirillum*), or sulfate or nitrate (the bacterium *Desulfovibrio*) (**Fig. 2A**).

In order to take advantage of their hydrogen-oxidizing neighbors, hydrogen-producing bacteria have evolved to grow in close proximity to the hydrogen-oxidizing bacteria or archaea. **Figure 2B** shows a section through a syntrophic community within a wastewater sludge granule, where anaerobic catabolism must degrade sewage components to

TABLE 1

Syntrophy: reactions producing and consuming H_2.

Genus	Reaction	$\Delta G^{\circ\prime}$* (kJ/mol)	ΔG** (kJ/mol)		
H_2 reduces an oxidant (HCO_3^-, SO_4^-, or NO_3^-)					
Methanospirillum	$4H_2 + HCO_3^- + H^+ \rightarrow CH_4 + 3H_2O$	–136			
Desulfovibrio	$4H_2 + SO_4^- + H^+ \rightarrow HS^- + 4H_2O$	–152			
Desulfovibrio	$4H_2 + NO_3^- + 2H^+ \rightarrow NH_4^+ + 3H_2O$	–600	**Oxidant for syntrophy:**		
Syntrophic catabolism producing H_2			**HCO_3^-**	**SO_4^-**	**NO_3^-**
Syntrophus	Benzoate + $7H_2O \rightarrow$ 3 acetate + HCO_3^- + $3H^+$ + $3H_2$	+70	–25	–33	–45
Syntrophomonas	Butyrate + $2H_2O \rightarrow$ 2 acetate + H^+ + $2H_2$	+49	–5	–13	–18

*$\Delta G^{\circ\prime}$ is the standard free energy change at pH 7 when the concentration of reactants and products are the same.

**$\Delta G'$ was the measured free energy change at pH 7 when syntrophic metabolism stopped.

Sources: Michael McInerney et al. 2007. *PNAS* **104**:7600; Ralf Cord-Ruwisch et al. 1988. *Arch. Microbio1.* **149**:350; Bradley E. Jackson and Michael J. McInerney. 2002. *Nature* **415**:454.

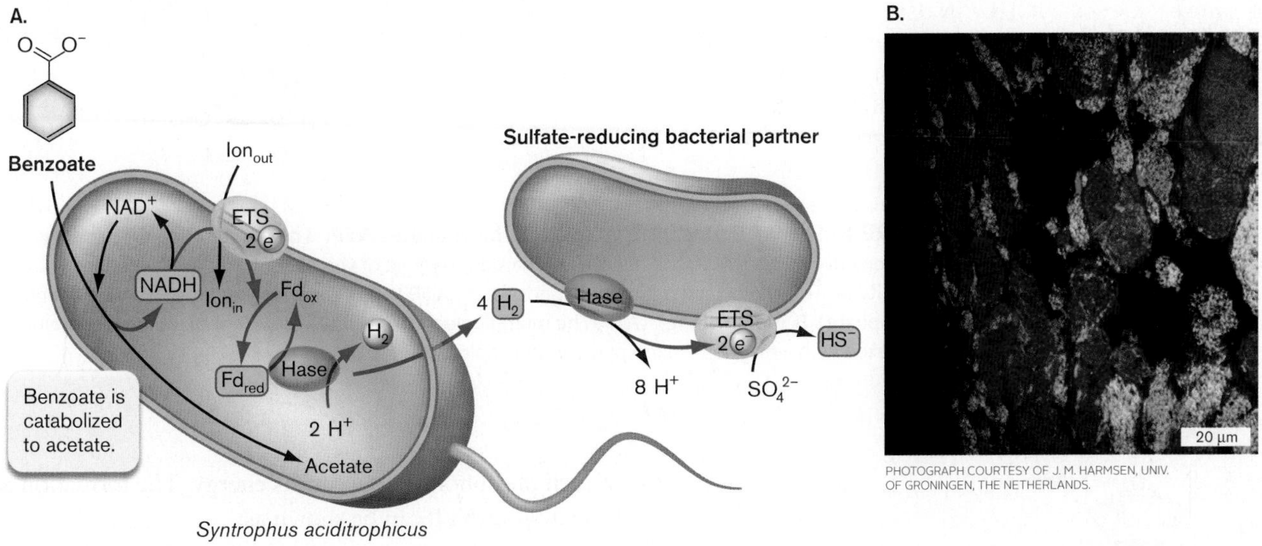
A.
Benzoate
Ion$_{out}$
NAD$^+$
ETS
2 e^-
NADH
Ion$_{in}$
Fd$_{ox}$
Fd$_{red}$
Hase
H_2
2 H$^+$
Acetate
Benzoate is catabolized to acetate.
Syntrophus aciditrophicus
Sulfate-reducing bacterial partner
4 H_2
Hase
ETS
2 e^-
HS$^-$
8 H$^+$
SO_4^{2-}
B.
20 µm

PHOTOGRAPH COURTESY OF J. M. HARMSEN, UNIV. OF GRONINGEN, THE NETHERLANDS.

FIGURE 2 ■ Syntrophy between *Syntrophus aciditrophicus* and a sulfate-reducing bacterium, *Desulfovibrio*. **A.** *Syntrophus aciditrophicus* catabolizes benzoate to acetate, transferring electrons back onto the hydrogen ions removed from the substrate. The H_2 formed must donate electrons to an electron transport system (ETS), despite a positive ΔG value (reverse electron transfer). Rapid removal of H_2 by the sulfate-reducing partner (*Desulfovibrio sp.*) provides energy, making the net ΔG value negative. **B.** Syntrophic community of a wastewater sludge granule, labeled by fluorescence in situ hybridization (FISH). The sludge includes colonies of H_2-producing bacteria (red fluorescent label), H_2-consuming methanogens (yellow) and other archaea (green).

CO_2. The granule is stained with fluorescent markers for the H_2-producing bacteria (red) and their partner methanogen (yellow). In order for the syntrophy to proceed, the H_2 from the catabolizing bacteria must be transferred quickly to the methanogenic partner, by processes that are not yet understood. Genomic analysis of *Syntrophus aciditrophicus* reveals genes for flagellar motility, suggesting that flagellar motility helps the cell find its partner in order to form an association. Microscopy reveals intercellular nanotubes that may transmit electrons from the H_2 producer to the partner (discussed in Chapter 14).

Under laboratory conditions, McInerney and postdoctoral fellow Jessica Sieber (**Fig. 1**) demonstrated syntrophic growth in the presence of defined concentrations of reactants and products. Their calculations showed that cells grew with negative values of free energy change. Yet even in the presence of the syntrophic partner, these anaerobic pathways have free energy changes that are remarkably low (about –20 kJ/mol), often below the theoretical minimum required to generate ATP (about –60 kJ /mol). Such extremely low energy levels require special arrangements called "reverse electron transport," in which electrons are transferred from one carrier to the next despite a positive ΔG value (requiring energy input). The electron transfer must be coupled with a compensating reaction that spends enough energy to yield a net negative ΔG value (discussed further in Chapter 14).

Syntrophic partnerships are considered "thermodynamic extremophiles" because they grow at such low ΔG values. Nonetheless, these organisms are of global importance because they play an essential role in recycling the small-molecule products of anaerobic catabolism. Furthermore, they present exciting industrial applications. For instance, the methanogenic partnership between H_2-producing bacteria and *Methanococcus maripaludis* might be used to attack unextractable oil residues and convert the carbon into methane for recovery as natural gas. Other syntrophic interactions may be used to increase the efficiency of sewage mineralization during water treatment.

RESEARCH QUESTION

How could you use radioisotope labeling to demonstrate that a syntrophic partnership of two species catabolizes benzoate? What controls would you run?

McInerney, Michael J., Lars Rohlin, Housna Mouttaki, UnMi Kim, Rebecca S. Krupp, et al. 2007. The genome of *Syntrophus aciditrophicus*: Life at the thermodynamic limit of microbial growth. *PNAS* **104**:7600.

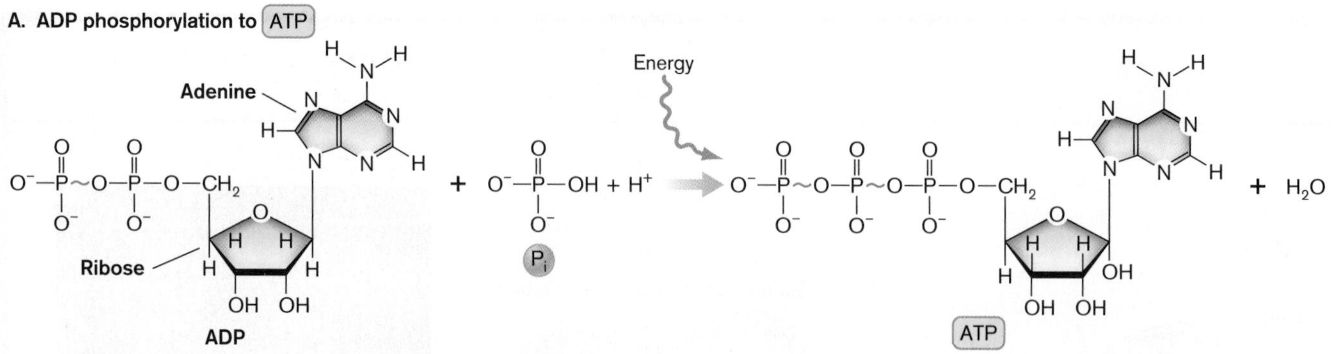
A. ADP phosphorylation to ATP
Adenine
Ribose
ADP
Energy
Pi
+ H2O
ATP

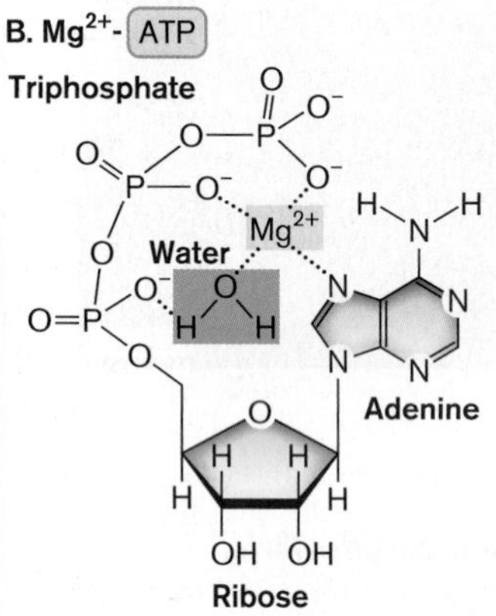
B. Mg2+-ATP
Triphosphate
Mg2+
Water
Adenine
Ribose

FIGURE 13.9 ■ ADP plus inorganic phosphate makes ATP. The reaction requires energy input (positive ΔG) because the negatively charged oxygens of the phosphates are forced into proximity. **A.** The chemical reaction phosphorylating ADP (adenosine diphosphate) to ATP (adenosine triphosphate). **B.** Model of Mg^{2+}-ATP. The multiple negative charges of ATP are stabilized by being complexed with a magnesium ion plus a water molecule.

Under physiological conditions, ATP always forms a complex with Mg^{2+} (**Fig. 13.9B**). The magnesium cation partly neutralizes the negative charges of the ATP phosphates, stabilizing the structure in solution. Most enzyme-binding sites for ATP actually bind Mg^{2+}-ATP. This is one reason magnesium is an essential nutrient for all living cells.

ADP phosphorylation to ATP. During cell metabolism, ATP is generated by **phosphorylation**, the condensation of inorganic phosphate with adenosine diphosphate (ADP):

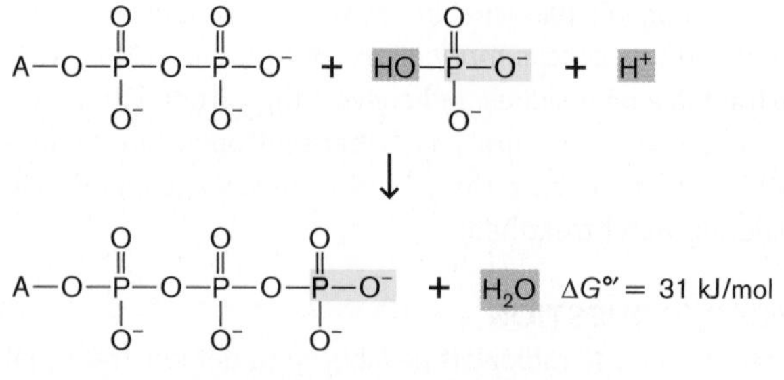
ΔG°′ = 31 kJ/mol

The phosphorylation of ADP to form ATP requires energy input (positive ΔG).

Why does ATP formation require energy? The inorganic phosphate has four oxygen atoms that share a negative charge. When phosphate reacts with another phosphate to form a bond, the charged oxygens of adjacent phosphates are forced into proximity, despite charge repulsion. The negatively charged oxygens in ATP show charge repulsion, associated with a large negative ΔG of hydrolysis. Hydrolysis of each phosphoryl group yields energy. The formation and hydrolysis of ATP can be shown as:

$$\text{A–P\textasciitilde P} + H^+ + P_i \rightleftharpoons \text{A–P\textasciitilde P\textasciitilde P} + H_2O$$

where ~ designates each energy-storing phosphoryl group and P_i designates inorganic phosphate; or, in more concise shorthand:

$$\text{ADP} + P_i \rightleftharpoons \text{ATP} + H_2O$$

ATP is a "medium size" energy carrier, since the cell contains many phosphorylated molecules that yield greater energy upon hydrolysis. **Table 13.3** summarizes examples of free energy change associated with hydrolysis of various phosphoryl groups. As we will see, phosphoryl group hydrolysis with $\Delta G°'$ values larger than –31 kJ/mol can yield energy that is stored by ATP formation. ATP hydrolysis can yield energy to phosphorylate molecules at $\Delta G°'$ values smaller than –31 kJ/mol.

ATP transfers energy. ATP can transfer energy to cell processes in three different ways: hydrolysis releasing phosphate; hydrolysis releasing pyrophosphate (diphosphate); and phosphorylation of an organic molecule. Each process serves different functions in the cell.

- **Hydrolysis releasing phosphate.** The **hydrolysis** of ATP at the terminal phosphate consumes H_2O to produce ADP and P_i, releasing energy. The energy released by ATP hydrolysis can be transferred to a coupled reaction of biosynthesis, such as building an amino acid. The two reactions are coupled by an enzyme with binding sites specific for ATP and the substrate.
- **Hydrolysis releasing pyrophosphate.** ATP can hydrolyze at the middle phosphate, releasing pyrophosphate (PP_i). The pyrophosphate usually hydrolyzes shortly

TABLE 13.3

Hydrolysis of phosphoryl groups: values of $\Delta G°'$ (pH 7.0, 25°C).

Reaction of hydrolysis	$\Delta G°'$ (kJ/mol)
Glucose 6-P + $H_2O \rightarrow$ glucose + P_i	–14
Fructose 1,6-bis-P + $H_2O \rightarrow$ fructose 6-P + P_i	–16
ATP + $H_2O \rightarrow$ ADP + P_i + H^+	–31
ADP + $H_2O \rightarrow$ AMP + P_i + H^+	–33
ATP + $H_2O \rightarrow$ AMP + PP_i (pyrophosphate) + H^+	–46
PP_i (pyrophosphate) + $H_2O \rightarrow$ 2 P_i + H^+	–19
1,3-Bis-P-glycerate + $H_2O \rightarrow$ 3-P-glycerate + P_i	–49
Phosphoenolpyruvate + $H_2O \rightarrow$ pyruvate + P_i	–62

afterward to make $2P_i$. Overall, the release of $2P_i$ from ATP yields approximately twice as much energy as the release of $1P_i$. Pyrophosphate release and subsequent hydrolysis drives a reaction strongly forward, because twice as much energy would be required to reverse the reaction. Pyrophosphate is released in reactions that must avoid reversal—for example, the incorporation of nucleotides into growing chains of RNA.

- **Phosphorylation of an organic molecule.** ATP can transfer its phosphate to the hydroxyl group of a molecule such as glucose to activate the substrate for a subsequent rapid reaction. No inorganic phosphate appears, and no water molecule is consumed:

$$\text{ATP} + \text{glucose} \rightarrow \text{ADP} + \text{glucose 6-P}$$

Some enzymes catalyze ATP transfer of phosphate to activate sugar molecules for catabolism. Other enzymes couple the phosphorylation of a sugar to its transport across the cell membrane; consequently, these enzymes make up the **phosphotransferase system (PTS)**. The PTS enzymes play a critical role in determining which nutrients from the environment some microbes can acquire and catabolize (discussed in Chapter 4).

Thought Questions

13.4 When ATP phosphorylates glucose to glucose 6-phosphate, what is the net value of $\Delta G°'$? What if ATP phosphorylates pyruvate? Can this reaction go forward without additional input of energy? (See **Table 13.3**.)

13.5 Linking an amino acid to its cognate tRNA is driven by ATP hydrolysis to AMP (adenosine monophosphate) plus pyrophosphate. Why release PP_i instead of P_i?

ATP produced by glucose catabolism. A large number of ATPs can be formed by coupling ATP synthesis to the step-by-step breakdown and oxidation of a food molecule such as glucose. In theory, complete oxidation of glucose through respiration can produce as many as 38 ATP molecules. The overall $\Delta G°'$ of the coupled reactions is:

$$C_6H_{12}O_6 + 6H_2O + 6O_2 \rightarrow 12H_2O + 6CO_2 \quad \Delta G°' = -2{,}878 \text{ kJ/mol}$$

$$38[\text{ADP} + P_i \rightarrow \text{ATP} + H_2O] \quad \Delta G°' = 38 \times (+31 \text{ kJ/mol})$$

$$\text{Net } \Delta G°' = -1{,}700 \text{ kJ/mol}$$

The difference in $\Delta G°'$ for the coupled reactions is the energy lost as heat and entropy—in this case, –1,700 kJ/mol (–2,878 kJ/mol + 1,178 kJ/mol). Thus, the maximal efficiency of energy capture by ATP is about 40%, a level that may be approached by highly efficient systems such as mitochondria. When ΔG values are corrected for cellular concentrations of reactants and products, the actual efficiency appears to be greater than 50%. Under conditions such as low oxygen concentration, much smaller amounts of ATP are made per molecule of glucose. For comparison, the efficiency of a typical machine, such as an internal combustion engine, is about 20%.

Thought Question

13.6 In the microbial community of the bovine rumen, the actual ΔG value has been calculated for glucose fermentation to acetate:

$$C_6H_{12}O_6 + 2H_2O \rightarrow 2C_2H_3O_2^- + 2H^+ + 4H_2 + 2CO_2 \quad \Delta G = -318 \text{ kJ/mol}$$

If the actual ΔG for ATP formation is +44 kJ/mol and each glucose fermentation yields four molecules of ATP, what is the thermodynamic efficiency of energy gain? Where does the lost energy go?

Note that besides ATP, other nucleotides carry energy. Guanosine triphosphate (GTP) provides energy for ribosome elongation of proteins. And the phosphodiester bonds of all four nucleotide triphosphates, as well as their corresponding deoxyribonucleotide triphosphates, carry energy for their own incorporation into RNA and DNA, respectively.

NADH Carries Energy and Electrons

Another major energy carrier is **nicotinamide adenine dinucleotide** (**NADH**, reduced; **NAD⁺**, oxidized). Unlike ATP, NADH carries energy associated with two electrons that reduce a substrate. NADH carries two or three times as much energy as ATP, depending on cell conditions. It is used to carry electrons from breakdown products of glucose.

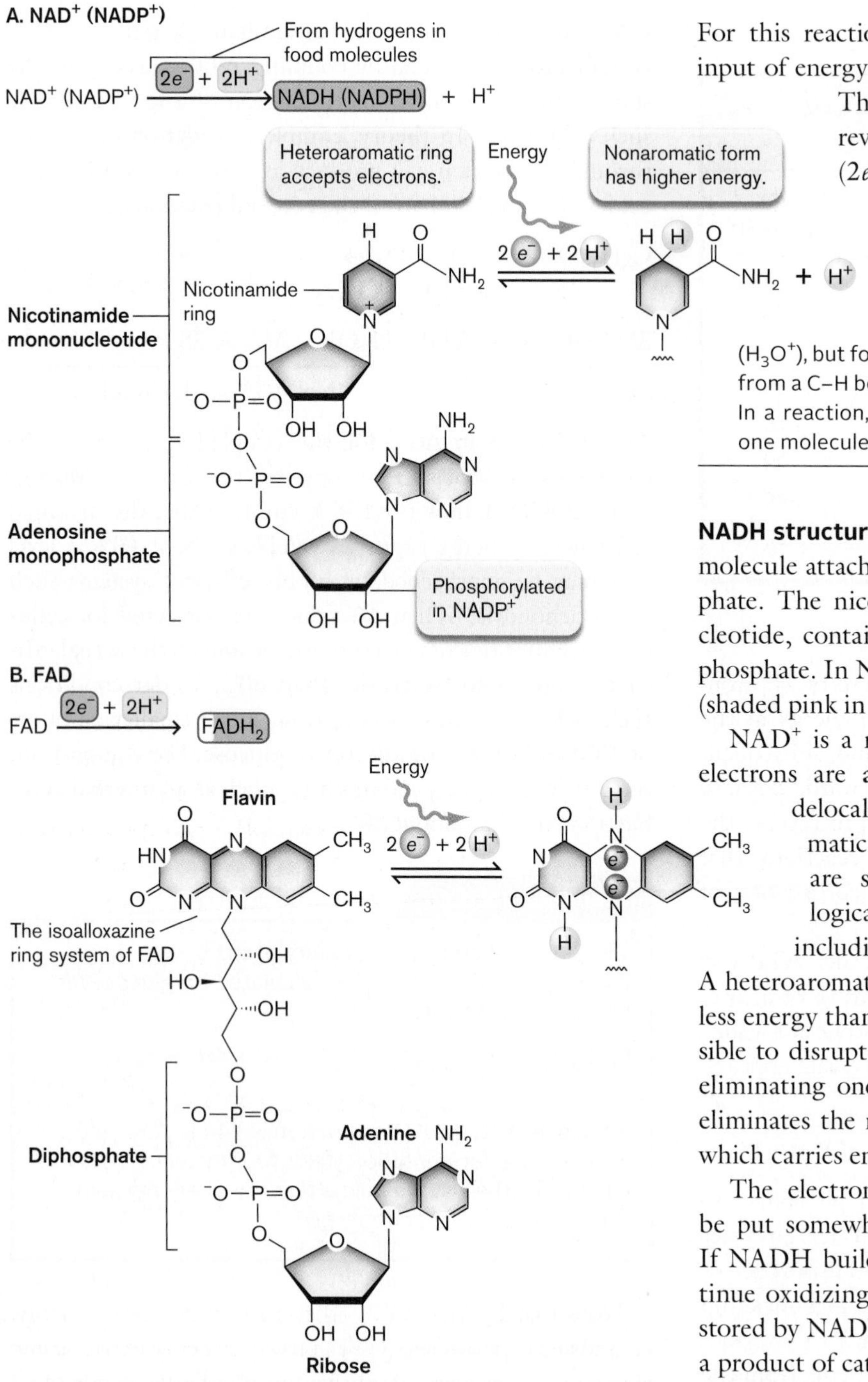
A. NAD+ (NADP+)
From hydrogens in food molecules
2e + 2H+
NAD+ (NADP+) → NADH (NADPH) + H+
Heteroaromatic ring accepts electrons.
Energy
Nonaromatic form has higher energy.
Nicotinamide ring
Nicotinamide mononucleotide
Adenosine monophosphate
Phosphorylated in NADP+
B. FAD
FAD → FADH2
Flavin
The isoalloxazine ring system of FAD
Diphosphate
Adenine
Ribose

FIGURE 13.10 ■ Reduction of NAD⁺ and FAD. A. In the reduction of NAD⁺, the nicotinamide ring (shaded pink) loses a double bond as two electrons are gained from an electron donor. Two hydrogen atoms are consumed; one bonds to NADH, while the other ionizes. **B.** In FAD, the flavin isoalloxazine ring system gains two electrons associated with two hydrogens.

Its oxidized form, NAD⁺, receives two electrons ($2e^-$) plus a hydrogen ion (H^+) from a food molecule; a second H^+ from the food molecule enters the solution. Overall, reduction of NAD⁺ consumes two hydrogen atoms to make NADH:

$$NAD^+ + 2H^+ + 2e^- \rightarrow NADH + H^+ \qquad \Delta G^{\circ\prime} = +62 \text{ kJ/mol}$$

For this reaction, $\Delta G^{\circ\prime}$ is positive; therefore, it requires input of energy from the catabolism of the food molecule. The reduced energy carrier NADH can then reverse this reaction by donating two electrons ($2e^-$) to another molecule, regenerating NAD⁺.

Note: A hydrogen ion, or "proton" (H^+), does not exist free in solution. In water, a H^+ combines with H_2O to form a hydronium ion (H_3O^+), but for clarity we use H^+. An atom of hydrogen removed from a C–H bond consists of a proton (H^+) plus an electron (e^-). In a reaction, the proton and electron may be transferred to one molecule or to separate molecules.

NADH structure and function. NAD⁺ consists of an ADP molecule attached to nicotinamide instead of a third phosphate. The nicotinamide mononucleotide, like a ribonucleotide, contains a nitrogenous base attached to a sugar phosphate. In NAD⁺, the nicotinamide has a ring structure (shaded pink in **Fig. 13.10A**) that forms a stable cation.

NAD⁺ is a relatively stable structure because the ring electrons are **aromatic**; that is, the bonding electrons delocalize around the ring, as in benzene. Aromatic rings that contain noncarbon atoms are said to be "heteroaromatic." Many biologically active molecules are heteroaromatic, including adenine and other nucleotide bases. A heteroaromatic ring is stable, but its disruption requires less energy than the disruption of benzene. Thus, it is possible to disrupt the ring by adding two electrons with H^+ eliminating one double bond. The donation of electrons eliminates the ring's aromatic nature, generating NADH, which carries energy in an amount useful for cell reactions.

The electrons transferred to NADH eventually must be put somewhere else, onto the next electron acceptor. If NADH builds up in a cell, no NAD⁺ remains to continue oxidizing food molecules. One way that the energy stored by NADH can be spent is to transfer $2H^+ + 2e^-$ onto a product of catabolism. For example, in ethanolic fermentation to make wine or beer, NADH reduces pyruvate to ethanol. In this case, however, the energy is lost to the cell. Alternatively, NADH can transfer its electrons to one of a series of electron carrier molecules known as the **electron transport system (ETS)**. Electron transport within bacteria can actually be used to generate electricity for commercial power (discussed in Chapter 14). Examples of electron transfer reactions are shown in the "tower of power" in **Table 13.4**. For example, adding two electrons ($2e^-$) to NAD⁺ to make NADH has a highly negative value of $E^{\circ\prime}$ (positive $\Delta G^{\circ\prime}$), which means that the reaction requires energy input.

TABLE 13.4

Standard reduction potentials.*

Electron acceptor	→	electron donor	$E^{\circ\prime}$ (mV)	$\Delta G^{\circ\prime}$ (kJ)
$2H^+ + 2e^-$	→	H_2	–420	+81
$NAD^+ + 2H^+ + 2e^-$	→	$NADH + H^+$	–320	+62
$FAD + 2H^+ + 2e^-$	→	$FADH_2$	–220	+42
$FMN + 2H^+ + 2e^-$	→	$FMNH_2$	–190	+37
Menaquinone + $2H^+ + 2e^-$	→	Menaquinol	–74	+14
Fumarate + $2H^+ + 2e^-$	→	Succinate	+33	–6
Ubiquinone + $2H^+ + 2e^-$	→	Ubiquinol	+110	–21
$NO_3^- + 2H^+ + 2e^-$	→	$NO_2^- + H_2O$	+420	–81
$½O_2 + 2H^+ + 2e^-$	→	H_2O	+820	–158

*For a more extensive list, see Table 14.1.

Note: Electron transport is driven by the reduction potential *E* (volts). The reduction potential *E* is related to free energy change: $\Delta G = -nFE$, where *n* is the number of electrons and *F* is Faraday's constant, 96.5 kJ/(V · mol). Reduction potentials, electron transport, and proton motive force are discussed further in Chapter 14.

The Electron Transport System

The electron transport system, ETS (or electron transport chain), stores energy from electron transfer as ion gradients across the membrane of the cell or an organelle. The ETS enables production of ATP. The ETS includes a series of proteins and small organic molecules that can be reduced and cyclically reoxidized. At the end of the series of oxidation-reduction transfers, the electrons are transferred to a **terminal electron acceptor** whose product leaves the cell. For example, as a terminal electron acceptor, molecular oxygen (O_2) is reduced to H_2O.

The reaction of O_2 reduction to H_2O is coupled to oxidation of NADH or another reduced energy carrier:

$$NADH + H^+ \rightarrow NAD^+ + 2H^+ + 2e^- \quad \Delta G^{\circ\prime} = -62 \text{ kJ/mol}$$
$$½O_2 + 2H^+ + 2e^- \rightarrow H_2O \quad \Delta G^{\circ\prime} = -158 \text{ kJ/mol}$$

$$NADH + H^+ + ½O_2 \rightarrow NAD^+ + H_2O \quad \Delta G^{\circ\prime} = -220 \text{ kJ/mol}$$

The total energy spent by NADH oxidation through the ETS is –220 kJ/mol (–62 kJ/mol – 158 kJ/mol). This energy is converted to transmembrane proton potential (composed of the H^+ concentration difference plus the charge difference across the membrane). The proton potential drives nutrient transport, motility, and synthesis of ATP. The ETS and the proton potential are discussed in detail in Chapter 14.

Many microbes can reduce terminal electron acceptors other than O_2, such as nitrate (NO_3^-) (see **Table 13.1**). Humans, unlike most microbes, have no alternative electron acceptors; thus, we need to breathe oxygen. When no terminal electron acceptor is available to accept electrons, NADH builds up to unfavorably high levels, leaving no more NAD^+ available to be reduced. When muscle tissues run out of oxygen, they transfer electrons from NADH back to the products of glucose catabolism, producing lactic acid. The buildup of lactic acid drives down pH, and the muscle cramps.

Other energy carriers that transfer electrons. Different steps of metabolism utilize different but related energy carriers. For example, NADPH differs from NADH only in its extra phosphate attached to the 2′ carbon of adenine nucleotide; the amount of energy carried is the same. Some enzymes can utilize both NADPH and NADH, whereas other enzymes use only one or the other.

Another related energy carrier is **flavin adenine dinucleotide (FAD)**, in which flavin substitutes for nicotinamide. The flavin nucleotide includes a ring structure whose aromaticity is eliminated by its receiving two electrons (**Fig. 13.10B**). The redox function of the flavin isoalloxazine ring system is similar to that of NADH:

$$FAD + 2H^+ + 2e^- \rightarrow FADH_2$$

Like NADH, $FADH_2$ donates $2e^-$ to an electron acceptor. $FADH_2$ is a weaker electron donor than NADH, but when $FADH_2$ is combined with a strong electron acceptor such as O_2, electron transfer occurs and significant energy is released ($\Delta G^{\circ\prime} = -42 - 158 = -200$ kJ/mol).

Why do different kinds of reactions use different energy carriers?

- **Different redox levels.** Food molecules may have more or fewer electrons (level of reduction/oxidation) than those associated with the cell structure. For example, lipids are more highly reduced than glucose. Thus, lipid catabolism requires a greater proportion of electron-accepting energy carriers (such as NAD^+ or $NADP^+$) than does glucose catabolism, and it makes relatively few ATP molecules directly. A combination of energy carriers with different redox states enables cells to balance their overall redox potential while transferring energy.
- **Different amounts of energy.** Biochemical reactions yield different amounts of energy—that is, different values of ΔG. Suppose a reaction can provide more than enough energy to generate ATP from ADP (31 kJ), but not quite enough to generate NADH from NAD^+ (62 kJ). An example is the conversion of succinate to fumarate in the TCA cycle (see Section 13.6). This reaction provides the energy to reduce FAD to $FADH_2$, whose oxidation by O_2 can yield two molecules of ATP. Thus, the use of $FADH_2$ enables the cell to make more efficient use of its food than if generation of ATP or NADH were the only choices.
- **Regulation and specificity.** Specific energy carriers can direct metabolites into different pathways serving different functions. For example, in many bacteria NADH is directed into the ETS, whereas NADPH, the 2′-phosphorylated form of NADH, is directed into biosynthesis of cell components such as amino acids and lipids.

The concentrations and reduction level of energy carriers provide much information on the state of a cell, such as the effects of environmental stress on cell metabolism. But energy carriers such as NADH undergo rapid turnover (interconversion with NAD^+). How can NADH and ATP concentrations be observed within living cells? One method uses nuclear magnetic resonance (NMR) spectroscopy (discussed in eTopic 13.1). The use of NMR to observe living cells led to the development of magnetic resonance imaging (MRI) to observe the entire human body.

Enzymes Catalyze Metabolic Reactions

In living cells, each reaction must occur only as needed, in the right amount at the right time. The rate of a reaction is determined by the **activation energy (E_a)**, the input energy needed to generate the high-energy transition state on the way to products (**Fig. 13.11**). Most biochemical reactions have an activation energy that exceeds the average kinetic energy of the reactant molecules colliding. Thus, no matter how negative the ΔG, the reaction will proceed at a significant rate only when the activation energy is lowered by interaction with a catalyst, an agent that participates in a reaction without being consumed.

Biological reactions are catalyzed by **enzymes**, structures composed of protein (or, in some cases, RNA) that bind substrates of a specific reaction. The enzyme lowers the activation energy by bringing the substrates in proximity to one another and by correctly orienting them to react. In some cases, enzymes provide a reactive amino acid residue to participate in a transition state between reactants and products. Microbial enzymes have growing importance in industry; they are used for food production, fabric treatment, and drug therapies (discussed in Chapter 16).

Enzymes couple specific energy-yielding reactions (such as those of glucose breakdown) with the cell's reactions requiring energy, such as making ATP. An example of coupled reactions is shown in **Figure 13.12**. The substrate phosphoenolpyruvate (PEP), a phosphorylated breakdown product of glucose, is converted to pyruvate while the phosphate is added to ADP to generate ATP:

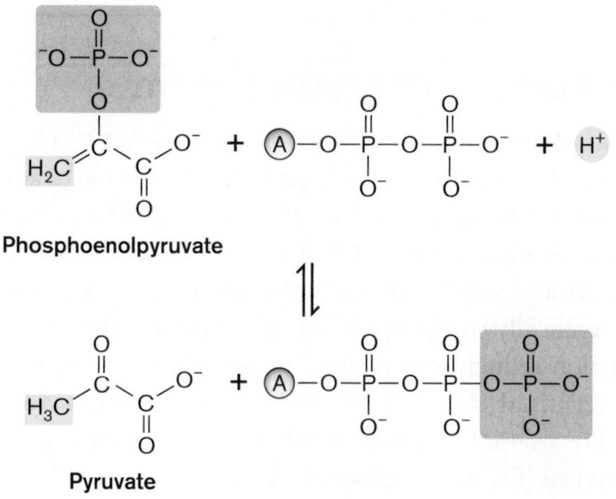

Phosphoenolpyruvate
Pyruvate

The $\Delta G^{\circ\prime}$ of phosphate cleavage from PEP is −62 kJ/mol, whereas the $\Delta G^{\circ\prime}$ of ATP formation is only +31 kJ/mol. Thus, the net $\Delta G^{\circ\prime}$ is negative (−31 kJ/mol), and the reaction goes forward to pyruvate. But a high activation energy makes the reaction extremely slow; it essentially never occurs on the timescale of life. The reaction occurs only when the two reacting substrates (PEP and ADP) are coupled by the enzyme pyruvate kinase (**Fig. 13.12A**). Pyruvate kinase has specific binding sites for each of its substrates: PEP and ADP. ADP and PEP are brought together by the enzyme and positioned so as to lower the activation energy of phosphate transfer.

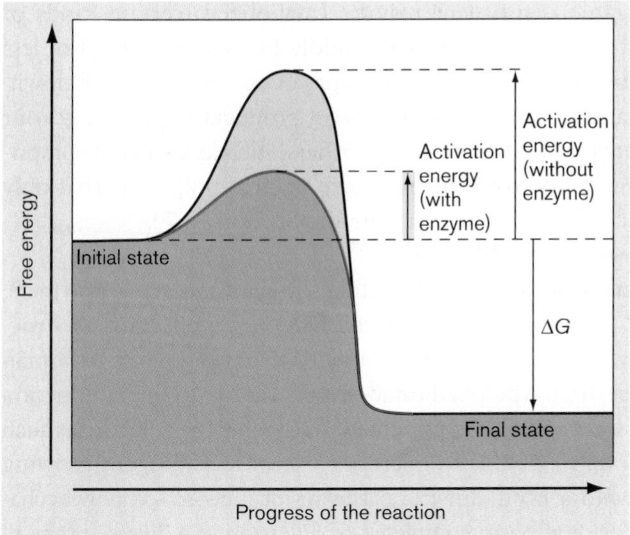

Free energy
Initial state
Activation energy (with enzyme)
Activation energy (without enzyme)
ΔG
Final state
Progress of the reaction

FIGURE 13.11 ■ Enzymes lower the activation energy of the transition state. In the presence of an enzyme, the activation energy of the reaction is decreased, allowing rapid conversion of reactants to products.

In addition to the catalytic site, the enzyme has an **allosteric site** (a regulatory site distinct from the substrate-binding site) for its activator, fructose 1,6-bisphosphate. The difference between a substrate-binding site and an allosteric site can be seen in the molecular model based on X-ray crystallography (**Fig. 13.12B**). The allosteric site is found at a distance from the substrate-binding site, but its interaction with the regulator, fructose 1,6-bisphosphate, alters the conformation of the entire enzyme, increasing the rate of reaction. As we will see, fructose 1,6-bisphosphate is a central intermediate of glycolysis; thus, it makes sense that as this molecule builds up, it activates pyruvate kinase to remove products farther down the chain of catabolic reactions, maintaining the steady flow of metabolites.

Thought Question

13.7 What would happen to the cell if pyruvate kinase catalyzed PEP conversion to pyruvate but failed to couple this reaction to ATP production?

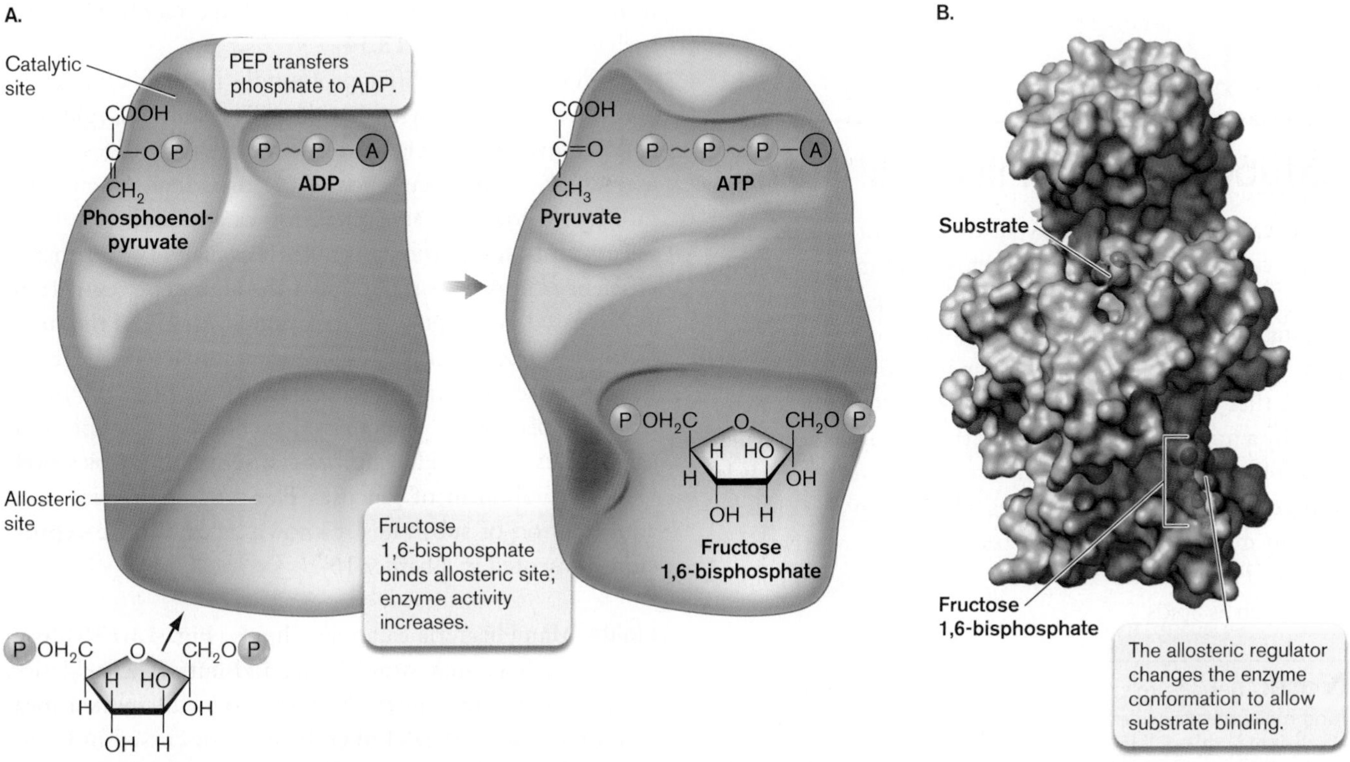

A.
Catalytic site
PEP transfers phosphate to ADP.
COOH
C—O P
CH2
Phosphoenol-pyruvate
P~P—A
ADP
COOH
C=O
CH3
Pyruvate
P~P~P—A
ATP
Allosteric site
Fructose 1,6-bisphosphate binds allosteric site; enzyme activity increases.
P OH2C O CH2O P
H HO
H OH
OH H
Fructose 1,6-bisphosphate
Fructose 1,6-bisphosphate
B.
Substrate
Fructose 1,6-bisphosphate
The allosteric regulator changes the enzyme conformation to allow substrate binding.

FIGURE 13.12 ■ The enzyme pyruvate kinase. A. Pyruvate kinase catalyzes the transfer of a phosphoryl group from PEP to ADP, generating pyruvate and ATP. The enzyme possesses separate binding sites for substrates and allosteric effectors. **B.** The molecular model of pyruvate kinase, based on a crystal structure of the enzyme bound to a substrate analog and an allosteric effector, fructose 1,6-bisphosphate. (PDB code: 1A3W)

Note that pyruvate kinase is actually named for its reverse reaction: the use of ATP to phosphorylate pyruvate to PEP—perhaps because the activity of the purified enzyme was originally observed in the reverse direction. Enzymes can catalyze both forward and reverse reactions. The predominant direction of catalysis depends on the concentrations of substrates and products (which determine ΔG) and on allosteric regulators.

To Summarize

- **Catabolic pathways organize the breakdown of large molecules** in a series of sequential steps coupled to reactions that store energy in small carriers such as ATP and NADH.
- **ATP and other nucleotide triphosphates store energy** in the form of phosphodiester bonds.
- **NADH, NADPH, and $FADH_2$** each store energy associated with an electron pair that carries reducing power.
- **Enzymes catalyze reactions by lowering the ΔG** required to reach the transition state. They couple energy transfer reactions to specific reactions of biosynthesis and cell function.

13.4

Catabolism: The Microbial Buffet

In the early twentieth century, it was thought that microbes could catabolize a limited subset of naturally occurring organic molecules, such as sugars. Molecules not known to be catabolized were termed "xenobiotics," especially if they were "synthetic" products of human industry. We have since found that virtually any organic molecule can be catabolized by a microbe that has evolved the appropriate enzymes. Catabolism by microbes in our own digestive system helps us digest complex food molecules. Catabolism also plays key roles in microbial disease; for example, the causative agent of acne, *Propionibacterium acnes,* degrades skin cell components such as lipids.

Note: Besides releasing energy, the breakdown of organic food molecules provides substrates for biosynthesis. Biosynthesis is covered in Chapter 15.

Substrates for Catabolism

Here we present several important classes of catabolic substrates (**Fig. 13.13**). While in principle virtually any organic constituent may be catabolized, certain kinds of substrates are used more rapidly because they require less activation energy or fewer types of enzymes to break down. Many of these substrates form products important to our nutrition and technology. Others, such as aromatic components of petroleum, are environmental pollutants that only bacteria and fungi can degrade (Section 13.7).

Carbohydrates. Carbohydrates (sugars and sugar polymers, or polysaccharides) (**Fig. 13.13A**), are important as structural components of cells, and for their central role in human digestion. Microbial catabolism is central to our production of food and drink (presented in Chapter 16). Glucose as such is rarely available to microbes, except to pathogens growing within a host. But the pathways of glucose, or polysaccharides catabolism complete the digestion of a diverse range of food molecules found in the environment. For example, the cellulose of plant cell walls, the starch of potatoes, the pectin of fruit—all contain sugar chains that are broken down by microbial enzymes, first to short chains (oligosaccharides), then to two-sugar units (disaccharides), and then to monosaccharides such as glucose or fructose. In some sugars, the aldehyde is replaced by a hydroxyl (sorbitol, mannitol) or a carboxylate (gluconate, glucuronate). Polysaccharides are hydrolyzed to products that enter central catabolic pathways such as glycolysis (**Fig. 13.14**).

Different species differ profoundly in their abilities to digest particular polysaccharides. The starches (glucans) and pectins (pectic acids) are the most widely digested. In a complex environment such as the intestinal lumen, with numerous potential substrates for catabolism, organisms select substrates on the basis of their availability and energy efficiency. Substrates are selected through gene regulation, as discussed in Chapter 10. For example, in *E. coli* the sugar lactose induces transcription of genes that encode beta-galactosidase (*lacZ*) and lactose permease (*lacY*). But in the presence of glucose, a preferred carbon source, *lac* transcription is halted. Halting *lac* transcription enables preferential catabolism of glucose. The process of prioritized consumption of substrates is known as **catabolite repression** (presented in Chapter 10).

Lipids. Many bacteria catabolize lipids (**Fig. 13.13B**) from sources such as milk, animal fats, and nuts. The oxidation of lipid catabolites causes the rancid odor of spoiled meat or butter (issues of food microbiology, discussed in Chapter 16). Microbes catabolize lipids by hydrolysis to glycerol and fatty acids (**Fig. 13.14**). Glycerol, a three-carbon triol, can be considered a three-carbon sugar; it commonly enters catabolism as an intermediate of glycolysis. Alternatively, other pathways break down glycerol to acetate. Fatty acids,

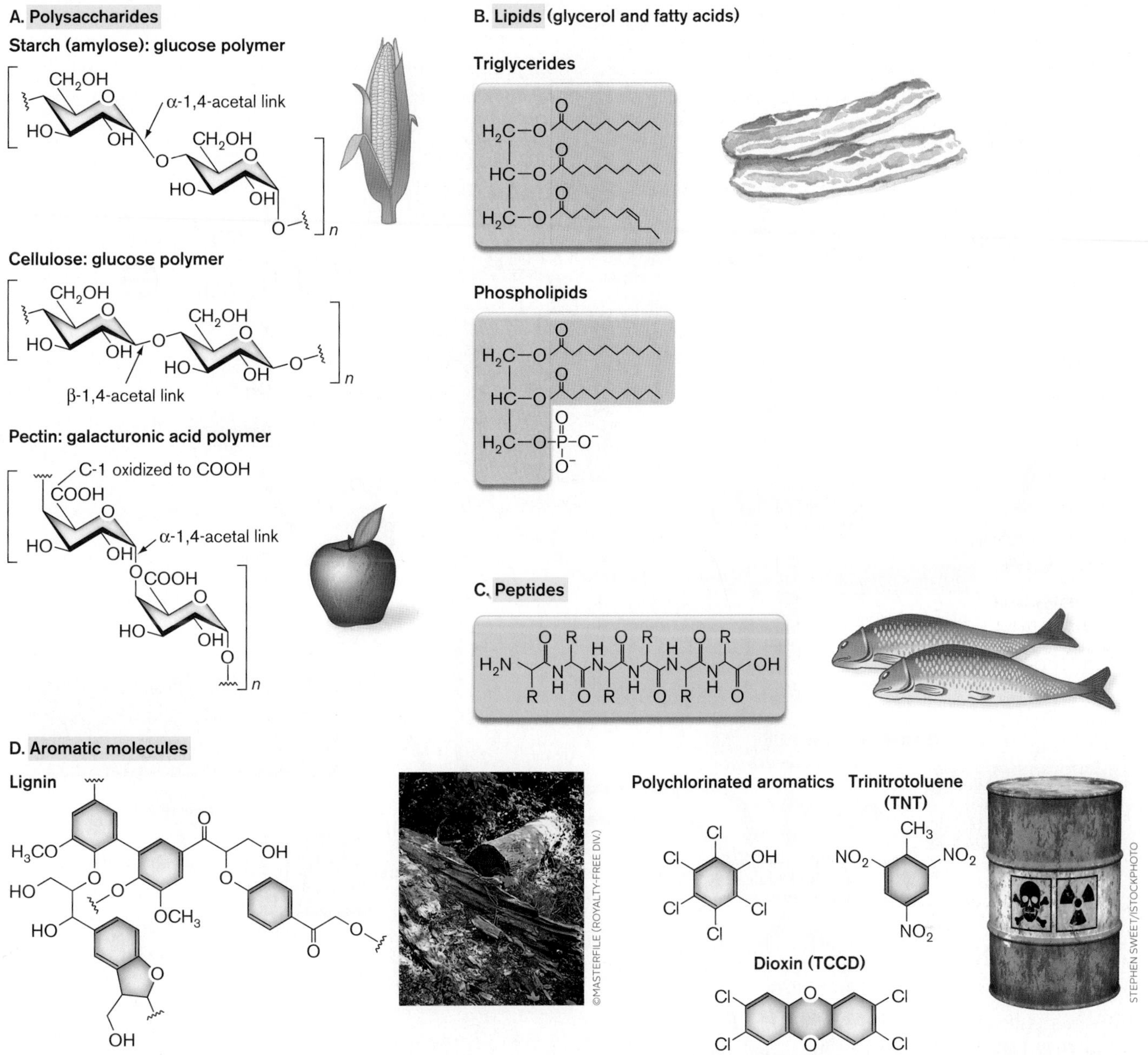
A. Polysaccharides
Starch (amylose): glucose polymer
CH2OH
O
α-1,4-acetal link
HO
OH
CH2OH
O
HO
OH
n
Cellulose: glucose polymer
β-1,4-acetal link
Pectin: galacturonic acid polymer
C-1 oxidized to COOH
COOH
α-1,4-acetal link
B. Lipids (glycerol and fatty acids)
Triglycerides
H2C–O
HC–O
Phospholipids
C. Peptides
H2N
OH
D. Aromatic molecules
Lignin
H3CO
OCH3
OH
Polychlorinated aromatics
Trinitrotoluene (TNT)
CH3
NO2
Cl
Dioxin (TCCD)
©MASTERFILE (ROYALTY-FREE DIV.)
STEPHEN SWEET/ISTOCKPHOTO

FIGURE 13.13 ▪ Complex carbon sources for catabolism. A. Polysaccharides such as starch, cellulose, and pectin are hydrolyzed to glucose. **B.** Lipids are broken down to acetate. **C.** Peptides are hydrolyzed to amino acids and then broken down to acetate, amines, and other molecules. **D.** Complex aromatic molecules such as lignins and halogenated aromatic pollutants are broken down to acetate and other molecules.

more highly reduced than glycerol, undergo oxidative breakdown by the fatty acid degradation (FAD) pathway, forming acetyl groups. These acetyl groups enter the TCA cycle when a terminal electron acceptor is available; alternatively, they enter fermentation or anaerobic syntrophy.

Peptides. We commonly think of proteins as essential parts of a cell. However, when present in excess, proteins can be catabolized to provide energy. Initially, proteins are broken into peptides (**Fig. 13.13C**) by sequence-specific proteases. Peptides are then hydrolyzed to individual amino acids (**Fig. 13.14**). Some pathogens catabolize specific amino acids as part of the disease process. For example, *Legionella pneumophila,* the cause of legionellosis, catabolizes threonine. Threonine catabolism is required for the pathogen to grow within macrophages, a type of white blood cells. Thus, threonine catabolism may be a target for new drugs against *L. pneumophila.*

Specific enzymes catalyze the early steps in the degradation of each amino acid until products are formed that

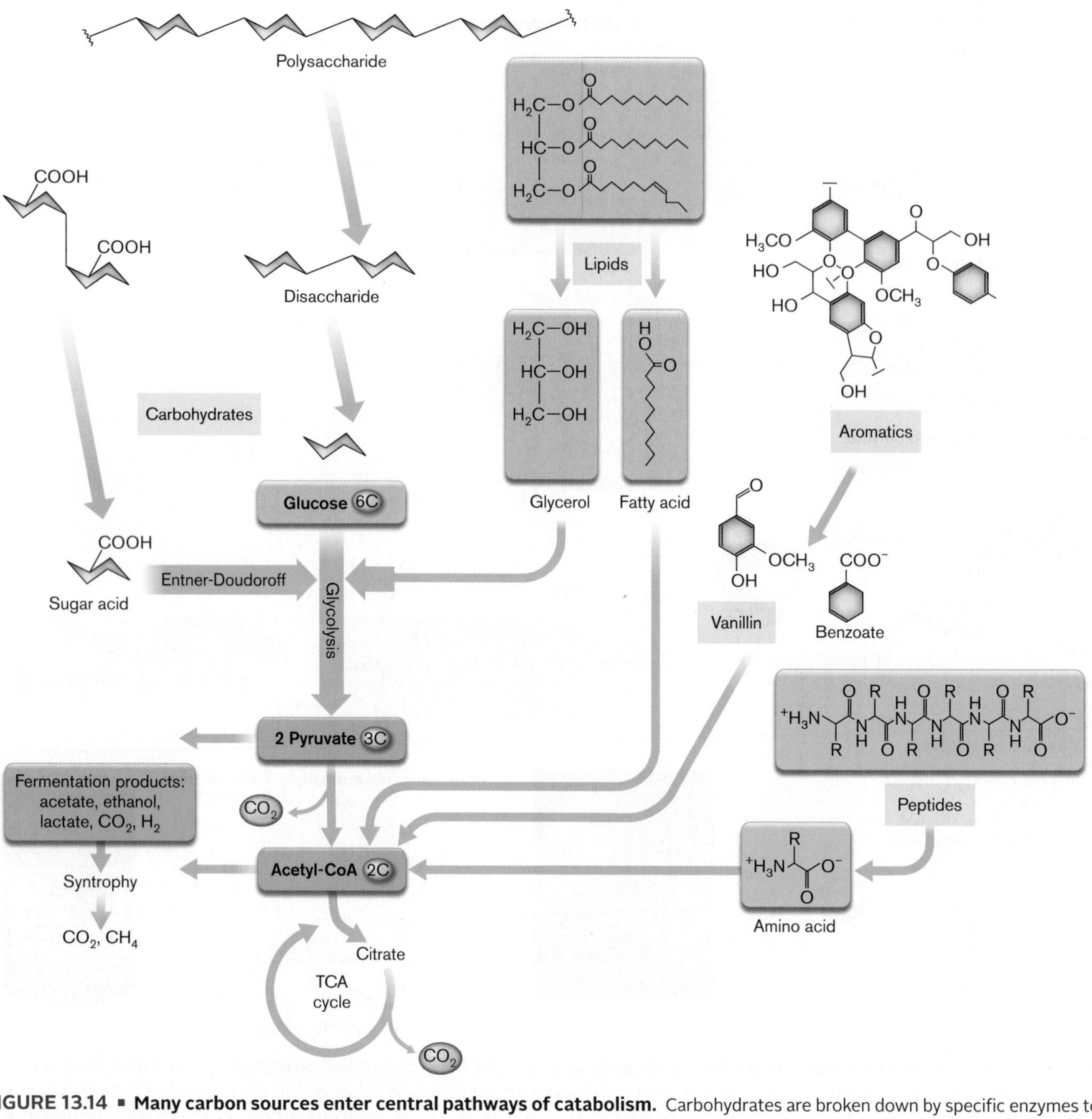

Polysaccharide
COOH
COOH
Disaccharide
Carbohydrates
Lipids
Glycerol
Fatty acid
Glucose 6C
Aromatics
Sugar acid
Entner-Doudoroff
Glycolysis
Vanillin
Benzoate
2 Pyruvate 3C
Fermentation products: acetate, ethanol, lactate, CO_2, H_2
CO_2
Peptides
Acetyl-CoA 2C
Syntrophy
CO_2, CH_4
Amino acid
Citrate
TCA cycle
CO_2

FIGURE 13.14 ■ Many carbon sources enter central pathways of catabolism. Carbohydrates are broken down by specific enzymes to disaccharides and then to monosaccharides such as glucose. Glucose and sugar acids are converted to pyruvate, which releases acetyl groups. Acetyl groups or acetate are also the breakdown products of fatty acids, amino acids, and complex aromatic plant materials such as lignin.

can enter common pathways of carbohydrate catabolism or the TCA cycle. The initial step of amino acid degradation is one of two kinds: decarboxylation (removal of CO_2) to produce an amine, or deamination (removal of NH_3) to produce a carboxylic acid. Carboxylic acids are degraded through the TCA cycle. Amine products, such as cadaverine and putrescine, are often excreted, causing the noxious odor of decomposing flesh.

Aromatic molecules. Aromatic compounds are more difficult to digest than sugars because of the exceptional stability of aromatic ring structures. Yet many bacteria metabolize benzene derivatives, and even polycyclic aromatic molecules, either partly or all the way to CO_2. A particularly important aromatic substance found in nature is **lignin** (**Fig. 13.13D**), which forms the key structural support of trees and woody stems. A discouragingly complex molecule, lignin is made from six-carbon sugars converted to benzene rings, with ether connections that are difficult for enzymes to break down. Fungi and soil bacteria catabolize lignin to oxidized benzene derivatives such as benzoate and vanillin. Further breakdown produces acetyl-CoA, which enters the TCA cycle.

Today, the environment contains increasing amounts of human-made aromatic compounds, produced for herbicides and other industrial uses, that are highly toxic pollutants. These include halogenated aromatics, such as polychlorinated biphenyls, the source of highly toxic dioxins. Halogenated aromatics turn out to be catabolized by a number of soil microbes, which are promising candidates for bioremediation of polluted environments (described in Section 13.7). Other types of aromatic molecules catabolized by microbes are polycyclic aromatic hydrocarbons (PAHs). The multiple fused rings of PAH compounds may take years to biodegrade in natural environments. PAHs are found in petroleum, as well as in coal tar products such as creosote, used to preserve wood (**Fig. 13.15**).

How do we discover the reactions of microbial catabolism for substrates as complex as creosote? **Figure 13.15** shows an example of data obtained from ^{13}C-label tracer molecules observed by ^{13}C nuclear magnetic resonance (NMR). The creosote component acenaphthene was labeled with ^{13}C at a single carbon (position 1) so that its NMR resonance signal would appear as a single peak. Then, as microbes degraded the sample, various peaks appeared, which were identified as various oxidized products of acenaphthene. By piecing together many such experiments, it is possible to map the full degradation pathway of catabolism, and to combine these data with genetic experiments to identify the enzymes.

Products of Catabolism

What is the ultimate fate of all the bits of organic carbon chewed up by microbial catabolism? The answer has myriad consequences for human nutrition and environmental cycling, as well as food and industrial production. From **Table 13.1**, recall two major forms of catabolism: **fermentation**, in which all the electrons from organic substrates are put back onto the organic products; and **respiration**, in which the electrons removed are ultimately transferred to an inorganic electron acceptor such as oxygen. In a third case, **photoheterotrophy**, bacteria gain energy from light while using organic carbon substrates for catabolism and/or biosynthesis. Photoheterotrophy is discussed in Chapter 14.

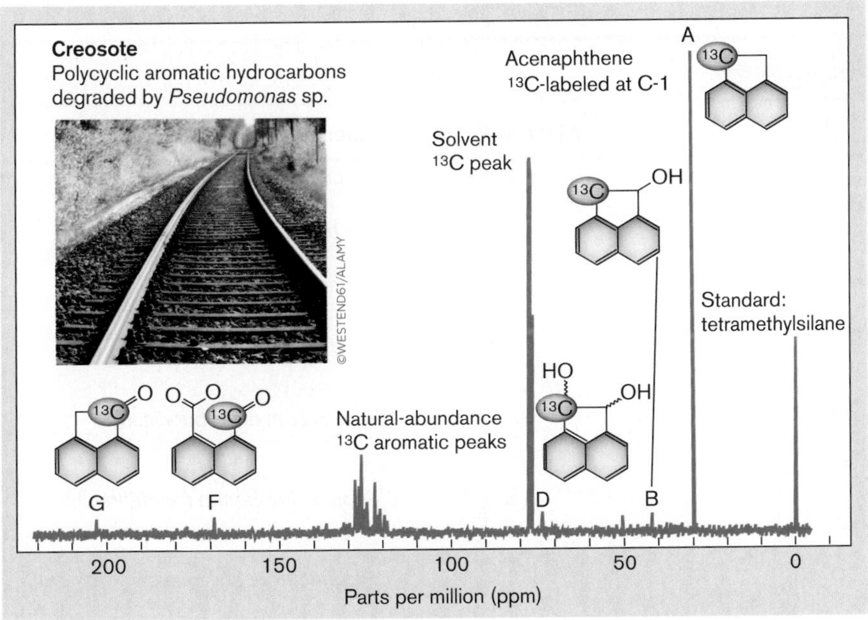
Creosote
Polycyclic aromatic hydrocarbons degraded by *Pseudomonas* sp.
©WESTEND61/ALAMY
Acenaphthene
13C-labeled at C-1
A
Solvent
13C peak
OH
Standard:
tetramethylsilane
HO
OH
Natural-abundance
13C aromatic peaks
O
G
F
D
B
200
150
100
50
0
Parts per million (ppm)

FIGURE 13.15 ■ Isotope labeling reveals pathway of creosote catabolism. Railroad ties (inset) are a source of the pollutant creosote, which contains a mixture of polycyclic aromatic compounds such as acenaphthene. ^{13}C-labeled acenaphthene is included in a creosote sample undergoing catabolism by *Pseudomonas* sp. strain A2279. After 7 days, the oxidized digestion products are revealed by characteristic resonance peaks in the ^{13}C NMR spectrum. *Source:* Modified from Sergey A. Selifonov et al. 1998. *Appl. Environ. Microbiol.* **64**:1447.

Fermentation. First identified by Louis Pasteur as *la vie sans air* ("life without air"), fermentation is the partial breakdown of organic food without net transfer of electrons to an inorganic terminal electron acceptor. Thus, fermentation occurs without oxygen. Fermentation has a negative ΔG, owing to breakdown of a large molecule to several smaller products, which are usually more stable as well. Examples of fermentation pathways include ethanolic fermentation producing alcoholic beverages, and lactate fermentation by lactic acid bacteria producing cheese and yogurt. (More examples are discussed below; also see **Table 13.5**).

Respiration. Respiration combines catabolic breakdown of organic molecules with electron transfer to a terminal electron acceptor such as oxygen. Respiration yields far more energy from catabolism than does fermentation. For humans, respiration is synonymous

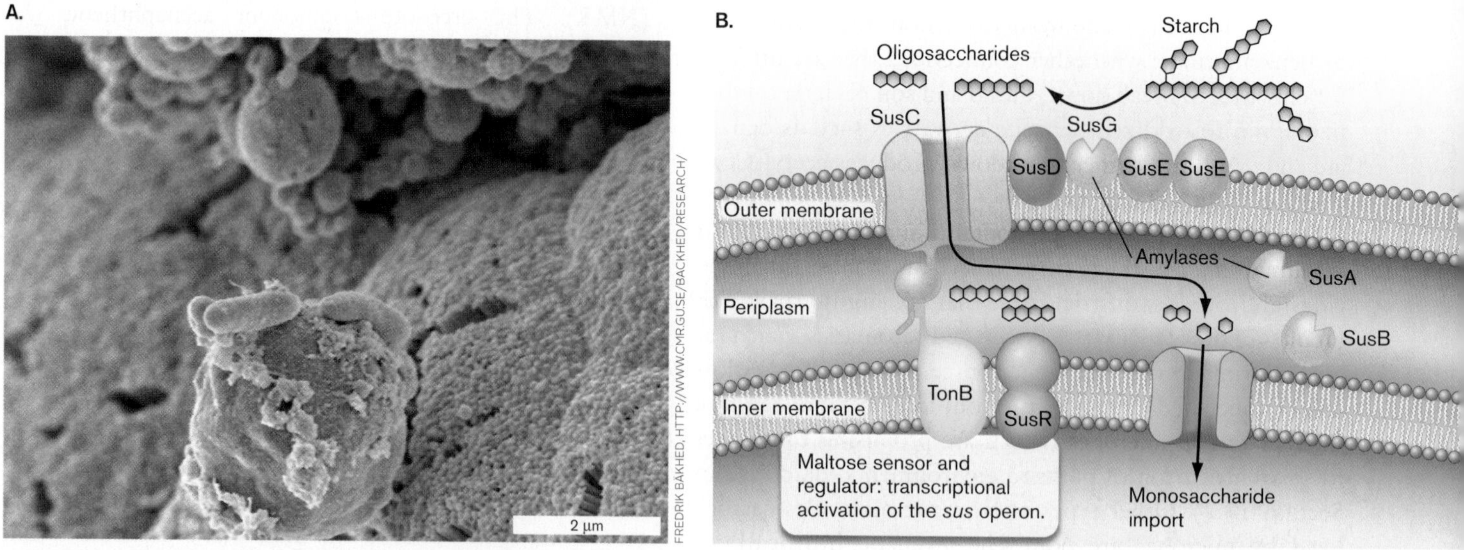
A.
2 μm
FREDRIK BAKHED, HTTP://WWW.CMR.GUSE/BACKHED/RESEARCH/
B.
Starch
Oligosaccharides
SusC
SusG
SusD
SusE
SusE
Outer membrane
Amylases
SusA
Periplasm
SusB
TonB
Inner membrane
SusR
Maltose sensor and
regulator: transcriptional
activation of the *sus* operon.
Monosaccharide
import

FIGURE 13.16 ▪ Polysaccharide catabolism by intestinal bacteria. **A.** *Bacteroides thetaiotaomicron* (colorized yellow) attached to a starch food particle within mouse intestine. **B.** Starch degradation enzymes and transporters within *Bacteroides* envelope. Extracellular enzymes degrade starch to oligosaccharides for transport across the outer membrane by Sus proteins. Periplasmic enzymes cleave the oligosaccharides to disaccharides and monosaccharides for transport across the inner membrane to the cytoplasm. **C.** Human gut bacteria, *Bacteroides plebeius,* have obtained genes from marine bacteria enabling digestion of marine seaweed polysaccharides such as porphyran. Genes encode enzymes (yellow) and Sus-like transport proteins (orange). Percentages indicate gene similarity between species. *Source:* Part B modified from Nicole Koropatkin et al. 2012. *Nat. Rev. Microbiol.* **10**:323; part C modified from Jan-Hendrik Hehemann et al. 2010. *Nat. Rev. Microbiol.* **464**:908.

TABLE 13.5

Fermentation reactions in bacteria (examples).

Reaction	ATP produced	Species of bacteria
$C_6H_{12}O_6 \rightarrow 2CH_3CH_2OH + 2CO_2$ [ethanolic]	2	*Zymomonas* sp.
$C_6H_{12}O_6 \rightarrow 2CH_3CH_2OCOO^- + 2H^+$ [lactate]	2	*Lactobacillus acidophilus*
$C_6H_{12}O_6 \rightarrow CH_3CH_2OH + CO_2 + CH_3CH_2OCOO^- + H^+$ [heterolactic]	2	*Leuconostoc mesenteroides*
$C_6H_{12}O_6 \rightarrow$ succinate, 2-oxoglutarate, acetate, ethanol, formate, lactate, CO_2, H_2 [mixed-acid; equation unbalanced]	Varies	*Escherichia coli*
$C_6H_{12}O_6 \rightarrow$ butanol, acetone, butyric acid, isopropanol, CO_2 [equation unbalanced]	Varies	*Clostridium acetobutylicum*
$2CH_3CH_2OCOO^-$ (lactate) $\rightarrow$ $CH_3(CH_2)COO^-$ (propionate) $+ CH_3COO^- + CO_2 + 2H^+$	3	*Propionibacterium freudenreichii*
$2C_2H_2$ (acetylene) $+ 3H_2O \rightarrow CH_3CH_2OH + CH_3COO^- + H^+$	1	*Pelobacter acetylenicus*
2 Citrate$^{3-} + H^+ \rightarrow$ 2 succinate$^{2-} + CH_3COO^- + 2CO_2$	1	*Providencia rettgeri*
CH_3CHNH_2COOH (alanine) $+ 2NH_2CH_2COOH$ (glycine) $+ 2H_2O \rightarrow$ $3CH_3COO^- + 3NH_4^+ + CO_2$ [Stickland reaction]	3	*Clostridium sporogenes*
$COOH(CH_2)_2COO^-$ (succinate) $\rightarrow CH_3(CH_2)COO^-$ (propionate) $+ CO_2$ $[Na^+_{in} \rightarrow Na^+_{out}]$	0	*Propionigenium modestum*

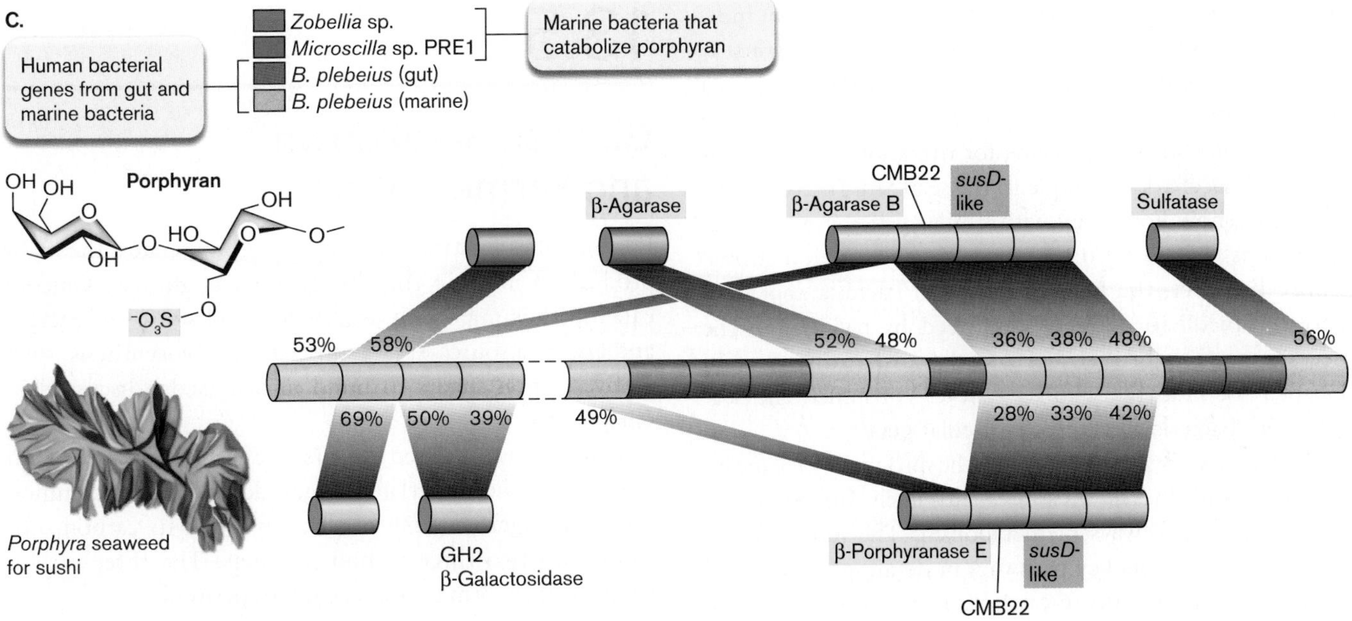
C.
Human bacterial genes from gut and marine bacteria
Zobellia sp.
Microscilla sp. PRE1
B. plebeius (gut)
B. plebeius (marine)
Marine bacteria that catabolize porphyran
Porphyran
β-Agarase
β-Agarase B
CMB22
susD-like
Sulfatase
53% 58%
52% 48%
36% 38% 48%
56%
69% 50% 39%
49%
28% 33% 42%
Porphyra seaweed for sushi
GH2
β-Galactosidase
β-Porphyranase E
susD-like
CMB22

with breathing; but in the absence of O_2, many microbes use alternative electron acceptors, such as nitrate or sulfate. The use of a terminal electron acceptor other than O_2 is called **anaerobic respiration**. Thus, microbes—unlike humans and other animals—have the capacity for anaerobic respiration.

Catabolism: Cell and Molecular Biology

In natural ecosystems, how do cells manage to break down complex carbon sources? Free-living bacteria and fungi may catabolize thousands of different carbon sources, each requiring specific transporters and enzymes for initial breakdown. Most of these, particularly complex plant constituents, are not digestible by animals, whose genomes fail to encode the necessary enzymes. The only polysaccharides that human enzymes can digest are starch, lactose, and sucrose. Yet human breast milk includes a mixture of complex milk oligosaccharides, including branched chains of *N*-acetylglucosamine, sialic acid, and fucose. Amazingly, our own breast milk has evolved to support bacteria with specific catabolic abilities to colonize the infant gut.

It is well known that animals such as cattle require microbes living within their digestive tracts to ferment cellulose from grasses or wood. We now know that humans, too, require related bacteria to digest a variety of plant-derived fibers. For example, both cattle and humans contain anaerobes of the Gram-negative genus *Bacteroides* and the Gram-positive genus *Ruminococcus* (originally named for the bovine rumen). By evolving a symbiosis with microbes, animals avoid the need to acquire new catabolic genes in their own genome. So in effect, the human gut microbes are a functional part of the human body. The microbial genomes are functionally part of the human metagenome, the total sequence of genomes of a community of organisms. Medical conditions such as obesity may depend in part on the catabolic activities of gut biota.

How do our gut bacteria actually handle complex food sources? In the first line of attack, bacterial cells such as *Bacteroides* species adhere to large food particles, which are a mass of tangled molecules (**Fig. 13.16A**). Since bacteria have solid cell walls, incapable of phagocytosis, a bacterium cannot take up a large food particle, but it can secrete enzymes to break down the fibers into short oligosaccharides (**Fig. 13.16B**). The outer membrane-inserted enzyme shown in the figure (SusG) is part of a multiprotein system called Sus for "starch utilization system." Other genes encode the oligosaccharide outer membrane transporter (SusC and SusD) and the periplasmic amylases that break down the oligosaccharides (SusA and SusB). The monosaccharide products are then transported across the inner membrane into the cytoplasm, for glycolysis.

Genome analysis reveals hundreds of Sus homologs, called Sus-like systems, that target different polysaccharides; these comprise, for example, 18% of the genome of *Bacteroides thetaiotaomicron*. In the example shown

(**Fig. 13.16C**), intestinal bacteria were analyzed from individuals with a long dietary history of eating *Porphyra* seaweed (nori, used to wrap sushi). Their intestinal bacteria include a strain of *Bacteroides plebeius* whose genes encode enzymes and Sus-like proteins for digestion of the sulfonated polysaccharides unique to the seaweed. Interestingly, the genes show homology with similar genes from marine bacteria associated with the seaweed (*Zobellia* and *Microscilla*). In effect, the human-associated bacteria appear to have "learned" to digest the seaweed by picking up the necessary genes through horizontal transfer (discussed in Chapter 9). Humans who have never consumed the seaweed lack bacteria with these particular genes.

As we saw in **Figure 13.14**, the products of catabolism of many diverse substrates ultimately funnel into a few common pathways of metabolism. The remainder of this chapter presents key pathways in detail: glycolysis and other pathways of glucose catabolism, the TCA cycle, and the catechol pathway of benzoate catabolism. These pathways play key roles in medical microbiology and in industrial fields such as bioremediation.

To Summarize

- **Carbohydrates, or polysaccharides,** are broken down to disaccharides, and then to monosaccharides. Sugars and sugar derivatives, such as amines and acids, are catabolized to pyruvate.
- **Pyruvate and other intermediary products of sugar catabolism** are fermented, or they are further catabolized to CO_2 and H_2O through the TCA cycle (in the presence of a terminal electron acceptor) or to CO_2, H_2, and CH_4 through fermentation and syntrophy.
- **Lipids and amino acids** are catabolized to glycerol and acetate, as well as other metabolic intermediates.
- **Aromatic compounds such as lignin and benzoate derivatives** are catabolized to acetate through different pathways, such as the catechol pathway.
- **Fermentation and respiration** complete the process of catabolism. In fermentation, the catabolite is broken down to smaller molecules without an inorganic electron acceptor. Respiration requires an inorganic terminal electron acceptor such as O_2 or nitrate. In photoheterotrophy, catabolism is supplemented by light absorption.
- **Catabolism of complex substrates requires enzymes and transporters** of specialized multi-protein Sus-like systems. Different bacteria contain homologous systems that are adapted to different carbon sources.

13.5

Glucose Breakdown and Fermentation

For microbes feasting on complex carbohydrates, the main product of most of their breakdown is glucose. Glucose catabolism is important as a widespread source of energy, and also as a source of key substrates for biosynthesis, such as five-carbon sugars to build nucleic acids (discussed in Chapter 15).

Glucose and related sugars are catabolized through a series of phosphorylated sugar derivatives. A common theme in sugar catabolism is the split of a six-carbon substrate into two three-carbon products. The three-carbon products may form two molecules of pyruvate:

$$C_6H_{12}O_6 \rightarrow 2C_3H_4O_3 + 4H\ (2\ NADH + 2H^+)$$

Under anaerobiosis—the prevailing condition of many microbial habitats—the pyruvate obtained from sugar breakdown must be converted to compounds that receive electrons from NADH, in order to restore the electron-accepting form NAD^+. Different microbes reduce pyruvate to different end products of fermentation. Alternatively, through respiration, NADH may reduce an electron acceptor such as oxygen or nitrate, allowing pyruvate to feed into the TCA cycle (discussed in Section 13.6).

Note: The carboxylic acid intermediates of metabolism exist in equilibrium with their dissociated, or ionized, forms, identified by the suffix "-ate." For example, lactic acid dissociates to lactate; acetic acid dissociates to acetate. We use the "-ate" terms for acids whose ionized form predominates under typical cell conditions (around pH 7).

To catabolize glucose, bacteria and archaea use three main routes to pyruvate (**Fig. 13.17**):

- **Glycolysis**, or the **Embden-Meyerhof-Parnas (EMP) pathway**, in which glucose 6-phosphate isomerizes to fructose 6-phosphate, ultimately forming two molecules of pyruvate. EMP is used by many bacteria, eukaryotes, and archaea. From each glucose, the pathway generates net two ATP and two NADH.
- **Entner-Doudoroff (ED) pathway**, in which glucose 6-phosphate is oxidized to 6-phosphogluconate, a phosphorylated sugar acid. Alternatively, sugar acids may be

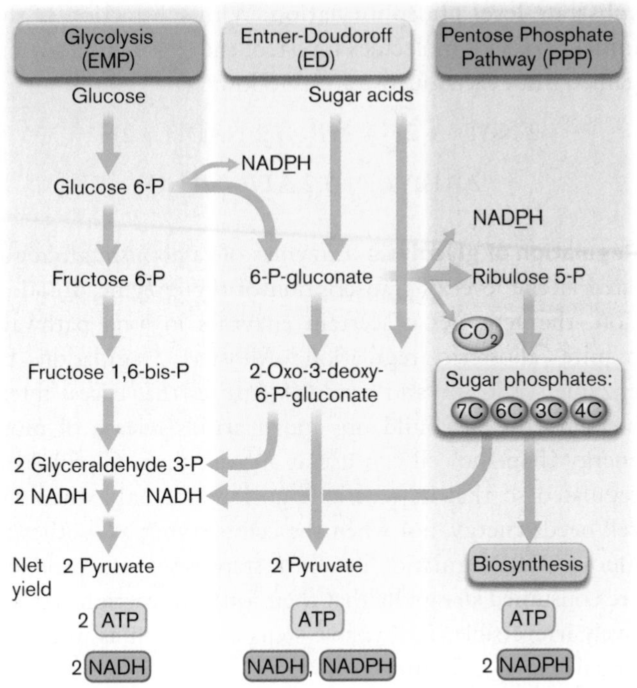
Glycolysis (EMP)
Entner-Doudoroff (ED)
Pentose Phosphate Pathway (PPP)
Glucose
Sugar acids
NADPH
Glucose 6-P
NADPH
Fructose 6-P
6-P-gluconate
Ribulose 5-P
CO_2
Fructose 1,6-bis-P
2-Oxo-3-deoxy-6-P-gluconate
Sugar phosphates:
7C 6C 3C 4C
2 Glyceraldehyde 3-P
2 NADH
NADH
Net yield
2 Pyruvate
2 Pyruvate
Biosynthesis
2 ATP
ATP
ATP
2 NADH
NADH, NADPH
2 NADPH

FIGURE 13.17 ■ From glucose to pyruvate: three pathways. The Embden-Meyerhof-Parnas (EMP) pathway of glycolysis, the Entner-Doudoroff (ED) pathway, and the pentose phosphate pathway (PPP) each catabolize carbohydrates by related but different routes.

converted directly to 6-phosphogluconate. Sugar acids often derive from intestinal mucus, and the ED pathway is essential for enteric bacteria to colonize the intestinal epithelium. The ED pathway generates only one ATP, one NADH, and one NADPH.

- **Pentose phosphate pathway (PPP)**, also known as the pentose phosphate shunt, in which glucose 6-phosphate is oxidized to 6-phosphogluconate, and then decarboxylated to a five-carbon sugar (pentose), ribulose 5-phosphate. The PPP produces sugars of three to seven carbons, which serve as precursors for biosynthesis or are converted to pyruvate as needed. The PPP generates one ATP plus two NADPH, the reducing cofactor most commonly associated with biosynthesis.

Glycolysis: The EMP Pathway

The **Embden-Meyerhof-Parnas (EMP) pathway**, or **glycolysis**, is the form of glucose catabolism most commonly studied in introductory biology; it is central for animals and plants, as well as many bacteria. In the EMP pathway, one molecule of D-glucose undergoes stepwise breakdown to two molecules of pyruvic acid (or its anion, pyruvate) (**Fig. 13.18**). Glucose breaks down in two stages. In the first stage, the glucose molecule is primed for breakdown by two steps of sugar phosphorylation by ATP. Each ATP phosphotransfer step spends Gibbs energy, as we saw for glucose 6-phosphate formation in **Table 13.3**. The phosphoryl groups tag two sides of the glucose for splitting into two three-carbon molecules of glyceraldehyde 3-phosphate (G3P).

In the second stage, each glyceraldehyde 3-phosphate is oxidized by NAD^+ through steps leading to pyruvate. Each conversion of glyceraldehyde 3-phosphate to pyruvate forms two molecules of ATP—one from dephosphorylation of the substrate, and one from addition of inorganic phosphate. Since 2 ATP were spent originally to phosphorylate glucose, the net gain of energy carriers is two molecules of NADH plus two molecules of ATP.

Most of the conversion steps are associated with a small change in energy—so small that the sign of ΔG depends on the concentrations

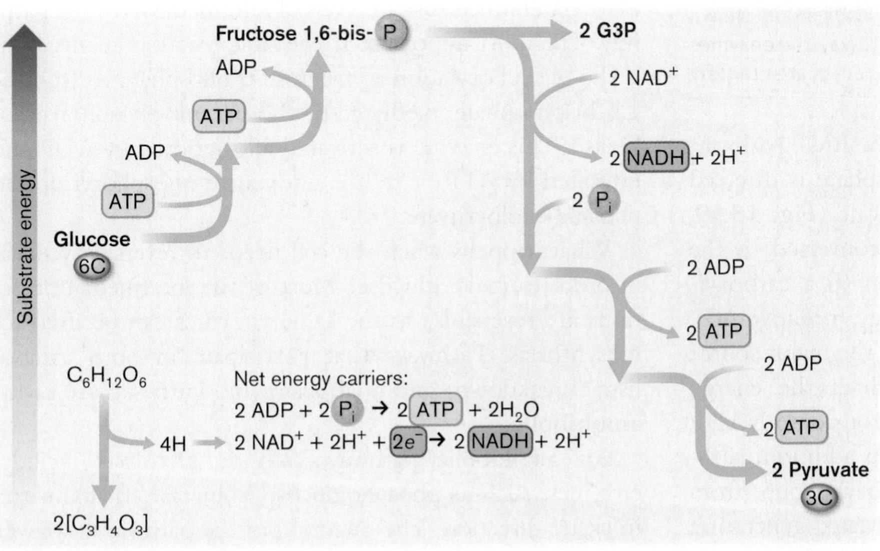
Substrate energy
Fructose 1,6-bis- P
2 G3P
ADP
ATP
2 NAD^+
ADP
2 NADH + $2H^+$
ATP
2 P_i
Glucose
6C
2 ADP
2 ATP
2 ADP
$C_6H_{12}O_6$
Net energy carriers:
2 ADP + 2 P_i → 2 ATP + $2H_2O$
4H → 2 NAD^+ + $2H^+$ + $2e^-$ → 2 NADH + $2H^+$
2 ATP
2 Pyruvate
3C
$2[C_3H_4O_3]$

FIGURE 13.18 ■ Substrate energy changes during the Embden-Meyerhof-Parnas pathway (glycolysis). Glucose is activated through two substrate phosphorylations by ATP. The breakdown of glucose to two molecules of pyruvate is coupled to net production of two ATP and two NADH. Phosphoryl groups are shown as P.

of substrates or products; thus, some individual steps are reversible. In the cytoplasm, however, as intermediate products form they are quickly consumed by the next step, so the pathway flows in one direction. The direction of flow is determined by the key irreversible reactions that consume ATP. These steps drive the pathway by spending energy.

Phosphorylation and splitting of glucose. In the first stage of the EMP pathway, the six-carbon sugar is activated by two phosphorylation steps (**Fig. 13.19**, left). The first phosphoryl group (phosphate) is added at carbon 6 of glucose. (In some species of bacteria, the first phosphoryl group is added by phosphoenolpyruvate instead of ATP, but the net effect is the same.) The next enzyme-catalyzed step, rearrangement of glucose 6-phosphate to fructose 6-phosphate, involves no significant change in energy but prepares the sugar to receive the second phosphoryl group, forming fructose 1,6-bisphosphate.

The sugar then splits into two three-carbon sugars (trioses), each tagged with one of the two phosphates. The splitting of this molecule has a favorable entropy change (ΔS), but its chemical change (ΔH) is unfavorable, largely canceling out the energy yield. The two triose phosphates—glyceraldehyde 3-phosphate (G3P) and dihydroxyacetone phosphate (DHAP)—have nearly the same energy, so an enzyme interconverts them reversibly. Interconversion is necessary because only G3P proceeds further in the pathway.

Note: In Chapters 13–16, every substrate conversion shown requires catalysis by an enzyme. For the EMP pathway, the enzymes are shown, but for other pathways, the enzyme names are omitted.

ATP generation. In the second stage of the EMP pathway, each three-carbon glyceraldehyde 3-phosphate is directed into an energy-yielding pathway to pyruvate (**Fig. 13.19**, right). First, the aldehyde (R-CHO) is converted to the carboxylate ($R\text{-}COO^- + H^+$). Conversion to a carboxylate releases a substantial amount of energy (negative ΔG). This oxidation of the aldehyde represents the major source of energy in glycolysis—the step at which the energy obtained is used to transfer a pair of electrons onto NAD^+, forming NADH (with an ionized H^+). In addition, sufficient energy is released to add a phosphoryl group (from inorganic phosphate, P_i) to the carboxylate, generating 1,3-bisphosphoglycerate.

In subsequent steps, the added phosphoryl group will be transferred to ADP, yielding the one net ATP generated per pyruvate (two per glucose). The transfer of a phosphoryl group from an organic substrate to make ATP is called **substrate-level phosphorylation**. With subtraction of the initial two ATP molecules invested, the net energy carriers gained from each glucose are as follows:

$$2\ NAD^+ + 2e^- + 2H^+ \rightarrow 2\ NADH + 2H^+$$

$$2\ ADP + P_i \rightarrow 2\ ATP + 2H_2O$$

Regulation of glycolysis. Enzymes of catabolism are regulated at the level of transcription of the enzyme. In addition, the activities of certain enzymes in long pathways require allosteric regulation. Allosteric regulation by enzyme substrates and products ensures that excess intermediates do not build up, and it avoids release of more energy than the cell can use at a given time. Glycolysis is regulated so that its reactions go forward only when the cell needs energy, not when the cell is trying to synthesize glucose. The regulation occurs at steps where the products are consumed so rapidly that their forward reaction is effectively irreversible. Irreversible steps are shown as unidirectional arrows in **Figure 13.19**.

Thought Question

13.8 Some bacteria make an enzyme, dihydroxyacetone kinase, that phosphorylates dihydroxyacetone to dihydroxyacetone phosphate (DHAP). How could this enzyme help the cell yield energy?

Enzymes that catalyze irreversible steps are regulated so as to maintain consistent levels of intermediates in the pathway. The most important irreversible reaction in glycolysis is the phosphorylation of fructose 6-phosphate to fructose 1,6-bisphosphate, mediated by the enzyme phosphofructokinase. This enzyme is activated allosterically by ADP and inhibited by ATP or by the alternative phosphoryl donor, phosphoenolpyruvate.

What happens when the cell needs to reverse glycolysis in order to make glucose? Most of the intermediate reactions are reversible, so the same enzymes can be used for biosynthesis. Pathways that participate in both catabolism (breakdown) and anabolism (biosynthesis) are called **amphibolic**.

An amphibolic pathway such as glycolysis includes enzymes, such as phosphoglucose isomerase, that can run in either direction. The direction of the pathway at a given time is determined by key enzymes that operate only in the catabolic direction or in the anabolic direction. For example, in glycolysis the ATP phosphorylation of fructose 6-phosphate is catalyzed by the enzyme phosphofructokinase, whereas in biosynthesis this step is reversed by a

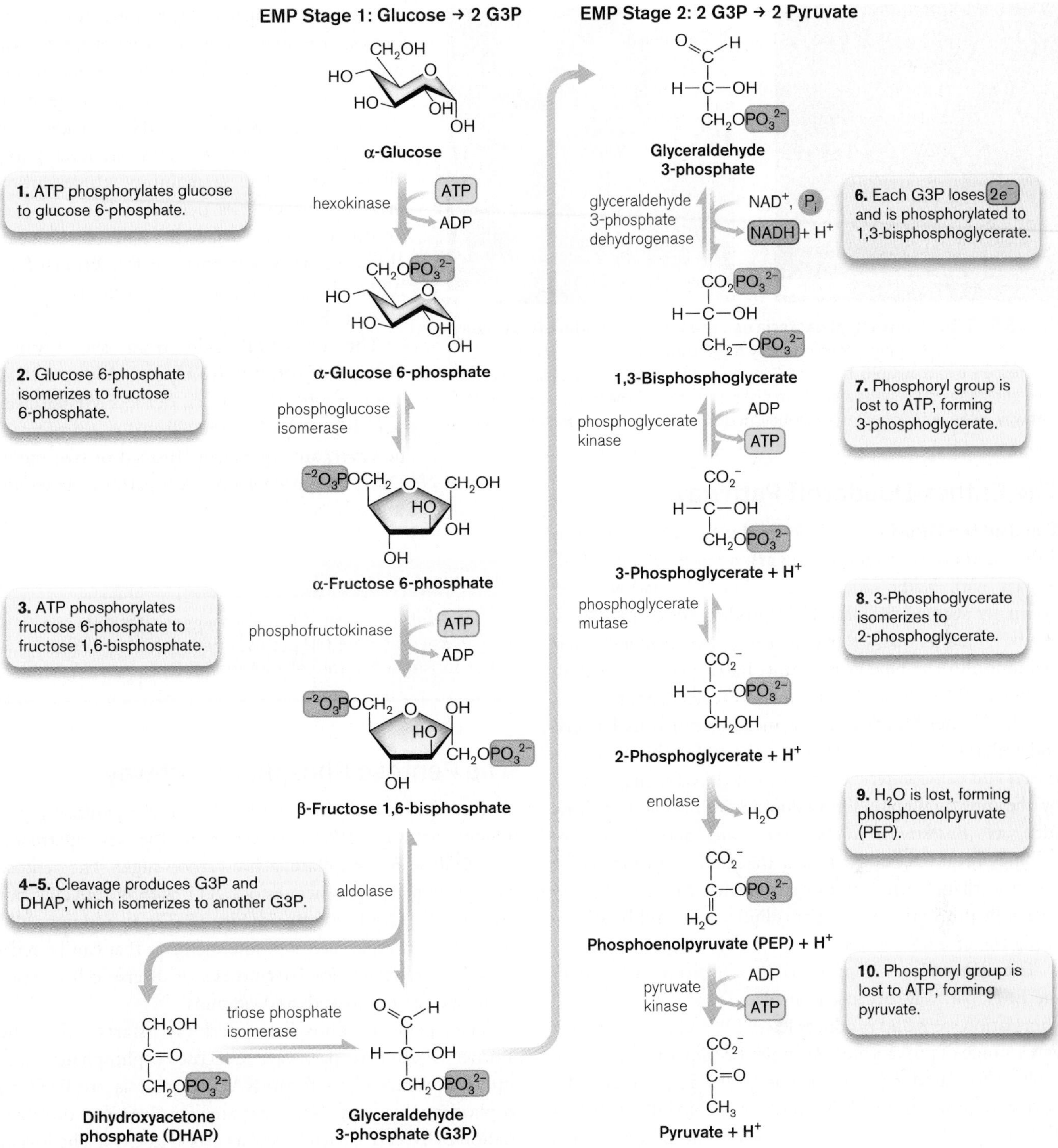

EMP Stage 1: Glucose → 2 G3P
α-Glucose
1. ATP phosphorylates glucose to glucose 6-phosphate.
hexokinase
ATP
ADP
α-Glucose 6-phosphate
2. Glucose 6-phosphate isomerizes to fructose 6-phosphate.
phosphoglucose isomerase
α-Fructose 6-phosphate
3. ATP phosphorylates fructose 6-phosphate to fructose 1,6-bisphosphate.
phosphofructokinase
ATP
ADP
β-Fructose 1,6-bisphosphate
4–5. Cleavage produces G3P and DHAP, which isomerizes to another G3P.
aldolase
triose phosphate isomerase
Dihydroxyacetone phosphate (DHAP)
Glyceraldehyde 3-phosphate (G3P)
EMP Stage 2: 2 G3P → 2 Pyruvate
Glyceraldehyde 3-phosphate
glyceraldehyde 3-phosphate dehydrogenase
NAD+, Pi
NADH + H+
6. Each G3P loses 2e− and is phosphorylated to 1,3-bisphosphoglycerate.
1,3-Bisphosphoglycerate
phosphoglycerate kinase
ADP
ATP
7. Phosphoryl group is lost to ATP, forming 3-phosphoglycerate.
3-Phosphoglycerate + H+
phosphoglycerate mutase
8. 3-Phosphoglycerate isomerizes to 2-phosphoglycerate.
2-Phosphoglycerate + H+
enolase
H2O
9. H2O is lost, forming phosphoenolpyruvate (PEP).
Phosphoenolpyruvate (PEP) + H+
pyruvate kinase
ADP
ATP
10. Phosphoryl group is lost to ATP, forming pyruvate.
Pyruvate + H+

FIGURE 13.19 ■ Embden-Meyerhof-Parnas pathway (glycolysis).

different enzyme, fructose bisphosphatase. Instead of regenerating ATP, fructose bisphosphatase removes the second phosphoryl group as inorganic phosphate—a step yielding energy and thus driving the whole pathway in reverse (toward biosynthesis of sugar). The two enzymes are regulated differently; the catabolic enzyme phosphofructokinase is activated by ADP, a signal of energy need, whereas the biosynthetic enzyme is inhibited by such signals.

A.
B.
10 μm

PAUL COHEN, U. RHODE ISLAND AND TYRRELL CONWAY, U. OKLAHOMA

COURTESY OF TYRRELL CONWAY

FIGURE 13.20 ■ Intestinal bacteria use the Entner-Doudoroff pathway. **A.** Intestinal *E. coli* (orange) feed primarily on gluconate from mucus secretions (fluorescence micrograph). **B.** Tyrrell Conway, at the University of Oklahoma, used genomics and genetic analysis to dissect the role of the Entner-Doudoroff pathway in the enteric bacterial catabolism of sugar acids from intestinal mucus.

The Entner-Doudoroff Pathway

The **Entner-Doudoroff (ED) pathway** offers a slightly different route to catabolize sugars, as well as sugar acids (sugars with acidic side chains). The ED pathway was originally studied for its role in production of the Mexican beverage *pulque,* or "cactus beer," by *Zymomonas* fermentation of the blue agave plant. Later, genomic analysis by Tyrrell Conway and colleagues revealed genes encoding the Entner-Doudoroff enzymes in numerous bacteria and archaea. In the human colon, the ED pathway enables *E. coli* and other enteric bacteria to feed on mucus secreted by the intestinal epithelium (**Fig. 13.20**). Some gut flora, such as *Bacteroides thetaiotaomicron,* actually induce colonic production of the mucus that they consume. These bacteria that "farm" intestinal mucus enhance human health by preventing pathogen colonization, and by stimulating the immune system.

The ED pathway appears to have evolved earlier than the EMP pathway, because it involves fewer substrate phosphorylation steps and produces less ATP, and it is found in a wider range of prokaryotes. As in the EMP pathway, glucose is phosphorylated to glucose 6-phosphate (**Fig. 13.21**). The next step, however, involves oxidation by $NADP^+$ at carbon 1, with loss of two hydrogens to form 6-phosphogluconate. Gluconate is a sugar acid found in intestinal mucus; it can be phosphorylated to enter the ED pathway.

The hydrogens and electrons from glucose 6-phosphate are transferred to $NADP^+$ instead of NAD^+ as in the EMP pathway. This step differs from the EMP pathway in two respects: The carrier used is $NADP^+$ instead of NAD^+, and the electron transfer occurs early, without a second ATP-consuming phosphorylation step. When the six-carbon substrate is eventually split into two three-carbon products, one of the three-carbon products is glyceraldehyde 3-phosphate, which enters the second stage of glycolysis (see **Fig. 13.19**). NADH is made, and two ATP are made by substrate-level phosphorylation. The remaining three-carbon product, however, is pyruvate. This one-step production of pyruvate short-circuits the catabolic pathway, missing the formation of an ATP. The unused potential energy is released as waste heat.

The net ATP gain from the Entner-Doudoroff pathway is only one ATP per molecule of glucose, half that of the EMP pathway (see **Fig. 13.17**). The electrons transferred, however, are equivalent: Instead of two molecules of NADH, the Entner-Doudoroff pathway generates one NADH and one NADPH.

Thought Question

13.9 Explain why the ED pathway generates only one ATP, whereas the EMP pathway generates two. What is the consequence for cell metabolism?

The Pentose Phosphate Pathway

A third pathway of glucose catabolism is the **pentose phosphate pathway (PPP)**, which forms the key intermediate **ribulose 5-phosphate**, a five-carbon sugar. The pentose phosphate pathway generates one ATP with no NADH, but two NADPH for biosynthesis (**Fig. 13.22**). In addition, PPP generates a complex series of intermediates that can be redirected as substrates for biosynthesis of diverse cell components such as amino acids and vitamins.

The pentose phosphate pathway starts like the Entner-Doudoroff pathway: Glucose 6-phosphate gives up two electrons to form NADPH and is oxidized to 6-phosphogluconate. The next step involves a second oxidation by $NADP^+$, with loss of a carbon as CO_2. The loss of CO_2 generates the five-carbon sugar ribulose 5-phosphate (hence the pathway name "pentose phosphate"). In succeeding steps, pairs of sugars, such as sedoheptulose 7-phosphate and glyceraldehyde 3-phosphate, exchange short carbon chains, giving rise to sugar phosphates of various lengths—for example, ribose 5-phosphate and erythrose 4-phosphate, which are precursors of purines and aromatic amino acids, respectively. Alternatively, if these

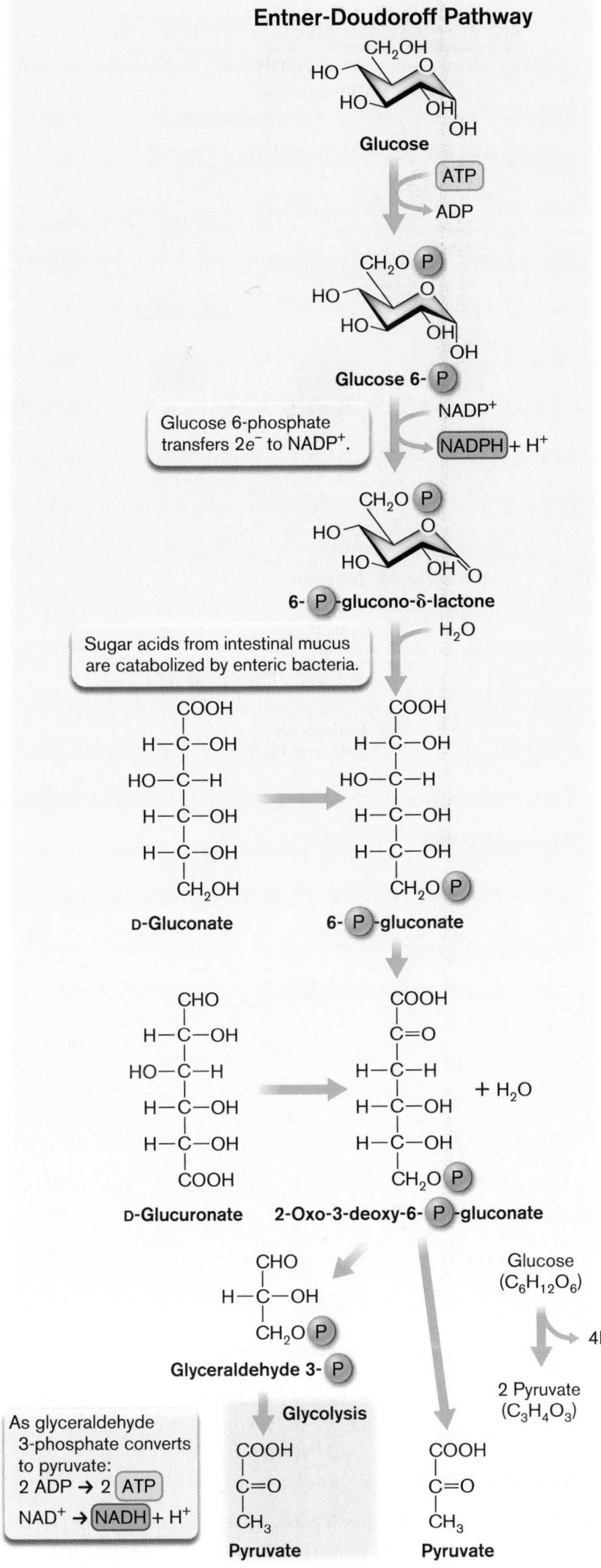
Entner-Doudoroff Pathway
Glucose
ATP
ADP
Glucose 6-P
NADP+
NADPH + H+
Glucose 6-phosphate transfers 2e− to NADP+.
6-P-glucono-δ-lactone
H2O
Sugar acids from intestinal mucus are catabolized by enteric bacteria.
D-Gluconate
6-P-gluconate
D-Glucuronate
2-Oxo-3-deoxy-6-P-gluconate
+ H2O
Glyceraldehyde 3-P
Glycolysis
Glucose (C6H12O6)
4H
2 Pyruvate (C3H4O3)
Net energy carriers:
ADP + Pi → ATP + H2O
NADP+ + 2H+ + 2e− → NADPH + H+
NAD+ + 2H+ + 2e− → NADH + H+
As glyceraldehyde 3-phosphate converts to pyruvate:
2 ADP → 2 ATP
NAD+ → NADH + H+
Pyruvate
Pyruvate

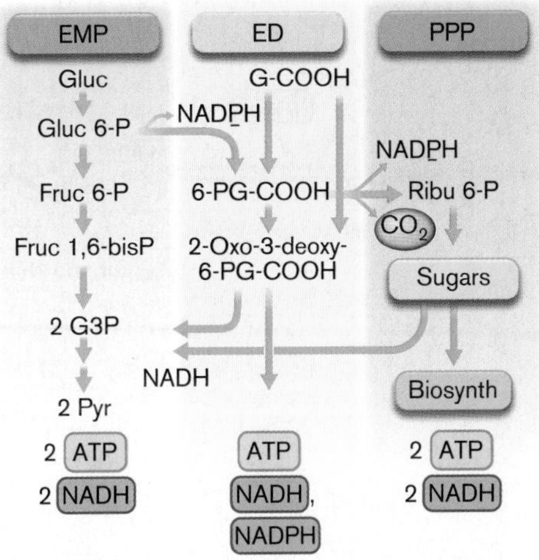
EMP
ED
PPP
Gluc
G-COOH
Gluc 6-P
NADPH
NADPH
Fruc 6-P
6-PG-COOH
Ribu 6-P
CO2
Fruc 1,6-bisP
2-Oxo-3-deoxy-6-PG-COOH
Sugars
2 G3P
NADH
Biosynth
2 Pyr
2 ATP
2 NADH
ATP
NADH,
NADPH
2 ATP
2 NADH

FIGURE 13.21 ■ Entner-Doudoroff pathway. Glucose 6-phosphate is oxidized to 6-phosphogluconate, with one pair of electrons transferred to NADPH. The 6-phosphogluconate is dehydrated and cleaved to form one pyruvate plus one glyceraldehyde 3-phosphate (G3P) that enters the EMP pathway to pyruvate.

routes to biosynthesis are not taken, the intermediates convert to fructose 6-phosphate and reenter the EMP pathway, where ATP and NADH are produced.

Fermentation Completes Catabolism

None of the pathways from glucose to pyruvate constitutes a completed pathway of catabolism, because NADH and NADPH remain to be recycled. In the absence of oxygen or other electron acceptors, heterotrophic cells must transfer the hydrogens from NADH + H^+ back onto the products of pyruvate, forming partly oxidized fermentation products with the same redox level (balance of O and H) as the original glucose had (**Table 13.5**). For example, glucose may be fermented to two molecules of lactic acid (**lactate fermentation**) or two molecules of ethanol plus two CO_2 (**ethanolic fermentation**) or one lactic acid, one ethanol, and one CO_2 (**heterolactic fermentation**). These fermentation products are then excreted from the cell. The large quantities of substrate consumed in fermentation generate large amounts of waste products to be excreted. These bacterial wastes are actually useful to human "fermentation industries" such as the production of alcoholic beverages (ethanolic fermentation) or cheese (lactate fermentation).

Why do fermenting bacteria give up such large quantities of waste products that retain usable energy? The reason

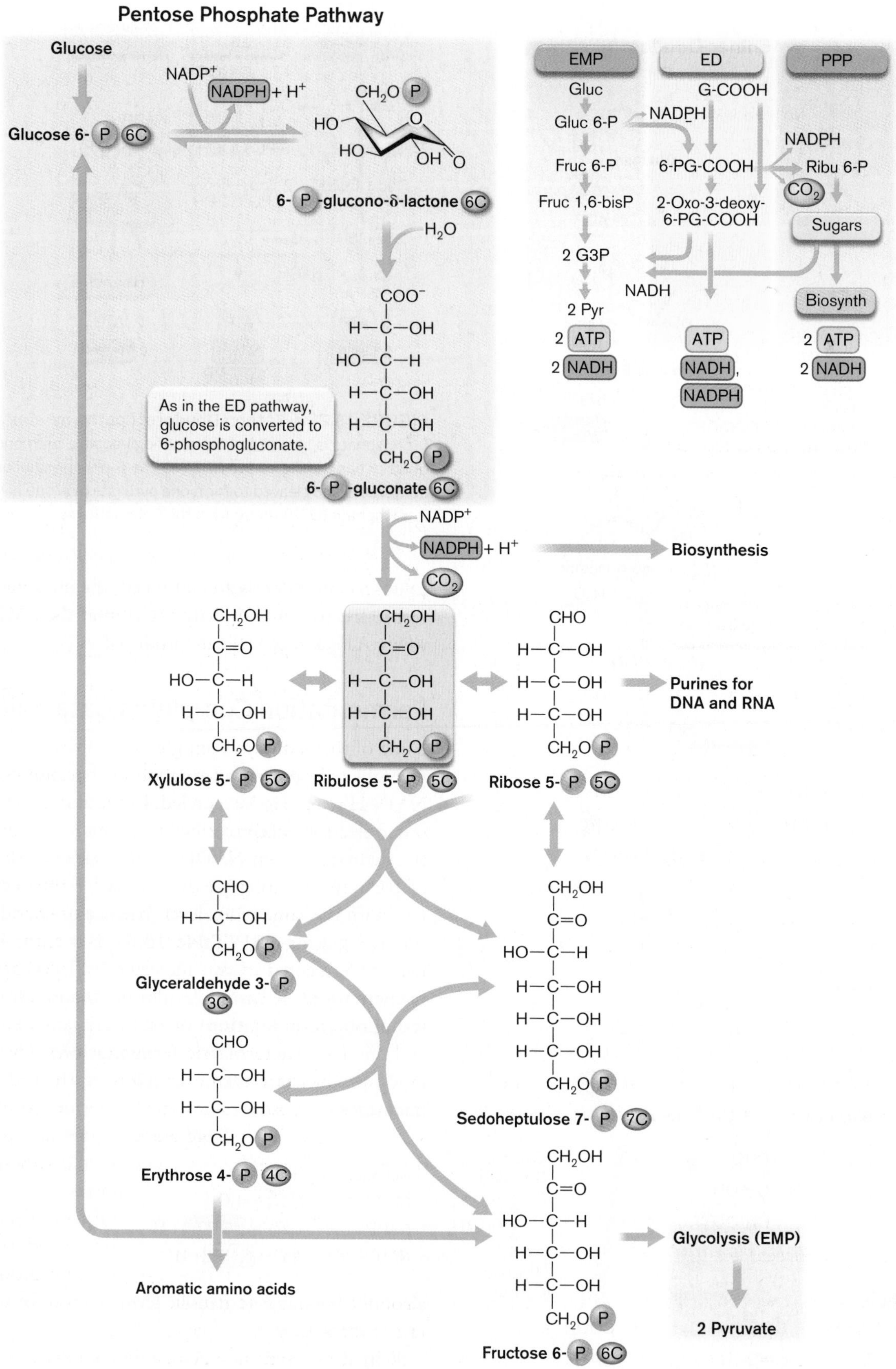
Pentose Phosphate Pathway
Glucose
NADP+
NADPH + H+
Glucose 6-P 6C
6-P-glucono-δ-lactone 6C
H2O
COO−
6-P-gluconate 6C
As in the ED pathway, glucose is converted to 6-phosphogluconate.
NADP+
NADPH + H+
CO2
Biosynthesis
EMP
ED
PPP
Gluc
G-COOH
Gluc 6-P
NADPH
Fruc 6-P
6-PG-COOH
Ribu 6-P
CO2
Fruc 1,6-bisP
2-Oxo-3-deoxy-6-PG-COOH
Sugars
2 G3P
NADH
2 Pyr
Biosynth
2 ATP
2 NADH
ATP
NADH, NADPH
2 ATP
2 NADH
Xylulose 5-P 5C
Ribulose 5-P 5C
Ribose 5-P 5C
Purines for DNA and RNA
Glyceraldehyde 3-P 3C
Sedoheptulose 7-P 7C
Erythrose 4-P 4C
Aromatic amino acids
Fructose 6-P 6C
Glycolysis (EMP)
2 Pyruvate

FIGURE 13.22 ■ Pentose phosphate pathway of glucose catabolism. Like the Entner-Doudoroff pathway, the pentose phosphate pathway forms 6-phosphogluconate. One CO_2 is released, and two NADPH are produced for biosynthesis. The pathway can generate ribose 5-phosphate for purine synthesis or erythrose 4-phosphate to synthesize aromatic amino acids.

is that in the absence of oxygen (or another electron acceptor), the fermentation products cannot yield energy. Most fermentation pathways do not generate ATP beyond that produced by substrate-level phosphorylation, the direct transfer of a phosphate group from an organic phosphate to ADP. An example is the final step of glycolysis, catalyzed by pyruvate kinase (see **Fig. 13.19**). Much of the energy available from glucose remains unspent or is lost as heat. Nevertheless, fermentation is essential for microbes in environments such as anaerobic soil or animal digestive tracts. Even aerated cultures of bacteria start fermenting once their demand for oxygen exceeds the rate of oxygen dissolving in water. Microbes compensate for the low efficiency of fermentation by consuming large quantities of substrate and excreting large quantities of fermentation products. An advantage of fermentation is that the rapid accumulation of acids or ethanol can inhibit growth of competitors.

Mixed-acid fermentation. Numerous pathways have evolved to dispose of the waste in different forms (**Fig. 13.23**). *E. coli* ferments by a combination of routes known collectively as **mixed-acid fermentation**, forming acetate, formate, lactate, and succinate, as well as ethanol, H_2, and CO_2. Hydrogen (H_2) and carbon dioxide (CO_2) are the main gases passed by the human colon.

Colonic H_2 plus CO_2 can yield further energy for methanogenic microbes, through conversion to methane (discussed in Chapter 14). Excess hydrogen and methane gases can cause problems for medical procedures such as colonoscopy. For colonoscopy, the colon may need to be flushed with a carbohydrate solution, which bacteria may ferment, releasing hydrogen. When a polyp is removed from the colon by electrocautery (high-frequency electric current), the gas may ignite, causing an explosion. Colonic explosion can be avoided by flushing out the gases before electrocautery.

During mixed-acid fermentation, the proportions of products vary with pH. Low pH favors ethanol and lactate (which minimize acidification) in preference to formate and acetate. *Clostridium* species produce alcohols (butanol, isopropanol); *Porphyromonas gingivalis,* a cause of periodontal disease, produces short-chain acids (propionate, butyrate).

Many fermentation products share key intermediates, such as acetyl-CoA. Acetyl-CoA is a versatile two-carbon intermediate, the "Lego block" of metabolism. The molecule consists of an acetyl group esterified to **coenzyme A (CoA)** (**Fig. 13.24**), an important coenzyme whose discovery won Fritz Lipmann the 1953 Nobel Prize in Physiology or Medicine, along with Hans Krebs. CoA has a thiol (SH) that exchanges its hydrogen for an acyl group, thus activating the molecule for transfer in various metabolic pathways.

For example, the enzyme pyruvate formate lyase splits pyruvate to form acetyl-CoA plus formate:

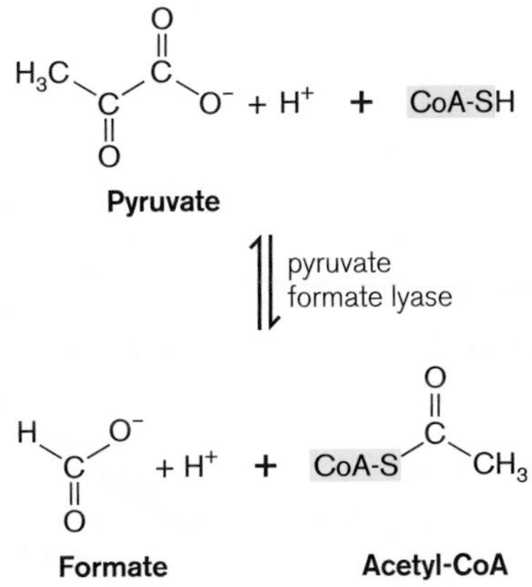
Pyruvate
+ H⁺
CoA-SH
pyruvate
formate lyase
Formate
+ H⁺
CoA-S
Acetyl-CoA

The SH group of CoA is indicated to emphasize its role in accepting the acetyl group to form acetyl-CoA. The hydrogen ($H^+ + e^-$) of the SH group is transferred onto the carboxyl carbon of pyruvate, generating formate.

Acetyl-CoA can be converted to various fermentation products by several different pathways. The simplest is an exchange of water with CoA, forming acetate:

$$CH_3CO—S—CoA + H_2O \rightleftharpoons CH_3COO^- + H^+ + HS—CoA$$

Incorporation of water restores the hydrogen to the thiol of CoA, and the OH to acetate. The acetate thus formed may then be excreted by the cell.

Note that the excreted acids and alcohols are readily recovered by cells when an electron acceptor becomes available for oxidation or when these fermentation products are needed as building blocks for biosynthesis. Alternatively, the excreted "wastes" may be utilized by other species capable of metabolizing them further. For example, the H_2 released by gut bacteria through mixed-acid fermentation can be oxidized to water by the gastric pathogen *Helicobacter pylori.* Alternatively, H_2 plus CO_2 can be used by gut methanogens to yield energy with release of methane.

In soil ecosystems, various other fermentations have evolved to utilize available substrates (see **Table 13.5**). *Clostridium* species generate butanol and other solvents of great industrial value. Other species ferment pairs of amino acids, in a type of mechanism called the **Stickland reaction**. A Stickland reaction avoids generating H_2 (a loss that would waste reduction potential) and produces as much as 3 ATP. By contrast, the succinate fermentation of *Propionigenium modestum* is an anaerobic reaction that proceeds with too small a ΔG value to generate a single ATP. Instead, the decarboxylation step (catalyzed by methylmalonyl-CoA decarboxylase) is coupled to the pumping of sodium ions across the cell membrane. The sodium pump requires a smaller energy input than ATP synthesis. As the gradient of sodium ions builds up, it can drive ATP synthesis through a sodium-driven ATP synthase,

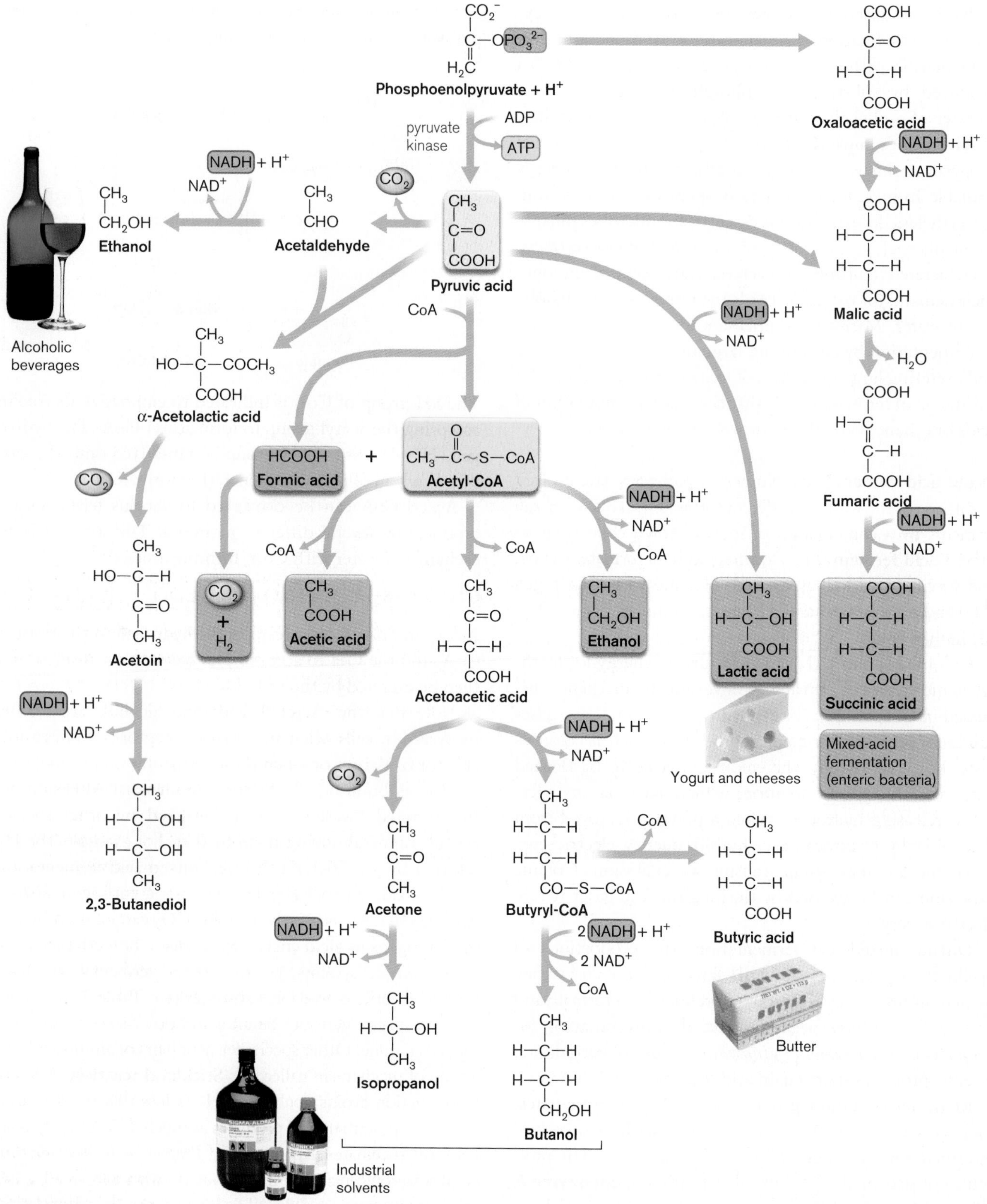

Phosphoenolpyruvate + H+
pyruvate kinase
ADP
ATP
Oxaloacetic acid
NADH + H+
NAD+
Malic acid
H2O
Fumaric acid
Succinic acid
Mixed-acid fermentation (enteric bacteria)
Pyruvic acid
CO2
Acetaldehyde
Ethanol
Alcoholic beverages
α-Acetolactic acid
CoA
HCOOH
Formic acid
Acetyl-CoA
CO2 + H2
Acetic acid
Acetoacetic acid
Ethanol
Lactic acid
Yogurt and cheeses
Acetoin
2,3-Butanediol
Acetone
Isopropanol
Butyryl-CoA
2 NADH + H+
2 NAD+
Butanol
Butyric acid
BUTTER
Butter
Industrial solvents

FIGURE 13.23 ■ Fermentation pathways. Alternative pathways from pyruvate and phosphoenolpyruvate to end products, many of which we use for food or industry. Different species conduct different portions of the pathways shown.

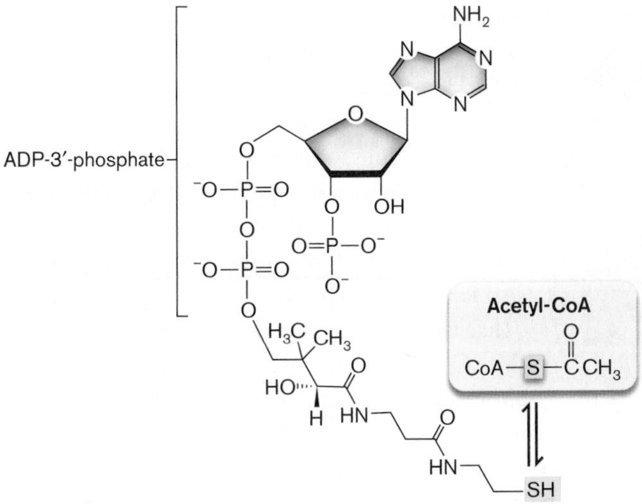
ADP-3′-phosphate
Acetyl-CoA
CoA—S—CCH₃
SH

FIGURE 13.24 ■ Structure of coenzyme A. The thiol (SH) forms an ester link with the COOH of acetic acid, generating acetyl-CoA.

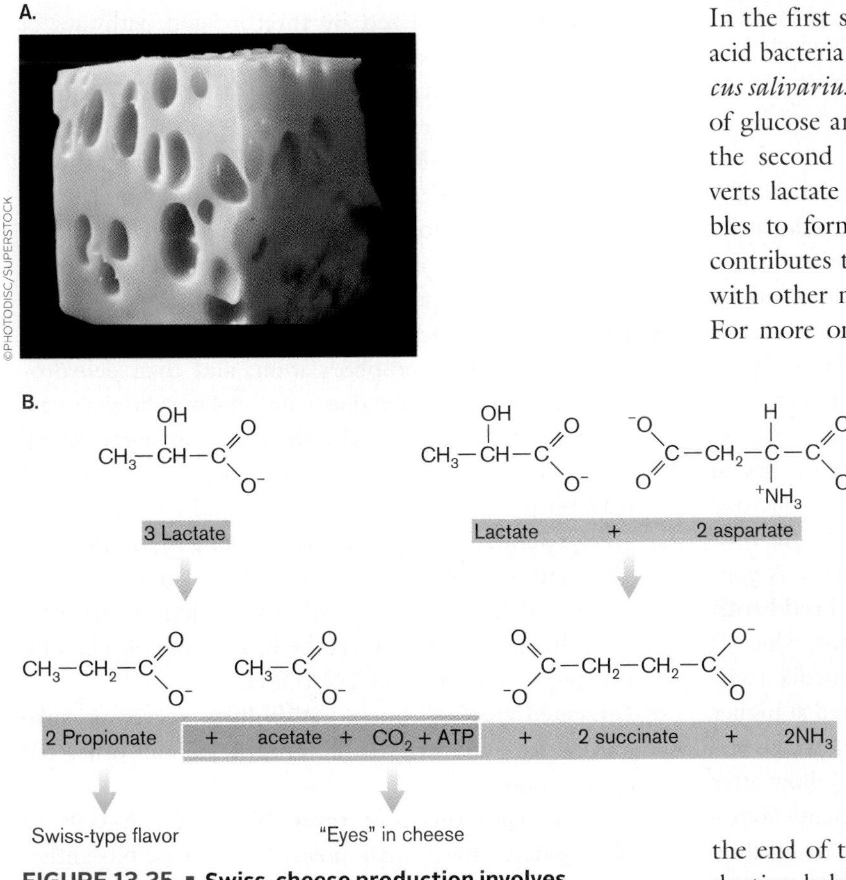
A.
B.
3 Lactate
Lactate + 2 aspartate
2 Propionate + acetate + CO2 + ATP + 2 succinate + 2NH3
Swiss-type flavor
"Eyes" in cheese

FIGURE 13.25 ■ Swiss-cheese production involves fermentation of lactate to propionate. A. Swiss cheese (Emmentaler). **B.** Lactate is fermented by *Propionibacterium freudenreichii* to propionate, acetate, and CO_2. Concurrent fermentation of lactate and aspartate generates additional CO_2, increasing the size and number of eyes.

homologous to the famous proton-driven ATP synthase (discussed in Chapters 4 and 14).

Food and Industrial Applications

The "waste products" of fermentation retain much of their organic structure and food value. Their carbon-hydrogen bonds retain the ability to reduce oxygen, and their organic structures can be used for biosynthesis. Thus, fermentation products have proved extremely useful in human culture and technology. For thousands of years, ethanolic fermentation by yeast has been used to produce wine and beer, while lactate fermentation has been used to produce yogurt and cheese (Chapter 16). The minor product butyric acid (butyrate) lends taste to butter.

Fermentations involving amino acids generate small amounts of products with intense odors and flavors. Products of amino acid catabolism confer some of the distinctive flavors of fermented foods and beverages, such as cheese. Swiss cheese production involves two fermentation stages. In the first stage, at high temperature, thermophilic lactic acid bacteria such as *Lactobacillus helveticus* and *Streptococcus salivarius* ferment the milk sugar lactose, a disaccharide of glucose and galactose, generating mainly lactic acid. In the second stage, *Propionibacterium freudenreichii* converts lactate to propionate, acetate, and CO_2, which bubbles to form the "eyes" (**Fig. 13.25**). The propionate contributes to the distinctive flavor of Swiss cheese, along with other minor side products that have intense flavors. For more on Swiss cheese fermentation, see **eTopic 13.2**. Microbial food production is discussed in Chapter 16.

In the chemical industry, microbial fermentation produces industrial solvents such as butanol and acetone. Acetone production had historic impact during World War I, when Britain needed a source of acetone to manufacture gunpowder. Acetone and butanol were produced as fermentation products by *Clostridium acetobutylicum*, a bacterium identified by the biochemist Chaim Weitzmann. Weitzmann was a Russian-born Jew who sought a Jewish homeland in Palestine. At the end of the war, Weitzmann's process for acetone production helped earn him the British government's support for the national homeland of Israel, where the biochemist later became the country's first president.

Today, microbial fermentation produces many "commodity chemicals," such as ethanol, butanol, and glycerol.

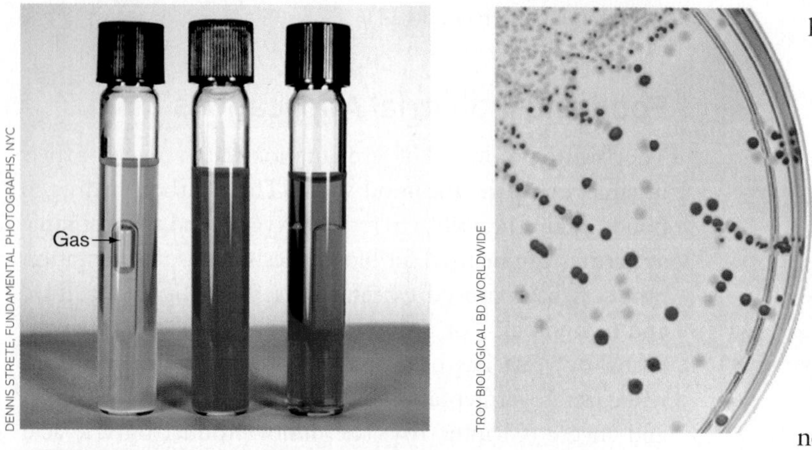
A. Phenol red broth test
B. Sorbitol MacConkey agar
Gas
DENNIS STRETE, FUNDAMENTAL PHOTOGRAPHS, NYC
TROY BIOLOGICAL BD WORLDWIDE

FIGURE 13.26 ■ Clinical tests based on fermentation. **A.** Phenol red broth test. Organisms are cultured in a tube of broth containing phenol red. The dye turns yellow when protonated at low pH. Phenol red is orange-red at neutral pH, yellow at lower pH (acid), and red at higher pH (base). A small inverted tube (Durham tube) collects gaseous fermentation products (CO_2 and H_2). **Left to right:** *Escherichia coli* gives acidic fermentation products (yellow) and gas in Durham tube; *Alcaligenes faecalis* does not ferment, tube turns red without gas; uninoculated control, red. **B.** Sorbitol fermentation test for pathogen *E. coli* O157:H7. White colonies (strain O157:H7) fail to ferment sorbitol, unlike red colonies (nonpathogenic *E. coli*).

Compared to industrial alternatives, such as production from petroleum, microbial culture is environmentally desirable and energy efficient. Furthermore, microbes are "smart" producers of small but complex pharmaceuticals such as vitamins, amino acids, and antibiotics, molecules that would require many steps of organic synthesis. Applications of microbial biosynthesis are discussed further in Chapters 15 and 16.

Another important application of fermentation lies in diagnostic microbiology. To quickly identify the microbe causing a disease and prescribe an effective antibiotic, hospitals use rapid and inexpensive biochemical tests. A general test for fermentation pathways is the **phenol red broth test** (**Fig. 13.26A**). Phenol red is a pH indicator, which is orange-red at neutral pH. It turns yellow in media acidified by fermentation acids (below pH 6.8) and red at higher pH (above pH 7.4). A culture of *Escherichia coli,* which ferments quickly, turns phenol red to a medium yellow after 24 hours. *Alcaligenes faecalis,* meanwhile, ferments poorly on sugar but converts peptides in the broth to alkaline amines; this culture turns deep red.

More specific tests depend on the microbe's ability to ferment specific sugars. Different species possess different enzymes able to convert different sugars and sugar derivatives to glucose, which then enters the common fermentation pathway. A well-known example is the use of sorbitol MacConkey agar to test for *E. coli* O157:H7, a lethal pathogen contaminating beef and vegetables. The O157:H7 strain of *E. coli* has a set of virulence genes absent from nonpathogenic *E. coli* strains present among our normal colon biota. But the pathogen also happens to lack genes present in normal *E. coli,* such as a gene encoding an enzyme to ferment sorbitol. Thus, failure to ferment sorbitol indicates a high probability that the strain is *E. coli* O157:H7. On sorbitol MacConkey agar, bacteria that ferment sorbitol during growth produce acids. The acidity causes a dye to turn red (the opposite of the phenol red test). Failure to ferment sorbitol (observed as pale colonies) indicates high probability of *E. coli* O157:H7 (**Fig. 13.26B**).

To Summarize

- **Glucose is catabolized** by three related pathways of stepwise degradation of glucose to pyruvate. In the absence of oxygen, completion of catabolism requires generation of fermentation products.
- **In the Embden-Meyerhof-Parnas (EMP) pathway,** glucose is activated by two substrate phosphorylations, and then cleaved to two three-carbon sugars. Both sugars eventually are converted to pyruvate. The pathway produces 2 ATP and 2 NADH.
- **In the Entner-Doudoroff (ED) pathway,** glucose is activated by one phosphorylation, and then dehydrogenated to 6-phosphogluconate. 6-Phosphogluconate is cleaved to pyruvate and a three-carbon sugar, which enters the EMP pathway to form pyruvate. The ED pathway produces 1 ATP, 1 NADH, and 1 NADPH.
- **In the pentose phosphate pathway (PPP),** 6-phosphogluconate is converted to a series of sugars, each containing three to seven carbons—including fructose 6-phosphate, which reenters the EMP pathway. The PPP may produce 1 ATP and 2 NADPH.
- **Intermediates of sugar catabolism serve as substrates for biosynthesis** of amino acids and other cell components.
- **Glucose catabolism is reversible,** enabling cells to build glucose from small molecules. Glucose biosynthesis uses some enzymes of glycolysis in reverse, but also requires enzymes that bypass irreversible steps in the catabolic pathway.
- **Fermentation is the completion of catabolism** without the electron transport system and a terminal electron acceptor. The electrons from NADH are restored to

pyruvate or its products in reactions that generate fermentation products, including alcohols and carboxylates, as well as H_2 and CO_2. Energy is stored in the form of ATP. Fermentation has applications in food, industrial, and diagnostic microbiology.

13.6

The Tricarboxylic Acid (TCA) Cycle

The products of sugar breakdown can be catabolized to CO_2 and H_2O through the **tricarboxylic acid (TCA) cycle**. The TCA cycle generates molecules of NADH and $FADH_2$ which donate electrons to an electron transport system (ETS) with a terminal electron acceptor such as O_2 (discussed in Chapter 14).

The TCA cycle is also known as the Krebs cycle, named for Hans Krebs (1900–1981), who shared the 1953 Nobel Prize in Physiology or Medicine with Fritz Lipmann. Krebs and his colleagues at Sheffield University, England, studied catabolism by observing the oxidizing activities of crude enzyme preparations from sources such as pigeon breast muscle, beef liver, and cucumber seeds. In all of these animal and plant tissues, the TCA cycle is conducted by mitochondria, using virtually the same process as their bacterial ancestors used.

Glucose catabolism connects with the TCA cycle through the breakdown of pyruvate to acetyl-CoA and CO_2. Recall from Section 13.4 that acetyl-CoA is also generated from many other catabolic pathways, including those for breakdown of fatty acids, amino acids, and even benzoate derivatives. Regardless of its source, acetyl-CoA enters the TCA cycle by condensing with the four-carbon intermediate oxaloacetate to form citrate (**Fig. 13.27**). Citrate undergoes two steps of oxidative decarboxylation, in which CO_2 is released and two hydrogens with electrons are transferred to make NADH + H^+ or $FADH_2$. The TCA cycle, in whole or in part, is found in all microbial species except for degenerately evolved pathogens that are dependent on host metabolism, such as *Treponema pallidum*, the cause of syphilis (**eTopic 13.3**).

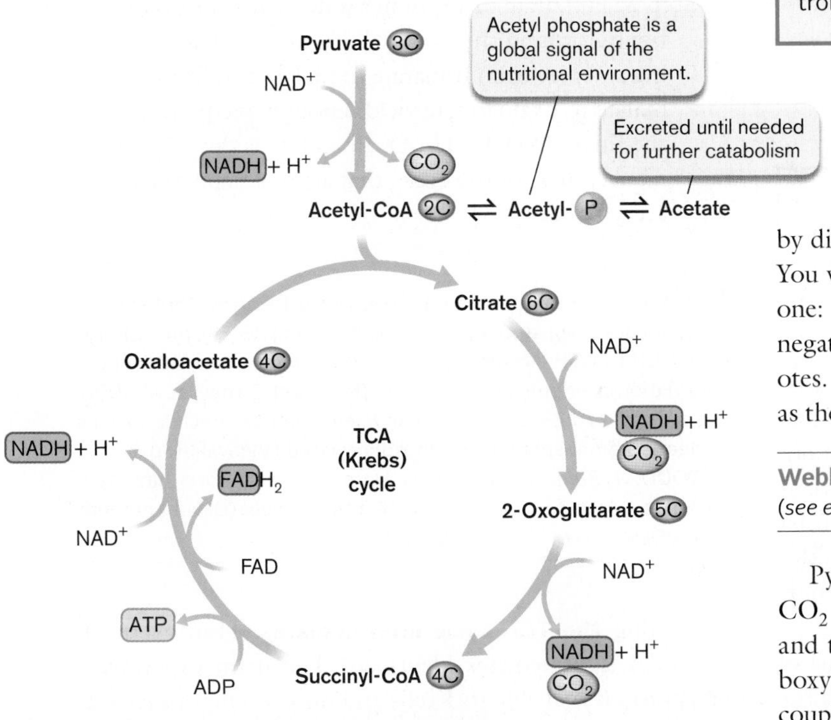
Pyruvate 3C
Acetyl phosphate is a global signal of the nutritional environment.
NAD+
Excreted until needed for further catabolism
NADH + H+
CO2
Acetyl-CoA 2C
Acetyl-P
Acetate
Citrate 6C
NAD+
Oxaloacetate 4C
NADH + H+
CO2
NADH + H+
FADH2
TCA (Krebs) cycle
2-Oxoglutarate 5C
NAD+
FAD
NAD+
ATP
NADH + H+
CO2
ADP
Succinyl-CoA 4C

FIGURE 13.27 ■ Acetyl-CoA feeds into the TCA cycle. Pyruvate undergoes oxidative decarboxylation and incorporates CoA to form acetyl-CoA. Depending on the state of the cell, either acetyl-CoA is converted to acetate for excretion or it enters the TCA cycle.

Thought Question

13.10 If a cell respiring on glucose runs out of oxygen and other electron acceptors, what happens to the electrons transferred from the catabolic substrates?

We present first the connecting step between pyruvate and acetyl-CoA, followed by the details of the TCA cycle. Bacteria and archaea use at least ten known variations on the TCA cycle, conducted by diverse species under various environmental conditions. You will no doubt be relieved to hear that we present only one: the Krebs pathway, which is found in most Gram-negative and Gram-positive bacteria, and in most eukaryotes. Other pathways are detailed in online resources such as the KEGG Pathway Database.

Weblinks KEGG (Kyoto Encyclopedia of Genes and Genomes) (*see ebook*)

Pyruvate is converted to acetyl-CoA through removal of CO_2 and transfer of $2e^-$ onto NAD^+. The removal of CO_2 and transfer of two electrons is known as oxidative decarboxylation. The oxidative decarboxylation of pyruvate, coupled to CoA incorporation, is performed by an unusually large multisubunit enzyme called the **pyruvate dehydrogenase complex**, or **PDC**. PDC is a key component of metabolism in bacteria and mitochondria, the first molecular player to direct sugar catabolism into respiration. In human mitochondria, defects in PDC affect organs that have a high metabolic rate, such as heart and brain, causing myocardial malfunction and heart failure, and neurodegeneration. A structural model for PDC, and the details of its reaction mechanism, are shown in **eTopic 13.4**.

The overall reactions catalyzed by PDC are:

$$CH_3COCOO^- + H^+ + HS\text{-}CoA \rightarrow$$
$$CH_3CO\text{-}S\text{-}CoA + CO_2 + 2H^+ + 2e^-$$
$$NAD^+ + 2H^+ + 2e^- \rightarrow NADH + H^+$$

The removal of stable CO_2 yields energy for the electron transfer to NADH. The protons dissociated from pyruvate and from the thiol (SH) of CoA ($2H^+$ in total) are balanced by the net gain of protons by NADH + H^+.

The activity of PDC is increased by high concentrations of its substrates (CoA and NAD^+) and inhibited by its products (acetyl-CoA and NADH). The product acetyl-CoA may enter one of several pathways. In *E. coli,* when glucose is plentiful, acetyl-CoA is mostly converted to acetate via the intermediate acetyl phosphate (**Fig. 13.27**). Acetyl phosphate is a global signal molecule that indicates to the cell the quantity and quality of carbon source available. As glucose decreases, the cell starts to take back acetate, converting it back to acetyl-CoA for entry into the TCA cycle. At the level of gene expression, PDC responds to environmental conditions. As would be expected, PDC gene expression is repressed by carbon starvation and by low levels of oxygen.

Thought Question

13.11 Compare the reactions catalyzed by pyruvate dehydrogenase and pyruvate formate lyase (see Section 13.5). What conditions favor each reaction, and why?

Acetyl-CoA Enters the TCA Cycle

Recall from the discussion of glucose catabolism that acetyl-CoA is a key intermediate of various fermentations (see **Fig. 13.23**). But when a strong electron acceptor is available, acetyl-CoA can enter the TCA cycle to transfer its electrons to electron carriers (**Figs. 13.27** and **13.28**). First, the acetyl group condenses with oxaloacetate, a four-carbon dicarboxylate (double acid). The condensation forms citrate, a six-carbon tricarboxylate. An advantage of intermediates with two or more acidic groups is that the concentration of the fully protonated form is extremely low; thus, the molecule is unlikely to be lost from the cell by diffusion across the membrane, as are monocarboxylic acids, such as acetate.

Through the rest of the cycle, citrate loses two carbons as CO_2 by a series of reactions that transfer increments of energy to 3 NADH, $FADH_2$, and ATP. Each reaction step couples energy-yielding to energy-storing events.

Step 1. As the acetyl group condenses with oxaloacetate, the hydrolysis of acetyl-CoA consumes a molecule of H_2O to restore HS-CoA. The removal of HS-CoA yields energy to incorporate acetate into oxaloacetate, forming citrate.

Step 2. Citrate undergoes two rearrangements (with little energy change) to form isocitrate. Isocitrate then undergoes oxidative decarboxylation. As we saw for pyruvate, removal of CO_2 yields energy to transfer $2H^+ + 2e^-$ to form NADH + H^+, producing 2-oxoglutarate (alpha-ketoglutarate).

Step 3. 2-Oxoglutarate undergoes decarboxylation to release CO_2 and make another NADH + H^+. In this case, CoA is incorporated, making succinyl-CoA.

Step 4. Succinyl-CoA releases CoA, yielding energy to phosphorylate ADP to ATP. To form fumarate, $2H^+ + 2e^-$ are transferred to FAD to form $FADH_2$—a reaction involving negligible free energy change. FAD is reduced instead of NAD^+ because electron donation from succinyl-CoA does not yield enough energy to reduce NAD^+.

Step 5. Fumarate incorporates water across its double bond, forming the hydroxy acid malate. The increasing stability from fumarate to malate and from malate to oxaloacetate yields enough energy to form the final NADH + H^+. Oxaloacetate is the original intermediate of the cycle, that again accepts the next acetyl-CoA.

Note: Some textbooks state that succinyl-CoA synthetase, the enzyme catalyzing step 4 in the TCA cycle (Fig. 13.28), phosphorylates GDP to GTP. According to the primary literature, ADP phosphorylation predominates in *E. coli* (Margaret Birney et al. 1996. *J. Bacteriol.* **178**:2883), whereas in *Pseudomonas* species, various nucleotide diphosphates are phosphorylated (Vinayak Kapatral et al. 2000. *J. Bacteriol.* **182**:1333). Human mitochondria have two enzymes, which form ATP and GTP, respectively (David Lambeth et al. 2004. *J. Biol. Chem.* **279**:36621).

Observing the TCA cycle intermediates. How were all the TCA intermediates identified? The main experimental approach available to Krebs and his contemporaries was to guess at dozens of short-chain acids known to exist in cells, and then to add each individually to an enzyme preparation and test for TCA cycle activity. In aerobic organisms, the TCA cycle is tightly tied to respiration, so an

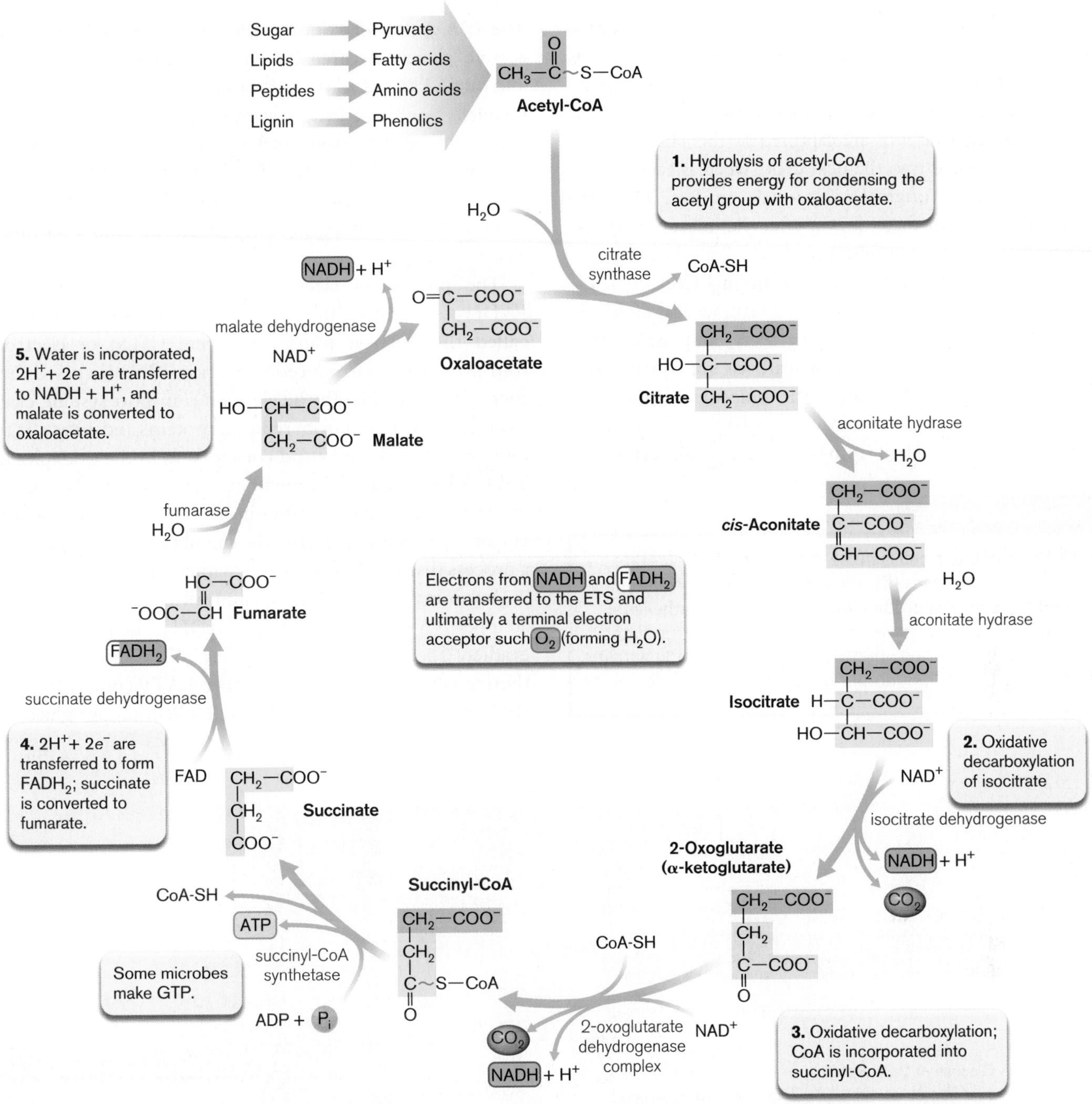
Sugar
Pyruvate
Lipids
Fatty acids
Peptides
Amino acids
Lignin
Phenolics
CH3—C~S—CoA
Acetyl-CoA
1. Hydrolysis of acetyl-CoA provides energy for condensing the acetyl group with oxaloacetate.
H2O
citrate synthase
CoA-SH
NADH + H+
O=C—COO−
CH2—COO−
Oxaloacetate
malate dehydrogenase
NAD+
5. Water is incorporated, 2H+ + 2e− are transferred to NADH + H+, and malate is converted to oxaloacetate.
HO—CH—COO−
CH2—COO− Malate
CH2—COO−
HO—C—COO−
Citrate CH2—COO−
aconitate hydrase
H2O
CH2—COO−
cis-Aconitate C—COO−
CH—COO−
fumarase
H2O
Electrons from NADH and FADH2 are transferred to the ETS and ultimately a terminal electron acceptor such O2 (forming H2O).
HC—COO−
−OOC—CH Fumarate
H2O
aconitate hydrase
FADH2
succinate dehydrogenase
CH2—COO−
Isocitrate H—C—COO−
HO—CH—COO−
2. Oxidative decarboxylation of isocitrate
4. 2H+ + 2e− are transferred to form FADH2; succinate is converted to fumarate.
FAD
CH2—COO−
CH2
COO−
Succinate
NAD+
isocitrate dehydrogenase
2-Oxoglutarate (α-ketoglutarate)
NADH + H+
CO2
CoA-SH
Succinyl-CoA
ATP
CH2—COO−
CH2
C~S—CoA
O
Some microbes make GTP.
succinyl-CoA synthetase
ADP + Pi
CoA-SH
CH2—COO−
CH2
C—COO−
O
CO2
2-oxoglutarate dehydrogenase complex
NAD+
NADH + H+
3. Oxidative decarboxylation; CoA is incorporated into succinyl-CoA.

FIGURE 13.28 ▪ The tricarboxylic acid (TCA) cycle. Acetyl-CoA derived from pyruvate and other catabolic pathways enters the TCA cycle. Blue highlighting shows the fate of labeled acetate incorporated into the TCA cycle. (1) The acetyl group condenses with four-carbon oxaloacetate to produce citrate, a tricarboxylic acid. (2) Citrate rearranges to isocitrate, which is decarboxylated and transfers $2H^+ + 2e^-$ to form $NADH + H^+$. (3) 2-Oxoglutarate is decarboxylated and transfers $2H^+ + 2e^-$ to form $NADH + H^+$ while incorporating CoA to form succinyl-CoA. (4) Succinate is symmetrical; thus, the "old" and "new" carbons are now indistinguishable. Succinate transfers $2H^+ + 2e^-$ to form $FADH_2$, forming fumarate. (5) Water is incorporated, and $2H^+ + 2e^-$ are transferred to form $NADH + H^+$, forming oxaloacetate, ready to take up another acetyl group.

increase in uptake of oxygen signaled a TCA intermediate. A major experimental advance was the use of tracer isotopes such as ^{14}C, discovered by Martin Kamen in 1940 (see **eTopic 15.1**).

Radioisotope tracers answered an important question about the TCA cycle. As the two acetyl carbons cycle through, which two carbons of each intermediate are removed as CO_2? Which carbons are lost first: the acetyl carbons or those of the original oxaloacetate? This question was answered by use of substrates radiolabeled with ^{14}C (**Fig. 13.28**, highlighted green). In the experiment, bacteria are fed ^{14}C-radiolabeled acetate, which enters the TCA cycle. The radiolabeled carbons are captured by the TCA intermediates and retained into the next cycle, whereas two COOH carbons from the original oxaloacetate are lost. Thus, the four-carbon intermediate does not recycle intact, but breaks down and re-forms each time it passes through the cycle.

Loss of the second CO_2 forms succinyl-CoA, which is converted to succinate. Succinate is a symmetrical molecule; thus, the former identities of the acetyl and oxaloacetate moieties are now erased, and the carbons are equally likely to disappear from either half of the molecule. Further conversion steps regenerate oxaloacetate—half its carbon from the acetyl group and half from the original oxaloacetate.

Thought Question

13.12 Suppose a cell is pulse-labeled with ^{14}C-acetate (fed the ^{14}C label briefly, then "chased" with unlabeled acetate). Can you predict what will happen to the level of radioactivity observed in isolated TCA intermediates? Plot a curve showing your predicted level of radioactivity as a function of the number of rounds of the cycle.

The TCA cycle and oxidative phosphorylation. In all, each acetate generates three NADH molecules, one $FADH_2$, and one ATP; and all the carbons from pyruvate (ultimately from glucose) have been released as waste CO_2. From the standpoint of the carbon skeleton, the glucose breakdown is now complete. But do we have a completed metabolic pathway? No, because all of the NADH and $FADH_2$ need to be recycled by donating their electrons onto a terminal electron acceptor.

The process of electron transfer from NADH and $FADH_2$ is mediated by a series of cell membrane proteins called the electron transport system (ETS) or electron transport chain (introduced in Section 13.3; discussed further in Chapter 14). In the electron transport system, electrons are transferred from reduced proteins and cofactors to more oxidized proteins and cofactors, as in the examples in **Table 13.4**. Some of the membrane proteins use the energy of electron transfer to pump protons, generating a gradient of hydrogen ions across the membrane (**Fig. 13.29**)—a process discussed in detail in Chapter 14.

Assuming a theoretical maximum yield of 3 ATP generated per NADH and 2 ATP per $FADH_2$, the hydrogen ion gradient then drives the membrane ATP synthase to synthesize as many as 34 ATP. Another 4 ATP come from glucose breakdown and the TCA cycle (38 total per glucose).

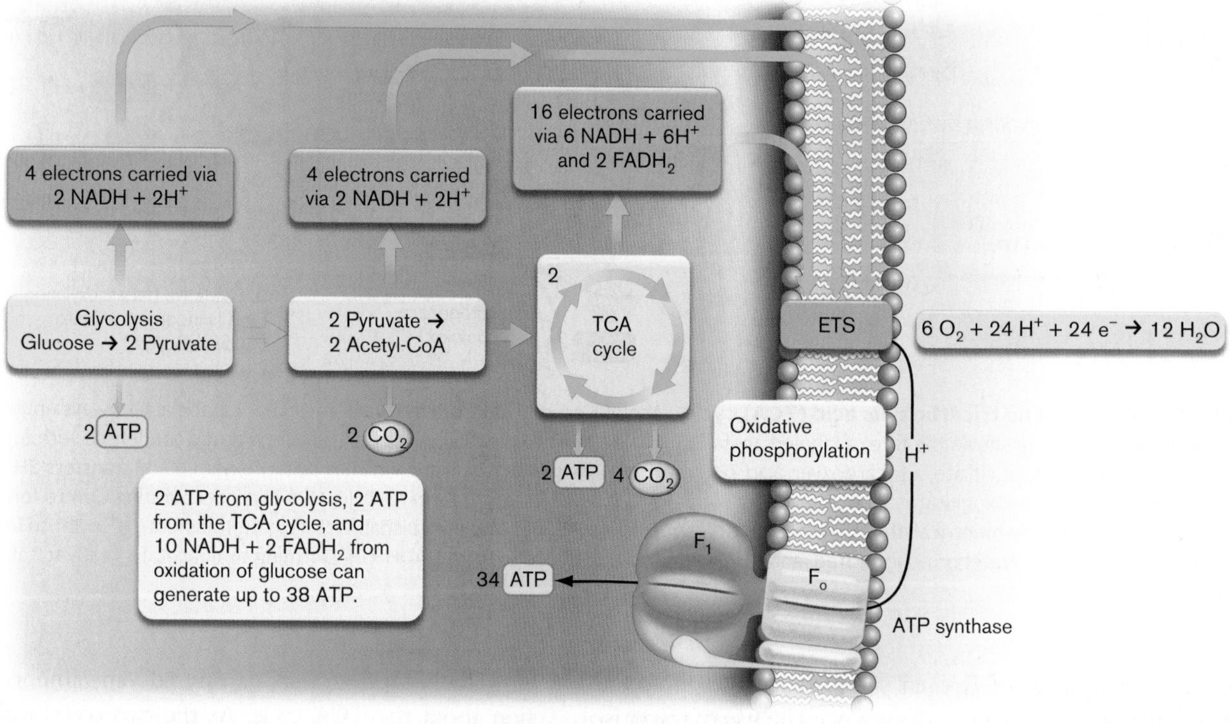
4 electrons carried via 2 NADH + 2H⁺
4 electrons carried via 2 NADH + 2H⁺
16 electrons carried via 6 NADH + 6H⁺ and 2 FADH₂
Glycolysis Glucose → 2 Pyruvate
2 Pyruvate → 2 Acetyl-CoA
2
TCA cycle
ETS
6 O₂ + 24 H⁺ + 24 e⁻ → 12 H₂O
2 ATP
2 CO₂
2 ATP 4 CO₂
Oxidative phosphorylation
H⁺
2 ATP from glycolysis, 2 ATP from the TCA cycle, and 10 NADH + 2 FADH₂ from oxidation of glucose can generate up to 38 ATP.
F₁
34 ATP
F₀
ATP synthase

FIGURE 13.29 ■ Complete oxidation of glucose. Glucose catabolism generates ATP through substrate-level phosphorylation and through the electron transport system pumping H^+ ions to drive the ATP synthase. The complete oxidative breakdown of glucose to CO_2 and H_2O could theoretically generate up to 38 ATP. Under actual conditions, the number is smaller.

Under actual conditions, however, bacteria make less ATP; about 20 ATP per glucose are made by a well-aerated culture of *E. coli*. Bacterial cells make trade-offs for flexibility, spending energy to maintain a stable proton potential during extreme changes in external pH and redox levels.

The overall process of electron transport and ATP generation is termed **oxidative phosphorylation**. The overall process of oxidative catabolism from substrate breakdown to oxidative phosphorylation is a form of **respiration**. The overall equation for the aerobic respiration of glucose is:

$$C_6H_{12}O_6 + 6H_2O + 6O_2 \rightarrow 12H_2O + 6CO_2$$

Glucose respiration can generate a relatively large number of ATPs per glucose—far more than fermentation can. In bacteria, however, the actual number of ATPs generated varies widely with availability of carbon source and oxygen. For example, as oxygen decreases in the environment, the ability to oxidize NADH decreases, so the cell may make only 1 or 2 ATP per NADH (discussed in Chapter 14). In addition, the enzymes of the TCA cycle are regulated extensively by substrate activation and product inhibition, and their expression is induced by high levels of oxygen and glucose.

Glyoxylate bypass. What happens when glucose is scarce and cells need carbon both for energy and for biosynthesis? Bacteria may catabolize lipids instead of glucose, breaking down the fatty acids into acetyl-CoA for the TCA cycle. But oxygen may be limited, and carbons would be lost as CO_2. Some bacteria can switch to using a modified TCA cycle called the **glyoxylate bypass** (**Fig. 13.30A**). The glyoxylate bypass allows lung pathogens such as *Pseudomonas aeruginosa* and *Mycobacterium tuberculosis* to catabolize fatty acids, via their breakdown to acetyl-CoA.

The glyoxylate bypass consists of two enzymes that divert isocitrate to glyoxylate, and then incorporate a second acetyl-CoA to form malate. The malate can then regenerate oxaloacetate to complete the bypass cycle, donating $2e^-$ to form NADH. The net reaction is:

$$2\ \text{Acetyl-CoA} + \text{oxaloacetate} + \text{NAD}^+ \rightarrow \text{succinate} + \text{malate} + 2\ \text{CoA} + \text{NADH} + \text{H}^+$$

Alternatively, malate or oxaloacetate can be diverted into biosynthesis of glucose (gluconeogenesis)—a pathway that reverses much of glucose catabolism (Chapter 15). Most bacteria need sugar biosynthesis to build their cell walls.

A.

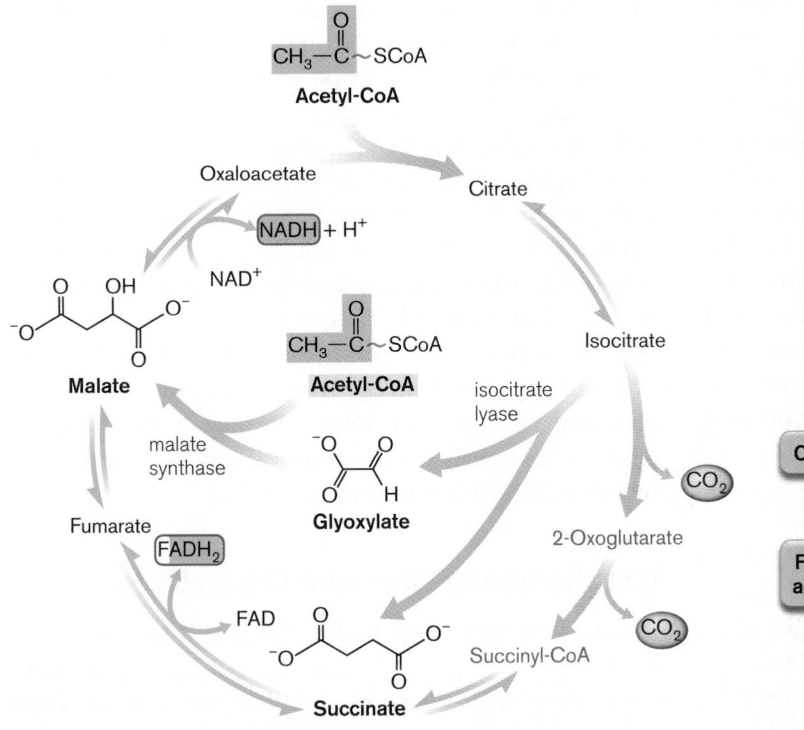

CH3—C~SCoA
Acetyl-CoA
Oxaloacetate
Citrate
NADH + H+
NAD+
Malate
CH3—C~SCoA
Acetyl-CoA
Isocitrate
isocitrate lyase
malate synthase
Glyoxylate
CO2
Fumarate
FADH2
2-Oxoglutarate
FAD
CO2
Succinyl-CoA
Succinate

B. *M. tuberculosis* in macrophages

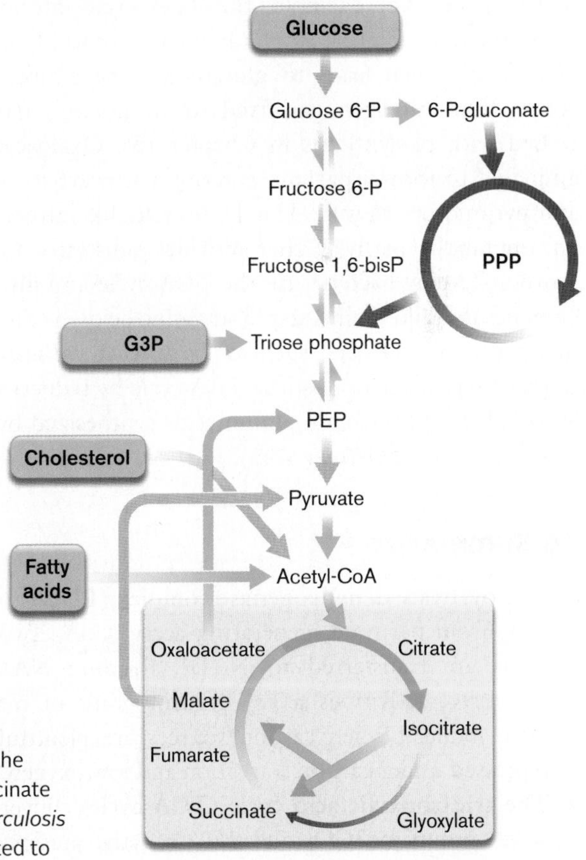

Glucose
Glucose 6-P
6-P-gluconate
Fructose 6-P
PPP
Fructose 1,6-bisP
G3P
Triose phosphate
PEP
Cholesterol
Pyruvate
Fatty acids
Acetyl-CoA
Oxaloacetate
Citrate
Malate
Isocitrate
Fumarate
Succinate
Glyoxylate

FIGURE 13.30 ▪ The glyoxylate bypass. A. Instead of releasing two CO_2, the glyoxylate bypass incorporates a second molecule of acetyl-CoA, producing succinate plus malate. **B.** The glyoxylate bypass in the metabolism of *Mycobacterium tuberculosis* growing within mouse macrophages. Some malate and oxaloacetate are converted to pyruvate and phosphoenolpyruvate (PEP), respectively, for biosynthesis of glucose. For glucose biosynthesis, some glycolytic enzymes act in reverse.

The glyoxylate bypass cuts out all loss of CO_2 and electron transfer to energy carriers, with the exception of one $FADH_2$ and possibly one NADH from malate to oxaloacetate. Thus, limited energy is released, but two carbons can be diverted to biosynthesis. **Figure 13.30B** shows the pathogen *Mycobacterium tuberculosis* using the glyoxylate bypass. A mystery of the disease tuberculosis is the ability of *M. tuberculosis* to persist for long periods, growing slowly within macrophages. What carbon source and metabolism does the pathogen use? The question was addressed by labeling infected cells with carbon sources enriched for the isotope ^{13}C. The intracellular bacteria were then isolated by cell fractionation (discussed in Chapter 3) and analyzed for ^{13}C isotope enrichment of various components. It turns out that persistent intracellular *M. tuberculosis* catabolizes host lipids via the glyoxylate bypass, diverting much of the carbon to build sugars and amino acids for bacterial cell growth. Thus, key enzymes of the glyoxylate bypass, and their regulators, offer targets for new antibiotics.

The TCA cycle for amino acid biosynthesis. Analysis of pathway evolution indicates that the TCA cycle originally evolved to provide substrates to build amino acids (Chapter 15). For example, the TCA cycle intermediate 2-oxoglutarate (alpha-ketoglutarate) is aminated to form glutamate, which leads to glutamine. The amine group comes from nitrogen gas fixed to ammonium ion (discussed with biosynthesis in Chapter 15). Oxaloacetate is aminated to form aspartate, entering pathways to purines and pyrimidines as well. The TCA cycle, like glycolysis, is an amphibolic pathway that provides substrates for biosynthesis. Many bacteria use the TCA cycle and glycolytic enzymes to build their sugars and amino acids, as discussed in Chapter 15. Others, such as *Treponema pallidum,* the cause of syphilis, have lost the TCA cycle by reductive evolution. They must obtain amino acids synthesized by their host organism (see **eTopic 13.3**).

To Summarize

- **The pyruvate dehydrogenase complex (PDC)** removes CO_2 from pyruvate, generating acetyl-CoA. Two electrons are transferred to NAD^+, forming NADH + H^+. PDC activity is a key control point of metabolism, induced when carbon sources are plentiful, and repressed under carbon starvation and low oxygen.
- **The tricarboxylic acid cycle (TCA cycle)** converts the acetyl group to $2CO_2$ and $2H_2O$ in the presence of a terminal electron acceptor such as O_2 to receive the electrons associated with the hydrogen atoms.
- **Acetyl-CoA enters the TCA cycle by condensing with oxaloacetate to form citrate.** A series of enzymes sequentially removes carbon dioxide and water molecules and generates 3 NADH, $FADH_2$, and ATP. Each reaction step couples energy-yielding to energy-storing events.
- **NADH and $FADH_2$** transfer electrons to the electron transport system and ultimately the terminal electron acceptor, such as O_2. Electron transport generates a transmembrane proton potential that powers the membrane-embedded ATP synthase to make ATP.
- **Respiration** consists of substrate catabolism plus oxidative phosphorylation, the process of electron transport and ATP generation.
- **The glyoxylate bypass** provides a way to gain limited energy from the TCA cycle while avoiding CO_2 loss; thus the bypass diverts intermediates to sugar biosynthesis.

13.7

Aromatic Pollutants

Many carbon sources for catabolism contain aromatic forms such as the benzene ring. Natural sources of aromatic carbon include lignin from wood, and phenanthrenes (three fused rings) from petroleum. Aromatic components of petroleum cause much of the environmental damage during oil spills such as the *Deepwater Horizon* spill in 2010. Industrial aromatic compounds, such as nitrate explosives, aniline dyes, and the solvent toluene, often pollute soil. To remove aromatic pollutants from water and soil, we depend on microbial catabolism. Aromatic molecules are notoriously difficult to catabolize because of the stability of the benzene ring. The benzene ring breaks down slowly, but over time, bacteria and fungi catabolize a wide range of aromatic molecules. These microbial processes have major industrial and environmental importance.

Catabolism of Benzene Derivatives

As we saw earlier (**Fig. 13.14**), fungi and bacteria catabolize lignin and petroleum to form single-ring aromatic compounds such as benzoate, vanillin, and phenols. These products are further catabolized to acetyl-CoA, which enters the TCA cycle. Benzoate, activated to benzoyl-CoA, has a central role in the catabolism of aromatic molecules, comparable to that of glucose in polysaccharide catabolism. Bacteria such as *Pseudomonas* and *Rhodococcus* species degrade benzoate and related molecules aerobically or anaerobically. Anaerobic degradation takes much longer, but it is critical because the volume of anaerobic habitat (such as soil) greatly exceeds that of oxygenated habitat.

Aerobic benzene catabolism. Benzene and related aromatic compounds, such as toluene (methylbenzene), chlorobenzoate, and nitrobenzene, can be catabolized via sequential oxidation steps, requiring the presence of an ETS that terminates with O_2 (**Fig. 13.31**). Early in the pathways, enzymes must remove substituents such as chlorides or nitrates. The methyl group of toluene is oxidized to carboxylate ($RCOO^-$), whose removal then drives a key breakdown step. Aerobic degradation commonly proceeds through the intermediate catechol, a benzene ring bearing two adjacent hydroxyl groups. Each benzene derivative is converted to catechol by a specific dioxygenase, an enzyme that coordinately oxygenates two adjacent ring carbons. Next follows ring cleavage and breakdown to pyruvate (forming NADH) and to acetyl-CoA, which may enter the TCA cycle and respiration.

The intermediate catechol then undergoes another key oxidation by a catechol dioxygenase, which adds two more oxygens while cleaving the ring. In different bacterial species, the enzyme may oxidize either the 1,2 positions or the 2,3 positions, as shown in **Figure 13.31**. Typical products are succinyl-CoA and acetyl-CoA, which enter the TCA cycle, completing breakdown to CO_2. The details of benzene catabolism vary among bacterial species, and many diverse mechanisms continue to be discovered. For research examples, see **eTopic 13.5**.

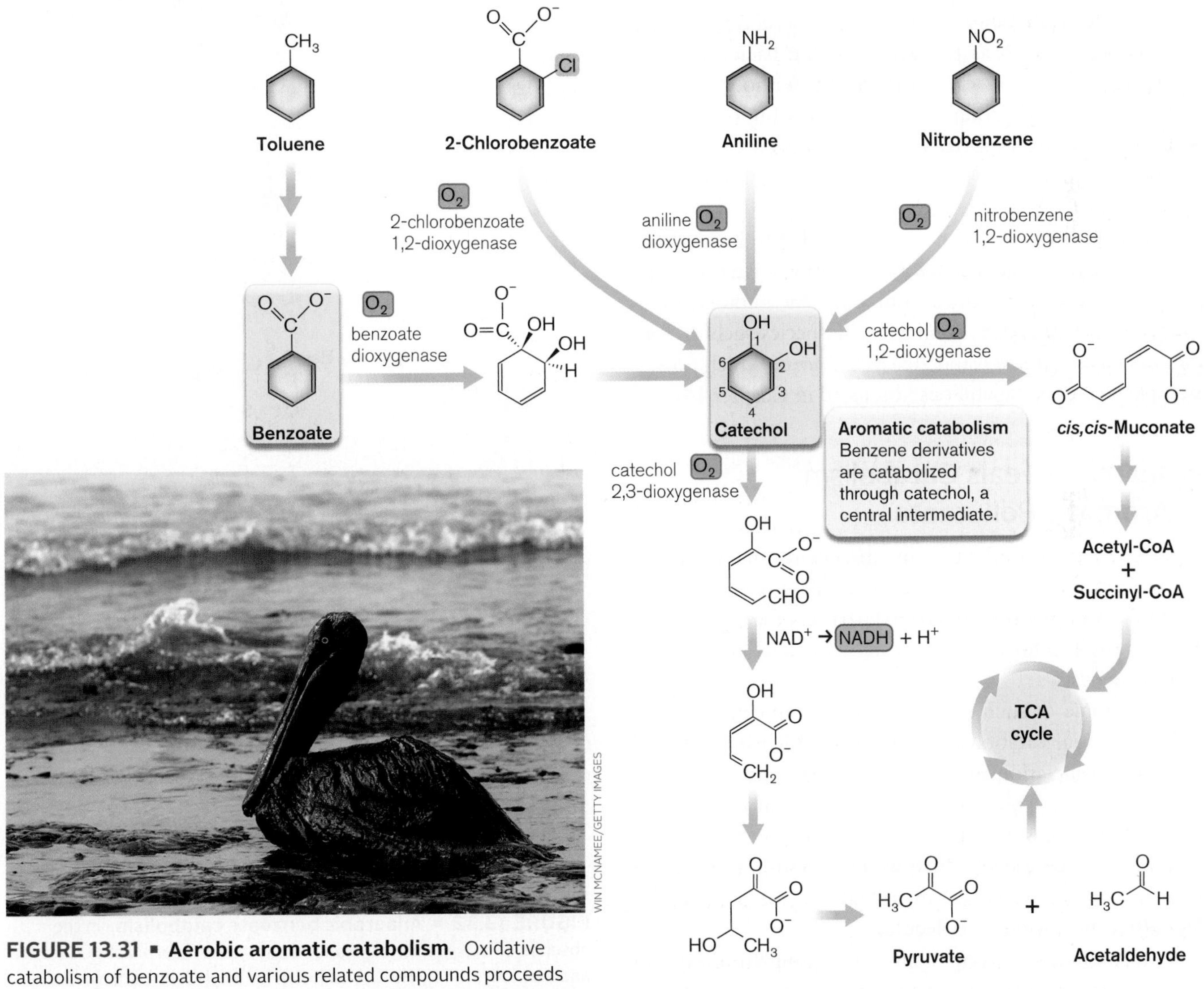
Toluene
2-Chlorobenzoate
Aniline
Nitrobenzene
2-chlorobenzoate 1,2-dioxygenase
aniline dioxygenase
nitrobenzene 1,2-dioxygenase
benzoate dioxygenase
Benzoate
catechol 1,2-dioxygenase
Catechol
cis,cis-Muconate
Aromatic catabolism
Benzene derivatives are catabolized through catechol, a central intermediate.
catechol 2,3-dioxygenase
NAD+ → NADH + H+
Acetyl-CoA
+
Succinyl-CoA
TCA cycle
Pyruvate
Acetaldehyde
WIN MCNAMEE/GETTY IMAGES

FIGURE 13.31 ■ Aerobic aromatic catabolism. Oxidative catabolism of benzoate and various related compounds proceeds through catechols. Catechols are degraded through several alternative pathways to the TCA cycle. Steps requiring oxidation are marked O_2. **Inset:** Bird contaminated by petroleum from an offshore wellhead.

Anaerobic benzene catabolism. A challenge for biodegradation is that many pollution sources reach deep underground, where the soil is anoxic. Thus, oxygen is unavailable to conduct the conversions shown in **Figure 13.32**, particularly formation of the carboxylate ($RCOO^-$) and the introduction of hydroxyl groups. How can microbes catabolize benzene and benzoate derivatives without oxygen?

Early stages of benzene and naphthalene catabolism involve the incorporation of CO_2, forming the carboxylate group of benzoate. Because CO_2 is a very weak oxidant, these reactions require input of energy by hydrolysis of ATP. The carboxylate is then activated by HS-CoA, forming benzoyl-CoA, the same key intermediate as for aerobic benzene catabolism.

Anaerobically, the benzoyl-CoA must use reducing energy from NADH to hydrogenate its ring carbons, thus breaking the aromaticity. Some bacteria spend ATP as well, whereas others, such as the iron-reducing bacterium *Geobacter metallireduscens*, can break the ring without spending ATP. The hydroxyl groups that enable shifting of the double bond positions are introduced by incorporation of H_2O. Catabolism continues, forming three acetyl groups activated by HS-CoA, plus one CO_2.

Where do bacteria get the energy needed to break down the aromatic rings under anaerobic conditions? Research suggests that the energy comes from anaerobic photosynthesis or anaerobic respiration, conducted by species such as *Rhodopseudomonas palustris* and *Azoarcus evansii*, which possess multiple metabolic capabilities (discussed in Chapter 14).

Research Reveals Catabolism of Aromatic Pollutants

The rise of toxic pollutants in our environment poses a growing challenge for remediation. To understand the metabolism of potential microbial partners for bioremediation, research requires sophisticated chemical and genetic tools of analysis.

An example of such research is the bioinformatic analysis of bacterial catabolism of nicotine. Nicotine is found in high concentration in waste tobacco material from cigarette processing. This is one of many drug molecules that can contaminate soil and water supplies. Another is caffeine from coffee waste, which we presented in the chapter opener. How do we find a microbial enzyme pathway for degrading these unique molecules?

At Shanghai Jiao Tong University, Hongzhi Tang and colleagues characterized the pathway of nicotine degradation in *Pseudomonas putida* S16, a strain that efficiently degrades

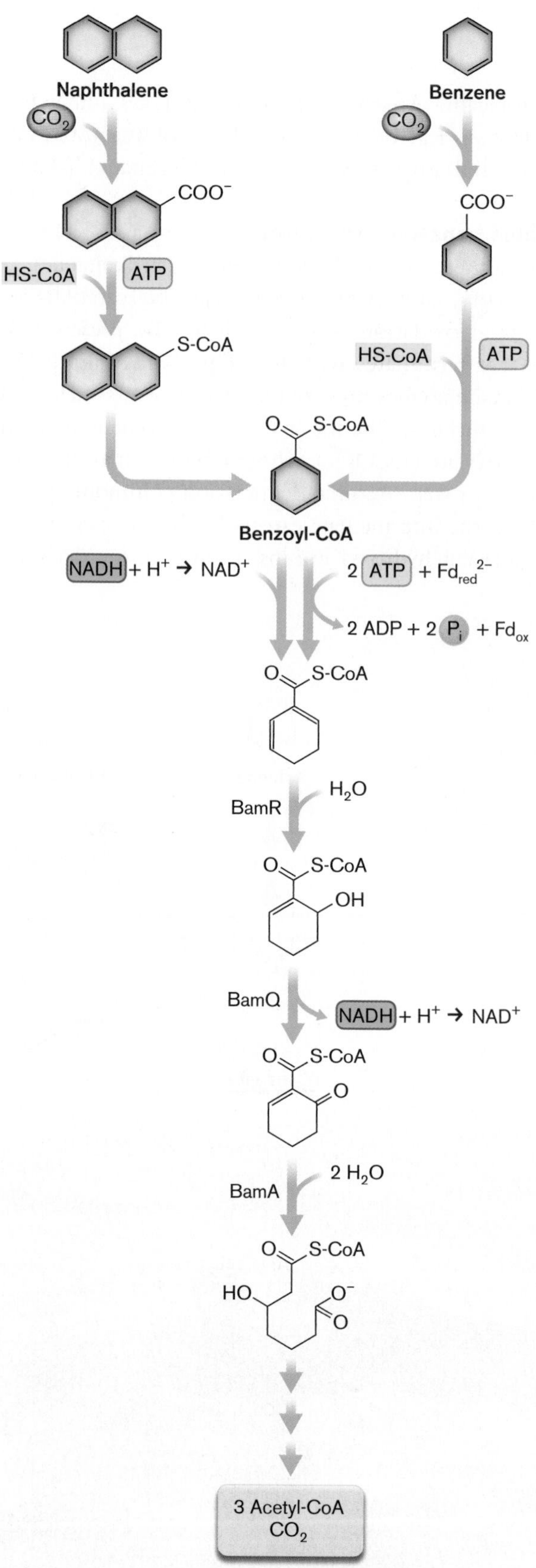

Naphthalene
Benzene
CO2
CO2
COO⁻
COO⁻
HS-CoA
ATP
S-CoA
HS-CoA
ATP
O
S-CoA
Benzoyl-CoA
NADH + H⁺ → NAD⁺
2 ATP + Fd red 2−
2 ADP + 2 Pi + Fd ox
O
S-CoA
BamR
H2O
O
S-CoA
OH
BamQ
NADH + H⁺ → NAD⁺
O
S-CoA
O
BamA
2 H2O
O
S-CoA
HO
O⁻
O
3 Acetyl-CoA
CO2

FIGURE 13.32 ▪ Anaerobic benzoate catabolism. In the absence of oxygen, benzene and naphthalene incorporate CO_2 and are activated by HS-CoA to form benzoyl-CoA. Benzoyl-CoA breakdown requires several steps of hydrolysis (H_2O) and of reduction (H) by carriers such as ferredoxin and NADH. The final three acetyl-CoA molecules may undergo fermentation. *Source:* Modified from Georg Fuchs et al. 2011. *Nat. Rev. Microbiol.* **9**:803.

nicotine (**Fig. 13.33**). Experiments with isotope labeling and gene complementation helped identify some enzymes that catalyze steps in the degradation pathway. Additional enzymes were identified by sequence comparison with the sequences of other organisms known to degrade nicotine. **Figure 13.33B** presents the species whose nicotine-degrading enzymes were shown to be homologs of genes in *P. putida*. The *P. putida* genes for nicotine catabolism were found to lie within a genomic island whose sequence analysis shows signs of horizontal transfer from other species (discussed in Chapter 9). For comparison, the homologs from five species are shown. While the sequences of homologous enzymes (orthologs) from different species match well, their linear order varies among the species.

Do such catabolic genes function together as a unit? To test this question, the genomic island from *P. putida* was moved into *E. coli*, an organism unable to degrade nicotine. An *E. coli* strain containing the nicotine catabolism genes on a plasmid gained the ability to degrade nicotine. This finding is important because it points to the possibility of engineering an industrial strain for remediation of nicotine-contaminated waste. For further discussion of industrial microbiology, see Chapter 16.

To Summarize

- **Catabolism of aromatic molecules** by bacteria and fungi recycles lignin and other important substances within ecosystems. Toxic pollutants are also degraded.
- **Benzoate undergoes aerobic catabolism to catechol.** The catechol ring is cleaved, generating acetyl-CoA, which enters the TCA cycle.
- **Anaerobic catabolism of benzoate** involves activation by HS-CoA and reduction by NADH.
- **Polycyclic aromatic hydrocarbons (PAHs)** are degraded slowly by microbes. PAH catabolism has exciting potential applications for bioremediation of toxic pollutants.
- **Aromatic catabolism by environmental microbes** is important for remediation of waste contaminated by drug molecules. To identify catabolic microbes, researchers use tools such as isotope labeling and bioinformatics.

A. Nicotine catabolism

N
N
CH2
Nicotine

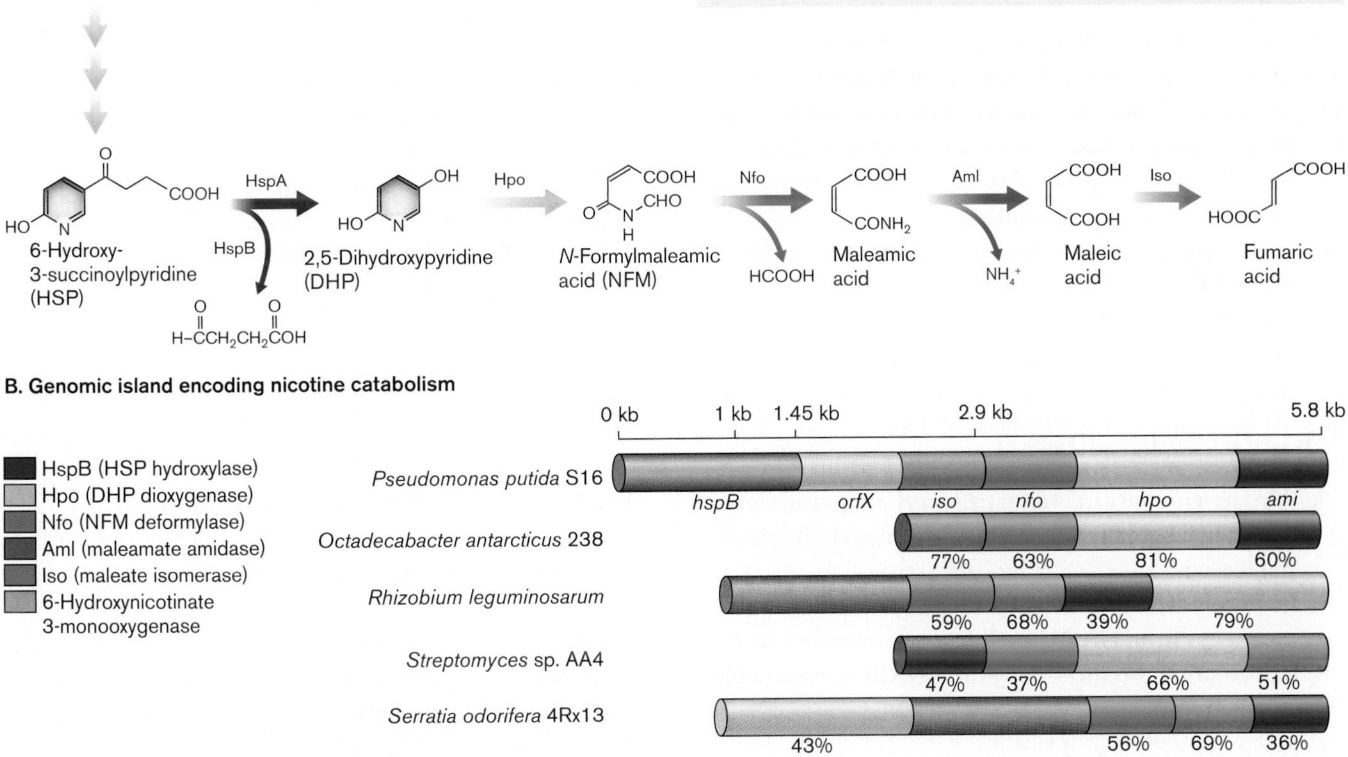
O
COOH
HO
N
6-Hydroxy-
3-succinoylpyridine
(HSP)
HspA
HspB
H–CCH2CH2COH
OH
2,5-Dihydroxypyridine
(DHP)
Hpo
COOH
CHO
N-Formylmaleamic
acid (NFM)
Nfo
HCOOH
COOH
CONH2
Maleamic
acid
Aml
NH4+
COOH
COOH
Maleic
acid
Iso
COOH
HOOC
Fumaric
acid
B. Genomic island encoding nicotine catabolism
0 kb
1 kb
1.45 kb
2.9 kb
5.8 kb
HspB (HSP hydroxylase)
Hpo (DHP dioxygenase)
Nfo (NFM deformylase)
Aml (maleamate amidase)
Iso (maleate isomerase)
6-Hydroxynicotinate
3-monooxygenase
Pseudomonas putida S16
hspB
orfX
iso
nfo
hpo
ami
Octadecabacter antarcticus 238
77%
63%
81%
60%
Rhizobium leguminosarum
59%
68%
39%
79%
Streptomyces sp. AA4
47%
37%
66%
51%
Serratia odorifera 4Rx13
43%
56%
69%
36%

FIGURE 13.33 ■ Nicotine catabolism: bioinformatics. **A.** Proposed pathway for nicotine catabolism. **B.** Genomic island encoding enzymes for nicotine catabolism, sequenced from five different species. Genes are assigned putative functions based on sequence similarity to genes from related species. *Source:* Hongzhi Tang et al. 2012. *Sci. Rep.* **2**:377.

Concluding Thoughts

In this chapter we have seen how energy-yielding reactions enable living cells to generate order from disorder. In fact, the result of cell metabolism is to increase entropy (chaos) in the universe as a whole, even while complexity (order) increases in the living cell. Living organisms acquire energy by transferring energy from reactions with negative ΔG to cell-building reactions with positive ΔG. The fundamental energy-yielding pathways of Earth's biosphere are those of photosynthesis, and the biomass generated through photosynthesis provides materials for catabolism. Much of human civilization has been built on harnessing microbial catabolism for waste treatment, food production, and biotechnology.

All forms of metabolism involve chemical exchange of electrons through oxidation and reduction. In Chapter 14 we focus on the transport of electrons between donor and acceptor molecules, through membrane-embedded complexes that generate ion gradients across the membrane. This transport of electrons is fundamental to respiration, the oxidative completion of catabolism to CO_2 and water. It also underlies the autotrophic means of building biomass: the gain of energy from light (phototrophy) and from mineral oxidation (lithotrophy).

CHAPTER REVIEW

Review Questions

1. Why must the biosphere continually take up energy from outside? Why can't all the energy be recycled among organisms, like the fundamental elements of matter?
2. Explain how a biochemical reaction can be driven by a change in enthalpy, ΔH. Explain how a different reaction can be driven by a change in entropy, ΔS. In each case, explain the role of the free energy change, ΔG.
3. Why do some biochemical reactions release energy only above a threshold temperature?
4. How do organisms determine which of their catabolic pathways to use? How does catabolism depend on environmental factors?
5. Beer is produced by yeast fermentation of grain to ethanol. Why must the process of beer production be anaerobic? Why are such large quantities of ethanol produced with a relatively small production of yeast biomass?
6. Explain the three different routes to catabolize glucose to pyruvate. Why is it necessary to start by spending one or two molecules of ATP?
7. Explain how the TCA cycle incorporates an acetyl group. How are the two CO_2 molecules removed?
8. Compare and contrast aerobic and anaerobic processes of benzoate catabolism.

Thought Questions

1. Why are glucose catabolism pathways ubiquitous, even in bacterial habitats where glucose is scarce? Give several reasons.
2. In glycolysis, explain why bacteria have to return the hydrogens from NADH back onto pyruvate to make fermentation products. Why can't NAD^+ serve as a terminal electron acceptor, like O_2?
3. Why does catabolism of benzene derivatives yield less energy than sugar catabolism? Why is benzene-derivative catabolism nevertheless widespread among soil bacteria?
4. Why do environmental factors regulate catabolism? For example, why are amino acids decarboxylated at low pH? Cite other examples.

Key Terms

activation energy (508)
adenosine triphosphate (ATP) (501)
allosteric site (509)
amphibolic (518)
anabolism (493)
anaerobic respiration (515)
aromatic (506)
calorimeter (497)
catabolism (492)
catabolite repression (510)
chemo- (492)
chemolithotrophy (492)
chemoorganotrophy (492)
coenzyme A (CoA) (523)
electron acceptor (501)
electron donor (501)
electron transport system (ETS) (506)
Embden-Meyerhof-Parnas (EMP) pathway (516, 517)
energy (494)
energy carrier (501)
enthalpy (ΔH) (496)
Entner-Doudoroff (ED) pathway (516, 520)
entropy (ΔS) (494, 496)
enzyme (508)
ethanolic fermentation (521)
fermentation (513)
flavin adenine dinucleotide (FAD) (507)
Gibbs free energy change (ΔG) (495)
glycolysis (516)
glyoxylate bypass (531)
heterolactic fermentation (521)
heterotrophy, heterotroph (492)
hydrolysis (504)
joule (J) (497)
lactate fermentation (521)
lithotrophy (492)
lignin (512)
mixed-acid fermentation (523)
nicotinamide adenine dinucleotide (NADH, NAD^+) (505)
organotrophy (492)
oxidative phosphorylation (531)
pentose phosphate pathway (PPP) (517, 520)
phenol red broth test (526)
phosphorylation (504)
phosphotransferase system (PTS) (505)
photo- (492)
photoheterotrophy (513)
photosynthesis (492)
phototrophy (492)
pyruvate dehydrogenase complex (PDC) (527)
respiration (513, 531)
ribulose 5-phosphate (520)
Stickland reaction (523)
substrate-level phosphorylation (518)
syntrophy (500)
terminal electron acceptor (507)
tricarboxylic acid (TCA) cycle (527)

Recommended Reading

Atlas, Ronald M. 1995. Petroleum biodegradation and oil spill bioremediation. *Marine Pollution Bulletin* **31**:178–182.

Bäckhed, Fredrik, Ruth E. Ley, Justin L. Sonnenburg, Daniel A. Peterson, and Jeffrey I. Gordon. 2005. Host-bacterial mutualism in the human intestine. *Science* **307**:915–920.

Brown, Stacie A., Kelli L. Palmer, and Marvin Whiteley. 2009. The host as a growth medium. *Nature Reviews. Microbiology* **6**:657–666.

Bunge, Michael, Lorenz Adrian, Angelika Kraus, Matthias Opel, Wilhelm G. Lorenz, et al. 2003. Reductive dehalogenation of chlorinated dioxins by an anaerobic bacterium. *Nature* **421**:357–360.

Eisenreich, Wolfgang, Thomas Dandekar, Jürgen Heesemann, and Werner Goebel. 2010. Carbon metabolism of intracellular bacterial pathogens and possible links to virulence. *Nature Reviews. Microbiology* **8**:401–412.

Fraser, Claire M., Steven J. Norris, George M. Weinstock, Owen White, Granger G. Sutton, et al. 1998. Complete genome sequence of *Treponema pallidum*, the syphilis spirochete. *Science* **281**:375–388.

Fuchs, Georg, Matthias Boll, and Johann Heider. 2011. Microbial degradation of aromatic compounds—From one strategy to four. *Nature Reviews. Microbiology* **9**:803–816.

Head, Ian M., D. Martin Jones, and Wilfred F. M. Röling. 2006. Marine microorganisms make a meal of oil. *Nature Reviews. Microbiology* **4**:173–182.

Jackson, Bradley E., and Michael J. McInerney. 2002. Anaerobic microbial metabolism can proceed close to thermodynamic limits. *Nature* **415**:454–456.

Jeon, C. O., W. Park, P. Padmanabhan, C. DeRito, J. R. Snape, et al. 2003. Discovery of a bacterium, with distinctive dioxygenase, that is responsible for in situ biodegradation in contaminated sediment. *Proceedings of the National Academy of Sciences USA* **100**:13591–13596.

Koropatkin, Nicole M., Elizabeth A. Cameron, and Eric C. Martens. 2012. How glycan metabolism shapes the human gut microbiota. *Nature Reviews. Microbiology* **10**:323–335.

Ladas, Spiros D., George Karamanolis, and Emmanuel Ben-Soussan. 2007. Colonic gas explosion during therapeutic colonoscopy with electrocautery. *World Journal of Gastroenterology* **13**:5295–5298.

Maurer, Lisa M., Elizabeth Yohannes, Sandra S. Bon-Durant, Michael Radmacher, and Joan L. Slonczewski. 2005. pH regulates genes for flagellar motility, catabolism, and oxidative stress in *Escherichia coli* K-12. *Journal of Bacteriology* **187**:304–319.

McInerney, Michael J., Jessica R. Sieber, and Robert P. Gunsalus. 2011. Microbial syntrophy: Ecosystem-level biochemical cooperation. *Microbe* **6**:479–485.

Peekhaus, Norbert, and Tyrrell Conway. 1998. What's for dinner? Entner-Doudoroff metabolism in *Escherichia coli*. *Journal of Bacteriology* **180**:3495–3502.

Sonnenburg, Justin L., Jian Xu, Douglas D. Leip, Chien-Huan Chen, Benjamin P. Westover, et al. 2005. Glycan foraging in vivo by an intestine-adapted bacterial symbiont. *Science* **307**:1955–1959.

Tang, Hongzhi, Yuxiang Yao, Lijuan Wang, Hao Yu, Yiling Ren, et al. 2012. Genomic analysis of *Pseudomonas putida*: Genes in a genome island are crucial for nicotine degradation. *Scientific Reports* **2**:377.

Xu, Jian, Magnus K. Bjursell, Jason Himrod, Su Deng, Lynn K. Carmichael, et al. 2003. A genomic view of the human–*Bacteroides thetaiotaomicron* symbiosis. *Science* **299**:2074–2075.

A.

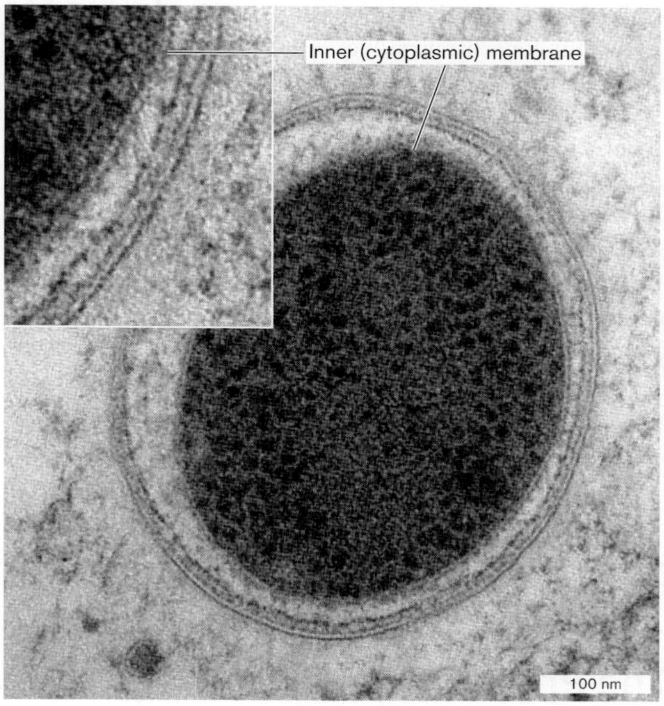
Inner (cytoplasmic) membrane
100 nm
YANPING LIU ET AL. 2006. J. GASTROENTEROL. 41:569–574. © SPRINGER-VERLAG 2006.

B.

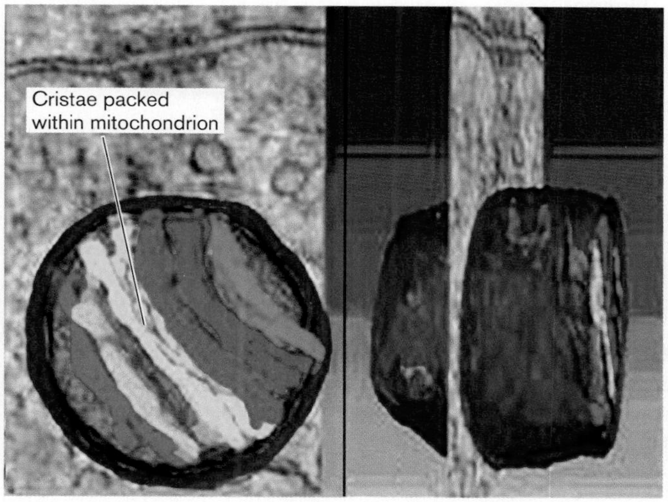
Cristae packed within mitochondrion
GUY A. PERKINS ET AL. 2010. J. NEUROSCI. 30:1015

FIGURE 14.3 ▪ Respiratory membranes. **A.** Electron transport occurs in the inner (cytoplasmic) membrane of *Helicobacter pylori.* The cytoplasmic membrane and cell wall are surrounded by periplasm and outer membrane. **B.** Mitochondrion within the brain stem neuron of a cat, modeled by cryo-EM tomography, shown with one EM section through the cell. The mitochondrial outer membrane (blue) encloses inner membrane organized in pockets called cristae (other colors).

Bacteria such as *Shewanella* and *Geobacter* species donate electrons to metals instead of to oxygen. These bacteria have extra cytochromes. The outer membrane cytochromes help metal-reducing bacteria gain access to insoluble forms of metal—or to the surface of a fuel cell electrode (see **Special Topic 14.2**). In other bacteria, such as the nitrite oxidizer *Nitrospira,* the respiratory membranes form intracytoplasmic pockets called "lamellae."

Respiratory membranes similar to those of bacteria are found within our own mitochondria (**Fig. 14.3B**). Mitochondrial respiration uses only O_2 as a terminal electron acceptor. The ETS proteins are embedded in folds of the mitochondrial inner membrane called "cristae" (singular, crista). The inner membrane separates the inner mitochondrial space from the intermembrane space (between the inner and outer membranes). The mitochondrial inner membrane, including its electron transport proteins, evolved from the cell membrane of an endosymbiotic bacterial ancestor. Among modern bacteria, the closest relatives of mitochondria are the rickettsias, obligate intracellular parasites that cause serious diseases, such as Rocky Mountain spotted fever. Mitochondrial evolution is discussed further in Chapter 17.

Electron carrier molecules include proteins and small organic cofactors bound to the proteins. The protein components of an ETS were first discovered in the 1930s at Cambridge University by Russian scientist David Keilin (1887–1963). Keilin was an entomologist who studied insect mitochondria. The mitochondrial inner membranes contained proteins called **cytochromes**, which were named for their deep colors, typically red to brown. The colors derive from absorption of visible light due to the relatively small energy transitions of electrons in a cytochrome.

In prokaryotes, cytochromes are often found in the cell membrane. **Figure 14.4** shows the absorbance spectrum of

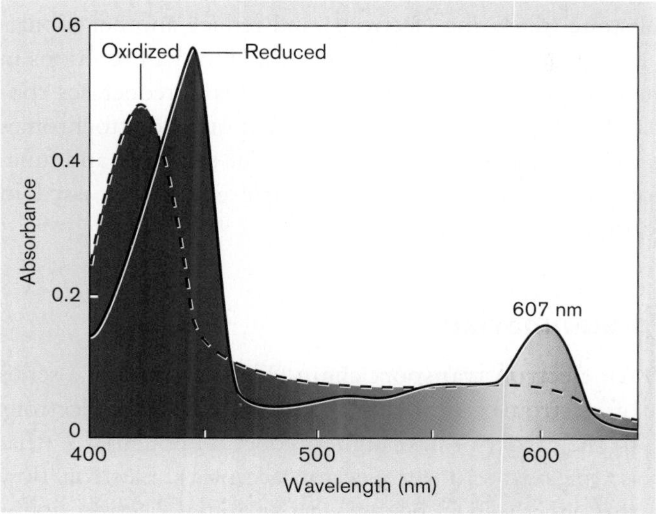
Oxidized
Reduced
607 nm
Absorbance
0.6
0.4
0.2
0
400
500
600
Wavelength (nm)

FIGURE 14.4 ▪ Light absorbance spectrum of a cytochrome. Absorption peaks shift between the oxidized and reduced forms of a cytochrome from the electron transport system of *Haloferax volcanii,* a halophilic archaeon.

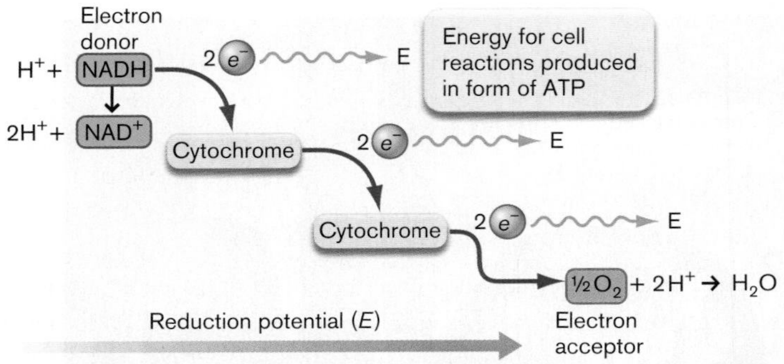

Electron donor
H⁺+ NADH
2H⁺+ NAD⁺
2e⁻
E
Energy for cell reactions produced in form of ATP
Cytochrome
2e⁻
E
Cytochrome
2e⁻
E
½O₂ + 2H⁺ → H₂O
Reduction potential (E)
Electron acceptor

FIGURE 14.5 ■ Electron transport system. Keilin's model for electron transfer through cytochromes. Each cytochrome in turn receives electrons from a stronger electron donor and transfers them to a stronger electron acceptor.

a cytochrome from the cell membrane of *Haloferax volcanii*, a halophilic archaeon isolated from the Dead Sea. In the reduced cytochrome, light absorption peaks in the blue range (440 nm) and in the red (607 nm). Upon oxidation, however, the cytochrome loses its absorption peak at 607 nm, and the 440-nm peak shifts to a shorter wavelength. Different species of bacteria make many different cytochromes, with different absorption peaks, but all cytochromes show spectral change with change in reduction state.

A membrane ETS typically includes several different cytochromes with different reduction potentials. Keilin proposed that the cytochromes pass electrons sequentially from each protein complex to the next-stronger electron acceptor, with each step providing a small amount of energy to the organism (**Fig. 14.5**). The electron transport proteins are called **oxidoreductases** because they oxidize one substrate (removing electrons) and reduce another (donating electrons). Thus, they couple different half reactions in the electron tower (see **Table 14.1**). Oxidoreductases consist of multiple-protein complexes that include cytochromes as well as noncytochrome proteins. The structure and function of ETS oxidoreductase complexes are discussed in Section 14.3.

To Summarize

- **An electron transport chain (ETS)** consists of a series of electron carriers that sequentially transfer electrons to the carrier of next-higher reduction potential E (that is, the next-stronger electron acceptor). Electron flow through the ETS begins with an initial electron donor from outside the cell and ultimately transfers all electrons to a terminal electron acceptor.
- **The reduction potential E** for a complete redox reaction must be positive to yield energy for metabolism. The standard reduction potential $E°'$ assumes that all reactant concentrations equal 1 M, at pH 7.
- **Concentrations of electron donors and acceptors** in the environment influence the actual reduction potential E experienced by the cell.
- **The ETS is embedded in a membrane that separates two compartments.** Two aqueous compartments must be separate to maintain an ion gradient generated by the ETS.
- **The ETS is composed of protein complexes and cofactors.** Protein complexes called oxidoreductases include cytochromes and noncytochrome proteins. Cytochromes are colored proteins whose absorbance spectrum shifts when there is a change in redox state.

14.2

The Proton Motive Force

The sequential transfer of electrons from one ETS protein to the next yields energy to pump ions (in most cases H^+) across the membrane. Proton pumping generates a **proton motive force** (or **proton potential**) composed of the H^+ concentration difference plus the charge difference across the membrane. The proton motive force (PMF) drives many different cellular processes, including ATP synthesis, nutrient transport, and flagellar rotation. Bacteria swim using rotary motors powered by a proton current.

The ETS Pumps Protons

The ion pumped by most ETS complexes is H^+. In water, H^+ never occurs as such, because it combines with a water molecule to form hydronium ion, H_3O^+. Current data, however, are consistent with the passage of hydrogen nuclei (protons) through the proton pumps of the electron transport system. Within the pump complex, the proton associates with one chemical group after another; for example, H^+ may combine with the amine of an amino acid (RNH_2) to form an ammonium ion (RNH_3^+), and then transfer to a different proton-accepting group within the protein.

Note: In this book we use "hydrogen ions" and "H^+" interchangeably with "protons."

The transfer of H^+ through a proton pump generates an H^+ concentration difference across the membrane. Since H^+ carries a positive charge, the proton transfer also generates a charge difference across the membrane. The H^+

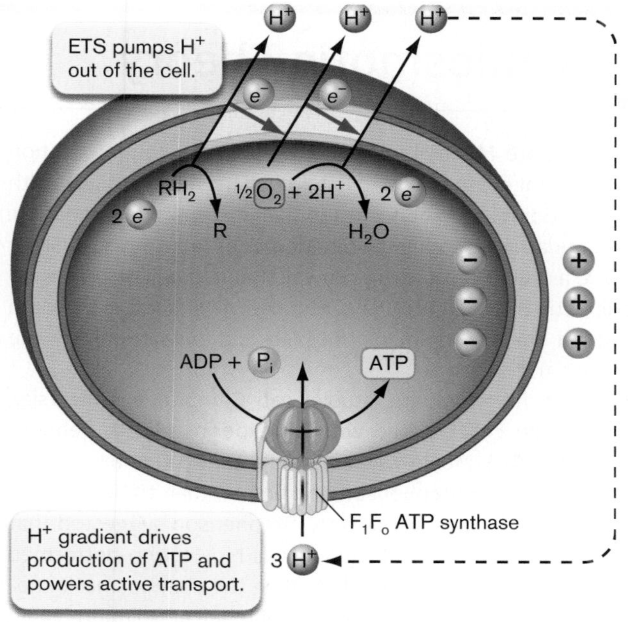
ETS pumps H+ out of the cell.
H+
e−
RH2
R
½O2 + 2H+
H2O
2 e−
ADP + Pi
ATP
F1Fo ATP synthase
3 H+
H+ gradient drives production of ATP and powers active transport.

FIGURE 14.6 ▪ An ETS generates proton motive force (Δp). An electron transport system may pump protons out of the cell. The resulting electrochemical gradient of protons (proton motive force) drives conversion of ADP to ATP through ATP synthase. It also drives nutrient transport, flagellar rotation, and other processes (see Fig. 14.9).

concentration difference (ΔpH) plus the charge difference ($\Delta\psi$) make a proton potential (Δp), also called a proton motive force (PMF). The proton potential (in volts) equals the electrochemical proton gradient (in joules) divided by Faraday's constant. The proton potential stores energy to drive protons back across the membrane through devices such as the ATP synthase and the flagellar motor. In pathogens, the proton potential drives drug efflux pumps that confer resistance to antibiotics such as tetracycline.

The discovery of the proton potential radically changed the field of biochemistry. Early in the twentieth century, Keilin and other scientists knew that the energy acquired by electron transport proteins was used to make ATP, but they did not know how. Most were convinced that electron transport was somehow directly coupled to ATP synthesis. The actual means of coupling was discovered by a student of Keilin's: Peter Mitchell (1920–1992). In 1961, Mitchell proposed an astonishing explanation for the coupling of electron transport to ATP synthesis (**Fig. 14.6**). The chemiosmotic theory states that the energy from electron transfer between membrane proteins is used to pump protons across the membrane, accumulating a higher H^+ concentration in the compartment outside. The pump generates the proton motive force Δp, which stores energy that can be used to make ATP (discussed in Section 14.3).

At first, many biologists found it difficult to understand how a proton potential could be coupled to ATP synthesis at an enzyme complex separate from the ETS, at a distant location on the membrane. The explanation is that the H^+ concentration gradient and charge difference exist everywhere on the membrane separating the two compartments. Because the membrane is impermeable to hydrogen ions, the H^+ current can flow back only through a proton-driven complex such as the membrane ATP synthase (discussed in Section 14.3). In effect, the proton potential is a "proton battery," analogous to the electron potential of an electrical battery. The concept of cellular electricity proved controversial among scientists, who devised many experiments to test it before it was accepted (see **Special Topic 14.1**).

Δp Includes $\Delta\psi$ and ΔpH

When protons are pumped across the membrane, energy is stored in two different forms—the separation of charge, or electrical potential (because H^+ carries a positive charge), and the gradient of H^+ concentration, or pH difference—as shown in **Figure 14.7**. Either form (or both) can drive

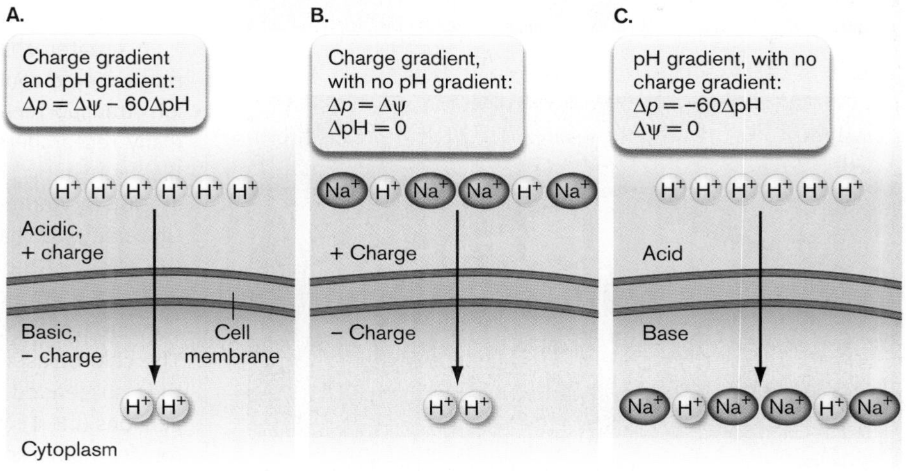
A.
Charge gradient and pH gradient: Δp = Δψ − 60ΔpH
H+ H+ H+ H+ H+ H+
Acidic, + charge
Basic, − charge
Cell membrane
H+ H+
Cytoplasm
B.
Charge gradient, with no pH gradient: Δp = Δψ ΔpH = 0
Na+ H+ Na+ Na+ H+ Na+
+ Charge
− Charge
H+ H+
C.
pH gradient, with no charge gradient: Δp = −60ΔpH Δψ = 0
H+ H+ H+ H+ H+ H+
Acid
Base
Na+ H+ Na+ Na+ H+ Na+

FIGURE 14.7 ▪ Electrical potential and pH difference. Proton motive force drives protons into the cell (arrow). **A.** The proton motive force Δp is composed of the transmembrane electrical potential $\Delta\psi$ (the charge difference), plus the transmembrane pH difference ΔpH (the chemical concentration gradient of H^+). **B.** If the pH inside and outside the cell is equal, then $\Delta p = \Delta\psi$. **C.** If the electrical charge inside and out is equal, then $\Delta p = -60\Delta$pH.

Special Topic 14.1: Testing the Chemiosmotic Theory

In 1961, as a young researcher at the University of Edinburgh, Peter Mitchell (**Fig. 1A**) first developed the chemiosmotic theory. He developed the idea of vectorial metabolism—that metabolic energy could be stored in a directional form, in an ion gradient. The energy stored in a gradient of ions across a membrane could be interconverted with the energy of chemical interactions, such as oxidation of glucose. Through vectorial metabolism, the oxidation-reduction reactions of the electron transport system could drive the pumping of protons across a membrane. The proton gradient could then drive protons back through the F_1F_o ATP synthase, driving synthesis of ATP.

But Mitchell's colleagues rejected his ideas. Researchers had spent many years discovering the molecular "intermediates" of metabolism, and they expected to find one or two more between the ETS and the ATP synthase. Furthermore, most biochemists lacked sufficient training in physics to appreciate Mitchell's ideas, and Mitchell's communication skills did not cross the gap. So Mitchell used his family wealth to leave Edinburgh and form the Glynn Research Foundation. His foundation supported a bioenergetics laboratory at Glynn House, a country estate in Cornwall that Mitchell restored and equipped (**Fig. 1B**). At Glynn, Mitchell developed a research program to test various principles and predictions of the chemiosmotic theory.

During his graduate studies at Cambridge University, Mitchell had collaborated with an experimentalist, Jennifer Moyle (**Fig. 1A**). Moyle was a student of the Cambridge biochemist Marjorie Stephenson, the first woman to be selected as a Fellow of the Royal Society. Stephenson suggested that Moyle work with Mitchell, and it was Moyle who performed most of the early experiments testing the chemiosmotic theory. Moyle continued working with Mitchell at Edinburgh, and she moved with him to Glynn, where they worked together for the next 30 years.

In the 1960s, Mitchell and Moyle devised several experiments to test the chemiosmotic theory. A key requirement was to show that the ETS generates a proton potential. Moyle showed that respiration of mitochondria is associated with proton efflux: Mitochondria isolated from rat livers were exposed to oxygen, causing efflux of hydrogen ions. The H^+ efflux occurred as electrons were transferred across the ETS and hydrogen ions were expelled by the proton pumps. In Moyle's experiments, the number of protons extruded per electron transferred down the ETS was consistent with the chemiosmotic model. Similar results were obtained with vesicles made first from mitochondrial membranes and later from the membranes of chloroplasts and bacteria.

A greater challenge was to demonstrate that a proton motive force could, in fact, drive the ATP synthase and other ion transporters without any undiscovered intermediate. Many researchers attempted to show this, often with conflicting results. The first successful experiment was reported in 1966 by André Jagendorf, a colleague from Johns Hopkins University who had visited Mitchell at Glynn. Jagendorf tested the effect of a proton gradient imposed on spinach chloroplasts (**Fig. 2A**). Chloroplasts contain ATP synthase directed outward (that is, driven by proton flow from inside to outside). The chloroplasts were partly opened by osmotic shock, and then suspended in medium containing concentrated hydrogen ions (pH 4). With the osmotic balance restored, the chloroplast membranes closed again, with increased $[H^+]$ trapped inside. When the pH outside was raised to pH 8.3—that is, lower $[H^+]$—protons from the acidic interior of the vesicles flowed out through the ATP synthase, and ADP and inorganic phosphate were converted to ATP. Jagendorf interpreted this result as consistent with Mitchell's chemiosmotic theory for the mechanism of phosphorylation.

A.

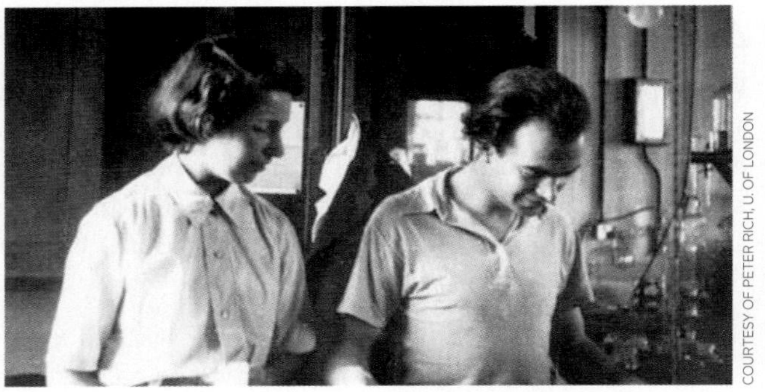
COURTESY OF PETER RICH, U. OF LONDON

B.

COURTESY OF PETER RICH, U. OF LONDON

FIGURE 1 ■ Peter Mitchell and Jennifer Moyle proposed and tested the chemiosmotic theory. A. Peter Mitchell and Jennifer Moyle conducting biochemistry research at Cambridge University. **B.** Glynn House, the Regency mansion in Cornwall, England, that Mitchell restored for his laboratory, is shown here in 1990 during a conference in Mitchell's honor, attended by five Nobel Prize winners.

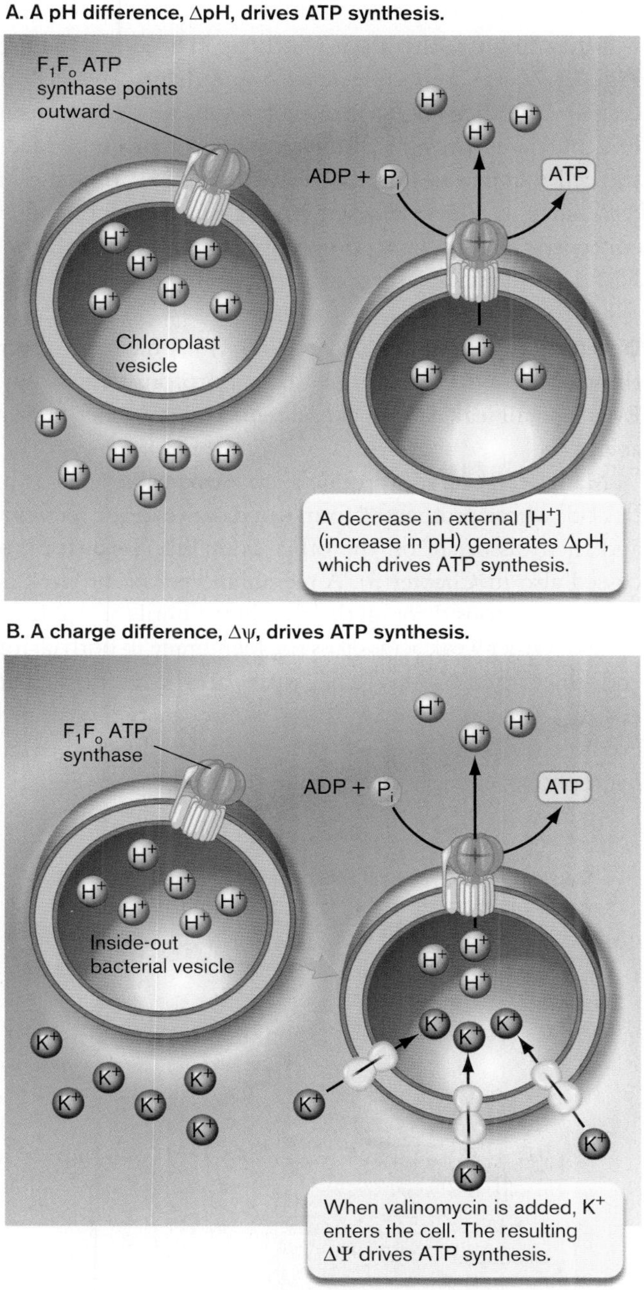
A. A pH difference, ΔpH, drives ATP synthesis.
F1Fo ATP synthase points outward
ADP + Pi
ATP
H+
Chloroplast vesicle
A decrease in external [H+] (increase in pH) generates ΔpH, which drives ATP synthesis.
B. A charge difference, Δψ, drives ATP synthesis.
F1Fo ATP synthase
ADP + Pi
ATP
H+
Inside-out bacterial vesicle
K+
When valinomycin is added, K+ enters the cell. The resulting ΔΨ drives ATP synthesis.

FIGURE 2 ■ Either ΔpH or Δψ drives ATP synthesis. **A.** A pH difference imposed across the chloroplast inner membrane drives a proton current through an outwardly directed ATP synthase. **B.** A charge difference generated by K^+ influx drives a proton current through ATP synthase.

Other researchers showed that ATP synthesis could be driven by a charge difference across a vesicle membrane (**Fig. 2B**). An artificial electrical potential (Δψ) was applied by loading vesicles with potassium ions (K^+). The potassium ions were conducted across the membrane by the ionophore valinomycin, a small cyclic peptide that binds to an ion and solubilizes it in the membrane. Valinomycin specifically binds K^+, while its hydrophobic side chains solubilize it in the membrane; thus the molecule conducts K^+ across the membrane down its concentration gradient (see **Fig. 2B**). In the experiment shown, the vesicles are "inside-out" bacterial membrane vesicles, in which the ATP synthase points outward instead of inward. The K^+ influx adds positive charge, which drives H^+ out through ATP synthase, catalyzing formation of ATP. Thus, Δψ drives formation of ATP.

These experiments, however, could not definitively rule out the existence of an undetected component of the membrane that somehow energized ATP synthase. A more compelling experiment was performed in 1975 by Efrem Racker and colleagues at Cornell University. Racker developed a process to make artificial vesicles called "liposomes," composed of purified phospholipids in the absence of any membrane proteins. He then purified a proton pump called bacteriorhodopsin from a halophilic archaeon, a *Halobacterium* species. Bacteriorhodopsin acts as a single-protein proton pump driven by light absorption. The bacteriorhodopsin was combined with phospholipids and with ATP synthase purified from mitochondria to obtain "reconstituted liposomes" with functional bacteriorhodopsin and mitochondrial ATP synthase. When the liposomes were exposed to light, the bacteriorhodopsin pumped protons and the ATP synthase made ATP. Thus, a proton pump and a proton-driven ATP synthase from two completely different organisms could work together in a membrane, connected solely by the proton motive force.

In 1978, Mitchell received the Nobel Prize in Chemistry, and the scientific community fully embraced the role of ion currents in metabolism. In later life, when a colleague disagreed with his views, Mitchell would turn off his hearing aid. Today the chemiosmotic principle is firmly established as the basis of energy transduction in respiration, lithotrophy, and photosynthesis. Furthermore, the proton potential drives numerous cellular devices, such as drug efflux pumps that enable bacteria to expel antibiotics.

RESEARCH QUESTION

Suppose you have a strain of pathogenic *E. coli* that expels an antibiotic via the AcrAB-TolC efflux pump. What experiment could provide evidence that efflux is driven by a proton current?

Seeger, Marcus A., André Schiefner, Thomas Eicher, François Verrey, Kay Diedrichs, et al. 2006. Structural asymmetry of AcrB trimer suggests a peristaltic pump mechanism. *Science* **313**:1295.

Δp-dependent cell processes. Both forms of energy contribute to the Δp (proton potential):

- **The electrical potential** ($\Delta\psi$) arises from the separation of charge between the cytoplasm (more negative) and the solution outside the cell membrane (more positive). For many bacteria, this "battery" potential is about –50 to –150 mV.
- **The pH difference** (ΔpH) is the log ratio of external to internal chemical concentrations of H^+. For example, if the bacterial internal pH is 7.5 and the external pH is 6.5, the ΔpH is 1.0, and the ratio of $[H^+]_{out}$ to $[H^+]_{in}$ is 10. A ΔpH of 1.0 corresponds to a proton potential of approximately –60 mV when the temperature is 25°C.

The relationship between electrical and chemical components of the proton potential Δp (in mV) is given by:

$$\Delta p = \Delta\psi - (2.3RT/F)\Delta\text{pH}$$

or approximately:

$$\Delta p = \Delta\psi - 60\Delta\text{pH}$$

For cells grown at neutral pH, all three terms (Δp, $\Delta\psi$, and -60ΔpH) usually have a negative value, meaning that their force drives protons inward from outside.

In living cells, the relative contributions of $\Delta\psi$ and -60ΔpH vary depending on other sources of charge difference and protons, as shown in **Figure 14.7**. Both the $\Delta\psi$ and ΔpH components of Δp are influenced by other factors besides ETS proton transport. For example:

- **The charge difference $\Delta\psi$** includes charges on other ions, such as K^+ and Na^+. These ions are pumped or exchanged by ion-specific membrane transport proteins.
- **The pH difference ΔpH** is affected by metabolic generation of acids, such as fermentation acids, or by pH changes outside the cell. Permeant acids can run down the gradient through the membrane, collapsing the ΔpH to zero.

Overall, the cell uses its various membrane pumps and metabolic pathways to adjust and maintain its proton motive force at a size sufficient to drive ATP synthesis but not so great as to disrupt the membrane.

Thought Question

14.3 What do you think happens to $\Delta\psi$ as the cell's external pH increases or decreases? What could happen to the Δp of bacteria that are swallowed and enter the extremely acidic stomach?

How are the Δp and its two components actually measured? In larger cells, such as those of eukaryotes, a microelectrode can be inserted into the cell to measure reduction potentials directly. Most bacterial cells are too small to have electrodes inserted into them, but $\Delta\psi$ and ΔpH can be measured by the uptake of molecules whose ability to permeate the membrane depends on their ionization state (**eTopic 14.2**).

What about other ions, such as the sodium ions in **Fig. 14.7**; can their concentration gradient drive cell processes? A gradient of Na^+ concentration can drive transport of a nutrient through a transporter protein embedded in the cell membrane (discussed in Chapter 4). In some bacteria, such as *Vibrio cholerae* (the cause of cholera), an Na^+ concentration gradient can power ATP synthesis by an Na^+-dependent ATP synthase.

Dissipation of proton motive force. What happens if something disrupts the membrane so that protons leak through? Even a small leak can quickly dissipate all the energy stored as ΔpH.

Many fermentation products are weak acids, such as acetic acid, whose protonated form can dissolve in the membrane and then dissociate on the other side (**Fig. 14.8**, top) (discussed also in Chapter 3). A membrane-permeant weak acid conducts protons through the membrane until the ΔpH is dissipated. Other weak acids cross the membrane in both charged and uncharged forms. The weak acid 2,4-dinitrophenol (DNP) dissociates to an anion whose charge is relatively evenly distributed around the molecule; thus, it remains hydrophobic enough to penetrate the membrane (**Fig. 14.8**,

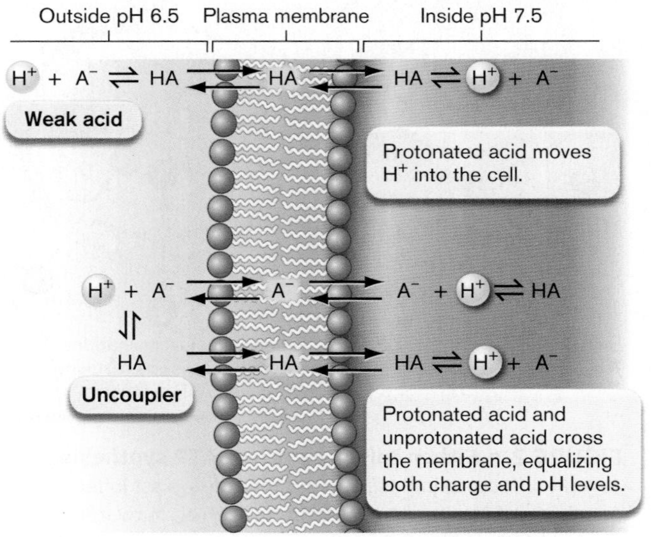
Outside pH 6.5
Plasma membrane
Inside pH 7.5
H+ + A− ⇌ HA
HA
HA ⇌ H+ + A−
Weak acid
Protonated acid moves H+ into the cell.
H+ + A−
A−
A− + H+ ⇌ HA
HA
HA
HA ⇌ H+ + A−
Uncoupler
Protonated acid and unprotonated acid cross the membrane, equalizing both charge and pH levels.

FIGURE 14.8 ▪ Membrane-permeant weak acids and uncouplers. A weak acid crosses the membrane in the protonated form and then dissociates, acidifying the cell. An uncoupler crosses the membrane in both protonated and unprotonated forms, cyclically acidifying the cell and dissipating the charge difference (electrical potential, $\Delta\psi$).

bottom). Because both protonated (uncharged) and unprotonated (negatively charged) forms cross the membrane, DNP can cyclically bring protons into the cell and collapse both $\Delta\psi$ and ΔpH. Such molecules are called **uncouplers** because they uncouple electron transport from ATP synthesis through the membrane ATP synthase. Uncouplers are highly toxic to bacteria, as well as to human cells. All cells need to maintain ion gradients and potentials.

Δ*p* Drives Many Cell Functions

Besides ATP synthesis, Δ*p* drives many cell processes directly, such as the rotation of flagellar motors (**Fig. 14.9**). Proton transport is also coupled to transport of ions such as K^+ or Na^+ through parallel transport (symport) or oppositely directed transport (antiport), as discussed in Chapters 3 and 4. Ion flux then drives uptake of nutrients such as amino acids or efflux of molecules such as antibacterial drugs. Drug efflux pumps are a major problem in hospitals, where the pump proteins are encoded by genes carried on plasmids that spread among virulent strains.

Thought Question

14.4 Suppose that de-energized cells of *E. coli* (Δ*p* = 0) with an internal pH 7.6 are placed in a solution at pH 6. What do you predict will happen to the cell's flagella? What does this demonstrate about the function of Δ*p*?

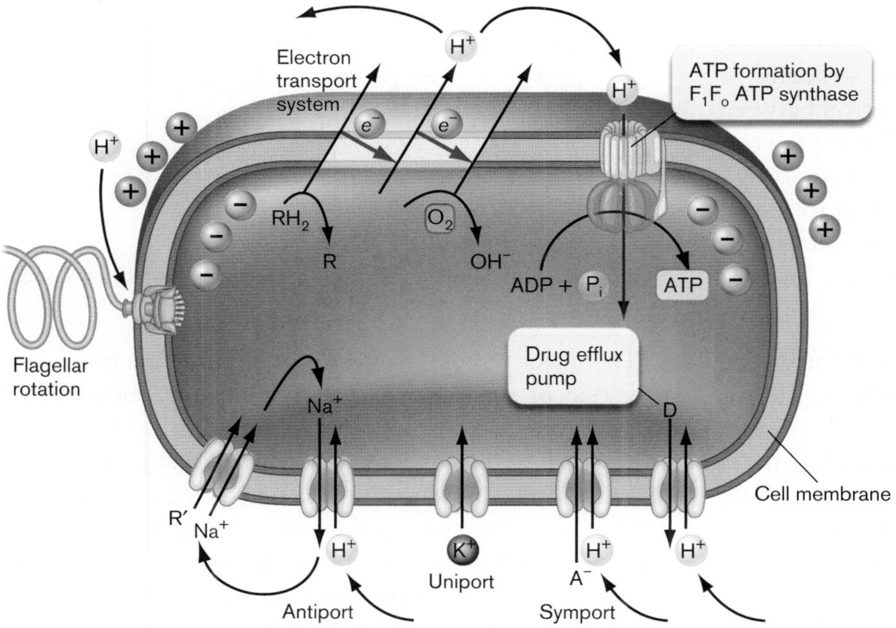

Electron transport system
H+
H+
ATP formation by F1F0 ATP synthase
e−
e−
H+
RH2
R
O2
OH−
ADP + Pi
ATP
Flagellar rotation
Drug efflux pump
Na+
D
Cell membrane
R′
Na+
H+
K+
H+
H+
Uniport
A−
Antiport
Symport

FIGURE 14.9 ■ Processes driven by the proton motive force. Processes powered by proton potential include ATP synthesis through the F_1F_o ATP synthase, flagellar rotation, uptake of nutrients, and efflux of toxic drugs. "R" refers to an organic nutrient.

To Summarize

- **The ETS complexes generate a proton motive force (Δ*p*).** The force is usually directed inward, driving protons into the cell.
- **The proton potential drives ATP synthase** and other functions, such as ion transport and flagellar rotation.
- **The proton potential (Δ*p*, measured in millivolts)** is composed of the electrical potential ($\Delta\psi$) and the hydrogen ion chemical gradient (ΔpH): $\Delta p = \Delta\psi - 60\Delta pH$.
- **Uncouplers are molecules taken up by cells in both protonated and unprotonated forms.** Uncouplers can collapse the entire proton potential, thus uncoupling respiration from ATP synthesis.

14.3

The Respiratory ETS and ATP Synthase

In the respiratory electron transport system (ETS), a series of carrier molecules harvests the reducing potential of electrons in small steps. The respiratory ETS most commonly presented is that of the mitochondrial inner membrane, in which NADH and $FADH_2$ transfer electrons ultimately to O_2, producing H_2O. Microbes also use many alternative electron donors and acceptors. For example, when *Salmonella enterica* infects the gut epithelium, macrophages attempt to kill them by secreting toxic reactive oxygen compounds. But one compound formed is tetrathionate ($S_4O_6^{2-}$), which *S. enterica* in the gut can use as a terminal electron acceptor.

Cofactors Allow Small Energy Transitions

ETS proteins such as cytochromes associate electron transfer with small, reversible energy transitions. The small energy transitions are mediated by **cofactors**, small molecules that associate with the protein. **Figure 14.10** shows the protein structure of a cytochrome from the pathogen *Pseudomonas aeruginosa*. The protein, cytochrome *c*, transfers electrons for nitrate respiration. Its reddish

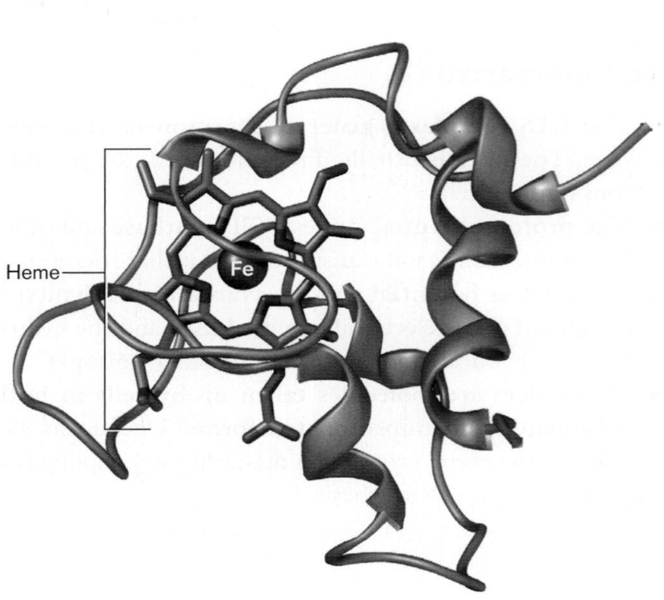
Heme
Fe

FIGURE 14.10 ▪ **Cytochrome *c*.** Portion of the protein structure of cytochrome *c* containing one heme cofactor, from the pathogen *Pseudomonas aeruginosa*. (PDB code: 2PAC)

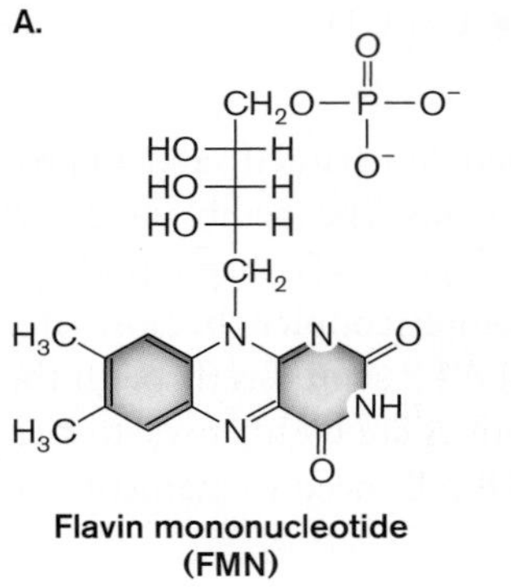
A.
HO
HO
HO
H
H
H
NH
Flavin mononucleotide (FMN)

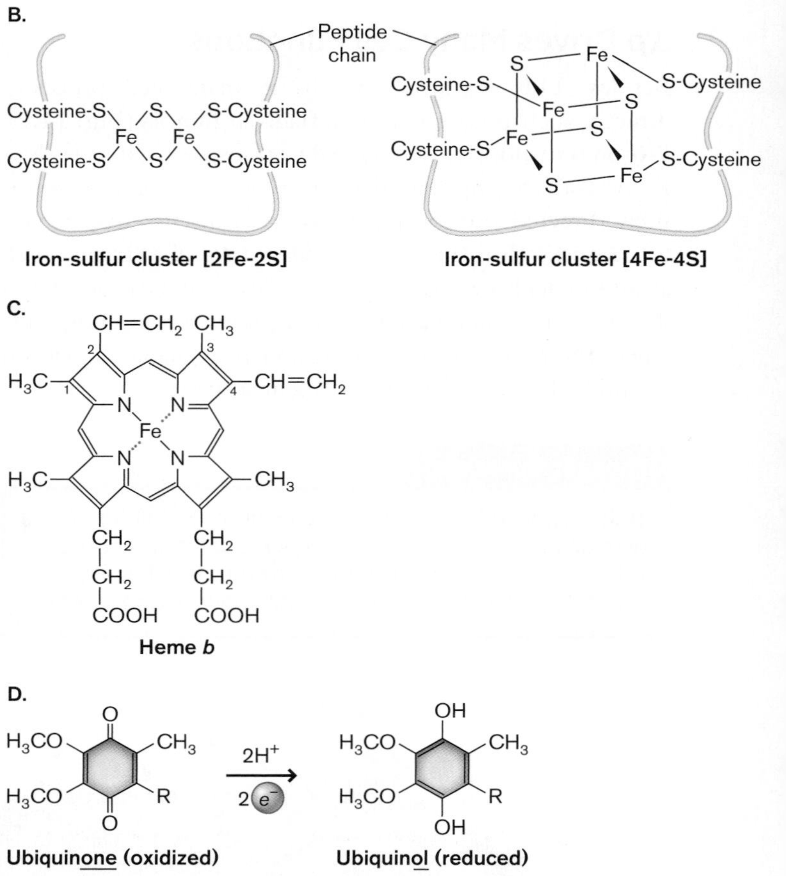
B.
Peptide chain
Cysteine-S
S-Cysteine
Fe
Iron-sulfur cluster [2Fe-2S]
Iron-sulfur cluster [4Fe-4S]
C.
CH=CH2
CH3
COOH
Heme b
D.
H3CO
CH3
R
OH
2H+
2e−
Ubiquinone (oxidized)
Ubiquinol (reduced)

FIGURE 14.11 ▪ **Reaction centers for electron transport.** **A.** Flavin mononucleotide (FMN). **B.** Iron-sulfur clusters: [2Fe-2S] and [4Fe-4S]. **C.** Heme *b*. The side chains of the ring vary among hemes, yielding different levels of redox potential. **D.** Ubiquinone, which is reduced to ubiquinol.

color derives from the buried cofactor called **heme**, a ring of conjugated double bonds surrounding an iron ion (Fe^{2+} or Fe^{3+}). The heme plays a key role in acquiring and transferring electrons, with an Fe^{2+}/Fe^{3+} transition. A metal-reducing bacterium such as *Geobacter* may make over a hundred different types of cytochromes in its envelope, where electrons accumulate until the bacterium finds an oxidized metal to accept them.

Cofactors such as heme allow small, reversible redox changes. Other examples include flavin mononucleotide (FMN) and ubiquinone (**Fig. 14.11**). The structure of each cofactor must allow transition of an electron between closely spaced energy levels to avoid "spending it all in one place." If all energy were spent in one transition, most of it would be lost as heat instead of being converted to several small processes, such as pumping H^+ across the membrane.

Small energy transitions typically involve these kinds of molecular structures:

- **Metal ions such as iron or copper**, coordinated (and hence held in place) with amino acid residues. Iron is often coordinated by sulfur atoms of cysteine residues in the protein; examples shown in **Figure 14.11B** are [2Fe-2S] and [4Fe-4S]. Transition metals make useful electron carriers because their outer electron shell has several closely spaced energy levels, facilitating small energy transitions.
- **Conjugated double bonds and heteroaromatic rings**, such as the nicotinamide ring of $NAD^+/NADH$, also provide narrowly spaced energy transitions. Membrane-soluble carriers such as **quinones** (reduced to **quinols**) allow even smaller energy transitions than does $NAD^+/NADH$.

The major protein complexes of electron transport each have one or more redox centers containing either metal ions or conjugated double bonds or both. In FMN, the conjugated double bonds allow a small energy transition (**Fig. 14.11A**). In iron-sulfur clusters—[2Fe-2S] and [4Fe-4FS]—the metal atoms provide the site for a small energy transition, its size dependent on the cluster's connections within the associated protein (**Fig. 14.11B**). The heme group found in cytochromes and other oxidoreductases contains extensive conjugated double bonds, coordinated around the metal Fe^{3+} (**Fig. 14.11C**). Reduction by a transferred electron converts the iron to Fe^{2+}. The branching of

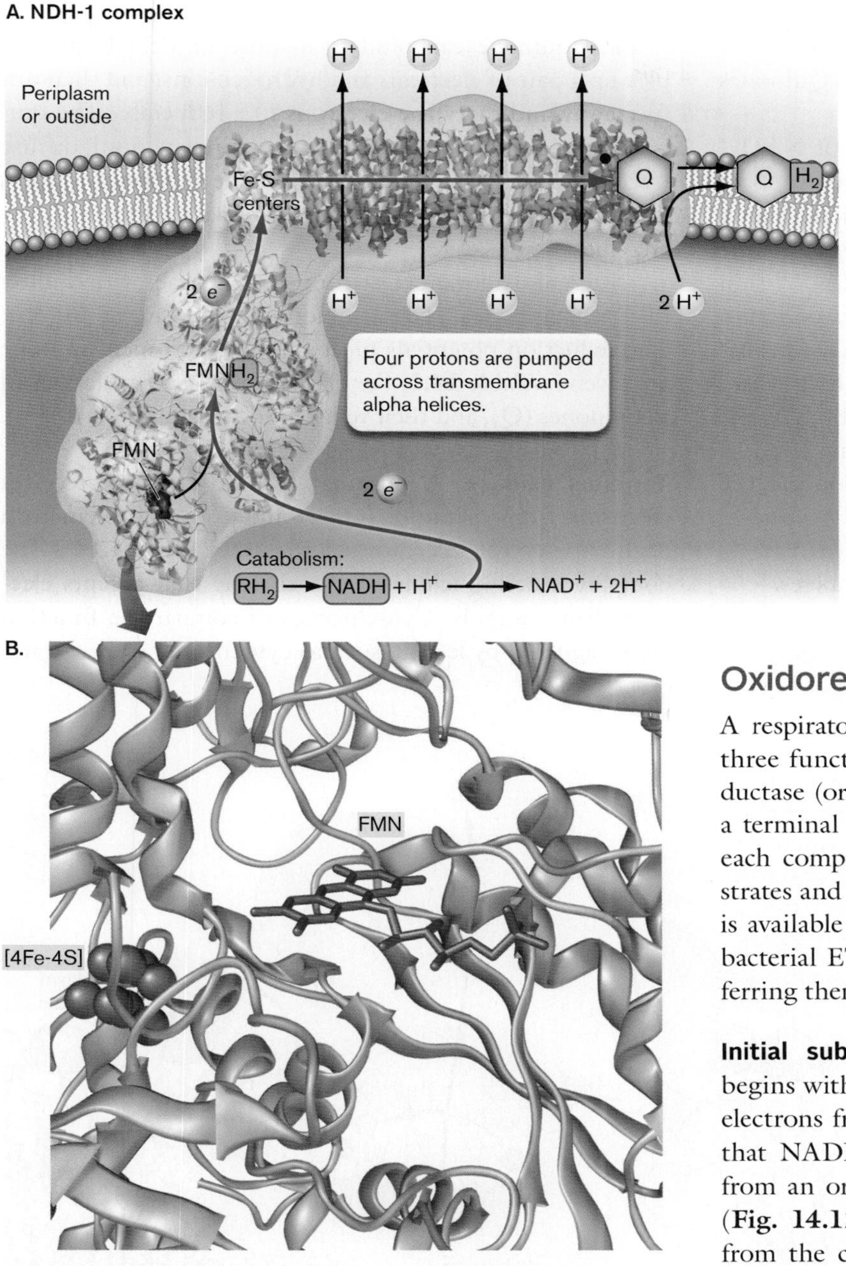
A. NDH-1 complex
Periplasm
or outside
H+
Fe-S
centers
Q
Q H2
2 e-
FMNH2
FMN
H+
2 H+
Four protons are pumped
across transmembrane
alpha helices.
2 e-
Catabolism:
RH2 NADH + H+ NAD+ + 2H+
B.
FMN
[4Fe-4S]

FIGURE 14.12 ▪ NADH:quinone oxidoreductase complex (NDH-1). A. NDH-1 transfers two electrons from NADH onto FMN ($FMNH_2$), through several Fe-S centers, and ultimately onto a quinone, Q (forming quinol, QH_2). The energy from oxidizing NADH is coupled to pumping $4H^+$ across the cell membrane. H^+ ions whose flow increases proton potential are highlighted yellow. **B.** Within the NDH-1 complex, FMN lies adjacent to the first iron-sulfur center, [4Fe-4S]. (PDB code: 1NOX) *Source:* Part A modified from R. Efremov and L. Sazanov. 2010. *Nature* **476**:414; part B modified from L. A. Sazanov and P. Hinchliffe. 2006. *Science* **311**:1430–1436.

side chains on the ring varies among different hemes, altering the magnitude of $E^{\circ\prime}$ for the redox couple Fe^{3+}/Fe^{2+}.

Some electron carriers are small molecules that associate loosely with a protein complex and then come off to diffuse freely throughout the membrane. Mobile electron carriers include quinones such as ubiquinone (**Fig. 14.11D**), which are reduced to quinols:

$$Q + 2H^+ + 2e^- \rightarrow QH_2$$

Quinols carry electrons and protons laterally within the membrane between the proton-pumping protein complexes of the ETS. After transferring their electrons to the next protein complex, they revert to quinones, capable of accepting electrons again.

Note: *Dehydrogenases, reductases,* and *oxidases* are all **oxidoreductases**, which oxidize one substrate (remove electrons) and reduce another (donate electrons). Oxidoreductases that accept electrons from NADH or $FADH_2$ are also called dehydrogenases, because their reaction releases hydrogen ions.

Oxidoreductase Protein Complexes

A respiratory electron transport system includes at least three functional components: an initial substrate oxidoreductase (or dehydrogenase), a mobile electron carrier, and a terminal oxidase. Microbes make alternative versions of each component using alternative electron-donating substrates and terminal electron acceptors, depending on what is available in the environment. Here we present a typical bacterial ETS receiving electrons from NADH and transferring them to oxygen.

Initial substrate oxidoreductase. A respiratory ETS begins with an initial oxidoreductase that receives a pair of electrons from an organic substrate such as NADH. Note that NADH forms by receiving two electrons plus $2H^+$ from an organic product of catabolism (designated RH_2) (**Fig. 14.12A**). The $2H^+$ are ultimately balanced by $2H^+$ from the cytoplasm combining with O_2 (or another terminal electron acceptor) at the end of the ETS. The two electrons ($2e^-$) from NADH enter an ETS protein complex embedded in the membrane.

NADH donates electrons to NADH dehydrogenase (NADH:quinone oxidoreductase, NDH-1) (**Fig. 14.12A**). In bacteria, the NDH-1 complex includes 14 different subunits; in mitochondria there are 46, forming one of the most elaborate membrane complexes known. A bacterial NDH-1 complex includes the cofactor FMN, as well as several iron-sulfur clusters—typically 7[4Fe-4S] and 2[2Fe-2S]. The cofactors "hand off" electrons to each other through adjacent connections; see, for example, the placement of FMN and the first [4Fe-4S] within the peptide coils of NDH-1 (**Fig. 14.12B**). Each electron from NADH travels through FMN and the iron-sulfur series. At the end of the chain, the electrons and $2H^+$ from solution are transferred to a quinone, which is thus reduced to a quinol. Quinones are designated Q; and quinols, QH_2.

Within the NDH-1 complex, the oxidation of NADH and reduction of Q to QH_2 yields energy to pump up to $4H^+$ across the membrane. A crystallographic model shows four apparent proton channels through transmembrane alpha helices of the protein subunits (**Fig. 14.12A**). Within each channel, a hydrogen ion hops along a series of amino acid residues. The H^+ translocation is driven by a conformational change in the alpha helices throughout the protein, arising from the initial two-electron reduction by NADH. The hydrogen ions pumped across the membrane contribute to the proton potential Δp. In human mitochondria, the NADH dehydrogenase (aka complex I) is critical for health; genetic defects in complex I are associated with diseases such as Parkinson's disease and some forms of diabetes.

Note that the $4H^+$ pumped across the membrane are distinct from the $2H^+$ acquired by the quinone ($Q \rightarrow QH_2$). In some halophilic bacteria, $4Na^+$ are pumped instead of $4H^+$.

Note: In our figures, protons that cross the membrane by the end of the ETS (and thus contribute to Δp) are highlighted yellow.

Not all substrate oxidoreductases pump protons. For example, *E. coli* has an alternative NADH dehydrogenase (NDH-2) that transfers two electrons to Q without pumping additional protons across the membrane. (The unused energy is lost as heat.) NDH-2 functions during rapid growth, when the cell must limit its proton potential to avoid membrane breakdown. Other complexes, such as succinate dehydrogenase, transfer electrons from substrates in a reaction that lacks sufficient energy to pump extra protons.

Quinone pool. A quinone can receive $2e^-$ from the substrate oxidoreductase, along with $2H^+$ from solution, to balance the negative charges, yielding a quinol (**Fig. 14.12A**). The quinols diffuse within the membrane and carry reduction energy to other ETS components. After transferring $2e^-$ to the next protein complex, $2H^+$ are released. Usually the $2H^+$ released are on the opposite side of the membrane from where $2H^+$ were originally picked up (**Fig. 14.13**). Thus, besides electron transfer, a quinol may contribute two protons to the transmembrane proton potential. The reoxidized carriers then recycle back as quinones.

Each quinone can bind to a substrate dehydrogenase, pick up a pair of electrons and hydrogen ions, and then diffuse away and carry the electrons to a reductase. The quinones and quinols, referred to as the **quinone pool**, diffuse freely within the membrane. Thus, the quinones/quinols are able to transfer electrons between many different redox enzymes.

Different oxidoreductase complexes interact with slightly different quinones, such as ubiquinone and menaquinone. The reduction potentials of ubiquinone and menaquinone are given in **Table 14.1**. For clarity, this chapter refers to all as quinones (Q), and their reduced forms as quinols (QH_2).

Terminal oxidase. A terminal oxidase complex receives electrons from a quinol (QH_2) and transfers them to a terminal electron acceptor, such as O_2 (**Fig. 14.13**). The complex usually includes a cytochrome that accepts electrons from quinols. Cytochromes of comparable function are designated by letters, such as cytochrome *b* (*E. coli* and

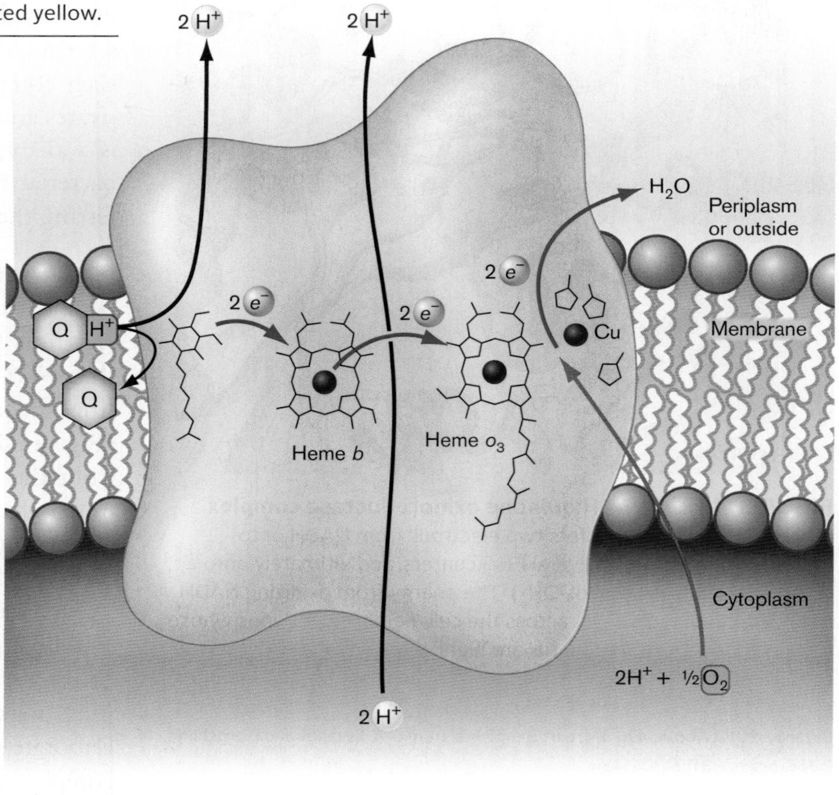

2 H+
2 H+
H2O
Periplasm or outside
2 e−
2 e−
2 e−
Q
H+
Q
Cu
Membrane
Heme b
Heme o3
Cytoplasm
2H+ + ½O2
2 H+

FIGURE 14.13 ▪ Cytochrome *bo* quinol oxidase complex. Each quinol (QH_2) transfers two electrons to heme *b*. The two H^+ from each quinol are expelled to the periplasm (or outside the bacterial cell). The $2e^-$ from the quinols are transferred to heme o_3—a step coupled to pumping of $2H^+$ across the membrane. At heme o_3, $2e^-$ combines with $2H^+$ plus an oxygen atom from O_2 to form water. *Source:* Based on the structure determined by Jeff Abramson et al. 2000. *Nat. Struct. Biol.* **7**:910–917.

mitochondria) and cytochrome *c* (mitochondria). The cytochrome is bound to an oxidase complex containing a series of electron-transferring carriers: two iron-centered hemes and three copper atoms. This unique center couples electron transfer and proton pumping.

The *E. coli* cytochrome *bo* quinol oxidase consists of cytochrome *b* plus oxidase complex *o*. The cytochrome *b* subunit receives two electrons from a quinol ($QH_2 \rightarrow Q$) and releases the $2H^+$ out to the periplasm. Each electron from quinol travels through the two hemes of the oxidase complex. Because the two quinol hydrogens originated from $2H^+$ in the cytoplasm (see **Fig. 14.12**) and $2H^+$ were released outside by a quinol, there is a net increase of $2H^+$ outside the cell. In addition, the transfer of $2e^-$ between the two oxidase hemes is coupled to pumping of $2H^+$ from the cytoplasm across to the periplasm (**Fig. 14.13**).

The second heme of the oxidase (heme o_3) acts to reduce an atom of oxygen from O_2. Each oxygen atom receives two electrons and combines with two protons ($2H^+$) from the cytoplasm to form H_2O. The $2H^+$ consumed balances the $2H^+$ released by catabolism to make NADH + H^+.

Note that the oxidase is conventionally shown as obtaining two electrons from the cytoplasm and donating them to an atom of an oxygen molecule ($\frac{1}{2}O_2$). A full reaction cycle of cytochrome *bo* quinol oxidase actually puts four electrons from two quinols (originally 2 NADH) onto O_2, taking up $4H^+$ from the cytoplasm to make two molecules of H_2O.

Besides cytochrome *bo* quinol oxidase, bacteria express different terminal oxidases that differ with respect to the ratio of cytoplasmic protons pumped to electrons transferred. For example, the alternative cytochrome *bd* oxidase reduces O_2 to water but pumps no extra protons. Although it pumps no protons, cytochrome *bd* oxidase can bind O_2 at much lower concentrations and donate electrons, completing the respiratory circuit. Thus, the *bd* oxidase enables *E. coli* to respire within low-oxygen habitats such as the mammalian intestine.

ETS Pathways for Organotrophy

As we have seen, an ETS for respiration on organic substrates includes at least three phases of electron transfer: (1) Electrons from an organic substrate are donated to an initial oxidoreductase; (2) the electrons are transferred to a quinone, which is reduced to a quinol; and (3) the quinol electrons are transferred to a terminal oxidase while its $2H^+$ are released outside. Both enzymes and quinones show considerable complexity and diversity among different species in different environments. **Figure 14.14** ▶ summarizes one example of a complete ETS for oxidation of NADH by $\frac{1}{2}O_2$ in the inner membrane of *E. coli*.

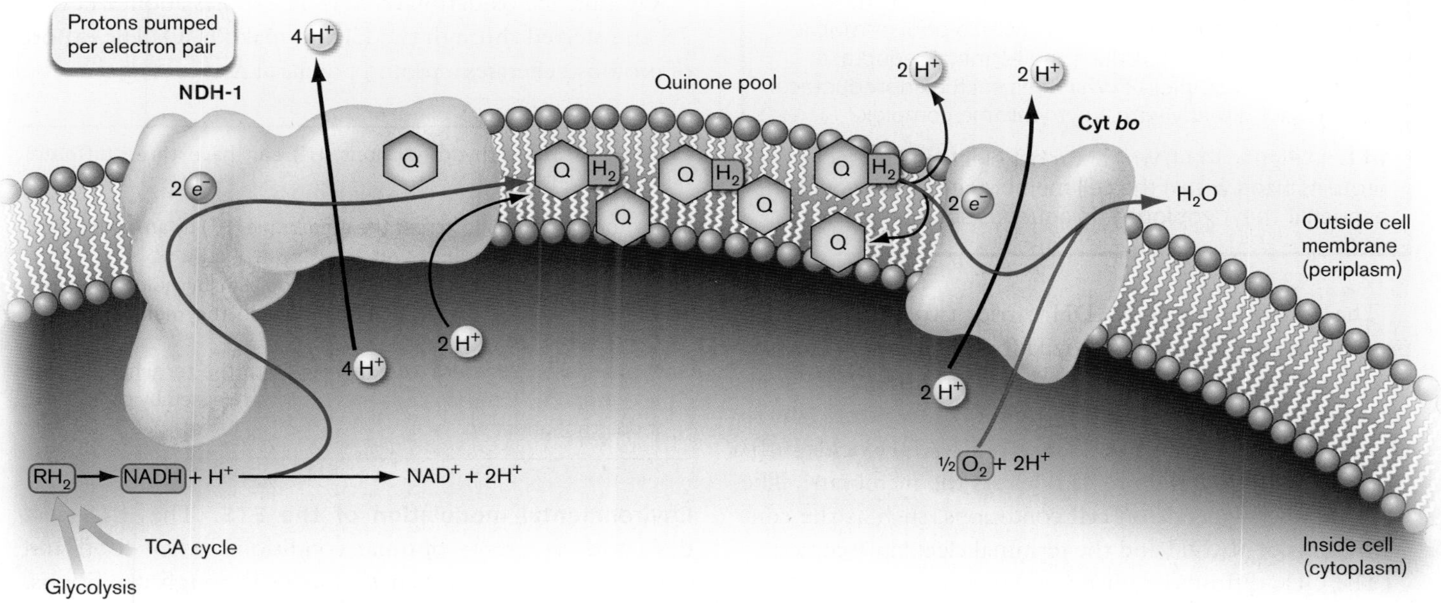
Protons pumped per electron pair
NDH-1
4 H+
Quinone pool
2 H+
2 H+
Cyt bo
Q
QH2
2e−
H2O
Outside cell membrane (periplasm)
2 H+
4 H+
2 H+
RH2 → NADH + H+ → NAD+ + 2H+
½O2 + 2H+
TCA cycle
Glycolysis
Inside cell (cytoplasm)

FIGURE 14.14 ▪ **A bacterial ETS for aerobic NADH oxidation.** In *E. coli*, electrons from NDH-1 are transferred to quinones, generating quinols, which transfer electrons onto cytochrome *bo* (Cyt *bo*). For each NADH oxidized, up to $8H^+$ may be pumped across the membrane. ▶

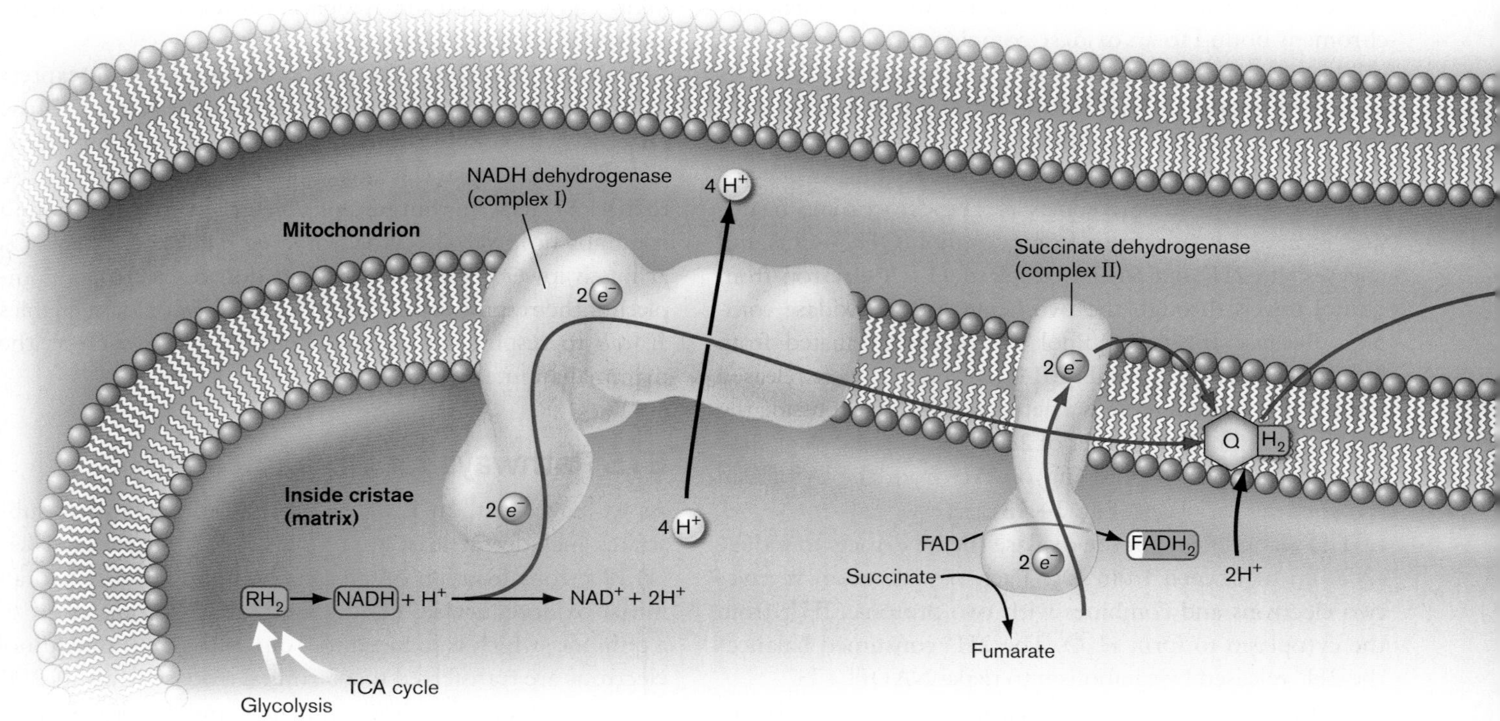

NADH dehydrogenase (complex I)
4 H+
Mitochondrion
Succinate dehydrogenase (complex II)
2 e−
2 e−
Q H2
Inside cristae (matrix)
2 e−
4 H+
FAD
FADH2
2 e−
2H+
Succinate
RH2
NADH + H+
NAD+ + 2H+
Fumarate
TCA cycle
Glycolysis

FIGURE 14.15 ■ Mitochondrial electron transport. In addition to NADH dehydrogenase and a terminal cytochrome oxidase, mitochondria possess ubiquinol:cytochrome *c* oxidoreductase, which provides an intermediate electron transfer step. As a result, mitochondrial membranes export 10–12 H^+ per NADH.

Thought Question

14.5 In Figure 14.14, what is the advantage of the oxidoreductase transferring electrons to a pool of mobile quinones, which then reduce the terminal reductase (cytochrome complex)? Why does each oxidoreductase not interact directly with a cytochrome complex?

14.6 In Figure 14.14, why are most electron transport proteins fixed within the cell membrane? What would happen if they "got loose" in aqueous solution?

The entering carrier NADH carries two electrons with protons obtained from catabolized food molecules. The two electrons ($2e^-$) from NDH-1 and two protons ($2H^+$) from solution are transferred onto Q (quinone), converting it to QH_2 (quinol). The transfer of $2e^-$ from NADH yields sufficient energy to pump up to $4H^+$ across the membrane. The exact number depends on cell conditions, such as the concentrations of NADH and the terminal electron acceptor.

The QH_2 diffuses within the membrane until it reaches a terminal oxidase complex, such as the cytochrome *bo* quinol oxidase. The $2H^+$ from QH_2 are released outside the cell, while the $2e^-$ enter the oxidase, reducing the two hemes. The two electrons join $2H^+$ from the cytoplasm, combining with an oxygen atom to make H_2O. The reaction is coupled to pumping of $2H^+$ across the membrane plus a net increase of $2H^+$ outside through redox reactions.

Overall, the oxidation of NADH exports about $8H^+$ per $2e^-$ transferred through the ETS to make H_2O. The export of protons generates a proton potential Δp.

Note: Protons (hydrogen ions, H^+) can have three different fates in the ETS:

1. Protons are pumped across the membrane (H^+) by an oxidoreductase complex. Contributes Δp.
2. Protons are consumed from the cytoplasm by quinone/quinol, while other protons are released outside the membrane (H^+). Contributes Δp.
3. Protons are consumed by combining with the terminal electron acceptor (O_2). If the loss balances protons released by catabolism, it does not affect Δp.

Environmental modulation of the ETS. The ETS just described represents optimal conditions, when food and oxygen are unlimited. What happens in tough conditions, when food (that is, electron donors) or oxygen is scarce? Under varying environmental conditions, bacteria adjust the efficiency of their ETS by expressing alternative oxidoreductases. For example, at low concentrations of oxygen (microaerophilic conditions, discussed in Chapter 5)

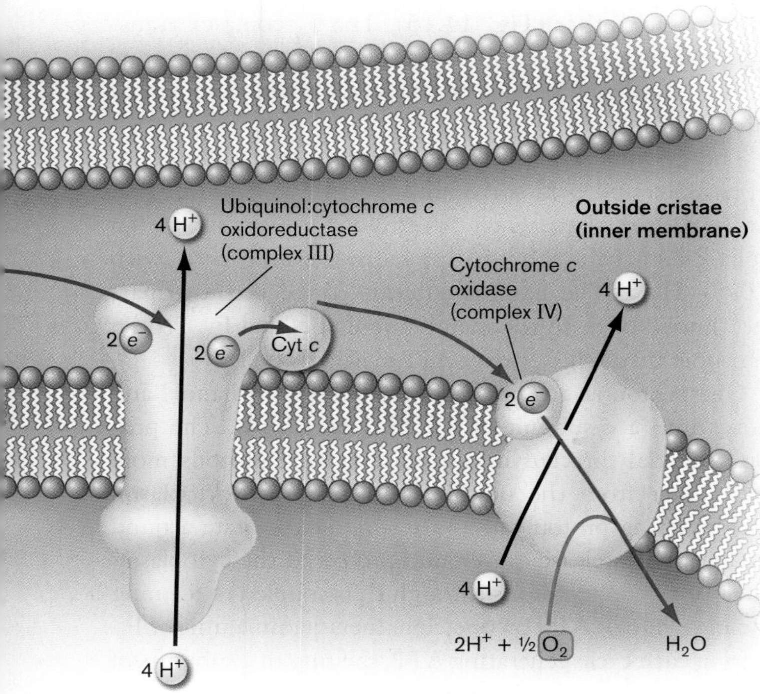

Ubiquinol:cytochrome c oxidoreductase (complex III)
Outside cristae (inner membrane)
Cytochrome c oxidase (complex IV)
4 H+
2 e−
Cyt c
2H+ + ½ O2
H2O

the reduction potential *E* is decreased. So the ETS may be unable to reduce O_2 to H_2O while pumping four protons. Instead, *E. coli* uses a different cytochrome quinol oxidase that has higher affinity for oxygen but pumps fewer protons. Thus, the bacteria will gain less energy, but they will still be able to grow. Environmental regulation of ETS is discussed further in **eTopic 14.3**.

Bacteria also have alternative oxidoreductases to serve different electron donors and acceptors. Some enzymes take electrons from donors lacking the potential of NADH; for example, succinate dehydrogenase catalyzes the one step of the TCA cycle that yields $FADH_2$, a step that provides not quite enough energy to produce NADH (discussed in Chapter 13). Succinate dehydrogenase is the only TCA enzyme embedded in the membrane as an ETS component. When O_2 concentration is so low as to be thermodynamically unavailable, some bacteria can use other electron acceptors, such as nitrate (anaerobic respiration, discussed in Section 14.4). Yet another alternative is to oxidize inorganic electron donors (lithotrophy, discussed in Section 14.5).

Mitochondrial respiration. In contrast to bacteria, mitochondria have only a single ETS, optimized for a relatively uniform intracellular environment (**Fig. 14.15**). Protected by the eukaryotic cytoplasm, mitochondria do not need to use alternative versions of their ETS to respire under different conditions. Instead, a set of just four electron-carrying complexes has evolved so as to maximize the energy obtained from NADH, minimizing energy lost as heat. This mitochondrial ETS is nearly universal throughout the cells of animals, plants, and most eukaryotic microbes.

The mitochondrial ETS has homologs (proteins encoded by genes with a common ancestor) of bacterial ETS components, including NADH dehydrogenase, succinate dehydrogenase, and cytochrome *c* oxidase. Nevertheless, the mitochondrial ETS differs from that of *E. coli* in these respects:

- **An intermediate cytochrome oxidase complex transfers electrons.** Besides NADH dehydrogenase (complex I) and cytochrome *c* oxidase (complex IV), mitochondria show an intermediate step of electron transfer to ubiquinol:cytochrome *c* oxidoreductase (complex III, shown in **Fig. 14.15**). The intermediate electron transfer step pumps an additional $2H^+$. Another $2e^-$ and $2H^+$ come from succinate dehydrogenase (complex II), which forms $FADH_2$ through the TCA cycle.
- **The mitochondrial ETS pumps more protons per NADH.** As many as 10–12 protons may be pumped per NADH, in contrast to 2–8 protons in *E. coli*.
- **Homologous complexes have numerous extra subunits.** Mitochondria have evolved additional nonhomologous subunits specific to eukaryotes. For example, the homologous subunits of cytochrome *c* oxidoreductase are enveloped by a series of eukaryotic-specific proteins.

The Proton Potential Drives ATP Synthesis

The proton potential drives synthesis of ATP by the membrane ATP synthase, also known as the F_1F_o ATP synthase. The proton-driven synthesis of ATP completes the cycle of oxidative phosphorylation, in which hydrogen ions pumped by the ETS drive phosphorylation of ADP to ATP by the ATP synthase. The same ATP synthase can use the proton potential generated by lithotrophy (Section 14.5) or by phototrophy (Section 14.6). Surprisingly, despite the homology of ATP synthase across all forms of life, the complex is a target for antibiotics. For example, the ATP synthase of *Mycobacterium tuberculosis,* the cause of tuberculosis, is inhibited specifically by bedaquiline, an antibiotic approved in 2012 for patients whose disease resists all other drugs.

The F_1F_o ATP Synthase

The F_1F_o ATP synthase is a protein complex highly conserved in the bacterial cell membrane, the mitochondrial inner membrane, and the chloroplast thylakoid membrane. An elegant molecular machine, the ATP synthase

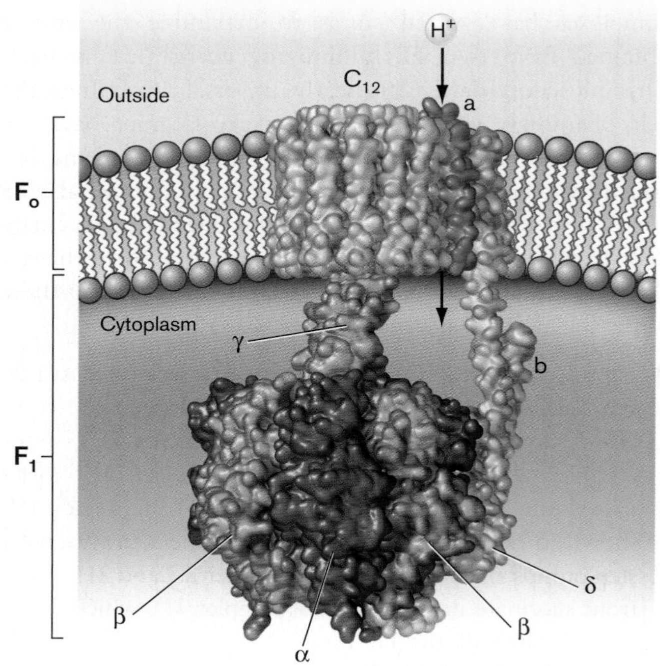
H+
C12
Outside
a
Fo
Cytoplasm
γ
b
F1
δ
β
α
β

FIGURE 14.16 ▪ Bacterial membrane-embedded ATP synthase (F_1F_o ATP synthase). The F_1F_o complex plugged into the bacterial plasma membrane. The three pairs of alpha and beta subunits rotate around the gamma axle, catalyzing formation and hydrolysis of ATP. The ring of c subunits rotates while translocating three protons. (PDB codes: 1B9U, 1C17, 1E79, 2CLY)

is composed of two complexes, F_o and F_1 that rotate relative to each other (**Fig. 14.16**). The F_o complex translocates protons across the membrane. Twelve c subunits form a cylinder embedded in the membrane, stabilized by subunits a and b.

The F_1 complex consists of six alternating subunits of types alpha and beta surrounding a gamma subunit that acts as a drive shaft. Each "third" of the F_1 complex (an alpha plus a beta subunit) interconverts ADP + P_i with ATP + H_2O. The gamma subunit connects the tripartite "knob" of F_1 to the membrane-embedded F_o. Proton transport through F_o drives ATP synthesis by F_1.

One proton at a time enters the subunit-a channel and moves into a c subunit of F_o (**Fig. 14.17** ▶). The proton potential directed inward ensures that protons more often enter from the outside than from the cytoplasm. Each entering proton causes a c subunit to rotate around the axle, with release of a bound proton to the cytoplasm. The flux of three protons through F_o is coupled to forming one molecule of ATP by one alpha-beta-gamma unit of F_1. During each cycle generating ATP, the ring of c subunits of the F_o rotor rotate in the membrane one-third of one turn relative to the F_1 in the cytoplasm.

Note that the generation of ATP is completely reversible, so ATP hydrolysis by F_1 can pump protons back through

FIGURE 14.17 ▪ H^+ flux drives ATP synthesis. **A.** Three protons enter c subunits of the F_o complex. The number of c subunits varies from 8–15 among bacterial species; 12 are shown here. **B.** The ring of c subunits rotates one-third turn relative to the F_1. Flux of three protons through F_o is coupled to F_1 converting ADP + P_i to ATP. ▶

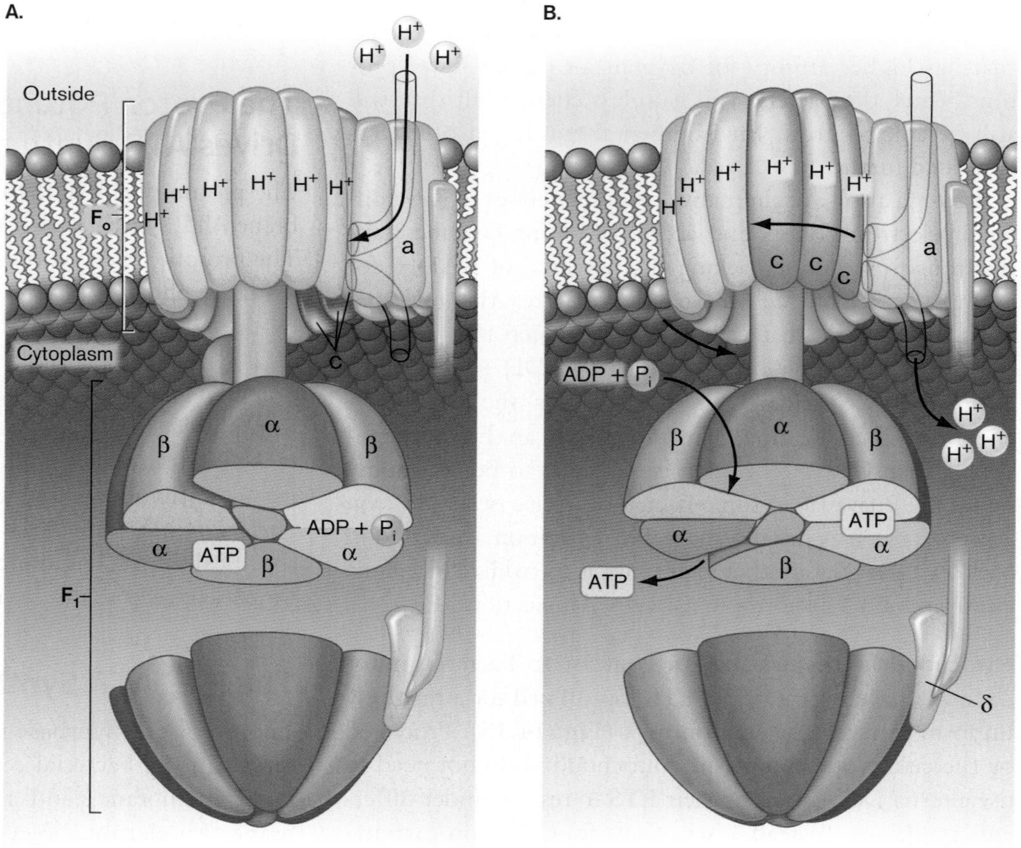
A.
Outside
H+
Fo
a
c
Cytoplasm
β
α
ADP + Pi
ATP
F1
B.
c
ADP + Pi
ATP
δ

F_o across the membrane. In the absence of a proton potential, a high ATP concentration can drive the F_o in reverse, actually pumping protons to generate Δp. This reversal of ATP synthase is used, for example, by *Enterococcus faecalis,* intestinal Gram-positive cocci that generate ATP mainly by fermentation. *E. faecalis* can operate the membrane-embedded F_1F_o ATP synthase in reverse, consuming ATP and thus generating a proton potential for nutrient uptake and ion transport.

How do alkaliphilic bacteria make ATP and maintain a proton potential when growing in alkaline lakes up to pH 11? At high external pH, the large inverted ΔpH would be expected to eliminate the proton potential; nevertheless, alkaliphiles make ATP. Terry Krulwich, at Mount Sinai School of Medicine, investigates the unusual properties of ATP synthase in alkaliphiles (discussed in **eTopic 14.4**).

Thought Question

14.7 Would *E. coli* be able to grow in the presence of an uncoupler that eliminates the proton potential supporting ATP synthesis?

Na^+ Pumps: An Alternative to H^+ Pumps

While a proton potential provides primary energy storage for most species, some bacteria generate an additional potential of sodium ions. A sodium motive force (ΔNa^+) is analogous to the proton motive force in that it includes the electrical potential $\Delta\psi$ plus the sodium ion concentration gradient (log ratio of the Na^+ concentration difference across the membrane). For extreme halophilic archaea, which grow in concentrated NaCl, the sodium potential entirely substitutes for the proton potential to drive ATP synthesis. These "haloarchaea" make use of the high external Na^+ concentration to store energy in the form of a sodium potential.

In some bacteria, an ETS oxidoreductase pumps Na^+ instead of H^+. For example, the proton-pumping NADH dehydrogenase can be supplemented by an NADH dehydrogenase that pumps Na^+ out of the cell. This primary sodium pump is found in many pathogens, including *Vibrio cholerae* (the cause of cholera) and *Yersinia pestis* (the cause of bubonic plague). These pathogens use the sodium-rich circulatory fluids of their hosts to store energy in a sodium potential.

To Summarize

- **Electron carriers** containing metal ions and/or conjugated double-bonded ring structures are used for electron transfer. For example, cytochrome *bo* quinol oxidase has two hemes and three copper ions.
- **A substrate dehydrogenase** receives a pair of electrons from a particular reduced substrate, such as NADH. NADH dehydrogenase (NADH:quinone oxidoreductase) typically has an FMN carrier and nine iron-sulfur clusters.
- **Quinones receive electrons** from the substrate dehydrogenase and become reduced to quinols. Typically, a quinol receives $2H^+$ from the cytoplasm, and then releases $2H^+$ across the membrane upon transfer of $2e^-$ to an electron acceptor complex.
- **Protons are pumped** by substrate dehydrogenases (oxidoreductases) and terminal oxidases. The number of protons pumped by a bacterial ETS is determined by environmental conditions, such as the concentrations of substrate and terminal electron acceptor.
- **Protons are consumed** by combining with the terminal electron acceptor, such as combining with oxygen to make H_2O.
- **The proton potential drives ATP synthesis** through the membrane-embedded F_1F_o ATP synthase. Three protons drive each F_1F_o cycle, synthesizing one molecule of ATP. Some bacteria use a similar ATP synthase driven by Na^+.

14.4

Anaerobic Respiration in Organotrophs

Nearly all multicellular animals and plants require electron transport to oxygen. Likewise, some bacteria are called **obligate aerobes** because they grow only using O_2 as a terminal electron acceptor; examples include important nitrogen-fixing bacteria, such as *Sinorhizobium meliloti* and *Azotobacter vinelandii.* However, other bacteria and archaea use a wide range of terminal electron acceptors, including metals, oxidized ions of nitrogen and sulfur, and chlorinated organic molecules. This **anaerobic respiration** generally occurs in environments where oxygen is scarce, such as in wetland soil and water, and within the human digestive tract.

Electron Acceptors and Donors

An organotroph such as *E. coli* may possess several different terminal oxidoreductases to reduce alternative electron

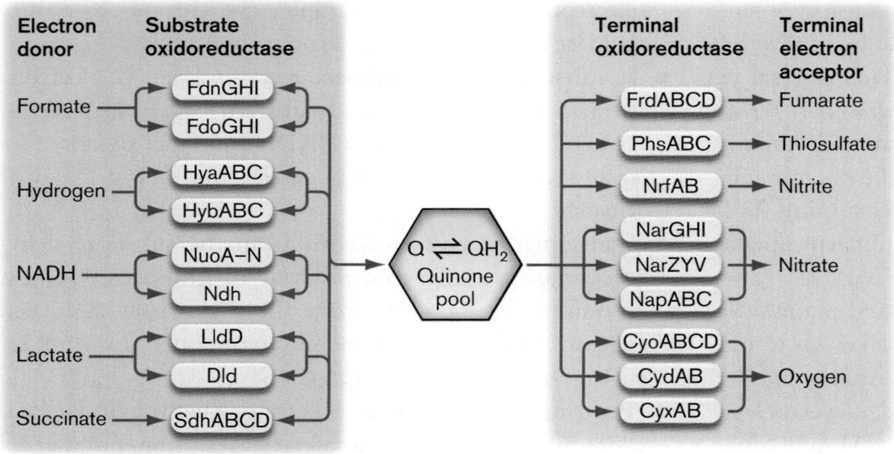
Electron donor
Substrate oxidoreductase
Terminal oxidoreductase
Terminal electron acceptor
Formate
FdnGHI
FdoGHI
Hydrogen
HyaABC
HybABC
NADH
NuoA–N
Ndh
Lactate
LldD
Dld
Succinate
SdhABCD
Q ⇌ QH2
Quinone pool
FrdABCD
Fumarate
PhsABC
Thiosulfate
NrfAB
Nitrite
NarGHI
NarZYV
NapABC
Nitrate
CyoABCD
CydAB
CyxAB
Oxygen

FIGURE 14.18 ▪ Alternative electron donors and electron acceptors. *Escherichia coli* can oxidize various foods (electron donors) while reducing various electron acceptors. Each electron donor may utilize alternative substrate oxidoreductases, depending on the environmental conditions. Each substrate oxidoreductase sends electrons through the quinone pool to the terminal oxidoreductases.

acceptors (**Fig. 14.18**). These enzymes, although comparable to cytochrome quinol oxidases, are conventionally termed "reductases" to emphasize reduction of the alternative electron acceptor. Some of the electron acceptors are inorganic, such as nitrate (NO_3^-) reduced to nitrite (NO_2^-), or NO_2^- reduced to NO (nitric oxide). Others are organic products of catabolism; for example, the TCA cycle intermediate fumarate can be reduced to succinate. Organic electron acceptors play important roles in food decomposition. For example, the substance trimethylamine oxide, used by fish as an osmoprotectant against sea salt, is reduced by bacteria to trimethylamine—the main cause of the "fishy" smell.

At the top of the ETS, alternative dehydrogenases (or substrate oxidoreductases) receive electrons from different organic electron donors, as well as from molecular hydrogen (H_2). The enzymes "connect" to various terminal oxidoreductases through the pool of quinones. Note that the various electron donors differ greatly in their reduction potential and hence in their capacity to generate proton potential (see **Table 14.1**). In a given environment, bacateria use the strongest electron donor and the strongest electron acceptor available. The best donor and best acceptor usually induce the expression of genes encoding their respective redox enzymes. For example, in the presence of nitrate, genes encoding nitrate reductase are expressed. At the same time, nitrate represses the expression of reductases for poorer electron acceptors, such as fumarate. (The mechanisms of gene induction and repression are discussed in Chapter 10.)

Nitrogen and Sulfur: Oxidized Forms

Oxidized forms of nitrogen and sulfur, usually anions such as nitrate or sulfate, can accept electrons from many species of soil and water bacteria. The nitrogen series offers an abundant source of strong electron acceptors. Reduction of oxidized states of nitrogen for energy yield is called **dissimilatory denitrification**. Dissimilatory nitrate reduction to ammonium (DNRA) contributes to respiration, whereas assimilatory reduction of nitrate generates ammonium ion for fixation into biomass (discussed in Chapters 15 and 22). Some pathogens are denitrifiers, such as *Neisseria meningitidis* (a cause of meningitis) and *Brucella* species that cause brucellosis in cattle, sheep, and dogs.

Oxidized forms of nitrogen. In the dissimilatory nitrogen redox series, a given oxidation state can serve as an acceptor in one redox couple but as a donor in the next. The redox couples are summarized here:

$$\underset{\text{Nitrate}}{NO_3^-} \xrightarrow{2e^-} \underset{\text{Nitrite}}{NO_2^-} \xrightarrow{e^-} \underset{\text{Nitric oxide}}{NO} \xrightarrow{e^-} \underset{\text{Nitrous oxide}}{\tfrac{1}{2}N_2O} \xrightarrow{e^-} \underset{\text{Nitrogen gas}}{\tfrac{1}{2}N_2}$$

Each nitrogen state requires a specialized reductase, such as nitrate reductase or nitrite reductase, to receive electrons from the ETS. During the reactions, oxygen atoms are removed and combined with protons to form water. The full series of nitrate reduction to N_2 plays a crucial role in producing the nitrogen gas of Earth's atmosphere (discussed in Chapter 22).

An alternative option for many soil bacteria, such as *Bacillus* species, is to reduce nitrite to ammonium ion (NH_4^+):

$$NO_2^- + 8H^+ + 6e^- \rightarrow NH_4^+ + 2H_2O$$

Dissimilatory nitrate and nitrite reduction to ammonium ion can increase the soil pH. The high pH will precipitate metals such as iron.

The presence of a particular reductase can be used as a diagnostic indicator for clinical isolates of bacteria. For example, the chemical test for nitrate reduction is a key step in the diagnosis of *Neisseria gonorrhoeae*, the causative agent of gonorrhea. *N. gonorrhoeae* happens to lack the terminal reductase for nitrate; hence, it tests negative, whereas several closely related species test positive.

Respiration using oxidized forms of nitrogen is widespread among bacteria and archaea. Most eukaryotes must breathe oxygen, but some eukaryotic microbes respire on nitrate and nitrite. In anaerobic soil, many yeasts and filamentous fungi can reduce nitrate to nitrite, and nitrite to nitrous oxide. In the digestive tracts of termites and of ruminant vertebrates, many fungi and protists grow anaerobically.

Oxidized forms of sulfur. In a similar series, oxidized forms of sulfur serve as electron acceptors for bacteria that synthesize appropriate reductases:

$$\underset{\text{Sulfate}}{SO_4^{2-}} \xrightarrow{2e^-} \underset{\text{Sulfite}}{SO_3^{2-}} \xrightarrow{2e^-} \underset{\text{Thiosulfate}}{\tfrac{1}{2}S_2O_3^{2-}} \xrightarrow{2e^-} \underset{\text{Elemental sulfur}}{S^0} \xrightarrow{2e^-} \underset{\text{Hydrogen sulfide}}{H_2S}$$

Their redox potentials are generally lower than those for oxidized nitrogen forms. Nevertheless, sulfate and sulfite receive electrons from many kinds of electron donors, including acetate, hydrocarbons, and H_2. Sulfate-reducing bacteria and archaea are widespread in the ocean, from the arctic waters to submarine thermal vents. The ubiquity of sulfate reduction may be related to the high sulfate content of seawater, in which SO_4^{2-} is the most common anion after chloride.

Dissimilatory Metal Reduction

An important class of anaerobic respiration involves the reduction of metal cations, or **dissimilatory metal reduction**. The term "dissimilatory" indicates that the metal reduced as a terminal electron acceptor is excluded from the cell. This is in contrast to minerals reduced for the purpose of incorporation into cell components (assimilatory metal reduction). Metal-reducing bacteria offer the intriguing prospect of making electricity in a fuel cell (see **Special Topic 14.2**). The metals most commonly reduced through anaerobic respiration are iron ($Fe^{3+} \rightarrow Fe^{2+}$) and manganese ($Mn^{4+} \rightarrow Mn^{2+}$), but virtually any metal with multiple redox states can be reduced or oxidized by some bacteria.

Metal electron acceptors require a specific oxidoreductase (or reductase). Metal ions in solution interact directly with reductases, but oxidized metals such as Fe^{3+} are often barely soluble in water. How does a bacterial reductase interact with an insoluble source of metal outside the cell? Soil bacteria may "shuttle" electrons to extracellular metals using quinone-like degradation products of lignin, a complex aromatic substance that forms the bulk of wood and woody stems. Lignin degradation products are also called "humics" because of their presence in humus, the organic components of soil (discussed in Chapter 21). Humics accumulate in anaerobic environments, where decomposition is slow. Alternatively, bacteria may contact insoluble metals using surface-bound cytochromes—and even electron-conducting "nanowires" made of protein (**Special Topic 14.2**).

Anoxic or low-oxygen environments, such as the bottom of a lake or wetland sediment, offer a series of different electron acceptors (**Fig. 14.19**). The stronger electron acceptors will be consumed, in turn, by species that have the terminal oxidases to use them. For example, as oxygen grows scarce, nitrate is reduced to nitrogen gas ($NO_3^- \rightarrow N_2$) by denitrifying bacteria. As nitrate is used up, other species reduce manganese ($Mn^{4+} \rightarrow Mn^{2+}$), iron ($Fe^{3+} \rightarrow Fe^{2+}$), and sulfate ($SO_4^{2-} \rightarrow H_2S$). At the bottom of the lake or sediment, carbon dioxide is reduced to methane by methanogenesis. Methanogenesis is conducted by archaea (see Section 14.5). Microbial reduction of minerals plays a critical role in every ecosystem and participates in the geochemical cycling of elements throughout Earth's biosphere (discussed in Chapter 22).

A particularly interesting genus of metal-reducing bacteria is *Geobacter*, which reduces metals such as iron and manganese, as well as uranium ($U^{6+} \rightarrow U^{4+}$). Reduced uranium is insoluble in water, precipitating as uranite (UO_2). *G. metallireducens* can reduce uranium and respire on simple organic substrates such as acetate, providing a way to remediate uranium-contaminated water. The U.S. Department of Energy has used uranium-reducing bacteria for remediation at Rifle, Colorado, where uranium contamination has threatened the Colorado River. In this process, acetate is pumped into the water table, where *G. metallireducens* oxidizes the acetate to CO_2 by reducing U^{6+} to U^{4+}. The reduced uranium then precipitates out of the water into the soil, where it can be collected and removed, while the cleansed water flows through.

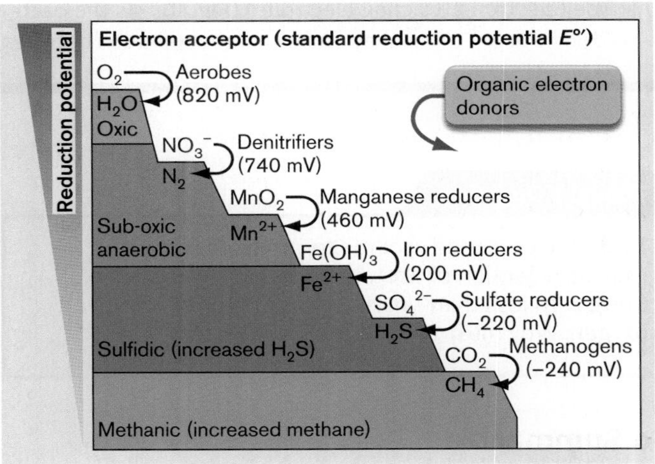

Reduction potential
Electron acceptor (standard reduction potential E°′)
O2
H2O
Aerobes (820 mV)
Oxic
Organic electron donors
NO3−
N2
Denitrifiers (740 mV)
MnO2
Mn2+
Manganese reducers (460 mV)
Sub-oxic anaerobic
Fe(OH)3
Fe2+
Iron reducers (200 mV)
SO4 2−
H2S
Sulfate reducers (−220 mV)
Sulfidic (increased H2S)
CO2
CH4
Methanogens (−240 mV)
Methanic (increased methane)

FIGURE 14.19 ■ Anaerobic respiration in a lake water column. As each successive terminal electron acceptor is used up, its reduced form appears. The next-best electron acceptor is then used, generally by a different species of microbe.

Special Topic 14.2: Bacterial Electric Power

In natural biofilms, bacteria transfer electrons to metals and even between neighboring bacteria. What if we could harness bacterial electron transfer to power our own electrical devices? Kenneth Nealson (**Fig. 1A**) and students at the University of Southern California are trying to do just that.

The concept of bacterial electric power is surprisingly recent. Up to 20 years ago, it was thought that bacteria could reduce only soluble ions such as nitrate. But Nealson, like the famous nineteenth-century microbial ecologist Sergei Winogradsky, was fascinated by the metal transformations he observed in wetlands and lake sediments. Nealson was particularly intrigued by the high levels of reduced manganese (Mn^{2+}) found in Lake Oneida, New York. He reasoned that only microbial activity could account for so much reduced manganese, compared to other lakes where manganese is in the oxidized form (Mn^{4+}, as the insoluble mineral MnO_2). In 1988, Nealson and his colleague Charles Myers discovered a bacterium that could respire anaerobically by donating electrons to MnO_2, releasing soluble Mn^{2+}. The bacterium, named *Shewanella oneidensis* MR-1 (for "manganese reducer") contains a large number of different cytochromes. The cytochromes help the bacterium donate electrons to different kinds of metals, even insoluble forms such as iron and cobalt embedded within clay.

How can bacteria reduce a substance outside their cell envelopes? Nealson and co-workers found evidence for several mechanisms, such as the shuttling of electrons from outer membrane cytochrome complexes, and the extension of electron-conducting "nanowires" made of protein. The nanowires help bacteria form a biofilm that connects them to each other and connects their cytochromes to the metal electron acceptor. In a fuel cell, the bacteria form a biofilm on the anode (electron-attracting electrode) (**Fig. 1B**). As the bacteria oxidize organic fuel, they cause charge separation between electrons and hydrogen ions. The electrons then pass within a current to the cathode.

The "fuel" for the cell can be a mixture of organic substances derived from any kind of food waste or sewage. Organic waste includes small organic molecules such as lactate, acetate, and even formaldehyde. The biofilm bacteria on the anode can remove hydrogens from these organic molecules, separating the electrons and hydrogen ions (**Fig. 1C**). The hydrogen ions migrate through a polymer membrane, whereas the electrons enter the anode leading to an electrical wire. The remaining carbon and oxygen atoms of the fuel are released as CO_2. The process is similar to natural respiration, except that instead of molecular oxygen, the electron acceptor is an electrode made of graphite. To complete the circuit, the electrons from the wire current ultimately react with oxygen and hydrogen ions to form water, as in aerobic respiration.

So far, microbial fuel cells have been able to generate milliamps of current, enough to drive small devices, such as clocks and marine data sensors. Microbial fuel cells offer a clean way to make electricity, with inexpensive fuel. The main challenge is generating larger currents. Many researchers are working to ramp up the currents and make the dream of bacterial electricity a reality.

RESEARCH QUESTION

How might you test the question of whether bacteria require protein nanowires to generate electricity in a fuel cell?

El-Naggar, Mohamed Y., Greg Wanger, Kar Man Leung, Thomas D. Yuvzinsky, Gordon Southam, et al. 2010. Electrical transport along bacterial nanowires from *Shewanella oneidensis* MR-1. *Proceedings of the National Academy of Sciences USA* **107**:18127.

Thought Question

14.8 The proposed scheme for uranium removal requires injection of acetate under highly anoxic conditions, with less than 1 part per million (ppm) dissolved oxygen. Why must the acetate be anoxic?

To Summarize

- **Anaerobic terminal electron acceptors**, such as nitrogen and sulfur oxyanions, oxidized metal cations, and oxidized organic substrates, accept electrons from a specific reductase complex of an ETS.
- **Nitrate** is successively reduced by bacteria to nitrite, nitric oxide, nitrous oxide, and ultimately nitrogen gas. Alternatively, nitrate and nitrite may be reduced to ammonium ion, a product that alkalinizes the environment.
- **Sulfate** is successively reduced by bacteria to sulfite, thiosulfate, elemental sulfur, and hydrogen sulfide. Sulfate reducers are especially prevalent in seawater.
- **Oxidized metal ions** such as Fe^{3+} and Mn^{4+} are reduced by bacteria in soil and aquatic habitats—a form of anaerobic respiration also known as dissimilatory metal reduction.
- **The geochemistry of natural environments** is shaped largely by anaerobic bacteria and archaea.

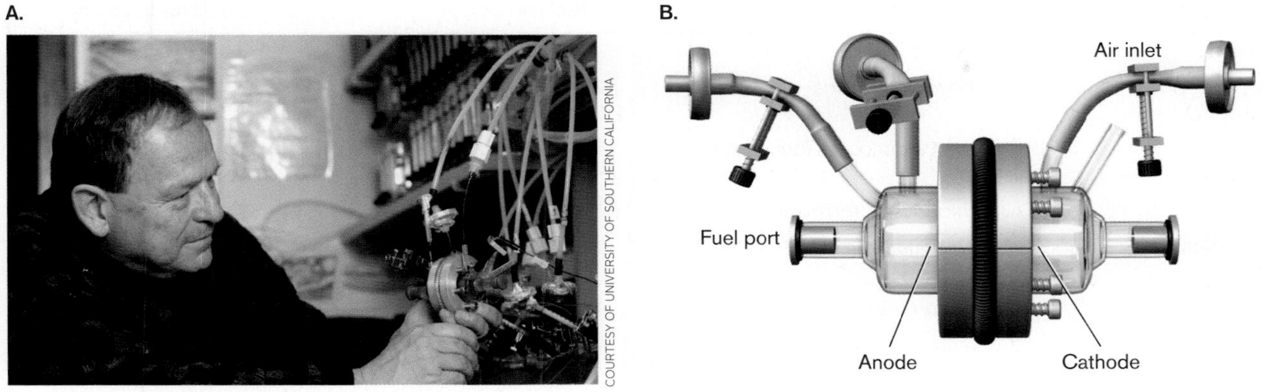
A.
COURTESY OF UNIVERSITY OF SOUTHERN CALIFORNIA
B.
Air inlet
Fuel port
Anode
Cathode

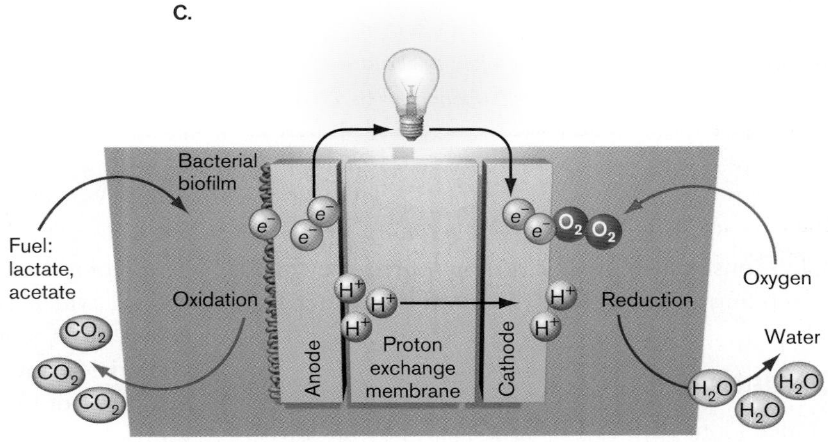
C.
Bacterial biofilm
Fuel: lactate, acetate
Oxidation
CO_2
CO_2
CO_2
e^-
H^+
Anode
Proton exchange membrane
Cathode
O_2
Oxygen
Reduction
Water
H_2O

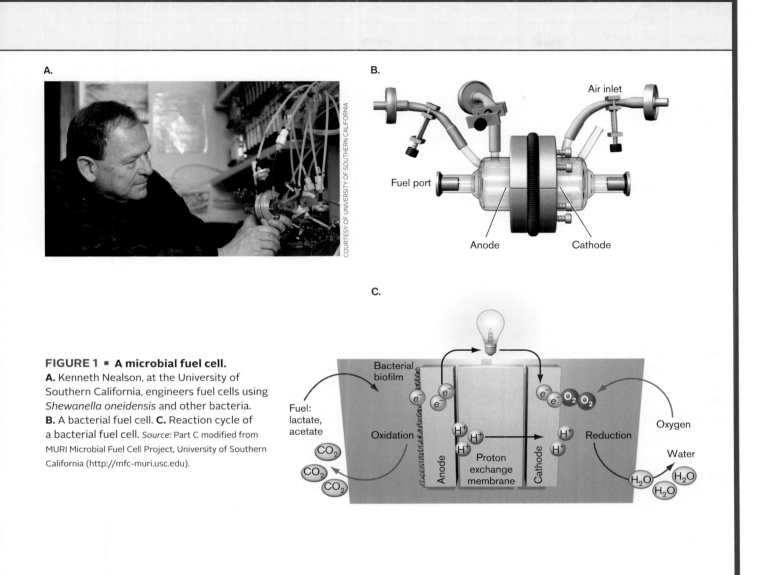

FIGURE 1 ▪ A microbial fuel cell. **A.** Kenneth Nealson, at the University of Southern California, engineers fuel cells using *Shewanella oneidensis* and other bacteria. **B.** A bacterial fuel cell. **C.** Reaction cycle of a bacterial fuel cell. *Source:* Part C modified from MURI Microbial Fuel Cell Project, University of Southern California (http://mfc-muri.usc.edu).

14.5

Lithotrophy and Methanogenesis

Many reduced minerals and single-carbon compounds can serve as electron donors for an ETS, in the energy-yielding form of metabolism known as **lithotrophy** (or **chemolithotrophy**). Some organotrophs have alternative oxidoreductases that conduct lithotrophy by oxidizing H_2 or Fe^{2+}; for example, the gastric pathogen *Helicobacter pylori* oxidizes hydrogen gas released by mixed-acid-fermenting bacteria in the colon. Other bacteria are "obligate lithotrophs"; that is, they oxidize only inorganic molecules. Nearly all lithotrophs are bacteria or archaea. Thus, bacteria and archaea fill many key niches in ecosystems that eukaryotes cannot.

Lithotrophy includes many kinds of electron donors, from metals and anions to single-carbon groups (**Table 14.2**). Each type of electron donor, such as H_2, NH_4^+, or Fe^{2+}, requires a specialized electron-accepting oxidoreductase. Most inorganic substrates other than H_2 are relatively poor electron donors compared to organic donors such as glucose. Therefore, the terminal electron acceptor is usually a strong oxidant, such as O_2, NO_3^-, or Fe^{3+}. Obligate lithotrophs (organisms that conduct only

TABLE 14.2

Lithotrophy: electron donors and acceptors.

Type of lithotrophy	Species example	Electron donor	Electron acceptor
Hydrogenotrophy	*Aquifex aeolicus*	$H_2 \rightarrow 2H^+ + 2e^-$	$O_2 \rightarrow H_2O$
Sulfate reduction (hydrogenotrophy)	*Desulfovibrio vulgaris*	$H_2 \rightarrow 2H^+ + 2e^-$	$SO_4^{2-} \rightarrow S^O, HS^-$
Methanogenesis (hydrogenotrophy)	*Methanocaldococcus jannaschii*	$H_2 \rightarrow 2H^+ + 2e^-$	$CO_2 \rightarrow CH_4$
Iron oxidation	*Acidithiobacillus ferrooxidans*	$Fe^{2+} \rightarrow Fe^{3+} + e^-$	$O_2 \rightarrow H_2O$
Ammonia oxidation (nitrosification)	*Nitrosomonas europaea*	$NH_3 \rightarrow NO_2^-$	$O_2 \rightarrow H_2O$
Nitrification	*Nitrobacter winogradskyi*	$NO_2^- \rightarrow NO_3^-$	$O_2 \rightarrow H_2O$
Anammox	"*Candidatus* Kuenenia stuttgartiensis"	$NH_3 + 4H^+ + 4e- \rightarrow N_2$	$NO_2 \rightarrow N_2$
Methylotrophy	*Methylococcus capsulatus*	$CH_4 \rightarrow CO_2 + 4H^+ + 4e^-$	$O_2 \rightarrow H_2O$
Carboxidotrophy	*Carboxydothermus hydrogenoformans*	$CO \rightarrow CO_2 + e^-$	$H_2O \rightarrow H_2$
Sulfide oxidation	*Sulfolobus solfataricus*	$HS^- \rightarrow S^O + H^+ + 2e^-$	$O_2 \rightarrow H_2O$
Sulfur oxidation	*Acidithiobacillus thiooxidans*	$S^O + 2H^+ + 4e^- \rightarrow H_2SO_4$	$O_2 \rightarrow H_2O$

lithotrophy) consume no organic carbon source; they build biomass by fixing CO_2 (a process discussed in Chapter 15).

Iron Oxidation

Reduced metal ions such as Fe^{2+} and Mn^{2+} provide energy through oxidation by O_2 or NO_3^-. Bacteria perform these lithotrophic reactions in soil where weathering exposes reduced minerals. They generate metal ions with higher oxidation states (such as Fe^{3+} or Mn^{4+}), which other bacteria use for anaerobic respiration. Environments such as ponds and wetlands that experience frequent shifts between oxygen availability and oxygen depletion are likely to host a variety of metal-oxidizing lithotrophs, as well as metal-reducing anaerobic heterotrophs. The roles of lithotrophy in ecology are discussed further in Chapters 21 and 22.

An example of lithotrophy is iron oxidation by the bacterium *Acidithiobacillus ferrooxidans* (**Fig. 14.20**). These bacteria oxidize Fe^{2+} (ferrous ion) using the reduction potential between Fe^{3+}/Fe^{2+} and O_2/H_2O:

	$E^{o\prime}$ (pH 2)	$G^{o\prime}$ (pH 2)
$2Fe^{2+} \rightarrow 2Fe^{3+} + 2e^-$ (pH 2)	–(+770) mV	+149 kJ/mol
$\frac{1}{2}O_2 + 2H^+ + 2e^- \rightarrow H_2O$ (pH 2)	+1,100 mV	–212 kJ/mol
$2Fe^{2+} + \frac{1}{2}O_2 + 2H^+ \rightarrow 2Fe^{3+} + H_2O$	+330 mV	–63 kJ/mol

The removal of electrons from Fe^{2+} requires an input of energy, which is compensated for by the larger yield of energy from reducing oxygen to water. The net reduction potential is small; thus, *A. ferrooxidans* must cycle large quantities of iron in order to grow. The reaction goes forward only at low pH, such as that of acid mine drainage where other microbes oxidize sulfur to sulfuric acid. The bacterial cell must keep its cytoplasmic pH considerably higher (pH 6.5) by transmembrane exchange of H^+ with cations, and inverting the electrical potential $\Delta\psi$ (positive inside). Thus, the proton potential Δp exists entirely in the form of ΔpH.

Unlike organic electron donors, metals such as iron must be accessed from insoluble particles outside the cell. In *A. ferrooxidans*, the electrons are collected from iron outside by an outer membrane cytochrome c_2. The cytochrome is associated with a periplasmic protein called rusticyanin. Rusticyanin collects electrons while excluding the potentially toxic metal from the cell. The electrons from rusticyanin can be transferred to an inner membrane cytochrome complex that reduces oxygen to water.

For iron oxidation, the ΔG of reaction is too small to pump protons. So how do cells obtain enough energy to synthesize ATP or to form NADPH for biosynthesis? An alternative pathway directs electrons to an NADPH oxidoreductase, which generates NADPH for carbon dioxide

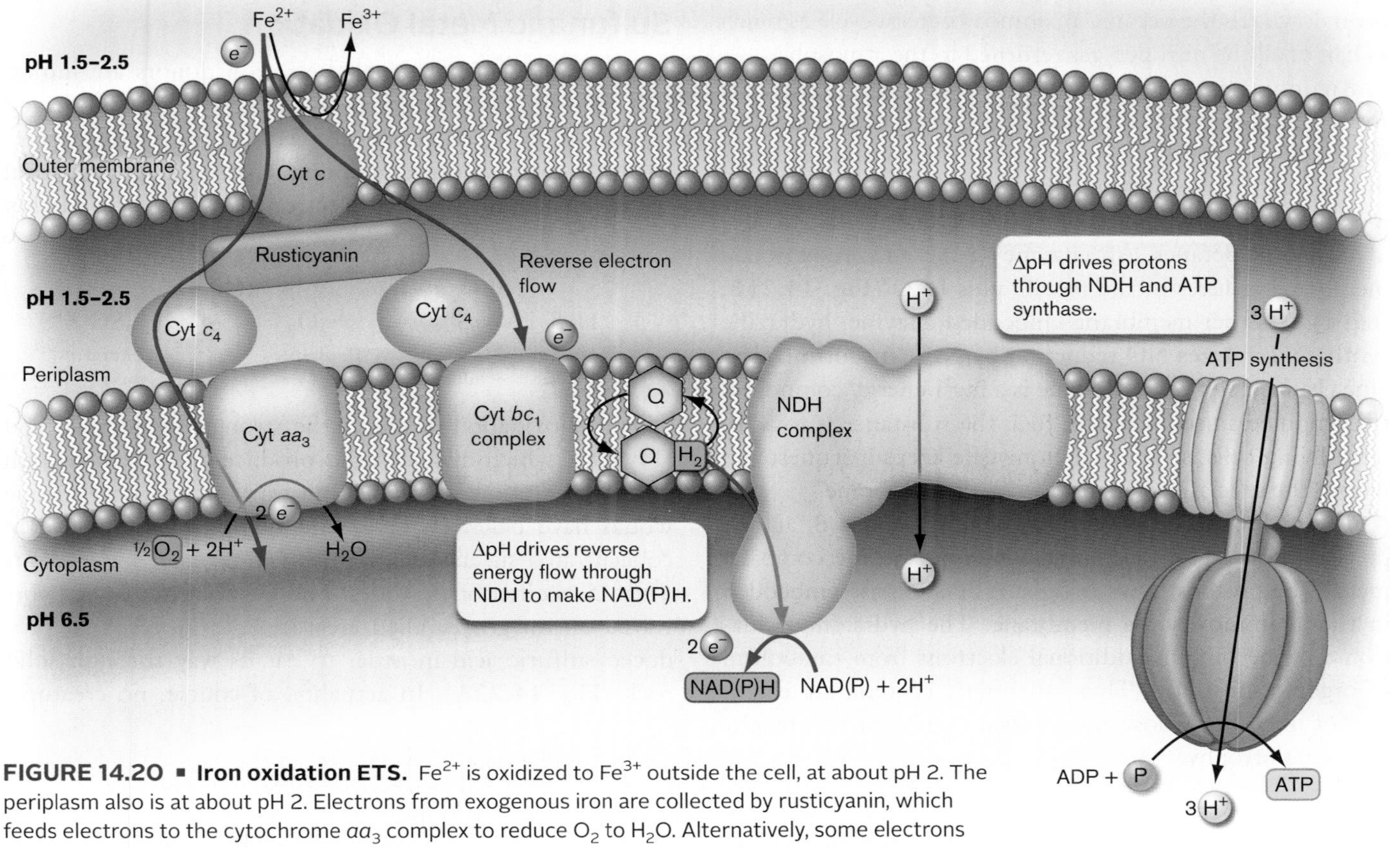

pH 1.5–2.5
Fe^{2+}
Fe^{3+}
e^-
Outer membrane
Cyt c
Rusticyanin
pH 1.5–2.5
Cyt c_4
Cyt c_4
Reverse electron flow
e^-
Periplasm
Cyt aa_3
Cyt bc_1 complex
Q
Q H_2
NDH complex
H^+
ΔpH drives protons through NDH and ATP synthase.
3 H^+
ATP synthesis
2 e^-
½O_2 + $2H^+$
H_2O
Cytoplasm
pH 6.5
ΔpH drives reverse energy flow through NDH to make NAD(P)H.
H^+
2 e^-
NAD(P)H
NAD(P) + $2H^+$
ADP + P
ATP
3 H^+

FIGURE 14.20 ■ Iron oxidation ETS. Fe^{2+} is oxidized to Fe^{3+} outside the cell, at about pH 2. The periplasm also is at about pH 2. Electrons from exogenous iron are collected by rusticyanin, which feeds electrons to the cytochrome aa_3 complex to reduce O_2 to H_2O. Alternatively, some electrons from rusticyanin reduce the cytochrome bc_1 complex. ΔpH drives reverse electron flow through NADPH oxidoreductase to form NADPH. Modified from Raquel Quatrini et al. 2009. *BMC Genomics* **10**:394.

fixation and biosynthesis. The pathway requires **reverse electron flow**. In reverse electron flow, an electron donor reduces an ETS with an unfavorable reduction potential, requiring input of energy. The energy in this case comes from the large ΔpH across the inner membrane (pH 6.5 inside, and pH 2 outside). Reverse electron flow occurs in some kinds of lithotrophy, in syntrophy (discussed in Chapter 13), and in phototrophy (discussed in Section 14.6).

Nitrogen Oxidation

A kind of lithotrophy essential for the environment is oxidation of nitrogen compounds:

$$\underset{\text{Ammonium}}{NH_4^+} \xrightarrow{\frac{1}{2}O_2} \underset{\text{Hydroxylamine}}{NH_2OH} \xrightarrow{O_2} \underset{\text{Nitr\textit{ous} acid (nitrite)}}{HNO_2} \xrightarrow{\frac{1}{2}O_2} \underset{\text{Nitr\textit{ic} acid (nitrate)}}{HNO_3}$$

Reduced forms of nitrogen, such as ammonium ion derived from fertilizers, support growth of **nitrifiers**, bacteria that generate nitrites or nitrates (forming nitrous acid or nitric acid, respectively). Acid production can degrade environmental quality. In soil treated with artificial fertilizers, nitrifiers decrease the ammonium ions obtained from such fertilizer and produce toxic concentrations of nitrites, which leach into groundwater. Still, nitrifiers can be useful in sewage treatment, where they eliminate ammonia that would harm aquatic life (discussed in Chapter 22).

Ammonia/ammonium and nitrite are relatively poor electron donors compared to organic molecules. Thus, their oxidation through the ETS pumps fewer protons, and the bacteria must cycle relatively large quantities of substrates to grow. As we saw for iron oxidation, reduced cofactors such as NADPH must be obtained through reverse electron flow involving redox carriers with different amounts of energy.

What happens to ammonium ion from detritus that accumulates in anaerobic regions, such as the bottom of a lake? Surprisingly, NH_4^+ can yield energy through oxidation by nitr<u>i</u>te produced from nitr<u>a</u>te respiration:

$$NH_4^+ + NO_2^- \rightarrow N_2 + 2H_2O \qquad \Delta G^{\circ\prime} = -357 \text{ kJ/mol}$$

Under conditions of high ammonium and extremely low oxygen, this metabolism supports ample growth of bacteria. Known as the **anammox reaction**, anaerobic ammonium oxidation plays a major role in wastewater treatment, where it eliminates much of the ammonium ion from sewage breakdown. In the oceans, anammox bacteria cycle as much as half of all the nitrogen gas returned to the atmosphere.

Anammox is conducted by planctomycetes, irregularly shaped bacteria with unusual membranous organelles that fill much of the cell (**Fig. 14.21A**). The central compartment is called the anammoxosome. The anammoxosome membrane is composed of unusual ladder-shaped lipids called ladderanes. An enzyme of the anammoxosomal membrane reduces nitrite to NO plus H_2O (**Fig. 14.21B**; step 1). Another membrane-embedded enzyme, hydrazine synthase, catalyzes NO reduction by NH_4^+ to form hydrazine (N_2H_4); step 2. Hydrazine is a high-energy compound that engineers use for rocket fuel; the substance is toxic to most living cells, so the planctomycete keeps it sequestered within the specialized anammoxosome membrane.

The hydrazine is further oxidized to N_2; step 3, and its protons are released within the anammoxosome compartment. The protons then drive an ATP synthase embedded in the anammoxosome membrane. The hydrazine oxidation enzyme obtains additional electrons from catabolism of organic substrates. Thus, anammox bacteria are examples of microbes whose metabolism combines lithotrophy and organotrophy.

Sulfur and Metal Oxidation

Major sources of lithotrophic electron donors are minerals containing reduced sulfur, such as hydrogen sulfide and sulfides of iron and copper. As we saw for nitrogen compounds, each sulfur compound that undergoes partial oxidation may serve as an electron donor and be further oxidized.

$$\underset{\text{Hydrogen sulfide}}{H_2S} \xrightarrow{\frac{1}{2}O_2} \underset{\text{Elemental sulfur}}{S^0} \xrightarrow{\frac{1}{2}O_2} \underset{\text{Thiosulfate}}{\tfrac{1}{2}S_2O_3^{2-}} \xrightarrow{O_2 + H_2O} \underset{\text{Sulfuric acid}}{H_2SO_4}$$

Sulfur oxidation produces the strong acid sulfuric acid (H_2SO_4), which dissociates to produce an extremely high H^+ concentration. In the early twentieth century, no one would have believed that living organisms could live in concentrated sulfuric acid, much less produce it. The *Star Trek* science fiction episode "The Devil in the Dark" portrayed an imaginary alien creature, the Horta, that produced sulfuric acid in order to eat its way through solid rock (**Fig. 14.22A**). In actuality, of course, no creatures

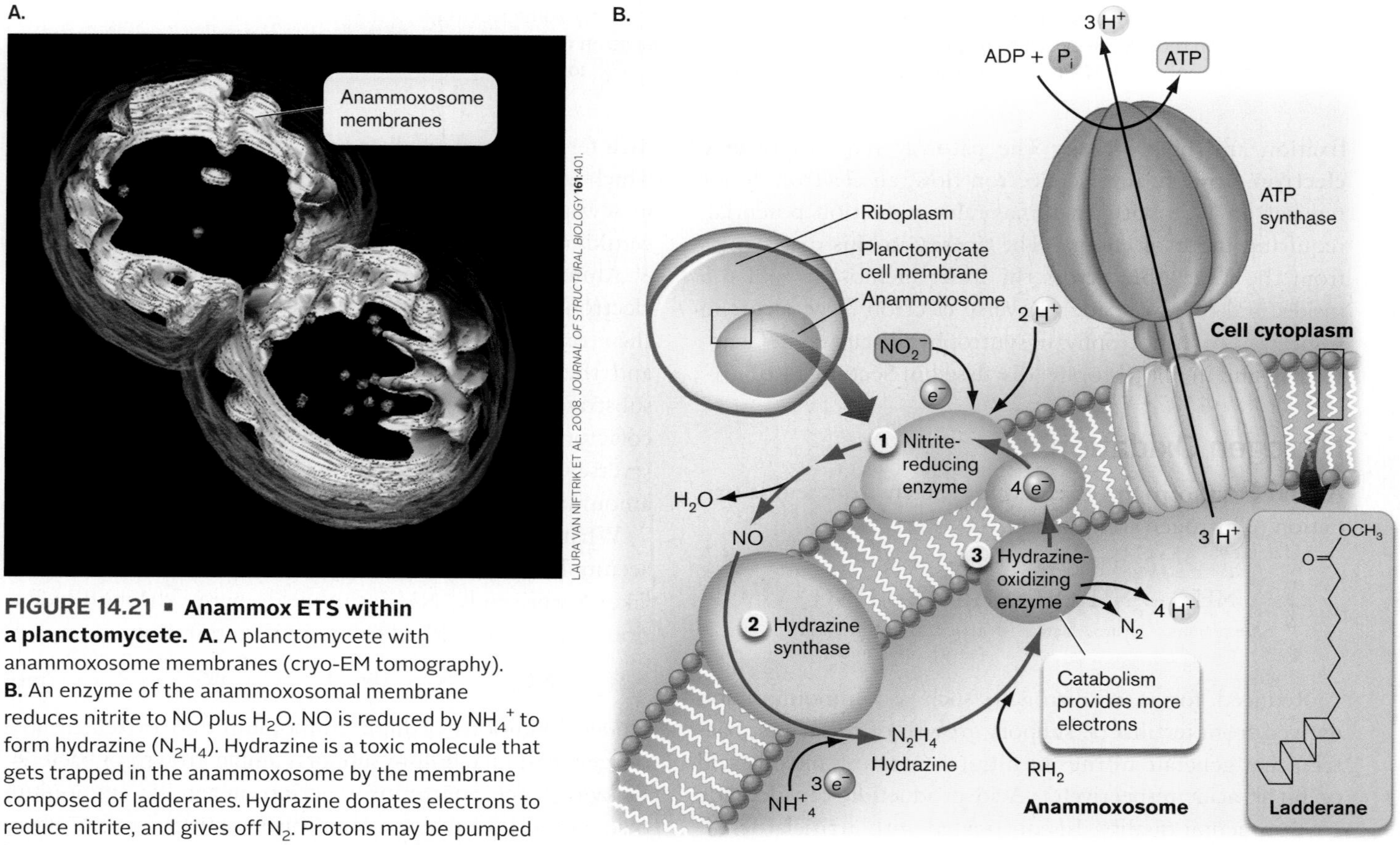
A.
Anammoxosome membranes
LAURA VAN NIFTRIK ET AL. 2008. JOURNAL OF STRUCTURAL BIOLOGY 161:401.
B.
3 H+
ADP + Pi
ATP
ATP synthase
Riboplasm
Planctomycate cell membrane
Anammoxosome
2 H+
Cell cytoplasm
NO2−
e−
1 Nitrite-reducing enzyme
4 e−
H2O
NO
3 H+
3 Hydrazine-oxidizing enzyme
4 H+
N2
2 Hydrazine synthase
Catabolism provides more electrons
N2H4
Hydrazine
RH2
3 e−
NH4+
Anammoxosome
OCH3
O
Ladderane

FIGURE 14.21 ■ Anammox ETS within a planctomycete. **A.** A planctomycete with anammoxosome membranes (cryo-EM tomography). **B.** An enzyme of the anammoxosomal membrane reduces nitrite to NO plus H_2O. NO is reduced by NH_4^+ to form hydrazine (N_2H_4). Hydrazine is a toxic molecule that gets trapped in the anammoxosome by the membrane composed of ladderanes. Hydrazine donates electrons to reduce nitrite, and gives off N_2. Protons may be pumped by the hydrazine synthase complex.

A. ***Star Trek*: an imaginary creature produces H_2SO_4**

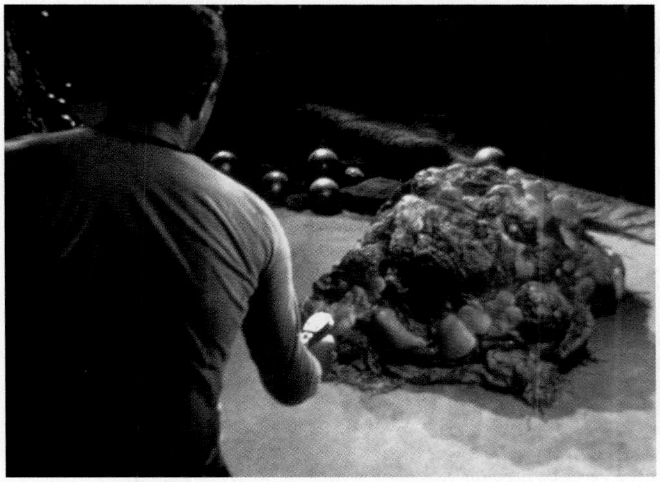

PARAMOUNT/CBS PHOTO ARCHIVE

B. A hot spring supports sulfur-oxidizing archaea

© GETTY IMAGES

C. ***Sulfolobus acidocaldarius***

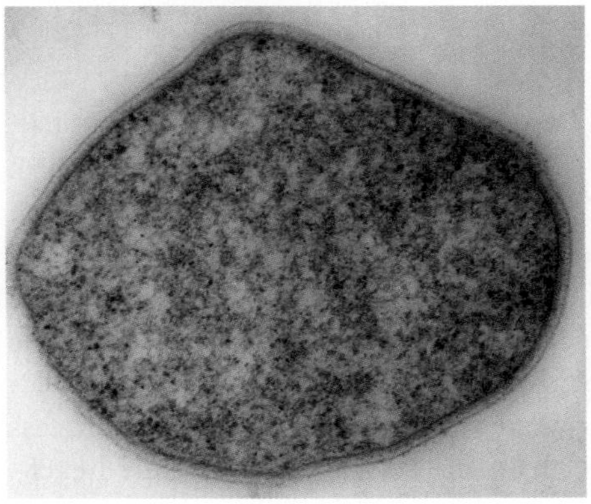

COURTESY OF KENNETH STEDMAN, PORTLAND STATE UNIVERSITY

FIGURE 14.22 ■ Organisms produce sulfuric acid: science and science fiction. A. In the *Star Trek* episode "The Devil in the Dark," starship officers encounter an imaginary creature, called the Horta, that tunnels through rock by producing sulfuric acid. **B.** Volcanic rocks and hot springs support growth of sulfur-oxidizing thermophilic archaea such as *Sulfolobus* species, whose growth at pH 2 colors the rocks. **C.** *Sulfolobus acidocaldarius* grows as irregular spheres about 1 μm in diameter (TEM).

beyond the size of a microbe are yet known to grow in sulfuric acid. But archaea such as *Sulfolobus* species oxidize hydrogen sulfide to sulfuric acid and grow at pH 2, often in hot springs at near-boiling temperatures (**Fig. 14.22B** and **C**). *Sulfolobus* makes irregular Horta-shaped cells without even a cell wall to maintain shape; how it protects its cytoplasm from disintegration is not yet understood.

Microbial sulfur oxidation can cause severe environmental acidification, eroding concrete structures and stone monuments. The problem is compounded by sulfur oxidation in the presence of iron, such as in iron mine drainage. For example, the sulfur-oxidizing archaeon *Ferroplasma acidarmanus* was discovered in the huge abandoned mine Iron Mountain, in northern California, by Katrina Edwards of the Woods Hole Oceanographic Institution and colleagues from the University of Wisconsin–Madison. *Ferroplasma* oxidizes ferrous sulfide (FeS_2) with ferric iron (Fe^{3+}) and water:

$$FeS_2 + 14Fe^{3+} + 8H_2O \rightarrow 15Fe^{2+} + 2SO_4^{\ 2-} + 16H^+$$

This reaction generates large quantities of sulfuric acid. The acidity of the mine water is near pH 0—one of the most acidic environments found on Earth. As *Ferroplasma* grows, it forms biofilms of thick streamers in the mine drainage, which poison aquatic streams.

Anaerobic reactions between sulfur and iron cause hidden hazards for human technology, such as the corrosion of steel in underwater bridge supports. Anaerobic corrosion was long considered a mystery, since iron was known to rust by means of spontaneous oxidation by O_2. In anaerobic conditions, however, sulfur-reducing bacteria can corrode iron (**Fig. 14.23**). In one pathway, the bacteria reduce elemental sulfur (S^0) with H_2 to hydrogen sulfide (H_2S). H_2S then combines with iron metal (Fe^0), which gives up two electrons to form Fe^{2+}, precipitating as iron sulfide (FeS). The displaced $2H^+$ combine with the $2e^-$ from iron, regenerating H_2—now available to reduce sulfur once again. In an alternative mechanism, bacteria use Fe^0 to reduce sulfate directly to FeS. These damaging processes may resemble the iron-based metabolism of Earth's most ancient life-forms.

Nevertheless, like the imaginary Horta that ended up helping miners with their excavations, acid-producing microbes are now used to supplement commercial mining. Lithotrophs such as *Acidithiobacillus ferrooxidans* oxidize

A.

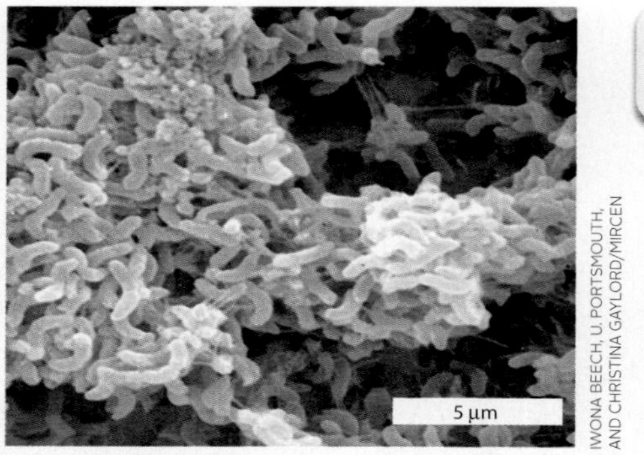
5 μm
IWONA BEECH, U. PORTSMOUTH, AND CHRISTINA GAYLORD/MIRCEN

B.

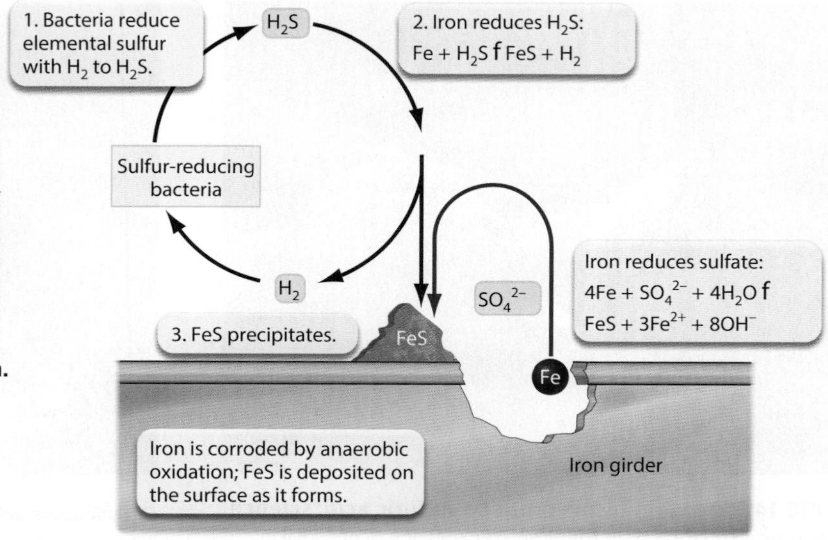
1. Bacteria reduce elemental sulfur with H_2 to H_2S.
H_2S
2. Iron reduces H_2S:
$Fe + H_2S$ f $FeS + H_2$
Sulfur-reducing bacteria
Iron reduces sulfate:
$4Fe + SO_4^{2-} + 4H_2O$ f
$FeS + 3Fe^{2+} + 8OH^-$
H_2
SO_4^{2-}
3. FeS precipitates.
FeS
Fe
Iron is corroded by anaerobic oxidation; FeS is deposited on the surface as it forms.
Iron girder

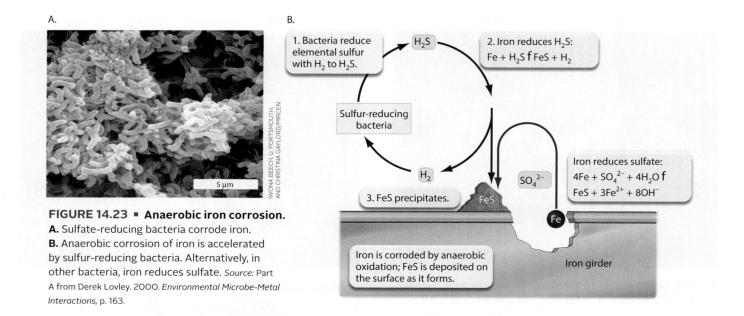

FIGURE 14.23 ▪ Anaerobic iron corrosion. **A.** Sulfate-reducing bacteria corrode iron. **B.** Anaerobic corrosion of iron is accelerated by sulfur-reducing bacteria. Alternatively, in other bacteria, iron reduces sulfate. *Source:* Part A from Derek Lovley. 2000. *Environmental Microbe-Metal Interactions*, p. 163.

sulfides of iron and copper found in minerals such as chalcopyrite ($CuFeS_2$), chalcocite (Cu_2S), and covellite (CuS). The oxidation of Cu^+ to Cu^{2+}, as well as the acidification resulting from production of sulfate, dissolves the metal from the rock. Oxidation of Cu^+ can occur either aerobically with O_2 or anaerobically with NO_3^- present in the soil. Other metals oxidized by *A. ferrooxidans* include selenium, antimony, molybdenum, and uranium. The process of metal dissolution from ores is called **leaching**. Leaching of minerals has been a part of mining since ancient times, long before the existence of microbes was known. Today, over 10% of the copper supply in the United States is provided by microbial leaching from ores too low in copper to smelt directly (**Fig. 14.24**). A similar process is being developed to mine gold, and special strains of *A. ferrooxidans* are being engineered to optimize gold recovery.

Hydrogenotrophy Uses H_2 as an Electron Donor

The use of molecular hydrogen (H_2) as an electron donor is called **hydrogenotrophy** (**Table 14.3**). An example is oxidation of H_2 by sulfur to form H_2S, performed by *Pyrodictium brockii,* an archaeon that grows at thermal vents

A. A copper mine **B. *Acidithiobacillus ferrooxidans***

ROBERT HARDING/ALAMY
0.2 μm
COURTESY OF L. E. MURR

FIGURE 14.24 ▪ Copper mining. **A.** Bingham Canyon copper mine near Salt Lake City, Utah, where copper is leached from low-grade ores by *Acidithiobacillus ferrooxidans* (**B**), a Gram-negative rod that oxidizes copper and iron sulfides (TEM).

TABLE 14.3

Hydrogenotrophy: examples.

Specific reaction	General description
$O_2 + 2H_2 \rightarrow 2H_2O$	Aerobic oxidation of H_2
Fumarate + $H_2 \rightarrow$ succinate	Organic + $H_2 \rightarrow$ organic
$2CO_2 + 4H_2 \rightarrow CH_3COOH + 2H_2O$	Mineral + $H_2 \rightarrow$ organic
$2H^+ + SO_4^{2-} + 4H_2 \rightarrow H_2S + 4H_2O$	Mineral + $H_2 \rightarrow$ mineral
$CO_2 + 4H_2 \rightarrow CH_4 + 2H_2O$	Methanogenesis

above 100°C. Hydrogen gas is a stronger electron donor than most organic foods, so it can be oxidized with the full range of electron acceptors. It is available in anaerobic communities as a fermentation product.

Hydrogenotrophy is hard to categorize. Because H_2 is inorganic, oxidation by O_2 is considered lithotrophy; yet many species that oxidize H_2 with molecular oxygen are organotrophs that also catabolize organic foods. When hydrogen reduces an organic electron acceptor, such as fumarate, the process may be considered either fermentation or anaerobic respiration. When hydrogen reduces a mineral such as sulfur or sulfate, the process is anaerobic lithotrophy.

A form of hydrogenotrophy with enormous potential for bioremediation is **dehalorespiration**, in which halogenated organic molecules serve as electron acceptors for H_2 (**Fig. 14.25A**). Chlorinated molecules such as chlorobenzenes, perchloroethene, and polyvinyl chloride (known as PVC) are highly toxic environmental pollutants. In dehalorespiration, the chlorine is removed as chloride anion and replaced by hydrogen—a reaction requiring input of two electrons from H_2. Many soil bacteria conduct dehalorespiration; a particularly unusual cell wall–less bacterium was discovered that dechlorinates chlorobenzene, a highly stable aromatic molecule (**Fig. 14.25B**).

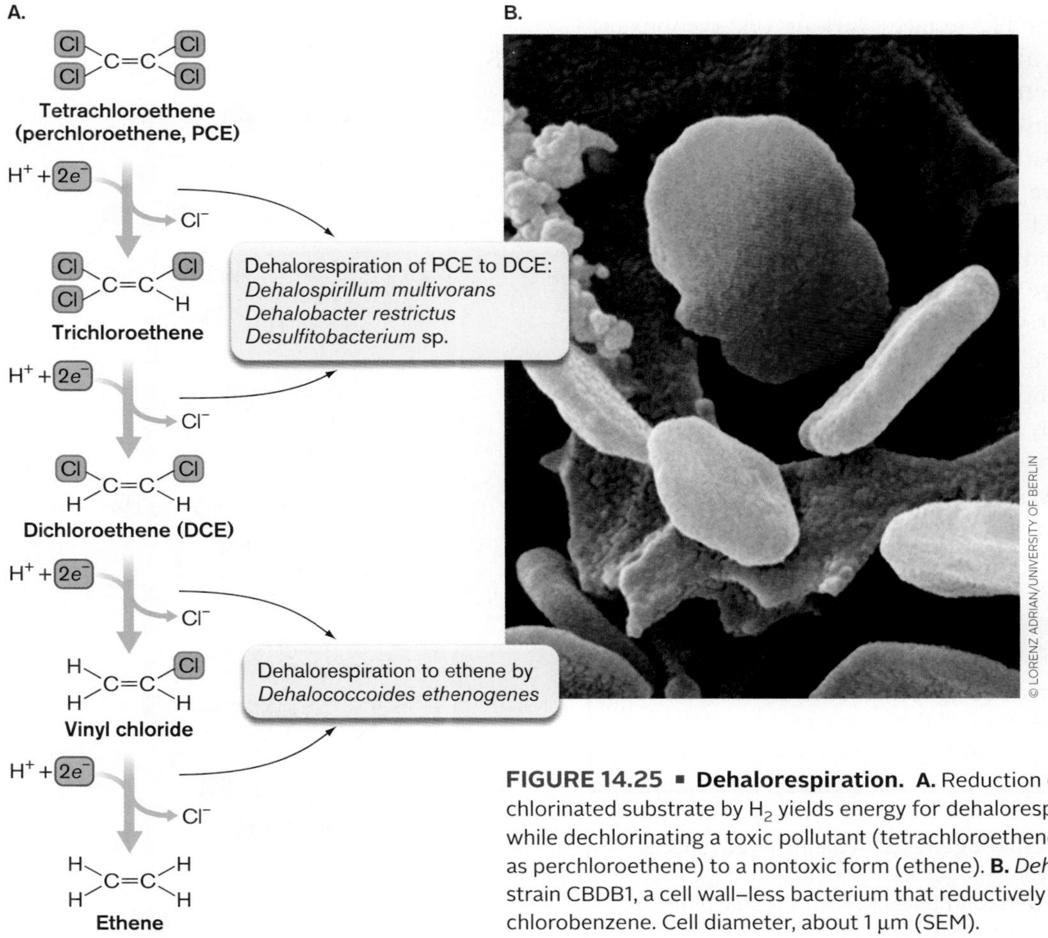
A.
Tetrachloroethene (perchloroethene, PCE)
H+ + 2e−
Cl−
Trichloroethene
Dehalorespiration of PCE to DCE:
Dehalospirillum multivorans
Dehalobacter restrictus
Desulfitobacterium sp.
H+ + 2e−
Cl−
Dichloroethene (DCE)
H+ + 2e−
Cl−
Vinyl chloride
Dehalorespiration to ethene by
Dehalococcoides ethenogenes
H+ + 2e−
Cl−
Ethene
B.
© LORENZ ADRIAN/UNIVERSITY OF BERLIN

FIGURE 14.25 ■ Dehalorespiration. A. Reduction of a chlorinated substrate by H_2 yields energy for dehalorespiring bacteria while dechlorinating a toxic pollutant (tetrachloroethene, also known as perchloroethene) to a nontoxic form (ethene). **B.** *Dehalococcoides* strain CBDB1, a cell wall–less bacterium that reductively dechlorinates chlorobenzene. Cell diameter, about 1 µm (SEM).

Methanogenesis

Hydrogen is such a strong electron donor that it can even reduce the highly stable carbon dioxide to methane. Reduction of CO_2 and other single-carbon compounds, such as formate, to methane is called **methanogenesis**. Methanogenesis supports a major group of archaea known as **methanogens**, many of which grow solely by autotrophy, generating methane.

The simplest form of methanogenesis involves hydrogen reduction of CO_2:

$$CO_2 + 4H_2 \rightarrow CH_4 + 2H_2O \qquad E^{\circ\prime} = 180 \text{ mV}$$

Because both sides of the equation contain a weak electron acceptor (CO_2, H_2O) and a strong electron donor (H_2, CH_4), it was surprising that such a reaction could yield energy for growth. But the presence of sufficient carbon dioxide and hydrogen supports enormous communities of methanogens. Such conditions prevail wherever bacteria grow by fermentation and their gaseous products are trapped, such as in landfills, where methane can be harvested as natural gas, as well as in the digestive systems of cattle and humans. Variants of methanogenesis also include pathways by which H_2 reduces various one-carbon and two-carbon molecules, such as methanol, methylamine, and acetic acid. Diverse methanogens are found in all kinds of environments (discussed in Chapter 19).

A simplified pathway of methanogenesis from CO_2 is shown in **Figure 14.26**. The CO_2 undergoes stepwise hydrogenation, and each oxygen is reduced to water. The increasingly reduced carbon is transferred through a series of unique cofactors (methanofuran, tetrahydromethanopterin, coenzyme M-SH). Three of the hydrogenation steps involve a membrane ETS complex that includes the carrier coenzyme F_{420}, whose reaction generates a proton potential. Note, however, that the final step of methane production generates a transmembrane sodium potential (ΔNa^+), which drives ATP synthesis by a sodium-powered ATP synthase embedded in the cell membrane. Further details of methanogenesis are presented in Chapter 19.

Thought Question

14.9 Hydrogen gas is so light that it rapidly escapes from Earth. Where does all the hydrogen come from to be used for hydrogenotrophy and methanogenesis?

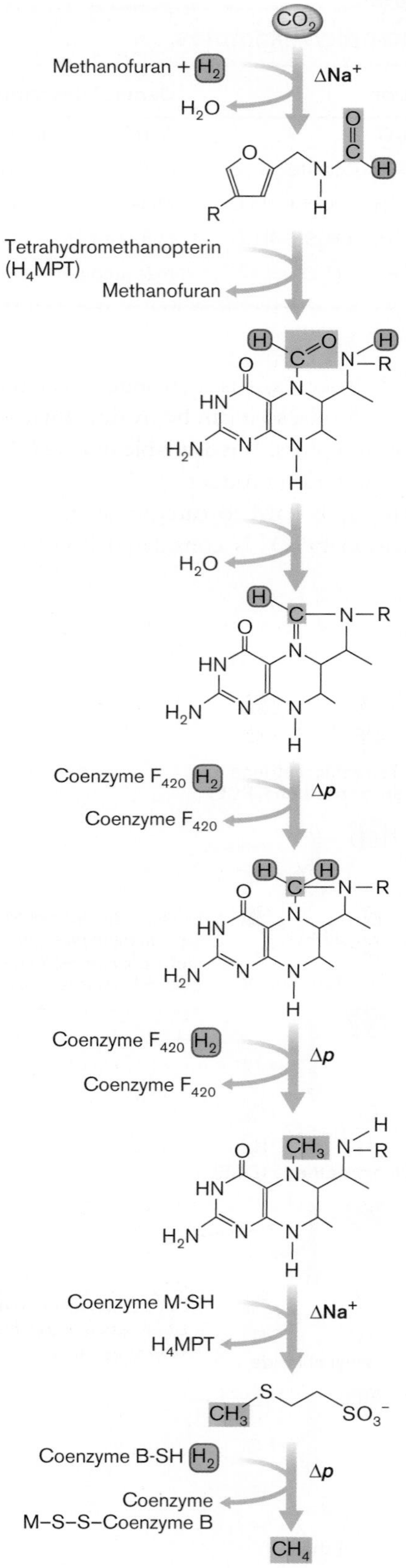

CO2
Methanofuran + H2
ΔNa+
H2O
Tetrahydromethanopterin (H4MPT)
Methanofuran
H2O
Coenzyme F420 H2
Δp
Coenzyme F420
Coenzyme F420 H2
Δp
Coenzyme F420
Coenzyme M-SH
ΔNa+
H4MPT
Coenzyme B-SH H2
Δp
Coenzyme M-S-S-Coenzyme B
CH4

FIGURE 14.26 ■ Methanogenesis. A methanogen reduces carbon dioxide with hydrogen, generating methane (CH_4). The incorporation of hydrogen contributes to both a proton potential and a sodium potential (discussed in Chapter 19).

Methylotrophy

Many forms of catabolism release reduced single-carbon molecules such as methanol (CH_3OH) or methylamine (CH_3NH_2). Oxidation of single-carbon molecules via an ETS is called **methylotrophy**, a form of metabolism conducted by many soil bacteria. In addition, the methane released by methanogenesis provides a niche for **methanotrophy**, a form of methylotrophy in which bacteria and archaea oxidize methane. In deep-ocean sediment, the activity of methanogens is so high that enormous quantities of methane become trapped on the seafloor in the form of water-based crystals known as methane hydrates. If all the methane were to be released at once, it would greatly accelerate global warming.

How is this methane recycled? Until recently, it was thought that most methanotrophy required O_2; indeed, many bacteria are aerobic methanotrophs. However, the deep-ocean sediments contain other species of bacteria and archaea that oxidize methane using nitrite or sulfate. This anaerobic methane oxidation may be critical for the global carbon cycle, as it suggests a mechanism for removal of deep sea methane (discussed in Chapters 21 and 22).

To Summarize

- **Lithotrophy (chemolithotrophy)** is the acquisition of energy by oxidation of inorganic electron donors.
- **Reverse electron flow** powered by ΔP can generate NADPH.
- **Sulfur oxidation** includes oxidation of H_2S to sulfur or to sulfuric acid by sulfur-oxidizing bacteria, often accompanied by iron oxidation. Sulfuric acid production leads to extreme acidification.
- **Nitrogen oxidation** includes successive oxidation of ammonia to hydroxylamine, nitrous acid, and nitric acid. Anammox, the oxidation of ammonium ion by nitrite, returns half the ocean's N_2 to the atmosphere.
- **Hydrogenotrophy** uses hydrogen gas as an electron donor. Hydrogen (H_2) has sufficient reducing potential to donate electrons to nearly all biological electron acceptors, including chlorinated organic molecules (through dehalorespiration).
- **Methanogenesis** is the oxidation of H_2 by CO_2, releasing methane. Methanogenesis is performed only by the methanogen group of archaea.
- **Methylotrophs** use O_2, nitrite, or sulfate to oxidize single-carbon compounds such as methane, methanol, or methylamine. A class of methylotrophs called **methanotrophs** specifically oxidize methane.

14.6

Phototrophy

On Earth today, the ultimate source of electrons driving metabolism is phototrophy, the harnessing of photoexcited electrons to power cell growth. Every year, photosynthesis converts more than 10% of atmospheric carbon dioxide to biomass, most of which then feeds microbial and animal heterotrophs. Most of Earth's photosynthetic production, especially in the oceans, comes from microbes (discussed in Chapter 21).

In phototrophy, the energy of a photoexcited electron is used to pump protons. Different kinds of phototrophy include the bacteriorhodopsin proton pump; the single-cycle chlorophyll-based photosystems I and II; and the double-cycle Z pathway of oxygenic photosynthesis in cyanobacteria and chloroplasts. Diverse kinds of phototrophic bacteria are presented in Chapter 18.

Retinal-Based Proton Pumps

In most ecosystems, the dominant source of carbon and energy is photosynthesis based on chlorophyll. At the same time, many halophilic archaea and marine bacteria supplement their metabolism with a simpler, more ancient form of phototrophy based on a single-protein light-driven proton pump containing the pigment retinal. Many varieties of the retinal-based proton pump have evolved, known as **bacteriorhodopsin** in haloarchaea and as **proteorhodopsin** in bacteria. We present bacteriorhodopsin as a relatively simple form of phototrophy, followed by the more complex ETS-based photolysis in bacteria and chloroplasts.

Bacteriorhodopsin and proteorhodopsin. Bacteriorhodopsin is a small membrane protein commonly found in halophilic archaea (or haloarchaea) such as *Halobacterium salinarum,* a single-celled archaeon that grows in evaporating salt flats containing concentrated NaCl (discussed in Chapter 19). For many years, bacteriorhodopsin-like proton pumps were thought to be limited to extreme halophilic archaea. As bacterial genomes were sequenced, however, homologs of the protein appeared in several species of proteobacteria; the homologs were termed proteorhodopsin. The proteorhodopsin genes appear to have entered bacteria by horizontal transfer from halophilic archaea. In 2005, Oded Béjà and colleagues from Israel, Austria, Korea, and the United States surveyed the genomes of unculturable bacteria from the upper waters of the Mediterranean and

Red seas. They found that 13% of the marine bacteria contain proteorhodopsins, accounting for a substantial—and previously unrecognized—fraction of marine phototrophy.

Bacteriorhodopsin absorbs light with a broad peak in the green range; thus, the organisms containing large amounts of bacteriorhodopsin reflect blue and red, appearing purple. The protein consists of seven hydrophobic alpha helices surrounding a molecule of **retinal**, the same cofactor bound to light-absorbing opsins in the vertebrate retina (**Fig. 14.27A**). In bacteriorhodopsin, the retinal is attached to the nitrogen end of a lysine residue (**Fig. 14.27B**).

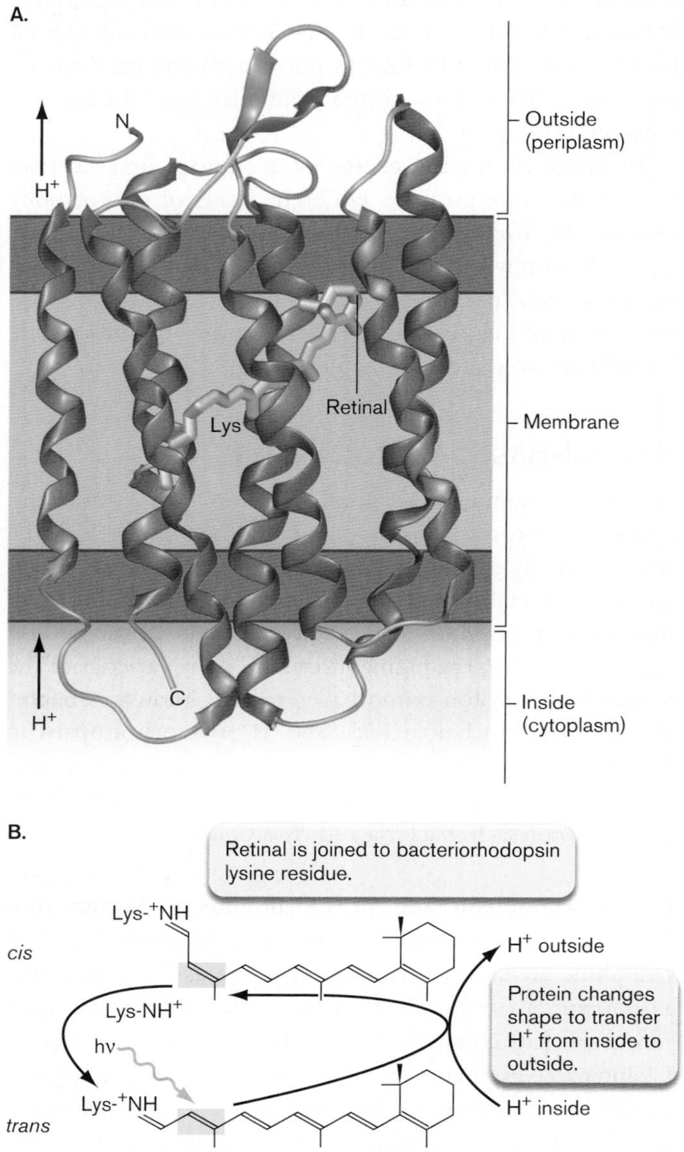
A.
N
H+
Outside
(periplasm)
Retinal
Lys
Membrane
C
H+
Inside
(cytoplasm)
B.
Retinal is joined to bacteriorhodopsin lysine residue.
Lys-+NH
cis
H+ outside
Lys-NH+
hv
Protein changes shape to transfer H+ from inside to outside.
Lys-+NH
trans
H+ inside

FIGURE 14.27 ■ The light-driven cycle of bacteriorhodopsin. **A.** Bacteriorhodopsin contains seven alpha helices that span the membrane in alternating directions and surround a molecule of retinal, which is linked to a lysine residue. (PDB code: 1FBB) **B.** A photon (hv) is absorbed by retinal, which shifts the configuration from *trans* to *cis*. The cycle of excitation and relaxation back to the *trans* form is coupled to pumping of $1H^+$ from the cytoplasm across the membrane.

Retinal has a series of conjugated double bonds that absorb visible light. Upon absorbing a photon, an electron in one of the double bonds is excited to a higher energy level. This process is called photoexcitation. As the electron falls back to the ground state, the double bond shifts position from *trans* (substituents pointing opposite) to *cis* (substituents pointing in the same direction). This change in shape of the retinal alters the conformation of the entire protein, causing it to pick up a proton from the cytoplasm. Eventually, the retinal switches back to its original *trans* configuration. The reversion to *trans* is coupled to the release of a proton from the opposite end of the protein facing outside of the cell. Thus, photoexcitation of bacteriorhodopsin is coupled to the pumping of one H^+ across the membrane.

The proton gradient generated by bacteriorhodopsin drives ATP synthesis by a typical F_1F_o ATP synthase. Light capture by bacteriorhodopsin supplements, but does not replace, catabolism for energy and heterotrophy for carbon source. The combination of light absorption and heterotrophy is called **photoheterotrophy**.

Purple membrane captures light rays. One problem every phototroph needs to solve is how to "capture" light rays. For chemotrophs, food molecules diffuse in solution and can be picked up by receptors for transport into a cell. In phototrophy, however, a photon impinges on one point of the cell, where it either is absorbed or passes through. Thus, the only way to absorb a high percentage of photons is to spread light-absorbing pigments over a wide surface area. To maximize light absorption, *Halobacterium salinarum* archaea pack their entire cell membranes with bacteriorhodopsin. The protein forms trimers that pack in hexagonal arrays, forming the "purple membrane" (**Fig. 14.28**).

Although the bacteriorhodopsin cycle is much simpler than chlorophyll-based photosynthesis (discussed next), it nevertheless illustrates several principles that apply to more complex forms of phototrophy:

- A photoreceptor absorbs light, causing excitation of an electron to a higher energy level, followed by return to the ground state.
- To maximize light collection, large numbers of photoreceptors are packed throughout a membrane.
- The photocycle (absorption and relaxation of the light-absorbing molecule) is coupled to energy storage in the form of a proton gradient.

FIGURE 14.28 ■ Bacteriorhodopsin purple membrane. Trimers of bacteriorhodopsin (monomers shown red, blue, and green) are packed in hexagonal arrays, forming the "purple membrane." (PDB code: 1MOL) *Source:* Purple Membrane: Theoretical Biophysics Group, VMD Image Gallery, NIH Resource for Macromolecular Modeling and Bioinformatics.

A. Roger Stanier

B. Germaine Cohen-Bazire

C. *Merismopedia sp.*

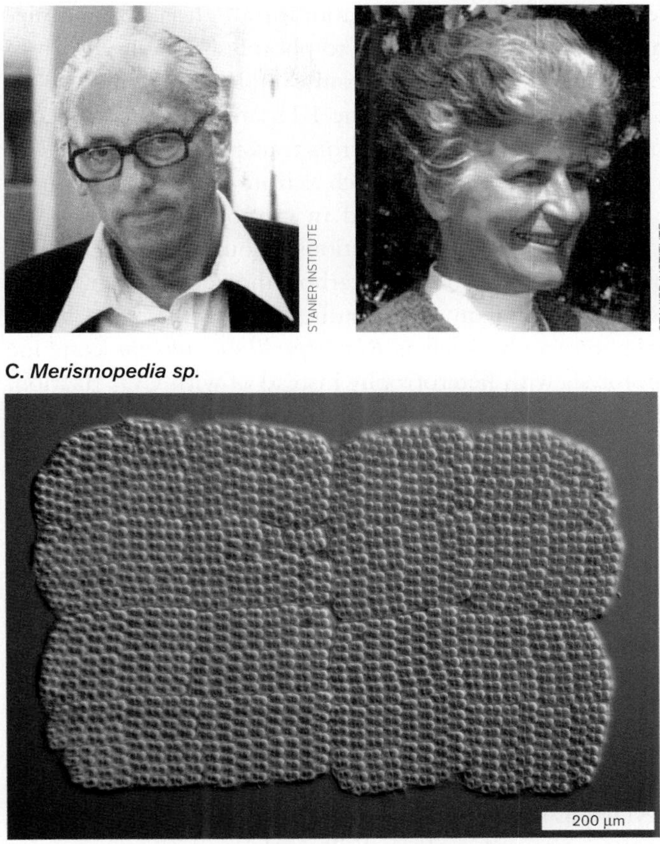
200 μm

STANIER INSTITUTE

STANIER INSTITUTE

WIM VAN EGMOND/VISUALS UNLIMITED, INC.

FIGURE 14.29 ■ Roger Stanier and Germaine Cohen-Bazire studied cyanobacterial photosynthesis. A. Roger Stanier pioneered the study of cyanobacterial physiology. **B.** Germaine Cohen-Bazire performed the first studies of genetic regulation of bacterial photosynthesis. **C.** Green colonies of the cyanobacterium *Merismopedia* sp.

Note: Distinguish among these terms of phototrophy:

- **Photoexcitation** means light absorption that raises an electron to a higher energy state, as in bacteriorhodopsin.
- **Photoionization** means light absorption that causes electron separation.
- **Photolysis** means light absorption coupled to splitting a molecule.
- **Photosynthesis** means photolysis with CO_2 fixation and biosynthesis.

Chlorophyll Photoexcitation and Photolysis

Cyanobacteria and chloroplasts, as well as other kinds of bacteria, obtain energy by photoexcitation of chlorophylls. Figuring out their "light reactions"—the fundamental source of energy for Earth's biosphere—was one of the most exciting projects of the twentieth century. Among hundreds of important contributors, we note two major figures: the married couple Roger Stanier (1916–1982) and Germaine Cohen-Bazire (1920–2001) (**Fig. 14.29A** and **B**). Stanier, a Canadian microbial physiologist at UC Berkeley, clarified the nature of cyanobacteria as phototrophic prokaryotes distinct from eukaryotic algae, and he helped distinguish the water-based photosynthesis of cyanobacteria from the sulfide metabolism of purple bacteria. Cohen-Bazire was a French bacterial geneticist who had studied *lac* operon regulation with Nobel laureate Jacques Monod at the Pasteur Institute in Paris. Cohen-Bazire applied her genetics skills to phototrophs, and she conducted the first genetic analysis of photosynthesis in purple bacteria and cyanobacteria.

Cyanobacteria, the only oxygen-producing bacteria, appear green, like algae or plants (**Fig. 14.29C**). Their green color arises from their chlorophyll, which absorbs blue and red but reflects green. Cyanobacteria include a wide range of species, such as the ocean's major producers, the submicroscopic *Prochlorococcus marinus,* barely visible under a light microscope. Other cyanobacteria have cells as large as eukaryotic algae and form complex developmental structures with important symbiotic associations (see Chapters 18 and 21). Cyanobacteria are among the most successful and diverse groups of life on Earth. Cyanobacteria, and the chloroplasts of algae and plants, produce all the oxygen available for aerobic life.

Overview of photolysis. The energy for photosynthesis derives from the photoexcitation of a light-absorbing pigment. Photoexcitation leads to photolysis, the light-driven separation of an electron from a molecule coupled to an ETS. The components of the ETS are often homologous to those of respiratory electron transport, and they share common electron carriers such as iron-sulfur clusters.

In plant chloroplasts and in cyanobacteria, photolysis is known as the "light reactions," coupled to the "light-independent reactions" of carbon dioxide fixation. Note, however, that many of the sulfur- or organic-based bacterial phototrophs, such as *Rhodospirillum rubrum,* combine photolysis with heterotrophy instead of with CO_2 fixation. At the same time, lithotrophic bacteria such as *Nitrospira* and *Acidithiobacillus* fix CO_2 using energy from mineral oxidation instead of photolysis. In this chapter we focus on photolysis as it occurs in cyanobacteria. We discuss CO_2 fixation along with other biosynthetic pathways later, in Chapter 15.

In ETS-based photosynthesis, photoexcitation leads to separation of an electron from a donor molecule such as H_2O or H_2S. Each electron is then transferred to an ETS, whose components show common ancestry with respiratory ETS proteins. The ETS generates a proton potential and the reduced cofactor NADPH (instead of consuming NADH, as in respiration). The proton potential drives ATP synthesis through an F_1F_o ATP synthase, similar to the one for respiration.

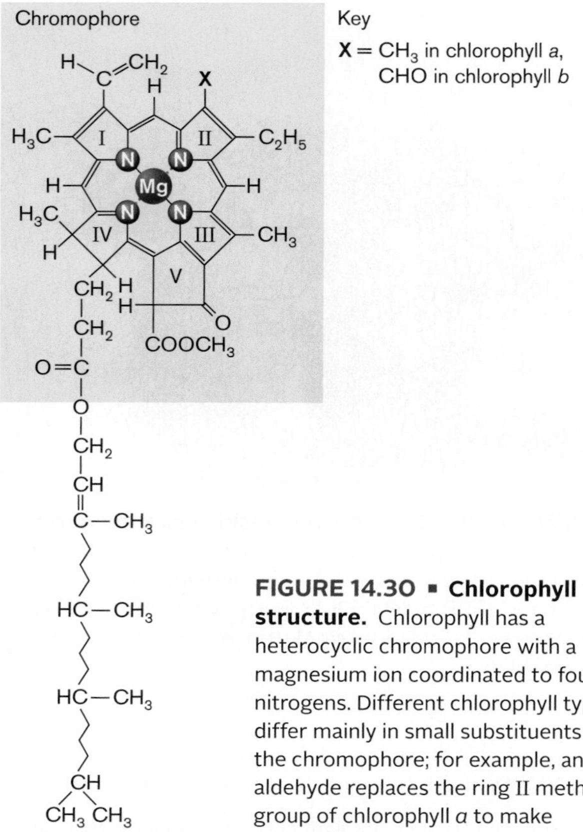
Chromophore
Key
X = CH_3 in chlorophyll *a*,
CHO in chlorophyll *b*
I
II
III
IV
V
Mg
N
X
C_2H_5
CH_3
H_3C
COOCH$_3$

Chlorophyll

FIGURE 14.30 ■ Chlorophyll structure. Chlorophyll has a heterocyclic chromophore with a magnesium ion coordinated to four nitrogens. Different chlorophyll types differ mainly in small substituents of the chromophore; for example, an aldehyde replaces the ring II methyl group of chlorophyll *a* to make chlorophyll *b*.

Chlorophylls absorb light. The main light-absorbing pigments are **chlorophylls**. Each type of chlorophyll contains a characteristic **chromophore**, a light-absorbing electron carrier. The chlorophyll chromophore consists of a heteroaromatic ring complexed to a magnesium ion (Mg^{2+}) (**Fig. 14.30**). As we saw for ETS electron carriers, metal ions and aromatic molecules offer electrons with relatively narrow energy transitions. The chromophore absorbs a photon through a reversible energy transition, such that the chlorophyll can alternate between excited and ground states.

Chlorophyll molecules differ slightly in their substituent groups around the ring; for example, chlorophyll *a* of chloroplasts has a methyl group in ring II, whereas chlorophyll *b* has an aldehyde. Both chlorophylls *a* and *b* are made by chloroplasts and by cyanobacteria, their nearest bacterial relatives. Because they absorb red and blue, they reflect the middle range of the spectrum and so appear green (**Fig. 14.31A**).

By contrast, the chlorophylls of anaerobic phototrophs, or "purple bacteria," such as *Rhodobacter* and *Rhodospirillum,* absorb most strongly in the far-red (infrared) and, in some cases, ultraviolet range (**Fig. 14.31B**). Their chlorophylls are specifically named **bacteriochlorophylls**. The purple bacteria grow in pond water or sediment. Bacteriochlorophyll absorption over an extended range of wavelengths helps capture light missed by the cyanobacteria and algae at the water's surface. In purple bacteria, bacteriochlorophylls are supplemented by accessory pigments called **carotenoids**, which absorb light of green wavelengths and transfer the energy to bacteriochlorophyll. The combination of green-absorbing carotenoids and infrared-absorbing bacteriochlorophylls makes cultures appear deep purple or brown.

The infrared radiation absorbed by bacteriochlorophylls is too weak to permit splitting H_2O to produce oxygen. Thus, purple bacteria are limited to photolysis of H_2S and small organic molecules; many are photoheterotrophs. On the other hand, infrared rays are available in water below the oxygenic phototrophs absorbing red and blue. Thus, anaerobic phototrophs grow at depths where their more high-powered oxygenic relatives do not.

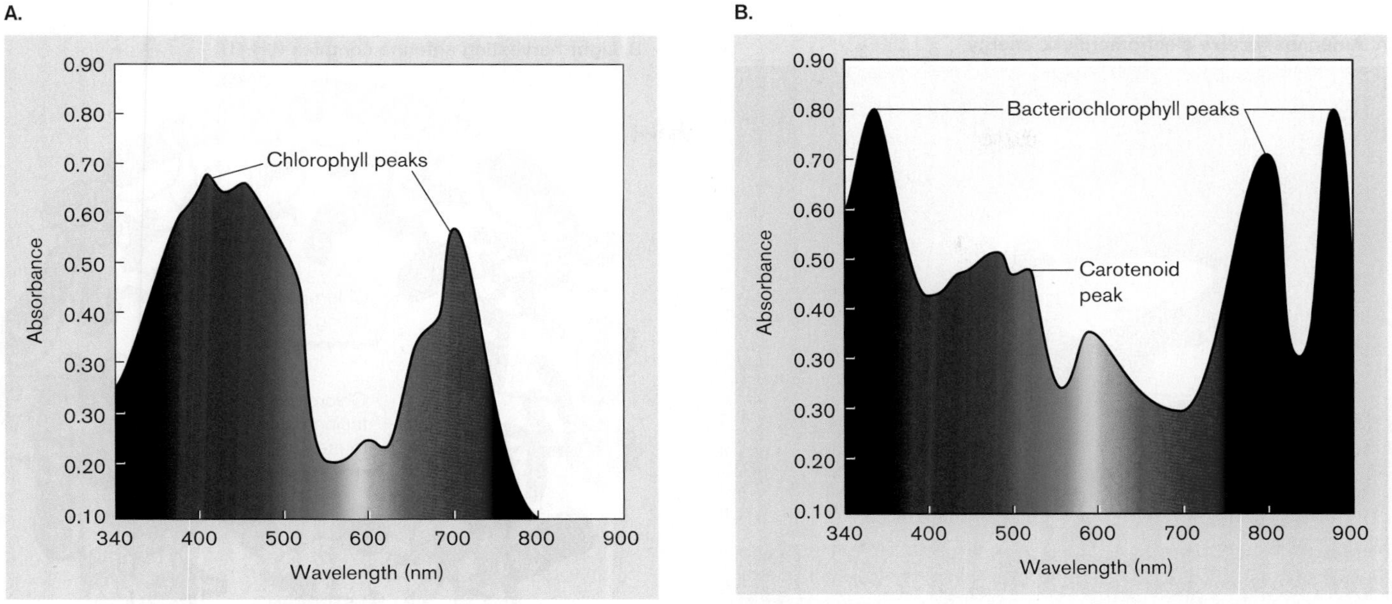
A.
Chlorophyll peaks
Absorbance
0.90
0.80
0.70
0.60
0.50
0.40
0.30
0.30
0.20
0.10
340
400
500
600
700
800
900
Wavelength (nm)
B.
Bacteriochlorophyll peaks
Carotenoid peak
Absorbance
0.90
0.80
0.70
0.60
0.50
0.40
0.30
0.30
0.20
0.10
340
400
500
600
700
800
900
Wavelength (nm)

FIGURE 14.31 ■ **Absorbance spectra of photosynthetic pigments. A.** Absorption by chloroplasts, including chlorophyll *a*, chlorophyll *b*, and carotenoid accessory pigments. The middle range (green) is reflected. **B.** Absorption by purple photosynthetic bacteria, including bacteriochlorophyll and carotenoids. The main absorption is infrared (range 750–900 nm); therefore, the bacteria appear purple or brown.

Note: Distinguish among these classes of photopigments:

- **Bacteriorhodopsin** is a retinal-containing proton pump.
- **Chlorophyll** is a charge-separating photopigment (usually referring to chloroplasts and cyanobacteria).
- **Bacteriochlorophyll** is a charge-separating chlorophyll of anaerobic purple and green bacteria (also known generically as chlorophyll).
- **Carotenoid** is an accessory pigment that absorbs the midrange of light wavelengths but does not directly conduct photolysis.

Antenna complex and reaction center. Light photons cannot be transported or concentrated in a compartment. Instead, they must be captured by absorption. The larger the array of absorptive molecules, the more photons will be captured.

To maximize light collection, many molecules of chlorophyll are grouped in an **antenna complex**. These antenna complexes are arranged like a satellite dish within the plane of the membrane in an elaborate cluster around accessory proteins (**Fig. 14.32A,B**). The clusters then associate in a ring around the **reaction center (RC)**, the protein complex in which chlorophyll photoexcitation connects to the ETS (**Fig. 14.32C**). Throughout the complex of bacteriochlorophylls and accessory pigments, whichever pigment molecule happens to be in the right place at the right time captures the photon. The energy from the photon then transfers at random from one chromophore to the next, until it arrives at the reaction center for electron transfer to the ETS. Other kinds of bacteria have other forms of antenna complex, such as the "phycobilisome" of cyanobacteria (shown later in **Fig. 14.36**).

In purple bacteria and cyanobacteria, the efficiency of photon uptake is increased by the extensive backfolding of the photosynthetic membranes in oval pockets stacked like pita breads (**Fig. 14.33**). These oval pockets are called **thylakoids**. The extensive packing of thylakoids gives an incident photon hundreds of chances to meet a chlorophyll at just the right angle for absorption.

The thylakoids are connected by tubular extensions, so that there exists one interior space, the **lumen**, separated topologically from the regular cytoplasm, or **stroma**. Protons are pumped from the stroma across the thylakoid membrane into the lumen. The F_1F_o complex is embedded in the thylakoid, where it makes ATP using the proton current running through it into the cytoplasm. The F_1 knob of ATP synthase appears to face "outward" in photosynthetic organelles (as opposed to "inward" in respiratory chains). In each case, however, the proton current and ATP motor face in the same direction with respect to the cytoplasm (stroma). The proton potential is more negative in the cytoplasm (stroma), thus drawing protons through the ATP synthase to generate ATP.

The photolytic electron transport system. In photolysis, the absorption of light by chlorophyll or bacteriochlorophyll drives the separation of an electron. The chlorophyll

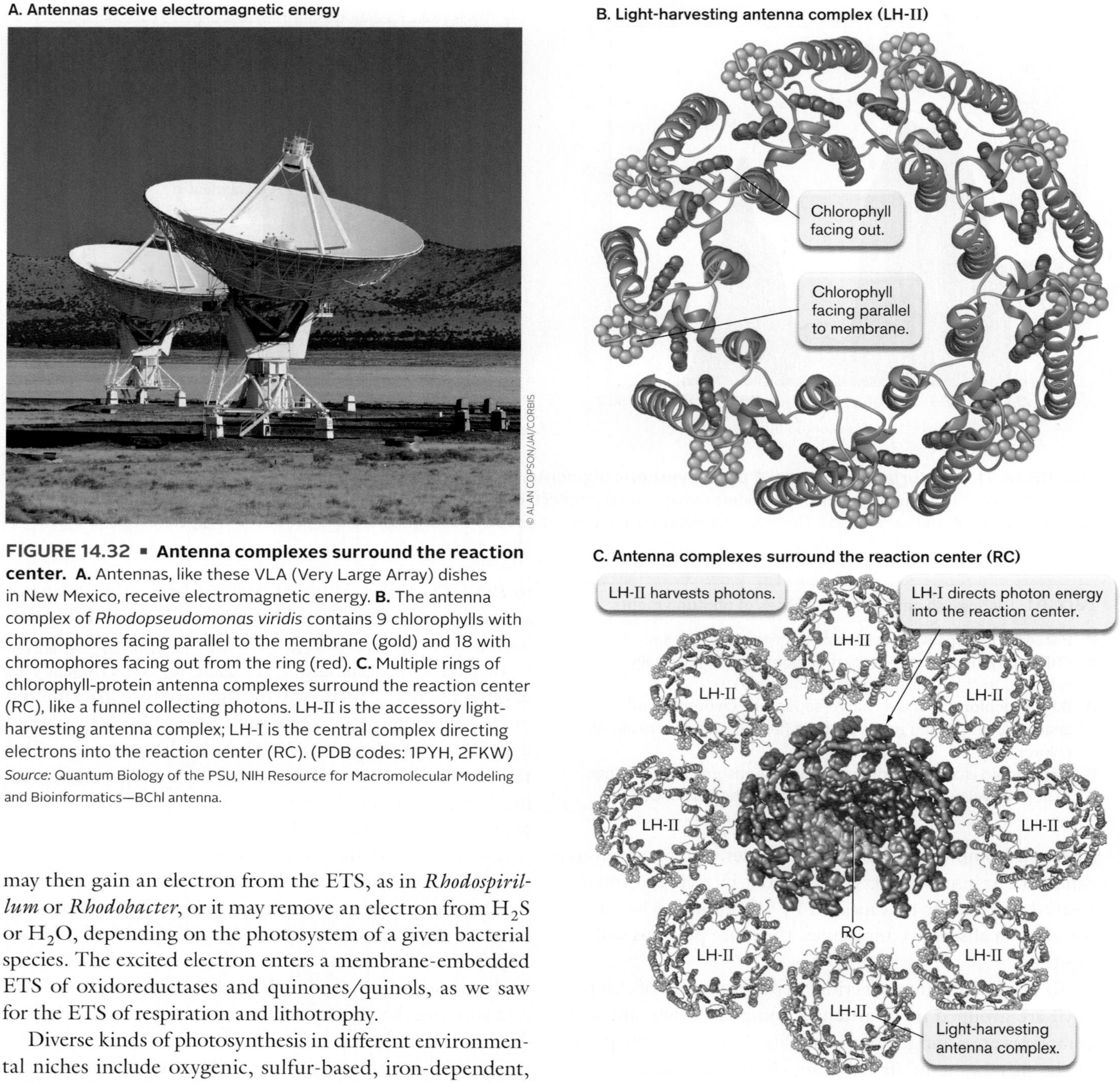
A. Antennas receive electromagnetic energy
© ALAN COPSON/JAI/CORBIS
B. Light-harvesting antenna complex (LH-II)
Chlorophyll facing out.
Chlorophyll facing parallel to membrane.
C. Antenna complexes surround the reaction center (RC)
LH-II harvests photons.
LH-I directs photon energy into the reaction center.
LH-II
RC
Light-harvesting antenna complex.

FIGURE 14.32 ■ Antenna complexes surround the reaction center. **A.** Antennas, like these VLA (Very Large Array) dishes in New Mexico, receive electromagnetic energy. **B.** The antenna complex of *Rhodopseudomonas viridis* contains 9 chlorophylls with chromophores facing parallel to the membrane (gold) and 18 with chromophores facing out from the ring (red). **C.** Multiple rings of chlorophyll-protein antenna complexes surround the reaction center (RC), like a funnel collecting photons. LH-II is the accessory light-harvesting antenna complex; LH-I is the central complex directing electrons into the reaction center (RC). (PDB codes: 1PYH, 2FKW)
Source: Quantum Biology of the PSU, NIH Resource for Macromolecular Modeling and Bioinformatics—BChl antenna.

may then gain an electron from the ETS, as in *Rhodospirillum* or *Rhodobacter*, or it may remove an electron from H_2S or H_2O, depending on the photosystem of a given bacterial species. The excited electron enters a membrane-embedded ETS of oxidoreductases and quinones/quinols, as we saw for the ETS of respiration and lithotrophy.

Diverse kinds of photosynthesis in different environmental niches include oxygenic, sulfur-based, iron-dependent, and even heterotrophic photolysis. Nevertheless, all forms of photolysis share a common design:

1. **Antenna system.** The antenna system (**Fig. 14.25B,C**) maximizes photon capture. A phototrophic antenna system is a large complex of chlorophylls that captures photons and transfers their energy amongst the photopigments until it reaches a reaction center. A complex of chlorophylls and accessory pigments in the photosynthetic membrane collects photons. Energy from each photoexcited electron is transferred among antenna pigments and eventually to the reaction center.
2. **Reaction center complex.** In the reaction center, the photon energy is used to separate an electron from chlorophyll. The electron is replaced by one from a small molecule such as H_2S (**photosystem I**, or PS I) or from the ETS (**photosystem II**, or PS II).

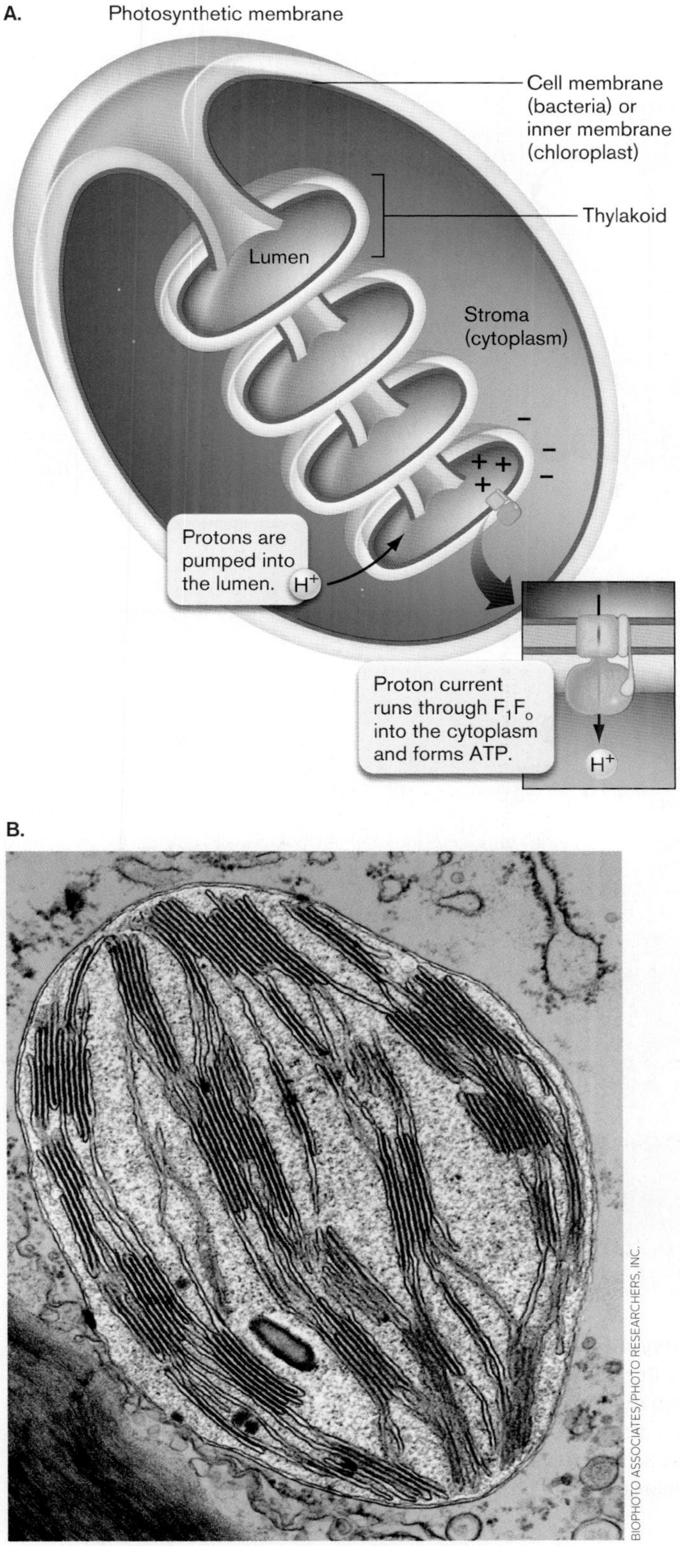
A.
Photosynthetic membrane
Cell membrane (bacteria) or inner membrane (chloroplast)
Thylakoid
Lumen
Stroma (cytoplasm)
Protons are pumped into the lumen.
H+
Proton current runs through F_1F_o into the cytoplasm and forms ATP.
H+
B.
BIOPHOTO ASSOCIATES/PHOTO RESEARCHERS, INC.

FIGURE 14.33 ▪ Photosynthetic membranes. A. The photosynthetic membranes of bacteria and chloroplasts appear as hollow disks with tubular interconnections. The disks are called thylakoids. Topologically, the membrane separates the cytoplasm (stroma) from the interior space (lumen). **B.** TEM section of a chloroplast, showing the stacked thylakoids.

3. **Electron transport system.** Each photoexcited electron enters an ETS. In PS I, electrons separated from H_2O or H_2S are transferred to $NADP^+$ to form NADPH. In PS II, the electron separated from bacteriochlorophyll is replaced by an electron returned from the ETS. In the **oxygenic Z pathway** (H_2O photolysis), electrons flow from PS II into PS I, ultimately releasing O_2 from H_2O.
4. **Energy carriers.** In PS I, electrons are used to make NADPH. In PS II, electron transfer provides energy to pump protons and drive the synthesis of ATP. The Z pathway makes both NADPH and ATP, which are used to fix CO_2.

Photosystems I and II

The steps of photolysis and electron transport occur in three different kinds of systems, in different classes of bacteria:

- **Anaerobic photosystem I** receives electrons associated with hydrogens from H_2S, HS^-, or H_2, or even from reduced iron (Fe^{2+}). Found in chlorobia ("green sulfur" bacteria) and in chloroflexi (filamentous green bacteria).
- **Anaerobic photosystem II** returns an electron from the ETS to bacteriochlorophyll. Found in alphaproteobacteria, "purple nonsulfur" bacteria, and other proteobacteria.
- **Oxygenic Z pathway** includes homologs of photosystems I and II. Two pairs of electrons are received from two water molecules to generate O_2. Found in cyanobacteria and in the chloroplasts of green plants.

The components of photosystems I and II (PS I and PS II) share common ancestry. Each system runs anaerobically, producing sulfur or oxidized organic by-products, but not O_2. Each photosystem shows more recent homology with the respective PS I and PS II components of the oxygenic Z pathway (so called because the electron path through a diagram of the two photosystems traces a Z). The Z pathway ultimately generates O_2—the source of most of the oxygen we breathe.

Note: In photolysis, each quantum of light excites a single electron. Some ETS components, such as the quinones/quinols ($Q \rightarrow QH_2$) actually process two of these electrons in completing their redox cycle. The single-electron intermediate states of quinones are not shown in Figures 14.35–14.37.

Photosystem I in chlorobia. In bacteria such as *Chlorobium* species, the reaction center (RC) contains bacteriochlorophyll P840, named for its peak absorption at 840 nm;

that is, the near infrared, actually beyond the range humans can see. But P840 and the chlorophylls of the antenna complex also absorb light over shorter wavelengths, the range of 400–550 nm.

The *Chlorobium* antenna complex consists of a membrane compartment called a chlorosome (**Fig. 14.35**). A single chlorosome may contain 200,000 molecules of bacteriochlorophyll, harvesting photons with nearly 100% efficiency; it is so sensitive that some chlorobia actually harvest thermal radiation from deep-sea thermal vents.

When any one bacteriochlorophyll absorbs a photon, the energy transfers amongst the photopigments until it reaches the PS I reaction center (**Fig. 14.34**). The photon yields sufficient energy to donate the high-potential electron to a high-potential quinone: phylloquinone/phylloquinol (PQ). Phylloquinol donates the electron to **ferredoxin**, an FeS protein. Ferredoxin transfers the electron to the enzyme ferredoxin-NAD^+ reductase; two of these electrons then reduce NAD^+ or the energetically equivalent $NADP^+$. The reduced carrier (NADH or NADPH) provides reductive energy for CO_2 fixation and biosynthesis (discussed in Chapter 15).

Chlorobium is a true autotroph, fixing CO_2 for biosynthesis (Chapter 15). However, electrons from sulfur or hydrogen gas supplement the phototrophic ETS. The electron flow in bacteria using PS I generates a net proton gradient by consuming H^+ inside and generating H^+ outside the cell, thus providing proton motive force to drive ATP synthesis.

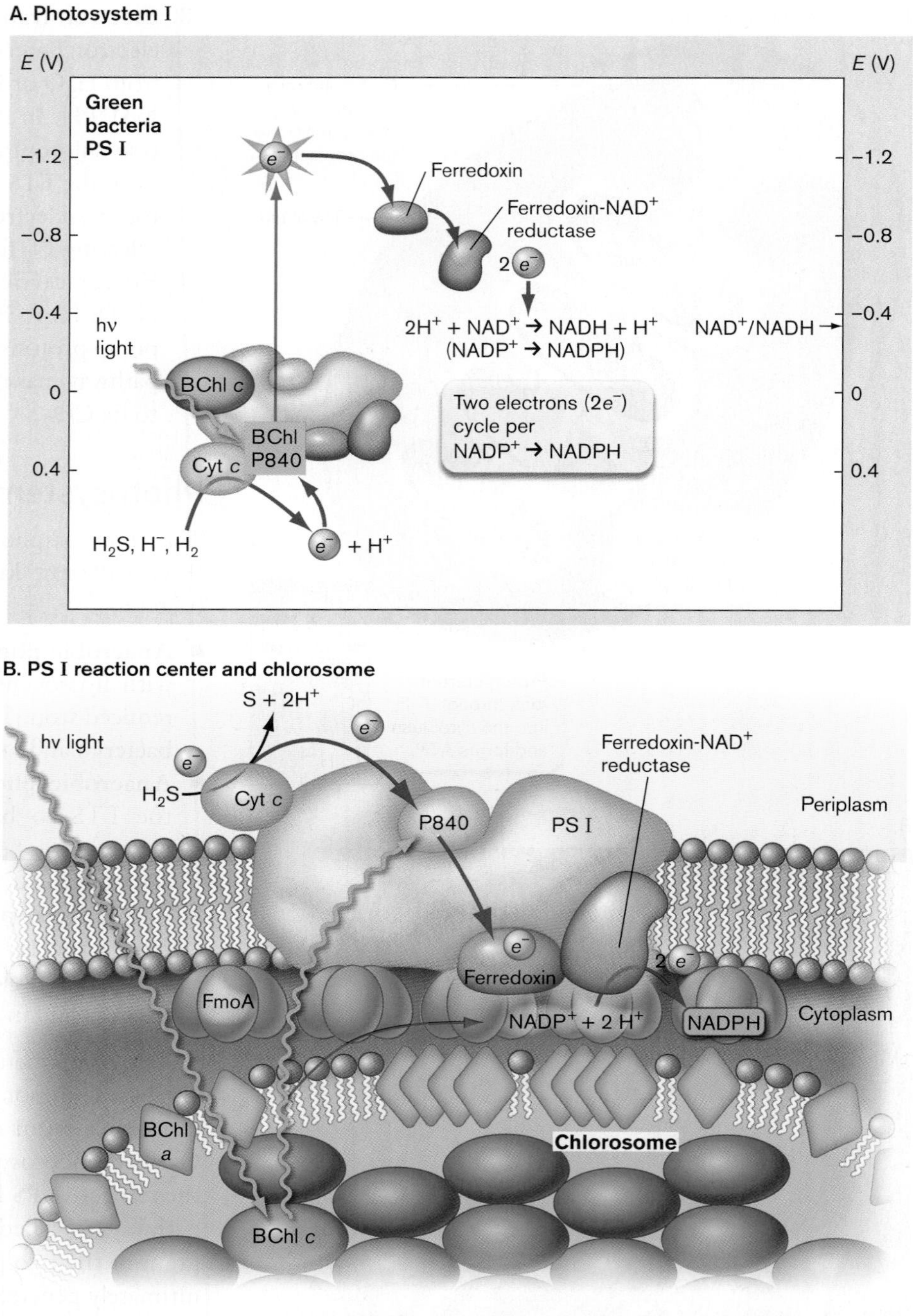
A. Photosystem I
E (V)
E (V)
Green bacteria PS I
–1.2
–0.8
–0.4
0
0.4
e⁻
Ferredoxin
Ferredoxin-NAD⁺ reductase
2e⁻
hv light
2H⁺ + NAD⁺ → NADH + H⁺
(NADP⁺ → NADPH)
NAD⁺/NADH
BChl c
BChl P840
Cyt c
Two electrons (2e⁻) cycle per NADP⁺ → NADPH
H₂S, H⁻, H₂
e⁻ + H⁺
B. PS I reaction center and chlorosome
S + 2H⁺
e⁻
hv light
e⁻
H₂S
Cyt c
Ferredoxin-NAD⁺ reductase
P840
PS I
Periplasm
e⁻
Ferredoxin
2e⁻
FmoA
NADP⁺ + 2 H⁺
NADPH
Cytoplasm
BChl a
Chlorosome
BChl c

FIGURE 14.34 ▪ Photosystem I separates electrons from sulfides and organic molecules. A. In green sulfur bacteria photoexcitation of P840 transfers e^- to a quinone (phylloquinone, PQ), at high reduction potential *E*. From PQ the electron is transferred to ferredoxin (FD). Ferredoxin is oxidized by ferredoxin-NAD^+ reductase (FNR), donating the $2e^-$ to NAD^+ (or the energetically equivalent $NADP^+$) to form NADH (or NADPH). **B.** The chlorosome antenna complex transfers photon energy to the PS I reaction center.

Photosystem II in alphaproteobacteria. Phototrophic alphaproteobacteria such as *Rhodospirillum rubrum* and *Rhodopseudomonas palustris* are typically found in wetlands and streams, where they capture light not used by other phototrophs. The antenna complex (LH) was shown in **Fig. 14.32**. The peak wavelength absorbed by bacteriochlorophyll P870 lies so far into the infrared (800–1,100 nm) that the photon energy is insufficient to reduce NAD(P) to NAD(P)H. The electrons from P870 are transferred by low-potential quinols to a terminal cytochrome oxidoreductase (**Fig. 14.35**). The quinols pass their electrons to cytochromes, while moving $2H^+$ across the membrane.

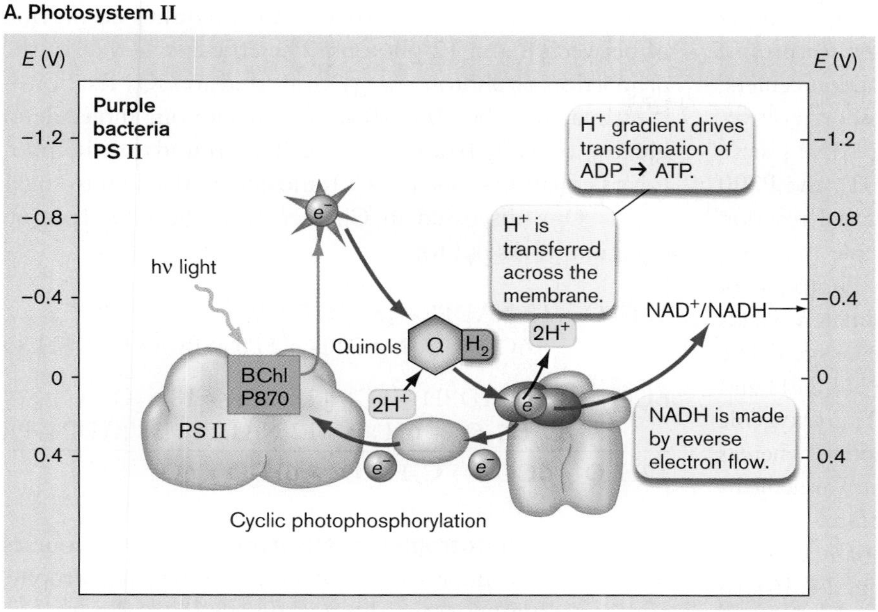
A. Photosystem II
E (V)
E (V)
Purple bacteria PS II
−1.2
−0.8
−0.4
0
0.4
−1.2
−0.8
−0.4
0
0.4
H+ gradient drives transformation of ADP → ATP.
H+ is transferred across the membrane.
hν light
e−
Quinols
Q
H2
2H+
2H+
NAD+/NADH
BChl P870
PS II
e−
e−
NADH is made by reverse electron flow.
Cyclic photophosphorylation

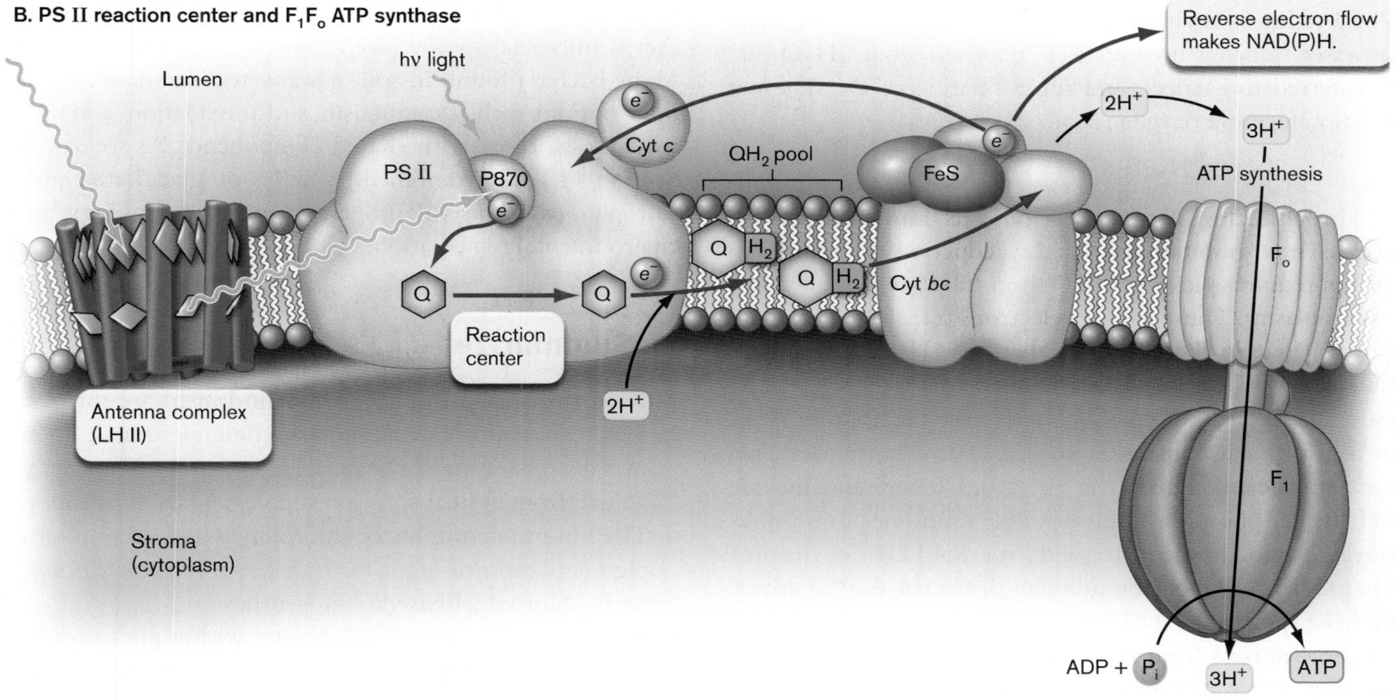
B. PS II reaction center and F1Fo ATP synthase
Reverse electron flow makes NAD(P)H.
Lumen
hν light
e−
Cyt c
QH2 pool
e−
FeS
2H+
3H+
ATP synthesis
PS II
P870
e−
Q
H2
Q
H2
Q
Q
e−
Cyt bc
Fo
Reaction center
2H+
Antenna complex (LH II)
F1
Stroma (cytoplasm)
ADP + Pi
3H+
ATP

FIGURE 14.35 ■ Photosystem II separates an electron from bacteriochlorophyll. **A.** In purple bacteria, Bchl P870 donates an energized electron to a quinone (Q). Two of these donated electrons complete the conversion of quinone to quinol (QH_2). Electrons flow through cytochrome *bc*, coupled to pumping of protons. The proton potential drives synthesis of ATP. The cytochrome *bc* complex transfers the electrons back to P870. **B.** PS II reaction center.

When the electrons reach cytochrome *c*, they flow back to bacteriochlorophyll, where they can be reexcited by photon energy from the antenna complex. Because the electron path traces back to its source, the ETS of photosystem II leading to ATP synthesis is called **cyclic photophosphorylation**.

Since the reduction potential is too small to reduce $NADP^+$ to NADPH, photosystem II requires **reverse electron flow**. As we saw for iron oxidation (Section 14.5), in reverse electron flow an electron donor reduces an ETS with an unfavorable reduction potential, requiring input of energy. Purple bacteria obtain this energy by spending ATP to increase the proton potential, or from a pathway outside photolysis, such as catabolism of organic compounds. The organic compounds also provide substrates for biosynthesis. Thus, purple bacteria are photoheterotrophs.

Thought Question

14.10 Suppose you discover bacteria that require a high concentration of Fe^{2+} for photosynthesis. Can you hypothesize what the role of Fe^{2+} may be? How would you test your hypothesis?

Oxygenic photolysis. The "Z" pathway of photolysis found in cyanobacteria and chloroplasts combines key features of both PS I and PS II (**Fig. 14.36**). Both reaction centers, however, contain chlorophylls that absorb at shorter wavelengths (higher energy) than those of the respective purple or green homologs: P680 instead of P870 (PS II), and P700 instead of P840 (PS I). Thus, the cyanobacterial reaction centers can split water—a highly stable molecule that cannot be photolyzed by anaerobic phototrophs. Photolysis of water requires greater energy input, but ultimately yields greater energy overall, and produces molecular oxygen. The energy potentials are high enough to generate NADPH and fix CO_2 into biomass (presented in Chapter 15). Oxygenic phototrophs dominate the shallow water depths, whereas anaerobes grow at lower depths, using light at wavelengths unused by cyanobacteria and algae near the surface.

Cyanobacteria harvest light via an exceptionally efficient antenna complex called the phycobilisome (**Fig. 14.36**). In the PSII reaction center, the photoexcitation of chlorophyll P680 yields enough energy to split H_2O. The entire cycle of water splitting involves $4H^+$ removed from $2H_2O$, $4e^-$ transferred to carriers, and the formation of O_2. The net reaction forming oxygen is:

$$2H_2O \rightarrow 4H^+ + 4e^- + O_2$$

Through the ETS, the electrons are transferred to quinones. As in respiration, each $2e^-$ reduction of quinone to quinol requires pickup of $2H^+$ from the stroma (equivalent to cytoplasm). Thus, the four electrons transferred generate a net change in the proton gradient of $4H^+$. Furthermore, the energy of electron transfer to cytochrome *bf* enables pumping of an additional $2H^+$ across the membrane. Thus, in all, for each conversion of $2H_2O$ to O_2 the net protons transferred across the thylakoid membrane include $4H^+$ (water photolysis) plus $4 \times 2H^+$ (quinones to quinols) through cytochrome *bf*, to yield a total of $12H^+$ for the proton gradient. The proton gradient drives the ATP synthase to make approximately 3 ATP per O_2 formed.

The electrons from PS II do not cycle back to the PS II reaction center, as they do in purple bacteria. Instead they are transferred to PS I by a protein called plastocyanin. The energy of the electron transferred by plastocyanin is augmented through absorption of a second photon by the chlorophyll of PS I. Subsequent electron flow through ferredoxin can now generate NADH or NADPH. Some of the electron flow instead cycles back to cytochrome *bf*, where it contributes to pumping protons.

The overall equation for energy yield of oxygenic photolysis can be represented as:

$$2H_2O + 2\ NADP^+ + 3\ [ADP + P_i] \rightarrow O_2 + 2\ [NADPH + H^+] + 3\ ATP + 3H_2O$$

To release one O_2 and fix one CO_2 requires absorption of between 8 and 12 photons. The efficiency—that is, the proportion of photon energy converted to CO_2 fixation—is estimated to be 20%–30%. To generate one molecule of glucose by CO_2 fixation, we need six rounds of the photolysis equation—one per CO_2 molecule "fixed" into sugar ($C_6H_{12}O_6$) (discussed in Chapter 15). The CO_2 fixation equation works out to:

$$12H_2O + 12\ NADP^+ + 18\ [ADP + P_i] \rightarrow 6O_2 + 12\ [NADPH + H^+] + 18\ ATP + 18H_2O$$

$$\mathbf{6CO_2} + 12\ [NADPH + H^+] + 18\ ATP + 18H_2O \rightarrow \mathbf{C_6H_{12}O_6} + 6H_2O + 12\ NADP^+ + 18\ [ADP + P_i]$$

$$12H_2O + \mathbf{6CO_2} \rightarrow \mathbf{C_6H_{12}O_6} + 6H_2O + 6O_2$$

Facultative Phototrophy, or Mixotrophy. Biology courses often give the impression that heterotrophy, lithotrophy, and photosynthesis are discrete metabolic pathways utilized by completely different organisms. In fact, however, a microbial species rarely depends on a single option. Many bacteria found in soil or water use alternative metabolic options such as catabolism, sulfur oxidation, and photosynthesis, often with shared components. As we saw in Section 14.1, the combined pathways of a facultative phototroph possess useful properties for biotechnology and environmental remediation.

To Summarize

- **Bacteriorhodopsin** and **proteorhodopsin** are forms of a light-driven proton pump that contains retinal, found in haloarchaea and in bacteria, respectively. The energy gained from light absorption supplements heterotrophy.
- **The antenna complex** of chlorophylls and other photopigments captures light for transfer to the reaction center in chlorophyll-based photosynthesis.
- **Thylakoids** are folded membranes within phototrophic bacteria or chloroplasts. The membranes extend the area for chlorophyll light absorption, and they separate two compartments to form a proton gradient.
- **Photosystem I** obtains electrons from H_2S or HS^-. The electrons are transferred through an ETS to form NADH or NADPH.
- **Photosystem II** transfers an electron through an ETS and pumps H^+ to generate ATP. An electron ultimately returns to bacteriochlorophyll through cyclic photophosphorylation. Reverse electron flow generates NADPH or NADH.
- **The oxygenic Z pathway in cyanobacteria and chloroplasts** includes homologs of photosystems I and II.

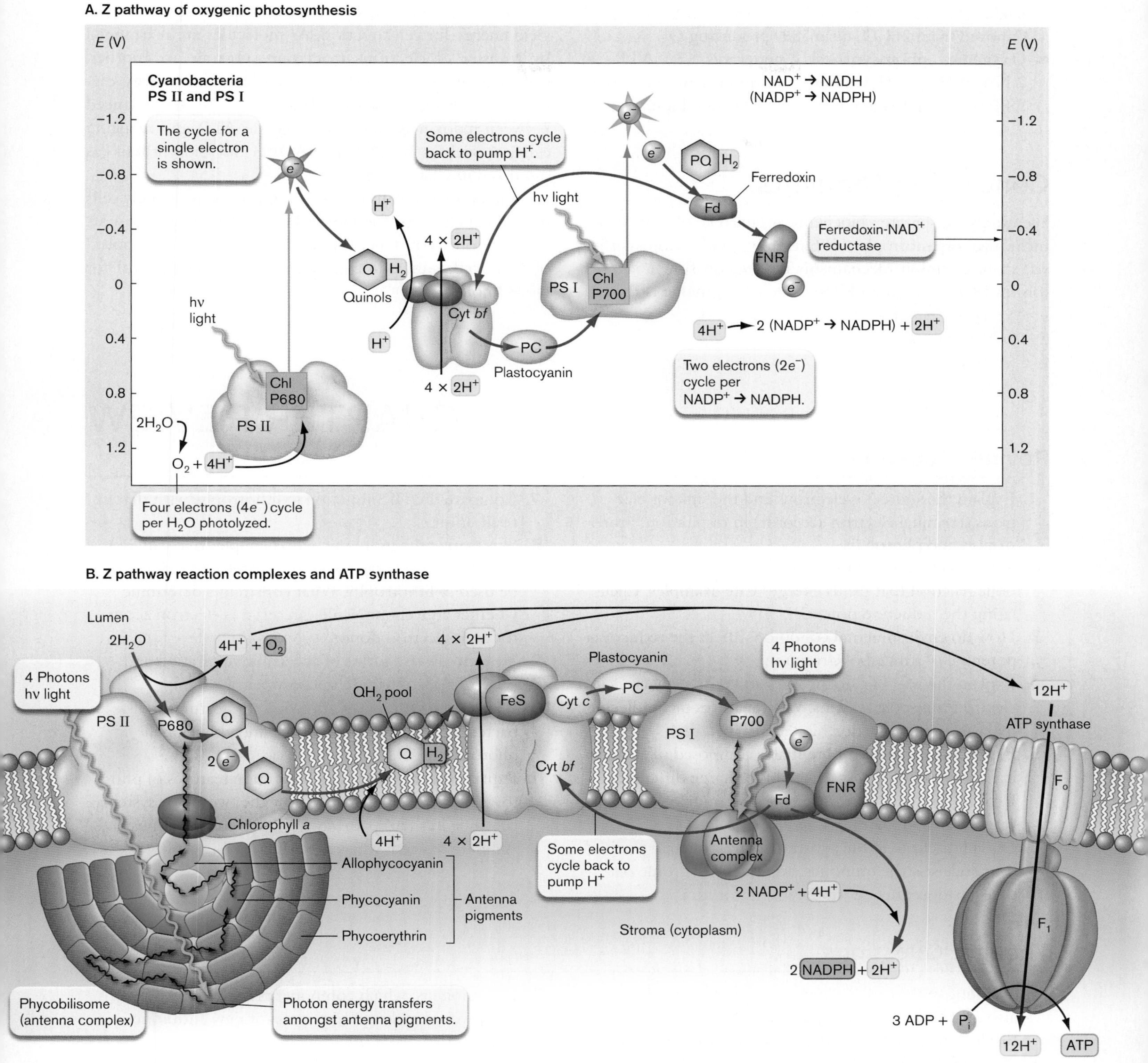
A. Z pathway of oxygenic photosynthesis
E (V)
E (V)
Cyanobacteria
PS II and PS I
−1.2
−0.8
−0.4
0
0.4
0.8
1.2
The cycle for a single electron is shown.
Some electrons cycle back to pump H+.
NAD+ → NADH
(NADP+ → NADPH)
e−
PQ H2
Ferredoxin
Fd
hν light
H+
4 × 2H+
Q H2
Quinols
Cyt bf
PS I
Chl P700
FNR
Ferredoxin-NAD+ reductase
hν light
H+
PC
Plastocyanin
4 × 2H+
4H+ → 2 (NADP+ → NADPH) + 2H+
Two electrons (2e−) cycle per NADP+ → NADPH.
Chl P680
PS II
2H2O
O2 + 4H+
Four electrons (4e−) cycle per H2O photolyzed.
B. Z pathway reaction complexes and ATP synthase
Lumen
2H2O
4H+ + O2
4 × 2H+
Plastocyanin
4 Photons hν light
4 Photons hν light
12H+
ATP synthase
PS II
P680
Q
QH2 pool
FeS
Cyt c
PC
P700
PS I
2 e−
Q
Q H2
Cyt bf
e−
Fd
FNR
Fo
Chlorophyll a
4H+
4 × 2H+
Antenna complex
Allophycocyanin
Phycocyanin
Phycoerythrin
Antenna pigments
Some electrons cycle back to pump H+
2 NADP+ + 4H+
Stroma (cytoplasm)
F1
2 NADPH + 2H+
Phycobilisome (antenna complex)
Photon energy transfers amongst antenna pigments.
3 ADP + Pi
12H+
ATP

FIGURE 14.36 ▪ Oxygenic photosynthesis in cyanobacteria and chloroplasts. **A.** Each H_2O is photolyzed via the Z pathway of PS II and PS I. The $2e^-$ from each water ($4e^-$ in all) are transferred to the quinone pool, which transfers $4e^-$ to cytochrome *bf*. Cytochrome *bf* pumps $4 \times 2\ H^+$ across the membrane and transfers $4e^-$ via plastocyanin to a PS I containing chlorophyll P700. A second photon excites P700, enabling transfer of e^- to ferredoxin ($4e^-$ per O_2 formed), and from there to $NADP^+ \rightarrow NADPH$. **B.** The Z pathway within a photosynthetic membrane. *Source:* Part B based on crystallographic data from thermophilic cyanobacteria, from Genji Kurisu et al. 2003. *Science* **302**:1009.

Eight photons are absorbed, and two electron pairs are removed from $2H_2O$, ultimately producing O_2.

- **Oxygenic photosynthesis generates 3 ATP + 2 NADPH** per $2H_2O$ photolyzed and O_2 produced. The ATP and NADPH are used to fix CO_2 into biomass.

Concluding Thoughts

Seemingly disparate biochemical means of nutrition, including organotrophy, lithotrophy, and photosynthesis, share common mechanisms of electron flow and proton transfer through an ETS. A recurring theme is that all forms of metabolism involve electron transfer reactions that yield energy for cell function. As molecules are rearranged by transfer of electrons from one substrate to another, energy is provided to form ion gradients and energy carriers. The energy carriers always need to be balanced between redox-neutral carriers, such as ATP, and reducing carriers, such as NADH or NADPH, for biosynthesis (as discussed in Chapter 15).

Whatever the means of obtaining energy, ultimately cells must spend energy for biosynthesis. Chapter 15 presents how microbes construct the fundamental "nuts and bolts" of their cells—including products surprisingly useful for biotechnology.

CHAPTER REVIEW

Review Questions

1. Explain the source of electrons and the sink for electrons (terminal electron acceptor) in respiration, lithotrophy, and photolysis.
2. How do bacteria combine redox couples for a metabolic reaction that yields energy? Cite examples, calculating the reduction potential.
3. How do environmental conditions affect the reduction potential of a metabolic reaction?
4. Explain the role of cytochromes and redox cofactors in electron transport systems. What features of a molecule make it useful for redox biochemistry?
5. Explain how a proton potential is composed of a chemical concentration difference plus a charge difference. Explain how each component of Δp can drive a cellular reaction.
6. Explain the role of the substrate dehydrogenase (oxidoreductase), the quinones, and the cytochrome oxidase (oxidoreductase) in the respiratory ETS.
7. Compare the ETS function in lithotrophy with that in respiration.
8. Summarize the inorganic redox couples that can be used in anaerobic respiration and those that can be used in lithotrophy. What constraints determine whether a given molecule can serve as electron acceptor or as electron donor?
9. How do diverse forms of anaerobic respiration and lithotrophy contribute to ecosystems?
10. Explain the differences and common features of bacteriorhodopsin phototrophy and chlorophyll phototrophy.
11. Explain the differences and common features of photosystems I and II. Explain how the two photosystems combine in the Z pathway. Why can the Z pathway generate oxygen, whereas PS I and PS II cannot?

Thought Questions

1. The lung pathogen *Pseudomonas aeruginosa*, which also grows in soil, can respire aerobically or else anaerobically using nitrate. Under what conditions would *P. aeruginosa* use each form of metabolism? What part of its ETS would need to change to accommodate the different forms?
2. What environments favor oxygenic photosynthesis, versus sulfur phototrophy? Explain.
3. In pathogens, which components of the ETS do you think would make good targets for new antibiotics, and why?
4. Devise a form of energy-yielding metabolism in which fumarate is converted to succinate; and a different form, in which succinate is converted to fumarate. Explain why the two reactions are reasonable, and try on the Internet to find actual organisms that obtain energy through these reactions.

Key Terms

anaerobic respiration (559)
anammox reaction (564)
antenna complex (575, 576)
bacteriochlorophyll (574, 575)
bacteriorhodopsin (571, 575)
carotenoid (575)
chemolithoautotroph (542)
chemolithotrophy (542, 563)
chemoorganotrophy (542)
chlorophyll (574, 575)
chromophore (574)
cofactor (551)
cyclic photophosphorylation (579)
cytochrome (545)
dehalorespiration (569)
dissimilatory denitrification (560)
dissimilatory metal reduction (561)
electron acceptor (541)
electron donor (541)
electron transport system (ETS) (electron transport chain) (540, 541, 577)
ferredoxin (578)
heme (552)
hydrogenotrophy (568)
leaching (568)
lithotrophy (540, 542, 563)
lumen (575)
methanogen (570)
methanogenesis (570)
methanotrophy (571)
methylotrophy (571)
nitrifier (565)
obligate aerobe (559)
organotrophy (540, 542)
oxidoreductase (546, 553)
oxygenic Z pathway (577)
photoexcitation (573)
photoheterotrophy (572)
photoionization (573)
photolysis (573)
photosynthesis (573)
photosystem I (577)
photosystem II (577)
phototrophy (540, 542)
proteorhodopsin (571)
proton potential (proton motive force) (546)
quinol (552)
quinone (552)
quinone pool (554)
reaction center (RC) (575, 576)
redox couple (542)
respiration (541)
retinal (572)
reverse electron flow (565, 579)
standard reduction potential ($E°$) (542)
stroma (575)
thylakoid (575)
uncoupler (551)

Recommended Reading

Andries, Koen, Peter Verhasselt, Jerome Guillemont, Hinrich W. H. Göhlmann, Jean-Marc Neefs, et al. 2005. A diarylquinoline drug active on the ATP synthase of *Mycobacterium tuberculosis. Science* **307**:223–227.

Beatty, J. Thomas, Jörg Overmann, Michael T. Lince, Ann K. Manske, Andrew S. Lang, et al. 2005. An obligately photosynthetic bacterial anaerobe from a deep-sea hydrothermal vent. *Proceedings of the National Academy of Sciences USA* **102**:9306–9310.

Ferreira, Kristina N., Tina M. Iverson, Karim Maghlaoui, James Barber, and So Iwata. 2004. Architecture of the photosynthetic oxygen-evolving center. *Science* **303**:1831–1838.

Flores, Enrique, and Antonia Herrero. 2010. Compartmentalized function through cell differentiation in filamentous cyanobacteria. *Nature Reviews. Microbiology* **8**:39–50.

Ishii, Shun'ichi, Takefumi Shimoyama, Yasuaki Hotta, and Kazuya Watanabe. 2008. Characterization of a filamentous biofilm community established in a cellulose-fed microbial fuel cell. *BMC Microbiology* **8**:6.

Jones, Shari A., Fatema Z. Chowdhury, Andrew J. Fabich, April Anderson, Darrel M. Schreiner, et al. 2009. Respiration of *Escherichia coli* in the mouse intestine. *Infection and Immunity* **75**:4891–4899.

Kashefi, Kazem, Jason M. Tor, Kelly P. Nevin, and Derek R. Lovley. 2001. Reductive precipitation of gold by dissimilatory Fe(III)-reducing bacteria and archaea. *Applied and Environmental Microbiology* **67**:3275–3279.

Kuenen, J. Gijs. 2008. Anammox bacteria: From discovery to application. *Nature Reviews. Microbiology* **6**:320–326.

Lower, Brian H., Liang Shi, Ruchirej Yongsunthon, Timothy C. Droubay, David E. McCready, et al. 2007. Specific bonds between an iron oxide surface and outer membrane cytochromes MtrC and OmcA from *Shewanella oneidensis* MR-1. *Journal of Bacteriology* **189**:4944–4952.

Pfeffer, Christian, Steffen Larsen, Jie Song, Mingdong Dong, Flemming Besenbacher, et al. 2012. Filamentous bacteria transport electrons over centimetre distances. *Nature* **491**:218–221.

Philippot, Laurent. 2005. Denitrification in pathogenic bacteria: For better or worst? *Trends in Microbiology* **13**:191–192.

Preiss, Laura, Adriana L. Klyszejko, David B. Hicks, Jun Liu, Oliver J. Fackelmayer, et al. 2013. The c-ring stoichiometry of ATP synthase is adapted to cell physiological requirements of alkaliphilic *Bacillus pseudofirmus* OF4. *Proceedings of the National Academy of Sciences USA* **110**:7874–7879.

Raghoebarsing, Ashna A., Arjan Pol, Katinka T. van de Pas-Schoonen, Alfons J. P. Smolders, Katharina F. Ettwig, et al. 2006. A microbial consortium couples anaerobic methane oxidation to denitrification. *Nature* **440**:918–921.

Reguera, Gemma, Kevin D. McCarthy, Teena Mehta, Julie S. Nicoll, Mark T. Tuominen, et al. 2005. Extracellular electron transfer via microbial nanowires. *Nature* **435**:1098–1101.

Strous, Marc, Eric Pelletier, Sophie Mangenot, Thomas Rattei, Angelika Lehner, et al. 2006. Deciphering the evolution and metabolism of an anammox bacterium from a community genome. *Nature* **440**:790–794.

Winter, Sebastian E., Parameth Thiennimitr, Maria G. Winter, Brian P. Butler, Douglas L. Huseby, et al. 2010. Gut inflammation provides a respiratory electron acceptor for *Salmonella*. *Nature* **467**:426–429.

CHAPTER 15
Biosynthesis

15.1 Overview of Biosynthesis

15.2 CO_2 Fixation: The Calvin Cycle

15.3 CO_2 Fixation: Diverse Pathways

15.4 Biosynthesis of Fatty Acids and Polyketides

15.5 Nitrogen Fixation

15.6 Biosynthesis of Amino Acids and Nitrogenous Bases

15.7 Biosynthesis of Tetrapyrroles

How do microbes build their cells? Some bacteria and archaea build themselves entirely from carbon dioxide and nitrogen gas plus a few salts. From these simple molecules, microbial enzymes construct amino acids, the cell wall and envelope, and all the machinery of the cell. Many microbes produce products with valuable properties, such as antibiotics. In the laboratory, we can engineer microbial biosynthesis to make cloned proteins, pesticides, and industrial reagents.

Once the key elements of carbon and nitrogen are incorporated into small molecules, more complex structures are built by many enzymes in intricate pathways. How do cells organize their biosynthesis to build precisely the forms they need? How do bacteria avoid wasting energy on excess production? These questions are answered by the use of radioisotope tracers, genetic analysis of mutants, and the decoding of genomes. When we discover a new species, its genome reveals its capacities for biosynthesis, including hints of new pharmaceuticals.

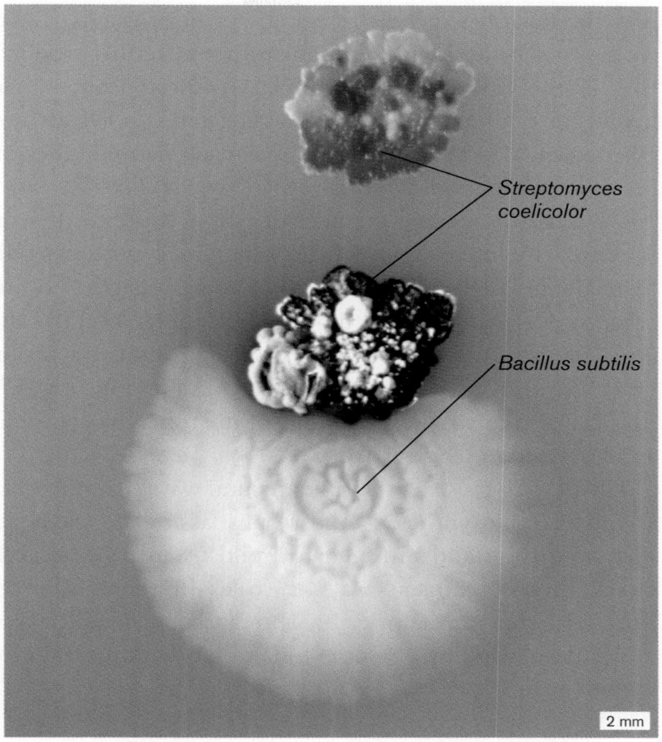

Streptomyces coelicolor
Bacillus subtilis
2 mm

CURRENT RESEARCH highlight

Microbes make antibiotics. Bacteria synthesize thousands of molecules that build their cells. Different products are made depending on the conditions—such as the presence of a competing species. *Streptomyces coelicolor* bacteria detect the presence of *Bacillus subtilis*—and to decrease the competitor's growth, they synthesize an antibiotic, actinorhodin. To analyze the products of *S. coelicolor,* Jeramie Watrous and colleagues collected mass spectra (spectra of product fragment masses) from the surface of the bacterial colony. The mass spectra were obtained by a technique called "nanoDESI," in which a tiny sample of the colony is taken up by capillary action and vacuum aspiration into the mass spectrometer. The masses of the ionized fragments of the bacterial products were analyzed by computer and identified using information from the bacterial genome. The data reveal a network of products made by *S. coelicolor* grown in isolation (red colony) or adjacent to *B. subtilis* (dark purple colony). *S. coelicolor* synthesizes the red pigment prodigiosin whether or not the competitor *B. subtilis* is present, but makes the blue actinorhodin only when *B. subtilis* is present. *Source:* Jeramie Watrous et al. 2012. *PNAS* **109**:E1743–E1752.

In Chapters 13 and 14 we learned how microbes use chemical reactions to gain energy, storing it in ion gradients and in small molecules such as ATP and NADPH. Chapter 15 shows how the microbes spend this energy for biosynthesis. Today we can use bacterial biosynthesis to manufacture products that in the past required artificial synthesis from petroleum. For example, products in a class called polyhydroxyalkanoates (a form of polyester) are made naturally by soil bacteria such as *Ralstonia eutropha* (**Fig. 15.1A**). *R. eutropha* synthesizes storage granules of poly-3-hydroxybutyrate. A related molecule, poly-4-hydroxybutyrate, is used commercially by Tepha Medical Devices to manufacture surgical sutures that are absorbed by the body (**Fig. 15.1B**). The molecule is synthesized by an engineered industrial strain of *E. coli*, in a reaction that consumes ATP (**Fig. 15.1C**). Additional industrial applications of microbial biosynthesis are described in Chapter 16.

Chapter 15 presents the fundamental ways that microbes accomplish biosynthesis. Autotrophs assimilate elements such as carbon and nitrogen to build useful carbon skeletons; these organisms play key roles in the food web. Microbial enzyme factories build remarkably complex biomolecules, including vitamins and antibiotics important for human health.

15.1

Overview of Biosynthesis

Biosynthesis is the building of complex biomolecules, also known as **anabolism**, the reverse of catabolism. **Figure 15.2** presents an overview of anabolism, or biosynthesis, and how it relates to catabolic pathways we saw in Chapter 13. Some microbes synthesize all their organic components from minerals such as carbonate and nitrate, while others must obtain essential molecules from their environment. Biosynthesis requires:

- **Essential elements.** Biosynthesis requires carbon, oxygen, hydrogen, nitrogen, and other essential elements. Carbon is obtained either through CO_2 fixation (autotrophy) or through acquisition of organic molecules made by other organisms (heterotrophy). Autotrophs assemble carbon and water into small molecules such as acetyl-CoA that serve as substrates or building blocks for the cell. Heterotrophs break down molecules such as sugars and peptides in order to synthesize acetyl-CoA and other key substrates. Besides carbon, biosynthesis

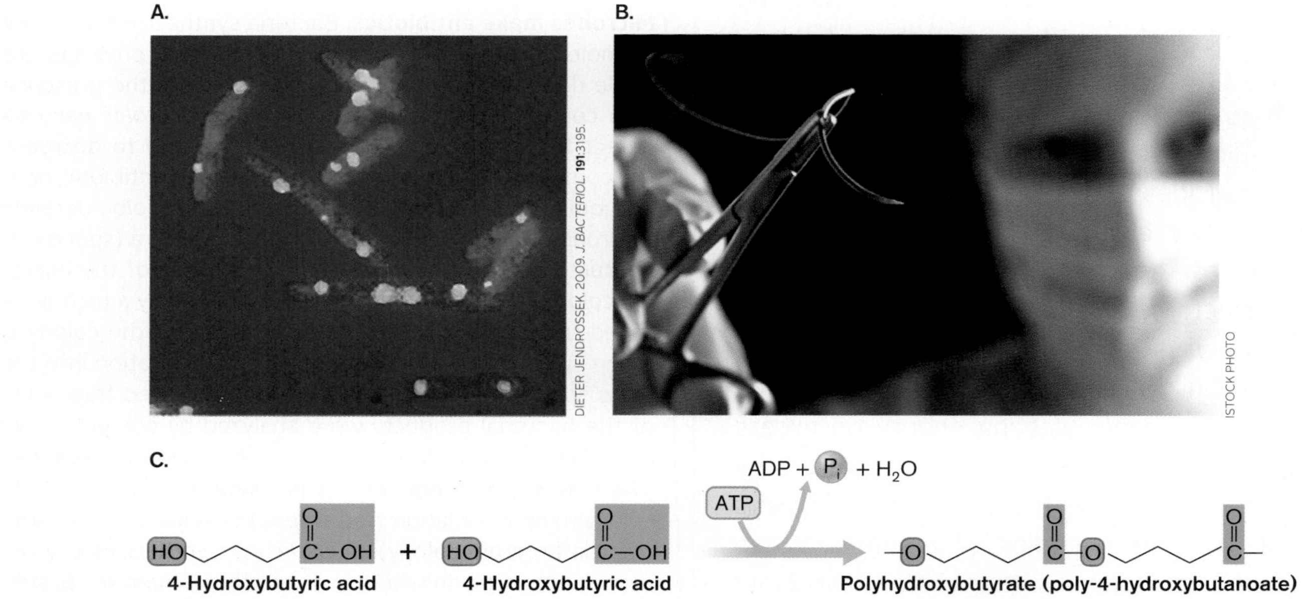
A.
DIETER JENDROSSEK. 2009. J. BACTERIOL. 191:3195.
B.
ISTOCK PHOTO
C.
ADP + Pi + H2O
ATP
HO C—OH + HO C—OH
4-Hydroxybutyric acid
4-Hydroxybutyric acid
Polyhydroxybutyrate (poly-4-hydroxybutanoate)

FIGURE 15.1 ■ Bacterial polyester: poly-4-hydroxybutyrate. A. *Ralstonia eutropha* bacteria form storage granules of a polyhydroxyalkanoate, poly-3-hydroxybutyrate. **B.** TephaFLEX surgical sutures, which are absorbed by the body. Sutures are manufactured from poly-4-hydroxybutyrate, synthesized by engineered *E. coli.* **C.** Multiple units of 4-hydroxybutyric acid condense to form poly-4-hydroxybutyrate. The reaction consumes ATP.

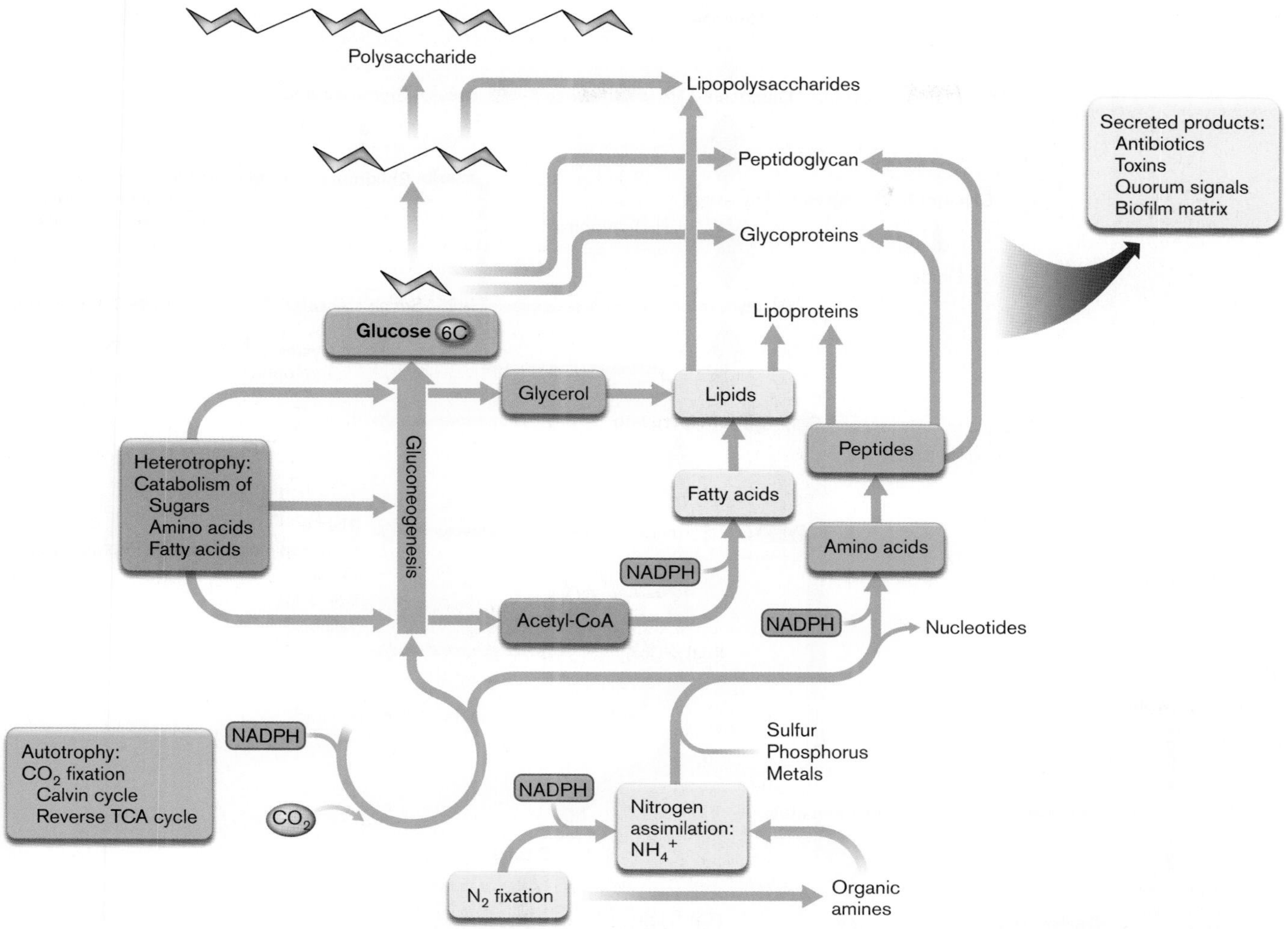
Polysaccharide
Lipopolysaccharides
Secreted products:
Antibiotics
Toxins
Quorum signals
Biofilm matrix
Peptidoglycan
Glycoproteins
Lipoproteins
Glucose 6C
Glycerol
Lipids
Peptides
Heterotrophy:
Catabolism of
Sugars
Amino acids
Fatty acids
Gluconeogenesis
Fatty acids
NADPH
Amino acids
Acetyl-CoA
NADPH
Nucleotides
NADPH
Sulfur
Phosphorus
Metals
Autotrophy:
CO_2 fixation
Calvin cycle
Reverse TCA cycle
NADPH
CO_2
Nitrogen
assimilation:
NH_4^+
N_2 fixation
Organic
amines

FIGURE 15.2 ■ Biosynthesis: an overview. Microbes obtain carbon skeletons through CO_2 fixation or through catabolism of compounds formed by other microbes. Nitrogen is obtained by N_2 fixation or by uptake of nitrates or organic amines. All biosynthesis requires energy from ATP and from reducing cofactors such as NADPH.

must assimilate nitrogen and sulfur for proteins, phosphorus for DNA, and metals for metal-containing enzymes.

- **Reduction.** Biosynthesis usually reduces the substrate, by hydrogenation and by removing oxygen. Cell components such as lipids and amino acids are more reduced than substrates such as CO_2 and acetate, so their biosynthesis requires a reducing agent such as NADPH.
- **Energy.** Building complex structures, with or without reduction, requires energy. Biosynthetic enzymes spend energy by coupling their reactions to the hydrolysis of ATP, the oxidation of NADPH, or the flux of ions down a transmembrane ion gradient.

Substrates for Biosynthesis

Where do the building blocks for biosynthesis come from? Many substrates arise from glucose catabolism and the tricarboxylic acid (TCA) cycle, central catabolic pathways discussed in Chapter 13 (**Fig. 15.3**). For example, the succinyl-CoA molecules used to build the tetrapyrrole ring of vitamin B_{12} come directly from the TCA cycle, while the glycines derive from 3-phosphoglycerate, an intermediate of glycolysis. Glycerol 3-phosphate provides the glyceride backbone of lipids. Pyruvate provides the backbone of several amino acids with aliphatic side chains, whereas erythrose 4-phosphate contributes to the ring structures of aromatic amino acids. Other amino acids derive their

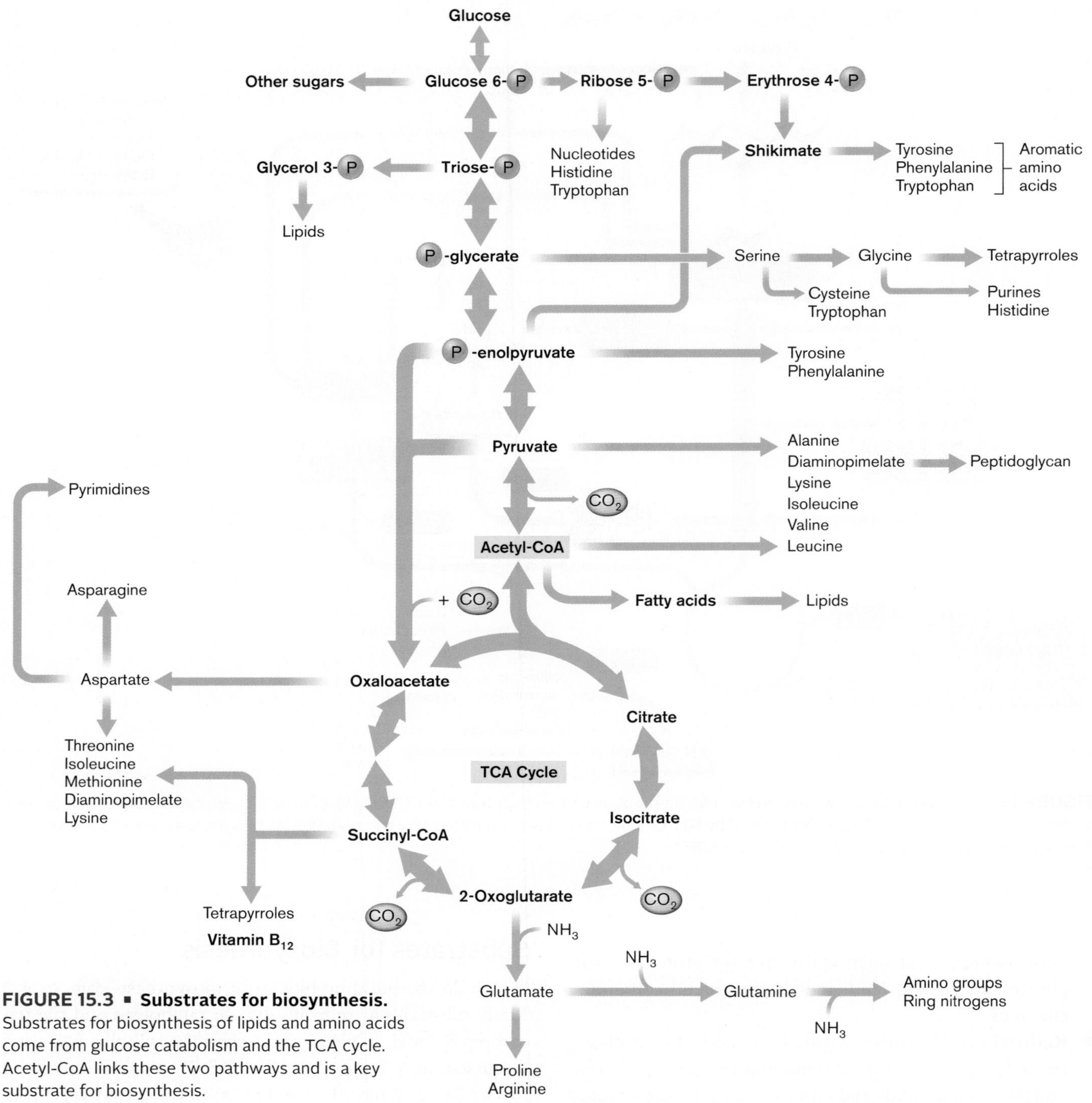

Glucose
Other sugars
Glucose 6-P
Ribose 5-P
Erythrose 4-P
Nucleotides
Histidine
Tryptophan
Shikimate
Tyrosine
Phenylalanine
Tryptophan
Aromatic amino acids
Glycerol 3-P
Triose-P
Lipids
P-glycerate
Serine
Glycine
Tetrapyrroles
Cysteine
Tryptophan
Purines
Histidine
P-enolpyruvate
Tyrosine
Phenylalanine
Pyruvate
Alanine
Diaminopimelate
Peptidoglycan
Lysine
Isoleucine
Valine
Leucine
CO_2
Acetyl-CoA
Pyrimidines
Asparagine
+ CO_2
Fatty acids
Lipids
Aspartate
Oxaloacetate
Citrate
Threonine
Isoleucine
Methionine
Diaminopimelate
Lysine
TCA Cycle
Succinyl-CoA
Isocitrate
2-Oxoglutarate
CO_2
CO_2
Tetrapyrroles
Vitamin B_{12}
NH_3
NH_3
Glutamate
Glutamine
Amino groups
Ring nitrogens
NH_3
Proline
Arginine

FIGURE 15.3 ▪ Substrates for biosynthesis. Substrates for biosynthesis of lipids and amino acids come from glucose catabolism and the TCA cycle. Acetyl-CoA links these two pathways and is a key substrate for biosynthesis.

carbon skeleton from TCA cycle intermediates oxaloacetate and 2-oxoglutarate, which incorporate nitrogen in the form of ammonium ion (NH_4^+).

Note that glucose catabolism and the TCA cycle are reversible; some autotrophs can synthesize entire sugar molecules, starting with CO_2 and working up through the reverse TCA cycle and reverse glycolysis. Thus, many common metabolites are available both to autotrophs and to heterotrophs, as well as to microbes of mixed metabolism.

Biosynthesis Spends Energy

Biosynthesis costs energy in several ways. The organism's genome must maintain the DNA that encodes all the enzymes that catalyze all the steps of the pathway; a mutation in any one enzyme may negate the entire process. Enzymes couple synthetic reactions to reactions releasing energy. The genomic and energetic costs lead microbes to evolve several strategies to control these costs:

- **Regulation.** Biosynthesis is regulated at several levels by the enzyme products. In this way, microbes avoid making more than they need under given environmental conditions. Regulation occurs at the levels of transcription and translation of enzymes, as well as through feedback inhibition of enzyme activity.
- **Genome degeneration.** Each species makes an evolutionary "choice" whether to maintain the expense of a particular assimilatory or biosynthetic pathway, such as N_2 fixation, or to lose the pathway and become dependent on other species in the environment. Parasitic microbes that grow only inside a host cell show extensive loss of biosynthetic genes.
- **Secondary products.** Free-living bacteria often produce secondary products that are not essential nutrients but enhance nutrient uptake or inhibit competitors. Many secondary products are antibiotics, such as streptomycin, produced by the actinomycete *Streptomyces coelicolor.*

First we present the pathways by which autotrophs assimilate carbon and nitrogen into simple building blocks, such as acetyl-CoA. Then we will present pathways of construction of the key parts of the cell, such as fatty acids and amino acids. Finally, we will consider exciting recent discoveries in the biosynthesis of secondary products such as antibiotics.

15.2

CO_2 Fixation: The Calvin Cycle

The fundamental significance of **carbon dioxide fixation** by green plants and algae was recognized in the early twentieth century. "Fixation" refers to the covalent incorporation of a small molecule into larger biochemical material. This incorporation of carbon requires tremendous energy input, as well as a large degree of reduction to incorporate hydrogen atoms.

The majority of the biomass on Earth consists of carbon fixed by chloroplasts and bacteria through the **reductive pentose phosphate cycle**, which recycles a pentose phosphate intermediate. This cycle is also known as the **Calvin cycle**, for which Melvin Calvin (1911–1997) was awarded the 1961 Nobel Prize in Chemistry. The mechanism of the pentose phosphate cycle was solved by Calvin with colleagues Andrew Benson and James Bassham, at UC Berkeley. The importance of the Calvin cycle to the global ecosystem can scarcely be overestimated; it plays a major role in removing atmospheric CO_2. The Calvin cycle's responsiveness to CO_2, temperature, and other factors must be considered in all models of global warming.

Note: The Calvin cycle is also known as the **Calvin-Benson cycle**, the **Calvin-Benson-Bassham cycle**, or the **CBB cycle**. This book uses the terms "Calvin cycle" and "CBB cycle."

The Calvin cycle is performed by several categories of bacteria (**Fig. 15.4** and **Table 15.1**). Photoautotrophs such

A. B. C.

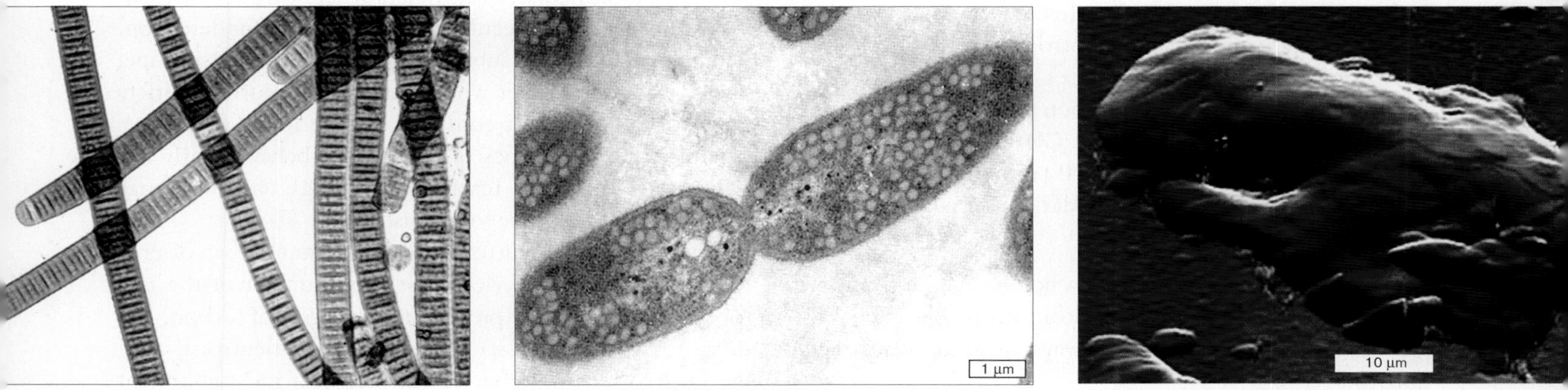
1 µm
10 µm

MICHAEL CLAYTON, U. WISCONSIN, MADISON — MCKENZIE AND KAPLAN, UT-MEDICAL SCHOOL, HOUSTON — S. N. TAN AND M. CHEN. 2012. *HYDROMETALLURGY* **119–120**: 87–94

FIGURE 15.4 ■ Phototrophs and chemolithoautotrophs use the Calvin cycle. **A.** *Oscillatoria,* a filamentous cyanobacterium, fixes CO_2 through oxygenic photosynthesis. Filaments are about 8 µm wide (light micrograph). **B.** *Rhodobacter sphaeroides,* purple photoheterotrophs with spherical photosynthetic membranes (TEM). **C.** The chemolithoautotroph *Acidithiobacillus ferrooxidans* corrodes concrete while oxidizing sulfur and iron. The bacteria use the Calvin cycle to fix CO_2 (TEM).

TABLE 15.1

Carbon dioxide fixation pathways.

Pathway	Organisms in which pathways occur: Bacteria	Archaea	Eukaryotes
Photoautotrophs, photoheterotrophs, and lithoautotrophs (Section 15.2)			
Calvin cycle	Cyanobacteria; purple phototrophs; lithotrophs	Rubisco homologs appear, but their function is unclear	Chloroplasts
Lithoautotrophic bacteria and archaea (Section 15.3)			
Reductive (reverse) TCA cycle	Green sulfur phototrophs (*Chlorobium*); thermophilic epsilonproteobacteria	Hyperthermophilic sulfur oxidizers (*Thermoproteus* and *Pyrobaculum*)	Anaplerotic reactions fix CO_2 to regenerate TCA intermediates
Reductive acetyl-CoA pathway	Anaerobes: acetogenic bacteria and sulfate reducers	Methanogens; other anaerobes	None known
3-Hydroxypropionate cycle	Green phototrophs (*Chloroflexus*)	Aerobic sulfur oxidizers (*Sulfolobus*)	None known

as cyanobacteria—and all plant chloroplasts—fix CO_2 using the Calvin cycle. So do some photoheterotrophs, such as *Rhodobacter sphaeroides*, when light is available; in the dark, they use preformed substrates (heterotrophy). Lithoautotrophs such as *Acidithiobacillus ferrooxidans* fix CO_2 using energy from metal oxidation.

- **Oxygenic phototrophic bacteria (cyanobacteria) and the chloroplasts of algae and multicellular plants fix CO_2** by the Calvin cycle coupled to oxygenic photosynthesis. The Calvin cycle is also referred to as the "dark reactions," or "light independent reactions," of oxygenic photosynthesis. Cyanobacteria are believed to generate the majority of oxygen gas in Earth's atmosphere.
- **Facultatively anaerobic purple bacteria**, including sulfur oxidizers and photoheterotrophs such as *Rhodospirillum* and *Rhodobacter*, use the Calvin cycle to fix CO_2. These bacteria also obtain carbon through catabolism.
- **Lithoautotrophic bacteria** fix CO_2 through the Calvin cycle, using NADPH and ATP provided by the oxidation of minerals. Mineral oxidation actually consumes oxygen.

To date, the Calvin cycle has not been found among archaea. Archaea and some lithoautotrophic bacteria fix carbon by several different pathways, as described in Section 15.3.

Calvin Cycle Intermediates

Early in the twentieth century, biochemists tried to figure out the mechanism of CO_2 fixation, believing that agricultural photosynthesis could be made more efficient. With the tools then available, however, researchers had no hope of sorting out the intermediate products through which CO_2 was fixed. Elucidating the pathways of CO_2 fixation required development of two fundamental tools: tracer radioisotopes, which are specific compounds labeled with radioactivity; and paper chromatography, a means of separating labeled compounds based on differential migration in a solvent. Both isotope labeling and chromatographic separation remain key tools of biochemical analysis today, enhanced by modern developments.

Carbon isotope labeling. As a means of tracking the conversion of intermediates within a biochemical pathway, specific atoms within molecules can be "labeled" for detection. Atoms are labeled by substituting an uncommon isotope, a version of an element whose number of neutrons differs from that of the most common isotope found in nature. Chemically, the isotopes of an element behave nearly the same in biochemical reactions, although enzymes may show a slight preference for one isotope.

The number of neutrons added to the number of protons (atomic number) yields the mass number of the isotope. For example, the predominant isotope of carbon, ^{12}C (carbon-12), contains six protons and six neutrons, adding up to a mass number of 12. The next most abundant isotope, ^{13}C, contains seven neutrons. An isotopic label needs to be detected on the basis of its physical properties, such as its mass (by mass spectroscopy) or its nuclear magnetic resonance (by NMR spectroscopy). The most sensitive property for detection is radioactive decay. Radioactive decay requires an unstable isotope (radioisotope), such as ^{14}C.

In the 1930s, when Calvin started his research, few useful radioisotopes were known for biological molecules. The discovery of ^{14}C by Martin Kamen (1913–2002) in 1940 revolutionized biochemistry, enabling the discovery of all kinds of cellular metabolism (**eTopic 15.1**).

Paper chromatography separates intermediates of CO_2 fixation. When ^{14}C-labeled substrates became available in the 1940s, a method was needed to isolate and characterize the unknown intermediates. The technique of **paper chromatography** was developed by Calvin for his studies of carbon fixation. In paper chromatography, a mixture of chemicals is spotted on a special paper, one end of which is immersed in a carrier solvent (**Fig. 15.5A**). The solvent is then drawn up through the paper by capillary action. Because the various chemical components of the solution differ in their solubility, they travel in the solvent at different rates. Afterward, the paper is dried and turned at a right angle. A second solvent is then added and drawn through in the crosswise direction. In the second solvent, the chemicals travel at different rates than in the first solvent, and thus are separated in two dimensions. The chance of two different kinds of molecules migrating the same distance in both solvents is very low.

Figure 15.5B shows one of Calvin's original chromatographic separations. The alga *Chlorella* was exposed to light so that it could conduct photosynthesis. $^{14}CO_2$ was added. Then, samples of the alga were killed at specific times by being plunged into boiling alcohol. Five seconds after the addition of $^{14}CO_2$, the main spots showing radioactivity correspond to 3-phosphoglycerate (PGA) and glyceraldehyde 3-phosphate (G3P), early intermediates of CO_2 fixation. Thirty seconds after $^{14}CO_2$ was added, radiolabel is detected in subsequent products, such as amino acids. The chromatography data illustrate the extraordinary speed of CO_2 fixation, while revealing a glimpse of its earliest intermediates. At even earlier times (2 seconds), only PGA shows radiolabel.

Figure 15.5C shows the portion of the cycle deduced from the data: CO_2 is incorporated to make PGA, and then is hydrogenated to G3P by an unknown reducing agent (later shown to be NADPH). In this particular experiment, the key intermediate that first assimilates CO_2 remains undetected. The key intermediate was later identified as the five-carbon sugar ribulose 1,5-bisphosphate.

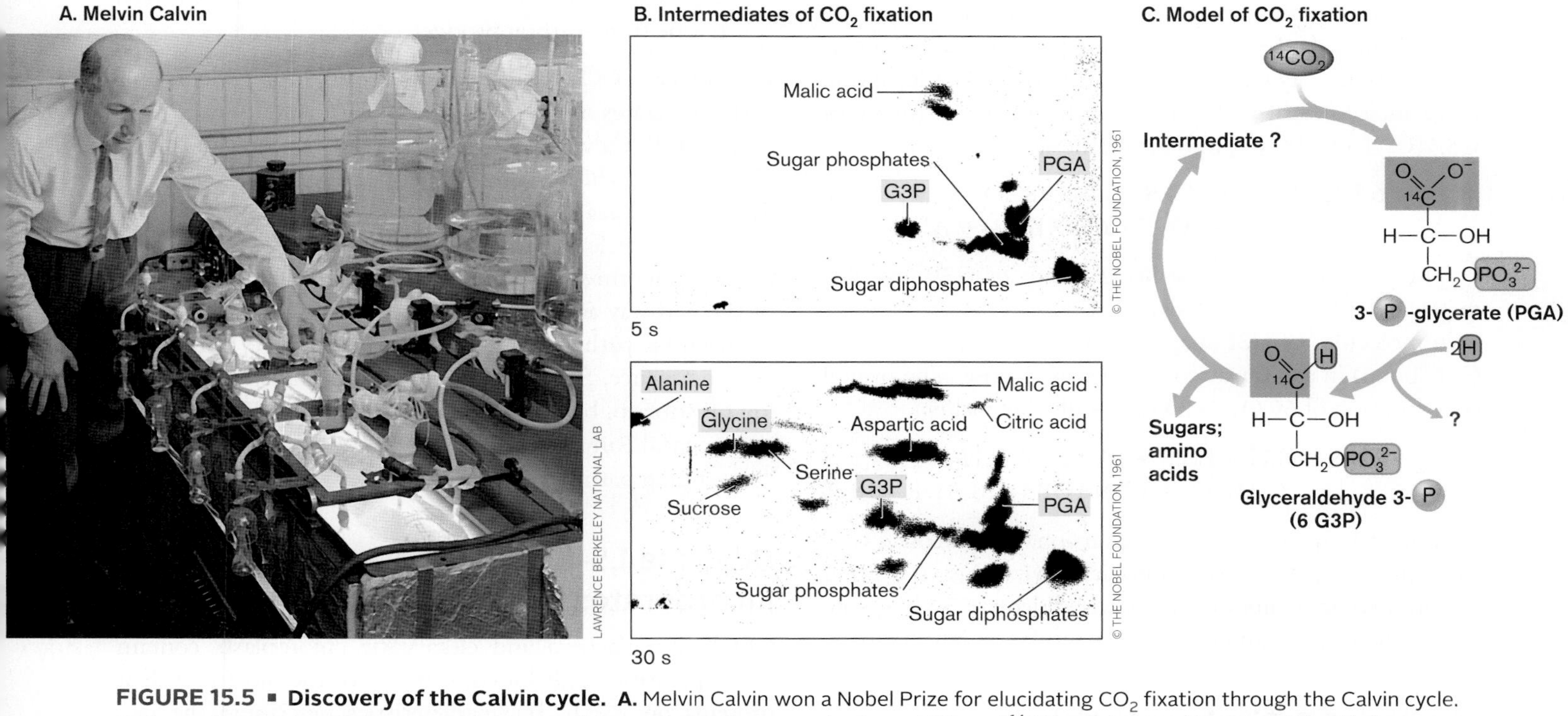
A. Melvin Calvin
LAWRENCE BERKELEY NATIONAL LAB
B. Intermediates of CO2 fixation
Malic acid
Sugar phosphates
PGA
G3P
Sugar diphosphates
5 s
© THE NOBEL FOUNDATION, 1961
Alanine
Malic acid
Glycine
Aspartic acid
Citric acid
Serine
G3P
Sucrose
PGA
Sugar phosphates
Sugar diphosphates
30 s
C. Model of CO2 fixation
14CO2
Intermediate ?
3-P-glycerate (PGA)
2H
?
Sugars; amino acids
Glyceraldehyde 3-P (6 G3P)

FIGURE 15.5 ■ Discovery of the Calvin cycle. A. Melvin Calvin won a Nobel Prize for elucidating CO_2 fixation through the Calvin cycle. **B.** Paper chromatography reveals early intermediate products. Five seconds after addition of $^{14}CO_2$ to photosynthesizing *Chlorella* (top panel), the main spots showing radioactivity contain 3-phosphoglycerate (PGA) and glyceraldehyde 3-phosphate (G3P). Thirty seconds after $^{14}CO_2$ addition (bottom panel), radiolabel has entered subsequent products, such as amino acids. **C.** Model of CO_2 fixation suggested by the chromatogram of ^{14}C-labeled intermediates.

Thought Question

15.1 Propose a simple experiment to reveal the key intermediate to receive CO_2.

Since Calvin's time, chromatographic separation has been developed into ever more sophisticated forms. High-performance liquid chromatography (HPLC) provides high-resolution separation of chemical products through columns packed with beads of various physical properties.

Overview of the Calvin Cycle

Decades of biochemical and genetic experiments established the details of CO_2 fixation in the Calvin cycle. An overview is shown in **Figure 15.6**; greater detail appears in **Figure 15.8**.

In each "turn" of the cycle, one molecule of CO_2 is condensed (combined, forming a new C–C bond) with the five-carbon sugar ribulose 1,5-bisphosphate. The resulting six-carbon intermediate splits into two molecules of 3-phosphoglycerate (PGA). The fixed CO_2 ultimately ends up as a carbon of glyceraldehyde 3-phosphate (G3P), reduced by $2H^+ + 2e^-$ from NADPH + H^+. Recall from Chapter 13 that NADPH is a phosphorylated derivative of NADH commonly associated with biosynthesis.

How does the fixed CO_2 become one "corner" of a glucose molecule? For every three turns of the cycle, fixing three molecules of CO_2, the cycle feeds one molecule of G3P ($C_3H_5O_3–PO_3^{2-}$) into biosynthesis:

$$3CO_2 + 6\ NADPH + 6H^+ + 9\ ATP + 9H_2O \rightarrow C_3H_5O_3–PO_3^{2-} + 6\ NADP^+ + 9\ ADP + 8\ P_i$$

The H_2O and the phosphoryl group of G3P are ultimately recycled during biosynthetic assimilation of G3P. Two molecules of G3P may condense (that is, form a new C–C bond) in a pathway to synthesize glucose. The overall condensation of $6CO_2 \rightarrow 2$ G3P $\rightarrow$ glucose yields:

$$6CO_2 + 12\ NADPH + 12H^+ + 18\ ATP + 18H_2O \rightarrow C_6H_{12}O_6 + 12\ NADP^+ + 18\ ADP + 18\ P_i$$

Alternatively, instead of conversion to glucose, G3P can enter biosynthesis of amino acids, vitamins, and other essential components of cells. Glyceraldehyde 3-phosphate is the fundamental unit of carbon assimilation into biomass.

Ribulose 1,5-Bisphosphate Is Reduced and Regenerated

Each "turn" of the Calvin cycle has three main phases: carboxylation and splitting into two three-carbon (3C) intermediates, reduction of two molecules of PGA to two molecules of G3P, and regeneration of ribulose 1,5-bisphosphate (see **Fig. 15.6**):

1. **Carboxylation and splitting: 6C → 2[3C].** Ribulose 1,5-bisphosphate condenses with CO_2 and H_2O, mediated by ribulose 1,5-bisphosphate carbon dioxide reductase/oxidase, generally referred to by the acronym **Rubisco**. Rubisco generates a six-carbon intermediate, which immediately hydrolyzes (splits into two parts by incorporating H_2O). The split produces two molecules of PGA, one of which contains the CO_2 fixed by this cycle.
2. **Reduction of PGA to G3P.** The carboxyl group of each PGA molecule is phosphorylated by ATP. The phosphorylated carboxyl group is then hydrolyzed and reduced by NADPH, forming G3P.
3. **Regeneration of ribulose 1,5-bisphosphate.** Of every six G3P, resulting from three cycles of fixing CO_2, five G3P enter a complex series of reactions (including hydrolysis of 3 ATP) to regenerate three molecules of ribulose 1,5-bisphosphate. The net conversion of five G3P molecules to three molecules of ribulose 1,5-bisphosphate releases $2H_2O$, restoring two of the $3H_2O$ fixed with $3CO_2$. The remaining sixth G3P exits the cycle, available to be used in the biosynthesis of sugars and amino acids. Thus, three fixed carbons lead to one three-carbon product.

Overall, each CO_2 fixed sends one carbon into biosynthesis and regenerates one ribulose 1,5-bisphosphate.

The full Calvin cycle occurs only in bacteria and in chloroplasts, which evolved from bacteria. It has not been found in archaea or in the cytoplasm of eukaryotes. Thus, the Calvin cycle appears to have evolved after the divergence of the three domains of life. The relative uniformity of the pathway across species, compared with the diversity of anaerobic pathways of CO_2 fixation, supports the view of late emergence. Some archaeal genomes do show a homolog of Rubisco, but this enzyme does not fix CO_2. Instead, the archaeal Rubisco forms PGA out of adenine, through a "scavenging pathway" of catabolism.

Ribulose 1,5-Bisphosphate Incorporates CO_2

CO_2-fixing bacteria and eukaryotic chloroplasts contain large quantities of Rubisco; it is one of the most prevalent proteins on Earth. Its characteristics determine the rate and efficiency of CO_2 fixation. The structure of Rubisco is highly conserved across bacterial and chloroplast domains. It consists of two types of subunits, designated small and large (**Fig. 15.7A**). The large (L) subunit contains

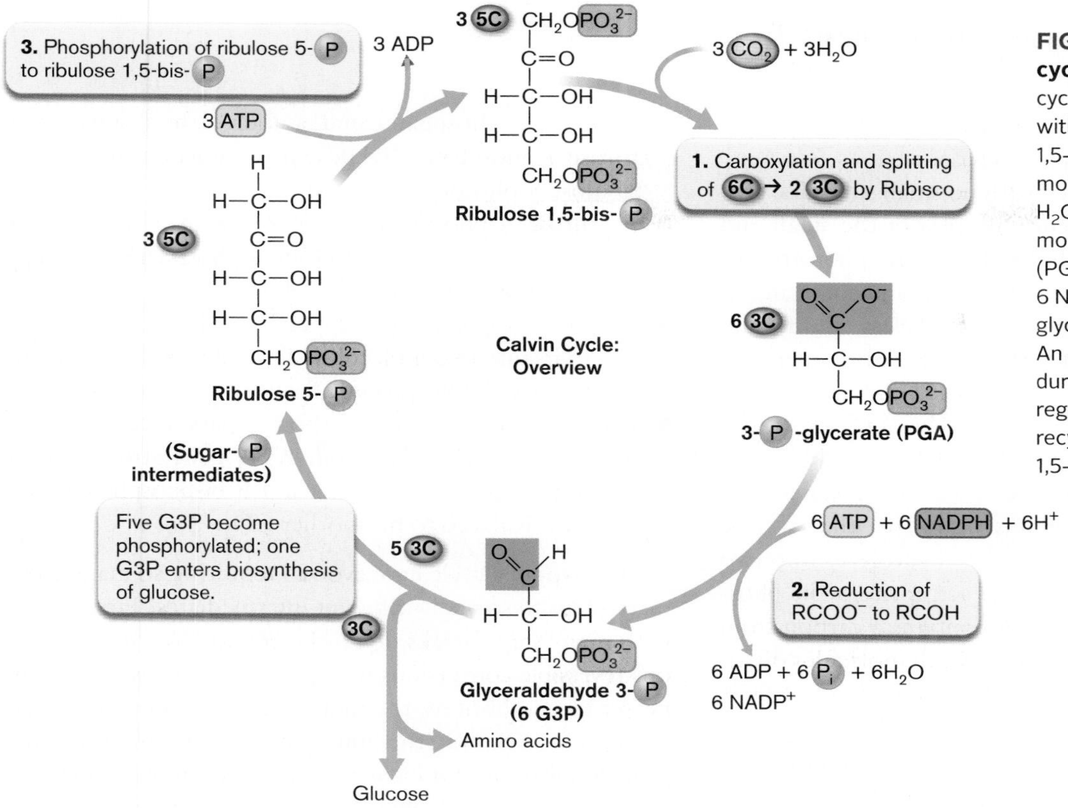
3. Phosphorylation of ribulose 5-P to ribulose 1,5-bis-P
3 ADP
3 ATP
3 5C
Ribulose 1,5-bis-P
3 CO2 + 3H2O
1. Carboxylation and splitting of 6C → 2 3C by Rubisco
Ribulose 5-P
Calvin Cycle: Overview
6 3C
3-P-glycerate (PGA)
(Sugar-P intermediates)
Five G3P become phosphorylated; one G3P enters biosynthesis of glucose.
5 3C
3C
6 ATP + 6 NADPH + 6H+
2. Reduction of RCOO⁻ to RCOH
Glyceraldehyde 3-P (6 G3P)
6 ADP + 6 Pi + 6H2O
6 NADP+
Amino acids
Glucose

FIGURE 15.6 ▪ The Calvin cycle: overview. The Calvin cycle condenses CO_2 and H_2O with the intermediate ribulose 1,5-bisphosphate. Overall, three molecules each of CO_2 and of H_2O are fixed and split into six molecules of 3-phosphoglycerate (PGA), which are reduced by 6 NADPH (with 6 ATP) to six glyceraldehyde 3-phosphate (G3P). An additional 3 ATP are consumed during sugar exchange reactions to regenerate three molecules of the recycled 5C intermediate ribulose 1,5-bisphosphate.

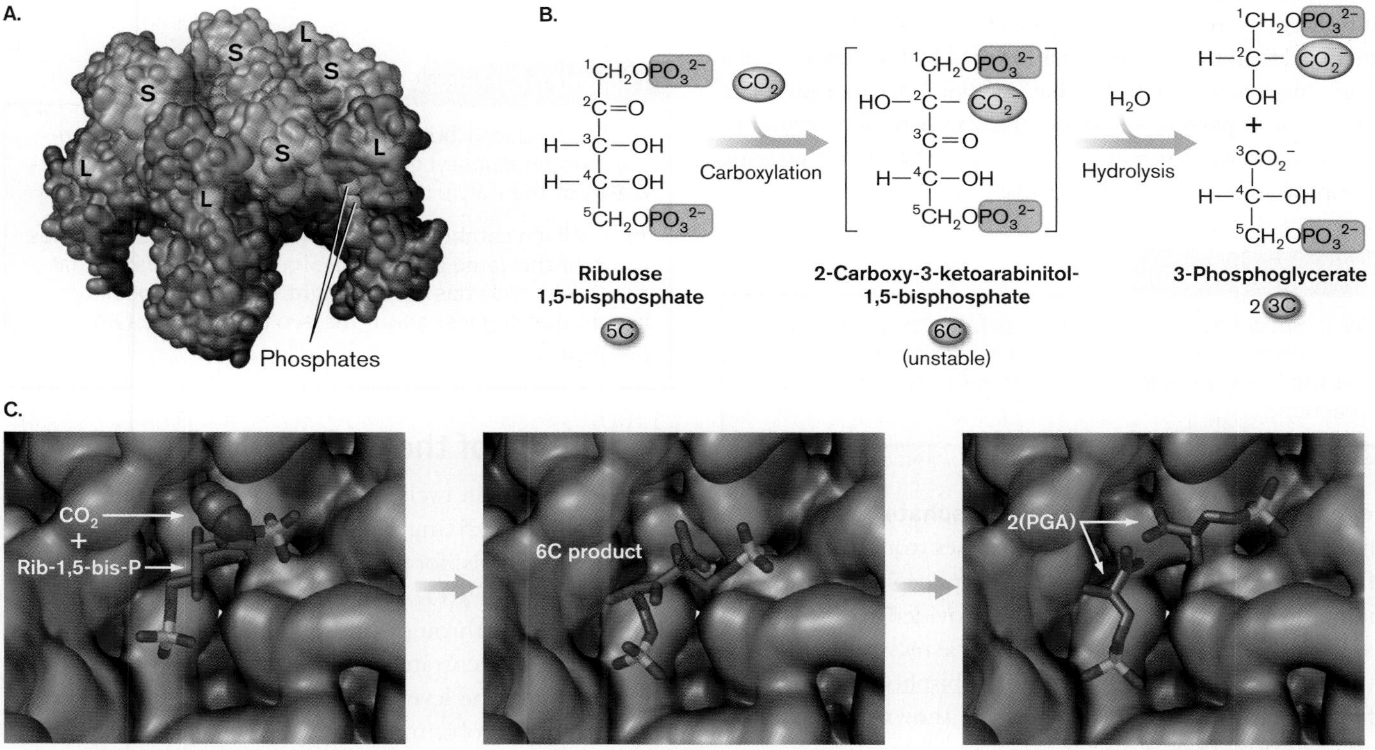
A.
S
L
Phosphates
B.
CO2
Carboxylation
H2O
Hydrolysis
Ribulose 1,5-bisphosphate
5C
2-Carboxy-3-ketoarabinitol-1,5-bisphosphate
6C
(unstable)
3-Phosphoglycerate
2 3C
C.
CO2
+
Rib-1,5-bis-P
6C product
2(PGA)

FIGURE 15.7 ▪ The mechanism of Rubisco. A. Rubisco from *Alcaligenes eutrophus* consists of eight large subunits (L) crowned by eight small subunits (S). Only the upper four L and S subunits are shown. On each L subunit, a catalytic site contains two phosphates (orange), which compete with the substrates for binding. (PDB code: 1BXN) **B.** Rubisco adds CO_2 to ribulose 1,5-bisphosphate to give an unstable six-carbon intermediate bound to the enzyme. The bound intermediate hydrolyzes to form two molecules of 3-phosphoglycerate (PGA). Both PGA molecules enter the cellular pool of PGA and behave equivalently in the rest of the cycle. **C.** The mechanism of CO_2 fixation at the catalytic site. CO_2 adds to the ketone, generating the 6C intermediate, which splits into two molecules of PGA. The two phosphates appear to act as anchors for the molecule(s) throughout CO_2 assimilation and cleavage. (PDB code: 1RUS) the active (catalytic) site. The function of the small (S) subunit remains unclear; some bacteria, such as *Rhodospirillum rubrum,* have a Rubisco with no small subunits. Different species contain different multiples of the small and large subunits: eight (in lithotrophs, green phototrophs, and some algae), six (in plant chloroplasts and some algae), or two (in purple phototrophs). Nevertheless, the fundamental mechanism of CO_2 fixation in all these organisms appears to be similar.

Mechanism of Rubisco. Initially, Rubisco adds CO_2 to ribulose 1,5-bisphosphate to give an unstable six-carbon intermediate that remains bound to the enzyme (**Fig. 15.7B** and **C**). The intermediate hydrolyzes into two molecules of PGA, each held at the active site by its phosphate. Only one molecule of PGA contains a carbon from the newly fixed CO_2; nevertheless, as both molecules dissociate from the enzyme they enter the cellular pool of PGA and behave equivalently in the rest of the cycle.

Rubisco has a high affinity for CO_2, and the typical concentration of Rubisco active sites (one per large subunit, six per complex) in plant chloroplasts is 4 mM—about 500 times greater than the concentration of CO_2. Thus, considerable energy is invested in CO_2 absorption. Yet the efficiency of carbon fixation by Rubisco is lowered by the existence of a competing reaction with O_2 that leads to 2-phosphoglycolate instead of 3-phosphoglycerate. Researchers are trying to engineer a more efficient oxygen-resistant Rubisco enzyme to improve efficiency of crop yields.

Thought Question

15.2 Speculate on why Rubisco catalyzes a competing reaction with oxygen. Why might researchers be unsuccessful in attempting to engineer Rubisco without this reaction?

Regeneration of ribulose 1,5-bisphosphate. Overall, each glyceraldehyde 3-phosphate (G3P) arises from three rounds of CO_2 fixation by ribulose 1,5-bisphosphate. Thus, to maintain the cycle, for each G3P provided to biosynthesis, five other molecules of G3P must be recycled into three five-carbon molecules of ribulose 1,5-bisphosphate. Details of regenerating the intermediate are shown in the lower half of **Figure 15.8**.

The regeneration pathway is summarized here:

- Two molecules of G3P condense to form the six-carbon sugar fructose 6-phosphate.
- Fructose 6-phosphate condenses with a third G3P. The nine-carbon molecule splits to form a five-carbon sugar (xylulose 5-phosphate) and a four-carbon sugar (erythrose 4-phosphate). Xylulose 5-phosphate rearranges to ribulose 5-phosphate.
- Erythrose 4-phosphate condenses with a fourth G3P (rearranged to dihydroxyacetone 3-phosphate) to make a seven-carbon sugar (sedoheptulose 7-phosphate).
- The seven-carbon sugar condenses with the fifth G3P and splits into two molecules of the ribulose 5-phosphate (via xylulose 5-phosphate and ribose 5-phosphate).
- Each of the three five-carbon sugars receives a second phosphoryl group from ATP, generating ribulose 1,5-bisphosphate. Each ribulose 1,5-bisphosphate is now ready for Rubisco to fix another CO_2.

Why would a cycle have evolved requiring so many enzymatic steps to so many different intermediates? First, a cycle with many steps breaks down the energy flow into numerous reversible conversions with near-zero values of ΔG. The nearer to equilibrium, the more energy is conserved in the conversion. Second, the multiple different intermediates provide substrates for biosynthesis. For example, some molecules of erythrose 4-phosphate and ribose 5-phosphate are withdrawn from the cycle to build aromatic amino acids and nucleotides.

Thought Questions

15.3 Why does ribulose 1,5-bisphosphate have to contain two phosphoryl groups, whereas the other intermediates of the Calvin cycle contain only one?

15.4 Which catabolic pathway (see Chapter 13) includes some of the same sugar-phosphate intermediates that the Calvin cycle has? What might these intermediates in common suggest about the evolution of the two pathways?

Regulation of the Calvin Cycle

How is the Calvin cycle organized and regulated? Expression of the CO_2-fixing enzymes varies with CO_2 concentration, light levels (for phototrophs), and temperature. The concentration of CO_2 is a special problem because CO_2 diffuses readily through phospholipid membranes. Thus, cells cannot concentrate this substrate across the cell membrane to reach the level needed to drive Rubisco. The gas concentration problem is solved by some species through enzymatic conversion of CO_2 to bicarbonate (HCO_3^-), which is trapped in the cytoplasm, unable to leak out of the cell membrane. This enzyme system is called the **carbon-concentrating mechanism (CCM)**. Other bacteria use alternative CO_2-fixing systems adapted to different CO_2 concentrations.

FIGURE 15.8 ■ The Calvin cycle in detail. The Calvin cycle assimilates three CO_2, forming six 3-phosphoglycerate (PGA) reduced to glyceraldehyde 3-phosphate (G3P), and converts five G3P into three ribulose 1,5-bisphosphate. Labels "3C," "5C," "6C," and so on indicate the number of carbon atoms per molecule.

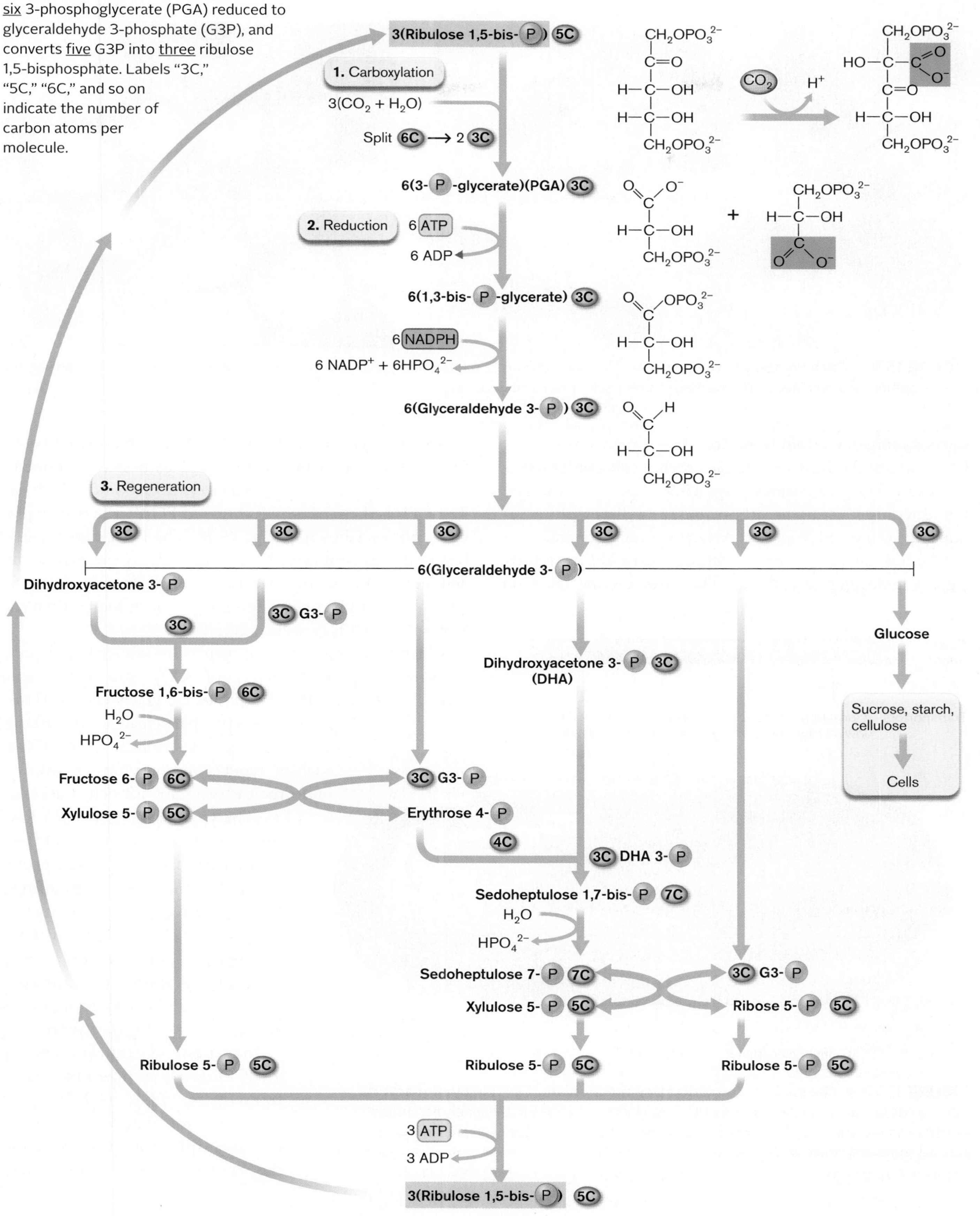
3(Ribulose 1,5-bis-P) 5C
1. Carboxylation
3(CO2 + H2O)
Split 6C ⟶ 2 3C
CO2
H+
6(3-P-glycerate)(PGA) 3C
2. Reduction
6 ATP
6 ADP
6(1,3-bis-P-glycerate) 3C
6 NADPH
6 NADP+ + 6HPO4 2−
6(Glyceraldehyde 3-P) 3C
3. Regeneration
6(Glyceraldehyde 3-P)
Dihydroxyacetone 3-P
3C G3-P
Fructose 1,6-bis-P 6C
H2O
HPO4 2−
Fructose 6-P 6C
Xylulose 5-P 5C
3C G3-P
Erythrose 4-P
4C
Dihydroxyacetone 3-P 3C
(DHA)
3C DHA 3-P
Sedoheptulose 1,7-bis-P 7C
Sedoheptulose 7-P 7C
Xylulose 5-P 5C
3C G3-P
Ribose 5-P 5C
Ribulose 5-P 5C
Glucose
Sucrose, starch, cellulose
Cells
3 ATP
3 ADP
3(Ribulose 1,5-bis-P) 5C

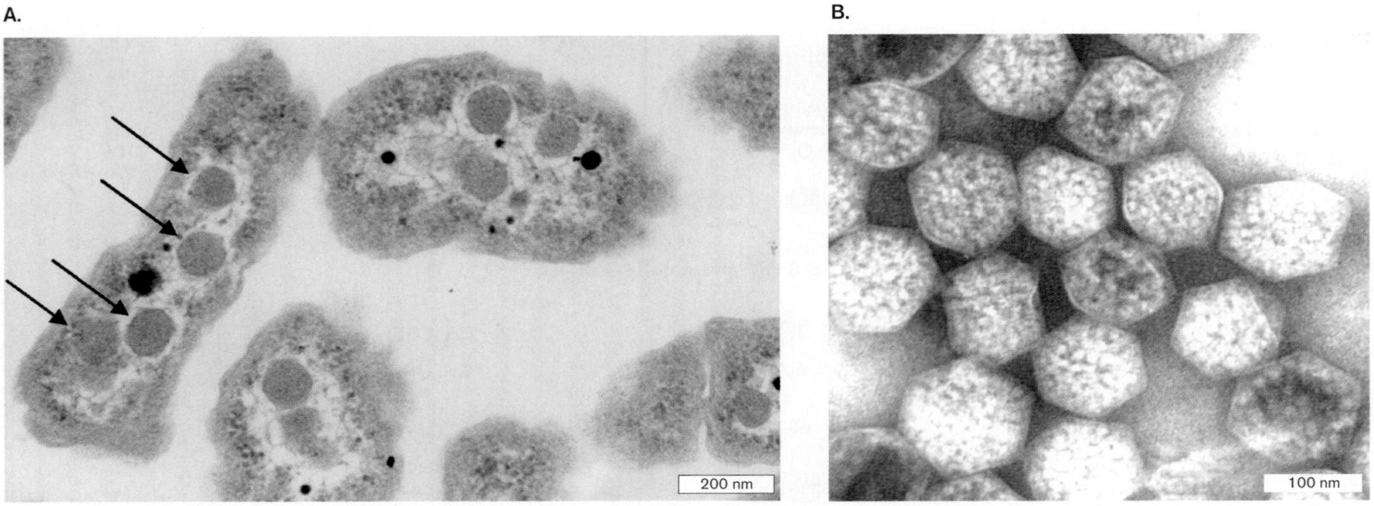
A.
B.
200 nm
100 nm

ADAPTED FROM TSAI, ET AL. 2007. *PLOS BIOLOGY* **5**: E144. MICROGRAPHS COURTESY OF SABINE HEINHORST.

FIGURE 15.9 ■ Carboxysomes. A. *Halothiobacillus neapolitanus,* a sulfur-oxidizing lithoautotroph. Thin section (TEM) showing polyhedral carboxysomes (arrows). **B.** Isolated carboxysomes, packed with Rubisco complexes (TEM).

Carboxysomes contain Rubisco. Many organisms that fix CO_2 contain the Rubisco complex within subcellular structures called **carboxysomes** (**Fig. 15.9**). Carboxysomes are found within CO_2-fixing lithotrophs, as well as within cyanobacteria and chloroplasts. A carboxysome consists of a polyhedral shell of protein subunits surrounding tightly packed molecules of Rubisco. The carboxysome takes up bicarbonate (converted from CO_2) by an unknown process. Once inside the carboxysome, the bicarbonate is immediately converted to CO_2 by the enzyme carbonic anhydrase. The CO_2 is then fixed by Rubisco to PGA—the first step of CO_2 fixation (shown in **Fig. 15.8**). The PGA exits the carboxysome to complete the Calvin cycle in the cytoplasm. Mutant strains of bacteria lacking carboxysomes can fix CO_2 only at high concentration (5%), much higher than the atmospheric CO_2 concentration (0.037%).

When CO_2 levels decline, genes are induced to express transmembrane transporters of CO_2 and bicarbonate (HCO_3^-) (**Fig. 15.10**). The inducible genes encode transport systems that participate in the carbon-concentrating mechanism. The CCM enables rapid concentration of HCO_3^- into the carboxysome for dehydration to CO_2. The CCM genes encode a low-affinity CO_2 transporter (NdhF4), a high-affinity CO_2 transporter (NdhF3), and two high-affinity HCO_3^- transporters (CmpABCD and SbtA). The low-affinity CO_2 transporter is effective only when CO_2 concentration is high. By contrast, the high-affinity transporters are induced by CO_2 starvation.

Induction of transporters by low CO_2 levels is shown by reverse transcriptase PCR (RT-PCR), a technique for detection of minute quantities of mRNA using the enzyme reverse transcriptase

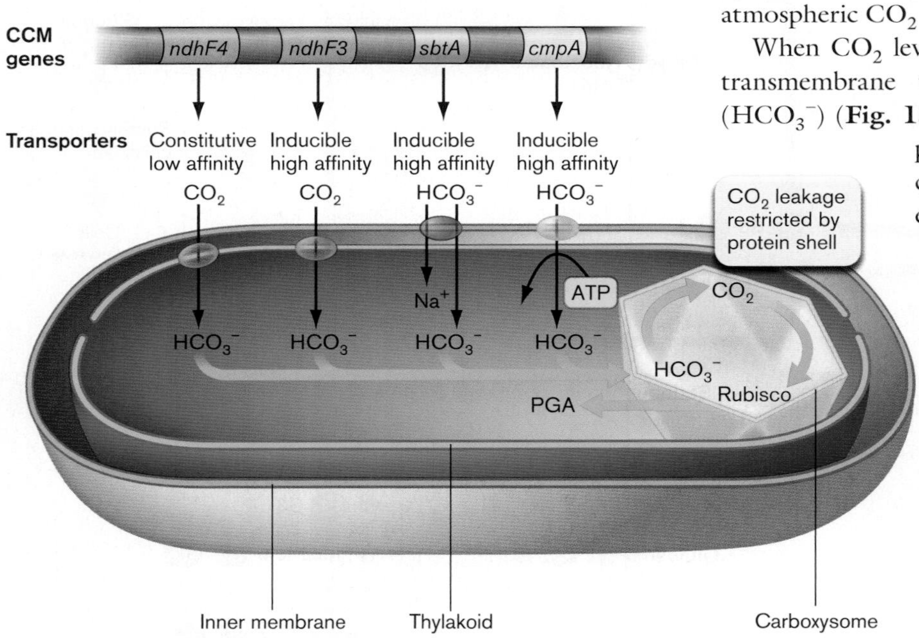
CCM genes
ndhF4
ndhF3
sbtA
cmpA
Transporters
Constitutive low affinity
Inducible high affinity
Inducible high affinity
Inducible high affinity
CO2
CO2
HCO3−
HCO3−
CO2 leakage restricted by protein shell
Na+
ATP
CO2
HCO3−
HCO3−
HCO3−
HCO3−
HCO3−
Rubisco
PGA
Inner membrane
Thylakoid
Carboxysome

FIGURE 15.10 ■ The carbon-concentrating mechanism (CCM) in cyanobacteria. CO_2 and HCO_3^- are brought into the cell by transport complexes in the inner membrane or thylakoid membranes of *Synechocystis* species. The HCO_3^- is concentrated in the carboxysome and converted to CO_2 by carbonic anhydrase. *Source:* Model from Dean Price and colleagues, Australian National University at Canberra; Patrick McGinn et al. 2003. *Plant Physiol.* **132**:218.

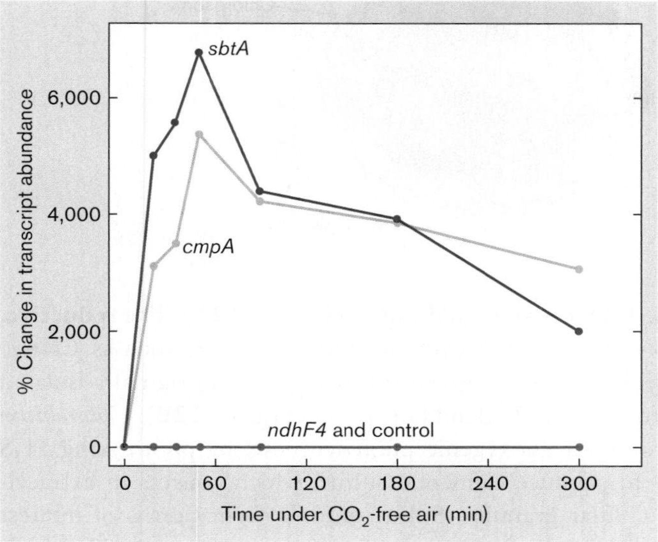

sbtA
cmpA
ndhF4 and control
% Change in transcript abundance
6,000
4,000
2,000
0
0
60
120
180
240
300
Time under CO2-free air (min)

FIGURE 15.11 ■ Time course of CCM gene transcription. When CO_2 is limiting, *Synechocystis* genes encoding the carbon uptake complexes show induced transcription. Induction is measured by reverse transcriptase PCR (RT-PCR). The *sbtA* and *cmpA* genes show especially high induction, whereas the constitutive system (*ndhF4*) shows no induction.

followed by the polymerase chain reaction, discussed in Chapter 12 and illustrated in Fig. 12.11. In RT-PCR, the RNA transcripts are reverse-transcribed to DNA; the DNA is then amplified by PCR to generate a detectable product. RNA transcripts are obtained for genes expressed before and after CO_2 limitation. The *sbtA* and *cmpA* transcripts, encoding HCO_3^- transporters, show particularly strong induction by CO_2 (**Fig. 15.11**). By contrast, the constitutive low-affinity CO_2 transporter *ndhF4* shows no induction.

Alternative CO_2 fixation systems. Instead of carboxysomes, some phototrophs possess alternative systems of CO_2 fixation adapted to different levels of CO_2. For example, the genome of the purple phototroph *Rhodobacter sphaeroides* includes two unlinked operons, each encoding a different set of CO_2 fixation genes, called form I and form II. Form I and form II each encode all the key enzymes, such as Rubisco. When CO_2 is limiting, form I enzymes work best; when CO_2 levels are saturating, form II enzymes work best.

To Summarize

- **The Calvin cycle** fixes CO_2 by reductive condensation with ribulose 1,5-bisphosphate. The Calvin cycle is used by cyanobacteria and chloroplasts and by some lithotrophs and photoheterotrophs.
- **^{14}C radiolabel and paper chromatography** were first used to identify intermediates of the Calvin cycle.
- **Rubisco** catalyzes the condensation of CO_2 with ribulose 1,5-bisphosphate. The six-carbon intermediate immediately splits into two molecules of 3-phosphoglycerate (PGA), which are activated by ATP and reduced by NADPH to glyceraldehyde 3-phosphate (G3P).
- **One of every six G3P is converted to glucose** or amino acids. The other five molecules of G3P undergo reactions to regenerate ribulose 1,5-bisphosphate.
- **Carboxysomes sequester and concentrate CO_2** for fixation by the Calvin cycle. CO_2 levels regulate gene expression for carboxysome transporters and the Calvin cycle.

15.3

CO_2 Fixation: Diverse Pathways

We have seen how oxygenic phototrophs and some lithoautotrophs fix CO_2 by the Calvin cycle. Other bacteria, however, fix CO_2 by alternative means (see **Table 15.1**). From the standpoint of evolution and phylogenetic distribution, the most ancient pathways are the reductive, or reverse, TCA cycle of anaerobic phototrophs and the reductive acetyl-CoA pathway of methanogens. Diverse kinds of CO_2 fixation play essential roles in soil, aquatic, and wetland ecosystems, as well as in the digestive systems of animals such as cattle.

The Reductive, or Reverse, TCA Cycle

Portions of the TCA cycle (discussed in Section 13.6) are found in all major groups of organisms. Most of the individual reaction steps of the TCA cycle are reversible, allowing the assimilation of small amounts of CO_2. Furthermore, all organisms, including humans, fix small amounts of CO_2 that regenerate TCA cycle intermediates. These regeneration steps are called **anaplerotic reactions**. Common anaplerotic reactions are the formation of oxaloacetate from phosphoenolpyruvate (PEP), catalyzed by PEP carboxylase; and the formation of malate from pyruvate, catalyzed by malic enzyme.

In some anaerobic bacteria and archaea, the entire TCA cycle functions in "reverse," reducing CO_2 to generate

A.

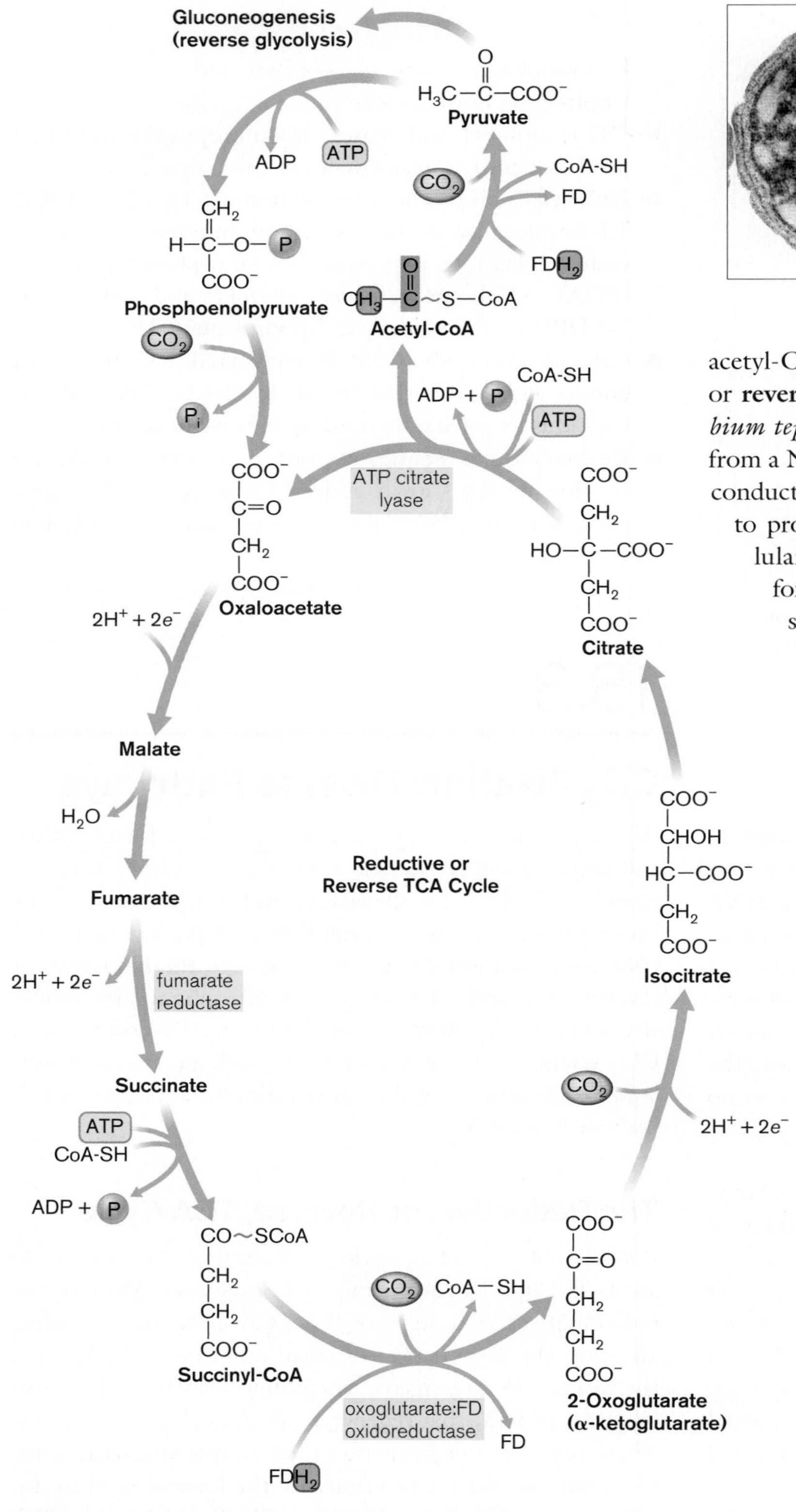

Gluconeogenesis (reverse glycolysis)
Pyruvate
ATP
ADP
CO2
CoA-SH
FD
FDH2
Phosphoenolpyruvate
Acetyl-CoA
CO2
Pi
CoA-SH
ADP + P
ATP
ATP citrate lyase
Oxaloacetate
Citrate
2H+ + 2e−
Malate
H2O
Fumarate
Reductive or Reverse TCA Cycle
Isocitrate
2H+ + 2e−
fumarate reductase
Succinate
CO2
2H+ + 2e−
ATP
CoA-SH
ADP + P
Succinyl-CoA
CO2
CoA−SH
2-Oxoglutarate (α-ketoglutarate)
oxoglutarate:FD oxidoreductase
FD
FDH2

B.

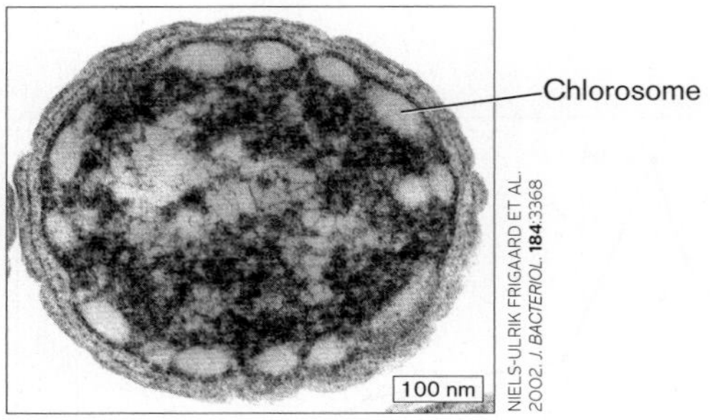

Chlorosome
100 nm
NIELS-ULRIK FRIGAARD ET AL. 2002. J. BACTERIOL. 184:3368

FIGURE 15.12 ▪ The reductive, or reverse, TCA cycle for CO_2 fixation. **A.** Anaerobic phototrophs and archaea use the reverse TCA cycle to fix carbon into biomass. CO_2 is fixed by several intermediates, including succinyl-CoA, 2-oxoglutarate, and acetyl-CoA. Reduction (addition of $2H^+ + 2e^-$) is performed by NADPH or NADH and by reduced ferredoxin (FDH_2). **B.** *Chlorobium tepidum,* a green sulfur bacterium that fixes CO_2 by the reductive TCA cycle (TEM). Chlorosome membranes contain the reaction centers of photolysis.

acetyl-CoA and build sugars (**Fig. 15.12A**). The **reductive,** or **reverse, TCA cycle** is used by bacteria such as *Chlorobium tepidum,* a green sulfur bacterium originally isolated from a New Zealand hot spring (**Fig. 15.12B**). *Chlorobium* conducts anoxygenic photosynthesis by photolyzing H_2S to produce elemental sulfur, which collects in extracellular granules. Sulfur granule formation is of interest for the development of a process to remove H_2S from sulfide-generating industries such as refining of coal and oil. The reductive TCA cycle is also used by epsilonproteobacteria of hydrothermal vent communities and by sulfur-reducing archaea, such as *Thermoproteus* and *Pyrobaculum.*

Reversal of the TCA cycle uses four or five ATP to fix four molecules of CO_2 and generates one molecule of oxaloacetate. The enzymes used are the same as for the "forward" cycle, except for three key enzymes that spend ATP to drive the reaction in reverse: ATP citrate lyase, 2-oxoglutarate:ferredoxin oxidoreductase, and fumarate reductase. For example, ATP citrate lyase catalyzes the cleavage of citrate into acetyl-CoA and oxaloacetate. CO_2 assimilation requires reduction by NADPH. The reductive TCA cycle is believed to be the most ancient means of CO_2 fixation and amino acid biosynthesis, the original cycle of biomass generation in the ancestors of all three living domains.

From CO_2 to acetyl-CoA, the reverse TCA cycle essentially reverses the overall cycle of catabolism:

$$2CO_2 + 2\ ATP + 8H^+ + 8e^- + H\text{-}SCoA \rightarrow CH_3CO\text{-}SCoA + 3H_2O + 2\ ADP + 2\ P_i$$

The coenzyme A is recycled as the acetyl group enters biosynthesis. For example, acetyl-CoA can assimilate another CO_2 with reduction to pyruvate. Pyruvate then builds up to glucose and related sugars by **gluconeogenesis** (reverse glycolysis), with expenditure of ATP and NADPH. Alternatively, acetyl-CoA can enter biosynthetic pathways to produce fatty acids or amino acids.

Unlike the Calvin cycle, the reverse TCA cycle provides several different molecular intermediates that can assimilate CO_2. Here are the key steps:

- Succinyl-CoA assimilates CO_2 to form 2-oxoglutarate (alpha-ketoglutarate).
- 2-Oxoglutarate assimilates CO_2 to form isocitrate.
- Acetyl-CoA (produced by the reverse TCA cycle) assimilates CO_2 to form pyruvate.

Each CO_2 assimilation requires one or more reduction steps. 2-Oxoglutarate is reduced by the protein ferredoxin (FDH_2). Ferredoxin also mediates acetyl-CoA reduction to pyruvate. Other reduction steps may be accomplished by NADPH or NADH.

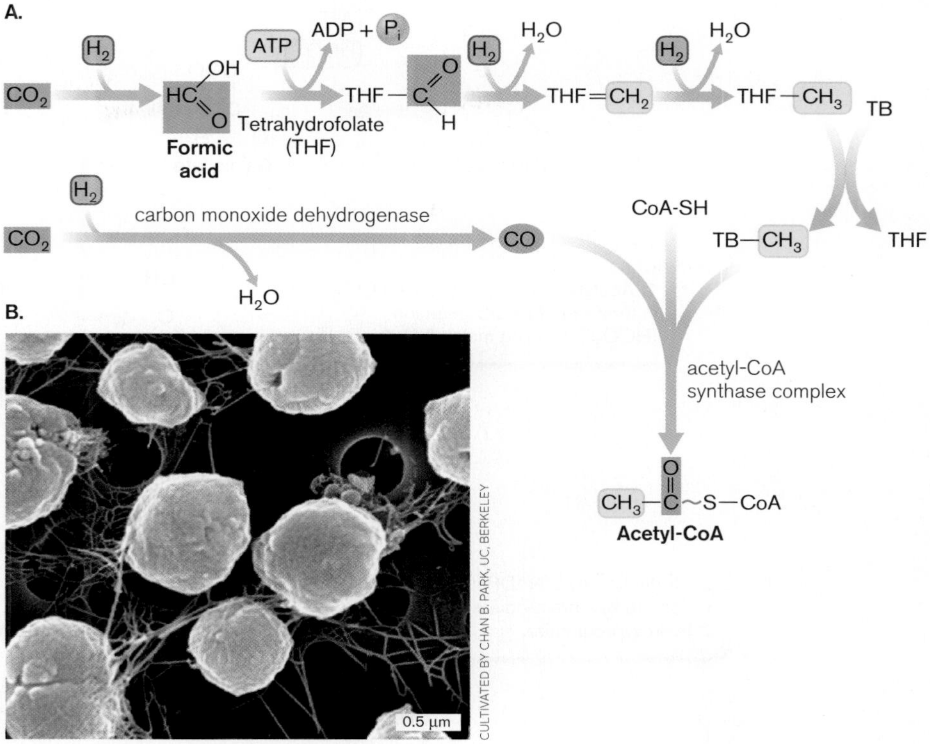
A.
CO_2
H_2
OH
HC
O
Formic acid
ATP
ADP + P_i
THF
Tetrahydrofolate (THF)
O
C
H
H_2
H_2O
THF=CH_2
H_2
H_2O
THF—CH_3
TB
THF
TB—CH_3
CoA-SH
CO_2
H_2
carbon monoxide dehydrogenase
CO
H_2O
acetyl-CoA synthase complex
CH_3—C~S—CoA
Acetyl-CoA
B.
0.5 µm
CULTIVATED BY CHAN B. PARK, UC BERKELEY

FIGURE 15.13 ■ The reductive acetyl-CoA pathway of CO_2 fixation. A. The first CO_2 enters the linear pathway by reduction to formate and is transferred onto the cofactor tetrahydrofolate (THF). Following reduction, the methyl group is transferred to a vitamin B_{12}–like cofactor (TB). The second CO_2 is reduced to carbon monoxide (CO) by the enzyme carbon monoxide dehydrogenase, and then incorporated into acetyl-CoA. **B.** *Methanocaldococcus jannaschii,* a thermophilic marine methanogen that fixes CO_2 by the reductive acetyl-CoA pathway (SEM). Methanogens use a cofactor called methanopterin (MPT) instead of THF.

The Acetyl-CoA Pathway

Yet another route for CO_2 assimilation into acetyl-CoA and pyruvate is the **reductive acetyl-CoA pathway** (**Fig. 15.13**). In the acetyl-CoA pathway, two CO_2 molecules are condensed through converging pathways to form the acetyl group of acetyl-CoA. The acetyl-CoA pathway is used by anaerobic soil bacteria, such as *Clostridium thermoaceticum,* and by most autotrophic sulfate reducers, such as *Desulfobacterium autotrophicum.* It also provides the main source of biomass for methanogens. Most of the CO_2 absorbed by methanogens is used to yield energy by reducing CO_2 to methane (discussed in Chapter 14). However, 5% of the CO_2 absorbed by a methanogen enters the acetyl-CoA pathway for biosynthesis, generating the entire biomass of the organism.

The substrates and ultimate products of the acetyl-CoA pathway are the same as those of the reverse TCA cycle, except that the reducing agent is H_2 instead of NADPH. The acetyl-CoA pathway is slightly more efficient than the TCA cycle, requiring one less ATP:

$$2CO_2 + ATP + 4H_2 + H\text{-}SCoA \rightarrow CH_3CO\text{-}SCoA + 3H_2O + ADP + P_i$$

Nevertheless, the order of the pathway and its intermediate compounds are completely different from those of the TCA cycle. The reductive acetyl-CoA pathway is linear, with no recycled intermediates.

Reduction of the first CO_2. The first CO_2 enters the linear pathway by reduction to formate. The formate is transferred onto a complex carrier cofactor. In bacteria, the cofactor is tetrahydrofolate (THF), a reduced form of **folate**. Folate (folic acid) is a heteroaromatic cofactor; it is an essential vitamin required by many organisms, including humans. A different cofactor, methanopterin (MPT), carries the formate in methanogenic archaea. In either case, the formate carbon is reduced in successive steps to a methyl group ($-CH_3$).

Reductive CO_2 incorporation into acetyl-CoA. The second CO_2 is reduced to carbon monoxide (CO) by the enzyme carbon monoxide dehydrogenase. This same enzyme is used in reverse reaction by some lithotrophs to gain energy from carbon monoxide as an electron donor. For fixation, the CO is condensed with the methyl group carried by the vitamin B_{12}–like cofactor TB to form acetyl-CoA. The acetyl-CoA then enters pathways of biosynthesis, as it does when formed by the reverse TCA cycle.

A.

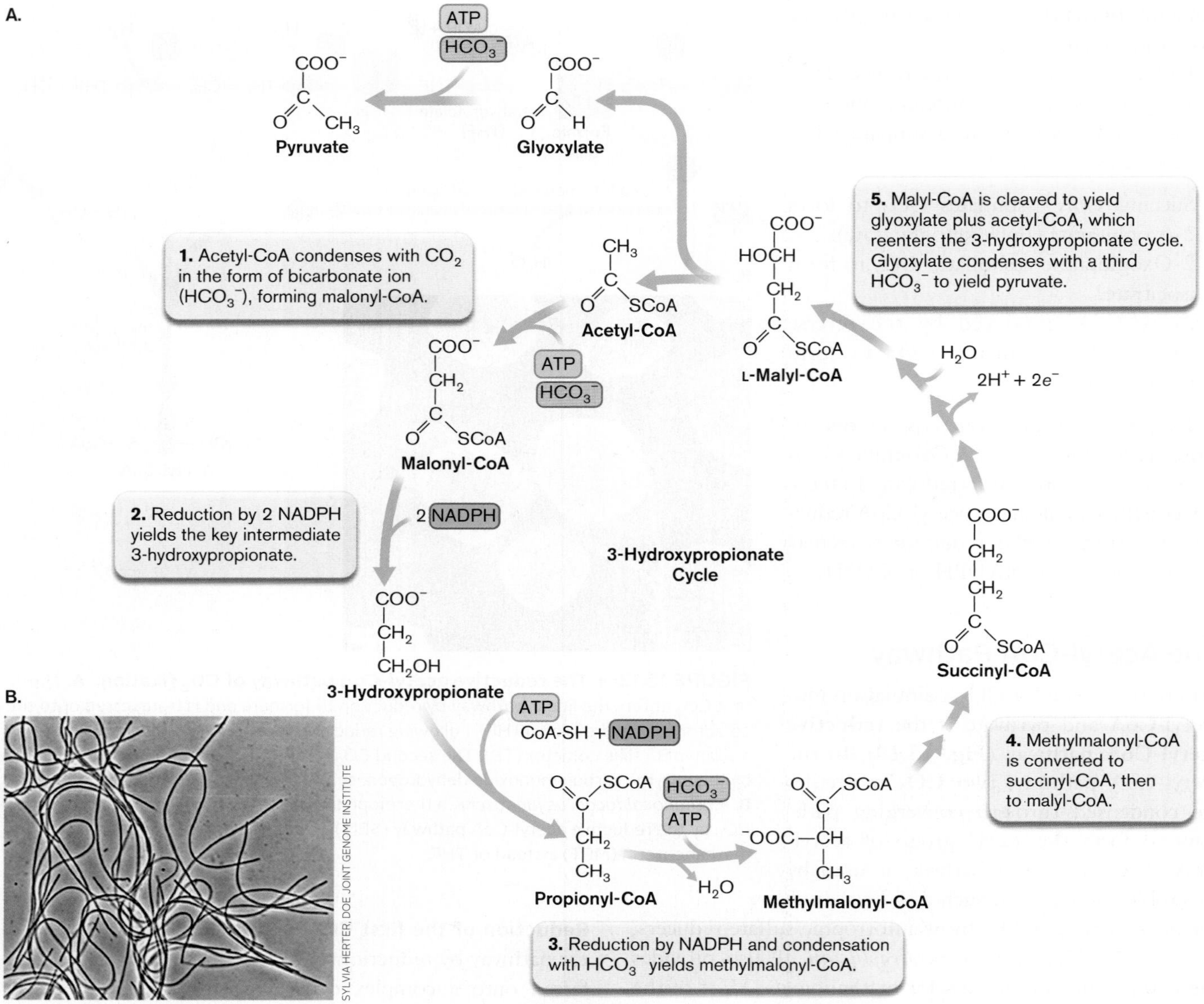
ATP
HCO_3^-
Pyruvate
Glyoxylate
1. Acetyl-CoA condenses with CO_2 in the form of bicarbonate ion (HCO_3^-), forming malonyl-CoA.
Acetyl-CoA
L-Malyl-CoA
5. Malyl-CoA is cleaved to yield glyoxylate plus acetyl-CoA, which reenters the 3-hydroxypropionate cycle. Glyoxylate condenses with a third HCO_3^- to yield pyruvate.
H_2O
$2H^+ + 2e^-$
Malonyl-CoA
2. Reduction by 2 NADPH yields the key intermediate 3-hydroxypropionate.
2 NADPH
3-Hydroxypropionate Cycle
Succinyl-CoA
3-Hydroxypropionate
B.
ATP
CoA-SH + NADPH
4. Methylmalonyl-CoA is converted to succinyl-CoA, then to malyl-CoA.
Propionyl-CoA
Methylmalonyl-CoA
3. Reduction by NADPH and condensation with HCO_3^- yields methylmalonyl-CoA.
SYLVIA HERTER, DOE JOINT GENOME INSTITUTE

C.

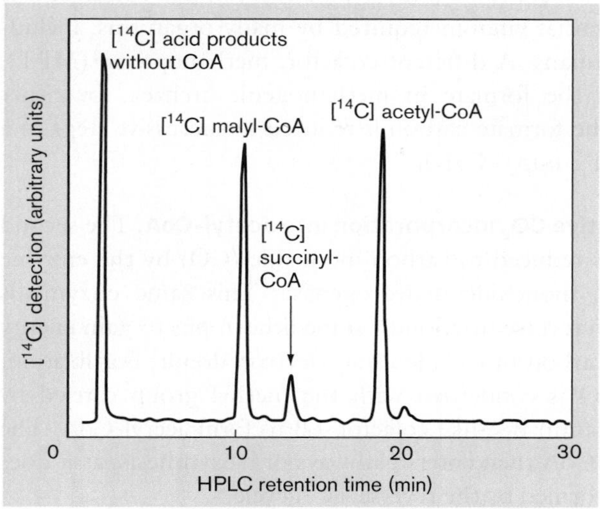
[14C] acid products without CoA
[14C] malyl-CoA
[14C] acetyl-CoA
[14C] succinyl-CoA
[14C] detection (arbitrary units)
0
10
20
30
HPLC retention time (min)

FIGURE 15.14 ▪ The 3-hydroxypropionate cycle of CO_2 fixation. **A.** In the 3-hydroxypropionate cycle, acetyl-CoA condenses with CO_2 in the form of bicarbonate ion (HCO_3^-), forming malonyl-CoA (1). Reduction by 2 NADPH forms the key intermediate 3-hydroxypropionate (2). Condensation with a third NADPH (3) and a second HCO_3^- forms methylmalonyl-CoA. Step 4 converts methylmalonyl-CoA to malyl-CoA. In step 5, malyl-CoA is cleaved to yield glyoxylate plus acetyl-CoA, which reenters the 3-hydroxypropionate cycle. Glyoxylate condenses with a third HCO_3^- to form pyruvate. **B.** *Chloroflexus,* a green phototroph that performs the 3-hydroxypropionate cycle. **C.** Products of [^{14}C] acetate are incorporated into the 3-hydroxypropionate cycle. After 1 minute, samples were separated by high-performance liquid chromatography (HPLC). The nonpolar coenzyme A esters of each acid separate in the HPLC solvent.
Source: Part A modified from Silke Friedmann et al. 2006. *J. Bacteriol.* **188**:2646; part C modified from Sylvia Herter et al. 2002. *J. Biol. Chem.* **277**:20277.

The fixation of CO_2 and release of methane by methanogens is a major concern for cattle farming. Within the rumen, bacteria convert cattle feed into CO_2 and H_2, and methanogens divert a significant portion of these gases into methane. The methane then escapes through burping and flatulence. If methanogen growth could be prevented, the CO_2 and H_2 could be fixed by other microbes into fermentation acids assimilated by the rumen. Beef production would then release less of the greenhouse gas methane.

The 3-Hydroxypropionate Cycle

The green photoheterotroph *Chloroflexus aurantiacus* fixes CO_2 yet another way: by the **3-hydroxypropionate cycle** (**Fig. 15.14**). *Chloroflexus* bacteria are filamentous thermophiles that absorb light at longer wavelengths than cyanobacteria and often grow below cyanobacteria in thick mats of biofilm. The 3-hydroxypropionate cycle is also used by archaea, such as the thermoacidophile *Acidianus brierleyi* and the aerobic sulfur oxidizer *Sulfolobus metallicus*.

In the 3-hydroxypropionate cycle, CO_2 is fixed by acetyl-CoA into 3-hydroxypropionate (**Fig. 15.14A**). Acetyl-CoA condenses with hydrated CO_2 (bicarbonate ion, HCO_3^-) and is reduced by 2 NADPH to 3-hydroxypropionate. Condensation with a second molecule of HCO_3^- forms methylmalonyl-CoA, followed by several intermediates to malyl-CoA. These intermediates were detected by [^{14}C] acetate incorporation into the metabolism of *Chloroflexus* (**Fig. 15.14C**). Malyl-CoA releases a molecule of acetyl-CoA to renew the cycle, plus glyoxylate. Glyoxylate condenses with a third HCO_3^- to form pyruvate, which enters standard routes for biosynthesis of sugars and amino acids. Thus, in all, three molecules of CO_2 are fixed into one molecule of pyruvate, which serves as a substrate for biosynthesis.

To Summarize

- **The reductive, or reverse, TCA cycle** fixes CO_2 in some anaerobes and archaea.
- **Anaplerotic reactions** in all organisms regenerate TCA cycle intermediates by fixing CO_2.
- **The acetyl-CoA pathway** in anaerobic bacteria and methanogens fixes CO_2 by condensation to form acetyl-CoA.
- **The 3-hydroxypropionate cycle** in *Chloroflexus* and in some hyperthermophilic archaea fixes CO_2 in a cycle generating the intermediate 3-hydroxypropionate.

15.4

Biosynthesis of Fatty Acids and Polyketides

The biosynthesis of cell parts begins with small, versatile substrates such as acetyl-CoA, a product of catabolic pathways such as the TCA cycle and benzoate catabolism (presented in Section 13.6). A key pathway of biosynthesis from acetyl-CoA generates fatty acids. Fatty acids condense with glycerol to form the phospholipids of cell membranes; they also form the lipid components of envelope proteins and are converted to enzyme cofactors. Fatty acid biosynthesis provides targets for antimicrobial drugs such as triclosan, an inhibitor of enoyl-ACP reductase. Related condensation pathways yield materials such as polyesters and polyketide antibiotics, such as tetracycline.

The repeating form of fatty acids, polyketides, and polyesters enables long-chain molecules to be synthesized with a relatively small number of different enzymes. By contrast, other cell components, such as amino acids, have more complex nonrepeating structures whose biosynthetic enzymes require long operons (presented in Section 15.6).

Fatty Acids Are Built from Repeating Units

Fatty acids exemplify the construction of cell components from repeating units (**Fig. 15.15**). The construction of molecules based on repeating units requires a cyclic pathway that feeds its products back repeatedly as substrates for further synthesis. The advantage of a cyclic process is that large polymers can be made with a limited number of enzymes; for example, the mycolic acids of *Mycobacterium tuberculosis* are extended to lengths of over 100 carbons.

While fatty acid structures used by cells include many different forms (discussed in Chapter 3), their synthesis always begins with the successive addition of two-carbon acetyl groups. Thus, most fatty acids contain an even number of carbon atoms.

The fatty acid synthase complex. The cyclic process of fatty acid synthesis is managed by the **fatty acid synthase complex**. The complex contains all the enzymes and component binding proteins bound together in proximity, so that all steps occur in one place without "losing" the unfinished molecule. The main bacterial version of this complex is designated FASII. All of the components are essential

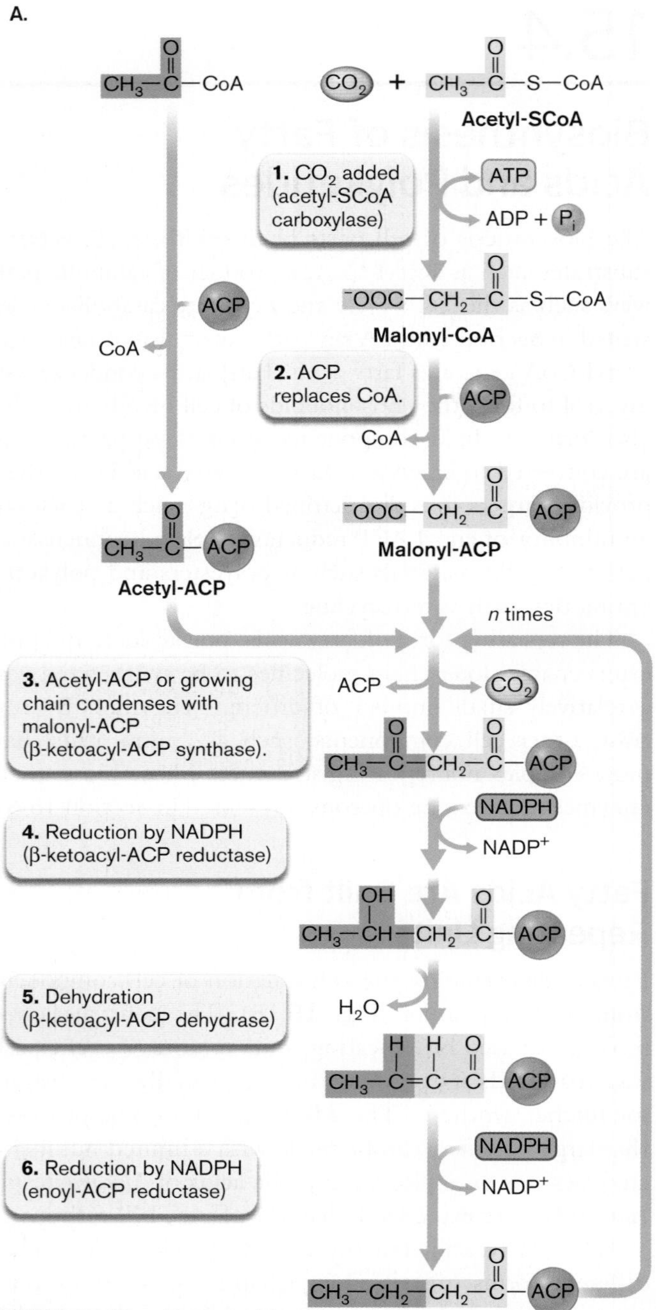
A.
CH3–C–CoA
CO2 + CH3–C–S–CoA
Acetyl-SCoA
1. CO2 added (acetyl-SCoA carboxylase)
ATP
ADP + Pi
⁻OOC–CH2–C–S–CoA
Malonyl-CoA
ACP
CoA
2. ACP replaces CoA.
ACP
CoA
⁻OOC–CH2–C–ACP
Malonyl-ACP
CH3–C–ACP
Acetyl-ACP
n times
3. Acetyl-ACP or growing chain condenses with malonyl-ACP (β-ketoacyl-ACP synthase).
ACP
CO2
CH3–C–CH2–C–ACP
NADPH
4. Reduction by NADPH (β-ketoacyl-ACP reductase)
NADP+
OH
CH3–CH–CH2–C–ACP
5. Dehydration (β-ketoacyl-ACP dehydrase)
H2O
CH3–C=C–C–ACP
NADPH
6. Reduction by NADPH (enoyl-ACP reductase)
NADP+
CH3–CH2–CH2–C–ACP

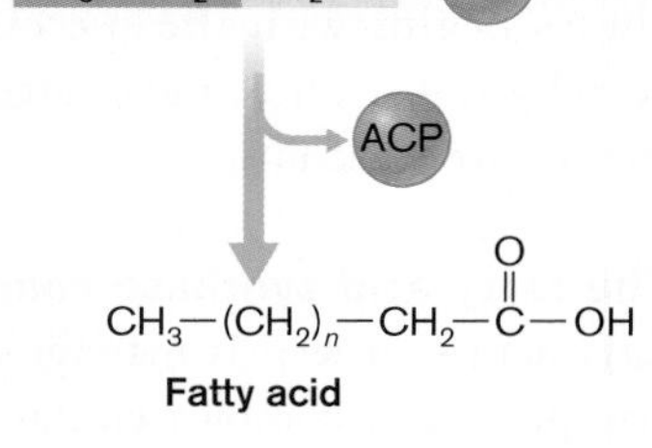
ACP
CH3–(CH2)n–CH2–C–OH
Fatty acid

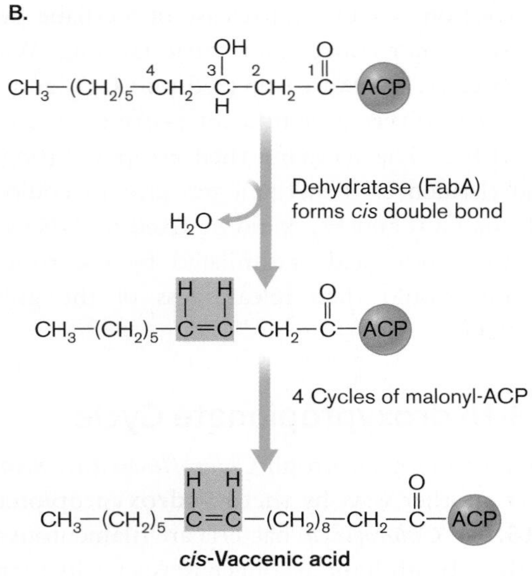
B.
CH3–(CH2)5–CH2–C–CH2–C–ACP
Dehydratase (FabA) forms cis double bond
H2O
CH3–(CH2)5–C=C–CH2–C–ACP
4 Cycles of malonyl-ACP
CH3–(CH2)5–C=C–(CH2)8–CH2–C–ACP
cis-Vaccenic acid

FIGURE 15.15 ▪ Fatty acid biosynthesis. **A.** Stepwise elongation of a saturated fatty acid. Units of acetyl-CoA are carboxylated to malonyl-CoA, and then successively condense to form the long chain of a fatty acid. Each two-carbon unit requires reduction by 2 NADPH. **B.** An alkene "kink" can be left unsaturated during chain extension. A special dehydratase enzyme forms the double bond between the third and fourth carbons, rather than the second and third. In this position, the double bond escapes reduction as the chain lengthens through subsequent cycles of malonyl addition.

for viability and thus are potential drug targets. Analogous multienzyme complexes are used by actinomycete bacteria to synthesize other long-chain products, such as polyketide antibiotics.

Activation of acetyl-CoA. Most acetyl groups within the cell are "tagged" with coenzyme A (CoA), a cofactor that directs acetate into catabolic pathways such as the TCA cycle. To redirect acetyl groups into fatty acid biosynthesis (**Fig. 15.15**), acetyl-CoA molecules are first tagged at the "back end" by condensation with CO_2 (**Fig. 15.15A**, step 1), catalyzed by acetyl-CoA carboxylase. The addition of CO_2 in this step does not count as carbon fixation, because the CO_2 added does not permanently add to the carbon skeleton. Instead, like phosphate or CoA, the CO_2 exists to be displaced when its function is no longer required. The CO_2-tagged acetyl-CoA is called malonyl-CoA. The coenzyme A is replaced by **acyl carrier protein (ACP)**, making malonyl-ACP (step 2).

Malonyl-ACP condenses with the growing chain. In step 3 (**Fig. 15.15A**), malonyl-ACP hooks onto the head of an acetyl-ACP or a longer-growing chain. The process of hooking onto the new unit involves displacement of CO_2 from the back end of malonyl-ACP, which replaces ACP from the front end of the growing chain. The growing chain now contains a ketone from the former acetyl group. The ketone must be reduced to CH_2 by 2 NADPH, gaining $4H^+ + 4e^-$ (steps 4–6). The chain is now fully hydrogenated (saturated).

Once hydrogenated, the chain is ready to take on the next malonyl-ACP, which, as before, loses CO_2 and replaces the ACP from the front of the growing chain (return to step 3). Successive addition can continue many times to build a saturated fatty acid.

Unsaturation. Certain fatty acids require unsaturated "kinks" in the chain to serve a structural purpose, such as to increase the fluidity of the membrane. Such a kink is formed by an unsaturated double bond between adjacent carbons (an alkene). Unsaturation can be generated during a cycle of elongation (**Fig. 15.14**, step 5). During the first four cycles of acetyl addition, an alkene bond forms between the second and third carbons of the fatty acid. After the fifth two-carbon addition, however, a special dehydratase enzyme generates the alkene double bond between the third and fourth carbons (**Fig. 15.15B**). This alkene fails to be hydrogenated further. Instead, several additional malonyl units are added, generating a long-chain fatty acid with a kink of a *cis* double bond.

Regulation of fatty acid synthesis. The synthesis of fatty acids consumes enormous quantities of reducing energy; thus, it must be regulated closely to avoid waste. From a structural standpoint, production of fatty acids incorporated into membranes must be balanced with growth of the cytoplasm. Furthermore, the many different variants of fatty acids are regulated in response to particular environmental needs. In *E. coli,* key points of regulation include the following:

- **Acetyl-CoA carboxylase represses its own transcription.** Transcription of an operon encoding two subunits of acetyl-CoA carboxylase (AccB, AccC) is repressed by one of its subunits, protein AccB. As AccB increases in concentration, it binds the promoter of the *accBC* operon, repressing further transcription. Thus, initiation of fatty acid biosynthesis (**Fig. 15.15A**, step 1) is always limited by the amount of AccB and AccC enzyme subunits that are present.
- **Starvation blocks fatty acid biosynthesis.** Starvation for carbon sources blocks fatty acid biosynthesis through the "stringent response." Blockage is mediated by the polyphosphorylated nucleotide ppGpp (guanosine tetraphosphate), the global regulator of the stringent response (discussed in Section 10.3).
- **Temperature regulates fatty acid composition.** Bacterial fatty acid composition is regulated by environmental factors such as temperature. In *E. coli,* low temperature favors unsaturated fatty acids because they are less rigid and maintain membrane flexibility. Low temperature induces expression of the gene *fabA* encoding the dehydratase enzyme that desaturates the fatty acid bond. As the dehydratase activity is increased, more unsaturated fatty acids are made.

Thought Question

15.5 For a given species, uniform thickness of a cell membrane requires uniform chain length of its fatty acids. How do you think chain length may be regulated?

Cyclic pathways of elongation comparable to fatty acid biosynthesis are used to build other kinds of polymers for energy storage. For example, many bacteria synthesize polyesters such as polyhydroxybutyrate (see **Fig. 15.1**). The term "polyester" indicates the multiple ester groups that are formed by repeated esterification of the carboxylic acid group of the chain with the hydroxyl group of a new alkanoate unit. Polyesters are insoluble in water, so they collect as storage granules within the bacterial cell, ready to release stored energy when needed. Polyester granules are synthesized by human pathogens such as *Legionella pneumophila,* the cause of legionellosis. Polyester storage helps *L. pneumophila* survive in water sources such as air-conditioning units. Commercially, polyesters produced by soil bacteria such as *Ralstonia eutropha* are used to manufacture biodegradable surgical sutures (see **Fig. 15.1B**).

Polyketide Antibiotics

The biosynthesis of fatty acids provides a model for study of the related biosynthesis of a major family of pharmaceuticals, the polyketides. A well-known polyketide is erythromycin, a broad-spectrum antibiotic frequently prescribed for pneumonia and other serious bacterial infections (**Fig. 15.16**). Polyketides are commonly produced by actinomycetes, particularly species of *Streptomyces* (eTopic 15.2). Antibiotics are produced as **secondary metabolites**, or **secondary products**—molecules that are not essential for survival but can enhance competition with other bacteria in a crowded natural environment.

The synthesis of polyketide antibiotics involves cyclic elongation by an enormous enzyme complex called a **modular enzyme**, which consists of multiple modules that add similar but nonidentical units to a growing chain. Like fatty acids, polyketides are built by the successive condensation of malonyl-ACP units, but each malonyl group carries a unique extension, or R group, instead of a CO_2. In erythromycin, the R groups are all methyl groups. In other polyketides, the R groups range from hydroxyl groups and amides to aromatic rings, some of which undergo secondary reactions and interconnections.

How is the order of different R groups determined? The modular enzyme contains a series of active sites, or modules, each of which catalyzes the addition of one of a series

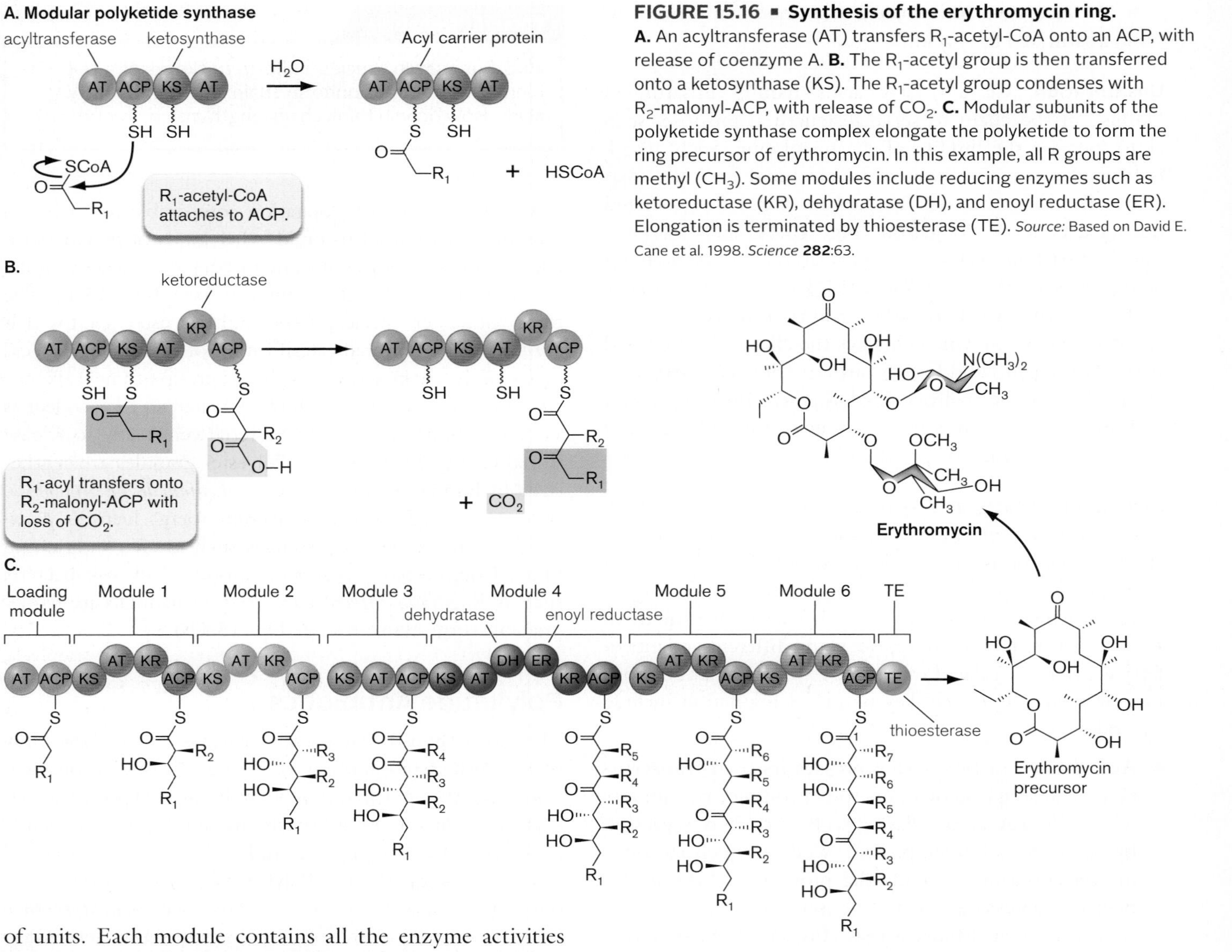
A. Modular polyketide synthase
acyltransferase
ketosynthase
Acyl carrier protein
AT
ACP
KS
SH
SCoA
H2O
HSCoA
R1-acetyl-CoA attaches to ACP.
B.
ketoreductase
KR
R1-acyl transfers onto R2-malonyl-ACP with loss of CO2.
CO2
C.
Loading module
Module 1
Module 2
Module 3
Module 4
Module 5
Module 6
TE
dehydratase
enoyl reductase
DH
ER
thioesterase
Erythromycin
Erythromycin precursor
N(CH3)2
OCH3

FIGURE 15.16 ■ Synthesis of the erythromycin ring. **A.** An acyltransferase (AT) transfers R_1-acetyl-CoA onto an ACP, with release of coenzyme A. **B.** The R_1-acetyl group is then transferred onto a ketosynthase (KS). The R_1-acetyl group condenses with R_2-malonyl-ACP, with release of CO_2. **C.** Modular subunits of the polyketide synthase complex elongate the polyketide to form the ring precursor of erythromycin. In this example, all R groups are methyl (CH_3). Some modules include reducing enzymes such as ketoreductase (KR), dehydratase (DH), and enoyl reductase (ER). Elongation is terminated by thioesterase (TE). *Source:* Based on David E. Cane et al. 1998. *Science* **282**:63.

of units. Each module contains all the enzyme activities needed to add the unit and recognizes a unit with one specific R group. The modular polyketide synthase may consist of a single protein with a series of domains, or it may consist of a complex of several protein subunits. Either the multidomain enzyme or the complex acts as an assembly line to generate the specific polyketide.

Figure 15.16A and **B** show the chain extension synthesis of a polyketide. The initial two domains of enzyme 1 are acyltransferase (AT) and an acyl carrier protein (ACP) domain, analogous to the ACP of fatty acid biosynthesis but specialized to accept one component of the polyketide. The acyltransferase transfers the acyl group (R_1-acetyl) from R_1-acetyl-CoA onto the ACP, with release of CoA (**Fig. 15.16A**). The R_1-acetyl is then transferred to a ketosynthase domain (KS).

Meanwhile, a malonyl group with its own R group (R_2-malonyl) has been transferred by a second AT onto the second ACP domain (**Fig. 15.16B**). The R_1-acetyl group then leaves KS and condenses with R_2-malonyl-ACP, releasing CO_2. The R_1R_2 acyl chain subsequently undergoes a series of extensions by R-malonyl groups (**Fig. 15.16C**). For erythromycin, all R groups are methyl; other polyketides have R groups that are more complex. The initial AT and ACP domains constitute a "loading module" for the first acyl group, whereas subsequent sets of domains constitute "extender modules" that each add a different R-malonyl group. Extender modules can include secondary activities such as dehydratase (DH; removes OH) and ketoreductase (KR; hydrogenates a ketone to OH). In principle, the modular approach can construct a limitless range of products with diverse antibiotic properties.

Elongation of the polyketide is terminated by thioesterase (TE), which hydrolyzes the thioester bond to the final ACP. This polyketide chain is an erythromycin precursor. The precursor requires additional enzymes (not shown)

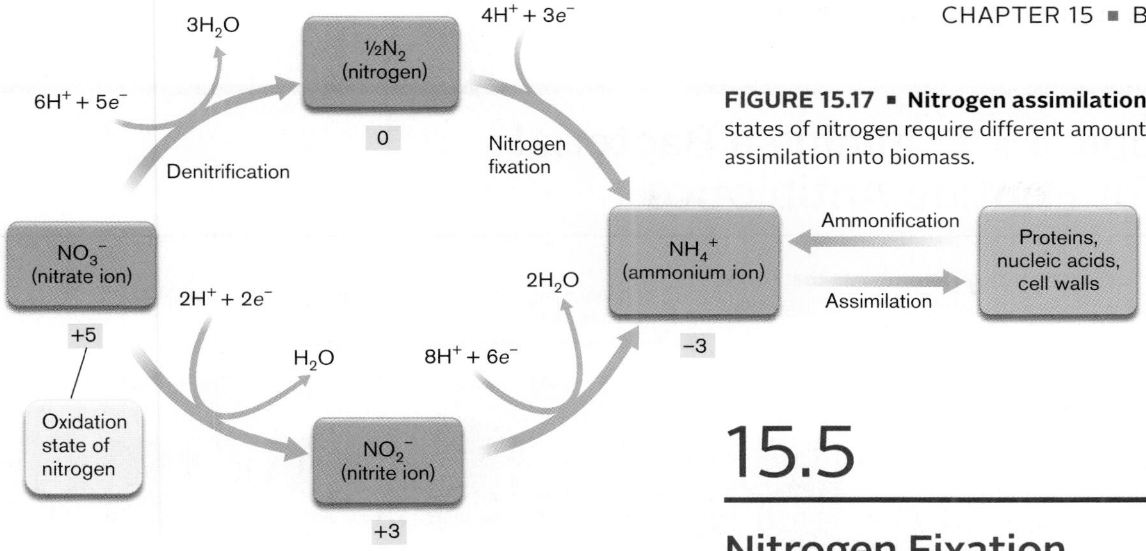
3H2O
½N2
(nitrogen)
4H+ + 3e−
6H+ + 5e−
0
Nitrogen
fixation
Denitrification
NO3−
(nitrate ion)
2H+ + 2e−
2H2O
NH4+
(ammonium ion)
Ammonification
Assimilation
Proteins,
nucleic acids,
cell walls
+5
H2O
8H+ + 6e−
−3
Oxidation
state of
nitrogen
NO2−
(nitrite ion)
+3

FIGURE 15.17 ■ Nitrogen assimilation. Different oxidation states of nitrogen require different amounts of reducing energy for assimilation into biomass.

to add extra components, such as sugars, to complete the erythromycin.

The concept of the modular enzyme is of growing importance to understanding antibiotic synthesis and bioengineering new products. Another class of antibiotics consists of polypeptides formed on modular enzymes instead of a ribosome, called nonribosomal peptides. The biosynthesis of a nonribosomal peptide antibiotic, vancomycin, is presented in **eTopic 15.3**. Modular enzyme biosynthesis of peptide antibiotics offers exciting prospects for the design of new drugs (see **Special Topic 15.1**).

To Summarize

- **Fatty acid biosynthesis** involves successive condensation of malonyl-ACP groups formed from acetyl groups tagged with acyl carrier protein (ACP) and a carboxylate. Each successive malonyl group is transferred onto the growing acyl chain, with release of CO_2.
- **The growing acyl chain is hydrogenated.** Each added unit is hydrogenated by two molecules of NADPH unless an unsaturated kink is required.
- **Some fatty acids are partly unsaturated.** An unsaturated kink may be generated by a special dehydratase, which generates the alkene double bond between the third and fourth carbons.
- **Fatty acid biosynthesis is regulated** by the levels of acetyl-CoA carboxylase and by the stringent response to carbon starvation. Bond saturation is regulated by temperature and other environmental factors.
- **Polyesters** for energy storage are synthesized by cyclic elongation of polyhydroxyalkanoates.
- **Polyketide antibiotics are synthesized by modular enzymes.** Nonribosomal peptide antibiotics are also made by modular enzymes.

15.5

Nitrogen Fixation

The synthesis of amino acids, cell walls, and other cofactors requires an additional element we have not yet considered—namely, nitrogen. In principle, nitrogen should be more accessible to cells than carbon is, since nitrogen gas (N_2) comprises more than three-quarters of our atmosphere. In fact, however, N_2, with its triple bond, is one of the most stable molecules in nature. An enormous input of energy is required to reduce nitrogen to ammonia for assimilation into carbon skeletons. Early in evolution, all cells may have fixed their own N_2, but today only certain species of bacteria and archaea retain the ability. All other organisms depend on reduced or oxidized forms of nitrogen, which ultimately derive from N_2. Thus, all living organisms, directly or indirectly, depend on N_2-fixing prokaryotes within the biosphere (discussed in Chapters 21 and 22).

Note: Bacteria and archaea play essential roles in global cycling of nitrogen, sulfur, and phosphorus. The geochemical cycling of these elements is discussed in Chapter 22.

Nitrogen Assimilation

To assimilate into biomass, nitrogen must be fully reduced to ammonia (NH_3), which at pH 7 is mostly protonated to ammonium ion (NH_4^+). Unlike carbon, nitrogen rarely appears in oxidized form in complex biomolecules. Inorganic forms, such as nitric oxide (NO), are used as defense mechanisms against invading pathogens (see Chapter 23) or as signaling molecules. In macromolecules, however, virtually all the nitrogen is reduced; organic compounds containing oxidized nitrogen are generally toxic.

While N_2 is the ultimate source and sink of biospheric nitrogen, several oxidized or reduced forms occur in the environment (**Fig. 15.17**). Most free-living bacteria can acquire nitrate (NO_3^-) or nitrite (NO_2^-) for reduction to ammonium ion, although they repress these energy-expensive pathways when ammonium ion is present.

Special Topic 15.1: Mining a Bacterial Genome for Peptide Antibiotics

Bacteria synthesize a vast array of peptide products that serve development, communication, and combat. These peptides have unique modifications that depart from the standard 20 amino acid residues. Some peptides are translated by ribosomes and then modified, whereas others, the nonribosomal peptides, are synthesized entirely by modular enzymes comparable to those that make polyketides. Nonribosomal peptides include powerful antibiotics such as vancomycin and daptomycin, and immunosuppressant drugs such as cyclosporine. Vancomycin biosynthesis is presented in **eTopic 15.3**.

As resistance evolves, in our arms race with pathogens, we always need more new antibiotics. The stendomycins are an emerging class of lipopeptides (peptides attached to lipid) that have antifungal activity (**Fig. 1**). Where do we find them? We can mine the genomes of the limitless strains of bacteria in our environment, which undergo their own arms races with each other. The challenge is (1) to detect previously unknown antibiotics within cells—a chemical "needle in the haystack"—and (2) to reveal the genes encoding their biosynthetic enzymes. The genes can then be cloned in an engineered fermenter strain (discussed in Chapter 16).

So which clues come first—the peptide or the biosynthetic genes? Either the peptide or the gene sequence may offer a clue, which then serves to interrogate the other side. A multistep interactive approach between peptide and gene was devised by Pieter Dorrestein and colleagues at UC San Diego (**Fig. 2**). The approach is called peptidogenomics. **Figure 3** outlines the peptidogenomic approach that identified the structure and biosynthetic genes for stendomycin I–VI, an antifungal agent produced by the actinomycete *Streptomyces hygroscopicus*.

Stendomycin was the first antibiotic identified in the bacterium *S. hygroscopicus* by a version of mass spectrometry called MALDI-TOF (matrix-associated laser desorption/ionization time-of-flight) mass spectrometry. In the MALDI technique, a UV laser beam is trained on a sample embedded in a chemical matrix. A thin portion of sample plus matrix is "desorbed"—that is, ionized—and the ions are taken up into the mass spectrometer. The "time of flight" into the detector offers a measure of the size of the ionized molecules.

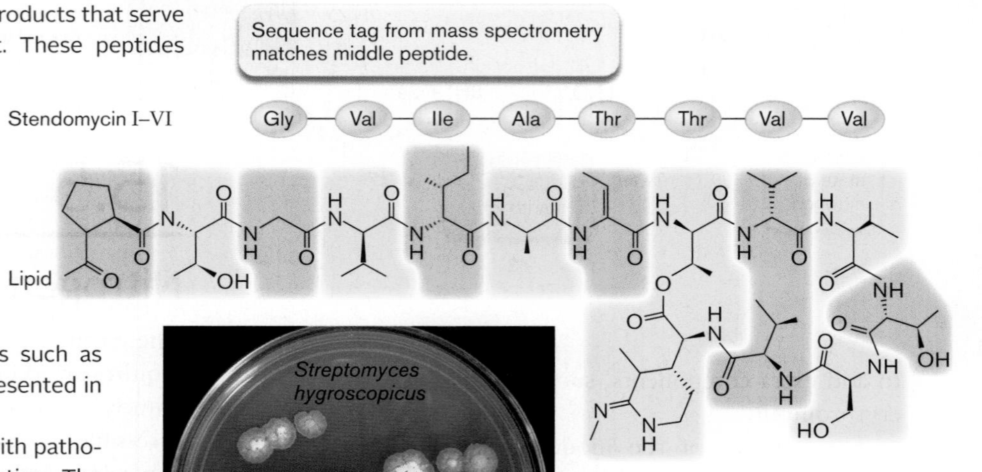
Sequence tag from mass spectrometry matches middle peptide.
Stendomycin I–VI
Gly
Val
Ile
Ala
Thr
Thr
Val
Val
Lipid

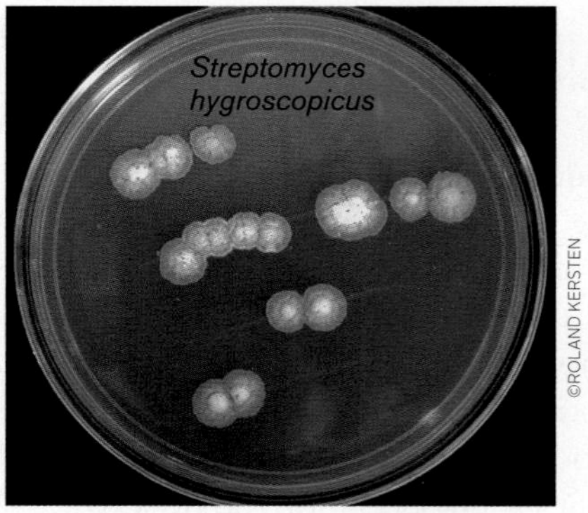
Streptomyces hygroscopicus
©ROLAND KERSTEN

FIGURE 1 ■ A nonribosomal peptide antibiotic, stendomycin I–VI, is synthesized by *Streptomyces hygroscopicus.*

COURTESY PIETER DORRESTEIN

FIGURE 2 ■ Pieter Dorrestein, at UC San Diego, discovers new antibiotic peptides by peptidogenomics.

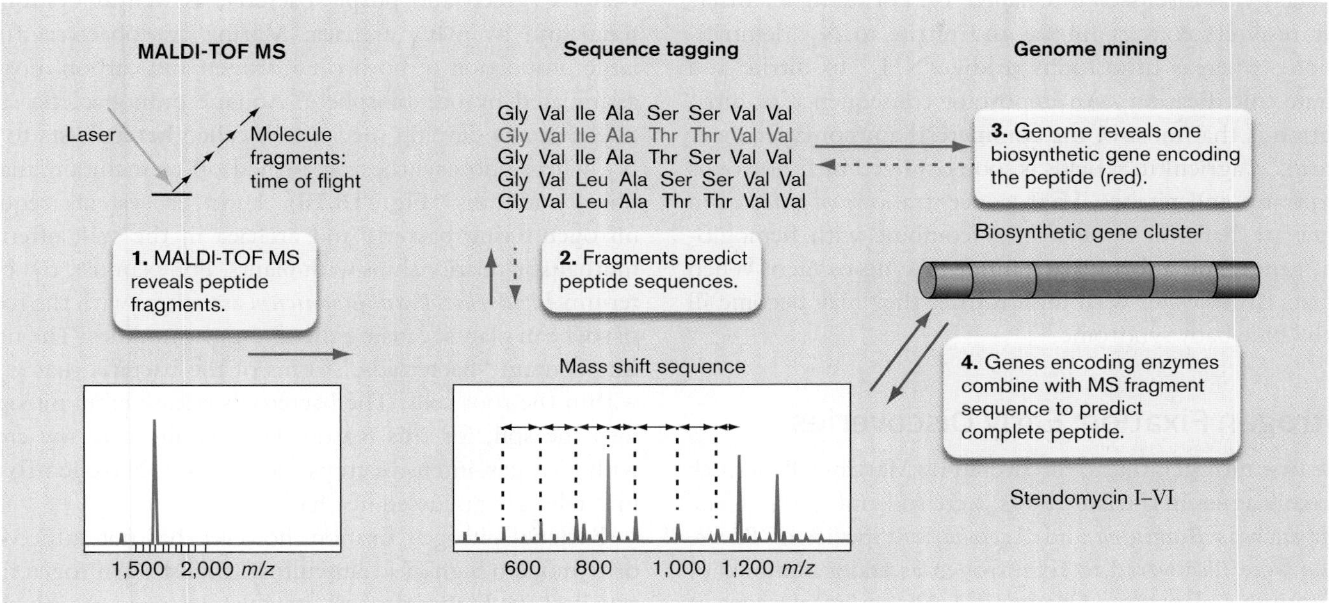
MALDI-TOF MS
Sequence tagging
Genome mining
Laser
Molecule fragments: time of flight
Gly Val Ile Ala Ser Ser Val Val
Gly Val Ile Ala Thr Thr Val Val
Gly Val Ile Ala Thr Ser Val Val
Gly Val Leu Ala Ser Ser Val Val
Gly Val Leu Ala Thr Thr Val Val
3. Genome reveals one biosynthetic gene encoding the peptide (red).
Biosynthetic gene cluster
1. MALDI-TOF MS reveals peptide fragments.
2. Fragments predict peptide sequences.
Mass shift sequence
4. Genes encoding enzymes combine with MS fragment sequence to predict complete peptide.
Stendomycin I–VI
1,500 2,000 m/z
600 800 1,000 1,200 m/z

FIGURE 3 ■ Peptidogenomics combines MALDI-TOF mass spectrometry (MS) with sequence analysis of biosynthetic gene clusters to reveal structure of novel antibiotics. m/z = mass-to-charge ratio (mass per charge).

For Dorrenstein's work, the MALDI laser was aimed at a matrix-embedded colony of *S. hygroscopicus*. Fragments of thousands of molecules were generated, among them components with sizes characteristic of certain uniquely modified amino acids that appear only in nonribosomal peptides, such as dehydroalanine. In addition, the fragments predicted the existence of a peptide containing eight amino acids, whose sequence did not match any translated sequence within the *S. hygroscopicus* genome. The data were consistent with eight slightly different versions of the peptide, called a sequence tag (**Fig. 3**).

The peptide sequence tag was then used to query the genome for genes encoding modular enzymes that might catalyze the synthesis of the amino acids in the peptide. Previous bioinformatic studies have characterized many families of modular enzymes that evolved from common ancestors. In seeking signature patterns for such genes, a cluster of genes for modular enzymes was identified that appeared likely to synthesize one version of the peptide: Gly-Val-Ile-Ala-Thr-Thr-Val-Val. This peptide proved to be the backbone of the lipopeptide stendomycin I–VI (**Fig. 1**). Note that in the mature lipopeptide, other enzymes have catalyzed various modifications, such as an ester linkage between the carboxyl terminus and the hydroxyl group of a threonine residue.

The full structure of stendomycin I–VI was identified and confirmed by back-and-forth testing between models predicted by the mass spectrometry fragments and by the gene cluster analysis. The method is now being used to identify many other peptide products. We can only hope that our rate of drug discovery outpaces the rate of resistance arising in pathogens.

RESEARCH QUESTION

The genomic cluster shown in **Figure 3** encodes three modular enzymes for stendomycin biosynthesis, one of which synthesizes the tag peptide. What do the other two enzymes synthesize? How could you test this question using peptidogenomics?

Roland D. Kersten, Yu-Liang Yang, Yuquan Xu, Peter Cimermancic, Sang-Jip Nam, et al. A mass spectrometry–guided genome mining approach for natural product peptidogenomics. 2011. *Nature Chemical Biology* **7**:794–802.

Even nitrogen-fixing legume symbionts, such as *Rhizobium,* can utilize nitrate.

In natural environments, most potential sources of nitrogen are subject to competition from dissimilatory metabolism in which the molecule is oxidized or reduced for energy, as discussed in Chapter 14. For example, anaerobic respirers convert nitrate and nitrite to N_2 (denitrification), whereas lithotrophs oxidize NH_4^+ to nitrite and nitrate (nitrification). An important consequence of nitrification is that most of the commercial ammonia fertilizer spread on agricultural fields is soon oxidized by lithotrophs to nitrates and nitrites. High concentrations of nitrates in water are harmful because they combine with hemoglobin, generating a form that cannot take up oxygen. When infants drink water with high nitrite, they may become ill with "blue baby syndrome."

Nitrogen Fixation: Early Discoveries

The first nitrogen fixers, discovered by Martinus Beijerinck and colleagues in the late 1800s, were soil and wetland bacteria such as *Beggiatoa* and *Azotobacter.* Species of *Rhizobium* were discovered to fix nitrogen as endosymbionts of leguminous plants (see Chapter 21). For several decades, it was believed that N_2 was fixed only by a few special bacteria in the soil or in symbiotic plant bacteria. But in the 1940s, Martin Kamen and colleagues noticed that phototrophs such as *Rhodospirillum rubrum* produce hydrogen gas—a known by-product of nitrogen fixation. Nitrogen fixation was hard to demonstrate reliably in the laboratory because at that time researchers did not know that in *R. rubrum* nitrogen fixation is repressed by the presence of alternative nitrogen sources, such as ammonia. When traces of ammonia and other nitrogen sources were eliminated, Kamen's student Herta Bregoff found that these photosynthetic bacteria actually fix nitrogen.

The fixation of nitrogen was confirmed by experiments showing uptake of the heavy isotope ^{15}N into biomass. Nitrogen is now known to be fixed by most phototrophic bacteria (green and purple bacteria, as well as cyanobacteria) and by many archaea. Marine cyanobacteria fix a large proportion of both the nitrogen and carbon dioxide assimilated by our biosphere. Aquatic cyanobacteria such as *Anabaena* develop special cells called **heterocysts** to fix N_2, where photosynthesis is turned off to maintain anaerobic conditions (**Fig. 15.18**). Land ecosystems require nitrogen-fixing bacteria and archaea in the soil, often in mutualistic relationships with plants. For example, the bacterium *Bradyrhizobium japonicum* associates with the roots of soybean plants, causing them to grow nodules. The nodules contain "bacteroids," forms of the bacteria that grow within the root cells. The bacteroids release extra nitrogen into the soil; for this reason, farmers alternate soy crops with nitrogen-intensive crops such as corn. Nitrogen-fixing mutualism is discussed in Chapter 21.

Bacterial nitrogen fixation, however, has not sufficed to drive modern high-yield agriculture. Industrial nitrogen fixation uses the **Haber process**, in which nitrogen gas is hydrogenated by methane (natural gas) to form ammonia, under extreme heat and pressure. Scientists estimate that the Haber process reduces more atmospheric N_2 than all of Earth's other processes combined. While this amount of nitrogen represents a small fraction of the atmosphere, the environmental effects are huge, polluting waterways and causing marine areas of hypoxia called "dead zones" (discussed in Chapters 21 and 22). The anthropogenic (human-caused) influx of nitrogen into the biosphere is an astonishing example of the influence of human society on our planet.

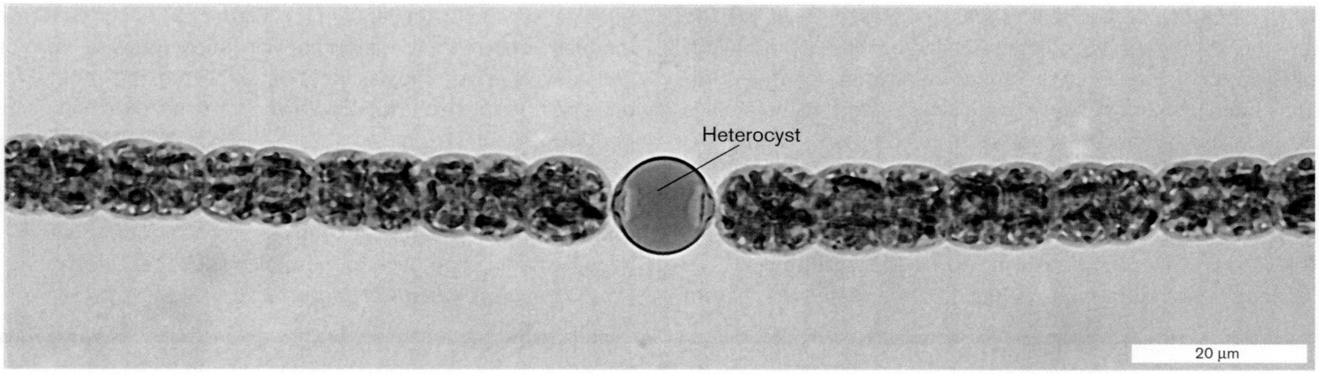
Heterocyst
20 μm

RALF WAGNER, U. DUSSELDORF; MARK SCHNEEGURT, WICHITA STATE; AND CYANOSITE

FIGURE 15.18 ▪ Nitrogen fixation requires specialized structures. *Anabaena spiroides,* a filamentous cyanobacterium, segregates N_2 fixation in heterocysts, cells that do not photosynthesize and thus maintain anoxic conditions.

The Mechanism of Nitrogen Fixation

In living cells, nitrogen fixation is an enormously energy-intensive process. The mechanism is largely conserved across species:

$$N_2 + 8H^+ + 8e^- + 16\ ATP \rightarrow 2NH_3 + H_2 + 16\ ADP + 16\ P_i$$

The electrons are donated by NADH, H_2, or pyruvate, or obtained through photosynthesis. The total energy investment includes approximately 3 ATP-equivalents per $2e^-$, plus 16 ATP molecules as shown. That makes 12 + 16 = 28 ATP in all—a large part of the energy gained from oxidation of glucose. The production of H_2 is surprising, as it consumes extra ATP. Hydrogen loss results from initiating the cycle of nitrogen reduction (discussed shortly). Some bacteria have secondary reactions to reclaim the lost hydrogen with part of its lost energy.

Note: Although nitrogen fixation is commonly represented as producing ammonia (NH_3), under most conditions of living cells the predominant form is protonated ammonium ion (NH_4^+).

Nitrogenase reaction mechanism. The overall conversion of nitrogen gas to two molecules of ammonia is catalyzed in four cycles by **nitrogenase**, an enzyme highly conserved in nearly all nitrogen-fixing species. Like Rubisco, nitrogenase probably evolved once in a common ancestor of all nitrogen-fixing organisms.

The mechanism of nitrogenase is of intense interest to agricultural scientists because of the potential benefits of improving efficiency of plant growth and of extending nitrogen-fixing symbionts to nonleguminous plants such as corn. In 1960, scientists at the DuPont Laboratory first isolated the nitrogenase complex from a bacterium, *Clostridium pasteurianum*. They measured its activity by incorporating heavy-isotope nitrogen, $^{15}N_2$, into $^{15}NH_3$. Since then, the detailed structure of nitrogenase has been solved (**Fig. 15.19**).

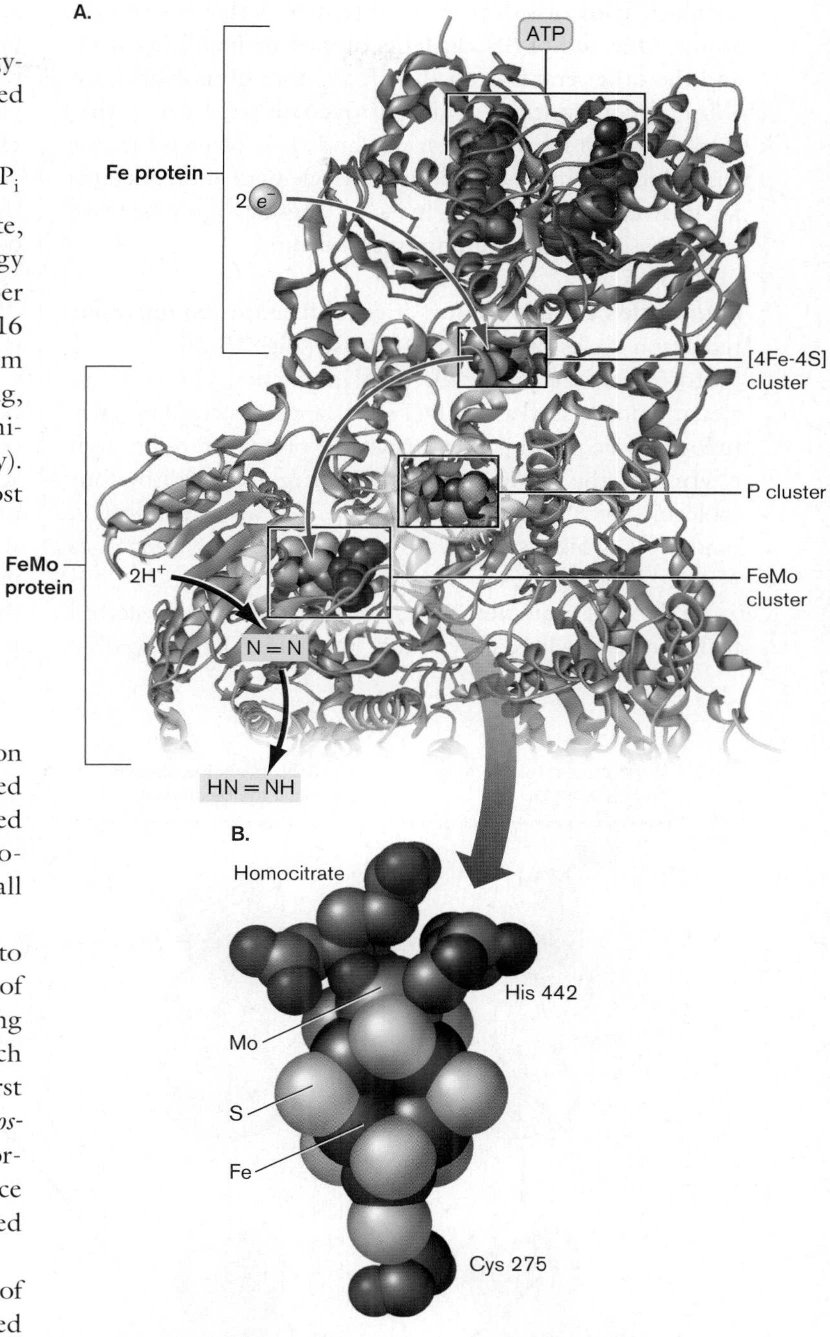
A.
ATP
Fe protein
2 e⁻
[4Fe-4S] cluster
P cluster
FeMo protein
2H⁺
FeMo cluster
N = N
HN = NH
B.
Homocitrate
His 442
Mo
S
Fe
Cys 275

FIGURE 15.19 ■ Structure of the nitrogenase complex. **A.** In each active site of the Fe protein–FeMo protein complex, the Fe protein binds ATP and receives electrons from the electron donors. The electrons are subsequently channeled down through the [4Fe-4S] cluster and the P cluster (Fe-S cluster) to the FeMo cluster, where they reduce the N_2. (PDB code: 1N2C) **B.** The metal cluster (Fe_7S_9Mo) is held in place by coordination with a molecule of homocitrate plus two amino acid residues of nitrogenase: His 442 and Cys 275.

The active nitrogenase complex includes two kinds of subunits encoded by different genes: Fe protein (shaded green in **Fig. 15.19**), containing a [4Fe-4S] center; and FeMo protein (shaded cyan), containing iron and molybdenum. The Fe protein contains a typical [4Fe-4S] structure to facilitate electron transfer. As we learned in Chapter 14, metal atoms help to transfer electrons because their orbitals are closely spaced and transitions between these orbitals involve relatively small amounts of energy. The electrons are funneled through a second Fe-S cluster (the P cluster) to the FeMo (iron-molybdenum) cluster. The FeMo cluster is an unusual structural characteristic of nitrogenase, in which trios of sulfur atoms alternate with trios of iron atoms. One end of the cluster is capped with another iron, and the other end is capped with an atom of molybdenum (Mo). A consequence of the nitrogenase structure is that most nitrogen-fixing organisms (and their plant hosts, for leguminous symbionts) require the element molybdenum for growth. Some bacteria make an alternative nitrogenase that substitutes vanadium for molybdenum.

Four cycles of reduction. Nitrogen fixation requires four reduction cycles through nitrogenase (**Fig. 15.20**).

To initiate the first cycle of N_2 reduction, $2e^-$ from an electron donor such as NADH or H_2 are transferred by a ferredoxin to Fe protein. The reduced Fe protein transfers each electron to the FeMo center, with energy supplied by four molecules of ATP (**Fig. 15.20**, step 1). The FeMo protein binds $2H^+$, which is reduced to H_2 (step 2). Only then does N_2 bind to the active site, by displacing the H_2 (step 3).

In the next reduction cycle, two electrons are transferred to Fe protein, where they reduce the iron. The reduced Fe protein transfers each electron to the FeMo center, near the binding site for N_2. The electron transfer requires expenditure of 2 ATP per electron, or 4 ATP per $2e^-$. Two hydrogen ions join the N_2 and receive the two electrons, forming HN═NH.

A third pair of electrons enters the Fe protein, which then joins another $2H^+$, reducing N_2 to $H_2N–NH_2$. The cycle again requires hydrolysis of 4 ATP. A final cycle of electron transfer and H^+ uptake reduces $H_2N–NH_2$ to $2NH_3$. At typical cytoplasmic pH values (near pH 7), NH_3 is protonated to NH_4^+.

The loss of H_2 during nitrogen fixation is puzzling because it represents lost energy in a highly energy-intensive process. The actual fate of the H_2 varies among species. *Klebsiella pneumoniae,* a Gram-negative organism in which nitrogen regulation has been much studied, gives off the H_2 without further reaction. On the other hand, *Azotobacter* uses an irreversible hydrogenase enzyme to convert H_2 back to $2H^+$ and recover the transferred electrons. In leguminous rhizobial symbionts such as *Sinorhizobium* species, H_2

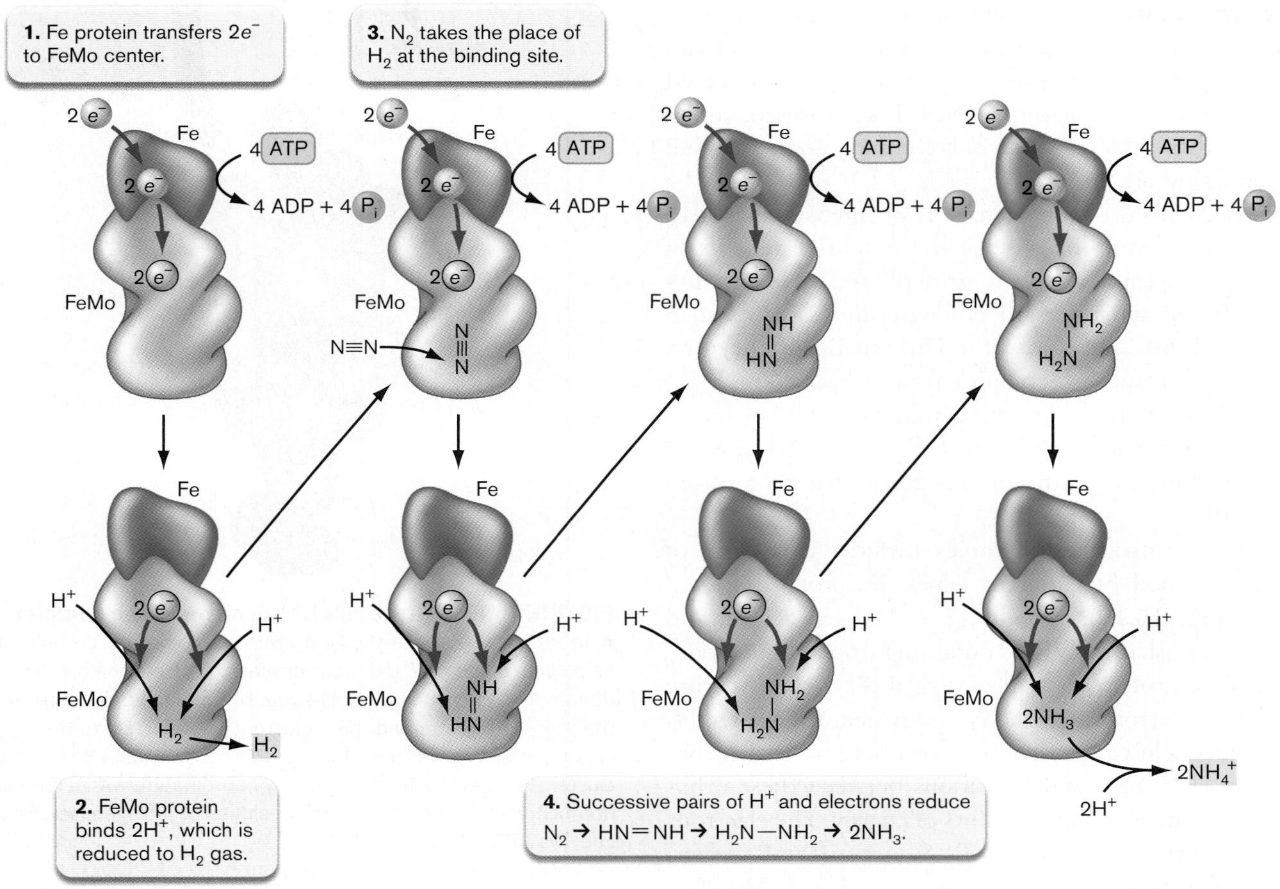
1. Fe protein transfers 2e⁻ to FeMo center.
3. N₂ takes the place of H₂ at the binding site.
2e⁻
Fe
4 ATP
4 ADP + 4 Pi
FeMo
N≡N
NH
HN
NH₂
H₂N
H⁺
H₂
2NH₃
2NH₄⁺
2H⁺
2. FeMo protein binds 2H⁺, which is reduced to H₂ gas.
4. Successive pairs of H⁺ and electrons reduce N₂ → HN=NH → H₂N—NH₂ → 2NH₃.

FIGURE 15.20 ■ Nitrogen fixation by nitrogenase. The enzyme nitrogenase successively reduces nitrogen by electron transfer, ATP hydrolysis, and H^+ incorporation. The nitrogenase complex includes two classes of subunits encoded by different genes: Fe protein (shaded green) and FeMo protein (shaded blue).

recovery varies surprisingly among strains and can affect the efficiency of plant growth.

The process of nitrogen fixation is expensive, both in terms of protein synthesis and in terms of reducing energy and ATP. Even bacteria capable of nitrogen fixation repress the process in the presence of other nitrogen sources, such as nitrate (NO_3^-), nitrite (NO_2^-), ammonia (NH_3), and nitrogenous organic molecules acquired from dead cells.

Anaerobiosis and N_2 Fixation

The reductive action of nitrogenase is extremely sensitive to oxygen because of the large reducing power needed to make NH_4^+. Thus, cells can fix nitrogen only in an anaerobic environment. This is no problem for anaerobes. But oxygenic phototrophs such as cyanobacteria face an obvious problem, as do bacterial symbionts of oxygenic plants. Aerobic and oxygenic organisms have developed several solutions to the problem of fixing nitrogen in aerobic environments:

- **Protective proteins.** Aerobic *Azotobacter* species synthesize protective proteins that stabilize nitrogenase and prevent attack by oxygen. For rhizobia, on the other hand, their plant hosts produce leghemoglobin (named for "legume" plants), a form of hemoglobin that sequesters oxygen away from the bacteria.
- **Temporal separation of photosynthesis and N_2 fixation.** Some species of cyanobacteria fix nitrogen only at night, when photosynthesis does not occur and thus no oxygen is made.
- **Specialized cells for N_2 fixation.** Filamentous cyanobacteria develop specialized nitrogen-fixing cells called heterocysts (see **Fig. 15.18**). The heterocysts lose their photosynthetic capacity entirely and specialize in nitrogen fixation. Heterocyst development is directed by a complex genetic program induced by nitrogen starvation. In natural environments, the heterocysts leak organic acids that attract heterotrophic bacteria, which use up all the oxygen around the heterocyst, generating ideal anaerobic conditions for nitrogen fixation.

A. Sydney Kustu

SYDNEY KUSTU, UC BERKELEY

B. Low NH_4^+ and low O_2 up-regulate nitrogenase expression

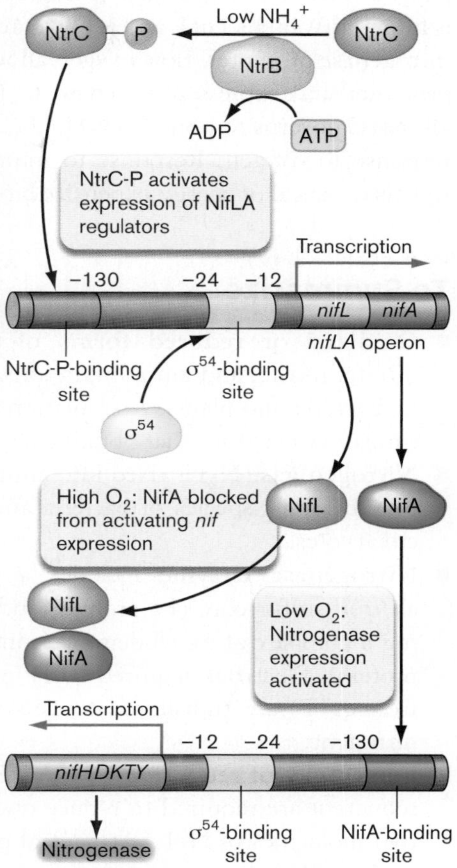
NtrC
P
Low NH4+
NtrC
NtrB
ADP
ATP
NtrC-P activates expression of NifLA regulators
Transcription
−130
−24
−12
nifL
nifA
nifLA operon
NtrC-P-binding site
σ54-binding site
σ54
High O2: NifA blocked from activating nif expression
NifL
NifA
NifL
NifA
Low O2: Nitrogenase expression activated
Transcription
−12
−24
−130
nifHDKTY
Nitrogenase
σ54-binding site
NifA-binding site

FIGURE 15.21 ▪ Regulation of nitrogen fixation. A. Sydney Kustu has made important discoveries about NtrC and other forms of nitrogen regulation in the Gram-negative soil bacterium *Klebsiella pneumoniae.* **B.** When cellular levels of NH_4^+ are low, NtrC is phosphorylated and activates expression of *nifLA*. Expression is coactivated by the nitrogen starvation sigma factor, sigma-54. The NifA protein (and sigma-54) activates expression of *nif* genes encoding nitrogenase. When oxygen levels are high, however, nitrogen fixation cannot occur, so NifA is blocked by binding NifL. (Numbers refer to positions upstream of the translation site.)

Molecular Regulation of N_2 Fixation

Nitrogen fixation costs substantial energy, and therefore the process is highly regulated. Oxygen and NH_4^+ availability regulate expression of the *nif* genes (*nifHDKTY*) encoding nitrogenase and other nitrogen-fixation proteins. Regulation of *nif* is mediated by several molecular regulators, including a nitrogen starvation sigma factor (sigma-54) and the NtrB-NtrC two-component signal transduction system (discussed in Section 10.7). The NtrB-NtrC system has been dissected in *Klebsiella pneumoniae* by Sydney Kustu and colleagues at UC Berkeley (**Fig. 15.21A**).

The NtrC protein regulates nitrogenase expression in response to NH_4^+ concentration (**Fig. 15.21B**). When NH_4^+ is low (thus the cell needs to fix nitrogen), NtrC is phosphorylated by NtrB. The NtrC-P phosphoprotein now binds its DNA site to activate expression of *nifLA,* making NifL and NifA proteins. Expression of *nifLA* is coactivated by sigma-54. Sigma-54 and the NifA protein together activate expression of *nifHDKTY* to make nitrogenase. Thus, low NH_4^+ concentration turns on nitrogen fixation. When cellular levels of NH_4^+ are high, no activators bind the *nifHDKTY* promoter, and the nitrogen fixation genes are not expressed.

Nitrogen fixation cannot occur with oxygen, so high oxygen levels block nitrogenase expression. When oxygen is high, NifA binds NifL and is prevented from binding the nitrogenase promoter. Low oxygen allows NifA to bind the promoter and express nitrogenase to fix nitrogen. Overall, NtrC governs response to NH_4^+, whereas NifA governs response to oxygen. Response to multiple environmental signals is typical of energy-expensive biosynthetic pathways.

To Summarize

- **Oxidized or reduced forms of nitrogen**, such as nitrate, nitrite, and ammonium ions, can be assimilated by bacteria and plants. Assimilation competes with dissimilatory reactions that obtain energy.
- **Nitrogen gas (N_2)** is fixed into ammonium ion (NH_4^+) only by some species of bacteria and archaea, never by eukaryotes.
- **Nitrogenase enzyme** includes a protein containing an iron-sulfur core (Fe protein) and a protein containing a complex of molybdenum, iron, and sulfur (FeMo protein). Electrons acquired by Fe protein (with energy from ATP) are transferred to FeMo protein to reduce nitrogen.
- **Four cycles of reduction by NADPH** or an equivalent reductant are required to reduce one molecule of N_2 to two molecules of NH_3. At neutral pH, NH_3 is protonated to NH_4^+.
- **Oxygen inhibits nitrogen fixation.** Bacteria and plants have various means of separating nitrogen fixation from aerobic respiration, such as heterocyst development or temporal separation.
- **Regulation of nitrogen fixation** by nitrogen and oxygen levels occurs via NtrC and sigma-54 regulation of transcription.

15.6

Biosynthesis of Amino Acids and Nitrogenous Bases

Where do microbes obtain amino acids to make their proteins and cell walls, as well as nitrogenous bases to synthesize DNA and RNA? When possible, microbes obtain these molecules from their environment through membrane-embedded transporters. But competition for such valuable nutrients is high, especially for free-living microbes in soil or water. Most free-living microbes and plants have the ability to make all the standard amino acids and bases of the genetic code, as well as nonstandard variants used for cell walls and transfer RNAs. We can use microbial biosynthesis of amino acids for industrial production of food supplements, for ourselves and for farm animals.

Amino Acid Synthesis

Like fatty acid biosynthesis, synthesis of amino acids and nitrogenous bases requires the input of large amounts of reducing energy. These compounds pose additional challenges because of their unique and diversified forms, which cannot be made by the cyclic processes that generate molecules from repeating units. Synthesis of complex, asymmetrical molecules such as amino acids requires many different conversions, each mediated by a different enzyme. Nevertheless, some economy is gained by an arrangement of branched pathways in which early intermediates are utilized to form several products (**Fig. 15.22**). For example, oxaloacetate is converted to aspartate, which can be converted to four other amino acids.

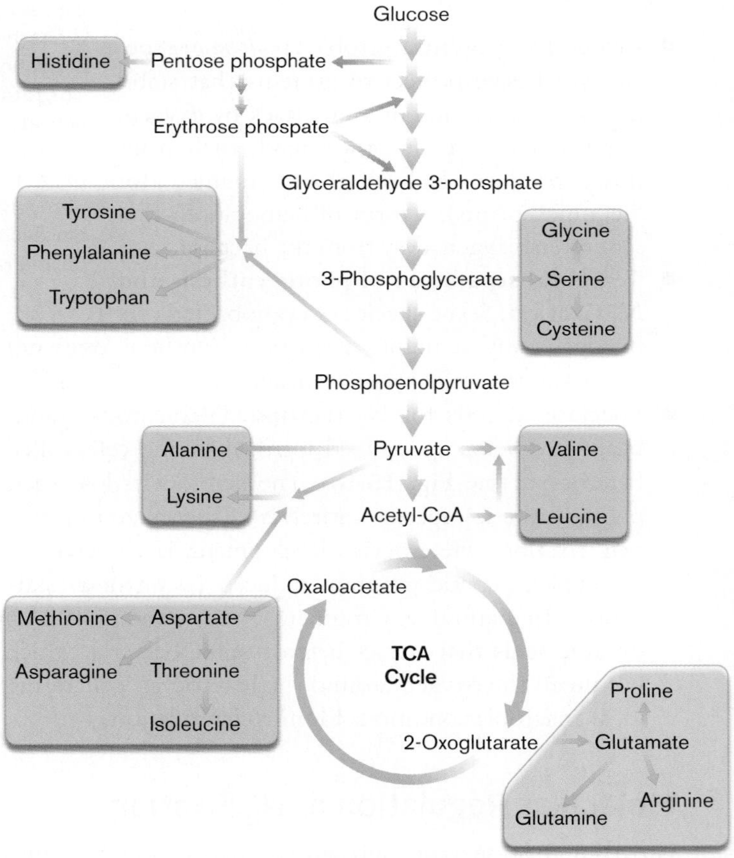
Glucose
Histidine
Pentose phosphate
Erythrose phospate
Glyceraldehyde 3-phosphate
Tyrosine
Phenylalanine
Tryptophan
Glycine
3-Phosphoglycerate
Serine
Cysteine
Phosphoenolpyruvate
Alanine
Pyruvate
Valine
Lysine
Acetyl-CoA
Leucine
Oxaloacetate
Methionine
Aspartate
Asparagine
Threonine
Isoleucine
TCA
Cycle
Proline
2-Oxoglutarate
Glutamate
Arginine
Glutamine

FIGURE 15.22 ■ Major pathways of amino acid biosynthesis. Some amino acids arise from key metabolic intermediates, whereas others must be synthesized out of other amino acids.

The carbon skeletons of amino acids arise from diverse intermediates of metabolism (see **Fig. 15.22**). As in fatty acid biosynthesis, precursor molecules are channeled into amino acid biosynthesis by specialized cofactors and reducing energy carriers such as NADPH. Note that certain amino acids arise directly from key metabolic intermediates (for example, glutamate from 2-oxoglutarate), whereas others must be synthesized from preformed amino acids (for example, glutamine, proline, and arginine from glutamate). Some amino acids can arise from more than one source; for example, leucine and isoleucine can be made from succinate as well as from pyruvate.

It has been hypothesized that the amino acids arising in just one or two steps from central intermediates are more ancient in cell evolution than those requiring more complex pathways. Five of these "ancient" amino acids—glutamate, aspartate, valine, alanine, and glycine—are the same as those detected in meteorites, whose composition resembles that of prebiotic Earth (**Fig. 15.23**). These same five amino acids also appear in early-Earth simulation experiments in which methane, ammonia, and water are heated under reducing conditions and subjected to electrical discharge. Thus, we speculate that the first amino acids that early cells evolved to make were the same as those that arose spontaneously in the prebiotic chemistry of our planet.

A more subtle effect of evolution has been to adjust the amino acid composition of proteins based on the energetic cost of biosynthesis. Proteins that are secreted or that project outside the cell cannot be recycled; their amino acids are ultimately lost to the cell. Thus, secreted and externally projecting proteins have evolved to contain the "cheaper" amino acids—that is, the amino acids whose synthesis requires spending fewer molecules of ATP and NADPH.

©WWW.METEORITES.COM.AU

FIGURE 15.23 ■ The Murchison meteorite. Fragments of a meteorite that fell in Murchison, Australia, in 1969 were shown to contain the five fundamental amino acids of biosynthetic pathways (glutamate, aspartate, valine, alanine, and glycine).

This effect can be seen in the diagram of a bacterial flagellum and its attached motor, which contains extracellular as well as cytoplasmic components, colored on a scale based on biosynthetic "expense" (**Fig. 15.24**). The external and secreted components favor less expensive amino acids, compared to the cytoplasmic components.

Assimilation of NH_4^+

Unlike sugars and fatty acids, amino acids must assimilate another key ingredient: nitrogen. The NH_4^+ produced by N_2 fixation or by nitrate reduction is the key source of nitrogen for biosynthesis. But NH_4^+ always exists in equilibrium with NH_3, which is toxic to cells. Moreover, NH_3 travels freely through membranes, making it difficult to store NH_4^+ within a cell. Deprotonation to NH_3 increases as pH rises; by pH 9.2, deprotonation reaches 50%. Even at neutral pH, a very small equilibrium concentration of NH_3 can drain NH_4^+ out of the cell. Thus, cells avoid storing high levels of NH_4^+; instead, the fixed nitrogen is incorporated immediately into organic products.

2-Oxoglutarate and glutamate condense with NH_4^+. In most bacteria, the main route for NH_4^+ assimilation is by condensing NH_4^+ with 2-oxoglutarate to form glutamate, or with glutamate to form glutamine (**Fig. 15.25**). Three key enzymes interconvert these substrates:

- **Glutamate dehydrogenase (GDH)**, actually named for its reverse activity, condenses NH_4^+ with 2-oxoglutarate to form glutamate. The condensation requires reduction by NADPH.
- **Glutamine synthetase (GS)** condenses a second NH_4^+ with glutamate to form glutamine—a process driven by spending one ATP.
- **Glutamate synthase (GOGAT, glutamine:2-oxoglutarate aminotransferase)** converts 2-oxoglutarate plus glutamine into two molecules of glutamate. Different variants of this enzyme use different reducing agents: NADPH, NADH, or ferredoxin. The NADPH variant is shown.

Through these reactions, all three substrates exchange amino groups readily. High nitrogen levels induce GDH to take up NH_4^+, but repress GS and GOGAT. GS has a higher affinity than GDH for NH_4^+. Low nitrogen levels induce GS (to make glutamine) and GOGAT (to convert some glutamine to glutamate as needed).

Both glutamate and glutamine contribute an amine, as well as their carbon skeletons, to the synthesis of other amino acids in the biosynthetic "tree." The transfer of ammonia between two metabolites such as glutamate and

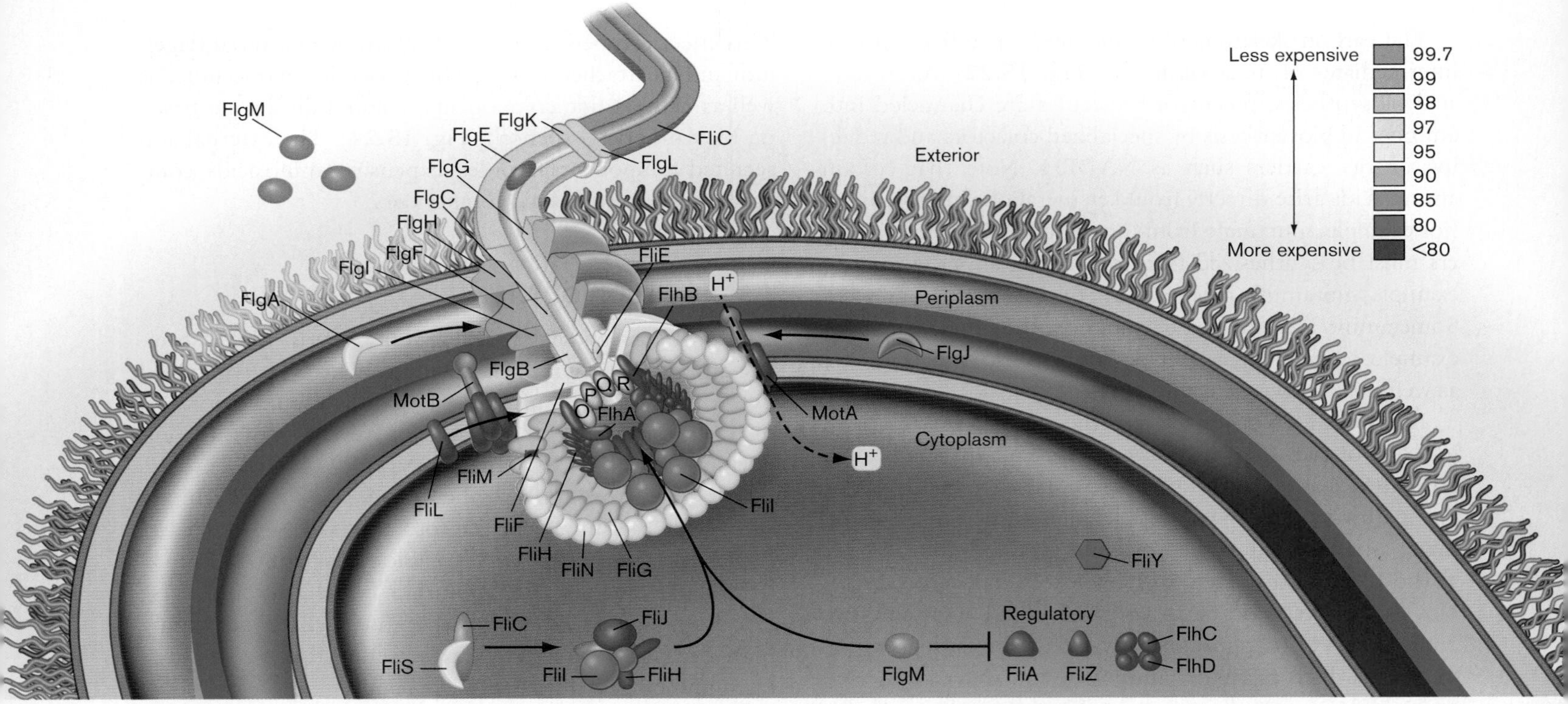

FlgM
FlgK
FlgE
FliC
FlgL
FlgG
FlgC
FlgH
FlgF
FlgI
FlgA
FliE
FlhB
H+
Exterior
Periplasm
FlgJ
FlgB
MotB
P
Q
R
O
FlhA
MotA
H+
Cytoplasm
FliM
FliL
FliF
FliH
FliN
FliG
FliI
FliY
FliC
FliJ
FliS
FliI
FliH
Regulatory
FlgM
FliA
FliZ
FlhC
FlhD
Less expensive
More expensive
99.7
99
98
97
95
90
85
80
<80

FIGURE 15.24 ▪ External proteins and proteins extending outside the cell use less expensive amino acids. Diagram of flagellar proteins (FlgE, FlgK, FlgL, Flg, FliC) attached to the motor complex. The external and secreted structures (FlgE, FliC, FlgM) are composed of less energy-expensive amino acids. Modified from Daniel Smith and Matthew Chapman, 2010. *Mbio* **1**: e00131

glutamine is called **transamination**. Transamination also occurs between many other pairs of amino acids and metabolic intermediates. For example, glutamate transfers NH_3 to oxaloacetate, making aspartate and 2-oxoglutarate; the reaction can also be reversed. In another example, valine transfers an amine group to pyruvate, generating alanine and 2-ketoisovalerate.

Thought Question

15.6 Suggest two reasons why transamination is advantageous to cells.

The cellular levels of glutamate and glutamine act as indicators of nitrogen availability. When mineral forms such as ammonium and nitrate are absent, some microbes seek organic sources of nitrogen in their environment. During nitrogen starvation, NtrC is phosphorylated and induces expression of glutamine synthetase as well as nitrogenase. NtrC phosphate also induces a high-affinity ammonia transporter and transporters for organic sources of nitrogen, such as amino acids, oligopeptides, and cell wall fragments containing amino sugars. All these molecules can be "scavenged" to obtain nitrogen for biosynthesis.

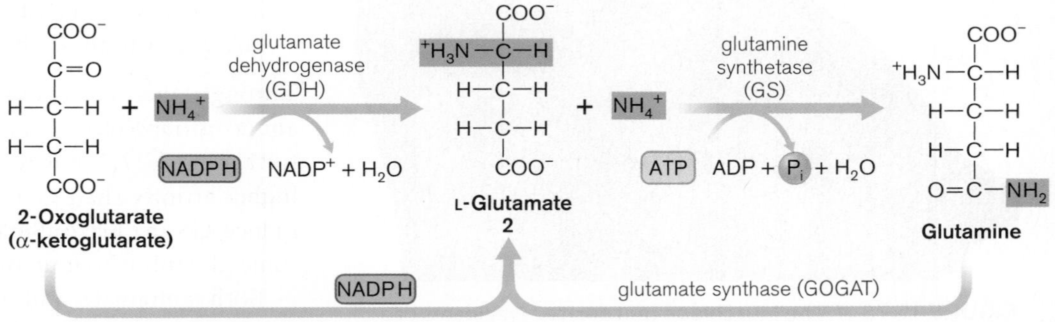

glutamate dehydrogenase (GDH)
NADPH
NADP+ + H2O
2-Oxoglutarate (α-ketoglutarate)
L-Glutamate 2
glutamine synthetase (GS)
ATP
ADP + Pi + H2O
Glutamine
NADPH
glutamate synthase (GOGAT)

FIGURE 15.25 ▪ Assimilation of NH_4^+ into glutamate and glutamine. The key TCA cycle intermediate 2-oxoglutarate (alpha-ketoglutarate) incorporates one molecule of NH_4^+ at the ketone to form glutamate. Glutamine can combine with oxoglutarate to produce two molecules of glutamate. These reactions provide sources of nitrogen to feed other pathways of amino acid synthesis.

Building Complex Amino Acids

Seven "fundamental" amino acids have relatively simple biosynthetic pathways. Others require longer pathways involving numerous enzymes. One amino acid produced by a complex pathway is arginine (**Fig. 15.26**). Bacterial arginine biosynthesis generally involves about a dozen different enzymes distributed among four to eight operons. Operon expression is regulated by an arginine repressor that binds the promoter of each operon to prevent transcription in the presence of sufficient arginine for cell needs. In some

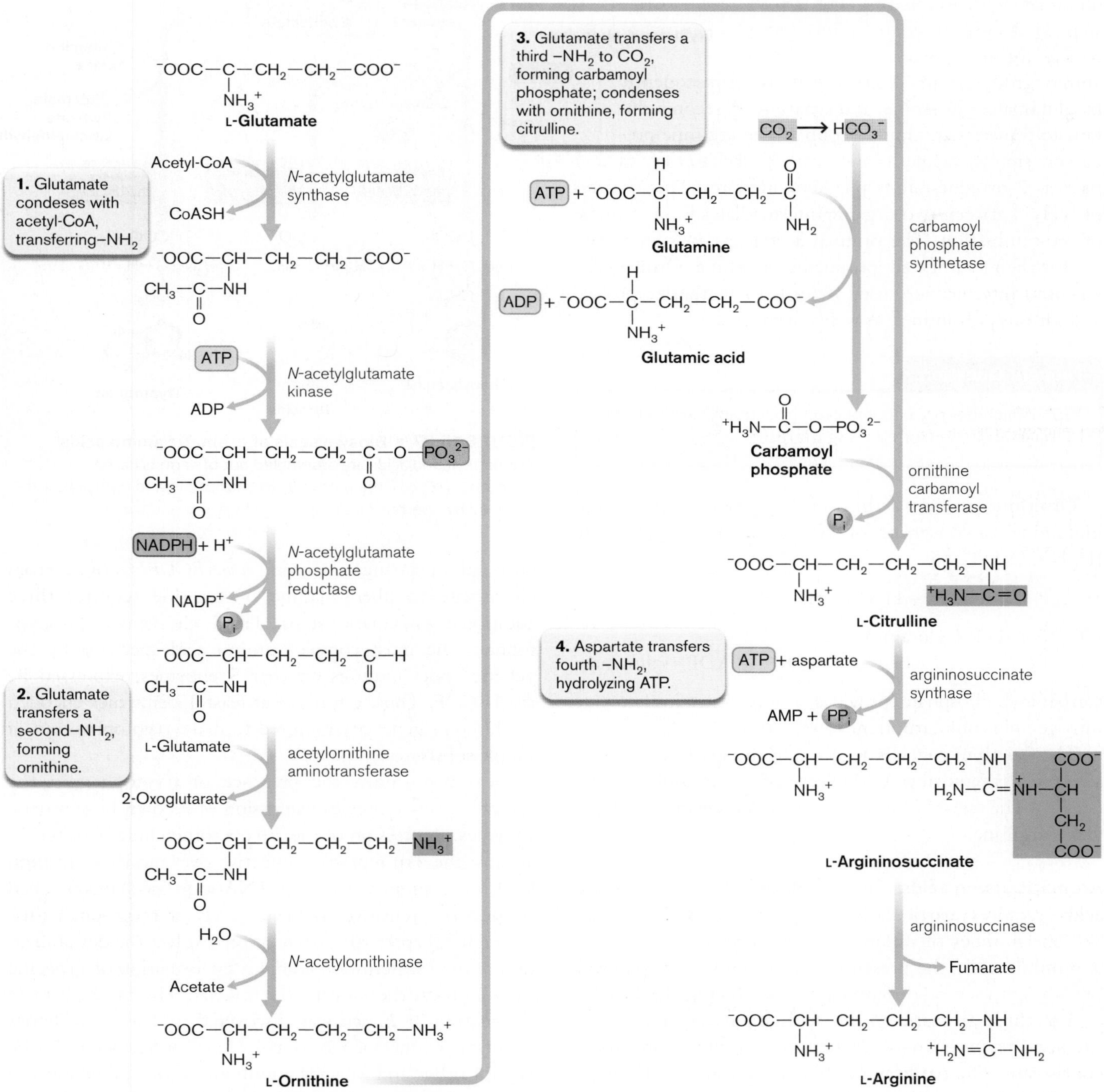

L-Glutamate
1. Glutamate condeses with acetyl-CoA, transferring–NH2
Acetyl-CoA
CoASH
N-acetylglutamate synthase
ATP
ADP
N-acetylglutamate kinase
NADPH + H+
NADP+
Pi
N-acetylglutamate phosphate reductase
2. Glutamate transfers a second–NH2, forming ornithine.
L-Glutamate
2-Oxoglutarate
acetylornithine aminotransferase
H2O
Acetate
N-acetylornithinase
L-Ornithine
3. Glutamate transfers a third –NH2 to CO2, forming carbamoyl phosphate; condenses with ornithine, forming citrulline.
CO2 → HCO3−
ATP
Glutamine
carbamoyl phosphate synthetase
ADP
Glutamic acid
Carbamoyl phosphate
ornithine carbamoyl transferase
Pi
L-Citrulline
4. Aspartate transfers fourth –NH2, hydrolyzing ATP.
ATP + aspartate
AMP + PPi
argininosuccinate synthase
L-Argininosuccinate
argininosuccinase
Fumarate
L-Arginine

FIGURE 15.26 ■ Arginine biosynthesis in *E. coli*. In arginine biosynthesis, two glutamate condense with acetyl-CoA, and nitrogen is donated by glutamate and glutamine.

species, the arginine repressor also activates enzymes of arginine catabolism, making the excess amino acid available as a carbon source.

Arginine biosynthesis. Arginine synthesis begins with the condensation of glutamate with acetyl-CoA (**Fig. 15.26**, step 1), transferring an amino group to the arginine precursor. Three more amino groups ($–NH_2$) are transferred subsequently by glutamate, glutamine, and aspartate. A second glutamate transfers an amino group to the arginine precursor (step 2), while its own carbon skeleton cycles back as 2-oxoglutarate (alpha-ketoglutarate). This transfer of NH_4^+ between two organic intermediates is an example of transamination. The original acetyl group from acetyl-CoA is then hydrolyzed, producing ornithine. Ornithine is a central intermediate, used by cells to synthesize proline and various polyamines, as well as arginine.

Thought Question

15.7 Which energy carriers (and how many) are needed to make arginine from 2-oxoglutarate?

Ornithine receives a third amino group (step 3) from glutamine via conversion of CO_2 to carbamoyl phosphate [$H_2N\text{-}CO(PO_4)^{2-}$]:

$$CO_2 + H_2O \rightarrow H^+ + HCO_3^-$$

$$HCO_3^- + ATP + \text{glutamine} \rightarrow H_2N\text{-}CO(PO_4)^{2-} + ADP + \text{glutamate}$$

Carbamoyl phosphate provides a way to assimilate one nitrogen plus one carbon into a structure.

The fourth nitrogen is acquired from aspartate in a two-step process requiring ATP release of pyrophosphate (PP_i) (step 4). The release of fumarate, a TCA cycle intermediate, yields arginine.

Aromatic amino acids. The complexity of aromatic amino acids requires particularly energy-expensive biosynthesis. The number of different enzymes required, however, is minimized by the presence of a common core pathway branching to several different amino acids (**Fig. 15.27**).

The three aromatic amino acids—phenylalanine, tyrosine, and tryptophan—each require a common precursor, chorismate. The pathway to chorismate starts with simple glycolytic intermediates (phosphoenolpyruvate and erythrose 4-phosphate), but its synthesis requires 10 enzymes encoded in a single operon, *aroABCDEFGHKL*. From chorismate to phenylalanine or tyrosine requires three additional enzymatic steps. From chorismate to tryptophan, the most complex amino acid specified by the genetic code, requires another 5 enzymes, expressed by *trpABCDE*. Thus, a total of at least 15 enzymes encoded by different genes are required to make tryptophan out of simple substrates.

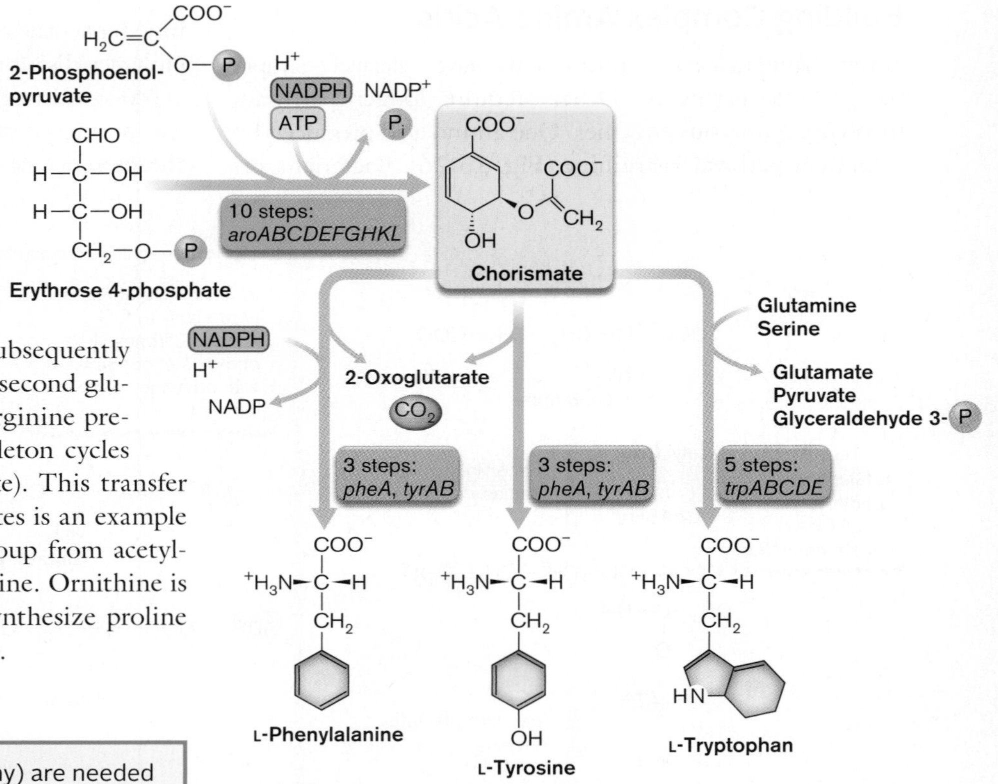
2-Phosphoenol-pyruvate
Erythrose 4-phosphate
H+
NADPH
NADP+
ATP
Pi
10 steps: aroABCDEFGHKL
Chorismate
NADPH
H+
NADP
2-Oxoglutarate
CO2
Glutamine
Serine
Glutamate
Pyruvate
Glyceraldehyde 3-P
3 steps: pheA, tyrAB
3 steps: pheA, tyrAB
5 steps: trpABCDE
L-Phenylalanine
L-Tyrosine
L-Tryptophan

FIGURE 15.27 ■ Biosynthesis of aromatic amino acids. Aromatic amino acids are assembled out of various carbon skeletons. In *E. coli,* the pathways to chorismate and tryptophan are encoded by operons of contiguous genes, *aro* and *trp.*

Not surprisingly, the presence of tryptophan in the cell effectively represses expression of its own biosynthetic enzymes. Repression of the *trp* operon includes two mechanisms: the Trp repressor, effective over moderate to high levels of tryptophan; and an RNA loop mechanism called attenuation, sensitive to lower levels of tryptophan (discussed in Chapter 10). Attenuation involves the destabilization of the transcription complex by formation of a specific stem loop on the nascent *trp* mRNA. The mechanism of attenuation in *E. coli* was discovered in 1981 by Charles Yanofsky, winner of a National Medal of Science in 2003. Several other amino acids commonly show attenuation in bacteria—particularly histidine, threonine, phenylalanine, valine, and leucine.

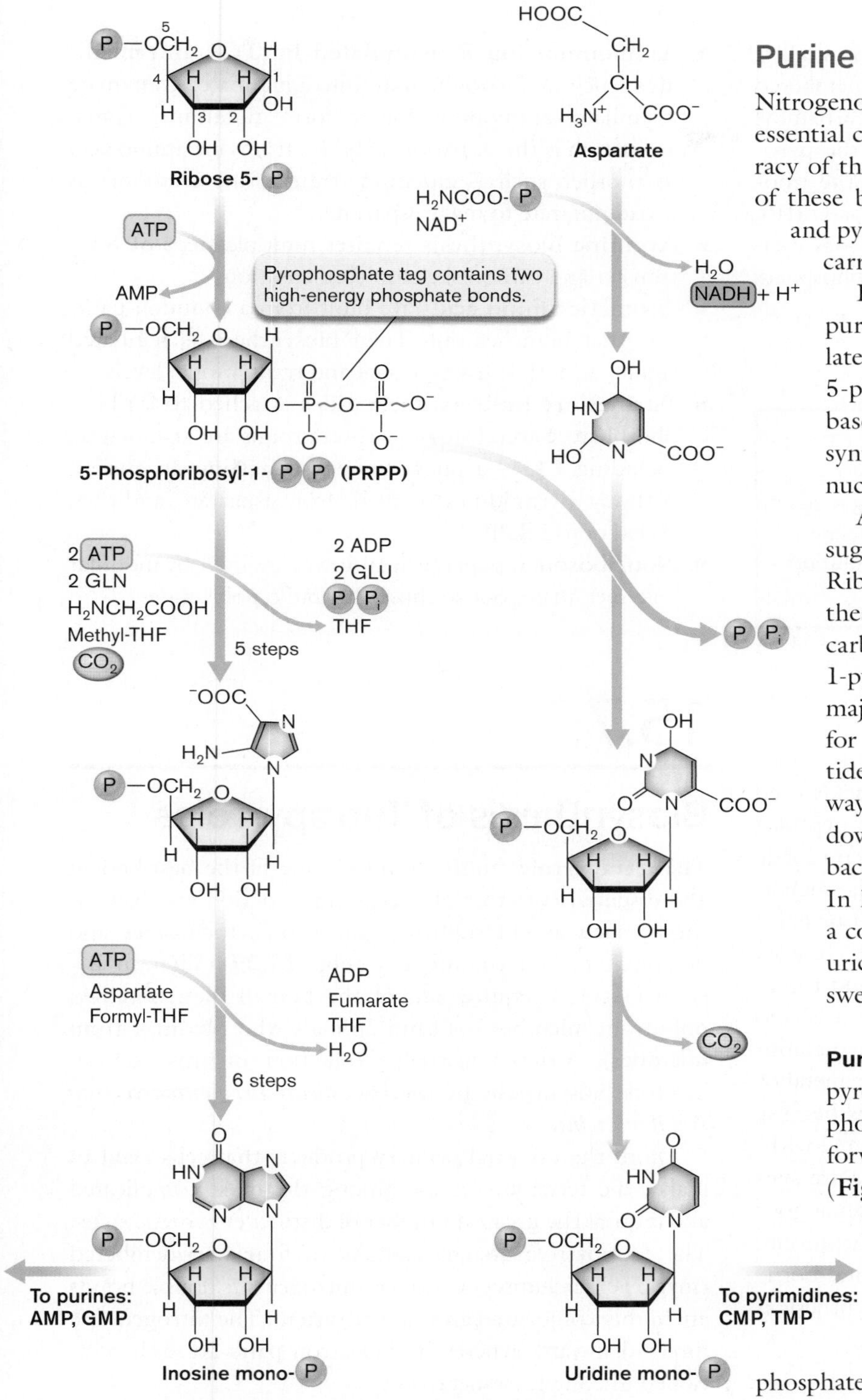

Ribose 5-P
ATP
AMP
Pyrophosphate tag contains two high-energy phosphate bonds.
5-Phosphoribosyl-1-P P (PRPP)
2 ATP
2 GLN
H2NCH2COOH
Methyl-THF
CO2
2 ADP
2 GLU
P Pi
THF
5 steps
ATP
Aspartate
Formyl-THF
ADP
Fumarate
THF
H2O
6 steps
To purines: AMP, GMP
Inosine mono-P
Aspartate
H2NCOO-P
NAD+
H2O
NADH + H+
P Pi
CO2
To pyrimidines: CMP, TMP
Uridine mono-P

FIGURE 15.28 ■ Biosynthesis of purines and pyrimidines. Both purines and pyrimidines are built onto ribose 5-phosphate. Ribose 5-phosphate is directed into purine and pyrimidine biosynthesis by the activating step of ATP → AMP, converting the sugar to 5-phosphoribosyl-1-pyrophosphate (PRPP). The purine ring is built up out of successive additions of amines, one-carbon units, plus a formyl group from formyl tetrahydrofolate (formyl-THF). The purine ring is modified to yield AMP and GMP. Likewise, the pyrimidine ring is modified to yield CMP or TMP.

Purine and Pyrimidine Synthesis

Nitrogenous bases include purines and pyrimidines, the essential coding components of DNA and RNA. The accuracy of the entire genetic code requires accurate synthesis of these bases. Besides their role in nucleic acids, purine and pyrimidine nucleotides such as ATP serve as energy carriers.

Bases are built onto ribose 5-phosphate. The purine and pyrimidine bases are not built as isolated units; rather, they are constructed on a ribose 5-phosphate substrate, forming a nucleotide (sugar-base-phosphate). Note that ribonucleotides are synthesized first and then converted to deoxyribonucleotides by enzymatic removal of the 2′ OH.

As we have seen, ribose 5-phosphate and related sugars participate in numerous metabolic pathways. Ribose 5-phosphate is directed into nucleotide synthesis when it is tagged with pyrophosphate at the carbon 1 (C-1) position, forming 5-phosphoribosyl-1-pyrophosphate (PRPP) (**Fig. 15.28**). PRPP is a major metabolic intermediate, the starting point for the synthesis of purine and pyrimidine nucleotides. PRPP is also produced by "scavenger" pathways, in which excess nitrogenous bases are broken down and recycled. Its synthesis is regulated by feedback inhibition to avoid overproduction of purines. In humans, overproduction of PRPP leads to gout, a condition in which the purine breakdown product uric acid precipitates in the joints, causing painful swelling of wrists and feet.

Purine synthesis. The hydrolysis of PRPP releases pyrophosphate, thus spending two high-energy phosphate bonds. This irreversible reaction drives forward the subsequent reactions to make purine (**Fig. 15.28**, left). To construct the purine, the pyrophosphate at carbon 1 (C-1) is replaced by an amine from glutamine. The purine is built up by addition of a series of single-carbon groups (formyl, methyl, and CO_2) alternating with nitrogens from glutamine and carbamoyl phosphate. The number of single-carbon assimilations, including assimilation of CO_2, is striking. It may reflect the ancient origin of the purine synthetic pathway, which evolved in CO_2-fixing autotrophs.

The first purine constructed is inosine monophosphate. Inosine is the "wobble" purine found at the third position of tRNA anticodons. Inosine monophosphate is subsequently converted to adenosine monophosphate (AMP) or guanosine monophosphate (GMP).

Pyrimidine synthesis. The pyrimidine is built by a slightly different route (**Fig. 15.28**, right). First, the six-membered pyrimidine ring forms from aspartate plus carbamoyl phosphate. The pyrimidine ring then displaces the pyrophosphate of PRPP, attaching by a nitrogen to the ribosyl carbon 1. The first pyrimidine built is uracil (as UMP), which can be converted to cytosine or thymine—as cytidine monophosphate (CMP) or thymidine monophosphate (TMP), respectively.

Thought Questions

15.8 Why are purines synthesized onto a sugar-base-phosphate?

15.9 Why are the ribosyl nucleotides synthesized first, and then converted to deoxyribonucleotides as necessary? What does this order suggest about the evolution of nucleic acids?

Nonribosomal Peptide Antibiotics

In addition to the standard amino acids used by ribosomes, there exist hundreds of different amino acids used by cells for functions such as the cross-bridges of peptidoglycan (discussed in Chapter 3). Some of these are made by one-step modification of standard amino acids, such as epimerization (altering the chirality from L to D) or halogenation with chlorine or fluorine. Others have complex carbon skeletons, including unsaturated bonds and three-membered rings. Both standard and nonstandard amino acids are used by actinomycetes to build secondary metabolites with antimicrobial activity. These secondary metabolites are called **nonribosomal peptide antibiotics** because their peptide backbone is constructed by an enzyme without ribosomes. Nonribosomal peptide antibiotics are synthesized by modular "assembly-line" enzymes, analogous to those that build polyketides. An example is vancomycin, the main antibiotic used against the lethal hospital-borne pathogen *Clostridium difficile*. Vancomycin biosynthesis is presented in eTopic 15.3.

To Summarize

- **Amino acid biosynthesis** requires numerous different enzymes to catalyze many unique conversions. Structurally related amino acids branch from a common early pathway.
- **Metabolic intermediates** from glycolysis and the TCA cycle initiate amino acid biosynthetic pathways.
- **Ammonium ion is assimilated by TCA intermediates**, such as 2-oxoglutarate into glutamate. Glutamate assimilates ammonium ion to form glutamine. Transamination is the donation of NH_4^+ from one amino acid to another, such as glutamine transferring ammonia to 2-oxoglutarate to make aspartate.
- **Arginine biosynthesis** requires multiple steps of NH_3 transfer and carbon skeleton condensation.
- **Aromatic amino acids** are built from a common pathway that branches out. Their biosynthesis is regulated tightly at both transcriptional and translational levels.
- **Purines are built as nucleotides** attached to a ribose phosphate. Several single-carbon groups are assimilated, including CO_2—a phenomenon suggesting an ancient pathway. Pyrimidines are made from aspartate, and then added onto PRPP.
- **Nonribosomal peptide antibiotics** are built by modular enzymes analogous to those that build polyketides.

15.7

Biosynthesis of Tetrapyrroles

The **tetrapyrrole** family includes some of the best-known and essential cofactors of energy transduction and biosynthesis, such as chlorophylls, hemes of cytochromes and hemoglobin, and vitamin B_{12} (**Fig. 15.29**). Vitamin B_{12} is an essential requirement of the human diet, obtained only from microbes (or from animals who obtain it from microbes). Most industrial production of this vitamin involves biosynthesis by the bacterium *Propionibacterium freudenreichii*.

Of all the essential primary products that cells need to make, the tetrapyrroles are among the most complicated and require the greatest number of distinct enzymatic steps. The tetrapyrrole system consists of four five-membered rings, each including an atom of nitrogen. Its double bonds are highly conjugated around the system. The nitrogens are directed inward, where their electron pairs are well positioned to complex a metal ion.

Tetrapyrroles arise from condensation of four pyrrole rings. The five-membered pyrrole ring has three substituent groups, one of which ($-CH_2NH_2$) serves as a linker in condensation of the tetrapyrrole. The individual pyrrole rings are generated by one of two different pathways (**Fig. 15.30**), found in different organisms: the glutamate pathway (in most bacteria, as well as in plants) and the glycine-succinate pathway (in purple bacteria, animals, and

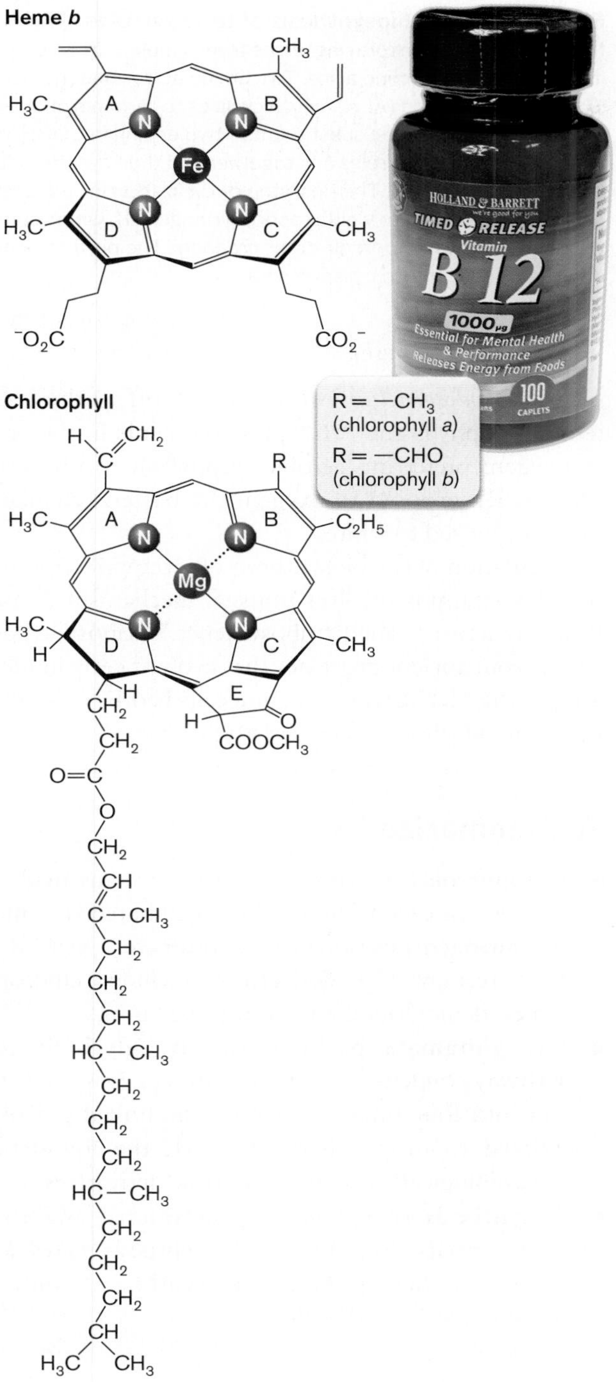
Heme b
Chlorophyll
R = —CH3 (chlorophyll a)
R = —CHO (chlorophyll b)
HOLLAND & BARRETT
TIMED RELEASE
Vitamin
B 12
1000μg
Essential for Mental Health & Performance
Releases Energy from Foods
100 CAPLETS

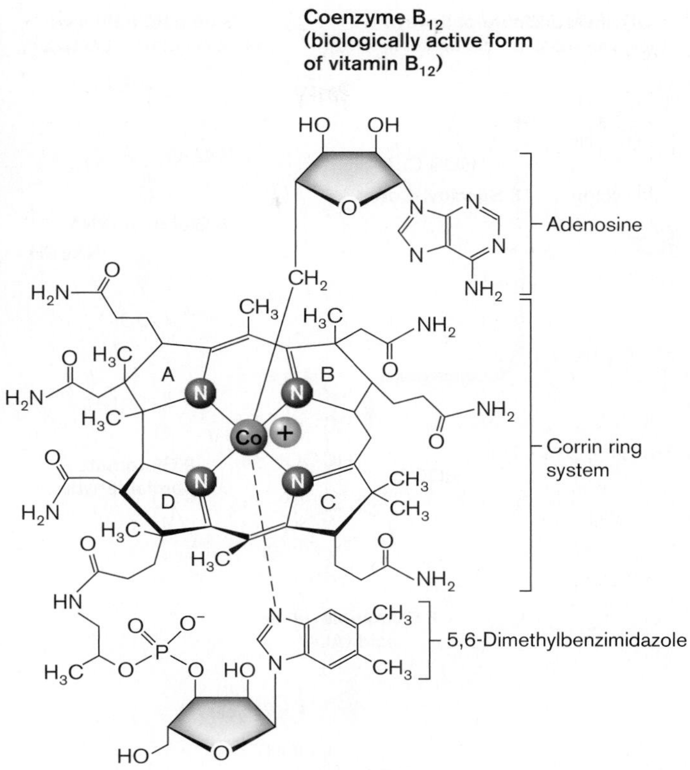
Coenzyme B12 (biologically active form of vitamin B12)
Adenosine
Corrin ring system
5,6-Dimethylbenzimidazole

FIGURE 15.29 ■ Tetrapyrroles. Tetrapyrrole derivatives include hemes, chlorophylls, and coenzyme B_{12} (the active form of vitamin B_{12}). Different variations of the ring enable coordination to different metal atoms.

yeast). The glutamate pathway is unusual in that it starts from a glutamine residue attached to a transfer RNA, generally used in protein synthesis. In either case, condensation and rearrangement generates the linear molecule 5-aminolevulinic acid (ALA). Two molecules of ALA cyclize to form the ring of porphobilinogen.

Once the pyrroles are formed, an enzyme condenses four of them in a chain, removing each amine as NH_4^+. The four linked pyrroles are conventionally labeled A through D. When the chain cyclizes, an unusual isomerization occurs in which the D pyrrole is inverted. The mechanism of the D-ring isomerization is unknown, but it imparts a critical asymmetry to the molecule as a whole. This first cyclized tetrapyrrole is called uroporphyrinogen III.

Uroporphyrinogen III provides the foundation of most biologically active tetrapyrroles. By inspecting functional tetrapyrroles, we can see the subtle modifications that distinguish them (see **Fig. 15.29**). Chlorophyll and coenzyme B_{12} (the active form of vitamin B_{12}, in which adenosine replaces cyanide) each have an extra-long side chain at the D ring, leading to completely different functional groups. The ring systems of both heme and chlorophyll are more unsaturated than the original ring system found in uroporphyrinogen III; the extra conjugated double bonds facilitate energy transitions in these electron carriers. The cobalamin core of coenzyme B_{12} has a more hydrogenated D ring and one less carbon (a carbon is lost in the connection between rings A and D).

The synthesis of tetrapyrroles is highly regulated by nutritional needs. For example, purple photoheterotrophs

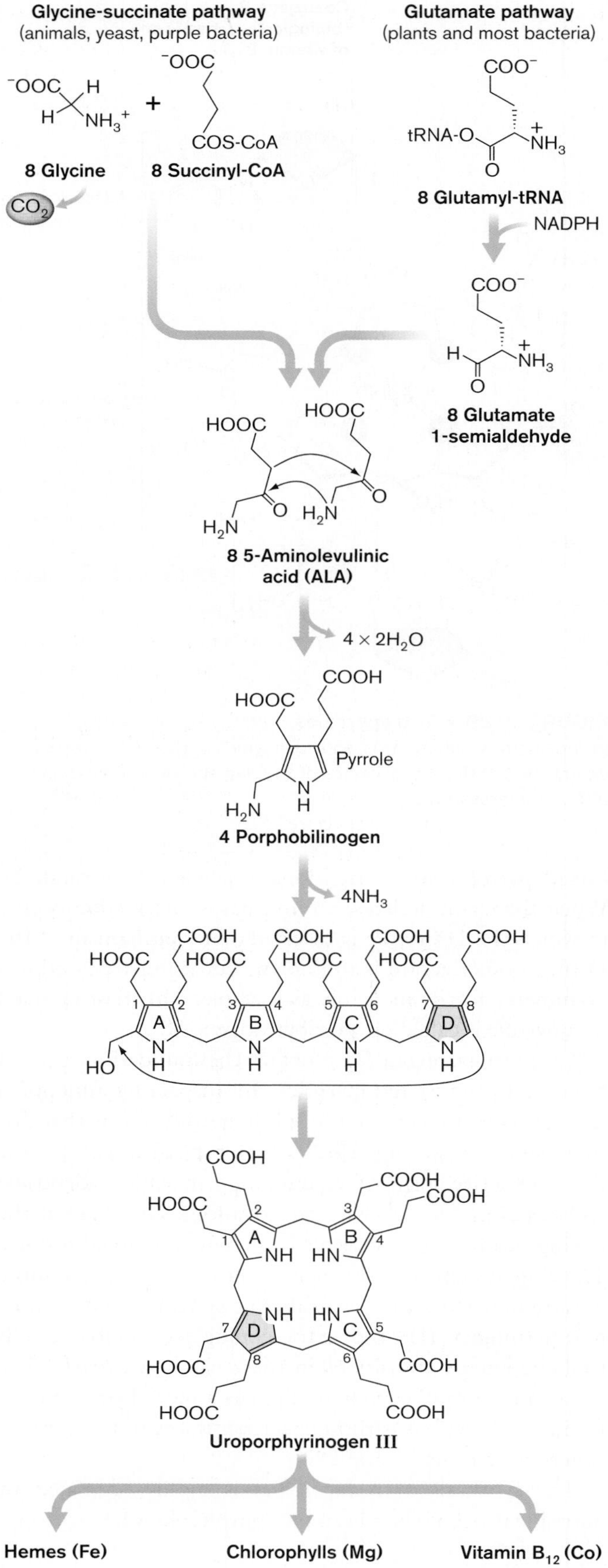

Glycine-succinate pathway
(animals, yeast, purple bacteria)
Glutamate pathway
(plants and most bacteria)
8 Glycine
8 Succinyl-CoA
CO2
8 Glutamyl-tRNA
NADPH
8 Glutamate
1-semialdehyde
8 5-Aminolevulinic acid (ALA)
4 × 2H2O
Pyrrole
4 Porphobilinogen
4NH3
Uroporphyrinogen III
Hemes (Fe)
Chlorophylls (Mg)
Vitamin B12 (Co)

FIGURE 15.30 ■ Biosynthesis of tetrapyrroles. The fundamental tetrapyrrole ring arises from simple TCA intermediates and amino acids. Two amino acids condense to form 5-aminolevulinic acid (ALA), which cyclizes to form porphobilinogen, a pyrrole ring with three substituents—two ending in COOH, one in NH_2. Four of the pyrroles link together and then cyclize to form the corrin ring system. The final step of cyclization includes reversal of the D-ring linkage to yield uroporphyrinogen III, which enters various pathways to form different products. The products contain different metal ions (in parentheses).

such as *Rhodospirillum rubrum* switch off synthesis of bacteriochlorophylls and other photopigments in the presence of oxygen, preferring aerobic respiration to phototrophy. Thus, when exposed to oxygen, the bacteria change color from purple-red to white.

Regulation of the biosynthesis of tetrapyrroles and other complex vitamins involves unusual mechanisms, based on RNA structure, called "riboswitches." Riboswitches may derive from ancient processes that evolved early in life's history, in the "RNA world" before cells had DNA or protein. For more on riboswitches, see **eTopic 15.4**.

To Summarize

- **Tetrapyrroles** are conjugated ring systems made up of pyrroles. In each pyrrole, a five-membered ring includes one nitrogen positioned to coordinate a central metal ion. Tetrapyrrole derivatives include chlorophylls, hemes, hemoglobin, and cyanocobalamin.
- **The glutamate pathway or the glycine-succinate pathway** condenses eight molecules of 5-aminolevulinic acid into four linked pyrroles. The linked pyrroles are cyclized to uroporphyrinogen III, the foundation of most biologically active tetrapyrrole derivatives.
- **Biosynthesis** is regulated by nutritional and environmental needs. Tetrapyrrole biosynthesis takes lots of energy, so these products are synthesized only when necessary.

Concluding Thoughts

In living cells, no enzymatic pathway occurs in isolation. All pathways of energy acquisition and biosynthesis occur together, sharing common substrates and products. For example, acetyl-CoA is produced by glycolysis or by fatty acid degradation; it is then utilized by pathways that synthesize fatty acids, amino acids, bases, and tetrapyrroles. Tetrapyrrole biosynthesis requires products of the TCA cycle. At the same time, the TCA cycle is needed to provide energy through catabolism. The microbial cell regulates these competing needs on an extremely rapid timescale: Within seconds of the disappearance or appearance of a nutrient, such as an amino acid, its biosynthesis is turned on or shut off.

A given microbe needs to build as much biomass as it can by the most efficient means available. For a pathogen growing within a host, this means utilizing preformed host compounds for rapid growth and replication. Free-living microbes, however, face intense competition for organic nutrients; so, in addition, they need to fix essential elements into readily accessible forms. Many soil and marine microbes build the most complicated biological molecules out of single-carbon and single-nitrogen sources, although they scavenge preformed organic substrates when available.

Microbial metabolism has many applications in food preparation and preservation, and in industrial production of antibiotics and other pharmaceuticals. The principles of food and industrial microbiology are discussed in Chapter 16. Later, in Chapters 21 and 22, we will see how microbial metabolism within populations and communities contributes to global cycles of Earth's biosphere.

CHAPTER REVIEW

Review Questions

1. What are the sources of substrates for biosynthesis? From what kinds of pathways do they arise?
2. How do microbial species economize by synthesizing only the products they need? Cite long-term as well as short-term mechanisms.
3. Compare and contrast the different cycles of carbon dioxide fixation. What classes of organisms conduct each type?
4. How was carbon isotope labeling used to identify intermediates of the Calvin cycle?
5. How is ribulose 1,5-bisphosphate consumed and re-formed through the Calvin cycle? What key products emerge to form sugars and amino acids?
6. How do oxygenic phototrophs maintain CO_2 at sufficient levels to conduct the Calvin cycle?
7. Explain the cyclic process of chain extension in fatty acid biosynthesis. Explain the generation of occasional unsaturated "kinks" in the chain.
8. Explain the different kinds of regulation of fatty acid biosynthesis.
9. What are the different environmental sources of nitrogen? Explain how and why microbes use these different sources.
10. Explain the process by which nitrogenase converts N_2 to $2NH_4^+$. Why is H_2 formed?
11. Explain the different ways that microbes maintain anaerobic conditions for nitrogenase.
12. Explain the molecular basis for regulation of nitrogen fixation and nitrogen scavenging.
13. Compare and contrast the general scheme of the biosynthesis of amino acids with that of fatty acids.
14. Outline the interconversions of 2-oxoglutarate, glutamate, and glutamine that provide nitrogen for amino acid biosynthesis.
15. Compare and contrast the processes of purine and pyrimidine biosynthesis. What is the role of the sugar ribose in each case?
16. How does the biosynthesis of tetrapyrroles generate diverse products from common small organic substrates?

Thought Questions

1. Why do some soil microbes fix N_2, whereas others depend on available nitrate, ammonium ion, or organic nitrogen? What environmental conditions would favor each strategy?
2. Disease-causing bacteria vary widely in their ability to synthesize amino acids. What kinds of pathogens would be likely to make their own amino acids, and what kinds would not?
3. *Mycoplasma genitalium,* an organism growing in human skin, lacks the ability to synthesize fatty acids. How do you think it makes its cell membrane? How could you test this?

Key Terms

acyl carrier protein (ACP) (602)
anabolism (586)
anaplerotic reaction (597)
biosynthesis (586)
Calvin cycle (Calvin-Benson cycle, Calvin-Benson-Bassham cycle, CBB cycle) (589)
carbon-concentrating mechanism (CCM) (594)
carbon dioxide fixation (589)
carboxysome (596)
fatty acid synthase complex (601)
folate (599)
gluconeogenesis (598)
glutamate dehydrogenase (GDH) (613)
glutamate synthase (glutamine:2-oxoglutarate aminotransferase) (GOGAT) (613)
glutamine synthetase (GS) (613)
Haber process (608)
heterocyst (608)
3-hydroxypropionate cycle (601)
modular enzyme (603)
nitrogenase (609)
nonribosomal peptide antibiotic (618)
paper chromatography (591)
reductive acetyl-CoA pathway (599)
reductive pentose phosphate cycle (589)
reductive (reverse) TCA cycle (598)
Rubisco (592)
secondary metabolite (secondary product) (603)
tetrapyrrole (618)
transamination (614)

Recommended Reading

Gollnick, Paul, Paul Babitzke, Alfred Antson, and Charles Yanofsky. 2005. Complexity in regulation of tryptophan biosynthesis in *Bacillus subtilis. Annual Review of Genetics* **39**:47–68.

Herter, Sylvia, Jan Farfsing, Nasser Gad'On, Christoph Rieder, Wolfgang Eisenreich, et al. 2001. Autotrophic CO_2 fixation by *Chloroflexus aurantiacus*: Study of glyoxylate formation and assimilation via the 3-hydroxypropionate cycle. *Journal of Bacteriology* **183**:4305–4316.

Iancu, Cristina V. H., Jane Ding, Dylan M. Morris, D. Prabha Dias, Arlene D. Gonzales, et al. 2007. The structure of isolated *Synechococcus* strain WH8102 carboxysomes as revealed by electron cryotomography. *Journal of Molecular Biology* **372**:764–773.

Kamen, Martin. 1985. *Radiant Science, Dark Politics. A Memoir of the Nuclear Age.* University of California Press, Berkeley.

Kerfeld, Cheryl A., William B. Greenleaf, and James N. Kinney. 2010. The carboxysome and other bacterial microcompartments. *Microbe* **5**:257–263.

Rehm, Bernd H. A. 2010. Bacterial polymers: Biosynthesis, modifications and applications. *Nature Reviews. Microbiology* **8**:578–592.

Schirmer, Andreas, Rishali Gadkari, Christopher D. Reeves, Fadia Ibrahim, Edward F. DeLong, et al. 2005. Metagenomic analysis reveals diverse polyketide synthase gene clusters in microorganisms associated with the marine sponge *Discodermia dissoluta. Applied and Environmental Microbiology* **71**:4840–4849.

Xu, Ying, Roland D. Kersten, Sang-Jip Nam, Liang Lu, Abdulaziz M. Al-Suwailem, et al. 2012. Bacterial biosynthesis and maturation of the didemnin anticancer agents. *Journal of the American Chemical Society* **134**:8625–8632.

Yang, Yu-Liang, Yuquan Xu, Paul Straight, and Pieter C. Dorrestein. 2009. Translating metabolic exchange with imaging mass spectrometry. *Nature Chemical Biology* **5**:885–887.

Zhang, Yong-Mei, Stephen W. White, and Charles O. Rock. 2006. Inhibiting bacterial fatty acid synthesis. *Journal of Biological Chemistry* **281**:17541–17544.

Zimmer, Daniel P., Eric Soupene, Haidy L. Lee, Volker F. Wendisch, Arkady B. Khodursky, et al. 2000. Nitrogen regulatory protein C–controlled genes of *Escherichia coli*: Scavenging as a defense against nitrogen limitation. *Proceedings of the National Academy of Sciences USA* **97**:14674–14679.

CHAPTER 16

Food and Industrial Microbiology

16.1 Microbes as Food

16.2 Fermented Foods: An Overview

16.3 Acid- and Alkali-Fermented Foods

16.4 Ethanolic Fermentation: Bread and Wine

16.5 Food Spoilage and Preservation

16.6 Industrial Microbiology

Microbes have nourished humans for centuries, generating cheese, bread, wine and beer, tempeh, and soy sauce. And yet, from the moment of harvest, we humans compete with microbes for our food. Microbes from the food's surface or from the air colonize food, and their uncontrolled growth can render the food rancid or putrid. Historically, the need for food storage and preservation has led to practices such as drying, salting, smoking, and adding spices, all of which retard microbial growth.

Today the principles of food microbiology extend to a growing field of industrial microbiology. Industrial microbiology includes the development of microbial products such as antibiotics and enzymes, as well as transgenic microbes that produce human proteins such as insulin and growth hormones. Many products derive from extremophiles, organisms adapted to extreme environments, whose enzymes can withstand industrial conditions. Industrial microbiology faces special challenges with regard to microbial growth conditions, scaled-up fermentation, genetic optimization of production, and safety testing.

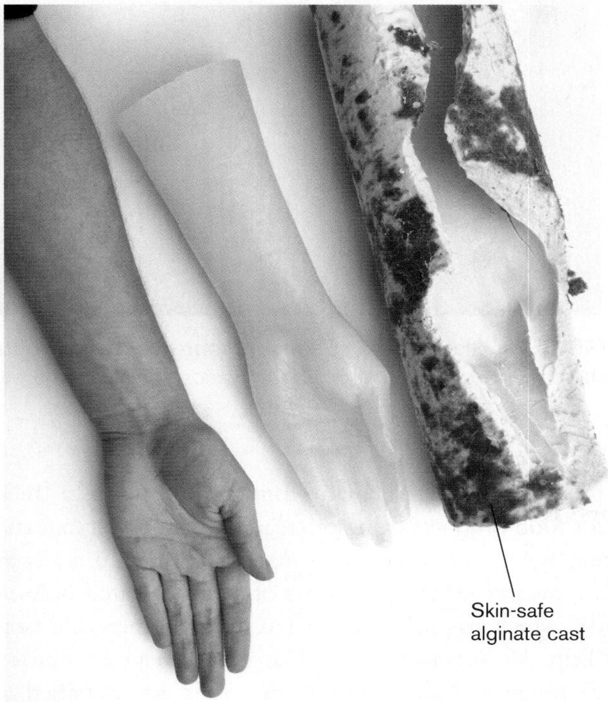

Skin-safe alginate cast

CURRENT RESEARCH highlight

Modeling cast from microbes. Alginate is a sugar-acid polymer marketed for life casting, dental impressions, and waterproofing material. Alginate is harvested from giant kelp, but a new start-up company, Progenesis Technologies, is developing exceptionally pure alginates from genetically engineered bacteria. The original bacteria, *Pseudomonas aeruginosa,* produced alginate as virulence factors for biofilms. The engineered bacteria will safely produce alginate pure enough for food products, for medical applications including absorbent sanitary pads, and for military use as radioisotope decontamination agents. In 2011, Progenesis Technologies opened a new laboratory in Huntington, West Virginia, as the state's first locally owned and operated genetic engineering company. *Source:* Scenics & Science/Alamy.

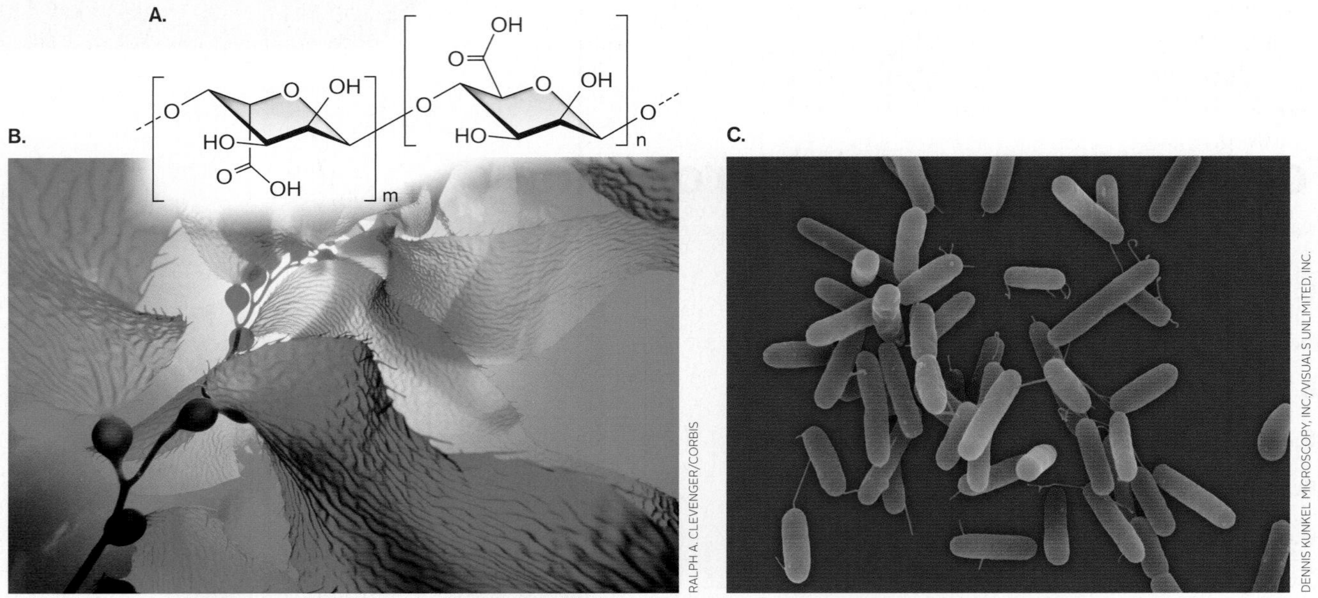
A.
B.
C.
OH
O
OH
HO
OH
m
n
RALPH A. CLEVENGER/CORBIS
DENNIS KUNKEL MICROSCOPY, INC./VISUALS UNLIMITED, INC.

FIGURE 16.1 ▪ Giant kelp and *Pseudomonas* bacteria produce industrial alginate. **A.** Alginate polysaccharide. **B.** Giant kelp, *Macrocystis pyrifera*. **C.** *Pseudomonas aeruginosa*, an alginate-producing bacterium that is under commercial development.

The product alginate has many commercial uses, as a thickener for food products from ice cream to faux caviar, a textile printing agent, and a modeling material for dental materials and life casting. Alginate consists of a carboxylated polysaccharide, which has long been harvested commercially from giant kelp, *Macrocystis pyrifera* (**Fig. 16.1**). Kelps are "brown algae," simple colonial phototrophs that are classified as microbial eukaryotes (presented in Chapter 20). Now a start-up company, Progenesis Technologies, is developing a new microbial source of alginate: *Pseudomonas* bacteria. The bacteria produce alginate in purer forms, better suited for sophisticated medical and military applications (see Current Research Highlight). Alginate is just one of a mushrooming number of products in what we call the **fermentation industry**.

The fermentation industry reaches back to 5000 BCE, the date of pottery jars excavated from a Neolithic mud-brick kitchen in the Zagros Mountains of modern Iran (**Fig. 16.2A**). The jars contain residue showing chemical traces of grapes fermented to wine. Besides making wine, Neolithic people leavened bread, another staple food requiring microbial growth. For thousands of years, microbial fermentation has been a daily part of human life and commerce.

Today, microbial products are made by bacteria grown in giant fermentation vessels (**Fig. 16.2B**). Industrial "fermentation" generally uses respiratory metabolism to maximize microbial growth. Microbial growth drives companies such as DuPont Industrial Biosciences, now earning $400 million in annual revenues from engineering microbial enzymes for industrial use. DuPont's Genencor Science program delivers 250 products ranging from contact lens cleansing agents to fashion finishes for

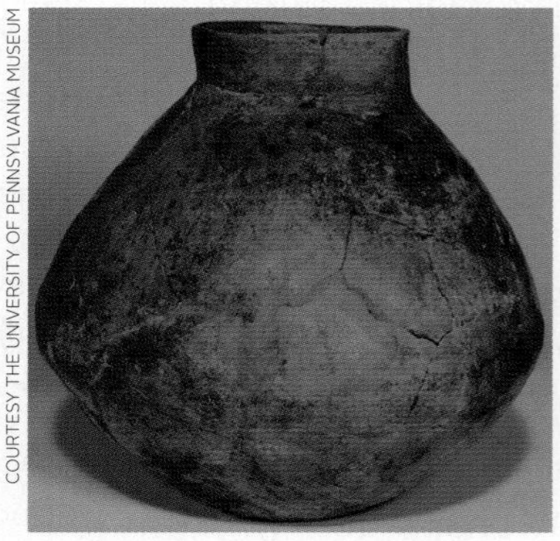
A.
COURTESY THE UNIVERSITY OF PENNSYLVANIA MUSEUM

B.
COURTESY OF GENEFERM BIOTECHNOLOGY CO., LTD.

FIGURE 16.2 ▪ The fermentation industry: then and now. **A.** A Neolithic jar that contained wine, from 5000 BCE. Excavated at Hajji Firuz Tepe, Iran. **B.** An industrial fermentation apparatus for growing microbes to produce enzyme products, at GeneFerm Biotechnology Co., Ltd., in Taipei, Taiwan.

denim at manufacturing centers around the world. Another company, Novozymes, based in Denmark, is the world's largest discoverer and manufacturer of enzymes, mostly through microbial fermentation. Novozymes marketed over 700 products in 2011.

The fermentation industry originated from long-standing microbial associations with human food. Until the relatively recent invention of steam-pressure sterilization, all foods contained live microbes. Microbial metabolism could spoil food—or it could improve the food by adding flavor, preserving valuable nutrients, and preventing growth of pathogens. As we learned in Chapter 1, one of the first great microbiologists, Louis Pasteur, began his career as a chemist investigating fermentation in winemaking. In Chapter 16 we see how the many kinds of microbial biochemistry presented in Chapters 13–15 give rise to the characteristics of food so familiar to us—from the taste of cheese and chocolate to the rising of bread dough and the physiological effects of alcoholic beverages. Industrial research continues to improve food through food biochemistry, discovering the molecular basis for the flavor and texture of microbial foods; food preservation, eliminating undesirable microbial decay by preservative methods; and food engineering, improving the quality, shelf life, and taste of microbial foods.

Beyond food, the commercial applications of microbes extend to industrial microbiology, the use of microbes to generate useful products of all kinds. Industrial microbiology generates enzymes, chemical feedstocks, fuels, and pharmaceuticals. New products are developed through genomic engineering of "industrial microbes" to produce them. The field of **synthetic biology** (discussed in Chapter 12) could someday yield microbes that act as nanomachines, perhaps even within the human body. Science fiction writers imagine even more amazing applications, such as bacteria growing cables for space elevators in the novel *The Highest Frontier* (2011). While such applications seem fanciful at present, in the past century few imagined that microbial products would now account for a multibillion-dollar industry.

16.1 Microbes as Food

Certain kinds of microbial bodies have long been eaten as food, especially the fruiting bodies of fungi and the fronds of marine algae. Single-celled algae and cyanobacteria are also used as food supplements. These microbes can provide important sources of protein, vitamins, and minerals.

Edible Fungi

In the children's classic *Homer Price,* by Robert McCloskey, settlers on the Ohio frontier save themselves from starvation when they discover "forty-two pounds of edible fungus, in the wilderness a-growin'." Fungal **fruiting bodies**, multicellular reproductive structures that generate spores, are commonly known as mushrooms (see Chapter 20). Mushrooms offer a flavorful source of protein and minerals. The protein content of edible mushrooms can be as high as 25% dry weight, comparable to that of whole milk (percent dry weight), and includes all essential dietary amino acids.

Mushrooms contributed to the survival of preindustrial humans, while killing those unlucky enough to consume varieties that were poisonous. Less than 1% of mushrooms are poisonous, but those few are deadly, such as the amanita, or "destroying angel," which produces toxins including the RNA polymerase inhibitor alpha-amanitin. People who ingest the amanita typically die of liver failure. Other kinds of mushrooms actually invite consumption in order to disperse their spores. For example, the underground truffles prized in European cooking produce odorant molecules that attract animals to dig them up. Truffle odorants include sex pheromones such as androstenol, found in human perspiration.

Many varieties of edible mushrooms are farmed and marketed for human fare. **Figure 16.3** shows the culturing of *Agaricus bisporus,* the mushroom variety most commonly sold in the United States as button mushrooms and portobellos. Button mushrooms are harvested at an early stage, whereas portobellos are harvested later, when the

KELLY IVORS

FIGURE 16.3 ■ Commercial mushroom production. Farming of *Agaricus bisporus* mushrooms on horse manure compost, by Penn State graduate student Kelly Ivors.

gills are fully exposed and some of the moisture has evaporated. The decrease in moisture in portobellos concentrates the flavor and gives them a dense, meaty texture; the mushrooms are often served in gourmet sandwiches as a vegetarian alternative to hamburger. *Agaricus* culture was first developed around 1700 in France, where the mushrooms were grown in underground caves. In modern mushroom farms, *Agaricus* mushrooms are cultured on composted horse manure or chicken manure in chambers controlled for temperature and humidity.

A.
B.
COURTESY OF TAO NATURE INTERNATIONAL
KLAUS ARRAS/STOCKFORD

FIGURE 16.4 ■ Nori production for sushi. A. Nori grows from nets in seawater. **B.** Toasted sheets of nori wrap rice, vegetables, and fish to make sushi.

Other mushroom varieties from China and Japan are grown on logs or wooden blocks. Wood-grown mushrooms include the strong-flavored *Lentinula edodes* (black forest mushroom, or shiitake), *Pleurotus* (oyster mushrooms), and *Flammulina velutipes* (enoki mushrooms), with long, thin, white stalks and delicate flavor.

Edible Algae

Several kinds of seaweed (marine algae) are cultivated, most notably in Japan. The red alga *Porphyra,* a eukaryotic "true alga" (discussed in Chapter 20), forms large multicellular fronds cultured for **nori** (**Fig. 16.4**). Nori is best known for its use in wrapping rice, fish, and vegetables to form sushi.

For nori production, the red algae are grown as "seeds," or starter cultures, in enclosed tanks. The starter cultures are then distributed on nets in a protected coastal area, usually an estuary. The cultures grow until they hang heavy from the nets, when they are harvested for processing into sheets. The sheets are toasted, turning dark green.

Edible Bacteria and Yeasts

Since prehistoric times, consumption of single-celled fungi or yeasts has provided a supplemental source of protein and vitamins. Yeasts such as *Saccharomyces* species were grown to high concentration in fermented milks and grain beverages. Traditional beers contained only a low percentage of alcohol with a thick suspension of nutrient-rich yeasts. Especially important was the content of vitamin B_{12}, an essential substance for the human diet that cannot be obtained from plant sources.

Few bacteria are edible as isolated organisms, mainly because their small cells contain a relatively high proportion of DNA and RNA. The high nucleic acid content is a problem because nucleic acids contain purines, which the human digestive system converts to uric acid. Because we humans lack the enzyme urate oxidase, the uric acid cannot be metabolized. Consumption of more than 2 grams per day of nucleic acids causes uric acid to precipitate, resulting in painful conditions such as gout and kidney stones. For this reason, most bacteria cannot be consumed in quantity, except as a minor component of food mass, as in fermented foods.

An exception is the cyanobacterium *Spirulina,* whose purine content is low enough to include as a modest part of the human diet (**Fig. 16.5A**). *Spirulina* consists of spiral-shaped cells that grow photosynthetically in freshwater. *Spirulina* is sold as a food additive rich in protein, vitamin B_{12}, and minerals. It also contains antioxidant substances that may prevent cancer. *Spirulina* is grown with illumination in special ponds lined for food production. The final product is collected and vacuum-dried to form a dark green powder of flour-like consistency.

In the food industry, *Spirulina* is classified as a form of **single-celled protein**, a term for edible microbes of high food value. Other kinds of single-celled protein include eukaryotes such as yeast and algae. The twentieth century saw the development of single-celled protein as a food source for impoverished populations. The yeast *Saccharomyces cerevisiae* was grown for protein by Germany during World War I, using inexpensive molasses for culture; and during World War II, *Candida albicans* was grown on paper mill wastes. Single-celled protein was later promoted by Western countries as a food for rapidly expanding populations of developing countries. This idea inspired the science fiction film *Soylent Green* (1973), in which people of a future overpopulated Earth are forced to eat "Soylent," a food supposedly based on soybeans and single-celled protein, though its true source is recycled humans (**Fig. 16.5B**).

A.

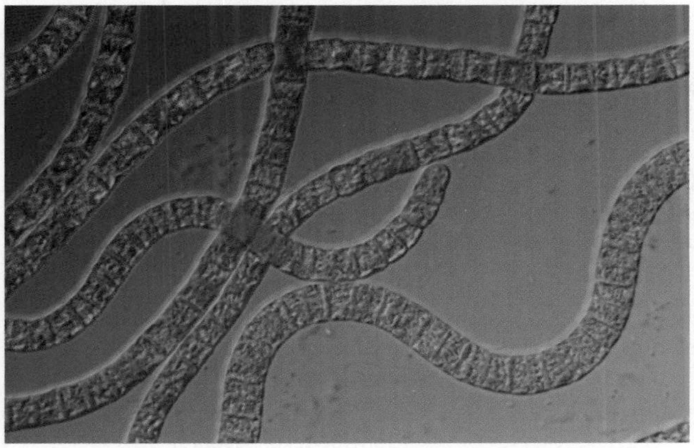

B.

FIGURE 16.5 ■ Single-celled protein. A. *Spirulina* filaments are processed into protein-rich food supplements. (LM, 250×) **B.** Scene from *Soylent Green* (1973), a science fiction film about a future Earth whose overgrown population is forced to eat a form of food called Soylent, supposedly based on soybeans and single-celled protein.

To Summarize

- **Fungal fruiting bodies** such as mushrooms and truffles are consumed as a protein-rich food.
- **Edible algae** include nori (toasted red algae), used to wrap sushi, and kelp (brown algae), which provides food additives such as alginate.
- ***Spirulina*** is an edible cyanobacterium, a source of single-celled protein. Most bacteria, however, are inedible in isolation because of their high concentration of nucleic acids.
- **Yeasts** have been grown as an economical protein and vitamin supplement.

16.2

Fermented Foods: An Overview

Virtually all human cultures have developed varieties of **fermented foods**, food products that are modified biochemically by microbial growth. The purposes of food fermentation include the following:

- **To preserve food.** Certain microbes, particularly the lactobacilli, metabolize only a narrow range of nutrients before their waste products build up and inhibit further growth. Typically, the waste fermentation products that limit growth are carboxylic acids, ammonia (alkaline), or alcohol. Buildup of these substances renders the product stable for much longer than the original food substrate.
- **To improve digestibility.** Microbial action breaks down fibrous macromolecules and tenderizes the product, making it easier for humans to digest. Meat and vegetable products are tenderized by fermentation.
- **To add nutrients and flavors.** Microbial metabolism generates vitamins, particularly vitamin B_{12}, as well as flavor molecules such as esters and sulfur compounds.

Different societies have devised thousands of different kinds of fermented foods. Examples are given in **Table 16.1**. Fermented foods that are produced commercially include dairy products such as cheese and yogurt, soy products such as miso (from Japan) and tempeh (from Indonesia), vegetable products such as sauerkraut and kimchi, and various forms of cured meats and sausages. Alcoholic beverages are made from grapes and other fruits (wine), grains (beer and liquor), and cacti (tequila). Other kinds of foods require microbial treatment for special purposes, such as leavening by yeast (for bread) or cocoa bean fermentation (for chocolate). Besides commercial production, numerous fermented products are homemade by traditional methods thousands of years old. Such products are known as "traditional fermented foods." Occasionally, a traditional fermented food enters commercial production and becomes widespread. For example, soy sauce, a traditional Japanese product, was marketed by the Kikkoman company and achieved global distribution in the twentieth century.

The nature of fermented foods depends on the quality of the fermented substrate, as well as on the microbial species and the type of biochemistry performed. Traditional fermented foods usually depend on **indigenous microbiota**—that is, microbes found naturally in association with the food substrate; or on starter cultures derived

TABLE 16.1

Fermented foods and beverages.

Product (origin)	Description	Microbial genera
Acidic fermentation of dairy products, meat, and fish		
Buttermilk (Asia, Europe)	Bovine milk, lactic fermented	*Lactococcus*
Yogurt (Asia, Europe)	Bovine milk, lactic fermented and coagulated	*Lactobacillus, Streptococcus*
Kefir (Russia)	Bovine or sheep's milk, mixed fermentation, acidic with some alcohol	*Lactobacillus, Streptococcus,* yeasts, others
Sour cream (Asia, Europe)	Bovine cream, lactic fermented	*Lactococcus*
Cheese (Asia, Europe)	Milk (bovine, sheep, or goat), lactic fermented, coagulated, and pressed; in some cases cooked; mold ripened (spiked or coated)	Acid fermentation: *Lactobacillus, Streptococcus, Propionibacterium*; mold ripening: *Penicillium*
Sausage (Asia, Europe)	Ground beef and/or pork encased with starter culture, lactic fermented, dried, or smoked	*Lactobacillus, Pediococcus, Staphylococcus,* others
Fermented fish (Africa, Asia)	Many kinds of fish, mixed fermentation, acid and amines produced	Unknown
Acidic fermentation of vegetables		
Tempeh (Indonesia)	Soybean cakes, fungal fermentation	*Rhizopus oligosporus*
Miso (Japan)	Soy and rice paste, fungal fermentation	*Aspergillus*
Soy sauce (China)	Extract of soy and wheat, fungal fermentation, brined, bacterial fermentation	*Aspergillus,* followed by halotolerant bacteria
Kimchi (Korea)	Cabbage, peppers, and other vegetables, with fish paste, brined; container is buried	*Leuconostoc,* other bacteria
Sauerkraut (Europe)	Cabbage, fermented, making lactic and acetic acids, ethanol, and CO_2	*Leuconostoc, Pediococcus, Lactobacillus*
Pickled foods (Asia)	Cucumbers, carrots, fish; brined, then fermented	*Leuconostoc, Pediococcus, Lactobacillus*
Kenkey (Western Africa)	Maize, fermented, wrapped in banana leaves and cooked	Unknown
Chocolate (South America)	Cocoa beans, soaked and fermented before processing to chocolate	*Lactobacillus, Bacillus, Saccharomyces*
Alkaline fermentation		
Pidan (China, Japan)	Duck eggs, coated in lime (CaO), aged, producing ammonia and sulfur odorants	*Bacillus*
Natto (China, Japan)	Whole soybeans, fermented	*Bacillus natto*
Dawadawa (Africa)	Locust beans, fermented	*Bacillus*
Ogiri (Africa)	Melon seed paste, fermented	*Bacillus*
Leavened bread dough		
Yeast breads (Asia, Europe)	Ground grain, dough leavened by yeast	*Saccharomyces*
Sourdough (Egypt)	Ground grain, dough leavened by starter culture from previous dough	*Saccharomyces, Torulopsis, Candida*
Injera (Ethiopia)	Ground teff grain, dough leavened and fermented 3 days by organisms from the grain	*Candida*
Alcoholic fermentation		
Wine (Asia, Europe)	Grape juice, yeast fermented, followed by malolactic fermentation	*Saccharomyces, Oenococcus*
Beer (Asia, Europe, Africa)	Barley and hops, yeast fermented	*Saccharomyces*
Sake (Japan)	Rice extract, yeast fermented	*Saccharomyces*
Tequila (Mexico)	Blue agave, yeast fermented and distilled	*Saccharomyces*
Whiskey (United Kingdom)	Barley or other grains or potatoes, fermented and distilled	*Saccharomyces*

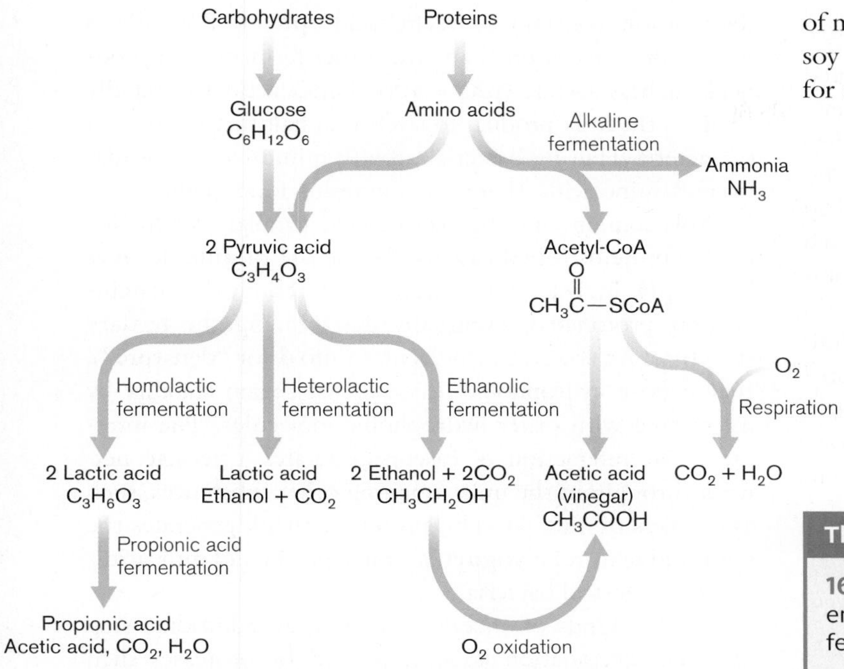
Carbohydrates
Proteins
Glucose
$C_6H_{12}O_6$
Amino acids
Alkaline
fermentation
Ammonia
NH_3
2 Pyruvic acid
$C_3H_4O_3$
Acetyl-CoA
$CH_3C(=O){-}SCoA$
O_2
Homolactic
fermentation
Heterolactic
fermentation
Ethanolic
fermentation
Respiration
2 Lactic acid
$C_3H_6O_3$
Lactic acid
Ethanol + CO_2
2 Ethanol + $2CO_2$
CH_3CH_2OH
Acetic acid
(vinegar)
CH_3COOH
$CO_2 + H_2O$
Propionic acid
fermentation
Propionic acid
Acetic acid, CO_2, H_2O
O_2 oxidation

FIGURE 16.6 ▪ Major chemical conversions in fermented foods.

from a previous fermentation, as in yogurt or sourdough fermentation. Commercial food-fermenting operations use highly engineered microbial strains to inoculate their cultures, although in some cases indigenous microbiota still participate. For example, wines and cheeses aged in the same caves for centuries often include fermenting organisms that persist in the air and the containers used.

Major classes of fermentation reactions are summarized in **Figure 16.6**. The most common conversions involve anaerobic fermentation of glucose, as discussed in Chapter 13. Glucose is fermented to lactic acid (**lactic acid fermentation**) in cheeses and sausages, primarily by lactic acid bacteria such as *Lactobacillus*. A second-stage fermentation of lactic acid to propionic acid (**propionic acid fermentation**) by *Propionibacterium* generates the special flavor of Swiss and related cheeses. Some kinds of vegetable fermentation, as in sauerkraut, involve production of lactic acid and CO_2, with small amounts of acetic acid and ethanol. This **heterolactic fermentation** is conducted primarily by *Leuconostoc*. Fermentation to ethanol plus carbon dioxide without lactic acid (**ethanolic fermentation**) is conducted by yeast during bread leavening and production of alcoholic beverages.

In some food products, particularly those fermented by *Bacillus* species, proteolysis and amino acid catabolism generate ammonia in amounts that raise pH (**alkaline fermentation**). For example, alkaline fermentation forms the soybean product natto. Other products require the growth of mold, such as the mold-spiked Roquefort cheese and the soy product tempeh. Mold growth requires some oxygen for aerobic respiration. Respiration must be limited, however, to avoid excessive decomposition of food substrate and loss of food value.

Note that the conversions cited here include only the major reactions in achieving the food product. In addition, thousands of minor or secondary reactions occur, some of which produce tiny amounts of potent odorants and flavors. While these flavor molecules have less nutritional consequence than the main fermentation products, they provide the complex, "sophisticated" taste for which fine cheeses, wines, and soy products are known.

Thought Questions

16.1 Why do the lipid components of food experience relatively little breakdown during anaerobic fermentation?

16.2 Why does oxygen allow excessive breakdown of food, compared with anaerobic processes?

To Summarize

- **Anaerobic fermentation of food** enhances preservation, digestibility, nutrient content, and flavor.
- **Acidic fermentations** lead to organic acid fermentation products, such as lactate and propionate.
- **Alkaline fermentations** produce ammonia and break down proteins to peptides.
- **Ethanolic fermentation** produces ethanol and carbon dioxide.
- **Lipids are relatively stable** under anaerobic conditions of fermentation.

16.3

Acid- and Alkali-Fermented Foods

Many food fermentations produce acids or bases. An acid or base serves as an effective preservative because the pH change is unlikely to be reversed, and because animal or plant bodies grown at near-neutral pH are unlikely to support growth of acidophiles or alkaliphiles, which grow at extreme pH conditions.

Acidic Fermentation of Dairy Products

The conversion of milk to solid or semisolid fermented products dates far back in human civilization. The practice of milk fermentation probably arose among herders who collected the milk of their pack animals but had no way to prevent the rapid growth of bacteria. The milk had to be stored in a portable container such as the stomach of a slaughtered animal. After hours of travel, the combined action of lactic acid–producing bacteria and stomach enzymes caused the coagulation of milk proteins into **curd**. The curd naturally separated from the liquid portion, called **whey**. Both curds and whey can be eaten, as in the nursery rhyme "Little Miss Muffet." The curds, however, are particularly valuable for their concentrated protein content.

Curd formation. A **cheese** is any milk product from a mammal (usually cow, sheep, or goat) in which the milk protein coagulates to form a semisolid curd. Curd formation results from two kinds of processes: acidification, usually as a result of the microbial production of lactic acid; and treatment by proteolytic enzymes such as rennet. The curd may then be separated and processed to varying degrees, depending on the type of cheese.

How does milk coagulate? The major organic components of cow's milk are milk fat (about 4% unless skimmed), protein (3.3%), and the sugar lactose (4.7%). Milk starts out at about pH 6.6, very slightly acidic. At this pH, the milk proteins are completely soluble in water; otherwise they would clog the animal's udder as the milk came out. Fermentation generally begins with bacteria such as *Lactobacillus* and *Streptococcus* (**Fig. 16.7**). As bacteria ferment lactose to lactic acid, the pH starts to decline. The dissociation constant of lactic acid (pK_a = 3.9) allows greater deprotonation than with other fermentation products, such as acetate (pK_a = 4.8). Thus, lactic acid rapidly acidifies the milk product to levels that halt further growth of bacteria. Halting bacterial growth minimizes the oxidation of amino acids, thereby maintaining food quality.

Milk contains micelles (suspended droplets) of hydrophobic proteins called caseins. As the pH of milk declines below pH 5, the acidic amino acid residues of caseins become protonated, eventually destabilizing the tertiary structure. As the casein molecules unfold (or "denature"), they expose hydrophobic residues that regain stability by interacting with other hydrophobic molecules. The intermolecular interaction of caseins generates a gel-like network throughout the milk, trapping other substances, such as droplets of milk fat. This protein network generates the semisolid texture of **yogurt**, a simple product of milk acidified by lactic acid bacteria.

In most kinds of cheese formation, an additional step of casein coagulation is accomplished by proteases such as rennet. Rennet derives from the fourth stomach of a calf, although modern versions are made by genetically engineered bacteria. Calf rennet includes two proteolytic enzymes: chymosin and pepsin. Chymosin specifically cleaves casein into two parts, one of which is charged and water-soluble, the other hydrophobic. The hydrophobic portion forms a firmer curd than intact casein and results in the harder texture of solid cheeses. The water-soluble portion, about one-third of the total casein, enters the whey and is lost from the curd. Processing of some cheese varieties includes high temperature, which denatures even the whey protein, so it is retained in the curd.

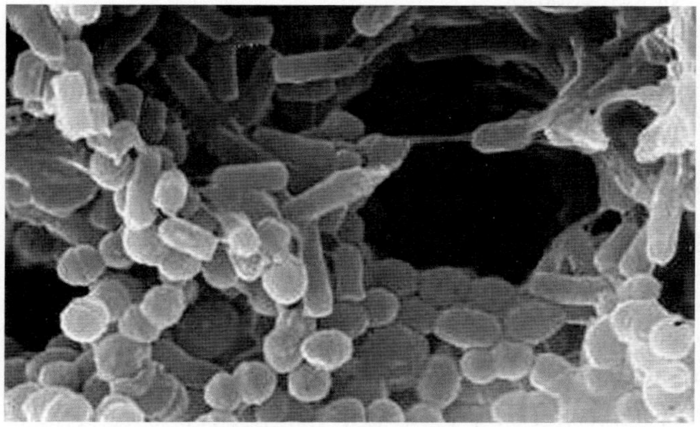

SYLVIE LORTAL, *HANDBOOK OF FOOD AND BEVERAGE FERMENTATION TECHNOLOGY*, 2004, CRC PRESS.

FIGURE 16.7 ▪ Bacterial community within Emmentaler cheese. Bacterial species include *Lactobacillus helveticus* (rods, 2.0–4.0 μm in length) and *Streptococcus thermophilus* (cocci). SEM.

Varieties of cheese. An extraordinary number of cheese varieties have been devised (**Fig. 16.8**). These fall into several categories based on particular steps in their production.

- **Soft, unripened cheeses**, such as cottage cheese and ricotta, are coagulated by bacterial action, without rennet. The curd is cooked slightly, and the whey is partly drained, but their water content is 55% or greater. These cheeses spoil easily; there are no steps of aging, or **ripening**.
- **Semihard, ripened cheeses**, such as Muenster and Roquefort, include rennet for firmer coagulation, and the curd is cooked down to a water content of 45%–55%. The cheese is aged for several months.
- **Hard cheeses**, such as Swiss cheese and cheddar, are concentrated to even lower water content. Extra-hard varieties, such as Parmesan and Romano, have a water content as low as 20%. These cheeses are aged for many months, even several years.

A.
B.
C.
D.
TEUBNER/GETTY IMAGES
MAXIMILIAN STOCK LTD/JUPITER IMAGES
CHRISTINA PETERS/JUPITER IMAGES
STEVEN MARK NEEDHAM/JUPITER IMAGES

FIGURE 16.8 ▪ Cheese varieties. **A.** Cottage cheese, an unripened perishable cheese. **B.** Emmentaler Swiss cheese, with eyes produced by carbon dioxide fermentation. **C.** Feta cheese, a soft cheese from goat's milk, preserved in brine. **D.** Roquefort, a medium-hard cheese ripened by spiking with *Penicillium roqueforti.*

- **Brined cheeses**, such as feta, are permeated with brine (concentrated salt), which limits further bacterial growth and develops flavor. Harder cheeses, such as Gouda, may be brined at the surface.
- **Mold-ripened cheeses** are inoculated with mold spores that germinate and grow during the ripening, or aging, process to contribute texture and flavor. The mold may be inoculated on the surface, to form a crust (as in Brie and Camembert), or it may be spiked deep into the cheese (as in blue cheese or Roquefort).

Cheese production. Commercial production of cheese involves a standard series of steps (**Fig. 16.9**). At each of these steps, choices of treatment lead to very different varieties. Key steps are illustrated for the example of Gouda cheese in **Figure 16.10**.

In the first step, the milk is filtered to remove particulate objects, such as straw, and microfiltered or centrifuged to remove potentially pathogenic bacteria and spores. Most modern production includes flash pasteurization (brief heating to 72°C; discussed in Chapter 5), although some traditional cheeses continue to be made from unpasteurized milk. Unpasteurized milk in cheese has been linked to illness, particularly from *Listeria,* bacteria that grow at typical refrigeration temperatures.

The fermenting microbes are added as a **starter culture**. The starter was traditionally derived from a sample of previous fermentation, in which case the flora are undefined. Commercial cheese production now uses defined species. In all but the soft cheeses, bacterial coagulation and curd formation are supplemented by rennet or by genetically engineered proteases.

The solid curd is then cut, or **cheddared** (hence the name "cheddar" cheese). The finer the pieces, the more whey can be pressed out and the harder the cheese produced. Curd is then heat treated, with or without the whey; if whey is included, more protein is retained. Brining at this stage leads to a salty cheese, such as feta.

The pressed curd is then shaped into a mold, which determines the ultimate shape of the cheese. Before ripening

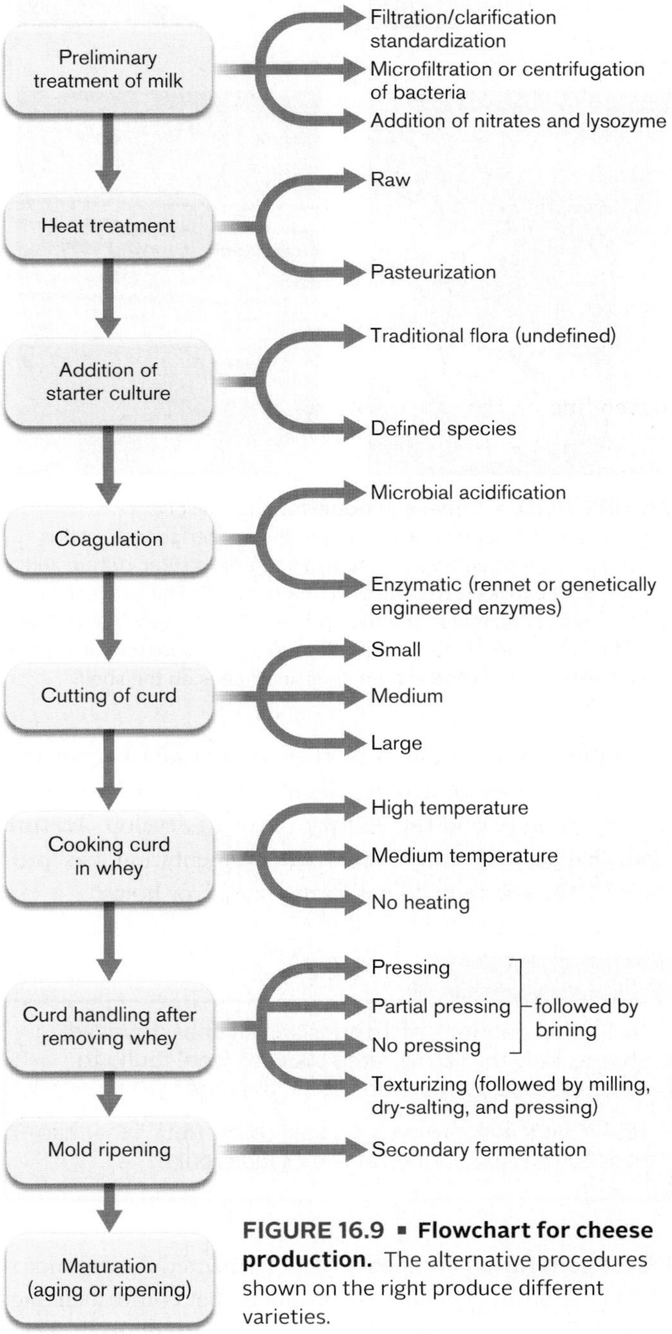
Preliminary treatment of milk
Filtration/clarification standardization
Microfiltration or centrifugation of bacteria
Addition of nitrates and lysozyme
Heat treatment
Raw
Pasteurization
Addition of starter culture
Traditional flora (undefined)
Defined species
Coagulation
Microbial acidification
Enzymatic (rennet or genetically engineered enzymes)
Cutting of curd
Small
Medium
Large
Cooking curd in whey
High temperature
Medium temperature
No heating
Curd handling after removing whey
Pressing
Partial pressing
No pressing
followed by brining
Texturizing (followed by milling, dry-salting, and pressing)
Mold ripening
Secondary fermentation
Maturation (aging or ripening)

FIGURE 16.9 ▪ Flowchart for cheese production. The alternative procedures shown on the right produce different varieties.

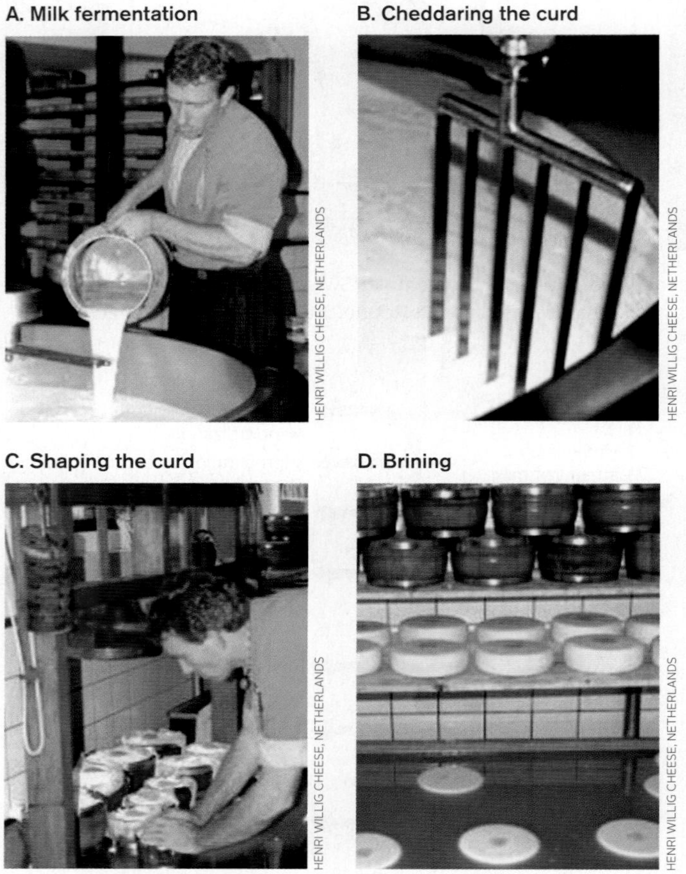
A. Milk fermentation
HENRI WILLIG CHEESE, NETHERLANDS
B. Cheddaring the curd
HENRI WILLIG CHEESE, NETHERLANDS
C. Shaping the curd
HENRI WILLIG CHEESE, NETHERLANDS
D. Brining
HENRI WILLIG CHEESE, NETHERLANDS

FIGURE 16.10 ▪ Cheese production. Gouda cheese is produced at the Henri Willig factory in the Netherlands. **A.** Milk is poured into a fermentation tub with a bacterial starter culture and rennet. **B.** The milk curd is cut, or "cheddared." **C.** The curds are shaped into round molds and then pressed to remove whey. **D.** The solidified curds are floated in brine to form the characteristic rind of Gouda cheese. The cheese then dries and ripens on the shelf.

(or aging), the cheese may be floated in brine to generate a rind; or it may be coated or spiked with a *Penicillium* mold. The ripening period then allows flavor to develop. Texture also changes; for example, where fermentation has produced CO_2, the trapped gas forms "eyes," or holes.

Thought Questions

16.3 In an outbreak of listeriosis from unpasteurized cheese, only the refrigerated cheeses were found to cause disease. Why would this be the case?

16.4 Cow's milk contains 4% lipid (butterfat). What happens to the lipid during cheese production?

Flavor generation in cheese. In all fermented foods, microbial metabolism generates by-products that confer a characteristic aroma and flavor. In some cases, particular species confer distinctive flavors; for example, *Propionibacterium* ferments lactate or pyruvate to propionate, a distinctive flavor component of Swiss cheese. All bacteria generate a surprising range of side reactions, forming trace products that confer distinctive flavors. For example, while *Lactobacillus* converts most of the lactose to lactic acid, a small fraction of the pyruvate is converted to acetoin, acetaldehyde, or acetic acid, which contribute flavor. Most amino acids are retained intact, but traces are converted to flavorful alcohols, esters, and sulfur compounds (**Fig. 16.11**). For example, methanethiol (CH_3SH) contributes to the desirable flavor of cheddar cheese. Lipids are not significantly metabolized by *Lactobacillus,* but in mold-ripened cheeses such as Camembert, *Penicillium* oxidizes a small amount of the lipids to flavorful methylketones, alcohols, and lactones.

Acid Fermentation of Vegetables

Many kinds of vegetable products are based on microbial fermentation. Commercial products marketed globally include pickles, soy sauce, and sauerkraut. Other products provide staple foods for particular nations or regions, such as Indonesian tempeh and Korean kimchi.

Soy fermentation. Soybeans offer one of the best sources of vegetable protein and are indispensable for the diet of millions of people, particularly in Southeast Asia. In North America, soy products are important for vegetarian consumption and as a milk substitute, as well as for animal feed. But soybeans also contain substances that decrease their nutritive value. Phytate, or inositol hexaphosphate, chelates minerals such as iron, inhibiting their absorption by the intestine. Lectins are proteins that bind to cell-surface glycoproteins within the human body. At high concentration, soybean lectins may upset digestion and induce autoimmune diseases. Protease inhibitors interfere with chymotrypsin and trypsin, thus decreasing the amount of protein that can be obtained from soy-based food.

All of these drawbacks of soybeans are diminished by microbial fermentation, while the protein content remains comparable to that of the unfermented bean (40%). A variety of fermented soy foods have been developed. Most soy fermentation involves mold growth, supplemented by bacteria that contribute vitamins, including vitamin B_{12}.

A major fermented soy product is **tempeh**, a staple food of Indonesia, the world's fourth-most-populous country, as well as of other countries in Southeast Asia (**Fig. 16.12**). Tempeh consists of soybeans fermented by *Rhizopus oligosporus,* a common bread mold. Besides decreasing the negative factors of soy, the mold growth breaks down proteins into more digestible peptides and amino acids. During World War II, tempeh was fed to American prisoners

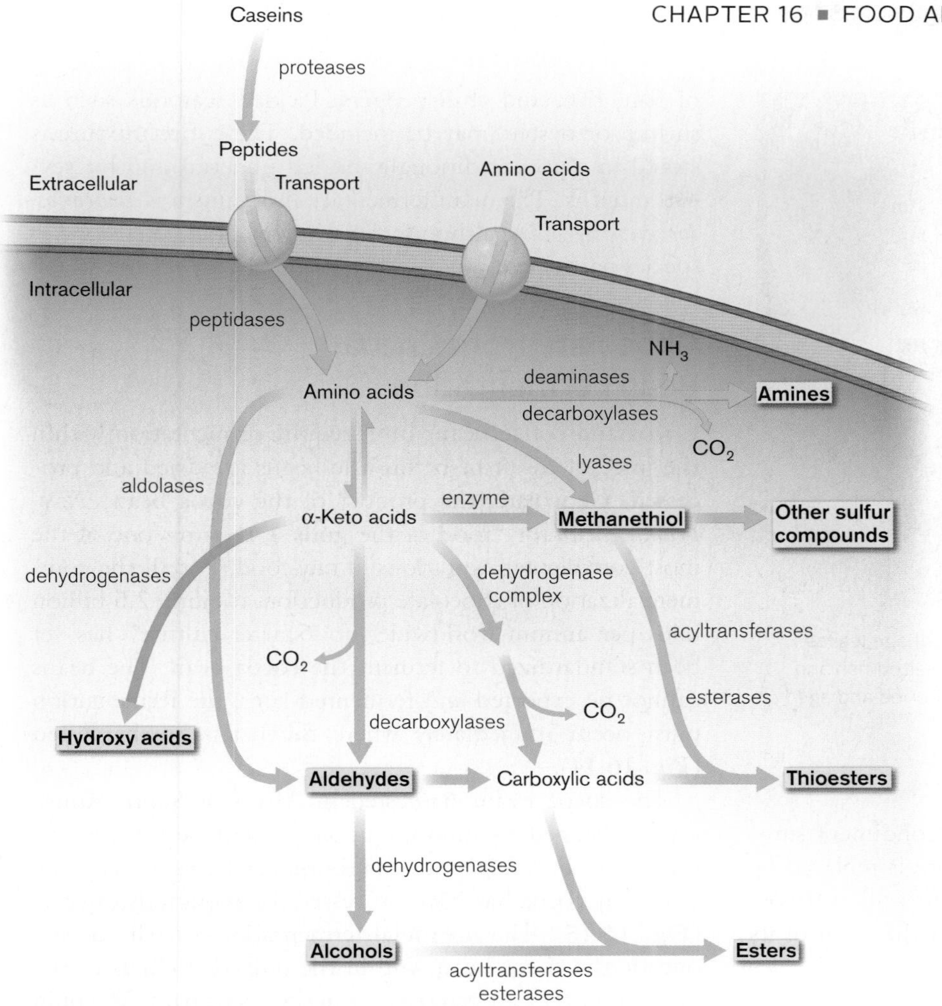
Caseins
proteases
Peptides
Extracellular
Transport
Amino acids
Transport
Intracellular
peptidases
NH3
Amino acids
deaminases
decarboxylases
Amines
CO2
lyases
aldolases
α-Keto acids
enzyme
Methanethiol
Other sulfur compounds
dehydrogenases
dehydrogenase complex
acyltransferases
CO2
esterases
CO2
Hydroxy acids
decarboxylases
Aldehydes
Carboxylic acids
Thioesters
dehydrogenases
Alcohols
Esters
acyltransferases
esterases

FIGURE 16.11 ▪ Flavor generation from amino acid catabolism. Casein catabolism generates flavor molecules (highlighted). Extracellular enzymes break down casein into peptides and amino acids, which are taken into the bacterial cell by membrane transporters. The amino acids are fermented to volatile alcohols and esters. In some cases, they combine with sulfur to form methanethiol and other sulfur-containing odorants characteristic of cheese.

of war held by the Japanese. The tempeh was later credited with saving the lives of prisoners whose dysentery and malnutrition had impaired their ability to absorb intact proteins.

Tempeh is commonly produced in home-based factories in Indonesia. The soybeans are soaked in water overnight, allowing initial fermentation by naturally present lactic acid bacteria; in some cases, a crude "starter" may be introduced from the water of soybeans soaked previously. This prefermentation allows bacterial generation of vitamins and produces mild acid that promotes growth of mold. The soaked beans are then hulled, cooked, and cooled to room temperature for inoculation with *R. oligosporus* spores or with a previous tempeh culture. The inoculated beans are wrapped in banana leaves or in perforated plastic bags and then allowed to incubate for 2 days. The mold grows as a white mycelium that permeates the beans, joining them into a solid cake. The final product has a mushroom-like taste and is often served fried or grilled like a hamburger.

Other soy products undergo acidic fermentation by the mold *Aspergillus oryzae*. The Japanese condiment **miso** is made from ground soy and rice, salted and fermented for 2 months by *A. oryzae*.

A. B.

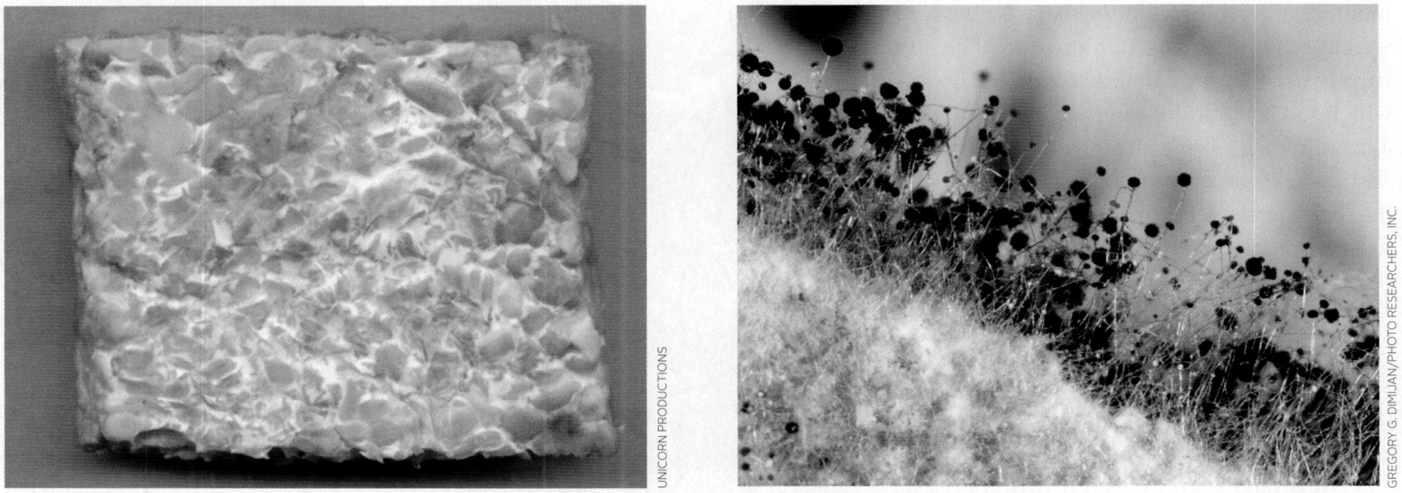
UNICORN PRODUCTIONS
GREGORY G. DIMIJIAN/PHOTO RESEARCHERS, INC.

FIGURE 16.12 ▪ Tempeh, a mold-fermented soy product. **A.** Fried tempeh. **B.** *Rhizopus oligosporus* mold, used to make tempeh.

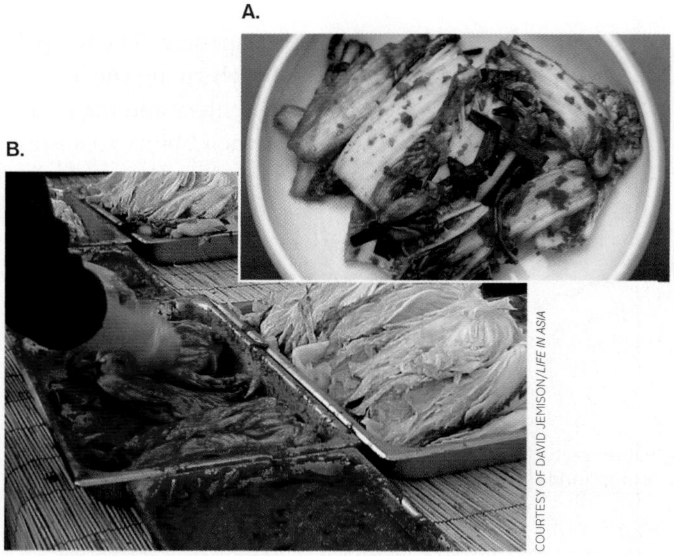

ALAMY

COURTESY OF DAVID JEMISON/*LIFE IN ASIA*

FIGURE 16.13 ▪ Kimchi. A. To make kimchi, cabbage leaves are layered alternating with chili paste containing salted fish and vegetables. **B.** The salted layers are packed to be buried and aged for 2 months.

Soy sauce is made from jiang, a Chinese condiment similar to miso in which the rice starter culture is replaced by wheat. The fermentation generates glutamic acid, a flavor-enhancing compound known popularly in the form of its salt—monosodium glutamate, or MSG.

Fermentation of cabbage and other vegetables. Various leaf vegetables are fermented by traditional societies, originally as a means of storage over the winter months. In Europe and North America, the best-known fermented products include sauerkraut and pickles. Sauerkraut production involves heterolactic fermentation by *Leuconostoc mesenteroides.* In heterolactic fermentation, each fermented sugar molecule yields lactic acid, as well as ethanol and carbon dioxide. The culture is used to inoculate shredded cabbage, which is layered in alternation with salt. The salt helps limit the number of species and the extent of microbial growth. A similar brine-enhanced fermentation process is used to pickle cucumbers, olives, and other vegetables.

An important food based on brine-fermented cabbage is Korean **kimchi** (**Fig. 16.13**). Kimchi is prepared from Chinese cabbage, salted and layered with radishes, peppers, onions, and other vegetables. The vegetables are covered with a paste of fish, rice, and chili peppers. Pickled seafoods such as shrimp or oysters may be included. The entire mixture is stored in a pot, traditionally buried underground for several months. The main fermentation organism is *Leuconostoc mesenteroides,* although *Streptococcus* and *Lactobacillus* species participate.

Chocolate from Cocoa Bean Fermentation

Cocoa and coffee beans both require fermentation within the juice of the fruit before the beans are dried and processed. Chocolate, the product of the cocoa bean, *Theobroma cacao* (or "food of the gods"), requires one of the most complex fermentations of any food. For all the commercialization of chocolate production, totaling 2.5 billion kilos per annum worldwide, no "starter culture" has yet been standardized to ferment the cocoa bean. The beans cannot be exported and fermented later; the fermentation must occur immediately where the beans are harvested (**Fig. 16.14**).

The cocoa beans harvested in Africa or South America are heaped in mounds upon plantain leaves for fermentation by indigenous microorganisms, essentially the same way cocoa has been processed for thousands of years (**Fig. 16.15**). The microbial fermentation actually occurs outside the cocoa bean, within the pulp that clings to the beans after they are removed from the cocoa fruit. The pulp contains approximately 15% sugars and pectin (a branched polysaccharide), 2% citric acid, plus a rich supply of amino acids and minerals. These nutrients support growth of many kinds of microbes. Brazilian microbiologist Rosane

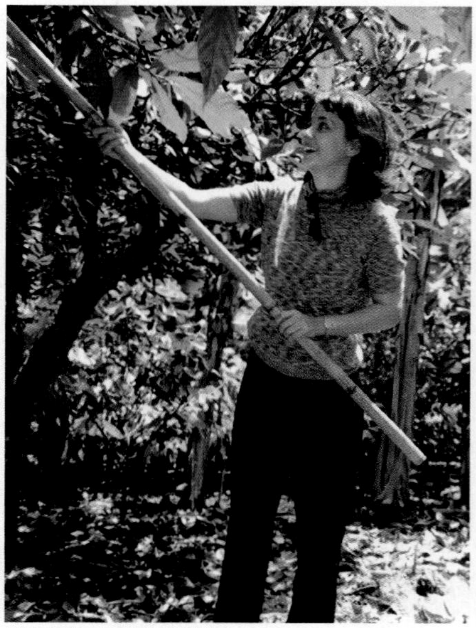

COURTESY OF ROSANE SCHWAN

SUPERSTOCK

FIGURE 16.14 ▪ Cocoa beans. A. Rosane Schwan harvests cocoa beans for her research on cocoa fermentation. **B.** Cocoa fruit, showing beans encased in mucilage.

A.

ICCO

B.

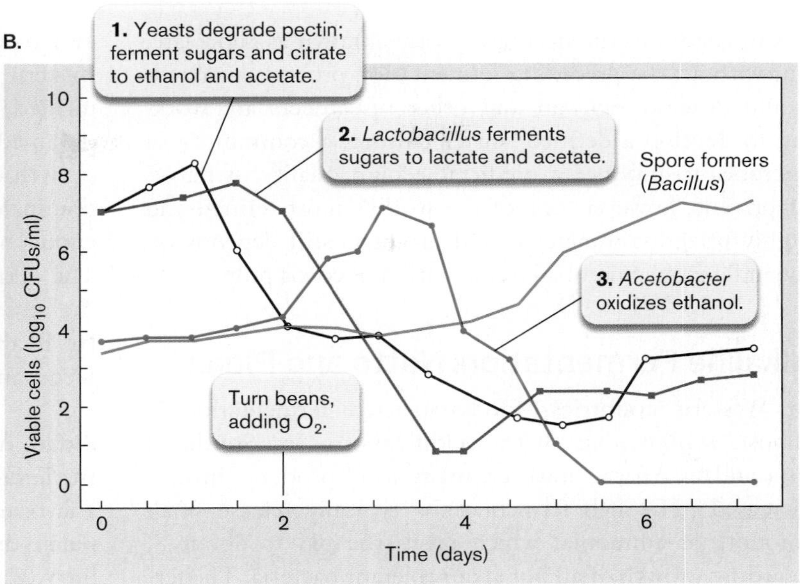
1. Yeasts degrade pectin; ferment sugars and citrate to ethanol and acetate.
2. *Lactobacillus* ferments sugars to lactate and acetate.
Spore formers (*Bacillus*)
3. *Acetobacter* oxidizes ethanol.
Turn beans, adding O_2.
Viable cells ($\log_{10}$ CFUs/ml)
10
8
6
4
2
0
0
2
4
6
Time (days)

FIGURE 16.15 ▪ Microbial succession during cocoa pulp fermentation. **A.** Heap of beans covered by plantain leaves. Fruit pulp ferments, liquefies, and drains away, while the beans acidify and turn brown. **B.** Yeasts ferment citrate to alcohol; then, lactic acid bacteria convert sugars to lactate. With aeration, acetic acid bacteria oxidize ethanol to CO_2. CFUs = colony-forming units. *Source:* Data in part B from R. Schwan.

Schwan (**Fig. 16.14A**), of the University of Lavras, Brazil, analyzed the microbial community. She defined three stages of succession: yeasts, lactic acid bacteria, and acetic acid bacteria (**Fig. 16.15B**).

Yeasts (anaerobic). Citric acid acidifies the pulp (pH 3.6). The acidity favors growth of yeasts, including *Candida, Kloeckera,* and *Saccharomyces.* The yeasts consume sugars and citric acid, increasing pH. They also degrade pectin into glucose and fructose, allowing the pulp to liquefy. As the liquefied pulp drains, yeast ferments the remaining sugars to ethanol, CO_2, and mixed acids such as acetate. The reactions release heat, increasing the temperature and accelerating metabolism. But the acetate, a permeant acid (discussed in Chapter 13) crosses the yeast cell membrane, where it lowers cell pH and inhibits growth. The ethanol and acetate also penetrate the bean embryo, killing the cells and releasing enzymes that generate key flavor molecules of chocolate.

Lactic acid bacteria (anaerobic). The consumption of citric acid by yeast increases the bean pH to pH 4.2, encouraging growth of lactic acid bacteria, such as *Lactobacillus plantarum. Lactobacillus* species convert sugars to lactic acid, as well as acetate and CO_2. As fermentable substrates disappear, however, lactic acid bacteria are inhibited.

Acetic acid bacteria (aerobic). After 2 days, the beans are turned over and mixed periodically to permit access to oxygen. Aerobic acetic acid bacteria such as *Acetobacter* now oxidize the ethanol and acids to CO_2. The consumption of acids neutralizes undesirable acidity. Heat is released, increasing the temperature to as high as 50°C. Oxygen penetrates the bean, oxidizing key components such as polyphenols. Polyphenol oxidation generates the brown color of cocoa and contributes flavor.

After fermentation and pulp drainage, the beans are dried and roasted. The roasting process completes the transformation of cocoa substances that contribute flavor. Cocoa liquor and cocoa butter are extracted from the beans and then recombined with sugar and other components to make "cocoa mass" (**Fig. 16.16**). The cocoa mass is stirred for several days to achieve a smooth texture; then

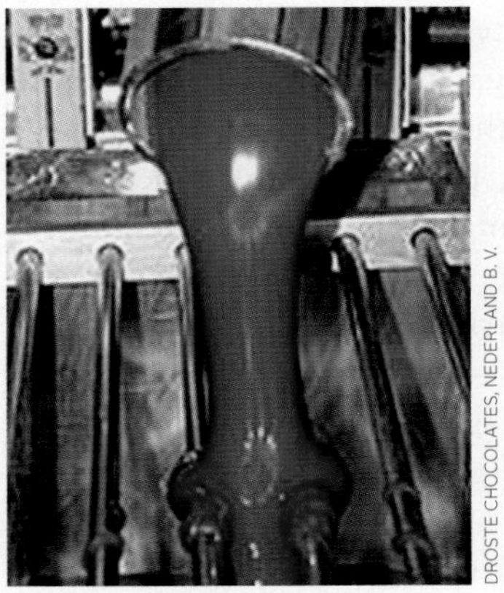
DROSTE CHOCOLATES, NEDERLAND B. V.

FIGURE 16.16 ▪ Chocolate manufacture. Cocoa mass contains cocoa butter and liquor extracted from the cocoa beans, mixed with sugar and other ingredients.

it is molded into the decorative forms known as chocolate. But without the preceding fermentation process, no flavor would develop. Schwan and other researchers are working to develop a defined starter culture, a community of microbes to produce a predictable high quality of flavor. At present, however, one of the world's most refined and highly prized commercial food products still depends on the indigenous microbial fermentation of cocoa pulp.

Alkaline Fermentation: Natto and Pidan

In Western countries, food-associated fermentation is almost synonymous with acidification. In Southeast Asia and in Africa, however, many food products involve increased pH. Such fermentations typically release small amounts of ammonia, which raises the pH to about 8, retarding growth of all but alkali-tolerant bacteria. The fermenting bacteria are usually *Bacillus,* aerobic species tolerant of moderate alkali and capable of extensive proteolysis and amino acid decomposition. The fermentation needs to be controlled to limit the loss of protein content, but the end result is a highly stable food product.

Natto. The Japanese soybean product **natto** is prepared by a process similar to that for tempeh. Soybeans are washed, pre-fermented, and cooked briefly before incubation with the starter organism *Bacillus natto.* The fermenting beans are incubated in a shallow, ventilated container at a slightly raised temperature (40°C).

B. natto secretes numerous extracellular enzymes, including proteases, amylases, and phytases. These enzymes decrease the undesirable components of soy, such as phytates and lectins, while liberating more easily digestible peptides and amino acids. Some of the amino acids are deaminated, generating ammonia; a well-ventilated natto chamber allows most of this gas to escape. In addition, *B. natto* synthesizes extracellular polymers such as polyglutamate (a peptide chain consisting exclusively of glutamic acid residues). Polyglutamate generates long elastic strings that bind the beans together. The stretching of these strings from chopsticks is considered a sign of a good natto (**Fig. 16.17A**).

Alkali-fermented vegetables. In Africa, numerous vegetable products are based on alkaline fermentation predominantly by *Bacillus* species. An example is dawadawa, a paste of fermented locust beans common in western Africa. The locust beans are washed and supplemented with potash, or potassium hydroxide, originally obtained from wood ashes. This addition of alkali retards growth of bacteria other than *Bacillus* species, which predominate at higher pH. The beans are fermented by indigenous bacteria (bacteria already present in the beans). The fermented beans are sun-dried, releasing most of the ammonia, and pounded into cakes for storage. Similar alkali-fermented vegetables include ogiri, from melon seeds, and ugba, from oil beans.

Pidan. An ancient means of preserving eggs has produced the famous Chinese delicacy pidan, or "century egg," now a favorite at dim sum restaurants (**Fig. 16.17B**). To make pidan, duck eggs are covered with a mixture of brewed tea, lime (CaO), and sodium carbonate (Na_2CO_3). The lime and sodium carbonate react to form NaOH, which penetrates the eggshells, raising pH and coagulating the egg white proteins. The eggs are buried in mud for several months, during which time the combined action of alkali and *Bacillus* fermentation generates dark colors and interesting flavors.

Thought Question

16.5 In traditional fermented foods, without pure starter cultures, what determines the kind of fermentation that occurs?

A.

COURTESY OF MATT WEGENER

B.

ELEANOR NAKAMA-MITSUNAGA, HONOLULU STAR-BULLETIN

FIGURE 16.17 ▪ Alkali-fermented foods. **A.** Natto consists of soybeans fermented by *Bacillus natto.* The fermentation generates long strings of polyglutamate. **B.** Pidan, or "century egg," consists of duck eggs coagulated by sodium hydroxide and fermented by *Bacillus* species. One egg is cut open, revealing the transformed yolk, which develops a greenish color.

To Summarize

- **Milk curd** forms by lactic acid fermentation and rennet proteolysis, rendering casein insoluble. The cleaved peptides coagulate to form a semisolid curd. The main fermentative organisms are lactic acid bacteria.
- **Cheese varieties** include unripened cheese, semihard and hard cheeses that are cooked down and ripened, brined cheeses, and mold-ripened cheeses.
- **Cheese flavors** are generated by minor side products of fermentation, such as alcohols, esters, and sulfur compounds.
- **Soy fermentation** to tempeh and other products improves digestibility and decreases undesirable soy components such as phytates and lectins. The fermentative agent of tempeh is the bread mold *Rhizopus oligosporus.*
- **Vegetables** are fermented and brined to make sauerkraut and pickles. Cabbage and supplementary foods are fermented and brined to make kimchi.
- **Cocoa fermentation** for chocolate requires complex fermentation of cocoa beans within the fruit pulp, including anaerobic fermentation by yeast and lactic acid bacteria, and aerobic respiration by *Acetobacter* species.
- **Alkali-fermented vegetables** include the soy product natto, the egg product pidan, and the locust bean product dawadawa. The main fermenting organisms are *Bacillus* species.

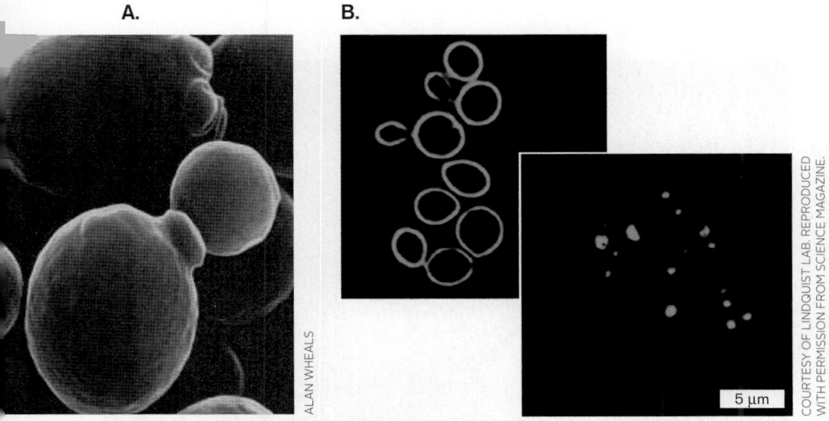
A.
B.
5 µm
ALAN WHEALS
COURTESY OF LINDQUIST LAB. REPRODUCED WITH PERMISSION FROM SCIENCE MAGAZINE.

FIGURE 16.18 ■ Baker's yeast, the "champion" fermenter. **A.** *Saccharomyces cerevisiae* cells budding; some show bud scars (SEM). **B.** *S. cerevisiae* is used to study the function of human proteins such as alpha-synuclein (green fluorescence), which plays a role in Parkinson's disease. Yeast cells engineered to express one copy of the gene (left panel) show the protein normally within their cell membrane. Two gene copies (right panel) cause the protein to clump and kill the cells.

16.4

Ethanolic Fermentation: Bread and Wine

Some of our most nutritionally significant and culturally important foods, including most bread and alcoholic beverages, require ethanolic fermentation by yeast fungi. Ethanolic fermentation converts pyruvate to ethanol and carbon dioxide:

$$C_3H_4O_3 \rightarrow CH_3CH_2OH + CO_2$$

The most prominent yeast used is *Saccharomyces cerevisiae,* known as baker's yeast or brewer's yeast (**Fig. 16.18**). A hardy organism, *S. cerevisiae* easily survives on a grocery shelf for home use and is genetically tractable for fundamental research. The yeast has been studied since the time of Pasteur, who used it to prove the biological basis of fermentation (presented in Chapter 1). *S. cerevisiae* today is a major model system of cell biology, yielding the molecular secrets of human cancer and other diseases.

Bread making depends on carbon dioxide to generate air spaces that **leaven** the dough, making its substance easier to chew and digest. The small amount of ethanol produced is eliminated during baking. For alcoholic beverages, however, ethanol is the key product, accompanied by carbon dioxide bubbles for "fizz," known as carbonation.

Bread Dough Is Leavened by Microbial CO_2

Bread is made in many different forms (**Fig. 16.19**) and from diverse kinds of flour, or ground grain. The earliest

GUY CALI AND ASSOCIATES/STOCKFORD

FIGURE 16.19 ■ Yeast bread. Many varieties of bread are made.

breads probably arose from grain mush naturally contaminated by yeast. Later, bread makers learned to include yeast left over from wine or beer production as a starter culture.

Yeast bread production. The preparation of all forms of yeast bread requires the same fundamental steps. A starter culture of yeast is included in the dough. The yeast can be commercial baker's yeast, or it can be **sourdough** starter, an undefined microbial population derived from a previous batch of dough. Analysis of sourdough shows mainly yeasts and lactobacilli, which release acids that favor the growth of the yeasts. The dough is kneaded to develop a fine network of air pockets and allowed to rise, expanding with production of the carbon dioxide gas (**Fig. 16.20**). The finest-textured breads are made from wheat flour, which contains gluten, a protein complex that forms a fine molecular network supporting the rising dough.

Modern bread sold commercially is produced on an industrial scale. The dough is made in huge vats and then cut into regular chunks that are placed into the bread pans. The loaves are baked under hot-air convection, a system that greatly decreases the baking time. In the United States, most commercial bread is sliced mechanically—an invention that dates back to 1912. Sliced bread requires preservative chemicals to prevent growth of mold on the exposed interior.

Thought Question

16.6 Compare and contrast the role of fermenting organisms in the production of cheese and bread.

SUSAN C. BOURGOIN/FOOD PIX/JUPITER IMAGES

FIGURE 16.20 ■ Making bread. As yeast fermentation generates carbon dioxide gas, the dough rises.

Injera: extended fermentation. Most kinds of bread involve only a short fermentation period, just long enough to produce enough gas for leavening. A prolonged fermentation, with more extensive microbial activity, occurs in the dough for an Ethiopian bread called **injera** (**Fig. 16.21**). The high microbial content provides a substantial source of vitamins not found in quick-rising breads.

Injera is made from teff (*Eragrostis tef*), a grain with small, round kernals that have high protein and lack gluten. Teff grows in arid regions, and is now being cultivated in the United States for gluten-free products. Because teff lacks gluten, it cannot rise as much as wheat flour, but it makes a kind of flatbread. The dough is spread into a wide pancake, and the organisms present in the grain and air are allowed to ferment it for 3 days. The fermentation includes a succession of species, usually dominated by the yeast *Candida*.

A.

JIM SUGAR/CORBIS

B.

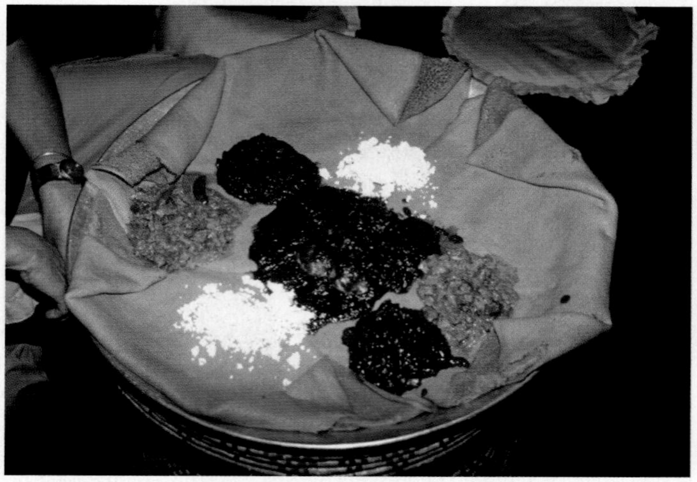
ALAMY

FIGURE 16.21 ■ Injera. A. After 3 days of fermentation, injera dough is baked in a ceramic pan upon a "Mirte" charcoal stove. **B.** Injera forms an edible tablecloth for a variety of Ethiopian foods.

The extended fermentation generates a complex range of by-products that confer exceptional flavors in the baked product. In Ethiopia, injera forms the basis of an entire meal, served with other food items placed upon it as an edible tablecloth. Diners wrap samples of each food in a fold of injera and consume them together.

Alcoholic Beverages: Beer and Wine

Ethanolic fermentation of grain or fruit was important to early civilizations because it provided a drink free of waterborne pathogens. Traditional forms of beer also provided essential vitamins in the unfiltered yeasts.

Ethanol is unique among fermentation products in that it provides a significant source of caloric intake, but it is also a toxin that impairs mental function. A modest level of ethanol enters the human circulation naturally from intestinal flora, equivalent to a fraction of a drink per day. The human liver produces the enzyme alcohol dehydrogenase, which detoxifies ethanol. This enzyme in a healthy liver can metabolize small amounts of alcohol without harm. However, excess alcohol consumption can overload the liver's capacity for detoxification and permanently damage the liver and brain.

Beer: alcoholic fermentation of grain. Beer production is one of the most ancient fermentation practices and is depicted in the statuary of ancient Egyptian tombs dated to 5,000 years ago (**Fig. 16.22A**). The earliest Sumerian beers were made from bread soaked in water and fermented. Today, most beer is produced commercially by fermenting barley using giant vats (**Fig. 16.22B**). Production of high-quality beer involves complex processing with many steps, including germination of barley grains, mashing in water and cooking, and introduction of hops for flavor (eTopic 16.1).

First the barley grains must germinate; that is, the seed embryos must start to grow. The germinating embryo makes enzymes needed to break down the barley starch to maltose (disaccharide) and glucose. Most of the sugars are fermented by yeast to ethanol and carbon dioxide. However, minor side products contribute flavors—or unpleasant off-flavors if present in too great an amount (**Fig. 16.23**). For example, off-flavors may result from the presence of small amounts of oxygen that oxidize some ethanol to acetaldehyde.

The yeast ferment most sugars to ethanol, but a small fraction is drawn off to make amino acids via TCA cycle intermediates, as discussed in Chapter 15. The 2-oxo acids of the TCA cycle are analogous to pyruvate, with the methyl group replaced by extended carbon chains (R group). A tiny amount of the 2-oxo acids is converted to long-chain alcohols, which add desirable flavor to beer.

Thought Question

16.7 Compare and contrast the role of low-concentration by-products in the production of cheese and beer.

Wine: alcoholic fermentation of fruit. The fermentation of fruit gives rise to wine, another class of alcoholic products of enormous historical and cultural significance. Grapes produce the best-known wines, but wines and distilled liquors are also made from apples, plums, and other fruits. The key difference between fermentation of fruits and fermentation of grains is the exceptionally high monosaccharide content in fruits. Grape juice, for example, can contain concentrations of glucose and fructose as high as 15%. The availability of simple sugars allows yeast to begin fermenting immediately, with no need for preliminary breakdown of long-chain carbohydrates, as in the malting and mashing of beer.

Most modern wine production uses strains of the grape *Vitis vinifera*. The grapes are

A.

EGYPTIAN MUSEUM, CAIRO/BORROMEO/ART RESOURCE, NY

B.

COURTESY OF HOWSTUFFWORKS.COM

FIGURE 16.22 ■ Beer production: ancient and modern. A. Making beer in ancient Egypt, circa 3000 BCE. The mash is stirred in earthen jars. **B.** Fermentors in a modern brewery.

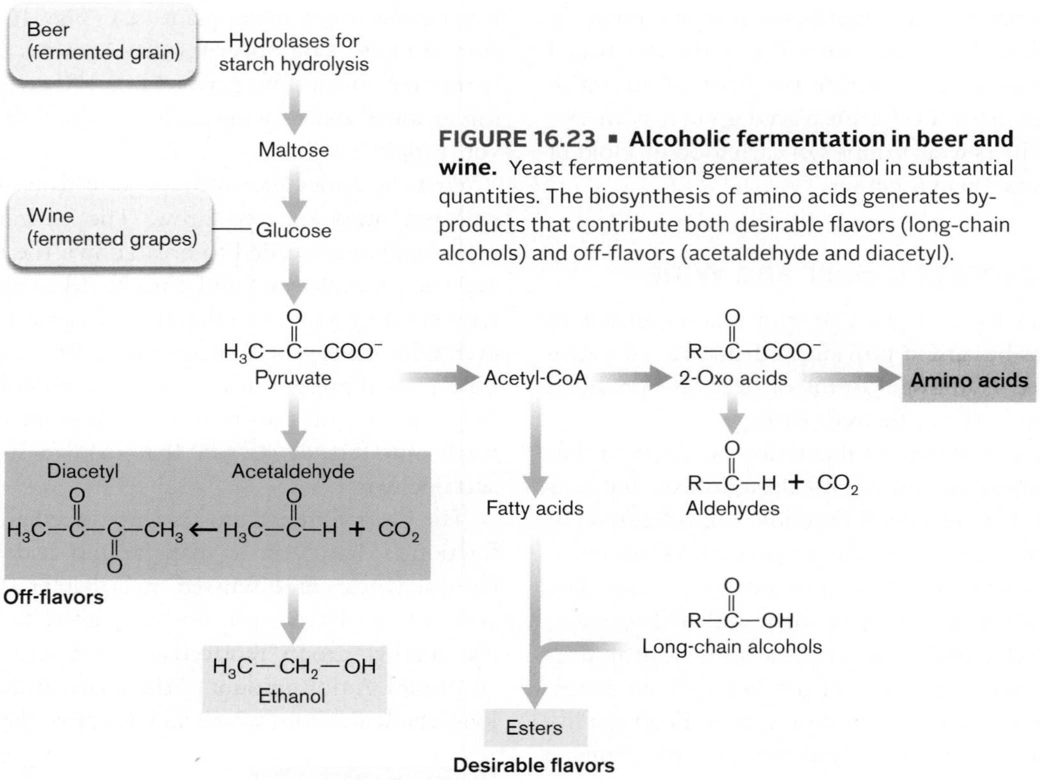
Beer
(fermented grain)
Hydrolases for starch hydrolysis
Maltose
Wine
(fermented grapes)
Glucose
H3C—C—COO−
Pyruvate
Acetyl-CoA
R—C—COO−
2-Oxo acids
Amino acids
Diacetyl
Acetaldehyde
H3C—C—C—CH3 ← H3C—C—H + CO2
Off-flavors
Fatty acids
R—C—H + CO2
Aldehydes
R—C—OH
Long-chain alcohols
H3C—CH2—OH
Ethanol
Esters
Desirable flavors

FIGURE 16.23 ■ Alcoholic fermentation in beer and wine. Yeast fermentation generates ethanol in substantial quantities. The biosynthesis of amino acids generates by-products that contribute both desirable flavors (long-chain alcohols) and off-flavors (acetaldehyde and diacetyl).

crushed to release juices, usually in the presence of antioxidants such as sulfur dioxide (**Fig. 16.24**). For white wine, the skins are removed before juice is fermented. For red wine, the skins are included in early fermentation to extract the red and purple anthocyanin pigments, as well as phenolic flavor compounds. The first few days of fermentation are dominated by indigenous species of yeast naturally present on the grapes, such as *Kloeckera* and *Hanseniaspora* species. Commercial producers usually inoculate with standard *Saccharomyces cerevisiae,* whose population dominates the late stage of fermentation (6–20 days). Yeast growth ends once the ethanol level reaches about 15%; to achieve higher alcohol content, distillation is required.

After fermentation, the wine is drained, or "racked," from the sediment of grape and yeast material, the lees. The liquid may be further clarified by centrifugation. Then it is stored for 2–3 weeks in tanks or barrels. During storage, a second stage of fermentation may be performed, called **malolactic fermentation**. Malolactic fermentation is needed to decrease the acidity from malic acid (found in grapes). Malic acid is converted to lactic acid, with a higher dissociation constant (a weaker acid). The L-malate is decarboxylated to L- or D-lactate:

$$^{-}HOOC—CH_2—CHOH—COOH^{-} \rightarrow CH_3—CHOH—COOH + CO_2$$

The wine is seeded with *Oenococcus oeni* bacteria, which ferment L-malate (deprotonated L-malic acid).

As in beer production, yeast fermentation of wine produces numerous minor products contributing flavor, such as long-chain alcohols and esters. At the same time, overgrowth of yeast or the growth of undesired species can produce excess amounts of these compounds, such as sulfides and phenolics, giving rise to off-flavors. Some undesired species require oxygen exposure, whereas others can grow during storage and bottling. The balance of microbial populations is challenging to control and has a major role in determining the quality of a given wine vintage.

To Summarize

- **Bread is leavened** by yeasts conducting limited ethanolic fermentation, producing enough carbon dioxide gas to expand the dough.
- **Injera** bread dough undergoes more extensive fermentation by indigenous organisms and, as a result, generates multiple flavors.
- **Beer** requires alcoholic fermentation of grain. Barley grains are germinated, allowing enzymes to break down the starch to maltose for yeast fermentation.

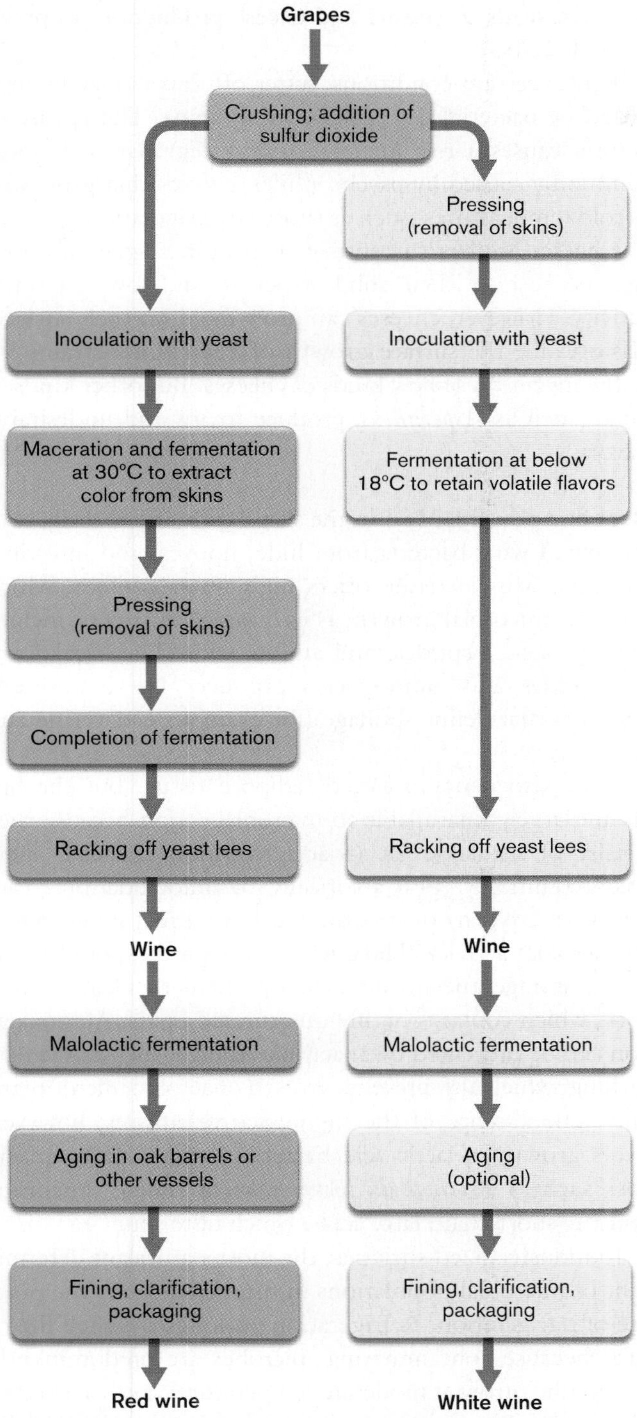
Grapes
Crushing; addition of sulfur dioxide
Pressing (removal of skins)
Inoculation with yeast
Inoculation with yeast
Maceration and fermentation at 30°C to extract color from skins
Fermentation at below 18°C to retain volatile flavors
Pressing (removal of skins)
Completion of fermentation
Racking off yeast lees
Racking off yeast lees
Wine
Wine
Malolactic fermentation
Malolactic fermentation
Aging in oak barrels or other vessels
Aging (optional)
Fining, clarification, packaging
Fining, clarification, packaging
Red wine
White wine

FIGURE 16.24 ▪ Production of red and white wines. Left: For red wine, the grapes are fermented with the skins at a temperature that increases extraction of color and tannins. **Right:** For white wine, the skins are removed before addition of yeast starter, and the temperature is kept low to retain volatile flavors. Both kinds of wine usually undergo malolactic fermentation by *Oenococcus oeni,* special lactic acid bacteria that consume malic acid.

- **Secondary products of grain fermentation**, such as long-chain alcohols and esters, generate the special flavors of beer.
- **Wine** derives from alcoholic fermentation of fruit, most commonly grapes. The grape sugar (glucose) is fermented by yeast to alcohol. A secondary product, malate, undergoes malolactic fermentation by *Oenococcus oeni* bacteria.

16.5 Food Spoilage and Preservation

We humans have always competed with microbes for our food. When early humans killed an animal, microbes commenced immediately to consume its flesh. Because meat perished so fast, it made economic sense to share the kill immediately and consume it all as soon as possible. Vegetables might last longer, but eventually they succumbed to mold and rot. Later societies developed preservation methods, such as drying, smoking, and canning, that enabled humans to survive winters and dry seasons on stored food.

Modern food preservation depends on **antimicrobial agents** (chemical substances that either kill microbes or slow their growth), as well as physical preservative measures, such as treatment with heat and pressure. These general principles of microbial control are described in Chapter 5. Here we focus on microbial contamination and food preservation from the perspective of the food industry.

Food Spoilage and Food Contamination

After food is harvested, several kinds of chemical changes occur. Some begin instantly, whereas others take several days to develop. Some changes, such as meat tenderizing, may be considered desirable; others, such as putrefaction, render food unfit for consumption. The major classes of food change include:

- **Enzymatic processes.** Following the death of an animal, its flesh undergoes proteolysis by its own enzymes. Limited proteolysis tenderizes meat. Plants after harvest undergo other changes; for example, in harvested corn the sugar rapidly converts to starch. That is why vegetables taste sweetest immediately after harvest.
- **Chemical reactions with the environment.** The most common abiotic chemical reactions involve oxidation by

air—for example, lipid autooxidation, which generates rancid odors.

- **Microbiological processes.** Microbes from the surface of the food begin to consume it—some immediately, others later in succession—generating a wide range of chemical products. In meat, internal organs of the digestive tract are an important source of microbial decay.

Microbial activity can aid food production, but it can also have various undesirable effects. Two different classes of microbial effects are distinguished: food spoilage and food contamination with pathogens.

Food spoilage refers to microbial changes that render a product obviously unfit or unpalatable for consumption. For example, rancid milk or putrefied meats are unpalatable and contain metabolic products that may be deleterious to human health, such as oxidized fatty acids or organic amines. Even in these cases, however, the definition of spoilage depends partly on cultural practice. What is sour milk to one person may be buttermilk to another; what one society considers spoiled meat, another may consider merely aged.

Different pathways of microbial metabolism lead to different kinds of spoilage. Sour flavors result from acidic fermentation products, as in sour milk. Alkaline products generate bitter flavor. Oxidation, particularly of fats, causes **rancidity**, whereas general decomposition of proteins and amino acids leads to **putrefaction**. The particularly noxious odors of putrefaction derive from amino acid breakdown products that often have apt names, such as the amines cadaverine and putrescine and the aromatic product skatole.

"Food contamination," or **food poisoning**, refers to the presence of microbial pathogens that cause human disease—for example, rotaviruses that cause gastrointestinal illness. Pathogens usually go unnoticed as food is consumed, because their numbers are very low, and they may not even grow in the food. Even the freshest-appearing food may cause serious illness if it has been contaminated with a small number of pathogens.

How Food Spoils

Different foods spoil in different ways, depending on their nutrient content, the microbial species, and environmental factors such as temperature. **Table 16.2** summarizes common forms of spoilage.

Dairy products. Milk and other dairy products contain carbon sources, such as lactose, protein, and fat. In fresh milk, the nutrient most available for microbial catabolism is lactose, which commonly supports anaerobic fermentation to sour milk. Fermentation by the right mix of microbes, however, leads to yogurt and cheese production, as previously described.

Under certain conditions, bitter off-flavors may be produced by bacterial degradation of proteins. The release of amines causes a rise in pH. Protein degradation is most commonly caused by psychrophiles, species that grow well at cold temperatures, such as those of refrigeration.

Cheeses are less susceptible than milk to general spoilage, because of their solid structure and lowered water activity. However, cheeses can grow mold on their surface. Historically, the surface growth of *Penicillium* strains led to the invention of new kinds of cheeses. But other kinds of mold, such as *Aspergillus,* produce toxins and undesirable flavors.

Meat and poultry. Meat in the slaughterhouse is easily contaminated with bacteria from hide, hooves, and intestinal contents. Muscle tissue offers high water content, which supports microbial growth, as well as rich nutrients, including glycogen, peptides, and amino acids. The breakdown of peptides and amino acids produces the undesirable odorants that define spoilage (for example, cadaverine and putrescine).

Meat also contains fat, or adipose tissue, but the lipids are largely unavailable to microbial action because they consist of insoluble fat (triacylglycerides). Instead, meat lipids commonly spoil abiotically by autooxidation (reaction with oxygen) of unsaturated fatty acids, independent of microbial activity. Thus, when meats are exposed to air during storage, they turn rancid—particularly meats such as pork, which contains highly unsaturated lipids. Autooxidation can be prevented by anaerobic storage, such as vacuum packing, which also prevents growth of aerobic microorganisms. The absence of the suppressed organisms, however, favors growth of lactic acid bacteria and facultative anaerobes such as *Brochothrix thermosphacta.* These organisms generate short-chain fatty acids, which taste sour.

In industrialized societies, the most significant determinant of microbial populations in meat spoilage is the practice of refrigeration. Refrigeration prolongs the shelf life of meat because contaminating microbes are predominantly mesophilic (grow at moderate temperatures, as discussed in Chapter 5). But ultimately, the few psychrotrophs initially present do grow; typically, these are *Pseudomonas* species. The pseudomonads are also favored by the low pH of meat (pH 5.5–7.0), which results from the accumulation of lactic acid in the muscle.

Seafood. Fish and other seafood contain substantial amounts of protein and lipids, as well as amines such as trimethylamine oxide. Fish spoils more rapidly than meat

TABLE 16.2

Food spoilage (examples).

Food product	Signs of spoilage	Microbial cause
Dairy products		
Milk	Sour flavor	Lactic acid bacteria produce lactic and acetic acids.
Milk	Coagulation	Lactic acid bacteria produce proteases that destabilize casein and lower pH, causing coagulation.
Milk	Bitter flavor	Psychrophilic bacteria degrade proteins and amino acids.
Cheese	Open texture, fissures	Lactic acid bacteria produce carbon dioxide.
Cheese	Discoloration and colonies	Molds such as *Penicillium* and *Aspergillus* grow on the cheese.
Meat and poultry		
Meat and poultry	Rancid flavor	Psychrotrophic bacteria produce fatty acids that become oxidized.
Meat and poultry	Putrefaction	*Pseudomonas* and other aerobes degrade amino acids, producing amines and sulfides.
Meat and poultry	Discolored patches	Molds such as *Mucor* and *Penicillium* grow on the surface.
Eggs	Pink or greenish egg white	*Pseudomonas* and related bacteria grow on albumin, producing water-soluble pigments.
Eggs	Sulfurous odor	Bacterial growth on albumin releases hydrogen sulfide.
Seafood		
Fish	Fishy smell	Anaerobic psychrophiles such as *Photobacterium* convert trimethylamine oxide to trimethylamine.
Fish	Odor of putrefaction	*Pseudomonas* and other Gram-negative species degrade amino acids, producing amines and sulfides.
Shellfish	Odor of putrefaction	*Vibrio* and other marine bacteria decompose the protein.
Fruits, vegetables, and grains		
Plants before harvest	Rotting or wilting	Plant pathogens, most commonly fungi such as *Alternaria, Aspergillus,* and *Penicillium.*
Stored plant foods	Rotting or wilting	Molds or bacteria produce degradative enzymes, such as pectinases and cellulases.
Apples, pears, cherries	Geosmin off-flavor	*Penicillium* mold.
Peeled oranges	Discoloration and off-flavor	*Enterobacter* and *Pseudomonas* spp.
Pasteurized fruit juices	Medicine-like phenolic off-flavor	Acid- and heat-tolerant spore former, *Alicyclobacillus* sp., produces 2-methoxyphenol (guaiacol).
Bread	Ropiness	*Bacillus* spp. grow, forming long filaments.
Bread	Red discoloration	*Serratia marcescens.*

and poultry for several reasons. First, fish do not thermoregulate, and they inhabit relatively low-temperature environments. Because fish grow in low-temperature environments, their surface microorganisms tend to be psychrotrophic and thus grow well under refrigeration. In addition, marine fish contain high levels of the osmoprotectant trimethylamine oxide, which bacteria reduce to trimethylamine, a volatile amine that gives seafood its "fishy" smell. Finally, the rapid microbial breakdown of proteins and amino acids leads to foul-smelling amines and sulfur compounds, such as hydrogen sulfide and dimethyl sulfide.

Thought Question

16.8 Why would bacteria convert trimethylamine oxide (TMAO) to trimethylamine? Would this kind of spoilage be prevented by exclusion of oxygen?

Plant foods. Fruits, vegetables, and grains spoil differently from animal foods because of their high carbohydrate content and their relatively low water content. The low water content of plant foods usually translates into considerably longer shelf life than for animal-based foods. Carbohydrates favor microbial fermentation to acids or alcohols that limit further decomposition, and this microbial action can be managed to produce fermented foods, as described in Section 16.2.

Plant pathogens rarely infect humans but may destroy the plant before harvest. Most plant pathogens are fungi, although some are bacteria, such as *Erwinia* species. Historically, plant pathogens have caused major agricultural catastrophes, such as the Irish potato famine, caused by a fungus-like pathogen. Plant pathogens continue to devastate local economies and cause shortages worldwide; for example, the witches'-broom fungus *Crinipellis perniciosa* causes a fungal disease of cocoa trees that has drastically cut Latin American cocoa production.

After harvest, various molds and bacteria can soften and wilt plant foods by producing enzymes that degrade the pectins and celluloses that give plants their structure. In general, the more processed the food, the greater the opportunities for spoilage. For example, citrus fruits generally last for several weeks, but peeled oranges are susceptible to spoilage by Gram-negative bacteria.

Baked bread usually resists spoilage, except for surface molds. In rare cases, however, improperly baked bread can show contamination. The appearance of red bread, caused by the red bacterium *Serratia marcescens,* is believed to have been the source of the "blood" observed in communion bread during a Catholic mass in the Italian town of Bolsena in 1263—an event that became known as the Miracle of Bolsena.

Pathogens Contaminate Food

Intestinal pathogens spread readily because many pathogens can be transmitted through food without any outward sign that the food is spoiled. The U.S. Centers for Disease Control and Prevention (CDC) estimates that there are 76 million cases of gastrointestinal illness a year in this country, usually spread through water or food. Thus, one in four Americans experiences gastrointestinal illness in a given year. In 2013, under President Obama, the Food and Drug Administration implemented new controls and rules as part of the 2011 Food Safety Modernization Act, the largest reform of the U.S. food safety laws in 70 years. The new laws require, for example, that crop irrigation water be free of pathogens and that agricultural workers have access to bathroom facilities.

An example of food contamination is the 2008 outbreak of *Salmonella enterica* from peanut products. Peanuts contaminated at one processing plant led to an epidemic that sickened 700 people across the United States (**Fig. 16.25A**). The first cases of *Salmonella* infection were reported to the CDC on September 1, 2008. Most infected individuals developed diarrhea, fever, and abdominal cramps 12–72 hours after infection, and symptoms lasted 4–7 days. Over the next 6 months, cases were reported from nearly all U.S. states. The curve of the outbreak (cases rising and then falling) followed the profile of a single-source epidemic, in which all infections are ultimately traced back to one source. (Epidemics are discussed in Chapter 28.)

What was the original source of the widespread outbreak? The CDC researchers compared the food intake histories of ill persons against matched controls. They found a statistical association between illness and intake of peanut butter, eventually narrowed to a specific brand of peanut butter sold to institutions. As the epidemic grew, cases emerged in which the contaminated food product was crackers filled with peanut butter cream. Ultimately, the peanut butter and cream were traced back to peanuts from a single factory in Georgia. At the food plant, the source of *Salmonella* contamination could not be identified, but the plant records showed that product samples had tested positive for *Salmonella*. Instead of discarding the product, the plant had retested the samples until they "tested negative." Numerous health violations were cited, including gaps in the walls and dirt buildup throughout the plant.

Nevertheless, *Salmonella* bacteria are very common pathogens. Did all of the cases of illness result from a common strain? The CDC used DNA analysis to show that all patients carried a common strain of *S. enterica* serovar Typhimurium. (A serovar is a strain whose surface proteins elicit a distinctive immune response.) The strain was identified by analyzing its genomic DNA cleaved by restriction endonucleases (see Section 7.6). Each restriction endonuclease cleaves DNA at sequence-specific positions. Strains that differ at key restriction sites generate cleavage fragments of differing length, which are separated by pulsed-field electrophoresis (**Fig. 16.25B**). In electrophoresis, applied voltage causes DNA fragments to migrate different distances according to size; the pulsed field optimizes separation of the largest sizes. The distance each fragment moves is visualized as a band in the gel. The band pattern, or "fingerprint," of *Salmonella* DNA from infected patients showed the same fragment lengths as *Salmonella* DNA from the peanut butter sample (labeled "(1)" in **Fig. 16.25B**). The band pattern from this peanut butter sample (1) was different from that of another infected peanut butter sample (2);

A.

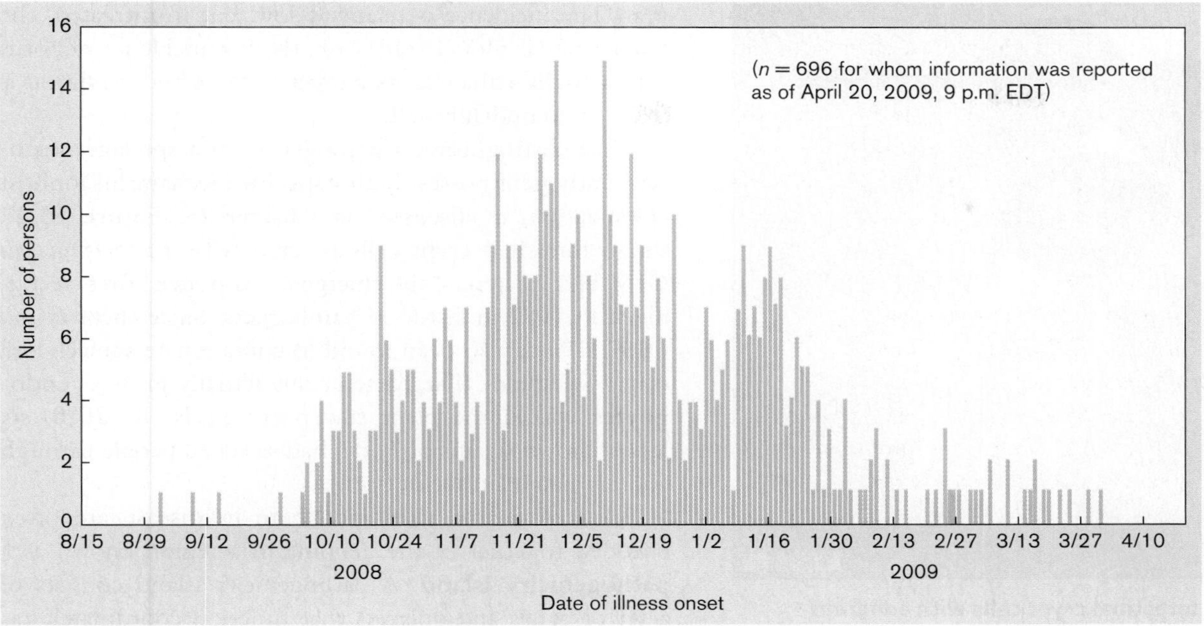
16
14
12
10
8
6
4
2
0
Number of persons
(n = 696 for whom information was reported as of April 20, 2009, 9 p.m. EDT)
8/15 8/29 9/12 9/26 10/10 10/24 11/7 11/21 12/5 12/19 1/2 1/16 1/30 2/13 2/27 3/13 3/27 4/10
2008
2009
Date of illness onset

B.

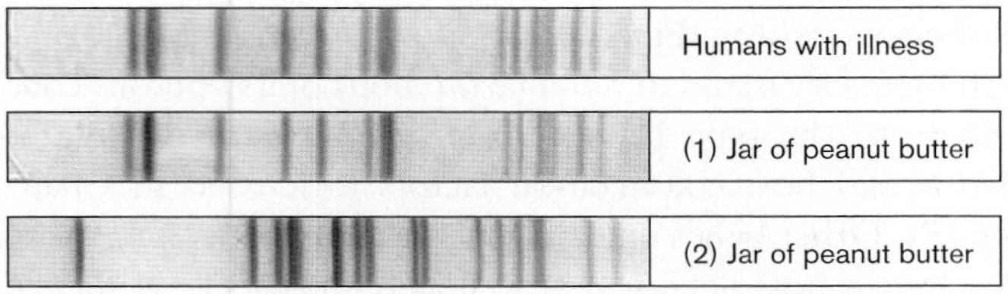
Humans with illness
(1) Jar of peanut butter
(2) Jar of peanut butter

FIGURE 16.25 ▪ *Salmonella enterica* outbreak from contaminated peanut butter. **A.** Infected individuals reported to the CDC from September 1, 2008, through April 20, 2009. **B.** Electrophoretic separation of restriction-digest DNA fragments from bacteria strains isolated from a human with illness. (1) Peanut butter containing the same strain; (2) peanut butter containing a different strain of *S. enterica*. *Source:* www.cdc.gov.

the bacterium in (2) proved unrelated to the *Salmonella* outbreak.

This case illustrates several troubling features of food contamination in modern society. It shows the consequence of a food production plant's failure to follow regulations, and the failure of health inspection to enforce them. The contaminated product shipped out to a diverse array of institutions such as schools, and secondary producers such as cookie manufacturers, who incorporated the peanut butter cream ingredient. The bacteria then remained viable in contaminated food products for many months, sickening people long after the contamination event had occurred.

Food-Borne Pathogens Emerge from Environment and Agriculture

Food-borne pathogens typically arise from a range of sources; see, for example, the diagram of transmission of *Listeria monocytogenes*, a psychrotrophic pathogen that invades the cells of the intestinal epithelium, causing listeriosis (**Fig. 16.26**). (Psychrotrophic organisms grow optimally at moderate temperatures but also grow slowly at

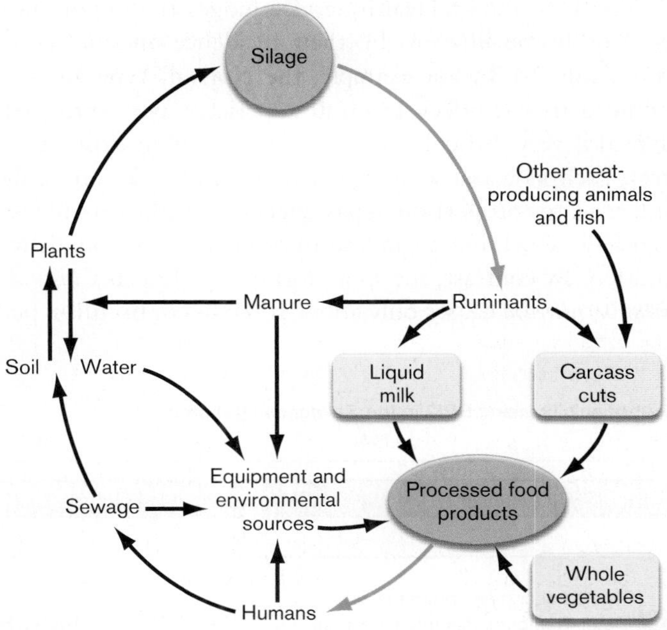
Silage
Other meat-producing animals and fish
Plants
Manure
Ruminants
Soil
Water
Liquid milk
Carcass cuts
Equipment and environmental sources
Processed food products
Sewage
Whole vegetables
Humans

FIGURE 16.26 ▪ Transmission of *Listeria monocytogenes*. Transmission occurs through various routes, including passage through food products. Colored arrows indicate transmission of disease.

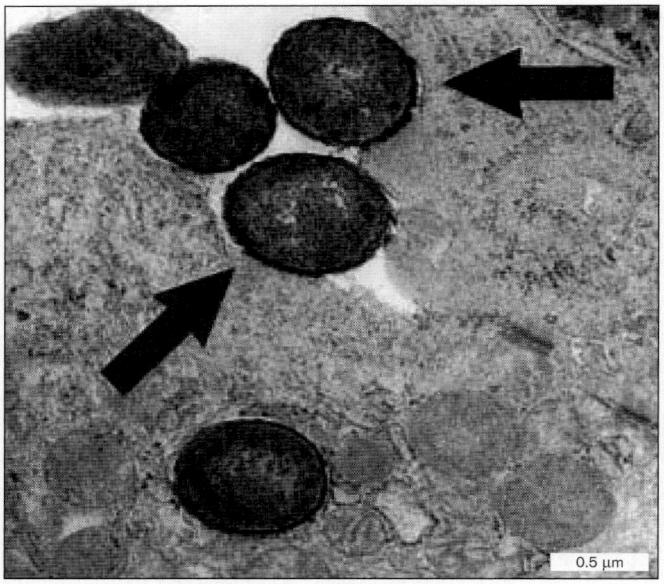
0.5 µm

FIGURE 16.27 ▪ Intestinal crypt cells with adherent bacteria, *Escherichia coli* strain O157:H7. A gnotobiotic (germ-free) piglet was infected with the bacteria (arrows). (TEM)

lower temperatures, typically 0°C–30°C.) *Listeria* can be transferred from soil and feed to cattle, whose manure then cycles it back to soil. From cattle, the pathogen contaminates milk and meat, where it can eventually infect human consumers. Because *Listeria* is a psychrotroph, it outcompetes other food-borne bacteria under refrigeration.

The U.S. Public Health Service judges the importance of food-borne illnesses by their incidence and/or severity (**Table 16.3**). For example, the Norwalk-type viruses, or noroviruses, infect 180,000 Americans per year; their spread is very difficult to control, especially in close quarters, such as a cruise ship, where an outbreak can easily infect a large proportion of passengers. The course of illness is usually short, but it can lead to complications from dehydration. By contrast, the spore-forming pathogen *Clostridium botulinum* causes only about 100 cases of botulism per year. The incidence is relatively low, but if untreated, the fatality rate is 50%. In this case, the low incidence of botulism actually enhances its danger because the condition is likely to go undiagnosed.

What distinguishes a pathogen from a spoilage organism? Pathogens possess highly specific mechanisms for host colonization, as discussed in Chapter 25. **Figure 16.27** shows intestinal crypt cells covered with *Escherichia coli* O157:H7 bacteria, an emergent pathogen first recognized in 1982 in fast-food hamburgers. Since then, *E. coli* O157:H7 has also been found to contaminate spinach and other vegetables. The bacteria can actually grow as **endophytes** within the plant transport vessels. By 2010, six lesser-known strains of *E. coli* had sickened people through contaminated lettuce or beef.

Bacterial factors that contribute to disease are often encoded together in the genome in a region known as a **pathogenicity island**. A pathogenicity island consists of a set of genes and operons that function coordinately (as discussed in Chapter 9). The colocalization of the genes enables transfer of virulence capability to other species as pathogens evolve. **Figure 16.28** shows an example of a pathogenicity island in *Salmonella*. Four of its operons contribute to the type III secretion complex, which secretes toxins and host colonization factors (discussed in Chapter 25). Other genes encode outer membrane proteins that counteract host defenses, as well as regulators of virulence gene expression.

The most dangerous consequence of infection by food-borne pathogens is the production of a potentially fatal toxin. For example, *E. coli* O157:H7 infection of the intestines can be overcome, but the bacteria produce Shiga toxin, which can destroy the kidneys. In cases of adult botulism, *Clostridium botulinum* does not usually grow within the patient; the botulinum toxin comes from bacteria that grew previously in improperly sterilized food. The botulinum toxin has a highly specific effect, inhibiting synaptic vesicle fusion in the terminals of peripheral motor neurons

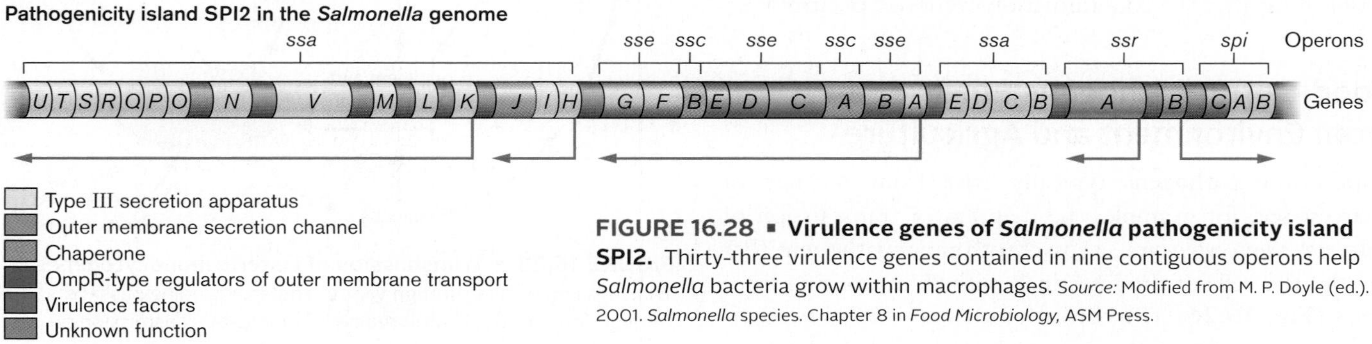
Pathogenicity island SPI2 in the Salmonella genome
ssa
sse ssc
sse
ssc sse
ssa
ssr
spi
Operons
U T S R Q P O N V M L K J I H G F B E D C A B A E D C B A B C A B
Genes
Type III secretion apparatus
Outer membrane secretion channel
Chaperone
OmpR-type regulators of outer membrane transport
Virulence factors
Unknown function

FIGURE 16.28 ▪ Virulence genes of *Salmonella* pathogenicity island SPI2. Thirty-three virulence genes contained in nine contiguous operons help *Salmonella* bacteria grow within macrophages. *Source:* Modified from M. P. Doyle (ed.). 2001. *Salmonella* species. Chapter 8 in *Food Microbiology*, ASM Press.

TABLE 16.3

Food-borne pathogens in the United States.[a]

Pathogen	Incidence and transmission	Course of illness
Norovirus (Norwalk and Norwalk-like viruses)	Most common cause of diarrhea; also called "stomach flu" (no connection with influenza). 180,000 cases per year are estimated. Transmitted mainly by virus-contaminated food and water. Infection rates are highest under conditions of crowding in close quarters, such as inside a ship or a nursing home.	Disease lasts 1 or 2 days. Includes vomiting, diarrhea, and abdominal pain; headache and low-grade fever may occur.
Salmonella	Most common food-borne cause of death; more than 1 million cases per year; estimated 600 deaths per year. Transmission nearly always through food—raw, undercooked, or recontaminated after cooking, especially eggs, poultry, and meat; also contaminates dairy products, seafood, fruits, and vegetables.	Causes gastrointestinal disease that includes diarrhea, fever, and abdominal cramps lasting 4–7 days. Fatal cases are most common in immunocompromised patients.
Campylobacter	More than 1 million cases of campylobacteriosis per year; estimated 100 deaths per year. Grows in poultry without causing symptoms. Transmission is mainly through raw and undercooked poultry; contaminates half of poultry sold. Occurs less often in dairy products or in foods contaminated after cooking.	In humans, usually causes severe bloody diarrhea, fever, and abdominal cramps lasting 7 days. Fatal cases are most common in immunocompromised patients.
Escherichia coli O157:H7	An emerging pathogen, first recognized in a hamburger outbreak in 1982; now known to infect 73,000 people yearly, including 60 deaths per year. Grows in cattle without causing symptoms. Transmitted through ground beef; also through unpasteurized cider and from produce, where it grows as an endophyte.	In humans, usually causes severe bloody diarrhea and abdominal cramps lasting 5–10 days. About 5% of patients, especially children and elderly, develop hemolytic uremic syndrome, in which the red blood cells are destroyed and the kidneys fail.
Clostridium botulinum	Causes about 100 cases per year of botulism, with a 50% fatality rate if untreated. *C. botulinum* grows in improperly home-canned foods, more rarely in commercially canned low-acid foods and improperly stored leftovers such as baked potatoes. Spores occur in honey, endangering infants under 2 years of age.	Botulinum toxin from growing bacteria causes progressive paralysis, with blurred vision, drooping eyelids, slurred speech, difficulty swallowing, and muscle weakness. Infant botulism causes lethargy and impaired muscle tone, leading to paralysis.
Listeria monocytogenes	*Listeria* bacteria grow in animals without causing symptoms. Animal feces may contaminate water, which is then used to wash vegetables. Transmission occurs mainly through vegetables washed in contaminated water and through soft cheeses. *Listeria* is psychrotrophic, growing at refrigeration temperatures.	Listeriosis involves fever, muscle aches, and sometimes gastrointestinal symptoms. In pregnant women, symptoms may be mild but lead to serious complications for the unborn child.
Shigella	*Shigella* infects about 18,000 people a year in the United States; in developing countries, *Shigella* infections are endemic in most communities. Transmission occurs through fecal-oral contact or from foods washed in contaminated water.	Shigellosis involves gastrointestinal symptoms such as diarrhea, fever, and stomach cramps, usually lasting 7–10 days. Complications are rare.
Staphylococcus aureus	*S. aureus* is best known as the cause of skin infections transmitted through open wounds. However, *S. aureus* can also be transmitted through high-protein foods such as ham, dairy products, and cream pastries.	*S. aureus* causes toxic shock syndrome. Can also cause food poisoning via preformed toxins.
Toxoplasma gondii	*T. gondii* is a parasite believed to infect 60,000 people annually, most with no symptoms. In a few cases, serious disease results. *T. gondii* is transmitted through contact with feces of infected animals, particularly cats, or through contaminated foods such as pork.	Toxoplasmosis causes mild flu-like symptoms; but in pregnant women, its transmission to the unborn child can lead to severe neurological defects, including death. Neurological complications also occur in immunocompromised patients.
Vibrio vulnificus	*V. vulnificus* is a free-living marine organism that contaminates seafood or open wounds. About 40 cases per year are reported.	*V. vulnificus* can infect the bloodstream, causing septic shock. Mainly threatens people with preexisting conditions such as liver disease.

[a]Ten major food-borne pathogens highlighted by the U.S. Public Health Service (USPHS).

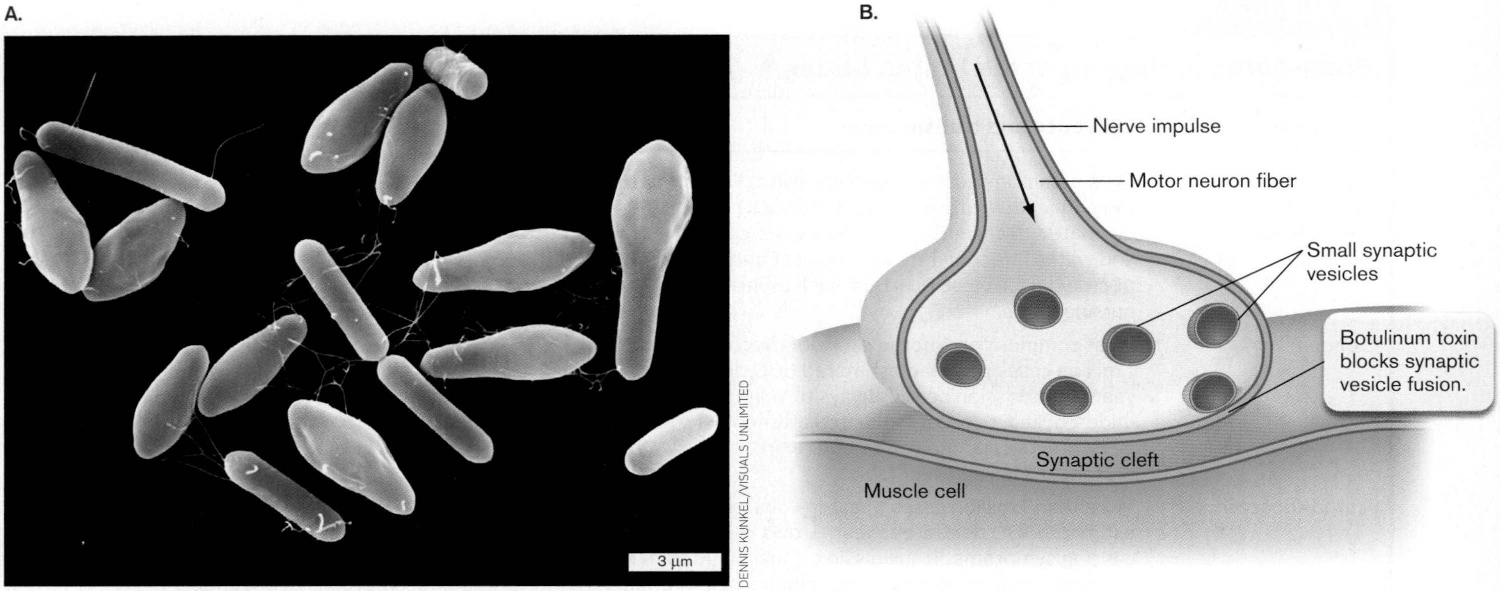
A.
3 μm
DENNIS KUNKEL/VISUALS UNLIMITED
B.
Nerve impulse
Motor neuron fiber
Small synaptic vesicles
Botulinum toxin blocks synaptic vesicle fusion.
Synaptic cleft
Muscle cell

FIGURE 16.29 ■ *Clostridium botulinum* produces botulinum toxin. **A.** Club-shaped morphology of *C. botulinum* cells containing endospores. **B.** Botulinum toxin inhibits synaptic vesicle fusion in the terminal of a peripheral motor neuron, preventing activation of the muscle cell.

(**Fig. 16.29**). Synaptic inhibition prevents activation of muscle cells, causing flaccid paralysis. Microbial toxins are discussed further in Chapter 25.

Food Preservation

Cultural practices and cuisines have long evolved to limit food spoilage. Such practices include cooking (heat treatment), addition of spices (chemical preservation), and fermentation (partial microbial digestion). In modern commercial food production, spoilage and contamination are prevented by numerous methods based on fundamental principles of physics and biochemistry that limit microbial growth (discussed in Chapter 5).

Physical means of preservation. Specific processes that preserve food based on temperature, pressure, or other physical factors include:

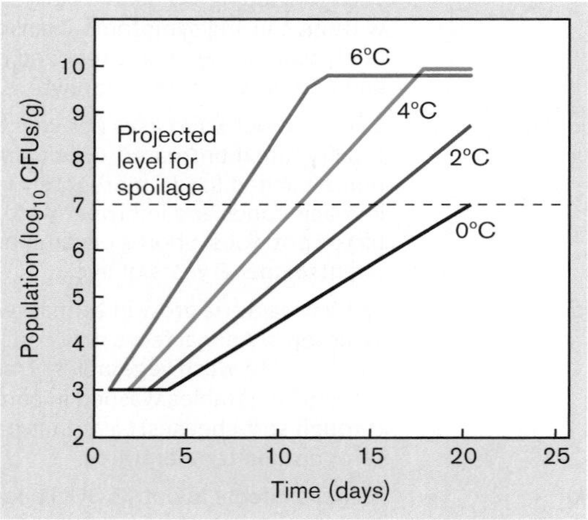
6°C
4°C
2°C
0°C
Projected level for spoilage
Population (log_{10} CFUs/g)
Time (days)
10
9
8
7
6
5
4
3
2
0
5
10
15
20
25

FIGURE 16.30 ■ Bacterial growth in ground beef. The growth rate of total aerobic bacteria in ground beef declines at lower storage temperatures. CFUs = colony-forming units.
Source: Modified from M. P. Doyle (ed.). 2001. Meat, poultry, and seafood. Chapter 5 in *Food Microbiology*, ASM Press.

- **Dehydration and freeze-drying.** Removal of water prevents microbial growth. Water is removed either by application of heat or by freezing under vacuum (known as **freeze-drying**, or **lyophilization**). Drying is especially effective for vegetables and pasta. The disadvantage of drying is that some nutrients are broken down.
- **Refrigeration and freezing.** Refrigeration temperature (typically –2°C to 16°C) slows microbial growth, as shown in an experiment comparing bacterial growth in ground beef at different temperatures (**Fig. 16.30**). Nevertheless, refrigeration also selects for psychrotrophs, such as *Listeria*. Freezing halts the growth of most microbes, but preexisting contaminant strains often survive to grow again when the food is thawed. This is why deep-frozen turkeys, for example, can still cause *Salmonella* poisoning, especially if the interior is not fully thawed before roasting.
- **Controlled or modified atmosphere.** Food can be packed under vacuum or stored under atmospheres with decreased oxygen or increased CO_2. Controlled atmospheres limit abiotic oxidation, as well as microbial growth. For example, CO_2 storage is particularly effective for extending the shelf life of apples.
- **Pasteurization.** Invented by Louis Pasteur, pasteurization is a short-term heat treatment designed to decrease

microbial contamination with minimal effect on food value and texture (discussed in Chapter 5). For example, milk is commonly pasteurized at 63°C for 30 minutes, followed by quick cooling to 4°C. Pasteurization is most effective for extending the shelf life of liquid foods with consistent, well-understood microbial flora, such as milk and fruit juices.

- **Canning.** In canning, the most widespread and effective means of long-term food storage, food is cooked under pressure to attain a temperature high enough to destroy endospores (typically 121°C). Commercial canning effectively eliminates microbial contaminants, except in very rare cases. The main drawback of canning is that it incurs some loss of food value, particularly that of labile biochemicals such as vitamins, as well as loss of desirable food texture and taste.
- **Ionizing radiation.** Exposure to ionizing radiation, known as food irradiation, effectively sterilizes many kinds of food for long-term storage. The main concerns about food irradiation are its potential for unknown effects on food chemistry and the hazards of the irradiation process itself for personnel involved in food processing. Nevertheless, irradiation has proved highly effective at eliminating pathogens that would otherwise cause serious illness.

Often, two or more means of preservation are used in combination, such as acid treatment and refrigeration. For example, **Figure 16.31** shows results of a typical experiment measuring the effect of pH on microbial survival in refrigerated food—in this case, *E. coli* O157:H7 in Greek eggplant salad. Note the critical threshold pH required to decrease bacterial counts. At pH 4.0, about the pH of lemon juice, the bacteria show a steep exponential death curve, whereas at pH 4.5, the bacteria remain viable for many days.

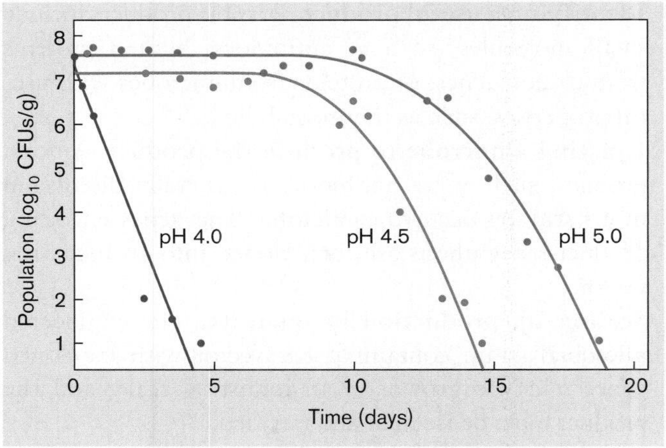
Population (log_{10} CFUs/g)
Time (days)
pH 4.0
pH 4.5
pH 5.0
8
7
6
5
4
3
2
1
0
5
10
15
20

FIGURE 16.31 ■ Bacteria die sooner at lower pH. Survival curves of *E. coli* O157:H7 in eggplant salad stored at 5°C at pH 4.0, pH 4.5, and pH 5.0. CFUs = colony-forming units. *Source:* Modified from Panagiotis N. Skandamis and George-John E. Nychas. 2000. *Appl. Environ. Microbiol.* **66**:1646.

How does the food industry know how long to treat food for sterilization? Several measurements indicate the efficiency of heat killing. The D-value (decimal reduction time) was described in Chapter 5. Two additional measures are 12D, the amount of time required to kill 10^{12} spores (or reduce a population 12 logs); and the z-value, the increase in degrees Celsius needed to lower the D-value to 1/10 of the time. If, for example, D_{100} (the D-value at 100°C) and D_{110} (the D-value at 110°C) for a given organism are 20 minutes and 2 minutes, respectively, then $12D_{100}$ equals 240 minutes (that is, 20 minutes × 12), and the z-value is 10°C (because a 10°C increase in temperature reduced the D-value to 1/10, from 20 minutes to 2 minutes). These measurements are determined empirically for each organism. The values are extremely important to the canning industry, which must ensure that canned goods do not contain spores of *Clostridium botulinum,* the previously described anaerobic soil microbe that causes the paralyzing food-borne disease botulism (see **Fig. 16.29**).

Because the tastes of certain foods suffer if they are overheated, z-values and 12D-values are used to adjust heating times and temperatures to achieve the same sterilizing result. Take the following example, where D_{121} is 10 minutes and $12D_{121}$ is 120 minutes. Sterilizing at 121°C for 120 minutes might result in food with a repulsive taste, whereas decreasing the temperature and extending the heating time may yield a more palatable product. The D- and z-values are used to adjust conditions for sterilization at a lower temperature. If D_{121} is 15 minutes (the time needed to kill 90% of cells) and the z-value is known to be 10°C (the temperature change needed to change D-value tenfold), then decreasing temperature by 10°C, to 111°C, will mean D_{111} is 150 minutes (10 × D_{121}). Therefore, $12D_{111}$ is 1,800 minutes. Sterilization may take longer, but food quality is likely to remain high, because the sterilizing temperature is lower.

Chemical means of preservation. Many kinds of chemicals are used to preserve foods. Major classes of chemical preservatives include:

- **Acids.** While microbial fermentation can preserve foods by acidification, an alternative approach is to add acids directly. Organic acids commonly used to preserve food include benzoic acid, sorbic acid, and propionic acid. The acids are generally added as salts: sodium benzoate, potassium sorbate, sodium propionate. These acids act by crossing the cell membrane in the protonated form and then releasing their protons at the higher intracellular pH. For this reason, they work best in foods that already have moderate acidity (pH 5–6), such as dried fruits and processed cheeses.

- **Esters.** The esters of organic acids often show antimicrobial activity whose basis is poorly understood. Examples include fatty acid esters and parabens (benzoic acid esters). They are used to preserve processed cheeses and vegetables.
- **Other organic compounds.** Numerous organic compounds, both traditional and synthetic, have antimicrobial properties. For example, cinnamon and cloves contain the benzene derivative eugenol, a potent antimicrobial agent.
- **Inorganic compounds.** Inorganic food preservatives include salts, such as phosphates, nitrites, and sulfites. Nitrites and sulfites inhibit aerobic respiration of bacteria, and their effectiveness is enhanced at low pH. These substances, however, may have harmful effects on humans; nitrites can be converted to toxic nitrosamines, and sulfites cause allergic reactions in some people.

Thought Question

16.9 Is it possible for physical or chemical preservation methods to completely eliminate microbes from food? Explain.

To Summarize

- **Food spoilage** refers to chemical changes that render food unfit for consumption. Food spoils through degradation by enzymes within the food, through spontaneous chemical reactions, and through microbial metabolism.
- **Food contamination, or food poisoning,** refers to the presence of microbial pathogens that cause human disease, or toxins produced by microbial growth. Food harvesting, food processing, and shared consumption are all activities that spread pathogens.
- **Dairy products** can be soured by excessive fermentation or made bitter by bacterial proteolysis.
- **Meat and poultry** are putrefied by decarboxylating bacteria, which produce amines with noxious odors.
- **Fish and other seafood** spoil rapidly because their unsaturated fatty acids rapidly oxidize; they harbor psychrotrophic bacteria that grow under refrigeration; and their trimethylamine oxide is reduced by bacteria to the fishy-smelling trimethylamine.
- **Vegetables** spoil by excess growth of bacteria and molds. Plant pathogens destroy food crops before harvest.
- **Food preservation** includes physical treatments, such as freezing and canning, as well as the addition of chemical preservatives, such as benzoates and nitrites.

16.6

Industrial Microbiology

The production and preservation of food is only one field of **industrial microbiology**, the commercial exploitation of microbes. Industrial microbiology is commonly understood to include a broad range of commercial products derived from microbes, including vaccines and clinical devices; industrial solvents and pharmaceuticals; bioinformatic analysis of genomes; and genetically modified plants and animals using microbial vectors.

Other important fields of industrial microbiology are wastewater treatment, bioremediation, and environmental management, covered in Chapters 21 and 22.

Industrial Microbiology Aims for Commercial Success

The practical application of a microbial product or device may arise out of an industrial laboratory, or it may be conceived by a research scientist with the aim of meeting a compelling need in society. For example, **Special Topic 16.1** profiles a biotechnology company formed to develop innovative products for the prevention, testing, and treatment of tuberculosis. In all cases, however, the key goal is success in the marketplace—that is, to generate a product that achieves adoption by customers in preference to alternative technologies. The product's sales must cover the costs of raw materials and production and (in a for-profit company) generate a profit for the shareholders. Success requires:

- **Identifying a useful product.** Possible products include small molecules, such as antibiotics; human proteins from cloned genes; or proteins from microbes with useful properties, such as thermostability.
- **Isolating a microbe to produce the product.** A novel product, such as an antibiotic, is generally discovered in a naturally occurring microbe. The genes encoding product biosynthesis are then cloned into an industrial vector.
- **Scaling up production in quantity.** The engineered microbial strain containing the vector with its cloned genes must be grown on an industrial scale, and the product must be isolated and purified.
- **Developing a business plan.** The scientist-entrepreneur must obtain partners skilled at industrial management, finance, and marketing. Patents must be filed to protect intellectual property rights.

TABLE 16.4

Commercial products from microbes.

Class of product	Product	Producer microbe[a]
Agricultural products	Gibberellin (plant hormone)	*Fusarium moniliforme* (fungus)
	Fungicides	*Coniothyrium minitans* (fungus)
	Insecticides (live pathogens)	*Bacillus thuringiensis*
Enzymes	Alpha-amylase	*Bacillus subtilis*
	Amyloglucosidase	*Aspergillus niger* (fungus)
	Lactase (beta-galactosidase)	*Kluyveromyces lactis* (fungus)
	Lipases	*Candida cylindraceae* (yeast)
	Alkaline protease	*Aspergillus oryzae* (fungus)
Food supplements	L-Lysine	*Brevibacterium lactofermentum*
	L-Tryptophan	*Klebsiella aerogenes*
	Monosodium glutamate (MSG)	*Corynebacterium ammoniagenes*
	Vitamin B_{12}	*Pseudomonas denitrificans*
	Vitamin C (ascorbic acid)	*Acetobacter suboxidans*
Fuels and solvents	Acetone	*Clostridium* spp.
	Butanol	*Clostridium acetobutylicum*
	Glycerol	*Zygosaccharomyces rouxii* (yeast)
	Methane (natural gas)	Methanogens (archaea)
Organic acids	Acetic acid	*Acetobacter xylinum*
	Citric acid	*Aspergillus niger* (fungus)
	Lactic acid	*Lactobacillus delbruckii*
Pharmaceuticals: antibiotics	Streptomycin	*Streptomyces griseus*
	Penicillin	*Penicillium chrysogenum* (fungus)
	Erythromycin	*Saccharopolyspora erythraea*
	Tetracycline	*Streptomyces aureofaciens*
Pharmaceuticals: other drugs	Lysergic acid (LSD, hallucinogen)	*Claviceps paspali* (fungus)
	Cyclosporine (immunosuppressant)	*Trichoderma polysporum* (fungus)
	Steroids	*Arthrobacter* spp.
Pharmaceuticals: cloned human proteins	Insulin	Recombinant *Escherichia coli*
	Interferon	Recombinant *E. coli* or *Saccharomyces cerevisiae* (yeast)
	Human growth hormone	Recombinant *E. coli*
	CC10 lung development protein	Recombinant *E. coli*
	Antibodies	Baculovirus-infected caterpillars
	Cancer regulators	Baculovirus-infected caterpillars

[a]Bacteria, unless stated otherwise.

Source: M. J. Waites. 2001. *Industrial Microbiology*, Blackwell Science, Table 4.1, pp. 76–77.

- **Safety and efficacy testing.** Human consumption or consumer use requires many levels of clinical testing for approval by government agencies.
- **Effective marketing.** The benefits of the new product must be communicated effectively to convince customers of its superiority to current products or processes.

Failure at any of the preceding tasks spells doom for the product. Thus, a prudent business plan includes having multiple alternative products in development. Although the failure rate of new products is high, all the products we use had to overcome these risks.

Molecular Products from Human or Microbial Sources

Increasingly, microbes are grown to produce a commercially valuable chemical substance, such as a vitamin, an industrial solvent, or an enzyme (**Table 16.4**). Each product

Special Topic 16.1: Companies Take On Tuberculosis

In the 1990s, newspaper headlines announced the resurgence of tuberculosis (TB) in American cities. Carol Nacy (**Fig. 1**) was then a chief scientific officer at a small biotechnology firm, following a research career at the Walter Reed Army Institute of Research and a term as president of the American Society for Microbiology. In 1996, Nacy was invited by the National Institutes of Health to assist in a review of tuberculosis research grant support to U.S. universities. She was surprised to discover the lack of attention to TB, a disease that is the world's number one killer of women aged 15–44 and the leading killer of men, after traffic accidents. The United States spends $1 billion yearly to treat 13,000 incident cases; yet the standard antibiotics for tuberculosis were developed before 1970, and the main diagnostic test available (the tuberculin skin test) dates to 1880, the time of Robert Koch. The best available vaccine—the Bacille Calmette-Guérin (BCG) live, attenuated vaccine—is only 50% effective.

Despite the clear need, tuberculosis has been of little interest to industrial research or development. So Nacy decided to found two new companies to address TB: a nonprofit medical research company to develop vaccines and conduct clinical trials, and a for-profit company to develop innovative drugs and treatment devices.

© JOANNE LAWTON/WASHINGTON BUSINESS JOURNAL

FIGURE 1 ▪ Carol Nacy developed two companies to fight tuberculosis.

A nonprofit company develops vaccines. The nonprofit company Aeras Global TB Vaccine Foundation, directed by Jerald Sadoff, has received over $200 million in grants from the Bill & Melinda Gates Foundation—more than doubling the amount spent worldwide on TB vaccines. Additional funding has come from the U.S. Centers for Disease Control and Prevention and from the government of Denmark. The Aeras foundation established a clinical trial site in Cape Town, South Africa. South Africa has the highest rate of pediatric TB in the world: over 500 cases per 100,000 persons. Aeras trained South African investigators and support staff in the art of clinical trials, while helping community health care workers to serve the local community. They helped test and counsel the community for HIV, a major risk factor for tuberculosis, and they vaccinated thousands of babies with BCG vaccine.

By 2009, Aeras had 12 different experimental TB vaccines in clinical and preclinical trials. The experimental vaccines are based on leading-edge principles of molecular biology, such as a recombinant vaccinia virus expressing antigenic proteins from *Mycobacterium tuberculosis*; recombinant fusion proteins from *M. tuberculosis*; and a mutant strain of the bacterium deleted for genes encoding two transcriptional sigma factors. An advanced "DNA vaccine" was developed consisting of a recombinant DNA vector designed to express antigenic proteins within host cells. The fundamental principles behind these experimental vaccines are introduced in Chapters 11 and 12.

A for-profit company develops drugs and devices. Nacy's for-profit company, Sequella, targeted innovative ideas with high risk but also high potential to improve performance, rapidity, and safety of diagnosing and treating TB infections. Nacy and her cofounders scanned the academic community for novel ideas that had succeeded against TB in "proof of principle" animal models. The most promising ideas were developed for improved antibiotics, rapid and less invasive tests for TB exposure, and devices to measure the extent of pulmonary infection. Because any one idea had a high risk of failure, multiple prospects were pursued in each category.

Several Sequella products have since reached advanced clinical trials. Their lead product is the transdermal patch test, a diagnostic transdermal patch to distinguish between active

requires a gene or operon of genes encoding either the product itself or the enzymes for the product's biosynthesis. We may distinguish between two fundamentally different sources of products: cloned genes from human, animal, or plant sources; and native microbial products, often from newly discovered species in extreme environments. Cloned human genes typically encode a protein of valuable function in the human body. For example, the

TB and prior TB infection or BCG vaccination (**Fig. 2**). Prior to the transdermal patch, it was difficult to determine whether an individual who tested positive for TB antibodies actually had active infection or had simply been exposed to the bacillus in the past. The transdermal product allows antigens to penetrate the skin without needle injection, and it produces results more reliable than those of the standard tuberculin skin test.

For treatment of TB, several possible new antibiotics are in the pipeline. The most promising antibiotic, SQ109, was discovered by high-throughput screening of a chemical library, in collaboration with Clifton Barry at the National Institutes of Health. The chemical library consisted of over 60,000 analogs of a known TB antibiotic, ethambutol (**Fig. 3**). Ethambutol is part of the current standard course of treatment for TB, whose 6-month time course has a poor compliance rate. It is hoped that improved drugs will shorten the time course and improve compliance, thereby decreasing the appearance of drug-resistant strains.

The analog molecules were selected for their common diamine core, with different combinations of side chains. The 60,000 compounds in the library were subjected to combinatorial screening, a mathematically intensive analysis based on numerous tests. Of the compounds tested, 2,796 showed activity against *M. tuberculosis* in the test tube. The 69 best compounds were tested for cytotoxicity in tissue culture, activity in TB-infected macrophages, and activity in infected animals. The compound with the greatest efficacy and fewest side effects was SQ109, a molecule with an unusual cage-like side group of three fused rings. SQ109 is now in human clinical trials. With these products, Sequella has positioned itself to develop more effective vaccines, diagnostics, and treatments to alleviate the global burden of tuberculosis—and to make a handsome profit for investors.

A.

B.

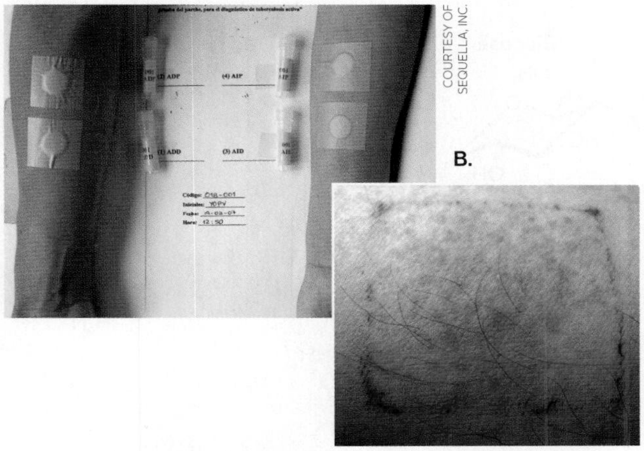
COURTESY OF SEQUELLA, INC.
COURTESY OF SEQUELLA, INC.

FIGURE 2 ■ The transdermal patch reveals active TB infection. A. The transdermal patch administers an antigen test reagent without requiring needle injection. **B.** A positive test result for active TB.

RESEARCH QUESTION

What challenges may face SQ109 in human trials? What problems may arise during human trials that do not appear when the drug is tested in tissue cultures?

Ma, Zhenkun, Christian Lienhardt, Helen McIlleron, Andrew J. Nunn, and Xiexiu Wang. 2010. Global tuberculosis drug development pipeline: The need and the reality. *Lancet* **375**:2100–2109.

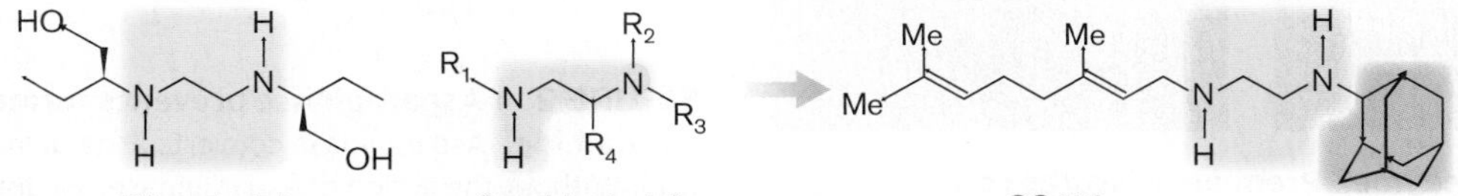
HO
H
N
N
H
OH
R1
R2
N
N
R3
H
R4
Me
Me
Me
N
H
N
H

Ethambutol **Combinatorial library of 63,238 diamines** **SQ109**

FIGURE 3 ■ A new antibiotic for TB is obtained by screening ethambutol analogs. A combinatorial library of compounds containing the ethambutol diamine core (yellow) was screened for antibacterial effect against *Mycobacterium tuberculosis.* The most promising agent screened was SQ109, with an unusual carbon-cage side group (pink).

Clarassance company produces the recombinant human protein CC10, a lung development protein that is often deficient in the lungs of premature infants. The recombinant protein, produced and purified from a recombinant bacterium, can be provided to the infant to reduce lung inflammation.

Cloning gene products in recombinant organisms is discussed in Chapter 12. In this chapter we discuss obtaining

Special Topic 16.2: Microbial Enzymes Make Money

How can we clean our clothes with less pollution, and keep their colors bright? And how can we remove carcinogens from our food? Ingenious answers lie in the microbial genomes. Microbial genes encode enzymes with properties worth hundreds of billions of dollars. But it takes a sophisticated research company to discover enzymes, develop them, and make a profit.

One of the foremost enzyme producer companies today is Novozymes. Founded in 1925, the company produced insulin and trypsin by extraction from animal tissues—a laborious, expensive process. In 1963, Novozymes produced its first product of microbial fermentation: the detergent enzyme Alcalase. In the 1980s the company took up genetic modification (recombinant DNA) to develop dozens of profitable microbial enzymes for cleaning, such as Lipolase, a lipase (fat cleavage) from the fungus *Thermomyces lanuginosus*; and Protamex, a protease (protein cleavage) from *Bacillus* bacteria. These enzymes break down and solubilize various components of food stains. Active at room temperature, these enzymes precisely remove the most common food stains while avoiding the energy-expensive heating of water. They also avoid the need for phosphate detergents, which pollute waterways.

Some cleaning enzymes actually improve the appearance of clothing. The cellulase Carezyme, for example, hydrolyzes

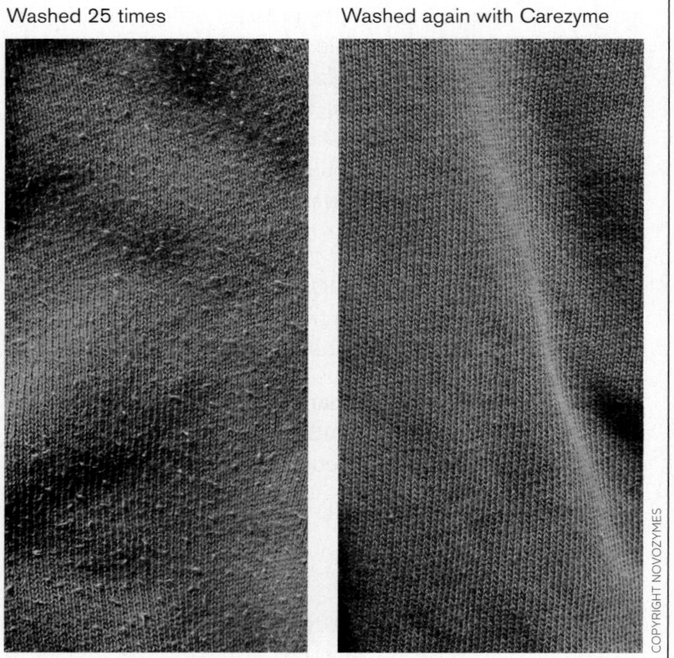
Washed 25 times
Washed again with Carezyme
COPYRIGHT NOVOZYMES

FIGURE 1 ▪ Carezyme® Premium restores color. During cleaning, the cellulase cleaves frayed ends of material, removing the fuzzy appearance and improving the garment's color.

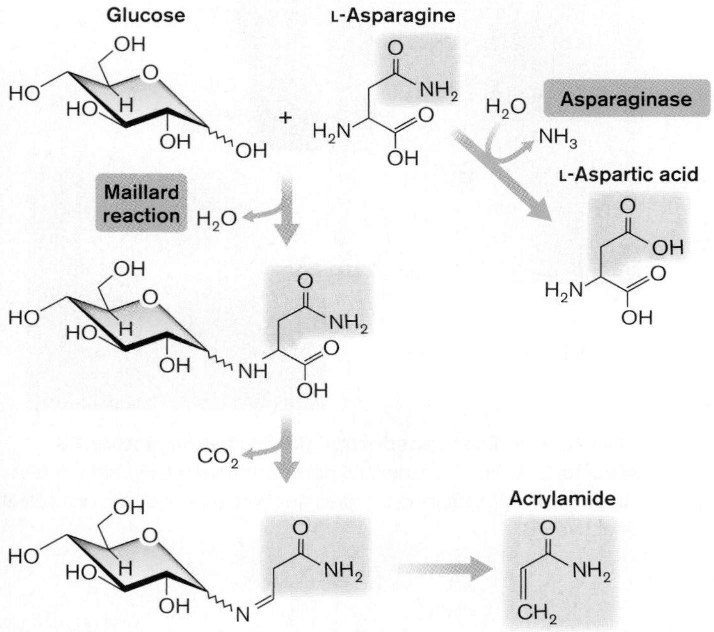
Glucose
L-Asparagine
+
Asparaginase
H_2O
NH_3
L-Aspartic acid
Maillard reaction
H_2O
CO_2
Acrylamide

FIGURE 2 ▪ Asparaginase prevents formation of acrylamide. Asparaginase converts asparagine into aspartic acid. Without the action of asparaginase, asparagine combines with sugar at high temperature (the Maillard reaction) and then releases acrylamide.

products from native microbial sources. For example, the Danish company Novozymes markets over 700 microbial enzymes for products ranging from laundry detergents to carcinogen-eliminating food additives (discussed in **Special Topic 16.2**). Novozymes also produces numerous enzymes as biocatalysts for green chemistry. "Green chemistry" refers to environmentally friendly procedures for reactions in organic chemistry, typically using water solution in place of petroleum-derived organic solvents.

Bioprospecting. To identify new microbial products, chemical companies screen thousands of microbial strains from diverse ecosystems. The search for organisms with potential commercial applications is called **bioprospecting**. Bio-short, frayed ends of cellulose that project from a cotton garment after many cleanings (**Fig. 1**). Cotton consists of crystalline fibers of cellulose (glucose polysaccharide, discussed in Chapter 13). Cutting the frayed ends removes the fuzzy appearance, restoring the solid color of the garment. But why does this powerful enzyme not destroy the entire garment? The enzyme has been engineered for high specificity: It cleaves only the highly disordered stray fibrils, not the ordered cellulose of the intact fabric. The enzyme was originally isolated from a species of *Bacillus* bacteria that uses it to catabolize plant material.

Novozymes also develops enzymes with exciting applications for food production. In 2002, food scientists reported in *Nature* that high-temperature cooking of many foods results in formation of acrylamide, a carcinogen. The acrylamide forms in foods such as bread crusts and crackers, French fries, and coffee. It was shown to arise from a reaction between sugar and amino acid, particularly the amino acid asparagine (**Fig. 2**). First the asparagine condenses with the sugar by a mechanism known as the Maillard reaction. The molecule then rearranges, releasing acrylamide.

How can the acrylamide production be prevented? Novozymes scientists reasoned that most of the acrylamide production could be avoided by pretreatment of the food with an enzyme, asparaginase, which deaminates asparagine to aspartic acid. They obtained the asparaginase gene from a strain of *Aspergillus* fungi. An industrial organism was engineered to produce the enzyme in quantity, and marketed as Acrylaway. **Figure 3** outlines the use of Acrylaway to avoid acrylamide production in French fries. The enzyme is added to potato strips during the dipping step, before the potatoes are dried and fried. Because the enzyme is completely "natural," it introduces no new toxin. Industrial trials completed in 2012 confirmed that Acrylaway effectively protects many kinds of food from carcinogenic acrylamide.

RESEARCH QUESTION

The Novozymes product Acrylaway solves a particular problem for a particular food industry. Search on the Internet for another industrial problem that might be solved by a microbial enzyme. Outline the steps you might take to develop such an enzyme product.

Stadler, Richard H., Imre Blank, Natalia Varga, Fabien Robert, Jörg Hau, et al. 2002. Food chemistry: Acrylamide from Maillard reaction products. *Nature* **419**:449–450.

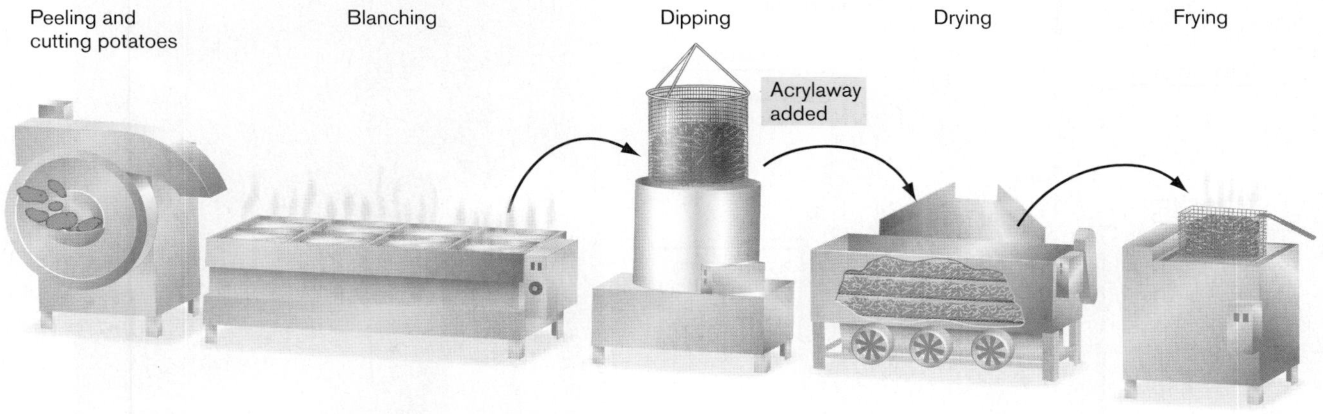
Peeling and cutting potatoes
Blanching
Dipping
Drying
Frying
Acrylaway added

FIGURE 3 ■ Asparaginase in French fries. Asparaginase added at the dipping step prevents acrylamide formation in French fries.

prospecting can be done anywhere, from one's backyard to Yellowstone National Park. Unique ecosystems are the most promising sources of previously unknown microbial strains with valuable properties. Extreme environments, such as the hot springs of Yellowstone, are particularly promising because their products may tolerate conditions of temperature or pH required for industrial use. Psychrophiles from extremely cold environments, such as Antarctica, are a useful source for enzymes that do not require heating for activity, such as the Stainzyme laundry detergent enzyme. Unfortunately, many such unique ecosystems are endangered by pollution, human-introduced invasive species, and global climate change.

An important aspect of bioprospecting is "mining the genome." Once a promising source strain is obtained, its genome is sequenced to investigate the genetic sequences that encode the useful product and regulate its expression. The operons encoding the product (or enzymes for its production) can then be cloned for optimal production (see Chapter 12). The cloned genes are transferred into an **industrial strain**, a strain whose growth characteristics are well studied and optimized for industrial production. An industrial strain must possess the following attributes:

- **Genetic stability and manipulation.** The industrial strain must reproduce reliably, without major DNA rearrangements. It must also have an efficient gene transfer system by which vectors can introduce genes of interest into its genome.
- **Inexpensive growth requirements.** Industrial strains must grow on low-cost carbon sources with minimal special needs, such as vitamins, and at easily maintained conditions of temperature and gases.
- **Safety.** Industrial strains must be nonpathogenic and must not produce toxic by-products.
- **High level of product expression.** The strain or recombinant vector must possess an efficient gene expression system to generate the desired product as a high proportion of its cell mass.
- **Ready harvesting of product.** Either the product must be secreted by the cell or, if the product is intracellular, the cells must be easily breakable to liberate the product.

Species for industrial strains include the bacterium *Bacillus subtilis*, the yeast *Candida utilis*, and the filamentous fungus *Aspergillus niger*. Each of these species is safe; grows to high density on inexpensive carbon sources, such as molasses; and expresses desired products at high concentration.

Thought Question

16.10 Why would different industrial strains or species be used to express different kinds of cloned products?

Fermentation Systems

Commercial success requires optimizing every detail of the fermentation system. "Fermentation" in industrial terms refers not just to anaerobic metabolism, but to all means of growth of microbes on an industrial scale. In an **industrial fermentor**, the growth vessel and all its environmental supports, such as temperature control and oxygenation, must be scaled up to thousands of liters (**Fig. 16.32**). This

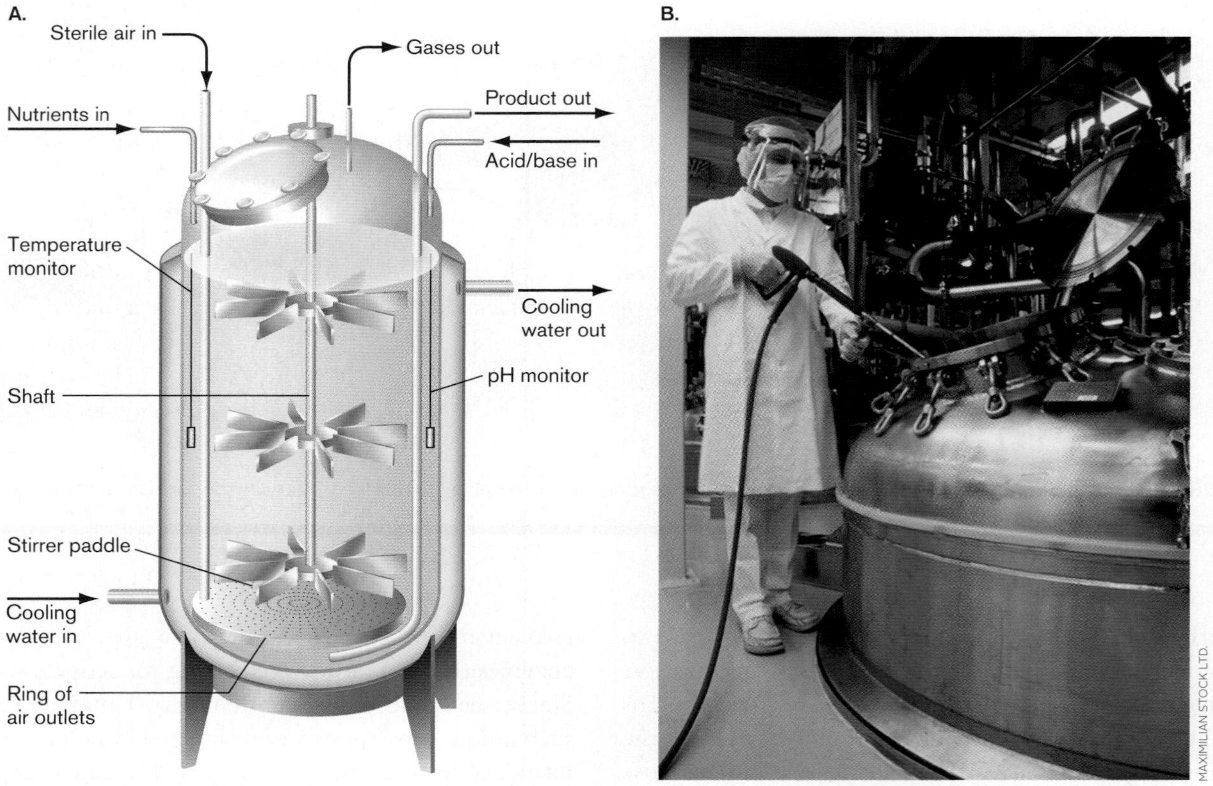
A.
Sterile air in
Gases out
Product out
Nutrients in
Acid/base in
Temperature monitor
Cooling water out
pH monitor
Shaft
Stirrer paddle
Cooling water in
Ring of air outlets
B.
MAXIMILIAN STOCK LTD.

FIGURE 16.32 ■ An industrial fermentor. A. Industrial production of microbial products requires scaled-up culture of the production microorganism. **B.** An industrial fermentor.

increase in scale generates many problems of quality control, such as maintaining uniform temperature, pH, and oxygenation throughout the vessel and minimizing foaming of the culture liquid. A small change in any of the growth factors can impact production costs and profit margin. Another major concern is to avoid contamination by other organisms.

The fermentor is the core of the first half of industrial production, known as **upstream processing** (**Fig. 16.33**, top). Upstream processing refers to the culturing of the industrial microbe to produce large quantities of product. All aspects of the process must be controlled to maximize the final concentration of product, which in most cases peaks at a specific time in the microbial growth cycle. Following microbial growth, the culture must be harvested and the product purified. These processes constitute **downstream processing** (**Fig. 16.33**, bottom). The first step of downstream processing is to separate the microbial cells from the culture fluid, by centrifugation or by filtration. Next, the **primary recovery** of product follows one of two different pathways, depending on whether the product is maintained within the cells or secreted into the culture fluid. Many kinds of subsequent purification and finishing steps are necessary before the product has acceptable quality for its desired use. Again, failure of any detail can render the entire product unusable.

Products designed for human internal use face formidable hurdles in clinical testing and, finally, approval by the appropriate regulatory agency, such as the U.S. Food and Drug Administration (FDA). After millions of dollars are invested in process development, the product may still fail this final test and never come to market. Not surprisingly, a pharmaceutical company must research thousands of potential products before achieving one that makes a profit. The consumer cost inevitably includes the development costs not only of the one successful product, such as recombinant insulin, but also of all the products that failed.

Despite all the hurdles to overcome, companies such as Novozymes sell hundreds of products and make billion-dollar profits. Examples of commercially successful products are described in **Special Topic 16.2**.

Production in Plant or Animal Host Systems

Some kinds of products require subtle kinds of processing that occur correctly only within eukaryotic cells. For example, protein products may require posttranslational modifications such as glycosylation (attachment of polysaccharide chains). In such cases, the expressed genes must be transferred from a bacterial or viral vector into an animal, fungal, or plant tissue culture or to a transgenic organism. The microbially transformed organism may itself be the industrial product. Here we present an example of microbial production involving the plant vector host *Agrobacterium tumefaciens*. A second example, that of an insect caterpillar virus, is presented in eTopic 16.2.

***Agrobacterium tumefaciens* engineers plants.** Bacteria and viruses provide the vectors for transferring genes into multicellular organisms, as discussed in Chapter 12. A bacterium of major industrial importance is *Agrobacterium tumefaciens*, a tumor-inducing plant pathogen that conducts natural genetic engineering on dicot plants (**Fig. 16.34**). Long before scientists invented "recombinant DNA," *A. tumefaciens* had evolved a gene transfer system by which it induces infected plant cells to generate food molecules to feed the pathogen. This highly efficient gene transfer system is readily modified to insert genes conferring traits of interest, such as herbicide resistance, into plant genomes.

Tumorigenic strains of *A. tumefaciens* possess a special plasmid for engineering plant cells, called the **Ti plasmid** (tumor-inducing plasmid). The Ti plasmid is transferred into a plant cell through a process mediated by bacterial proteins, similar to conjugation (see Section 9.2). The Ti plasmid includes the *vir* operon, which encodes virulence genes, as well as T-DNA (transferable DNA), a sequence of genes that will be transferred to the host plant and recombined into its genome. T-DNA encodes tumor induction genes, as well as enzymes for biosynthesis of a carbon and nitrogen source called an opine. Opines are specialized amino acids made by a one-step synthesis from arginine, typically by amination of a central metabolite such as pyruvate or 2-oxoglutarate. A given strain of *A. tumefaciens* typically provides one type of opine synthesis enzyme and has the ability to metabolize the corresponding opine.

The *vir* gene products from the Ti plasmid detect the presence of a plant host and stimulate plasmid transfer (see **Fig. 16.34**, left). In the cell envelope, VirA protein detects a chemical signal from a wounded plant, which is capable of being infected. VirA then activates VirG to induce expression of other *vir* genes, encoding endonucleases VirD1 and VirD2. The VirD endonucleases cleave the left and right ends of the T-DNA and direct its transfer into the plant cell. Within the plant cell nucleus, the T-DNA becomes integrated into the plant genome, where it induces opine production. The opine-producing cells proliferate, forming a tumor.

For industrial use (**Fig. 16.34**, right), a recombinant strain of *A. tumefaciens* has the Ti plasmid divided into two separate plasmids. One contains the *vir* operon conducting DNA transfer; the other contains T-DNA with its

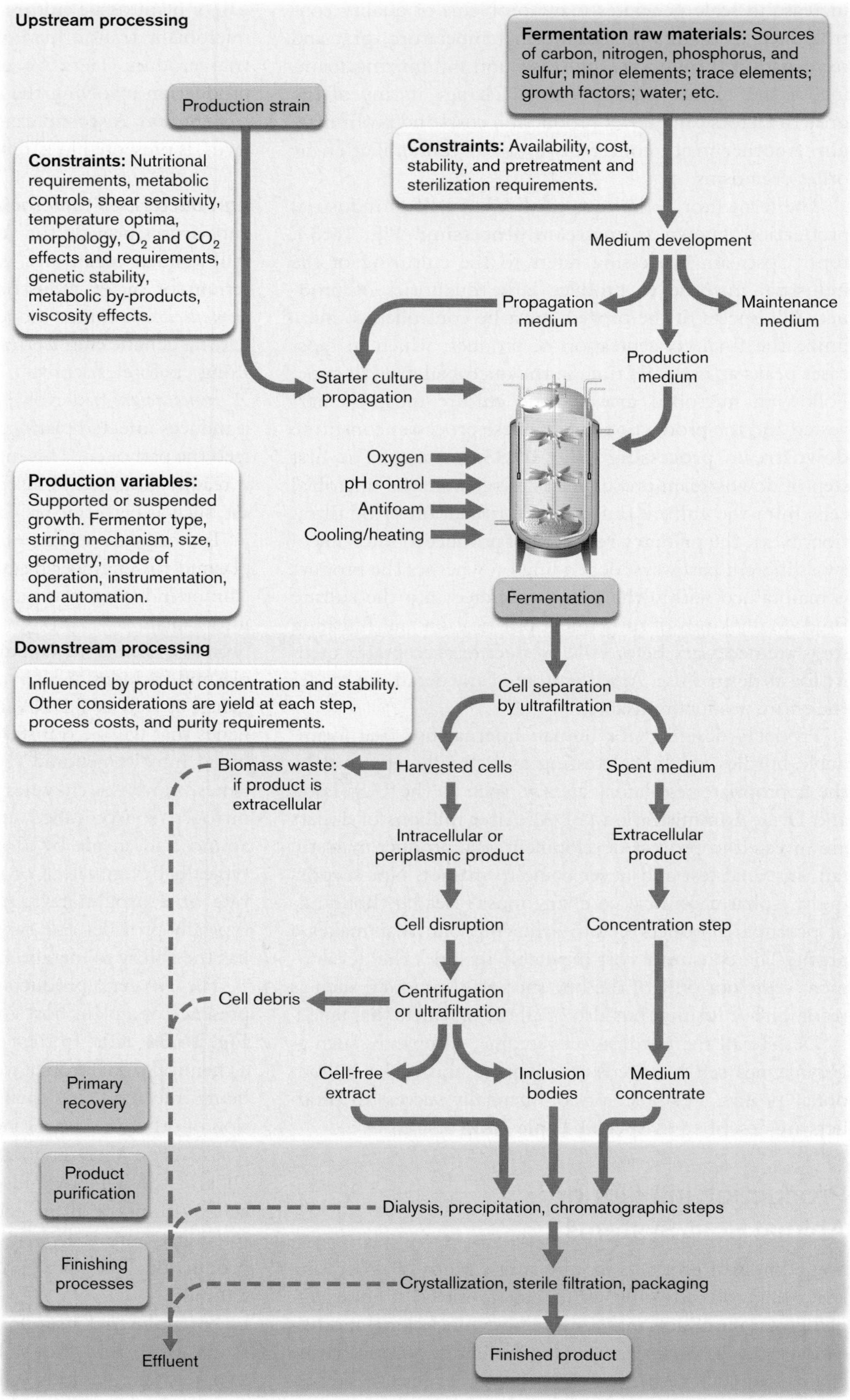
Upstream processing
Fermentation raw materials: Sources of carbon, nitrogen, phosphorus, and sulfur; minor elements; trace elements; growth factors; water; etc.
Production strain
Constraints: Availability, cost, stability, and pretreatment and sterilization requirements.
Constraints: Nutritional requirements, metabolic controls, shear sensitivity, temperature optima, morphology, O_2 and CO_2 effects and requirements, genetic stability, metabolic by-products, viscosity effects.
Medium development
Propagation medium
Maintenance medium
Production medium
Starter culture propagation
Oxygen
pH control
Antifoam
Cooling/heating
Production variables: Supported or suspended growth. Fermentor type, stirring mechanism, size, geometry, mode of operation, instrumentation, and automation.
Fermentation
Downstream processing
Influenced by product concentration and stability. Other considerations are yield at each step, process costs, and purity requirements.
Cell separation by ultrafiltration
Biomass waste: if product is extracellular
Harvested cells
Spent medium
Intracellular or periplasmic product
Extracellular product
Cell disruption
Concentration step
Cell debris
Centrifugation or ultrafiltration
Cell-free extract
Inclusion bodies
Medium concentrate
Primary recovery
Product purification
Dialysis, precipitation, chromatographic steps
Finishing processes
Crystallization, sterile filtration, packaging
Effluent
Finished product

FIGURE 16.33 ▪ Details of upstream and downstream processing. Top: Upstream processing involves engineering of the microbial strain and large-scale growth to generate the product. **Bottom:** Downstream processing involves product concentration and purification.

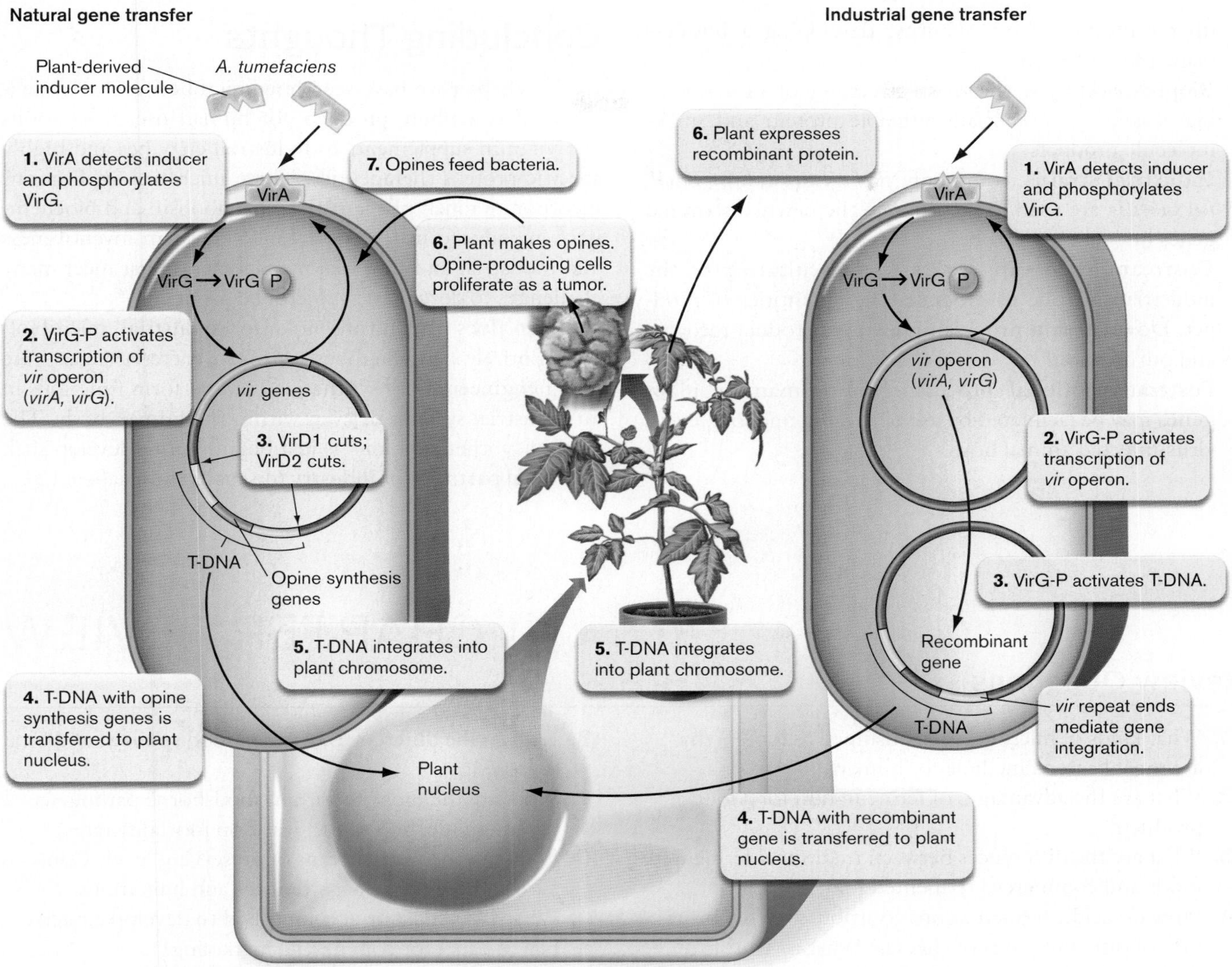
Natural gene transfer
Industrial gene transfer
Plant-derived inducer molecule
A. tumefaciens
1. VirA detects inducer and phosphorylates VirG.
VirA
VirG → VirG P
2. VirG-P activates transcription of vir operon (virA, virG).
vir genes
3. VirD1 cuts; VirD2 cuts.
T-DNA
Opine synthesis genes
4. T-DNA with opine synthesis genes is transferred to plant nucleus.
5. T-DNA integrates into plant chromosome.
6. Plant makes opines. Opine-producing cells proliferate as a tumor.
7. Opines feed bacteria.
Plant nucleus
6. Plant expresses recombinant protein.
5. T-DNA integrates into plant chromosome.
1. VirA detects inducer and phosphorylates VirG.
vir operon (virA, virG)
2. VirG-P activates transcription of vir operon.
3. VirG-P activates T-DNA.
Recombinant gene
T-DNA
vir repeat ends mediate gene integration.
4. T-DNA with recombinant gene is transferred to plant nucleus.

FIGURE 16.34 ■ *Agrobacterium tumefaciens:* a natural gene transfer vector for plants. Left: *A. tumefaciens* transfers T-DNA containing opine synthesis genes into a plant, which then produces opines to feed the bacteria. **Right:** *A. tumefaciens* can be engineered to transfer T-DNA containing a recombinant gene of interest into the plant genome. A recombinant strain of *A. tumefaciens* has the Ti plasmid divided into two separate plasmids: one containing the *vir* operon conducting DNA transfer, the other containing T-DNA with most of its genes substituted by a desired recombinant gene.

left and right ends intact but most of its genes substituted by a desired recombinant gene. Upon infection, the virulence system induces transfer of the T-DNA to the plant cell without any tumor-inducing genes, allowing genomic integration of the recombinant gene without tumor induction or opine production.

Thought Question

16.11 Why would an herbicide resistance gene be desirable in an agricultural plant? What long-term problems might be caused by microbial transfer of herbicide resistance genes into plant genomes?

To Summarize

- **Industrial microbiology** includes the production of vaccines and clinical devices, industrial solvents and pharmaceuticals, and genetically modified plants and animals.
- **Microbial product molecules** may be indigenous, or they may be cloned from nonmicrobial sources. Thermophiles and psychrophiles are particularly important sources of new strains with potentially interesting new properties.
- **Microbial products must be competitive** with alternative technologies. Developing a competitive new molecular product requires identifying a useful molecule, isolating and engineering a strain to produce it, scaling

up for production in quantity, developing a business plan, and testing for safety.

- **Bioprospecting** is the mass screening of new microbial strains for potentially valuable protein and small-molecule products.
- **Industrial strains**, commonly *Escherichia coli* or *Bacillus subtilis*, are used to incorporate the newly discovered genes into a host microbe.
- **Upstream processing** refers to the culturing of the industrial microbe to produce large quantities of product. **Downstream processing** involves product recovery and purification.
- **Posttranscriptional processing of human or plant genes** may be facilitated by use of a transgenic source or a virus-infected animal host.

Concluding Thoughts

In this chapter we have seen how microbes from our environment contribute products for human use, from foods and vitamin supplements to industrial enzymes and highly specific protein therapeutics. These microbial applications incorporate much of the natural metabolism and biochemistry introduced in Chapters 13–15. Human inventiveness and industrial and financial management must meet many challenges to develop new microbial products.

What does the future hold for industrial microbiology? Entirely new microbes—viruses, bacteria, algae—are being engineered or "synthesized" to perform functions in an industrial system, or even within the human body. The field of synthetic biology is just beginning to develop such microbial partners for industry (discussed in Chapter 12).

CHAPTER REVIEW

Review Questions

1. What kinds of microbes are consumed as food? Why are most bacteria inedible for humans?
2. What are the advantages of fermentation for a food product?
3. What are the differences between traditional fermented foods and commercial fermented foods?
4. How do acidic fermentations contribute to the formation of different kinds of cheeses? What is the role of different kinds of metabolism performed by different microbial species?
5. Compare and contrast acidic and alkaline fermentation processes. What different kinds of foods are produced?
6. Compare and contrast the role of ethanolic fermentation in bread making and winemaking.
7. How do relatively minor fermentation reactions contribute to the flavor of food?
8. Explain the differences between food spoilage and food poisoning.
9. What are the most important food-borne pathogens, based on infection rates? Based on mortality rates?
10. What are the major means of preserving food? Compare and contrast their strengths and limitations.
11. What tasks must be accomplished to develop a microbial product for commercial marketing?
12. What are the sources of potential new microbial products? What is the difference between a source strain and an industrial strain?
13. Explain upstream processing and downstream processing.
14. Explain why genes encoding industrially useful products may be transferred into plant or animal systems for production. What are the roles of microbes in these systems?

Thought Questions

1. In cheese production, how do different kinds of fermenting microbes generate different flavors?
2. If you were a food safety regulator, which pathogen on the list in **Table 16.3** would you consider your top priority? Defend your answer based on factors such as numerical incidence of infection, severity of disease, and economic losses due to illness.
3. Suppose you undertake industrial production of a recombinant glycoprotein for human therapy. Would you produce your product in bacteria, in yeast, or in caterpillars? Explain the advantages and limitations of your choice.

Key Terms

alkaline fermentation (629)
antimicrobial agent (641)
bioprospecting (654)
cheddared (631)
cheese (630)
curd (630)
downstream processing (657)
endophyte (646)
ethanolic fermentation (629)
fermentation industry (624)
fermented food (627)
food poisoning (642)
food spoilage (642)
freeze-drying (648)
fruiting body (625)
heterolactic fermentation (629)
indigenous microbiota (627)
industrial fermentor (656)
industrial microbiology (650)
industrial strain (656)
injera (638)
kimchi (634)
lactic acid fermentation (629)
leaven (637)
lyophilization (648)
malolactic fermentation (640)
miso (633)
natto (636)
nori (626)
pathogenicity island (646)
primary recovery (657)
propionic acid fermentation (629)
putrefaction (642)
rancidity (642)
ripening (630)
single-celled protein (626)
sourdough (638)
starter culture (631)
synthetic biology (625)
tempeh (632)
Ti plasmid (657)
upstream processing (657)
whey (630)
yogurt (630)

Recommended Reading

Centers for Disease Control and Prevention. 2008. Outbreak of *Listeria monocytogenes* infections associated with pasteurized milk from a local dairy—Massachusetts, 2007. *Morbidity and Mortality Weekly Report* **57**:1097–1100.

Centers for Disease Control and Prevention. 2009. Multistate outbreak of *Salmonella* infections associated with peanut butter and peanut butter–containing products—United States, 2008–2009. *Morbidity and Mortality Weekly Report* **58**:1–6.

Chang, Shu-Ting, and Philip G. Miles. 2004. *Mushrooms: Cultivation, Nutritional Value, Medicinal Effect, and Environmental Impact.* 2nd ed. CRC Press, New York.

Doyle, Michael P., and Robert L. Buchanan (eds.). 2013. *Food Microbiology: Fundamentals and Frontiers.* 4th ed. ASM Press, Washington, DC.

Giraffa, Giorgio. 2004. Studying the dynamics of microbial populations during food fermentation. *FEMS Microbiological Reviews* **28**:251–260.

Hui, Y. H., Lisbeth Meunier-Goddick, Åse S. Hansen, Jytte Josephsen, Wai-Kit Nip, et al. (eds.). 2004. *Handbook of Food and Beverage Fermentation Technology.* Marcel Dekker, New York.

Marilley, L., and M. G. Casey. 2004. Flavours of cheese products: Metabolic pathways, analytical tools and identification of producing strains. *International Journal of Food Microbiology* **90**:139–159.

Mills, David A., Helen Rawsthorne, C. Parker, D. Tamir, and K. Makarova. 2005. Genomic analysis of *Oenococcus oeni* PSU-1 and its relevance to winemaking. *FEMS Microbiological Reviews* **29**:465–475.

Schwan, Rosane F. 1998. Cocoa fermentations conducted with a defined microbial cocktail inoculum. *Applied Environmental Microbiology* **64**:1477–1483.

Schwan, Rosane F., and Alan E. Wheals. 2004. The microbiology of cocoa fermentation and its role in chocolate quality. *Critical Reviews in Food Science and Nutrition* **44**:205–221.

Steinkraus, Keith H. (ed.). 1995. *Handbook of Indigenous Fermented Foods.* 2nd ed. Marcel Dekker, New York.

SOURCE: DUBILIER

Nicole Dubilier, head of the Symbiosis Group at the Max Planck Institute for Marine Microbiology.

AN INTERVIEW WITH:

Nicole Dubilier

Marine Animals with Bacterial Symbionts

Nicole Dubilier, a marine microbiologist, is the head of the Symbiosis Group at the Max Planck Institute for Marine Microbiology in Bremen, Germany. She earned her PhD in marine zoology from the University of Hamburg in 1992 and then did a postdoctoral fellowship at Harvard University. Since 1997, Dubilier and her students have studied symbiotic bacteria of marine animals from hydrothermal vents, cold seeps, whale falls, and shallow-water coastal sediments.

How did you decide to make a career in marine microbiology?

I did not originally want to become a microbiologist. I thought microbes were uninteresting organisms, too small to be observed with the naked eye and therefore not important and fascinating like animals such as dolphins. In high school I decided to become a marine biologist because I loved the ocean, having spent all my summers on Fire Island, New York. I thought that it would be brilliant to have a career where I could spend half my day diving and the other half doing fascinating experiments in the lab.

It wasn't until the end of my university studies that I began to realize how interesting microbiology is, in part because of the discovery of hydrothermal vents in the deep sea and the slew of papers that came out in the early 1980s describing exotic vent inhabitants such as giant tube worms without a mouth or gut that are "fed" by symbiotic bacteria that they host.

Why did you undertake your study of symbiosis between marine animals and bacteria?

I did my PhD at the University of Hamburg in Germany with Olav Giere, working on the adaptations that allow marine worms to live in oxygen-poor and sulfide-rich mudflats. I struggled with developing the kind of fascination and interest that I felt was essential for becoming a good scientist and devoting my life to a career in research. After earning my PhD in 1992, I gave myself one more year to figure out if I wanted to stay in science. Olav and I agreed that I would use molecular methods to identify the bacterial symbionts of small gutless worms from coral reef sediments that Olav had been working on.

I was fortunate to be able to do this postdoc at Harvard University with Colleen Cavanaugh, one of the first scientists to discover the hydrothermal vent symbioses. I really took to the molecular methods that Colleen and other environmental microbiologists began using in the late 1980s. These methods were so powerful in yielding insights into the biology and evolutionary ecology of uncultivable microbes.

During my 2 years in Colleen's lab, I discovered my passion and fascination for microbial symbioses and am extremely grateful that I was able to find "my calling." What I love about the symbioses I work on are the remarkable habitats they occur in. I get to go on research cruises with deep-diving submersibles and explore the exotic beauty of hydrothermal vents. Equally dazzling are the coral reef islands where we collect our gutless worms, like those on the Great Barrier Reef or in the Caribbean. And when I am out in the field (which is only once a year—much too seldom for my taste), I consider myself very lucky to actually be paid to do my work.

What challenges did you face in studying these symbioses?

When I started studying the small oligochaete worms, we did not know how many symbiont species they have. At the time, it was assumed that most hosts had only one or two different types of symbionts, but morphological analyses of the oligochaete worms indicated they might have more. In the early 1990s when I began working on these worms, we were using manual methods to sequence the symbionts' 16S rRNA genes, and it took me nearly 10 years to fully characterize the true diversity of the bacterial symbionts. Thanks to intensive sequencing efforts and imaging methods, such as fluorescence in situ hybridization, that allow us to visualize bacteria based on their 16S rRNA genes, we now know that these worms have five to six different types of symbionts.

Another challenge is that so far, no one has been able to cultivate the symbionts from our hosts. This means that we have to regularly go on excursions and research cruises to collect the worms and mussels we work on. I love this part of my work; I find it physically satisfying to experience these organisms and their environments with all of my senses. I also enjoy the spontaneity of working in the field and having to continually improvise because things rarely go as originally planned. When you are at sea or on a small coral reef island, you can't order something you forgot or quickly replace a piece of equipment that breaks down, so you learn to be inventive.

How did graduate students and collaborators contribute to your project?

After my postdoc with Colleen, I joined the Max Planck Institute for Marine

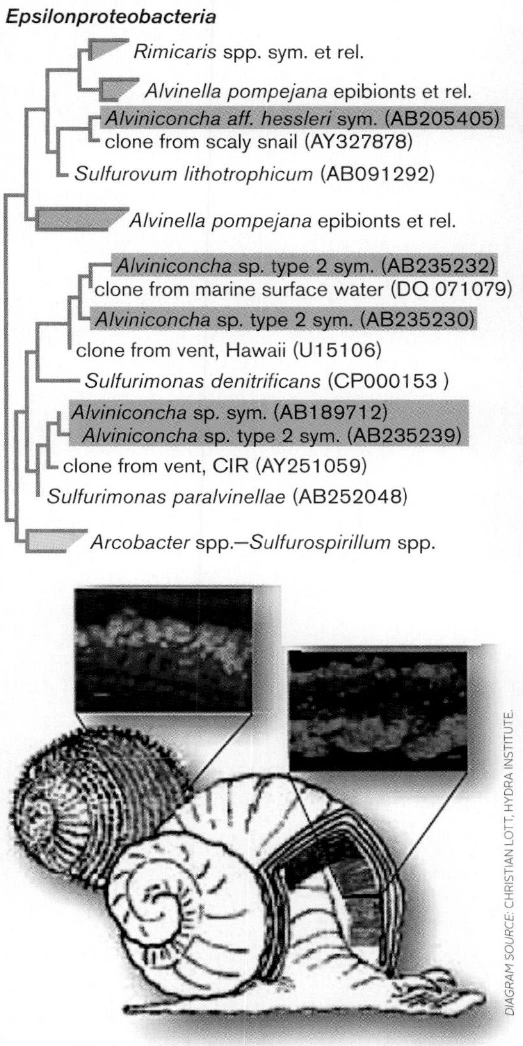
Epsilonproteobacteria
Rimicaris spp. sym. et rel.
Alvinella pompejana epibionts et rel.
Alviniconcha aff. hessleri sym. (AB205405)
clone from scaly snail (AY327878)
Sulfurovum lithotrophicum (AB091292)
Alvinella pompejana epibionts et rel.
Alviniconcha sp. type 2 sym. (AB235232)
clone from marine surface water (DQ 071079)
Alviniconcha sp. type 2 sym. (AB235230)
clone from vent, Hawaii (U15106)
Sulfurimonas denitrificans (CP000153)
Alviniconcha sp. sym. (AB189712)
Alviniconcha sp. type 2 sym. (AB235239)
clone from vent, CIR (AY251059)
Sulfurimonas paralvinellae (AB252048)
Arcobacter spp.—Sulfurospirillum spp.
DIAGRAM SOURCE: CHRISTIAN LOTT, HYDRA INSTITUTE.

Phylogeny tree of bacterial symbionts of the deep-sea gastropod *Alviniconcha.*
Source: Hidetoshi Urakawa et al. 2005. *Environ. Microbiol.* 7:750.

Microbiology in Bremen, Germany. Our institute is quite unique in the field of marine microbiology because we have scientists working on a broad range of topics, covering everything from biogeochemistry to bioinformatics, so I can almost always find someone who has the expertise I need. The contributions of my graduate students and postdocs have been the other critical contribution to the research in my group. I have a team of highly motivated and gifted young scientists, and I greatly enjoy working with them and helping them learn how to use their skills to become successful researchers.

Alviniconcha gastropods held by Dubilier.

What key technologies have contributed to your research?

Molecular methods have revolutionized my ability to understand the biology, ecology, and evolution of the symbiotic bacteria I work on. The "omics" (genomics, transcriptomics, proteomics, and metabolomics) provide us with a wealth of information about how symbionts gain energy, carbon, and other nutrients from the environment, and how they interact with each other and with their hosts. We are using new imaging methods, such as NanoSIMS, a high-resolution imaging technique, to identify the chemical and isotopic composition of individual cells, and to link the identity of different symbionts to their function. We also profit immensely from rapid advances in deep-sea technology and our collaboration with the MARUM research center in Bremen, which has remotely operated vehicles (ROVs, or unmanned submersibles). ROVs work for extended periods on the seafloor at great depths deploying the instruments we need to study the deep-sea environment and collect our symbiotic animals.

Do marine symbiotic bacteria have practical applications?

Many symbiotic bacteria produce bioactive compounds that have pharmaceutical potential. For example, bacterial symbionts of sponges and bryozoans produce antimalarial and anticancer compounds, or metabolites with antimicrobial and antifungal activity. We are excited because we have just discovered that one of the symbionts we are working on produces insecticidal toxins. Our symbiont toxins are highly similar to toxins used by the bacterial symbionts of parasitic nematodes to kill the insects they prey on.

How does your family relate to your work?

I have been together with my husband since I began my PhD, and he has always understood how important my work is to me and has supported me throughout my career by accepting that I often work long hours and travel a lot. And when our son was born, my husband shared equally in his upbringing, despite the demands of his own career as an orthopedist. Our son is now 14, and he just wandered into the room while I was writing this, so I asked him how he relates to my work. His answer was: "Well I'm very proud of you, and I think it's cool that you're a marine biologist." I wonder if one day he'll say he thinks it's cool that I'm a marine microbiologist.

CHAPTER 17
Origins and Evolution

17.1 Origins of Life
17.2 Early Metabolism
17.3 Microbial Phylogeny and Gene Transfer
17.4 Adaptive Evolution
17.5 Microbial Species and Taxonomy
17.6 Symbiosis and the Origin of Mitochondria and Chloroplasts

Microbial life appeared as early as 3.8 billion years ago, soon after our planet Earth formed out of dust from the young Sun. Since then, microbes have evolved into forms adapted to diverse ways of life, from psychrophiles beneath the ice of Antarctica to anaerobes in the human colon. Descendants of those early microbes include all living plants and animals, including ourselves. How did microbes originate and evolve? How did their evolving metabolism shape the chemistry of Earth's crust and atmosphere?

Today we watch microbial evolution as it happens—in the laboratory, and even within the human body. We trace the descent of microbes through their genomes, and discover their amazing array of abilities. Microbes evolve along a branching tree of life, and also through gene transfer between distant branches. Many microbes evolve fascinating partnerships with other living things, such as bacteria that fix nitrogen within plant cells. Other microbes evolve as pathogens at the expense of their host and cause disease. And commercial laboratories speed up microbial evolution to develop new drugs and vaccines.

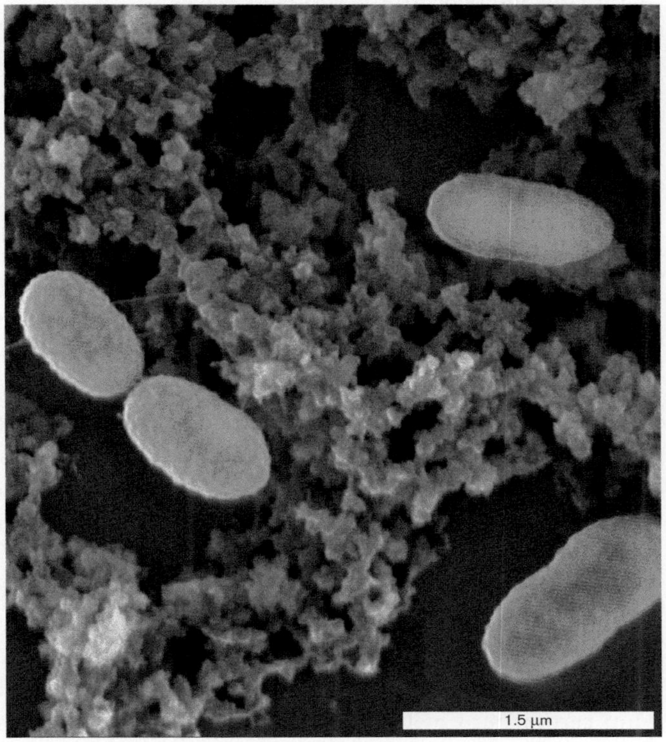
1.5 µm

CURRENT RESEARCH highlight

Dairy bacterial evolution. *Lactococcus lactis,* a major fermenter of the dairy industry, is the state microbe of Wisconsin. *L. lactis* subspecies *cremoris* is shown amidst milk solids coagulated during twelve hours of fermentation (colorized SEM). The *L. lactis* strains used by the dairy industry have evolved genetic sequences that differs from that of strains naturally occurring on plants. Herwig Bachmann and colleagues at NIZO food research, Netherlands, conducted experimental evolution to test how plant-derived *L. lactis* strains could evolve to ferment milk effectively for dairy production. In 2012, the researchers reported that following 1,000 generations grown in milk, the bacteria showed increased fitness and had evolved traits comparable to those observed in commercial dairy strains. The mechanisms of evolution included loss of insertion elements and gain of mutator traits which increase the rate of genetic change. Thus, the evolutionary adaptation of bacteria to the dairy niche can be observed experimentally. *Source:* Herwig Bachmann, et al. 2012. *Genome Research* **22:**115. Micrograph courtesy of I. Ayala-Hernández, et al. 2010. *Food Hydrocolloids* **23:**1299.

For centuries, observers of the natural world have wondered where life came from. As early as 1802, the naturalist Erasmus Darwin, grandfather of Charles Darwin, wrote:

> Organic life beneath the shoreless waves
> Was born and nurs'd in ocean's pearly caves;
> First forms minute, unseen by spheric glass,
> Move on the mud, or pierce the watery mass;
> These, as successive generations bloom,
> New powers acquire and larger limbs assume.

Thus, nineteenth-century biologists developed the idea that all living organisms had evolved from microbes, perhaps even from cells too small to be seen with the "spheric glass" of a microscope. Even without modern tools of isotope analysis and genetics, thoughtful observers recognized the overwhelming commonalities among all living cells, such as the membrane-enclosed compartment of cytoplasm and common metabolic pathways such as sugar metabolism. Today, lines of evidence from geology, biochemistry, and genetics overwhelmingly support the microbial origin of life.

What did early life look like? The earliest forms of life for which we have clear fossil evidence are bacterial communities called **stromatolites** (**Fig. 17.1A**). A stromatolite is a bulbous mass of sedimentary layers of limestone (calcium carbonate, $CaCO_3$) accreted by microbes over years, even centuries. Within the outer layers, microbes grow as a "microbial mat," a kind of biofilm. The outermost layers of the mat contain oxygenic phototrophs, such as diatoms and filamentous cyanobacteria, that exude bubbles of oxygen. A few millimeters below the surface, red light supports bacteria photolyzing H_2S to sulfate, which is then reduced by still lower layers of sulfate-reducing bacteria. Stromatolites today survive mainly in isolated pools whose high salt concentration excludes predators, as in Hamlin Pool, Shark Bay, Australia. But 3 billion years ago, without predators, stromatolites covered shallow seas all over Earth. Fossil stromatolites appear in ancient rock formations such as the 3.4-billion-year-old Strelley Pool Chert, a sedimentary formation in Pilbara Craton, Australia (**Fig. 17.1B**). Their layers preserve the wavy form of the microbial mats, remarkably similar to living stromatolites.

Questions about life's origin have long sparked controversy. In eighteenth-century Europe, the idea of "spontaneous generation" of microbes was considered so dangerous that priests and monks performed scientific experiments to disprove it (described in Chapter 1). The priests ultimately won their scientific point: All microbes today have "parents"—that is, preexisting microbes. But these experiments did not address the origin of the first living cells or how early life gave rise to modern species. This concept was so controversial that in the early twentieth century, laws forbade teaching it in American public schools. In 1929, such a law in Tennessee was put to the test by high school teacher John Scopes in what became the famous Scopes trial (**Fig. 17.2**). According to a 14-year-old student who testified at the trial, Scopes taught that "there was a little germ of one cell organism formed, and this organism kept evolving . . . and from this was man." A large body of evidence now supports the view that microbes living more than 3 billion years ago gave rise to all animals and plants, as well as to all modern microbes.

Chapter 17 explores the evidence from geochemistry and molecular biology for the nature of the earliest cells,

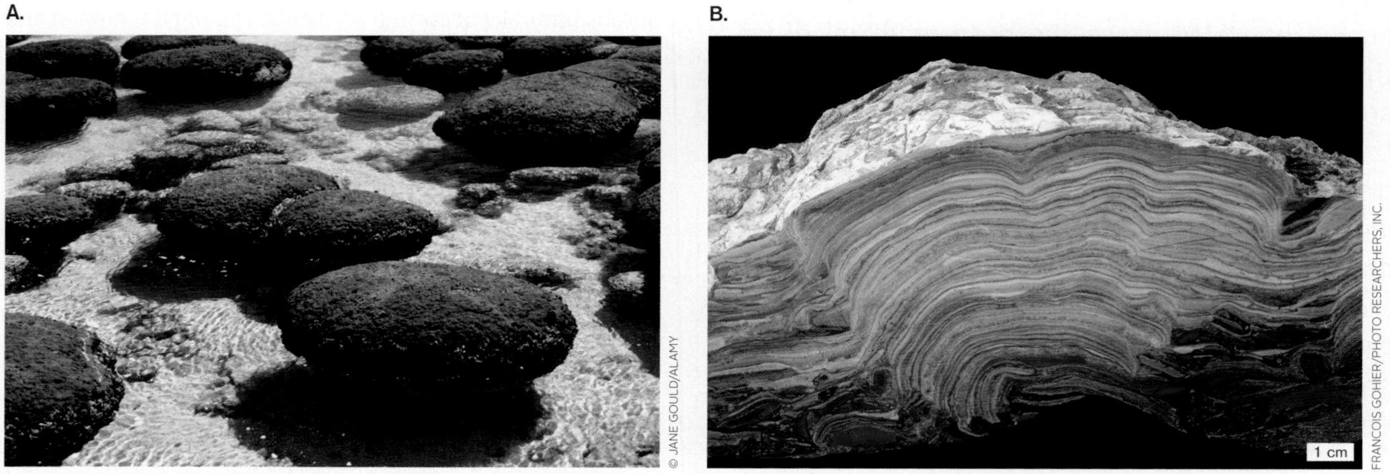
A.
B.
© JANE GOULD/ALAMY
1 cm
FRANCOIS GOHIER/PHOTO RESEARCHERS, INC.

FIGURE 17.1 ■ Stromatolites: an ancient life-form. A. Cyanobacteria form colonial stromatolites, the present-day structures believed to most closely resemble the earliest forms of life on Earth. Shark Bay, Western Australia. **B.** Section through a 3.4-billion-year-old fossil stromatolite from the Strelley Pool Chert, Pilbara Craton, Australia.

T.T.MARTIN. HEADQUARTERS
ANTI-EVOLUTION LEAGUE
"THE CONFLICT"-"HELL & THE HIGH SCHOOL"
BRYAN'S
BOOKS

HULTON-DEUTSCH COLLECTION/CORBIS

FIGURE 17.2 ■ Objecting to microbial ancestry. In 1925, outside the Scopes trial in Dayton, Tennessee, demonstrators opposed the teaching that all life—including humans—evolved from a microbe.

as well as the challenges in interpreting data from so long ago. We show how molecular techniques reveal deep similarities among all life-forms, such as the core macromolecular apparatus of DNA, RNA, and proteins. We watch the mechanisms of microbial evolution emerge from laboratory evolution experiments, such as the famous long-term evolution experiment on *E. coli* conducted by Richard Lenski (see the Part 1 interview).

We focus on three major concerns:

- The origin of life on Earth and the nature of the earliest cells.
- The divergence of microbes from common ancestors, modified by gene transfer and symbiosis.
- The mechanisms of microbial evolution, as it occurs in nature and in the laboratory.

17.1 Origins of Life

Before the first cells could evolve, several fundamental conditions were required:

- **Essential elements.** Because all life on Earth is composed of molecules, the origin of life required the fundamental elements that compose organic molecules.
- **Continual source of energy.** The generation of life requires continual input of energy, which ultimately is dissipated as heat. The main source of energy for life is nuclear fusion reactions within the Sun.
- **Temperature range permitting liquid water.** Above 150°C, life's macromolecules fall apart; below the freezing point of water, metabolic reactions cease. Maintaining the relatively narrow temperature range conducive to life depends on the nature of our Sun, our planet's distance from the Sun, and the heat-trapping capacity of our atmosphere.

Elements of Life

For life to arise and grow, elements such as carbon and oxygen needed to be available on Earth. The planet Earth coalesced during formation of the solar system 4.5 Gyr (gigayears, or billions of years) ago. Central to the solar system is our Sun, a "yellow" star of medium size and surface temperature (5770 K). The Sun's surface temperature generates electromagnetic radiation across the spectrum, peaking in the range of visible light. As we learned in Chapter 13, the photon energies of visible light are sufficient to drive photosynthesis but not so energetic that they destroy biomolecules. Thus, the stellar class of our Sun makes organic life possible.

Note: In geological description, a billion years (10^9) is a gigayear, or Gyr. A million years (10^6) is a megayear, or Myr.

The Sun's surface temperature and luminosity are generated by nuclear fusion reactions in which hydrogen nuclei fuse to form helium nuclei. (Be careful to distinguish nuclear reactions, involving nuclei, from chemical reactions, involving electrons.) Besides hydrogen and helium, 2% of the solar mass consists of heavier elements, such as carbon, nitrogen, and oxygen, as well as traces of iron and other metals—elements that compose Earth, including its living organisms. Where did these heavier elements come from? To answer this question, we must look to other stars in the universe at different stages of their development (**Fig. 17.3**).

Elements of life formed within stars. Throughout the universe, young stars such as our Sun fuse hydrogen to form helium. As stars age, they use up all their hydrogen. With hydrogen gone, the aging star contracts and its temperature rises, enabling helium nuclei to fuse, forming carbon (see **Fig. 17.3**). Carbon drives a cyclic nuclear reaction, the CNO (carbon-nitrogen-oxygen) cycle, to form isotopes of nitrogen and oxygen. Subsequent nuclear reactions generate heavier elements through iron (Fe). Thus, the major elements of biomolecules were formed within stars that aged before our solar system was born.

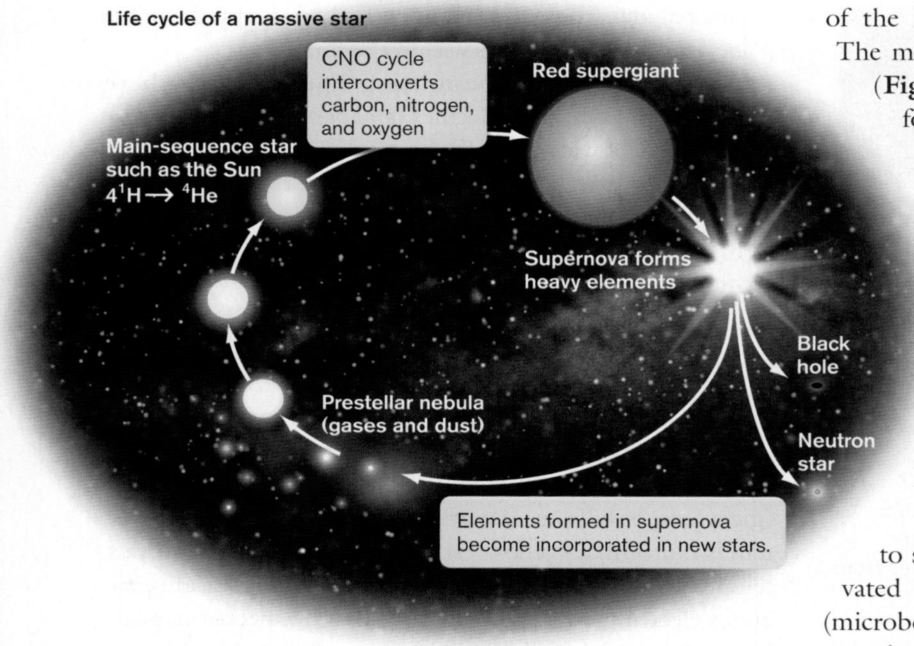
Life cycle of a massive star
CNO cycle interconverts carbon, nitrogen, and oxygen
Red supergiant
Main-sequence star such as the Sun
4¹H → ⁴He
Supernova forms heavy elements
Black hole
Prestellar nebula (gases and dust)
Neutron star
Elements formed in supernova become incorporated in new stars.

FIGURE 17.3 ■ Stellar origin of atomic nuclei that form living organisms. In young stars, hydrogen nuclei fuse to form helium. In older stars, fusion of helium forms carbon, nitrogen, oxygen, and all the heavier elements up to iron. Massive stars explode as supernovas, spreading all the elements of the periodic table across space. These elements are picked up by newly forming stars, such as our own Sun.

The later nuclear reactions of aging stars generate heavier nuclei, up to iron. The aging star expands, forming a red giant (see **Fig. 17.3**). When a star of sufficient mass expands (a supergiant), it explodes as a **supernova**. The explosion of a supernova generates in a brief time all the heaviest elements and ejects the entire contents of the star at near light speed. Billions of years before our Sun was born, the first stars aged and died, spreading all the elements of the periodic table across the universe. Some of these elements coalesced with our Sun and formed the planets of our solar system. In effect, all life on Earth is made of stardust, the remains of stars long gone.

Thought Question

17.1 What would have happened to life on Earth if the Sun were a different stellar class, substantially hotter or colder than it is?

Elemental composition of Earth. When our solar system formed, individual planets coalesced out of matter attracted by the force of gravity. Because of Earth's small size, most of the hydrogen gas escaped Earth's gravity very early. The most abundant dense component of Earth was iron (**Fig. 17.4**). Much of Earth's iron sank to the center to form the core. The core is surrounded by a mantle, composed primarily of iron combined with less dense crystalline minerals such as silicates of iron and magnesium: $(Fe,Mg)_2SiO_4$. The mantle is coated by Earth's thin outer crust. The crust is composed primarily of silicon dioxide, SiO_2, also known as quartz or chert. Crustal rock includes smaller amounts of numerous minerals, including the carbonates and nitrates that provided the essential elements for life. Overall, the crust shows a redox gradient, reducing in the interior and oxidizing at the surface.

The crust provides a habitat for microbes down to surprising depths, such as within gold mines excavated to 3 kilometers (km). Some endolithic microbes (microbes living within rock) metabolize by oxidizing electron donors generated through decay of radioactive metals. The discovery of endolithic organisms deep in the crust was of great interest to NASA scientists seeking life on Mars. Of the planets, Mars most closely resembles Earth in geology and distance from the Sun, and its crust might provide a habitat similar to Earth's.

The outer surface of the crust supports the remainder of the **biosphere**, the sum total of all life on Earth. The biosphere generates oxidants (electron acceptors), most notably O_2. Oxygen-breathing organisms can live only on the outer surface, where O_2 is produced by photosynthesis.

Earth's atmosphere. From the crust and the mantle of early Earth, volcanic activity released gases such as carbon dioxide and nitrogen, which formed Earth's first atmosphere, while volcanic water vapor formed the ocean. The composition of this first atmosphere, before life evolved, looked much like that of Mars: thin, about 1% as dense as that of Earth today, consisting primarily of CO_2. But unlike Mars, Earth developed living organisms that filled the atmosphere with gaseous N_2 and O_2 and that continue to produce these gases today. Organisms also produce CO_2, as well as fixing it into biomass. Some CO_2 and N_2 arise from geological sources such as volcanoes, but their contribution is small compared to that of biological cycles (discussed in Chapter 22). The overall composition of Earth's atmosphere is determined by living organisms, primarily microbes.

Temperature. Another important aspect of Earth's habitat, determined by the atmospheric density and composition, is temperature. Atmospheric gases absorb light and

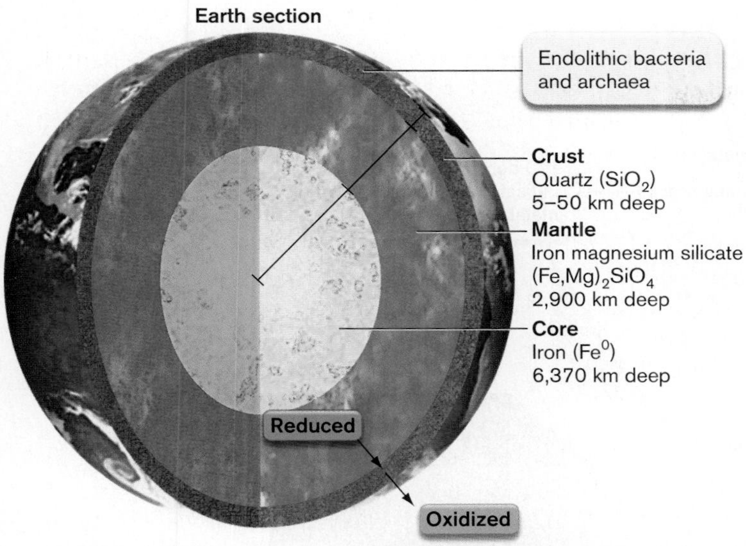
Earth section
Endolithic bacteria and archaea
Crust
Quartz (SiO_2)
5–50 km deep
Mantle
Iron magnesium silicate
$(Fe,Mg)_2SiO_4$
2,900 km deep
Core
Iron (Fe^0)
6,370 km deep
Reduced
Oxidized

FIGURE 17.4 ■ Geological composition of Earth. The cross section of Earth shows the core, the mantle, and the thin outer crust. The core and mantle are rich in iron; oxygen content increases toward the crust. The crust is composed primarily of silicates such as quartz (SiO_2). Crustal rock supports endolithic bacteria and archaea.

convert the energy to heat, raising the temperature of the surface and atmosphere. This rise in temperature is known as the **greenhouse effect**. Because carbon dioxide is an especially potent greenhouse gas, the CO_2-rich atmosphere of early Earth could have heated the planet to temperatures approaching that of Venus, eliminating the possibility of life. Instead, microbial consumption of CO_2 and generation of nitrogen and oxygen gases limited Earth's surface temperature to an average of 13°C. The cooling effect may have led to an ice age, possibly reversed by rising methane from methanogens. One way or another, the history of Earth's atmosphere is intimately related to the history of microbial evolution.

Geological Evidence for Early Life

The first period of Earth's existence, ranging from 4.5 to 3.8 Gyr ago, is designated the **Hadean eon** (**Fig. 17.5**), named for Hades, the ancient Greek world of the dead. During the Hadean eon, repeated bombardment by meteorites vaporized the oceans, which then cooled and recondensed. Meteor bombardment may have killed off incipient life more than once before living microbes finally became established. Still, scientists speculate on whether some forms of life might have survived Hadean conditions, perhaps growing 3 km below the Earth's surface. Like the dead spirits imagined by the Greeks to have populated Hades, Earth's earliest cells might have reached deep enough within the crust that they were protected from the heat and vaporization occurring at the surface.

In 2009, a belt of rock in Ungava, Québec, was dated by radioisotope decay at 4.3 Gyr ago, near the time that Earth formed. Some researchers find evidence for early life in this rock, but overall the evidence for life earlier than 3.8 Gyr ago remains speculative.

The Archaean eon. The earliest geological evidence for life that is generally accepted dates to 3.8–2.5 Gyr ago, in the **Archaean eon** (**Fig. 17.5**). In the Archaean, meteor bombardment was less frequent, and the Earth's crust had become solid. The Archaean marked the first period with stable oceans containing the key ingredient of life: liquid water. Water is a key medium for life because it remains liquid over a wide range of temperatures and because it dissolves a wide range of inorganic and organic chemicals. Rock strata dating to the Archaean eon reveal the first evidence of living organisms and their metabolic processes.

Note: The term "Archaean" refers to the earliest geological eon when life existed, whereas "archaeal" is the adjective referring to the taxonomic domain Archaea. The domain Archaea (originally "Archaebacteria") was named by Carl Woese based on his theory that the species of this domain most closely resembled the earliest life-forms of the Archaean eon. In fact, early life may have encompassed diverse traits later associated with archaea, bacteria, and eukaryotes.

How and when did living cells arise out of inert materials? Without a time machine to take us back 4 Gyr, we must rely on evidence from Earth's geology. Interpreting geology is a challenge because most forms of evidence for early life are indirect and subject to multiple interpretations. The farther back in time, the more change has occurred to the rock strata and the greater the difficulties are. One way to meet this challenge, however, is to compare the results from different kinds of evidence (**Table 17.1**). If two or more kinds of evidence (such as microfossils and isotope ratios) point to life in the same location, the conclusion is strengthened.

Stromatolites. Fossil stromatolites are layers of carbonate or silicate rock that resemble modern living stromatolites. Stromatolites grow in shallow marine water as mats of cyanobacteria. The cyanobacteria secrete mucus that traps sediment, which precipitates calcium carbonate from the water, cementing the sediment into hard rock. Fossil stromatolites date as early as 3.4 Gyr ago (see **Fig. 17.1**). These rock formations appear remarkably similar to the layered forms of stromatolites today. The ancient rock, however, is too deformed to reveal the detailed structure of cells, and the biological origin of such fossils is questioned by some researchers.

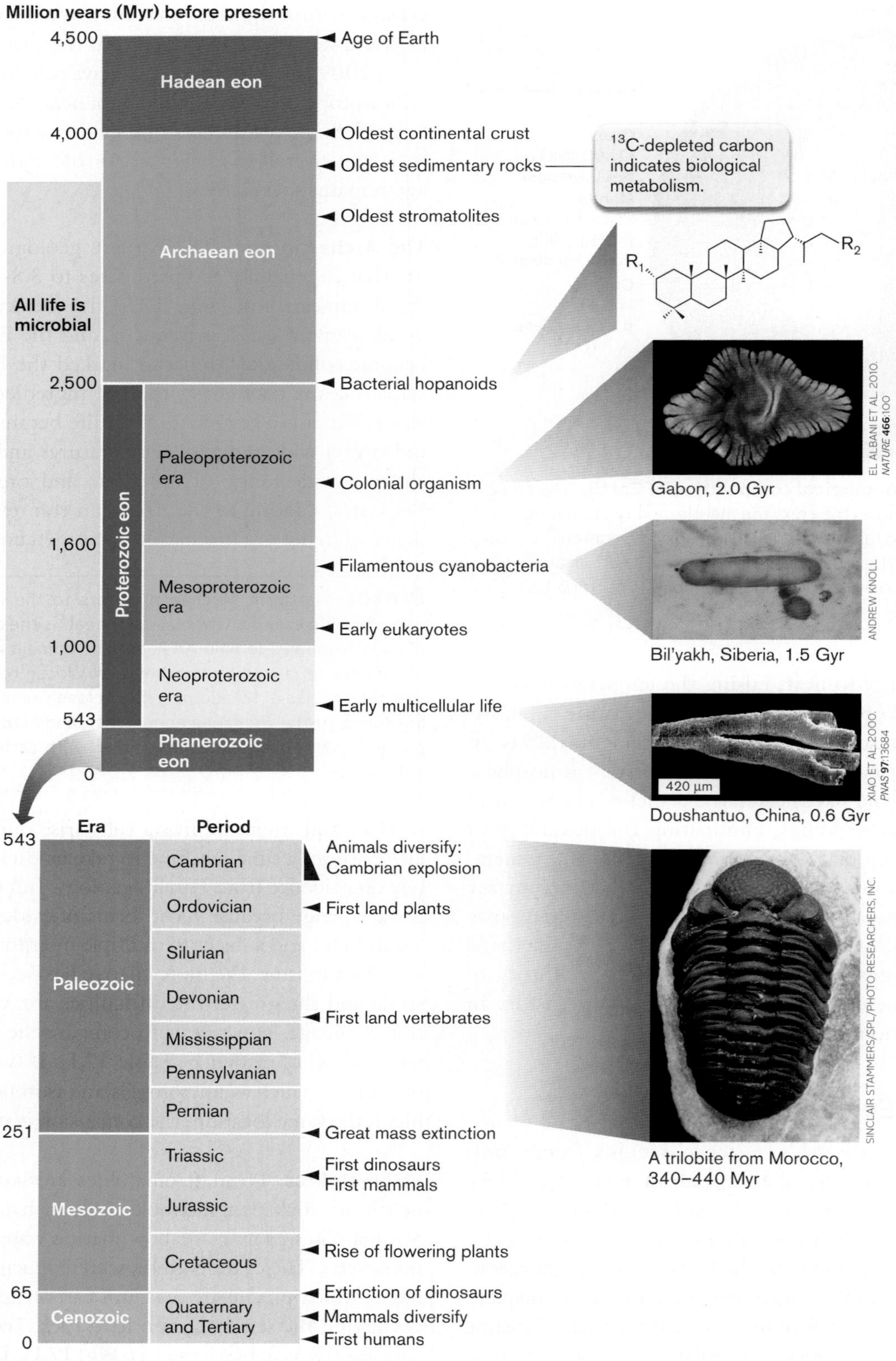
Million years (Myr) before present
4,500
Hadean eon
4,000
Archaean eon
All life is microbial
2,500
Proterozoic eon
Paleoproterozoic era
1,600
Mesoproterozoic era
1,000
Neoproterozoic era
543
Phanerozoic eon
0
Age of Earth
Oldest continental crust
Oldest sedimentary rocks
^{13}C-depleted carbon indicates biological metabolism.
Oldest stromatolites
Bacterial hopanoids
Colonial organism
Filamentous cyanobacteria
Early eukaryotes
Early multicellular life
R_1
R_2
Gabon, 2.0 Gyr
EL ALBANI ET AL. 2010. NATURE 466:100
Bil'yakh, Siberia, 1.5 Gyr
ANDREW KNOLL
420 µm
Doushantuo, China, 0.6 Gyr
XIAO ET AL. 2000. PNAS 97:13684
Era
Period
543
Paleozoic
Cambrian
Ordovician
Silurian
Devonian
Mississippian
Pennsylvanian
Permian
251
Mesozoic
Triassic
Jurassic
Cretaceous
65
Cenozoic
Quaternary and Tertiary
0
Animals diversify: Cambrian explosion
First land plants
First land vertebrates
Great mass extinction
First dinosaurs
First mammals
Rise of flowering plants
Extinction of dinosaurs
Mammals diversify
First humans
A trilobite from Morocco, 340–440 Myr
SINCLAIR STAMMERS/SPL/PHOTO RESEARCHERS, INC.

FIGURE 17.5 ■ Geological evidence for early life. The geological record shows evidence of microbial life early in Earth's history, 3 Gyr before the first multicellular forms.

TABLE 17.1

Geological evidence of early life.

Type of evidence	Advantages	Limitations
Stromatolites		
Layers of phototrophic microbial communities grew and died, their form filled in by calcium carbonate or silica.	Fossil stromatolites can be observed in the oldest rock of the Archaean eon. Their distinctive shapes resemble those of modern living stromatolites.	Some layered formations attributed to stromatolites have been shown to be generated by abiotic (nonbiological) processes.
Microfossils		
Early microbial cells decayed, and their form was filled in by calcium carbonate or silica. The size and shape of microfossils resemble those of modern cells.	Microfossils are visible and measurable under a microscope, offering direct evidence of cellular form.	Microscopic rock formations require subjective interpretation. Some formations may result from abiotic processes.
Isotope ratios		
Microbes fix $^{12}CO_2$ more readily than $^{13}CO_2$. Thus, limestone depleted of ^{13}C must have come from living cells. Similarly, sulfate reduced by sulfate-respiring bacteria shows depletion of ^{34}S compared with ^{32}S.	Isotope ratios offer objective measurement of a highly reproducible physical quantity. They provide the best evidence for dating the earliest life. Isotope ratios generated by key biochemical reactions can calibrate the time line of phylogenetic trees.	We cannot prove absolutely that no abiotic process could generate a given isotope ratio. Isotope ratios tell us nothing about the form of early life or how it evolved.
Biosignatures		
Certain organic molecules found in sedimentary rock are known to be formed only by certain microbes. These molecules are used as biosignatures.	Biosignatures such as hopanoids are complex molecules specific to bacteria.	A biosignature thought specific to one kind of organism may be discovered in others. In the oldest rocks, organic biosignatures are eliminated entirely by metamorphic processes.
Oxidation state		
The oxidation state of metals such as iron and uranium indicates the level of O_2 available when the rock formed. Banded iron formations (BIFs) suggest oxidation by microbial phototrophs that intermittently produced oxygen.	Oxidized metals offer evidence of microbial processes even in highly deformed rock.	It is hard to rule out abiotic causes of oxidation. If biological processes were the cause, the kind of metabolism is not revealed.

Microfossils. The most convincing evidence for early microbial life is the visual appearance of **microfossils**, microscopic fossils in which minerals have precipitated and filled in the form of ancient microbial cells (**Fig. 17.6**). Microfossils are dated by the age of the rock formation in which they are found, which in turn is based on evidence such as radioisotope decay. Convincing microfossils need to show regular 3D patterns of cells that cannot be ascribed to abiotic (nonbiological) causes.

The earliest convincing microfossils are dated at 2.0 Gyr ago. Microfossils dated to 2.0 Gyr ago include filamentous prokaryotes in the Gunflint Formation, Ontario, Canada (**Fig. 17.6A**). The Gunflint outcrops consist of chert, a kind of silicate formed by precipitation from an ancient sea. The sea was rich in carbonates and reduced iron, a good combination for redox metabolism. Some of the fossils resemble the form of filamentous iron-metabolizing bacteria today, such as *Leptothrix* species (**Fig. 17.6B**). Other microfossils dated at 2.0 Gyr ago resemble colonial cyanobacteria (compare **Fig. 17.6C** with **17.6D**). More recent strata, dated at 1.2 Gyr ago, contain larger fossil cells comparable to those of modern eukaryotes such as algae (**Fig. 17.6E** and **F**).

If life existed in more ancient times, such as the Archaean eon, where are the microfossils? Archaean rock is

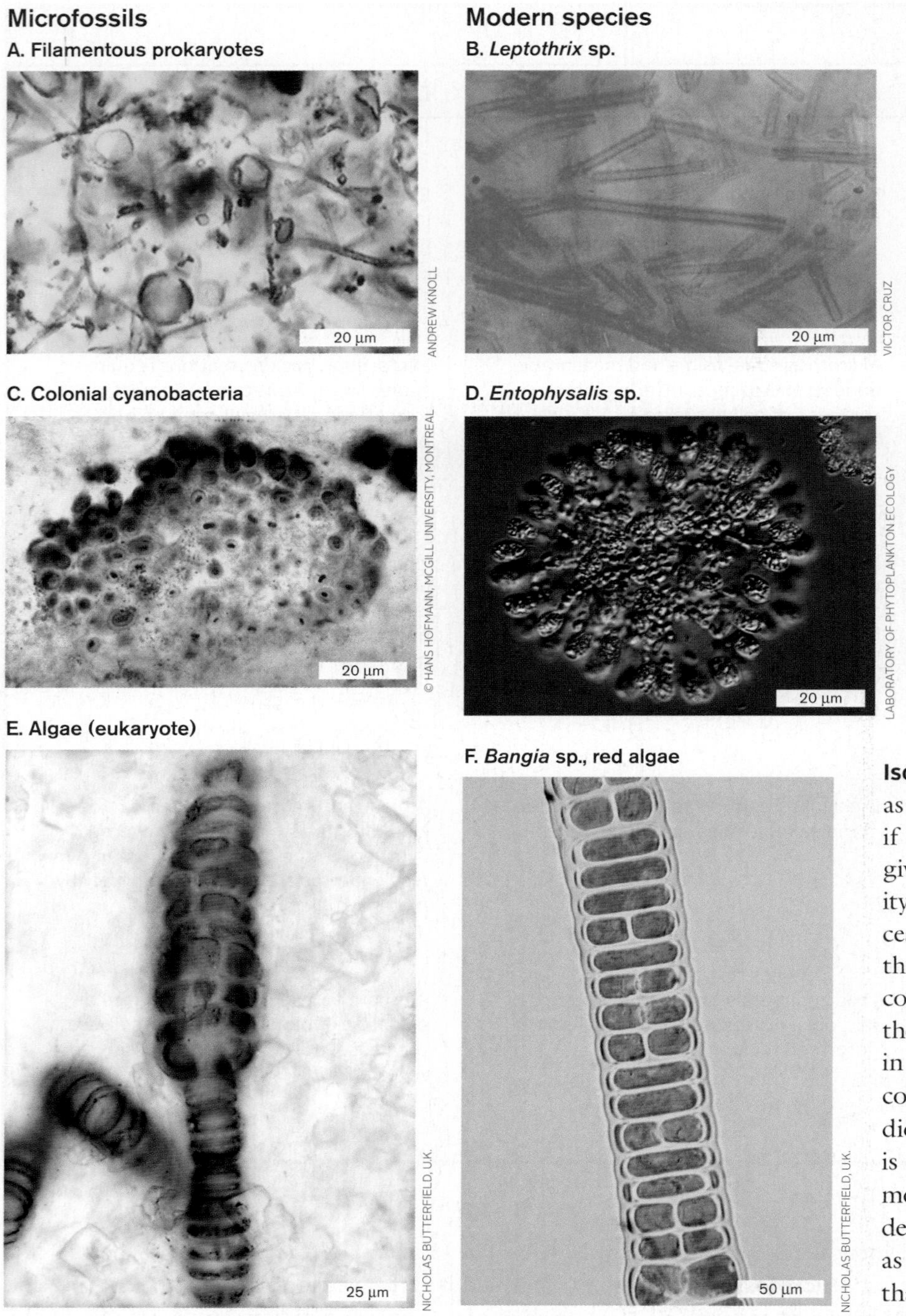

Microfossils
A. Filamentous prokaryotes
20 μm
ANDREW KNOLL
Modern species
B. Leptothrix sp.
20 μm
VICTOR CRUZ
C. Colonial cyanobacteria
20 μm
© HANS HOFMANN, MCGILL UNIVERSITY, MONTREAL
D. Entophysalis sp.
20 μm
LABORATORY OF PHYTOPLANKTON ECOLOGY
E. Algae (eukaryote)
25 μm
NICHOLAS BUTTERFIELD, U.K.
F. Bangia sp., red algae
50 μm
NICHOLAS BUTTERFIELD, U.K.

FIGURE 17.6 ■ Microfossils compared with modern bacteria. A. Filamentous prokaryotes, 2.0 Gyr old, from Gunflint Formation, Ontario, Canada. **B.** Modern *Leptothrix* filamentous bacteria. **C.** Colonial cyanobacteria, about 2.0 Gyr old, from Belcher Islands, Canada. **D.** Modern *Entophysalis* cyanobacteria. **E.** Filamentous algae, 1.2 Gyr old, from arctic Canada. **F.** Modern red algae, a eukaryote, *Bangia* sp.

metamorphic, greatly modified by temperature and pressure. The macroscopic contours of stromatolites can be identified, but microfossil interpretation is highly controversial. For example, microfossils of cyanobacteria dated to 3.85 Gyr by William Schopf in the early 1990s were accepted and described in many textbooks (**Fig. 17.7**). The 3.85-Gyr-old fossils have since been reinterpreted by Martin Brasier and colleagues as nonbiogenic artifacts (caused by abiotic processes). The form of the proposed Archaean microfossils is less regular and convincing than the form of later specimens, particularly when observed at different angles not shown in the original publication.

Another kind of evidence is that of a **biosignature**, or **biological signature**, a chemical indicator of life. Biosignatures have been found that are even earlier than the oldest fossils. Their significance is limited, however, as it is hard to rule out nonbiogenic explanations, so researchers seek additional, corroborating evidence based on independent principles.

Isotope ratios. An **isotope ratio** may serve as a biosignature in fossil rock (**Fig. 17.8A**) if the ratio between certain isotopes of a given element is altered by biological activity. Enzymatic reactions, unlike abiotic processes, are so selective for their substrates that their rates may differ for molecules containing different isotopes. For example, the carbon-fixing enzyme Rubisco, found in chloroplasts, preferentially fixes CO_2 containing ^{12}C rather than ^{13}C. The carbon dioxide fixed into microbial cells eventually is converted to calcium carbonate in sedimentary rock. Thus, the calcium carbonate deposited by CO_2-fixing autotrophs (such as cyanobacteria) shows lower ^{13}C content than calcium carbonate deposited by abiotic processes (**Fig. 17.8B**). The difference, $\delta^{13}C$, is defined by the fractional difference (in parts per thousand) between the $^{13}C/^{12}C$ ratios in a sample versus a standard inorganic rock:

$$\delta^{13}C = \frac{^{13}C/^{12}C\ (\text{experimental}) - {}^{13}C/^{12}C\ (\text{standard})}{^{13}C/^{12}C\ (\text{standard})} \times 1{,}000$$

Typical $\delta^{13}C$ values are shown in **Figure 17.8B**. Organisms on land and sea show $\delta^{13}C$ values of –10 to –30 parts per thousand, representing a significant ^{13}C depletion through carbon fixation. A comparable $\delta^{13}C$ is observed in fossil fuels, which formed from plant and animal bodies decomposed by bacteria.

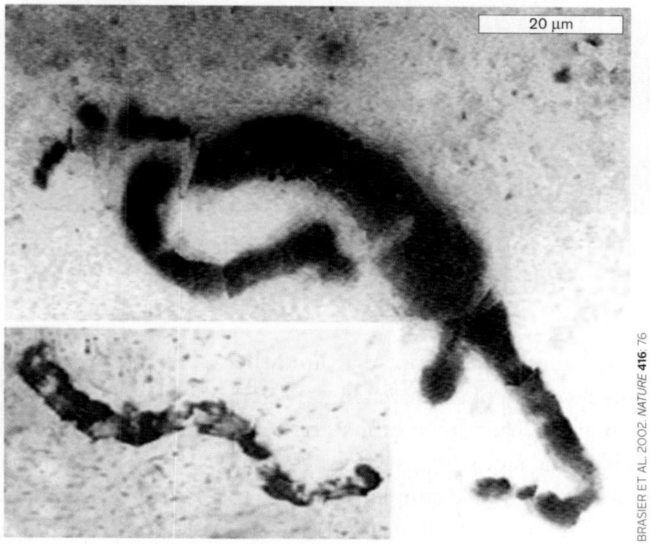
20 μm
BRASIER ET AL. 2002. NATURE 416:76

FIGURE 17.7 ■ Microfossils or artifacts? Structures originally identified as cyanobacterial microfossils from Western Australia, dated to 3.85 Gyr ago. Further testing indicated that the structures are nonbiological artifacts.

The most ancient mineral samples showing a substantial $\delta^{13}C$ (about −18 parts per thousand) are graphite granules in the Isua rock bed of West Greenland, dated at 3.7 Gyr old. The graphite granules were analyzed by Minik Rosing, a native Greenlander at the Danish Lithosphere Centre (**Fig. 17.8A**). The graphite grains derive from microbial remains buried within sediment that subsequently metamorphosed, driving out the water content but leaving behind the telltale carbon. By contrast, carbonate rock of nonbiogenic origin from the same formation shows a $\delta^{13}C$ near zero.

Bacterial hopanoids. A different kind of biosignature is given by organic molecules specific to a particular lifeform. Certain organic molecules may last within rock for hundreds of millions of years. A particularly durable class of molecules consists of membrane lipids. Recall from Chapter 3 that bacterial cell membranes contain steroid-like molecules called hopanoids (**Fig. 17.8C**). A hopanoid consists of four or five fused rings of hydrocarbon with side groups that vary depending on the bacterial species. The hopanoid derivative 2-methylhopane is found in sedimentary rock of the Hamersley Basin of Western Australia, dated to 2.5 Gyr ago. This biosignature offers evidence that some kind of bacteria existed by the end of the Archaean eon.

A. Minik Rosing

BBC

B. ^{13}C isotope depletion

C. 2-Methylhopane

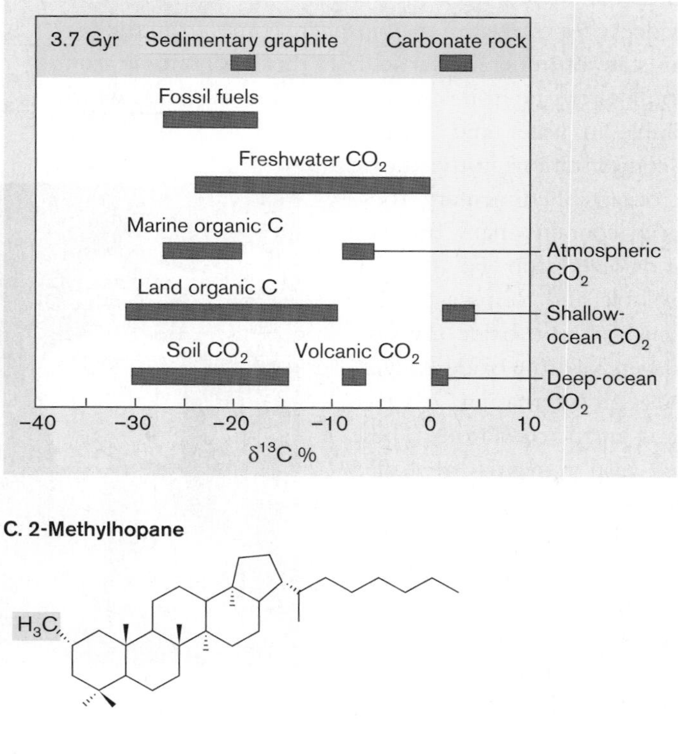
3.7 Gyr
Sedimentary graphite
Carbonate rock
Fossil fuels
Freshwater CO2
Marine organic C
Atmospheric CO2
Land organic C
Shallow-ocean CO2
Soil CO2
Volcanic CO2
Deep-ocean CO2
−40
−30
−20
−10
0
10
δ13C %
H3C

FIGURE 17.8 ■ Biosignatures of early life. A. Minik Rosing (left), in Greenland, shows the Isua rocks whose carbon isotope ratios indicate photosynthesis at 3.8 Gyr ago. **B.** ^{13}C isotope depletion (negative $\delta^{13}C$) occurs in biomass as a result of the Calvin cycle. Negative $\delta^{13}C$ is observed at 3.7 Gyr in sedimentary graphite, which may derive from sedimented phototrophs. Little or no isotope depletion is seen in carbonate rock, which has no biological origin. **C.** 2-Methylhopane, a biological signature of bacteria, is found in rock strata dated at 2.5 Gyr ago. The 2-methyl group is highlighted.

A.

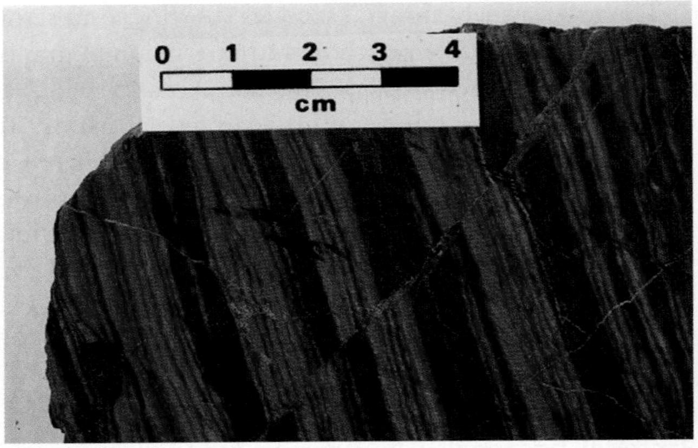
0 1 2 3 4
cm
KURT KONHAUSER

B.

KURT KONHAUSER

FIGURE 17.9 ▪ Banded iron formations. A. Banded iron formation in ancient sedimentary rock. Its main component is chert, a form of quartz (silicon dioxide, SiO_2) with layers colored red by iron oxide (Fe_2O_3). **B.** The BHP Billiton Iron Ore mine at Newman, Western Australia.

Banded Iron Formations Reveal Oxidation by O_2

An extraordinary event in the planet's history was the evolution of the first oxygenic phototrophs: cyanobacteria that split water to form O_2. The entry of O_2 into Earth's biosphere is often portrayed as a sudden event that would have been disastrous to microbial populations lacking defenses against its toxicity. In fact, geological evidence shows that oxygen arose gradually in the oceans, starting about 2 Gyr ago, and may have arisen and disappeared numerous times before reaching a high, steady-state level in our atmosphere. The mechanism of the oxygen fluctuation is unknown, but it must have occurred through cycles of aerobic and anaerobic microbial metabolism.

Evidence for oxygen in the biosphere comes from the oxidation state of minerals, particularly those containing iron. The bulk of crustal iron is in the reduced form (Fe^{2+}), which is soluble in water and reached high concentrations in the anoxic early oceans. Sedimentary rock, however, contains many fine layers of oxidized iron (Fe^{3+}), which is insoluble and forms a precipitate, such as iron oxide (Fe_2O_3). The layers of iron oxide suggest periods of alternating oxygen-rich and anoxic conditions. These layered iron minerals are called **banded iron formations (BIFs)** (**Fig. 17.9A**). A common form of banded iron consists of gray layers of silicon dioxide (SiO_2) alternating with layers colored red by iron oxides and iron oxyhydroxides [$FeO_x(OH)_y$]. Banded iron formations are widespread around the world and provide our major sources of iron ore (**Fig. 17.9B**).

Banded iron formations are often found in rock strata containing signs of past life such as ^{13}C depletion and other biomarkers. For example, the Isua formations (Greenland) and Hamersley formations (Western Australia), which both show ^{13}C depletion, also contain extensive banded iron. Calculations show that the layers of oxidized iron could result from biological metabolism involving iron oxidation. One possibility is that the iron was oxidized by chemolithotrophs, using molecular oxygen produced by cyanobacteria. The Archaean and early Proterozoic eons experienced fluctuating levels of molecular oxygen in the atmosphere. These fluctuations could have led to oscillating levels of iron oxide, thus producing bands in the sediment, as microbes used up all the oxygen.

Alternatively, Dianne Newman, at the California Institute of Technology, proposes that the iron oxides arose

A.

COURTESY OF DIANNE NEWMAN

B.

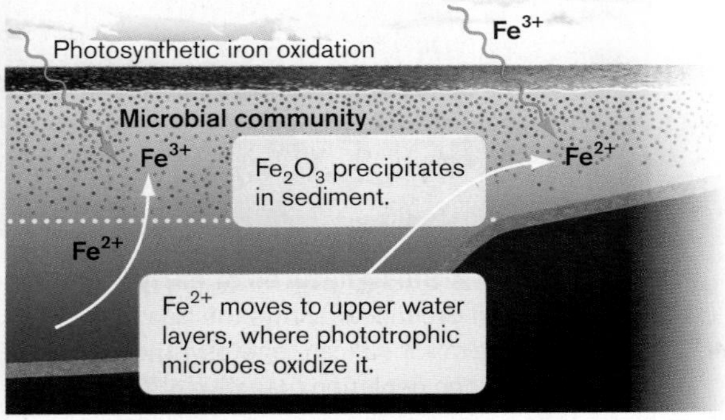
Photosynthetic iron oxidation
Fe^{3+}
Microbial community
Fe^{3+}
Fe_2O_3 precipitates in sediment.
Fe^{2+}
Fe^{2+}
Fe^{2+} moves to upper water layers, where phototrophic microbes oxidize it.

FIGURE 17.10 ▪ Iron phototrophy. A. Dianne Newman at the California Institute of Technology proposes that early iron phototrophs caused the iron oxide deposition generating banded iron formations. **B.** Photosynthetic oxidation of Fe^{2+} to Fe^{3+} may have generated sedimentary layers of Fe_2O_3 and $FeO_x(OH)_y$.

directly from anaerobic photosynthesis, in which the reduced iron served as the electron donor (**Fig. 17.10**). In iron phototrophy, light excites an electron from Fe^{2+}, oxidizing the ion to Fe^{3+}, while the excited electron cycles through an electron transport system to yield energy (discussed in Chapter 14). Newman discovered iron phototrophy in modern purple bacteria such as *Rhodopseudomonas palustris*. In the ancient Earth, photosynthetic oxidation of Fe^{2+} to Fe^{3+}, or "photoferrotrophy," could have occurred in cycles until the marine iron was all oxidized, generating sedimentary layers of iron oxides and iron oxyhydroxides. Today, photoferrotrophs are abundant in deep anoxic lakes that resemble our model of the anoxic Archaean ocean.

By 2.3 Gyr ago, the prevalence of oxidized iron and other minerals indicates the steady rise of oxygen from photosynthesis in Earth's atmosphere. All the dissolved Fe^{2+} from the ocean floor was oxidized, leaving the oceans in the iron-poor state that persists today. As the most efficient electron acceptor, molecular oxygen enabled the evolution of aerobic respiratory bacteria. Aerobic bacteria gave rise to mitochondria, which enabled the evolution of eukaryotes and, ultimately, multicellular organisms (**Fig. 17.11**).

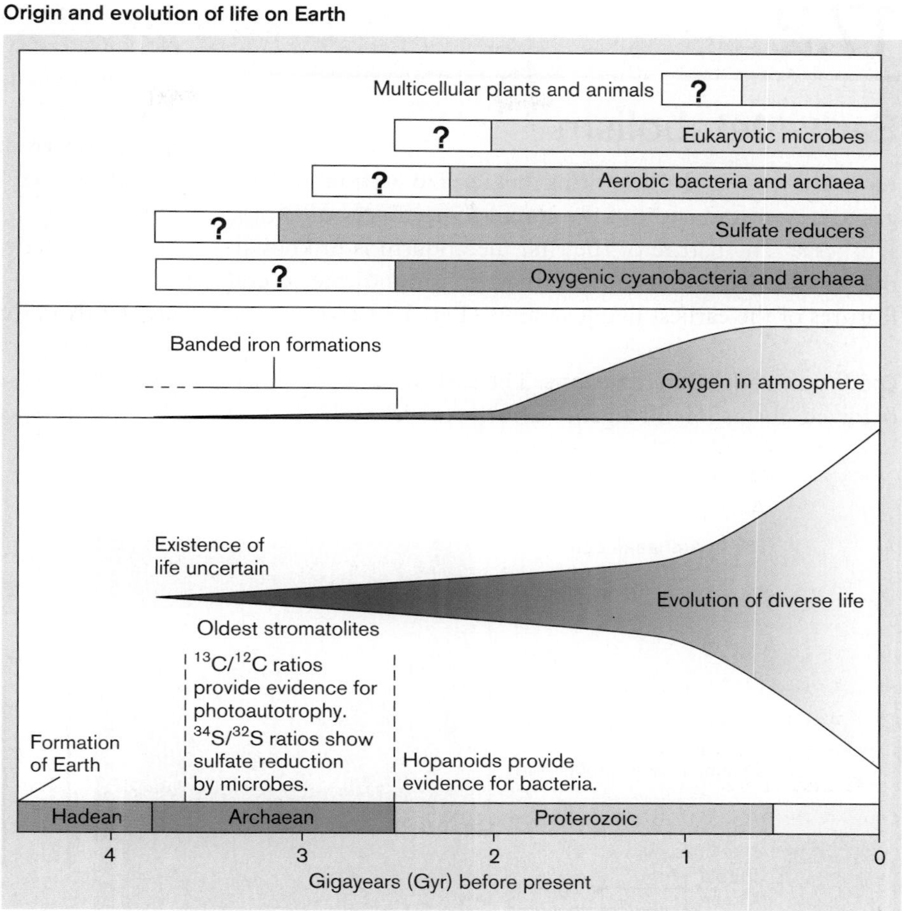
Origin and evolution of life on Earth
Multicellular plants and animals
?
Eukaryotic microbes
Aerobic bacteria and archaea
Sulfate reducers
Oxygenic cyanobacteria and archaea
Banded iron formations
Oxygen in atmosphere
Existence of life uncertain
Evolution of diverse life
Oldest stromatolites
$^{13}C/^{12}C$ ratios provide evidence for photoautotrophy.
$^{34}S/^{32}S$ ratios show sulfate reduction by microbes.
Formation of Earth
Hopanoids provide evidence for bacteria.
Hadean
Archaean
Proterozoic
4
3
2
1
0
Gigayears (Gyr) before present

FIGURE 17.11 ■ Proposed time line for the origin and evolution of life. The planet Earth formed during the Hadean eon (about 4.5 Gyr ago). The environment was largely reducing until cyanobacteria pumped O_2 into the atmosphere. When the O_2 level reached sufficient levels (about 0.6 Gyr ago), multicellular animals and plants evolved. Question marks designate periods when evidence for given life-forms is uncertain.

Remarkably, modern cells are still composed primarily of reduced molecules, highly reactive with oxygen—a relic of the time when our ancestral cells evolved in the absence of oxygen. The conditions under which such cells may have evolved can be simulated in the laboratory—conditions under which some of life's most common molecules, such as adenine and simple amino acids, form spontaneously (presented earlier in Special Topic 1.1). These early-Earth simulation experiments can never prove the actual conditions under which life began, but they can suggest testable models with intriguing implications.

To Summarize

- **Elements of life** were formed through nuclear reactions within stars that exploded into supernovas before the birth of our own Sun.
- **Reduced molecules** compose Earth's interior. Oxidized minerals are found only near the surface. Early Earth had no molecular oxygen (O_2).
- **Archaean rocks show evidence for life** based on fossil stromatolites, isotope ratios, and chemical biosignatures. Fossil stromatolites appear in chert formations formed 3.4 Gyr ago. Isotope ratios for carbon indicate photosynthesis at 3.7 Gyr ago and sulfate reduction at 3.47 Gyr ago. Bacterial hopanoids appear at 2.5 Gyr ago.
- **Microfossils of filamentous and colonial prokaryotes** date to 2.0 Gyr ago. At 1.2 Gyr ago, larger fossil cells resemble those of modern eukaryotes.
- **Banded iron formations** reflect the cyclic increase and decrease of oxygen produced by cyanobacteria and consumed through reaction with reduced iron. After all the ocean's iron was oxidized, oxygen increased gradually in the atmosphere.

17.2

Early Metabolism

How did the earliest life-forms metabolize without oxygen gas to respire and without the complex machinery of photosynthesis? The nature of the first metabolism is unknown, but geochemistry and modern metabolism indicate several features of the earliest living systems (**Fig. 17.12A**).

Oxidation-reduction reactions. The early oceans contained oxidized forms of nitrogen, sulfur, and iron that could interact with reduced minerals from the crust. For example, nitrate (NO_3^-) or sulfate (SO_4^{2-}) could be reduced by hydrogen gas to yield energy (hydrogenotrophy; discussed in Chapter 14). The oxidized molecules were generated by reactions driven by ultraviolet radiation, which penetrated the atmosphere in the absence of the ozone layer. Sulfur isotope ratios ($^{34}S/^{32}S$) suggest the growth of sulfate-reducing bacteria as early as 3.47 Gyr ago.

Light-driven ion pumps. A simple light-driven pump, such as the bacteriorhodopsin of haloarchaea (halophilic archaea), could have conducted the first kind of phototrophy. The absorption spectrum of bacteriorhodopsin matches the

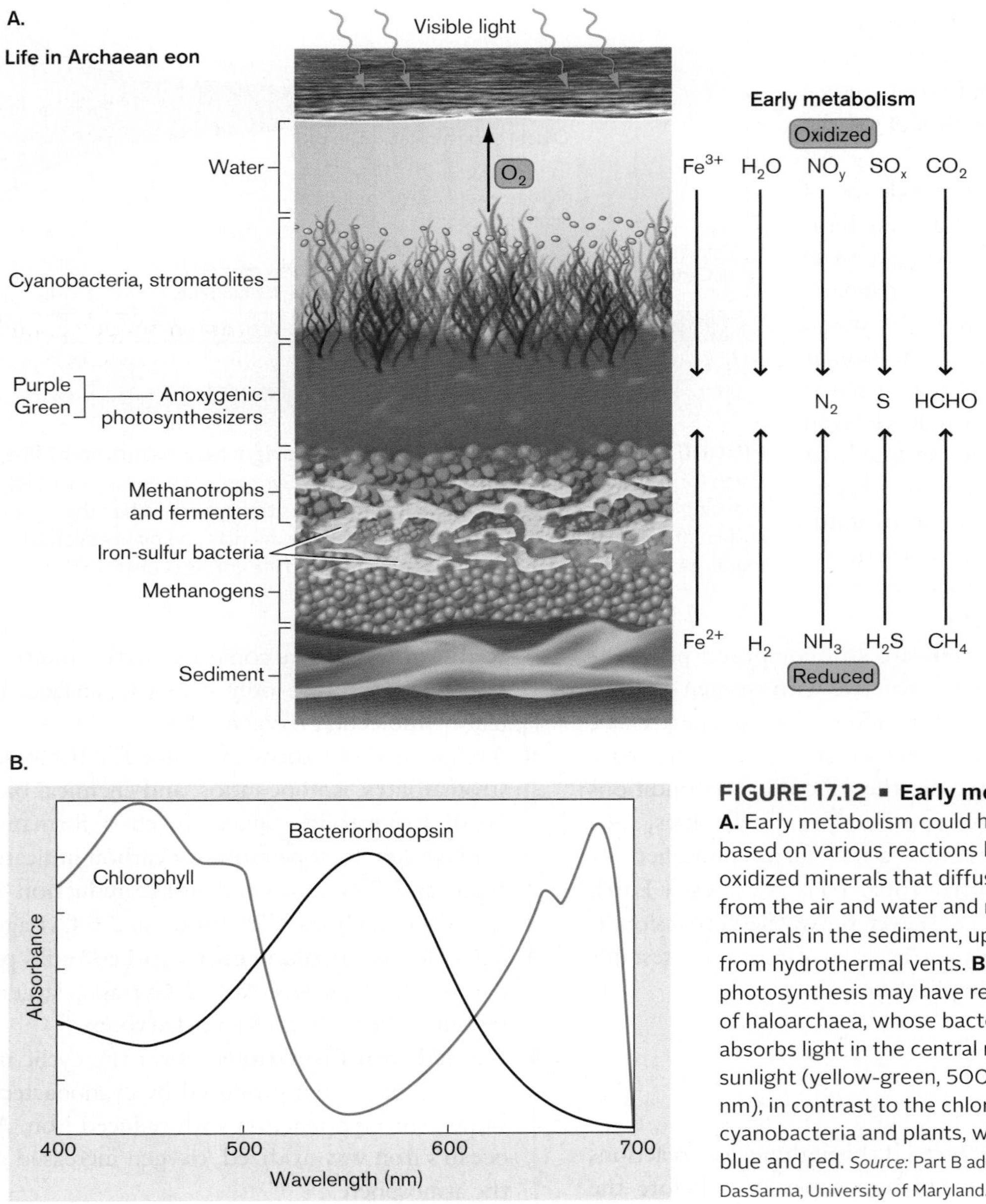
A.
Life in Archaean eon
Visible light
Water
O_2
Cyanobacteria, stromatolites
Purple
Green
Anoxygenic photosynthesizers
Methanotrophs and fermenters
Iron-sulfur bacteria
Methanogens
Sediment
Early metabolism
Oxidized
Fe^{3+} H_2O NO_y SO_x CO_2
N_2 S HCHO
Fe^{2+} H_2 NH_3 H_2S CH_4
Reduced
B.
Bacteriorhodopsin
Chlorophyll
Absorbance
400
500
600
700
Wavelength (nm)

FIGURE 17.12 ▪ Early metabolism. **A.** Early metabolism could have been based on various reactions between oxidized minerals that diffuse down from the air and water and reduced minerals in the sediment, upwelling from hydrothermal vents. **B.** Early photosynthesis may have resembled that of haloarchaea, whose bacteriorhodopsin absorbs light in the central range of sunlight (yellow-green, 500–600 nm), in contrast to the chlorophyll of cyanobacteria and plants, which absorbs blue and red. *Source:* Part B adapted from Shil DasSarma, University of Maryland.

peak of solar radiation reaching the upper layers of ocean (**Fig. 17.12B**), in contrast to the chlorophylls of cyanobacteria and plants, which absorb the outer ranges of blue and red. Perhaps cyanobacteria evolved in the presence of haloarchaea, filling the unexploited photochemical niches.

Methanogenesis. Climate models of early Earth suggest a methane atmosphere, produced by methanogenic archaea. Methanogenesis involves reaction of H_2 and CO_2 producing CH_4 and H_2O. Methanogens show highly divergent genomes—a finding that suggests early evolution of their common ancestor. Their biochemistry and evolution (discussed in Chapter 19) are consistent with proposed models of ancient life.

Models for the First Cells

Various models have been proposed to explain how the first living cells originated from nonliving materials, and how they replicated and evolved. Models for early life attempt to address the following questions: In what kind of environment did the first cells form? What kind of metabolism did the first cells use to generate energy? What was their hereditary material?

Three kinds of models attempt to explain how nonliving materials gave rise to a cell. The **prebiotic soup** model proposes that **abiotic** (nonliving) reactions involving simple reduced chemicals such as ammonia and methane could have assembled complex organic molecules of life. These complex macromolecules eventually acquired the apparatus needed for self-replication and membrane compartmentalization. **Metabolist models** propose that central components of intermediary metabolism, including using the TCA cycle to generate amino acids, arose from self-sustaining chemical reactions based on inorganic chemicals. These abiotic reactions then acquired self-replicating macromolecules and membranes. Finally, the **RNA world** is a model of early life in which RNA performed all the informational and catalytic roles of today's DNA and proteins. The concept of an RNA world draws upon the emerging sequences of genomes, which reveal thousands of catalytic and structural RNAs.

The prebiotic soup. In the mid-twentieth century, biochemists Aleksandr Oparin, at Moscow University, and Stanley Miller and Harold Urey, at the University of Chicago, showed that organic building blocks of life such as amino acids could arise abiotically out of a mixture of water and reduced chemicals, including CH_4, NH_3, and H_2 (**Fig. 17.13**). The mixture was subjected to an electrical discharge, similar to the lightning discharges that arise from volcanic eruptions, which would have been common in the late Hadean or early Archaean eon. The chemical reaction produced fundamental amino acids such as glycine and alanine. Similar experiments by Juan Oró, at the University of Houston, showed the formation of adenine by condensation of ammonia and hydrogen cyanide.

The same amino acids and nucleobases arising from early-Earth simulation are also found in meteorites, which are believed to retain the chemistry of the early solar system "frozen" in time. But unlike the chemical experiments, the meteorites show a remarkable predominance of left-handed mirror forms (L-enantiomers) of the amino acids—the same L form utilized in the protein synthesis of all life

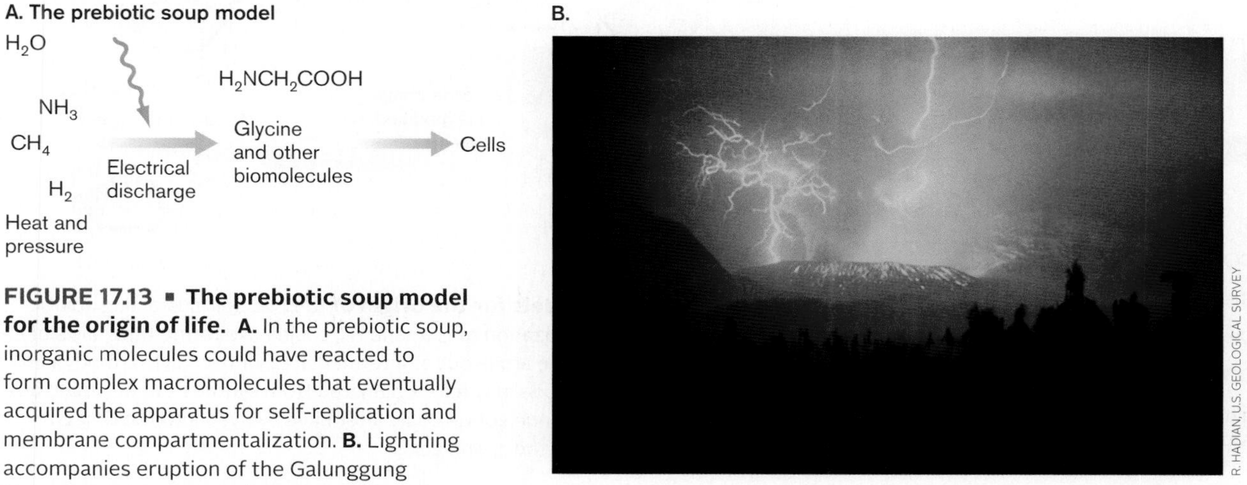
A. The prebiotic soup model
H₂O
NH₃
CH₄
H₂
Heat and pressure
Electrical discharge
H₂NCH₂COOH
Glycine and other biomolecules
Cells
B.
R. HADIAN, U.S. GEOLOGICAL SURVEY

FIGURE 17.13 ■ The prebiotic soup model for the origin of life. A. In the prebiotic soup, inorganic molecules could have reacted to form complex macromolecules that eventually acquired the apparatus for self-replication and membrane compartmentalization. **B.** Lightning accompanies eruption of the Galunggung Volcano in West Java, Indonesia, in 1982. The first biomolecules may have formed as a result of lightning triggered by volcanic eruption.

on Earth. Some researchers hypothesize that organic compounds formed in outer space were restricted to the L form by an unknown process and then "seeded" the propagation of L-form compounds in Earth's prebiotic soup.

The original model conditions for the prebiotic soup assumed that molecules in the early Archaean ocean were largely reduced, with little or no oxygen present in the atmosphere. More recent geochemical evidence suggests that the early ocean actually included oxidized forms of nitrogen, sulfur, and iron that arose through reactions driven by ultraviolet radiation, which penetrated the atmosphere in the absence of the ozone layer. These oxidized minerals could have reacted with the reduced crustal minerals, releasing energy to drive production of more complex biomolecules.

Forming the first cell, or "proto-cell," must have required enclosing the first biochemical reactants within a membrane-like compartment. Jack Szostak and colleagues at Harvard Medical School investigate how such compartments can arise spontaneously from fatty acid glycerol esters. The fatty acid derivatives are "amphipathic"; that is, they possess both hydrophobic portions that associate together and hydrophilic portions that associate with water. In water, the fatty acid derivatives collect in "micelles"—small, round aggregates in which the hydrophobic portions associate in the interior and the hydrophilic portions associate with water. Under certain conditions, micelles can aggregate to form hollow vesicles of membrane. Szostak showed how the vesicles can take up molecules such as RNA, suggesting a primitive cell-like form. These spontaneous processes of membrane formation offer models for how the first living cells may have arisen.

Metabolist models. Other models attempt to explain the origin of biosynthesis on the basis of CO_2 fixation, the fundamental metabolism of all life today. Proponents of a metabolist model, including Harold Morowitz, at George Mason University, and Günter Wächterhäuser, at University of Regensburg, propose that CO_2-based metabolism originated through self-sustaining reactions (**Fig. 17.14A**). Simulation experiments suggest that such premetabolic reactions could have been catalyzed by metal sulfides prevalent in the early ocean. For example, abiotic polymerization of CO_2 into TCA cycle intermediates may be catalyzed by FeS.

How did simple inorganic reactions lead to biochemical cycles involving more complex biopolymers, such as nucleic acids and proteins? A major challenge is to explain the genetic code by which nucleic acid codons are assigned to unique amino acids. The genetic code may have arisen from an earlier pretranslational mechanism of amino acid biosynthesis (**Fig. 17.14B**). In this mechanism, proposed by Morowitz and colleagues, each amino acid was originally synthesized from a TCA cycle acid complexed to a dinucleotide. The dinucleotide later evolved into the first two nucleotides of the codon specifying the amino acid. The proposed link between a specific dinucleotide and each amino acid would explain certain features of the genetic code, such as the fact that most amino acids specified by

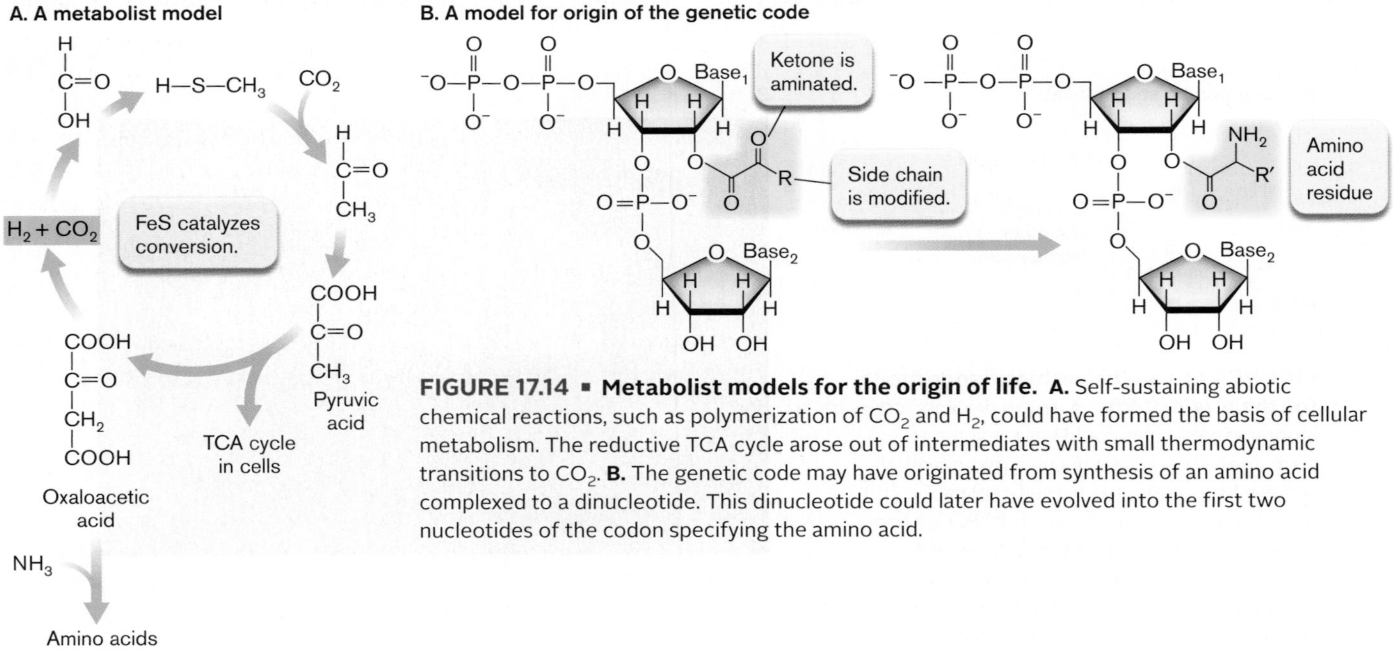
A. A metabolist model
H—S—CH3
CO2
H2 + CO2
FeS catalyzes conversion.
Pyruvic acid
TCA cycle in cells
Oxaloacetic acid
NH3
Amino acids
B. A model for origin of the genetic code
Base1
Ketone is aminated.
Side chain is modified.
Base2
Amino acid residue

FIGURE 17.14 ▪ Metabolist models for the origin of life. **A.** Self-sustaining abiotic chemical reactions, such as polymerization of CO_2 and H_2, could have formed the basis of cellular metabolism. The reductive TCA cycle arose out of intermediates with small thermodynamic transitions to CO_2. **B.** The genetic code may have originated from synthesis of an amino acid complexed to a dinucleotide. This dinucleotide could later have evolved into the first two nucleotides of the codon specifying the amino acid.

a codon starting with the same nucleotide are synthesized from the same TCA cycle acid (discussed in Chapter 15).

The RNA world. Neither the prebiotic soup model nor the metabolist models account for the evolution of macromolecules that encode complex information, such as nucleic acids and proteins. A candidate for life's first "informational molecule" is RNA. RNA is a relatively simple biomolecule, with only 4 different "letters," compared to the 20 standard amino acids of proteins. Its purine base adenine arises spontaneously from ammonia and carbon dioxide under conditions believed to resemble those of the Archaean eon. Its ribose sugar is a fundamental building block of living cells, with key roles in numerous biochemical pathways, such as the Calvin cycle. Most exciting, in 2009, John Sutherland and co-workers discovered an organic reaction pathway that generates a sugar-base ribonucleotide in the absence of any enzyme. This discovery suggests a way that the first RNA molecules could have formed before there were living cells.

For several reasons, RNA is a better candidate than DNA for the earliest information molecule. Compared to DNA, RNA requires less energy to form and degrade. RNA's pyrimidine base uracil is formed early by biochemical pathways; only later is it converted to the thymine used by DNA.

Most important, RNA molecules have been shown to possess catalytic properties analogous to those of proteins. Catalytic RNA molecules are called **ribozymes**. The first ribozyme, discovered by Nobel laureate Thomas Cech in the protist *Tetrahymena,* can splice introns in mRNA. Other ribozymes actually catalyze synthesis of complementary strands of RNA, suggesting a model for early replication of RNA chromosomes. The most elaborate example of catalytic RNA is found in the ribosome. In the ribosome, X-ray crystallography reveals that the key steps of protein synthesis, such as peptide bond formation, are actually catalyzed by the RNA components, not proteins (discussed in Chapter 8). The ribosomal proteins possess relatively little catalytic function; their main role seems to be protection and structural support of the RNA.

Could RNA molecules have composed the earliest cells? In 2009, Tracey Lincoln and Gerald Joyce at the Scripps Research Institute devised a system in which two RNA ribozymes catalyze each other's synthesis. The ribozyme model system suggests how, in the earliest cells, RNA might have fulfilled the key functions that are today filled by DNA and proteins, including information storage, replication, and catalysis (**Fig. 17.15**). This model is known as the "RNA world."

The prominent function of RNA in the ribosome, one of life's most ancient and conserved molecular machines, suggests a model for the transition from an RNA world to the modern cell. In the ribosome, the actual steps of catalysis, such as forming the peptide bond, are conducted by RNA subunits acting as ribozymes. The ribozymes are stabilized by protein subunits. Thomas Cech, Sidney Altman, and colleagues propose that the earliest RNA components of cells evolved by acquiring proteins to enhance stability. The proteins helped prevent the tendency of RNA to hydrolyze (come apart in reaction with water). As cells evolved, their peptide components increased through natural selection, and the RNA subunits may have shrunk by reductive evolution (the evolutionary loss of unneeded parts). A few complexes, such as the ribosome, still maintain their ribozymes; for others, perhaps all that remains are one or two nucleotides. Dinucleotide cofactors such as NADH persist in enzymes today, perhaps representing vestigial remnants of the RNA world. For more on RNA evolution, and surprising applications for medicine, see **eTopic 17.1**.

The RNA-world model explains the central role of RNA throughout the history of living cells. Yet the role of RNA offers little clue as to the origins of cell compartmentalization and metabolism. No single model of life's origin yet addresses all the requirements for a living cell—metabolism, membrane compartmentalization, and hereditary material. Each model does, however, offer important insights into the evolutionary history and mechanisms of life today.

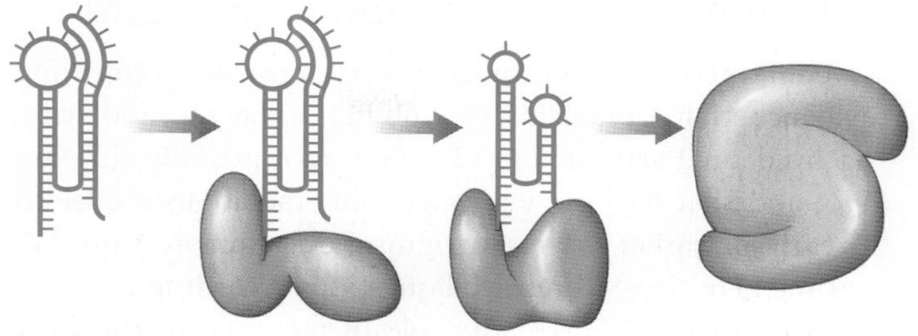
Ribozyme
Ribonucleoprotein enzyme
Protein enzyme with nucleotide coenzyme

FIGURE 17.15 ■ From the RNA world to proteins. Earliest cells may have been composed of RNA enzymes (ribozymes). As the RNA cells evolved, ribozymes acquired protein subunits that eventually assumed most of their catalytic functions. Remnants of the original RNA may persist as nucleotide cofactors such as NADH.

Thought Question

17.2 Outline the strengths and limitations of each model of the origin of living cells. Which aspects of living cells does each model explain?

Unresolved Questions about Early Life

Overall, geology and biochemistry provide compelling evidence that organisms resembling today's cyanobacteria lived on Earth at least 2.5 Gyr ago, possibly 3.7 Gyr ago, and that bacteria with anaerobic metabolism evolved as early or earlier. Many intriguing questions remain. We outline here three unresolved questions regarding the temperature of early Earth, the role of methane in the early atmosphere, and the actual source of Earth's first cells.

Thermophile or psychrophile? The apparent existence of life so soon after Earth cooled suggests a thermophilic origin. Thermophily is supported by the fact that in the domains Bacteria and Archaea, the deepest-branching species (that is, species that diverged the earliest from others in the domain) are thermophiles. Such organisms could have thrived at hydrothermal vents, which offer a continuous supply of H_2S and carbonates.

On the other hand, after meteoric bombardment abated, early Earth should have become glacially cold. In the Archaean, solar radiation was 20%–30% less intense than it is today, and the thin CO_2 atmosphere was insufficient to increase the temperature by a greenhouse effect. A colder habitat would support psychrophiles. Psychrophiles might have had an advantage in an RNA world, given the thermal instability of RNA compared to DNA and proteins.

A world of methane? Among Earth's earliest life-forms were methanogens (methane producers). Methanogens are one of the most widely divergent groups of organisms and persist today in environments ranging from anaerobic sediment to the human intestine. Methanogenesis requires only carbon dioxide and hydrogen gas, which would have been plentiful in early anoxic sediment. The production of methane, an extremely potent greenhouse gas, could have greatly increased Earth's temperature during the Archaean eon. Thus, methanogenesis could explain how Earth escaped the permanent freeze of Mars. Overheating would have been halted when the oxygen gas produced by cyanobacteria enabled growth of methanotrophs, bacteria that oxidize methane. The decline of methane and the rise of CO_2 then would have brought about relative thermal stability.

The debate over the temperature and climate of early Earth has interesting implications for Earth today, when we again face the prospect of massive global climate change. Human agriculture favors explosive growth of methanogens, which threaten to accelerate global warming faster than the biosphere can moderate it. Understanding the climate of early Earth may help us better understand and manage our own climate, discussed in Chapter 22.

Origin on Earth or elsewhere? Emerging evidence from fossils and geochemistry has inexorably pushed back the earliest known dates for several kinds of metabolism closer to 3.7 Gyr ago (see **Fig. 17.11**). This implies that as soon as Earth cooled to a temperature suitable for life, all the fundamental components of cells evolved almost immediately. How could life, with all its diverse kinds of metabolism, have arisen so quickly?

The idea that life-forms originated elsewhere and "seeded" life on Earth is called **panspermia**. Theories of panspermia remain highly speculative. One hypothesis is that microbial cells originated on Mars and were then carried to Earth on meteorites. As the solar system formed, Mars would have cooled sooner than Earth, and as a smaller planet, its weaker gravity would have generated less bombardment by meteorites. Martian rocks ejected into space by meteor impact have reached Earth, and calculations based on simulated space habitats show that microbes could survive such a journey. A Martian origin, however, gains us only about half a billion years; it does not really explain the origin of life's complexity and diversity. Did life-forms come from still farther away, perhaps borne on interstellar dust from some other solar system?

A more likely explanation is that early evolution on Earth occurred much faster than later evolution. For example, RNA viruses such as influenza and HIV mutate and evolve much faster than modern cells (discussed in Chapter 11). If early cells with RNA genomes mutated as fast as viruses, they would have evolved and diverged faster than cellular organisms that we know today.

Thought Question

17.3 Suppose that a NASA rover were to discover living organisms on Mars. How might such a find shed light on the origin and evolution of life on Earth?

To Summarize

- **Early metabolism** involved anaerobic oxidation-reduction reactions. Likely forms of early metabolism include sulfate respiration, light-driven ion pumps, iron phototrophy, and methanogenesis.
- **Prebiotic soup models** propose that the fundamental biochemicals of life arose spontaneously through condensation of reduced inorganic molecules.
- **Metabolist models** propose that components of intermediary metabolism arose from self-sustaining chemical reactions that connected nucleotides with amino acids, forming the basis of the genetic code.

- **The RNA-world model** proposes that in the first cells, RNA performed all the informational and catalytic roles of today's DNA and proteins.
- **Thermophile or psychrophile?** Classic models of early life assume thermophily, but Earth may actually have been cold when the first cells originated.
- **A world of methane?** If the first cells were methanogens, methane production could have led to the first greenhouse effect, warming the Earth and enabling evolution of other kinds of life.
- **Origin on Earth or elsewhere?** Isotope ratios suggest the presence of complex metabolism by 3.7 Gyr ago, shortly after Earth cooled (3.9 Gyr ago). Simpler cells existing before 3.9 Gyr ago may have evolved much faster than life today. A more speculative possibility is that life first evolved on another planet.

17.3

Microbial Phylogeny and Gene Transfer

The unifying assumption of modern biology is genetic relatedness, or molecular phylogeny. Phylogeny generates a series of branching groups of related organisms called **clades**. Each clade is a **monophyletic group**—that is, a group of species that share a common ancestor not shared by any species outside the clade. Each monophyletic group then branches into smaller monophyletic groups and ultimately species. The full description of branching divergence of a species is called its **phylogeny**. This description of phylogeny however must be modified to include the contributions of **horizontal gene transfer** (discussed next).

Divergence through Mutation and Natural Selection

Populations of organisms diverge from each other through several fundamental mechanisms of evolution. These include:

- **Random mutation.** DNA sequences change through rare mistakes (in bacteria and archaea, typically one out of a million base pairs) as the chromosome replicates. Replication errors result in mutation. Most mutations are neutral; that is, they have no effect on gene function.
- **Natural selection and adaptation.** In a given environment, natural selection favors organisms that produce greater numbers of offspring in that environment (discussed in Section 17.4). Genes encoding traits under selection pressure may show mutation frequencies much higher or lower than those generated by the random mutation rate. Natural selection enables a population to adapt to a changing environment.
- **Reductive evolution (degenerative evolution).** In the absence of selection for a trait, the genes encoding the trait accumulate mutations without affecting the organism's reproductive success. Because mutations that decrease function are more common than mutations that improve function, accumulating mutations without selection pressure leads to decline and ultimately loss of the trait. The loss or mutation of DNA encoding unselected traits is called **reductive evolution** (or **degenerative evolution**).

Random mutations with neutral effects that are not subject to selection tend to accumulate at a steady rate over generations because the error rate of the DNA replication machinery stays about the same. The resulting "genetic drift" causes sequences in separate populations to diverge over time. The constancy of mutation rate (within limits) provides a tool for us to measure the time of divergence of species based on their DNA sequences.

Thought Question

17.4 What kinds of DNA sequence changes have no effect on gene function?

Molecular Clocks Are Based on Mutation Rate

An important conceptual advance of the twentieth century was that information contained in a macromolecule such as protein or DNA could measure the history of a species. The temporal information contained in a macromolecular sequence is called a **molecular clock**. Molecular clocks have revolutionized our understanding of the emergence of all living organisms, including human beings. The first molecular clocks, based on protein sequencing and DNA hybridization, were developed in the 1960s. Subsequently, the sequences of ribosomal RNA (rRNA) were used by Carl Woese to reveal the divergence of three domains of life. The rRNA sequence is particularly useful because ribosome structure and function are highly similar across all organisms. Genomic sequences now offer many other genes to measure divergence at different levels of classification.

A molecular clock is based on the acquisition of new random mutations in each round of DNA replication. In

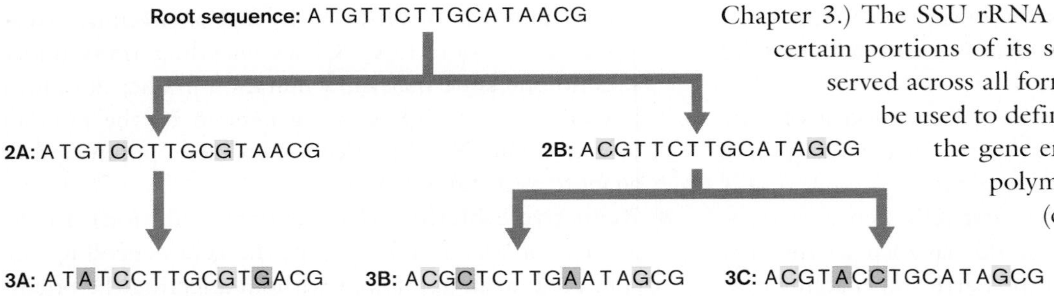
Root sequence: ATGTTCTTGCATAACG
2A: ATGTCCTTGCGTAACG
2B: ACGTTCTTGCATAGCG
3A: ATATCCTTGCGTGACG
3B: ACGCTCTTGAATAGCG
3C: ACGTACCTGCATAGCG

FIGURE 17.16 ▪ The molecular clock. As genetic molecules reproduce, the number of mutations (shaded in yellow and green) accumulated at random is proportional to the number of generations and thus the time since divergence. In each sequence designation (for example, "2A"), the number indicates the generation and the letter identifies a specific strain.

Figure 17.16, each offspring in generation 2 acquires two new mutations; their sequences now differ by 25%. In the next generation, each individual propagates the earlier mutant sequences while acquiring two more random mutations. Strain 3A now differs by 50% from strains 3B and 3C, which differ by 25% from each other. (Actual mutation frequencies, of course, are much lower—about one base per million per generation.) The key assumption is that the chromosomes of offspring acquire a consistent number of random mutations from their parents, and therefore the number of sequence differences between two species should be proportional to the time of divergence between them.

In practice, the molecular clock works best for a particular gene sequence with the following features:

- **The gene has the same function across all species compared.** That is, all versions are orthologous; they have not evolved to serve different functions. Functional difference may lead to different rates of change.
- **The generation time is the same for all species compared.** Shorter generation times (more frequent reproductive cycles) lead to overestimates of the overall time of divergence because of the increased opportunity for DNA mutation.
- **The average mutation rate remains constant among species and across generations.** If different species mutate at different rates, species with more rapid rates of mutation will appear to have diverged over a longer time than is actually the case.

Genes that show the most consistent measures of evolutionary time encode components of the transcription and translation apparatus, such as ribosomal RNA and proteins, tRNA, and RNA polymerase. The most widely used molecular clock is the gene encoding the **small-subunit rRNA (SSU rRNA)**. The SSU rRNA is also known by its sedimentation coefficient: 16S rRNA (bacteria) or 18S rRNA (eukaryotes). (Sedimentation coefficients are discussed in Chapter 3.) The SSU rRNA is particularly useful because certain portions of its sequence are remarkably conserved across all forms of life. These portions can be used to define primers to amplify DNA of the gene encoding the rRNA, using the polymerase chain reaction, or PCR (discussed in Section 7.6). The gene sequence lying between a pair of primers will show greater variation, allowing distinction between different clades. PCR can be used to amplify genes even from a mixture of uncultured organisms.

Use of a molecular clock requires the alignment of homologous sequences in divergent species or strains (**Fig. 17.17**). Alignment is the correlation of portions of two gene sequences that diverged from a common ancestral sequence (homologous sequence). The process requires assumptions and decisions about base substitutions, as well as insertions and deletions. For example, in **Figure 17.17** the best alignment requires us to assume that three bases were lost from the fourth sequence (or else inserted into the ancestor of the other sequences). The relative differences among the sequences can be used to propose a tree

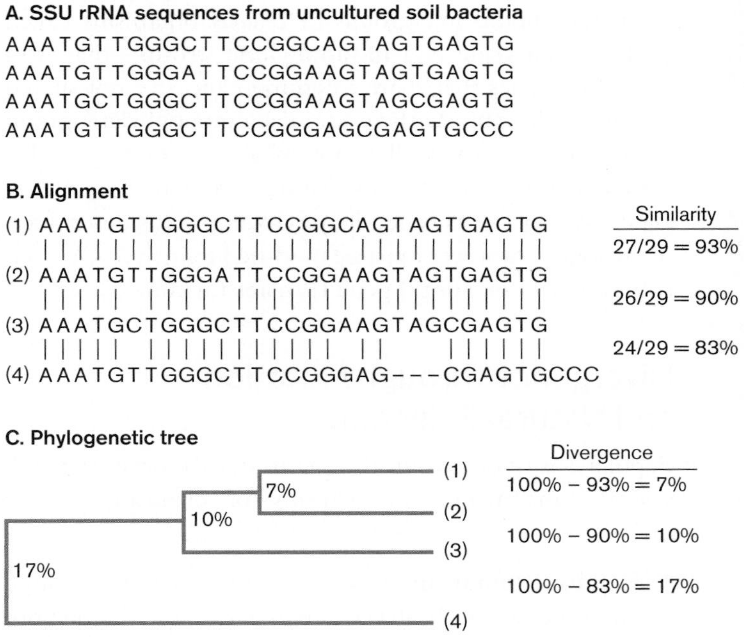
A. SSU rRNA sequences from uncultured soil bacteria
AAATGTTGGGCTTCCGGCAGTAGTGAGTG
AAATGTTGGGATTCCGGAAGTAGTGAGTG
AAATGCTGGGCTTCCGGAAGTAGCGAGTG
AAATGTTGGGCTTCCGGGAGCGAGTGCCC
B. Alignment
Similarity
(1) AAATGTTGGGCTTCCGGCAGTAGTGAGTG
27/29 = 93%
(2) AAATGTTGGGATTCCGGAAGTAGTGAGTG
26/29 = 90%
(3) AAATGCTGGGCTTCCGGAAGTAGCGAGTG
24/29 = 83%
(4) AAATGTTGGGCTTCCGGGAG---CGAGTGCCC
C. Phylogenetic tree
7%
10%
17%
(1)
(2)
(3)
(4)
Divergence
100% – 93% = 7%
100% – 90% = 10%
100% – 83% = 17%

FIGURE 17.17 ▪ DNA sequence alignment. **A.** SSU rRNA sequences from different organisms can be aligned at homologous regions. **B.** The best alignment is that which minimizes mismatches. **C.** A possible phylogenetic tree of divergence of the four sequences.

of divergence (**Fig. 17.17C**). In the tree, the length of each branch is proportional to the number of differences between two sequences. The number of differences, or divergence, is given by 100% minus the percent similarity between the aligned sequences.

In practice, a much larger amount of sequence with multiple differences is needed to calculate a phylogenetic tree. All trees are based on probability, with ambiguities depending on the assumptions of how sequences change. The data are calculated by computer programs, which may yield different results depending on their assumptions. Standard assumptions include the following:

- **Minimum number of changes.** The best alignment between two sequences is that which assumes the smallest number of mutational changes.
- **Functional sequences change more slowly.** Sequences that encode essential catalytic portions of the gene product are maintained by selection pressure and change more slowly than portions of the molecule without essential function.
- **Third-base codon positions show more random change.** In a protein-coding gene, many codons have multiple anticodons that differ at the third base, so third-base changes are least likely to change the amino acid. Thus, third-base nucleotides offer the best molecular clock information.

Phylogenetic Trees

Once homologous sequences are aligned, the frequency of differences between them can be used to generate a **phylogenetic tree** that estimates the relative amounts of evolutionary divergence between the sequences. If divergence rate over time, or mutation rate, is the same for all sequences compared, then divergence data can be used to infer the length of time since two species shared a common ancestor.

A phylogenetic tree is a model. All phylogenetic trees depend on complex mathematical analysis to measure degrees of divergence and propose a tree that most probably connects present sequences with their common ancestor. Because we can never know the exact tree with certainty, different types of calculations may lead to slightly different results. Two common approaches are:

- **Maximum parsimony.** Evolutionary distances are computed for all pairs of taxa based on the numbers of nucleotide or amino acid substitutions between them. A proposed common ancestor, or "ancestral state," is reconstructed. All possible trees comparing relative time of divergence from the common ancestor are computed. The "best fit" tree is defined as the one requiring the fewest mutations to fit the data (that is, the one that is most "parsimonious"). A limitation of parsimony reconstruction is that more than one tree may produce results consistent with the data.
- **Maximum likelihood.** For each possible tree, one calculates the likelihood (probability) that such a tree would have produced the observed DNA sequences. The probability of given mutations is based on complex statistical calculations. Maximum-likelihood methods require large amounts of computation but obtain the most information from the data, usually generating one tree or a small set of probable trees.

A portion of a computed phylogenetic tree is shown in **Figure 17.18**. The sequence data were obtained from PCR-amplified sequences of SSU rRNA from isolates in Obsidian Pool, a thermal spring at Yellowstone National Park. The sequences were analyzed by Susan Barns and

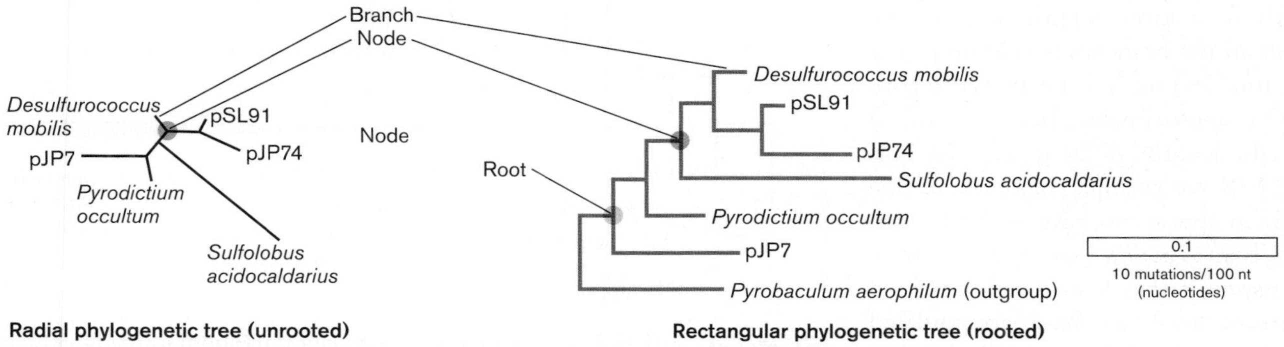
Branch
Node
Desulfurococcus mobilis
pSL91
pJP74
pJP7
Pyrodictium occultum
Sulfolobus acidocaldarius
Node
Root
Pyrobaculum aerophilum (outgroup)
0.1
10 mutations/100 nt (nucleotides)
Radial phylogenetic tree (unrooted)
Rectangular phylogenetic tree (rooted)

FIGURE 17.18 ■ Phylogenetic trees: rooted and unrooted. Comparison of rRNA sequences was used to generate a phylogenetic tree for thermophiles isolated from Obsidian Pool in Yellowstone National Park. **Left:** The radial phylogenetic tree shows the degrees of relatedness among taxa based on the number of molecular substitutions on each branch (see scale bar). **Right:** The rectangular tree shows the same divergence distances, lined up in parallel. This rectangular tree includes a root, a distance to an outgroup organism. A rooted tree indicates the position of a common ancestor. *Source:* Susan M. Barnes et al. 1996 *PNAS* **93**: 9188.

colleagues at Indiana University in the laboratory of Norman Pace, a pioneering investigator of extreme habitats. Note that some of the Obsidian Pool samples are known species, such as the thermophile *Pyrodictium occultum,* whereas others are uncharacterized organisms given alphanumeric designations, their existence known only by the rRNA sequences reported here. In fact, any microbial habitat surveyed by PCR, including a "soap scum" biofilm on a shower curtain, will yield previously unknown species (**Special Topic 17.1**).

The Obsidian Pool rRNA sequences were aligned as shown in **Figure 17.17**, and divergence distances were analyzed by maximum likelihood. In each tree, the total distance (in terms of base-substitution frequency) between any two sequences is approximated by the distance from each sequence and its branch point, or **node** (**Fig. 17.18**). Phylogenetic trees can be drawn in different ways. The tree at left in **Figure 17.18** is a radial tree, in which branches indicate distance outward from their nodes. The tree at right shows the same divergence distances from nodes, drawn as a rectangular tree. In a rectangular tree, all branches run in parallel. A rectangular tree helps compare divergent branches, although it takes up more space than a radial tree, especially when large numbers of organisms are compared.

The tree at left in **Figure 17.18** is "unrooted"; that is, it shows only the relative distances between different sequences, without indicating which of these diverged earliest from the common ancestor. To obtain the **root of a tree** (highlighted yellow in the figure), a sequence must be compared from well outside the selected cluster of organisms. The root is the position from which the most different sequence (the "outgroup," in this case *Pyrobaculum aerophilum*) first diverged from the rest.

If the length of each branch corresponds to a given length of time, we expect all the branches to add up to the same total length. In some trees this condition is approximated, but in microbial trees the lengths differ greatly. In **Figure 17.18**, for example, *Sulfolobus acidocaldarius* appears to have evolved much more than *Desulfurococcus fermentans.* The reason is that *S. acidocaldarius* and its recent ancestors have accumulated mutations faster. In every tree, some lineages accumulate mutations faster than others. The difference in rate arises from differences in mutation rate and from differences in generation time between organisms whose sequences are compared. Thus, our molecular clocks are inevitably distorted, especially for distantly diverged organisms with disparate mutation rates.

A phylogenetic tree can compare any set of organisms, even from multiple habitats. **Figure 17.19** shows the divergence of selected Gammaproteobacteria, a clade of Gram-negative bacteria (presented in Chapter 18). The tree is based on a set of protein sequences of "housekeeping" genes—that is, genes encoding functions essential for all cells, and usually transmitted vertically (parent to offspring). *Escherichia, Shigella,* and *Salmonella* are closely related genera of intestinal bacteria, including pathogenic strains and commensal strains (normally present). The genus *Klebsiella* includes species that cause pneumonia. *Photorhabdus luminescens* is a nematode bacterium that helps its host parasitize insects. *Yersinia pestis* causes bubonic plague. *Photobacterium profundum* is a marine barophile (high pressure) growing in a deep-sea trench. The tree is rooted by *Shewanella,* a genus of metal reducers used to construct fuel cells. Overall, we find that bacteria of bewildering diversity, from human pathogens to deep-ocean dwellers, diverged relatively recently from a common ancestor.

One group, the genus *Buchnera,* appears to have diverged much faster than the other bacteria. *Buchnera* species are intracellular endosymbionts of aphids. The intracellular symbionts have undergone reductive evolution, losing many functions supplied by their host. Intracellular endosymbionts typically mutate much faster than free-living

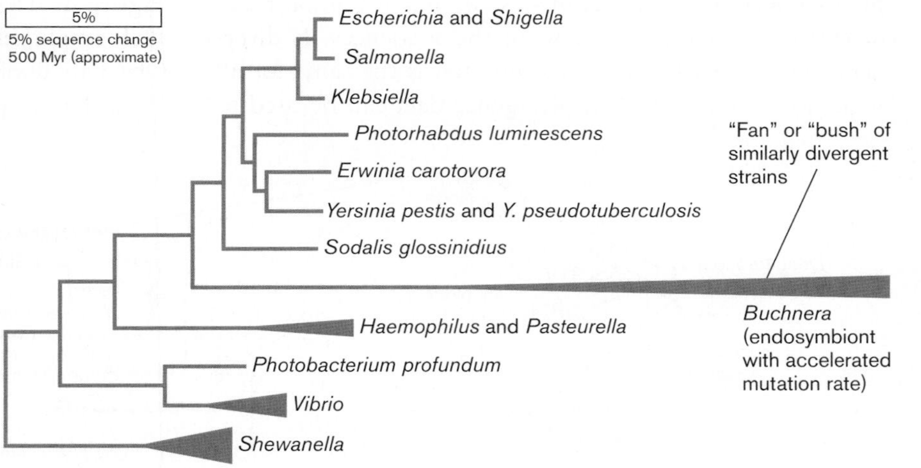
5%
5% sequence change
500 Myr (approximate)
Escherichia and Shigella
Salmonella
Klebsiella
Photorhabdus luminescens
Erwinia carotovora
Yersinia pestis and Y. pseudotuberculosis
Sodalis glossinidius
"Fan" or "bush" of similarly divergent strains
Haemophilus and Pasteurella
Buchnera (endosymbiont with accelerated mutation rate)
Photobacterium profundum
Vibrio
Shewanella

FIGURE 17.19 ■ Phylogeny of selected Gammaproteobacteria. The phylogenetic tree was derived from concatenated sequences of highly conserved "housekeeping" proteins. The scale bar corresponds to 5% amino acid sequence divergence. *Source:* Modified from Morgan Price et al. 2008. *Genome Biol.* **9**:R4 and from Fabia Battististuzzi et al. 2004. *BMC Evol. Biol.* **4**:44.

bacteria. In addition, they undergo intense selection pressure for adaptation to their obligate host. Endosymbiosis is discussed further in Section 17.6.

Buchnera and several other lineages show a widening branch. The widening branch indicates a "fan" or "bush" of strains equally distant from a common node (branch point). The cause of "bushy" branches is debated; some researchers argue that all branches become bushy once enough strains have been sequenced.

How do we calibrate a phylogenetic tree—that is, relate the number of mutations to the time since divergence? We need an external measure of time. A tree can be calibrated if some kind of fossil evidence or geological record exists to confirm at least one branch point of the tree. But for microbes, such fossil calibrations remain speculative. One convincing method of calibration is to correlate the divergence of microbial species growing only within particular host species with the divergences of their hosts, based on the host fossil record. For example, the exceedingly rapid divergence of *Buchnera* species (**Fig. 17.19**) can be calibrated on the basis of their host insects. Such calculations, however, reveal vastly divergent rates of evolutionary change in different bacterial taxa.

Could a life-form mutate even faster than *Buchnera*? Some RNA viruses, such as HIV, mutate so fast that they form multiple strains within one host (discussed in Chapter 11).

Thought Question

17.5 What are the major sources of error in constructing phylogenetic trees?

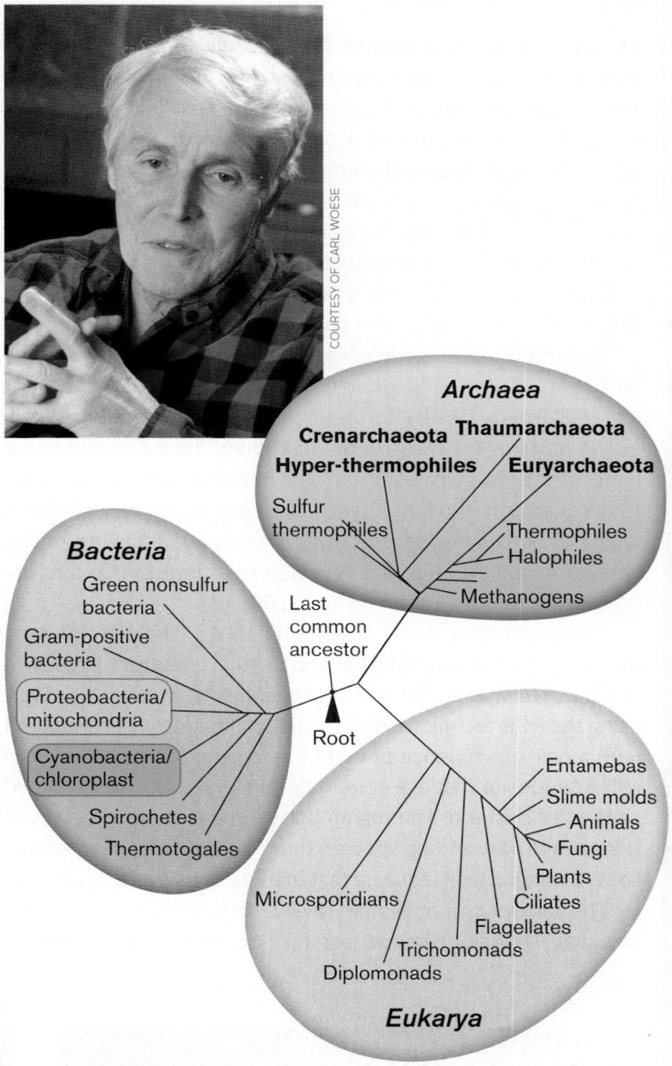
COURTESY OF CARL WOESE
Archaea
Crenarchaeota
Thaumarchaeota
Hyper-thermophiles
Euryarchaeota
Sulfur thermophiles
Thermophiles
Halophiles
Methanogens
Bacteria
Green nonsulfur bacteria
Gram-positive bacteria
Proteobacteria/ mitochondria
Cyanobacteria/ chloroplast
Spirochetes
Thermotogales
Last common ancestor
Root
Entamebas
Slime molds
Animals
Fungi
Plants
Ciliates
Flagellates
Trichomonads
Diplomonads
Microsporidians
Eukarya

FIGURE 17.20 ■ Three domains of life. Carl Woese (1928–2012) (inset) used SSU rRNA sequencing to reveal three equally distinct domains of life: Bacteria, Eukarya (eukaryotes), and Archaea.

Divergence of Three Domains of Life

Carl Woese first used SSU rRNA phylogeny to reveal the existence of a third kind of life, Archaea, roughly as distant from bacteria as from eukaryotes (**Fig. 17.20**). The three fundamental groups of life-forms—Archaea, Bacteria, and Eukarya—are termed **domains**. How was an entire domain of life missed in the past? Many archaea grow in habitats previously thought inhospitable for life; for example, *Thermoplasma* species grow at 60°C and pH 2 (**Fig. 17.21**). Others, such as methanogens and halophiles, were long known to microbial ecologists, but without tools for genetic analysis they were simply classified among bacteria.

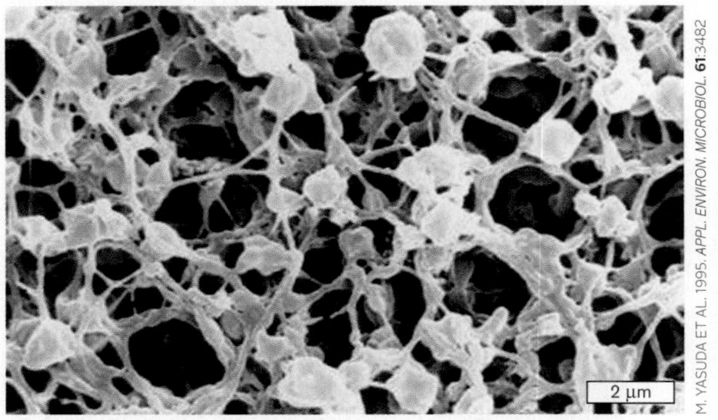
2 µm
M. YASUDA ET AL. 1995. APPL. ENVIRON. MICROBIOL. 61:3482

FIGURE 17.21 ■ *Thermoplasma.* This archaeon lives at 60°C at pH 2—with no cell wall, only a cell membrane.

Rooting the tree of life. Where is the root of the tree, the position of the last universal common ancestor of all

Special Topic 17.1: Phylogeny of a Shower Curtain Biofilm

The deep sea and the tropical forest receive much attention as habitats in which to discover new exotic life-forms. But novel microorganisms can be discovered closer to home—in the soil, on the roots of grass, in our own digestive tract. Even within our homes lurk microbial communities that include potentially dangerous pathogens. Such organisms were discovered by Norman Pace (**Fig. 2B**) and colleagues at San Diego State University and at the University of Colorado, taking a break from their submarine studies to explore a domestic aquatic habitat: the "soap scum" biofilm on a shower curtain (**Fig. 1**).

Domestic water sources are well known as a source of pathogens such as *Legionella pneumophila,* the cause of legionellosis, a deadly form of pneumonia. Pathogens persist in biofilms growing within plumbing and air-conditioning lines. Lesser-known reservoirs for microbes are the biofilms that collect on household surfaces such as shower curtains, which are often neglected during cleaning. Viable cells of the shower curtain biofilm are revealed by fluorescence microscopy using the DAPI stain for DNA (**Fig. 1B**).

Pace's students Scott Kelley and Ulrike Theisen obtained biofilm from a shower curtain and used it to extract DNA. The biofilm DNA was amplified for sequences transcribed to rRNA using the PCR technique. The PCR amplification was conducted with a pair of sequence primers based on portions of the SSU rRNA sequence that are extremely conserved across all known species of bacteria. The region between the primers, however, is known to vary among species; thus, the DNA amplified should show distinguishing features that identify the source organism.

The shower curtain biofilm revealed a wide range of species, including 117 unique sequences for SSU rRNA. The phylogeny shows how shower curtain samples relate to known bacteria, offering clues as to how they might interact with humans. The most abundant groups of species include two genera of alphaproteobacteria: *Sphingomonas* and *Methylobacterium. Sphingomonas* is named for its outer membrane content of sphingolipids, lipids containing an amide link to fatty acids. The phylogeny of the shower curtain sphingomonads is shown in **Figure 2A**. *Sphingomonas* species are known to grow in a wide range of soil and water habitats and are occasionally isolated as yellow colonies from natural water supplies. They catabolize complex carbon sources such as dibenzofuran and hexachlorocyclohexane and thus are of interest for bioremediation of environmental pollutants. Some species, however—particularly *S. paucimobilis* (highlighted in **Fig. 2**)—cause opportunistic infections in immunocompromised individuals, including bacteremia, peritonitis, and abscesses.

Species of the next-most-abundant genus, *Methylobacterium,* are versatile heterotrophs known for their ability to metabolize single-carbon sources such as methanol and methylamine. Species grow in soil and water, as well as within plant tissues; they are also isolated from automobile air-conditioning systems, printing paper machines, and dental unit water lines. Their pink color may be the source of the pink color commonly observed in shower biofilms. Some species, such as *M. extorquens* and *M. zatmanii,* cause illness in immunocompromised individuals, including pneumonia, skin ulcers, and bacteremia.

Growing numbers of immunocompromised individuals care for themselves at home, where they need to control their own microbiological exposure. Thus, it is of concern to discover opportunistic pathogens in a home setting. The authors

A.

© DENA DIGILIO BETZ

B.

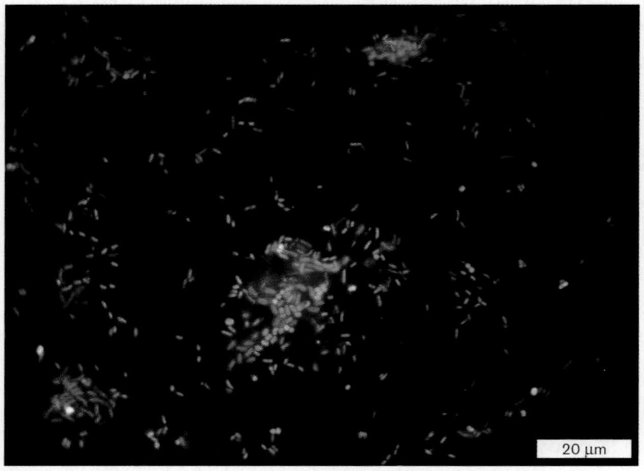
20 μm

SCOTT T. KELLEY ET AL. 2004. *APPL. ENVIRON. MICROBIOL.* **70**:4187

FIGURE 1 ▪ Bacterial diversity on a shower curtain. **A.** Shower curtain "soap scum" consists of a biofilm. **B.** Biofilm from a shower curtain is visualized by fluorescence microscopy using DAPI stain.

conclude with a reminder that "exposure can be minimized by regular cleaning or by changing shower curtains."

RESEARCH QUESTION

In the shower curtain biofilm, what do you think is the source of the bacterial species found growing there? How might you test your hypothesis? What tools would you need, and what controls would you perform?

Kelley, Scott T., Ulrike Theisen, Largus T. Angenent, Allison St. Amand, and Norman R. Pace. 2004. Molecular analysis of shower curtain biofilm microbes. *Applied and Environmental Microbiology* **70**:4187–4192.

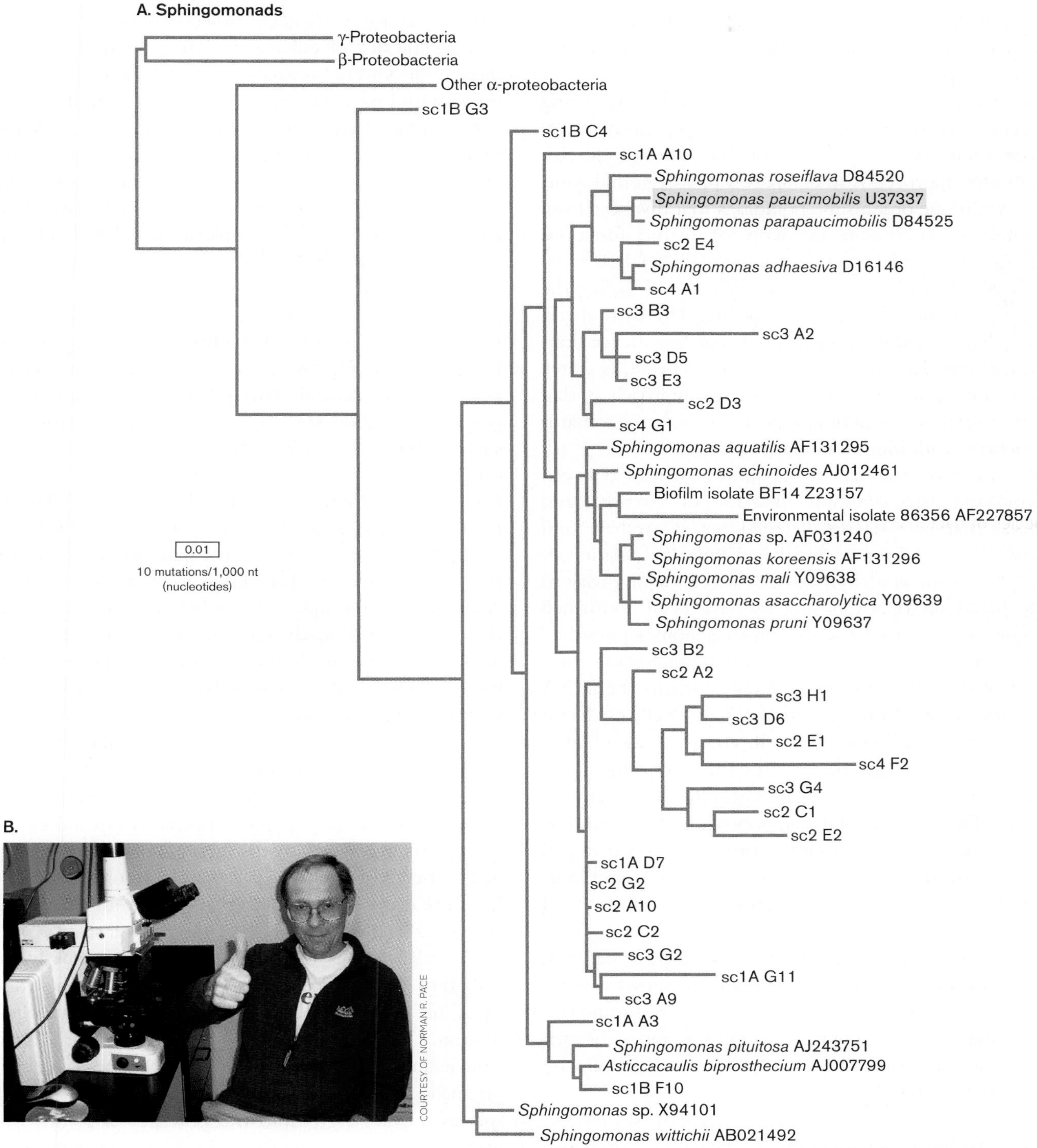
A. Sphingomonads
γ-Proteobacteria
β-Proteobacteria
Other α-proteobacteria
sc1B G3
sc1B C4
sc1A A10
Sphingomonas roseiflava D84520
Sphingomonas paucimobilis U37337
Sphingomonas parapaucimobilis D84525
sc2 E4
Sphingomonas adhaesiva D16146
sc4 A1
sc3 B3
sc3 A2
sc3 D5
sc3 E3
sc2 D3
sc4 G1
Sphingomonas aquatilis AF131295
Sphingomonas echinoides AJ012461
Biofilm isolate BF14 Z23157
Environmental isolate 86356 AF227857
Sphingomonas sp. AF031240
Sphingomonas koreensis AF131296
Sphingomonas mali Y09638
Sphingomonas asaccharolytica Y09639
Sphingomonas pruni Y09637
sc3 B2
sc2 A2
sc3 H1
sc3 D6
sc2 E1
sc4 F2
sc3 G4
sc2 C1
sc2 E2
sc1A D7
sc2 G2
sc2 A10
sc2 C2
sc3 G2
sc1A G11
sc3 A9
sc1A A3
Sphingomonas pituitosa AJ243751
Asticcacaulis biprosthecium AJ007799
sc1B F10
Sphingomonas sp. X94101
Sphingomonas wittichii AB021492
0.01
10 mutations/1,000 nt
(nucleotides)
B.
COURTESY OF NORMAN R. PACE

FIGURE 2 ▪ ***Sphingomonas* phylogeny from a shower curtain biofilm.** **A.** Phylogeny of sphingomonad bacteria was based on SSU RNA gene sequence comparison. **B.** Norman R. Pace characterized the first sequence phylogeny of thermophiles from high-temperature environments. His laboratory has since characterized the microbial diversity of other environments, such as a shower curtain biofilm.

life-forms? Which of the three domains (Bacteria, Archaea, Eukarya) first diverged from the other two? The question has profound importance for biology because the research community bases its "model systems" for study on their commonalities with organisms of importance to humans. For example, investigators of intron splicing in archaea argue that archaea represent a model system for related processes in complex eukaryotes such as humans.

The root, however, can be found only by measuring divergence relative to an outside group of organisms. Since no "outgroup" exists for the tree of all life, how is the tree rooted? One approach is to compare a pair of homologous genes within one organism—homologs that diverged from a common ancestral gene and acquired distinct functions (paralogs). The pair of paralogs chosen for analysis must have diverged within the common ancestral cell before the divergence of the three domains of life. The early divergence is determined by comparing the pair's sequence similarity in many different organisms and constructing a tree of gene phylogeny. A limitation of this approach is that genes with different functions have been subjected to natural selection in addition to random variation.

Most phylogenetic data so far indicate a root between Bacteria and the common ancestor of Archaea and Eukarya; that is, Bacteria diverged before Archaea and Eukarya diverged from each other. While the position of the root remains controversial, the three major divisions of life based on rRNA phylogeny have been confirmed by sequence data from the numerous genomes published to date. Furthermore, structural and physiological comparison of the three domains largely confirms the rRNA tree (**Table 17.2**). Note first that all living cells on Earth share profound similarities. All cells consist of membrane-enclosed compartments that shelter the same fundamental apparatus of cell production: the DNA-RNA-protein machine. The fundamental components of this machine appear to have evolved before the three domains diverged from their last universal common ancestor. From a molecular standpoint, all cells on Earth appear more similar than they are different.

Nonetheless, important differences emerge between each pair of domains; indeed, each domain shows distinctive traits absent or scarce in the other two. We summarize here the major features common to most members of each domain. Various classes and species within each domain are explored in Chapters 18–20.

Archaea, bacteria, and eukaryotes. Of the three domains, the eukaryotes stand out as having a nucleus and other complex membranous organelles (see **Table 17.2**). Eukaryotic organelles include mitochondria and chloroplasts, which evolved from internalized bacteria. Bacteria and archaea possess no nucleus and have relatively simple intracellular membranes. Their size is limited by diffusion across the cell membrane, with occasional exceptions, such as the "giant bacterium" *Epulopiscium fishelsoni*. The larger size and complexity of eukaryotic cells generally require the most high-powered sources of energy, such as aerobic respiration and oxygenic photosynthesis, although some protists conduct fermentation. By contrast, the prokaryotes (bacteria and archaea) employ a wider range of metabolic alternatives, including lithotrophy and anaerobic respiration. Finally, eukaryotic plants and animals have attained a degree of multicellular complexity unknown in the prokaryotic domains.

On the other hand, eukaryotes share key traits with archaea that distinguish both from bacteria. The core information machinery of eukaryotes more closely resembles that of archaea. The two domains share closely related components of the central DNA-RNA-protein machine: RNA polymerase, ribosomes, and transcription factors. Even such hallmarks of eukaryotes as intragenic introns, the splicing machinery, and the "RNA interference" regulatory complexes are found in archaea. This explains why rRNA trees place archaea closer to eukaryotes than to bacteria.

At the same time, eukaryotes share fundamental structures with bacteria that differ from those in archaea. Archaea possess unique cell membrane components, such as their ether-linked lipids. Outside the archaea, ether-linked membrane lipids are found in only a few deep-branching bacterial species that share habitat (and exchange genes) with hyperthermophilic archaea. Only the domain Archaea includes species capable of growth in the most "extreme" environments of temperature (above 110°C or below –20°C) and pH (below pH 1), although other archaeal species grow well at mesophilic temperatures in soil or water. Perhaps the most striking distinction of archaea is the complete absence of archaeal pathogens of animals or plants. Even the many methanogens that live within animal digestive tracts have never been shown to cause disease.

Overall, the three-domain phylogeny divides life usefully into three distinctive groups. Yet the tree also shows signs of gene flow unaccounted for by monophyletic descent. The eukaryotes contain mitochondria derived from assimilated bacteria whose genomes persist within the organelle. And pathogenic bacteria such as *Agrobacterium* species transfer DNA into the genomes of plants. Moreover, sequenced genomes reveal evidence of gene transfer between bacteria and archaea sharing high-temperature habitats. What if the tree of life is not strictly monophyletic?

TABLE 17.2

Three domains of life.

Characteristic	Traits of living organisms		
	All cells on Earth resemble each other in these traits:		
Chromosomal material	Double-stranded DNA		
RNA transcription	Common ancestral RNA polymerase		
Translation	Common ancestral rRNAs and elongation factors		
Protein	Common ancestral functional domains		
Cell structure	Aqueous cell compartment bounded by a membrane		
	Comparison of domains:		
	Bacteria	**Archaea**	**Eukarya**
	Archaea resemble bacteria in these traits:		
Cell volume	1–100 μm^3 (usually)		1–10^6 μm^3
DNA chromosome	Circular (usually)		Linear
DNA organization	Nucleoid		Nucleus with membrane
Gene organization	Multigene operons		Single genes
Metabolism	Denitrification, N_2 fixation, lithotrophy, respiration, and fermentation		Respiration and fermentation
Multicellularity	Simple		Simple or complex
		Archaea resemble eukaryotes in these traits:	
Intron splicing	Introns are rare	Introns are common	
RNA polymerase	Bacterial homologs	Eukaryotic homologs	
Transcription factors	Bacterial homologs	Eukaryotic homologs	
Ribosome sensitivity to chloramphenicol, kanamycin, and streptomycin	Sensitive	Resistant	
Translation initiator	Formylmethionine	Methionine (except mitochondria use formylmethionine)	
Cell wall	Peptidoglycan	Pseudopeptidoglycan or other polymer; or protein S-layer	
	Bacteria resemble eukaryotes and differ from archaea in these traits:		
Methanogenesis	No	Yes	No
Thermophilic growth	Up to 90°C	Up to 120°C	Up to 80°C
Photosynthesis	Many species; bacteriochlorophyll (proteorhodopsin derived from archaea)	Haloarchaea only; bacteriorhodopsin	Many species; chlorophyll (bacterial origin)
Light absorption	Red and blue (chlorophyll absorption)	Green (central range of solar spectrum; no chlorophyll)	Red and blue (chloroplasts of bacterial origin)
Membrane lipids (major)	Ester-linked fatty acids	Ether-linked isoprenoid	Ester-linked fatty acids
Pathogens infecting animals or plants	Many pathogens	No pathogens	Many pathogens

Horizontal Gene Transfer

In retrospect, the very first demonstration of hereditary material—the transformation of avirulent *Pneumococcus* to virulence in 1928—involved **horizontal gene transfer** (or lateral gene transfer). Horizontal gene transfer is the acquisition of a piece of DNA from another cell, as distinguished from **vertical gene transfer**, the transmission of an entire genome from parent to offspring. Among bacteria and archaea, DNA is transferred horizontally by plasmids, transposable elements, and bacteriophages, as well as through the process of transformation, as discussed in Chapter 9. For example, drug resistance genes are transferred from harmless human-associated bacteria to pathogens. Another example is the transfer of genes encoding light-driven proton pumps (bacteriorhodopsin) from halophilic archaea into marine bacteria. Such transfer events are relatively rare, occurring perhaps once in a million generations. But over time, the number of such "rare" events can accumulate.

How can we tell when a genome contains DNA sequence "transferred" from a distant relative? One sign of horizontal transfer is a DNA sequence whose GC/AT ratio (proportion of GC and AT base pairs) differs from the rest of the genome. A surprising proportion of genomic DNA can show "spikes" of GC content that differ from the GC/AT ratio of neighboring sequences. These regions of anomalous GC/AT ratio indicate origin elsewhere, even from species of a different domain. In some archaea, particularly the hyperthermophiles *Pyrococcus* and *Aeropyrum*, 10%–20% of the genes appear to come from bacteria that share their high-temperature environment. On the other hand, the thermophilic bacterium *Thermotoga maritima* shows a number of genes transferred from archaea.

Between more closely related taxa, transfer occurs even faster. For example, the *Escherichia coli* genome acquired about 18% of its genes from closely related species after its relatively recent divergence from the close relative *Salmonella enterica*. Some medically important genera, such as *Neisseria* (which causes gonorrhea and meningitis), are particularly "recombinogenic." Rapid gene exchange enables pathogens to avoid the host immune system by expressing novel proteins not recognized by host antibodies (discussed in Chapter 23). Furthermore, the genomes of eukaryotes include many genes acquired from bacteria, most of them from the endosymbiotic ancestors of their mitochondria and chloroplasts (discussed in Section 17.6)

How does horizontal gene transfer affect microbial phylogeny? In 1999, Ford Doolittle, at Dalhousie University, redrew the standard tree of life with a bewildering array of cross-cutting lineages to show how actual phylogeny combines horizontal and vertical transfer (**Fig. 17.22**). Perhaps

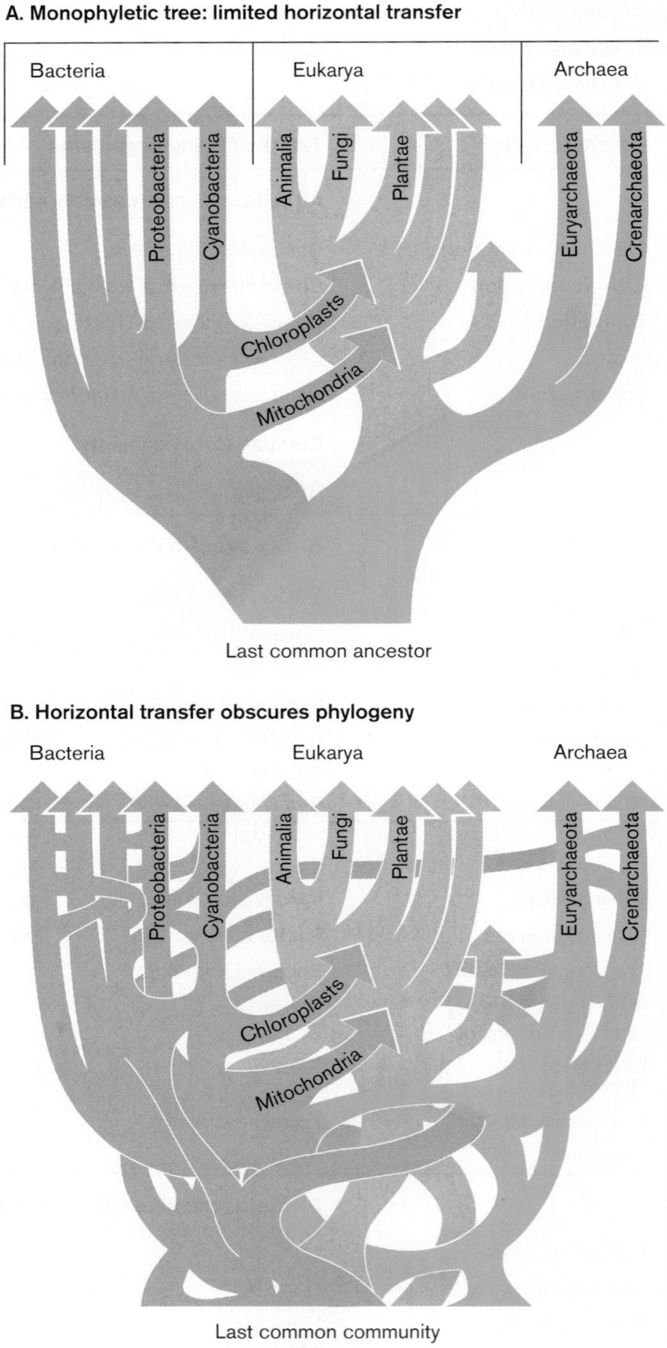
A. Monophyletic tree: limited horizontal transfer
Bacteria
Eukarya
Archaea
Proteobacteria
Cyanobacteria
Animalia
Fungi
Plantae
Euryarchaeota
Crenarchaeota
Chloroplasts
Mitochondria
Last common ancestor
B. Horizontal transfer obscures phylogeny
Bacteria
Eukarya
Archaea
Proteobacteria
Cyanobacteria
Animalia
Fungi
Plantae
Euryarchaeota
Crenarchaeota
Chloroplasts
Mitochondria
Last common community

FIGURE 17.22 ■ Vertical and horizontal gene transfer. **A.** The traditional view of phylogeny holds that the vast majority of gene transfer is vertical and that most lineages are monophyletic. Only occasional transfers occur, such as that of mitochondria and chloroplasts from bacteria into eukaryotes. **B.** An alternative view argues that horizontal gene transfer is so prevalent among microbes that it obscures monophyletic distinctions between taxa. *Source:* Modified from W. Ford Doolittle. 1999. *Science* **284**:2124.

the "last common ancestor" was actually a last common community of diverse life-forms that contributed different parts of our genetic legacy.

Thought Question

17.6 What are the limits of evidence for horizontal gene transfer in ancestral genomes? What alternative interpretation might be offered?

Reconciling Vertical and Horizontal Gene Transfer

One approach to sorting out vertical and horizontal gene transfer was proposed by James Lake and colleagues at UCLA and developed further by Doolittle. Lake's approach assumes that some classes of genes nearly always transmit vertically—particularly "informational genes," which specify products essential for transcription and translation. Informational genes include RNA polymerase, as well as ribosomal RNAs such as SSU rRNA, and elongation factors. Informational genes need to interact directly in complex ways with large numbers of cellular components; thus, their capacity for horizontal transfer is limited. On the other hand, "operational genes" are those whose products govern metabolism, stress response, and pathogenicity. Operational genes function with relative independence from other cell components and consequently move more easily among distantly related organisms, particularly organisms that share the same habitat. An important category of such movable genes is virulence factors. Pathogens show extensive horizontal transfer of virulence genes and genes encoding resistance to host defenses and antibiotics.

Groups of such genes are often transferred together on plasmids or on **genomic islands**, regions of DNA of foreign origin that confer special properties, such as virulence or nitrogen fixation (discussed in Section 9.7). Examples of a virulence plasmid and a genomic island are found in the deadly pathogen *E. coli* O157:H7 (eTopic 17.2). But in some cases, the group of genes transferred is so large that it makes up a substantial chunk of an organism's genome. For example, the Aquificales group of thermophilic bacteria includes a large section derived from Proteobacteria similar to *E. coli*—so large that the Aquificales phylogeny remains unclear.

A balanced view of vertical and horizontal gene transfer is shown in **Figure 17.23**. The row of arrows designates vertical transfer of the bulk of the genomes of the two diverging species. Black lines represent vertically transferred genes, such as ribosomal RNA; gray indicates genes with more flexible function. At various levels, colored lines indicate horizontal transfer, where genes enter each lineage by processes such as conjugation or transformation. Gray lines peel out, indicating loss of a gene by mutation. Overall, the vertical lineage persists for most of the core informational genes (black lines), while horizontally acquired genes enter the genome. The persistence of genes showing vertical inheritance in nearly all life (such as SSU rRNA) may reflect the phylogeny of the organism as a whole, despite the other genes that are transferred horizontally. But to fully assess phylogeny, we must sequence entire genomes.

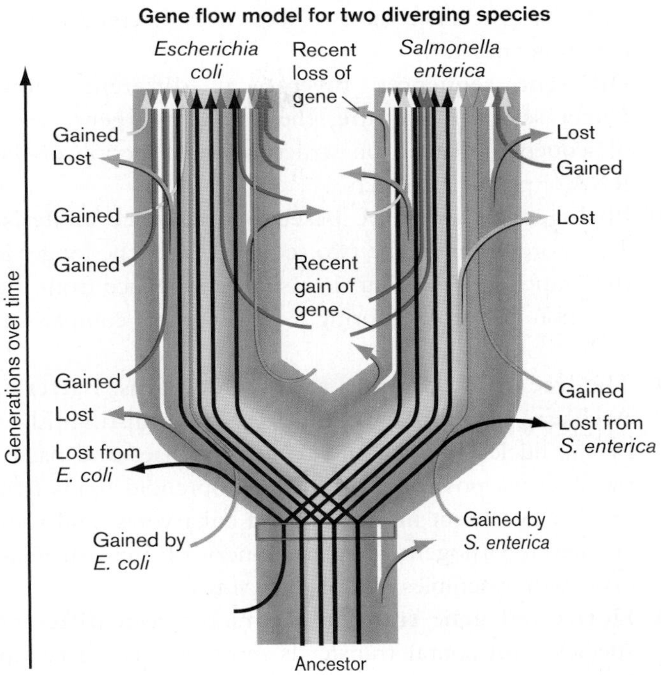

Gene flow model for two diverging species
Escherichia coli
Recent loss of gene
Salmonella enterica
Generations over time
Gained
Lost
Gained
Gained
Recent gain of gene
Gained
Lost
Lost from E. coli
Gained by E. coli
Lost
Gained
Lost
Gained
Lost from S. enterica
Gained by S. enterica
Ancestor

FIGURE 17.23 ■ Genes enter and leave genomes. Both rRNA trees and whole-genome trees consistently reflect monophyletic descent (black lineages) down to the genus level. Below the genus level, monophyletic descent may be obscured by high rates of horizontal transfer (colored lines entering, gray lines departing). White spaces indicate genes lost by reductive evolution.

To Summarize

- **Phylogeny is the divergence of related organisms.** Organisms diverge through random mutation, natural selection, and reductive evolution.
- **Molecular clocks are based on mutation rate.** Given a constant mutation rate and generation time, the degree of difference between two DNA sequences correlates

with the time since the two sequences diverged from a common ancestor.

- **Different sequences diverge at different rates.** Under selection pressure, the actual divergence rates of sequences depend on structure and function of the RNA or protein products.
- **Phylogenetic trees are based on sequence analysis.** The more different the two sequences are, the longer is the branch representing time since divergence from the common ancestor. Rooting a tree requires comparison with an outgroup.
- **The tree of life diverges to three domains: Bacteria, Archaea, and Eukarya.** Eukaryotes are distinguished by the nucleus, which is lacking in archaea and bacteria. Archaea possess ether-linked isoprenoid lipids that are rare or absent in bacteria and eukaryotes, and they are never pathogens. The machinery of archaeal gene expression resembles that of eukaryotes.
- **Horizontal gene transfer occurs between different species.** Horizontal transfer is most frequent between closely related species or between distantly related species that share a common habitat. Informational genes with many molecular interactions transfer vertically, whereas operational genes that function independently of other components are more likely to transfer horizontally.
- **Horizontal gene transfer is important for adaptation to new environments and for pathogenesis.** Gene transfer among pathogenic and nonpathogenic strains leads to emergence of new pathogens.

17.4

Adaptive Evolution

In discussing phylogeny, we focused on the accumulation of random mutations that cause genome sequences to diverge at a steady rate. The rate of nonselected sequence changes enables us to measure evolutionary time. But how does evolution enable survival in new environments? Adaptive evolution requires **natural selection**. In natural selection, the genetic variants that come into being by random mutation differ in their chance of survival. Those variants that survive to leave more offspring undergo "selection"; that is, their traits are overrepresented in the next generation. Over many generations, the descendants develop traits very different from those of their ancestors.

How can we study adaptive evolution—especially the evolution of different species, which occurs over millions of years? We have several kinds of evidence:

- **Genome analysis.** Comparing gene sequences, both within and between genomes, enables us to track how organisms adapted in the past.
- **Strongly selective environments.** Environments under intensive selective pressure, such as antibiotic exposure in hospitals and high-temperature habitats, reveal rapid evolution.
- **Experimental evolution.** Experimental strategies can now reveal evolution in the laboratory, enabling us to test predictive models.

Genome Analysis

As discussed in Section 17.3, the sequence of genomes reveals descent over time based on steady accumulation of mutations. But other mutations undergo selection that provides a cell with new functions. A common way this occurs is by **gene duplication**.

As cells replicate their DNA over many generations (discussed in Chapter 9), occasionally the DNA polymerase will make a duplicate copy of a gene. With duplicate copies of a gene, one copy may acquire mutations that detract from its function without detriment to the organism because the "backup" copy still functions. But suppose the mutated copy gains additional mutations that actually alter its function, providing a new function to the cell. For example, a gene encoding a transporter for one sugar may now encode a transporter that better "fits" a different sugar. The two **paralogous genes**, or **paralogs**, have evolved to serve different functions. Paralogs are a major source of raw material that contributes to new functions arising through evolution. We can detect paralogs in a genome, through sequence relatedness; for example, the genome of the hyperthermophilic bacterium *Thermotoga maritima* shows paralogous ABC transporters for lactose, cellobiose, mannose, and xylose, among others. Evolution of paralogs provides an important way for organisms to enhance fitness in a complex environment.

Does a population ever lose some of its paralogs? Certain environments select for loss of many genes—a phenomenon called **degenerative**, or **reductive**, **evolution**. The selection for loss is "positive" because organisms save energy by avoiding replication and expression of unneeded genes. Intracellular pathogens and obligate symbionts often lose large chunks of their genome, by processes such as intracellular recombination of the chromosome. The lost genes typically encoded functions supplied by the host, such as capture of nutrients and generation of energy. Lost genes can be identified by comparison with free-living species. For example, *Treponema pallidum,* the cause of syphilis, has a genome of 1.1 million base pairs, which is about a quarter the size of the *E. coli* genome. *T. pallidum* has lost all of its enzymes used in the TCA cycle, respiration, and amino acid biosynthesis. These genes are present in the genomes of free-living treponemes.

Surprisingly, dilute nutrient-poor natural environments also select for gene reduction in certain species. The marine cyanobacterium *Prochlorococcus,* a major source of global photosynthesis, has lost numerous genes that confer little advantage in the open ocean. Missing genes encode transporters for sugars and amino acids (which are scarce in the ocean) as well as genes for flagellar motility and pili. More surprising, *Prochlorococcus* species also lack catalase, despite their high output of hydrogen peroxide, which requires catalase to detoxify. Their hydrogen peroxide is detoxified by catalase-positive bacteria attracted to the oxygen produced by *Prochlorococcus.*

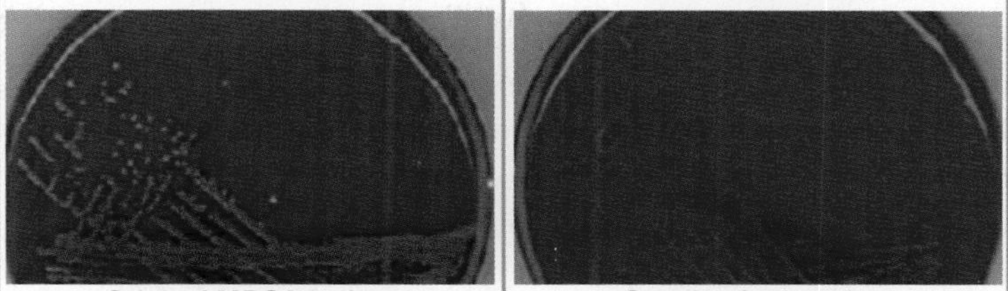
MRSA bacteria isolated from patient
Before antibiotics
After antibiotics

Antibiotics	Original MRSA isolate	Small-colony variant
Vancomycin	Partly resistant	Resistant
Rifampin	Sensitive	Resistant
Ciprofloxacin	Sensitive	Resistant
Linezolid	Sensitive	Partly resistant

FIGURE 17.24 ■ Evolution of antibiotic resistance in MRSA. Left: The initial MRSA isolate infecting the patient showed partial resistance to vancomycin but was sensitive to rifampin, ciprofloxacin, and linezolid. **Right:** Following exposure to these antibiotics, a new strain was isolated that grew more slowly (the small-colony variant) but was resistant to all the antibiotics. *Source:* Wei Gao et al. 2010. *PLoS Pathogens* **6**: e1000944.

Strongly Selective Environments

Evolution requires many generations, but certain microbes may undergo 40 generations in a day. When such microbes are under strong selection pressure, such as the presence of an antibiotic, evolution may occur surprisingly fast. In the case shown in **Fig. 17.24**, a hospitalized patient was infected by MRSA, a deadly strain of *Staphylococcus aureus* that had already evolved resistance to drugs such as methicillin. The patient was an elderly man with a weakened immune system, unable to eliminate even small populations of an opportunistic pathogen such as MRSA. The original culture from the patient's blood showed bacteria sensitive to four antibiotics. But prolonged exposure resulted in natural selection for resistance. Ultimately, following a total of 12 weeks' exposure to linezolid (one of the last antibiotics effective against MRSA), the bacteria isolated showed resistance to three other antibiotics plus partial resistance to linezolid.

In **Figure 17.24**, note that the latest drug-resistant strain actually made smaller colonies with or without the drug—the "small-colony variant." In other words, natural selection yielded a population that was the "fittest" under a particular environmental condition—the presence of linezolid in an immunocompromised host—where even a slow-growing strain could persist. In the absence of the drug, the original strain would have outcompeted the small-colony variant.

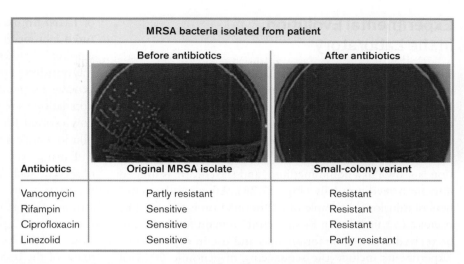

In the case just described, what was the molecular basis of the multidrug resistance? Researchers obtained DNA from the patient's original MRSA, as well as the later small-colony variant. The two genomes were sequenced and compared. Just three point mutations in the small-colony variant accounted for the drug resistance, as well as the retarded growth. One of these mutations derepressed a stress regulon, causing accumulation of the stress signal ppGpp (guanosine tetraphosphate). The ppGpp stress regulon includes expression of many protective genes that enable a cell to survive in the presence of antibiotics. But the cost to the cell of expressing this regulon is a slower growth rate; like a community under "terror alert," the cell's normal everyday processes are slowed by the demands of the stress response.

The MRSA example illustrates two key points:

- **The "fittest" trait depends on the environment in which selection occurs.** The presence of an antibiotic selects individuals that are resistant, despite slower growth (small-colony size). Without the antibiotic, the faster-growing individuals prevail.
- **Rapid adaptive evolution may be enhanced by disabling regulation.** In this case, the ppGpp stress regulon was derepressed, enabling stress responses that normally would be turned off because they inhibit cell growth.

Experimental Evolution in the Laboratory

The process of evolution can be studied in the laboratory—an approach known as **experimental evolution** or **laboratory evolution**. In a typical evolution experiment, the experimenter founds a population of bacteria or other organism that is then grown under a chosen set of environmental conditions. In one typical setup, the population is grown in a liquid medium, and the culture is diluted into fresh medium daily (**Fig. 17.25**). After regular numbers of dilutions, a sample of the population is frozen. The bacteria in this "frozen fossil record" remain alive and can be revived at any time for analysis and use in experiments. Experiments include the sequencing of genomic DNA of clones isolated from these frozen time points. With laboratory evolution experiments, we can test predictions of how new traits appear and use genome sequences to examine the mutations that led to the new traits. Evolution experiments can also include many replicate populations, making it possible to examine the repeatability of evolution.

A landmark experiment on evolution in the laboratory was performed by Richard Lenski at Michigan State University (see the Part 1 interview). In 1988, Lenski began a long-term evolution experiment when he founded 12 populations of *E. coli* from a single clone. The 12 populations were cultured in a medium in which growth was limited by a small amount of glucose. Every 24 hours, 1% of each population was transferred to fresh medium and then grew for about 6.7 generations per day. Every 500 generations, samples of each population were frozen for later study. The populations have now evolved for more than 50,000 generations, and daily dilutions are continuing in Lenski's lab.

During the experiment, many changes have been observed, such as the evolution of bacteria that can grow faster than the ancestor and form larger cells. The most striking change was first observed after generation 33,000. One of the 12 populations suddenly became much denser (**Fig. 17.25**). Examination of this population revealed that many of the bacteria had evolved a remarkable new phenotype: They were able to grow aerobically on citrate—a trait found in other species but not found in *E. coli*. The bacteria are called citrate plus, or Cit^+.

Since citrate has been available in the medium from the beginning, why did the Cit^+ mutation take so long to evolve? By testing *E. coli* from earlier generations that he had stored in a freezer, Lenski was able to show that over the first 31,000 generations, no bacteria could metabolize citrate; but then a rare mutation occurred that enabled citrate uptake and catabolism. Over a few generations, the Cit^+ cells had increased their proportion of the population, until they became the predominant strain.

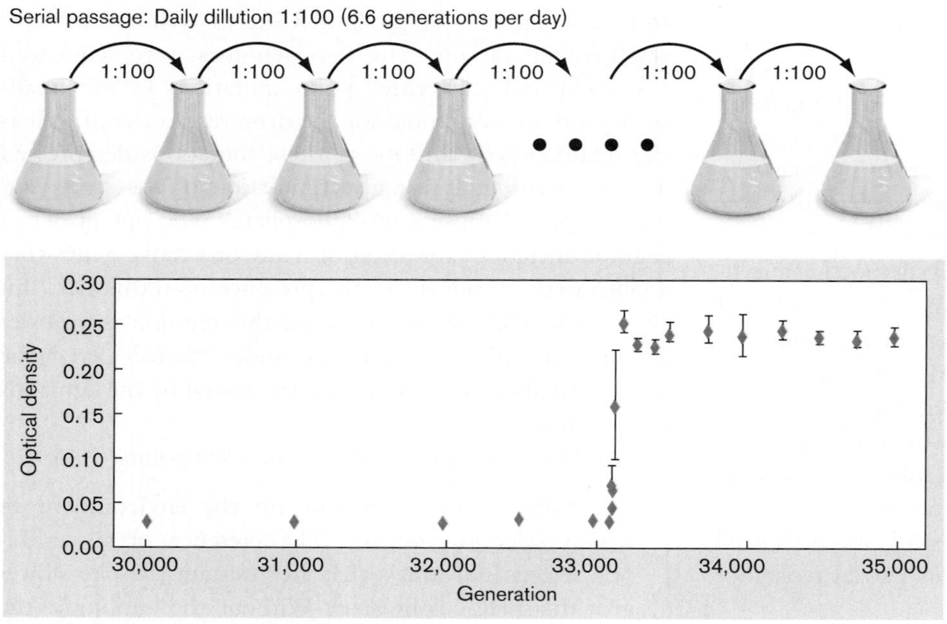
Serial passage: Daily dillution 1:100 (6.6 generations per day)
1:100
Optical density
Generation
0.30
0.25
0.20
0.15
0.10
0.05
0.00
30,000
31,000
32,000
33,000
34,000
35,000

FIGURE 17.25 ■ Evolution of an aerobic, citrate-utilizing strain of *E. coli*. The bacteria were evolved by serial transfer in a glucose-limited growth medium. The medium also contained citrate to allow the bacteria to take up iron. After 33,000 generations (estimated cell doublings), the bacterial population suddenly grew to a much greater density because of the rise to high frequency of a strain that had evolved the ability to grow on the citrate at about 31,000 generations. *Source:* Adapted from Zachary Blount et al. 2008. *PNAS* **105**:7899.

But was the overall phenotype really so sudden? Could any *E. coli* cell mutate to catabolize citrate? Lenski proposed that the earlier generations had acquired some kind of "potentiating" mutations that somehow enabled *E. coli* to gain the Cit^+ mutation. He called this the "historical contingency" hypothesis. Lenski and his students went into the population's fossil record and isolated clones from various time points. They then tested the ability of these clones, as well as of the ancestral strain, to mutate to Cit^+. They found that only clones isolated from time points after 20,000 generations reevolved the new Cit^+ phenotype; the original strain did not. This result supported the historical contingency model.

To gain additional information about the early and late mutations,

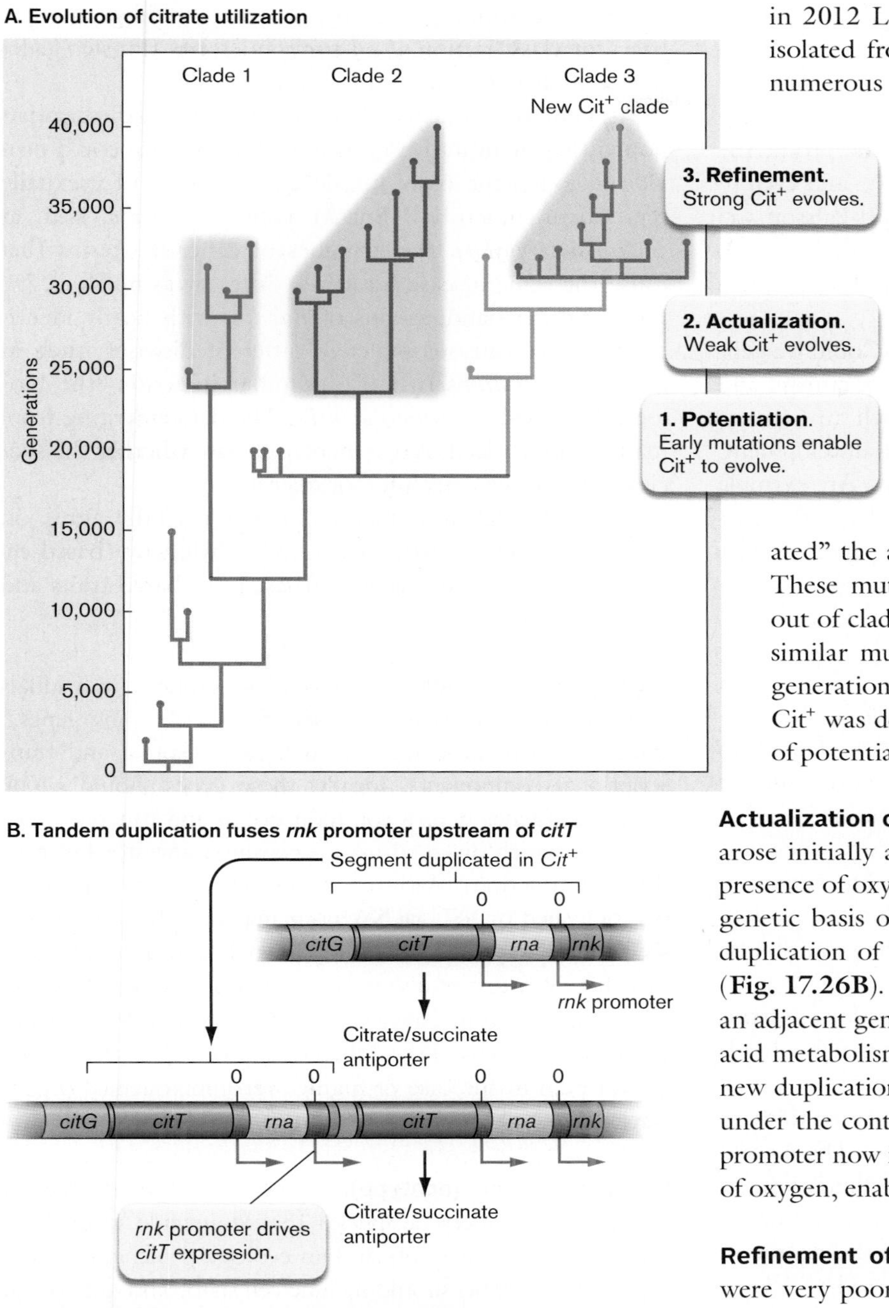
A. Evolution of citrate utilization
Clade 1
Clade 2
Clade 3
New Cit+ clade
40,000
35,000
30,000
25,000
20,000
15,000
10,000
5,000
0
Generations
3. Refinement.
Strong Cit+ evolves.
2. Actualization.
Weak Cit+ evolves.
1. Potentiation.
Early mutations enable Cit+ to evolve.
B. Tandem duplication fuses rnk promoter upstream of citT
Segment duplicated in Cit+
citG
citT
rna
rnk
rnk promoter
Citrate/succinate antiporter
O
rnk promoter drives citT expression.

FIGURE 17.26 ■ Evolution of citrate utilization in *E. coli*. **A.** One or more of the mutations accumulated over 30,000 generations potentiated evolution of the Cit$^+$ trait. The Cit$^+$ trait first appeared in about generation 31,000. Subsequent mutations refined the phenotype, increasing the rate of citrate utilization. Ball symbols indicate genomes from the population's history that were sequenced for analysis. **B.** The Cit$^+$ mutation. After 31,000 generations, a tandem duplication event placed a copy of the *rnk* promoter, which directs expression when oxygen is present, upstream of the *citT* gene encoding a citrate transporter that is normally expressed only without oxygen. This new *rnk-citT* module allows the CitT citrate transporter to be expressed during aerobic metabolism, giving the bacteria access to the citrate present in the medium. *Source:* Adapted from Zachary Blount et al. 2012. *Nature* **489**:513.

in 2012 Lenski sequenced the genomes of 29 clones isolated from the population's frozen fossil record at numerous time points through 40,000 generations (**Fig. 17.26**). The strains were measured for citrate utilization, and for ability to evolve Cit$^+$. The data fit the following model of stages of evolving a new trait: (1) potentiation to achieve useful mutations; (2) actualization of a novel mutant phenotype; and (3) refinement, or increasing the degree of the phenotype.

Potentiating mutations. During the first 31,000 generations, early mutations occurred in some cells that did not confer citrate catabolism but somehow "potentiated" the ability of cells to gain a Cit$^+$ mutation later. These mutations led to the appearance of Cit$^+$ cells out of clade 3 (**Fig. 17.26A**) and to the appearance of similar mutations in "replayed" evolution from late-generation clade 3 strains. Thus, the potentiation of Cit$^+$ was demonstrated statistically, although the basis of potentiation remains unclear.

Actualization of the novel phenotype. The Cit$^+$ mutation arose initially as a weak ability to transport citrate in the presence of oxygen, and therefore to respire on citrate. The genetic basis of this mutation was found to be a tandem duplication of a gene encoding a citrate transporter, *citT* (**Fig. 17.26B**). The duplication also includes a promoter for an adjacent gene, *rnk*, which encodes a regulator of nucleic acid metabolism, expressed in the presence of oxygen. The new duplication rewires how *citT* is regulated by placing it under the control of a copy of the *rnk* promoter. The *rnk* promoter now initiates transcription of *citT* in the presence of oxygen, enabling the cell to respire on citrate.

Refinement of the phenotype. The earliest Cit$^+$ clones were very poor at growing on the citrate in the medium. But later clones evolved by mutations that increased the rate of citrate utilization—that is, by evolutionary refinement of the Cit$^+$ trait. The researchers showed that this improvement was due to an increase in the copy number of the duplicated segment that produces the *rnk-citT* module, which increased the number of expressed copies of *citT*. This improved growth on citrate eventually led to the Cit$^+$ clones coming to dominate the population. Interestingly, though, even the "refined" Cit$^+$ cells never fully took over the cultures. A proportion of cells always remained that had adapted to limiting glucose by means that did not involve using citrate.

It remains to be seen how well the above model explains evolution in other situations, and what alternative models exist. Nevertheless, laboratory evolution is already used widely in industrial microbiology (discussed in Chapter 16). Evolution can be used to increase the productivity of strains producing commercial products—and even to develop new products. For example, Bernhard Palsson and colleagues at UC San Diego used laboratory evolution to evolve a strain of *Streptomyces* to produce high concentrations of a novel antibiotic against MRSA.

Suppose we take evolution to an extreme: Could we generate in vitro all the possible mutants of an organism, and set them against each other all at once? Such an extreme approach is, in fact, used for commercial development, through what is called **directed evolution**. An example of directed evolution is the development of hyperthermophilic forms of a xylanase enzyme used for paper production (**Special Topic 17.2**).

To Summarize

- **Adaptive evolution** requires natural selection.
- **Genome analysis** tracks how organisms adapted in the past. Gene duplications provide the opportunity to evolve paralogous genes with different functions.
- **Degenerative (reductive) evolution** occurs when unneeded genes are lost from the genome. The organism saves energy by avoiding their replication and expression.
- **Strongly selective environments**, such as antibiotic exposure, reveal rapid evolution.
- **Selective pressure depends on the particular environment.** A trait favored in one environment may be disadvantageous in another.
- **Experimental evolution in the laboratory enables** us to test hypotheses about natural selection. The Lenski long-term evolution experiment shows potentiation, to achieve useful mutations; actualization of a novel mutant phenotype; and refinement, or increasing the degree of the phenotype. Laboratory evolution is used widely in industrial development.

17.5

Microbial Species and Taxonomy

What is a species? Among eukaryotes, a species is defined by the principle that members of different species do not normally interbreed. Thus, the failure to interbreed is the traditional property defining species. Bacteria and archaea, however, reproduce asexually, so interbreeding is not a basis for classification. And some microbes transfer genes horizontally between distantly related clades.

As genome sequence data became available, scientists hoped that quantitative measures of divergence could provide a consistent basis for defining species of asexually reproducing microbes. But in some organisms, such as *Helicobacter pylori,* the genomes of different strains that cause the same disease (gastritis) differ by as much as 7%. On the other hand, strains of *Bacillus* with nearly identical genomes cause completely different diseases, such as anthrax (*B. anthracis*) and caterpillar infection (the biological pesticide *B. thuringiensis*). The data emerging from many genomes leads researchers to debate whether the species concept lacks meaning for microbes.

Amid the debates, microbiologists generally agree on the importance of two perspectives: phylogeny (based on DNA relatedness) and ecology (based on shared traits and ecological niche).

Phylogenetic relatedness. A species is a group of individuals that share relatedness of a key set of "housekeeping genes," typically informational genes such as ribosomal and transcriptional components. Ideally, these genes should all be orthologs (genes with a common origin and function), not paralogs (which diverged from a common ancestor but now differ in function). Within a genus, species that cannot be distinguished by SSU rRNA alone may be defined by analysis of multiple genes. For example, analysis of multiple gene loci effectively distinguishes *Neisseria meningitidis* (the cause of meningitis) from *Neisseria lactamica,* a harmless resident of the nasopharynx. The multigene approach has proved successful even in the case of highly "recombinogenic" organisms known to acquire and rearrange genes readily.

Ecological niche (ecotype). Besides a high degree of genomic relatedness, a species should include individuals that share common traits and an ecological niche, or "ecotype." Shared traits should include cell shape and nutritional requirements, and there should be a common habitat and life history (for example, causing the same disease). By these criteria, highly divergent strains of *Helicobacter pylori* causing gastritis make up one species, whereas *Bacillus anthracis* (anthrax) and *Bacillus thuringiensis* (caterpillar infection) are different species despite their highly similar genomes.

A working definition of species. While the debate goes on, many microbiologists accept the following criteria for a working definition of a microbial species:

- **DNA hybridization ≥70%.** When DNA from two different genomes is denatured and mixed together, the strands reanneal to form a hybrid helix. If the proportion of hybridization is 70% or greater, the two organisms are usually considered the same species.

- **SSU rRNA similarity ≥97%.** Two organisms with 97% or greater similarity in SSU rRNA sequence are considered to share the same genus. The larger percentage reflects the relatively high conservation of rRNA sequence, compared to other genes.
- **Average nucleotide identity (ANI) of orthologs ≥95%.** If whole genomes are available and annotated, we can define all the orthologous genes (genes of the same function) shared by a pair of strains. If two strains share 95% or greater ANI for these genes, they are the same species.

Most well-known species meet all three of these criteria, and hard-to-classify organisms meet at least one. The working definition classifies nearly all known bacteria in a way that reconciles phylogeny with ecotype. For example, *Helicobacter pylori* genomes with a common gastric pathology show exceptional variation due to horizontal gene transfer, even including their SSU rRNA genes. Nevertheless, a set of orthologs can be defined that share 95% ANI.

At the same time, new questions arise from the availability of multiple sequenced genomes for a single species. Suppose that every time we sequence a new isolate from nature, or from a clinical specimen, we find that every new genome sequenced has a few new genes absent from previously sequenced isolates. How, then, do we define the gene map? This situation would be unheard of for animals and plants, in which the gene map is fixed by Mendelian recombination. It is common, however, for bacteria such as *Bacillus cereus,* a Gram-positive soil bacterium and food pathogen (**Fig. 17.27**). For each new genome sequenced, several hundred new genes are identified that are absent from all the other genomes so far. In a pathogen, some of these new genes could be involved in pathogenesis, with profound implications for medical therapy and vaccine development.

Should we now attempt to define bacterial and archaeal genomes not just by a single sequenced genome, but by the sum total of all expected genes in all possible isolates? This theoretical total is called the **pan-genome**. The pan-genome includes genes present in all sequenced genomes of a species (known as its **core genome**) plus "accessory genes" present in one or more sequenced isolates. But how do we estimate the size of a pan-genome for a given species? A statistical model can predict the chance of finding new genes every time we sequence another genome. A few species, such as the anthrax agent *Bacillus anthracis,* have a "closed" pan-genome that appears to be defined by a relatively small number of natural isolates. But for *Bacillus cereus* (**Fig. 17.27**), after the first ten genomes, each new genome still reveals another 250 genes. Amazingly, the *B. cereus* pan-genome appears to be infinite in size. This is called an "open" pan-genome. Other species with open pan-genomes include the pneumonia pathogen *Streptococcus pneumoniae,* the marine cyanobacterium *Prochlorococcus marinus,* and the halophilic archaeon *Haloquadratum walsbyi.* Open pan-genomes may well predominate in nature; thus, most bacterial and archaeal species have access to their core genome plus an uncountable number of possible accessory genes.

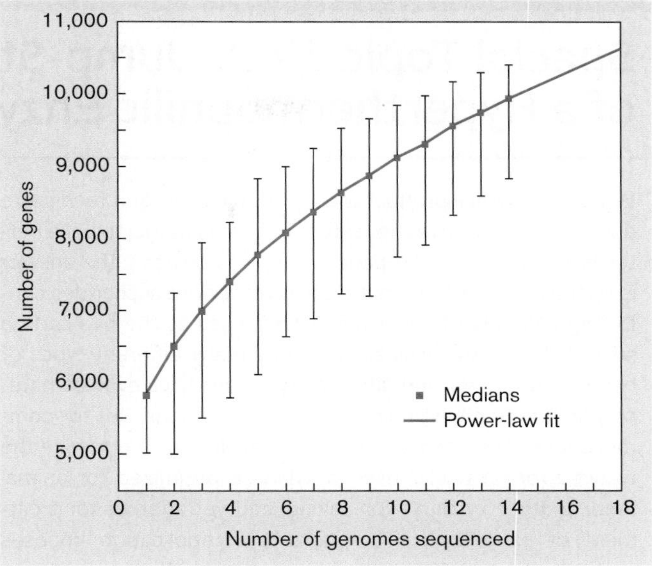
11,000
10,000
9,000
8,000
7,000
6,000
5,000
Number of genes
0
2
4
6
8
10
12
14
16
18
Number of genomes sequenced
Medians
Power-law fit

FIGURE 17.27 ■ Pan-genome of *Bacillus cereus*. The median number of genes found is shown as a function of the number of *B. cereus* genomes sequenced. Error bars indicate the range of gene number found upon sequencing different combinations of genomes. *Source:* Modified from Hervé Tettelin et al. 2008. *Curr. Opin. Microbiol.* **12**:472.

Classification and Nomenclature

Defining a species is part of the task of **taxonomy**, the description of distinct life-forms and their organization into different categories with shared traits. Taxonomy includes **classification**, the recognition of different classes of life; **nomenclature**, the naming of different classes; and **identification**, the recognition of the class of a given microbe isolated in pure culture. Taxonomy is critical for every microbiological pursuit, from diagnosing a patient's illness to understanding microbial ecosystems.

Classification generates a hierarchy of **taxa** (groups of related organisms; singular, taxon) based on successively narrow criteria. The fundamental basis of modern taxonomy is DNA sequence similarity, but the use of DNA arises within a long historical tradition of phenotypic description that shapes the views and practice of microbial taxonomy.

Special Topic 17.2: Jump-Starting Evolution of a Hyperthermophilic Enzyme

What commercial product is used to make chicken feed more digestible, to improve the texture of cotton fabric, to bake better bread, and to make paper using less bleach? The answer is xylanase, an enzyme that cleaves plant polysaccharides containing a variety of sugar monomers, such as the five-carbon sugar xylose. Xylanases are made by many different types of bacteria and fungi, and different xylanases have evolved naturally to function best under different conditions. But for commercial use, we need to "fine-tune" evolution to optimize the enzyme for particular uses: a xylanase optimized for animal feed, another xylanase for baking, and yet another for treatment of paper pulp. The paper pulp application imposes particularly demanding conditions of high pH and high temperature, and requires an enzyme that is stable above 100°C.

To design types of xylanases and other enzymes for industrial conditions, the Verenium Corporation in San Diego developed a technology of directed evolution. As introduced in Chapter 12, directed evolution "jump-starts" the evolution process by generating a large number of possible mutations—in some cases, the entire collection of single mutations possible in a given gene. Large-scale screening then "selects" for the mutant strains with optimal activity. While directed evolution may involve more than one round of mutation and screening, it avoids the need for hundreds or thousands of rounds of growth and selection.

To generate a hyperthermophilic xylanase, Verenium scientists first undertook "discovery," a process of collecting and testing genes encoding xylanase from many naturally occurring bacteria and fungi from all over the world. The most promising xylanase producers were collected from an alkaline hot spring of the Russian Kamchatka Peninsula, including bacteria such as *Bacillus* species and fungi such as *Trichoderma*. The scientists identified 200 different xylanases having promising levels of activity and stability. Both bacterial and fungal xylanases showed homology—members of a group called "family 11," which shares a common ancestor and related peptide structure.

From the hot-spring collection, several genes were selected that encoded the best thermophilic xylanases active at high pH. The top genes were used for "gene assembly," in which parental genes are recombined to form a highly diverse set of daughter genes (**Fig. 1**). In effect, the gene assembly process jump-starts evolution by forming in one generation a collection of recombinations that in nature would require many generations to occur by chance.

In gene assembly, the parental genes were cleaved by restriction endonucleases into several fragments (using methods described in Chapters 7 and 12). The gene fragments were then shuffled like a deck of cards by recombination in the test tube, generating a library of many different daughter genes (**Fig. 1**). For example, six parental genes cleaved into five fragments could recombine (assuming one fragment of each type) into $5^6 = 15{,}625$ different recombinants. The library was then screened in *E. coli* for activity of all the expressed enzymes, using a robotic screening system. Verenium scientists selected the gene encoding the xylanase that was most active at the highest temperatures, up to 76°C.

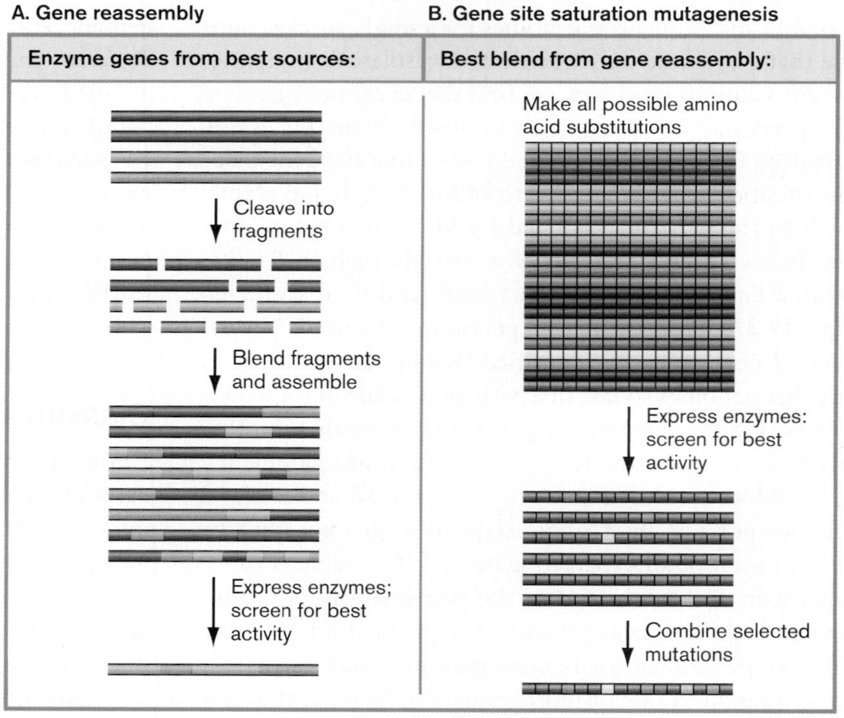
A. Gene reassembly
B. Gene site saturation mutagenesis
Enzyme genes from best sources:
Best blend from gene reassembly:
Cleave into fragments
Blend fragments and assemble
Express enzymes; screen for best activity
Make all possible amino acid substitutions
Express enzymes: screen for best activity
Combine selected mutations

FIGURE 1 ▪ Directed evolution by gene reassembly and in vitro mutagenesis. A. Candidate genes are cleaved, and then the fragments are blended and assembled (gene reassembly). **B.** The best product of gene reassembly is subjected to gene site saturation mutagenesis (GSSM), in which all possible amino acid substitutions are made. The mutants are screened and subjected to a combinatorial process that generates all possible mutant combinations.

The product of gene assembly was then subjected to a secondary method of directed evolution called "gene site saturation mutagenesis" (GSSM). GSSM was performed in collaboration with another lab—that of Claire Dumon and Harry Gilbert at Newcastle University, England. The GSSM method also imitates nature, but it yields in one "generation" a pool of all possible single-codon mutations that, in nature, might occur, but only over many generations. For GSSM, the evolved thermophilic xylanase gene was copied by PCR using primers with 64-fold degeneracy—that is, short DNA sequences that replace each codon in the gene sequence with all 64 possible codons (**Fig. 1**). For example, in a gene of 300 codons, making all the codon substitutions with 20 different amino acids at each position results in 300 × 20 = 6,000 different single-substitution mutations. The mutant genes were cloned in an expression vector and transformed into bacteria. To ensure that greater than 95% of the library was sampled, 70,000 different colonies were screened for xylanase activity and thermostability. Clones were selected that increased enzyme stability at 76°C from 10% to 50%.

The thermostable mutant genes were then assembled and combined by a multisite combinatorial assembly process that generates all possible combinations of the single-codon substitutions. **Figure 2** shows a distribution of the number of combined mutations in combinatorial library isolates with enzyme activity at temperatures as high as 86°C. The combinatorial process ultimately yielded an enzyme (**Fig. 3**) that is active at even higher temperatures, up to 101°C. A compelling result is that none of the mutations found would have been predicted from crystal structure to increase thermostability; they arose only from the screen of all possible mutants and, in combination, yield dramatic enhancement.

The hyperthermophilic xylanase is now used for paper pulp manufacture, enabling the pulp to be processed with much lower levels of bleach, which is toxic to the environment. Thus, directed evolution in the test tube jump-started the development of a product to enhance industrial sustainability.

RESEARCH QUESTION

In **Figure 2**, why do you think the distribution of combined-mutation isolates is shifted for larger numbers of mutations in the hyperthermophilic isolates? What does this observation predict about the evolution of hyperthermophiles?

Dumon, Claire, Alexander Varvak, Mark A. Wall, James E. Flint, Richard J. Lewis, et al. 2008. Engineering hyperthermostability into a GH11 xylanase is mediated by subtle changes to protein structure. *Journal of Biological Chemistry* **283**:22557–22564.

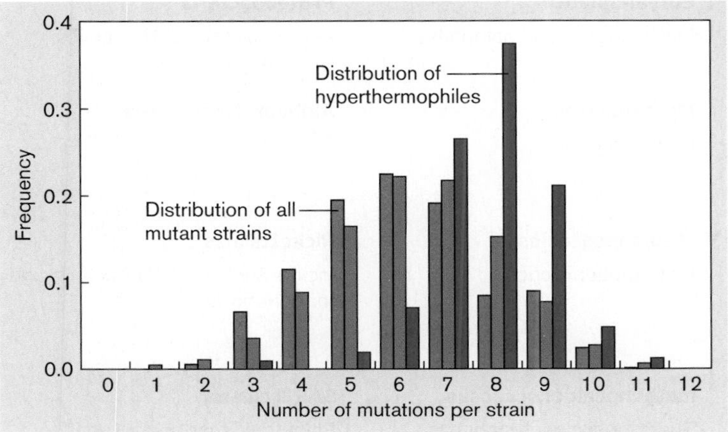
0.4
0.3
0.2
0.1
0.0
Frequency
Distribution of
hyperthermophiles
Distribution of all
mutant strains
0 1 2 3 4 5 6 7 8 9 10 11 12
Number of mutations per strain

FIGURE 2 ▪ **Distribution of mutants in the combinatorial library.** The frequencies of combination strains are plotted as a function of number of mutations per strain. Blue bars represent the frequencies of hyperthermophiles; red bars represent the distribution of all combination strains. *Source:* Modified from Claire Dumon et al. 2008. *J. Biol. Chem.* **283**:22557–22564, fig. 2.

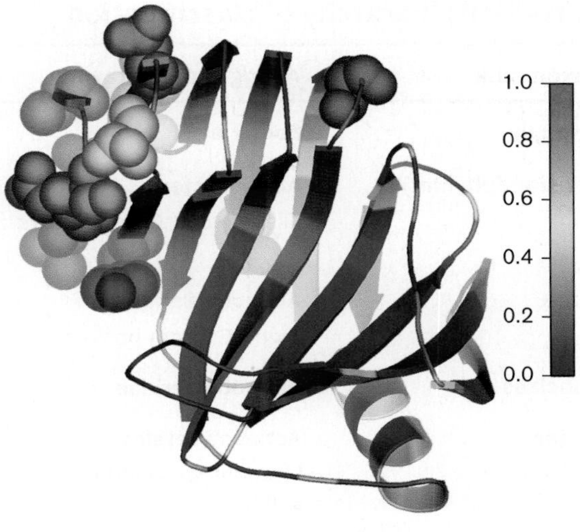
1.0
0.8
0.6
0.4
0.2
0.0

FIGURE 3 ▪ **Hyperthermophilic xylanase.** The crystallographic model shows the position of mutations conferring enhanced thermostability. Amino acid color represents the proportion of the GSSM combinatorial library strains that include a substitution at that position. The amino acid substitutions that were combined in the best hyperthermophilic xylanase are shown as space-filled atoms. *Source:* Modified from Claire Dumon et al. 2008. *J. Biol. Chem.* **283**:22557–22564, fig. 5.

Historically, taxa have been defined and named on the basis of a combination of genetic and phenotypic traits. The nomenclature of microbes remains surprisingly fluid: As new traits are identified and the genetic sequence of a microbe is established, species are all too frequently renamed. Current information on microbial names is available through online databases such as the List of Prokaryotic Names with Standing in Nomenclature.

Weblinks List of Prokaryotic Names with Standing in Nomenclature (LPSN) (*see ebook*)

Traditional classification designates levels of taxonomic hierarchy, or rank, such as phylum, class, order, and family. Some levels of rank are designated by certain suffixes—for example, "-ales" (order) and "-aceae" (family). The ultimate designation of a type of organism is that of **species**, which includes the capitalized **genus name** (group of closely related species) followed by the uncapitalized **species name**—for example, *Streptomyces coelicolor*.

The conventions of nomenclature and levels of rank vary widely across different taxa, as shown in **Table 17.3**. Species with a long history of study are given many levels of rank in the literature, whereas species discovered recently show relatively few, regardless of genetic diversity. *S. coelicolor* is a well-studied member of the actinomycetes, filamentous Gram-positive bacteria that produce many complex natural products, including antibiotics. The actinomycetes are subdivided into classes, orders, and families, as well as sublevels of the traditional ranks. Compare the actinomycetes with a species of a more recently described class, the hyperthermophilic methanogen *Methanocaldococcus jannaschii* (formerly *Methanococcus jannaschii*). *M. jannaschii* is defined only by the minimal levels of rank, without subclasses or suborders.

In 2002, a group of mostly unculturable marine bacteria, the SAR11 cluster, was found to comprise approximately 25% of all marine microbial cells. SAR11 was originally defined by its SSU rRNA sequence of uncultured environmental samples. Surprisingly, the rRNA sequence places these marine oligotrophs in the order Rickettsiales, most of whose known members are obligate intracellular parasites, such as *Rickettsia,* the cause of Rocky Mountain spotted fever—and also the endosymbiont ancestor of our mitochondria. The SAR11 cluster has not been classified below the rank of order, except for one species cultured in 2002: *Pelagibacter ubique.*

TABLE 17.3

Taxonomic hierarchy of classification.

Taxon rank	A long-studied taxon	A less-studied taxon	An emerging taxon
Domain	**Bacteria**	**Archaea**	**Bacteria**
Division (phylum)	**Actinobacteria** Filamentous; Gram-positive	**Euryarchaeota** Methanogens and halophiles	**Proteobacteria** Purple bacteria and relatives; Gram-negative
Class	**Actinobacteria** High GC; Gram-positive	**Methanococci** Methanogens	**Alphaproteobacteria** Gram-negative
Subclass	**Actinobacteridae**		
Order	**Actinomycetales** Filamentous; acid-fast stain	**Methanococcales** Methanogenic cocci	**Rickettsiales** Includes intracellular bacteria and mitochondria
Suborder	**Streptomycineae**		
Family	**Streptomycetaceae** Filamentous; hyphae produce spores	**Methanocaldococcaceae** Thermophilic methanogens	**SAR11 cluster** Planktonic marine bacteria, mostly uncultured
Genus	***Streptomyces***	***Methanocaldococcus***	***Pelagibacter***
Species (date first described)	***S. coelicolor* (1908)**	***M. jannaschii* (1984)**	***P. ubique* (2002)**

Note: Taxonomic categories generally have two forms: formal and informal. The formal term is capitalized, with a latinized suffix: Actinomycetales, *Pseudomonas, Micrococcus.* The informal term is lowercase, in some cases with an anglicized ending; and informal references to genera are not italicized: actinomycetes, pseudomonads, micrococci.

Emerging Clades: Unclassified and Uncultured Bacteria

As we discover new microorganisms, how do we decide what to call them? The term **emerging** is used to refer to an organism recently discovered or described. If the new organism causes a disease, it is called an "emerging pathogen." An emerging organism that cannot be cultured may be known only by its habitat and its small-subunit rRNA sequence. Such organisms require a culture method and a description of essential cell structure and metabolism before designation as new species. "Incompletely described" emerging organisms are designated as follows:

- **Unclassified/uncultured organism.** An unclassified organism, or uncultured organism, is assigned to a taxonomic rank based on SSU rRNA sequence but has not yet been grown in pure culture—for example, an "uncultured actinomycete."
- **Environmental sample.** An environmental sample is designated by its habitat and assigned a rank based on SSU rRNA. Examples of environmentally defined actinomycetes in the NCBI database include "oil-degrading bacterium AOB1" and "glacier bacterium FJS11."
- **Candidate species.** A cultured organism with some physiological characterization beyond DNA sequence may be published with a provisional status of **candidate species**, designated by the prefatory term "*Candidatus.*" For example, "*Candidatus* Nostocoida limicola" was published as "a filamentous bacterium from activated sludge."

Nongenetic Categories for Medicine and Ecology

Genetic relatedness is the standard for classifying and naming organisms in all fields of biology. At the same time, several nongenetic systems of categorization serve a practical purpose in certain fields. These systems include:

- **Phenotypic categories for identification.** Categories such as pigmentation or cell shape (rod or coccus) may have minor genetic significance but are useful for practical identification of organisms isolated from field or clinical sources.
- **Ecological categories.** In ecology, the niche filled by an organism may be more important than its phylogeny. For example, cyanobacteria and sequoia trees both fill the role of photosynthetic producer of biomass. Both are primary producers—a trophic category of organisms that feed other organisms in the food web. Ecological categories are discussed further in Chapter 21.
- **Disease categories.** In medical microbiology, microorganisms are categorized according to the type of disease they cause or the host organ system they inhabit. For example, *Mycoplasma pneumoniae* and influenza virus are both pulmonary pathogens, whereas *Escherichia coli* and *Bacteroides thetaiotaomicron* are both normal members of the gut microbiome. Disease categories are discussed further in Chapters 25 and 26.

Defining a Species

A commonly accepted set of names for species and higher taxa is essential for research and communication. The accepted rules for naming species and taxa have been determined by the International Committee on Systematics of Prokaryotes (ICSP). The ICSP establishes minimal criteria for designating species, genera, classes, and other taxa of bacteria and archaea. To establish a new species, a previously unknown form of microbe must be isolated and grown in pure culture; this cultured organism is known as an **isolate**. The isolate's unique genetic and phenotypic traits are published, with a proposed species name designated "*Candidatus*" for **candidate species**. An example of a candidate species is "*Candidatus* Nitrosopumilus maritimus," a marine archaeon that oxidizes ammonia, first isolated in 2005. The candidate species becomes accepted as an official species upon publication in the *International Journal of Systematic and Evolutionary Microbiology,* the official journal of record for novel prokaryotic taxa.

All accepted taxonomic categories and species descriptions are compiled in *Bergey's Manual of Determinative Bacteriology,* a multivolume reference work of prokaryotic taxonomy. Up-to-date taxonomy of bacteria and archaea, as well as eukaryotes, is maintained in the "Taxonomy" browser of the interactive National Center for Biotechnology Information (NCBI) online database, supported by the U.S. government.

In recent years, the standard practice for accepting and naming species has been overwhelmed by the number of new isolates, as well as uncultured microbes, reported in the literature. Many uncultured microbes are known only by their SSU rRNA sequences. Increasingly, however, uncultured organisms have large portions of their genomes sequenced and assembled using **metagenomics** (presented in Chapters 7 and 21). Metagenomics is the sequencing of multiple genomes in an environmental community. The community can be sampled from any kind of habitat, such as the human digestive tract or the Sargasso Sea.

Identification

Once a species has been described and classified, we need a way to identify future members of the species isolated from natural environments. Identification poses special difficulties with microbes, which, by definition, are invisible to the unaided eye. Even under the electron microscope, thousands of divergent species may possess similar shape and form.

Because the definition of a species is based on its genetic sequence, the most consistent and straightforward way to identify an isolate is to sequence part or all of its genome or to attempt hybridization of its DNA with a labeled sequence probe (discussed in Section 7.6). In clinical practice, such DNA-based methods are used increasingly. Nevertheless, even in the clinical lab—as well as in field and environmental microbiology—it is convenient to narrow down the possibilities using various easily determined traits, such as cell shape, staining properties, and metabolic reactions. Thus, practical identification is based on a combination of phylogeny (relatedness based on DNA sequence divergence) and phenetic, or phenotypic, traits.

A common strategy of practical identification is the **dichotomous key**, in which a series of yes/no decisions successively narrows down the possible categories of species. **Figure 17.28** shows an example of a dichotomous key used to identify filamentous bacteria isolated from a wastewater treatment facility. Filamentous bacteria are important in wastewater because their entangled filaments interfere with the settling of bacteria out of the treated water. Note that in the key shown, most of the traits are phenotypic, such as cell size and motility. Most of the branch decisions have two choices, although one juncture (cell motility and shape) has four. The key identifies some organisms down to the species level (*Sphaerotilus natans*), others to the genus level (*Flexibacter*), and others only to numbered samples whose characterization remains incomplete (0914, 021N).

Thought Question

17.7 What would be the identification of a straight, nonsheathed, Gram-negative bacterium that has sulfur granules, is motile, and is 1.0 μm wide? What would happen if you assigned the bacterium a width of 0.9 μm?

A disadvantage of the dichotomous key is that it requires a series of steps, each of which takes time. In the clinical setting, time is critical in identifying a potential pathogen and prescribing appropriate treatment. An alternative means of identification is the probabilistic indicator. A **probabilistic indicator** is a battery of biochemical tests performed simultaneously on an isolated strain. The indicator requires a predefined database of known bacteria from a well-studied habitat, such as Gram-negative bacteria from the human intestinal tract. To build such a database, numerous strains of each species must be isolated in pure culture and tested for all the traits in the database. The fraction of isolates that test positive for each trait, for each species, is noted in the database as the probability of obtaining a positive result. The database can be used to identify a specimen isolated from a patient (discussed in Chapter 28). But, like the dichotomous key, the probabilistic indicator works only if the isolate coincides with a member of the predefined database.

To Summarize

- **Microbial species** are defined by sequence similarity of vertically transmitted genes such as SSU rRNA sequences and multiple orthologous genes. The species definition should be consistent with the ecological niche or pathogenicity.
- A **pan-genome** includes core genes possessed by all isolates of a species plus accessory genes found in some isolates but not others. A pan-genome may be open (infinite number of genes) or closed (finite set of available genes).
- **Taxonomy** is the description and organization of lifeforms into classes (taxa). Taxonomy includes classification, nomenclature, and identification.
- **Classification** is traditionally based on a hierarchy of ranks. Groups of organisms long studied tend to have many ranks, whereas recent isolates have few.
- **DNA sequence relatedness** defines microbial taxa. Below genus level, however, the definition of bacterial species can be problematic.
- **Emerging taxa** are types of organisms recently discovered or described. They may be uncultured, and their phylogeny may be uncertain.
- **Practical identification** is based on phenotypic and genetic traits. Methods of identification include the dichotomous key and the probabilistic test battery. Both methods assume a predefined set of organisms.

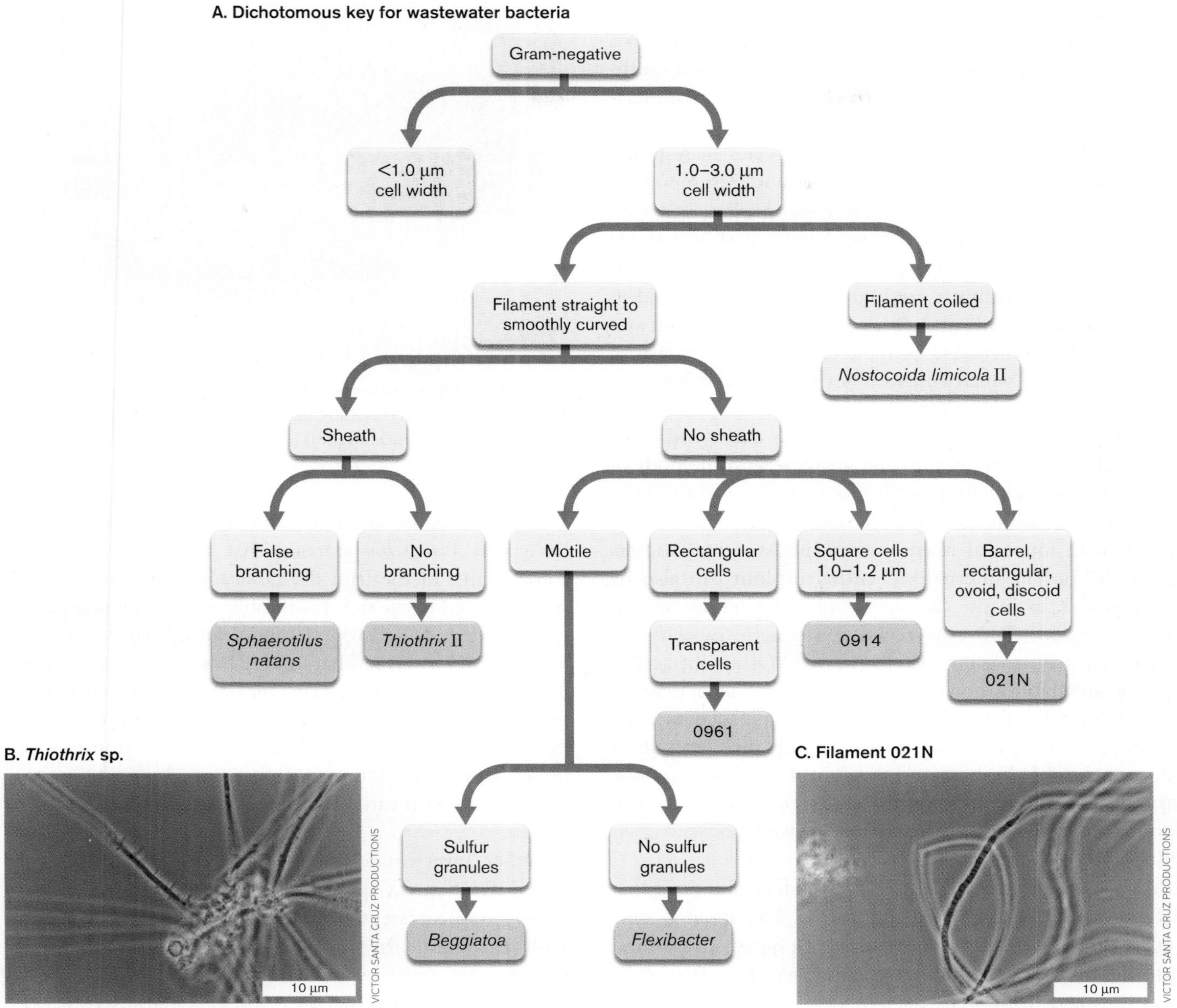
A. Dichotomous key for wastewater bacteria
Gram-negative
<1.0 μm cell width
1.0–3.0 μm cell width
Filament straight to smoothly curved
Filament coiled
Nostocoida limicola II
Sheath
No sheath
False branching
No branching
Motile
Rectangular cells
Square cells 1.0–1.2 μm
Barrel, rectangular, ovoid, discoid cells
Sphaerotilus natans
Thiothrix II
Transparent cells
0914
021N
0961
Sulfur granules
No sulfur granules
Beggiatoa
Flexibacter
B. Thiothrix sp.
10 μm
VICTOR SANTA CRUZ PRODUCTIONS
C. Filament 021N
10 μm
VICTOR SANTA CRUZ PRODUCTIONS

FIGURE 17.28 ■ Dichotomous key for wastewater bacteria. **A.** A section of a dichotomous key for identifying Gram-negative bacteria from wastewater. The designations 0961, 0914, and 021N refer to unidentified species. **B.** *Thiothrix* species, a filamentous wastewater bacterium. **C.** Filament 021N, an unclassified organism that forms starburst filaments. *Source:* A. modified from Santa Cruz Database Central, Wastewater.

17.6

Symbiosis and the Origin of Mitochondria and Chloroplasts

So far, we have largely considered single species in isolation. In fact, however, all organisms evolve in the presence of other kinds of species, with whom they share interactions, both positive and negative. A major engine of evolution is **symbiosis**, the intimate association of two unrelated species. The ecology and behavioral adaptations of microbial symbiosis are discussed in Chapter 21; here we focus on the role of symbiosis in evolution of cells.

Evolution of Endosymbiosis

The word "symbiosis" is popularly understood to mean **mutualism**, a relationship in which both partners benefit and may absolutely require each other. Biologists, however, recognize **parasitism**, in which one partner is harmed, as a relationship equally as intimate as mutualism; both mutualism and parasitism are forms of symbiosis. Intimate relationships between species, either negative or positive, lead to **coevolution**, the evolution of two species in response to one another, showing parallel phylogeny. An important bacterial mutualism is that of nitrogen fixation, in which rhizobia form intracellular "bacteroids" within legume plants that cannot fix nitrogen on their own. Both rhizobia and their plant hosts are highly evolved to respond to each other chemically and develop the nitrogen-fixing system. Another remarkable example of coevolution is that of leaf-cutter ants, which cultivate both fungal and bacterial partners (discussed in eTopic 17.3).

The most intimate kind of symbiosis is **endosymbiosis**, in which one partner population grows within the body of another organism. Endosymbiosis includes communities of microbes within the digestive tracts of animals, such as the human intestinal microbiome (discussed in Chapters 21 and 23). The internalized **endosymbiont** can also be intracellular, as in the case of rhizobial bacteroids within legume tissues. Rhizobia retain the genetic capacity for independence, growing readily in soil. Other intracellular endosymbionts, however, become wholly dependent on their host cells. Pathogenic endosymbionts, such as chlamydias, evolve specialized traits enabling their growth at the expense of the host and evasion of the host immune system. But endosymbionts also undergo drastic reductive evolution, evolving ever-deeper interdependence with their host cells.

A simple example of intracellular endosymbiosis is that of the alga *Chlorella* growing within *Paramecium bursaria* (**Fig. 17.29**). The algae conduct photosynthesis and provide nutrients to the paramecium, which in turn shelters the algae from predators and viruses. The relationship is highly specific—only certain species of algae and paramecia participate—and the algal growth is limited to a population that avoids harming the host. This relationship may give clues to how the bacterial ancestor of chloroplasts began its intracellular existence.

Despite this intimate relationship, *Chlorella* retains its ability to multiply outside the paramecium; thus this symbiosis is reversible. Moreover, under conditions of starvation in the absence of light, the paramecium may start to digest its endosymbionts as prey. Thus, the nature of the symbiosis (mutualistic or predatory) depends on the environment.

Many invertebrate animals, often themselves parasites of animals or plants, possess obligate bacterial endosymbionts.

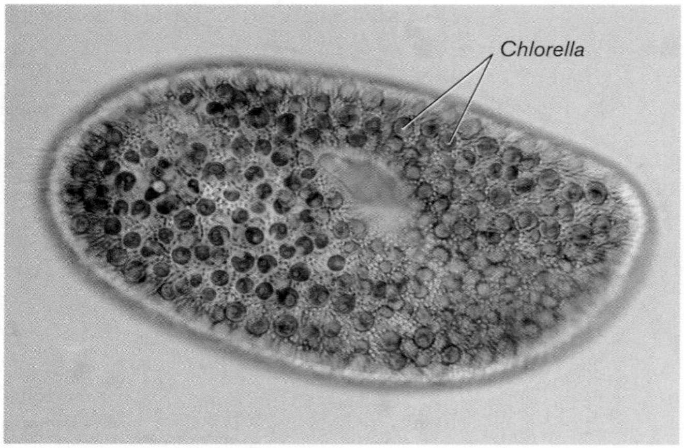
Chlorella
WIM VAN EGMOND/VISUALS UNLIMITED

FIGURE 17.29 ▪ Endosymbiosis. *Paramecium bursaria,* a ciliate protist with endosymbiotic *Chlorella* algae. Cell length of *Paramecium*: 100–150 μm (LM).

In some cases the endosymbiont is a parasite, such as the bacterium *Wolbachia pipientis* that grows within cells of some strains of the fruit fly, *Drosophila*. *Wolbachia* bacteria are related to the rickettsias, obligate intracellular pathogens carried by arthropods and transmitted to humans. The *Wolbachia* strains that infect *Drosophila* cells are transmitted only through egg cells; they cannot exist outside the insect. The bacteria have evolved ways to manipulate *Drosophila* reproduction to enhance their own transmission; for example, they feminize infected males so as to produce eggs, which carry *Wolbachia* into the next generation of flies.

Other invertebrate endosymbionts are mutualists. In fact, 15% of insect species depend on intracellular bacteria to produce essential nutrients, such as certain amino acids or vitamins. In these mutualisms, both partners have lost essential traits by reductive evolution, and each now requires the partner species to provide the lost function. The degree of genome reduction is most extreme in the intracellular partner, which can be seen as heading in the same evolutionary direction that led to mitochondria and chloroplasts.

A surprising number of human invertebrate parasites, such as filarial nematodes and Anopheles mosquitoes, carry bacterial endosymbionts required for host growth. This discovery has exciting implications for treatment of parasitic diseases. Invertebrate parasites such as filarial nematode worms invade human lymph nodes, causing forms of disease (filariasis) that are notoriously difficult to treat. Few anti-nematode compounds are sufficiently selective for worm metabolism versus human metabolism, since both are eukaryotic animals. A form of filariasis is elephantiasis, in which a limb expands with edema due to blockage of lymph ducts by worms (**Fig. 17.30**). Filariasis afflicts more than 120 million people worldwide, largely in the Indian subcontinent and in Africa.

The filarial nematode *Brugia malayi* harbors *Wolbachia* endosymbionts (**Fig. 17.31**). These *Wolbachia* strains differ from those that parasitize insects. *Wolbachia* may have entered the nematode originally as a pathogen or parasite, and then persisted because of its metabolic contributions to the host. The nematode endosymbiont strains are mutualists; their presence is required for the nematode's embryonic development. The bacteria are found within tissue layers beneath the nematodes' skin and within the uterine tubes of females, where they enter the developing offspring (**Fig. 17.31**). When human patients infected by the nematodes are treated with antibiotics such as tetracycline, the bacteria disappear from worm tissues. The worm burden gradually decreases, and no offspring are produced. Antibacterial antibiotics eliminate the worms sooner and more completely than does treatment with anti-nematode agents.

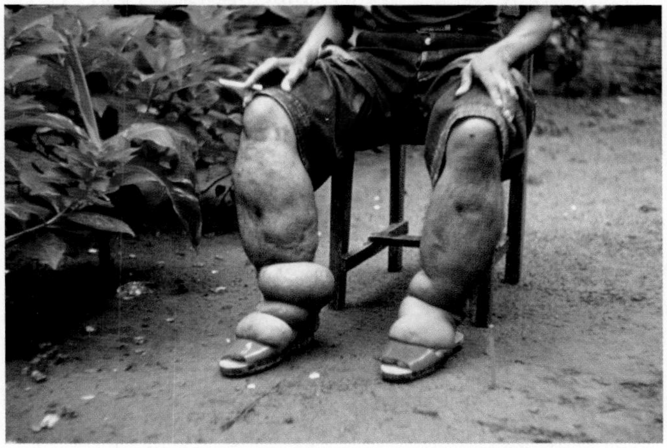
CDC

FIGURE 17.30 ■ Filariasis. Patient suffering from filariasis, a form also known as elephantiasis.

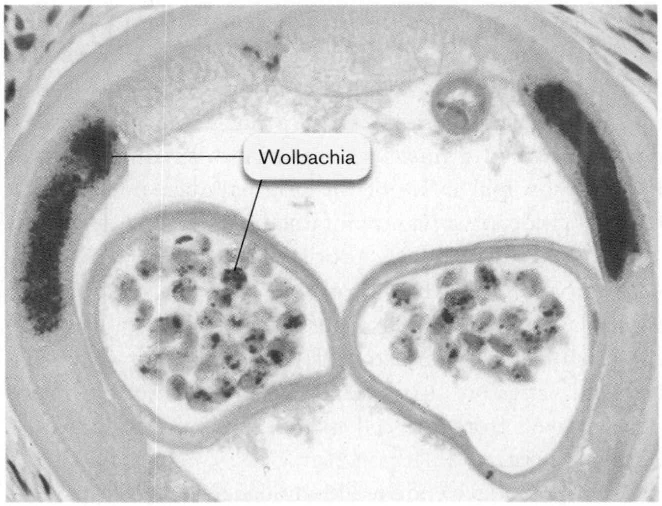
Wolbachia
K. PFARR ET AL. 2005. PLOS MED. 2: E110

FIGURE 17.31 ■ The filarial endosymbiont *Wolbachia*. Cross section of the nematode *Brugia malayi*, showing *Wolbachia* bacteria (stained pink) within the worm's dermis and uterine tubes.

The genome of a *Wolbachia* strain from a filarial nematode reveals extensive reductive evolution. With barely a million base pairs, the *Wolbachia* genome has lost many metabolic pathways. It retains glycolysis and the TCA cycle but has lost the pathways for biosynthesis of all amino acids and most vitamins. It nonetheless retains pathways to make purines, pyrimidines, and the coenzymes riboflavin and FAD—essential pathways lost by its host nematode. Overall, *Wolbachia* appears to be evolving into an organelle of its host, like the ancestors of mitochondria and chloroplasts.

Mitochondria and Chloroplasts

As Lynn Margulis and colleagues have shown, the assimilation of endosymbionts as mitochondria and chloroplasts played a central role in the evolution of eukaryotes (**Fig. 17.32**). Many similar endosymbioses are known today, such as the free-living algae *Chlorella* acquired by the protist predator *Paramecium bursaria* (see **Fig. 17.29**). Like *Wolbachia,* mitochondria evolved from a bacterium related to the rickettsias. The mitochondrial ancestor must have entered the eukaryotic lineage as, or shortly after, the eukaryotes diverged from archaea, since all known eukaryotes retain mitochondria or vestigial remnants of mitochondrial genomes (discussed in Chapter 20). Mitochondria provide the cell with the essential functions of electron transport and respiration. The electron transport system (ETS) is found in the mitochondrial inner membrane, believed to derive from the cell membrane of the ancestral bacterium. The outer membrane may derive from the invaginating membrane of the host cell that originally engulfed the endosymbiont.

Chloroplasts arose from cyanobacteria at some point before the divergence of red and green algae (discussed in Chapter 20). A model for cyanobacterial uptake can be seen in the protist *Glaucocystophyta,* which independently (and much later) took up cyanobacterial endosymbionts. The endosymbionts of *Glaucocystophyta* retain cell walls and some metabolism. Like mitochondria, chloroplasts possess inner and outer membranes, believed to derive from the ancestral endosymbiont and host, respectively. Photosynthetic complexes are located in the thylakoid membranes, similar to those of modern cyanobacteria.

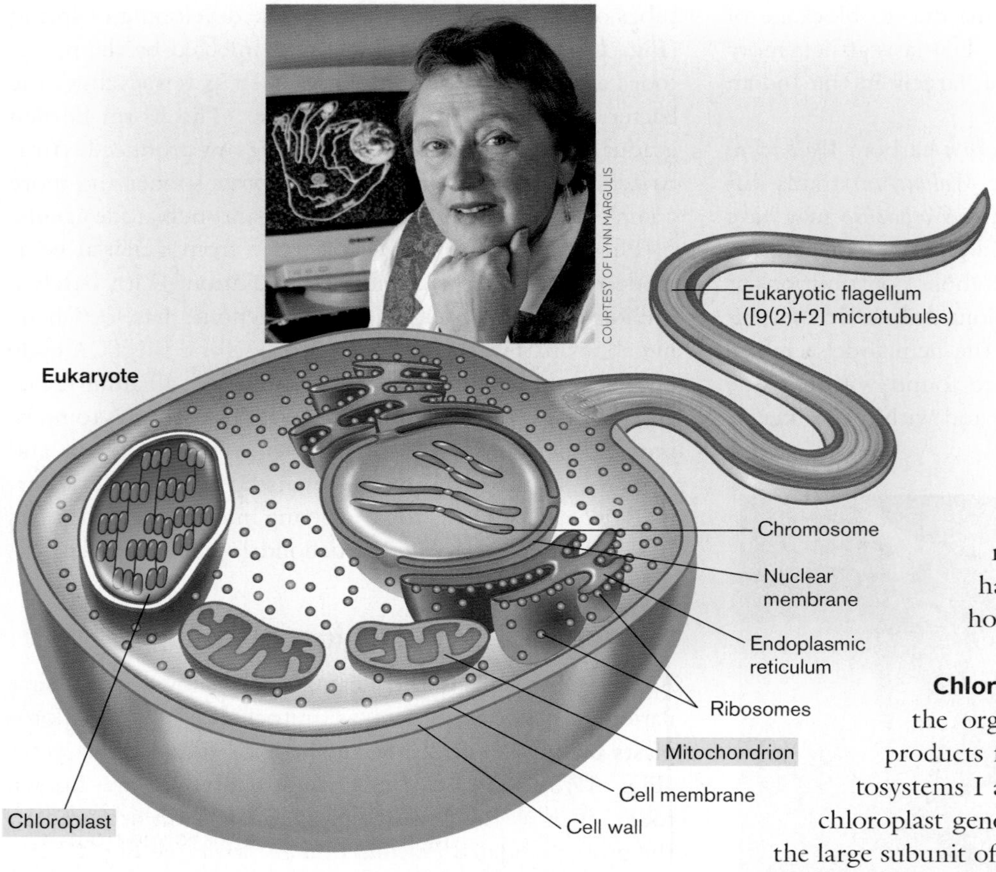
COURTESY OF LYNN MARGULIS
Eukaryote
Eukaryotic flagellum
([9(2)+2] microtubules)
Chromosome
Nuclear
membrane
Endoplasmic
reticulum
Ribosomes
Mitochondrion
Cell membrane
Cell wall
Chloroplast

FIGURE 17.32 ▪ Endosymbiotic cells evolved into mitochondria and chloroplasts. Eukaryotic cells contain mitochondria and chloroplasts, organellar remnants of ancient endosymbioses. **Inset:** Lynn Margulis (1938–2011).

The genomes of mitochondria and chloroplasts both show extreme reduction (**Fig. 17.33**)—even more extreme than that of any known endosymbiotic bacteria. The few genes that remain in the organellar genome include remnants of the central transcription-translation apparatus, such as rRNA and tRNAs, as well as a handful of genes whose products are essential for survival of the host cell: for respiration (mitochondria) or photosynthesis (chloroplasts).

Mitochondrial genome. In human mitochondria, the mitochondrial genome encodes key parts of the respiratory chain, including subunits of NADH dehydrogenase, cytochrome oxidase, and ATP synthase (**Fig. 17.33A**). Mutations in these key genes lead to serious diseases; for example, damage to mitochondrial genes of respiration is associated with motor neuron disease, parkinsonism, and forms of ataxia.

But thousands of genes encoding ETS subunits, as well as other essential parts of mitochondria, have migrated from the mitochondrion to the nucleus. The nuclear acquisition probably occurred through accidental copying of mitochondrial genes into the nuclear genome. Reductive evolution then occurred, faster in the mitochondrial copy because of the faster mutation rate. Some of these nuclear-acquired mitochondrial genes show tissue-specific expression, resulting in different mitochondrial types associated with different tissues. Thus, some mitochondrial defects are actually inherited through the nuclear genome. The mitochondria have evolved as integral parts of the host cell.

Chloroplast genome. In chloroplasts, the organellar genome encodes essential products for photosynthesis, including photosystems I and II and the ATP synthase. The chloroplast genome shown in **Fig. 17.33B** retains the large subunit of Rubisco, whereas the gene encoding the small subunit has migrated to the nucleus.

Remarkably, the process of **symbiogenesis**, the generation of new symbiotic associations, continues in a number of protist species. Protist-algae, also known as "secondary symbiont" algae, result from symbiogenesis in which an alga (containing a chloroplast) was engulfed by an ancestral protist. The protist cell contains the degenerate remains of the algal endosymbiont (**Fig. 17.34**). The algal mitochondrion was lost through reductive evolution, and the nucleus shrank to a nucleomorph, the vestigial remains of a nucleus containing a small amount of the chromosomal DNA of the original algal nucleus. But the algal chloroplast was maintained, "enslaved" by its new host. The result is a new protist-alga species, *Guillardia theta,* capable of both phototrophy and heterotrophy. Its chloroplast has a double membrane derived from the original cyanobacterial ancestor of the chloroplast and from the algal ancestor, surrounded by another double membrane derived from the cell membranes of the alga and a secondary host.

Other secondary endosymbiont algae, such as kelps, diatoms, and dinoflagellates, are discussed in Chapter 20. In some species, tertiary symbiosis has been documented, in which a secondary-endosymbiont alga has been swallowed in turn by another protist.

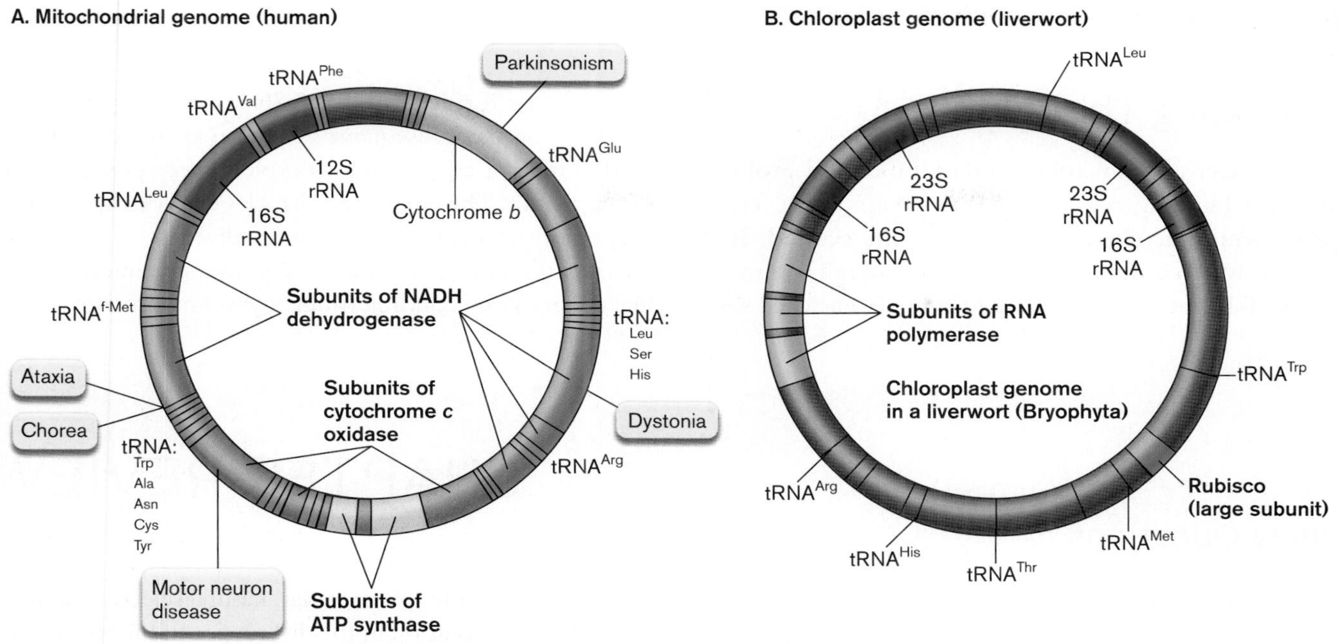
A. Mitochondrial genome (human)
tRNA^Phe
tRNA^Val
Parkinsonism
tRNA^Glu
12S rRNA
tRNA^Leu
16S rRNA
Cytochrome b
tRNA^f-Met
Subunits of NADH dehydrogenase
tRNA: Leu Ser His
Ataxia
Chorea
Subunits of cytochrome c oxidase
Dystonia
tRNA: Trp Ala Asn Cys Tyr
tRNA^Arg
Motor neuron disease
Subunits of ATP synthase
B. Chloroplast genome (liverwort)
tRNA^Leu
23S rRNA
23S rRNA
16S rRNA
16S rRNA
Subunits of RNA polymerase
tRNA^Trp
Chloroplast genome in a liverwort (Bryophyta)
tRNA^Arg
Rubisco (large subunit)
tRNA^His
tRNA^Thr
tRNA^Met

FIGURE 17.33 ▪ Genomes of mitochondria and chloroplasts. A. The mitochondrial genome retains large- and small-subunit rRNAs, five tRNA genes, plus subunits of the respiratory electron transport chain. Bubbles indicate human diseases associated with mitochondrial defects. **B.** The chloroplast genome retains large- and small-subunit rRNAs (23S and 16S), several tRNA and RNA polymerase genes, plus Rubisco (large subunit) and components of photosystems I and II (dispersed around the circle, not labeled in the figure).

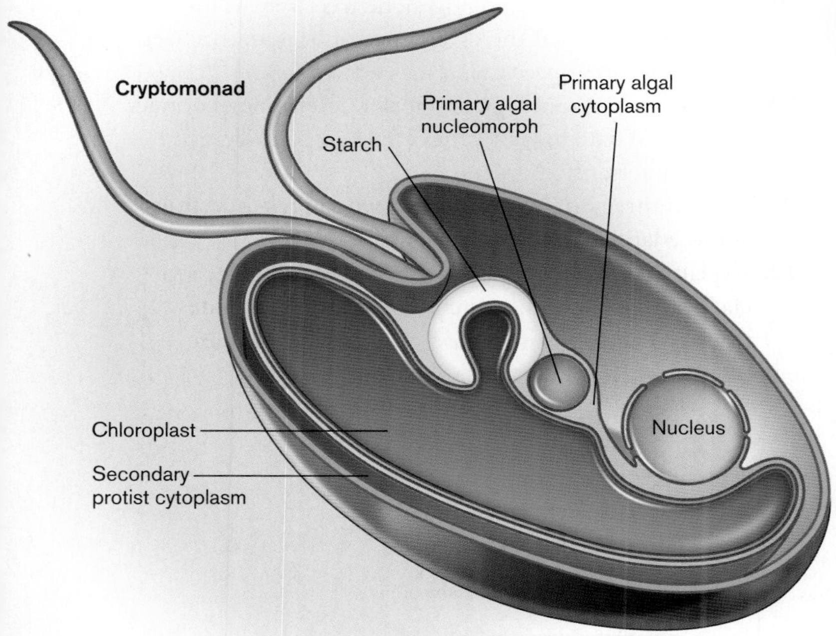
Cryptomonad
Primary algal cytoplasm
Primary algal nucleomorph
Starch
Chloroplast
Nucleus
Secondary protist cytoplasm

FIGURE 17.34 ▪ A protist-alga results from secondary endosymbiosis. The protist *Guillardia theta* contains the primary algal chromosome surrounded by the remains of the primary algal cytoplasm and nucleus, which has shrunk to a nucleomorph.
Source: Modified from Paul R. Gilson. 2001. *Genome Biol.* **2**:1022.

Thought Question

17.8 Besides mitochondria and chloroplasts, what other kinds of entities within cells might have evolved from endosymbionts?

To Summarize

- **Symbiosis is the intimate association of two unrelated species.** A symbiosis in which both partners benefit is called **mutualism**. If one partner benefits while harming the other, the symbiosis is called **parasitism**.
- **Symbiotic partners undergo coevolution**, the evolution of two species in response to one another. Coevolution involves reductive (degenerative) evolution, in which each partner species loses some functions that the other partner provides.
- **An endosymbiont lives inside a much larger host species.** Many microbial cells harbor endosymbiotic bacteria whose metabolism yields energy for their hosts.
- **Many invertebrates harbor endosymbiotic bacteria.** The bacteria are required for host survival and, in some cases, for pathology caused by a parasitic invertebrate.
- **Mitochondria evolved from endosymbionts.** The ancestor of mitochondria was an alphaproteobacterium related to rickettsias.
- **Chloroplasts evolved from endosymbionts.** The chloroplast ancestor was a cyanobacterium.

Concluding Thoughts

Over 3.85 Gyr ago, microbial communities had evolved, their strata building rock layers that persist today. From these or other microbes, all subsequent life evolved. It is hard to say which is more astonishing: the overall commonalities of all living cells, including membrane-enclosed support systems for genomes of 3.85 billion years of shared ancestry, or the subsequent evolution of organisms with vastly different adaptations to exploit every possible niche of our planet. The next three chapters explore these diverse adaptations: Chapter 18, bacterial diversity; Chapter 19, archaeal diversity; and Chapter 20, diversity among microbial eukaryotes, including fungi, algae, and protists.

CHAPTER REVIEW

Review Questions

1. What was the composition of Earth's early crust and atmosphere? What processes changed their composition to that found today?
2. What kinds of evidence support the presence of life in the Archaean eon? What are the advantages and limitations of each kind of evidence?
3. What kinds of metabolism are believed to have existed in Archaean life? What kinds of evidence support their existence?
4. Compare and contrast three models for the origin of the first cells. Which features of life does each model explain, and which features are unexplained?
5. Explain the roles of classification, nomenclature, and identification for microbial taxonomy.
6. Why is the definition of species in bacteria and archaea more problematic than for eukaryotes? What is generally considered the present basis for defining prokaryotic species?
7. Discuss the roles of mutation, natural selection, and reductive evolution in the divergence of microbial species. Cite specific examples.
8. Explain the basis of a "molecular clock" for measuring microbial evolution. What fundamental properties must be met by a gene to function as a molecular clock? What are the limitations of a molecular clock?
9. Explain the basis of a phylogenetic tree. Why is the fundamental tree at the divergence of bacteria, archaea, and eukaryotes unrooted?
10. How does horizontal gene transfer determine genomic content? What kinds of genes are likely to undergo horizontal transfer?
11. Explain three different ways that we can test questions about adaptive evolution.
12. Explain how endosymbiosis can lead to obligate association. Explain how reductive evolution and gene transfer lead to the evolution of organelles that are inseparable from host cells.

Thought Questions

1. How convincing are the microfossils in **Figure 17.6**? What criteria do you think would define a microfossil?
2. In the phylogeny shown here, where are the root and the outgroup? How does the outgroup organism differ from the others? Which two organisms are the most closely related? Which node represents the last common ancestor of *Neisseria* and *Haemophilus*? Which genome has evolved much faster than the others, and why?

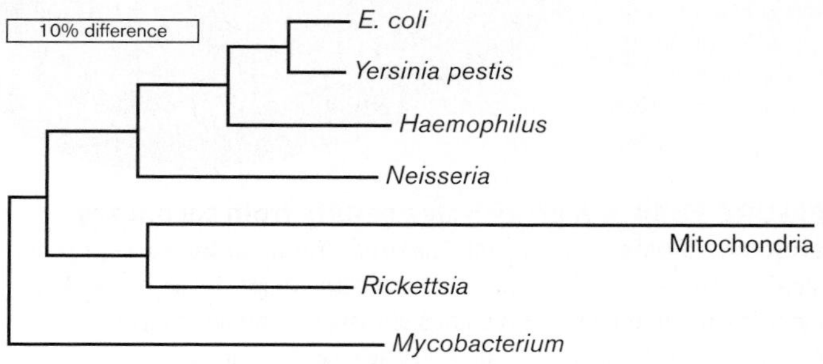

10% difference
E. coli
Yersinia pestis
Haemophilus
Neisseria
Mitochondria
Rickettsia
Mycobacterium

3. Design an evolution experiment in the laboratory to evolve a bacterium that breaks down a dangerous pollutant such as dioxin. How would you select the starting organism, and what experimental steps would you perform?

Key Terms

abiotic (677)
Archaean eon (669)
banded iron formation (BIF) (674)
biosignature (biological signature) (672)
biosphere (668)
candidate species (701)
clade (681)
classification (697)
coevolution (704)
core genome (697)
degenerative evolution (681, 692)
dichotomous key (702)
directed evolution (696)
domain (685)
emerging (701)
endosymbiont (704)
endosymbiosis (704)
experimental evolution (694)
gene duplication (692)
genomic island (691)
genus name (700)
greenhouse effect (669)
Hadean eon (669)
horizontal gene transfer (681, 690)
identification (697)
isolate (701)
isotope ratio (672)
laboratory evolution (694)
metabolist model (677)
metagenomics (701)
microfossil (671)
molecular clock (681)
monophyletic group (681)
mutualism (703)
natural selection (692)
node (684)
nomenclature (697)
pan-genome (697)
panspermia (680)
paralogous gene (paralog) (692)
parasitism (704)
phylogenetic tree (683)
phylogeny (681)
prebiotic soup (677)
probabilistic indicator (702)
reductive evolution (681, 692)
ribozyme (679)
RNA world (677)
root of a tree (684)
small-subunit (SSU) rRNA (682)
species (700)
species name (700)
stromatolite (666)
supernova (668)
symbiogenesis (706)
symbiosis (703)
taxon (697)
taxonomy (697)
vertical gene transfer (690)

Recommended Reading

Achtman, Mark, and Michael Wagner. 2008. Microbial diversity and the genetic nature of microbial species. *Nature Reviews. Microbiology* **6**:431–440.

Barns, Susan M., Charles F. Delwiche, Jeffery D. Palmer, and Norman R. Pace. 1996. Perspectives on archaeal diversity, thermophily and monophyly from environmental rRNA sequences. *Proceedings of the National Academy of Sciences USA* **93**:9188–9193.

Blount, Zachary D., Jeffrey E. Barrick, Carla J. Davidson, and Richard E. Lenski. 2012. Genomic analysis of a key innovation in an experimental *Escherichia coli* population. *Nature* **489**:513–518.

Blount, Zachary D., Christina Z. Borland, and Richard E. Lenski. 2008. Historical contingency and the evolution of a key innovation in an experimental population of *Escherichia coli. Proceedings of the National Academy of Sciences USA* **105**:7899–7906.

Brasier, Martin D., Owen R. Green, Andrew P. Jephcoat, Annette K. Kleppe, Martin J. Van Kranendonk, et al. 2002. Questioning the evidence for Earth's oldest fossils. *Nature* **416**:76–81.

Charusanti, Pep, Nicole L. Fong, Harish Nagarajan, Alban R. Pereira, Howard J. Li, et al. 2012. Exploiting adaptive laboratory evolution of *Streptomyces clavuligerus* for antibiotic discovery and overproduction. *PLoS ONE* **7**:e33722.

Currie, Cameron R., Michael Poulsen, John Mendenhall, Jacobus J. Boomsma, and Johan Billen. 2006. Coevolved crypts and exocrine glands support mutualistic bacteria in fungus-growing ants. *Science* **311**:81–83.

Fraser, Christophe, Eric J. Alm, Martin F. Polz, Brian G. Spratt, and William P. Hanage. 2009. The bacterial species challenge: Making sense of genetic and ecological diversity. *Science* **322**:741–746.

Gevers, Dirk, Frederick M. Cohan, Jeffrey G. Lawrence, Brian G. Spratt, Tom Coenye, et al. 2005. Re-evaluating prokaryotic species. *Nature Reviews. Microbiology* **3**:733–739.

Hanage, William P., Christophe Fraser, and Brian G. Spratt. 2005. Fuzzy species among recombinogenic bacteria. *BMC Biology* **3**:6.

Hanczyc, Martin M., Shelly M. Fujikawa, and Jack W. Szostak. 2003. Experimental models of primitive cellular compartments: Encapsulation, growth, and division. *Science* **302**:618–622.

Jiao, Yongqin, Andreas Kappler, Laura R. Croal, and Dianne K. Newman. 2005. Isolation and characterization of a genetically-tractable photoautotrophic Fe(II)-oxidizing bacterium, *Rhodopseudomonas palustris* strain TIE-1. *Applied and Environmental Microbiology* 71:4487–4496.

Lapierre, Pascal, and J. Peter Gogarten. 2009. Estimating the size of the bacterial pan-genome. *Trends in Genetics* **25**:107–110.

Lincoln, Tracey A., and Gerald F. Joyce. 2009. Self-sustained replication of an RNA enzyme. *Science* **323**:1229–1232.

Moran, Nancy A., John P. McCutcheon, and Atsushi Nakabachi. 2008. Genomics and evolution of heritable bacterial symbionts. *Annual Reviews in Genetics* **42**:165–190.

Ochman, Howard. 2005. Genomes on the shrink. *Proceedings of the National Academy of Sciences USA* **102**:11959–11960.

Salzberg, Steven L., Julie C. Dunning Hotopp, Arthur L. Delcher, Mihai Pop, Douglas R. Smith, et al. 2005. Serendipitous discovery of *Wolbachia* genomes in multiple *Drosophila* species. *Genome Biology* **6**:R23.

Tettelin, Hervé, David Riley, Ciro Cattuto, and Duccio Medini. 2008. Comparative genomics: The bacterial pan-genome. *Current Opinion in Microbiology* **12**:472–477.

Wernegreen, Jennifer J. 2005. Endosymbiosis: Lessons in conflict resolution. *PLoS Biology* **2**:307–311.

CHAPTER 18

Bacterial Diversity

18.1 Bacterial Diversity at a Glance

18.2 Cyanobacteria: Oxygenic Phototrophs

18.3 Firmicutes and Actinobacteria (Gram-Positive)

18.4 Proteobacteria (Gram-Negative)

18.5 Deep-Branching Gram-Negative Phyla

18.6 Spirochetes: Sheathed Spiral Cells with Internalized Flagella

18.7 Chlamydiae, Planctomycetes, and Verrucomicrobia: Irregular Cells

Bacteria vary tremendously in their cell structure and metabolism. They include heterotrophs, phototrophs, and lithotrophs—and some species can be classified as all three. There are obligate aerobes, anaerobes, and microaerophiles. Their cell shapes include rods, cocci, spirals, and budding forms. Ecologically, bacteria include mutualists, pathogens, and organisms that cannot be cultured in our laboratories. New species are continually discovered in soil and water, in our homes, and even within our own bodies.

How do we make sense of the thousands of different kinds of bacteria? Chapter 18 introduces the major kinds of bacteria about which we know the most. These include the Gram-positive rods and cocci, the Gram-negative proteobacteria, anaerobes, phototrophs, and spirochetes. There are planctomycetes, intracellular parasitic chlamydias, and deep-branching thermophiles. Even now, all the species we know represent but a tiny fraction of the diverse bacterial species growing in nature.

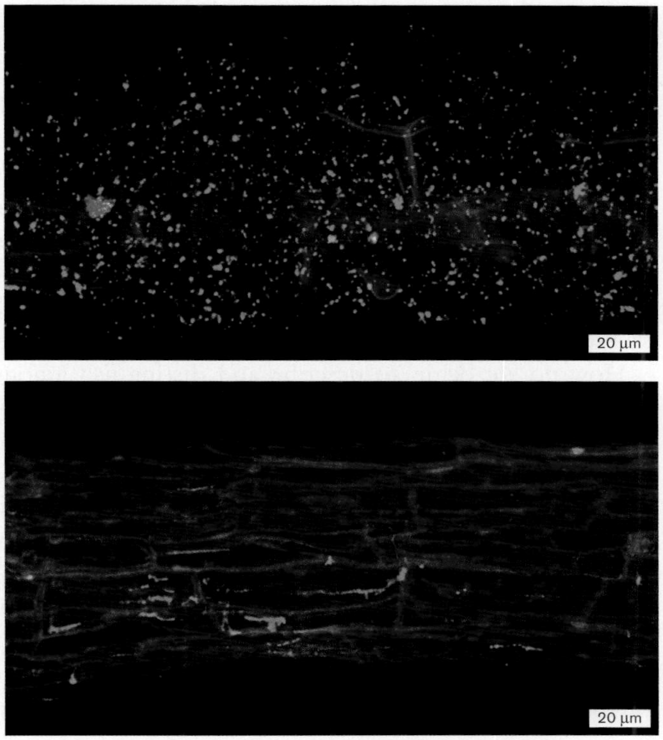
20 µm
20 µm

CURRENT RESEARCH **highlight**

Fluorescence reveals bacterial hitchhikers. Many unknown microbes grow in association with plants—and when we eat plants, we eat their microbes. A fluorescent reporter (green) linked to a PCR probe reveals bacteria growing on the roots of the model plant *Arabidopsis thaliana.* The bacteria include species of Bacteroidetes (top) and Actinobacteria (bottom). The data were obtained in 2012 by Davide Bulgarelli and colleagues at the Max Planck Institute for Plant Breeding Research, Cologne, Germany, using a taxon-specific probe for ribosomal RNA (rRNA). When the probe DNA hybridizes with rRNA, its attached enzyme catalyzes the formation of fluorophores. This method is called catalyzed reporter deposition fluorescence in situ hybridization (CARD-FISH). Why the plant hosts the Bacteroidetes and Actinobacteria is unknown; one hypothesis is that the Actinobacteria produce antibiotics that protect the plant from pathogens. *Source:* Davide Bulgarelli et al. 2012. *Nature* **488**:91–95.

Bacteria evolve a bewildering array of life-forms that colonize every habitat on Earth, from the dunes of the Sahara desert to the ice-buried Lake Vostok of Antarctica. Every day we discover new species never before recorded—not just from exotic environments, but from habitats much closer to home. In 2012, the Human Microbiome Project Consortium reported a landmark survey of microbial communities throughout the human body. Habitats such as the nostrils, tongue, and digestive tract were sampled from more than 200 healthy individuals (**Fig. 18.1**). From each sample, the DNA of every possible microorganism was sequenced. The genome sequences reveal extraordinary diversity, including new species alongside familiar inhabitants of humans. Individual humans show surprisingly different profiles of microbial species. For example, the tongues swabbed from 172 different individuals show very different proportions of firmicutes and proteobacteria (**Fig. 18.1**). In the future, we aim to link specific human-associated microbes with health issues ranging from obesity to aging.

How do we begin to describe and distinguish among all the kinds of bacteria on Earth? Even in familiar habitats the vast majority of species remain unknown, and new deep-branching ones continue to be discovered. At the same time, new approaches to genomics, microscopy, and culturing clarify our view of major groups of bacteria. Chapter 18 surveys major bacterial taxa that are of physiological, ecological, and medical importance. Our treatment emphasizes phylogeny (evolutionary relatedness of taxa), as well as key traits of a taxon, such as the Gram-positive cell wall of Firmicutes and the oxygenic photosynthesis of cyanobacteria. We introduce the key roles of microbes in communities. Microbial communities and ecology are explored further in Chapter 21, and global cycling is discussed in Chapter 22.

For each major taxonomic group, we describe a few key species to represent the spectrum of diversity. Additional genera and species are referenced in an expanded table found in Appendix 4. For further exploration, we recommend the **National Center for Biotechnology Information (NCBI) Taxonomy Database**, a public online resource funded by the U.S. government that includes the taxonomy of biological species reported in the literature and an ever-growing database of sequenced genomes.

Weblinks National Center for Biotechnology Information (NCBI) *(see ebook)*

18.1

Bacterial Diversity at a Glance

To survey bacterial diversity in one chapter is like touring all the countries of a continent in a single day. Like countries, bacterial taxa have complex traits and histories, and often contested borders. But overall, bacteria share major traits in common. We review these common traits of bacteria and then go on to explore their differences.

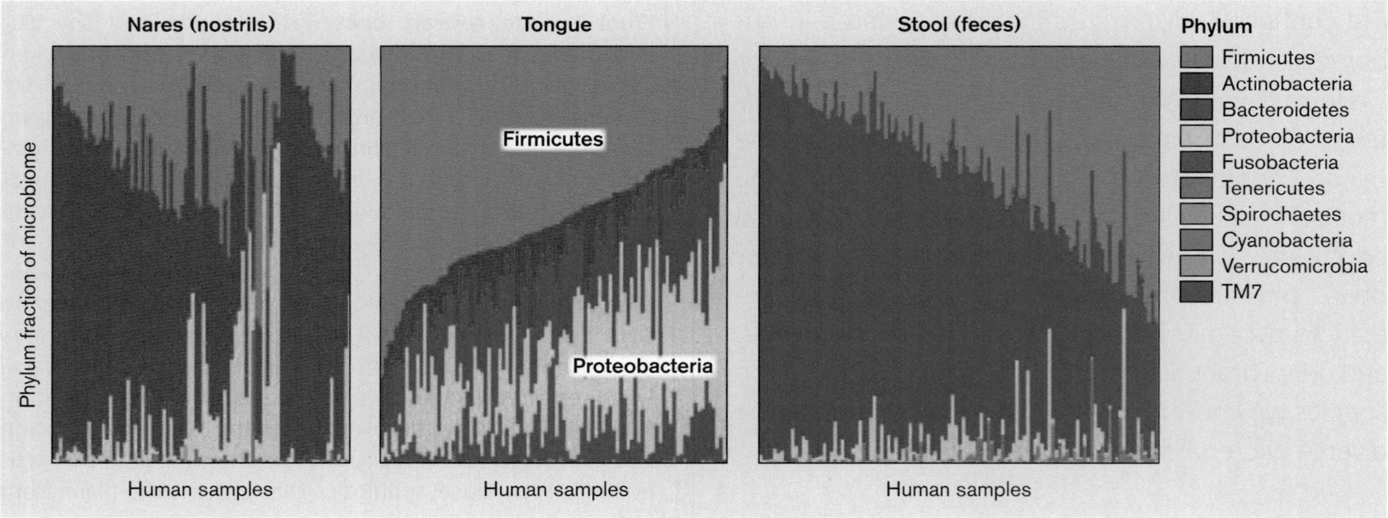
Nares (nostrils)
Tongue
Stool (feces)
Phylum
Firmicutes
Actinobacteria
Bacteroidetes
Proteobacteria
Fusobacteria
Tenericutes
Spirochaetes
Cyanobacteria
Verrucomicrobia
TM7
Phylum fraction of microbiome
Firmicutes
Proteobacteria
Human samples
Human samples
Human samples

FIGURE 18.1 ▪ Microbiomes of parts of the human body. Microbial communities were sampled from multiple individuals. The bacterial categories (phyla) were identified by DNA sequencing. *Source:* Modified from Fig. 2 of Human Microbiome Project Consortium. 2012. *Nature* **486**:207.

Bacteria: Common Traits and Diverging Phylogeny

Chapter 17 summarized the differences and similarities of the three major domains—Bacteria, Archaea, and Eukarya (see Table 17.2). A profound common feature of bacteria is their central apparatus for gene expression, particularly their RNA polymerase and their ribosomal RNAs and translation factors. Bacterial gene expression complexes differ more from those of Archaea or Eukarya than those of Archaea or Eukarya differ from each other. This subtle point of molecular biology has a profound consequence for human medicine and agriculture: It underlies the selective activity of many antibiotics, such as streptomycin, that attack only bacteria without affecting animals or plants.

Another trait distinguishing bacteria from archaea and eukaryotes is that most bacterial cells possess a cell wall of peptidoglycan (discussed in Chapter 3). Peptidoglycan is composed of disaccharide-peptide chains that can cross-link in three dimensions; key enzymes that build the peptide links are blocked by antibiotics such as penicillin and vancomycin. Some archaea possess analogous sugar-peptide structures called "pseudopeptidoglycan" (discussed in Chapter 19), but their structural details and antibiotic sensitivity differ fundamentally from those of bacterial peptidoglycan. Eukaryotes such as fungi and plants have cell walls of polysaccharides such as cellulose and chitin (discussed in Chapter 20).

In bacteria, variant forms of peptidoglycan distinguish different species. For instance, the Gram-positive pathogen *Staphylococcus aureus* has cell wall peptides cross-linked by pentaglycine (a chain of five glycine residues). Some species, such as mycoplasmas, lack peptidoglycan altogether, but they arose by reductive evolution from bacteria that possess it.

The phylogeny of known bacteria is presented in **Figure 18.2**. Not shown are numerous uncultivated bacteria that branch from all parts of this tree, known only from

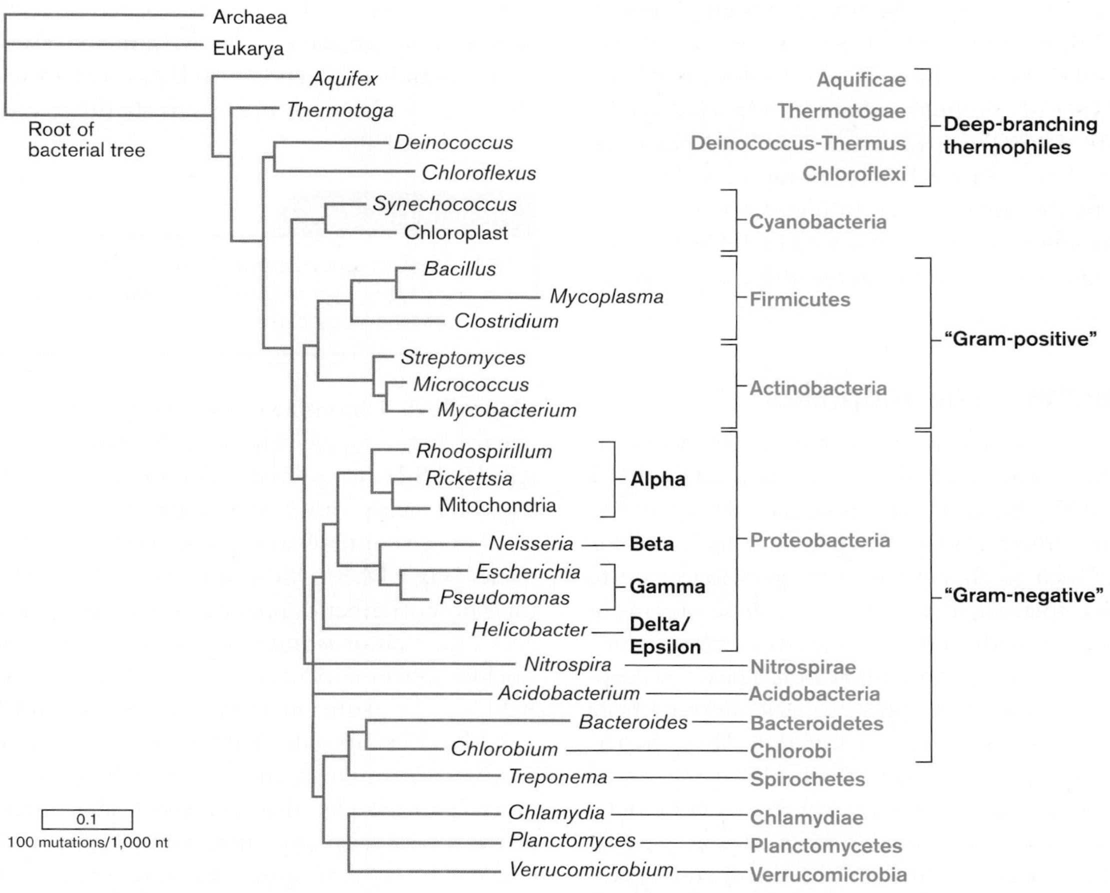
Archaea
Eukarya
Root of bacterial tree
Aquifex
Thermotoga
Deinococcus
Chloroflexus
Synechococcus
Chloroplast
Bacillus
Mycoplasma
Clostridium
Streptomyces
Micrococcus
Mycobacterium
Rhodospirillum
Rickettsia
Mitochondria
Alpha
Neisseria
Beta
Escherichia
Pseudomonas
Gamma
Helicobacter
Delta/ Epsilon
Nitrospira
Acidobacterium
Bacteroides
Chlorobium
Treponema
Chlamydia
Planctomyces
Verrucomicrobium
Aquificae
Thermotogae
Deinococcus-Thermus
Chloroflexi
Deep-branching thermophiles
Cyanobacteria
Firmicutes
Actinobacteria
"Gram-positive"
Proteobacteria
"Gram-negative"
Nitrospirae
Acidobacteria
Bacteroidetes
Chlorobi
Spirochetes
Chlamydiae
Planctomycetes
Verrucomicrobia
0.1
100 mutations/1,000 nt

FIGURE 18.2 ■ **Bacterial phylogeny.** A phylogenetic tree of representative Bacteria based on 16S rRNA sequence comparison. The tree is rooted with respect to Archaea and Eukarya. Bold labels correspond to major groups in Table 18.1. Phylum names are lettered blue.

their DNA sequence. The names of well-studied phyla are lettered in blue. A **phylum** (plural, **phyla**) is defined as a group of organisms sharing a common ancestor that diverged early from other bacteria, based on small-subunit rRNA (SSU rRNA) sequence (discussed in Chapter 17). Increasingly, whole-genome data are supplementing and refining our definition of bacterial clades. Phyla and other major divisions are also defined on the basis of historical convention and consensus of the research community.

A bacterial phylum comprises species that share key traits as well as ancestry. While sharing key traits, the member species often show remarkable diversity in other ways. Some phyla, such as Cyanobacteria, share a unique form of metabolism (oxygenic photosynthesis), yet have evolved many diverse cell shapes and grow in diverse habitats. Other phyla, such as Spirochetes, share a unique cell structure while diverging in habitat and metabolism.

Table 18.1 introduces the major groups of bacteria that appear most frequently in the literature. **An expanded version of this table, including many important species, appears in Appendix 4B.** In addition, more than 100 other "candidate" bacterial phyla have been proposed, and deep-branching clades continue to be discovered. For example, in 2004 a new phylum, TG-1, was described comprising many species of cellulose-digesting bacterial symbionts of protists within the termite gut. The TG-1 phylum is studied as a model for plant cellulose conversion to biofuels.

The sections that follow offer a "one-day tour" of the seven major groups of Bacteria presented in Table 18.1. We will discuss some of these groups in more detail in Sections 18.2–18.7.

Deep-Branching Thermophiles

Which kinds of bacteria were the first to diverge from the ancestral archaea and eukaryotes—and may retain more traits shared with Archaea? The bacteria that appear to have diverged earliest (called "deep-branching") include extremophiles such as *Aquifex aeolicus* (growing at up to 95°C at marine thermal vents). However, these organisms also show rapid growth and high mutation rates, which may have accelerated their molecular clock. Thus, the deep-branching thermophiles may appear to have diverged from other bacteria earlier than they in fact did. These hyperthermophilic bacteria also share their high-temperature habitats with Archaea—a possible source of horizontal gene flow (see Chapter 17). For example, *Aquifex* species have the archaeal trait of ether-linked membrane lipids. The genes encoding these "archaeal" lipids appear to have entered bacteria by horizontal transfer from archaea to bacteria sharing the high-temperature habitat.

Aquificae and Thermotogae. Most members of the thermophilic phyla Aquificae and Thermotogae are hydrogenotrophs, oxidizing hydrogen gas with molecular oxygen to make water. *Aquifex pyrophilus,* a flagellated rod, was first discovered by extremophile microbiologist Karl Stetter at Regensburg University in a submarine hydrothermal vent north of Iceland (for interview, see **eTopic 18.1**). Stetter named the bacterium *Aquifex,* Latin for "water maker," because it oxidizes hydrogen to form water. *Aquifex* species are obligate autotrophs, fixing CO_2 into biomass using the reverse TCA cycle. Other genera of the phylum Aquificae oxidize small organic molecules or sulfides, and form filamentous mats in hydrothermal springs. *Thermocrinis ruber* grows at 82°C–88°C as a mat of pink filamentous streamers in the outflow channel of Octopus Spring, in Yellowstone National Park.

Another deep-branching phylum of thermophiles (growing at 50°C–80°C) is Thermotogae. *Thermotoga maritima* is a sulfur-reducing respirer, originally isolated from a geothermal vent in Volcano, Italy. The cells have a loosely bound sheath, or "toga," for which the genus is named. Remarkably, nearly a quarter of the *T. maritima* genome appears to derive from archaea. This mosaic bacterial-archaeal genome reflects extensive horizontal gene transfer between organisms sharing a common high-temperature habitat.

Thought Question

18.1 What taxonomic questions are raised by the apparent high rate of gene transfer between archaea and thermophilic bacteria?

Chloroflexi. Chloroflexi bacteria are filamentous photoheterotrophs, supplementing heterotrophy with photosystem II (PS II) to generate ATP (discussed in Chapter 14). Together with other thermophiles, Chloroflexi species form massive microbial mats in the hot springs of Yellowstone (**Fig. 18.3**). Most species of *Chloroflexus* contain their photosynthetic apparatus within membranous organelles called **chlorosomes**. The chlorosome contains bacteriochlorophylls *c* and *d* (Bchl *c* and Bchl *d*), complexed with Bchl *a* at the plasma membrane (presented in Chapter 14).

Chloroflexi are informally called "green nonsulfur bacteria" to distinguish them from Chlorobi, a phylum of green phototrophs that are strict anaerobes (presented below, under deep-branching Gram-negative phyla). Some Chloroflexi appear green because their bacteriochlorophylls absorb primarily blue and red, allowing transmission of green. Other species, however, appear red or yellow owing to accessory pigments.

TABLE 18.1

Bacterial diversity.* (Expanded table appears in Appendix 4B.)

Deep-branching thermophiles. Thermophilic bacteria that diverged early from archaea and eukaryotes. Many genes transferred laterally from archaea.

Aquificae. Hyperthermophiles (70°C–95°C). Oxidize H_2. *Aquifex.*

Chloroflexi. Filamentous phototrophs, often with chlorosomes. *Chloroflexus aurantiacus.*

Deinococcus-Thermus. Radiation-resistant species and thermophiles. *Deinococcus radiodurans.*

Thermotogae. Thermophiles (55°C–100°C). Anaerobic heterotrophs. *Thermotoga* sp.

Cyanobacteria. Oxygenic photoautotrophs with thylakoid membranes; share ancestry with chloroplasts.

- **Chroococcales.** Square colonies based on two division planes.
- **Gloeobacterales.** Lack thylakoids; conduct photosynthesis in cell membrane.
- **Nostocales.** Filamentous chains with N_2-fixing heterocysts. Some grow symbiotically with corals or plants. *Nostoc* sp.
- **Oscillatoriales.** Filamentous chains with motile hormogonia (short chains). *Oscillatoria* sp.
- **Pleurocapsales.** Globular colonies; reproduce through baeocytes.
- **Prochlorales.** Tiny single cells, elliptical or spherical (0.5 μm). *Prochlorococcus* sp.

Firmicutes and Actinobacteria (Gram-positive). Peptidoglycan multiple layers, cross-linked by teichoic acids.

Firmicutes. Low-GC, Gram-positive rods and cocci.

- **Bacillales.** Aerobic or facultative anaerobes. *Bacillus subtilis.*
- **Clostridiales.** Anaerobic rods. *Clostridium botulinum* and *Clostridium tetani.*
- **Lactobacillales.** Non-spore formers. Facultative anaerobes. Ferment, producing lactic acid. *Lactobacillus lactis.*
- **Tenericutes or Mollicutes.** Lack cell wall; require animal host. *Mycoplasma pneumoniae.*

Actinobacteria. High-GC, Gram-positive bacteria with moderate salt tolerance.

- **Actinomycetales.**

  Actinomycetaceae. Filamentous, producing aerial hyphae and spores. Actinomycetes such as *Streptomyces coelicolor.*

  Corynebacteriaceae. Irregularly shaped rods. *Corynebacterium diphtheriae.*

  Micrococcaceae. Small airborne cocci. *Micrococcus luteus.*

  Mycobacteriaceae. Exceptionally complex cell walls. *Mycobacterium tuberculosis.*

  Propionibacteriaceae. Propionic acid fermentation. *Propionibacterium shermanii.*
- **Bifidobacteriales.** Ferment without gas. *Bifidobacterium* sp.

Proteobacteria. Gram-negative bacteria with diverse cell forms and metabolism.

Alphaproteobacteria (class)

- **Caulobacterales.** Aquatic oligotrophs; alternate stalk and flagellum. *Caulobacter crescentus.*
- **Rhizobiales.** Plant mutualists and pathogens. *Sinorhizobium meliloti.*
- **Rhodobacterales, Rhodospirillales.** Flagellated photoheterotrophs.
- **Rickettsiales.** Includes intracellular parasites; related to mitochondria. *Rickettsia rickettsii* causes Rocky Mountain spotted fever.
- **Sphingomonadales.** Heterotrophs and photoheterotrophs; some opportunistic pathogens. *Sphingomonas* sp.

Betaproteobacteria (class)

- **Burkholderiales.** *Burkholderia pseudomallei* causes melioidosis.
- **Neisseriales.** Diplococci. *Neisseria gonorrhoeae* causes gonorrhea; *N. meningitidis* causes meningitis.

Gammaproteobacteria (class)

- **Acidithiobacillales.** Lithotrophs. *Acidithiobacillus ferrooxidans* oxidizes iron and sulfur.
- **Aeromonadales.** Aquatic heterotrophs such as *Aeromonas hydrophila.*
- **Enterobacteriales.** Facultative anaerobes; colonize the human colon.
- **Legionellales.** *Legionella pneumophila* causes legionellosis pneumonia.
- **Pseudomonadales.** Rods; aerobic or respire on nitrate; catabolize aromatics. *Pseudomonas aeruginosa* infects lungs in cystic fibrosis patients.
- **Thiotrichales.** Lithotrophs and heterotrophs.
- **Vibrionales.** Marine heterotrophs. *Vibrio cholerae* causes cholera.

Deltaproteobacteria (class)

- **Bdellovibrionales.** Periplasmic predators.
- **Desulfobacterales.** Reduce sulfate.
- **Myxococcales.** Gliding bacteria that form fruiting bodies.

Epsilonproteobacteria (class)

- **Campylobacterales.** Spirillar pathogens. *Campylobacter jejuni* causes gastroenteritis. *Helicobacter pylori* causes gastritis.

Deep-branching Gram-negative phyla.

Acidobacteria. Acidophiles and thermophiles. "*Candidatus* Chloracidobacterium" species.

Bacteroidetes. Obligate anaerobes. Heterotrophs that feed on diverse carbon sources. *Bacteroides thetaiotaomicron.*

Chlorobi. Green sulfur-oxidizing phototrophs. *Chlorobium tepidum.*

Fusobacteria. Gram-negative anaerobic bacteria found in septicemia and in skin ulcers. *Fusobacterium nucleatum.*

Nitrospirae. Oxidize nitrite; aerobic or facultative. *Nitrospira marina.*

Spirochetes (Spirochaeta). Narrow, coiled cells with axial filaments, encased by sheath. Polar flagella beneath sheath double back around cell.

- **Spirochaetales.** Aquatic free-living; endosymbionts and pathogens (width <0.5 μm; length 10–20 μm).

  Borrelia. *B. burgdorferi* causes Lyme disease, transmitted by ticks.

  Hollandina. Termite gut endosymbionts.

  Leptospira. L-shaped animal pathogens; cause leptospirosis.

  Spirochaeta. Aquatic, free-living heterotrophs.

  Treponema. *T. pallidum* causes syphilis.

Chlamydiae, Planctomycetes, and Verrucomicrobia. Irregular cells lacking peptidoglycan, with subcellular structures analogous to those of eukaryotes.

Chlamydiae. Intracellular cell wall–less pathogens of animals or protists. *Chlamydia trachomatis* causes sexually transmitted disease and trachoma (eye infection).

Planctomycetes. Nucleoid has double membrane analogous to eukaryotic nuclear membrane and other intracytoplasmic subcellular membrane compartments. *Brocadia* species anaerobically oxidize ammonium and release N_2 (anammox reaction). *Pirellula* species inhabit marine sediment.

Verrucomicrobia. Stalk-like appendages contain actin filaments. Aquatic oligotrophs. *Prosthecobacter* sp.

*Bulleted terms are representative orders within the phylum (unless stated otherwise).

Deinococcus-Thermus. The phylum Deinococcus-Thermus possesses a unique structural trait: the substitution of L-ornithine for diaminopimelic acid in the peptidoglycan cross-bridge. *Thermus* species (growing at 70°C–75°C) are heterotrophs commonly isolated from hot tap water. *Deinococcus* species, however, are not thermophilic. *D. radiodurans* bacteria resist extremely high doses of ionizing radiation (discussed in Chapters 5 and 9). An otherwise ordinary heterotroph, *D. radiodurans* was originally isolated from cans of meat supposedly sterilized with a radiation dose of several megarads. *Deinococcus* bacteria are found suspended in air, both indoors and high in the atmosphere; they resist drought. *Deinococcus* species, unlike *Thermus* species, have a thick cell wall that stains Gram-positive, as well as an S-layer.

Cyanobacteria

The **Cyanobacteria** (Section 18.2) are a phylum of profound importance for all ecosystems. Cyanobacteria conduct photosynthesis by splitting water and releasing oxygen (O_2). Some cyanobacteria are poisoned by hydrogen sulfide (H_2S), but other species can adapt to use H_2S when available.

Cyanobacteria and chloroplasts (eukaryotic organelles that evolved from cyanobacteria) are the only life-forms that produce oxygen gas. These bacteria have a unique two-photosystem apparatus for oxygenic photosynthesis arranged in lamellar arrays of membranes called thylakoids (discussed in Chapter 14). Unique to cyanobacteria and chloroplasts are the chlorophylls, most notably chlorophylls *a* and *b*; these molecules are distinct from the "bacteriochlorophylls" used by non-oxygenic bacterial phototrophs. Cyanobacteria share a common metabolism, yet the cell shape and organization of cyanobacterial species show an immense range of different forms, including chains of cells, square arrays, and globular colonies. Many cyanobacteria grow in seawater or freshwater, whereas others grow in mutualism with a eukaryotic host.

Gram-Positive Bacteria

Two major phyla, **Firmicutes** and **Actinobacteria**, are called the "Gram-positive bacteria" because most members stain Gram-positive (Section 18.3). A few Firmicutes (the mycoplasmas) fail to stain Gram-positive because they lack cell walls, and some Actinobacteria possess a thick waxy coat that excludes the Gram stain. Most Firmicutes and Actinobacteria have an exceptionally thick cell wall, with several layers of peptidoglycan threaded

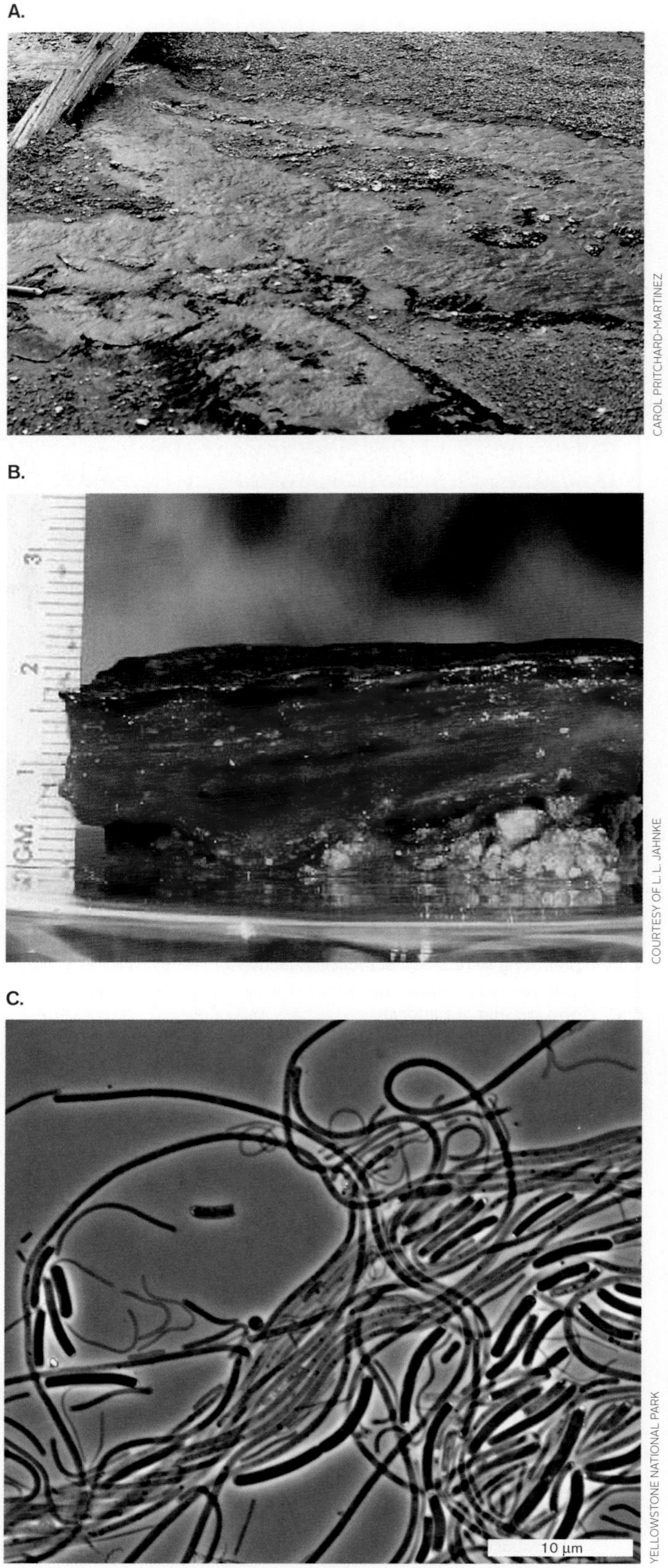
A.
CAROL PRITCHARD-MARTINEZ
B.
CM
1
2
3
COURTESY OF L. L. JAHNKE
C.
10 µm
YELLOWSTONE NATIONAL PARK

FIGURE 18.3 ▪ A deep-branching thermophile: *Chloroflexus*. **A.** This hot-spring bacterial mat in Yellowstone National Park contains *Chloroflexus* species and other thermophilic bacteria. **B.** Bacterial mat section showing layers of *Chloroflexus*. **C.** *Chloroflexus* bacteria (light micrograph).

by supporting molecules such as teichoic acids or mycolic acids. The thick, reinforced cell wall is what retains the Gram stain (discussed in Chapter 2). In addition, most Gram-positive species possess a well-developed S-layer of protein with glycan strands. By contrast, in other bacterial phyla the S-layer is either absent or present in diminished form.

Firmicutes and Actinobacteria differ genetically in their **GC content**—that is, the proportion of their genomes consisting of guanine-cytosine base pairs (as opposed to adenine-thymine). Firmicutes (low GC) generally grow as well-defined rods or cocci, isolated or in simple filaments consisting of cells that divide but remain attached end to end. Many Firmicutes form **endospores**, inert heat-resistant spores that can remain viable for thousands of years. Endospores are the most durable type of spore formed by bacteria. Actinobacteria have a relatively high GC content, and are known as the "high-GC Firmicutes."

Actinobacteria include the **actinomycetes** (order Actinomycetales), which undergo complex life cycles forming filamentous hyphae and arthrospores. Other groups closely related to actinomycetes grow as isolated rods and cocci, often with variable shape, such as the corynebacteria. Actinomycete relatives include the well-known causative agents of tuberculosis (*Mycobacterium tuberculosis*) and leprosy (*M. leprae*).

Proteobacteria

Proteobacteria are called "Gram-negative" because their single layer of peptidoglycan fails to retain the Gram stain (Section 18.4). Species of Proteobacteria have a complex outer membrane consisting of lipopolysaccharides (LPS) and porins. All Proteobacteria have in common the LPS outer membrane, but they show an immense range of shape (rods, cocci, spirals, and filaments) and metabolism (heterotrophy, lithotrophy, and anaerobic phototrophy). Most species are aerobic, facultative, or microaerophilic (requiring a low level of O_2; inhibited at higher levels).

Proteobacteria include five major classes, named Alphaproteobacteria, Betaproteobacteria, Gammaproteobacteria, Deltaproteobacteria, and Epsilonproteobacteria. (Some sources separate the Greek letter from the name "Proteobacteria"—for example, "Alpha Proteobacteria," "Beta Proteobacteria," and so on.) Proteobacteria include famous model organisms and pathogens, such as *Escherichia coli, Salmonella enterica,* and *Yersinia pestis.* Many are human or animal symbionts, either as mutualists or as pathogens—including *Rickettsia,* the genus most closely related to mitochondria.

Deep-Branching Gram-Negative Bacteria

Several deep-branching phyla stain Gram-negative but diverge distantly from the Proteobacteria. They show diverse metabolism and morphology (Section 18.5). Most members of the phyla **Bacteroidetes** and **Chlorobi** are obligate anaerobes. Within Bacteroidetes, *Bacteroides* species ferment complex carbohydrates, serving as the major mutualists of the human gut. By contrast, the closely related Chlorobi species are anaerobic "green sulfur" phototrophs that photolyze sulfides or hydrogen. The **Nitrospirae** largely resemble proteobacteria in form. Most oxidize nitri̲te (NO_2^-) to nitra̲te (NO_3^-). Nitrite oxidation is a lithotrophic conversion essential for ecosystems (see Chapter 22).

Other Gram-negative clades are referred to as **emerging**—that is, recently defined or characterized. **Acidobacteria** is an emerging phylum of soil bacteria; and the phylum **Fusobacteria** includes human pathogens.

Spirochetes (Spirochaeta)

The **Spirochetes** (Section 18.6) have evolved a unique and complex cellular form of a flexible, extended spiral, resembling an old-style telephone cord. The cytoplasm and cell membrane are contained within an outer membrane called the sheath. Between the sheath and the cell membrane extend flagella doubled back from each pole. The rotation of these flagella is coordinated so as to twist and flex the helical body, generating motility and chemotaxis. Spirochetes include many free-living forms in aquatic systems, as well as digestive endosymbionts and pathogens.

Chlamydiae, Planctomycetes, and Verrucomicrobia

Three phyla of bacteria have unusual cell shapes: **Chlamydiae**, **Planctomycetes**, and **Verrucomicrobia** (Section 18.7). Most members of these groups have lost their peptidoglycan cell walls, but they show complex structural adaptations and developmental forms. The Chlamydiae (best-known genus *Chlamydia*) are intracellular parasites that lose most of their cell envelope during intracellular growth. The replicating parasites generate multiple spore-like "elementary bodies" that escape to infect the next host.

Planctomycetes, by contrast, are free-living aquatic bacteria, with stalked cells that reproduce by budding. Each planctomycete cell contains an extra double membrane surrounding its nucleoid, analogous to a eukaryotic nuclear membrane, though it evolved independently.

Verrucomicrobia, also free-living, are bacteria with wart-like protruding structures containing actin.

Thought Questions

18.2 Which taxonomic groups in **Table 18.1** stain Gram-positive and which stain Gram-negative? Which group contains both Gram-positive and Gram-negative species? For which groups is the Gram stain undefined, and why?

18.3 Which groups of bacterial species share common structure and physiology within the group? Which groups show extreme structural and physiological diversity?

Note: Formal names for taxa such as phyla, orders, and genera are capitalized (phylum Cyanobacteria, genus *Streptococcus*). Informal forms are lowercase roman (cyanobacteria, streptococci), as are adjectival forms (cyanobacterial, streptococcal).

The seven groups outlined in **Table 18.1** represent those bacteria of greatest historical interest and those best characterized by microbiologists. Following our brief tour of the bacterial domain, we now explore some of the diverse species within the major groups. Additional species are compiled in **Appendix 4**.

To Summarize

- **Deep-branching thermophiles** such as *Aquifex* and *Thermotoga* species share traits and habitats with thermophilic archaea. They show extensive transfer of archaeal genes. *Deinococcus* species are highly resistant to ionizing radiation. *Chloroflexus* species are thermophilic photoheterotrophs.
- **Cyanobacteria** conduct oxygenic photosynthesis. Species vary widely in their cell shape and ecological niche.
- **Firmicutes and Actinobacteria** have a thick cell wall and generally stain Gram-positive.
- **Proteobacteria and other Gram-negative taxa** have a thin cell wall and an LPS-containing outer membrane. They show exceptionally diverse metabolism and ecological adaptations.
- **Spirochetes** have flexible, spiral-shaped cells with complex intracellular architecture.
- **Chlamydiae, Planctomycetes, and Verrucomicrobia** have irregularly shaped cells with complex intracellular form and development.
- **New isolates** continually reveal previously unknown taxa of bacteria.

Note: The main organizing principle used in Chapters 18–20 is that of phylogeny based on DNA relatedness. Characteristic traits described for each branch apply to the majority of its known species, though many exceptions have evolved, such as members of Spirochetes that lack spiral form.

18.2

Cyanobacteria: Oxygenic Phototrophs

What produces all the oxygen we breathe? All of the oxygen gas in Earth's atmosphere comes from Cyanobacteria (or Cyanophyta) and from plant chloroplasts that evolved from a cyanobacterial ancestor. The phylum is named for the blue phycocyanin accessory pigments possessed by some genera, giving them a bluish tint. Cyanobacteria commonly appear green because of the predominant blue and red absorption by chlorophylls *a* and *b* (see **Table 18.2**). Essentially the same chlorophylls are found in plant chloroplasts, which evolved from internalized cyanobacteria (discussed in Chapter 17). Some cyanobacteria, however, appear red because of the accessory pigment phycoerythrin, which absorbs blue-green light in a range missed by cyanobacterial chlorophylls. Cyanobacteria are the only prokaryotes that use both photosystems I and II, photolyzing water to produce oxygen, as explained in Chapter 14. Under anoxic conditions, most cyanobacteria can also photolyze hydrogen, reduced sulfur compounds, and organic compounds.

The success of oxygenic phototrophy is shown by its fundamental similarity across all members of Cyanobacteria, despite their variety of cell form, behavior, and genetic diversity. Even two cyanobacterial genera such as *Synechococcus* and *Anabaena* share only 37% of their genes. Cyanobacteria are found in all habitats, from tropical soils to Antarctica. Cyanobacteria include species edible for humans, in products such as *Spirulina* salad or *Nostoc* soup.

How does the oxygenic photosynthesis of cyanobacteria compare with the non-oxygenic photosynthesis of thermophilic Chloroflexi, or of sulfur-based phototrophs among the Proteobacteria? **Table 18.2** summarizes key features of phototrophy in clades throughout the bacterial domain, including those presented in this chapter. All light-harvesting complexes descend from one of two ancestral sources: the chlorophyll/bacteriochlorophyll with electron transport (PS I, PS II) or the proteorhodopsin proton pump. These photosystems have diverged through vertical inheritance as well as by horizontal transfer between different clades.

TABLE 18.2

Phototrophic bacteria.

Taxon	Energy generation	Cell structure and photopigments	Absorption spectrum (whole cell)	Reaction center
Chloroflexi "Green nonsulfur" *Chloroflexus*	$+O_2$ Heterotrophy $-O_2$ Photoheterotrophy or use reduced sulfur	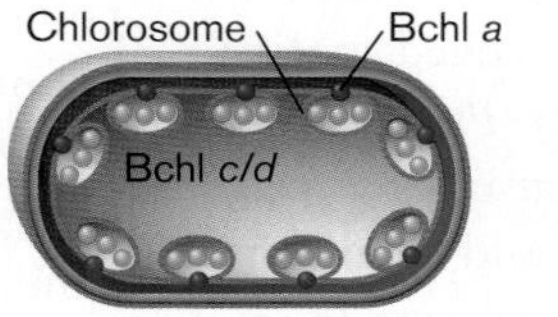 Chlorosome Bchl *a* Bchl *c/d*	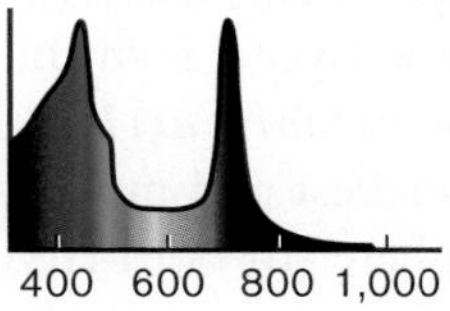 400 600 800 1,000	PS II
Cyanobacteria "Blue-greens" *Anabaena*	$+O_2$ Oxygenic phototrophy $-O_2$ Photolithoautotrophy on reduced sulfur	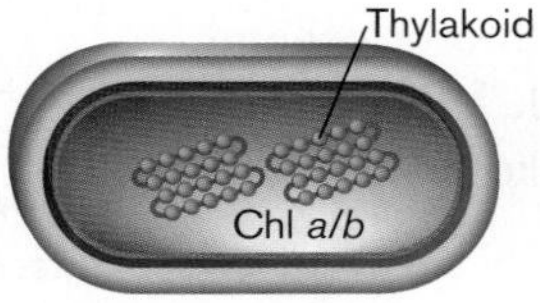 Thylakoid Chl *a/b*	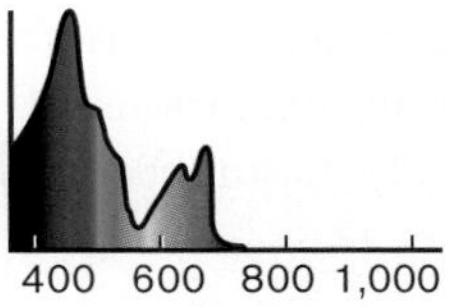 400 600 800 1,000	PS I and PS II
Firmicutes "Sun bacteria" *Heliobacterium*	$+O_2$ $-O_2$ Photoheterotrophy	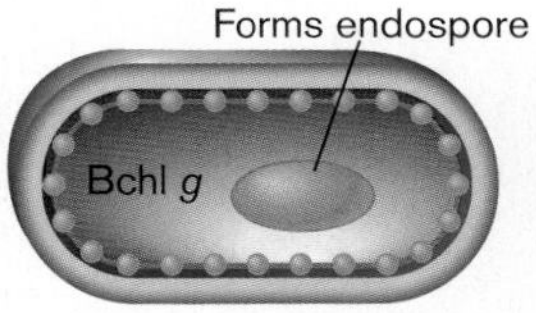 Forms endospore Bchl *g*	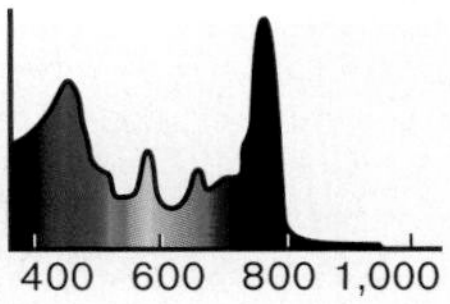 400 600 800 1,000	PS I
Proteobacteria "Purple nonsulfur" (Alpha and Beta classes) Bchl *a* *Rhodospirillum*	$+O_2$ Heterotrophy $-O_2$ Photoheterotrophy	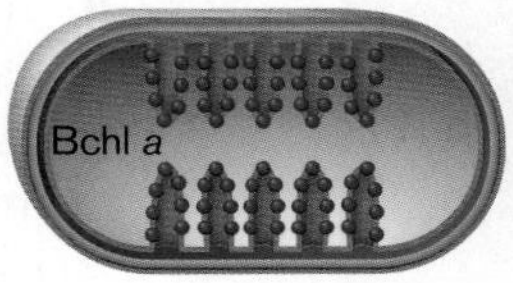 Bchl *a*	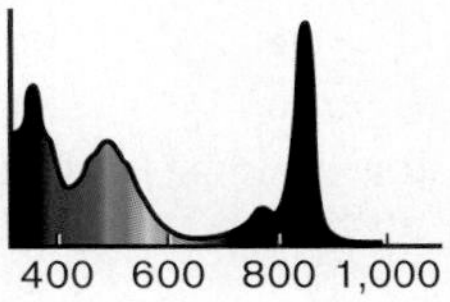 400 600 800 1,000	PS II
Bchl *b* *Blastochloris*	$+O_2$ Heterotrophy $-O_2$ Photoheterotrophy	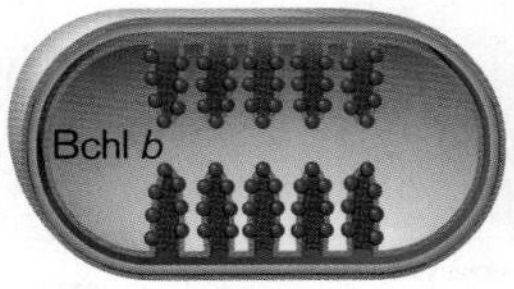 Bchl *b*	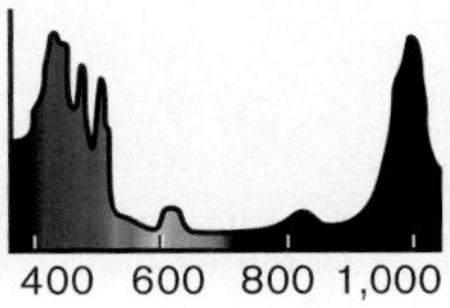 400 600 800 1,000	PS II
"Purple sulfur" (Gamma class) *Chromatium*	$+O_2$ $-O_2$ Photolithoautotrophy on reduced sulfur	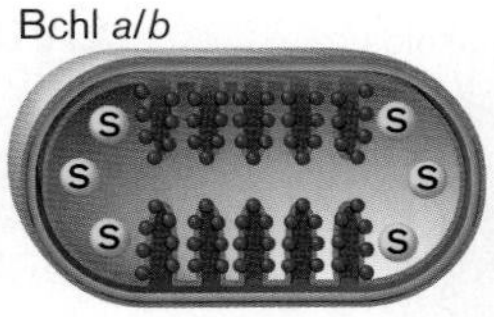 Bchl *a/b* S	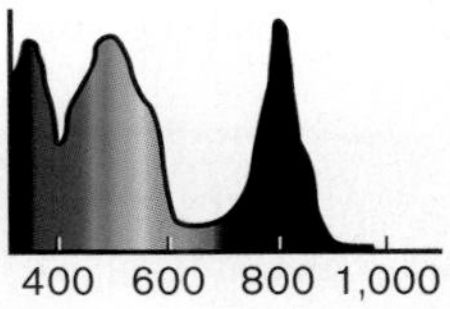 400 600 800 1,000	PS II
"Proteorhodopsin" (Alpha and Gamma classes) *Pelagibacter* (SAR11) SAR86	$+O_2$ Heterotrophy $-O_2$ Photoheterotrophy	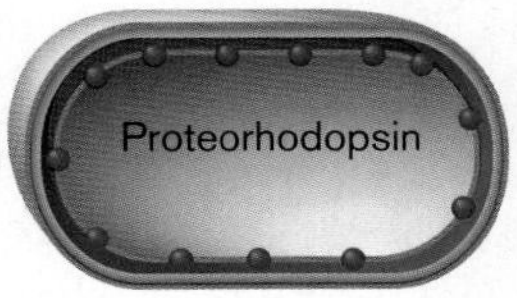 Proteorhodopsin	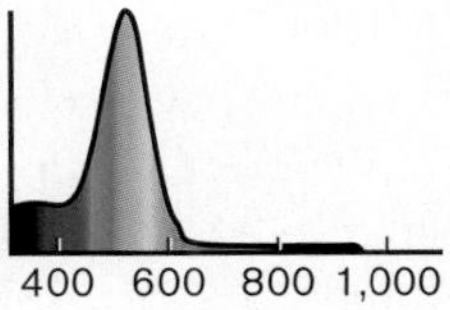 400 600 800 1,000	PR
Chlorobi "Green sulfur" *Chlorobium*	$-O_2$ only Photolithoautotrophy on reduced sulfur	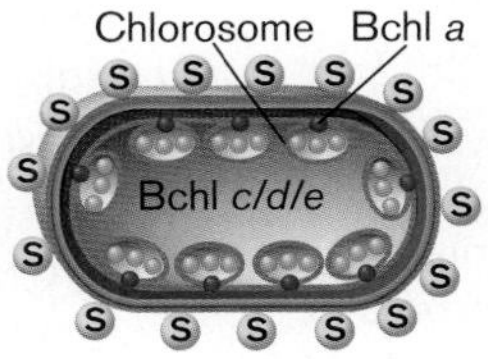 Chlorosome Bchl *a* Bchl *c/d/e* S	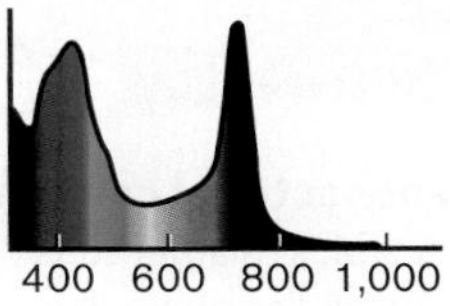 400 600 800 1,000	PS I

Sources: J. Overman and F. Garda-Pichet. 2001. The phototrophic way of life. In *Prokaryotes.* 2002. Springer-Verlag: Oded Beja et al. 2001. *Nature* **411**:786–789.

Cyanobacterial Cell Structure

The photosynthetic apparatus of cyanobacteria is organized within thylakoids, pockets of membrane resembling flattened spheres packed with reaction centers. The thylakoids may be distributed through the cell, as in filamentous genera such as *Nostoc* (**Fig. 18.4A**), or they may encircle the cell in concentric layers, as in the single-celled marine species *Prochlorococcus marinus* (**Fig. 18.4B**). *Prochlorococcus* is one of the smallest and most abundant oxygen producers in the biosphere. In both large cells and small, the thylakoids are completely separate from the plasma membrane, unlike the attached chlorosomes of Chloroflexi and the plasma membrane extensions of "purple" Proteobacteria (see **Table 18.2**). Cyanobacterial thylakoids resemble the thylakoids of eukaryotic chloroplasts; they are the most complex and specialized form of photosynthetic apparatus.

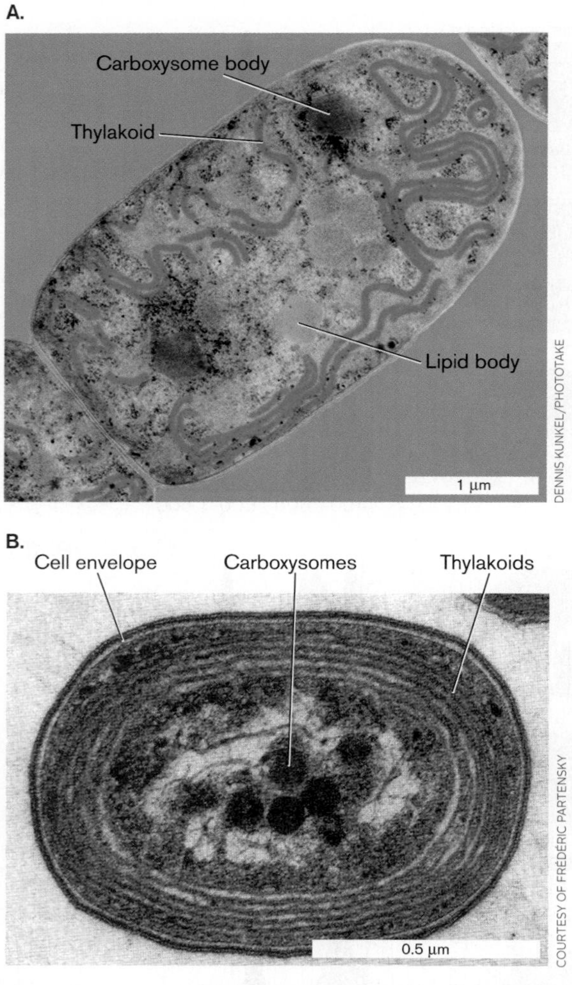
A.
Carboxysome body
Thylakoid
Lipid body
1 µm
DENNIS KUNKEL/PHOTOTAKE
B.
Cell envelope
Carboxysomes
Thylakoids
0.5 µm
COURTESY OF FRÉDÉRIC PARTENSKY

FIGURE 18.4 ■ Cyanobacterial cell structure. A. Intracellular organelles of *Nostoc*, a typical filamentous cyanobacterium (colorized TEM). **B.** Intracellular organelles of *Prochlorococcus*, a prochlorophyte cyanobacterium, the smallest known phototroph. *Prochlorococcus* accounts for 40%–50% of marine phototrophic biomass.

Cyanobacteria have several other subcellular structures (**Fig. 18.4**). **Carboxysomes** (also known as polyhedral bodies) are rich in the enzyme Rubisco, and they fix CO_2 (discussed in Chapter 15). Cyanobacteria store energy-rich compounds in lipid bodies. To maintain height in the water column and thus access to sunlight, cyanobacteria have **gas vesicles**, whose buoyancy enables cells to float. Their external structures include a thick peptidoglycan cell wall, similar to that of Gram-positive cells, plus several external layers that vary with different species. Many species move by "gliding," a form of motility whose mechanism is poorly understood.

Besides fixing CO_2, most cyanobacteria fix N_2. Because nitrogen fixation requires the absence of oxygen, cyanobacteria have to solve the problem of maintaining anaerobic biochemistry while producing huge quantities of highly toxic O_2. Different species solve this problem in different ways:

- Formation of specialized nitrogen-fixing cells called **heterocysts**. Heterocysts exclude oxygen.
- Temporal separation, alternating between photosynthesis during daylight and nitrogen fixation at night.
- Accumulation of large aggregates of cells in which the interior becomes sufficiently anaerobic for nitrogenase to function, while the exterior continues oxygenic photosynthesis.
- Symbiosis with microbes or plants that consume oxygen or otherwise maintain anoxic conditions.

Filamentous and Colonial Cyanobacteria

Cyanobacteria have evolved several major categories of form. Single-celled forms include *Synechococcus* and *Prochlorococcus,* the most abundant phototrophs in the oceans. Other genera, such as *Oscillatoria,* generate long filaments of hundreds or even thousands of cells (**Fig. 18.5**). *Oscillatoria* cells are stacked like plates, wider than they are long. To disseminate their cells beyond the biofilm, the filaments produce **hormogonia** (singular, **hormogonium**), short motile chains of three to five cells. Other filamentous genera, such as *Nostoc,* arrange their filaments in balls of mucilage, possibly to protect themselves from grazing (**Fig. 18.6A**). Most filamentous species develop heterocysts to fix nitrogen (**Fig. 18.6B**). The function of heterocysts is discussed in Chapter 15.

Under environmental stress, such as light limitation or phosphate starvation, filamentous cyanobacteria such as *Anabaena* form specialized spore cells called **akinetes**. An

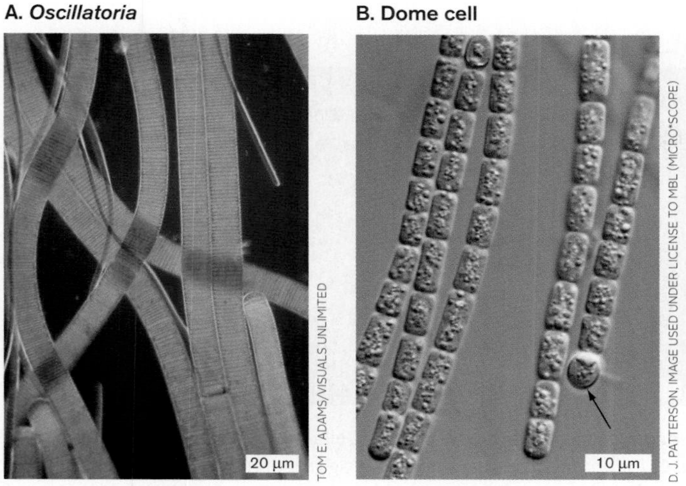
A. Oscillatoria
B. Dome cell
20 µm
10 µm
TOM E. ADAMS/VISUALS UNLIMITED
D. J. PATTERSON, IMAGE USED UNDER LICENSE TO MBL (MICRO*SCOPE)

FIGURE 18.5 ■ Filamentous cyanobacteria. A. *Oscillatoria* forms large filaments whose cells are wider than their length. **B.** Each filament terminates in a dome cell (arrow).

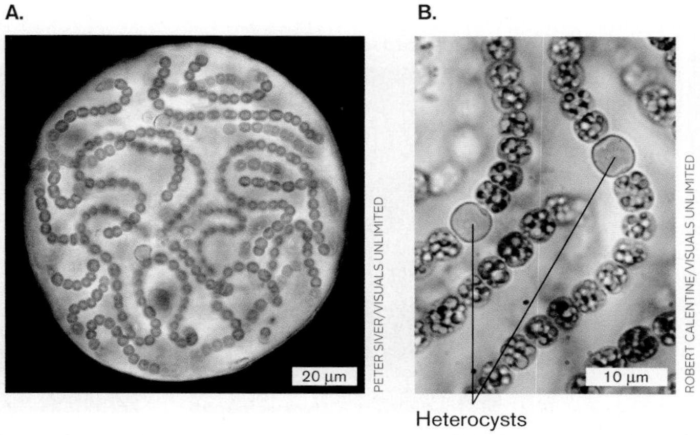
A.
B.
20 µm
10 µm
Heterocysts
PETER SIVER/VISUALS UNLIMITED
ROBERT CALENTINE/VISUALS UNLIMITED

FIGURE 18.6 ■ Clumped filamentous cyanobacteria. A. *Nostoc* forms filaments that clump together in large balls of mucilage. **B.** *Nostoc* makes heterocysts.

akinete forms as a long, oval cell adjacent to a heterocyst, where it stores nitrogen and develops a thickened envelope. Like other types of spores, akinetes resist desiccation and remain viable for long periods. **Table 18.3** compares the properties of akinetes with those of other spore types (which are discussed under Firmicutes and other taxa). Akinetes lie dormant but viable until improved conditions permit germination and growth of new vegetative filaments.

TABLE 18.3

Spore types in bacteria.

Spore type	Bacteria that produce the spore	Initiation of spore formation	Formation of the spore	Properties of the spore
Akinete	Filamentous cyanobacteria	Light limitation Cold temperature Phosphate starvation	An akinete develops next to a heterocyst, as a large oval cell with a multilayered envelope.	Desiccation and cold resistant Viable for decades
Arthrospore	Actinomycetes	Carbon starvation Phosphate starvation	At the tip of an aerial mycelium, cells undergo vegetative division and pinch off as arthrospores.	Desiccation resistant Heat resistant
Elementary body	Chlamydias	Completion of intracellular life cycle	Intracellular chlamydia reticular bodies replicate and then develop into elementary bodies with cross-linked outer membrane proteins. Elementary bodies survive outside the host cell.	Survives outside host Desiccation resistant
Endospore	Firmicutes	Carbon starvation Nitrogen starvation Phosphate starvation Low pH Peptide antibiotics	Individual cell develops mother cell and forespore. The forespore develops into an endospore with a spore coat reinforced by keratin and calcium dipicolinate.	Highly heat and desiccation resistant Viable for centuries
Myxospore	Myxobacteria	Nutrient starvation Heat shock Glycerol Dimethyl sulfoxide	Myxobacteria aggregate to form a fruiting body in which vegetative cell division forms a mass of myxospores.	Desiccation resistant UV resistant

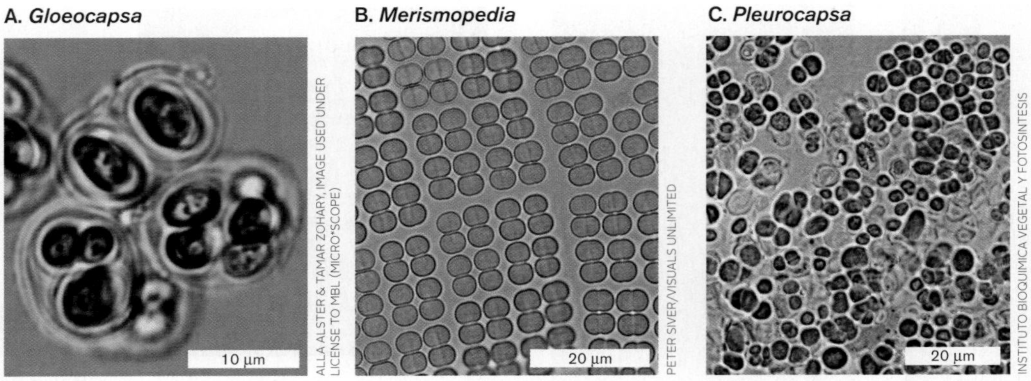
A. Gloeocapsa
B. Merismopedia
C. Pleurocapsa
10 µm
20 µm
20 µm
ALLA ALSTER & TAMAR ZOHARY, IMAGE USED UNDER LICENSE TO MBL (MICRO*SCOPE)
PETER SIVER/VISUALS UNLIMITED
INSTITUTO BIOQUIMICA VEGETAL Y FOTOSINTESIS

FIGURE 18.7 ■ Single-celled and colonial cyanobacteria. **A.** *Gloeocapsa* is surrounded by mucus. Cells grow as single cells, doublets, or quartets. **B.** *Merismopedia* forms extended quartets, octets, and so on. **C.** *Pleurocapsa* forms enormous aggregates that release baeocytes.

In lake water, akinete germination may cause toxic blooms of *Anabaena*.

A filamentous cyanobacterium important for coastal areas is *Lyngbya*, a common source of toxic blooms. *Lyngbya* filaments, known as "giant *Lyngbya*," can overgrow ponds, coastal corals, and sea grass, killing fish and sickening humans.

Thought Question

18.4 What are the relative advantages and disadvantages of propagation by hormogonia, as compared with akinetes?

Nonfilamentous species divide to form small groups or larger colonies (**Fig. 18.7**). *Gloeocapsa* and *Chroococcus* form doublets or quartets, encased in a thick protective mucous slime (**Fig. 18.7A**). Others, such as *Merismopedia*, continue cell division in two planes, extending to form long, square sheets of attached cells (**Fig. 18.7B**). Colonial genera such as *Myxosarcina* and *Pleurocapsa* reproduce by multiple fission, forming large cell aggregates (**Fig. 18.7C**). As the aggregate matures, some cells continue to divide and release single cells called baeocytes. Each baeocyte reproduces and develops into a new cell aggregate. The aggregate group maintains anoxic conditions at the center for nitrogen fixation.

Thought Question

18.5 What are the relative advantages and disadvantages of the different strategies for maintaining separation of nitrogen fixation and photosynthesis?

Cyanobacterial Communities

What niches do Cyanobacteria fill in ecosystems? In the ocean, cyanobacteria are the main primary producers (discussed in Chapter 21). Environmental change causes the colonial cyanobacterium *Trichodesmium* to form giant blooms visible from outer space, covering many square kilometers of ocean surface (**Fig. 18.8**). Such a bloom can be triggered by an influx of iron carried by wind from a dust storm blowing off the Sahara desert. The concentrated *Trichodesmium* converts nitrogen into forms that promote growth of eukaryotic algae, which prove toxic to consumers such as fish and manatees.

In salt marshes and sand flats, cyanobacteria participate in multilayered microbial mats, such as the section shown in **Figure 18.9**, cut from the sand flats of Great Sippewissett Salt Marsh, on Cape Cod, Massachusetts. The high concentration of sulfides in the sediment supports growth

HORN POINT LABORATORY, THOMAS C. MALONE

FIGURE 18.8 ■ Marine cyanobacterial bloom. A ship plows through an ocean bloom of *Trichodesmium*.

Cyanobacteria and diatoms
Purple sulfur proteobacteria
Long-wavelength purple sulfur bacteria
3 cm
JÖRG OVERMANN AND FERRAU GARCIA-PICHEL

FIGURE 18.9 ▪ Cyanobacteria in microbial mats. Cutaway through a multilayered microbial mat from the sand flats of Great Sippewissett Salt Marsh (Cape Cod, Massachusetts). Cyanobacteria and diatoms form the upper green layer, above layers of purple sulfur proteobacteria.

of high populations of sulfur phototrophs. Typically, cyanobacteria and eukaryotic algae such as diatoms form the upper green layer. Below, the purple layer consists of "purple sulfur" proteobacteria, whose photopigments (primarily Bchl *a*) absorb at longer wavelengths (discussed in Section 18.5). The pale-colored layer below the purple layer consists of proteobacteria with Bchl *b*, which absorbs farther into the infrared.

Cyanobacteria share many kinds of mutualism with animals, plants, fungi, and protists. Sponges growing on coral reefs may harbor endosymbiotic cyanobacteria that provide the sponges with nutrients from photosynthesis. The products of photosynthesis supplement the nutrients obtained by the sponges from their filter feeding, and these extra nutrients greatly augment the sponge growth rate within the competitive coral reef environment. Sponge symbionts produce many pharmaceutically active compounds.

To Summarize

- **Cyanobacteria** are the only oxygenic prokaryotes. They conduct photosynthesis in thylakoids, fix CO_2 in carboxysomes, and maintain buoyancy using gas vesicles. They exhibit gliding motility.
- **Single-celled cyanobacteria** such as *Prochlorococcus* are among the smallest and most abundant phototrophic producers in the oceans.
- **Filamentous cyanobacteria** such as *Nostoc* and *Oscillatoria* are common in freshwater lakes. Filaments form heterocysts to fix nitrogen, and reproduce by hormogonia or by akinetes.
- **Colonial cyanobacteria** such as *Myxosarcina* produce large cell aggregates with an anaerobic core for nitrogen fixation. The colonies reproduce through baeocytes.
- **Symbiotic associations** of cyanobacteria occur with animals, fungi, and plants.

18.3

Firmicutes and Actinobacteria (Gram-Positive)

Which bacteria have the toughest, thickest cell walls? Most bacteria of the phylum Firmicutes, meaning "tough skin" bacteria, have thick peptidoglycan cell walls that retain the Gram stain (discussed in Chapter 3). The thick cell wall helps exclude antibiotics and antibacterial agents from competitors in the environment. Members of a related phylum, Actinobacteria, also stain Gram-positive. As shown in **Table 18.1**, we define Firmicutes as the phylum of Gram-positive bacteria that contains "low-GC" species (having a low ratio of GC/AT base pairs), whereas Actinobacteria are the "high-GC" species. Species in both phyla have thick cell walls reinforced by teichoic acids, cross-threading phosphodiester chains of glycerol and ribitol (discussed in Chapter 3).

Many firmicutes, such as *Bacillus* and *Clostridium* species, survive unfavorable environmental conditions by forming durable endospores. Non-spore formers such as *Lactobacillus* and *Streptococcus* may have evolved from a common Firmicutes ancestor that formed endospores. In many cases, the machinery to form endospores is discovered in the genomes of firmicutes previously thought to be non-spore formers, such as *Carboxydothermus* sp., soil bacteria that oxidize CO (carbon monoxide) to CO_2. On the other hand, actinobacteria of the order Actinomycetales (actinomycetes), such as *Streptomyces,* do not form endospores, but they develop filaments dispersing arthrospores (see **Table 18.3**).

Firmicutes Include Endospore-Forming Rods

Endospore-forming bacteria are common in soil and air because their spore forms resist desiccation and can remain viable in a dormant state for thousands of years. The best-known orders are Bacillales (mainly aerobic respirers) and Clostridiales (obligate anaerobes). Both groups include species of environmental and economic importance, as well as causative agents of well-known diseases.

Bacillales, genus *Bacillus*. *Bacillus* was one of the first bacterial genera to be classified in the nineteenth century (**Fig. 18.10**). Colonies soon appear on a nutrient agar plate exposed to air. *Bacillus* species can be isolated from soil or food by suspending a sample in water and heating at 80°C for half an hour. Vegetative cells (that is, cells undergoing binary fission) and non-spore formers are killed at that temperature. The remaining endospores will germinate and grow on a beef broth agar plate at 25°C–30°C. *Bacillus* species isolated in this way include over a thousand characterized strains, all but a few of them harmless to humans.

The large, rod-shaped **vegetative cells** (growing and replicating form) of *Bacillus* species are easily stained and visualized. A species of particular scientific importance is *Bacillus subtilis*, the best-studied Gram-positive organism, a "model system" for Firmicutes. The *B. subtilis* genome sequence reveals a large number of transporters for carbon sources and many secretory complexes for industrially important enzymes and drug resistance proteins. It also reveals several integrated prophages (integrated phage genomes). Phage genomes contribute to bacterial evolution by transferring genes between different strains and species, as discussed in Chapter 17.

B. subtilis is used as a model system to study stress response. Enormous changes in protein expression accompany starvation and general stress conditions. As the vegetative cells run short of nutrients on an agar plate, they begin a program to "sporulate"—that is, develop inert endospores (**Fig. 18.10A**). The life cycle of endospore production involves a coordinated developmental plan, in which the cell divides near the pole instead of at the cell equator (**Fig. 18.10B**). The polar compartment develops as the **forespore**, directed by unique regulatory proteins such as SpoIIIE. The larger compartment, called the **mother cell**, provides DNA and nutrients to the growing forespore and disintegrates after release of the mature endospore. When the released endospore encounters favorable conditions of moisture and nutrients, it germinates and restarts vegetative growth. The full sporulation cycle is discussed in Chapter 4.

A *Bacillus* species of great economic importance is *B. thuringiensis*, the most successful biological control agent yet produced (**Fig. 18.11**). *B. thuringiensis* was originally discovered in 1901 by Japanese bacteriologist Shigetane Ishiwata as a cause of disease in silkworms. The organism proved easy to culture on agar-based medium, and its spores are now applied as an insecticide against the gypsy moth caterpillar. During sporulation,

A.

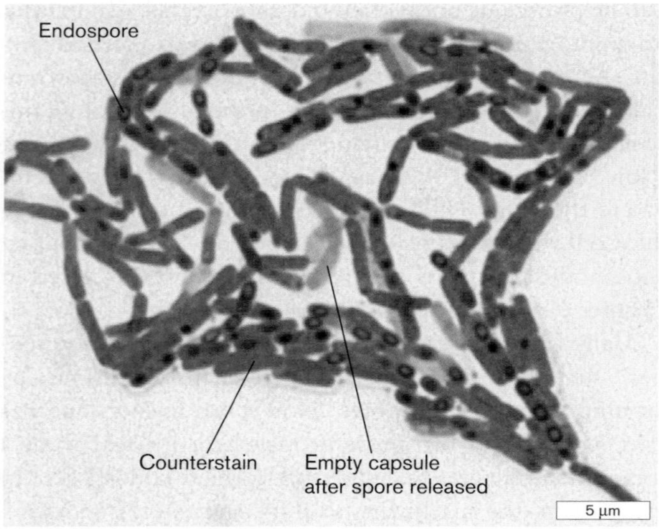
Endospore
Counterstain
Empty capsule
after spore released
5 μm
CDC/COURTESY OF LARRY STAUFFER, OREGON STATE PUBLIC HEALTH LABORATORY

B.

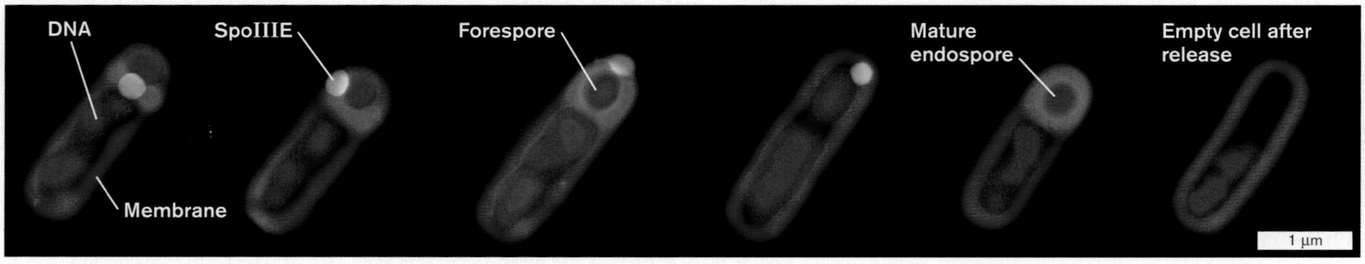
DNA
SpoIIIE
Forespore
Mature
endospore
Empty cell after
release
Membrane
1 μm
PNAS COVER, DEC. 1999/MARC SHARP AND KIT POGLIANO, UC SAN DIEGO

FIGURE 18.10 ■ *Bacillus* species: Gram-positive endospore formers. **A.** Gram-stained *Bacillus* species, sporulating culture. Endospores stain green. Malachite green stain. **B.** Correlated fluorescence imaging of membrane migration, protein translocation, and chromosome localization during *B. subtilis* sporulation. Membranes were stained with red fluorescent FM4-64. Chromosomes were localized with the blue fluorescent nuclear counterstain DAPI. The small, green fluorescent patches indicate the localization of a green fluorescent protein (GFP) gene fusion to SpoIIIE, a protein essential for both initial membrane fusion and forespore engulfment. Progression of the engulfment is shown from left to right.

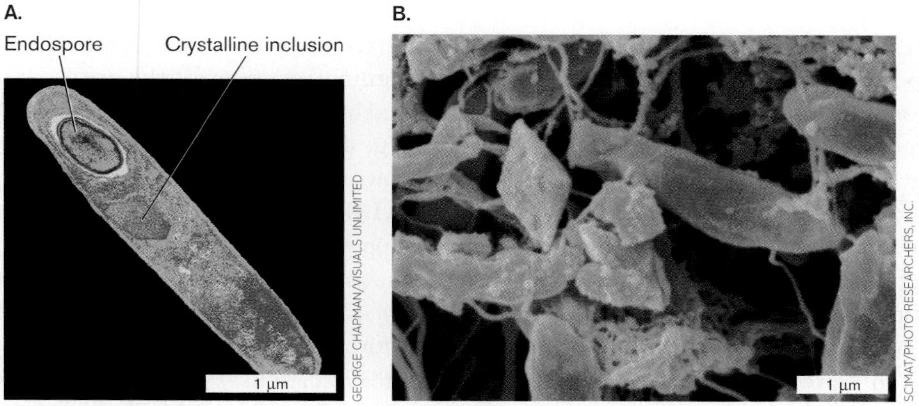
A.
Endospore
Crystalline inclusion
1 µm
GEORGE CHAPMAN/VISUALS UNLIMITED
B.
1 µm
SCIMAT/PHOTO RESEARCHERS, INC.

FIGURE 18.11 ■ *Bacillus thuringiensis:* the biological insecticide. **A.** Sporulating cell of *B. thuringiensis,* showing crystalline inclusion of insecticidal toxin (colorized TEM). **B.** Toxin crystals (colorized blue) embedded in *B. thuringiensis* colony.

B. thuringiensis generates a diamond-shaped crystal adjacent to its endospore (**Fig. 18.11A**). The crystal contains an insecticidal protein known as delta endotoxin. The crystals are ultimately released from the cells (**Fig. 18.11B**). The toxin is activated only at high pH in the digestive tracts of insect larvae; thus, it is safe for animals with acidic digestive tracts.

Nevertheless, controversy arose when the gene encoding the toxin was introduced transgenically into plants, with the aim of producing insect-resistant crops. The concern is that the *Bt* gene might escape into wild plants and endanger beneficial or harmless species of moths and butterflies.

Many *Bacillus* species are extremophiles, growing at high pH (*B. alkalophilus*), at high temperature (*B. thermophilus*), or in high salt (*B. halodurans*). Some species combine alkaliphily with thermophily, as in the case of the "alkalithermophile" *B. alkalophilus.* Genomic research has focused on the surprisingly subtle differences that distinguish an extremophile from a closely related mesophile. For example, thermostability of proteins can be determined by comparing the protein sequences of a thermophile with those of a related mesophile. The thermophilic sequence shows specific patterns of amino acid residues that confer thermostability. Such amino acid substitutions provide useful information for industrial engineering of enzymes.

Clostridiales, genus *Clostridium.* The anaerobic family of spore formers includes the genus *Clostridium,* species of which cause botulism (*C. botulinum*) and tetanus (*C. tetani*) (**Fig. 18.12A**). The botulism toxin, botulinum, or "Botox," is famous for its therapeutic use to relax muscle spasms and smooth wrinkles in skin (**Fig. 18.12B**) (discussed in Chapter 26). Other species, such as *C. acetobutylicum,* have economic importance as producers of industrial solvents such as butanol and acetone (for industrial microbiology, see Chapter 16). The butanol pathway is an example of the diverse fermentative strategies found among clostridia. Unlike *Bacillus* (a monophyletic clade with a common ancestor), the *Clostridium* group is polyphyletic, representing many clades that branch among different genera.

The best-known *Clostridium* species, such as *C. botulinum* and *C. tetani,* sporulate in a form distinct from that of *Bacillus* (see **Fig. 18.12A**). The growing *Clostridium* endospore swells the end of the cell, forming a "drumstick" appearance. Mature clostridial spores are generally less heat resistant than the spores of *Bacillus* and thus more difficult to isolate in pure culture from natural environments. Nevertheless, *Clostridium* spores are found in soil and water, ready to germinate and grow when the environment becomes anoxic. Many harmless species are found in the human colon, particularly in infants. Pathogenic *C. botulinum* can grow within the colon of very young infants, and infant botulism has been implicated in some cases of sudden infant death syndrome.

Another species, *Clostridium difficile,* is a life-threatening intestinal pathogen, resistant to most antibiotics. *C. difficile* grows in patients treated with antibiotics that eliminate normal

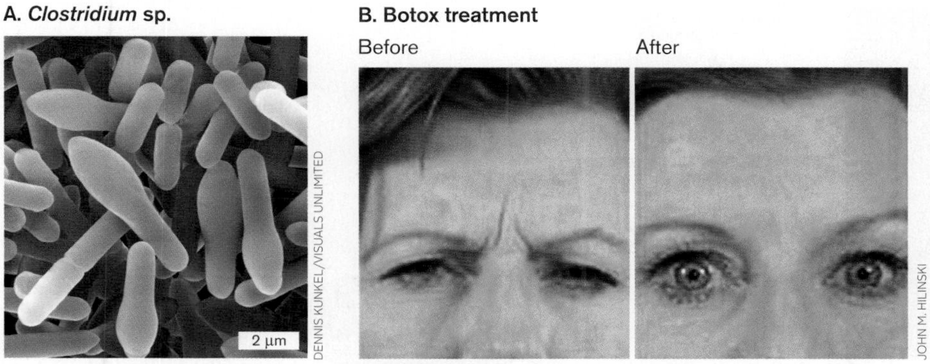
A. Clostridium sp.
2 µm
DENNIS KUNKEL/VISUALS UNLIMITED
B. Botox treatment
Before
After
JOHN M. HILINSKI

FIGURE 18.12 ■ *Clostridium* species: spore-forming anaerobes. **A.** As *Clostridium* cells sporulate, the endospore swells, forming a characteristic "drumstick" appearance. **B.** The deadly botulinum toxin (Botox) from *C. botulinum* is used to relax muscle spasms. This woman was unable to open her eyelids fully (left frame)—a condition that was cured by injection of Botox (right frame).

enteric bacteria, allowing growth of the pathogen. In 2013, a new way to eliminate *C. difficile* was reported: a bile salt analog was found to prevent germination of clostridial endospores. The bile salt analog prevented disease in mice, and it will be explored for developing antigermination therapies in humans.

Related to *Clostridium* are the Heliobacteriaceae, or "Sun bacteria," the only known phototrophs in Firmicutes (see **Table 18.2**). Their name derives from their yellowish color, caused by the shift of their red-peak absorbance into the infrared, which allows transmission of red light plus green light (perceived as yellow). The heliobacteria are photoheterotrophs, with a PS I reaction center providing a modest boost of energy to supplement heterotrophic metabolism of simple organic compounds. Most heliobacterial cells are elongated rods with peritrichous flagella, and they form endospores—the only known endospore-forming phototrophs.

Note: Distinguish *Heliobacterium* from *Helicobacter*, a helical-shaped species of the Gammaproteobacteria.

Variant sporulation and "live birth." The order Clostridiales includes species of exceptionally large bacteria that grow only within the digestive tract of specific animal hosts. These species have evolved intriguing variations of the endospore former life cycle, as revealed by Esther Angert and colleagues at Cornell University (**Fig. 18.13A**). *Metabacterium polyspora* (size 15–20 µm) grows throughout the digestive tract of guinea pigs (**Fig. 18.13B**). Endospores ingested from feces germinate in the upper intestine but rarely undergo binary fission. Instead, the growing cell forms forespores at both poles (**Fig. 18.13C**). The forespores actually multiply within the mother cell to form several endospores, which are released in the colon before defecation.

An even larger enteric endosymbiont is *Epulopiscium fishelsoni*, found in the digestive tract of surgeonfish (**Fig. 18.14**). These bacteria are large enough to be seen by eye—about the size of the period at the end of this sentence. As in *M. polyspora*, Angert showed that *E. fishelsoni* reproduction is synchronized with the digestive cycle of its host, but it has gone even further in transformation of the sporulation cycle. Binary fission is eliminated; the cell must fission at both poles. Each polar fission generates an intracellular daughter cell that grows to nearly the full length of the mother cell. From two to seven intracellular offspring ultimately emerge in "live birth" from the mother cell, which then disintegrates.

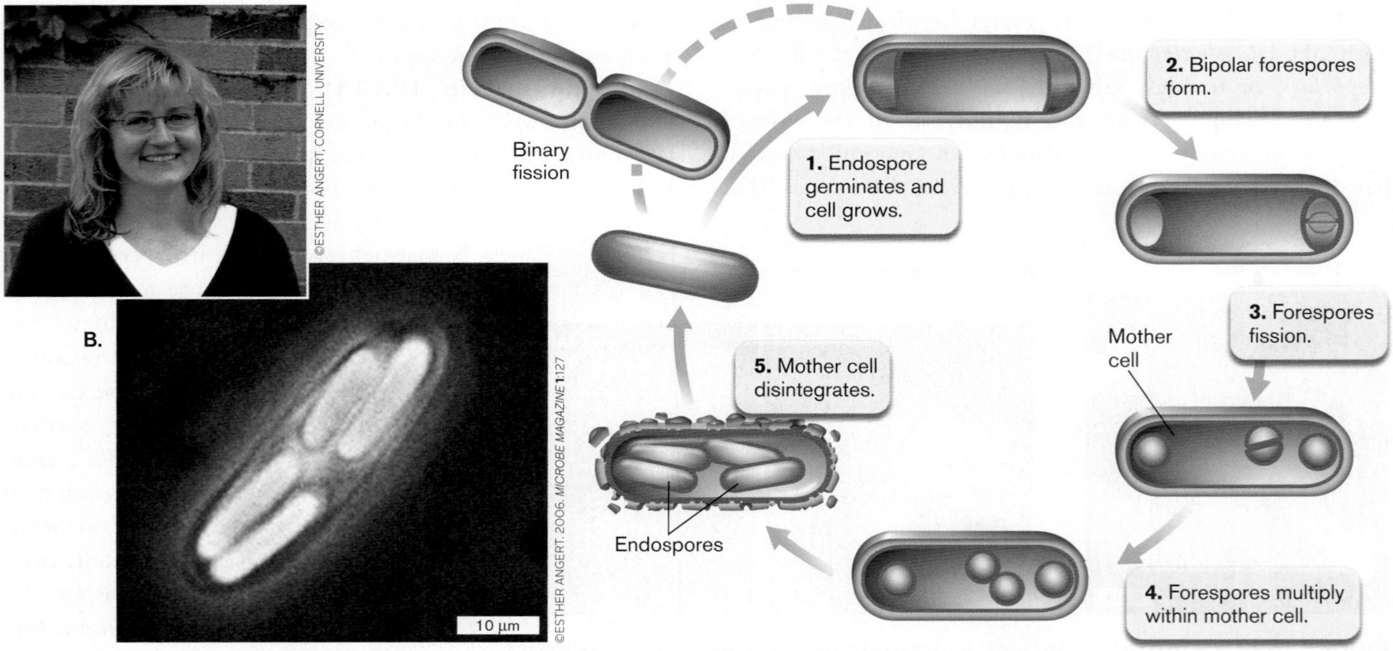
A.
©ESTHER ANGERT, CORNELL UNIVERSITY
B.
10 µm
©ESTHER ANGERT. 2006. MICROBE MAGAZINE 1:127
C. Multiple endospore formation; limited binary fission
Binary fission
1. Endospore germinates and cell grows.
2. Bipolar forespores form.
3. Forespores fission.
Mother cell
4. Forespores multiply within mother cell.
5. Mother cell disintegrates.
Endospores

FIGURE 18.13 ▪ Multiple endospore formation. A. Esther Angert, at Cornell University, characterized unusual forms of sporulation and reproduction in exceptionally large firmicute bacteria. **B.** *Metabacterium polyspora* forms multiple endospores (phase-contrast LM). **C.** A forespore forms at each pole. Forespores fission and multiply within the mother cell and then are released. Germinated cells undergo limited or no binary fission.

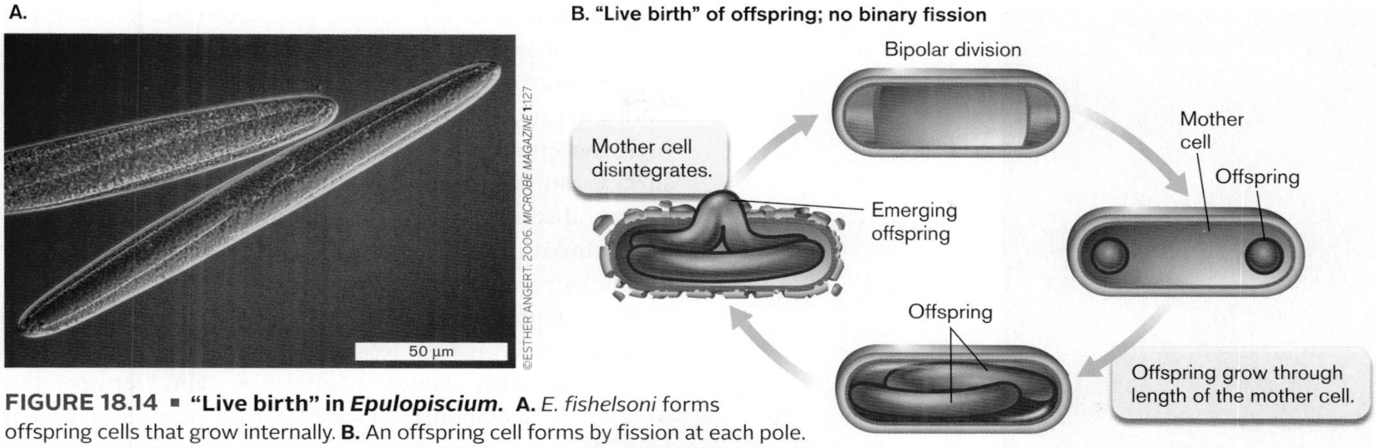
A.
50 µm
©ESTHER ANGERT. 2006. MICROBE MAGAZINE 1:127
B. "Live birth" of offspring; no binary fission
Bipolar division
Mother cell
Offspring
Offspring grow through length of the mother cell.
Offspring
Emerging offspring
Mother cell disintegrates.

FIGURE 18.14 ■ "Live birth" in *Epulopiscium*. **A.** *E. fishelsoni* forms offspring cells that grow internally. **B.** An offspring cell forms by fission at each pole. The cells grow internally until released. No binary fission occurs outside the mother cell.

Non-Spore-Forming Firmicutes

Which Firmicutes do not form spores? Both Bacillales and Clostridiales, as well as other Gram-positive orders, include many non-spore-forming rods and cocci. In these taxa, the endospore-forming system was probably lost by reductive evolution. Non-spore-forming firmicutes include human pathogens such as *Listeria* and *Streptococcus* species, as well as lactic acid bacteria that are important for food production.

Listeria species are facultative anaerobic bacilli, named for the British surgeon Joseph Lister (1827–1912), who was the first to promote antisepsis during surgery. They include enteric pathogens, such as *L. monocytogenes*, that contaminate cheese and sauerkraut (discussed in Chapter 16). Unlike other food-associated organisms, *Listeria* grows at temperatures as low as 4°C. Under preindustrial conditions of food preparation, *L. monocytogenes* was generally outcompeted by other flora. The era of refrigeration led to the emergence of *Listeria* as the cause of listeriosis, a severe gastrointestinal illness that can progress to the nervous system. *L. monocytogenes* cells are taken up by macrophages into phagocytic vesicles, but they avoid digestion and escape the vesicles. The bacteria then multiply as they travel through the host cytoplasm, generating "tails" of actin (**Fig. 18.15** ▶). The actin tails eventually project the cells of *Listeria* out of the original host cell and enable it to penetrate a neighboring host cell.

Lactic acid bacteria. The lactic acid bacteria (order Lactobacillales) are aerotolerant (capable of growth in the presence of oxygen), though they do not use oxygen to respire. Most lactic acid bacteria are obligate fermenters; that is, they generate ATP by substrate-level phosphorylation (discussed in Chapter 13). They ferment primarily by converting sugars to lactic acid (a fermentation pathway discussed in Chapter 13). As the acid builds up, the pH decreases until it halts bacterial growth; thus, the carbon source retains much of its food value for human consumption. This is the basis of yogurt and cheese production. *Lactococcus* and *Lactobacillus* species are extremely important for the dairy industry (discussed in Chapter 16). Other common genera of lactic acid bacteria include *Leuconostoc,* which often spoils meat, and *Pediococcus,* found in sauerkraut and fermented bean products, as well as meat products such as sausage.

The shape of lactate-producing bacteria varies among species, from long, thin rods to curved rods and cocci. Most lactic acid bacteria have fastidious growth requirements and need many amino acids and vitamins. They can be isolated from pasture grasses incubated anaerobically in moderate acid (pH 5). The human intestinal flora include species of lactic acid bacteria. Certain species, particularly *Lactobacillus acidophilus,* are believed to play a positive role in human health by inhibiting the growth of pathogens. For this reason *L. acidophilus* may be ingested as a probiotic therapy.

***Staphylococcus* and *Streptococcus*.** The staphylococci are facultative aerobic cocci that grow in clusters (**Fig. 18.16A** and **B**), often packed in hexagonal arrays (see panel A). They include common skin flora such as *Staphylococcus epidermidis.* The staphylococci are generally salt tolerant, and their fermentation generates short-chain fatty acids that inhibit growth of skin pathogens. Certain species, however, are themselves serious pathogens. *Staphylococcus aureus* causes impetigo and toxic shock syndrome, as well as pneumonia, mastitis, osteomyelitis, and other diseases (discussed in Chapter 26). It is a major cause of nosocomial (hospital-acquired) infections, especially contamination of surgical

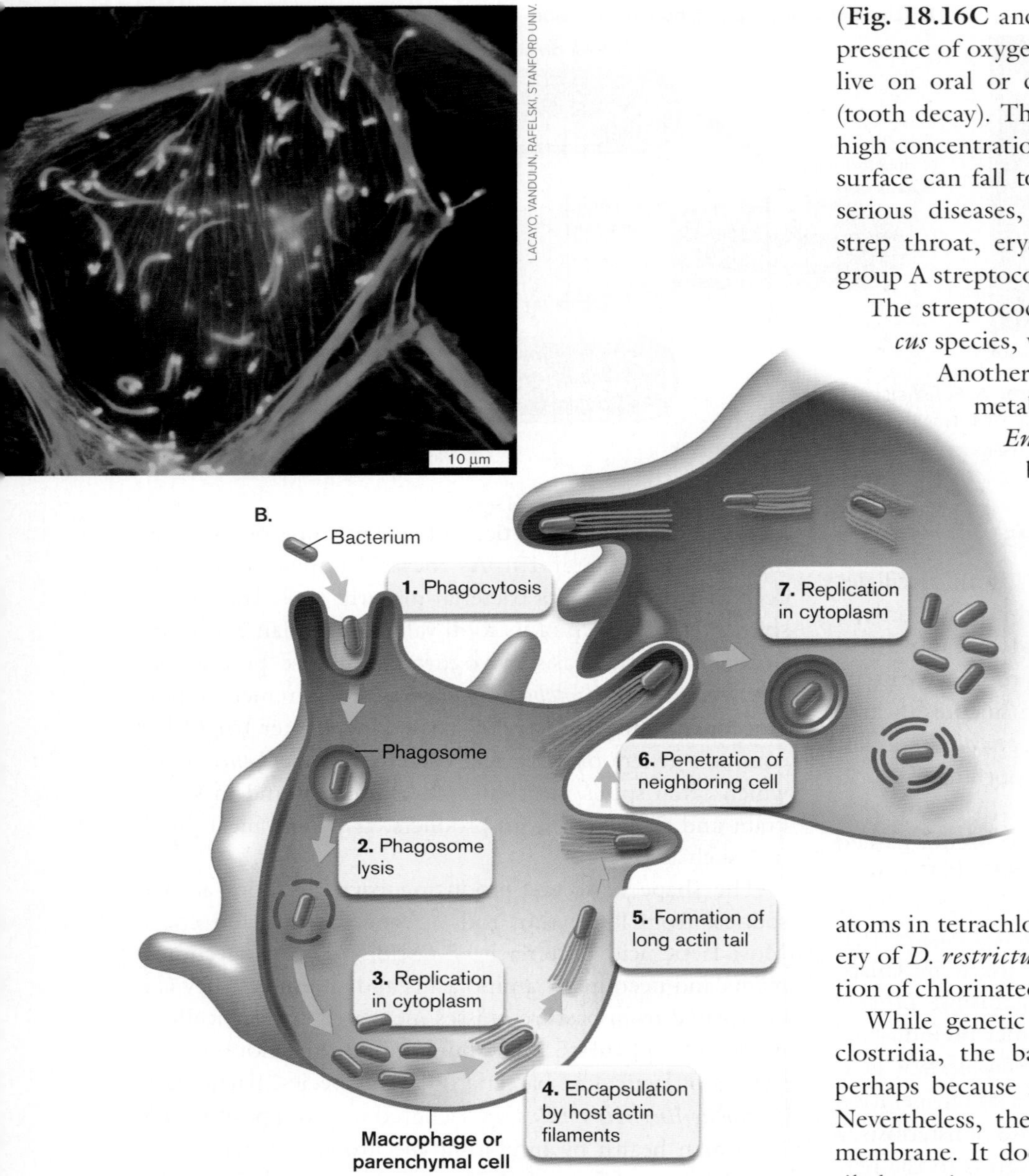
A.
LACAYO, VANDUIJN, RAFELSKI, STANFORD UNIV.
10 µm
B.
Bacterium
1. Phagocytosis
Phagosome
2. Phagosome lysis
3. Replication in cytoplasm
4. Encapsulation by host actin filaments
5. Formation of long actin tail
6. Penetration of neighboring cell
7. Replication in cytoplasm
Macrophage or parenchymal cell

FIGURE 18.15 ■ *Listeria monocytogenes:* intracellular pathogen travels on tails of actin. **A.** Fluorescent antibodies mark the tails of polymerized actin (green) behind the *Listeria* cell bodies (red) traveling within an infected macrophage. **B.** Invading bacteria encapsulate themselves in actin. Actin tails propel them through the host cytoplasm and out through the cell membrane to invade a neighboring cell. *Source:* Part B from U. South Carolina, Microbiology and Immunology On Line.

wounds. The most dangerous strains, now resistant to most known antibiotics, are termed MRSA (methicillin-resistant *S. aureus*).

Streptococcus species generally form chains instead of clusters, because their cells divide in a single plane (**Fig. 18.16C** and **D**). They are aerotolerant (grow in the presence of oxygen) but metabolize by fermentation. Many live on oral or dental surfaces, where they cause caries (tooth decay). Their fermentation of sugars produces such high concentrations of lactic acid that the pH at the tooth surface can fall to pH 4. *Streptococcus* species cause many serious diseases, including pneumonia (*S. pneumoniae*), strep throat, erysipelas, and scarlet fever (*S. pyogenes* or group A streptococci).

The streptococci are less salt tolerant than *Staphylococcus* species, which tolerate as much as 5%–15% NaCl. Another genus whose size and fermentative metabolism resembles that of streptococci is *Enterococcus*. *E. faecalis* is a common member of the intestinal flora, and related strains are enteric pathogens.

Anaerobic dechlorinators. Some firmicutes from the soil show promising abilities to degrade chlorinated pollutants, such as dry-cleaning solvents that are biodegraded very slowly in the environment. The chlorinated molecules are reduced as alternative electron acceptors. *Dehalobacter restrictus,* a flagellated rod, was isolated as an anaerobe capable of respiring by donating electrons to chlorine atoms in tetrachloroethene (**Fig. 18.17**). Thus, the discovery of *D. restrictus* has promising potential for bioremediation of chlorinated pollutants.

While genetic analysis places *D. restrictus* among the clostridia, the bacterium actually stains Gram-negative, perhaps because its peptidoglycan layer is relatively thin. Nevertheless, the species shows no Gram-negative outer membrane. It does possess a thick S-layer of hexagonally tiled proteins, typical of Gram-positive bacteria.

Mycoplasmas lack a cell wall. The mycoplasmas are cell wall-less bacteria, classified as **Tenericutes** or Mollicutes (named in Latin "soft skin"). Mycoplasms have completely lost their cell wall and S-layer through reductive evolution, retaining only their cell membrane. Presumably, the loss of these energy-expensive structures enhanced the reproductive rate of cells in a protected host environment. The Mollicutes comprise many genera of flexible wall-less cells that maintain a shape through some kind of cytoskeleton (**Fig. 18.18A** and **B**). On agar they form colonies of a characteristic "fried-egg" appearance (**Fig. 18.18C**).

The best-known genus of Mollicutes is *Mycoplasma,* although its species now appear to fall in several distantly

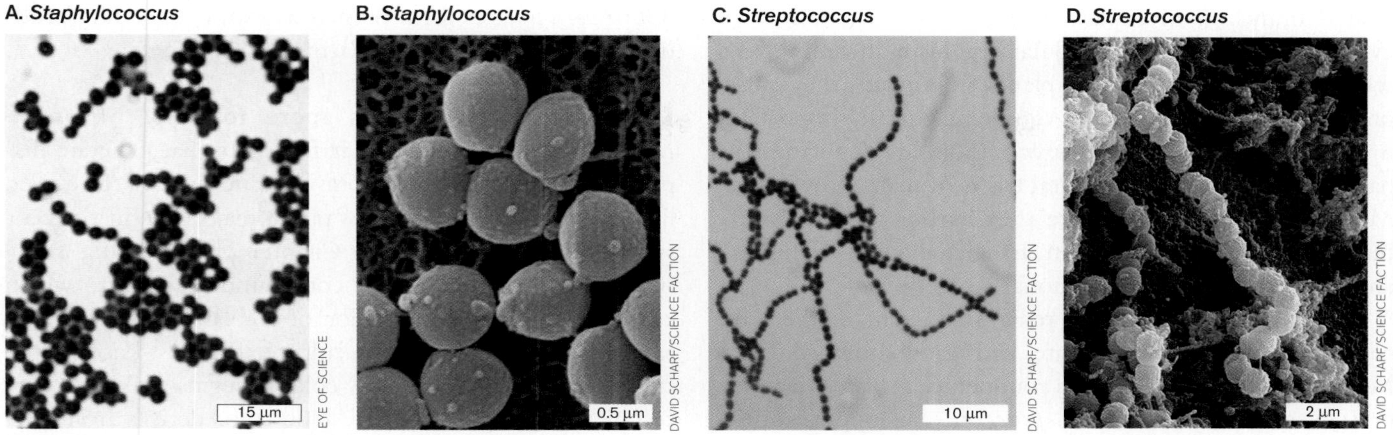
A. Staphylococcus
B. Staphylococcus
C. Streptococcus
D. Streptococcus
15 μm
EYE OF SCIENCE
0.5 μm
DAVID SCHARF/SCIENCE FACTION
10 μm
DAVID SCHARF/SCIENCE FACTION
2 μm
DAVID SCHARF/SCIENCE FACTION

FIGURE 18.16 ▪ Staphylococci and streptococci. **A.** *Staphylococcus* species (Gram stain). **B.** *Staphylococcus* species (colorized SEM). **C.** *Streptococcus* species (Gram stain). **D.** *Streptococcus* species (colorized SEM).

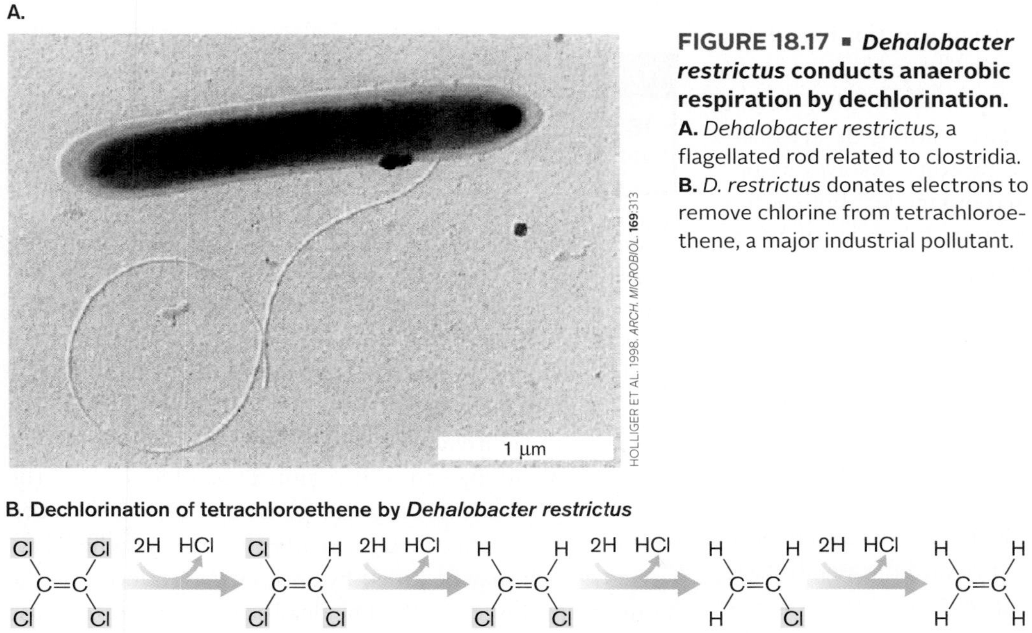
A.
1 μm
HOLLIGER ET AL. 1998. ARCH. MICROBIOL. 169:313
B. Dechlorination of tetrachloroethene by Dehalobacter restrictus
2H HCl
2H HCl
2H HCl
2H HCl
Cl Cl C=C Cl Cl
Cl H C=C Cl Cl
H H C=C Cl Cl
H H C=C H Cl
H H C=C H H

FIGURE 18.17 ▪ *Dehalobacter restrictus* conducts anaerobic respiration by dechlorination. **A.** *Dehalobacter restrictus,* a flagellated rod related to clostridia. **B.** *D. restrictus* donates electrons to remove chlorine from tetrachloroethene, a major industrial pollutant.

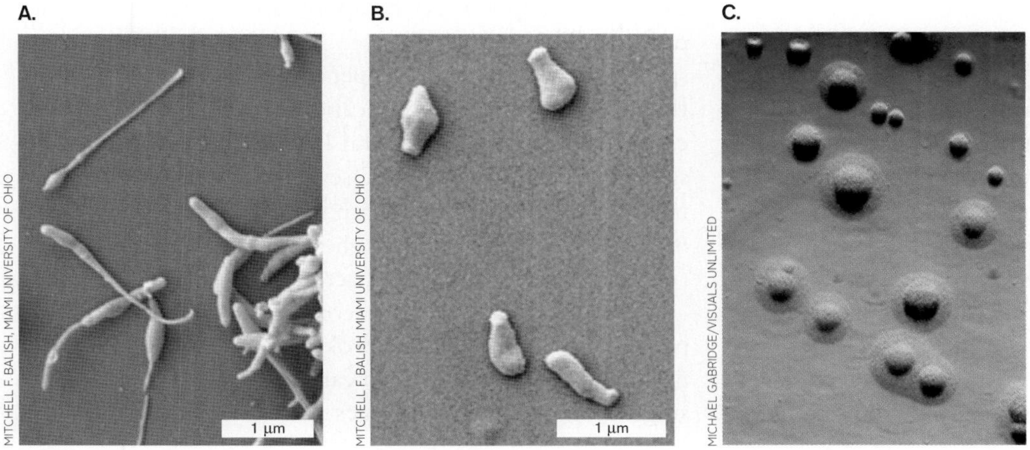
A.
B.
C.
1 μm
MITCHELL F. BALISH, MIAMI UNIVERSITY OF OHIO
1 μm
MITCHELL F. BALISH, MIAMI UNIVERSITY OF OHIO
MICHAEL GABRIDGE/VISUALS UNLIMITED

FIGURE 18.18 ▪ Mycoplasmas: parasites without cell walls. **A.** *Mycoplasma penetrans* cells have an elongated tip used for attachment to the host (SEM). **B.** *M. mobile* (SEM). **C.** Mycoplasmas cultured on agar show a "fried-egg" colony shape.

related branches. Mycoplasmas are found as parasites of every known class of multicellular organism, including vertebrates, insects, and vascular plants; in humans, they cause pneumonia and meningitis. Another medically important mycoplasma is *Ureaplasma urealyticum,* an opportunistic pathogen inhabiting the genital tracts of men and women. Most individuals are unaware they harbor the organism, but *U. urealyticum* has been associated with urethritis, amniotic infections, and pulmonary infections.

Some *Mycoplasma* species remain adherent to the host cell, whereas others penetrate and grow intracellularly. Most mycoplasma cells have a rounded cell shape with one or two extended tips. In *M. penetrans,* an opportunistic pathogen infecting AIDS patients, the cell's attachment tip is coated with adhesion molecules that enable attachment to a host cell surface (**Fig. 18.18A**). The attachment tip penetrates deep into epithelial tissues. By contrast, the fish pathogen *M. mobile* has no attachment tip, but glides along a surface at a rate of seven cell lengths per second (**Fig. 18.18B**). The basis of mycoplasma motility is not understood, although it may involve gliding or cytoskeletal contraction. In some species, motility involves a specialized cell tip called the "terminal organelle," but species that lack this organelle are equally motile. Species of *Spiroplasma* have a spiral shape and undergo corkscrew motion; the basis for this motion is also unknown.

Mycoplasma genomes are among the smallest in known cellular organisms. They also lack biosynthetic pathways for amino acids and phospholipids, which instead must be acquired from the host. Mycoplasmas have unique nutritional requirements, such as cholesterol, a membrane component typical of eukaryotes but rare for prokaryotes. For these reasons, mycoplasmas are difficult to grow in pure culture, although they readily infect tissue cultures. In fact, mycoplasma contamination of tissue culture is so prevalent that it has compromised major studies of cancer and AIDS.

Actinomycetes Form Multicellular Filaments

Which group of Firmicutes forms complex multicellular filaments? The phylum Actinobacteria, the "high-GC Gram-positives," comprises several orders, including Actinomycetales and Bifidobacteriales. Members of the order Actinomycetales are "actinomycetes." Actinomycetes include filamentous spore formers such as *Streptomyces* and *Frankia,* as well as short-chain organisms such as *Mycobacterium,* and irregularly shaped cells such as *Corynebacterium.* An emerging group of marine actinomycetes such as *Salinispora* are isolated from sediment and from sponges. *Salinispora* species form numerous exotic secondary products that show promise as pharmaceutical agents.

***Streptomyces:* filamentous spore formers.** *Streptomyces* bacteria form multicellular filaments that generate dispersible spores. The streptomycete life cycle is discussed in Chapter 4. Streptomycetes play a major role in the ecosystems of soil (discussed in Chapter 21). Decaying *Streptomyces* cells produce the compound **geosmin**, which causes the characteristic odor of soil and can affect the taste of drinking water. In culture, the best-known species, *S. coelicolor* (Latin, "sky color"), forms strikingly blue colonies (**Fig. 18.19A**). The blue color derives from several pigments, including actinorhodin, a polyketide antibiotic. Other species produce filaments that are red, orange, green, or gray, depending on their distinctive products, many of which are antibiotics.

Thought Question

18.6 Why would *Streptomyces* produce antibiotics targeting other bacteria?

Streptomyces species are obligate aerobes, requiring access to air to complete their life cycle. When a *Streptomyces* spore germinates, it extends **vegetative mycelia**, branched filaments that grow into the substrate. Some of the filaments then grow upward into the air, where they develop into **aerial mycelia**. The aerial mycelia in some species grow in tightly coiled spirals (**Fig. 18.19B** and **C**). As the mycelial colony runs out of nutrients, older cells of the filament age and lyse, releasing nutrients that are absorbed by the younger cells. The nutrients also attract other scavengers, which may be killed by antibiotics produced by the aging streptomycete cells. The dead scavengers, too, release nutrients that feed the growing tip of the mycelium.

As mycelia mature, they fragment into smaller cells called **arthrospores**. Arthrospores are vegetative cells, not dormant like endospores (see **Table 18.3**). The arthrospores separate and are dispersed by the wind, enabling them to colonize a new location. Streptomycete mycelia can be obtained from natural habitats by burying a glass slide in soil and then waiting several days for spores to germinate, covering the slide with mycelia. They are challenging to isolate in pure culture, however, because their coiled filaments trap cells of other bacteria.

The *S. coelicolor* genome contains over 8 million base pairs—one of the largest prokaryotic genomes. Streptomycete chromosomes are linear with special "telomeres," single-stranded end sequences that double back to form

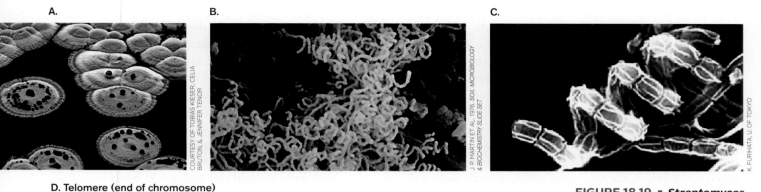

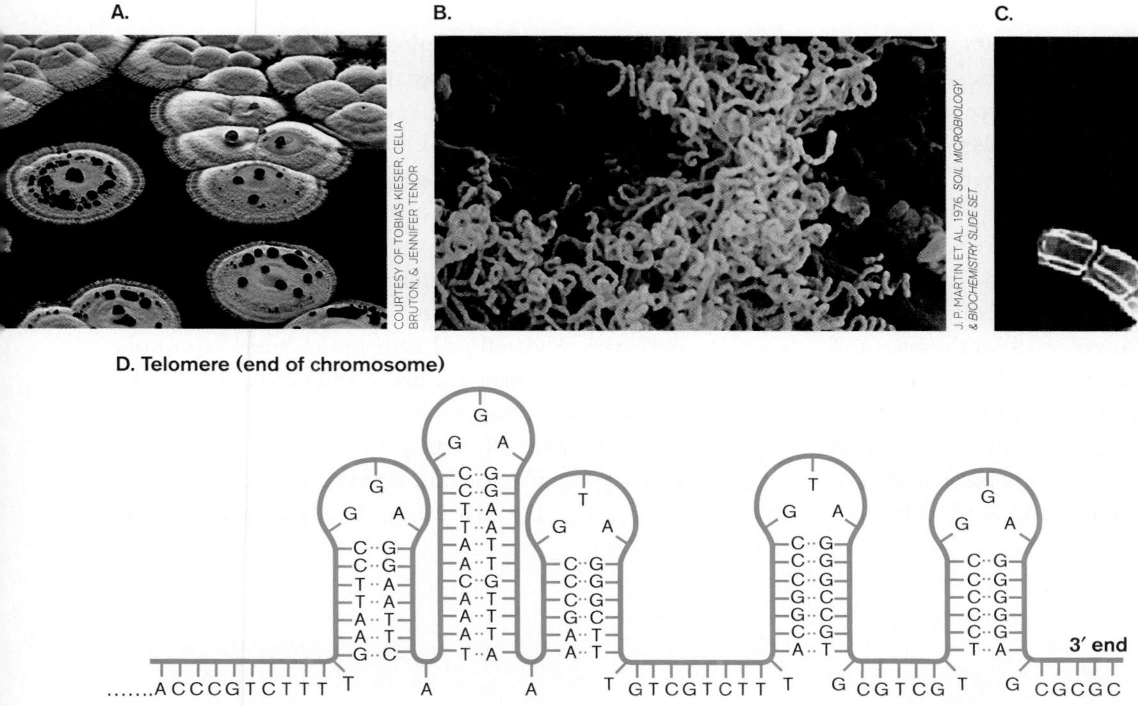
A.
B.
C.
COURTESY OF TOBIAS KIESER, CELIA BRUTON, & JENNIFER TENOR
J. P. MARTIN ET AL. 1976, SOIL MICROBIOLOGY & BIOCHEMISTRY SLIDE SET
K. FURIHATA, U. OF TOKYO
D. Telomere (end of chromosome)
3′ end
.......ACCCGTCTTT
GTCGTCTT
CGTCG
CGCGC

FIGURE 18.19 ▪ *Streptomyces* bacteria. **A.** Colonies of *S. coelicolor* show sky-blue mycelia. **B.** *Streptomyces* cells form coiled filaments (filament width approx. 0.5 μm; SEM). **C.** Close-up of a coiled filament, showing individual cells (SEM). **D.** Hairpin-looped telomere end of the linear chromosome of *S. griseus*. *Source:* Part D from *Prokaryotes*, Springer.

hairpin loops (**Fig. 18.19D**). Much of the lengthy genome of a streptomycete encodes catabolism of a rich array of diverse organic components of decaying plant and animal matter, including even lignin. Other genes encode extensive operons for production of diverse secondary products (see Chapter 15), including antibiotics. More than half the antibiotics currently used in medicine derive from *Streptomyces* species.

Note: Distinguish the order Actinomycetales from the genus *Actinomyces* within this order. Distinguish *Streptomyces*, filamentous rod-shaped actinomycetes, from nonactinomycete *Streptococcus*, Gram-positive cocci that form short, unbranched chains.

Actinomycetes associated with animals and plants. Many marine actinomycetes associate with sponges, where they produce antibiotics that may help the sponge resist pathogens. Pharmaceutical companies now investigate these sponge symbionts for new drugs. Some *Streptomyces* species maintain a mutualism with leaf-cutter ants. The ants culture the bacteria on special organs to produce antibiotics against parasites of their fungal gardens (discussed in eTopic 17.3). A few actinomycetes are animal pathogens; for example, *Actinomyces* species cause actinomycosis, a form of skin abscesses in humans and cattle.

Actinomycetes have many mutually beneficial associations with plants. For example, *Frankia* associates with the alder tree, which develops orange-yellow-colored nodules on its roots to fix nitrogen (**Fig. 18.20**). *Frankia* species live as **endophytes**, endosymbionts of vascular plants. This nitrogen fixation mutualism is analogous to the better-known symbiosis between legumes and rhizobia (discussed under Alphaproteobacteria, in Section 18.5). Other *Frankia* species benefit wheat by conferring resistance to major fungal pathogens, such as the fungus that causes "take-all" disease, as well as to certain parasites and insects. Still other *Frankia* species grow as saprophytes, consuming dead leaf litter. A few actinomycetes cause plant diseases. For example, potato scab disease is caused by *Streptomyces scabies*.

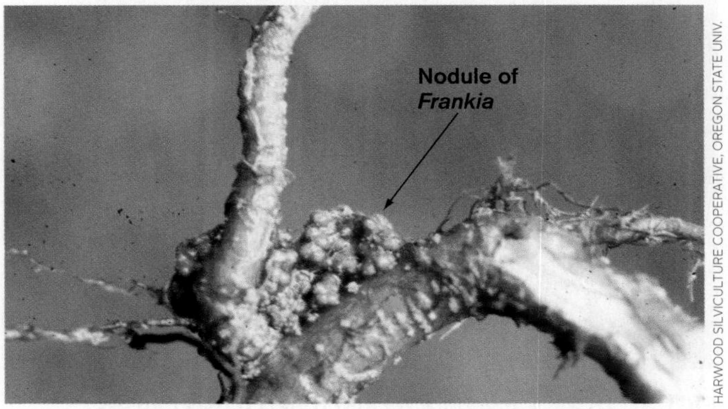
Nodule of
Frankia
HARWOOD SILVICULTURE COOPERATIVE, OREGON STATE UNIV.

FIGURE 18.20 ▪ *Frankia* species associate endophytically with plants. Root of alder tree (*Alnus glutinosa*) bears the orange-yellow-colored nodules (arrowhead) containing *Frankia*.

Nonmycelial Actinobacteria: *Mycobacterium* and *Corynebacterium*. These actinobacteria share the thick acid-fast cell wall of *Streptomyces* but lack the mycelial lifestyle. Two genera that cause dreaded diseases are *Mycobacterium* (**Fig. 18.21**) and *Corynebacterium*. Both genera have thick cell envelopes containing **mycolic acids** and phenolic glycolipids. The mycolic acids of *M. tuberculosis* are extremely diverse and include some of the longest-chain acids known, up to 90 carbons (**Fig. 18.21C**). The mycolic acids are linked to arabinogalactan, a polymer of arabinose and galactose built on the peptidoglycan (discussed in Chapter 3). The mycolyl-arabinogalactan-peptidoglycan complex forms a waxy coat that impedes the entry of nutrients through porins and thus limits growth rate, but it also protects the bacterium from host defenses and antibiotics. For this reason, the cure of tuberculosis requires an exceptionally long course of antibiotic therapy.

Mycobacterium includes the species *M. tuberculosis* and *M. leprae,* as well as lesser-known pathogens such as *M. ulcerans.* Cells of *M. tuberculosis* can be detected by the **acid-fast stain** as tiny rods associated with sloughed cells in sputum (**Fig. 18.21A**). In the acid-fast stain, cells are penetrated with a dye that is retained under treatment with acid alcohol (discussed in Chapter 2). The acid-fast property is associated with unusual cell wall lipids, such as mycolic acids. *M. tuberculosis* bacteria are challenging to culture; they form crinkled colonies after 2 weeks of growth on agar-based media (**Fig. 18.21B**).

The closely related species *M. leprae* causes the disfiguring disease leprosy (**Fig. 18.22A**). The species has one of

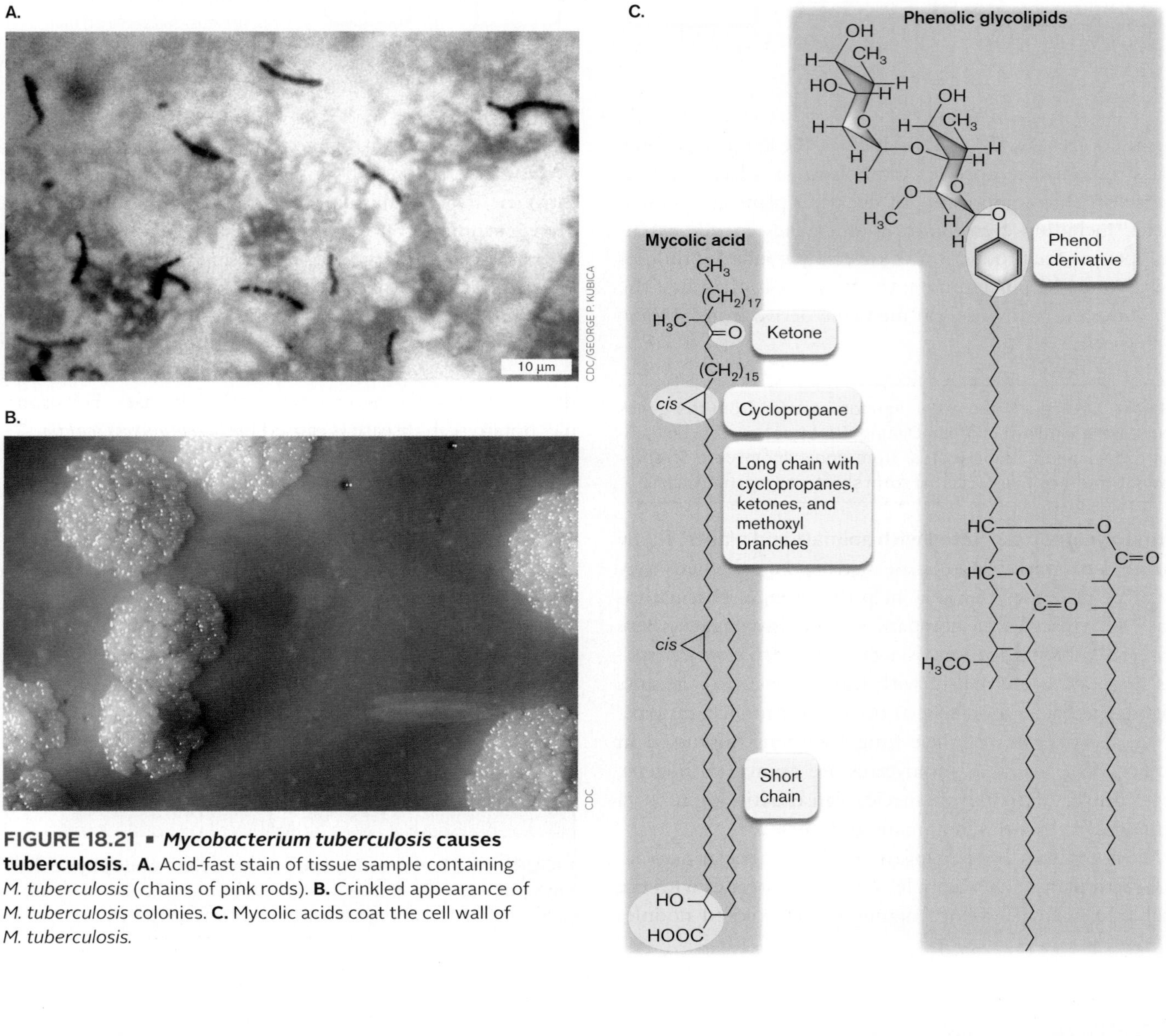
A.
10 μm
CDC/GEORGE P. KUBICA
B.
CDC
C.
Phenolic glycolipids
Phenol derivative
Mycolic acid
Ketone
Cyclopropane
Long chain with cyclopropanes, ketones, and methoxyl branches
Short chain
cis
HO
HOOC
H_3CO

FIGURE 18.21 ■ *Mycobacterium tuberculosis* causes tuberculosis. **A.** Acid-fast stain of tissue sample containing *M. tuberculosis* (chains of pink rods). **B.** Crinkled appearance of *M. tuberculosis* colonies. **C.** Mycolic acids coat the cell wall of *M. tuberculosis.*

the longest known doubling times of any pathogen (about 14 days), and it can take a year to grow enough cells in the laboratory for observation. Lower temperature is required; for this reason, leprosy attacks the extremities (hands and feet, which have lower temperature). Culture on artificial media is impossible; the bacteria can be grown only within low-temperature animals, such as armadillos, or within genetically immunodeficient mice.

The genomes have been sequenced for both *M. tuberculosis* and *M. leprae*. *M. tuberculosis* has surprisingly few recognizable pathogenicity genes but a large number of environmental stress components, including 16 environmental sigma factors (discussed in Chapter 8), as well as 250 genes for its complex lipid metabolism. Over half the *M. leprae* genome consists of pseudogenes, homologs of *M. tuberculosis* genes undergoing reductive evolution (**Fig. 18.22B**). Thus, *M. leprae* appears to be an evolving pathogen "caught in the act" of losing many genes no longer needed in its sheltered host environment. How it lost the need for so many genes preserved in *M. tuberculosis* remains a mystery. *M. leprae* causes disease worldwide, including 250 cases of leprosy annually in the United States. In 2013, a new test was approved that reveals leprosy infection a year before symptoms appear—allowing cure with antibiotics before nerve damage is irreversible.

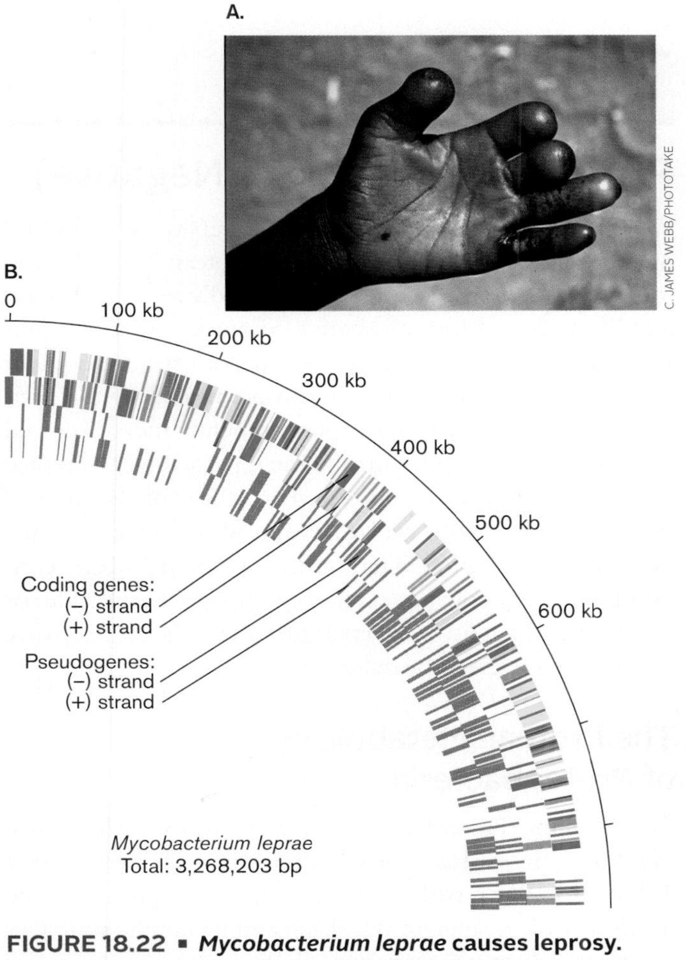
A.
C. JAMES WEBB/PHOTOTAKE
B.
0
100 kb
200 kb
300 kb
400 kb
500 kb
600 kb
Coding genes:
(–) strand
(+) strand
Pseudogenes:
(–) strand
(+) strand
Mycobacterium leprae
Total: 3,268,203 bp

FIGURE 18.22 ■ *Mycobacterium leprae* causes leprosy. **A.** Hand disfigured by leprosy. **B.** The genome of *M. leprae* shows a high content of decaying pseudogenes (gray bars), most of which correspond to functional genes in the genome of *M. tuberculosis*.

Mycobacteria also include a much larger number of harmless commensals, such as *M. smegmatis*, isolated from human skin. Species of mycobacteria can be isolated from soil and water, as well as from various animal sources. Their culture is difficult because of their slow growth rates, but isolation can be enhanced by treatment with a base (NaOH or KOH) at concentrations that kill most other bacteria.

Note: Distinguish *Mycobacterium*, rods whose cell walls contain mycolic acids, from *Mycoplasma*, firmicutes lacking cell walls, related to *Bacillus* and *Clostridium*.

Irregularly shaped actinomycetes. Several nonmycelial actinomycetes show unusual cell shapes. Members of the genus *Corynebacterium* include soil bacteria, as well as pathogens such as *C. diphtheriae*, the cause of the lung disease diphtheria. *Corynebacterium* species grow as irregularly shaped rods, which may divide by a "half-snapping" mechanism in which one side of the cell remains attached like a hinge (**Fig. 18.23A**). Related soil bacteria include the genera *Nocardia* and *Rhodococcus*.

Soil bacteria of the genus *Arthrobacter* exhibit an unusual cell cycle in which coccoid stationary-phase cells sprout into rods, which eventually run out of nutrients and revert to the coccoid form. The growing rods form irregular branched filaments (**Fig. 18.23B**). An *Arthrobacter*

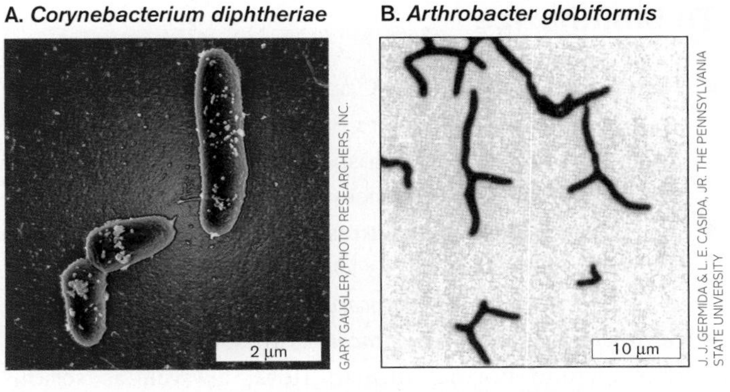
A. Corynebacterium diphtheriae
2 µm
GARY GAUGLER/PHOTO RESEARCHERS, INC.
B. Arthrobacter globiformis
10 µm
J. J. GERMIDA & L. E. CASIDA, JR. THE PENNSYLVANIA STATE UNIVERSITY

FIGURE 18.23 ■ Irregularly shaped actinomycetes: *Corynebacterium* and *Arthrobacter*. **A.** *Corynebacterium diphtheriae* (cells 1–2 µm in length) divides by snapping off one side while remaining attached at the other; the result is a typical V shape or "Chinese letter" arrangement (colorized SEM). **B.** *Arthrobacter globiformis* cultures form coccoid cells in stationary phase. With added nutrients, the coccoid cells grow out as rods.

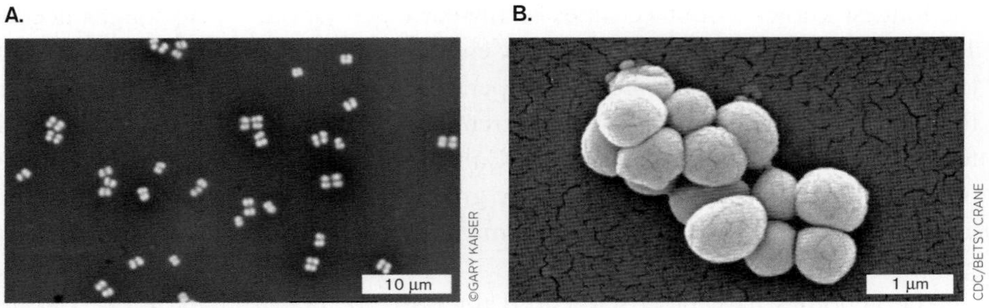
A.
10 μm
©GARY KAISER
B.
1 μm
CDC/BETSY CRANE

FIGURE 18.24 ■ *Micrococcus* species grow in tetrads or sarcinae. **A.** *M. luteus* growing in tetrads (negative or indirect stain with nigrosin). **B.** *M. luteus* bacteria grow in clumps when cultured on a solid substrate (SEM).

species was discovered conducting anaerobic respiration by reduction of hexavalent chromium (Cr^{6+}), a toxic metal pollutant, to a less toxic oxidation state. *Arthrobacter* now shows potential as an agent of bioremediation of hexavalent chromium.

Micrococcaceae. Relatives of *Arthrobacter* are nonfilamentous cocci such as *Micrococcus. M. luteus* is one of the most widespread of soil bacteria, appearing readily as yellow colonies on agar plates exposed to air. Historically, the genus *Micrococcus* in the family Micrococcaceae was classified with *Staphylococcus,* but its DNA sequence data now place *Micrococcus* and most of the Micrococcaceae within the actinomycetes.

Micrococci are aerobic heterotrophs, commonly isolated from air and dust, although their habitat of choice is human skin. Micrococci grow in square or cuboid formations, dividing in two or three planes (**Fig. 18.24**); cuboid clusters are known as **sarcinae** (singular, **sarcina**). *M. luteus* is harmless to humans and grows well at room temperature, so it makes an excellent laboratory organism for observation by students.

To Summarize

- **Firmicutes**, the low-GC Gram-positive bacteria, include endospore-forming genera such as *Bacillus* and *Clostridium*. The cycle of endospore formation was probably present in the common ancestor of this phylum.
- **Nonsporulating firmicutes** include pathogenic rods such as *Listeria,* as well as food-producing bacteria such as *Lactobacillus* and *Lactococcus.*
- ***Staphylococcus* and *Streptococcus*** are Gram-positive cocci that include normal human flora, as well as serious pathogens causing toxic shock syndrome, pneumonia, and scarlet fever.
- **Mycoplasmas** belong phylogenetically to Firmicutes but lack the cell wall and S-layer. They have flexible cytoskeletons and show ameboid motility. Mycoplasma species cause diseases such as meningitis and pneumonia.
- **Actinobacteria** (order Actinomycetales) include mycelial spore-forming soil bacteria, such as the actinomycete *Streptomyces.* Other actinobacteria have irregularly shaped cells, such as *Corynebacterium* species that cause diphtheria.
- **Mycobacteria** are actinobacterial rods whose cell envelope contains a diverse assemblage of complex mycolic acids. Mycobacterial species cause tuberculosis and leprosy. They stain acid-fast.

18.4

Proteobacteria (Gram-Negative)

Which phylum shows the greatest diversity of form and metabolism? Proteobacteria—the "protean," or "many-formed," bacteria—have exceptionally diverse metabolism and lifestyles, yet they all share a common structure: their Gram-negative cell envelope, which consists of an outer membrane, peptidoglycan cell wall permeated by the periplasm, and inner membrane (plasma membrane) (discussed in Chapter 3). The outer membrane is packed with receptor proteins and porins, comprising two-thirds of the mass of the membrane. Porins evolve so as to admit nutrients while excluding antibiotics. The outer membrane lipids contain long sugar polymer extensions (lipopolysaccharide, or LPS). In pathogens, LPS repels phagocytosis and has toxic effects when released by dying cells.

The Protean Metabolism of Proteobacteria

The phylum Proteobacteria is defined to include five monophyletic classes, labeled Alpha through Epsilon (see **Table 18.1**). Even within each closely related group, we see nearly as wide a range of cell shape and metabolic strategies as we see in Proteobacteria as a whole.

A closer look at proteobacterial metabolism shows that processes that at first seem very different actually connect

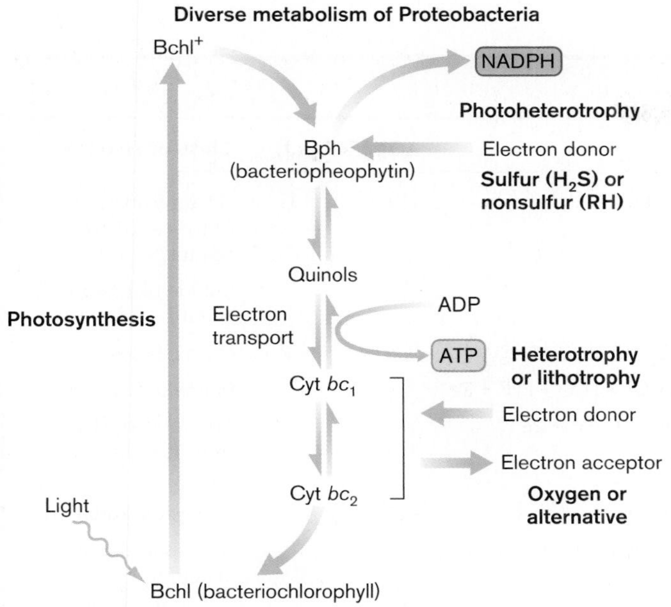
Diverse metabolism of Proteobacteria
Bchl+
NADPH
Photoheterotrophy
Bph
(bacteriopheophytin)
Electron donor
Sulfur (H2S) or
nonsulfur (RH)
Quinols
ADP
Photosynthesis
Electron
transport
ATP
Heterotrophy
or lithotrophy
Cyt bc1
Electron donor
Electron acceptor
Cyt bc2
Oxygen or
alternative
Light
Bchl (bacteriochlorophyll)

FIGURE 18.25 ■ The protean metabolism of Proteobacteria. In many Gram-negative species, metabolic diversity arises through minor "add-ons" of biochemical modules such as light absorption by bacteriochlorophyll, use of sulfide or organic electron donors, and use of oxygen or alternative (anaerobic) electron acceptors.

through linked modules (**Fig. 18.25**) (metabolism is discussed in detail in Chapter 14). Many proteobacteria can oxidize H_2 (hydrogenotrophy) and small organic acids, as well as complex organic molecules with aromatic rings. Some "photoheterolithotrophs" are able to carry out nearly all the fundamental classes of metabolism, depending on environmental conditions, such as availability of light, oxygen, and nutrients. An example is *Rhodopseudomonas palustris,* the species of Alphaproteobacteria characterized and sequenced by Caroline Harwood at the University of Washington (see Figure 14.2 and **eTopic 14.1**). In such a "protean" organism, the core of all proteobacterial energy acquisition is a respiratory chain of electron donors and acceptors. Electrons may enter the chain from a photoexcited chlorophyll, from organic electron donors such as sugars or benzoates, or from a mineral electron donor such as reduced sulfur or iron (lithotrophy). In facultative anaerobes, the electron acceptor may be a mineral such as nitrate or sulfate (anaerobic respiration). The capabilities of a given organism depend on which oxidoreductases its genome encodes.

Photoheterotrophy. Light-supplemented heterotrophy occurs in Alpha-, Beta-, and Gammaproteobacteria. A remarkable diversity of light absorption niches has evolved from a common ancestor of photosystems I and II (see **Table 18.2**). The various bacteriochlorophylls of Proteobacteria generally peak in two ranges, in the blue and in the red or infrared. Bacteriochlorophyll *b,* in *Blastochloris viridis,* peaks well beyond 1,000 nm. Species with different photopigments often grow together in stratified layers of sediment or wetland—the infrared absorbers below, where they capture the longer wavelengths "left over" from the shorter-wavelength absorbers above.

The prevalence of homologous photosystems suggests that the common ancestor of the Gammaproteobacteria was a photoheterotroph. At the same time, other evidence supports horizontal transfer of photosystems among proteobacterial branches. The evidence is particularly strong in the case of **proteorhodopsin**, a homolog of the retinal protein light-driven proton pump first characterized in halophilic archaea (discussed in Chapters 14 and 19). The proteorhodopsin light pump in *Pelagibacter,* a marine relative of *Rickettsia,* absorbs green and yellow (wavelength 500–600 nm), the midrange of solar radiation reaching Earth's surface. Different species have proteorhodopsins with very different absorption ranges, apparently adapted to different niches in the marine ecosystem.

In the literature, proteobacterial phototrophs are historically called "purple bacteria." The actual colors of all the phototrophs range from purple-red through yellow and brown. "Purple sulfur bacteria" are species that photolyze reduced forms of sulfur, such as H_2S, HS^-, S^0, or $S_2O_3^{2-}$ (thiosulfate), whereas "purple nonsulfur bacteria" photolyze H_2 or use photosystem II for cyclic photophosphorylation (discussed in Chapter 14). However, it turns out that many purple bacteria also use photosystem II. Most proteobacterial phototrophs conduct photosynthesis only in the absence of oxygen, reverting to aerobic heterotrophy when oxygen is present. Some, however, are obligate anaerobes. Still others, found in marine environments, surprisingly photolyze sulfide or organic compounds only in the presence of oxygen. Making sense of this extraordinary diversity is the job of marine and aquatic microbiologists, particularly those calculating global cycles of carbon and oxygen (discussed in Chapters 21 and 22).

Thought Question

18.7 Why might genes for the proteorhodopsin light-powered proton pump be more likely to transfer horizontally than the bacteriochlorophyll-based photosystems PS I and PS II?

Lithotrophy. The use of inorganic electron donors recurs throughout the phylogeny of bacteria, with remarkable

TABLE 18.4

Lithotrophy and methylotrophy (examples).

Class/phylum	Example species	Reaction	$\Delta G^{\circ\prime}/2e^-$ (kJ)	Class of reaction
Alphaproteobacteria	*Nitrobacter winogradskyi*	$NO_2^- + \frac{1}{2}O_2 \longrightarrow NO_3^-$	–54	Nitrite oxidation (nitrification to nitrate)
	Paracoccus pantotrophus	$H_2S + 2O_2 \longrightarrow 2H^+ + SO_4^{2-}$		Sulfide oxidation to sulfate
	Roseobacter litoralis	$CO + H_2O \longrightarrow CO_2 + 2H^+ + 2e^-$		CO oxidation
	Hyphomicrobium	$CH_3OH + O_2 \longrightarrow CO_2 + H_2O + 2H^+ + 2e^-$		Methylotrophy
Betaproteobacteria	*Nitrosomonas europaea*	$NH_3 + O_2 \longrightarrow HNO_2^- + 3H^+ + 2e^-$		Ammonia oxidation (nitrification to nitrite)
	Ralstonia eutropha	$2H_2 + \frac{1}{2}O_2 \longrightarrow 2H_2O$	–237	Hydrogen oxidation
	Thiobacillus denitrificans	$3S^0 + 4NO_3^- \longrightarrow 3SO_4^{2-} + 2N_2$		Anaerobic sulfur oxidation
Gammaproteobacteria	*Acidithiobacillus ferrooxidans*	$4FeS_2 + 15O_2 + 14H_2O \longrightarrow 4Fe(OH)_3 + 16H^+ + 8SO_4^{2-}$	–164	Iron-sulfur oxidation
	Acidithiobacillus thiooxidans	$2S^0 + 3O_2 + 2H_2O \longrightarrow 2SO_4^{2-} + 4H^+$	–196	Sulfur oxidation
	Methylococcus capsulatus	$CH_4 + 2O_2 \longrightarrow HCO_3^- + H^+ + H_2O$	–203	Methanotrophy
	Beggiatoa alba	$2H_2S + O_2 \longrightarrow 2S^0 + 2H_2O$	–210	Sulfide oxidation
	Chromatium	$4Fe^{2+} + CO_2 + 11H_2O + h\nu \longrightarrow 4Fe(OH)_3 + [CH_2O] + 8H^+$		Iron phototrophy (photoferrotrophy)
Deltaproteobacteria	*Desulfovibrio*	$4S^0 + 4H_2O \longrightarrow SO_4^{2-} + 3HS^- + 5H^+$	–11.3 (at pH 8)	Sulfur oxidation
Nitrospirae	*Nitrospira*	$NO_2^- + \frac{1}{2}O_2 \longrightarrow NO_3^-$	–54	Nitrite oxidation (nitrification to nitrate)
Planctomycetes	*Brocadia anammoxidans*	$NH_4^+ + NO_2^- \longrightarrow N_2 + 2H_2O$	–238	Anaerobic ammonium oxidation (anammox)

diversity among Proteobacteria (**Table 18.4**). The ability to oxidize or reduce minerals evolved multiple times in different clades by modification of the electron transport chain. Many lithotrophic reactions generate acid, which conveniently breaks down rock containing additional reduced minerals. Bacterial and archaeal lithotrophy has tremendous significance for global cycling of elements in the biosphere (discussed in Chapter 22). Some lithotrophs are autotrophs, fixing carbon dioxide into biomolecules by the Calvin cycle (in carboxysomes) or the reverse TCA cycle. Others have alternative pathways of heterotrophy when organic foods are available. An interesting kind of lithotrophy performed by many Proteobacteria is carbon monoxide (CO) oxidation (**Special Topic 18.1**). Oxidizing CO is important because this toxic substance is a by-product of animal and plant metabolism.

We now survey species of the major classes of Proteobacteria.

Alphaproteobacteria: Photoheterotrophs, Methylotrophs, and Endosymbionts

The class Alphaproteobacteria includes photoheterotrophs and heterotrophs, as well as bacteria metabolizing single-carbon compounds. A remarkable array of intracellular mutualists and pathogens arises in this group.

Photoheterotrophs. The Alpha class photoheterotrophs are generally unicellular. Their cell shapes range from flagellated spirilla to rounded rods (*Rhodobacter sphaeroides*) (**Fig. 18.26A**), wide spirals (*Rhodospirillum rubrum*),

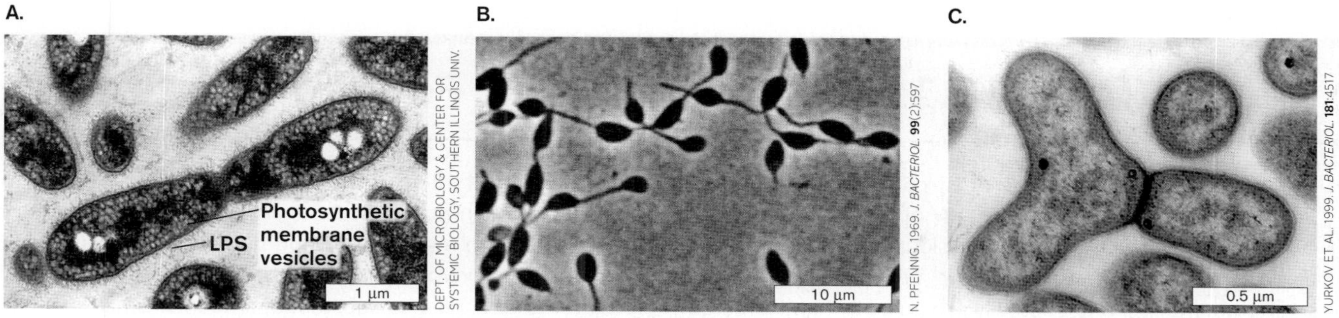
A.
B.
C.
Photosynthetic membrane vesicles
LPS
1 µm
DEPT. OF MICROBIOLOGY & CENTER FOR SYSTEMIC BIOLOGY, SOUTHERN ILLINOIS UNIV.
10 µm
N. PFENNIG. 1969. J. BACTERIOL. 99(2):597
0.5 µm
YURKOV ET AL. 1999. J. BACTERIOL. 181:4517

FIGURE 18.26 ■ Alphaproteobacterial photoheterotrophs. **A.** *Rhodobacter sphaeroides*, showing intracellular photosynthetic membranes and outer membrane LPS filaments (TEM). **B.** *Rhodomicrobium vannielii* with stalked cells (phase contrast). **C.** *Citromicrobium* species, an aerobic photoheterotroph, forms highly pleomorphic shapes, including the Y shape (TEM).

and stalked cells (*Rhodomicrobium*) (**Fig. 18.26B**). The TEM of *R. sphaeroides* (**Fig. 18.26A**) shows that the cell is packed with photomembranes (membranes containing the photosynthetic complex) arranged in folds or tubes invaginated from the cytoplasmic membrane. These intracellular photomembranes expand the surface area for photon capture, but during growth with oxygen, the photomembranes disappear. The outer surface of the cell shows LPS filaments extending from the outer membrane.

The metabolism of *Rhodospirillum rubrum* shifts drastically with oxygen. Anaerobically, *R. rubrum* bacteria grow dark red because they synthesize membrane containing Bchl *a*, which absorbs green and infrared and reflects primarily red. With oxygen, however, *R. rubrum* fails to synthesize Bchl *a*, and the cells grow white as their metabolism switches to straight heterotrophy. Heterotrophy is supplemented by oxidizing small molecules such as carbon monoxide.

Alpha photoheterotrophs requiring oxygen for phototrophy (but not photolyzing water for O_2) were found unexpectedly in genomic surveys of marine bacteria. Originally thought to be straight heterotrophs, these organisms revealed genes encoding bacteriochlorophyll. Their aerobic heterotrophy was shown to be driven by light absorption. The reason for the oxygen requirement is unclear, since their photosystems do not involve oxygen. Examples of O_2-requiring phototrophs have since been found in most of the proteobacterial clades, and species have been isolated from all major habitats, including freshwater, marine, and soil ecosystems. Some show unusual cell morphology, such as the Y shape of *Citromicrobium* (**Fig. 18.26C**).

The aerobic photoheterotroph *Erythromicrobium ramosum* can reduce toxic metal compounds such as tellurite ion (TeO_3^{2-}). The cells generate tellurium crystals that take up 30% of their cell weight, a trait that we might use to remove tellurite from liquid waste. Other toxic metals reduced by aerobic photoheterotrophs include selenium and arsenic. Thus, these obscure bacteria now have a promising future in remediation of metal-contaminated industrial wastes.

Aquatic and soil oligotrophs. Alphaproteobacteria also include many nonphototrophic heterotrophs of soil and water. Many aquatic and soil heterotrophs are oligotrophs adapted to extremely low nutrient concentrations; an example is *Caulobacter crescentus*, the stalk-to-flagellum organism discussed in Chapters 3 and 4. Oligotrophic bacteria often have unusual extended shapes enhancing nutrient uptake, such as the starlike cell aggregates of *Seliberia stellata*. In *Seliberia* species, the individual tightly coiled rods generate oval or spherical reproductive cells by a budding process. The budding reproductive cells germinate into rods, which then form new aggregates.

Some heterotrophs common in soil are pathogens. *Brucella* spp. are intracellular pathogens of animals that can also infect humans. The soil pathogen *Granulibacter bethesdensis* is associated with chronic granulomatous disease, an inherited disorder of the phagocyte oxidase system that leaves patients susceptible to infection. Other pathogens are carried by insects or animal hosts; for example, *Bartonella henselae* causes cat scratch fever.

Methylotrophy and methanotrophy. Methylotrophy is the ability of an organism to oxidize reduced single-carbon compounds such as methanol, methylamine, or methane. The Alphaproteobacteria include several genera of methylotrophs, which are found in all environments, including soil, freshwater, and the ocean. Most methylotrophs can grow on both single-carbon and organic compounds. One such versatile genus, *Methylobacterium*, is equally at home in soil and water, on plant surfaces, and as a contaminant of facial creams and purified water for silicon chip manufacture. Other species are restricted to single-carbon compounds, incapable of metabolizing organic compounds with carbon-carbon bonds. Methylotrophs that grow solely on methane (CH_4) are called **methanotrophs**.

Methane-oxidizing bacteria of both Alpha and Gamma classes contribute to aquatic ecosystems, serving as major food sources for zooplankton. They eliminate much of the

Special Topic 18.1: Carbon Monoxide: Food for Bacteria?

Carbon monoxide (CO) is the most common cause of fatal poisoning in industrial countries. A product of incomplete combustion as in automobile emissions, inhaled CO kills by binding hemoglobin and preventing uptake of oxygen. But small amounts of CO are actually produced by human metabolism, as well as by plant photosynthesis and marine photochemistry. What happens to all this CO?

In natural environments, CO serves as an electron donor, which some bacteria can oxidize for food:

$$CO + H_2O \longrightarrow CO_2 + 2H^+ + 2e^-$$

The CO is oxidized by water to CO_2, transferring two electrons to an ETS to yield energy. Thus, bacteria may play a previously unrecognized role in cycling CO and preventing toxic buildup in a local environment.

Where in nature do we find CO? Major sources are anthropogenic (human-made), such as automobiles and home furnaces that burn fossil fuels, wood-burning stoves, and forest clearing (**Fig. 1**). Volcanic eruptions and deep-sea hydrothermal vents also release substantial amounts. Small amounts of CO arise from living processes such as plant metabolism. Some bacteria that live in association with plant leaves or roots can assimilate the CO either as a sole energy source (lithotrophy) or as a lithotrophic supplement to organic respiration. The CO may donate electrons to molecular oxygen, or to anaerobic electron acceptors such as nitrate. Some nitrogen-fixing bacteria, such as *Bradyrhizobium* species, metabolize CO as part of their legume endosymbiosis: The intracellular bacteroids oxidize plant-produced CO to CO_2, which the plant can then fix into biomass.

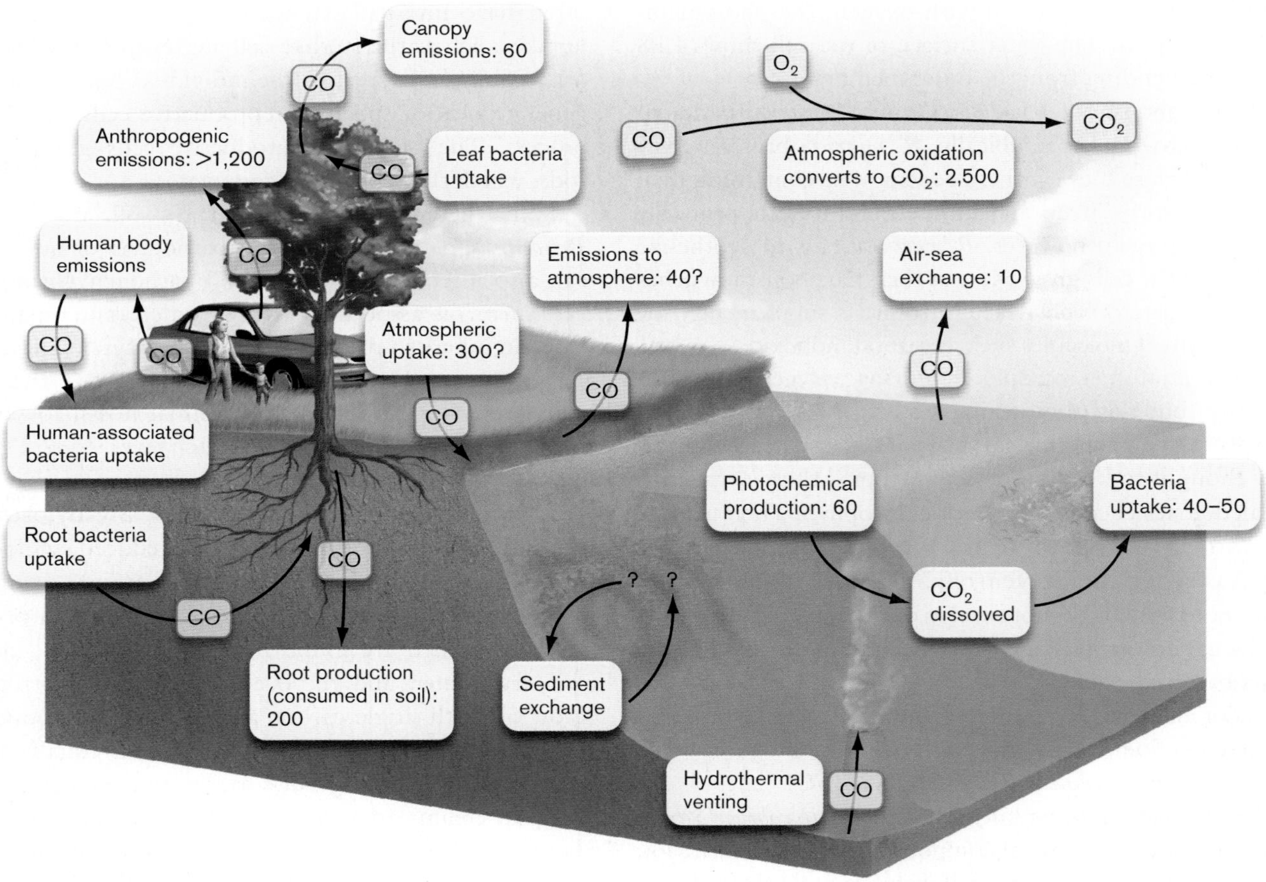
Canopy emissions: 60
CO
O_2
CO_2
Anthropogenic emissions: >1,200
Leaf bacteria uptake
CO
CO
Atmospheric oxidation converts to CO_2: 2,500
Human body emissions
CO
Emissions to atmosphere: 40?
Air-sea exchange: 10
CO
CO
Atmospheric uptake: 300?
CO
CO
CO
Human-associated bacteria uptake
Photochemical production: 60
Bacteria uptake: 40–50
Root bacteria uptake
CO
CO_2 dissolved
? ?
CO
Root production (consumed in soil): 200
Sediment exchange
Hydrothermal venting
CO

FIGURE 1 ■ Carbon monoxide emission and uptake. CO is produced in large amounts by oxidation in the atmosphere and by human burning of organic fuels. Biological sources of CO include plant and animal metabolism, and microbial processes. Much of this CO is consumed by bacteria. Human-associated bacteria consume CO from human metabolism. Units of CO are teragrams per year (10^{12} g/yr). *Source:* Modified from Gary M. King and Caroline F. Weber. 2007. *Nat. Rev. Microbiol.* **5**:107.

Human metabolism generates CO through breakdown of the heme portion of hemoglobin, generating a concentration of up to 5 parts per million, about five times the level of atmospheric CO. These low concentrations of CO serve as neurotransmitters and immune system regulators. But human CO may also supplement the metabolism of slow-growing pathogens such as *Mycobacterium tuberculosis.* Mycobacterial CO oxidation has been proposed as a target for anti-tuberculosis therapy.

In marine environments, a major source of CO emission is coastal water, where sunlight degrades dissolved organic matter from decayed plants and algae. These light-driven reactions, called photodegradation, yield CO as a by-product, building up surprisingly high levels in the water. More than 80% of this CO was found to be oxidized by microbial processes, but which organisms are involved? John D. Tolli (currently professor of biology at Southwestern College) and colleagues, while at the Woods Hole Oceanographic Institution, investigated this question by isolating CO-oxidizing bacteria from the water at Vineyard Sound, Massachusetts. They used a technique called substrate-tracking microautoradiography (STAR) for uptake of radiolabeled CO by bacterial enzymes. The STAR technique was combined with fluorescent in situ hybridization (FISH), in which fluorescent-labeled DNA from known CO-oxidizing organisms is hybridized to rRNA from newly isolated bacteria (a technique described in Appendix 3). The STAR-FISH technique allowed simultaneous assay of colonies for (1) ^{14}C-CO uptake (STAR), (2) total cellular DNA using the fluorophore DAPI, and (3) rRNA hybridization (FISH) with known bacterial probes. The probe sequences used were for *Roseobacter,* a known CO-oxidizing bacterium of the Alphaproteobacteria clade. Thus, CO-oxidizing colonies were identified, and their relationship to *Roseobacter* was determined by comparing DNA sequences.

Figure 2 shows the phylogeny of some of the isolates Tolli found, based on SSU rRNA sequence comparison. Some isolates turned out to be known species of Proteobacteria; others showed novel sequences never before reported (those with prefix "JT"). The strongest CO oxidizers (JT-01, JT-08, JT-22) fell within the *Roseobacter* clade. *Roseobacter* species are oval-shaped bacteria that grow in rosettes and, in some cases, produce pink colonies. Abundant in coastal water, they are aerobic anoxygenic phototrophs; that is, they use only sulfide or H_2 for photosynthesis and do not produce O_2, but for unknown reasons their growth requires O_2. Four other JT isolates showed gammaproteobacterial DNA sequences distantly related to *Pseudomonas aeruginosa,* a soil bacterium and opportunistic pathogen. These diverse marine bacteria may play a key role in environmental CO removal, contributing to the health of the marine food chain.

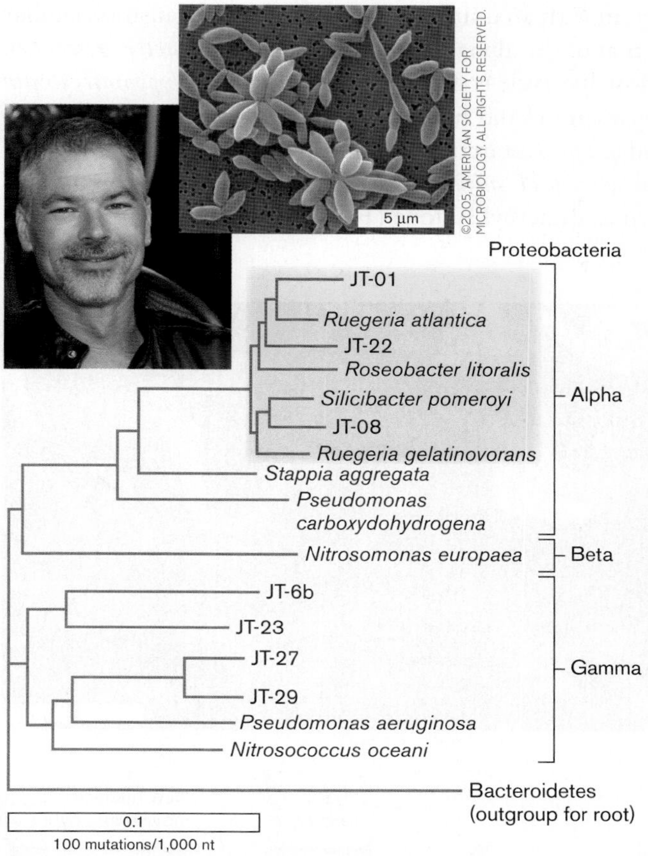
5 µm
©2005, AMERICAN SOCIETY FOR MICROBIOLOGY. ALL RIGHTS RESERVED.
Proteobacteria
JT-01
Ruegeria atlantica
JT-22
Roseobacter litoralis
Silicibacter pomeroyi
JT-08
Ruegeria gelatinovorans
Stappia aggregata
Pseudomonas carboxydohydrogena
Alpha
Nitrosomonas europaea
Beta
JT-6b
JT-23
JT-27
JT-29
Pseudomonas aeruginosa
Nitrosococcus oceani
Gamma
Bacteroidetes (outgroup for root)
0.1
100 mutations/1,000 nt

FIGURE 2 ▪ Phylogeny of coastal CO-oxidizing bacteria. CO-oxidizing bacteria were isolated from Vineyard Sound, Massachusetts. The phylogenetic tree was calculated by comparing SSU rRNA sequences of newly described CO-oxidizing isolates (prefix "JT") and known CO-oxidizing species. The *Roseobacter* clade is highlighted. **Insets:** John Tolli, professor of biology at Southwestern College (*left*); *Roseobacter* species (*right*) (SEM). *Source:* Modified from John D. Tolli et al. 2006. *Appl. Environ. Microbiol.* **72**:1966. Inset from Jesper B. Bruhn et al. 2005. *Appl. Environ. Microbiol.* **71**:7263.

RESEARCH QUESTION

How could you devise an enrichment culture technique to isolate CO-oxidizing microbes from soil samples?

Tolli, John D., S. M. Sievert, and C. D. Taylor. 2006. Unexpected diversity of bacteria capable of carbon monoxide oxidation in a coastal marine environment, and contribution of the *Roseobacter*-associated clade to total CO oxidation. *Applied and Environmental Microbiology* **72**:1966–1973.

methane produced by methanogens before it reaches the atmosphere, where it has a potent greenhouse effect (see Chapter 22).

An interesting methylotroph is *Hyphomicrobium,* a bacterium with an unusual stalk-to-flagellum transition similar to that of the alphaproteobacterium *Caulobacter crescentus,* whose life cycle is described in Chapter 3. *Hyphomicrobium* species are found in environments as diverse as wastewater sludge and Antarctic island soil. For example, the Antarctic soil species *H. sulfonivorans* metabolizes sulfur compounds such as dimethyl sulfone (**Fig. 18.27**). Like *Caulobacter,* a flagellated cell of *Hyphomicrobium* species may lose its flagellum to form a stalk. But unlike *Caulobacter, Hyphomicrobium* then forms a daughter cell from the opposite end of the stalk! As the DNA replicates, one daughter nucleoid must migrate all the way through the stalk to reach the daughter cell. The daughter cell forms a flagellum and septates, separating from the parent.

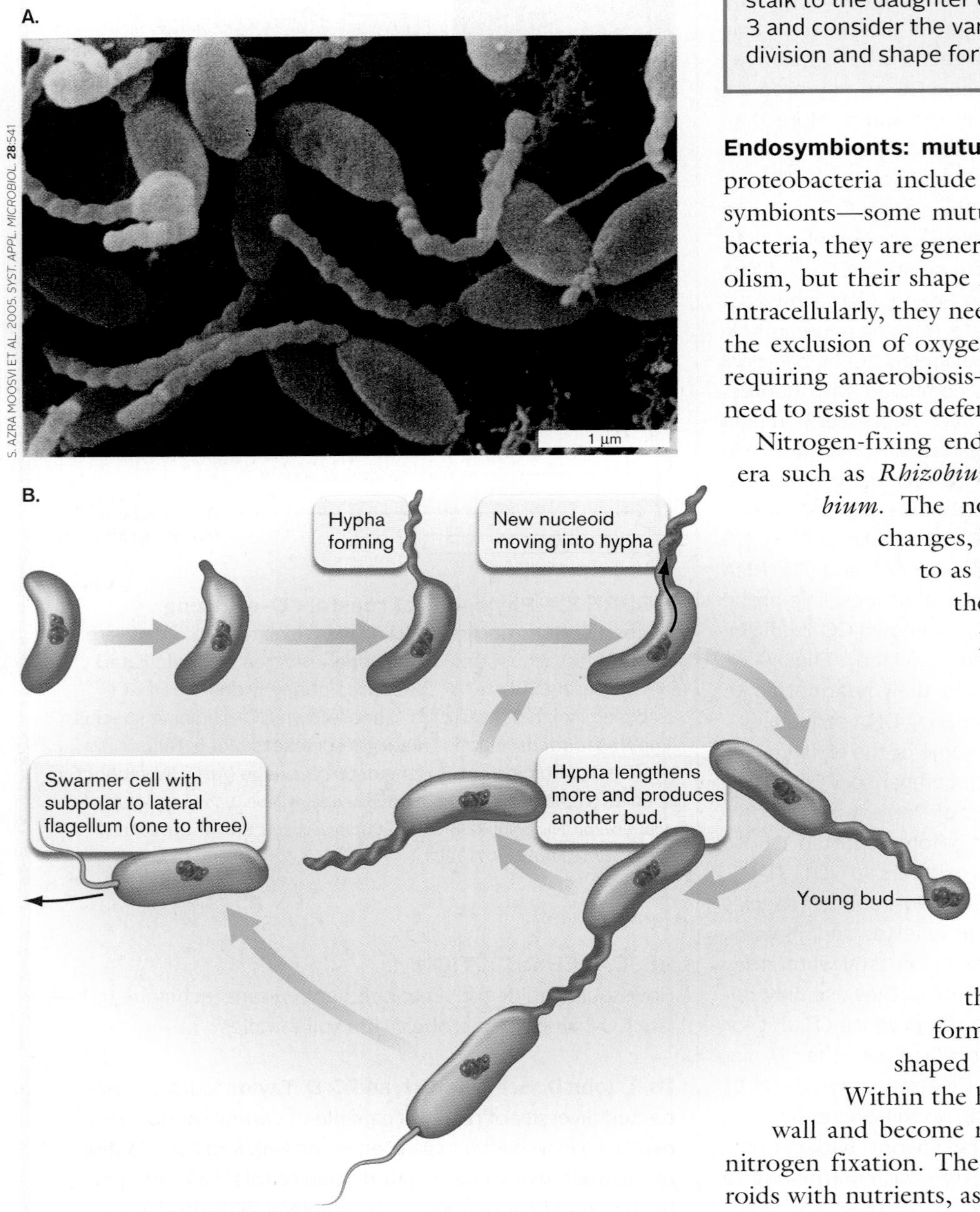
A.
S. AZRA MOOSVI ET AL. 2005. SYST. APPL. MICROBIOL. 28:541
1 µm
B.
Hypha forming
New nucleoid moving into hypha
Swarmer cell with subpolar to lateral flagellum (one to three)
Hypha lengthens more and produces another bud.
Young bud

FIGURE 18.27 ■ *Hyphomicrobium sulfonivorans* from Antarctic soil. A. These bacteria with coiled stalks catabolize dimethyl sulfone. **B.** Life cycle of *Hyphomicrobium* species.

Thought Question

18.8 Can you hypothesize a mechanism for migration of the daughter nucleoid of *Hyphomicrobium* through the stalk to the daughter cell? For possibilities, see Chapter 3 and consider the various molecular mechanisms of cell division and shape formation.

Endosymbionts: mutualists and parasites. The Alphaproteobacteria include many highly evolved intracellular symbionts—some mutualists, others parasites. As isolated bacteria, they are generally rod-shaped with aerobic metabolism, but their shape is transformed within the host cell. Intracellularly, they need to solve special problems, such as the exclusion of oxygen from nitrogen fixation, a process requiring anaerobiosis—or, in the case of pathogens, the need to resist host defenses.

Nitrogen-fixing endosymbionts of plants include genera such as *Rhizobium, Bradyrhizobium,* and *Sinorhizobium.* The nomenclature has undergone many changes, but the species are generally referred to as rhizobia. Rhizobia can live freely in the soil, but they prefer to colonize plants, usually legumes such as peas or alfalfa, where they form distinctive nodule structures. Each bacterial species colonizes a particular host range for infection. Complex chemosensory processes attract the bacteria to the surface of the host root, where their presence induces root hairs to form curls around the bacteria (**Fig. 18.28**). The curling of the root hairs enables the bacteria to form infection threads, chains of rod-shaped bacteria that invade the root cells. Within the host cells, the bacteria lose their cell wall and become rounded **bacteroids**, specialized for nitrogen fixation. The host plant cells provide the bacteroids with nutrients, as well as protective components such as **leghemoglobin**, an oxygen-binding protein that maintains anaerobiosis within infected cells. Leghemoglobin turns the nodule interior pink (**Fig. 18.28A**).

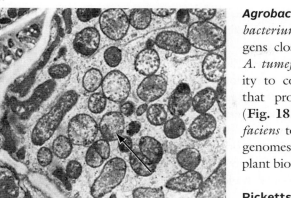

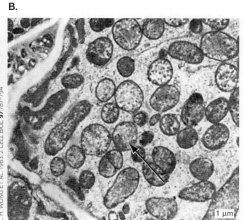

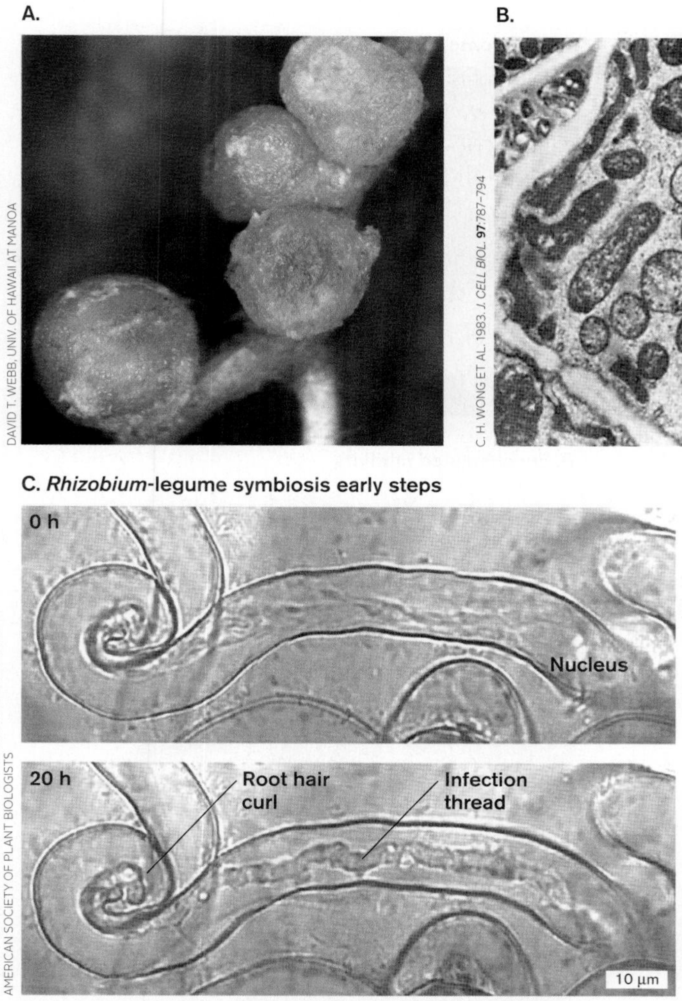
A.
B.
DAVID T. WEBB, UNIV. OF HAWAII AT MANOA
C. H. WONG ET AL. 1983. J. CELL BIOL. 97:787–794
1 µm
C. Rhizobium-legume symbiosis early steps
0 h
Nucleus
20 h
Root hair curl
Infection thread
10 µm
JOELLE FOURNIER ET AL. 2008. PLANT PHYSIOL. 148:1985. © 2008 AMERICAN SOCIETY OF PLANT BIOLOGISTS

FIGURE 18.28 ▪ Rhizobia: legume endosymbionts. **A.** Legume nodules cut open to show pink regions where the plant cells produce leghemoglobin to maintain anaerobic conditions for bacteroid nitrogen fixation. **B.** Within the host cells, *Sinorhizobium meliloti* cells form bacteroids (arrow) that fix nitrogen while receiving nutrients from the plant. **C.** Clover root hair curls around infecting *S. meliloti.* The bacteria enter the curl and grow down the root hair as an infection thread that penetrates the legume cells, enabling the bacteria to colonize in the form of bacteroids. Micrographs show times 0 and 20 hours postinfection.

Agrobacterium. The genus *Agrobacterium* includes plant pathogens closely related to the rhizobia. *A. tumefaciens* is known for its ability to convert host cells to a form that produces tumors called galls (**Fig. 18.29**). The ability of *A. tumefaciens* to insert its DNA into plant genomes has made it a major tool for plant biotechnology (see Chapter 16).

Rickettsias: intracellular pathogens and mitochondria. Short coccoid rods, rickettsias lack flagella and can grow only within a host cell. The best-known rickettsia is *Rickettsia rickettsii,* the cause of Rocky Mountain spotted fever, a disease spread by ticks throughout the United States. *Rickettsia* species parasitize human endothelial cells (**Fig. 18.30**). The bacteria induce phagocytosis and then dissolve the phagocytic vesicle and escape into the cytoplasm. Some rickettsias propel themselves through the host cell by polymerizing cytoplasmic actin behind them—a process similar to that of *Listeria* (described under Firmicutes). The actin tails eventually project outward as

NIGEL CATTLIN/PHOTO RESEARCHERS, INC.

FIGURE 18.29 ▪ Plant gall induced by *Agrobacterium tumefaciens.* Crown gall on chrysanthemum plant.

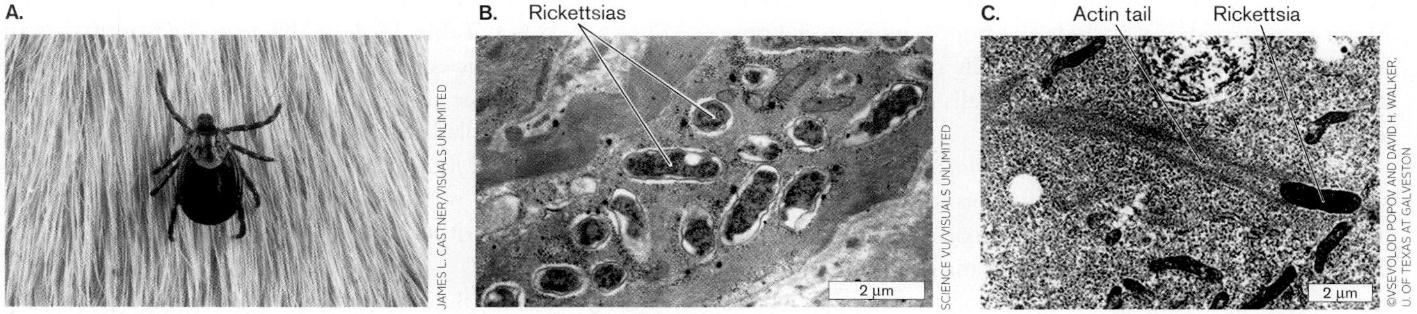
A.
JAMES L. CASTNER/VISUALS UNLIMITED
B. Rickettsias
2 µm
SCIENCE VU/VISUALS UNLIMITED
C. Actin tail Rickettsia
2 µm
©VSEVOLOD POPOV AND DAVID H. WALKER, U. OF TEXAS AT GALVESTON

FIGURE 18.30 ▪ Rickettsias are obligate intracellular parasites. **A.** The Rocky Mountain tick carries *Rickettsia rickettsii,* the cause of Rocky Mountain spotted fever. **B.** *Rickettsia* species parasitize human endothelial cells (TEM). **C.** The rickettsias propel themselves through the host cell by polymerizing cytoplasmic actin behind them (TEM).

filopodia, extensions of host cytoplasm and membrane that protect the bacteria from host defenses while enabling them to invade adjacent cells.

Genetic analysis shows that the rickettsia clade includes mitochondria. All eukaryotic mitochondria appear to be descendants of an ancient rickettsial parasite whose respiratory apparatus ultimately became essential to power eukaryotic cells. These free-living species, such as *Pelagibacter ubique* (isolate SAR11), contain proteorhodopsin (discussed in Chapter 14). It was unexpected to find a nonparasitic marine "rickettsia," let alone one with a form of phototrophy.

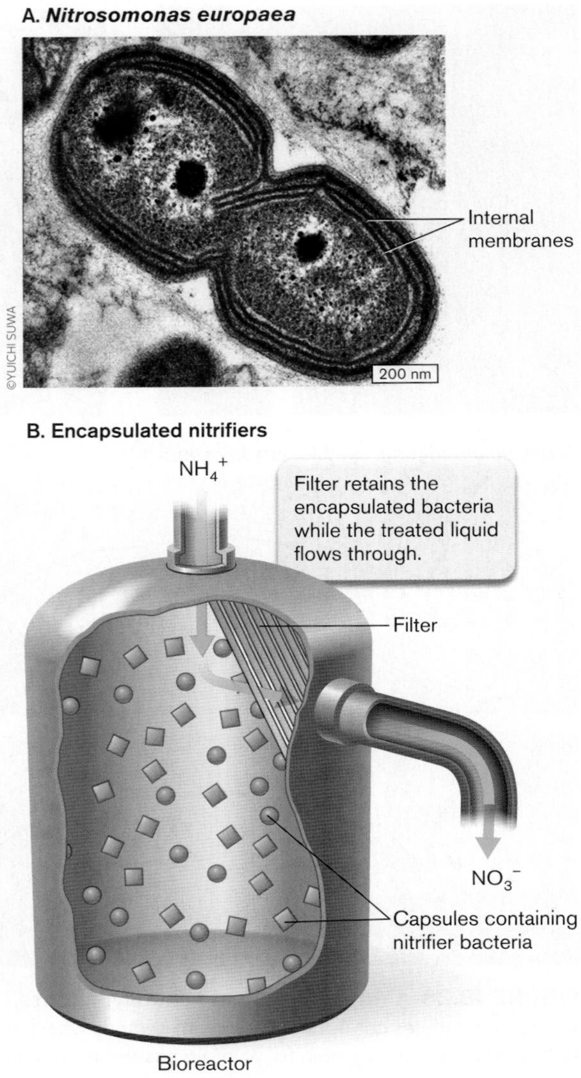
A. Nitrosomonas europaea
Internal membranes
©YUICHI SUWA
200 nm
B. Encapsulated nitrifiers
NH_4^+
Filter retains the encapsulated bacteria while the treated liquid flows through.
Filter
NO_3^-
Capsules containing nitrifier bacteria
Bioreactor

FIGURE 18.31 ■ Nitrifiers. A. *Nitrosomonas europaea,* of the Betaproteobacteria, oxidizes ammonia to nitrite (TEM). Internal membranes contain the electron transport complexes. **B.** Wastewater treatment uses nitrifier bacteria to remove ammonia.

Betaproteobacteria: Photoheterotrophs, Lithotrophs, and Pathogens

The Betaproteobacteria include photoheterotrophs such as *Rhodocyclus,* as well as a diverse range of lithotrophs (see **Table 18.4**). Several important pathogens infect humans and other animals.

Lithotrophs: nitrifiers and sulfur oxidizers. An important group of nitrogen lithotrophs consists of the <u>**nitrifiers**</u> (**Fig. 18.31A**). Nitrifiers oxidize ammonia (NH_3) to nitrite (NO_2^-) or nitrite to nitrate (NO_3^-). Typically, different species conduct the two reactions separately while coexisting in soil and water. Nitrifiers are of enormous economic and practical importance for wastewater treatment because they decrease the reduced nitrogen content of sewage. Special systems have been developed to retain nitrifier bacteria behind filters as one stage of water treatment. In the system shown in **Figure 18.31B**, nitrifying bacteria are encapsulated in pellets to retain them within the bioreactor while the treated water flows through a filter.

The most commonly isolated ammonia oxidizers are *Nitrosomonas* species of the Beta class. *Nitrosomonas* cells conduct electron transport through extensive internal membranes that are either stacked or invaginated. There are related genera, such as *Nitrosolobus* and *Nitrosovibrio,* some with peritrichous or polar flagella. One ammonia-oxidizing genus of the Gamma group has been found: *Nitrosococcus.* Major nitrite oxidizers include the nonproteobacterial phylum Nitrospirae. Although genetically outside the Proteobacteria, the unusual spiral-shaped cells of Nitrospirae have a Gram-negative cell envelope.

Bacteria of the genus *Acidithiobacillus* (recently reassigned from class Beta to Gammaproteobacteria) oxidize iron or sulfur (**Fig. 18.32A**). Iron-oxidizing bacteria commonly form the brown stains found inside plumbing. Most species are short rods or vibrios (comma-shaped). *Acidithiobacillus* and other sulfur-oxidizing genera can undergo a number of different reactions oxidizing H_2S to S^0 and S^0 to SO_4^{2-} (see **Table 18.4**). Sulfate production makes an environment acidic enough to erode stone monuments and the interior surface of concrete sewer pipes. Sulfur oxidation is often coupled to oxidation of iron, $Fe^{2+} \longrightarrow Fe^{3+}$. The bacterium *A. ferrooxidans* is known for its role in acidification of mine water, which contributes to leaching of iron, copper, and other minerals (**Fig. 18.32B**).

Pathogens. The Beta class includes aerobic heterotrophic cocci such as *Neisseria*. The cocci of *Neisseria* species form

A.
5 µm
P. ROMANO ET AL. 2001. J. CHEM. TECH. BIOTECH. 76:723
B.
PHOTO COURTESY OF KENNECOTT UTAH COPPER, SALT LAKE CITY, UT

FIGURE 18.32 ▪ *Acidithiobacillus:* iron oxidizers. **A.** *A. ferrooxidans* leaches copper and iron from molybdenite ore (SEM). The hexagonal object is a molybdenite crystal. **B.** In a copper mine, *Acidithiobacillus* species can oxidize copper ores, leaching the copper into solution for retrieval.

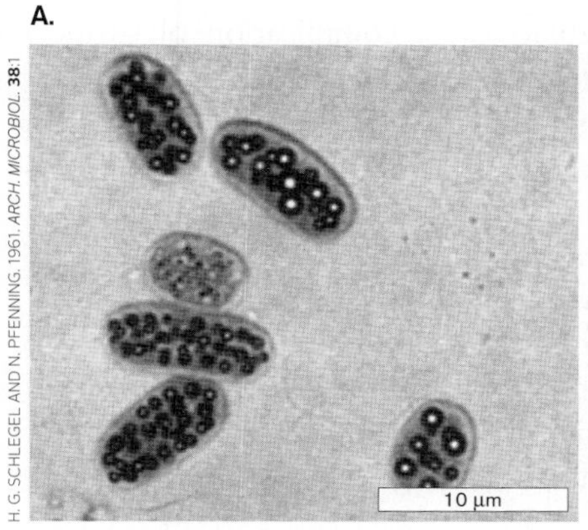
A.
10 µm
H. G. SCHLEGEL AND N. PFENNING. 1961. ARCH. MICROBIOL. 38:1

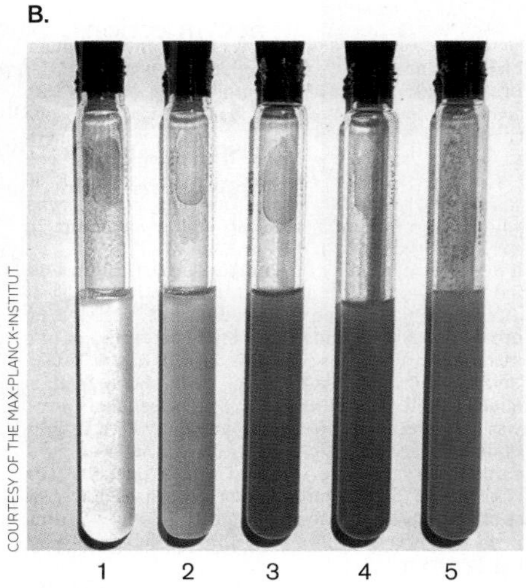
B.
COURTESY OF THE MAX-PLANCK-INSTITUT
1
2
3
4
5

FIGURE 18.33 ▪ *Chromatium:* sulfur and iron phototrophs. A. *Chromatium* forms single-flagellated rods full of sulfur granules. **B.** Photoferrotrophy. Under illumination, color develops over time (1–5) as a *Chromatium* isolate oxidizes Fe^{2+} to Fe^{3+}.

distinctive pairs known as **diplococci**. Most *Neisseria* species are harmless commensals of the nasal or oral mucosa, but *N. gonorrhoeae* causes the sexually transmitted disease gonorrhea. *N. gonorrhoeae* is actually a microaerophile, requiring a narrow range of oxygen concentration; it has fastidious growth requirements, necessitating cultivation on special blood-based medium. A related organism, *N. meningitidis,* may be carried asymptomatically by a quarter of the human population, but occasionally it causes meningitis, which can be fatal. Other neisserias, such as *N. sicca,* are easily isolated from skin and often presented as an unknown for identification by undergraduates.

Other members of the Beta class are important animal and plant pathogens, such as *Burkholderia*. *B. cepacia* was originally isolated from onions as a cause of bulb rot; it is now known to be a major opportunistic invader of the lungs of cystic fibrosis patients.

Gammaproteobacteria: Photolithotrophs, Enteric Flora, and Pathogens

The Gammaproteobacteria are well known for the Enterobacteriaceae family of facultative anaerobes found in the human colon, including important pathogens. In the environment, marine species catabolize complex organic pollutants, such as petroleum hydrocarbons. Gammaproteobacteria also include unusual phototrophs, some of which oxidize iron and nitrite.

Sulfur and iron phototrophs. The Gamma group phototrophs, such as *Chromatium* species (**Fig. 18.33A**), mainly utilize sulfide and produce sulfur, which is deposited as intracellular granules visible within the cytoplasm. Their phototrophy is entirely anaerobic. Some of the Gamma group are true autotrophs and do not use organic substrates. Some of these actually conduct phototrophy using iron (Fe^{2+}) to donate electrons. Iron phototrophy, or "photoferrotrophy" (**Fig. 18.33B**), is considered an intriguing possibility for metabolism of early life (discussed in Chapter 17). *Thiocapsa* uses NO_2^- and was the first nitrogen-based phototroph discovered.

Sulfur lithotrophs. The sulfur-oxidizing genus *Beggiatoa* was one of the first kinds of lithotrophs described by pioneer microbial ecologist Sergei Winogradsky (discussed in Chapter 1). *Beggiatoa* species oxidize H_2S to elemental sulfur, which collects as sulfur granules within the periplasm. *Beggiatoa* also stores carbon in cytoplasmic granules of polyhydroxybutyrate—a common strategy among proteobacteria. The cells of *Beggiatoa* grow as extended filaments with visible sulfur granules, forming biofilms on sulfide-rich sediment. Another sulfur oxidizer, the genus *Thioploca*, forms thick biomats in aquatic sediment. The bacteria grow as long filaments that glide through hollow tubes full of mucous secretions. **Figure 18.34** shows filaments of *Thioploca* sp. emerging from a sheath.

Enteric fermenters and respirers. The family Enterobacteriaceae, facultative anaerobes of the Gamma subdivision, include some of the most intensively studied species of all bacteria. Species are readily isolated from the contents of the human digestive tract and easily grown on laboratory media based on human food. The best-known species of Enterobacteriaceae—indeed, the best studied of all bacterial species—is the model organism *Escherichia coli.* Some strains of *E. coli* grow normally in the human intestine, feeding on our mucous secretions and producing vitamins, such as vitamin K. But other strains, such as *E. coli* O157:H7, cause serious illness. A large proportion of the world's children die of *E. coli*–related intestinal illness before the age of 5.

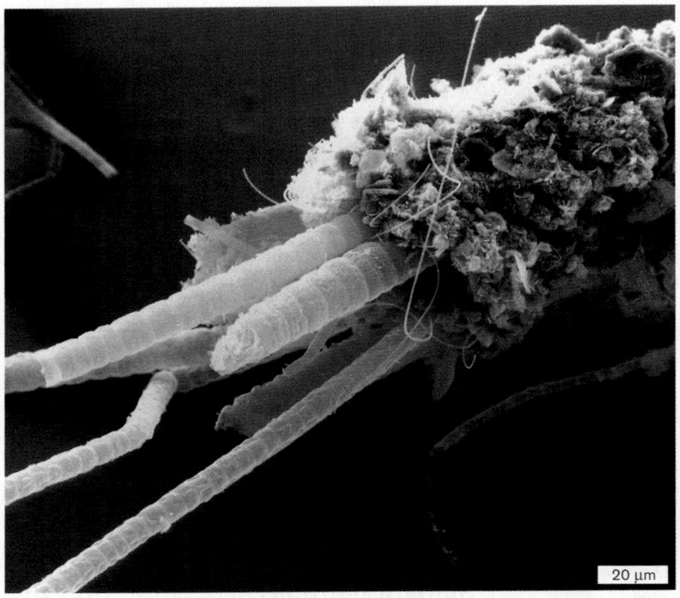
20 µm
ROLF SCHMALJOHANN ET AL. 2001. MAR. ECOL. PROG. SER. 211:27

FIGURE 18.34 ▪ *Thioploca* filaments emerging from a sheath. *Thioploca* species oxidize sulfur in marine sediment, using oxygen or nitrate; this sample is from the northeastern Arabian Sea off Pakistan.

The Enterobacteriaceae are Gram-negative rods, although as nutrients diminish, their size dwindles almost to a coccoid form. They grow singly, in chains, or in biofilms. Many species are motile, with numerous flagella. Most strains grow well with or without oxygen, by either respiration (aerobic or anaerobic) or fermentation. They ferment rapidly on carbohydrates, generating fermentation acids, ethanol, and gases (CO_2 plus H_2) in varying proportions, depending on the species. Their presence in the intestine supports the growth of organisms utilizing these gases, including methanogens (discussed in Chapter 19). Many strains form biofilms. Biofilm formation explains the persistence of drug-resistant infections, such as those associated with urinary catheters in long-term hospital patients.

Thought Question

18.9 Why do you think it took many years of study to realize that *Escherichia coli* and other Proteobacteria can grow as a biofilm?

Because they are easy to cultivate and have been studied extensively in clinical laboratories, many genera of Enterobacteriaceae are familiar to students in introductory microbiology laboratory courses. A common laboratory exercise is to distinguish *Enterobacter* and *Klebsiella* species from *E. coli* by their fermentation to the pH-neutral product butanediol, which tests positive in the Voges-Proskauer test. (Fermentation is discussed in Chapter 13.) *Enterobacter* species occur more frequently in freshwater streams than in the human body, although a few species colonize the intestine and cause illness.

Proteus mirabilis and *P. vulgaris* cause bladder and kidney infections, particularly as a complication of surgical catheterization. *Proteus* species are heavily flagellated and display a remarkable **swarming** behavior (**Fig. 18.35**). In response to an environmental signal, the flagellated rods grow into long-chain swarmer cells. The swarmers gather together, forming "rafts" that swim together and grow into a complex biofilm.

Besides symbionts of animals, Enterobacteriaceae include plant pathogens, such as *Erwinia carotovora* and related species. *Erwinia* species cause wilts, galls, and necrosis of a wide variety of plants, including bananas, tomatoes, and orchids.

Related facultative rods in soil and water conduct anaerobic respiration by donating electrons from organic substrates to a variety of metals, such as iron and magnesium. *Shewanella oneidensis* and other metal reducers are used to make electricity in fuel cells (discussed in Chapter 14).

Aerobic rods. Closely related to Enterobacteriaceae are several genera of rod-shaped bacteria that are obligate respirers. Many are obligate aerobic respirers (requiring O_2 for growth), although some can use alternative electron acceptors such as nitrate. These genera metabolize an extraordinary range of natural compounds, including aromatic derivatives of lignin; thus, they have important roles in natural recycling and soil turnover.

The Pseudomonadaceae are a large and amorphous group. As the "pseudo" prefix suggests, their taxonomic unit is poorly defined and includes species whose DNA sequence has necessitated their reassignment to other groups. The Gamma class pseudomonads, such as *Pseudomonas aeruginosa* and *P. fluorescens,* respire on oxygen or nitrate and are vigorous swimmers with single or multiple polar flagella. *P. aeruginosa* can swim throughout a standard agar plate (much to the chagrin of students attempting to isolate colonies). Nevertheless, under appropriate environmental conditions, pseudomonad cells give up their motility and develop biofilms (**Fig. 18.36**) (discussed in Chapter 4). Biofilms of *P. aeruginosa* cause lethal infections of the pulmonary lining of cystic fibrosis patients.

Some pseudomonad pathogens have the unusual ability to infect both plants and animals. For example, *P. aeruginosa* commonly infects plants as well as humans. *P. fluorescens* infects seedlings and causes rotting of citrus fruit, while it also appears as an opportunistic pathogen of immunocompromised cancer patients. Other pseudomonad species, however, are harmless residents of soil or sewage.

Legionella pneumophila is a well-publicized pathogen related to the pseudomonads. Incapable of growth on sugars, *L. pneumophila* requires oxygen to respire on amino acids. The organism exhibits an unusual dual lifestyle, alternating between intracellular growth within human macrophages and intracellular growth within freshwater amebas (**Fig. 18.37**). Growth within amebas facilitates transmission through aerosols into the human lung.

A.

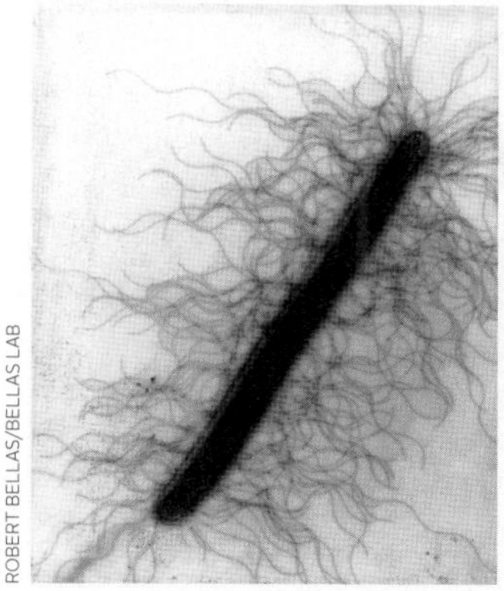
ROBERT BELLAS/BELLAS LAB

B.

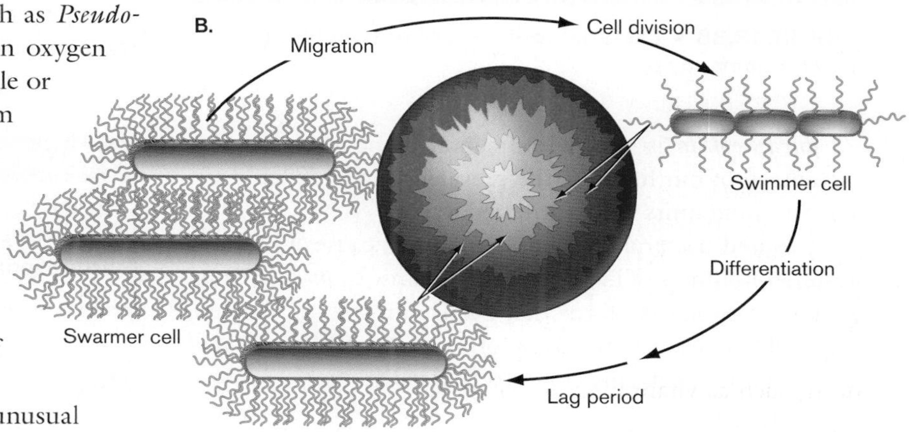
Migration
Cell division
Swimmer cell
Differentiation
Lag period
Swarmer cell

FIGURE 18.35 ■ The enteric rod *Proteus mirabilis:* isolated swimmer or cooperative swarmer. A. A thickly flagellated swarmer cell (TEM). Cell length can reach 20 μm. **B.** Waves of migration of *P. mirabilis* through blood agar.

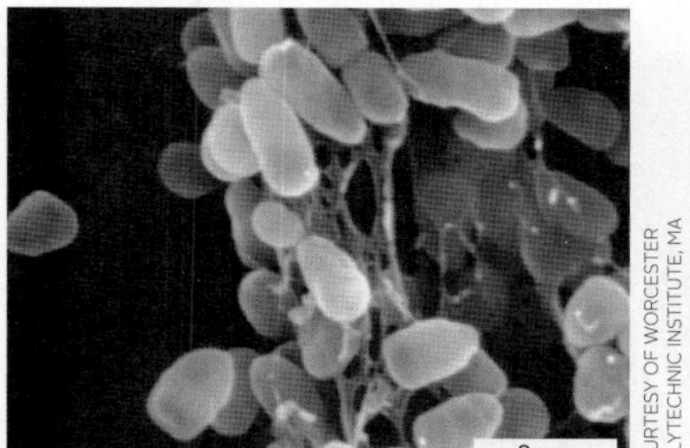
2 μm
COURTESY OF WORCESTER POLYTECHNIC INSTITUTE, MA

FIGURE 18.36 ■ *Pseudomonas* species form biofilms. *P. fluorescens* biofilm on a plant surface.

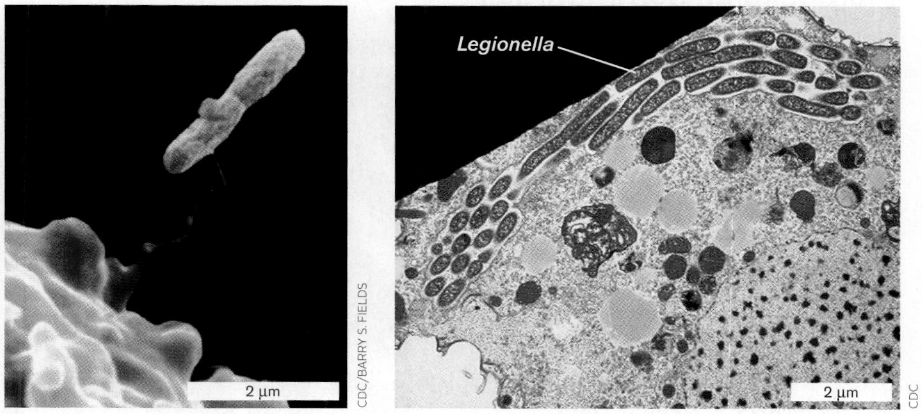
A.
B.
2 μm
CDC/BARRY S. FIELDS
Legionella
2 μm
CDC

FIGURE 18.37 ■ *Legionella pneumophila* colonizes an ameba. A. *L. pneumophila* cell caught by an ameba's pseudopod (colorized SEM). **B.** *L. pneumophila* cells have colonized the ameba (TEM).

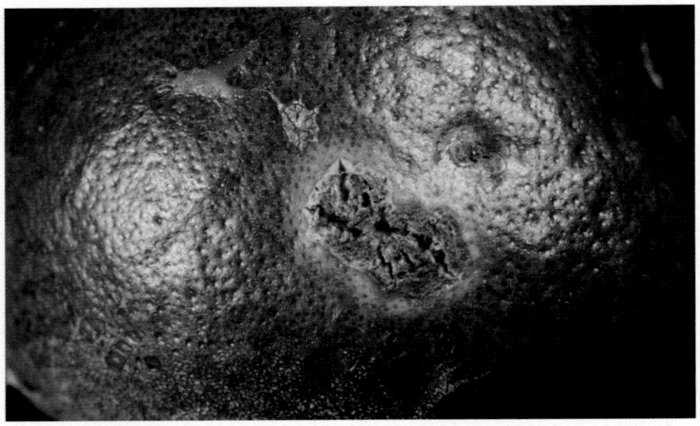

NIGEL CATTLIN/SCIENCE SOURCE

FIGURE 18.38 ■ Bacterial spot disease of tomato, caused by *Xanthomonas* sp.

L. pneumophila is an environmental pathogen that takes advantage of our lifestyle (the prevalence of large-scale air-conditioning units with unfiltered water).

A related bacterium, *Coxiella burnetii,* causes Q fever, a respiratory illness of livestock and humans. *C. burnetii* converts to a spore-like form that persists in soil. *C. burnetii* resembles a rickettsia in some of its strategy of pathogenesis, but it is genetically closer to *Legionella*.

Plant pathogens. Gammaproteobacteria include important plant pathogens such as *Xanthomonas* species. *Xanthomonas* is a flagellate rod that colonizes a wide range of agricultural plants, such as tomatoes (**Fig. 18.38**), potatoes, onions, broccoli, and citrus fruits. The disease may mottle the leaves and fruit, and it may spread throughout the plant.

Deltaproteobacteria: Lithotrophs and Multicellular Communities

The Deltaproteobacteria include important sulfur and iron reducers such as the fuel-cell bacterium *Geobacter metallireducens* (discussed in Chapter 14). Other species have complex life cycles that include multicellular developmental forms. The best-studied example is the myxobacteria. Myxobacteria have exceptionally large genomes, such as that of *Sorangium cellulosum* (12.6 Mb).

Myxobacteria. The myxobacteria, such as *Myxococcus xanthus,* are free-living soil bacteria that can grow as isolated cells or come together to form a differentiated structure for the purpose of reproduction (**Fig. 18.39**). When nutrients are plentiful, the myxobacteria grow and divide as individual cells. As nutrients run out, the cells begin to attract each other, moving into parallel formations. Myxobacteria have no flagella, but they move along a surface by using a form of motility called gliding.

The formations develop into a bulging mass, and the aggregating cells coalesce to form a **fruiting body** (**Fig. 18.40**). Fruiting-body formation also occurs in slime molds, a class of eukaryotic microbe (discussed in Chapter 20). The myxobacteria, however, have evolved their process independently. Within the *Myxococcus* fruiting body, durable spherical cells, called **myxospores**, develop. Ultimately, the myxospores are released to be carried on the wind to a more favorable location.

Thought Question

18.10 Compare and contrast the formation of cyanobacterial akinetes, firmicute endospores, actinomycete arthrospores, and myxococcal myxospores.

Note: Distinguish my<u>x</u>obacteria from *My<u>c</u>obacterium,* the Gram-positive genus that includes the causative agents of tuberculosis and leprosy.

Bdellovibrios parasitize bacteria. Bacteria can be parasitized or preyed on by smaller bacteria. The Delta class includes *Bdellovibrio* species, which attack proteobacterial host cells. The structure of the "attack cell" is a small, comma-shaped rod with a single flagellum. The attack cell attaches to the envelope of its host and then penetrates into the periplasm, where it uses host resources to grow

A. B.

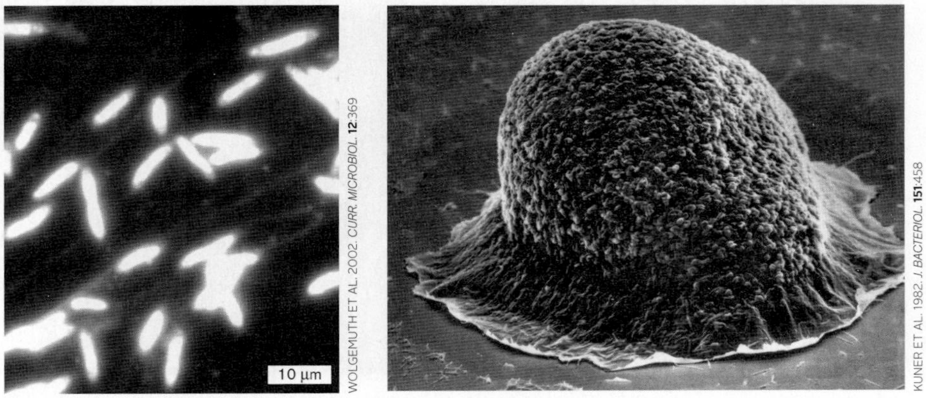
10 µm
WOLGEMUTH ET AL. 2002. CURR. MICROBIOL. 12:369
KUNER ET AL. 1982. J. BACTERIOL. 151:458

FIGURE 18.39 ■ Myxobacteria. A. *Myxococcus xanthus* cells glide while producing trails of slime (dark-field LM). **B.** Upon starvation, cells come together to generate a fruiting body (100–200 µm across) packed with spherical myxospores (SEM).

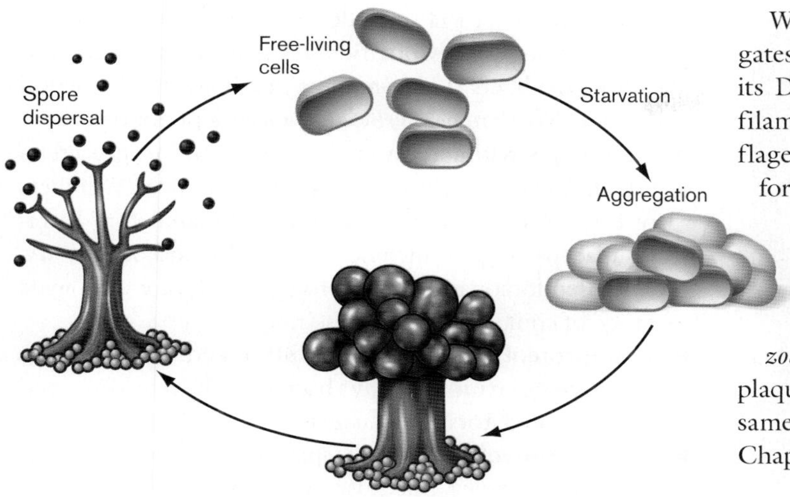
Free-living cells
Starvation
Spore dispersal
Aggregation
Fruiting body

FIGURE 18.40 ▪ The life cycle of *Myxococcus xanthus.* Starving myxobacteria glide toward each other to aggregate. The aggregation generates a fruiting body with bulges packed with small, spherical myxospores. *Source:* Dale Kaiser.

(**Fig. 18.41**). The growing cell produces enzymes that cross the inner membrane to degrade host macromolecules and make their components available to the bdellovibrio. The entire host cell loses its shape and becomes a protective incubator for the predator. This stage is called the bdelloplast.

Within the periplasm, the invading bdellovibrio elongates as a spiral filament while replicating several copies of its DNA. When most nutrients have been exhausted, the filament septates into multiple short cells. The cells develop flagella, and the bdelloplast bursts, releasing the newly formed attack cells (**Fig. 18.41B**).

Bdellovibrios can be isolated from sewage, soil, or marine water—any environmental source of Gram-negative prey bacteria. Different species infect *E. coli* and *Pseudomonas,* as well as *Agrobacterium* and *Rhizobium* in their free-living state. They can be cultured as plaques on a top-agar plate containing host bacteria—the same procedure used to isolate bacteriophages (discussed in Chapter 6).

Epsilonproteobacteria: Microaerophilic Helical Pathogens

Epsilonproteobacteria include the genera *Campylobacter* and *Helicobacter. Helicobacter pylori* is well known today as the causative agent of gastritis and stomach ulcers (**Fig. 18.42**). Yet, until recently, most microbiologists believed that bacteria could not live in the acidic stomach. The discovery of *H. pylori* as the cause of gastritis earned Barry Marshall and J. Robin Warren, of the University of Western Australia, the 2005 Nobel Prize in Physiology or Medicine.

H. pylori and related species grow primarily on the stomach epithelium, at about pH 6, which is less acidic than the

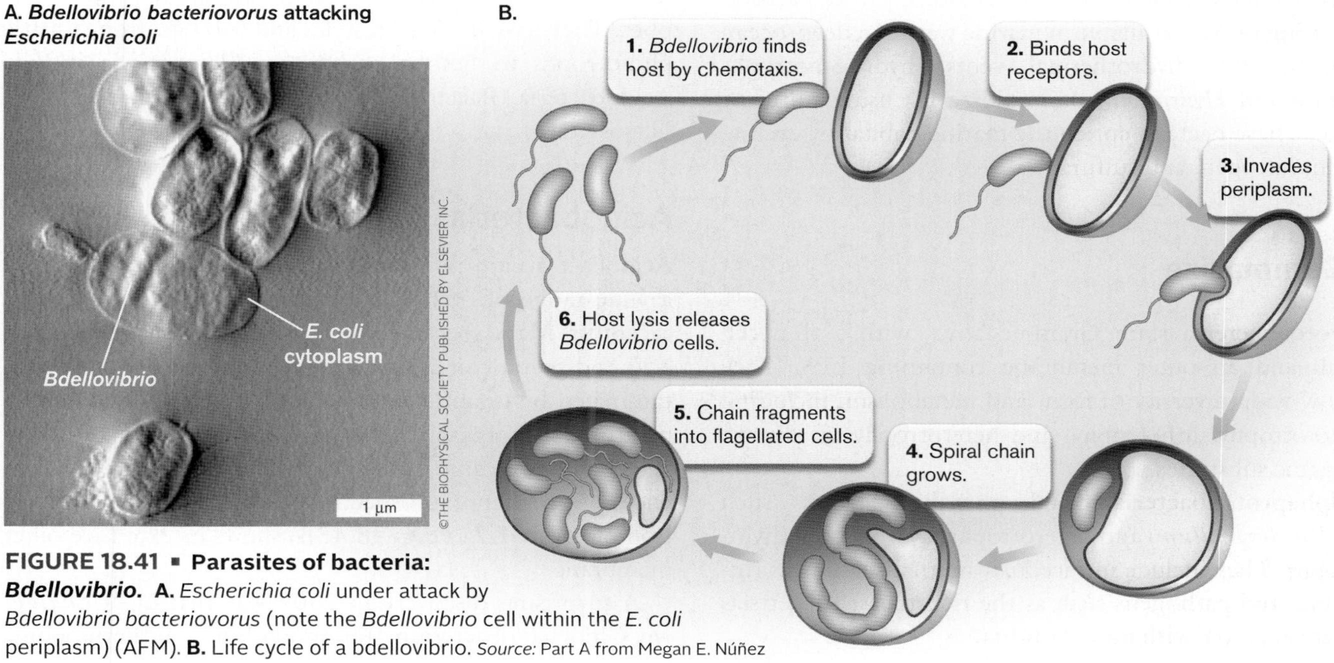
A. *Bdellovibrio bacteriovorus* attacking *Escherichia coli*
E. coli cytoplasm
Bdellovibrio
1 µm
©THE BIOPHYSICAL SOCIETY PUBLISHED BY ELSEVIER INC.
B.
1. *Bdellovibrio* finds host by chemotaxis.
2. Binds host receptors.
3. Invades periplasm.
4. Spiral chain grows.
5. Chain fragments into flagellated cells.
6. Host lysis releases *Bdellovibrio* cells.

FIGURE 18.41 ▪ Parasites of bacteria: *Bdellovibrio.* **A.** *Escherichia coli* under attack by *Bdellovibrio bacteriovorus* (note the *Bdellovibrio* cell within the *E. coli* periplasm) (AFM). **B.** Life cycle of a bdellovibrio. *Source:* Part A from Megan E. Núñez et al. 2003. *Biophys. J.* **84**:3379.

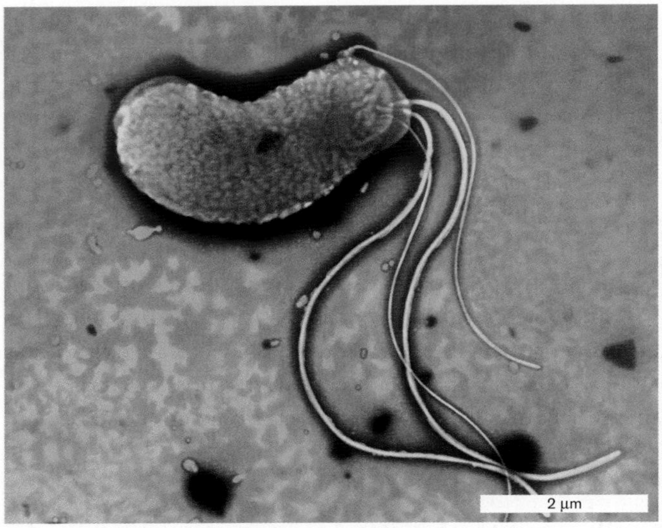
2 µm

CAMR/A. BARRY DOWSETT/PHOTO RESEARCHERS, INC.

FIGURE 18.42 ■ *Helicobacter*, a neutralophile growing within the acidic stomach. *H. pylori*, a short spirillum with unusual knobbed flagella projecting from one end.

gastric contents (pH 2–4). The bacteria bury themselves in the epithelial layer and neutralize their acidic surroundings by making urease enzyme, which converts urea to ammonia and carbon dioxide. *Helicobacter* species form wide spiral cells (spirilla) with an unusual grouping of flagella at one end. The metabolism of *H. pylori* is microaerophilic, requiring a low level of oxygen. The bacteria can be isolated through biopsy of the gastric mucosa.

Several groups of related Epsilonproteobacteria are sulfur oxidizers and sulfur reducers, found in marine and aquatic habitats. *Thiovulum* species oxidize sulfides aerobically, using oxygen available in marine water. In deep-ocean sediment, near hydrothermal vents, hydrogenotrophs *Nautilia* and *Hydrogenimonas* oxidize H_2 using sulfur or nitrate. These bacteria enrich the marine habitat by cycling carbon, nitrogen, and sulfur.

To Summarize

- **Proteobacteria** stain Gram-negative, with a thin cell wall and an outer membrane containing LPS. They show wide diversity of form and metabolism, including phototrophy, lithotrophy, and heterotrophy on diverse organic substrates.
- **Alphaproteobacteria** include photoheterotrophs (such as *Rhodospirillum*) and heterotrophs, as well as methylotrophs. They include intracellular mutualists such as rhizobia, and pathogens such as the rickettsias. Rickettsias share ancestry with mitochondria.
- **Betaproteobacteria** include photoheterotrophs (*Rhodocyclus*), as well as nitrifiers (*Nitrosomonas*) and iron-sulfur oxidizers (*Thiobacillus*). Pathogenic diplococci include *Neisseria gonorrhoeae,* the cause of gonorrhea.
- **Gammaproteobacteria** include sulfur and iron bacteria (*Acidithiobacteria*, *Chromatium*), as well as members of Enterobacteriaceae found in the human colon. Intracellular pathogens include *Salmonella* and *Legionella*. The pseudomonads, aerobic rods, can respire on a wide range of complex organic substrates.
- **Deltaproteobacteria** include sulfur and iron reducers (*Geobacter*), fruiting-body bacteria (*Myxococcus*), and bacterial predators (*Bdellovibrio*).
- **Epsilonproteobacteria** are spirillar pathogens such as *Helicobacter pylori,* the cause of gastritis.

18.5

Deep-Branching Gram-Negative Phyla

Are there other phyla besides Proteobacteria that possess an outer membrane and stain Gram-negative? Some Gram-negative bacteria branch more deeply from the Proteobacteria than the five Greek-lettered classes do from each other, and they are thus classified as separate phyla. Most members of these phyla (though not all) are obligate anaerobes. They have diverse lifestyles and habitats, from aquatic phototrophs to human pathogens. Major phyla include Acidobacteria, Bacteroidetes, Chlorobi, Fusobacteria, and Nitrospirae.

Acidobacteria

Acidobacteria are abundant in soils, where they metabolize a wide range of organic substrates using diverse electron acceptors. Many species grow in extreme conditions such as acid and metal contamination—for example, in soils contaminated by uranium mining. Others grow at high temperature, such as isolate K22 (**Fig. 18.43**) obtained from the Taupo Volcanic Zone, New Zealand, by GNS Science, a geoscience prospecting company. The thin section of isolate K22 (**Fig. 18.43B**) shows its extensive outer membrane.

A surprising discovery in 2007 was that of "*Candidatus* Chloracidobacterium thermophilum," a thermophilic

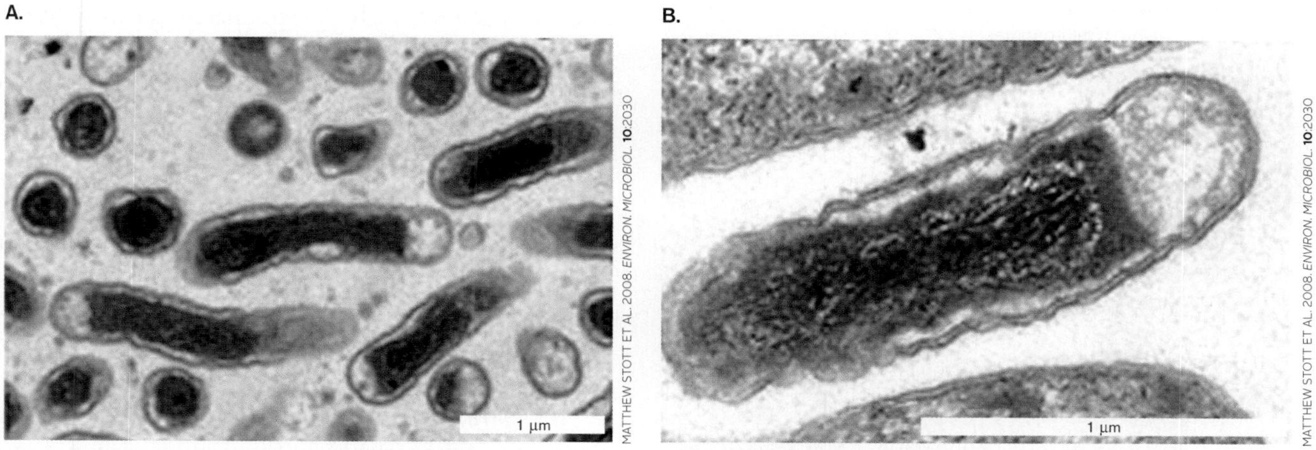
A.
B.
1 μm
1 μm
MATTHEW STOTT ET AL. 2008. ENVIRON. MICROBIOL. 10:2030
MATTHEW STOTT ET AL. 2008. ENVIRON. MICROBIOL. 10:2030

FIGURE 18.43 ▪ Acidobacteria isolate K22. A. A thermophilic acidophile, showing a Gram-negative outer membrane. **B.** Thin section of isolate K22 (TEM).

aerobic phototroph. Isolated from Octopus Spring, at Yellowstone National Park, this bacterium is a photoheterotroph that supplements catabolism with light absorption by photosystem I (PS I). It grows chlorosomes similar to those of Chloroflexi (Section 18.1). These distantly related organisms may have shared photosynthesis genes via horizontal gene transfer.

Bacteroidetes

The phylum Bacteroidetes includes genera such as *Bacteroides* and *Flavobacterium*. *Bacteroides* species, such as *B. fragilis* and *B. thetaiotaomicron*, are the major inhabitants of the human colon (**Fig. 18.44**). Their envelope polysaccharides help the bacteria evade the immune system. *Bacteroides* species grow anaerobically under extremely low oxygen concentrations such as those found in the human intestine (1 ppm), yet they can actually use the oxygen they find; thus they have been called "nanoaerobes."

Their main source of energy is fermentation of a wide range of sugar derivatives from plant material, compounds that are indigestible by humans and potentially toxic. *Bacteroides* converts these substances into simple sugars and fermentation acids, some of which are absorbed by the intestinal epithelium. Thus, *Bacteroides* species serve two important functions for their host: They break down potential toxins in plant foods, and their fermentation products make up as much as 15% of the caloric value we obtain from food. A third benefit of *Bacteroides* is their ability to remove side chains from bile acids, enabling the return of bile acids to the hepatic circulation. In effect, our gut bacteria including *Bacteroides* constitute a functional organ of the human body.

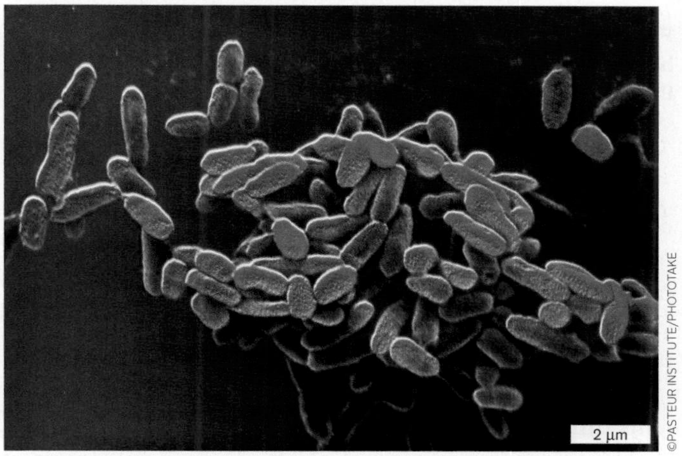
2 μm
©PASTEUR INSTITUTE/PHOTOTAKE

FIGURE 18.44 ▪ *Bacteroides fragilis* cells colonize the human colon. *B. fragilis* causes intestinal infections (colorized SEM).

Bacteroides species cause trouble, however, when they reach parts of the body not designed to host them. During abdominal surgery, bacteria can escape the colon and invade the surrounding tissues. The displaced bacteria can form an abscess, a localized mass of bacteria and pus contained in a cavity of dead tissue. The interior of the abscess is anaerobic and often impenetrable to antibiotics.

Chlorobi

The Chlorobi (mainly *Chlorobium* species) are known informally as "green sulfur bacteria." While they are genetically close to *Bacteroides*, their metabolism is surprisingly different. Chlorobi are strict photolithotrophs, using PS I to split electrons from H_2, H_2S, or other reduced sulfur

A.

B.

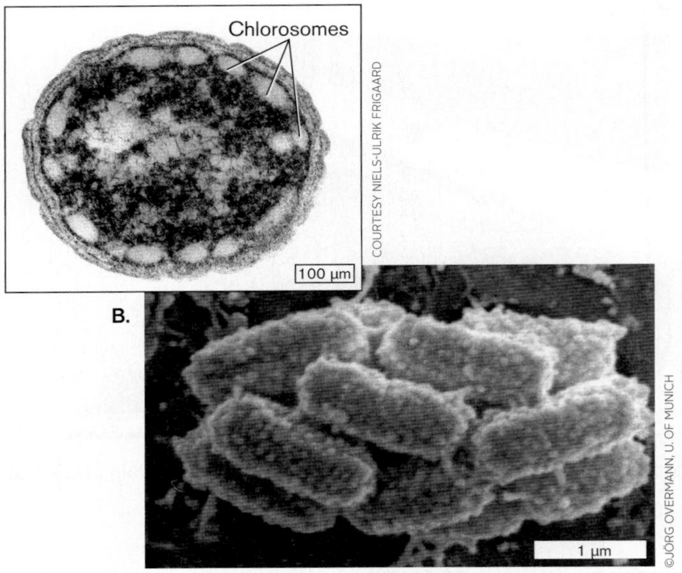
Chlorosomes
100 µm
COURTESY NIELS-ULRIK FRIGAARD
1 µm
©JÖRG OVERMANN, U. OF MUNICH

FIGURE 18.45 ▪ *Chlorobium:* Gram-negative "green bacteria." **A.** *Chlorobium tepidum* cell containing chlorosomes. **B.** *Chlorobium* sp. cells covered with sulfur globules; cells form a cluster surrounding a nonphototrophic symbiotic bacterium.

compounds (**Fig. 18.45**; see also **Table 18.2**). During the oxidation of sulfide, *Chlorobium* species deposit elemental sulfur extracellularly, forming attached sulfur globules (**Fig. 18.45B**). In contrast, the sulfur granules of Gammaproteobacteria such as *Chromatium* are deposited internally.

Chlorobium species require extensive membrane systems full of photopigments for light absorption. The chlorophyll reaction centers of *Chlorobium* are contained within chlorosomes associated with the cytoplasmic membrane (**Fig. 18.45A**), similar to chlorosomes of the deep-branching phylum Chloroflexi. The photopigment of *Chlorobium* is predominantly Bchl *c*, which absorbs in the blue (460 nm) and near-infrared (750 nm), thus reflecting the middle range, brownish green.

Note: Distinguish phylum Chlorobi (sulfur photoautotrophs) from phylum Chloroflexi (thermophilic filamentous photoheterotrophs). Both taxa are phototrophs containing green photopigment complexes arranged in chlorosomes, but they are deeply divergent genetically.

Fusobacteria

The Fusobacteria are an emerging taxon of surprisingly virulent pathogens. *Fusobacterium nucleatum* was identified from dental plaque; this bacterium is the second most frequent cause of human abscesses (after *Bacteroides*). Within biofilms such as dental plaque, *Fusobacterium* species form carbohydrate bridges with numerous other kinds of bacteria, such as spirochetes, proteobacteria, and firmicutes, as well as eukaryotic pathogens such as fungi. These distantly related biofilm partners have transferred exceptional numbers of genes to the *F. nucleatum* genome.

Nitrospirae

The phylum Nitrospirae consists of Gram-negative spiral bacteria that oxidize nitrite ion to nitrate (NO_2^- to NO_3^-). Their cell structure resembles that of the Proteobacteria, although their phylogenetic branch is deep enough for assignment to a separate phylum. Most species, such as *Nitrospira* spp. (see **Table 18.1**), are true lithotrophs or autotrophs, fixing carbon in the form of carbon dioxide or carbonate using carboxysomes. *Nitrospira* species are generally found in freshwater or salt water. Their removal of excess nitrite makes a key contribution to aquatic ecosystems.

Another important genus is *Leptospirillum,* which includes acidophilic iron oxidizers. *Leptospirillum* spp. are also strict autotrophs, fixing carbon by using Fe^{2+} as their electron donor and O_2 as the electron acceptor. Their metabolism generates acid, contributing to acid mine drainage in iron mines at Iron Mountain, California, where they grow in massive pink biofilms.

To Summarize

- **Acidobacteria** are Gram-negative soil bacteria, including many acidophiles.
- **Bacteroidetes** are anaerobes that ferment complex plant materials in the human colon. They may enter body tissues through wounds and cause abscesses.
- **Chlorobi** are green sulfur phototrophs, obligate anaerobes incapable of heterotrophy.
- **Fusobacteria** are pathogens that cause septicemia and skin ulcers.
- **Nitrospirae** are Gram-negative spiral bacteria that oxidize nitrite to nitrate (*Nitrospira*).

18.6

Spirochetes: Sheathed Spiral Cells with Internalized Flagella

Which bacterial cells form tightly coiled spirals? Spirochetes (or Spirochaeta) is a unique phylum of heterotrophic bacteria. While different species of spirochetes conduct a broad range of heterotrophy, from aerobic to anaerobic, all

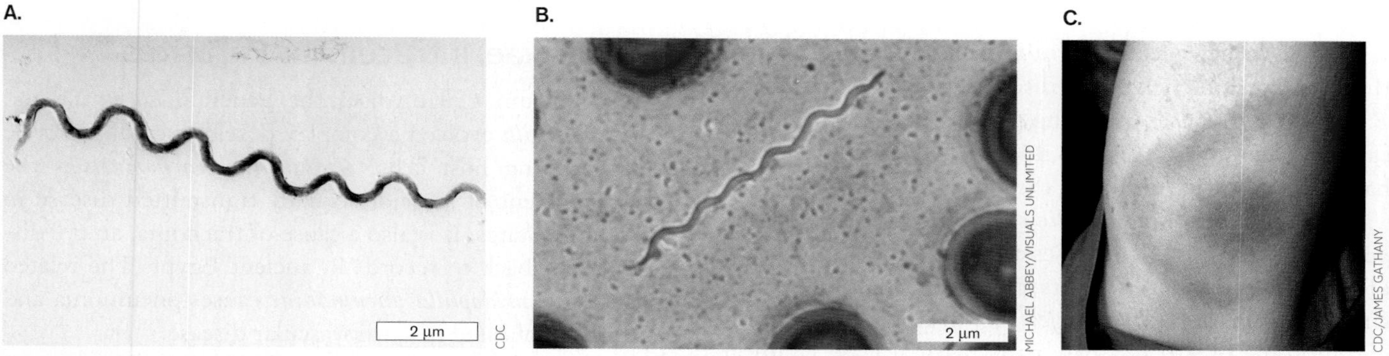
A.
2 µm
CDC
B.
2 µm
MICHAEL ABBEY/VISUALS UNLIMITED
C.
CDC/JAMES GATHANY

FIGURE 18.46 ▪ Spirochetes. **A.** *Treponema pallidum,* the causative agent of syphilis (TEM). **B.** *Borrelia recurrentis* cell in a blood film (stained micrograph). The organism causes relapsing fever. **C.** Lesion caused by tick-borne infection by *Borrelia burgdorferi,* a cause of Lyme disease.

spirochetes share a distinctive cell structure consisting of a long, tight spiral that is quite flexible (**Fig. 18.46**). In many species, the spiral is so thin that its width cannot be resolved by bright-field microscopy, and the organisms can pass through a filter of pore size 2 µm.

Cell Structure of Spirochetes

The spirochete cell is surrounded by a thick outer sheath of lipopolysaccharides and proteins. The spirochete sheath is similar to a proteobacterial outer membrane, except that the periplasmic space completely separates the sheath from the plasma membrane. At each end of the cell, one or more polar flagella extend and double back around the cell body within the periplasmic space (**Fig. 18.47**). The periplasmic flagella (axial fibrils) rotate on proton-driven motors, as do regular flagella; but because they twine back around the cell body, their rotation forces the entire cell to twist around, corkscrewing through the medium.

This corkscrew motion of the cell body turns out to have a physical advantage in highly viscous environments, such as human mucous secretions or agar culture medium. Few nonspirochetes can swim through agar at standard culture concentrations (1.5% agar); thus, growth within agar provides a way to isolate anaerobic spirochetes from environmental sources. When an environmental sample is inoculated into agar, other bacteria grow and concentrate at the injection point, whereas spirochetes migrate outward in a "veil" through the agar.

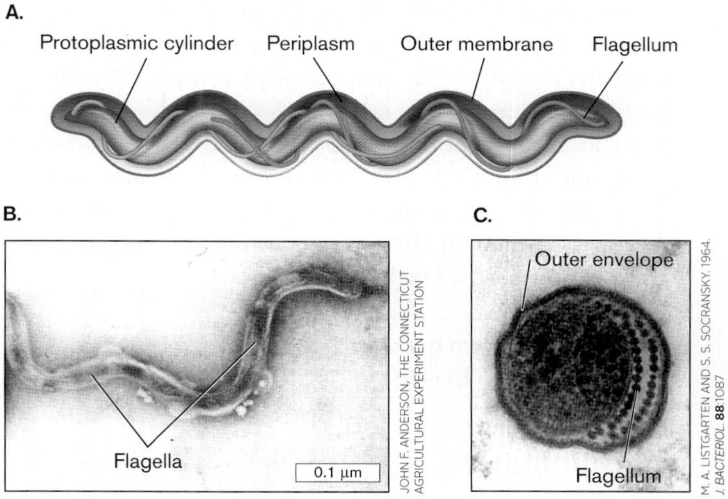
A.
Protoplasmic cylinder
Periplasm
Outer membrane
Flagellum
B.
Flagella
0.1 µm
JOHN F. ANDERSON, THE CONNECTICUT AGRICULTURAL EXPERIMENT STATION
C.
Outer envelope
Flagellum
M. A. LISTGARTEN AND S. S. SOCRANSKY. 1964. J. BACTERIOL. 88:1087

FIGURE 18.47 ▪ Spirochete structure. **A.** Spirochete cell structure, showing the arrangement of periplasmic flagella. **B.** Spirochete with two periplasmic flagella, isolated from mouse blood (TEM). **C.** Cross section through a spirochete, showing outer envelope and flagella (axial fibrils) (TEM). The unidentified spirochete was obtained from a human lesion of acute necrotizing ulcerative gingivitis.

Spirochete Diversity

Spirochetes grow in a wide range of habitats, from ponds and streams to the digestive tracts of animals. Aquatic systems carry free-living sugar fermenters of the genus *Spirochaeta*. A particularly interesting spirochete community is found in the termite gut, where the organisms form elaborate symbiotic associations with protists and assist in the digestion of cellulose.

The best-known spirochetes are those that cause human and animal diseases. The causative agent of the sexually transmitted disease syphilis is the spirochete *Treponema pallidum* (**Fig. 18.46A**). The cell of *T. pallidum* is too narrow (0.2 µm) to visualize by bright-field microscopy; instead, dark-field, fluorescence, or electron microscopy must be used. While *T. pallidum* is a highly virulent pathogen, other members of the genus *Treponema* are normal residents of the human and animal oral, intestinal, and genital regions.

T. pallidum is not culturable in the laboratory, but its genome sequence reveals much about its physiology. The genome of 1.1 million base pairs (presented in **eTopic 13.3**) is highly degenerate, lacking nearly all components of biosynthesis and of the TCA cycle; its only system for ATP production is glycolysis.

The spirochete genus *Borrelia* includes two pathogens that cause serious tick-borne diseases in the United States. *B. recurrentis* causes relapsing fever, and *B. burgdorferi* causes Lyme disease (**Fig. 18.46B** and **C**). *Borrelia* species have been cultured, although they grow very slowly.

A unique feature of *Borrelia* species is their multipartite genome. Each species possesses a linear main chromosome of less than a million base pairs, plus a number of linear and circular plasmids. *B. burgdorferi,* for example, has a linear chromosome of 910,725 base pairs, with at least 17 linear and circular plasmids that total an additional 533,000 base pairs. The reason for this unusual fragmentation of *Borrelia* genomes is unknown.

Spirochetes include important pathogens of animals, such as *Leptospira,* the cause of leptospirosis, a form of nephritis (kidney inflammation) with complications in the liver and other organs. *Leptospira* cells are known for the peculiar L shape of the cell, as each end of the spirochete turns out at an angle. The genus *Treponema* also includes the causative agent of digital dermatitis in cattle and sheep. As in other bacterial phyla, however, the pathogenic species are far outnumbered by harmless flora.

Note: Distinguish the spirochete *Leptospira* from the nitrite-oxidizing proteobacterium *Leptospirillum.*

To Summarize

- **The spirochete cell** is a tight coil, enclosed by a sheath and periplasmic space containing periplasmic flagella.
- **Spirochete motility** occurs by a flexing motion caused by rotation of the periplasmic flagella, propagated the length of the coil.
- **Spirochetes grow in diverse habitats.** Some are free-living fermenters in water or soil. Others are pathogens, such as *Treponema pallidum,* the cause of syphilis. Still others are endosymbionts of an animal digestive tract, such as the termite gut.

18.7

Chlamydiae, Planctomycetes, and Verrucomicrobia: Irregular Cells

Several related phyla of bacteria have either no cell walls or diminished cell walls. These organisms evolved independently of the mycoplasmas, and their alternative cell forms show diverse environmental adaptations.

Chlamydiae: Intracellular Parasites

In the phylum Chlamydiae, the genera *Chlamydia* and *Chlamydophila* evolved a complex developmental life cycle of parasitizing host cells. *Chlamydia trachomatis* is the causative agent of a major sexually transmitted disease in the United States. It is also a cause of trachoma, an eye disease dating back to records in ancient Egypt. The related species *Chlamydophila pneumoniae* causes pneumonia and has been implicated in cardiovascular disease.

Chlamydiae alternate between two developmental stages with different functions: elementary bodies and reticulate bodies (**Fig. 18.48A**). The form of chlamydia transmitted outside host cells is called an **elementary body**. Like endospores, elementary bodies are metabolically inert, with a compacted chromosome. While lacking a cell wall, they possess an outer membrane whose proteins are cross-linked by disulfide bonds, making a tough coat that provides osmotic stability. The elementary body adheres to a host cell surface and is endocytosed (**Fig. 18.48B**).

To reproduce, the elementary body must transform itself into a **reticulate body**, named for the netlike appearance of its uncondensed DNA. The reticulate body has an active metabolism and divides rapidly, but outside the cell it is incapable of infection and vulnerable to osmotic shock. To complete the infection cycle, therefore, reticulate bodies must develop into new elementary bodies before exiting the host. When the host cell lyses, the elementary bodies are released to infect new cells. Chlamydiae infect a wide range of host cell types, from respiratory epithelium to macrophages.

Planctomycetes: A Nucleus-like Compartment

Another cell wall–less group, the Planctomycetes, evolved largely as free-living organisms. Planctomycetes are oligotrophs, heterotrophs requiring nutrients at extremely low concentration. They grow in aquatic, marine, and saline environments. Their mechanism of osmoregulation remains poorly understood.

Planctomycete cells possess multiple internal membrane compartments of unknown function (**Fig. 18.49A**). An internal membrane just inside the cell membrane divides the cytoplasm into concentric portions. A double membrane surrounds the entire nucleoid, analogous to the double membrane surrounding the eukaryotic nucleus, a remarkable example of independent analogous evolution between bacteria and the eukaryotes. The actual function of the nucleoid-surrounding membrane varies with different species; for example, in anammox planctomycetes,

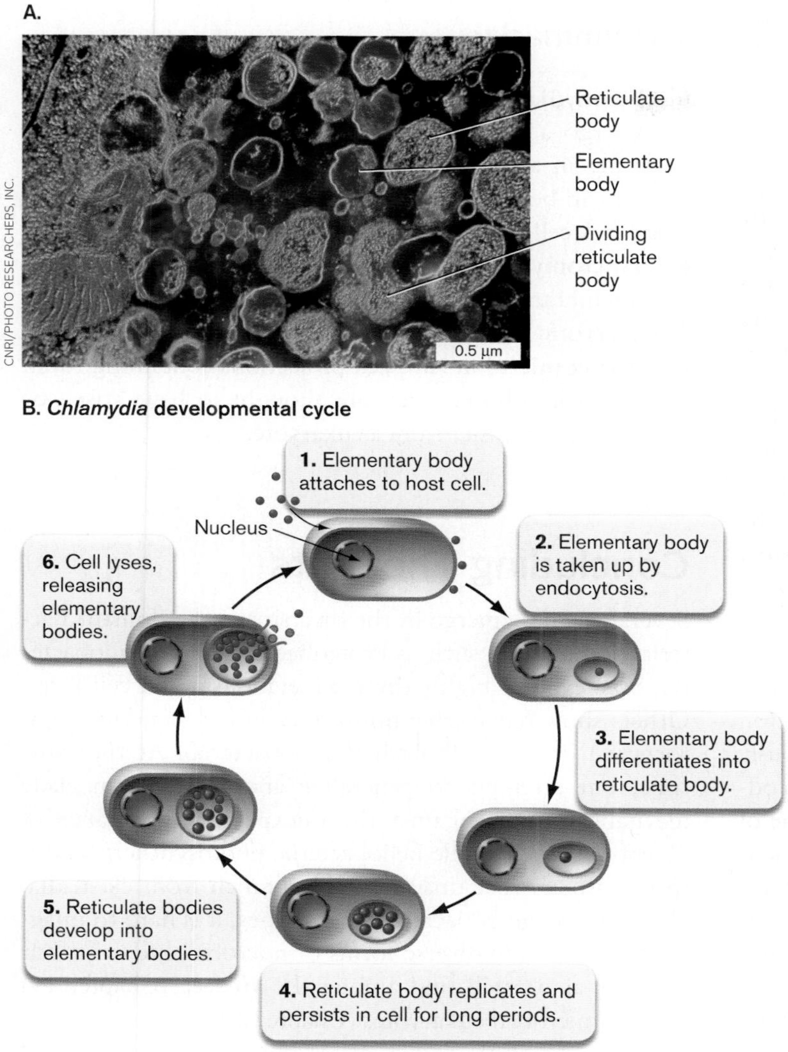
A.
Reticulate body
Elementary body
Dividing reticulate body
0.5 µm
CNRI/PHOTO RESEARCHERS, INC.
B. Chlamydia developmental cycle
1. Elementary body attaches to host cell.
Nucleus
2. Elementary body is taken up by endocytosis.
3. Elementary body differentiates into reticulate body.
4. Reticulate body replicates and persists in cell for long periods.
5. Reticulate bodies develop into elementary bodies.
6. Cell lyses, releasing elementary bodies.

FIGURE 18.48 ■ *Chlamydia* life cycle. A. *Chlamydia trachomatis* multiplying within a human cell (colorized TEM). Infected cell, equivalent to step 5 in part (B), contains reticulate bodies (yellow) growing and dividing, as well as newly formed elementary bodies (red). **B.** *Chlamydia* species persist outside the host as a spore-like "elementary body" that is metabolically inactive and contains compacted DNA. Upon endocytosis, the elementary body avoids lysosomal fusion and develops into a "reticulate body" in which the DNA, now uncondensed, has a reticular (netlike) appearance. The reticulate body replicates within the host cytoplasm and then develops into new elementary bodies that are released when the cell lyses.

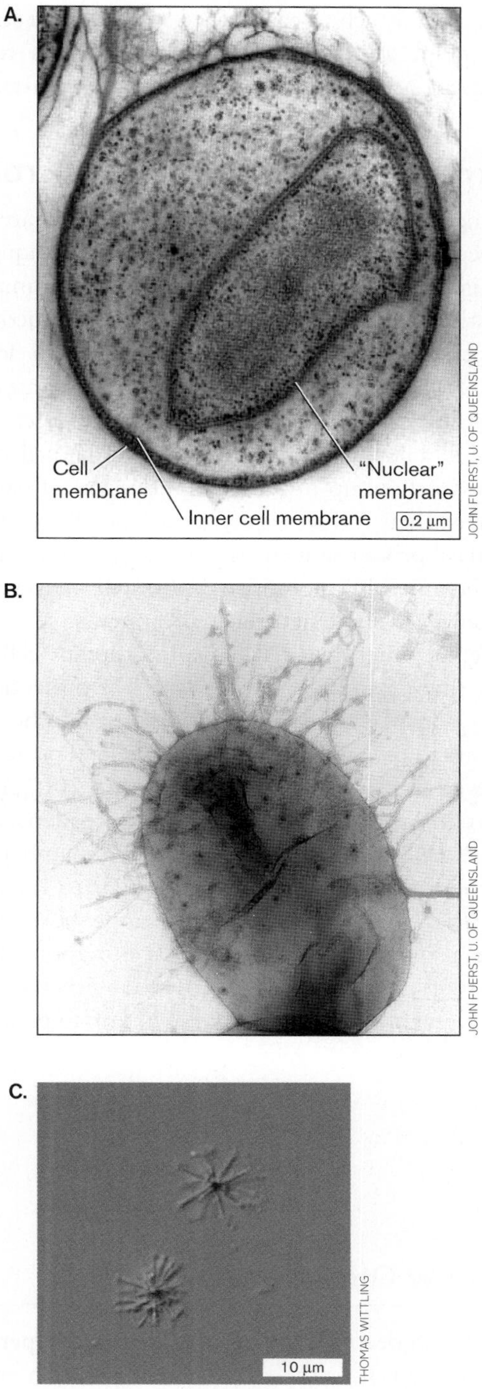
A.
Cell membrane
Inner cell membrane
"Nuclear" membrane
0.2 µm
JOHN FUERST, U. OF QUEENSLAND
B.
JOHN FUERST, U. OF QUEENSLAND
C.
10 µm
THOMAS WITTLING

FIGURE 18.49 ■ *Planctomyces:* bacteria with a "nuclear membrane." A. Section through *Gemmata obscuriglobus* showing membrane compartmentalization (TEM). The DNA is contained within a double membrane analogous to a eukaryotic nuclear membrane. **B.** A swarmer cell of *Planctomyces* sp. with multiple flagella (TEM). **C.** *Planctomyces bekefii* cells form stalks that attach to each other at the center to yield a starlike form. (LM)

a single membrane enclosing the nucleoid contains the complexes conducting the anammox reaction (anaerobic ammonium oxidation; discussed in Chapter 14).

Like eukaryotic protists (discussed in Chapter 20), the planctomycetes have flexible cell bodies that can assume diverse forms. For example, some *Planctomyces* species have rotary flagella (**Fig. 18.49B**), whereas *P. bekefii* cells have stalks that attach to each other to generate a starlike aggregate (**Fig. 18.49C**). Most planctomycetes reproduce by budding, a strategy typical of eukaryotic yeasts.

Verrucomicrobia: Wrinkled Microbes

Verrucomicrobia, "wrinkled microbes," are irregularly shaped bacteria found in a wide variety of aquatic and terrestrial environments, as well as in the mammalian gastrointestinal tract (see **Table 18.1**). Most Verrucomicrobia are oligotrophs, growing heterotrophically in low-salt habitats. Some are ectosymbionts of protists, attached to the cell surface of the eukaryote, where they eject harpoon-like objects. They are rarely cultured, and until recently they failed to show up in PCR amplification of natural isolates because their rDNA sequences are poorly amplified by the standard primer sequences. Using appropriate primer pairs, Verrucomicrobia comprise 5% of all surveyed microbial sequences in some natural environments.

Verrucomicrobia have peptidoglycan cell walls, but their shape is dominated by wart-like projections containing a poorly characterized cytoskeleton. The cytoskeleton appears to contain tubulin, a cytoskeletal protein previously believed to exist only in eukaryotes. In 2002, genes encoding tubulin were found in the partly sequenced genome of *Prosthecobacter dejongeii,* a free-living member of Verrucomicrobia. The genes appear so similar to those of eukaryotes that they must have undergone horizontal transfer from a eukaryotic genome. This horizontal transfer of a eukaryotic trait may be contrasted with the independent development of a nucleus-like structure in planctomycetes.

To Summarize

- **Chlamydiae** are obligate intracellular parasites that undergo a complex developmental progression, culminating in a spore-like form called an elementary body that can be transmitted outside the host cell. Chlamydiae lack cell walls.
- **Planctomycetes** lack cell walls, and they have evolved a membrane enclosing the nucleoid, analogous to the eukaryotic nuclear membrane.
- **Verrucomicrobia** have cell projections containing tubulin. Their tubulin genes are thought to have arisen by horizontal transfer from a eukaryote.

Concluding Thoughts

Several themes emerge in the diversity of the domain Bacteria. Some phyla, such as Proteobacteria and Actinobacteria, have evolved highly diverse metabolism and cell form. Others show remarkable uniformity in cell structure (Spirochetes) or in metabolism (Cyanobacteria). At the same time, our attempts to generalize about any given clade inevitably run up against the unexpected appearance of exceptions, such as the heliobacteria, photosynthetic endospore formers that unaccountably branch from clostridia. Given the range of bacterial phenotypes, it is hard to imagine that yet more diverse forms of microbial life exist; but they do, as we will find among the archaea (Chapter 19) and the microbial eukaryotes (Chapter 20).

CHAPTER REVIEW

Review Questions

1. Which deep-branching phyla include hyperthermophiles? Why is the actual branch position of these groups controversial?
2. Compare and contrast Chloroflexi and Cyanobacteria with respect to habitat, cell structure, and means of photosynthesis.
3. Compare and contrast the colonial and filamentous cyanobacteria with respect to life cycle and means of nitrogen fixation.
4. Name and describe three genera of Firmicutes that form endospores. Discuss the difference between single and multiple spore formers. Explain the existence of non-spore-forming species of Firmicutes.
5. Compare and contrast firmicute endospores and actinomycete arthrospores with respect to their means of production, their resistance properties, and their dispersal mechanisms.
6. Describe the diverse kinds of metabolism available to different species of Proteobacteria. Explain how it is possible for some species to perform many different kinds of energy-gaining metabolism.

7. What do species of Bacteroidetes and Chlorobi have in common, and how do they differ?
8. Explain the structure and mechanism of motility typical in species of Spirochetes.
9. Discuss the properties of as many intracellular parasites as you recall from various phyla. What do they have in common, and how do they differ?
10. Describe the unique cell structure of *Planctomyces.* Explain how the traits of planctomycete cells appear analogous to aspects of eukaryotic cells.
11. How do microbiologists seek to find previously unknown kinds of bacteria? What techniques reveal unknown microorganisms?

Thought Questions

1. Why are the Proteobacteria metabolically diverse? How is it possible that so many closely related organisms use such different molecules to yield energy?
2. How do you think *Mycobacterium tuberculosis* manages to grow despite its thick envelope screening out most nutrients?
3. For motility, what are the relative advantages of external flagella versus the flexible spiral cells of spirochetes?
4. Why do different species of microbes grow together in layered biofilms (a) on a sand flat and (b) on the surface of human teeth?

Key Terms

acid-fast stain (732)
Acidobacteria (717)
Actinobacteria (716)
actinomycete (717)
aerial mycelium (730)
akinete (720)
arthrospore (730)
bacteroid (740)
Bacteroidetes (717)
carboxysome (720)
Chlamydiae (717)
Chlorobi (717)
chlorosome (714)
Cyanobacteria (716)
diplococcus (743)
elementary body (752)
emerging (717)
endophyte (731)
endospore (717)
Firmicutes (716)
forespore (724)
fruiting body (746)
Fusobacteria (717)
gas vesicle (720)
GC content (717)
geosmin (730)
heterocyst (720)
hormogonium (720)
leghemoglobin (740)
methanotroph (737)
methylotrophy (737)
mother cell (724)
mycolic acid (732)
myxospore (746)
nitrifier (742)
Nitrospirae (717)
phylum (714)
Planctomycetes (717)
Proteobacteria (717)
proteorhodopsin (735)
reticulate body (752)
sarcina (734)
Spirochetes (717)
swarming (744)
Tenericutes (Mollicutes) (728)
vegetative cell (724)
vegetative mycelium (730)
Verrucomicrobia (717)

Recommended Reading

Anderson, Gregory G., Joseph J. Palermo, Joel D. Schilling, Robyn Roth, John Heuser, et al. 2003. Intracellular bacterial biofilm-like pods in urinary tract infections. *Science* **301**:105–107.

Andersson, Siv G. E., Alireza Zomorodipour, Jan O. Andersson, Thomas Sicheritz-Pontén, U. Cecilia M. Alsmark, et al. 1998. The genome sequence of *Rickettsia prowazekii* and the origin of mitochondria. *Nature* **396**:133–140.

Angert, Esther. 2006. Beyond binary fission: Some bacteria reproduce by alternative means. *Microbe* **1**:127–131.

Beatty, J. Thomas, Jörg Overmann, Michael T. Lince, Ann K. Manske, Andrew S. Lang, et al. 2005. An obligately photosynthetic bacterial anaerobe from a deep-sea hydrothermal vent. *Proceedings of the National Academy of Sciences USA* **102**:9306–9310.

Bryant, Donald A., and Niels-Ulrik Frigaard. 2006. Prokaryotic photosynthesis and phototrophy illuminated. *Trends in Microbiology* **14**:488–496.

Campbell, Barbara J., Annette Summers Engel, Megan L. Porter, and Ken Takai. 2006. The versatile ε-proteobacteria: Key players in sulphidic habitats. *Nature Reviews. Microbiology* **4**:458–468.

Chouari, Rakia, Denis Le Paslier, Patrick Daegelen, Philippe Ginestet, Jean Weissenbach, et al. 2003. Molecular evidence for novel planctomycete diversity in a municipal wastewater treatment plant. *Applied and Environmental Microbiology* **69**:7354–7363.

Dover, Lynn G., A. M. Cerdeno-Tarraga, M. J. Pallen, J. Parkhill, and Gurdyal S. Besra. 2004. Comparative cell wall core biosynthesis in the mycolated pathogens, *Mycobacterium tuberculosis* and *Corynebacterium diphtheriae. FEMS Microbiological Reviews* **28**:225–250.

Frigaard, Niels-Ulrik, Asuncion Martinez, Tracy J. Mincer, and Edward F. DeLong. 2006. Proteorhodopsin lateral gene transfer between marine planktonic Bacteria and Archaea. *Nature* **439**:847–850.

Geissinger, Oliver, Daniel P. R. Herlemann, Erhard Mörschel, Uwe G. Maier, and Andreas Brune. 2009. The ultramicrobacterium "*Elusimicrobium minutum*" gen. nov., sp. nov., the first cultivated representative of the termite group 1 phylum. *Applied and Environmental Microbiology* **75**:2831–2840.

Giovannoni, Stephen J., H. James Tripp, Scott Givan, Mircea Podar, Kevin L. Vergin, et al. 2005. Genome streamlining in a cosmopolitan oceanic bacterium. *Science* **309**:1242–1245.

Howerton, Amber, Manomita Patra, and Ernesto Abel-Santos. 2013. A new strategy for the prevention of *Clostridium difficile* infections. *Journal of Infectious Disease*. [Epub ahead of print.] doi: 10.1093/infdis/jit068.

Human Microbiome Project Consortium. 2012. Structure, function and diversity of the healthy human microbiome. *Nature* **486**:207–214.

Jenkins, Cheryl, Ram Samudrala, Iain Anderson, Brian P. Hedlund, Giulio Petroni, et al. 2002. Genes for the cytoskeletal protein tubulin in the bacterial genus *Prosthecobacter. Proceedings of the National Academy of Sciences USA* **99**:17049–17054.

Jensen, Paul R., Philip G. Williams, Dong-Chan Oh, Lisa Zeigler, and William Fenical. 2007. Species-specific secondary metabolite production in marine actinomycetes of the genus *Salinispora. Applied and Environmental Microbiology* **73**:1146–1152.

Kindaichi, Tomonori, Tsukasa Ito, and Satoshi Okabe. 2004. Ecophysiological interaction between nitrifying bacteria and heterotrophic bacteria in autotrophic nitrifying biofilms as determined by microautoradiography-fluorescence in situ hybridization. *Applied and Environmental Microbiology* **70**:1641–1650.

King, Gary M., and Carolyn F. Weber. 2007. Distribution, diversity and ecology of aerobic CO-oxidizing bacteria. *Nature Reviews. Microbiology* **5**:107–118.

Masaki, Toshihiro, Jinrong Qu, Justyna Cholewa-Waclaw, Karen Burr, Ryan Raaum, et al. 2013. Reprogramming adult Schwann cells to stem cell-like cells by leprosy bacilli promotes dissemination of infection. *Cell* **152**:51–67.

Moosvi, S. Azra, Ian R. McDonald, David A. Pearce, Donovan P. Kelly, and Ann P. Wood. 2005. Molecular detection and isolation from Antarctica of methylotrophic bacteria able to grow with methylated sulfur compounds. *Systematic and Applied Microbiology* **28**:541–554.

Schlieper, Daniel, María A. Oliva, José M. Andreu, and Jan Löwe. 2005. Structure of bacterial tubulin BtubA/B: Evidence for horizontal gene transfer. *Proceedings of the National Academy of Sciences USA* **102**:9170–9175.

Venter, J. Craig, Karin Remington, John F. Heidelberg, Aaron L. Halpern, Doug Rusch, et al. 2004. Environmental genome shotgun sequencing of the Sargasso Sea. *Science* **304**:66–74.

Wang, Jenny, Cheryl Jenkins, Richard I. Webb, and John A. Fuerst. 2002. Isolation of *Gemmata*-like and *Isosphaera*-like planctomycete bacteria from soil and freshwater. *Applied and Environmental Microbiology* **68**:417–422.

Wu, Dongying, Philip Hugenholtz, Konstantinos Mavromatis, Rüdiger Pukall, Eileen Dalin, et al. 2009. A phylogeny-driven genomic encyclopaedia of Bacteria and Archaea. *Nature* **462**:1056–1060.

Wu, Martin, Qinghu Ren, A. Scott Durkin, Sean C. Daugherty, Lauren M. Brinkac, et al. 2005. Life in hot carbon monoxide: The complete genome sequence of *Carboxydothermus hydrogenoformans* Z-2901. *PLoS Genetics* **1**:e65.

CHAPTER 19
Archaeal Diversity

19.1 Archaeal Traits

19.2 Crenarchaeota across the Temperature Range

19.3 Thaumarchaeota: Symbionts and Ammonia Oxidizers

19.4 Methanogens

19.5 Haloarchaea

19.6 Thermophilic and Acidophilic Euryarchaeota

19.7 Deeply Branching Divisions

Archaea are the most ecologically diverse of the three domains. Diverse species grow in all soil and water habitats, in symbiosis with animals and plants, and in extreme environments that exclude bacteria and eukaryotes. Archaea include hyperthermophiles inhabiting the hottest environments on Earth, as well as Arctic and Antarctic psychrophiles growing beneath sea ice and in subglacial lakes. Many archaea have adapted to extreme pressure, acidity, or salinity. Haloarchaea grow in concentrated brine, coloring salt lakes red. Other species have forms of metabolism, such as methanogenesis, that are unique to archaea. Their proposed resemblance to the earliest life-forms earned this group the name Archaea.

Archaea in moderate environments participate in microbial communities, often growing with bacteria in multispecies biofilms. Marine archaea include free-living autotrophs, members of methane-eating syntrophies, and symbionts of fish and sponges. Methanogens grow in all kinds of anaerobic environments, from wetland soil to our own digestive tract. Yet surprisingly, the archaeal domain lacks pathogens. No archaeon has yet been shown to cause disease.

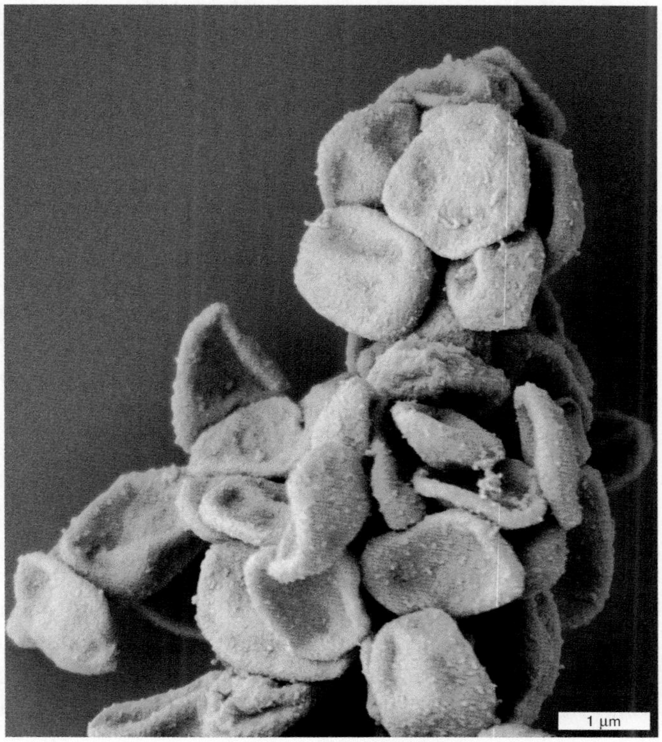

1 μm

CURRENT RESEARCH **highlight**

Salt-loving haloarchaea. *Haloferax volcanii* grows in concentrated salt habitats, such as the Great Salt Lake and the Dead Sea. Its unusual concave cells have no cell wall, only an S-layer. The lack of a cell wall allows two cells to fuse and exchange cytoplasmic material. While these archaea are prokaryotes (lack a nucleus), they can exchange genes through cell fusion and recombination between their nucleoids—a process remarkably similar to that of sexual exchange in eukaryotes. In 2012, Uri Gophna and co-workers showed that two species of *Haloferax*—*H. volcanii* and *H. mediterranei*—which differ by more than a tenth of their genomes, can recombine with each other, generating interspecies hybrids. *Source:* Jerry Eichler, Ben Gurion University of the Negev, Israel, http://lifeserv.bgu.ac.il/wb/jeichler/

In 2012, an archaeon was the subject of an unusual grant award from the Bill & Melinda Gates Foundation. Shiladitya DasSarma at the University of Maryland School of Medicine proposed to use haloarchaea—archaea that grow in concentrated salt—as a delivery vehicle for recombinant vaccines against typhoid bacteria. The recombinant haloarchaea vaccine can be embedded in salt crystals (**Fig. 19.1**)—a cheap and convenient delivery method for developing countries where typhoid is widespread. Vaccine delivery is a surprising application for a member of the diverse range of organisms known as archaea.

Overall, species of archaea grow at a wider range of temperature and other environmental factors than either bacteria or eukaryotes. We find archaea throughout soil and water, and within the human digestive tract. Other species grow in alkaline soda lakes, in extremely acidic mine runoff streams, or in superheated hot springs. Extremophiles are of interest to biotechnology companies because their exceptionally thermostable enzymes can catalyze reactions under industrially useful conditions, such as high salt or acidity or temperatures above the boiling point of water.

How do archaea in "moderate" habitats interact with bacteria and eukaryotes? Many archaea join bacteria within biofilms in soil and on marine particles. Archaea may cooperate with eukaryotes—for example, by colonizing the surface of plant roots. Even in these shared habitats, species of archaea show unique traits, such as energy-yielding methanogenesis and cyclic diether membranes, found only in the archaeal domain, not in bacteria or eukaryotes. Perhaps the most compelling unique feature of archaea, from the human point of view, is the complete absence of known archaeal pathogens of animals or plants. A few reports find methanogens associated with periodontal disease, but even in these cases, no causal relationship has been demonstrated.

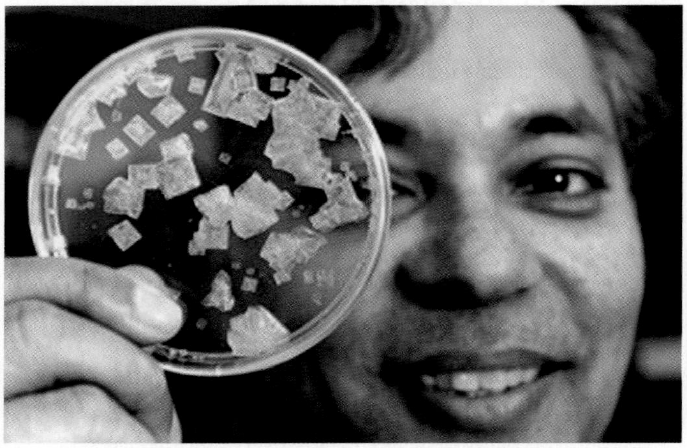

PHOTO AVAILABLE FROM DASSARMA

FIGURE 19.1 ■ Haloarchaea embedded in salt crystals can deliver a recombinant vaccine. The haloarchaeon *Halobacterium* NRC-1 can be engineered to express an antigen. The vaccine retains potency when the halophilic organisms are embedded in NaCl. Crystals are held here by Shiladitya DasSarma, University of Maryland School of Medicine.

In Chapter 19 we explore the diversity of archaea. We emphasize the unique structures and metabolism of archaeal cells. While surveying their major taxonomic categories, we introduce research techniques used to study organisms in extreme environments and those with unique forms of metabolism, such as methanogenesis. New species of archaea continue to be discovered; for further exploration, we recommend the online **National Center for Biological Information (NCBI) Taxonomy Database**.

19.1

Archaeal Traits

What unique traits distinguish archaea from other domains of life? Archaea share key traits with bacteria (such as the lack of a nucleus) but other traits with eukaryotes (such as transcription factors). And key traits such as isoprenoid membrane lipids are unique to archaea. Here we summarize key traits of archaea and then outline the major phylogenetic divisions.

Archaeal Cell Structure and Metabolism

Distinctive features of archaea, sometimes called "archaeal signatures," include components of the cell membrane and envelope and certain metabolic pathways to gain energy (**Table 19.1**).

Isoprenoid membranes. The most distinctive structure of archaea is their membrane (**Fig. 19.2**). The membrane lipids of archaea differ profoundly from those of bacteria and eukaryotes, with the exception of a few thermophilic bacteria that show evidence of horizontal transfer of archaeal genes conferring synthesis of archaeal-type lipids. Archaeal lipid structure includes:

- **L-Glycerol.** Archaeal membrane lipids incorporate L-glycerol, rather than the mirror-symmetrical form D-glycerol, which is used by bacteria and eukaryotes. The two chiral forms show similar thermal stability, but their biochemistry requires different enzymes, and thus they represent a deep divergence in ancestry. In some archaea the glycerol is extended by six carbons, forming nonitol (nine OH groups).

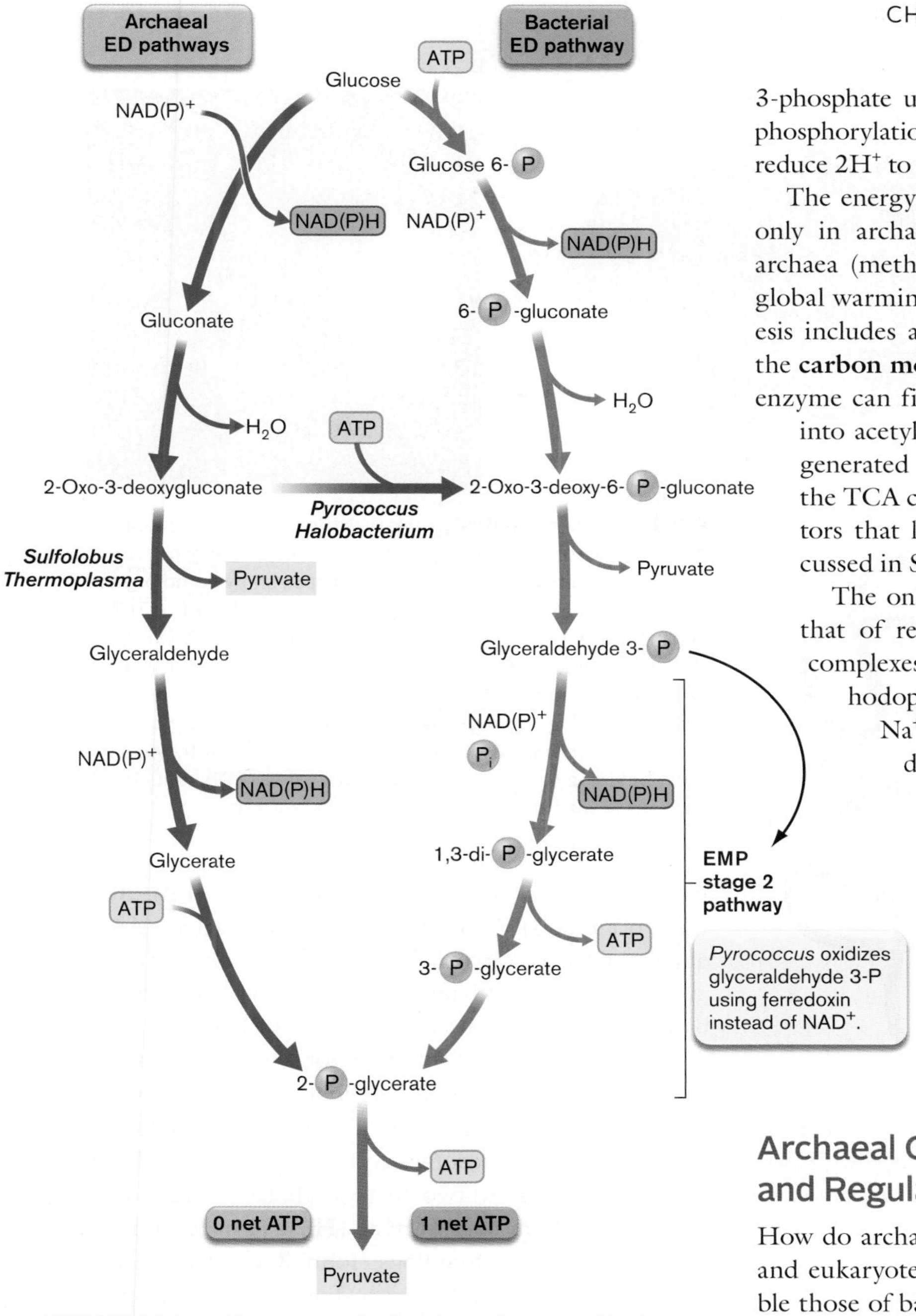
Archaeal ED pathways
Bacterial ED pathway
ATP
Glucose
NAD(P)+
Glucose 6-P
NAD(P)H
NAD(P)+
NAD(P)H
6-P-gluconate
Gluconate
H2O
H2O
ATP
2-Oxo-3-deoxygluconate
2-Oxo-3-deoxy-6-P-gluconate
Pyrococcus
Halobacterium
Sulfolobus
Thermoplasma
Pyruvate
Pyruvate
Glyceraldehyde
Glyceraldehyde 3-P
NAD(P)+
Pi
NAD(P)+
NAD(P)H
NAD(P)H
Glycerate
1,3-di-P-glycerate
EMP stage 2 pathway
ATP
ATP
3-P-glycerate
Pyrococcus oxidizes glyceraldehyde 3-P using ferredoxin instead of NAD+.
2-P-glycerate
ATP
0 net ATP
1 net ATP
Pyruvate

FIGURE 19.4 ■ Glucose catabolism in archaea. *Sulfolobus* and *Thermoplasma* species catabolize glucose to pyruvate by a modified ED pathway without phosphorylating glucose and produce no net ATP. *Halobacterium* species phosphorylate 2-oxo-3-deoxygluconate and produce 1 net ATP by the EMP stage 2 pathway. *Pyrococcus furiosus* oxidizes glyceraldehyde 3-phosphate using ferredoxin instead of NAD^+ and avoids phosphorylation.

thus channeling into the "standard" ED pathway. The ED pathway generates one molecule of pyruvate and one molecule of 3-phosphoglycerate, which produces one net ATP through the second stage of the EMP pathway. A unique variant of the EMP pathway is seen in the vent thermophile *Pyrococcus furiosus,* which oxidizes glyceraldehyde 3-phosphate using ferredoxin instead of NAD and avoids phosphorylation. The reduced ferredoxin is then used to reduce $2H^+$ to H_2 in an energy-yielding reaction.

The energy-yielding process of **methanogenesis** occurs only in archaea. Methane production by methanogenic archaea (methanogens) makes a growing contribution to global warming, as discussed in Chapter 22. Methanogenesis includes a unique pathway of carbon fixation, called the **carbon monoxide reductase pathway** because the key enzyme can fix CO as well as CO_2. A CO group is fixed into acetyl-CoA by condensation with a methyl group generated by methanogenesis. Acetyl-CoA then enters the TCA cycle. Methanogenesis also uses unique cofactors that largely replace NAD, FAD, and FMN (discussed in Section 19.4).

The only known form of phototrophy in archaea is that of retinal-based ion pumps. Membrane protein complexes known as **bacteriorhodopsin** and halorhodopsin contain retinal and pumps for H^+ or Na^+ (discussed in Chapter 14). Bacteriorhodopsins are found in haloarchaea (formerly called halobacteria) such as *Halobacterium halobium*; the species was named before it was known to be an archaeon (discussed in Section 19.5). Genes encoding light-absorbing proton pumps have been transferred to many marine archaea and bacteria. By contrast, the chlorophyll-based phototrophy found in bacteria and plants is completely unknown in archaea.

Archaeal Genome Structure and Regulation

How do archaeal genomes compare with those of bacteria and eukaryotes? The genomes of archaea generally resemble those of bacteria in size and gene density, and genes of related function are often arranged in operons, like those of bacteria. Nevertheless, certain features of genome structure are unique to archaea, and some traits more closely resemble those of eukaryotes than those of bacteria.

A distinctive feature of archaeal chromosome function is the "reverse gyrase" enzyme found in hyperthermophiles. For comparison, all bacteria and eukaryotes, as well as mesophilic archaea, have gyrase to maintain their DNA in an "underwound" negatively supercoiled state (presented in Chapter 7). But hyperthermophilic archaea with reverse gyrase maintain positive supercoiling to stabilize their DNA. The positive superturns "overwind" the DNA, preventing the helix from melting into separate strands at high temperature.

A.

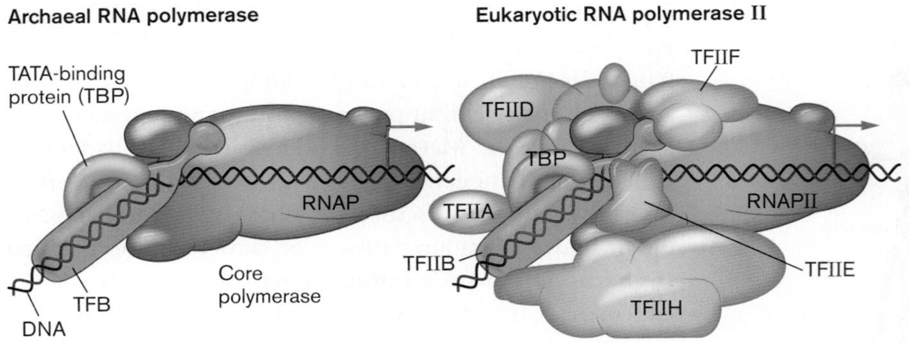
Archaeal RNA polymerase
Eukaryotic RNA polymerase II
TATA-binding protein (TBP)
RNAP
Core polymerase
TFB
DNA
TFIIF
TFIID
TBP
TFIIA
RNAPII
TFIIB
TFIIE
TFIIH

B.

JOHN REEVE

C.

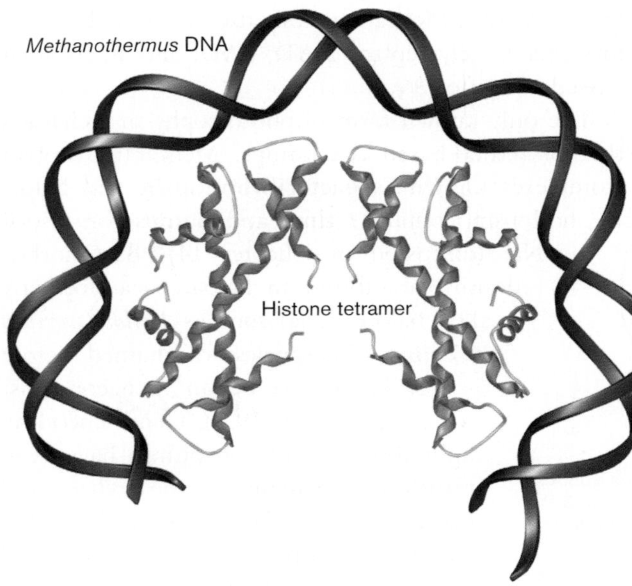
Methanothermus DNA
Histone tetramer

FIGURE 19.5 ▪ Central genetic molecules of Archaea resemble those of Eukarya. A. The core RNA polymerase (RNAP) subunits of the archaeon *Sulfolobus* show homology to those of the eukaryotic RNA polymerase II (RNAPII). The TATA-binding protein (TBP, green) and transcription factor B (TFB, pink) also show homology to eukaryotic counterparts. **B.** John Reeve (left), and a graduate student at the Ohio State University, study the molecular biology of methanogens using an anaerobic glove box. **C.** *Methanothermus fervidus* DNA binds to histone tetramers that show homology to the histones of eukaryotes. (PDB code: 1A7W)
Source: Kathryn A. Bailey et al. 2002. *J. Biol. Chem.* **277**:9293.

A unique feature of archaeal RNA is the distinctive modified bases in their tRNA molecules. In particular, the guanosine analog archaeosine (7-formamidino-7-deazaguanosine) is used by nearly all archaea but no bacteria or eukaryotes. Other unusual tRNA bases, such as queuosine, are found only in bacteria and eukaryotes, not in archaea.

The genes encoding most archaeal proteins contain uninterrupted coding sequence, as in bacteria; but certain tRNA gene sequences are interrupted by introns (nontranslated sequence), similar to the tRNA introns found in eukaryotes. Furthermore, the archaeal apparatus for DNA and RNA polymerases, transcription factors, and protein synthesis show remarkable similarity to those of eukaryotes. **Figure 19.5A** compares the components of an archaeal RNA polymerase (from *Sulfolobus*) with those of eukaryotic RNA polymerase II. The archaeal polymerase possesses two transcription factors (regulatory protein components) that are found in eukaryotes: TATA-binding protein (TBP, a subunit of TFIID) and transcription factor B (TFIIB, which in archaea is designated TFB). By contrast, bacteria have no homologs of these factors. A consequence of the eukaryotic-like transcription and translation in archaea is that archaea are resistant to antibacterial antibiotics that target transcription and translation.

Another eukaryotic structure for which archaeal homologs were discovered is that of **histones**, the fundamental packaging proteins of DNA. The histone complex found in eukaryotic chromosomes contains a histone $(H3 + H4)_2$ tetramer flanked by two histone (H2A + H2B) dimers. The archaeal homologs form an $(H3 + H4)_2$ tetramer, with no (H2A + H2B) homologs. John Reeve and colleagues at the Ohio State University (**Fig. 19.5B**) showed that isolated DNA of specific sequences could be bound and curved around a histone tetramer from the methanogen *Methanothermus fervidus* (**Fig. 19.5C**). The DNA has AT-rich sequences that specifically fit the histone complex. The key sequences of the histones (the ones that bind to DNA) show homology to eukaryotic histones. Histones have since been found in many species of archaea.

Phylogeny of Archaea

How have archaea diverged from the common ancestor and evolved into the diverse forms we find today? For several reasons, the phylogeny of Archaea is a challenge to define.

Most archaea are uncultured and are known solely through small-subunit (SSU) rRNA sequence or through metagenomics (the genome sequences of a microbial community). Many of their genomes are highly "recombinogenic." For example, two different samples of *Ferroplasma acidarmanus* show 99% identical SSU rRNA, yet their overall genomes differ by 22%, implying extensive horizontal gene exchange. Finally, deeply branching clades such as the Ancient Archaeal Group (AAG) have such divergent SSU rRNA that their sequence fails to amplify with "standard" PCR primers based on the most highly conserved regions of the gene. We suspect that many deeply branching groups remain unknown, as yet undetected by our current molecular tools.

The domain Archaea includes three phyla, also called divisions, that have been extensively characterized (**Fig. 19.6**). These phyla are **Crenarchaeota**, **Thaumarchaeota**, and **Euryarchaeota**. Major taxa are outlined in **Table 19.2**. **An expanded version of this table appears in Appendix 4B.**

Note: The phylogeny of archaea undergoes frequent revision, as new taxa are discovered and novel genomes are sequenced. For example, the taxa now designated Thaumarchaeota were originally classified with the Crenarchaeota, until multilocus genome comparison revealed the deeper division between the two clades.

Crenarchaeota. The Crenarchaeota include a substantial proportion of soil, marine, and benthic (marine sediment) microbial communities, including associations with plants and animals. Understanding these organisms is important to assess their contribution to the global carbon cycle.

A fascinating group of crenarchaeotes is the thermophiles. Many thermophilic crenarchaeotes metabolize sulfur, either by anaerobic reduction (such as by H_2 to form H_2S) or by aerobic oxidation (by O_2 to form sulfuric acid). Anaerobic sulfur metabolizers include moderate thermophiles (growth range about 60°C–80°C), as well as hyperthermophiles (90°C–120°C). Many of the hyperthermophiles are also barophiles, growing under high pressure

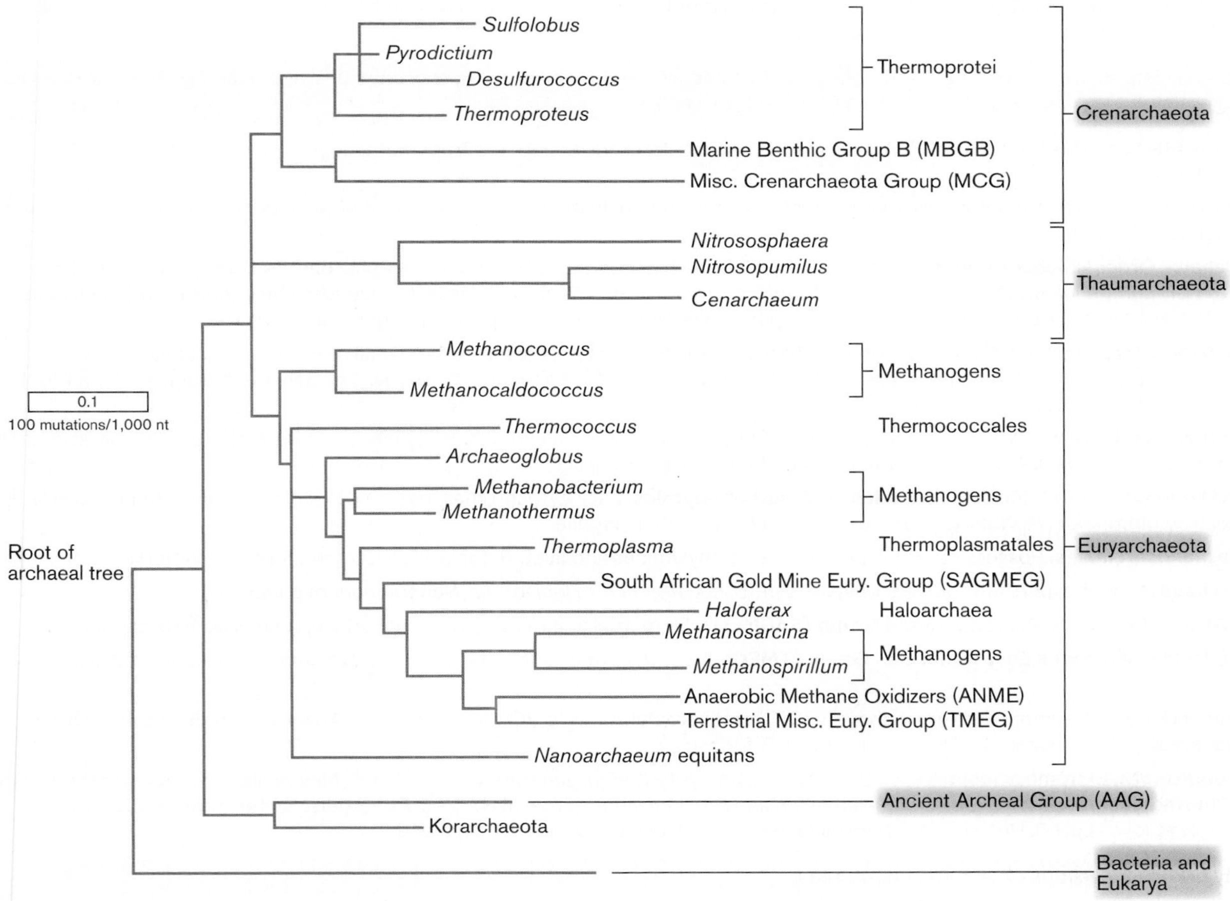

Sulfolobus
Pyrodictium
Desulfurococcus
Thermoproteus
Thermoprotei
Crenarchaeota
Marine Benthic Group B (MBGB)
Misc. Crenarchaeota Group (MCG)
Nitrososphaera
Nitrosopumilus
Cenarchaeum
Thaumarchaeota
0.1
100 mutations/1,000 nt
Methanococcus
Methanocaldococcus
Methanogens
Thermococcus
Thermococcales
Archaeoglobus
Methanobacterium
Methanothermus
Methanogens
Root of archaeal tree
Thermoplasma
Thermoplasmatales
Euryarchaeota
South African Gold Mine Eury. Group (SAGMEG)
Haloferax
Haloarchaea
Methanosarcina
Methanospirillum
Methanogens
Anaerobic Methane Oxidizers (ANME)
Terrestrial Misc. Eury. Group (TMEG)
Nanoarchaeum equitans
Ancient Archeal Group (AAG)
Korarchaeota
Bacteria and Eukarya

FIGURE 19.6 ▪ **Archaeal divergence.** Major clades, divisions, and metagenomic groups of archaea. Divergence was based approximately on SSU rRNA and genomic sequences.

TABLE 19.2

Archaeal diversity.* (Expanded table appears in Appendix 4B.)

Crenarchaeota. Tetraether membranes surrounded by S-layer. Wide range of growth temperatures.

Marine Benthic Group B (MBGB). Anoxic marine sediments: deep-sea methane hydrates, hydrothermal vents, coastal regions.

Miscellaneous Crenarchaeote Group (MCG). Widely divergent crenarchaeotes in diverse soil and water.

Thermoprotei. Many thermophiles in hot springs and marine vents. Also includes marine mesophiles and psychrophiles. *Thermosphaera.*

- **Caldisphaerales.** Thermoacidophilic heterotrophs grow in hot springs. *Caldisphaera* spp.
- **Desulfurococcales.** Anaerobic sulfur reduction with organic electron donors. Irregularly shaped cells with glycoprotein S-layer; no cell wall (*Aeropyrum pernix, Desulfurococcus fermentans*). *Ignicoccus islandicus. Pyrodictium abyssi, Pyrodictium occultum, Pyrolobus fumarii* grow at marine thermal vents, up to 110°C. Unique periplasmic space contains membrane vesicles.
- **Psychrophilic marine crenarchaeotes (uncharacterized).** *Marine water,* deep sea, and Antarctica.
- **Sulfolobales.** Aerobic acidophiles; moderate thermophiles. Oxidize H_2S to H_2SO_4. *Sulfolobus, Sulfurisphaera, Acidianus.*
- **Thermoproteales.** *Pyrobaculum, Thermoproteus, Vulcanisaeta.*

Thaumarchaeota. Tetraether membranes include crenarchaeol. Marine and soil archaea that oxidize NH_3 with O_2. Important marine sources of nitrates for phytoplankton, and of methylphosphonate (CH_3-PO_3^{2-}), which bacteria convert to methane. Includes members of Marine Group 1.1a.

- **Cenarchaeales.** Sponge symbionts; grow at 10°C. *Cenarchaeum symbiosum.*
- **Nitrosopumilales.** Ammonia-oxidizing archaea (AOA), marine and soil. *Nitrosopumilus maritimus, Nitrososphaera gargensis.*
- **Psychrophilic marine thaumarchaeotes (uncharacterized).** *Marine water,* deep sea, and Antarctica. Anaerobic heterotrophs, sulfate reducers, nitrite-reducing methanotrophs.

Euryarchaeota. Metabolism includes methanogenesis, halophilic photoheterotrophy, and sulfur and hydrogen oxidation; acidophiles and alkaliphiles. Methanogens and halophiles have rigid cell walls.

Anaerobic Methane Oxidizers (ANME). Anoxic marine sediments. Oxidize methane from methanogens, in syntrophy with sulfate-reducing bacteria.

Archaeoglobi. Order **Archaeoglobales**. Hyperthermophiles; sulfate oxidation of H_2 or organic hydrogen donors; reverse methanogenesis. *Archaeoglobus fulgidus.*

Haloarchaea. Order **Halobacteriales**. Halophiles; grow in brine (concentrated NaCl). Conduct photoheterotrophy, with light-driven H^+ pump and Cl^- pump. *Haloarcula, Halobacterium, Haloferax* grow in salterns and salt lakes. *Haloquadra* (*Haloquadratum*) are square-shaped. *Halorubrum lacusprofundi* is an Antarctic psychrophile. *Natronococcus* spp. are alkaliphiles in soda lakes.

Methanogens (four classes). Generate methane from CO_2 and H_2, formate, acetate, other small molecules; strict anaerobes. Pseudopeptidoglycan or sulfated chondroitin cell walls. Grow in anaerobic soil, water, or animal digestive tract. Deep-sea psychrophiles generate methane hydrates.

- **Methanobacteriales.** Lack cytochromes; reduce CO_2, formate, or methanol with H_2 by electron bifurcation. *Methanobrevibacter smithii* and *Methanosphaera stadtmanae* inhabit human digestive tract.
- **Methanomicrobiales, Methanococcales, and Methanopyrales.** Lack cytochromes; reduce CO_2, formate, or methanol with H_2 by electron bifurcation. *Methanocaldococcus jannaschii,* vent thermophile.
- **Methanosarcinales.** Possess cytochromes; reduce methylamines and acetate (as well as CO_2 and formate) with H_2.

Nanoarchaeota. Vent hyperthermophiles; obligate symbionts attached to *Ignicoccus. Nanoarchaeum equitans.*

South African Gold Mine Euryarchaeotal Group (SAGMEG). Terrestrial and marine deep subsurface. Anaerobic heterotrophs.

Terrestrial Miscellaneous Euryarchaeotal Group (TMEG). Terrestrial deep subsurface and surface soils, marine and freshwater sediments.

Thermococci. Hyperthermophiles (grow above 100°C) and barophiles (up to 200 atm pressure). Anaerobes; reduce sulfur. Order **Thermococcales:** *Thermococcus, Pyrococcus abyssi, P. furiosus.*

Thermoplasmata. Extreme acidophiles; oxidize sulfur from pyrite (FeS_2), generating sulfuric acid. Mesophiles or moderate thermophiles. Order **Thermoplasmatales:** *Ferroplasma acidiphilum* and *F. acidarmanus* grow at 37°C–50°C. Oxidize sulfur from FeS_2, generating ambient pH as low as pH 0. No cell wall. *Thermoplasma acidophilum* grows at 59°C and pH 2.

Deeply branching groups of uncultured Archaea.

Ancient Archaeal Group (AAG). Vent hyperthermophiles and benthic subsurface communities.

Korarchaeota. Hyperthermophiles in Yellowstone hot springs and in deep-sea thermal vents.

Marine Hydrothermal Vent Group (MHVG). Vent hyperthermophiles.

*Bulleted terms are representative orders (unless stated otherwise).

at hydrothermal vents on the ocean floor; an example is *Pyrodictium abyssi.* Yet other crenarchaeotes are mesophiles (growing at moderate temperatures) and even psychrophiles (growing at temperatures below 20°C), found in Antarctic lakes. Overall, the Crenarchaeota encompass the widest range of growth temperature of any division of life.

Thaumarchaeota. The phylum Thaumarchaeota (originally classified with Crenarchaeota) includes many mesophilic heterotrophs and sulfur oxidizers in soil and water. Some are key symbionts of marine invertebrate animals such as deep-sea sponges. Another important group are the ammonia oxidizers (AOA), which oxidize ammonia to nitrite, thus playing a major role in the nitrogen cycle (discussed in Chapters 21 and 22).

Euryarchaeota. The phylum Euryarchaeota also includes members throughout soil and water, and associated with plants and animals. They show a greater range of metabolism than the Crenarchaeota. The most highly divergent group of Euryarchaeota is the methanogens. Methanogens serve a key energetic role in ecosystems by offering an anaerobic mechanism to remove excess H_2 and other small-molecule reductants. Despite their common energetic pathway, methanogens branch deeply among other euryarchaeotic groups, and they show a wide range of cell shape and environmental adaptations.

One branch related to methanogens is the Haloarchaea, extreme halophiles that are the only form of life to grow in concentrated brine (NaCl). Most haloarchaea, such as *Halobacterium* species NRC-1, are photoheterotrophs that can supplement their metabolism with light-driven ion pumps. Still other genera, such as *Archaeoglobus,* show mixed physiology, including components of methanogenesis linked to other pathways, such as glycolysis. The euryarchaeotes include mesophiles as well as thermophiles, such as the order Thermococcales, but few psychrophiles have been found so far.

Euryarchaeotes include extreme acidophiles, as well as extreme alkaliphiles. With respect to pH, the euryarchaeotes show the widest range of any clade, from *Ferroplasma* (the order Thermococcales), growing at pH 0, to *Natronococcus,* growing at pH 10.

Note: In Table 19.2, certain archaeal names contain a "bacteria" component—for example, the genus *Halobacterium.* Such organisms were known and named as bacteria before 1977, when the category "archaea" was first defined.

Emerging phyla. Besides the major phyla Crenarchaeota, Thaumarchaeota, and Euryarchaeota, metagenomic analysis continues to reveal new clades that branch near the root of the tree. The Ancient Archaeal Group (AAG) was discovered in hydrothermal vents and deep benthic sediment. These organisms failed to be detected by PCR primers based on known SSU rRNA gene sequence. Thus, their detection required reverse transcription of rRNA from uncultured cells. The reverse transcripts showed that the rRNA sequence of AAG organisms differs by 20% from the most conserved regions of other known archaeal rRNA.

As more genomes have been sequenced, other proposed deeply branching lineages have since been assigned to known clades. The vent hyperthermophiles known as Korarchaeota appear to branch with the Ancient Archaeal Group. The **Nanoarchaeota** are tiny hyperthermophiles that grow as obligate symbionts of a crenarchaeote, *Ignicoccus.* Genomic analysis now suggests that the Nanoarchaeota are a fast-evolving lineage (like other symbionts with reduced genomes) and that they are euryarchaeotes, branching from the Thermococcales group.

Thought Question

19.1 If two deeply diverging clades each show a wide range of growth temperature, what does this suggest about the evolution of thermophily or psychrophily?

To Summarize

- **Archaeal membranes are composed of L-glycerol diether or tetraether lipids**, with isoprenoid side chains that may include cross-links or pentacyclic rings.
- **Many archaea have no peptidoglycan cell wall, but an S-layer of protein.** Some methanogens have cell walls of pseudomurein.
- **Glucose is catabolized by variants of the ED pathway.** Other metabolic pathways found in archaea include methanogenesis and retinal-associated light-driven ion pumps such as bacteriorhodopsin.
- **Central genetic functions of archaea resemble those of eukaryotes**, as seen in the structure of DNA and RNA polymerases and of histone-like DNA-binding proteins.
- **The most studied phyla or divisions of archaea are Crenarchaeota, Thaumarchaeota, and Euryarchaeota.** Crenarchaeota include sulfur hyperthermophiles, mesophiles, and psychrophiles. Thaumarchaeota include ammonia oxidizers, marine invertebrate symbionts, and others. Euryarchaeota include methanogens, halophiles, and acidophiles.
- **Metagenomics reveals uncultured organisms.** We continue to discover deeply branching clades of archaea by designing new PCR primers, reverse-transcribing community rRNA, and sequencing metagenomes.

19.2

Crenarchaeota across the Temperature Range

How were the first crenarchaeotes discovered and described? The name Crenarchaeota means "scalloped archaea," derived from the amorphous scalloped cell shapes of the first hyperthermophiles discovered. The first members described were hyperthermophiles such as *Thermosphaera* species (**Table 19.2**). Since then, however, numerous mesophiles and psychrophiles have been identified in soil and marine habitats.

©INGA SPENCE/VISUALS UNLIMITED

FIGURE 19.7 ■ Thermophiles colonize a hot spring. Morning Glory Pool, a hot spring in Yellowstone National Park, supports thermophilic archaea and bacteria.

Habitats for Thermophiles

Thermophiles and hyperthermophiles commonly grow in hot springs and geysers, such as those of Yellowstone National Park (**Fig. 19.7**) or the Solfatara volcanic area near Naples, Italy. A hot spring occurs where water seeps underground above a magma chamber, which heats the water to near boiling. The heated water expands and is forced upward through fissures, coming out in a heated spring. In a geyser, the water is heated under pressure. As the water escapes upward, it turns into steam, which expands and jets upward, falling into a heated pool. These heated pools and their surrounding edges generate extreme ranges of temperature, mineral content, and acidity. They support a diverse range of microbial life, including thermophilic cyanobacteria and firmicutes, as well as archaea.

Several features of hot springs and geysers are important for thermophiles:

- **Reduced minerals.** The heated water dissolves high concentrations of sulfides and other reduced minerals. When the water emerges and cools, the minerals precipitate. These reduced minerals serve as rich energy sources for autotrophs.
- **Low oxygen content.** At higher temperatures, the oxygen concentration of water declines. Therefore, hyperthermophiles tend to be anaerobic, although there are important exceptions, such as the aerobic sulfur oxidizer *Sulfolobus*.
- **Steep temperature gradients.** The temperature of the water falls dramatically within a short distance from the source, forming a steep gradient. Different species of thermophiles are adapted to different temperatures and grow in separate patches at the different temperatures, causing a variegated pattern.
- **Acidity.** Some hot-spring environments show extreme acidity. The acidity results from oxidation of sulfur or iron in reactions that generate strong inorganic acids such as sulfuric acid (H_2SO_4).

A special subcategory of volcanic hot-spring habitats is that of submarine hydrothermal vents on the ocean floor. The vent thermophiles must evolve to grow at high pressure under several kilometers of ocean. Pressure increases by approximately 100 atm per kilometer of ocean depth. Organisms that grow only at high pressure are called **barophiles** (discussed in Chapter 5).

Desulfurococcales: Reducing Sulfur from Hot Springs

Species of the order **Desulfurococcales** show distinctive cell structures and forms of metabolism (**Fig. 19.8A**; **Table 19.3**). All possess elaborate S-layers but lack a cell wall; their diphytanylglycerol membranes contain a combination of diethers and tetraethers. Many take advantage of the high temperatures that increase the thermodynamic favorability of sulfur redox reactions. An example is *Desulfurococcus fermentans* (**Fig. 19.8A**), a flagellated coccoid cell isolated from hot springs; it grows optimally at 85°C. *D. fermentans* respires anaerobically by reducing elemental sulfur (S^0) to sulfide (HS^-). Sulfur reduction is coupled to oxidation of small organic molecules such as sugars.

Another flagellated coccus, *Ignicoccus islandicus*, has an unusual cell architecture (**Fig. 19.8B**). *Ignicoccus* has an

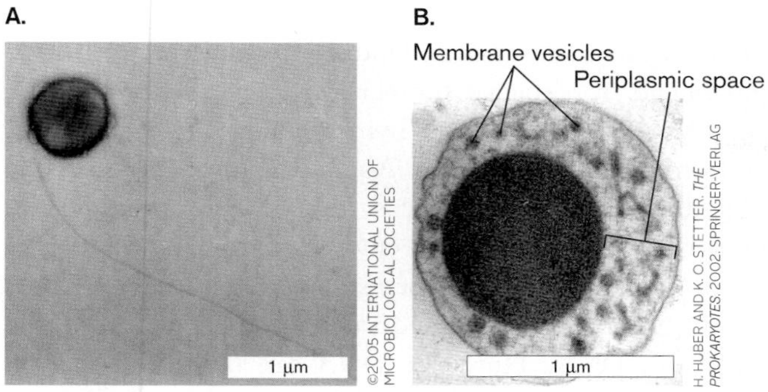
A.
B.
Membrane vesicles
Periplasmic space
1 μm
1 μm
©2005 INTERNATIONAL UNION OF MICROBIOLOGICAL SOCIETIES
H. HUBER AND K. O. STETTER. THE PROKARYOTES. 2002. SPRINGER-VERLAG

FIGURE 19.8 ▪ Hyperthermophilic crenarchaeotes. **A.** *Desulfurococcus fermentans* (shadow EM). **B.** *Ignicoccus islandicus* (TEM). Its unique periplasmic space contains membrane vesicles. *Source:* A. Perevalova et al. 2005. *Int. J. Syst. Evol. Microbiol.* **55**: 995.

outer membrane surrounding its cytoplasmic membrane, with a large periplasmic space between them. The periplasmic space contains membrane-enclosed vesicles of unknown function. The evolution of this compartment, similar to the extra membranes of the bacterial genus *Planctomyces* (see Chapter 18), suggests a model for an intermediate stage of evolution of the eukaryotic nucleus. *I. islandicus,* unlike *Desulfurococcus,* is a marine organism, growing at temperatures as high as 98°C. It is a lithotroph, oxidizing hydrogen with sulfur:

$$H_2 + S^0 \longrightarrow H_2S$$

Most of the cultured species of Desulfurococcales are obligate anaerobes. An exception is *Aeropyrum pernix,* one of the first archaea to have its genome sequenced. *A. pernix* is an aerobic heterotroph, respiring with O_2 on complex compounds during growth at 70°C–100°C.

Barophilic Vent Hyperthermophiles

The most extreme hyperthermophiles are barophiles adapted to grow near hydrothermal vents at the ocean floor. The high pressure beneath several kilometers of ocean allows water to remain liquid at temperatures above 100°C; the highest known temperature for growth of an organism is 121°C.

A common feature of thermal vents is the **black smoker** (**Fig. 19.9**). A black smoker is a chimneylike structure resulting from the upwelling of seawater superheated by an undersea magma chamber. As in a geyser aboveground, the heated water is forced upward through a small opening. Because the thermal vent is under steam pressure, the water can reach temperatures of over 400°C, enabling it to dissolve high concentrations of minerals such as iron II sulfide (FeS). When the rising water escapes, however, it immediately cools, depositing iron sulfide around the edge of the vent chimney and precipitating iron sulfide particles that cloud the water—hence the term "black smoker." While no organism can grow at 400°C, various species of archaea are adapted to grow in the range of 100°C–120°C, where the vent stream meets the seawater and minerals precipitate (**Fig. 19.9B**).

TABLE 19.3

Hyperthermophilic Crenarchaeota.

Representative species	Growth temperature (°C)	Growth pH	Cell shape	Metabolism
Aeropyrum pernix	70–100°	pH 5–9	Cocci	O_2 respiration
Desulfurococcus fermentans	78–87°	pH 6	Flagellated cocci	Anaerobic S^0 respiration or fermentation
Ignicoccus islandicus	70–98°	pH 5–7	Flagellated cocci with periplasmic space	Anaerobic lithotrophy, S^0 oxidation of H_2
Pyrodictium abyssi	80–110°	pH 5–7	Disks linked by cannulae	Anaerobic fermentation
Pyrodictium brockii	85–110°	pH 5–7	Disks linked by cannulae	Anaerobic oxidation of H_2 by S^0 or $S_2O_3^{2-}$ or fermentation
Sulfolobus solfataricus	50–87°	pH 2–4	Irregular cocci	O_2 respiration on S^0, producing H_2SO_4
Thermosphaera aggregans	65–90°	pH 5–7	Flagellated cocci in aggregates	Anaerobic fermentation

A.

MICHAEL PERFIT, U. OF FLORIDA, AND NOAA VENTS PROGRAM

B.

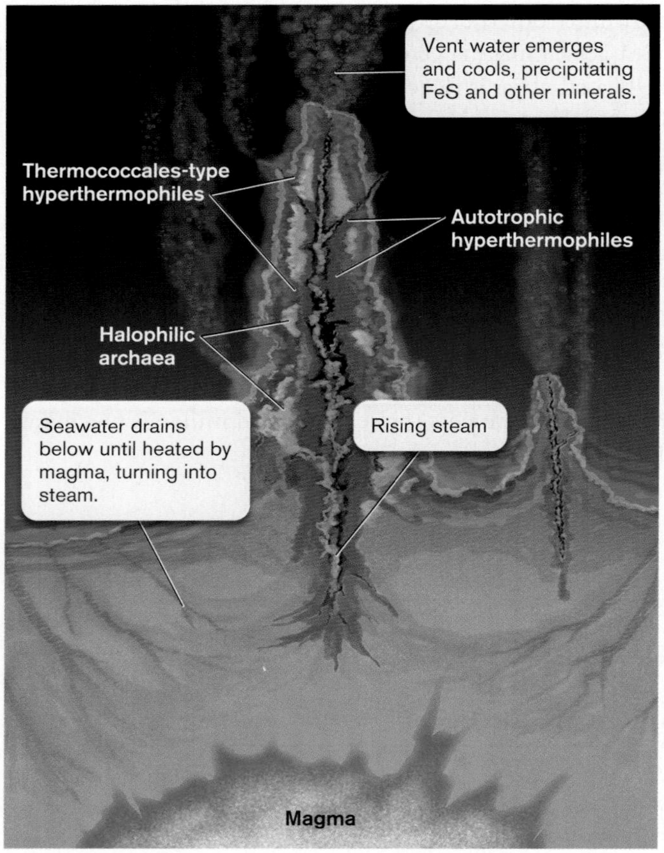
Vent water emerges and cools, precipitating FeS and other minerals.
Thermococcales-type hyperthermophiles
Autotrophic hyperthermophiles
Halophilic archaea
Seawater drains below until heated by magma, turning into steam.
Rising steam
Magma

FIGURE 19.9 ■ Extreme temperature and pressure: black smoker vents. A. Black smoker vents with steam escaping from "chimneys" of sulfide minerals at the Juan de Fuca Ridge on the ocean floor. **B.** Different parts of the smoker vent system support different classes of archaea.

How do we study organisms under such extreme conditions? To study hyperthermophiles from black smoker vents requires specialized equipment. The isolation of such organisms is a challenge because their habitats endanger our own survival. Undersea vent systems must be approached by a special submersible device with a robotic arm. An example is the Environmental Sample Processor from the Monterey Bay Aquarium Research Institute (**Fig. 19.10A**). The robotic system samples temperature and other properties of fluid emerging from a black smoker hydrothermal vent. It can then sample organisms for study. An advanced version of this device can actually process the organism's DNA. Thus, the DNA can be obtained from vent-adapted microbes that could not survive transfer to a laboratory at sea level. The robotic sample processor is supported by NASA as a model for a future space probe to explore one of Jupiter's moons, Europa, considered a possible source of extraterrestrial life.

Organisms that do survive transport to sea level must nonetheless be maintained at high pressure and temperature to ensure viability. In the laboratory, we must maintain all devices for microscopy and cultivation under pressure and at high temperature. Organisms are cultured in a pressurized cell such as the "Deep Aquarium" (**Fig. 19.10B**). The culture must be provided with reduced minerals and gases needed for the growth of vent microbes.

Vent-adapted crenarchaeotes include *Pyrodictium abyssi* (**Fig. 19.11**), *P. occultum*, and *P. brockii*; the latter is named for Thomas Brock of the University of Wisconsin–Madison, a pioneering researcher of hyperthermophiles. For energy, *Pyrodictium* species reduce sulfur to H_2S, either with molecular hydrogen or with organic compounds. A membrane-embedded sulfur-reducing complex and a proton-translocating ATP synthase have been isolated from *P. abyssi*. The complexes are extremely heat stable, exhibiting a temperature optimum of 100°C.

Pyrodictium species grow as flat, disk-shaped cells that can be as thin as 0.1 μm. The cells contain a periplasm and outer membrane with an S-layer that is coated with zinc sulfide, presumably precipitated from the vent minerals. The cell disks are interconnected by periplasmic extensions called **cannulae** (singular, **cannula**). The cannulae can extend to more than 0.1 mm, forming complex networks of connections (**Fig. 19.11**); in liquid culture, the networks grow into white balls up to 10 mm in diameter. Cryo-electron tomography of a *Pyrodictium* cell shows that the cannulae bridge the periplasm between cells, but not the cytoplasm. The cannulae may enable *Pyrodictium* cells to share nutrients and maintain a biofilm, while keeping their cellular identity distinct with separated cytoplasm.

What happens when a *Pyrodictium* cell divides? Cells of *P. abyssi* generate new cannulae as they undergo fission (**Fig. 19.12**). Some of the new cannulae form as loops

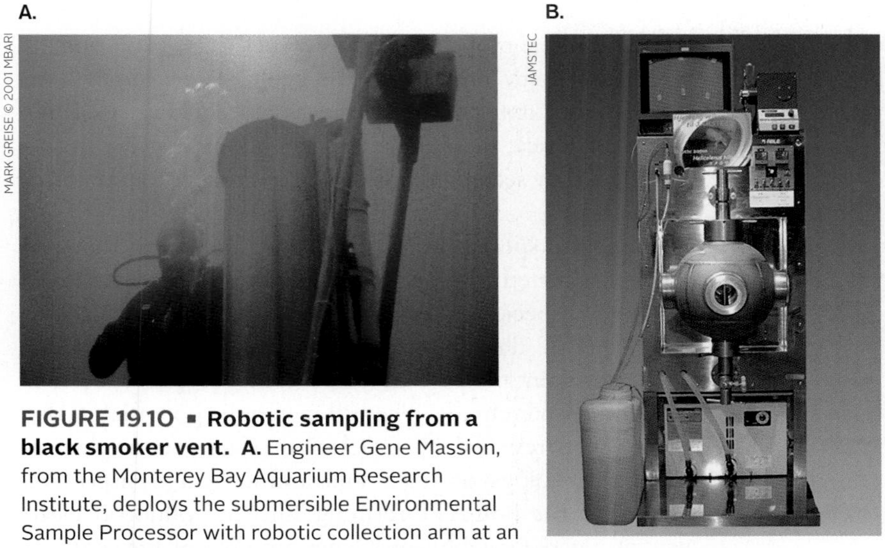
A.
B.
MARK GREISE © 2001 MBARI
JAMSTEC

FIGURE 19.10 ▪ Robotic sampling from a black smoker vent. A. Engineer Gene Massion, from the Monterey Bay Aquarium Research Institute, deploys the submersible Environmental Sample Processor with robotic collection arm at an ocean site off the coast of Maine. **B.** The Deep Aquarium, a pressurized device for cultivation of vent organisms. *Source:* A. Courtesy of Japan Agency for Marine-Earth Science and Technology; B. Courtesy of Monterey Bay Aquarium Research Institute.

connecting the two daughter cells while simultaneously pushing the two cells apart. In this fashion, the cell division process expands the cell network.

Note: Two genera of vent thermophiles have similar names but only distant genetic relatedness: *Pyrodictium abyssi,* a crenarchaeote; and *Pyrococcus abyssi,* a euryarchaeote.

The intricate cell network of *Pyrodictium abyssi* is an example of a single-species biofilm. Other forms of biofilms are found at hydrothermal vents. *Thermosphaera aggregans* forms colonies so tightly bound that they cannot be dissociated by protease treatment or sonication (**Fig. 19.13**). Multispecies biofilms of hyperthermophiles line the chimneys of black smoker vents.

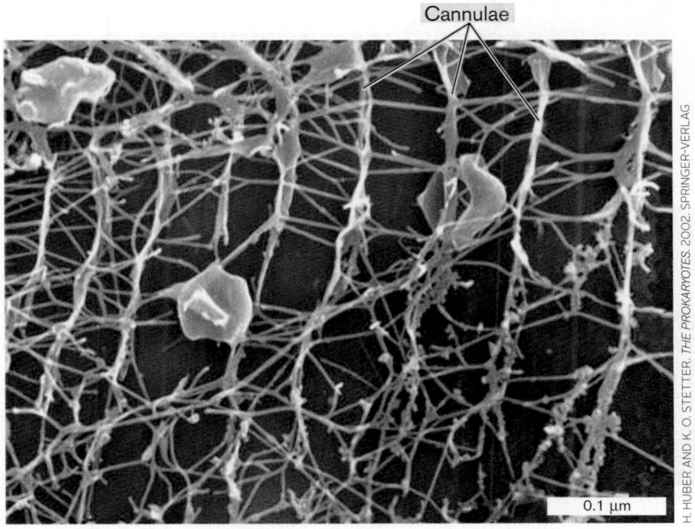
Cannulae
0.1 μm
H. HUBER AND K. O. STETTER. THE PROKARYOTES. 2002. SPRINGER-VERLAG

FIGURE 19.11 ▪ *Pyrodictium abyssi* growing as networks of cells linked by cannulae. SEM.

Thought Question

19.2 What might be the advantages of flagellar motility for a hyperthermophile living in a thermal spring or in a black smoker vent? What would be the advantages of growth in a biofilm?

Sulfolobales: High Temperature and Extreme Acid

Can some archaea grow in extreme acid, as well as high temperature? The crenarchaeote order **Sulfolobales** includes species that respire by oxidizing sulfur (instead of reducing it like *Desulfurococcus* does). These organisms, such as *Sulfolobus* species, grow at 80°C–90°C within hot springs and solfataras (volcanic vents that emit only gases). Ken Stedman and colleagues at Portland State University study *Sulfolobus solfataricus,* a species that grows at 80°C and pH 2

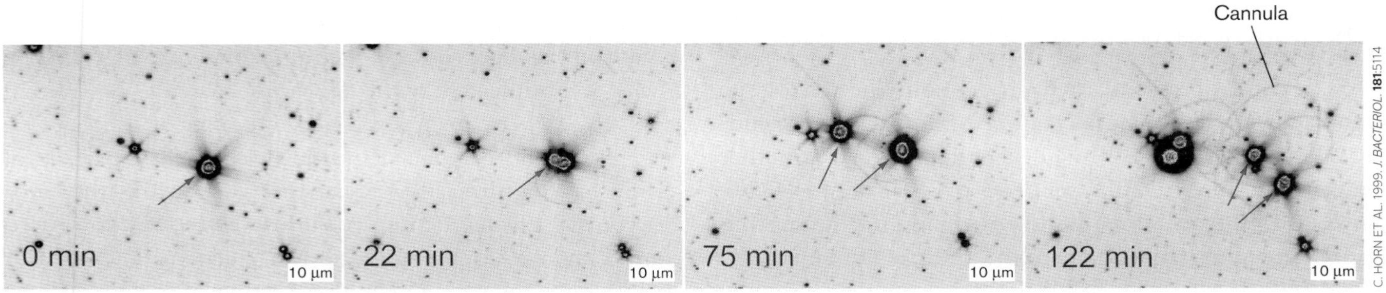
Cannula
0 min
22 min
75 min
122 min
10 μm
C. HORN ET AL. 1999. J. BACTERIOL. 181:5114

FIGURE 19.12 ▪ *Pyrodictium abyssi* undergoing cell division. Cells of *P. abyssi* generate new interconnecting cannulae as they divide.

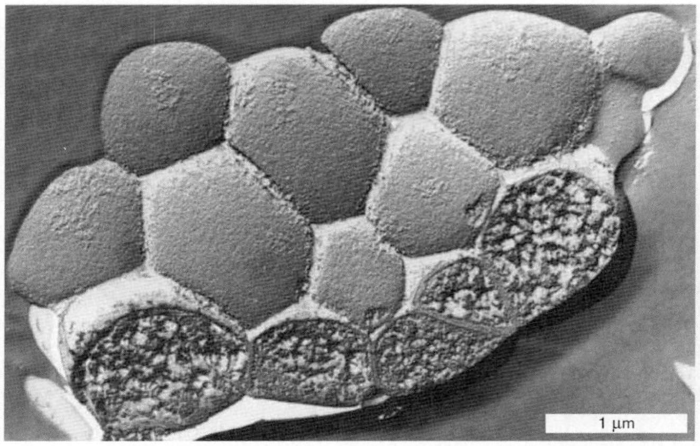
1 µm

H. HUBER AND K. O. STETTER. *THE PROKARYOTES*. 2002. SPRINGER-VERLAG.

FIGURE 19.13 ▪ Hyperthermal biofilms. Freeze-etched sample of *Thermosphaera aggregans* reveals a dense aggregate of cells (SEM).

A.

IMAGE COURTESY OF KEN STEDMAN

B.

1 µm

OLIVER MECKES/NICOLE OTTAWA

FIGURE 19.14 ▪ Isolating *Sulfolobus* at Yellowstone. **A.** A researcher holds a collecting tube at length to obtain samples from a steaming spring, Rabbit Creek, at Yellowstone National Park. **B.** *Sulfolobus* species grow at 80°C at pH 2–3.

(**Fig. 19.14**). *Sulfolobus* species oxidize organic compounds with oxygen, or they oxidize S^0 or H_2S to sulfuric acid:

$$2S^0 + 3O_2 + 2H_2O \longrightarrow 2H_2SO_4 \longrightarrow 4H^+ + 2SO_4^{\,2-}$$

As a result of sulfuric acid production, the pH of the organism's surroundings falls to pH 2–3, effectively excluding all but acidophiles. While other archaea grow at higher temperatures (up to 120°C) or lower pH (below pH 0), *Sulfolobus* is of interest as a "double extremophile," requiring both high temperature and extreme acidity simultaneously.

Sulfolobus cells have a membrane composed mainly of tetraethers with cyclopentane rings (see **Fig. 19.2E**). Tetraether membranes are commonly seen in acidophilic thermophiles, probably because they are exceptionally impermeable to protons. Remarkably, *Sulfolobus* has no cell wall, but only an S-layer of glycoprotein (**Fig. 19.15A**). Like all archaea, these organisms are nonpathogenic to animals, yet they secrete toxins deadly to competitor strains of *Sulfolobus*.

Sulfolobus species are not obligate autotrophs; they can also grow heterotrophically on sugars or amino acids. In fact, many species are easily cultured in tryptone broth at 80°C, pH 3. Their internal pH is typically pH 6.5; thus, they maintain more than three units of pH difference across their membrane. The full metabolic potential of this organism is revealed by annotation of its genome, which reveals homologs of sugar and amino acid transporters, as well as the non-ATP-forming Entner-Doudoroff pathway of glucose catabolism. Genes are present for enzymes to oxidize HS^- and $S_2O_3^{\,2-}$, as well as S^0. The main redox carrier for respiration appears to be ferredoxin (instead of NADH, which is relatively unstable at high temperature).

Thought Question

19.3 What problem with cell biochemistry is faced by acidophiles that conduct heterotrophic metabolism?

Because *Sulfolobus* species are readily cultured, their metabolism and ecology have been studied extensively, revealing unexpected features. Some species show multiple origins of replication of their DNA; for example, *S. solfataricus* and *S. acidocaldarius* each use three active replication origins. The presence of multiple origins is yet another feature of archaeal DNA management that resembles that of eukaryotes.

Another interesting discovery in *Sulfolobus* was that of archaeal viruses. *Sulfolobus* species are attacked by a number of viruses, including fuselloviruses (**Fig. 19.15B**). Fuselloviruses generally resemble bacteriophages in size and function, but their capsid is spindle-shaped, forming a cone at each end. Spindle-shaped capsids are common in archaeal viruses, but never seen in bacteriophages.

How does a virus infect an archaeon? The process of a viral infection has been observed in *S. solfataricus* (**Fig. 19.16**). Cells were infected with *Sulfolobus* turreted icosahedral virus (STIV), a virus isolated from a boiling acid hot spring in Yellowstone National Park. Each icosahedral particle of STIV has 12 turret-like projections, as shown in the cryo-EM image reconstruction in **Figure 19.16A**. Mature virions contain a double-stranded DNA genome coated with lipid within the capsid. The thin section of an infected cell (**Fig. 19.16B**) shows progeny virions packed into a hexagonal array, portions of which poke through the host cell's S-layer

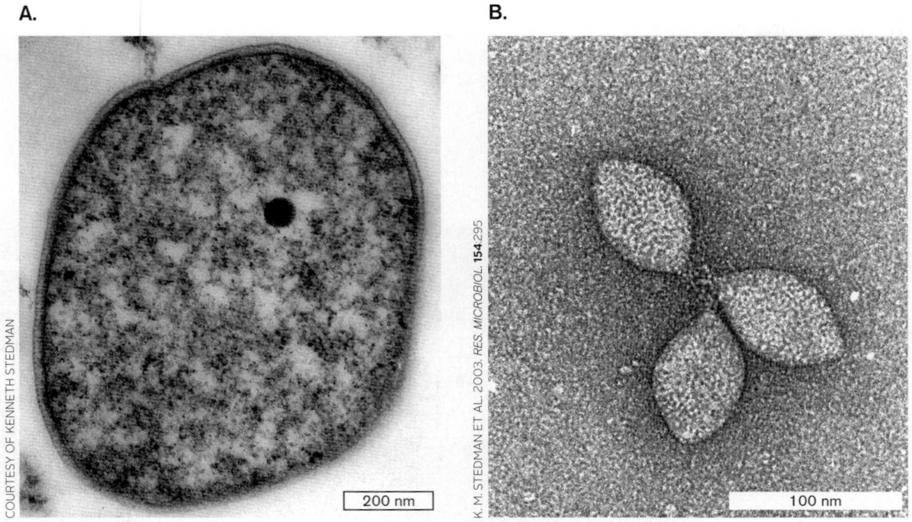
A.
B.
200 nm
100 nm

FIGURE 19.15 ■ *Sulfolobus*. **A.** *Sulfolobus* cell with thick S-layer. **B.** Fusellovirus, a double cone–shaped virus that infects *S. solfataricus*.

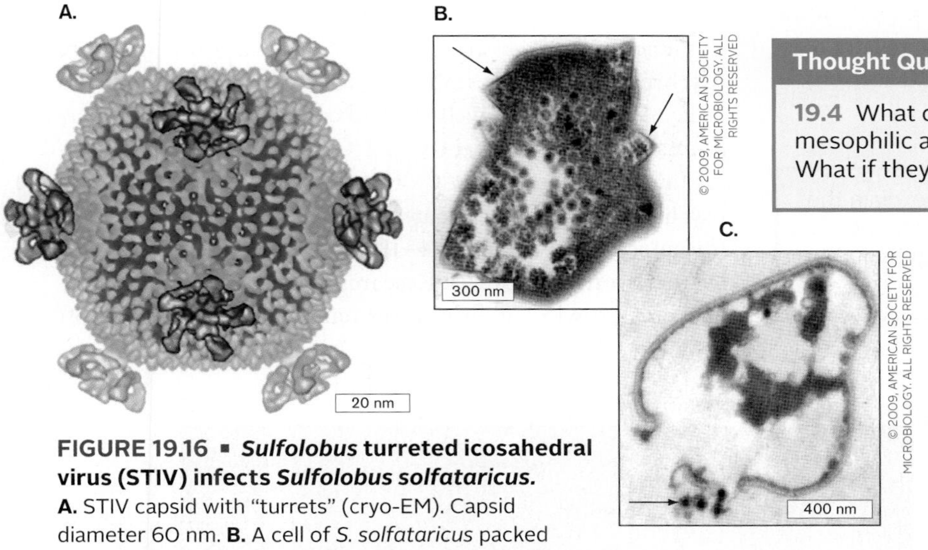
A.
B.
C.
20 nm
300 nm
400 nm

FIGURE 19.16 ■ *Sulfolobus* turreted icosahedral virus (STIV) infects *Sulfolobus solfataricus*. **A.** STIV capsid with "turrets" (cryo-EM). Capsid diameter 60 nm. **B.** A cell of *S. solfataricus* packed with hexagonal array of STIV particles. Arrows point to pyramidal bulges where virus arrays poke through a breach in the S-layer. **C.** Empty cell membrane and S-layer following lysis and viral release. Arrow points to released virus particles. *Source:* Part A from George Rice et al. 2004. *PNAS* **101**:7716. Parts B and C from Susan K. Brumfield et al. 2009. *J. Virol.* **83**:5964.

in pyramidal forms. After lysis (**Fig. 19.16C**), the S-layer complex is all that remains of the empty cell. The S-layer appears surprisingly intact, showing that it provides a sturdy cell covering perhaps comparable in strength to a cell wall.

This lytic cycle resembles lytic and fast-release cycles of bacterial and eukaryotic viruses, although the capsid "turrets" and the pyramidal bulges of the lysing cell are unique to archaea. The structure of a capsid protein component shows surprising homology to both bacterial and eukaryotic viral proteins. This finding supports the hypothesis that modern viruses derive from an ancient reproductive form predating the divergence of the three domains of life (discussed in Chapter 11).

Numerous other archaeal viruses have been discovered in high-temperature environments, including icosahedral, tailed, and filamentous forms. All known archaeal viruses have genomes consisting of double-stranded DNA, suggesting that only double-stranded DNA is stable enough for virus particles to persist at high temperature. In 2012, however, a possible archaeal RNA-dependent RNA transcriptase (discussed in Chapter 11) was identified from a Yellowstone hot spring. This finding suggests that there may be positive-strand RNA viruses that infect hyperthermophilic archaea.

Thought Question

19.4 What conclusions might be drawn if viruses of mesophilic archaea are found to have RNA genomes? What if they all have DNA genomes only?

Besides Sulfolobales, the class Thermoprotei includes other orders of thermoacidophiles. Caldisphaerales is an order of thermoacidophiles first isolated from a Philippine hot spring at Mount Makiling. Members of Caldisphaerales, such as *Caldisphaera*, typically grow at pH 3 up to 80°C. Unlike *Sulfolobus* species, *Caldisphaera* species are anaerobes or microaerophiles, tolerating only low concentrations of oxygen. They grow by fermentation or anaerobic respiration. Another order, Thermoproteales, includes hyperthermoacidophiles isolated from marine vents; thus, they survive extreme pressure, as well as heat and acid! *Thermoproteus* species grow at temperatures up to 97°C and at pH values lower than pH 3. Their rod-shaped cells are less than 0.3 μm in length—one of the smallest cell types known. Their metabolism is autotrophic, gaining energy by reducing sulfur with H_2 to H_2S.

Mesophiles and Psychrophiles

Because archaea were first isolated from extreme habitats, it came as a surprise when SSU rRNA probes revealed crenarchaeotes in moderate habitats throughout the biosphere. Many crenarchaeotes—perhaps the majority in

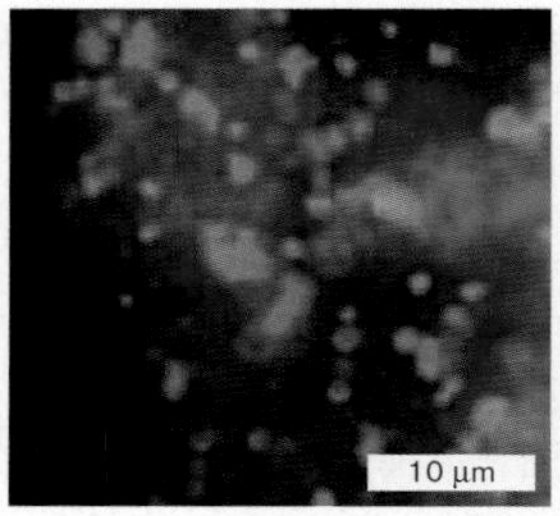
10 µm

FIGURE 19.17 ■ Crenarchaeotes growing on the surface of a tomato root. Crenarchaeotes were detected by fluorescence in situ hybridization (FISH, explained in Appendix 3). Red fluorescence arises from a fluorophore-DNA probe that hybridizes to an rRNA sequence specific to Crenarchaeota.

H. SIMON, OREGON HEALTH & SCIENCE UNIVERSITY, BEAVERTON, OR

nature—grow in water or soil, or in association with plants. For example, **Figure 19.17** shows crenarchaeotes growing on the surface of a tomato root. Others are marine mesophiles. The first marine mesophiles were found in 1992 by Edward DeLong and colleagues in samples from the Pacific Ocean. A survey in 2001 of marine archaea at the Hawaii Ocean Time-series Station found high numbers of crenarchaeotes (**Fig. 19.18**). The abundance of crenarchaeotes varied according to season and increased with depth, typically comprising 40% of the total microbial population at depths of 1,000 meters, where temperatures are cold. Some of the organisms identified as crenarchaeotes were later reassigned as thaumarchaeotes (discussed in the next section).

Thought Question

19.5 What hypothesis might you propose to explain the reason for the time and depth distribution of marine Crenarchaeota seen in **Figure 19.18**? How might you test your hypothesis?

RICARDO CAVICCHIOLI. 2006. *NAT. REV. MICROBIOL.* **4**:331

FIGURE 19.19 ■ Ace Lake, Antarctica. Psychrophilic archaea and other microbes grow at low temperatures, including the coldest habitats known.

Genetic surveys also reveal crenarchaeotes that are psychrophiles—species adapted to low temperatures. Ace Lake, Antarctica, is a well-studied cold habitat with temperatures in the range of 14°C–24°C and bottom anoxic layers never warmer than 2°C (**Fig. 19.19**). The lake supports psychrophilic crenarchaeotes, as well as euryarchaeotes (discussed in Sections 19.4–19.6). Alison Murray and colleagues from the Desert Research Institute found crenarchaeotes growing at even lower temperatures in sea ice off

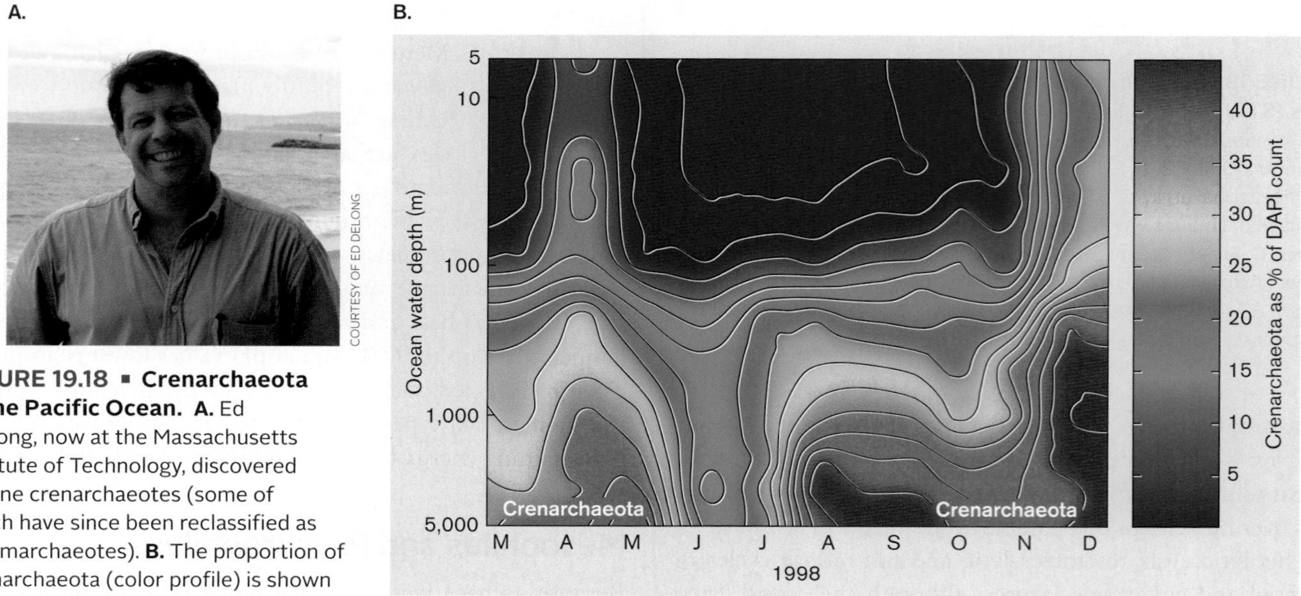
A.
B.
COURTESY OF ED DELONG
Ocean water depth (m)
5
10
100
1,000
5,000
Crenarchaeota
Crenarchaeota
M A M J J A S O N D
1998
40
35
30
25
20
15
10
5
Crenarchaeota as % of DAPI count

FIGURE 19.18 ■ Crenarchaeota in the Pacific Ocean. A. Ed DeLong, now at the Massachusetts Institute of Technology, discovered marine crenarchaeotes (some of which have since been reclassified as thaumarchaeotes). **B.** The proportion of Crenarchaeota (color profile) is shown as a function of depth and season, measured at the Hawaii Ocean Time-series Station. Crenarchaeotes were identified by FISH (explained in Appendix 3). A fluorescein-labeled DNA probe specific to Crenarchaeota was hybridized to a cellular rRNA sequence; DNA was detected using the DNA-binding fluorophore DAPI. *Source:* Part B from Markus Karner et al. 2001. *Nature* **409**:507.

A.
B.
C.
COURTESY OF ALISON MURRAY
©2003 ALISON MURRAY
COURTESY OF ALISON MURRAY

FIGURE 19.20 ■ Collecting psychrophiles in Antarctica. **A.** Alison Murray, from the Desert Research Institute, studies psychrophilic microbes in Antarctica. **B.** Microbes are collected off Bonaparte Point, Antarctica, from seawater samples down to 45 meters. **C.** Antarctic seawater samples are filtered to concentrate psychrophilic archaea and bacteria.

Antarctica (**Fig. 19.20**). Organisms were collected from ice and seawater at a temperature of −1.8°C; the water remains liquid below zero because of the high salt concentration.

Other crenarchaeotes are adapted to high pressure as well as cold. At the ocean floor, at temperatures of about 2°C, benthic strata show crenarchaeotes that include anaerobic heterotrophs, sulfate reducers, and anaerobic methanotrophs (methane oxidizers). Methanotrophs are particularly crucial in recycling the methane produced by seafloor methanogens (euryarchaeotes, discussed in Section 19.4).

To Summarize

- **Crenarchaeote species are found in a wide range of extreme and moderate habitats.**
- **Habitats for hyperthermophiles** include hot springs and submarine hydrothermal vents. Vent organisms are barophiles as well as thermophiles. Anaerobic hyperthermophilic acidophiles include Caldisphaerales and Thermoproteales.
- **Desulfurococcales includes diverse thermophiles.** Most are anaerobes that use sulfur to oxidize hydrogen or organic molecules.
- ***Pyrodictium* species show unusual networked cells.** Disk-shaped cells are interconnected by cytoplasmic bridges called cannulae.
- ***Sulfolobus* species are aerobic thermoacidophiles.** *Sulfolobus* species oxidize sulfur or H_2S to sulfuric acid, and they catabolize organic compounds.
- **Oceans, soil, plant roots, and animals** provide habitats for mesophilic and psychrophilic crenarchaeotes. Psychrophiles in marine sediment play a key role in recycling methane hydrates produced by methanogens.

19.3

Thaumarchaeota: Symbionts and Ammonia Oxidizers

The first mesophilic crenarchaeotes were identified by the sequence of their SSU rRNA. But more recent analysis combined the sequences of multiple genes, such as those encoding ribosomal proteins. This more finely tuned analysis showed that some of the organisms branched deeply from the rest. These deep-branching archaea were reclassified as a new clade, the Thaumarchaeota. Thaumarchaeotes synthesize a distinctive tetraether lipid called crenarchaeol, containing a six-membered cyclic ring (**Fig. 19.21**). Crenarchaeol was originally named for "Crenarchaota" but now appears to be found mainly in those organisms reclassified as Thaumarchaeota. Thaumarchaeota include intriguing symbionts of marine sponges, as well as ammonia oxidizers that perform a key role in cycling nitrogen.

Marine Sponge Symbionts

Some of the psychrophilic thaumarchaeotes found by DeLong and colleagues live as endosymbionts of marine

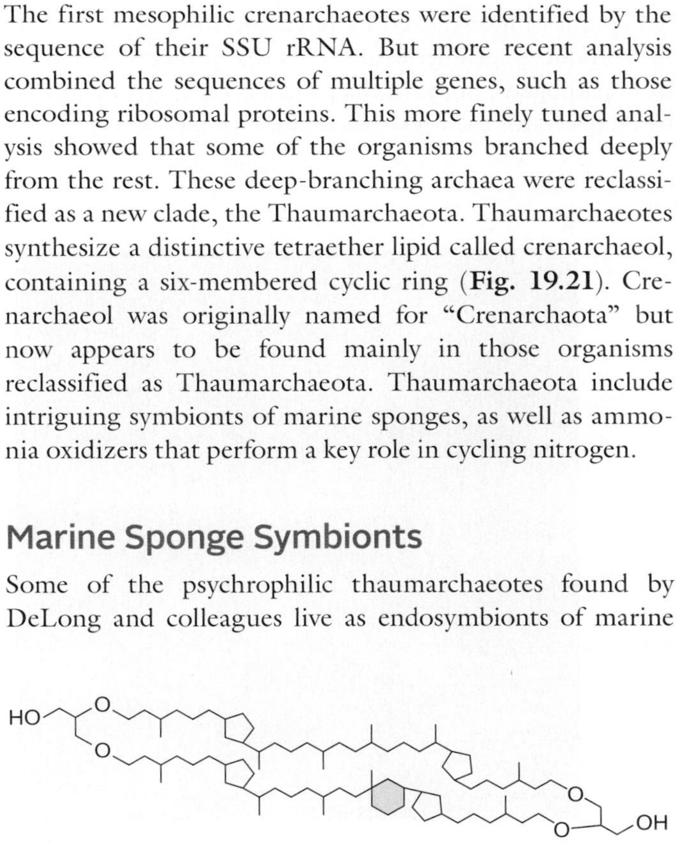
HO
O
O
O
O
OH

FIGURE 19.21 ■ Crenarchaeol: a biosignature for crenarchaeotes and thaumarchaeotes. Crenarchaeol is a diphytanylglycerol diether containing a six-membered cyclic ring.

A. B.

E. F. DELONG. 2003. *ASM NEWS* **69**(10)

PRESTON, MONTEREY BAY AQUARIUM RESEARCH INSTITUTE, MOSS LANDING, CA

FIGURE 19.22 ▪ Symbiosis between a thaumarchaeote and a sponge. **A.** *Cenarchaeum symbiosum* inhabits the sponge *Axinella mexicana.* **B.** Differential fluorescent staining of *C. symbiosum* present in sponge tissue reveals that archaeal DNA (stained with DAPI, green) is segregated from host cell ribosomes (FISH with an rRNA probe, red).

animals such as the sponge. The thaumarchaeote *Cenarchaeum symbiosum* inhabits the sponge *Axinella mexicana* (**Fig. 19.22A**). Like other marine thaumarchaeotes, *C. symbiosum* has yet to be grown in pure culture. But the presence of the microbes is shown by the fluorescence of a DNA probe that hybridizes to sequences specific to *C. symbiosum* (**Fig. 19.22B**). How the microbe benefits its sponge host is unknown, but the sponge and its endosymbionts can be cocultured in an aquarium for many years. A hypothesis investigated by researchers is that *C. symbiosum* produces antimicrobial agents that protect the sponge from pathogens. Some of the products of *C. symbiosum* are being tested for their pharmacological properties.

Ammonia-Oxidizing Thaumarchaeotes Cycle Global Nitrogen

From a global standpoint, the most significant thaumarchaeotes discovered are the ammonia oxidizing archaea (AOA) such as those of the order Nitrosopumilales. Ammonia-oxidizing archaea gain energy by aerobically oxidizing ammonia to nitrite:

$$2NH_3 + 3O_2 \longrightarrow 2NO_2^- + 2H_2O + 2H^+$$

This lithotrophic reaction yields redox energy, enabling the microbe to fix CO_2 for biomass (discussed in Chapter 14). The reaction plays a key role in the global nitrogen cycle, as

A. B.

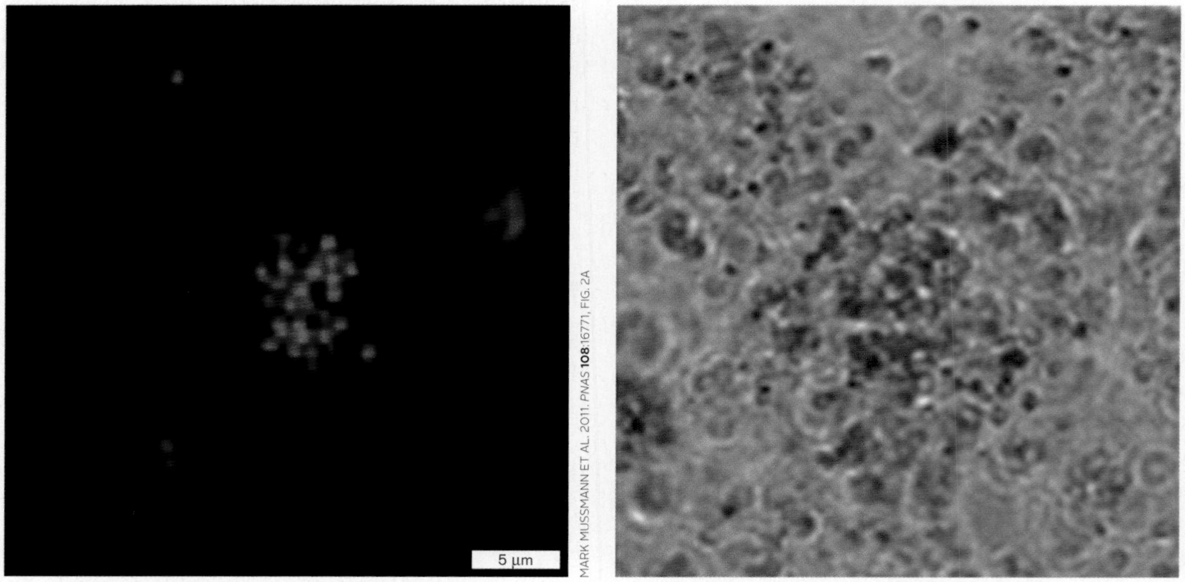
5 µm

MARK MUSSMANN ET AL. 2011. *PNAS* **108**:16771, FIG. 2A

MARK MUSSMANN ET AL. 2011. *PNAS* **108**:16771, FIG. 2B

FIGURE 19.23 ▪ Ammonia-oxidizing Thaumarchaeota in nitrogen-rich industrial sludge. **A.** FISH probe hybridizes with rRNA sequence specific to thaumarchaeotes. **B.** Light micrograph.

the first step of returning organic nitrogen to atmospheric N_2 (discussed in Chapters 21 and 22). It also provides a major source of nitrite for marine phytoplankton.

The first known ammonia-oxidizing thaumarchaeote, *Nitrosopumilus maritimus,* was isolated from a marine aquarium in 2005 by David Stahl and colleagues at the University of Washington, Seattle. In 2006, Ann Pearson and colleagues from Harvard University and Scripps Institution of Oceanography showed that AOA are widespread throughout the oceans. Since then, AOA have been found throughout marine and aquatic water, as well as soil. They appear in industrial waste sludge, revealed by fluorescent DNA probes (**Fig. 19.23**). While bacteria also oxidize ammonia, in many habitats the AOA are the dominant oxidizers.

Nitrosopumilus and related AOA of Marine Group 1 oxidize ammonia at extremely low concentrations, thus quickly removing a substance toxic to other members of the ecosystem. In some habitats, they perform most of the recycling of ammonia, such as the ammonia excreted by fish in aquaria. Ammonia-oxidizing thaumarchaeotes continue to be discovered in a growing range of habitats (see **Special Topic 19.1**).

Further metagenomic studies of soil and water reveal additional novel kinds of Thaumarchaeota, including thermophiles. Overall, the clade may turn out to be as diverse as the Crenarchaeota.

To Summarize

- **Thaumarchaeotes include symbionts of marine sponges.** Emerging thaumarchaeotes include mesophiles, psychrophiles, and thermophiles in many environments.
- **Ammonia-oxidizing archaea (AOA)** contribute to global nitrogen cycling in water and soil.

19.4

Methanogens

The third major branch of known archaea is Euryarchaeota, the "broad-ranging archaea" (**Table 19.2**). The euryarchaeotes are dominated by the **methanogens**, a vast array of species, all of which derive energy through reactions producing methane. Biogenic methane formation has actually been observed since 1776, when the Italian priest Carlo Campi and physicist Alessandro Volta (1745–1827) investigated bubbles of "combustible air" from a wetland lake. Volta wrote:

> Being in a little boat on Lake Maggiore, and passing close to an area covered with reeds, I started to poke and stir the bottom with my cane. So much air emerged that I decided to collect a quantity in a large glass container. . . . This air burns with a beautiful blue flame.

Volta noted that methane arose from wetlands containing water-saturated decaying plant material. In 1882, the German medical researcher Hermann von Tappeiner (1847–1927) combined plant materials with ruminant stomach contents (a source of methanogens) and showed that both components were essential to produce methane. During the late nineteenth century, in England, methane was collected from manure and sewage and used as a fuel for street lamps.

Different Paths to Methane

Different species generate methane from different substrates, such as CO_2 and H_2 or small organic compounds, generally fermentation products of bacteria. Each pathway of methanogenesis generates a small free energy change ($\Delta G^{\circ\prime}$) that is just enough to drive processes of carbon fixation and biosynthesis. All types of methanogenesis are poisoned by molecular oxygen and therefore require extreme anaerobiosis. Major substrates and reactions include:

Carbon dioxide: $CO_2 + 4H_2 \longrightarrow CH_4 + 2H_2O$
Formic acid: $4CHOOH \longrightarrow CH_4 + 3CO_2 + 2H_2O$
Acetic acid: $CH_3COOH \longrightarrow CH_4 + CO_2$
Methanol: $4CH_3OH \longrightarrow 3CH_4 + CO_2 + 2H_2O$
Methylamine: $4CH_3NH_2 + 2H_2O \longrightarrow 3CH_4 + CO_2 + 4NH_3$
Dimethyl sulfide: $2(CH_3)_2S + 2H_2O \longrightarrow 3CH_4 + CO_2 + 2H_2S$

In CO_2 reduction, the most common form of methanogenesis, carbon dioxide plus molecular hydrogen combine to form water and methane. CO_2-reducing methanogens are autotrophs, growing solely on CO_2 and H_2 with a source of nitrogen and other minerals. Methanogenesis from other carbon sources, such as formate, acetate, or methanol, is heterotrophic, and generates CO_2 as a product in addition to methane. The nitrogen-containing substrate methylamine generates ammonia in addition to methane and CO_2, whereas the sulfur-containing substrate dimethyl sulfide generates hydrogen sulfide.

Special Topic 19.1: Eating Ammonia: Thaumarchaeotes

We commonly think of ammonia as a cleaning agent with an off-putting odor. In marine environments, excreted ammonia can build up to high levels, requiring oxidation by archaea and bacteria. The thaumarchaeotes thus help balance the ecology of intertidal pools and beaches. But most ammonia-oxidizing archaea (AOA) are inhibited by levels of ammonia above 2 mM. Could some AOA oxidize ammonia at higher concentrations, preventing buildup of levels toxic for animals and plants?

Tatsunori Nakagawa and colleagues at Nihon University, Japan, used enrichment culture (described in Chapter 4) to hunt for new kinds of AOA in the seafloor of Tanoura Bay. Using sterile tubes, they obtained samples of sand from the seafloor. The samples were serially diluted in medium containing ammonium sulfate and adjusted to pH 8, a pH high enough for significant deprotonation of ammonium ion (NH_4^+) to ammonia (NH_3). During the enrichment culture, ammonia was progressively consumed and converted to nitrite (NO_2^-) (**Fig. 1**). In the culture, the gene encoding ammonia oxidase (*amoA*) was identified by PCR amplification. The ammonia-oxidizing archaea (AOA) were identified by fluorescence in situ hybridization (FISH) using a fluorophore attached to DNA that hybridizes to rRNA of *Nitrosopumilus* species (**Fig. 2**). The same microbial sample showed very little hybridization to a bacterial probe (**Fig. 2C**). Thus, the culture was highly enriched for AOA, although bacteria could never be removed completely; the researchers propose that AOA require an unknown product of cocultured bacteria.

A culture was obtained consisting of 99% of a single archaeal isolate. The isolate's SSU rRNA sequence showed close relatedness to *Nitrosopumilus maritimus,* as well as more distant relatedness to other thaumarchaeotes and crenarchaeotes (**Fig. 3**). The new isolate was designated NM25, a candidate species of *Nitrosopumilus*. But, despite the relatedness to other AOA organisms, NM25 shows an exceptional tolerance to ammonia. While other cultured AOA are inhibited by 2-mM ammonia, isolate NM25 grows at ammonia concentrations as high as 25 mM. Thus, NM25 may play an especially important role by oxidizing ammonia at levels too toxic for other organisms. Once the ammonia concentration is decreased, other environmental archaea and bacteria may remove the traces that remain.

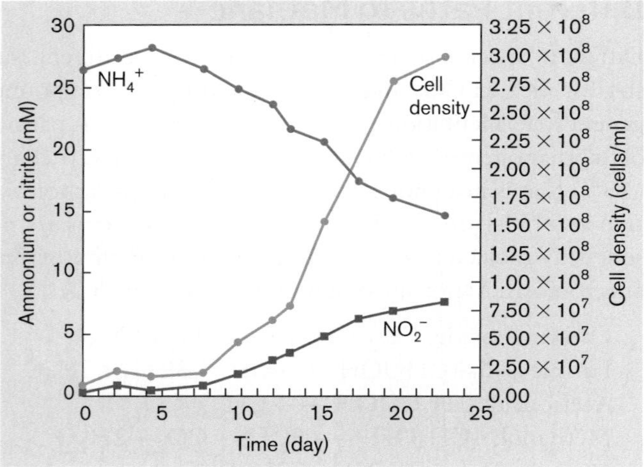

30
25
20
15
10
5
0
NH4+
Cell density
NO2−
0
5
10
15
20
25
Time (day)
Ammonium or nitrite (mM)
3.25 × 10^8
3.00 × 10^8
2.75 × 10^8
2.50 × 10^8
2.25 × 10^8
2.00 × 10^8
1.75 × 10^8
1.50 × 10^8
1.25 × 10^8
1.00 × 10^8
7.50 × 10^7
5.00 × 10^7
2.50 × 10^7
0.00
Cell density (cells/ml)

FIGURE 1 ■ Enrichment culture for ammonia oxidizers. Ammonia-oxidizing microbes from eelgrass seafloor samples convert ammonia to nitrite.

RESEARCH QUESTION

Could you design an enrichment culture for ammonia-oxidizing microbes from the soil where you live? How would your experiment be modified from that of Nakagawa?

Matsutani, Naoki, Tatsunori Nakagawa, Kyoko Nakamura, Reiji Takahashi, Kiyoshi Yoshihara, et al. 2011. Enrichment of a novel marine ammonia-oxidizing archaeon obtained from sand of an eelgrass zone. *Microbes and Environments* **26**:23–29.

Note that only a narrow range of substrates supports methanogenesis. For unknown reasons, most methanogens lack the vast array of alternative pathways found in soil bacteria such as *Rhodopseudomonas* or *Streptomyces*. Thus, methanogens generally require close association with bacterial partners to provide their substrates—a relationship called **syntrophy** (discussed in Chapter 14). Syntrophic methanogens usually grow in habitats with minimal resource flux, where hydrogen and carbon dioxide gases can be trapped for their use. Removal of these gases then enhances bacterial metabolism.

Many kinds of methanogens can be cultured in the laboratory. Their culture poses special challenges because the reactions of methanogenesis are halted by oxygen. Thus, most methanogens are strict anaerobes, although a few that tolerate oxygen have been found. To exclude oxygen, all growth and manipulation of cultures are performed in an anaerobic chamber. Furthermore, methanogens that use

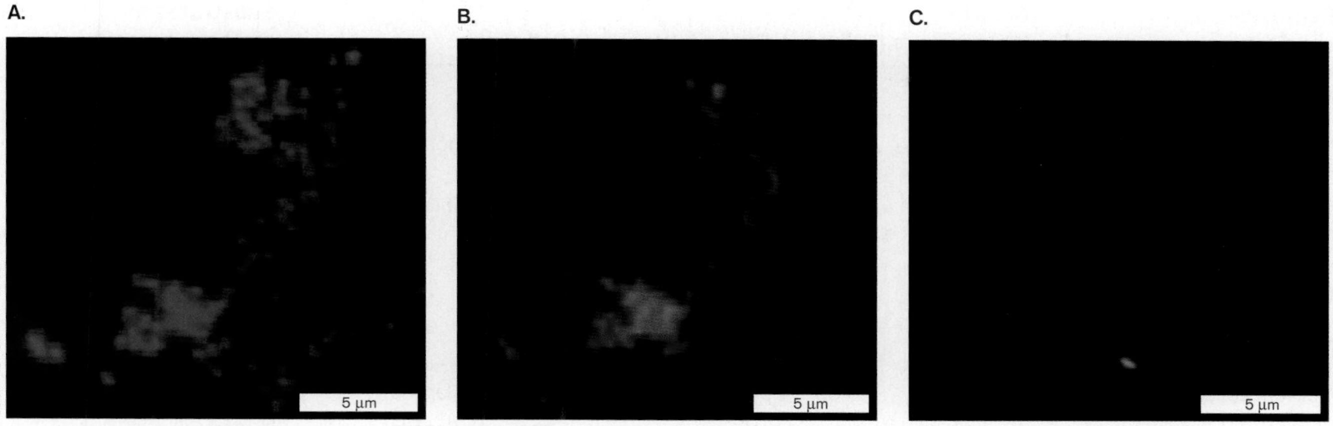
A.
B.
C.
5 µm
5 µm
5 µm

COURTESY OF PRIYA DASSARMA

FIGURE 2 ■ Ammonia-oxidizing archaea detected in enrichment culture. A. Nucleoid DNA of all archaea and bacteria is revealed by DAPI fluorescent stain. **B.** AOA archaea revealed by FISH using a DNA probe encoding a gene for an ammonia oxidation enzyme. **C.** Bacteria revealed by FISH using a probe specific to bacterial genomes.

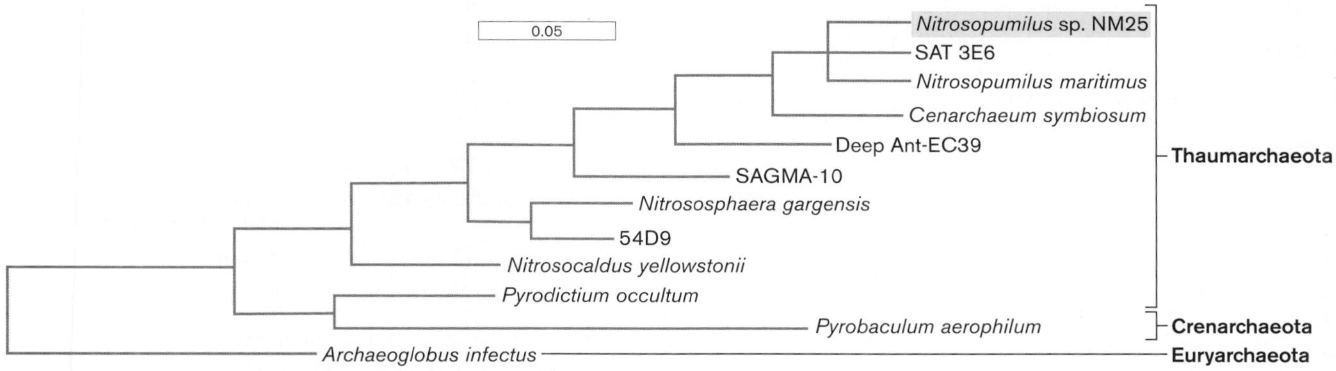
0.05
Nitrosopumilus sp. NM25
SAT 3E6
Nitrosopumilus maritimus
Cenarchaeum symbiosum
Deep Ant-EC39
SAGMA-10
Nitrososphaera gargensis
54D9
Nitrosocaldus yellowstonii
Pyrodictium occultum
Pyrobaculum aerophilum
Archaeoglobus infectus
Thaumarchaeota
Crenarchaeota
Euryarchaeota

FIGURE 3 ■ Phylogeny of ammonia-oxidizing thaumarchaeote NM25. The phylogenetic tree is based on comparison of SSU rRNA sequences. Strain NM25 shows close relatedness to *Nitrosopumilus* species, and more distant relatedness to other thaumarchaeotes, to crenarchaeotes, and to a euryarchaeote.

CO_2 and H_2 as substrates must receive a steady supply of these gases. Even more challenging is the culture of hyperthermophilic methanogens, such as *Methanopyrus,* a vent hyperthermophile that grows at scorching temperatures up to 122°C. For these organisms, we must maintain both high temperature and high pressure.

Thermophiles, mesophiles, and psychrophiles. Methanogens include four classes, of which five major orders are shown in **Table 19.4**. The amount of divergence within an order can be nearly as great as that between orders. Particularly striking is the wide range of growth temperatures among closely related species. Thermophiles and even hyperthermophiles branch from closely related mesophiles; for example, the order Methanobacteriales includes *Methanobrevibacter ruminantium* (37°C–39°C), *Methanobacterium thermoautotrophicum* (50°C–75°C), and *Methanothermus fervidus* (60°C–97°C).

TABLE 19.4

Methanogens.

Representative species	Growth temperature (°C)	Growth pH	Cell shape	Substrates for methanogenesis
Methanobacteriales				
Methanobrevibacter ruminantium	37–39°	pH 6–9	Chains of short rods	H_2 and CO_2
Methanobacterium thermoautotrophicum	50–75°	pH 7–8	Filaments of long rods	H_2 and CO_2, formate
Methanothermus fervidus	60–97°	pH 6–7	Rods	H_2 and CO_2
Methanococcales				
Methanococcus vannielii	20–40°	pH 7–9	Cocci with flagella	H_2 and CO_2
Methanocaldococcus jannaschii	48–94°	pH 6–7	Cocci with flagella	H_2 and CO_2
Methanomicrobiales				
Methanoculleus olentangii	30–50°	pH 6–8	Cocci	Acetate, complex nutrients
Methanospirillum hungatei	20–45°	pH 6–7	Spirilla	Acetate
Methanopyrales				
Methanopyrus kandleri	84–122°	pH 6–8	Long rods (2–14 μm)	H_2 and CO_2
Methanosarcinales				
Methanosarcina barkeri	20–50°	pH 5–7	Aggregates of cocci	H_2 and CO_2, methanol, methylamine, acetate
Methanosaeta concilii	10–45°	pH 6–8	Filaments of rods	Acetate
Methanohalophilus zhilinae	45°	pH 8–10	Cocci	Methanol, methylamine, dimethyl sulfide

Source: Harald Huber and Karl O. Stetter. 2002. *The Prokaryotes*, Springer-Verlag.

Methanogens Show Diverse Cell Forms

Despite their metabolic similarity, methanogens display an astonishing diversity of form, perhaps as diverse as the entire domain of bacteria (**Fig. 19.24**). For example, *Methanocaldococcus jannaschii* cells grow as cocci with numerous flagella attached to one side, while *Methanosarcina mazei* forms peach-shaped cocci lacking flagella. *Methanothermus fervidus* cells are short, fat rods without flagella, and *Methanobacterium thermoautotrophicum* grows as elongate rods reminiscent of the bacterial genus *Bacillus*. Still others, such as *Methanospirillum hungatei,* form wide spirals. The morphological diversity of methanogens may be explained in part by their rigid cell walls, which can maintain a distinctive shape.

The composition of methanogen cell walls is much more diverse than that of bacteria. *Methanobacterium* species have a cell wall composed of pseudopeptidoglycan, a structure in which chains of alternating amino sugars are linked by peptide cross-bridges analogous to those of peptidoglycan. By contrast, *Methanosarcina* species have a cell wall composed of sulfated polysaccharides. The genera *Methanomicrobium* and *Methanococcus* have protein cell walls.

Filamentous methanogens form chains of large cells similar to those of filamentous cyanobacteria. Filamentous methanogens, such as *Methanosaeta,* perform a key function in the treatment of sewage waste (**Fig. 19.25**). In waste treatment, the raw sewage first undergoes aerobic

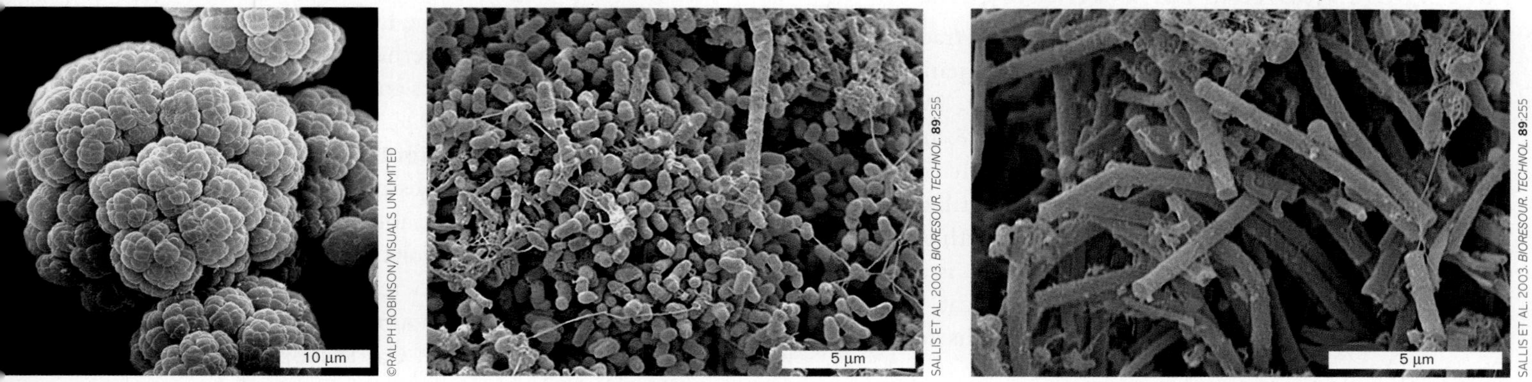
A. *Methanosarcina mazei*
10 µm
©RALPH ROBINSON/VISUALS UNLIMITED
B. *Methanothermus fervidus*
5 µm
SALLIS ET AL. 2003. *BIORESOUR. TECHNOL.* **89**:255
C. *Methanobacterium thermoautotrophicum*
5 µm
SALLIS ET AL. 2003. *BIORESOUR. TECHNOL.* **89**:255

FIGURE 19.24 ■ Methanogens show a wide range of shapes. A. *Methanosarcina mazei,* a lobed coccus form lacking flagella (SEM). **B.** *Methanothermus fervidus,* a short bacillus (SEM). **C.** *Methanobacterium thermoautotrophicum,* an elongate bacillus (SEM).

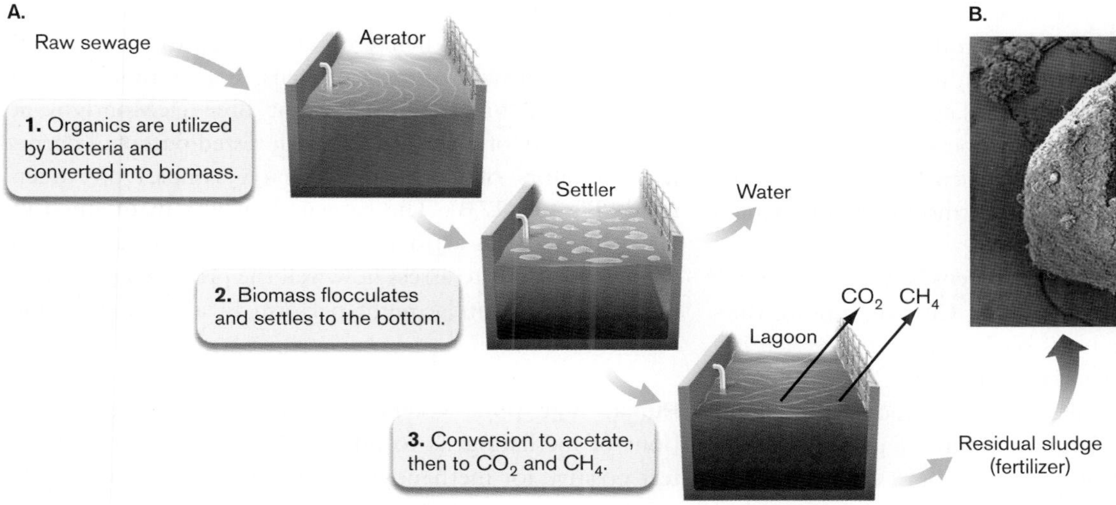
A.
Raw sewage
Aerator
1. Organics are utilized by bacteria and converted into biomass.
Settler
Water
2. Biomass flocculates and settles to the bottom.
CO_2 CH_4
Lagoon
3. Conversion to acetate, then to CO_2 and CH_4.
Residual sludge (fertilizer)

B.
200 µm
SALLIS ET AL. 2003. *BIORESOUR. TECHNOL.* **89**:255

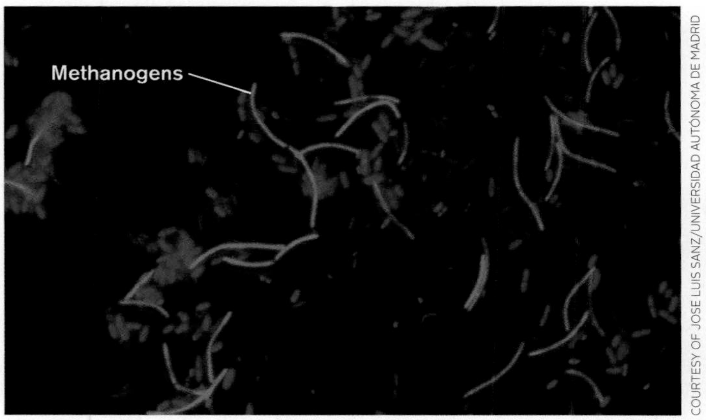
C.
Methanogens
COURTESY OF JOSE LUIS SANZ/UNIVERSIDAD AUTÓNOMA DE MADRID

FIGURE 19.25 ■ Filamentous methanogens bind bacterial communities in waste treatment. A. Raw sewage is aerated and decomposed by bacteria, followed by anaerobic incubation. Under anaerobiosis, the bacterial waste products are converted to methane and CO_2. The bacteria are packed together by filamentous methanogens to form sludge. **B.** Sludge forms granules packed with bacteria and methanogens. **C.** Filamentous methanogens entangle bacteria, forming "flocs" that settle in the wastewater tank. Methanogens fluoresce red with a DNA probe for Methanomicrobiales (7–10 µm in length), while other wastewater microbes fluoresce blue (DAPI stain).

respiration by bacteria, converting the organic materials to small molecules such as CO_2 and acetate, followed by anaerobic digestion of the remainder. The bacteria performing anaerobic decomposition become trapped in filaments of *Methanosaeta,* which convert bacterial fermentation products such as acetate into methane and CO_2. The methanogenic filaments serve a key function by trapping bacteria into granules that settle out from the liquid.

Anaerobic Habitats for Methanogens

Methanogens grow in soil, in animal digestive tracts, and in marine floor sediment. A major methanogenic environment is the anaerobic soil of wetlands (**Fig. 19.26A**). The wetlands that generate the most methane are typically disturbed or artificial wetlands, particularly rice paddies, which contain high levels of added fertilizer that bacteria convert to the substrates used by methanogens. From the standpoint of the organism, methane is an incidental by-product, but this product has great significance for our biosphere as a greenhouse gas (discussed in Chapter 22).

Methanogenesis in soil and landfills. Chinese farmers have found that one way to decrease methane production is to drain the soils used for rice production. Changsheng Li and colleagues at the University of New Hampshire showed in experimental plots that drained soil produces less methane. The drained soil receives more oxygen, which blocks methanogenesis. Draining soil also stimulates rice root development and accelerates decomposition of organic matter in the soil to release nitrogen. Sufficiently dry rice paddy soil can actually become a sink for atmospheric methane, thanks to methane-oxidizing bacteria.

Another major source of methanogens is landfills. Landfills such as those outside New York City are among the largest human-made structures on Earth. They are rich in organic wastes, which bacteria ferment to CO_2, H_2, and short-chain organic molecules that methanogens convert to methane. As a result, large amounts of gas can build up and spontaneously combust, causing explosions. To avoid explosions, the methane needs to be piped out. In some cases, the gas can be collected and used to generate electricity.

As global temperatures increase, the largest release of methane may come from thawing permafrost and retreating glaciers (**Fig. 19.26B**). Thawing sediments release large quantities of methane from accelerated methanogenesis. In some cases the methane bubbles up so fast that it melts holes in the ice sheets covering arctic lakes.

Digestive methanogenic symbionts. Numerous methanogens also grow within the digestive fermentation chambers of animals such as termites and cattle (**Fig. 19.27A**) (discussed in Chapter 21). Termites have to aerate their mounds continually in order to remove methane; when rainfall temporarily clogs the mound, the mound can be ignited by lightning and explode. Cattle support methanogenesis within their rumen and reticulum; a common veterinary trick is to insert a cannula and light a flame on the escaping gas. Bovine methanogenesis diverts carbon from meat production, and it makes a significant contribution to global methane.

Research on rumen microbiology focuses on attempts to suppress growth of methanogens. For example, Steven Ragsdale and his graduate student Bree DeMontigny at the University of Nebraska–Lincoln tested potential chemical inhibitors of methanogenesis on samples of rumen fluid (**Fig. 19.27B**). The rumen fluid was maintained in vials that serve as artificial rumens. The artificial rumens allow researchers to assess how well the proposed methane-blocking compounds might perform in an actual bovine rumen.

Human digestion. Recent research reveals key contributions of methanogens to human digestion. Most humans test positive for methanogens, primarily the species *Methanobrevibacter smithii* and *Methanosphaera stadtmanae*. These archaea constitute about 10% of gut anaerobes. *M. smithii* increases the efficiency of digestion by consuming excess reduced products (formate and H_2) from bacterial mixed-acid fermentation (discussed in Chapter 13). High levels of H_2 inhibit bacterial NADH dehydrogenases and thus decrease the proton potential for ATP production. *Methanosphaera stadtmanae* consumes methanol, a potentially toxic by-product of the bacterial degradation of pectin, a major fruit polysaccharide. Thus, methanogens may enhance the growth of human enteric bacteria.

If methanogens influence the fermentation efficiency of gut flora, do they affect the caloric content we obtain from food? In 2006, Buck Samuel and Jeffrey Gordon at Washington University in St. Louis conducted an experiment to test the

A. B.

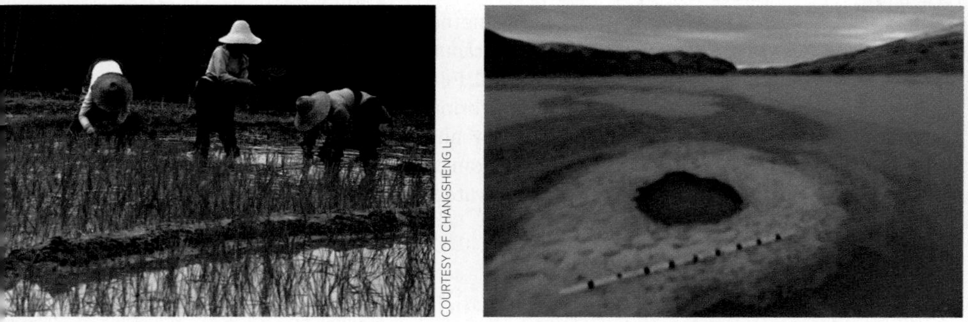

COURTESY OF CHANGSHENG LI

KATEY M. WALTER ANTHONY. 10.1038/NGEO1480. *NATURE GEOSCIENCE*, SUPPLEMENTARY FIGURE 2.

FIGURE 19.26 ▪ Terrestrial habitats for methanogens. A. Chinese farmers plant rice in waterlogged soil, an ideal anaerobic habitat for methanogens. **B.** Methane gas bubbles up from a lake at a retreating glacier in West Greenland.

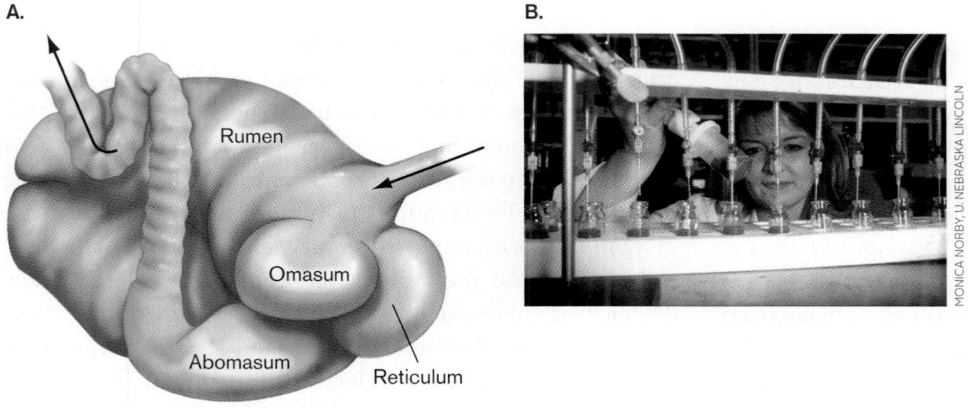
A.
B.
Rumen
Omasum
Abomasum
Reticulum
MONICA NORBY, U. NEBRASKA LINCOLN

FIGURE 19.27 ▪ Methane production in the bovine rumen. A. Methanogens grow with the community of fermentative bacteria in the rumen and reticulum of ruminating animals. **B.** Bree DeMontigny, a graduate student in animal science who worked with Stephen Ragsdale at the University of Nebraska–Lincoln, is shown here pipetting rumen fluid samples into vials that serve as "artificial rumens." The vials were then incubated and assayed for the amount of methane formed. Ragsdale is now at the University of Michigan–Ann Arbor.

effect of methanogenesis on energy balance in gnotobiotic (germ-free) mice. The gnotobiotic mice were colonized with *Bacteroides thetaiotaomicron,* an anaerobic bacterium that ferments complex polysaccharides from the plants we consume. The mice were colonized in the presence or absence of *Methanobrevibacter smithii,* which consumes H_2 through methanogenesis. Mice colonized with both *B. thetaiotaomicron* and *M. smithii* were found to make more fat after consuming the same quantity of food than those colonized solely with *B. thetaiotaomicron.* This result contrasts with the case of the bovine rumen, in which methanogenesis draws off so much carbon that it decreases fermentation efficiency. The results suggest interesting hypotheses for future study of weight control in humans.

COURTESY OF THE U.S. GEOLOGICAL SURVEY

FIGURE 19.28 ▪ Methane hydrate. Methane gas hydrates, a crystalline form of methane and water produced by benthic methanogens, are a potential source of fuel.

Methane hydrates on the seafloor. Psychrophilic marine methanogens grow at or beneath the seafloor. These seabed methanogens generate large volumes of methane that seep up slowly from the sediment. Under the great pressure of the deep ocean, the methane becomes trapped as **methane gas hydrates**, which are crystalline materials in which methane molecules are surrounded by a cage of water molecules (**Fig. 19.28**). The methane hydrates accumulate in vast quantities. Methane hydrates are of interest as a potential source of natural gas. But if a large part of this methane were to be released, it could greatly accelerate global warming.

Fortunately, much of the methane produced by seafloor methanogens is oxidized to CO_2 by **anaerobic methane oxidizers (ANME)** of the groups ANME-1, -2, and -3. The ANME euryarchaeotes oxidize methane in syntrophy with sulfur-reducing bacteria such as *Desulfurococcus* (discussed in Chapter 18). The bacteria reduce sulfate (abundant in seawater) to sulfide. Coupled together, the reactions of methane oxidation and sulfate reduction have the negative value of ΔG needed to drive metabolism for both kinds of organisms.

Yet another group of euryarchaeotes actually reverses part of methanogenesis. *Archaeoglobus fulgidus,* a deep-sea vent hyperthermophile, oxidizes methyl groups back to CO_2 using the methanogenic cofactor methanopterin (MPT) (see Section 19.6). The methyl groups are oxidized by reduction of sulfate. *Archaeoglobus* is the only archaeon known to reduce sulfate without a bacterial partner.

Biochemistry of Methanogenesis

Methanogenesis is extremely important for our environment, as the process releases a potent greenhouse gas. Thus, much effort goes into better understanding how methanogenesis works. Knowledge of methanogenic biochemistry may help us develop controls, such as methanogen-specific antibiotics to minimize methane output from cattle.

Cofactors for methanogenesis. The process of methanogenesis uses a series of specific cofactors to carry each carbon from CO_2 (or other substrates) as it becomes progressively reduced by hydrogen. The hydrogen atoms also require redox carriers. Most of the cofactors are unique to methanogens, although the general structural types resemble those of redox cofactors we saw in Chapter 14 (**Fig. 19.29**). For example, cofactor F_{420} is a heteroaromatic molecule (containing nitrogens in the aromatic rings) that undergoes a redox transition by acquiring or releasing two hydrogens—a transition similar to that of the nicotinamide ring of NADH.

Methanogenesis from CO_2. The formation of methane from CO_2 and H_2 (**Fig. 19.30**) is technically a form of anaerobic respiration in which H_2 is the electron donor and CO_2 is the terminal electron acceptor (discussed in Chapter 14). The process fixes CO_2 onto the cofactor methanofuran (MFR) and then passes the carbon stepwise from one cofactor to the next, each time losing an oxygen to form water or gaining a hydrogen carried by another cofactor.

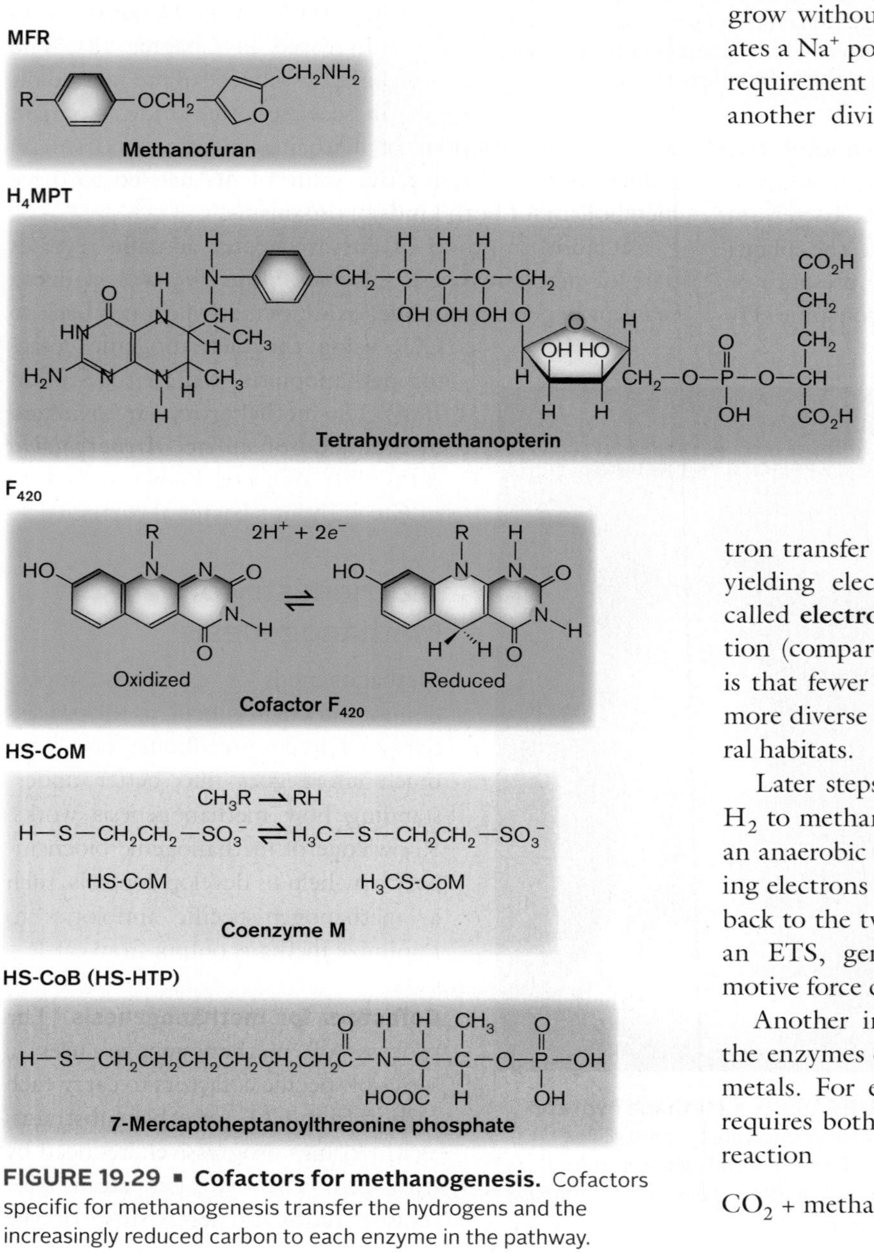
MFR
Methanofuran
H_4MPT
Tetrahydromethanopterin
F_{420}
$2H^+ + 2e^-$
Oxidized
Reduced
Cofactor F_{420}
HS-CoM
HS-CoM
H_3CS-CoM
Coenzyme M
HS-CoB (HS-HTP)
7-Mercaptoheptanoylthreonine phosphate

FIGURE 19.29 ▪ Cofactors for methanogenesis. Cofactors specific for methanogenesis transfer the hydrogens and the increasingly reduced carbon to each enzyme in the pathway.

The first step in the conversion of CO_2 and H_2 to methane is fixing the carbon onto methanofuran (see **Fig. 19.30**). This requires placement of protons onto the oxygen to form water. The mechanism of methanogenesis depends on the available concentration of H_2. High-H_2 environments, such as oil wells and sewage sludge, favor genera such as *Methanosarcina* (**Fig. 19.30A**). The high H_2 concentration allows electron donation to CO_2 from an electron transport system (ETS). The ETS may provide energy from a sodium potential (ΔNa^+). Most methanogens require Na^+ for growth, unlike bacteria, many of which can grow without sodium. Methanogenesis ultimately generates a Na^+ potential that drives ATP synthesis. The sodium requirement of methanogens is something they share with another division of Euryarchaeota, the halophiles (discussed in Section 19.5).

Other methanogens such as *Methanobacterium* and *Methanococcus* use much lower concentrations of H_2 (<1/10,000 atm) common in soil or water. At lower concentrations of substrate, the free energy (ΔG) of reaction becomes less favorable (discussed in Chapters 13 and 14). So the methanogen needs to couple CO_2 reduction to an energy-spending reaction, the reduction of CoM-S-S-CoB (**Fig. 19.30B**). This coupling of an energy-spending electron transfer (CO_2 reduction via ferredoxin) to an energy-yielding electron transfer (CoM-S-S-CoB reduction) is called **electron bifurcation**. The cost of electron bifurcation (compared to the ETS used with high H_2, panel A) is that fewer ATPs can be made; but it allows growth of more diverse methanogens, in a much wider range of natural habitats.

Later steps in methanogenesis reduce the carbon with H_2 to methane. In the final step, CoM-S-S-CoB serves as an anaerobic terminal electron acceptor for an ETS accepting electrons from H_2. Overall, H_2 reduces CoM-S-S-CoB back to the two cofactors HS-CoM and HS-CoB through an ETS, generating proton motive force. The proton motive force drives ATP synthase.

Another important feature of methanogenesis is that the enzymes catalyzing each step require several transition metals. For example, the hydrogenase that reduces F_{420} requires both nickel and iron. The enzyme catalyzing the reaction

$$CO_2 + \text{methanofuran} + 3H \longrightarrow \text{CHO-methanofuran} + H_2O$$

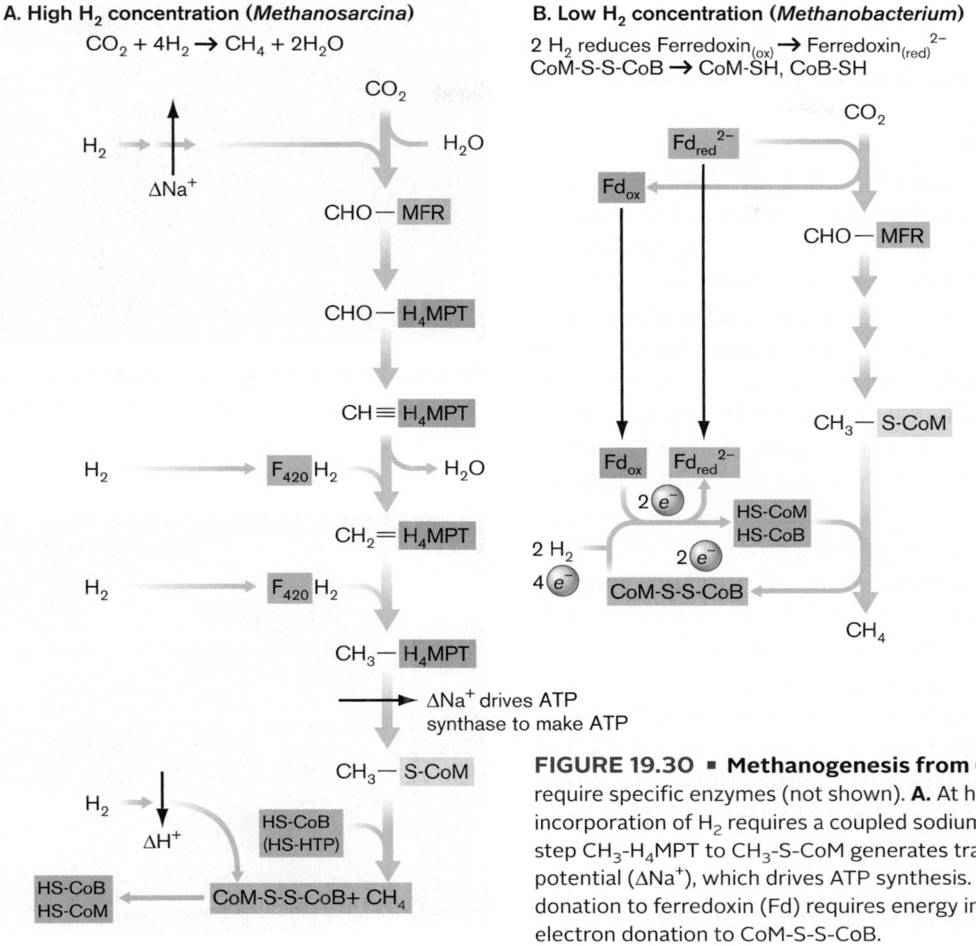
A. High H_2 concentration (*Methanosarcina*)
$CO_2 + 4H_2 \rightarrow CH_4 + 2H_2O$
CO_2
H_2
ΔNa^+
H_2O
CHO— MFR
CHO— H_4MPT
CH≡H_4MPT
H_2
F_{420} H_2
H_2O
CH_2=H_4MPT
H_2
F_{420} H_2
CH_3— H_4MPT
ΔNa^+ drives ATP synthase to make ATP
CH_3— S-CoM
H_2
ΔH^+
HS-CoB (HS-HTP)
HS-CoB
HS-CoM
CoM-S-S-CoB+ CH_4
B. Low H_2 concentration (*Methanobacterium*)
2 H_2 reduces $Ferredoxin_{(ox)} \rightarrow Ferredoxin_{(red)}^{2-}$
CoM-S-S-CoB → CoM-SH, CoB-SH
CO_2
Fd_{red}^{2-}
Fd_{ox}
CHO— MFR
CH_3— S-CoM
Fd_{ox}
Fd_{red}^{2-}
2 e^-
HS-CoM
HS-CoB
2 H_2
2 e^-
4 e^-
CoM-S-S-CoB
CH_4

FIGURE 19.30 ■ Methanogenesis from CO_2 and H_2. All steps require specific enzymes (not shown). **A.** At high H_2: The initial incorporation of H_2 requires a coupled sodium potential (ΔNa^+). The step CH_3-H_4MPT to CH_3-S-CoM generates transmembrane sodium potential (ΔNa^+), which drives ATP synthesis. **B.** At low H_2: Electron donation to ferredoxin (Fd) requires energy input from coupled electron donation to CoM-S-S-CoB.

requires either molybdenum or tungsten, depending on the species; some species have two alternative enzymes, depending on which metal is available. Another metal, cobalt, is required for a B_{12}-related cofactor that participates in methane production from methanol and methylamines.

Thought Question

19.6 What do the multiple metal requirements suggest about how and where the early methanogens evolved?

Methanogenesis from acetate. Methane production from acetate is particularly important for wastewater treatment, where most of the substrate consists of short-chain bacterial fermentation products. The acetate methanogenesis pathway is not fully understood, but the initial incorporation of H_2 probably requires a coupled gradient of sodium ion, as it does for CO_2 methanogenesis. The two carbons from acetate enter different pathways, which eventually converge:

$$CH_3\text{-S-CoM} + \text{HS-CoB} \longrightarrow CH_4 + \text{CoM-S-S-CoB}$$

To Summarize

- **Methanogens gain energy through redox reactions that generate methane** from CO_2 and H_2, formate, and acetate. They require association with bacterial species that generate the needed substrates.
- **Methanogens have rigid cell walls of diverse composition in different species**, including pseudopeptidoglycan, protein, and sulfated polysaccharides.
- **Species of methanogens show a wide range of different shapes**, including rods (single or filamentous), cocci (single or clumped), and spirals.
- **Methanogens inhabit anaerobic environments** such as wetland soil, marine benthic sediment, and animal digestive organs.
- **Biochemical pathways of methanogenesis** involve transfer of the increasingly reduced carbon to cofactors that are unique to methanogens.

19.5

Haloarchaea

In the saturated salt water of Utah's Great Salt Lake or Israel's Dead Sea, swimmers float with their heads well above the dense brine, but few living things can grow (**Fig. 19.31**). The main inhabitants of high-salt environments are extremely halophilic archaea, members of the class **Haloarchaea**. Most haloarchaea belong to the order **Halobacteriales**, which was named before the archaea were classified as distinct from bacteria. The order Halobacteriales includes species of haloarchaea that diverge among the methanogens (see **Fig. 19.6**). The halophiles, however, do not conduct methanogenesis, but grow as photoheterotrophs, using light energy to drive a retinal-based ion pump to establish a proton potential. Their photopigments color salterns, brine pools that are evaporated to mine salt (**Fig. 19.32**). Salted foods such as meat and fish and salt-cured hides can be spoiled by haloarchaea.

The halophilic archaea require at least 1.5 M NaCl or equivalent ionic strength, and most grow optimally at near saturation (about 4.3 M, seven times the concentration of seawater). Most are colored red by bacterioruberin, which protects cells from damage by light (**Fig. 19.33**). Thus, their growth colors their habitat red.

ALISON WRIGHT

FIGURE 19.31 ▪ A high-salt environment. A swimmer floats with her head up in the highly concentrated brine of the Dead Sea, in Israel.

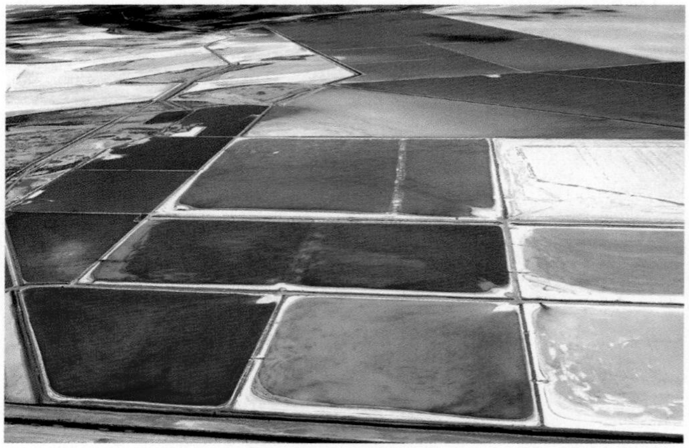
COURTESY OF DON GREEN PHOTOGRAPHY AND MORTON SALT

FIGURE 19.32 ▪ A saltern for salt production. Aerial view of a solar saltern facility in Grantsville, Utah.

A.

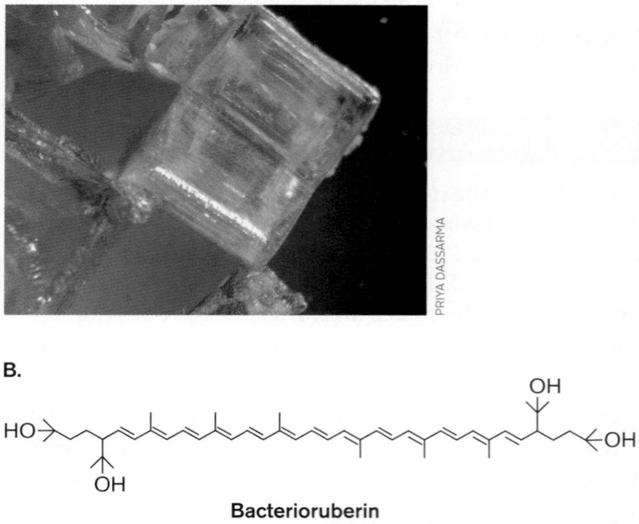
PRIYA DASSARMA
B.
HO
OH
OH
OH
Bacterioruberin

FIGURE 19.33 ▪ Red pigmentation. A. Brine inclusions in a salt crystal colored by haloarchaea. **B.** The red pigment bacterioruberin protects cells from damage by light.

Haloarchaeal Form and Physiology

Halophilic microbes need a way to maintain turgor pressure—that is, to avoid cell shrinkage as cytoplasmic water runs down the osmotic gradient toward external high salt. Most bacterial halophiles compensate for high external salt by uptake or synthesis of other kinds of osmolytes, such as small organic molecules. Haloarchaea, however, adapt to high external NaCl by maintaining a high intracellular concentration of potassium chloride (about 4-M KCl). Potassium ion concentrations are moderately high in most microbial cells (commonly 200 mM), but the exceptionally high KCl concentration within haloarchaea requires major physiological adaptations:

- **High GC content of DNA.** High salt concentration decreases the fidelity in base pairing of DNA. Because the triple hydrogen bonds of GC pairs hold more strongly than AT pairs, the exceptionally high GC content of haloarchaea (above 60% for most species) may protect their DNA from denaturation in high salt.
- **Acidic proteins.** Most haloarchaeal proteins are highly acidic, with an exceptional density of negative charges at their surface. The high negative charge maintains a layer of water in the form of hydrated potassium ions (K^+). This unusual hydration layer keeps the acidic proteins soluble in the high salt within the cytoplasm.

The high salt content of haloarchaea provides a convenient way to lyse cell contents for analysis: Simply transfer cells into low-salt buffer, and they fall apart owing to osmotic shock. This technique provides a quick way for beginning students to isolate DNA. This and other techniques have made the organism *Halobacterium* a model system for molecular biology education (**Special Topic 19.2**).

The properties of diverse halophiles are presented in **Table 19.5**. The haloarchaea are generally uniform with respect to temperature range (mesophilic). With respect to pH, two categories are found: halophiles that grow at neutral pH, and those that are alkaliphiles, growing in soda lakes above pH 9.

With respect to cell shape, haloarchaea display considerable diversity. Some species form symmetrical rods, as in *Halobacterium* NRC-1, the model system studied by Shiladitya DasSarma and colleagues at the University of Maryland. Other species form pleomorphic cells, flattened like pancakes; and still others form regular cocci (*Haloferax dombrowskii*) (**Fig. 19.34A**). A few species grow as flattened squares, the only microbial cells known to be square (*Haloquadratum walsbyi*) (**Fig. 19.34B** and **C**). The basis for maintaining the various shapes is unknown. In oligotrophic (low-nutrient) environments, it is likely that the greater surface-to-volume ratio gives flattened cells a competitive edge in obtaining nutrients.

The cell envelopes of most haloarchaea contain rigid cell walls of glycoprotein, as do some of the methanogens. Most species possess flagella for phototaxis and **gas vesicles** for maintaining their buoyancy in the upper layer of the water column. Unlike phospholipid vesicles, gas vesicles are made entirely of protein, and they are filled with air.

TABLE 19.5

Halophilic archaea.

Representative species	NaCl range (M)	Growth temperature (°C)	Growth pH	Cell shape	Location of isolate
Haloarcula quadrata	2.7–4.3	53°	pH 7	Square flat; pleomorphic	Sabkha, Sinai, Egypt
Haloarcula valismortis	3.5–4.3	40°	pH 7.5	Pleomorphic rods	Salt pools, Death Valley, CA
Halobacterium salinarum	3.0–5.2	35–50°	pH 7	Rods	Salted cow hide
Halococcus morrhuae	2.5–5.2	30–45°	pH 7	Cocci	Dead Sea, Israel
Haloferax volcanii	1.5–5.2	40°	pH 7	Pleomorphic dish-shaped	Dead Sea, Israel
Halorubrum lacusprofundi	1.5–5.2	1–44°	pH 7	Long rods (12 µm)	Deep Lake, Antarctica
Natronococcus occultus	1.4–5.2	30–45°	pH 9.5	Cocci	Lake Magadi, Kenya
Natronomonas pharaonis	2.0–5.2	45°	pH 9–10	Rods	Wadi Natrun, Egypt

Sources: Aharon Oren, The order Halobacteriales, *The Prokaryotes*, Springer-Verlag; *Halorubrum lacusprofundi* growth temperature from Ricardo Cavicchioli. 2006. *Nat. Rev. Microbiol.* **4**:331.

A.

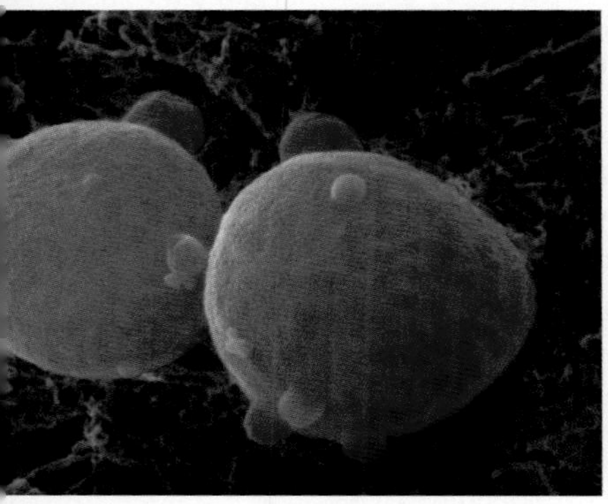

EYE OF SCIENCE/SCIENCE SOURCE

B. Square haloarchaea

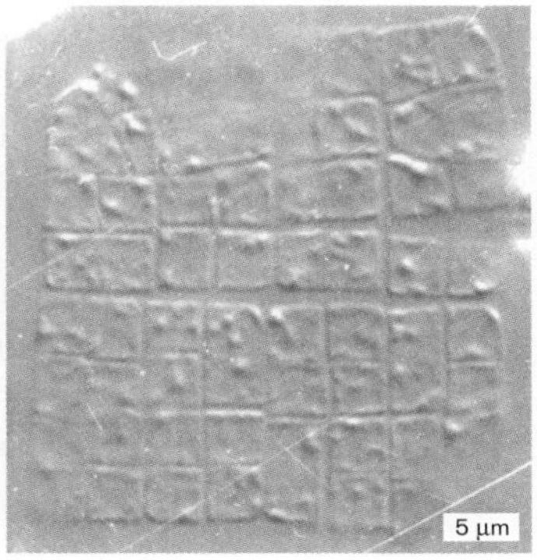
5 µm

M. KESSEL AND Y. COHEN. 1982. *J. BACTERIOL.* **150**:851

C. *Haloquadratum walsbyi*

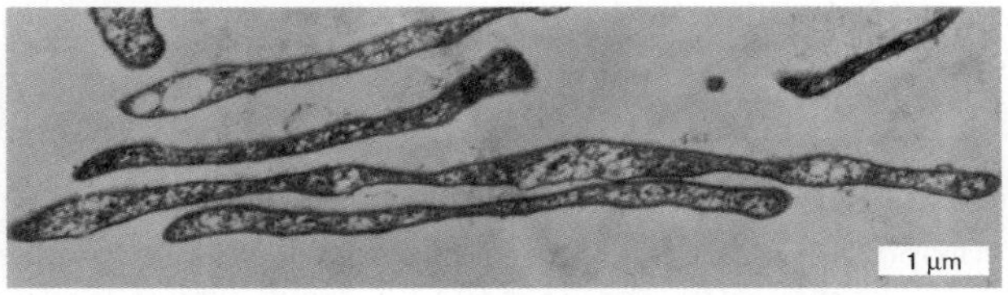
1 µm

W. STOECKENIUS. 1981. *J. BACTERIOL.* **148**:352

FIGURE 19.34 ■ Diverse haloarchaea. A. *Haloferax dombrowskii* (SEM). **B.** Square haloarchaea (*Haloquadratum walsbyi*) during their sixth round of cell division, which alternates in vertical and horizontal directions. (Photomicrograph, Nomarski optics.) **C.** Cross-sectioned cells of square haloarchaea show their thinness (TEM).

Special Topic 19.2: Haloarchaea in the Classroom

A halophilic archaeon, *Halobacterium* NRC-1, is used as a teaching tool for precollege education. Microbes relate to many precollege content areas, including health, food production and preservation, environmental science, genetics, and biotechnology. But many of the bacteria traditionally used by educators cause occasional illness in immunocompromised individuals, such as children with diabetes. An advantage of *Halobacterium* is that the archaeon has never caused a documented illness in humans.

Halobacterium grows readily at room temperature on inexpensive broth medium supplemented with concentrated NaCl, which excludes most contaminants. Growth produces attractive colonies with mutants that show interesting color phenotypes (**Fig. 1**). For example, colonies with sectored colors demonstrate gene mutation. And *Halobacterium* cells lyse readily in water for instant extraction of DNA, as well as other cellular components, such as gas vesicles and purple membrane.

For use in high schools, educator Priya DasSarma has developed a series of kits for studying *Halobacterium* and leads workshops to train teachers (**Fig. 2**). DasSarma and her husband, Shiladitya DasSarma, at the University of Maryland, have studied *Halobacterium* for two decades, developing the tools of genetics and molecular biology to investigate haloarchaeal physiology, including sequencing its genome. Now Priya DasSarma has developed sets of hands-on laboratory exercises that enable students to frame experimental questions and test them. Available from Carolina Biological Supply Company, the exercises in these kits teach a wide range of concepts in microbiology, such as exponential growth and competition, motility, antibiotic sensitivity and resistance, and genetic transformation and complementation.

In DasSarma's workshop, students and teachers learn to use the microbe for inquiry-based teaching by asking funda-

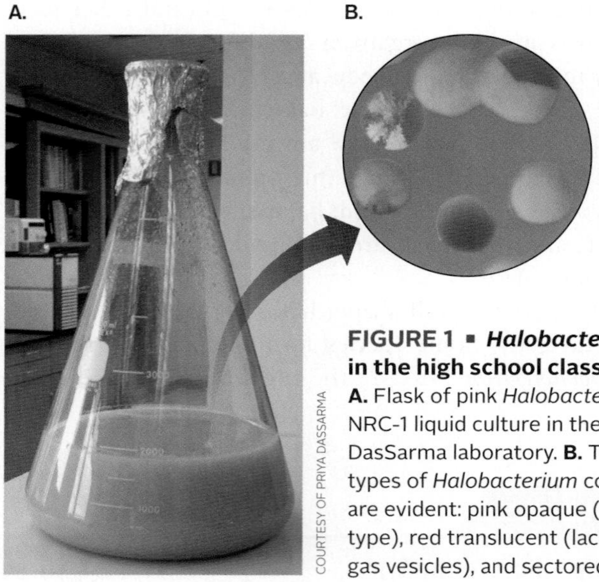
A.
B.
COURTESY OF PRIYA DASSARMA

FIGURE 1 ■ *Halobacterium* in the high school classroom. **A.** Flask of pink *Halobacterium* NRC-1 liquid culture in the DasSarma laboratory. **B.** Three types of *Halobacterium* colonies are evident: pink opaque (wild type), red translucent (lacking gas vesicles), and sectored.

Different Hypersaline Habitats Support Different Haloarchaea

Not all hypersaline (high-salt) habitats are alike. They differ with respect to pH, temperature, and the presence of other minerals, such as magnesium ion. Different kinds of hypersaline habitats support different species of haloarchaea (**Table 19.5**), as well as halophilic bacteria. Major types of habitats include:

- **Thalassic lakes.** Thalassic lakes (Greek *thalassa,* meaning "ocean"), such as the Great Salt Lake, in Utah, contain saturated salts with essentially the same ionic proportions as the ocean: Na^+ and Cl^- (NaCl) predominate, followed by Mg^{2+}, K^+ and SO_4^{2-}. Thalassic lakes (also called brine lakes) support genera such as *Halobacterium.* Antarctic brine lakes support growth of cold-adapted halophiles, such as *Halorubrum lacusprofundi,* at temperatures as low as –18°C.
- **Athalassic lakes.** Athalassic lakes, such as the Dead Sea in Israel, contain higher proportions of magnesium ions. These habitats favor genera such as *Haloarcula,* which require 100-mM Mg^{2+} for growth and grow best at magnesium concentrations above 1 M.
- **Solar salterns.** These are artificial pools of brine, or saturated NaCl, that evaporate in sunlight, precipitating halite (salt crystals) for commercial production. Commercial evaporation pools for salt actually benefit from the red microbes, whose light absorption accelerates heating and evaporation.
- **Brine pools beneath the ocean.** Undersea brine pools collect near geothermal vents, as in the Gulf of Mexico, Mediterranean Sea, and Red Sea. These hypersaline regions contain salts brought up by vent water. They support hyperthermophilic halophiles, most of which have not yet been cultured.

COURTESY OF PRIYA DASSARMA

FIGURE 2 ■ High school teachers use *Halobacterium*. Educator Priya DasSarma (right) leads a workshop including teachers Janet Kovach (left) and Christina K. Schwalm. The workshop trains teachers to use *Halobacterium* in the classroom.

mental questions about what the students can see. Why does one genetic form of *Halobacterium* appear beautiful pink, whereas a closely related form appears deep red? The pink, milky appearance is caused by light scattering from intracellular gas vesicles and light absorption by red carotenoid pigments. A mutant strain lacking gas vesicles shows the bright red pigment. On agar plates, why are some *Halobacterium* colonies pink and red, sectored like a pie? Red sectors arise from spontaneous gas vesicle mutations in a fraction of cells within a colony; the mutants' descendants grow outward in a red wedge.

The gas vesicles keep the cells afloat in their natural habitat, such as the Great Salt Lake. But DasSarma's research lab has found a medical application for the gas vesicle genes, which can be fused to a gene expressing an antigen. The recombinant microbe may then serve as a vaccine. In the teaching lab, students can share the excitement of this research by isolating *Halobacterium* DNA. The cells easily lyse in distilled water; then the DNA is spooled out for genetic analysis. The availability of a completely sequenced genome permits a wide range of teaching applications.

RESEARCH QUESTION

Cells from a pink colony of *Halobacterium* can be collected in a tabletop centrifuge—but when resuspended, the cells appear dark red. What might have happened to the cells? Can you propose an experiment to test your hypothesis?

Stuart, Elizabeth S., Fazeela Morshed, Marinko Sremac, and Shiladitya DasSarma. 2004. Cassette-based presentation of SIV epitopes with recombinant gas vesicles from halophilic archaea. *Journal of Biotechnology* **114**:225–237.

- **Alkaline soda lakes** such as Lake Magadi in Kenya, where carbonate salts drive the pH above pH 9. These lakes support alkaliphilic haloarchaea such as *Halobacterium*.
- **Underground salt deposits,** which contain micropockets of salt-saturated water where halophiles may survive for thousands of years.

Retinal-Based Photoheterotrophy

Most haloarchaea are photoheterotrophs, in which light-directed energy acquisition supplements respiration on complex carbon sources. Respiration occurs with oxygen or anaerobically with nitrate. A typical habitat for haloarchaea starts out as a pool with moderate salt content, containing a range of bacterial and archaeal species with varied tolerance to salt. As the pool evaporates, the salt concentrates, and the various bacteria die and lyse, releasing cell components that the haloarchaea can metabolize. The halophiles supplement their utilization of organic substrate energy by using light-driven ion pumps.

Light is captured by retinal-containing transmembrane proteins called **bacteriorhodopsin** (**BR**) and the chloride pump **halorhodopsin** (**HR**) (**Fig. 19.35**). (The role of light-driven ion pumps in phototrophy is discussed in Chapter 14.) The proton pump bacteriorhodopsin was named to indicate that it is a prokaryotic version of the better-known eukaryotic retinal rhodopsins; the name was chosen before the domain Archaea was recognized. The chloride pump halorhodopsin was discovered and named in reference to the chloride ion (a halide). The two proteins are homologs with similar structure and mechanism. In the cell membrane, they form complexes that aggregate in patches called purple membrane.

Each bacteriorhodopsin proton pump contains seven alpha helices that traverse the membrane, surrounding a

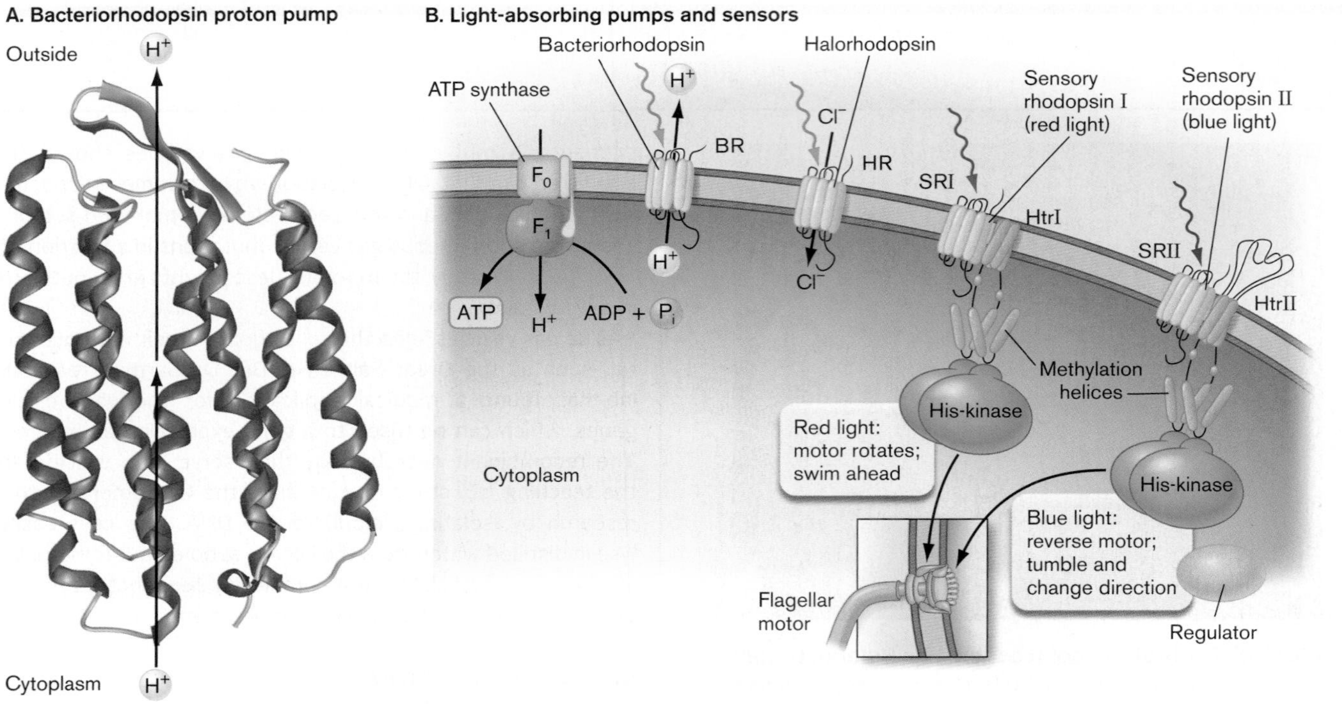
A. Bacteriorhodopsin proton pump
Outside
H+
Cytoplasm
H+
B. Light-absorbing pumps and sensors
Bacteriorhodopsin
Halorhodopsin
ATP synthase
H+
BR
Cl−
HR
Sensory rhodopsin I (red light)
Sensory rhodopsin II (blue light)
SRI
HtrI
SRII
HtrII
F0
F1
H+
Cl−
ATP
H+
ADP + Pi
Methylation helices
His-kinase
Red light: motor rotates; swim ahead
Cytoplasm
His-kinase
Blue light: reverse motor; tumble and change direction
Flagellar motor
Regulator

FIGURE 19.35 ▪ Light-driven ion pumps and sensors. **A.** Bacteriorhodopsin absorbs light and pumps H^+ out of the cell (PDB code: 1QKO), whereas light-activated halorhodopsin pumps chloride into the cell (not shown). **B.** Rhodopsin family molecules in the membrane of *Halobacterium*: bacteriorhodopsin (BR), light-driven proton extrusion; halorhodopsin (HR), light-driven chloride intake; sensory rhodopsins I and II (SRI and SRII), with their signal transduction proteins HtrI and HtrII. The sensory transduction proteins phosphorylate and dephosphorylate a protein that regulates the direction of rotation of the flagellar motor. ▶

buried molecule of retinal, the same light-absorbing molecule found in photoreceptors of the human retina. Light absorption triggers a conformational change that enables a proton from the cytoplasmic face to be picked up by an aspartate residue. The proton is then transferred stepwise through several other amino acid residues in the protein, leading to release of the proton outside the cell. The net result of proton transfer by bacteriorhodopsin is generation of a proton motive force that can run a proton-driven ATP synthase, storing energy as ATP (**Fig. 19.35B**▶).

Halorhodopsin has a similar structure, in which chloride (instead of H^+) is pumped <u>into</u> the cell instead of outward. Because chloride is negatively charged, this chloride transport contributes to the proton motive force. Light-driven chloride pumps are unique to haloarchaea.

Haloarchaea also possess homologs of bacteriorhodopsin that serve as sensory devices: the sensory rhodopsins I and II. These proteins, too, each have seven alpha helices containing retinal. When the sensory rhodopsins absorb light, they signal the cell to swim using its flagella (discussed in Chapter 3). The activated sensory rhodopsin I directs the cell to swim toward red light, the optimum range for photosynthesis. Sensory rhodopsin II is activated to reverse the flagellar motor and make the cell swim away from blue and ultraviolet light, which causes photooxidative damage to DNA. The rhodopsins signal through the chemotaxis machinery to the flagellar motor, using histidine kinase enzymes that phosphorylate regulatory proteins (see **Fig. 19.35B**). This response to light, or phototaxis, is an important model for archaeal molecular regulation.

The photosynthesis and phototaxis systems can be seen in the context of the overall metabolic map of *Halobacterium* NRC-1 (**Fig. 19.36**). In the metabolic map, the proton-translocating ATP synthase is driven by the proton potential generated through bacteriorhodopsin. Alternatively, a proton potential can be generated by the respiratory chain, through consumption of organic foods. Glucose is catabolized by the semiphosphorylated Entner-Doudoroff pathway, in which the intermediate 2-oxo-3-deoxygluconate is formed and then phosphorylated, producing one net molecule of ATP.

The *Halobacterium* genome reveals homologs for many genes that encode transporters of organic food molecules such as sugars and amino acids. In addition, the genome contains a number of cation and anion transporter genes. Some transporters move in needed molecules such as

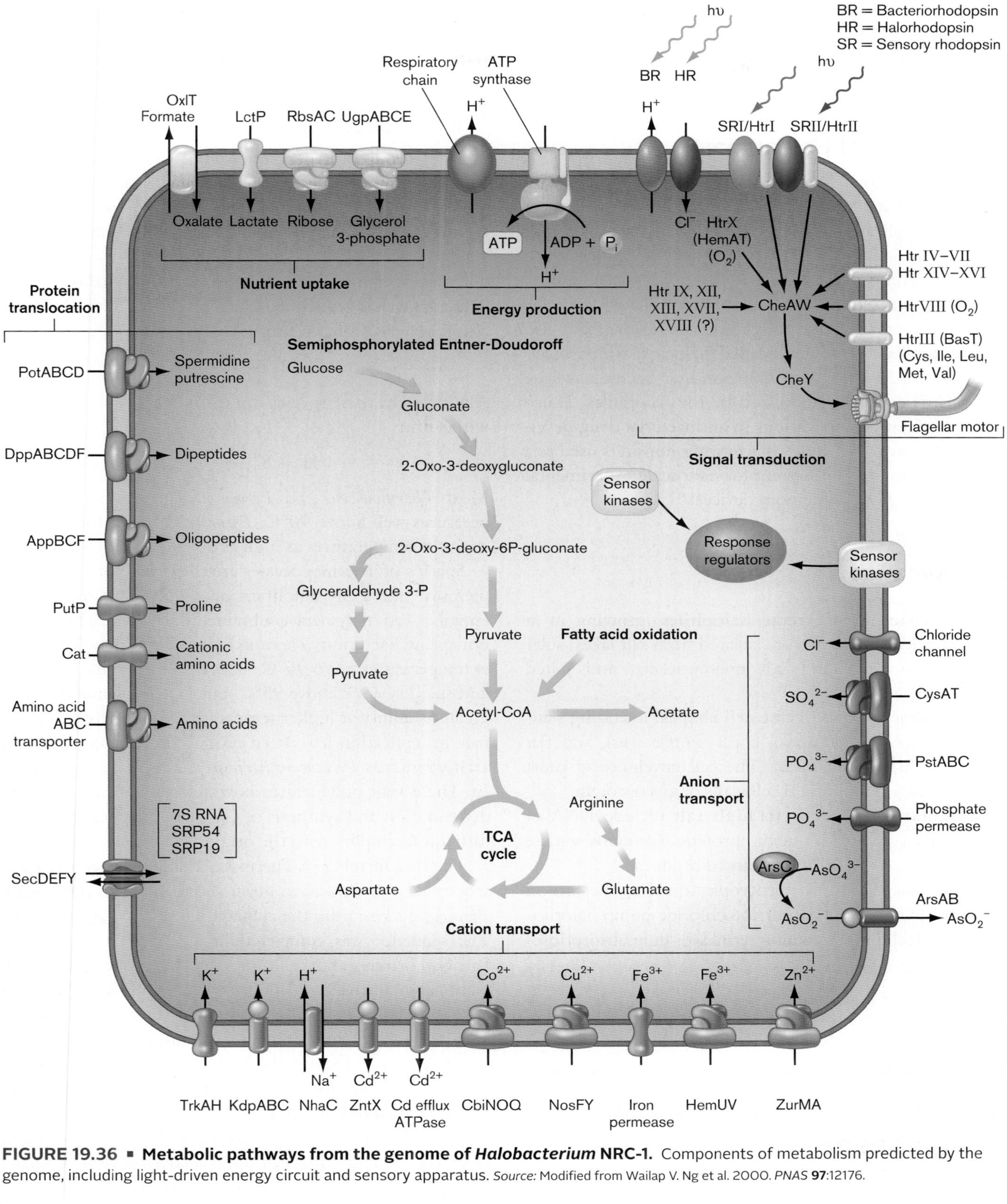
BR = Bacteriorhodopsin
HR = Halorhodopsin
SR = Sensory rhodopsin
hυ
hυ
Respiratory chain
ATP synthase
BR HR
OxlT
Formate
LctP
RbsAC
UgpABCE
H+
H+
SRI/HtrI
SRII/HtrII
Oxalate
Lactate
Ribose
Glycerol 3-phosphate
ATP
ADP + Pi
H+
Cl−
HtrX (HemAT) (O2)
Htr IV–VII
Htr XIV–XVI
Nutrient uptake
Energy production
Protein translocation
Htr IX, XII, XIII, XVII, XVIII (?)
CheAW
HtrVIII (O2)
HtrIII (BasT) (Cys, Ile, Leu, Met, Val)
Semiphosphorylated Entner-Doudoroff
PotABCD
Spermidine putrescine
Glucose
CheY
Gluconate
Flagellar motor
DppABCDF
Dipeptides
2-Oxo-3-deoxygluconate
Signal transduction
Sensor kinases
AppBCF
Oligopeptides
2-Oxo-3-deoxy-6P-gluconate
Response regulators
Sensor kinases
Glyceraldehyde 3-P
PutP
Proline
Pyruvate
Fatty acid oxidation
Cat
Cationic amino acids
Chloride channel
Cl−
Pyruvate
Amino acid ABC transporter
Amino acids
Acetyl-CoA
Acetate
SO4 2−
CysAT
PO4 3−
PstABC
Anion transport
7S RNA
SRP54
SRP19
Arginine
PO4 3−
Phosphate permease
TCA cycle
SecDEFY
ArsC
AsO4 3−
Aspartate
Glutamate
ArsAB
AsO2−
AsO2−
Cation transport
K+ K+ H+ Co2+ Cu2+ Fe3+ Fe3+ Zn2+
Na+ Cd2+ Cd2+
TrkAH
KdpABC
NhaC
ZntX
Cd efflux ATPase
CbiNOQ
NosFY
Iron permease
HemUV
ZurMA

FIGURE 19.36 ▪ **Metabolic pathways from the genome of *Halobacterium* NRC-1.** Components of metabolism predicted by the genome, including light-driven energy circuit and sensory apparatus. *Source:* Modified from Wailap V. Ng et al. 2000. *PNAS* **97**:12176.

phosphate, whereas others pump out harmful substances such as arsenate. Efflux of inorganic toxins is particularly critical for growth in evaporating brine, where the concentrations of trace metals are increasing.

Applications of Haloarchaea

Haloarchaeal proteins and other cell components have found applications in research and industry. The most promising application is that haloarchaeal gas vesicles can be genetically engineered to serve as a vaccine (mentioned at the beginning of the chapter). Gas vesicles containing proteins with fused antigen peptides are purified from the cells and used to elicit antibody response. Haloarchaea also produce antibacterial agents called halocins, which accelerate the death and lysis of surrounding bacteria, whose components can be metabolized by the halophiles. Halocins have potential applications in antibacterial drug development. In nanotechnology, bacteriorhodopsin is used as a photoswitch in a model system for nanoscale experimental electronic circuits and memory devices.

To Summarize

- **Haloarchaea are extreme halophiles**, growing in at least 1.5 M NaCl. They are isolated from salt lakes, solar salterns, underground salt micropockets, and salted foods.
- **Haloarchaea show diverse cell shapes**, including slender rods (*Halobacterium*), cocci (*Halococcus*), and flat squares (*Haloquadratum*). The cell envelopes of most haloarchaea contain rigid cell walls of glycoprotein.
- **Molecular adaptations to high salt** include DNA of high GC content and acidic proteins (proteins with a high number of negatively charged residues).
- **Retinal-based photoheterotrophy** involves the proton pump bacteriorhodopsin or the chloride pump halorhodopsin. Both pumps contain retinal for light absorption.

19.6

Thermophilic and Acidophilic Euryarchaeota

Besides the methanogens and halophiles, the phylum Euryarchaeota includes a number of hyperthermophiles that superficially resemble the crenarchaeotes, although their genetic divergence from these organisms is deep. Some of the euryarchaeotes are acidophiles that grow under the most extreme acidic conditions on Earth.

Thermococcales

A major group of hyperthermophilic euryarchaeotes is the order Thermococcales, including genera such as *Thermococcus* and *Pyrococcus* (**Fig. 19.37A**). These genera are most commonly isolated from submarine solfataras, volcanic vents that emit only gases. Most are anaerobes fermenting complex carbon sources, such as peptides or carbohydrates. Their growth is accelerated by the use of elemental sulfur as a terminal electron acceptor for anaerobic respiration:

$$2H^+ + 2e^- + S^0 \longrightarrow H_2S$$

Alternatively, these species can oxidize molecular hydrogen with sulfur:

$$H_2 + S^0 \longrightarrow H_2S$$

Most *Thermococcus* and *Pyrococcus* species grow at temperatures well above 90°C; *Pyrococcus woesei,* for example, grows at temperatures as high as 105°C.

Species of Thermococcales are the source of "vent polymerases" for PCR amplification. They now replace the enzyme Taq polymerase obtained from the Yellowstone hot-spring bacterium *Thermus aquaticus,* which grows only at temperatures up to 90°C. The Taq enzyme has only limited stability at or above 95°C, but DNA polymerases with greater stability at higher temperatures and increased accuracy in replication have been produced from vent-dwelling archaea such as *Pyrococcus furiosus* and *Thermococcus litoralis.* These vent polymerases now allow higher-temperature denaturation and synthesis of GC-rich sequences that were difficult to amplify with the original enzyme.

The first member of Thermococcales whose genome was sequenced was *Pyrococcus abyssi.* (Remember to distinguish *Pyrococcus abyssi* from the crenarchaeote *Pyrodictium abyssi.*) The sequence was completed in 1999 by Daniel Prieur, Patrick Forterre, and colleagues at Genoscope, the French National Sequencing Center. The genome of *Pyrococcus abyssi* reveals an especially high number of eukaryotic homologs for DNA replication, transcription, and protein translation. Examples include the Eukarya-like primase, helicase, and endonuclease for generation of Okazaki fragments in the lagging strand of DNA synthesis. At the same time, the genome also shows bacterial homologs for cell division and DNA repair. Thus, *P. abyssi* offers a striking view of mixed heritage from the common ancestor of the three domains.

Pyrococcus species possess enzymes conducting most of the classic conversions of the EMP pathway of glycolysis,

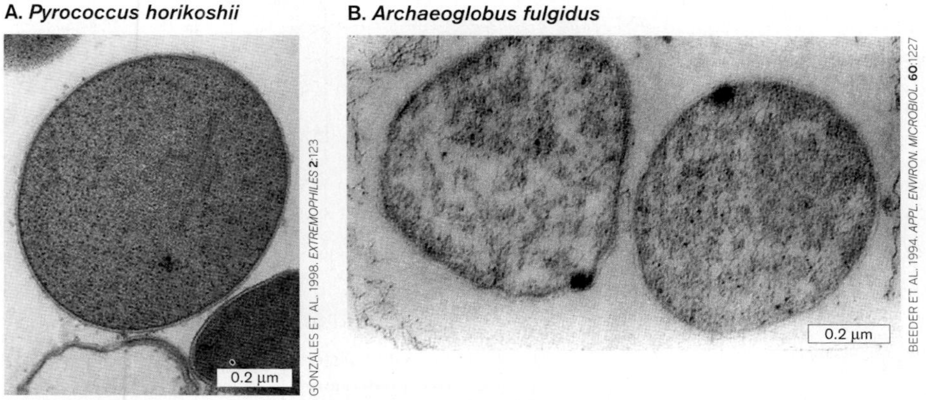
A. *Pyrococcus horikoshii*
B. *Archaeoglobus fulgidus*
0.2 µm
0.2 µm
GONZALES ET AL. 1998. *EXTREMOPHILES* **2**:123
BEEDER ET AL. 1994. *APPL. ENVIRON. MICROBIOL.* **60**:1227

FIGURE 19.37 ▪ Hyperthermophilic Euryarchaeota. A. *Pyrococcus horikoshii*, isolated from a hydrothermal vent at the Okinawa Trough (TEM). **B.** *Archaeoglobus fulgidus* (TEM), isolated from hot water in the North Sea oil field.

but several steps use enzymes unrelated to those in bacteria. Examples include two ADP-dependent enzymes—glucokinase and phosphofructokinase—as well as the phosphate-independent enzyme glyceraldehyde 3-phosphate ferredoxin oxidoreductase. The glyceraldehyde 3-phosphate conversion is unique in that it sidesteps the use of inorganic phosphate, which is required at this step by bacterial glycolysis (shown earlier, **Fig. 19.4**).

Pyrococcus and other members of Thermococcales have enzymes requiring tungsten, a metal rarely required outside the archaea. (Tungsten is present in elevated concentrations at hydrothermal vents.) Another unusual feature of *Pyrococcus* is that both proton motive force generation and the ATP synthase appear to involve Na^+/H^+ antiport. This dependence on sodium is a trait shared with the methanogens and the halophiles.

Thought Question

19.7 Compare and contrast the metabolic options available for *Pyrococcus* and for the crenarchaeote *Sulfolobus*.

Archaeoglobus Reverses Part of Methanogenesis

While many archaea reduce sulfur, only one group is known to reduce sulfate ion (SO_4^{2-}) without a bacterial partner. The order Archaeoglobales includes *Archaeoglobus fulgidus*, a hyperthermophile with unusual metabolic characteristics. **Figure 19.37B** shows *A. fulgidus* isolated from water at 75°C beneath an oil rig in the North Sea.

The genome of *A. fulgidus* reveals an exceptionally complex array of metabolic pathways, including an acetyl-CoA degradation pathway that reverses part of methanogenesis (**Fig. 19.38**). *A. fulgidus* uses several enzymes and cofactors, such as methanopterin (MPT), present in methanogens. But instead of generating methane from CO_2, they run the reactions in reverse, converting the methyl group of acetate to CO_2. The energy required is gained through fermentation of carbohydrates and through anaerobic respiration with sulfate. The sulfate is successively reduced to sulfite and sulfide. When organic substrates are available, sulfate reduction yields more energy than methanogenesis.

Thermoplasmatales

The order Thermoplasmatales includes thermophiles and acidophiles with no cell wall and no S-layer, but only a plasma membrane. How they maintain their cells against extreme conditions is poorly understood. An example is the moderate thermophile isolated from self-heating coal refuse piles, *Thermoplasma acidophilum*, growing at 59°C and pH 2. The cells have flagella and are motile, but it is unclear how the flagella maintain a torque against the membrane without a rigid envelope for support. The metabolism of *T. acidophilum*, like that of *Pyrococcus*, is based on S^0 respiration of organic molecules.

The genome sequence of *T. acidophilum* contains just 1.5 million bp—one of the smallest known for a free-living organism. It shows substantial evidence of horizontal transfer from the crenarchaeote *Sulfolobus*, a hyperthermophile that shares the same range of habitats. Horizontal transfer between distant relatives turns out to be common among the hyperthermophiles, causing some controversies in their classification.

Another acidophilic genus of the order Thermoplasmatales is *Ferroplasma* (**Fig. 19.39**). *Ferroplasma* species are found in mines containing iron pyrite ore (FeS_2). Using dissolved Fe^{3+} as an oxidizing agent in the presence of H_2O, they oxidize the sulfur to sulfuric acid. The chemical equation is:

$$FeS_2 + 14Fe^{3+} + 8H_2O \longrightarrow 15Fe^{2+} + 2SO_4^{2-} + 16H^+$$

The reaction generates pH values below pH 0 (1-M H^+). This degree of acidity is enough to dissolve a metal shovel within a day.

The amorphous cells of *Ferroplasma acidarmanus* grow in biofilms that form long streamers into water draining

FIGURE 19.38 ▪ Metabolism of *Archaeoglobus fulgidus*. *A. fulgidus* gains energy by sulfate respiration on small organic molecules such as lactate. This process drives a unique acetyl-CoA degradation pathway involving reversal of methanogenesis. *Source:* H. P. Klenk et al. 1997. *Nature* **390**: 364.

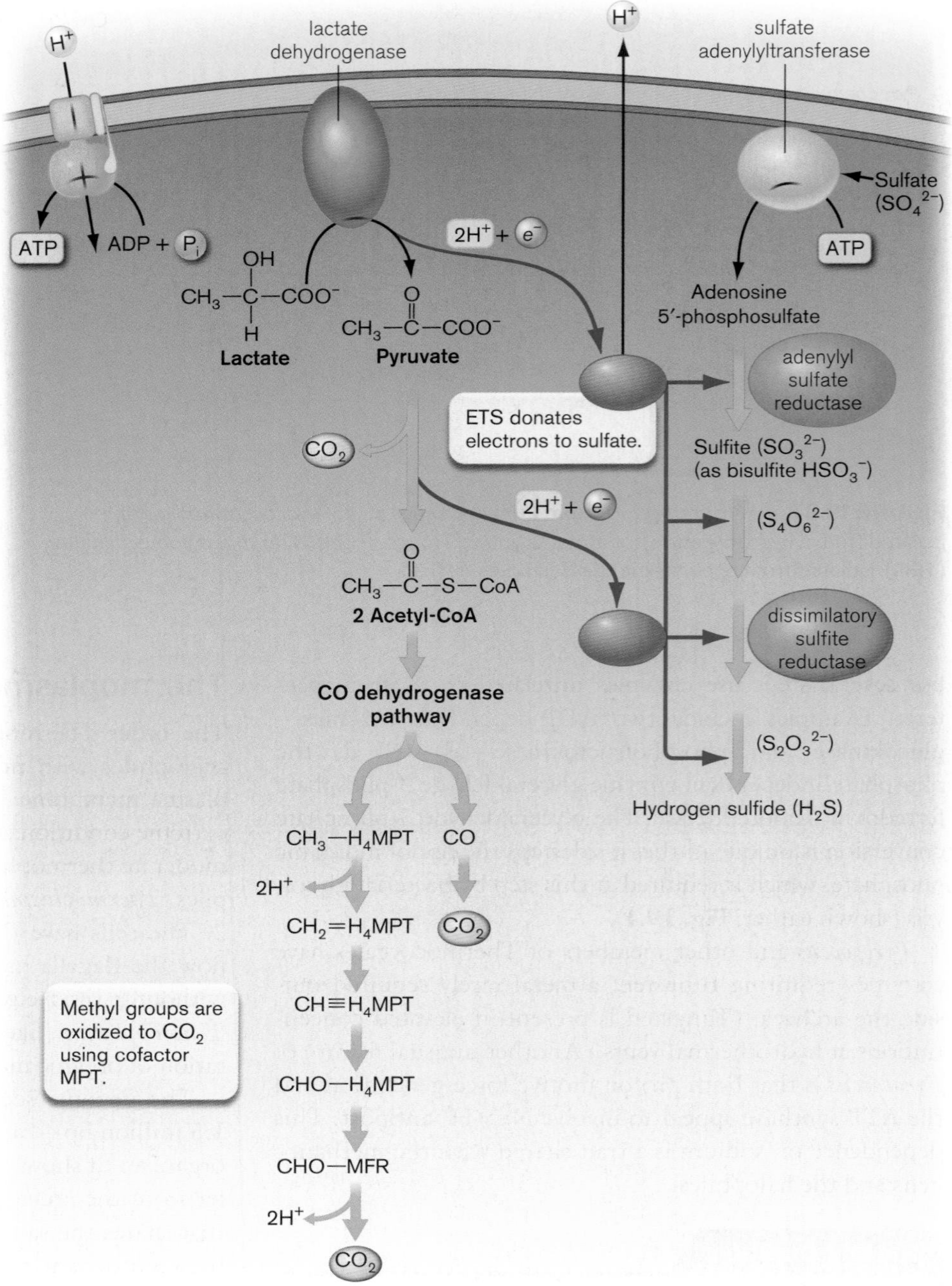

H^+
ATP
ADP + P_i
lactate dehydrogenase
OH
CH_3—C—COO^-
H
Lactate
O
CH_3—C—COO^-
Pyruvate
$2H^+ + e^-$
H^+
sulfate adenylyltransferase
Sulfate (SO_4^{2-})
ATP
Adenosine 5′-phosphosulfate
adenylyl sulfate reductase
ETS donates electrons to sulfate.
CO_2
Sulfite (SO_3^{2-}) (as bisulfite HSO_3^-)
$2H^+ + e^-$
($S_4O_6^{2-}$)
O
CH_3—C—S—CoA
2 Acetyl-CoA
dissimilatory sulfite reductase
CO dehydrogenase pathway
($S_2O_3^{2-}$)
Hydrogen sulfide (H_2S)
CH_3—H_4MPT
CO
$2H^+$
CH_2=H_4MPT
CO_2
Methyl groups are oxidized to CO_2 using cofactor MPT.
CH≡H_4MPT
CHO—H_4MPT
CHO—MFR
$2H^+$
CO_2

FIGURE 19.39 ▪ The extreme acidophile *Ferroplasma*. **A.** *F. acidiphilum* grows at pH 0 (TEM). **B.** Streamers of *F. acidarmanus* anchored to deposits of pyrite within the Iron Mountain mine in California. The stream is about a meter across, its water about pH 0. This level of acidity will dissolve a metal shovel in a day.

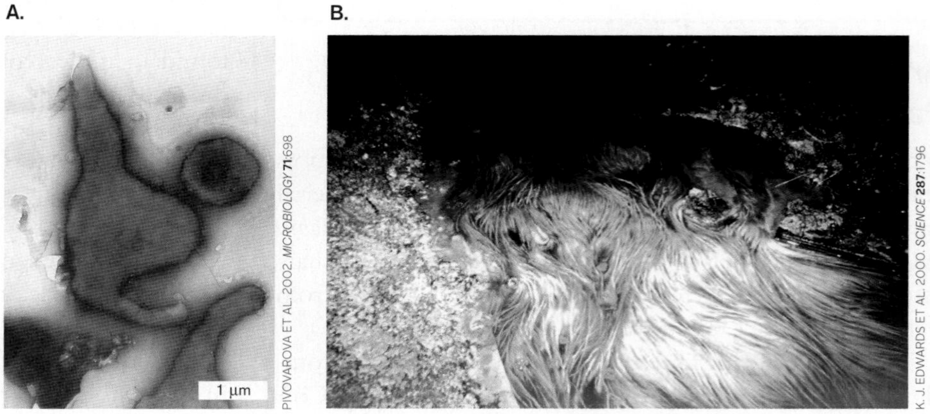

A.
B.
1 µm
PIVOVAROVA ET AL. 2002. *MICROBIOLOGY* **71**:698
K. J. EDWARDS ET AL. 2000. *SCIENCE* **287**:1796

from the mine (**Fig. 19.39B**). The oxidative disintegration of iron-bearing ores can be useful for leaching of minerals, but it also causes acid mine drainage into aquatic systems (discussed in Chapter 22).

Thought Question

19.8 Compare and contrast sulfur metabolism in *Pyrococcus* and *Ferroplasma.*

Nanoarchaeota

The smallest known euryarchaeotes are the Nanoarchaeota, recognized so far for a single species, *Nanoarchaeum equitans* (**Fig. 19.40**). The organism consists of exceptionally small cells that are obligate symbionts of the crenarchaeote *Ignicoccus.* The *N. equitans* cell in **Figure 19.40** is attached to *I. hospitalis* by a membrane bridge (black arrow). The host cell may harbor up to four of the smaller cells. Both host and symbiont genomes have been sequenced, revealing extensive coevolution of the two.

The *N. equitans* genome is exceptionally small (under 500 kb) and shows evidence of rapid degenerative evolution. The diminished genome is typical of a dependent organism that has lost numerous genes for functions now provided by a host. It remains unclear, however, whether *N. equitans* is parasitic on *Ignicoccus* or if it makes a metabolic contribution to its host (which is able to grow without the attached *N. equitans*).

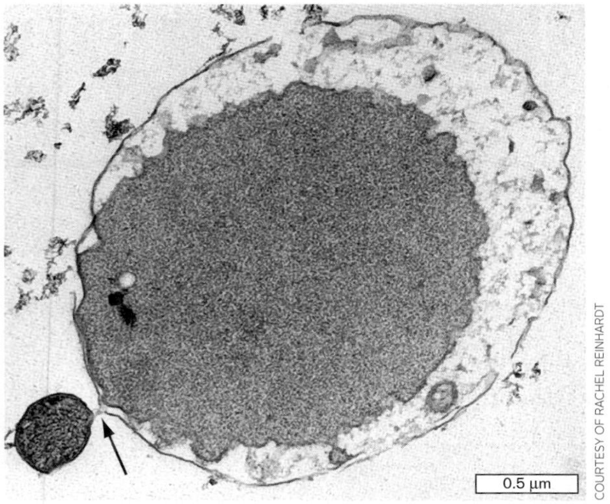
0.5 μm
COURTESY OF RACHEL REINHARDT

FIGURE 19.40 ▪ *Nanoarchaeum equitans*, a small obligate symbiont attached to *Ignicoccus hospitalis*. TEM.

To Summarize

- **Thermococcales includes hyperthermophiles and acidophiles.** Most species use sulfur to oxidize complex organic substrates.
- ***Pyrococcus* and *Thermococcus* are the source of vent polymerases** for the polymerase chain reaction (PCR).
- **Tungsten is commonly used by enzymes** of Thermococcales species.
- ***Archaeoglobus* species reduce sulfate.** The methyl group of acetate is oxidized to CO_2 by a pathway using methanogenic cofactors.
- **Thermoplasmatales includes extreme acidophiles.** *Ferroplasma* oxidizes iron pyrite ore (FeS_2) in a process that generates concentrated sulfuric acid, causing acid mine drainage.
- ***Nanoarchaeum equitans* is a tiny obligate symbiont that grows attached to *Ignicoccus hospitalis.***

19.7

Deeply Branching Divisions

We continue to identify emerging clades of archaea, such as the Thaumarchaeota, through PCR-amplified rDNA probes. Some of the sequences of emerging clades diverge more deeply than the divergence between Crenarchaeota and Euryarchaeota. Most such samples are uncultivated, and little is known of them other than their rRNA sequence. Nevertheless, new techniques of genome sequencing enable us in some cases to read most of a genome from a single cell. The genomes of such mystery strains may provide clues as to their needs for cultivation.

A deeply branching division is the Ancient Archaeal Group (AAG) of hyperthermophiles. AAG includes the Korarchaeota vent hyperthermophiles isolated by Susan Barns and Norman Pace. The organisms have not been isolated in pure culture, but in 2008 the genome of one korarchaeote was sequenced from an enriched mixed culture at 80°C–90°C at Obsidian Pool, Yellowstone National Park. The organism, provisionally named "*Candidatus* Korarchaeum cryptofilum," grows in long, thin filaments less than 200 nm wide. Its genes suggest that it gains energy mainly from anaerobic peptide fermentation.

Concluding Thoughts

Overall, species of archaea inhabit a wider range of environments than either bacteria or eukaryotes—from extreme heat to extreme cold, from high pH to extreme acid, as well as temperate environments such as the oceans. Crenarchaeotes are found in all soil and water habitats, and as endosymbionts of marine animals. Archaeal metabolism includes unique capacities, such as methanogenesis. Methanogens are found in all soil and water environments, as well as the human digestive tract. Yet we probably know less about the actual scope of Archaea than we do about the other two domains because so many members remain uncultured. As of this writing, we recognize three major divisions: the crenarchaeotes, including sulfur-metabolizing hyperthermophiles, as well as marine mesophiles and psychrophiles; the thaumarchaeotes, including ammonia oxidizers and marine endosymbionts; and the euryarchaeotes, including methanogens, halophiles, and sulfur-metabolizing extremophiles. New deeply branching isolates continue to be discovered, potentially representing entire new divisions. Much of what we call Archaea remains to be explored.

CHAPTER REVIEW

Review Questions

1. What distinctive structures are seen in the archaeal cell membrane and envelope?
2. Which aspects of archaeal genetics resemble the genetics of bacteria, and which aspects have more in common with eukaryotes?
3. Compare and contrast diverse members of the Crenarchaeota and Thaumarchaeota.
4. Outline the genetic phylogeny and key traits of these groups of archaea: Haloarchaea, methanogens, Thermococcales, Thermoplasmatales.
5. What are some specific physiological adaptations found in hyperthermophiles? Psychrophiles? Extreme acidophiles?
6. Outline three different specific types of mutualism involving an archaeal symbiont.
7. Explain how different archaea contribute to cycling of nitrogen and sulfur in ecosystems.
8. Explain what is known and what is unknown about the following groups of archaea: marine pelagic crenarchaeotes; marine benthic anaerobes; soil and plant root–associated crenarchaeotes. What kinds of experiments may reveal additional traits of these organisms?

Thought Questions

1. What approaches can you use to discover previously unknown deeply branching groups of archaea? Explain the strengths and limitations of each method.
2. Why do you think we have found no archaea that are pathogens of animals or plants?
3. Why do you think methanogens appear in many branches among different groups, whereas haloarchaea branch as a single group?

Key Terms

anaerobic methane oxidizers (ANME) (781)
bacteriorhodopsin (BR) (761, 787)
barophile (766)
black smoker (767)
cannula (768)
carbon monoxide reductase pathway (761)
Crenarchaeota (763)
Desulfurococcales (766)
Electron bifurcation (782)
Euryarchaeota (763)
gas vesicle (785)
Haloarchaea (784)
Halobacteriales (784)
halorhodopsin (HR) (787)
histone (762)
isoprenoid (759)
methane gas hydrate (781)
methanogen (775)
methanogenesis (761)
Nanoarchaeota (765)
pseudomurein (760)
pseudopeptidoglycan (760)
Sulfolobales (769)
syntrophy (776)
tetraether (759)
Thaumarchaeota (763)

Recommended Reading

Barns, Susan M., Charles F. Delwiche, Jeffrey D. Palmer, and Norman R. Pace. 1996. Perspectives on archaeal diversity, thermophily and monophyly from environmental rRNA sequences. *Proceedings of the National Academy of Sciences USA* **93**:9188–9193.

Biddle, Jennifer F., Julius S. Lipp, Mark A. Lever, Karen G. Lloyd, Ketil B. Sørensen, et al. 2006. Heterotrophic Archaea dominate sedimentary subsurface ecosystems off Peru. *Proceedings of the National Academy of Sciences USA* **103**:3846–3851.

Bolduc, Benjamin, Daniel P. Shaughnessy, Yuri I. Wolf, Eugene Koonin, Francisco F. Roberto, et al. 2012. Identification of novel positive-strand RNA viruses by metagenomic analysis of archaea-dominated Yellowstone hot springs. *Journal of Virology* **86**:5562–5573.

Brochier-Armanet, Céline, Bastien Boussau, Simonetta Gribaldo, and Patrick Forterre. 2008. Mesophilic crenarchaeota: Proposal for a third archaeal phylum, the Thaumarchaeota. *Nature Reviews. Microbiology* **6**:245–252.

Brumfield, Susan K., Alice C. Ortmann, Vincent Ruigrok, Peter Suci, Trevor Douglas, et al. 2009. Particle assembly and ultrastructural features associated with replication of the lytic archaeal virus *Sulfolobus* turreted icosahedral virus. *Journal of Virology* **83**:5964–5970.

Cavicchioli, Ricardo. 2006. Cold-adapted archaea. *Nature Reviews. Microbiology* **4**:331–343.

DasSarma, Priya, Regie C. Zamora, Jochen A. Müller, and Shiladitya DasSarma. 2012. Genome-wide responses of the archaeon *Halobacterium* sp. Strain NRC-1 to oxygen limitation. *Journal of Bacteriology* **194**:5530. doi:10.1128/JB.01153-12.

Edwards, Katrina J., Philip L. Bond, Thomas M. Gihring, and Jillian F. Banfield. 2000. An archaeal extreme acidophile involved in acid mine drainage. *Science* **287**:1796–1799.

Elkins, James G., Mircea Podar, David E. Graham, Kira S. Makarova, Yuri Wolf, et al. 2008. A korarchaeal genome reveals insights into the evolution of the Archaea. *Proceedings of the National Academy of Sciences USA* **105**:8102–8107.

Golyshina, Olga V., and Kenneth N. Timmis. 2005. *Ferroplasma* and relatives, recently discovered cell wall-lacking archaea making a living in extremely acid heavy metal-rich environments. *Environmental Microbiology* 7:1277–1288.

Karner, Markus B., Edward F. DeLong, and David M. Karl. 2001. Archaeal dominance in the mesopelagic zone of the Pacific Ocean. *Nature* **409**:507–510.

Karr, Elizabeth A., Joshua M. Ng, Sara M. Belchik, W. Matthew Sattley, Michael T. Madigan, et al. 2006. Biodiversity of methanogenic and other archaea in the permanently frozen Lake Fryxell, Antarctica. *Applied and Environmental Microbiology* **72**:1663–1666.

Kaster, Anne-Kristin, Johanna Moll, Kristian Parey, and Rudolf K. Thauer. 2011. Coupling of ferredoxin and heterodisulfide reduction via electron bifurcation in hydrogenotrophic methanogenic archaea. *Proceedings of the National Academy of Sciences USA* **108**:2981–2986.

Könneke, Martin, Anne E. Bernhard, José R. de la Torre, Christopher B. Walker, John B. Waterbury, et al. 2005. Isolation of an autotrophic ammonia-oxidizing marine archaeon. *Nature* **437**:543–546.

Lepp, Paul W., Mary M. Brinig, Cleber C. Ouverney, Katherine Palm, Gary C. Armitage, et al. 2006. Methanogenic Archaea and human periodontal disease. *Proceedings of the National Academy of Sciences USA* **101**:6176–6181.

Pernthaler, Annelie, Anne E. Dekas, C. Titus Brown, Shana K. Goffredi, Tsegereda Embaye, et al. 2008. Diverse syntrophic partnerships from deep-sea methane vents revealed by direct cell capture and metagenomics. *Proceedings of the National Academy of Sciences USA* **105**:7052–7057.

Pester, Michael, Christa Schleper, and Michael Wagner. 2011. The Thaumarchaeota: An emerging view of their phylogeny and ecophysiology. *Current Opinion in Microbiology* **14**:300–306.

Ruepp, Andreas, Werner Graml, Martha-Leticia Santos-Martinez, Kristin K. Koretke, Craig Volker, et al. 2000. The genome sequence of the thermoacidophilic scavenger *Thermoplasma acidophilum. Nature* **407**:508.

Samuel, Buck S., and Jeffrey I. Gordon. 2006. A humanized gnotobiotic mouse model of host–archaeal–bacterial mutualism. *Proceedings of the National Academy of Sciences USA* **103**:10011–10016.

Sapra, Rajat, Karine Bagramyan, and Michael W. W. Adams. 2003. A simple energy-conserving system: Proton reduction coupled to proton translocation. *Proceedings of the National Academy of Sciences USA* **100**:7545–7550.

Sauder, Laura A., Katja Engel, Jennifer C. Stearns, Andre P. Masella, Richard Pawliszyn, et al. 2011. Aquarium nitrification revisited: Thaumarchaeota are the dominant ammonia oxidizers in freshwater. *PLoS ONE* **6**:e23281.

Stams, Alfons J. M., and Caroline M. Plugge. 2009. Electron transfer in syntrophic communities of anaerobic bacteria and archaea. *Nature Reviews. Microbiology* 7:568–577.

CHAPTER 20
Eukaryotic Diversity

20.1 Phylogeny of Eukaryotes

20.2 Fungi

20.3 Algae

20.4 Amebas and Slime Molds

20.5 Alveolates: Ciliates, Dinoflagellates, and Apicomplexans

20.6 Trypanosomes and Metamonads

The domain Eukarya encompasses a breathtaking range of size and shape, from giant whales and sequoias to microbial fungi, algae, and protists. Some eukaryotic microbes are single cells as small as the smallest bacteria.

Fungi include multicellular forms such as mushrooms, as well as unicellular yeasts and filamentous *Penicillium* and *Neurospora.* Fungal filaments interconnect the roots of forest plants, forming a vast underground network of nutrients. Algae conduct photosynthesis using chloroplasts. Algae include broad sheets of kelp, as well as unicellular phytoplankton. Many algae turn out to be secondary endosymbionts—protists whose ancestors engulfed preexisting algae whole, only to assimilate them and utilize their prey's chloroplasts.

Protists include amebas, zigzag-swimming euglenas with their chloroplasts, paramecia with hundreds of cilia, and stalked vorticellae whose rings of cilia draw prey toward the mouth. Most protists are free-living predators in the food web of soil and water. But some are deadly parasites, such as the trypanosome of sleeping sickness and the plasmodium of malaria.

1 µm

ALISON R. TAYLOR

CURRENT RESEARCH highlight

Algae form miniature sculptures. Coccolithophores (or coccoliths) such as *Coccolithus pelagicus* are single-celled algae that produce intricate scales of calcium carbonate. Coccoliths are among the ocean's most abundant phototrophs, causing massive blooms that can be seen by satellite. Their calcium carbonate plates dissolve in acid; thus, their survival is endangered by ocean acidification caused by rising atmospheric CO_2. In 2011, Alison Taylor and colleagues showed how *C. pelagicus* uses voltage-gated H^+ channels to regulate its cytoplasmic pH. Increased H^+ concentration in the cytoplasm elevates an electrical potential difference across the plasma membrane. The elevated potential activates the channel to export H^+ and thereby protects the coccolithophore's calcification from acidity.

Source: Alison Taylor et al. 2011. *PLoS Biology* **9:** e1001085.

When we think of eukaryotes, we think first of plants and animals consisting of complex multicellular bodies. Macroscopic multicellular eukaryotes provide most of the food we consume, from breakfast cereal to beef. But eukaryotes also include vast numbers of microbes, such as protists (algae and protozoa). While we rarely eat these directly, marine and aquatic protists form much of the vast food web supporting fish and other metazoan seafood, a major food source of protein. A drop of pond water reveals tiny flagellates and algae, such as a desmid with its symmetrical leaf-shaped cell (**Fig. 20.1**). The green desmid has chloroplasts to fix carbon, contributing biomass to the food web. Other protists, such as euglenas, conduct photosynthesis but also swim using their flagella—and they phagocytose organic nutrients for respiration. Indeed, many marine protists have dual roles as producer and consumer.

Of the three domains of life, Eukarya shows the greatest range of size and shape. In Chapter 20 we explore the diversity of eukaryotic microbes, as essential partners in ecosystems and as parasites that cause devastating pathology. Recall that the structure of eukaryotic cells is defined by the presence of the nucleus and other membrane-enclosed organelles, which enable eukaryotic cells to grow a thousandfold larger than those of prokaryotes (for review, see Appendix 2). Despite their extraordinary range of form, the metabolism of eukaryotes is less diverse than that of either bacteria or archaea. Most eukaryotes conduct either oxygenic photosynthesis or heterotrophy—or both. All have descended from an ancestral cell that engulfed a bacterial endosymbiont, giving rise to mitochondria, the source of aerobic respiration. And all phototrophs descended from a cell that engulfed the bacterial ancestor of chloroplasts.

This chapter presents the phylogeny of major groups of eukaryotes, including the branching of animals and plants from the microbial family tree. We explore the form and function of fungi, amebas and slime molds, algae, and various classes of parasites. **Table 20.1** summarizes representative clades of microbial eukaryotes. **For a more extensive table, see Appendix 4.**

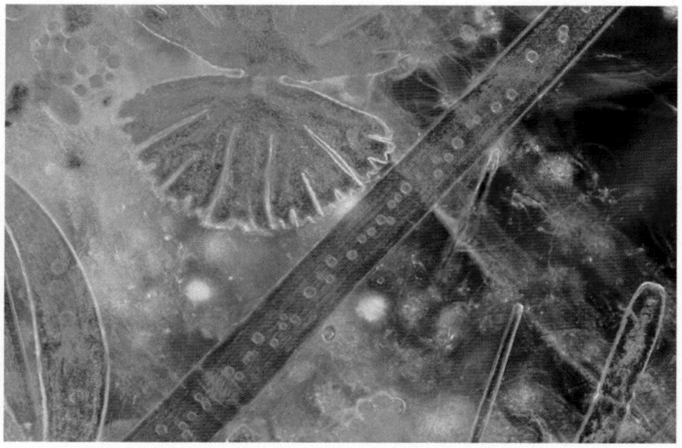

FIGURE 20.1 ■ Desmids and filamentous algae in a drop of pond water. Algae have chloroplasts that conduct photosynthesis.

20.1

Phylogeny of Eukaryotes

Our view of the domain Eukarya, including its relation to Bacteria and Archaea, has undergone several transitions over the past century, as discussed in Chapter 1. Classifying eukaryotes has always been a challenge, for several reasons. Complex eukaryotic cells frequently lose structures through reductive (degenerative) evolution. Thus, for example, a clade originally defined by possession of flagella often includes members that lack flagella. In addition, superficially similar forms of organisms have evolved independently in distantly related taxa; this is called convergent evolution. For example, the "water molds" that grow on aquarium fish superficially resemble fungi, but they actually evolved in a clade that includes brown algae and diatoms.

Another challenge for classification is the size and complexity of eukaryotic genomes, which has delayed the completion of genome sequences. Genomes are key because the SSU rRNA (small-subunit rRNA) sequences of eukaryotes fail to distinguish major clades as clearly as they do in bacteria or in archaea. Emerging genome sequences of microbial eukaryotes now show the relatedness of the major clades. While large questions remain, researchers are approaching a consensus view of eukaryotic descent—including the branch that gave rise to our own human species.

Historical Overview of Eukaryotes

For most of human history, life was understood in terms of macroscopic multicellular eukaryotes: animals (creatures that move to obtain food) and plants (rooted organisms that grow in sunlight). Fungi, which lack photosynthesis, were nonetheless considered a form of plant because they grow on the soil or other substrate. Thus, **mycology**, the study of fungi, was often included with botany, the study of plants. But the basis of fungal growth was poorly understood, and its mystery was often associated with magic. For example, people were mystified by the sudden growth of mushrooms in a ring, which they called a "fairy ring" (**Fig. 20.2A**). The mushrooms actually arise as fruiting bodies from the tips of fungal hyphae (filaments of cells)

TABLE 20.1

Eukaryotic microbial diversity.* (Expanded table appears in Appendix 4D.)

Opisthokonta (fungi and metazoan animals). Single flagellum on reproductive cells. Includes multicellular animals.

Metazoa (animals). Multicellular organisms with motile cells and body parts. Includes colonial animals, both invertebrates and vertebrates. *Homo sapiens.*

Choanoflagellata. Single flagellum with collar of microvilli. Resemble sponge choanocytes. Possible link to common ancestor of multicellular animals.

Eumycota (fungi). Cells form hyphae with cell walls of chitin.

Ascomycota. Fruiting bodies form asci containing haploid ascospores. Includes opportunistic pathogens and bread-making yeasts. *Penicillium, Aspergillus, Saccharomyces cerevisiae, Stachybotrys.* Lichens are a mutualism between an ascomycete and green algae (*Trebouxia*) or cyanobacteria (*Nostoc*).

Basidiomycota. Basidiospores form primary and secondary mycelia; generate mushrooms. May be edible (*Lycoperdon*) or toxic (*Amanita*). Plant pathogens (*Ustilago maydis* causes corn smut). Human pathogens (*Cryptococcus neoformans*).

Chytridiomycota. Zoospores (motile gametes) with a single flagellum resemble the gametes of animals. Saprophytes or anaerobic rumen fungi. *Allomyces.* Frog pathogens (*Batrachochytrium dendrobatidis*), bovine rumen digestive endosymbionts (*Neocallomastix*).

Glomeromycota. Mutualists of plant roots, forming arbuscular mycorrhizae, filamentous networks that share nutrients with and among diverse plants.

Microsporidia. Single-celled parasites that inject a spore through a tube into a host cell, causing microsporidiosis. *Encephalitozoon* species. Commonly infect AIDS patients.

Zygomycota. Nonmotile gametes grow toward each other and fuse to form the zygote (zygospore). Saprophytes or insect parasites. Some form mycorrhizae.

Viridiplantae (primary endosymbiotic algae and plants). Includes algae and multicellular plants. Chloroplasts arose from a primary endosymbiont.

Charophyta (stoneworts). Multicellular algae with rhizoids that adhere to sediment. Form crust of calcium carbonate, giving the name "stoneworts."

Chlorophyta (green algae). Chlorophyll *a* confers green color. Inhabit upper waters.

- **Unicellular with paired flagella.** *Chlamydomonas, Volvox* (colonial).
- **Multicellular.** *Ulva* grow in sheets; *Spirogyra* form chains; *Cymopolia* forms calcified stalks with filaments.
- **Picoeukaryotes.** *Ostreococcus* and *Micromonas* are unicellular algae <3 μm in diameter.
- **Siphonous algae.** *Caulerpa* species consist of a single cell with multiple nuclei, growing to indefinite size.

Rhodophyta (red algae). Phycoerythrin obscures chlorophyll, colors the algae red. Absorption of blue-green light enables colonization of deeper waters. *Porphyra* forms sheets edible by humans; *Mesophyllum* forms coralline algae, hardened by calcium carbonate crust; resembles coral.

Amoebozoa (amebas and slime molds). Lobe-shaped (lobose) pseudopods driven by sol-gel transition of actin filaments.

Amebas. Unicellular. No microtubules to define shape. Life cycle is primarily asexual. Predators in soil or water. Giant free-living amebas (*Amoeba proteus*); parasites (*Entamoeba histolytica*).

Mycetozoa. Slime molds. Upon starvation, amebas aggregate to form a fruiting body, which undergoes meiosis and produces spores. Cellular slime molds (*Dictyostelium*); plasmodial slime molds (*Physarum polycephalum*). Some slime-mold amebas may generate flagella in an aqueous habitat.

Rhizaria (amebas with filament-shaped pseudopods). Filament-shaped (filose) pseudopods. Some species have a test (shell) of silica or other inorganic materials.

Possess flagella or pseudopods (Cercozoa); form spiral tests (Foraminifera); form thin pseudopods called filopodia (Radiolaria).

Alveolata (having cortical alveoli). Cortex contains flattened vesicles called alveoli, reinforced below by lateral microtubules.

Ciliophora. Common aquatic predators. Reproduce by conjugation, in which micronuclei are exchanged; then regenerate macronucleus for gene expression. Covered with cilia (*Paramecium*); mouth ringed with cilia (*Vorticella*); suctorians (*Acineta*).

Dinoflagellata. Secondary or tertiary endosymbiont algae, from engulfment of red algae or diatoms. Cortical alveoli contain stiff plates. Pair of flagella, one wrapped around the cell. Free-living aquatic (*Peridinium*); zooxanthellae, endosymbionts of coral (*Symbiodinium*).

Apicomplexa (formerly Sporozoa). Parasites with complex life cycles. Lack flagella or cilia; possess apical complex for invasion of host cells. Vestigial chloroplasts. *Plasmodium falciparum* causes malaria; *Toxoplasma gondii* causes feline-transmitted toxoplasmosis; *Cryptosporidium parvum* is a waterborne opportunistic parasite.

Heterokonta (having pair of dissimilar flagella). Paired flagella of dissimilar form, one much shorter than the other.

Diatoms (Bacillariophyceae); kelps (Phaeophyceae); golden algae (Chrysophyceae); coccolithophores (Prymnesiophyceae); water molds (Oomycetes).

Euglenozoa or Discicristata (having disk-shaped cristae). Usually possess a deep feeding groove. Disk-shaped cristae of mitochondria.

Free-living flagellates (Euglenida and Jakobida); trypanosomes (*Trypanosoma brucei*, sleeping sickness; *T. cruzi*, Chagas' disease).

Metamonada (vestigial mitochondria). Parasitic or symbiotic flagellates. Mitochondria and Golgi degenerated through evolution.

Human intestinal parasites (*Giardia intestinalis*, or *G. lamblia*); symbionts of termite gut (*Pyrsonympha*).

Bulleted terms are representative groups within the phylum.

A.

HTTP://WWW.UGAURBANAG.COM/CONTENT/FAIRY-RING

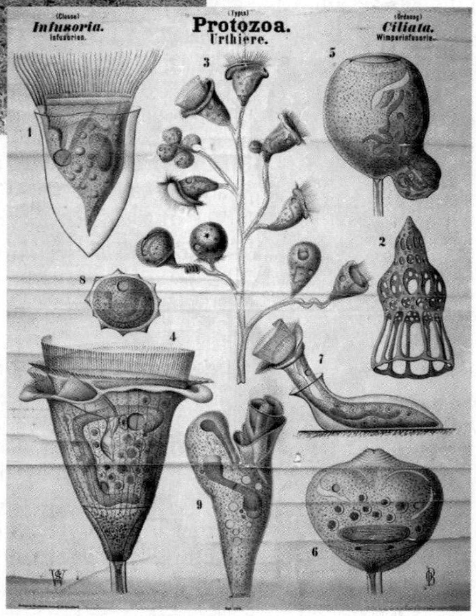
Infusoria.
Protozoa.
Urthiere.
Ciliata.

B.

THE MBLWHOI LIBRARY

FIGURE 20.2 ■ Traditional views of fungi and protozoa (protists). **A.** Basidiomycete mushrooms growing in a "fairy ring." **B.** A nineteenth-century depiction of ciliated protists, by Rudolf Leuckart.

that propagate from a single spore and extend radially underground.

In the eighteenth and nineteenth centuries, microscopists came to recognize microscopic forms of fungi such as filamentous hyphae and unicellular yeasts. Other unicellular life-forms, such as amebas and paramecia, were motile and appeared more like microscopic animals. These animal-like organisms were called **protozoa** (singular, **protozoan**) (**Fig. 20.2B**). The cellular dimensions of protozoa were typically ten- to a hundredfold larger than those of bacteria, and their form and motility offered intriguing subjects for observation. So did single-celled phototrophs such as diatoms and dinoflagellates, which were called **algae** (singular, **alga**). Algae were thought of as unicellular plants, although simple multicellular forms were known. The unicellular and microscopic forms of fungi, protozoa, and algae came to be included in the subject of microbiology.

Discoveries in physiology led us to redefine these organisms. For example, the motile organisms defined as protozoa often contain chloroplasts and fit the classification of algae. On the other hand, slime molds, originally classified with fungi, show form and motility more typical of protozoa. By the mid-twentieth century, naturalists classified protozoa, unicellular algae, and undifferentiated colonial forms as **protists**. Researchers including Herbert Copeland, Robert Whittaker, and Lynn Margulis attempted to refine the definition of "protist" to better distinguish microbial life-forms. Today, molecular phylogeny shows that protists comprise several clades equally distant from each other as from animals and plants (**Fig. 20.3**). In the terminology used today:

- **Protist** refers to single-celled and colonial eukaryotes other than fungi. Protists include many diverse clades of algae and protozoa.
- **Protozoa** are protists that are single-celled heterotrophs. They include environmental consumers, as well as medically important parasites such as *Giardia*.

The phylogeny of microbial eukaryotes has only recently emerged from DNA sequences (**Fig. 20.3**). Eukaryotic genomes are typically severalfold larger than those of bacteria and archaea, and 50%–90% of their DNA consists of noncoding sequences. Thus, eukaryotic genomes take longer to sequence and are more challenging to annotate than those of prokaryotes.

Furthermore, the evolution of eukaryotes includes multiple events of **endosymbiosis**, in which an engulfed cell evolved into an essential organelle. An endosymbiotic incorporation of a proteobacterium by the ancestor of all eukaryotes gave rise to mitochondria. Later incorporation of a cyanobacterium by the ancestor of plants and algae gave rise to chloroplasts. Much later, several lineages of protists took up chloroplast-bearing algae, which now show varying stages of evolution as organelles. Algae of the lineage derived from a single endosymbiotic event are **primary endosymbionts** (the Viridiplantae in **Fig. 20.3**), while the protists that later incorporated algae are **secondary endosymbionts** (green asterisks in **Fig. 20.3**).

Although uncertainties remain, a consensus view of eukaryotic phylogeny has emerged based on a combination of DNA sequence comparison, protein trees, and the appearance of specific gene fusions and deletions. This consensus phylogeny is shown in **Figure 20.3**. The distinct clades of protists include amebas and slime molds (Amoebozoa); amebas with needlelike pseudopods (Rhizaria); ciliates and dinoflagellates (Alveolata); oomycetes, brown algae, and diatoms (Heterokonta); and other groups, such as Metamonada.

We will now summarize key traits of each major group. Diversity within each group is explored in the sections that follow (compiled in **Table 20.1**).

Note: Medical textbooks also cover invertebrate animal parasites such as worms and mites as eukaryotic agents of disease, although they are not considered microbes.

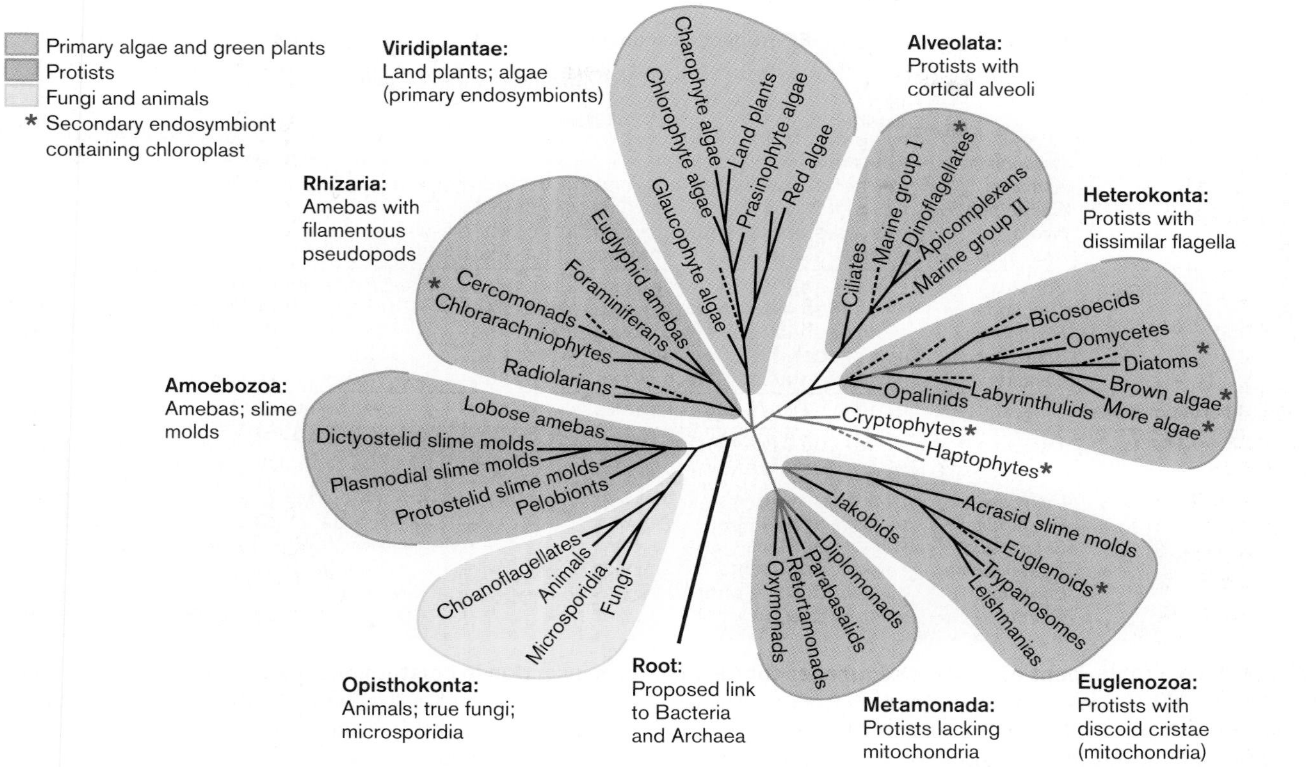
Primary algae and green plants
Protists
Fungi and animals
* Secondary endosymbiont containing chloroplast
Viridiplantae: Land plants; algae (primary endosymbionts)
Charophyte algae
Chlorophyte algae
Glaucophyte algae
Land plants
Prasinophyte algae
Red algae
Alveolata: Protists with cortical alveoli
Ciliates
Marine group I
Dinoflagellates*
Apicomplexans
Marine group II
Heterokonta: Protists with dissimilar flagella
Bicosoecids
Oomycetes
Diatoms*
Brown algae*
More algae*
Labyrinthulids
Opalinids
Cryptophytes*
Haptophytes*
Rhizaria: Amebas with filamentous pseudopods
Euglyphid amebas
Foraminiferans
*Cercomonads
Chlorarachniophytes
Radiolarians
Amoebozoa: Amebas; slime molds
Lobose amebas
Dictyostelid slime molds
Plasmodial slime molds
Protostelid slime molds
Pelobionts
Choanoflagellates
Animals
Microsporidia
Fungi
Opisthokonta: Animals; true fungi; microsporidia
Root: Proposed link to Bacteria and Archaea
Oxymonads
Retortamonads
Parabasalids
Diplomonads
Jakobids
Metamonada: Protists lacking mitochondria
Acrasid slime molds
Euglenoids*
Trypanosomes
Leishmanias
Euglenozoa: Protists with discoid cristae (mitochondria)

FIGURE 20.3 ■ Phylogeny of eukaryotes based on DNA sequence data. Dotted lines denote emerging little-known strains. Green asterisks denote organisms that have secondary endosymbiont chloroplasts. Gray branches indicate less certain lineage. *Source:* Modified from Sandra L. Baldauf. 2003. *Science* **300**:1703.

Opisthokonts: Animals and Fungi

Where do humans and other multicellular animals fit into this taxonomy? The position of animals ("metazoa") among the eukaryotic microbial clades is of interest because it suggests which contemporary microbes most closely resemble our own cells. The degree of relatedness can help define microbial model systems for probing key questions of human cell biology. For example, the baker's yeast *Saccharomyces cerevisiae* shares so much of its genetic machinery with humans that it provides a model for cancer, cell trafficking defects, and even the basis of human aging.

As we saw in Chapter 17, the first measure of species relatedness is based on DNA sequences encoding ribosomal RNA, particularly small-subunit rRNA. For eukaryotes, however, rRNA sequences yield ambiguous results, in part because of differing rates of change among different clades. So researchers have searched the genomes for stronger clues to relative divergence—that is, which clade diverged earliest from the others. A particularly strong clue would be the appearance of a gene insertion or deletion unique to a particular clade. Such a sequence indicates that all species containing the gene insertion or deletion must have diverged after their common ancestor branched from the outgroup.

Surprisingly, genomic analysis relates animals more closely to fungi (such as yeasts) than to motile microbial eukaryotes (such as paramecia or amebas). The true **fungi**, or **Eumycota**, are heterotrophs, either single-celled or growing in nonmotile filaments of cells called hyphae. Nevertheless, animals and fungi share several key gene insertions and deletions, as discovered by Sandra Baldauf and colleagues at the University of York. Baldauf focused on a short sequence insertion in a key gene encoding a protein translation factor, elongation factor 1α (EF-1α) (**Fig. 20.4**). The inserted DNA, encoding 12 amino acids, appears in all sequenced genomes of animals and fungi and is absent from all plants and most protists. It is present in organisms, such as microsporidians, that were previously considered protists but whose genomes now indicate classification under fungi. The EF-1α gene sequence, as well as several others, supports the grouping of animals and fungi (including microsporidians) together in one clade (see **Fig. 20.3**).

Animal and fungal cells also share a structural feature distinguishing them from protists: the presence of an unpaired flagellum. Both animals and fungi include species whose life cycle has a uniflagellar stage, in contrast to other microbial eukaryotes whose flagella are paired (for example, euglenas). In the case of humans, the uniflagellar stage is the spermatozoan. Similarly, some species of fungi generate uniflagellar reproductive cells called **zoospores**. As

EF-1α peptide sequence alignment

Group	Organism	197	Insertion within EF gene	265
Opisthokonts: Animals	*Homo*	PISGWNGDNMLEPSAN-MPWFKG	WKVTRKDG-----NASG	TTLLEALDCILPPTRPTDKPLRLPL
	Drosophila	PISGWHGDNMLEPSTN-MPWFKG	WEVGRKEG-----NADG	KTLVDALDAILPPARPTDKALRLPL
	Acmaea	PISGYNGDNMLEKSPN-MPWYKG	WKVEQKDDKGNASTVTG	DTLTQALDSIQPPKRPTDKALRLPL
	Caenorhabditis	PISGFNGDNMLEVSSN-MPWFKG	WAVERKEG-----NASG	KTLLEALDSIIPPQRPTDRPLRLPL
	Hydra	PVSGWHGDNMIEPSPN-MSWYKG	WEVEYKDTG----KHTG	KTLLEALDNIPLPARPSSKPLRLPL
	Anemonia	PISGWHGDNMLEKSDK-MPWWNG	FELFNKSQG---SKTG	TTLFDGLDDINVPSRPTDKALRLPL
Opisthokonts: Fungi	*Sordaria*	PISGFNGDNMLEASTN-CPWYKG	WEKETKAG-----KSTG	KTLLEAIDAIEQPKRPTDKPLRLPL
	Triochoderma	PISGFNGDNMLTPSTN-CPWYKG	WEKETKAG-----KFTG	KTLLEAIDSIEPPKRPTDKPLRLPL
	Histoplasma	PISGFEGDNMIEPSPN-CTWYKG	WNKETASG-----KSSG	KTLLDAIDAIEPPTRPTDKPLRLPL
	Schizosacch.	PVSGFQGDNMIEPTTN-MPWYQG	WQKETKAG-----VVKG	KTLLEAIDSIEPPARPTDKPLRLPL
	Candida	PISGWNGDNMIEPSTN-CPWYKG	WEKETKSG-----KVTG	KTLLEAIDAIEPPTRPTDKPLRLPL
Opisthokonts: Microsporidians	*Glugea*	PISGYLGINIVEKGDK-FEWFKG	WKPV-SGA-----GDSI	FTLEGALNSQIPPPRPIDKPLRMPI
Plants	*Triticum*	PISGFEGDNMIERSTN-LDWYKG	-----------------	PTLLEALDQINEPKRPSDKPLRLPL
	Arabidopsis	PISGFEGDNMIERSTN-LDWYKG	-----------------	PTLLEALDQINEPKRPSDKPLRLPL
Protists	*Dictyostelium*	PISGWNGDNMLERSDK-MEWYKG	-----------------	RTLLEALDAIVEPKRPHDKPLRIPL
	Plasmodium	PISGFEGDNLIEKSDK-TPWYKG	-----------------	RTLIEALDTNQPPKRPYDKPLRIPL
	Entamoeba	PISGFQGDNMIEPSTN-MPWYKG	-----------------	PTLIGALDSVTPPERPVDKPLRLPL
	Trypanosoma	PISGWQGDNMIEKSEK-MPWYKG	-----------------	PTLLEALDMLEPPVRPSDKPLRLPL
	Euglena	PISGWNGDNMIEASEN-MGWYKG	-----------------	LTLIGALDNLEPPKRPSDKPLRLPL
	Giardia	PTSGWTGDNIMEKSDK-MPWYKG	-----------------	PCLIDAIDGLKAPKRPTDKPLRLPI

FIGURE 20.4 ▪ Alignment of peptide sequences for EF-1α reveals insertion in opisthokont species. A DNA insertion encoding 11–17 amino acid residues appears in the EF-1α sequence of animals, fungi, and microsporidians (known collectively as opisthokonts) but not in plants or most protists. This finding suggests that opisthokonts have a common ancestor that diverged from the eukaryote lineage before divergence of the plants and non-opisthokont protists. *Source:* Modified from Sandra L. Baldauf. 2003. *Science* **300**:1703.

a whole, the clade including single-flagellum members is termed "opisthokont," based on the Greek words meaning "backward-pointing pole," because the flagellum points backward like a barge pole.

Among opisthokonts, the microbes that diverged most recently from animals (600 million years ago) appear to be the choanoflagellates. Genetic studies of choanoflagellates reveal several genes found only in animals. The prefix "choano-," meaning "funnel," refers to the collar of filaments surrounding the flagellum. The collared cells of choanoflagellates closely resemble the choanocyte cells of colonial sponges, an ancient form of animal (**Fig. 20.5**). Furthermore, choanoflagellates form colonies that crudely resemble the colonial structure of sponges. Thus, choanoflagellates may represent a "missing link" between animals and the microbial eukaryotes.

Note: Eukaryotic flagella are whiplike organelles composed of microtubules and surrounded by a membrane; their action is powered by ATP along the entire filament. Distinguish them from bacterial and archaeal flagella, which are rotary helical filaments composed entirely of protein subunits; their rotation is powered at the base by proton motive force.

Fungi (Eumycota) consist of cells with chitinous cell walls that grow in chains called **hyphae** (singular, **hypha**). Fungi range from single-celled organisms, such as yeasts, to complex multicellular forms such as mushrooms. The deepest-branching clade of fungi, Chytridiomycota, produces the uniflagellar zoospores that define opisthokonts. Other fungi, however, generate nonmotile reproductive cells, a result of reductive evolution.

Several taxa that historically were grouped with fungi (for example, slime molds) are now classified genetically as protists. Slime molds generate populations of cells that migrate into a unified structure called a **fruiting body** to make reproductive cells. Slime molds are now grouped with amebas (Amoebozoa, discussed in Section 20.4). Water molds, which are plant and animal pathogens of the class Oomycetes, are now recognized as heterokont protists (see **eTopic 20.1**).

Algae Evolved by Engulfing Phototrophs

Algae are commonly defined as single-celled plants and simple multicellular plants lacking true stems, roots, and leaves. Algal cells contain **chloroplasts**, membrane-enclosed organelles of photosynthesis that evolved from a cyanobacterium (**Fig. 20.6**). In Earth's biosphere, algae plus bacterial phototrophs feed all marine and aquatic ecosystems, producing the majority of oxygen and biomass available for Earth's consumers.

How did ancestral algae get their chloroplasts? The "green plants" (Viridiplantae) include primary

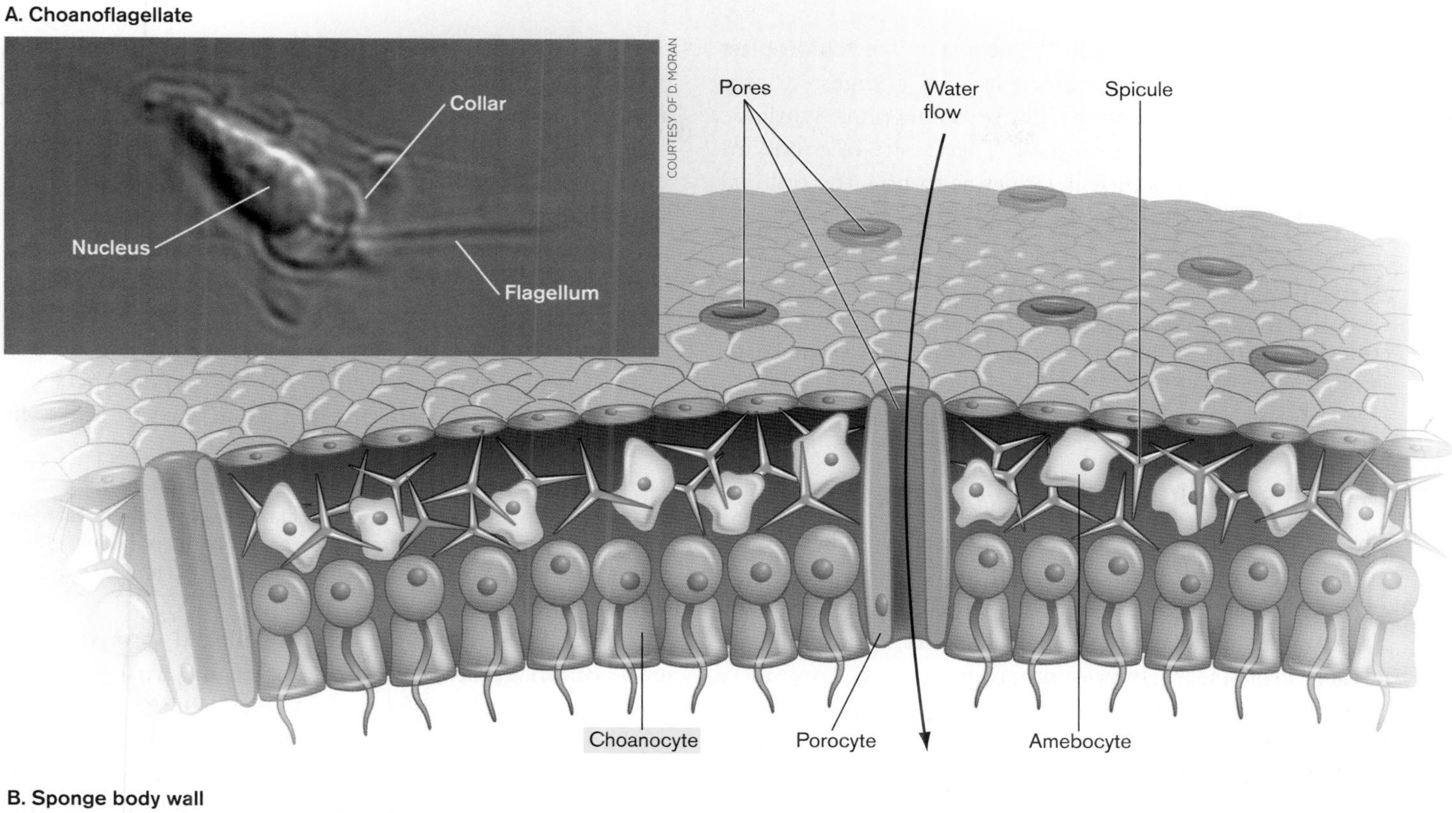
A. Choanoflagellate
Collar
Nucleus
Flagellum
COURTESY OF D. MORAN
Pores
Water flow
Spicule
Choanocyte
Porocyte
Amebocyte
B. Sponge body wall

FIGURE 20.5 ■ Choanoflagellates resemble sponge choanocytes. A. A choanoflagellate, *Acanthocorbis unguiculata* (cell length 6 μm). **B.** Sponge choanocytes resemble choanoflagellates. Within the sponge, choanocytes assist the circulation of water and the uptake of nutrients.

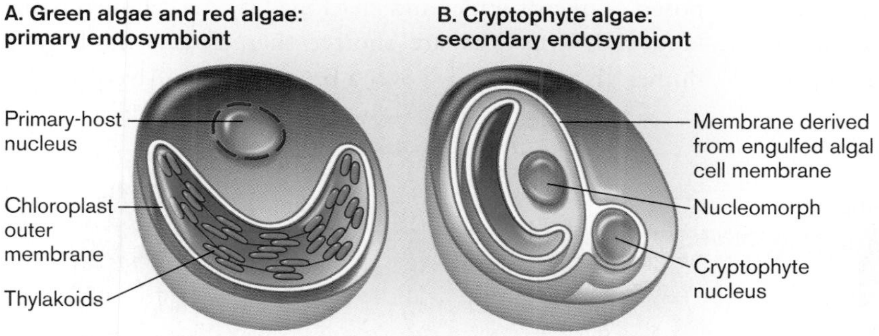
A. Green algae and red algae: primary endosymbiont
B. Cryptophyte algae: secondary endosymbiont
Primary-host nucleus
Chloroplast outer membrane
Thylakoids
Membrane derived from engulfed algal cell membrane
Nucleomorph
Cryptophyte nucleus

FIGURE 20.6 ■ Chloroplast evolution: primary and secondary endosymbiosis. A. Green algae (Chlorophyta) and red algae (Rhodophyta) contain chloroplasts (green) that evolved from engulfed cyanobacteria. Primary-host cytoplasm is colored yellow; primary-host nuclear membrane is colored brown. **B.** Cryptophyte algae contain chloroplast (green), vestigial primary-host cytoplasm (yellow), and vestigial nucleus or nucleomorph (brown) from the engulfed primary endosymbiont.

endosymbiont algae descended from a common ancestor containing a chloroplast. The chloroplasts of primary endosymbionts are enclosed by two membranes (**Fig. 20.6A**): the inner membrane (from the ancestral phototroph's cell membrane) and the outer membrane (from the host cell membrane as it enclosed its prey). Both **green algae** (**chlorophytes**) and **red algae** (**rhodophytes**) are primary endosymbionts. Their chloroplasts diverged from their common ancestor to utilize pigments absorbing different ranges of the light spectrum. Genetically, the green algae are the most closely allied with plants.

Surprisingly, several other taxa traditionally considered to be algae turn out to be secondary endosymbionts derived from protist hosts. The symbiotic history is most evident in the cryptophyte algae (**Fig. 20.6B**), which still retain a vestigial nucleus, or **nucleomorph**, derived from the engulfed cell. Their chloroplast is surrounded by two extra membranes, one from the primary endosymbiont (engulfed alga) cell membrane and one from the secondary host. Other secondary endosymbiont algae include the **chrysophytes**, such as kelps and diatoms. The **dinoflagellates**, photoheterotrophs of the protist clade Alveolata, are secondary or tertiary endosymbionts, descended from a flagellate that consumed one or more types of algae.

Dinoflagellates also engage in "kleptoplasty," or "chloroplast stealing," in which the chloroplast of a digested prey is retained long enough to derive some photosynthetic energy, but ultimately consumed. The variety of endosymbiosis among protists provides clues as to how the original chloroplast evolved within the ancestral algae.

Note: "Algae" may refer to primary algae, or "true algae," as well as secondary endosymbiont algae that derive from protist clades. Fungi (the "true fungi," or Eumycota) are opisthokonts. Several fungus-like organisms have been reclassified within protist clades.

Protists Include Many Divergent Clades

The protists actually include several distantly related categories of eukaryotes (see **Table 20.1**). All protists are heterotrophs, commonly predators or parasites, although many also conduct photosynthesis as secondary endosymbionts (algae). Protists are important producers and consumers in marine, aquatic, and soil food webs. In ecology, phototrophic protists are termed "phytoplankton" and heterotrophs are termed "zooplankton," although many in fact are "mixotrophs" that act as both producers and consumers.

Amebas are unicellular organisms of highly variable shape that form **pseudopods**, locomotory extensions of cytoplasm enclosed by the cell membrane. Their size can reach several millimeters, and they can eat small invertebrates. There are two major groups of amebas: Amoebozoa and Rhizaria (see **Table 20.1**). The Amoebozoa, the most familiar kind of amebas, have lobed pseudopods, pseudopods that extend lobes of cytoplasm through cytoplasmic streaming. Most lobed amebas are free-living in aquatic habitats, but some cause human diseases such as meningitis. Still other kinds of lobed amebas are cellular slime molds, in which individual amebas converge to form a fruiting body. Amebas of the second group, Rhizaria, have thin, filamentous pseudopods, often radially arranged like a star, as in heliozoan amebas. Some Rhizaria, such as the foraminiferans, form inorganic shells called tests. Fossil foraminiferan tests are common in rock formations derived from ancient seas. Foraminiferan shells formed the white cliffs of Dover in Britain and the stone used to build the Egyptian pyramids.

Note: DNA sequencing led to reclassification of the filamentous amebas as Rhizaria (formerly Cercozoa, in the previous edition of this book). "Cercozoa" now denotes a subgroup under Rhizaria.

Alveolates (Alveolata) include ciliated protists (ciliates), dinoflagellates, and apicomplexans. An example is the ciliated protist *Paramecium* (**Fig. 20.7A**). Alveolates are known for their complex outer covering, or cortex. The cortex contains networks of vesicles called **cortical alveoli** (**Fig. 20.7B**). Alveoli store calcium ion and in some species grow protective plates. Organisms equipped with paired flagella or cilia are known as **flagellates** and **ciliates**, respectively. **Flagella** (singular, **flagellum**) and **cilia** (singular, **cilium**) are essentially equivalent organelles composed of microtubules and enveloped by the cell membrane (**Fig. 20.7C**). Cilia are shorter than flagella and more numerous, and usually cover a broad surface. Alveolata also

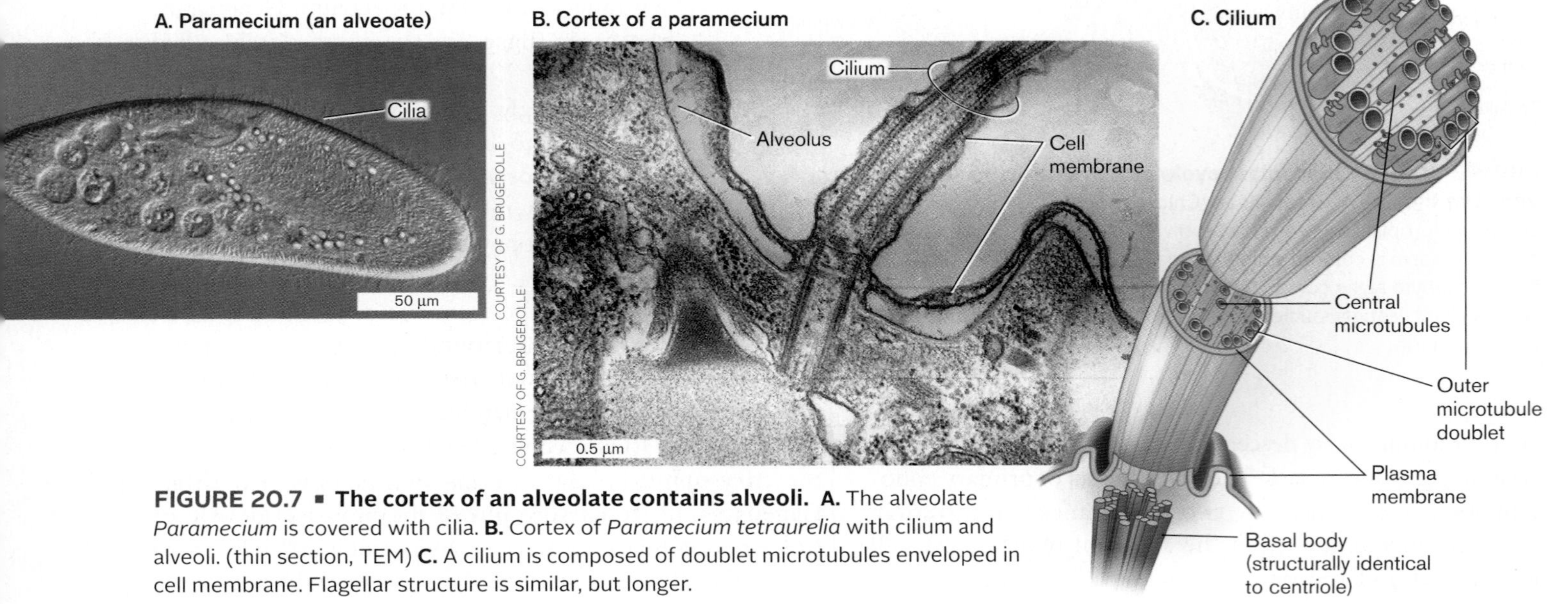
A. Paramecium (an alveoate)
Cilia
50 µm
COURTESY OF G. BRUGEROLLE
B. Cortex of a paramecium
Cilium
Alveolus
Cell membrane
0.5 µm
COURTESY OF G. BRUGEROLLE
C. Cilium
Central microtubules
Outer microtubule doublet
Plasma membrane
Basal body (structurally identical to centriole)

FIGURE 20.7 ■ The cortex of an alveolate contains alveoli. A. The alveolate *Paramecium* is covered with cilia. **B.** Cortex of *Paramecium tetraurelia* with cilium and alveoli. (thin section, TEM) **C.** A cilium is composed of doublet microtubules enveloped in cell membrane. Flagellar structure is similar, but longer.

includes a major group of parasites that no longer possess flagella, the **apicomplexans**. A well-known apicomplexan parasite is *Plasmodium falciparum*, which causes malaria.

Heterokonts (Heterokonta) are named for their pairs of differently shaped flagella (see **Table 20.1**). Heterokonts are thus distinguished from opisthokonts, which possess a single unpaired flagellum (if any). Flagellated heterokonts possess two flagella of unequal length and different structure. They include many voracious zooplankton. Nonflagellated heterokonts include the oomycetes, or water molds (formerly classified with fungi), as well as diatoms and kelps (secondary endosymbiotic algae with chloroplasts). **Diatoms** are single cells with unique bipartite shells that fit together like a petri dish. The shells of diatoms form an infinite variety of different patterns for different species. **Kelps**, also known as brown algae, extend multicellular sheets floating at the water's surface. The kelp known as sargassum weed is famous for its growth in the Sargasso Sea.

The Euglenozoa include free-living protists such as euglenas, as well as parasites showing extensive evolutionary reduction. Most Euglenozoa have mitochondria with distinctive disk-shaped cristae (membrane pockets). Parasitic Euglenozoa include the trypanosomes that cause sleeping sickness and Chagas' disease. A related clade, Metamonada, includes the waterborne parasite *Giardia intestinalis* (*G. lamblia*). In metamonads the mitochondria have lost their genomes and degenerated into **mitosomes** (discussed in Section 20.6).

A. B.

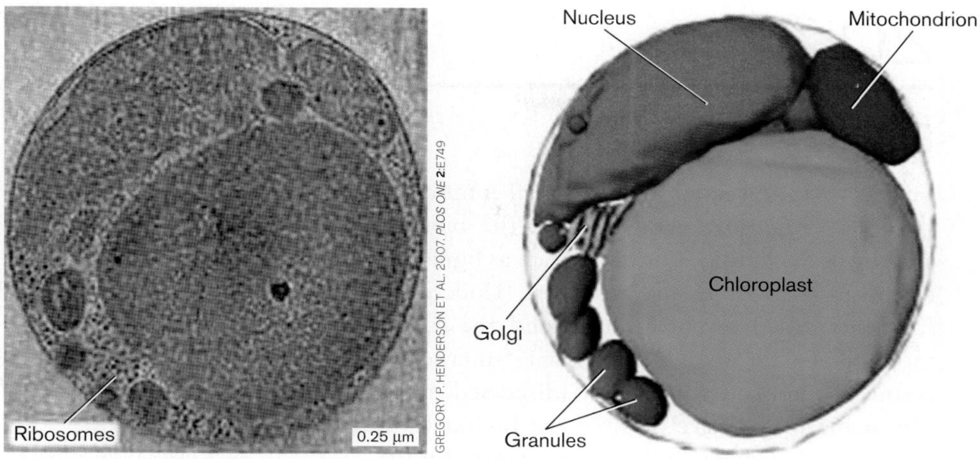
Ribosomes
0.25 µm
GREGORY P. HENDERSON ET AL. 2007. PLOS ONE 2:E749
Nucleus
Mitochondrion
Chloroplast
Golgi
Granules

FIGURE 20.8 ■ A tiny eukaryote still has organelles. *Ostreococcus tauri*, visualized by cryo-EM **(A)** and in 3D by tomography **(B)**.

Emerging Eukaryotes

Are there still eukaryotes that we haven't yet discovered? Genes from natural communities continually reveal new species of microbial eukaryotes in previously unknown divisions. Many of the new isolates are single cells as small as bacteria, designated nanoeukaryotes (3–10 µm) and picoeukaryotes (0.5–3 µm). For example, the picoeukaryote *Ostreococcus tauri* is a green alga less than 2 µm across (**Fig. 20.8**). The tiny alga consists of a flat, disk-shaped cell containing one or two of each type of organelle (a mitochondrion, a chloroplast, a stack of Golgi) packed tightly within the small volume. The genome of *O. tauri* is also downsized; at only 8 Mb, it is barely twice that of the bacterium *Escherichia coli*.

Genetic analysis shows that similar miniaturized eukaryotes branch deeply from all groups in the phylogenetic tree, potentially doubling the known number of major eukaryotic taxa. Furthermore, our metagenomic analysis reveals eukaryotes in environments previously believed to be restricted to bacteria and archaea, ranging from Antarctic sea ice to the anaerobic submarine sediments, and the hyperacidic Tinto River in Spain. In every new habitat tested, new deep-branching clades of eukaryotes emerge, with new implications for ecology and global cycling (discussed in Chapters 21 and 22). The wealth of new genomic data is reshaping our understanding of the domain Eukarya.

To Summarize

- **Opisthokonta** includes true fungi (Eumycota) and multicellular animals (Metazoa), as well as certain kinds of protists.
- **Viridiplantae** includes green plants and primary endosymbiont algae.
- **Protist groups** include Amoebozoa, the unshelled amebas with lobe-shaped pseudopods; Rhizaria, the filopodial amebas; Alveolata, ciliates and flagellates with complex cortical structure; Heterokonta, the kelps, diatoms, and flagellates with nonequivalent paired flagella; and Euglenozoa and Metamonada, primarily parasites.
- **Many protists are phototrophs as well as heterotrophs**, based on secondary or tertiary endosymbiosis derived from engulfed algae.
- **Metagenomic analysis reveals new clades of microbial eukaryotes**. New kinds of microbial eukaryotes emerge from all habitats.

20.2

Fungi

Fungi provide essential support for all communities of multicellular organisms. Fungi recycle the biomass of wood and leaves, including substances such as lignin, which other organisms may be unable to digest. Underground fungal filaments called mycorrhizae extend the root systems of most plants, forming a nutritional "internet" that interconnects the plant community (discussed in Chapter 21). Mycorrhizae may have inspired the fictional underground tree network depicted in the film *Avatar* (2009).

Within animal digestive systems, fungi ferment plant materials. On the other hand, pathogenic fungi infect plants and animals, and they contribute to the death of immunocompromised human patients. Still other fungi produce antibiotics such as penicillin, as well as food products such as wine and cheeses (discussed in Chapter 16).

Shared Traits of Fungi

Most fungi share these distinctive traits:

- **Absorptive nutrition.** Most fungi cannot ingest particulate food, as do protists, because their cell walls cannot part and re-form, like the flexible pellicles of amebas and ciliates. Instead they secrete digestive enzymes and then absorb the broken-down molecules from their environment.
- **Hyphae.** Most fungi grow by extending multinucleate cell filaments called hyphae (**Fig. 20.9A**). As a hypha extends, its nuclei divide mitotically without cell division, generating a multinucleate cell. Hyphae grow by cytoplasmic extension and branching. A branched mass of extending hyphae is called a **mycelium** (plural, **mycelia**).
- **Cell walls contain chitin.** Chitin is an acetylated amino-polysaccharide of immense tensile strength, stronger than steel (**Fig. 20.9B**). Its strength derives from multiple hydrogen bonds between fibers. Chitinous cell walls enable fungi to penetrate plant or animal cells, including tough materials such as wood. Inhibitors of chitin synthesis, such as the polyoxins and nikkomycins, are used as antibiotics against fungal infections.
- **Membranes contain ergosterol.** Ergosterol is an analog of cholesterol not found in animals or plants. Ergosterol is so distinctive to fungi that its presence can be used as a measure of fungal content in plant material such as grains. Inhibitors of ergosterol biosynthesis, such as the triazoles, are used to treat fungal infections. Another antifungal agent, nystatin, specifically binds ergosterol and forms membrane pores that leak K^+ ions.

Fungal Hyphae Absorb Nutrients

How do fungal hyphae grow and form colonies of "mold"? A fungal hypha expands at the tip. Cytoplasmic expansion is driven by turgor pressure against the chitin cell wall—the force that enables fungi to penetrate tough materials such as wood. Fungal hyphae can extend as fast as half a centimeter per hour.

The cytoplasmic turgor pressure is regulated by uptake of hydrogen ions in exchange for potassium ions (**Fig. 20.10A**). Loss of turgor pressure—for example, by puncture of the hypha—leads to accelerated K^+ uptake and

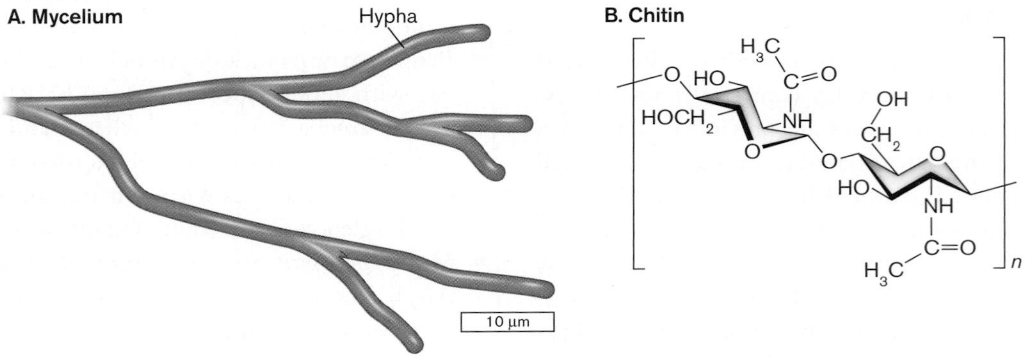
A. Mycelium
Hypha
10 μm
B. Chitin
H3C
C=O
HO
O
HOCH2
NH
O
O
OH
CH2
O
HO
NH
C=O
H3C
n

FIGURE 20.9 ▪ Fungi grow hyphae with cell walls of chitin. A. Fungal hyphae extend and form branches, generating a mycelium. **B.** Chitin consists of beta-linked polymers of *N*-acetylglucosamine.

water influx, restoring turgor. Molecules that cause loss of K^+, such as nystatin, serve as antifungal agents.

At the hypha's growing tip, turgor pressure pushes the cell membrane forward, and the membrane expands by incorporating vesicles generated from the endoplasmic reticulum (as seen in the micrograph, **Fig. 20.10B**). The ER and mitochondria store Ca^{2+}, whose release triggers vesicle fusion with the plasma membrane. The fused vesicles provide phospholipids and proteins to extend the membrane surface area as the cytoplasm expands.

Just behind the hypha's growing tip lies its absorption zone (see **Fig. 20.10A**). The absorption zone takes in nutrients from the surrounding medium, such as the cytoplasm of an invaded animal cell. Behind the absorption zone, the older part of the hypha collects and stores nutrients. As the storage zone expands, the nucleus divides multiple times. Septa form across the hypha, partly compartmentalizing the cytoplasm. As the older part of the hypha ages, its tubular form begins to lyse, releasing cell constituents. This aging part of the hypha is called the senescence zone (not shown).

As hyphae grow, branches extend from their sides. The hyphae branch and extend radially, forming the mycelium. The mycelium forms the characteristic round, fuzzy colony of a fungus or "mold." On a substrate such as wood or agar, mycelia grow in two forms: aerial mycelium, which extends out into the air; and surface mycelium, which grows into and along the surface of the substrate.

Unicellular Fungi

Despite the advantages of multicellular hyphae, some fungi are unicellular, known as **yeasts**. Yeast forms evolved in many different fungal taxa. The yeast *Saccharomyces*

A.

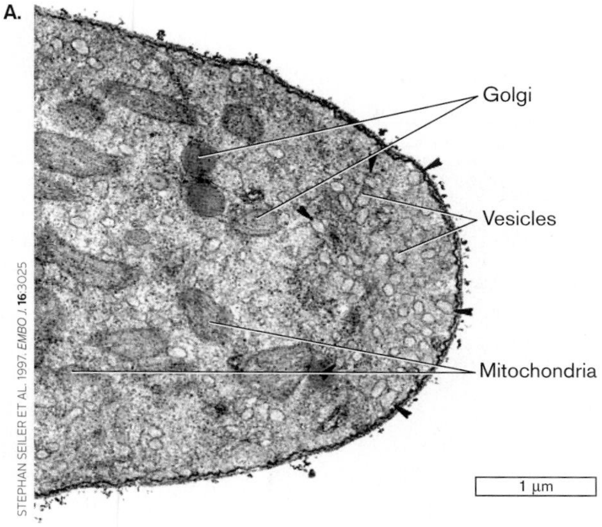

Golgi
Vesicles
Mitochondria
1 μm
STEPHAN SEILER ET AL. 1997. EMBO J. 16:3025

B.

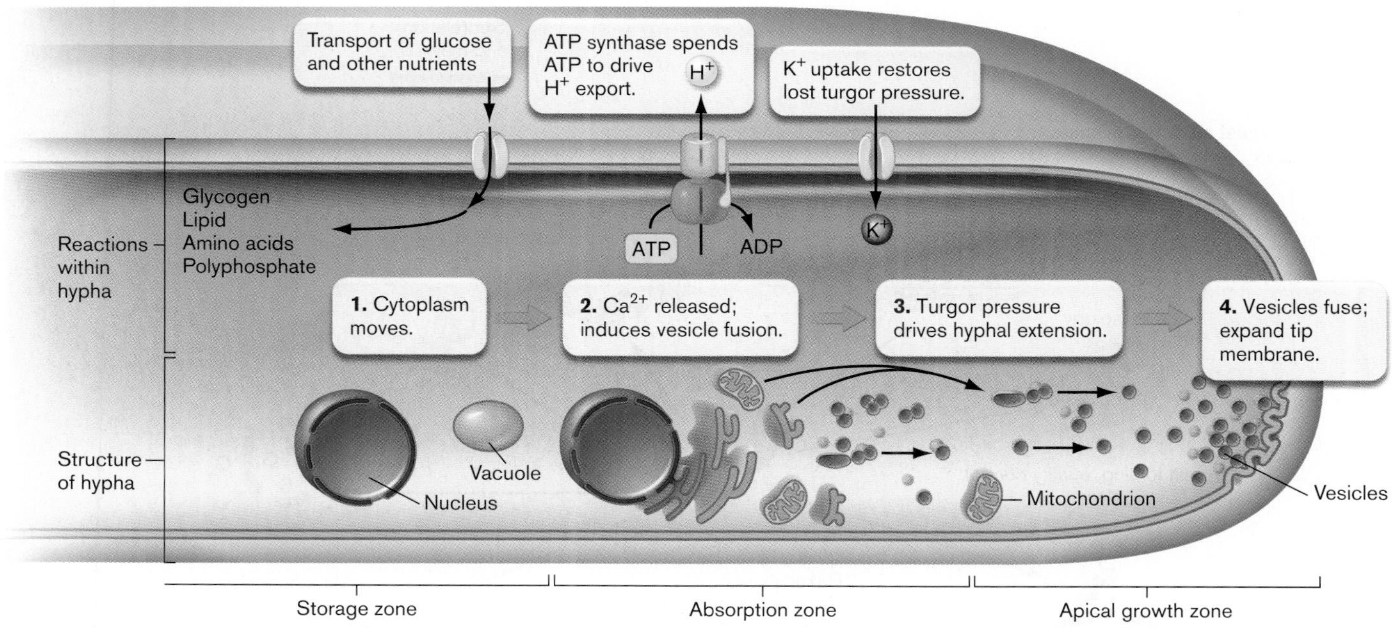

Transport of glucose and other nutrients
ATP synthase spends ATP to drive H+ export.
H+
K+ uptake restores lost turgor pressure.
Glycogen
Lipid
Amino acids
Polyphosphate
Reactions within hypha
ATP
ADP
K+
1. Cytoplasm moves.
2. Ca2+ released; induces vesicle fusion.
3. Turgor pressure drives hyphal extension.
4. Vesicles fuse; expand tip membrane.
Structure of hypha
Vacuole
Nucleus
Mitochondrion
Vesicles
Storage zone
Absorption zone
Apical growth zone

FIGURE 20.10 ▪ Cellular basis of hyphal extension. A. Section through the growing tip of a hypha (TEM). Vesicles collect at the tip, where they fuse into the cell membrane, enabling extension. **B.** The absorption zone takes in nutrients. Cytoplasm moves toward the tip of the apical growth zone, driven by turgor pressure. Turgor pressure is regulated by H^+ export and K^+ uptake. Ca^{2+} released by the ER and mitochondria induces vesicles to fuse and to expand the plasma membrane at the growing tip.

cerevisiae (**Fig. 20.11A**) is used to leaven bread and to brew wine and beer (for more on food microbiology, see Chapter 16). *S. cerevisiae* reproduces by **budding**, in which mitosis of the mother cell generates daughter cells of smaller size. The mother cell acquires a bud scar where the smaller one pinched off (**Fig. 20.11A**). After generating a limited number of buds, the mother cell senesces and dies. Thus, yeasts provide a unicellular model system for the process of aging.

Other yeasts, such as *Candida albicans,* are important members of human vaginal flora but can cause opportunistic infections. Some are important opportunistic pathogens, occurring frequently in AIDS patients; for example, *Pneumocystis jirovecii* (formerly *P. carinii*) is a yeast-form

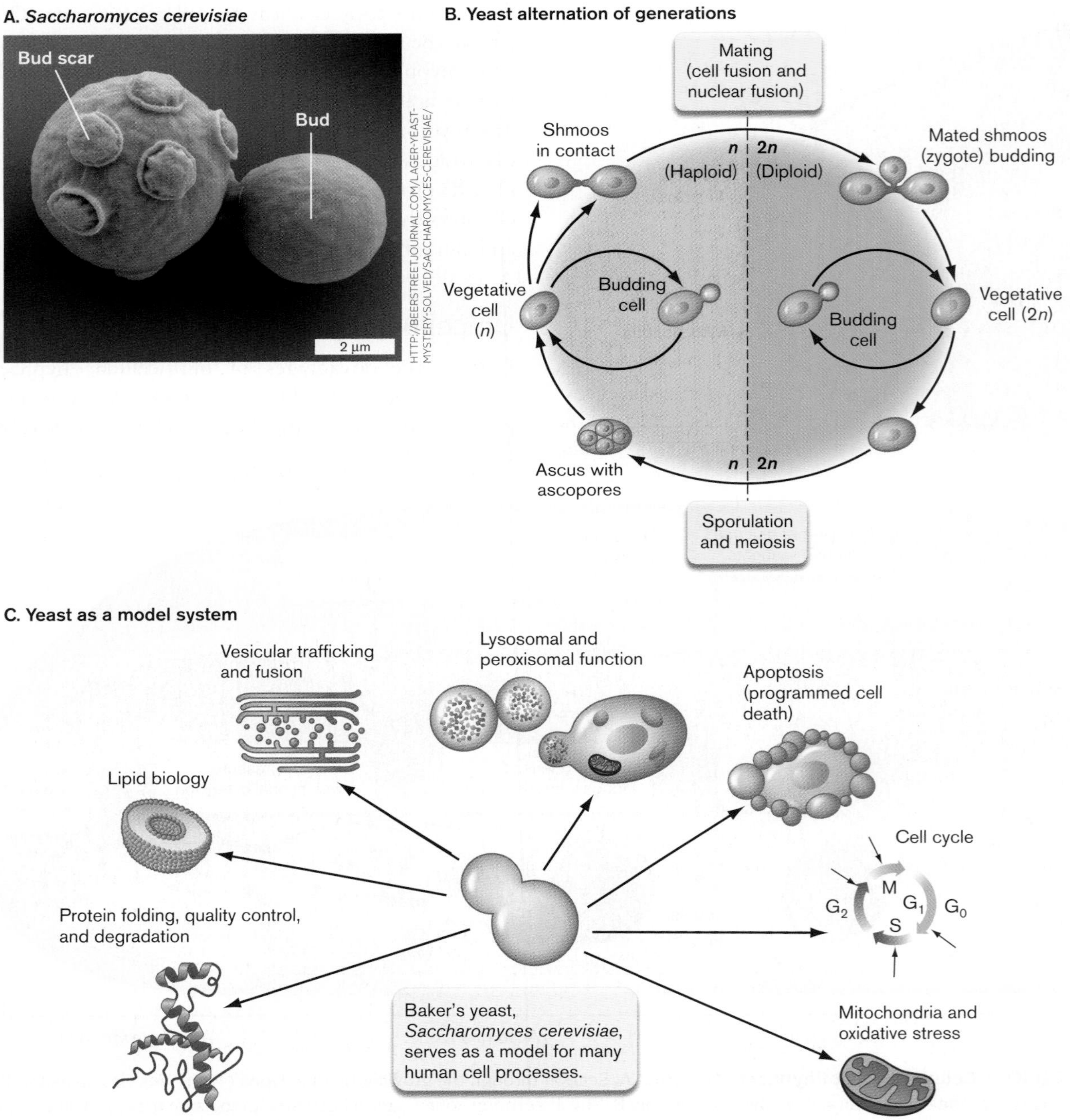
A. Saccharomyces cerevisiae
Bud scar
Bud
2 µm
HTTP://BEERSTREETJOURNAL.COM/LAGER-YEAST-MYSTERY-SOLVED/SACCHAROMYCES-CEREVISIAE/
B. Yeast alternation of generations
Mating (cell fusion and nuclear fusion)
Shmoos in contact
n 2n
(Haploid) (Diploid)
Mated shmoos (zygote) budding
Vegetative cell (n)
Budding cell
Budding cell
Vegetative cell (2n)
Ascus with ascopores
n 2n
Sporulation and meiosis
C. Yeast as a model system
Vesicular trafficking and fusion
Lysosomal and peroxisomal function
Apoptosis (programmed cell death)
Lipid biology
Cell cycle
M G1 G0 G2 S
Protein folding, quality control, and degradation
Baker's yeast, Saccharomyces cerevisiae, serves as a model for many human cell processes.
Mitochondria and oxidative stress

FIGURE 20.11 ■ Yeasts are nonmycelial fungi. A. *Saccharomyces cerevisiae,* or baker's yeast, reproduces by budding (SEM; cell size 3–6 µm). Upper cell shows six bud scars. **B.** In the life cycle of *S. cerevisiae,* haploid cells reproduce many generations by budding. **C.** *S. cerevisiae* serves as a model for human cell processes. *Source:* Part C modified from Vikram Khurana and Susan Lindquist. 2010. *Nat. Rev. Neurosci.* **11**:436–449.

ascomycete, whereas *Cryptococcus neoformans* is a yeast-form basidiomycete (discussed shortly). Pathogens such as *Candida albicans* can grow either as single cells (yeast form) or as mycelia; these are known as "dimorphic" fungi. The yeast form occurs normally in the mucosa, but germination of mycelia leads to disease. A very different dimorphic fungal pathogen is *Blastomyces dermatitidis,* the cause of blastomycosis, a type of pneumonia. *B. dermatitidis* forms a mycelium in culture and in soil environments, but it grows as a yeast within the infected lung.

Yeast as a model organism for research. The yeast *Saccharomyces cerevisiae* provides a major research subject for eukaryotic biology (**Fig. 20.11C**). Its cells grow rapidly, in haploid and diploid forms, and offer convenient genetic recombination and transformation. The yeast genome of 6,000 genes includes many human homologs, such as actin and the *ras* proto-oncogene (gene involved in cancer). In 2001, the Nobel Prize in Physiology or Medicine was awarded to yeast researchers Leland Hartwell and Paul Nurse for their discovery of key molecules of the eukaryotic cell cycle whose defects lead to carcinogenesis. Often entire networks of proteins interacting in yeast have human homologs; thus, *S. cerevisiae* has been called a "single-celled human." For a surprising application of the yeast model to studying disease of the brain, see **Special Topic 20.1**.

Thought Question

20.1 Why would yeasts remain unicellular? What are the relative advantages and limitations of hyphae?

Yeast reproductive cycles. Some yeasts are asexual, whereas others can undergo sexual **alternation of generations** (**Fig. 20.11B**). This life cycle alternates between generation of a haploid population, with a single copy of each chromosome (n), and a diploid population, with a diploid chromosome number ($2n$). The haploid form develops gametes to fertilize each other, making a $2n$ zygote. After vegetative (nonsexual) divisions, the $2n$ form undergoes meiosis, regenerating the haploid form. (The process of meiosis is reviewed in Appendix 2.) Alternation of generations allows an organism to respond genetically to environmental change by reassorting its genes through meiosis, and by recombining them through fertilization. Gene reassortment and recombination provide new genotypes, some of which may increase survival in the changed environment.

In baker's yeast (*Saccharomyces cerevisiae*), haploid spores divide and proliferate by mitosis, forming a haploid mycelium. Under environmental signals such as starvation, mating factors induce the haploid cells to differentiate into gamete forms called "shmoos" (see **Fig. 20.11B**). Gametes of two different mating types fuse, and their nuclei combine to form a zygote. In the diploid generation, the zygote divides mitotically, generating a population of diploids that appear superficially similar to haploid cells. Under stress, particularly desiccation, the diploids undergo meiosis to reassort their genes for combinations that may better survive the changed environment. Meiosis generates an **ascus** (plural, **asci**) that contains four haploid spores.

Many fungi and protists undergo modified versions of alternation of generations, utilizing a wide variety of haploid and diploid structures to accomplish essentially the same genetic tasks. In many fungi, the haploid form predominates; for example, ascomycetes such as *Aspergillus* and *Neurospora* form mainly haploid mycelia. In some fungi, no sexual reproduction has been observed, probably because the inducing conditions are unknown. Species that lack a known sexual cycle are called **mitosporic fungi**, also known as "imperfect" fungi. Mitosporic species are found in many different clades. An example is the famous *Penicillium* mold, an ascomycete, from which we discovered the antibiotic penicillin.

Thought Question

20.2 Why would some fungi conduct asexual reproduction under most conditions? What are the advantages and limitations of sexual reproduction?

Mycelia, Mushrooms, and Mycorrhizae

Different species of fungi show vastly different forms, from the familiar mushrooms (fruiting bodies that can weigh several pounds) to the mycelia of pathogens and the symbiotic partners of algae in lichens. Major clades of fungi include Chytridiomycota, Zygomycota, Ascomycota, and Basidiomycota.

Note: The major groups of fungi are also known by names with the alternative suffix "-etes": Chytridiomycetes, Zygomycetes, Ascomycetes, Basidiomycetes.

Chytridiomycota: motile zoospores. The deepest-branching clade of fungi is Chytridiomycota (the chytrids), which share with animals and choanoflagellates the motile, flagellated reproductive form known as a zoospore. The zoospore form has been lost by other fungi.

Chytrid species include bovine rumen inhabitants whose hyphae penetrate tough plant material, facilitating digestion. An example is *Neocallomastix* species

Special Topic 20.1: Yeast: A Single-Celled Human Brain?

The yeast *Saccharomyces cerevisiae* provides a model for human diseases such as cancer because it shares so many homologs of our genes. Most of these genes encode fundamental parts of cells, such as actin and cell growth regulators. But of course the single-celled fungus lacks the differentiated parts and connectors of human cells such as neurons. So yeast could not serve as a model for complex diseases of the brain. Or could it?

Neurodegenerative diseases can result from disorders of cell function, such as lysosomal storage and degradation. A model of Alzheimer's disease holds that a protein called beta-amyloid gets cleaved to a peptide that is secreted; the secreted peptide returns to the cell by endocytosis. The endocytosed beta-amyloid peptide then interferes with intracellular trafficking by an unknown mechanism. To investigate the mechanism of beta-amyloid toxicity, Susan Lindquist at the Whitehead Institute for Biomedical Research (**Fig. 1**) developed a yeast model.

First, Lindquist and her students constructed a yeast strain that expresses beta-amyloid peptide from a plasmid. (We describe methods to construct such strains in Chapter 12.) In this yeast strain, the beta-amyloid gets secreted, but it returns through endocytosis to the endoplasmic reticulum (ER) for trafficking—just as it does in a human cell. So what happens to the yeast? The yeast grows more slowly because beta-amyloid disrupts ER trafficking (**Fig. 2**). The controls—a yeast protein expressed on a plasmid, and a plasmid vector expressing no protein—grow normally.

The disruption of trafficking by beta-amyloid was consistent with its proposed role in Alzheimer's disease. But to strengthen the connection—and to reveal other parts of the process—Lindquist used her yeast model to screen for yeast

B. P. TRIVEDI. 2012. *NATURE* **482**: 294

FIGURE 1 ▪ Susan Lindquist, Whitehead Institute for Biomedical Research. Lindquist develops yeast models of human disease.

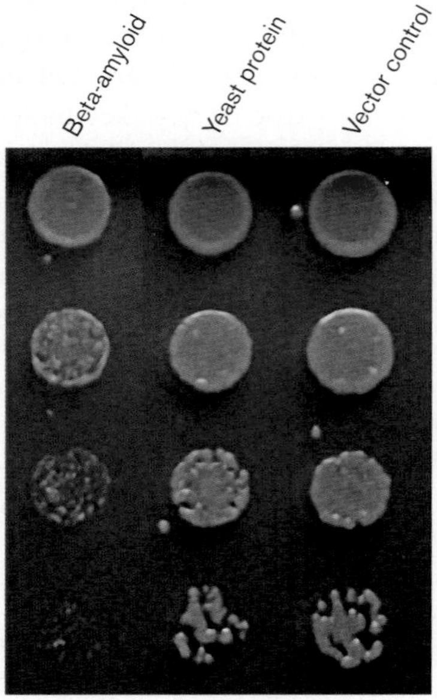

Beta-amyloid
Yeast protein
Vector control

SEBASTIAN TREUSCH ET AL. 2011. *SCIENCE* **334**: 1241

FIGURE 2 ▪ Yeast expresses human beta-amyloid peptide. Yeast cultures were serially diluted (vertical direction) and spread on agar to grow patches. A plasmid expressing beta-amyloid (left) causes slower growth (thinner patches) than a plasmid expressing a normal yeast protein or no protein (vector control) (middle and right, respectively).

(**Fig. 20.12A**). Unlike most fungi, *Neocallomastix* is an obligate anaerobe whose mitochondria have evolved into **hydrogenosomes**, organelles that ferment carbohydrates in a pathway generating H_2. Hydrogenosomes are a unique adaptation of certain anaerobic fungi and protists.

Other chytrids are aerobic animal pathogens. **Figure 20.12B** shows the skin of a frog infected by the chytridiomycete *Batrachochytrium dendrobatidis*. *B. dendrobatidis* has caused a widespread die-off of frogs in Central and South America, in an epidemic associated with global warming. The mycelium of *B. dendrobatidis* grows within the frog skin, producing capsules full of diploid zoospores called zoosporangia. Each zoosporangium protrudes through the skin surface, ready to expel zoospores in search of a new host.

proteins that could overcome beta-amyloid interference. So she transformed her yeast model strain with a library of yeast genes that were "overexpressed" (that is, expressed on a plasmid at levels severalfold greater than normal). She identified 23 suppressors, genes whose overexpression suppressed the effect of beta-amyloid and restored normal yeast growth. The suppressors also restored normal trafficking of a reporter protein, a fluorescent YFP fusion protein expressed by the yeast genome (**Fig. 3**). The control cells showed reporter protein fluorescence localized normally to the vacuole (**Fig. 3A**). Beta-amyloid expression prevented trafficking to the vacuole (**Fig. 3B**), but expression of the suppressor protein YAP1802 restored movement to the vacuole (**Fig. 3C**).

Are these suppressors of the yeast model relevant to humans? Six of the suppressor genes turned out to have homologs in the human genome already identified as possible risk factors for Alzheimer's disease. The YAP1802 human homolog is a protein called PICALM. Lindquist tested whether PICALM overexpression could protect rat brain neurons from beta-amyloid toxicity. She provided PICALM to the neurons by expression on a lentiviral vector—that is, a gene therapy vector derived from HIV virus (discussed in Chapter 11). Remarkably, PICALM restored normal growth to neurons that had taken up beta-amyloid. For the future, Lindquist and her colleagues are using the yeast model to reveal additional human genes linked to Alzheimer's disease, as well as possible targets for chemotherapy and gene therapy.

RESEARCH QUESTION

How would you go about using the yeast model to identify additional human genes involved in Alzheimer's disease?

Treusch, Sebastian, Shusei Hamamichi, Jessica L. Goodman, Kent E. S. Matlack, Chee Yeun Chung, et al. 2011. Functional links between Ab toxicity, endocytic trafficking, and Alzheimer's disease risk factors in yeast. *Science* **334**:1241–1245.

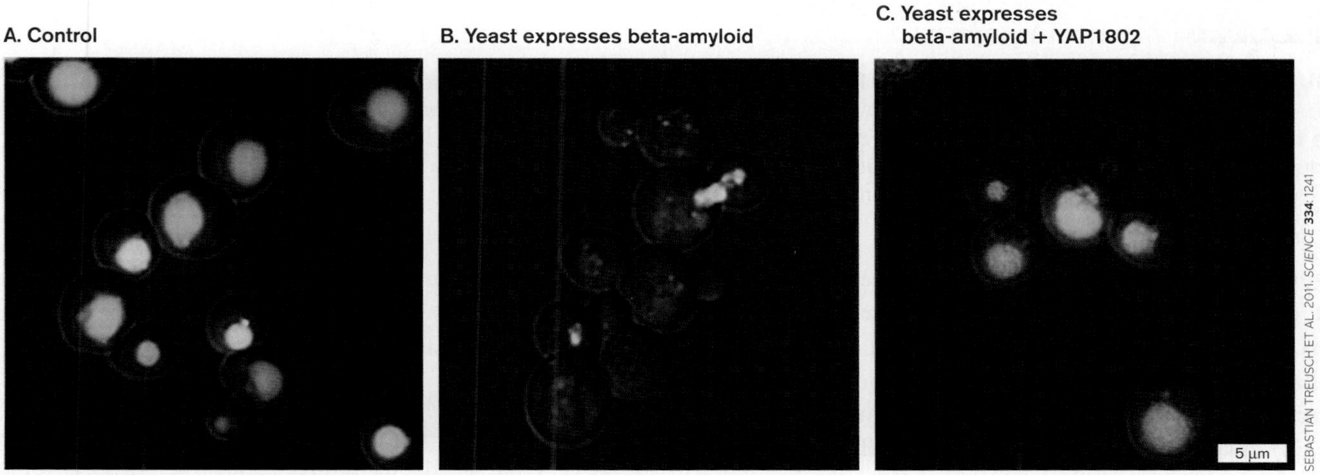
A. Control
B. Yeast expresses beta-amyloid
C. Yeast expresses beta-amyloid + YAP1802
5 μm
SEBASTIAN TREUSCH ET AL. 2011. SCIENCE 334: 1241

FIGURE 3 ▪ Beta-amyloid disrupts traffic of yeast protein. A. A yeast protein fused to yellow fluorescent protein (YFP) is trafficked normally to the vacuole (fluorescence microscopy). **B.** Beta-amyloid expression interferes with transport. **C.** YAP1802 expression restores transport.

The life cycle of a chytrid includes both haploid (gametophyte) and diploid (sporophyte) mycelia (**Fig. 20.12C**). Haploid mycelia produce motile gametes that detect each other by sex-specific attractants. The gametes fuse to produce a motile zygote. The zygote forms a cyst, a cell with arrested metabolism that can persist for long periods. In a favorable environment, the cyst germinates to form a diploid mycelium, or sporophyte. The sporophyte generates zoosporangia full of zoospores. There are two alternative forms of zoosporangia: those that produce diploid zoospores, which form cysts and regenerate the diploid mycelium; and those that undergo meiosis to produce haploid zoospores. The haploid zoospores generate a haploid mycelium (gametophyte) capable of producing haploid gametes.

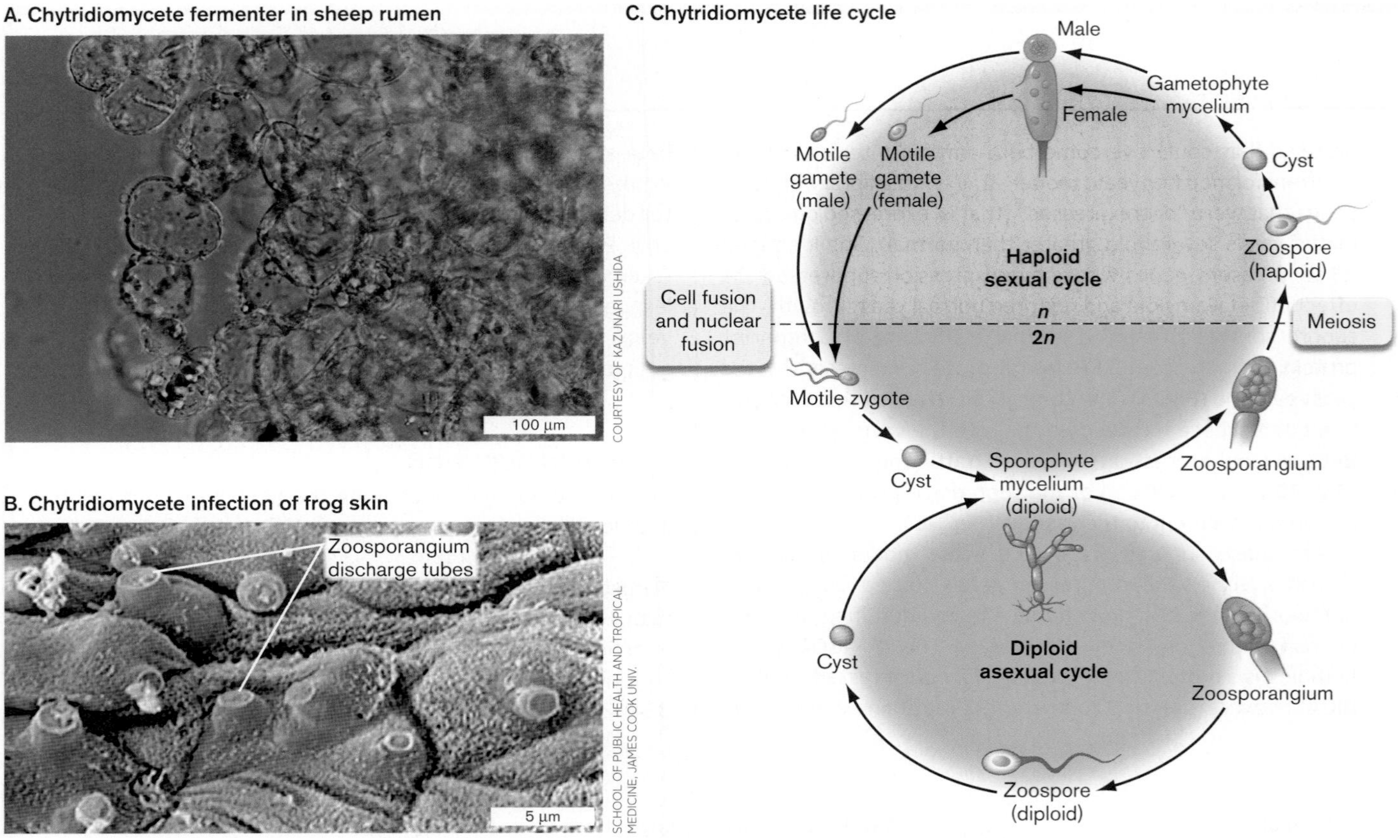
A. Chytridiomycete fermenter in sheep rumen
100 μm
COURTESY OF KAZUNARI USHIDA
B. Chytridiomycete infection of frog skin
Zoosporangium
discharge tubes
5 μm
SCHOOL OF PUBLIC HEALTH AND TROPICAL MEDICINE, JAMES COOK UNIV.
C. Chytridiomycete life cycle
Male
Female
Gametophyte
mycelium
Cyst
Zoospore
(haploid)
Motile
gamete
(male)
Motile
gamete
(female)
Haploid
sexual cycle
Cell fusion
and nuclear
fusion
n
2n
Meiosis
Motile zygote
Cyst
Sporophyte
mycelium
(diploid)
Zoosporangium
Diploid
asexual cycle
Cyst
Zoosporangium
Zoospore
(diploid)

FIGURE 20.12 ▪ Chytridiomycete form and life cycle. **A.** Chytrids such as *Neocallomastix* species ferment complex plant material for ruminant animals such as cattle and sheep. **B.** A pathogenic chytrid, *Batrachochytrium dendrobatidis,* infects the skin of a frog. Frog skin cells are penetrated by discharge tubes of diploid zoosporangia about to release zoospores. **C.** Life cycle of a chytrid. The diploid mycelium produces motile zoospores that form cysts in a poor environment. Alternatively, the diploid mycelium undergoes meiosis to form a haploid mycelium (gametophyte) that produces motile gametes.

Zygomycota: nonmotile sporangia. The **zygomycetes** and other nonchytridiomycete fungi generate nonmotile spores. Nonmotile spores require transport by air or water, or ballistic expulsion (expulsion under pressure) from a spore-bearing organ, called the **sporangium** (plural, **sporangia**). A common zygomycete is the bread mold *Rhizopus* (**Fig. 20.13A**). Most zygomycetes, such as *Mucor* species, are soil molds that decompose plant material or other fungi or the droppings of animals (**Fig. 20.13B**). These modest molds fill important niches in all terrestrial ecosystems.

The life cycle of a zygomycete parallels that of a chytrid in its alternative options of haploid (n) and diploid ($2n$) forms (**Fig. 20.13C**). The mechanics differ, however, owing to the lack of motile gametes. The haploid spore (sporangiospore) is disseminated through air currents. A sporangiospore does not directly undergo sexual reproduction; it grows into a haploid mycelium. The haploid mycelium then forms special hyphae whose tips differentiate into gamete cells. The gametes cannot separate from the filament; instead, two gamete-bearing hyphae must grow toward each other in order to fuse and form a **zygospore**. The zygospore undergoes meiosis and generates the sporangium, a haploid structure that releases **sporangiospores**.

Thought Question

20.3 What are the advantages and limitations of motile gametes, as compared to nonmotile spores?

Ascomycota: mycelia with paired nuclei. The **ascomycete** fungi are famous in the history of science, as well as in the culinary arts (**Fig. 20.14**). The bread mold *Neurospora* was used by George Beadle and Edward Tatum in

Thought Question

20.4 Compare the life cycle of an ascomycete (**Fig. 20.14C**) with that of a chytridiomycete (**Fig. 20.12C**). How are they similar, and how do they differ?

A. Bread mold, *Rhizopus*

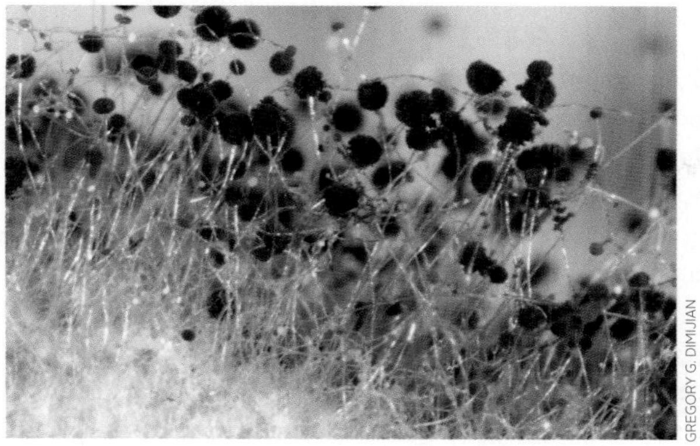
GREGORY G. DIMIJIAN

B. *Mucor* diploid hyphae form zygospores

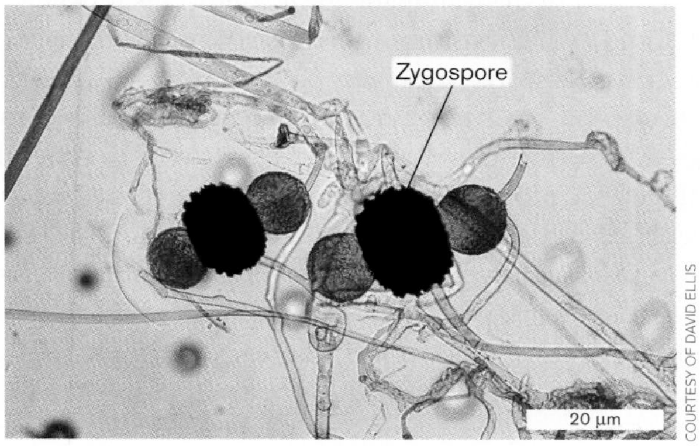
Zygospore
20 µm
COURTESY OF DAVID ELLIS

C. Zygomycete life cycle

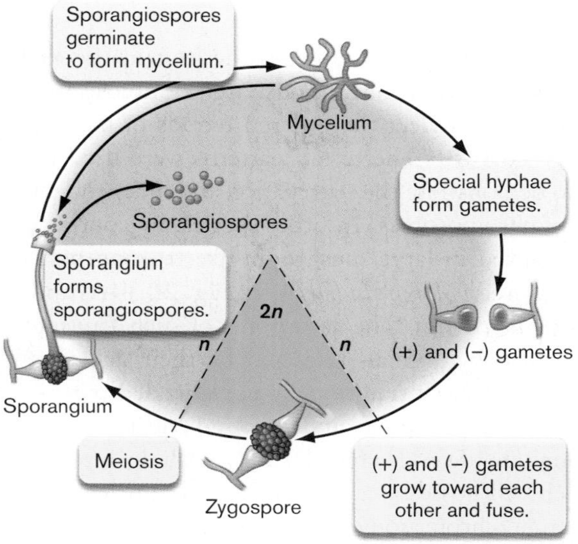
Sporangiospores germinate to form mycelium.
Mycelium
Special hyphae form gametes.
Sporangiospores
Sporangium forms sporangiospores.
2n
n
n
(+) and (–) gametes
Sporangium
Meiosis
Zygospore
(+) and (–) gametes grow toward each other and fuse.

FIGURE 20.13 ▪ Zygomycete fungi form nonmotile sporangia. **A.** *Rhizopus* (bread mold) haploid sporangia contain sporangiospores. **B.** Diploid hyphae of *Mucor* species terminate in zygospores. **C.** The life cycle of zygomycetes involves primarily haploid mycelia. Special hyphae form gametes at their tips. Gametes of different mating types fuse to form the diploid zygospore. The zygospore undergoes meiosis, regenerating haploid cells that form sporangia. The sporangia release sporangiospores, which germinate to form new mycelia.

A. Morel (an ascomycete fruiting body)

ED RESCHKE/PETER ARNOLD

B. Asci containing ascospores

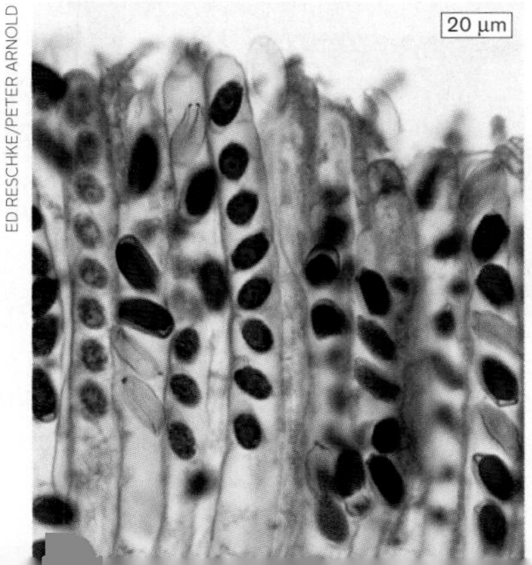
ED RESCHKE/PETER ARNOLD
20 µm

C. Ascomycete life cycle

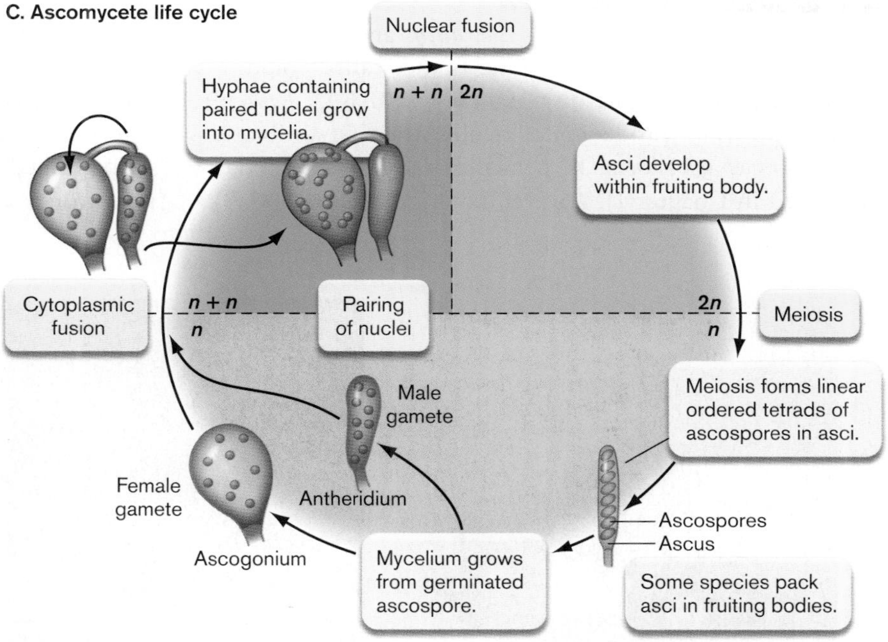
Nuclear fusion
Hyphae containing paired nuclei grow into mycelia.
n + n
2n
Asci develop within fruiting body.
Cytoplasmic fusion
n + n
n
Pairing of nuclei
2n
n
Meiosis
Male gamete
Meiosis forms linear ordered tetrads of ascospores in asci.
Female gamete
Antheridium
Ascospores
Ascus
Ascogonium
Mycelium grows from germinated ascospore.
Some species pack asci in fruiting bodies.

FIGURE 20.14 ▪ Ascomycetes produce large fruiting bodies. **A.** The culinary delicacies known as morels are fruiting bodies of the species *Morchella hortensis.* The dark pits of the morel are lined with asci. **B.** Ascomycete asci containing ascospores (stained red). **C.** The life cycle of an ascomycete alternates between the diploid and haploid forms. The diploid mycelium produces asci, within which the haploid ascospores are formed.

the 1940s to formulate the one gene–one protein theory. In *Neurospora,* meiosis produces pods (asci) of **ascospores** aligned in rows that reflect the ordered tetrads of meiotic division (**Fig. 20.14B**). The tetrad patterns were used by geneticists to demonstrate the segregation and independent assortment of chromosomes. In other species, by contrast, the asci are packed in large mushroom-like fruiting bodies known as morels (*Morchella hortensis*; **Fig. 20.14A**) and truffles (*Tuber aestivum*). The ascospores of such fruiting bodies are spread by animals attracted by their delicious flavor. Human collectors traditionally use muzzled pigs to detect and unearth the famous underground truffles.

The ascomycete life cycle (**Fig. 20.14C**) includes a phase in which each cell possesses a pair of separate nuclei, one from each parent (chromosome number is designated $n + n$). The "dikaryotic" (paired-nuclei) phase is generated by haploid mycelia in which male and female reproductive structures fuse, followed by migration of all the male nuclei into the female structure. The paired nuclei then undergo several rounds of mitotic division while migrating into the growing mycelium. In the mycelial tips, the paired nuclei finally fuse (becoming $2n$) and the mycelial tips develop into asci. Each ascus then undergoes meiosis in which the haploid products segregate in the same order that the meiotic chromosomes separated.

Some ascomycetes, such as *Aspergillus* and *Penicillium* species, form small asexual fruiting bodies called conidiophores for airborne spore dispersal (**Fig. 20.15A**). *Penicillium* is known for producing penicillin, the first antibiotic in widespread use; forms of penicillin are still used today (discussed in Chapter 1). *Aspergillus* (**Fig. 20.15A** and **B**) is a growing medical problem as an opportunistic pathogen of immunocompromised patients. *Aspergillus* can produce toxins (called mycotoxins) such as aflatoxin. Aflatoxin poisoning commonly affects livestock, and in some cases agricultural workers; the toxin causes liver damage, immunosuppression, and cancer.

Conidiophore-forming ascomycetes such as *Aspergillus* and *Stachybotrys* are the major form of mold associated with dampness in human dwellings; for example, they caused massive damage to homes flooded in the wake of Hurricane Katrina in 2005 (**Fig. 20.15C**). The flooding of homes full of drywall made ideal conditions for the growth of mold, which commonly consists of airborne ascomycete mycelia. Mold grew not only on materials submerged, but also on the surface above, exposed to water-saturated air, up to 3 feet above the flood line (the highest level submerged). Unfortunately, most homeowner insurance policies covered damage only "up to the flood line."

Many ascomycetes are pathogens of animals or plants. For example, *Microsporum* and *Trichophyton* species cause ringworm skin infection, whereas *Magnaporthe oryzae* causes rice blast, the most serious disease of cultivated rice. Other species, however, are beneficial symbionts of plants, including crop plants such as beans, cucumbers, and cotton. *Trichoderma* species grow on the roots, or in some cases within the vascular tissue, of the plant. The fungi share nutrients with the plant, and they induce plant defenses against pathogens. *Trichoderma* even has a commercial use in cloth processing; the fungus is used to make "stonewashed jeans," as its cellulase enzymes partly digest the cotton.

A.
B.
C.
DAVID M. PHILLIPS/VISUALS UNLIMITED
SCIENCE PHOTO LIBRARY
JOAN SLONCZEWSKI

FIGURE 20.15 ■ ***Ascomycyte molds.*** **A.** *Aspergillus* forms a microscopic asexual fruiting structure called a conidiophore, containing spores in its spherical tip. **B.** Colony of *Aspergillus nidulans* on an agar plate. **C.** Black mold (such as *Stachybotrys*) grows above the flood line on a kitchen wall. An undergraduate volunteer shows the presence of mold in New Orleans, 6 months after flooding caused by Hurricane Katrina.

A. *Amanita*

©FRED STEVENS

B. *Aleuria*

©BRUCE FUHRER

FIGURE 20.16 ■ Mushrooms and other basidiomycetes form large, complex fruiting bodies. A. *Amanita phalloides* makes one of the most dangerous toxins known: alpha-amanitin, an inhibitor of RNA polymerase II. **B.** *Aleuria* mushrooms releasing spores from their gills, to be carried on the wind.

Basidiomycota: cells with paired nuclei form mushrooms. The **basidiomycetes** form large, intricate fruiting bodies known as "true" mushrooms (**Fig. 20.16**). Mushrooms produce some of the world's deadliest poisons, such as alpha-amanitin, which inhibits RNA polymerase II. Alpha-amanitin is produced by the amanita, or "destroying angel" (**Fig. 20.16A**), a taste of which is usually fatal. (The amanita's own RNA polymerase is insensitive to the toxin.) Other mushroom species include some of the world's most prized culinary delights, such as the portobello. Many grow in soil, while others, such as *Piptoporus,* grow on tree bark. Some mushrooms have evolved elaborate insect-attracting structures and odors, such as the "starfish stinkhorn," with its ring of bright red horns.

The mushroom itself is only the fruiting body of the basidiomycete, whose life cycle involves transitions among n, $n + n$, and $2n$ (**Fig. 20.17A**) similar to those of ascomycetes (see **Fig. 20.14C**). In the basidiomycete, however, the

A. Basidiomycete (mushroom) life cycle

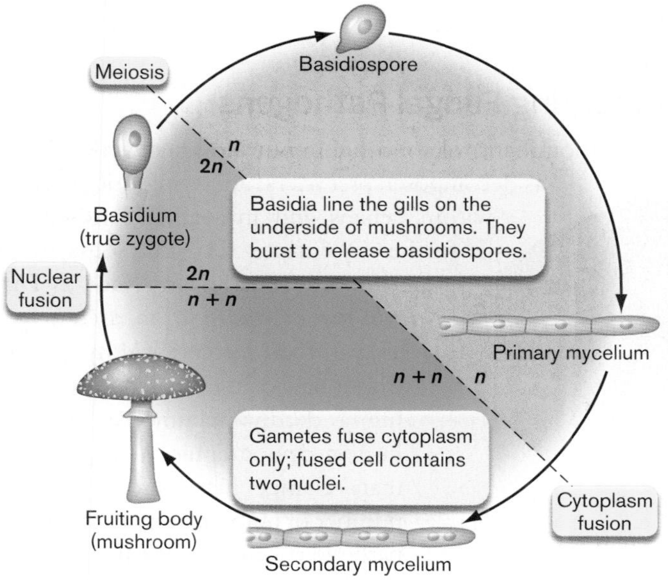
Meiosis
Basidiospore
n
2n
Basidium (true zygote)
Basidia line the gills on the underside of mushrooms. They burst to release basidiospores.
Nuclear fusion
2n
n + n
Primary mycelium
n + n
n
Gametes fuse cytoplasm only; fused cell contains two nuclei.
Cytoplasm fusion
Fruiting body (mushroom)
Secondary mycelium

B.

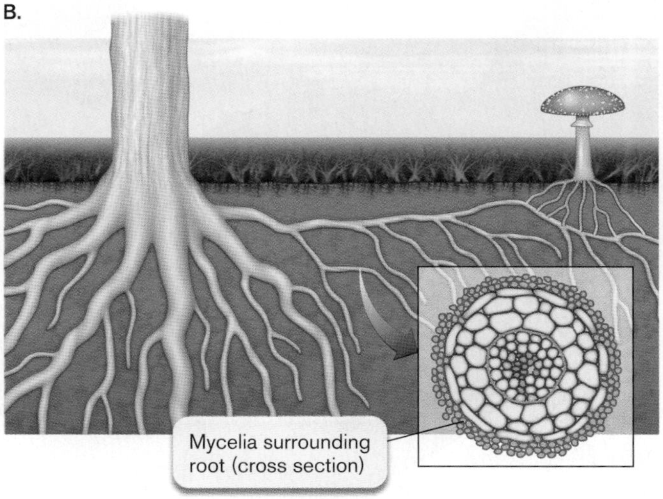
Mycelia surrounding root (cross section)

FIGURE 20.17 ■ Mushroom life cycle. A. Haploid basidiospores generate primary mycelium underground, where they form gametes. Gametes of opposite mating types fuse their cytoplasm only, forming secondary mycelium. The parental nuclei remain separate throughout many generations of mitosis during development of the fruiting body (mushroom). As the mushroom matures, the basidia undergo nuclear fusion and meiosis, forming progeny basidiospores. **B.** The underground secondary mycelia of some mushrooms form mycorrhizae with tree roots. Mycorrhizae enhance and extend the absorptive power of the tree roots, while obtaining plant sugars for the fungus.

fruiting body consists largely of cells with paired nuclei ($n + n$). A few of the paired nuclei fuse to form diploid cells ($2n$) called basidia (singular, basidium), which line the gills of the mushroom. The basidia undergo meiosis to form haploid **basidiospores** (n). Some types of basidia can release basidiospores under pressure, whereas other basidiospores are transmitted by wind (see **Fig. 20.16B**).

The basidiospores germinate to form underground mycelium. This haploid "primary mycelium" generates gametes that ultimately fuse to form $n + n$ "secondary mycelium." The primary and secondary mycelia may radiate underground, unseen, until their tips generate mushrooms aboveground, at points approximately equidistant from the origin. The result is a mysterious "fairy ring" of mushrooms (see **Fig. 20.2A**). In a forest, these invisible underground hyphae of basidiomycetes and zygomycetes contribute **mycorrhizae** (singular, **mycorrhiza**) that gain sugars from trees while extending the tree root systems (**Fig. 20.17B**).

Glomeromycota form arbuscular mycorrhizae. A remarkable phylum of fungi is the Glomeromycota, all of which are obligate mutualists of plants. These fungi, such as *Glomus* species, form extensive networks of filamentous connections with plant roots similar to the mycorrhizae formed by basidiomycetes; but unlike the basidiomycete mycorrhizae, the Glomeromycota form **arbuscular mycorrhizae**, in which the fungal filaments actually penetrate plant cells in a most intimate symbiosis (**Fig. 20.18**). Mycorrhizae expand the roots' absorptive capacity, while obtaining plant sugars for the fungus. More than 90% of all land plants, including trees, depend on these fungal interconnections, which share nutrients among many unrelated plants, as well as the fungi (discussed in Chapter 21).

In arbuscular mycorrhizae, a fungal hypha grows between plant cells without breaching the plant cell wall (**Fig. 20.18B**). As the hypha branches into the plant cytoplasm, the plant cell wall invaginates to accommodate the branch, while maintaining a "periarbuscular space" between the plant cell wall and the fungal plasma membrane. This highly regulated invaginating branch is called an "arbuscule"—hence the term "arbuscular mycorrhiza." Arbuscular formation is regulated by plant hormones called strigolactones. The arbuscule expands the surface area to exchange sugars from the plant for ammonium and phosphate from the fungus.

A.

B.

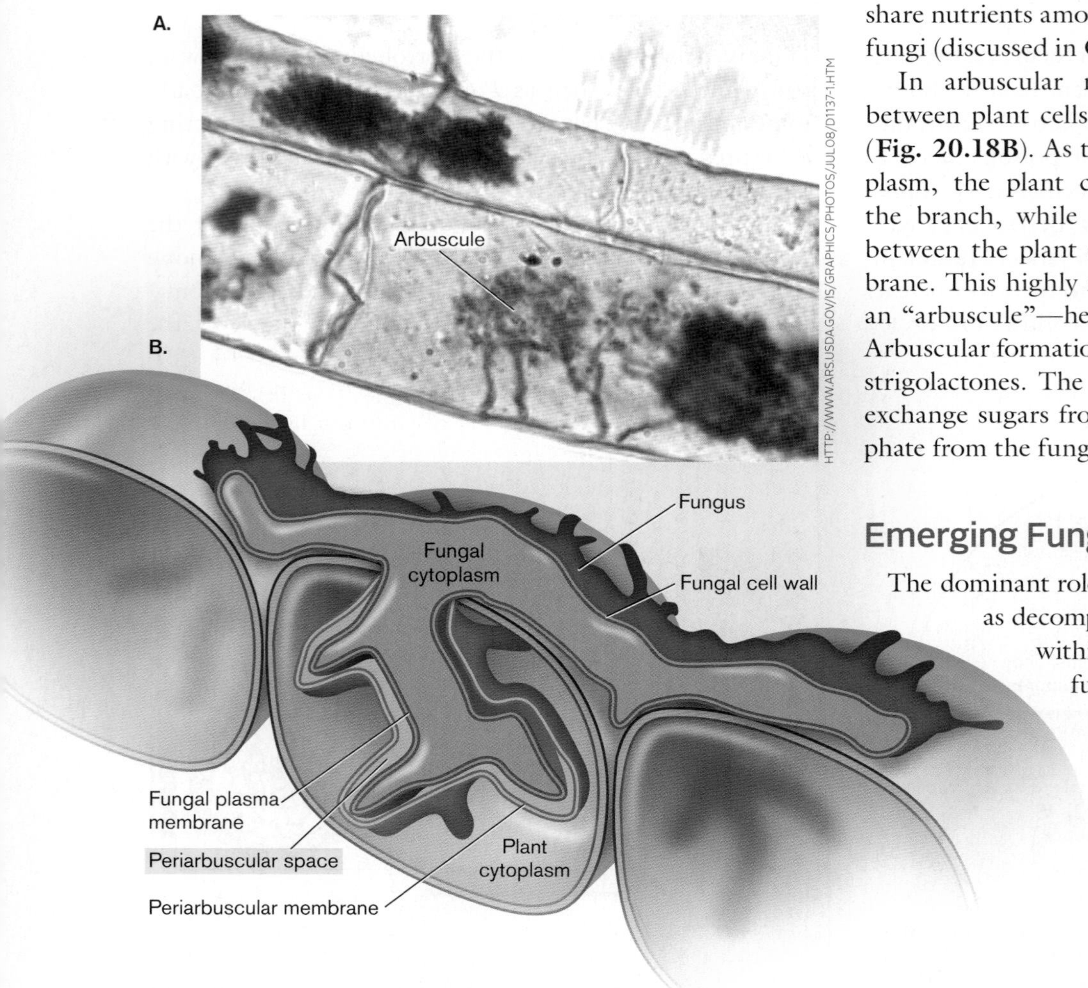
Arbuscule
Fungus
Fungal cytoplasm
Fungal cell wall
Fungal plasma membrane
Periarbuscular space
Plant cytoplasm
Periarbuscular membrane

HTTP://WWW.ARS.USDA.GOV/IS/GRAPHICS/PHOTOS/JUL08/D1137-1.HTM

FIGURE 20.18 ▪ Arbuscular mycorrhiza. A. Glomeromycota fungi invade corn root cells, as part of a mutualistic symbiosis to exchange nutrients. **B.** The fungal hypha grows between the plant cells, and then into a plant cell to form an arbuscule. The arbuscule expands surface area for exchange while maintaining the plant cell wall intact. *Source:* Part B modified from Martin Parniske. 2008. *Nat. Rev. Microbiol.* **6**:763.

Emerging Fungal Pathogens

The dominant role of fungi in our biosphere is positive—as decomposers, recyclers, and symbiotic partners within lichens and mycorrhizae (discussed further in Chapter 21). But some fungi are important pathogens, such as *Histoplasma capsulatum*, an ascomycete fungus that infects healthy people who inhale contaminated dust, causing a deadly pneumonia. And as human demographics shifts and climate change increases, a growing number of human, animal, and plant pathogens emerge (**Table 20.2**). For immunocompromised patients, especially the elderly confined to hospitals, a growing threat is *Aspergillus* sp., which can colonize the

TABLE 20.2

Emerging fungal pathogens.

Species	Phylum	Host	Disease
Aspergillus fumigatus	Ascomycota	Humans (immunocompromised)	Aspergillosis of lung, and elsewhere in the body
Cryptococcus neoformans	Basidiomycota	Humans (immunocompromised)	Meningitis and meningoencephalitis
Encephalitozoon intestinalis	Microsporidia	Humans with AIDS	Microsporidiosis (intestinal)
Histoplasma capsulatum	Ascomycota	Humans, dogs, cats	Histoplasmosis (lung)
Pneumocystis jirovecii	Ascomycota	Humans (immunocompromised)	Pneumocystis pneumonia
Stachybotrys chartarum	Ascomycota	Humans	Black mold disease; respiratory damage
Exserohilum rostratum	Ascomycota	Humans	Wound and skin infections; contaminated injections
Batrachochytrium dendrobatidis	Chytridiomycota	Amphibians (frogs and toads)	Chytridiomycosis
Fusarium solani	Ascomycota	Sea turtles (loggerheads)	Hatch failure
Geomyces destructans	Ascomycota	Brown bats	White nose disease
Puccinia graminis	Basidiomycota	Wheat	Wheat stem rust
Magnaporthe oryzae	Ascomycota	Rice	Rice blast disease
Nosema species	Microsporidia	Honeybees	Colony collapse disorder

lung and other tissues. In 2012, contaminated steroid injections led to an outbreak of infections by *Exserohilum rostratum* and other previously rare opportunists. Other fungal pathogens cause massive mortality of animals and plants.

Several parasitic organisms originally classified as protozoa because of their superficial appearance have now been shown to be fungi or fungus-related, based on their genome sequence and biochemistry. The reclassification has important consequences for research, taxonomy, and therapy. An example is the ascomycete *Pneumocystis jirovecii*. The organism was first described in 1909, when it was thought to be a life stage of a trypanosome causing Chagas' disease (discussed in Section 20.6). When the organism was recognized as a distinct species of protists infecting animals, it was named *Pneumocystis carinii*. In the 1970s, a strain of *Pneumocystis* was renamed *P. jirovecii* for causing pneumonia in immunocompromised humans. In the 1980s, the rise of AIDS led to a sudden increase in infections. Sequencing the organism's genome revealed it to be an ascomycete fungus. As a result, the organism's name was called into question by fungus researchers; its developmental forms and biochemistry were reevaluated, and medical research was redirected to culture and treat the organism based on fungal physiology.

A major clade of parasites now shown to be fungi is Microsporidia. Microsporidia have small genomes with many degenerate genes; their mitochondria have lost their DNA and are nonfunctional. Microsporidia form spores as small as a few micrometers in size, which can infect animal cells. The microsporidian spore extrudes a specialized invasion complex called the polar tube that penetrates the host cell, typically a macrophage. In humans, microsporidians are opportunistic pathogens such as the intestinal parasite *Encephalitozoon intestinalis,* emerging with the rise of AIDS and the growth of elderly and immunocompromised populations.

Other taxa originally assigned as fungi, based on superficial appearance, are now classified as protists, based on genetic similarity. The **oomycetes**, or "water molds," were originally classed as fungi because of their fungus-like filaments, which infect plants and animals. Oomycete *Phytophthora infestans* devastated Irish potato crops in the 1840s, causing the Great Irish Famine, in which a million people died. Today, a related species, *Phytophthora ramosum,* causes "sudden oak death," a disease killing tens of thousands of oaks and other trees in the western United States. Oomycetes are discussed further in **eTopic 20.1**.

To Summarize

- **Fungi form hyphae with cell walls of chitin.** Hyphae absorb nutrients from decaying organisms or from infected hosts. Some fungi remain unicellular; these are called yeasts, or mitosporic fungi.
- **Chytridiomycete fungi have motile zoospores.** Motile reproductive forms are a trait shared with animals. Flagellar motility has been lost by other fungi through reductive evolution.
- **Zygomycete fungi form haploid mycelia.** Hyphal tips differentiate into gametes and grow toward each other to undergo sexual reproduction. Some zygomycetes form mycorrhizae that connect the roots of plants.
- **Ascomycete fungal mycelia form paired nuclei.** Within the hyphal cells, the paired nuclei fuse, followed by meiosis and development of ascospores. Some ascomycetes form fruiting bodies called conidiophores.
- **Basidiomycete fungi form mushrooms.** Cells with paired nuclei (secondary mycelium) form large fruiting bodies called mushrooms. The paired nuclei fuse to form the diploid basidium, which generates haploid basidiospores. The basidiospores develop underground hyphae or mycorrhizae that interconnect plant roots.
- **Glomeromycota form arbuscular mycorrhizae.** These fungi form intimate mutualistic networks of connections with plant roots, exchanging minerals for plant sugars.
- **Emerging fungal pathogens threaten humans, plants, and animals.**

20.3

Algae

Algae are primary producers in all ecosystems, most crucially aquatic and marine habitats. In aquatic and marine ecology, the algae, together with photosynthetic bacteria, are known as **phytoplankton** (see Chapter 21). All algae possess chloroplasts.

The primary endosymbiotic algae, or "true algae," are products of a single ancestral endosymbiosis that also gave rise to land plants. The biochemistry and cell structures of true algae and land plants are similar; thus, the study of photosynthesis in the alga *Chlorella* by Martin Kamen, Melvin Calvin, and others has advanced our understanding of plant physiology (discussed in Chapter 15).

Other photosynthetic eukaryotes, or secondary endosymbiotic algae, arose from protists that engulfed a primary or secondary endosymbiont. Secondary endosymbionts show "mixotrophic" nutrition, involving both phototrophy and heterotrophy. For example, dinoflagellate "algae" are voracious predators on smaller protists. The heterokont algae (diatoms, coccolithophores, and kelps) are covered in this section. Dinoflagellates are covered under alveolates (Section 20.5).

Primary Endosymbiotic Algae

The primary endosymbiotic algae include two major clades: Chlorophyta, or green algae; and Rhodophyta, or red algae—although not all members of each group appear green or red, respectively. Rhodophyta that appear red have a secondary pigment called phycoerythrin, in addition to green chlorophyll.

Green algae (Chlorophyta). Many green algae are unicellular. An important model system for genetics and phototaxis is *Chlamydomonas reinhardtii,* a unicellular chlorophyte common in freshwater systems as well as Antarctic pools. The genetics of cell cycle regulation in *C. reinhardtii* provides clues to tumor formation in humans.

C. reinhardtii has a symmetrical pair of flagella, a common pattern for green and red algae and their gametes (**Fig. 20.19**). The alga swims forward by bending its flagella back toward the cell, like a breast stroke. *Chlamydomonas* cells are mostly haploid, reproducing by asexual cell division (**Fig. 20.20**). For sexual reproduction, opposite mating types fuse to form a zygote, which loses flagella and grows a spiny protective coat. The zygote undergoes meiosis to regenerate haploid cells.

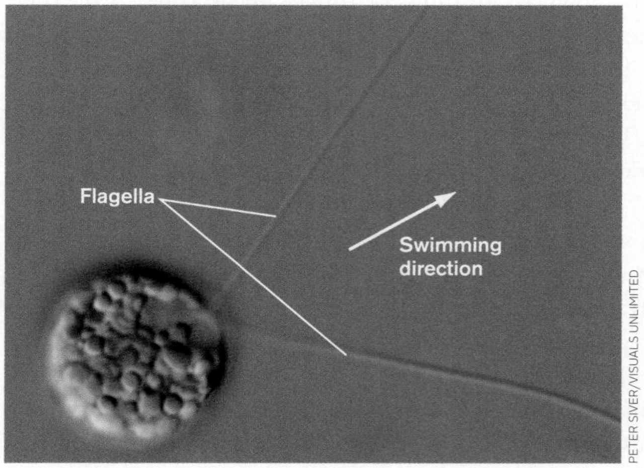
Flagella
Swimming direction
PETER SIVER/VISUALS UNLIMITED

FIGURE 20.19 ■ A single-celled green alga: *Chlamydomonas reinhardtii.* *Chlamydomonas* has a green chloroplast and grows as a photoautotroph (cell diameter 10–20 μm). A symmetrical pair of flagella pulls the cell forward.

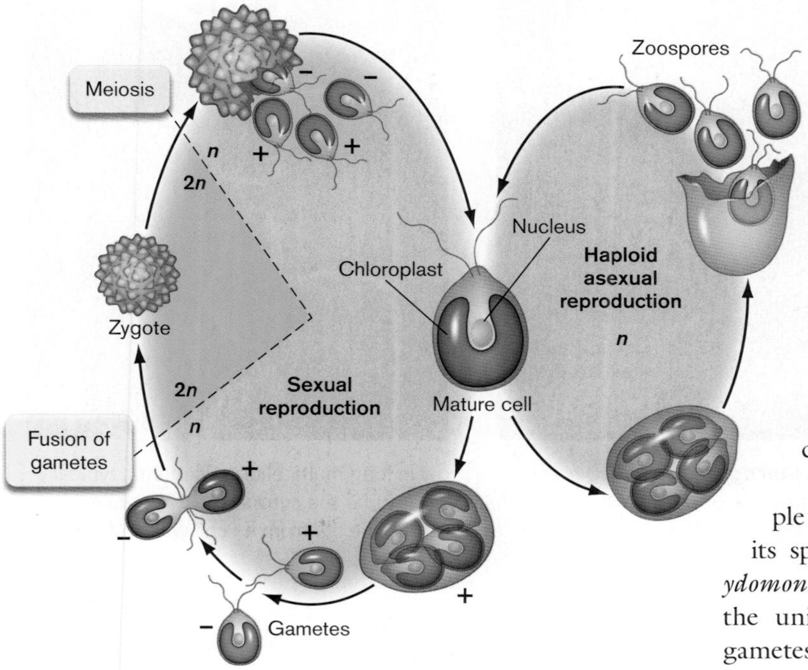

Meiosis
n
2n
+
+
−
−
Zoospores
Nucleus
Haploid
asexual
reproduction
n
Chloroplast
Zygote
2n
Sexual
reproduction
Mature cell
Fusion of
gametes
n
+
−
+
+
−
Gametes

FIGURE 20.20 ▪ ***Chlamydomonas* alternation of generations.** *Chlamydomonas* reproduces asexually as haploid cells. Alternatively, the alga generates gametes that fuse into a zygote that immediately undergoes meiosis to regenerate haploid cells.

The cell ultrastructure of *Chlamydomonas* is typical of algal cells (**Fig. 20.21**). The nucleus is cupped by a single chloroplast, which is surrounded by a double membrane. The double membrane indicates a primary endosymbiont; the inner membrane derives from the ancestral bacterium, and the outer membrane derives from the engulfing host. Within the chloroplast is a pyrenoid, an organelle that concentrates bicarbonate (HCO_3^-) and converts it to CO_2 for fixation. The pyrenoid is surrounded by one or more starch bodies for energy storage. The starch is broken down to sugars as needed, followed by glycolysis and respiration in the mitochondria. Osmolarity is maintained by the contractile vacuole. The *Chlamydomonas* cell is encased in a cell wall composed predominantly of glycoprotein. Other green algae have cellulose cell walls similar to those of plants.

Other algae form long filaments of cells. An example is *Spirogyra*, a common pond dweller known for its spiral chloroplasts (**Fig. 20.22A**). As with *Chlamydomonas*, the haploid form predominates; but unlike the unicellular alga, *Spirogyra* forms neither flagellated gametes nor zoospores. Instead, its sexual reproduction requires alignment of two filaments of opposite mating type (**Fig. 20.22B**). Cells conjugate (form cytoplasmic bridges) between the two filaments. The "male" gametes are those whose cytoplasm inserts through the conjugation bridge to join that of the "female" gamete. As the gametes fuse, a row of empty cell walls is left behind (**Fig. 20.22C**). The zygote eventually hatches and germinates a new chain of cells.

Some marine algae grow in undulating sheets. An example familiar to beach bathers is the "sea lettuce" *Ulva*

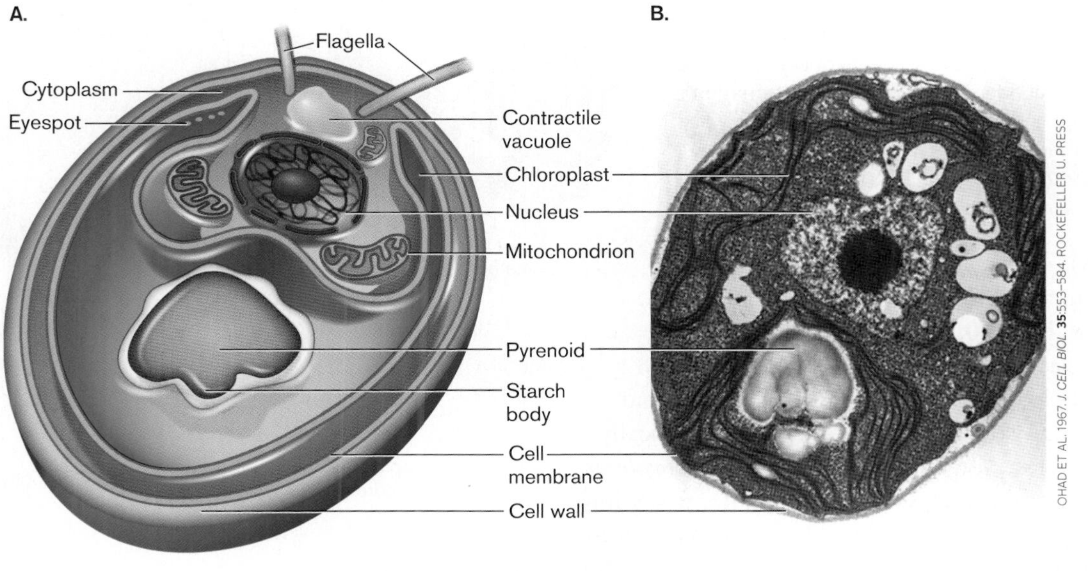

A.
B.
Flagella
Cytoplasm
Eyespot
Contractile
vacuole
Chloroplast
Nucleus
Mitochondrion
Pyrenoid
Starch
body
Cell
membrane
Cell wall
OHAD ET AL. 1967. J. CELL BIOL. 35:553–584. ROCKEFELLER U. PRESS

FIGURE 20.21 ▪ **Cell structure of *Chlamydomonas reinhardtii*. A.** The cell membrane is surrounded by a cell wall of cellulose and glycoproteins. The single chloroplast fills much of the cell and wraps around the nucleus. Within the chloroplast lies a pyrenoid, a structure for concentrating bicarbonate ion for conversion to CO_2. The pyrenoid is surrounded by starch bodies that store high-energy compounds. A contractile vacuole maintains constant osmotic pressure. **B.** Electron micrograph of *C. reinhardtii* shows the nucleus, chloroplast, pyrenoid, and other organelles.

A. B. C.

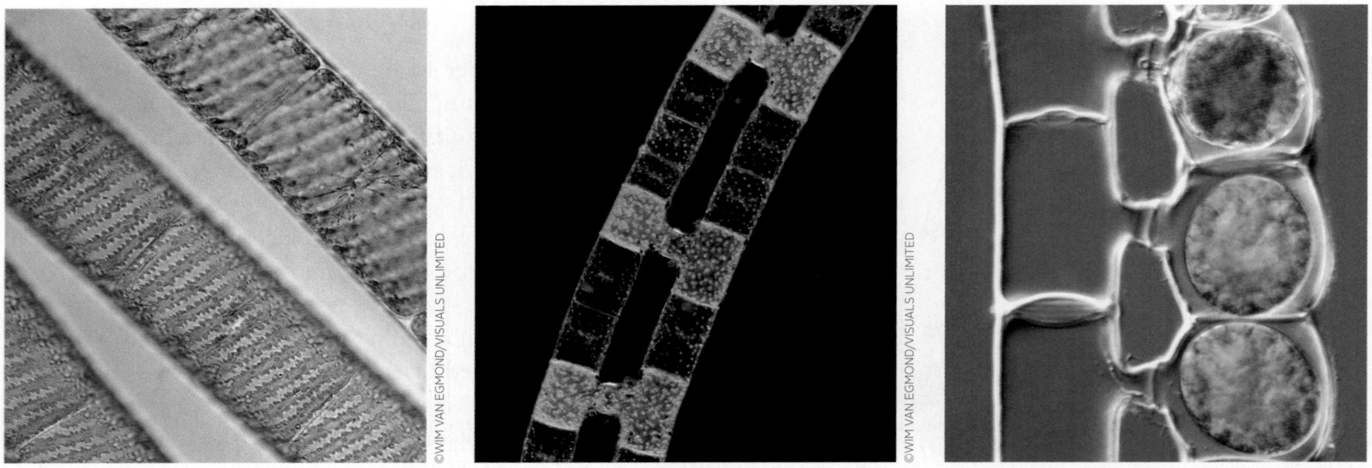

©WIM VAN EGMOND/VISUALS UNLIMITED
©WIM VAN EGMOND/VISUALS UNLIMITED
M. I. WALKER

FIGURE 20.22 ▪ Filaments with spiral chloroplasts: *Spirogyra.* **A.** *Spirogyra* species grow in long multicellular filaments, typically 25 μm in width and several centimeters long. Each cell contains one or more chloroplasts that spiral around the cytoplasm. **B.** Sexual reproduction involves conjugation between cells of two mating types. The cytoplasm from each male cell exits from its cell wall and enters the female cell. **C.** Gamete fusion is complete, generating zygotic spores.

A.

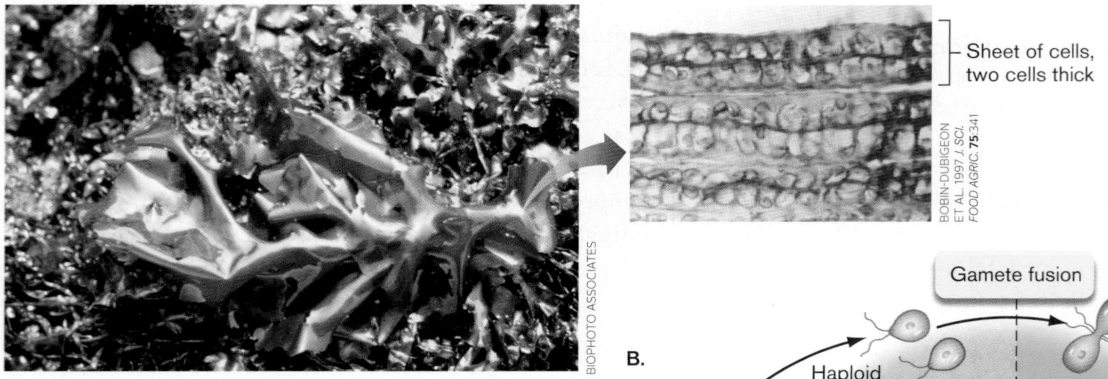

Sheet of cells, two cells thick
BOBIN-DUBIGEON ET AL. 1997 J. SCI. FOOD AGRIC. 75:341
BIOPHOTO ASSOCIATES

B.

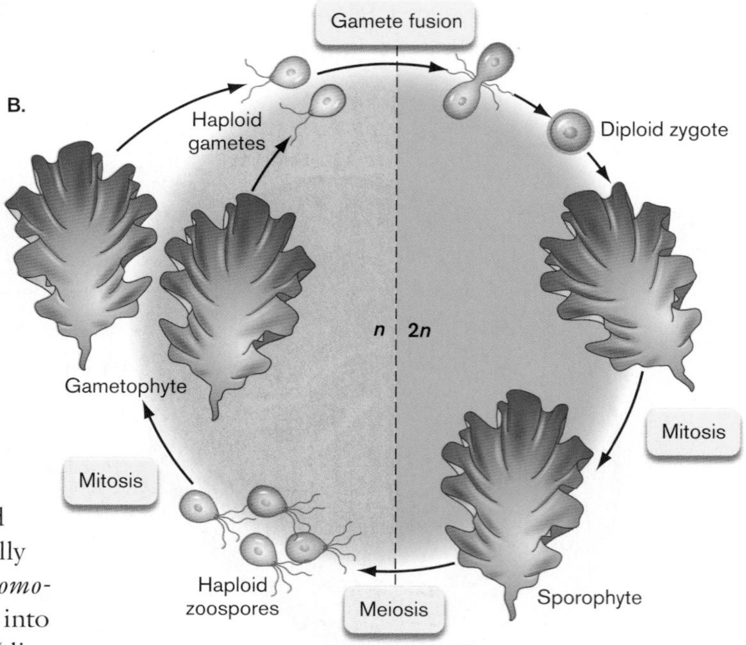

Gamete fusion
Haploid gametes
Diploid zygote
n
2n
Gametophyte
Mitosis
Mitosis
Haploid zoospores
Meiosis
Sporophyte

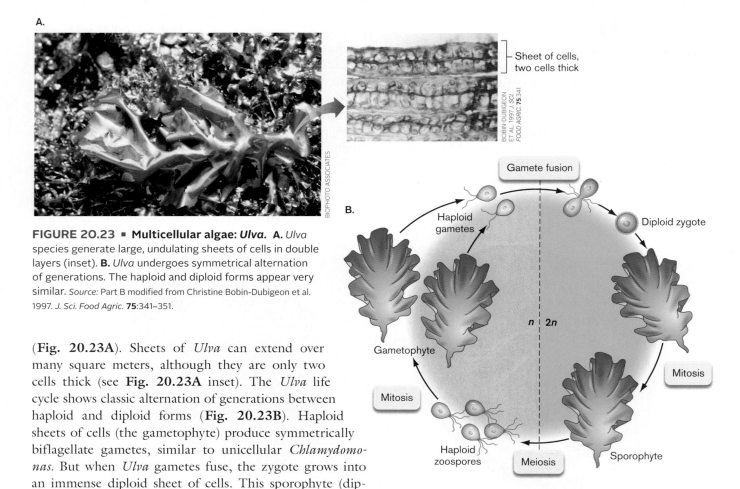

FIGURE 20.23 ▪ Multicellular algae: *Ulva.* **A.** *Ulva* species generate large, undulating sheets of cells in double layers (inset). **B.** *Ulva* undergoes symmetrical alternation of generations. The haploid and diploid forms appear very similar. *Source:* Part B modified from Christine Bobin-Dubigeon et al. 1997. *J. Sci. Food Agric.* **75**:341–351.

(**Fig. 20.23A**). Sheets of *Ulva* can extend over many square meters, although they are only two cells thick (see **Fig. 20.23A** inset). The *Ulva* life cycle shows classic alternation of generations between haploid and diploid forms (**Fig. 20.23B**). Haploid sheets of cells (the gametophyte) produce symmetrically biflagellate gametes, similar to unicellular *Chlamydomonas*. But when *Ulva* gametes fuse, the zygote grows into an immense diploid sheet of cells. This sporophyte (diploid multicellular body) appears similar in form to the gametophyte. The sporophyte eventually undergoes meiosis, releasing haploid zoospores with two pairs of flagella. The zoospores undergo mitosis and regenerate the gametophyte.

Marine and freshwater algae show many complex and beautiful forms. Colonial algae such as *Volvox* (**Fig. 20.24A**) generate geodesic spheres of biflagellate cells similar to those of *Chlamydomonas*. Their flagella point outward from the

A. *Volvox* B. *Cymopolia*

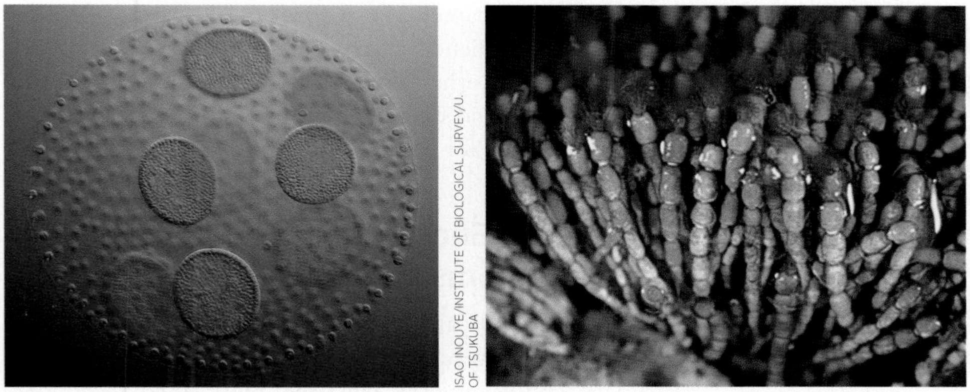

FIGURE 20.24 ▪ Diverse green algae. **A.** *Volvox* forms colonies of cells similar to those of *Chlamydomonas*; the colony may grow to a size of 1–3 mm. Progeny colonies form within the larger sphere until the sphere bursts, releasing them. **B.** *Cymopolia* forms calcified stalks of cells, sprouting filaments that conduct photosynthesis.

A. B.

FIGURE 20.25 ▪ Red algae: *Porphyra*. **A.** *Porphyra* forms large, red multicellular sheets. **B.** The sheets are harvested and toasted, known as "nori." Nori are used to wrap sushi, a Japanese delicacy.

colony, propelling it forward and drawing nutrients across its surface. Each cell of *Volvox* connects by cytoplasmic bridges to five or six of its neighbors. The colony reproduces by generating daughter colonies within the sphere, which grow until the outer sphere falls apart, liberating the daughters.

Other green algae, such as *Cymopolia,* form long calcified stalks punctuated by tufts of green filaments, where most of the photosynthesis occurs (**Fig. 20.24B**). Algae of yet another group, the siphonous algae, are known for forming long, siphon-like tubes and fronds without cell partitions—in effect, giant organisms that are unicellular. Originally grown for aquariums, the siphonous species *Caulerpa taxifolia* has become a notorious pest in the Mediterranean Sea and off the coast of California. *C. taxifolia* is exceptionally cold resistant and produces toxins that prevent grazing.

Red algae (Rhodophyta). Red algae, or rhodophytes, are colored red by the photopigment phycoerythrin. Phycoerythrin absorbs efficiently in the blue and green range, which green algae fail to absorb. Because the red algae absorb wavelengths missed by the green algae, they can colonize deeper marine habitats below the green algal populations.

Rhodophytes include unicellular, filamentous, and multicellular forms. Several kinds are human food sources. *Porphyra* forms large sheets that are harvested in Japan for use as nori for wrapping sushi, a delicacy that includes rice, vegetables, and uncooked fish (**Fig. 20.25**). Red algae contain valuable polymers called sulfated polygalactans (sugar polymers with sulfate side chains). Sulfated polygalactans include agar, used to solidify microbial growth media; agarose, a processed sugar derivative used to form electrophoretic gels; and carrageenan, an additive used in processed foods.

Rhodophytes such as *Plocamium* form delicate branched filaments. Others, the **coralline algae**, calcify their fronds into hardened shapes similar to corals (**Fig. 20.26**). The calcified bodies of corallines grow at greater depths than corals, and they often form the foundation for coral reefs.

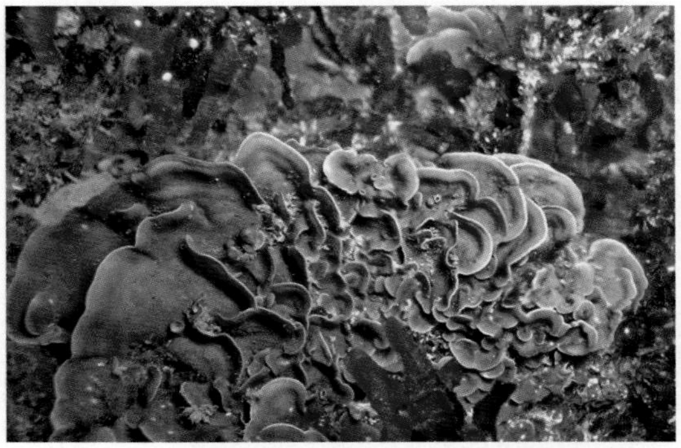

FIGURE 20.26 ▪ Coralline algae. *Mesophyllum* forms calcified fronds that provide a foundation for coral reefs.

Thought Question

20.5 How are coralline algae able to grow at greater depths than coral?

Algae serve as symbiotic partners for many important living systems. For example, certain algae grow in intimate association with fungi to form unified structures called **lichens**, important colonizers of dry and cold habitats (discussed in Chapter 21). A form of algal-fungal ground cover similar to lichens is **cryptogamic crust**, common on desert soil. Still other algae grow within the cells of paramecia, hydras, and corals, providing photosynthetic nutrition in exchange for protection.

Secondary Endosymbiotic Algae: Diatoms and Kelps

Many kinds of algae, called **secondary endosymbionts**, arose from ancestral protists that once engulfed a primary endosymbiotic alga (see **Fig. 20.6**). Secondary

A.

DARLYNE A. MURAWSKI/NATIONAL GEOGRAPHIC/GETTY IMAGES

FIGURE 20.27 ■ Diatoms. A. Diatoms of various species. **B.** Life cycle of a centric diatom, *Thalassiosira* sp. Each vegetative cell division (lower right) requires formation of an in-fitting frustule half. Successive divisions result in progressive decrease in size (lower left). At a critical size limit, the diatom must undergo meiosis to form eggs and sperm (upper left). As gametes fuse, they form an auxospore (upper right), which regenerates a frustule of the original size.

B.

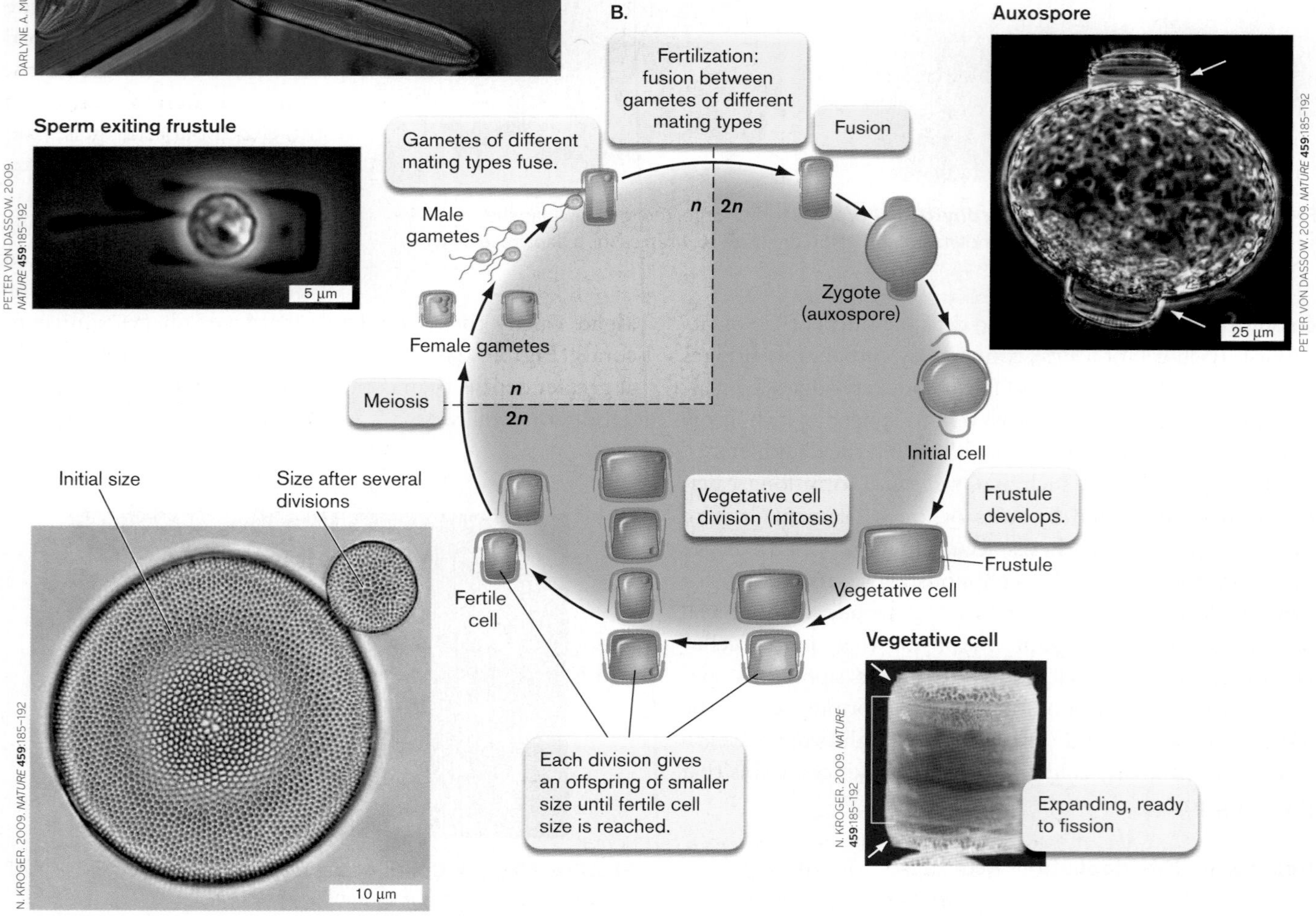
Auxospore
25 µm
PETER VON DASSOW, 2009. NATURE 459:185–192
Fertilization: fusion between gametes of different mating types
Fusion
Gametes of different mating types fuse.
Sperm exiting frustule
5 µm
PETER VON DASSOW, 2009. NATURE 459:185–192
Male gametes
n
2n
Zygote (auxospore)
Female gametes
Meiosis
n
2n
Initial cell
Initial size
Size after several divisions
Vegetative cell division (mitosis)
Frustule develops.
Frustule
Vegetative cell
Fertile cell
Vegetative cell
Each division gives an offspring of smaller size until fertile cell size is reached.
Expanding, ready to fission
10 µm
N. KROGER, 2009. NATURE 459:185–192
N. KROGER, 2009. NATURE 459:185–192

endosymbiotic algae show two traits that distinguish them from primary endosymbionts:

- **More than two membranes surround the chloroplast.** The extra membranes derive from the cell membrane of the engulfed alga.
- **Metabolism of secondary endosymbionts includes heterotrophy.** By contrast, the primary endosymbionts are near-obligate autotrophs, catabolizing only the simplest substrates, such as acetate.

Several major groups of secondary endosymbiotic algae are heterokonts (stramenopiles). These include the diatoms (Bacillariophyceae); the brown algae (Phaeophyceae), such as kelps; the golden algae (Chrysophyceae), mainly pale-colored flagellates; the yellow-green algae (Xanthophyceae); and other less studied forms of phytoplankton. Flagellated species of heterokont algae always show a pair of differently shaped flagella. Typically, one flagellum is brush-like with side branches, while the other is shorter, often wrapped around the cell, its function uncertain.

Diatoms (Bacillariophyceae). Diatoms are unicellular algae found ubiquitously in aquatic and marine waters (**Fig. 20.27**). They conduct a fifth of all photosynthesis on Earth, and they fix as much biomass as all the terrestrial rain forests. A diatom grows a unique kind of bipartite shell called a **frustule**. The frustule is composed of silica (cross-linked silicon dioxide, SiO_2). The silicate frustules protect diatoms from many kinds of predators. Diatoms are nonetheless consumed by flagellates and amphipods (shrimplike invertebrates) and are infected by viruses. Frustules of decomposed diatoms eventually sediment on the ocean floor, where they build sedimentary rock strata more than a kilometer thick. This "diatomaceous earth" is used in insulation material and in toothpaste. Diverse species of diatoms are highly sensitive to environmental factors such as pH, and the frequency of their shells in sediment can be used to track a lake's environmental history.

Frustules of different species form an extraordinary range of shapes with intricate pore formations (**Fig. 20.27A**). The shapes fall into two classes: centric, with radial symmetry; or pennate, with bilateral symmetry. All, however, pose a unique challenge to cell division: As the diatom grows and fissions, each daughter cell receives one parental half of the frustule while forming a new half fitting within the parental half, like the bottom dish of a petri plate (**Fig. 20.27B**). Thus, each generation results in an inexorable decline in size of the organism. As its cell size reaches a critical point, the diatom must undergo meiosis to generate gametes. Centric diatoms form egg cells and flagellated sperm (as shown in **Fig. 20.27B**), whereas pennate diatoms form equivalent ameboid gamete cells. When the gametes fuse, they form a special kind of zygote called an auxospore. The auxospore generates a frustule of the same size as the original diatom.

Coccolithophores (Prymnesiophyceae). Coccolithophores, known as "coccoliths," are haptophyte algae superficially similar to diatoms in that their cells have a solid mineral exoskeleton (**Fig. 20.28**). Instead of silicate, however, their exoskeleton is composed of calcium carbonate ($CaCO_3$). The calcium carbonate grows in multiple plates (unlike the unitary exoskeleton of a diatom). The plates may extend radially in all directions, such as the "trumpet" shapes of *Discosphaera tubifera* (**Fig. 20.28**), or form multiple layers of flat oval shapes encasing the cell, as in *Emiliana huxleyi*. The calcium carbonate plates protect the tiny cell from some predators. Coccoliths such as *E. huxleyi* generate huge milky blooms that appear in NASA satellite images. The blooms are dissipated by predation or by virus infection.

Coccoliths are increasingly recognized as major players in the ocean's carbon cycle (discussed also in Chapter 22). They sequester large amounts of carbon from CO_2 into their carbonate shells. Unfortunately, their calcium carbonate is sensitive to acidification caused by the global rise in CO_2. For this reason, intensive research is focused on understanding the response of coccoliths such as *E. huxleyi* to pH change. In 2011, a large-scale study conducted

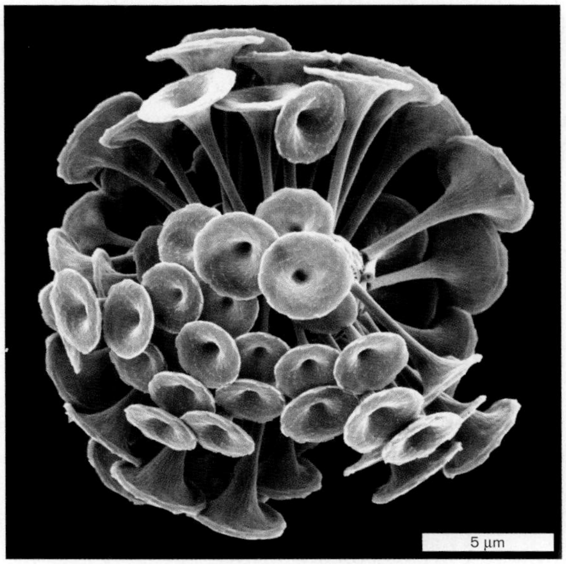
5 µm

FIGURE 20.28 ▪ A coccolithophore, *Discosphaera tubifera*. The tiny cell is surrounded by a much larger volume of trumpet-shaped coccoliths. From the Alboran Sea, western Mediterranean.

Sargassum weed
VISUALS UNLIMITED

FIGURE 20.29 ■ Kelp forests. *Sargassum natans* forms the basis of the Sargasso Sea. The brown alga forms stalks with leaflike blades and round gas bladders to keep the alga afloat. Kelp forests support animals such as the sea turtle.

by several European universities concluded that increasing CO_2 levels correlate with a decline in the overall mass of marine coccoliths. At the same time, however, certain strains of *Emiliana* were shown to grow despite low pH, so many uncertainties remain.

Brown algae (Phaeophyceae). Brown algae, such as kelps, possess vacuoles of leucocin, an oily lipid for energy storage whose color gives the organism a brown or yellow tint. Kelps are familiar to ocean bathers as the long, dark brown blades that root near the beach until the surf rips the blades off and tosses them ashore. Kelps support important communities of multicellular organisms known as kelp forests. The most famous kelp forests are those of the unrooted **sargassum weeds**, which float on the Sargasso Sea. Sargassum consists of stalks with photosynthetic blades and round gas bladders to keep the organism afloat (**Fig. 20.29**). Sargassum supports a complex food web of invertebrate and vertebrate animals, including worms, crabs, fish, and sea turtles.

To Summarize

- **Chlorophyta (green algae)** absorb red and blue light and grow near the top of the water column. Green algae include unicellular, filamentous, and sheet forms.
- **Rhodophyta (red algae)** have the accessory photopigment phycoerythrin, which absorbs green and longer-wavelength blue light, enabling growth at greater depths. Red algae include species of diverse forms, many of which are edible for humans.
- **Secondary endosymbiotic algae** are derived from protists that had engulfed primary symbiotic algae. They are mixotrophs, combining phototrophy and heterotrophy.
- **Diatoms are heterokonts with silicate shells called frustules.** Diatoms replicate by an unusual division cycle generating successively smaller frustules.
- **Kelps are heterokonts that grow in long, sheetlike fronds.** Kelps play an important role in the ecology of the open ocean, as well as the ecology of marine beaches.

20.4

Amebas and Slime Molds

The ameba (alternative spelling, "amoeba") is familiar to most of us as an apparently amorphous form of microscopic life, capable of engulfing and consuming prey in a dramatic fashion (**Fig. 20.30**). While the ameba's shape is exceptionally variable, the pseudopods, or "false feet," that it extends, far from being amorphous, are complex structures that undertake highly controlled and specific movements. Most amebas are free-living predators in soil or water, engulfing prey by **phagocytosis**. They extend in size up to 5 mm, large enough to phagocytose bacteria, algae, ciliates, smaller amebas, and even invertebrates such as rotifers. A few are dangerous parasites of humans or animals. Furthermore, free-living amebas can harbor bacterial pathogens such as *Legionella pneumophila*, which contaminates water supplies and air ducts. The bacteria cause legionellosis, an often fatal form of pneumonia. The host ameba enables the pathogen's persistence and transmission to human hosts.

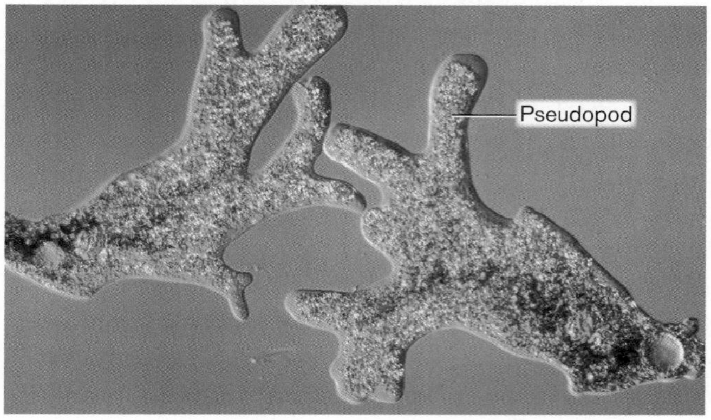
Pseudopod
M. I. WALKER/PHOTO RESEARCHERS, INC.

FIGURE 20.30 ■ Amebas. *Amoeba proteus* moves by extending its pseudopods.

Free-living amebas such as *Acanthamoeba* species are common predators in the soil microbial community. They cause problems when they contaminate contact lens cleaning solutions, causing keratitis (infection of the cornea). Wearers of reusable contact lenses have a keratitis infection rate of two per thousand.

Thought Question

20.6 What might happen when an ameba phagocytoses algae?

Note: The taxonomy of amebas and slime molds remains problematic, with diverse views as to the number of clades, their relatedness, and their degree of divergence.

Pseudopod Motility

Ameba-like cell forms occur in many different species. Some species persist as an ameba throughout all or most of the life cycle. Others can convert to flagellated forms, particularly when the habitat fills with water—a low-viscosity condition favoring flagellar motility. Still other protist species, such as dinoflagellates, never become fully amebic, but they can extend a pseudopod to engulf prey.

Different kinds of amebas have different kinds of pseudopods. "Classic" amebas (Amoebozoa) have lobe-shaped pseudopods (**Fig. 20.30**). Lobe-shaped pseudopods have the most variable shape. A different form is the sheetlike pseudopod, or **lamellar pseudopod**. Lamellar pseudopods are extended by dinoflagellates. Similar lamellar pseudopods are generated by human white blood cells such as leukocytes. Finally, needlelike pseudopods, or filopodia, are thin extensions reinforced by microtubules. Filopodia are made by the Rhizaria amebas, some of which have tests (shells), such as Foraminifera (spiral shells) and Radiolaria (radial form).

The extension of lobe-shaped and lamellar pseudopods has been studied closely for its relevance to human white blood cells (for more on white blood cells as host defenses, see Chapter 23). The mechanism of pseudopod motility remains poorly understood, but it is known to involve a sol-gel transition between cortical cytoplasm (just beneath the cell surface) and the cytoplasm of the deeper interior (**Fig. 20.31**). The tip of a pseudopod contains a gel of polymerized actin beneath its cell membrane. From the center of the ameba, liquid cytoplasm (sol) containing actin subunits streams forward along microtubular "tracks," powered by ATP hydrolysis. The actin subunits stream into the pseudopod, where they polymerize, forming a gel. The gel region grows, pushing the membrane forward and extending the pseudopod. As the gel is pushed backward, it resolubilizes to continue the cycle.

Amebas can have one nucleus or multiple nuclei. They are usually haploid and reproduce asexually by nuclear mitosis, without dissolution of the nuclear membrane, followed by fission of the cytoplasm. Some species do have reproductive alternatives, such as cyst formation, gamete fusion and meiosis, and even growth of flagella in a favorable habitat.

Thought Question

20.7 What kind of habitat would favor a flagellated ameba?

Ameba genetics is poorly understood, but at least one ameba genome has been sequenced—that of the intestinal parasite *Entamoeba histolytica*. The sequence contains 20 Mb of DNA in 14 chromosomes; some of these are linear, whereas others are circular. Closely related strains show considerable variation in organization, suggesting that ameba genomes undergo extensive rearrangement.

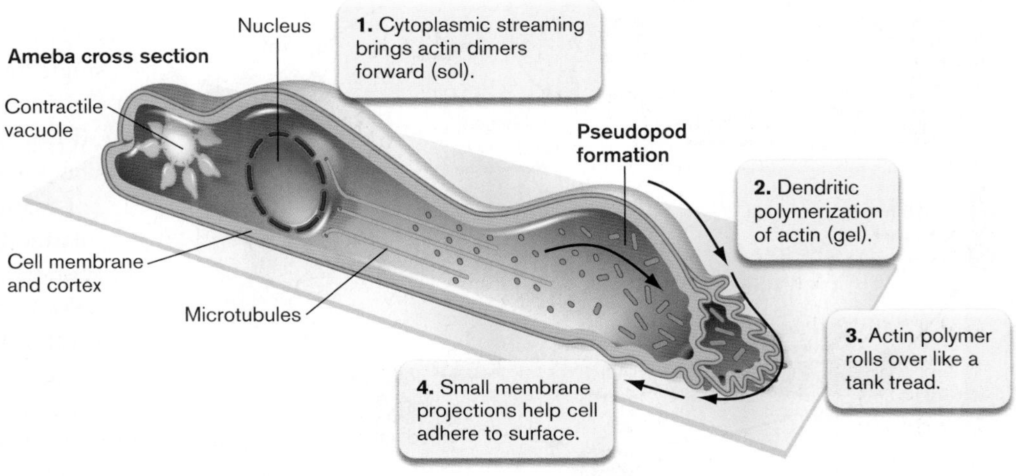
Ameba cross section
Nucleus
1. Cytoplasmic streaming brings actin dimers forward (sol).
Contractile vacuole
Pseudopod formation
2. Dendritic polymerization of actin (gel).
Cell membrane and cortex
Microtubules
3. Actin polymer rolls over like a tank tread.
4. Small membrane projections help cell adhere to surface.

FIGURE 20.31 ■ Pseudopod motility. A pseudopod extends by flow of liquid cytoplasm (sol state) followed by actin polymerization (gel state). As actin polymerizes, the cell rotates down toward the substrate like a tank tread.

Slime Molds

Some amebas conduct a life cycle in which thousands of individuals (all members of one species) aggregate into a complex differentiated fruiting body (**Fig. 20.32A**). Such an organism is called a **slime mold**. Slime molds, as the name implies, were originally classified with fungi because their fruiting bodies superficially resemble fungal reproductive forms. There are two kinds of slime molds: **cellular slime molds**, which form from ameboid cells that aggregate into a multicellular "slug"; and **plasmodial slime molds**, in which an ameba undergoes mitosis without cell division, forming a multinucleate single cell.

A well-studied example of a cellular slime mold is *Dictyostelium discoideum,* historically an important model system for multicellular development (**Fig. 20.32**). *D. discoideum* amebas are relatively small, about 10 μm, but large enough to consume bacteria. They can be cocultured on a plate with *Escherichia coli.* As the haploid amebas consume bacteria, they divide asexually until their food runs out. At this point, a few amebas begin to emit the aggregation signal molecule cyclic AMP (cAMP). An ameba emitting cyclic AMP attracts other amebas nearby, which move toward the center and begin emitting cyclic AMP as well. Successive waves of cyclic AMP continue to attract thousands of amebas to the center, where they pile on top of each other to form a slug as long as 1 mm. The slug then migrates, attracted by light and warmth, to find an appropriate place to form a fruiting body and disperse its spores.

The slug at last differentiates into a fruiting body, a spherical sporangium supported on a stalk of largely empty cells that emerges from a basal disk. The sporangium then releases spores (also called cysts), which are dispersed on air currents and can remain viable for several years. When a spore detects chemical signals from bacteria, it germinates an ameba to feed on them.

Note that the entire reproductive cycle just described is asexual; the amebas and their differentiated structures remain haploid throughout. *D. discoideum* amebas do have a sexual alternative (illustrated in **Fig. 20.32B**), in which cells of opposite mating type can fuse to form a diploid zygote and then undergo meiosis, restoring haploid amebas.

By contrast, plasmodial slime molds, such as *Physarum polycephalum,* develop from a single diploid ameba. As the ameba grows, its nuclei multiply, forming a **plasmodium** (plural, **plasmodia**), a giant multinucleate cell that can spread over an area of many square centimeters. Out of the plasmodium arise fruiting bodies whose sporangia undergo meiosis, producing haploid spores. Two spores then fertilize each other to form a diploid ameba that can develop again into a plasmodium. A large plasmodium can occasionally be seen as a yellow mass of slime spreading over decaying wood.

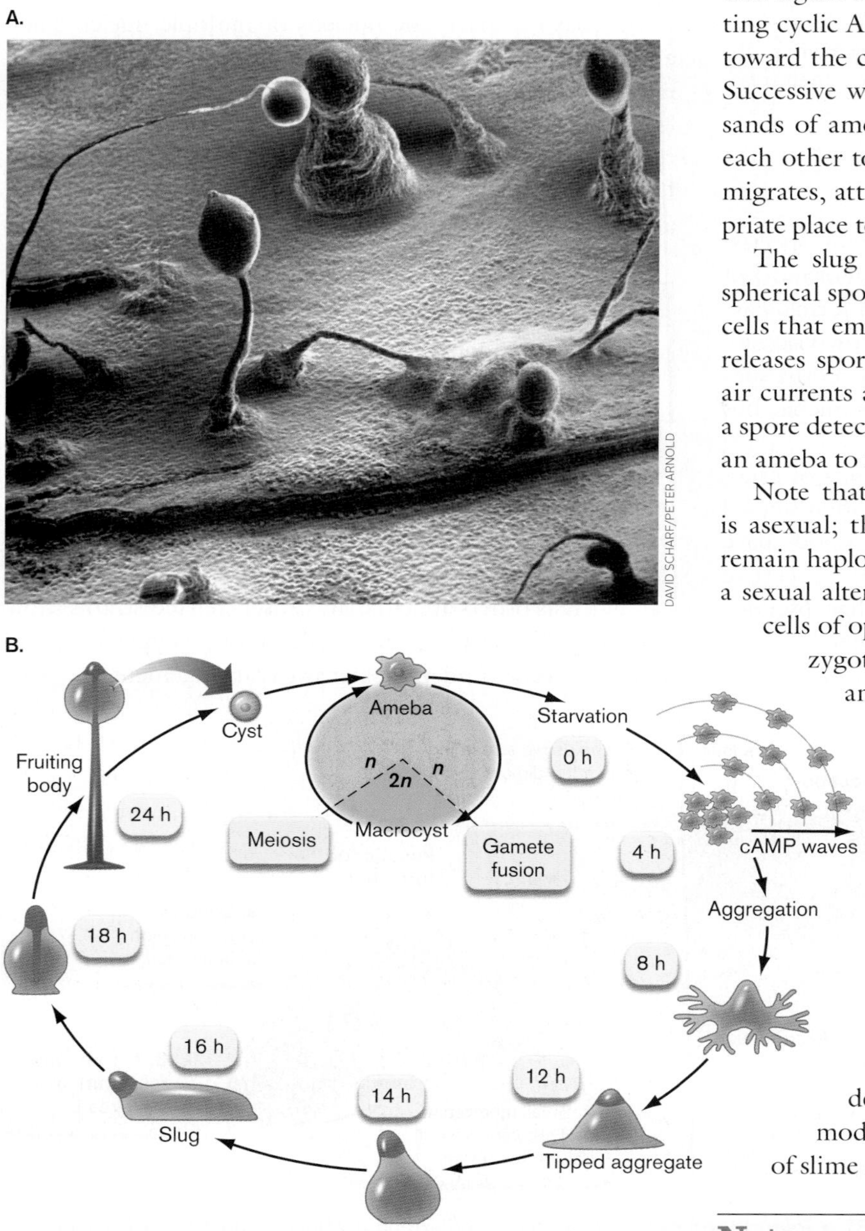
A.
DAVID SCHARF/PETER ARNOLD
B.
Ameba
Cyst
Starvation
Fruiting body
n
n
2n
0 h
24 h
Meiosis
Macrocyst
Gamete fusion
4 h
cAMP waves
Aggregation
18 h
8 h
16 h
12 h
14 h
Slug
Tipped aggregate

FIGURE 20.32 ■ A cellular slime mold: *Dictyostelium discoideum.* **A.** Fruiting bodies of *D. discoideum* (composite light micrograph). **B.** Life cycle of *D. discoideum.*

Note: Distinguish the term "plasmodium" (a large multinucleate cell) from the genus *Plasmodium* (an apicomplexan parasite, such as *Plasmodium falciparum,* which causes malaria).

A. Radiolarian tests

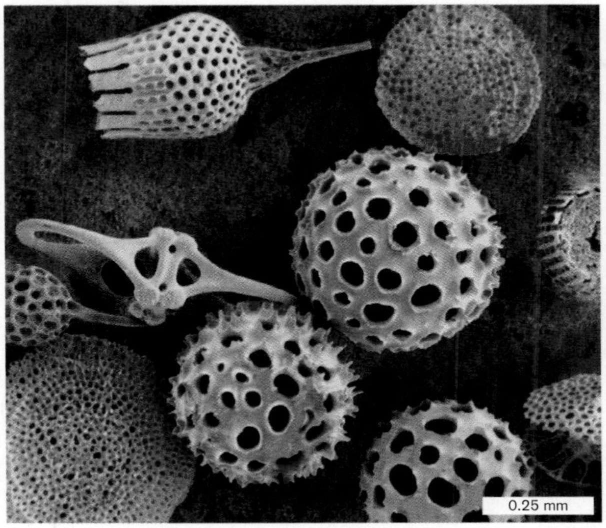
0.25 mm
DENNIS KUNKEL MICROSCOPY, INC.

B. A foraminiferan

500 µm
C. HEMLEBEN/©CUSHMAN FOUNDATION FOR FORAMINIFERAL RESEARCH, INC.

FIGURE 20.33 ■ Filamentous and shelled amebas. **A.** Shells (tests) of radiolarians (colorized SEM). **B.** A live foraminiferan, *Globigerinella aequilateralis* (dark-field).

Filamentous and Shelled Amebas

Amebas of a distinct clade (Rhizaria) form needlelike pseudopods. Most of these amebas are encased by mineral shells called tests (**Fig. 20.33A**). A major group of shelled amebas is the **radiolarians**, whose shells are made of silica perforated with numerous holes through which pseudopods appear to radiate in all directions. Many different kinds of radiolarians exist today, and many can be recognized in fossil rock. A second group of shelled amebas is the **foraminiferans** (**Fig. 20.33B**). The foraminiferans, or forams, generate shells of calcium carbonate as chambers laid down in helical succession. Their pseudopods all extend from one opening in the most recent chamber.

Radiolarians and forams grow in marine and aquatic habitats. Their shells make up a large part of reef formations, sedimentary rock, and beach sand. Forams, in particular, are used in geological surveys as indicators of petroleum deposits.

To Summarize

- **Amebas** move using pseudopods. In different species, pseudopods are lobe-shaped, lamellar, or filamentous (filopodia).
- **Cytoplasmic streaming** through cycles of actin polymerization and depolymerization drives the extension and retraction of pseudopods.
- **Slime molds** show an asexual reproductive cycle in which a fruiting body produces spores. In cellular slime molds, amebas aggregate to form a slug. In plasmodial slime molds, a single ameba develops into a multinucleate cell.
- **Radiolarians** have silicate shells penetrated by filamentous pseudopods.
- **Foraminiferans** have calcium carbonate shells with helical arrangement of chambers. The most recent chamber opens to extend filamentous pseudopods.

20.5

Alveolates: Ciliates, Dinoflagellates, and Apicomplexans

The Alveolata include voracious predators such as the ciliated protists (**Fig. 20.34A**). Alveolates are named for the flattened vacuoles called alveoli (singular, alveolus) within their outer cortex (**Fig. 20.34B**; see also **Fig. 20.7A**). Some alveoli contain plates of stiff material, such as protein, polysaccharide, or minerals. Besides alveoli, most alveolate protists possess other kinds of cortical organelles, such as extrusomes for delivery of enzymes or toxins, bands of microtubules for reinforcement, and whiplike cilia or flagella. The alveolate cell form is highly structured, in contrast to the amorphous shape of amebas. Major groups of alveolates include ciliates, dinoflagellates, and apicomplexans.

Ciliates

A diverse group of alveolates known as Ciliophora, or ciliates, possess large numbers of cilia, short projections containing [9(2)+2] microtubules. Their whiplike action is

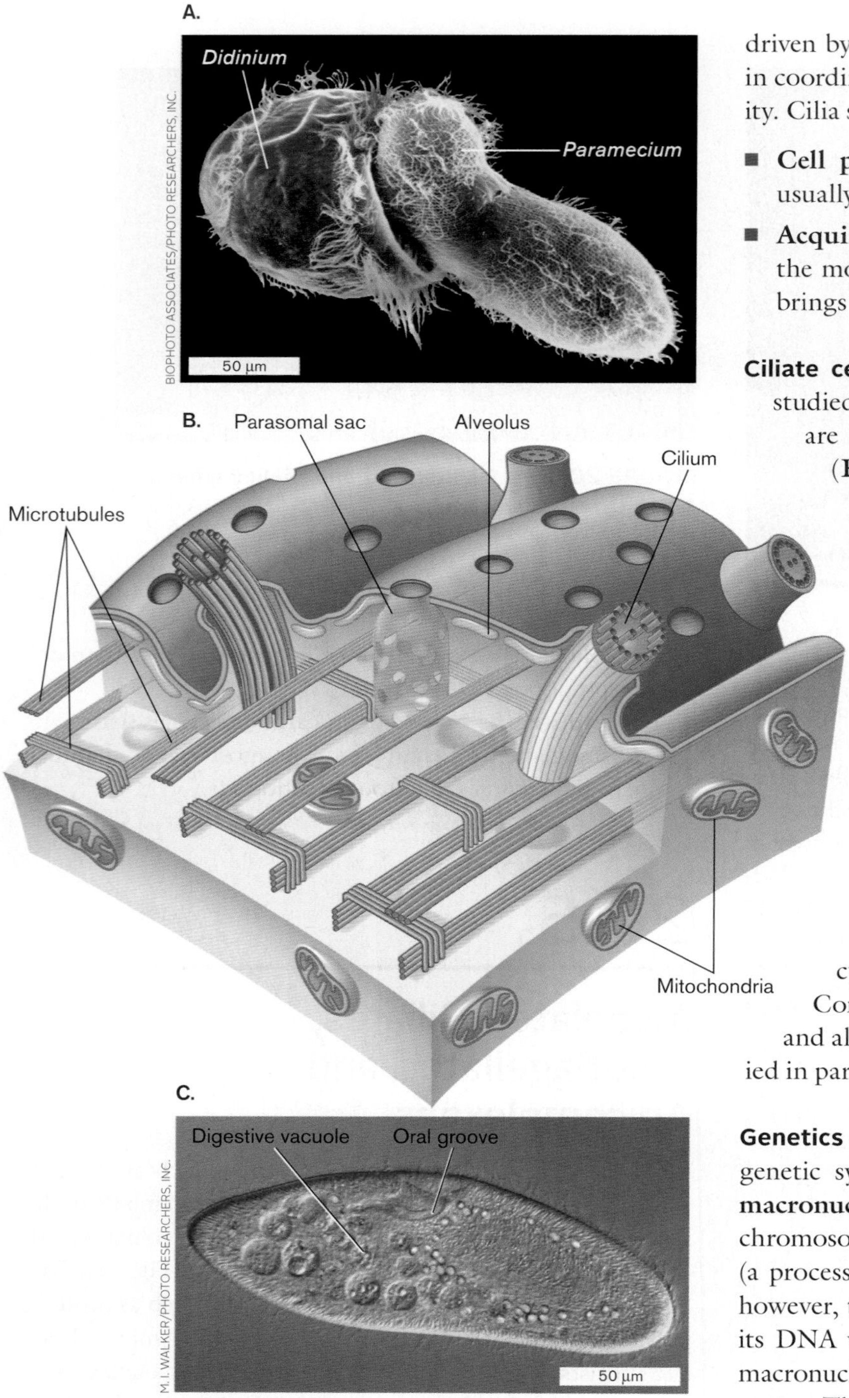
A.
Didinium
Paramecium
50 µm
BIOPHOTO ASSOCIATES/PHOTO RESEARCHERS, INC.
B.
Parasomal sac
Alveolus
Cilium
Microtubules
Mitochondria
C.
Digestive vacuole
Oral groove
50 µm
M. I. WALKER/PHOTO RESEARCHERS, INC.

FIGURE 20.34 ■ Ciliated protists. A. *Didinium* consuming *Paramecium* (SEM). **B.** Cortical structure of a ciliate. Beneath the outer membrane lie flattened sacs of fluid called alveoli. The cilia, composed of [9(2)+2] microtubules, are rooted in a complex network of lateral microtubules. Parasomal sacs take up nutrients and form endocytic vesicles. **C.** A paramecium has digestive vacuoles and an oral groove for ingestion (colorized phase contrast).

driven by ATP (for review, see Appendix 2). The cilia beat in coordinated waves that maximize the efficiency of motility. Cilia serve two functions:

- **Cell propulsion.** Coordinated waves of beating cilia, usually covering the cell surface, propel the cell forward.
- **Acquiring food.** By generating water currents into the mouth of the cell, a ring of cilia around the mouth brings food into the cell.

Ciliate cell structure. *Paramecium* is one of the most studied ciliates. Paramecia feed on bacteria and in turn are consumed by larger ciliates, such as *Didinium* (**Fig. 20.34A**). They can also take up smaller particles through endocytosis by specialized pores in their cortex, called parasomal sacs (**Fig. 20.34B**).

The cell structure of *Paramecium* includes an **oral groove** for uptake of food driven by the beating cilia (**Fig. 20.34C**). Once ingested through the oral groove, a food particle travels within the digestive vacuole in a circuit around the cell. The digestive vacuole ultimately empties into the cytoproct, a specialized vacuole for the discharge of waste outside the cell. Paramecia maintain osmotic balance by means of a **contractile vacuole**, a vacuole that withdraws water from the cytoplasm to shrink it or contracts to expand it. Contractile vacuoles are widespread among protists and algae, but their mode of action has been most studied in paramecia.

Genetics and reproduction. Most ciliates have a complex genetic system involving one or more **micronuclei** and **macronuclei**. The micronucleus contains a diploid set of chromosomes that undergoes meiosis for sexual exchange (a process reviewed in Appendix 2). For gene expression, however, the micronucleus generates hundreds of copies of its DNA within a macronucleus. The DNA copies in the macronucleus are rearranged and fragmented to small segments. The small DNA segments generate a large number of "telomeres," chromosome ends. In humans, telomere shortening is associated with aging; thus, ciliate macronucleus formation provides a model system for study of human aging (**eTopic 20.2**).

In ciliates, only macronuclear genes are transcribed to RNA and translated to protein. When a ciliate reproduces asexually, the micronucleus undergoes mitosis, whereas the macronucleus divides by a different mechanism that is poorly understood. Cell division occurs across the long

A. Conjugating paramecia

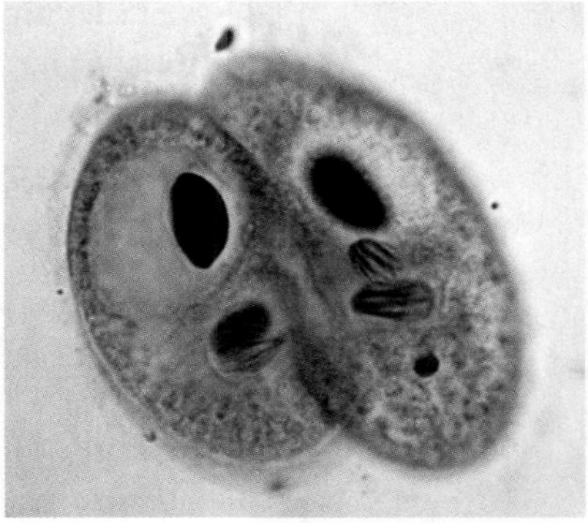
©MICHAEL ABBEY/VISUALS UNLIMITED

B. Conjugation between ciliates

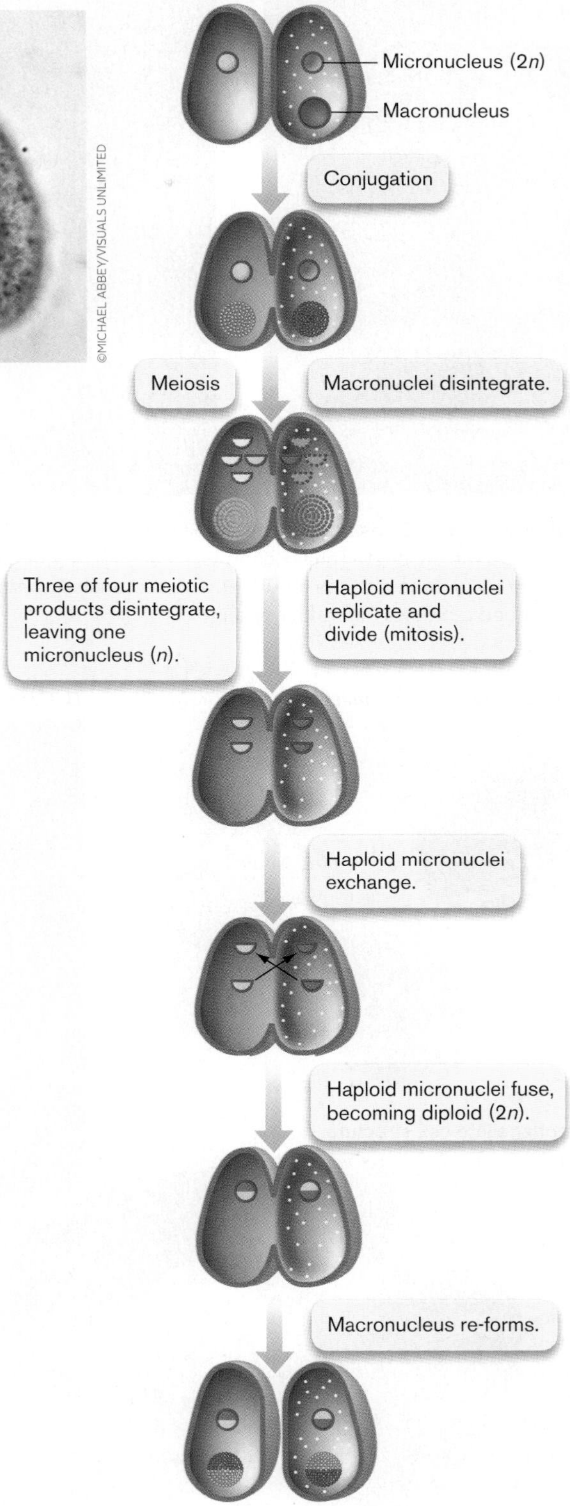
Micronucleus (2n)
Macronucleus
Conjugation
Meiosis
Macronuclei disintegrate.
Three of four meiotic products disintegrate, leaving one micronucleus (n).
Haploid micronuclei replicate and divide (mitosis).
Haploid micronuclei exchange.
Haploid micronuclei fuse, becoming diploid (2n).
Macronucleus re-forms.

FIGURE 20.35 ■ Conjugation. A. Two paramecia conjugating (light micrograph). **B.** In conjugation, two paramecia of opposite mating type form a cytoplasmic bridge. The $2n$ micronucleus of each cell undergoes meiosis. Each macronucleus, as well as three out of four meiotic products, disintegrates. The haploid micronuclei undergo mitosis, forming two daughter micronuclei. Daughter nuclei from each cell are exchanged across the cytoplasmic bridge and then fuse with their respective counterparts, restoring $2n$ micronuclei. The cells come apart, and each micronucleus generates a new macronucleus.

axis, necessitating generation of a new oral groove for the posterior daughter cell and a new cytoproct for the anterior daughter cell—again, a process poorly understood.

Most ciliates are diploid and never produce haploid gamete cells. Instead, their sexual reproduction involves exchange of micronuclei. The two ciliates of a mating pair exchange haploid micronuclei by **conjugation**. In conjugation, two cells of opposite mating type form a cytoplasmic bridge and exchange their nuclear products of meiosis (**Fig. 20.35**). While the cells connect, the micronucleus of each cell undergoes meiosis to form four haploid nuclei. Three out of four of the haploid nuclei disintegrate, as does the entire macronucleus. The haploid micronuclei then undergo mitosis, and one of each daughter nuclei is exchanged across the cytoplasmic bridge. Each transferred nucleus then fuses with its haploid counterpart, restoring diploidy. The two cells come apart, and each recombined micronucleus generates a new macronucleus.

Thought Questions

20.8 Compare and contrast the process of conjugation in ciliates and bacteria (see Chapter 9).

20.9 For ciliates, what are the advantages and limitations of conjugation, as compared with gamete production?

Stalked ciliates. Some ciliates adhere to a substrate and use their cilia primarily to obtain prey. **Stalked ciliates** such as *Stentor* and *Vorticella* have a ring of cilia surrounding a large mouth (**Fig. 20.36A**). The ciliary beat is specialized to draw large currents of water and whatever prey it carries. Stalked ciliates are commonly found in pond sediment and in wastewater during biological treatment by microbial digestion, where they are attached to flocs of filamentous bacteria.

Another group of stalked ciliates, the suctorians (**Fig. 20.36B**), possess cilia for only a short period after a daughter cell is released by the stalked cell. The daughter cell swims by ciliary motion until it finds a good habitat in which

A. ***Stentor*** **(stalked ciliate)**

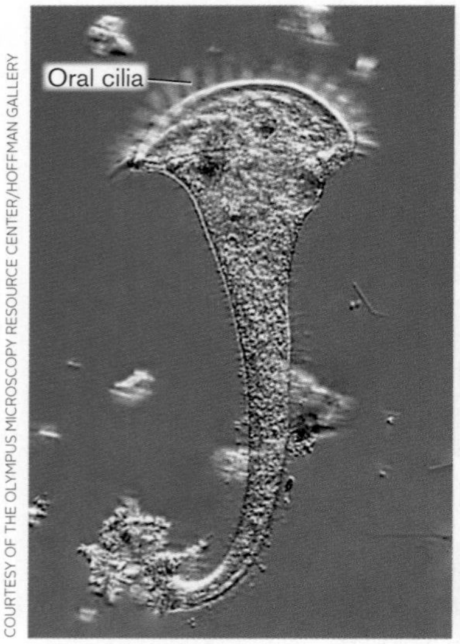
Oral cilia
COURTESY OF THE OLYMPUS MICROSCOPY RESOURCE CENTER/HOFFMAN GALLERY

B. ***Acineta*** **(suctorian)**

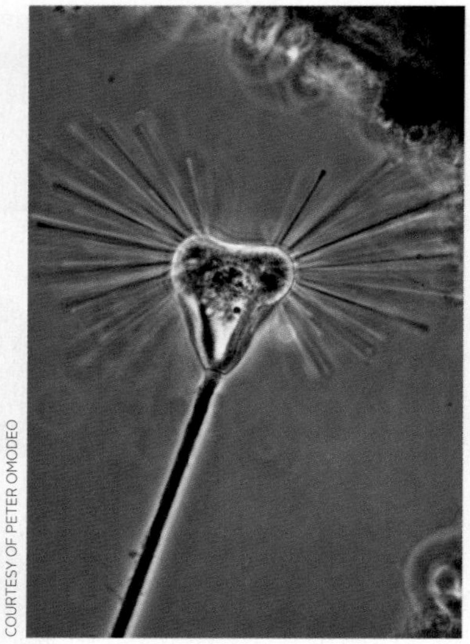
COURTESY OF PETER OMODEO

FIGURE 20.36 ■ Stalked ciliates. A. *Stentor,* a ciliate with a flexible stalk, 1.5–2.0 mm in length (phase contrast). The oral ring of cilia generates currents drawing food into the mouth. **B.** The suctorian *Acineta* replaces cilia with knobbed tentacles (light micrograph).

to settle, whereupon its cilia are replaced by knobbed tentacles similar to the filopodia of shelled amebas. Suctorians prey on swimming ciliates such as paramecia.

Dinoflagellates Are Phototrophs and Predators

The dinoflagellates (Dinoflagellata) are a major group of marine phytoplankton, essential to marine food webs. Like ciliates, they are highly motile, but instead of numerous short cilia, dinoflagellates possess just two long flagella, one of which wraps along a crevice encircling the cell (**Fig. 20.37A**). Some dinoflagellates possess elaborate hornlike extensions (**Fig. 20.37B**). The cell extensions increase the range of nutrient uptake, and they may deter predation.

Dinoflagellates are secondary or tertiary algal endosymbionts. They have a chloroplast derived from a red alga, which in some species was later replaced by a heterokont alga, itself a secondary

A. ***Gymnodinium*** **(dinoflagellate)**

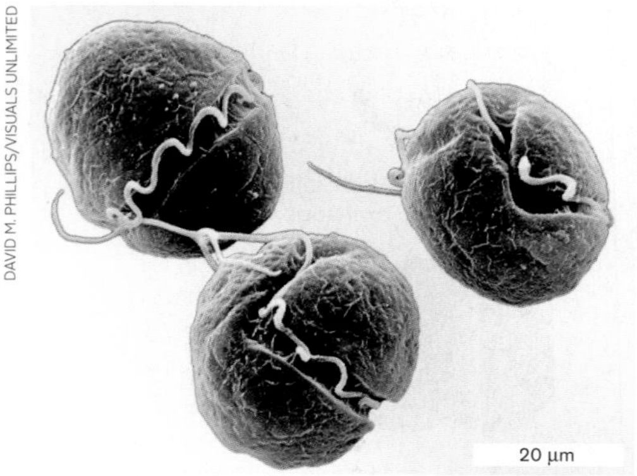
DAVID M. PHILLIPS/VISUALS UNLIMITED
20 µm

B. ***Ceratium*** **spp.**

©DENNIS KUNKEL/VISUALS UNLIMITED
100 µm

C. Dinoflagellate cell structure

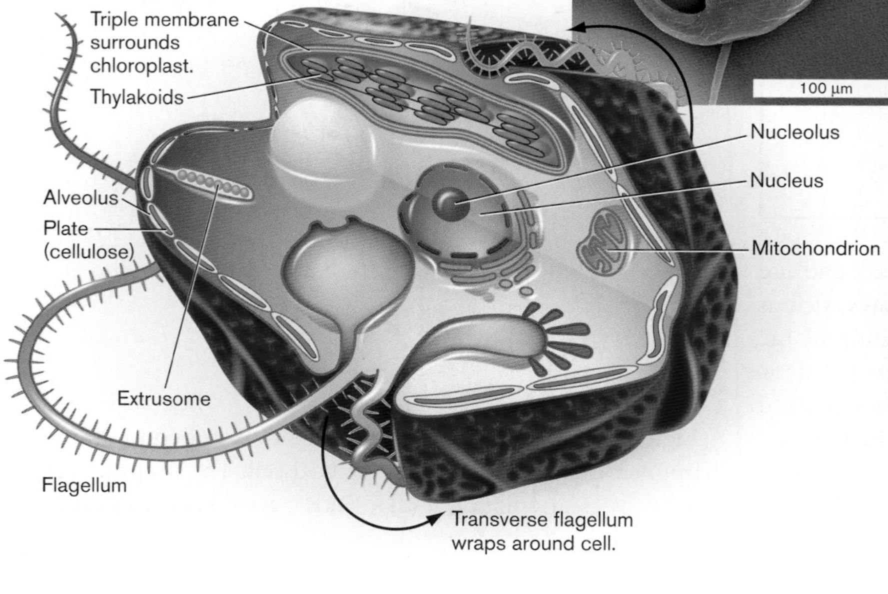
Triple membrane surrounds chloroplast.
Thylakoids
Nucleolus
Nucleus
Alveolus
Plate (cellulose)
Mitochondrion
Extrusome
Flagellum
Transverse flagellum wraps around cell.

BILL BACHMAN/PHOTO RESEARCHERS, INC.

D. Red tide

FIGURE 20.37 ■ Dinoflagellates. A. *Gymnodinium* sp., a dinoflagellate, with one of its two flagella wrapped around the cell (colorized SEM). **B.** *Ceratium* species, dinoflagellates with "horns." **C.** Diagram of a dinoflagellate. Protective plates of protein or calcified polysaccharides are formed within cortical alveoli. The chloroplast is surrounded by a triple membrane. **D.** "Red tide," caused by a bloom of dinoflagellates.

endosymbiont (**Fig. 20.37C**). Some dinoflagellates possess carotenoid pigments that confer a red color. Blooms of red dinoflagellates cause the famous red tide, which may have inspired the biblical story of the plague in which water turns to blood (**Fig. 20.37D**). Dinoflagellates release toxins that can be absorbed by shellfish, poisoning consumers months or years later.

The armor-plated appearance of a dinoflagellate results from its stiff alveolar plates, composed of cellulose (**Fig. 20.37C**). The complex outer cortex includes various extrusomes (organelles that extrude a defensive substance) and endocytic pores, as well as a species-specific pattern of alveolar plates. Dinoflagellates supplement their photosynthesis by predation, extending a special type of pseudopod to engulf prey. Some dinoflagellates have evolved to lose their chloroplasts altogether, becoming obligate predators or parasites.

Some dinoflagellates inhabit other organisms as endosymbionts, providing sugars from photosynthesis in exchange for a protected habitat. Their hosts include shelled amebas, sponges, sea anemones, and, most important, reef-building corals. Coral endosymbionts, known as **zooxanthellae** (singular, **zooxanthella**), are vital to reef growth. The zooxanthellae are temperature sensitive, and their health is endangered by global warming. Rising temperatures in the ocean lead to coral bleaching (the expulsion of zooxanthellae), after which the coral dies.

Apicomplexans Are Specialized Parasites

Apicomplexans include many human parasites, such as the intestinal parasite *Cyclospora,* which infected hundreds of people in the United States in 2013. Apicomplexan cells have an **apical complex**, a highly specialized structure that facilitates entry of the parasite into a host cell. Another important apicomplexan is *Toxoplasma gondii,* a parasite commonly carried by cats and transmissible to humans, where it can harm a developing fetus. Like the ciliates and dinoflagellates, apicomplexans possess an elaborate cortex composed of alveoli, pores, and microtubules. But as parasites, apicomplexans have undergone extensive reductive evolution, losing their flagella or cilia. They possess a unique organelle called the apicoplast, derived by genetic reduction from an endosymbiotic chloroplast. No capacity for photosynthesis remains, but the apicoplast provides one essential function in fatty acid metabolism.

The best-known apicomplexan is *Plasmodium falciparum,* the main causative agent of **malaria**, the most important parasitic disease of humans worldwide. **Figure 20.38** shows red blood cells in the early "ring stage" of infection, and in the "schizont" stage which bursts, releasing progeny parasites. *P. falciparum* is carried by mosquitoes, which transmit the parasite to humans when the insect's proboscis penetrates the skin. The disease is endemic in areas inhabited by 40% of the world's population; it infects hundreds of millions of people and kills more than a million African children each year.

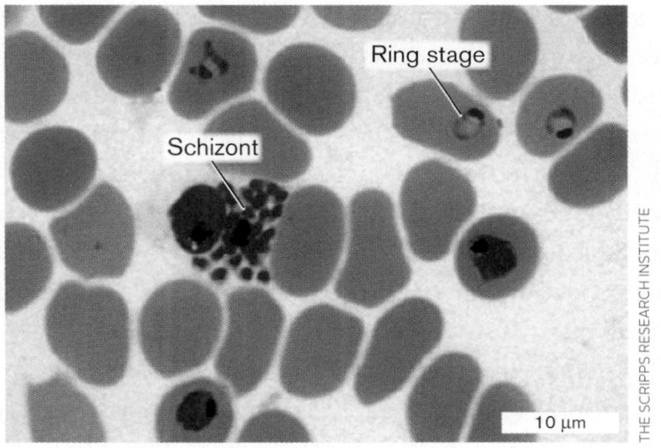
Ring stage
Schizont
10 µm
THE SCRIPPS RESEARCH INSTITUTE

FIGURE 20.38 ■ *Plasmodium falciparum,* a cause of malaria. Red blood cells infected with *P. falciparum,* which is stained purple with a dye that interacts with DNA (light micrograph). One late-stage infected blood cell (schizont) can be seen bursting, unleashing parasites on surrounding cells.

The transmitted parasites invade the liver and then develop into the **merozoite** form that invades red blood cells (**Fig. 20.39**). The merozoite first contacts a red blood

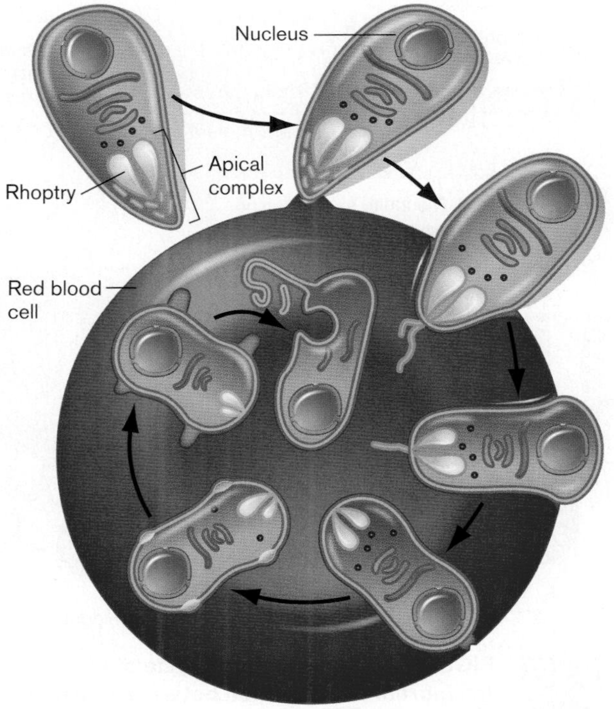
Nucleus
Apical complex
Rhoptry
Red blood cell

FIGURE 20.39 ■ Merozoite form of *Plasmodium falciparum* invades a red blood cell. The apical complex facilitates invasion and then dissolves as the merozoite transforms into an intracellular form.

cell with its apical complex. The apical complex contains secretory organelles called rhoptries that inject enzymes to aid entry of the parasite. The cone-like tip of the apical complex penetrates the host cell, enabling secretion of lipids and proteins that facilitate invasion. Eventually, the entire merozoite enters the host cell, leaving no traces of the parasite on the host cell surface. Thus, the internalized parasite becomes invisible to the immune system until its progeny burst out.

P. falciparum acquires resistance rapidly and no longer responds to drugs such as quinine that nearly eliminated the disease half a century ago. The life cycle and molecular properties of *P. falciparum* have been studied extensively for clues to aid in the development of new antibiotics and vaccines. The elaborate life cycle of *P. falciparum* and other parasites involves several common features:

- **Schizogony**, mitotic reproduction of a diploid form (in the mammalian host) to achieve a large population within a host tissue. Usually, the nuclei multiply first, followed by separation of individual nucleated cells.
- **Gamogony**, the meiosis and differentiation of haploid cells into male and female gametes capable of fertilization.
- **Mitotic reproduction** and development of the diploid forms (within the insect) into a spore-like form transmissible to the next host.

In the case of malaria (**Fig. 20.40**), whip-shaped sporozoites injected by the mosquito invade the liver, where they undergo schizogony (nuclear multiplication followed by cell separation). The cell products of schizogony are called merozoites. The merozoites from the liver then invade red blood cells, where they feed on hemoglobin. An early infected blood cell appears as a "ring stage" (**Fig. 20.40**). The parasite multiplies, filling the host cell, now called a "schizont." The schizont bursts, liberating progeny merozoites that invade another round of red cells. The bursting of red blood cells also releases cell fragments that trigger the cyclic fevers characteristic of malaria.

Some of the merozoites in the bloodstream undergo meiosis, generating pre-gamete cells, or gametocytes. The gametocytes multiply and mature (gamogony), and then they are acquired by bloodsucking mosquitoes. In the mosquito's midgut, the gametocytes develop into female eggs and flagella-like male cells. The male cells fertilize the egg cells, and the resulting zygotes undergo repeated mitosis and differentiate into sporozoites, which enter the salivary gland for transmission to the next human host.

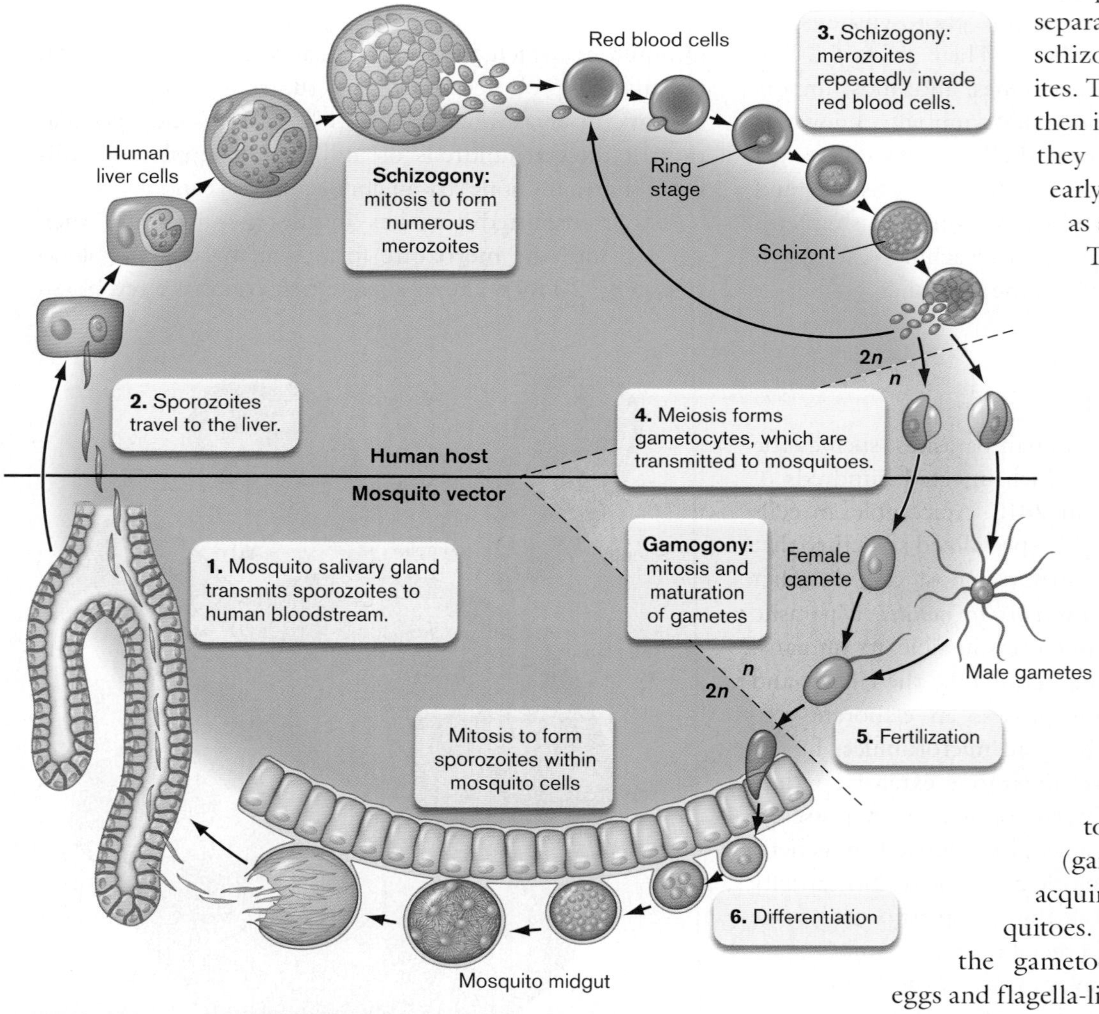
Red blood cells
3. Schizogony: merozoites repeatedly invade red blood cells.
Human liver cells
Schizogony: mitosis to form numerous merozoites
Ring stage
Schizont
2n
n
2. Sporozoites travel to the liver.
4. Meiosis forms gametocytes, which are transmitted to mosquitoes.
Human host
Mosquito vector
Gamogony: mitosis and maturation of gametes
Female gamete
1. Mosquito salivary gland transmits sporozoites to human bloodstream.
Male gametes
n
2n
5. Fertilization
Mitosis to form sporozoites within mosquito cells
6. Differentiation
Mosquito midgut

FIGURE 20.40 ■ Malaria: cycle of *Plasmodium falciparum* transmission between mosquito and human.

The nuclear genome of *P. falciparum* consists of 24 Mb contained in 14 chromosomes. Sequence annotation and expression studies predict 5,300 protein-encoding open reading frames (ORFs), comparable to the number in a yeast genome. The parasite has lost many genes encoding enzymes and transporters while expanding its repertoire of proteins involved in antigenic diversity. In addition, the parasite contains two smaller nonnuclear genomes: that of its mitochondria and that of the chloroplast-derived apicoplast.

How can we treat malaria and eradicate the disease? The malarial genome reveals promising targets for drug design. For example, the fatty acid biosynthesis occurring within the apicoplast is targeted by triclosan and other antibiotics. Other promising targets for antimalarial drugs are the unique proteases required to digest hemoglobin within the *P. falciparum* food vacuole.

To Summarize

- **Ciliates are covered with numerous cilia.** Cilia provide motility and help capture prey. Ciliates undergo complex reproductive cycles involving exchange of micronuclei through conjugation.
- **Dinoflagellates are phototrophic predators.** Dinoflagellates are tertiary endosymbiotic algae. Their alveoli contain calcified plates; they have paired flagella, one of which is used for propulsion. Predation occurs by extension of a lamellar pseudopod.
- **Apicomplexans are parasites that penetrate host cells.** The apicoplast is a specialized organ for cell penetration. Apicomplexans such as *Plasmodium falciparum* conduct complex life cycles within mammalian and arthropod hosts.

20.6

Trypanosomes and Metamonads

The group Euglenida includes flagellated protists such as *Euglena,* with chloroplasts arising from secondary endosymbiosis. Like other algal protists, the euglenas combine photosynthesis and heterotrophic nutrition. The Euglenida, however, also include a group of obligate parasites called **trypanosomes**. Trypanosomes consist of an elongated cell with a single flagellum. The trypanosome has a unique organelle called the "kinetoplast," consisting of a mitochondrion containing a bundle of multiple copies of its circular genome, usually placed near the base of the flagellum.

Trypanosomes cause some of the most gruesome and debilitating conditions known to humanity, such as leishmaniasis (**Fig. 20.41A**). *Leishmania major* (**Fig. 20.41B**) causes skin infections that may enter the internal organs. If untreated, leishmaniasis can lead to swelling and decay of the extremities (**Fig. 20.41A**) and eventually death. Carried by sand flies, *Leishmania* infects 1.5 million people annually, in South America, Africa and the Middle East, and southern Europe. *Leishmania* often infects Americans serving in Iraq; for this reason, returning veterans from Iraq are permanently restricted from donating blood.

Another major disease caused by trypanosomes is trypanosomiasis, also known as African sleeping sickness. A major killer of humans and livestock, in some sub-Saharan countries trypanosomiasis causes the second largest number of deaths after AIDS. The parasite, *Trypanosoma brucei* (**Fig. 20.41C**), is carried by the tsetse fly. *T. brucei* multiplies in the bloodstream of the host animal,

A. Leishmania infection
COURTESY OF CDC/D. S. MARTIN
B. Leishmania major
10 μm
©DENNIS KUNKEL
C. Trypanosoma brucei
Trypanosome
5 μm
EYE OF SCIENCE/PHOTO RESEARCHERS, INC.

FIGURE 20.41 ■ Trypanosomes.
A. Patient suffering from *Leishmania* infection (leishmaniasis). **B.** Cluster of *L. major* undergoing schizogony within the sand fly. **C.** *Trypanosoma brucei,* seen here among red blood cells, cause of African sleeping sickness.

Special Topic 20.2: The Trypanosome: A Shape-Shifting Killer

The trypanosome *Trypanosoma brucei* has extraordinary abilities to change its surface proteins to indefinitely thwart the immune response. But this system of surface protein variation requires an enormous genetic repertoire, which must be activated only when needed, within the mammalian host bloodstream. At other stages of its life cycle, the trypanosome has very different needs. So the parasite differentiates among at least six differently shaped forms, from insect to mammalian host. Each form has different biochemical properties enabling survival and growth in the particular host organ environment. How does the trypanosome know when to differentiate? The control points offer chances for drug treatment.

An example of such a control point is the transition from the mammalian bloodstream to the insect salivary gland, where the trypanosome becomes transmissible to the next host (**Fig. 1**). In the bloodstream, most trypanosomes assume the "slender" form, which multiplies to large numbers while undergoing antigenic variation. The slender form, however, cannot survive in the tsetse fly. So a fraction of the slender forms differentiate into the "stumpy" form, so called for the wider appearance of the cell. The stumpy form does not proliferate, because it is arrested in its cell cycle; however, unlike the slender form, it can survive in the tsetse fly. When the stumpy form is sucked up by the fly, it must differentiate again, into another form ("procyclic"), which can multiply in the very different host environment of the fly's midgut. The procyclic form then differentiates into the epimastigote and finally the metacyclic form, which expresses the antigenically variable proteins needed to grow within the bloodstream of the next mammalian host.

What signals induce the stumpy form to differentiate into the procyclic form? An experiment showed that a drop in temperature from 37°C (human body temperature) to 20°C (approximate temperature of the tsetse fly in the evening)

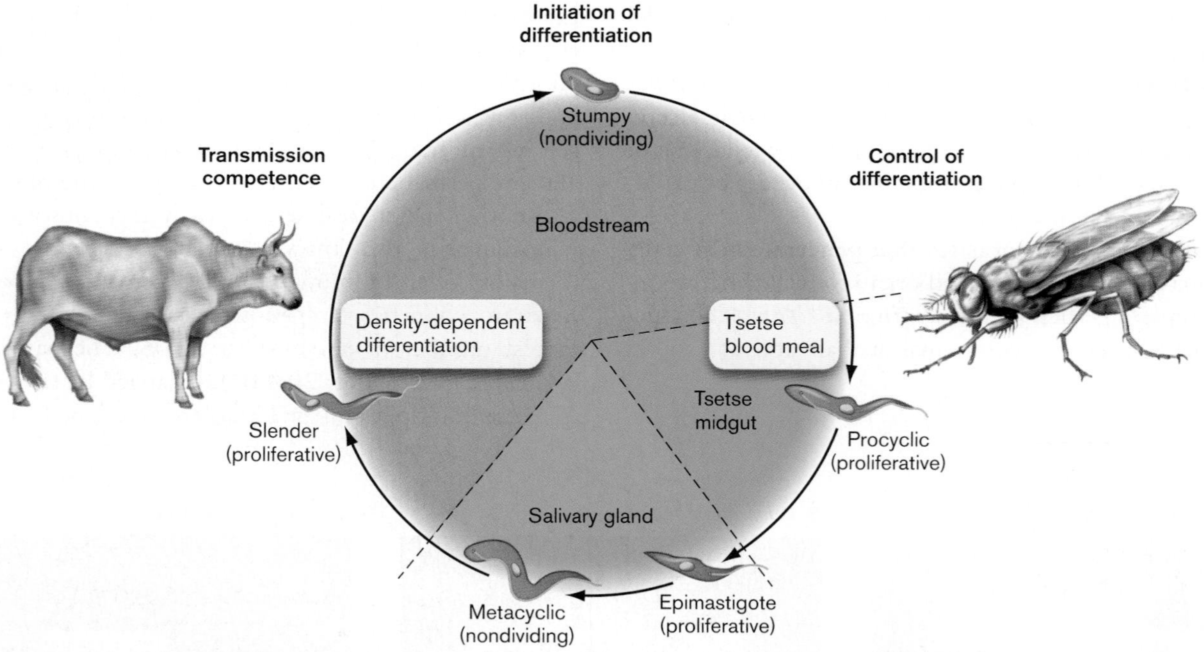
Initiation of differentiation
Stumpy (nondividing)
Transmission competence
Control of differentiation
Bloodstream
Density-dependent differentiation
Tsetse blood meal
Tsetse midgut
Slender (proliferative)
Procyclic (proliferative)
Salivary gland
Metacyclic (nondividing)
Epimastigote (proliferative)

FIGURE 1 ■ Stages of the trypanosome life cycle. The "slender" form of the trypanosome proliferates in the bloodstream, but only the "stumpy" form can survive in the tsetse salivary gland. The "stumpy" form must differentiate to the "procyclic" form and other forms that proliferate within the insect midgut.

causing repeated cycles of proliferation and fever that ultimately lead to death, if untreated. This trypanosome is known for its extraordinary degree of antigenic variation. Its genome includes 200 different active versions of its variant surface glycoprotein (VSG), the antigen inducing the immune response, as well as 1,600 different "silent" versions that can recombine with "active" VSG to make further variations. In effect, the trypanosome overwhelms the host immune system by continually generating new antigenic forms until the host repertoire of antibodies is exhausted. In order to infect its human host, the trypanosome needs to interconvert among several different forms could induce differentiation of the stumpy form to the procyclic form. But what components of the stumpy form process the signal? This question was addressed by Keith Matthews and colleagues at the University of Edinburgh. Matthews compared the transcriptomes (the gene expression levels throughout the genome) of two different strains of trypanosomes, one of which was defective for differentiation. The gene expression levels were compared by microarray analysis (discussed in Chapter 10). The microarray analysis revealed two genes expressed only in the differentiation-positive strain, encoding cell-surface proteins PAD1 and PAD2, named for "proteins associated with differentiation."

The roles of the PAD1 and PAD2 genes in differentiation were tested by immunofluorescence microscopy. First, a mixture of stumpy and slender forms was prepared under phase contrast (**Fig. 2A**). The fixed forms were then treated with PAD1-specific antibody, followed by a fluorescent secondary antibody. Fluorescence (**Fig. 2B**) showed signal throughout the surface of the stumpy forms but not the slender forms. Thus, PAD1 protein is expressed only in the stumpy form; it may be required for differentiation of the slender form to the stumpy form.

What happens at the next transition, to the procyclic form within the tsetse fly? Other experiments showed that PAD1 disappeared in the procyclic form, but there was increased expression of a related protein, PAD2. Furthermore, the increased expression and cellular distribution of PAD2 specifically correlated with the temperature shift that activates differentiation. At the higher temperature of the mammalian bloodstream (37°C), PAD2 protein (green) was confined to the flagellar pocket of the trypanosome (**Fig. 3A**); at the lower temperature of the insect environment (20°C), PAD2 was increased and distributed around the surface of the stumpy cell (**Fig. 3B**). The timing of PAD2 activation is consistent with its requirement for differentiation from the stumpy form to the procyclic form. Thus, PAD1 and PAD2 proteins play critical roles in trypanosome differentiation and suggest targets for antimicrobial therapy.

RESEARCH QUESTION

What pattern of expression might you expect to find for a different PAD protein expressed during differentiation of a metacyclic form into a slender form? At what temperature would you expect normal expression? How might this information provide a useful drug target for human therapy?

Dean, Samuel, Rosa Marchetti, Kiaran Kirk, and Keith R. Matthews. 2009. A surface transporter family conveys the trypanosome differentiation signal. *Nature* **459**:213–217.

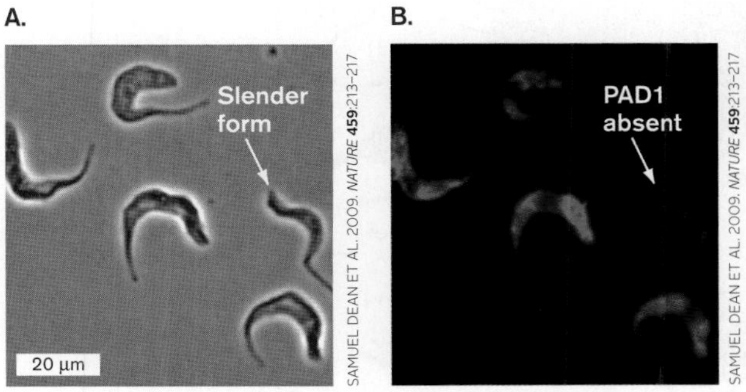
A.
B.
Slender form
20 µm
SAMUEL DEAN ET AL. 2009. NATURE 459:213–217
PAD1 absent
SAMUEL DEAN ET AL. 2009. NATURE 459:213–217

FIGURE 2 ■ **PAD1 expression in stumpy and slender forms. A.** Under phase-contrast microscopy, the slender form is distinguished from the stumpy form. **B.** Immunofluorescence shows PAD1 protein (red) in the stumpy form but not in the slender form.

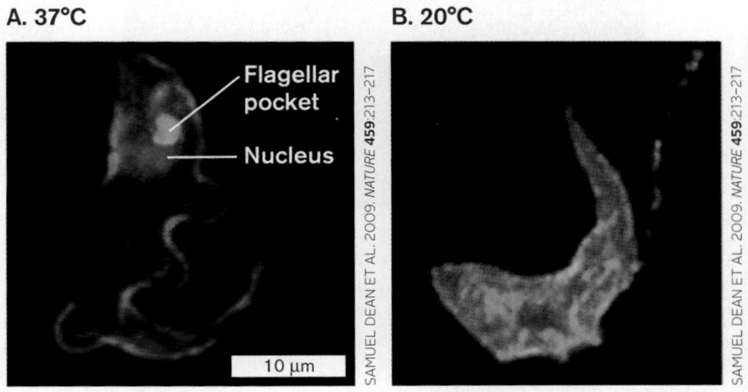
A. 37°C
B. 20°C
Flagellar pocket
Nucleus
10 µm
SAMUEL DEAN ET AL. 2009. NATURE 459:213–217
SAMUEL DEAN ET AL. 2009. NATURE 459:213–217

FIGURE 3 ■ **Temperature shift alters PAD2 expression. A.** At 37°C, PAD2 immunofluorescence (green) is confined to the flagellar pocket of the stumpy form. DAPI fluorescence indicates nuclear DNA (blue), and alpha-tubulin immunofluorescence indicates the cytoplasm (red). **B.** At 20°C, PAD2 is increased and distributed throughout the cell.

of its life cycle. The molecular basis of conversion offers targets for drug therapy. An example of research on the trypanosome life cycle is described in **Special Topic 20.2**.

A related trypanosome, *T. cruzi,* is carried by reduviid bedbugs; it causes Chagas' disease, a debilitating infection of the heart and other internal organs. Chagas' disease is prevalent in South and Central America, and global warming is expected to expand its range north.

Another important group of parasites and symbionts is Metamonada. The metamonad parasites include the diplomonads (named for their double nuclei), such as *Giardia intestinalis* (*G. lamblia*), a common intestinal parasite

(**Fig. 20.42**). *Giardia* is a frequent nemesis of day-care centers, but it also occurs in aquatic streams visited by bears and other wildlife. *Giardia* occasionally contaminates community water supplies in the United States and has become endemic in major Russian cities. *Giardia* and other metamonads are noted for their anaerobic metabolism and their degenerated organelles, reflecting their adaptation to the anaerobic intestinal environment.

The giardial life cycle alternates between two major forms: the trophozoite and the dormant cyst. The trophozoite (**Fig. 20.42B**) has two nuclei (and therefore, 4*n* chromosomes). There are eight pairs of flagella, and an "adhesive disk" enabling the parasite to adhere to the intestinal epithelium. The cell body contains no Golgi, and its mitochondria have degenerated to "mitosomes." Mitosomes lack mitochondrial genomes but retain some proteins expressed by former mitochondrial genes that have migrated to the giardial nuclear genome. When the trophozoite experiences stress conditions, such as high levels of bile and a high pH, the organism encysts (**Fig. 20.42C**). The cyst detaches from the intestine and is expelled from the host. It remains dormant until ingestion by a new host, where stomach acid triggers differentiation into a trophozoite.

Concluding Thoughts

The microbial eukaryotes, including fungi, algae, and many kinds of protists, serve many diverse roles in our ecosystems. A survey of protists may leave an impression that most protist clades exist primarily to parasitize humans. However, recent genetic surveys based on DNA sequence analysis of environmental communities suggest the existence of twice as many protist clades in nature as we have studied to date. Most of these unknown clades have no direct connection with humans, yet they may fill crucial niches in the ecosystems on which human existence depends. The remaining chapters in Part 4 emphasize the interconnections among many kinds of microbes that form the foundations of Earth's biosphere.

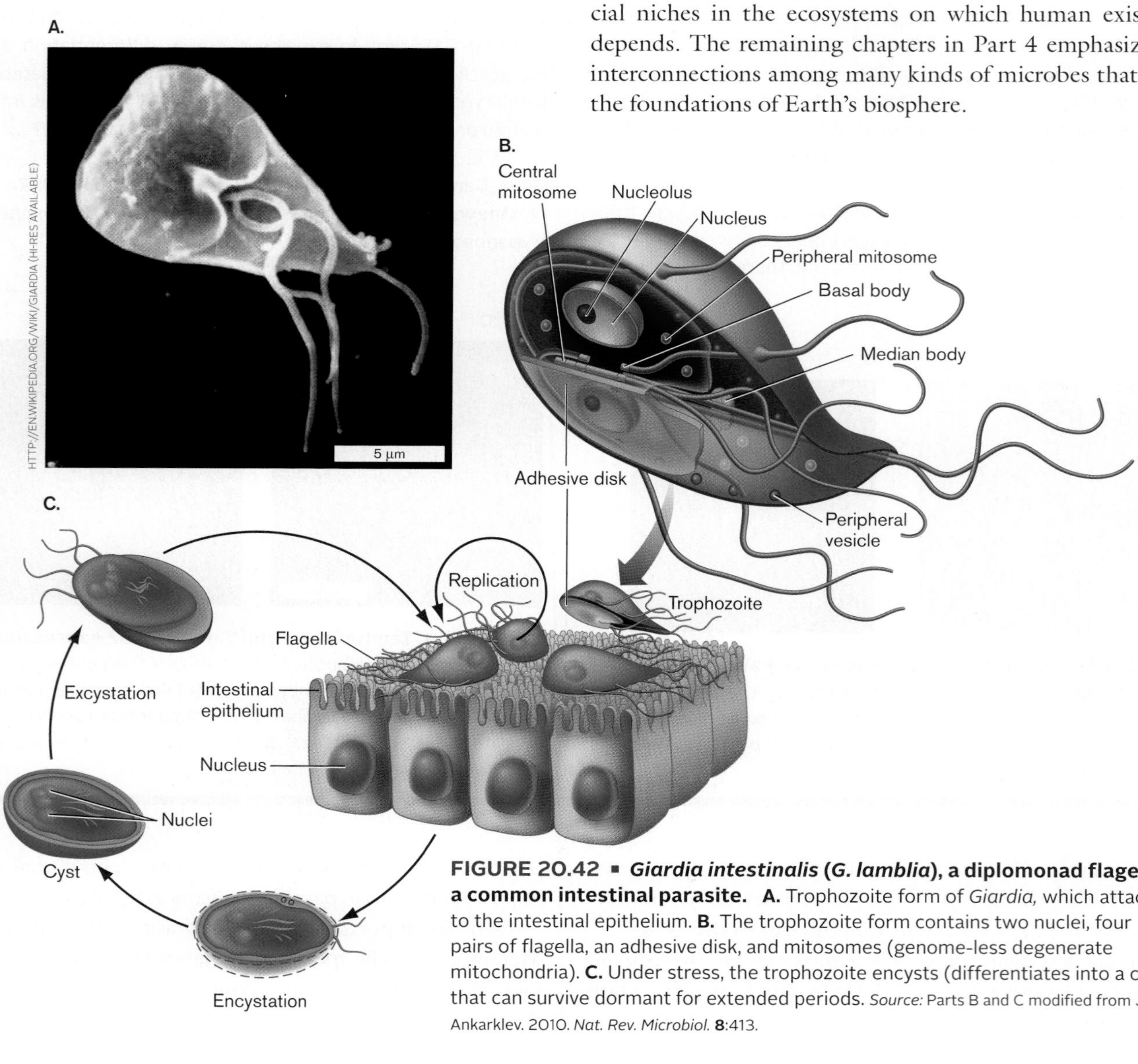
A.
HTTP://EN.WIKIPEDIA.ORG/WIKI/GIARDIA (HI-RES AVAILABLE)
5 μm
B.
Central mitosome
Nucleolus
Nucleus
Peripheral mitosome
Basal body
Median body
Adhesive disk
Peripheral vesicle
Trophozoite
C.
Replication
Flagella
Excystation
Intestinal epithelium
Nucleus
Nuclei
Cyst
Encystation

FIGURE 20.42 ■ *Giardia intestinalis* (*G. lamblia*), a diplomonad flagellate, a common intestinal parasite. **A.** Trophozoite form of *Giardia,* which attaches to the intestinal epithelium. **B.** The trophozoite form contains two nuclei, four pairs of flagella, an adhesive disk, and mitosomes (genome-less degenerate mitochondria). **C.** Under stress, the trophozoite encysts (differentiates into a cyst) that can survive dormant for extended periods. *Source:* Parts B and C modified from Johan Ankarklev. 2010. *Nat. Rev. Microbiol.* **8**:413.

CHAPTER REVIEW

Review Questions

1. Discuss the evidence for the branching of fungi and animals within one clade, the Opisthokonta, which is distinct from algae and protists.
2. How do primary symbiont algae differ from secondary and tertiary symbiont algae? Compare with respect to cell structure and nutritional options.
3. Compare and contrast the molecular basis of motility in amebas and ciliates. Cite particular species.
4. Compare and contrast the traits of these heterokont protists: kelps, diatoms, and dinoflagellates. Compare with respect to cell structure, colony organization, and nutritional options.
5. Summarize the key traits of fungi. What do fungi have in common with protists, and how do they differ?
6. Outline the life cycles of the major phyla of fungi: Chytridiomycota, Zygomycota, Ascomycota, and Basidiomycota. Explain their ecological significance.
7. Compare and contrast the traits of green and red algae (Chlorophyta and Rhodophyta, respectively).
8. Outline the life cycle of the slime mold *Dictyostelium discoideum*. Compare and contrast its features with that of fungi that produce fruiting bodies, such as basidiomycetes.
9. Outline the complex parasitic life cycles of an apicomplexan parasite and a trypanosome. Cite evidence of reductive evolution, as well as of evolution of elaborate specialized structures to facilitate the parasite's life cycle.

Thought Questions

1. Compare and contrast eukaryotic microbes that have inorganic shells or plates. What is their composition, and how do they grow?
2. Explain mixotrophy. Why are so many marine eukaryotes mixotrophs?
3. Why do eukaryotes show such a wide range of cell size? What selective forces favor large cell size, and what favors small cell size?
4. Do eukaryotic parasites have genomes that are larger or smaller than those of free-living organisms? Explain.

Key Terms

alga (800, 802)
alternation of generations (809)
alveolate (804)
ameba (804)
apical complex (831)
apicomplexan (805)
arbuscular mycorrhizae (816)
ascomycete (812)
ascospore (814)
ascus (809)
basidiomycete (815)
basidiospore (816)
budding (808)
cellular slime mold (826)
chloroplast (802)
chrysophyte (803)
ciliate (804)
cilium (804)
conjugation (829)
contractile vacuole (828)
coralline alga (821)
cortical alveolus (804)
cryptogamic crust (822)
diatom (805, 823)
dinoflagellate (803)
endosymbiosis (800)
Eumycota (801)
flagellate (804)
flagellum (804)
foraminiferan (827)
fruiting body (802)
frustule (823)
fungus (801)
gamogony (832)
green alga (chlorophyte) (803)
heterokont (805)
hydrogenosome (810)
hypha (802)
kelp (805)
lamellar pseudopod (825)
lichen (822)
macronucleus (828)
malaria (831)
merozoite (831)
micronucleus (828)
mitosome (805)
mitosporic fungus (809)
mycelium (806)
mycology (798)
mycorrhizae (816)
nucleomorph (803)
oomycete (817)
oral groove (828)
phagocytosis (824)
phytoplankton (818)
plasmodial slime mold (826)
plasmodium (826)
primary endosymbiont (800)
protist (800)
protozoan (800)

pseudopod (804)
radiolarian (827)
red alga (rhodophyte) (803, 821)
sargassum weed (824)
schizogony (832)
secondary endosymbiont (800, 822)
slime mold (826)
sporangiospore (812)
sporangium (812)
stalked ciliate (829)
trypanosome (833)
yeast (807)
zoospore (801)
zooxanthella (831)
zygomycete (812)
zygospore (812)

Recommended Reading

Amaral-Zettler, Linda A., Felipe Gomez, Erik R. Zettler, Brendan G. Keenan, Ricardo Amils, et al. 2002. Eukaryotic biodiversity and physiology at acidic extremes: Spain's Tinto River. Eukaryotic diversity in Spain's River of Fire. *Nature* **417**:137.

Ankarklev, Johan, Jon Jerlström-Hultqvist, Emma Ringqvist, Karin Troell, and Staffan G. Svärd. 2010. Behind the smile: Cell biology and disease mechanisms of *Giardia* species. *Nature Reviews. Microbiology* **8**:413–422.

Armbrust, E. Virginia. 2009. The life of diatoms in the world's oceans. *Nature* **459**:185–192.

Armbrust, E. Virginia, John A. Berges, Chris Bowler, Beverley R. Green, Diego Martinez, et al. 2004. The genome of the diatom *Thalassiosira pseudonana*: Ecology, evolution, and metabolism science. *Science* **306**:79–86.

Baldauf, Sandra L. 2003. The deep roots of eukaryotes. *Science* **300**:1703–1706.

Baum, Jake, Anthony T. Papenfuss, Buzz Baum, Terence P. Speed, and Alan F. Cowman. 2006. Regulation of apicomplexan actin-based motility. *Nature Reviews. Microbiology* **4**:621–628.

Beaufort, L., I. Probert, T. de Garidel-Thoron, E. M. Bendif, D. Ruiz-Pino, et al. 2011. Sensitivity of coccolithophores to carbonate chemistry and ocean acidification. *Nature* **476**:80–83.

Fisher, Matthew C., Daniel. A. Henk, Cheryl J. Briggs, John S. Brownstein, Lawrence C. Madoff, et al. 2012. Emerging fungal threats to animal, plant and ecosystem health. *Nature* **484**:186–194.

Gardner, Malcolm J., Shamira J. Shallom, Jane M. Carlton, Steven L. Salzberg, Vishvanath Nene, et al. 2002. The genome sequence of the human malaria parasite *Plasmodium falciparum*. *Nature* **419**:498–511.

Haldar, Kasturi, Sophien Kamoun, N. Luisa Hiller, Souvik Bhattacharje, and Christiaan van Ooij. 2006. Common infection strategies of pathogenic eukaryotes. *Nature Reviews. Microbiology* **4**:922–931.

Harman, Gary E., Charles R. Howell, Ada Viterbo, Ilan Chet, and Matteo Lorito. 2004. *Trichoderma* species—opportunistic, avirulent plant symbionts. *Nature Reviews. Microbiology* **2**:43–56.

Henderson, Gregory P., Lu Gan, and Grant J. Jensen. 2007. 3-D ultrastructure of *O. tauri*: Electron cryotomography of an entire eukaryotic cell. *PLoS One* **8**:e749.

Kronstad, James W., Rodgoun Attarian, Brigitte Cadieux, Jaehyuk Choi, Cletus A. D'Souza, et al. 2011. Expanding fungal pathogenesis: *Cryptococcus* breaks out of the opportunistic box. *Nature Reviews. Microbiology* **9**:193–203.

Lew, Roger R. 2011. How does a hypha grow? The biophysics of pressurized growth in fungi. *Nature Reviews. Microbiology* **9**:509–518.

Parfrey, Laura W., and Laura A. Katz. 2010. Dynamic genomes of eukaryotes and the maintenance of genomic integrity. *Microbe* **5**:156–163.

Parniske, Martin. 2008. Arbuscular mycorrhiza: The mother of plant root endosymbioses. *Nature Reviews. Microbiology* **8**:763–775.

Pounds, J. Alan, Martin R. Bustamante, Luis A. Coloma, Jamie A. Consuegra, Michael P. L. Fogden, et al. 2007. Widespread amphibian extinctions from epidemic disease driven by global warming. *Nature* **439**:161–167.

Wilson, Richard A., and Nicholas J. Talbot. 2009. Under pressure: Investigating the biology of plant infection by *Magnaporthe oryzae*. *Nature Reviews. Microbiology* **7**:185–196.

CHAPTER 21
Microbial Ecology

21.1 Metagenomes—and Beyond
21.2 Functional Ecology
21.3 Symbiosis
21.4 Marine and Aquatic Microbes
21.5 Soil and Subsurface Microbes
21.6 Plant Microbial Communities
21.7 Animal Microbial Communities

Metagenomes—the DNA of microbial communities—reveal microbes in all habitats on Earth, from Antarctic lakes to the acid drainage of iron mines. Microbes evolve to grow in the air, water, and soil that surround us. In the ocean, vast quantities of microbes produce the biomass that ultimately feeds fish and humans. In forests and fields, microbes are major consumers; they decompose the majority of plant material, generating fertile soil. Diverse microbes largely determine the quality of soil, air, and water for human life. Microbial communities form functional components of all plants and animals, including human beings. Human microbial communities offer opportunities for medical therapy, from probiotics to fecal transplants.

For their plant and animal hosts, microbes provide nutrition, enhance development, and even control behavior. Microbes form elaborate symbiotic systems such as mycorrhizae, underground networks of fungal hyphae that support and connect the roots of plants throughout forest and field. Rhizobia fix nitrogen for legumes, whereas algae photosynthesize for coral reefs. Anaerobes digest complex plant fibers for termites, cattle, and even humans. Chapter 21 explores how microbes interact with each other and with their many diverse habitats on Earth.

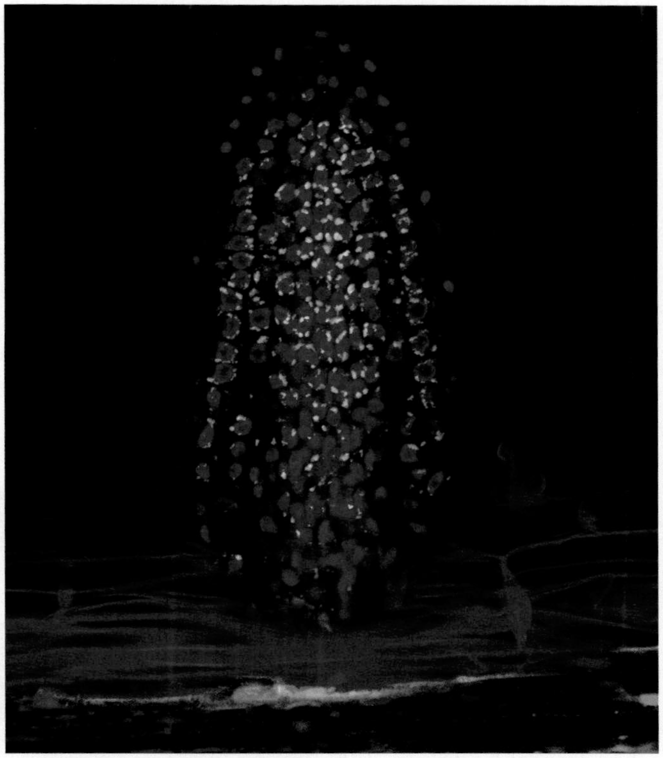

SARAH LEBEIS/U. OF NORTH CAROLINA, DEPT. OF ENERGY JOINT GENOME INSTITUTE

CURRENT RESEARCH highlight

Bacterial communities protect plants. Communities of bacteria colonize a new root of the plant *Arabidopsis thaliana.* The bacteria (green) and the root cell nuclei (blue) are visualized by fluorescence in situ hybridization (FISH) with probes hybridized to rRNA, using confocal laser scanning microscopy. In 2012, Derek Lundberg, Sarah Lebeis, and colleagues analyzed the metagenome of *Arabidopsis* endophytes, or plant-associated bacteria. Endophytes include species of Actinobacteria and Proteobacteria that may help the plant resist pathogens and survive environmental stresses such as high salt or temperature. Distinct clades of bacteria grow on the root surface or within the plant tissues. Communities of endophytic bacteria and fungi form a functional part of all plants—including those we and other animals consume as food. *Source:* Derek Lundberg et al. 2012. *Nature* **488**:86.

When honeybees share food through their mouthparts, they offer more than nectar and signal molecules: They also share their microbial communities (**Fig. 21.1**). Microbes have evolved to colonize every habitat of our biosphere, including soil, water, air, and the bodies of plants and animals. We can analyze the microbial communities within animals such as honeybees through **metagenomics**, the sequencing of DNA directly from a microbial community. The metagenomes of honeybee gut communities show a consistent set of microbe types, including unique types of lactobacilli (lactic acid bacteria) as well as bee-specific isolates of Gammaproteobacteria and Betaproteobacteria. How and why do bees maintain these gut microbiota? Microbial ecologists hypothesize that the bees' bacteria aid digestion and provide antibiotics to counter pathogens. The bees, their nectar-providing plants, and their microbial communities form a balanced ecosystem.

Chapter 21 explores how microbial communities function in their ecosystem. Microbial communities critically impact other organisms in all habitats, from oceans and forests to the interstices of rock (**Fig. 21.2**). Microbes recycle organic material in aquatic and terrestrial ecosystems, providing resources for plants and animals; and deep below Earth's surface, microbes shape the rock of Earth's crust. Our agricultural productivity depends on the positive and negative effects of microbes in the soil and air. Next, Chapter 22 will show how microbial cycling of the elements shapes the global biosphere as a whole. Microbial cycles of elements through our soil, air, and water increasingly face the impact of human technology. To preserve our planet's health, we will need to know far more about microbial ecology.

Microbes fill unique niches in ecosystems. "No man is an island," nor is any microbe. All organisms evolve within an **ecosystem** consisting of populations of species plus their habitat or environment. A **population** is a group of individuals of one species living in a common location. The sum of all the populations of different species constitute a **community**.

Lactobacillus johnsonii
5 µm

HTTP://WWW.MICHIGANBEES.ORG/2012/TROPHALLAXIS/

M. CARINA AUDISIO, 2010. *MICROBIOLOGICAL RESEARCH* **166**:1.

FIGURE 21.1 ■ Honeybees share food containing microbes. Food shared by bees contains beneficial gut bacteria. Inset: *Lactobacillus johnsonii*, lactic acid bacteria isolated from gut of honeybees (SEM).

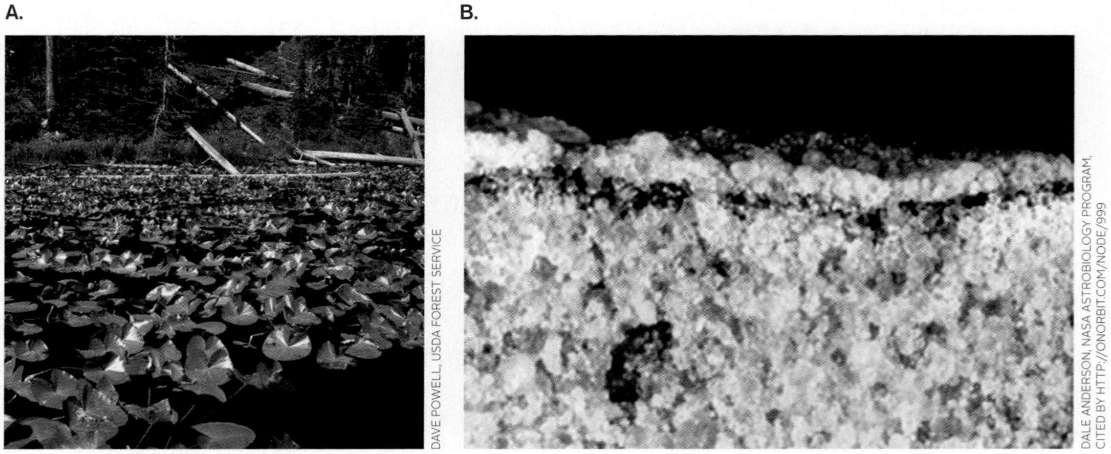
A.
B.

DAVE POWELL, USDA FOREST SERVICE

DALE ANDERSON, NASA ASTROBIOLOGY PROGRAM, CITED BY HTTP://ONORBIT.COM/NODE/999

FIGURE 21.2 ■ Microbes shape all environments. **A.** Microbes recycle organic material in aquatic and terrestrial habitats, providing resources for plants and animals. **B.** Endolithic microbes (black and green) grow within sandstone rock, Battleship Promontory, Antarctica.

Within a community, each population of organisms fills a specific **niche**. The niche is a set of conditions enabling an organism to grow and reproduce, including its habitat, resources, and relations with other species of the ecosystem. For example, the niche of *Synechococcus,* a cyanobacterium, is that of a free-living marine organism that fixes CO_2 into biomass while producing molecular oxygen that is used by swarms of respiring bacteria. The habitat of *Synechococcus* is the upper water layer of the ocean; its biomass provides food for protist predators, which in turn feed invertebrates and fish.

The essential role of microorganisms in all ecosystems was formulated by the Dutch microbiologist Cornelius B. van Niel (1897–1985). Van Niel was the first to show that bacteria in the soil and water can conduct photosynthesis without producing oxygen, using electron donors such as H_2S instead of H_2O. This surprising discovery revealed one of many kinds of metabolism unique to microbes, unknown in plants or animals. Other unique forms of microbial metabolism (discussed in Chapters 13–15) include nitrogen fixation by bacteria and archaea, and the degradation of lignin by bacteria and fungi. Moreover, microbial metabolism provides ecosystems with their sole source of key elements such as sulfur, phosphorus, and iron.

Unique microbial metabolism provides unique roles for microbes in ecosystems. Van Niel expressed this principle in two hypotheses of microbial ecology:

- **Every molecule existing in nature can be used as a source of carbon or energy by a microorganism somewhere.** Any cell component or product, such as amino acids and H_2S, can participate in some kind of energy-yielding reaction. We now extend this principle to include human-made compounds entering the environment. If an energy-yielding reaction exists, some microbe will evolve to use it.
- **Microbes are found in every environment on Earth.** Every possible habitat for life supports microbes. In fact, the largest part of the biosphere (below Earth's surface) is inhabited solely by microbes.

So how can we test these hypotheses for a given ecosystem, such as a freshwater lake or a human digestive tract? We raise these questions:

- **In a given ecosystem, what microbes exist?** To find the microbes that function in an ecosystem is a challenge, especially for those whose distribution is sparse. Two complementary approaches to identify members of microbial communities are **metagenomic** analysis and **enrichment culture**.
- **What functions do microbes provide?** Microbes cycle essential nutrients through a food web. They also serve more complex functions that we are just beginning to discover, such as defending host organisms from pathogens, and even modulating animal development and behavior.

Note: Microbial ecology is an interdisciplinary enterprise that uses methods presented throughout this book, such as fluorescence microscopy (Chapter 2), enrichment culture (Chapter 4), metagenome sequencing (Chapter 7), isotope labeling (Chapter 13), and molecular phylogeny (Chapter 17). Additional methods, such as gel electrophoresis and FISH (fluorescence in situ hybridization) are described in Appendix 3.

21.1

Metagenomes—and Beyond

Since microbes are microscopic, how do we identify individual organisms and distinguish them from their surroundings? Recall that the great eighteenth-century taxonomist Carolus Linnaeus called certain microbes "*Chaos*" because he thought we would never be able to clearly distinguish one from another. In the nineteenth century, Robert Koch developed techniques of pure culture, and Sergei Winogradsky developed enrichment culture—methods that isolate microbial species and reveal their metabolic traits (discussed in Chapters 1 and 4). **New ways to culture bacteria previously thought unculturable are described in Section 21.4.**

Even the most advanced methods of culture miss vast numbers of microbes that are unculturable—at least, so far. Many of these unculturable species emerge through DNA sequencing of environmental metagenomes (introduced in Chapter 7). In 1991, one of the first to sequence genes from environmental samples was Norman Pace, then at the University of Indiana, who cloned ribosomal RNA genes from plankton filtered from the Pacific Ocean. The small subunit (SSU) rRNA sequence became the standard for identifying environmental taxa (discussed in Chapter 17). In 1998, Jo Handelsman and colleagues used shotgun sequencing to analyze large portions of genomes from a soil microbial community. Handelsman coined the term "metagenome" to refer to the sum of genomes obtained directly from a community. In 2011, an international consortium sequenced the metagenome of the human gut **microbiome** (host-associated microbial community) from

TABLE 21.1

Metagenomic analysis: examples.

Microbial target community	Discovery	Sequencing approach	Reference
Marine picoplankton (microbes of size 0.2–2.0 μm) from filtered water of the Pacific Ocean	Identified 15 uncultured species of cyanobacteria and proteobacteria	Cloned SSU rRNA genes; Sanger dideoxy sequencing gels	Thomas M. Schmidt et al. 1991. *J. Bacteriol.* **173**:4371
Soil from an agricultural research station, Madison, WI	Unexpected diversity of uncultured firmicutes and proteobacteria	Genome libraries in bacterial artificial chromosomes; ABI automated gel sequencer	Michelle Rondon et al. 2000. *Appl. Environ. Microbiol.* **66**:2541
Pink biofilm growing at pH 1 in acid mine drainage, Iron Mountain, CA	Near-complete genomes of *Leptospirillum* and *Ferroplasma*	Genome libraries in plasmids; ABI capillary sequencer	Gene Tyson et al. 2004. *Nature* **428**:37
Marine microbes (size 0.1–3.0 μm) from filtered Sargasso Sea surface water	1,800 genomic species; 148 new bacterial taxa; surprisingly many bacteria contain proteorhodopsin and are photoheterotrophs	Genome libraries in plasmids; ABI capillary sequencer	J. Craig Venter et al. 2004. *Science* **304**:66
Indoor urban air, from a shop in Singapore, compared to soil and water	Indoor air contains relatively low bacterial diversity; includes opportunistic pathogens	Genome libraries in plasmids; ABI capillary sequencer	Susannah Tringe et al. 2008. *PLoS One* **3**:e1862
Bacteria from leaf-cutter-ant gut fungus gardens, Gamboa, Panama	Novel carbohydrate-digesting enzymes that may enhance biofuel production	PCR amplified and cloned library; Roche 454 pyrosequencer	Garret Suen et al. 2010. *PLoS Genet.* **6**:e1001129
Rumen microbiome of fistulated cows fed switchgrass	15 novel bacterial genomes; 27,700 carbohydrate-digesting enzymes	Genome libraries; Illumina sequencer	Matthias Hess et al. 2011. *Science* **331**:463
Human gut microbiomes of 22 individuals from 4 countries	Microbial genomes vary with human age; taxa clusters are shared by all human subjects	Genome libraries; ABI capillary sequencer	Manimozhiyan Arumugam et al. 2011. *Nature* **473**:174
Soil of Amazon rain forest, compared with converted agricultural soil	Agricultural conversion leads to large-scale decrease of microbial diversity	Bar code PCR amplification and Roche 454 pyrosequencer	Jorge Rodrigues et al. 2013. *PNAS* **110**:988

22 individuals in four countries. Examples of metagenome studies, with research references, are listed in **Table 21.1**.

Metagenomic sequencing is a powerful tool, but it poses challenges far beyond those of sequencing a single intact genome. The huge quantities of data require advanced sequencing methods and computational algorithms for analysis. At the same time, new methods of microbial culture reveal species that even metagenomics can miss. Here we present the power and pitfalls of metagenomics—and beyond, the new emerging tools of microbial ecology.

Sampling the Environment

To sequence a metagenome requires a series of steps, each of which presents important choices (**Fig. 21.4**). First, we need to obtain appropriate DNA samples. Sampling requires that we define a **target community** to study. Extreme environments pose safety considerations; for example, the acid mine drainage site sampled by Jillian Banfield's group has acidity strong enough to dissolve an iron shovel (**Fig. 21.3A**). Nonetheless, Banfield sequenced community DNA including the genome of the hyperacidophile *Leptospirillum* (**Fig. 21.3B**). Other target communities require interaction with a host animal, such as the rumen of a fistulated cow (**Fig. 21.3C** and **D**).

Sampling the target community. Sequencing a metagenome requires several steps (**Fig. 21.4**). The cells of the target community must be separated from their surroundings without loss of DNA (**Fig. 21.4**, step 1). For example,

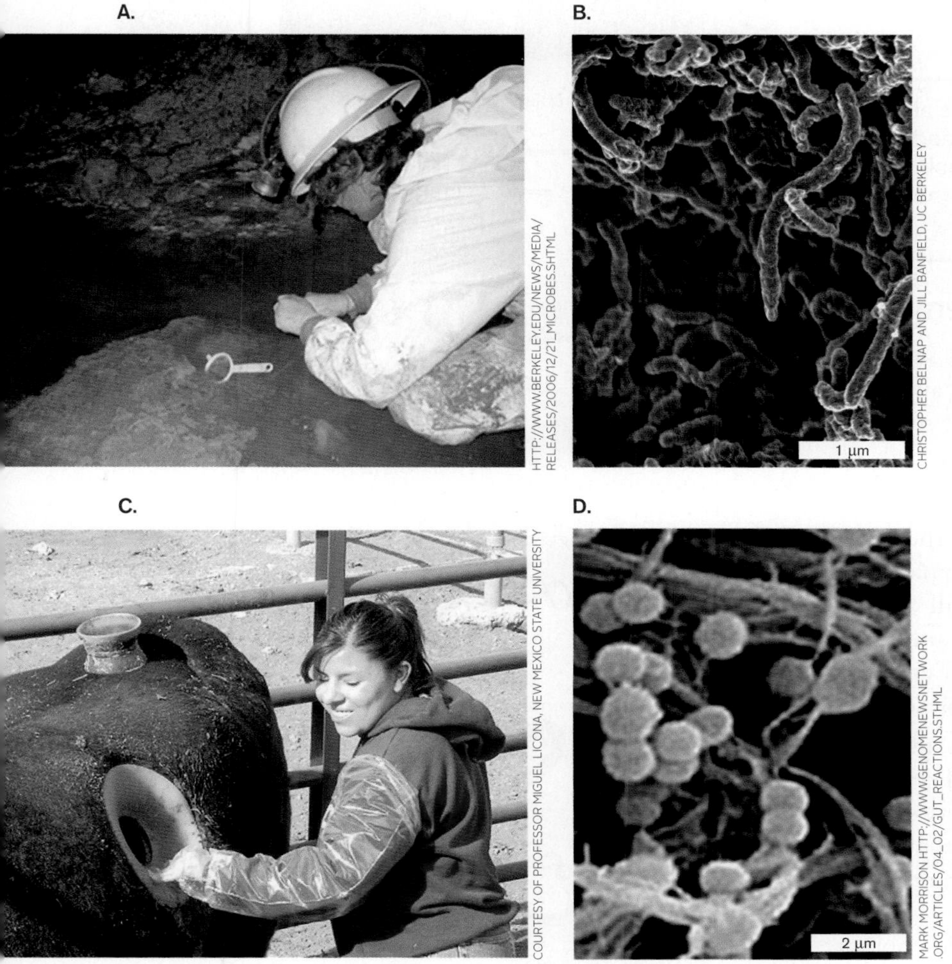
A.
B.
C.
D.
1 µm
2 µm
HTTP://WWW.BERKELEY.EDU/NEWS/MEDIA/RELEASES/2006/12/21_MICROBES.SHTML
CHRISTOPHER BELNAP AND JILL BANFIELD, UC BERKELEY
COURTESY OF PROFESSOR MIGUEL LICONA, NEW MEXICO STATE UNIVERSITY
MARK MORRISON HTTP://WWW.GENOMENEWSNETWORK.ORG/ARTICLES/04_02/GUT_REACTIONS.SHTML

FIGURE 21.3 ■ Sampling a target community of microbes. **A.** Jillian Banfield samples "pink slime" of archaea growing in acid mine drainage at Iron Mountain, California, one of the largest Superfund cleanup sites. **B.** *Leptospirillum* sp., bacteria whose genome emerged out of Banfield's metagenomic analysis. **C.** An undergraduate student samples rumen contents from a fistulated (cannulated) cow. The closable opening (see the plug sitting on the cow's back) does not harm the animal. **D.** *Ruminococcus albus* bacteria digest plant fibers within the rumen of a cow.

sampling of a soil community requires removal of humic acids (derivatives of the wood component lignin), which inhibit DNA polymerases. Suppose the target community inhabits a host plant or animal; what additional separation is required? We must separate the DNA of the target community from the DNA of the host, which can contaminate the DNA pool to be sequenced. Thus, we need a distinctive separation protocol to dislodge host-associated microbes from their host, before the host nuclei break open.

Filtering the sample. A different problem arises when sampling a large, dilute target community, such as an ocean. Which community interests you—the protists and small invertebrates? the bacteria and archaea? the viruses? If you sample the entire community, one fraction may yield most of the DNA but overwhelm other community members. To manage the complexity, you may decide to **filter** or fractionate the target community (**Fig. 21.4**, step 2). Filtering can be physical—for example, by pore size or by cell sorting through flow cytometry. But bacteria, protists, and even viruses may overlap considerably in size. Alternatively, once DNA has been sequenced, "computational filtration" can eliminate DNA from excluded sources such as protists, when you are focusing on bacteria. However, computational filtering may also exclude sequences of interest simply because they are new and match nothing reported before in the databases.

Isolating DNA. Once separated from the physical habitat, the cells of the target community must be opened in such a way that all of the DNA is released with minimal breakage of the strands (**Fig. 21.4**, step 3). We can lyse the cells by "bead beating" or by sonication (discussed in Chapter 2). The DNA can then be purified by precipitation with phenol and ethanol, or by binding to special filters. But—unlike the analysis of a single-species genome—for a metagenome we must account for different species that possess different kinds of enzyme inhibitors, as well as envelope, sheath, and S-layers of diverse composition.

How can we ensure that our protocol will be optimal for all the thousands of species in the community? In fact, there is no single best way to extract metagenomic DNA. Different researchers argue for one of two main approaches:

- **Use a single, universally applied method of DNA extraction for all target communities.** If all research groups sample metagenomes using a common DNA extraction method, then results may be compared across all projects.
- **Use multiple DNA extraction methods for each target.** If a research group uses multiple DNA extraction methods to sample one target community, then they have the best chance to maximize coverage of all the microbial genomes in the sample.

Thought Question

21.1 Suppose you are conducting a metagenomic analysis of soil sampled from different parts of a wetland. Your budget for soil analysis covers only ten sample analyses. Would you use one DNA extraction method, or multiple methods?

Metagenome Sequencing and Assembly

Once the DNA samples are obtained, how do we sequence them? In most cases, the sample DNA must be amplified (**Fig. 21.4**, step 4) followed by high-throughput sequencing (step 5). The choice of sequencing method depends on several factors.

- **DNA concentration.** How concentrated are the DNA samples relative to other cell components? Samples such as a human tissue biopsy, or groundwater, will contain DNA in such small quantities that it must be amplified. Amplification is achieved by cloning or PCR, either of which can introduce sequence errors and sample loss.
- **Sample diversity.** Species in the community may differ in their fractional representation by several orders of magnitude. How important are the community's rarest members? This question is hard to answer without first taking a trial run at the genome to establish a rarefaction curve (discussed shortly).
- **Functional interest.** Is our aim a complete description of the target community, such as the microbiota inhabiting the human stomach? Or do we focus on a narrower goal, such as identifying soil actinomycete genes that produce novel antibiotics? A narrower focus allows quicker analysis of the DNA—unless those organisms are among the community's rarer members.

Sequencing technologies. Different methods of DNA sequencing (**Fig. 21.4**, step 5) differ with respect to the amount of PCR amplification, the length of the sequence read, and the accuracy of the sequence. The methods generally available in 2012 are compared in **Table 21.2**. The ABI capillary technology based on Sanger dye termination provides the most accurate sequence, but the lower cost of Roche 454 pyrosequencing can prove more economical, if sufficient replicates are read. The thousandfold greater volume of data provided by Illumina sequencing expands the feasible size of projects, but the shorter reads are more challenging to assemble.

Assembling genomes. Once the DNA sequences are obtained, how do we assemble them into genomes (**Fig. 21.4**, step 6)? Recall that the genome of a single organism requires **assembly** of overlapping fragments

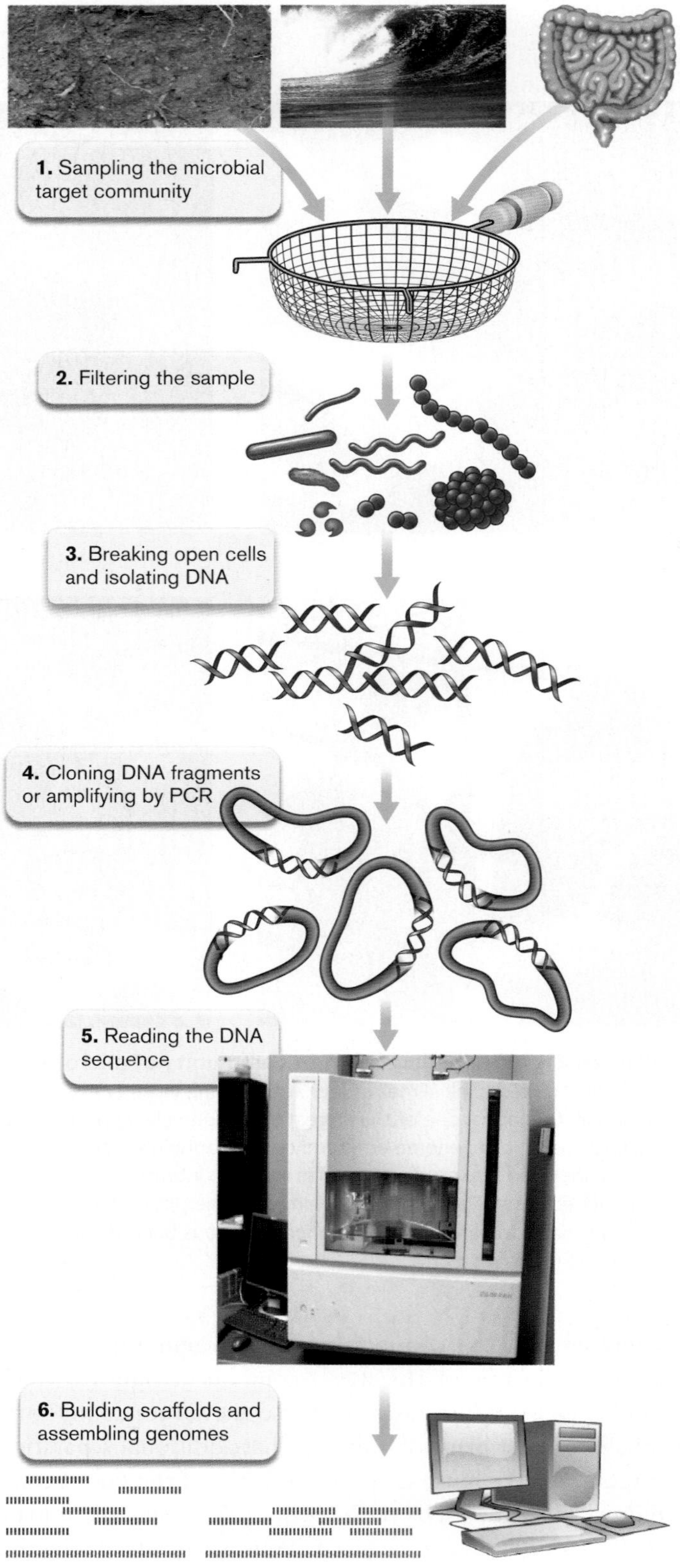
1. Sampling the microbial target community
2. Filtering the sample
3. Breaking open cells and isolating DNA
4. Cloning DNA fragments or amplifying by PCR
5. Reading the DNA sequence
6. Building scaffolds and assembling genomes

FIGURE 21.4 ▪ Sequencing a metagenome. Select a target community to sample, from a habitat such as soil, water, or host plant or animal. Remove the sampled microbes from their environment (step 1) and filter the desired class of organisms (step 2). Lyse the cells and isolate pure, intact DNA (step 3). Copy the DNA, either by cloning or by PCR amplification (step 4). Read the DNA sequence using a high-throughput sequencer (step 5). Use computational software to build scaffolds and assemble genomes (step 6).
Source: John Wooley et al. 2010 *PLoS Computational Biology* **6:**e1000667.

TABLE 21.2

High-throughput sequencing methods.

Sequencer	Read length (base pairs)	Technology	+ Advantages – Limitations
ABI 3730	600–900	Sanger dideoxy chain termination by fluorescent dyes; template requires cloned genome fragments (shotgun sequencing).	+ Highest accuracy; longest reads for assembly – Cloning bias; some genes toxic to vector
Roche 454	400–500	PCR-amplified template on bead in well. Polymerase extension generates pyrophosphate, sulfurase makes ATP, and luciferase produces light indicating nucleotide addition.	+ PCR amplification replaces cloning; avoids cloning bias; lower cost for high-volume sequence – Less accurate than ABI Sanger; long mononucleotide repeats cannot be read
Solexa Illumina	75–100	PCR-amplified template; sequencing by synthesis; thousands of samples per instrument channel.	+ High-volume yield (thousandfold higher than Sanger) – Short reads are harder to assemble; less accurate than ABI Sanger

into **contigs**, regions of contiguous DNA sequence without gaps. The fragments and contigs resemble a jigsaw puzzle with thousands of pieces, to be accomplished by advanced computation programs (discussed in Chapter 7). But a metagenome requires assembly of thousands of jigsaw puzzles, whose pieces are all jumbled together. In practice, metagenomes rarely yield complete genomes of individual species. In one example, the acid mine community sequenced by Gene Tyson, Jill Banfield, and colleagues (**Fig. 21.3A**; **Table 21.1**) yielded near-complete genomes of two organisms: *Leptospirillum* sp. and *Ferroplasma acidarmanus*. These genome assemblies were possible because the extreme environment supported only a few species.

A highly diverse community, the cow rumen microbiome sequenced by Matthias Hess and colleagues, yielded 15 partial genomes ranging from 60% to 93% estimated completeness (**Fig. 21.5A**). The completeness of each genome was estimated from the number of "core genes"

A.

Genome bin	Genome size (Mb)	Phylogenetic order	Estimated completeness
AFa	2.87	Spirochaetales	92.98%
AMa	2.21	Spirochaetales	91.23%
AIa	2.53	Clostridiales	90.10%
AGa	3.08	Bacteroidales	89.77%
AN	2.02	Clostridiales	78.50%
AJ	2.24	Bacteroidales	75.96%
AC2a	2.07	Bacteroidales	75.96%
AWa	2.02	Clostridiales	75.77%
AH	2.52	Bacteroidales	75.45%
AQ	1.91	Bacteroidales	71.36%
AS1a	1.75	Clostridiales	70.99%
APb	2.41	Clostridiales	64.85%
BOa	1.67	Clostridiales	64.16%
ADa	2.99	Myxococcales	62.13%
ATa	1.87	Clostridiales	60.41%

B.

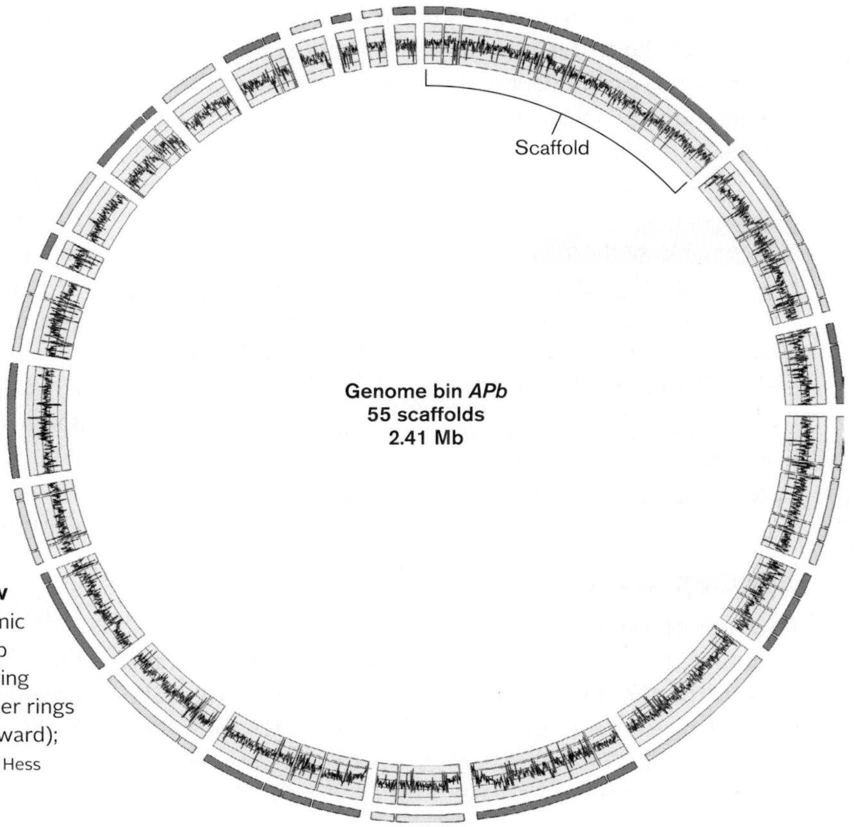
Scaffold
Genome bin APb
55 scaffolds
2.41 Mb

FIGURE 21.5 ■ Partial genomes assembled from a cow rumen microbes. A. Genome bins were matched to taxonomic groups based on comparison with reference sequences. **B.** Map of genome bin *APb,* a member of the order Clostridiales, showing assembled scaffolds. Gaps represent incomplete regions. Inner rings indicate fold coverage by sequence reads (backward and forward); red lines indicate 25-fold coverage. *Source:* Modified from Matthia Hess et al. 2011. *Science* **331**:463.

identified. Core genes are defined as those genes found in nearly all members of a given taxonomic order, such as Clostridiales for the group of genomes (genome bin) *APb* (highlighted row). The fraction of the core genes found divided by those expected for the order gives an estimate of completeness of the genome sequence.

The partial genomes that Hess found, however, account for only a tiny fraction of the species present, out of 268 gigabases (billions of base pairs) sequenced. Each partial genome assembled from the rumen community represents a **bin**—that is, a set of sequence reads from closely related members of one taxonomic unit, showing a given level of similarity. Since no two individual organisms possess exactly the same sequence, the investigator must decide how much difference to allow in defining a taxon. **Binning**, the sorting of sequences into taxonomic bins, requires further computational analysis. Factors for computation may include, for example, similarity of base composition and similarity to reference database sequences.

Each partial genome is composed of **scaffolds**. A scaffold is composed of large segments of DNA sequence pieced together from contigs (**Fig. 21.5B**). Between scaffolds, there usually remain unsequenced gaps. Inside the wheel of scaffolds, the jagged trace represents the degree of **coverage**—that is, the number of times a given region was sequenced; multiple copies of overlapped sequence represent a reliable assembly. But the remainder of the rumen sequences form unattached scaffolds and singleton reads out of thousands of unknown genomes. And these unattached sequences encode tens of thousands of novel enzymes for carbohydrate digestion—important for the biofuel industry, which aims to maximize plant conversion to fuel.

Thought Question

21.2 Suppose you plan to sequence a marine metagenome for the purpose of estimating carbon dioxide fixation and release, to improve our model for global climate change. Do you focus your resources on assembling as many complete genomes as possible, or do you focus on identifying all the community's enzymes of carbon metabolism?

Species Diversity

How can we identify all of the taxa present in a community, including those that do not yield full genomes? The standard method for identifying species is to compare the SSU rRNA genes—that is, genes encoding 16S rRNA for bacteria and archaea, or 18S rRNA for eukaryotes (discussed in Chapter 17). SSU rRNA gene similarity is used to define **operational taxonomic units (OTUs)**, a working metagenomic definition of "species." The use of SSU rRNA sequence does have limitations, particularly in that a given species' genome may possess multiple copies—between 1 and 16 copies for bacterial 16S rRNA genes, and even wider copy variation for the 18S rRNA genes of microbial eukaryotes. Furthermore, in some cases SSU rRNA genes evolve faster than other genes in a genome, or even show signs of horizontal gene transfer. Thus, analysis of metagenomes may need to include additional "housekeeping genes," that is, genes expressed constitutively for fundamental cell functions. Housekeeping genes, such as those encoding translation factors, show orthologs in a wide range of species.

Nevertheless, SSU rRNA remains the best starting point to reveal the general categories of microbes found in a community; their relative abundance in a community; and the overall diversity of the target, compared to other communities. But first, the investigator must show how much of the actual diversity the sampled SSU rRNA reads represent. The degree of diversity sampled can be estimated by a **rarefaction curve** (**Fig. 21.6A**). A rarefaction curve plots the number of different OTUs (species) found, as a function of increasing sample size. If resampling continues to reveal new OTUs, then the community diversity has not yet been fully sampled. On the other hand, if the number of OTUs reaches a plateau, then the community diversity may be well represented. For example, **Figure 21.6A** shows rarefaction curves for metagenomes from indoor air samples (Air-1, Air-2) compared to those from nearby samples of soil and river water (Soil-1, Soil-2, Water). The curves for soil and water show a continual increase in OTUs with resampling, whereas the indoor air samples approach a plateau after about 2,500 clones. Thus, the indoor air samples appear to contain less diversity than the outdoor samples, and the indoor diversity is well represented by the air samples.

For the same study, **Figure 21.6B** compares the relative distribution of proteobacterial taxa found in each of the metagenomes compared. Both of the air samples show a high proportion of Caulobacterales (blue bars), including species known for opportunistic infections in "clean" hospital settings. The water sample shows a majority of Rhodobacterales (yellow bar). The soil samples show a broader distribution of taxa than either air or water.

Functional Annotation

Once we have our metagenomic DNA sequence reads organized into scaffolds and partial genomes, how do we "call" the genes? There is no single method to ensure that we recognize all the actual genes encoding functional

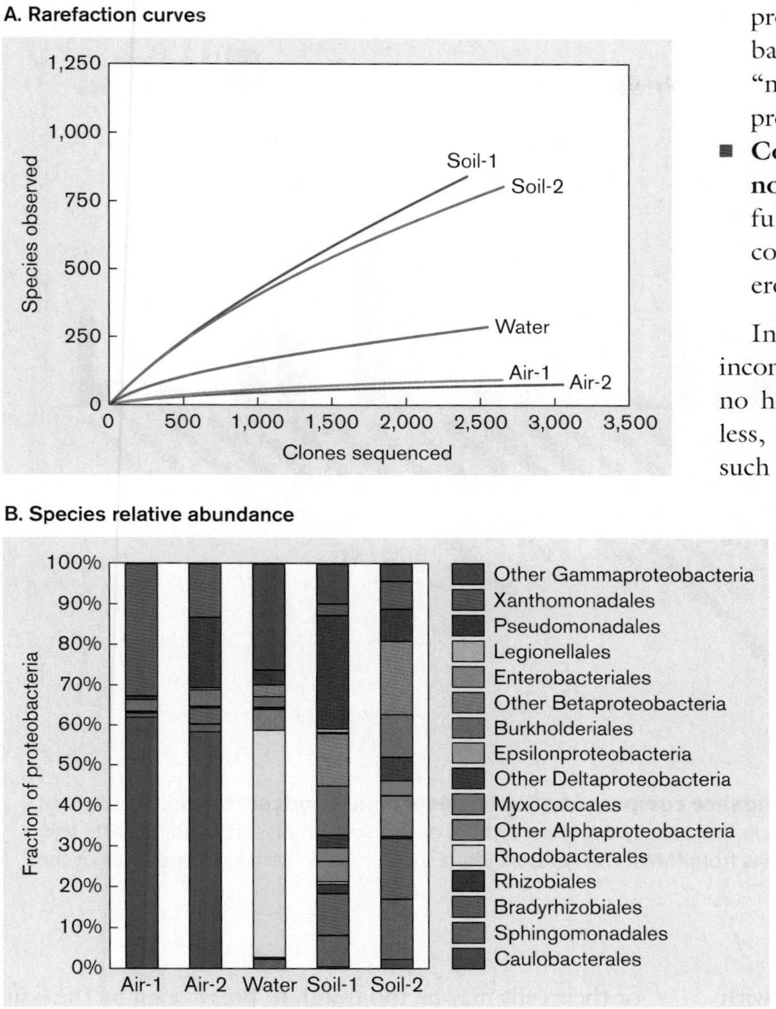
A. Rarefaction curves
Species observed
1,250
1,000
750
500
250
0
Soil-1
Soil-2
Water
Air-1
Air-2
0
500
1,000
1,500
2,000
2,500
3,000
3,500
Clones sequenced
B. Species relative abundance
Fraction of proteobacteria
100%
90%
80%
70%
60%
50%
40%
30%
20%
10%
0%
Air-1
Air-2
Water
Soil-1
Soil-2
Other Gammaproteobacteria
Xanthomonadales
Pseudomonadales
Legionellales
Enterobacteriales
Other Betaproteobacteria
Burkholderiales
Epsilonproteobacteria
Other Deltaproteobacteria
Myxococcales
Other Alphaproteobacteria
Rhodobacterales
Rhizobiales
Bradyrhizobiales
Sphingomonadales
Caulobacterales

FIGURE 21.6 ■ Diversity in metagenomes. Indoor air, outdoor soil, and river water were sampled from an urban environment of Singapore. **A.** Rarefaction curves show the number of species, or OTUs, found as a function of the number of clones sequenced for the metagenome. **B.** For each target community, the bar heights represent the proportion of proteobacteria in various orders. *Source:* Modified from Susannah Tringe et al. 2008. *PLoS One* **3**:e1862.

products—or that we don't mistakenly define some noncoding sequences as genes (false positives). Many computational tools are used to call genes. Some approaches include:

- **Gene structure.** The six reading frames are searched for start and stop codons that bracket sequences of appropriate length to encode proteins (open reading frames, or ORFs). To be transcribed and translated to a functional product, each ORF must have an upstream promoter sequence, as well as a ribosomal start site (discussed in Chapters 8 and 10).
- **Homologs and motifs.** The metagenomic reads may match homologs of related genes from databases of previously sequenced organisms. In addition, the databases include short, recurring peptide sequences called "motifs" that are common to functional classes of proteins.
- **Comparison with previously processed metagenomes.** Metagenome comparison is particularly useful in the clinical setting, where diagnostic tests aim to compare the microbiomes of diseased patients with a reference collection of "normal" human microbiomes.

In practice, all of these gene-calling approaches are incomplete because they miss truly novel genes for which no homologs or motifs exist in the databases. Nonetheless, functional annotation can yield surprising results—such as the discovery from Craig Venter's Sargasso Sea metagenome that abundant marine proteobacteria contain proteorhodopsins homologous to the archaeal light-harvesting proton pump bacteriorhodopsin (discussed in Chapter 14). As many as 80% of marine bacteria have proteorhodopsin, and thus appear to be photoheterotrophs—a finding of importance for global measures of photosynthesis.

Functional genes offer a basis for comparing two microbial communities. An example of a comparative overview of functional genes is shown in **Fig. 21.7**. This figure compares the abundance of functional gene categories between microbes from desert soil and nondesert soil. The most striking difference between the two is that the desert soil genomes contain far fewer genes for virulence and defense, such as antibiotic production and resistance. This finding confirms that microbes in deserts are more isolated than those in moist soil communities; thus, the survival of desert microbes depends less on competition with fellow microbes than on the ability to tolerate extreme conditions of dryness. On the other hand, the desert genomes encode a greater number of products for protein metabolism. This finding suggests that growth under deprivation may require a greater range of metabolic options for growth of cell biomass.

Beyond the Metagenome

Metagenomics is our most powerful tool to identify the microbes that inhabit an ecosystem. Yet serious limitations remain. In a recent study of soil in an apple orchard, reported in 2012, Jo Handelsman and colleagues compared the sampling rate of a metagenomic screen versus plate culture. For isolation of the soil community, they used a sophisticated culture medium tailored to support

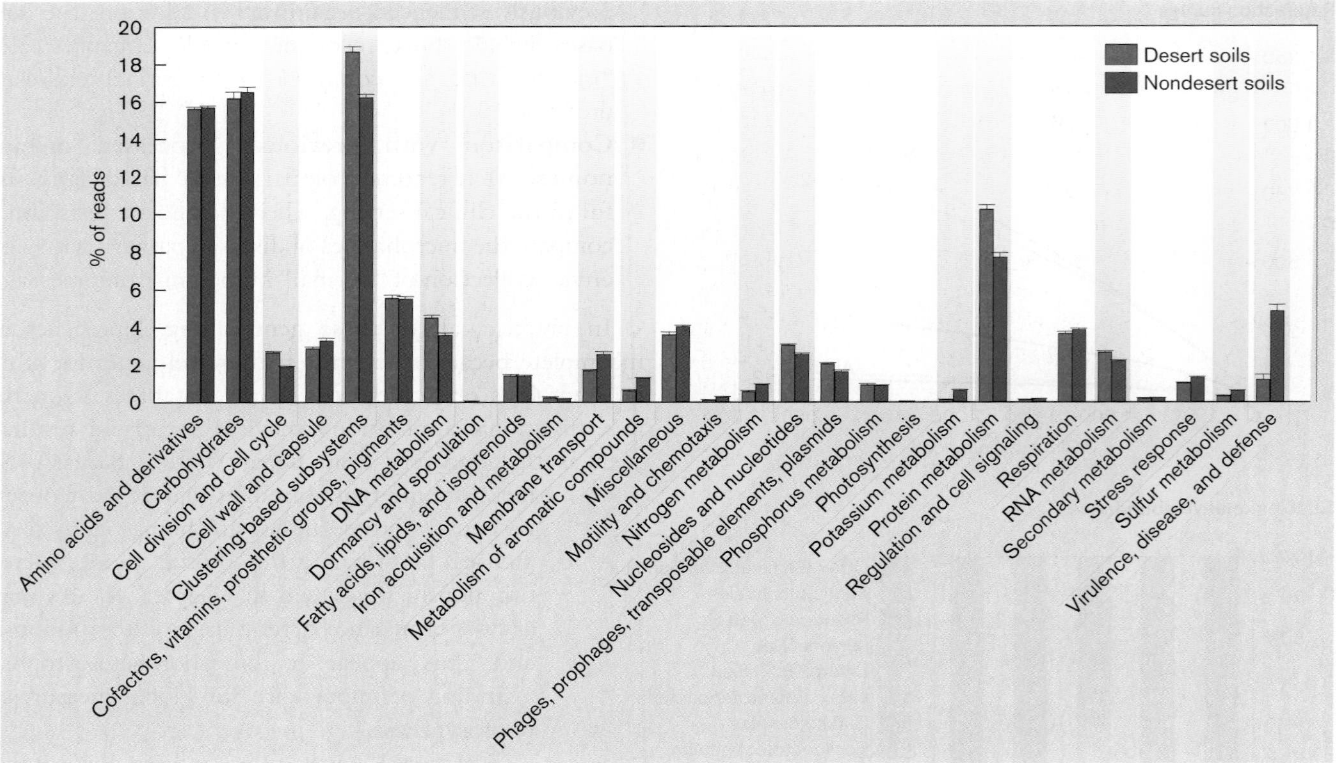
% of reads
0
2
4
6
8
10
12
14
16
18
20
Desert soils
Nondesert soils
Amino acids and derivatives
Carbohydrates
Cell division and cell cycle
Cell wall and capsule
Clustering-based subsystems
Cofactors, vitamins, prosthetic groups, pigments
DNA metabolism
Dormancy and sporulation
Fatty acids, lipids, and isoprenoids
Iron acquisition and metabolism
Membrane transport
Metabolism of aromatic compounds
Miscellaneous
Motility and chemotaxis
Nitrogen metabolism
Nucleosides and nucleotides
Phages, prophages, transposable elements, plasmids
Phosphorus metabolism
Photosynthesis
Potassium metabolism
Protein metabolism
Regulation and cell signaling
Respiration
RNA metabolism
Secondary metabolism
Stress response
Sulfur metabolism
Virulence, disease, and defense

FIGURE 21.7 ▪ Functional gene categories: relative abundance compared between desert and nondesert soils. Highlighted categories denote significant differences in abundance of genes in functional categories. Sixteen desert soil samples from hot deserts and Antarctic cold deserts were compared with seven non-desert soils from forests and grasslands. (Error bars represent standard error of the mean.) *Source:* Modified from Noah Fierer. 2012. *PNAS* **109**:21390, fig. 3.

growth of bacteria from the rhizosphere (associated with plant roots) and exclude fast-growing fungi. From the same locations, a metagenome was obtained by pyrosequencing. To their surprise, more than 60% of the cultured bacteria failed to appear in the metagenome.

How could these cultured organisms be missed in the metagenome? The researchers hypothesize that the cultured organisms represent organisms of the "rare biosphere"—that is, species of such low abundance in the soil that the metagenomic screen does not pick them up. But upon culturing, these rare organisms are favored by the sudden provision of concentrated nutrients—a condition that inhibits the more abundant soil organisms, many of which are oligotrophs (discussed in Chapter 4). These "copiotrophs" or "weed organisms" may normally be rare but prevail when nutrients suddenly appear.

Other reasons why metagenomes may miss community members include:

- **Failure of the filters and DNA isolation.** Organisms may unintentionally be lost by the filtering procedures, or their cells may be too tough to break open by the lysis procedure.
- **Failure of the PCR primers.** The "universal primers" for SSU rRNA, in fact, miss many species whose SSU rRNA sequences vary beyond the range of the primer sequences. It is hard for us to know how widespread such organisms are.

Beyond metagenomics, how do we know which genes of the genomes are actually utilized by the organisms of the target community? New technologies are emerging to sample not just the DNA, but the RNA and protein pools of a community. **Metatranscriptomics** is the study of the RNA transcripts obtained from an environmental community (described in Section 10.9). The "metatranscriptome" gives a snapshot of gene expression activity of a community at a given point in time. A metatranscriptome of a marine water community is shown in Section 21.3 (**Fig. 21.17**). **Metaproteomics** is the study of protein data obtained from environmental samples using techniques of proteomics (discussed in Chapters 3 and 5).

To Summarize

- **A metagenome is the sum total of all DNA sequenced from an environmental sample.** The environment may be any natural or human-made environment, or the microbiome (host-associated microbes) of a plant or animal.
- **Samples are obtained from a target community of a defined environment.** Sampling requires separating the microbes from their physical environment, filtering the organisms of interest, breaking open the cells, and purifying the DNA. Different sampling procedures may lead to different views of a metagenome.
- **The metagenomic DNA is sequenced.** High-throughput sequencing methods, such as Sanger capillary sequencing, pyrosequencing, and Illumina sequencing by synthesis, generate large amounts of data, which require computational analysis.
- **The sequence reads are assembled into scaffolds.** Scaffolds are assembled into partial genomes defined by taxonomic binning.
- **Species diversity of a metagenome may be estimated by a rarefaction curve.**
- **Functional annotation offers clues as to the ecological functions contributed by the microbes.**
- **Emerging technologies address the limits of metagenomics.** Metagenomes miss unknown numbers of rare species, some of which may actually be cultured. Gene expression and function are investigated through metatranscriptomics and metaproteomics.

21.2 Functional Ecology

How do microbes contribute to ecosystems? The van Niel hypotheses imply that there is an extraordinary variety of species using different energy-yielding reactions. The energy acquired by microbes, as well as the elements they assimilate into biomass, then circulate throughout the ecosystem. Other functions that microbes provide include detoxifying wastes and protecting partners from pathogens. As endosymbionts of animals and plants, microbes can modulate the host's development and behavior.

What a microbe gives to and takes from its ecosystem depends on its genome (as revealed by metagenomic studies) and its environment. Each organism's genome provides a set of metabolic enzymes available to interconvert molecules. For example, the genomes of pseudomonads encode enzymes for catabolizing complex aromatic molecules (benzene derivatives), such as lignins from the wood of dead trees. Thus, the pseudomonad can recycle woody carbon into the ecosystem. And the pseudomonad's genes themselves become an environmental resource for other microbes, via horizontal transfer or transformation, expanding the genetic potential of the microbial "pan-genome" (discussed in Chapter 17).

The environment's physical and chemical factors, such as temperature, nutrient supplies, oxygen availability, and pH, all influence a microbe's metabolic options. For example, if H_2 is available in the presence of O_2, *Rhodopseudomonas palustris* will oxidize hydrogen to water. On the other hand, if O_2 is absent but light is available, *R. palustris* will use anaerobic photosynthesis, splitting H_2S to make sulfur. Often, metabolic options also require metabolic activity from other organisms in the community. Fermenting bacteria provide the H_2 that *R. palustris* oxidizes.

All organisms depend, directly or indirectly, on the presence of other organisms. Cooperation by organisms may be incidental, as in the case of hydrogen-oxidizing bacteria using H_2 from fermenters; or it may involve mutualism, a highly evolved partnership in which two species have coevolved to support each other (discussed in Section 21.3).

Carbon Assimilation and Dissimilation: The Food Web

The interactions between microbes and their ecosystems include two common roles of metabolic input and output, often referred to as assimilation and dissimilation, respectively. Most of these metabolic processes were discussed in Chapters 13, 14, and 15, but ecology offers a different perspective.

Assimilation refers to processes by which organisms acquire an element, such as carbon from CO_2, to build into cells. When the environment lacks organic compounds containing an element such as nitrogen or phosphorus, microbes may assimilate the element from mineral sources. Common kinds of assimilation include carbon dioxide fixation and nitrogen fixation. Organisms that produce biomass from inorganic carbon (usually CO_2 or bicarbonate ions) are called **primary producers**. Producers are a key determinant of productivity for other members of the ecosystem.

Dissimilation is the process of breaking down organic nutrients to inorganic minerals such as CO_2 and NO_2^-, usually through oxidation. Microbial dissimilation releases minerals for uptake by plants, and it provides the basis of wastewater treatment (discussed in Chapter 22). But microbial dissimilation can decrease habitat quality by removing organic nitrogen. When soil bacteria break down amines (RNH_2) to ammonium ion (NH_4^+), nitrifying bacteria such as *Nitrosomonas* oxidize the ammonium to nitrite and nitrate. These highly soluble anions are then washed out of soil into the ground water.

This chapter covers microbial assimilation and dissimilation of carbon and nitrogen in association with the plants and animals of an ecosystem. The cycles of other key elements, and their effects on the global biosphere, are explored in Chapter 22.

Thought Question

21.3 From Chapters 13–15, give examples of microbial metabolism that fit patterns of assimilation and dissimilation.

The major interactions among organisms in the biosphere are dominated by the production and transformation of **biomass**, the bodies of living organisms. To obtain energy and materials for biomass, all organisms participate in **food webs** (**Fig. 21.8**). A food web describes the ways in which various organisms consume each other, as well as each other's products. Levels of consumption are called **trophic levels**. Organisms at each trophic level consume biomass of organisms from another level, usually closer to producers in the food web. At each trophic level, the fraction of biomass retained by the consumer is small; most is released as CO_2, through respiration to provide energy.

Every food web depends on primary producers for two things:

- **Absorbing energy from outside the ecosystem.** A key source of energy is sunlight, and the producers are phototrophs.
- **Assimilating minerals into biomass.** The biomass of producers is then passed on to subsequent trophic levels.

The majority of carbon in Earth's biosphere is assimilated by oxygen-producing phototrophs such as cyanobacteria, algae, and plants. Certain important ecosystems are founded on lithotrophs—for example, the hydrothermal vent communities, in which bacteria oxidize hydrogen sulfide to fix CO_2, capturing both gases as they well up from Earth's crust. The vent communities use oxygen generated by phototrophs from above.

In addition to producers, all ecosystems include **consumers**, which acquire nutrients from producers and ultimately dissimilate biomass by respiration, returning carbon back to the atmosphere (see **Fig. 21.8A**). Consumers constitute several trophic levels based on their distance from the primary producers. The first level of consumers, generally called **grazers**, directly feed on producers. Grazers usually convert 90% of the producer carbon back to CO_2 through respiratory metabolism, yielding energy. The next level of consumers, often called **predators**, feed on the grazers, again converting 90% back to atmospheric CO_2. In microbial ecosystems, the trophic relationships are often highly complex, because a given species may act as both producer and consumer.

At each trophic level, some of the organisms die, and their bodies are consumed by **decomposers**, returning carbon and minerals back to the environment for use by producers. Decomposers are all microbes (fungi or bacteria). Decomposers have particularly versatile digestive enzymes

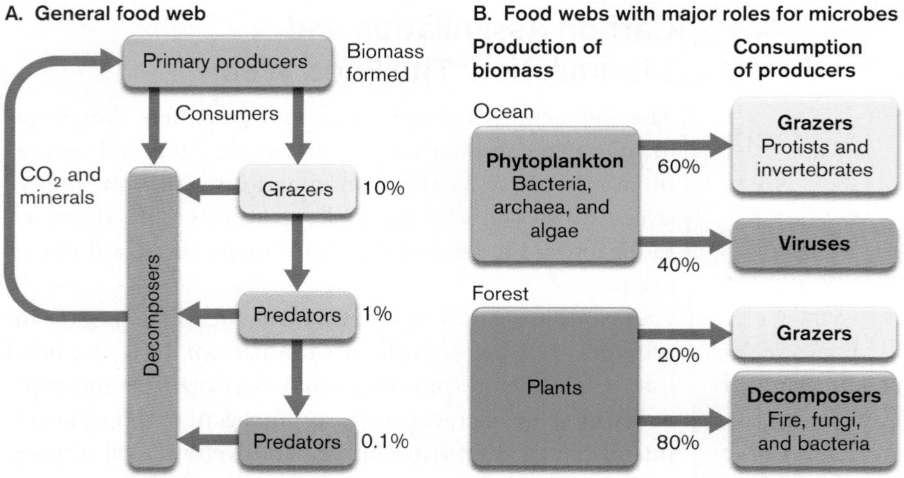
A. General food web
Primary producers
Biomass formed
Consumers
CO_2 and minerals
Grazers
10%
Decomposers
Predators
1%
Predators
0.1%
B. Food webs with major roles for microbes
Production of biomass
Consumption of producers
Ocean
Phytoplankton
Bacteria, archaea, and algae
60%
Grazers
Protists and invertebrates
40%
Viruses
Forest
Plants
20%
Grazers
80%
Decomposers
Fire, fungi, and bacteria

FIGURE 21.8 ▪ Microbes within food webs. A. Biomass production. Primary producers fix CO_2 into biomass, grazers (primary consumers) feed on primary producers, and predators consume the grazers and smaller predators. Decomposers consume dead bodies of organisms from every level, recirculating their minerals to the producers. Percentages indicate the fraction of original CO_2 converted to biomass at each trophic level. **B.** In different habitats, the relative significance of the various trophic components varies. In marine ecosystems, the primary producers are bacteria, archaea, and algae. Viruses break down both producer and consumer microbes. In a forest, the major producers are trees, much of whose biomass is poorly digested by grazers; 80% of consumption occurs by fungal and bacterial decomposition. In some forests, the major decomposer is fire.

capable of breaking down complex molecules such as lignin. Without decomposers, carbon and minerals needed by phototrophs would be locked away by ever-increasing mounds of dead biomass. Instead, all biomass is recycled somewhere in the biosphere. As we learned in Chapter 13, the energy gained by ecosystems can be cycled in part, but all is eventually lost as heat.

In the function of ecosystems, phylogenetic distinctions among the domains bacteria, archaea, and eukaryotes are of less significance than the biological and biochemical consequences of an organism's presence in the community—that is, what the organism produces or consumes. Thus, in this chapter we place greater emphasis on trophic roles than on phylogenetic distinctions.

The relative significance of microbial and multicellular producers and consumers varies considerably among different habitats. A major difference appears between marine and terrestrial ecosystems (**Fig. 21.8B**). In the oceans, most CO_2 fixation and biomass production are performed by the smallest inhabitants, phototrophic bacteria. The main consumers are protists and viruses. Viruses are the most numerous replicating forms in the ocean—and they lyse most marine cells before any multicellular predators get a chance to consume them.

In terrestrial ecosystems, by contrast, the major primary producers and fixers of CO_2 are multicellular plants. Plants generate **detritus**, discarded biomass such as leaves and stems, that requires decomposition by fungi and bacteria. While viruses are important, multicellular consumers such as worms and insects play a much greater role.

The differences between the food webs of ocean and dry land explain why most of the food we harvest from the ocean consists of predators at the higher trophic levels (fish), whereas most food harvested on land consists of producers and first-level consumers (plants and herbivores). Fish depend on a vast, invisible food web of microbes, with several trophic levels that dissipate the assimilated carbon. Thus, the numbers of fish remain limited, despite the seemingly huge volume of ocean. Marine fish populations are especially vulnerable to overharvesting and to any environmental change that disrupts the food web.

Oxygen and Other Electron Acceptors

The availability of oxygen and other electron acceptors is the most important factor that determines how nutrients containing carbon, nitrogen, and sulfur are assimilated and dissimilated. Examples of aerobic and anaerobic metabolism are presented in **Table 21.3**. In aerobic environments, microbes use molecular oxygen as an electron acceptor to respire on organic compounds (abbreviated CHO)

TABLE 21.3

Aerobic and anaerobic metabolism.

Element	Oxidized by O_2 (lithotrophy)	Reduced by CHO* or H_2 (anaerobic respiration)
Nitrogen	$NH_3 + O_2 \rightarrow NO_3^-$	$NO_3^- + CHO \rightarrow N_2$
Manganese	$Mn^{2+} + O_2 \rightarrow Mn^{4+}$	$Mn^{4+} + CHO \rightarrow Mn^{2+}$
Iron	$Fe^{2+} + O_2 \rightarrow Fe^{3+}$	$Fe^{3+} + CHO \rightarrow Fe^{2+}$
Sulfur	$H_2S + O_2 \rightarrow SO_4^{2-}$	$SO_4^{2-} + CHO \rightarrow H_2S$
Carbon	$CH_4 + O_2 \rightarrow CO_2$	$CO_2 + H_2 \rightarrow CH_4$

*CHO = organic material.

produced by other organisms (discussed in Chapters 13 and 14). Aerobic respiration on organic compounds is highly dissimilatory in that it tends to break compounds down to CO_2. Microbes also use oxygen to respire on reduced minerals such as NH_3, H_2S, and Fe^{2+} (lithotrophy). Lithotrophy is coupled to CO_2 fixation and is therefore assimilatory metabolism.

In anaerobic environments, such as deep soil, microbes use minerals such as Fe^{3+} and NO_2^- to oxidize organic compounds supplied by other organisms (anaerobic respiration). Other anaerobes use these electron acceptors to oxidize reduced minerals. Anaerobic environments allow much slower rates of assimilation and dissimilation than occur in the presence of oxygen. The total volume of anaerobic microbial communities, however, far exceeds that of the oxygenated biosphere.

Temperature, Salinity, and pH

Other abiotic factors can profoundly affect the environment for microbes and other members of the food chain, either directly or by impacting other factors. The effects are most significant for **extremophiles**, species that grow in environments considered extreme by human standards (**Table 21.4**), discussed in Chapter 5. Temperature limits the rate of metabolism. Higher temperatures found in hot springs (80–100°C) enable the fastest growth rates measured (doubling times as short as 10 minutes for some hyperthermophiles). On the other hand, temperature limits the oxygen concentration in water, so hyperthermophiles often reduce sulfur instead.

High salt concentration limits the growth of microbes adapted to freshwater conditions. By contrast, many microbial species have adapted to high salinity (they are called halophiles). The haloarchaea, for instance, bloom

TABLE 21.4

Extremophiles.

Class of extremophile	Typical environmental conditions for growth
Acidophile	Acidic environments at or below pH 3
Alkaliphile	Alkaline environments at a range of pH 9–14
Barophile	High pressure, usually at ocean floor, from 200 to 1,000 atm
Endolith	Within rock crystals down to a depth of 3 km
Halophile	High salt, typically above 2-M NaCl
Hyperthermophile	Extreme high temperature, above 80°C
Oligotroph	Low carbon concentration, below 1 ppm
Psychrophile	Low temperature, below 15°C
Thermophile	Moderately high temperature, 50°C–80°C
Xerophile	Desiccation, water activity below 0.8

in population as a body of water shrinks and becomes hypersaline.

Acidity is important geologically because a high concentration of hydrogen ions accelerates the release of reduced minerals from exposed rock. Extreme acidity is often produced by lithotrophs (discussed in Section 14.5). The increasing acidity releases minerals to be oxidized, while excluding acid-sensitive competitors. Other kinds of habitats, such as soda lakes, show extreme alkalinity resulting from high sodium carbonate.

As our understanding of ecosystems deepens, we discover other remarkable ways that organisms benefit each other. For example, plant-associated bacteria and fungi enhance the uptake of nutrients and protect the host from pathogens. Animal-associated microbes protect the host from pathogens, enhance digestion, and modulate the immune system. These processes are detailed in the next section, and in the sections on plant and animal microbial communities (Sections 21.6 and 21.7).

To Summarize

- **Microbial populations fill unique niches in ecosystems.** Microbes are ubiquitous; they occupy every niche and environment on Earth. Every chemical reaction that may yield free energy can be utilized by some kind of microbe.
- **Microbes fix or assimilate essential elements into biomass**, which recycles within ecosystems. Important elements, such as nitrogen, are fixed solely by bacteria and archaea.
- **Primary producers fix single-carbon units, usually CO_2.** Microbial primary producers include algae, cyanobacteria, and lithotrophs.
- **Consumers and viruses break down the bodies of producers,** generating CO_2 and releasing heat energy. Dissimilation is the process of breaking down nutrients to inorganic minerals such as CO_2 and NO_3^-, usually through oxidation.
- **Decomposers such as fungi and bacteria release nutrients from dead organisms.**
- **Microbial activity depends on levels of oxygen, carbon, nitrogen, and other essential elements**, as well as environmental factors such as temperature, salinity, and pH. The largest volume of our biosphere contains anaerobic bacteria and archaea. Beneath Earth's surface, most metabolism is anaerobic.

21.3

Symbiosis

One of the most fascinating results of evolution is how organisms adapt to the presence of others. Some relationships of microbes occur at a distance; for example, the oxygen gas released by marine cyanobacteria is breathed by organisms around the globe (discussed in Chapter 22). But other kinds of interactions require close association between microbes of different species, or between a microbial population and a host plant or animal. For instance, cyanobacterial heterocysts (nitrogen-fixing cells) attract swarms of heterotrophic bacteria whose respiration consumes oxygen, maintaining low-oxygen conditions required for nitrogen fixation.

The most intimate association between organisms of different species is called **symbiosis** (plural, **symbioses**). Symbiotic associations include a full range of both positive and negative relationships (**Table 21.5**). Whether positive or negative, both partners evolve in response to each other.

Mutualism Involves Partner Species That Require Each Other

The most highly evolved forms of symbiosis involve partner species that require each other for survival—a phenomenon termed **mutualism**. Mutualism can involve two or more

TABLE 21.5

Types of symbiotic associations involving microbial species.

Type of interaction	Effects of interaction	Example
Mutualism	Two organisms grow in an intimate species-specific relationship in which both partner species benefit and may fail to grow independently.	Lichens consist of fungi and algae (in some cases cyanobacteria) growing together in a complex layered structure. Each species requires the presence of the other.
Synergism	Both species benefit through growth, but the partners are easily separated and either partner can grow independently of the other.	Human colonic bacteria ferment, releasing H_2 and CO_2, which methanogens convert to methane. The methanogens gain energy, and the bacteria benefit energetically from the removal of their fermentation products.
Commensalism	One species benefits, while the partner species neither benefits nor is harmed.	In wetlands, *Beggiatoa* bacteria oxidize H_2S for energy. Removal of H_2S enables growth of other microbes for whom H_2S is toxic. The other microbes are not known to benefit *Beggiatoa.*
Amensalism	One species benefits by harming another. The relationship is nonspecific.	In the soil, *Streptomyces* bacteria secrete antibiotics that lyse other species, releasing their cell contents for *Streptomyces* to consume.
Parasitism	One species (the parasite) benefits at the expense of the other (the host). The relationship is usually obligatory for the parasite.	*Legionella pneumophila,* the cause of legionellosis, parasitizes amebas in natural aquatic habitats. Within the human lung, *L. pneumophila* parasitizes macrophages.

microbial partners. It can also involve one or more microbial partners with a plant or animal host.

In some cases, both partners absolutely require each other; in other cases, one or the other is incapable of growing alone. The mutually beneficial relationship is maintained by numerous genetic responses that regulate each partner, avoiding damage to the other. Mutualisms such as nitrogen-fixing rhizobia within legumes can have enormous practical applications (discussed in Section 12.6).

Microbial partners: lichens. A highly evolved form of mutualism is the **lichen** (**Fig. 21.9**). Lichens consist of an intimate symbiosis between a fungus and an alga or cyanobacterium—sometimes both. The symbiosis requires

A.

JOAN SLONCZEWSKI

B.

JOAN SLONCZEWSKI

C. Lichen cross section

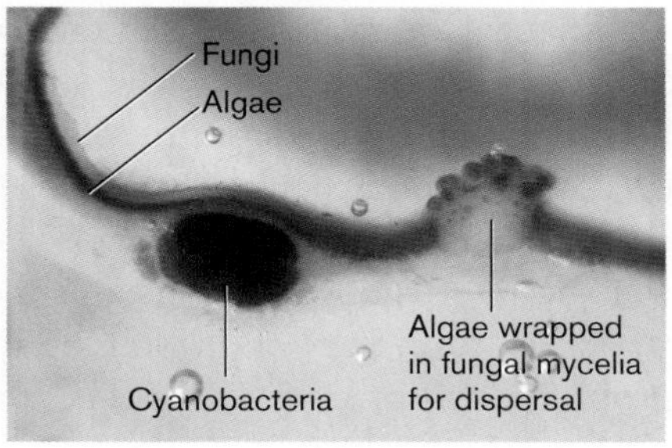
Fungi
Algae
Cyanobacteria
Algae wrapped in fungal mycelia for dispersal

STEPHEN & SYLVIA SHARNOFF, *LICHENS OF N. AMERICA*

FIGURE 21.9 ■ Lichens. **A.** A tombstone at Kenyon College cemetery encrusted with lichens (pale green) and mosses (dark green), a nonvascular plant. **B.** Close-up of the lichens in part (A). **C.** Section through a lichen (*Lobaria pulmonaria*) shows fungal, algal, and bacterial symbionts.

A. Lichen ground cover

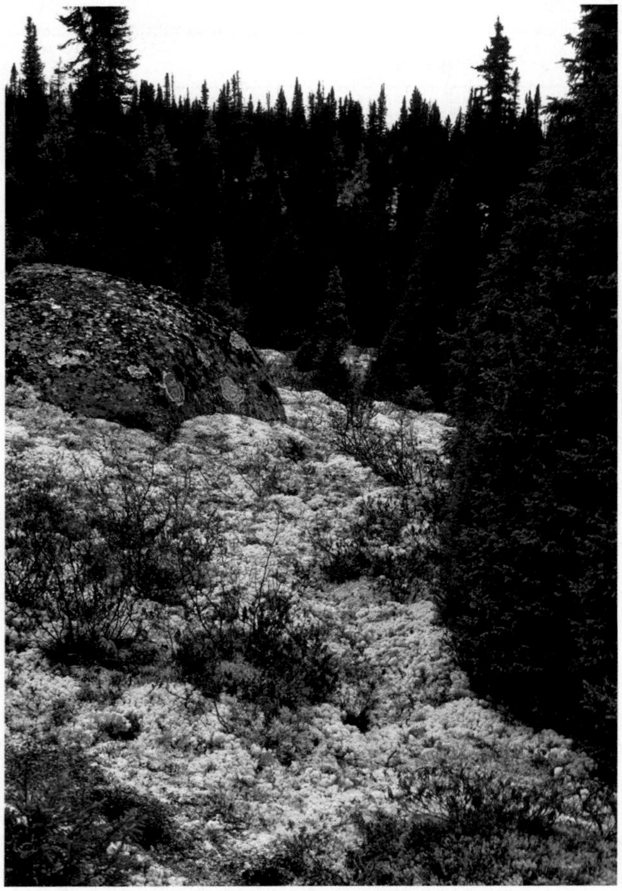

STEPHEN & SYLVIA SHARNOFF

B. Cryptogamic crust

STEPHEN & SYLVIA SHARNOFF

FIGURE 21.10 ■ Lichens within ecosystems. Boreal forest ecosystems and native cultures depend on lichens. **A.** Lichens cover most of the ground in boreal forests. **B.** Cryptogamic crust in Montana wilderness.

compatible partner species. The alga or bacterium provides photosynthetic nutrition, while the fungus provides minerals and protection. Lichens grow very slowly, but they tolerate extreme desiccation.

Lichens show a surprising variety of form. Different species may form a flat crust, branched filaments, or leaflike lobes. A cross section of the leaflike lichen *Lobaria pulminaria* reveals a layer of algae (green) covered by fungal mycelium (white), which protects the algae from ultraviolet light damage (**Fig. 21.9C**). In addition to the algae, this lichen includes patches of cyanobacteria, which fix nitrogen. Thus, the fungus, algae, and cyanobacteria form a three-way mutualism. For dispersal, the lichen forms asexual clumps of algae wrapped in fungal mycelium. The clumps flake off and are carried by wind to new locations.

Lichens are essential for boreal (northern) forests, where they cover the majority of the ground and provide food for grazing animals (**Fig. 21.10A**). In winter, lichens are a major food source for caribou, which dig beneath the snow to obtain them. In desert regions of the western and southwestern United States, lichens form part of a complex ground cover called **cryptobiotic soil** or **cryptogamic crust** (**Fig. 21.10B**). (*Crypto* means "hidden.") Cryptobiotic soil includes a patchy association of lichens, mosses, and nonlichenous algae, cyanobacteria, and fungi. The crust takes many years to grow a centimeter thick and is easily destroyed by bikes or hiking boots. The crust plays a vital role in desert ecosystems by holding moisture, protecting plant seeds for germination, and preventing erosion.

Microbial partners with an animal host. A particularly complex metabolic mutualism is that of termites, whose digestive tract contains bacteria that catabolize wood polysaccharides such as cellulose. The termite feeds on wood and is completely dependent on its symbiotic bacteria.

Within the termite digestive organ (called the hindgut), its wood-digesting bacteria form highly complex associations with protists such as *Mixotricha paradoxa* (**Fig. 21.11**), a large ciliate (as long as half a millimeter). The *Mixotricha* protist contains several kinds of bacterial endosymbionts. The protist takes up termite-ingested wood particles by phagocytosis, and then the protist's intracellular bacteria digest the wood polysaccharides. *Mixotricha* also has organelles that appear to be vestigial remnants of another endosymbiont, diminished by reductive evolution. On the protist's surface, four kinds of bacteria are attached. Two of the attached species are spirochetes, one significantly larger than the other. The spirochetes extend from the protist's cell membrane; they are flagellated, and their flagellar motility propels the protist cell. Two other species of "anchor bacteria" are Gram-negative rods attached to knobs of the protist surface. All members of the partnership have evolved an obligate relationship that provides food energy for the symbionts.

The relationship among microbial symbionts within the termite gut community generates a complex series of

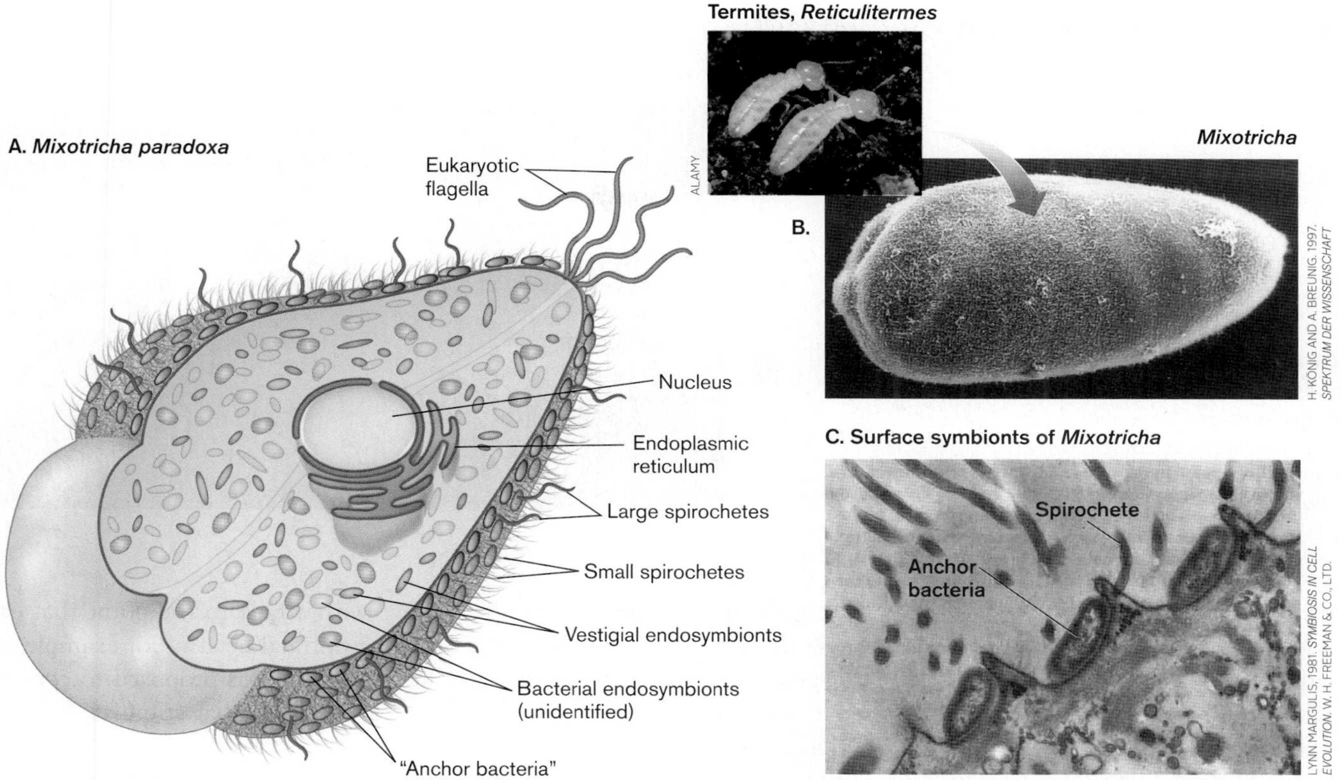
Termites, Reticulitermes
ALAMY
A. Mixotricha paradoxa
Eukaryotic flagella
Nucleus
Endoplasmic reticulum
Large spirochetes
Small spirochetes
Vestigial endosymbionts
Bacterial endosymbionts (unidentified)
"Anchor bacteria"
Mixotricha
B.
H. KÖNIG AND A. BREUNIG, 1997. SPEKTRUM DER WISSENSCHAFT
C. Surface symbionts of Mixotricha
Spirochete
Anchor bacteria
LYNN MARGULIS, 1981. SYMBIOSIS IN CELL EVOLUTION. W. H. FREEMAN & CO., LTD.

FIGURE 21.11 ▪ *Mixotricha paradoxa:* a multiple symbiont. A. The protist *Mixotricha paradoxa* possesses attached spirochetes (large and small species), "anchor bacteria," and two kinds of bacterial endosymbionts. **B.** Soldier termites, *Reticulotermes flavipes*, contain *Mixotricha* (SEM) and other gut endosymbionts that digest wood cellulose. **C.** TEM section through pellicle, including anchor bacteria and attached spirochetes.

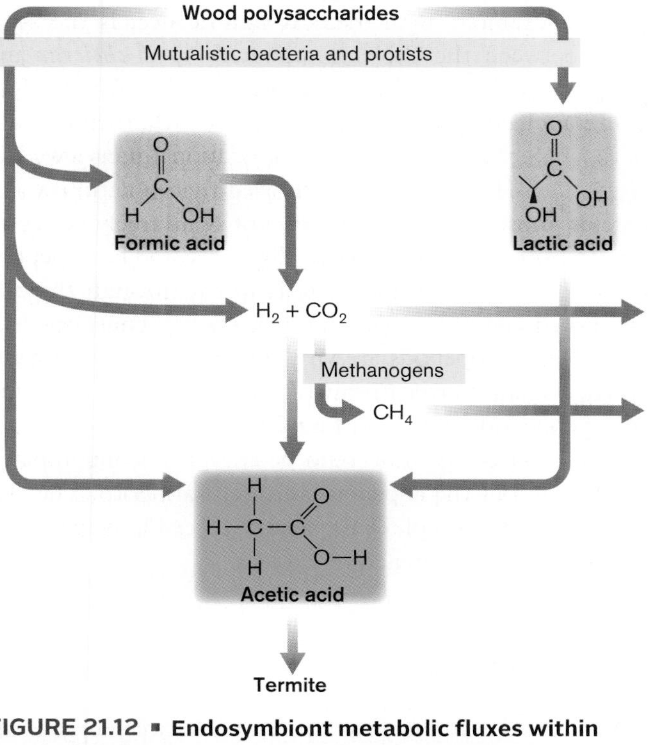
Wood polysaccharides
Mutualistic bacteria and protists
Formic acid
Lactic acid
H2 + CO2
Methanogens
CH4
Acetic acid
Termite

FIGURE 21.12 ▪ Endosymbiont metabolic fluxes within termite hindgut. Bacteria and their protist symbionts ferment wood polysaccharides to lactic acid, formic acid, acetic acid, H_2, and CO_2 within the hindgut of a termite, *Reticulitermes santonensis.* Some hydrogen is lost from the gut, some is converted to methane, and some is converted to acetic acid. Acetic acid is absorbed through the outer lining and feeds the termite.

metabolic fluxes (**Fig. 21.12**). This kind of metabolic cooperation is called **syntrophy**, which means "feeding together." For syntrophy, the fluxes within the community must balance energetically with a negative value of ΔG, as they would for a single free-living organism (discussed in Chapter 13). Most commonly, one species produces a product that is consumed by the second species.

In the simplified model shown in **Figure 21.12**, the wood polysaccharides are converted by bacteria to fermentation acids, such as acetic acid, which is absorbed by the termite. (The termite's metabolism then oxidizes the acetate to CO_2.) Other products of the termite bacterial fermentation include CO_2 and H_2, which can be converted to methane by methanogenic archaea. In some termites the H_2 builds up to levels as high as 30%. The termite microbial mutualism is being studied as a model system for production of hydrogen biofuel.

Digestive mutualisms are also important for mammals such as cattle (discussed in Section 21.6). Even we humans derive about 15% of our caloric intake from bacterial fermentation of plant fibers that we cannot otherwise digest.

Symbiosis Involves Varying Degrees of Cooperation and Parasitism

Symbiosis between organisms involves a range of interdependence, from obligate mutualism (cooperation) to obligate parasitism (see **Table 21.5**). In a gut community,

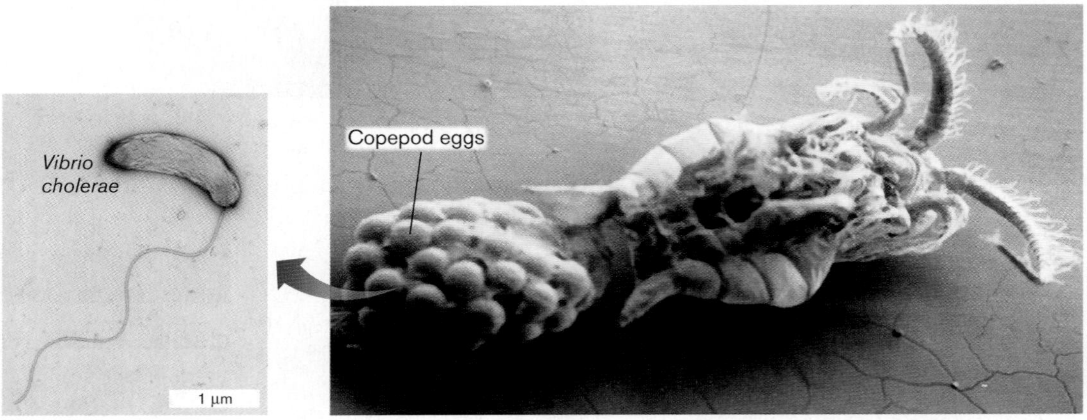
Vibrio cholerae
1 µm
Copepod eggs

FIGURE 21.13 ■ Copepod that carries *Vibrio cholerae.* Copepod (1 to 2 mm) with case of eggs to be "hatched" by *V. cholerae* bacteria (TEM).

GOPAL MURTI/PHOTO RESEARCHERS

COURTESY OF RITA COLWELL

some of the microbial members may enhance each other's growth, but they can also grow independently. Their optional cooperation is called **synergism**, in which both species benefit but can grow independently and show less specific cellular communication. For example, human colonic bacteria produce fermentation products that colonic methanogens metabolize to methane. The methanogens gain energy, and the fermenting bacteria benefit energetically through the steady-state removal of their end products.

In other cases, one species derives benefit from another without return; for example, some wetland bacteria derive benefit from *Beggiatoa* because *Beggiatoa* bacteria oxidize H_2S, which inhibits growth of other species. An interaction that benefits one partner only is called **commensalism**. Commensalism is difficult to define in practice, since "commensal" microbes often provide a hidden benefit to their host. For example, gut bacteria such as *Bacteroides* species were considered commensals until it was discovered that their metabolism aids human digestion.

An interaction that harms one partner nonspecifically, without an intimate symbiosis, is called **amensalism**. An example of amensalism is actinomycete production of antimicrobial peptides that kill surrounding bacteria. The dead bacterial components are then catabolized by the actinomycete.

Finally, **parasitism** is an intimate relationship in which one member (the parasite) benefits while harming a specific host. Many microbes have evolved specialized relationships as parasites, including intracellular parasitic bacteria such as the rickettsias, which cause diseases such as Rocky Mountain spotted fever.

The distinction between mutualist and parasite is often subtle. Lichens consist of a mutualistic association between fungus and algae, but environmental change can convert the fungus to a parasite. On the other hand, parasitic microbes may coevolve with a host to the point that each depends on the other for optimal health. For example, the high incidence of human allergies is proposed to correlate with lack of exposure to parasites that stimulate development of the immune system.

Multiple partner species can form a complex web of positive and negative dependence. An example is the leaf-cutter-ant symbiosis with fungi, which includes a fungal mutualist, a fungal parasite, and a bacterial mutualist that counteracts the parasite (presented in eTopic 17.3). Another example, with medical significance, is the symbiosis between the cholera pathogen *Vibrio cholerae* and copepods. This relationship was discovered by Anwar Huq, from Bangladesh, working with Rita Colwell, then director of the U.S. National Science Foundation (interviewed in eTopic 1.1). Huq and Colwell showed that in natural water systems, most *V. cholerae* cells do not swim freely but colonize the surfaces of copepods (**Fig. 21.13**). The copepods actually depend on these bacteria to eat through the chitin of their egg cases, releasing their young. Thus, bacterial mutualists of copepods are virulent pathogens of humans, whose cholera diarrhea returns the organism to the water full of copepods. But Huq found that drinking water contaminated by *V. cholerae* could be cheaply decontaminated by filtering out the copepods through sari cloth. The elucidation of this complex microbial partnership reaped benefits for people threatened by cholera.

To Summarize

- **Symbiosis** is an intimate association between organisms of different species.
- **Mutualism** is a form of symbiosis in which each partner species benefits from the other. The relationship is usually obligatory for growth of one or both partners.

- **Lichens** are a mutualistic community of algae or cyanobacteria with fungi. Lichens are essential producers for dry soil habitats.
- **Termite gut mutualists** include cellulose-digesting bacteria and protists. A major by-product is hydrogen gas.
- **Syntrophy** is a metabolic association between two species, requiring both partners in order to complete the metabolism with a negative value of ΔG.
- **Parasitism** is a form of symbiosis in which one species grows at the expense of another, usually much larger, host organism.
- **Interactions of multiple species can include both mutualism and parasitism.**

21.4

Marine and Aquatic Microbes

What microbial communities inhabit the oceans? Oceans cover more than two-thirds of Earth's surface, reaching depths of several kilometers and forming an immense habitat (**Fig. 21.14**). Both oceans and freshwater support huge quantities of bacteria and algae, which drive vast ecosystems. Marine and freshwater microbes have enormous human impact, from the maintenance of fisheries to the production of biofuels. Algae are being engineered to produce biofuels, at rates 30-fold more energy-efficient than those of terrestrial biofuel plants. A science fiction scenario, *The Highest Frontier* (2011), imagines an orbital space habitat powered by an outer shell of marine microbial phototrophs.

Marine water is distinguished by its salt concentration, averaging 3.5%. The major ions are Na^+ and Cl^-, with significant levels of sulfate and iodide. Marine salt concentration is high enough to prevent growth of many aquatic and terrestrial bacteria, such as *E. coli*. Nevertheless salt-tolerant organisms such as *Vibrio cholerae* grow well over a broad range of salt concentration.

Marine Habitats

Marine waters vary considerably with respect to temperature, pressure, light penetration, and concentration of organic matter. In the open ocean, the water column (known as the **pelagic zone**) is subdivided into distinct regions (see **Fig. 21.14**):

- **Neuston (about 10 μm).** The neuston is the air-water interface. Although extremely thin, the neuston layer contains the highest concentration of microbes. Many algae and protists have evolved so as to "hang" from the layer of surface tension.
- **Euphotic zone (100–200 m).** The **euphotic zone**, or **photic zone**, is the upper part of the water column, which receives light for phototrophs. In the open ocean, the euphotic zone extends down a couple of hundred meters, whereas at the **coastal shelf** (<200 meters to floor), a higher concentration of silt and organisms decreases the photic zone to as little as 1 meter.

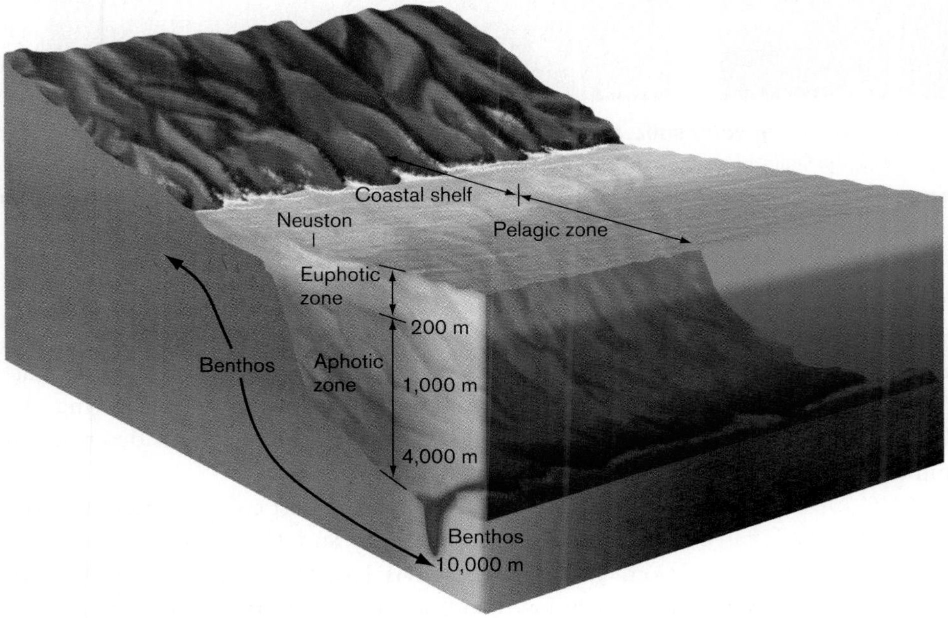
Coastal shelf
Neuston
Pelagic zone
Euphotic zone
200 m
Benthos
Aphotic zone
1,000 m
4,000 m
Benthos
10,000 m

FIGURE 21.14 ■ Regions of marine habitat. The marine habitat subdivides into several categories. The coastal shelf region is defined as the water extending from the shoreline out to a depth of 200 meters. The pelagic zone (open ocean) includes several depth regions: the neuston, the microscopic interface between water and air; the euphotic zone of light penetration, where phototrophs can grow, down to 100–200 meters; the aphotic zone, which supports only heterotrophs and lithotrophs; and the benthos, at the ocean floor.

Special Topic 21.1: Cleaning Up the *Deepwater* Oil Spill

On the night of April 20, 2010, an explosion ripped through the *Deepwater Horizon* oil rig in the Gulf of Mexico. The explosion and subsequent collapse of the rig caused leaks in the oil pipeline, 1.5 kilometers below the ocean surface. For the next 4 months, until the well was capped, about a million gallons of oil per day leaked out into the ocean, and smaller amounts continued to leak thereafter. The oil impacted marine life throughout the gulf, and damaged coastal ecosystems of Florida, Louisiana, and Mississippi.

What happened to all the oil? And could microbes remove it? Crude oil includes thousands of components, ranging from methane and short-chain hydrocarbons to complex aromatic molecules, which different microbes catabolize at different rates. Researchers Molly Redmond and David Valentine, from UC Santa Barbara, arrived to find out what was eating the gulf oil (**Fig. 1**).

Redmond and Valentine sampled microbes from the seawater during several months that followed the spill. They analyzed 16S rRNA libraries from water contaminated by the plume of oil rising from the leak, as well as from uncontaminated gulf seawater (**Fig. 2**). Nonplume (uncontaminated) water shows a broad range of diverse taxa, including Proteobacteria, Cyanobacteria, and Bacteroidetes. The plume of oil contamination, however, showed a marked shift to particular taxa: the genus *Colwellia* (blue-gray bars), a benthic gammaproteobacterium named for Rita Colwell in 1988; and a novel clade of the order Oceanospirillales, which the researchers designated DWH Oceanospirillales (bright blue bars). These two taxa made up the bulk of organisms sampled from the plume in the months of May and June, respectively. Another genus emerging in June, *Cycloclasticus* (green bars), was known for degrading cyclic aromatic hydrocarbons.

The researchers hypothesized that the plume-enriched bacteria such as *Colwellia* were consuming oil components and incorporating the carbon into their biomass. To test the hypothesis, they incubated seawater with added methane,

A. B.

experimentation_ON

HTTP://WWW.LIVESCIENCE.COM/14154-METHANE-MICROBES-DEEPWATER-HORIZON-OIL-SPILL-BTS.HTML

HTTP://WWW.NSF.GOV/NEWS/NEWS_SUMM.JSP?ORG=NSF&CNTN_ID=121845&PREVIEW=FALSE

FIGURE 1 ■ Sampling oil-eating bacteria from the *Deepwater* spill. A. David Valentine (right)and John Kessler, wearing respirators, on a ship near the pipeline leaking oil into the Gulf of Mexico. **B.** Molly Redmond analyzes a water sample for oil-eating bacteria.

- **Aphotic zone.** Below the reach of light, in the aphotic zone, only heterotrophs and lithotrophs can grow.
- **Benthos.** The benthos includes the region where the water column meets the ocean floor, as well as sediment below the surface. Organisms that live in the benthos, such as those of thermal vent communities, are called **benthic organisms**.

Another important determinant of marine habitat is the **thermocline**, a depth at which temperature decreases steeply and water density increases. A thermocline typically exists in an unmixed region. At the thermocline, a population of heterotrophs will peak, feeding on organic matter that settles from above. An experiment showing the thermocline, conducted by undergraduates, is described in **eTopic 21.1**.

Coastal regions show the highest concentration of nutrients and living organisms, and the least light penetration. The open ocean is largely oligotrophic (having an extremely low concentration of nutrients and organisms).

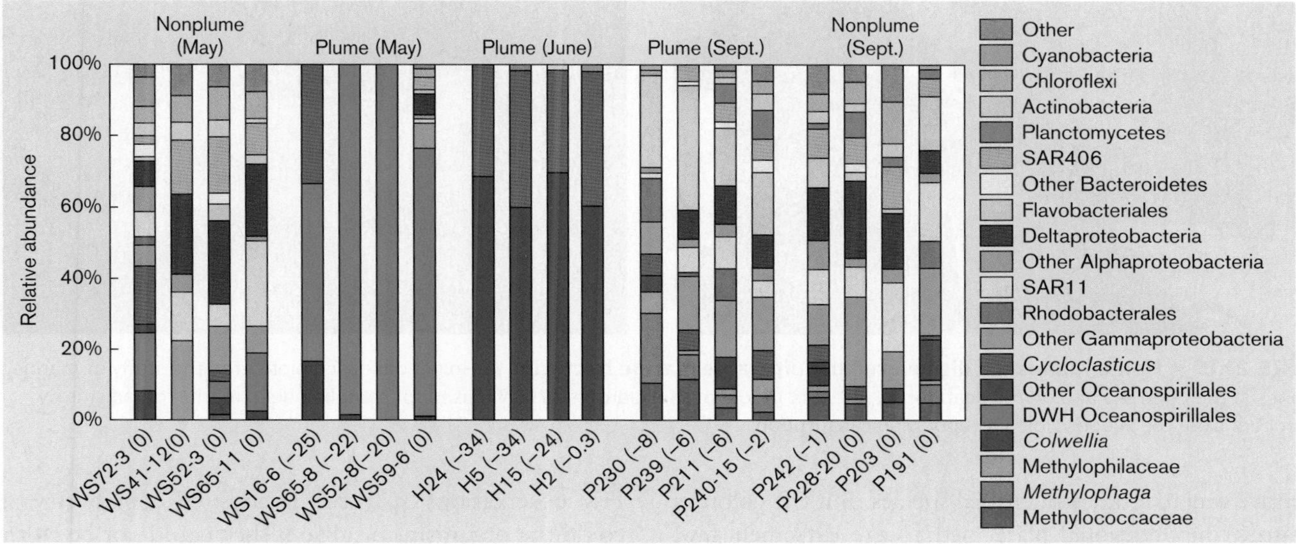
Nonplume (May)
Plume (May)
Plume (June)
Plume (Sept.)
Nonplume (Sept.)
100%
80%
60%
40%
20%
0%
Relative abundance
WS72-3 (0)
WS41-12(0)
WS52-3 (0)
WS65-11 (0)
WS16-6 (–25)
WS65-8 (–22)
WS52-8(–20)
WS59-6 (0)
H24 (–34)
H5 (–34)
H15 (–24)
H2 (–0.3)
P230 (–8)
P239 (–6)
P211 (–6)
P240-15 (–2)
P242 (–1)
P228-20 (0)
P203 (0)
P191 (0)
Other
Cyanobacteria
Chloroflexi
Actinobacteria
Planctomycetes
SAR406
Other Bacteroidetes
Flavobacteriales
Deltaproteobacteria
Other Alphaproteobacteria
SAR11
Rhodobacterales
Other Gammaproteobacteria
Cycloclasticus
Other Oceanospirillales
DWH Oceanospirillales
Colwellia
Methylophilaceae
Methylophaga
Methylococcaceae

FIGURE 2 ■ Bacterial taxa: relative abundance in 16S rRNA clone libraries. Samples were obtained from the plume (oil spreading through seawater) and from nonplume seawater. During May and June, major taxa shifted to oil-consuming *Colwellia* and Oceanospirillales bacteria. By September, after much of the plume had dissipated, the microbial distribution appeared more similar to that of the nonplume water. *Source:* Modified from Molly C. Redmond and David L. Valentine. 2012. *PNAS* **109**:20292–20297, fig. 1.

propane, or benzene labeled with a heavy atom, ^{13}C. They then isolated metagenomic DNA and separated "heavy" from "light" DNA by density-gradient centrifugation. Clone libraries from the density-separated DNA pools showed that *Colwellia* incorporated ^{13}C from propane and benzene, whereas methylotrophic (methyl-oxidizing) taxa of Methylococcaceae incorporated heavy methane. Thus, *Colwellia* appears to be an early consumer of petroleum components, whereas the role of DWH Oceanospirillales remains inconclusive.

Overall, researchers believe that microbial catabolism rapidly consumed the most digestible components of the leaked oil, such as methane and hydrocarbons. But large quantities of more complex aromatics remain on the seafloor, and throughout the ecosystem. These, too, will ultimately be catabolized, but they will take much longer, perhaps years, to dissipate.

RESEARCH QUESTION

Why might the researchers' experiments have failed to demonstrate oil uptake by DWH Oceanospirillales bacteria? What possibilities do you suggest?

Redmond, Molly C., and David L. Valentine. 2012. Natural gas and temperature structured a microbial community response to the *Deepwater Horizon* oil spill. *PNAS* **109**:20292–20297.

The concentration of heterotrophic microorganisms determines the **biochemical oxygen demand** (**BOD**; also called **biological oxygen demand**), the amount of oxygen removed from the water by aerobic respiration. Normally, the open ocean has such a low concentration of organisms that the BOD is extremely low; therefore, the dissolved oxygen content is high. This explains why enough oxygen reaches the ocean floor to serve lithotrophs and methanotrophs. The BOD rises, however, when there is excess sewage or petroleum such as that spilled by the *Deepwater Horizon* oil rig in 2010. Microbial consumption of these wastes at first deprives fish of oxygen, but the process is the only way the ocean recovers (see **Special Topic 21.1**).

Culturing the Uncultured

To nineteenth-century microbiologists, the oceans appeared virtually free of bacteria. Trained in the tradition of Robert Koch (discussed in Chapter 1), microbiologists attempted to isolate marine bacteria by plate culture, considered the

A. Holger Jannasch   B. Submersible *Alvin*

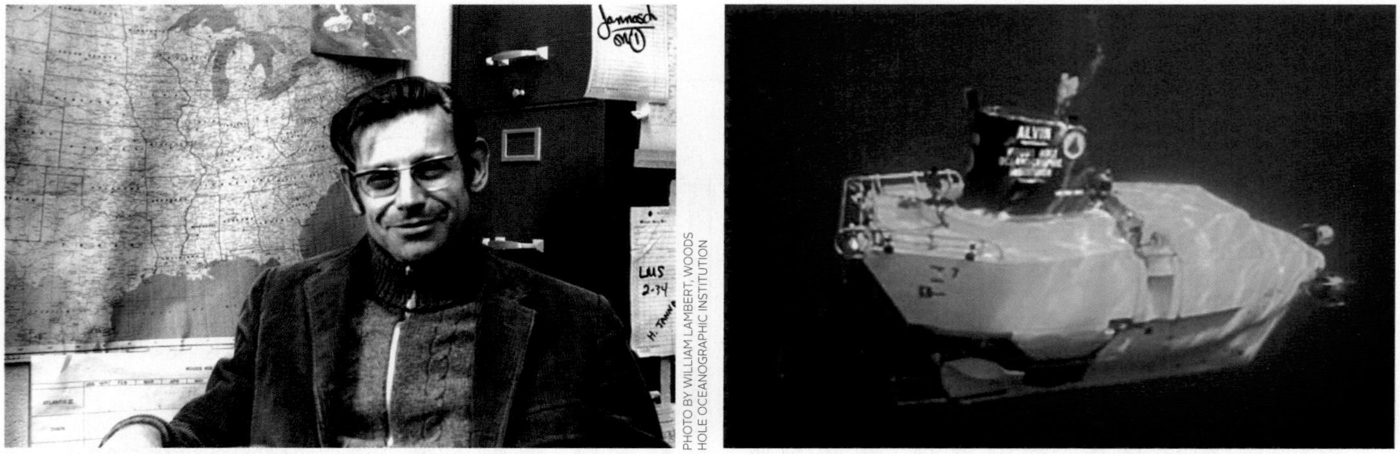

PHOTO BY WILLIAM LAMBERT, WOODS HOLE OCEANOGRAPHIC INSTITUTION

COURTESY OF WOODS HOLE OCEANOGRAPHIC INSTITUTION

FIGURE 21.15 ■ Holger Jannasch discovered unculturable marine bacteria. A. Holger Jannasch pioneered the study of marine microbes in the open sea and on the seafloor. **B.** A recent model of the submersible *Alvin* used for sampling deep marine organisms by researchers from the Woods Hole Oceanographic Institution.

definitive way to study a microbial species. But the colonies that grew on traditional plate media were extremely few. Then, in 1959, the German microbial ecologist Holger Jannasch (1927–1998) (**Fig. 21.15A**) showed that many more bacteria could be seen by light microscopy than could be grown on plates. Later, with colleagues at the Woods Hole Oceanographic Institution, Jannasch went on to study life at the ocean floor, using the famous submersible vessel *Alvin* (**Fig. 21.15B**).

Jannasch was one of the first to show that most marine microorganisms could not be cultured in the laboratory. His observations sparked decades of controversy. How could the organisms be alive if they could not be cultured? In 1985, Rita Colwell (**Fig. 21.16A**), then at the University of Maryland, introduced the concept of **viable but nonculturable (VBNC)**, referring to environmental organisms that metabolize but could not be cultured. In some cases, previously cultured bacteria, such as the cholera pathogen *Vibrio cholerae,* could enter such a state (**Fig. 21.16B**)—a phenomenon of great importance for medicine. In other cases, environmental organisms were found in an unculturable state, with no known means of culture. Today we

A.

COURTESY OF RITA COLWELL

B.

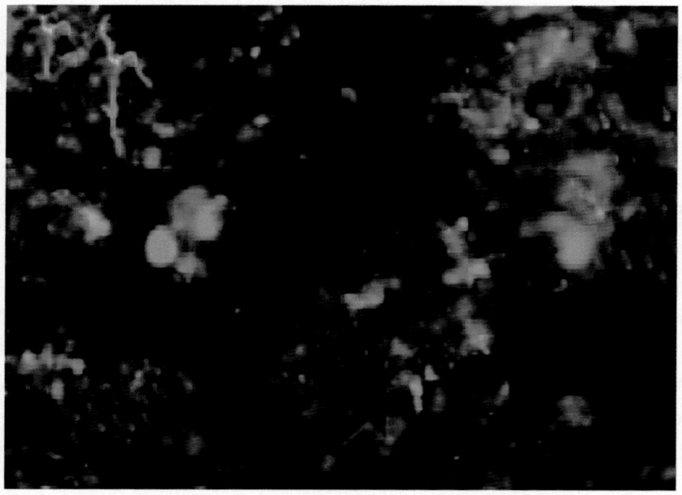

MUNIRUL ALAM ET AL. 2007. *PNAS* **104**:17801–17806, FIG. 3D

FIGURE 21.16 ■ Uncultured bacteria. A. Rita Colwell, University of Maryland, former director of the National Science Foundation. **B.** Uncultured cells of *Vibrio cholerae* in a biofilm (green fluorescence). Fluorescence microscopy.

term such environmental organisms **uncultured**—with the understanding that their culture requirements may yet be discovered.

In fact, emerging culture methods now enable many uncultured organisms to grow in the laboratory. Often microbial growth requires hidden synergy or mutualism with other organisms in the natural environment. For example, in 2008 Erik Zinser and colleagues at the University of Tennessee–Knoxville showed that the cyanobacterium *Prochlorococcus,* the ocean's most abundant oxygenic phototroph, can be cultured in the presence of a "helper bacterium" that catalyzes hydrogen peroxide, a by-product of oxygenic photosynthesis. The helpers produce catalase, which *Prochlorococcus* has lost through degenerative evolution, enabling more efficient growth of the tiny phototroph.

Another exciting approach to culturing is to mimic the natural environment in which organisms are found. In 2002, Kim Lewis and colleagues at Northeastern University attempted to isolate bacteria from intertidal sediment using a growth chamber perfused with seawater. The sampled bacteria were embedded in agar, over which the seawater flowed. As much as 40% of the microbes inoculated formed colonies in the agar—far more than the typical fraction attempted under standard culture conditions (roughly 0.1%). Some of the colonies persisted through passage onto standard agar plates, but most failed to isolate as "pure" cultures of a single species. They could grow only in intimate contact with microbes of other species, providing unknown growth factors. In 2011, Lewis showed that some of these shared growth factors are siderophores. Siderophores are secreted molecules that bind iron and make it available to cells in the vicinity. These findings suggest that intimate symbiosis, rather than single-species growth, may be the more common condition for microbes in nature.

Marine Metagenomes

Cultured or not, the marine microbial communities are now emerging through metagenomes. Craig Venter's pioneering expedition in 2007 identified 25,000 different microbial species per liter of seawater, from ocean water samples taken from around the globe. The most abundant strains fell under the newly discovered genus *Pelagibacter,* or "pelagic zone bacteria." These Alphaproteobacteria, surprisingly, are related to rickettsias, which are obligate intracellular parasites (discussed in Chapter 18). Proteorhodopsin enables these marine heterotrophs to make ATP while starving for organic nutrients, which are scarce in the open ocean. Further studies confirm that over half the species of marine bacteria are Proteobacteria, in contrast with the earlier view that cyanobacteria were the most abundant marine bacteria.

As discussed in the previous section, even metagenomes miss community members that may fill important niches. To find them, researchers set out to explore expressed genes—a "metatranscriptome." In 2008, Sallie Chisholm and Edward DeLong from the Massachusetts Institute of Technology and Penn State sequenced the mRNAs expressed by marine bacteria obtained from a Hawaiian research station. The RNA molecules were polyadenylated (a string of adenines was added) using an enzyme that favors mRNA and excludes ribosomal RNAs because of their high degree of secondary structure (intramolecular folding). The resulting RNA pool, enriched for expressed mRNA, was reverse-transcribed to DNA (called cDNA for "complementary DNA"). The cDNA was then amplified and sequenced.

The sequences of the cDNA (representing expressed genes) were compared to those amplified from metagenomic DNA of the same microbial community. The relative abundance of expressed and metagenomic sequences gives a measure of expression levels of all genes, which are plotted in declining rank order (**Fig. 21.17**). The most highly expressed genes included those encoding functional

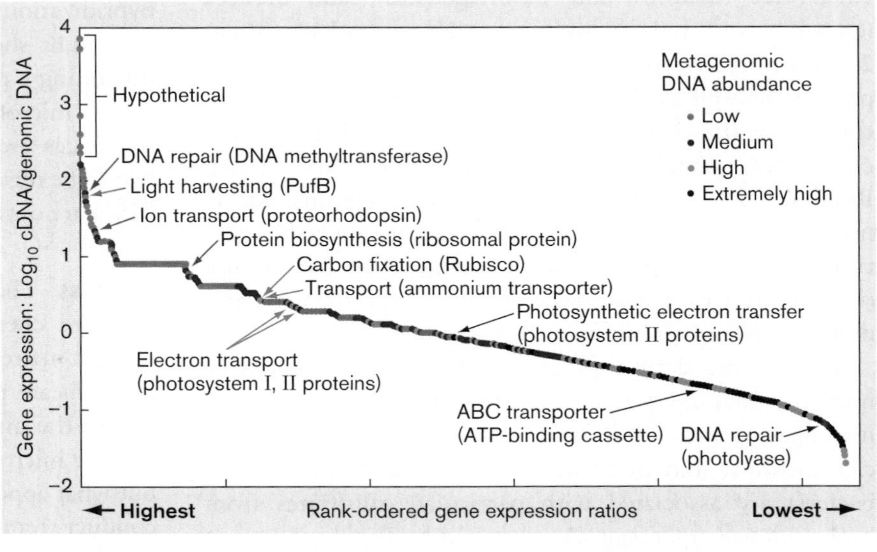

Gene expression: Log10 cDNA/genomic DNA
4
3
2
1
0
−1
−2
Hypothetical
DNA repair (DNA methyltransferase)
Light harvesting (PufB)
Ion transport (proteorhodopsin)
Protein biosynthesis (ribosomal protein)
Carbon fixation (Rubisco)
Transport (ammonium transporter)
Photosynthetic electron transfer (photosystem II proteins)
Electron transport (photosystem I, II proteins)
ABC transporter (ATP-binding cassette)
DNA repair (photolyase)
Metagenomic DNA abundance
Low
Medium
High
Extremely high
Highest
Rank-ordered gene expression ratios
Lowest

FIGURE 21.17 ■ Genes expressed in a marine microbial community. Expression of each gene is measured as RNA (copied to cDNA) divided by the gene's relative abundance in the metagenome. The Y axis represents the ratio cDNA/metagenomic DNA for all expressed genes, plotted in rank order (from highest to lowest). The relative abundance of the DNA sequence within the metagenome is color-coded. The genes most highly expressed tend to be rarest in the genome (colored red).

elements of photosynthesis, such as light-harvesting proteins and Rubisco. Genes for DNA repair were also highly expressed, presumably to correct UV damage in the open ocean. But the most highly expressed genes were some of the rarest in the metagenomes (colored red in **Fig. 21.17**). And 40% of all the cDNA (expressed genes) had no counterpart in the genomic DNA (not shown in the figure). Thus, expression analysis does indeed reveal organisms undetected in the metagenome. It also suggests that vast ranges of organisms may remain unreached by either method.

Beyond DNA, microbial ecology addresses organisms in their relationships with their habitat and with each other. We will now introduce approaches to measure microbial populations and their interactions within marine communities. These approaches have enormous practical applications, from the management of fisheries to the assessment of greenhouse gas fluxes (a topic pursued in Chapter 22).

Measuring Planktonic Communities

In marine science, the term **plankton** refers to organisms that float passively in water. Microbiologists use the term loosely, as microbial "plankton" include motile bacteria and protists. Marine phytoplankton (microbial phototrophs) produce a substantial part of the world's oxygen and consume much of the atmospheric CO_2.

Microbial plankton, or **microplankton**, include numerous members of the three domains discussed in Chapters 18, 19, and 20: bacteria (cyanobacteria and proteobacteria), eukaryotes (diatoms and dinoflagellates), and archaea (crenarchaeotes and euryarchaeotes). **Nanoplankton** (about 2–20 μm) arbitrarily include smaller algae and flagellated protists, as well as filamentous cyanobacteria. **Picoplankton** (about 0.2–2 μm) consist of bacteria and the smaller eukaryotes, including the smallest living cells known. Besides these size classes of cells, the term "femtoplankton" may refer to marine viruses, the smallest detectable particles capable of reproduction. Note however that the use of these prefixes (nano-, pico-, femto-) is approximate, and does <u>not</u> refer to size units such as nanometers.

Despite the definition of "plankton," not all marine microbes float independently. Many form biofilms on colonial algae such as kelps, or on suspended inorganic particles known as **marine snow**. As many as half of all marine bacteria are associated with particulate substrates from broken-down organisms. As observed by Farooq Azam at the Scripps Institution of Oceanography, "Seawater is an organic matter continuum, a gel of tangled polymers with embedded strings, sheets, and bundles of fibrils and particles, including living organisms, as 'hotspots.'" Thus, while the ocean as a whole has a low average nutrient concentration, marine waters include a suspension of concentrated tangles of nutrient-rich substrates. Collectively, their global volume is so large that their sedimentation rate influences calculations of Earth's carbon cycle.

Population size can be estimated in different ways: the number of individual reproductive units, the total organic biomass available to consumers, or the rate of productivity or assimilation of key nutrients, such as carbon dioxide. All ways of measuring these quantities present challenges; there is no single right method, but different approaches answer different questions.

Fluorescence microscopy. Marine microbes of all sizes, even viruses, can be detected and counted under the microscope using a DNA-intercalating fluorescent dye. Recall that fluorescence enables detection even of particles whose size is below the resolution limit defined by the wavelength of light (discussed in Chapter 2). Fluorescence microscopy with DNA-binding fluorophores is used to detect planktonic microbes (**Fig. 21.18A**). The fluorescent dye DAPI reveals extremely small bacteria and algae (large spots in the figure) and viruses (small spots). DAPI-stained samples can be quantified and used to map plankton distribution through space and time (**Fig. 21.18B**).

How do we distinguish different kinds of microbes in a mixed sample? Differential detection of taxonomic classes requires a fluorophore attached to a DNA probe, a short sequence that can hybridize only to DNA or RNA of a particular taxon. This technique is called fluorescence in situ hybridization (FISH); for example, the Current Research Highlight shows a FISH micrograph of diverse bacteria inhabiting a plant root. In ecology research, FISH usually targets microbial rRNA instead of DNA because the rRNA sequences are present in 10,000-fold greater copy number. (The methods of FISH and the more sensitive CARD-FISH are outlined in Appendix 3.)

Biomass. The amount of biomass can be determined by standard chemical assays of protein and other forms of organic matter. Net biomass of a population, however, does not indicate productivity within an ecosystem, because it misses the amount of carbon cycled through respiration. Marine microbes have an extremely rapid rate of turnover, but what appears to be a small population may nonetheless conduct tremendous rates of carbon fixation into biomass, which is rapidly consumed by the next trophic level.

Incorporation of radiolabeled substrates. The measured amount of biomass does not indicate the rate of biomass

A.

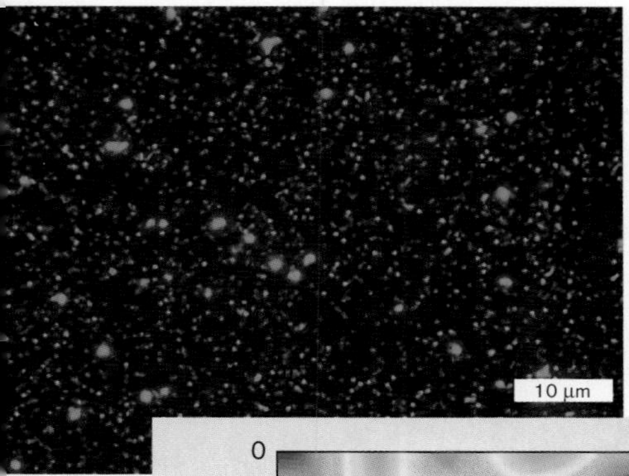
10 μm
THE OCEANIC MICROBIAL OBSERVATORY

B.

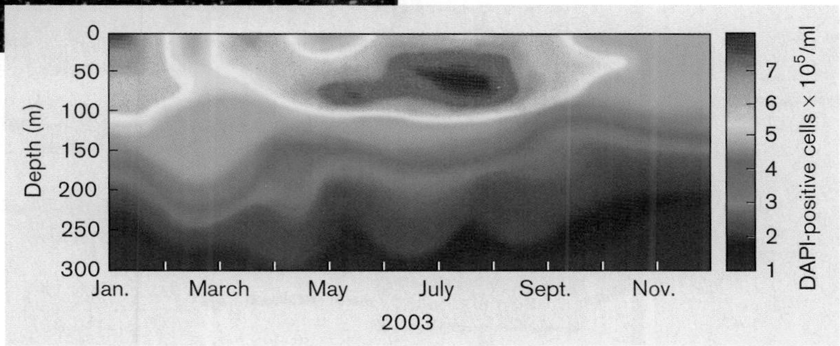
Depth (m)
0
50
100
150
200
250
300
Jan.
March
May
July
Sept.
Nov.
2003
7
6
5
4
3
2
1
DAPI-positive cells × 10^5/ml

FIGURE 21.18 ■ Fluorescence observation of picoplankton. **A.** Picoplankton, mainly bacteria, observed at a sampling station near Bermuda (fluorescence micrograph). The bacteria (large spots) and viruses (small spots) are stained with DAPI, a fluorescent dye that intercalates DNA. **B.** Concentration of picoplankton as a function of depth and time over the course of a year. Picoplankton were observed by fluorescence using DAPI DNA stain.

production. The rate of production can be estimated by the cells' incorporation of a radiolabeled substrate. Uptake of $^{14}CO_2$ indicates the rate of carbon fixation—a highly significant property for the study of global warming. Uptake of ^{14}C-thymidine measures the rate of DNA synthesis, an indicator of the rate of cell division (example shown in eTopic 21.1). These procedures can be conducted in seawater under native conditions, without requiring laboratory cultivation.

A limitation of these methods is that addition of the radiolabeled substrate may raise a nutrient concentration to artificially high levels that distort the naturally occurring rates of activity. In the case of labeled thymidine, another limitation is that not all growing cells incorporate exogenous thymidine into their DNA.

Planktonic Food Webs

Marine plankton interact with each other as producers and consumers in a food web. A simplified outline of the food web in the open ocean is shown in **Figure 21.19A**. The diagram of a food web is often called a "spaghetti diagram" because so many trophic interactions cross each other in various directions. The phototrophic producers are known as **phytoplankton**. Examples of phytoplankton include the cyanobacteria *Prochlorococcus* (predominating in the upper photic layer) and *Synechococcus* (in a slightly deeper layer reached by longer wavelengths of light). In addition, there are myriad forms of algae, such as diatoms and dinoflagellates.

Bacteria and protists are consumed by larger protists, which feed small invertebrates, which feed larger invertebrates and, ultimately, vertebrates such as fish.

All levels of microbial plankton undergo intense predation by viruses. The degree of viral predation is difficult to measure, but recent studies indicate that cell lysis by viruses breaks down about half of microbial biomass (discussed in Chapter 6). Virus particles represent a significant sink for carbon and nitrogen. They accelerate the return of minerals to producers and necessitate a larger base of producers to sustain the ecosystem. Some marine viruses are highly host specific, infecting only certain species of dinoflagellates or cyanobacteria. Their presence selects for diverse communities containing numerous scattered species. Other viruses attack many hosts—and they transfer genes from one host to another, such as the genes encoding photosystems. Thus, marine viruses are a dominant force determining community species distribution and genome content.

Thought Question

21.4 How do viruses select for increased diversity of microbial plankton?

Ocean ecosystems contain a large number of trophic levels, of which the top consumers include fish and humans. Because each trophic level spends 90% of its food intake for energy, ecosystems require a huge lower foundation to sustain the highest-level consumers. This is one reason why fisheries worldwide are now in danger of running out of fish for human consumption: Fish are being harvested faster than the ecosystem can replace them.

Note that actual food webs are far more complex than the one shown in **Figure 21.19A**. One source of complexity is that many "algal protists," such as chrysophytes (golden algae) and dinoflagellates, are actually **mixotrophs**, organisms that both fix CO_2 through photosynthesis and catabolize organic compounds from outside. Mixotrophy offers the opportunity to grow at night, without light, and to acquire scarce minerals, such as iron, from prey. Mixotrophs include secondary endosymbiont algae (discussed in Chapters 17 and 20), such as kelp and sargassum weed, as well as protists containing cyanobacterial or algal endosymbionts. A scheme for a food web that includes mixotrophs is shown in **Figure 21.19B**. The scheme is

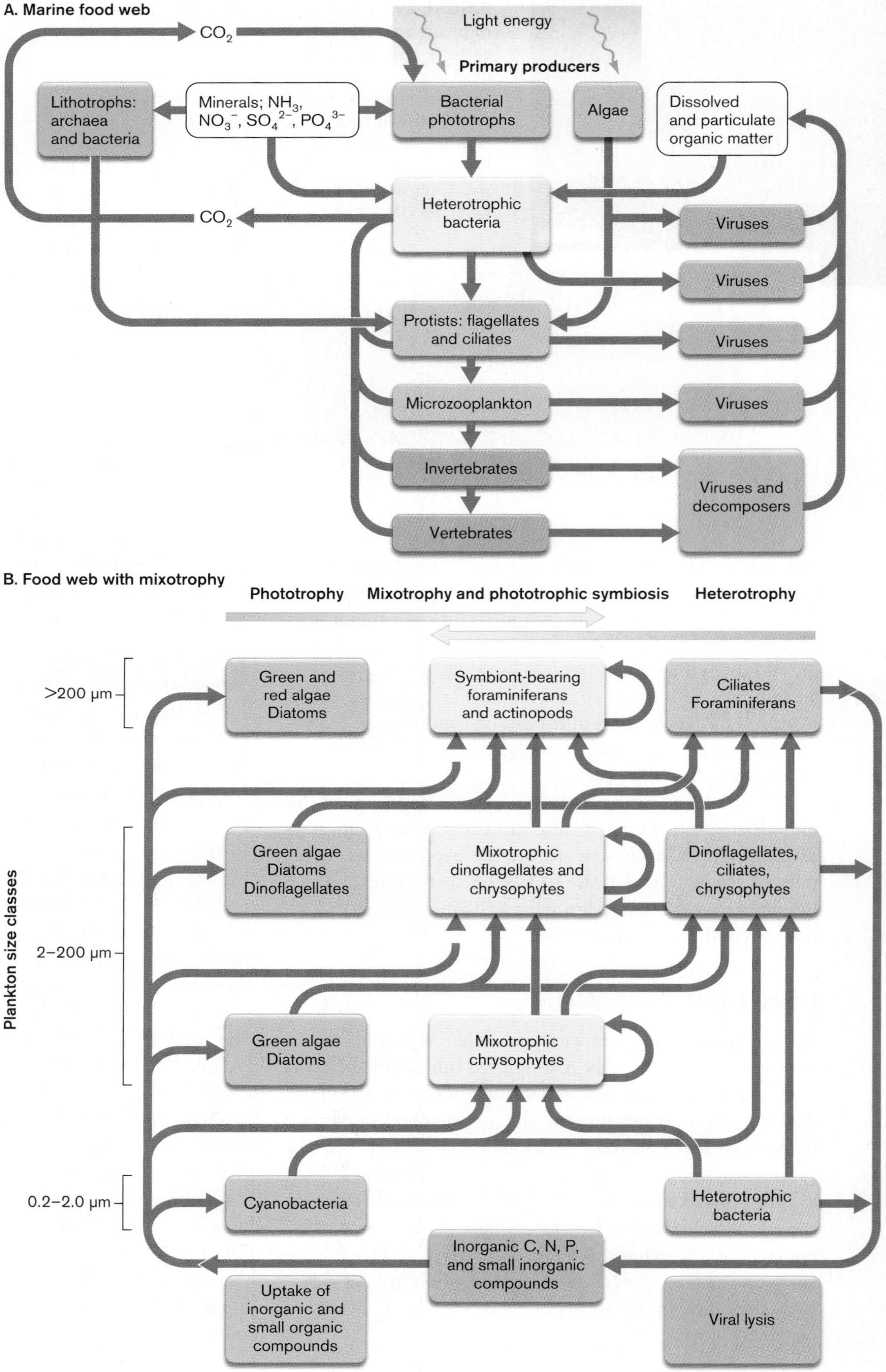
A. Marine food web
Light energy
Primary producers
CO_2
Lithotrophs: archaea and bacteria
Minerals; NH_3, NO_3^-, SO_4^{2-}, PO_4^{3-}
Bacterial phototrophs
Algae
Dissolved and particulate organic matter
Heterotrophic bacteria
CO_2
Viruses
Viruses
Protists: flagellates and ciliates
Viruses
Microzooplankton
Viruses
Invertebrates
Viruses and decomposers
Vertebrates
B. Food web with mixotrophy
Phototrophy
Mixotrophy and phototrophic symbiosis
Heterotrophy
Plankton size classes
>200 μm
2–200 μm
0.2–2.0 μm
Green and red algae Diatoms
Symbiont-bearing foraminiferans and actinopods
Ciliates Foraminiferans
Green algae Diatoms Dinoflagellates
Mixotrophic dinoflagellates and chrysophytes
Dinoflagellates, ciliates, chrysophytes
Green algae Diatoms
Mixotrophic chrysophytes
Cyanobacteria
Heterotrophic bacteria
Inorganic C, N, P, and small inorganic compounds
Uptake of inorganic and small organic compounds
Viral lysis

FIGURE 21.19 ■ The pelagic marine food web. A. In the marine water column, the vast majority of carbon transfer occurs among microbes. Arrows lead from organisms consumed and point to their consumers. The primary producers are phototrophic bacteria and algae, with a smaller contribution from lithotrophic bacteria and archaea. Grazers include heterotrophic bacteria and protists. Predators include protist flagellates and ciliates. About 40% of bacterial and protist biomass is degraded by viral lysis. Multicellular organisms cycle a relatively small fraction of biomass. **B.** A more complex view of the marine food web includes mixotrophs (organisms that combine phototrophic and heterotrophic nutrition) and symbiotic associations between protists and phototrophic bacteria or algae.

further complicated by the existence of elaborate mutualistic associations between different kinds of protists, algae, and bacteria.

Finally, remember that numerous unique kinds of microbes exist only as symbionts of macroscopic organisms, including invertebrates, fish, and marine mammals. Microbial communities within plants and animals are discussed in detail in Sections 21.5 and 21.6.

The Ocean Floor

The ocean floor (benthos) experiences extreme pressure beneath several kilometers of water. Most organisms that live there are pressure-dependent species, known as **barophiles** (discussed in Chapter 5). Barophiles require high pressure for growth (200–1,000 atm), failing to grow when cultured at sea level. Cold temperatures (about 2°C) select for **psychrophiles** (cold-adapted species), whose rate of growth is relatively slow. In the absence of light, heterotrophic bacteria depend on detritus from above as their carbon source; but by the time any material reaches the benthos, much of the organic carbon has been depleted by prior consumers. A metagenome of the benthos (deep ocean) reported in 2011 by Emiley Eloe, Douglass Bartlett, and colleagues at UC San Diego revealed primarily Proteobacteria, Bacteroidetes, and Planctomycetes. Their genomes encode a large number of heavy-metal transport and resistance proteins, which help cells survive toxic metals upwelling from the ocean floor. These metal resistance genes may have practical applications for bioremediation.

Among the first to investigate life on the ocean floor, in the 1960s, were Jannasch and colleagues at the Woods Hole Oceanographic Institution. Using the submersible research vessel *Alvin,* they discovered entirely new forms of benthic bacteria and archaea. Some of these microbes grow in thick biofilms at black smoker **thermal vents**, or **hydrothermal vents**, such as those of Guaymas Basin, in the Gulf of California (**Fig. 21.20**) (discussed in Chapter 19). The clouds rising from the thermal vent are minerals precipitating as the superheated solution meets cold seawater. The reduced minerals from the vents support entire ecosystems of lithotrophic bacteria, including mutualists of uniquely evolved giant clams and worms. Many vent-dependent microbes are **thermophiles**, adapted to high temperatures (discussed in Chapter 5). Many are hyperthermophilic archaea such as *Pyrodictium occultum,* growing at temperatures above 100°C, which are reached only at high pressure (1,000 atm; discussed in Chapter 19).

The ocean floor provides reduced inorganic minerals such as iron sulfide (FeS) and manganese (Mn^{2+}) that can combine with dissolved oxygen to drive lithotrophy (discussed in Chapter 14). Lithotrophic metabolism by microbes in the sediment generates a permanent voltage potential between the reduced sediment and the oxidizing water. As a result, the entire benthic interface between floor and water acts as a charged battery that can actually generate electricity.

The benthic redox gradient is enhanced dramatically at hydrothermal vents, where volcanic activity causes upwelling of hydrogen sulfide (H_2S), H_2, and carbonates (**Fig. 21.20A**). At a hydrothermal vent, these reduced minerals are brought up by seawater that seeps through the sediment until it reaches a magma pool, where it becomes superheated and rises to the surface as steam. Sulfate-reducing bacteria reduce sulfate from seawater with H_2 upwelling from the vent fluids, to form H_2S. As the H_2S rises, it is oxidized by sulfur-oxidizing bacteria such as *Thiomicrospira.* Nearby sediment (anaerobic) supports methanogens, and the methane they produce seeps up into the oxygenated water where it is oxidized by methanotrophs.

The high concentrations of sulfides near thermal vents have selected for a remarkable evolution of invertebrate species that feed through mutualistic associations with H_2S-oxidizing bacteria. The H_2S oxidizers fix carbon from the carbonates, generating organic metabolites that feed their animal hosts—species of worms, anemones, and giant clams, all closely related to surface-dwelling species that would be poisoned by H_2S. The tube worm *Riftia* is colored bright red by a pigment carrying H_2S and O_2 in its circulatory fluid (**Fig. 21.20B**). The worm has evolved such a complete dependence on its symbionts that it has lost its own digestive tract.

Surprisingly, microbial communities related to those at thermal vents are found in unheated benthic habitats known as **cold seeps**. A cold seep is a region of the ocean floor where methane or petroleum leaks up through a rock fissure, or from methane hydrates (crystalline methane-water deposits) formed by methanogens (discussed in Chapter 19). Cold-seep organisms are adapted to high pressure, but unlike vent organisms, they are psychrophiles. Bacteria adapted to cold seeps and benthos break down the petroleum leaked from incidents such as the *Deepwater Horizon* oil rig blowout (**Special Topic 21.1**). Cold-seep endosymbionts within animals are presented in the Part 2 interview with Nicole Dubilier, and discussed in eTopic 21.2.

Freshwater Microbial Communities

Many features of marine habitats also apply to freshwater aquatic systems, such as lakes and rivers. Freshwater habitats, of course, contain much lower salt, usually less than 0.1%. The study of aquatic systems is called limnology.

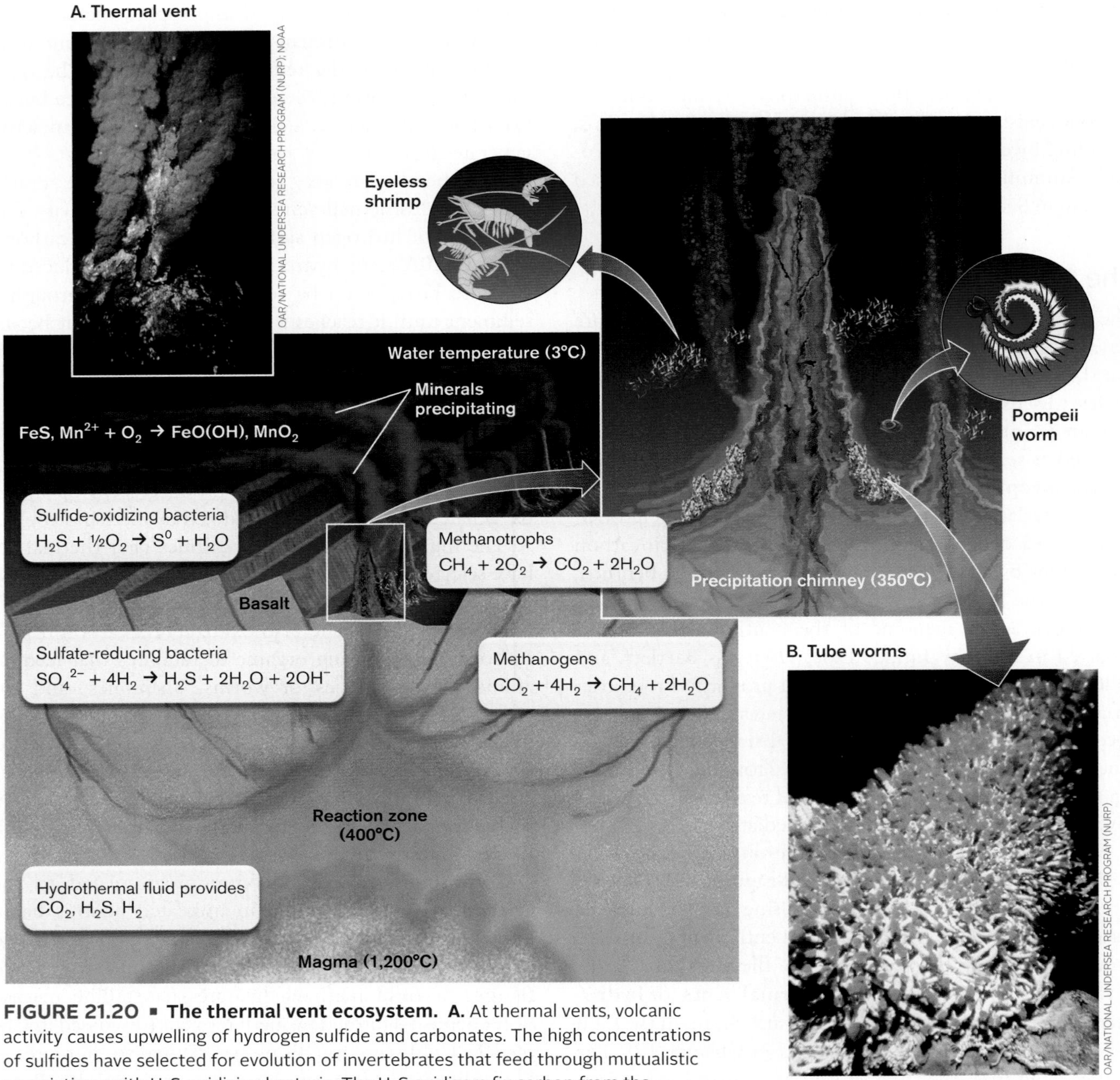
A. Thermal vent
OAR/NATIONAL UNDERSEA RESEARCH PROGRAM (NURP); NOAA
Eyeless shrimp
Pompeii worm
Water temperature (3°C)
Minerals precipitating
FeS, Mn^{2+} + O_2 → FeO(OH), MnO_2
Sulfide-oxidizing bacteria
H_2S + ½O_2 → S^0 + H_2O
Methanotrophs
CH_4 + $2O_2$ → CO_2 + $2H_2O$
Precipitation chimney (350°C)
Basalt
Sulfate-reducing bacteria
SO_4^{2-} + $4H_2$ → H_2S + $2H_2O$ + $2OH^-$
Methanogens
CO_2 + $4H_2$ → CH_4 + $2H_2O$
B. Tube worms
Reaction zone (400°C)
Hydrothermal fluid provides CO_2, H_2S, H_2
Magma (1,200°C)
OAR/NATIONAL UNDERSEA RESEARCH PROGRAM (NURP)

FIGURE 21.20 ▪ The thermal vent ecosystem. A. At thermal vents, volcanic activity causes upwelling of hydrogen sulfide and carbonates. The high concentrations of sulfides have selected for evolution of invertebrates that feed through mutualistic associations with H_2S-oxidizing bacteria. The H_2S oxidizers fix carbon from the carbonates, generating organic metabolites that feed their animal hosts. **B.** Vent bacteria within the gills of tube worms provide energy from oxidation of hydrogen sulfide. *Source:* Part A from Raina Maier et al. 2000. *Environmental Microbiology.* Academic Press. Reprinted by permission of Elsevier, Ltd.

Like the marine water column, the profile of a lake water column is defined by depth, aeration, and temperature (**Fig. 21.21**).

Large, undisturbed lakes are usually **oligotrophic**, with dilute concentrations of nutrients and microbes (**Fig. 21.21A**). The warm upper water layer above the thermocline is called the epilimnion. The epilimnion water is warm, well mixed, and oxygenated relative to the lower layers of water, which are isolated from the atmosphere. It supports oxygenic phototrophs such as algae and cyanobacteria. Below the epilimnion lies a thermocline, a steep transition zone to colder, denser water below.

A. Oligotrophic lake (dilute nutrients)

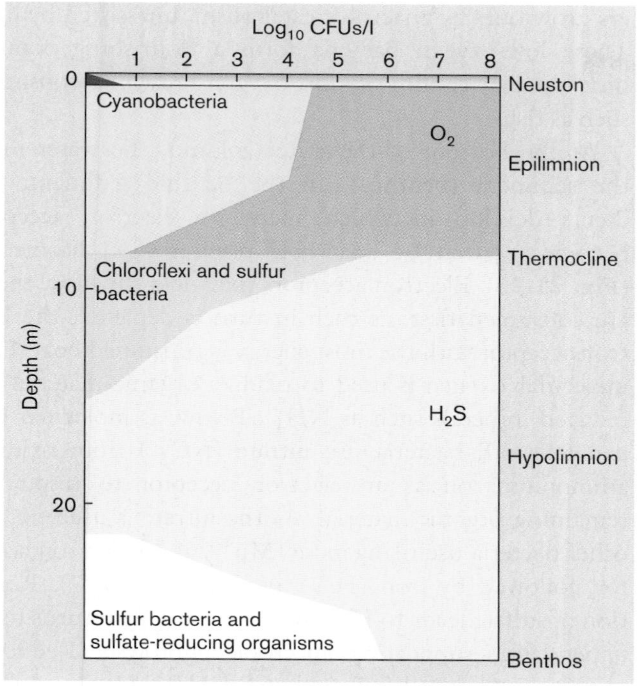
Log10 CFUs/l
1 2 3 4 5 6 7 8
0
10
20
Depth (m)
Cyanobacteria
O_2
Chloroflexi and sulfur bacteria
H_2S
Sulfur bacteria and sulfate-reducing organisms
Neuston
Epilimnion
Thermocline
Hypolimnion
Benthos

B. Eutrophic lake (rich nutrients)

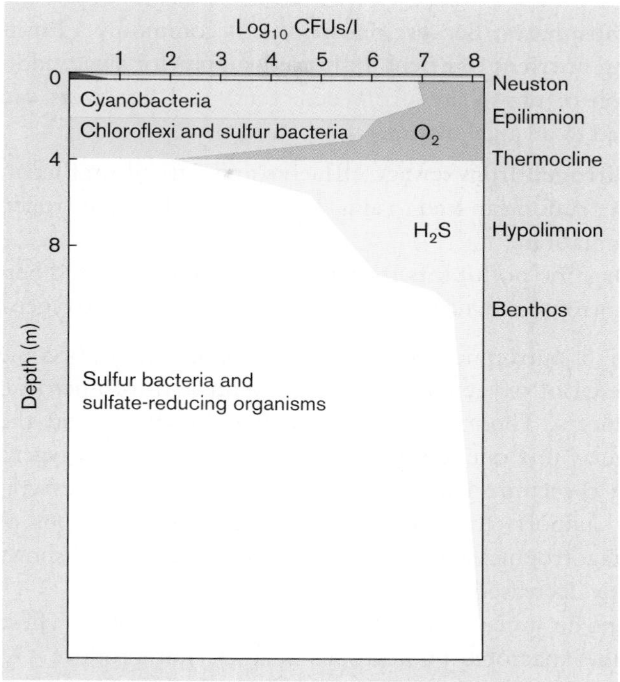
Log10 CFUs/l
1 2 3 4 5 6 7 8
0
4
8
Depth (m)
Cyanobacteria
Chloroflexi and sulfur bacteria
O_2
H_2S
Sulfur bacteria and sulfate-reducing organisms
Neuston
Epilimnion
Thermocline
Hypolimnion
Benthos

C. Algal bloom

© JOHN SEEWER/AP

FIGURE 21.21 ■ Aquatic microbial metabolism in lakes. **A.** Depth profile of an oligotrophic lake. The oxygenated zone (epilimnion) reaches to the thermocline, below which only anaerobes grow. CFU = colony-forming unit. **B.** Depth profile of a eutrophic lake. The oxygenated epilimnion is much shallower, and microbial concentrations are tenfold higher than in an oligotrophic lake. **C.** Algal bloom from fertilizer runoff fills a channel into the reservoir at Grand Lake St. Marys State Park, Ohio, in 2010.

The epilimnion reaches only about 10 meters in depth, much shallower than the marine euphotic zone. In a lake, the depth of light penetration varies greatly, depending on the concentration of microbes and particulate matter. At the edge of the lake, where the upper layer becomes shallow enough for rooted plants, is the **littoral zone**. In the deeper water, below the epilimnion, lies the hypolimnion, a region that becomes anoxic.

A lake that receives large concentrations of nutrients, such as runoff from agricultural fertilizer or septic systems, becomes **eutrophic** (**Fig. 21.21B**). In a eutrophic lake, the nutrients support growth of algae to high densities, causing an **algal bloom** (**Fig. 21.21C**). The algal bloom is rapidly consumed by heterotrophic bacteria, whose respiration removes all the oxygen. This oxygen loss causes the anoxic hypolimnion to reach nearly up to the surface of the lake.

But all fish and invertebrates need access to the epilimnion for oxygen. In a eutrophic lake, fish die off owing to lack of oxygen, which heterotrophic microbes have consumed. Such a lake is said to have a high level of biochemical oxygen demand (BOD). For example, in 2010, fertilizer runoff caused eutrophication of Grand Lake St. Marys in Ohio. The lake was covered with thick green algae that released toxins, depleted oxygen, ruined boat hulls, and prevented human use of the lake.

Common causes of eutrophication include:

- **Phosphates.** Because phosphorus is commonly a **limiting nutrient** (nutrient in shortest supply) for algae, addition of phosphates from detergents and fertilizers can lead to an algal bloom.
- **Nitrogen** from sewage effluents and agricultural fertilizer runoff can lead to algal blooms by relieving nitrogen limitation.
- **Organic pollutants** from sewage effluents overfeed heterotrophic bacteria, depleting the epilimnion of oxygen.

In a eutrophic lake, the lower layers have become depleted of oxygen as a result of overgrowth of microbial producers. Thermal stratification may break up, and the lake may mix one to several times per year, thus reoxygenating the entire lake. A permanently eutrophic lake typically supports ten times the microbial concentrations of an oligotrophic lake (see **Fig. 21.21A** and **B**), but shows greatly decreased animal life.

Anoxic water supports only anaerobic microbes. These include anaerobic phototrophs that do not produce O_2. Enough light may penetrate to support anaerobic H_2S-oxidizing phototrophs such as *Chlorobium* and *Rhodopseudomonas*. Although H_2S photolysis provides less energy than oxygenic H_2O photolysis, these bacteria have evolved to use chlorophylls whose spectrum extends into the infrared (**Fig. 21.22**). Light in the infrared portions of the spectrum cannot be used by oxygenic phototrophs, because the photon energy is insufficient to split water. The less efficient H_2S photolyzers, however, can harness the energy of red and infrared radiation (discussed in Chapter 14). H_2S-oxidizing phototrophs overlap metabolically with anaerobic heterotrophs and lithotrophs that reduce oxidized minerals. Some bacteria, such as *Rhodospirillum rubrum*, can grow anaerobically with or without light; others grow only by anaerobic catabolism, unassisted by light. These low-oxygen bacteria form a flourishing community, but they cannot support oxygen-breathing consumers such as fish.

At the bottom of the water column, the water meets the sediment (benthos). In the benthic sediment, gradients develop in which successive electron acceptors become reduced by anaerobic respirers and lithotrophs (**Fig. 21.23**). Electron acceptors that yield the most energy are consumed first; as each in turn is depleted, the electron acceptor with the most energy is consumed next. First, molecular oxygen is used to oxidize organic material and reduced minerals such as NH_4^+. Below, as molecular oxygen falls off, bacteria use nitrate (NO_3^-) from oxidized ammonium ion as an electron acceptor to respire on remaining organic material. As the nitrate is used up, still other bacteria use manganese (Mn^{4+}) as an electron acceptor, followed by iron (Fe^{3+}) and sulfate (SO_4^{2-}). Reduction of sulfate leads to H_2S, which eventually returns to the upper layers supporting anaerobic photolysis. Reduction of CO_2 by H_2 produces methane (CH_4). Methane collects below and sometimes ignites when it escapes to the surface. Methane from freshwater lakes and streams is emerging as a major contributor to global warming (discussed in Chapter 22).

Note: Certain minerals and organic molecules occur in equilibrium between ionized and un-ionized states over the range of pH typical of most common habitats (pH 4–9). Examples include ammonia (NH_3), which protonates to ammonium ion (NH_4^+), and organic acids such as acetic acid (CH_3COOH), which deprotonates to acetate (CH_3COO^-). In this chapter we refer to the form most prevalent at pH 7 unless stated otherwise.

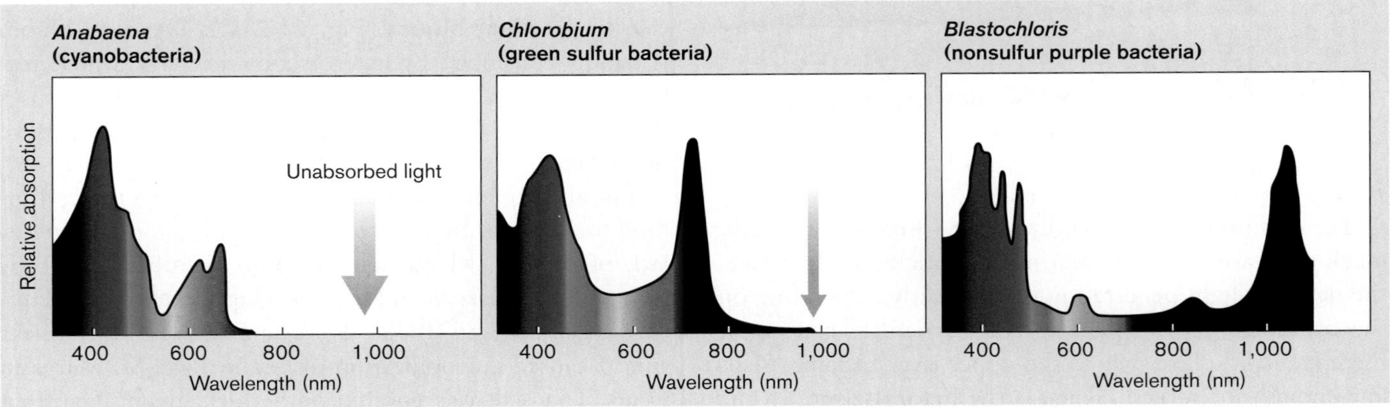
Anabaena
(cyanobacteria)
Chlorobium
(green sulfur bacteria)
Blastochloris
(nonsulfur purple bacteria)
Relative absorption
Unabsorbed light
400
600
800
1,000
Wavelength (nm)

FIGURE 21.22 ■ Absorbance spectra of chlorophyll from lake phototrophs. In the upper waters, algae and cyanobacteria absorb primarily blue and red. Below, where red has been absorbed by microorganisms above, anaerobic phototrophs such as *Blastochloris* absorb infrared (wavelengths beyond 750 nm).

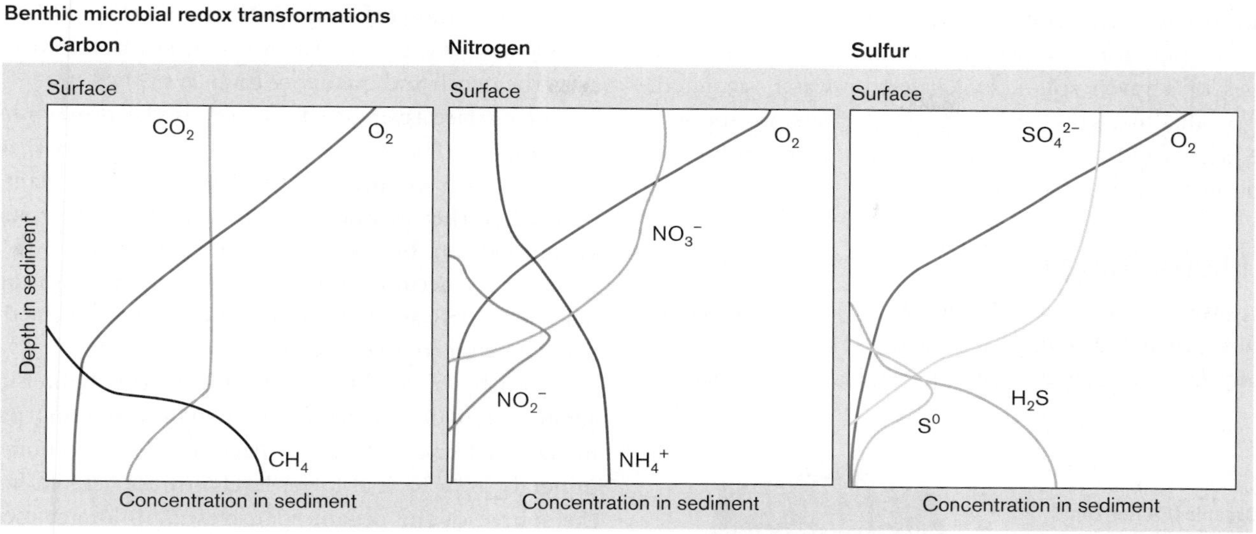
Benthic microbial redox transformations
Carbon
Nitrogen
Sulfur
Surface
CO_2
O_2
CH_4
Depth in sediment
Concentration in sediment
NO_3^-
NO_2^-
NH_4^+
SO_4^{2-}
S^0
H_2S

FIGURE 21.23 ■ **Redox gradients in the benthic sediment.** At the top of the sediment interface with the water column, minerals are first oxidized lithotrophically by O_2; then, as O_2 declines, the oxidized minerals are used as alternative electron acceptors for anaerobic respiration.

Throughout the water column, protists consume algae and bacteria, while fungi decompose detritus. They also interact with invertebrates and fish as parasites. Other important consumers are the viruses, which lyse about half of microbial populations in lakes, as they do in the ocean. Viruses limit the number of microbes and keep the water clear enough for light to penetrate.

To Summarize

- **The euphotic zone of the ocean is the upper part of the water column that receives light for phototrophs.** Below, in the aphotic zone, only heterotrophs and lithotrophs can grow. The benthos includes the region where the water column meets the ocean floor, as well as sediment below the surface.
- **Most marine microbes cannot be cultured or require special culturing methods.** Some can be cultured using conditions that mimic their natural habitat, and allow intimate association with other species that provide growth factors such as siderophores.
- **Plankton are small floating organisms, including swimming microbes.** Phytoplankton are phototrophs such as cyanobacteria and algae. Microbial consumers include protists and viruses. Many marine protists are mixotrophs (producers and consumers in one).
- **Picoplankton include bacteria and small microbial eukaryotes.** They are measured by fluorescence microscopy. Their biochemical rate of production is measured by uptake of radiolabeled nutrients.
- **Benthic microbes are barophiles.** The seafloor supports psychrophiles, whereas hydrothermal vents support thermophiles. Vents and cold seeps support sulfur-oxidizing bacteria, sulfur-reducing bacteria, methanogens, and methanotrophs. Bacteria that oxidize H_2S and methane feed symbiotic animals such as tube worms.
- **Aquatic freshwater lakes have stratified water levels.** As depth increases, minerals become increasingly reduced. Anaerobic forms of metabolism predominate, with the more favorable alternative electron acceptors used in turn.
- **Lakes may be oligotrophic or eutrophic.** Eutrophic lakes may show such high biochemical oxygen demand (BOD) that the oxygen concentration falls to levels too low to support vertebrate life.

21.5

Soil and Subsurface Microbes

In contrast to the ocean, where the base of producers is almost entirely microbial, the major producers of terrestrial ecosystems are macroscopic plants. Plants vary greatly in size and form, from mosses and bryophytes to prairie grasses and forest trees, but most terrestrial plants are rooted in **soil**. Soil is a complex mixture of decaying organic and mineral matter that feeds vast communities of microbes—arguably the most complex microbial ecosystem on our planet. And soil-based agriculture is the major source of food for our planet's human inhabitants. The qualities of a given soil—oxygenated or water saturated, acidic or alkaline, salty or fresh, nutrient-rich or -poor—define what food can be grown and whether the human community will eat or starve.

Soil Microbiology

The general structure of soil (**Fig. 21.24**) includes a series of layers called "horizons" that arise as a result of soil-forming factors such as rainfall, temperature variation, wind, and biological activity. Note that the soils of different habitats, such as prairie, forest, and desert, vary greatly as to the depth and quality of each layer.

The surface layer of soil we see is the organic horizon (O horizon). The organic horizon consists of dark, organic detritus, such as shreds of leaves fallen from plants. The detritus of the organic horizon is in the earliest stages of decomposition by microbes, primarily fungi and bacteria such as actinomycetes. Early-stage decomposition is defined loosely as a state in which the origin of the detritus may be still recognizable.

Beneath the organic horizon lies the lighter-colored aerated horizon (A horizon), in which organic particles in more advanced stages of decomposition combine with minerals from rock at lower levels. In the aerated horizon, the source of the organic particles is no longer recognizable, and decomposers have broken down some of the more difficult-to-digest plant structural components, such as lignin (a complex aromatic polymer found in wood; discussed in the next section). This partly decomposed material is often sold by garden stores as peat or topsoil.

In well-drained soil, both the organic and aerated horizons are full of oxygen, as well as nutrients liberated by the decomposers and used by plants. Soil consists of a complex assemblage of organic and inorganic particles (**Fig. 21.25**).

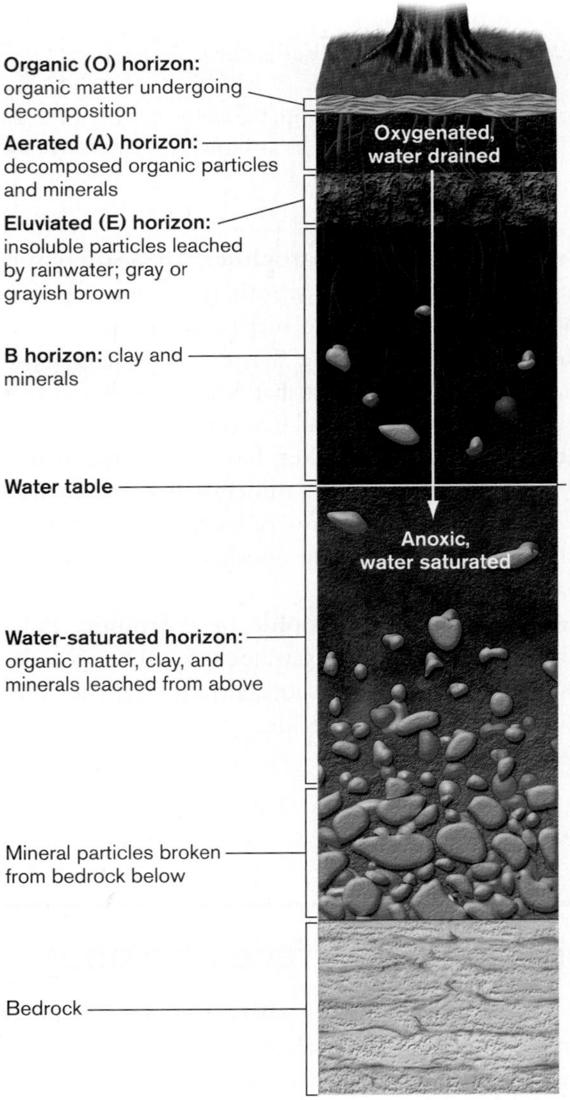
Organic (O) horizon: organic matter undergoing decomposition
Aerated (A) horizon: decomposed organic particles and minerals
Eluviated (E) horizon: insoluble particles leached by rainwater; gray or grayish brown
B horizon: clay and minerals
Water table
Water-saturated horizon: organic matter, clay, and minerals leached from above
Mineral particles broken from bedrock below
Bedrock
Oxygenated, water drained
Anoxic, water saturated

FIGURE 21.24 ■ The soil profile. Soil forms layers in which decomposing organic material predominates at the top, and minerals toward the bottom, at bedrock. The top layers are aerated, providing heterotrophs with access to O_2, whereas the bottom layers are water saturated and anaerobic.

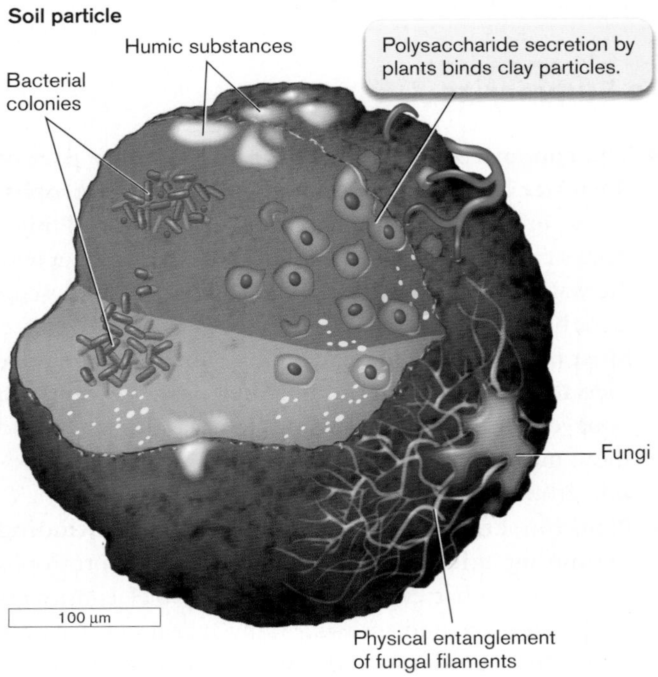
Soil particle
Humic substances
Polysaccharide secretion by plants binds clay particles.
Bacterial colonies
Fungi
100 μm
Physical entanglement of fungal filaments

FIGURE 21.25 ■ Microbes in soil and rock. Soil particles support growth of complex assemblages of microbes. A soil particle contains bacterial colonies, biofilm associations, and microbes associated with fungi and plant roots.

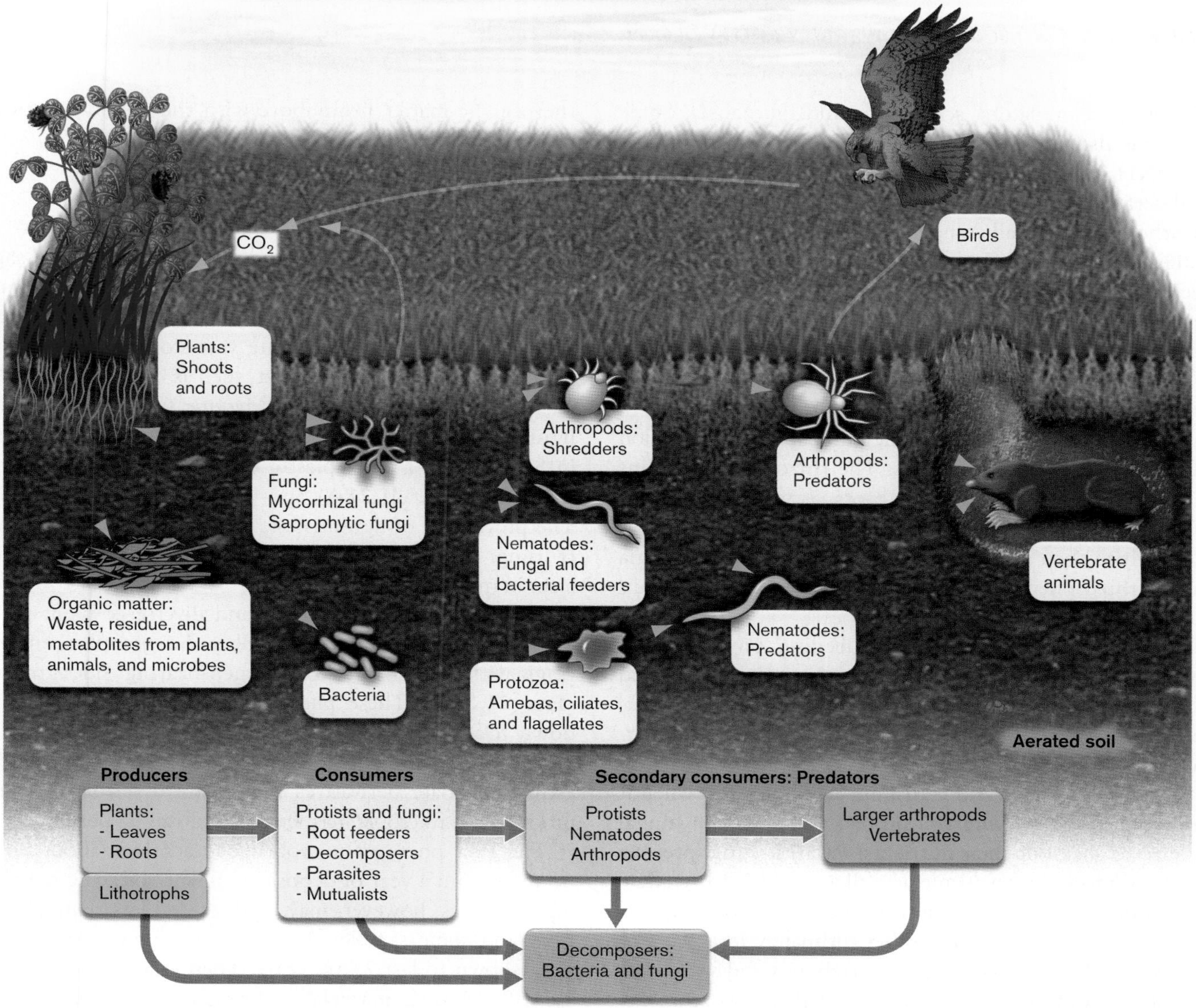
CO2
Birds
Plants:
Shoots
and roots
Arthropods:
Shredders
Arthropods:
Predators
Fungi:
Mycorrhizal fungi
Saprophytic fungi
Nematodes:
Fungal and
bacterial feeders
Vertebrate
animals
Organic matter:
Waste, residue, and
metabolites from plants,
animals, and microbes
Nematodes:
Predators
Bacteria
Protozoa:
Amebas, ciliates,
and flagellates
Aerated soil
Producers
Consumers
Secondary consumers: Predators
Plants:
- Leaves
- Roots
Lithotrophs
Protists and fungi:
- Root feeders
- Decomposers
- Parasites
- Mutualists
Protists
Nematodes
Arthropods
Larger arthropods
Vertebrates
Decomposers:
Bacteria and fungi

FIGURE 21.26 ▪ The soil food web (aerated zone). Plants are the major producers, although some production also occurs from lithotrophs such as ammonia oxidizers. Detritus from plants is decomposed by fungi and bacteria, which feed protists and small invertebrates such as nematodes. Protists and small invertebrates are consumed by larger invertebrates and vertebrate animals.

Between the soil particles are air spaces that provide access to oxygen, allowing aerobic respiration. Each particle of soil supports miniature colonies, biofilms, and filaments of bacteria and fungi that interact with each other and with the roots of plants.

Below the aerated horizon, the eluviated horizon (E horizon) experiences periods of water saturation from rain (see **Fig. 21.24**). Rainwater leaches (dissolves and removes) some of the organic and mineral nutrients from the upper layers. Below the eluviated horizon lie increasing proportions of minerals and rock fragments broken off from bedrock below. These lower water-saturated layers form the **water table**. This anoxic, water-saturated region contains mainly lithotrophs and anaerobic heterotrophs.

The soil layers finally end at bedrock, a source of mineral nutrients such as carbonates and iron. Interestingly, bedrock is permeated with microbes. Core samples show that crustal rock as deep as 3 km down contains **endoliths**, bacteria growing between crystals of solid rock. What energy source feeds microbes trapped within rock? For some endoliths, a surprising answer may be the radioactive decay of uranium. Uranium-238 decay generates hydrogen radicals that combine to form hydrogen gas. The hydrogen gas combines with CO_2 from carbonate rock, providing an electron donor and carbon source for methanogens and other endolithic lithotrophs.

The Soil Food Web

The top horizons of soil feature a food web of extraordinary complexity (**Fig. 21.26**). The major producers are green plants, whose leaves generate detritus and whose root systems feed predators, scavengers, and mutualists. Some carbon is also fixed by lithotrophs oxidizing reduced nitrogen (NH_3), hydrogen sulfide (H_2S), and iron (Fe^{2+}). In well-aerated soil, however, the proportion of carbon fixed by lithotrophs is small compared to that fixed by plants.

Plant matter is decomposed by many species of fungi, such as *Mycena* species, and by bacteria such as the actinomycetes. The actinomycetes include *Streptomyces,* a genus famous for the production of antibiotics and for generating chemicals whose odors give soil its characteristic smell (**Fig. 21.27A**). Besides fallen leaves, another source of organic matter from plants is the **rhizosphere**, the region of soil surrounding plant roots. The rhizosphere contains proteins and sugars released by roots, as well as sloughed-off plant cells. These materials feed large numbers of bacteria, which then cycle minerals back to the plant. Bacteria in the rhizosphere may also discourage growth of plant pathogens.

In the aerated zone, bacteria feeding on leaf detritus and root exudates are then preyed on by protists and nematodes. Many complex interactions occur. The nematode *Heterorhabditis bacteriophora* (**Fig. 21.27B**) carries symbiotic *Photorhabdus luminescens* bacteria, which it alternately consumes and transmits as a mutualist when it infects an insect. *Vampirella* protists drill holes in fungal hyphae to suck out their nutrients. Parasitic fungi prey on plants or invertebrates; some actually capture and strangle nematodes. Mycorrhizal fungi extend the absorptive surface area of plant roots. Microbes ultimately feed invertebrates, which then feed larger invertebrates and vertebrate predators. Some predators, such as earthworms and burrowing animals, enhance the soil quality by turning over the matter, aerating the soil particles and helping to mix the organic matter from above with the mineral particles from below.

Decomposition of Lignin to Humus

A critical role of fungal decomposers (also known as **saprophytes**) is the breakdown of extremely complex structural components of vascular plants such as grass and trees. Trees, in particular, accumulate vast stores of biomass in forms that are difficult to digest, such as lignin. Fungal and bacterial decomposers possess enzyme systems to degrade lignin and other complex components of plants (presented in Section 13.7). Examples of decomposers include white rot fungi (**Fig. 21.28A**) and actinomycete soil bacteria. The prevalence of lignin is one reason that decomposition by fungi plays a much larger role in terrestrial ecosystems than in marine ecosystems.

Lignin is a highly complex and diverse covalent polymer composed of interlinked phenolic groups (benzene rings with OH or related oxygen-bearing side groups) (**Fig. 21.28B**). These phenolic polymers can be broken down relatively rapidly to smaller units, some containing only a single phenol or benzoic acid. One of the most interesting benzoate derivatives is vanillin, a flavor molecule produced by microbial fermentation of vanilla beans.

The first phase of microbial degradation (about 50%) occurs within a year of deposition in the soil. The remaining phenolics, however, may be degraded less than 5% per year; and some samples dated by ^{14}C isotope ratios have been shown to last 2,000 years. These phenolic molecules are called **humic material** or **humus**. Because of its slow degradation, humic material provides a steady slow-release supply of nutrients for plant growth. But forests whose rate

A. *Streptomyces griseus*

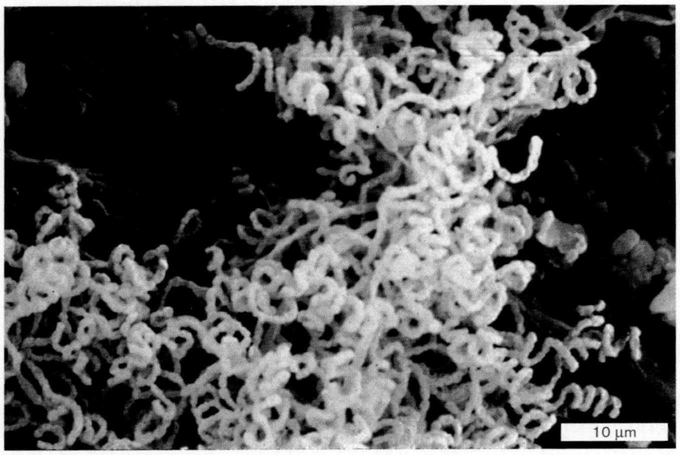
10 µm

REPRINTED FROM *THE SOIL BIOLOGY PRIMER* WITH PERMISSION. ©2000 SOIL AND WATER CONSERVATION SOCIETY

B. Nematode with symbiotic bacteria

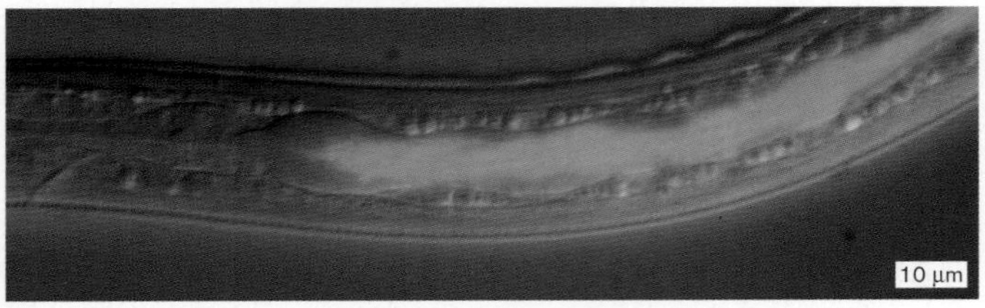
10 µm

HTTP://WWW.WORMBOOK.ORG/CHAPTERS/WWW_GENOMESHBACTERIOPHORA/GENOMESHBACTERIOPHORA.HTML

FIGURE 21.27 ■ Soil microbes. A. Hyphae of actinomycete bacteria, *Streptomyces griseus* (filament width 1 µm, SEM). *Streptomyces* bacteria give the soil its characteristic odor. **B.** The nematode *Heterorhabditis bacteriophora* carries luminescent *Photorhabdus luminescens* bacteria, which it alternately consumes and transmits to a host insect.

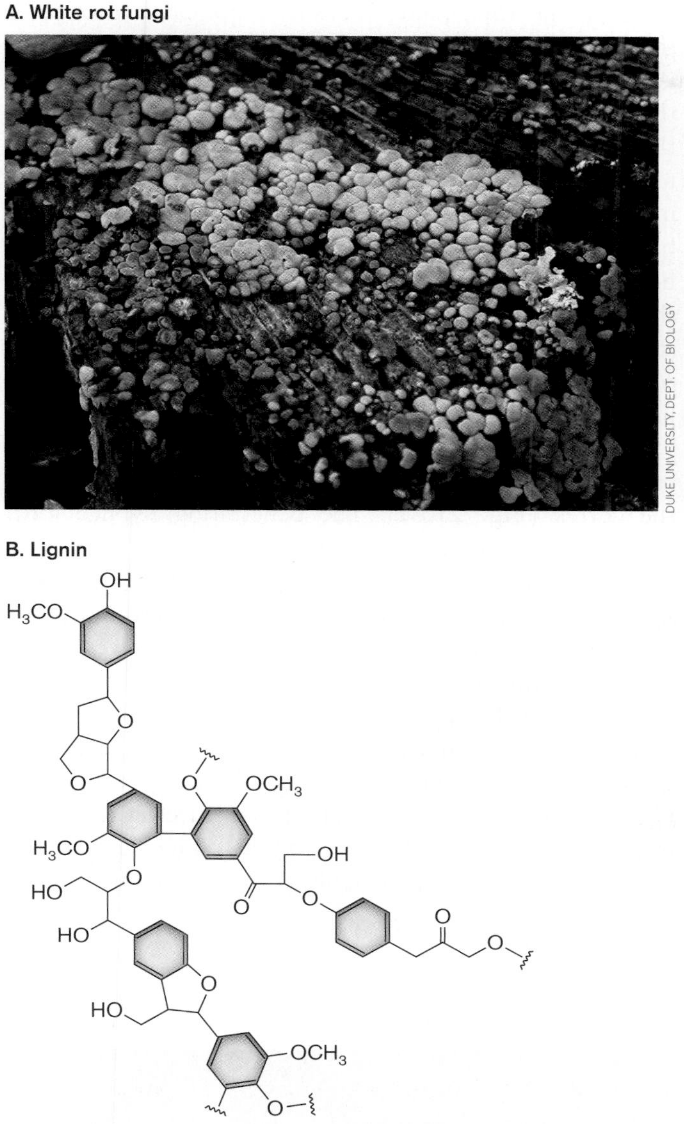
A. White rot fungi
DUKE UNIVERSITY, DEPT. OF BIOLOGY
B. Lignin

FIGURE 21.28 ■ Microbial formation of humus. A. *Xylobolus frustulatus*, a white rot fungus, growing on a willow log. The fungus degrades lignin. **B.** Lignin is a complex organic polymer that is a component of wood and bark—one of the most abundant polymers on Earth.

of microbial decomposition is particularly low—for example, the New Jersey Pine Barrens—depend on fire to clear the mounting layers of humus and return its minerals to the ecosystem.

Lignin is also a major constituent of newsprint and other kinds of paper disposed of by composting or in landfills. Because of the low degradation rate, particularly in anoxic soil, newsprint in landfills can maintain its structure for decades. Archaeology of landfills outside New York City reveals readable headlines from papers dating to the 1940s.

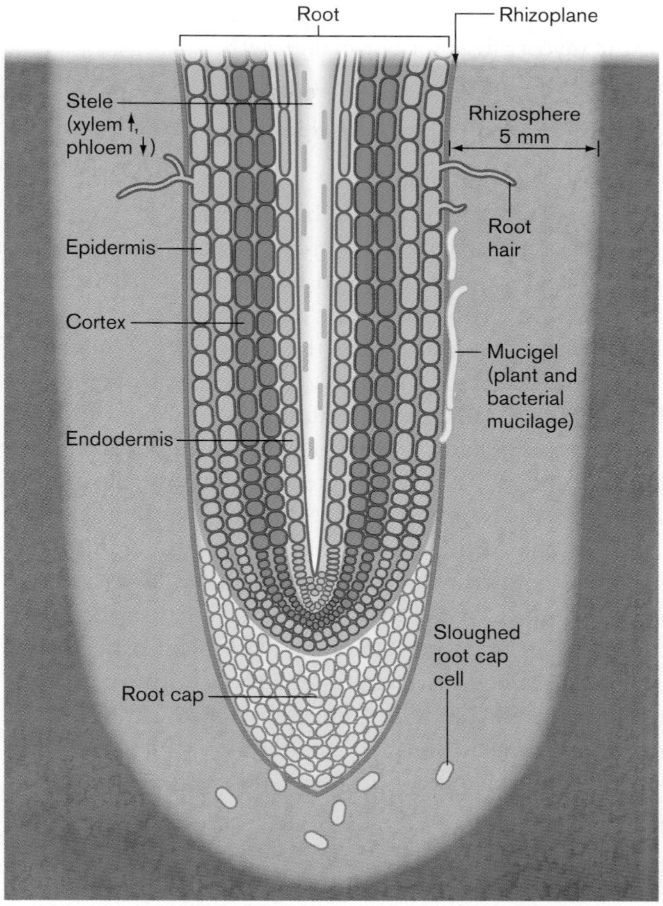
Root
Rhizoplane
Stele
(xylem ↑,
phloem ↓)
Rhizosphere
5 mm
Root
hair
Epidermis
Cortex
Mucigel
(plant and
bacterial
mucilage)
Endodermis
Sloughed
root cap
cell
Root cap

FIGURE 21.29 ■ Plant roots offer special habitats for microbes. The rhizoplane is the region of soil directly contacting the plant root surface. The rhizosphere is the soil outside the rhizoplane that receives substances from the root, such as mucilage, sloughed cells, and exudates.

Microbes Associated with Roots

The presence of plant roots provides yet another level of complexity to soil communities (**Fig. 21.29**). Plant roots influence the surrounding soil by taking up nutrients and by secreting organic substances and molecules that modulate their surroundings. The environment surrounding a plant root can be further subdivided into two categories: the **rhizoplane**, the root surface; and the **rhizosphere**, the region of soil outside the root surface but still influenced by plant exudates (materials secreted by the plant). Particular bacterial species are adapted to these environments. For example, in anaerobic wetland soil, the rhizoplane and rhizosphere of plant roots provide oxygen for methanotrophs that oxidize methane produced by methanogens.

The rhizoplane and rhizosphere also provide the environment for symbiotic fungi that generate mycorrhizae. At least 80% of plants in nature, including 90% of forest trees, require mycorrhizae for optimal growth.

Mycorrhizae: The Fungal Internet

The function and significance of mycorrhizae for plant growth is just beginning to be understood. **Mycorrhizae** (singular, **mycorrhiza**; from *myco*, "fungal," and *rhiza*, "root") consist of fungal mycelia that associate intimately with the roots of plants, extending access to minerals while obtaining in return the energy-rich products of plant photosynthesis. Mycorrhizae were first discovered in the 1880s by German truffle hunters who sought to cultivate the prized delicacy, the fruiting body of an ascomycete (for review of fungi, see Chapter 20). The propagation of truffles was investigated by mycologist Albert Frank at the Landwirtschaftliche Hochschule Berlin. To Frank's surprise, he found that the truffles extended their mycelia far beyond the site of the fruiting body, and that the mycelia formed an impenetrable tangle with plant roots. Frank called the tangled mycelia "fungus-roots" or mycorrhizae. A century later, we are beginning to appreciate that these mysterious fungal-root tangles offer a vast interconnected network for exchange of nutrients among fungi and many different plants, like an internet connecting countless sites.

Two different kinds of mycorrhizae are observed: ectomycorrhizae and endomycorrhizae. **Ectomycorrhizae** colonize the rhizoplane, the surface of plant rootlets—the most distal part of plant roots (**Fig. 21.30**). The fungal mycelia never penetrate the root cells. They form a thick mantle surrounding the root and growing between the root cells, then extend long mycelia away from the root to absorb nutrients. Numerous kinds of fungi form ectomycorrhizae, including ascomycetes, such as truffles; and basidiomycetes, known by their mushrooms, such as stinkhorns. Plants grown with ectomycorrhizae invest less of their body mass in roots and more in the aboveground stems and leaves—an important consideration for agriculture, in which the aboveground plant is usually the part harvested.

Endomycorrhizae form a more intimate association, in which the fungal hyphae penetrate plant cells deep within the cortex (**Fig. 21.31**). The penetrating hyphae form knobbed branches that resemble microscopic "trees," or arbuscules, within the root cells. Some of the hyphae form specialized vesicles within the plant that store nutrients. Another name for this kind of mycorrhizae is **vesicular-arbuscular mycorrhizae (VAM)**.

Endomycorrhizae are more specialized than ectomycorrhizae. They comprise a relatively small number of fungal species, such as members of the Glomeromycota genus *Glomus*, and they show obligate dependence on their host plants. Their presence in nature and their importance in the ecosystem, however, are actually greater. Endomycorrhizal

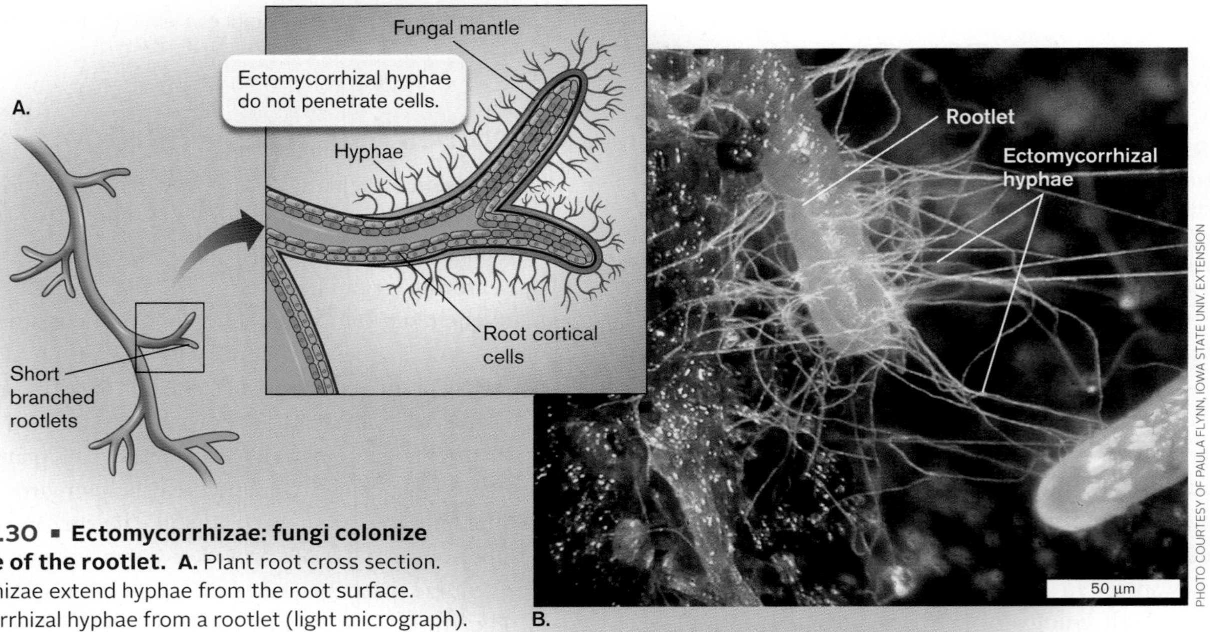
A.
Fungal mantle
Ectomycorrhizal hyphae do not penetrate cells.
Hyphae
Root cortical cells
Short branched rootlets
Rootlet
Ectomycorrhizal hyphae
50 µm
B.
PHOTO COURTESY OF PAULA FLYNN, IOWA STATE UNIV. EXTENSION

FIGURE 21.30 ▪ Ectomycorrhizae: fungi colonize the surface of the rootlet. A. Plant root cross section. Ectomycorrhizae extend hyphae from the root surface. **B.** Ectomycorrhizal hyphae from a rootlet (light micrograph).

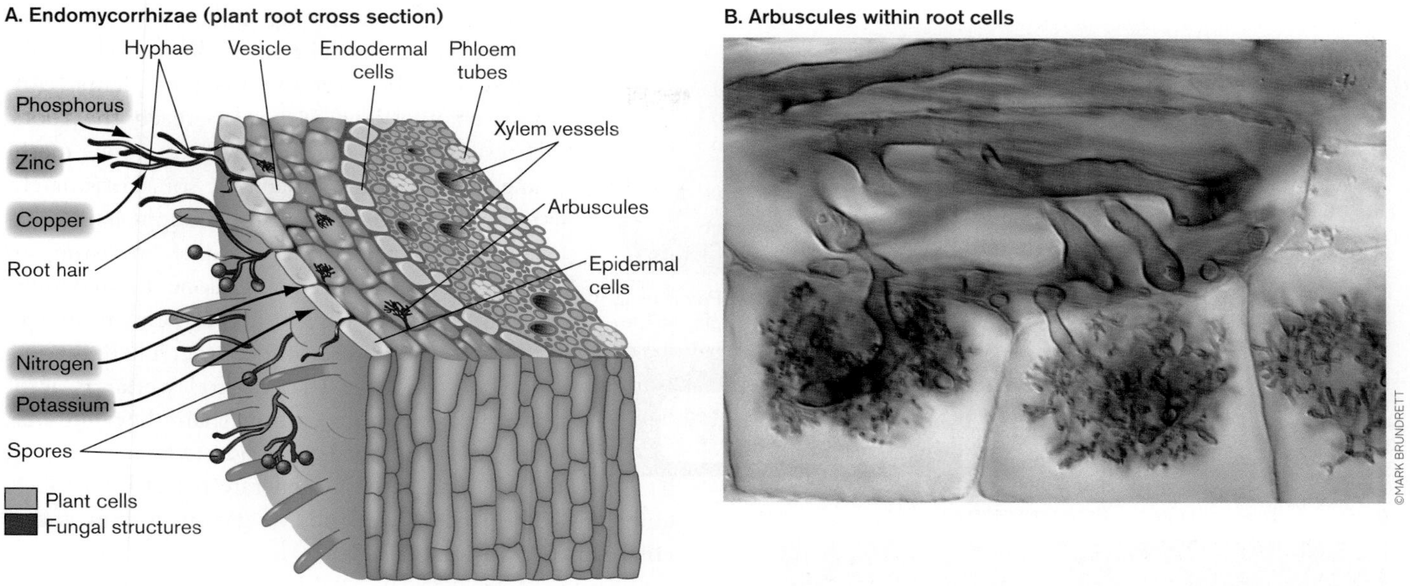
A. Endomycorrhizae (plant root cross section)
Hyphae
Vesicle
Endodermal cells
Phloem tubes
Xylem vessels
Arbuscules
Epidermal cells
Phosphorus
Zinc
Copper
Root hair
Nitrogen
Potassium
Spores
Plant cells
Fungal structures
B. Arbuscules within root cells
©MARK BRUNDRETT

FIGURE 21.31 ▪ Endomycorrhizae: fungi invade root cells, forming arbuscules. **A.** Plant root cross section. Endomycorrhizal hyphae penetrate cells deep within the root cortex. **B.** The penetrating hypha forms an arbuscule within a root cell.

species exist entirely underground (they do not form mushrooms), and they completely lack sexual cycles. They may acquire 25% of the photosynthetic product of their hosts in exchange for tremendous expansion of access to soil resources.

Mycorrhizae greatly enhance the plant's uptake of water, as well as minerals such as nitrogen and phosphorus. In addition, the hyphae sequester toxins, and they actually distribute organic substances from one plant to another. Mycorrhizae may join many different plants, even different species, in a vast, nutrient-sharing network.

Thought Question

21.5 Design an experiment to test the hypothesis that the presence of mycorrhizae enhances plant growth in nature.

Wetland Soils

So far, we have considered the interface between ground and water (the benthic sediment beneath oceans and lakes) and the interface between ground and air (aerated soil). An interesting case that combines the two is wetlands (**Fig. 21.32A**). A **wetland** is defined as a region of land that undergoes seasonal fluctuations in water level, so that sometimes the land is dry and oxygenated, and at other times water saturated and anaerobic. Wetlands provide many crucial functions; for example, the Everglades filter much of the water supply for Florida communities.

Wetland soil that undergoes such periods of anoxic water saturation is known as **hydric soil**. Hydric soil is characterized by "mottles," patterns of color and paleness (**Fig. 21.32B**). The reddish brown portions (for example, surrounding a plant root) result from oxidized iron (Fe^{3+}). The gray portions indicate loss of iron in its water-soluble reduced form (Fe^{2+}) generated by anaerobic respiration.

Soil becomes anoxic when the rate of oxygen diffusion is too low to support aerobic metabolism. Anaerobic metabolism allows much lower rates of production than metabolism in the presence of oxygen because anaerobes use oxidants of lower redox potential and limited quantity, such as sulfate and nitrate. Many kinds of anaerobic bacteria inhabit wetlands. For example, denitrifiers (bacteria using nitrate to oxidize organic food) remove nitrate from water before it enters the water table—one of the ways that wetlands protect our water supply.

The alternative oxidants (electron acceptors) are always in limited supply, so further catabolism occurs by fermentation. Fermentation allows only incomplete breakdown of food molecules, generating a rich and diverse supply of nutrients for a variety of consumers, including aerobic organisms when the water recedes. Thus, despite its lower overall productivity, anaerobic soil contributes to

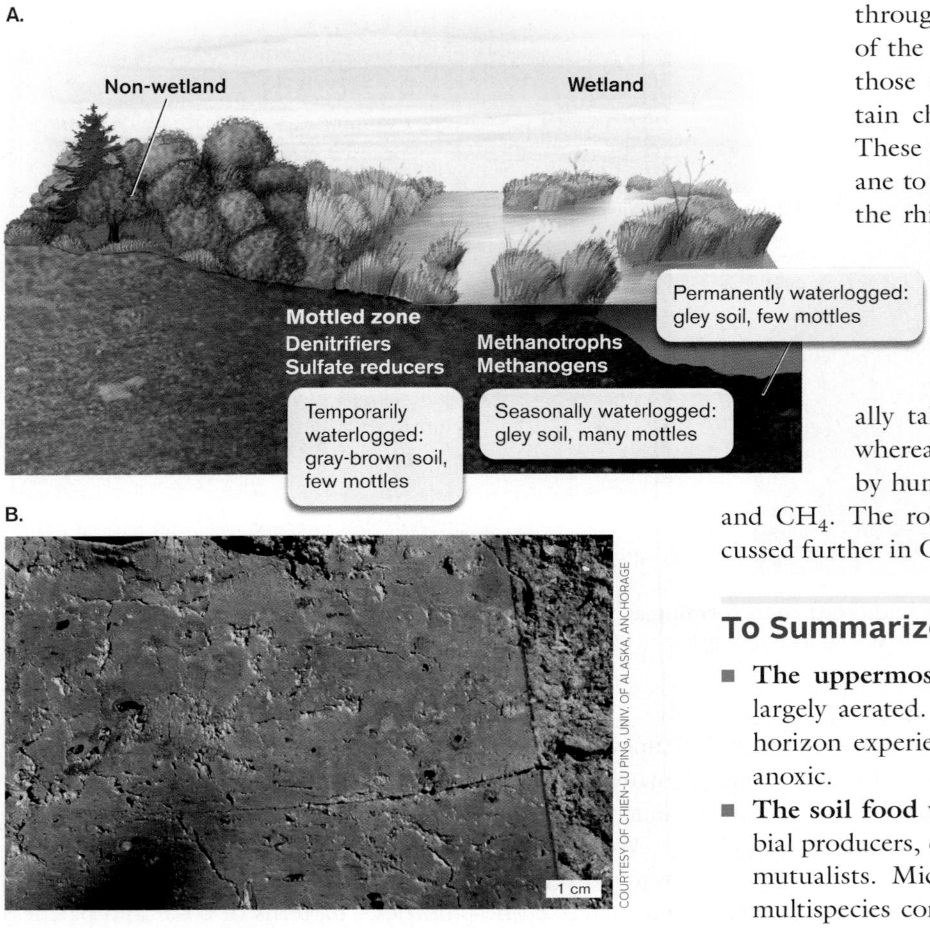
A.
Non-wetland
Wetland
Permanently waterlogged: gley soil, few mottles
Mottled zone
Denitrifiers
Sulfate reducers
Methanotrophs
Methanogens
Temporarily waterlogged: gray-brown soil, few mottles
Seasonally waterlogged: gley soil, many mottles
B.
1 cm
COURTESY OF CHIEN-LU PING, UNIV. OF ALASKA, ANCHORAGE

FIGURE 21.32 ▪ Wetland soil. A. A true wetland experiences periods of water saturation alternating with dry soil. Soil that is water saturated becomes anaerobic because O_2 diffuses slowly through water. **B.** Alternating periods of water saturation and dryness give the soil a mottled color. The red-orange-colored portions around root holes result from oxidized iron (Fe^{3+}), whereas the gray portions ("gley") indicate that water has washed iron away after reduction (Fe^{2+}).

the nutritional diversity of wetland ecosystems. The relatively slow rate of decomposition can lead to accumulation of high levels of organic carbon, particularly rich for plant growth.

The anoxic conditions of wetlands also favor methanogenesis. Methanogenesis is performed solely by archaea (discussed in Chapters 14 and 19). Methanogenesis occurs when fermenting bacteria generate H_2, CO_2, and other one- or two-carbon substrates that methanogens convert to methane. Methane is a more potent greenhouse gas than CO_2, and although the current methane concentration in our atmosphere is low, it is rising exponentially.

In water-filled anaerobic soils, particularly those of rice fields, significant quantities of methane escape to the air through air-conducting channels in the roots of the rice plant. The roots of rice plants, like those of other wetland vascular plants, contain channels for oxygen to reach the roots. These same channels, however, allow methane to escape from anaerobic soil. Fortunately, the rhizosphere of the roots supports methanotrophs that oxidize methane, so research is being done to maximize methanotroph activity and minimize the release of methane. There is some evidence that natural wetlands actually take in more carbon than they put out, whereas disturbed wetlands (wetlands altered by human activity) generate net efflux of CO_2 and CH_4. The role of wetlands in global cycling is discussed further in Chapter 22.

To Summarize

- **The uppermost horizons of soil** consist of detritus, largely aerated. Below the aerated layers, the eluviated horizon experiences water saturation. Lower layers are anoxic.
- **The soil food web** includes a complex range of microbial producers, consumers, predators, decomposers, and mutualists. Microbial mats and biofilms are complex multispecies communities of bacteria, fungi, and other microbes.
- **Fungi decompose lignin**, a complex aromatic tree component that is challenging to digest. Lignin decomposition forms humus.
- **Fungi form symbiotic associations** with plant roots called mycorrhizae. Mycorrhizae transport soil nutrients among many different kinds of plants.
- **Wetland soils** alternate aerated (dry) with anoxic (water-saturated) conditions. Anoxic wetland soil favors methanogenesis. Wetland soils are among the most productive ecosystems.

21.6

Plant Microbial Communities

Plant tissues provide unique habitats for microbes. The microbial partners may grow as mutualists, commensals, or parasites. Plants coevolve with their associated microbes, some of which can grow nowhere else. In other cases, microbes evolve alternative lifestyles—for example, as helpful endosymbionts of plants but as pathogens of animals. When we eat plants, we also consume all their endophytic microbes.

Endophytes Grow within Plants

Plant tissues contain a complex vascular system for nutrient transport, in which phloem tubes conduct photosynthesized sugars down from the leaves, and xylem tubes bring water and minerals up from the roots. Unlike the sterile blood vessels of animals, plant transport vessels are normally colonized by endophytic microbes, or **endophytes** (**Fig. 21.33**). Endophytes include fungi and bacteria, most of which have alternative lifestyles in the soil or as pathogens of animals.

Plants tolerate endophytes because they confer substantial benefits. For example, prairie grasses called fescue, grazed by cattle in the southeastern United States, host fungal epiphytes. The fungi, *Neotyphodium coenophialum,* produce alkaloids that deter insect predators, pathogens, and root-feeding nematodes. Unfortunately, some of the alkaloids poison cattle, but agricultural scientists have engineered fungal strains that still protect the plant from pathogens while allowing cattle to graze.

Human pathogens such as *E. coli* O157:H7 and *Salmonella enterica* can grow endophytically in crop plants such as spinach and alfalfa (**Fig. 21.33A**). Endophytic pathogens pose a problem for the food industry because the bacteria cannot be "washed off" from raw produce. On the other hand, endophytes such as *Stenotrophomonas* species have potentially valuable uses. *Stenotrophomonas* bacteria produce enzymes that deter many plant pathogens. The bacteria also absorb and concentrate toxic metals such as arsenic, suggesting a possible use for bioremediation of metal-contaminated soil. They produce promising antibiotics and proteases for cleansing agents. Other kinds of endophytes protect plants from heat, salt, and drought—contributions of growing importance as our global climate changes.

Rhizobia Fix Nitrogen for Legumes

A form of bacteria-plant mutualism critical for agriculture is that of nitrogen fixation by **rhizobia** (singular, **rhizobium**), a group of soil-dwelling Alphaproteobacteria (discussed in Chapter 18). Major rhizobial genera include *Rhizobium, Bradyrhizobium,* and *Sinorhizobium.* Rhizobia, associated with legumes such as peas and beans, fix more nitrogen than the plants absorb from soil, actually increasing the soil's nitrogen content. For this reason, farmers often alternate crops such as corn with soybeans to restore nitrogen to the soil. The rhizobial bacteria develop specialized forms within plant cells, called **bacteroids**.

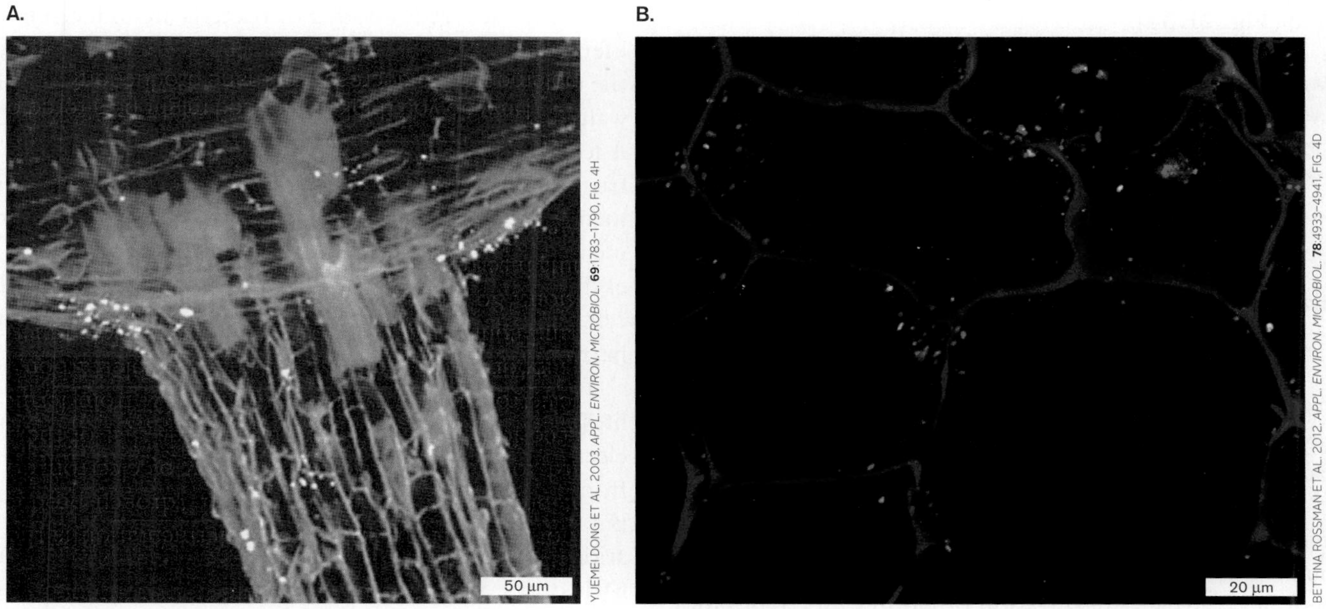
A.
50 μm
YUEMEI DONG ET AL. 2003. APPL. ENVIRON. MICROBIOL. 69:1783–1790, FIG. 4H
B.
20 μm
BETTINA ROSSMAN ET AL. 2012. APPL. ENVIRON. MICROBIOL. 78:4933–4941, FIG. 4D

FIGURE 21.33 ▪ Endophytic bacteria within plants. A. *Salmonella enterica* bacteria labeled with GFP fluorescence colonize an alfalfa root. **B.** Proteobacteria growing between cells of a banana seedling. FISH with fluorescent probes reveals Alphaproteobacteria (green), other Proteobacteria (red), and plant cell walls (violet).

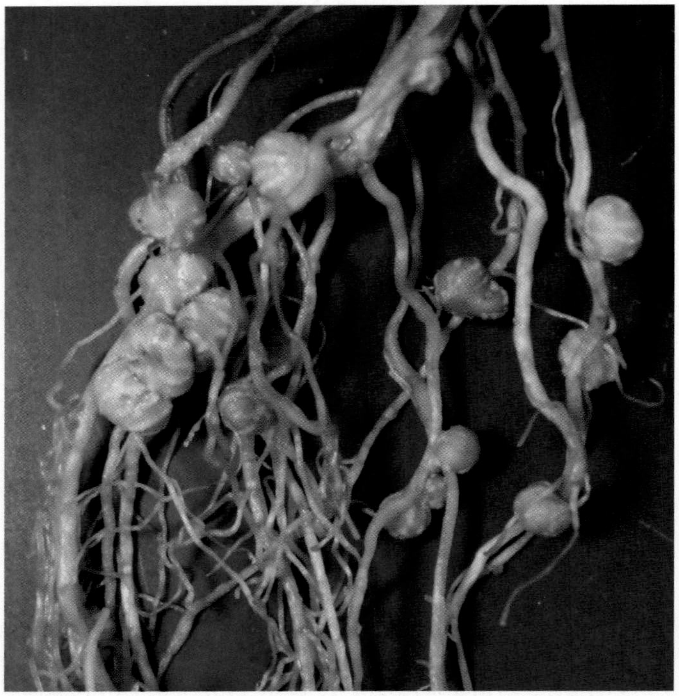

FIGURE 21.34 ■ *Rhizobium* nodules on pea plant roots. Rhizobia induce legume roots to form nitrogen-fixing nodules.

Bacteroids lack cell walls and are unable to reproduce; their function is specialized for nitrogen fixation. The rhizobial infection of root hairs induces the formation of nodules within which the nitrogen-fixing bacteroids are sequestered (**Fig. 21.34**).

Thought Question

21.6 How do you think symbiotic rhizobia reproduce? Why do bacteroids develop if they cannot proliferate?

The initiation, development, and maintenance of the rhizobia-legume symbiosis poses intriguing questions of genetic regulation. How does the association begin? How do host plant and bacterium recognize each other as suitable partners? The legume exudes signal molecules called **flavonoids** into its rhizosphere. Flavonoids resemble steroid hormones such as estrogen and have similar effects on animals; they are also called phytoestrogens. The flavonoids are detected by rhizobial bacteria, which respond by chemotaxis, swimming toward the root surface. Flavonoids then induce bacterial expression of Nod factors, molecules composed of chitin with lipid attachments. Nod factors communicate with the host plant and help establish species specificity between bacterium and host.

Thought Question

21.7 High levels of nitrate or ammonium ion corepress the expression of Nod factors (see **Figure 21.35**). What is the biological advantage of Nod regulation?

The entry of the bacteria into the host involves a fascinating interplay between bacterial and plant cells whose mechanism remains largely unknown (**Fig. 21.35A**). First, a bacterium is attracted by flavonoids to the surface of a root hair extended by a root epidermal cell (**Fig. 21.35A**, step 1). The bacterial Nod factor induces the root hair to curl around it and ultimately surround the bacterium with plant cell envelope (steps 2 and 3). The bacterium then induces growth of a tube poking into the plant cell (step 4). The tube growth is directed by the plant nucleus, which migrates toward the plant cortex (step 5). As the tube grows, bacteria proliferate, forming a column of cells that projects down the tube (steps 6 and 7). This column of cells is known as the **infection thread**. The infection thread can be visualized by light microscopy (**Fig. 21.35B**).

As the infection thread develops, signals from the bacteria induce the cortical cells (below the epidermis) to prepare to receive the bacteria. The bacteria induce further tube formation into the cortical cells and continue penetration as they grow. The penetration of the infection thread into the cortex is shown in fluorescence micrographs (**Fig. 21.35C**) in which the bacteria are engineered to express a fluorescent protein.

The cortical cells invaded by bacteria are induced to proliferate in an organized manner, forming nodules. Within the nodules, most of the infecting bacteria differentiate into wall-less bacteroids that will fix nitrogen (**Fig. 21.36A**). A few bacteria fail to differentiate; their fate is unclear. The bacteroids remain sequestered within a sac of plant-derived membrane known as the **symbiosome**. The symbiosome membrane contains special transporters that mediate the exchange of nutrients between the bacteroid and its host cell, sustaining bacteroid metabolism while preventing harm to the host.

From the plant cytoplasm, the bacteroid receives catabolites such as malate, which enter the TCA cycle and donate electrons for respiration. The oxygen for respiration comes from the plant's photosynthesis, regulated by leghemoglobin to maintain levels low enough to allow nitrogen fixation (**Fig. 21.37**). Nitrogen fixation consumes about a fifth of the plant's photosynthetic products. As discussed in Chapter 15, nitrogen is fixed by the nitrogenase enzyme into ammonium ions:

$$N_2 + 10H^+ + 8e^- \rightarrow 2NH_4^+ + H_2$$

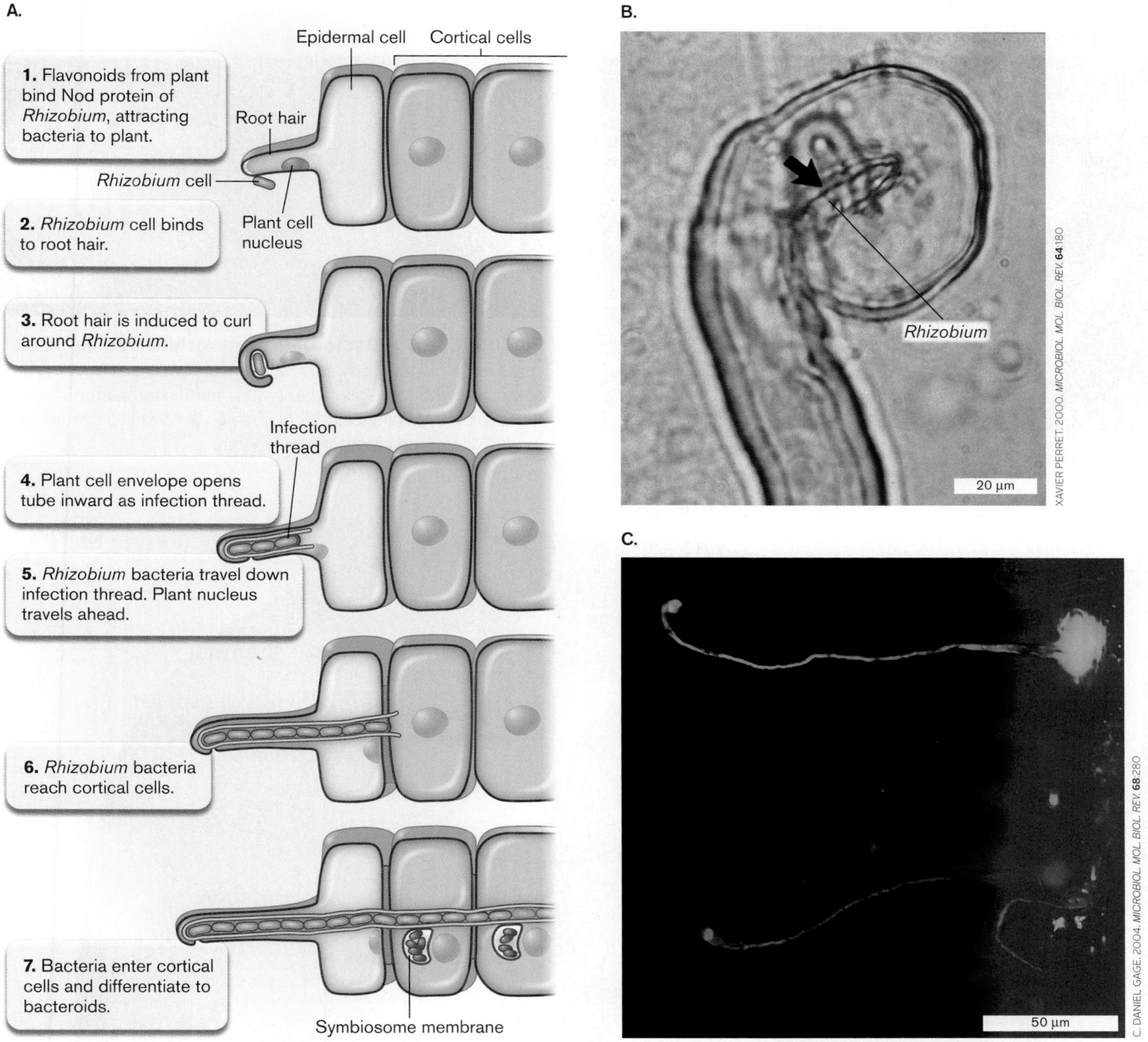
A.
Epidermal cell
Cortical cells
1. Flavonoids from plant bind Nod protein of Rhizobium, attracting bacteria to plant.
Root hair
Rhizobium cell
2. Rhizobium cell binds to root hair.
Plant cell nucleus
3. Root hair is induced to curl around Rhizobium.
Infection thread
4. Plant cell envelope opens tube inward as infection thread.
5. Rhizobium bacteria travel down infection thread. Plant nucleus travels ahead.
6. Rhizobium bacteria reach cortical cells.
7. Bacteria enter cortical cells and differentiate to bacteroids.
Symbiosome membrane
B.
Rhizobium
20 μm
XAVIER PERRET. 2000. MICROBIOL. MOL. BIOL. REV. 64:180
C.
50 μm
C. DANIEL GAGE. 2004. MICROBIOL. MOL. BIOL. REV. 68:280

FIGURE 21.35 ■ The infection thread. A. Rhizobia are attracted to the legume by chemotaxis toward exuded flavonoids. A bacterium induces an epidermal root hair to curl around it and take it up into the infection thread, a tube of plant cell wall material. The thread eventually penetrates cortical cells, where the bacteria lose their cell walls and become nitrogen-fixing bacteroids. **B.** Root hair curling around *Rhizobium,* and the formation of an infection thread (light microscopy). **C.** Infection threads invading the cortex (fluorescence microscopy). Bacteria express either DsRed (pink) or green fluorescent protein (green).

The reaction requires expenditure of 8 NADPH or NADH plus 16 ATP, which are generated by respiration. But respiration requires oxygen, which poisons nitrogenase. Thus, oxygen needs to be delivered to the bacteroid only as needed, and in an amount just enough to run respiration. The oxygen is sequestered and brought to the bacteroid by **leghemoglobin**, an iron-bearing plant protein related to blood hemoglobin. Leghemoglobin in plant cytoplasm colors the nodule pink (**Fig. 21.36B**).

Overall, the nitrogen fixation symbiosis is kept in balance by several regulatory mechanisms. The presence of ammonium or nitrate ions inhibits symbiosis and nitrogen fixation. The bacteroids cannot synthesize their own amino acids; instead, they must provide ammonium to the plant cytoplasm for assimilation into amino acids, some of which cycle back to the bacteroid. But what is known of regulation is dwarfed by the unanswered questions: How is the infection thread formed? How does the plant allow

A. Bacteroids within plant cells

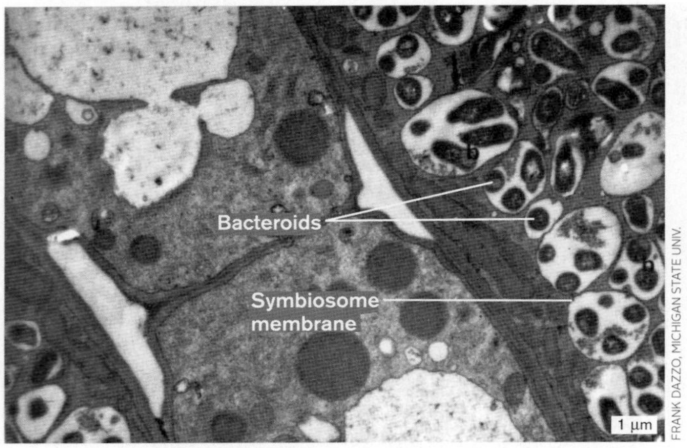
Bacteroids
Symbiosome membrane
1 μm
FRANK DAZZO, MICHIGAN STATE UNIV.

B. Nodules contain leghemoglobin

STEPHANIE BATCHELET

FIGURE 21.36 ▪ Bacteroids form within nodules. A. *Rhizobium* cells within plant cytoplasm lose their cell walls and become bacteroids, contained within the plant-generated symbiosome membrane. Bacteroid-containing cells form nodules. **B.** Within a sectioned nodule, the pink color indicates leghemoglobin.

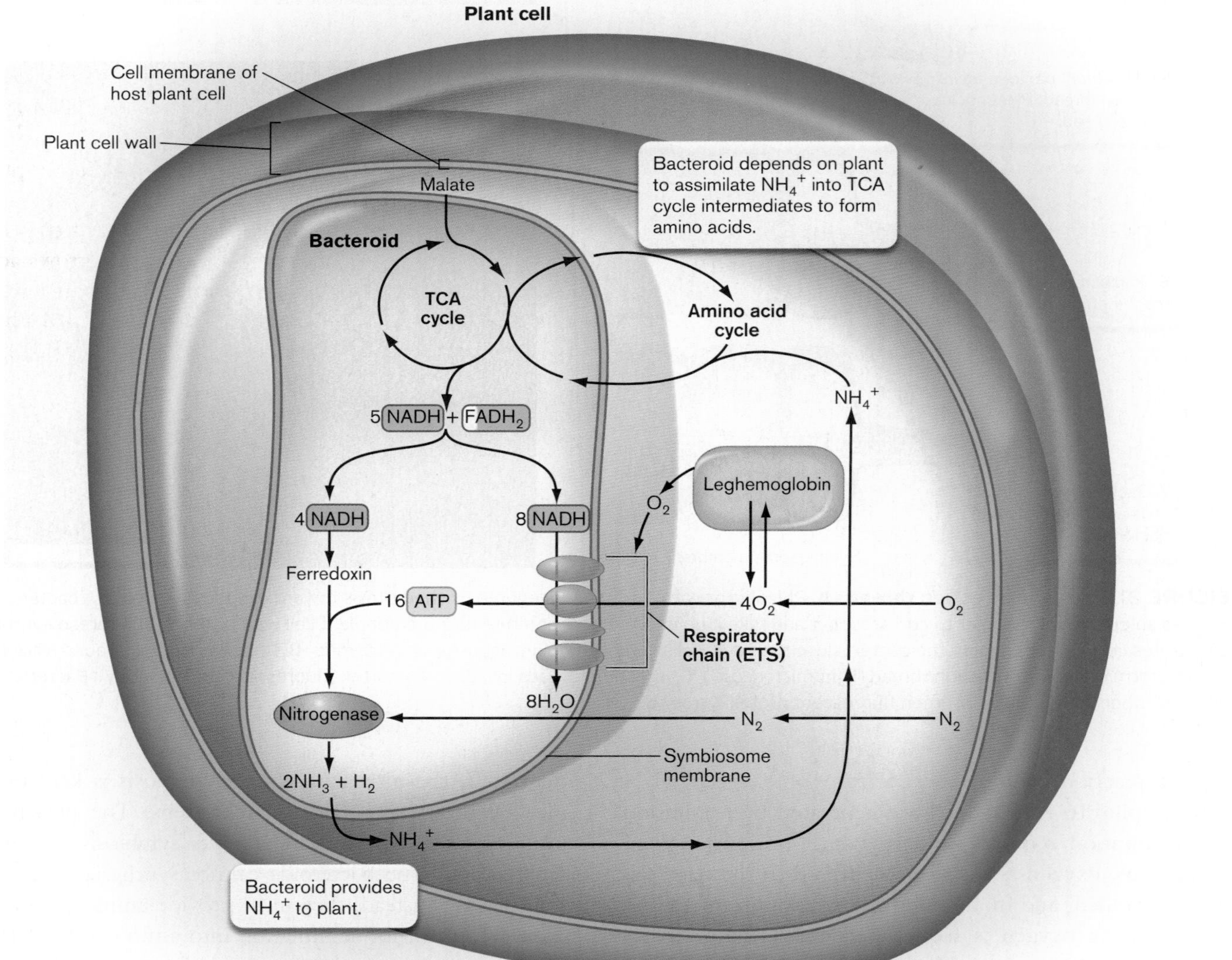
Plant cell
Cell membrane of host plant cell
Plant cell wall
Malate
Bacteroid depends on plant to assimilate NH_4^+ into TCA cycle intermediates to form amino acids.
Bacteroid
TCA cycle
Amino acid cycle
NH_4^+
5 NADH + $FADH_2$
Leghemoglobin
O_2
4 NADH
8 NADH
Ferredoxin
16 ATP
$4O_2$
O_2
Respiratory chain (ETS)
$8H_2O$
Nitrogenase
N_2
N_2
Symbiosome membrane
$2NH_3 + H_2$
NH_4^+
Bacteroid provides NH_4^+ to plant.

FIGURE 21.37 ▪ Energy and oxygen regulation during nitrogen fixation. The bacteroid receives photosynthetic products from the plant, such as malate and oxygen, to generate ATP for nitrogen fixation. The amount of oxygen is regulated closely by leghemoglobin. The bacteroid provides nitrogen (fixed as ammonium ion) to the plant cell.

A. Virus infection of tulips B. Crown gall tumor on rose stem C. Anthracnose fungus on white oak leaf

COURTESY OF KATHY NICHOLS
ALAMY
PAULA FLYNN, IOWA STATE EXTENSION SERVICE

FIGURE 21.38 ■ Plant diseases range from innocuous to devastating. **A.** Striped tulips result from a virus. **B.** Crown gall tumor on rose stem, caused by *Agrobacterium tumefaciens.* **C.** White oak leaf spotted by anthracnose fungus.

infection while preventing uncontrolled growth of bacteria? What determines how much of the plant's photosynthetic products are harvested by the bacteria? Why is hydrogen gas released, and how can this loss of potential energy be prevented? What limits the host specificity of rhizobia to legumes, and can it be extended to other crop plants, such as corn? Research on these questions is critical for agriculture.

Plant Pathogens

We have seen how many bacteria and fungi interact positively with plants, but many other species act as pathogens. In any environment, pathogens are always outnumbered by the vast community of neutral or helpful microbes. Nevertheless, when a pathogen does colonize a plant, its growth can have effects ranging from minimal to devastating (**Fig. 21.38**). A relatively harmless plant virus was associated with a famous historical phenomenon: the sixteenth-century tulip craze in the Netherlands. The virus caused streaking of tulip petals (**Fig. 21.38A**), a pattern much admired by tulip fanciers. Other viruses, however, can cause devastating blights and epidemics. (For more on viruses, see Chapters 6 and 11.)

A pathogenic relative of rhizobia is the bacterium *Agrobacterium tumefaciens,* whose DNA transforms plant cells to form crown gall tumors (**Fig. 21.38B**). *A. tumefaciens* has an unusually broad host range, and its natural genetic transformation system has been applied widely for commercial plant engineering (discussed in Chapter 16). The tumors remain largely confined and have relatively little effect on plant growth. Other bacterial pathogens, particularly species of *Erwinia* and *Xanthomonas,* severely damage plants.

The most common plant pathogens are fungi. Fungal diseases such as anthracnose (**Fig. 21.38C**) cause substantial losses in agriculture, affecting cucumbers, tomatoes, and other vegetables. Dutch elm disease, which has wiped out nearly all the native elms of the United States, is caused by the fungus *Ophiostoma novo-ulmi.* The fungus is carried by bark beetles, which bore into the xylem, damaging the plant's transport vessels and allowing access for fungal spores.

Some fungal pathogens generate specialized structures to acquire nutrients from plants. As a hypha grows across the plant epidermis, its tip can penetrate the plant cell wall, followed by ingrowth of a bulbous extension called a **haustorium** (plural, **haustoria**) (**Fig. 21.39**). The haustorium

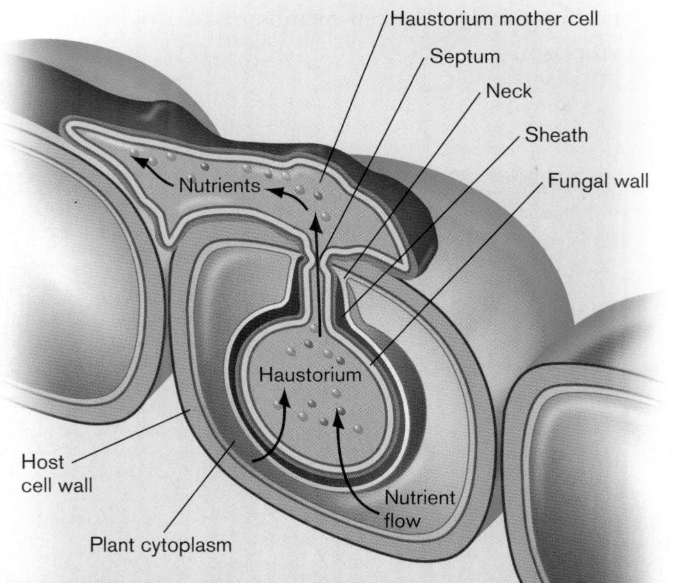
Haustorium mother cell
Septum
Neck
Sheath
Fungal wall
Nutrients
Haustorium
Host cell wall
Nutrient flow
Plant cytoplasm

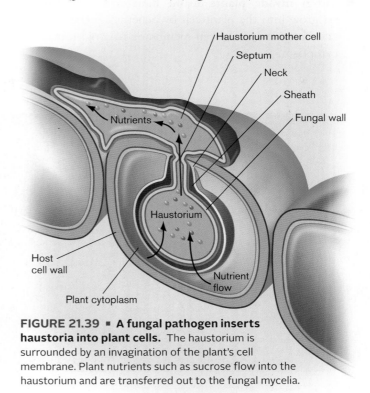

FIGURE 21.39 ■ A fungal pathogen inserts haustoria into plant cells. The haustorium is surrounded by an invagination of the plant's cell membrane. Plant nutrients such as sucrose flow into the haustorium and are transferred out to the fungal mycelia.

never penetrates the plant cell membrane, thus avoiding leakage and loss of plant cytoplasm. Instead, it causes the membrane to invaginate, while expanding into the volume of the plant cell. The haustorium takes up nutrients such as sucrose, generated by adjacent chloroplasts. Depending on the species of fungus, haustorial parasitism can lead to mild growth retardation, or it can rapidly kill the plant.

Thought Question

21.8 Compare and contrast the processes of plant infection by rhizobia and by fungal haustoria.

To Summarize

- **Endophytes** are bacteria or fungi that grow within plant transport vessels, conferring benefits such as resistance to pathogens.
- **Rhizobia induce legume roots** to nodulate for nitrogen fixation. The bacteria enter the root as infection threads. Some of the bacteria enter root cells and develop into nitrogen-fixing bacteroids.
- **Bacteroids** gain energy from plant cell respiration but must remain anaerobic. The anaerobic state is protected by the oxygen-binding protein leghemoglobin.
- **Plant pathogens** include bacteria, fungi, and viruses. Some pathogens only mildly affect the plant, whereas others cause devastation.
- **Fungi invade plants using haustoria.** The haustoria grow into the plant cell by penetrating the cell wall and invaginating the cell membrane to avoid leakage of cytoplasm.

21.7

Animal Microbial Communities

Animals, like plants, have microbial communities on their surfaces or within certain internal organs. Many microbes have beneficial effects, such as enhancing digestion or generating protective substances in the skin. A relatively small proportion of animal-associated microbes cause disease. We discuss here just a few examples of particular significance for ecology and veterinary science. Human microbiota are discussed in Chapter 23. Pathogenesis and disease in humans are discussed in Chapters 25 and 26.

Zooxanthellae in Corals and Sea Anemones

Many invertebrate animals acquire endosymbiotic algae, most commonly dinoflagellates. These endosymbionts are called **zooxanthellae** (singular, **zooxanthella**). The algae receive protection from predators, while the animal receives photosynthetic products. Examples include anemones, clams, and corals (**Fig. 21.40A**). Coral zooxanthellae are extremely important to the biosphere because healthy coral is required for reef formation and much of the biological productivity of coastal shelf ecosystems. The slight rise in temperature that has occurred from global warming has already caused severe problems with **coral bleaching**, in which the algal symbionts die or are expelled. The coral turns white and soon dies, unless its symbionts return.

The most common algal partners are members of the dinoflagellate genus *Symbiodinium* (**Fig. 21.40B**). A given

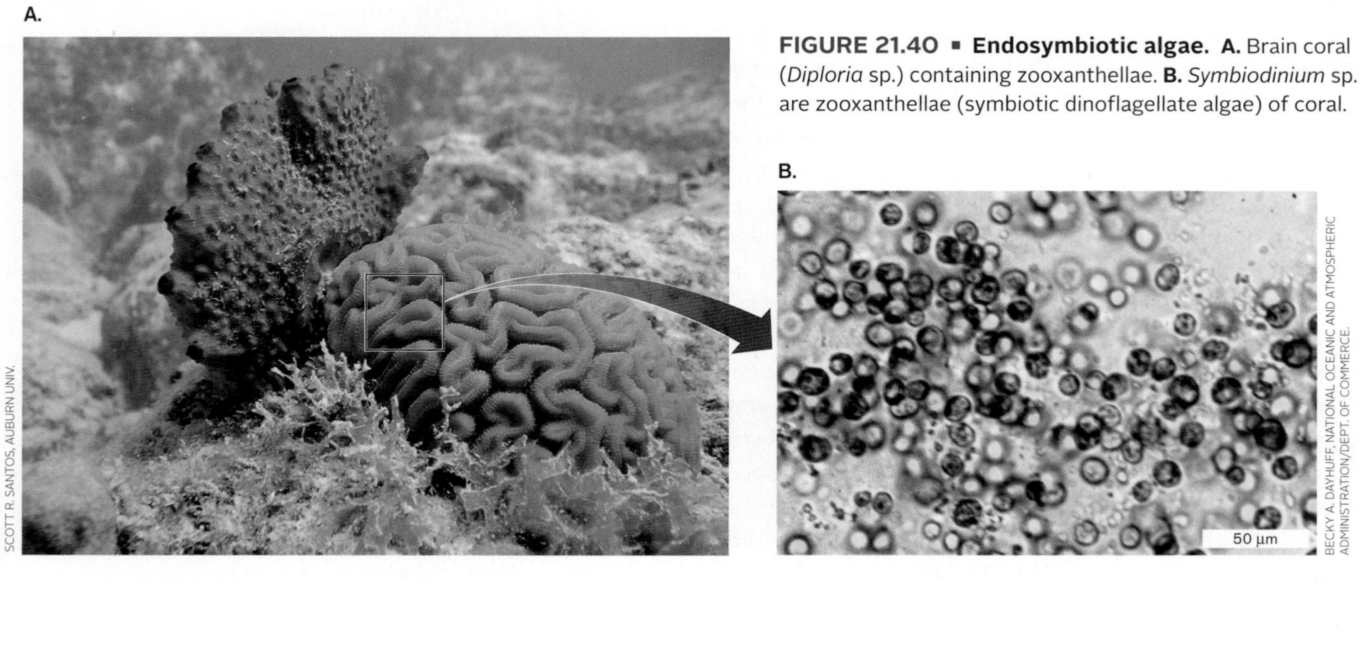
A.
B.
50 µm
SCOTT R. SANTOS, AUBURN UNIV.
BECKY A. DAYHUFF, NATIONAL OCEANIC AND ATMOSPHERIC ADMINISTRATION/DEPT. OF COMMERCE.

FIGURE 21.40 ■ Endosymbiotic algae. A. Brain coral (*Diploria* sp.) containing zooxanthellae. **B.** *Symbiodinium* sp. are zooxanthellae (symbiotic dinoflagellate algae) of coral.

coral or anemone may harbor several different species of *Symbiodinium,* which show different preferences for light or shade and different tolerances for temperature change. Studies of coral bleaching due to temperature increase suggest that corals containing diverse species of symbionts are more likely to survive, because one of their species may happen to be resistant to a rise in temperature.

A. Sponge containing actinomycetes

NICOLE WEBSTER, AUSTRALIAN INSTITUTE OF MARINE SCIENCE

B. Actinomycete isolated from sponge

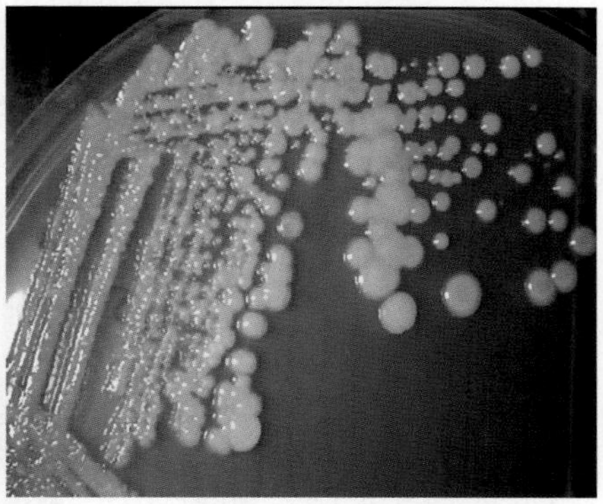

PHOTO BY MATTHEW ANDERSON, RUSSELL HILL'S LAB, U. OF MARYLAND BIOTECHNOLOGY INSTITUTE

FIGURE 21.41 ■ Sponge bacterial communities. A. The sponge *Rhopaloeides odorabile* contains actinomycetes that produce novel antimicrobial compounds. **B.** A sponge-associated actinomycete growing on an agar plate.

Sponge Communities

Marine sponges contain large numbers of associated bacteria, which may comprise up to 40% of the actual biomass of the animal. Some sponge-associated bacteria are consumed as food, but others remain associated permanently. For example, Russell Hill and colleagues at the University of Maryland, Baltimore, characterized the bacterial species associated with the sponge *Rhopaloeides odorabile,* which grows off the Australian Great Barrier Reef (**Fig. 21.41**). Using 16S rRNA probes, they identified novel species of deltaproteobacteria, planctomycetes, flavibacteria, and green sulfur and nonsulfur proteobacteria. Some of these bacteria can fix carbon for the sponge in regions of low oxygen, where the algae cannot photosynthesize. Other bacteria produce antimicrobial substances that defend the sponge from pathogens. Sponge bacteria, therefore, may offer a wealth of previously unknown antibiotics.

Digestive Communities

Vertebrate animals harbor highly complex and nutritionally important communities of gut microbes. Digestive chambers, such as the bovine rumen and the human colon, support thousands of species of bacteria, protists, and archaea. The human colon contains numerous fermenters and methanogens, some of which feed on intestinal mucus, whereas others digest complex plant fibers that our intestinal lining cannot, thus providing up to 15% of our caloric intake.

The bovine rumen microbial community. The most extensively studied digestive communities are those of ruminants, such as cattle, sheep, llamas, and caribou. Throughout most of human civilization, ruminants have provided protein-rich food, textile fibers, and mechanical work. A historical reference is the biblical injunction to consume an animal that "is cleft-footed and chews the cud"—that is, "ruminates," or redigests its food in the fermentation chamber known as the **rumen** (**Fig. 21.42**).

The microbial community of the rumen enables herbivores to acquire nutrition from complex plant fibers that the animal could not otherwise digest. From a genomic standpoint, such an arrangement makes evolutionary sense. If the animal had to digest all the diverse polysaccharide chains encountered in nature, its own genome would have to encode a wide array of different enzyme systems. Instead, the ruminant relies on diverse microbial species to conduct various kinds of digestion. Organisms that partly digest a substrate (for instance, converting sugars to lactate) provide a substrate such as short-chain acids that the host animal can absorb and digest to completion by aerobic respiration.

Thought Question

21.9 Why does ruminant fermentation provide food molecules that the animal host can use? How is the animal able to obtain nourishment from waste products that the microbes could not use?

The bovine gut system has four chambers (**Fig. 21.42A**). The rumen initially digests the feed and then passes it to the reticulum. The reticulum breaks the feed into smaller pieces and traps indigestible objects, such as stones or nails. After

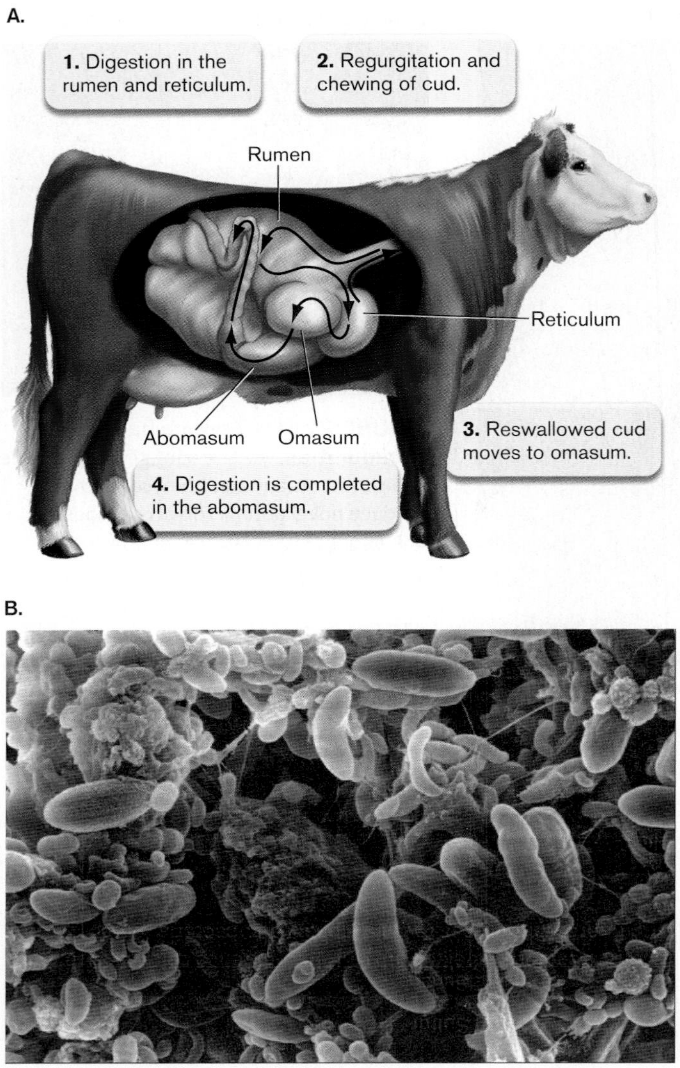
A.
1. Digestion in the rumen and reticulum.
2. Regurgitation and chewing of cud.
Rumen
Reticulum
Abomasum
Omasum
3. Reswallowed cud moves to omasum.
4. Digestion is completed in the abomasum.
B.
MIE UNIV., FACULTY OF BIORESOURCES

FIGURE 21.42 ■ The bovine rumen. A. The rumen is the largest of four chambers in the bovine stomach. **B.** Bacteria and fungi growing within the rumen.

initial digestion, feed is regurgitated for rechewing and then returned to the rumen, by far the largest of the chambers. In the rumen, feed is broken down to small particles and fermented slowly by thousands of species of microbes. The partially digested feed passes to the omasum, which absorbs water and short-chain acids produced by fermentation. The abomasum then decreases pH and secretes enzymes to digest proteins before sending its contents to the colon for further nutrient absorption and waste excretion.

While rumen digestion has been studied since the 1830s, major advances in understanding rumen fermentation were first achieved in 1966 by Robert Hungate (1906–2004) at UC Davis, who pioneered techniques of anaerobic microbiology. An example of Hungate's methods still in use today is that of obtaining anaerobic cultures from a **fistulated**, or **cannulated, cow**—that is, a cow in which an artificial connection is made between the rumen and the animal's exterior (see **Fig. 21.3B**). The cow is unharmed by the fistula and rumen sampling.

Metagenomics coupled with bioenergetics studies show how different microbes fill different niches in ruminal metabolism (**Fig. 21.43**). Cattle grown on relatively poor forage (that is, forage high in complex plant content) show a high proportion of ruminal fungi, the chytridiomycetes (discussed in Chapter 20). Chytridiomycete mycelia appear on ruminal food particles, and their motile zoospores—formerly mistaken for protists—swim through rumen fluid. By contrast, cattle fed a high cellulose diet, such as hay, grow faster and show cellulolytic bacteria such as *Ruminococcus albus* and *Fibrobacter flavefaciens.*

Metabolism of cellulolytic bacteria requires the presence of small amounts of branched-chain fatty acids. The branched-chain acids turn out to be produced by amino acid fermenters such as *Megasphaera elsdenii* and *Peptostreptococcus anaerobius.* However, too much degradation of amino acids can lead to overproduction of ammonia, poisoning the animal; thus, the protein content of cattle feed must be limited.

Another problem with ruminal fermentation is the frequent production of H_2 and CO_2. Hydrogen production is hard to avoid because the quantity of electron donors (reduced food molecules) greatly exceeds that of the available electron acceptors. The H_2 and CO_2 from fermentation support methanogens, wasting valuable carbon from feed and contributing a substantial part of global methane emissions. So much methane is formed by the rumen that a cannula inserted into the rumen liberates enough of the gas to light a flame.

Thought Question

21.10 How do you think cattle feed might be altered or supplemented to decrease methane production?

A concern with modern cattle rearing is the shift in feed from hay to grain, whose higher content of starch leads to more rapid digestion and faster growth of the animal. Unfortunately, rapid starch digestion favors fermenters such as *Prevotella* species and *E. coli,* which generate higher acid levels and gases, leading to "starch bloat." Furthermore, rumen acidity selects for acid-resistant pathogens such as the *E. coli* strain O157:H7.

Other vertebrate digestive communities. Many kinds of vertebrate animals besides ruminants have gut microbial

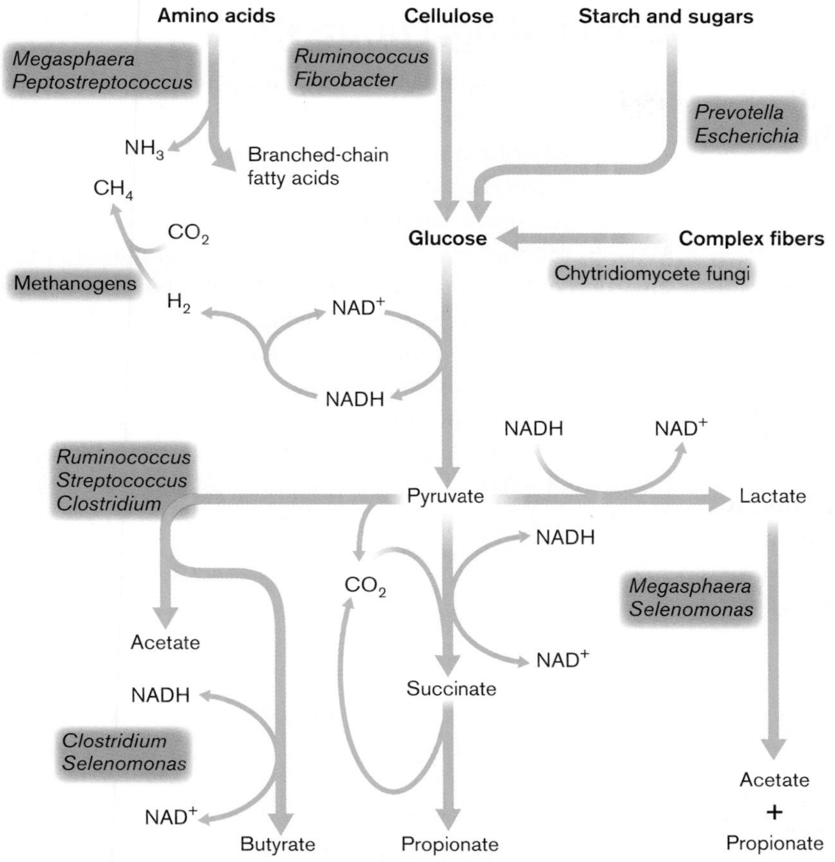

Amino acids
Cellulose
Starch and sugars
Megasphaera
Peptostreptococcus
Ruminococcus
Fibrobacter
Prevotella
Escherichia
NH3
Branched-chain fatty acids
CH4
CO2
Glucose
Complex fibers
Chytridiomycete fungi
Methanogens
H2
NAD+
NADH
NADH
NAD+
Ruminococcus
Streptococcus
Clostridium
Pyruvate
Lactate
NADH
CO2
Megasphaera
Selenomonas
Acetate
NAD+
NADH
Succinate
Clostridium
Selenomonas
Acetate
+
NAD+
Butyrate
Propionate
Propionate

FIGURE 21.43 ▪ Ruminal metabolism. Various microbes participate in digesting food, ultimately producing short-chain acids that are absorbed by the bovine gut epithelium. *Source:* Based on J. B. Russell and J. L. Rychlik. 2001. *Science* **292**:1119–1122.

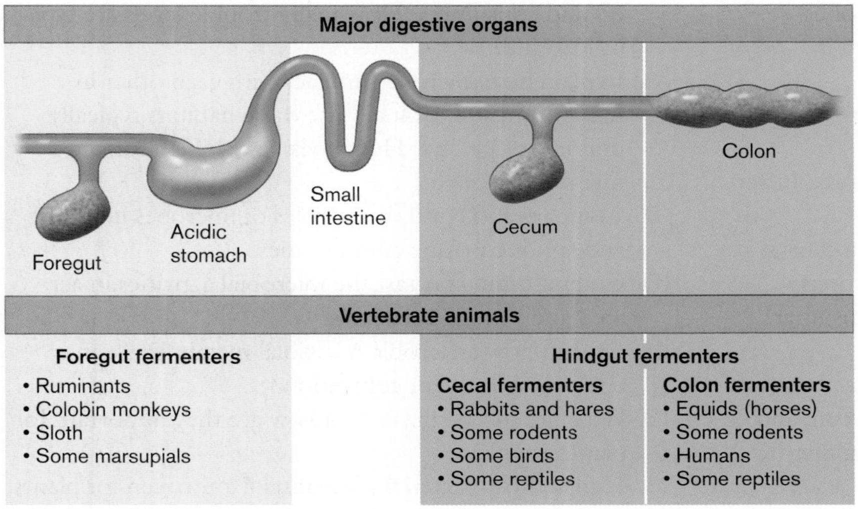

Major digestive organs
Colon
Small intestine
Cecum
Acidic stomach
Foregut
Vertebrate animals
Foregut fermenters
• Ruminants
• Colobin monkeys
• Sloth
• Some marsupials
Hindgut fermenters
Cecal fermenters
• Rabbits and hares
• Some rodents
• Some birds
• Some reptiles
Colon fermenters
• Equids (horses)
• Some rodents
• Humans
• Some reptiles

FIGURE 21.44 ▪ Vertebrate gut fermentation. Vertebrate animals vary in their location of microbial fermentation. Fermenting microbes may be active mainly in the foregut (rumen), before the stomach, as in cattle; in the cecum, a digestive organ following the small intestine of rabbits and rodents; or in the colon (large intestine), as in humans. *Source:* Modified from William Karasov and Hannah Carey. 2009. *Microbe* **4**:323.

communities whose fermentation contributes to host nutrition (**Fig. 21.44**). Cattle are considered "foregut" fermenters because their fermenting microbes receive the feed before the host's small intestine has digested and absorbed the best nutrients. Other foregut fermenters include colobin monkeys, and marsupials such as kangaroos. Kangaroos differ from cattle in that their foregut retention time is much shorter, so there is less time for action of slow-growing microbes such as methanogens. Thus, kangaroos produce little methane, in contrast to cattle. Switching from cattle to kangaroo meat has been proposed in Australia as a way to decrease greenhouse gas emissions.

Other animals, including humans, conduct microbial fermentation in the "hindgut" region (in humans, the colon). Hindgut fermentation favors bacteria capable of digesting complex plant materials that pass undigested through the small intestine. For example, the prominent *Bacteroides* species ferment mucopolysaccharides, pectin, and arabinogalactan, among many others. *Bacteroides* species release oligosaccharides (short-chain sugar polymers), which are fermented further by *Bifidobacterium* species. In addition, bacterial biosynthesis may contribute essential amino acids absorbed by the colon. The human colon also supports methanogens, such as *Methanobrevibacter smithii*.

Besides nutrition, human microbiota interact positively with our immune system and generate defenses against pathogens, as do plant endophytes. For example, *Bifidobacterium* species down-regulate human secretion of cytokines, molecules that cause inflammation in ulcerative colitis. Other colonic bacteria produce bacteriocins, molecules that prevent pathogens from adhering to the gut lining. Gut bacteria even interact with the brain. In mice, probiotic gut bacteria secrete neurotransmitters that improve brain development and function. In humans, abnormalities of the gut microbiome are associated with encephalopathy leading to psychiatric disorders. Human microbiota are discussed further in Chapter 23.

To Summarize

- **Animals harbor microbial communities.** Microbes populate the skin, where they provide protection from pathogens. Other species grow in the animal's digestive system, where they enhance nutrition.
- **Corals and sea anemones harbor zooxanthellae**, algae that provide products of photosynthesis in exchange for a protected habitat.
- **The rumen of ruminant animals is a complex microbial digestive chamber.** Rumen microbes, including bacteria, protists, and fungi, digest complex plant materials. The microbial digestion generates short-chain fatty acids that are absorbed by the intestinal lining.
- **Foregut and hindgut fermentation contribute to digestion** in many vertebrate animals, including humans.

Concluding Thoughts

In this chapter we have seen how microbes colonize a vast array of habitats, ranging from the deserts and oceans to the roots of plants and the digestive tracts of animals. Microbes are found in the upper layers of our planet's atmosphere, down to the deepest rock strata we can reach. Wherever found, microbes both respond to and modify the environment that surrounds them. While this chapter has focused on food webs in local habitats, Chapter 22 takes a global perspective of microbial ecology and its roles in the cycling of Earth's essential elements, such as nitrogen and iron. We will see how microbes interact with global climate change and examine the evidence for existence of microbial life beyond Earth.

CHAPTER REVIEW

Review Questions

1. How do we analyze a metagenome of a microbial community? What are the advantages and pitfalls of metagenomics, compared to culturing microbes?
2. What unique functions do microbes perform in ecosystems?
3. Explain the difference between carbon assimilation and dissimilation.
4. What kinds of microbial metabolism are favored in aerated environments? In anoxic environments?
5. What are examples of microbial producers in ecosystems? Include phototrophs as well as lithotrophs. Can a microbe be both a producer and a consumer? Explain.
6. Explain the microbial relationships in various forms of symbiosis, including mutualism, commensalism, and parasitism. Outline an example of each, detailing the contributions of each partner.
7. Compare and contrast the marine food web with the soil food web. What kinds of organisms are the producers and consumers? How many trophic levels are typically found?
8. Explain how microbes interact with each other in marine and soil habitats. Are these habitats typically uniform or patchy? How does "patchiness" affect microbial growth?
9. Compare and contrast the roles of microbes in photic and aphotic marine communities.
10. Compare and contrast the microbial activities in aerated and waterlogged soils.
11. Explain how anaerobic microbial metabolism can enrich soil for plant cultivation.
12. What are mycorrhizae, and how are they important for plant growth?
13. Explain how bacterial mutualists fix nitrogen for plants.

Thought Questions

1. Explain what you can learn about a marine microbial community from a metagenome, as compared with a metatranscriptome. How and why might the two approaches yield different results?
2. Explain, with specific examples, what mutualism and parasitism have in common, and how they differ. Explain an example of a relationship that combines aspects of both.
3. The photic zone and the benthic zone pose challenges and opportunities for marine microbes. What challenges do they have in common, and how do they differ?

Key Terms

algal bloom (867)
amensalism (856)
assembly (844)
assimilation (849)
bacteroid (877)
barophile (865)
benthic organism (858)
bin, binning (846)
biochemical oxygen demand (biological oxygen demand) (BOD) (859)
biomass (850, 862)
cannulated cow (884)
coastal shelf (857)
cold seep (865)
commensalism (856)
community (840)
consumer (850)
contig (845)
coral bleaching (882)
coverage (846)
cryptobiotic soil (cryptogamic crust) (854)
decomposer (850)
detritus (851)
dissimilation (849)
ecosystem (840)
ectomycorrhizae (874)
endolith (871)
endomycorrhizae (874)
endophyte (877)
enrichment culture (841)
euphotic zone (857)
eutrophic (867)
extremophile (851)
filter (843)
fistulated cow (884)
flavonoid (878)
food web (850)
grazer (850)
haustorium (881)
humic material (humus) (872)
hydric soil (875)
hydrothermal vent (865)
infection thread (878)
leghemoglobin (879)
lichen (853)
lignin (872)
limiting nutrient (868)
littoral zone (867)
marine snow (862)
metagenomics, metagenomic (840, 841)
metaproteomics (848)
metatranscriptomics (848)
microbiome (841)
microplankton (862)
mixotroph (863)
mutualism (852)
mycorrhizae (874)
nanoplankton (862)
niche (841)
oligotrophic (866)
operational taxonomic unit (OTU) (846)
parasitism (856)
pelagic zone (857)
photic zone (857)
phytoplankton (863)
picoplankton (862)
plankton (862)
population (840)
predator (850)
primary producer (849)
psychrophile (865)
rarefaction curve (846)
rhizobium (877)
rhizoplane (873)
rhizosphere (872, 873)
rumen (883)
saprophyte (872)
scaffold (846)
soil (869)
symbiosis (852)
symbiosome (878)
synergism (856)
syntrophy (855)
target community (842)
thermal vent (865)
thermocline (858)
thermophile (865)
trophic level (850)
uncultured (861)
vesicular-arbuscular mycorrhizae (VAM) (874)
viable but nonculturable (VBNC) (860)
water table (871)
wetland (875)
zooxanthella (882)

Recommended Reading

Azam, Farooq, and Francesca Malfatti. 2007. Microbial structuring of marine ecosystems. *Nature Reviews. Microbiology* **5**:782–792.

DeLong, Edward F. 2009. The microbial ocean from genomes to biomes. *Nature* **459**:200–212.

Dubilier, Nicole, Claudia Bergin, and Christian Lott. 2008. Symbiotic diversity in marine animals: The art of harnessing chemosynthesis. *Nature Reviews. Microbiology* **6**:725–740.

Fierer, Noah, and Robert B. Jackson. 2006. The diversity and biogeography of soil bacterial communities. *Proceedings of the National Academy of Sciences USA* **103**:626–631.

Flint, Harry J., Edward A. Bayer, Marco T. Rincon, Raphael Lamed, and Bryan A. White. 2008. Polysaccharide utilization by gut bacteria: Potential for new insights from genomic analysis. *Nature Reviews. Microbiology* **6**:121–131.

Gage, Daniel J. 2004. Infection and invasion of roots by symbiotic, nitrogen-fixing rhizobia during nodulation of temperate legumes. *Microbiology and Molecular Biology Reviews* **68**:280–300.

Huq, Anwar, Mohammed Yunus, Syed Salahuddin Sohel, Abbas Bhuiya, Michael Emch, et al. 2010. Simple sari filtration is sustainable and continues to protect villagers from cholera in Matlab, Bangladesh. *mBio* **1**:e00034-10.

Hurwitz, Bonnie L., and Matthew B. Sullivan. 2013. The Pacific Ocean Virome (POV): A marine viral metagenomic dataset and associated protein clusters for quantitative viral ecology. *PLoS One* **8**:e57355.

Johnson, Zackary I., Erik R. Zinser, Allison Coe, Nathan P. McNulty, E. Malcolm, et al. 2006. Niche partitioning among *Prochlorococcus* ecotypes along ocean-scale environmental gradients. *Science* **311**:1737–1740.

Kaeberlein, T., K. Lewis, and S. S. Epstein. 2002. Microorganisms in pure culture in a simulated natural environment. *Science* **296**:1127–1129.

Karasov, William H., and Hannah V. Carey. 2009. Metabolic teamwork between gut microbes and hosts. *Microbe* **4**:323–328.

Karl, David M. 2007. Microbial oceanography: Paradigms, processes, and promise. *Nature Reviews. Microbiology* **5**:759–769.

Kiers, E. Toby, Marie Duhamel, Yugandhar Beesetty, Jerry A. Mensah, Oscar Franken, et al. 2011. Reciprocal rewards stabilize cooperation in the mycorrhizal symbiosis. *Science* **333**:880–882.

Moran, Nancy A., Allison K. Hansen, J. Elijah Powell, and Zakee L. Sabree. 2012. Distinctive gut microbiota of honey bees assessed using deep sampling from individual worker bees. *PLoS One* 7:e36393.

Ottesen, Elizabeth A., Curtis R. Young, John M. Eppley, John P. Ryan, Francisco P. Chavez, Christopher A. Scholin, and Edward F. DeLong. 2013. Pattern and synchrony of gene expression among sympatric marine microbial populations *Proc. Natl. Acad. Sci. USA* **110:**E488–E497.

Parniske, Martin. 2008. Arbuscular mycorrhiza: The mother of plant root endosymbioses. *Nature Reviews. Microbiology* **6**:763–775.

Rodrigues Jorge L. M., Vivian H. Pellizari, Rebecca Mueller, Kyunghwa Baek, Ederson da C. Jesus, et al. 2013. Conversion of the Amazon rainforest to agriculture results in biotic homogenization of soil bacterial communities. *Proceedings of the National Academy of Sciences USA* **110**:988–993.

Rusch, Douglas B., Aaron L. Halpern, Granger Sutton, Karla B. Heidelberg, Shannon Williamson, et al. 2007. The Sorcerer II global ocean sampling expedition: Northwest Atlantic through eastern tropical Pacific. *Public Library of Science Biology* **5**(3):e77.

Sato, Tomoyuki, Yuichi Hongoh, Satoko Noda, Satoshi Hattori, Sadaharu Ui, et al. 2009. *Candidatus* Desulfovibrio trichonymphae, a novel intracellular symbiont of the flagellate *Trichonympha agilis* in termite gut. *Environmental Microbiology* **11**:1007–1015.

CHAPTER 22

Microbes in Global Elemental Cycles

22.1 Biogeochemical Cycles
22.2 The Carbon Cycle and Bioremediation
22.3 The Hydrologic Cycle and Wastewater Treatment
22.4 The Nitrogen Cycle
22.5 Sulfur, Phosphorus, and Metals
22.6 Astrobiology

Microbes throughout the biosphere recycle carbon, nitrogen, sulfur, and other elements essential for all life. Through their biochemical transformations, diverse microbial activities largely determine the quality of soil, air, and water for human life. Today, all of these geochemical cycles are altered profoundly by human activity. The burning of fossil fuels, which were generated millions of years ago by subterranean microbes, releases quantities of carbon dioxide too great to be absorbed by marine bacteria and algae. Growing rice production increases the release of methane by methanogens that thrive in submerged rice paddies. Both carbon dioxide and methane are "greenhouse gases," which lead to global warming. Will we humans learn to manage the perturbations of our own biosphere, from pollution to global warming?

Microbes can help us manage environmental change, both globally and locally, through bioremediation. From local wastewater treatment and bioremediation of polluted soil, to the control of greenhouse gases, microbes are our hidden partners on Earth. And Earth's microbial cycles lead us to wonder whether biospheres exist on other worlds. Our neighbor planet Mars shows tantalizing signs of water and the possibility of hidden microbial life.

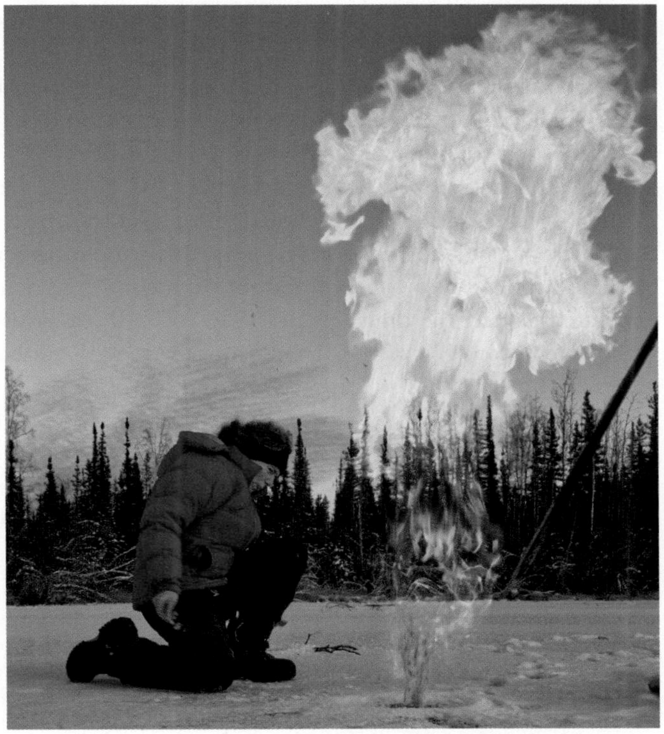

CURRENT RESEARCH **highlight**

Permafrost burning. Arctic permafrost, ground that stays frozen all year, stores large quantities of partly decomposed plant matter (organic-horizon soil). As global temperatures rise, the permafrost melts, and microbial metabolism activates. Bacterial fermentation releases CO_2 and H_2, which methanogens convert to methane. Katey Anthony at the University of Alaska Fairbanks demonstrates how bubbles of methane beneath the ice burst into the atmosphere, where they can ignite into flame. In 2012, Andrew McDougall and colleagues at the University of Victoria (British Columbia), modeled both the response of permafrost carbon release and its feedback effect on global warming. They project that permafrost could release more than a quarter of its carbon stores by the year 2100, and that this amount could add another 1.5°C to the global temperature by the year 2300. *Source:* Andrew MacDougall et al. 2012. *Nature Geoscience* **5**:719.

Two billion years ago ancient cyanobacteria began to photolyze water and produce molecular oxygen, as discussed in Chapter 17. Oxygen is a powerful oxidant, lethal to most life in that anoxic era, when all living organisms were anaerobic microbes. Most anaerobic species must have gone extinct, except for the lucky few that evolved mechanisms of antioxidant protection. Since then, microbes have shaped our biosphere by releasing oxygen, by fixing nitrogen and returning it to the atmosphere, and by fixing and producing carbon dioxide. These and countless other microbial processes have made Earth's atmosphere what it is, and they continue their roles in its homeostasis.

Yet, during the past century human technology has upended key aspects of the global biospheric chemistry that microbes had balanced for so long. Half the nitrogen in the biosphere now comes from anthropogenic (human-generated) sources. From the years 1985 to 2012, each year was hotter than average—a trend consistent with global warming by greenhouse gases. The rapid increase of the greenhouse gases carbon dioxide (CO_2), methane (CH_4), and nitrous oxide (N_2O) correlates with the industrial age (**Fig. 22.1**). These gas levels from thousands of years ago are measured from air bubbles trapped in ice in Antarctica. The gases are generated by bacteria, methanogens, and other life-forms; but human technology accelerates their release. Burning petroleum releases CO_2 from a product that bacteria and plants took millions of years to form. At the same time, phototrophic bacteria and plants fix much of the global carbon dioxide into biomass. But the added human output of CO_2 has outpaced the rate of plant and microbial CO_2 fixation. Global warming accelerates methane release by methanogens. And bacteria oxidize sewage nitrogen to N_2O.

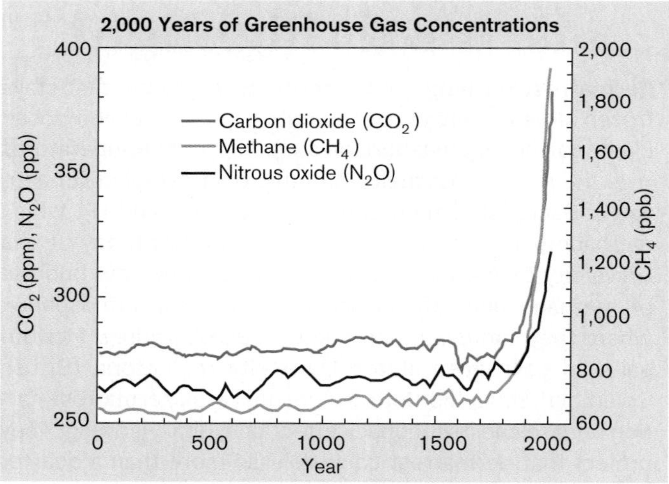
2,000 Years of Greenhouse Gas Concentrations
Carbon dioxide (CO2)
Methane (CH4)
Nitrous oxide (N2O)
CO2 (ppm), N2O (ppb)
CH4 (ppb)
400
350
300
250
2,000
1,800
1,600
1,400
1,200
1,000
800
600
0
500
1000
1500
2000
Year

FIGURE 22.1 ▪ Global greenhouse gases. Atmospheric levels of CO_2, methane, and nitrous oxide since 1750 were measured in Antarctic ice cores. The recent increase of these greenhouse gases accompanies the rise in fossil fuel burning and fertilizer-intensive agriculture. Various microbes produce and consume these gases. *Source:* Modified from U.S. Global Change Research Program. 2009. *Global Climate Change Impacts in the United States.* Cambridge University Press.

Concern over the increasing greenhouse gases led the United Nations, meeting in Kyoto in 1997, to adopt the Kyoto Protocol for reduction of industrial emissions of CO_2, CH_4, and other gases that contribute to global warming. Over 160 nations adopted the protocol, although the United States declined, contending that it posed a disproportionate economic burden on developed nations. Will the human-induced global warming cause mass extinctions of a majority of Earth's species—as did the rise of ancient cyanobacteria? Or can we use our knowledge of microbial ecology to channel microbial activities into recovering the balance—for example, by increasing microbial CO_2 fixation?

Throughout most of this book, we present microbial biochemistry in the context of growth of individual organisms. Here in Chapter 22, we show how the collective metabolic activities of microbial populations contribute to global cycles of elements throughout Earth's biosphere. We consider the ways that we humans enlist microbes for **bioremediation**, the use of microbial metabolism to reverse pollution and restore chemical cycles—for example, the bioremediation of a contaminated aquifer at Aberdeen Proving Ground, Maryland (see **Special Topic 22.1**). Finally, we explore how our awareness of global microbiology has renewed our quest for life beyond Earth.

22.1

Biogeochemical Cycles

Microbes have exceptional abilities to interconvert molecules. Nearly any kind of molecule can be metabolized by some species somewhere if the reaction provides energy or building blocks for biomass. But even microbes face the limits of the actual elements themselves: No microbe, nor any other living creature, can convert one element to another. Living organisms conduct chemical reactions, not nuclear reactions—although microbes that live deep underground may obtain chemical energy from hydrogen ions generated by uranium decay (discussed in Chapter 21).

Microbes Cycle Essential Elements

Because organisms cannot form their own elements, they need to get them from their environment. While small amounts of matter enter the biosphere from outer space, and gases continually escape Earth, the rate of these changes is tiny compared to the rates of living processes. So, organisms acquire their elements either from nonliving components of their environment, such as by fixing atmospheric CO_2, or from other organisms, by grazing, predation, or decomposition. Furthermore, all organisms recycle their components back to the biosphere. The partners in this recycling include **abiotic** entities such as air, water, and minerals, as well as **biotic** entities such as predators and decomposers. Collectively, the metabolic interactions of microbial communities with the biotic and abiotic components of their ecosystems are known as **biogeochemistry** or **geomicrobiology**.

TABLE 22.1

Typical elemental composition of Gram-negative bacteria.

Element	Dry weight (%)
Carbon	50
Oxygen	20
Nitrogen	14
Hydrogen	8
Phosphorus	3
Sulfur	1
Potassium	1
Sodium	1
Calcium	0.5
Magnesium	0.5
Chlorine	0.5
Iron	0.2
All others	0.3

Which elements need to be recycled and made available for life? In most organisms, six elements predominate: carbon, oxygen, nitrogen, hydrogen, phosphorus, and sulfur (discussed in Chapter 4). **Table 22.1** summarizes the elemental composition typical of Gram-negative bacteria. In other microbes, elemental proportions vary with species and growth conditions. Some species include inorganic components such as the silicate shells of diatoms or the sulfur granules of phototrophic bacteria. All of these elements flow in **biogeochemical cycles** of nutrients throughout the biotic and abiotic components of the biosphere. The environmental levels of key elements can limit biological productivity. For example, the concentration of iron limits the marine populations of phytoplankton, which depend on iron supplied by wind currents. Other elements, such as zinc and copper, are **micronutrients**, nutrients required in still smaller amounts. Micronutrients, too, must be cycled, although their flux is challenging to measure.

Sources and Sinks of Essential Elements

Biogeochemical cycles include both biological components (such as phototrophs that consume CO_2 and heterotrophs that release CO_2) and geological components (such as volcanoes that release CO_2 and oceans that absorb CO_2). The major parts of the biosphere containing significant amounts of an element needed for life are called **reservoirs** of that element (**Table 22.2**). Each reservoir acts both as a **source** of that element for living organisms and as a **sink** to which the element returns. For example, the ocean is an important reservoir for carbon (CO_2 in equilibrium with HCO_3^-, bicarbonate ion). The ocean's carbon cycles rapidly; thus, the ocean serves as both a source and a sink for carbon.

As elements cycle from sources to sinks, microbial metabolism generates a series of redox changes (discussed in Chapter 14). The major oxidation states of carbon, nitrogen, and sulfur are summarized in **Table 22.3**. Biospheric carbon can be found as CH_4 generated by methanogens (completely reduced, −4), as CO_2 produced by respiration and fermentation (completely oxidized, +4), or as one of various intermediate states of oxidation. Nitrogen is excreted by many organisms in its most reduced form, ammonia (NH_3), which is protonated to ammonium ion (NH_4^+). Ammonium is oxidized by lithotrophic bacteria through several stages to nitrite (NO_2^-) and nitrate (NO_3^-), which serve as terminal electron acceptors for anaerobic respiration. Sulfides (H_2S, HS^-) serve as electron donors for respiration and sulfur phototrophy, whereas oxidized forms, such as sulfite (SO_3^{2-}) and sulfate (SO_4^{2-}), serve in anaerobic respiration.

Thought Question

22.1 Why is oxidation state critical for the acquisition, usability, and potential toxicity of cycled compounds? Cite examples based on your study of microbial metabolism.

How do we study microbial cycling on a global scale? How do we figure out whether ecosystems are net sources or sinks of CO_2? Does microbial activity enhance or limit availability of nitrogen? Rates of flux of elements in the biosphere are very difficult to measure, yet the questions have enormous political and economic implications.

TABLE 22.2

Global reservoirs of carbon, nitrogen, and sulfur.[a]

Reservoir	Carbon	Rate of cycling	Nitrogen	Rate of cycling	Sulfur	Rate of cycling
Atmosphere	700 (CO_2)	Fast	3,900,000 (N_2)	Slow	0.0014 (SO_2, H_2S)	Fast
Ocean						
Biomass	4	Fast	0.5	Fast	0.15	Fast
Organic molecules	2,100	Fast	300	Fast	–	–
Inorganic molecules	38,000 (HCO_3^-, CO_3^{2-})	Fast	20,000 (N_2) 690 (NO_3^-, NO_2^-, NH_4^+)	Slow Fast	1,200,000 (SO_4^{2-})	Slow
Land						
Biomass	500	Fast	25	Fast	8.5	Fast
Organic matter (soil)	1,200	Fast	110	Slow	16	Fast
Crust (below land and ocean)	120,000,000	Slow	770,000	Slow	18,000,000	Slow
Fossil fuel (oil, coal, natural gas)	13,000	Fast				

Source: R. Maier et al. 2000. *Environmental Microbiology.* Academic Press, San Diego, CA.

[a]Units are 10^9 metric tons, or 10^{12} kg.

TABLE 22.3

Oxidation states of cycled compounds.

Oxidation state	Carbon		Nitrogen		Sulfur	
−4	CH_4	Methane				
−3			NH_3, NH_4^+	Ammonia, ammonium ion		
−2	CH_3OH, $(CH_2)_n$	Methanol, hydrocarbon	$H_2N\text{-}NH_2$	Hydrazine	H_2S, HS^-	Sulfides
−1			NH_2OH	Hydroxylamine		
0	$(CH_2O)_n$	Carbohydrate	N_2	Nitrogen	S^0	Elemental sulfur
+1			N_2O	Nitrous oxide		
+2	CHOOH	Formic acid	NO	Nitric oxide	$S_2O_3^{2-}$	Thiosulfate
+3			HNO_2, NO_2^-	Nitrous acid, nitrite ion		
+4	CO_2, HCO_3^-	Carbon dioxide, bicarbonate ion	NO_2	Nitrogen dioxide	SO_3^{2-}	Sulfite
+5			HNO_3, NO_3^-	Nitric acid, nitrate ion		
+6					H_2SO_4, SO_4^{2-}	Sulfuric acid, sulfate

To measure environmental carbon, nitrogen, and other elements, various methods are used. These methods fall under the following categories:

- **Chemical and spectroscopic analysis.** Bulk quantities of CO_2, nitrates, and other chemicals can be determined by sophisticated chemical instrumentation. Atmospheric CO_2 is measured by infrared absorption spectroscopy, applied to samples from towers such as those of the NASA FLUXNET study (**Fig. 22.2A**). Gas chromatography is used to separate and quantify various gases, including oxygen, nitrogen, sulfur dioxide, and carbon monoxide. Mass spectrometry detects extremely small quantities of different molecules, even distinguishing between elemental isotopes.
- **Radioisotope incorporation.** The influx and efflux of CO_2 can be measured by the uptake of ^{14}C-labeled substrates in a small, controlled model ecosystem called a **mesocosm**. Alternatively, CO_2 flux can be measured with radioisotope tracers in the field using a field chamber.
- **Stable isotope ratios.** Some enzyme reactions show a preference for one isotope over another, such as ^{14}N versus ^{15}N. For example, denitrifiers (bacteria that metabolize nitrate) strongly prefer the ^{14}N isotope, leaving behind nitrate enriched in ^{15}N. The $^{14}N/^{15}N$ ratio is measured using mass spectrometry. Measuring nitrogen isotope ratios can indicate whether denitrifiers could have conducted metabolism in the sample.

To Summarize

- **Microbes cycle essential elements in the biosphere.** Many key cycling reactions are performed only by bacteria and archaea.
- **Elements cycle between organisms and abiotic sources and sinks.** The most accessible source of carbon and nitrogen is the atmosphere. Earth's crust stores large amounts of key elements, but their availability to organisms is limited.
- **Environmental flux of elements is measured through chemistry.** Methods include infrared spectroscopy and mass spectrometry, gas chromatography, radioisotope incorporation, and the measurement of isotope ratios.

A.

B.

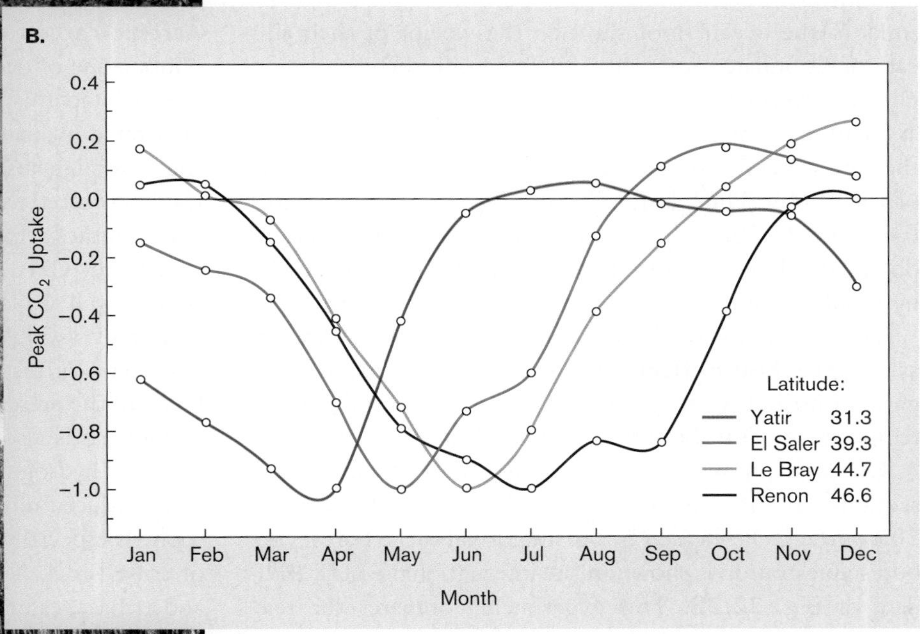
Peak CO_2 Uptake
0.4
0.2
0.0
−0.2
−0.4
−0.6
−0.8
−1.0
Jan
Feb
Mar
Apr
May
Jun
Jul
Aug
Sep
Oct
Nov
Dec
Month
Latitude:
Yatir 31.3
El Saler 39.3
Le Bray 44.7
Renon 46.6

AUSTIN CARY MEMORIAL FOREST, U. FLORIDA, GAINESVILLE

FIGURE 22.2 ▪ Measuring flux of elements in the biosphere. **A.** A U.S. FLUXNET tower at Austin Cary Memorial Forest in Florida is used for atmospheric CO_2 sampling as part of a global effort to monitor carbon flux. **B.** Peak CO_2 uptake by forests, recorded across the seasons by FLUXNET towers at Yatir, Israel; El Saler, Spain; Renon, Italy; Le Bray, France. *Source:* Kadmiel Maseyk, 2013. *FluxLetter* **5**:15

22.2

The Carbon Cycle and Bioremediation

The foundation of all food webs involves influx and efflux of carbon. The major reservoirs of carbon are shown in **Table 22.2**. In theory, carbonate rock forms the largest reservoir of carbon. But Earth's crust is the source least accessible to the biosphere as a whole. Crustal rock provides carbon only to organisms at the surface and to subsurface microbes that grow extremely slowly. Thus, subsurface carbon turnover is very slow. The carbon reservoir that cycles most rapidly is that of the atmosphere, a source of CO_2 for photosynthesis and lithotrophy. The atmosphere also acts as a sink for CO_2 produced by heterotrophy and by geological outgassing from volcanoes.

The atmospheric reservoir is much smaller than other sources, such as the oceans, crustal rock, and fossil fuels. For this reason, the industrial burning of fossil fuels has perturbed the balance between atmospheric CO_2 and larger reservoirs, such as the ocean. The ocean actually absorbs a good part of the extra CO_2, where the carbon is eventually converted to carbonates. In addition, marine phototrophs such as diatoms and coccolithophores trap a substantial amount of carbon in biomass. A portion of their biomass sinks to the ocean floor through the weight of their silicate or carbonate exoskeletons. But despite these buffering effects of the ocean, atmospheric CO_2 continues to increase at an annual rate of 2 parts per million (ppm). This is twice the rate observed in the past century, during the decade of the 1960s. Thus, atmospheric CO_2 is rising at an ever faster rate. In 2013, the Mauna Loa Observatory in Hawaii observed a level of CO_2 at 400 ppm—believed to be the highest level on Earth for the past three million years.

The CO_2 traps solar radiation as heat—a process known as the **greenhouse effect**. CO_2 is one of several greenhouse gases contributing to global warming, the overall rise in temperature of our biosphere over the past hundred years.

On land, terrestrial plants, particularly forest trees, sequester significant amounts of carbon—perhaps 10%–20% of the CO_2 released by burning fossil fuels. Forest carbon sequestration is shown in the international FLUXNET data of **Fig. 22.2B**. This experiment compares the seasonal patterns of net CO_2 uptake (the negative values on the Y axis) for forests at four different latitudes. It shows that the higher the latitude of the forest, the later in summer its CO_2 uptake peaks. Such measurements provide the basis for modeling global climate change, and for negotiating agreements to "trade" pollution for forest growth.

In 2013, another important carbon sink was discovered by Karina Clemmensen, Björn Lindahl, and colleagues at Uppsala BioCenter, Sweden: the **mycorrhizae** (mycorrhizal fungi) associated with forest plant roots (presented in Chapter 21). Clemmensen showed that in some forests the roots and fungi can sequester as much as 22 kilograms of carbon per square meter of forest soil, which may be 70% of the total carbon sequestered. Thus, mycorrhizal fungi play an important role in minimizing release of the greenhouse gas CO_2.

Note, however, that being a greenhouse gas does not make CO_2 inherently "bad" for the environment. In fact, if heterotrophic production of CO_2 were to cease altogether, phototrophs would run out of CO_2 in roughly 300 years, despite the vast quantities of carbon present in the ocean and crust. Thus, both CO_2 fixers and heterotrophs need each other for a continuous cycle.

Carbon Cycles Depend on Oxygen

The global cycle of carbon in the biosphere is closely linked to the cycles of oxygen and hydrogen, elements to which most carbon is bonded (**Fig. 22.3**). Overall, carbon cycles between carbon dioxide (CO_2) and various reduced forms of carbon, including biomass (living material). Note that the results of carbon cycling differ greatly, depending on the presence of molecular oxygen.

Aerobic carbon cycling. Aerated ecosystems include the photic zone of oceans and the oxygenated surface of terrestrial habitats (discussed in Chapter 21) (**Fig. 22.3**, top). In an aerated (or oxic) habitat, such as the marine photic zone, the ecosystem absorbs enough light for the rate of photosynthesis to exceed the rate of heterotrophy. Microbial and plant photosynthesis fixes CO_2 into biomass, designated by the shorthand $[CH_2O]$. Phototrophs include bacteria and protists, as well as plants. Aerobic CO_2 fixation is accompanied by release of O_2. The O_2 is then used by heterotrophs (such as bacteria, protists, and animals) to convert $[CH_2O]$ back to CO_2. In the presence of light, a net excess of O_2 is released.

Biomass is also produced through lithotrophy—the oxidation of hydrogen, hydrogen sulfide, ferrous iron (Fe^{2+}) and other reduced minerals, and even carbon monoxide. Lithotrophy is especially prominent in soil and in weathered areas of crustal rock. Lithotrophy is performed solely by bacteria and archaea, essential microbial partners in these ecosystems.

Anaerobic carbon cycling. Anoxic environments support lower rates of biomass production than do oxygen-rich environments because they depend on oxidants of lower redox potential and limited quantity, such as Fe^{3+}. Anaerobic conversion of CO_2 to biomass is done mainly by bacteria and

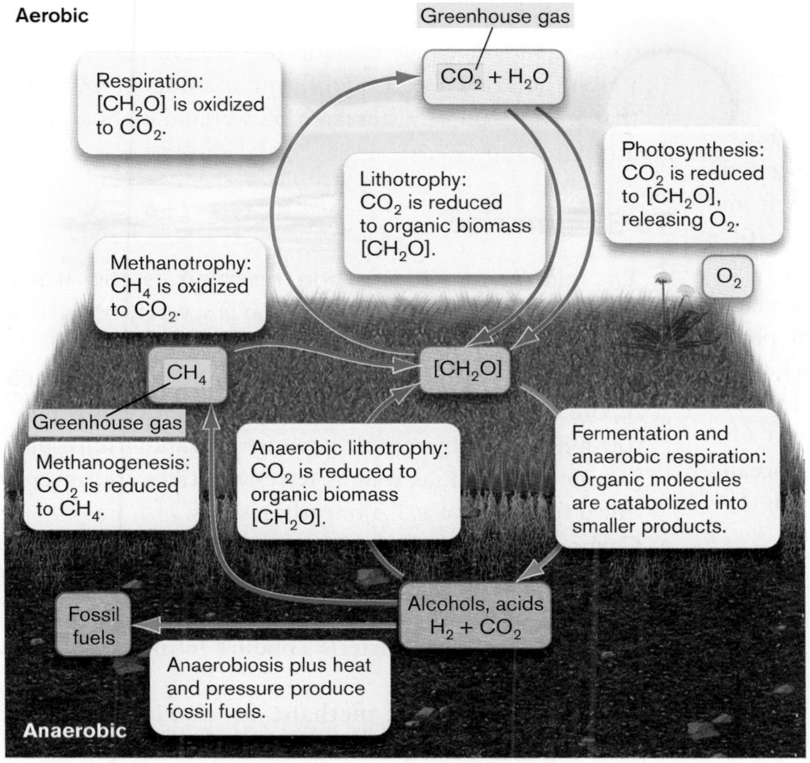
Aerobic
Greenhouse gas
$CO_2 + H_2O$
Respiration: $[CH_2O]$ is oxidized to CO_2.
Lithotrophy: CO_2 is reduced to organic biomass $[CH_2O]$.
Photosynthesis: CO_2 is reduced to $[CH_2O]$, releasing O_2.
O_2
Methanotrophy: CH_4 is oxidized to CO_2.
CH_4
$[CH_2O]$
Greenhouse gas
Methanogenesis: CO_2 is reduced to CH_4.
Anaerobic lithotrophy: CO_2 is reduced to organic biomass $[CH_2O]$.
Fermentation and anaerobic respiration: Organic molecules are catabolized into smaller products.
Fossil fuels
Alcohols, acids $H_2 + CO_2$
Anaerobiosis plus heat and pressure produce fossil fuels.
Anaerobic

FIGURE 22.3 ▪ The carbon cycle: aerobic and anaerobic. Aerobic and anaerobic conversions of carbon. Blue = reduction of carbon; red = oxidation of carbon; orange = fermentation; $[CH_2O]$ = organic biomass. In an aerobic environment (top), photosynthesis generates molecular oxygen (O_2), which enables the most efficient metabolism by heterotrophs, methanotrophs, and lithotrophs. In an anaerobic environment (bottom), photosynthesis generates only oxidized minerals, which support limited anaerobic respiration. Fermentation generates organic carbon products, as well as CO_2 and H_2. In the absence of oxygen, methanogens convert carbon dioxide (CO_2) and molecular hydrogen (H_2) to methane (CH_4), one of the most potent greenhouse gases.

archaea. Vast, permanently anaerobic habitats extend several kilometers below Earth's surface, encompassing greater volume than the rest of the biosphere put together (**Fig. 22.3**, bottom). In these habitats, endolithic bacteria inhabit the interstices of rock crystals (discussed in Chapter 21).

In soil and water, anaerobic metabolism includes fermentation of organic carbon sources, as well as respiration and lithotrophy with alternative electron acceptors such as nitrate, ferric iron (Fe^{3+}), and sulfate. Anaerobic decomposition by microbes is one stage in the formation of fossil fuels such as oil and natural gas (primarily methane). In soil, anoxic conditions (extremely low in O_2) favor incomplete breakdown of organic material. This characteristic of anaerobic soil actually enriches ecosystems, particularly those of wetlands, which undergo periodic cycles of aeration and hydration. The partly decomposed matter becomes available for further decomposition with oxygen. Anoxic environments near the surface also favor production of methane from the H_2, CO_2, and other fermentation products of anaerobes.

A major source of concern for global greenhouse gases is the methane hydrates accumulating in deep marine sediments, generated by huge benthic communities of methanogens (discussed in Chapters 19 and 21). Warming of methane hydrate releases gaseous methane, which quickly rises to the atmosphere. Geological evidence suggests that rapid methane release accompanied the retreat of the glaciers during ice age transitions. A rapid methane release today could accelerate global warming.

Fortunately, much of the methane hydrates are oxidized by microbial mats of sulfate-reducing bacteria and anaerobic methane-oxidizing archaea (ANME, discussed in Section 19.4). The sulfate reducers plus the methane oxidizers conduct a syntrophic reaction, for which the overall ΔG value is negative:

$$CH_4 + SO_4^{2-} \rightarrow HCO_3^- + HS^- + H_2O$$

In this metabolism, methane is the initial electron donor oxidized by the ANME partner, and sulfate is the terminal electron acceptor reduced by the bacteria. The high sulfate concentration of marine water drives this reaction in anoxic sediment, where all O_2 has been consumed by microbes that oxidize upwelling reduced minerals (such as sulfide oxidizers at thermal vents and cold seeps, discussed in Chapter 21).

In 2012, researchers reported a different source of marine methane, from aerated water: the aerobic ammonia-oxidizing archaea, such as *Nitrosopumilus* (discussed in Chapter 19). William W. Metcalf and colleagues at the University of Illinois at Urbana-Champaign showed that *Nitrosopumilus* sp. conduct reactions that degrade toxic phosphonates, organic compounds containing a direct carbon-phosphorus bond. The reactions release methylphosphonate, a compound that many kinds of bacteria convert to methane in order to acquire the scarce phosphate for phospholipids and nucleic acids. In addition, the sequence of the *Nitrosopumilus* enzyme for methylphosphonate production revealed homologs in the Global Ocean Sampling Expedition (GOS) metagenome—for example, in the bacterium *Pelagibacter*. Thus, both archaeal and bacterial methylphosphonate may be sources of methane from aerated habitats.

The Global Carbon Balance

What determines atmospheric levels of CO_2? The global balance of biological CO_2 fixation and release largely determines the level of CO_2. Since the beginning of the industrial age, however, the release of CO_2 has accelerated significantly. The major part of this increase comes from the combustion of **fossil fuels**, which adds about 6×10^{15} grams of carbon annually to the atmosphere. Fossil fuels are the product of microbial anaerobic digestion of plant and animal remains, reduced to hydrocarbons by the pressure and heat of Earth's crust. When burned as fuel, carbon that had accumulated over millions of years is rapidly returned to the atmosphere as CO_2. Some of the CO_2 flux is compensated by increased CO_2 fixation and ocean absorption, but about a tenth of the carbon remains in the atmosphere.

Another factor in the increase of atmospheric CO_2 is microbial decomposition in the soil (discussed in Chapter 21). Microbes release carbon through respiration (aerobic and anaerobic), as well as fermentation in anoxic strata, and methanogenesis. A major concern today is the increased rate of microbial activity in arctic permafrost as temperatures increase. The arctic permafrost is estimated to store more than twice the amount of carbon that is in the atmosphere. As permafrost melts, so do methane hydrates from methane produced earlier by methanogens (see the Current Research Highlight at the start of the chapter). The microbial release of methane and CO_2 will then increase the rate of global warming, which in turn thaws the permafrost faster.

The contribution of ocean ecosystems to carbon flux is even more difficult to measure. Metagenomic surveys are revealing previously unrecognized communities of marine phototrophs, such as submicroscopic cyanobacteria, the prochlorophytes (*Prochlorococcus* species). In the deep ocean, similar studies reveal new methane-oxidizing archaea (ANME) that play a critical role in removing benthic methane hydrates.

Studies of carbon and oxygen flux based on isotope ratios suggest a much greater level of biological production than earlier estimates suggested. The oceans conduct more than half of the global biological uptake of carbon from the atmosphere. This kind of information is crucial for predicting the global effects of changes in atmospheric CO_2.

From a political standpoint, ecosystems such as forests that act as carbon sinks by fixing carbon into stable biomass are considered desirable because they lessen the rate of CO_2 input into the atmosphere. Ecosystems that act as sources of carbon dioxide may be viewed with disfavor because they contribute to global warming. But what if CO_2-generating ecosystems provide other environmental benefits? For example, wetlands are among Earth's most productive ecosystems, supporting vast amounts of plant and animal life, although they also release significant amounts of CO_2 and methane. An example of research on wetland carbon cycles is shown in eTopic 22.1.

To Summarize

- **The most accessible carbon reservoir is the atmosphere (CO_2).** Atmospheric carbon is severely perturbed by the burning of fossil fuels.
- **Cyanobacteria and other phytoplankton cycle much of the CO_2 in the biosphere.** Oxygen released by phototrophs is used by aerobic heterotrophs and lithotrophs. Marine plankton, as well as terrestrial trees and mycorrhizal fungi, serve as major carbon sinks.
- **Carbon cycling is linked to the cycling of hydrogen and oxygen.**
- **Anaerobic environments cycle carbon through bacteria and archaea.** Bacteria conduct fermentation and anaerobic respiration.
- **Methanogens release methane in anoxic water and soil.** In oxygenated water, methane may be released by bacteria via methylphosphonate. Methane is oxidized to CO_2 by methane-oxidizing bacteria and archaea.
- **Microbial decomposition returns CO_2 to the atmosphere.** Microbial decomposition is the main source of accelerated CO_2 flux from the soil, especially from melting permafrost.

22.3

The Hydrologic Cycle and Wastewater Treatment

The fate and distribution of complex carbon compounds are largely functions of the **hydrologic cycle**, or **water cycle**, the cyclic exchange of water between atmospheric water vapor and Earth's ecosystems (**Fig. 22.4A**). A vast reservoir of water is supplied by the ocean. In the hydrologic cycle, water precipitates as rain, which is drawn by gravity into groundwater, rivers, lakes, and ultimately the ocean. All along this route, of course, evaporation returns water to the air. Human communities interact with the hydrologic cycle by drawing water for drinking and other purposes, and by returning wastewater. Before wastewater can be safely returned to the hydrologic cycle, organic contaminants must be removed. Key parts of that treatment are performed by microbes.

A. Hydrologic cycle

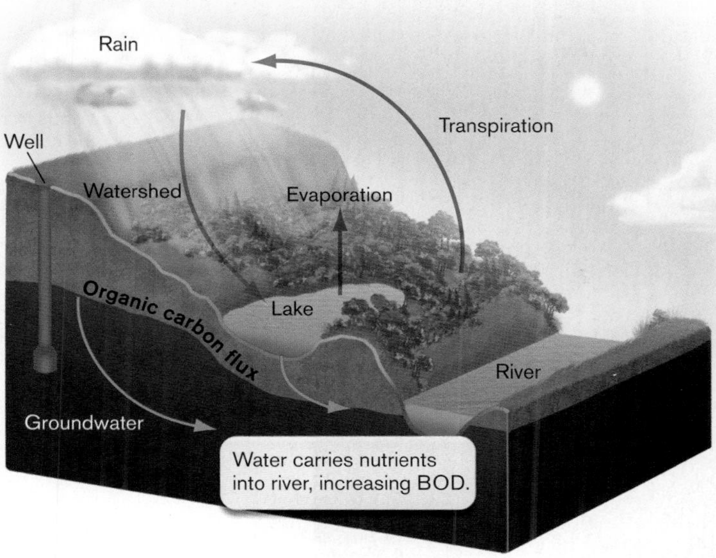
Rain
Transpiration
Well
Watershed
Evaporation
Organic carbon flux
Lake
River
Groundwater
Water carries nutrients into river, increasing BOD.

B. BOD measurement

JOEL GORDON, PITTSFIELD WASTEWATER TREATMENT PLANT

C. BOD analyzer

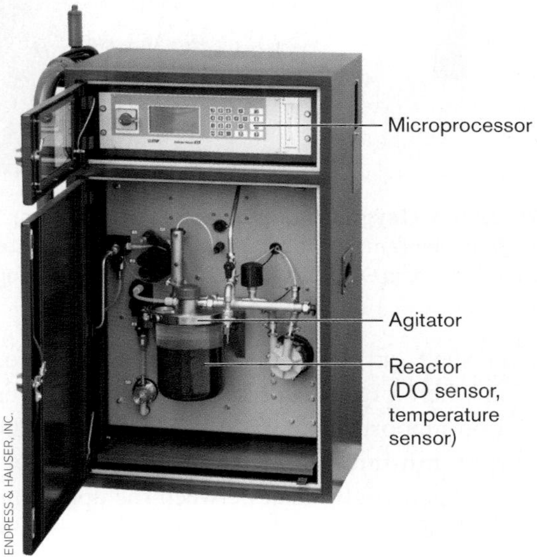
Microprocessor
Agitator
Reactor (DO sensor, temperature sensor)
ENDRESS & HAUSER, INC.

FIGURE 22.4 ■ The hydrologic cycle interacts with the carbon cycle. **A.** The hydrologic cycle carries bacteria and organic carbon into groundwater and aquatic systems. **B.** Bottled water samples are measured for dissolved oxygen over time; the rate of decrease of dissolved oxygen indicates biochemical oxygen demand (BOD). The rate of decrease of dissolved oxygen in water samples is approximately proportional to the concentration of organic matter available for respiration. **C.** A microprocessor-controlled BIOX-1010 BOD analyzer measures rate of respiration. Water samples are mixed with a concentrated microbial biomass, and a dissolved-oxygen (DO) sensor measures small rates of oxygen decrease over time.

Biochemical Oxygen Demand

What determines the health of an aquatic ecosystem? As discussed in Chapter 21, a major factor in the health of ecosystems is the balance between the level of oxygen and the levels of reduced organic nutrients. Organic contamination destabilizes marine, aquatic, and terrestrial ecosystems. Water passing through the ground and aquatic ecosystems carries organic carbon material from humus, sewage, and fertilizer runoff. A sudden influx of rich carbon substrates accelerates respiration by aquatic microbes. Microbial respiration then competes with that of fish, invertebrates, and amphibians for the limited supply of oxygen dissolved in water, raising the **biochemical oxygen demand (BOD)**, also called the biological oxygen demand (BOD). The higher the concentration of organic substances, the higher the BOD arising from microbial oxygen consumption.

High BOD can cause a massive die-off of fish and other aquatic animals. Thus, a routine part of monitoring the health of lakes and streams is the measurement of BOD. A standard value of BOD is defined by measuring the rate of oxygen uptake in a water sample by a defined set of heterotrophic bacteria (**Fig. 22.4B**). Oxygen uptake is observed in a BOD analyzer (**Fig. 22.4C**), which detects dissolved oxygen in water. The rate of decrease of dissolved oxygen measured by the BOD analyzer is approximately proportional to the amount of dissolved organic matter available for respiration. Note, however, that the BOD in a natural environment will depend on the microbes actually present, as well as the plants and animals competing for oxygen and other resources.

Until recently, BOD was considered a local issue, affecting the health of lakes and rivers in a community. But today we recognize huge impacts of rising BOD in the oceans (**Fig. 22.5**). In the oceans, oxygen levels are high near the surface, where phototrophs release oxygen, but organic nutrients are so scarce that respiration is limited. Oxygen is also high in the deep benthos, because most organic nutrients have been consumed, and because the sheer volume of water can hold dissolved oxygen in large amounts. But near

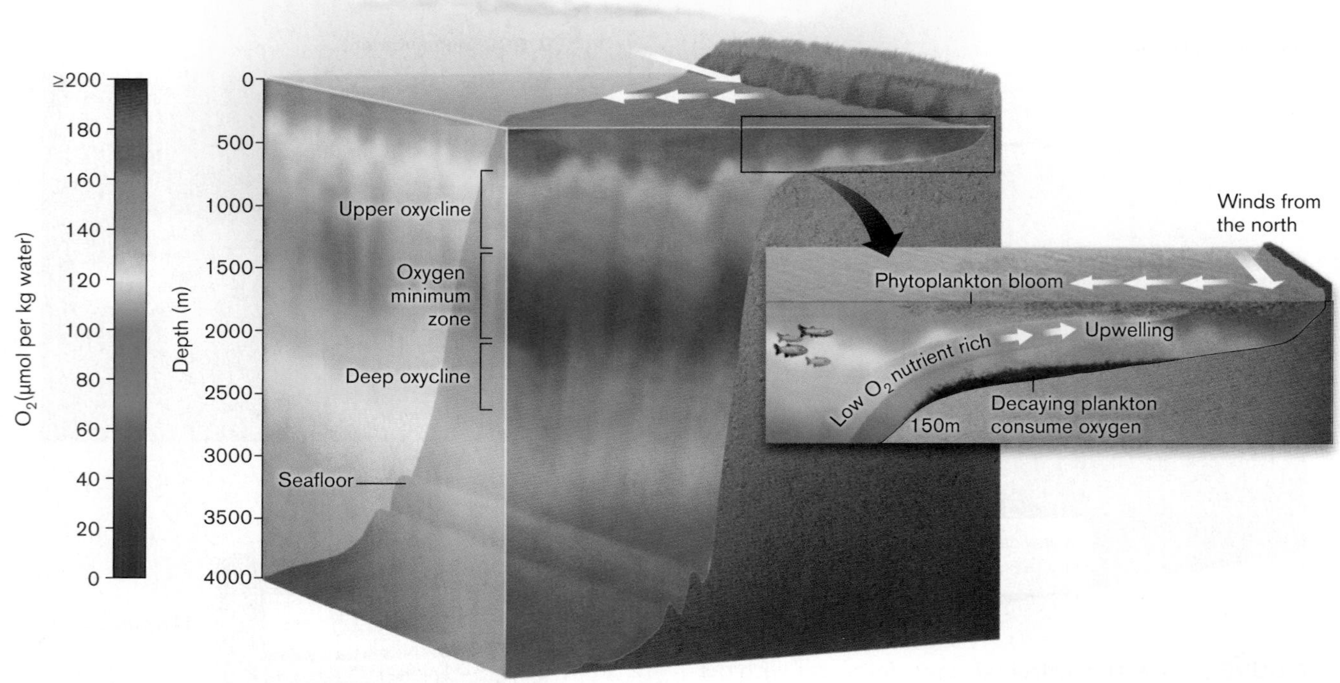
≥200
180
160
140
120
100
80
60
40
20
0
O_2(µmol per kg water)
Depth (m)
0
500
1000
1500
2000
2500
3000
3500
4000
Upper oxycline
Oxygen minimum zone
Deep oxycline
Seafloor
Winds from the north
Phytoplankton bloom
Upwelling
Low O_2 nutrient rich
150m
Decaying plankton consume oxygen

FIGURE 22.5 ■ Oxygen minimum zone in the ocean. The oxygen minimum zone (OMZ) occurs in the region of the coastal shelf where nutrients from below meet the oxygen produced by phototrophs above. In this region, microbial respiration consumes oxygen faster than the organic nutrients. The OMZ can expand upward when phytoplankton blooms decay, consuming the surface oxygen.

the coastal shelf, currents may carry sediment up to a middle region where organic nutrients meet the oxygen. This combination supports rapid bacterial respiration. The result is an **oxygen minimum zone (OMZ)**, a region of low or near-zero oxygen sandwiched between the upper and lower oxygenated layers. Above and below the anoxic water, there is a steep oxycline (gradient of oxygen concentration). A well-known oxygen minimum zone occurs off the coast of Oregon, where crabs and other animals are found dying as they try to escape asphyxiation.

Oxygen minimum zones are worsened by increasing temperatures, which accelerate respiration, and by influx of sewage and agricultural waste. The zone expands upward toward the surface and downward to the sediment, trapping crabs and fish. Today, large regions of ocean have become **dead zones**, or **zones of hypoxia**, devoid of most fish and invertebrates. A major dead zone is a region in the Gulf of Mexico off the coast of Louisiana where the Mississippi River releases about 40% of the U.S. drainage to the sea (**Fig. 22.6A**). Over its long, meandering course, which includes inputs from the Ohio and Missouri rivers, the Mississippi builds up high levels of organic pollutants, as well as nitrates from agricultural fertilizer. When these nitrogen-rich substances flow rapidly out to the gulf in the spring, they lift the nitrogen limitation on algal growth and feed massive algal blooms. The algal population then crashes, and their sedimenting cells are consumed by heterotrophic bacteria. The heterotrophs use up the available oxygen, causing **hypoxia**. Hypoxia kills off the fish, shellfish, and crustaceans over a region equivalent in size to the state of New Jersey.

In 2010 the Gulf of Mexico dead zone was expanded by the unprecedented spill of oil from the *Deepwater Horizon* oil well blowout (**Fig. 22.6B**). The offshore oil platform exploded, releasing millions of barrels of oil into the Gulf over 3 months. The leaked oil killed wildlife throughout the gulf, causing unprecedented environmental damage. Workers tried to contain the spill, but ultimately most of the oil was consumed by marine bacteria. Only bacteria and archaea possess the enzymes needed to degrade the complex organic mixture of petroleum, which includes waxy aromatic compounds such as paraffins (discussed in Chapter 21). While consumption by oil-eating bacteria was important for breaking down the pollution, their respiration raised BOD levels throughout the gulf, thus lowering oxygen available to wildlife. Furthermore, studies of earlier oil spills, such as the *Exxon Valdez* oil spill in 1989, show that certain oil components linger in the environment for decades, including toxins such as polycyclic aromatic hydrocarbons (PAHs). Further research is needed to increase the microbial degradation of pollutants.

Dead zones now occur along the coasts of industrial and developing countries throughout the world, including, for example, India and Australia. They deplete habitat for much marine life—for example, forcing sharks to swim out

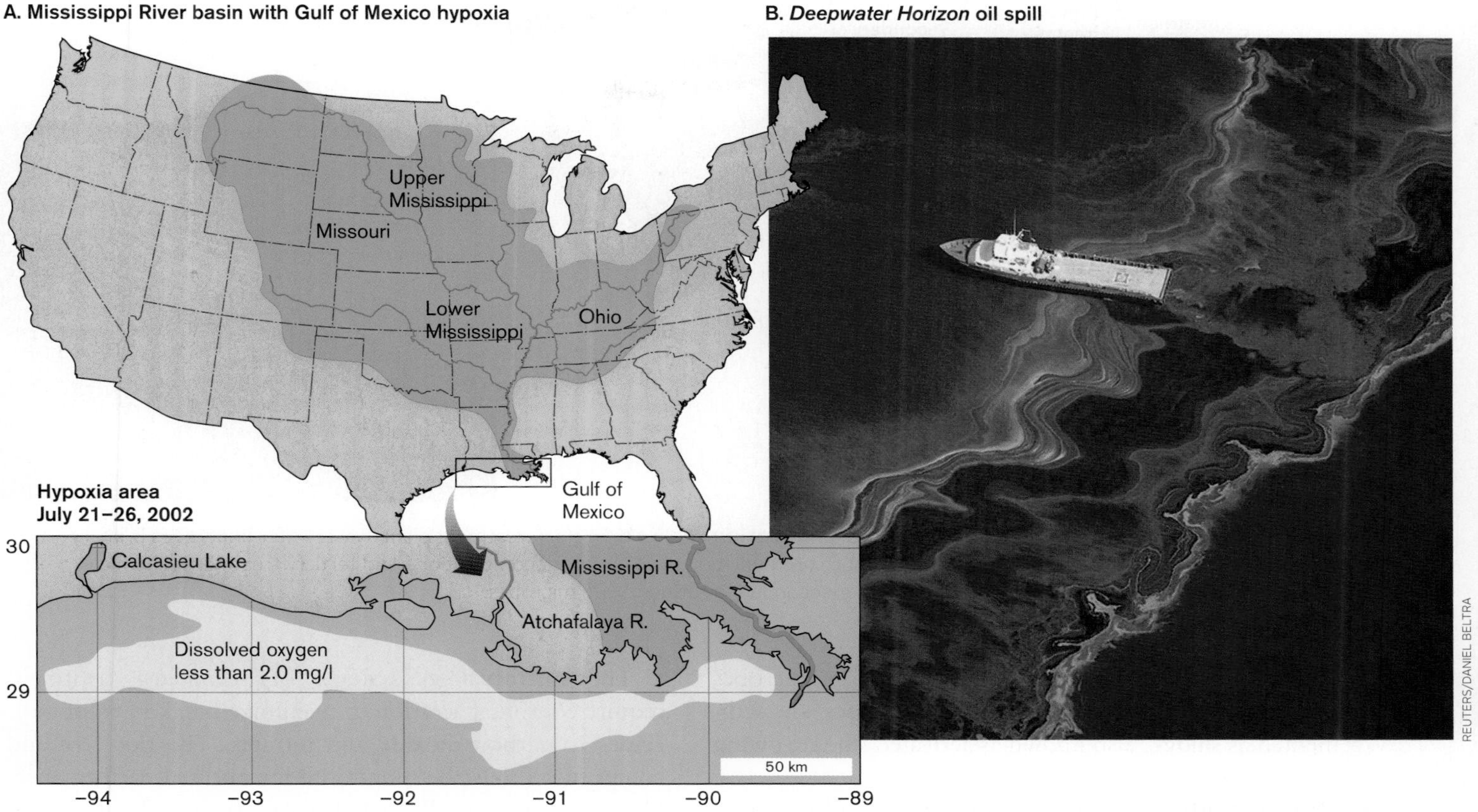
A. Mississippi River basin with Gulf of Mexico hypoxia
Upper Mississippi
Missouri
Lower Mississippi
Ohio
Hypoxia area
July 21–26, 2002
Gulf of Mexico
30
29
Calcasieu Lake
Mississippi R.
Atchafalaya R.
Dissolved oxygen less than 2.0 mg/l
50 km
−94
−93
−92
−91
−90
−89
B. Deepwater Horizon oil spill
REUTERS/DANIEL BELTRA

FIGURE 22.6 ■ The dead zone in the Gulf of Mexico. A. Map of the Mississippi River drainage, which empties into the Gulf of Mexico. Every summer, drainage high in organic carbon and nitrogen causes algal blooms, leading to hypoxia and death of fish. **B.** Aerial view of the oil leaked from the *Deepwater Horizon* oil wellhead in the Gulf of Mexico, 2010. *Source:* A: National Oceanographic and Atmospheric Administration.

of hypoxic regions and nearer the shore. Dead zones contribute to the crash of fisheries worldwide, removing a critical food resource for human populations. Plans to clean up dead zones face daunting costs.

To avoid such dead zones, release of industrial pollutants such as petroleum must be prevented. In addition, all the communities throughout the river drainage areas must treat their wastes to eliminate nitrogenous wastes before disposal. There are two common approaches to community wastewater treatment, both of which involve microbial partners: wastewater treatment plants and wetland filtration. Both approaches depend on microbes to remove organic carbon and nitrogen from water before it returns to aquatic systems and ultimately the ocean.

Wastewater Treatment

In industrialized nations, all municipal communities use some form of **wastewater treatment** (**Fig. 22.7**). The purpose of wastewater treatment is to decrease the BOD and the level of human pathogens before water is returned to local rivers. The treatment process includes microbial metabolism. A modern wastewater plant can convert sewage into water that exceeds all government standards for humans to drink.

The wastewater treatment plant is the final destination for all household and industrial liquid wastes passing through the municipal sewage system. A typical plant includes the following stages of treatment:

- **Preliminary treatment** consists of screens that remove solid debris, such as sticks, dead animals, and feminine hygiene items.
- **Primary treatment** includes fine screens and sedimentation tanks that remove insoluble particles. The particles eventually are recombined with the solid products of wastewater treatment to form what is known as **sludge**. The sludge ultimately is used for fertilizer or landfill.
- **Secondary treatment** consists primarily of microbial ecosystems that decompose the soluble organic content of wastewater, by aerobic and anaerobic respiration. The nutrient removal process may include biological removal of nitrogen and phosphorus. If included at this point, nitrogen is removed by nitrifying bacteria that oxidize ammonium, and phosphorus is removed

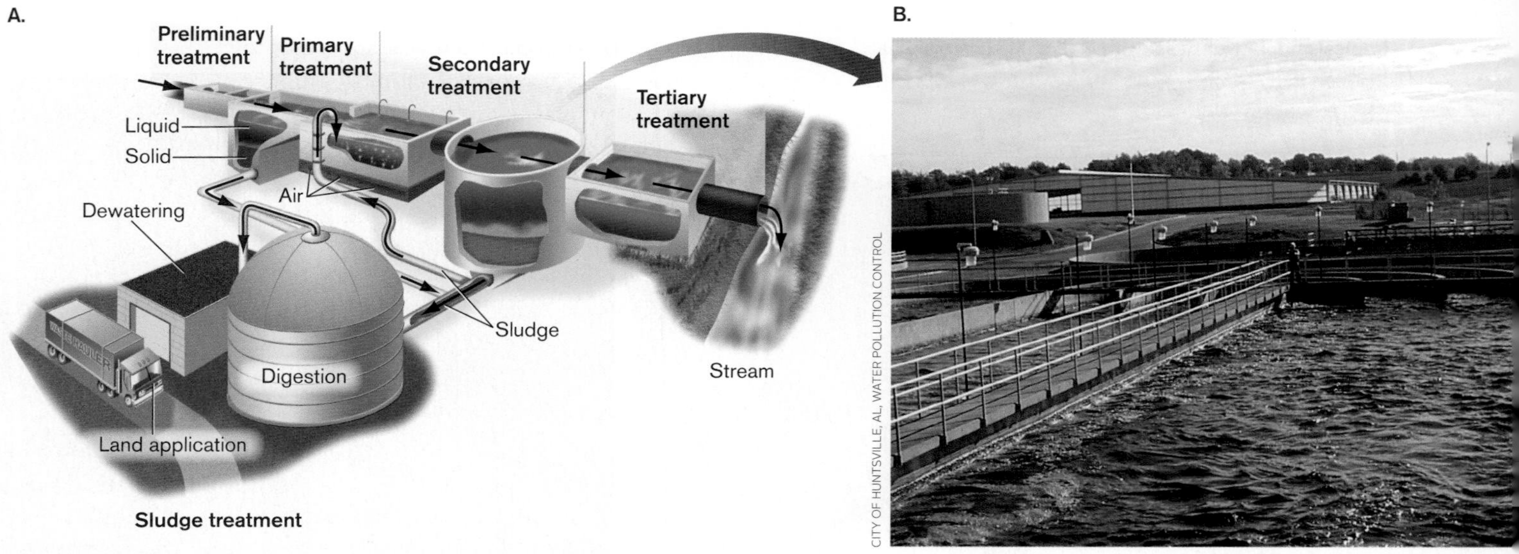
A.
Preliminary treatment
Primary treatment
Secondary treatment
Tertiary treatment
Liquid
Solid
Air
Dewatering
Sludge
Stream
Digestion
Land application
Sludge treatment
B.
CITY OF HUNTSVILLE, AL, WATER POLLUTION CONTROL

FIGURE 22.7 ■ Wastewater treatment with bioremediation. A. In a municipal treatment plant, wastewater undergoes primary treatment (filtering and settling), secondary treatment (bioremediation by microbial decomposition), and tertiary treatment (chemical treatments including chlorination). **B.** Aeration basin for secondary treatment with microbes.

by polyphosphate-accumulating bacteria. The microbes form particulate **flocs** of biofilm. The flocs are sedimented as sludge, also known as activated sludge, owing to their microbial activity.

- **Tertiary (advanced) treatment** includes filtration of particulates from the microbial flocs of secondary treatment, and may include chemical processes to decrease nitrogen and phosphorus. The final step involves disinfection to eliminate pathogens, usually by a chemical process such as chlorination. The treated water is then returned to local aquatic waterways.

The microbial ecosystems of secondary treatment require continual aeration to maximize breakdown of molecules to carbon dioxide and nitrates. The floc size and composition must be monitored for optimal performance. Floc microbes typically include bacilli such as *Zoogloea, Flavobacterium,* and *Pseudomonas,* as well as filamentous species such as *Nocardia* (**Fig. 22.8A**). Optimal treatment depends on the ratio of filamentous to single-celled bacteria: enough filaments to hold together the flocs for sedimentation, but not so many as to trap air and cause flocs to float and foam, preventing sedimentation.

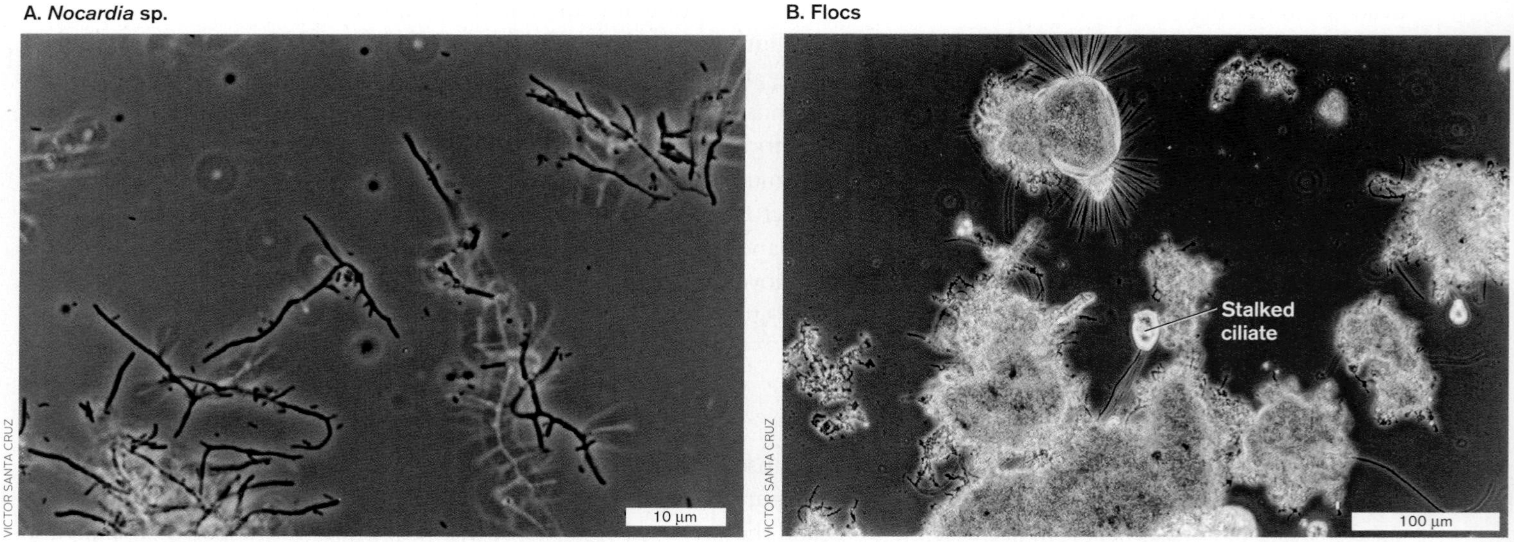
A. *Nocardia* sp.
VICTOR SANTA CRUZ
10 μm
B. Flocs
Stalked ciliate
VICTOR SANTA CRUZ
100 μm

FIGURE 22.8 ■ Microbes in wastewater bioremediation. A. Filamentous *Nocardia* sp. bacteria from flocs formed during secondary treatment (light micrograph). **B.** Flocs with a stalked ciliate, which preys on bacteria (light micrograph).

Besides bacteria, the ecosystem of activated sludge includes filamentous methanogens (discussed in Chapter 19), which metabolize short-chain molecules such as acetate within the anaerobic interior of flocs. Thus, wastewater treatment generates methane, often in quantities that can be recovered as fuel.

In addition, the bacteria are preyed upon by protists such as stalked ciliates (**Fig. 22.8B**), swimming ciliates, and amebas, as well as invertebrates such as rotifers and nematodes. The predators serve the valuable function of limiting the numbers of planktonic single-celled bacteria, enabling the bulk of the biomass to be removed by sedimentation.

Thought Question

22.2 What would happen if wastewater treatment lacked microbial predators? Why would the result be harmful?

Wastewater treatment plants are remarkably effective at converting human wastes to ecologically safe water and ultimately human drinking water. The plants are, however, impractical for purifying the runoff from large agricultural operations. For large-scale alternatives to treatment plants, communities and agricultural operations are looking to wetland restoration. Much of our current water supply is already filtered and purified by natural wetlands, such as the Florida Everglades (**Fig. 22.9A**). Wetlands remove nitrogen through the action of denitrifying bacteria. In wetlands, rainwater and river water trickle slowly through vast stretches of soil, where microbial conversion acts as the foundation of a macroscopic ecosystem including trees and vertebrates. Thus, much of the carbon and nitrogen is fixed into valuable biomass. In the wetlands of the Everglades, the remaining water filters slowly through limestone into underground aquifers, which ultimately provide water through wells to human communities.

Some agricultural operations are building artificial wetlands to replace treatment plants. **Figure 22.9B** shows a hog farm where a series of terraced wetlands was built to drain liquefied manure. The wetlands were found to produce fewer odors and to remove organics more efficiently and at a lower cost than do traditional filtration plants.

A much greater challenge is that of industrial runoff, in which toxic wastes from factory chemistry enter an aquifer, at levels endangering human health. These chemicals may poison the treatment plants designed for human effluents. Large regions of land surrounding such a plant may require remediation that is expensive or impractical. And traces of pollutants such as chlorinated aromatics reach every spot on the globe; for example, dioxins are found in the tissues of Antarctic penguins.

How can we remediate industrial pollutants before they poison the local countryside and spread throughout the globe? One way is to harness naturally occurring bacteria for bioremediation. The bacterial community's reaction rate can be enhanced through selection by enrichment culture (described in Chapter 4). In the culture, organisms from the polluted site are cultivated repeatedly in the presence of added pollutant, and then tested for activity using radiolabeled substrates. An example of successful bioremediation is the case of Aberdeen Proving Ground (**Special Topic 22.1**).

A. Everglades

B. Artificial wetlands

EVERGLADES NATIONAL PARK

TIM MCCABE, USDA/NRCS

FIGURE 22.9 ▪ Water filtration by wetlands. A. The marshes of the Everglades act as natural filters that bioremediate water entering the aquifers of southern Florida. **B.** A filtering system installed by Steve Kerns on his hog farm in Taylor County, Iowa. A series of hillside terraces form wetlands containing bacteria that metabolize hog manure and wastewater.

Special Topic 22.1: Bioremediation of Weapons Waste

Since 1917, the Aberdeen Proving Ground has been a site for experimental manufacture and testing of weapons and ammunition; chemical weapons were stored there until 2006. Located near the mouth of the Susquehanna River, where it enters Chesapeake Bay, the proving ground releases chlorinated pollutants that threaten a wide area of nature preserves and historical sites. A plume of waste, including highly toxic compounds such as chlorinated eth<u>e</u>nes and eth<u>a</u>nes, has invaded the surrounding wetland. How can this site be cleaned up and made safe?

Michelle Lorah, Emily Majcher, and colleagues at the U.S. Geological Survey (USGS) are enlisting indigenous bacteria to do the job. As a demonstration project, they built a "reactive mat," a microbial mat containing cultures of bacteria that catabolize chlorinated hydrocarbons. An example of such bacteria presented in Chapter 14 was *Dehalobacter* sp., which remove chloride from chlorinated molecules by oxidizing H_2—a process called "dehalorespiration," a form of hydrogenotrophy in which the halogenated hydrocarbon actually serves as the electron acceptor. Lorah's team used enrichment culture to identify a consortium (mixed population) of species that rapidly reduce halogenated hydrocarbons by removing chloride and substituting a hydrogen atom. The team obtained samples from the Aberdeen wetland and then cultured the bacterial community in the presence of chlorinated hydrocarbons. The consortium samples were "fed" chlorinated hydrocarbons plus lactate or H_2 as an electron donor, and incubated anaerobically. An enriched consortium was identified, WBC-2, which could break down 90% of a dichloroethene within a month. The consortium contained a variety of dechlorinating bacteria, such as *Dehalococcoides* species.

Lorah's team installed a microbial mat containing consortium WBC-2 on sediment where a contaminated plume of groundwater seeped into the wetland at Aberdeen (**Fig. 1**). The aim was to filter the chlorinated molecules through the mat before they reached the water. The water above and below the mat was monitored for a period of a year, during which time more than 94% of the chloroethenes and chloroethanes were degraded. **Figure 2** shows an example of results from one month in which researchers measured the concentrations of these pollutants in water above and below the microbial mat. Overall, this pilot study demonstrates the practical capacity of a selectively enriched microbial mat to remediate heavily polluted groundwater.

RESEARCH QUESTION

What techniques might you use to study the microbial mat and determine which species are actually metabolizing the chlorinated molecules? Suppose you could isolate these species; what would be the advantages and disadvantages of doing so, as opposed to using the original consortium?

USGS

FIGURE 1 ▪ Bacterial reactive mat designed to dechlorinate pollutants in wetland at Aberdeen Proving Ground. The mat contained peat and compost augmented with the WBC-2 consortium of dechlorinating bacteria (inset, SEM).

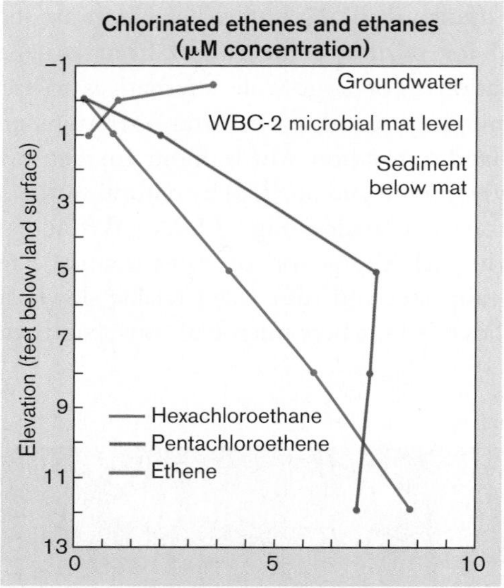
Chlorinated ethenes and ethanes
(μM concentration)
Groundwater
WBC-2 microbial mat level
Sediment
below mat
Elevation (feet below land surface)
Hexachloroethane
Pentachloroethene
Ethene
−1
1
3
5
7
9
11
13
0
5
10

FIGURE 2 ▪ Chlorinated ethenes and ethanes depleted by the WBC-2 microbial mat. The plume of water flowed up through the mat, where hexachloroethane and pentachloroethene concentrations decline. Concentration of the dechlorinated product ethene rises above the mat. Data for the month of June 2005, from the USGS study.

Emily H. Majcher, Michelle M. Lorah, Daniel J. Phelan, and Angela L. McGinty. 2009. Design and performance of an enhanced bioremediation pilot test in a tidal wetland seep, West Branch Canal Creek, Aberdeen Proving Ground, Maryland. U.S. Geological Survey Scientific Investigations Report 2009–5112.

To Summarize

- **The hydrologic cycle is the cyclic exchange of water between the atmosphere and the biosphere.** Water precipitates as rain, which enters the ground and ultimately flows to the oceans. Along the way, some of the water evaporates, returning to the atmosphere.
- **Water carries organic carbon that generates biochemical oxygen demand (BOD).** High BOD accelerates heterotrophic respiration and depletes oxygen needed by fish.
- **Oxygen minimum zones are hypoxic regions of ocean sandwiched between upper and lower oxygenated layers.** Low oxygen occurs at middle depth, where the oxygen from phototrophs meets upwelling organic nutrients.
- **Dead zones occur where sewage and agricultural runoff expand the oxygen minimum zones.** Dead zones exclude all aerobic life.
- **Wastewater treatment cuts down BOD.** Secondary treatment involves formation of flocs, microbial communities that decompose the soluble organic content.
- **Wetlands filter water naturally.** Wetland filtration helps purify groundwater entering aquifers.
- **Industrial effluents are highly toxic and can reach all parts of the globe.** Bioremediation with microbes may eliminate such toxins.

22.4

The Nitrogen Cycle

Besides carbon, oxygen, and hydrogen, another major element that cycles largely by microbial conversion is nitrogen (**Fig. 22.10**). The nitrogen cycle is notable in two respects:

- **Many oxidation states.** A larger number of oxidation states exist for biological nitrogen than for any other major biological element (see **Table 22.3**). Conversion between these oxidation states requires metabolic processes such as N_2 fixation, lithotrophy, and anaerobic respiration.
- **Dependence on prokaryotes.** Many steps of the nitrogen cycle require bacteria and archaea. Without bacteria and archaea, the nitrogen cycle would not exist.

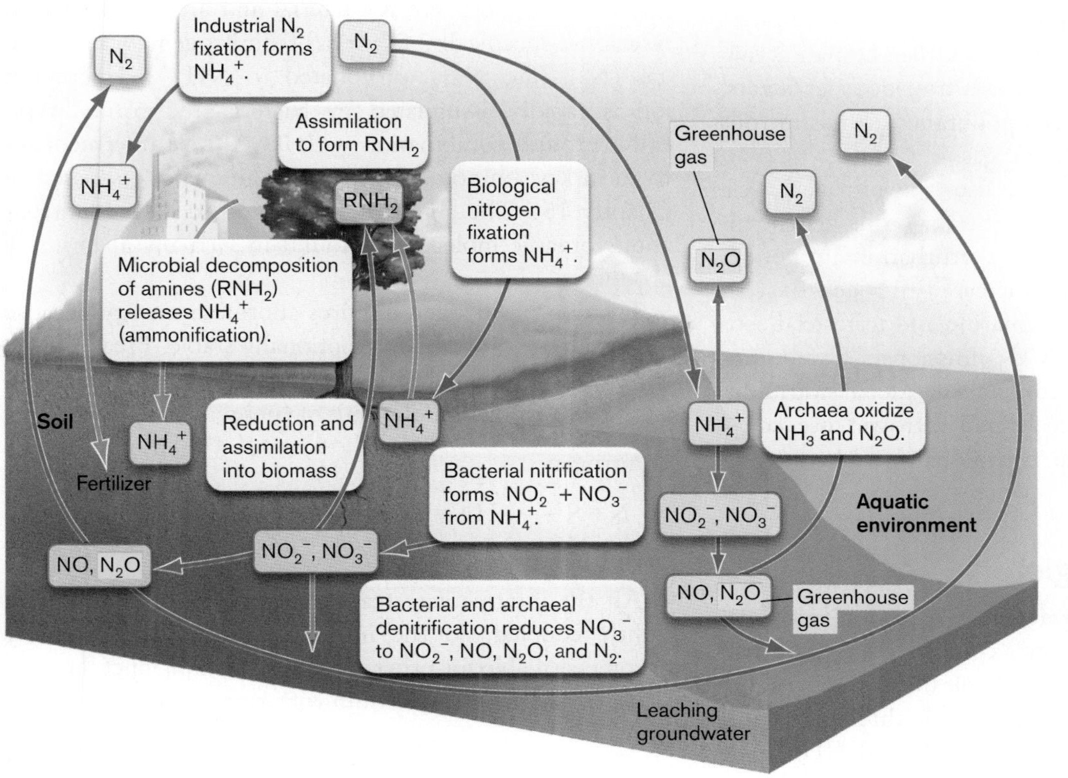

N_2
Industrial N_2 fixation forms NH_4^+.
N_2
Assimilation to form RNH_2
NH_4^+
RNH_2
Biological nitrogen fixation forms NH_4^+.
Greenhouse gas
N_2
N_2
N_2O
Microbial decomposition of amines (RNH_2) releases NH_4^+ (ammonification).
Soil
NH_4^+
Fertilizer
Reduction and assimilation into biomass
NH_4^+
Bacterial nitrification forms $NO_2^- + NO_3^-$ from NH_4^+.
NH_4^+
Archaea oxidize NH_3 and N_2O.
NO_2^-, NO_3^-
Aquatic environment
NO, N_2O
NO_2^-, NO_3^-
Bacterial and archaeal denitrification reduces NO_3^- to NO_2^-, NO, N_2O, and N_2.
NO, N_2O
Greenhouse gas
Leaching groundwater

FIGURE 22.10 ▪ The global nitrogen cycle. Bacteria and archaea interconvert forms of nitrogen throughout the biosphere. Blue = reduction; red = oxidation; orange = redox-neutral.

Thought Question

22.3 How many kinds of biomolecules can you recall that contain nitrogen? What are the usual oxidation states for nitrogen?

Sources of Nitrogen

Where is nitrogen found on Earth? As we saw for carbon, a significant amount of nitrogen is found in Earth's crust, in the form of ammonium salts in rock (see **Table 22.2**). Unlike crustal carbon, however, the nitrogen found in rock is largely inaccessible to microbes. The major accessible source of nitrogen is the atmosphere, of which dinitrogen gas (N_2) constitutes 79%. However, N_2 is a highly stable molecule that requires enormous input of reducing energy before assimilation is possible (see Chapter 15). Thus, for many natural ecosystems, and most forms of agriculture, nitrogen is the limiting nutrient for primary productivity.

Until recently in Earth's history, nitrogen was fixed entirely by nitrogen-fixing bacteria and archaea. In the twentieth century, however, the Haber process was invented for artificial nitrogen fixation to generate fertilizers for agriculture. The process was devised by German chemist Fritz Haber, who won the 1918 Nobel Prize in Chemistry. In the Haber process, N_2 is hydrogenated by methane under extreme heat and pressure. Today, the Haber process for producing fertilizers accounts for approximately 30%–50% of all nitrogen fixed on Earth. Other human activities, such as fuel burning and use of nitrogenous fertilizers, contribute to oxidized nitrogen pollutants such as nitrous oxide (N_2O), a potent greenhouse gas.

Because of extensive fertilizer use, the nitrogen cycle today is the most perturbed of the major biogeochemical cycles. The ultimate effects of perturbation are not yet clear. Some models show that increased nitrogen fixation could fertilize CO_2 fixation by marine and terrestrial ecosystems, partly decreasing net CO_2 emissions. The amount, however, will be too small to prevent global warming. Other models show that accelerated nitrogen use could limit nitrogen availability in the global biosphere, with unknown consequences.

The "Nitrogen Triangle"

The numerous oxidation states available for microbial metabolism of nitrogen generate a complex cycle of conversions. One way to organize the complexity is to envision a "nitrogen triangle" whose corners include three key forms: atmospheric $\mathbf{N_2}$; the reduced forms NH_3 and NH_4^+; and the oxidized forms NO_2^- and NO_3^- (**Fig. 22.11A**). At the base of the triangle, both reduced and oxidized forms of nitrogen are assimilated into biomass.

Thought Question

22.4 The nitrogen cycle has to be linked with the carbon cycle, since both contribute to biomass. How might the carbon cycle of an ecosystem be affected by increased input of nitrogen?

We will consider in turn the three sides of the nitrogen triangle:

- **N_2 fixation to ammonium (NH_4^+)**, a form assimilated into biomass by microbes and plants.
- **Ammonia oxidation, in which ammonia is oxidized aerobically to nitrite (NO_2^-) and nitrate (NO_3^-)**, which are also assimilated by microbes and plants. Aerobic oxidation to nitrite or nitrate is called nitrification. In the absence of oxygen gas, anaerobic ammonium oxidation by nitrite generates N_2—a reaction called anammox.
- **Denitrification of NO_2^- and NO_3^- back to N_2** (or, in carbon-rich habitats, reduction to NH_3).

Nitrogen fixation. The main avenue for entry of nitrogen into the biosphere is bacterial and archaeal **nitrogen fixation**—specifically, fixation of dinitrogen, or nitrogen gas (N_2), into NH_3, protonated to NH_4^+. Ammonium ion is rapidly assimilated by bacteria and plants, typically through combination with TCA cycle intermediates to form key amino acids, such as glutamate (discussed in Chapter 15). The ability to assimilate NH_4^+ into nitrogenous organic molecules is found in virtually all primary producers.

Fixation of nitrogen requires enormous energy because the triple bond of N_2 is exceptionally stable. Breaking the triple bond to generate ammonia requires a series of reduction steps involving high input of energy:

Dinitrogen

$$N{\equiv}N \xrightarrow[2H^+ + 2e^-]{} HN{=}NH \xrightarrow[2H^+ + 2e^-]{} H_2N{-}NH_2 \xrightarrow[2H^+ + 2e^-]{} 2NH_3$$

All the intermediate reactions of nitrogen fixation are tightly coupled, so the intermediate compounds rarely become available for other uses. The final product, ammonia, exists in ionic equilibrium with ammonium ion:

$$NH_3 + H_2O \rightleftharpoons NH_4^+ + OH^-$$

A. The nitrogen triangle

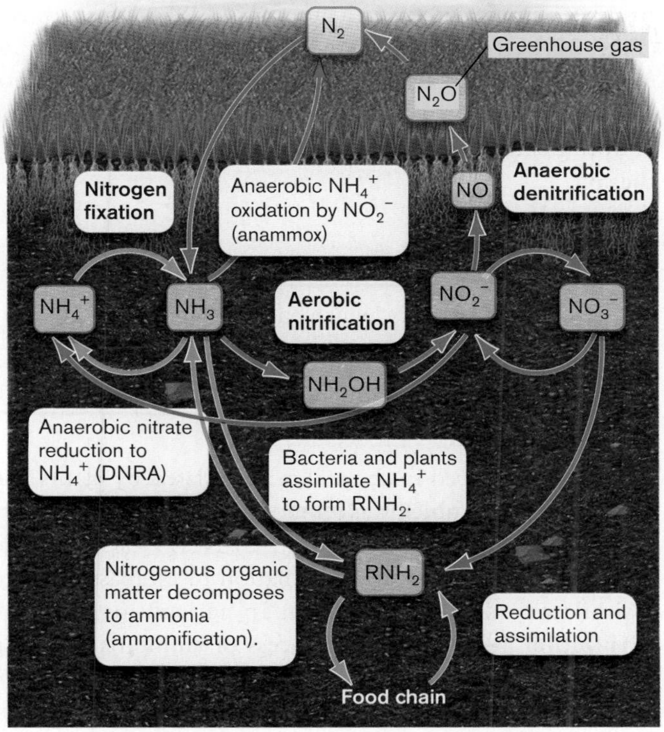
N_2
Greenhouse gas
N_2O
NO
Anaerobic denitrification
Nitrogen fixation
Anaerobic NH_4^+ oxidation by NO_2^- (anammox)
NH_4^+
NH_3
Aerobic nitrification
NO_2^-
NO_3^-
NH_2OH
Anaerobic nitrate reduction to NH_4^+ (DNRA)
Bacteria and plants assimilate NH_4^+ to form RNH_2.
Nitrogenous organic matter decomposes to ammonia (ammonification).
RNH_2
Reduction and assimilation
Food chain

B. *Nitrobacter winogradskyi*

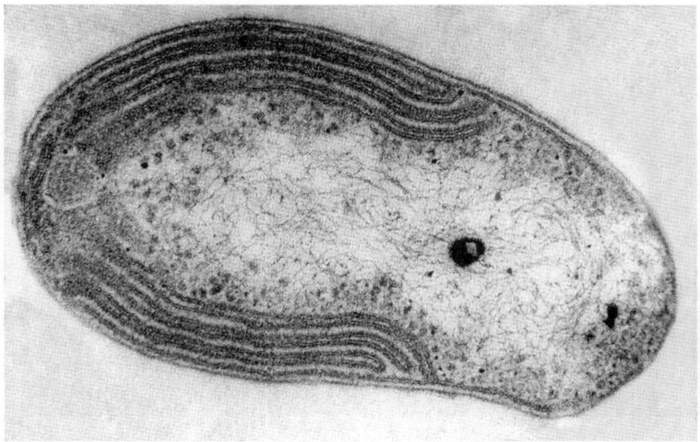
SCIENTECMATRIX.COM

FIGURE 22.11 ■ The nitrogen cycle: fixation, nitrification, denitrification. A. The "nitrogen triangle" consists of nitrogen fixation and assimilation, reductive dissimilation of nitrate (denitrification; blue), and oxidation (nitrification; red). Denitrification includes production of the potent greenhouse gas nitrous oxide (N_2O). Assimilation into biomass is often reductive; virtually all nitrogen in biomolecules is highly reduced. Oxidation of ammonia generates nitrites and nitrates, whose runoff can pollute water supplies. **B.** *Nitrobacter winogradskyi* oxidizes nitrite to nitrate. Folded layers of membrane contain the electron transport complexes. (Cell length 0.5–1.0 µm, TEM.)

Half the ammonia is protonated at about pH 9.3, so at near-neutral pH, most exists in the protonated form, NH_4^+.

Nitrogen fixation is catalyzed by the enzyme nitrogenase (discussed in Chapter 15). The highly reductive reactions of nitrogenase require exclusion of oxygen; yet they also require tremendous input of energy, usually provided by aerobic metabolism. Therefore, nitrogen-fixing bacteria generally isolate anaerobic nitrogen fixation from aerobic metabolism by one of several mechanisms, such as the heterocysts of cyanobacteria (discussed in Chapter 15).

Given the energy expense and the need to exclude oxygen, many species of bacteria, as well as all eukaryotes, have lost the nitrogen fixation pathway by degenerative evolution. But all ecosystems, both marine and terrestrial, include some species of bacteria and archaea that fix N_2 into ammonia. Nitrogen-fixing bacteria in the soil include obligate anaerobes, such as *Clostridium* species, and facultative Gram-negative enteric species of *Klebsiella* and *Salmonella,* as well as obligate respirers such as *Pseudomonas.* In oceans and in freshwater systems, cyanobacteria are the major nitrogen fixers. After fixation, all these organisms assimilate the reduced nitrogen into essential components of their cells. Within an ecosystem, nitrogen fixers ultimately make the reduced nitrogen available for assimilation by nonfixing microbes and plants, either directly through symbiotic association (such as that of rhizobia and legumes, discussed in Chapter 21) or indirectly through predation (marine cyanobacteria) and decomposition (soil bacteria).

If nitrogen fixation is ubiquitous in soil and water, then why is nitrogen limiting? Nitrogen fixation is extremely energy intensive; thus, the rate of fixation usually fails to meet the potential demand of other members of the ecosystem. The one exception is legume symbiosis with rhizobia, which provide ample nitrogen for their hosts.

In agriculture, symbiotic nitrogen fixation by rhizobial bacteria increases the yield of crops such as soybeans (**Fig. 22.12A**). To enhance colonization by nitrogen fixers, farmers apply molecules called isoflavonoids, which mimic the natural plant-derived attractants for rhizobia.

Nitrification. Free ammonia in soil or water is quickly oxidized for energy by **nitrifiers**, bacterial species that possess enzymes for oxidation of ammonia to nitrite (NO_2^-), or of nitrite to nitrate (NO_3^-). This process is called **nitrification**. Nitrification of ammonia is a form of lithotrophy, an energy generation pathway involving oxidation of minerals. The

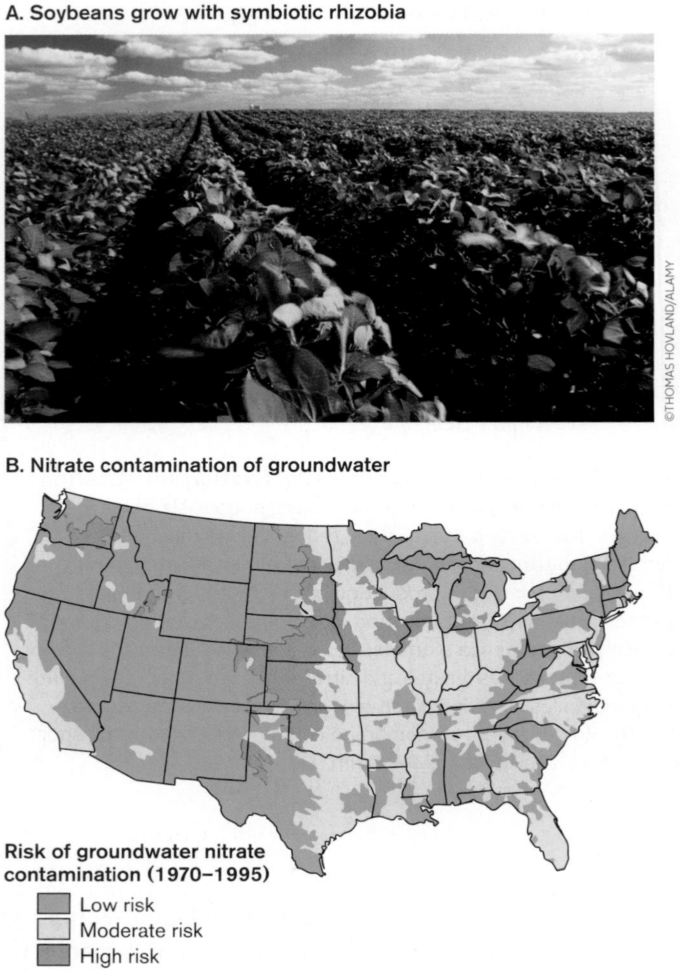
A. Soybeans grow with symbiotic rhizobia
©THOMAS HOVLAND/ALAMY
B. Nitrate contamination of groundwater
Risk of groundwater nitrate contamination (1970–1995)
Low risk
Moderate risk
High risk

FIGURE 22.12 ■ Agricultural benefits and consequences of microbial nitrogen metabolism. A. Soybean field in Maine. Rhizobial nitrogen fixation enhances growth of soybeans and other major crops. **B.** Nitrate in drinking water is especially prevalent in agricultural regions of the United States. Ammonification of fertilizer, followed by nitrification, generates nitrate and nitrite. Nitrate and nitrite runoff from oxidized nitrogenous fertilizers pollutes streams and groundwater.

nitrification pathway generates the red base of the triangle in **Figure 22.11A**. The pathway of nitrification includes:

$$\begin{array}{ccccccc} \text{Ammonia} & & & & \text{Nitrite} & & \text{Nitrate} \\ NH_3 & \longrightarrow & NH_2OH & \longrightarrow & NO_2^- & \longrightarrow & NO_3^- \\ & +\tfrac{1}{2}O_2 & & +O_2 & + & +\tfrac{1}{2}O_2 & \\ & & & & H_2O + H^+ & & \end{array}$$

The pathway includes two separate energy generation mechanisms, utilized by different species: (1) ammonia through NH_2OH to nitrite, and (2) nitrite to nitrate. Typically, soil contains both kinds of microbes living together—a collaboration ensuring complete conversion to nitrate. Nitrifying genera include *Nitrosomonas,* which oxidizes ammonia to nitrite, as well as *Nitrobacter* (**Fig. 22.11B**) and *Nitrospira,* which oxidize nitrite to nitrate. Nitrite can also serve as an electron donor for photosynthesis by *Thiocapsa* species. Note that the production of both nitrite and nitrate generates acid, which can acidify the soil.

Thought Question

22.5 In the laboratory, which bacterial genus would likely grow on artificial medium including NH_2OH as the energy source: *Nitrosomonas* or *Nitrobacter*?

Nitrate produced in the soil is assimilated by plants and bacteria nearly as quickly as ammonium ion, although extra energy is needed to reduce nitrate to NH_4^+ for incorporation into biomass. Nitrate assimilation to biomass is called **assimilatory nitrate reduction**. In agriculture, intensive fertilization generates a large excess of ammonia resulting from **ammonification**, the breakdown of organic nitrogen releasing ammonia. Ammonia is oxidized rapidly by lithotrophic bacteria (nitrifiers). The excess ammonia leads to a buildup of nitrites and nitrates, which are highly soluble in water, and they readily diffuse into aquatic systems. Aquatic nitrate reacts with organic compounds to form toxic nitrosamines. Nitrate influx also relieves the nitrogen limit on algae, causing algal blooms and raising BOD. Chronic nitrate influx leads to eutrophication and die-off of fish.

Consumption of nitrate in drinking water can lead to methemoglobinemia, a blood disorder in which hemoglobin is inactivated. Methemoglobinemia occurs in infants whose stomachs are not yet acidic enough to inhibit growth of bacteria that convert nitrate to nitrite. Nitrite oxidizes the iron in hemoglobin, eliminating its capacity to carry oxygen. The failure to carry oxygen leads to a bluish appearance, one cause of "blue baby syndrome." Nitrite-induced blue baby syndrome is a problem in intensively cultivated agricultural regions, such as Kansas and Nebraska (**Fig. 22.12B**).

Agricultural runoff also contributes to oxygen minimum zones and dead zones, such as those off the Oregon coast and in the Gulf of Mexico. The amines in sewage are removed as ammonia (ammonification), and the ammonia is oxidized by lithotrophs. The major marine ammonia oxidizers include the Thaumarchaeota, such as *Nitrosopumilus* (discussed in Chapter 19).

Denitrification. N_2 is regenerated by anaerobic respiration (**Fig. 22.11A**), in which an oxidized form of nitrogen such as nitrate or nitrite receives electrons from organic electron donors. Bacteria and archaea reduce nitrate through a series of decreased oxidation states back to atmospheric nitrogen:

$$\underset{\text{Nitrate}}{2NO_3^-} \xrightarrow[4H^+ + e^-]{} \underset{\text{Nitrite}}{2NO_2^-} + 2H_2O \xrightarrow[4H^+ + 2e^-]{} \underset{\text{Nitric oxide}}{2NO} + 2H_2O \xrightarrow[2H^+ + 2e^-]{}$$

$$\underset{\text{Nitrous oxide}}{N_2O} + H_2O \xrightarrow[2H^+ + 2e^-]{} \underset{\text{Dinitrogen}}{N_2} + H_2O$$

In terms of elemental flux through ecosystems, nitrate and nitrite reduction are known as **denitrification** or **dissimilatory nitrate reduction**. (In contrast, assimilatory nitrate reduction incorporates the nitrogen into biomass.) Anaerobic respirers in soil or water use an oxidized form of nitrogen as an alternative electron acceptor in the absence of O_2 (discussed in Chapter 14). All types of nitrogen-based anaerobic respiration are repressed in the presence of oxygen, a more favorable electron acceptor; therefore, denitrification is limited to anoxic habitats.

In undisturbed environments, the products of nitrate respiration rarely build up to levels that harm the ecosystem. Heavy fertilization, however, causes buildup of excess nitrate and so increases the environmental rate of denitrification. During this process, some of the nitrogen escapes as nitrous oxide gas (N_2O). A highly potent greenhouse gas, N_2O generates 200 times the warming effect of CO_2; thus, relatively small amounts of N_2O can make a disproportionate contribution to global warming. Furthermore, N_2O in the upper atmosphere reacts catalytically with ozone, depleting the ozone layer. Thus, the atmospheric effects of bacterial denitrification are a serious concern in agricultural and waste treatment processes.

Nitrous oxide also builds up in marine dead zones, where the nitrates from ammonium oxidation are subsequently used as electron acceptors. Syed Wajih Naqvi and colleagues at the National Institute of Oceanography in Goa, India, investigated denitrification in the zone of hypoxia off the coast of India (**Fig. 22.13A**). The study revealed unexpectedly high levels of N_2O production at a series of stations within the zone. In an experiment, introduction of nitrate into water samples from the dead zone led quickly to production of nitrite and N_2O (**Fig. 22.13B**), a process caused by denitrifying bacteria. Studies in 2012 suggest additional N_2O production by ammonia-oxidizing archaea such as *Nitrosopumilus*. Thus, bacterial denitrification and archaeal ammonia oxidation in polluted ocean waters may contribute significantly to global warming.

Dissimilatory nitrate reduction to ammonia. Most environmental nitrate and nitrite are reduced via N_2O as described previously. Certain conditions, however, favor the alternate route of dissimilatory nitrate reduction to ammonia (DNRA) (see **Fig. 22.11A**). Nitrate reduction to ammonia is a form of anaerobic lithotrophy or hydrogenotrophy in which nitrate serves as an electron acceptor and hydrogen gas (H_2) is the electron donor:

$$NO_3^- + 4H_2 + H^+ \longrightarrow NH_3 + 3H_2O$$

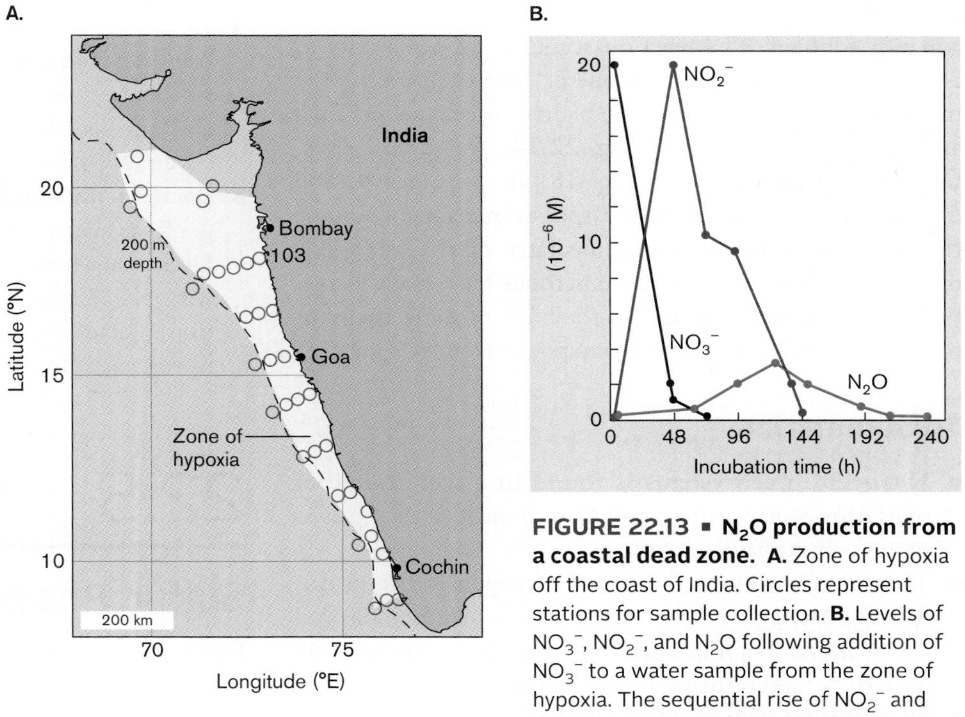
A.
India
Bombay
103
200 m
depth
Goa
Zone of
hypoxia
Cochin
200 km
Latitude (°N)
20
15
10
70
75
Longitude (°E)
B.
20
10
0
(10⁻⁶ M)
NO2−
NO3−
N2O
0
48
96
144
192
240
Incubation time (h)

FIGURE 22.13 ■ N_2O production from a coastal dead zone. A. Zone of hypoxia off the coast of India. Circles represent stations for sample collection. **B.** Levels of NO_3^-, NO_2^-, and N_2O following addition of NO_3^- to a water sample from the zone of hypoxia. The sequential rise of NO_2^- and N_2O indicates metabolism of denitrifying bacteria.

Bacteria reduce nitrate to ammonia mainly in anaerobic environments rich in organic carbon and H_2 generated by fermentation, but low in reduced nitrogen, such as sewage sludge and stagnant water. Another carbon-rich habitat favoring this pathway is the rumen, the digestive tract of cattle, goats, and other ruminant animals (see Chapter 21). Ruminants consume grasses whose cellulose requires digestion by mutualistic microbes. Ruminants also depend on their digestive microbes to assimilate nitrate into ammonia and synthesize amino acids.

Anaerobic ammonium oxidation (anammox). For many years, denitrification was considered the main way that nitrogen compounds return N_2 to the atmosphere. In 2003, two research groups from Europe and Costa Rica, led by Tage Dalsgaard and Marcel Kuypers, showed that in anoxic deep-sea water the major source of N_2 is bacteria conducting anaerobic ammonium oxidation by nitrite (the anammox reaction, discussed in Chapter 14):

$$NH_4^+ + NO_2^- \longrightarrow N_2 + 2H_2O$$

In anammox, ammonium ion serves as electron donor, and nitrite serves as anaerobic electron acceptor—an unusual combination of two different oxidation states of the same key element, nitrogen. The reaction is a kind of anaerobic lithotrophy.

Anammox by bacteria has now been observed in all kinds of anaerobic habitats, including terrestrial soil and aquatic sediment. The reaction accounts for a majority of all N_2 returned to the atmosphere. Among bacteria, the main anammox contributors are planctomycete genera such as *Kuenenia* and *Scalindua* (**Fig. 22.14**). Anammox planctomycetes (discussed in Chapter 18) are an unusual kind of cell wall–less bacteria with a special interior membrane that segregates the toxic intermediates of the anammox reaction. Promising anammox microbes have been identified from wastewater sludge, in the hope of using them for more effective removal of excess nitrogen from wastewater.

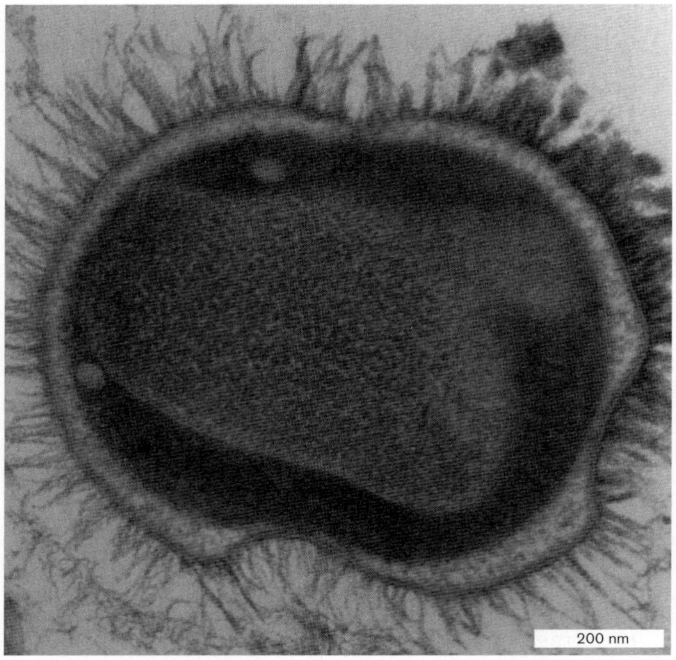
200 nm

L. VAN NIFTRIK AND M. S. M. JETTEN. 2012. *MICROBIOL. MOL. BIOL. REV.* **76**:585–596, FIG. 6D.

FIGURE 22.14 ▪ Anammox bacterium: *Scalindua*. The planctomycete *Scalindua* has an interior compartment to undergo anammox lithotrophy (discussed in Chapter 14). Pili surround the envelope.

To Summarize

- **Nitrogen in ecosystems is found in a wide range of oxidation states.** Interconversion of most of these states requires bacteria or archaea.
- **The main source and sink of nitrogen is the atmosphere.** N_2 is fixed into NH_4^+ by some bacteria and archaea. Denitrifying bacteria reduce NO_3^- successively back to N_2 and return it to the atmosphere.
- **Nitrogen fixation is conducted by symbiotic bacteria in association with specific plants.** Legume-associated nitrogen fixation is critical for agriculture.
- **Nitrification ($NH_4^+ \longrightarrow NO_2^- \longrightarrow NO_3^-$)** is aerobic oxidation of ammonia to nitrite and nitrate. Nitrification yields energy for lithotrophic bacteria in soil and water, and consumes oxygen, thus expanding marine oxygen minimum zones.
- **In anoxic zones containing high concentrations of organics, bacteria and archaea reduce nitrogen.** Bacteria conduct anaerobic respiration, reducing nitrate to N_2 and nitrous oxide (N_2O), a greenhouse gas. In the deep ocean, NO_3^- may be reduced by hydrogen gas to NH_3.
- **Bacteria and archaea oxidize ammonia.** When oxygen is available, bacteria and archaea oxidize ammonia. Under anaerobic conditions, bacteria oxidize NH_4^+ with NO_3^-, generating nitrogen gas (the anammox reaction). Bacterial anammox returns a substantial fraction of nitrogen gas to the atmosphere.

22.5

Sulfur, Phosphorus, and Metals

Besides carbon and nitrogen, many other elements participate in biochemical cycles that have important consequences for the biosphere, as well as for human environments. Sulfur undergoes a "triangle" of redox conversions analogous to that of nitrogen. Phosphorus, unlike other biological elements, is generally assimilated in the oxidized state, and phosphate is often a limiting nutrient for plants. Iron cycles in complex interactions with sulfur and phosphate. Beyond these macronutrients, many toxic metals in the environment, such as mercury and arsenic, are either bioactivated or detoxified by microbes.

The Sulfur Cycle

Sulfur is a major component of biomass, including proteins and cofactors. Like carbon and nitrogen, sulfur can be assimilated in either mineral or organic form. At the same time, assimilation competes with dissimilation. Reduced and oxidized forms of sulfur (comparable to those of nitrogen; see **Table 22.3**) offer electron donors and acceptors for dissimilatory reactions that generate energy. Reduced forms of sulfur, such as H_2S, can also participate in photolysis through reactions analogous to those of H_2O. Oxidized forms, such as sulfate, serve as anaerobic electron acceptors. As with nitrogen, most of these redox reactions are performed solely by bacteria and archaea. The biochemistry of sulfur cycling has important environmental consequences, such as the corrosion of concrete and iron.

In the ocean, sulfate is the second most common anion after chloride. Marine sulfate turns over slowly and constitutes an essentially limitless supply. Thus, sulfur is rarely a limiting nutrient. Marine algae release various reduced forms of sulfur, particularly dimethyl sulfide [$(CH_3)_2S$], which causes part of the "salty sea smell." Dimethyl sulfide enters the atmosphere, where it becomes oxidized to aerosols that may help water condense and form clouds. In the atmosphere, the overall amount of sulfur (mainly sulfur dioxide) is small. Nevertheless, some sulfur compounds generate toxic effects, as well as acidic pollution. Thus, both biochemical and industrial sources of atmospheric sulfur are of concern.

Competing assimilatory and dissimilatory sulfur reactions in the biosphere form a "sulfur triangle" including H_2S, S^0, and SO_4^{2-} (**Fig. 22.15A**). The oxidation and reduction pathways are analogous to those of nitrogen. Bacterial sulfur metabolism includes additional options of anaerobic phototrophy, such as H_2S photolysis. In an aquatic system, H_2S arises from spring waters welling up from the sediment and from decomposition of detritus. In decomposition, anaerobic respirers such as *Desulfovibrio* species convert sulfate to sulfur, then to H_2S. As H_2S rises

A. The sulfur triangle

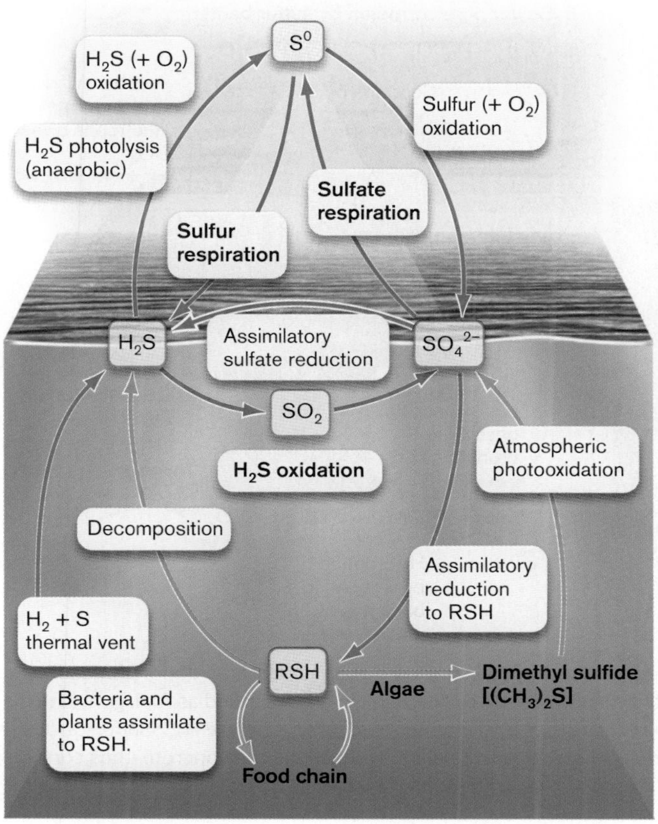

S⁰
H₂S (+ O₂) oxidation
Sulfur (+ O₂) oxidation
H₂S photolysis (anaerobic)
Sulfate respiration
Sulfur respiration
H₂S
Assimilatory sulfate reduction
SO₄²⁻
SO₂
H₂S oxidation
Atmospheric photooxidation
Decomposition
Assimilatory reduction to RSH
H₂ + S thermal vent
RSH
Algae
Dimethyl sulfide [(CH₃)₂S]
Bacteria and plants assimilate to RSH.
Food chain

B. White sulfur bacteria

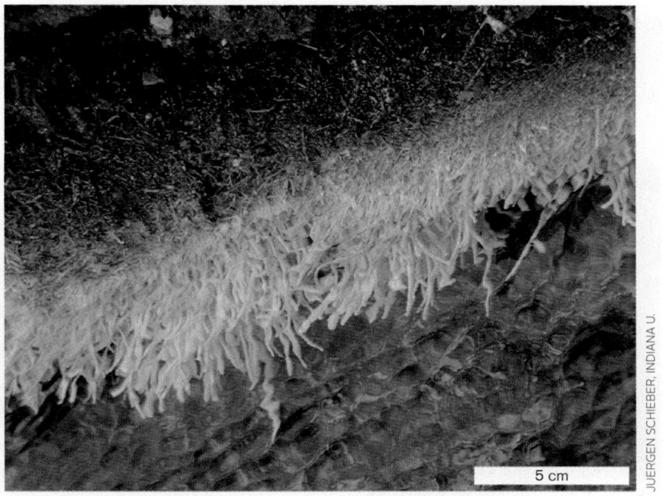

5 cm
JUERGEN SCHIEBER, INDIANA U.

FIGURE 22.15 ■ The sulfur cycle. A. The "sulfur triangle." With oxygen, bacteria and archaea oxidize H_2S to sulfur dioxide, and then to sulfate. Anaerobically, H_2S may be photolyzed to sulfur. Sulfate serves as an electron acceptor for anaerobic respiration, or it may be assimilated into biomass, with reduced sulfur groups (RSH). Elemental sulfur may be oxidized to sulfate by lithotrophs. **B.** Microbial mat of white, sulfur-oxidizing bacteria (probably *Beggiatoa*) growing at a sulfide spring.

to the oxygenated surface water, it is readily oxidized by sulfur-oxidizing bacteria such as *Acidithiobacillus* and *Beggiatoa*. *Beggiatoa* species are known as "white sulfur bacteria" because they form white mats (**Fig. 22.15B**), their appearance due to the sulfur granules generated by sulfide oxidation. Microbial sulfur oxidation is helpful for environments because it removes H_2S, which is highly toxic to most nonsulfur bacteria and plants. Because of the toxicity of H_2S, autotrophs more readily assimilate SO_4^{2-} into biomass, despite the extra energy needed to reduce SO_4^{2-} to the thiol form found in proteins.

If light is available, an alternate route for H_2S oxidation is photolysis by anaerobic phototrophs such as *Rhodopseudomonas* species. Some phototrophic bacteria further oxidize the S^0 generated to sulfate. In some sulfur-rich lakes, such as the Russian Lake Sernoye, underground springs pump so much H_2S that most of the sulfur is photolytically converted to S^0, which forms up to 5% of the sediment. This elemental sulfur can be mined commercially.

In thermal vent ecosystems, several sulfur-based reactions drive metabolism. For example, thermal vent archaea such as *Pyrodictium* species use sulfur to oxidize hydrogen gas to H_2S:

$$H_2 + S^0 \longrightarrow H_2S$$

This type of sulfur-based hydrogenotrophy is enhanced by the vent conditions of extreme pressure and temperature (100°C)—conditions under which elemental sulfur exists in a molten state more accessible to microbes than solid sulfur.

Decomposition of biomass generates various organic sulfur compounds, many of which are volatile. Odors of certain microbial sulfur products contribute to the smell of rotting eggs, but others enhance the taste of cheeses (discussed in Chapter 16).

Sulfur is reduced by sulfur-reducing bacteria in many subsurface environments, such as beneath deposits of oil and coal (**Fig. 22.16A**). Oil and coal contain sulfur in the form of SO_4^{2-} and provide a carbon source for sulfate respirers (sulfate-reducing bacteria that respire using sulfate as a terminal electron acceptor). The products of sulfate respiration include S^0 and organic sulfur. When the fuel is burned, these forms of sulfur cause severe pollution. Before burning, however, the S^0 and organic sulfur can be oxidized by sulfur-oxidizing bacteria. Fuel processors are now turning to sulfur-oxidizing microbes for experimental use in "desulfuration," the removal of sulfur from coal.

One habitat that exhibits the entire range of sulfur oxidation states is a sewer pipe. In a concrete sewer pipe, the

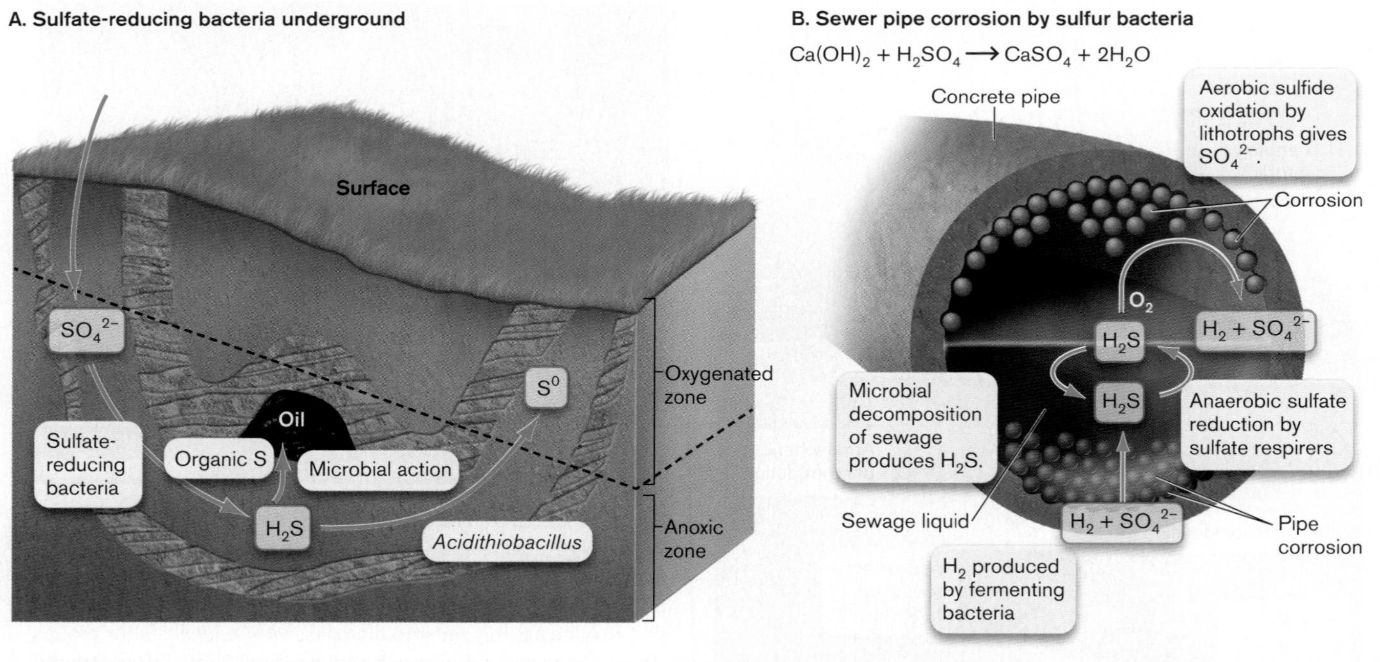
A. Sulfate-reducing bacteria underground
Surface
SO_4^{2-}
S^0
Oil
Sulfate-reducing bacteria
Organic S
Microbial action
H_2S
Acidithiobacillus
Oxygenated zone
Anoxic zone
B. Sewer pipe corrosion by sulfur bacteria
$Ca(OH)_2 + H_2SO_4 \longrightarrow CaSO_4 + 2H_2O$
Concrete pipe
Aerobic sulfide oxidation by lithotrophs gives SO_4^{2-}.
Corrosion
O_2
H_2S
$H_2 + SO_4^{2-}$
Microbial decomposition of sewage produces H_2S.
H_2S
Anaerobic sulfate reduction by sulfate respirers
Sewage liquid
$H_2 + SO_4^{2-}$
Pipe corrosion
H_2 produced by fermenting bacteria

FIGURE 22.16 ▪ Consequences of the sulfur cycle. A. Oil- or coal-bearing geological strata provide rich electron donors for sulfate-reducing bacteria (sulfate respirers). Eventually, various forms of sulfur contaminate the oil or coal, which, when burned as fuel, generates first SO_2 and ultimately sulfuric acid (acid rain). **B.** In a sewer pipe, sulfate-reducing bacteria (sulfate respirers) generate H_2S. Sulfur-oxidizing bacteria then oxidize H_2S to sulfate in the form of sulfuric acid. The sulfuric acid reacts with calcium hydroxide in the concrete, thus corroding the interior surface of the pipe.

alternation between anaerobic and oxygenated sulfur biochemistry causes severe corrosion (**Fig. 22.16B**). The microbial decomposition of sewage yields large quantities of toxic H_2S, which then volatilizes to high levels that endanger sewer workers. The H_2S is then oxidized to sulfuric acid by *Acidithiobacillus ferrooxidans,* a bacterium that colonizes the surface of the concrete. The sulfuric acid (H_2SO_4) decreases the pH at the concrete surface to pH 2. In the concrete surface, sulfuric acid converts calcium hydroxide to soluble calcium sulfate. Over several years, this corrosion can eat away half the thickness of a sewer pipe.

Sulfur metabolism shows important connections with metabolism of metals. For example, sulfur-oxidizing bacteria such as *A. ferrooxidans* oxidize iron as well (discussed shortly).

Thought Question

22.6 Compare and contrast the cycling of nitrogen and sulfur. How are the cycles similar? How are they different?

The Phosphate Cycle

Phosphorus is a fundamental element of nucleic acids, phospholipids, and phosphorylated proteins. Unlike sulfur and nitrogen, which are found in several different oxidation states, phosphorus is cycled mainly in the fully oxidized state of phosphate (**Fig. 22.17**). The absence of fully reduced phosphorus in ecosystems may be due to the fact that reduced phosphorus (phosphine, PH_3) undergoes spontaneous combustion in the presence of oxygen. Nevertheless, some anaerobic decomposers use phosphate as a terminal electron acceptor, reducing it to phosphine. In marshes and graveyards, where extensive decomposition occurs, phosphine emanates from the ground, where it ignites with a green glow. Microbial respiration of phosphate might be the cause of such "ghostly" apparitions.

Where is phosphate available? Although phosphate is abundant in Earth's crust, its availability in ecosystems is limited by its tendency to precipitate with calcium, magnesium, and iron ions. Thus, dissolved phosphate in water and soil is often a limiting nutrient for productivity. In natural ecosystems, the available phosphate is taken up rapidly by bacteria and phytoplankton, and then consumed by grazers and predators and dispersed by decomposers.

Marine water is extremely limited for phosphate, because of the distance from sediment minerals. Thus, the genomes of marine phototrophs encode many systems for acquiring phosphate from organic sources. Some actually acquire phosphorus in a partly reduced form, such as phosphite (PO_3^{3-}) or phosphonate (HPO_3^{2-}), both of which are available in marine water. In addition, marine phototrophs show an unusual ability to substitute nonphosphorus lipids for phospholipids, thus cutting their phosphorus needs by half. For example, the cyanobacteria *Prochlorococcus, Synechococcus,* and *Trichodesmium* can replace membrane phospholipids with sulfonated lipids. Similarly, algae such as *Thalassiosira* species can replace phosphatidylcholine with betaine, which contains no phosphate but includes a carboxylate group.

Another microbial response to phosphate scarcity is to replace membrane phosphates with organic phosphonates, in which the phosphorus atom bonds directly to carbon ($R–PO_3^{2-}$), instead of the easily hydrolyzed phosphoester bond. These organic phosphonates, however, are cleaved by marine archaea, the ammonia oxidizers such as *Nitrosopumilus.* The archaea release methylphosphonate, which many bacteria can cleave to obtain phosphate. Phosphate

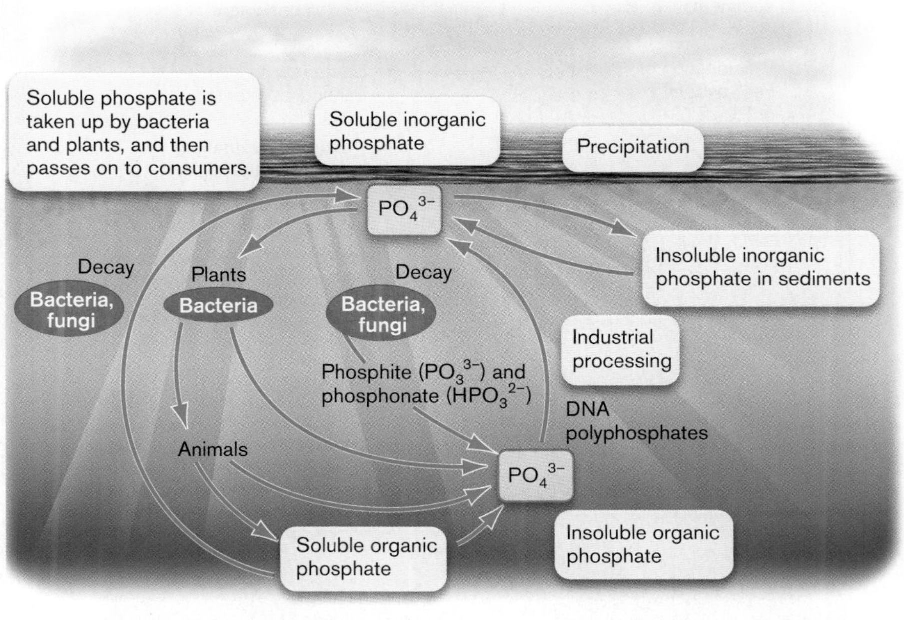
Soluble phosphate is taken up by bacteria and plants, and then passes on to consumers.
Soluble inorganic phosphate
Precipitation
PO_4^{3-}
Insoluble inorganic phosphate in sediments
Decay
Bacteria, fungi
Plants
Bacteria
Decay
Bacteria, fungi
Industrial processing
Phosphite (PO_3^{3-}) and phosphonate (HPO_3^{2-})
DNA polyphosphates
Animals
PO_4^{3-}
Soluble organic phosphate
Insoluble organic phosphate

FIGURE 22.17 ▪ The phosphate cycle. Phosphorus in the biosphere occurs entirely in the form of inorganic or organic phosphate. Most phosphate precipitates as insoluble salts in sediment. The small amount of soluble phosphate is taken up by plants and bacteria, which may then be taken up by consumers. Decomposers return phosphate to the environment.

scavenging from methylphosphonate releases methane, thus interacting with the carbon cycle.

In agriculture, phosphate is often added as a fertilizer. Phosphate fertilizer is obtained by treating calcium phosphate rock with sulfuric acid, producing calcium sulfate (gypsum) and phosphoric acid. Excess phosphate from fertilizer or industry may drain into streams and lakes where phosphorus is the limiting element. The sudden influx of phosphate causes an algal bloom. The overgrowth of algae leads to overgrowth of heterotrophs, depletion of oxygen, and destruction of the food chain.

Thought Question

22.7 Compare and contrast the cycling of nitrogen and phosphorus. How are the cycles similar? How are they different?

The Iron Cycle

Why do organisms need iron? As a micronutrient, iron forms a negligible part of biomass but is essential for growth (discussed in Chapter 4). Organisms require iron as a cofactor for enzymes and an essential component of oxygen carrier molecules such as hemoglobin. Other common micronutrients required for enzymes and cofactors are zinc, copper, and selenium.

Iron in soil and sediment. Iron is a major component of Earth's crust, and substantial quantities are present in most soil and aquatic sediment. Yet the availability of iron to organisms is limited by its extremely low solubility in the oxidized form. In microbial biochemistry, the oxidized form, ferric iron (Fe^{3+}), interconverts with the reduced form, ferrous iron (Fe^{2+}). In the presence of oxygen, iron metal (Fe^0) "rusts" and is therefore available to organisms mainly as Fe^{3+}. Ferric iron is especially insoluble at high pH, precipitating with hydroxide ions as ferric hydroxide [$Fe(OH)_3$] or with phosphate ions as ferric phosphate ($FePO_4$) (**Fig. 22.18A**).

In iron mines, where pyrite (FeS_2) is exposed to air, spontaneous oxidation releases sulfuric acid:

$$2FeS_2 + 7O_2 + 2H_2O + 4H^+ \longrightarrow 2Fe^{2+} + 4H_2SO_4$$

This reaction drives the growth of lithotrophs such as *Acidithiobacillus ferrooxidans*, which greatly increases the rate of sulfuric acid production, leading to acid mine drainage.

A. The iron cycle

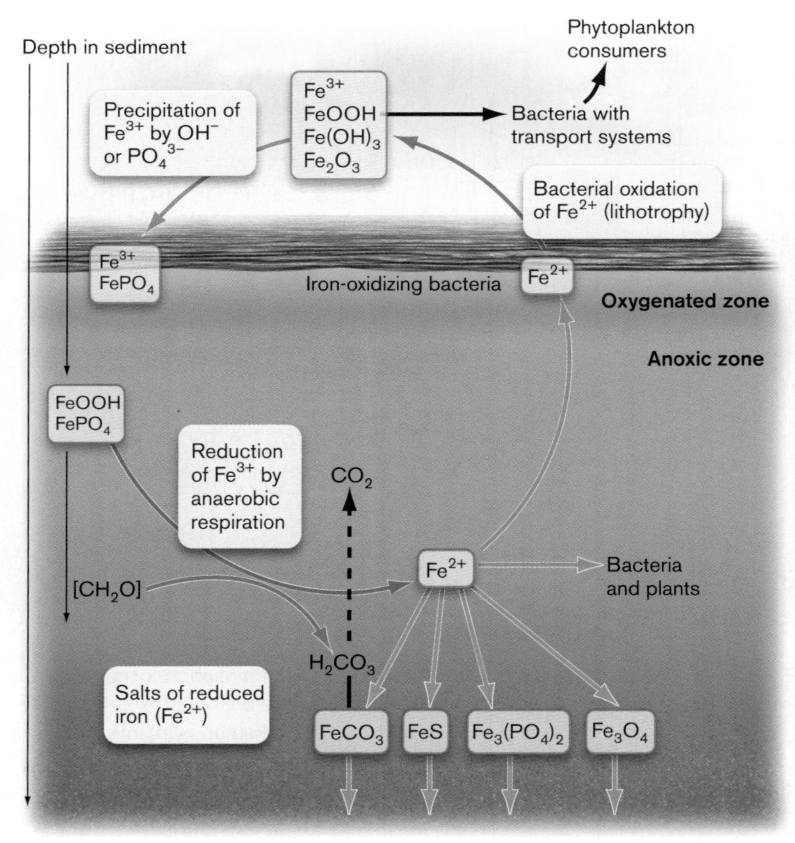
Depth in sediment
Phytoplankton consumers
Precipitation of Fe3+ by OH− or PO43−
Fe3+ FeOOH Fe(OH)3 Fe2O3
Bacteria with transport systems
Bacterial oxidation of Fe2+ (lithotrophy)
Fe3+ FePO4
Iron-oxidizing bacteria
Fe2+
Oxygenated zone
Anoxic zone
FeOOH FePO4
Reduction of Fe3+ by anaerobic respiration
CO2
Fe2+
Bacteria and plants
[CH2O]
H2CO3
Salts of reduced iron (Fe2+)
FeCO3
FeS
Fe3(PO4)2
Fe3O4

B. Iron precipitation due to lithotrophy

BRYAN ALLEN

FIGURE 22.18 ■ The iron cycle. **A.** Ferric iron (oxidized iron, Fe^{3+}) precipitates with hydroxide or phosphate. Only bacteria can assimilate Fe^{3+}. In anoxic sediment, bacterial respiration reduces Fe^{3+} to Fe^{2+}, a more soluble form available to plants. **B.** Home plumbing shows signs of lithotrophic oxidation of Fe^{2+} to Fe^{3+} (orange precipitate).

THOMAS R. FLETCHER/ALAMY

FIGURE 22.19 ▪ Mine drainage enters a stream. Lithotrophic oxidation of reduced iron yields the orange material polluting this stream in Preston County, West Virginia.

Where acid mine drainage enters an aquatic system, the reduced iron is oxidized by lithotrophs to ferric hydroxide, forming an orange precipitate, seen in home plumbing (**Fig. 22.18B**). Iron mine drainage causes severe pollution of streams (**Fig. 22.19**).

In anoxic habitats, such as benthic sediment, Fe^{3+} is largely reduced to Fe^{2+} by anaerobic respiration. Reduction leads to loss of the reddish color, generating gray-colored sediment, as in wetland soil (discussed in Chapter 21). The reduction of Fe^{3+} to Fe^{2+} is one of the enriching contributions of anaerobic sediment to aquatic wetlands and coastal estuaries, because reduced iron is more available to plants and bacteria. Oxidized iron, in aerobic soils, can be taken up only by bacteria. Bacteria synthesize special iron uptake systems, including molecules called **siderophores** that bind ferric ion outside, then are taken up by the cell (see Section 4.2). Bacterial iron then becomes available to consumers in the ecosystem.

Iron cycling often connects with the sulfur cycle in ways that can prove unfortunate for human engineering. The most serious problem is anaerobic corrosion of iron. Iron corrodes spontaneously in the presence of oxygen. In anoxic environments, however, little spontaneous corrosion occurs. Instead, iron is corroded by sulfate-reducing bacteria (**Fig. 22.20**). Bacteria such as *Desulfovibrio* and *Desulfobacter* can grow on the iron surface, forming an anaerobic biofilm. Within the biofilm, iron is oxidized to Fe^{2+}, and sulfate dissolved in the water is reduced to ferrous sulfide:

$$4Fe^0 + SO_4^{2-} + 4H_2O \longrightarrow FeS + 3Fe^{2+} + 8OH^-$$

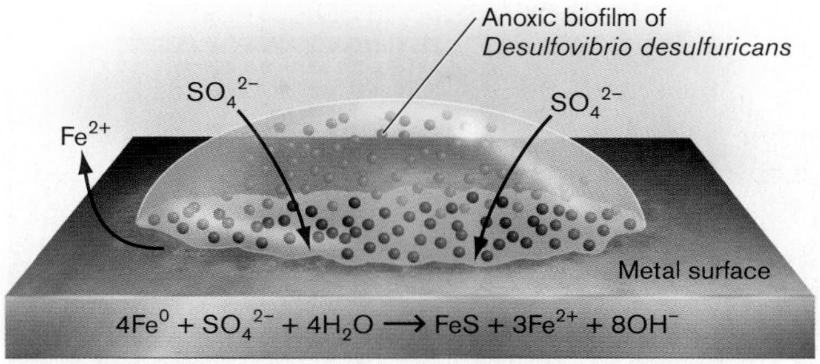
Anoxic biofilm of
Desulfovibrio desulfuricans
SO_4^{2-}
SO_4^{2-}
Fe^{2+}
Metal surface
$4Fe^0 + SO_4^{2-} + 4H_2O \longrightarrow FeS + 3Fe^{2+} + 8OH^-$

FIGURE 22.20 ▪ Anaerobic corrosion. Iron-sulfur bacteria convert Fe^O to FeS. FeS flakes off, exposing iron to further oxidation and rapid corrosion. *Source:* Modified from Hang T. Dinh et al. 2004. *Nature* **427**:829–832.

The FeS flakes away, exposing more iron to react with water, resulting in cyclic corrosion.

Marine iron. In most ocean water, iron is extremely scarce. This is because the benthic sediment containing iron is so distant that its iron is largely inaccessible. The main source of marine iron is eolian (wind-borne) dust from dry land, such as windstorms from the Sahara desert. Most of the wind-borne iron is particulate, oxidized, and unavailable to eukaryotic phytoplankton (algae). Thus, bacteria that acquire and reduce ferric iron provide the main entry of iron into the ecosystem. This may explain why so many marine algae are mixotrophs: Although their photosynthesis can fix plenty of carbon for biomass, they consume bacteria as a source of iron. Another source of marine iron, discovered in 2013, is hydrothermal vents.

Iron is thus a limiting nutrient for marine phytoplankton. As discussed in Chapter 21, when higher quantities of a limiting nutrient enter an ecosystem, the limited populations rapidly increase. The rise of algal populations following iron addition was demonstrated in an "iron fertilization" experiment conducted in 2007 by Phillip Boyd and colleagues at the University of Otago, New Zealand. Several thousand kilograms of ferrous sulfate ($FeSO_4$) was released in the Southern Ocean, near Antarctica. Within a week after the iron release, there was a bloom of phytoplankton (**Fig. 22.21**). The dominant phytoplankton in the bloom were diatoms of the species *Fragilariopsis kerguelensis.* The diatom bloom was so large that it was detected by a NASA satellite as a region of increased color reflected by chlorophyll.

The dramatic effects of iron fertilization led some researchers to propose that increasing iron throughout the oceans could cause blooms of diatoms removing CO_2 from the atmosphere. If a sufficiently large proportion of the diatoms were to escape predation and fall to the ocean

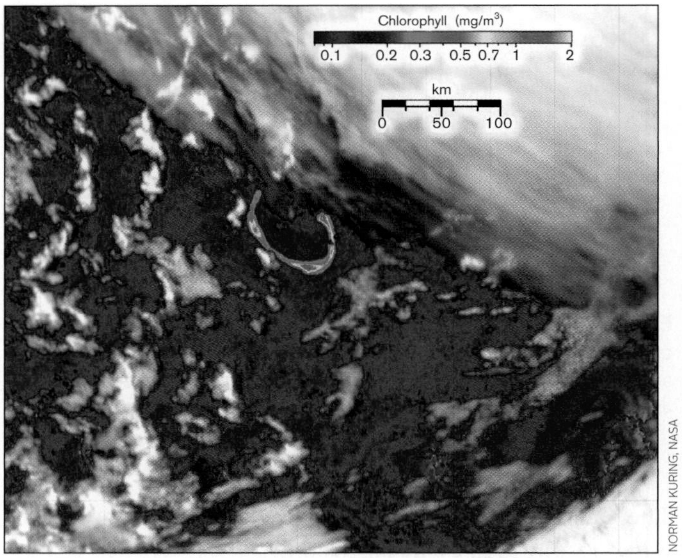
Chlorophyll (mg/m³)
0.1 0.2 0.3 0.5 0.7 1 2
km
0 50 100
NORMAN KURING, NASA

FIGURE 22.21 ■ Iron-induced phytoplankton bloom in the Southern Ocean. Chlorophyll concentration was determined from satellite measurements of color, using the Sea-viewing Wide Field-of-view Sensor (SeaWiFS). The phytoplankton bloom was observed several days after iron sulfate was introduced. White regions indicate cloud cover. *Source:* Phillip W. Boyd et al. 2007. *Science* **315**:612–617.

floor, they would effectively remove their carbon from circulation. An alternative outcome, however, might be eutrophication, causing an ecological collapse comparable to a dead zone.

In 2009, a major test of iron fertilization was conducted by the LOHAFEX project of the National Institute of Oceanography of India and the Alfred Wegener Institute for Polar and Marine Research of Germany. The LOHAFEX experiment (*loha* meaning "iron" in Hindi; *fex* for "fertilization experiment") was opposed by many environmental groups, who considered it a form of pollution, but it was ultimately carried out. In the experiment, 20,000 kg of iron sulfate was introduced in the Southern Ocean, causing a bloom across 300 km^2. The bloom consisted of coccoliths, algae with plates of calcium carbonate instead of silicate (discussed in Chapter 20). But the coccoliths attracted amphipods, shrimplike predators that soon consumed the algae, recycling most of their carbon back to CO_2.

While iron fertilization is unlikely to meet our need for CO_2 removal, microbiology offers other approaches to offset the perturbation of Earth's carbon cycle by human technology. For example, bacteria and algae may be engineered to provide biofuels, such as hydrogen gas, whose use yields energy without releasing CO_2.

Our view of Earth's biosphere increasingly is moving toward environmental management—the concept that wilderness as such no longer exists, but only a biosphere to be managed for better or worse. Microbes will have many roles as partners in environmental management.

Other Metals in the Environment

Another concern for environmental management is that of toxic metals and metalloids such as mercury and arsenic. Besides iron, numerous trace metals interact with bacterial species as either electron donors or acceptors (**Table 22.4**). Bacterial conversion of metals can either produce toxic species or remove toxic metals from ecosystems. For example, chromium-6 [Cr(VI)] is an extremely toxic pollutant that can be reduced by soil bacteria to the much less toxic Cr(III). Microbial metal cycling is discussed further in eTopic 22.2.

How can bioremediation remove toxic minerals from groundwater? An example is the bioremediation of arsenic, a critical water contaminant in densely populated countries such as India, Bangladesh, and Cambodia. Arsenic causes cancers and crippling skin disorders. Arsenic and other metals can be removed from groundwater via bioremediation. Bhaskar Sengupta and colleagues at Queen's University, Belfast, developed a bioreactor technology for subterranean arsenic removal, or SAR (**Fig. 22.22**). This process takes advantage of the fact that different oxidation states of arsenic differ in their solubility. For the bioreactor, groundwater is cycled through an oxygen pump and returned to the source. Naturally occurring lithotrophic bacteria oxidize arsenite (AsO_3^{3-}, water-soluble) to arsenate (AsO_4^{3-}, insoluble). Bacteria also oxidize iron and manganese to their less soluble forms, as discussed in Chapter 14. The negatively and positively charged ions precipitate together, effectively removed from the water.

Multiple Limiting Factors Modulate Complex Ecosystems

A common assumption of terrestrial and aquatic ecology is that a single nutrient, such as phosphorus or iron, is limiting for a given ecosystem. The requirement for more than one limiting factor is known as **resource colimitation**. In highly oligotrophic marine water, populations often depend on multiple limiting factors. For example, in the Baltic Sea both nitrogen and phosphorus are present in such low concentrations that they must be added together in order to stimulate a phytoplankton bloom. In another example, the North Atlantic is limited for both phosphorus and iron. Addition of both P and Fe stimulates growth of nitrogen-fixing cyanobacteria.

In some cases, different species are limited by different nutrients. In the LOHAFEX experiment, the Southern

TABLE 22.4

Microbial metabolism of metals.

Metal	Major conversions	Microbial genera (examples)	Effects within environment
Manganese, Mn	$Mn^{2+} \longrightarrow Mn^{4+}$	*Hyphomicrobium, Arthrobacter*	Mn is a trace element required for enzymes
	$Mn^{4+} \longrightarrow Mn^{2+}$	*Geobacter, Pseudomonas*	Anaerobic respiration in sediment
Mercury, Hg	$Hg^{2+} \longrightarrow Hg^{0}$	*Acidithiobacillus*	Hg^{0} volatilizes; little harm
	$Hg^{2+} \longrightarrow (CH_3)Hg^{+}$	*Desulfovibrio*	$(CH_3)Hg^{+}$ is a severe neurotoxin; accumulates at higher trophic levels, such as fish
Arsenic, As	$AsO_3^{3-} \longrightarrow AsO_4^{3-}$	*Alcaligenes, Pseudomonas*	Removal of poisonous AsO_3^{3-} via use as terminal e^- acceptor, converting to insoluble AsO_4^{3-}
	$AsO_4^{3-} \longrightarrow AsO_2^{-}$	*Bacillus, Chrysiogenes, Pyrobaculum*	AsO_4^{3-} (arsenate) used as terminal e^- acceptor
	$AsO_4^{3-} \longrightarrow (CH_3)_3As$	*Candida* (a fungus), *Scopulariopsis* (a fungus)	Methylarsines. Poisonous; inhalation of moldy wallpaper with arsenic pigment
Chromium, Cr	$CrO_4^{2-} \longrightarrow Cr^{3+}$	*Aeromonas, Arthrobacter, Desulfovibrio*	CrO_4^{2-} [Cr(VI)] is mutagenic and carcinogenic; as e^- acceptor, reduced to Cr^{3+} [Cr(III)], less toxic
Vanadium, V	$VO_3^{-} \longrightarrow VO(OH)$	*Veillonella, Desulfovibrio, Clostridium*	V is a trace element for some nitrogenases and invertebrate blood pigments
			VO_3^{-} (vanadate) is oxidized as an e^- donor
Selenium, Se	$Se^{0} \longrightarrow SeO_3^{2-}$	*Bacillus, Micrococcus*	Se is a trace element; small amounts in food help remove mercury. Larger amounts are toxic
Uranium, U	$UO_2^{2+} \longrightarrow UO_2$	*Veillonella, Shewanella, Geobacter*	UO_2^{2+} [U(VI)], soluble; is respired to uranium dioxide, UO_2 [U(IV)], insoluble; used for cleanup of radioactive uranium

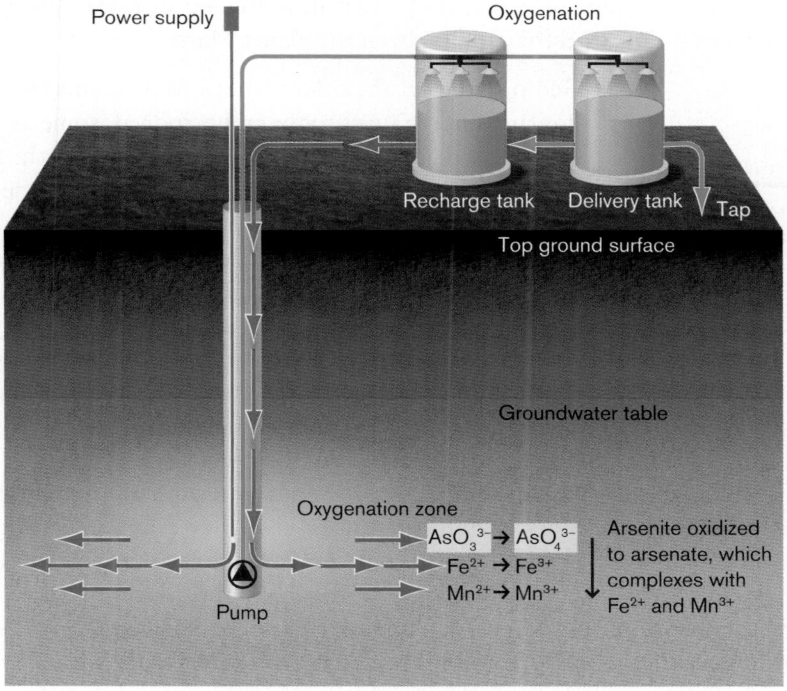
Power supply
Oxygenation
Recharge tank
Delivery tank
Tap
Top ground surface
Groundwater table
Oxygenation zone
AsO3 3- → AsO4 3-
Fe2+ → Fe3+
Mn2+ → Mn3+
Arsenite oxidized to arsenate, which complexes with Fe2+ and Mn3+
Pump

FIGURE 22.22 ▪ Subterranean arsenic removal. Bioremediation of arsenic-contaminated groundwater by lithotrophic bacteria. Water cycles through oxygenation tanks back to the source, where bacteria oxidize arsenite, iron, and manganese. The oxidized form of arsenic, which is arsenate, precipitates with iron and manganese for convenient removal.

Ocean diatoms were limited for both iron and silicate; but coccoliths, whose shells are calcium carbonate rather than silicate, required only added iron to bloom.

To Summarize

- **Sulfate is abundant in marine water.**
- **Oxidized and reduced forms of sulfur are cycled in ecosystems.** Sulfate and sulfite serve as electron acceptors for respiration. Hydrogen sulfide serves as an electron donor. Sulfur cycling participates in acid mine drainage and pipe erosion.
- **Phosphorus cycles primarily in the fully oxidized form (phosphate).** Phosphate limits growth of phototrophic bacteria and algae in some aquatic and marine systems.
- **Iron cycles in oxidized and reduced forms.** Oxidized iron (Fe^{3+}) serves as a terminal electron acceptor in anaerobic soil and water. Reduced iron (Fe^{2+}) from rock is oxidized through weathering or mining. Bacterial lithotrophy accelerates iron oxidation, leading to acidification.
- **Metal toxins can be metabolized by bacteria.** Bacterial metabolism may either increase or decrease toxicity. Metabolic conversion to an insoluble form, such as arsenite to arsenate, offers an effective means of bioremediation.
- **Marine habitats show resource colimitation.** Multiple resources may be limiting for the phytoplankton community or for different populations.

22.6 Astrobiology

As we come to appreciate the ubiquitous contributions of microbes to shaping our planet, in all its diverse habitats, increasingly we wonder whether microbes exist on worlds beyond Earth. Is Earth unique in supporting life, or have living cells evolved as well on Mars or Venus or on Jupiter's planet-sized moons? Astrobiology is the study of life in the universe, including its origin and possible existence outside Earth. The discovery of life beyond Earth would arguably be the most significant advance in science since a human set foot on the moon.

If the same physical and chemical laws govern the universe everywhere, then it is hard to suppose that only one of the billions of stars would have a planet supporting life. On the other hand, we have no idea how many planets have been capable of developing and sustaining a biosphere. As Isaac Asimov said, "There are two possibilities. Maybe we're alone. Maybe we're not. Both are equally frightening."

If life exists elsewhere, is it built on the same fundamental elements as ours? Many lines of evidence suggest that the biochemistry of life elsewhere would resemble that of Earth. Terrestrial life is founded on macroelements in the first two rows of the periodic table, including carbon, nitrogen, oxygen, phosphorus, and sulfur (see **Table 22.1**; for the periodic table, see Appendix 1). The valence numbers of these elements from the middle of the periodic table enable them to form complex molecular structures with strong covalent bonds. The same fundamental molecules that appear in early-Earth simulation experiments, such as adenine and glycine, also appear in meteorites. Thus, we suspect that the fundamental building blocks for biochemistry are universal. On the other hand, if life could indeed be founded on some other basis, how would we recognize it?

If life is found on other planets, it will almost certainly include microbes. In fact, a case can be made that the majority of biospheres in our galaxy would consist entirely of microbial life, since microbes inhabit a wider range of conditions than do multicellular plants and animals. Even on Earth itself, the largest bulk of our biosphere—including deep sediments and rock strata—consists of microbial ecosystems.

Did Mars Once Support a Biosphere?

For several reasons, the most studied candidate for extraterrestrial life has been the planet Mars:

- **Geology.** Of all the solar planets, Mars seems the most similar to Earth in its topography; indeed, some areas of Mars remarkably resemble desertscapes on Earth. Martian rock contains the fundamental elements needed for life.
- **Day and year length.** Mars has a day length similar to that of Earth, and a year only twice as long as ours.
- **Temperature.** The average temperature on Mars is 220 K (–53°C), too cold for most biochemistry on Earth, but its temperature rises above freezing at the equator. By contrast, the torrid heat of Venus (460°C) would exclude stable macromolecules.
- **Atmosphere.** The overall atmospheric pressure on Mars, 6 millibars (mbar), is barely a hundredth that of Earth (1,013 mbar) and lacks molecular oxygen. Thus, aerobes could not grow. But the Martian atmosphere does include carbon dioxide, actually at 20 times the CO_2 content on Earth, so there would be plenty for photoautotrophic production of biomass.

- **Water.** Surface water freezes out of the Martian atmosphere, without existing as a liquid. But mineral formations suggest that liquid water flowed in the past. Liquid water may yet exist deep underground, supporting life-forms similar to the endoliths of Earth's crustal rock.

The existence of liquid water is a key question because on Earth, wherever liquid water exists—even brine (concentrated salt) at –20°C—there we find microbial life. NASA's Mars rovers, most recently the *Curiosity* rover in 2012, have found more water than expected. Water condenses and freezes out of the Martian atmosphere, generating ice clouds and snow. The Martian soil composition showed high levels of perchlorate, a chemical that attracts water to form liquid solution and can support the metabolism of some Earth microbes.

Other evidence for liquid water in the past on Mars comes from geological formations surveyed by the *Mars Reconnaissance Orbiter* in 2007. The orbiter mapped sedimentary deposits in Mars's Jezero crater (**Fig. 22.23**). The crater formations reveal flow patterns typical of a river delta flowing into a lake. Claylike mineral deposits (false-colored green in **Fig. 22.23**) may have trapped organic compounds needed for life. The layered patterns are best explained by a model involving fluid flow, such as the flow of water.

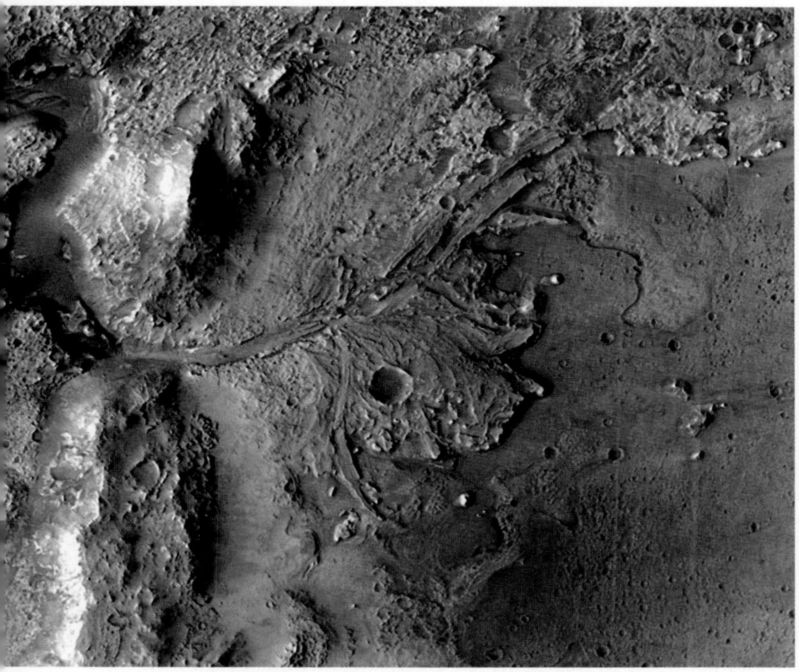
BETHANY EHLMANN, NASA/JPL/JHUAPL/MSSS/BROWN UNIVERSITY

FIGURE 22.23 ■ Evidence of past water on Mars. Sedimentary deposits in the Mars Jezero crater, mapped by the *Mars Reconnaissance Orbiter* in 2007. The flow patterns resemble a river delta, best explained by a model involving the flow of liquid water. Green indicates claylike mineral deposits.

If life once existed on Mars, what became of it? Two main possibilities are considered:

- **Life developed and existed until the planet froze.** Under this scenario, life originated much as it did on Earth, during a time of heavy bombardment from space and outgassing of nitrogen and carbon dioxide. Unfortunately, however, life failed to generate sufficient atmosphere for a greenhouse effect to sustain temperate conditions. No oxygenic organisms produced molecular oxygen and an ozone layer.
- **Life developed and still exists underground.** In the absence of an ozone layer, the Martian surface is sterilized by cosmic radiation. Nevertheless, microbes may yet exist deep underground, similar to the endolithic prokaryotes on Earth. Endoliths metabolize within interstitial water and are protected from cosmic radiation.

Do Biosignatures Indicate Life?

If microbes do exist on Mars, they should be detectable. But the detection of unknown life, even on Earth, remains a challenge. The advent of PCR sequence detection of species based on ribosomal genes has revealed thousands of unknown species, most of which cannot be grown or recognized by other methods. Other species may well be missed if their rRNA sequences fail to amplify with our probes. On Mars, assuming life evolved independently of Earth life-forms, we would have no way to define sequence probes for detection.

Instead, researchers try to define **biosignatures**, chemical and physical signs that could only have been formed by life. Most of the proposed biosignatures are based on types of evidence for life on Earth, either as fossils of ancient life (discussed in Chapter 17) or as signs of current life in extreme habitats. Proposed biosignatures include the following:

- **Microfossils.** The mineralization of microbial cells leads to the formation of structures that can be visualized under a microscope. The cell fossils must be sufficiently distinct to establish that no abiotic process could have formed them.
- **Isotope ratios.** Certain biochemical reactions preferentially use one isotope of an atom over another; for example, Rubisco, the key enzyme of carbon dioxide fixation, uses ^{12}C in preference to ^{13}C. Thus, photosynthetic use of ^{12}C can decrease the $^{12}C/^{13}C$ ratio in subsequent carbonate deposits. Isotope ratios of nitrogen, oxygen, and sulfur are also used as biosignatures.
- **Mineral deposits.** Certain mineral formations are observed to be caused only by microbial activity. For example, insoluble manganese oxides such as Mn_2O_3 are almost always the result of microbial oxidation of

reduced manganese. Reduced manganese ions are very stable, and their abiotic oxidation rate is extremely slow.

- **Metabolic activity.** Samples of soil can be incubated with radioactive tracer substances such as $^{14}CO_2$ and tested for metabolic conversion or incorporation into biomass. It must be established that the conversion could not have occurred abiotically and that no organisms from Earth were present.

Various kinds of evidence for life on Mars have been reported, such as possible microfossils within a Martian meteorite that landed in Antarctica. Metabolic activity was tested in samples obtained by the NASA *Viking* lander in 1975. As of this writing, however, no evidence has proved conclusive for active or past living microorganisms on Mars.

Could Mars Be Terraformed to Support Earth Life?

If no life exists—or if it exists only in the form of microbes deep underground—should we consider human intervention, or **terraforming**, to make Mars habitable for life from Earth? Scenarios for terraforming, long explored in science fiction, are now receiving serious thought among space scientists. Terraforming Mars would require increasing the temperature and air pressure. The temperature of the atmosphere might be increased by release of greenhouse gases. In addition, sufficient carbon dioxide and nitrogen gases must be made available from the Mars surface rock. In principle, microbes from Earth could be seeded to grow and generate an atmosphere containing nitrogen, oxygen, and CO_2.

The dilemmas and consequences of terraforming are depicted in the novel *Red Mars* (1993) by Kim Stanley Robinson. In favor of terraforming, it is argued that Mars offers enormous natural resources of potential benefit for humanity, especially as terrestrial resources are used up. Human settlements on Mars would be a major step forward for space exploration. On the other hand, it is argued that the planet Mars is a natural monument, a place with its own right to exist as such. It should be allowed to remain in its natural state for future generations to appreciate. As a practical matter, terraforming remains unfeasible for the near future. For example, the amount of chlorofluorocarbons required to raise Martian temperature is calculated to be 100 times greater than our global capacity to produce such substances.

Does Europa Have an Ocean?

Farther out in the solar system, surprising candidates for microbial life are the moons of Jupiter. In 2000, the *Galileo* space probe passed several of Jupiter's moons, including Ganymede, Callisto, and Europa (**Fig. 22.24**). While their distance from the Sun results in extreme cold, these bodies receive extra heat from friction generated by tidal forces from the giant planet Jupiter. In the case of Europa, the tidal forces have been calculated to provide enough heat to liquefy water without boiling it off. Furthermore, measurement of Europa's magnetic field suggests that its composition includes a dense iron core surrounded by 15% water. Most of the water must be locked in ice, but tidal heating could melt enough water for an underlying ocean of brine. On Earth, similar brine lakes beneath the ice of Antarctica harbor halophilic archaea that grow at –20°C.

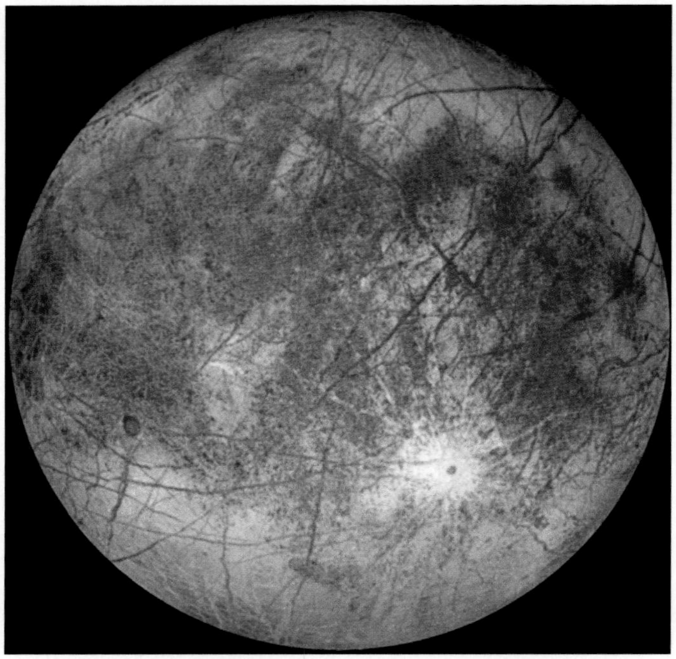

NASA/DRL

FIGURE 22.24 ■ Jupiter's moon Europa. Does life exist beneath Europa's ice, in a salty sea?

If such oceans do exist, how could they support life without photosynthesis? Photosynthesis is impossible at Jupiter's distance from the Sun. However, an alternative source of chemical energy might be the influx of charged particles accelerated in Jupiter's magnetic field. Charged particles entering Europa's ice can react with water to form hydrogen peroxide (H_2O_2). The H_2O_2 then breaks down, releasing molecular oxygen. Oxygen reaching the brine layer below could combine with electron donors from crustal vents and power metabolism. Alternatively, molecular hydrogen (H_2) could be generated from water ionized by decay of radioisotopes. A similar source of H_2 for life has been proposed to occur on Earth in rock strata several kilometers below the surface, where it may support lithotrophs. On Europa, the hydrogen could then combine with oxygen gas or other oxidants to power life. These schemes are highly speculative—but tantalizing enough to encourage future NASA missions to take a closer look at Europa and its sister moons.

Does Life Exist on Planets of Distant Stars?

The past decade of astronomy has seen an extraordinary growth in our knowledge of solar systems beyond our own. Now that we know that so many other stars possess planets, we can only wonder whether they also possess biospheres of life-forms we would recognize.

In recent years, several hundred "extrasolar" planets have been detected around distant stars, including some of low mass comparable to that of Earth. Could we ever hope to detect signs of life on an extrasolar planet? A possible means of detection is suggested by William Sparks and colleagues at the Space Telescope Science Institute in Baltimore, and collaborating institutions. They show that light scattered by marine phytoplankton exhibits a property called "circular polarization." Circular polarization arises from substances that are homochiral—that is, present in only one of two mirror forms, such as the L- and D-forms of an amino acid. Homochirality is a strong biosignature, typical of proteins and metabolites. If someday we have telescopes capable of detecting light from distant planets, circular polarization might provide evidence of life.

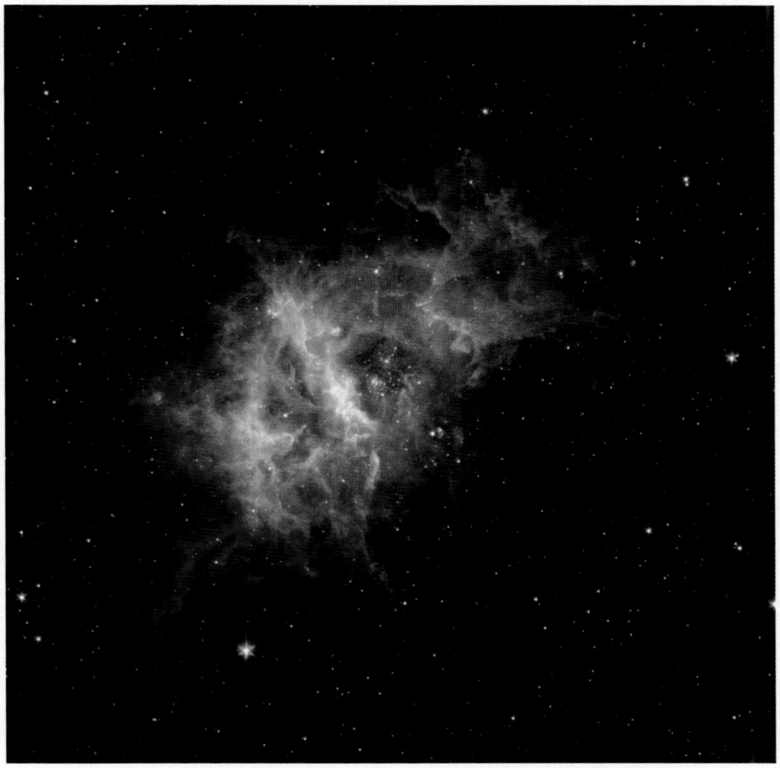
NASA

FIGURE 22.25 ▪ A stellar nursery. Nebula RCW-49 contains more than 300 newborn stars, from which NASA's Spitzer Space Telescope detected spectroscopic signals of common organic constituents of life (infrared photograph).

Figure 22.25 shows an infrared photograph through the Spitzer Space Telescope of nebula RCW-49, a gaseous cloud full of newborn stars. Recall from Chapter 17 that as stars form, they take up dust of supernovas that includes all the elements needed to form biomolecules. Spectroscopic observation of RCW-49 reveals stars surrounded by disks coalescing into planets. The planetary disks contain icy particles full of organic molecules such as methanol, glycine, and ethylene glycol, a reduced form of sugar. Could there be biospheres in the making?

To Summarize

- **Astrobiology is the study of life in the universe, including possible habitats outside Earth.**
- **The search for extraterrestrial life is based on methods similar to those used to seek early life on Earth.** Evidence includes chemical and physical biosignatures, isotope ratios, microfossils, and metabolic activity.
- **Mars is the planet whose geology most closely resembles that of Earth.** Geological features strongly support the past existence of flowing water, a prerequisite for microbial life.
- **Jupiter's moon Europa is proposed as another possible site for life.** Europa is bathed in a sea of brine similar to terrestrial habitats for halophiles.

Concluding Thoughts

Our observations of distant molecules, as well as our analysis of meteorites (discussed in Chapter 17), suggest that the biomolecules of Earth fit into a universal pattern of interstellar chemistry. Whether life exists elsewhere or we are alone, we must remember that our Earth is the only place we know of at this time that can support humans and the forms of life we require for our own survival. The survival of our entire biosphere depends on our microbial partners cycling key elements and acquiring energy to drive the food web. For the first time in history, human technology now rivals the ability of microbes to alter fundamental cycles of biogeochemistry. But to manage and moderate our alterations—for our own survival and that of the biosphere—our fate still depends on the microbes.

CHAPTER REVIEW

Review Questions

1. Identify the major sources and sinks of carbon, nitrogen, and sulfur. Which sources recycle rapidly, and why?
2. Explain how the carbon cycle differs in oxygenated and anoxic environments.
3. Explain two different chemical methods of measuring the environmental levels of carbon and nitrogen.
4. Outline the hydrologic cycle. Explain the role of biochemical oxygen demand (BOD) in water quality and how it may be perturbed by human pollution.
5. Outline the functions of a wastewater treatment plant. Include the phases of primary, secondary, and tertiary treatment. Explain the roles of microbes in these phases of water treatment.
6. Outline the main transformations of the nitrogen cycle. Which reactions are carried out only by microbes?
7. Outline the main transformations of the sulfur cycle. Which features are comparable to the nitrogen cycle, and which are unique to the sulfur cycle?
8. For the iron cycle, explain aerobic and anaerobic processes of microbial transformation. Explain how microbial iron transformation may be linked to sulfur transformation.
9. Explain how bacteria may convert metals to toxic forms. Alternatively, explain how bacteria may detoxify metal-polluted sediment.
10. Offer several arguments for and against the existence of life on planets beyond Earth.

Thought Questions

1. How does influx of nitrogen-rich fertilizers to soil ecosystems increase the rate of CO_2 efflux from wetlands?
2. In the past 50 years, most of the wetlands off the coast of Louisiana have been destroyed. How is wetland destruction related to formation of the dead zone in the Gulf of Mexico? How could the dead zone be revived, and why would the cost of restoration be projected to be billions of dollars?
3. What nutrients are limiting in the marine benthos, and why?

Key Terms

abiotic (891)
ammonification (906)
anammox (908)
assimilatory nitrate reduction (906)
biochemical oxygen demand (BOD) (897)
biogeochemical cycle (891)
biogeochemistry (891)
bioremediation (890)
biosignature (917)
biotic (891)
dead zone (898)
denitrification (907)
dissimilatory nitrate reduction (907)
floc (900)
fossil fuel (896)
geomicrobiology (891)
greenhouse effect (894)
hydrologic cycle (896)
hypoxia (898)
mesocosm (893)
micronutrient (891)
mycorrhizae (894)
nitrification (905)
nitrifier (905)
nitrogen fixation (904)
oxygen minimum zone (OMZ) (898)
reservoir (891)
resource colimitation (914)
siderophore (913)
sink (891)
sludge (899)
source (891)
terraforming (918)
wastewater treatment (899)
water cycle (896)
zone of hypoxia (898)

Recommended Reading

Arrigo, Kevin R. 2009. Marine microorganisms and global nutrient cycles. *Nature* **437**:349–355.

Canfield, Donald E., Alexander N. Glazer, and Paul G. Falkowski. 2010. The evolution and future of Earth's nitrogen cycle. *Science* **330**:192–193.

Clemmensen, Karina E., Adam Bahr, Otso Ovaskainen, Anders Dahlberg, Alf Ekblad, et al. 2013. Roots and associated fungi drive long-term carbon sequestration in boreal forest. *Science* **339**:1615–1618.

Dinh, Hang T., Jan Kuever, Marc Mussmann, Achim W. Hassel, Martin Stratmann, et al. 2004. Iron corrosion by novel anaerobic microorganisms. *Nature* **427**:829–832.

Gruber, Nicolas, and James N. Galloway. 2008. An Earth-system perspective of the global nitrogen cycle. *Nature* **451**:293–296.

Johnston, Andrew W. B., Jonathan D. Todd, and Andrew R. J. Curson. 2012. Microbial origins and consequences of dimethyl sulfide. *Microbe* **4**:181–185.

Maranger, Roxanne, D. F. Bird, and Neil M. Price. 1998. Iron acquisition by photosynthetic marine phytoplankton from ingested bacteria. *Nature* **396**:248–251.

Montzka, S. A., E. J. Dlugokencky, and J. H. Butler. 2011. Non-CO_2 greenhouse gases and climate change. *Nature* **476**:43–50.

Naqvi, Syed Wajih A., D. A. Jayakumar, P. V. Narvekar, H. Naik, V. V. S. S. Sarma, et al. 2000. Increased marine production of N_2O due to intensifying anoxia on the Indian continental shelf. *Nature* **408**:346–349.

Okubo, Chris H., and Alfred S. McEwen. 2007. Fracture-controlled paleo-fluid flow in Candor Chasma, Mars. *Science* **315**:983–985.

Pollard, Raymond T., Ian Salter, Richard J. Sanders, Mike I. Lucas, C. Mark Moore, et al. 2009. Southern Ocean deep-water carbon export enhanced by natural iron fertilization. *Nature* **457**:577–581.

Sohm, Jill A., Eric A. Webb, and Douglas G. Capone. 2011. Emerging patterns of marine nitrogen fixation. *Nature Reviews. Microbiology* **9**:499–508.

Sparks, William B., James Hough, Thomas A. Germer, Feng Chen, Shiladitya DasSarma, et al. 2009. Detection of circular polarization in light scattered from photosynthetic microbes. *Proceedings of the National Academy of Sciences USA* **106**:7816–7821.

Van Mooy, Benjamin A. S., Helen F. Fredricks, Byron E. Pedler, Sonya T. Dyhrman, David M. Karl, et al. 2009. Phytoplankton in the ocean use non-phosphorus lipids in response to phosphorus scarcity. *Nature* **458**:69–72.

Wright, Jody J., Kishori M. Konwar, and Steven J. Hallam. 2012. Microbial ecology of expanding oxygen minimum zones. *Nature* **10**:381–394.

Wu, Jingfeng, William Sunda, Edward A. Boyle, and David M. Karl. 2000. Phosphate depletion in the western North Atlantic Ocean. *Science* **289**:759–762.

COURTESY OF FERRIC FANG

Ferric Fang, University of Washington School of Medicine.

AN INTERVIEW WITH:

Ferric Fang

Molecular Microbiology Dissects a Pathogen

Ferric Fang has taught at the University of Washington School of Medicine for 12 years, and previously at UC San Diego and the University of Colorado. He also oversees the Clinical Microbiology Laboratory at Harborview Medical Center in Seattle, where he works on molecular methods to improve the diagnosis of bacterial infections. He is editor in chief of the journal *Infection and Immunity* and a member of the American Academy of Microbiology. In studies of pathogenic *Salmonella* species, his research group discovered a mechanism that enables a pathogen to acquire new genes expanding its ability to infect a mammalian host.

Why did you decide to study microbiology?

After a series of unsuccessful summer student research projects, I initially concluded that I was not meant to be a scientist. I instead went to medical school with the intention of becoming a clinician-teacher. However, while fulfilling the requirements of an infectious disease fellowship, I unexpectedly fell in love with research. I hasten to add that it was not love at first sight, but rather a gradual progression from flirtation to intrigue to passionate obsession.

How did you choose to study *Salmonella*? What makes this organism interesting?

Salmonella enterica is versatile, genetically tractable, important for human health, and rapidly growing. *Salmonella* interacts with nearly all components of mucosal and systemic immunity, and can therefore be used to model many types of host-pathogen interaction. *Salmonella* remains an important threat to human health, causing numerous food-borne outbreaks, life-threatening enteric fever, and devastating systemic infections in people coinfected with HIV. Ingenious scientists who have studied *Salmonella* for many decades have developed powerful methods for analysis and genetic manipulation, so I can spend most of my time doing interesting experiments instead of trying to invent tools. Finally, if you are an impatient person, as I am, the importance of rapid growth cannot be overemphasized. *Salmonella* divides approximately every 20 minutes in rich medium. If nutritional sources weren't limiting, one *Salmonella* bacterium happily dividing could exceed the mass of Earth in less than 2 days. One can get experimental results very quickly with *Salmonella*.

What is "xenogeneic silencing," and how did you discover it? Could this be considered a bacterial "immune system"?

"Xenogeneic silencing" is a mechanism by which the nucleoid-associated DNA-binding protein H-NS selectively silences horizontally acquired genes by targeting sequences with high AT content (that is, GC content lower than that of most of the genome). My laboratory's discovery of xenogeneic silencing arose from a casual conversation at the end of a lab meeting. My postdoc, Will Navarre, and I were trying to understand the implications of AT-rich DNA recognition by the H-NS protein. We initially envisaged xenogeneic silencing as a bacterial "immune system" to silence the expression of AT-rich foreign genes, and this may, in fact, be its original function.

Will identified H-NS binding sites in the *Salmonella* chromosome using chromatin immunoprecipitation on microarray (ChIP-on-chip) analysis. He then discovered a striking correlation between AT content and H-NS binding. We have since come to appreciate how this unusual means of genetic regulation might facilitate the acquisition of DNA by horizontal gene transfer. Since unregulated expression is a major barrier to gene transfer, xenogeneic silencing provides a mechanism by which foreign genes can be transcriptionally silenced to prevent their deleterious effects on the cell, thereby promoting their acquisition and subsequent integration into existing regulatory networks. Xenogeneic silencing provides an important insight into how bacterial evolution occurs and helps explain why horizontally acquired DNA tends to be AT rich: Such sequences are better tolerated and able to be incorporated by the recipient.

What current research question are you pursuing, and how do students participate?

We are interested in mechanisms of innate immunity and the strategies used by

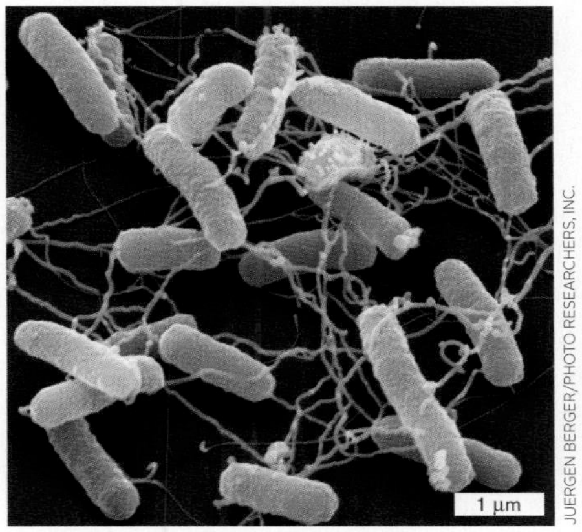
1 µm
JUERGEN BERGER/PHOTO RESEARCHERS, INC.

Salmonella enterica.

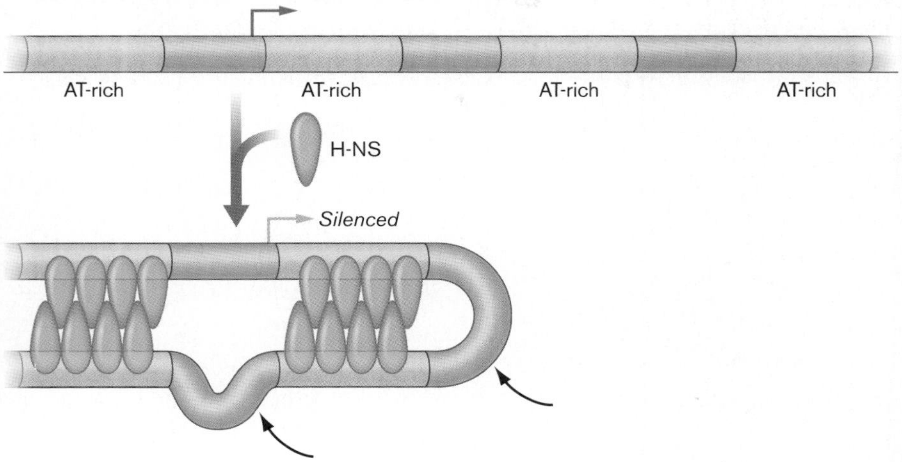
AT-rich
AT-rich
AT-rich
AT-rich
H-NS
Silenced

Bridging model for H-NS binding to DNA.

pathogens to avoid or resist host defenses. Current projects focus on virulence gene regulation, targets of host-derived nitric oxide and reactive oxygen species, bacterial iron metabolism, and typhoid pathogenesis. Students have made significant contributions to each of these projects. For instance, in a study of sigma factors (global regulators of transcription) my first graduate student, Traci Testerman, discovered that the alternative sigma factor RpoE is highly expressed in stationary phase. I was rather skeptical at the time, but Traci pursued this line of inquiry, which eventually led to an unexpected and important regulatory linkage among the RpoE, RpoH, and RpoS sigma factors.

I try to achieve a balanced approach with students so that they can get help when they really need it but also have freedom to explore questions on their own. Oswald Avery used to say, "Disappointment is my daily bread," and frustration is one of the major engines of scientific progress. This means that I have to allow students to take a crack at solving problems before jumping in myself. Over the years, my students have taught me a lot.

Does your work have implications for patient care and public health?

Ann Larson and I recently developed a PCR-based assay to detect toxigenic *Clostridium difficile* from stool specimens. This has improved our ability to detect *C. difficile* infections by 30% and is directly benefiting patients at our hospital. Much of my research would be considered basic rather than applied, but any research leading to a better understanding of the natural world has the potential to be applied for the good of society.

What is your most interesting project volunteering for the American Society for Microbiology?

The ASM is among the best professional societies in the world. My greatest pleasure in chairing the ASM General Meeting came from the opportunity to work with the many dedicated volunteers who put the meeting together. Currently I am involved with ASM Journals as the editor in chief of *Infection and Immunity*. I am enjoying the chance to work with a talented group of editors and staff to improve the journal and make it more relevant and useful in the Internet age.

What advice do you have for today's students?

Rather than trying to get into a "hot" field, I recommend pursuing something that is extremely interesting to you. Ten years down the road, you'll be surprised to find that everyone regards you as an expert! As Niels Bohr once said, "An expert is a person who has made all the mistakes that can be made in a very narrow field." Also don't be afraid to have multiple pots on the fire; you never know which line of investigation is going to turn out to be the most rewarding. Last but not least, don't forget that science is a team sport. Science represents the cumulative efforts of countless thousands of scientists over time; it is rather silly that we place so much emphasis on individual achievements in science when everyone's work is so dependent on that of others.

How does your family relate to your work? Do you have interesting pursuits outside of science?

My family has always been my highest priority. I have heard a number of scientists say that their secret is having the most patient and supportive spouse in the world, but it is actually true in my case. My kids seem intent on distancing themselves from science as far as possible (my daughter is an artist at Harvard, and my son is a classical guitarist at the Eastman School of Music). My major passion outside of family and work is music, especially jazz. If only I had a more consistent high register on the trumpet, we wouldn't be having this conversation!

CHAPTER 23

Human Microbiota and Innate Immunity

23.1 Human Microbiota: Location and Shifting Composition

23.2 Risks and Benefits of Microbiota

23.3 Overview of the Immune System

23.4 Barbarians at the Gate: Innate Host Defenses

23.5 The Acute Inflammatory Response

23.6 How Phagocytes Detect and Kill Microbes

23.7 Interferon, Natural Killer Cells, and Toll-like Receptors

23.8 Complement's Role in Innate Immunity

23.9 Fever

The human body teems with microbial hitchhikers. Most are harmless as long as they stay where they belong, such as on the skin or in the intestines, and many are beneficial to their unsuspecting hosts. Consequently, organisms that are part of the normal human microbiota typically exist in a symbiotic relationship with us. Of course, the human body is also under constant attack from microbial invaders that inhabit the surrounding environment. Fortunately, we have developed a series of barriers and elaborate fail-safe mechanisms that keep normal microbiota and invading pathogens at bay. Obstacles such as skin and stomach acid, which are considered to be nonspecific defenses, will repel most microorganisms. But for those microbes able to breach these barriers, there await adaptive and innate immune defenses.

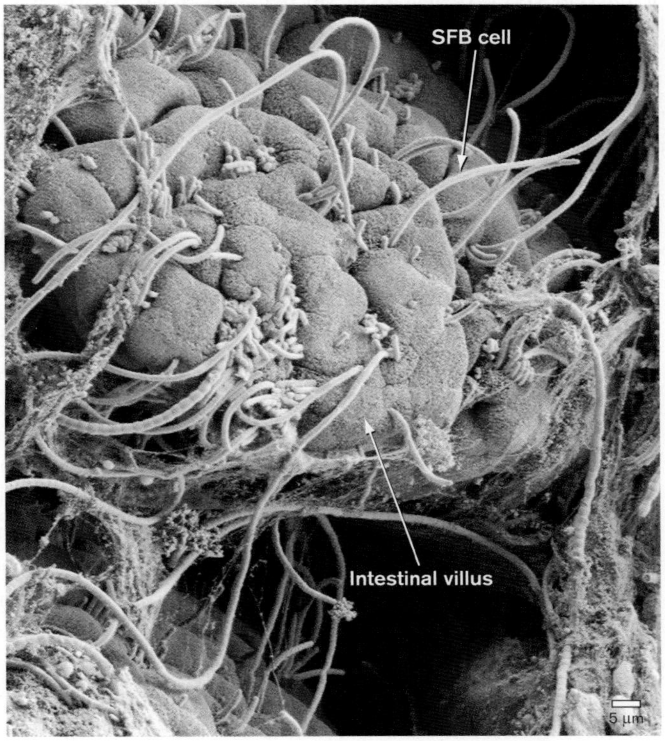

SFB cell
Intestinal villus
5 μm

CURRENT RESEARCH highlight

Microbiota "tune" the immune system. The trillions of microbes that populate our gut are called our microbiota. We harbor these bacteria for many reasons, but one of the most compelling involves their role in shaping our immune system. The specific gut microbes that help develop a host's immune system are actually tailored to the host. For instance, mice lacking microbiota fail to develop a normal immune system. Infusing these unfortunate mice with human gut microbiota is no help, but immunity develops normally in mice transplanted with mouse microbiota. One group of bacteria that shape mouse immunity consists of segmented filamentous bacilli (SFBs), shown here attached by holdfasts to mouse small intestine. SFBs are spore-forming, Gram-positive bacteria (15–75 μm in length) that stimulate production of a type of white blood cell called T_H17 lymphocytes. T_H17 cells are essential for a balanced immune system and crucial to our survival as a species. Thus, without SFBs or their equivalent in human microbiota, we might cease to exist. *Source:* Lauren Gravitz. 2012. *Nature* **485**:S12–S13. Photo by Alice Liang (NYU) and Doug Wei (Carl Zeiss, Inc.); colorized by Eric Roth (NYU).

Imagine two vacationing tourists, unknown to each other, each visiting the Alabama Gulf of Mexico coast at the same time. Both decide to take a swim in the gulf. One, a 12-year-old girl, has a cut on her leg caused by the sharp edge of a scooter. The other, a man 66 years of age, recently cut his arm on a nail protruding from a wooden handrail. Within moments of entering the gulf, several small Gram-negative microbes invade both of their bodies through those cuts. Four days later, the young girl is heading back to Birmingham in the back of the family car, oblivious to the battle recently waged in her bloodstream. At the same time, the man lies dead of an aggressive blood infection.

Both swimmers were attacked by the same pathogen, *Vibrio vulnificus.* Why did one live and the other die? As is the case for most healthy people, a variety of nonspecific, innate immune factors present in the girl's body killed the invading pathogen before it could multiply. The man, on the other hand, was an alcoholic with liver disease. He was deficient in several of those defense mechanisms—mechanisms that made the difference between life and death. What are these powerful innate factors? How can they be nonspecific, able to attack many different microbes without ever previously encountering any of them? And why do they kill the invaders and not our own cells?

Before exploring questions of immunity and self-defense, we must understand that each of us is a self-contained ecosystem, home to numerous microbes inhabiting all sorts of body niches. The day-to-day presence of these microbial guests both shapes and sharpens our immune system, but it is a benefit that comes with considerable risk.

23.1

Human Microbiota: Location and Shifting Composition

From the moment of our birth to the time of our death, we are constantly populated by microbes. In fact, our bodies carry ten times as many bacterial cells as human cells. Colonization occurs where our body meets the external environment (for example, mouth, skin, and parts of the genitourinary tract). More sequestered body sites are sterile. For example, most internal organs, blood, and cerebrospinal fluid should not harbor any bacteria. The presence of microbes at these sites usually means an infection is under way. Bacteria routinely found at nonsterile body sites (such as the intestine) were previously called **commensal organisms** ("commensal" comes from the Latin for "to share a table"). Strictly speaking, a commensal organism derives benefit from the host, but the host is unaffected. However, many organisms originally called commensal were later shown to also benefit the host in relationships described as mutualistic (see Section 17.6).

We now understand that bacteria colonizing our bodies may be as important collectively as a kidney or liver. For instance, Gary Siuzdak and his colleagues at the Scripps Research Institute showed that our body cells are continually bathed in metabolites produced by gut microbes. It is now thought that these metabolites, which circulate in our blood, significantly and favorably influence human health and development. The consortium of colonizing microbes has been dubbed the human **microbiota** or **microbiome**.

Metagenomic strategies (described in Chapter 7) are being employed to sequence all the genomes of our resident microbes. Remember, the majority of species that make up our microbiota are unknown and have never been grown in the lab. The collective genome of all microorganisms living in or on the human body is called a **metagenome**. We have gained tremendous insight into host-microbe relationships from the various microbiome projects under way. For instance, comparisons of microbiomes from the guts of lean and obese mice have shown vast differences in microbial species. The data suggest that different microbiota influence the efficiency of calorie harvesting from the diet and alter how derived energy is stored. Indeed, a microbiome study of sets of human twins in which one from each set was obese and the other lean found dramatic differences in their gut microbiomes.

Weblinks Human Microbiome Project (*see ebook*)

In this section we describe some well-known members of the human microbiota and explain why they populate different body sites. **Figure 23.1** illustrates human body sites (skin, respiratory, digestive, and genitourinary tracts) colonized by bacteria and provides examples of the bacteria found there. Realize that the various species comprising normal microbiota at a given body site exist as a biofilm consortium, chemically communicating with each other and their host.

Table 23.1 lists the four prominent ecosystems of humans, along with the number of microorganisms that typically inhabit them (called the **bioburden**). The table includes the ratio of aerobes to anaerobes and the initial origins of resident species. Note that the species composing our microbiota will vary throughout life as a result of continual reseeding from the environment. Each body site

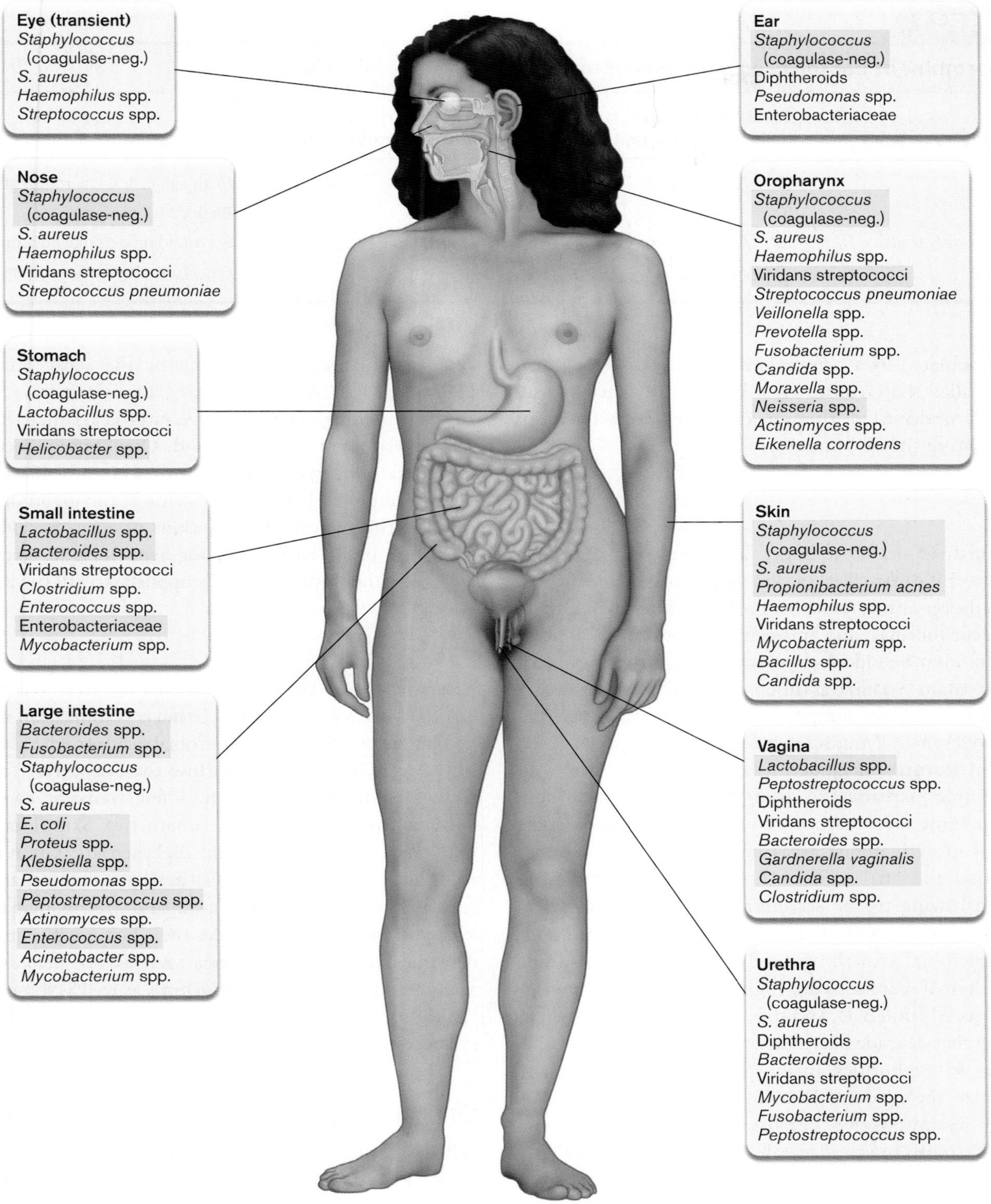
Eye (transient)
Staphylococcus (coagulase-neg.)
S. aureus
Haemophilus spp.
Streptococcus spp.
Nose
Staphylococcus (coagulase-neg.)
S. aureus
Haemophilus spp.
Viridans streptococci
Streptococcus pneumoniae
Stomach
Staphylococcus (coagulase-neg.)
Lactobacillus spp.
Viridans streptococci
Helicobacter spp.
Small intestine
Lactobacillus spp.
Bacteroides spp.
Viridans streptococci
Clostridium spp.
Enterococcus spp.
Enterobacteriaceae
Mycobacterium spp.
Large intestine
Bacteroides spp.
Fusobacterium spp.
Staphylococcus (coagulase-neg.)
S. aureus
E. coli
Proteus spp.
Klebsiella spp.
Pseudomonas spp.
Peptostreptococcus spp.
Actinomyces spp.
Enterococcus spp.
Acinetobacter spp.
Mycobacterium spp.
Ear
Staphylococcus (coagulase-neg.)
Diphtheroids
Pseudomonas spp.
Enterobacteriaceae
Oropharynx
Staphylococcus (coagulase-neg.)
S. aureus
Haemophilus spp.
Viridans streptococci
Streptococcus pneumoniae
Veillonella spp.
Prevotella spp.
Fusobacterium spp.
Candida spp.
Moraxella spp.
Neisseria spp.
Actinomyces spp.
Eikenella corrodens
Skin
Staphylococcus (coagulase-neg.)
S. aureus
Propionibacterium acnes
Haemophilus spp.
Viridans streptococci
Mycobacterium spp.
Bacillus spp.
Candida spp.
Vagina
Lactobacillus spp.
Peptostreptococcus spp.
Diphtheroids
Viridans streptococci
Bacteroides spp.
Gardnerella vaginalis
Candida spp.
Clostridium spp.
Urethra
Staphylococcus (coagulase-neg.)
S. aureus
Diphtheroids
Bacteroides spp.
Viridans streptococci
Mycobacterium spp.
Fusobacterium spp.
Peptostreptococcus spp.

FIGURE 23.1 ▪ Examples of normal microbiota present at various colonizing sites. Highlighted organisms are common inhabitants of the location. *Staphylococcus* species are differentiated by their differing abilities to produce coagulase, an enzyme that coagulates serum. *Staphylococcus aureus* is generally coagulase-positive (there are some coagulase-negative mutants), whereas other species of staphylococci normally found on skin are coagulase-negative. Some organisms—*Streptococcus pneumoniae,* for example—are pathogens but can be found as normal microbiota in some individuals.

TABLE 23.1

The prominent bacterial ecosystems of humans.

Microbial reservoirs	Total "bioburden" (colony-forming units)	Aerobe-to-anaerobe ratio	How acquired
Skin	10^{4-6}/sweat gland	1:10	Birth canal; oral environments
Mouth	10^{6-8}	1:10	Birth canal; caregiver
Genitourinary tract	10^{8-9}/vagina or ureter	1:100	Surrounding external environment
Intestine	$10^{11}/cm^3$	1:1,000	Baby formula; mother; caregiver

is also populated by a consortium of bacterial and human viruses, called the **virome**. Most human-associated viruses are poorly understood, though some may interact with us by stimulating the immune system.

Skin

The human adult, on average, is covered with 2 square meters (over 9 square feet) of skin (**epidermis**) populated by 10^{12} microorganisms. As with all colonized body sites, there are resident (normal) and transient members of the microbiota. But even a resident microbe exhibits diversity in that different strains colonize at different times. Thus, a human's microbiome does not comprise a static population but represents a vibrantly changing mix of strains and species.

Several features of epidermis make it difficult to colonize. Large expanses of skin are subject to drying, although some areas harbor enough moisture to support microbial growth; moist areas include scalp, ear, armpit, and genital and anal regions. The skin has an acidic pH (pH 4–6) owing to the secretion of organic acids by oil and sweat glands. As noted in Section 5.5, organic acids inhibit microbial growth by lowering bacterial cytoplasmic pH. Epidermal secretions are high in salt and low in water activity (see Section 5.4), and they contain enzymes such as lysozyme that degrade bacterial peptidoglycan.

Despite these hurdles, many species of bacteria manage to colonize the epidermal habitat. Most of these organisms are Gram-positive because Gram-positive bacteria tend to be more resistant to salt and dryness. *Staphylococcus epidermidis,* various *Bacillus* species, and yeasts such as *Candida* are common examples of normal skin microbes. Though normal, they are not always innocuous. One member of the skin's resident microbiota, the Gram-positive, anaerobic rod *Propionibacterium acnes,* causes acne, a very visible plague of adolescence. Curiously, this is the same genus (but different species) that makes Swiss cheese (see Section 16.3). Increased hormonal activity during the teen years stimulates oil production by the sebaceous glands (**Fig. 23.2**). *P. acnes* readily degrades the triglycerides in this oil, turning them into free fatty acids that then promote inflammation of the gland. One consequence of the inflammatory response is the formation of a blackhead, a plug of fluid and keratin that forms in the gland duct. The result is the typical skin eruptions of acne. Because of its microbial basis, treatments for acne include tetracycline (oral) or clindamycin (topically applied) to kill the bacteria.

Eye

The eye is exposed to the outside environment, so is it heavily colonized? Actually, it's not, because colonization is inhibited by the presence of antimicrobial factors such as lysozyme in the tears that continually rinse the eye surface (conjunctiva). Despite this protection, a few transient commensal bacteria can be found on the conjunctiva. Skin flora such as *Staphylococcus epidermidis* and diphtheroids (Gram-positive rods that look like clubs), as well as some Gram-negative rods such as *Escherichia coli, Klebsiella,* and *Proteus,* manage to at least temporarily make the eye their home without causing damage. Occasionally, bacteria such as *Streptococcus pneumoniae* and *Haemophilus influenzae,* as well as some viruses, can cause an ocular disease known as pinkeye, in which the eye becomes reddened with a watery discharge.

Oral and Nasal Cavities

When do we acquire a microbiome? Within hours after birth, a human infant's mouth becomes colonized with nonpathogenic *Neisseria* species (Gram-negative cocci), *Streptococcus, Actinomyces, Lactobacillus* (all Gram-positive), and some yeasts. These organisms come from the environment surrounding the newborn, such as the mother's skin and garments. As teeth emerge in the newborn, the anaerobic space between teeth and gums supports the growth of anaerobes, such as *Prevotella* and *Fusobacterium*

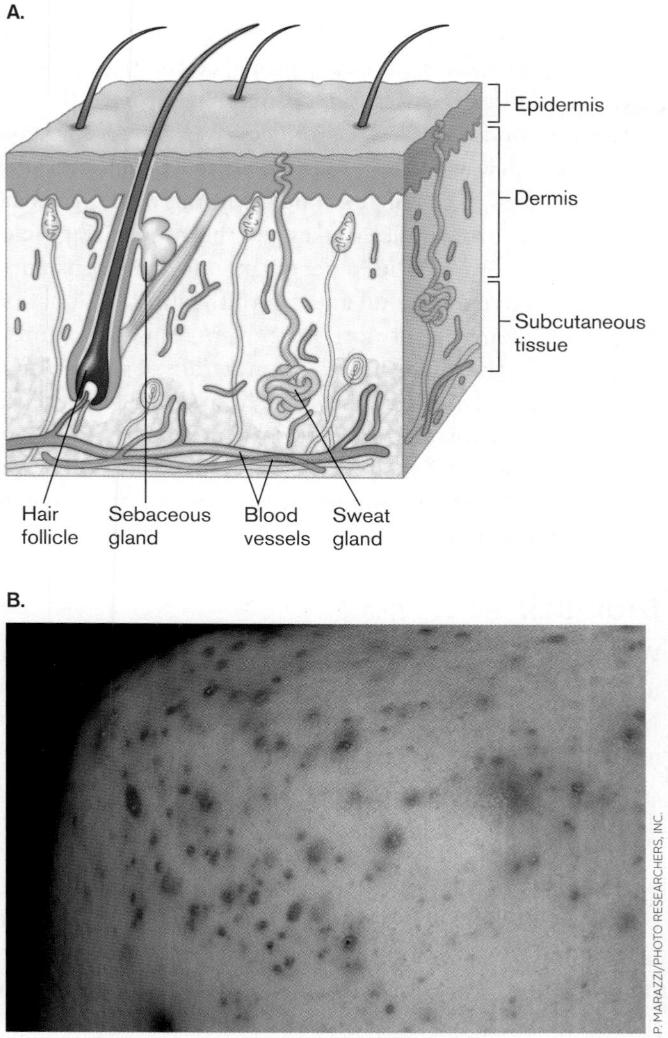
A.
Epidermis
Dermis
Subcutaneous tissue
Hair follicle
Sebaceous gland
Blood vessels
Sweat gland
B.
P. MARAZZI/PHOTO RESEARCHERS, INC.

FIGURE 23.2 ▪ Microbiology of skin and the development of acne. **A.** The location of sebaceous glands in skin. **B.** Acne.

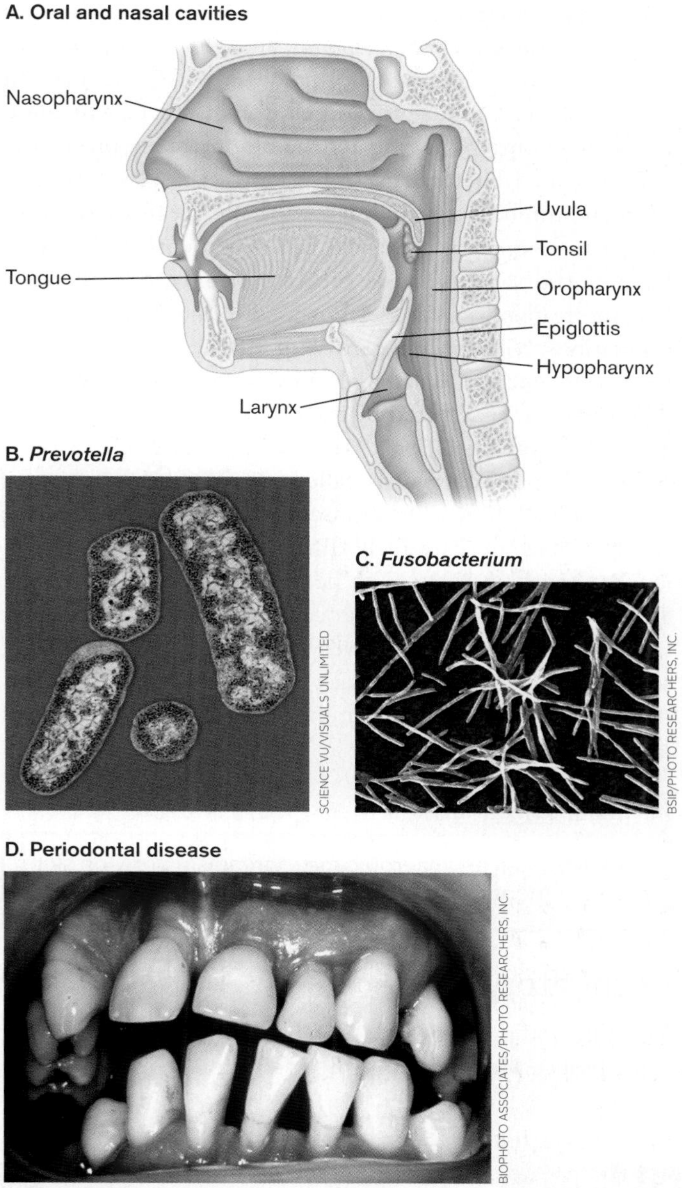
A. Oral and nasal cavities
Nasopharynx
Tongue
Larynx
Uvula
Tonsil
Oropharynx
Epiglottis
Hypopharynx
B. Prevotella
SCIENCE VU/VISUALS UNLIMITED
C. Fusobacterium
BSIP/PHOTO RESEARCHERS, INC.
D. Periodontal disease
BIOPHOTO ASSOCIATES/PHOTO RESEARCHERS, INC.

FIGURE 23.3 ▪ Examples of anaerobic oral microbiota and periodontal disease. **A.** Structures of the oral and nasal cavities. **B.** *Prevotella* (colorized TEM; each cell approx. 2 μm long). **C.** *Fusobacterium* (SEM; each cell approx. 1–5 μm long). **D.** Symptoms of periodontal disease include red, swollen gums, bleeding gums, gum shrinkage, and teeth drifting apart.

(**Fig. 23.3B** and **C**). Whatever the organism, colonizers of the oral cavity must be able to adhere to surfaces, like teeth and gums, to avoid mechanical removal and flushing into the acidic stomach. The teeth and gingival crevices are colonized by over 500 species of bacteria.

Organisms like *Streptococcus mutans,* which attaches to tooth enamel, and *Streptococcus salivarius,* which binds gingival surfaces, form a glycocalyx that enables them to firmly adhere to oral surfaces and to each other. They are but two of the microbes that lead to dental plaque formation. The acidic fermentation products of these organisms demineralize teeth and cause dental caries (tooth decay).

Important microbial habitats of the throat include the **nasopharynx**, which is the area leading from the nose to the oral cavity, and the **oropharynx**, which lies between the soft palate and the upper edge of the epiglottis (**Fig. 23.3A**). Organisms like *Staphylococcus aureus* and *Staphylococcus epidermidis* populate these sites. The nasopharynx and oropharynx can also harbor relatively harmless streptococci such as *Streptococcus salivarius* and *Streptococcus oralis,* as well as *Streptococcus mutans,* a major cause of tooth decay. Other oropharyngeal organisms include a large number of diphtheroids and the small Gram-negative rod *Moraxella catarrhalis.* Within the tonsillar crypts (small pits along the tonsil surface) lie anaerobic species such as *Prevotella, Porphyromonas,* and *Fusobacterium.*

The oral microbiota are normally harmless, but they can cause disease. Dental procedures, for instance, will often cause these organisms to enter the bloodstream, producing what is called **bacteremia** (infection of the bloodstream). Normal immune mechanisms typically clear these transient bacteremias quite easily; but in patients who have a heart mitral valve prolapse (heart murmur), microbes can become trapped in the defective valve and form bacterial vegetations. Vegetations consist of a large number of bacterial cells encased within glycocalyx (a polysaccharide or peptide polymer secreted by the organism) and fibrin (produced by clotting blood). Because the onset of disease is often insidious (slow), it is called subacute bacterial endocarditis (**Fig. 23.4**). Once ensconced within a vegetation, the microbes are extremely difficult to kill with antibiotics. Previously, patients with heart valve defects were given antibiotics prior to dental procedures. Immunocompetent individuals, however, are at very low risk of valve infection, so prophylactic treatment with antibiotics before a dental procedure is no longer recommended unless the patient is immunocompromised.

Thought Question

23.1 How can an anaerobic microorganism grow on skin or in the mouth, both of which are exposed to air?

Respiratory Tract

The lungs and trachea do not harbor long-term resident microbiota. Many organisms entering the nasopharynx become trapped in the nose by cilia that beat toward the pharynx. The microbes are thus propelled toward the acidic stomach and death. Microorganisms that slip into the trachea are trapped by mucus produced by ciliated epithelial cells lining airways. Cilia usher the microbes up and away from the lungs. The ciliated mucous lining of the trachea, bronchi, and bronchioles makes up the **mucociliary elevator** (**Fig. 23.5**). The mucociliary elevator constantly sweeps foreign particles up and out of the lung. This action is extremely important for preventing respiratory infections. When the mucociliary elevator fails, as when it is covered with tar from years of heavy smoking, or when it is overwhelmed by the inhalation of too many infectious microbes, infections such as the common cold (for example, rhinovirus) or pneumonia (for example, *Streptococcus pneumoniae*) can result.

Stomach

We have known for a hundred years that the stomach contents are acidic and that gastric acidity can kill bacteria. Just how important that acidity is for protection against microbes is illustrated by the infection caused by *Vibrio cholerae,* the causative agent of cholera. Cholera is a severe diarrheal disease endemic to many of the poorer countries of the world. The toxin produced by the organism acts on the intestinal lining to cause voluminous diarrhea—as much as 10 liters a day. Death can occur rapidly as a result of dehydration and shock. Although cholera actually affects the intestines, not the stomach, the bacteria must survive passage through the stomach to reach the intestines.

In the middle of the twentieth century, the U.S. government was concerned that troops dispatched to endemic

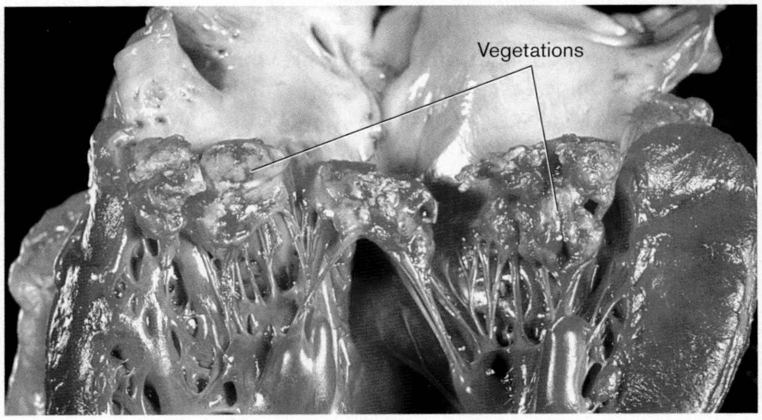
Vegetations

FIGURE 23.4 ▪ Gross pathology of subacute bacterial endocarditis involving the mitral valve. The left ventricle of the heart has been opened to show mitral valve fibrin vegetations due to infection. These growths are not present in a normal heart.

FIGURE 23.5 ▪ Mucociliary elevator. Movement of these hairlike cilia ushers particles up and out of the trachea and lungs (colorized SEM; diameter between 0.5 and 1 μm).

countries could develop cholera and become incapacitated in large numbers, so the U.S. Army decided to test potential cholera vaccines for their efficacy. The vaccines were living cultures of genetically weakened cholera bacteria that could not cause serious disease. Living strains were preferred because they were expected to grow in the gastrointestinal tract and stimulate natural mucosal immunity (discussed later and in Section 24.3). But after huge numbers of virulent organisms were administered to healthy volunteers (as a control group), very few contracted cholera. This result was confusing because hundreds of thousands of malnourished people in India easily contract the disease.

Why did volunteers given live cholera generally fail to develop the disease? Variations in resistance to cholera were related to differences in stomach acidity. The organism *V. cholerae* is extremely sensitive to low pH; even a pH of 4 will readily kill it. Healthy volunteers had stomach pH values well below 4, so *V. cholerae* was easily killed. Malnourished people, however, suffer from hypochlorhydria (decreased stomach acid). This permissive environment gives ingested microbes time to enter the less acidic intestine, where they can thrive and cause disease. Cholera epidemics in which thousands of people die, even today, typically occur among poor, malnourished populations. Because the U.S. populace is better nourished, cholera is not a major problem in the United States.

Although the stomach contents are very acidic, the mucous lining of the stomach is much less so. It is there that some bacteria can take refuge. In fact, the stomach harbors a diverse microbiota as detected using cultural and molecular techniques. A classic stomach pathogen is *Helicobacter pylori*. Although this organism has a remarkable ability to resist acidic pH (it survives at pH 1 using the enzyme urease to generate ammonia, which neutralizes acid), it will not grow in strong acid pH conditions. It can grow and divide, however, in the mucous lining of the stomach, where the pH is closer to 5 or 6 (**Fig. 23.6**). The U.S. Centers for Disease Control and Prevention (CDC) estimates that *Helicobacter* colonizes the stomachs of half the world's population. Most of the time, *Helicobacter* does not cause any apparent problem, but on occasion, the organism can produce gastric ulcers and even cancer.

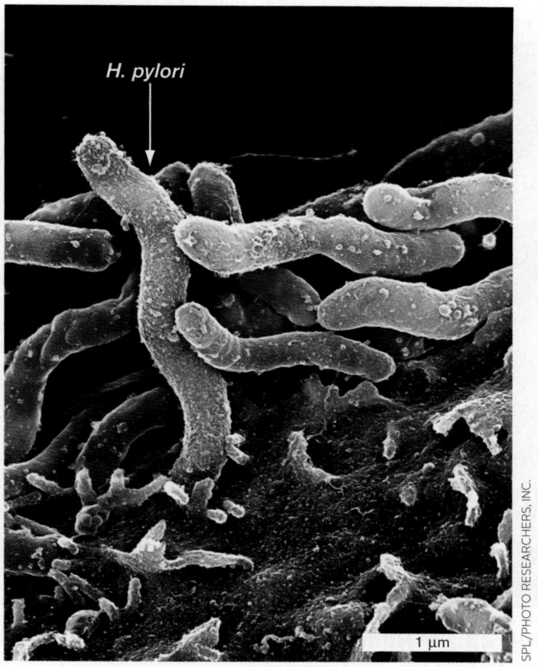
H. pylori
1 µm
SPL/PHOTO RESEARCHERS, INC.

FIGURE 23.6 ▪ *Helicobacter pylori* defies stomach acidity. *Helicobacter pylori* (colorized SEM) growing in the mucus on stomach epithelium. *H. pylori* bacteria attach to gastric epithelial cells and induce specific changes in cellular function, such as an increase in the expression of laminin receptor 1, a protein associated with malignancy. Peptic ulcers can result.

Intestine

The intestine is an extremely long tube consisting of several sections, each of which supports the growth of different bacterial species (see **Fig. 23.1**). Pancreatic secretions (at pH 10) enter the intestine at a point just past the stomach and raise the intestinal pH to about pH 8, which immediately relieves the acid stress placed on microbes. The relatively high pH and bile content of the duodenum and jejunum allow colonization by only a few resident, mostly Gram-positive, bacteria (enterococci, lactobacilli, and diphtheroids). These particular Gram-positive organisms possess a bile salt hydrolase that helps them grow in the presence of intestinal bile salts. The distal parts of the human intestine (ileum and colon) have a slightly acidic pH (pH 5–7) and a lower concentration of bile salts—conditions that support a more vibrant ecosystem. The intestine actually contains roughly 10^9–10^{11} bacteria per gram of feces (**Fig. 23.7**). This generally anaerobic environment is populated by both anaerobic and facultative microbes in a ratio of 1,000 anaerobes to one facultative organism. Why is the intestinal lumen anaerobic? The small amount of oxygen that diffuses from the intestinal wall into the lumen is immediately consumed by the facultative microbes, which render the environment anaerobic.

A well-studied model anaerobe, *Bacteroides thetaiotaomicron,* benefits us by breaking down the complex carbohydrates we eat. Much of this organism's genome is dedicated to taking many of the complex carbohydrates we eat and breaking them down into products that can be absorbed by the body. We humans actually absorb 15%–20% of our daily caloric intake in this way. Another

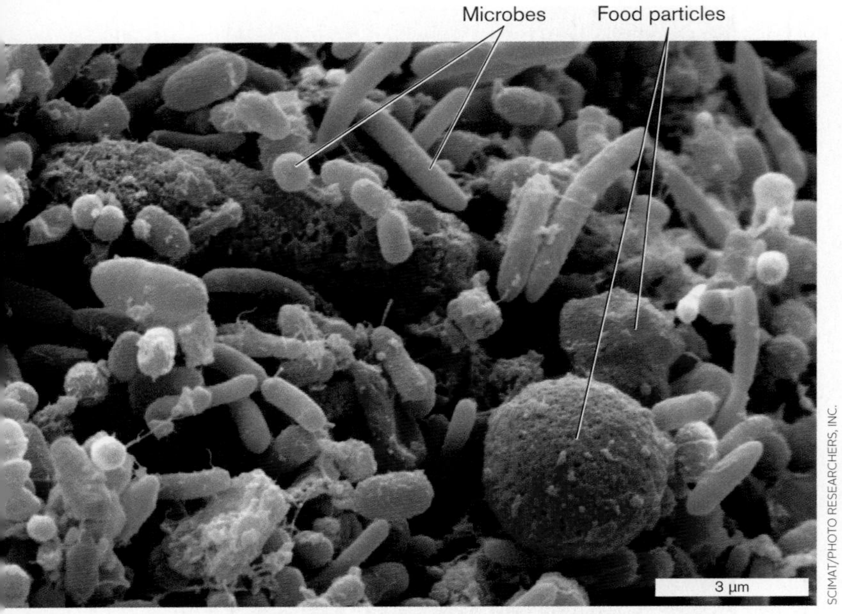
Microbes
Food particles
3 µm
SCIMAT/PHOTO RESEARCHERS, INC.

FIGURE 23.7 ■ Human feces showing food particles and resident bacteria. Colorized SEM.

important group of anaerobes consists of the segmented filamentous bacilli (SFBs) mentioned in the chapter opener. These spore formers are important for developing a healthy mouse immune system. Although SBFs have not been found in human intestines, research strongly suggests another organism with similar immune enhancing properties exists in our intestines, too.

Only recently have we started documenting which microbes inhabit our intestines (mainly because we cannot culture them). Culture-independent techniques like ribotyping (a bacterial classification based on 16S RNA sequences) and metagenomic sequencing have revealed the identity, diversity, and variation among species of the human microbiota. For example, of 395 unique "phylotypes" identified in humans, 60% were unknown at the species level, 80% have never been cultured, and huge differences in the population of microbes were observed between individuals.

All organisms normally residing in the intestine are innocuous and considered normal biota. But if they accidentally escape into nearby tissues, they can cause serious disease. Intestinal bacteria, encompassing 400–500 different species, have special talents that allow them to bind and colonize the intestinal cell wall. One reason why such a large and eclectic mix of species is supported is that different bacteria attach to different host cell receptors. Another reason is that many different food sources are available to support diverse groups of microbes.

Organisms that inhabit the intestine include facultative anaerobes in the family Enterobacteriaceae, of which *E. coli* is a member; and many "strict" anaerobes, such as *Bacteroides* (Gram-negative rods), *Peptostreptococcus* (Gram-positive cocci), and *Clostridium* (Gram-positive, spore-forming rods). In fact, the vast majority of our intestinal biota consists of the phyla Bacteroidetes and Firmicutes, of which *Bacteroides* species and *Clostridium* species, respectively, are members. Besides bacteria, other common and innocuous inhabitants are the yeast *Candida albicans* and protozoa such as *Trichomonas hominis* and *Entamoeba hartmanni.* The composition of intestinal microbiota among humans is a continuum of many different combinations of microbes. At one end of the spectrum are microbiomes dominated by *Bacteroides* while at the other end are communities dominated by *Prevotella* (another Gram-negative anaerobic species).

Does diet have an effect on our intestinal microbiota? The large intestine contains at least 500 bacterial phylotypes, as well as methanogenic archaea. Species diversity is due in part to the varied nutrients present in the intestine. But because diet influences which types of nutrients are available, a person's diet strongly influences the composition of intestinal microbiota. A recent study has shown, for example, that people with high protein and animal fat diets had fecal communities enriched with the *Bacteroides*-dominant enterotype, while those with carbohydrate-rich diets had fecal communities dominated by *Prevotella.* Human breast milk contains special complex carbohydrates that our own intestinal cells cannot digest, but the bacteria there can. Thus, human breast milk evolved to encourage specific populations of microbiota. The effects of different microbiota on human health are under study. For instance, a connection between obesity and the presence of methanogens in the intestine appears likely.

In addition to ingested foods, the mucus layer covering epithelial tissues is also recognized as an important source of carbohydrates in the intestine. Mucus is a complex gel of glycoproteins and glycolipids whose sugar substituents include *N*-acetylglucosamine, *N*-acetylgalactosamine, galactose, fucose, sialic acids, and lesser amounts of glucuronate, galacturonate, and gluconate. Interestingly, *E. coli* grows poorly in feces, but grows well in mucus by catabolizing gluconate via the Entner–Doudoroff pathway (see Section 13.5). Thus, *E. coli* doesn't grow on what we eat but on what we secrete.

Despite this buffet-like assortment of food, many species still end up vying for the same nutrient, so competition for food is fierce. Perturbing the balance of power with antibiotics, diet, or stress can lead to disease. Although we have considerable knowledge of the gastrointestinal microbiota,

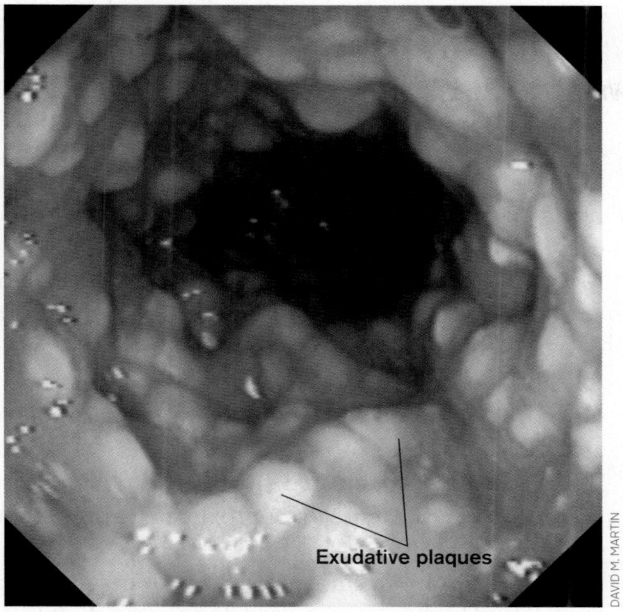
Exudative plaques
DAVID M. MARTIN

FIGURE 23.8 ▪ Pseudomembranous enterocolitis caused by *Clostridium difficile*. The organism is often a part of the normal intestinal microbiota, but antibiotic treatment, especially following surgery, can kill off competing microbes, leaving *C. difficile* to grow unabated. A toxin produced by the organism growing at the epithelial surface damages and kills host cells, causing exudative plaques to form on the intestinal wall. The small plaques eventually coalesce to form a large pseudomembrane that can slough off into the intestinal contents.

there remains much to learn. For example, humans and microbes sense each other's secreted chemical signals and alter gene expression in response. How do these chemical "conversations" impact our health?

Thought Questions

23.2 Why do many Gram-positive microbes that grow on the skin, such as *Staphylococcus epidermidis*, grow poorly or not at all in the gut?

23.3 How might normal biota escape the intestine and cause disease at other body sites?

Probiotics. As already noted, the microbial composition of the intestine is complex but balanced. It is complex because many species reside there. It is balanced because these species manage to coexist without eliminating each other. The rival organisms help prevent infection by helping to develop our immune system, by competing with pathogenic species for nutrients and actually contributing to nutrition (for example, *E. coli* makes vitamin B_{12}, whereas other organisms help digest complex substrates).

Emotional stress, a change in diet, or antibiotic therapy can throw off that balance and lead to poor digestion or disease, such as pseudomembranous enterocolitis (**Fig. 23.8**). Some people favor restoring the natural microbial balance by orally ingesting living microbes such as the lactobacilli present in yogurt. Taking these supplements, called **probiotics** (from the Greek meaning "for life"), is thought to restore balance to the microbial community and return the host to good health. The most commonly used probiotic genera are *Lactobacillus* and *Bifidobacterium* (**Fig. 23.9**). The potential mechanisms by which they improve intestinal health include competitive bacterial interactions (normal biota prevent growth of pathogenic bacteria), production of antimicrobial compounds, and immunomodulation (an

A.

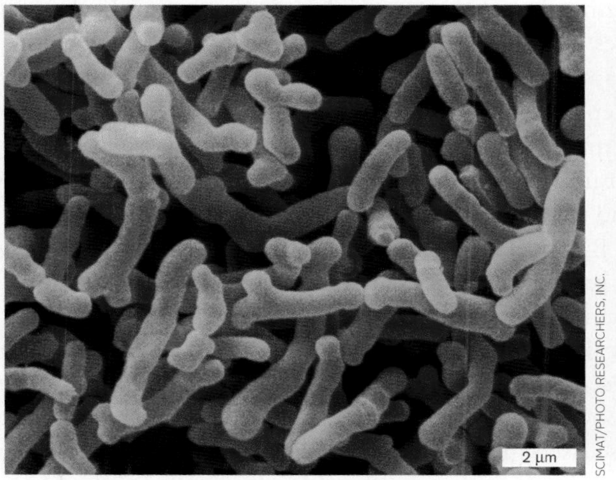
2 µm
SCIMAT/PHOTO RESEARCHERS, INC.

B.

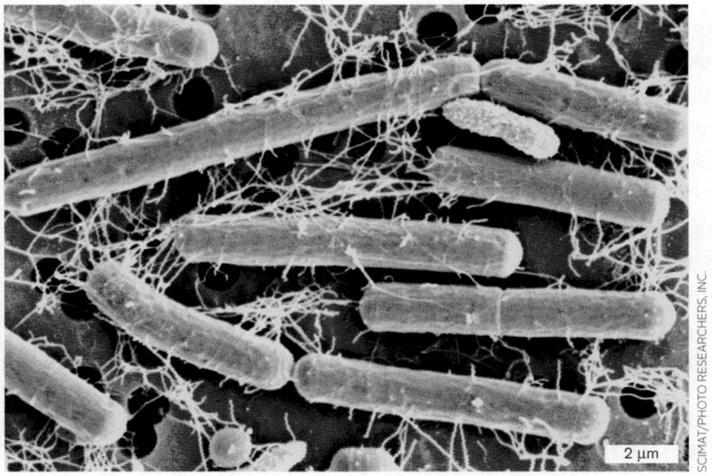
2 µm
SCIMAT/PHOTO RESEARCHERS, INC.

FIGURE 23.9 ▪ Commonly used probiotic microorganisms. A. *Bifidobacterium* (note the Y shape of some cells). **B.** *Lactobacillus acidophilus.* Colorized SEMs.

ability to change the activity of cells of the immune system). Probiotics are now being used to treat several gastrointestinal disorders, including inflammatory bowel disease (IBD). An extreme form of probiotics is the fecal transplant, where the intestinal microbiome of a healthy person is transferred to a relative with a severe intestinal disease such as colitis. Restoring a "normal" microbiome in this way has successfully "cured" IBD or ulcerative colitis patients that did not respond to antibiotics.

Thought Question

23.4 Why can the colon be considered a fermenter?

Genitourinary Tract

Much of the genitourinary tract is normally free from microbes. These areas include the kidneys, which remove waste products from the blood; ureters, which remove urine from the kidneys; and the urinary bladder, which holds urine until it is excreted. The distal urethra, however—because of its proximity to the outside world—normally contains *Staphylococcus epidermidis*, *Enterococcus* species, and some members of Enterobacteriaceae. Some of these organisms can cause bladder disease (known as urinary tract infection, or UTI) if they make their way into the bladder (for example, via catheterization).

The large surface area and associated secretions of the female genital tract make it a rich environment for microbes. Composition of the vaginal microbiota changes with the menstrual cycle, owing to changing nutrients and pH. The mildly acidic nature of vaginal secretions (approximately pH 4.5) discourages the growth of many microbes. As a result, the acid-tolerant *Lactobacillus acidophilus* is among the most populous vaginal species. Healthy women appear to fall into two broad categories: 30% have lactobacillus as the primary member of their vaginal microbiota, while 70% have mixed species with few lactobacilli. Women in the latter group appear to be more susceptible to sexually transmitted infections.

The balance between different species that comprise vaginal microbiota is crucial to preventing infection. Antibiotic therapy to treat an infection anywhere in the body can also affect the vaginal microbiota. The resulting imbalance can allow infection by *Candida albicans* (a yeast infection), which is not susceptible to antibiotics designed to kill bacteria.

To Summarize

- The **normal microbiota** present on body surfaces constantly changes.
- **Normal microbiota can cause disease** if microbes breach the surfaces they colonize and gain access to the circulation or deeper tissues.
- **Normal microbiota help prevent disease** by inhibiting colonization by pathogens.
- **Skin microbiota** consists primarily of Gram-positive microbes, including *Propionibacterium acnes*, which can cause acne.
- **Microbe-free areas of the body** include the lungs, cerebrospinal fluid, and bladder. The eyes are generally microbe-free, although some microbes can be present for short periods.
- **Oral and nasal surfaces** are colonized by aerobic and anaerobic microbes.
- **The intestine** is populated by 10^9–10^{11} microbes per gram of feces in a ratio of approximately 1,000 anaerobes to one aerobe.
- **Vaginal microbiota** influence susceptibility to sexually transmitted infections.

23.2

Risks and Benefits of Microbiota

It is enticing to portray colonizing bacteria as hostile armies encamped at our body's gates (also called portals of entry), just waiting for the opportunity to invade. But bacteria do not possess hostile intent, only the need to find food. The microbes of our normal microbiota, unlike true pathogens, have not evolved tools to pass beyond their resident niche. The needs of commensal organisms are met at the colonizing site. However, accidental penetration beyond these sites can cause disease. A cancerous lesion in the colon, for example, can provide a passageway for microbes to enter deeper tissues and cause disease. *Bacteroides fragilis*, a harmless anaerobe in the gut, can invade tissues through surgical wounds, causing abscesses that persist (and even lead to gangrene) after abdominal surgery (**Fig. 23.10**).

By and large, the body's barriers of defense work well to prevent incursions by microbiota or, in the event of an incursion, to kill the invader. These defenses sometimes break down in persons with an immune system compromised by medical treatments (such as anticancer drugs) or by disease (such as a deficiency in complement factors, discussed in Section 23.3). Such a person is described as a **compromised host** and can be repeatedly infected by normal biota. Organisms causing disease in this situation are called **opportunistic pathogens**.

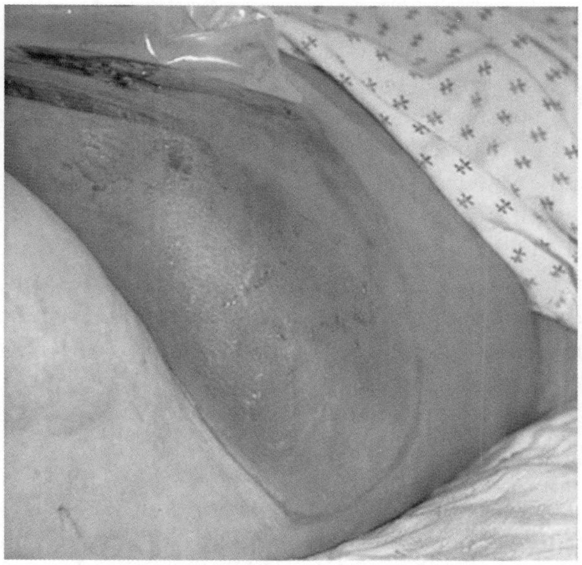
ABSTR HEMATOL ONCOL © 2003 CLIGGOTT PUBLISHING, DIVISION OF SCP COMMUNICATIONS

FIGURE 23.10 ▪ Anaerobic gas gangrene of the abdominal wall caused by *Bacteroides*. This infection was an unfortunate result of bowel surgery. Organisms escaped from the intestine and initiated infection in the abdominal wall.

Microbiome species are often beneficial (mutualistic). They can interfere with the colonization of pathogens by competing for attachment receptors on host cells, competing for food sources, and synthesizing antimicrobial compounds such as fermentation end products. For example, lactic acid made by lactobacilli keeps the vaginal pH acidic, which dissuades colonization by various pathogenic microbes. In another example, a polysaccharide produced by *B. fragilis* stimulates production of an anti-inflammatory cytokine that will prevent gastrointestinal colitis caused by *Helicobacter hepaticus*. **Cytokines** are small, secreted host proteins that bind to cells of the immune system and regulate their function. (Cytokines are discussed in Sections 23.5 and 24.6.)

Commensal organisms also enhance function of the immune system. Bacterial proteins such as catalase can act as **immunomodulins**. Immunomodulins made by normal biota growing on mucosal surfaces modify the secretion of host proteins, such as cytokines and tumor necrosis factor, which influence the immune response. Another potential benefit of commensal organisms is their development as vaccine delivery systems (eTopic 23.1).

Some toxic bacterial products also have benefits. **Enterotoxins**, for example, are proteins produced by some Gram-negative pathogens that cause diarrhea. But they may also protect against colorectal cancer by activating membrane calcium channels in intestinal epithelial cells. Increasing membrane conductance for calcium turns on an antiproliferative pathway that provides resistance to colon cancer.

The benefit that microbiota offer their hosts is especially apparent in gnotobiotic animals. A **gnotobiotic animal** is an animal that is germ-free or one in which all the microbial species present are known. Developing a gnotobiotic colony of animals involves delivering offspring by cesarean section under aseptic conditions in an isolator. The newborn is moved to a separate isolator where all entering air, water, and food are sterilized. Once gnotobiotic animals are established, the colony is maintained by normal mating between the members. Germ-free animals, however, often have poorly developed immune systems, lower cardiac output, and thin intestinal walls, and they are more prone to infection by pathogens. The lesson from these animals is that the presence of microbiota continually challenges the immune system and keeps it active. In the chapter opening, Current Research highlight, Gnotobiotic mice were used to identify the role of segmented filamentous bacilli in the development of mouse immunity.

To Summarize

- **Opportunistic pathogens** infect only compromised hosts.
- **Benefits of commensal microbes** include interfering with pathogen colonization, the production of immunomodulatory proteins, and the potential use as vaccine delivery vehicles.
- **Gnotobiotic animals** are germ-free or colonized by a known set of microbes.

23.3

Overview of the Immune System

Because we are surrounded by and host numerous bacteria, how is it we are not constantly infected? Here we begin our discussion of the many layers of protection designed to prevent infection and disease. As you will see, numerous physical and chemical barriers provide an effective first line of defense against infection by invading microbes. As such, they play a critical role in managing our resident ecosystems. But these barriers are not unbreachable. Organisms can still slip through. Consequently, humans, as well as other mammals, have a more aggressive defense called the immune system. The **immune system** is an integrated system of organs, tissues, cells, and cell products that differentiates self from nonself and neutralizes potentially pathogenic organisms or substances. This complex collection of cells and soluble proteins is capable of responding to nearly any foreign molecular structure.

Innate and Adaptive Immunity

There are two broad types of immunity: **innate immunity** (often called **nonadaptive immunity**) and **adaptive immunity** (discussed in detail in Chapter 24). Innate and adaptive immunity have several key differences. Innate immunity includes physical barriers such as skin, chemical barriers such as stomach acid, and relatively nonspecific cellular responses to infection that engage if the physical and chemical barriers are breached. The cellular innate responses are triggered by microbial structures such as peptidoglycan and lipopolysaccharides. Innate immunity is essentially "hardwired" into the body. It is present at birth, so that mechanisms of innate immunity exist before the body ever encounters a microbe. The protection afforded by innate immunity is nonspecific, capable of blocking or attacking many different types of foreign substances and organisms.

In contrast to innate immunity, adaptive immunity is designed to react to very specific structures called **antigens**. An antigen is any chemical, compound, or structure foreign to the body that elicits an immune response. Adaptive immune responses to a specific antigen do not occur until the body "sees" that antigen. Once activated, adaptive immune mechanisms can recognize at least 10^{10} different antigenic structures and specifically launch a directed attack against each one. Once such an attack has been activated, the organism keeps a "memory" of the exposure in the form of specific memory cells, and an encounter with the microbe years later will reactivate the memory cells specific to the antigen.

The two types of immunity are illustrated by the response of the immune system to infection by the microorganism *Neisseria gonorrhoeae,* which causes the sexually transmitted disease gonorrhea. A component of the innate immune response is **complement**, composed of several soluble protein factors constantly present in the blood. Within moments of an initial infection, complement proteins form holes in the bacterial membrane, thereby killing the microbe (see Section 23.8). On the other hand, the adaptive immune response will generate specific antibodies made to the cells of *N. gonorrhoeae* that escaped the innate mechanisms. These antibodies, however, are not made until well after the organism infects a person. Together, innate and adaptive immunity can help ward off disease caused by this organism. Unfortunately, *N. gonorrhoeae* infection does not generate good "memory," so reinfection with this organism is possible.

It is important to know that the innate and adaptive immune systems are interconnected. Antibodies made by the adaptive immune system will trigger parts of the innate immune system, such as the complement cascade we will discuss later. Likewise, activation of innate resistance mechanisms will cause the release of small immunomodulatory peptides (cytokines) that influence the type and strength of adaptive immunity brought to bear. In military terms, it is similar to the army coordinating its actions with those of the air and naval forces.

Infection versus Disease

Contact with an infectious agent does not guarantee that a person will actually contract the disease. If the number of infecting organisms is small and the immune system (innate and adaptive) is effective, the individual may not develop disease (see the chapter introduction), although a person known to have been exposed to certain microorganisms will be treated with antibiotics as a preventive measure. Chapter 26 more fully discusses the difference between being infected and having a disease.

Any microbe that launches a successful attack on a human or other animal and causes disease must first breach the host's physical and chemical barriers to gain entrance to the body. It must then survive the innate defense mechanisms and begin to multiply. Finally, the microbe must

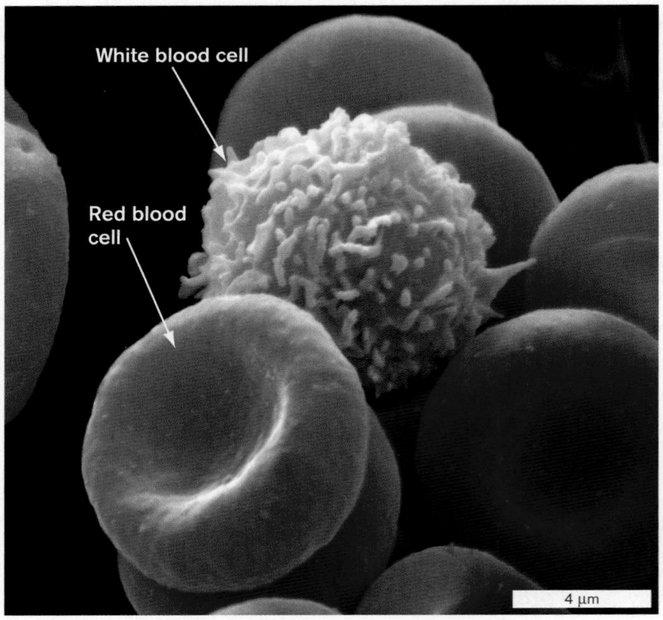
White blood cell
Red blood cell
4 µm

FIGURE 23.11 ▪ Red and white blood cells. This colorized scanning electron micrograph illustrates the relative sizes and 3D morphologies of these cell types.

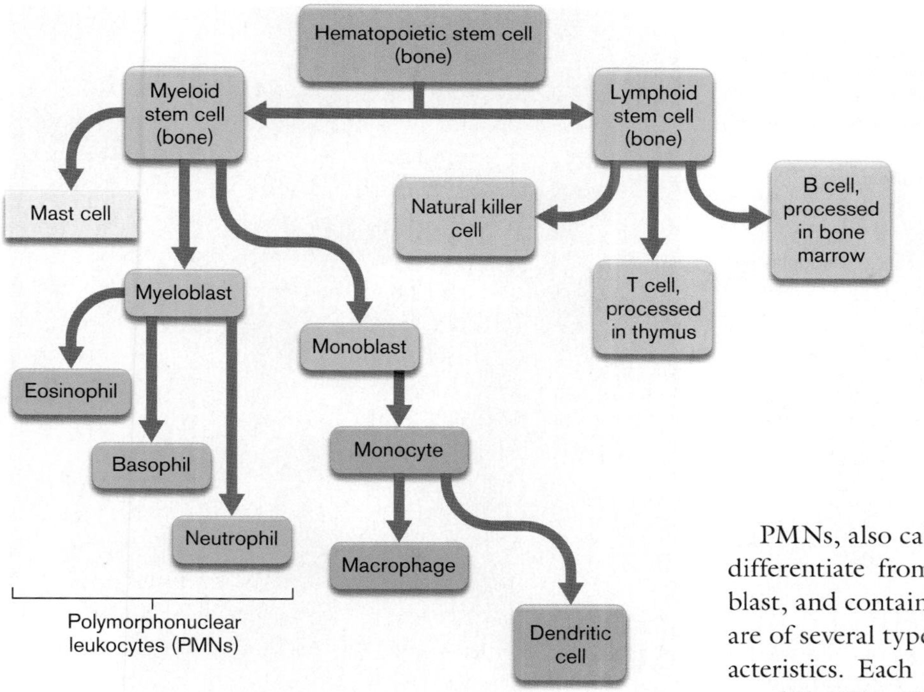
Hematopoietic stem cell (bone)
Myeloid stem cell (bone)
Lymphoid stem cell (bone)
Mast cell
Natural killer cell
B cell, processed in bone marrow
T cell, processed in thymus
Myeloblast
Monoblast
Eosinophil
Basophil
Monocyte
Neutrophil
Macrophage
Dendritic cell
Polymorphonuclear leukocytes (PMNs)

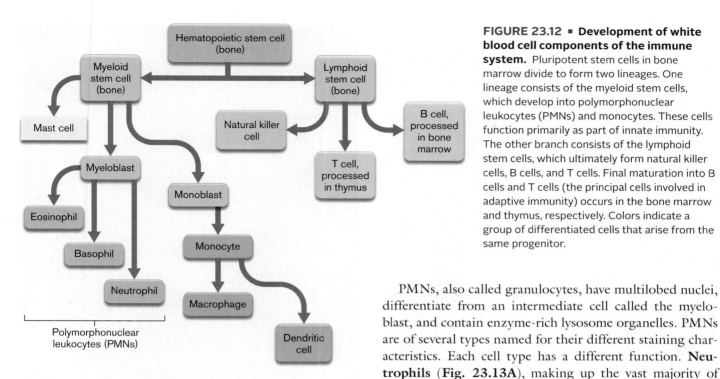

FIGURE 23.12 ▪ Development of white blood cell components of the immune system. Pluripotent stem cells in bone marrow divide to form two lineages. One lineage consists of the myeloid stem cells, which develop into polymorphonuclear leukocytes (PMNs) and monocytes. These cells function primarily as part of innate immunity. The other branch consists of the lymphoid stem cells, which ultimately form natural killer cells, B cells, and T cells. Final maturation into B cells and T cells (the principal cells involved in adaptive immunity) occurs in the bone marrow and thymus, respectively. Colors indicate a group of differentiated cells that arise from the same progenitor.

surmount the last line of defense, adaptive immunity, which begins to respond as the microbe struggles to overcome innate immune defenses. The rest of this chapter will discuss the various innate defense mechanisms. Adaptive immunity is discussed in Chapter 24.

Although innate and adaptive immunity are often treated as separate entities, certain kinds of cells and organs play a role in both types of immunity. We thus introduce various cells and organs of the immune system as a whole before focusing on innate immunity. Our coverage of innate immunity will include physical and chemical barriers to infection, inflammation and nonspecific killing through phagocytosis, neutrophil extracellular traps, interferon, natural killer cells, and complement.

Cells of the Immune System

Blood is composed of red blood cells, white blood cells (also generally known as leukocytes), and platelets (**Fig. 23.11**). The many types of white blood cells are formed by the differentiation of stem cells produced in bone marrow (**Fig. 23.12**). Among these white blood cells are various components of innate immunity that differentiated from myeloid stem cells. These differentiated cells include:

- Polymorphonuclear leukocytes (PMNs)
- Monocytes
- Macrophages
- Dendritic cells
- Mast cells

PMNs, also called granulocytes, have multilobed nuclei, differentiate from an intermediate cell called the myeloblast, and contain enzyme-rich lysosome organelles. PMNs are of several types named for their different staining characteristics. Each cell type has a different function. **Neutrophils** (**Fig. 23.13A**), making up the vast majority of white cells in the blood, can engulf microbes by phagocytosis. Phagocytosis involves the extrusion of pseudopods that attach to, and envelop the pathogen, which ends up in a phagosome vacuole. The phagocyte then kills the

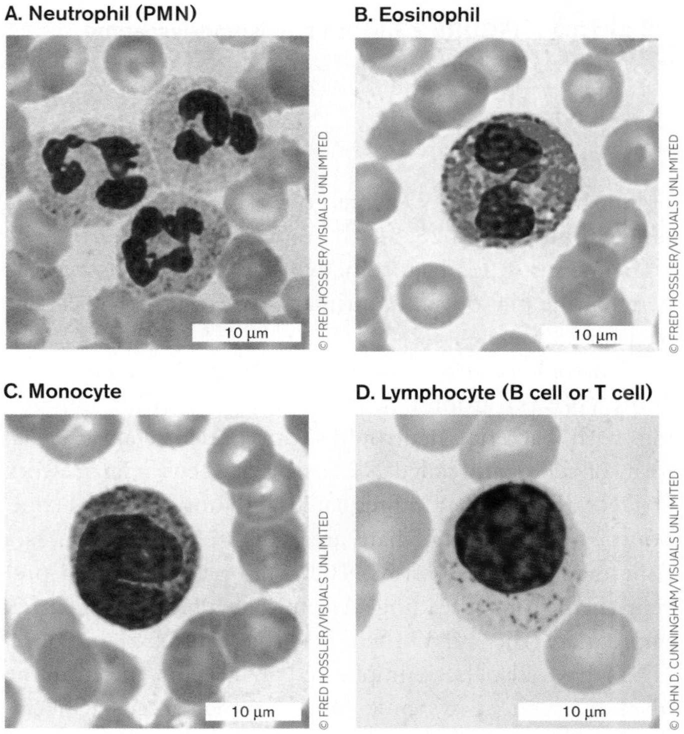
A. Neutrophil (PMN)
B. Eosinophil
C. Monocyte
D. Lymphocyte (B cell or T cell)
10 µm
10 µm
10 µm
10 µm
© FRED HOSSLER/VISUALS UNLIMITED
© FRED HOSSLER/VISUALS UNLIMITED
© FRED HOSSLER/VISUALS UNLIMITED
© JOHN D. CUNNINGHAM/VISUALS UNLIMITED

FIGURE 23.13 ▪ Types of white blood cells. A. Neutrophil (PMN). **B.** Eosinophil. **C.** Monocyte. **D.** Lymphocyte (B cell or T cell).

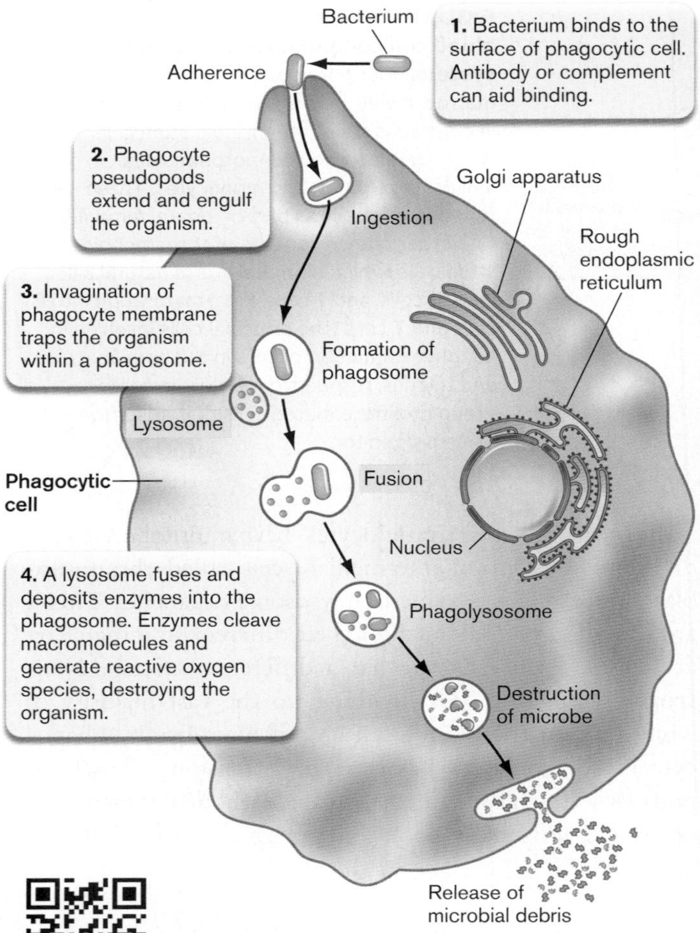
Bacterium
1. Bacterium binds to the surface of phagocytic cell. Antibody or complement can aid binding.
Adherence
2. Phagocyte pseudopods extend and engulf the organism.
Golgi apparatus
Ingestion
Rough endoplasmic reticulum
3. Invagination of phagocyte membrane traps the organism within a phagosome.
Formation of phagosome
Lysosome
Phagocytic cell
Fusion
Nucleus
4. A lysosome fuses and deposits enzymes into the phagosome. Enzymes cleave macromolecules and generate reactive oxygen species, destroying the organism.
Phagolysosome
Destruction of microbe
Release of microbial debris

FIGURE 23.14 ■ Phagosome-lysosome fusion.

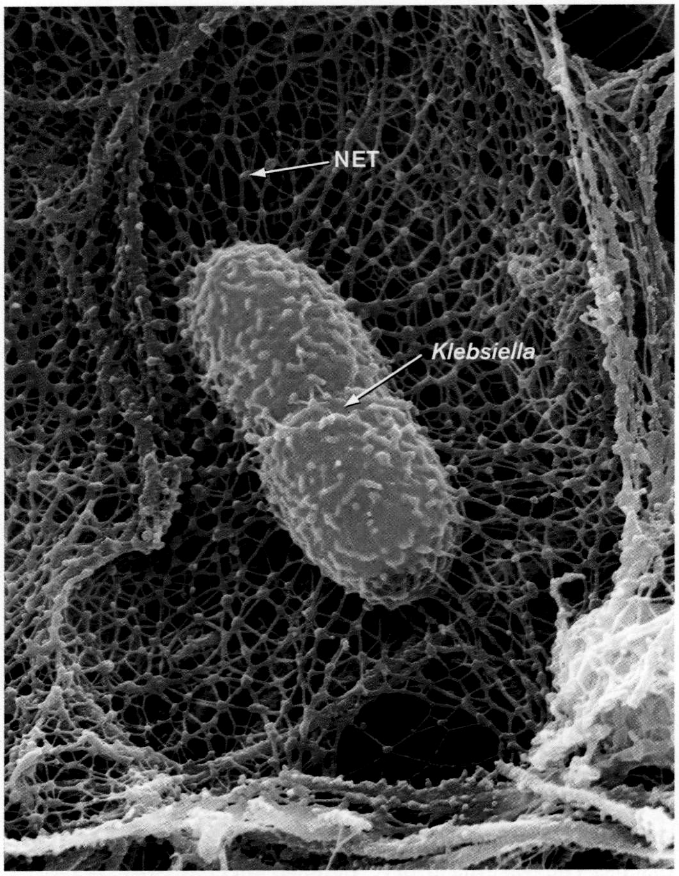
NET
Klebsiella

HTTP://JCB-BIOWRITES.RUPRESS.ORG/2010/11/NEW-ISSUE-NOVEMBER-1ST.HTML
HTTP://JCB.RUPRESS.ORG/CONTENT/191/3.COVER-EXPANSION

FIGURE 23.15 ■ Neutrophil extracellular trap. An infected mouse lung shows a *Klebsiella pneumoniae* bacterium (colorized; size approx. 1 μm) snared in a neutrophil extracellular trap (green), a web of decondensed chromatin released by neutrophils to catch and kill pathogens.

organism by fusing enzyme-gorged lysosomes with the **phagosomes** (**Fig. 23.14**). Enzymes spilling from the lysosome into the phagosome will destroy various components of the microbe and, ultimately, the microbe itself (see Section 23.6).

Neutrophils also "throw" **neutrophil extracellular traps (NETs)** around nearby pathogens. After interacting with bacteria, neutrophils can undergo an unusual form of cell death called NETosis that spews a latticework of DNA (chromatin) impregnated with antimicrobial compounds into the immediate area (**Fig. 23.15**). Much as a fisherman's net traps fish, NETs trap pathogens and prevent them from spreading. Antimicrobial compounds that impregnate the NET then kill the captured microbes. NETs have also been implicated in certain autoimmune diseases, such as systemic lupus erythematosus (**Special Topic 23.1**).

Thought Question

23.5 **Figure 23.15** shows how a neutrophil extracellular trap can ensnare a nearby pathogen. What bacterial structure might blunt the microbicidal effect of NETs?

Basophils, which stain with basic dyes, and **eosinophils** (**Fig. 23.13B**), which stain with the acidic dye eosin, do not phagocytose microbes or throw NETs, but release products, such as major basic protein, that are toxic to the microbe. These two types of blood cells also release chemical mediators (called vasoactive agents) that affect the diameter and permeability of blood vessels (the significance of which is discussed in Section 23.5). Eosinophils play a major role in killing multicellular parasites like helminth worms. **Mast cells** are similar to basophils in structure but differentiate in a lineage separate from PMNs. Unlike PMNs, mast cells are residents of connective tissues and mucosa and do not circulate in the bloodstream. Basophils and mast cells also contain high-affinity receptors for a class of antibody called immunoglobulin E (IgE) associated with allergic responses (detailed in Section 24.8).

Note: Because neutrophils constitute the vast majority of PMNs, the terms "neutrophil" and "PMN" are often used interchangeably.

Monocytes (**Fig. 23.13C**) are white blood cells with a single nucleus (not multilobed like a PMN); they engulf (phagocytose) foreign material. Monocytes circulating in the blood can migrate out of blood vessels into various tissues and differentiate into **macrophages** and **dendritic cells** (**Fig. 23.16**). Macrophages are phagocytic and form a major part of the amorphous **reticuloendothelial system**, which is widely distributed throughout the body. The reticuloendothelial system is a group of cells with the ability to take up and sequester particles. In addition to macrophages, the system is composed of specialized endothelial cells lining the sinusoids of the liver, spleen, and bone marrow, as well as reticular cells of lymphatic tissue (macrophages) and of bone marrow (fibroblasts). The function of macrophages and the reticuloendothelial system is to phagocytose microorganisms.

Macrophages are the cells most likely to make first contact with invading pathogens. They have two functions. As part of innate immunity, they kill invaders directly. Protrusions from the macrophage surface extend and clasp nearby bacteria, pulling them into the cell (**Fig. 23.16A**). Once ingested by the macrophage, the bacteria are destroyed. Subsequently, as the first step in adaptive immunity, the remnants are processed (degraded) into smaller peptides (called antigens), which are then presented on the macrophage cell surface. Specific white blood cells called T cells (a type of lymphocyte) can bind the displayed antigens and become activated as part of the adaptive immune response. Thus, macrophages are considered **antigen-presenting cells** (see Chapter 24).

Dendritic cells (**Fig 23.16B**) are also antigen-presenting cells. Present in the spleen and lymph nodes, they, like macrophages, can take up, process, and present small antigens on their cell surface. Dendritic cells are different from macrophages in structure and in the fact that dendritic cells primarily take up small soluble antigens from their surroundings rather than through phagocytosis and the degradation of whole bacteria.

A. Macrophage engulfing bacteria

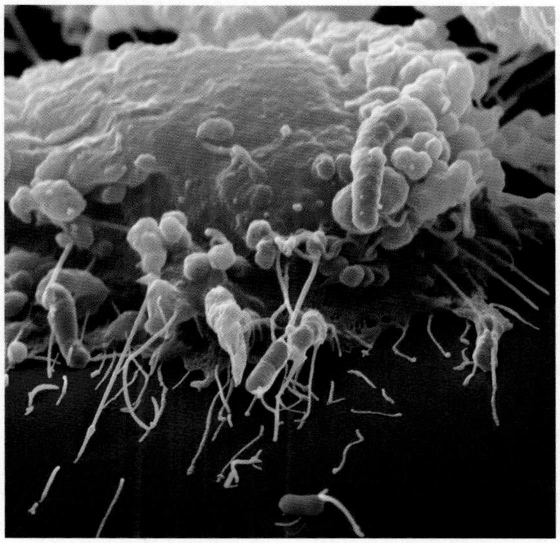

B. Dendritic cell

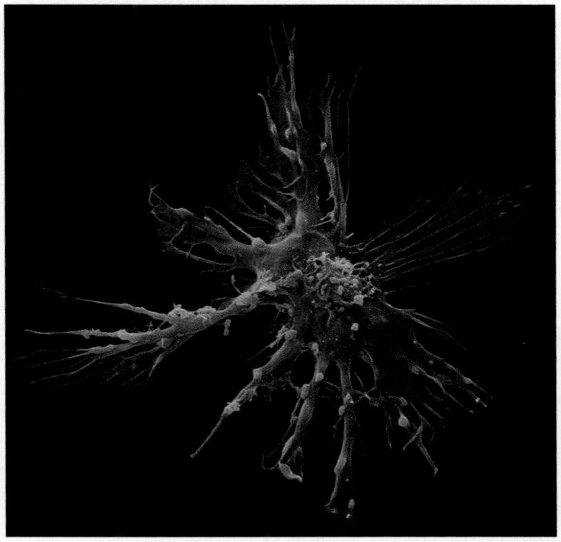

FIGURE 23.16 ▪ The innate immune system depends on white blood cells called macrophages and dendritic cells. **A.** Membrane protrusions from a macrophage (20 μm long) detecting and engulfing bacteria (pink; *E. coli*, 1.5 μm long). Colorized SEM. **B.** Dendritic cell (colorized SEM).

Platelets are small, puzzle piece–shaped cell fragments (they lack a nucleus) that derive from megakaryocytes, cells that differentiate from the precursors of red blood cells. Platelets, which circulate in the bloodstream, are required for efficient blood clotting. When activated by damaged endothelial cells (cells lining blood vessels), platelets clump, become trapped by fibrin, and form plugs that stop bleeding. Several factors produced by platelets also help wound

Special Topic 23.1: Are NETs a Cause of Lupus?

It started with slight joint pain in her hands. Then, within weeks, Annette developed searing pain in all her joints. Blood tests revealed that Annette had systemic lupus erythematosus (SLE), an autoimmune disease in which the immune system can mistakenly attack any part of the body. "Lupus" (Greek for "wolf") comes from a butterfly-shaped facial rash that resembles the pattern of fur on a wolf's face (**Fig. 1**). In addition to joint pain and skin rashes, SLE patients may have heart arrhythmias, abdominal pain, difficulty breathing, fatigue, fever, and swollen lymph nodes—depending on which parts of the body are under attack.

An important test used to diagnose lupus looks for antinuclear antibodies (ANAs) in the patient's blood. Antibodies are small proteins that usually bind substances foreign to the body (Chapter 24). Lupus patients, however, are autoreactive, so they make antibodies to their own DNA and other "self" proteins.

The mechanisms underlying SLE are complex and not fully defined. One component that seems important is the production of type I interferons. In addition to their role in inhibiting viral replication, type I interferons increase the lytic potential of natural killer cells (Section 23.7) and stimulate the development of cytotoxic lymphocytes (via T_H1 cells, as discussed in Chapter 24). Both cell types can cause tissue damage. Type I interferons can also lower the "threshold" for making autoreactive antibodies. But what leads to increased interferon production in lupus patients?

Recent evidence suggests that a unique subset of neutrophils present in all lupus patients can initiate a cycle of pronounced interferon production via NET formation. **Figure 2** illustrates that these special neutrophils, called low-density granulocytes (LDGs), more easily undergo NETosis, as compared to typical healthy control neutrophils or to normal-density neutrophils from lupus patients. NETosis generates an ample source of autoantigen, such as chromatin, to which autoantibodies can be made, and helps propagate tissue damage, especially to blood vessel endothelial cells, by NET-entangled peroxidases and proteases.

But what stimulates interferon production? Autoantibodies that bind to antimicrobial peptides like LL-37 embedded in NETs promote transport of DNA into circulating plasmacytoid dendritic cells, a primary source of interferon-alpha. The subsequent interaction between DNA and Toll-like receptor 9 (TLR9 binds to CpG sequences in DNA) on these cells stimulates interferon-alpha production. Interferon-alpha not only enhances natural killer (NK) cell and cytotoxic lymphocyte activity, but also amplifies NET release by other LDGs (**Fig. 3**). The end result is a cycle of inflammation and tissue damage that contributes to the symptoms of SLE. Stopping NET release by LDGs could be a way to lessen the symptoms of SLE.

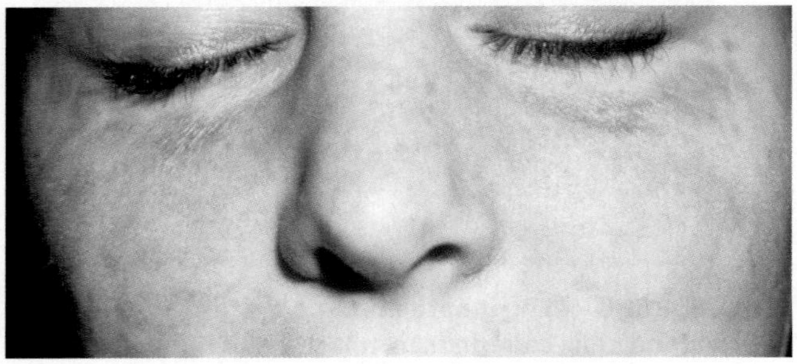

DR. KEN GREER/VISUALS UNLIMITED, INC.

FIGURE 1 ▪ Characteristic butterfly rash of lupus.

RESEARCH QUESTION

How would you determine whether NET formation by LDGs really stimulates interferon alpha production by plasmacytoid dendritic cells? Design an experiment using components of the system and wells of a microtiter plate.

Villanueva, E., S. Yalavarthi, C. C. Berthier, J. B. Hodgin, R. Khandpur, et al. 2011. Netting neutrophils induce endothelial damage, infiltrate tissues, and expose immunostimulatory molecules in systemic lupus erythematosus. *J. Immunol.* **187**:538–552.

repair. Recent evidence also shows that platelets are part of the innate immune system. Invading bacteria can bind platelets and trigger the release of antimicrobial peptides (Section 23.4). Bacterially activated platelets can also induce NET formation by neutrophils.

A **white blood cell (WBC) differential** is a test that physicians often order to help them diagnose an infection. **Table 23.2** presents general guidelines for interpreting a WBC differential, indicating which cell types increase or decrease in response to infections with bacteria, viruses, or parasites (protozoa or worms). Notice that total WBC counts increase in each case, but the type of WBC that increases in number differs with the agent.

Lymphoid Organs

Lymphocytes (**Fig. 23.13D**), which are the main participants in adaptive immunity, are present in blood at about 2,500 cells per microliter, accounting for over one-third of all peripheral white blood cells. However, an individual lymphocyte spends most of its life within specialized solid tissues (lymphoid organs) and enters the bloodstream only

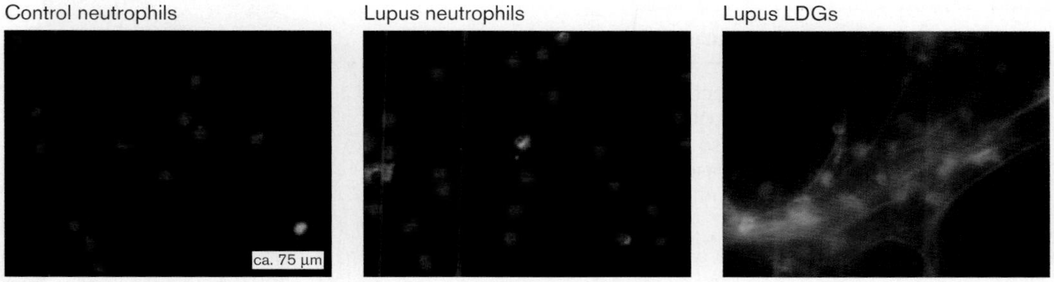
Control neutrophils
Lupus neutrophils
Lupus LDGs
ca. 75 μm

FIGURE 2 ▪ Circulating lupus low-density granulocytes undergo increased NETosis. Representative images of control neutrophils, lupus normal-density neutrophils, and lupus low-density granulocytes (LDGs) isolated from peripheral blood. Cells were fixed to coverslips and stained for nuclei and NETs. Blue structures are fluorescently stained nuclei (Hoescht stain); NETs are stained green by a fluorochrome-tagged antibody that binds to a protein (elastase) embedded in NET strands. Even without provocation, lupus LDGs release NETs.

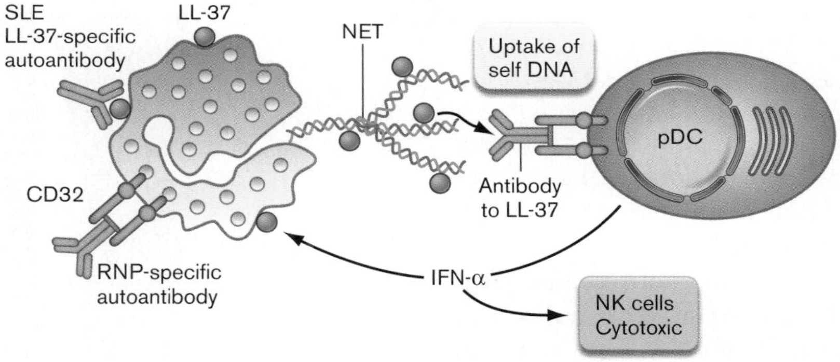
SLE
LL-37-specific
autoantibody
LL-37
NET
Uptake of
self DNA
pDC
CD32
Antibody
to LL-37
RNP-specific
autoantibody
IFN-α
NK cells
Cytotoxic

FIGURE 3 ▪ NET stimulation of IFN-alpha. In systemic lupus erythematosus (SLE), autoantibodies specific for ribonucleoproteins (RNPs) and the antimicrobial peptide LL-37 accelerate the release of NETs by neutrophils. Next, autoantibodies that bind to antimicrobial peptides (LL-37) present in NETs promote DNA transport into plasmacytoid dendritic cells (pDCs). DNA binding to Toll-like receptor 9 (TLR9) in vacuole membranes stimulates the production of interferon-α (IFN-α). IFN-α, in turn, enhances LL-37 surface expression on neutrophils, which amplifies NET release induced by autoantibodies. The progressive increase in proteases and peroxidases released by NETs and the activation of natural killer (NK) cells and cytotoxic lymphocytes enhances tissue damage. (CD32 is a membrane receptor for antibodies.) *Source:* Modified from Alberto Mantovanti et al. 2011. *Nat. Rev. Immunol.* **11**:519–531.

periodically, where it migrates from one place to another, surveying tissues for possible infection or foreign antigens. Consequently, most lymphocytes are found in the lymph nodes or spleen. No more than 1% of the total lymphocyte population circulates in the blood at any one time.

The tissues of the immune system, where the great majority of lymphocytes are found, are classified as primary or secondary lymphoid organs or tissues, depending on their function (**Fig. 23.17**). The primary lymphoid organs and tissues are where immature lymphocytes mature into antigen-sensitive B cells and T cells. **B cells**, which ultimately produce antibodies (see Chapter 24), develop in bone marrow tissue, whereas **T cells**, which modulate various facets of adaptive immunity, develop in the thymus, an organ located above the heart.

The secondary lymphoid organs serve as stations where lymphocytes can encounter antigens. These encounters lead to the differentiation of B cells into antibody-secreting plasma cells, and of T cells into antigen-specific helper cells, as will be discussed in Chapter 24. The spleen is an example of a secondary lymphoid organ. It is designed to filter blood directly and detect microorganisms. Macrophages in

TABLE 23.2

Guidelines for interpreting white blood cell (WBC) counts.*

	Normal	Elevated During
Total WBC count	4,500–11,000 per mm^3	Bacterial infection (12,000–30,000 per mm^3); also viral, and parasitic infections, and allergy
Differential:		
Neutrophils	54%–62%	Bacterial infection (also high immature forms, band cells)
Eosinophils	1%–3%	Parasitic infection
Basophils	0%–0.75%	Allergy
Lymphocytes	25%–33%	Viral infection
Monocytes	3%–7%	

*These are general guidelines. Individual organisms or certain noninfectious medical conditions or immunological defects can alter the findings. A blank space indicates little change in the parameter.

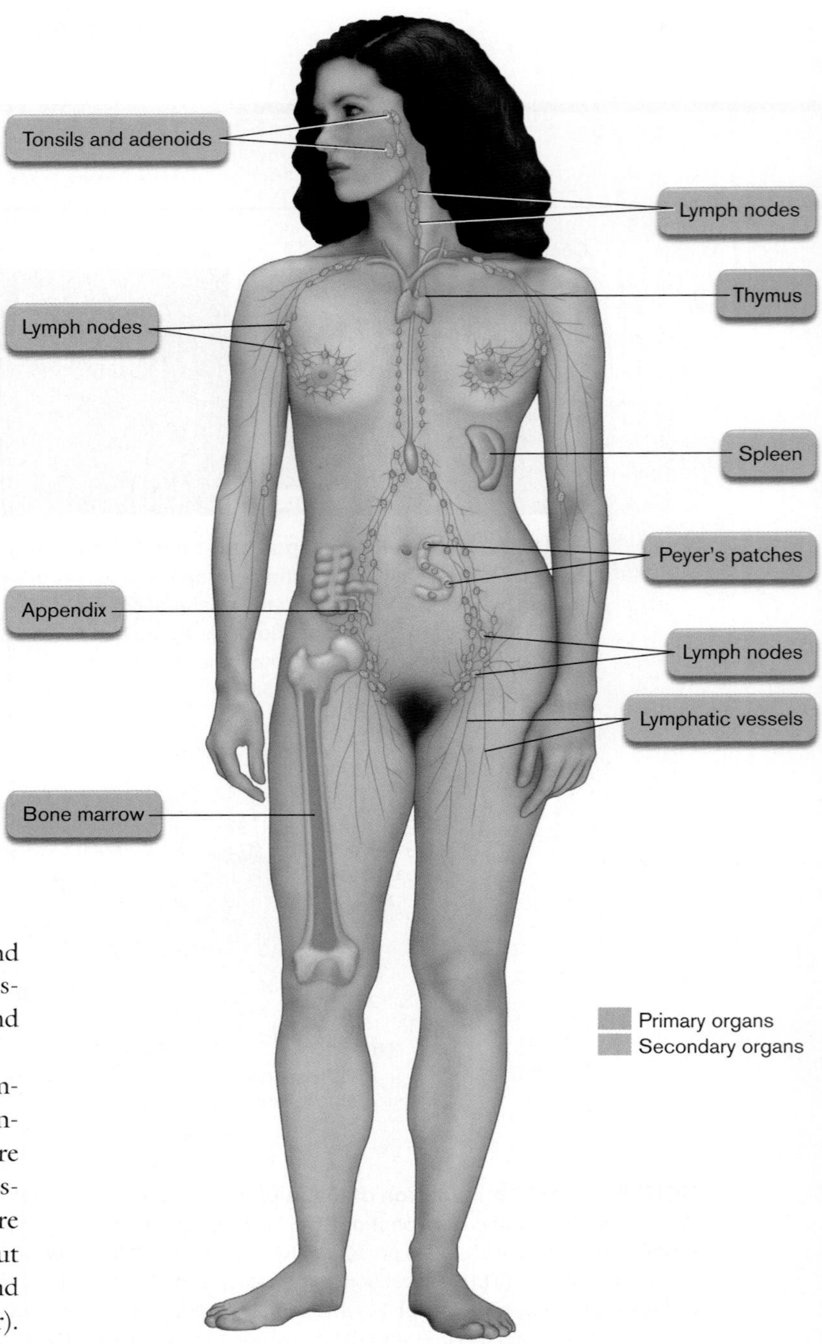
Tonsils and adenoids
Lymph nodes
Thymus
Lymph nodes
Spleen
Peyer's patches
Appendix
Lymph nodes
Lymphatic vessels
Bone marrow
Primary organs
Secondary organs

FIGURE 23.17 ■ **Lymphoid organs.**

the spleen engulf these organisms and destroy them, and then migrate to other secondary lymphoid organs and present pieces of the microbe (called antigens) to the B and T cells, which then become activated.

The **lymph nodes** are another kind of secondary lymphoid organ. Lymph nodes are arranged to trap organisms from local tissues, not from blood. Lymph nodes are situated at various sites in the body where lymphatic vessels converge (for example, under the armpits). There are also lymphoid tissues in the mucosal regions of the gut and respiratory tracts (for example, Peyer's patches and gut-associated lymphoid tissue, or GALT, discussed later). Other secondary lymphoid organs are the tonsils, adenoids, and appendix.

To Summarize

- **The immune system** consists of both innate and adaptive mechanisms that recognize and eliminate pathogens.
- **Innate immunity** includes physical barriers and some cellular responses to various microbial structures.
- **Adaptive immunity** is a cellular response to specific structures (antigens) in which a memory of exposure is produced.
- **Myeloid bone marrow stem cells** differentiate to form cells of the innate immune system—namely, phagocytic PMNs, monocytes, macrophages, antigen-presenting dendritic cells, and mast cells. **Platelets** are derived from a different cell line.
- **Lymphoid stem cells** differentiate into natural killer cells (part of the innate immune system) and lymphocytes (cells of the adaptive immune system). Lymphocytes are classified as B cells, which ultimately produce antibodies, and T cells, which regulate adaptive immunity.

23.4

Barbarians at the Gate: Innate Host Defenses

When defending a castle in medieval times, the first lines of defense included physical barriers (the castle wall), chemical barriers (boiling oil tossed onto invaders trying to scale the wall), and finally hand-to-hand combat once the wall was breached. Similarly, the body's initial defenses against infectious disease are composed of physical, chemical, and cellular barriers designed to prevent a pathogen's access to host tissues. Although generally described as nonspecific, some innate defense systems are more specific than others.

Physical Barriers to Infection

The first line of defense against any potential microbial invader (either commensal or pathogenic) occurs where parts of the body interface with the environment. These interfaces (skin, lung, gastrointestinal tract, genitourinary tract, and oral cavities) have similar defense strategies, although each has unique characteristics. A defense common to all host surfaces involves **tight junctions**—watertight adhesions that link adjacent epithelial cells at mucosal surfaces and endothelial cells lining blood vessels. Tight junctions prevent bacteria, and even host cells, from moving between internal and external host compartments. The "glue" that holds tight junctions together is a series of interconnecting glycoprotein molecules (**Fig. 23.18**).

Skin. Few microorganisms can penetrate skin, because of the thick keratin armor produced by closely packed cells called keratinocytes. Keratin protein is a hard substance (hair and fingernails are made of it) that is not degraded by known microbial enzymes. An oily substance (sebum) produced by the sebaceous glands also covers and protects the skin. Its slightly acidic pH inhibits bacterial growth. Competition between species also limits colonization by pathogens, and microorganisms that manage to adhere are continually removed by the constant shedding of outer epithelial skin layers.

Other, more specialized cells just under the skin can recognize microbes managing to slip through the physical barrier. They are part of a consortium of cells called **skin-associated lymphoid tissue (SALT)**. **Langerhans cells** make up a significant portion of SALT. They are specialized dendritic cells that can phagocytose microbes. Once a lymphoid Langerhans cell has ingested a microbe, the cell migrates by ameboid movement to nearby lymph nodes and "presents" parts of the microbe to the immune system to activate antimicrobial immunity.

Note: Do not confuse phagocytic Langerhans cells with the pancreatic "islets of Langerhans," which secrete insulin.

Mucous membranes. Mucosal surfaces form the largest interface (200–300 square meters) between the human host and the environment (the intestine alone is 7–8 meters, or 23–26 feet, long). Mucous membranes in general present a containment problem. They must be selectively permeable in order to exchange nutrients, as well as to export products and waste components. At the same time, they must constitute a barrier against invading pathogens. Mucosal membranes are covered with special tightly knit epithelial layers that support this barrier function. The mucus secreted from stratified squamous epithelial cells

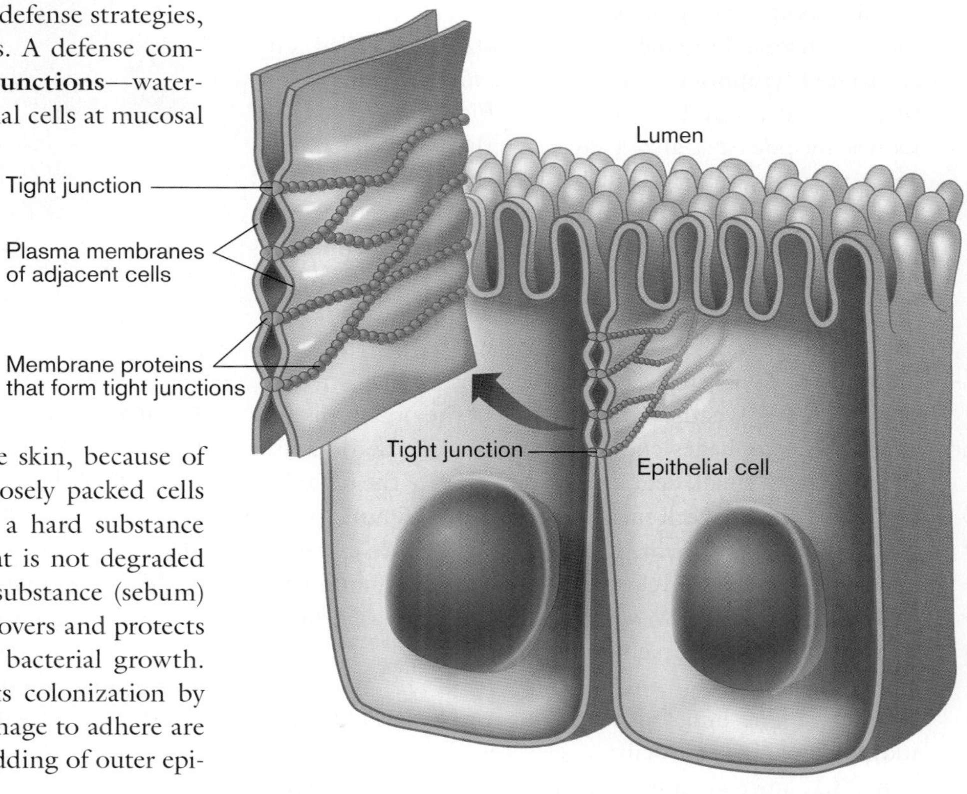
Lumen
Tight junction
Plasma membranes of adjacent cells
Membrane proteins that form tight junctions
Tight junction
Epithelial cell

FIGURE 23.18 ▪ Tight junctions hold adjacent cells together. Tight junctions are formed by an interconnected series of glycoproteins (blowup). By tightly linking adjacent cell membranes together, tight junctions produce a barrier through which bacteria, viruses, and other host cells cannot easily pass.

coats mucosal surfaces and traps microbes. Compounds within the mucus can serve as a food source for some commensal organisms, but other secreted compounds—like the enzymes lysozyme, which cleaves cell wall peptidoglycan, and lactoperoxidase, which produces superoxide radicals—can kill an organism trapped in the mucus.

Semispecific innate immune mechanisms are also associated with mucosal surfaces. Host cells, even epithelial cells, in mucosa have evolved mechanisms to distinguish harmless compounds from microorganisms. Patterns of conserved structures on microbes, called **microbe-associated molecular patterns (MAMPs)**, are recognized by host cell surface receptors such as various Toll-like receptors (discussed in Section 23.7) and CD14. Once a MAMP has been recognized, the host cell sends out chemicals that can activate immune system cells involved with innate and adaptive immune mechanisms.

Note: MAMPs were previously called PAMPs, for "pathogen-associated molecular patterns." Because the structures recognized are also present on nonpathogenic bacteria and viruses, the term was changed to MAMP.

Like skin, the gastrointestinal system possesses an innate mucosal immune system, in this case called **gut-associated lymphoid tissue (GALT)**. GALT includes tonsils, adenoids, and Peyer's patches (**Fig. 23.19A**). These tissues include specialized **M cells** that dot the intestinal surface and are wedged between epithelial cells. "M" stands for "microfold," which describes their appearance (**Fig. 23.19B**). These are fixed cells that take up microbes (microbiota or pathogens) from the intestine and release them, or pieces of them, into a pocket formed on the opposite, or basolateral, side of the cell. Other cells of the innate immune system, such as macrophages (which migrate through tissues), gather here and collect the organisms that emerge. Macrophages engulf and try to kill the organism and then place small, degraded components of the organism on the macrophage cell surface where other immune system cells can recognize them. As a result, M cells are extremely important for the development of mucosal immunity to pathogens. However, M cells can also serve as a portal for some pathogens to gain entry to the body.

The lungs. The lungs also have a formidable defense. In addition to the mucociliary elevator discussed in Section 23.1, microorganisms larger than 100 μm become trapped by hairs and cilia lining the nasal cavity and trigger a forceful expulsion of air from the lungs (a sneeze). The sneeze is designed to clear the organism from the respiratory tract. Organisms that make it to the alveoli are met by phagocytic cells called **alveolar macrophages**. These cells can ingest and kill most bacteria.

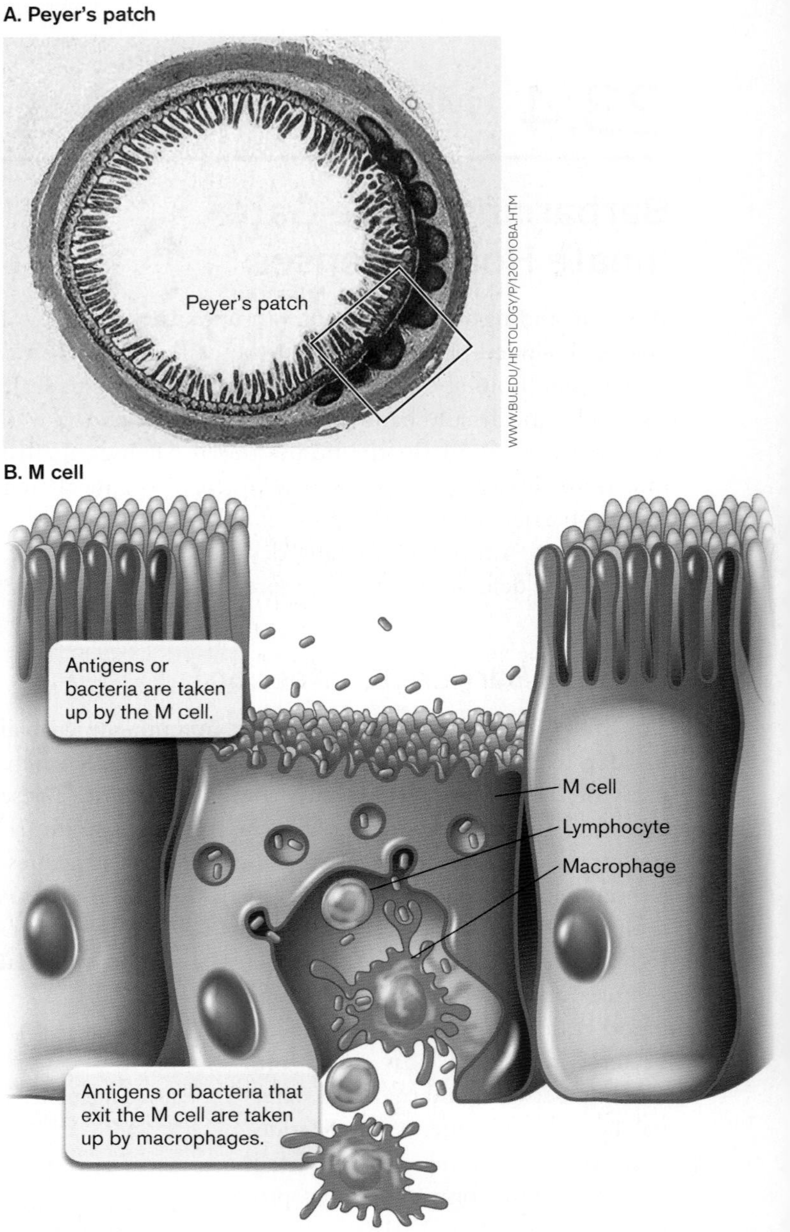
A. Peyer's patch
Peyer's patch
WWW.BU.EDU/HISTOLOGY/P/12001OBA.HTM
B. M cell
Antigens or bacteria are taken up by the M cell.
M cell
Lymphocyte
Macrophage
Antigens or bacteria that exit the M cell are taken up by macrophages.

FIGURE 23.19 ▪ Gut-associated lymphoid tissue (GALT). **A.** A Peyer's patch located on the intestine. **B.** Diagram of an M cell (microfold cell).

Another important factor that prevents lung infections is an epithelial membrane protein called **cystic fibrosis transmembrane conductance regulator (CFTR)**. CFTR is a membrane chloride channel that regulates chloride movement across the membrane, an essential part of hydrating mucus in healthy individuals. Localization of the protein at the lung cell surface is depicted in **Figure 23.20**. Cystic fibrosis patients have a defective CFTR and are much more susceptible to lung infections, especially those caused by *Pseudomonas aeruginosa*. The *P. aeruginosa* strains posing the

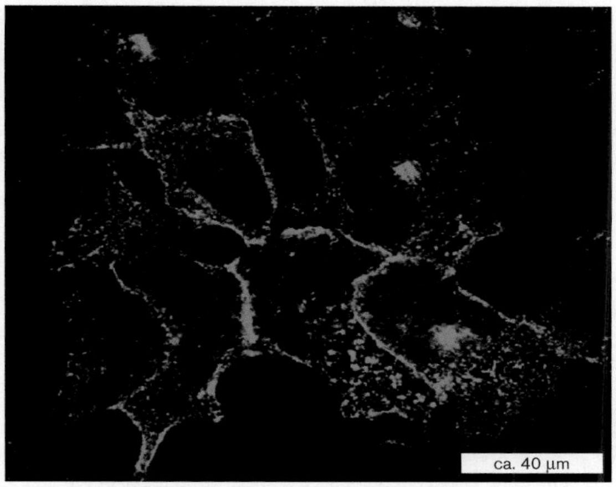
ca. 40 µm

AFFINITY BIOREAGENTS

FIGURE 23.20 ▪ Plasma membrane location of the cystic fibrosis gene product (CFTR). Confocal laser scanning fluorescence microscopy was used to image living epithelial cells of the human airway. The cells were transfected with DNA expressing a cystic fibrosis transmembrane conductance regulator (CFTR) protein tagged with green fluorescent protein (CFTR-GFP). Note that the fluorescence is heaviest at the edge of the cell, indicating that the CFTR-GFP protein is localized at the cell surface.

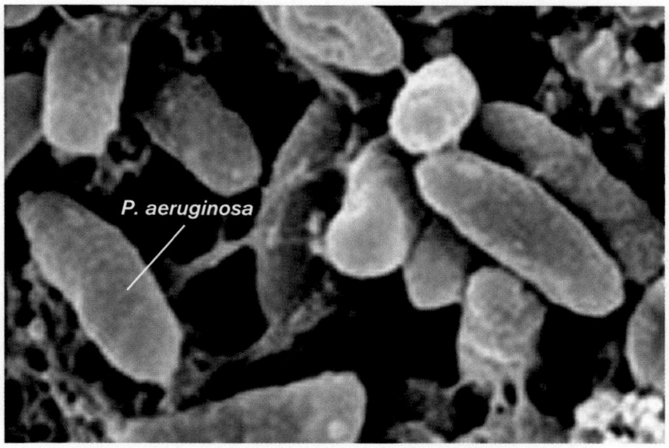
P. aeruginosa

SEBASTIEN VILAIN

FIGURE 23.21 ▪ Mucoid strains of *Pseudomonas aeruginosa* that infect cystic fibrosis patients. The organism is shown growing in a biofilm with strings of dehydrated exopolysaccharide (EPS) connecting cells. This EPS material gives colonies of these strains a very mucoid appearance. (Cells 2–3 µm long; SEM.)

most serious threat produce a thick slimy material (alginate) that retards other lung clearance mechanisms (**Fig. 23.21**). How CFTR helps protect lungs from infection is not clear. One hypothesis is that normal CFTR is a receptor for *P. aeruginosa* lipopolysaccharide (LPS), which enables endocytosis and killing of the microbe by phagolysosome mechanisms. Another hypothesis is that normal CFTR helps maintain an airway pH optimal for the activity of secreted antimicrobial peptides. Either way, CFTR is part of the innate barrier guarding against lung infection.

Chemical Barriers to Infection

Some examples of chemical barriers were mentioned previously, such as the acidic pH of the stomach, lysozyme in tears, and generators of superoxide. In addition, a variety of human cells generate small antimicrobial, cationic (positively charged) peptides called defensins (**Table 23.3**). These antimicrobial peptides are important components of innate immunity against microbial infections.

Defensins range in length from 29 to 47 amino acids and are found in mammals, birds, amphibians, and plants. Defensins (and other antimicrobial peptides) destroy an invading microbe's cytoplasmic membrane and are effective against Gram-positive and Gram-negative bacteria, fungi, and even some viruses (such as those with membranes, like HIV). To kill, the peptides must bind the negatively charged outer membrane lipopolysaccharides of Gram-negative bacteria and move into the periplasm. Defensins are then pulled into the cytoplasmic membrane of either Gram-positive or Gram-negative bacteria by the transmembrane electrical potential, about –150 mV in bacteria (that is, the cell interior is more negative than the exterior). The peptides

TABLE 23.3

Categories of natural antimicrobial peptides.

Class (Examples)	Major presence in humans (Source)
Alpha defensins (-1, -2, -3, -4[a])	Neutrophils (Stored in granules)
Alpha defensins (-5, -6)	Paneth cells, small intestine (Stored in granules)
Cathelicidins (LL-37, hCAP18)	Neutrophils (Secreted)
Histatins	Saliva (Secreted)
Beta defensins (HBD-1, -2)	Epithelia (Secreted)
Kinocidins (tPMP,[b] PF-4)	Platelets (Secreted)
Other species	
Maganins	Frogs
Protegrins	Pigs
Indolicidin	Cattle

[a]Alpha defensins are named alpha defensin-1, alpha defensin-2, etc.

[b]tPMP = thrombin-induced platelet microbicidal protein.

assemble into channels that destroy the cytoplasmic membrane barrier, killing the bacterial cell. Defensins generally do not affect eukaryotic cells, because these cells have a very low membrane potential (–15 mV) across their plasma membrane. Antimicrobial peptides are produced by many human cells, including cells of the skin, lungs, genitourinary tract, and gastrointestinal tract (**Fig. 23.22**). Specific defensins produced by different animals may partially explain pathogen-host specificity (see **eTopic 23.2**).

One form of vertebrate defensins (alpha) are stored in membrane-enclosed granules within neutrophils and in Paneth cells in the small intestine (**Fig. 23.22A**). When stimulated, these cells **degranulate** (release their granule contents) by fusing their granule membranes to cytoplasmic or vacuolar membranes, dumping their contents into the surroundings or into phagocytic vacuoles, where the alpha defensins can destroy engulfed microbes (**Fig. 23.22B**). In contrast, the beta defensins are not stored in cytoplasmic granules. The synthesis of beta defensins is activated only after contact with bacteria or their other products. Other cationic antimicrobial peptides are listed in **Table 23.3**. Neutrophil cathelicidins are discussed in **eTopic 23.3**.

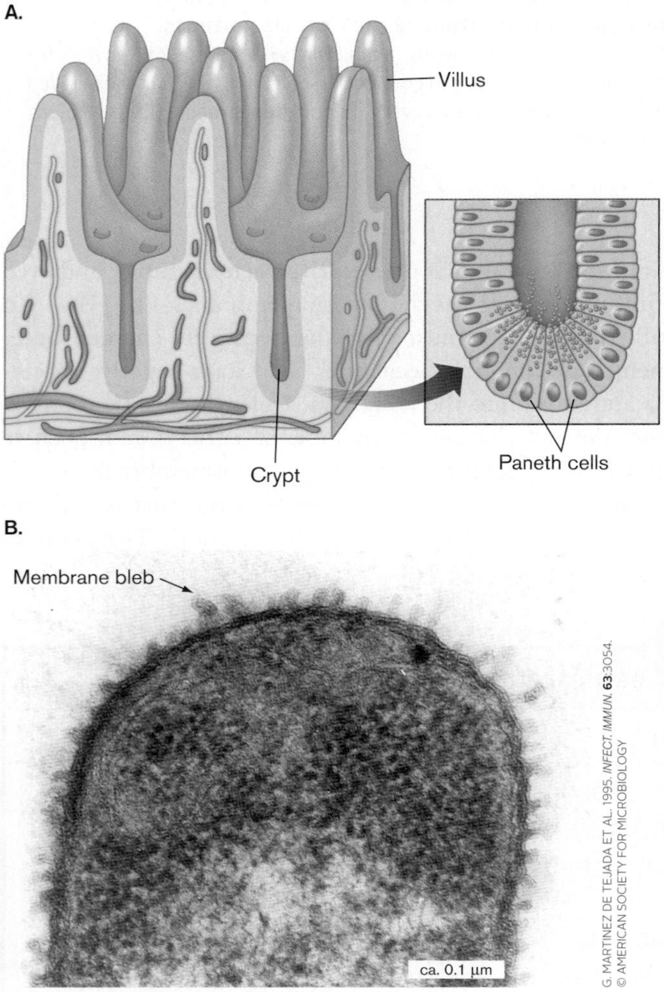
A.
Villus
Crypt
Paneth cells
B.
Membrane bleb
ca. 0.1 µm
G. MARTINEZ DE TEJADA ET AL. 1995. *INFECT. IMMUN.* **63**:3054.
© AMERICAN SOCIETY FOR MICROBIOLOGY

FIGURE 23.22 ■ Defensins. A. Certain defensins are produced in the crypts of the intestine. The crypts contain granule-rich Paneth cells (blowup) that discharge their granules into the crypt lumen in response to the entry of bacteria or as a result of food-related stimulation by acetylcholine. **B.** Effect of cationic peptides on *E. coli* O111. Polymyxin B is a small cationic peptide antibiotic that mimics the action of defensins. In this micrograph, polymyxin B causes blebs of membrane to ooze from the surface of the cell.

Thought Question

23.6 Why do defensins have to be so small? Do defensins kill normal microbiota?

To Summarize

- **Skin defenses** against invading microbes include closely packed keratinocytes and a SALT lymphoid system made up largely of phagocytic Langerhans cells.
- **Mucous-membrane defenses** involve secreted enzymes, cytokines, and GALT tissues, such as Peyer's patches, that contain phagocytic M cells.
- **M cells** in gut-associated lymphoid tissues sample bacterial cells at their surface and release pieces of them to immune system cells.
- **Phagocytic alveolar macrophages** inhabit lung tissues, contributing to nonspecific defense.
- **Chemical barriers against disease** include cationic defensins, acid pH in the stomach, and superoxide produced by certain cells.
- **Microbe-associated molecular patterns (MAMPs)** are recognized by Toll-like receptors (TLRs) found on many host cells, such as macrophages. Binding triggers release of chemical signal molecules that activate innate and adaptive immune mechanisms.

23.5

The Acute Inflammatory Response

The boil shown in **Figure 23.23** is an inflammatory response triggered by infection with the organism *Staphylococcus aureus*. Inflammation is a critical innate defense in the war between microbial invaders and their hosts. It

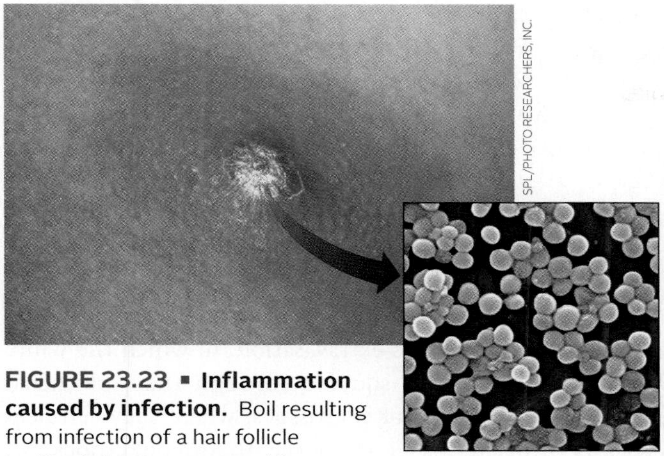
SPL/PHOTO RESEARCHERS, INC.
© MEDISCAN/CORBIS

FIGURE 23.23 ▪ Inflammation caused by infection. Boil resulting from infection of a hair follicle by *Staphylococcus aureus* (size 0.5–1.0 μm; blowup, colorized SEM).

provides a way for phagocytic cells (such as neutrophils) normally confined to the bloodstream to gain access to infected sites within tissues. Movement of these cells out of blood vessels is called **extravasation** or diapedesis, which we discuss shortly. Once at the infection site, the neutrophils begin engulfing microbes. The white pus associated with an infection is teeming with these white blood cells. The five cardinal signs of inflammation, first described over 2,000 years ago, are: redness, warmth, pain, swelling, and altered function at the affected site.

Although many things can trigger inflammation, our focus is how microbes cause the response. The process begins with the infection itself. Microorganisms introduced into the body—for example, on a wood splinter—will begin to grow and produce compounds that damage host cells (**Fig. 23.24** ▶). Resident macrophages that wander into the infected area engulf these organisms and then release inflammatory mediators (chemoattractants) that "call" for more help. These mediators include **vasoactive factors** such as leukotrienes, platelet-activating factor, and prostaglandins, which act on blood vessels of the microcirculation, increasing blood volume and capillary permeability to help deliver white blood cells to the area. In addition, small protein molecules called cytokines are secreted, diffusing to the vasculature and stimulating the expression of specific receptors (selectins) on the endothelial cells of capillaries and venules.

Vasoactive Factors and Cytokines Initiate Extravasation

How do neutrophils pass through blood vessel walls? The passage of neutrophils through vascular walls (extravasation, or diapedesis; **Fig. 23.25**) requires a relaxation of endothelial cell adhesion. The process is initiated by vasoactive factors released by macrophages; these factors increase vascular permeability and stimulate vasodilation.

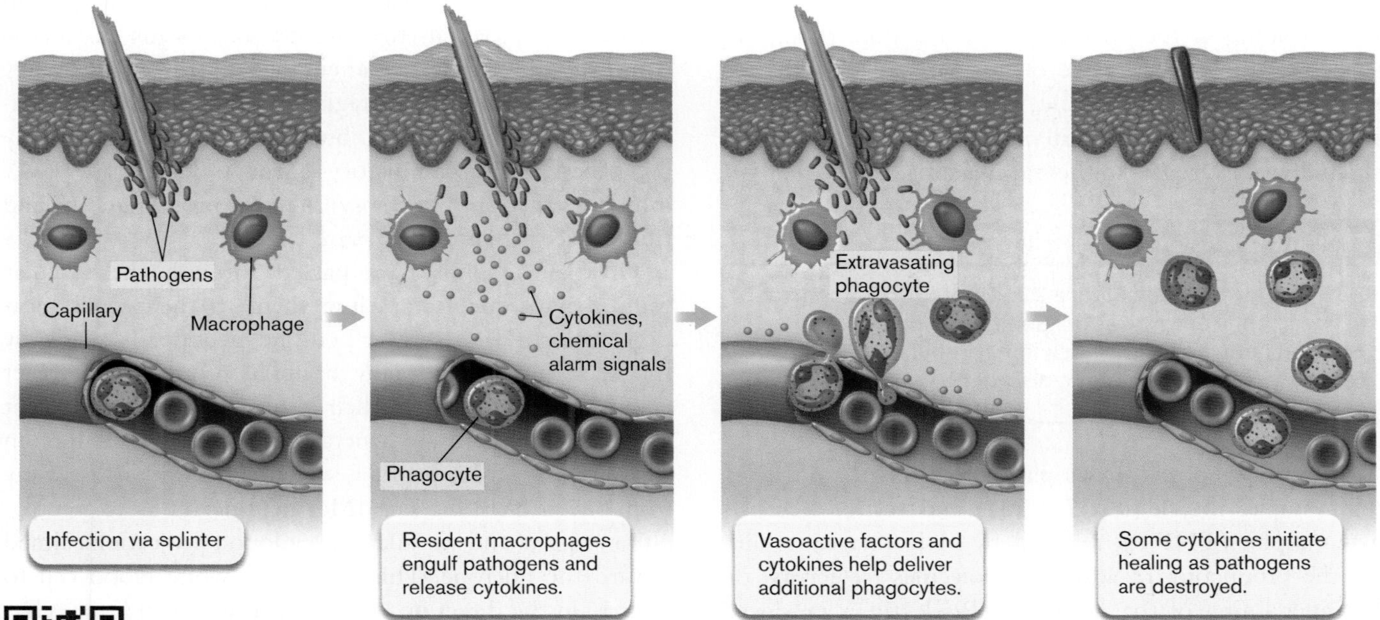
Pathogens
Capillary
Macrophage
Infection via splinter
Cytokines, chemical alarm signals
Phagocyte
Resident macrophages engulf pathogens and release cytokines.
Extravasating phagocyte
Vasoactive factors and cytokines help deliver additional phagocytes.
Some cytokines initiate healing as pathogens are destroyed.

FIGURE 23.24 ▪ Basic inflammatory response. Neutrophils (a type of phagocyte) circulate freely through blood vessels and can squeeze between cells in the walls of a capillary (extravasation) to the site of infection. They then engulf and destroy any pathogens they encounter. ▶

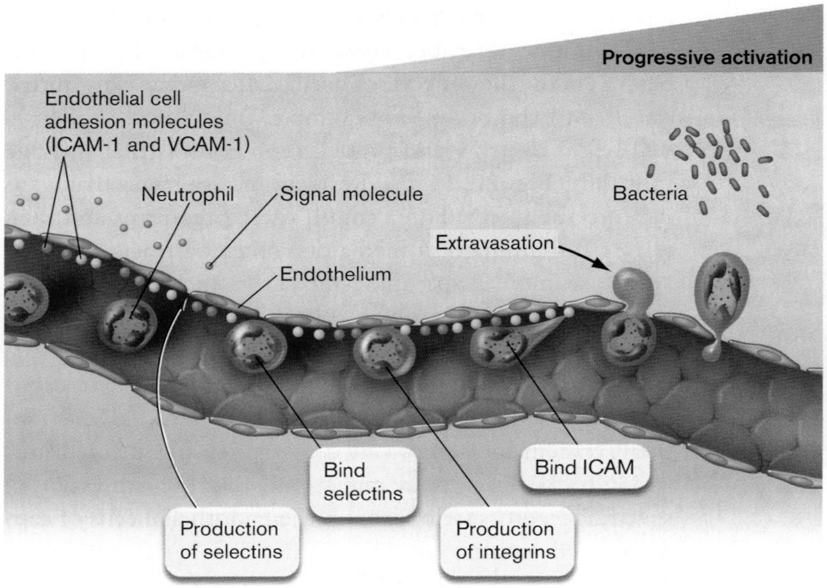

Progressive activation
Endothelial cell adhesion molecules (ICAM-1 and VCAM-1)
Neutrophil
Signal molecule
Bacteria
Extravasation
Endothelium
Bind selectins
Bind ICAM
Production of selectins
Production of integrins

FIGURE 23.25 ■ Mechanism of extravasation. Extravasation is the process by which neutrophils move from the bloodstream into surrounding tissues. Signal molecules produced by damaged tissue cells induce the production of selectins (produced early in the process) on the surfaces of endothelial cells, and integrins on the surface of the white blood cell (selectins and integrins are not shown). Selectins capture neutrophils traveling through blood vessels. Neutrophils begin to roll along the vessel wall, and the integrins on their surface lock onto the endothelial cell's adhesion molecules (ICAM-1 or VCAM-1). The neutrophils are progressively activated while rolling and ultimately squeeze through the wall between endothelial cells (extravasation).

Vasodilation slows blood flow and, as a result, increases blood volume in the affected area. The more permeable vessel also allows the escape of plasma into tissues. Both events cause localized swelling, redness, and heat. Vasoactive factors also stimulate local nerve endings, causing pain, which draws awareness to the infected area.

Thought Question

23.7 As illustrated in **Figure 23.25**, integrin is important for neutrophil extravasation. Some individuals, however, produce neutrophils that lack integrin. What is the likely consequence of this genetic disorder?

Many kinds of cytokines play a role in inflammation. For example, **interleukin 1 (IL-1)** and **tumor necrosis factor alpha (TNF-α)**, released by macrophages, stimulate the production of adhesion molecules (**selectins**) on the inner lining of the capillaries. P-selectin is produced first, followed by E-selectin. The selectins snag neutrophils zooming by in the bloodstream, slow them down, and cause them to roll along the endothelium (see **Fig. 23.25**). Rolling neutrophils that encounter inflammatory mediators are activated to produce and display integrin adhesion molecules on their surface. The integrins on the neutrophils lock onto endothelial adhesion molecules ICAM-1 (intracellular adhesion molecule 1) and VCAM-1 (vascular cell adhesion molecule 1). The binding of cell adhesion molecules stops the neutrophils from rolling and initiates extravasation, in which the white blood cells squeeze through the endothelial wall and into the tissues, where they can help macrophages attack the invading microbes.

Damaged tissue cells in the area of inflammation release **bradykinin**, a nine-amino-acid polypeptide that helps loosen the tight junctions between endothelial cells to promote extravasation (**Fig. 23.26**). Bradykinin also triggers degranulation of mast cells. The histamine released from mast cells further loosens the endothelial cell junctions, increasing vascular permeability. More fluid enters tissues and accumulates (edema). Even though tight junctions are loosened, to actually pass between endothelial cells, activated neutrophils need the enzyme sialidase to temporarily break carbohydrate linkages that hold tight junctions together.

What causes the pain of inflammation? Bradykinin induces capillary cells to make prostaglandins that cause pain by stimulating nerve endings in the area. A key enzyme involved in prostaglandin synthesis is cyclooxygenase (COX). Aspirin, ibuprofen, and the popularly prescribed anti-inflammatory agent Celebrex are COX inhibitors that prevent the synthesis of prostaglandins and thus reduce inflammatory pain.

Once neutrophils have passed through the vascular wall, chemotactic factors lure them to the proper location. One of these, fMet-Leu-Phe peptide, is made by the bacterium itself. Many bacterial proteins have fMet (*N*-formylmethionine) as their N-terminal amino acid, but mammalian proteins, in general, do not. Bacteria will often cleave off the fMet peptide, which can then diffuse away from the bacterium. The fMet peptide binds to neutrophil receptors and stimulates pseudopod projections aimed toward the microbe. This causes the white blood cell to migrate in the direction of the infection. In addition, **chemokine** peptides (IL-8 and MCP-1) produced by damaged tissues can serve as chemoattractants for these white blood

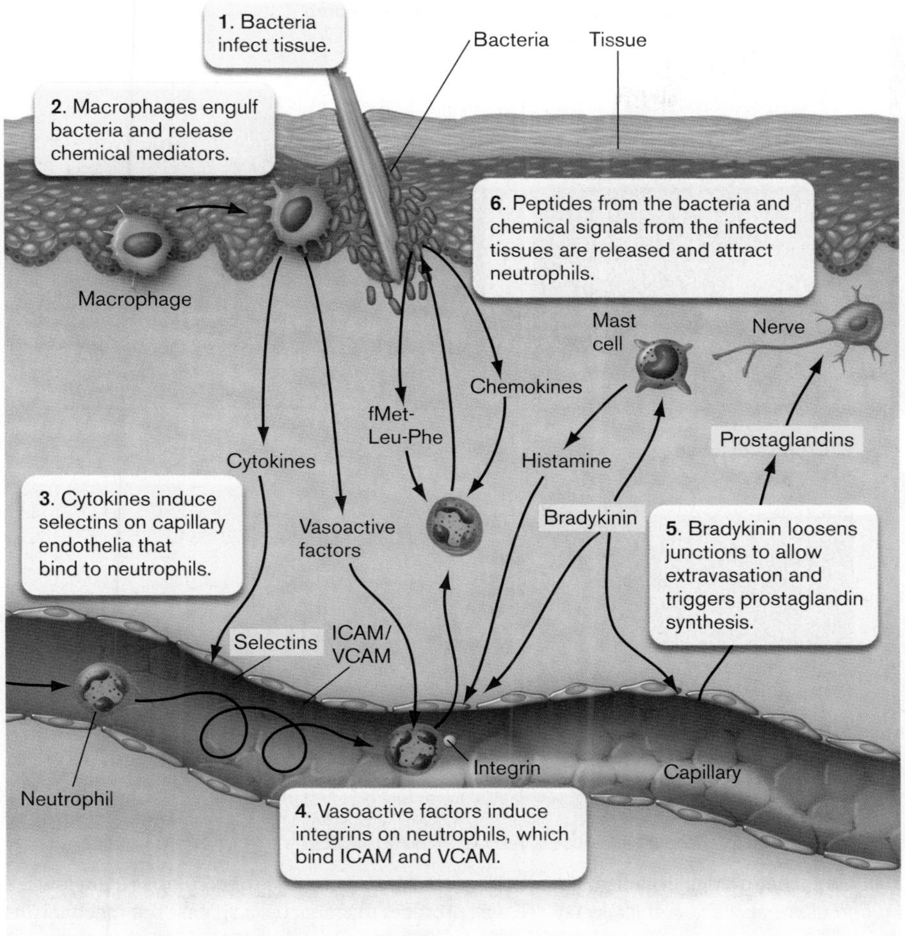
1. Bacteria infect tissue.
Bacteria
Tissue
2. Macrophages engulf bacteria and release chemical mediators.
6. Peptides from the bacteria and chemical signals from the infected tissues are released and attract neutrophils.
Macrophage
Mast cell
Nerve
Chemokines
fMet-Leu-Phe
Prostaglandins
Cytokines
Histamine
3. Cytokines induce selectins on capillary endothelia that bind to neutrophils.
Vasoactive factors
Bradykinin
5. Bradykinin loosens junctions to allow extravasation and triggers prostaglandin synthesis.
Selectins
ICAM/VCAM
Integrin
Capillary
Neutrophil
4. Vasoactive factors induce integrins on neutrophils, which bind ICAM and VCAM.

FIGURE 23.26 ▪ Summary of the inflammatory response.

cells. Once phagocytes arrive at the site of infection, they begin devouring microbes. Realize, however, that much of the damage caused by an infection is not due directly to the microbe but is the result of the body's reaction to its presence.

Thought Question

23.8 What happens to all the neutrophils that enter a site of infection once the infection has resolved?

Chronic Inflammation Causes Permanent Damage

Inflammation that persists over months or years is called **chronic inflammation** and is provoked by the long-term presence of a causative stimulus. Chronic inflammation inevitably causes permanent tissue damage, even though the body attempts repair. The causes of chronic inflammation are many. For example, infectious organisms such as *Mycobacterium tuberculosis, Actinomyces bovis,* and various protozoan parasites can avoid or resist host defenses (**Fig. 23.27A**). As a result, they persist at the site and continually stimulate the basic inflammatory response. Nonliving, irritant material like wood splinters, inhaled asbestos particles, or surgical implants can also cause chronic inflammation. Autoimmune diseases are another important cause. Autoimmunity (reaction against self) occurs when there is a failure to regulate some aspect of adaptive immunity (see Chapter 24). As a result, the immune system recognizes a part of the body as foreign (not self) and begins to react against it. Rheumatoid arthritis is one example of the body attacking itself.

Whatever causes chronic inflammation, there is continual recruitment of macrophages and lymphocytes from the circulation. The body may attempt to "wall off" the site of inflammation by forming a **granuloma**. A granuloma begins as an aggregation of the mononuclear inflammatory cells surrounded by a rim of lymphocytes. The body then deposits fibroconnective tissue around the lesion, causing tissue hardening known as fibrosis.

Several forms of granulomas are shown in **Figure 23.27**. Mycobacteria resist killing by macrophages and can be found living within these cells (**Fig. 23.27A**). Resistance to host defenses can produce long-term chronic infections and the production of granulomas. For example, skin infections caused by *Mycobacterium marinum* will produce skin granulomas (**Fig. 23.27B**), while *M. tuberculosis* can cause liver granulomas (**Fig. 23.27C**). Another disease thought to involve granuloma formation is Crohn's disease, which commonly manifests as abdominal pain, frequent bowel movements, and rectal bleeding. Crohn's disease has been attributed to an autoimmune reaction possibly activated in some way by intestinal microbiota. In this case, the intestinal bacteria are thought to cause a chronic inflammation resulting in characteristic granulomas (**Fig. 23.27D** and **E**).

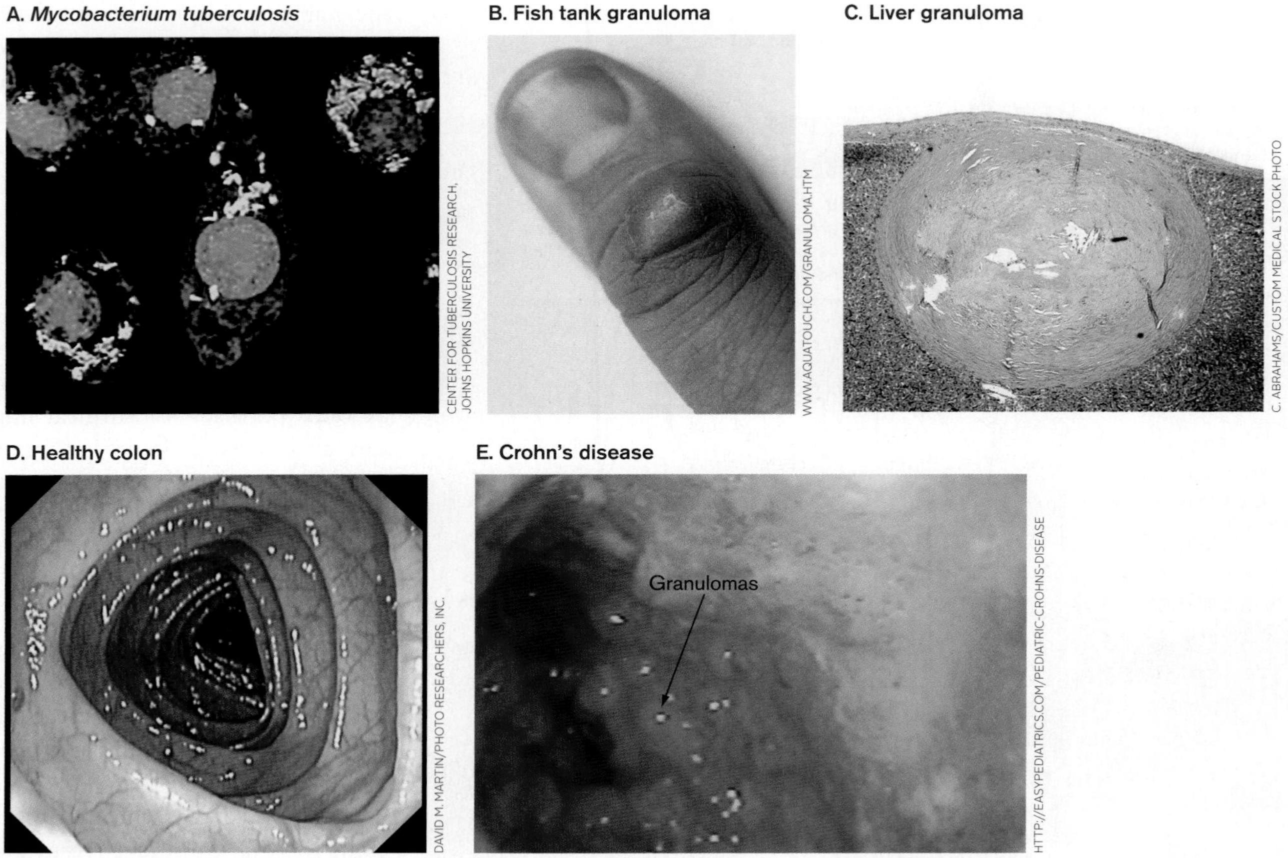
A. *Mycobacterium tuberculosis*
CENTER FOR TUBERCULOSIS RESEARCH, JOHNS HOPKINS UNIVERSITY
B. Fish tank granuloma
WWW.AQUATOUCH.COM/GRANULOMA.HTM
C. Liver granuloma
C. ABRAHAMS/CUSTOM MEDICAL STOCK PHOTO
D. Healthy colon
DAVID M. MARTIN/PHOTO RESEARCHERS, INC.
E. Crohn's disease
Granulomas
HTTP://EASYPEDIATRICS.COM/PEDIATRIC-CROHNS-DISEASE

FIGURE 23.27 ▪ Chronic inflammation. A. The fluorescent orange organisms shown here are *Mycobacterium tuberculosis* (2 μm long; fluorescence microscopy) within macrophages in a tuberculosis abscess. A thick, waxy cell wall protects mycobacteria against the mechanisms used by macrophages to destroy microorganisms. As a result, they survive for prolonged periods within macrophages. The continual stimulation of an inflammatory response leads to chronic inflammation. **B.** Fish tank granuloma. *Mycobacterium marinum* can cause tuberculosis-like infection in fish. The organism can accidentally enter the human body through an open wound or abrasion during cleaning of an aquarium containing infected fish. The infection is first noted as a lesion that heals very slowly on the hand or forearm. Once it does heal, it often forms a granuloma at the site that contains live organisms. **C.** Cross section of liver showing a necrotizing granuloma caused by *M. tuberculosis* that spread through the bloodstream to this site after escaping the lung. **D.** Healthy colon. **E.** Intestinal granulomas of Crohn's disease.

To Summarize

- **Inflammation** begins with extravasation that moves neutrophils from the bloodstream into infected tissues.
- **Macrophages** in tissues engulf microorganisms and release vasoactive factors that increase vascular permeability, cytokines that stimulate the production of selectin receptors, and chemoattractant molecules that call in neutrophils.
- **Extravasation** is the process by which neutrophils pass between cells of the endothelial wall. Once out of the circulation, the neutrophils travel to the site of infection.
- **Bradykinin** causes the release of prostaglandins, which produce pain in the affected area.
- **Chronic inflammation** results from the persistent presence of a foreign object.

23.6

How Phagocytes Detect and Kill Microbes

Phagocytosis (briefly described in Section 23.3) is an effective nonspecific immune response. However, it would be disastrous if white blood cells indiscriminately phagocytosed host as well as nonhost structures. Fail-safe controls built into our immune defenses prevent this from happening. Without these restraints, the mammalian immune system would constantly attack itself. When one of these fail-safe mechanisms fails, autoimmune diseases can arise.

Phagocytes Recognize Alien Cells and Particles

For phagocytosis to proceed, macrophages and neutrophils must first recognize the surface of a particle as foreign (**Fig. 23.28**). When a phagocyte surface interacts with the surface of another body cell, the phagocyte becomes temporarily paralyzed (inept at pseudopod formation). Paralysis allows the phagocyte to evaluate whether the other cell is friend or foe, self or nonself. Self-recognition involves glycoproteins located on the white blood cell membrane binding inhibitory glycoproteins present on all host cell membranes. The inhibitory glycoprotein on human cells is called CD47. Because invading bacteria lack these inhibitory surface molecules, they can readily be engulfed.

Although many bacteria, such as *Mycobacterium* or *Listeria* species, are easily recognized and engulfed by phagocytosis (**Fig. 23.28A–C**), others, such as *Streptococcus pneumoniae,* possess polysaccharide capsules that are too slippery for pseudopods to grab (**Fig. 23.28D**). This is where innate immunity and adaptive immunity join forces. Adaptive immunity produces anticapsular antibodies that aid the innate immune mechanism of phagocytosis through a process known as **opsonization** (**Fig. 23.29**). The anticapsular antibodies coat the surface of the bacterium, leaving the tail end of the antibodies, called the Fc region (see Section 24.3), pointing outward. These bacteria are said to be "opsonized." The Fc regions of these antibodies are recognized and bound by specific receptors on phagocyte cell surfaces. As a result, the antibodies stick the opsonized bacteria to phagocytic Fc receptors, which allow the phagocyte to more easily engulf the invader.

Oxygen-Independent and -Dependent Killing Pathways

During phagocytosis, the cytoplasmic membrane of the phagocyte flows around, and then engulfs, the bacterium, producing an intracellular phagosome, as described earlier. Subsequent phagosome-lysosome fusion (producing a phagolysosome; see **Fig. 23.14**) results in both oxygen-independent and oxygen-dependent killing pathways. Mechanisms independent of oxygen include enzymes like lysozyme to destroy the cell wall, compounds such as lactoferrin to sequester iron away from the microbe, and defensins, small cationic antimicrobial peptides (described in Section 23.4).

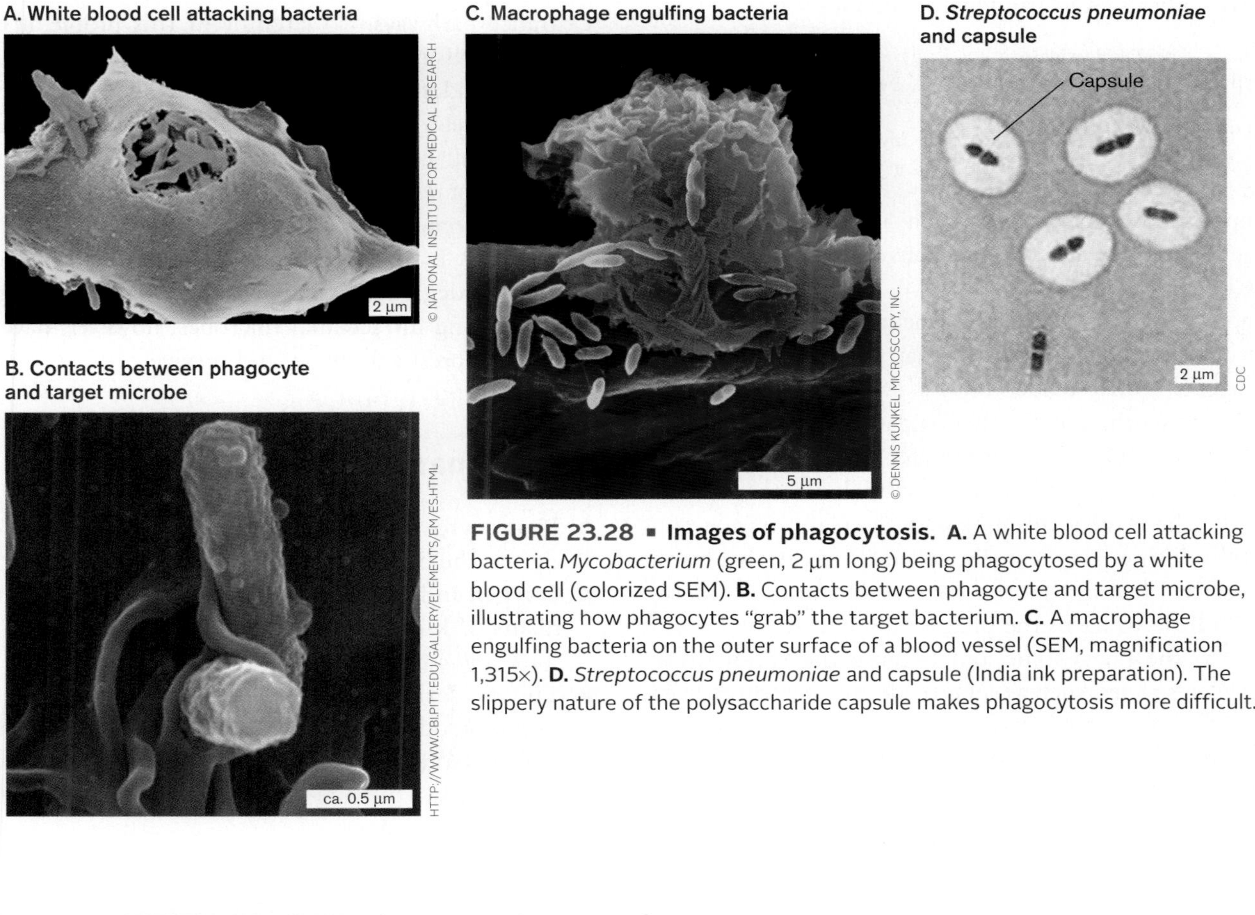
A. White blood cell attacking bacteria
2 µm
© NATIONAL INSTITUTE FOR MEDICAL RESEARCH
B. Contacts between phagocyte and target microbe
ca. 0.5 µm
HTTP://WWW.CBI.PITT.EDU/GALLERY/ELEMENTS/EM/ES.HTML
C. Macrophage engulfing bacteria
5 µm
© DENNIS KUNKEL MICROSCOPY, INC.
D. Streptococcus pneumoniae and capsule
Capsule
2 µm
CDC

FIGURE 23.28 ▪ **Images of phagocytosis. A.** A white blood cell attacking bacteria. *Mycobacterium* (green, 2 µm long) being phagocytosed by a white blood cell (colorized SEM). **B.** Contacts between phagocyte and target microbe, illustrating how phagocytes "grab" the target bacterium. **C.** A macrophage engulfing bacteria on the outer surface of a blood vessel (SEM, magnification 1,315×). **D.** *Streptococcus pneumoniae* and capsule (India ink preparation). The slippery nature of the polysaccharide capsule makes phagocytosis more difficult.

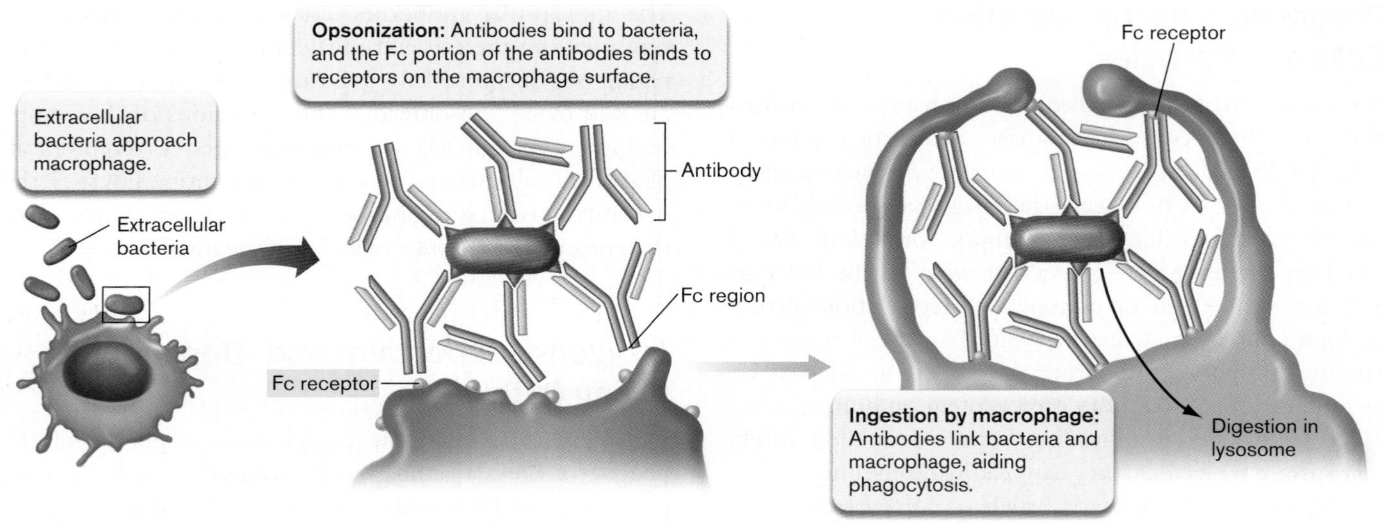

Opsonization: Antibodies bind to bacteria, and the Fc portion of the antibodies binds to receptors on the macrophage surface.
Extracellular bacteria approach macrophage.
Extracellular bacteria
Antibody
Fc region
Fc receptor
Fc receptor
Ingestion by macrophage: Antibodies link bacteria and macrophage, aiding phagocytosis.
Digestion in lysosome

FIGURE 23.29 ▪ Opsonization. Opsonization is a process that facilitates phagocytosis. Here, macrophage Fc receptors bind to the Fc region of antibodies binding to bacteria.

Oxygen-dependent mechanisms are activated through the Toll-like receptors (discussed in Section 23.7). Oxygen-dependent mechanisms kill through the production of various oxygen radicals. NADPH oxidase, myeloperoxidase, and nitric oxide synthetase in the phagosome membrane are extremely important. NADPH oxidase yields superoxide ion ($^{\bullet}O_2^-$), hydrogen peroxide, and, ultimately, hydroxyl radicals ($^{\bullet}OH$) and hydroxide ions (OH^-).

$$NADPH + 2O_2 \xrightarrow{\text{NADPH oxidase}} NADP^+ + H^+ + 2\,^{\bullet}O_2^- \xrightarrow{\text{superoxide dismutase}}$$

$$H_2O_2 + Cl^- \xrightarrow{\text{myeloperoxidase}} OH^- + HOCl$$

Myeloperoxidase, present only in neutrophils, converts hydrogen peroxide and chloride ions to hypochlorous acid (HOCl).

Macrophages, mast cells, and neutrophils also generate reactive nitrogen intermediates that serve as potent cytotoxic agents. Nitric oxide (NO) is synthesized from arginine by NO synthetase. Further oxidation of NO by oxygen yields nitrite (NO_2^-) and nitrate (NO_3^-) ions. All of these reactive oxygen species attack bacterial membranes and proteins. These mechanisms greatly increase oxygen consumption during phagocytosis, called the **oxidative burst**. The reactive chemical species formed during the oxidative burst do little to harm the phagocyte because the burst is limited to the phagosome and because the various reactive oxygen species, such as superoxide, are very short-lived. Although phagocytes are very good at clearing infectious agents, many bacteria have developed ways to outsmart this aspect of innate immunity.

Autophagy and Intracellular Pathogens

Intracellular pathogens that grow in eukaryotic cytoplasm can be a serious problem for the host. Many pathogens, such as *Mycobacterium tuberculosis*, the cause of tuberculosis, enter the host cell in ways that bypass endosome formation or, if they do enter via an endosome, can escape from that compartment. These pathogens block normal host cell clearance pathways. To circumvent this problem, eukaryotic host cells (not just phagocytes) use a process that normally degrades damaged organelles (called **autophagy**) to clear themselves of intracellular pathogens. During autophagy, the cell constructs a double membrane around the organism (or damaged organelle). This structure, called the autophagosome, sequesters the microbe from the nutrient-rich cytosol. Lysosomes then fuse with the autophagosome, depositing degradative enzymes that digest the organism. Ever-adapting intracellular microbes, however, have found ways to suppress autophagy and survive.

To Summarize

- **Phagocytosis** is selective for particles recognized as foreign to the body.
- **Oxygen-independent and oxygen-dependent mechanisms of killing** are initiated by fusion between a lysosome and a bacteria-containing phagosome.
- **The oxidative burst**, a large increase in oxygen consumption during phagocytosis, results in the production of superoxide ions, nitric oxide, and other reactive oxygen species.

- **Autophagy** is a process by which intracellular bacteria can be sequestered from the cytoplasm (via an autophagosome) and killed following fusion with a lysosome.

23.7

Interferon, Natural Killer Cells, and Toll-like Receptors

When a community is threatened by a thief, a neighbor who has been robbed alerts others to take precautions. And, if the thief is discovered, police are called to arrest him. Similarly, interferon peptides and natural killer cells are two innate mechanisms of defense that, respectively, warn healthy host cells of a nearby infection and seek out and then destroy cells already infected. Toll-like and NOD-like receptors, on the other hand, are tantamount to burglar alarm systems that activate upon encountering an intruder.

Interferons Are Host Specific, Not Virus Specific

In 1957, it was discovered that cells exposed to inactivated viruses produce at least one soluble factor that can "interfere" with viral replication when applied to newly infected cells. The term **interferon** was coined to represent these molecules. Actually, several different macromolecules interfere with viral replication. Interferons are low-molecular-weight cytokines (14–20 kDa) produced by many eukaryotic cells in response to intracellular infection. The action of interferons is usually species specific (interferon from mice will not work on human cells) but virus nonspecific (human interferon will help protect against both poliovirus and influenza virus, for example).

There are two general types of interferons, which differ in the receptors they bind and the responses they generate. Type I interferons have high antiviral potency; they consist of IFN-alpha (IFN-α), IFN-beta (IFN-β), and IFN-omega (IFN-ω). Type II interferon, IFN-gamma (IFN-γ), has more of an immunomodulatory function that is discussed in Chapter 24.

Type I interferons can bind to specific receptors on uninfected host cells and render those cells resistant to viral infection. The host cell becomes resistant because interferon induces the intracellular production of two classes of proteins. One class encompasses double-stranded RNA-activated endoribonucleases that can cleave viral RNA. Proteins in the second class are protein kinases that phosphorylate and inactivate eukaryotic initiation factor 2 (eIF2), which is required to translate viral RNA. These mechanisms affect both RNA and DNA viruses because protein synthesis is required for the propagation of all viruses. However, RNA viruses are better than DNA viruses at inducing interferon. Type I interferons are used to treat certain viral infections (for example, hepatitis C).

Type II interferon functions by activating various white blood cells—for example, macrophages, natural killer cells, and T cells—to increase the number of **major histocompatibility complex (MHC)** antigens on their surfaces. MHC proteins are important for recognizing self and for presenting foreign antigens to the adaptive immune system. They are discussed more fully in Chapter 24.

Natural Killer Cells Recognize Infected Cells and Cancer Cells

Body cells that are infected or cancerous can be a major problem for the host. Infected cells can harbor the pathogenic microbe, hiding it from the immune system. Cancer cells, on the other hand, can take over and kill the host. A class of lymphoid cells called natural killer (NK) cells identifies and handles these situations. NK cells are formed in bone marrow (see **Fig. 23.12**). Instead of killing microbes, the mission of these cells is to destroy host cells that harbor microorganisms or that have been transformed into cancer cells (**Fig. 23.30**). Natural killer cells recognize changes in cell-surface proteins of infected or cancer cells (MHC class I molecules; see Section 24.6) and then degranulate to release chemicals that kill those cells.

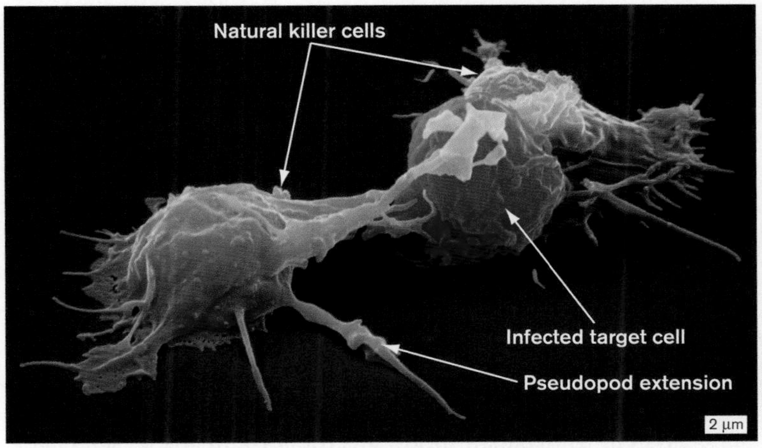
Natural killer cells
Infected target cell
Pseudopod extension
2 µm

EYE OF SCIENCE/PHOTO RESEARCHERS, INC.

FIGURE 23.30 ▪ Natural killer cells. Natural killer (NK) cells attack eukaryotic cells infected by microbes, not the microbes themselves. Perforin produced by the NK cell punctures the membranes of target cells, causing them to burst. (colorized SEM)

Natural killer cells recognize their targets in two basic ways. One involves MHC class I molecules, whereas the other utilizes Fc receptors. A normal host cell displays two classes of MHC molecules on the outside of the cell membrane. MHC I is an indicator of "self." (MHC II molecules are discussed in Chapter 24.) NK cells have specific receptors that bind to self MHC I molecules on the surfaces of other cells in the body. An NK cell that "touches" a self MHC I–containing cell from the same person will not attack that cell. However, if a host cell lacks MHC class I molecules, NK cells perceive the target as foreign and a potential threat. Host cells can lose their MHC molecules during infection or as a result of malignant transformation. When an NK cell encounters a host cell lacking these markers, the NK cell inserts a pore-forming protein (**perforin**) into the membrane of the target cell, through which cytotoxic enzymes are delivered.

Natural killer cells also contain Fc receptors on their cell surface. The second killing mechanism, called **antibody-dependent cell-mediated cytotoxicity (ADCC)**, is activated when the Fc receptor on the NK cell links to an antibody-coated host cell. The part of an antibody that does not bind to a target molecule (antigen) is called the Fc region. Why would host cells be coated with antibodies? During their replication, many viruses place viral proteins in the membrane of the infected cell. Antibodies to those viral proteins will coat the compromised cell, tagging it for ADCC. Once the compromised cell has been targeted, it is killed by the NK cell in the same way as described for cells lacking MHC—namely, by insertion of a perforin molecule. This killing mechanism is another example of cooperation between innate immunity (NK cells) and adaptive immunity (antibody-producing lymphocytes).

Thought Question

23.9 If NK cells can attack infected host cells coated with antibody, why won't neutrophils?

Toll-like Receptors Recognize Microbe-Associated Molecular Patterns

The faster the body can detect the presence of pathogens, the more quickly it can begin to deal with them. The more quickly it deals with them, the better the outcome of an infection. It takes time, however, for the adaptive immune system to make antibody specific for a microbe (see Chapter 24). All the while, the pathogen can grow and cause disease. Fortunately, bacteria and viruses possess unique structures that immediately tag them as foreign. Structures such as peptidoglycan, flagellin, and lipoteichoic acids are not normally present in tissues unless bacteria or viruses are present. These structures have microbe-associated molecular patterns (MAMPs) that can be recognized by Toll-like receptors present on various host cell types (**Table 23.4**).

First discovered in insects and named Toll receptors, **Toll-like receptors (TLRs)** are evolutionarily conserved cell-surface glycoproteins present on the cells of many eukaryotic genera. They are transmembrane proteins with an extracellular Toll/interleukin 1 receptor domain (TIR domain). Humans have numerous Toll-like receptors, each of which recognizes different MAMPs present on pathogenic microorganisms, making them an innate defense mechanism with some degree of specificity. For example, TLR2 binds to lipoarabinomannan from mycobacteria, zymosan from yeasts, lipopolysaccharide (LPS) from spirochetes, and peptidoglycan. TLR4, on the other hand, binds LPS, as well as host proteins released at sites of infection (for example, heat-shock protein 60). CD14, another host cell surface protein, serves as a coreceptor for LPS. Note that these receptors bind fragments of structures after they are released from the microbe. They don't interact with the whole organism.

Note: The term "Toll gene" came from Christiane Nüsslein-Volhard's 1985 exclamation, "That's crazy!" (in German, "Das ist ja toll!") when shown the underdeveloped posterior of a mutant fruit fly. Thus, the gene was dubbed "Toll." Toll-like receptors in mammals display sequence similarities to the Toll genes involved with insect embryogenesis.

Once bound to their ligands, the TLRs trigger an intracellular regulatory cascade via their TIR domain, causing the host cell to release proteins called cytokines that diffuse away from the site, bind to receptors on various cells of the immune system (see **Table 23.4**), and direct them to engage the invader. Cytokines are discussed further in Chapter 24. The cells that respond to cytokines can be part of innate immunity, adaptive immunity, or both. TLR recognition of MAMPs can also trigger autophagy in infected cells. Bruce Beutler and Jules Hoffman shared the 2011 Nobel Prize in Physiology or Medicine for their work on the role of TLRs in immunity.

NOD-like receptors. While TLRs are important sensors of external MAMPs (or of MAMPs in endosomes), **NOD-like receptor (NLR)** proteins are important cytoplasmic sensors of MAMPs (**Table 23.4**). NLRs are structurally similar to a family of plant proteins called NODs (nucleotide-binding oligomerization domain) that provide resistance to pathogens. In mammals, NLRs are part of large, intracellular complexes of proteins called **inflammasomes**. When bound to a MAMP, NLRs trigger a signal

TABLE 23.4

Examples of Toll-like receptors and NOD-like receptors.

Receptor	MAMPs recognized	Source	Host cells	Location
TLR1	Lipopeptides	Bacteria	Monocytes/macrophages, dendritic cells, B cells	Cell surface
TLR2	Glycolipids, lipoteichoic acids	Bacteria	Monocytes/macrophages, dendritic cells, mast cells	Cell surface
TLR3	Double-stranded RNA	Viruses	Dendritic cells, B cells	Cell compartment
TLR4	Lipopolysaccharide, heat-shock proteins	Bacteria	Monocytes/macrophages, dendritic cells, mast cells, intestinal epithelium	Cell surface
TLR5	Flagellin	Bacteria	Monocytes/macrophages, dendritic cells, intestinal epithelium	Cell surface
TLR6	Diacyl lipopeptides	*Mycoplasma*	Monocytes/macrophages, mast cells, B cells	Cell surface
TLR9	Unmethylated CpG residues in DNA	Bacteria	Monocytes/macrophages	Cell compartment
NOD 1	Component of Gram-negative peptidoglycan	Gram-negative bacteria	Many cell types	Inflammasome
NOD 2	Peptidoglycan component	Bacteria	Macrophages, dendritic cells, epithelia of lung and GI tract	Inflammasome
NLRP-3	Peptidoglycan	Bacteria	Many cell types	Inflammasome
NLRP-4	Flagellin, CpG, ATP, dsRNA	Bacteria	Many cell types	Inflammasome

pathway in inflammasomes different from that used by TLRs. The result, though, is the same; namely, cytokines are made that stimulate inflammation and activate adaptive immune mechanisms. In general, TLRs sense extracellular pathogens, whereas NLRs sense cytoplasmic pathogens.

To Summarize

- **Interferons are species-specific molecules** that can nonspecifically interfere with viral replication (type I) and modulate the immune system (type II).
- **Natural killer (NK) cells** are a class of white blood cells that destroy cancer cells or cells harboring microorganisms.
- **Natural killer cells target host cells** that have lost MHC class I receptors as a result of infection or cancer, or host cells that are coated with antibody (antibody-dependent cell-mediated cytotoxicity, ADCC).
- **Natural killer cells kill** by inserting perforin pores into the membranes of target cells.
- **Toll-like receptors (TLRs)** on host cells and **NOD-like receptors** recognize different microbe-associated molecular patterns (MAMPs).

23.8

Complement's Role in Innate Immunity

WBCs engulf and kill pathogens, but can simple serum proteins also kill microbes? Yes, a series of 20 serum proteins (complement factors) that make up the complement cascade can also attack bacterial invaders. Complement was first discovered as a heat-labile component of blood that enhances (or complements) the killing effect of antibodies on bacteria. Several complement factors are proteases that sequentially form and cleave other complement factors. (The liver is the main source of complement proteins.) Once a complement cascade is triggered, several things happen. Pores are inserted into bacterial membranes, causing cytoplasmic leaks, while pieces of some complement proteins attract WBCs and facilitate phagocytosis (opsonization).

Complement Activation Pathways

The three routes to complement activation are officially known as the classical pathway, the alternative pathway, and the lectin pathway. The classical complement pathway depends on antibody, so it is part of adaptive rather than innate immunity and is discussed in Section 24.7. The lectin pathway requires the synthesis of mannose-binding lectin by the liver in response to macrophage cytokines. Lectin coats the surface of invading microbes and activates complement without antibody. The lectin pathway connects to the classical complement pathway. We focus here in Chapter 23 on the alternative pathway because it is a well-characterized part of innate resistance. Like the lectin pathway, the alternative pathway is a nonspecific defense mechanism (that is, it does not require antibody for activation). The alternative complement pathway can attack invading microbes long before a specific immune response can be launched.

One goal of the complement cascade is to insert pores into target microbial membranes. The pores destroy membrane integrity, and thus kill the cell. The alternative complement pathway begins with the complement factor C3. In blood, C3 slowly cleaves into C3a and C3b (**Fig. 23.31**, step 1). C3b, under normal circumstances, is rapidly degraded—a process that thwarts inadvertent complement activation. However, if C3b meets LPS on an invading Gram-negative microbe, the bound C3b becomes stable and binds another factor, designated factor B (step 2), and makes factor B susceptible to cleavage by yet another protein, factor D (step 3). The resulting complex, called C3bBb, has two roles. It is a C3 convertase that can quickly cleave more C3 to amplify the cascade and is changed by another serum protein (properdin) into what is called C5 convertase (step 4). From this point on, all complement pathways are identical. C5 convertase cleaves C5 in serum to C5a and C5b (step 5), and C5b then forms a prepore complex by binding to C6 and C7 (step 6). The resulting C5bC6C7 complex binds to membranes. Finally, C8 and C9 factors join in to form the **membrane attack complex (MAC)**, becoming a destructive pore in the outer membrane of the Gram-negative cell (step 7).

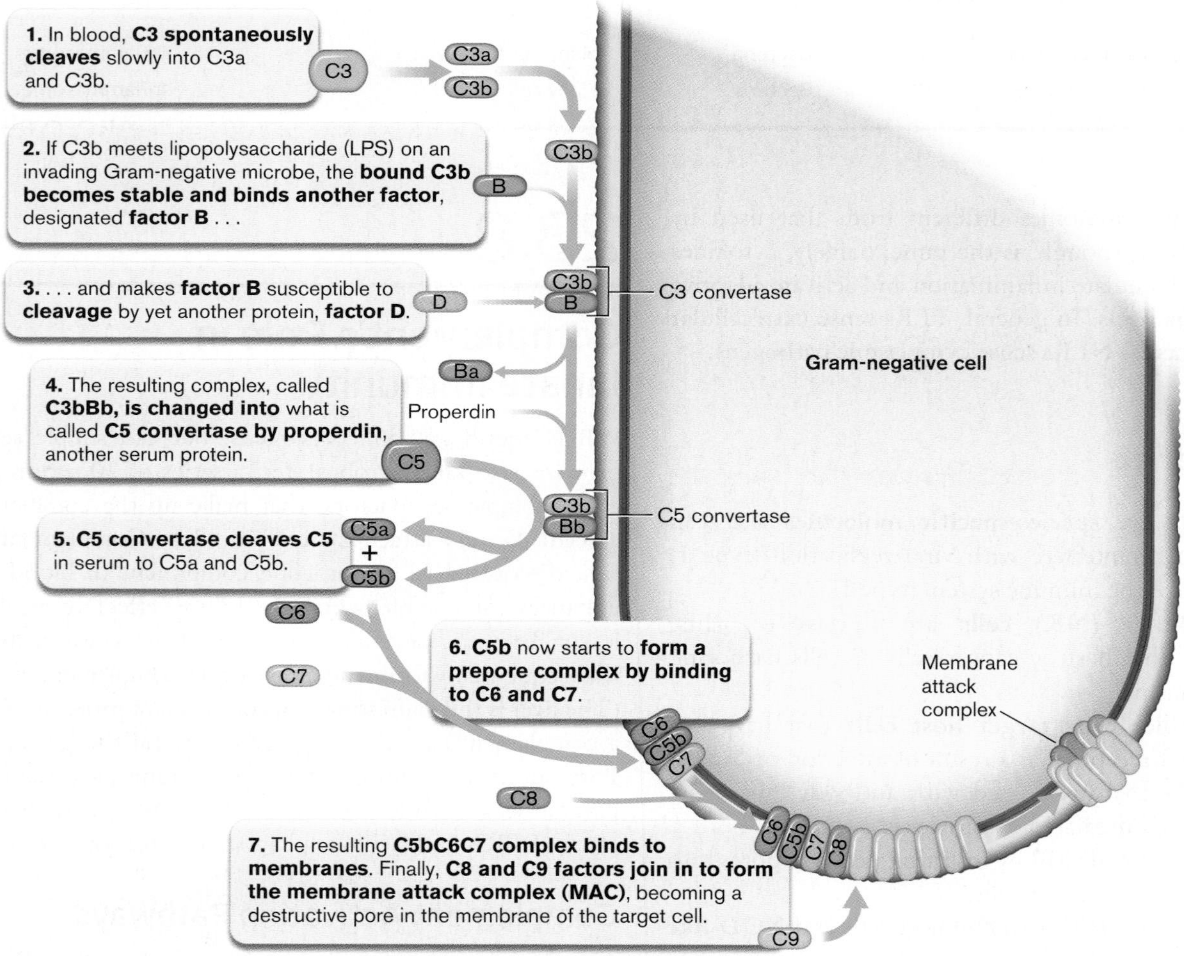
1. In blood, C3 spontaneously cleaves slowly into C3a and C3b.
C3
C3a
C3b
2. If C3b meets lipopolysaccharide (LPS) on an invading Gram-negative microbe, the bound C3b becomes stable and binds another factor, designated factor B . . .
B
3. . . . and makes factor B susceptible to cleavage by yet another protein, factor D.
D
C3 convertase
Gram-negative cell
Ba
4. The resulting complex, called C3bBb, is changed into what is called C5 convertase by properdin, another serum protein.
Properdin
C5
Bb
C5 convertase
5. C5 convertase cleaves C5 in serum to C5a and C5b.
C5a
+
C5b
C6
C7
6. C5b now starts to form a prepore complex by binding to C6 and C7.
Membrane attack complex
C8
7. The resulting C5bC6C7 complex binds to membranes. Finally, C8 and C9 factors join in to form the membrane attack complex (MAC), becoming a destructive pore in the membrane of the target cell.
C9

FIGURE 23.31 ■ The alternative complement pathway. Although called "alternative," this complement cascade is part of the first-line innate defense.

Thought Question

23.10 **Figure 23.31** shows how the complement cascade can destroy a bacterial cell. Factor H is a blood protein that regulates complement activity. Factor H binds host cells and inhibits complement from attacking our cells by accelerating degradation of C3b and C3bBb. How could bacteria take advantage of factor H?

Lysozyme (present in serum) enters through outer membrane pores of Gram-negative bacteria and cleaves peptidoglycan, making the cell membrane more susceptible to the membrane attack complex. Gram-positive bacteria are resistant to complement because they lack an outer membrane (no LPS to efficiently start the cascade) and have a thick peptidoglycan layer that hinders access of complement components. Even in the absence of LPS, however, there are ways to activate complement that involve antigen-antibody complexes (see Section 24.7). Even eukaryotic cells infected with viruses and coated with antibody can activate complement via the antibody-dependent classical complement.

Other Roles for Complement Peptides in Innate Immunity

Factor C3b, in addition to initiating the complement cascade, is a potent opsonin. An **opsonin** is any factor that can promote phagocytosis (related to opsonization, described earlier). PMNs (neutrophils) have specific C3b receptors on their surface. Thus, when C3b binds to a bacterial cell surface, it tags that cell and makes it easier for PMNs to grab and engulf the organism.

The complement fragments C5a (which forms part of the prepore complex) and C3a have many roles in immune function. They are **anaphylatoxins**, which trigger degranulation of vasoactive factors such as histamine from endothelial cells, mast cells, or phagocytes. They can also stimulate chemotaxis of immune cells. C5a and C3a bind to separate receptors on WBC plasma membranes and activate separate signal cascade pathways. C5a triggers Ca^{2+} release from intracellular stores and stimulates the actin polymerization needed for cell migration. Conversely, C3a triggers Ca^{2+} influx, which facilitates extravasation. These peptide mediators also stimulate the release of certain cytokines, such as IL-4 from monocytes and mast cells, that prepare capillary endothelium for the rolling and adhesion of neutrophils. C5a will also up-regulate P-selectin and ICAM-1. Thus, some complement factors not only help directly destroy target bacterial cells, but also facilitate phagocytosis and contribute directly to inflammation.

Acute-Phase Reactants, Complement, and Heart Disease

As noted earlier, inflammation is associated with the production of various cytokines by macrophages. Some of these cytokines (such as IL-1, TNF-α, and IL-6) travel to the liver, where they stimulate synthesis of several so-called acute-phase reactant proteins, including **C-reactive protein**. Named for its ability to activate complement, C-reactive protein is an acute-phase reactant that will bind to components of bacterial cell surfaces but not to host cell membranes (**Fig. 23.32**). Once tied to the bacterial cell surface, C-reactive protein will bind complement factor C1q of the classical complement pathway (Chapter 24), ultimately converting factor C3 to C3b, propagating the complement cascade. C-reactive protein accelerates C3b production at the bacterial surface, where it can do the most damage.

Elevated levels of C-reactive protein have been linked to an increased risk of cardiovascular disease. The hypothesis is that fatty deposits in the arteries trigger inflammation in the area of deposition. Inflammation then triggers an increase in C-reactive protein, which can be measured in blood as an indicator of cardiovascular disease. Taking an aspirin a day may reduce the risk of heart attack by preventing inflammation and, in turn, reducing levels of C-reactive protein. Whether C-reactive protein has a direct role in cardiovascular disease is not known.

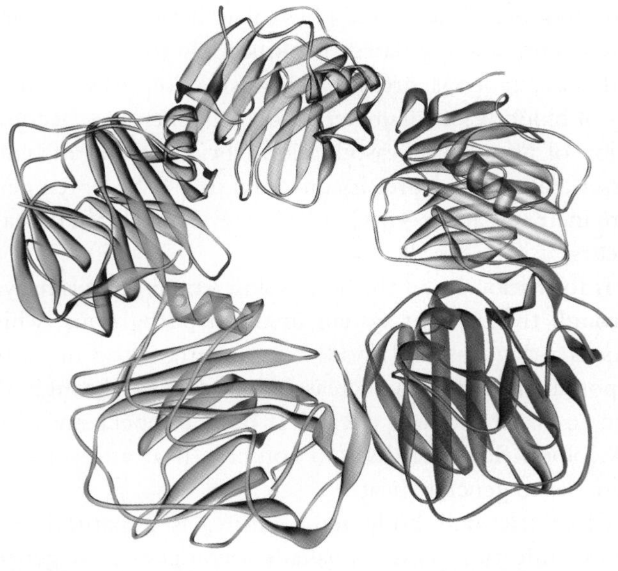

KENNETH EWARD/BIOGRAFX/PHOTO RESEARCHERS, INC.

FIGURE 23.32 ■ C-reactive protein. The 3D structure of pentameric human C-reactive protein (CRP) is unique. CRP is synthesized as a 206-amino-acid polypeptide that folds to form a flattened jelly roll structure, which then assembles into a radially symmetrical pentamer that circulates in serum. This structure can bind bacterial cell surfaces and C3 complement factor.

To Summarize

- **Complement** is a series of 20 proteins naturally present in serum.
- **Activation of the complement cascade** results in a pore being introduced into target membranes.
- **The three pathways for activation** are the classical, alternative, and lectin pathways.
- **The alternative activation pathway** begins when complement factor C3b is stabilized by interaction with the LPS of an invading microbe.
- **The cascade of protein factors**, C3b ⟶ B ⟶ factor D ⟶ properdin ⟶ C5 ⟶ C6 ⟶ C7 ⟶ C8 and C9, results in the formation of a membrane attack complex in target membranes.
- **C-reactive protein** in serum is activated when bound to microbial structures and will convert C3 to C3b, which can start the complement cascade.

23.9

Fever

What is fever and why is it a good thing? In a healthy individual, body temperature is kept constant within a very small range (36°C–38°C), despite large differences in surrounding temperature and physical activity. Body temperature is normally regulated by blood flow through the skin and subcutaneous areas. Vasoconstriction (the tightening of blood vessel diameter) allows the increased accumulation of heat, while vasodilation (the loosening of blood vessel diameter) secures its quick release. If body temperature increases above 42°C–43°C, irreversible brain damage occurs.

Information about the body's skin temperature is relayed through the nervous system to the hypothalamus, which acts as a thermostat. If body temperature is too high, the hypothalamus directs increased blood flow through the skin to accelerate heat release. If body temperature is too low, blood flow decreases to conserve heat and shivering increases to generate heat.

Fever (elevated body temperature) is a natural reaction to infection. Fever is usually accompanied by general symptoms, such as sweating, chills, sensation of cold, and other subjective sensations. Substances that cause fever are known as **pyrogens**. Pyrogens fall into two classes: exogenous and endogenous. Exogenous pyrogens are those that originate outside the body, such as bacterial toxins. Endogenous pyrogens, on the other hand, are formed by the body's own cells in response to an outside stimulus (such as a bacterial toxin).

Exogenous pyrogens generally cause fever by inducing the release of endogenous pyrogens such as interferon, tumor necrosis factor, IL-6, and other cytokines. Pyrogens cross the blood-brain barrier and bind to neurons in the thermoregulatory center of the anterior hypothalamus. Cytokine-receptor interaction stimulates the production of phospholipase A2, an enzyme required to make prostaglandins. Prostaglandin E2 is made and changes the responsiveness of thermosensitive neurons. In other words, prostaglandin E2 turns up the body's thermostat.

Because the ideal growth temperature for many microbes is 37°C, elevated temperature can place the organism outside its "comfort zone" of growth. There is also evidence that fever reduces iron availability to bacteria (cytokine release causes an increase in iron storage protein). Slower growth of the pathogen "buys" the immune system time to subdue the infection before it is too late. Consequently, interventions that reduce a moderate fever caused by infection may be counterproductive to a speedy recovery.

Thought Question

23.11 If increased fever limits bacterial growth, why do bacteria make pyrogenic toxins?

To Summarize

- **The hypothalamus** acts as the body's thermostat.
- **Exogenous and endogenous pyrogens** elevate body temperature by stimulating prostaglandin production.
- **Prostaglandins** change the responsiveness of thermosensitive neurons in the hypothalamus.

Concluding Thoughts

The human body plays host to many species of microbes (our microbiota), many of which contribute to our health and well-being. The various "hardwired" innate immune mechanisms described here in Chapter 23 keep our microbiota at bay and provide an effective first line of defense against potential pathogens. The next chapter addresses what happens when microbes breach these innate defenses. Unlike the general protective mechanisms just discussed, the system of adaptive immunity generates a molecular defense specifically tailored to a given pathogen.

CHAPTER REVIEW

Review Questions

1. Name some sterile body sites.
2. What body sites are colonized by normal microbiota?
3. Under what circumstances can commensal organisms cause disease?
4. Why are commensal organisms beneficial to the host?
5. Name and describe various types of innate immunity.
6. What are probiotics? How do they help maintain health?
7. How do the lungs avoid being colonized?
8. What is a gnotobiotic animal?
9. Describe GALT and SALT.
10. Describe some chemical barriers to infection.
11. Discuss the different types of white blood cells.
12. What is a lymphoid organ?
13. Outline the process of inflammation.
14. Explain why phagocytes do not indiscriminately phagocytose body cells.
15. What is interferon?
16. Describe antibody-dependent cell-mediated cytotoxicity.
17. How does complement kill bacteria?
18. Why might fever be helpful in fighting infection?

Thought Questions

1. Is it common for microbial pathogens to pass the placental barrier (called transplacental transmission) and infect the fetus? Consider papillomavirus, *Listeria monocytogenes, Escherichia coli*, HIV, *Treponema pallidum, Neisseria gonorrhoeae*, and *Staphylococcus aureus*.
2. The vagina contains competing, commensal microbes that contribute to the health of the organ. So, is "cleaning" the vagina by douching actually unhealthy and likely to lead to an increased chance of infection (vaginosis)?
3. Why have microbes not altered their structures to avoid being recognized by Toll-like receptors?

Key Terms

adaptive immunity (936)
alveolar macrophage (944)
anaphylatoxin (957)
antibody-dependent cell-mediated cytotoxicity (ADCC) (954)
antigen (936)
antigen-presenting cell (939)
autophagy (952)
B cell (941)
bacteremia (930)
basophil (938)
bioburden (926)
bradykinin (948)
C-reactive protein (957)
chemokine (948)
chronic inflammation (949)
commensal organism (926)
complement (936)
compromised host (934)
cystic fibrosis transmembrane conductance regulator (CFTR) (944)
cytokine (935)
defensin (945)
degranulate (946)
dendritic cell (939)
enterotoxin (935)
eosinophil (938)
epidermis (928)
extravasation (947)
gnotobiotic aånimal (935)
granuloma (949)
gut-associated lymphoid tissue (GALT) (944)
immune system (935)
immunomodulin (935)
inflammasome (954)
innate immunity (936)
interferon (953)
interleukin 1 (IL-1) (948)
Langerhans cell (943)
lymph node (942)
lymphocyte (940)
M cell (944)
macrophage (939)
major histocompatibility complex (MHC) (953)
mast cell (938)
membrane attack complex (MAC) (956)
metagenome (926)
microbe-associated molecular pattern (MAMP) (944)
microbiome (926)
microbiota (926)
monocyte (939)
mucociliary elevator (930)
nasopharynx (929)
neutrophil (937)
neutrophil extracellular trap (NET) (938)
NOD-like receptor (NLR) (954)
nonadaptive immunity (936)
opportunistic pathogen (934)

opsonin (957)
opsonization (951)
oropharynx (929)
oxidative burst (952)
perforin (954)
phagosome (938)
platelet (939, 942)
probiotic (933)
pyrogen (958)
reticuloendothelial system (939)
selectin (948)
skin-associated lymphoid tissue (SALT) (943)
T cell (941)
tight junction (943)
Toll-like receptor (TLR) (954)
tumor necrosis factor alpha (TNF-α) (948)
vasoactive factor (947)
virome (928)
white blood cell (WBC) differential (940)

Recommended Reading

Chung, Hachung, Sunje J. Pamp, Jonathan A. Hill, Neeraj K. Surana, Sanna M Edelman, et al. 2012. Gut immune maturation depends on colonization with a host-specific microbiota. *Cell* **149**:1578–1593.

Clemente, Jose C., Luke K. Ursell, Laura Wegener Parfrey, and Rob Knight. 2012. The impact of the gut microbiota on human health: An integrative view. *Cell* **148**:1258–1270.

Delgado, Monica A., and Vojo Deretic. 2009. Toll-like receptors in control of immunological autophagy. *Cell Death and Differentiation* **16**:976–983.

Harris, Kristina, Amira Kassis, Geneviève Major, and Chieh J. Chou. 2012. Is the gut microbiota a new factor contributing to obesity and its metabolic disorders? *Journal of Obesity* **2012**:879151.

Hathaway, Lucy J., and Jean-Pierre Kraehenbuhl. 2000. The role of M cells in mucosal immunity. *Cellular and Molecular Life Science* **57**:323–332.

Hewetson, J. T. 1904. The bacteriology of certain parts of the human alimentary canal and of the inflammatory processes arising therefrom. *British Medical Journal* **2**:1457–1460.

Kaplan, Mariana J., and Marko Radic. 2012. Neutrophil extracellular traps: Double-edged swords of innate immunity. *Journal of Immunology* **189**:2689–2695.

Kumar, Sushil, Harshad Ingle, Durbaka Vijaya Raghava Prasad, and Himanshu Kumar. 2012. Recognition of bacterial infection by innate immune sensors. *Critical Reviews in Microbiology* **39**(3): 229–246.

Moore, Aimee M., Christian Munck, Morton Sommer, and Gautam Dantus. 2011. Functional metagenomic investigations of the human intestinal microbiota. *Frontiers in Microbiology* **2**:188.

Niedergang, Florence, and Jean-Pierre Kraehenbuhl. 2000. Much ado about M cells. *Trends in Cell Biology* **10**:137–141.

Ogawa, Michinaga, and Chihiro Sasokawa. 2006. Bacterial evasion of the autophagic defense system. *Current Opinion in Microbiology* **9**:62–68.

Orvedahl, Anthony, and Beth Levine. 2009. Eating the enemy within: Autophagy in infectious diseases. *Cell Death and Differentiation* **16**:57–69.

Skeldon, Alexander, and Maya Saleh. 2011. The inflammasomes: Molecular effectors of host resistance against bacterial, viral, parasitic, and fungal infections. *Frontiers in Microbiology* **2**:15.

Stetson, Daniel B., and Ruslan Medzhitov. 2006. Type I interferons in host defense. *Immunity* **25**:373–381.

Stuart, Lynda M., and R. Alan B. Ezekowitz. 2005. Phagocytosis: Elegant complexity. *Immunity* **22**:539–550.

van Sorge, Nina M., and Kelly S. Doran. 2012. Defense at the border: The blood-brain barrier versus bacterial foreigners. *Future Microbiology* 7: 383–394.

Wikoff, William R., Andrew T. Anfora, Jun Liu, Peter G. Schultz, Scott A. Lesley, et al. 2009. Metabolomics analysis reveals large effects of gut microflora on mammalian blood metabolites. *Proceedings of the National Academy of Sciences USA* **106**:3698–3703.

Wu, Gary D., Jun Chen, Christian Hoffmann, Kyle Bittinger, Ying-Yu Chen, et al. 2011. Linking long-term dietary patterns with gut microbial enterotypes. *Science* **334**:105–108.

Zhou, Wuding. 2012. The new face of anaphylatoxins in immune regulation. *Immunobiology* **217**:225–234.

CHAPTER 24

The Adaptive Immune Response

24.1 Overview of Adaptive Immunity

24.2 Immunogenicity

24.3 Antibody Structure and Diversity

24.4 Primary and Secondary Antibody Responses

24.5 Genetics of Antibody Production

24.6 T cells Link Antibody and Cellular Immune Systems

24.7 Complement as Part of Adaptive Immunity

24.8 Hypersensitivity and Autoimmunity

Memory extends beyond the brain. Adaptive immunity also depends on a kind of memory—one that does not rely on nerves and neurons. Our immune system's memory relies on special lymphocytes in the circulation called memory B cells and T cells that form during an infection and "remember" it for years. Although a single member of these cells can recall exposure to only one part of an infecting agent, collectively memory cells remember every microbe that has managed to breach our innate defenses. Armed with this "knowledge", they circulate throughout the body like tiny sentries, ready to detect and quickly respond to a second attack.

From birth, the immune system is able to adapt and recognize billions of possible foreign antigens. However, a system this flexible can also attack itself. Here in Chapter 24 we demystify how adaptive immunity develops, regulates itself, and neutralizes potential threats.

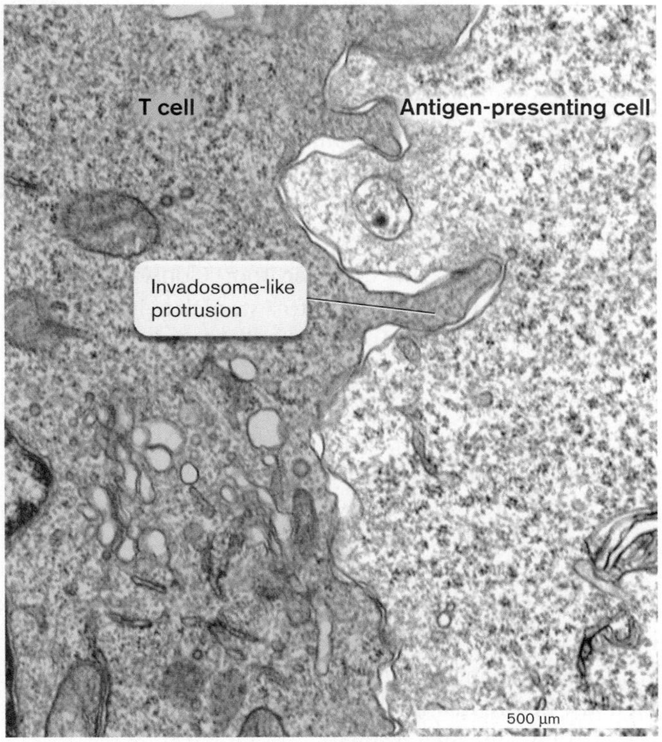

T cell
Antigen-presenting cell
Invadosome-like protrusion
500 µm

CURRENT RESEARCH **highlight**

T cells probe antigen-presenting cells for information. Lymphocytes called T cells are important regulators and enforcers of adaptive immunity. To become activated, T cells must recognize foreign molecules called antigens. However, T cells recognize foreign antigens only if the antigen is presented to them on another host cell called an antigen-presenting cell (APC). The APC engulfs microbes, breaks them into tiny pieces (antigens), and transports the antigens to their cell surface. T cells then scan the surfaces of different APCs to find the right foreign antigens. The photomicrograph shows a T cell (red) extending an invadosome-like protrusion (ILP) to probe the surface of a blood vessel endothelial cell (also an APC). The intimate cell-cell contacts initiated by ILPs appear to be required for T cells to recognize their matched antigen. ILPs, then, are the sensory organelles of immune surveillance. *Source:* Sharpe, A. H. and C. V. Carman. 2012. *J. Immunol.* **188:** 3686–3699.

David Philip Vetter was born in 1971 without an immune system. Better known as the "bubble boy," he was the only human known to live in a plastic, germ-free bubble for his entire 12 years of life (1971–1984) (**Fig. 24.1A**). His predicament stemmed from a genetic disease known as severe combined immunodeficiency (SCID). In the most serious form of SCID (caused by a defective cytokine receptor), the patient harbors no T cells and has dysfunctional (or sometimes no) B cells, so the body cannot launch a meaningful immune defense against any invading microbe—whether pathogen or normal microbiota. Because David's older brother had died of SCID before David was born, physicians were alerted that David might have the disorder too. Consequently, David was transferred to a sterile environment within seconds after birth to await a bone marrow transplant. Water, air, food, diapers, clothes—all were disinfected with special cleaning agents before entering his sterile plastic bubble. He was handled only through special plastic gloves attached to the wall. He lived for 12 years physically isolated in this plastic, sterile environment, venturing out only in a NASA-designed sterile space suit. A bone marrow transplant was attempted in 1984, but his defective immune system failed to protect him from Epstein-Barr virus, which went undetected in the transplanted cells. He died just before his thirteenth birthday.

A.

COURTESY OF TEXAS CHILDREN'S HOSPITAL

B.

COURTESY OF UCL NEWS

FIGURE 24.1 ■ Living without an immune system. A. David Vetter, the bubble boy, inside his environmental bubble. The tube behind him is a port that was used to introduce sterile food, clothes, and other items. **B.** Adrian J. Thrasher, University College London, conducted a successful gene therapy trial on children with severe combined immunodeficiency (SCID).

In 2004, another dramatic attempt was made to cure the disease using stem cells. Stem cells are undifferentiated, and capable of changing into many different cell types, including B cells and T cells (see Section 23.3). A team of scientists headed by Dr. Adrian Thrasher of University College London (**Fig. 24.1B**) removed stem cells from the bone marrow of four SCID children who lacked the gene for the cytokine receptor. They inserted the gene encoding the normal cytokine receptor into a severely defective leukemia virus that can infect cells without causing disease. The debilitated leukemia virus served as a gene vector to deliver the normal cytokine receptor gene into the patient's stem cells. Once the genetic material entered the nucleus, the healthy copy of the gene began to function and the corrected stem cells were reintroduced into the patients. After this gene replacement therapy, all four children started making T cells with the correct receptor and produced functional B cells. The patients developed an immune system and were discharged.

SCID dramatically illustrates the importance of our immune system and how fragile its development is. All it takes is a small defect in a single gene to subvert the entire process.

Chapter 24 begins by describing the two types of adaptive immunity and the factors that influence the immunogenicity of foreign proteins, lipids, and so on. We will explore how B cells ultimately differentiate into plasma cells and make antibodies (defining what is called humoral, or circulating, immunity) and how certain T-cell lymphocytes develop to directly kill infected host cells in what is called cell-mediated or cellular immunity. You will also learn that another type of T cell controls the balance between humoral and cell-mediated responses to a given infection. Ultimately, you will appreciate that adaptive immunity is a major reason the human race still exists.

24.1

Overview of Adaptive Immunity

Recall from Chapter 23 that the immune system has both nonadaptive and adaptive mechanisms. Nonadaptive (innate) immune mechanisms are present from birth, whereas adaptive immunity develops as the need arises. For instance, adaptive immunity against malaria does not develop until the individual has encountered the plasmodial parasite that causes the disease. The adaptive immune response—**adaptive immunity**—is a complex, interconnected, and cross-regulated defense network.

Note: The terms "adaptive immune response" and "immune response" are often used interchangeably.

Two types of adaptive immunity are recognized: humoral immunity and cell-mediated immunity. In **humoral immunity**, **antibodies** are produced that directly target microbial invaders. The term "humoral" means "related to body fluids." Thus, antibodies are proteins that circulate in the bloodstream and recognize foreign structures called antigens. An **antigen** (also called an **immunogen**) is any molecule that will, when introduced into a person, elicit the synthesis of antibodies that specifically bind the antigen. Antigens stimulate B cells (B lymphocytes) to differentiate into antibody-producing cells. **Cell-mediated immunity**, the second type of adaptive immunity, employs teams of T cells (T lymphocytes) that recognize antigens and then destroy host cells infected by the microbe possessing the antigen. In truth, the humoral and cellular immune responses are intertwined, each relying on some facet of the other to work efficiently. T cells serve a central role in adaptive immunity by determining whether humoral or cell-mediated mechanisms predominate in response to a specific antigen.

What triggers an immune response, and how long does it take? Adaptive immunity develops over a 3- to 4-day period after exposure to an invading microbe. The immune system does not recognize the whole microbe, but innumerable tiny pieces of it. Each small segment of an antigen that is capable of eliciting an immune response is called an **antigenic determinant** or **epitope**. Many single-protein antigens are recognized when the larger antigen is broken into smaller segments upon being phagocytosed. Even distinct tertiary (3D) shapes within a protein may be counted as antigenic determinants if they produce a specific response. This happens when stretches of amino acids far removed from each other in a protein's primary sequence align side by side in 3D space after folding (**Fig. 24.2**). Such a 3D structure may be recognized by the immune system as a single entity or antigen. Besides proteins, other structures in the cell, such as complex polysaccharides, can have linear and 3D epitopes. So the immune response to a microbe is really a composite of thousands of B-cell responses to different epitopes. The response to each individual epitope is **clonal**; that is, it gives rise to a population of cells that originate from a single B cell. This means that each clone of B cells will target a unique epitope.

The humoral immune response requires several cell types and cell-to-cell interactions. What are those interactions, and where do they take place? As illustrated in **Figure 24.3**, the process begins with an infection somewhere in the body. Dendritic cells and macrophages patrolling the area gather up the foreign antigens and present them on their cell surface. Phagocytic cells that degrade large antigens into smaller antigenic determinants and place those determinants on their cell surface are called **antigen-presenting cells (APCs)**. Many types of cells can be antigen-presenting. They include "professional" APCs, such as macrophages (monocytes), mast cells, or dendritic cells; and "nonprofessional" APCs (for example, endothelial cells or fibroblasts) under the right circumstances. Because dendritic cells are so important to the immune response, their discoverer, Ralph Steinman (1943–2011), was awarded the Nobel Prize in Physiology or Medicine in 2011 (**Fig. 24.3B**).

Once decorated with antigen, the professional antigen-presenting cell travels to secondary lymphoid organs (lymph nodes) where B cells and T cells await. Specific

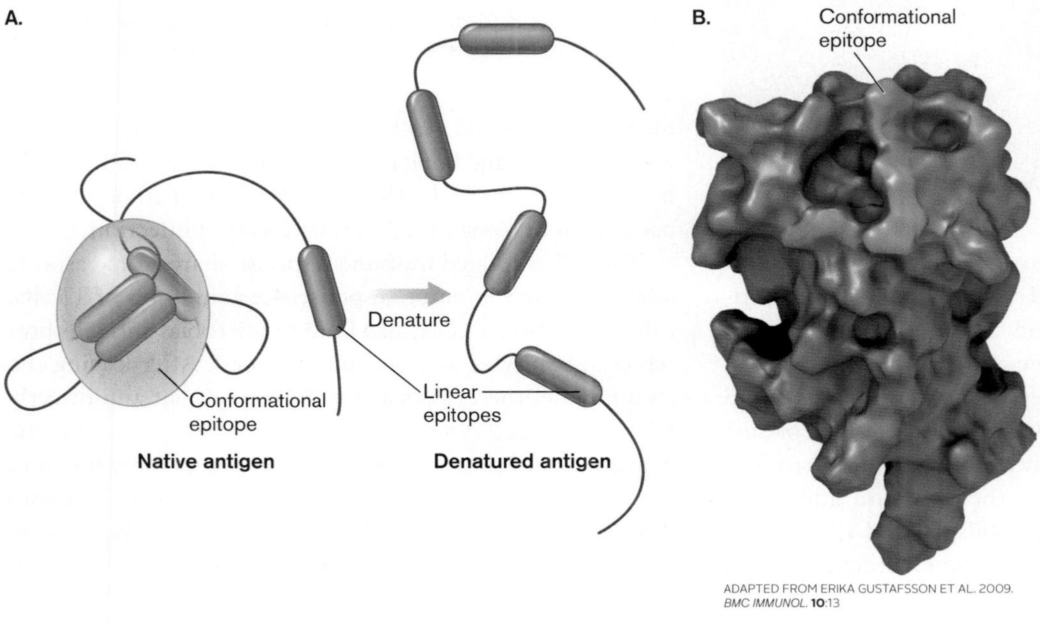
A.
Denature
Conformational epitope
Linear epitopes
Native antigen
Denatured antigen
B.
Conformational epitope
ADAPTED FROM ERIKA GUSTAFSSON ET AL. 2009. BMC IMMUNOL. 10:13

FIGURE 24.2 ■ Antigens and epitopes. A. Native proteins fold into a 3D shape, where several regions separated in the linear sequence can reside next to each other to form a conformational epitope. Denaturing the protein with the detergent sodium dodecyl sulfate (SDS) and reducing agents like dithiothreitol (to remove disulfide bonds) will unfold the protein and separate the various amino acid stretches that formed the conformational epitope. **B.** A 3D protein structure showing, in red, four amino acids that form a conformational epitope.

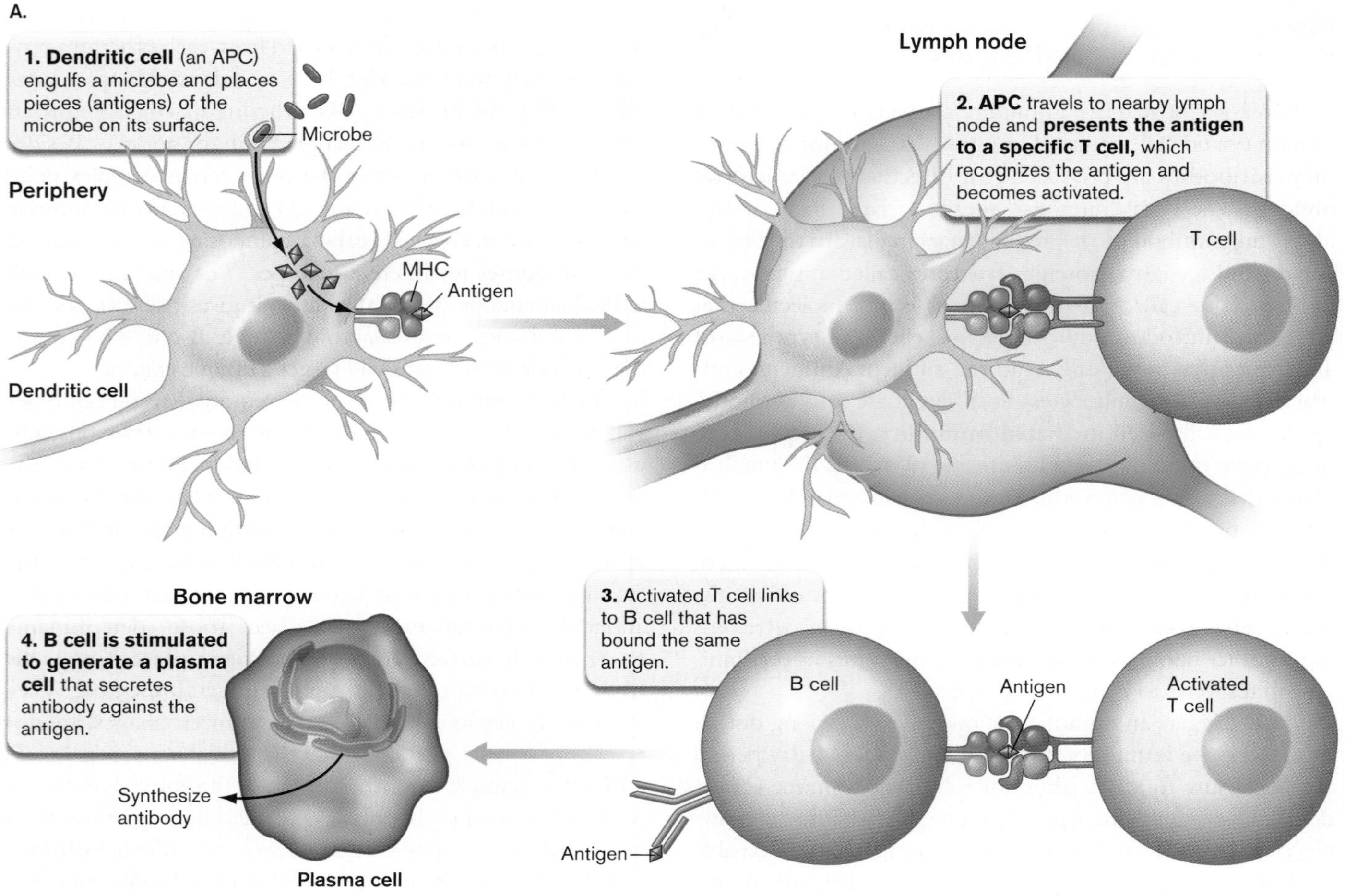
A.
1. Dendritic cell (an APC) engulfs a microbe and places pieces (antigens) of the microbe on its surface.
Microbe
Periphery
MHC
Antigen
Dendritic cell
Lymph node
2. APC travels to nearby lymph node and presents the antigen to a specific T cell, which recognizes the antigen and becomes activated.
T cell
Bone marrow
4. B cell is stimulated to generate a plasma cell that secretes antibody against the antigen.
Synthesize antibody
Plasma cell
3. Activated T cell links to B cell that has bound the same antigen.
B cell
Antigen
Activated T cell
Antigen

B.

ZACH VEILLEUX/ROCKEFELLER UNIVERSITY

FIGURE 24.3 ■ Cell-cell interactions involved in making antibody. A. There are two basic cell-cell interactions required to make an antibody: (1) An APC (dendritic cell) presents antigen to a helper T cell, and (2) the activated T cell links to and activates a B cell bound to the same antigen. The two interactions can be broken down into four steps. **B.** Ralph Steinman received the 2011 Nobel Prize in Physiology or Medicine for his discovery of dendritic cells.

T cells in the node then link to the antigen presented on the APC and become activated T cells. One type of activated T cell then binds to and activates a lymph node B cell that encountered the same antigen. That interaction authorizes B cells to generate plasma cells able to pump out large amounts of specific antibody. (Note that plasma cells are <u>not</u> B cells.) The activated B cells also produce memory B cells that remember the exposure and stand ready to quickly generate plasma cells, should the antigen be encountered months or years later. Once formed, the memory B cells and most plasma cells leave the lymph node and migrate to the bone marrow. Other plasma cells remain in the lymph node. In the bone marrow, memory B cells patiently wait to be called to future sites of infection.

The cell-mediated immune response shares some aspects of the humoral immune response. Certain types of T cells, called cytotoxic T cells, also bind to microbial antigens presented on an APC and become activated. After leaving the lymph node, the cytotoxic T cell can seek out and directly kill any host cell infected with the microbe. The cytotoxic T cell recognizes an infected cell that places antigens from the invading microbe on their host cell surface. Besides directly killing infected cells, cytotoxic T cells also synthesize and secrete growth factors called cytokines (Section

23.2) that incite nearby macrophages to indiscriminately attack cells in the local area. Cytokines are discussed in Section 24.6. Thus, cellular immunity, in general, is critical for dealing with intracellular pathogens such as viruses, whereas humoral immunity is most effective against extracellular bacterial pathogens like *Streptococcus pneumoniae,* one cause of pneumonia.

As we proceed through this chapter, we will reveal in layers how the immune system functions, with each layer building on the previous one. We will periodically return to a particular aspect of the immune response—for example, B-cell differentiation into plasma cells—to integrate seemingly distinct parts of the immune system into a unified concept of immunity.

Thought Question

24.1 Two different stretches of amino acids in a single protein form a 3D antigenic determinant. Will the specific immune response to that 3D antigen also respond to one of the two amino acid stretches alone?

To Summarize

- **An antigen** can elicit an antibody response. An antigen usually consists of many different epitopes (antigenic determinants), each of which binds to a different, specific antibody.
- **Humoral immunity** against infection is the result of antibody production originated by B cells.
- **Cell-mediated (cellular) immunity** involves a type of lymphocyte called T cells, which control antibody production and can directly kill host cells.

24.2

Immunogenicity

Immunogenicity measures the effectiveness by which an antigen elicits an immune response. One antigen can be more immunogenic than another. For example, proteins are the strongest antigens, but carbohydrates can also elicit immune reactions. Nucleic acids and lipids are usually weaker antigens, in part because these molecules are very flexible and present a variable 3D structure that does not easily interact with antibodies. Nucleic acids and lipids are also weak antigens because they are both made of relatively uniform repeating units. Proteins are more effective antigens for three reasons: They form a variety of shapes, they maintain their tertiary structure, and they are made of many different amino acids that can be assembled in many different combinations. These features provide stronger interactions with antibodies in the bloodstream and enable better recognition by lymphocytes, the cellular workhorses of the immune system.

Several other factors contribute to the immunogenicity of proteins (see eTopic 24.1). For example, the larger the antigen, the more likely it is that phagocytic cells will "see" and engulf it. This is important because, as noted earlier, an immune response cannot occur until phagocytic antigen-presenting cells (such as macrophages and dendritic cells) first engulf large antigens and degrade them, presenting the epitopes on their cell surface. The immune response begins when a B cell binds to a foreign peptide and a T lymphocyte binds to the same foreign peptide displayed on the surface of an antigen-presenting cell. These events happen independently.

The presentation of antigens on APCs forces the antigen to be placed on a membrane surface protein structure called the **major histocompatibility complex**, or **MHC** (discussed later). The more tightly an antigen can bind to these MHC surface proteins, the more immunogenic it is. The stronger the binding, the easier it is for T cells to recognize the complex.

Each specific antigen shows a different **threshold dose** needed to generate an optimal response. A dose higher or lower than that threshold will not generate as strong an immune response. Lower doses activate only a few B cells, whereas exceedingly high doses of antigen can cause **B-cell tolerance**, a state in which B cells have been overstimulated to the point that they do not respond to subsequent antigen exposures and make antibody. Tolerance is part of the reason your immune system does not react against your own protein antigens.

As you might expect, the body must regulate the immune system carefully so that a response is not leveled against itself. In effect, the immune system must become "blind" to its own antigens; as a result, the host will often be blind to foreign antigens that resemble epitopes of its own cells. Therefore, the more complex the foreign protein is, the more likely it will possess antigenic determinants that a lymphocyte can recognize as nonself. The farther an antigen is from "self," the greater its immunogenicity will be.

Immunological Specificity

Our earliest clues into the nature of immunity came from smallpox. Smallpox is a devastating disease that caused enormous suffering and killed millions of people in the seventeenth and eighteenth centuries (Chapter 1 and **Fig. 24.4**). There was no cure, and the only available preventive treatment was to take dried material from the lesions of a previous smallpox sufferer, place it on a healthy person, and hope the person survived. Those who survived were protected from subsequent bouts of smallpox but were still susceptible to other diseases. This early observation gave rise to the idea of **immunological specificity**, which means that an immune response to one antigen is not effective against a different antigen. In other words, the immune response to smallpox will not protect someone against the plague bacillus (*Yersinia pestis*), which is antigenically different from the smallpox virus.

While immunological specificity is important, it is not absolute. As described in Chapter 1, an English country physician named Edward Jenner (see Fig. 1.19B) in the late eighteenth century (long before viruses were discovered) learned to protect townsfolk from deadly smallpox disease (caused by variola virus) by inoculating them with scrapings from lesions produced by cowpox (a tamer disease caused by vaccinia virus; see **Fig. 24.4C**). This story illustrates that an immune reaction against one organism or virus may be sufficient to protect against an antigenically related, if not identical, organism. The technique of exposing individuals to "tame" microbes to protect them against pathogens, now generally called **vaccination**, has been used to protect humans against many bacterial and viral pathogens (**Table 24.1**). Most vaccinations today involve administering crippled (live, attenuated) strains of the pathogenic microbe or inactivated microbial toxins (for example, diphtheria toxin).

Cross-protection, in which immunization against one microbe protects against a second, will work only if two proteins critical to the pathogenesis of the two different microorganisms share key antigenic determinants. No cross-protection occurs if these determinants differ significantly. A good example is the common cold, which is caused by hundreds of closely related rhinovirus strains (rhinitis, a runny nose, is one of the symptoms of this viral disease). Infection with one strain will not immunize the victim against a second strain. The reason is that the structures of rhinovirus proteins that attach to the ICAM-1 surface protein on host cells differ dramatically between different strains of rhinovirus (**Fig. 24.5A**). Antibodies called neutralizing antibodies, which bind to the attachment protein on one strain of rhinovirus, will prevent infection by that

A. Smallpox patient

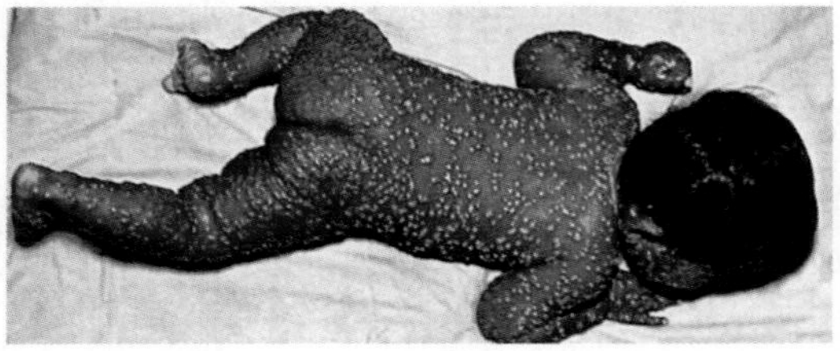
CDC

B. Variola major

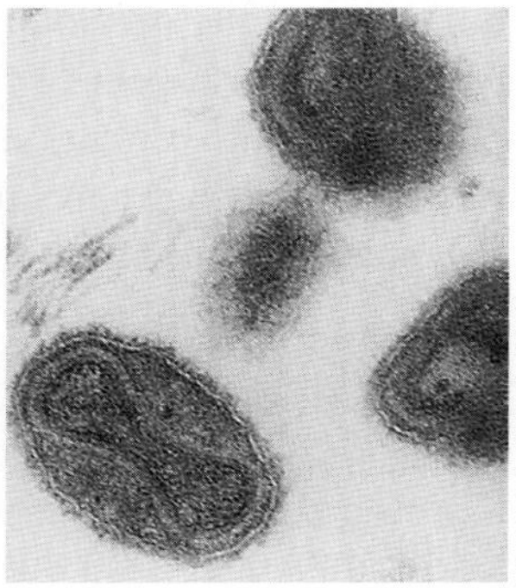
REPRODUCED WITH PERMISSION
OF THE ASM MICROBELIBRARY

C. Vaccinia

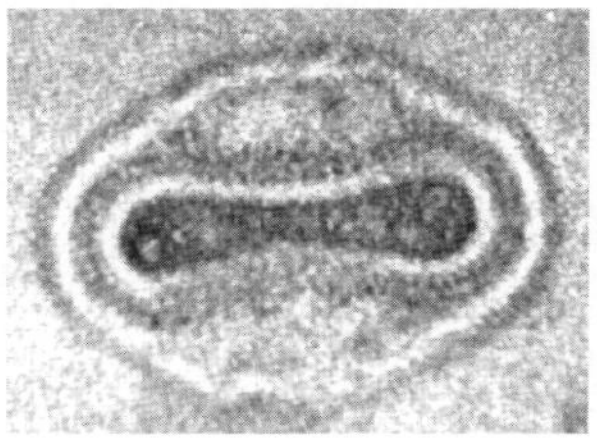
©1999 AMERICAN SOCIETY
FOR MICROBIOLOGY

FIGURE 24.4 ▪ Immunological specificity is the basis of vaccination. A. Smallpox patient covered with white pox pustules. **B.** The smallpox virus, variola major (300 nm long, TEM). The photo shows the dumbbell-shaped, membrane-enclosed nucleic acid core. **C.** The vaccinia virus that causes cowpox (360 nm long, EM). Edward Jenner recognized the similarity between the deadly smallpox and less severe cowpox diseases and used cowpox scrapings to vaccinate humans against smallpox.

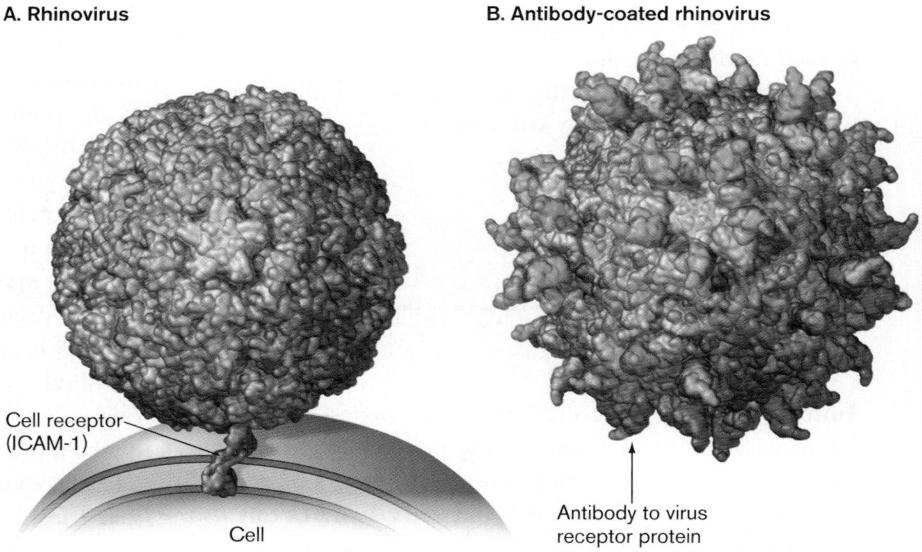
A. Rhinovirus
B. Antibody-coated rhinovirus
Cell receptor
(ICAM-1)
Cell
Antibody to virus
receptor protein

FIGURE 24.5 ▪ Antibodies prevent rhinovirus attachment to cell receptors. **A.** The complex rhinovirus capsid is pictured here attaching to the cell-surface molecule ICAM-1 (intercellular adhesion molecule, shown in reddish brown). (PDB code: 1rhi) **B.** Rhinovirus coated with protective (neutralizing) antibodies (green) that block the ICAM-1 receptors on the virus. As a result, the virus fails to attach to and infect the host cell. (PDB code: 1RVF)

TABLE 24.1

Vaccines against viral and bacterial pathogens.

Disease	Vaccine	Vaccination recommended for:
Viral diseases		
Chickenpox	Attenuated strain (will still replicate)	Children 12–18 months
Hepatitis A	Inactivated virus (will not replicate)	Children 12 months
Hepatitis B	Viral antigen	Newborns
Influenza	Inactivated virus or antigen	Everyone, after 6 months old, yearly
Measles, mumps, rubella (MMR)	Attenuated viruses; MMR combined vaccine	Children 12 months
Polio	Inactivated (injection, Salk)	Children 2–3 months
Rabies	Inactivated virus	Persons in contact with wild animals
Yellow fever	Attenuated virus	Military personnel
Bacterial diseases		
Anthrax	*Bacillus anthracis,* toxin components; unencapsulated strain	Agricultural and veterinary personnel; key health care workers
Cholera	Killed *Vibrio cholerae,* toxin components	Travelers to endemic areas
Diphtheria	Toxoid (inactivated toxin)	Children 2–3 months
Pertussis	Acellular *Bordetella pertussis*	Children 2–3 months
Tetanus	Toxoid	Children 2–3 months
Haemophilus influenzae type b (Hib) (meningitis)	Bacterial capsular polysaccharide	Children under 5 years
Lyme disease	*Borrelia burgdorferi,* lipoproteins OspA and OspC surface antigens	Canines; human vaccine discontinued
Meningococcal disease	*Neisseria meningitidis,* bacterial capsular polysaccharides	Children >2 years; adults >50 years
Pneumococcal pneumonia	*Streptococcus pneumoniae,* bacterial capsular polysaccharides	Children; adults >50 years
Tuberculosis (*Mycobacterium tuberculosis*)	Attenuated *Mycobacterium bovis* [BCG (Bacille Calmette-Guérin) vaccine]	Exposed individuals
Typhoid fever	Killed *Salmonella* Typhi	Individuals in endemic areas
Typhus	Killed *Rickettsia prowazekii*	Medical personnel in endemic areas; scientists; discontinued

strain (**Fig. 24.5B**) but will not bind an antigenically distinct ICAM-1 receptor protein from a different strain. A key to one lock will not work in a different lock.

Thought Question

24.2 How does a neutralizing antibody that recognizes a viral coat protein prevent infection by the associated virus?

Antigens and Immunogens

Antigens that, by themselves, can elicit antibody production are called **immunogens**. Molecules of molecular weight less than 1,000, however, are generally not immunogenic, because they do not bind MHC molecules. Nevertheless, these small molecules, called **haptens** (from the Greek word meaning "to fasten" and the German word for "stuff"), will elicit the production of specific antibodies if they are covalently attached to a larger carrier protein or other molecule (**Fig. 24.6**). Haptens can be thought of as small, incomplete antigens. An example of a hapten is the antibiotic penicillin, a serious cause of immune hypersensitivity reactions in some individuals (see Section 24.8).

Consider the following; a protein carrier like bovine serum albumin (BSA) injected into a mouse elicits antibodies that react against BSA (**Fig. 24.6**). BSA antigen is an

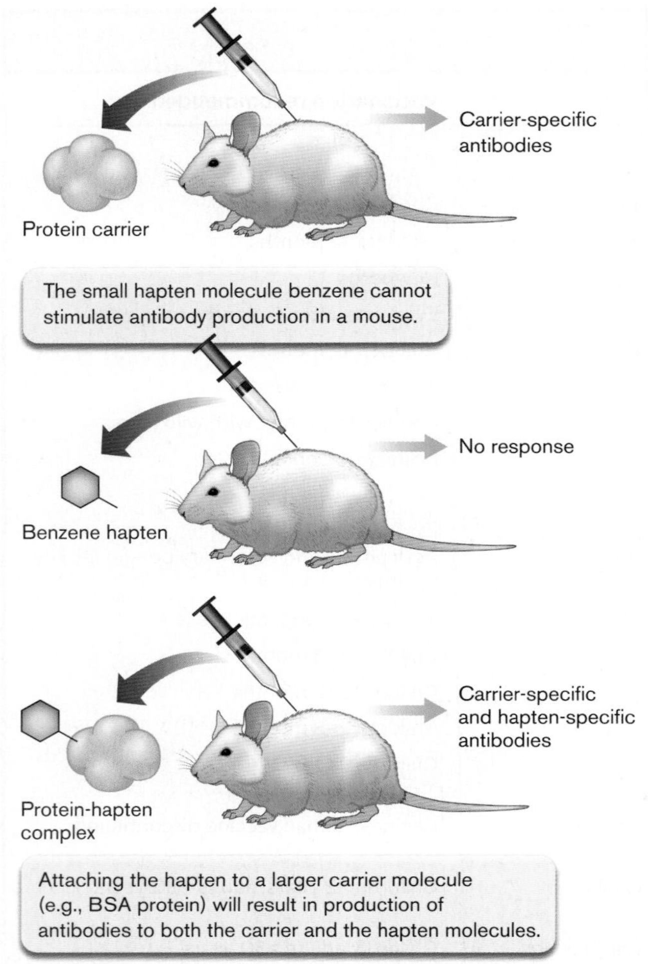
Carrier-specific antibodies
Protein carrier
The small hapten molecule benzene cannot stimulate antibody production in a mouse.
No response
Benzene hapten
Carrier-specific and hapten-specific antibodies
Protein-hapten complex
Attaching the hapten to a larger carrier molecule (e.g., BSA protein) will result in production of antibodies to both the carrier and the hapten molecules.

FIGURE 24.6 ▪ This basic schematic shows how haptens can elicit antibody production.

immunogen because it elicits an immune response to itself. In contrast, a mouse injected with the hapten benzene sulfate fails to produce antibodies to benzene sulfate. However, when the hapten chemical is attached to BSA and the hapten-BSA complex is injected, the mouse produces not only antibodies that react to the carrier (BSA), but also other antibodies that react against the benzene hapten. The reason for this carrier effect will become evident later in the chapter, when we describe how antigens are processed by cells of the immune system. Thus, antigens include immunogens that elicit an immune response by themselves, and haptens that must be attached to an immunogen in order to generate an immune response. Karl Landsteiner uncovered the antigen-immunogen-hapten relationship in the early 1900s, along with discovering the ABO blood group system that first defined the concept of immunological specificity (eTopic 24.2).

Thought Question

24.3 The attachment proteins of different rhinovirus strains all bind to ICAM-1. How can all these proteins be immunologically different if they find the same target (ICAM-1)? Why won't antibodies directed against one rhinovirus strain block the attachment of other rhinovirus strains?

To Summarize

- **Proteins are better immunogens** than nucleic acids and lipids because proteins have more diverse chemical forms.
- **Antigen-presenting cells (APCs)**, such as phagocytes, degrade microbial pathogens and present distinct pieces on their cell-surface MHC proteins.
- **Immunological specificity** means that antibody made to one epitope will not bind to different epitopes (although some weak cross-binding to similar epitopes can happen; for example, antibody to cowpox virus will bind to a similar epitope on smallpox virus).
- **A hapten** is a small compound that must be conjugated to a larger carrier antigen to elicit the production of an antibody.

24.3

Antibody Structure and Diversity

What are antibodies, and why are they important? Antibodies, also called **immunoglobulins**, are members of the larger immunoglobulin superfamily of proteins. Antibodies are the keys to immunological specificity. They are glycoproteins made by the body in response to an antigen. Members of the immunoglobulin superfamily of proteins have in common a 110-amino-acid domain with an internal disulfide bond. The immunoglobulin superfamily includes antibodies and other important binding proteins, such as the major histocompatibility proteins and B-cell receptors described later.

Note: Although antibodies are immunoglobulins, not all immunoglobulins are antibodies.

Like miniature "smart bombs," antibody immunoglobulins individually circulate through blood, ignoring all antigens except those for which they were designed. When

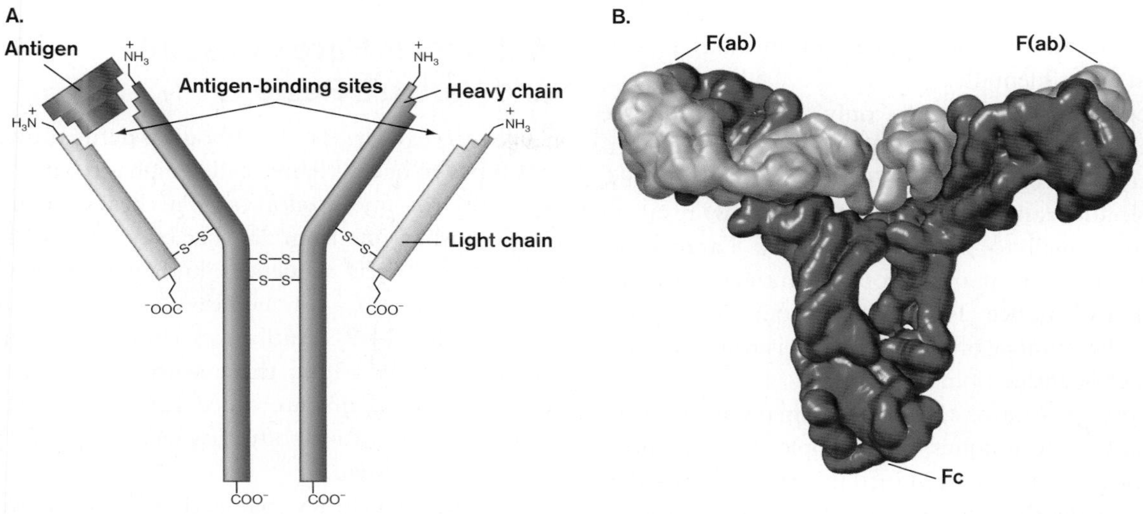
A.
Antigen
Antigen-binding sites
Heavy chain
Light chain
B.
F(ab)
F(ab)
Fc

FIGURE 24.7 ▪ Basic antibody structure. A. Each antibody contains two heavy chains and two smaller, light chains held together by disulfide bonds. The Y-shaped structure contains two antigen-binding sites, one at each arm of the molecule. The two antigen-binding sites are formed by the amino-terminal regions of the heavy- and light-chain pairs. **B.** The 3D structure of an antibody. The heavy chains are shown in blue, the light chains in yellow. The F(ab) regions represent the antigen-binding sites. The Fc portion points downward and is used to attach the antibody to different cell-surface molecules. (PDB code: 1R70)

an antibody finds its antigenic match, it binds to the antigen and initiates several events designed to destroy the target. Antibodies, in addition to being free-floating, are strategically situated on the surfaces of B cells, where they enable the B lymphocytes to recognize specific antigens (B-cell receptors are discussed in Section 24.4).

A typical antibody consists of four polypeptide chains. There are two large **heavy chains** and two smaller **light chains** (**Fig. 24.7**). The four polypeptides combine to form a Y-shaped tetrameric structure held together by disulfide bonds. Two bonds connect the two identical heavy chains to each other. One light chain is then attached near its carboxyl end to the middle of each heavy chain by a single disulfide bond. The antigen-binding sites are formed at the amino-terminal ends of the light and heavy chains. One antibody molecule possesses two identical antigen-binding sites, one on each "arm" of the molecule.

Antibody Structure Enables Immunoprecipitation

A typical antibody molecule has two antigen-binding sites, each of which binds identical antigens. Because of this ability to bind to more than one antigen molecule, antibodies can cross-link antigens in solution, ultimately forming complexes that are too large to remain soluble, so they fall out of solution (**Fig. 24.8**). The phenomenon, called **immunoprecipitation**, is normally observed only in vitro,

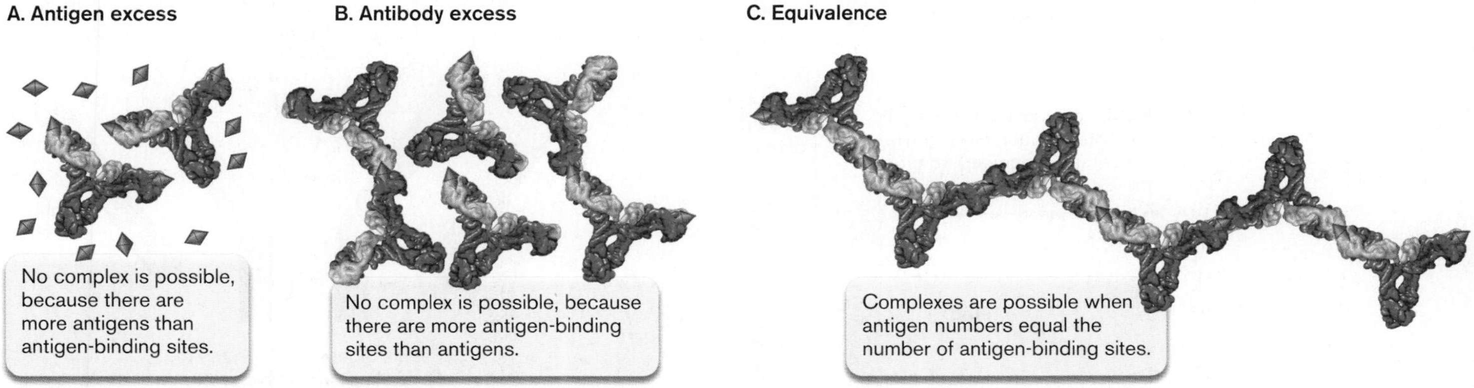
A. Antigen excess
B. Antibody excess
C. Equivalence
No complex is possible, because there are more antigens than antigen-binding sites.
No complex is possible, because there are more antigen-binding sites than antigens.
Complexes are possible when antigen numbers equal the number of antigen-binding sites.

FIGURE 24.8 ▪ Basis of immunoprecipitation. Only when the numbers of epitopes and antigen-binding sites are roughly equivalent **(C)** will a large complex form and fall out of solution.

where the concentration of antigen and antibody can be manipulated experimentally.

Immunoprecipitation occurs only with appropriate ratios of antigen and antibody molecules. Too many antigen molecules (antigen excess; **Fig. 24.8A**) or too few antigen molecules (antibody excess; **Fig. 24.8B**) result in complexes too small to immunoprecipitate. Large complexes are formed only at an appropriate antigen/antibody ratio, called **equivalence** (**Fig. 24.8C**). Equivalence is the point where the number of antigenic sites is roughly equal to the number of antigen-binding sites.

Immunoprecipitation is the basis for many experimental immunological techniques. For example, the concentration of an antigen can be determined in vitro, or antibodies can be used to identify and remove specific antigens from a complex mixture because the specific antigen-antibody complex precipitates. Appendix 3 describes some experimental uses for immunoprecipitation, such as radial immunodiffusion and western blotting.

Antibodies Have Constant and Variable Regions

There are five classes of antibodies, defined by five different types of heavy chains, called alpha (α), mu (μ), gamma (γ), delta (δ), and epsilon (ε). The heavy-chain classes are distinguished one from another by regions of highly conserved amino acid sequences, known as **constant regions** (denoted C_H or C_L for the heavy and light chains, respectively) (**Fig. 24.9**). Antibodies containing gamma heavy chains are called IgG; those with alpha, delta, mu, and epsilon heavy chains are called IgA, IgD, IgM, and IgE, respectively. Each antibody class serves a specific purpose in the immune system.

In contrast to heavy chains, there are only two classes of light chains—kappa (κ) and lambda (λ)—which are defined by their own constant regions (C_L). A single antibody of any heavy-chain class (for example, IgG) may contain two kappa light chains or two lambda light chains, but never

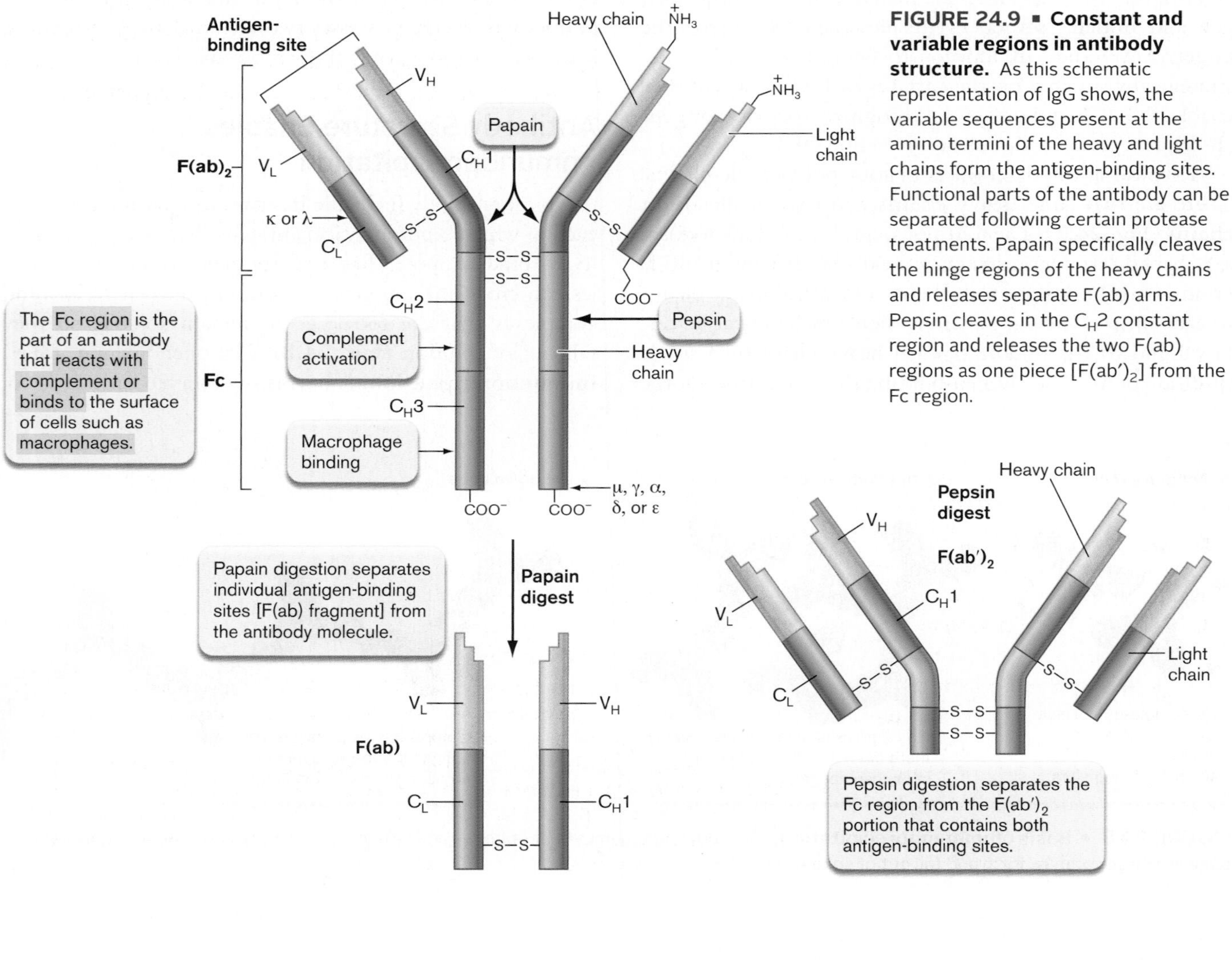
Antigen-binding site
Heavy chain
$\overset{+}{N}H_3$
V_H
$\overset{+}{N}H_3$
Papain
Light chain
$F(ab)_2$
V_L
C_H1
κ or λ
C_L
S–S
S–S
S–S
S–S
COO^-
C_H2
Pepsin
The Fc region is the part of an antibody that reacts with complement or binds to the surface of cells such as macrophages.
Complement activation
Heavy chain
Fc
C_H3
Macrophage binding
μ, γ, α, δ, or ε
COO^-
COO^-
Papain digestion separates individual antigen-binding sites [F(ab) fragment] from the antibody molecule.
Papain digest
V_L
V_H
F(ab)
C_L
C_H1
S–S
Heavy chain
Pepsin digest
V_H
$F(ab')_2$
C_H1
V_L
Light chain
C_L
S–S
S–S
S–S
S–S
Pepsin digestion separates the Fc region from the $F(ab')_2$ portion that contains both antigen-binding sites.

FIGURE 24.9 ■ Constant and variable regions in antibody structure. As this schematic representation of IgG shows, the variable sequences present at the amino termini of the heavy and light chains form the antigen-binding sites. Functional parts of the antibody can be separated following certain protease treatments. Papain specifically cleaves the hinge regions of the heavy chains and releases separate F(ab) arms. Pepsin cleaves in the C_H2 constant region and releases the two F(ab) regions as one piece [$F(ab')_2$] from the Fc region.

one of each. Two-thirds of all antibody molecules carry kappa chains; the rest have lambda chains.

The antigen-binding part of an antibody is formed by highly variable amino acid sequences situated at the amino-terminal ends of the light and heavy chains. These **variable regions** are referred to as the V_L and V_H regions, respectively (**Fig. 24.9**). The rest of each immunoglobulin chain is composed of the highly conserved constant regions. C_H1, C_H2, and C_H3 denote the three different constant regions in each heavy chain. Section 24.5 discusses the genes that code for these regions and the combinatorial possibilities underlying the formation of different antibody molecules.

In addition to the two "arms" that bind antigen, every antibody contains a "tail" called the **Fc region** (**Fig. 24.9**). The Fc region is not involved in antigen recognition but is important for anchoring antibodies to the surface of certain host cells and for binding components of the complement system.

Thought Question

24.4 Can an $F(ab')_2$ antibody fragment prevent the binding of rhinovirus to the ICAM-1 receptor on host cells? And can an $F(ab')_2$ antibody fragment facilitate phagocytosis of a microbe?

Isotypes, Allotypes, and Idiotypes

Antibodies differ from each other in both large and small ways. We describe antibody diversity in terms of the amino acid differences that distinguish antibodies between mammalian species (**isotypes**), between different individuals within a species (**allotypes**), and between different antibodies within an individual (**idiotypes**).

"Isotype" refers to the amino acid differences in the constant regions of heavy-chain classes (what makes IgG different from IgE). These differences will be the same in all members of a species (for example, *Homo sapiens*). Antibody "allotype" refers to any amino acid sequence differences in either heavy or light chains that make IgG from one person different from IgG in another person. Idiotype differences are even more subtle. "Idiotype" refers to amino acid sequence differences between two molecules of the same isotype (IgG, for example) in a single person. Idiotype differences usually occur in the antigen-binding region of an antibody.

In real terms, consider two different people, John and Sherrie. The blood of both John and Sherrie carries the human IgG isotype. However, all of the IgG molecules circulating in John have allotypic amino acid differences from the IgG antibodies circulating in Sherrie (John and Sherrie have different IgG allotypes) (**Fig. 24.10**). At another level, the IgG molecules within Sherrie that bind to epitope A possess idiotypic differences from her IgG molecules that bind epitope B.

Isotype, allotype, and idiotype differences in antibodies are important because they reflect different levels of antibody diversity. Each individual has five antibody isotypes: IgG, IgA, IgD, IgE, and IgM. Each isotype in an individual has several allotypic differences compared to the same isotype in another individual. Finally, a single antibody allotype in a person will contain millions of idiotypic differences that reflect antigen specificity. Idiotypic differences, by and large, occur in the antigen-binding sites of the various antibodies of a single individual.

Note that antibodies are proteins too, and are themselves antigens. Thus, the amino acid differences found within a single class of antibody—IgG, for example—also represent different antigenic epitopes within that class. Isotypic, allotypic, and idiotypic differences, discussed already, define the epitopes in antibody molecules that can be detected using the immunological techniques described in Appendix 3.

Thought Question

24.5 What types of antibodies will IgG taken from Sherrie raise when injected into John (for example, anti-isotype, anti-allotype, or anti-idiotype)?

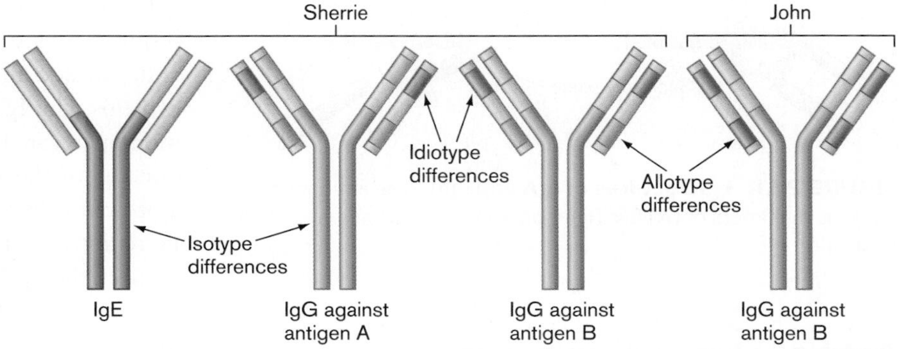
Sherrie
John
Idiotype differences
Allotype differences
Isotype differences
IgE
IgG against antigen A
IgG against antigen B
IgG against antigen B

FIGURE 24.10 ■ Isotype, idiotype, and allotype differences on antibodies. Colors indicate different amino acid sequences (or epitopes). Isotype differences occur in heavy-chain constant regions within the same individual (any IgE versus any IgG from Sherrie in this case). Idiotype differences occur in the antigen-binding regions in antibodies directed against different antigens (differences in Sherrie's IgG molecules directed against antigen A versus antigen B). Allotype differences occur in the light-chain or heavy-chain constant regions of different people (for example, differences in anti-B IgG antibodies taken from John compared to Sherrie).

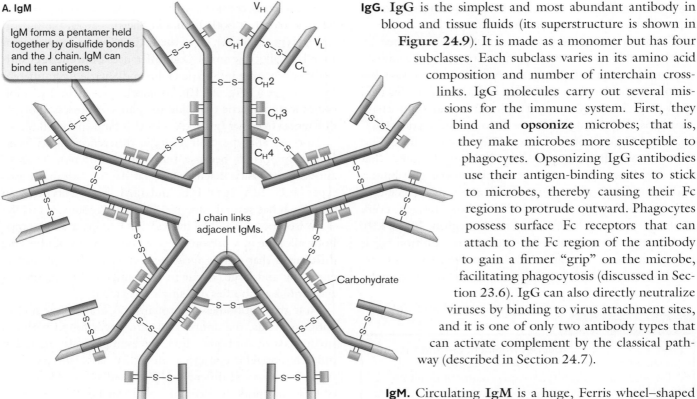

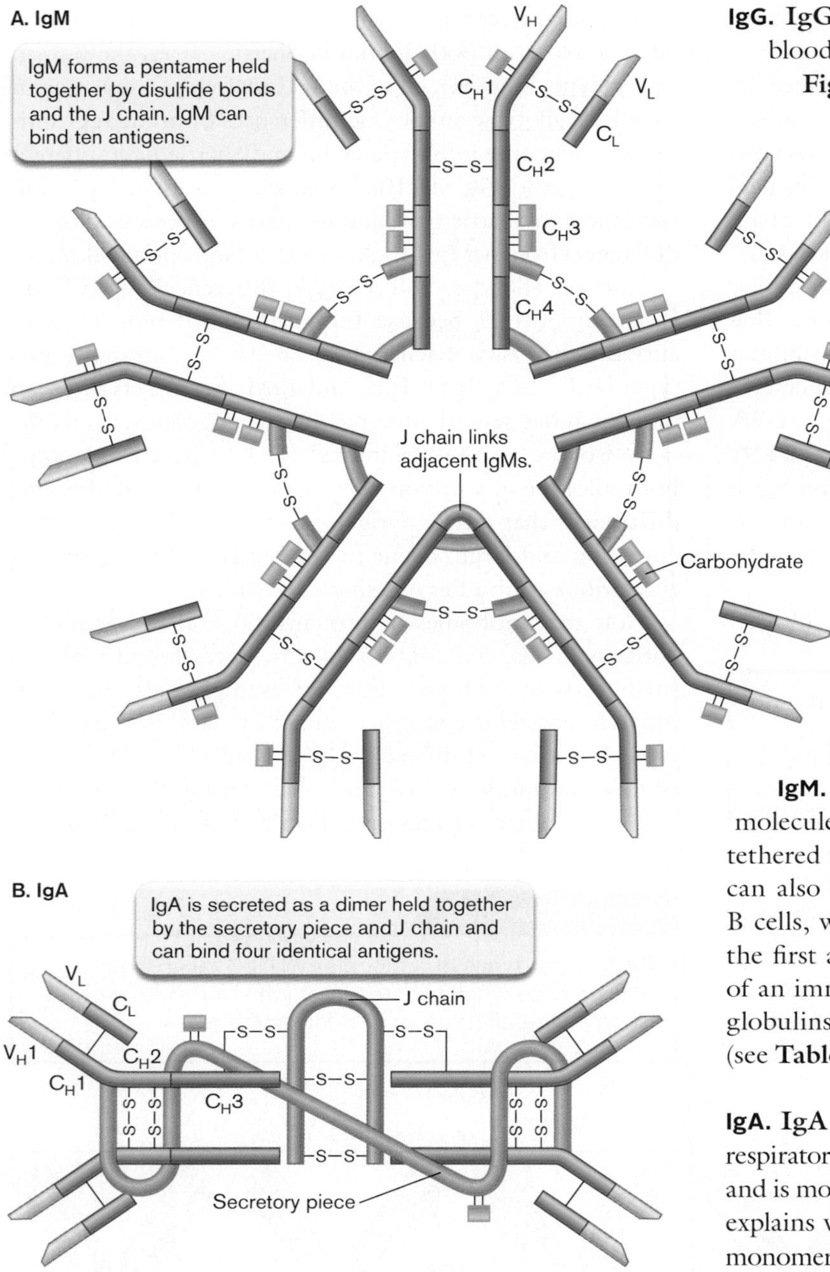
A. IgM
IgM forms a pentamer held together by disulfide bonds and the J chain. IgM can bind ten antigens.
V_H
V_L
C_H1
C_L
C_H2
C_H3
C_H4
S–S
J chain links adjacent IgMs.
Carbohydrate
B. IgA
IgA is secreted as a dimer held together by the secretory piece and J chain and can bind four identical antigens.
V_L
C_L
V_H1
C_H1
C_H2
C_H3
J chain
S–S
Secretory piece

FIGURE 24.11 ▪ **Structures of IgA and IgM.** The antibodies are made as multimers of **(A)** five (IgM) or **(B)** two (IgA) immunoglobulin molecules.

Antibody Isotype Functions and "Super" Structures

All antibody isotypes have the same basic structure. However, each isotype has a unique <u>super</u>structure (for example, monomer or dimer), and each is designed to carry out a different task. Some key properties of the five different immunoglobulin classes are listed in **Table 24.2**.

IgG. IgG is the simplest and most abundant antibody in blood and tissue fluids (its superstructure is shown in **Figure 24.9**). It is made as a monomer but has four subclasses. Each subclass varies in its amino acid composition and number of interchain cross-links. IgG molecules carry out several missions for the immune system. First, they bind and **opsonize** microbes; that is, they make microbes more susceptible to phagocytes. Opsonizing IgG antibodies use their antigen-binding sites to stick to microbes, thereby causing their Fc regions to protrude outward. Phagocytes possess surface Fc receptors that can attach to the Fc region of the antibody to gain a firmer "grip" on the microbe, facilitating phagocytosis (discussed in Section 23.6). IgG can also directly neutralize viruses by binding to virus attachment sites, and it is one of only two antibody types that can activate complement by the classical pathway (described in Section 24.7).

IgM. Circulating **IgM** is a huge, Ferris wheel–shaped molecule formed from five monomeric immunoglobulins tethered together by the J-chain protein (**Fig. 24.11B**). It can also be found in monomeric form on the surfaces of B cells, where it forms part of the B-cell receptor. IgM is the first antibody isotype detected during the early stages of an immune response. Unlike the smaller IgG immunoglobulins, IgM is so large that it cannot cross the placenta (see **Table 24.2**).

IgA. IgA is secreted across mucosal surfaces (linings of the respiratory, gastrointestinal, urinary, and reproductive tracts) and is most commonly found as a dimer (**Fig. 24.11A**). This explains why IgA can bind four molecules of antigen (each monomer can bind two molecules of antigen). The components of the IgA dimer are linked by disulfide bonds to a protein called the J chain, which joins two IgAs by their Fc regions. A sixth protein, the secretory piece, is wrapped around the IgA dimer during the secretion process. The secreted molecule, now called sIgA (secretory IgA), is found in tears, breast milk, and saliva, and on other mucosal surfaces. The molecule sIgA is important for mucosal immunity against pathogens that infect mucosal linings.

IgD. IgD and IgE, the last two antibody isotypes, are present at very low levels in the blood. IgD is a monomer that can neither bind complement nor cross the placenta. IgD molecules, however, are abundant on the surface of B cells.

TABLE 24.2

Properties of human immunoglobulins.

	IgG							
Property	**IgG1**	**IgG2**	**IgG3**	**IgG4**	**IgA**	**IgM**	**IgD**	**IgE**
Serum half-life (days)	21	20	7	21	6	10	3	2
% Total serum Ig	**70%**				15%–20%	5%–10%	0.2%	0.002%
Antigen-binding sites	2				2–4	5–10	2	2
Produced by fetus	Poor, if at all				Poor, if at all	Yes	?	Poor, if at all
Transmitted across placenta	Yes				No	No	No	No
Binds complement	Yes				No	Yes	No	No
Opsonizing	Yes				No	No	No	No
Binds mast cells	No				No	No	No	Yes

Attached to B-cell surfaces by Fc regions, IgD, along with monomeric IgM, can bind antigen and signal B cells to differentiate and make antibody.

IgE. IgE, also present in only trace amounts in the blood, is found more prominently bound to the surfaces of mast cells and basophils, where it has potent biological activity. Mast cells and basophils contain granules loaded with inflammatory mediators. The primary role of IgE is to amplify the body's response to invaders. Once secreted into serum, IgE attaches to mast cells (**Fig. 24.12A** and **B**), again by way of its Fc region, and, like a Venus flytrap, waits until its matched antigen binds to its antigen-binding site. When surface IgE molecules of two mast cells are cross-linked by antigen, a signal is sent internally that triggers degranulation (see Section 24.8). The subsequent release of histamine and other pharmacological mediators from these granules helps orchestrate the acute inflammation that takes place during early host responses to microbial infection (that is, while the antibody response is gearing up). The system also causes severe allergic hypersensitivities (such as anaphylaxis) and milder forms like hay fever (**Fig. 24.12C**).

To Summarize

- **Antibodies, or immunoglobulins**, are members of the immunoglobulin superfamily.
- **Antibodies are Y-shaped** molecules that contain two heavy chains and two light chains.

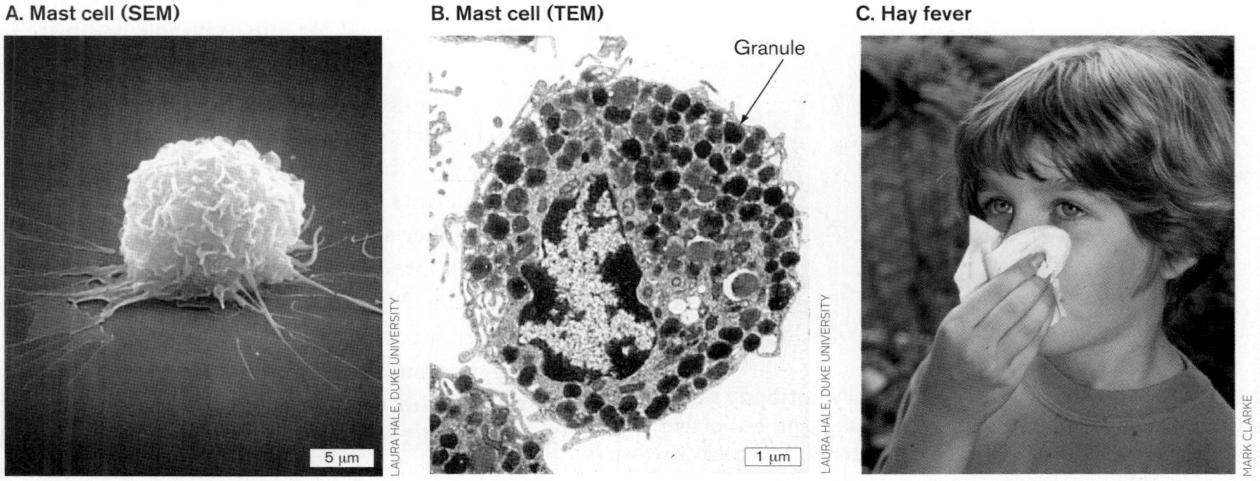
A. Mast cell (SEM)
5 µm
LAURA HALE, DUKE UNIVERSITY
B. Mast cell (TEM)
Granule
1 µm
LAURA HALE, DUKE UNIVERSITY
C. Hay fever
MARK CLARKE

FIGURE 24.12 ▪ **Mast cells. A.** Scanning EM of a mast cell. **B.** Transmission EM showing granules (arrow). **C.** Hay fever is the result of degranulation of IgE-coated mast cells, which release histamine and other pharmacological mediators.

- **There are five classes (isotypes) of heavy chains.** Each antibody isotype is defined by the structure of the heavy chain.
- **Each antibody molecule contains two antigen-binding sites.** Each binding site is formed by the hypervariable ends of a heavy- and light-chain pair.
- **The Fc portion** ("tail") of an antibody can bind to specific receptors on host cells. This binding is antigen independent.

24.4

Primary and Secondary Antibody Responses

Once you have been infected with a microorganism or have been given a vaccine, what happens? After a lag period of several days, antibodies begin to appear in the **serum** (the fluid that remains after blood clots). During the lag period, a series of molecular and cellular events causes a distinct subset of B cells (located in lymph nodes and spleen) to proliferate and differentiate into antibody-secreting **plasma cells** and **memory B cells**. Each B cell is genetically programmed to make antibodies to one antigen or epitope—a process called the **primary antibody response**, the events of which will be discussed later in the chapter.

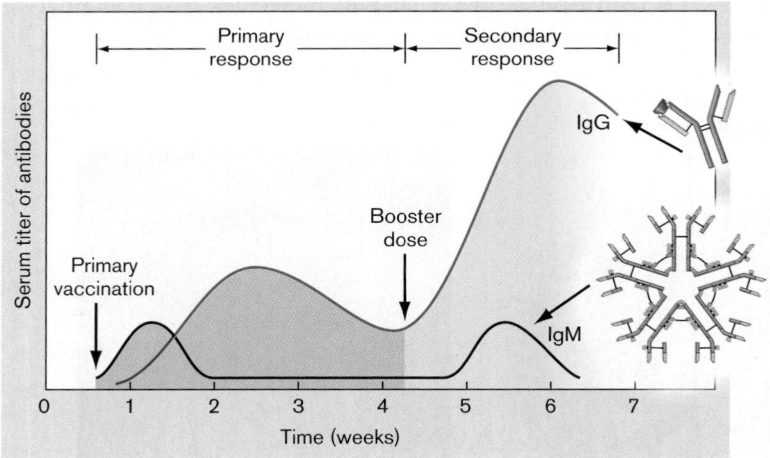

Primary response
Secondary response
Serum titer of antibodies
IgG
Booster dose
Primary vaccination
IgM
0
1
2
3
4
5
6
7
Time (weeks)

FIGURE 24.13 ■ Primary versus secondary antibody response. Primary vaccination or infection leads to the early synthesis of IgM followed by IgG. Reinfection or a second, booster dose of a vaccine results in a more rapid antibody response consisting mainly of IgG due to memory B cells formed during the primary response. Note that the time course and level of antibody made vary with the immunogen and the host.

A secondary exposure to the antigen, which can take place months or years after the initial encounter, will trigger a rapid, almost instantaneous increase in the production of antibodies and is called the **secondary antibody response** (**Fig. 24.13**). This quick response occurs thanks to the memory B cells formed during the primary response. Once stimulated, memory B cells rapidly differentiate into plasma cells and secrete antibody. Plasma cells are much larger than B cells because of an enormous increase in protein synthesis and secretion machinery.

The net result of the primary antibody response is the early synthesis and secretion of pentameric IgM molecules specifically directed against the antigen (also called the immunogen). Later during the primary response, a process known as **isotype switching** (**class switching**) occurs and the predominant antibody type produced becomes IgG rather than IgM (discussed shortly). Antibodies made during the primary response, while specific for the immunogen, are actually not of the highest affinity. Later responses by memory B cells increase antibody affinity (discussed in the next section).

As the immunogen is cleared from the body, the levels of both IgG and IgM decline because the plasma cells that produced them die. Plasma cells have an average life span of only 100 days. However, the immune system has been primed to respond more aggressively to the immunogen, should it encounter it again. A rapid response to a second encounter is possible because memory B cells produced during the primary response are maintained in lymph nodes and continue to divide. If a person later encounters the same antigen, these memory B cells quickly proliferate and differentiate into plasma cells with no lag phase. Thus, memory B cells quickly initiate the secondary antibody response (or anamnestic response, from the Greek *anamnesis,* meaning "remembrance"). Memory B cells comprise approximately 40% of the B-cell population.

During the secondary response, memory B cells that have undergone isotype switching from IgM to IgG become plasma cells that secrete copious amounts of IgG antibody. These antibodies have a higher specificity for the antigen than the antibodies produced during the primary response (see "Making Memory B Cells" in Section 24.5). The higher specificity results from hypermutation, resulting in some plasma cells that produce antibodies which bind their antigens more strongly than those produced by the clonal ancestor. Small amounts of IgM are also produced from the few memory cells that did not undergo isotype switching during the primary response.

The secondary antibody response is why vaccinations work. Before vaccinations, the only way to be protected from an infection by a given pathogen was to have already been infected by that pathogen. Memory cells made during the primary response will protect you against a second infection, but pathogens can do considerable harm during the lag phase of the primary response. To avoid this harm, an innocuous version of a pathogen, or a harmless piece of it, can be injected into a person to trigger a primary response without producing disease (or, at worst, producing only a mild form of the illness). Immunization thus primes the immune system to respond efficiently and without delay upon encountering the real pathogen. **Table 24.1** lists a variety of viral and bacterial diseases for which immunizations are available.

Thought Question

24.6 The mother of a newborn was found to be infected with rubella, a viral disease. Infection of the fetus could lead to serious consequences for the newborn. How could you determine whether the newborn was infected while in utero?

B Cells Differentiate into Plasma Cells by Clonal Selection

As mentioned earlier, each B cell circulating throughout the body or ensconced in a lymphoid organ is programmed to synthesize antibody that reacts with a single epitope (for example, a small portion of a protein). In a process called **clonal selection**, an invading antigen will inadvertently select which B-cell clone will proliferate to large numbers and differentiate into antibody-producing plasma cells or memory B cells. In this way, large amounts of antibody specific for the antigen are made. The mechanism of clonal selection begins with the antigen binding to a matching B cell (a B cell preprogrammed to bind to that antigen; see the discussion of B-cell receptors that follows) (**Fig. 24.14**).

Mature but naive B cells (those that have not previously encountered antigen) can produce only IgM and IgD, which have identical antigen specificities. These two antibody classes are displayed like tiny satellite dishes on the B-cell surface, anchored by their Fc regions through hydrophobic transmembrane segments. These surface antibodies are the keys to stimulating the proliferation and differentiation of B cells into antibody-producing plasma cells or memory B cells. Upon binding to its corresponding antigen via these surface antibodies, the B cell is said to become activated, whereby it multiplies and differentiates into a plasma cell that ultimately synthesizes only one antibody isotype (for example, IgG1 or IgA2). Clonal selection has begun. In addition to antigen binding, most B cells also require help from T cells to become plasma cells and memory B cells (discussed later).

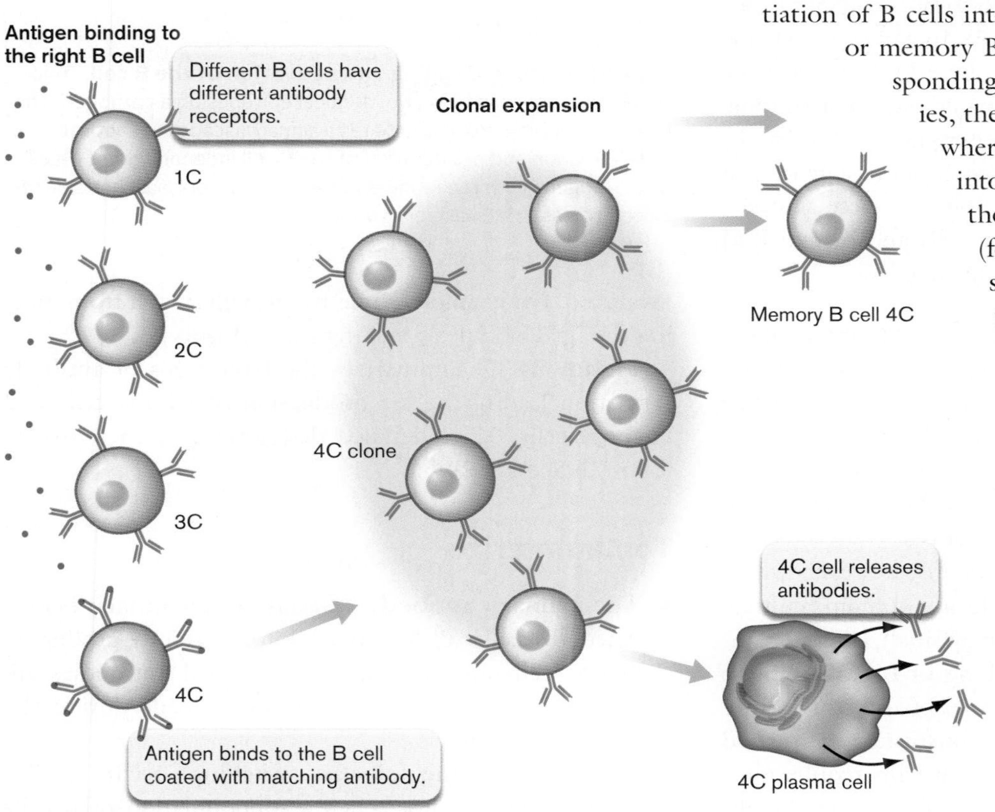
Antigen binding to the right B cell
Different B cells have different antibody receptors.
Clonal expansion
1C
2C
3C
4C
4C clone
Memory B cell 4C
4C cell releases antibodies.
Antigen binds to the B cell coated with matching antibody.
4C plasma cell

FIGURE 24.14 ■ Clonal selection theory. The B-cell population is composed of individuals that have specificity for different antigens. When a B cell contacts its cognate antigen, an intracellular signal is generated, leading to proliferation and differentiation of that clone (clonal expansion). Plasma cells and memory B cells result.

B-cell receptor. Each antibody bound to a B-cell membrane is associated with two other membrane proteins, called Igα and Igβ (these are not immunoglobulins, but they are designated "Ig" because they associate with the surface antibody). The complex is called the **B-cell receptor**, or **BCR** (**Fig. 24.15**). Each B cell may have upwards of 50,000 B-cell receptors. A microbe generally has multiple copies of the same epitope on its surface (think

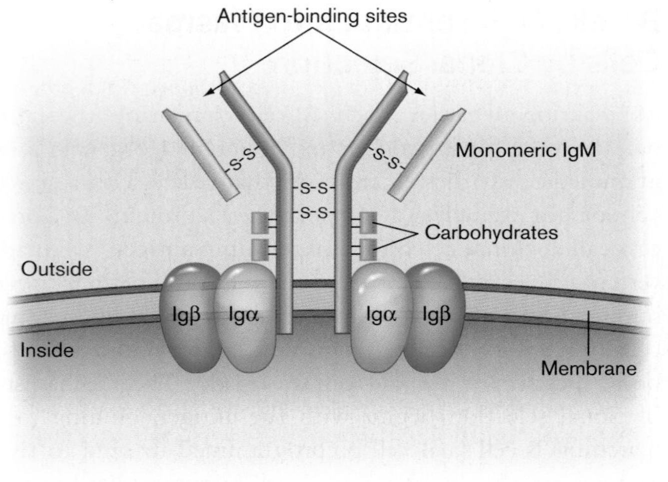
Antigen-binding sites
Monomeric IgM
S-S
S-S
S-S
S-S
Carbohydrates
Outside
Igβ
Igα
Igα
Igβ
Inside
Membrane

FIGURE 24.15 ▪ B-cell receptor. The B-cell receptor is formed as a complex consisting of a monomeric IgM plus Igα and Igβ in the membrane. Igα and Igβ are not immunoglobulins themselves, but associate with the antibody immunoglobulin part of the receptor.

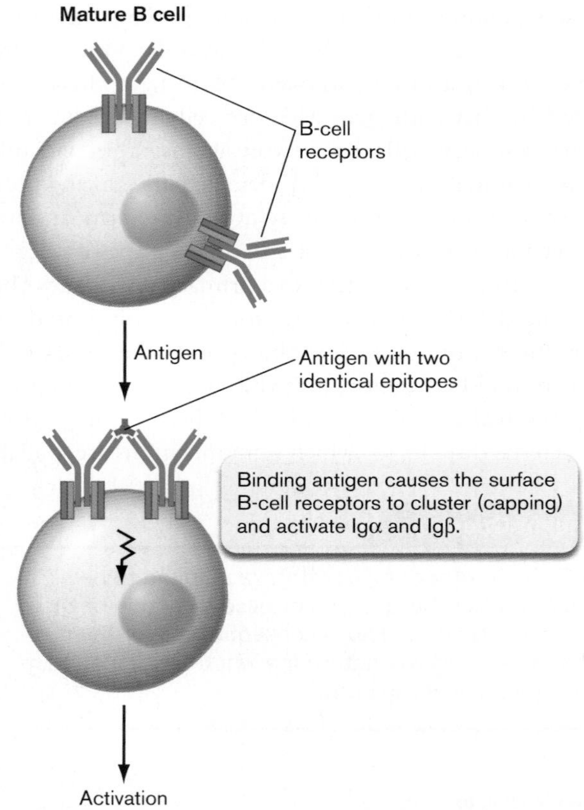
Mature B cell
B-cell receptors
Antigen
Antigen with two identical epitopes
Binding antigen causes the surface B-cell receptors to cluster (capping) and activate Igα and Igβ.
Activation

FIGURE 24.16 ▪ Capping and activation of the B cell. Two B-cell receptors can bind two identical epitopes on a pathogen. The resulting capping process initiates a signal cascade that activates differentiation and proliferation of the B cell independent of T-cell help. Antigens with repeating epitopes, such as polysaccharides, can directly cross-link B-cell receptors.

about a virus capsid). Each epitope can bind adjacent B-cell receptors on one B cell. Once bound, surface B-cell receptors begin to cluster in a process called capping. **Capping** activates Igα and Igβ to initiate a phosphorylation signal cascade directed into the nucleus (**Fig. 24.16**). In a phosphorylation cascade, a phosphate group donated by ATP is passed from one protein to another, usually ending up on, and activating, a transcriptional regulator. The transcription factors at the end of the BCR cascade stimulate transcription of genes that contribute to cellular proliferation. Some of these activated B cells can differentiate into plasma cells and secrete IgM antibody as part of the primary immune response.

T-cell-dependent and -independent antibody production. There are two routes by which antigens can stimulate B cells to differentiate into plasma cells. In one, called the T cell–independent route (described earlier), antigens that possess multiple repeating epitopes (for example, polysaccharide antigens) can directly cross-link B-cell receptors (the capping process)—a step necessary for triggering differentiation (see **Fig. 24.16**). Proteins, however, which are the largest group of antigens, do not contain multiple repeating units. Proteins possess many small, discrete, single epitopes, making the cross-linking of B-cell receptors difficult. B-cell responses to these types of antigens require help from specific T cells, and this constitutes the second route to B-cell activation. Thus, B cells usually require multiple signals to initiate a primary response. How T cells help foster B-cell activation will be discussed in Section 24.6. During B-cell activation, antibody heavy-chain isotype switching (class switching) can occur at the gene level. Class, or isotype, switching changes the class of antibodies produced. For example, some B cells will switch from making the IgM isotype to making the IgA isotype.

Figure 24.17 summarizes the basic steps of antibody formation leading to the production of plasma cells and memory cells. More detailed descriptions of these events follow in Section 24.5.

To Summarize

- **The primary antibody response** to an antigen begins when B cells differentiate into antibody-producing plasma cells and memory B cells. IgM antibodies are generally the first class of antibodies made during the primary response.
- **Isotype switching** (or class switching) occurs during the primary response when a subpopulation of B cells switches during differentiation from making IgM to making other antibody isotypes.
- **The secondary antibody response** occurs during subsequent exposures to an antigen and arises because

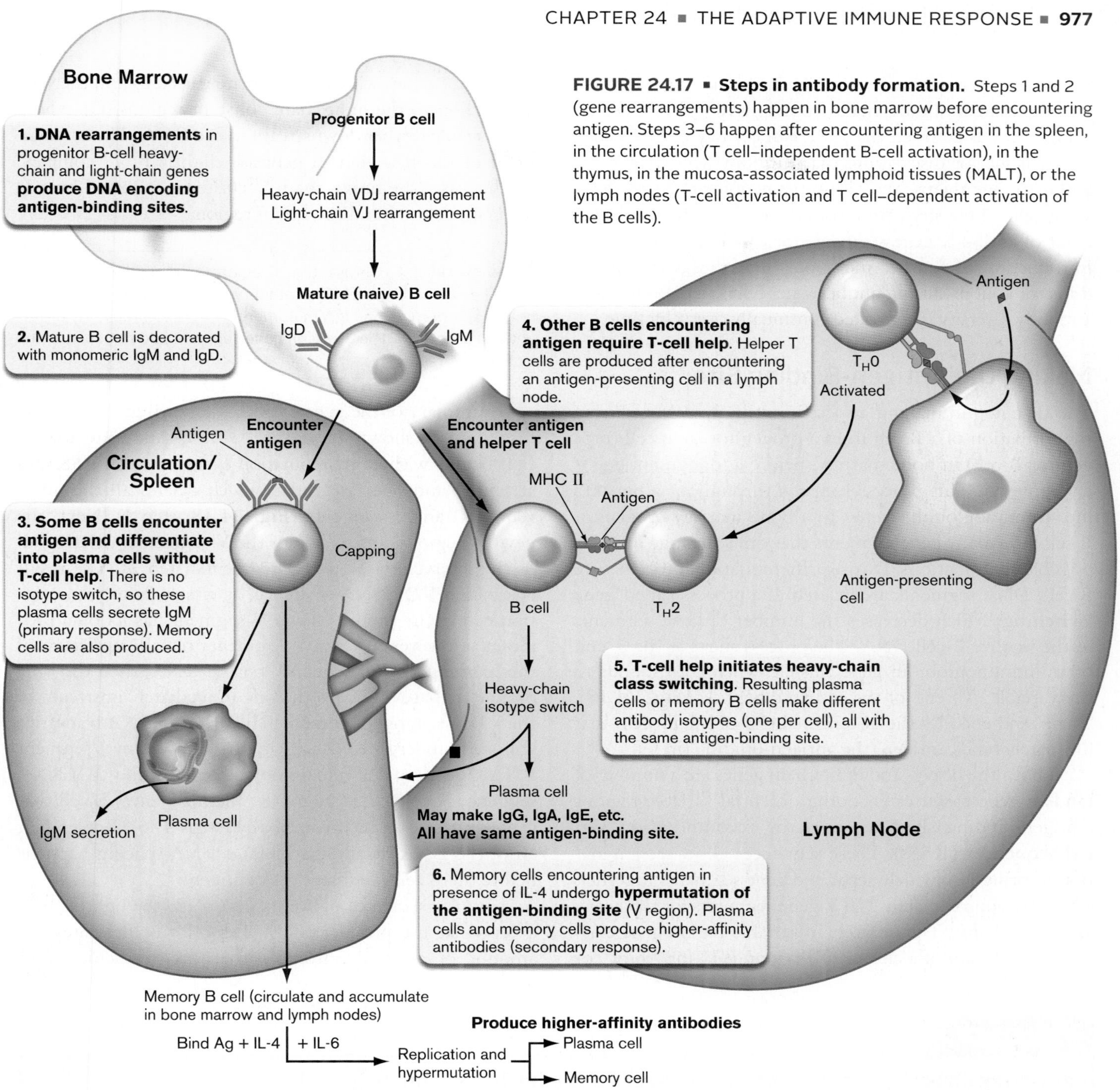

Bone Marrow
1. **DNA rearrangements** in progenitor B-cell heavy-chain and light-chain genes **produce DNA encoding antigen-binding sites**.
Progenitor B cell
Heavy-chain VDJ rearrangement
Light-chain VJ rearrangement
Mature (naive) B cell
2. Mature B cell is decorated with monomeric IgM and IgD.
IgD
IgM
Antigen
Encounter antigen
Circulation/ Spleen
3. **Some B cells encounter antigen and differentiate into plasma cells without T-cell help**. There is no isotype switch, so these plasma cells secrete IgM (primary response). Memory cells are also produced.
Capping
IgM secretion
Plasma cell
4. **Other B cells encountering antigen require T-cell help**. Helper T cells are produced after encountering an antigen-presenting cell in a lymph node.
Encounter antigen and helper T cell
MHC II
Antigen
B cell
T_H2
T_H0
Activated
Antigen
Antigen-presenting cell
Heavy-chain isotype switch
5. **T-cell help initiates heavy-chain class switching**. Resulting plasma cells or memory B cells make different antibody isotypes (one per cell), all with the same antigen-binding site.
Plasma cell
May make IgG, IgA, IgE, etc.
All have same antigen-binding site.
Lymph Node
6. Memory cells encountering antigen in presence of IL-4 undergo **hypermutation of the antigen-binding site** (V region). Plasma cells and memory cells produce higher-affinity antibodies (secondary response).
Memory B cell (circulate and accumulate in bone marrow and lymph nodes)
Bind Ag + IL-4
+ IL-6
Replication and hypermutation
Produce higher-affinity antibodies
Plasma cell
Memory cell

FIGURE 24.17 ■ Steps in antibody formation. Steps 1 and 2 (gene rearrangements) happen in bone marrow before encountering antigen. Steps 3–6 happen after encountering antigen in the spleen, in the circulation (T cell–independent B-cell activation), in the thymus, in the mucosa-associated lymphoid tissues (MALT), or the lymph nodes (T-cell activation and T cell–dependent activation of the B cells).

memory B cells are activated. IgG is the predominant antibody made.

- **Clonal selection** is the rapid proliferation of a subset of B cells during the primary or secondary antibody response.
- **A B-cell receptor** consists of a membrane-embedded antibody in association with the Igα and Igβ proteins. Binding of antigen to the B-cell receptor triggers B-cell proliferation and differentiation.

24.5

Genetics of Antibody Production

Before we discuss how T cells influence antibody production, let's jump ahead and look at how antibody diversity happens. It is estimated that each human can synthesize 10^{11} different antibodies. Given that each B cell displays antibodies to only one antigenic determinant, it follows that there are 10^{11} different B cells in the body. We have learned, however, that each person possesses only about 1,000 genes or gene segments involved in antibody formation. How are 10^{11} different antibodies made from only 10^3 genes? Susumu Tonegawa was awarded the 1987 Nobel Prize in Physiology or Medicine for discovering that antibody genes can move and rearrange themselves within the genome of a differentiating cell. Three steps are involved—namely, the rearrangement of antibody gene segments (or cassettes), the random introduction of somatic mutations, and the generation of different codons during antibody gene splicing. In humans, antibody diversity is generated continually over a lifetime.

Making the Antigen-Binding Site

The first step in making a specific antibody occurs during the formation of a B cell from a progenitor stem cell (progenitor B cell) in bone marrow, before a foreign antigen is encountered. This process happens throughout a person's life. Immunoglobulin genes in a bone marrow progenitor B cell have many gene segments that can rearrange in many possible combinations. During differentiation into a mature B cell, DNA segments are deleted in a process called **gene switching**, which decreases the number of gene segments in the mature B-cell DNA. The process starts at the 5′ end of an immunoglobulin gene cluster, which corresponds to the variable (V) end of the ultimate peptide (**Fig. 24.18**). The 5′ end encodes the N-terminal end of the protein that ultimately binds antigen (the antigen-binding site).

In both the heavy- and light-chain genes are a number of tandem gene cassettes encoding potential variable regions (antigen-binding sites) separated by **recombination signal sequences (RSSs)**. These sequences allow recombination to bring two widely separated gene segments together. There are approximately 170 V gene segments for the heavy and light chains. The light-chain variable-region gene cluster lies upstream of a cluster of J-region ("J" for "joint" or "joining") genes, which are ultimately used to join the variable region to the light-chain constant regions (C), whose genes reside farther downstream in the DNA. The arrangement of the heavy-chain genes is slightly more complex. In this case, the heavy-chain V cluster is followed by a D (diversity) cluster and then the J region.

Note: Do not confuse the J-region gene segments used to make heavy- and light-chain proteins with the J-chain protein that holds together IgM and IgA multimers. They are completely different and unrelated. The J-region gene segments do not encode the J chain.

A summary of the genetic processes leading to antibody formation is shown in **Figure 24.18**. Antibody formation begins with a recombination event between RSS sites at one D and one J segment, which deletes all the intervening D and J segments (**Fig. 24.18**, step 1). Next, the new DJ region joins to one of the V segments, deleting all of the intervening V and D segments (step 2). The result is a joined VDJ DNA sequence, at which point transcription can occur (step 3). Each V segment has its own promoter. However, if extra V segments remain upstream of the rearranged VDJ sequence on the DNA, then the only promoter that will fire is the one immediately upstream of the VDJ rearranged segment. The primary RNA transcript will then undergo RNA splicing to remove any J-segment RNA sequences that remain downstream of the VDJ RNA sequence (step 3). The result is a mature (naive) B cell that can synthesize a specific antibody (steps 4–6). Remember, all of the DNA recombination and RNA splicing happens before the B cell ever "sees" the antigen.

A similar sequence of events occurs for the light chains, except that the product is VJ. **Table 24.3** illustrates the amount of antibody diversity that can be achieved in

TABLE 24.3

Antibody diversity attributed to combinatorial joining in the human germ line.

	Number of:			
Chain type	**V regions**	**D regions**	**J regions**	**Number of combinations**
λ Light chains	30	0	4	$30 \times 4 = 120$
κ Light chains	40	0	5	$40 \times 5 = 200$
Heavy chains	100	27	6	$100 \times 27 \times 6 = 16{,}200$
Number of possible antibodies	16,200 heavy-chain combinations × 120 λ-chain combinations = 1.94×10^6			
	16,200 heavy-chain combinations × 200 κ-chain combinations = 3.24×10^6			
	$1.94 \times 10^6 + 3.24 \times 10^6 = 5.18 \times 10^6$ combinations			

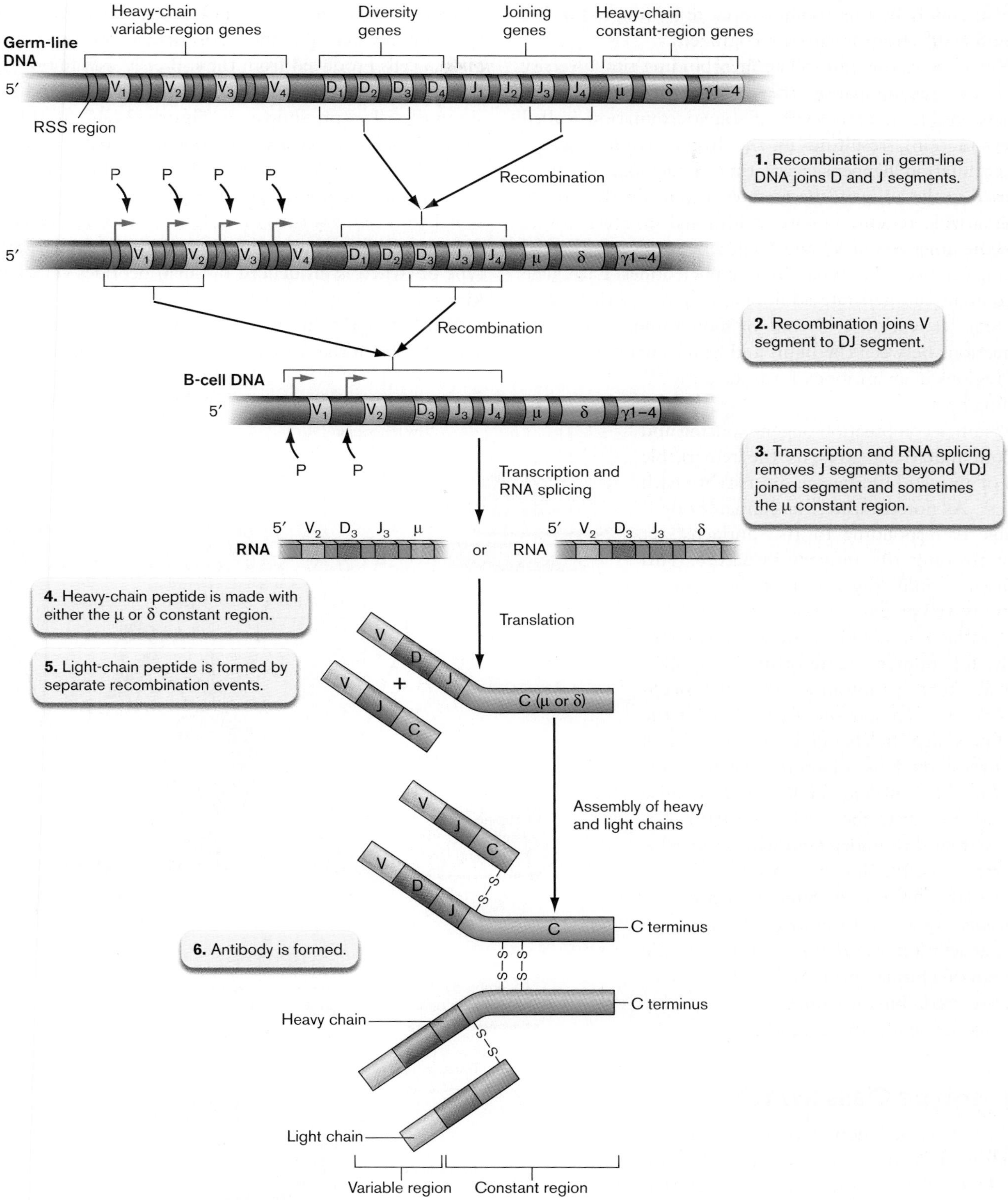
Heavy-chain variable-region genes
Diversity genes
Joining genes
Heavy-chain constant-region genes
Germ-line DNA
5′
RSS region
Recombination
1. Recombination in germ-line DNA joins D and J segments.
P
Recombination
2. Recombination joins V segment to DJ segment.
B-cell DNA
Transcription and RNA splicing
3. Transcription and RNA splicing removes J segments beyond VDJ joined segment and sometimes the μ constant region.
RNA
or
4. Heavy-chain peptide is made with either the μ or δ constant region.
5. Light-chain peptide is formed by separate recombination events.
Translation
C (μ or δ)
Assembly of heavy and light chains
C terminus
6. Antibody is formed.
Heavy chain
Light chain
Variable region
Constant region

FIGURE 24.18 ■ Formation of the VDJ regions of heavy chains. Note that only a small subset of the V, D, and J genes listed in Table 24.3 is shown in this model. RSS = recombination signal sequence.

humans simply by this combinatorial re-joining (a total of about 5×10^6 antigens can be recognized).

Where does the rest of antigen-binding-site diversity arise? After recombination, the V regions of the germ lines are susceptible to high levels of somatic mutation (called hypermutation), resulting in the hypervariable regions. Hypermutation happens every time a memory B cell is exposed to the antigen; the memory B cells divide and the hypervariable regions mutate. Additional diversity comes from the junctions of VJ and VDJ, where recombinational splicing can occur between different nucleotides. Each gene splice event can generate additional codons, so the resulting peptides will differ by one or more amino acids. The interactions between the light- and heavy-chain hypervariable regions in an antibody form the antigen-binding sites.

In sum, a combination of gene splicing and random mutations gives us the remarkable level of antigen-binding-site diversity we each possess. As noted earlier, the human body is capable of responding to 10^{11} antigens, yet there are only 10^8 antigens in nature. This apparent overkill suggests that the immune system is well prepared to cope with any possible antigen it could encounter. Unfortunately for humans, enterprising microbes, such as the trypanosomes that cause sleeping sickness, can stay one step ahead of the immune system by changing the structure of key surface antigens. Changing the antigenic structure of a protein renders useless those antibodies made to the previous structure.

It isn't hard to understand why multicellular organisms, like humans, need the capacity to make any one of billions of different antibodies quickly. Pathogens can undergo many generations of growth within the single life span of a human host. So humans have to generate recombinant clones of cells quickly to overcome rapidly dividing pathogens.

The Isotype Class Switch

We just described how a progenitor B cell becomes a mature (naive) B cell (see **Fig. 24.17**). The mature B cell (a B cell that has not yet "seen" the antigen but has already assembled its immunoglobulin VDJ binding site) produces both IgM and IgD B-cell receptors (so the naive B cell is referred to as IgM^+ IgD^+). The primary immune response begins when a mature (naive) B-cell receptor finds its matched antigen. In the early stages of the response, plasma cells produced from these B cells secrete only IgM, and eventually secrete IgM and IgD. As indicated in the summary diagram shown in **Figure 24.17**, if this activated B cell also receives appropriate signals from a certain type of T cell known as a **helper T cell (T_H cell)**, then further immunoglobulin isotype switching will occur; the switched B cell may then make IgG, IgA, or IgE, each with the same antigen recognition domain (variable region). The type of switch is influenced by small peptides called **cytokines** that are secreted by helper T cells.

What "flips" the switch? Notice in **Figures 24.18** and **24.19** that the constant-region gene segments defining

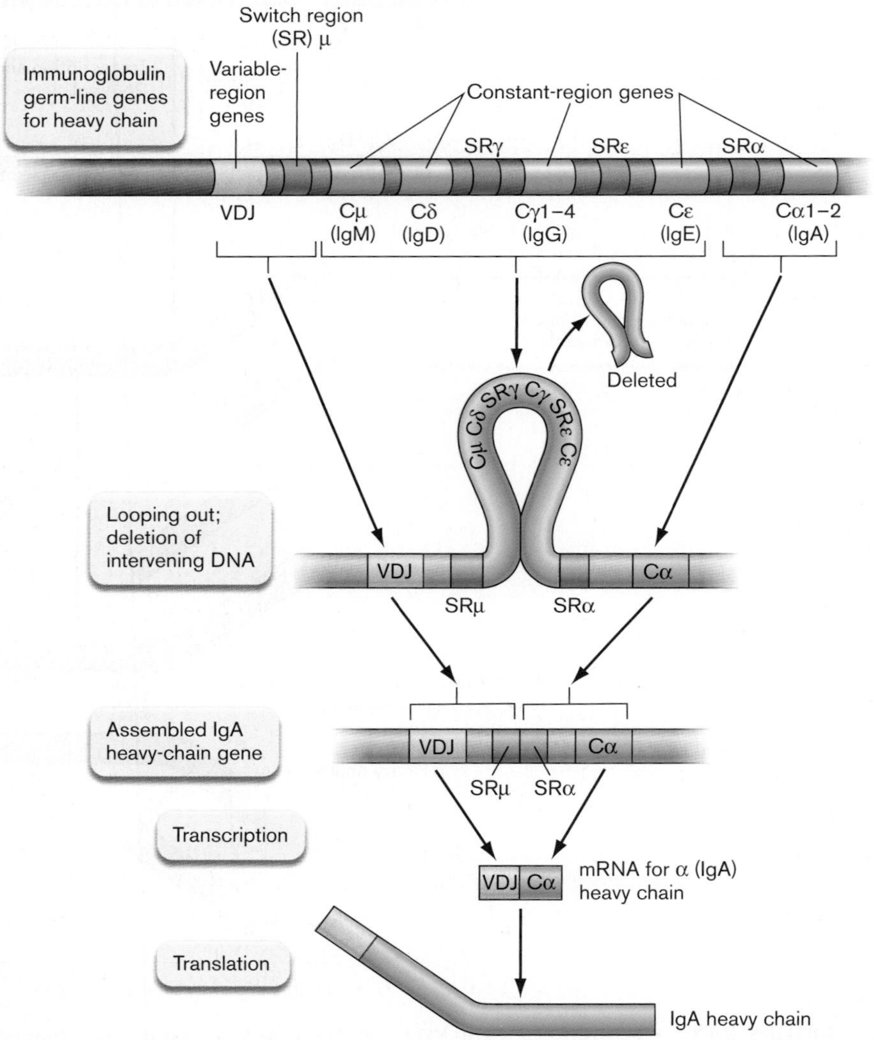
Switch region
(SR) μ
Immunoglobulin germ-line genes for heavy chain
Variable-region genes
Constant-region genes
SRγ
SRε
SRα
VDJ
Cμ (IgM)
Cδ (IgD)
Cγ1–4 (IgG)
Cε (IgE)
Cα1–2 (IgA)
Deleted
Cμ Cδ SRγ Cγ SRε Cε
Looping out; deletion of intervening DNA
VDJ
Cα
SRμ
SRα
Assembled IgA heavy-chain gene
VDJ
Cα
SRμ
SRα
Transcription
VDJ Cα
mRNA for α (IgA) heavy chain
Translation
IgA heavy chain

FIGURE 24.19 ▪ Heavy-chain class switching. As a B cell becomes activated, a switch in antibody isotype will occur. The switch involves recombination between isotype cassettes that brings one heavy-chain constant region (Cα in the example, for IgA) in tandem with a VDJ sequence.

different antibody heavy-chain classes are arranged in tandem after a VDJ region. The mechanism by which a B cell switches to make IgG (C regions gamma 1–4), IgE (C-epsilon), or IgA (C-alpha 1 and 2) is very similar to VDJ formation. Each constant segment, except delta, contains a repeating DNA base sequence called a **switch region**. Recombination between these switch regions will delete the intervening DNA between the VDJ region and one of the constant regions. The primary RNA transcript produced after this recombination event has all of its introns spliced out before translation. Because the VDJ region is the same regardless of which C_H gene is selected, the antibody produced will have the same antigenic specificity as the original IgM. However, the heavy-chain switch selection process is not random. The type of cytokine present at the time of the switch will influence which C_H gene is selected.

Note that any heavy-chain peptide (alpha, gamma, delta, and so on) can combine with any light chain (lambda or kappa). However, a single, mature B cell can make only one type of heavy chain and one type of light chain.

Thought Questions

24.7 B cells in early stages have both IgM and IgD surface antibodies, but the delta region has no switch region. Why does the delta region have no switch region?

24.8 Why do individuals with type A blood have anti-B and not anti-A antibodies?

Making Memory B Cells

Having encountered an antigen, the antigen-activated B cell will divide to make memory B cells, as well as plasma cells (**Figs. 24.14** and **24.17**). Memory B cells are B cells that have already undergone class switching (that is, they are committed to making IgG) but are very long-lived. Their long lives provide immunological memory. But why do they survive longer than regular B cells? Part of the reason is the production of what one might call an "anti-suicide" protein called Mcl-1. B cells that do not encounter their matched antigen eventually undergo a process of programmed cell death (apoptosis). However, if the B-cell receptor binds to its matched (cognate) antigen, then a signal is sent to the nucleus to make the Mcl-1 anti-apoptosis protein. The memory-B-cell line can now propagate for many years, guarding the person against microbial assault.

During their lifetimes, memory B cells, like other B cells, hypermutate the antigen-binding regions of antibody genes (the VDJ and VJ regions). These mutations increase the affinity of the antibody produced during the secondary response; that is, the hypermutation helps sharpen the antigen-binding site, making it even more specific toward a given microbe or antigen. Note, too, that as an immune response clears an infection, the antigen becomes scarce. This means that only B cells with the highest-affinity antibodies as part of their B-cell receptors will remain activated. This process is known as **affinity maturation**.

Vaccines and Immune System Compartments

Now that you know how antibodies are made, let's revisit vaccines. Vaccines come in two basic types: live, attenuated vaccines and killed-organism (or subunit) vaccines. What difference does it make which one you take? Don't they both generate immunity? Yes, but generally a live, attenuated vaccine is better because when a crippled but live microbe replicates at its normal body target site, an immune response most appropriate to that site develops. Take the case of the Salk (killed) and Sabin (attenuated) polio vaccines. Poliovirus typically enters the body through ingestion, replicates in the mucosa, moves to the regional lymph nodes, and produces a viremia. Eventually the virus can attack the central nervous system and cause paralysis.

The killed Salk vaccine, which is injected (inactivated polio vaccine, or IPV), generates an antibody response in the bloodstream capable of preventing the viremia and paralytic consequences of a natural infection, but it does not generate a mucosal immunity capable of preventing mucosal replication of the natural virus. Consequently, mild disease can result if wild poliovirus infects vaccinated people. These immunized people will also shed poliovirus in their feces, enabling the virus to spread to others who might not be vaccinated. In contrast, the live virus in the Sabin vaccine, which is administered orally (oral polio vaccine, or OPV), will replicate in the mucosal lymphatic system of the gut, where secretory IgA antibodies are best generated. Anti-polio IgA antibodies secreted in these mucosal areas will prevent wild-type poliovirus from replicating in the intestinal mucosa of a vaccinated person, so no virus is shed and there are no symptoms.

What accounts for this so-called **mucosal immunity**? Is the mucosal immune system a compartment distinct from the rest of the body? In large part, it is. An organism that infects the mucosa of the intestine, the lung, or urogenital tracts (whose lymphatic systems are collectively called mucosa-associated lymphoid tissues, or MALT) will cause infected tissues to generate a specific blend of cytokines at the site of infection that guide the heavy-chain switch of activated B cells toward IgA. In nonmucosal tissues, the cytokine mix generated during an infection encourages B cells to switch classes more toward IgG.

Mucosal immunity is even more compartmentalized than just described. Lymphocytes activated at one mucosal site—the intestine, for instance—will circulate through lymph but generally end up back in the original MALT, in this case the gut-associated lymphoid tissues (GALT). These lymphocytes are "imprinted" in some way during activation so that they home back to the original site of infection or vaccination. This homing "instinct" is thought to be due, in part, to chemokines (attractants), integrins (attachment molecules), and cytokines that are differentially expressed among different mucosal tissues. Compartmentalization within the mucosal immune system places constraints on the choice of vaccination route for inducing effective immune responses at desired sites. For instance, oral immunization generates effective immunity in the GALT but not in the lungs. This is fine in the case of poliovirus, because poliovirus doesn't affect the lungs. Nasal immunization, however, will elicit mucosal immunity in the lungs able to prevent influenza.

Thought Question

24.9 Why do immunizations lose their effectiveness over time?

To Summarize

- **Progenitor B cells** can make any antibody isotype but are programmed to bind only one epitope.
- **Mature (naive) B cells** have undergone VDJ rearrangement and make IgM and IgD surface antibodies (part of B-cell receptors).
- **Experienced B cells** have completed isotype switching, which occurs after naive B cells bind their target antigen. The B-cell receptors are made of a single antibody isotype that specifically binds one epitope.
- **During the primary antibody response**, IgM is the predominant class of antibody secreted.
- **Memory B cells** rapidly proliferate and differentiate into antibody-secreting plasma cells upon a second encounter with an antigen/epitope. Most of the antibody made during this secondary response is IgG.
- **Antibody diversity (idiotype)** occurs via a complex series of splicing events between adjacent DNA cassettes, as well as mutational events in DNA sequences encoding the hypervariable regions of heavy and light chains.
- **Mucosal immunity** is an immune system compartment mostly separate from the rest of the immune system.

24.6

T cells Link Antibody and Cellular Immune Systems

Do the two arms of the immune system (antibody-based and cell-mediated) know each other exist? Because different types of infections tilt the immune response one way or the other, the systems must communicate with each other to provide balance. T-cell lymphocytes, with integral roles in both antibody production and cell-mediated immunity, manage the balance.

Although derived from the same progenitor stem cell as B cells, T cells develop in the thymus (rather than in the bone marrow, where B cells develop), and they contain surface protein antigens different from those of B cells. T cells come in several varieties marked by the presence of different cell differentiation (CD) surface proteins and by the types of cytokines they produce (**Table 24.4**). Two critically important groups are helper T cells (already mentioned) and **cytotoxic T cells (T_C cells)**. Helper T cells display the surface antigen CD4, while cytotoxic T cells display CD8.

Helper T cells come in several models, but three types are central to adaptive immunity: T_H0 cells, a precursor to the other two types; T_H1 cells, which assist in the activation of cytotoxic T cells; and T_H2 cells, which stimulate B-cell differentiation into plasma cells. Cytotoxic T (T_C) cells are the "enforcers" of the cell-mediated immune response. They destroy the membranes of host cells infected with viruses or bacteria. The ratio between the number of T_H1 cells and T_H2 cells produced during an infection affects whether the immune response will lean more toward antibody (T_H2) or cell-mediated (T_H1) immune mechanisms. To be of any use, however, T cells must be activated by antigen.

T-Cell Activation Requires Antigen Presentation

T cells never bind free-floating antigen. T cells are activated only by antigen bound to another cell's surface. These other cells are collectively called **antigen-presenting cells (APCs)** because they present antigens to T cells (see Current Research Highlight). The APC surface proteins that hold and present the antigen are known as **major histocompatibility complex (MHC)** proteins (**Fig. 24.20A**). MHC proteins differ between species and among individuals within a species. They help determine whether a given

TABLE 24.4

Major classes of T cells.

T-cell type	Surface marker/ MHC restriction	Major function
Helper T cell		
T_H0 cell	CD4/Class II	Differentiate into T_H1 or T_H2
T_H1	CD4/Class II	Activate cytotoxic T cells
T_H2	CD4/Class II	B-cell helper
Cytotoxic T cell (T_C)	CD8/Class I	Kill virus-infected and cancer cells
Regulatory T cell (Treg)	CD4, CD25/Class II	Down-regulate immune response

[a]TGF = transforming growth factor.

antigen is recognized as coming from the host (a self antigen) or from another source (a foreign antigen), in a phenomenon called histocompatibility (hence the name "major histocompatibility complex"). The salient feature of all MHC molecules is that they bind antigen that has entered the host cell, and then present the antigen back on the cell surface. MHC molecules are critical to the immune system because the T cell, to be activated, must first recognize a foreign antigen attached to an MHC molecule on an APC.

Two classes of MHC molecules are found on cell surfaces. Both classes belong to the immunoglobulin family of proteins, but they are not immunoglobulins. **Class I MHC molecules** are found on all nucleated cells, whereas **class II MHC molecules** have a more limited distribution on professional APCs such as dendritic cells, B cells, and macrophages.

The surface CD proteins CD8 and CD4 mentioned earlier help T cells distinguish between MHC class I and class II molecules on antigen-presenting cells. The CD8 molecules on T_C cells selectively bind MHC class I, while CD4 molecules on T_H cells selectively bind MHC class II. Antigens presented on class I MHC molecules generally arise from intracellular pathogens, such as viruses and some bacteria that require cell-mediated immunity for resolution. In contrast, antigen peptides presented on class II MHC molecules originate from extracellular infections, resolved by antibody, and are recognized only by CD4 T_H cells.

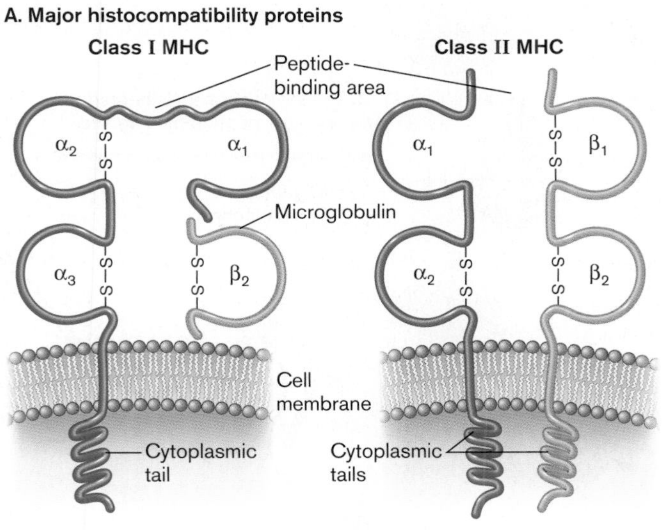
A. Major histocompatibility proteins
Class I MHC
Class II MHC
Peptide-binding area
α2
α1
α3
β2
Microglobulin
α1
β1
α2
β2
S–S
Cell membrane
Cytoplasmic tail
Cytoplasmic tails

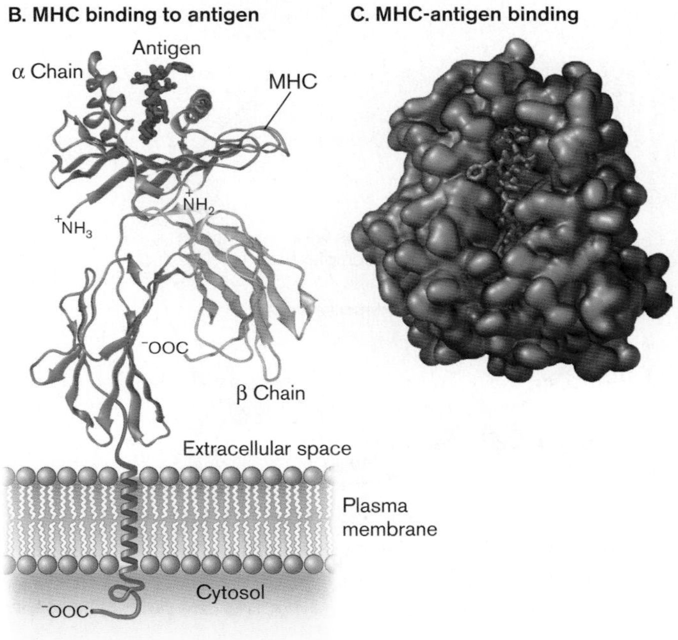
B. MHC binding to antigen
C. MHC-antigen binding
Antigen
α Chain
MHC
+NH3
NH2
⁻OOC
β Chain
Extracellular space
Plasma membrane
Cytosol
⁻OOC

FIGURE 24.20 ■ Major histocompatibility proteins. **A.** MHC class I molecules are composed of a 45-kDa chain and a small peptide called β_2-microglobulin (12 kDa). MHC class II molecules contain an alpha chain (30–34 kDa) and a beta chain (26–29 kDa). The peptide-binding regions of both classes show variability in amino acid sequence that yields different shapes and grooves. Peptide antigens nestle in the grooves and are held there awaiting interaction with T-cell receptors. CD8 T cells recognize antigen peptides associated with class I molecules, while CD4 T cells recognize peptides bound to class II molecules. **B.** Antigen binding to an MHC molecule. **C.** Top view of antigen (red) nestled in the MHC peptide-binding site. (PDB code: 1BII)

Antigen Processing and Presentation on APCs Occur by Two Paths

How are foreign antigens placed, or presented, on host cell surfaces? In the initial stages of an immune response, antigen-presenting cells internalize the pathogen, such as a virus or a bacterium. Inside the APCs, pathogen proteins are degraded into smaller peptides (epitopes). These epitopes are placed within MHC-binding clefts and transported back to the cell surface. Whether an antigen peptide binds to class I or class II MHC molecules generally depends on how the antigen initially entered the cell (**Fig. 24.21**).

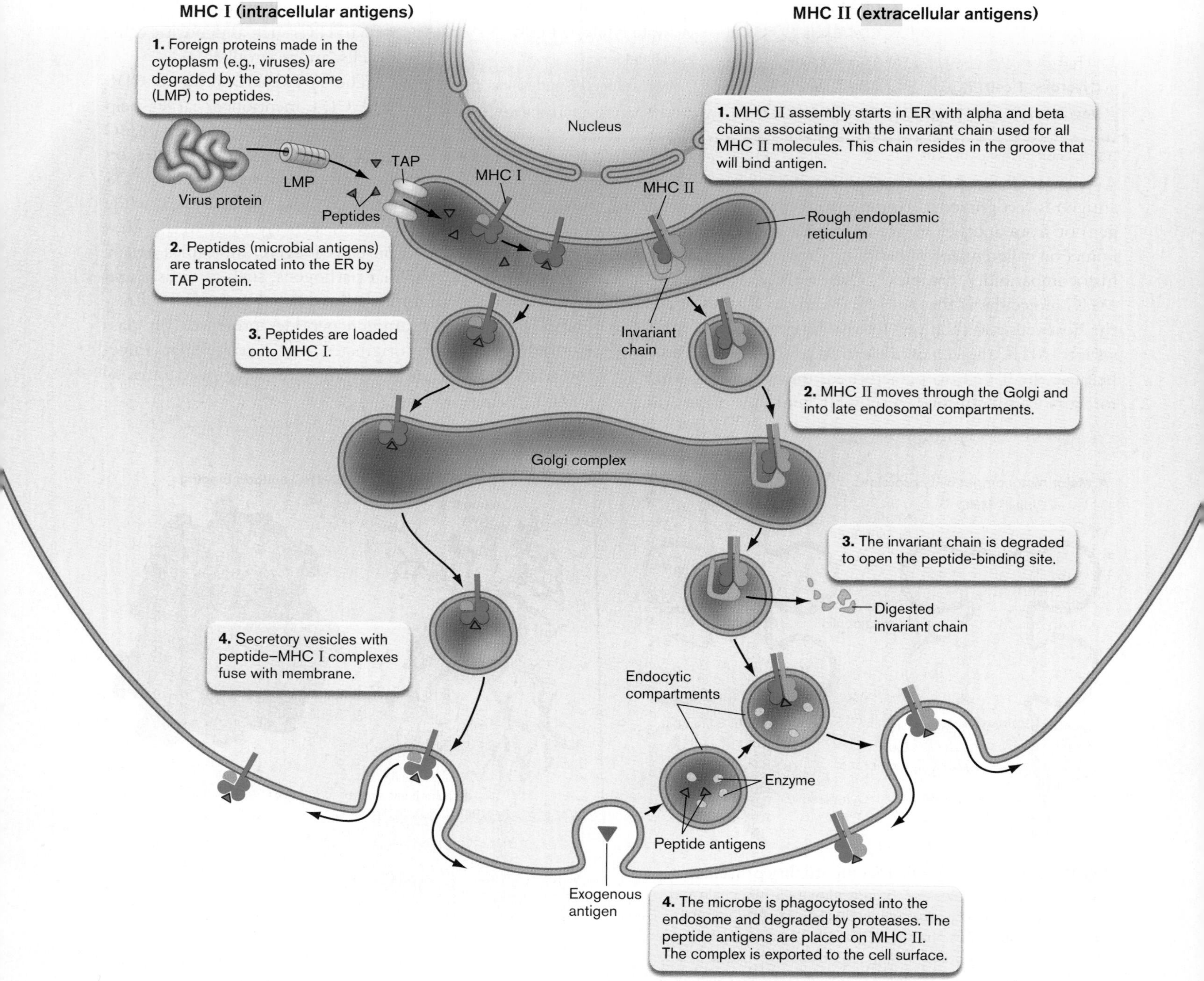
MHC I (intracellular antigens)
MHC II (extracellular antigens)
1. Foreign proteins made in the cytoplasm (e.g., viruses) are degraded by the proteasome (LMP) to peptides.
Virus protein
LMP
TAP
Peptides
MHC I
Nucleus
MHC II
1. MHC II assembly starts in ER with alpha and beta chains associating with the invariant chain used for all MHC II molecules. This chain resides in the groove that will bind antigen.
Rough endoplasmic reticulum
2. Peptides (microbial antigens) are translocated into the ER by TAP protein.
3. Peptides are loaded onto MHC I.
Invariant chain
2. MHC II moves through the Golgi and into late endosomal compartments.
Golgi complex
3. The invariant chain is degraded to open the peptide-binding site.
Digested invariant chain
4. Secretory vesicles with peptide–MHC I complexes fuse with membrane.
Endocytic compartments
Enzyme
Peptide antigens
Exogenous antigen
4. The microbe is phagocytosed into the endosome and degraded by proteases. The peptide antigens are placed on MHC II. The complex is exported to the cell surface.

FIGURE 24.21 ▪ Processing and presentation of antigens by antigen-presenting cells on class I and class II MHC proteins. **Left:** Microbial proteins made in the host cytoplasm are degraded and peptides are placed on MHC class I molecules in the endoplasmic reticulum (ER). **Right:** Microbial proteins made outside the cell are endocytosed (bottom), degraded in the endosome, and placed on MHC class II molecules. LMP = low-molecular-mass polypeptide component of the proteasome; TAP = transporter of antigen peptides.

Endogenous antigens, which are synthesized by viruses and intracellular bacteria as they grow within the cytoplasm of an APC, will attach to class I MHC molecules on the endoplasmic reticulum and are moved to the cell surface (**Fig. 24.21**, left). In contrast, exogenous antigens, which are produced outside of the APC (as are most bacterial antigens), will enter the cell via phagocytosis and attach to class II MHC molecules moved to the acidic phagosome or lysosome (**Fig. 24.21**, right). The MHC class II–peptide complex is then carried to the cell surface. Once the antigen is presented on the APC cell surface, T cells can interact with the antigen-MHC complex via the T-cell receptors (discussed next). Interactions between antigen-presenting cells and naive T cells occur within the lymph nodes, the spleen, or Peyer's patches in the gut. So APCs must make their way to those locations to generate an immune response.

Note: Pathogens are not the only items processed by APCs. Large proteins, carbohydrates, and other sizable molecules are engulfed, processed, and presented on surface MHC molecules.

The antigen-binding clefts of different MHC molecules differ markedly and have distinct binding affinities for different antigenic peptides (**Fig. 24.20B** and **C**).

T-Cell Receptors

T-cell receptors (TCRs) are the antigen-binding molecules (not immunoglobulins) present on the surfaces of T cells (**Fig. 24.22**). A TCR on helper T cells will bind only to antigens attached to MHC surface proteins on antigen-presenting cells, as noted earlier. However, the TCRs of cytotoxic T cells can bind viral antigens present on any virus-infected cell, whether or not that cell is normally considered an APC.

The T-cell receptor is composed of several transmembrane proteins. The part that recognizes antigen is composed of two molecules: alpha and beta. Much like the immunoglobulins, the alpha and beta proteins of the TCR are formed from gene clusters that undergo gene rearrangements. The diversity of TCR antigen receptors comes from random recombinations of various V, D, and J segments (analogous to, but different from, the immunoglobulin genes), and from variability in the precise joining of segments. There is no hypermutation, however.

The TCR alpha and beta proteins are found in a complex with four other peptides, which together form the CD3 complex. When stimulated by binding antigen, these ancillary CD3 complex proteins recruit and activate intracellular protein kinases and launch a phosphorylation cascade that triggers proliferation of the T cell and key immunological events that we describe later in the chapter.

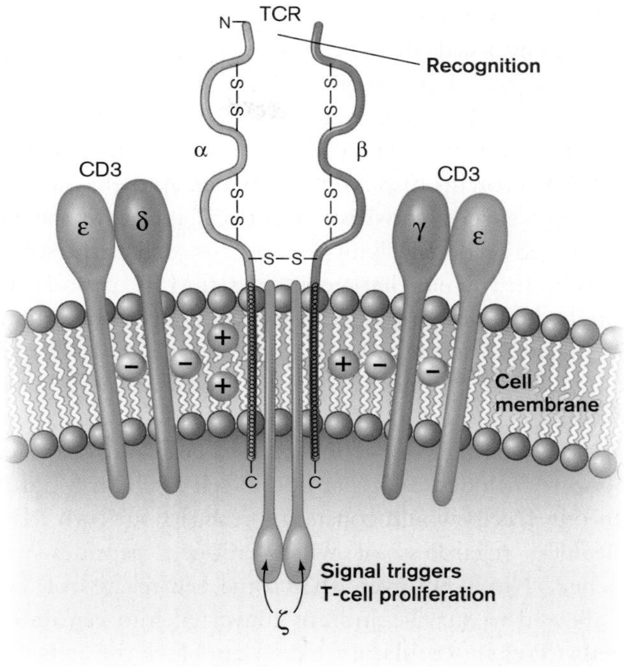
TCR
N
Recognition
α
β
CD3
CD3
ε
δ
γ
ε
S–S
Cell membrane
C
C
Signal triggers T-cell proliferation
ζ

FIGURE 24.22 ■ The T-cell receptor (TCR) and CD3 complex. T-cell receptor proteins are associated with CD3 proteins at the cell surface. Antigen binds to the alpha and beta subunits. The positive and negative charges holding the complex together come from amino acids in the peptide sequences. Once bound to antigen, the complex transduces a signal into the cell. This signal triggers T-cell proliferation.

Why don't your T cells see your body's antigens? Are they tolerant somehow? Actually, T cells that strongly bind self antigens are not made tolerant; they are killed in a process called T-cell education.

T-Cell Education and Deletion

Greek mythology tells of Narcissus, a young man punished by the gods for scorning the women who fell in love with him. One day Narcissus saw a beautiful face in a pool of water. Not knowing it was his own reflection, he fell in love, tumbled into the pool, and drowned. His inability to recognize himself is analogous to the danger posed by an immune system. Immune systems must be able to distinguish what is self (meaning antigens present in our own tissues) from what is not self. Otherwise our immune system would constantly attack our own cells. Innate immunity accomplishes this important task, in part, through the use of pattern recognition proteins such as the Toll-like and NOD-like receptors (Section 23.7). How adaptive immune mechanisms avoid recognizing self is equally important. At the outset of development, the immune system is fully capable of reacting against self.

One mechanism the body uses to avoid attacking itself is to delete any T cells that strongly react against self antigens. T cells undergo a two-stage selection or "education" process in the thymus to recognize self versus nonself. T cells bearing T-cell receptors (TCRs) that weakly recognize self MHC proteins displayed on thymus epithelial cells are allowed to survive (**positive selection**) and leave the thymus to seed secondary lymphoid organs such as the spleen. T cells in the thymus that recognize self MHC peptides too strongly, however, are killed, or deleted from the population (**negative selection**). Almost 95% of T cells entering the thymus die during these positive and negative selection processes.

This "weeding" out is important because if our T-cell repertoire included cells that bound self MHC too tightly, then our T cells would constantly react to our own MHC molecules, regardless of which antigen peptides were attached. Note, however, that some self-reactive T cells are allowed to survive and are converted into **regulatory T cells (Tregs)**. Regulatory T cells can block the activation of harmful, self-reactive (autoimmune) lymphocytes that escape deletion and enter the circulation.

Why is positive selection that promotes weak binding to MHC molecules important? The positive selection process is needed because T cells must be able to recognize self MHC proteins before the T-cell receptor can bind the MHC-associated antigen. Consequently, T cells from one individual normally do not recognize peptides placed on antigen-presenting cells from another individual—a property known as **MHC restriction**. CD4 T_H cells are class II MHC restricted, while CD8 T_C cells are class I restricted.

Rolf Zinkernagel (University of Zurich) and Peter Doherty (St. Jude Children's Research Hospital, Memphis) discovered this phenomenon in 1976 with a simple experiment. They demonstrated that mouse cytotoxic T cells would efficiently kill virus-infected target cells having an MHC I surface protein identical to that of the T cell, but failed to kill infected cells containing a different MHC I protein from a different strain of mouse. Thus, T-cell recognition has two components: self (the MHC molecule) and nonself (the antigen). This discovery won Zinkernagel and Doherty the 1996 Nobel Prize in Physiology or Medicine.

Why transplanted organs with one MHC type are prone to rejection by a recipient of a different MHC type is explained in eTopic 24.3.

Where are T cells "educated"? In contrast to negative selection for B cells, which occurs in bone marrow, T-cell education is limited to the thymus. But if the thymus expresses only thymus antigens, how can T cells that respond to antigens expressed on other host cells (for example, heart cells) be removed? The answer is a special gene activator called AIRE in thymus cells that stimulates the synthesis of all human proteins in small amounts. This expression is necessary to complete T-cell education within the thymus.

You might ask how someone whose thymus has been removed (done as a treatment for myasthenia gravis) can live if the organ is critical for T-cell maturation and for deleting self-reactive T cells. Actually, the thymus begins losing function shortly after birth, such that very little function remains in adults. Fortunately, some T-cell maturation can occur in secondary lymphoid tissues, which allows people with thymectomies to live relatively normal lives. Newborns, however, have not had sufficient time to populate their secondary extrathymic lymphoid organs with T cells.

APCs Activate T_H0 Helper T cells

Figure 24.23 summarizes the steps that lead to T-cell activation in the lymph node and highlights how activated T cells influence the two types of adaptive immunity: humoral (antibody) and cellular. The different classes of T cells, T_H and T_C, require different but interrelated activation programs. First we consider T_H cell activation. T_H cells require two molecular signals to become active. T-cell receptors on precursor T_H0 cells first recognize and link to antigen-MHC proteins on antigen-presenting cells (APCs) (**Fig. 24.23**, step 1a). In the case of T_H0 cells, the MHC molecules are class II.

The second signal needed to convert a T_H0 cell to a T_H1 or T_H2 cell involves a CD28 molecule present on the T_H0 cell surface binding to a B7 protein on the APC cell surface (step 1a). Different cytokines produced during infection will then convert activated T_H0 cells to T_H1 or T_H2 cells (step 2a). Mast cells, for instance, that engage parasites and allergens will secrete IL-4, which helps convert the activated T_H cell to T_H2. The activation of cytotoxic T cells (T_C cells) will be examined later.

T_H2 Cells Activate B Cells

During the early phase of a primary response, there is a type of B cell in the lymph node that can become activated after binding antigen if it also binds to complement C3 on a bacterium (Section 23.8), or if a microbe-associated molecular pattern (MAMP) binds to a Toll-like receptor on the B cell (Section 23.7). This activation does not result in heavy-chain class switching; only IgM is secreted by the resulting plasma cell. This is partly why the primary response starts with IgM (see **Fig. 24.13**).

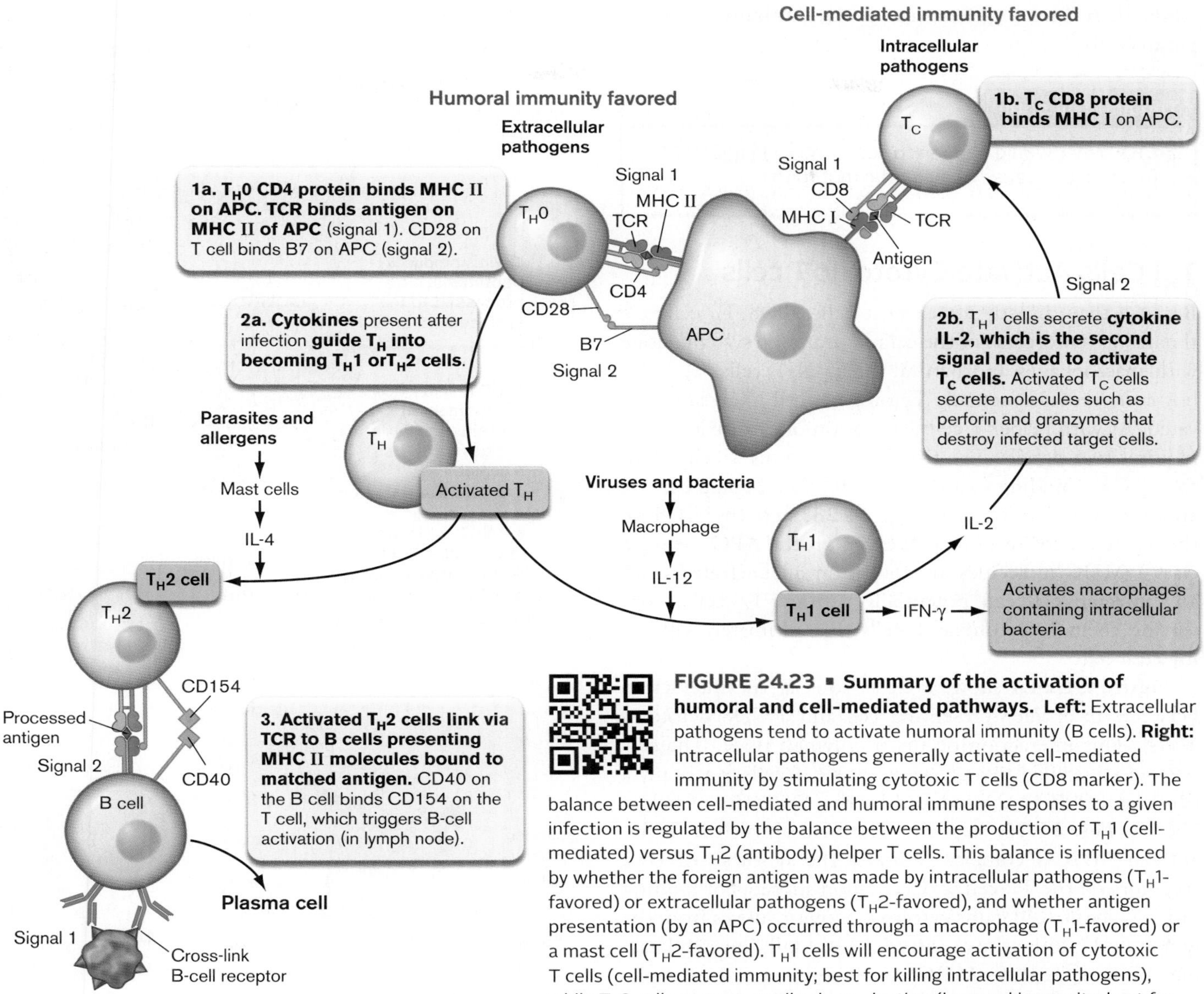

Cell-mediated immunity favored
Intracellular pathogens
1b. T_C CD8 protein binds MHC I on APC.
Humoral immunity favored
Extracellular pathogens
1a. T_H0 CD4 protein binds MHC II on APC. TCR binds antigen on MHC II of APC (signal 1). CD28 on T cell binds B7 on APC (signal 2).
T_H0
Signal 1
MHC II
TCR
Signal 1
CD8
MHC I
TCR
T_C
Antigen
CD4
Signal 2
CD28
B7
Signal 2
APC
2a. Cytokines present after infection guide T_H into becoming T_H1 or T_H2 cells.
2b. T_H1 cells secrete cytokine IL-2, which is the second signal needed to activate T_C cells. Activated T_C cells secrete molecules such as perforin and granzymes that destroy infected target cells.
Parasites and allergens
Mast cells
IL-4
T_H
Activated T_H
Viruses and bacteria
Macrophage
IL-12
T_H1
IL-2
T_H2 cell
T_H2
T_H1 cell
IFN-γ
Activates macrophages containing intracellular bacteria
CD154
Processed antigen
Signal 2
CD40
3. Activated T_H2 cells link via TCR to B cells presenting MHC II molecules bound to matched antigen. CD40 on the B cell binds CD154 on the T cell, which triggers B-cell activation (in lymph node).
B cell
Plasma cell
Signal 1
Cross-link B-cell receptor

FIGURE 24.23 ▪ Summary of the activation of humoral and cell-mediated pathways. Left: Extracellular pathogens tend to activate humoral immunity (B cells). **Right:** Intracellular pathogens generally activate cell-mediated immunity by stimulating cytotoxic T cells (CD8 marker). The balance between cell-mediated and humoral immune responses to a given infection is regulated by the balance between the production of T_H1 (cell-mediated) versus T_H2 (antibody) helper T cells. This balance is influenced by whether the foreign antigen was made by intracellular pathogens (T_H1-favored) or extracellular pathogens (T_H2-favored), and whether antigen presentation (by an APC) occurred through a macrophage (T_H1-favored) or a mast cell (T_H2-favored). T_H1 cells will encourage activation of cytotoxic T cells (cell-mediated immunity; best for killing intracellular pathogens), while T_H2 cells promote antibody production (humoral immunity; best for attacking extracellular pathogens). See text for more details. ▶

During the later stages of a primary response, however, a T_H2 helper T cell is required to activate most B cells after the B-cell receptors bind antigen (**Fig. 24.23**, step 3). This is called T-cell help. This help is specific to the antigen and triggers heavy-chain class switching. But how does a B cell gain specific T-cell help? The specific T_H2 cell must have a TCR able to bind the same antigen as the B cell. This way the T_H2 cell "knows" which B cell to help. As part of the B-cell receptor capping mechanism (antigen cross-linking of BCR), some antigen bound by B-cell receptors becomes internalized and processed to be presented back on the B cell's surface MHC receptor (**Fig. 24.23**, step 3). For instance, a T_H2 cell that was activated by dendritic cells presenting antigen A (step 1a) can also use its T-cell receptor to bind antigen A presented on a B-cell MHC II molecule (step 3). This contact allows CD40 on the B cell to bind CD154 on the T cell—an interaction that completes the activation of the B cell. The B cell will now differentiate into a plasma cell (**Fig. 24.23**, step 3).

During the secondary immune response, memory B cells also need T-cell help to become plasma cells, but direct contact with helper T cells is not required. Memory B cells that have antigen bound to their B-cell receptors can respond to the soluble IL-4 and IL-6 cytokines secreted by activated helper T cells without having direct contact with the T_H cell. IL-4 stimulates B-cell proliferation, while IL-6 directs differentiation into antibody-secreting plasma cells.

Thought Question

24.10 What would happen to someone lacking CD154 on T_H2 cells because of a genetic mutation?

T_H1 Cells Activate Cytotoxic T cells

Because they directly attack host cells, CD8 T_C cells are the major "enforcers" of the cellular immune system, along with macrophages and natural killer (NK) cells. T_C cells, however, are not actually cytotoxic until they are activated. As with helper T cells, activation of cytotoxic T cells requires two signals. The first signal involves T-cell receptors (TCRs) binding to antigen-MHC class I complexes on an antigen-presenting cell (**Fig. 24.23**, step 1b). CD8 on the T_C cell helps recognize MHC I on the APC. Because class I MHC molecules are found on all nucleated cells, any infected cell can potentially activate T_C cells, converting them to cytotoxic T cells that ultimately kill the infected cell.

Figure 24.24A shows a closer look at the interaction between an antigen-presenting cell and a CD8 cytotoxic T cell. This interaction occurs in a lymph node. The second signal needed to activate cytotoxic T cells is actually a cytokine called IL-2 produced by the T_H1 class of helper T cells (**Fig. 24.23**, step 2b). The first signal (the TCR–antigen–MHC I complex) initiates the synthesis of IL-2 receptors on the T_C-cell surface. Upon subsequent binding of IL-2, the T_C cell gains cytotoxic activity, moves from the lymph node to the site of infection, and kills any cell bearing the same peptide–MHC class I complex (for example, cells infected with the same virus that triggered T_C production) (**Fig. 24.24B**).

Cytotoxic T cells kill infected target cells by releasing the contents of their granules, which contain the proteins **granzyme** and **perforin**. Perforin produces a pore in the target cell membrane through which granzymes can enter. In the cytoplasm, granzymes cleave and activate caspase proteins in the infected cell. The activated caspases then trigger apoptosis and cell death. The infected host cell is sacrificed for the good of the whole animal, human or otherwise.

Why are cytotoxic T cells more effective than B cells and antibodies at clearing viral infections? A major reason is that intracellular pathogens (for example, viruses) hide inside host cells, where they are protected from antibody. Consequently, these pathogens are best killed when the harboring host cell is also sacrificed (via cellular immunity). Because all nucleated cells have class I MHC molecules, any infected cell can be recognized and killed by an appropriate T_C cell for the good of the host.

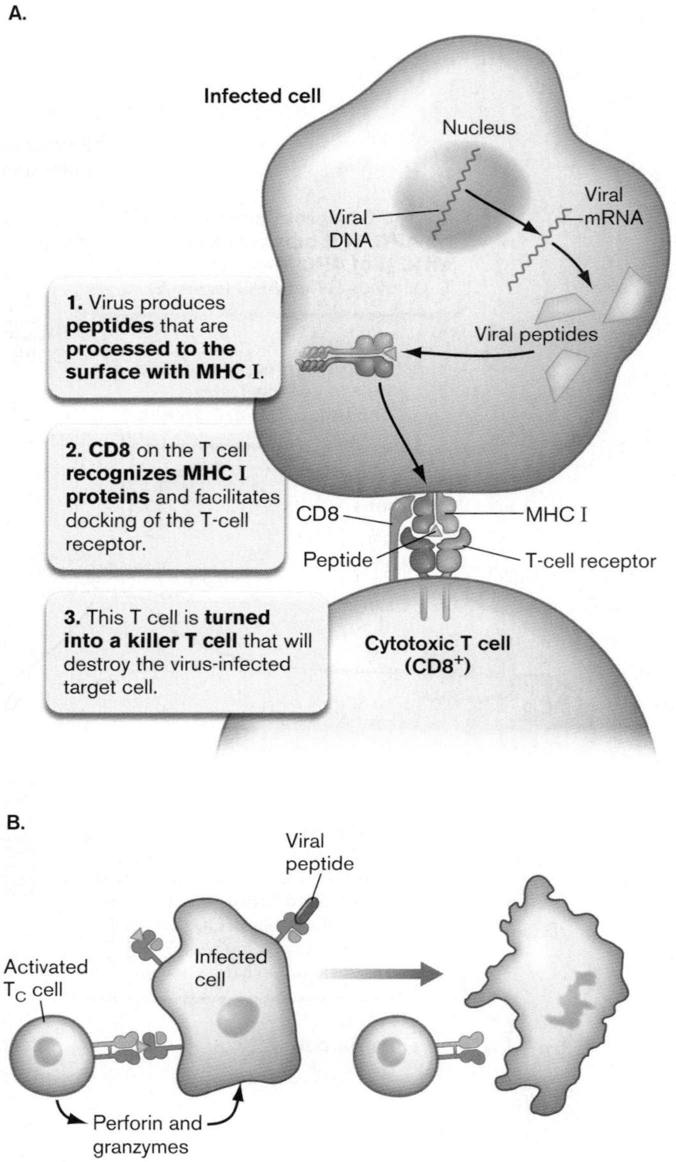
A.
Infected cell
Nucleus
Viral DNA
Viral mRNA
Viral peptides
1. Virus produces peptides that are processed to the surface with MHC I.
2. CD8 on the T cell recognizes MHC I proteins and facilitates docking of the T-cell receptor.
CD8
MHC I
Peptide
T-cell receptor
3. This T cell is turned into a killer T cell that will destroy the virus-infected target cell.
Cytotoxic T cell (CD8+)
B.
Viral peptide
Activated Tc cell
Infected cell
Perforin and granzymes

FIGURE 24.24 ■ Presentation of a viral antigen to a T cell and cytotoxic T-cell action. A. CD8 protein on a T cell directs the interaction between a T-cell receptor and a viral antigen bound to a class I MHC protein on an antigen-presenting cell. **B.** The activated T_C cell then leaves the lymph node and migrates to the site of infection, where it can recognize the viral peptide presented on the MHC class I receptor on an infected cell. This interaction authorizes the cytotoxic T cell to kill the infected cell.

Weblinks Activation of cytotoxic T cells (*see ebook*)

Superantigens Do Not Require Processing to Activate T cells

To stimulate T-cell responses, normal antigens require processing by antigen-presenting cells. Each peptide produced by antigen processing must be placed on an MHC molecule and presented to a cognate T cell. However, when an antigen is introduced into a host, only a few T cells have the proper TCR needed to recognize that antigen—in the range of 1–100 cells in a million. Proliferation of the T cells through antigen-dependent activation increases their number and increases the immune response to that antigen.

Some proteins, called **superantigens**, indiscriminately stimulate T cells by bypassing the normal route of antigen processing. In fact, recognition as an antigen is not even involved. Certain microbial toxins, such as staphylococcal toxic shock syndrome toxin (TSST), are actually superantigens. These proteins, illustrated in **Figure 24.25**, simultaneously bind to the outside of T-cell receptors on T cells and to the MHC molecules on APCs (for example, macrophages). This joining of the T cell and macrophage activates more T cells than a typical immune reaction does, and stimulates the release of massive amounts of inflammatory cytokines from both cell types.

The effect of superantigens can be devastating because cytokines like **tumor necrosis factor (TNF)** can overwhelm the host immune system regulatory network and cause severe damage to tissues and organs. The result is disease and sometimes death. Jim Henson, the creator of Kermit the Frog and other Muppet characters, died in 1990 from complications of pneumonia caused by a potent superantigen produced by *Streptococcus pyogenes.* This example is but one of many ways that overreaction by the immune system causes morbidity (disease) and mortality (death)—effects more pronounced than the direct effects of the pathogen involved.

CD4 Helper T cells and Cytokines Balance the Immune Response

Antigens presented on class I MHC molecules elicit only cell-mediated immunity (via CD8 T cells). Antigens presented on class II molecules, however, can actually stimulate both arms of the immune system by activating CD4

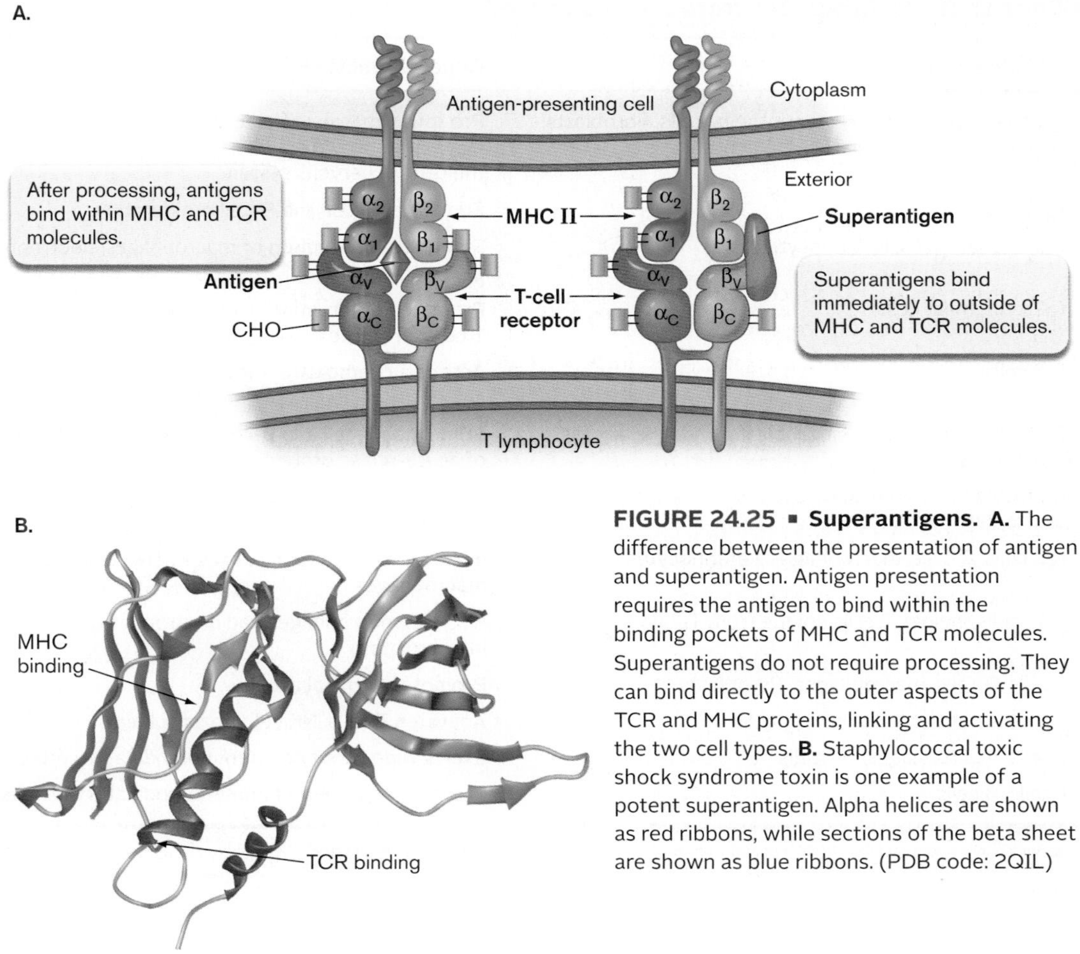
A.
Antigen-presenting cell
Cytoplasm
Exterior
After processing, antigens bind within MHC and TCR molecules.
MHC II
Superantigen
Antigen
T-cell receptor
CHO
Superantigens bind immediately to outside of MHC and TCR molecules.
T lymphocyte
B.
MHC binding
TCR binding

FIGURE 24.25 ■ **Superantigens. A.** The difference between the presentation of antigen and superantigen. Antigen presentation requires the antigen to bind within the binding pockets of MHC and TCR molecules. Superantigens do not require processing. They can bind directly to the outer aspects of the TCR and MHC proteins, linking and activating the two cell types. **B.** Staphylococcal toxic shock syndrome toxin is one example of a potent superantigen. Alpha helices are shown as red ribbons, while sections of the beta sheet are shown as blue ribbons. (PDB code: 2QIL) helper T cells. CD4 helper T cells do not kill host cells directly, but instead release various cytokines called lymphokines that do many things, including attract additional white cells to the area. **Table 24.5** lists a fraction of the many different cytokines produced by various cell types and their influence over the immune system. Activated T_H1 cells secrete cytokines that stimulate cytotoxic T_C cells (see **Fig. 24.23**, step 2b), as well as macrophages and natural killer cells. Activated T_H2 cells, on the other hand, trigger B cells to switch immunoglobulin classes and differentiate into plasma cells that secrete antibody (see **Fig. 24.23**, step 3).

But what determines whether T_H1 or T_H2 cells predominate during an infection? Part of the answer is found in the blend of cytokines generated by macrophages or mast cells during different types of infection. Infections by viruses and bacteria favor the production of T_H1 cells, whereas allergens and parasites favor differentiation to T_H2. Early in an infection, bacteria and viruses stimulate cells of the innate immune system (macrophages and natural killer cells) to produce cytokines that tilt development of T_H0 into T_H1 (one such cytokine is IL-12). During interactions with an allergen, on the other hand, mast cells produce IL-4, which promotes T_H0 differentiation into T_H2 cells (see **Fig. 24.23**, step 2a.

There is also cross-inhibition of T_H1 and T_H2 development. Cytokines such as interferon-gamma (IFN-γ) produced by T_H1 cells inhibit T_H2 development, while interleukins IL-4 and IL-10 secreted by T_H2 cells inhibit T_H1 differentiation. As a result, once the immune system starts to go down one pathway, the other pathway is held back. But this is not an all-or-none response. For instance, a viral infection will also result in T_H2 cells that stimulate antibody production. The tilt toward T_H1 or T_H2 predominance occurs because the body realizes which pathway, cell-mediated (T_H1) or humoral (T_H2), will more effectively clear a particular infection.

The importance of CD4 T cells in immunity is tragically illustrated by acquired immunodeficiency syndrome (AIDS). The human immunodeficiency virus (HIV) binds

TABLE 24.5

Select cytokines that modulate the immune response.*

Cytokine	Sample sources	General functions
IL-1	Many cell types, including endothelial cells, fibroblasts, neuronal cells, epithelial cells, macrophages	Pro-inflammatory; affects differentiation and activity of cells in inflammatory response; acts as endogenous pyrogen in the central nervous system
IL-2	T_H1 cells	Stimulates T-cell and B-cell proliferation
IL-3	T cells, mast cells, keratinocytes	Stimulates production of macrophages, neutrophils, mast cells, others
IL-4	T_H2 cells, mast cells	Promotes differentiation of CD4 and T cells into T_H2 helper T cells; promotes proliferation of B cells
IL-5	T_H2 cells	Acts as a chemoattractant for eosinophils; activates B cells and eosinophils
IL-6	T_H2 cells, macrophages, fibroblasts, endothelial cells, hepatocytes, neuronal cells	Stimulates T-cell and B-cell growth; stimulates production of acute-phase proteins
IL-8	Monocytes, endothelial cells, T cells, keratinocytes, neutrophils	Chemoattracts PMNs; promotes migration of PMNs through endothelium
IL-10	T_H2 cells, B cells, macrophages, keratinocytes	Inhibits production of IFN-γ, IL-1, TNF-α, IL-6 by macrophages
IL-17	T_H17 cells (helper T cells unique from T_H1 and T_H2)	Recruits macrophages and neutrophils to sites of inflammation
IFN-α/β	T cells, B cells, macrophages, fibroblasts	Promotes antiviral activity
IFN-γ	T_H1 cells, cytotoxic T cells, NK cells	Activates T cells, NK cells, macrophages
TNF-α	T cells, macrophages, NK cells	Exerts wide variety of immunomodulatory effects
TNF-β	T cells, B cells	Exerts wide variety of immunomodulatory effects

*IFN = interferon; IL = interleukin; NK = natural killer; PMN = polymorphonuclear leukocyte; TNF = tumor necrosis factor.

to CD4 molecules and is thus able to invade and infect CD4 T cells. As the disease progresses, the number of CD4 T cells declines below its normal level of about 1,000 per microliter. When the number of CD4 T cells drops below 400 per microliter, the ability of the patient to mount an immune response is jeopardized. The patient not only becomes hypersusceptible to infections by pathogens, but also becomes susceptible to infections by commensal organisms. Most AIDS patients actually die from these opportunistic infections.

Microbial Evasion of Adaptive Immunity

As efficient as our immune system is, numerous viral and bacterial pathogens have developed effective means for avoiding the adaptive immune response. Many viruses produce proteins that down-regulate production of class I MHC on infected cell surfaces. Down-regulation of MHC I will limit antigen presentation, since MHC I is needed for that process. On the other hand, losing MHC I should expose the infected host cell to natural killer cells because NK cells attack peers that lack MHC I (discussed in Section 23.7). To surmount this obstacle, for example, human cytomegalovirus places a decoy MHC I–like molecule on the surface of infected cells. These decoys are thought to bind inhibitory receptors on NK surfaces that block NK cell cytotoxicity.

Like viruses, bacteria are masters of illusion when it comes to the immune system. A major cause of gastric ulcers, *Helicobacter pylori,* expresses proteins from a cluster of pathogenicity genes that trigger apoptosis (programmed cell death) of T cells. Other bacteria have evolved mechanisms that interfere with signal transduction pathways controlling the expression of cytokines. For example, YopP from *Yersinia enterocolitica,* one cause of gastroenteritis, inhibits a specific signal transduction pathway needed to produce TNF, IL-1, and IL-8. Thus, *Yersinia* avoids the detrimental effects of those pro-inflammatory cytokines. Another method of immune avoidance is employed by various mycobacteria, some of which cause tuberculosis and leprosy (**Fig. 24.26**). These bacteria induce the production of anti-inflammatory cytokines, which dampen the immune response. *Mycobacterium*-infected macrophages produce IL-6, which inhibits T-cell activation, and IL-10, which down-regulates the production of MHC II molecules needed for specific activation of T cells. Fortunately for humans, in most cases the immune system catches on to these tricks and, through redundant humoral and cellular mechanisms, manages to resolve these infections.

Activated T_H1 Cells Also Activate Macrophages

Macrophages, too, must be activated to become highly effective killers of microbes. Activation of macrophages also requires two signals. One activation pathway involves interferon-gamma (IFN-γ) produced by nearby infected or damaged cells, followed by binding of the macrophage to lipopolysaccharide (LPS) or other microbial components via Toll-like or NOD-like receptors (TLRs or NLRs). Once activated, the macrophage becomes aggressive in terms of phagocytosis and increases production of numerous antimicrobial reactive oxygen intermediates.

As noted earlier, some bacterial pathogens, such as mycobacteria, the causative agents of tuberculosis and leprosy, grow primarily in the phagolysosomes of macrophages. There they are shielded from antibodies and cytotoxic T cells. Intracellular pathogens live in the usually hostile environment of the phagocyte either by inhibiting the fusion of lysosomes to the phagosomes in which they grow or by preventing the acidification of the vesicles needed to activate lysosomal proteases. However, a macrophage activated by a T_H1 cell can rid itself of such pathogens. Even unactivated macrophages are able to process some of the intracellular bacteria and place antigen from them on their class II MHC molecules. T_H1 cells then activate these macrophages though binding of their TCR molecules to the macrophage MHC II–antigen complex and by secreting interferon-gamma. The activated

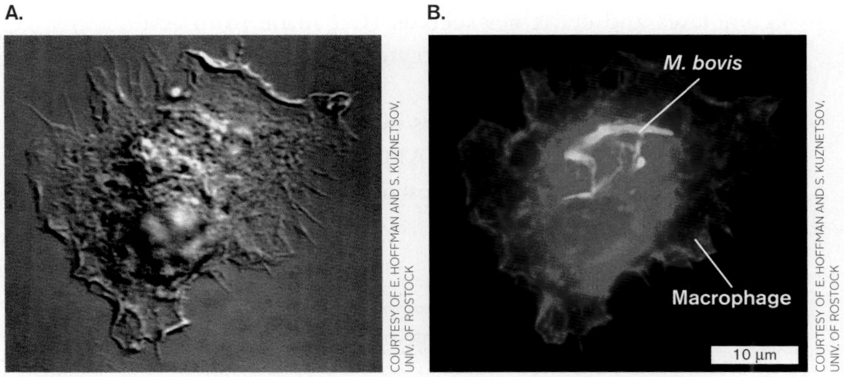
A.
COURTESY OF E. HOFFMAN AND S. KUZNETSOV, UNIV. OF ROSTOCK
B.
M. bovis
Macrophage
10 µm
COURTESY OF E. HOFFMAN AND S. KUZNETSOV, UNIV. OF ROSTOCK

FIGURE 24.26 ▪ *Mycobacterium bovis* growing within cultured macrophages (cell line J774). A. Differential interference contrast image of an infected macrophage. **B.** The same cell viewed by fluorescence microscopy. *M. bovis* is carrying green fluorescent protein. The macrophage was stained for F-actin (red) and nucleus (blue).

macrophages can then kill any bacteria that may be growing within them.

Given that activated macrophages are such effective assassins of microbial pathogens, why are macrophages not always kept in an active state? A major reason is that once activated, macrophages also damage nearby host tissue through the release of reactive oxygen radicals and proteases. Thus, effective killing of microbial pathogens comes at the expense of host tissue damage.

To Summarize

- **The major histocompatibility complex (MHC)** consists of membrane proteins with variable regions that can bind antigens. Class I MHC molecules are on all nucleated cells, while antigen-presenting cells contain both class I and class II MHC molecules.
- **Antigen-presenting cells (APCs)** such as dendritic cells present antigens synthesized during an intracellular infection on their surface class I MHC molecules, but place antigens from engulfed microbes or allergens on their class II MHC molecules.
- **Activation of a T_H0 cell** requires two signals: TCR/CD4 binding to an MHC II–antigen complex on an antigen-presenting cell, and CD20-B7 interaction.
- **T_H0 cell differentiation** to T_H1 or T_H2 cells is influenced by different cytokine "cocktails" secreted by macrophages and NK cells during an infection.
- **Activation of a B cell** into an antibody-producing plasma cell usually requires two signals: antigen binding to a B-cell receptor and binding to a T_H2 cell activated by the same antigen.
- **Activation of cytotoxic T cells** requires two signals: TCR/CD8 molecules that recognize MHC I–antigen complexes and cytokines such as IL-2 made from activated T_H1 cells. The activated cytotoxic T cells, in turn, destroy infected host cells.
- **Superantigens** stimulate T cells by directly linking TCR on a T cell with MHC on an APC without undergoing APC processing and surface presentation.

Thought Questions

24.11 Transplant rejection is a major consideration when transplanting most tissues, because host T_C cells can recognize allotypic MHC on donor cells. Why, then, are corneas easily transplanted from a donor to just about any other person?

24.12 Why does attaching a hapten to a carrier protein allow antihapten antibodies to be produced?

24.7

Complement as Part of Adaptive Immunity

Chapter 23 describes how complement, in what is called the alternative pathway, can attack invading microbes before an adaptive immune response is launched (see Section 23.8). Factor C3b binds to LPS and sets off a reaction cascade ending with a membrane attack complex (MAC) pore composed of C5b, C6, C7, C8, and C9 proteins (see Fig. 23.31). However, antibody made during the adaptive response to a pathogen offers another route to activate complement called the **classical complement pathway**. Dubbed "classical" because it was the first complement pathway to be discovered, it requires a few additional proteins before reaching C3, the linchpin factor connecting the two pathways.

The classical cascade begins when a complement C1 protein complex binds to the Fc region of an antibody bound to a bacterial or viral pathogen (**Fig. 24.27**). The bound C1 complex then cleaves two other complement factors, C2 and C4, not used in the alternative pathway. Two fragments, one each from C2 and C4, combine to form another protease, called C3 convertase, that cleaves C3 into C3a and C3b. C3b in the alternative pathway is stabilized by interacting with LPS. In the classical pathway, however, C3b combines with C3 convertase to make a C5 convertase (note that this C5 convertase is different from the C5 convertase formed in the alternative pathway; see Fig. 23.31). The subsequent steps leading to formation of a membrane attack complex (MAC) are the same as in the alternative pathway. As before, C5b binds to a target membrane and is joined by C6, C7, and C8. Multiple C9 proteins then assemble around the MAC and form a pore that compromises the integrity of the target cell.

In Section 23.8 we noted the existence of another complement activation pathway that does not require antigen-antibody complexes. The lectin activation pathway is initiated by lectins (produced by the liver), which recognize and coat the sugar structures that decorate the surfaces of infectious organisms. The lectin-coated polysaccharides trigger cleavage of factor C4 to C4a and C4b. C4b can cleave C2 and combine with one of the fragments to form C3 convertase. C3 convertase then continues through the classical activation pathway just described.

Weblinks Lectin complement activation pathway animation (see *ebook*)

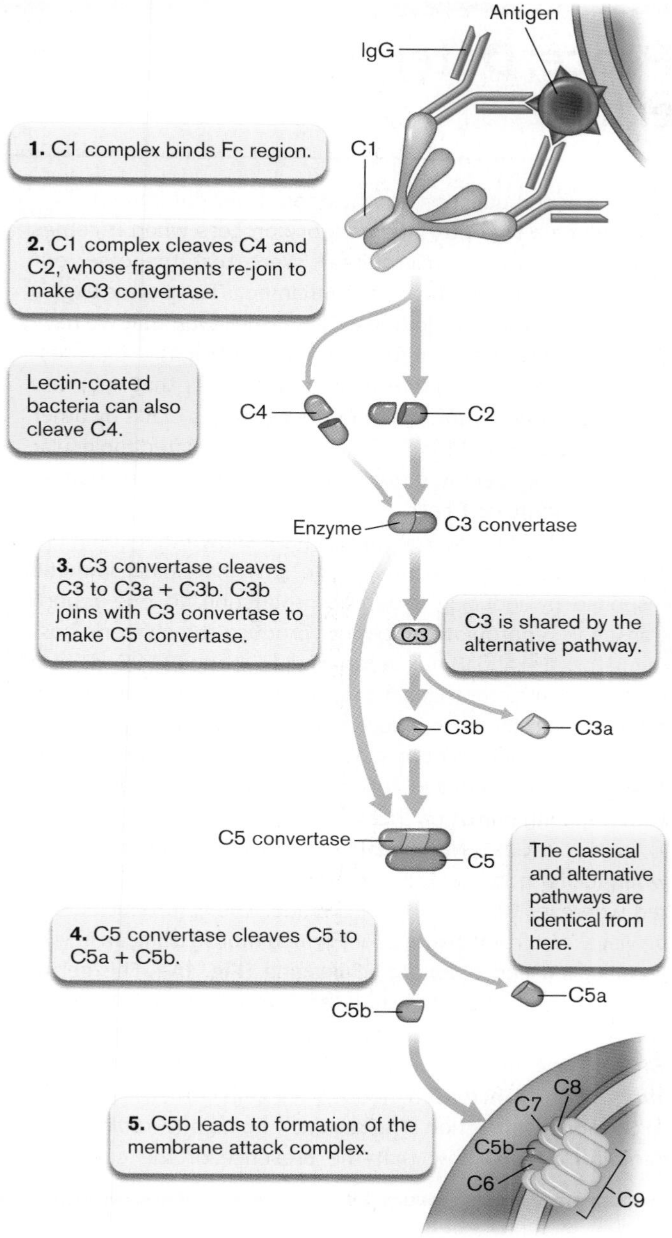
Antigen
IgG
1. C1 complex binds Fc region.
C1
2. C1 complex cleaves C4 and C2, whose fragments re-join to make C3 convertase.
Lectin-coated bacteria can also cleave C4.
C4
C2
Enzyme
C3 convertase
3. C3 convertase cleaves C3 to C3a + C3b. C3b joins with C3 convertase to make C5 convertase.
C3
C3 is shared by the alternative pathway.
C3b
C3a
C5 convertase
C5
The classical and alternative pathways are identical from here.
4. C5 convertase cleaves C5 to C5a + C5b.
C5a
C5b
5. C5b leads to formation of the membrane attack complex.
C7
C8
C5b
C6
C9

FIGURE 24.27 ▪ Classical complement cascade.

Why do we need complement with all the other immune responses at our disposal? The need becomes evident in individuals with complement deficiencies. These patients are extremely susceptible to recurrent septicemic (blood) infections by organisms such as *Neisseria gonorrhoeae* (the cause of gonorrhea) that normally do not survive forays into the bloodstream, because they are killed by complement. Why does reinfecting *Neisseria* not quickly succumb to all the other defenses, such as antibodies and cytotoxic T cells? It turns out that key antigens on the bacterial surface change shape over generations in a process called phase variation (see Sections 10.6 and 25.6). The new antigens are "invisible" to antibodies made against earlier versions.

Regulating Complement Activation

How do normal body cells prevent self-destruction following complement activation? The sequential assembly of the membrane attack complex provides several places where regulatory factors can intervene. One such factor is the host cell surface protein CD59. CD59 will bind any C5b-C8 complex trying to form in the membrane and prevent C9 from polymerizing. Thus, no pore is formed and the host cell is spared. (Complement will not normally attack uninfected or infected host cells, since both contain CD59.)

Another regulatory mechanism hinges on a normal serum protein called **factor H**. Factor H prevents the inadvertent activation of complement in the absence of infection. If factor H is unavailable, then the uncontrolled activation of complement can damage host cells despite the presence of CD59. To short-circuit the cascade, factor H binds to C3bBb, displaces Bb from the complex, and acts as a cofactor for factor I protease, which then cleaves C3b. Without C3b, the complement cascade stops. Some bacteria, such as certain strains of *Neisseria gonorrhoeae,* have learned to protect themselves from complement by binding factor H.

Microbiota can also assist host cells in resisting complement. The intestinal microbe *Bacteroides thetaiotaomicron* protects host cells from complement-mediated cytotoxicity by up-regulating a **decay-accelerating factor** present in host cell membranes. Decay-accelerating factor stimulates decay of complement factors and prevents their deposition at the cell surface. **Special Topic 24.1** describes another way *Bacteroides* modulates the immune system.

Other Roles for Complement Fragments

The C3a, C4a, and C5a cleavage fragments generated by the complement cascade do not participate in MAC formation but are important for amplifying the immune reaction. These peptides act as chemoattractants to lure more inflammatory cells into the area. C3b, which does participate in MAC formation, also can act as an opsonin when it is bound to a bacterial cell. An opsonin is a protein factor that can facilitate phagocytosis. Because phagocytes contain C3b receptors on their surfaces, it is easier for them to grab and engulf cells coated with C3b. Additional roles for complement fragments were discussed in Section 23.8.

Special Topic 24.1: An Uneasy Peace: Détente at the Microbiota-Intestine Interface

The epithelial surface of the intestinal mucosa comes in contact with an enormous amount of gut microbiota. So why doesn't the immune system kill these freeloaders? Or, for that matter, why isn't the intestinal tract damaged by chronic inflammation due to an immune response to the microbiota? The answers rely on a complex and well-regulated network of tolerance-inducing mechanisms present in the GALT (gut-associated lymphoid tissue) by which the inflammatory response is held in check.

First, why isn't the immune system activated to destroy the intestinal microbiota? Could our immune systems become tolerant to the presence of these organisms? Actually, there is an immune response. Intestinal dendritic cells sample commensal intestinal bacteria that inadvertently penetrate the epithelial layer and then, through antigen presentation in the regional mesenteric lymph nodes, induce IgA^{+} B cells that seed the mucosa with IgA plasma cells. Secretory IgA is transported through the epithelial layer into the mucosa, where it limits the penetration of commensal bacteria back through the epithelium and shapes the density of different bacterial species in the intestinal lumen. Thus, production of IgA is an important adaptation to the presence of commensal intestinal bacteria. Compartmentalization of the intestine's mucosal immune system allows a measured immune response toward commensals without needing the systemic immune system to become tolerant to these organisms. This is important because we must be able to mount a vigorous immune response in case an intestinal organism escapes to the bloodstream. Secreted IgA antibodies probably minimize systemic antibody responses to the microbiota by trapping bacterial antigens on mucosal surfaces, thereby decreasing movement into lymph nodes.

The second question is, why doesn't our gut immune system induce a vigorous inflammatory process when it comes in contact with gut microbiota? Were that the case, our intestines would be chronically inflamed. The answer to this question involves an additional $CD4^{+}$ T-cell type that we have discussed only briefly: **regulatory T cells (Tregs)**. Tregs present in our mucosa secrete a set of cytokines that suppress the production of pro-inflammatory cytokines made by other cells responding to MAMPs (microbiota-associated molecular patterns). Treg-secreted molecules also inhibit T_H1 responses involving cytotoxic T cells.

Several ingenious experiments have shown that commensal microbiota actually inhibit the gastrointestinal immune response by inducing Treg cell production. In one report, transgenic, gnotobiotic mice that produce the chimeric host regulator NF-kappaB–*lux* were used. NF-kappaB (NF-κB) is a transcription factor that activates production of inflammatory proteins. The factor is induced via innate immune pathways involving Toll-like receptors binding to MAMPs. The amount of NF-kappaB produced following an infection reflects the amount of inflammation that will result.

In this model, NF-kappaB was fused to luciferase, so when luciferin is injected into the mouse, light will be emitted wherever NF-kappaB–*lux* is made. The light emitted can be visualized from an intact, unharmed mouse using an in vivo technique called biophotonic imaging (**Fig. 1A**). The germ-free mouse on the left was given the pathogen *Salmonella enterica* serovar Typhimurium to stimulate inflammation, whereas the mouse on the right was fed *Bifidobacterium infantis* (which increased mouse T reg cells) for 3 weeks prior to *Salmonella* infection. The results 4 hours after *Salmonella* infection clearly show that the presence of the commen-

Thought Questions

24.13 Do bone marrow transplants in a patient with severe combined immunodeficiency (SCID) require immunosuppressive chemotherapy?

24.14 How can a stem cell be differentiated from a B cell at the level of DNA?

To Summarize

- **The classical pathway for complement activation** begins with an interaction between the Fc portion of an antibody bound to an antigen and C1 factor in blood. (The alternative pathway does not need antibody but begins when C3b binds to LPS on a bacterium.)
- **The Fc-C1 complex** reacts with C2 and C4, leading to production of C3b and a novel C5 convertase specific to the classical pathway.
- **After C5 convertase cleaves C5**, the classical pathway is the same as the alternative pathway, resulting in the formation of a membrane attack complex (MAC).
- **CD59 and/or factor H** prevents inappropriate MAC formation in host cells.

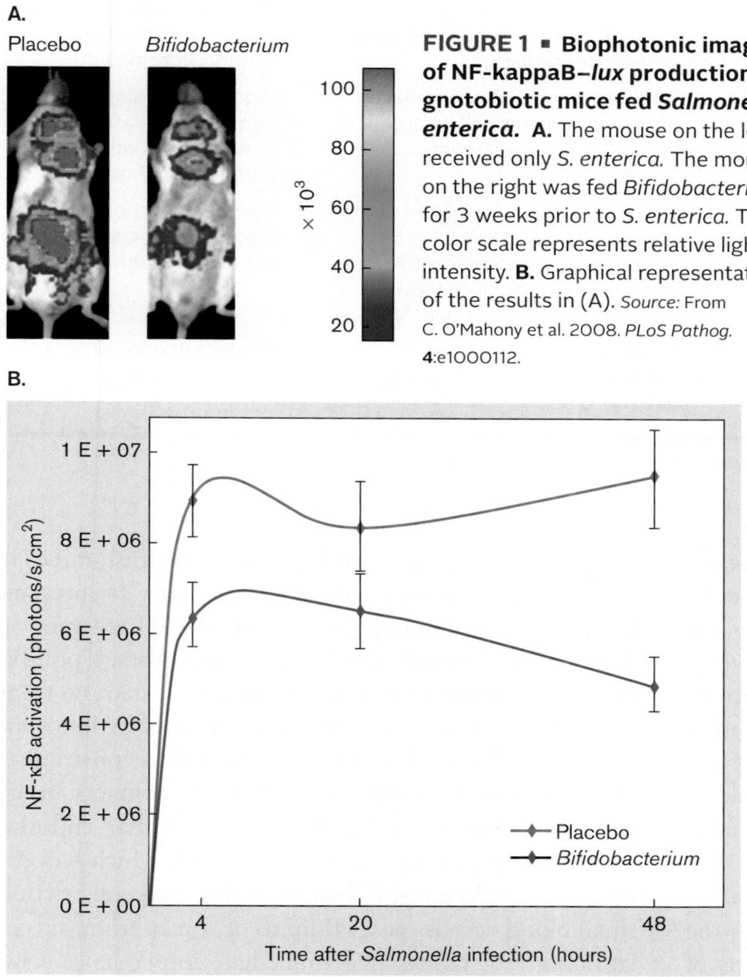
A.
Placebo
Bifidobacterium
100
80
60
40
20
× 10^3
B.
NF-κB activation (photons/s/cm^2)
1 E + 07
8 E + 06
6 E + 06
4 E + 06
2 E + 06
0 E + 00
4
20
48
Time after Salmonella infection (hours)
Placebo
Bifidobacterium

FIGURE 1 ▪ Biophotonic imaging of NF-kappaB–*lux* production in gnotobiotic mice fed *Salmonella enterica*. **A.** The mouse on the left received only *S. enterica.* The mouse on the right was fed *Bifidobacterium* for 3 weeks prior to *S. enterica.* The color scale represents relative light intensity. **B.** Graphical representation of the results in (A). *Source:* From C. O'Mahony et al. 2008. *PLoS Pathog.* **4**:e1000112.

sal organism *B. infantis* decreased the ability of *Salmonella* to activate NF-kappaB (**Fig. 1B**) and decreased the inflammatory response. The authors of the study further showed that feeding the gnotobiotic mice *B. infantis* increased the number of Treg cells produced, which reduced pro-inflammatory cytokine production.

The conclusion is that even a single commensal species can actually limit pro-inflammatory damage when the innate immune system is activated. It is not coincidental, then, that individuals with chronic inflammatory diseases often have decreased numbers of Treg cells. Indeed, a recent report shows that *Bifidobacterium* induces the production of Treg cells in humans too.

RESEARCH QUESTION

What effect might antibiotic treatment have on Treg production in a human? What, if any, consequence might develop?

Konieczna, P., D. Groeger, M. Ziegler, R. Frei, R. Ferstl, et al. 2012. *Bifidobacterium infantis* 35624 administration induces Foxp3 T regulatory cells in human peripheral blood: Potential role for myeloid and plasmacytoid dendritic cells. *Gut* **61**:354–366.

24.8

Hypersensitivity and Autoimmunity

Sometimes the immune system overreacts to certain foreign antigens and causes more damage than the antigen (microbe) alone might cause. Furthermore, some foreign antigens possess structures similar in shape to host structures and can trick the immune system into reacting against self. These immune miscues are called allergic hypersensitivity reactions, and the antigen causing the reaction is called an **allergen**. There are four types of hypersensitivity reactions (**Table 24.6**). The first three are antibody mediated; the fourth is cell mediated.

Case History: Type I Hypersensitivity

A bee stings a 9-year-old boy walking with his mother at the zoo. Within minutes, the boy begins sweating and itching. His chest then starts to tighten, and he has tremendous difficulty breathing. Terrified, he looks to his equally frightened mother for help.

TABLE 24.6

Summary of hypersensitivity reactions.

Type	Description	Time of onset	Mechanism[a]	Manifestations
I	IgE-mediated hypersensitivity	2–30 min	Ag induces cross-linking of IgE bound to mast cells with release of vasoactive mediators.	Systemic anaphylaxis, local anaphylaxis, hay fever, asthma, eczema
II	Antibody-mediated cytotoxic hypersensitivity	5–8 h	Ab directed against cell-surface antigens mediates cell destruction via ADCC or complement.	Blood transfusion reactions, hemolytic disease of the newborn, autoimmune hemolytic anemia
III	Immune complex–mediated hypersensitivity	2–8 h	Ag-Ab complexes deposited at various sites induce mast cell degranulation via Fc receptor; PMN degranulation damages tissue (localized reaction; **eTopic 24.4**).	Systemic reactions, disseminated rash, arthritis, glomerulonephritis
IV	Cell-mediated hypersensitivity	24–72 h	Memory T_H1 cells release cytokines that recruit and activate macrophages.	Contact dermatitis, tubercular lesions

[a]Ag = antigen; Ab = antibody; ADCC = antibody-dependent cell-mediated cytotoxicity; PMN = polymorphonuclear leukocyte.

This is a classic and severe example of **type I (immediate) hypersensitivity**, called **anaphylaxis**, in which smooth muscle contracts and capillaries dilate in response to the release of pharmacologically active substances. Type I hypersensitivity occurs within minutes of a <u>second</u> exposure to an allergen when the allergen reacts with IgE-coated mast cells. Recall that the primary role of IgE is to amplify the body's response to invaders by causing mast cells to release inflammatory mediators (see Section 24.3). Allergic individuals with type I hypersensitivity produce an excessive amount of IgE to the allergen. On initial exposure, an allergen elicits the production of IgE antibodies specific to the allergen (bee venom in this example). The Fc portions of these antibodies bind to Fc receptors on the surfaces of mast cells. IgE-coated mast cells are then said to be <u>sensitized</u>.

During a second exposure to the allergen, identical antigenic sites on the allergen bind to adjacent surface IgE molecules pointing out from the mast cell. The result is a bridge between adjoining binding sites (**Fig. 24.28**). This cross-linking inhibits **adenylate cyclase** in the mast cell. The resulting decrease in cellular cyclic adenosine monophosphate (cAMP) level causes mast cell granules to move to the cell surface, where they release their contents in a process called degranulation. Understanding this process is clinically important because anaphylaxis inhibitors such as epinephrine (also known as adrenaline) <u>stimulate</u> adenylate cyclase activity, stopping the anaphylaxis cascade in its tracks.

Mast cell degranulation releases chemicals with potent pharmacological activities. The most important of these is histamine, which binds to histamine receptors (H1 receptors) present on most body cells. Antihistamines have a structure similar to histamine and work by preventing histamine from binding (antagonist) the H1 receptor or by dampening the basal activity of H1 receptors. Upon binding to H1 receptors on smooth muscle, histamine triggers production of a signal molecule (inositol trisphosphate) that initiates smooth-muscle contraction to constrict small blood vessels. Histamine also weakens contacts between adhesion proteins (VE-cadherin) on vascular endothelial cells, causing gaps between cells through which blood fluids can seep. As a result, histamine-induced constriction of small blood vessels causes fluid to be forced from the circulation into the tissues. The immediate consequence is swelling (**edema**) in the joints and around the eyes and a rash (similar to hives), with burning and itching of the skin due to nerve involvement. In the clinical example involving the bee sting, the contraction of lung smooth muscles also led to breathing difficulties.

What we described was a severe type I allergic reaction caused by a bee sting. Type I hypersensitivity reactions do not usually involve the whole body. Most type I reactions are more localized and cause what is called atopic ("out of place") disease. Hay fever, or allergic rhinitis, is a common manifestation of atopic disease that can be caused by the inhalation of dust mite feces (**Fig. 24.29**), animal skin or hair (dander), and certain types of grass or weed pollens. This disease affects the eyes, nose, and upper respiratory tract. Atopic asthma, another form of type I hypersensitivity resulting from inhaled allergens, affects the lower respiratory tract and is characterized by wheezing and difficulty breathing.

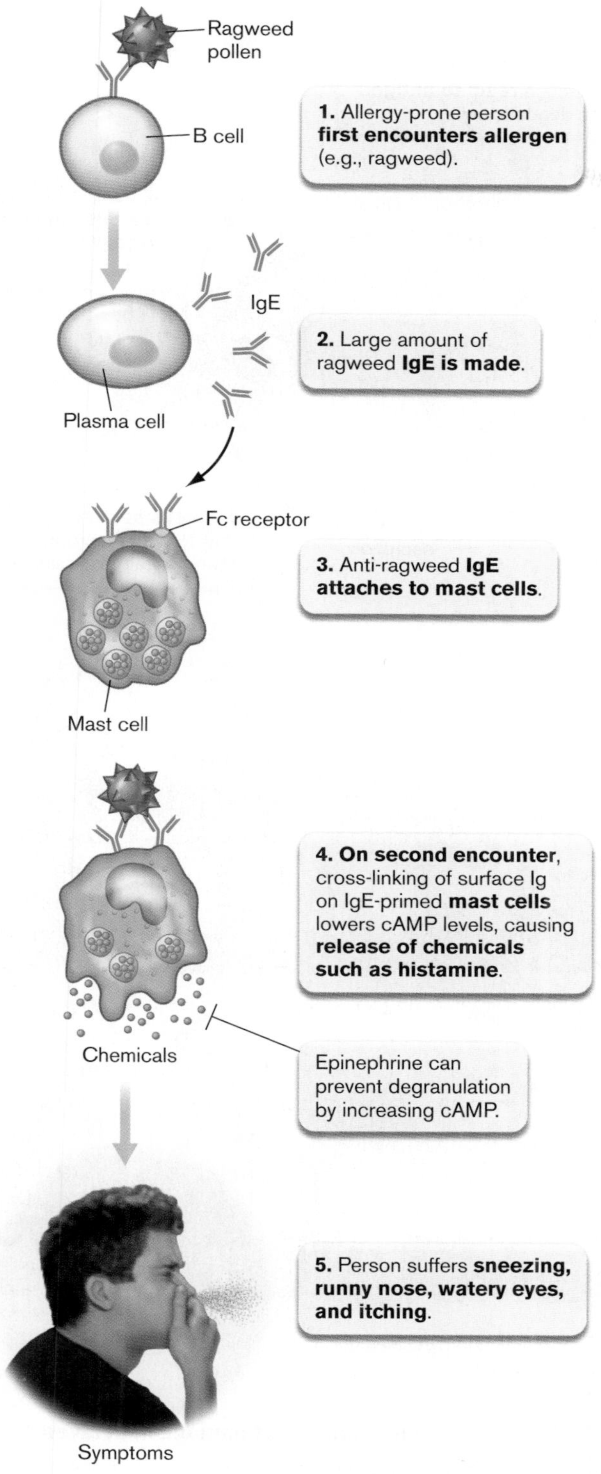
Ragweed pollen
B cell
1. Allergy-prone person **first encounters allergen** (e.g., ragweed).
IgE
Plasma cell
2. Large amount of ragweed **IgE is made**.
Fc receptor
3. Anti-ragweed **IgE attaches to mast cells**.
Mast cell
4. **On second encounter**, cross-linking of surface Ig on IgE-primed **mast cells** lowers cAMP levels, causing **release of chemicals such as histamine**.
Chemicals
Epinephrine can prevent degranulation by increasing cAMP.
5. Person suffers **sneezing, runny nose, watery eyes, and itching**.
Symptoms

FIGURE 24.28 ■ Events leading to type I hypersensitivity reactions.

The administration of antihistamines is an effective treatment of allergic rhinitis, but the more important chemical mediators of asthma are the leukotrienes produced during what is called **late-phase anaphylaxis**. In late-phase anaphylaxis, mast cells release chemotactic factors that call in eosinophils. Eosinophils entering the affected area produce large amounts of leukotrienes that,

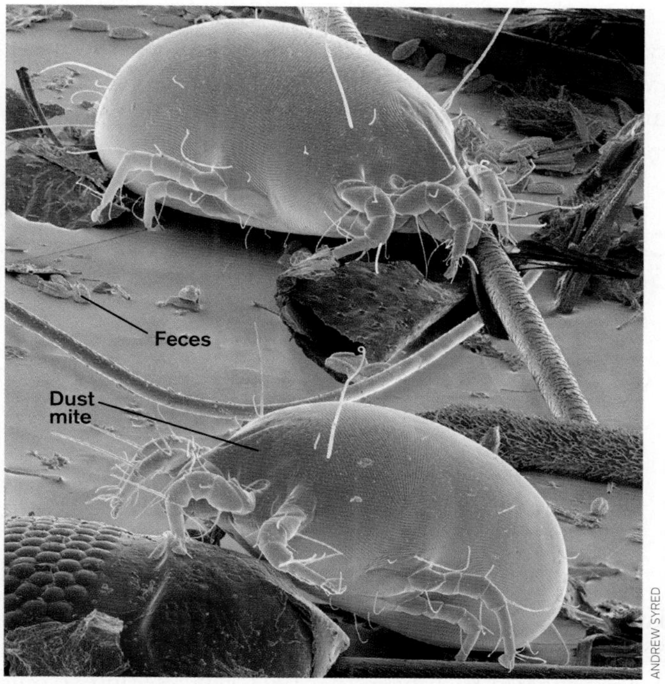
Feces
Dust mite
ANDREW SYRED

FIGURE 24.29 ■ Inhaled allergen. Dust mite (200–600 μm) and dust mite feces (colorized SEM).

like histamine, cause vasoconstriction and inflammation. At this point, antihistamines have little effect. Effective treatment includes inhaled steroids to minimize inflammation and bronchodilators (for example, albuterol) to widen bronchioles and facilitate breathing. One medication, Singulair, works by blocking leukotriene receptors in the lung. During severe asthma attacks, injected epinephrine is critical. Epinephrine will open airways to ease breathing through direct hormonal action.

People sensitized to allergens are not necessarily doomed to suffer with the allergy their entire life. A clinical treatment called **desensitization** can sometimes be used to prevent anaphylaxis. Desensitization involves injecting small doses of allergen over a period of months. This process is thought to produce IgG molecules that circulate, bind, and neutralize allergens before they contact sensitized mast cells. Desensitization has been useful in cases of asthma, food allergies, and bee stings, although results are variable and by no means guaranteed.

Another treatment for allergic asthma, omalizumab (Xolair), can actually prevent sensitization. Omalizumab is a monoclonal antibody (an antibody preparation of one antibody type that binds only one epitope). Administered by injection, omalizumab selectively binds IgE. Given to allergic patients once a month, the monoclonal antibody prevents IgE from binding to Fc receptors on mast cells, and thus prevents sensitization. A subsequent encounter with an allergen will not trigger an asthmatic attack.

Why don't all antigens generate type I hypersensitivity? It turns out that most antigens do not elicit high levels of IgE antibodies. It is unclear what gives certain antigens this capability and what makes different individuals prone to different allergies.

Case History: Type IV Hypersensitivity

The patient is an 8-year-old girl from Argentina. She recently received the BCG vaccine for tuberculosis before coming to the United States. Upon entering school, she is required to take a skin test for tuberculosis. She tries to refuse but is told it is a requirement and allows the nurse to apply the test to her arm. Three days later, the test site has a large red lesion and the skin is starting to slough (peel off).

Type IV hypersensitivity, also known as **delayed-type hypersensitivity (DTH)**, is the only class of hypersensitivity that is triggered by antigen-specific T cells. Because T cells have to react and proliferate to cause a response, it generally takes 24–48 hours before a reaction is noticed. This delay distinguishes type IV hypersensitivity from the more rapid antibody-mediated allergic reactions (types I–III). The girl in the case study was initially sensitized with BCG (an attenuated strain of *Mycobacterium bovis,* a close relative of *M. tuberculosis*) as a result of vaccination. Because the organism is intracellular, the vaccination produced a cell-mediated immunity, complete with preactivated memory T cells (similar to memory B cells). When she was reinoculated by the skin test, the memory T cells activated and elicited a localized reaction at the site of injection. Note: About 6 months after vaccination, the hypersensitivity usually diminishes and the tuberculin skin test becomes useful once again.

Type IV hypersensitivity develops in two stages. In the first stage (sensitization), antigen is processed and presented on cutaneous dendritic cells (called Langerhans cells). These APCs travel to the lymph nodes, where T_H0 cells can react to them as described earlier (see **Fig. 24.23**), generating activated T cells and a subset of memory T cells. On second exposure, two routes leading to DTH are possible. In the first pathway (**Fig. 24.30A**), memory T_H1 cells bind antigen that is complexed to class II MHC receptors (this happens at the site of infection) and release IFN-gamma, TNF-beta, and IL-2. These cytokines recruit macrophages and PMNs to the site and activate macrophages and natural killer cells to release inflammatory mediators that damage innocent, uninfected bystander host cells. In the second pathway (**Fig. 24.30B**), memory T_C cells recognize antigen on class I MHC receptors, become activated, and directly kill the host cell presenting the antigen. A hallmark of type IV hypersensitivity is that white blood cells, not serum, can transfer the sensitivity to a naive animal. This is because T cells, not serum antibody, cause the reaction.

Forms of DTH reactions include contact dermatitis (such as poison ivy) and allograft rejection. The antigens involved

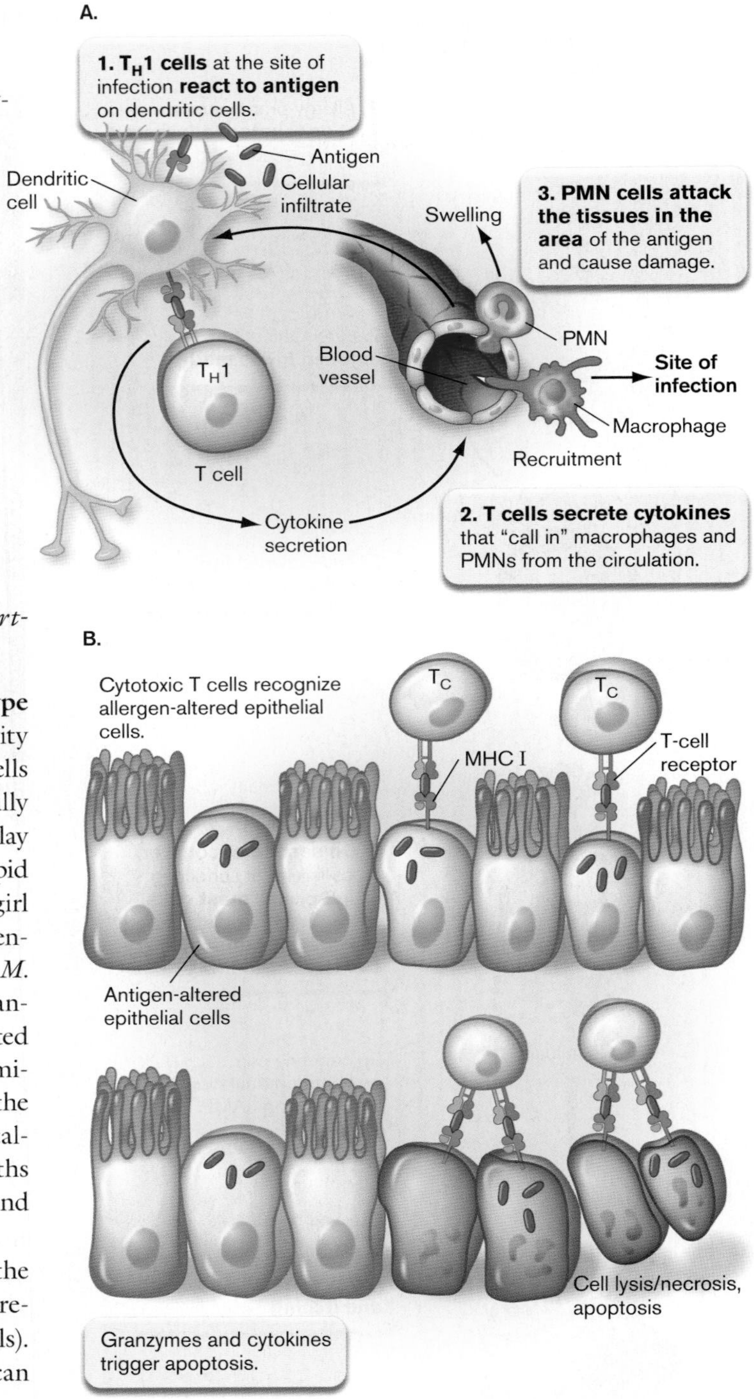
A.
1. TH1 cells at the site of infection react to antigen on dendritic cells.
Dendritic cell
Antigen
Cellular infiltrate
Swelling
3. PMN cells attack the tissues in the area of the antigen and cause damage.
PMN
Site of infection
Blood vessel
TH1
Macrophage
Recruitment
T cell
Cytokine secretion
2. T cells secrete cytokines that "call in" macrophages and PMNs from the circulation.
B.
Cytotoxic T cells recognize allergen-altered epithelial cells.
TC
TC
T-cell receptor
MHC I
Antigen-altered epithelial cells
Cell lysis/necrosis, apoptosis
Granzymes and cytokines trigger apoptosis.

FIGURE 24.30 ■ Mechanisms of damage in delayed-type hypersensitivity (type IV). **A.** T_H1 cells react to antigen on dendritic cells. **B.** Antigen-sensitized cytotoxic T cells recognize allergen haptens attached to proteins of other cells. This recognition triggers the release of granzymes and production of cytokines, which cause apoptosis of the target cell.

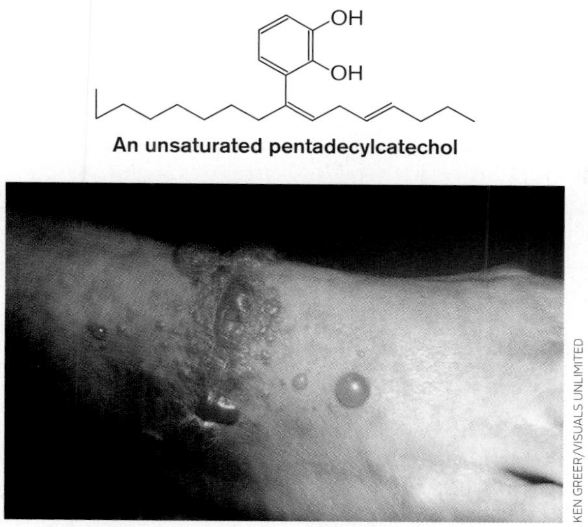
OH
OH
An unsaturated pentadecylcatechol
KEN GREER/VISUALS UNLIMITED

FIGURE 24.31 ■ Contact dermatitis. Pentadecylcatechols are chemicals present on the surface of poison ivy leaves. They are haptens that bind to proteins in dermal cells, where they can activate T cells. The result of the ensuing delayed-type hypersensitivity is contact dermatitis.

with contact dermatitis are usually small haptens that have to bind and modify normal host proteins to become antigenic (**Fig. 24.31**). In this case, the hapten-modified protein is processed, and fragments containing the bound hapten are presented on the surfaces of antigen-presenting cells.

There are two other types of hypersensitivity, both of which involve antigen-antibody complexes. **Type II hypersensitivity** occurs when antibody binds to host cell surface antigens. ABO blood group incompatibility, discussed in eTopic 24.2, is an example of type II hypersensitivity. **Type III hypersensitivity** happens when large complexes are formed between antibody and small soluble antigens. The complexes can trigger complement activation that leads to rash formation. Specific case histories and discussion of these types of hypersensitivity can be found in eTopic 24.4.

Autoimmunity: An Inability to Recognize Self

The ability to distinguish between self antigens and foreign antigens is crucial to our survival; without this ability, our immune system would constantly attack us from within (see eTopic 24.3). Normally, the body develops tolerance to self; occasionally, however, an individual loses immune tolerance against some self antigens and the body mounts an abnormal immune attack against its own tissues. The attack can involve antibodies or T cells and is called an **autoimmune response**. Autoimmune responses may or may not be associated with pathological changes (autoimmune disease). Almost 30% of the population will have an autoimmune antibody by age 65, but many will not exhibit disease. The mechanisms that lead to autoimmune disease are essentially hypersensitivity reactions.

An autoimmune response results in autoimmune disease when an autoantibody or autoimmune lymphocyte damages tissue components. Tissue damage can ensue when antibody to self activates antibody-dependent cell cytotoxicity (see Section 23.7). In this scenario, autoantibodies affixed to host cells can bind to NK cell Fc receptors and cause the NK cell to kill the tissue cell.

How might autoimmune antibodies be formed? One proposed mechanism starts with the occasional escape of self-reacting B cells from the negative selection process. Fortunately, self-reacting B cells that escape negative selection are usually not a problem, because the specific helper T cells needed to activate them were deleted from the T-cell population. How, then, can a self-reacting B cell be activated? The B-cell receptor on autoreactive B cells can, of course, take up and process a self antigen or a foreign antigen that mimics a self antigen. A foreign antigen that resembles self, however, will also contain one or more nonself epitopes that "piggyback" with the foreign epitope into the self-reacting B cell. As a result, the nonself epitope will also be presented on a B-cell MHC molecule. When this happens, T cells specific to the nonself epitope (these T cells were not deleted in the thymus) will bind the nonself epitope presented by the self-reacting B cell. T-cell binding activates the B cell to make antibody, but the antibody, and that of its descendant plasma cells, is directed to the self antigen, not the nonself antigen. Once made, the self-reacting antibodies begin attacking host antigen on host tissues.

A good example of autoimmunity caused by molecular mimicry is the M protein (pili) of *Streptococcus pyogenes,* a microbe that causes "strep throat" and the autoimmune disease rheumatic fever. M pili contain an epitope that resembles cardiac antigen, but the cardiac-like epitope is flanked by epitopes that are not related to the human host. The cardiac-like epitope can bind surface antibody on an escaped self-reacting B cell and be taken up. Because the nonself epitope and the heart-like epitope are contained within the same protein, the nonself epitope will "piggyback" its way into the B cell. The self epitope is processed and placed on the B-cell surface MHC. When this happens, T cells specific to the nonself epitope (M protein in this example) will recognize the nonself peptide and activate the B cell. Because the B cell is programmed to make antibody to the cardiac antigen, those self-reacting antibodies are made, begin to bind to cardiac tissue, and trigger autoimmune disease. In the case of *S. pyogenes,* cardiac tissue is damaged and rheumatic fever results. This similarity between the epitope in M protein and cardiac tissue is an example of antigenic mimicry.

Other autoimmune diseases involve the production of autoantibodies that bind and block certain cell receptors. Graves' disease (hyperthyroidism), for example, occurs

TABLE 24.7

Examples of autoimmune diseases.

Autoimmune disease	Autoantigen	Pathology
Type II hypersensitivity—mediated by antibody to cell-surface antigens		
Acute rheumatic fever	Streptococcal M protein, cardiomyocytes	Myocarditis, scarring of heart valves
Autoimmune hemolytic anemia	Rh blood group	Destruction of red blood cells by complement, phagocytosis
Goodpasture's syndrome	Basement membrane collagen	Pulmonary hemorrhage, glomerulonephritis
Graves' disease	Thyroid-stimulating hormone (TSH) receptor	Antibody stimulates T3, T4 production, hyperthyroidism
Myasthenia gravis	Acetylcholine receptor	Interrupts electrical transmission, progressive muscular weakness
Type III hypersensitivity—mediated by antibody complexes with small soluble antigens (immune complex)		
Systemic lupus erythematosus	DNA, histones, ribosomes	Arthritis, vasculitis, glomerulonephritis
Type IV hypersensitivity—mediated by antigen-specific T cells		
Type 1 diabetes	Pancreatic beta cell antigen	Beta cell destruction
Multiple sclerosis	Myelin protein	Demyelination of axons

when an autoantibody is made that binds to the thyroid-stimulating hormone (TSH) receptor. This receptor binding mimics actual TSH binding and continually stimulates the production of two hormones—triiodothyronine (T3) and thyroxine (T4)—which generally increases metabolic rate. The thyroid is not destroyed, but in fact enlarges to form a goiter. Examples of other important autoimmune diseases are listed in **Table 24.7**.

To Summarize

- **Allergens** cause the host immune system to overrespond or react against self.
- **Type I hypersensitivity** involves IgE antibodies bound to mast cells by the antibody Fc region. Binding of antigen to the mast cell–attached IgE causes mast cell degranulation. This type of hypersensitivity can occur within minutes of exposure.
- **Type IV hypersensitivity** (delayed-type hypersensitivity, or DTH) involves antigen-specific T cells. T_H1 cells release cytokines that activate macrophages and NK cells. T_C cells can directly kill cells that present the antigen. Reaction is seen within a few days of exposure.
- **Autoimmune disease** is caused by the presence of lymphocytes that can react to self.
- **Autoreactive B cells** (B cells that make antibody directed against self epitopes) can be activated if their BCR takes up a self-mimic epitope linked to a nonself epitope and presents the nonself epitope on the surface MHC. T cells that recognize the nonself epitope can then activate the B cell, which then secretes antibody against the self epitope.
- **Cytotoxic T cells can produce autoimmune disease** by killing host cells that make a self protein that closely resembles a foreign antigen.

Concluding Thoughts

Immunity is a remarkable feat of nature. The idea that any one person's immune system can recognize and respond to virtually any molecular structure and yet remain selectively "blind" to his or her own antigens is hard to comprehend. But even more remarkable, pathogenic microbes have managed ways to outmaneuver the interconnected redundancies and safeguards of the immune system. In some cases, the pathogen evolves to become less harmful and coexists with the host; indeed, pathogens in nature are far outnumbered by closely related strains that are harmless or even beneficial to their hosts. In other cases, however, evolution generates a never-ending arms race between pathogen and host. The next chapter describes some of the microbial strategies that contribute to the success of a pathogen.

CHAPTER REVIEW

Review Questions

1. Define "antigen," "epitope," "hapten," and "antigenic determinant."
2. What is the basic difference between humoral immunity and cellular immunity?
3. Why are proteins better immunogens than nucleic acids?
4. What makes IgA antibody different from IgG?
5. What is immunoprecipitation?
6. Explain isotypic, allotypic, and idiotypic differences in antibodies.
7. How is IgE involved in allergic hypersensitivity?
8. Discuss differences in the primary and secondary antibody responses.
9. Outline the basic steps that turn a B cell into a plasma cell.
10. What is isotype switching, and how is antibody diversity achieved?
11. What signals are needed to activate helper T cells?
12. What signals activate cytotoxic T cells?
13. Discuss the differences between transplant rejections of nucleated cells and the rejection of blood cells.
14. How do superantigens activate T cells?
15. Discuss the differences between the alternative and classical pathways of complement activation.
16. How does the host prevent membrane attack complexes from being formed in host cells?
17. Describe the development of type I hypersensitivity.
18. How does a B cell programmed to make an antibody against self become activated in the absence of specific T-cell help?

Thought Questions

1. Cytotoxic T cells lyse the membranes of host cells carrying viruses or bacteria. Why doesn't this lysis facilitate the spread of those organisms rather than help clear the infection?
2. Immunity raised by the poliovirus vaccine will prevent subsequent infection by poliovirus, but someone with an *S. aureus* infection (and the attendant immune response) can be reinfected many times by *S. aureus*. Why does the immune system work so well against infection by some pathogens (for example, poliovirus) but not others (for example, *Staphylococcus aureus*)?
3. (Refer to **eTopic 24.2**.) Why are blood group type O people "universal donors" (meaning they can donate their red blood cells to type O, type A, type B, or type AB individuals), but not universal acceptors of red blood cells from type O, A, B, or AB individuals? Which blood type person might be considered a universal recipient?
4. IgM is the first antibody produced during a primary immune response. Of the five different types of antibodies produced, why would the body want IgM to be the first?
5. Some pathogens, such as rubella virus, can cross the placental barrier from mother to the fetus and cause a dangerous infection. Which antibody isotype against rubella virus would you look for in a newborn to diagnose congenital rubella syndrome— IgG or IgM?

Key Terms

adaptive immunity (962)
adenylate cyclase (996)
affinity maturation (981)
allergen (995)
allotype (971)
anaphylaxis (996)
antibody (963)
antigen (963)
antigen-presenting cell (APC) (963, 982)
antigenic determinant (963)
autoimmune response (999)
B-cell receptor (975)
B-cell tolerance (965)
capping (976)
cell-mediated immunity (963)
class I MHC molecule (983)
class II MHC molecule (983)
class switching (974)
classical complement pathway (992)
clonal (963)
clonal selection (975)
constant region (970)
cytokine (980)
cytotoxic T cell (T_C cell) (982)
decay-accelerating factor (993)
delayed-type hypersensitivity (DTH) (998)
desensitization (997)
edema (996)
epitope (963)
equivalence (970)
factor H (993)
Fc region (971)

gene switching (978)
granzyme (988)
hapten (967)
heavy chain (969)
helper T cell (T_H cell) (980)
humoral immunity (963)
idiotype (971)
immediate hypersensitivity (996)
immunogen (963, 967)
immunogenicity (965)
immunoglobulin (968)
IgA (972)
IgD (972)
IgE (973)
IgG (972)
IgM (972)
immunological specificity (966)
immunoprecipitation (969)
isotype (971)
isotype switching (974)
late-phase anaphylaxis (997)
light chain (969)
major histocompatibility complex (MHC) (965, 982)
memory B cell (974)
MHC restriction (986)
mucosal immunity (981)
negative selection (986)
opsonize (972)
perforin (988)
plasma cell (974)
positive selection (986)
primary antibody response (974)
recombination signal sequence (RSS) (978)
regulatory T cell (Treg) (986, 994)
secondary antibody response (974)
serum (974)
superantigen (989)
switch region (981)
threshold dose (965)
tumor necrosis factor (TNF) (989)
type I hypersensitivity (996)
type II hypersensitivity (999)
type III hypersensitivity (999)
type IV hypersensitivity (998)
vaccination (966)
variable region (971)

Recommended Reading

Arrieta, Marie-Clair, and Brett B. Finlay. 2012. The commensal microbiota drives immune homeostasis. *Frontiers in Immunology* **3**:33. [Online.] http://www.frontiersin.org/Molecular_Innate_Immunity/10.3389/fimmu.2012.00033/full.

Bonneville, Marc, Rebecca L. O'Brien, and Willi Born. 2010. Gammadelta T cell effector functions: A blend of innate programming and acquired plasticity. *Nature Reviews. Immunology* **10**:467–478.

Christensen, Jeanette E., and Allan R. Thomsen. 2009. Coordinating innate and adaptive immunity to viral infection: Mobility is the key. *Acta Pathologica Microbiologica et Immunologica Scandinavica* **117**:338–355.

Cyster, Jason G. 2010. B cell follicles and antigen encounters of the third kind. *Nature Immunology* **11**:989–996.

Edelman, Sanna M., and Dennis L. Kasper. 2008. Symbiotic commensal bacteria direct maturation of the host immune system. *Current Opinion in Gastroenterology* **24**:720–724.

Joffre, Olivier P., Elodie Segura, Ariel Savina, and Sebastian Amigorena. 2012. Cross-presentation by dendritic cells. *Nature Reviews. Immunology* **12**:557–569.

Jost, Stephanie, and Marcus Altfeld. 2012. Evasion from NK cell-mediated immune responses by HIV-1. *Microbes and Infection* **14**:904–915.

Macpherson, Andrew J., and Emma Slack. 2007. The functional interactions of commensal bacteria with intestinal secretory IgA. *Current Opinion in Gastroenterology* **23**:673–678.

Noriega, Vanessa, Veronika Redmann, Thomas Gardner, and Domenico Tortorella. 2012. Diverse immune evasion strategies by human cytomegalovirus. *Immunologic Research* **54**:140–151.

O'Mahony, Caitlin, Paul Scully, David O'Mahony, Sharon Murphy, Francis O'Brien, et al. 2008. Commensal-induced regulatory T cells mediate protection against pathogen-stimulated NF-kappaB activation. *PLoS Pathogens* **4**:e1000112.

Palucka, Karolina, and Jacques Banchereau. 2012. Cancer immunotherapy via dendritic cells. *Nature Reviews. Cancer* **12**:265–277.

Robinson, Harriet L., and Rama Rao Amara. 2005. T-cell vaccines for microbial infections. *Nature Medicine* **11**:s25–s32.

Savino, Wilson. 2006. The thymus is a common target organ in infectious diseases. *PLoS Pathogens* **2**:472–483.

Spencer, Jo, Linda S. Klavinskis, and Louise D Fraser. 2012. The human intestinal IgA response; burning questions. *Frontiers in Immunology* **3**:108. [Epub.] doi:10.3389/fimmu.2012.00108.

Weill, Jean-Claude, Sandra Weller, and Claude-Agnès Reynaud. 2009. Human marginal zone B cells. *Annual Review of Immunology* **27**:267–285.

CHAPTER 25

Microbial Pathogenesis

25.1 Host-Pathogen Interactions

25.2 Virulence Factors and Pathogenicity Islands

25.3 Microbial Attachment: First Contact

25.4 Toxins Subvert Host Function

25.5 Deploying Toxins and Effectors

25.6 Surviving within the Host

25.7 Experimental Tools That Define Pathogenesis

Mammals have elaborate physical, chemical, and immunological defenses that protect against disease-causing microbes, or pathogens, yet every fortress has its weakness. Pathogenic microbes exploit those weaknesses, and the result is disease.

How does a pathogen differ from a commensal organism? The answer varies, depending on the pathogen. Some avoid phagocytosis by host cells, whereas others actively encourage it. Even more mysterious are pathogens that develop a latent, undetectable stage in the host, and then later emerge to cause disease. Some pathogens efficiently slay their hosts. Others persist for many years without killing. Which is the more successful pathogenic strategy?

The fundamental question of microbial pathogenesis is how an organism too small to be seen with the naked eye can kill a human that is 1 million times larger. Here in Chapter 25 we explore the strategies that different bacterial and viral pathogens use to accomplish this feat.

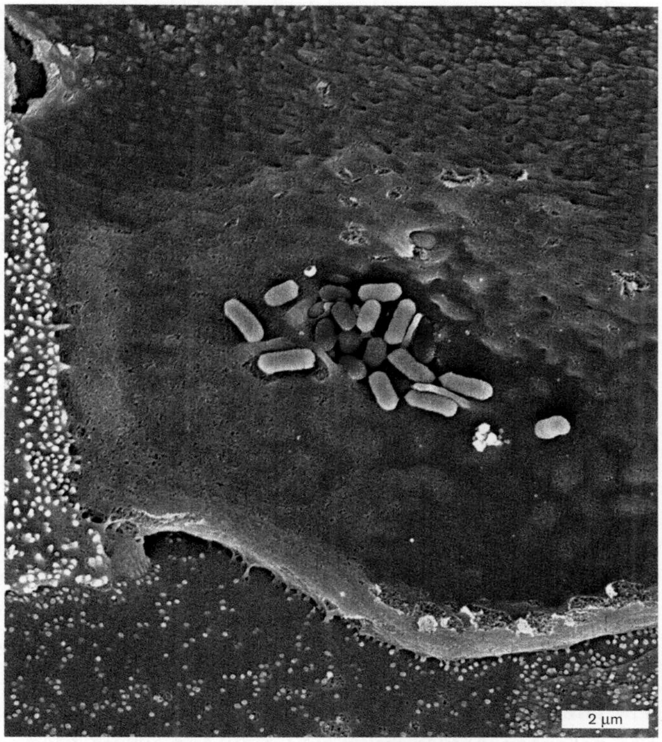

2 μm

A. TALHOUN ET AL. 2012. *CELL HOST MICROBE* **12**:645–656.

CURRENT RESEARCH **highlight**

***Salmonella* "morphs" intestinal host cells into "gateways."** *Salmonella enterica* (an important cause of typhoid fever) is ingested, crosses the gut epithelial layer, and uses macrophages as taxis to move through the host. *Salmonella* specifically targets antigen-sampling microfold (M) cells within intestinal Peyer's patches to transit the gut epithelium. Curiously, only a small number of these gateway M cells exist within Peyer's patches, but their numbers increase during *Salmonella* infection. How? This SEM image shows *Salmonella* converting an intestinal follicular-associated epithelial (FAE) cell (another cell type in Peyer's patches) into an M cell. *Salmonella* uses a syringe-like secretion system to "inject" the effector protein SopB into an FAE cell. SopB "tells" the FAE cell to become an M cell and effectively increases the number of gateways to deeper tissues. Bacteria attached to the FAE cell surface are artificially colored yellow, while the invading bacteria are red.

A 23-year-old Hispanic mother brought her 3-year-old daughter into the emergency room. The child was lethargic, had a fever of 40°C (104°F), and was having difficulty breathing. The mother explained that the family had arrived in the United States from El Salvador the previous week. The attending physician noted an extreme swelling of the child's cervical lymph nodes, giving the girl a thick, "bull-neck" appearance. She also noticed the beginnings of a membranous growth at the back of the child's throat that was starting to obstruct the trachea. The pseudomembrane was grayish in color and bled when scraped. When asked, the distraught mother admitted that the child had not received any vaccinations before arriving in the United States. Suspecting the nature of the child's illness, the physician granted immediate admission to the hospital and ordered administration of penicillin and a specific antitoxin. The results of a throat swab sent to the microbiology lab confirmed the physician's suspicion. The root of the child's disease was *Corynebacterium diphtheriae,* which causes diphtheria.

C. diphtheriae is a deadly pathogen able to kill humans by attacking the respiratory and cardiac systems. When untreated, death often comes by suffocation when the gray membrane completely covers the trachea. This pathogen is a Gram-positive, non-spore-forming rod identical in appearance to many common commensal throat microbes, such as the more docile *Corynebacterium striatum.* So, two Gram-positive rods, identical in appearance, are both found in the human throat. Why is one a killer and the other not? The child with diphtheria most likely came in contact with the pathogen before leaving El Salvador, where vaccinations in some areas are difficult to obtain. The question we ask in this chapter is not how the disease could have been prevented, but what genetic distinctions differentiate closely related disease-causing pathogens from innocuous nonpathogens. In the case of diphtheria, the difference between friend and foe is a bacteriophage genome embedded in the genome of the bacterium. This phage carries the gene for diphtheria toxin, whose properties will be described later.

Pathogens, such as *C. diphtheriae,* that kill their hosts or only transiently reside there must also be prepared for life outside the host. These microbes possess an alternate physiology that allows them to survive in nonhost environments such as a lake or soil. Other pathogens, however, are not as versatile and die when separated from their host. To survive, they must be passed directly from person to person. If the host dies before transmission, the pathogen dies with it. Thus, to kill or not to kill is an important question each pathogen must address. In this chapter we discuss various relationships that occur between pathogens and their hosts and the factors that contribute to **pathogenesis**, the process by which microbes cause disease in a host. The degree of harm that is caused depends on the mechanisms the pathogen has at its disposal.

25.1

Host-Pathogen Interactions

How long have we humans and our evolutionary ancestors suffered with infections? Millions of years, it turns out. Paleopathologists found evidence of brucellosis in a skeleton from an *Australopithecus africanus* male, a predecessor of *Homo sapiens* that lived over 2 million years ago. The vertebrae of this "person" exhibited damage characteristic of disease caused by a pathogenic species of *Brucella* (**Fig. 25.1**). Even though we have long been plagued by infectious diseases, the idea that these diseases can be caused by tiny living organisms invading our bodies became apparent only 150 years ago when Robert Koch discovered the microbial cause of anthrax.

To Catch a Pathogen

How are the microscopic agents of infectious disease discovered? The revelation that *Bacillus anthracis* causes anthrax led Koch to propose a set of steps, or postulates, (discussed in Section 1.3) needed to prove that a specific microbe causes a specific disease. Koch's postulates state that the organism must be present in every case of a disease, must be propagated in pure culture, must cause the same disease when inoculated into a naive host, and must be recovered from the newly diseased host.

Viruses, however, cannot grow in pure culture. Thus, viruses causing disease cannot meet Koch's postulates. To accommodate viral diseases, Koch's criteria were modified by Thomas Rivers in 1937 to include cultivating the agent in host cells (rather than in pure culture), proving that the agent passes through a 0.2-µm filter (bacteria do not), and demonstrating an immune response to the virus in patients. These steps were used in 2003 to rapidly discover the virus causing severe acute respiratory syndrome (SARS) and in 2013 to discover a related virus causing Middle Eastern Respiratory Syndrome (MERS). Both viruses were grown in a macaques monkey model, fulfilling Koch's postulates.

Fulfilling Koch's and Rivers's postulates remains the most persuasive evidence of causation, but there are some problems with this standard. Many agents cannot be cultured, and there may be no suitable animal model in which

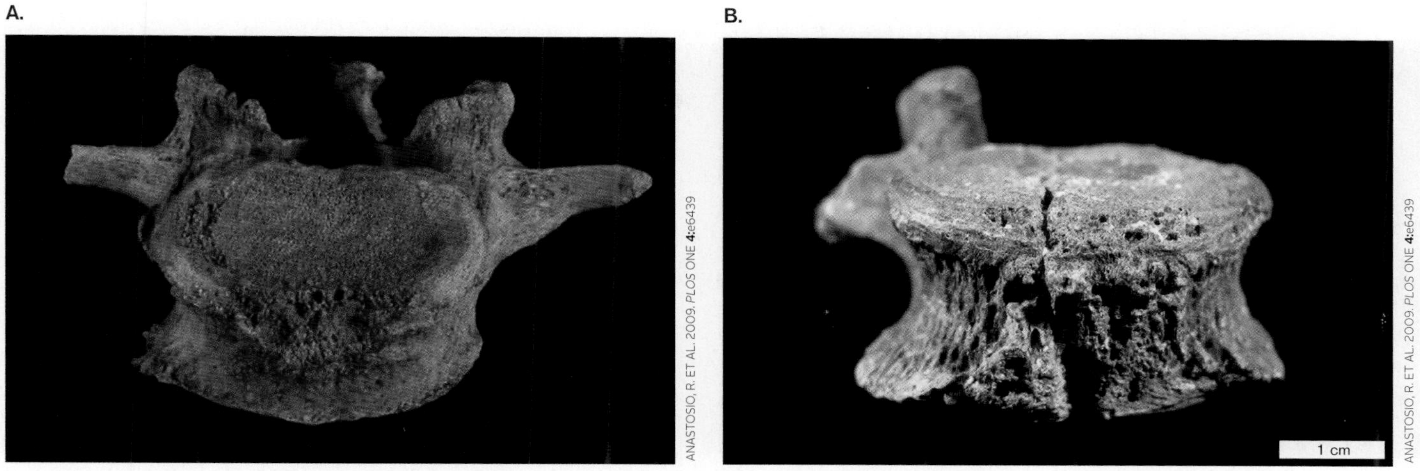
A.
B.
ANASTOSIO, R. ET AL. 2009. PLOS ONE 4:e6439
1 cm
ANASTOSIO, R. ET AL. 2009. PLOS ONE 4:e6439

FIGURE 25.1 ■ Paleopathological evidence of brucellosis in an ancestor of *Homo sapiens*. **A.** Modern human L4 vertebra affected by brucellosis. **B.** *Australopithecus africanus* L4, showing lesions consistent with brucellosis.

disease can be reproduced. In these situations we must resort to a statistical association between organism and disease based on the presence of the agent or its footprints (nucleic acid, antigen, and preferably, an immune response; see Chapter 28).

Knowing that some microbes cause disease while others do not raises an interesting question: What unique molecular mechanisms do disease-causing microbes use to cause disease?

The Language of Pathogenesis

Before discussing the microbial mechanisms of disease, we should first establish the vocabulary of infectious disease. The term **parasites**, in the broadest sense, includes bacteria, viruses, fungi, protozoa, and worms that colonize and harm their hosts. In practice, however, only disease-causing protozoa and worms are called parasites, whereas bacterial, viral, and fungal agents of disease are referred to as **pathogens**. Pathogens and parasites infect their animal and plant hosts in a variety of ways and enter into a variety of host-pathogen relationships, depending on the site of colonization. For example, organisms that live on the surface of a host are called **ectoparasites**. The fungus *Trichophyton rubrum*, one cause of athlete's foot, is an ectoparasite (**Fig. 25.2**). *Wuchereria bancrofti*, the worm parasite that causes elephantiasis, is an **endoparasite** because it lives inside the body (**Fig. 25.3**).

An **infection** occurs when a pathogen or parasite enters or begins to grow on a host, but the term "infection" does not necessarily imply overt disease. Any potential pathogen growing in or on a host is said to cause an infection, but that infection may be only transient if immune defenses kill the pathogen before noticeable disease results. Indeed,

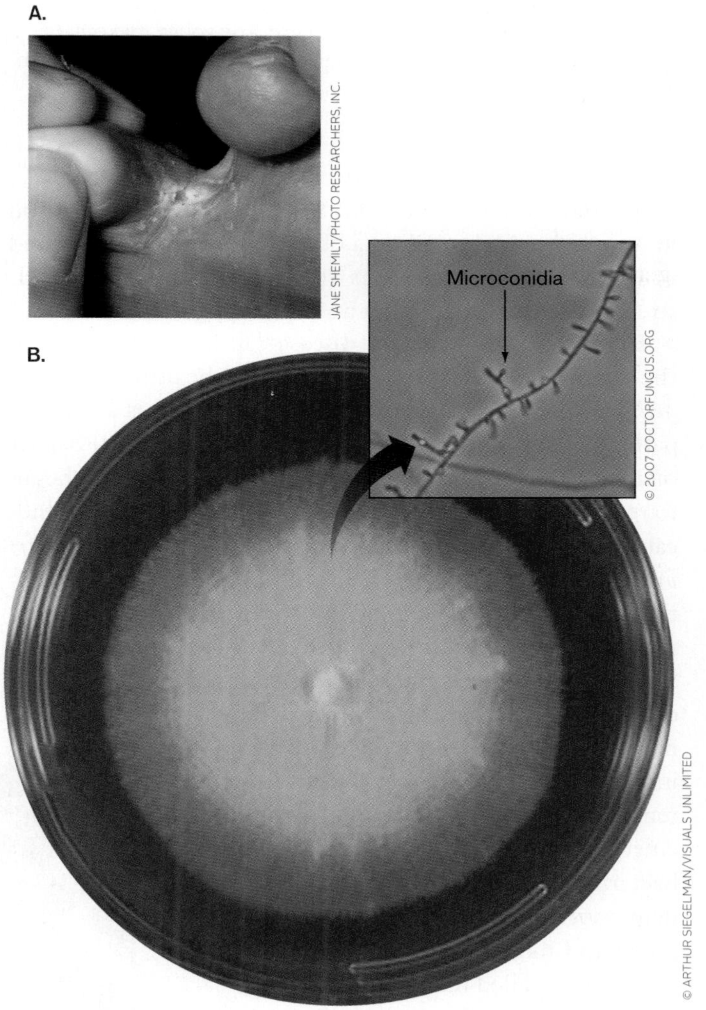
A.
JANE SHEMILT/PHOTO RESEARCHERS, INC.
B.
Microconidia
© 2007 DOCTORFUNGUS.ORG
© ARTHUR SIEGELMAN/VISUALS UNLIMITED

FIGURE 25.2 ■ An ectoparasite. **A.** Athlete's foot can be caused by the fungus *Trichophyton rubrum.* **B.** Colony morphology and microscopic, branching conidia (blowup) of *T. rubrum.* Conidia are asexual spores that grow on stalks called conidiophores (see Chapter 20).

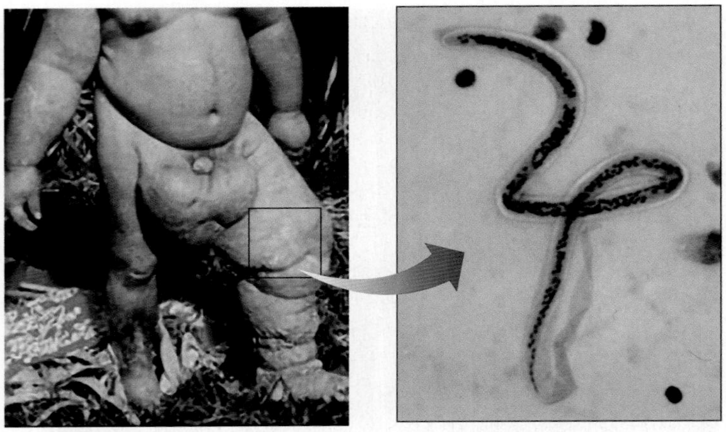

W. PETERS. 1992. *A COLOUR ATLAS OF ARTHROPODS IN CLINICAL MEDICINE.* ELSEVIER.

CDC

FIGURE 25.3 ■ An endoparasite. The disease filariasis, commonly known as "elephantiasis" for obvious reasons, is caused by the worm *Wuchereria bancrofti* (blowup), which enters the lymphatics and blocks lymphatic circulation. Adult worms are threadlike and measure 4–10 cm in length. The young microfilariae (shown) are approximately 0.5 mm in length. Though not a problem in the United States, *W. bancrofti* and elephantiasis are found throughout middle Africa, Asia, and New Zealand.

most infections go unnoticed. For example, every time you have your teeth cleaned by a dentist, your gums bleed and your resident oral microbes transiently enter the bloodstream, but you rarely suffer any consequences.

Primary pathogens are disease-causing microbes with the means to breach the defenses of a healthy host. *Shigella flexneri,* the cause of bacillary dysentery, is a primary pathogen. When ingested, it can survive the natural barrier of an acidic (pH 2) stomach, enter the intestine, and begin to replicate. **Opportunistic pathogens**, on the other hand, cause disease only in a compromised host. *Pneumocystis jirovecii* (previously *P. carinii*) is an opportunistic pathogen that causes life-threatening infections in AIDS patients, whose immune systems have been eroded (**Fig. 25.4A**). Some microbes even enter into a **latent state** during infection, in which the organism cannot be found by culture. Herpes virus, for instance, can enter the peripheral nerves and remain dormant for years, and then suddenly emerge to cause cold sores (**Fig. 25.4B**). The bacterium *Rickettsia prowazekii* causes epidemic typhus, but it can also enter a latent phase and months or years later cause a disease relapse called recrudescent typhus.

The term **pathogenicity** refers to an organism's ability to cause disease. It is defined in terms of how easily an organism causes disease (infectivity) and how severe that disease is (virulence). Pathogenicity, overall, is shaped by the genetic makeup of the pathogen. In other words, an organism is more—or less—pathogenic, depending on the tools at its disposal (such as toxins) and their effectiveness.

Virulence is a measure of the degree or severity of disease. For instance, Ebola virus and the closely related Marburg virus have case fatality rates near 70%, so they are highly virulent (**Fig. 25.5**). By contrast, rhinovirus, the cause of the common cold, is very effective at causing disease but almost never kills its victims, so it is highly infective but has a low virulence. Both organisms are pathogenic, but with one you live and the other you die.

One way to measure virulence is to determine how many bacteria or virions are required to kill 50% of an experimental group of animal hosts. This is called the **lethal dose 50% (LD_{50})**. A pathogen with a low LD_{50}, in which very few organisms (or viruses) are required to kill 50% of the hosts, is more virulent than one with a high LD_{50} (**Fig. 25.6**). For organisms that colonize but do not kill the host, the infectious dose needed to colonize 50% of the experimental hosts—that is, the **infectious dose 50% (ID_{50})**—can be measured. ID_{50} is measured by determining how many microbes are required to cause disease symptoms in half of an experimental group of hosts.

Although it might be possible to measure the infectious dose, rather than lethal dose, for a lethal pathogen, it is not typically done. LD_{50} gives a clear end point, so it is much easier to use when trying to determine the effectiveness of a given treatment (an antibiotic, for example) or to quantify the role of a given gene in pathogenesis.

Thought Question

25.1 **Figure 25.6** presents the association between LD_{50} and virulence. Is a microbe with an LD_{50} of 5×10^4 more or less virulent than a microbe with an LD_{50} of 5×10^7?

A. **B.**

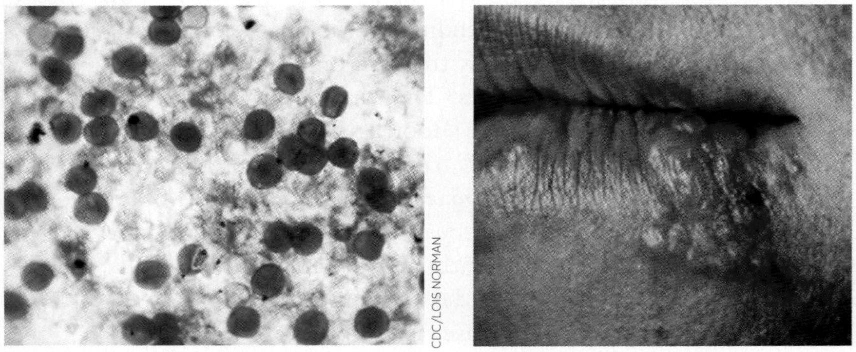

CDC/LOIS NORMAN CDC/HERMANN

FIGURE 25.4 ■ Opportunistic and latent infections. A. *Pneumocystis jirovecii* cysts (5–10 μm in diameter) in bronchoalveolar material. Notice that the fungi look like crushed Ping-Pong balls. **B.** Cold sore produced by a reactivated herpes virus hiding latent in nerve cells.

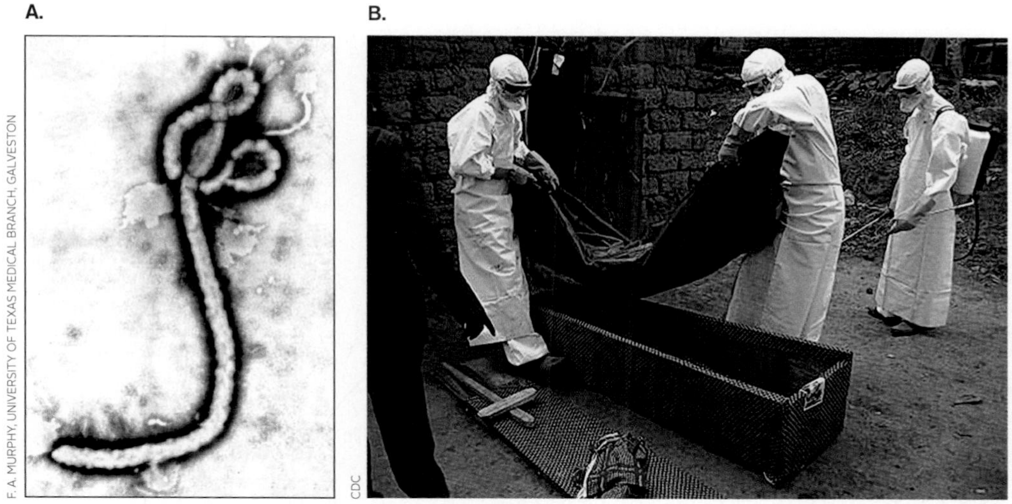
A.
B.
F. A. MURPHY, UNIVERSITY OF TEXAS MEDICAL BRANCH, GALVESTON
CDC

FIGURE 25.5 ▪ Highly virulent viruses. A. Ebola virus (approx. 1 μm long, TEM). **B.** The body of a victim of Marburg virus is placed in a coffin for safe burial in Angola. Marburg and Ebola cause hemorrhagic infections in which patients bleed from the mouth, nose, eyes, and other orifices. They have a 70%–80% mortality rate.

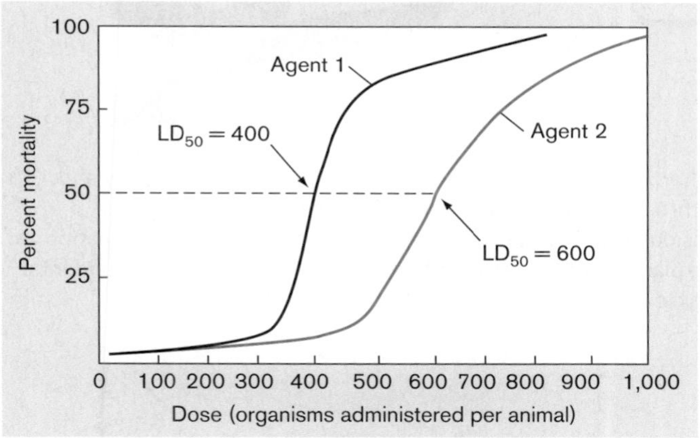
100
75
50
25
Percent mortality
Agent 1
Agent 2
$LD_{50} = 400$
$LD_{50} = 600$
0 100 200 300 400 500 600 700 800 900 1,000
Dose (organisms administered per animal)

FIGURE 25.6 ▪ Measurement of virulence. Each LD_{50} measurement requires infecting small groups of animals with increasing numbers of infectious agent and observing how many animals die. The number of microbes that kill half the animals is called the LD_{50} dose. In this example, agent 1 is more virulent than agent 2.

Infection Cycles Can Be Direct or Indirect

We tend to separate medical and environmental considerations when discussing pathogens, but greater insight can be gained from studying pathogens when both perspectives are considered. Most pathogens maintain a significant presence outside the human host. Some, such as the fungus *Histoplasma* (found in soil) or the bacterium *Legionella pneumophila* (found in natural water), find sanctuary in the natural environment. Many pathogens, such as the eastern equine encephalitis virus or avian flu virus, sustain their numbers in animals or insects. Keep this in mind as we discuss various infectious diseases, because many virulence factors actually evolved as adaptations within diverse nonhuman environments. Environment and evolution are integral factors to consider in the study of infectious disease.

Pathogens must pass somehow from one person or animal to another if a disease is to spread. The route of transmission an organism takes is called the **infection cycle**. A cycle of infection can be simple or complex (**Fig. 25.7**). Organisms that spread directly from person to person, such as rhinovirus or *Shigella,* have simple infection cycles. A person can also transmit some infectious agents to another person by contaminating food during preparation. Inanimate objects through which pathogens can be relayed to hosts are called **fomites** (for example, the tissue used to stifle a sneeze). More complex cycles often involve **vectors**, usually insects or ticks (arthropods), as intermediaries. Vectors carry infectious agents from one animal to another. A mosquito vector, for example, transfers the virus causing yellow fever from infected to uninfected individuals (**Fig. 25.8**) in what is called **horizontal transmission**. The mosquito can also bequeath this particular virus to its offspring via infected eggs in a form of **vertical transmission** called **transovarial transmission**. Although yellow fever is not a problem in the United States today, West Nile virus, a flavivirus closely related to yellow fever, currently claims several victims each year in this country.

Because insects and ticks are instrumental in transmitting pathogens, killing these arthropod vectors is an important way to halt the spread of disease. Interventions include spraying insecticide in a community during egg-hatching season or using other microbes as assassins "trained" to kill the vector. For example, *Bacillus thuringiensis* will kill many types of insects that carry infectious agents. More recently, an insect virus called baculovirus has been developed that kills the *Culex* mosquito vectors carrying the West Nile virus. The advantage of these vector-targeting microbes is that they do not kill other insects or animals, as do many chemical insecticides.

Another critical factor in an infection cycle is the "reservoir" of infection. A **reservoir** is an animal, bird, or insect that normally harbors the pathogen. In the case of yellow fever, the mosquito is not only the vector, but the reservoir as well, because the insect can pass the virus to future

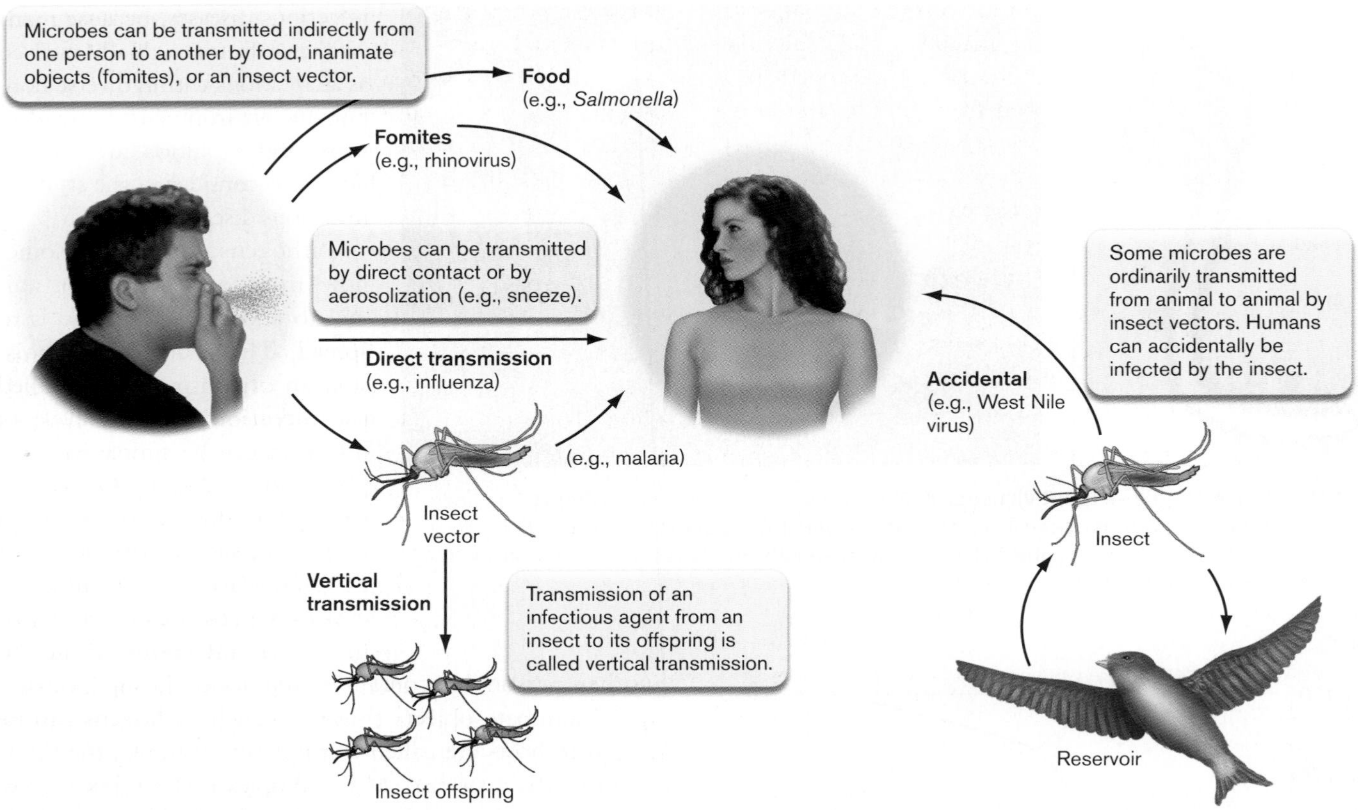

Microbes can be transmitted indirectly from one person to another by food, inanimate objects (fomites), or an insect vector.
Food
(e.g., Salmonella)
Fomites
(e.g., rhinovirus)
Microbes can be transmitted by direct contact or by aerosolization (e.g., sneeze).
Direct transmission
(e.g., influenza)
(e.g., malaria)
Insect vector
Vertical transmission
Transmission of an infectious agent from an insect to its offspring is called vertical transmission.
Insect offspring
Some microbes are ordinarily transmitted from animal to animal by insect vectors. Humans can accidentally be infected by the insect.
Accidental
(e.g., West Nile virus)
Insect
Reservoir

FIGURE 25.7 ■ **Infection cycles.** Infectious agents can be transmitted horizontally from one member of a species to another by a variety of means, such as food, fomites, aerosolization, direct contact, or arthropod (insect or tick) vector. Vertical transmission is the passage of a pathogen from parent to offspring. In insects, vertical transmission occurs because the egg itself is infected during formation. In humans vertical transmission occurs when the organism passes through the placenta to the fetus or to the newborn during birth. Accidental transmission happens when a host that is not part of the normal infection cycle unintentionally encounters that cycle.

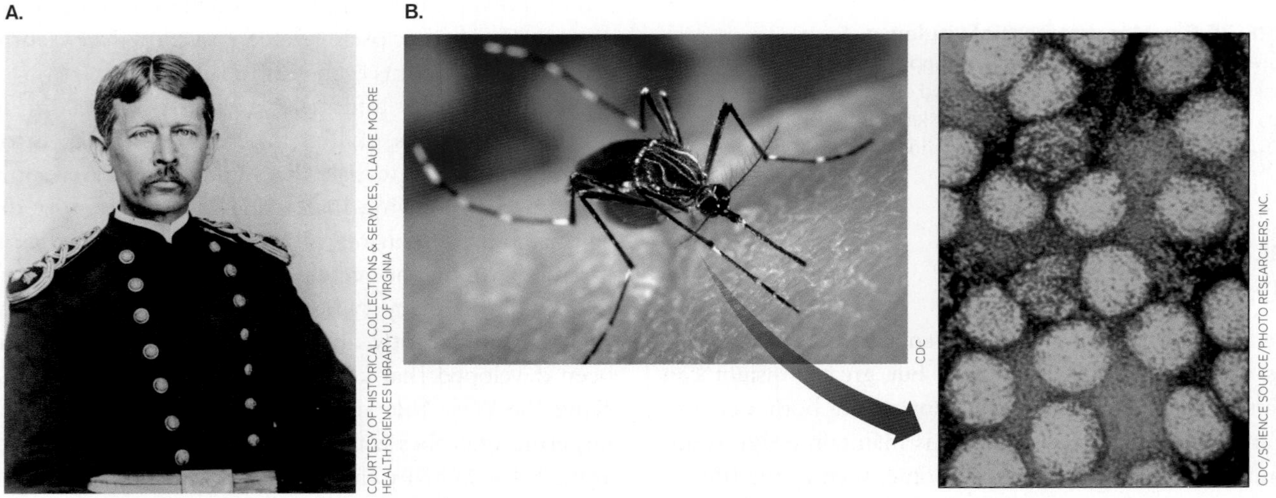

A.
B.
COURTESY OF HISTORICAL COLLECTIONS & SERVICES, CLAUDE MOORE HEALTH SCIENCES LIBRARY, U. OF VIRGINIA
CDC
CDC/SCIENCE SOURCE/PHOTO RESEARCHERS, INC.

FIGURE 25.8 ■ **Insect vector and yellow fever. A.** Walter Reed, a member of the University of Virginia School of Medicine class of 1869, proved in 1901 that the mosquito *Aedes aegypti* is the vector of transmission for yellow fever, a disease named for the jaundice produced by liver damage. **B.** *A. aegypti.* Yellow fever is caused by a flavivirus (inset, TEM) carried by this mosquito. The virus varies in size from 50 to 90 nm. Yellow fever remains endemic in the northern part of South America and in central Africa.

generations of mosquitoes through vertical transmission. The virus causing eastern equine encephalitis (EEE), however, uses birds as a reservoir. The microbe is normally a bird pathogen and is transmitted from bird to bird via a mosquito vector. The virus does not persist in the insect, but transmission by the insect vector keeps the virus alive by passing it to new avian hosts. Humans or horses entering geographic areas that harbor the disease (called endemic areas) can also be bitten by the mosquito. When this happens, they become accidental hosts and contract disease. The virus does not replicate to high titers in mammals, which means that horses and humans are poor reservoirs for it. EEE virus, however, does replicate to high numbers in the avian host.

Reservoirs are critically important for the survival of a pathogen and as a source of infection. If the eastern equine encephalitis virus had to rely on humans to survive, the virus would cease to exist because of limited replication potential and limited access to mosquitoes. Note that the reservoir of a given pathogen might not exhibit disease.

Even a simple infection cycle can become more complex. For example, rhinovirus can be spread person-to-person by a sneeze (airborne) or through the sharing of inanimate objects (fomites), such as contaminated utensils (fork, pen), towels, cloth handkerchiefs, and doorknobs. Handshaking is also an efficient means of transferring some pathogens. Imagine that one person in a city of 100,000 people has a cold and sneezes on her hands. Then, without washing them, she goes through the day shaking hands with 10 people, and each of those people shakes hands with another 10 people per day, and so on. If there were no repeat handshakes, and if none of the contacts washed their hands, it would take only 4 days to spread the virus throughout the population.

In this example, the entire populace of the city would eventually come in contact with the virus, but not everyone would actually contract disease. Additional factors influence whether the virus successfully replicates in a given individual.

Portals of Entry

How do infectious agents gain access to the body? Each organism is adapted to enter the body in different ways. Food-borne pathogens (for example, *Salmonella, E. coli, Shigella,* and rotavirus) are ingested by mouth and ultimately colonize the intestine. They have an oral portal of entry. Airborne organisms, in contrast, infect through the respiratory tract. Some microbes enter through the conjunctiva of the eye; others enter through the mucosal surfaces of the genital and urinary tracts. Agents that are transmitted only by mosquitoes or other insects enter their human hosts via the **parenteral route**, meaning injection into the bloodstream. Wounds and needle punctures can also serve as portals of entry for many microbes. Thus, shared needle use between drug addicts has been an important factor in the spread of HIV.

Immunopathogenesis of Infectious Diseases

Although we focus in this chapter on the mechanisms microbes use to cause disease (toxins, for example), it is often "friendly fire" by our immune system reacting to a pathogen that causes major tissue and organ damage. The immune response to any infection involves activating a complex network of cell types and soluble factors (discussed in Chapters 23 and 24) that may inadvertently damage the host to such a degree that it causes illness and even death. This collateral damage, or "immunopathology," is a calculated risk taken by the host in its haste to eradicate the pathogen. The term **immunopathogenesis** applies when the immune response to a pathogen is a contributing cause of pathology and disease.

The disease dengue hemorrhagic fever is a case in point. Caused by the dengue virus and transmitted by the *Aedes* mosquito, dengue fever manifests as a severe headache, muscle and joint pain, fever, and rash. The symptoms can also include abdominal pain, nausea, and vomiting. However, these symptoms are more a consequence of immunopathogenesis than they are a direct result of viral replication. Replication of the virus in host cells will lead to a massive activation of T cells ($CD4^+$ and $CD8^+$). The process triggers a cytokine cascade (some call it a "storm") that targets vascular endothelial cells and produces an endothelial "sieve" effect leading to fluid and protein leakage. Cytokines such as TNF-α, IL-1, and IFN-gamma (discussed in Chapters 23 and 24), along with many others that are released, contribute to inflammation and disease symptoms. So, to fully understand any infectious disease you must be aware of the pathogenic mechanisms wielded by the pathogen, but also realize that the disease symptoms may be due to immunopathogenesis.

To Summarize

- **Fulfilling Koch's and Rivers's postulates** is important for identifying the microbial cause of a new disease, but they must be supplemented with molecular or disease-tracking tools if one of the postulates cannot be fulfilled.

- **Infection does not equal disease.** The immunocompetence of the host and the virulence potential of the pathogen influence whether an infection causes disease.
- **Primary pathogens** have mechanisms that help the organism circumvent host defenses in a healthy host, whereas **opportunistic pathogens** cause disease only in a compromised host.
- **Pathogenicity** refers to the mechanisms a pathogen uses to produce disease and how efficient the organism is at causing disease, whereas **virulence** is a measure of disease severity.
- **Diseases can be spread** by direct or indirect contact between infected and uninfected persons/animals or by insect vectors.
- **Pathogens use portals of entry** best suited to their mechanisms of pathogenesis.
- **Immunopathogenesis** occurs when the immune system's response to an infection damages host cells and tissues.

25.2

Virulence Factors and Pathogenicity Islands

To cause disease, all pathogens must enter a host; find their unique niche; avoid, circumvent, or subvert normal host defenses; multiply; and eventually be transmitted to a new susceptible host. Pathogens can be distinguished from their avirulent counterparts by the presence of **virulence factors** that help accomplish these goals. Virulence factors, which are encoded by virulence genes, include toxins, attachment proteins, capsules, and other devices used to avoid host innate and adaptive immune systems. All of these factors enhance the disease-producing capability, or pathogenicity, of the pathogen. But how do we identify a virulence gene?

Molecular Koch's Postulates

Virulence is a measurable phenotype (Section 25.1). Genes encoding virulence factors are expressed just like those for other traits, but how do you find them? Identifying genes for a metabolic or biosynthetic pathway is relatively easy because it requires only a clear phenotype observed on an agar plate. If a gene involved with an amino acid biosynthetic pathway is defective, then the amino acid is not made and must be added to the medium or else the mutant will not grow. Finding a virulence gene is more difficult because the screen involves growth in a host. There is no in vitro way to identify a phenotype for virulence gene mutants. Virulence mutants can be recognized only if they fail to survive or cause disease in animals.

There have been numerous clever techniques developed over past decades to identify potential virulence genes. Examples are discussed in **eTopics 25.1** and **25.2**. Regardless of how the suspected virulence gene was found, it can be confirmed as having a role in virulence or pathogenicity only if it fulfills a set of "molecular Koch's postulates" originally formulated by Stanley Falkow, a preeminent infectious disease scientist. The postulates are as follows:

1. The phenotype under study should be associated with pathogenic strains of a species.
2. Specific inactivation of the suspected virulence gene(s) should lead to a measurable loss in virulence or pathogenicity. The gene(s) should be isolated by molecular methods.
3. Reversion or replacement of the mutated gene should restore pathogenicity.

A variety of other molecular questions are often asked. Answers to these questions can suggest the role of a virulence protein in pathogenesis. For example:

- Does moving the virulence gene into an avirulent strain impart a pathogenicity trait on the avirulent strain? (For instance, does moving a gene for attachment from a pathogen to a nonpathogen allow the nonpathogen to now attach to host cells?)
- Does the suspected virulence protein bind to important host proteins? Binding to a host protein could indicate the target of the microbial protein.
- Does the microbial protein resemble the sequence or structure of an important host protein? Such resemblance might indicate that the microbial protein mimics the function of the orthologous host protein.
- Does introducing the suspected virulence protein (or its encoding gene) into a host cell alter the host cell's physiology, or does a GFP-tagged version of the protein localize to a specific host cell compartment or organelle? A positive finding in either case could reveal a target for the suspected virulence protein.

In addition to the experimental approaches just described, there are bioinformatic ways to identify potential pathogenicity genes.

Pathogenicity Islands

Extensive sequencing efforts have allowed us to compare genomes of many pathogens and expose some "footprints" of their evolution. For example, in bacterial pathogens, most chromosomes are dotted with clusters of pathogenicity genes that encode virulence functions. These gene

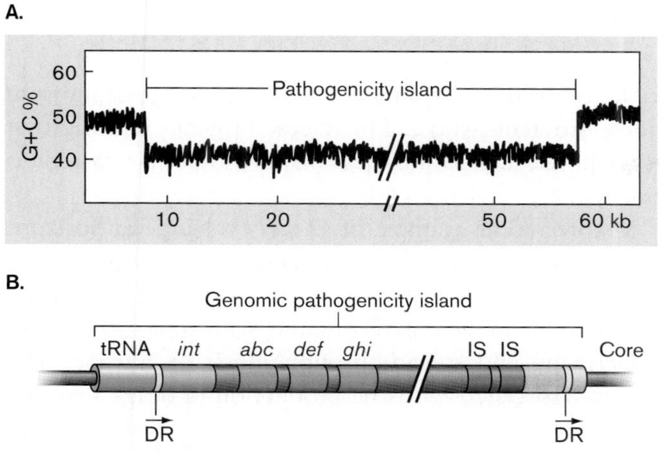
A.
G+C %
60
50
40
Pathogenicity island
10
20
50
60 kb
B.
Genomic pathogenicity island
tRNA
int
abc
def
ghi
IS
IS
Core
DR
DR

FIGURE 25.9 ■ Model pathogenicity island. **A.** The guanine + cytosine (GC) content of the island is different from that of the core genome. **B.** Schematic model of a pathogenicity island. The DNA block is linked to a tRNA gene and flanked by direct repeats (DR) that may be "footprints" of a transposon or viral-mediated transfer. The integrase gene (*int*) and insertion sequences (IS) may also be remnants of transposition.

clusters, called **pathogenicity islands**, can be considered the toolboxes of pathogens (originally discussed in Section 9.7). Many virulence genes reside in pathogenicity islands, although many others do not. Some virulence genes reside on plasmids (for example, the genes for the diarrhea-producing labile toxin of certain *E. coli* strains) or in phage genomes (such as the genes encoding the diphtheria toxin of *Corynebacterium diphtheriae*).

Most pathogenicity islands appear to have been horizontally transmitted (via conjugation or transduction; discussed in Section 9.2) from organisms long ago extinct into the ancestors of today's pathogens. Horizontal gene transfers move whole blocks of DNA (more than 10 kb) from one organism to another, placing the blocks directly in the chromosome in what is called a **genomic island** (see Chapter 9). If the island increases the "fitness" (virulence) of a microorganism (pathogen) that interacts with a host, it is called a pathogenicity island. Genomic islands generally reveal themselves by several anomalies that they possess with respect to the rest of the host genome:

- Genomic islands are often linked to a tRNA gene and generally have a GC/AT ratio very different from that of the rest of the chromosome (**Fig. 25.9A**). For example, a plot of GC content along the length of a chromosome may reveal that most of the genome has a 50% GC content. But somewhere in the middle, a 50-kb region sticks out on the graph, showing a content of 40%. This probably reflects the GC content of the microbe that donated the island. The reason tRNA genes are often the targets for insertion of pathogenicity islands is not known. One hypothesis is that the conserved secondary structure of tRNA genes facilitates integration by an integrase.
- Genomic islands are typically flanked by genes with homology to phage or plasmid genes (**Fig. 25.9B**). This arrangement is thought to reflect the transfer vector used to move the island from one organism to another.

But what do the pathogenicity genes do? Some genes encode molecular "grappling hooks," such as pili that attach to host cells. Once attached, microbes can secrete toxins that injure the host cell. Other bacteria wall themselves off to prevent damage by host inflammatory responses. Some bacterial pathogens are even capable of what could be called "host cell reprogramming." These organisms inject proteins directly into the host cell to disrupt normal signaling pathways. This reprogrammed target cell can be made to do one of several things; engulf the bacterium, "commit suicide" (undergo apoptosis), or engineer a tighter, more intimate attachment platform at the cell surface.

Table 25.1 lists several pathogenicity islands and their functions. It provides examples of pathogenicity islands

TABLE 25.1

Examples of pathogenicity islands.

Pathogenicity island	Function	Organism	Disease
HPI (high-pathogenicity island)	Iron uptake	*Yersinia* spp.	Plague, enterocolitis
VPI (*Vibrio* pathogenicity island)	Toxin production	*Vibrio cholerae*	Cholera
PAI III (pathogenicity island III)	Encoding adhesins	Uropathogenic *E. coli*	Urinary tract infection
SPI-1 and SPI-2 (*Salmonella* pathogenicity islands)	Type III secretion	*Salmonella enterica*	Gastroenteritis
SHI-1 and SHI-2 (*Shigella* islands)	Type III secretion	*Shigella flexneri*	Bloody diarrhea
YSA (*Yersinia* secretion apparatus)	Type III secretion	*Yersinia* spp.	Plague, enterocolitis
cag PI (cytotoxin-associated gene pathogenicity island)	Type IV secretion	*Helicobacter pylori*	Gastric ulcers, gastric cancer
icm/dot (intracellular multiplication)	Type IV secretion	*Legionella pneumophila*	Legionnaires' disease

present in different bacteria, names a key function of the products of the island, and identifies the disease caused. Various pathogenicity island functions will be described as the chapter proceeds.

Weblinks Virulence factors of pathogenic bacteria (*see ebook*)

Caught in the Act: Examples of Pathogen Evolution by Horizontal Gene Transfer

***Shigella* and *Escherichia*.** **Figure 25.10** compares the circular *Shigella flexneri* genome with the genomes of *E. coli* K-12 (a commonly used avirulent lab strain) and *E. coli* O157:H7 (a virulent strain of enterohemorrhagic *E. coli*). All three Gram-negative rods are closely related but differ greatly in terms of pathogenic potential and mechanisms. *S. flexneri* and *E. coli* O157:H7 cause bloody diarrhea, while *E. coli* K-12 has a commensal origin. DNA sequence analysis has revealed chromosome regions that are common among these organisms, as well as genomic islands specific to individual species and strains (**Fig. 25.10**). The core genes needed for sustaining growth (that is, for transcription, translation, replication, and so on) are common to all three organisms, whereas other genes may be present in only one. These unique genes and islands are thought to be the result of horizontal gene transfers originating from widely different genera. The degree of gene shuffling that was required to separate these otherwise similar bacteria is remarkable.

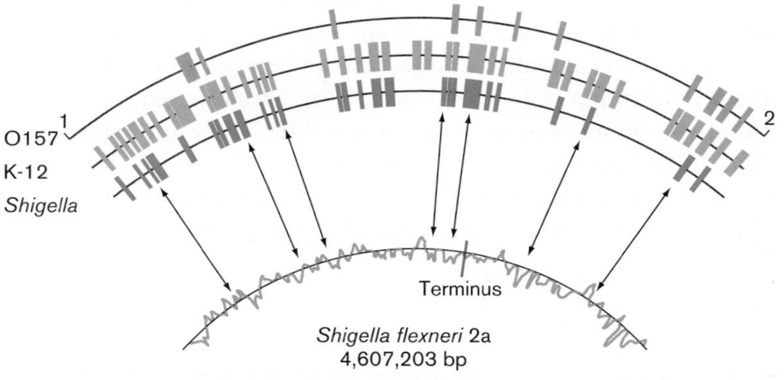
1
2
O157
K-12
Shigella
Terminus
Shigella flexneri 2a
4,607,203 bp

FIGURE 25.10 ■ Comparison of the *Shigella flexneri* 2a chromosome with chromosomes of *E. coli* K-12 and O157:H7 (EDL933). Segments of the three genomes between 1 Mb and 2 Mb are shown. Each gray line indicates DNA sequences shared among the organisms (O157, top arc; K-12, second arc; *Shigella,* third arc). Colored boxes depict genomic islands (including pathogenicity islands) present in each organism. The bottom arc illustrates the GC content of the *Shigella* genome. Each data point along the graph indicates GC content relative to AT content as averaged over a sliding 10-kb window. Note that major differences in *Shigella* GC content shown above or below the center line (indicating 50% GC content) often correlate with the genomic islands depicted in the third arc. Arrows indicate some islands with obvious correlation to GC content.

A more recent example of *E. coli* evolving via horizontal gene transfer involves the enterohemorrhagic strain O104:H4, which caused the frightening 2011 outbreak of diarrhea and hemolytic uremic syndrome that began in Germany and spread through much of Europe. The lethality associated with the production of Shiga toxin by this strain and its resistance against many antibiotics was remarkable. The presence of several new virulence traits, including antibiotic resistance, in this strain that are absent in its closest relative suggested the involvement of horizontal gene transfer in its evolution.

***Staphylococcus aureus*.** Another pathogen caught in the act of transferring virulence genes, this time in real time, is *Staphylococcus aureus,* a cause of boils, toxic shock syndrome, and many other infections. *Staphylococcus aureus* swaps pathogenicity genes by transduction. The *S. aureus* pan-genome possesses a large, 15- to 17-member family of chromosomal pathogenicity islands (SaPIs), most of which encode superantigens that cause superantigen-type diseases such as toxic shock syndrome (superantigens are discussed in Chapter 24). What is remarkable is that most SaPIs transfer between different strains with relative ease. Because of this mobility, SaPIs are widely distributed in the genomes of *S. aureus,* with many strains possessing two or more.

Interestingly, the genome organization of these elements is similar to that of a temperate phage. However, the SaPI phages are defective and do not form phage-like particles unless the strain is infected by certain types of helper phages that supply missing proteins. Once that happens, the elements are packaged into particles composed of phage proteins. The SaPI particles are released from the donor cell and can "infect" another strain of *S. aureus,* where the pathogenicity island is then incorporated into the recipient's genome. This is a clear example of horizontal gene transfer happening every day somewhere on the planet. Even more startling is that these phage particles can also transfer the SaPI genes with equal ease to *Listeria monocytogenes,* a completely different genus that causes food-borne infections.

Streptococcus agalactiae, another pathogen caught in the act of evolving via conjugation, is described in **eTopic 25.3**.

The following sections describe some of the specific tools that pathogens use to undermine the integrity of the body. From attachment, to toxins, to intracellular invasion, the infection process is like a chess match, with each side, human and microbe, trying to outmaneuver the other.

To Summarize

- **Fulfilling the molecular Koch's postulates** validates the identity of a virulence gene.
- **Pathogenicity islands** are DNA sequences within a species that are acquired by horizontal gene transfer from a different species.
- **Virulence genes** encode products that enhance the disease-causing ability of the organism. Many virulence genes can be found within pathogenicity islands, but some are located outside of an obvious genomic island or reside in plasmids.
- **Pathogenicity islands** contain distinct features, such as GC content and the remnants of phages or plasmids, that mark them as being different from the rest of the genome.
- **Gene transfer mechanisms** will, by horizontal transfer, move virulence factor genes and pathogenicity islands among bacterial strains and species.

25.3

Microbial Attachment: First Contact

Regardless of the disease, pathogens must reach a colonization site either through their own motility or by hitchhiking with a vector. Once at the site, the pathogen needs attachment mechanisms to stay there.

The human body has many ways to exclude pathogens. The lungs use a mucociliary elevator (or escalator; see Fig. 23.4) to rid themselves of foreign bodies, the intestine uses peristaltic action to ensure that its contents are constantly flowing, and the bladder uses contraction to propel urine through the urethra with tremendous force. How do bacteria ever manage to stay around long enough to cause problems? Like a person grasping a telephone pole during a hurricane, successful pathogens moving through the body manage to grab onto host cells and tenaciously hold on. Thus, the first step toward infection is attachment, also called adhesion. **Adhesin** is the general term for any microbial factor that promotes attachment.

Viruses attach to the host through their capsid or envelope proteins, which bind to the specific host cell receptors discussed in Chapters 6 and 11. Bacteria have a variety of similar strategies. They can use hairlike appendages called **pili** (also called **fimbriae**), whose tips contain receptors for mammalian cell-surface structures. Or they can use a variety of adherence proteins or other molecules (adhesins) that are not part of a pilus. Sometimes they use both. **Table 25.2** summarizes bacterial attachment strategies. Note that different pili can impart tissue specificity for attachment (for example, uropathogenic versus diarrheagenic *E. coli*).

Pili

Pili from different bacterial species are classified by sequence-based homologies. We will consider three groups in this chapter, called type I, type III, and type IV pili.

Type I pili are a group that, in general, adhere to mannose residues on host cell surfaces. Because adding free mannose will inhibit attachment of most type I pili, this binding is called "mannose sensitive." Another group of pili is called "mannose resistant" because the pili do not bind to mannose residues. There are at least two types of mannose-resistant pili. Members of one type, sometimes called **type III pili** (not to be confused with type III secretion), bind to red blood cells treated with tannic acid (to expose binding sites). The other, more commonly studied type is called **type IV pili**. Unlike type I pili, which simply stick out from the cell surface, type IV pili are more dynamic and are assembled on the cell surface through a very different pathway.

There are other types of attachment pili that do not fall neatly into one of these three groups. The primary classification of pili is now based largely on protein sequence information (deduced from DNA sequence) and may contradict earlier phenotype-based schemes. For instance, the pyelonephritis-associated pili (Pap) of the uropathogenic *E. coli* are mannose resistant, since they bind to a digalactoside on host surfaces (called the P-blood-group antigen). However, amino acid sequence homology suggests that Pap is very similar to type I pili. Note that pili, or other adhesins, can be virulence factors on some organisms but not on others.

Pilus Assembly

How bacteria assemble pili on their cell surfaces is an engineering marvel. The shafts of pili are cylindrical structures composed of identical pilin protein subunits. Several different proteins adorn the tip, including one at the very apex that binds to host receptors (**Fig. 25.11A**). How the tip protein manages to "hold on" in the face of tremendous shear forces (such as those exerted during urination) is described in eTopic 25.4. In addition to the structural components of pili, numerous other proteins collaborate to assemble the structure. Genes encoding a given pilin

TABLE 25.2

Specific attachments of bacteria to cell or tissue surfaces.

Bacterium	Adhesin	Host receptor	Attachment site	Disease
Streptococcus pyogenes	Protein F	Amino terminus of fibronectin	Pharyngeal epithelium	Sore throat
Streptococcus mutans	Glucan	Salivary glycoprotein	Pellicle of tooth	Dental caries
Streptococcus salivarius	Lipoteichoic acid	Unknown	Buccal epithelium of tongue	None
Streptococcus pneumoniae	Cell-bound protein	*N*-acetylhexosamine galactose disaccharide	Mucosal epithelium	Pneumonia
Staphylococcus aureus	Cell-bound protein	Amino terminus of fibronectin	Mucosal epithelium	Various
Neisseria gonorrhoeae	*N*-methylphenylalanine pili	Glucosamine galactose carbohydrate	Urethral/cervical epithelium	Gonorrhea
Enterotoxigenic *E. coli*	Type I fimbriae (pili)	Species-specific carbohydrate(s)	Intestinal epithelium	Diarrhea
Uropathogenic *E. coli*	Type I fimbriae (pili)	Complex carbohydrate	Urethral epithelium	Urethritis
Uropathogenic *E. coli*	P pili (pyelonephritis-associated pili)	P blood group	Upper urinary tract	Pyelonephritis
Bordetella pertussis	Pili ("filamentous hemagglutinin")	Galactose on sulfated glycolipids	Respiratory epithelium	Whooping cough
Vibrio cholerae	*N*-methylphenylalanine pili	Fucose and mannose carbohydrate	Intestinal epithelium	Cholera
Treponema pallidum	Peptide in outer membrane	Surface protein (fibronectin)	Mucosal epithelium	Syphilis
Mycoplasma	Membrane protein	Sialic acid	Respiratory epithelium	Pneumonia
Chlamydia	Unknown	Sialic acid	Conjunctival or urethral epithelium	Conjunctivitis or urethritis
Corynebacterium diphtheriae	Pili	Unknown	Pharyngeal epithelium	Diphtheria

A. FimH adhesin tip

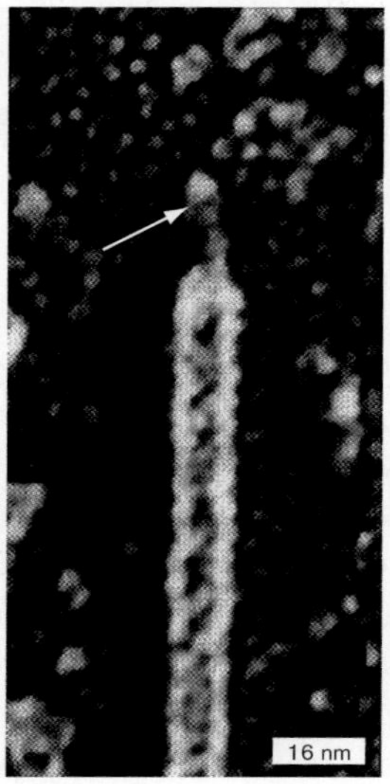
16 nm

C. HAL JONES ET AL. 1995. *PNAS* **92**:2081

B. Type I pilus gene cluster

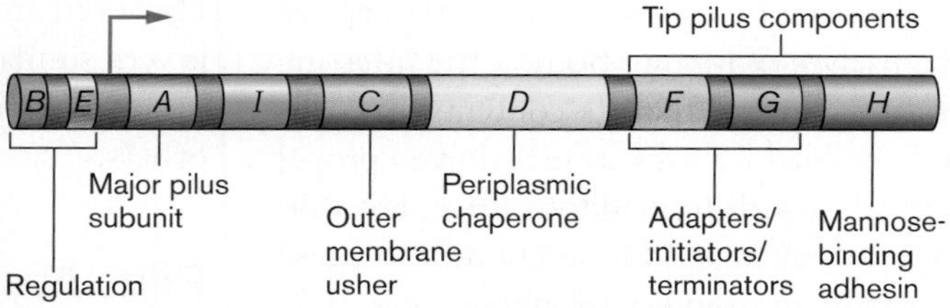
Tip pilus components
B
E
A
I
C
D
F
G
H
Regulation
Major pilus subunit
Outer membrane usher
Periplasmic chaperone
Adapters/ initiators/ terminators
Mannose-binding adhesin

FIGURE 25.11 ■ Attachment pili and encoding operon. **A.** High-resolution micrograph showing a type I pilus (TEM). The FimH adhesin at the tip (arrow) is the protein that binds to the cell receptor. **B.** Genetic organization of the type I gene cluster, which includes genes involved in pilus assembly. The genes are designated *fim A–I.*

protein and the associated assembly apparatus are typically arranged on the chromosome as an operon (**Fig. 25.11B**).

Assembly of type I pili. **Figure 25.12** illustrates the assembly of a type I pilus using uropathogenic *E. coli* Pap as a model. The mechanism is representative of other type I pili; only the names of the proteins will differ for each system. Protein components are secreted into the periplasm by the SecA-dependent general secretory system (discussed in Section 8.6). Once in the periplasm, the subunits are chaperoned by PapD to the membrane site of assembly, which is marked by the presence of the usher protein PapC. PapC proteins form channels in the outer membrane large enough to accommodate individual pilus subunits and, like an usher in a theater, direct the subunits to their proper places.

Chaperoning of the pilin building blocks by PapD is necessary to prevent pilin subunits from inadvertently assembling in the periplasm. As illustrated in **Figure 25.12**, appropriate assembly of pili at the usher site starts with the tip protein, PapG, which will ultimately bind to carbohydrates on host membranes after the pilus is complete. After PapG, the ushers add PapF and PapE, forcing PapG farther away from the surface. Then, identical PapA pilin subunits are strung together in a series to form the shaft. PapA subunits assemble by sequentially sharing a domain from each other, linking together like pieces of a jigsaw puzzle.

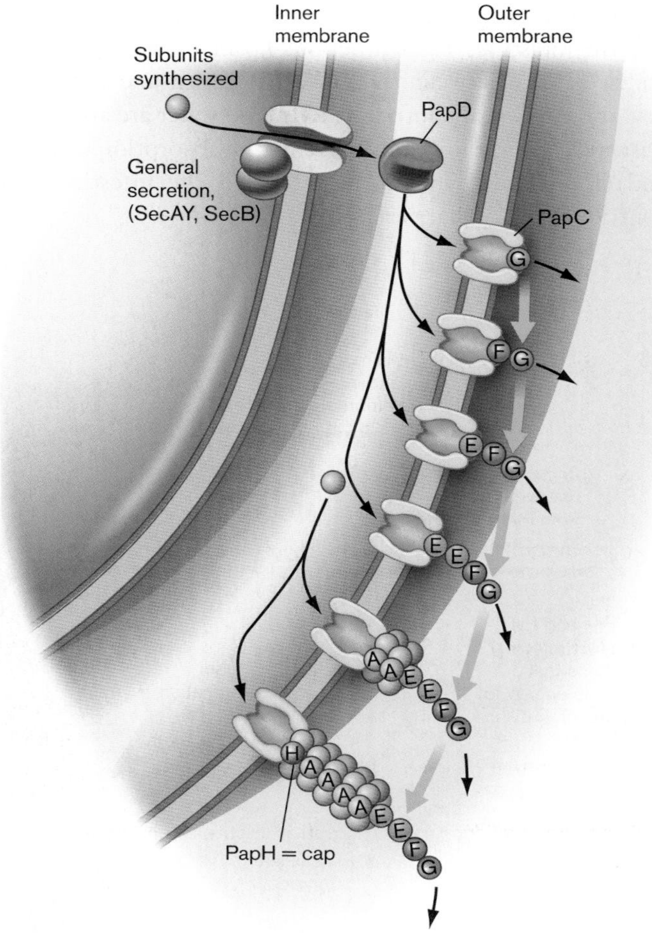
Inner membrane
Outer membrane
Subunits synthesized
PapD
General secretion, (SecAY, SecB)
PapC
G
F
E
A
H
PapH = cap

FIGURE 25.12 ■ **Assembly of type I pili.** The pyelonephritis-associated pilus (Pap) assembly illustrated is representative of other type I pili, such as in Fim in Figure 25.11. Only the names of the proteins will differ. Proteins are secreted by the Sec system to the periplasm, where they are chaperoned by PapD to the site of assembly. PapC, also called the usher, assembles the individual proteins in the proper order. Assembly starts with the tip protein, PapG, marked at the far right, which ultimately binds to carbohydrates on host membranes. The subunits fit together like pieces of a jigsaw puzzle. The arrow at the head of the elongating pilus indicates the direction of pilus growth.

Assembly of type IV pili. Other pili with an important role in pathogenesis are the type IV pili. Type IV pili are found in a broad spectrum of Gram-negative bacteria and share amino acid homology in their major pilin structure (**Fig. 25.13**). All type IV pili use similar secretion and assembly machinery involving at least a dozen proteins. One major difference between the assembly of type IV and type I pili is that type IV pilus proteins are never free in the periplasm; instead, they are transported directly from the cytoplasm through a channel in the outer membrane (**Fig. 25.13A**). Thus, type IV pilus assembly is SecA independent. Species with type IV pili include *Vibrio cholerae, Pseudomonas aeruginosa* (**Fig. 25.13B**), certain pathogenic strains of *E. coli, Neisseria meningitidis* (**Fig. 25.13C**), and *Neisseria gonorrhoeae*. The genes for type IV pili were duplicated and modified through evolution into a protein secretion mechanism designated type II secretion. Despite having evolved from genes encoding type IV pili, type II secretion systems export virulence proteins unrelated to pili (discussed in Section 25.5).

Type IV pili can actually make cells move because the assembly process involves the reiterative elongation and retraction of the pili. The process, called "twitching motility," occurs when the pilus elongates, attaches to a surface, and then depolymerizes from the base, which shortens the pilus and pulls the cell forward. This mechanism is akin to using a grappling hook to scale a building. The gliding motility of the slime mold *Myxococcus xanthus* is also due to type IV pili. The type IV pili of *Neisseria meningitidis,* shown in **Figure 25.13C**, are essential for crossing the blood-brain barrier and causing bacterial meningitis.

Recent evidence suggests that the retraction of type IV pili, at least in diarrhea-causing strains of *E. coli,* helps disrupt the tight junctions connecting adjacent cells that line the intestine. Tight junctions normally form a permeability barrier between the intestinal lumen and the intestine through which nutrients, ions, and water are absorbed. Disrupting tight junctions prevents the absorption of water and electrolytes, which accumulate in the intestine and contribute to the diarrhea.

Nonpilus Adhesins

Bacteria also carry afimbriate adhesins (proteins that aid in attachment but do not form pili) that mediate binding to host tissues (**Fig. 25.14**). Some examples include *Bordetella* pertactin (which binds to host cell integrin), *Streptococcus* protein F (which binds to fibronectin), *Streptococcus* M protein (which binds to fibronectin and complement regulatory factor H), and intimin of enteropathogenic *E. coli* (which binds to Tir; see Section 25.5). Many Gram-positive bacteria have surface-exposed proteins resembling pili that contain serine-rich repeats able to bind sialic acid or keratin.

Fimbriae (pili) often mediate the initial binding between bacterium and host, after which a more intimate attachment is formed by an afimbriate attachment protein. In the case of *Neisseria gonorrhoeae,* once the type IV pilus has attached to the surface of the mucosal epithelial cell, the filamentous pilus contracts, pulling the bacterium down onto the host cell membrane. Tight secondary interactions are then mediated by the neisserial Opa membrane proteins, another example of an afimbriate adhesin (Opa is named because of the opacity it adds to colony appearance).

MAM7, a promiscuous adhesin. Although many different bacterial adhesins are known, most are species specific, and many are not produced until after an infection starts, making these molecules less

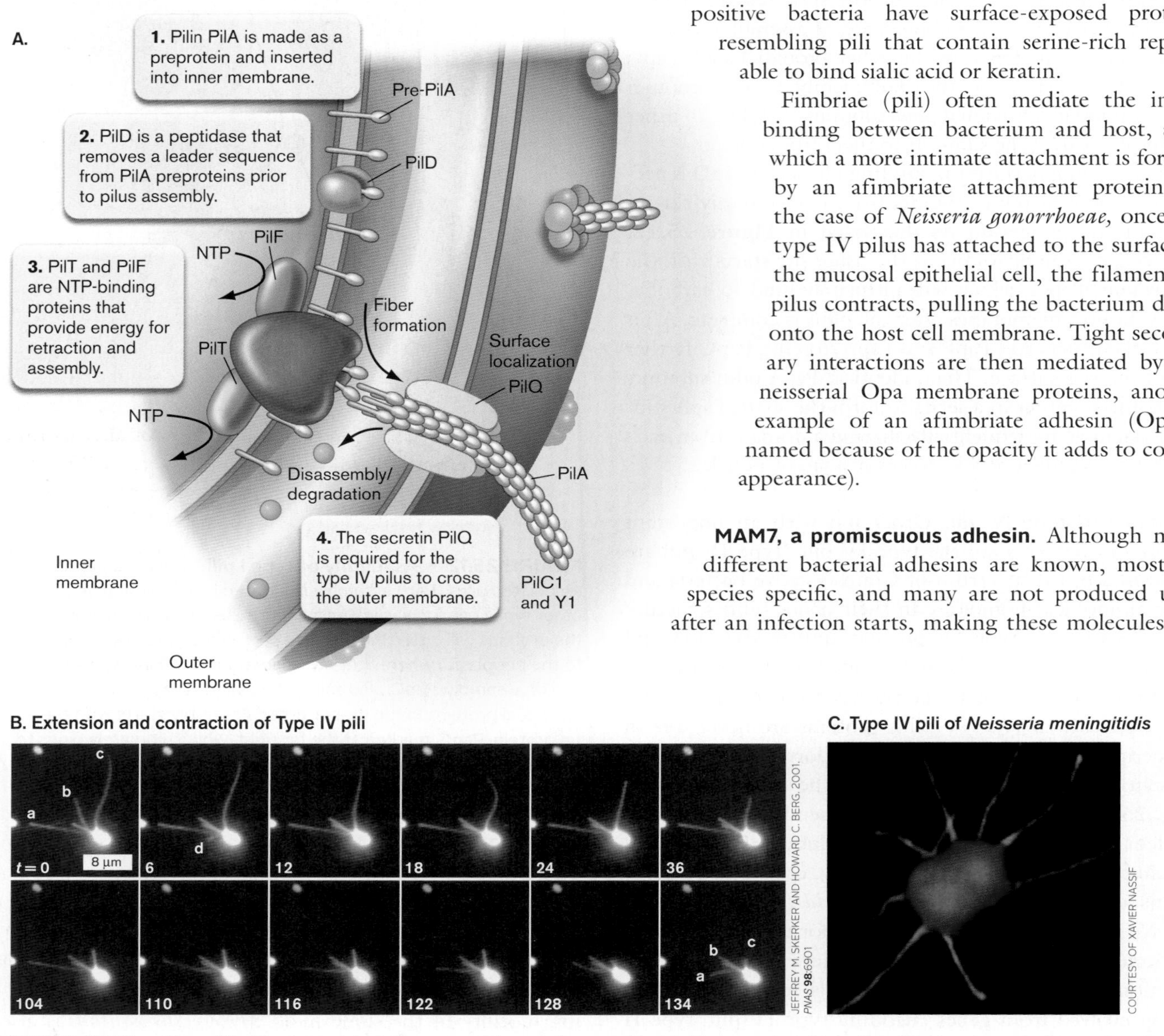
A.
1. Pilin PilA is made as a preprotein and inserted into inner membrane.
2. PilD is a peptidase that removes a leader sequence from PilA preproteins prior to pilus assembly.
3. PilT and PilF are NTP-binding proteins that provide energy for retraction and assembly.
4. The secretin PilQ is required for the type IV pilus to cross the outer membrane.
Pre-PilA
PilD
PilF
NTP
PilT
NTP
Fiber formation
Surface localization
PilQ
PilA
Disassembly/ degradation
PilC1 and Y1
Inner membrane
Outer membrane
B. Extension and contraction of Type IV pili
a
b
c
d
8 µm
t = 0
6
12
18
24
36
104
110
116
122
128
134
JEFFREY M. SKERKER AND HOWARD C. BERG. 2001. PNAS 98:6901
C. Type IV pili of Neisseria meningitidis
COURTESY OF XAVIER NASSIF

FIGURE 25.13 ■ Type IV pili. A. Model of pilus assembly and disassembly. In this example, PilA is the pilin protein, and PilC1 and Y1 form the attachment tip. Diameter of the filament approx. 6 nm. Assembly and disassembly require the hydrolysis of nucleoside triphosphate (NTP) and take place at the inner membrane, not in the periplasm. **B.** Photographic evidence of type IV pilus extension and retraction in cells of *Pseudomonas aeruginosa.* Filament b retracts; then filament d extends at 6 seconds and retracts. Filament c attaches briefly at its distal tip (note straightening at 24 seconds) and then begins to retract. Fluorescent microscopy. *t* = time in seconds. **C.** Type IV pili are essential for the interaction of *Neisseria meningitidis* with brain endothelial cells. Type IV pili are green in this SEM. Diplococcal cells approx. 1.6 µm in diameter.
Source: Part A modified from Bardy et al. 2003. *Microbiology* **149**:295–304.

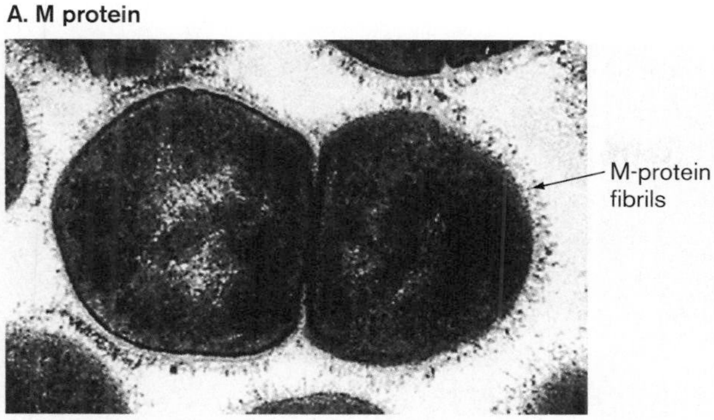
A. M protein
M-protein fibrils

MARIA FAZIO AND VINCENT A. FISCHETTI, ROCKEFELLER UNIVERSITY

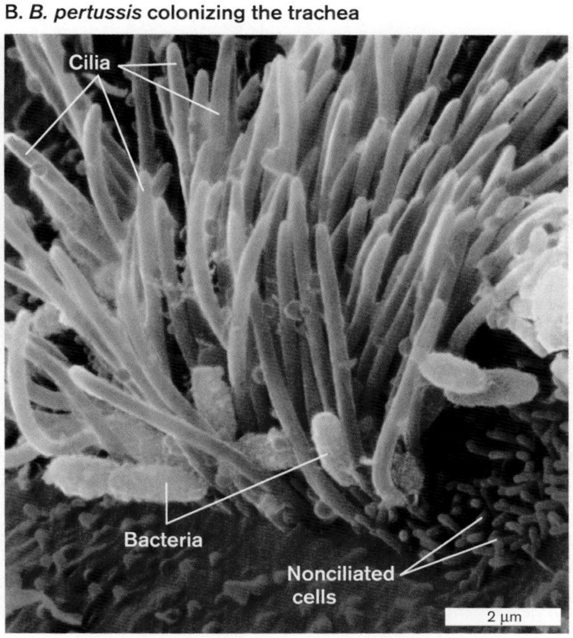
B. *B. pertussis* colonizing the trachea
Cilia
Bacteria
Nonciliated cells
2 μm

NIBSC/SCIENCE PHOTO LIBRARY

FIGURE 25.14 ■ Nonpilus adhesins. A. M-protein surface fibrils on *Streptococcus pyogenes* (TEM). Cells 0.5–1 μm in diameter. **B.** Colonization of tracheal epithelial cells by *Bordetella pertussis* (SEM). This organism uses a surface protein called pertactin, as well as a pilus called filamentous hemagglutinin (FHA) to bind bronchial cells.

suitable for initial binding to host cells. Could there be a single microbial adhesion molecule used as a nonspecific "first binder" by multiple pathogens? Kim Orth and colleagues at the University of Texas Southwestern Medical Center recently described one such adherence factor, called multivalent adhesion molecule 7 (MAM7). It is used by a variety of Gram-negative bacterial pathogens but not found in Gram-positives. MAM7 is an outer membrane protein containing seven so-called mammalian cell entry domains (initially described for *Mycobacterium tuberculosis*).

The mammalian cell entry domains of MAM7 establish protein-protein interactions with host cell fibronectin, as well as protein-lipid interactions with host membranes. MAM7 adhesins appear important during the early stages of an infection by providing an opportunity for stronger adhesin molecules to establish tighter contact. Once bound via MAM7, the specificity of a pathogen for a certain cell type must be mediated by other, strain-specific adhesion molecules. Orth's group has also shown that MAM7 could be an attractive general target for antimicrobials. They found that a nonpathogenic *E. coli* engineered to express MAM7 could bind to host cells and block attachment of several other Gram-negative pathogens, including *Yersinia* and *Vibrio* species. Once firmly bound to a host cell, each bacterial pathogen deploys its own set of toxins and effectors to subvert host cell functions, thereby establishing a growth niche for the pathogen.

Host susceptibility. Why are some people susceptible to certain infections while others are not? Part of the reason is immune competence, but another is receptor availability. Pathogens rely on key host surface structures such as gangliosides to recognize and attach to the correct host cell by the mechanisms just described. But a host species can evolve to become resistant to infection when the gene encoding the receptor (or receptor synthesis) mutates. The mutation could lead to complete loss of the protein or a change in its shape to prevent recognition or alter its function. An example is the T-cell surface protein CCR5, which acts as a receptor for HIV (see Section 11.4). Individuals with a genetic defect that eliminates CCR5 are resistant to HIV infection, so even without methods for preventing and curing HIV infection, humans would eventually evolve a level of resistance to HIV. Differences in attachment receptors also explain, at least in part, why some pathogens have broad host specificity while others more narrowly target their host. It can also explain aspects of tissue specificity.

Biofilms and Infections

As first discussed in Section 4.6, bacteria in most environments form organized, high-density communities of cells called biofilms that are embedded in self-produced exopolymer matrices. Biofilm development is an ancient prokaryotic adaptation that allows microorganisms to adhere to any surface, living or nonliving, and facilitates survival in hostile environments. Within a single biofilm you can find localized differences in the expression of surface molecules, antibiotic resistance, nutrient utilization, and virulence factors. Bacteria in biofilms also coordinate their behavior through cell-cell communication using secreted chemical signals.

Biofilms are important features in chronic infections found on oral, lung, and urogenital (bladder) tissues.

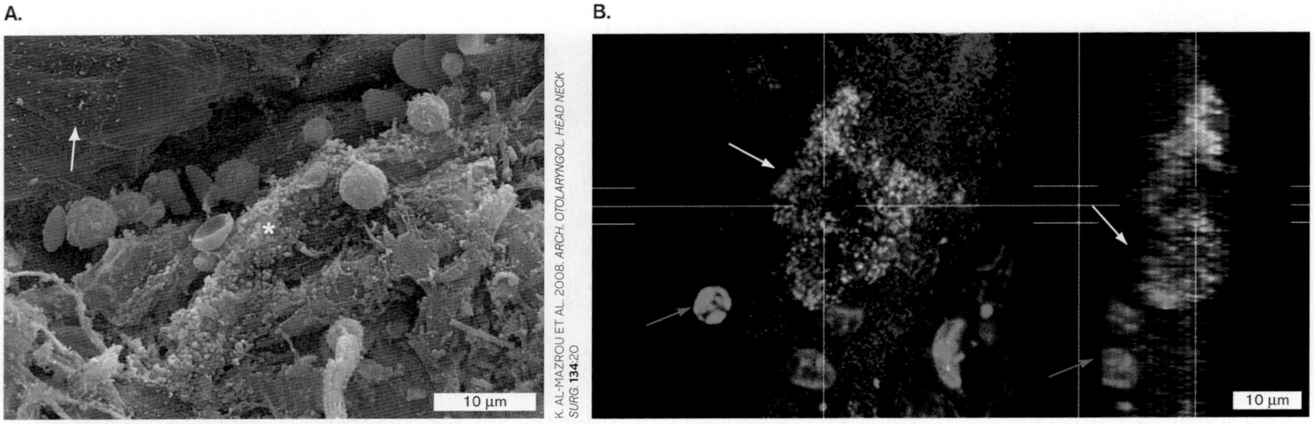
A.
B.
10 µm
10 µm

K. AL-MAZROU ET AL. 2008. *ARCH. OTOLARYNGOL. HEAD NECK SURG.* **134**:20

© 2009 BLACKWELL PUBLISHING, LTD.

FIGURE 25.15 ■ A bacterial biofilm infection. A. Scanning EM showing mixed bacteria (asterisk) and adherent biofilm on the surface epithelium (arrow) of an infected tonsil. **B.** Confocal micrograph (described in Chapter 2) showing biofilm clusters (white arrows) consisting of rods and cocci on the mucosa of a pediatric adenoid. Removal of the adenoid is a routine treatment for recurrent otitis media (middle ear infection). Specimens were treated with nucleic acid stains using the LIVE/DEAD *Bac*Light Bacterial Viability Kit, in which live bacteria stain green and dead bacteria stain red. Host inflammatory cells (red arrows) were also stained green, but their nuclei appear much larger than the bacteria. The mucosal surface (blue) was imaged using reflected light.

Pseudomonas aeruginosa causes a life-threatening, chronic lung infection in individuals with cystic fibrosis (CF). This microbe has been found growing as aggregates enclosed in a matrix within mucus from CF patients. It is thought that insufficient mucociliary clearance contributes to *P. aeruginosa* biofilm formation. Biofilms are also important in periodontitis (gum disease), indwelling catheter infections, infections of artificial heart valves, chronic urinary tract infections, recurrent tonsillitis, rhinosinusitis, chronic otitis media (middle ear infection), chronic wound infections, and osteomyelitis (bone infection). **Figure 25.15A** shows a scanning EM of a chronic infection of tonsil tissue from a pediatric patient. The fluorescent confocal image in **Figure 25.15B** shows a 3D view of a biofilm on adenoid tissue with live cells stained green and dead cells stained red. The chronic presence of these organisms in biofilms will continually stimulate innate immune mechanisms through interactions with Toll-like receptors and cause chronic inflammation as a result. A biofilm infection may linger for months, years, or even a lifetime. If the infection is associated with an implanted medical device (such as an artificial knee), the device may have to be replaced. Today, vascular catheter-related bloodstream infections are the most serious and costly health care–associated infections.

Biofilm infections are very important clinically because bacteria in biofilms exhibit tolerance to antimicrobial compounds and persistence in spite of sustained host defenses. Thus, biofilm infections are hard to cure. Tolerance to antibiotics may be caused by poor nutrient penetration through the exopolymer matrix into the deeper regions of the biofilm, leading to a stationary phase–like dormancy. Bacterial factors important to biofilm formation include type IV pili, quorum-sensing structural genes and regulators, and extracellular matrix synthesis. Interfering with cell-cell signaling is effective in preventing or limiting biofilm formation and may provide a target for new antimicrobial therapies.

To Summarize

- **Bacteria use pili and nonpilus adhesins** to attach to host cells.
- **Type I pili** produce a static attachment to the host cell, whereas **type IV pili** continually assemble and disassemble.
- **Nonpilus adhesins** are bacterial surface proteins, or other molecules, that can tighten interactions between bacteria and target cells.
- **Biofilms** play an important role in chronic infections by enabling persistent adherence and resistance to bacterial host defenses and antimicrobial agents.

25.4

Toxins Subvert Host Function

Following attachment, many microbes secrete protein toxins (called **exotoxins**) that kill host cells and unlock their nutrients (because dead host cells ultimately lyse). Bacterial pathogens have developed an impressive array of toxins that take advantage of different key host proteins or structures. All Gram-negative bacteria also possess a toxic compound called **endotoxin** (a part of lipopolysaccharide) that can hyperactivate host immune systems to harmful levels. Do not confuse <u>en</u>dotoxin with protein <u>ex</u>otoxins.

Microbial Exotoxins

Microbial exotoxins fall into nine categories based on their mechanisms of action (**Table 25.3**). These classes are summarized here, and several are illustrated in **Figure 25.16**.

- **Plasma membrane disruption.** Members of the first class, exemplified by the alpha (α) toxin of *Staphylococcus aureus*, form pores in host cell membranes and cause leakage of cell constituents (**Fig. 25.16A**).
- **Cytoskeleton alterations.** A second class of toxins can stimulate either actin polymerization or actin depolymerization. For example, *Vibrio cholerae* RTX toxin depolymerizes actin by cross-linking actin fibers. Modifying actin polymerization can alter cell shape or cause cell membranes to wrap around the pathogen, inviting the organism into the cytoplasm.
- **Protein synthesis disruption.** A third class, exemplified by diphtheria and Shiga toxins, targets eukaryotic ribosomes and destroys protein synthesis (**Fig. 25.16B**).
- **Cell cycle disruption.** These toxins either stop (*E. coli* cytolethal distending toxin, or CLDT) or stimulate (*Pasteurella multocida* toxin) host cell division.
- **Signal transduction disruption.** The fifth type of toxin subverts host cell second messenger pathways. Cholera toxin and *E. coli* ST (stable toxin), for instance, cause runaway synthesis of cyclic adenosine monophosphate (cAMP; described later) and cyclic guanosine monophosphate (cGMP) (**Fig. 25.16C**), respectively, in target cells. Elevated cAMP or cGMP levels, in turn, trigger critical changes in ion transport and fluid movement.
- **Cell-cell adherence.** These exotoxins are proteases that cleave proteins binding host cells to one another. One such toxin, exfoliative toxin of *Staphylococcus aureus*, breaks the connection between dermis and epidermis, giving victims the gruesome appearance of scalded skin.
- **Vesicle traffic.** The major toxin in this class (VacA of *Helicobacter pylori*) actually has several modes of action, depending on the host cell. The most visually striking effect is its ability to cause vacuolization, which is the fusion of numerous intracellular vesicles.
- **Exocytosis.** The eighth class of exotoxin includes two protease toxins that alter the movement of nerve cell cytoplasmic vesicles to membranes where they release neurotransmitters. One example is tetanus toxin, which cleaves host proteins required for exocytosis of the inhibitory neurotransmitter gamma-aminobutyric acid (GABA).
- **Superantigens.** Members of the ninth and last class of toxins activate the immune system without being processed by antigen-presenting cells (discussed in Section 24.6). The pyrogenic (fever-producing) toxins of *Staphylococcus aureus* (such as toxic shock syndrome toxin) and *Streptococcus pyogenes* are examples of superantigen toxins.

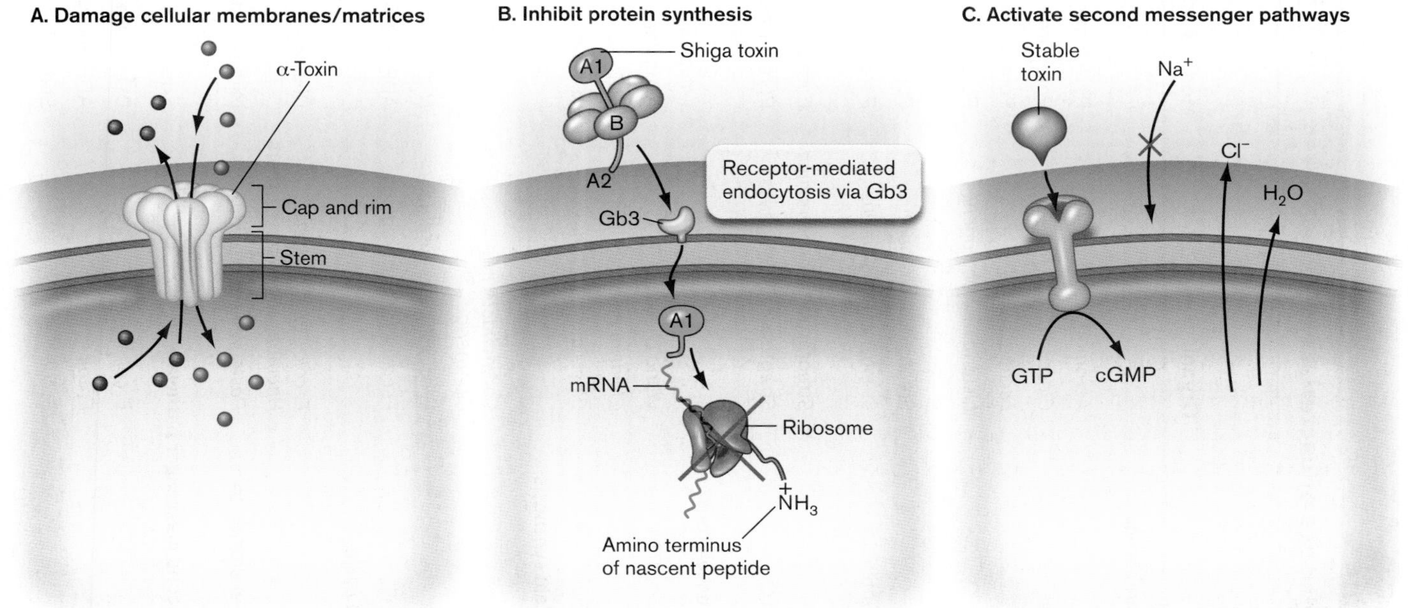
A. Damage cellular membranes/matrices
α-Toxin
Cap and rim
Stem
B. Inhibit protein synthesis
Shiga toxin
A1
B
A2
Gb3
Receptor-mediated endocytosis via Gb3
A1
mRNA
Ribosome
$\overset{+}{N}H_3$
Amino terminus of nascent peptide
C. Activate second messenger pathways
Stable toxin
Na^+
Cl^-
H_2O
GTP
cGMP

FIGURE 25.16 ■ **Three classes of microbial exotoxins.** These classes are defined by mode of action. **A.** Pore-forming toxins assemble in target membranes and cause leakage of compounds into and out of cells. **B.** Shiga toxin attaches to ganglioside Gb3, enters the cell, and cleaves 28S rRNA in eukaryotic ribosomes to stop translation. **C.** Enterotoxigenic *E. coli* heat-stable toxin affects cGMP production. The result is altered electrolyte transport—inhibition of Na^+ uptake and stimulation of Cl^- transport. In response to the resulting electrolyte imbalance, water leaves the cell.

TABLE 25.3

Characteristics of bacterial exotoxins.[a]

Toxin	Organism	Mode of action	Host target	Disease	Toxin implicated in disease[b]
Damage membranes					
Aerolysin	*Aeromonas hydrophila*	Pore former	Glycophorin	Diarrhea	(Yes)
Perfringolysin O	*Clostridium perfringens*	Pore former	Cholesterol	Gas gangrene[c]	(Yes)
Hemolysin[d]	*Escherichia coli*	Pore former	Plasma membrane	UTIs	(Yes)
Listeriolysin O	*Listeria monocytogenes*	Pore former	Cholesterol	Food-borne systemic illness, meningitis	Yes
Alpha toxin	*Staphylococcus aureus*	Pore former	Plasma membrane	Abscesses[c]	(Yes)
Panton-Valentine leukocidin	*Staphylococcus aureus*	Pore former	Plasma membrane	Abscesses, necrotizing pneumonia	(Yes)
Pneumolysin	*Streptococcus pneumoniae*	Pore former	Cholesterol	Pneumonia[c]	(Yes)
Streptolysin O	*Streptococcus pyogenes*	Pore former	Cholesterol	Strep throat, scarlet fever	Unknown
Disrupt cytoskeletons					
Vibrio cholerae RTX	*Vibrio cholerae*	Actin depolymerization	Cross-links actin fibers	Cholera	Unknown
C2 toxin	*Clostridium botulinum*	ADP-ribosyltransferase	Monomeric G-actin	Botulism	Unknown
Iota toxin	*Clostridium perfringens*	ADP-ribosyltransferase	Actin	Gas gangrene[c]	(Yes)
Inhibit protein synthesis					
Diphtheria toxin	*Corynebacterium diphtheriae*	ADP-ribosyltransferase	Elongation factor 2	Diphtheria	Yes
Shiga toxins	*E. coli/Shigella dysenteriae*	*N*-glycosidase	28S rRNA	HC and HUS	Yes
Exotoxin A	*Pseudomonas aeruginosa*	ADP-ribosyltransferase	Elongation factor 2	Pneumonia[c]	(Yes)
Disrupt cell cycle					
CLDT	*E. coli, Campylobacter, Haemophilus ducreyi*, others	DNase	DNA damage (triggers G_2 cell cycle arrest)	Diarrhea, chancroid, others	Unknown
Pasteurella multocida toxin	*Pasteurella multocida*	Mitogen (also activates Rho GTPases)	Nucleus (encourages cell division)	Wound infection	(Yes)
Activate second messenger pathways					
CNF	*E. coli*	Deamidase	Rho G proteins	UTIs	Unknown
LT	*E. coli*	ADP-ribosyltransferase	G proteins	Diarrhea	Yes
ST[d]	*E. coli*	Stimulates guanylate cyclase	Guanylate cyclase receptor	Diarrhea	Yes

EAST	*E. coli*	ST-like?	Unknown	Diarrhea	Unknown
Edema factor	*Bacillus anthracis*	Adenylate cyclase	ATP	Anthrax	Yes
Dermonecrotic toxin	*Bordetella pertussis*	Deamidase	Rho G proteins	Rhinitis	(Yes)
Pertussis toxin	*Bordetella pertussis*	ADP-ribosyltransferase	G protein(s)	Pertussis (whooping cough)	Yes
C3 toxin	*Clostridium botulinum*	ADP-ribosyltransferase	Rho G protein	Botulism	Unknown
Toxin A	*Clostridium difficile*	Glucosyltransferase	Rho G protein(s)	Diarrhea/PC	(Yes)
Toxin B	*Clostridium difficile*	Glucosyltransferase	Rho G protein(s)	Diarrhea/PC	Unknown
Cholera toxin	*Vibrio cholerae*	ADP-ribosyltransferase	G protein(s)	Cholera	Yes
Lethal factor	*Bacillus anthracis*	Metalloprotease	MAPKK1/MAPKK2	Anthrax	Yes
Cell-cell adherence					
Exfoliative toxins	*Staphylococcus aureus*	Serine protease, superantigen	Desmoglein, TCR and MHC II	Scalded skin syndrome[c]	Yes
Bacteroides fragilis toxin	Enterotoxigenic *Bacteroides fragilis*	Metalloprotease	E-cadherin (indirect?)	Diarrhea, inflammatory bowel disease	Yes
Vesicle traffic					
VacA	*Helicobacter pylori*	Large vacuole formation, apoptosis	Receptor-like protein tyrosine phosphatase, sphingomyelin	Gastric ulcers, gastric cancer	(Yes)
Exocytosis					
Neurotoxins A–G	*Clostridium botulinum*	Zinc metalloprotease	VAMP/synaptobrevin, SNAP-25 syntaxin	Botulism	Yes
Tetanus toxin	*Clostridium tetani*	Zinc metalloprotease	VAMP/synaptobrevin	Tetanus	Yes
Superantigens (activate immune response)					
Enterotoxins	*Staphylococcus aureus*	Superantigen	TCR and MHC II, medullary emetic center (vomit center)	Food poisoning[c]	Yes
Exfoliative toxins	*Staphylococcus aureus*	See "Cell-cell adherence" above			
Toxic shock syndrome toxin	*Staphylococcus aureus*	Superantigen	TCR and MHC II	Toxic shock syndrome[c]	Yes
Pyrogenic exotoxins	*Streptococcus pyogenes*	Superantigens	TCR and MHC II	Toxic shock syndrome, scarlet fever	Yes
Lethal factor	*Bacillus anthracis*	Metalloprotease	MAPKK1/MAPKK2	Anthrax	Yes

[a]**Abbreviations:** CLDT = cytolethal distending toxin; CNF = cytotoxic necrotizing factor; EAST = enteroaggregative *E. coli* heat-stable toxin; HC = hemorrhagic colitis; HUS = hemolytic uremic syndrome; LT = heat-labile toxin; MAPKK = mitogen-activated protein kinase kinase; MHC II = major histocompatibility complex class II; PC = antibiotic-associated pseudomembranous colitis; SNAP-25 = synaptosomal-associated protein; ST = heat-stable toxin; TCR = T-cell receptor; UTI = urinary tract infection; VAMP = vesicle-associated membrane protein.

[b]Yes = strong causal relationship between toxin and disease; (Yes) = role in pathogenesis has been shown in animal model or appropriate cell culture.

[c]Other diseases are also associated with the organism.

[d]Toxin is also produced by other genera of bacteria.

Two-subunit AB toxins. A common structural theme among many, but not all, bacterial toxins is that they have two subunits, usually called A and B. These two-subunit complexes are called AB toxins. The actual toxic activity in AB toxins resides within the A subunit. The role of the B subunit is limited to binding host cell receptors. Thus, the B subunit for each toxin delivers the A subunit to the host cell. Many AB toxins have five B subunits arranged as a ring, in the center of which is nestled a single A subunit (**Figs. 25.16B** and **25.17A**).

A major AB toxin subclass comprises toxins that have **ADP-ribosyltransferase** enzymatic activity. These enzyme toxins transfer the ADP-ribose group from an NAD molecule to a target protein (**Fig. 25.17B**). The ADP-ribosylated protein has an altered function. Sometimes the function is destroyed (for example, protein synthesis is destroyed by diphtheria toxin); other times the protein is locked into an active form insensitive to regulatory feedback control (for example, cAMP synthesis continues unchecked in the presence of cholera toxin).

The mechanisms of selected toxins representing the various classes are described in the following sections.

Alpha toxin. The hemolytic alpha toxin is produced by *Staphylococcus aureus,* an organism that causes boils and blood infections. Alpha toxin forms a transmembrane, oligomeric (seven-member) beta barrel pore in target cell membranes. It is easy to see how the resulting leakage of cell constituents and influx of fluid cause the target cell to burst. To form the pore, hydrophobic areas of each monomer face the lipids of the membrane, and hydrophilic residues face the channel interior. A completed pore and a cutaway view exposing the channel are illustrated in **Figure 25.18A** and **B**. Diagnostic microbiology laboratories visualize hemolysins (proteins that lyse red blood cells) such as alpha toxin by inoculating bacteria onto agar plates containing sheep red blood cells (**Fig. 25.18C**). The clear, yellow zones around the *S. aureus* colonies growing on blood agar indicate that the microbe secretes a hemolysin.

A.

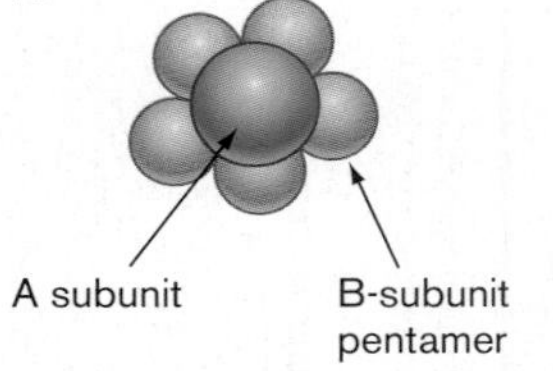
A subunit
B-subunit pentamer

B.

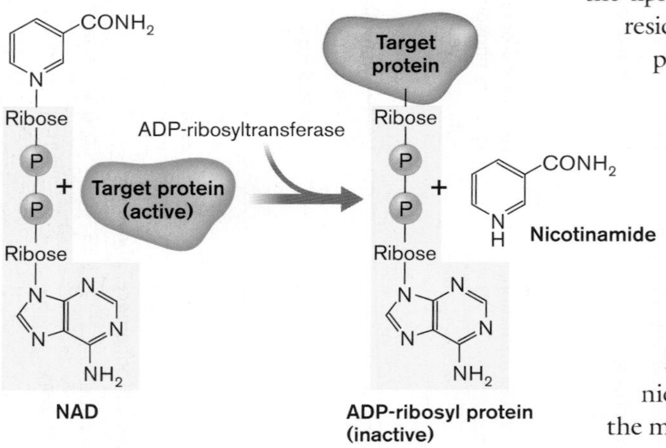
CONH2
N
Ribose
P
P
Ribose
N
N
N
N
NH2
NAD
+
Target protein (active)
ADP-ribosyltransferase
Target protein
Ribose
P
P
Ribose
NH2
ADP-ribosyl protein (inactive)
+
CONH2
N H
Nicotinamide

FIGURE 25.17 ■ AB toxins. **A.** A typical AB toxin consists of an A subunit and a pentameric B subunit joined noncovalently. **B.** Many AB toxins are ADP-ribosyltransferase enzymes that modify protein structure and function.

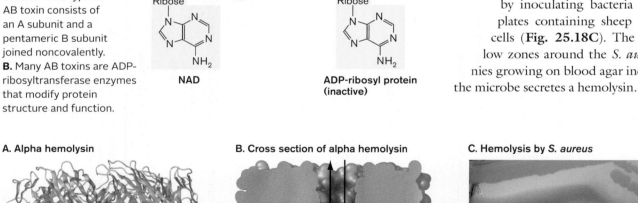

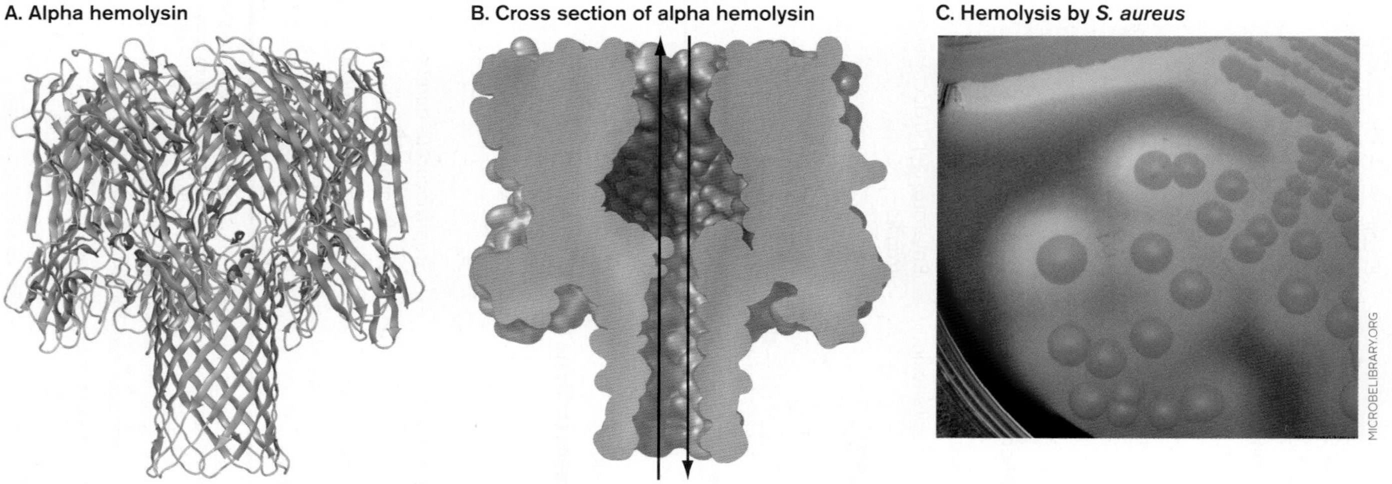
A. Alpha hemolysin
B. Cross section of alpha hemolysin
C. Hemolysis by S. aureus
MICROBELIBRARY.ORG

FIGURE 25.18 ■ Alpha hemolysin of *Staphylococcus aureus.* **A.** 3D figure of the pore complex comprising seven monomeric proteins. (PDB code: 7AHL) **B.** Cross section showing the channel. Arrows indicate movement of fluids through the pore. **C.** A blood agar plate inoculated with *S. aureus.* The alpha toxin is secreted by the organism and diffuses away from the producing colony. It forms pores in the red blood cells embedded in the agar, causing them to lyse, which is visible as a clear area surrounding each colony.

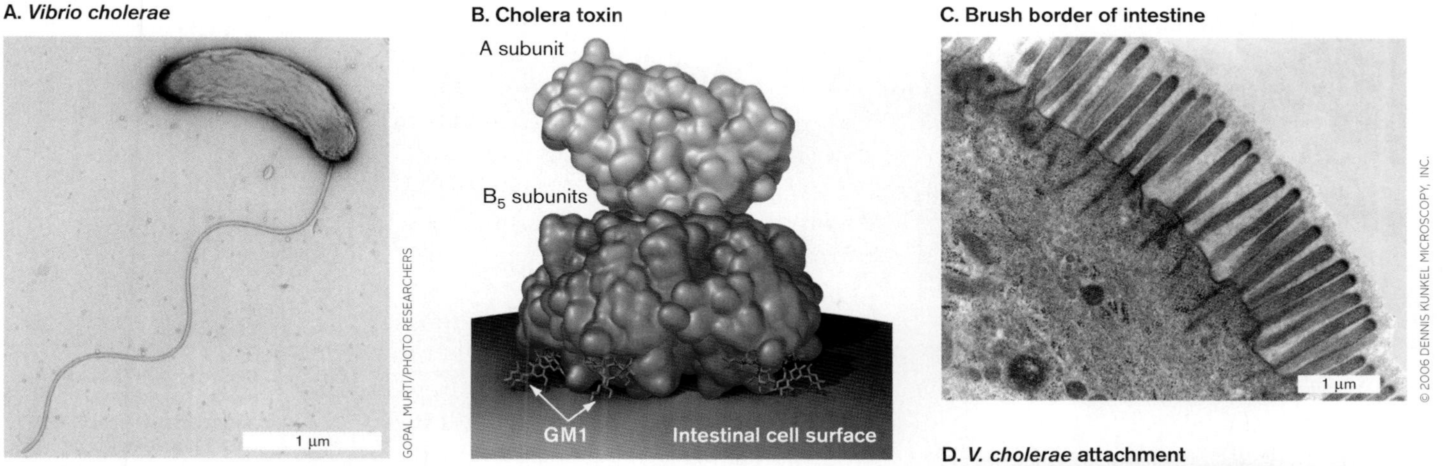
A. *Vibrio cholerae*
B. Cholera toxin
C. Brush border of intestine
1 μm
GOPAL MURTI/PHOTO RESEARCHERS
A subunit
B_5 subunits
GM1
Intestinal cell surface
1 μm
© 2006 DENNIS KUNKEL MICROSCOPY, INC.

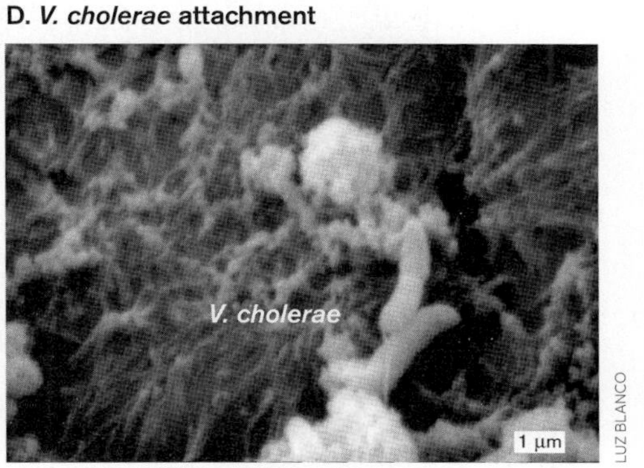
D. *V. cholerae* attachment
V. cholerae
1 μm
LUZ BLANCO

FIGURE 25.19 ▪ **Pathogenesis of cholera.** **A.** *Vibrio cholerae* (SEM). Note the slight curve of the cell and the presence of a single polar flagellum. **B.** 3D structure of cholera toxin binding ganglioside GM1 on the intestinal cell surface. (PDB code: 1S5F) **C.** Brush border of intestine (TEM). *V. cholerae* binds to the fingerlike villi on the apical surface. **D.** View of *V. cholerae* binding to the surface of a host cell (SEM). Note that *V. cholerae* does not invade the host cell.

Cholera and *E. coli* labile toxins. *Vibrio cholerae* (**Fig. 25.19A**) produces a severe diarrheal disease called cholera that generally afflicts malnourished people populating poor countries like Bangladesh, or countries in which access to clean water has been disrupted by war or natural disasters, such as the aftermath of the 2010 earthquake in Haiti. *V. cholerae* produces a gastrointestinal enterotoxin nearly identical to one produced by some strains of *E. coli* associated with what is known as "traveler's diarrhea." Enterotoxins specifically affect the intestine. The *E. coli* enterotoxin is called **labile toxin (LT)** because it is easily destroyed by heat. Cholera toxin (CT) and labile toxin are both AB toxins with identical modes of action, which is to increase the level of cAMP made inside the host cell.

Normally, intestinal transport mechanisms absorb NaCl and other ions (electrolytes), as well as water from food material moving through the intestine. These actions produce well-formed feces with very little water and salt content. CT and LT, however, reverse this process by secreting water and electrolytes into the intestinal lumen. After the bacteria attach to the cells lining the intestinal villi (**Fig. 25.19C** and **D**), they secrete their AB toxins (**Fig. 25.19B**). Both toxins have five B subunits arranged as a ring around a single A subunit. The B subunits bind to ganglioside GM1 on eukaryotic cell membranes and deliver the A subunit to the target cell (**Fig. 25.20**, steps 1–4). The A subunit possesses the toxic part of the molecule, an ADP-ribosyltransferase, which must be activated by the host.

The binding of CT or LT to GM1 triggers endocytosis and the formation of a toxin-containing vacuole, which is transported to the endoplasmic reticulum. During this time, the A subunit is cleaved by a host protease into two fragments, called A1 and A2, which are still held together by a disulfide bond. The reducing environment in the vacuole reduces that bond and frees the A1 peptide containing active ADP-ribosyltransferase into the endoplasmic reticulum, which exports the toxin into the cytoplasm.

The mission of the A1 peptide is to modify (that is, ADP-ribosylate) a membrane-associated GTPase (called a G protein or G factor) that binds to adenylate cyclase and controls its activity (**Fig. 25.20**, step 5). Human cells have two types of G-factor complexes that stimulate (G_s) or inhibit (G_i) adenylate cyclase, respectively, when bound to GTP (see **eTopic 25.5** for details). An intrinsic GTPase in each G factor hydrolyzes GTP and prevents continual stimulation of adenylate cyclase by G_s, or its inhibition by G_i. Cholera toxin (and *E. coli* labile toxin) ADP-ribosylate G_s and thereby inhibit the intrinsic GTPase activity (**Fig. 25.20**, step 5). As a result, adenylate cyclase is constantly stimulated to produce cAMP.

The high amount of cAMP stimulates a host protein kinase that activates various ion transport channels, including the cystic fibrosis transmembrane conductance regulator (CFTR), so named because a defect in this protein manifests as the lung disease cystic fibrosis. CFTR controls chloride transport in several cell types, including intestinal epithelia (discussed in Section 23.4). As a result of CFTR

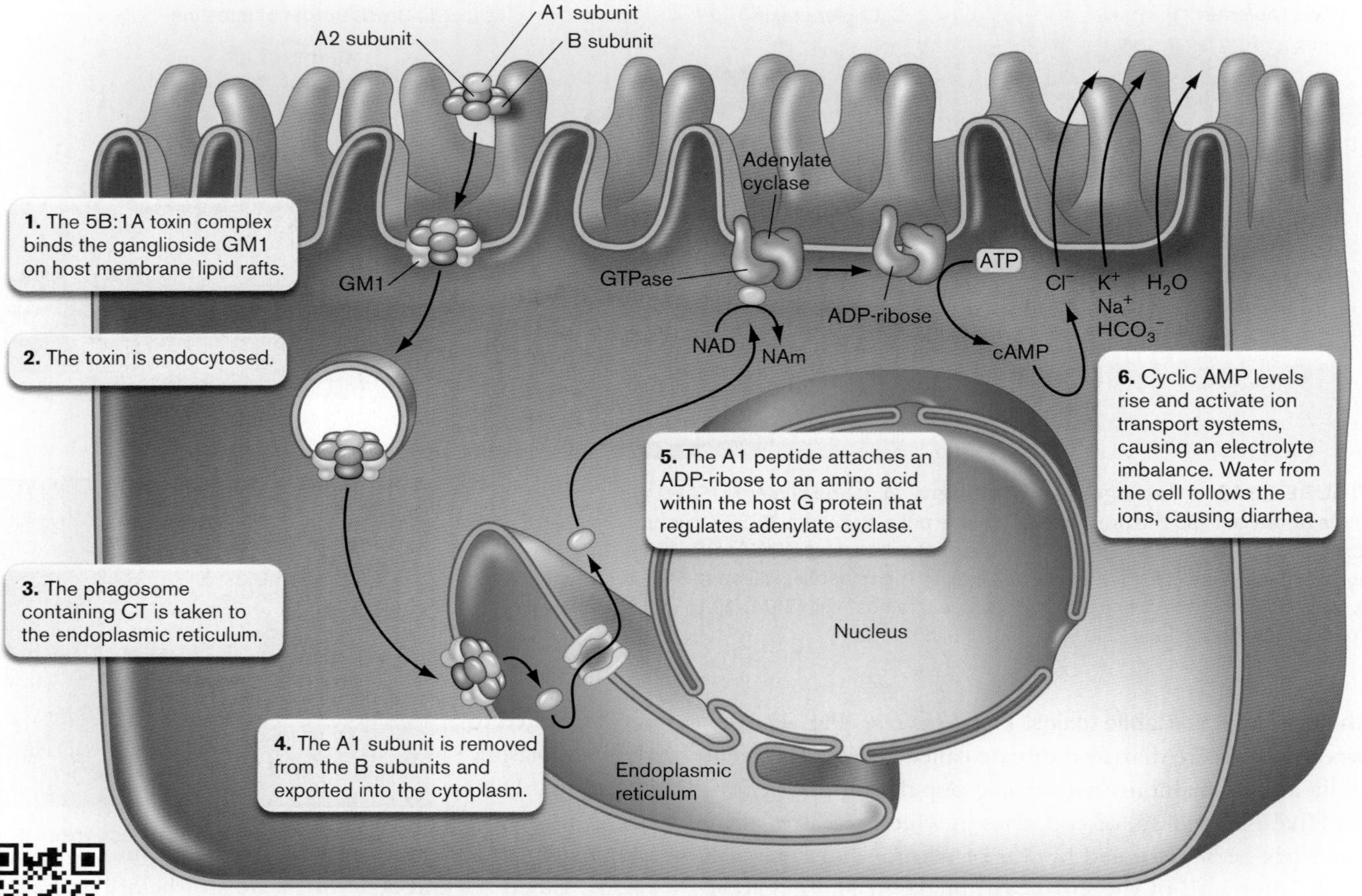
A1 subunit
A2 subunit
B subunit
1. The 5B:1A toxin complex binds the ganglioside GM1 on host membrane lipid rafts.
GM1
2. The toxin is endocytosed.
3. The phagosome containing CT is taken to the endoplasmic reticulum.
4. The A1 subunit is removed from the B subunits and exported into the cytoplasm.
Endoplasmic reticulum
Nucleus
5. The A1 peptide attaches an ADP-ribose to an amino acid within the host G protein that regulates adenylate cyclase.
Adenylate cyclase
GTPase
NAD
NAm
ADP-ribose
ATP
cAMP
Cl⁻ K⁺ H₂O
Na⁺
HCO₃⁻
6. Cyclic AMP levels rise and activate ion transport systems, causing an electrolyte imbalance. Water from the cell follows the ions, causing diarrhea.

FIGURE 25.20 ■ Cholera toxin mode of action. Delivery of cholera toxin into target cells and deregulation of adenylate cyclase activity. NAm = nicotinamide.

activation, chloride, sodium, and other ions leave the cell, and in an attempt to equilibrate osmolarity, water leaves as well. Because the affected cells line the intestine, the escaping water enters the intestinal lumen, leading to watery stools, or diarrhea (**Fig. 25.20**, step 6).

Thought Question

25.2 **Figure 25.20** illustrates how cholera toxin works to cause diarrhea. To develop a vaccine that generates protective antibodies, which subunit of cholera toxin should be used to best protect a person from the toxin's effects?

How is diarrhea a benefit to the microorganism? The more diarrhea is produced and expelled, the more organisms are made and disseminated throughout the environment. This increases the chance that another host will ingest the organism, ensuring survival of the species. Diarrhea can also benefit a pathogen by decreasing competition with other organisms as they are swept away. In fact, the vast majority of organisms found in the diarrheal fluids of cholera patients are *V. cholerae* bacteria.

Different pathogens have discovered alternative ways to alter host cAMP levels. The Gram-negative pathogen *Bordetella pertussis* causes a childhood respiratory infection called whooping cough, so named for the whooping sound made as a child tries to take a breath after a fit of coughing. This microbe also secretes an ADP-ribosylating toxin, but this one modifies G_i (the inhibitory factor that normally downregulates adenylate cyclase activity). ADP-ribosylation in this instance prevents that inhibition, with the result (again) of runaway cAMP synthesis. Pertussis toxin, however, is structurally different from cholera toxin. In addition, the organism makes a "stealth" adenylate cyclase that is secreted from the bacterium but remains inactive until it enters host cells, where it becomes active by binding the calcium-binding protein calmodulin. The resulting increase in cAMP levels causes inappropriate triggering of certain host cell signaling pathways and damages lung cells.

Many other bacterial toxins utilize ADP-ribosylation to target different host proteins. A classic example, discussed in **eTopic 25.6**, is diphtheria toxin, produced by *Corynebacterium diphtheriae*. This toxin kills cells by ADP-ribosylating protein synthesis elongation factor 2 (EF2). *C. diphtheriae* is a good example of a pathogen whose toxin gene (*dtx*) is part of a prophage genome integrated into the bacterial chromosome.

Thought Question

25.3 How might you experimentally determine whether a pathogen secretes an exotoxin? (*Hint:* You can learn more about how new toxins are found in **eTopic 25.7**.)

Shiga toxin: *Shigella* and *E. coli* O157:H7. *Shigella dysenteriae* and *E. coli* O157:H7 (also known as enterohemorrhagic *E. coli*) cause food-borne diseases whose symptoms include bloody diarrhea. These organisms produce an important toxin known as Shiga toxin (or Shiga-like toxin). The gene (*stx*) encoding Shiga toxin is part of a phage genome integrated into the bacterial chromosome. The toxin has five B subunits for binding and one A subunit imbued with toxic activity. The A subunit, upon entry, destroys protein synthesis by cleaving 28S rRNA in eukaryotic ribosomes. Strains that produce high levels of this toxin are associated with acute kidney failure known as hemolytic uremic syndrome.

Note: *Shigella dysenteriae* is the only *Shigella* species that produces Shiga toxin.

E. coli O157:H7 is a recently emerged pathogen. The organism can colonize cattle intestines without causing bovine disease; as a result, undetected bacteria can easily contaminate meat products following slaughter, as well as irrigation water from farms (see Chapter 28). The organism is sometimes referred to as the "Jack in the Box" microbe (or hamburger disease microbe if you are Canadian) because the first large U.S. outbreak was associated with fast-food hamburgers at a Washington State Jack in the Box restaurant in 1993. Both *E. coli* and *Shigella* have remarkable acid resistance mechanisms that rival that of the gastric pathogen *Helicobacter pylori*. This acid resistance makes these pathogens infectious at a very low infectious dose.

What activates *stx* expression? Iron availability is a key factor for inducing the expression of *stx* and many other virulence genes in pathogens. The body holds its iron tightly in proteins such as lactoferrin and transferrin. To an invading organism, the body is a very iron-poor environment. In the presence of low iron, expression of shiga toxin increases, the toxin kills host cells, and dead cells release their iron. Shiga toxin, then, offers a way to rob the host of its iron stores.

An intriguing question often pondered by scientists is, What roles do virulence factors play in the natural ecology of these bacteria? Surely factors such as Shiga toxin did not evolve after *Shigella* started infecting humans. William Lainhart, Gino Stolfa, and Gerald Koudelka from the State University of New York at Buffalo predicted that Shiga toxin was actually a natural defense against *Tetrahymena thermophila*, a ciliated protist that grazes on bacteria. Predatory ciliates are a major source of bacterial mortality in the environment. Confirming their hypothesis, Lainhart and his colleagues found that *Tetrahymena* was killed when cocultured with Shiga toxin–producing bacteria. They proposed that reactive oxygen species produced by *Tetrahymena* induce the SOS response in *Shigella*, which then activates phage replication and the expression of *stx* (SOS induction is discussed in Section 9.5). This is another example of altruistic behavior by members of a bacterial species in which some individuals sacrifice themselves (as a result of phage lysis) to save the population. For further discussion of the beneficial roles of viruses, see Chapter 6.

Thought Question

25.4 Would patients with iron overload (excess free iron in the blood) be more susceptible to infection?

Anthrax. A century ago, anthrax (caused by *Bacillus anthracis*; **Fig. 25.21A**) was mainly a disease of cattle and sheep. Humans acquired the disease only accidentally. Today we fear the deliberate shipment of *B. anthracis* through the mail (as happened in 2001) or its dispersion from the air ducts of heavily populated buildings. What makes this Gram-positive, spore-forming microbe so dangerous? In large part, its lethality is due to the secretion of a plasmid-encoded tripartite toxin. The core subunit of the toxin is called **protective antigen (PA)** because immunity to this protein protects hosts from disease. Protective antigen is made as a single peptide but then binds to the host cell surface (there are multiple receptors), where a human protease cleaves off a fragment (**Fig. 25.21B**). The remaining part of PA can self-assemble in the membrane to form heptameric (seven-membered) and octomeric (eight-membered) pores. The other two components of anthrax toxin—**edema factor (EF)** and **lethal factor (LF)**—bind to the PA rings and are carried into the cell (**Fig. 25.21C**). The complex is endocytosed, and the two proteins carried in are passed through the pore into the host cytoplasm. Proton motive force helps unfold the toxins and translocates them across the endocytic membrane. PA is the B

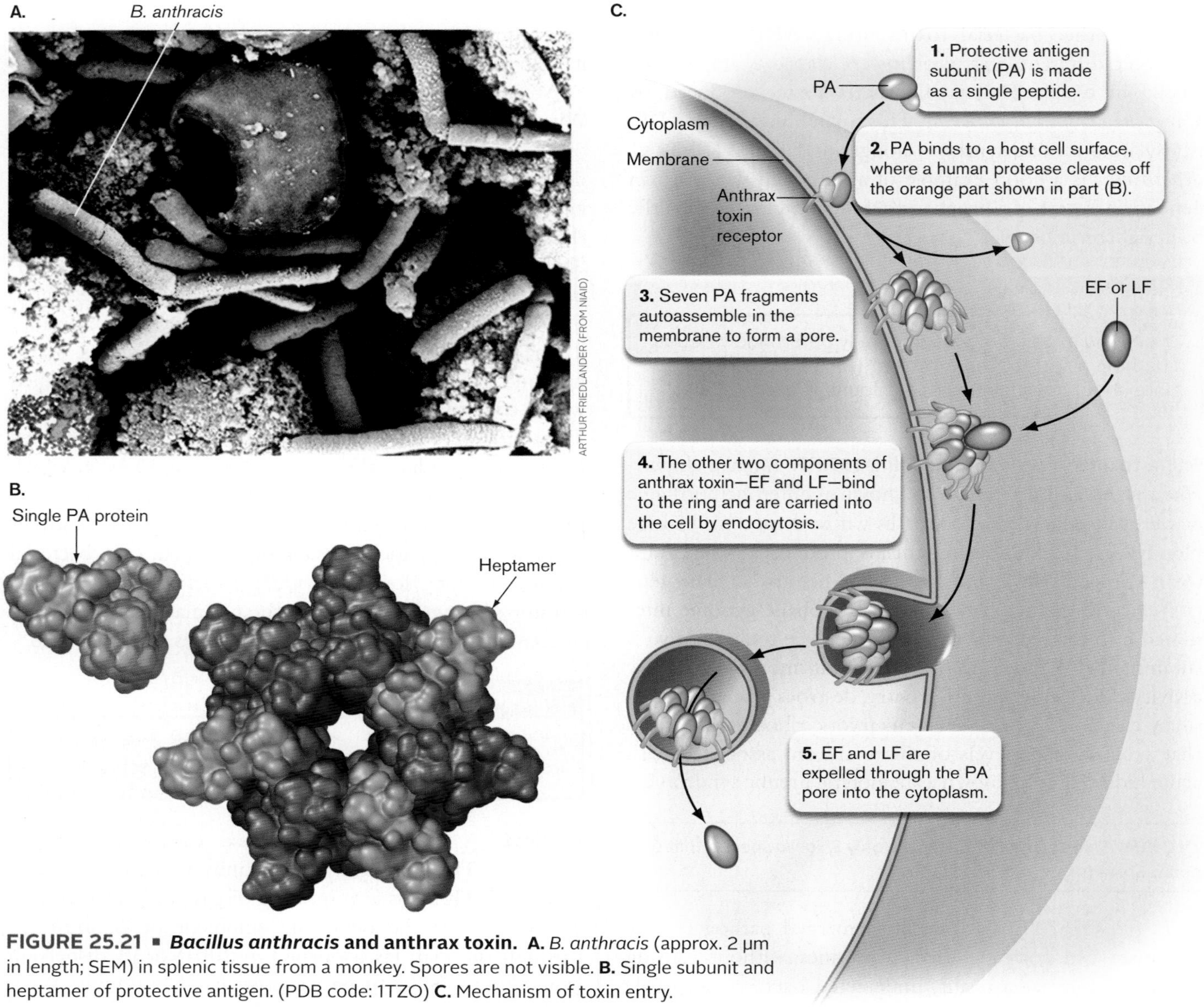
A.
B. anthracis
ARTHUR FRIEDLANDER (FROM NIAID)
B.
Single PA protein
Heptamer
C.
PA
1. Protective antigen subunit (PA) is made as a single peptide.
Cytoplasm
Membrane
2. PA binds to a host cell surface, where a human protease cleaves off the orange part shown in part (B).
Anthrax toxin receptor
3. Seven PA fragments autoassemble in the membrane to form a pore.
EF or LF
4. The other two components of anthrax toxin—EF and LF—bind to the ring and are carried into the cell by endocytosis.
5. EF and LF are expelled through the PA pore into the cytoplasm.

FIGURE 25.21 ■ *Bacillus anthracis* and anthrax toxin. **A.** *B. anthracis* (approx. 2 μm in length; SEM) in splenic tissue from a monkey. Spores are not visible. **B.** Single subunit and heptamer of protective antigen. (PDB code: 1TZO) **C.** Mechanism of toxin entry.

subunit for anthrax toxin while EF and LF represent different A subunits.

Edema factor and lethal factor are the toxic parts of anthrax toxin. Both are enzymes that attack the signaling functions of the cell. Edema factor is an adenylate cyclase that remains inactive until entering the cytoplasm, where it binds calmodulin. This binding activates adenylate cyclase, resulting in a huge production of cAMP, and inactivates calmodulin from its normal function in the cell.

Lethal factor is actually a protease that cleaves several host protein kinase kinases, each of which is part of a critical regulatory cascade affecting cell growth and proliferation. A protein kinase kinase is an enzyme that phosphorylates, and thereby activates, another protein kinase that can then phosphorylate one or more subsequent target proteins. One consequence of subverting these phosphorylation cascades is a failure to produce signals that recruit immune cells to fight the infection.

We have examined only a few of the many toxins employed by pathogens. Some of the others, including tetanus and botulism toxins, will be described in the next chapter. What should be apparent from our brief sampling is the evolutionary ingenuity that pathogens have used to try to tame the human host. A discussion of how new microbial toxins are discovered is found in **eTopic 25.7**.

Thought Question

25.5 Use an Internet search engine to determine what other toxins are related to the cholera toxin A subunit. (*Hint:* Start by searching on "protein" at PubMed to find the protein sequence; then use a BLAST program.)

Weblinks Anthrax tutorial (*see ebook*)

Endotoxin (LPS) Is Made Only by Gram-Negative Bacteria

Another important virulence factor common to all Gram-negative microorganisms is endotoxin present in the outer membrane (discussed in Chapter 3). Not to be confused with secreted exotoxins, endotoxin is an embedded part of the bacterial cell surface and an important contributor to disease. Endotoxin, otherwise called lipopolysaccharide (LPS), is composed of lipid A, core glycolipid, and a repeating polysaccharide chain known as the O antigen (**Fig. 25.22**). LPS molecules form the outer leaflet of the Gram-negative outer membrane (discussed in Chapter 3). As bacteria die, they release endotoxin. Endotoxin is a microbe-associated molecular pattern (MAMP) molecule that can bind to certain Toll-like receptors on macrophages or B cells and trigger the release of TNF-α, interferon, IL-1, and other cytokines (MAMPs, Toll-like receptors, and cytokines are discussed in Chapters 23 and 24). The release of these active agents causes a variety of symptoms, such as:

- Fever
- Activation of clotting factors, leading to disseminated intravascular coagulation
- Activation of the alternative complement pathway
- Vasodilation, leading to hypotension (low blood pressure)
- Shock due to hypotension
- Death when other symptoms are severe

The lipid A moiety of LPS possesses endotoxic activity (**Fig. 25.22**).

Note: Lipopolysaccharide O antigens forming part of the outer membrane are used to classify different strains of *E. coli* (for example, *E. coli* O157 versus *E. coli* O111), as well as other Gram-negative organisms.

The role of endotoxins can be seen in infections with the Gram-negative diplococcus *Neisseria meningitidis* (**Fig. 25.23A**), a major cause of bacterial meningitis. *N. meningitidis* has, as part of its pathogenesis, a septicemic phase in which the organism can replicate to high numbers in the bloodstream. The large amount of endotoxin present causes a massive depletion of clotting factors, which leads to internal bleeding, most prominently displayed to a physician as small pinpoint hemorrhages called **petechiae** on the patient's hands and feet (**Fig. 25.23B**). Capillary bleeding near the surface of the skin causes petechiae. One danger of treating massive Gram-negative sepsis with antibiotics is that the enormous release of endotoxin from dead bacteria could well kill the patient. Untreated Gram-negative sepsis is, however, almost always fatal, so its treatment, albeit risky, is imperative.

A proposed approach to prevent endotoxic shock is based on the knowledge that LPS must bind to Toll-like receptor TLR4 to cause endotoxic shock. What if we could neutralize TLR4 by antibody and prevent it from binding LPS during an infection? In a proof-of-principle experiment, antibody raised to TLR4 (anti-TLR4) was injected into mice. The antibody successfully blocked TLR4 receptors and protected the mice from *E. coli*–induced septic shock.

A. LPS membrane

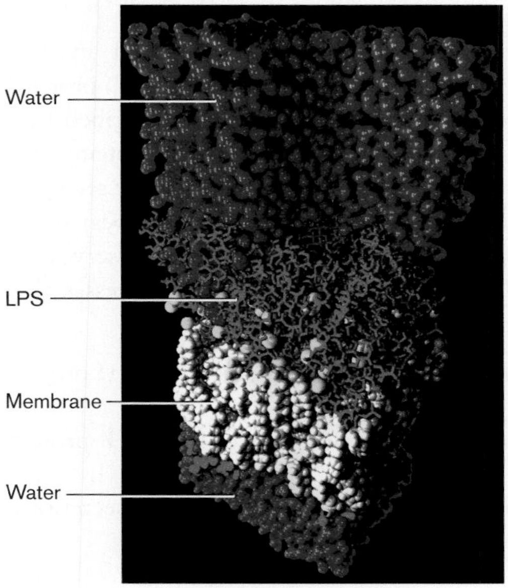
Water
LPS
Membrane
Water

PACIFIC NORTHWEST NATIONAL LABORATORY

B. Gram-negative bacterial endotoxin (lipopolysaccharide, LPS)

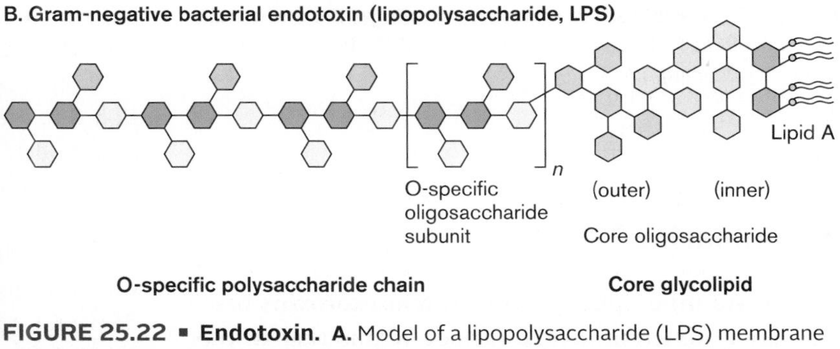
n
Lipid A
O-specific oligosaccharide subunit
(outer)
(inner)
Core oligosaccharide
O-specific polysaccharide chain
Core glycolipid

FIGURE 25.22 ■ Endotoxin. A. Model of a lipopolysaccharide (LPS) membrane of *Pseudomonas aeruginosa,* consisting of 16 lipopolysaccharide molecules (red) and 48 ethylamine phospholipid molecules (white). **B.** Basic structure of endotoxin, showing the repeating O-antigen side chain that faces out from the microbe and the membrane proximal core glycolipid and lipid A (contains endotoxic activity).

A.

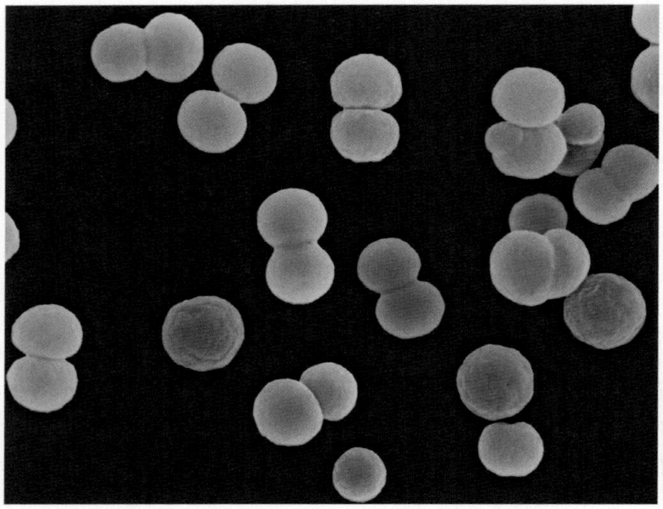

©DENNIS KUNKEL MICROSCOPY, INC.

B.

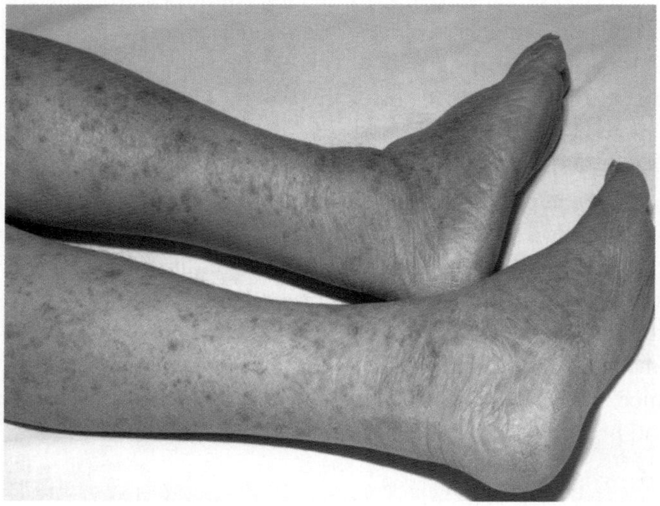

MEDISCAN/VISUALS UNLIMITED

FIGURE 25.23 ▪ Effect of *Neisseria meningitidis* endotoxin. **A.** *N. meningitidis* (cell 0.8–1 μm in diameter; SEM). **B.** Petechial rash caused by *N. meningitidis*.

To Summarize

- **There are nine categories of protein exotoxins** based on mode of action. These include toxins that disrupt membranes, inhibit protein synthesis, or alter host cell signal molecule synthesis, as well as superantigens and target-specific proteases.
- **Many bacterial toxins are two component, AB subunit toxins.** The B subunit promotes penetration through host cell membranes, whereas the A subunit has toxic activity.
- *Staphylococcus aureus* alpha toxin forms pores in host cell membranes.
- **Cholera toxin,** *E. coli* labile toxin, and pertussis toxin are AB toxins that alter host cAMP production by adding ADP-ribose groups to different G-factor proteins.
- **Shiga toxin** is an AB toxin that cleaves host cell 28S rRNA in host cell ribosomes.
- **Anthrax toxin** is a three-part AB toxin with one B subunit (protective antigen) and two different A subunits that affect cAMP levels (edema factor) and cleave host protein kinases (lethal factor).
- **Lipopolysaccharide (LPS)**, known as endotoxin, is an integral component of Gram-negative outer membranes and an important virulence factor that triggers massive release of cytokines from host cells. The indiscriminate release of cytokines can trigger fever, shock, and death.

25.5

Deploying Toxins and Effectors

A recurring theme among bacterial pathogens is the secretion of proteins that destroy, cripple, or subvert host target cells. The bacterial toxins described in Section 25.4 are secreted into the surrounding environment, where they float randomly until chance intervenes and they hit a membrane-binding site. However, many pathogens attach to tissue cells and inject bacterial proteins (called effectors) directly into the host cell cytoplasm. The proteins may not kill the cell, but they redirect host signaling pathways in ways that benefit the microbe.

Protein secretion pathways were introduced in Section 8.6, focusing on ATP-binding cassette (ABC) proteins as a model. Additional secretion models are described here in their critical role of delivering pathogenicity proteins such as toxins. A particularly interesting aspect of these secretory systems is that many of them evolved from, and bear structural resemblance to, other cellular structures that serve fundamental cellular functions. The molecular processes that are evolutionarily related to secretion include:

- Type IV pilus biogenesis (homologous to type II protein secretion)
- Flagellar synthesis (homologous to type III protein secretion)
- Conjugation (homologous to type IV protein secretion)

TABLE 25.4

Secretion systems for bacterial toxins.

Secretion type	Features	Examples
I	SecA dependent, one effector per system	*E. coli* alpha hemolysin, *Bordetella pertussis* adenylate cyclase
II	SecA dependent, similar to type IV pili	*Pseudomonas aeruginosa* exotoxin A, elastase, cholera toxin
III	SecA independent, syringe, related to flagella, multiple effectors secreted	*Yersinia* Yop proteins, *Salmonella* Sip proteins, enteropathogenic *E. coli* (EPEC) EspA proteins, TirA
IV	Related to conjugational DNA transfers, multiple effectors secreted	*B. pertussis* toxin, *Helicobacter* CagA
V	Autotransporter, SecA dependent to periplasm, self-transport through outer membrane, one effector per system	Gonococcal and *Haemophilus influenzae* IgA proteases
VI	Related to phage tails, single effector, harpoon mechanism	*Burkholderia* and *Vibrio cholerae* VgrG
VII	Unrelated to other systems	*Mycobacterium tuberculosis* Esx and Esp

Table 25.4 lists features of the seven export systems and examples of associated virulence effector proteins. Systems I, III, and IV will be described in more detail.

Type II Secretion Resembles Type IV Pilus Assembly

Cholera toxin, discussed in Section 25.4, is a well-known example of a toxin secreted by a **type II secretion system**. Type II secretion offers a clear example of how nature has modified the blueprints of one system to do a very different task. DNA sequence analysis has revealed that the genes used for type IV pilus biogenesis (see Section 25.3) were duplicated at some point during evolution and redesigned to serve as a protein secretion mechanism. Type IV pili have the unusual ability to extend and retract from the outer membrane—a property that produces the gliding motility of *Myxococcus* (see Section 4.7) and the twitching motility of *Neisseria* and *Pseudomonas*. As you might guess, assembly/disassembly of these appendages is quite complex.

Type II protein secretion mechanisms mirror this complexity. Proteins to be secreted first make their way, via the Sec-dependent general secretion pathway, to the periplasm, where they then encounter the appropriate type II secretion system. Because of this periplasmic "layover," the proteins are folded before secretion. Type II secretion systems use the pilus-like structure as a piston to ram the folded proteins through an outer membrane pore structure and into the surrounding void (**Fig. 25.24**). Piston action occurs via

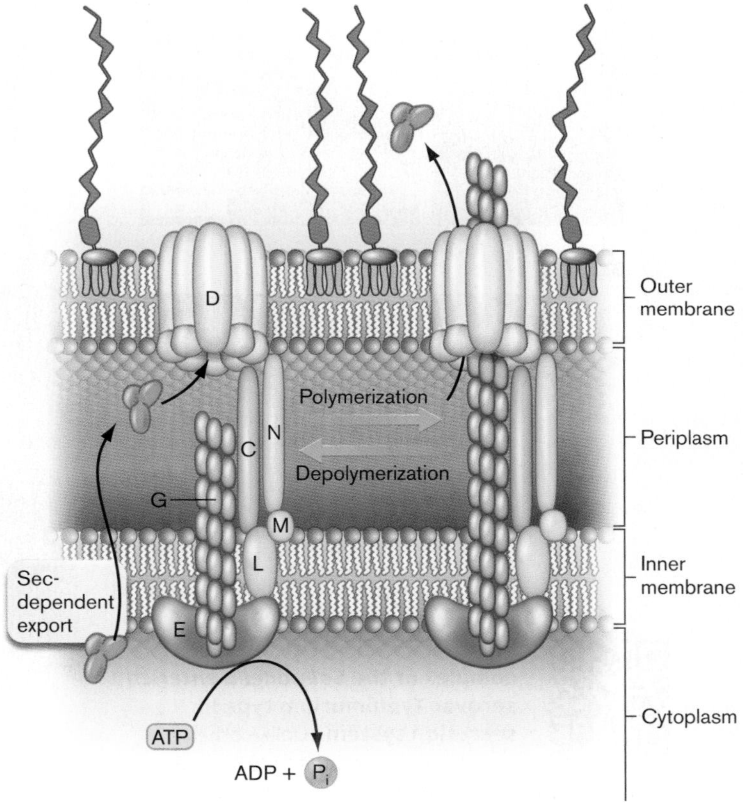

D
Polymerization
N
C
Depolymerization
G
M
L
Sec-dependent export
E
ATP
ADP + P_i
Outer membrane
Periplasm
Inner membrane
Cytoplasm

FIGURE 25.24 ■ Type II secretion. C, D, E, G, L, M, and N are protein components of the secretion system.

cyclic assembly and disassembly of pilus-like proteins, driving the pilus-like structure through an outer membrane pore and then retracting.

Type III Secretion Injects Effector Proteins into Host Cells

In the 1990s, it was discovered that the etiological agents of Black Death and various forms of diarrhea (caused by species of *Yersinia*, *Salmonella*, and *Shigella*) could somehow take the bacterial virulence proteins made in their cytoplasm and drive them directly into the eukaryotic cell cytoplasm without the protein ever getting into the extracellular environment. Direct delivery is a good idea because it eliminates the dilution that happens when a toxin is secreted into media. Another advantage of this strategy is that it avoids the need to tailor the toxin to fit a preexisting host receptor.

What kind of molecular machine can directly deliver cytoplasmic bacterial proteins into target cells? Research has shown that some microbes use tiny molecular syringes embedded in their membranes to inject proteins directly into the host cytoplasm; this mechanism of delivery is called a **type III secretion system (T3SS)**. **Figure 25.25** shows electron micrographs of type III secretion needles and a model of the complex. Genes encoding type III systems are actually related to flagellar genes, whose products export the flagellin proteins through the center of a growing flagellum (discussed in Section 3.7). It appears that flagellar genes were evolutionarily reengineered to encode proteins that act more like molecular syringes (**Fig. 25.25C**▶).

The bacterial virulence proteins secreted by type III systems subvert normal host cell signaling pathways, some of which cause dramatic rearrangements of host cytoskeleton at the cell membrane that lead to engulfment of the microbe (**Fig. 25.26**). The genes encoding type III systems

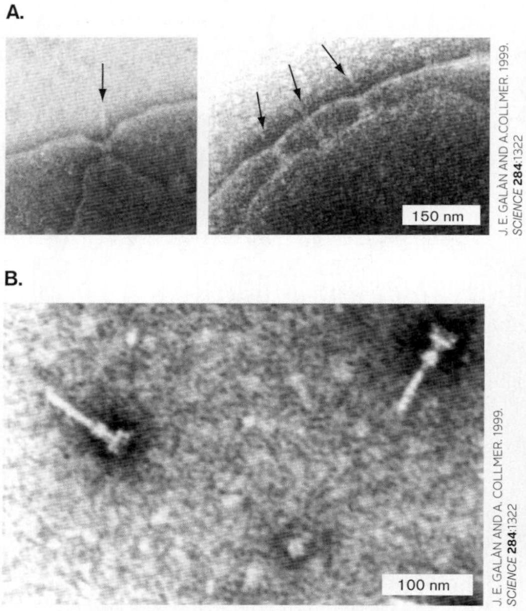
A.
150 nm
J. E. GALÁN AND A.COLLMER. 1999. SCIENCE 284:1322
B.
100 nm
J. E. GALÁN AND A. COLLMER. 1999. SCIENCE 284:1322

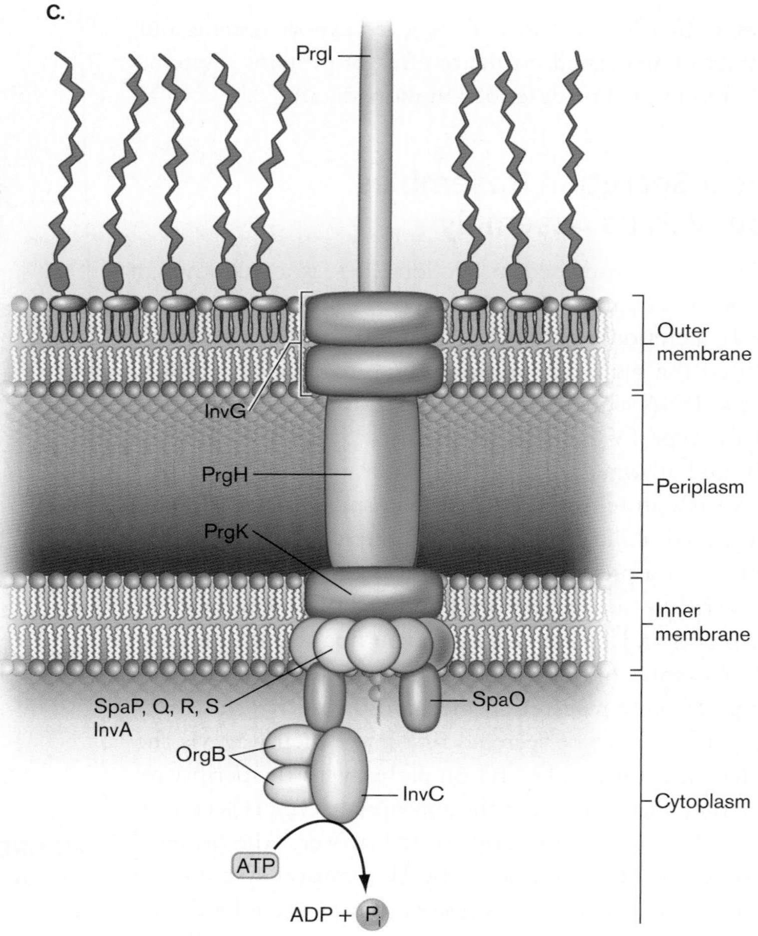
C.
PrgI
InvG
PrgH
PrgK
SpaP, Q, R, S
InvA
SpaO
OrgB
InvC
ATP
ADP + Pi
Outer membrane
Periplasm
Inner membrane
Cytoplasm

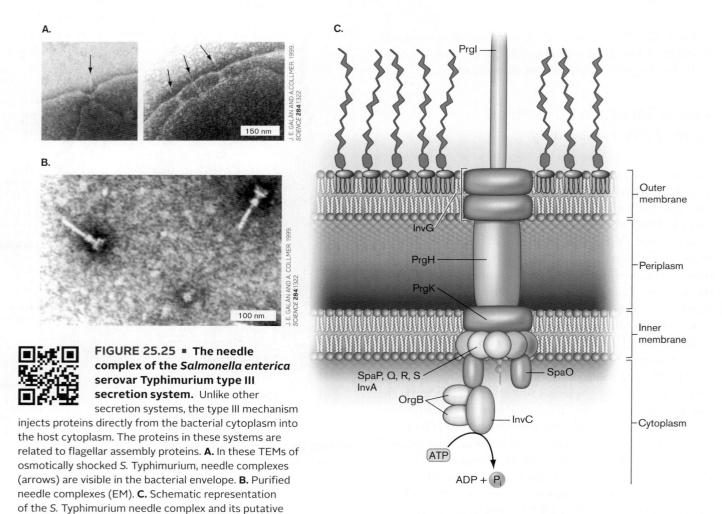

FIGURE 25.25 ▪ The needle complex of the *Salmonella enterica* serovar Typhimurium type III secretion system. Unlike other secretion systems, the type III mechanism injects proteins directly from the bacterial cytoplasm into the host cytoplasm. The proteins in these systems are related to flagellar assembly proteins. **A.** In these TEMs of osmotically shocked *S.* Typhimurium, needle complexes (arrows) are visible in the bacterial envelope. **B.** Purified needle complexes (EM). **C.** Schematic representation of the *S.* Typhimurium needle complex and its putative components. *Source:* Part C modified from Moat, Foster, and Spector. 2002. *Microbial Physiology,* 4th ed. © Wiley-Liss, Inc. ▶

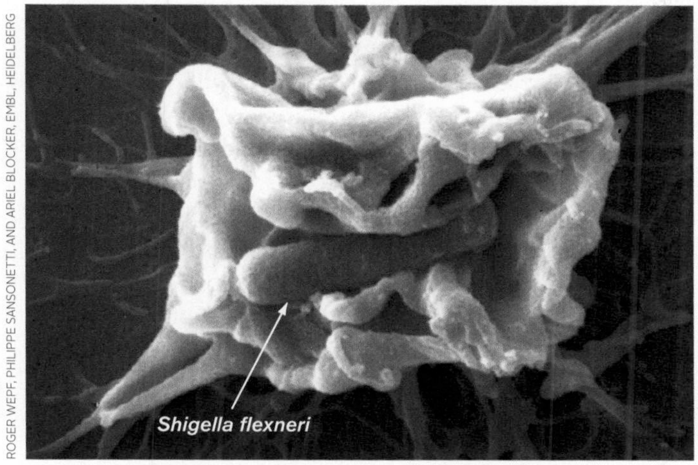
Shigella flexneri
ROGER WEPF, PHILIPPE SANSONETTI, AND ARIEL BLOCKER, EMBL, HEIDELBERG

FIGURE 25.26 ▪ *Shigella* invades a host cell ruffle produced as a result of type III secretion. *Shigella flexneri* entering a HeLa cell (SEM). The bacterium (small diameter approx. 1 μm) interacts with the host cell surface and injects (via its type III secretion apparatus) its invasin proteins, which choreograph a local actin-rich membrane ruffle at the host cell. The ruffle engulfs the bacterium and eventually disassembles, internalizing the bacterium.

are usually located within pathogenicity islands inherited from other microbial sources. Many bacterial pathogens use this type of secretion system, including plant pathogens such as *Pseudomonas syringae* (the cause of blight, a disease of many plants in which leaves or stems develop brown spots). Secretion is normally triggered by cell-cell contact between host and bacterium.

A recently proposed alternative to the injection model suggests that the T3SS needle complex is really a sensor of cell-cell contact and that the effector proteins are presituated on the bacterial membrane. Once tight cell-cell contact is sensed, the translocation of effectors across the host membrane is carried out by translocator proteins paired with the effector proteins on the bacterial membrane. There is evidence to support both models.

***E. coli* uses a T3SS to "inject" its own receptor into host cells.** Some microbes do not rely solely on the natural array of host receptors for attachment. Instead, these bacterial pathogens use a type III secretion system to insert their own receptors into target cells (**Fig. 25.27A**). One such group of enterprising pathogens is enteropathogenic *E. coli* (EPEC). Although EPEC uses pili to form an initial, loose attachment, the bacterium must ultimately establish a more intimate attachment. EPEC achieves this using an adhesion molecule on the bacterial cell called **intimin**. Intimin is a 94-kDa integral outer membrane bacterial protein needed for intimate adherence to host cells. However, there is no natural receptor for intimin on host membranes. As discovered by Brett Finlay and his colleagues in Vancouver, British Columbia, EPEC must deliver its own intimin receptor, a 57-kDa bacterial protein called Tir (for translocated

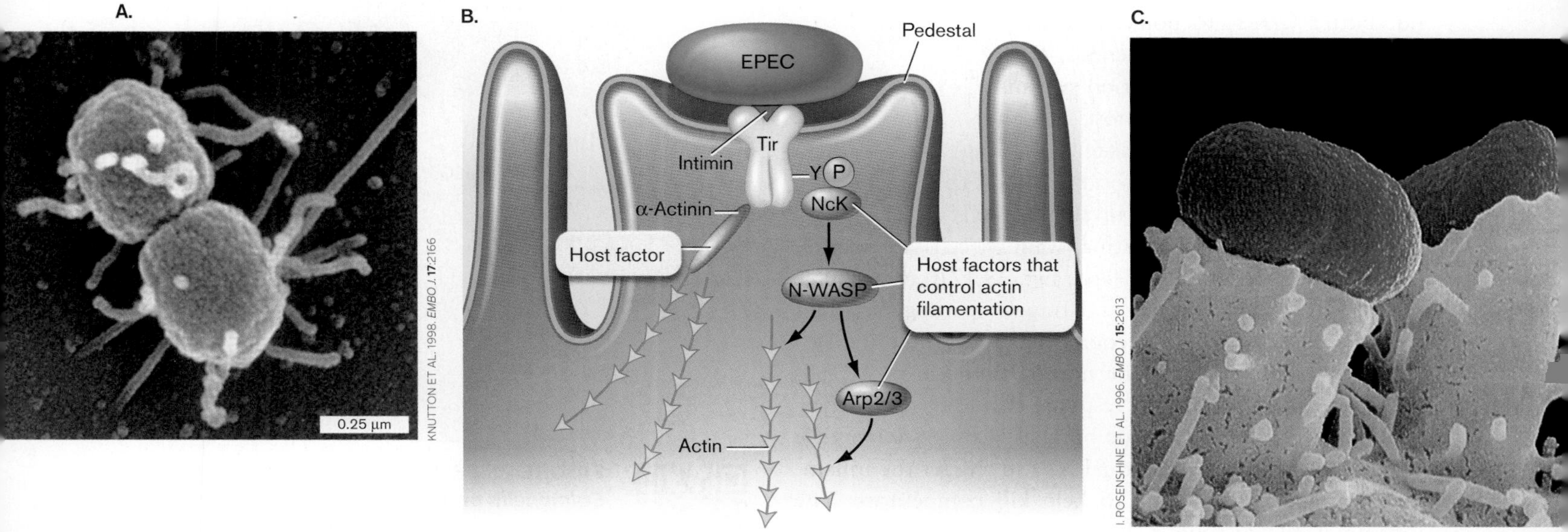
A.
0.25 μm
KNUTTON ET AL. 1998. EMBO J. 17:2166
B.
Pedestal
EPEC
Tir
Intimin
Y
P
NcK
α-Actinin
Host factor
Host factors that control actin filamentation
N-WASP
Arp2/3
Actin
C.
I. ROSENSHINE ET AL. 1996. EMBO J. 15:2613

FIGURE 25.27 ▪ *E. coli* type III secretion and cell-cell interaction. A. EspA filaments form a bridge between enteropathogenic *E. coli* (EPEC) and an epithelial cell during the early stages of attaching and effacing (A/E) lesion formation (SEM). These filaments are part of the type III secretion apparatus, functioning as a molecular syringe to inject proteins from pathogenic *E. coli* into the host cell. **B.** Model of cytoskeletal components within the EPEC pedestal. EPEC injects Tir protein into the host cell, where it moves to the membrane and acts as a receptor for intimin. Tir also communicates through phosphorylation with other host proteins to cause a change in actin cytoskeleton, which leads to pedestal formation. **C.** Pedestal formation (SEM).

intimin receptor) that the bacteria inject into the host cell. The genes that encode intimin, Tir, and the secretion apparatus are all part of an EPEC pathogenicity island.

Once injected and placed in the host membrane, Tir binds intimin on the bacterial surface (**Fig. 25.27B**). Think of Tir as a wall anchor you poke into a board in order to attach something to it. The result is a tighter, more intimate adherence between the bacterium and host cell surface required for infection to proceed. In addition, host protein kinases phosphorylate Tir at tyrosine residue 474. Phosphorylated Tir directly triggers a remarkable reorganization of host cellular cytoskeletal components (actin, alpha-actinin, ezrin, talin, and myosin light chain) such that a membrane "pedestal" is formed, raising the microbe up (**Fig. 25.27C**). The result of this attachment is the characteristic attaching and effacing (A/E) lesion, characterized by destruction of the microvilli and pedestal formation. By placing itself on a " pedestal," EPEC avoids engulfment and the perils of the phagolysosome.

Weblinks *E. coli* infection (*see ebook*)

***Salmonella* pathogenesis.** The Gram-negative bacterium *Salmonella enterica* is currently the most common bacterial food-borne pathogen in the United States (*Campylobacter* is second). Its transmission has been linked to everything from cantaloupe (2012) to peanut butter (2012) to poultry (pick a year). As part of its pathogenesis, *Salmonella* uses type III secretion systems to invade the eukaryotic host cell and become an intracellular parasite. Following ingestion of contaminated food or water, this bacterium attaches to and invades M cells that are interspersed along the intestinal wall. M cells are specialized intestinal epithelial cells (see Fig. 23.17B) that sample normal intestinal microbes and transfer pathogens across the epithelial barrier for recognition by the immune system. *Salmonella* subverts the normal function of M cells and causes an inflammatory response that leads to diarrhea. (Actually, a subset of "suicide bomber" cells of *Salmonella* Typhimurium invades intestinal cells to trigger the inflammatory response as a way to help kill competitors. The inflammatory neutrophils also make thiosulfate, which *Salmonella*, but not its competitors, can use as an alternative electron acceptor.)

During its evolutionary journey toward becoming a pathogen, *Salmonella* has acquired as many as 14 pathogenicity islands that are absent from related, but harmless, *E. coli* strains. Five of those islands can be found in all *S. enterica* serovars, but only two will be discussed here. *Salmonella* pathogenicity island 1 (SPI-1) encodes a type III protein secretion system that delivers a cocktail of at least 13 different protein toxins (called effector proteins) directly into the cytosol of host epithelial cells in the gut (**Fig. 25.28**▶). Inside epithelial cells, these effector proteins interfere with signal transduction cascades and modulate the host response. One mission of these effectors is to induce cytoskeletal rearrangements that cause ruffling of the eukaryotic membrane around the microbe (see **Fig. 25.26**). The membrane ruffling starts the process of engulfment. *Salmonella* induces this response as a way to avoid the normal endocytic process.

Once *Salmonella* enters an epithelial cell or a macrophage, it finds itself in a vacuole. In the normal course of events, an enzyme-packed lysosome would then fuse with the phagosome and release its contents in an effort to kill the invader. *Salmonella,* however, possesses a second pathogenicity island, called SPI-2, which subverts this host

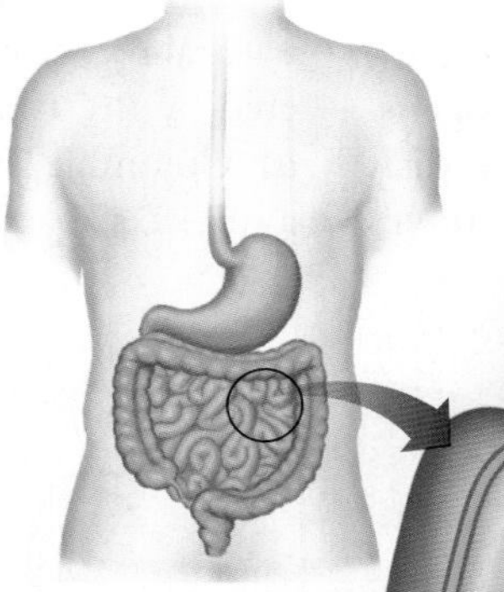

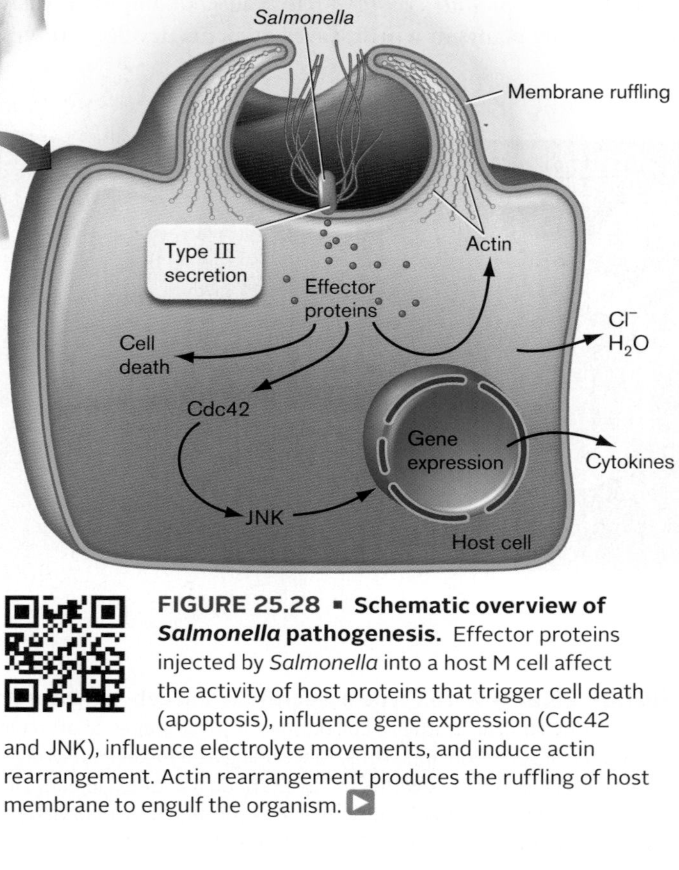
Salmonella
Membrane ruffling
Type III secretion
Actin
Effector proteins
Cell death
Cl^-
H_2O
Cdc42
Gene expression
Cytokines
JNK
Host cell

FIGURE 25.28 ▪ Schematic overview of *Salmonella* pathogenesis. Effector proteins injected by *Salmonella* into a host M cell affect the activity of host proteins that trigger cell death (apoptosis), influence gene expression (Cdc42 and JNK), influence electrolyte movements, and induce actin rearrangement. Actin rearrangement produces the ruffling of host membrane to engulf the organism. ▶

response. SPI-2 uses another type III secretion system to inject proteins that alter vesicle remodeling and trafficking, thereby reducing phagosome-lysosome fusion so that the intracellular bacteria are spared. Insofar as there are multiple recognized pathogenicity islands present in the genome, the pathogenesis of *Salmonella* is even more complex than just outlined.

Because of their importance to virulence, type III secretion systems are the subject of intensive research designed to exploit them as potential drug targets.

Weblinks *Salmonella* infection (*see ebook*)

Type IV Secretion Resembles Conjugation Systems

As described in Chapter 9, many bacteria can transfer DNA from donor to recipient cells via a cell-cell contact system known as conjugation (Section 9.2). The conjugation systems of some pathogens have been modified, through evolution, into new systems that transport proteins, or proteins plus DNA, directly into target cells. *Agrobacterium tumefaciens,* for example, uses its Vir system to transfer the tumor-producing Ti plasmid and some effector proteins into plant cells. The result is a plant cancer called crown gall disease. The bacterium that causes whooping cough in humans, *Bordetella pertussis,* also uses a type IV secretion system, to export pertussis toxin, but it simply exports the toxin, without injecting it into the host (**Fig. 25.29**). Type IV systems also differ with respect to whether the protein is taken directly from the cytoplasm, like CagA from *Helicobacter,* or from the periplasm, as with pertussis toxin. In the latter case, the SecA-dependent general secretory system first delivers the toxin to the periplasmic space.

Another unique toxin delivery system, type VI secretion, resembles a harpoon or blowgun (see **Special Topic 25.1**).

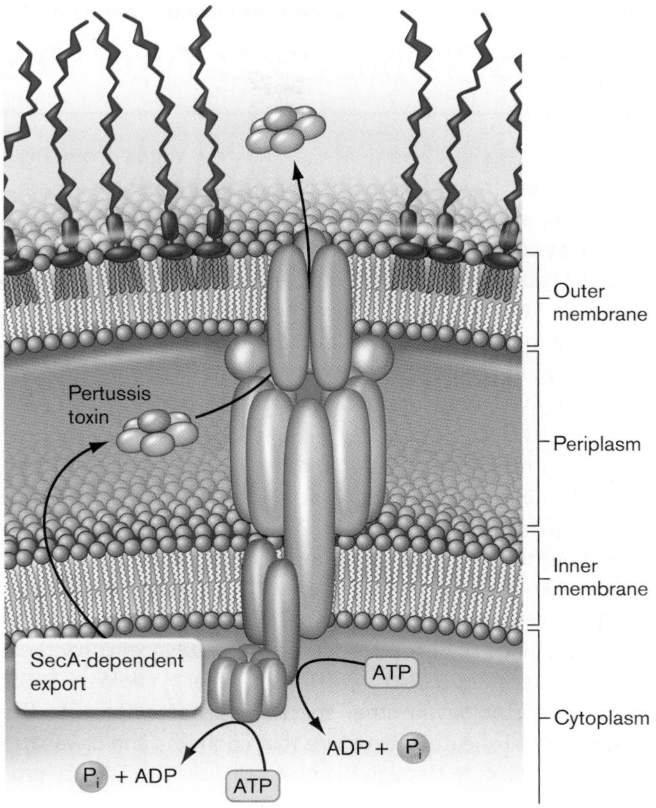
Outer membrane
Pertussis toxin
Periplasm
Inner membrane
SecA-dependent export
ATP
Cytoplasm
ADP + Pi
Pi + ADP
ATP

FIGURE 25.29 ■ Type IV secretion of pertussis toxin. Evolutionarily related to conjugation systems, this type IV system in *Bordetella pertussis* takes pertussis toxin from the periplasm (moved there by SecA-dependent transport) across the outer membrane.

Thought Question

25.6 Protein and DNA have very different structures. Why would a protein secretion system be derived from a DNA-pumping system?

To Summarize

- **Many pathogens** use specific protein secretion pathways to deliver toxins.
- **Type II secretion** systems use a pilus-like extraction/retraction mechanism to push proteins out of the cell.
- **Type III secretion** uses a molecular syringe to inject proteins from the bacterial cytoplasm into the host cytoplasm.
- **Type IV secretion** utilizes an entourage of proteins that resemble conjugation machinery to secrete proteins from either the cytoplasm or the periplasm.

25.6

Surviving within the Host

Once inside a host, how does a successful pathogen avoid detection and destruction? Many of the virulence factors in the pathogen's arsenal help the microbe escape or resist innate immune mechanisms. Others are dedicated to stealth—that is, hiding from the immune system. But before discussing how these organisms survive in a host, we must ask how the pathogen knows it is in a host.

Special Topic 25.1: Type VI Secretion: Poison Darts

Pygmies living in Central Africa and the Yagua from Peru both use an ancient weapon for hunting—namely, blowguns. Blowguns are carefully crafted, hollowed-out tubes designed to shoot curare-tipped poison darts that can penetrate and paralyze animal prey. What do blowguns and microbes have in common? Many Gram-negative microbes, such as *E coli, Vibrio cholerae, Pseudomonas,* and *Burkholderia,* make incredible blowgun-like devices that deliver either virulence proteins to eukaryotic prey, or antimicrobials into bacterial competitors to kill them.

These molecular weapons, called type VI secretion systems (T6SSs), comprise at least 13 proteins and are present in at least one copy in about 25% of all sequenced Gram-negative bacteria. The genomes of some bacteria, such as *Burkholderia,* carry five or six distinct T6SS gene clusters.

The components of T6SSs appear to be repurposed phage tail parts normally used to deliver phage DNA into bacterial host cells (see Fig. 6.13A). VipA/B proteins of *V. cholerae* T6SS, and their orthologs in other species, are homologous to the phage T4 tail sheath, a structure that contracts and drives the T4 needlelike core through the host cell wall. The T4 core provides a tube through which T4 DNA is delivered into the cytoplasm. Hcp proteins of T6SS are homologs of the phage T4 needle proteins, while VgrG proteins are structural homologs of the T4 spike complex that sits atop the T4 core. The T4 spike complex initially punctures the target bacterial cell surface. T6SSs actually resemble an inverted phage tail inside a bacterial cell. T6SSs translocate substrates from one cell directly into a target cell only after physical contact with the target.

Bacterial T6SSs are thought to kill or maim by translocating (firing) toxic effector molecules into bacterial competitors or into eukaryotic target cells. But do they actually fire like phage tails? Grant Jensen (California Institute of Technology; **Fig. 1A**) and John Mekalanos (Harvard University; **Fig. 1B**) wanted to find out.

The team and their colleagues used time-lapse fluorescence light microscopy to visually track what happens to a *Vibrio cholerae* T6SS sheath tagged with green fluorescent protein (VipA-GFP). As shown in **Figure 1C**, the sheaths of the type VI secretion system cycled between assembly, quick contraction, disassembly, and reassembly. Real-time video of this cycling can be seen at the link provided at the end of this feature.

Next, the group used cryo-EM imaging to visualize the T6SS in greater detail. The images showed straight tubular structures in two different conformations aligned perpendicular to the cytoplasmic membrane. The first structure was a long, thin, extended form that stretched almost the entire width of the cell (**Fig. 2A**). The second was a shorter, wider, contracted form

A.

COURTESY OF MARTIN PILHOFER

B.

COURTESY OF JOHN MEKALANOS

FIGURE 1 ■ Extension and contraction of type VI secretion system. Grant Jensen (standing, with Martin Pilhofer at the electron microscope) **(A)** and John Mekalanos **(B)** headed the study to visualize the T6SS mechanism of *Vibrio cholerae.* **C.** Time-lapse images showing the extension of the VipA-GFP structure from one side of the cell to another (arrows) followed by a contraction event and apparent disassembly of the contracted VipA-GFP structure.

C.

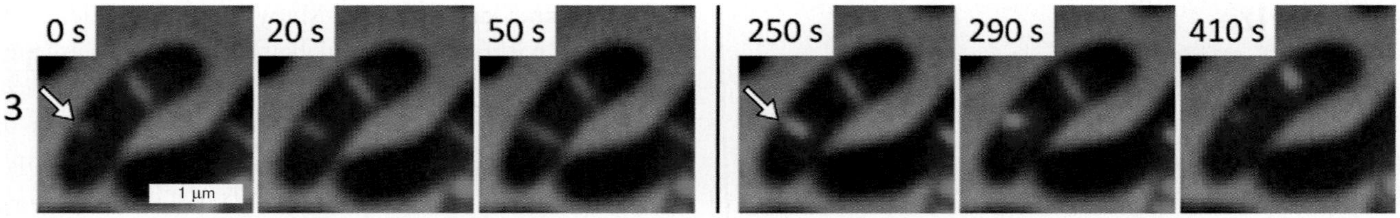
0 s
20 s
50 s
250 s
290 s
410 s
3
1 μm

NATURE 2012. **483:** 182–186

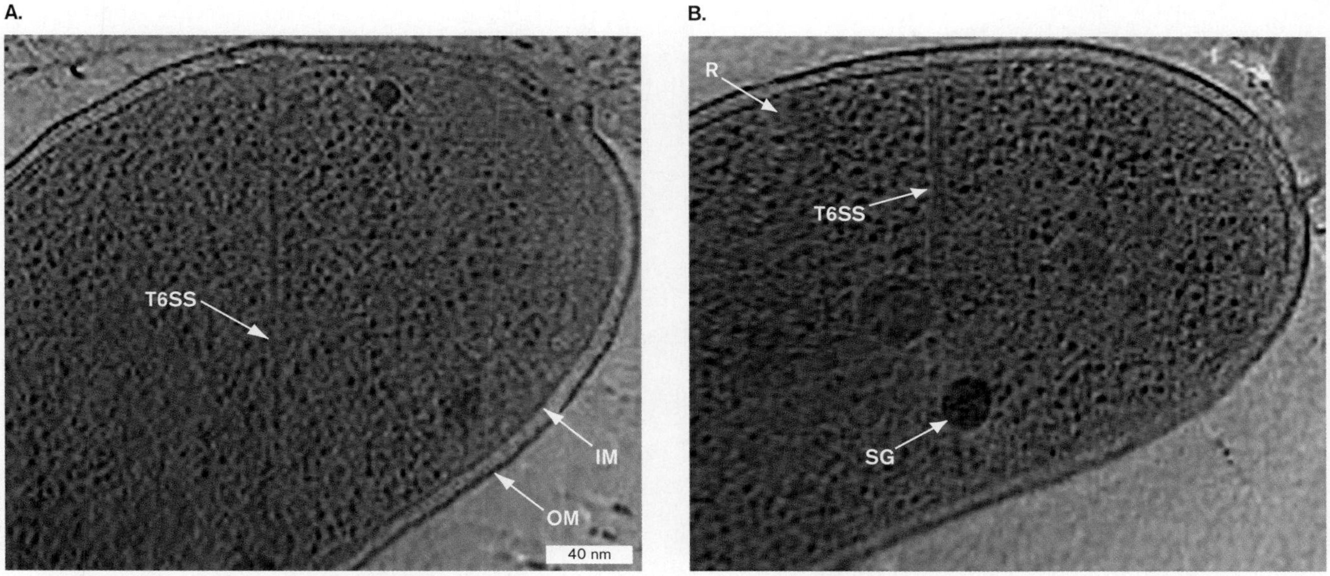
A.
T6SS
IM
OM
40 nm
B.
R
T6SS
SG

FIGURE 2 ■ Cryo-electron-tomographic imaging of T6SS structures inside intact cells. A. An extended T6SS structure. IM = inner membrane; OM = outer membrane. **B.** A contracted T6SS. R = putative ribosome; SG = polyphosphate storage granule.

(**Fig. 2B**). The researchers concluded that these structures corresponded to what was seen expanding and contracting in the fluorescent microscopic study. They further suggested that the rapid contraction (firing) of the type VI secretion system sheath provides the energy needed to shoot proteins out of the bacterium and into an adjacent target cell.

The VipA/B sheath and Hcp core proteins are depolymerized after firing by another T6SS gene product, called ClpV, a type of chaperone. The authors of the study propose, based on the rapid assembly and disassembly, that the disassembled Hcp and VipA/B monomers and dimers are reused to assemble a new microbial "blowgun." The model for T6SS assembly (loading), extension (ready to fire), contraction (firing), and disassembly is illustrated in **Figure 3** (next page).

But what do type VI secretion systems secrete? When engaged with eukaryotic cells, VgrG proteins appear to be the main, if not only, effector. VgrG proteins carry extended C-terminal regions that function in the eukaryotic host cell cytoplasm. For example, the C-terminal extension of *V. cholerae* VgrG1 carries an RtxA toxin domain that, after being fired into target cells, cross-links actin. VgrG1 from *Aeromonas hydrophila* induces host cell toxicity by ADP-ribosylating host cell actin, which ultimately triggers apoptotic cell death.

In addition to delivering VgrG effector "darts" into eukaryotic cells, the T6SSs have been characterized as mediators of competitive interbacterial interactions. The only antimicrobial T6SS effector characterized to date is the *Pseudomonas aeruginosa* protein Tse2, which appears to be the toxin component of a toxin immunity system (Chapter 10). Bacterial cells lacking Tsi2, the cognate immunity protein to Tse2, suffer arrested growth when Tse2 is expressed intracellularly.

With its role in virulence, its prevalence among Gram-negative bacterial genomes, and its predicted phage-like structure, the type VI secretion system has become an enticing object of research. Identification of pathogen-specific T6SS components and substrates may reveal novel targets for treating bacterial disease.

Weblinks Type VI secretion movie (*see ebook*)

RESEARCH QUESTION

Before searching for, or designing, a potential antimicrobial that would target one of the T6SS components, what experiment(s) could you do to determine whether a single antimicrobial compound could potentially target T6SS from many different organisms?

Basler, Marek, Martin Pilhofer, Gregory P. Henderson, Grant J. Jensen, and John J. Mekalanos. 2012. Type VI secretion requires a dynamic contractile phage tail-like structure. *Nature* **483**:182–186.

(continued)

Special Topic 25.1: *(continued)*

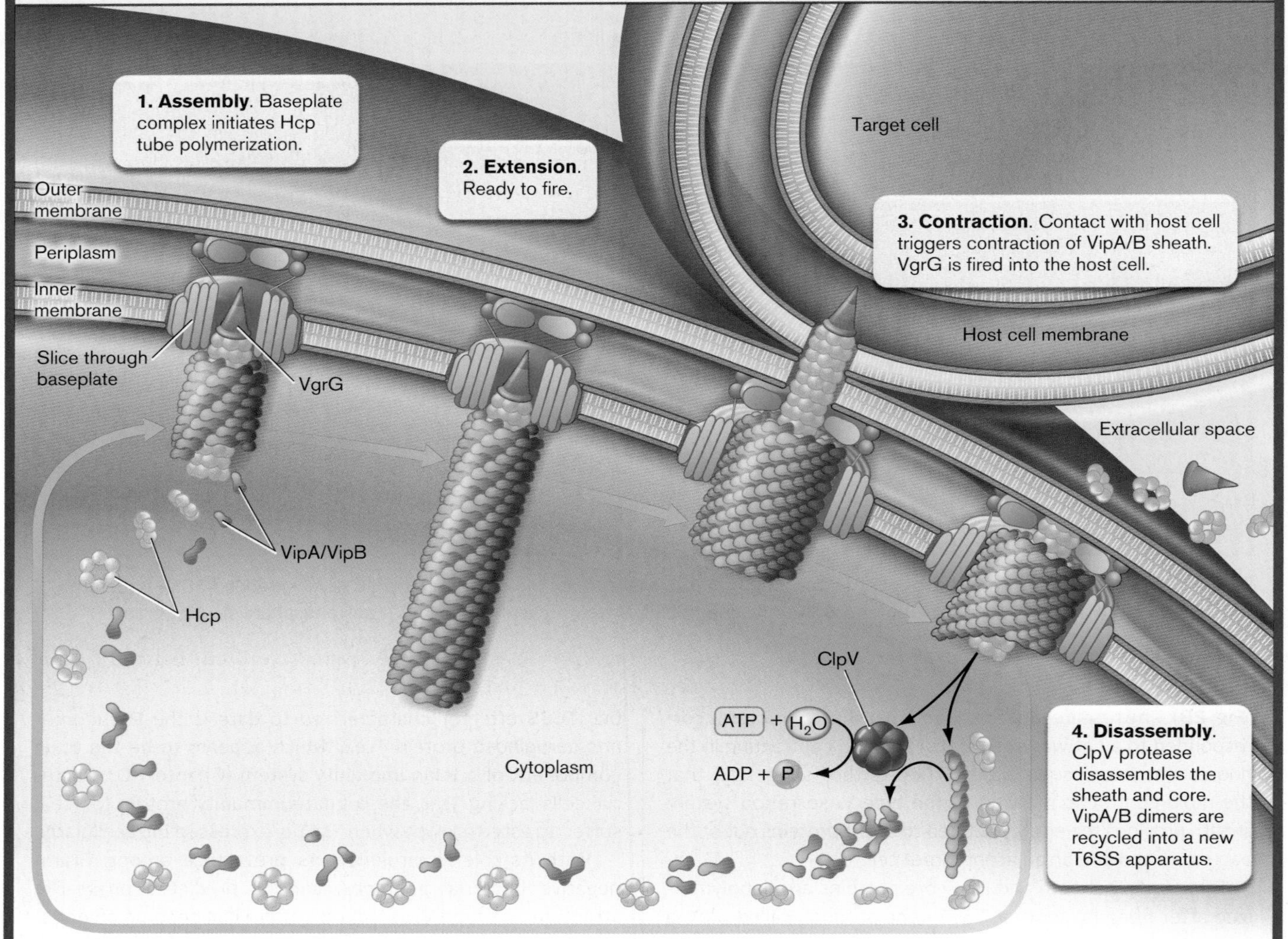

1. Assembly. Baseplate complex initiates Hcp tube polymerization.
2. Extension. Ready to fire.
Target cell
3. Contraction. Contact with host cell triggers contraction of VipA/B sheath. VgrG is fired into the host cell.
Outer membrane
Periplasm
Inner membrane
Slice through baseplate
VgrG
Host cell membrane
Extracellular space
VipA/VipB
Hcp
ClpV
ATP + H_2O
ADP + P
Cytoplasm
4. Disassembly. ClpV protease disassembles the sheath and core. VipA/B dimers are recycled into a new T6SS apparatus.

FIGURE 3 ■ Model of type VI secretion mechanism. This figure illustrates a Gram negative bacterium using a type VI secretion system to attack another Gram negative bacterium. A similar series of events takes place when a pathogen attacks a eukaryotic host cell. IM = inner membrane; OM = outer membrane; PG = peptidoglycan.

Where Am I?

Many pathogens can grow either outside or inside a host. To accommodate the needs imposed by these different environments, microbes employ alternate physiologies appropriate to each. Why make a type III secretion system if there are no host cells around? But how do microbial pathogens know whether they are in a host or in a pond? And what bacterial genes are expressed exclusively while in a host?

The same types of regulatory mechanisms that sense environmental conditions in a pond are used by the microbe to intuit its whereabouts in a host. That is, various sensing systems act in concert to recognize any specific environmental niche. Two-component signal transduction systems, discussed in Section 10.1, are used to monitor magnesium concentrations, which are characteristically low in a host cell vacuole. Other regulators measure pH, which will be acidic in the same vacuole. There are many other examples. The point is that there is no single in vivo sensor system. The various regulators collaborate to trigger the expression of virulence genes. The concentration of free iron, for example, which is typically very low in the host, is an important signal used to induce the synthesis of virulence proteins whose purpose is to liberate iron bound up in host proteins. However, regulators sensitive to other in vivo signals must also be activated to achieve a successful infection.

Cell-cell communication is also important during infections. *Pseudomonas aeruginosa,* for example, has at least two quorum-sensing systems that detect secreted autoinducers. As the number of bacteria in a given space increases, so, too, does the concentration of the chemical autoinducer. When the autoinducer reaches a critical concentration, it diffuses back into the bacterium or binds to a surface receptor and triggers the expression of bacterial target genes. Genes included in the *P. aeruginosa* quorum-sensing regulons encode *Pseudomonas* exotoxin A and other secreted proteins, such as elastase, phospholipase, and alkaline protease. Why would a pathogen employ quorum sensing to regulate virulence factors? One reason may be to prevent alerting the host that it is under attack before enough microbes can accumulate through replication. Tripping the host's alarms too early would make eliminating infection easy. However, waiting until a large number of bacteria have amassed before releasing toxins and proteases will increase the chance that the host can be overwhelmed.

Knowledge of quorum sensing has also provided an opportunity for treating disease. We know that bacteria can communicate with each other by secreting, and sensing, autoinducer chemicals. We also know that these molecules will accumulate in the growth environment and trigger collaborative responses, such as biofilm formation, among members of the population. A new chemical has been developed that interferes with quorum sensing and could serve as a new type of antibiotic. The compound, *N*-(2-oxocyclohexyl)-3-oxododecanamide, is an analog of the homoserine lactone autoinducers used by *P. aeruginosa*. The compound can bind, but not activate, the regulatory proteins LasR and RhlR. When tested in vitro, this compound reduced biofilm formation and decreased production of virulence factors.

Extracellular Immune Avoidance

This topic was first discussed in the chapters on host defense (Chapter 23) and immunology (Chapter 24). Many bacteria, such as *Streptococcus pneumoniae* and *Neisseria meningitidis,* produce a thick polysaccharide capsule that envelops the cell. Capsules help organisms resist phagocytosis in several ways. Recall that phagocytes must recognize bacterial cell-surface structures or surface-bound C3b complement factor to begin phagocytosis (see Section 23.6). Capsules cover bacterial cell wall components and mannose-containing carbohydrates that phagocytes normally use for attachment. The uniformity and slippery nature of capsule composition make it difficult for phagocytes to lock onto the bacterial cell.

But what about complement factor C3b, which can bind to the bacterial cell? Phagocytes have surface C3b receptors that latch on to C3b molecules. Capsules will envelop any C3b complement factor that binds to the bacterial surface, thereby hiding it from the phagocyte. Fortunately, immune defense mechanisms can eventually circumvent this avoidance strategy by producing opsonizing antibodies (IgG) against the capsule itself (see Section 24.3). The Fc regions of antibodies that bind to the capsule point away from the bacterium, so they are free to bind Fc receptors on phagocyte membranes. Binding of the Fc region to the phagocyte's Fc receptor triggers phagocytosis.

Pathogens can also use proteins on the cell surface to avoid phagocytosis. *Staphylococcus aureus* has a cell wall protein called **protein A** that binds to the Fc region of antibodies, hiding the bacteria from phagocytes. This works in two ways. Protein A can bind to the Fc region of an antibody before the antibody-binding site ever finds its bacterial target. Thus, the business end of the antibody—that is, the part with antigen-binding sites—is pointing away from the microbe. However, even if the antibody finds its antigen target on one bacterium, protein A from a second bacterium can bind to that Fc region, once again blocking phagocyte recognition.

Some microbes can trigger apoptosis in target host cells. Proteins made by the pathogen enter the host cell and trigger this programmed cell death. How does this help the pathogen? If a macrophage is targeted for self-destruction, it cannot destroy the microbe.

Another immune avoidance strategy used by microorganisms, both the extracellular and intracellular types, is to change their antigenic structure. Genes encoding flagella, pili, and other surface proteins often use site-specific gene inversions to express alternative proteins (for example, *Salmonella* phase variation) or slipped-strand mispairing to add or remove amino acids from a sequence (see Section 10.6). You can think of these processes as shape-shifting to avoid recognition.

Intracellular Pathogens

In an effort to escape both innate and humoral immune mechanisms (see Chapter 24), many bacterial pathogens, called **intracellular pathogens**, seek refuge by invading host cells. (Viruses, of course, are by definition intracellular pathogens.) Antibodies and phagocytic cells will not penetrate live host cells, so hiding there temporarily provides the pathogen safe harbor. Some bacteria dedicate their entire lifestyle to intracellular parasitism. *Rickettsia,* for example, for reasons unknown, will not grow outside a living eukaryotic cell. Other microbes, such as *Salmonella* and *Shigella,* are considered facultative intracellular pathogens. **Facultative intracellular pathogens** can live either inside host cells or free. We have already discussed how intracellular parasites get into cells, but how do they withstand intracellular attempts to kill them?

Once inside the phagosome, intracellular pathogens have three options to avoid being killed by a phagolysosome (**Fig. 25.30**). They can escape the phagosome, prevent phagosome-lysosome fusion, or survive the result.

Escape from the phagosome. The Gram-negative bacillus *Shigella dysenteriae* and the Gram-positive bacillus *Listeria monocytogenes,* both of which cause food-borne gastrointestinal disease, use hemolysins to break out of the phagosome vacuole before fusion. In this way, they completely avoid lysosomal enzymes. Once free in the cytoplasm, they are thought to enjoy unrestricted growth. Yet even in the cytoplasm, these microbes have found a way to redirect host cell function to their own ends.

A fascinating aspect of escaping the phagosome involves motility. *Shigella* and *Listeria* are both nonmotile at 37°C in vitro; however, they both move around inside the host cell, even though they have no flagella. How do they move? These species are equipped with a special device at one end of the cell that mediates host cell actin polymerization. The polymerizing actin, called a "rocket tail," propels the organism forward through the cell (**Fig. 25.31A**) until it reaches a membrane. The membrane is then pushed into an adjacent cell, where the organism once again ends up in a vacuole (**Fig. 25.30**, Fate 3). This strategy allows the microbe to spread from cell to cell without ever encountering the extracellular environment, where it would be vulnerable to attack. Actin motility is also a feature of some species of *Rickettsia* (**Fig. 25.32**), *Mycobacterium,* and *Burkholderia.*

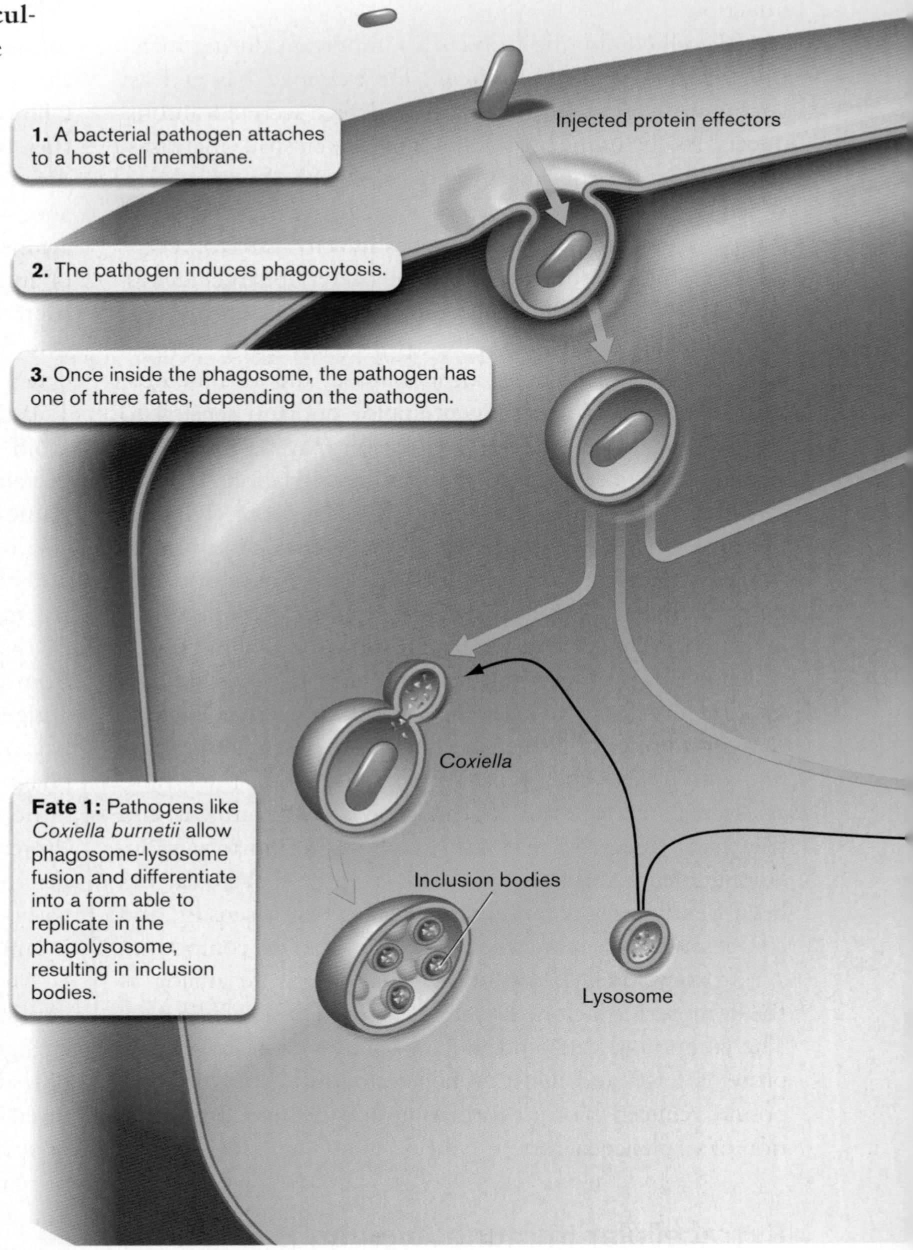
Injected protein effectors
1. A bacterial pathogen attaches to a host cell membrane.
2. The pathogen induces phagocytosis.
3. Once inside the phagosome, the pathogen has one of three fates, depending on the pathogen.
Coxiella
Fate 1: Pathogens like *Coxiella burnetii* allow phagosome-lysosome fusion and differentiate into a form able to replicate in the phagolysosome, resulting in inclusion bodies.
Inclusion bodies
Lysosome

FIGURE 25.30 ▪ Alternative fates of intracellular pathogens. Alternative fates of intracellular pathogens. Different pathogens have different strategies for surviving in a host cell. Some tolerate phagolysosome fusion (*Coxiella*), others prevent phagolysosome fusion (*Salmonella*), and still others escape the phagosome to replicate in the cytoplasm (*Shigella* and *Listeria*).

Inhibiting phagosome-lysosome fusion. Other intracellular pathogens avoid the hazard of lysosomal enzymes by preventing lysosomal fusion with the phagosome. *Salmonella, Mycobacterium, Legionella,* and *Chlamydia* are good examples. For example, *Legionella pneumophila* grows inside alveolar

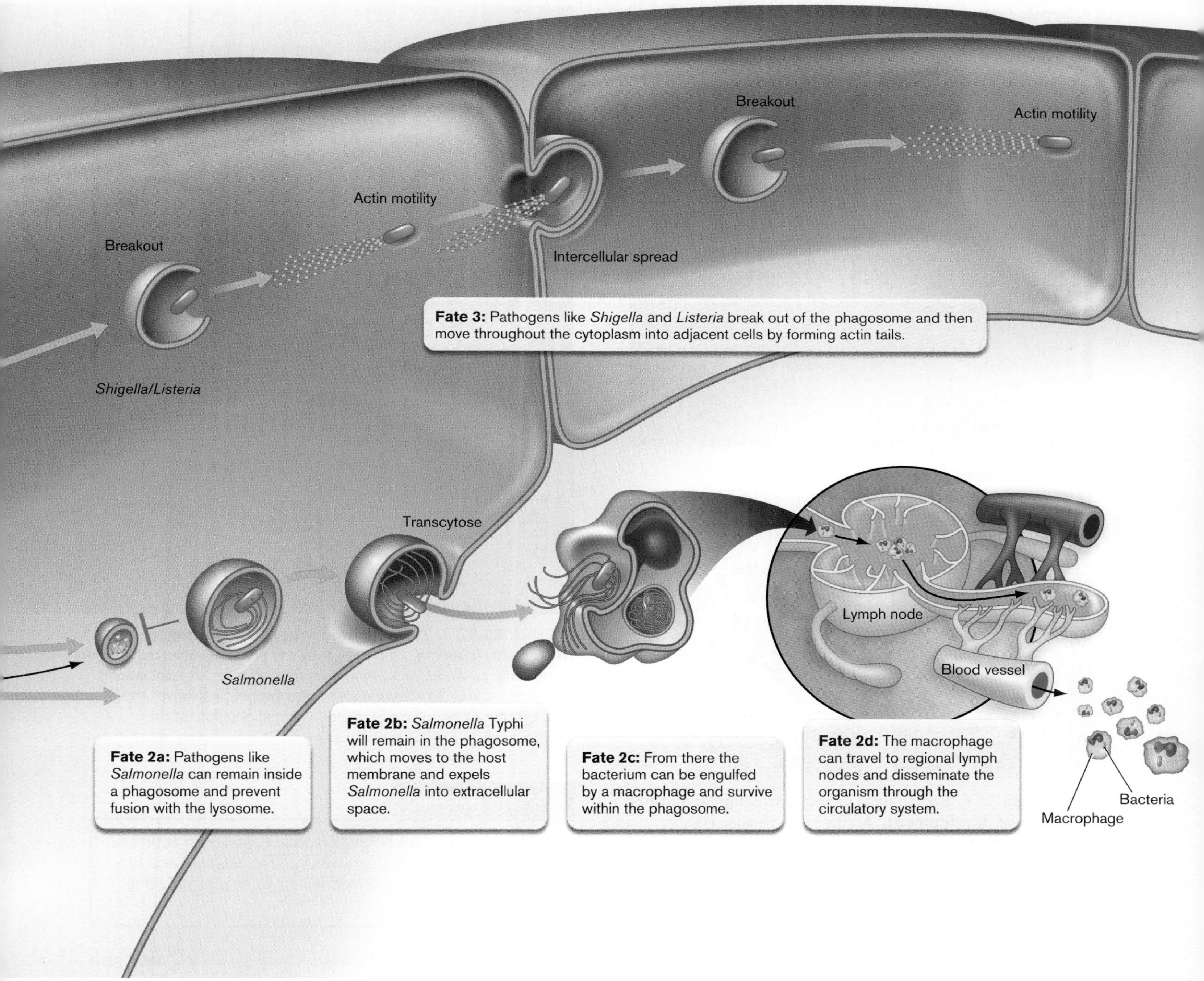
Breakout
Actin motility
Breakout
Actin motility
Intercellular spread
Fate 3: Pathogens like Shigella and Listeria break out of the phagosome and then move throughout the cytoplasm into adjacent cells by forming actin tails.
Shigella/Listeria
Transcytose
Salmonella
Lymph node
Blood vessel
Fate 2a: Pathogens like Salmonella can remain inside a phagosome and prevent fusion with the lysosome.
Fate 2b: Salmonella Typhi will remain in the phagosome, which moves to the host membrane and expels Salmonella into extracellular space.
Fate 2c: From there the bacterium can be engulfed by a macrophage and survive within the phagosome.
Fate 2d: The macrophage can travel to regional lymph nodes and disseminate the organism through the circulatory system.
Bacteria
Macrophage

macrophage phagosomes and produces the potentially fatal Legionnaires' disease, so named for the veterans group that suffered the first recognized outbreak in 1976. The organism, once inside a phagosome, uses a type IV secretion system to secrete proteins through the vesicle membrane and into the cytoplasm. These bacterial proteins interfere with the cell signaling pathways that cause phagosome-lysosome fusion. The result is that *L. pneumophila* can grow in a friendlier vesicle. Interestingly, *L. pneumophila* is actually a soil and water microbe. Its ability to survive inside macrophages evolved from its ability to survive in amebas, which serve as a natural reservoir for this pathogen.

Salmonella Typhi is another pathogen that prevents phagolysosome fusion, but this organism eventually leaves the

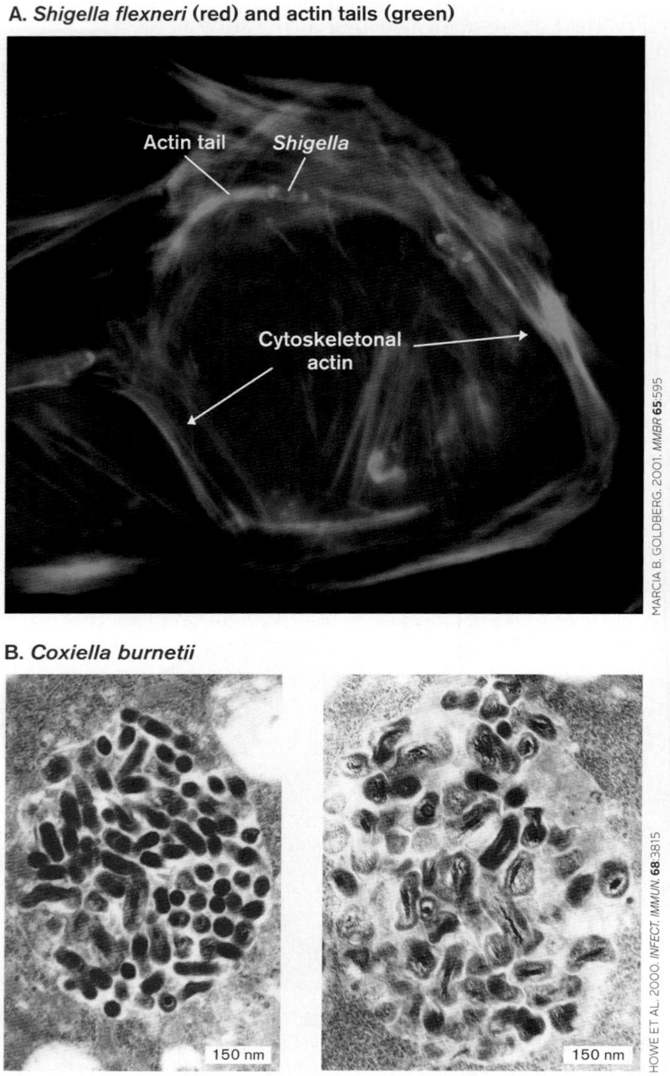
A. Shigella flexneri (red) and actin tails (green)
Actin tail
Shigella
Cytoskeletonal actin
MARCIA B. GOLDBERG. 2001. MMBR 65:595
B. Coxiella burnetii
150 nm
150 nm
HOWE ET AL. 2000. INFECT. IMMUN. 68:3815

FIGURE 25.31 ▪ Intracellular pathogens: *Shigella* motility and *Coxiella* development. **A.** Intracellular *Shigella flexneri* (fluorescence microscopy), fluorescently stained red (1 μm in length), are propelled through the host cytoplasm by actin tails, stained green. *Shigella* is a facultative intracellular pathogen. **B.** TEM image showing a typical vacuole in J774A.1 mouse macrophage cells infected with *Coxiella burnetii* at 2 hours (left) and 6 hours (right) postinfection. The organism, which prefers to live in the acidified vacuole, undergoes a form of differentiation that changes its shape and alters its interactions with the host cell. *Coxiella* is an obligate intracellular pathogen.

host cell (exocytosis) and enters extracellular tissue spaces where macrophages await (**Fig. 25.30**, Fate 2). The bacterium is once again engulfed, by a macrophage this time, but again survives inside the macrophage phagosome. The infected macrophage homes to a regional lymph node, enters the bloodstream, and, like a Trojan horse, disseminates *Salmonella* throughout the body.

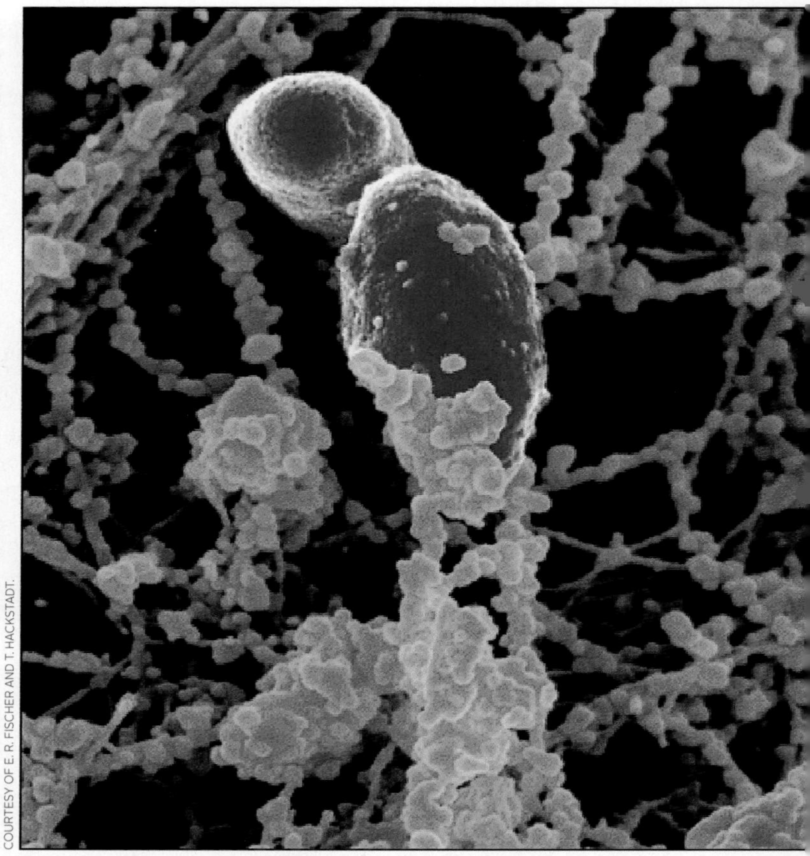
COURTESY OF E. R. FISCHER AND T. HACKSTADT.

FIGURE 25.32 ▪ The obligate intracellular pathogen *Rickettsia rickettsii*. This SEM shows *R. rickettsii* (blue, ca. 0.7 μm length), the cause of Rocky Mountain spotted fever, in association with host actin (gold). Several pathogens propel themselves though host cytoplasm by polymerizing host actin at one pole of the bacterial cell. *Source:* Damon W. Ellison et al. 2008. *Infect. Immun.* **76**:542–550.

Thought Question

25.7 **Figure 25.31A** shows *Shigella* forming an actin tail at one pole. Why do organisms such as *Shigella* and *Listeria* assemble actin-polymerizing proteins at only one pole?

Thriving under stress. In what could be called the "grin and bear it" strategy, some intracellular pathogens prefer the harsh environment of the phagolysosome. *Coxiella burnetii*, for example, grows well in the very acidic phagolysosome environment (**Fig. 25.31B**). This obligate intracellular organism (an organism that grows only inside another living cell) causes a flu-like illness called Q fever (query fever). The symptoms of Q fever include sore throat, muscle aches, headache, and high fever. It has a mortality rate of about 1%, so most people recover to good health. The organism allows phagosome-lysosome fusion because the acidic environment that results is needed for it to survive and grow.

Thought Question

25.8 How can you determine whether a bacterium is an intracellular parasite?

Why some bacteria are obligate intracellular pathogens is unclear. One intracellular bacterium, *Rickettsia prowazekii*, a cause of epidemic typhus, appears to be an "energy parasite" that can transport ATP from the host cytoplasm and exchange it for spent ADP in the bacterium's cytoplasm. But this does not explain its obligate intracellular status, since giving *Rickettsia* ATP outside a host does not allow the bacterium to grow. Other factors remain to be discovered.

Thought Question

25.9 Why might killing a host be a bad strategy for a pathogen?

Sleeping with the Enemy

As just described, many bacterial and, of course, viral pathogens find safe haven by growing inside host cells. However, from our knowledge of innate and adaptive immune responses it is not intuitively obvious why this is so. After all, infected cells present microbial antigens on their class I or class II MHC receptors to alert the innate and adaptive immune systems that the infected host cell must be killed to resolve the infection. In addition, pieces of intracellular microbes (flagella, LPS, peptidoglycan) will bind pattern recognition receptors (TLRs and NLRs) that activate intracellular inflammasomes. Inflammasomes trigger production of pro-inflammatory cytokines that mediate inflammation. So how do intracellular pathogens avoid destruction? It turns out that these invaders employ a variety of molecular tricks that misdirect the immune system much like a magician misdirects an audience. All of these strategies, summarized in **Figure 25.33**, are designed to buy the microbe more time to overwhelm the host.

Molecular mimicry. A variety of bacteria and viruses use mimicry to confuse the immune system (discussed in Section 24.8). In some cases, microbial proteins are made that look like cytokines or that bind to host cells and hitchhike via normal host trafficking to the nucleus, where the bacterial protein interferes with cytokine gene expression. One example involving *Shigella* is discussed in eTopic 25.8. These factors can manipulate the balance of helper T cells, for example, and send immunity down the wrong path for combating the microbe.

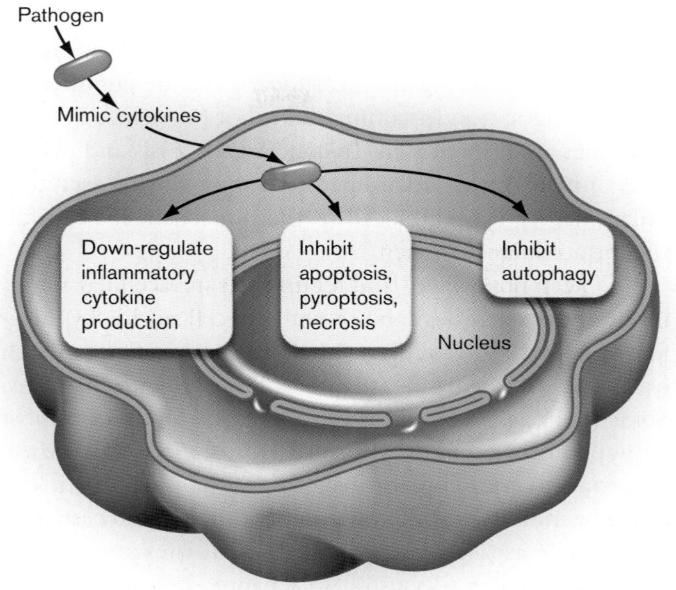
Pathogen
Mimic cytokines
Down-regulate inflammatory cytokine production
Inhibit apoptosis, pyroptosis, necrosis
Inhibit autophagy
Nucleus

FIGURE 25.33 ■ Summary of microbial strategies that misdirect the immune system. Bacteria and viruses produce different molecules that can mimic cytokines or transcriptional regulators, alter cytokine production, prevent programmed cell death, or inhibit autophagy.

Flipping cytokine profiles. Many pathogens defy death by growing inside macrophages that are "armed" with numerous antimicrobial weapons. Pathogens that can grow inside macrophages include bacteria such as *Salmonella*, *Yersinia*, *Listeria*, *Mycobacterium tuberculosis*, *Francisella tularensis*, and *Chlamydia*; as well as protozoans such as *Leishmania* and *Trypanosoma cruzi*. How do they survive? These, and many other pathogens, interrupt host cell signaling pathways designed to activate macrophage antimicrobial mechanisms. To perform this trick, most intracellular pathogens inhibit production of pro-inflammatory cytokines, such as TNF-α, IL-12, and IFN-gamma, but encourage production of anti-inflammatory cytokines like IL-10, TGF-beta, and IL-4.

Mycobacterium, *T. cruzi*, and *Leishmania* also down-regulate the expression of host membrane receptors for IFN-gamma (thereby inhibiting inflammation), and interfere with downstream regulators that activate the production of MHC class I proteins, which dulls host immune mechanisms. Reduced MHC I production means less antigen presentation, fewer activated T_H1 helper cells, and, thus, fewer cytotoxic T cells to attack the infected cell.

Stopping programmed cell death. When macrophages fail to eliminate infecting pathogens, they eventually give up and try to kill themselves and their stowaway microbial passengers through one of three programmed-cell-death pathways (apoptosis, necrosis, or pyroptosis) as a last resort to clear the infection. Apoptotic cells maintain membrane integrity during apoptotic death and are engulfed by nearby phagocytes. Inflammation is not provoked, because cytokines are not released. In contrast, necrosis and pyroptosis initiate rapid inflammatory responses by secreting inflammatory cytokines. All of these mechanisms can kill the intracellular pathogen.

To "keep hope alive" and retain their intracellular niche, intracellular microbes can prevent host cell suicide by interfering with the molecular signals that initiate host death programs, or they can activate pro-survival mechanisms. Some pathogens even synthesize microbial mimics of host anti-apoptotic proteins. *Yersinia enterocolitica* and *Mycobacterium tuberculosis,* for instance, prevent pyroptotic cell death by inhibiting inflammasome formation and caspase 1 activation. Caspase 1 is a protease that initiates pyroptosis.

Herpes viruses (discussed in Chapter 11), which cause cold sores, genital herpes, mononucleosis, and some cancers, can enter into a latent state. Latency requires shutting down viral replication and integrating viral DNA into a host genome. But why doesn't the virus trigger apoptosis in the host cell before becoming latent? Herpes viruses produce a number of small RNA molecules called microRNAs (miRNAs) that can interfere with the apoptosis program. One such miRNA prevents the translation of two host cell proteins required for cell death. The miRNAs bind to regions in the apoptosis mRNAs and promote their degradation. Once latent, the virus no longer produces any viral proteins, so the immune system of the infected individual cannot detect the infected cell.

Some intracellular pathogens (*Trypanosoma cruzi, Leishmania,* and *Yersinia pseudotuberculosis*) actually tilt macrophage suicide pathways toward apoptosis to promote pathogen dissemination (recall that apoptotic cells are engulfed by other phagocytes). Some intracellular pathogens both inhibit and activate host cell death—just not at the same time. Early in infection, *M. tuberculosis* (the cause of tuberculosis) orchestrates the inhibition of host cell suicide pathways to allow the organism to grow but then promotes suicide later as a way to disseminate.

Autophagy is a highly regulated mechanism by which eukaryotic cells form intracellular vesicles around damaged organelles to scavenge them for nutrients. Autophagy is also used as a universal innate defense mechanism to fight intracellular pathogens (**Fig. 25.34**). Autophagic vacuoles (autophagosomes) can encase these pathogens and deliver them to degradative lysosomes for destruction. Pathogen components are then sent to endosomes, where microbial structures are recognized by endosomal Toll-like receptors that trigger the innate immune system (Chapter 23).

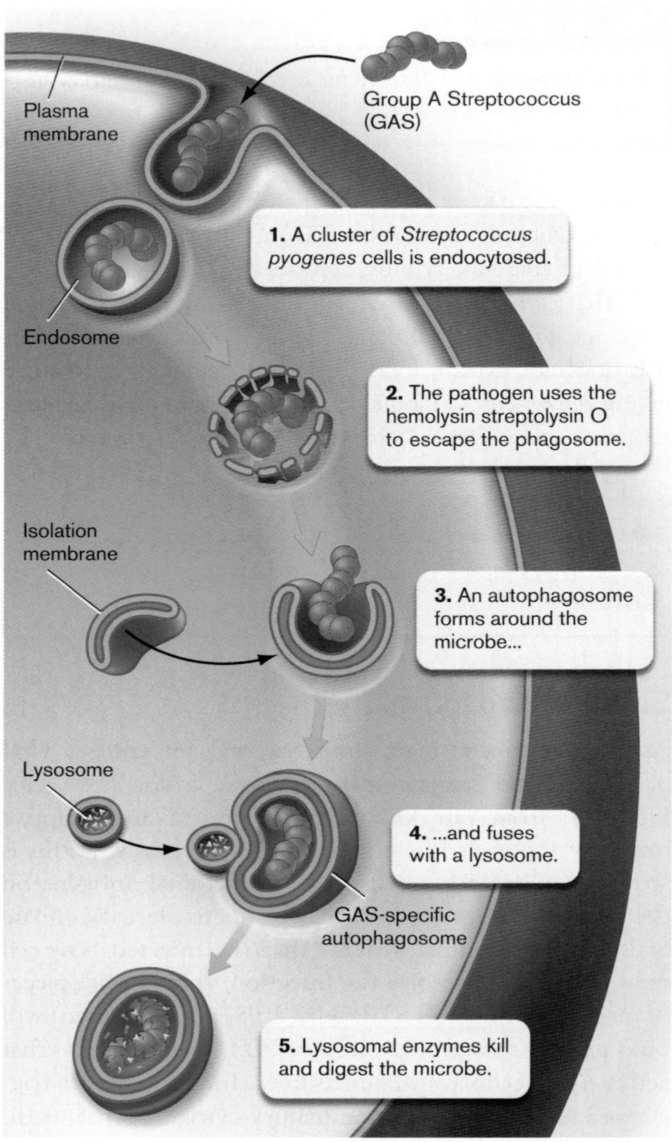
Plasma membrane
Group A Streptococcus (GAS)
1. A cluster of *Streptococcus pyogenes* cells is endocytosed.
Endosome
2. The pathogen uses the hemolysin streptolysin O to escape the phagosome.
Isolation membrane
3. An autophagosome forms around the microbe...
Lysosome
4. ...and fuses with a lysosome.
GAS-specific autophagosome
5. Lysosomal enzymes kill and digest the microbe.

FIGURE 25.34 ■ Autophagy as an innate immune mechanism. Cells in which a pathogen escapes the phagosome will try to form an intracellular vacuole (autophagosome) around the organism in a second attempt to kill it.

The microbial components (antigens) are also sent to cell compartments rich in major histocompatibility complex II (MHC II) molecules. MHC II molecules then rise to the cell surface and present the microbial antigens to the adaptive immune system as described in Chapter 24.

Intracellular pathogens, however, have evolved mechanisms that can prevent autophagy and prolong their survival. For example, the Nef protein of human immunodeficiency virus (HIV) and protein M2 of influenza prevent autophagosome formation by targeting beclin 1, a protein central to autophagosome production. Recent data

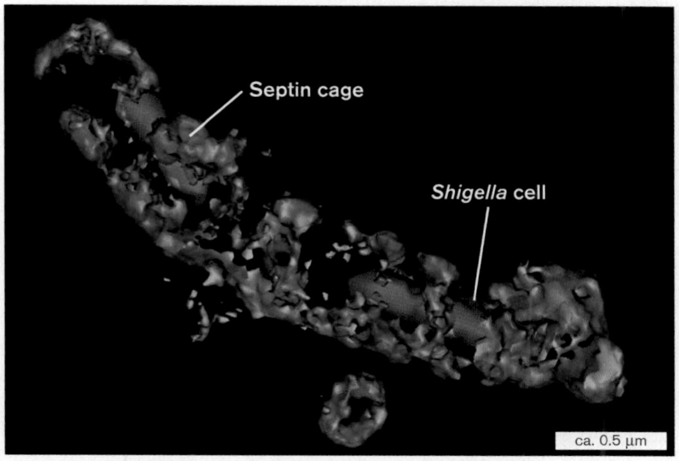
Septin cage
Shigella cell
ca. 0.5 μm
SERGE MOSTOWY ET AL. 2010. CELL HOST MICROBE. 8:433–444

FIGURE 25.35 ■ Entrapment of intracytoplasmic bacteria by septin cage-like structures. *Shigella* may move in the host cytoplasm by actin tails or may become trapped by a septin cage and marked for autophagy.

also indicate that RNA virus proteins interact with 35% of autophagy-associated proteins, suggesting that autophagy is widely targeted by pathogens.

Shigella, a cause of bacterial dysentery, avoids autophagy by making tails of polymerized host actin that propel the microbe through host cytoplasm (described earlier). This tactic may sometimes work, but the host can stop the bacterium from making actin tails by wrapping *Shigella* in septin filaments, as shown in the 3D rendering in **Figure 25.35**. Septin cages initially require actin to form, but then they inhibit actin polymerization. The septin-caged microbe is now trapped and marked for autophagy.

Redirecting host ubiquitylation signals. How do bacteria and viruses misdirect the immune system? One way is to target a regulatory mechanism common to many host systems—namely, ubiquitylation (ubiquitination). Ubiquitin is a highly conserved, 76-amino-acid polypeptide in eukaryotes that can be covalently attached to other proteins. Attachment involves an enzymatic cascade of three enzymes: E1, E2, and E3 (reviewed in **eTopic 8.3**). Depending on where a protein is ubiquitylated, the protein can be activated or, alternatively, tagged for destruction by the host proteasome. While there are only a few E1 and E2 enzymes, there are hundreds of E3 ubiquitin ligase enzymes that recognize and tag target proteins with polyubiquitin. There are also deubiquitylation enzymes that can reverse the process.

In contrast to their eukaryotic hosts, viral and bacterial pathogens lack ubiquitylation systems but have evolved E3 ligases and deubiquitylases that effectively subvert normal host ubiquitylation pathways (**Fig. 25.36**). Like someone changing signs on a highway, the viral enzymes cause host signaling systems, and the immune system, to veer off course.

There are several ways ubiquitylation normally directs the immune system. In the innate immune system, TLR pathway proteins are ubiquitylated as a way to induce production of inflammatory cytokines. When a MAMP binds to a TLR, ubiquitylation of signal induction proteins will activate the signal. In contrast, proteins that inhibit the pathway are marked by ubiquitin for destruction. The result is an activated signal pathway that leads to the formation of inflammatory cytokines (**Fig. 25.36**, step 1).

To subvert this system, some pathogens produce their own E3 ligases that divert the normal signal induction pathways. For example, rotavirus, a major cause of infant diarrhea, produces an E3 ligase that adds ubiquitin to activators of NF-kappaB, the transcription regulator that induces cytokine production (step 2). The ubiquitylated proteins are destroyed and NF-kappaB is not activated. Using a different strategy, the bacterial pathogen *Salmonella enterica* produces a deubiquitylase that removes ubiquitin from a normal inhibitor of NF-kappaB. The inhibitor is not degraded, which means that NF-kappaB is not activated and inflammatory cytokines are not made (step 3). In both instances, inflammation is minimized, allowing the pathogen to survive.

Ubiquitylation is an important mechanism for adaptive immunity too. For example, the process regulates when cell-surface MHC class I and II molecules are expressed in dendritic cells (**Fig. 25.37**). Before dendritic cells mature, MHC molecules are polyubiquitylated and degraded, limiting their placement on cell surfaces. After maturation, however, the MHC molecules are no longer ubiquitylated and, as a result, accumulate on the cell surface, making dendritic cells better equipped for antigen presentation. A number of viruses exploit this process by making viral E3 ligases that polyubiquitylate MHC proteins (**Fig. 25.37**). The MHC molecules that would present viral proteins to the immune system are degraded.

Some pathogens even make E3 ligases that ubiquitylate host proteins that are not normally tagged. One example is found in the interferon signal cascade. Interferons are secreted by virus-infected host cells, bind to interferon receptors on uninfected cells, and induce the expression of numerous genes that protect the uninfected cell from virus infection. Paramyxoviruses such as mumps and measles viruses produce E3 ligases that ubiquitylate key regulatory components (JAK-STAT) of the interferon signal cascade, which marks them for destruction. The host cell then becomes quite vulnerable to virus attack.

These examples demonstrate that ubiquitylation is a crucial part of immune regulation and a popular target of

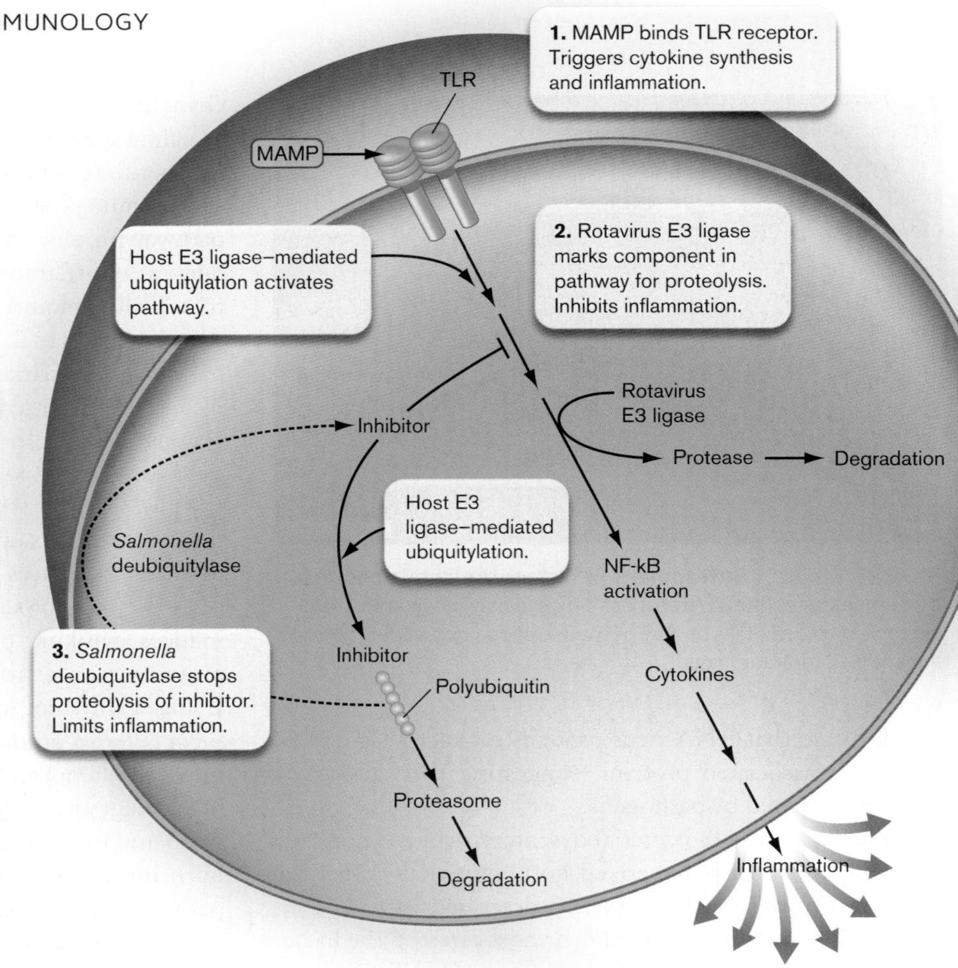
1. MAMP binds TLR receptor. Triggers cytokine synthesis and inflammation.
TLR
MAMP
Host E3 ligase–mediated ubiquitylation activates pathway.
2. Rotavirus E3 ligase marks component in pathway for proteolysis. Inhibits inflammation.
Inhibitor
Rotavirus E3 ligase
Protease
Degradation
Salmonella deubiquitylase
Host E3 ligase–mediated ubiquitylation.
NF-kB activation
3. Salmonella deubiquitylase stops proteolysis of inhibitor. Limits inflammation.
Inhibitor
Polyubiquitin
Cytokines
Proteasome
Degradation
Inflammation

FIGURE 25.36 ▪ Microbial E3 ligases and deubiquitylases alter innate immune systems. When a MAMP binds to a TLR receptor, a series of events, including a host E3 ligase–mediated ubiquitylation, activates the transcriptional regulator NF-kappaB, which activates transcription of inflammatory cytokine genes. The cytokines are secreted and initiate inflammatory processes. Microbial E3 ligases (for example, rotavirus E3 ligase) can ubiquitylate other components of the NF-kappaB activation pathway and mark them for destruction. As a result, cytokine synthesis is inhibited and inflammation is limited. *Salmonella* makes a deubiquitylase that removes polyubiquitin from an inhibitor of NF-kappaB activation. Deubiquitylation saves the inhibitor from destruction, allowing continued inhibition of NF-kappaB activation.

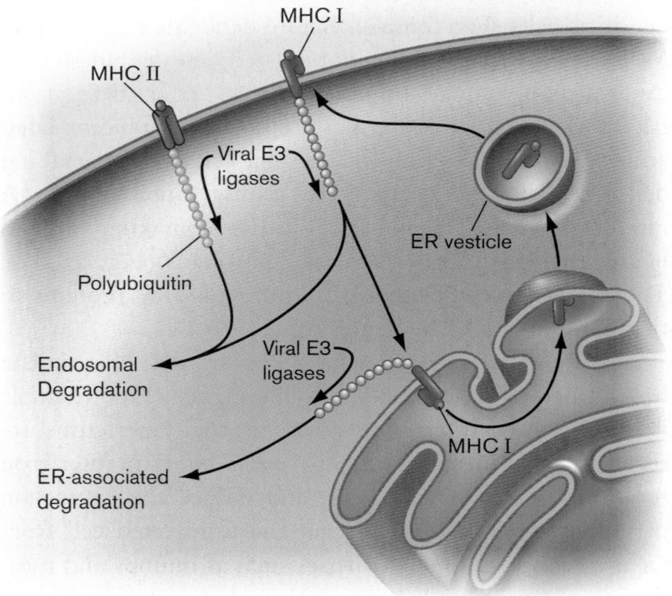
MHC I
MHC II
Viral E3 ligases
ER vesticle
Polyubiquitin
Endosomal Degradation
Viral E3 ligases
MHC I
ER-associated degradation

FIGURE 25.37 ▪ Microbial E3 ligases and deubiquitylases alter adaptive immune systems. MHC class I and class II molecules present antigens on host cell surfaces to helper T cells. Some viral E3 ligases can ubiquitylate the MHC molecules, marking them for degradation via endosomal pathways. Other E3 ligases can ubiquitylate MHC I molecules while in the endoplasmic reticulum (ER), leading to their degradation before being placed on the cell surface.

pathogens. Gaining an understanding of the ways ubiquitylation directs the immune system and how pathogens influence that direction could lead to new ways of enhancing host defense. Clearly, facultative and obligate intracellular pathogens have developed many ingenious strategies to take control of host cell defenses, but the host, for its part, has also evolved effective countermeasures.

Thought Question

25.10 You have discovered an E3 ligase that specifically modifies surface proteins of *Listeria monocytogenes.* Using what has been presented about toxin entry and protein secretion in Chapter 25, suggest some strategies for delivering this enzyme into host cells infected with this pathogen.

To Summarize

- **Two-component signal transduction systems** can regulate virulence gene expression in response to the host environment.
- **Quorum sensing** may prevent pathogens from releasing toxic compounds too early during infection.

- **Extracellular pathogens** evade the immune system by hiding in capsules, by changing their surface proteins, or by triggering apoptosis.
- **Intracellular bacterial pathogens** attempt to avoid the immune system by growing inside host cells. They use different mechanisms to avoid intracellular death.
- **Hemolysins** are used by certain pathogens to escape from the phagosome and grow in the host cytoplasm.
- **Actin tails** are used by some microbes to move within and between host cells.
- **Inhibiting phagosome-lysosome fusion** is one way pathogens can survive in phagosomes.
- **Molecular mechanisms for avoiding the immune system** include molecular mimicry, altering cytokine profiles, stopping programmed host cell death, interfering with autophagy, and redirecting ubiquitylation signals.
- **Specialized physiologies** enable some organisms to survive in the normally hostile environment of fused phagolysosomes.

25.7

Experimental Tools That Define Pathogenesis

Earlier (in Section 25.2) and in eTopics 25.1 and 25.2 we discussed how virulence genes in microbes can be identified using some clever selection techniques. But can we get a broader picture of an organism's pathogenesis without looking one gene at a time? What about pathogens that we cannot grow in vitro? Do they exist, and can bioinformatics help us figure out how to grow them in vitro? And how does a host respond to an infection? Can we use new molecular tools to examine the host's response? The next three examples explore these questions.

Helicobacter pylori, Gene Arrays, and Gastric Ulcers

Helicobacter pylori is a Gram-negative microbe, transmitted orally, that causes a variety of stomach diseases, including ulcers and cancer (**Fig. 25.38**). *H. pylori* has a genome size of 1.6 Mb, encoding 1,590 genes. The organism, however, is extremely difficult to grow in vitro and does not easily exchange genetic material. Therefore, classical approaches for studying its pathogenicity that rely on mutagenesis, clonal selection, and gene transfer are nearly impossible.

However, gene array technology, which can identify the repertoire of genes expressed under any condition (see Section 10.9), and signature-tagged mutagenesis (eTopic 25.1) have successfully identified numerous genes involved in the virulence of *Helicobacter*. A basic scheme of pathogenesis has been suggested as a result (**Fig. 25.39**). As with most microbial diseases, the first step is colonization. Following ingestion of the pathogen, *H. pylori* flagella propel the organism toward the mucosa (**Fig. 25.39**, step 1). As it approaches the mucosa, the pathogen produces intracellular and perhaps extracellular urease, which converts urea to CO_2 and ammonia, thereby decreasing the local acidity of the stomach (step 2). Two other enzymes, collagenase and mucinase, soften the mucous lining, which allows

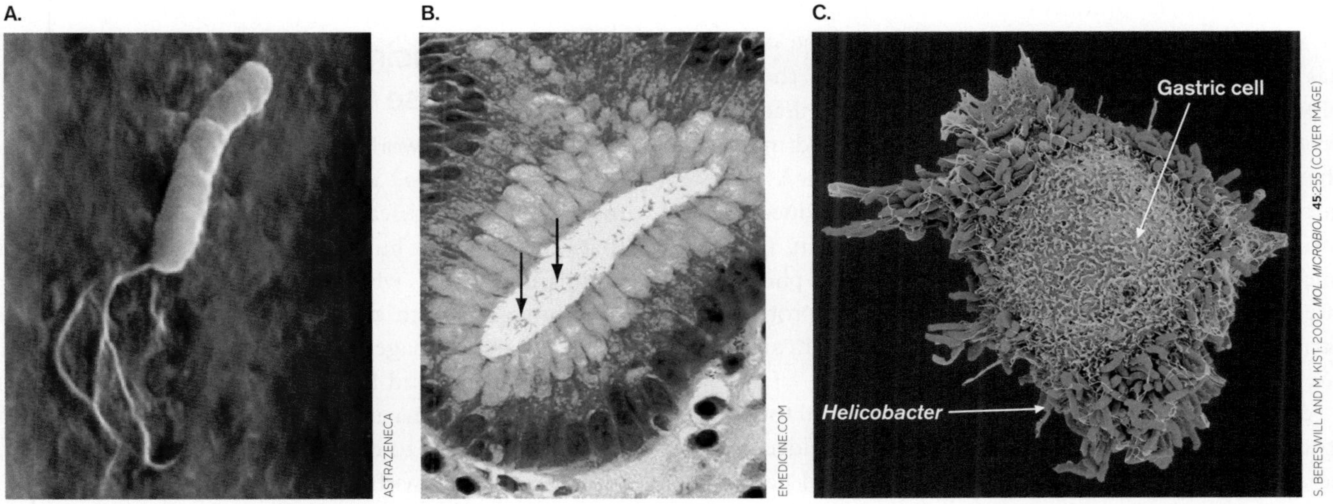
A.
B.
C.
ASTRAZENECA
EMEDICINE.COM
Gastric cell
Helicobacter
S. BERESWILL AND M. KIST. 2002. MOL. MICROBIOL. 45:255 (COVER IMAGE)

FIGURE 25.38 ▪ *Helicobacter* pathogenesis. A. *H. pylori* (SEM). Note the tuft of flagella at one pole. Cell length approx. 2 µm. **B.** *Helicobacter* (arrows) attached to gastric mucosa (SEM). **C.** *Helicobacter* (red) binding to human gastric epithelial cells (colorized SEM). Intimate binding facilitates translocation of the bacterial effector protein CagA into the gastric cell via type IV secretion.

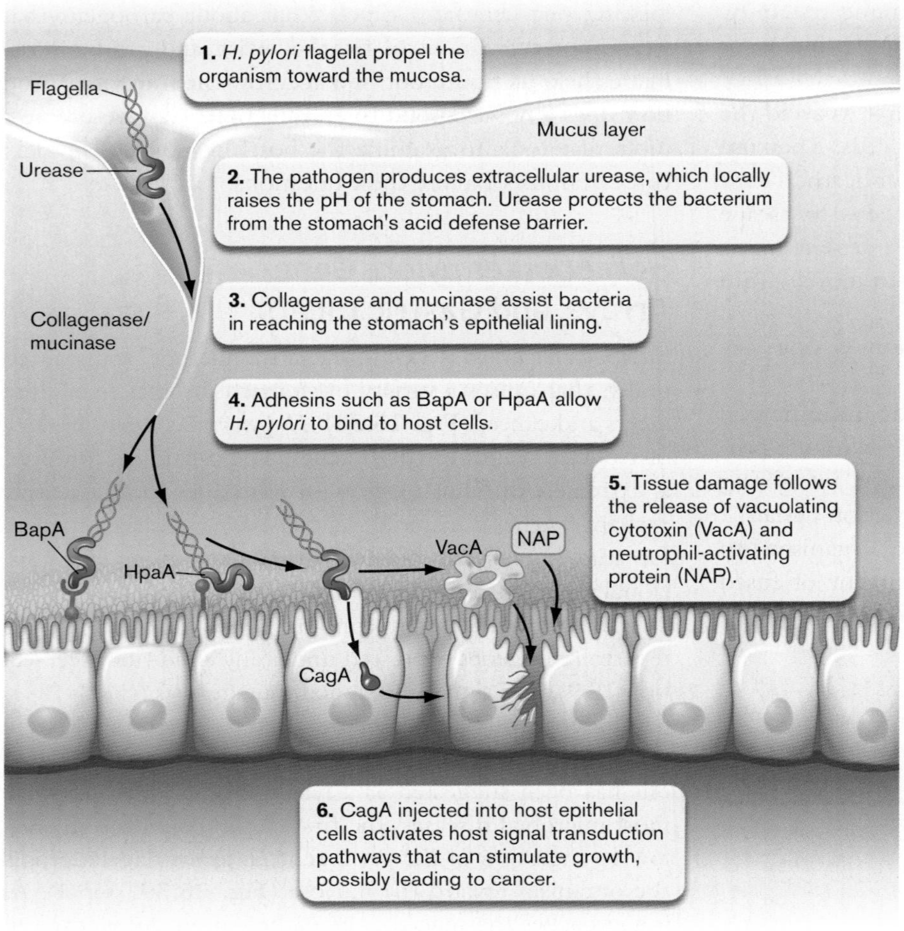
1. H. pylori flagella propel the organism toward the mucosa.
Flagella
Mucus layer
Urease
2. The pathogen produces extracellular urease, which locally raises the pH of the stomach. Urease protects the bacterium from the stomach's acid defense barrier.
3. Collagenase and mucinase assist bacteria in reaching the stomach's epithelial lining.
Collagenase/ mucinase
4. Adhesins such as BapA or HpaA allow H. pylori to bind to host cells.
5. Tissue damage follows the release of vacuolating cytotoxin (VacA) and neutrophil-activating protein (NAP).
BapA
HpaA
VacA
NAP
CagA
6. CagA injected into host epithelial cells activates host signal transduction pathways that can stimulate growth, possibly leading to cancer.

FIGURE 25.39 ■ Steps in *Helicobacter* colonization.

the bacteria to reach the stomach's epithelial lining (step 3). The epithelial lining is much less acidic than the lumen, so the organism can grow and divide. Once at the epithelium, *Helicobacter* produces various adhesins, such as BapA or HpaA, to bind host cells (step 4). Once the organism has adhered, tissue damage occurs following the release of vacuolating cytotoxin (VacA) and neutrophil-activating protein (NAP), which activates neutrophils and mast cells that further damage local tissue (step 5). VacA forms a hexameric pore in the host membrane and induces apoptosis (programmed cell death). Another protein, CagA, is injected into host epithelial cells, where it is phosphorylated; CagA then interacts with host signaling proteins and activates host signal transduction pathways that can stimulate growth, possibly leading to cancer (step 6).

As you can see, genomics has been a powerful tool in the study of *Helicobacter*. DNA sequence alone yielded a fairly detailed picture of *Helicobacter* physiology and pathogenesis based on homologies to known enzymes from other organisms (see Fig. 8.43). Several aspects of these models have been confirmed experimentally.

Whipple's Disease and DNA Sequencing

It is difficult to characterize a pathogen if you cannot even grow it in culture. The microorganism *Tropheryma whipplei* causes a gastrointestinal illness called Whipple's disease, whose symptoms include diarrhea, intestinal bleeding, abdominal pain, loss of appetite, weight loss, fatigue, and weakness. Though identified, the causative agent had never been grown outside of fibroblast cells, making it nearly impossible to study its physiology. Once the complete genome sequence of two strains of *T. whipplei* became available, however, Didier Raoult and his colleagues at the Marseille School of Medicine discovered how to feed this fastidious bacterium. The genome (less than 1 Mb) contained a great deal of functional information. Computer modeling studies used this information to discover that the organism lacks the machinery to make several amino acids. With this knowledge, the investigators were able to design a cell-free culture medium that supported the growth of three different strains of *T. whipplei,* as well as a new strain isolated from a heart valve. Similar applications of modern genomic techniques should lead to other successes in growing previously unculturable intracellular pathogens.

Deep Sequencing Reveals the Host Response to *Brucella*

Brucellosis is a worldwide zoonotic infectious disease caused by *Brucella,* a genus of fastidious Gram-negative rods. A zoonotic disease (zoonosis) is a disease of animals that can also affect humans. In humans, *Brucella* produces undulant fever in which the patient's temperature rises daily, going as high as 105°F, but falls at night. *Brucella* grows in macrophages that disseminate the organism to the liver, spleen, and bone marrow. Qianhong Liu at Jilin University in China used a technique called digital gene expression (DGE) to examine transcription in mouse macrophages infected with *Brucella melitensis.*

DGE technology uses beads covered with oligodeoxythymidine (oligo-dT) fragments to bind the poly-A tails of eukaryotic mRNA and pull down those mRNA molecules via centrifugation. The single-stranded RNAs are converted to complementary DNA (cDNA) by reverse transcriptase and then tagged with an oligonucleotide of known sequence. The tagged cDNAs are then cleaved into smaller pieces with a restriction endonuclease, and the tagged pieces are deep-sequenced using techniques described in Section 7.6. The number of times a cDNA from a given gene is sequenced is proportional to how many RNA transcripts of that gene were initially present.

Using this technology to compare transcripts from infected and uninfected macrophages, Liu and colleagues found that *B. melitensis* altered expression of almost 4,000 host genes. Consistent with the idea that *Brucella* is engulfed into macrophage endosomes and lives in a modified endoplasmic reticulum (ER) compartment, transcriptome analysis identified major changes in the expression of genes for the ER and lysosome. Genes involved with the apoptotic programmed-cell-death pathway were downregulated, saving *Brucella* from that fate, and the expression of many genes involved in NOD-like receptor, B-cell receptor, and Toll-like receptor signaling pathways of innate and adaptive immunity were diminished (Chapters 23 and 24). By examining the changes in host gene expression that take place during infection, some of which are orchestrated by the microbe, scientists hope to devise new ways to boost host resistance to infection.

To Summarize

- **Gene array technology** was used to predict mechanisms of *Helicobacter pylori* pathogenesis.
- **DNA sequencing** revealed growth requirements for the intestinal pathogen *Tropheryma whipplei*.
- **Deep sequencing** of transcripts from infected and uninfected host cells can expose a complex, yet orchestrated, response to infection.

Concluding Thoughts

This chapter has only scratched the surface of all that is known about microbial pathogenesis. It should be clear at this point that transmission, attachment, immune avoidance, and subversion of host signaling pathways are common goals of most successful pathogens, whether bacterial, fungal, viral, or parasitic. But even with all we know, there remains much to learn.

The remaining chapters deal with the basic principles used to diagnose disease and eradicate offending pathogens, but many pathogens remain hard to detect and difficult, if not impossible, to kill. So, the more we know about the mechanisms microbes use to cause disease, the better we will be at developing effective countermeasures.

As you will learn in the coming chapters, new pathogens are constantly emerging. Over the last few decades, we have seen the development of HIV, SARS, avian flu, hantavirus, West Nile virus, *E. coli* O157:H7, and the reemergence of flesh-eating streptococci, to name but a few. Can we ever stop pathogens from emerging? Probably not. For every countermeasure we develop, nature designs a counter-countermeasure. Our hope is that continuing research into the molecular basis of pathogenesis and antimicrobial pharmacology will keep us one step ahead.

CHAPTER REVIEW

Review Questions

1. Describe the differences between infection and disease; pathogenicity and virulence; LD_{50} and ID_{50}.
2. What is meant by direct versus indirect routes of infection?
3. What are the characteristics of a good reservoir for an infectious agent?
4. Name the various portals of entry for infectious agents, and a disease associated with each.
5. Describe the basic features of a pathogenicity island.
6. Explain various ways in which bacteria can attach to host cell surfaces.
7. Describe the basic steps by which pili are assembled on the bacterial cell surface. How do type I and type IV pili differ?
8. Explain the five broad categories of toxin mode of action.
9. What is ADP-ribosylation, and how does it contribute to pathogenesis?

10. Explain the differences between exotoxins and endotoxins.
11. Explain the mechanisms of secretion carried out by type II and type III protein secretion systems. What are the paralogous origins of these systems?
12. Describe the key features of *Salmonella* pathogenesis.
13. How can genomic approaches help identify pathogens in an infection?
14. What different mechanisms do intracellular pathogens use to survive within the infected host cell?
15. Describe different molecular strategies that microbes use to avoid the immune system.
16. How do bacteria determine whether they are in a host environment?
17. Discuss the relationships between ubiquitylation and intracellular pathogens.

Thought Questions

1. Why do new versions of swine and avian flu often originate in Asia?
2. How can you modify Koch's postulates to prove that a bacterial gene is a virulence factor?
3. How would you determine whether a particular pilus on group A streptococci (GAS) is required for the organism's pathogenesis? Use a tissue culture model.
4. Why have humans not developed resistance to microbial toxins?
5. You want to make a live oral vaccine for cholera. But, because *Vibrio cholerae* is an acid-sensitive organism, the person to be immunized would have to ingest large numbers of organisms. *E. coli*, on the other hand, is very acid resistant and able to survive stomach acidity for long periods of time. You think that moving the acid resistance system from *E. coli* to *V. cholerae* will solve this problem. Is there an ethical issue to consider when trying to move the acid resistance system from *E. coli* into *V. cholerae*?

Key Terms

adhesin (1013)
ADP-ribosyltransferase (1022)
autophagy (1042)
ectoparasite (1005)
edema factor (EF) (1025)
endoparasite (1005)
endotoxin (1018)
exotoxin (1018)
facultative intracellular pathogen (1038)
fimbria (1013)
fomite (1007)
genomic island (1011)
horizontal transmission (1007)
immunopathogenesis (1009)
infection (1005)
infection cycle (1007)
infectious dose 50% (ID_{50}) (1006)
intimin (1031)
intracellular pathogen (1037)
labile toxin (LT) (1023)
latent state (1006)
lethal dose 50% (LD_{50}) (1006)
lethal factor (LF) (1025)
opportunistic pathogen (1006)
parasite (1005)
parenteral route (1009)
pathogen (1005)
pathogenesis (1004)
pathogenicity (1006)
pathogenicity island (1011)
petechia (1027)
pilus (1013)
primary pathogen (1006)
protective antigen (PA) (1025)
protein A (1037)
reservoir (1007)
transovarial transmission (1007)
type I pilus (1013)
type II secretion system (1029)
type III pilus (1013)
type III secretion system (T3SS) (1030)
type IV pilus (1013)
vector (1007)
vertical transmission (1007)
virulence (1006)
virulence factor (1010)

Recommended Reading

Basler, Marek, Martin Pilhofer, Gregory P. Henderson, Grant J. Jensen, and John J. Mekalanos. 2012. Type VI secretion requires a dynamic contractile phage tail-like structure. *Nature* **483**:182–186.

Bierne, Hélène, and Pascale Cossart. 2012. When bacteria target the nucleus: The emerging family of nucleomodulins. *Cellular Microbiology* **14**:622–633.

Burrows, Lori L. 2005. Weapons of mass retraction. *Molecular Microbiology* **57**:878–888.

Casadevall, Arturo. 2008. Evolution of intracellular pathogens. *Annual Review of Microbiology* **62**:19–33.

Chen, John, and Richard P. Novick. 2009. Phage-mediated intergeneric transfer of toxin genes. *Science* **323**:139–141.

Dean, Paul, Marc Maresca, and Brendan Kenny. 2005. EPEC's weapons of mass subversion. *Current Opinion in Microbiology* **8**:28–34.

Dobrindt, Ulrich, Bianca Hochhut, Ute Hentschel, and Jörg Hacker. 2004. Genomic islands in pathogenic and environmental microorganisms. *Nature Reviews. Microbiology* **2**:414–424.

Groisman, Eduardo, and Josep Casadesús. 2005. The origin and evolution of human pathogens. *Molecular Microbiology* **56**:1–7.

Hall-Stoodley, Luanne, and Paul Stoodley. 2009. Evolving concepts in biofilm infections. *Cell Microbiology* **11**:1034–1043.

Hamon, Mélanie, Hélène Bierne, and Pascale Cossart. 2006. *Listeria monocytogenes*: A multifaceted model. *Nature Reviews. Microbiology* **4**:423–434.

Hsiao, Ansel, and Jun Zhu. 2009. Genetic tools to study gene expression during bacterial pathogen infection. *Advances in Applied Microbiology* **67**:297–314.

Ivarrson, Mattias E., Jean-Christophe Leroux, and Bastien Castagner. 2012. Targeting bacterial toxins. *Angewandte Chemie International Edition* **51**:4024–4045.

Jiang, Xiaomo, and Zhijian J. Chen. 2012. The role of ubiquitylation in immune defence and pathogen evasion. *Nature Reviews. Immunology* **12**:35–48.

Jin, Qi, Zhenghong Yuan, Jianguo Xu, Yu Wang, Yan Shen, et al. 2002. Genome sequence of *Shigella flexneri* 2a: Insights into pathogenicity through comparison with genomes of *Escherichia coli* K12 and O157. *Nucleic Acid Research* **30**:4432–4441.

Krachler, Anne Marie, Hyeilin Ham, and Kim Orth. 2011. Outer membrane adhesion factor multivalent adhesion molecule 7 initiates host cell binding during infection by Gram-negative pathogens. *Proceedings of the National Academy of Sciences USA* **108**:11614–11619.

Lainhart, William, Gino Stolfa, and Gerald B. Koudelka. 2009. Shiga toxin as a bacterial defense against a eukaryotic predator, *Tetrahymena thermophila*. *Journal of Bacteriology* **191**:5116–5122.

Lipkin, W. Ian. 2008. Pathogen discovery. *PLOS Pathogens* **4**:e1000002.

Liu, Qianhong, Wenyu Han, Changjiang Sun, Liang Zhou, Limin Ma, et al. 2012. Deep sequencing-based expression transcriptional profiling changes during *Brucella* infection. *Microbial Pathogenesis* **52**:267–277.

Malik-Kale, Preeti, Carrie E. Jolly, Stephanie Lathrop, Seth Winfree, Courtney Luterbach, et al. 2011. *Salmonella*—At home in the host cell. *Frontiers in Microbiology* **2**:125.

Mota, Luís Jaime, Isabel Sorg, and Guy R. Cornelis. 2005. Type III secretion: The bacteria-eukaryotic cell express. *FEMS Microbiology Letters* **252**:1–10.

Olsen, Randall J., Samuel A. Shelburne, and James M. Musser. 2009. Molecular mechanisms underlying group A streptococcal pathogenesis. *Cell Microbiology* **11**:1–12.

Pallen, Mark J., and Brendan W. Wren. 2007. Bacterial pathogenomics. *Nature* **449**:835–842.

Renesto, Patricia, Nicolas Crapoulet, Hiroyuki Ogata, Bernard La Scola, Guy Vestris, et al. 2003. Genome-based design of a cell-free culture medium for *Tropheryma whipplei*. *Lancet* **362**:447–449.

Rutherford, Steven T., and Bonnie L. Bassler. 2012. Bacterial quorum sensing: Its role in virulence and possibilities for its control. *Cold Spring Harbor Perspectives in Medicine* **2**:a012427.

Thi, P. Emily, Ulrike Lambertz, and Neil E. Reiner. 2012. Sleeping with the enemy: How intracellular pathogens cope with a macrophage lifestyle. *PLOS Pathogens* **8**:e1002551.

Thompson, Lucinda J., and Hilde de Reuse. 2002. Genomics of *Helicobacter pylori*. *Helicobacter* **7**:1–17.

Wangdi, Tamding, Sebastion E. Winter, and Andreas J. Baumler. 2012. Typhoid fever: "You can't hit what you can't see." *Gut Microbes* **3**: 88–92.

Wiles, Siouxsie, William P. Hanage, Gad Frankel, and Brian Robertson. 2006. Modeling infectious disease—Time to think outside the box? *Nature Reviews. Microbiology* **4**:307–312.

CHAPTER 26
Microbial Diseases

26.1 Characterizing and Diagnosing Microbial Diseases

26.2 Skin and Soft-Tissue Infections

26.3 Respiratory Tract Infections

26.4 Gastrointestinal Tract Infections

26.5 Genitourinary Tract Infections

26.6 Central Nervous System Infections

26.7 Cardiovascular System Infections

26.8 Systemic Infections

26.9 Immunization

In the U.S. Civil War, twice as many people died from infections as from battle wounds. Clearly, developments in medicine had not kept pace with innovations in rifle accuracy. Today, antibiotics can quell most wound infections, yet millions of people still die as a result of dangerous microbes that spread by food, air, body fluids, and even insects. How do clinicians recognize these many different types of infections? What key symptoms or laboratory tests help determine whether a person suffers from a viral or bacterial disease?

Microbial diseases are with us daily and are major contributors to global mortality. The emergence of new pathogens, increasing drug resistance, and threats of bioterrorism have heightened the need for investigations into microbial disease mechanisms and the body's ability to combat infectious agents. In addition, effective diagnostic algorithms are needed to quickly identify infectious diseases and prevent their spread. Chapter 26 explores the general classes and causes of microbial diseases and introduces the art of diagnosis.

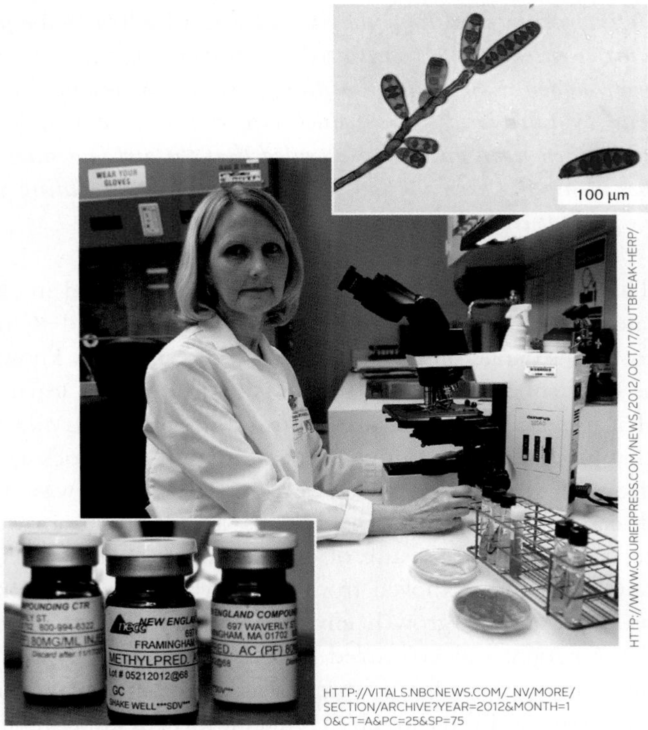

100 µm
METHYLPRED

AP PHOTO/THE CENTERS FOR DISEASE CONTROL

HTTP://WWW.COURIERPRESS.COM/NEWS/2012/OCT/17/OUTBREAK-HERP/

HTTP://VITALS.NBCNEWS.COM/_NV/MORE/SECTION/ARCHIVE?YEAR=2012&MONTH=10&CT=A&PC=25&SP=75

CURRENT RESEARCH highlight

Hunting a mysterious killer. In September 2012, a 51-year-old Baltimore woman developed an excruciating headache that penetrated into her face. This was followed by double vision, nausea, vertigo, and loss of muscle coordination. All are signs of meningitis, but a spinal tap provided no clues to the cause. At the same time, 600 miles away, Tonya Snyder (shown), a medical technologist at Vanderbilt University Medical Center, found something very unusual in one of her patients' spinal fluid samples—a fungus. That discovery helped investigators diagnose the very first case of meningitis linked to tainted steroids prepared by a Massachusetts compounding pharmacy. The solutions (left inset) were unknowingly contaminated with various fungi, such as *Exserohilum rostratum* (right inset), and injected directly into patient spinal columns to treat back pain. These fungi are harmless unless allowed to bypass the host's innate immune defenses. As of January 2013, there were 620 cases tied to these drugs, and 36 deaths.

In May, 2013, a visibly ill 16-year-old girl was taken to her pediatrician, where she complained of diarrhea, high fever (39.9°C; 103.8°F), and vomiting. She told the physician she had been healthy until 2 days prior. Upon physical exam, the doctor noticed that the girl had a much lower than normal blood pressure (76/48 mm Hg; normal range 90–130/60–90), a rapid heart rate (120 beats per minute; normal range 60–100), and an erythematous (red) rash on her trunk. Because of her deteriorating condition, the patient was admitted to the pediatric intensive care unit at the local hospital, and cultures were obtained. She was immediately given intravenous fluids and IV antibiotics, yet she died within a week. Patient history taken upon admission revealed that the girl had started her menstrual period 4 days before becoming ill, providing an important clue as to the cause of the disease.

This tragedy is part of a larger story that emerged in the late 1970s and early 1980s, when women started dying from this dangerous, new (emerging) disease. Now known as toxic shock syndrome, scientists and physicians learned that it is caused by certain strains of *Staphylococcus aureus* that produce a superantigen type of toxin (toxic shock syndrome toxin, or TSST; see Section 24.6). Why was the disease not recognized in previous decades? The answer turned out to be the use of one brand of superabsorbent tampons (since removed from the market). The tampons produced a rich growth environment for *S. aureus*. So, if the patient was colonized by a TSST-producing strain, huge amounts of toxin were released to circulate in the bloodstream. Even though the superabsorbant tampons are no longer available, toxic shock syndrome is still sometimes associated with menstruation, as in the case history. Today, however, we recognize that toxic shock syndrome is a possible consequence of any *S. aureus* infection and can occur in both men and women (most cases are now unrelated to menstruation or tampon use). We now have tools to diagnose it quickly.

In Chapter 25 we discussed the weapons that microbes use to cause disease (including TSST). But what of the diseases themselves? Microbial diseases are often presented from the microbe's point of view; for example, where in the body does *S. aureus* cause disease? Another approach, which we take here in Chapter 26, is to look at infections from the vantage point of the infected organ. We might ask, for example, which organisms cause vaginal infections versus lung infections, and how are they differentially diagnosed? Are the patient's complaints consistent with gastrointestinal tract disease or a lung infection? This approach is more attuned to the practices of health care workers and the fundamentals by which a physician interacts with a patient. An individual who goes to a clinician complaining of a fever, cough, and chest pain does not typically report having encountered a particular bacterium. The physician must determine from the patient that the disease is localized to the chest and then intuit that it may be a respiratory tract infection. Appropriate samples are then collected and sent to a clinical microbiology laboratory, where data are collected to confirm or dispute the presence of a specific etiological agent, such as *Streptococcus pneumoniae* or other lung pathogens.

We do not present an exhaustive compendium of microbial illnesses in this chapter, but a representative sampling of infections that illustrate key aspects of microbial disease. The material is arranged and presented so as to integrate your knowledge of microbiology and immunology within the framework of the practice of medicine and the study of infectious disease. The chapter closes by revisiting immunization (introduced in Chapters 23 and 24) in terms of herd immunity and the vaccine schedules used to prevent various diseases discussed here and in earlier chapters.

26.1

Characterizing and Diagnosing Microbial Diseases

What is the best way to classify microbial diseases? Although this chapter will classify representative microbial diseases by organ system, there are other very instructive ways to group and study microbial infections. These include the **organism** and **portal-of-entry** approaches. Each has clear benefits and pitfalls. The organism approach is useful when examining the different ways a given species can cause disease. For instance, *E. coli* can cause genitourinary tract infections, gastrointestinal infections, and meningitis. How do the various strains causing these diseases differ from one another? Are virulence factors present in the uropathogenic strain that are absent or different in the diarrhea-producing strain? (Virulence factors are discussed in Sections 25.2 and 25.3.)

Pathogens are also commonly classified by their route of infection (or portal of entry; see Chapter 23). In this approach, pathogens are classified as food-borne, airborne, blood-borne, or sexually transmitted. The portal-of-entry approach has its merits, but it sometimes seems at odds with the disease itself. *Salmonella enterica* serovar Typhi, for example, is categorized as a food-borne pathogen (the cause of typhoid fever). Food-borne pathogens typically have a fecal-oral route of infection and cause pronounced gastrointestinal disease. Yet *S.* Typhi does not actually cause gastrointestinal disease until very late in the disease process. After ingestion, it quickly enters the bloodstream from the intestine to produce a serious septicemia (blood infection) called enteric (typhoid) fever. At this stage, the organism cannot be found in feces. (Much later the organism can reenter the intestine via the gallbladder and cause mild gastrointestinal symptoms.)

The obvious value of knowing the route of infection is that society can design effective measures to avoid disease. Hand washing and proper food preparation, for instance, will effectively prevent the septicemia of typhoid fever. But not all organisms entering via the gastrointestinal tract cause gastrointestinal disease. Note that *S.* Typhi is different from *Salmonella enterica* serovar Enteritidis, a major cause of diarrhea. *S.* Enteritidis is also a food-borne pathogen, but it rarely enters the bloodstream.

The organ system approach has its drawbacks as well, because organisms that initially affect one organ system can disseminate to infect other organ systems. *Neisseria gonorrhoeae,* for example, initially infects the genitourinary tract, but can disseminate via the bloodstream, causing a blood infection. *Yersinia enterocolitica,* a cause of gastroenteritis, can disseminate from the intestine and cause abscesses elsewhere in the body. Yet we favor the organ system approach because it is illuminating to know how microbial diseases present themselves to physicians treating patients. The physician's examination starts by using a patient's symptoms to identify which organ system is affected, not by knowing the portal of entry. Your studies will benefit, however, from grouping the infectious agents we discuss also by portal of entry and taxonomy.

Patient Histories Provide Important Diagnostic Clues

A problem often faced by clinicians is that many infectious diseases display similar symptoms, making diagnosis difficult. *Vibrio cholerae* and enterotoxigenic *E. coli,* for instance, both produce diarrheal diseases characterized by cramps, lethargy, and liters of watery stool each day. Cholera, however, is not commonly seen in the United States, Canada, or other developed countries. Nevertheless, a clinician might suspect cholera if a patient recently traveled to or arrived from an endemic area (a specific geographic locale such as India or Bangladesh) where the disease is regularly observed. This travel information is not gained by examining the patient but comes from taking a "patient history."

Asking about a patient's hobbies is another critical part of taking a patient's history, because that information, too, can help diagnose a disease. For example, a person who is suffering from enlarged glands, fever, and headaches and who has recently gone rabbit hunting may have been exposed to the Gram-negative rod *Francisella tularensis* (**Fig. 26.1A**), an intracellular pathogen that infects various wild animals and is the cause of tularemia, an illness also known as "rabbit fever". A hunter with these symptoms could have accidentally infected a cut while cleaning the kill. Tularemia is an example of a **zoonotic disease**, an infection that normally affects animals but can be transmitted to humans.

Can knowing a person's occupation be important? Consider the case of a man with acute pneumonia. If the clinician discovers that the man is a sheep farmer, then *Coxiella*

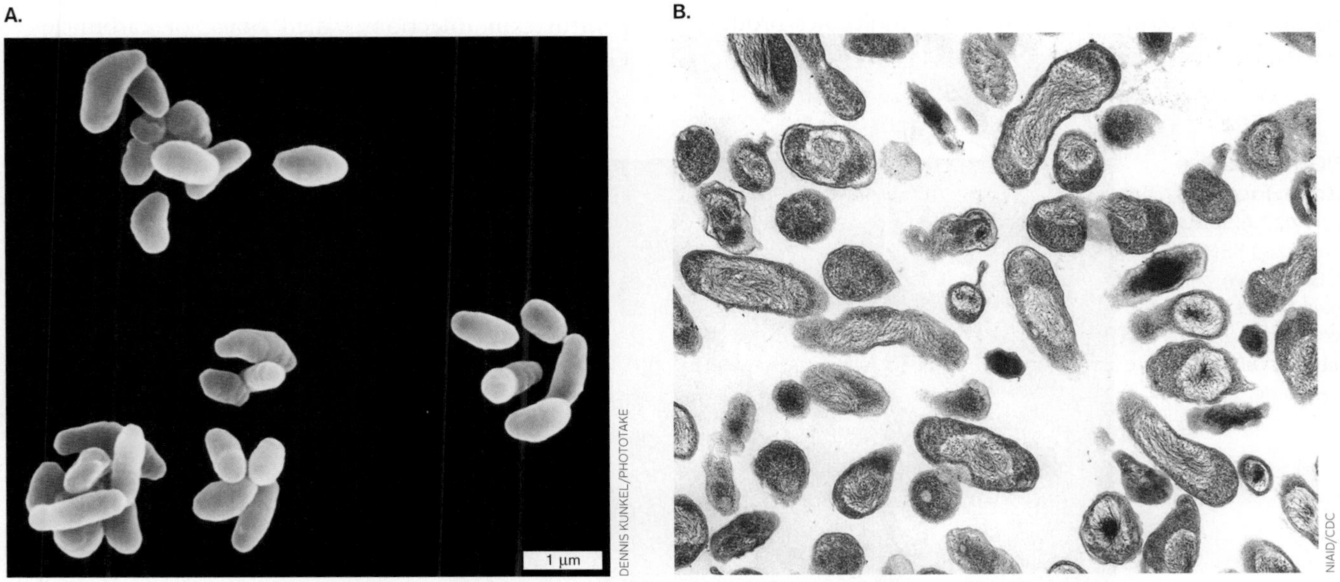
A.
B.
1 µm
DENNIS KUNKEL/PHOTOTAKE
NIAID/CDC

FIGURE 26.1 ▪ Examples of bacteria that cause zoonotic diseases. **A.** *Francisella tularensis* (colorized SEM), the cause of tularemia, a highly infectious disease spread usually via a tick vector but also through cuts. **B.** *Coxiella burnetii* (colorized TEM), the cause of Q fever. This irregularly shaped organism undergoes developmental stages that range in size from 0.2 to 1 µm.

burnetii, a pathogenic bacterial species that infects sheep and is shed in large quantities in placental materials, would be one possible cause of infection. The organism grows intracellularly only (it is an obligate intracellular pathogen). It infects sheep, cattle, and goats but does not usually cause clinical disease in these animals. *C. burnetii* is secreted in body fluids and shed in high numbers in the animal's amniotic fluids and placenta, heavily contaminating the soil. After the contaminated soil dries, the microbes become aerosolized whenever the dirt is disturbed. Humans (such as farmers) who inhale the dried particles can develop the lung infection called Q fever, another type of zoonotic disease (**Fig. 26.1B**).

What possible connection could there be between a woman with severe respiratory disease and her job handling animal hides in a leather factory? Knowing her occupation, a clinician might investigate whether she had contracted anthrax ("woolsorter's disease") from inhaling spores present on the animal hides.

Because patient histories are so helpful in diagnosing infectious diseases, we will present several case histories in the following sections and segue into discussions of various microbes that can infect each organ system. Key aspects of infectious diseases will be revealed as we proceed.

It should be stressed that further insight into pathogenesis and new innovations in diagnosis will come from the remarkable number of microbial genomes whose sequences have been completed, as well as from sequencing efforts that are still under way. A sampling of the pathogens whose genomes have already been sequenced is presented in eTopic 26.1. Many of the open reading frames (ORFs) identified have no known function and may eventually explain differences in host and tissue specificity and reveal new toxins or other pharmacologically active molecules produced by virulent microbes.

To Summarize

- **Pathogens can be classified** as food-borne, airborne, blood-borne, or sexually transmitted.
- **Understanding infectious disease** requires knowledge of the organ system, the portal of entry, and the infectious organism.
- **Patient histories** are vital in diagnosing microbial diseases.
- **Zoonotic diseases** are animal diseases accidentally transmitted to humans.

26.2

Skin and Soft-Tissue Infections

Skin infections range from simple boils to severe, complicated, so-called flesh-eating diseases (**Table 26.1**). Recall that the integrity of the skin, as well as the presence of normal skin microbiota, prevents infection. However, even minor insults to the skin (such as a paper cut) can result in infections, most of which are caused by the Gram-positive pathogen *Staphylococcus aureus.* Healthy individuals develop infections of the skin only rarely, whereas people with underlying immunosuppressive diseases such as diabetes are at much higher risk.

Boils

Staphylococcus aureus (**Fig. 26.2A**) is a common cause of painful skin infections called boils, or carbuncles. This Gram-positive organism, often a normal inhabitant of the

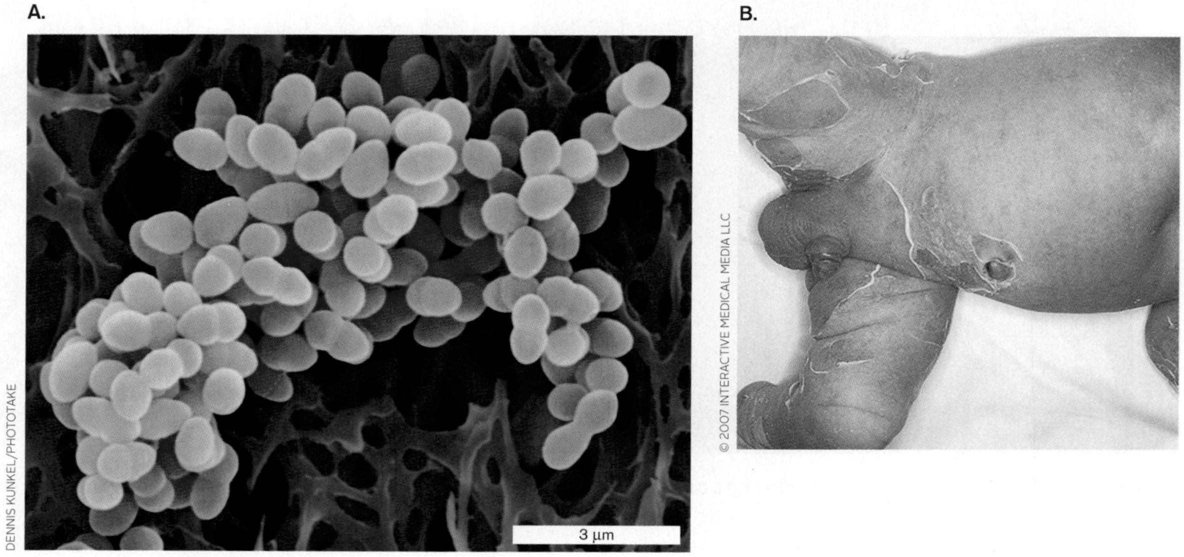
A.
B.
3 μm
DENNIS KUNKEL/PHOTOTAKE
© 2007 INTERACTIVE MEDICAL MEDIA LLC

FIGURE 26.2 ▪ *Staphylococcus aureus.* **A.** *S. aureus* (colorized SEM). **B.** Exfoliative toxin from some strains of *S. aureus* causes scalded skin syndrome.

TABLE 26.1

Common infectious diseases of the skin.

Disease	Symptoms	Etiological agent	Virulence factors
Bacterial			
Folliculitis	Boils	*Staphylococcus aureus* (G+ cocci); fibrin wall around abscess renders it poorly accessible to antibiotics	Coagulase, protein A, TSST, leukocidin, exfoliative toxin
Scalded skin syndrome	Peeling skin on infants, systemic toxin	*Staphylococcus aureus* (G+ cocci)	
Impetigo	Skin lesions on the face, mostly children	*Staphylococcus aureus* (G+ cocci)	
Scarlet fever	Sore throat, fever, rash	*Streptococcus pyogenes* (G+ cocci)	M-protein pili, C5a peptidase, hemolysin, pyrogenic toxins, others
Erysipelas	Skin lesions, usually facial, that spread to cause systemic infection	*Streptococcus pyogenes*	As for *S. pyogenes* (see above)
Cellulitis	Uncomplicated infection of the dermis	*S. aureus, S. pyogenes*	As for *S. pyogenes* (see above)
Necrotizing fasciitis	Rapidly progressive cellulitis	*S. aureus, S. pyogenes, Clostridium perfringens*	As for *S. pyogenes* (see above); a variety of toxins
Viral			
Rubella[a]	Discolored, pimply rash; mild disease unless congenital	Rubella virus [ssRNA(+)]	
Measles[a]	Severe disease, fever, conjunctivitis, cough, rash	Rubeola virus [ssRNA(−)]	V protein (interferes with interferon signaling)
Chickenpox[a]	Generalized discolored lesions	Varicella-zoster (dsDNA)	Glycoprotein B (fusion of viral and cellular membranes)
Shingles	Pain and skin lesions, usually on trunk in adults		Glycoprotein E (required for cell-cell fusion)
Smallpox[b]	Raised, crusted skin rash, highly contagious	Variola major (dsDNA)	SPICE (smallpox inhibitor of complement enzymes)
Warts[a]	Rapid growth of skin cells	Papillomaviruses (dsDNA)	E6 and E7 oncoproteins (Chapter 25)
Fungal			
Dermatomycosis	Dry, scaly lesions like athlete's foot	Dermatophytes	
Sporotrichosis	Granulomatous, pus-filled lesions; can disseminate to lungs or other organs	*Sporothrix schenckii*	
Blastomycosis	Granulomatous, pus-filled lesions; can disseminate to lungs or other organs	*Blastomyces dermatitidis*	BAD1 adherence
Candidiasis	Patchy inflammation of mouth (thrush) or vagina; can disseminate in immunocompromised patients	*Candida albicans*; *Candida glabrata*	Proteinase, phospholipase, Ssn6/Tup1 regulators, others
Aspergillosis	Infected wounds, burns, cornea, external ear	*Aspergillus* spp.	PacC/FOS1 regulators, gliotoxin
Zygomycosis	Oropharyngeal infections; affects mainly diabetic patients; can rapidly disseminate	*Mucor* and *Rhizopus* spp.	

[a]Vaccine is available against this agent.

[b]Vaccine is no longer in use, because disease has been eradicated. The exceptions are highly restricted laboratories.

nares (nostrils), can infect a cut or gain access to the dermis via a hair follicle. It possesses a number of enzymes that contribute to disease, including coagulase, which helps coat the organism with fibrin, thereby walling off the infection from the immune system and antibiotics. As a result, boils generally require surgical drainage, as well as antibiotic therapy. As noted earlier, some strains of *S. aureus* also produce toxic shock syndrome toxin, a superantigen that can lead to serious systemic symptoms. Recall that a superantigen links and activates antigen-presenting cells and T cells by binding to the outside of MHC class II receptors and T-cell receptor. Antigen recognition is not required. As a result, many different T cells become activated to release a flood of cytokines (see Chapters 24 and 25).

A particularly dangerous strain of *S. aureus*, called methicillin-resistant *S. aureus* (MRSA), has emerged. *S. aureus* infections are commonly treated with penicillin-like drugs, such as methicillin. MRSA has developed resistance to methicillin and many other penicillin-like drugs through a mutation that alters one of the proteins (called a penicillin-binding protein, PBP) involved in cell wall synthesis (see Chapter 27). Because methicillin is normally used as a first line of defense against staphylococcal infections, treatment failure of MRSA can be life-threatening, and alternative drugs, such as vancomycin, need to be used.

MRSA first appeared as an agent of nosocomial infections—infections that occur after a patient enters a hospital. However, these antibiotic-resistant organisms are no longer contained only in the hospital. Individuals who have not been in a hospital are being infected with MRSA in what are called "community-acquired infections." This is occurring at an epidemic rate in the United States (an incidence of approximately 20 per 100,000 population). Seventy percent of staphylococcal skin infections are now caused by MRSA. Thus, a physician can no longer assume that a patient walking into the office with a staphylococcal infection will respond to methicillin-like drugs such as oxacillin. The doctor must assume that it may be MRSA. As a result, treatment regimens around the country, and the world, are being forced to change. Today, vancomycin and linezolid are typically used to treat staph infections. Antibiotics are discussed in Chapter 27.

Other staphylococcal diseases are caused by toxin-producing strains in which the organism remains localized but the toxin disseminates. We have already mentioned TSST, but there are other toxins. For example, some strains of *S. aureus* produce a toxin called exfoliative toxin, which causes a blistering disease in children called staphylococcal scalded skin syndrome (**Fig. 26.2B**). Exfoliative toxin, like TSST, is a superantigen (described in Section 25.4). In addition, exfoliative toxin cleaves a skin cell adhesion molecule that, when inactivated, results in blisters.

Thought Question

26.1 Does *Staphylococcus aureus* have to disseminate through the circulation to produce the symptoms of scalded skin syndrome (SSS)? Explain why or why not.

Table 26.1 presents other common infections of the skin and soft tissues.

Case History: Necrotizing Fasciitis by "Flesh-Eating" Bacteria

*One weekend in June, Cassi was camping with her three children. She suffered a minor cut on her finger, which she bandaged properly. She also injured the left side of her body while playing sports with her kids. Not thinking much of either of her minor injuries, she went to bed. Two days later, Cassi was extremely ill. Her symptoms included vomiting, diarrhea, and a fever. She was also in severe pain where she had injured her side, and the area had begun to bruise (the skin was not broken). By the next day she could barely get out of bed, and by the end of the night she was having difficulty breathing and could not see. Her side began to leak fluid and blood. Cassi was admitted to the hospital in shock, with no detectable blood pressure. An infectious disease specialist diagnosed the problem as necrotizing fasciitis, and she was rushed into surgery. In an effort to save her life, about 7% of her body surface was removed. Because the large wound infection in her side would need to resolve before a skin graft could be performed to repair it, the hole in Cassi's body was left wide open (**Fig. 26.3B**). After nearly 3 months and several operations, Cassi recovered.*

What kind of organism can cause this type of devastating disease? The disease **necrotizing fasciitis**, also known as "flesh-eating disease," is rare and is often caused by the Gram-positive coccus *Streptococcus pyogenes* (**Fig. 26.3A**), a microbe normally associated with throat infections (pharyngitis). Although sometimes described as a recently emerging infectious disease, necrotizing fasciitis was first discovered in 1783, in France. Its incidence may have risen recently owing to the increased use of nonsteroidal, anti-inflammatory drugs (such as ibuprofen), which increase a person's susceptibility to infection. In this case history, Cassi probably had this organism on her skin when the injury to her side occurred. The injured area probably suffered an invisible microabrasion, providing a good growth

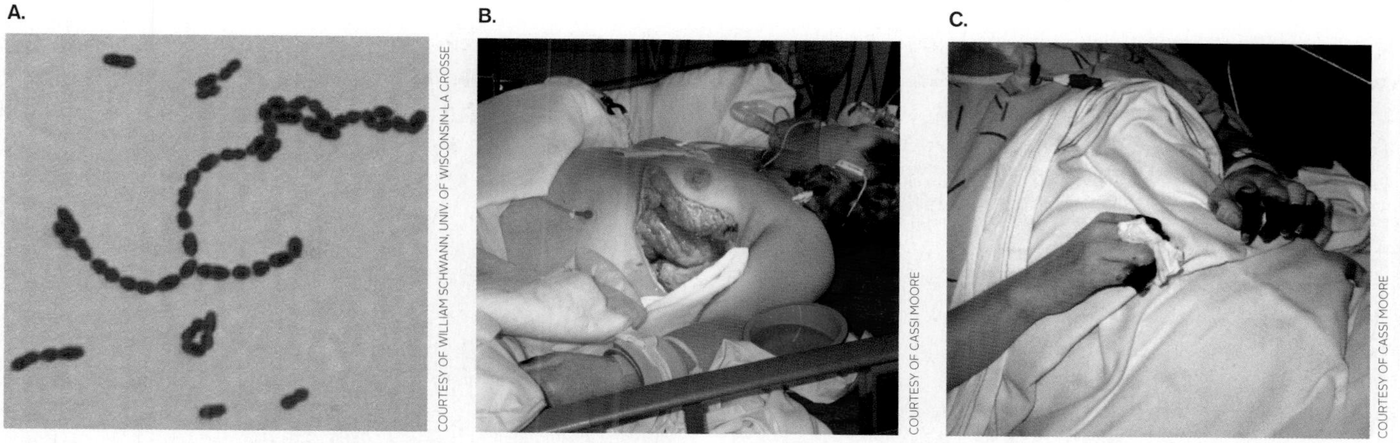
A.
B.
C.
COURTESY OF WILLIAM SCHWANN, UNIV. OF WISCONSIN-LA CROSSE
COURTESY OF CASSI MOORE
COURTESY OF CASSI MOORE

FIGURE 26.3 ■ Flesh-eating *Streptococcus pyogenes*. **A.** Gram stain of *Streptococcus pyogenes*. Each cell approx. 1 μm in diameter. **B.** Flesh removed from a patient in an effort to stop the spread of necrotizing fasciitis. **C.** Gangrene of the fingers caused by the vasopressors (agents that constrict blood vessels) given to maintain the patient's blood pressure.

environment for the organism, leading to the secretion of potent toxins and death of surrounding tissues.

Rapid, aggressive antibiotic treatment is required in these extreme cases, even before the clinical microbiology lab has had time to identify the organism. Therapy can include several antibiotics, such as clindamycin and metronidazole (which act against anaerobes and Gram-positive cocci) and gentamicin (a drug particularly effective against Gram-negative microbes). (Chapter 27 further discusses these and other antibiotics.) Often, however, antibiotic treatment of patients with necrotizing fasciitis is ineffective because of insufficient blood supply to the affected tissues.

A less aggressive but similar skin infection is **cellulitis**. Cellulitis is a non-necrotizing inflammation of the dermis that does not involve the fascia or muscles, but is characterized by localized pain, swelling, tenderness, erythema, and warmth. *S. pyogenes* is the most frequent cause of cellulitis in immunocompetent adults, but a number of other bacteria, including *Staphylococcus aureus*, Gram-negative bacilli, and anaerobes, can also cause this skin infection.

Streptococcus pyogenes wields many different virulence factors, including M protein (pilus-like), superantigen exotoxins, and secreted enzymes such as hyaluronidase and DNase (some of these are described in Chapter 25). The sources of many established virulence factor genes in *S. pyogenes* are the numerous prophages (phage genomes) integrated into the bacterial genome. Prophages constitute approximately 10% of the organism's genome. One study found that a soluble factor produced by human pharyngeal cells can facilitate activation of at least some of these phages and cause horizontal transfer of the associated virulence factors between strains of this pathogen. Prophage activation and phage production are discussed in Chapter 6.

Thought Question

26.2 Why would treatment of some infections require multiple antibiotics?

Weblinks Antibiotic treatment (*see ebook*)

Viral Diseases Causing Skin Rashes

Several viruses can produce skin rashes, although their route of infection is usually through the respiratory tract. Measles, for example (see Section 6.1), is a highly contagious viral infection caused by a paramyxovirus, whose hallmark symptom is skin rash (Fig. 6.3D). Also known as rubeola, the first signs are fever, cough, runny nose, and red eyes occurring 9–12 days after exposure. A few days later, spots (Koplik's spots) appear in the mouth, along with a sore throat. Then a skin rash develops that typically starts on the face and spreads down the body. The virus replicates in the lymph nodes and spreads to the bloodstream (viremia), where it can infect endothelial cells of the blood vessels. The rash occurs when T cells begin to interact with these infected cells.

Although skin rash is the main symptom of measles, infection can also cause respiratory symptoms and complications, including pneumonia, bronchitis, croup, and even a fatal encephalitis in immunocompromised patients. In the United States, measles has been almost completely eliminated by the measles, mumps, and rubella (MMR) multivalent vaccine. Worldwide, however, measles is still a serious problem in places where vaccinations are not routine.

Rubella virus, a togavirus, causes a maculopapular rash known as German measles, or three-day measles. The rash

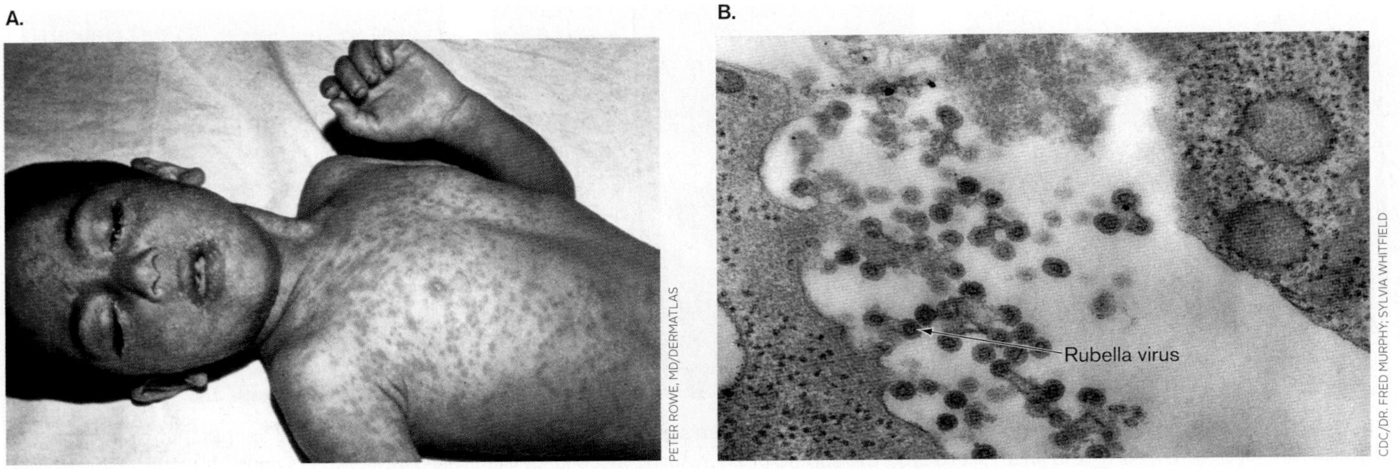
A.
PETER ROWE, MD/DERMATLAS
B.
Rubella virus
CDC/DR. FRED MURPHY; SYLVIA WHITFIELD

FIGURE 26.4 ■ German measles. A. Skin rash caused by rubella virus. **B.** Rubella virus budding from the cell surface to form an enveloped virus particle (approx. 50–70 nm; TEM).

is similar but less red than that of measles (**Fig. 26.4**). German measles is an infection that affects primarily the skin and lymph nodes and is usually transmitted from person to person by aerosolization of respiratory secretions. It is not dangerous in adults or children; the virus can, however, cross the placenta in a pregnant woman and infect her fetus. If the virus crosses the placenta within the first trimester, the result is congenital rubella syndrome, which can cause death or serious congenital defects in the developing fetus.

Other viruses affecting the skin, such as chickenpox, the related disease shingles, smallpox, and human papillomavirus, are included in **Table 26.1** and eTopic 26.2. The vaccination schedule for preventing these diseases is provided in **Table 26.8**.

To Summarize

- *Staphylococcus aureus* and *Streptococcus pyogenes* are common bacterial causes of skin infections. The organisms usually infect through broken skin.
- **Methicillin-resistant** *Staphylococcus aureus* (MRSA) has become an important cause of community-acquired staphylococcal infections.
- **Necrotizing fasciitis** is usually caused by *Streptococcus pyogenes,* but it can be the result of other infections.
- **Infections of the skin** can disseminate via the bloodstream to other sites in the body.
- **Rubeola and rubella viruses** infect through the respiratory tract, but their main manifestation is the production of similar maculopapular skin rashes.

26.3

Respiratory Tract Infections

Lung and upper respiratory tract infections are among the most common diseases of humans. Many different bacteria, viruses, and fungi are well adapted to grow in the lung. Successful lung pathogens come equipped with appropriate attachment mechanisms and countermeasures to avoid various lung defenses (such as alveolar macrophages). One reemerging bacterial pathogen, *Bordetella pertussis,* the cause of whooping cough, inhibits the mucociliary elevator by binding to lung cilia (eTopic 26.3). Although many microbes can infect the lung, most respiratory diseases are of viral origin, and most viral infections (such as the common cold) do not spread beyond the lung. Fortunately, viral diseases by and large are self-limiting and typically resolve within 2 weeks; however, the damage caused by a primary viral infection can lead to secondary infections by bacteria.

Bacterial infections of the lung, whether of primary or secondary etiology, require intervention. Today this means antibiotic therapy. Before the advent of antibiotics, the only recourse was to insert a tube into the patient's back to drain fluid accumulating in the pleural cavity around the lung (a pathological process known as pleural effusion). Unless released, the pressure on the lung will collapse the alveoli and make breathing difficult.

Bacterial infections of the lung that arise secondarily to viral disease occur, in part, because the patient dehydrates. The resulting increase in mucus viscosity hampers the mucociliary elevator (described in Section 23.1), as will growth of the bacterium itself. Because a compromised mucociliary elevator makes it difficult to expel the microbe, the patient's susceptibility to secondary bacterial infection increases. Hence, cold sufferers are advised to drink plenty of fluids, which help decrease the viscosity of mucus and consequently improve mucociliary elevator function.

Case History: Bacterial Pneumonia

*In March, an 80-year-old resident of a New Jersey nursing home had a fever accompanied by a productive cough with brown sputum (mucous secretions of the lung that can be coughed up). He reported to the attending physician that he had pain on the right side of his chest and suffered from night sweats. Blood tests revealed that his white blood cell (WBC) count was 14,000/µl (normal is 5,000–10,000/µl) with a makeup of 77% segmented forms (polymorphonuclear leukocytes, PMNs) and 20% bands (immature PMNs). The chest radiograph revealed a right upper lobe infiltrate with cavity formation (**Fig. 26.5A**). From this information, the clinician diagnosed pneumonia. Microscopic examination of the patient's sputum revealed Gram-positive cocci in pairs and short chains surrounded by a capsule (**Fig. 26.5B**). Bacteriological culture of his sputum and blood yielded* Streptococcus pneumoniae.

Pneumonia is a disease that can be caused by many different microbes (**Table 26.2**). The pneumococcus *Streptococcus pneumoniae* accounts for about 25% of community-acquired cases of pneumonia, but pneumococcal pneumonia occurs mostly among the elderly and immunocompromised, including smokers, diabetics, and alcoholics. A breakdown of pneumonia cases by causative organism is shown in **Figure 26.5C**.

The noses and throats of 30%–70% of a given population can contain *S. pneumoniae*. The microbe can be spread from person to person by sneezing, coughing, or other close, personal contact. Pneumococcal pneumonia may begin suddenly, with a severe shaking chill usually followed by high fever, cough, shortness of breath, rapid breathing, and chest pains.

Pneumococcal lung infection begins when the pneumococcus is aspirated into the lung. Once in the lung, the microbe grows in the nutrient-rich edema fluid of the alveolar spaces. Neutrophils and alveolar macrophages then arrive to try to stop the infection. They are called into the area from the circulation by chemoattractant chemokines released by damaged alveolar cells. The thick polysaccharide capsule of the pneumococcus, however, makes

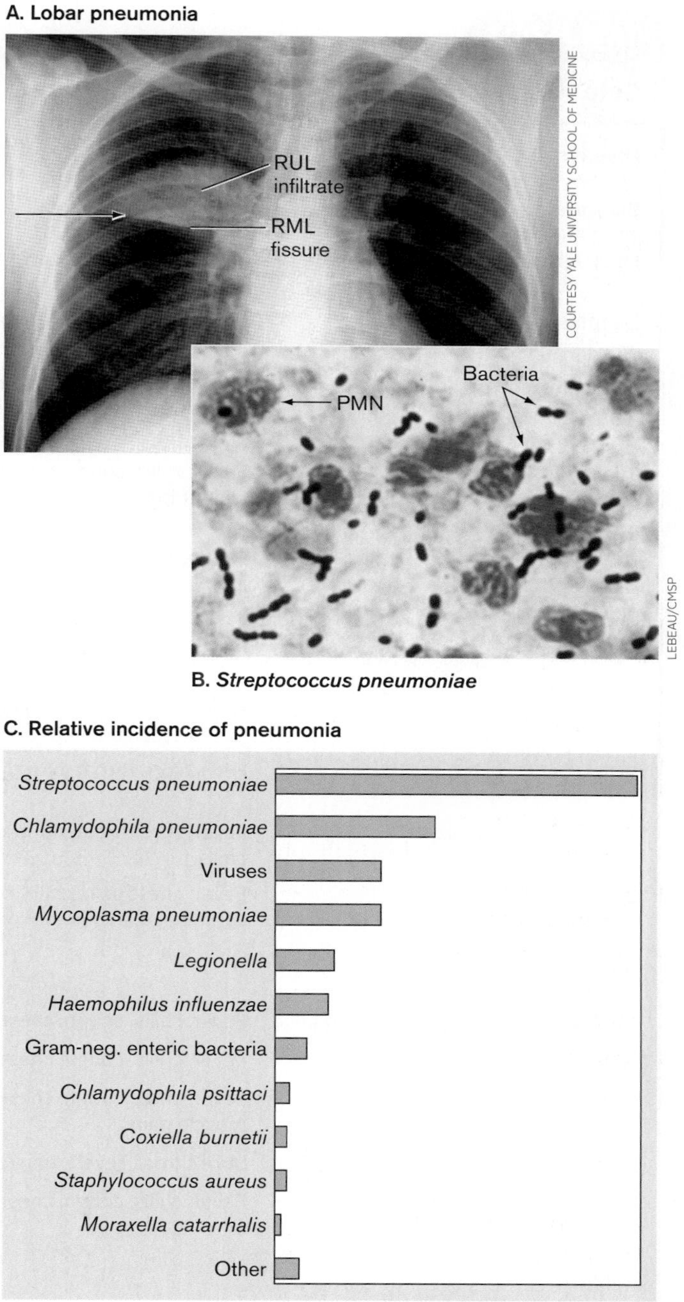
A. Lobar pneumonia
RUL infiltrate
RML fissure
COURTESY YALE UNIVERSITY SCHOOL OF MEDICINE
PMN
Bacteria
LEBEAU/CMSP
B. Streptococcus pneumoniae
C. Relative incidence of pneumonia
Streptococcus pneumoniae
Chlamydophila pneumoniae
Viruses
Mycoplasma pneumoniae
Legionella
Haemophilus influenzae
Gram-neg. enteric bacteria
Chlamydophila psittaci
Coxiella burnetii
Staphylococcus aureus
Moraxella catarrhalis
Other

FIGURE 26.5 ■ Pneumonia caused by *Streptococcus pneumoniae*. **A.** X-ray view of a patient with lobar pneumonia. Infiltrate in the right upper lobe (RUL) is caused by *S. pneumoniae.* The sharp lower border represents the upper boundary of the right middle lobe (RML) fissure (arrow). **B.** Micrograph of *S. pneumoniae.* Sputum sample showing numerous PMNs and extracellular diplococci in pairs and short chains. Bacteria range from 0.5 to 1.2 µm in diameter. **C.** Relative incidence of pneumonia caused by various microorganisms.

TABLE 26.2

Selected microbes that cause respiratory tract diseases.

Disease	Symptoms	Etiological agent
Bacterial		
Inhalation anthrax	Fever, muscle aches, hypotension, respiratory failure	*Bacillus anthracis* (G+ rod)
Diphtheria	Tracheal pseudomembrane	*Corynebacterium diphtheriae* (G+ rod)
Whooping cough	Fever, runny nose, sneezing, violent cough followed by inhalation "whoop"	*Bordetella pertussis* (G– coccobacillus) (see **eTopic 26.3**)
Tuberculosis	Fever, chills, cough, bloody sputum, fatigue, weight loss	*Mycobacterium tuberculosis* (acid-fast bacillus)
Pneumonia	Infects cystic fibrosis patients; fever, chills, cough, chest pain	*Pseudomonas aeruginosa* (G– rod)
	Fever, chills, cough, chest pain	*Streptococcus pneumoniae* (G+ diplococcus)
	Fever, sore throat, chest pain, runny nose (also associated with cardiovascular disease)	*Chlamydophila pneumoniae*
Psittacosis	Fever, sore throat, chest pain, runny nose	*Chlamydophila psittaci*
"Walking pneumonia"	Fever, sore throat, nonproductive cough, chills	*Mycoplasma pneumoniae* (wall-less microbe)
Legionnaires' disease (legionellosis)	Fever, chest pain, chills, cough, muscle pain, vomiting, diarrhea	*Legionella pneumophila* (G– rod)
Viral		
CMV disease	Fever, chills, cough, chest pain	Cytomegalovirus (CMV); dsDNA
RSV disease	Fever, chills, cough, chest pain	Respiratory syncytial virus (RSV); ssRNA (–)
Influenza	Fever, chills, cough, chest pain, sore throat, muscle pain	Influenza and parainfluenza viruses; ssRNA (–), segmented
	Sore throat, fever, runny nose	Adenovirus; dsDNA
Severe acute respiratory syndrome (SARS)	Fever, chills, cough, chest pain	SARS virus; ssRNA (+)
Fungal		
Aspergillosis	Affects lungs and sinuses; fever, chills, breathing difficulty	*Aspergillus* spp.
Histoplasmosis	Flu-like; pulmonary infiltrate	*Histoplasma capsulatum*
Coccidioidomycosis	Flu-like; pulmonary infiltrate	*Coccidioides immitis*
Blastomycosis	Chest pain, cough, skin lesion, pulmonary infiltrate	*Blastomyces dermatitidis*
Pneumocystosis	Infects AIDS patients; cough, fever, weight loss	*Pneumocystis jirovecii* (formerly *P. carinii*)

[a]BCG = Bacille Calmette-Guérin (a weakened strain of the bovine tuberculosis strain).

Virulence properties	Source/Transmission	Treatment	Vaccine
Peptide capsule; PA, LF, and EF toxins	Soil/Airborne	Ciprofloxacin	+, military
Diphtheria toxin	Humans/Airborne; localizes to nasopharynx	Penicillin	+
Tracheal cytotoxin, adenylate cyclase toxin, filamentous hemagglutinin (adhesin)	Humans/Respiratory droplets	Erythromycin	+
Cord factor, wax D, intracellular growth	Humans/Respiratory droplets	Combination therapy (rifampin, isoniazid, ethambutol, pyrazinamide)	+, BCG[a]
Exotoxin A, phospholipase C, exopolysaccharide, others	Water, soil/Inhalation or via spread via bloodstream from separate body site	Quinolones, aminoglycosides	
Capsule, pneumolysin	Humans/Inhalation	Macrolides, quinolones, ceftriaxone	Multivalent from capsule antigens
Obligate intracellular growth; prevents phagolysosome fusion	Humans/Person-to-person	Quinolones	
Obligate intracellular growth; prevents phagolysosome fusion	Bird droppings, dust/Inhalation	Tetracycline, erythromycin	
Adhesin tip	Humans/Inhalation, person-to-person	Erythromycin and other macrolides	
Intracellular growth, hemolysin, cytotoxin, protease	Water towers/Inhalation	Erythromycin	
	Humans/Saliva, tears, breast milk	Ganciclovir	Experimental
	Humans/Respiratory droplets	Treat symptoms; ribavirin	
	Humans (rarely from pigs, birds)/ Respiratory droplets	Zanamivir (Relenza)	+
	Humans/Respiratory droplets	Treat symptoms	+, military
	Humans/Respiratory droplets		
	Environment/Inhalation	Amphotericin B, voriconazole	
	Bird, chicken, bat droppings/ Inhalation	Amphotericin B, itraconazole	
	Environment/Inhalation	Amphotericin B, fluconazole	
	Environment/Inhalation	Amphotericin B, itraconazole	
	Environment/Inhalation	Trimethoprim sulfamethoxazole	

phagocytosis very difficult (see Section 25.6). In an otherwise healthy adult, pneumococcal pneumonia usually involves one lobe of the lungs; thus, it is sometimes called lobar pneumonia. The infiltration of PMNs and fluid lead to the typical radiological findings of diffuse, cloudy areas. In contrast, infants, young children, and elderly people more commonly develop an infection in other parts of the lungs, such as around the air vessels (bronchi), causing bronchopneumonia.

The white blood cell count in the case history is telling. The patient had an elevated WBC count (normal is 5,000–10,000/µl) and elevated band cells (normal is 0%–8%). These increases are indicative of a bacterial, not viral, infection. Neutrophils (PMNs), the front-line combatants against infection, rise in response to bacterial infections and are first released from bone marrow as immature band cells, whose presence is a sure sign of bacterial infection.

Several outbreaks of pneumococcal pneumonia have occurred over recent years in nursing homes, where numerous residents have been affected. These incidents underscore the importance of elderly people receiving the pneumococcal polysaccharide vaccine (PPSV) as a hedge against infection. While there are over 80 antigenic types of pneumococcal capsular polysaccharides, the injected vaccine contains only the 23 types that are most often associated with disease (Section 26.9). A vaccine that is formulated to respond to multiple antigens is called a "multivalent vaccine." The resulting immune response will protect vaccinated individuals against infection by those antigenic types. The vaccine is recommended for individuals over 65, as well as for those who are immunocompromised. The patient in this case history failed to receive the vaccine.

In addition to causing serious infections of the lungs (pneumonia), *S. pneumoniae* can invade the bloodstream (bacteremia) and the covering of the brain (meningitis). The death rates for these infections are about one out of every 20 who get pneumococcal pneumonia, about four out of 20 who get bacteremia, and six out of 20 who get meningitis. Individuals with special health problems, such as liver disease, AIDS (caused by HIV), or organ transplants, are even more likely to die from the disease, because of their compromised immune systems.

An emerging infectious disease problem throughout the United States and the world is the increasing resistance of *S. pneumoniae* to antibiotics. At least 30% of the strains isolated are already resistant to penicillin, the former drug of choice for treating the disease. Chapter 27 discusses why antibiotic resistance is on the rise for this and other microbes.

Case History: Disseminated Disease from a Fungal Lung Infection

*A 35-year-old male boxer named Tyrrell, recently admitted to a Maryland hospital, had been in good health until 6 months earlier, when he developed a chronic cough that produced blood-tinged white sputum. He also experienced flu-like symptoms, a decrease in appetite, and weight loss. One month prior to admission, he became so short of breath that he could no longer continue boxing. At that time, an X-ray taken in the emergency room showed right upper lobe infiltrate, indicating pneumonia (**Fig. 26.6A**). A tuberculosis skin test was negative. Tyrrell was given a prescription for the antibiotic azithromycin (a macrolide antibiotic commonly used to treat bacterial infections of the respiratory tract) and discharged. Despite antibiotic treatment, the cough never diminished, and he developed several painless subcutaneous nodules. The largest nodule was on his left leg, contained pus, caused pain, and eventually hindered walking (**Fig. 26.6B**). This symptom prompted his current admission to the hospital. The patient's history revealed that he installed home insulation for a living and had not traveled outside the area for the previous year. He complained of fevers, chills, night sweats, and a 10-kg (22-lb) weight loss. His right tibia was tender to the touch, indicating bone involvement. He denied having prior pneumonia, sinus infection, arthritis, hematuria (blood in the urine), numbness, or muscle weakness. He had no history of intravenous drug use and had been in a monogamous relationship for 4 years. His white cell count was 10,500/µl with a normal differential of PMNs and band cells.*

In this case, the expression of frankly purulent material (pus) from the left-leg nodule suggested that this was an infectious process. The infection in this patient probably started in the lung (clued by the cough), after which the organism disseminated throughout the body via the bloodstream. Fungus is a probable cause, given the chronic nature of the patient's symptoms. An alternative possibility would be tuberculosis, caused by the bacterium *Mycobacterium tuberculosis. M. tuberculosis* also causes a chronic lung infection and might have been suspected, except for the negative TB skin test. The tuberculin skin test involves injecting a small amount of mycobacterial antigen called PPD (purified protein derivative) under the skin of the lower arm. A person who has been infected with *M. tuberculosis* will exhibit a localized delayed-type hypersensitivity reaction at the site of injection, although this does not equate to currently active disease.

The most likely fungal causes of infection in this case history are the endemic mycoses, such as histoplasmosis,

A. Pneumonia infiltrate

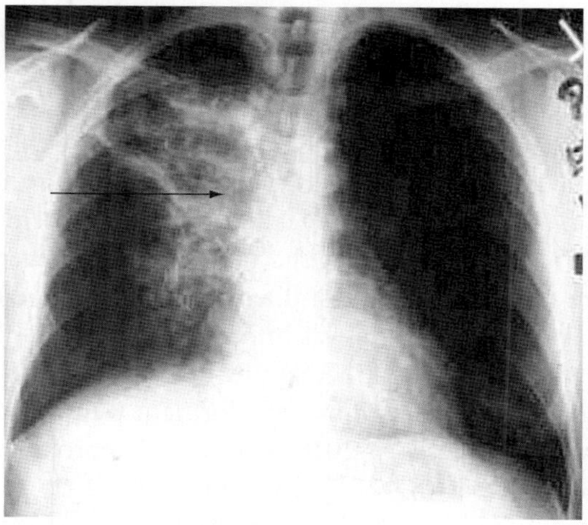

B. Leg lesion

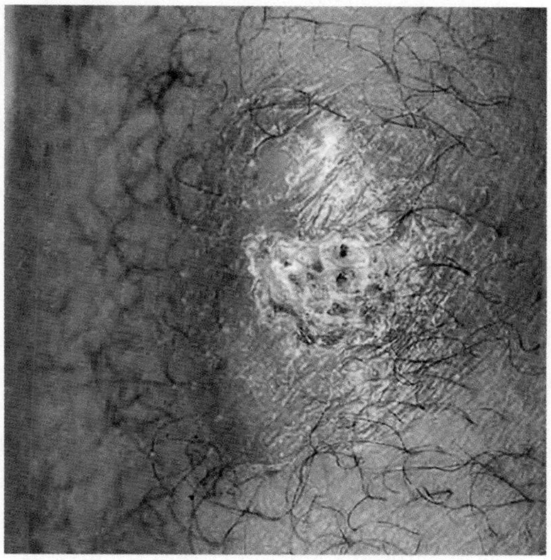

C. Colony of *Blastomyces dermatitidis*

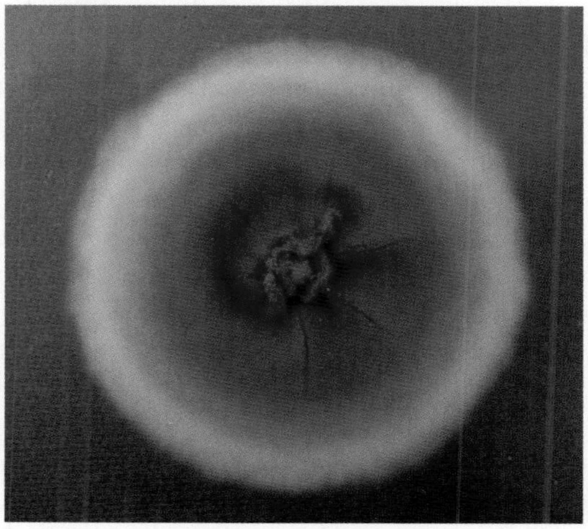

FIGURE 26.6 ▪ Pneumonia and metastatic disease caused by *Blastomyces dermatitidis*. **A.** Diffuse infiltrate in the right lung (arrow). **B.** Metastatic leg lesion at the tibia. **C.** Fungal colony of *B. dermatitidis*.

blastomycosis, and coccidioidomycosis. This patient had never traveled to the western United States, where coccidioidomycosis is endemic, so exposure to *Coccidioides* was ruled out. Histoplasmosis most commonly presents as a flu-like pulmonary illness, with erythema nodosum (tender bumps on skin) and arthritis (swollen joints) or arthralgia (joint pain), none of which the patient had. Blastomycosis can disseminate to the lung, skin, bone, and genitourinary tract, consistent with the pattern of organ involvement seen in this patient. *Cryptococcus,* an encapsulated yeast, is not the likely cause, since it typically requires an immunocompromised host to cause disease—not the case in this instance. (*Cryptococcus,* which causes cryptococcosis, is an opportunistic pathogen that commonly infects AIDS patients.) The most prevalent clinical form of cryptococcosis is meningoencephalitis, although disease can also involve the skin, lungs, prostate gland, urinary tract, eyes, myocardium, bones, and joints.

Amphotericin B, an antifungal agent (discussed in Section 27.8), was finally given to this patient. His fever lowered almost immediately, and the skin nodules diminished. After 2 weeks, a fungus was found in the cultures of the nodule biopsy, bronchoalveolar lavage (washes), and urine (**Fig. 26.6C**). This fungus was identified by a DNA probe as *Blastomyces dermatitidis,* confirming the diagnosis of blastomycosis.

B. dermatitidis is a dimorphic fungus that resides in the soil of the Ohio and Mississippi river valleys and the southeastern United States. The portal of entry is the respiratory tract, and infection is usually associated with occupational and recreational activities in wooded areas along waterways, where there is moist soil with a high content of organic matter and spores. The incubation period ranges from 21 to 106 days. This patient most likely inhaled conidia (fungal spores) from the soil while crawling underneath houses installing insulation. The physician learned that the patient used only a T-shirt to cover his mouth and nose—not an effective method of keeping spores from entering the respiratory tract. He should have worn a respirator.

Several critical features of this case help differentiate it from the preceding case of pneumococcal pneumonia. First, the initial macrolide antibiotic, azithromycin, should have killed most bacterial sources of infection. Second, the X-ray finding of diffuse infiltrate is more indicative of fungal lung infection than bacterial infection, which in a patient of this age would likely be confined to one lobe. The patient was young and in good health prior to the infection, making it unlikely to be pneumococcal pneumonia. The blood count was also a clue. Fungal infections do not usually cause an increase in WBCs or an increase in band cells. Finally, the

metastatic lesions (infectious lesions that develop at a secondary site away from the initial site of infection) on the leg were in no way consistent with *S. pneumoniae*. They arise when the organism moves through the bloodstream from the primary site of infection to another body site, where it can begin to grow. Many infectious diseases start out as a localized infection but end up disseminating throughout the body to cause metastatic lesions.

Note: The term "metastasis" means the spread of disease from one organ to another, noncontiguous organ. Only microbial infections and malignant cancer cells can metastasize.

Tuberculosis as a Reemerging Disease

Tuberculosis, caused by the acid-fast bacillus *Mycobacterium tuberculosis* (**Fig. 26.7** inset), was once considered of passing historical significance to physicians practicing in the developed world. In 1985, however, owing primarily to the newly recognized HIV epidemic and a growing indigent population, TB resurfaced, especially in inner-city hospitals. In 1991, highly virulent multidrug-resistant (MDR) strains of *M. tuberculosis* were reported. These strains not only produced fulminant (rapid-onset) and fatal disease among patients infected with HIV (the time between TB exposure to death is 2–7 months), but also proved highly infectious. Tuberculin skin test conversion rates of up to 50% were reported in exposed health care workers (the conversion rate today is less than 1%). A positive tuberculin skin test is seen as a delayed-type hypersensitivity reaction to *M. tuberculosis* proteins [called purified protein derivative (PPD)] injected under the skin. Note, however, that a positive tuberculin skin test does not signify active disease, but only that the person was infected at one time. The bacterium may have been killed by the immune system without having caused disease or may lie dormant waiting to reactivate. MDR-TB cases in 2010 represented approximately 1% of new TB cases (8,422). More information on mycobacteria structure can be found in Section 18.3 and on pathogenesis in Section 25.6.

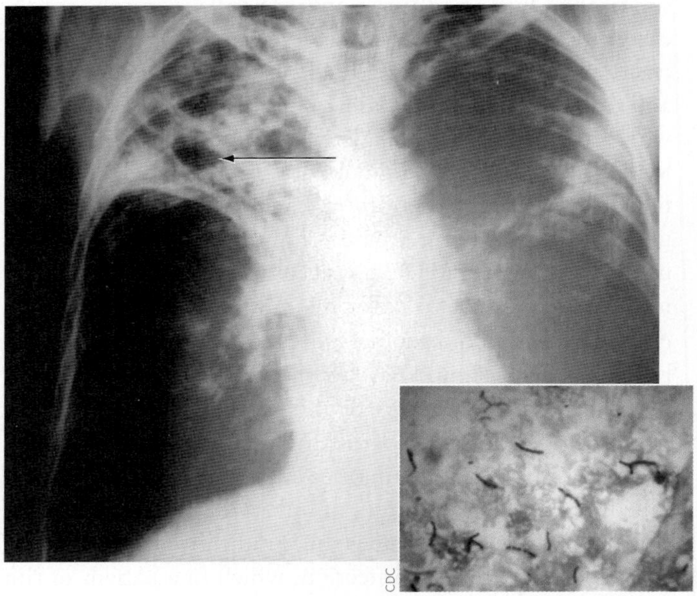
CDC

FIGURE 26.7 ▪ Calcified Ghon complex of tuberculosis. The arrow points to a complex in a patient's right upper lobe. Note the difference in appearance compared to Figure 26.6A. **Inset:** Acid-fast stain of *Mycobacterium tuberculosis.* (Cells ca. 2–3 μm length)

M. tuberculosis primarily causes a respiratory infection (pulmonary TB), but it can disseminate through the bloodstream to produce abscesses in many different organ systems (extrapulmonary TB). Extrapulmonary TB is sometimes called miliary TB because the size of the infected nodules (1–5 mm), called tubercles, approximates the size of millet seeds. Tubercles in the lung are filled with *M. tuberculosis* cells that drain into bronchial tubes and the upper respiratory tract. Aerosolization of respiratory secretions (coughing) directly spreads the tubercle bacilli from someone with active disease to an uninfected person (person-to-person transmission, no animal reservoir).

Once inhaled into the lung, there are three possible outcomes for the bacilli: They can die, they can produce primary disease, or they can become latent. In all three cases the bacteria are initially phagocytosed by alveolar macrophages and, if not killed, they survive ensconced within modified phagolysosomes. The bacilli can multiply in these vacuoles, kill the macrophage (via induced apoptosis), and then be released to infect other macrophages. Four to six weeks postinfection, an infected person can develop active primary disease marked by a productive cough that generates sputum. The patient also experiences fever, night sweats, and weight loss. Patients with active disease are contagious and produce a delayed-type hypersensitivity that makes them tuberculin-positive.

An alternative to primary disease is an asymptomatic latent tuberculosis infection (LTBI). Here the tubercle bacilli remain hidden in macrophages and may even become dormant. The delayed-type hypersensitivity response to the organism builds small, hard tubercles around these cells. Over time, the tubercles develop into caseous lesions that have a cheese-like consistency and can calcify into the hardened Ghon complexes seen on typical X-rays (**Fig. 26.7**).

The bacteria can remain dormant in these tubercles for many years. Even primary tuberculosis can resolve and become latent tuberculosis, unless the organisms were killed with antibiotics.

In a patient with LTBI, *M. tuberculosis* can sometimes, years after initial infection, overcome the imposed confinement of the immune system. The bacteria begin to multiply and cause **secondary**, or **reactivation, tuberculosis**. Secondary TB commonly occurs in immunocompromised people (such as HIV patients). The symptoms of secondary TB are more serious than those of primary TB and include severe coughing, greenish or bloody sputum, low-grade fever, night sweats, and weight loss. The gradual wasting of the body led to the older name for tuberculosis: "consumption."

Treatment of active TB is aggressive and involves a four-drug regimen, including isoniazid, rifampin, pyrazinamide, and ethambutol, given over a course of several months. MDR strains are defined as being resistant to two or more first-line drugs. Because of this resistance, MDR strains are treated with a nine-drug regimen. Even more dangerous are extensively drug-resistant tuberculosis (XDR-TB) strains, which are resistant to three or more second-line drugs in addition to two or more first-line drugs. These strains are almost untreatable.

Thought Question

26.3 Explain why patient noncompliance (failure to take drugs as directed) is thought to have led to XDR-TB.

Viral Diseases of the Lung

Numerous viruses can cause lung infections (see **Table 26.2**). Influenza virus and rhinovirus are described in Chapters 6 and 11, and compared in eTopic 26.4. The recently emerged infection SARS (severe acute respiratory syndrome) is discussed in Chapter 28. However, an important viral lung infection not discussed elsewhere is respiratory syncytial disease, caused by respiratory syncytial virus (RSV). A negative-sense, single-stranded-RNA, enveloped virus, RSV is the most common cause of bronchiolitis and pneumonia among infants and children under 1 year of age. Illness begins most frequently with fever, runny nose, cough, and sometimes wheezing. RSV is spread from respiratory secretions through close contact with infected persons or by contact with contaminated surfaces or objects. Infection can occur when the virus contacts mucous membranes of the eyes, mouth, or nose and possibly through the inhalation of droplets generated by a sneeze or cough. Unlike rubella or rubeola, which infect the respiratory tract and disseminate though the body, RSV remains localized in the lung.

The majority of children hospitalized for RSV infection are under 6 months of age. RSV can cause repeated infections throughout life, usually associated with moderate to severe cold-like symptoms. Severe lower respiratory tract disease may occur at any age, especially among the elderly or people with compromised cardiac, pulmonary, or immune systems. As yet, a vaccine to control this disease is not available.

Table 26.2 presents many other bacterial, fungal, and viral microbes that can cause respiratory tract infection. Be aware that very different diseases can produce similar symptoms. For instance, people constantly confuse influenza (the flu) with the common cold. Symptomatically, they may start out similarly, but there are telling differences. Influenza is characterized by fever, myalgia (muscle aches), pharyngitis (sore throat), and headache (viral infection discussed in Chapter 11). A runny nose is not one of the symptoms. The common cold, however, manifests as a runny nose, nasal congestion, sneezing, and throat irritation. No myalgia. The observant clinician will note the difference.

To Summarize

- **The mucociliary elevator** is a primary defense mechanism used by the lung to avoid infection.
- **Pneumonia** is caused by many microorganisms.
- **An elevated white cell count** in blood is an indicator of bacterial infection.
- **Pneumococcal vaccine** should be administered to the elderly because they are often immunocompromised.
- **Fungal agents** commonly cause long-term, chronic infections.
- **Tuberculosis** is an ancient yet reemerging bacterial disease with an increasing mortality rate caused by the development of multidrug-resistant strains, the susceptibility of HIV patients, and an increasing indigent population.
- **Localized infections in the lung can disseminate** via the bloodstream to form metastatic lesions at other body sites.
- **Respiratory syncytial virus** is one of several viruses that can cause lung disease, but it rarely disseminates.

26.4

Gastrointestinal Tract Infections

Nearly everyone has experienced diarrhea—a condition characterized by frequent loose bowel movements accompanied by abdominal cramps. Hundreds of millions of cases occur each year in the United States and are a major cause of death in developing countries. The loose stools usually result from inflammation (generally called gastroenteritis) due to viral growth, bacterial growth, or toxin production, causing large amounts of water and electrolytes to leave the intestinal cells and enter the intestinal lumen. As a result, the patient not only suffers diarrhea, but can become dangerously dehydrated. As with respiratory tract infections, most diarrheal disease is viral in origin, with rotavirus being the primary culprit. Among the bacteria, the Gram-negative rod *Salmonella enterica* serovar Typhimurium and the spiral-shaped or curved bacillus *Campylobacter* are the most frequent causes of self-limiting diarrheal disease.

The main symptoms of gastroenteritis, whether bacterial or viral, are watery diarrhea and vomiting. A more severe form of gastroenteritis is called dysentery, whose symptoms include abdominal pain, a persistent desire to empty the bowels, and diarrhea with passage of blood or mucus. Bacterial causes of dysentery include several *Shigella* species and some strains of *E. coli*. An ameba, *Entamoeba histolytica,* can also cause a form of the disease, called amebic dysentery. Features of bacterial dysentery are discussed in more detail later in this section.

Remarkably, *Staphylococcus aureus* causes gastrointestinal disease without ever producing infection. We have all heard of the local church picnic where scores of people become violently ill within hours of eating unrefrigerated potato salad. *S. aureus* is the usual cause of these disasters. Not all strains of *S. aureus* cause food poisoning, but certain strains can secrete enterotoxins into tainted foods such as pies, turkey dressing, or potato salad. After ingestion, the toxin travels to the intestine, where it enters the bloodstream and stimulates nerves leading to the vomit center in the brain. Because the toxin is preformed, symptoms occur quickly after ingestion. Within 2–6 hours, the poisoned patient will begin vomiting and may also experience diarrhea. The disease, though violent, is not life-threatening and usually resolves spontaneously within 24–48 hours. In contrast, diarrhea caused by infectious agents, such as *Salmonella enterica,* which must first grow in the victim, do not occur until 12–24 hours after ingestion, sometimes longer. A clinician noting quick onset of symptoms in a patient will immediately suspect staphylococcal food poisoning. Obviously, antibiotic treatment is not needed for staph food poisoning, but it may be indicated for other gastrointestinal infections.

It is important to recognize that the most important treatment for diarrhea is **rehydration therapy**. Vomiting and diarrhea can cause huge losses in water volume. Because water leaving tissues causes an osmotic imbalance, salts also leave the cells to try to re-establish that balance. The problem is that the resulting electrolyte imbalance can severely affect cardiovascular, respiratory, and renal systems. Consequently, replacing body fluids orally or by IV is imperative to prevent death. Rehydration solutions, therefore, must contain glucose as well as sodium and potassium salts in proper balance (for example, Pedialyte).

Antibiotics Are Often Inappropriate When Treating Diarrhea

Although antibiotic treatment of infectious gastroenteritis seems intuitive, it is rarely used. Most gastrointestinal infections are viral (for example, norovirus and rotavirus), so antibiotics are ineffective. Likewise, gastroenteritis caused by bacteria usually resolves spontaneously, without antibiotic treatment. However, severe systemic disease stemming from gastroenteritis can develop, often in the young or elderly. Diseases such as typhoid fever (*Salmonella* Typhi) or bacillary dysentery (*Shigella dysenteriae*) respond well to antibiotics.

In some cases, antibiotic treatment can actually trigger gastrointestinal disease. For example, the antibiotic clindamycin can kill most normal intestinal bacteria except the naturally resistant Gram-positive anaerobe *Clostridium difficile,* the causative agent of pseudomembranous enterocolitis. Unrestrained by microbial competition, *C. difficile* will grow in the intestine and produce specific toxins that can damage intestinal cells. The organism's growth leads to inflammation and the formation of pseudomembrane structures along the intestinal wall (refer to Fig. 23.8). Because they block the intestinal mucosa, pseudomembranes cause malabsorption of nutrients and water, which results in diarrhea. As the pseudomembrane enlarges, it begins to slough off and pass into the stool. Diagnosis of this disease involves PCR identification of the organism or immunological identification of the toxin in fecal samples.

Some patients suffer with recurrent *C. difficile* infections. It is not clear why, but one suggestion is that spores lodged in colon folds may escape clearance by peristalsis. Another hypothesis is that patients with recurrent disease have an impaired response to the *C. difficile* toxins. How do you prevent recurrences? In some instances, after vegetative *C. difficile* has been killed by antibiotic treatment, a procedure known as a fecal transplant (Section 23.1) can be used to restore a healthy gastrointestinal microbiota that prevents the recurrence of *C. difficile* disease.

How does the intestine restore its bacterial population after decimation by diarrhea or antibiotic treatment? Repopulation occurs, in part, by reingesting microbes from food. But recent research suggests that the much maligned and trivialized appendix is, among other things, an important reservoir of gut microbes that can seed the intestine and reestablish the microbiota. Once properly reestablished, gut microbiota are capable of fending off pathogens such as *C. difficile.* As evidence, researchers found that patients with appendectomies were more than twice as likely to develop repetitive infections with *C. difficile.*

Case History: Diarrhea and Dysentery

In April, a 6-year-old girl from Montgomery County, Pennsylvania, arrived at the ER with bloody diarrhea, a temperature of 39°C (102.2°F), abdominal cramping, and vomiting. Hospital admission occurred 5 days after a kindergarten field trip to the local dairy farm. When the parents were questioned about the child's activities during the trip, they said the child had purchased a snack while at the farm. Upon laboratory analysis, a fecal smear was positive for leukocytes, and isolation of organisms confirmed the presence of Gram-negative rods that produced Shiga toxins 1 and 2. In subsequent testing by pulsed-field gel electrophoresis, the isolate was indistinguishable from E. coli *O157:H7. By this time, the child had developed additional problems. Symptoms included puffy face and hands, as well as neurological abnormalities. Initial examination suggested renal failure, and laboratory analyses supported this diagnosis with thrombocytopenia (reduced blood platelet count) and hemolytic uremic syndrome (HUS; renal failure). The child was treated by fluid and electrolyte replacement (intravenous). Antibiotics were not administered.*

In this case history, the presence of leukocytes in a fecal smear is a sign that the intestinal pathogen may have invaded the epithelial mucosa of the intestine (or severely damaged it). Breaching this barrier sends out a chemical call to neutrophils, which then enter the area and, in an effort to kill the pathogen, also damage the intestinal cells. *Shigella dysenteriae, Salmonella enterica,* and enteroinvasive *E. coli* (EIEC) actually invade enterocytes and are considered intracellular pathogens. Enterohemorrhagic *E. coli* (EHEC), which also produces leukocytes and blood in stools, is not an intracellular parasite (it is not invasive), but causes damaging attachment and effacing lesions, described in Section 25.5, which destroy the mucosal epithelium. The resulting inflammation, in conjunction with damage caused by the Shiga toxin it produces, leads to blood and white cells in the stool. *E. coli* O157:H7, the etiological agent in the case history, is a common serotype of EHEC.

There are at least six different classes of pathogenic *E. coli* that differ in their repertoire of pathogenicity islands, plasmids, and virulence factors. They include EIEC and EHEC, already mentioned; enterotoxigenic *E. coli* (ETEC) and uropathogenic *E. coli* (UPEC), both described in Chapter 25; as well as enteropathogenic *E. coli* (EPEC) and enteroaggregative *E. coli* (EAEC). All but UPEC cause gastrointestinal disease. (A new diarrheagenic strain is described in **Special Topic 26.1**.) To tell these strains apart, each group has telltale O and H antigens that can be identified using serology.

Note: "O antigen" is part of the bacterium's LPS, while "H antigen" is flagellar protein. Thus, "O157:H7" denotes the specific version of LPS (O157) and flagellar protein (H7) found on *E. coli* O157:H7. Other pathogenic strains of *E. coli* have different O and H antigens.

Shiga toxin. *Shigella* and EHEC, the agent in the preceding case history, both produce toxins, called Shiga toxins 1 and 2, that are encoded by genes of bacteriophage genomes embedded in the bacterial chromosome. These toxins inhibit host protein synthesis and, in the process, damage endothelial cells in the intestine, kidney, and brain. Shiga toxin–induced death of vascular endothelial cells in the intestine causes the breakdown of blood vessel linings, followed by hemorrhage that manifests as bloody diarrhea.

Endothelial damage also triggers the formation of platelet-fibrin microthrombi (clots) that occlude blood vessels in the various organs, leading to two major syndromes: hemolytic uremic syndrome (HUS) and thrombotic thrombocytopenic purpura (TTP). HUS occurs when the microthrombi are limited to the kidney. The microclots clog the tiny blood vessels in this organ and cause decreased urine output, ultimately leading to kidney failure and death. In TTP, the clots occur throughout the circulation, causing reddish skin hemorrhages called petechiae and purpuras. Neurological symptoms (for example, confusion, severe headaches, and possibly coma) then arise from microhemorrhages in the brain. The hemorrhaging occurs because platelets needed for normal clotting have been removed from the circulation as they form the microthrombi. The decreased number of platelets is called thrombocytopenia.

The toxins, which are absorbed through the intestine and disseminated via the bloodstream, have five B subunits

Special Topic 26.1: Sprouts and an Emerging *Escherichia coli*

Starting in May 2011, people in Germany began dying from hemolytic uremic syndrome (HUS) (**Fig. 1**). As of June 3,228 cases and 35 deaths had been reported. By then a massive epidemiological hunt was under way for the cause of the disease and its source. The organism isolated in each case was a form of *E. coli* rarely seen before. Its LPS and flagellar serotype was O104:H4. The organism secreted type 2 Shiga toxin, which contributed to HUS but, strangely, did not produce the attaching and effacing proteins often associated with enterohemorrhagic *E. coli* (for instance, strain O157:H7). Unfortunately, this unique pathovar also harbored a plasmid that conveyed resistance to all penicillins, cephalosporins, and co-trimoxazole. Another feature was the presence of enteroaggregative adherence fimbriae (*aaf*), which allow *E. coli* to adhere like stacked bricks to host cells. Again, AAF factor is not usually seen in Shiga toxin–producing *E. coli* (called STEC), but it is a defining feature of enteroaggregative *E. coli* (EAEC).

One more odd thing about this HUS outbreak was that a majority of the patients (90%) were adults rather than children, the usual victims of EHEC or other STEC strains. All of this evidence plus extensive DNA sequence analysis indicate that the deadly O104:H4 strain is a genetic hybrid between enterohemorrhagic and enteroaggregative strains of *E. coli.* As a result, it is officially referred to as EAHEC O104:H4. What series of unfortunate events produced this "chimeric" pathogen? The *stx-2* gene is part of a prophage that can horizontally transfer between strains of *E. coli,* but the trafficking of the *aaf* and drug resistance characteristics is less apparent and under investigation.

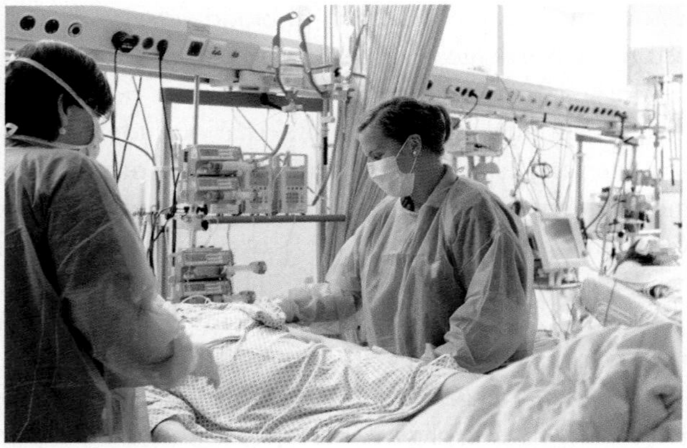

HTTP://WWW.SUEDDEUTSCHE.DE/WISSEN/EHEC-EPIDEMIC-BAHR-BEFUERCHTET-WEITERE-TODESFAELLE-1.1107576

FIGURE 1 ▪ Hemolytic uremic syndrome patient in Germany. This patient was being treated in the medical intensive care unit of the University Hospital Schleswig-Holstein in Lübeck. Note the biosafety precautions taken by the medical personnel (gloves, gown, and mask).

A particularly hideous aspect of disease produced by O104:H4 is that the prevalence of HUS symptoms in diarrheagenic patients far exceeds the usual rate for typical EHEC victims. But why? One hypothesis is that this variant produces high amounts of Shiga toxin (Stx-2) as compared to other STEC strains. Victor Gannon and his colleagues at the University of Lethbridge and the Public Health Agency of Canada demonstrated this in vitro using mitomycin C. Mitomycin C is a DNA intercalating agent that can trigger reactivation of prophages via the SOS response (Chapter 9) and, in so doing, it stimulates the expression of phage-encoded toxin genes. **Figure 2** shows that after mitomycin C treatment, the O104:H4 strain produced five to six times more Shiga toxin than other EHEC strains. It is thought that conditions in the gastrointestinal tract could induce a similar response.

The HUS outbreak started in May 2011. By early June the finger-pointing had begun. Germany initially blamed Spanish cucumbers, which led to a huge economic loss in Spain as foreign countries refrained from importing these vegetables. Later it was shown that Spain was not the source of this organism. Finally, the German agricultural minister of Lower Saxony announced that an organic farm near Uelzen, which produces a variety of sprouted foods, was the likely source of the *E. coli* outbreak (**Fig. 3**). Before the source was found, however, several tourists visiting Germany ingested the organism and returned to their home countries, including the United States and Canada, where they then developed HUS. Fortunately, as of this writing, no further major outbreaks caused by this pathogen have been reported.

RESEARCH QUESTION

Which features of the O104:H4 strain might be suitable targets for antimicrobial drug design? Explain why they might be suitable and what pitfalls there may be to using these drugs in a patient.

C. R. Laing, Y. Zhang, M. W. Gilmour, V. Allen, R. Johnson, et al. 2012. A comparison of Shiga-toxin 2 bacteriophage from classical enterohemorrhagic *Escherichia coli* serotypes and the German *E. coli* O104:H4 outbreak strain. *PLoS One* **7**:e37362. [Online.] http://www.plosone.org/article/info%3Adoi%2F10.1371%2Fjournal.pone.0037362.

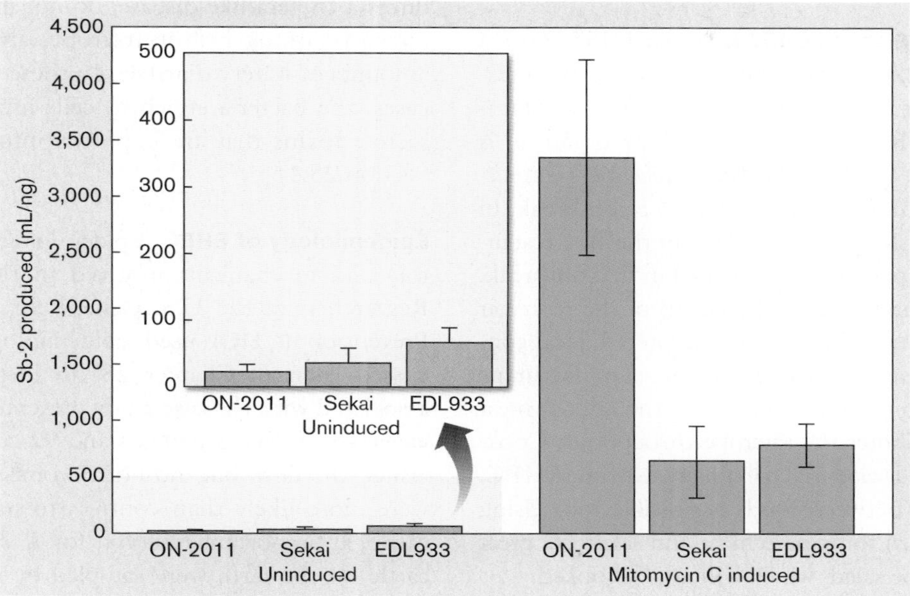
Sb-2 produced (mL/ng)
4,500
4,000
3,500
3,000
2,500
2,000
1,500
1,000
500
0
500
400
300
200
100
0
ON-2011
Sekai
EDL933
Uninduced
ON-2011
Sekai
EDL933
Uninduced
ON-2011
Sekai
EDL933
Mitomycin C induced

FIGURE 2 ■ **Shiga toxin production by *E. coli* pathovars.** Stx-2 production by *E. coli* O104:H4 outbreak-related strain ON-2011 and *E. coli* O157:H7 strains EDL933 and Sakai in uninduced and mitomycin C–induced states as measured by an Stx-2-specific ELISA. Error bars represent standard deviations from three independent replicates. Mitomycin C is a DNA intercalating agent that can trigger reactivation of prophages. Lytic replication can stimulate expression of phage-encoded toxin genes. *Source:* Adapted from Laing, C. R. et al. 2012. PLoS ONE **7**(5): e37362. © dpa.

A.

B.

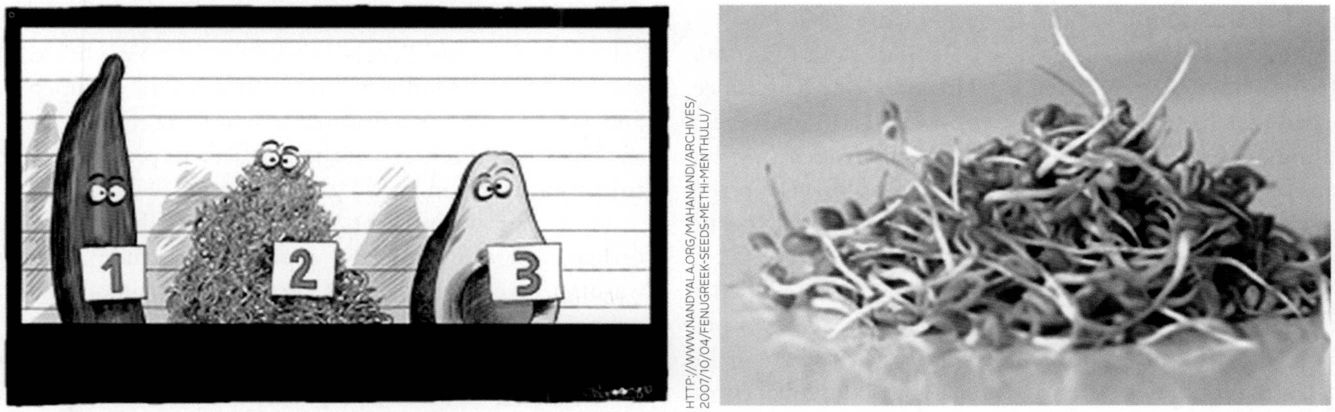
1
2
3

HTTP://JEFFREYHILL.TYPEPAD.COM/ENGLISH/2011/06/CARTOON-WHOS-GUILTY.HTML#.UN-7JXFYVT2

HTTP://WWW.NANDYALA.ORG/MAHANANDI/ARCHIVES/2007/10/04/FENUGREEK-SEEDS-METHI-MENTHULU/

FIGURE 3 ■ **What's the source? A.** Early on, the source of the new strain of *E. coli* was unclear, prompting this cartoon by Martin Sutovec depicting a police lineup composed of the suspects: cucumbers, sprouts, and avocados. **B.** Ultimately, fenugreek sprouts grown at an organic farm in Germany were identified as the source. These sprouts are commonly used in salads.

used to bind to target cell membranes, and one A subunit imbued with toxic activity (see Section 25.4). The A subunit, upon entry, destroys protein synthesis by cleaving an adenine from 28S rRNA in eukaryotic ribosomes.

Enterohemorrhagic *E. coli* (EHEC). *E. coli* O157:H7 is a recently emerged pathogen that can colonize cattle intestines without causing disease and, as a result, can contaminate meat products following slaughter. The organism is sometimes called the "Jack in the Box" microbe—a reference to the first documented large-scale U.S. outbreak, in 1993, linked to a Washington State Jack in the Box restaurant. As many as 600 people were sickened in that outbreak.

E. coli O157:H7 rarely affects the health of the reservoir animal. But when an infected steer is slaughtered, fecal contamination of the carcass can happen, despite manufacturers' considerable efforts to prevent it. Grinding the tainted meat into hamburger distributes the microbe throughout. Cooking burgers to 160°C is essential to kill any existing EHEC. Cross-contamination between foods is possible too. Using the same cutting board to prepare meat and salad is a great way to contaminate the salad, which will not be cooked.

Despite EHEC's common association with hamburger, vegetarians are not safe from this organism. During heavy rains, waste from a cattle farm can easily wash into nearby vegetable fields unless precautions are taken. If the cattle waste contains *E. coli* O157:H7, the crops are contaminated. One such outbreak occurred in 2006, when spinach from certain areas of California became contaminated with this pathogen, prompting a nationwide recall of bagged spinach and a month without spinach salad.

Early on, the remarkably low infectious dose of *E. coli* O157:H7 mystified researchers. However, we have since learned that *E. coli* has an impressive level of acid resistance, rivaling that of the gastric pathogen *Helicobacter pylori.* Acid resistance mechanisms permit *E. coli* to survive in the acidic stomach and enable a mere 10–100 individual organisms to cause disease.

As already noted, EHEC strains produce two toxins that are identical to the Shiga toxins produced by *Shigella* species. The toxins cleave host ribosomal RNA, thereby halting translation. As discussed earlier, one consequence of Shiga toxin is HUS. The development of HUS, as in the case history described, is a common consequence of *E. coli* O157:H7 infection. Unfortunately, HUS can be treated only with supportive care, such as blood transfusions and dialysis throughout the critical period until kidney function resumes. Antibiotic treatment can increase the release of Shiga toxins from the organisms and actually trigger HUS. Thus, antimicrobial therapy is not recommended.

In contrast to the case just described, many gastrointestinal infections do not produce fecal leukocytes or blood in the stool. Diarrheal diseases caused by *Vibrio cholerae* (cholera) or enterotoxigenic *E. coli* (ETEC), which produces a cholera-like disease, do not involve invasion of the intestinal lining by the microbe, and they yield copious amounts of watery diarrhea. In these two toxin-driven diseases, the bacteria attach to cells lining the intestine and secrete toxins that are imported into the target cells (see Section 25.4).

Epidemiology of EHEC. Epidemiology is the study of factors and mechanisms involved in the spread of disease. Researchers at the U.S. Centers for Disease Control and Prevention (CDC) used epidemiological principles (discussed later, in Chapter 28) to identify the risk factors associated with the case study presented earlier. They interviewed 51 infected patients and 92 controls (children who visited the farm but did not become ill). Infected patients were more likely than controls to have had contact with cattle, an important reservoir for *E. coli* O157:H7. All 216 cattle on the farm were sampled by rectal swab, and 13% yielded *E. coli* O157:H7 with a DNA restriction pattern indistinguishable from that isolated from the patients. This finding indicated that the cattle were the source of infection. Activities that promoted hand-to-mouth contact, such as nail biting and purchasing food from an outdoor concession, were more common among the children who contracted disease (fecal-oral route of infection). Furthermore, separate areas were not established for eating and interactions with farm animals. Visitors could touch cattle, calves, sheep, goats, llamas, chickens, and a pig while eating and drinking. Hand-washing facilities were unsupervised and lacked soap, and disposable hand towels were out of the children's reach. All of these situations provided opportunity for infection.

How can we prevent disease caused by enterohemorrhagic *E. coli*? Industry approaches include thorough washing of carcasses before processing, maintaining cold temperatures, and testing for possible contamination. In addition, the use of gamma irradiation to sterilize beef, spinach, and lettuce has been approved. In fact, a majority of U.S. hamburger beef is now irradiated, which has dramatically reduced the number of hamburger-related outbreaks of EHEC.

Type III secretion and diarrhea. Type III secretion was first described in Section 25.5, where we discussed the pathogenesis of *Salmonella enterica,* but these secretion systems are present in numerous Gram-negative pathogens, such as EHEC in our case history. Recall that type III protein secretion systems directly inject proteins from the cytoplasm of a bacterial pathogen into the cytoplasm of a target eukaryotic host cell. The system delivers proteins across three membranes—two for the Gram-negative bacterial pathogen and one for the target cell.

In addition to stimulating bacterial entry into host cells, bacterial proteins injected by type III transport systems cause host cells to secrete pro-inflammatory cytokines. The cytokines then "call in" inflammatory cells and alter ion transport through the epithelial membrane. Excessive export of ions such as chloride causes water to leave the cell in an attempt to equilibrate the internal and external ionic concentrations. The water entering the intestine results in diarrhea. Type III–translocated effector proteins induce many host cellular responses that contribute to intestinal inflammation during an infection.

Although most gastrointestinal disease is intestinal in locale, specialized microbes can also target the stomach, with its harsh acidic environment.

Case History: Ulcers—It's Not What You Eat

Gary was a 34-year-old accountant who had immigrated to Nebraska from Poland 7 years earlier. Since his teenage years, he had been bothered periodically by episodes of epigastric pain (pain around the stomach), nausea, and heartburn. Antacids usually alleviated the symptoms. Over the years, he received several courses of treatment with Tagamet or Pepcid to reduce acid secretion and provide relief. Recently, an upper-GI endoscopy had been performed, in which a long, thin tube tipped with a camera and light source was inserted into Gary's mouth and into his stomach. The view through the endoscope showed some reddened areas in the antrum (bottom part) of the stomach. The endoscope was also equipped with a small clawlike structure that obtained a small tissue sample from the lining of Gary's stomach. A urease test performed on the antral biopsy turned positive in 20 minutes. Histological examination of the biopsy confirmed moderate chronic active gastritis (inflammation of the stomach lining) and revealed the presence of numerous spiral-shaped organisms. Cultures of the antral biopsy were positive for Helicobacter pylori.

Painful and sometimes life-threatening gastric ulcers were for many years blamed on spicy foods and stress. These factors were believed to cause increased acid production that ate away at the stomach lining, even though the gastric mucosa is normally well protected from stomach acid, which can fall as low as pH 1.5. This protection argued against the model but was ignored. In the 1980s, after discovering odd, helical bacteria present in the biopsies of gastric ulcers, Australians J. Robin Warren and Barry Marshall (a medical intern at the Royal Perth Hospital at the time) (**Fig. 26.8A**) proposed that bacteria, not pepperoni, cause ulcers (**Fig. 26.8B**). Their hypothesis was viewed with skepticism and declared as heresy by the established medical community. Faced with disbelief bordering on ridicule, the young intern drank a vial of the helical organisms and waited. A week later he began vomiting and suffered other painful symptoms of gastritis. Barry Marshall could not have been happier. He had proved his point. We now know that this curly microbe causes the vast majority of stomach ulcers.

The discovery of *Helicobacter pylori* and its association with gastric ulcer disease led to an upheaval in gastroenterology. Prior to this discovery, treatment had focused on suppressing acid production, which did not provide long-term relief. Within 1 year after acid suppression therapy, up

A.

COURTESY OF BARRY MARSHALL

B.

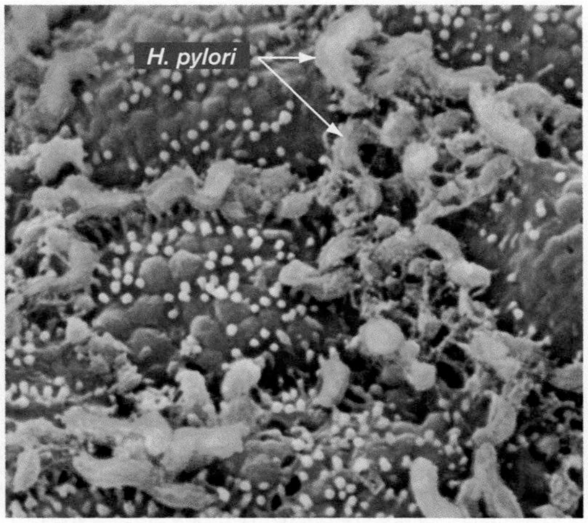
H. pylori

CNRI/PHOTO RESEARCHERS, INC.

FIGURE 26.8 ▪ A bacterial cause of gastric ulcers. A. Australian physician Barry Marshall was so sure he was right about the cause of stomach ulcers that he swallowed bacteria to prove his point. **B.** View of *Helicobacter pylori* in stomach crypts (colorized SEM). Bacteria are approx. 2 μm in length.

to 80% of patients suffer a relapse of their ulcer. Therapy now includes antimicrobial treatment to kill the bacteria and acid suppression therapy to prevent further inflammation while the ulcer heals. Warren and Marshall, who recovered from his gastritis, received the 2005 Nobel Prize in Physiology or Medicine for their groundbreaking work.

The exact mechanism by which *H. pylori* causes gastric ulcers is not known, although a variety of virulence factors have been described (see Section 25.7). The urease enzyme of *H. pylori* is a characteristic virulence feature. Urease converts urea, produced by the gastric lining, to ammonia and carbon dioxide. The ammonia neutralizes acid in and around the organism and allows the microbe to survive in the stomach. In an animal model, the urease enzyme of *H. pylori* was found to be important for bacterial colonization in the stomach. Treatment with urease inhibitors such as acetohydroxamic acid, however, does not lead to eradication of *H. pylori*. This failure is likely due to protection afforded the organism once it reaches the mucus layer that blankets the gastric epithelium (see **Fig. 26.8B**). The pH of this environment is closer to neutral, so urease may no longer be needed. Tools useful for diagnosing *H. pylori* include histology, rapid urease testing, and serology [for example, enzyme-linked immunosorbent assay (ELISA) to detect antibody to the CagA antigen].

Enzyme-linked immunosorbent assay (ELISA). ELISA is a common immunological tool to detect the presence, in serum, of antibodies to a specific organism—an indication of infection. The assay, described more fully in Section 28.4, is performed by coating wells in a plastic dish with an antigen (for example, *Helicobacter* CagA). Serum from the patient is then added to the well. If antibodies to CagA are present, they will bind to the CagA antigen. Unbound antibody is removed by washing, and a secondary antibody that binds human IgG is added to the well. These anti-antibodies have an enzyme linked to them. A sandwich is formed as follows: [plastic dish]–[CagA protein]–[anti-CagA antibody]–[anti-human IgG antibody]–[enzyme]. When substrate is added to the well, the enzyme acts on it to produce light or a chromogenic (colored) product. The more anti-CagA antibody present in the serum, the more light or product that is produced by the linked enzyme. Applications of ELISA and further details of the methodology involved are described in Section 28.4.

***Helicobacter pylori* and cancer.** Problems caused by *H. pylori* are not limited to gastric ulcers. The microbe has also been associated with gastric cancer. The evidence is not yet conclusive, but many individuals with gastric cancer are also colonized by *H. pylori*. The most compelling experimental proof is that gerbils infected with this organism develop gastric cancer. In addition, when the CagA protein of *H. pylori* is injected into gastric epithelial cells, it is phosphorylated on a tyrosine residue and activates a regulatory cascade that causes the gastric cell to proliferate (see Fig. 25.38). The connection between *H. pylori* and gastric cancer is sobering when you consider that *H. pylori* can be detected in about two-thirds of the world's population, especially in impoverished countries.

Rotavirus Is the Single Greatest Cause of Gastroenteritis

Many people wrongly think that most cases of diarrhea are caused by a bacterial agent. Actually, a virus, **rotavirus**, causes more intestinal disease than any bacterial species (see Table 6.1, Group III: dsRNA viruses). Rotavirus is highly infectious, spreading by the fecal-oral route; it is endemic around the globe; and it affects all age groups, although children between 6 and 24 months are most severely affected. It is estimated that by age 3, all children have had a rotavirus infection.

The incubation period is approximately 2 days, after which the victim commonly suffers frequent watery, dark green, explosive diarrhea. All of this may be accompanied by nausea, vomiting, and abdominal cramping. Severe dehydration and electrolyte loss due to the diarrhea will cause death unless supportive measures, such as fluid replacement, are undertaken. There is no cure, but most patients recover if rehydrated properly. Few deaths from rotavirus occur in the United States, but each year more than 600,000 children worldwide die from this viral diarrhea. The mortality and incidence of this disease are expected to decrease, however, because of a recently developed safe and effective vaccine (Section 26.9).

Protozoa Are Another Major Cause of Diarrheal Diseases

As we learned in Chapter 20, some protozoa (also called protists) cause serious human diseases. For instance, *Entamoeba histolytica* and *Cryptosporidium parvum* cause the diarrheal diseases amebic dysentery and cryptosporidiosis, respectively. In 2010, the CDC tallied 8,944 cases of cryptosporidiosis (up from 2,640 in 2005), a reportable disease in the United States. Two other amebas, *Naegleria* and *Acanthamoeba*, cause amebic meningoencephalitis. Because some species of *Acanthamoeba* can infect the eye, soft-contact wearers should take precautions to prevent contamination of their lenses.

A.

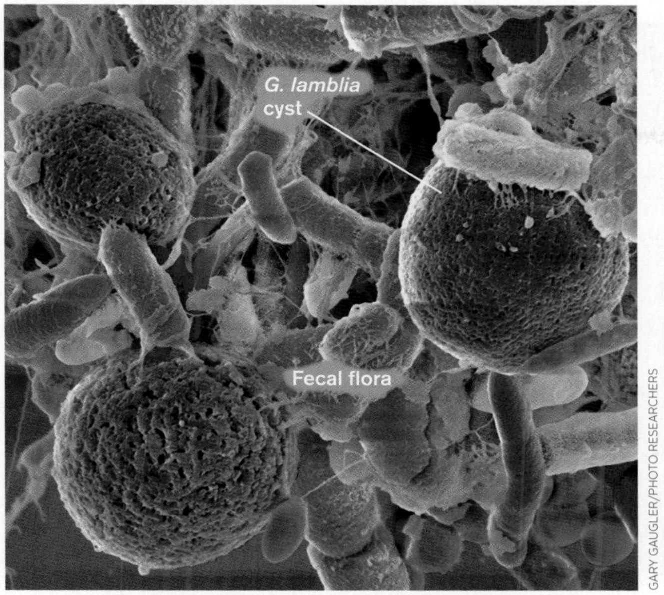
G. lamblia
cyst
Fecal flora
GARY GAUGLER/PHOTO RESEARCHERS

B.

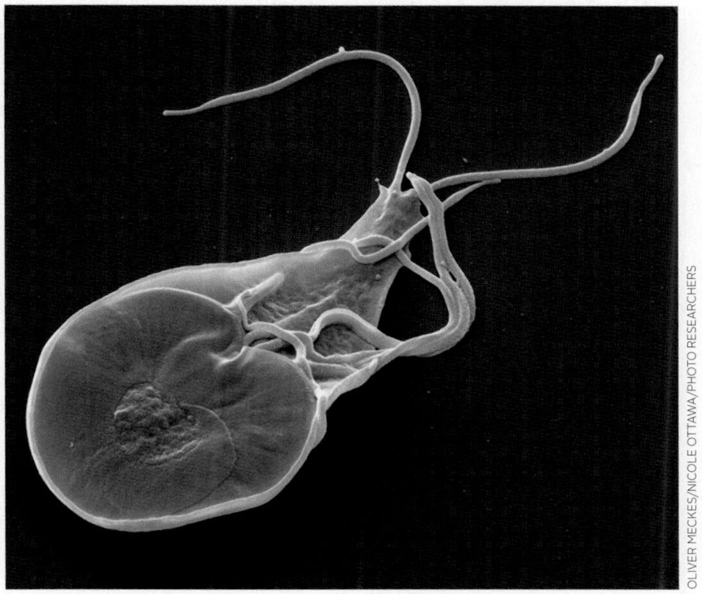
OLIVER MECKES/NICOLE OTTAWA/PHOTO RESEARCHERS

FIGURE 26.9 ▪ ***Giardia lamblia.*** This protist is a major cause of diarrhea in the world. **A.** Cysts (7–14 μm) present in fecal matter (colorized SEM). **B.** Trophozoite form (5–15 μm in length, colorized SEM).

The flagellated protozoan *Giardia lamblia* is a major cause of diarrhea throughout the world. In the United States alone, *G. lamblia* caused nearly 17,000 reported cases of giardiasis diarrhea in 2011 (over 19,000 in 2010) and likely caused thousands more that were not reported. *G. lamblia* enters a human or other host as a cyst present in drinking water contaminated by feces (**Fig. 26.9A**). Aside from humans, *G. lamblia* can be found in various rodents, deer, cattle, and even household pets. It is very infectious. Ingestion of as few as 25 cysts can lead to disease. Following ingestion, the hard, outer coating of the cyst is dissolved by the action of digestive juices to produce a trophozoite (**Fig. 26.9B**), which attaches itself to the wall of the small intestines and reproduces. Offspring quickly encyst and are excreted out of the host's body.

Asymptomatic carriers of *G. lamblia* are common; it has been estimated that anywhere from 1% to 30% of children in U.S. day-care centers are carriers. Disease usually manifests as greasy stools alternating between a watery diarrhea, loose stools, and constipation. However, some patients will experience explosive diarrhea. Diagnosis usually comes from observing the cysts or trophozoite forms of the protozoan in feces. Metronidazole is a drug often used to cure the disease. To prevent it in the first place, proper treatment of community water supplies is essential.

We have examined only a handful of the microbes that cause gastrointestinal infection. Others are listed in **Table 26.3** and are described in Chapter 20.

To Summarize

- **Diarrhea** leads to dehydration, for which fluid replacement is a critical treatment. Antibiotic treatment is usually not recommended.
- **Staphylococcal food poisoning** is not an infection. It is a toxigenic disease.
- **Antibiotic treatments** can sometimes cause gastrointestinal disease (for example, pseudomembranous enterocolitis by *Clostridium difficile*).
- **Bacteria that invade intestinal epithelial mucosal cells** lead to the presence of red and white blood cells in fecal contents. The culprits are intracellular pathogens such as *Shigella, Salmonella,* and EIEC.
- **Bacteria that do not invade intestinal cells** usually produce watery diarrhea. EHEC is an exception because the attachment and effacing lesions it produces result in bloody stools.
- **Bacterial toxins** produced by bacterial enteric pathogens can cause systemic symptoms.
- **The bacterium *Helicobacter pylori,*** a common cause of gastric ulcers, lives in the stomach and is highly acid resistant.
- **Rotavirus** is the single greatest cause of diarrhea worldwide.
- ***Giardia lamblia*** is a major protozoan cause of diarrhea worldwide.

TABLE 26.3

Selected microbes that cause diseases of the gastrointestinal tract.

Disease	Symptoms	Etiological agent	Virulence factors	Source	Treatment
Bacterial					
Staphylococcal food poisoning	Symptoms within 4 h of ingestion; nausea, vomiting, diarrhea	*Staphylococcus aureus* (G+)	Enterotoxin	Preformed toxin in foods	Supportive
Botulism	Symptoms begin quickly; flaccid paralysis	*Clostridium botulinum* (G+, anaerobe)	Neurotoxin	Preformed toxin in foods	Antiserum
Salmonellosis	Symptoms after 18 h; abdominal pain, diarrhea; invade intestinal M cells	*Salmonella enterica* [Gram-negative (G–)]	Type III secretion, intracellular growth	Chickens, other animals; fecal-oral route	Oral rehydration; antibiotics if severe
Typhoid fever	Headache, fever, chills, abdominal pain, rash (rose spots), hypotension, diarrhea in late stages	*Salmonella enterica* serovar Typhi (G–)	Type III secretion, intracellular growth, PhoPQ regulators, Vi antigen capsule	Human carriers (gallbladder reservoir); food, water	Quinolones
Traveler's diarrhea	Watery diarrhea	Enterotoxigenic *E. coli*	Labile and stable toxins	Humans; food, water	Oral rehydration
Gastroenteritis	Bloody diarrhea, HUS	Enterohemorrhagic *E. coli*	Intimin, Tir, type III secretion, Shiga toxin	Contaminated foods (hamburger) and crops	Oral rehydration; antibiotics if severe
Shigellosis	Bloody diarrhea, HUS	*Shigella* spp. (G–)	Shiga toxin, type III secretion, intracellular growth, actin-based motility, escape phagosome	Humans; fecal-oral route	Oral rehydration; antibiotics if severe
Cholera	Watery diarrhea	*Vibrio cholerae* (G–)	Cholera toxin, toxin-coregulated pili (TCPs), ToxR regulator	Human waste–contaminated water	Oral rehydration, antibiotics
Gastroenteritis	Diarrhea, blood in stool	*Vibrio parahaemolyticus* (G–)	Enterotoxin	Raw seafood	Self-limiting
Gastroenteritis	Watery diarrhea, nausea	*Clostridium perfringens* (G+)	Alpha toxin	Soil, food	Self-limiting
Pseudomembranous enterocolitis	Fever, abdominal pain, diarrhea, pseudomembrane in colon	*Clostridium difficile* (G+)	Cytotoxin, antibiotic resistance	Animals, normal microbiota	Vancomycin
Gastroenteritis	Fever, muscle pain, watery diarrhea, blood in stool, headache	*Campylobacter jejuni* (G–)	Cytotoxin, enterotoxin, adhesin	Poultry, unpasteurized milk	Erythromycin
Gastric ulcers	Abdominal pain, bleeding, heartburn	*Helicobacter pylori* (G–)	Adhesin, urease CagA, vacuolating toxin	?	Triple drug (omeprazole, clarithromycin, metronidazole)
Viral					
Stomach "flu"	Nausea, vomiting, diarrhea	Norovirus (Norwalk virus)	?	Fecal-oral route	Oral rehydration
		Rotavirus (most common cause)	?	Fecal-oral route	Oral rehydration

26.5

Genitourinary Tract Infections

Although the genital and urinary tracts are different organ systems, their close association in the body has led them to be grouped together when discussing infections.

Urinary Tract Infections

The urinary tract includes the kidneys, ureters, urinary bladder, and urethra. Infections anywhere along this route are called urinary tract infections (UTIs). UTIs are the second most common type of bacterial infection in humans, ranking in frequency just behind respiratory infections such as bronchitis or pneumonia. Along with numerous office visits, bladder infections and other UTIs result in 100,000 hospital admissions and $1.6 billion in medical expenses each year in the United States alone. Most sufferers are women. One in five women will get a urinary tract infection at some point in her life, and 20%–40% of those infected will develop recurrent infections.

Urine, as produced in the kidneys and stored in the bladder, is normally sterile. Microorganisms must be introduced into the bladder to cause infection. Active infection of the urinary tract occurs in one of three basic ways:

- **Infection from the urethra to the bladder.** This is the most common route. Bacteria residing along the superficial urogenital membranes of the urethra can ascend to the bladder. This is a more common occurrence in women than in men. Microorganisms can also be introduced into the bladder by means of mechanical devices such as catheters or cystoscopes that are passed through the urethra into the bladder.
- **Descending infection from the kidney.** Descending infection occurs when an infected kidney sheds bacteria that descend via the ureters into the bladder. Kidney infections arise when microorganisms are deposited in the kidneys from the bloodstream.
- **Ascending infection to the kidney.** In ascending infection, an established infection in the bladder ascends along the ureter to infect the kidney.

Thought Question

26.4 Why do you think most urinary tract infections occur in women?

Urine is bacteriostatic to most of the commensal organisms inhabiting the perineum and vagina, such as *Lactobacillus*, *Corynebacterium*, diphtheroids, and *Staphylococcus epidermidis*. In contrast, many Gram-negative organisms thrive in urine. As a result, most urinary tract infections are caused by facultative Gram-negative rods from the GI tract. The most common etiological agents of UTIs are:

- Certain serotypes of *E. coli* that comprise the uropathogenic *E. coli* (75% of all UTIs)
- *Klebsiella*, *Proteus*, *Pseudomonas*, *Enterobacter* (20%)
- *Staphylococcus aureus*, *Enterococcus*, *Chlamydia*, fungi, *Staphylococcus saprophiticus*, other (5%)

Case History: Classic Urinary Tract Infection

Lashandra was 24 years old and had been experiencing back pain, increased frequency of urination, and dysuria (painful urination) over the previous 3 days. She consulted her general practitioner, who requested that a midstream specimen of urine be examined. This was the first time Lashandra had ever suffered from persisting dysuria. Upon microscopic examination, the urine was found to contain more than 50 leukocytes per microliter and 35 red blood cells per microliter. No epithelial squamous cells were seen. The urine culture plated on agar medium yielded more than 10^5 colonies per milliliter (meaning more than 10^5 organisms per milliliter in the urine) of a facultative anaerobic Gram-negative bacillus capable of fermenting lactose.

The first question to ask in this case is whether the patient had a significant UTI. The purpose of the midstream urine collection is to provide laboratory data to make this determination. Even though urine in the bladder is normally sterile, urine becomes contaminated with normal microbiota that have adhered to the urethral wall. In a midstream collection, the patient urinates briefly, stops to position a collection jar, and resumes urinating to collect a midstream sample. This procedure minimizes the number of organisms in the sample by washing away organisms clinging to the urethra before actually collecting the sample. Nevertheless, the collected sample will still contain low numbers of organisms representing normal microbiota of the urethra. A diagnosis of UTI is made when the number of bacteria in the sample becomes greater than 10^5/ml. (However, that number can be as low as 1,000/ml in symptomatic women.) The patient in the case history is symptomatic and has sufficient numbers of bacteria in her urine to indicate a UTI.

The laboratory found the organism to be a Gram-negative bacillus that ferments lactose, suggesting *E. coli* as the likely culprit. Given that the normal habitat of *E. coli* is the gastrointestinal tract and this is the first UTI suffered by the patient, the infection is likely the result of an inadvertent introduction of the microbe into the urethra. The organism makes its way up the urethra and into the bladder. Another way the bladder can become infected is via a descending route from the kidney. Organisms from a kidney that was infected as a result of sepsis can descend along a ureter to cause a bladder infection.

Gram-negative rods not only thrive in urine but may be adapted to cause urinary tract infections through specialized pili. These pili have terminal receptors for glycolipids and glycoproteins present on urinary tract epithelial cells (**Fig. 26.10A**). Uropathogenic strains of *E. coli*, for example, typically have P-type pili, with a terminal receptor for the P antigen (a rather appropriate name for a bladder-specific virulence factor). The P antigen is a blood group marker found on the surface of cells lining the perineum and urinary tract; it is expressed by approximately 75% of the population. These individuals are particularly susceptible to UTIs. The P antigen can also be shed and found in vaginal and prostate secretions. When this happens, the secreted P antigens can be protective by acting as a decoy. Soluble P antigen can bind to the bacterial receptor and prevent binding of the organism to surface epithelium. Individuals most susceptible to UTIs are those who express P antigen on their cells but lack P antigen in their secretions. Some patients, usually women, are also susceptible to recurrent bladder infections. These are thought to be caused by uropathogenic *E. coli* that invade urinary tract epithelial cells and form compact intracellular biofilms (discussed in eTopic 26.5).

Urinary tract infections are among those most frequently acquired during a hospital stay (so-called nosocomial infections). In these cases, the causal organism is less likely to be *E. coli* and more likely to be another Gram-negative bacterium or *Staphylococcus*. Many UTIs resolve spontaneously, but others can progress to destroy the kidney or, via Gram-negative septicemia, the host. As a result, antibiotic therapy is recommended. In older patients, UTIs frequently show atypical symptoms, including delirium, which disappears when the UTI is treated.

Thought Question

26.5 Urine samples collected from six hospital patients were placed on a table at the nurses' station awaiting pickup from the microbiology lab. Several hours later, a courier retrieved the samples and transported them to the lab. The next day, the lab reported that four of the six patients had UTIs. Would you consider these results reliable? Would you start treatment based on these results?

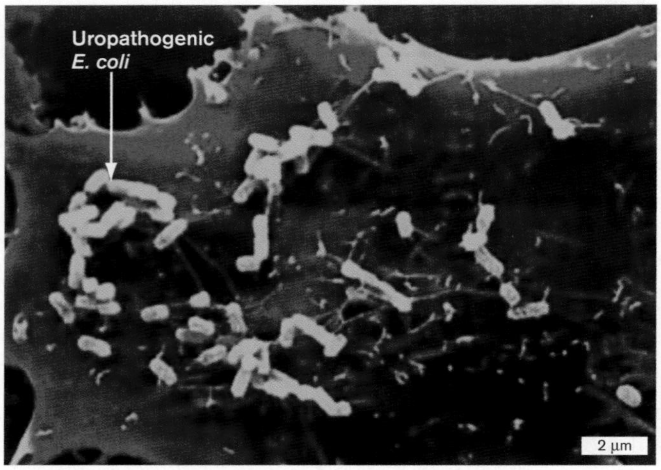
A.
Uropathogenic
E. coli
2 µm
COURTESY OF HILDE MERKERT, UNIVERSITY OF WUERZBURG

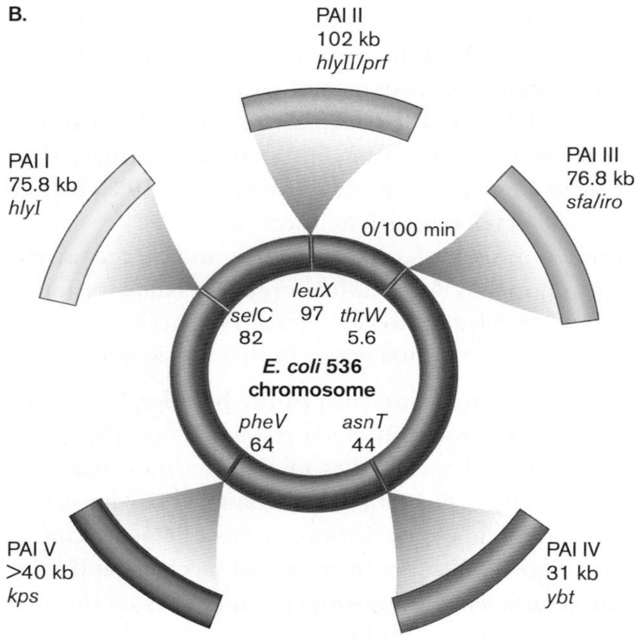
B.
PAI II
102 kb
hlyII/prf
PAI I
75.8 kb
hlyI
PAI III
76.8 kb
sfa/iro
0/100 min
leuX
97
selC
82
thrW
5.6
E. coli 536
chromosome
pheV
64
asnT
44
PAI V
>40 kb
kps
PAI IV
31 kb
ybt

FIGURE 26.10 ▪ Uropathogenic *E. coli*. A. Bladder cell with adherent uropathogenic *E. coli* (SEM). Bacteria are approx. 1 µm in length. **B.** Distribution of pathogenicity islands (PAIs) in uropathogenic *E. coli*. The location of each insert is given in map units within the circle representing the genome. The 0–100 map units are called centisomes. Each centisome is approx. 44 kb of DNA. Zero is arbitrarily placed at the *thr* (threonine) gene. The origin of replication on this map is near 82 centisomes. A chromosomal gene flanking the insert is also provided. The size of each island is provided above the insert. A key virulence gene for each island is listed.

What makes uropathogenic *E. coli* different from other *E. coli*? This is a question still under investigation, but genomic analysis has exposed five pathogenicity islands unique to these strains (**Fig. 26.10B**). The functions of these pathogenicity islands are still under investigation.

Sexually Transmitted Diseases

Sexually transmitted diseases (STDs) are defined as infections transmitted primarily through sexual contact, which may include genital, oral-genital, or anal-genital contact. The organisms or viruses involved are generally very susceptible to drying and require direct physical contact with mucous membranes for transmission. Because sex can take many forms in addition to intercourse, these microbes can initiate disease in the urogenital tract, rectum, or oral cavities. Condoms can prevent transmission, but do so, of course, only when used. Examples of common sexually transmitted diseases are listed in **Table 26.4**.

Case History: Secondary Syphilis

A pregnant 18-year-old woman came to the county urgent-care clinic with a low-grade fever, malaise, and headache. She was sent home with a diagnosis of influenza. She again sought treatment 7 days later, after she discovered a macular rash (flat, red) developing on her trunk, arms, palms of her hands, and soles of her feet. Further questioning of the patient when serology results were known revealed that 1 year earlier, she had had a painless ulcer on her vagina that healed spontaneously.

The vaginal ulcer, the long latent period, and secondary development of rash on the hands and feet described in the

TABLE 26.4

Microbes that cause sexually transmitted diseases.

Disease	Symptoms	Etiological agent	Virulence factors	Treatment	Reported cases
Gonorrhea	Purulent discharge, burning urination; can lead to sterility	*Neisseria gonorrhoeae* (G–)	Type IV pili, phase variation	Ceftriaxone	309,341[a]
Syphilis	1°: chancre; 2°: joint pain, rash; 3°: gummata, aneurism, central nervous system damage	*Treponema pallidum* (spirochete)	Motility	Penicillin	45,834[a]
Nongonococcal urethritis	Watery or mucoid urethral discharge, burning urination	*Chlamydia trachomatis*	Intracellular growth; prevents phagolysosome fusion	Azithromycin	1.3 million[a]
Trichomoniasis	Vaginal itching, painful urination, strawberry cervix	*Trichomonas vaginalis* (protozoan)	Cytotoxin	Metronidazole	149,000[a]
Chancroid	Painful genital lesion	*Haemophilus ducreyi* (G–)	?	Erythromycin	24[a]
HIV diagnoses ranging from asymptomatic to AIDS	For AIDS, fever, diarrhea, cough, night sweats, fatigue, opportunistic infections	HIV	gp120, Rev, Nef, and Tat proteins	Azidothymidine (AZT), protease inhibitors, zidovudine	35,741[a]
Genital herpes	Painful ulcer on external genitals, painful urination	Herpes simplex 2	Cell fusion protein, complement-binding protein, latency	Acyclovir, iododeoxyuridine	232,000[a] (45 million[b])
Genital warts	Warts on external genitals	Human papillomavirus	E6, E7 proteins	Vaccine now available	376,000[a] (20 million[b])

[a]CDC, reported cases for 2010.

[b]Total estimated current cases.

case history are classic symptoms of syphilis. Syphilis was recognized as a disease as early as the sixteenth century, but the organism responsible, the spirochete *Treponema pallidum,* was not discovered until 1905 (**Fig. 26.11A**) (Chapter 18 describes spirochete structure). The illness has several stages. The disease has often been called the great imitator because its symptoms in the second stage, as exhibited in the case history, can mimic many other diseases. The incubation stage can last from 2 to 6 weeks after transmission, during which time the organism multiplies and spreads throughout the body.

Primary syphilis is an inflammatory reaction at the site of infection called a **chancre** (**Fig. 26.11B**). About a centimeter in diameter, the chancre is painless and hard, and it contains spirochetes. Patients are usually too embarrassed to seek medical attention and, because it is painless, hope that it will just go away. It does go away after several weeks, and without scarring. The disease has now entered the primary latent stage. Over the next 5 years, symptoms may be absent, but at any time, as described in the case history, the infected person can develop the rash typical of **secondary syphilis** (**Fig. 26.11C**). The rash can be similar to rashes produced by many different diseases, which contributes to the "great imitator" label. The patient remains contagious in this stage. Some patients eventually progress over years to **tertiary syphilis**, in which a great number of symptoms can develop, affecting mainly the cardiovascular and nervous systems. The patient can develop dementia and eventually dies from the disease.

The presence of *T. pallidum* in tissues can be detected with fluorescent antibody, but the initial screen is usually serological (that is, patient serum is tested for antibodies). Antibiotics are useful for eradicating the organism, but there is no vaccine, and cure does not confer immunity.

The disease is particularly dangerous in pregnant women. The treponeme can cross the placental barrier and infect the fetus to cause **congenital syphilis**. At birth, infected newborns will have notched teeth (visible on X-rays), perforated palates, and other congenital defects. Women should be screened for syphilis as part of their prenatal testing to prevent these congenital infections.

Columbus and the New World theory of syphilis. An outbreak of syphilis that spread throughout Europe soon after Christopher Columbus and his crew returned from the Americas (1493) led to the theory that Columbus brought the treponeme to Europe from the New World. However, the theory that Columbus brought syphilis to Europe has been difficult to prove.

Scientists advocating either the New World origin or the Old World origin of syphilis have relied mainly on bone evidence to support their views. Treponemal diseases, such as syphilis and yaws (a milder form of disease caused by *T. pallidum* subspecies *pertenue*), leave characteristic marks on skeletons, usually in the tertiary phase of disease. So, if bones with signs of treponemal disease were found in Native Americans who died before Columbus arrived in America but were not found in pre-Columbian European remains, then case closed, Columbus did it. The problem is that paleontologists have reported evidence of treponemal diseases in the pre-Columbian skeletons of Native Americans and in those of Europeans.

The answer to this 500-year-old debate may be found in a recent study by Kristin Harper and colleagues, who used a phylogenetic approach to address the question. The group sequenced 26 geographically disparate strains of pathogenic *Treponema*. Of all the strains examined, the sexually transmitted syphilis-causing strains originated most recently and were more closely related to the nonsexually transmitted yaws-causing strains from South America, supporting the New World theory of syphilis. Thus, it seems the crew of the Columbus voyages brought smallpox and measles to America and returned to Europe with syphilis.

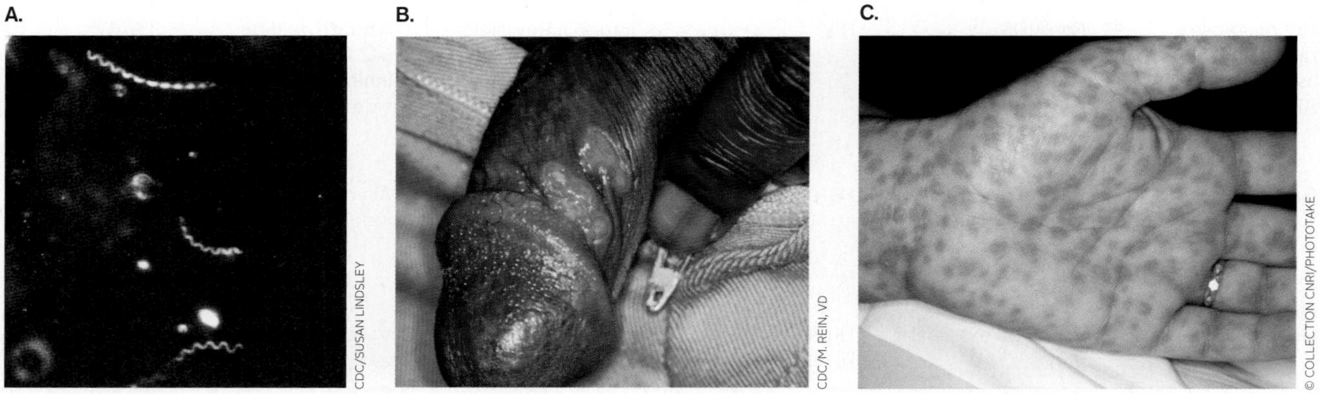
A.
B.
C.
CDC/SUSAN LINDSLEY
CDC/M. REIN, VD
© COLLECTION CNRI/PHOTOTAKE

FIGURE 26.11 ■ Syphilis. **A.** *Treponema pallidum* (dark-field microscopy). Organisms are 10–25 μm long. **B.** Chancre of primary syphilis. **C.** Rash of secondary syphilis.

The Tuskegee experiment. Unfortunately, much of what we know about syphilis is the result of the infamous Tuskegee experiment conducted in the 1930s in Alabama. The study was entitled "Untreated Syphilis in the Negro Male." Through dubious means and deception, a group of African-American males was enlisted in a study that promised treatment but whose real purpose was to observe how the disease progressed without treatment. Today, such experiments are barred, thanks to strict institutional review board (IRB) oversights in which human subjects must sign informed consent forms. An interesting treatise on the Tuskegee experiment can be found on the Internet at the National Center for Case Study Teaching in Science (search the site for "Bad Blood").

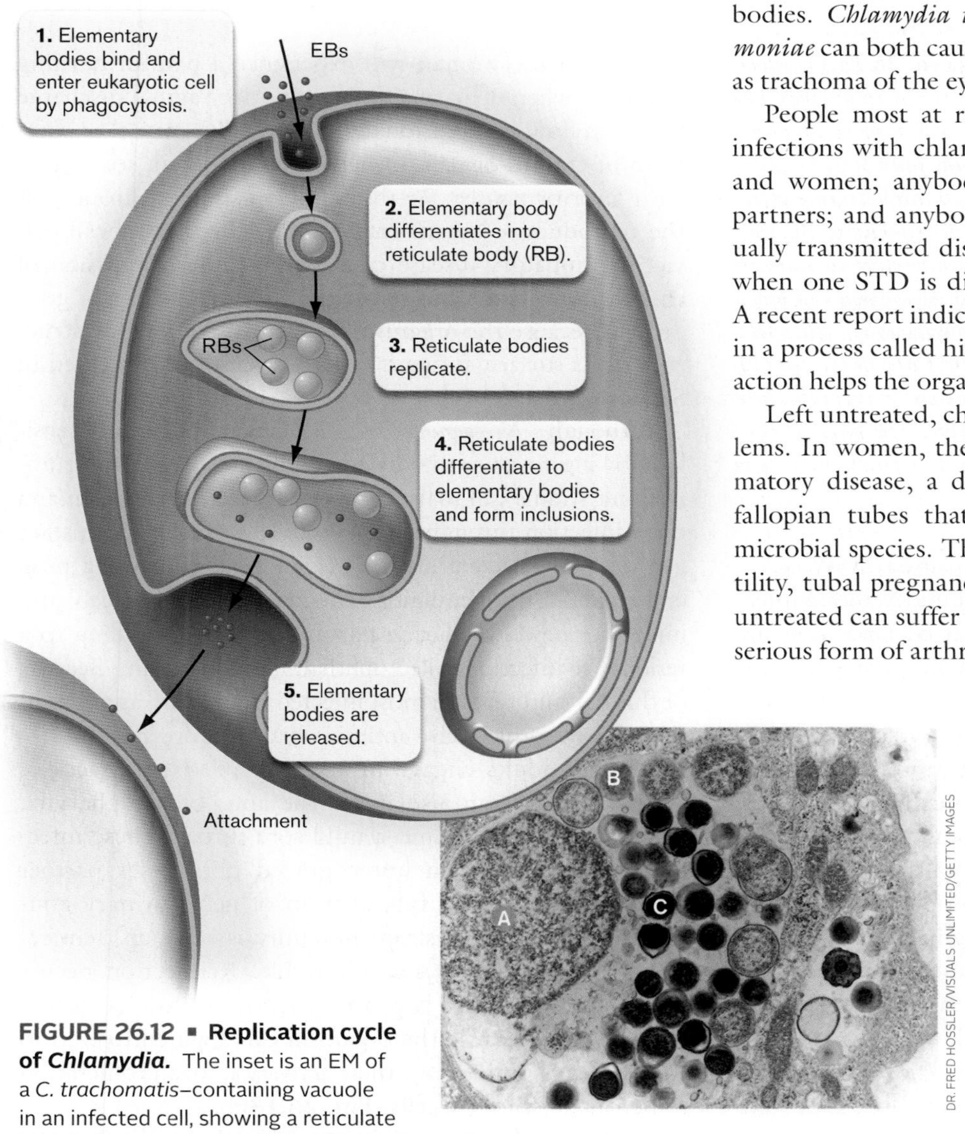
1. Elementary bodies bind and enter eukaryotic cell by phagocytosis.
EBs
2. Elementary body differentiates into reticulate body (RB).
RBs
3. Reticulate bodies replicate.
4. Reticulate bodies differentiate to elementary bodies and form inclusions.
5. Elementary bodies are released.
Attachment
A
B
C
DR. FRED HOSSLER/VISUALS UNLIMITED/GETTY IMAGES

FIGURE 26.12 ■ Replication cycle of *Chlamydia*. The inset is an EM of a *C. trachomatis*–containing vacuole in an infected cell, showing a reticulate body (A), an intermediate form between the elementary body and the reticulate body (B), and an elementary body (C).

Chlamydial Infections Are Often Silent

Chlamydia is the most frequently reported sexually transmitted infectious disease in the United States, according to the Centers for Disease Control and Prevention, but many people are completely unaware that they are infected. Three-fourths of infected women, for instance, have no symptoms.

The chlamydias are unusual Gram-negative organisms with a unique developmental cycle (Chapter 18 describes chlamydial morphology). They are obligate intracellular pathogens that start as a small, nonreplicating, infectious elementary body that enters target eukaryotic cells. Once inside vacuoles, they begin to enlarge into replicating reticulate bodies (**Fig. 26.12**). As the vacuole fills, the reticulate bodies divide to become new nonreplicating elementary bodies. *Chlamydia trachomatis* and *Chlamydophila pneumoniae* can both cause STDs, as well as other diseases, such as trachoma of the eye or pneumonia.

People most at risk of developing genitourinary tract infections with chlamydias are young, sexually active men and women; anybody who has recently changed sexual partners; and anybody who has recently had another sexually transmitted disease. The astute clinician knows that when one STD is discovered, others may also be present. A recent report indicates that *Chlamydia* can bind to sperm in a process called hitchhiking. It is thought that this interaction helps the organism spread to females.

Left untreated, chlamydia can cause serious health problems. In women, the organism can produce pelvic inflammatory disease, a damaging infection of the uterus and fallopian tubes that can be caused by several different microbial species. The damage produced can lead to infertility, tubal pregnancies, and chronic pelvic pain. Men left untreated can suffer urethral and testicular infections and a serious form of arthritis.

Case History: Gonorrhea

A 22-year-old mechanic saw his family doctor for treatment of painful urination and urethral discharge. The patient was sexually active, with three regular and several "one time–good time" partners. Physical examination was unremarkable except for prevalent urethral discharge. The discharge was Gram-stained and sent for culture. The Gram stain revealed many pus cells, some of which contained numerous

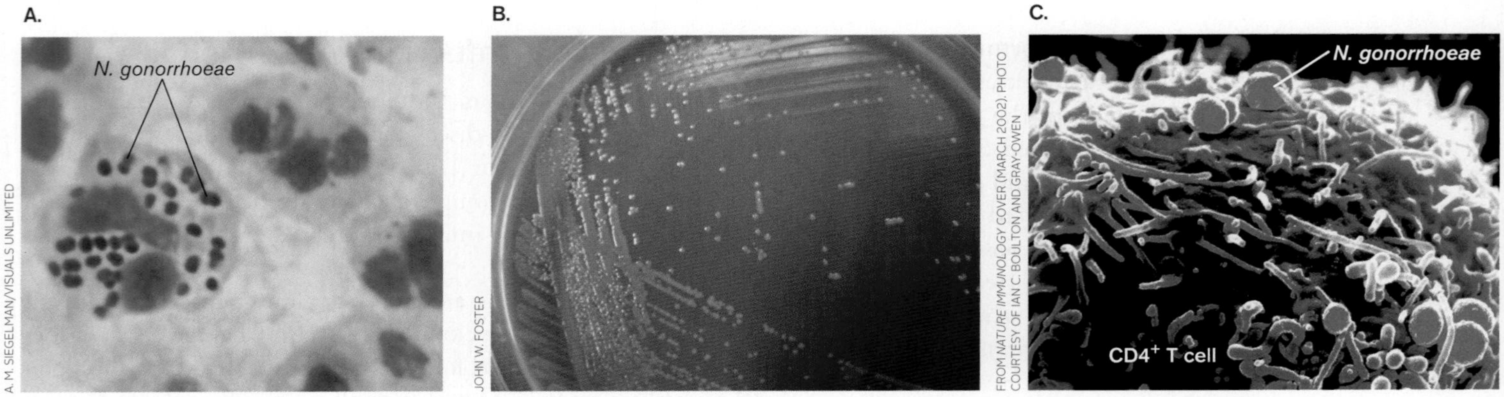
A.
N. gonorrhoeae
A. M. SIEGELMAN/VISUALS UNLIMITED
B.
JOHN W. FOSTER
C.
N. gonorrhoeae
CD4+ T cell
FROM NATURE IMMUNOLOGY COVER (MARCH 2002). PHOTO COURTESY OF IAN C. BOULTON AND GRAY-OWEN

FIGURE 26.13 ■ *Neisseria gonorrhoeae*. **A.** Within pus-filled exudates, the Gram-negative diplococci are found intracellularly inside PMNs. The intracellular bacteria (approx. 0.6–1 μm in diameter) in this case are no longer viable, having been killed by the antimicrobial mechanisms of the white cell. **B.** Colonies of *N. gonorrhoeae* growing on chocolate agar (agar plates containing heat-lysed red blood cells that turn the medium chocolate brown). **C.** *N. gonorrhoeae* binding to $CD4^+$ T cells (colorized SEM), inhibiting T-cell activation and proliferation, which may explain the ease of reinfection.

*phagocytosed Gram-negative diplococci (**Fig. 26.13A**). Blood was drawn for syphilis serology, which proved negative. The patient was given an intramuscular injection of ceftriaxone (250 mg), and oral tetracycline (500 mg, four times a day) was prescribed for 7 days. The bacteriology lab was able to recover the bacteria seen in the Gram-stained smear of the urethral discharge. The organism produced characteristic colonies on chocolate agar (agar plates containing heat-lysed red blood cells that turn the medium chocolate brown) (**Fig. 26.13B**). The case was subsequently reported to the state public health department. Upon his return visit, the patient's symptoms had resolved, and a repeat culture was negative. The patient confirmed that all three of his regular sexual contacts had been seen and evaluated at the STD clinic.*

The disease here is classic gonorrhea caused by *Neisseria gonorrhoeae*. A characteristic distinguishing *Neisseria* infections from *Chlamydia* infections is that bacterial cells are seen in gonorrheal discharges but not in chlamydial discharges. Gonorrhea has been a problem for centuries and remains epidemic in this country today. Symptoms generally occur 2–7 days after infection, but can take as long as 30 days to develop. Most infected men exhibit symptoms; only about 10%–15% do not. Symptoms include painful urination, yellowish white discharge from the penis, and in some cases swelling of the testicles and penis. The Greek physician Galen (AD 129–ca199) originally mistook the discharge for semen. This mistake led to the name *gonorrhea*, which means "flow of seed."

In contrast to men, most infected women (80%) do not exhibit symptoms and constitute the major reservoir of the organism. If they are asymptomatic, they have no reason to seek treatment and thus can spread the disease. When symptoms are present, they are usually mild. A symptomatic woman will experience a painful burning sensation when urinating and will notice vaginal discharge that is yellow or occasionally bloody. She may also complain of cramps or pain in her lower abdomen, sometimes with fever or nausea. As the infection spreads throughout the reproductive organs (uterus and fallopian tubes), pelvic inflammatory disease occurs (see earlier discussion of chlamydia). There is no serological test or vaccine for gonorrhea, because the organism frequently changes the structure of its surface antigens (phase variation is discussed in Section 10.6 and **eTopic 10.5**).

Although *N. gonorrhoeae* is generally serum sensitive, owing to its sensitivity to complement, certain serum-resistant strains can make their way to the bloodstream and carry infection throughout the body. As a result, both sexes can develop purulent arthritis (joint fluid containing pus), endocarditis, or meningitis. An infected mother can also infect her newborn during parturition (birth), leading to a serious eye infection called ophthalmia neonatorum. Because of this risk and because most infected women are asymptomatic, all newborns receive antimicrobial eyedrops at birth.

Because adults engage in a variety of sexual practices, *N. gonorrhoeae* can also infect the anus or the pharynx, where it can develop into a mild sore throat. These infections generally remain unrecognized until a sex partner presents with a more typical form of genitourinary gonorrhea. Because no lasting immunity is built up, reinfection with *N. gonorrhoeae* is possible. Reinfection occurs in part because there is phase variation in various surface antigens and because the organism can apparently bind to $CD4^+$ T cells, inhibiting their activation and proliferation to become memory T cells (**Fig. 26.13C**).

Over the decades, *N. gonorrhoeae* has incrementally developed resistance to antibiotics used in its treatment, but there has always been a new, effective drug ready to take the place of the old drug. In a frightening turn, that may no longer be the case. Dangerous new multidrug-resistant strains have appeared that are resistant to the last available drug—ceftriaxone. The alarm has been raised that new antibiotics must be developed if we are to prevent an uncontrollable explosion of cases of this already epidemic disease.

Thought Question

26.6 Given that *Neisseria gonorrhoeae* is exquisitely sensitive to ceftriaxone, why do you suppose the patient in this case was also treated with tetracycline? And why can one person be infected repeatedly with *N. gonorrhoeae*?

HIV Causes Sexually Transmitted and Blood-Borne Disease

Though HIV (human immunodeficiency virus) is believed to have originated around the year 1900, it was not discovered until 1981, when the virus caused the greatest pandemic of the late twentieth century. HIV remains a serious problem today, especially in Africa, where it is estimated that 10%–30% of the population is infected with HIV (prevalence in the United States is less than 1%). HIV has claimed the lives of almost 2 million people per year worldwide (about 20,000 per year in the United States). You may be surprised to learn that in the United States, women make up about 27% of new AIDS cases per year. The molecular biology and virulence of HIV are discussed in Chapters 11 and 25, and pathogenesis is covered in **eTopic 26.6**. This section focuses on the disease that HIV causes—namely, acquired immunodeficiency syndrome (AIDS).

HIV, a lentivirus in the retroviral family, is a prominent example of viruses that can be transmitted either sexually (vaginally, orally, anally, homosexually, or heterosexually) or through direct contact with body fluids, such as occurs with blood transfusion or the sharing of hypodermic needles by intravenous drug users. HIV is <u>not</u> transmitted by kissing, tears, or mosquito bites. It can, however, be transferred from mother to fetus through the placenta (transplacental transfer). **Figure 26.14A** shows a worldwide decrease in the number of newly infected HIV patients and deaths due to AIDS. The number of people living with HIV has increased, however, because of the development

A.

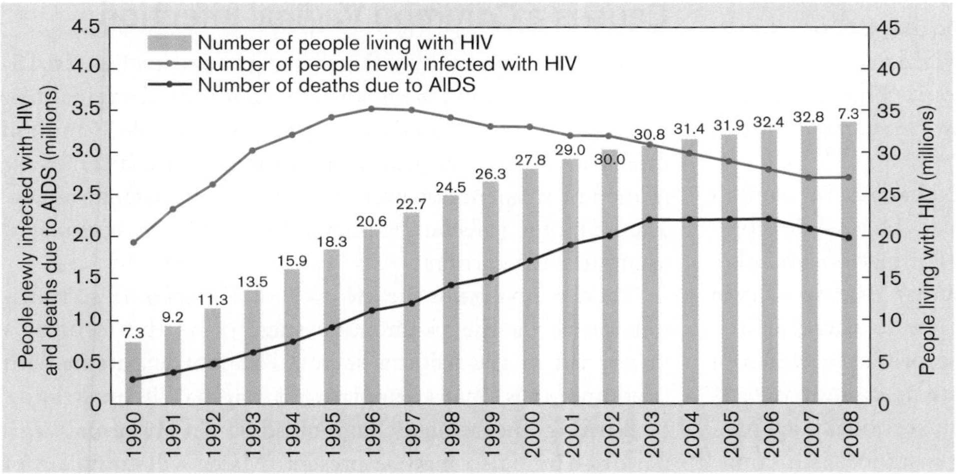
Number of people living with HIV
Number of people newly infected with HIV
Number of deaths due to AIDS
People newly infected with HIV and deaths due to AIDS (millions)
People living with HIV (millions)
4.5
4.0
3.5
3.0
2.5
2.0
1.5
1.0
0.5
0
45
40
35
30
25
20
15
10
5
0
7.3
9.2
11.3
13.5
15.9
18.3
20.6
22.7
24.5
26.3
27.8
29.0
30.0
30.8
31.4
31.9
32.4
32.8
7.3
1990
1991
1992
1993
1994
1995
1996
1997
1998
1999
2000
2001
2002
2003
2004
2005
2006
2007
2008

B.

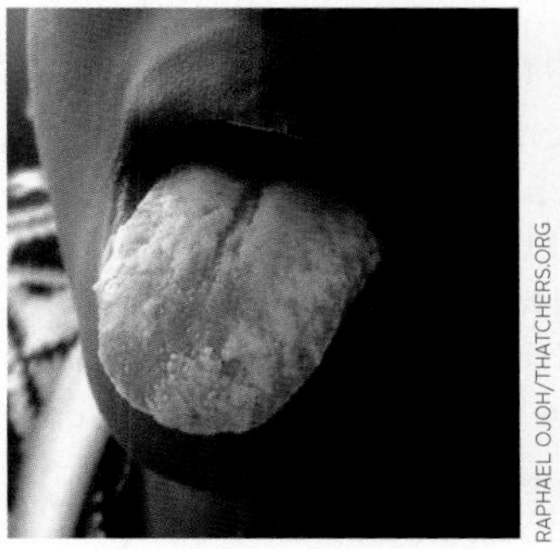
RAPHAEL OJOH/THATCHERS.ORG

C.

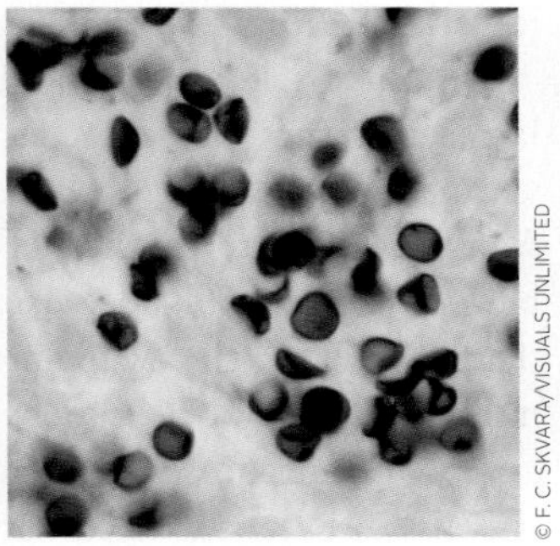
© F. C. SKVARA/VISUALS UNLIMITED

D.

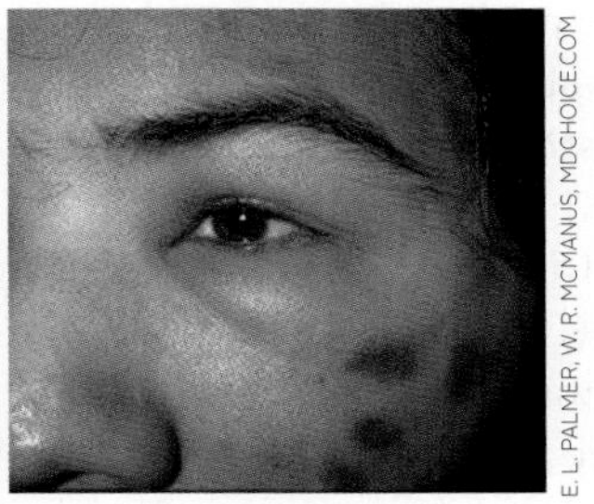
E. L. PALMER, W. R. MCMANUS, MDCHOICE.COM

FIGURE 26.14 ■ Acquired immunodeficiency syndrome. **A.** HIV infections and AIDS deaths worldwide. The number of people living with HIV continues to increase, yet the number of people newly infected with HIV and the number of deaths due to AIDS have decreased. **B.** Oral candidiasis (thrush). The white patches are caused by secondary infection by the yeast *Candida albicans*. **C.** *Pneumocystis jirovecii* infection of the lung. Note the cuplike appearance of the fungus, almost like crushed Ping-Pong balls. Organisms range from 2 to 6 µm in diameter. **D.** Kaposi's sarcoma (oval spots) and periorbital cellulitis infection.

of more effective antiviral treatments. The pathogenesis of HIV and the proportion of AIDS cases by sex and ethnicity are described in **eTopic 26.6**.

Having entered the bloodstream, HIV infects $CD4^+$ T cells and replicates very rapidly, producing a billion particles per day. The subsequent decrease in $CD4^+$ T cells leads to the disease symptoms collectively called AIDS. The symptoms of AIDS may take years to develop.

The early stage of the disease, previously known as AIDS-related complex (ARC), can include fever, headache, macular rash, weight loss, and the appearance of antibodies to HIV in serum. These early, relatively mild symptoms can manifest within a few months of infection, resolve within a few weeks, and then may recur. The debilitated immune system leads to secondary infections by the yeast *Candida albicans* (candidiasis) and related species (**Fig. 26.14B**).

For several years after infection the patient remains asymptomatic, but the virus will continue to replicate and destroy $CD4^+$ T cells (called chronic asymptomatic HIV). Initially the T cells are quickly replaced, but eventually the body cannot keep up and T-cell numbers fall. The disease progresses to advanced HIV disease (AIDS), marked by a significant depletion of the $CD4^+$ T-cell population. The CDC definition of AIDS includes all persons with a $CD4^+$ cell count of less than 200/µl. Once the $CD4^+$ T-cell population falls below 400 cells/µl, opportunistic infections begin to arise and disease processes begin. Opportunistic infections include pneumonia by *Mycobacterium avium-intracellulare* or *Pneumocystis jirovecii* (**Fig. 26.14C**), cryptococcal meningitis, *Histoplasma capsulatum* infection, and tuberculosis.

A patient with AIDS can exhibit changes in mental cognition, muscular action, and reflexes. These neurological changes appear coincident to inflammation and the demyelination of neurons. Cardiovascular disease is also common in AIDS patients. In addition to these afflictions, AIDS patients are more susceptible to cancers because their depressed immune systems cannot detect and destroy cancer cells generated by secondary agents. Kaposi's sarcoma (**Fig. 26.14D**), for instance, is a common cancer seen in AIDS patients that is caused by human herpes virus type 8 (HHV8). Kaposi's sarcoma originates in endothelial or lymphatic cells, but the resulting tumors can arise anywhere—gastrointestinal tract, mouth, lungs, skin, or brain.

Diagnosis of AIDS involves detecting anti-HIV antibodies and determining the $CD4^+$ T-cell count in a patient. An assay for HIV to determine viral load is done via quantitative PCR to detect HIV-specific genes such as *gag*, *nef*, or *pol*. Remember, a person who is HIV-positive does not necessarily have AIDS; the disease may take years to develop. A vaccine is not yet available to prevent AIDS, in part because the envelope proteins of the virus (see Fig. 11.24) typically change their antigenic shape. However, progression of the disease can be controlled by antiretroviral drugs that inhibit two HIV enzymes critical to the replication of the virus: reverse transcriptase and protease. Antiretroviral drugs are discussed further in Chapter 27.

New treatment regimens (HAART; see Chapter 27) have made HIV an increasingly survivable infection. Consider this: In the past, nearly all HIV-infected individuals eventually became ill and died from AIDS-related diseases, but in the United States at least half of, if not most, HIV-positive people now die from diseases unrelated to HIV (for example, heart attack). Because HIV infection is now considered a treatable disease, the CDC recommends routine screening for HIV.

Thought Question

26.7 Like the cause of plague, HIV is a blood-borne disease. Why, then, do fleas and mosquitoes fail to transmit HIV?

The Protozoan *Trichomonas vaginalis* Causes a Common Vaginal Infection

Trichomonas vaginalis is a flagellated protozoan (**Fig. 26.15**) that causes an unpleasant sexually transmitted vaginal disease called trichomoniasis. Approximately 2–3 million infections occur each year in the United States. It can occur in men or women; however, men are usually asymptomatic. Even among infected women, 25%–50% are considered asymptomatic carriers.

There is no cyst in the life cycle of *T. vaginalis*, so transmission is via the trophozoite stage only (the form of a protozoan in the feeding stage). The female patient with trichomoniasis may complain of vaginal itching and/or burning and a musty vaginal odor. An abnormal vaginal discharge also may be present. Males will complain of painful urination (dysuria), urethral or testicular pain, and lower abdominal pain.

Owing to colonization by lactobacilli (which produce large amounts of acidic lactic acid), the normal, healthy vagina has a pH of less than 4.5. However, since *T. vaginalis* feeds on bacteria, the pH of the vagina rises as the numbers of lactobacilli decrease. Definitive diagnosis requires demonstrating the flagellated protozoan in secretions by microscopy. PMNs, which are the primary host defense against the organism, are also usually present. Like giardiasis, this disease is treated with metronidazole.

DAVID M. PHILLIPS/VISUALS UNLIMITED

FIGURE 26.15 ▪ *Trichomonas vaginalis.* *T. vaginalis* is a protozoan that causes a common sexually transmitted disease (size 7 × 10 μm, SEM).

To Summarize

- **Urinary tract infections (UTIs)** can result from ascending (to the kidney) or descending (from the kidney) routes of infection. The most common route leading to bladder infection, however, is through the urethra.
- ***E. coli*** is the most common cause of UTIs.
- **Syphilis, gonorrhea, and chlamydia** are the most common sexually transmitted diseases.
- **A patient with one sexually transmitted disease** often has another sexually transmitted disease too.
- **Complement sensitivity** prevents dissemination by *Neisseria gonorrhoeae*. In contrast, *N. meningitidis*, a cause of meningitis, frequently disseminates in the bloodstream because it is complement resistant.
- **HIV depletion of CD4$^+$** T cells results in lethal secondary infections and cancers.
- ***Trichomonas vaginalis*** is a flagellated protozoan that causes a sexually transmitted vaginal disease. The reservoirs for this organism are the male urethra and female vagina.

26.6

Central Nervous System Infections

Microbes cannot gain easy access to the brain in large measure because of the **blood-brain barrier**, a filter mechanism that allows only selected substances into the brain. The blood-brain barrier works to our advantage when harmful substances, such as bacteria, are prohibited from entering. However, it works to our disadvantage when substances we want to enter the brain, such as antibiotics, are kept out. The barrier is not a single structure but a function of the way blood vessels, especially capillaries, are organized in the brain. Furthermore, the endothelial cells in those vessels have tight junctions that do not allow most compounds or microbes to cross. And yet, brain infections do occur.

Case History: Meningitis

*In April 2001, a 4-month-old infant from Saudi Arabia was hospitalized with fever, tender neck, and purplish spots (purpuric spots) on her trunk (**Fig. 26.16A**). Suspecting meningitis, the clinician took a cerebrospinal fluid (CSF) sample*

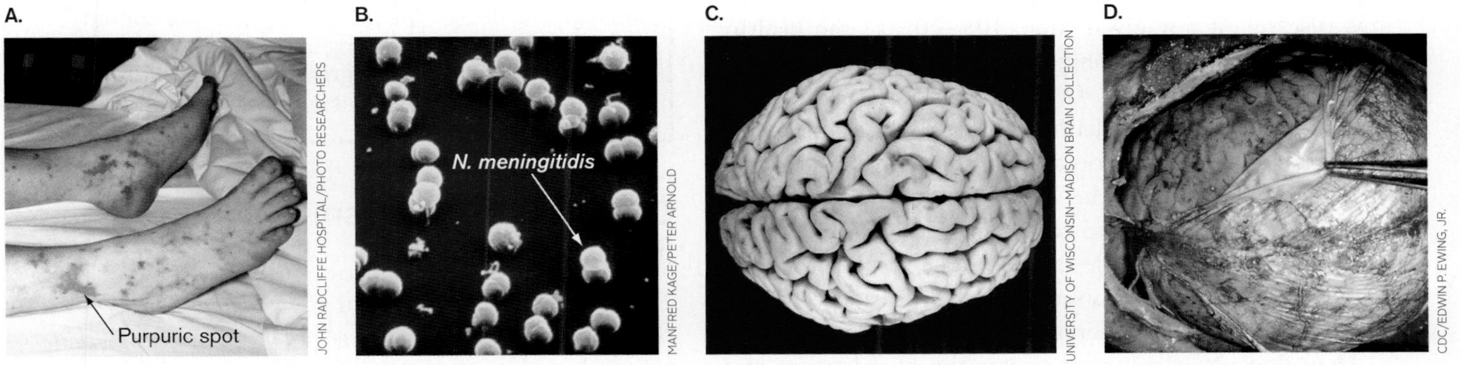
A.
Purpuric spot
JOHN RADCLIFFE HOSPITAL/PHOTO RESEARCHERS
B.
N. meningitidis
MANFRED KAGE/PETER ARNOLD
C.
UNIVERSITY OF WISCONSIN-MADISON BRAIN COLLECTION
D.
CDC/EDWIN P. EWING, JR.

FIGURE 26.16 ▪ Bacterial meningitis. Meningococcal disease is dreaded by parents and medical practitioners alike for its rapid onset and the difficulty of obtaining a timely and accurate diagnosis. **A.** Purpuric spots produced by local intravascular coagulation due to *Neisseria meningitidis* endotoxin. The rash in meningitis typically has petechial (small) and purpuric (large) components. **B.** *N. meningitidis* (diameter approx. 1 μm, SEM). **C.** Normal brain. **D.** Autopsy specimen of meningitis due to *Streptococcus pneumoniae*. Note the greening of the brain, compared with the pink normal brain.

and examined it by Gram stain. The smear revealed Gram-negative diplococci inside PMNs. The CSF was turbid with 900 leukocytes per microliter, and Neisseria meningitidis *was confirmed by culture. The child was treated with cefotaxime and made a full recovery. Her father, the person who brought her in, was clinically well. However, the meningococcus was isolated from his oropharynx, as well as from the throat of the patient's 2-year-old brother. Isolates from the patient, her father, and her brother were positive by agglutination with meningococcus A, C, Y, W135 polyvalent reagent. The father's vaccination certificate confirmed that he had received a quadrivalent meningococcal vaccine. All three isolates were sent to the WHO (World Health Organization) Collaborating Centre, which confirmed meningococcus serogroup W135. The three isolates were examined by pulsed-field gel electrophoresis and were found to be indistinguishable (that is, they had identical DNA restriction patterns).*

Meningitis is an inflammation of the meninges, the membranes that surround the brain and spinal cord. Meningitis can be either bacterial or viral in origin. Sinus and ear infections can extend directly to the meninges, while septicemic spread requires passage through the blood-brain barrier. Viral meningitis is serious but rarely fatal in people with a normal immune system. The symptoms generally persist for 7–10 days and then completely resolve. Bacterial meningitis is usually caused by *Streptococcus pneumoniae*, *Neisseria meningitidis*, or *Haemophilus influenzae*. Symptoms of bacterial meningitis can include sudden onset of fever, headache, neck pain or stiffness, painful sensitivity to strong light (photophobia), vomiting (often without abdominal complaints), and irritability. Prompt medical attention is extremely important because the disease can quickly progress to convulsions and death.

The meningococcus *N. meningitidis* (**Fig. 26.16B**) can colonize the human oropharynx, where it causes mild, if any, disease. At any given time, 10%–20% of the healthy population can be colonized and asymptomatic. The organism spreads directly by person-to-person contact or indirectly via droplet nuclei from sneezing or fomites. The problem arises when this organism enters the bloodstream. Unlike *N. gonorrhoeae*, the cause of gonorrhea, *N. meningitidis* is very resistant to complement, owing to its production of a polysaccharide capsule. The capsule allows the microbe to produce a transient blood infection (bacteremia) and reach the blood-brain barrier.

How does *N. meningitidis* enter the bloodstream and the brain? The organism initially uses type IV pili to adhere to and enter nasopharyngeal endothelial cells (type IV pili are discussed in Section 25.3). The bacteria then cross the endothelial cell layer by **transcytosis**, a process by which an internalized pathogen passes through a host cell to the opposite side. Ultimately, the pathogen passes into the capillary lumen and is swept away to the brain. Once in the brain, type IV pili again adhere to endothelial cells, but this time they trigger the recruitment of host proteins that destabilize intercellular junctions. The bacteria then slip between the loosened junctions and enter the cerebrospinal fluid. Once in the CSF, microbes can multiply almost at will. **Figure 26.16C** and **D** show the remarkable damage (greening) that *N. meningitidis* and other microbes, such as *S. pneumoniae*, cause in the brain.

Several antigenic types of capsules, called type-specific capsules, are produced by different strains of pathogenic *N. meningitidis*: types A, B, C, W135, and Y. Types A, C, Y, and W135 are usually associated with epidemic infections seen among people kept in close proximity, such as college students or military personnel. Type B meningococcus is typically involved in sporadic infections. Antibodies to these capsular antigens are used to classify the capsular types of the organisms causing an outbreak. Knowing the capsular type of organism involved in each case helps determine whether the disease cases are related and where the infection may have started.

Meningococcal meningitis is highly communicable. As a result, close contacts, such as the parents or siblings of any patient with meningococcal disease, should receive antimicrobial prophylaxis within 24 hours of diagnosis; a single dose of ciprofloxacin (a quinolone antibiotic, discussed in Section 27.4) can be given to adults, and 2 days of rifampicin given to children. Highly susceptible populations can be immunized with a vaccine containing four of the five capsular structures (type B is not immunogenic; Section 26.9).

Thought Question

26.8 Normal cerebrospinal fluid is usually low in protein and high in glucose. The protein and glucose content does not change much during a viral meningitis, but bacterial infection leads to greatly elevated protein and lowered glucose levels. What could account for this?

Case History: Botulism—It Is What You Eat

In June, a 47-year-old resident of Oklahoma was admitted to the hospital with rapid onset of progressive dizziness, blurred vision, slurred speech, difficulty swallowing, and nausea. Findings on examination included drooping eyelids, facial paralysis, and impaired gag reflex. He developed breathing difficulties and required mechanical ventilation. The patient reported that during the 24 hours before onset of symptoms, he had eaten home-canned green beans and a stew containing roast beef and potatoes. Analysis of the patient's stool detected botulinum type A toxin, but no Clostridium botulinum *organisms were found. The patient was hospitalized for 49 days, including 42 days on mechanical ventilation, before being discharged.*

Imagine a disease that causes complete loss of muscle function. Using secreted exotoxins (neurotoxins), two microbes cause lethal paralytic diseases. In one instance, the victim suffers a flaccid paralysis in which the muscles go limp, as in the case history just given, causing paralysis and respiratory difficulty. Voluntary muscles fail to respond to the mind's will because botulinum toxin interferes with neural transmission. The disease (called **botulism**) is typically food-borne and is caused by an anaerobic, Gram-positive, spore-forming bacillus named *Clostridium botulinum* (Chapter 18).

In striking contrast to botulism is tetanus, a very painful disease in which muscles continually and involuntarily contract (called tetany or spastic paralysis). Tetanus is caused by **tetanospasmin**, a potent exotoxin made by another anaerobic, Gram-positive spore-forming bacillus, called *Clostridium tetani* (**Fig. 26.17A**). Tetanospasmin interferes with neural transmission, but in contrast to botulism toxin, it causes excessive nerve signaling to muscles, forcing the victim's back to arch grotesquely while the arms flex and legs extend. The patient remains locked this way until death. Spasms can be strong enough to fracture the patient's vertebrae. **Figure 26.17B** shows the result of injecting a mouse's hind leg with just a tiny amount of tetanus toxin. In both botulism and tetanus, death can result from asphyxiation.

Botulism is typically caused by ingesting preformed toxin, although infected wounds or germination of ingested spores can also occasionally produce disease. The microbe germinates in the food and produces toxin. Because the organism is an anaerobe, an anaerobic environment must be present. Home-canning processes are designed to remove oxygen in the canned food, as well as sterilize the food; but if the sterilization process is incomplete, then surviving spores germinate and produce toxin. The toxin is susceptible to heat but will remain active in

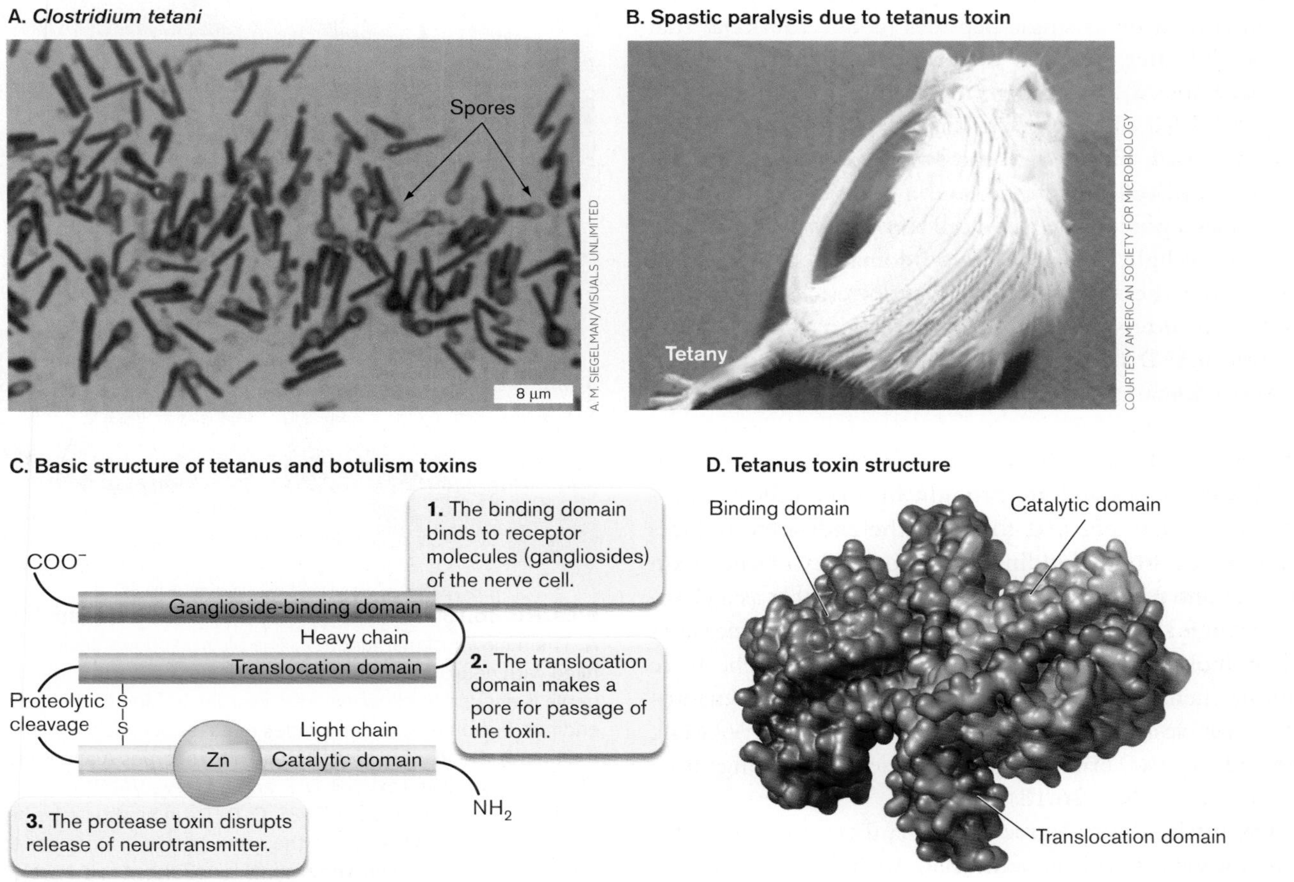
A. Clostridium tetani
Spores
8 µm
A. M. SIEGELMAN/VISUALS UNLIMITED
B. Spastic paralysis due to tetanus toxin
Tetany
COURTESY AMERICAN SOCIETY FOR MICROBIOLOGY
C. Basic structure of tetanus and botulism toxins
1. The binding domain binds to receptor molecules (gangliosides) of the nerve cell.
COO⁻
Ganglioside-binding domain
Heavy chain
Translocation domain
2. The translocation domain makes a pore for passage of the toxin.
Proteolytic cleavage
S
S
Light chain
Zn
Catalytic domain
NH2
3. The protease toxin disrupts release of neurotransmitter.
D. Tetanus toxin structure
Binding domain
Catalytic domain
Translocation domain

FIGURE 26.17 ■ **Tetanus and botulism toxins. A.** Photomicrograph of *Clostridium tetani* (cell length 4–8 µm). **B.** Mouse injected with tetanus toxin in left hind leg. **C.** Schematic diagram of tetanus and botulism toxins. **D.** 3D representation of the tetanus neurotoxin with the domains marked. (PDB code: 3BTA) improperly cooked food. After ingestion, the toxin is absorbed from the intestine. As in the case history given here, it is not unusual for the organism to be absent from stool samples. This is why serological identification of the toxin is important.

A rare form of botulism, called infant botulism or "floppy head syndrome," can occur when infants (not older children) are fed honey. Honey can harbor *C. botulinum* spores that can germinate in the gastrointestinal tract, after which the growing vegetative cells will secrete toxin.

Thought Question

26.9 Knowing the symptoms of tetanus, what kind of therapy would you use to treat the disease?

Toxin structure. Botulism and tetanus toxins share 30%–40% identity and have similar structures and nearly identical modes of action. Botulism and tetanus toxins are each composed of two peptides—a large, or heavy, fragment analogous to a B subunit of AB toxins and a small, or light, fragment analogous to an A subunit. Both toxins are initially made as single peptides (about 150 kDa) that are cleaved after secretion to form two fragments (heavy and light chains) that remain tethered by a disulfide bond (**Fig. 26.17C**). Each heavy chain includes the binding domain, which binds to receptor molecules (gangliosides) on the nerve cell membrane, and a translocation domain that makes a pore in the nerve cell through which the toxin passes. The light chains (catalytic domains) are proteases that disrupt the movement of exocytic vesicles containing neurotransmitters needed for contraction or relaxation. **Figure 26.17D** shows a 3D rendition of tetanus toxin with the three domains marked.

Mechanism of neurotoxin action. Both toxins bind to target membranes and are brought into the cell by endocytosis. The low pH that forms in the endosome reduces the disulfide bonds holding the two halves of the toxin together and causes the heavy chain to assemble as a channel through the membrane. That channel allows release of the proteolytic light chain into the cytoplasm. The toxic subunit then cleaves key host proteins, such as synaptobrevin (or vesicle-associated membrane protein, VAMP), involved in the exocytosis of vesicles containing neurotransmitters (**Fig. 26.18**).

With such a high degree of mechanistic similarity, why do tetanus and botulism toxins have such drastically different effects? The answer is based on where each toxin acts in the nervous system. Both toxins enter at peripheral nerve endings where nerve meets muscle. The disease botulism

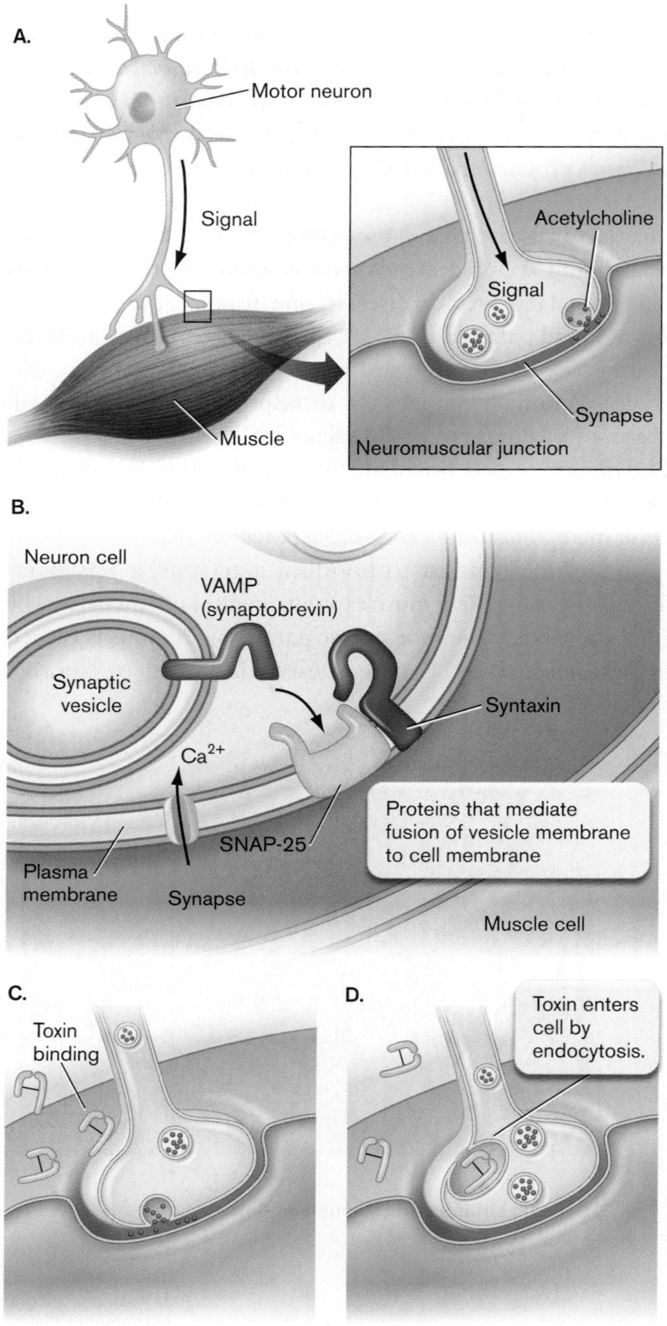
A.
Motor neuron
Signal
Muscle
Acetylcholine
Signal
Synapse
Neuromuscular junction
B.
Neuron cell
VAMP
(synaptobrevin)
Synaptic vesicle
Syntaxin
Ca^{2+}
Proteins that mediate fusion of vesicle membrane to cell membrane
SNAP-25
Plasma membrane
Synapse
Muscle cell
C.
Toxin binding
D.
Toxin enters cell by endocytosis.

FIGURE 26.18 ■ Mechanism of action of botulism toxin. **A.** The neuromuscular junction. The blowup shows vesicles filled with neurotransmitters. **B.** A series of proteins within the nerve is needed to allow synaptic vesicles to bind to the nerve endings. Fusion of the membranes releases acetylcholine into the neuromuscular junction. Botulism toxin types A and E cleave SNAP-25. Botulism toxins B, D, F, and G cleave VAMP. Botulism toxin C1 cleaves syntaxin and SNAP-25. **C, D.** Toxin binds via the heavy chain and is endocytosed into the nerve terminal. Once the toxin cleaves its target, the nerve terminal is no longer able to release acetylcholine.

results from the ingestion of preformed toxin in poorly prepared, contaminated foods. Whether the organism is also ingested is immaterial. Once botulism toxin has entered a peripheral nerve, it immediately cleaves peptides associated with the exocytic release of the neurotransmitter acetylcholine, such as VAMP, syntaxin, and synaptosomal-associated protein (SNAP-25) (**Fig. 26.18**). Without acetylcholine to activate nerve transmission, muscles will not contract and the patient is paralyzed.

Although botulism disease is now rare, the toxin has gained renewed interest in recent years. It is considered a select biological toxic agent of potential use to bioterrorists, but that is not where most interest lies. Because it can safely relax muscles in a localized area if injected in small doses, botulism toxin, or Botox, is used cosmetically by plastic surgeons to reduce facial wrinkles in some patients, and therapeutically by neurologists to treat migraine headaches.

Tetanus, in contrast to botulism, is not a food-borne disease. *Clostridium tetani* spores are introduced into the body by trauma (such as by stepping on the wrong end of a nail). Necrotic tissue then provides the anaerobic environment required for germination. Growing, vegetative cells release tetanospasmin, which enters the peripheral nerve cells at the site of injury. But rather than cleaving targets here, the toxin travels up axons in the direction opposite to nerve signal transmission until it reaches the spinal column, where it becomes fixed at the presynaptic inhibitory motor neuron. There the toxin also cleaves proteins like VAMP, but the function of vesicles in these nerves is to release inhibitory neurotransmitters that dampen nerve impulses. Tetanus toxin blocks release of the inhibitory neurotransmitters GABA and glycine into the synaptic cleft, leaving nerve impulses unchecked. As a result, impulses come too frequently and produce the generalized muscle spasms characteristic of tetanus. **Figure 26.19** illustrates the mechanism of action of tetanus toxin. A tetanus toxoid vaccine is available and is administered to children as part of the DTaP (diphtheria, tetanus, and acellular pertussis) vaccine (Section 26.9).

Thought Questions

26.10 **Figure 26.19** demonstrates that tetanus toxin has a mode of action (spastic paralysis) opposite to that of botulism toxin (flaccid paralysis). Since they have opposing modes of action, can botulism toxin be used to save a patient with tetanus?

26.11 How do the actions of tetanus toxin and botulinum toxin actually help the bacteria colonize or obtain nutrients?

26.12 If *Clostridium botulinum* is an anaerobe, how might botulism toxin get into foods?

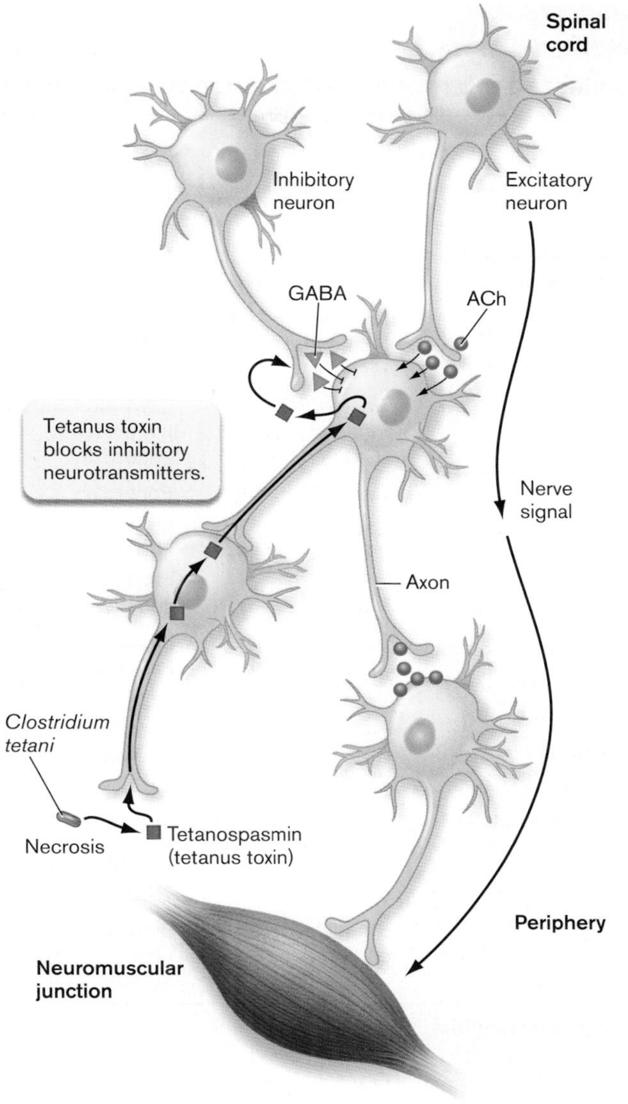
Spinal cord
Inhibitory neuron
Excitatory neuron
GABA
ACh
Tetanus toxin blocks inhibitory neurotransmitters.
Nerve signal
Axon
Clostridium tetani
Necrosis
Tetanospasmin (tetanus toxin)
Periphery
Neuromuscular junction

FIGURE 26.19 ■ Retrograde movement of tetanus toxin to an inhibitory neuron. Tetanus toxin enters the nervous system at the neuromuscular junction and travels retrogradely up the axons until reaching an inhibitory neuron located in the central nervous system. There it cleaves VAMP protein (Fig. 26.18B) associated with the exocytosis of vesicles containing inhibitory neurotransmitters. ACh = acetylcholine; GABA = gamma-aminobutyric acid.

Case History: Eastern Equine Encephalitis

In August, Mr. C brought his 21-year-old son, Rich, to a New Jersey emergency room. Rich appeared dazed and had trouble responding to simple commands. When questioned about his son's activities over the previous few months, Mr. C told the physician that Rich planned to enter veterinary school in the

TABLE 26.5

Selected microbes that cause meningitis or encephalitis.

Type of disease	Etiological agent	Virulence factors	Typical initial infection or source	Treatment	Vaccine
Bacterial (septic) meningitis	*Streptococcus pneumoniae* (G+)	Capsule, pneumolysin	Lung	Ampicillin	Multivalent, capsule
	Neisseria meningitidis (G–)	Capsule, IgA protease, endotoxin	Throat	Cephalosporin (3rd generation), ceftriaxone	Multivalent, capsule
	Haemophilus influenzae type B (G–)	Polyribitol capsule, IgA protease, endotoxin	Ear infection	Cephalosporin (3rd generation), ceftriaxone	Type b polysaccharide
	Other G– bacilli	Endotoxin	Septicemia		
	Group B streptococci (G+)	Sialic acid capsule, streptolysin, inhibition of alternate complement path	Neonate infected during parturition	Ampicillin	Capsular
	Listeria monocytogenes (G+)	Intracellular growth, PrfA regulator, actin-based motility	Mild GI disease of mother	Ampicillin plus gentamicin	
	Mycobacterium tuberculosis	Cord factor, wax D, intracellular growth	Lung	Combination therapy (rifampin, isoniazid, ethambutol, pyrazinamide)	BCG[c]
	Staphylococcus aureus (G+)	Coagulase, protein A, TSST, leukocidin	Septicemia	Methicillin, vancomycin	
	Staphylococcus epidermidis (G+)	Biofilm, slime	Complication of surgical procedure	Vancomycin, methicillin	
Aseptic meningitis[a]	Fungi (e.g., *Coccidioides, Cryptococcus*)		Sinusitis, direct spread to meninges	Ketoconazole, fluconazole	
	Amebas (e.g., *Naegleria*)		Swimming in contaminated waters	Amphotericin B	
	Treponema pallidum		Syphilis		
	Mycoplasmas	Adhesin tip	Respiratory	Erythromycin	
	Leptospira	Burrowing motility	Septicemia, water contaminated with animal urine	Erythromycin	Killed whole cell, animals only
Viral meningitis or encephalitis	Viruses (90% caused by enteroviruses)			Self-limiting	
Eastern equine encephalitis	EEE virus	?	Mosquito bite	None; often fatal	
West Nile disease	West Nile virus	?	Mosquito bite	Supportive therapy[b]	

[a]Aseptic meningitis is an inability to isolate bacterial sources by ordinary means.

[b]"Supportive therapy" means hospitalization, intravenous fluids, airway management, respiratory support, and prevention of secondary infections.

[c]BCG = Bacille Calmette-Guérin (a weakened strain of the bovine tuberculosis strain).

fall and had spent the month of July relaxing and sunning himself on New Jersey beaches. Aside from the shore, his favorite locale was a pond in a wooded area near a horse farm. On the afternoon prior to admission, Rich had become lethargic and tired. He returned home and went to bed. That evening, his father woke him for supper, but Rich was confused and had no appetite. By 11 p.m., Rich had a fever of 40.7°C (103.5°F) and could not respond to questions. A few hours later, when his father had trouble rousing him, he brought Rich to the ER. Over the next week, Rich's condition worsened to the point where his limbs were paralyzed. Two weeks later he died. Serum samples taken upon entering the hospital and a few days before he died showed a sixfold rise in antibody titer to eastern equine encephalitis (EEE) virus. Brain autopsy showed many small foci of necrosis in both the gray and white matter.

The EEE virus, a member of the family *Togaviridae,* is transmitted from bird to bird by mosquitoes. Horses contract the disease in the same way. Rich contracted it inadvertently while visiting the pond. The disease, an encephalitis, is often fatal (35% mortality) but fortunately rare. One reason human disease is rare is that the species of mosquito that usually transmits the virus between marsh birds does not prey on humans. But sometimes a human-specific mosquito bites an infected bird and then transmits EEE virus to humans. Another reason human disease is rare is that the virus generally does not fare well in the body. Many persons infected with EEE virus have no apparent illness because the immune system thwarts viral replication. However, as already noted, mortality is high in those individuals that do develop disease.

An interesting point in the case history is that diagnosis relied on detecting an increase in antibody titer to the virus. Typically, at the point that disease symptoms first appear, the body has not had time to generate large amounts of specific antibodies. After a week or so, often when the patient is nearing recovery (convalescence), antibody titers have risen manyfold. The rule of thumb is that a greater-than-fourfold rise in antibody titer between acute disease and convalescence (or in this case death) indicates that the patient has had the disease. While this knowledge could not help the patient in the case described here, it was valuable in terms of public health and prevention strategies.

Several bacterial, fungal, and viral causes of encephalitis and meningitis are listed in **Table 26.5**.

Prions Are Infectious Proteins

An unusual infectious agent called the prion has been implicated as the cause of a series of relatively rare but invariably fatal brain diseases (**Table 26.6**). Prions are infectious agents that do not have a nucleic acid genome. It seems that a protein alone can mediate an infection. The **prion** has been defined as a small proteinaceous infectious particle that resists inactivation by procedures that modify nucleic acids. The discovery that proteins alone can transmit an infectious disease came as a surprise to the scientific community. Diseases caused by prions are especially worrying, since prions resist destruction by many chemical agents and remain active after heating at extremely high temperatures. There have even been documented cases in which sterilized surgical instruments, originally used on a prion-infected person, still held infectious agent and transmitted the disease to a subsequent surgery patient.

How can a nonliving entity without nucleic acid be an infectious agent? Prions associated with human brain

TABLE 26.6

Prion diseases.

Disease	Susceptible animal	Incubation period	Disease characteristics
Creutzfeldt-Jakob (sporadic, familial, new variant [vCJD])	Human	Months (vCJD) to years	Spongiform encephalopathy (degenerative brain disease)
Kuru	Human	Months to years	Spongiform encephalopathy
Gerstmann-Straussler-Scheinker syndrome	Human	Months to years	Genetic neurodegenerative disease
Fatal familial insomnia	Human	Months to years	Genetic neurodegenerative disease with untreatable insomnia
Mad cow disease	Cattle	5 years	Spongiform encephalopathy
Wasting disease	Deer	Months to years	Spongiform encephalopathy

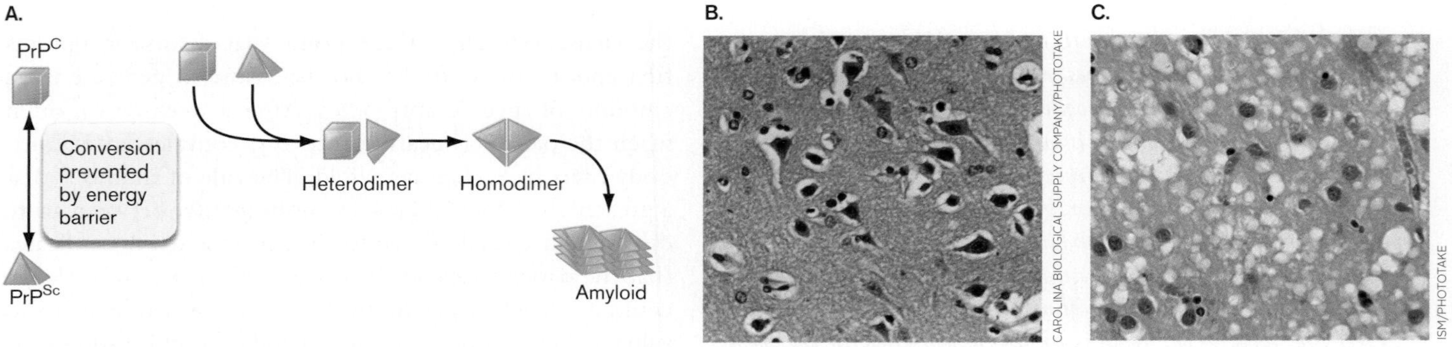
A.
PrP^C
Conversion prevented by energy barrier
PrP^Sc
Heterodimer
Homodimer
Amyloid
B.
CAROLINA BIOLOGICAL SUPPLY COMPANY/PHOTOTAKE
C.
ISM/PHOTOTAKE

FIGURE 26.20 ■ Spongiform encephalopathies. A. Refolding model of prion diseases. PrP is a brain protein that can take two forms: the normal form (PrP^C), which is a natural brain protein, and the prion form (PrP^{Sc}). When a prion form gains access to the brain (following ingestion or other means), it can cause the slow refolding of normal PrP^C proteins to form aggregates that damage the brain. **B.** Normal brain section. **C.** Section of brain taken from a CJD victim. Note the "Swiss cheese" appearance, indicating brain damage.

disease are thought to be aberrantly folded forms of a normal brain protein. The theory is that when a prion is introduced into the body and manages to enter the brain, it will cause normally folded forms of the protein to refold incorrectly (**Fig. 26.20A**). The improperly folded proteins fit together like Lego blocks to produce damaging aggregated structures within brain cells.

Some investigators suggest that living, infectious agents, such as *Spiroplasma* (a spiral-shaped genus whose members lack cell walls), can somehow precipitate a conformational change in the normal brain protein, converting it to an infectious prion protein. This proposal is not universally accepted, however, and has yet to be proved.

Prion diseases are often called **spongiform encephalopathies** because the postmortem appearance of the brain includes large, spongy vacuoles in the cortex and cerebellum. These are visible in a brain sample from a victim of one of these diseases, called Creutzfeldt-Jakob disease (CJD) (**Fig. 26.20B** and **C**). Most mammalian species appear to develop these diseases.

Since 1996, mounting evidence has pointed to a causal relationship between outbreaks in Europe of a disease in cattle called bovine spongiform encephalopathy (BSE, or "mad cow disease") caused by prions and a disease in humans called variant Creutzfeldt-Jakob disease (vCJD). Both disorders are invariably fatal. For humans to contract disease, prions from contaminated meat are ingested and penetrate the intestinal mucosa via the antigen-sampling M cells in Peyer's patches. The circulation then traffics the agent to the brain.

As of this writing, only four cases of BSE in cattle have been detected in the United States. Because of aggressive surveillance efforts in the United States and Canada (where only nine cases have been found), it is unlikely, but not impossible, that BSE will be a food-borne hazard to humans in this country. The CDC monitors the trends and current incidence of typical CJD (approximately 300 cases per year) and variant CJD (three cases total) in the United States (see eTopic 26.7). The worldwide incidence of CJD is one per 1 million population.

To Summarize

- ***Neisseria meningitidis*** is resistant to serum complement because it produces a type-specific capsule. This allows the organism to reach and then cross the blood-brain barrier.
- **A vaccine for *N. meningitidis*** is available, but none exists for *N. gonorrhoeae.*
- **Tetanospasmin** causes spastic paralysis.
- **Botulism toxin** causes flaccid paralysis.
- **Serological diagnosis of an infectious disease** is possible if specific pathogen antibody titer rises fourfold between the acute and convalescent stages.
- **Spongiform encephalopathies** are believed to be caused by nonliving proteins called prions.

26.7

Cardiovascular System Infections

Infections of the cardiovascular system include septicemia, **endocarditis** (inflammation of the heart's inner lining), pericarditis (inflammation of the heart's outer lining) and, possibly, atherosclerosis (the deposition of fatty substances along the inner lining of arteries; eTopic 26.8). These are all life-threatening diseases.

Septicemia is, by strict definition, the presence of bacteria or viruses in the blood. The presence of viruses is a condition called **viremia**, while bacteria in the circulation is more specifically called **bacteremia**. In practice, however, the terms "septicemia" and "bacteremia" are often used interchangeably. Septicemia can develop from a local infection situated anywhere in the body, although blood factors such as complement can nonspecifically kill many types of bacteria that enter the blood. Nevertheless, Gram-positives, Gram-negatives, aerobes, and anaerobes can all produce septicemia under the right conditions.

Endocarditis can be either viral or bacterial in origin. It can be a consequence of many bacterial diseases, such as brucellosis, gonorrhea, psittacosis, staphylococcal and streptococcal infections, candidiasis, and Q fever. Among the many viral causes are coxsackievirus, echovirus, Epstein-Barr virus, and HIV. Bacterial infections of the heart are always serious. Viral infections, although common, are rarely life-threatening in healthy individuals and are usually asymptomatic.

Case History: Bacterial Endocarditis

Elizabeth was 38 years old and had a history of mitral valve prolapse (a common congenital condition in which a heart valve does not close properly). She was recently admitted to the hospital complaining of fatigue, intermittent fevers for 5 weeks and headaches for 3 weeks—symptoms the physician recognized as possible indications of endocarditis. Elizabeth reported having a dental procedure a few weeks prior to the onset of symptoms. A sample of her blood placed in a liquid bacteriological medium grew Gram-positive cocci, which turned out to be Streptococcus mutans, *a member of the viridans streptococci. As a result, the diagnosis of bacterial endocarditis was confirmed. The patient began a 1-month course of intravenous penicillin G and gentamicin therapy and eventually recovered to normal health.*

Endocarditis (inflammation of the heart) is traditionally classified as acute or subacute, depending on the pathogenic organism involved and the speed of clinical presentation. Subacute bacterial endocarditis (SBE) has a slow onset with vague symptoms. It is usually caused by bacterial infection of a heart valve (**Fig. 26.21**). SBE infections are usually (but not always) caused by a viridans streptococcus from the oral microbiota (for example, *Streptococcus mutans,* a common cause of dental caries). "Viridans streptococci" is a general term used for commensal streptococci whose colonies produce green alpha hemolysis on blood agar ("viridans" from the Greek *viridis,* "to be green"). Most patients who develop infective endocarditis have mitral valve prolapse (90%), although this is frequently not the case when patients are intravenous drug abusers or have hospital-acquired (nosocomial) infections.

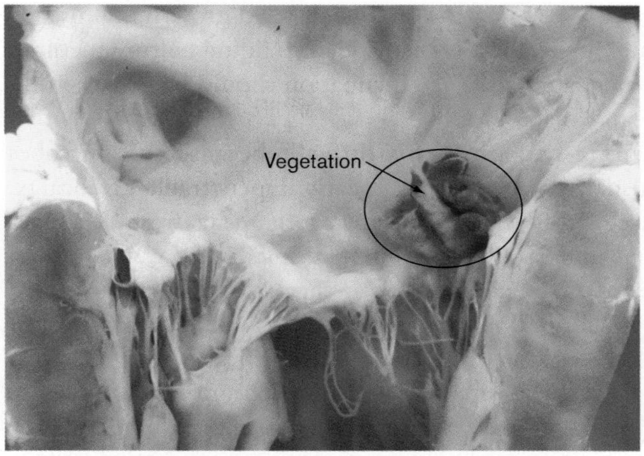
Vegetation

FIGURE 26.21 ■ View of bacterial endocarditis. Close-up of mitral valve endocarditis, showing vegetation (arrow).

As in the case history presented here, bacterial endocarditis can begin at the dentist's office. Following a dental procedure (such as tooth restoration) or even while brushing your teeth, oral bacteria can transiently enter the bloodstream and circulate. *S. mutans,* which is not normally a serious health problem, can become lodged onto damaged heart valves, grow as a biofilm, and secrete a thick glycocalyx coating that encases the microbes and forms a vegetation on the valve, damaging it further. If untreated, the condition can be fatal within 6 weeks to a year.

When more virulent organisms, such as *Staphylococcus aureus,* gain access to cardiac tissue, a rapidly progressive (acute) and highly destructive infection ensues. Symptoms of acute endocarditis include fever, pronounced valvular regurgitation (backflow of blood through the valve), and abscess formation.

Most patients with subacute bacterial endocarditis present with a fever that lasts several weeks. They also complain of nonspecific symptoms, such as cough, shortness of breath, joint pain, diarrhea, and abdominal or flank pain. Endocarditis is suspected in any patient who has a heart murmur and an unexplained fever for at least a week. It should also be considered in an intravenous drug abuser with a fever, even in the absence of a murmur. In either case, definitive diagnosis requires blood cultures that grow bacteria. Blood cultures involve taking samples of a patient's blood from two different locations (such as two different arms). Growth of the same organism in cultures taken from two body sites rules out inadvertent contamination with skin microbes, which would likely yield growth in only one culture. The blood is then added to liquid culture medium and incubated at 37°C. Incubation should be done aerobically and anaerobically.

Curing endocarditis is difficult because the microbes are usually ensconced in a nearly impenetrable glycocalyx. Consequently, eradicating microorganisms from the vegetations almost always requires hospitalization, where high doses of intravenous antibiotic therapy can be administered and monitored. Antibiotic therapy usually continues for at least a month, and in extreme cases, surgery may be necessary to repair or replace the damaged heart valve.

Patients with heart valve prolapse should not shy away from the dentist, however. As long as a healthy immune system is in place, the risk of infection is low. In fact, maintaining good oral hygiene can keep their risk of developing infectious endocarditis low. Immunocompromised patients, on the other hand, are at increased risk and should be treated prophylactically with oral penicillin or erythromycin 1 hour before a procedure to kill oral bacteria that enter the bloodstream. This prophylactic treatment can prevent the development of endocarditis in these patients.

Viruses such as adenovirus and some enteroviruses can also cause endocarditis, as well as a condition known as myocarditis, an inflammation of the heart muscle.

Thought Question

26.13 A patient presenting with high fever and in an extremely weakened state is suspected of having a septicemia. Two sets of blood cultures are taken from different arms. One bottle from each set grows *Staphylococcus aureus,* yet the laboratory report states that the results are inconclusive. New blood cultures are ordered. What would make these results inconclusive?

Malarial Parasites Feed on Blood

Malaria is the most devastating infectious disease known. Each year 300–500 million people develop malaria worldwide, and 1–3 million of these people, mostly children, die. Fortunately, the disease is relatively rare in the United States; only about 1,000 cases occur annually, and almost all are acquired as a result of international travel to endemic areas (**Fig. 26.22A**). Once again, this correlation illustrates the diagnostic importance of knowing a patient's history.

The disease is caused by four species of *Plasmodium*: *P. falciparum* (the most deadly), *P. malariae, P. vivax,* and *P. ovale*. The life cycle of *Plasmodium,* discussed in detail in Chapter 20, is complex and involves two cycles: an asexual erythrocytic cycle in the human and a sexual cycle in the mosquito (see Fig. 20.40).

In the erythrocytic cycle, the organisms enter the bloodstream through the bite of an infected female *Anopheles* mosquito (the mosquito injects a small amount of saliva containing *Plasmodium*). The haploid sporozoites travel immediately to the liver, where they undergo asexual fission to produce merozoites. Released from the liver, the merozoites attach to and penetrate red blood cells, where *Plasmodium* consumes hemoglobin and enlarges into a trophozoite. The protist nucleus divides, so that the cell, now called a schizont, contains up to 20 or so nuclei. The schizont then divides to make the smaller, haploid merozoites (**Fig. 26.22B**). The glutted red blood cell eventually lyses, releasing merozoites that can infect new red blood cells (see Fig. 20.39).

Weblinks Malaria animation (*see ebook*)

Sudden, synchronized release of the merozoites and red blood cell debris triggers the telltale symptoms of malaria: violent, shaking chills followed by high fever and sweating. The erythrocytic cycle, and thus the symptoms, repeats every 48–72 hours. After several cycles, the patient goes into remission lasting several weeks to months, after which there is a relapse.

Much of today's research focuses on why malarial relapse happens. Why does the immune system fail to eliminate the parasite after the first episode? When *Plasmodium* invades the red blood cells, it lines the blood cells with a protein, PfEMP1, which causes the parasite to stick to the sides of blood vessels. The parasite is thereby removed from circulation, but the protein cannot protect the parasite from patrolling macrophages, which eventually detect the invader and recruit other immune cells to fight it. So, during a malarial infection, a small percentage of each generation of parasites switches to a different version of PfEMP1 that the body has never seen before. In its new disguise, *Plasmodium* can invade more red blood cells and cause another wave of fever, headaches, nausea, and chills.

The antigenic shift happens when the parasite alters the PfEMP1 gene that is expressed. The body now has to repeat the recognition and attack responses all over again. The parasite has 60 cloaking genes, called *var,* that can be turned on and off individually, changing the organisms' antigenic structure, like a criminal repeatedly changing his disguise to elude police.

In April 2005, Australian scientists Alan Cowman, Brendan Crabb, and colleagues at the Walter and Eliza Hall Institute of Medical Research showed that *var* genes are

A.

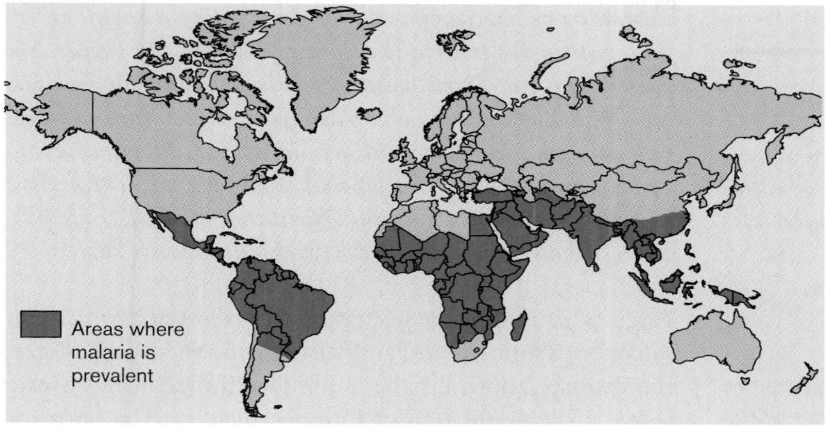
Areas where malaria is prevalent

B.

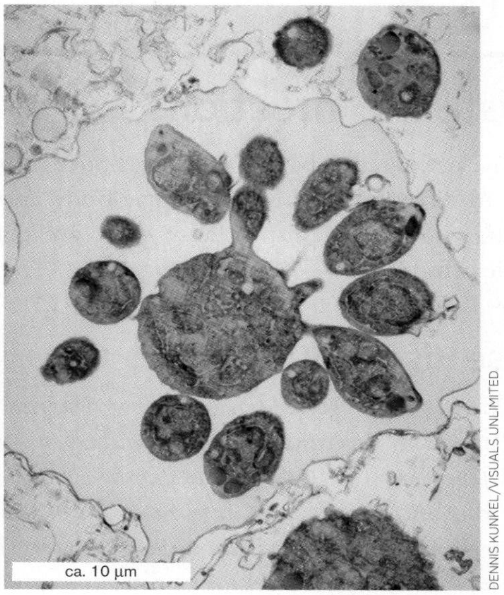
ca. 10 μm
DENNIS KUNKEL/VISUALS UNLIMITED

FIGURE 26.22 ■ Malaria is a major disease worldwide. A. Endemic areas of the world where malaria is prevalent. **B.** *Plasmodium falciparum* (colorized TEM). Schizont after completion of division. A residual body of the organism (yellow-green) is left over after division. The erythrocyte has lysed and only a ghost cell remains; no cytoplasm is seen surrounding the merozoites just being released. Free merozoites are seen outside the membrane.

regulated by chromosome packaging, which unwraps one gene to be expressed at a time and literally packs away the inactive genes. DNA can be encased so securely by some proteins that other proteins cannot access the nucleic acid for transcription—a process known as epigenetic silencing. Becoming immune to all the types of malaria can take upwards of 5 years and requires constant exposure; otherwise the immunity is lost. Many children do not live long enough to gain immunity to malaria in all its forms.

Diagnosis of malaria involves microscopic demonstration of the protist within erythrocytes (Wright stain) or serology to identify antimalarial antibodies. Treatment regimens include chloroquine or mefloquine, which kill the organisms in their erythrocytic asexual stages, and primaquine, effective in the exoerythrocytic stages. The chloroquine family of drugs acts by interfering with the detoxification of heme generated from hemoglobin digestion. Malaria parasites accumulate the hemoglobin released from red blood cells in plasmodial lysosomes, where digestion occurs. The parasites use the amino acids from hemoglobin to grow but find free heme toxic. To prevent eating itself to death (from accumulating too much heme), the organism detoxifies heme via polymerization, which produces a black pigment. Many antimalarial drugs, such as chloroquine, prevent polymerization by binding to the heme. As a result, the increased iron level (from heme) kills the parasite.

Chloroquine is given prophylactically to persons traveling to endemic areas. Unfortunately, *Plasmodium* has been developing resistance to these drugs, forcing the development of new ones. The antigenic shape-shifting carried out by this parasite has so far stymied development of an effective vaccine.

Bear in mind, we have described only selected organisms that cause cardiovascular infections; there are many more (for example, *Rickettsia typhi*). There is even evidence that *Chlamydophila pneumoniae* may have a role in coronary artery disease (see **eTopic 26.8**).

Note: Babesiosis is an emerging disease caused by a protozoan (*Babesia microti*) that, like *Plasmodium*, also infects red blood cells. *B. microti* is transmitted by the deer tick; the same insect that transmits the agent for Lyme disease. Typically, babesiosis is a mild, flu-like disease, but in its severe form can present with symptoms similar to malaria.

To Summarize

- **Blood cultures** are useful in diagnosing septicemia and endocarditis.
- **Endocarditis** can have acute or subacute onsets.
- **Subacute bacterial endocarditis** is usually an endogenous infection caused by *Streptococcus mutans.*
- **Malaria, caused by *Plasmodium* species,** manifests as repeated episodes of chills, fever, and sweating owing to the organism's ability to alter the antigenic appearance of its surface proteins and evade the immune response.

26.8

Systemic Infections

Many pathogenic bacteria can produce septicemia as a way to disseminate throughout the body and infect other organs. These organisms cause what are considered systemic infections.

Case History: The Plague

*A 25-year-old New Mexico rancher was admitted to an El Paso hospital because of a 2-day history of headache, chills, and fever (40°C; 104°F). The day before admission, he began vomiting. The day of admission, an orange-sized, painful swelling in the right groin area was noted (**Fig. 26.23A**). A lymph node aspirate and a smear of peripheral blood were reported to contain Gram-negative rods that exhibited bipolar staining (**Fig. 26.23B**). The patient's white blood cell count was 24,700/µl (normal is 5,000–10,000/µl), and his platelet count was 72,000/µl (normal is 130,000–400,000/µl). In the 2 weeks prior to becoming ill, the patient had trapped, killed, and skinned two prairie dogs, four coyotes, and one bobcat. The patient had cut his left hand shortly before skinning a prairie dog. PCR and typical biochemical testing of a Gram-negative rod isolated from blood cultures identified the organism as* Yersinia pestis, *the organism that causes plague. The patient received an antibiotic cocktail of gentamicin and tetracycline. He eventually recovered, after 6 weeks in intensive care.*

Plague is caused by the bacterium *Yersinia pestis,* which can infect both humans and animals. During the Middle Ages, the disease, known as the Black Death, decimated over a third of the population of Europe. Such was the horror it evoked that invading armies would actually catapult dead plague victims into embattled fortresses. This was probably the first case of biowarfare.

Contrary to popular belief, *Y. pestis* is present in the United States. The organism is endemic in 17 western states. It is normally transmitted from animal to animal, typically rodents like rats and even prairie dogs

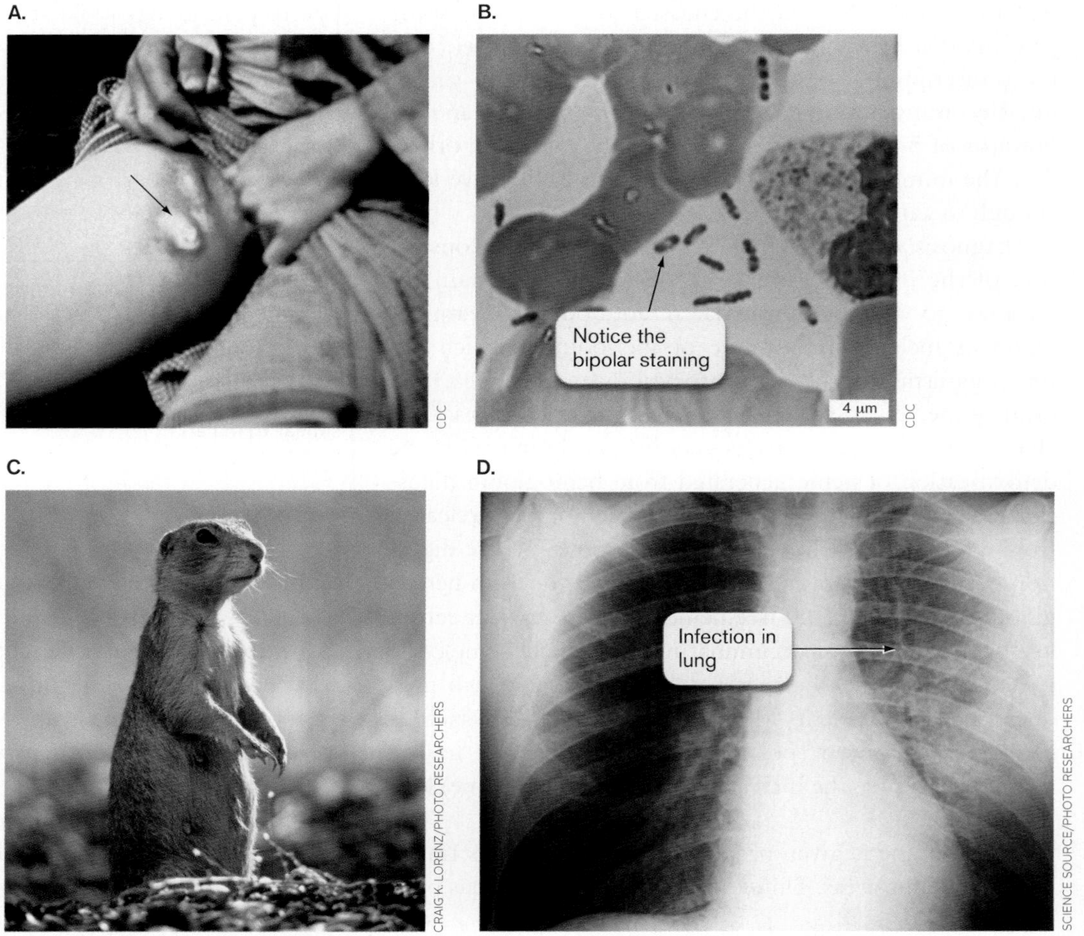
A.
CDC
B.
Notice the bipolar staining
4 µm
CDC
C.
CRAIG K. LORENZ/PHOTO RESEARCHERS
D.
Infection in lung
SCIENCE SOURCE/PHOTO RESEARCHERS

FIGURE 26.23 ▪ The plague. **A.** Classic bubo (swollen lymph node) of bubonic plague. **B.** *Yersinia pestis,* bipolar staining (length 1–3 µm). **C.** Prairie dogs are often hosts to fleas that carry plague bacilli. **D.** X-ray of pneumonic plague, showing bilateral pulmonary infection.

(**Fig. 26.23C**), by the bite of infected fleas. **Figure 26.24** illustrates the various infective cycles of the plague bacillus. Humans are not typically part of the natural infectious cycle. However, in the absence of an animal host, the flea can take a blood meal from humans and thereby transmit the disease to them. During the Middle Ages, urban rats venturing back and forth to the countryside became infected by the fleas of wild rodents that served as a reservoir. Upon returning to the city, the rat flea passed the organism on to other rats, which then died in droves. The rat fleas, deprived of their normal meal, were forced to feed on city dwellers, passing the disease on to them.

Individuals bitten by an infected flea or accidentally infected through a cut while skinning an infected animal first exhibit the symptoms of **bubonic plague**. Bubonic plague emerges as the organism moves from the site of infection to the lymph nodes, producing characteristically enlarged nodes called buboes (see **Fig. 26.23A**). From the lymph nodes, the pathogen can enter the bloodstream, causing **septicemic plague**. In this phase the patient can go into shock from the massive amount of endotoxin in the bloodstream. Neither bubonic nor septicemic plague is passed from person to person. As the organism courses through the bloodstream, however, it will invade the lungs and produce **pneumonic plague** (**Fig. 26.23D**), which can be easily transmitted from person to person through aerosol droplets generated by coughing. Pneumonic plague is the most dangerous form of the disease because it can kill quickly and spread rapidly through a population. Pneumonic plague is so virulent that an untreated patient can die within 24–48 hours. The organism is usually identified postmortem.

Y. pestis has numerous virulence factors. For instance, YadA is a surface adhesin that binds collagen. Another factor, the F1 protein capsule surface antigen, plays a part in blocking phagocytosis in mammalian hosts. Certain biofilms formed by *Y. pestis* are also important. An extracellular matrix synthesized by *Y. pestis* produces an

FIGURE 26.24 ▪ **The cycles of plague.** The sylvatic cycle occurs in the wild, where fleas transmit the organism between rodents. An accidental interaction with urban rats can trigger a similar urban cycle. Humans can be infected through contact with infected fleas coming from either cycle. Flea bite transmission initiates bubonic plague symptoms that can progress to pneumonic plague. Pneumonic plague is highly infectious, which can cause epidemic spread of the disease.

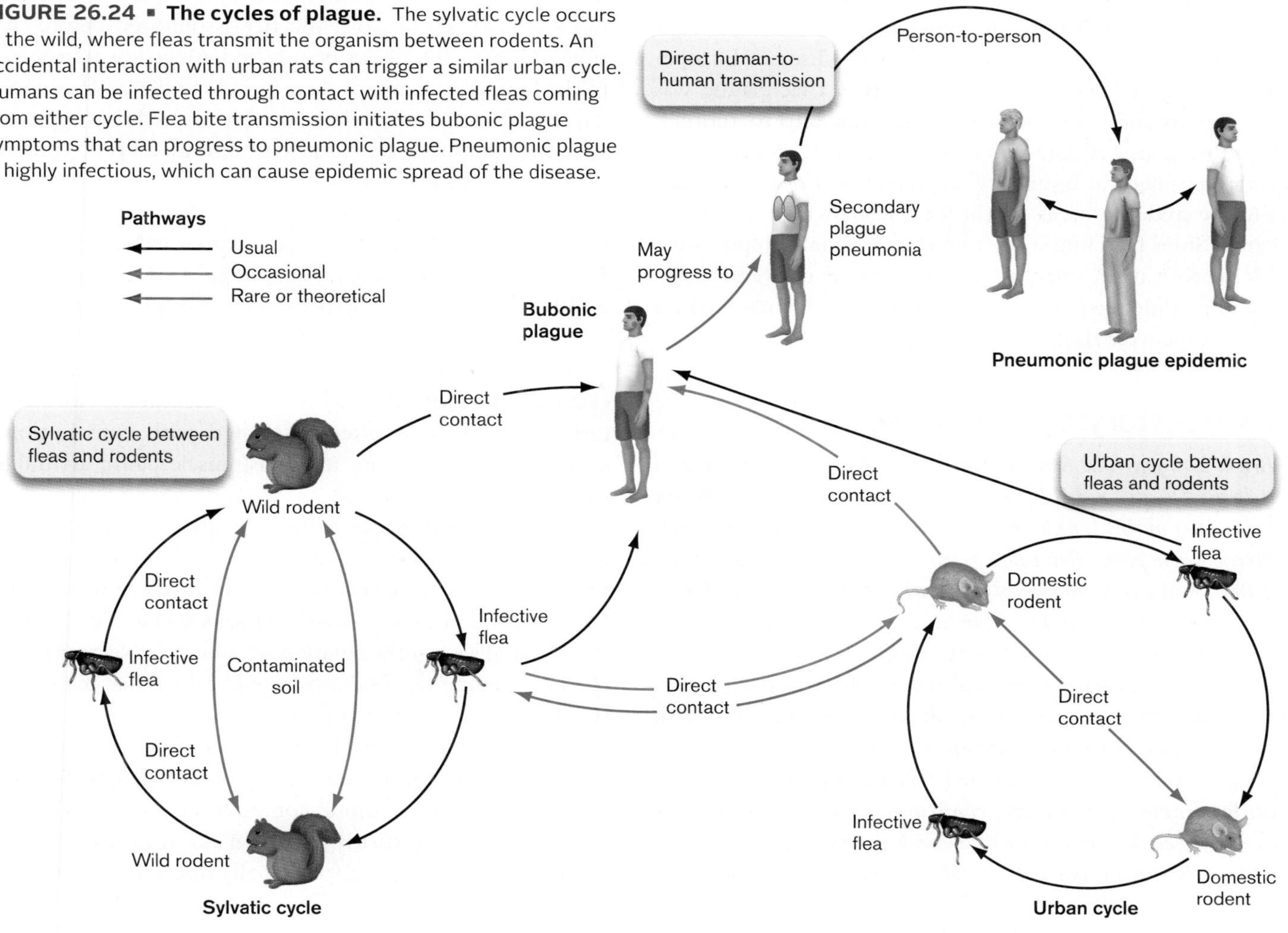

Pathways
Usual
Occasional
Rare or theoretical
Direct human-to-human transmission
Person-to-person
Secondary plague pneumonia
May progress to
Bubonic plague
Pneumonic plague epidemic
Direct contact
Sylvatic cycle between fleas and rodents
Wild rodent
Direct contact
Infective flea
Contaminated soil
Infective flea
Direct contact
Wild rodent
Sylvatic cycle
Direct contact
Direct contact
Urban cycle between fleas and rodents
Infective flea
Domestic rodent
Direct contact
Infective flea
Domestic rodent
Urban cycle

adherent biofilm in the flea midgut that contributes to flea-to-mammal transmission. The biofilm blocks flea digestion, making the flea feel "starved" even after a blood meal. Therefore, the flea jumps from host to host in a futile effort to feel full. As the flea tries to take a blood meal, the blockage causes the insect to regurgitate bacteria into the wound. This curious effect of *Y. pestis* on the insect vector is another unique aspect of how plague spreads so quickly.

Y. pestis also uses type III secretion systems to inject virulence proteins (YopB and YopD) into host cell membranes. Unlike *Salmonella,* which uses type III–secreted proteins to gain entrance into host cells, *Y. pestis* is not primarily an intracellular pathogen. Injection of the Yop proteins disrupts the actin cytoskeleton and so helps the organism evade phagocytosis. By evading phagocytosis, the organism avoids triggering an inflammatory response and produces massive tissue colonization.

Plague has disappeared from Europe; the last major outbreak occurred in 1772. The reason for its disappearance is not known, but it was probably the result of multiple factors, not the least of which was human intervention. Although it wasn't until the nineteenth century that doctors understood how germs could cause disease, Europeans recognized by the sixteenth century that plague was contagious and could be carried from one area to another. Beginning in the late seventeenth century, governments created a medical boundary, or *cordon sanitaire,* between Europe and the areas to the east from which epidemics came. Ships traveling west from the Ottoman Empire were forced to wait in quarantine before passengers and cargo could be unloaded. Those who attempted to evade medical quarantine were shot.

Case History: Lyme Disease

*Brad, a 9-year-old from Connecticut, developed a fever and a large (8-cm) reddish rash with a clear center (**erythema migrans**) on his trunk (**Fig. 26.25A**). He also had some left facial nerve palsy. Brad had returned a week previously from a Boy Scout camping trip to the local woods, where he did a lot of hiking. When asked by his physician, Brad admitted finding a tick on his stomach while in the woods but thinking little of it. The doctor ordered serological tests for* Borrelia burgdorferi *(the organism that causes Lyme disease),* Rickettsia rickettsii *(which produces Rocky Mountain spotted fever), and* Ehrlichia equi *(which causes ehrlichiosis). The ELISA test for* B. burgdorferi *came back positive, confirming a diagnosis of Lyme disease. The boy was given a 3-week regimen of doxycycline (a tetracycline derivative), which resolved the rash and palsy.*

Lyme arthritis was first reported in Lyme, Connecticut, in the 1970s, but the causative organism, *Borrelia burgdorferi,* was not identified until 1982. Since then, **Lyme disease** (aka **borreliosis**) has become the most common vector-borne illness in the United States and is considered an emerging infectious disease. The main endemic areas are the northeastern coastal area from Massachusetts to Maryland, Wisconsin and Minnesota, and northern California and Oregon. Lyme disease is also common in parts of Europe.

B. burgdorferi is a spirochete (**Fig. 26.25B**) transmitted to humans by ixodid ticks (hard ticks) (**Fig. 26.25C** and **D**). In the northeastern and central United States, where most cases occur, the deer tick *Ixodes scapularis* transmits the spirochete, usually during the summer months. In the western United States, *I. pacificus* is the tick vector.

During its nymphal stage (the stage after taking its first blood meal), *I. scapularis* is the size of a poppy seed. Its bite is painless, so it is easily overlooked. Infection takes place when the tick feeds because the spirochete is regurgitated into the host. However, the organism grows in the tick's digestive tract and takes about 2 days to make its way to the tick's salivary gland, so if the tick is removed before that time, the patient will not be infected. Once it is transferred to the human, the microbe can travel rapidly via the bloodstream to any area in the body, but it prefers to grow in skin, nerve tissue, synovium (joint lining), and the conduction system of the heart.

Lyme disease has three stages. Three general stages of Lyme disease are recognized. Stage 1 infections occur 3–30 days after the initial exposure. Approximately 75% of patients experience an erythema migrans rash, usually at the site of the tick bite, that varies in appearance but is classically erythematous with central clearing ("bull's-eye" rash). This stage is often associated with constitutional symptoms such as fever, myalgia (muscle pain), arthralgia (joint pain), and headache.

Stage 2 occurs weeks to months after the initial infection. In this stage, the patient can be quite ill with malaise, myalgia, arthralgia, or neurological or cardiac involvement. Common neurological manifestations include Bell's palsy (facial paralysis), inflammation of spinal nerve roots, and chronic meningitis. The most common cardiac manifestation is an irregular heart rhythm.

Stage 3 borreliosis occurs months to years later and can involve the synovium, nervous system, and skin, though skin involvement is more common in Europe than in North America. Arthritis occurs in the majority of previously untreated patients; it is usually intermittent, involves

A.

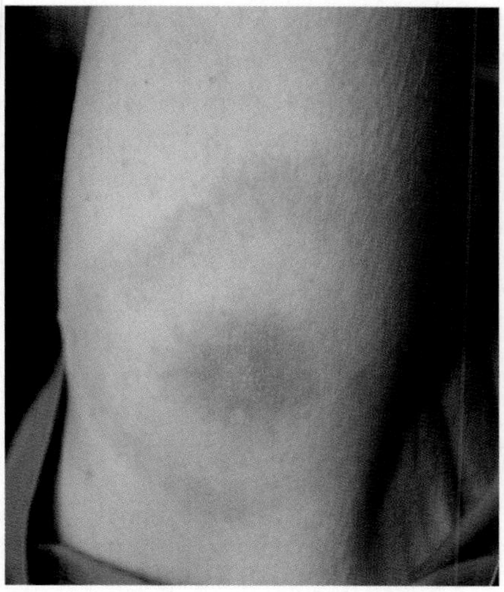
CDC

B.

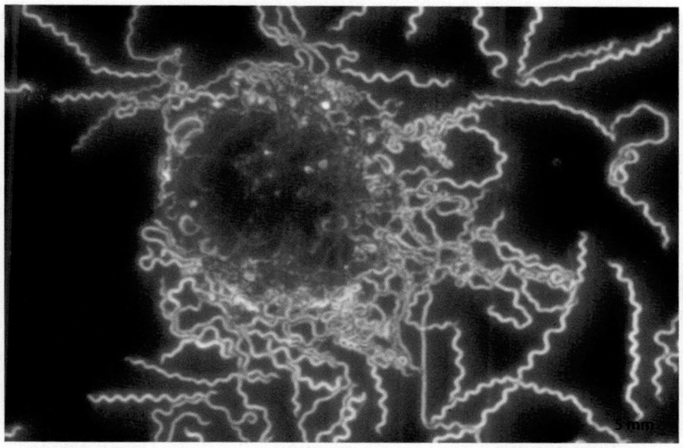
COURTESY JEFFREY NELSON

C.

DAVID M. PHILLIPS/PHOTO RESEARCHERS

D.

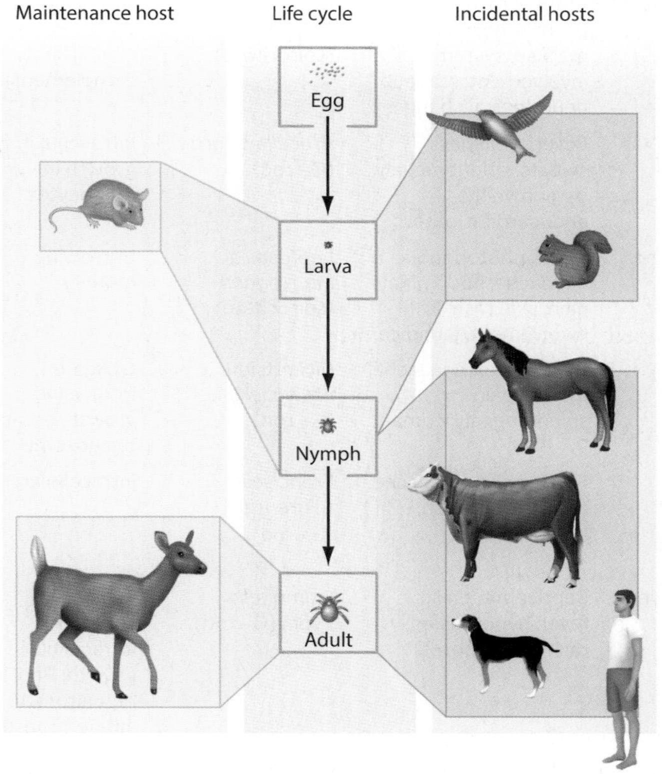
Maintenance host
Life cycle
Incidental hosts
Egg
Larva
Nymph
Adult

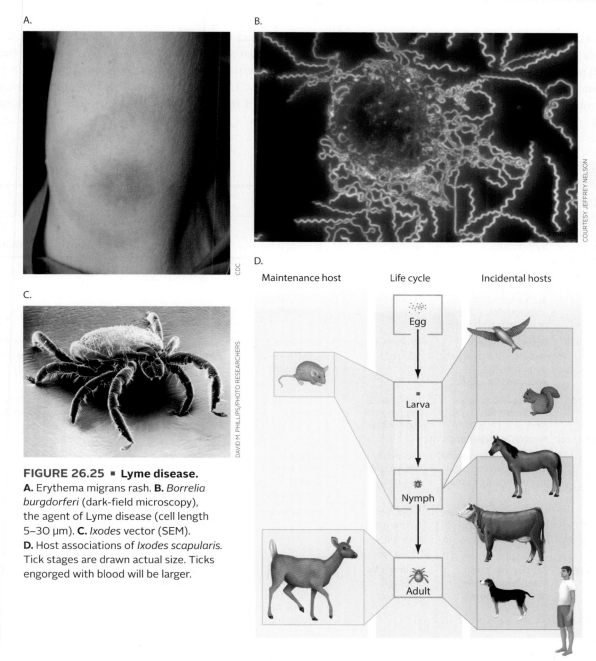

FIGURE 26.25 ▪ Lyme disease.
A. Erythema migrans rash. **B.** *Borrelia burgdorferi* (dark-field microscopy), the agent of Lyme disease (cell length 5–30 μm). **C.** *Ixodes* vector (SEM). **D.** Host associations of *Ixodes scapularis.* Tick stages are drawn actual size. Ticks engorged with blood will be larger.

the large joints, particularly the knee, and lasts from weeks to months in any given joint. Joint fluid analysis typically shows a WBC count of 10,000–30,000/μl. Late neurological involvement may include peripheral neuropathy and encephalopathy, manifested as memory, mood, and sleep disturbances.

Treatment with antibiotics is recommended for all stages of Lyme disease but is most effective in the early stages. Treatment for early Lyme disease (stage 1) is a course of doxycycline for 14–28 days. Lyme arthritis is typically slow to respond to antibiotic therapy. Despite antimicrobial drug treatment, patients with persistent active arthritis and persistently positive PCR tests may have incomplete microbial eradication and may be the most likely to benefit from repeated treatment with injected antibiotics. However, the disease in some people does not resolve at all—why is not known.

Curiously, 25% of people infected with *B. burgdorferi* never experience erythema migrans, and many infected individuals are also unsure of tick bites. Hence, patients with Lyme disease may present with arthritis as their first complaint. Although this makes diagnosis extremely difficult, knowing that the patient lives in or recently traveled to an endemic area can provide a critical clue.

Many other bacteria can cause septicemia and systemic illness (**Table 26.7**). Gram-negative organisms like *E. coli, Salmonella* Typhi, and *Francisella*, and Gram-positive microbes like *Staphylococcus aureus, Enterococcus*, and *Bacillus anthracis*, can grow in the bloodstream if they can gain entrance. Even anaerobes that are normal inhabitants of the intestine (for example, *Bacteroides fragilis*) can be lethal if they escape the intestine and enter the blood, as might happen following surgery. This is one reason surgical patients are given massive doses of antibiotics immediately before and after surgery.

TABLE 26.7

Selected microbes that cause systemic disease.

Disease	Symptoms	Etiological agent	Virulence properties	Source	Treatment	Vaccine
Lyme disease	Stage 1: rash; stage 2: chills, headache, malaise, systemic involvement; stage 3: neurological changes	*Borrelia burgdorferi* (spirochete)	Antigenic variation, OspE (binds complement)	Deer tick	Penicillin, tetracycline	No longer available
Brucellosis	Fever, weakness, sweats, splenomegaly, osteomyelitis, endocarditis, others	*Brucella abortis* (G– rod)	Intracellular, growth in monocytes	Animal products, unpasteurized milk	Doxycycline	Yes, for animals
Leptospirosis	Fever, photophobia, headache, abdominal pain, skin rash, liver involvement, jaundice	*Leptospira interrogans* (spirochete)	Burrowing motility	Urine of infected animals	Erythromycin, penicillin	Yes, for animals
Epidemic typhus	Chills, fever, headache, muscle pain, splenomegaly, coma	*Rickettsia prowazekii* (G– rod)	Obligate intracellular growth, escapes phagosome	Human louse, flying-squirrel flea	Tetracycline, chloramphenicol	
Tularemia	Fever, chills, headache, muscle pain, rash, bacteremia	*Francisella tularensis* (G– rod)	Intracellular	Rabbits, rodents, insect vectors	Gentamicin, streptomycin	Yes, but not currently in the U.S.
Typhoid fever	Septicemia, chills, fever, hypotension, rash (rose spots)	*Salmonella* Typhi (G– rod)	Type III secretion, intracellular growth, PhoPQ regulators, Vi antigen capsule	Gallbladder of human carrier	Ciprofloxacin, ceftriaxone	Vi antigen
	Septicemia, chills, fever, hypotension	*Salmonella choleraesuis* (G– rod)	Intracellular growth, invasin	Animals, poultry	Ceftriaxone	
Vibriosis	Serious with immunocompromised patients; fever, chills, multi-organ damage, death	*Vibrio vulnificus* (G– curved rod)	Cytolysin, capsule	Seawater, raw oysters	Tetracycline plus aminoglycoside	
Bubonic plague	Buboes (swollen lymph glands), high fever, chills, headache, cough, pneumonia, septicemia	*Yersinia pestis* (G– rod)	Intracellular growth, type III secretion of YOPs (*Yersinia* outer proteins), phospholipase D, toxin	Rodents, rodent fleas, human respiratory aerosol, potential bioterrorism agent	Streptomycin or tetracycline	Yes, but not available in the U.S.

Hepatitis Viruses Target the Liver

Hepatitis is a general term meaning inflammation of the liver. Hepatitis is caused by several viruses, including hepatitis A, B, and C viruses. Although these viruses are members of very different families, they all target the liver. We include them in this section on systemic infections because their infectious routes take them to the bloodstream before arriving at the liver.

Hepatitis A virus (HAV) is a single-stranded RNA picornavirus (**Fig. 26.26A**) that causes an acute infection spread person-to-person by the fecal-oral route, but hepatitis A can also result from eating undercooked shellfish collected from contaminated waters. The virus replicates in the intestinal endothelium and is disseminated via the bloodstream to the liver. After replicating in hepatocytes, the progeny enter the bile and are released into the small intestine, explaining why stools are so infectious. Though the virus has an early viremic stage after leaving the intestine, it is rarely transmitted by transfusion because the viremic stage is transient, ending after liver symptoms develop. In contrast, hepatitis B and C viruses produce persistent viremia and are readily transmitted by transfusion.

Many people who are infected with HAV are asymptomatic or exhibit very mild symptoms that include nausea, vomiting, diarrhea, low-grade fever, and fatigue. As the virus attacks the liver, patients may become jaundiced (from the accumulation of bilirubin in the skin), and their urine will turn dark brown. There is no specific treatment, but the disease usually lasts for only a few months and then resolves without establishing a carrier state. Disease can be prevented, however, if immunoglobulin is given to someone who has had contact with an infected individual. A hepatitis A vaccine containing inactivated virus (called HepA vaccine) is administered after 1 year of age. For those not vaccinated, frequent hand washing is important for preventing the spread of the disease because it interrupts the fecal-oral cycle.

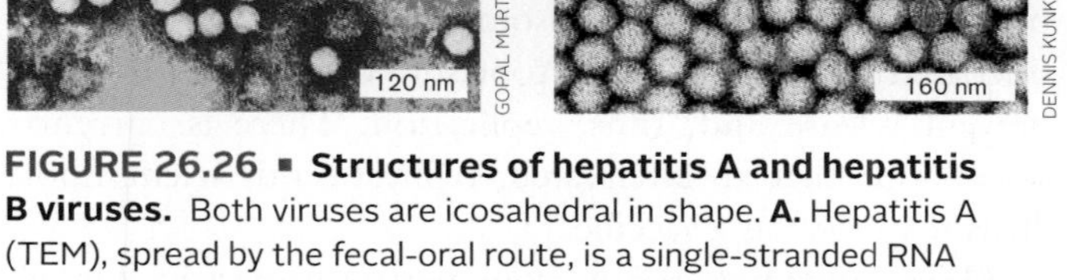
A.
120 nm
GOPAL MURTI/VISUALS UNLIMITED
B.
160 nm
DENNIS KUNKEL/VISUALS UNLIMITED

FIGURE 26.26 ▪ Structures of hepatitis A and hepatitis B viruses. Both viruses are icosahedral in shape. **A.** Hepatitis A (TEM), spread by the fecal-oral route, is a single-stranded RNA virus in the *Picornaviridae* family (30 nm in diameter). **B.** Hepatitis B (TEM) is an enveloped, double-stranded DNA virus in the *Hepadnaviridae* family (40 nm in diameter).

In contrast to HAV, hepatitis B virus (HBV) is a partially double-stranded circular DNA virus (family *Hepadnaviridae*) that causes diseases of varying severity. These include acute and chronic hepatitis, cirrhosis, and hepatocarcinoma. The virus also wears a membrane envelope donned when progeny viruses are released from infected cells. The virion coat protein, a surface antigen, is called HBsAg. The virus makes an excess amount of HBsAg, so it is sometimes extended as a tubular tail on one side of the virus particle and is often found in the blood of infected individuals in the form of noninfectious filamentous and spherical particles (**Fig. 26.26B**). The presence of HBsAg in blood is an indicator of HBV infection.

HBV is transferred primarily via blood transfusions, contaminated needles shared by IV drug users, and any human body fluid (saliva, semen, sweat, breast milk, tears, urine, or feces). It can even be transferred transplacentally to a fetus and can be sexually transmitted. Infection by HBV has two stages: a short-term acute phase and a long-term chronic phase that, if it extends beyond 6 months, may never resolve. Symptoms resemble those of the flu, but with jaundice and brown urine. Liver damage caused by HBV infection is due in large part to an efficient cell-mediated immune response. Cytotoxic T cells and natural killer cells cause immune lysis of infected liver cells. Over the long term, chronic hepatitis will lead to a scarred and hardened liver (cirrhosis), the only recourse being a liver transplant. Fortunately, about 90% of those infected are able to fight off infection and never proceed to the chronic stage. A HepB vaccine (made from recombinant HBsAg) is available. Its administration is recommended after birth, followed by booster shots administered by 2 months and 18 months of age.

Hepatitis C virus (HCV) causes another form of hepatitis. HCV is a single-stranded positive-sense, linear RNA virus with a lipid coat; it is a member of the *Flaviviridae* family. It is transmitted by blood transfusions and causes 90% of transfusion-related cases of hepatitis. It can also be transmitted by needle sticks, razor blades, tattooing, and, less frequently, by sex. Over 100 million people worldwide are infected with HCV. Screening for HCV (serology or PCR) is recommended for anyone who exhibits signs of hepatitis or practices the risky behaviors noted above. In addition, the CDC recommends that anyone born between

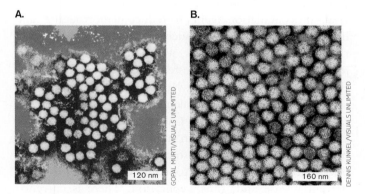

1945 and 1965 (baby boomers) be tested because, for unknown reasons, baby boomers are five times more likely to be infected than other adults.

Most HCV-infected individuals (80%) do not exhibit symptoms, and in those that do, symptoms may not appear for 10–20 years. At least 75% of patients who exhibit symptoms ultimately progress to chronic hepatitis requiring a liver transplant. Fortunately, infection can be detected using ELISA. Liver biopsies of HCV patients are used to determine the extent of liver damage, which in turn helps establish the stage of disease.

Prevention of HBV or HCV infection for health care personnel includes avoiding inadvertent needle sticks and, if such a stick should occur, the administration of immunoglobulin within 7 days. Chronic hepatitis can be treated with interferon, but more effective treatment now includes antiviral protease inhibitors that prevent the processing (by proteolysis) of an important HCV polyprotein. Though vaccines have been developed for HAV and HBV (Section 26.9), no vaccine is yet available for HCV. It is important to note that, since hepatitis viruses can be spread via contaminated blood products, all blood donations collected by the Red Cross and other agencies are tested for the presence of these viruses, as well as for HIV. Thus, the blood supply is safe.

Ebola—The Perfect Pathogen or Too Deadly for Its Own Good?

How would you define the perfect pathogen? Would it be an organism that can kill its host with terrifying ease and quickness? If so, Ebola virus would fit the description. Ebola virus, a lipid-enveloped, threadlike RNA virus (*Filoviridae*) (Fig. 25.5), was first associated with an outbreak of 318 cases of a hemorrhagic disease in Zaire in 1976. Of the 318 people who contracted the disease, 280 of them died within days. The disease was characterized by acute (rapid) onset of fever, severe muscle pains, horrible bleeding from multiple orifices (nose, mouth, anus, and vagina), and ultimately death. Also in 1976, 284 people in Sudan were infected with the virus, and 156 of them died. The Ebola virus has a frightening reputation. It spreads like wildfire through the body after infection, causing severe hemorrhagic fever, and typically kills 90% of its victims. Internal bleeding results in shock and acute respiratory distress, leading to death.

The symptoms of Ebola (and of a related disease caused by the Marburg virus) reflect subversion of the innate immune system coupled with uncontrolled viral replication, particularly in macrophages and dendritic cells. Ebola virus infection of these cells enhances production of proinflammatory cytokines, such as TNF-α, and inhibits stimulation of T-cell maturation by dendritic cells. Thus, Ebola infections stimulate inflammatory processes leading to tissue damage, but shut down early immune responses and prevent activation of adaptive immune responses, which allows unfettered viral replication.

Ebola viral proteins and their locations in the virion are shown in **Figure 26.27A**. Ebola VP35 protein is a component of the viral RNA polymerase complex, but it is also a potent inhibitor of host interferon (IFN) production. The cellular response to whatever IFN is made is inhibited by VP24, which blocks the nuclear accumulation of a regulatory protein called STAT1. STAT1 is critical to IFN-stimulated gene expression. These and other strategies allow rapid replication of the virus.

After replicating, Ebola offspring sprout from the cell surface in a mass of tangled threads (**Fig 26.27B**). These new virions go on to attack new cells, riddling blood vessels and organs with damage as they go. Rapid release of new virions involves subverting another host mechanism, tetherin, designed to slow viral spread. Paul Bates and his colleagues at the University of Pennsylvania discovered that the cellular protein tetherin essentially tethers mature virus particles inside a cell so that they are unable to spread. Tetherin is IFN induced and can restrict the spread of structurally diverse enveloped viruses, including HIV (thus, it is part of the innate response). Ebola glycoprotein, however, counteracts tetherin, so that nothing slows down viral spread.

The outlook for a patient infected with Ebola is dire. The incubation period is 4–16 days, and death occurs within 7–16 days. There are no effective drugs or vaccines. Administering blood plasma from people who have recovered, anticoagulation agents to reduce hemorrhaging, and interferon have had limited success. Chemical inhibitors of the host cathepsin proteases, which are required for viral replication, have been suggested as one possible antiviral therapy. Ebola enters a cell when the virus membrane glycoprotein attaches to host membranes. The virus is then taken up in an endosome. Cathepsins in the endosome cleave the viral glycoprotein and allow the virus membrane to fuse with the endosome membrane, releasing the uncoated virus into the cytoplasm. Cathepsin inhibitors prevent release and, thus, replication. There is currently some hope that an attenuated, replication-deficient Ebola virus may serve as a vaccine.

Ebola epidemics result from person-to-person contact or from inadvertent laboratory exposures. Fortunately, Ebola outbreaks are self-contained because the viruses kill

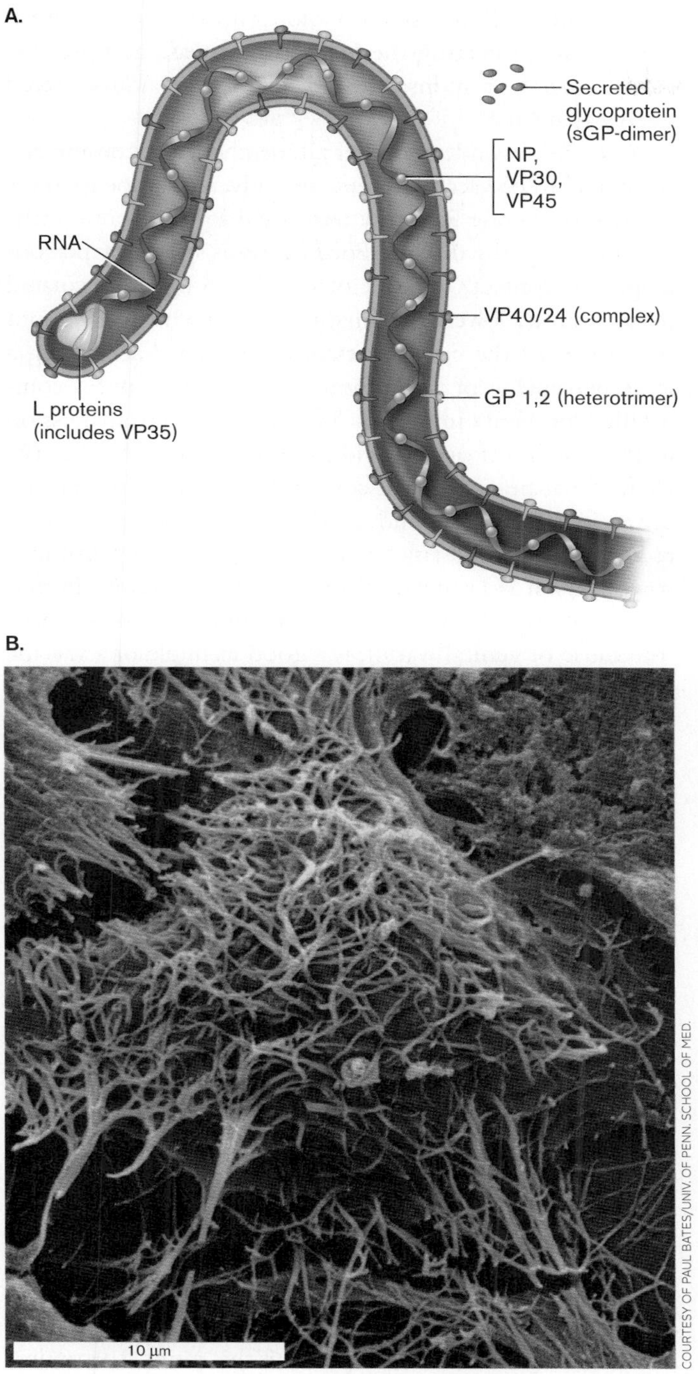
A.
Secreted
glycoprotein
(sGP-dimer)
NP,
VP30,
VP45
RNA
VP40/24 (complex)
GP 1,2 (heterotrimer)
L proteins
(includes VP35)
B.
10 µm
COURTESY OF PAUL BATES/UNIV. OF PENN. SCHOOL OF MED.

FIGURE 26.27 ■ **Ebola virion. A.** Composition of the virus. The ribonucleoprotein complex consists of the nucleoprotein (NP), the structural proteins VP30 and VP35, and the virion-associated RNA-dependent RNA polymerase (L proteins). The glycoprotein (GP-sGP) is an integral membrane protein that can be secreted. **B.** Threadlike Ebola virions budding from a cell (center). Ebola progresses so rapidly because it disables a host protein, called tetherin, that is normally used to limit virus release from infected cells. The result is the rapid release of massive numbers of virus particles that can then quickly spread infection to other organs and tissues.

their victims quickly. Death comes before the virus can be transmitted to a new host. Some scientists therefore argue that efficient and quick killing is not the mark of a perfect pathogen. The better pathogen lets its host linger to ensure a home for itself and more opportunity to disseminate. So where does Ebola go when it is not infecting humans? The natural ecology of these viruses is largely unknown, although an association with monkeys and/or bats as possible reservoirs is suggested.

Global health authorities became very concerned in 2009 about a report of Ebola virus transmission from a pig to a farmer in the Philippines. Given that the particular Ebola strain in question, called Ebola-Reston, has never caused disease in humans (the pig farmer developed antibodies to the strain, not disease), why should the World Health Organization be so concerned? While Ebola was thought to be carried only in humans and monkeys, it now appears that pigs represent a new reservoir for the virus. In addition, animals often play an important ecological role in the evolution of new viruses that can infect humans. For instance, different influenza viruses mix their genomes in birds (avian flu) and pigs (swine flu). Finding an Ebola virus in pigs, albeit one that is innocuous to humans, raises the possibility that a new emerging strain of Ebola-Reston could jump to humans and cause disease.

To Summarize

- **Septicemia** is caused by many Gram-positive and Gram-negative bacterial pathogens. It can start with the bite of an infected insect, introduction via a wound, escape from an abscess, or penetration of the mucosal epithelium by the pathogen (as through the intestine or vagina); it can lead to disseminated, systemic disease.
- **Plague** has sylvatic and urban infection cycles involving transmission between fleas and rats.
- ***Yersinia pestis*–infected flea bites** lead to bubonic plague. Bubonic plague can progress to septicemic and pneumonic stages.
- **Aerosolized respiratory secretions** will directly spread *Y. pestis* pneumonic plague from person to person (no insect vector).
- **Lyme disease** is caused by the spirochete *Borrelia burgdorferi,* which is transmitted from animal reservoirs to humans by the bite of *Ixodes* ticks.
- **There are three stages of Lyme disease**: stage 1, a bull's-eye rash (erythema migrans); stage 2, joint, muscle, and nerve pain; and stage 3, arthritis with WBCs in the joint fluid.

- **Hepatitis** is caused by several unrelated viruses; among them, HAV, HBV, and HCV account for most disease.
- **HAV** is transmitted by the fecal-oral route, does not establish chronic infection, and can be prevented by a vaccine.
- **HBV and HCV** can be transmitted by blood products (such as transfusions) and shared hypodermic needles and can lead to chronic hepatitis. Vaccine for HBV, but not HCV, is available.
- **Ebola virus** spreads from human to human and kills its victims quickly. Its viral proteins alter cytokine production and facilitate virus release from infected cells.

26.9

Immunization

Chapter 24 described the basic makeup of vaccines, explained how vaccines prevent infection, and listed many of the available vaccines (see Table 24.1). Now that we have described many of the organisms for which vaccines are available, more focus can be placed on how and when vaccines are delivered and whether vaccines are safe.

As noted in Table 24.1, some vaccines utilize killed organisms (examples are the hepatitis A vaccine and the Salk inactivated polio vaccine), while others contain attenuated microbes (BCG for tuberculosis or the Sabin live polio vaccine) or consist of purified components of an infectious agent (*Streptococcus pneumoniae* and *Haemophilus influenzae* type b capsular antigens). Some vaccines are injected in combination as polyvalent vaccines to control multiple diseases (for example, MMR for measles, mumps, and rubella), while others are given individually but may change from year to year (influenza viral capsid proteins).

Multiple vaccines can be given simultaneously and safely. Most vaccines are administered during childhood, when the diseases can be the most devastating. **Table 26.8** provides the current immunization schedule recommended for children and adolescents by the Centers for Disease Control and Prevention. Notice that all of these vaccines are given in multiple doses, called booster doses. The exception to this rule is the influenza vaccine, which is given in a single dose but changes every year. The reason for multiple doses, as noted in Chapter 24, is that secondary exposure to an antigen provides a more robust and long-lasting immunity. However, most vaccines are not administered until 2 months of age, because maternal antibody crossing through the placenta to the fetus (or to a newborn through breast milk) will persist for a short time in the newborn, temporarily protecting the baby from disease and possibly dampening the response to a vaccine antigen administered during that time.

You might wonder whether all members of a community must have to be vaccinated against a given microbe to lower the risk of disease for every individual in that community. It depends on the disease. For infections spread by person-to-person contact, the risk of disease to an unvaccinated person can be lowered dramatically even when only about two-thirds of the community is vaccinated. Vaccinating a large percentage of a community effectively conveys community (or herd) immunity by interrupting transmission of the disease. If one individual contracts the disease, the chance that he or she will come into contact with another unvaccinated person and transmit the disease is much reduced. Thus, the risk of disease to any single unvaccinated person is lessened as a result of community immunity. Gardasil, the vaccine against human papillomavirus (the cause of genital warts), is a good example of a vaccine that can provide herd immunity (See **eTopic 26.2**).

Herd immunity works well for diseases such as diphtheria, whooping cough (pertussis), measles, and mumps. However, herd immunity will not lower the risk of an unvaccinated person from contracting tetanus, which is not spread by person-to-person contact. *Clostridium tetani,* the agent whose toxin causes tetanus, is a ubiquitous soil organism transmitted through punctured skin. The risk of tetanus for an unvaccinated person doesn't change even if every other person in the community is vaccinated against tetanus.

The vast majority of people who receive vaccines suffer no, or only mild, reactions, such as fever or soreness at the injection site. Very rarely do more serious side effects, such as allergic reactions, occur. Vaccines are extremely safe. Unsettling, erroneous information circulates on the Internet purporting a link between vaccinations and other diseases, such as diabetes or autism, but no well-controlled scientific study supports these claims. The risks, including death, associated with these preventable infectious diseases are far greater than the minimal risk associated with being vaccinated against them. As proof of this point, the undervaccination of children in the United States in the 1980s and '90s has led to an increase in cases of whooping cough (27,550 in 2010), caused by *Bordetella pertussis.* Successful vaccination programs carried out in the United States have come close to eradicating many diseases once feared, such as polio, rubella, and diphtheria, and have dramatically lowered the morbidity and mortality of many others.

The current risk of infection for some diseases (because they are so rare) is lower than the risk of an adverse reaction to immunization. However, failing to vaccinate, as just

TABLE 26.8

Recommended childhood and adolescent immunization schedule, by vaccine and age–United States, 2009.*

Vaccine / Age ⟶	Birth	1 month	2 months	4 months	6 months	9 months	12 months	15–18 months	24 months	4–6 years	11–12 years
Hepatitis B[a]	HepB	HepB			HepB				HepB series		
Rotavirus			Rota	Rota	Rota						
Diphtheria, tetanus, acellular pertussis[b]			DTaP	DTaP	DTaP			DTaP		DTaP	Tdap
Haemophilus influenzae type b[c]			Hib	Hib	Hib[c]		Hib				
Inactivated poliovirus			IPV	IPV	IPV					IPV	
Measles, mumps, rubella							MMR			MMR	MMR
Varicella							Varicella			Varicella	
Meningococcal[d]							High risk				MCV4
Pneumococcal[e]			PCV13	PCV13	PCV13		PCV13		PPSV		
Influenza[f]					Influenza (yearly)						
Hepatitis A[g]							HepA series		HepA series		

*This schedule indicates the recommended ages for routine administration of currently licensed childhood vaccines, as of January 1, 2013.

■ Indicates age groups that warrant special effort to administer those vaccines not previously administered.

■ Range of recommended ages ■ Catch-up immunization ■ Assessment at age 11–12 years

[a]Hepatitis B vaccine (HepB). *At birth:* All newborns should receive monovalent HepB, administered soon after birth and before hospital discharge.

[b]Diphtheria and tetanus toxoids and acellular pertussis vaccine (DTaP). Tdap is a modified vaccine with lower doses of diphtheria and pertussis toxoids.

[c]*Haemophilus influenzae* type b conjugate vaccine (Hib).

[d]Meningococcal conjugate vaccine (MCV4). MCV4 should be administered to all children at age 11–12 years, as well as to unvaccinated adolescents at high school entry (age 15 years). Vaccine contains four types of capsules.

[e]Pneumococcal vaccine. The 13-valent pneumococcal conjugate vaccine (PCV) is recommended for all children aged 2–23 months and for certain children aged 24–59 months. The final dose in the series should be administered at age ≥12 months. Pneumococcal polysaccharide vaccine (PPSV) is a 23-valent vaccine recommended in addition to PCV for certain high-risk groups. PPSV is also recommended for people over 65.

[f]Inactivated influenza vaccine should be administered annually starting at age 6 months. Live, attenuated influenza vaccine should not be given until 2 years, and not to immunocompromised individuals.

[g]Hepatitis A vaccine (HepA). HepA is recommended for all children at age 1 year (12–23 months).

Source: http:www.cdc.gov/vaccines/schedules/downloads/child/0-6yrs-schedule-pr.pdf

mentioned, will result in a population susceptible to the microbe and a resurgence of serious disease.

Thought Question

26.14 How might you design/construct a more effective vaccine for an antigen (for instance, the *Yersinia pestis* F1 antigen, or hepatitis A) that would harness the power of a Toll-like receptor (TLR)? *Hint:* TLR5 from Table 23.5.

To Summarize

- **Vaccines** can be made from live, attenuated organisms; killed organisms; or purified microbe components.
- **Herd immunity** can help protect unimmunized persons from diseases transmitted person-to-person.
- **Serious side effects** very rarely result from immunizations.

Concluding Thoughts

This chapter has described key concepts of infectious disease using just a small sample of disease-causing pathogens. Other diseases are considered elsewhere in the book or on the accompanying website. These include anthrax, cholera, the common cold (caused by rhinovirus), dengue fever, diphtheria, gastric ulcers (*Helicobacter pylori*), herpes, AIDS, warts (human papillomavirus), influenza, salmonellosis, *Streptococcus agalactiae* infections, West Nile disease, and Whipple's disease (*Tropheryma whipplei*). The principal goal with this and the previous chapter was to illustrate how microbial metabolism can undermine the physiology of a human host. The next chapter will describe how humans fight back using pharmacology to sabotage the physiology of the infecting microbes.

CHAPTER REVIEW

Review Questions

1. How are pathogens classified by portal of entry?
2. Discuss some common skin infections.
3. What causes boils?
4. What are the symptoms of necrotizing fasciitis?
5. What is the difference between primary and secondary infections?
6. How do pneumococci avoid engulfment by phagocytes?
7. What causes the clouding seen in X-rays of infected lungs?
8. What are the key features of the pneumococcal vaccine?
9. Name the common fungal causes of lung disease.
10. What is a metastatic lesion?
11. Why is diarrhea watery?
12. What is the most common microbial cause of diarrhea?
13. List common bacterial agents that cause diarrhea.
14. Would you suspect *Salmonella* infection in a cluster of nauseated patients rushed to the hospital directly from a church picnic? Why or why not?
15. What is the significance of finding leukocytes in stool?
16. What is one reservoir of *E. coli* O157:H7?
17. How is a UTI diagnosed?
18. What is an important virulence determinant of uropathogenic *E. coli*?
19. What is the most common sexually transmitted disease?
20. How is gonorrhea different in men and women?
21. Why will *Neisseria gonorrhoeae* not usually disseminate in the bloodstream, while *N. meningitidis* will?
22. Name the major causes of bacterial meningitis. What are the two routes of infection?
23. If tetanus and botulism toxins have the same mode of action, why do they cause opposite effects on muscles?
24. What are some virulence factors of *Yersinia pestis*?

Thought Questions

1. Chickenpox is a disease of children and young adults caused by herpes virus 3 (varicella). It is characterized by a rash of fluid-filled vesicles that eventually become crusty. The rash starts on the trunk and spreads to the extremities. Illness usually resolves in 7–10 days, and the patient becomes immune to the disease. However, some individuals later in life (:60 years old) develop a painful disease called shingles caused by the same virus—even if they have never been reexposed to the virus. The lesions are tender, persistent vesicles that form on the skin. Propose a plausible explanation, considering the age of the shingles victims and the occurrence of severe pain in this illness, for how they contracted shingles and why the lesions are painful.
2. A 5-year-old male was brought to the emergency room by his grandmother, who had found the boy on the floor of her apartment covered in bloody, loose feces. Patient history revealed that the boy attended day care regularly. The diagnostic laboratory determined the etiological agent to be a Gram-negative rod. This organism is a facultative intracellular pathogen that escapes the host cell vacuole and moves in and between cells by actin polymerization. From your reading of Chapters 25 and 26, what do you think are the most likely genus and species? Why is the fact that the boy attended day care significant?
3. Why are urinary tract infections among the most commonly acquired nosocomial (hospital-acquired) infections?

4. Visit the MMWR (Morbidity and Mortality Weekly Report) page of the CDC website that provides the "Summary of Notifiable Diseases" (the URL as of this writing is www.cdc.gov/mmwr/mmwr_nd). Study the table that summarizes the monthly incidence of infections in the United States and compare the data for gonorrhea and Lyme disease. Explain the differences in monthly incidence of these two diseases, and then view the incidence of those diseases in your state.

Key Terms

bacteremia (1091)
blood-brain barrier (1083)
borreliosis (1096)
botulism (1085)
bubonic plague (1095)
cellulitis (1057)
chancre (1078)
chlamydia (1079)
congenital syphilis (1078)
endocarditis (1090)
erythema migrans (1096)
hepatitis (1099)
Lyme disease (1096)
metastatic lesion (1064)
necrotizing fasciitis (1056)
organism (1052)
pneumonic plague (1095)
portal-of-entry (1052)
primary syphilis (1078)
prion (1089)
reactivation tuberculosis (1065)
rehydration therapy (1066)
rotavirus (1072)
secondary syphilis (1078)
secondary tuberculosis (1065)
septicemia (1091)
septicemic plague (1095)
spongiform encephalopathy (1090)
tertiary syphilis (1078)
tetanospasmin (1085)
transcytosis (1084)
viremia (1091)
zoonotic disease (1053)

Recommended Reading

Chua, Caroline L., Graham Brown, John A. Hamilton, Stephen Rogerson, and Phillipe Boeuf. 2012. Monocytes and macrophages in malaria: Protection or pathology? *Trends in Parasitology* **29**:26–34.

Chun, Tae-Wook, and Antony S. Fauci. 2012. HIV reservoirs: Pathogenesis and obstacles to viral eradication and cure. *AIDS* **26**:1261–1268.

Coureuil, Mathieu, Olivier Join-Lambert, Herve Lecuyer, Sandrine Bourdoulous, Stefano Marullo, et al. 2012. Mechanism of meningeal invasion by *Neisseria meningitidis. Virulence* **3**:164–172.

Donaldson, David S., Atshushi Kobayashi, Hiroshi Ohno, Hideo Yagita, Ifor R. Williams, et al. 2012. M cell-depletion blocks oral prion disease pathogenesis. *Mucosal Immunology* **5**:216–225.

Kaletsky, Rachel L., Joseph R. Francica, Caroline Agrawal-Gamse, and Paul Bates. 2009. Tetherin-mediated restriction of filovirus budding is antagonized by the Ebola glycoprotein. *Proceedings of the National Academy of Sciences USA* **106**:2886–2891.

Hadjifrangiskou, Maria, and Scott J. Hultgren. 2012. What does it take to stick around? Molecular insights into biofilm formation by uropathogenic *Escherichia coli. Virulence* **3**:231–233.

Neemann, Kari, D. D. Eichele, P. W. Smith, R. Bociek, M. Akhtari, et al. 2012. Fecal microbiota transplantation for fulminant *Clostridium difficile* infection in an allogeneic stem cell transplant patient. *Transplant Infectious Disease* **14**:E161–E165.

Perrin, Agnès, Stéphane Bonacorsi, Ettiene Carbonnelle, Driss Talibi, Philippe Dessen, et al. 2002. Comparative genomics identifies the genetic islands that distinguish *Neisseria meningitidis,* the agent of cerebrospinal meningitis, from other *Neisseria* species. *Infection and Immunity* **70**:7063–7072.

Ruggiero, Paolo. 2012. *Helicobacter pylori* infection: What's new. *Current Opinion in Infectious Diseases* **25**:337–344.

Russell, David G. 2007. Who puts the tubercle in tuberculosis? *Nature Reviews. Microbiology* **5**:39–47.

Soon, J. M., P. Seaman, and R. N. Baines. 2012. *Escherichia coli* O104:H4 outbreak from sprouted seeds. *International Journal of Hygiene and Environmental Health* **216**:346–354.

Trevejo, Rosalie T., Margaret C. Barr, and Robert Ashley Robinson. 2005. Important emerging bacterial zoonotic infections affecting the immunocompromised. *Veterinary Research* **36**:493–506.

Zhou, Dongsheng, and Ruifu Yang. 2009. Molecular Darwinian evolution of virulence in *Yersinia pestis. Infection and Immunity* **77**:2242–2250.

Zinkernagel, Annelies S., and Victor Nizet. 2007. *Staphylococcus aureus*: A blemish on skin immunity. *Cell Host & Microbe* **1**:161–162.

CHAPTER 27
Antimicrobial Therapy

27.1 The Golden Age of Antibiotic Discovery
27.2 Fundamentals of Antimicrobial Therapy
27.3 Measuring Drug Susceptibility
27.4 Mechanisms of Action
27.5 Challenges of Drug Resistance
27.6 The Future of Drug Discovery
27.7 Antiviral Agents
27.8 Antifungal Agents

The discovery of antibiotics about 80 years ago has played a major role in increasing life expectancy throughout the world. Prior to 1918, the average life expectancy in the United States was 45–50 years. That number is now nearly 79 years, thanks in part to antibiotics. But antibiotics may soon become useless. Over several decades, antibiotics have been used indiscriminately to treat patients, and they are often included in animal feed—not to treat animals but to grow larger ones. These abuses, combined with the ability of bacteria to become antibiotic resistant, have led experts to predict an impending crisis in which the human race is vulnerable to infectious diseases we once thought had been conquered.

Important questions about chemotherapeutic agents will be addressed in this chapter, including: Why do antimicrobials inhibit the growth of bacteria, but not of humans or animals? What is antibiotic resistance? How do clinicians know which antibiotic to use to treat an infection? We will also discuss what makes a good antibiotic target, and how new antibiotics are discovered.

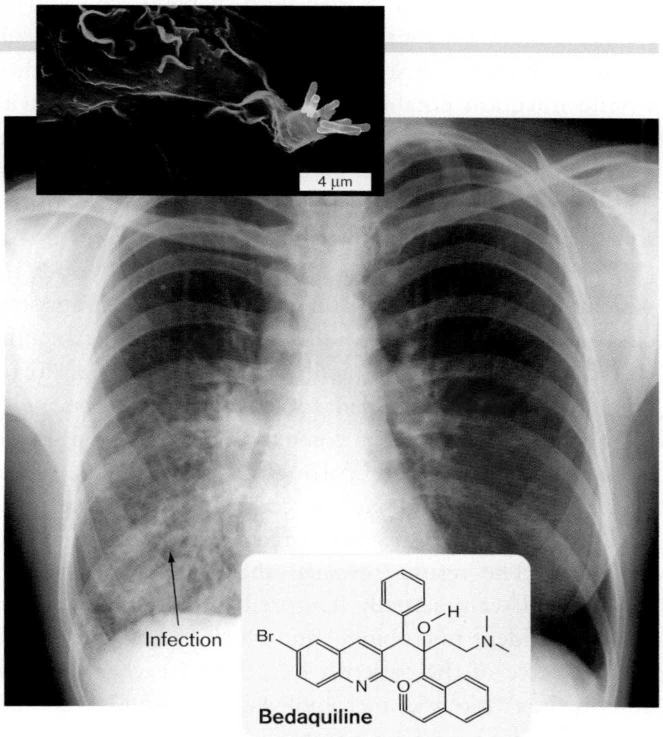

4 μm
Infection
Br
N
O
H
Bedaquiline

SIMON FRASER/SCIENCE PHOTO LIBRARY/SCIENCE

CURRENT RESEARCH highlight

A new weapon in our antibiotic arsenal. Almost 30% of Earth's humans harbor latent tuberculosis infections; that amounts to almost 2 billion people. The accompanying photos show the ravages of a pulmonary TB infection viewed via colorized X-ray, and an SEM of *Mycobacterium tuberculosis* infecting a lung macrophage. What makes the situation even more desperate is that of the 9 million new TB cases each year, 500,000 are multidrug-resistant (MDR). Unfortunately, the antibiotics for this pathogen have not changed for 40 years—until now. A new drug called bedaquiline (inset) was approved by the FDA in 2013 to treat MDR-TB. The antibiotic selectively targets the organism's energy-generating ATP synthase—a novel mode of action—and starves the pathogen of energy. The hope is that this new antibiotic can finally stem the rising tide of TB. Tempering that hope, however, is the knowledge that bedaquiline-resistant mutants of *M. tuberculosis* have already been isolated in the laboratory. *Source:* K. Andries et al. 2005. *Science* **307**:223–227.

Antibiotics undeniably have been a benefit to modern society. We live longer and contribute more to society than in the past, in large measure because of antibiotics. Infections we consider minor today often killed their victims just 50 or 60 years ago. But there has been a downside too. Consider the case of a 56-year-old man with diabetes who entered the hospital for a heart transplant. The operation went well, but 1 week after the surgery, he developed a severe chest wound infection notable for exuded pus. Treatment with methicillin, a penicillin derivative commonly used to treat infections, failed and the patient became comatose. The diagnostic microbiology laboratory ultimately identified the agent as a methicillin-resistant *Staphylococcus aureus* (MRSA), prompting the surgeon to immediately place the patient on intravenous vancomycin for 6 weeks. Vancomycin is an antibiotic, structurally different from methicillin, that can usually kill methicillin-resistant bacteria. Fortunately, this patient recovered; too often, they do not.

Where did the infection come from? Surprising as it may seem, this drug-resistant pathogen was a resident of the hospital itself. To find the source, nasal swabs were taken of all hospital personnel and screened for the presence of *S. aureus*. The results revealed that several members of the surgical team actually harbored this organism as part of their resident microbiota. But which individual was the actual source of the patient's infection? In what could be described as "forensic microbiology," each strain was subjected to pulsed-field gel electrophoresis, and the resulting genomic restriction patterns were compared with that of the isolate from the infected patient. A match pointed to the source. The unwitting culprit turned out to be the perfusionist who manipulated the tubing used for cardiopulmonary bypass.

Hospital-acquired, or **nosocomial**, infections are not unusual. As many as 5%–10% of all patients admitted to acute-care hospitals acquire nosocomial infections, resulting in over 80,000 deaths each year. The deaths are due, in part, to the poor health of the patient and in part to the antibiotic-resistant nature of bacteria lurking in hospitals. In fact, as many as 60%–70% of staph infections that develop in a hospital setting are the result of methicillin-resistant *S. aureus* (MRSA). A foreboding report from the United Kingdom finds that one in four nursing-home residents is colonized by MRSA. A recent study in the United States found that the noses of nearly one in five nonhospitalized people are also colonized by MRSA. This problem will only get worse.

We begin Chapter 27 with a discussion of the golden age of antibiotic discovery (1940–60) and move on to describe the basic concepts of antibiotics and their use and misuse. We will also delve into the ways genomic and proteomic approaches broaden our ability to search for new antibiotics and identify new antibiotic targets—all part of our attempt to stay one step ahead of evolving, antibiotic-resistant pathogens.

27.1

The Golden Age of Antibiotic Discovery

Antibiotics (from the Greek meaning "against life") are compounds produced by one species of microbe that can kill or inhibit the growth of other microbes. While we think of antibiotics as being a recent biotechnological development, their use has historical precedent. Ancient remedies called for cloths soaked with organic material to be placed on wounds to allow them to heal faster. This organic material likely contained "natural antibiotics" that killed bacteria and prevented further infection. The medicinal properties of molds were also recognized for centuries. Historical accounts refer to the ancient Chinese successfully treating boils with warm soil and molds scraped from cheeses, and in England a paste of moldy bread was a home remedy for wound infections up until the beginning of the twentieth century.

The modern antibiotic revolution began with the discovery of **penicillin** in 1928 by Sir Alexander Fleming (1881–1955). This discovery was actually a rediscovery, and was arguably one of the greatest examples of serendipity in science. Although Fleming generally receives the credit for discovering penicillin, a French medical student, Ernest Duchesne (1874–1912), originally discovered the antibiotic properties of *Penicillium* in 1896.

Duchesne observed that Arab stable boys at the nearby army hospital kept their saddles in a dark and damp room to encourage mold to grow on them. When asked why, they told him the mold helped heal saddle sores on the horses. Intrigued, Duchesne prepared a solution from the mold and injected it into diseased guinea pigs. All recovered. Although he submitted his work as a dissertation to the Pasteur Institute, it was ignored because he was young and unknown.

Penicillium was forgotten in the scientific community until Fleming rediscovered it one day in the late 1920s. Petri dishes were glass in those days and could be rewashed and sterilized. Fleming was preparing to wash a pile of old petri dishes he had used to grow the pathogen *Staphylococcus aureus*. He opened and examined each dish before tossing it into a cleaning solution. He noticed that one dish had grown contaminating mold, which in and of itself was not unusual in old plates, but all around the mold the staph bacteria had failed to grow (**Fig. 27.1A**). Fleming took a sample of the mold and found that it was from the penicillium family, later identified as *Penicillium notatum*. The mold appeared to have synthesized a chemical, now known as penicillin (**Fig. 27.1B**), which diffused through the agar, killing cells of *S. aureus* before they could form colonies. Fleming (**Fig. 27.1C**) presented his findings in 1929, but they raised little interest, since penicillin appeared to be unstable and would not remain active in the body long enough to kill pathogens.

As World War II began, an Oxford professor named Howard Florey (**Fig. 27.1D**), together with his colleague Ernst Chain, rediscovered Fleming's work, thought it held promise, and set about purifying penicillin. To their amazement, when the purified penicillin was injected into mice infected with staphylococci or streptococci, the majority of mice survived. Subsequent human trials also proved successful (**Fig. 27.1E**), and penicillin gained wide use, saving countless lives during the war. As a fitting tribute to serendipity, Fleming, Florey, and Chain received the 1945 Nobel Prize in Physiology or Medicine for their work. Duchesne's work, however, went unrecognized by the scientific community for decades.

The next landmark discovery in antibiotics was made by Gerhard Domagk (1895–1964), a German physician at the Bayer Institute of Experimental Pathology and Bacteriology, who investigated antimicrobial compounds in the 1930s (**Fig. 27.2A**). In 1935, Domagk's 6-year-old child was afflicted with a serious streptococcal infection induced by an innocent pinprick to the finger. The infection spread to the lymph nodes under her arm and became so severe that lancing and draining the pus 14 times did little to help. The only remaining alternative was to amputate the arm. Unfortunately, even this drastic measure would probably not save her life. Frustrated, Gerhard Domagk took what would appear to be drastic measures. He administered a dose of a red dye (prontosil) he was investigating

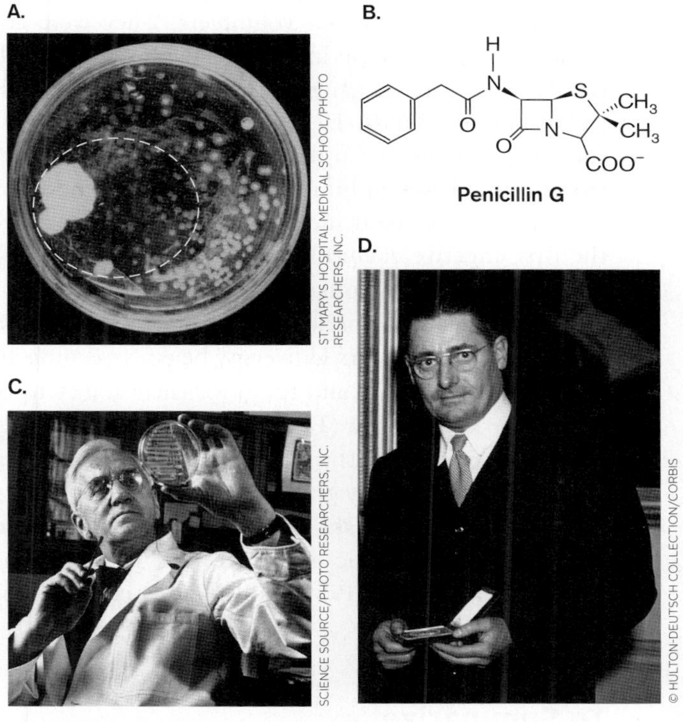
A.
B.
C.
D.
ST. MARY'S HOSPITAL MEDICAL SCHOOL/PHOTO RESEARCHERS, INC.
Penicillin G
SCIENCE SOURCE/PHOTO RESEARCHERS, INC.
© HULTON-DEUTSCH COLLECTION/CORBIS

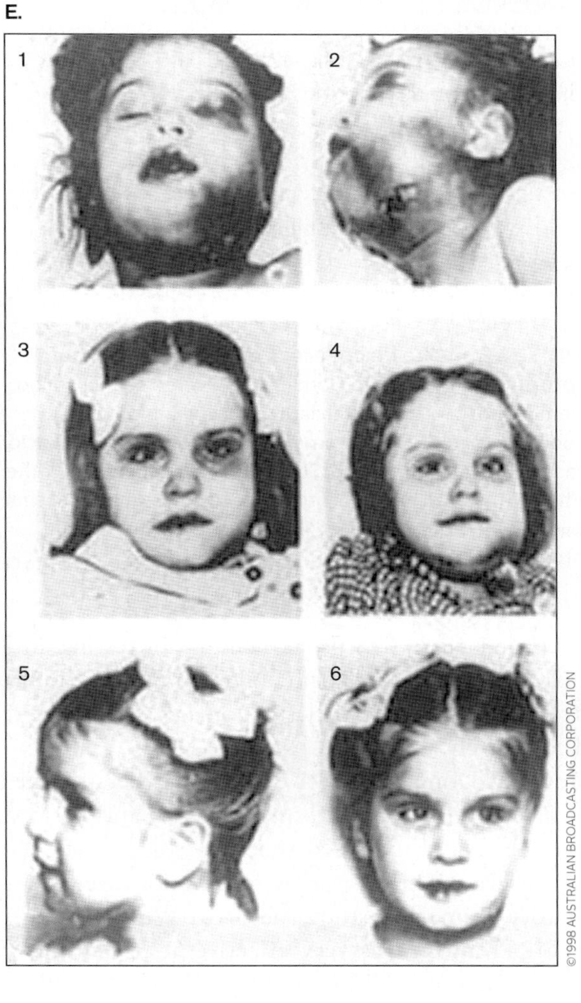
E.
1
2
3
4
5
6
©1998 AUSTRALIAN BROADCASTING CORPORATION

FIGURE 27.1 ■ The dawn of antibiotics. **A.** Alexander Fleming's photo of the dish with bacteria and penicillin mold. The ring highlights the area of decreased growth of *Staphylococcus aureus* colonies. **B.** The chemical structure of penicillin G. **C.** Alexander Fleming at work in his laboratory. **D.** Howard Florey. **E.** Pictures taken in 1942, shortly after the introduction of penicillin, show the improvement in a child suffering from an infection 4 days (panel 2) and 9 days (panel 4) after treatment. Panels 5 and 6 show her fully recovered.

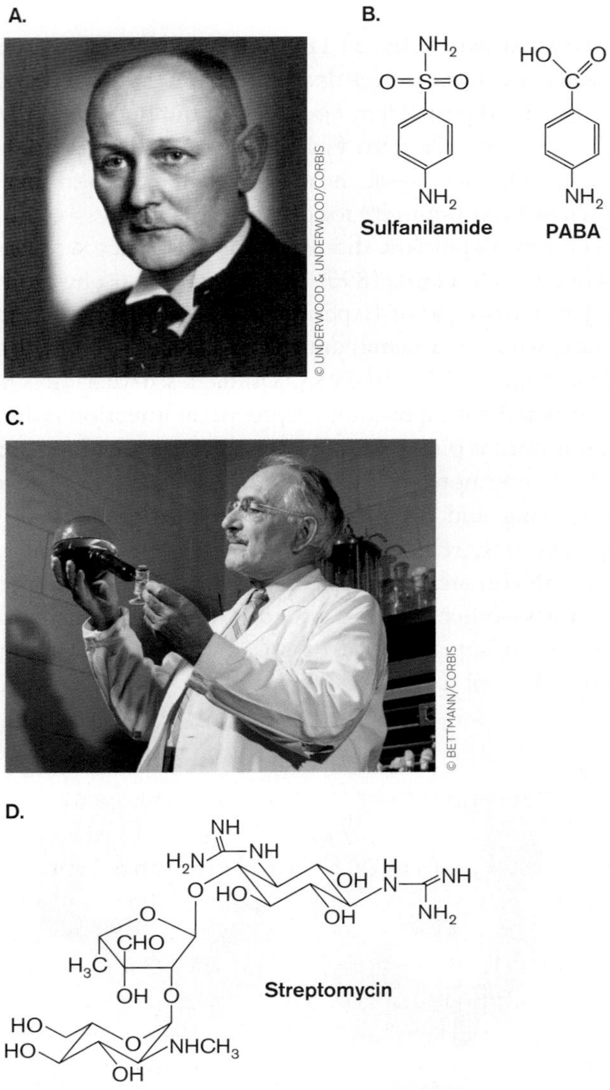
A.
B.
Sulfanilamide
PABA
© UNDERWOOD & UNDERWOOD/CORBIS
C.
© BETTMANN/CORBIS
D.
Streptomycin

FIGURE 27.2 ▪ The discoverers of sulfanilamide and streptomycin. **A.** Gerhard Domagk discovered sulfanilamide. **B.** Chemical structure of sulfanilamide, an analog of *para*-aminobenzoic acid (PABA), a precursor of the vitamin folic acid, which is necessary for growth. Sulfanilamide inhibits one of the enzymes that converts PABA into folic acid. **C.** Selman Waksman discovered streptomycin in 1944. **D.** Chemical structure of streptomycin.

that, on agar plates (the usual medium for testing antibiotics), had shown absolutely no ability to inhibit the growth of streptococcus. Nevertheless, Domagk's daughter recovered completely.

How did Domagk conceive of such a therapy when the conventional method of screening for antibiotic activity on agar plates indicated that this compound (Prontosil) was useless? The answer lay in his prior extensive use of laboratory animals to study the drug. When administered to mice, Prontosil was very effective at preventing infection. Had he not used live animals, Domagk would never have discovered that Prontosil was metabolized by the body into another compound, sulfanilamide, clearly lethal to the streptococcus. This finding led to an entire class of drugs, the **sulfa drugs**, that saved hundreds of thousands of lives, including that of Domagk's own child. The take-home message of this story is that an antibiotic's activity on a plate, or lack thereof, does not necessarily correlate with the drug's activity in a patient.

Sulfanilamide is an analog of *para*-aminobenzoic acid (PABA), a precursor of folic acid, a vitamin necessary for nucleic acid synthesis (**Fig. 27.2B**). Sulfanilamide and other sulfa drugs bind to and inhibit the enzyme that converts PABA to folic acid. Without folic acid to make nucleic acid precursors, the pathogen stops growing. Sulfa drugs inhibit bacterial growth without affecting human cells because folic acid is not synthesized by humans (it is a dietary supplement instead), and because bacteria do not transport folic acid (they must make it themselves).

In a dark turn, the pressing need for effective antibiotics during World War II dictated a change in the in vivo screening methods used by Domagk's German employer. Animal testing was abandoned in favor of human testing, but the humans were not volunteers. They were concentration camp prisoners intentionally infected with bacterial diseases, such as gangrene, and then treated with new chemical compounds. Domagk's possible participation in these experiments, even if unwilling, was a source of controversy that haunted him long after the war ended. Yet his contributions to medicine continued as he also developed the first effective chemotherapy for tuberculosis via Thiosemicarbazones and isoniazid, still used today.

During the same period of history, Selman Waksman (1888–1973) at Rutgers University began screening 10,000 strains of soil bacteria and fungi for their ability to inhibit growth or kill bacteria (**Fig. 27.2C**). In 1944, this herculean effort paid off with the discovery of streptomycin, an antibiotic produced by the actinomycete *Streptomyces griseus* (**Fig. 27.2D**). Waksman's discovery of streptomycin triggered the antibiotic gold rush that is still under way and earned him the 1952 Nobel Prize in Physiology or Medicine.

To Summarize

- **The importance of antibiotics** in treating disease was recognized in the early 1940s.
- **Some antimicrobial agents are initially inactive**, until converted by the body to an active agent.
- **Florey and Chain purified penicillin** and capitalized on Fleming's discovery of penicillin.

27.2

Fundamentals of Antimicrobial Therapy

Antibiotics comprise the vast majority of chemotherapeutic agents used to treat microbial diseases. As already noted, the term "antibiotic" originally referred to any compound produced by one species of microbe that could kill or inhibit the growth of other microbes. Today the term is also used for synthetic chemotherapeutic agents, such as sulfonamides, that are clinically useful but chemically synthesized. Many natural and synthetic compounds affect microbial growth, but their utility in a clinical setting is dictated by certain key characteristics.

Antibiotics Exhibit Selective Toxicity

As early as 1904, the German physician Paul Ehrlich (1854–1915) realized that a successful antimicrobial compound would be a "magic bullet" that selectively kills or inhibits the pathogen but not the host. This seemingly obvious premise was innovative at the time. Ehrlich made several discoveries based on this concept, the most celebrated of which was the arsenical compound known as Salvarsan. Salvarsan proved to be quite effective in killing the syphilis agent *Treponema pallidum* (this was long before penicillin was discovered). Syphilis, a sexually transmitted disease, had been untreatable and the source of considerable long-term suffering. Ehrlich's "magic bullet" concept is now known as **selective toxicity**. Although Ehrlich's selective toxicity concept was right, Salvarsan itself was not as selectively toxic as he thought. This arsenical compound harmed the host, but usually killed off the treponemes before killing the patient.

Selective toxicity is possible because key aspects of a microbe's physiology are different from those of eukaryotes. For example, suitable bacterial antibiotic targets include peptidoglycan, which eukaryotic cells lack, and ribosomes, which are structurally distinct between the Bacteria and Eukarya. Thus, chemicals like penicillin, which prevents peptidoglycan synthesis, and tetracycline, which binds to bacterial 30S ribosomal subunits, inhibit bacterial growth but are essentially invisible to host cells, since they do not interact with them at low doses.

While their intended targets are bacterial cells, some antibiotics, particularly at high doses, can interact with elements of eukaryotic cells and cause side effects that harm the patient. For example, chloramphenicol, a drug that targets bacterial 50S ribosomal subunits, can interfere with the development of blood cells in bone marrow—a phenomenon that may result in aplastic anemia (failure to produce red blood cells). The toxicity of an antibiotic can also depend on the age of the patient. Ciprofloxacin, for instance, can cause defects in human bone growth plates and should not be administered to children (Section 27.2). Problems can even arise if the drug does not directly impact mammalian physiology. For example, many people develop an extreme allergic sensitivity to penicillin, in which case the treatment of an infection may end up being worse than the infection itself. Physicians must be aware of these allergies and use alternative antibiotics to avoid harming their patients.

As a student of microbiology, you must be able to properly distinguish between the terms "drug susceptibility" and "drug sensitivity." A microbe is susceptible to the drug's action, but a human can develop an allergic sensitivity to the drug.

Antimicrobials Have a Limited Spectrum of Activity

No single antimicrobial drug affects all microbes. As a result, antimicrobial drugs are classified by the type of organisms they affect. Thus, we have antifungal, antibacterial, antiprotozoan, and antiviral agents. The term "antibiotic" is usually reserved for antimicrobial compounds that affect bacteria. Even within a group, one agent might have a very narrow **spectrum of activity**, meaning it affects only a few species, while another antibiotic inhibits many species. For instance, penicillin has a relatively narrow spectrum of activity, killing primarily Gram-positive bacteria. However, ampicillin is penicillin with an added amino group that allows the drug to more easily penetrate the Gram-negative outer membrane. As a result of this chemically engineered modification, ampicillin kills Gram-positive and Gram-negative organisms, giving it a broader spectrum of activity than penicillin has. There are antimicrobials as well that exhibit extremely narrow activities. One example is isoniazid, which is clinically useful only against *Mycobacterium tuberculosis,* the agent of tuberculosis. **eTopic 27.1** explores the spectrum of activity of select antibiotics.

As discussed in several prior chapters, we are increasingly aware that our natural microbiota contribute in important ways to human health and development. However, few studies have explored the possible impact of antibiotic use on host-microbiota interactions. We have known for decades that antibiotics—especially broad-spectrum antibiotics—can destroy the ecological balance of bacterial species in the gut (as well as at other body sites) and lead to gastrointestinal disease. Pathology can result when one species resistant to the antibiotic gains a growth advantage over various drug-susceptible species that ordinarily keep the pathogen in check (see the discussion of *Clostridium difficile* in Section 26.4). But what other effects might there be when the microbial balance of power in the intestine is disturbed? We may ultimately find that directing the toxicity of an antibiotic to a single bacterial species (while leaving all others alone) has benefits to human health that we do not currently understand.

Antibiotics Are Classified as Bacteriostatic or Bactericidal

Patients typically believe that all antibiotics kill their intended targets. This is a misconception. Many drugs simply prevent growth of the organism and let the body's immune system dispatch the intruding microbe. Thus, antimicrobials are also classified on the basis of whether or not they kill the microbe. An antibiotic is **bactericidal** if it kills the target microbe, whereas it is **bacteriostatic** if it merely prevents bacterial growth.

To Summarize

- **Antimicrobial agents** may be produced naturally or artificially.
- **Selective toxicity** refers to the ability of an antibiotic to attack a unique component of microbial physiology that is missing or distinctly different from eukaryotic physiology.
- **Antibiotic side effects** on mammalian physiology can limit the clinical usefulness of an antimicrobial agent.
- **Antibiotic spectrum of activity** is the range of microbes that a given drug affects.
- **Bactericidal antibiotics** kill microbes; **bacteriostatic antibiotics** inhibit microbial growth.

27.3

Measuring Drug Susceptibility

One critical decision a clinician must make when treating an infection is which antibiotic to prescribe for the patient. There are several factors to consider, including:

- **The relative effectiveness of different antibiotics on the organism causing the infection.** Making this determination requires learning whether the organism isolated from a specific patient has developed resistance to the drug.
- **The average attainable tissue levels of each drug.** An antibiotic may appear to work on an agar plate, but the concentration at which it affects bacterial growth may be too high to be safe in the patient. In fact, an important aspect of designing new antibiotics is to enhance the pharmacological activity of an existing drug—for example, modifying it so that the body does not break it down or quickly secrete it in urine.

Minimal Inhibitory Concentration

The in vitro effectiveness of an antimicrobial agent is determined by measuring how little of it is needed to stop growth. This amount is classically measured in terms of an antibiotic's **minimal inhibitory concentration (MIC)**, defined as the lowest concentration of the drug that will prevent the growth of an organism. But the MIC for any one drug will differ among different bacterial species. For example, the MIC of ampicillin needed to stop the growth of *Staphylococcus aureus* will be different from that needed to inhibit *Shigella dysenteriae*. The reasons that a drug may be more effective against one organism than another include the ease with which the drug penetrates the cell and the affinity of the drug for its molecular target.

So how do we measure MIC? As shown in **Figure 27.3**, an antibiotic is serially diluted along a row of test tubes containing nutrient broth. After dilution, the organism to be tested is inoculated at low, constant density into each tube, and the tubes are usually incubated overnight. Growth of the organism is seen as turbidity. In **Figure 27.3**, the tubes with the highest concentration of drug are clear, indicating no growth. The tube containing the MIC is the tube with the lowest concentration of drug that shows no growth. Note, however, that the MIC does not indicate whether a drug is bacteriostatic or bactericidal.

Thought Questions

27.1 **Figure 27.3** illustrates how MICs are determined. Test your understanding of how MICs are measured in the following example. The drug tobramycin is added to a concentration of 1,000 μg/ml in a tube of broth from which serial twofold dilutions were made. Including the initial tube (tube 1), there are a total of ten tubes. Twenty-four hours after all the tubes are inoculated with *Listeria monocytogenes*, turbidity is observed in tubes 6–10. What is the MIC?

27.2 What additional test performed on an MIC series of tubes will tell you whether a drug is bacteriostatic or bactericidal?

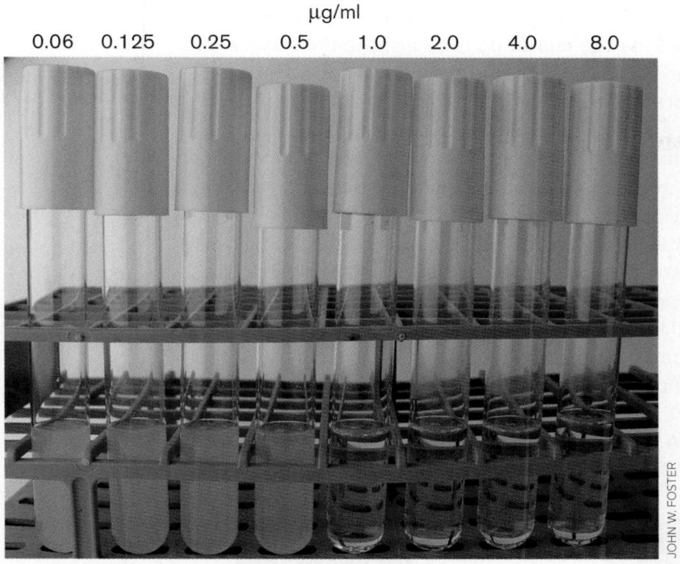
µg/ml
0.06
0.125
0.25
0.5
1.0
2.0
4.0
8.0
JOHN W. FOSTER

FIGURE 27.3 ■ Determining minimal inhibitory concentration (MIC). In this series of tubes, tetracycline was diluted serially starting at 8 µg/ml (far-right tube). Each tube was then inoculated with an equal number of bacteria. Turbidity indicates that the antibiotic concentration was insufficient to inhibit growth. The MIC in this example is 1.0 µg/ml.

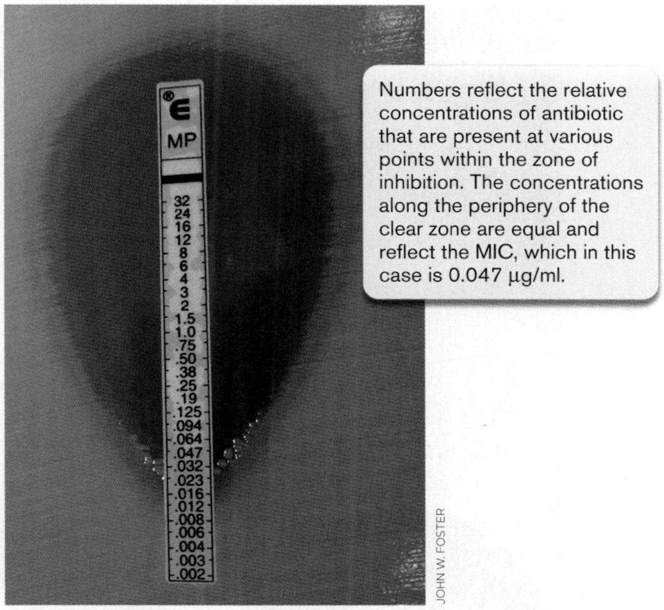
E
MP
32
24
16
12
8
6
4
3
2
1.5
1.0
.75
.50
.38
.25
.19
.125
.094
.064
.047
.032
.023
.016
.012
.008
.006
.004
.003
.002
Numbers reflect the relative concentrations of antibiotic that are present at various points within the zone of inhibition. The concentrations along the periphery of the clear zone are equal and reflect the MIC, which in this case is 0.047 µg/ml.
JOHN W. FOSTER

FIGURE 27.4 ■ An MIC strip test. The Etest (AB Biodisk) is a commercially prepared strip that produces a gradient of antibiotic concentration (in µg/ml) when placed on an agar plate. The MIC corresponds to the point where bacterial growth crosses the numbered strip.

MIC determinations are very useful for estimating a single drug's effectiveness against a single bacterial pathogen isolated from a patient, but they are not very practical when trying to screen 20 or more different drugs. Dilutions take time—time that the technician, not to mention the patient, may not have. The time required to evaluate antibiotic effectiveness can be reduced by using a strip test (like the Etest shown in **Fig. 27.4**) that avoids the need for dilutions. The strip, containing a gradient of antibiotic, is placed on an agar plate freshly seeded with a dilute lawn of bacteria. While the bacteria are trying to grow, the drug diffuses out of the strip and into the media. Drug emanating from the more concentrated areas of the strip will travel faster and farther through the agar than will drug from the less concentrated areas of the strip. Thus, the drug's effect (killing or inhibiting the growth of cells) will extend farther away from the strip at locations of high concentration than at locations of lower concentration. The result is a **zone of inhibition** where the antibiotic has stopped bacterial growth. The MIC is the point at which the elliptical zone of inhibition intersects with the strip.

Kirby-Bauer Disk Susceptibility Test

Although the strip test eliminates the time and effort needed to make dilutions, it would take 20 or more plates to test an equal number of antibiotics for just one bacterial isolate. Clinical labs can receive up to 100 or more isolates in one day, so individual MIC determinations are impractical. A simplified agar diffusion test, however, which can test 12 antibiotics on one plate, makes evaluating antibiotic susceptibility a manageable task.

Named for its inventors, the **Kirby-Bauer assay** uses a series of round filter paper disks impregnated with different antibiotics. A dispenser (**Fig. 27.5A**) delivers up to 12 disks simultaneously to the surface of an agar plate covered by a bacterial lawn. Each disk is marked to indicate the drug used. During incubation, the drugs diffuse away from the disks into the surrounding agar and inhibit growth of the lawn to different distances (**Fig. 27.5B–D**). The zones of inhibition vary in width, depending on the antibiotic used, the concentration of drug in the disk, and the susceptibility of the organism to the drug. The diameter of the zone correlates to the MIC of the antibiotic against the organism tested.

Correlations between MIC values and Kirby-Bauer zone sizes are made empirically. Every disk containing a given antibiotic is impregnated with a standard concentration of drug, and every antibiotic has a quantifiable MIC that will differ when tested against different bacterial species and strains. The outermost ring of the no-growth zone in a Kirby-Bauer disk test must, by definition, contain the minimal concentration of drug needed to prevent growth on agar. Thus, if species A and B have MIC values for penicillin

A.

JOHN W. FOSTER

FIGURE 27.5 ■ The Kirby-Bauer disk susceptibility test. A. Device used to deliver up to 12 disks to the surface of a Mueller-Hinton plate. The device is placed over the plate, and the plunger is depressed to deliver the disks. **B.** Disks impregnated with different antibiotics are placed on a freshly laid lawn of bacteria and incubated overnight. The clear zones around certain disks indicate growth inhibition. Shown are the results with normal *Staphylococcus aureus.* **C.** Results for methicillin-resistant *S. aureus* (MRSA). Note the lack of inhibition by the oxacillin disk (arrow). This strain is resistant to both methicillin and oxacillin because they are structurally similar. **D.** Results for *Streptococcus pneumoniae.* The brownish tint of the blood agar plates outside the zones of bacterial inhibition is caused by a hemolysin secreted by the lawn of pneumococci. C, chloramphenicol; CC, clindamycin; CZ, cefazolin; E, erythromycin; NOR, norfloxacin; OX, oxacillin; P, penicillin; RA, rifampin; SAM, sulbactam-ampicillin; SXT, sulfa-trimethoprim; TE, tetracycline; VA, vancomycin.

B. C. D.

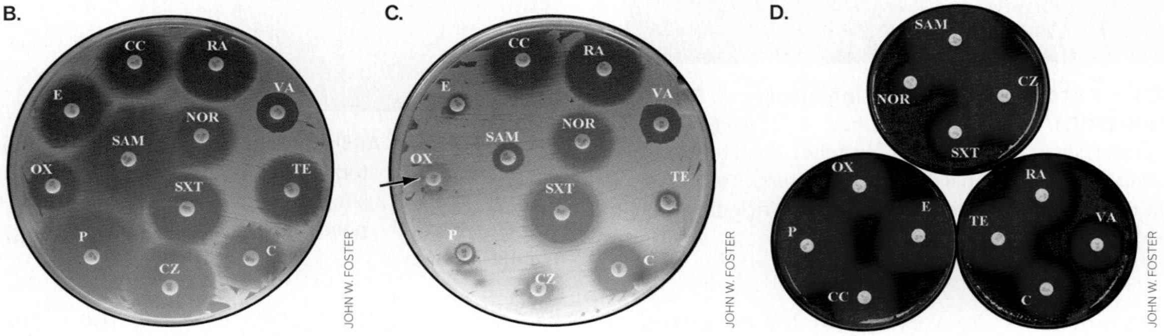
CC
RA
E
VA
NOR
SAM
OX
TE
SXT
P
C
CZ
JOHN W. FOSTER
CC
RA
E
VA
NOR
SAM
OX
SXT
TE
P
C
CZ
JOHN W. FOSTER
SAM
NOR
CZ
SXT
OX
E
RA
P
TE
VA
CC
C
JOHN W. FOSTER

of 4 μg/ml and 40 μg/ml, respectively, then species A will exhibit a proportionally larger zone of inhibition than species B in the disk test. A graph plotting MIC on one axis and zone diameter on the other provides the correlation.

After incubating agar plates in the Kirby-Bauer test, the diameters of the zones of inhibition around each disk are measured, and the results are compared with a table listing whether a zone is wide enough (meaning the MIC is low enough) to be clinically useful. **Table 27.1** shows susceptibility data for *Staphylococcus aureus.* The concentration of antibiotic used and the zone size that is considered clinically significant are correlated with the average attainable tissue level for each antibiotic. For the antibiotic to remain effective in vivo, it is important that the tissue concentration of the drug remain above the MIC; otherwise, invading bacteria will not be affected.

TABLE 27.1

Susceptibility results for *Staphylococcus aureus.*

		Zone of inhibition diameter (mm)		
Antibiotic	**Quantity in disk (μg)**	**Resistant**	**Intermediate**	**Susceptible**
Ampicillin	10	<12	12–13	>13
Chloramphenicol	30	<13	13–17	>17
Erythromycin	15	<14	14–17	>17
Gentamicin	10	≤12.5		>12.5
Streptomycin	10	<12	12–14	>14
Tetracycline	30	<15	15–18	>18

Correlating antibiotic MIC with tissue level. The average attainable tissue level for a drug depends on how quickly the antibiotic is cleared from the body via secretion by the kidney or destruction in the liver. It also depends on when side effects of the drug start to appear. The graph in **Figure 27.6** shows that as long as the concentration of the drug in tissue or blood remains higher than the MIC, the drug will be effective. The concentration can be kept at sufficient levels either by administering a higher dose, which runs the risk of side effects, or by giving a second dose at a time when the levels from the first dose have declined. This is why patients are told to take doses of some antibiotics four times a day and other antibiotics only once a day.

Thought Question

27.3 A patient with a bacterial lung infection was given the antibiotic shown in **Figure 27.6** and was told to take one pill twice a day. The pathogen is susceptible to this drug. Will the prescribed treatment be effective? Explain your answer.

To ensure reproducibility, the Kirby-Bauer test was standardized a half century ago. Reproducibility means that results from a laboratory in California will match those in Alabama, Ohio, or any other state. The following are standardizations used to make the test reproducible and easier.

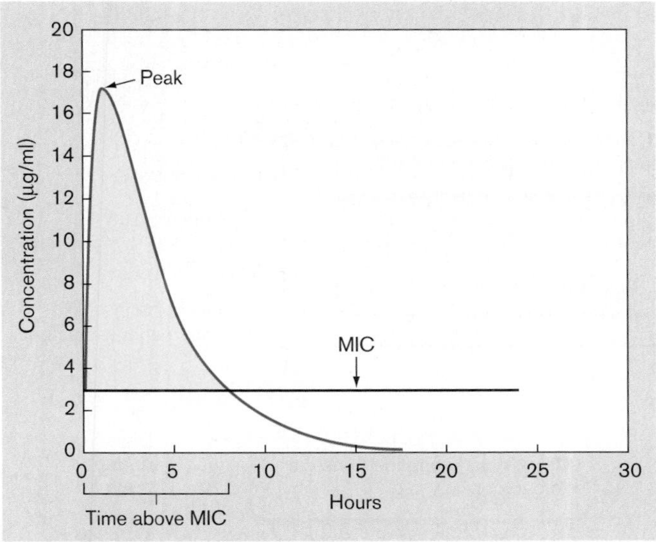
Peak
MIC
Concentration (μg/ml)
20
18
16
14
12
10
8
6
4
2
0
0
5
10
15
20
25
30
Hours
Time above MIC

FIGURE 27.6 ■ Correlation between MIC and serum or tissue level of an antibiotic. This graph illustrates the serum level of ampicillin over time. The important consideration here is how long the serum level of the antibiotic remains higher than the MIC. Once the concentration falls below the MIC, owing to destruction in the liver or clearance through the kidneys and secretion, the infectious agent fails to be controlled by the drug—in this case, 7–8 hours after the initial dose. To maintain a serum level higher than the MIC, a second dose would be taken. The shaded area of the curve represents time above MIC.

- **Size of the agar plate.** The plates used (150 mm) are larger than standard agar plates (100 mm) to accommodate more disks and to maintain sufficient distance between disks so that zones of inhibition do not overlap.
- **Depth of the media.** Antibiotics diffuse out of impregnated disks in not two, but three, dimensions. Because diffusion cannot occur very far downward in a thinly poured agar, the drug is forced to move more laterally. Thus, the zone of inhibition measured from a thinly poured agar plate will be larger than the zone from a thick agar plate.
- **Media composition.** Media composition can also affect results. For example, most common laboratory media contain *para*-aminobenzoic acid (PABA), which is used by the cell to make folic acid, a vitamin needed for purine and pyrimidine synthesis. Sulfonamides are analogs of PABA that competitively inhibit one of the enzymes needed to convert PABA to folic acid. Media containing PABA, such as standard nutrient agar, will flood the bacterial cell with PABA and limit the ability of the sulfa drugs to compete for the enzyme. Thus, even though the drug might be effective in vivo, on the agar plate there would be no zone of inhibition. The standardized medium used for the Kirby-Bauer test, called **Mueller-Hinton agar**, contains no PABA.
- **The number of organisms spread on the agar plate.** The more organisms there are placed on a plate, the smaller the zone of inhibition will be. This phenomenon is observed because there is a time lag between dropping a disk onto a plate and the diffusion of the antibiotic. The more organisms there are initially on a plate, the less time it takes for them to form visible growth; so, by the time the antibiotic gets to them, visible growth has already formed. As a result, a standard optical density solution of each organism is prepared, and a cotton swab is used to spread the entire agar surface.
- **Size of the disks.** A standard diameter of 6 mm means that all antibiotics start diffusing into agar at the same point.
- **Concentrations of antibiotics in the disks.** The higher the concentration of antibiotic in a disk, the faster the drug can diffuse through the plate and kill bacteria (or at least inhibit growth) before replicating cells are able to form visible growth. To avoid differences between labs, the concentration of each given drug impregnating a disk has been standardized.
- **Incubation temperature.** Incubation temperature will not affect growth and diffusion equally. To avoid differences, a temperature of 37°C is standard.

Thought Question

27.4 You are testing whether a new antibiotic will be a good treatment choice for a patient with a staph infection. The Kirby-Bauer test using the organism from the patient shows a zone of inhibition of 15 mm around the disk containing this drug. Clearly, the organism is susceptible. But you conclude from other studies that the drug would be ineffective in the patient. What would make you draw this conclusion?

To Summarize

- **The spectrum of an antibiotic and the susceptibility of the infectious agent** are critical points of information required before prescribing antibiotic therapy.
- **Minimal inhibitory concentration (MIC)** of a drug, when correlated with average attainable tissue levels of the antibiotic, can predict the effectiveness of an antibiotic in treating disease.
- **MIC is measured** using tube dilution techniques, but it can be approximated using the Kirby-Bauer disk diffusion technique.

27.4

Mechanisms of Action

As noted in Section 27.2, selective toxicity of an antibiotic depends on enzymes or structures unique to the bacterial target cell. The following aspects of a microbe's physiology are classic targets:

- Cell wall synthesis
- Cell membrane
- DNA synthesis
- RNA synthesis
- Protein synthesis
- Metabolism

Table 27.2 summarizes the general targets of common antibiotics. Chapters 3, 7, and 8 describe these cellular components and provide the basis for understanding how antibiotics work. The mechanisms of action for antibiotics affecting DNA, RNA, and protein synthesis are described in Chapters 7 and 8, so they receive only brief mention here.

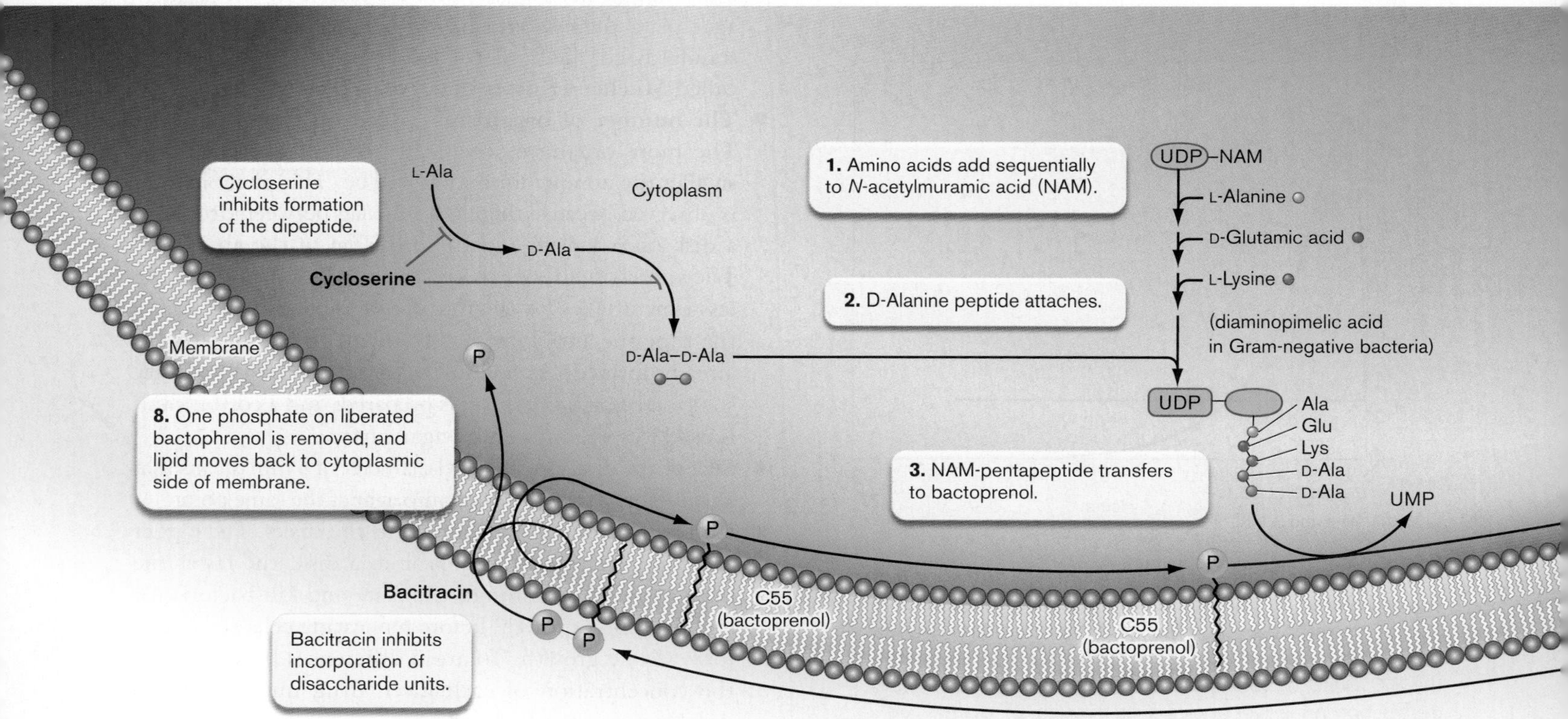
Cycloserine inhibits formation of the dipeptide.
L-Ala
Cytoplasm
D-Ala
Cycloserine
Membrane
P
D-Ala–D-Ala
1. Amino acids add sequentially to N-acetylmuramic acid (NAM).
UDP–NAM
L-Alanine
D-Glutamic acid
L-Lysine
2. D-Alanine peptide attaches.
(diaminopimelic acid in Gram-negative bacteria)
UDP
Ala
Glu
Lys
D-Ala
D-Ala
UMP
8. One phosphate on liberated bactophrenol is removed, and lipid moves back to cytoplasmic side of membrane.
3. NAM-pentapeptide transfers to bactoprenol.
Bacitracin
C55 (bactoprenol)
C55 (bactoprenol)
Bacitracin inhibits incorporation of disaccharide units.

FIGURE 27.7 ■ Peptidoglycan synthesis in a Gram-positive bacterium, and targets of antibiotics. Several small-molecular-weight compounds are sequentially joined to form a disaccharide unit that will be added to preexisting extracellular chains of this unit. Red lines indicate inhibition. Cycloserine inhibits ligation of the two D-alanines (step 2); bacitracin inhibits linking of the disaccharide units; vancomycin and the beta-lactams, such as penicillin, inhibit the peptide cross-linking of peptidoglycan side chains.

TABLE 27.2

Mechanisms of action of antimicrobial agents.

Target	Antibiotic examples
Cell wall synthesis	Penicillins, cephalosporins, bacitracin, vancomycin
Protein synthesis	Chloramphenicol, tetracyclines, aminoglycosides, macrolides, lincosamides
Cell membrane	Polymyxin, amphotericin, imidazoles (vs. fungi)
Nucleic acid function	Nitroimidazoles, nitrofurans, quinolones, rifampin; some antiviral compounds, especially antimetabolites
Intermediary metabolism	Sulfonamides, trimethoprim

Cell Wall Antibiotics

Bacterial cell walls are the basis for selective toxicity for some antibiotics because peptidoglycan does not exist in mammalian cells; thus, antibiotics that target the synthesis of these structures should selectively kill bacteria. The following case history illustrates the use of two cell wall–targeting antibiotics and also reveals how bacteria can evolve to escape destruction.

Case History: Meningitis

A 3-year-old child was brought to the emergency room crying, with a stiff neck and high fever. Gram stain of cerebrospinal fluid revealed Gram-positive cocci, generally in pairs. The diagnosis was meningitis. The physician immediately prescribed intravenous ampicillin. Unfortunately, the child's condition worsened, so antibiotic treatment was changed to a third-generation cephalosporin (which will better cross the blood-brain barrier). The patient began to improve within hours and was released after 2 days. A report from the clinical microbiology laboratory identified the organism as Streptococcus pneumoniae.

Both of the antibiotics used in this case kill bacteria by targeting cell wall synthesis. Synthesis of peptidoglycan (introduced in Chapter 3) is a complex process but is represented simply in **Figure 27.7**. Basically, sugar molecules called *N*-acetylglucosamine (NAG) and *N*-acetylmuramic

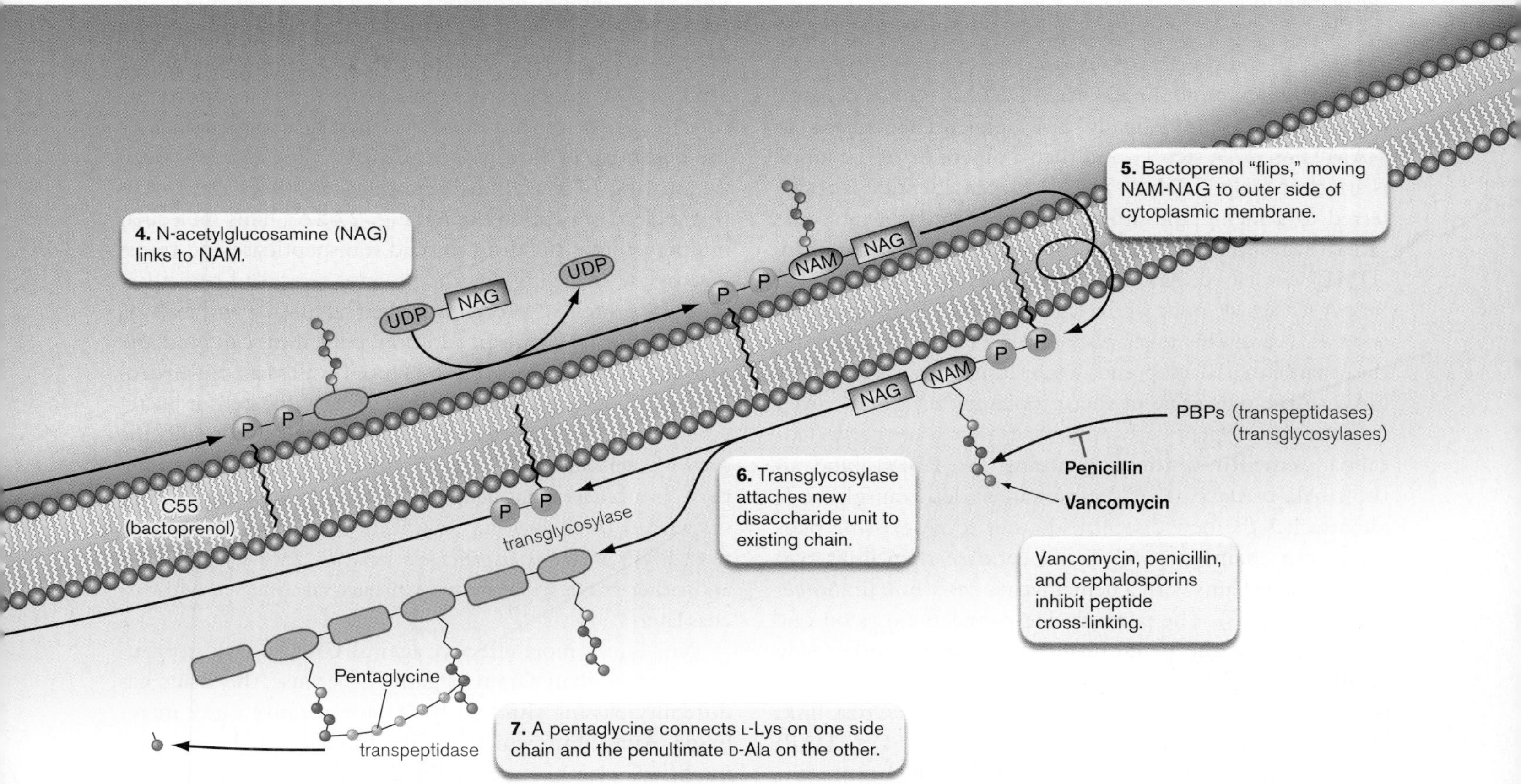

4. N-acetylglucosamine (NAG) links to NAM.
UDP
NAG
UDP
P
P
NAM
NAG
5. Bactoprenol "flips," moving NAM-NAG to outer side of cytoplasmic membrane.
C55 (bactoprenol)
transglycosylase
NAG
NAM
PBPs (transpeptidases) (transglycosylases)
Penicillin
Vancomycin
6. Transglycosylase attaches new disaccharide unit to existing chain.
Vancomycin, penicillin, and cephalosporins inhibit peptide cross-linking.
Pentaglycine
transpeptidase
7. A pentaglycine connects L-Lys on one side chain and the penultimate D-Ala on the other.

acid (NAM) are made by the cell and linked together by a **transglycosylase** enzyme into long chains assembled at the cell wall. *N*-acetylmuramic acid contains a short side chain of amino acids that is assembled enzymatically, not by a ribosome. Rigidity of the macromolecular structure, essential for maintaining cell shape, is achieved by cross-linking the side chains from adjacent strands. The enzyme **transpeptidase** (D-alanyl-D-alanine carboxypeptidase/transpeptidase) catalyzes the cross-link. **Figure 27.7** also indicates where several antibiotics target various stages of this assembly process.

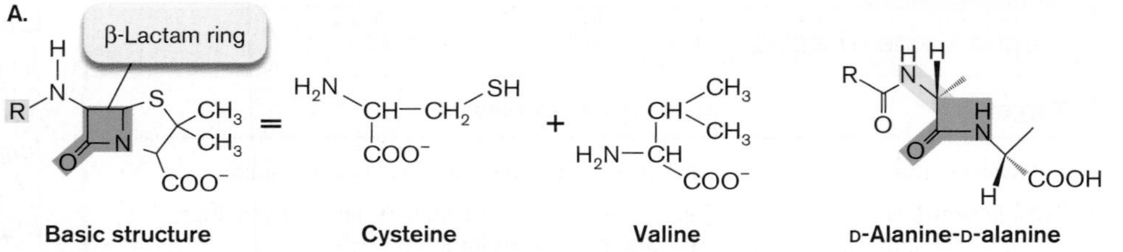
A.
β-Lactam ring
Basic structure
Cysteine
Valine
D-Alanine-D-alanine

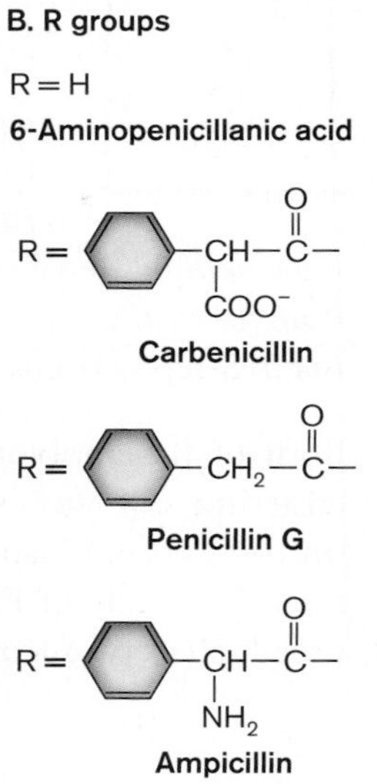
B. R groups
R = H
6-Aminopenicillanic acid
Carbenicillin
Penicillin G
Ampicillin

FIGURE 27.8 ■ The structure of penicillins. **A.** Penicillanic acid (R = H) is derived from cysteine and valine. Also shown is the D-alanine-D-alanine structure of peptidoglycan, which is structurally similar to the beta-lactam ring of penicillins (green shading). **B.** The R group highlighted in (A) can be any one of a number of different groups, some of which are shown here. Modifying this group changes the pharmacological properties and antimicrobial spectrum. Methicillin is penicillin G with $-OCH_3$ groups on the second and sixth carbons of the benzene ring.

Peptidoglycan is assembled outside the cell membrane. Before we can explain how cell wall antibiotics work, you need to know how the cell wall is made. Synthesis of peptidoglycan starts in the cytoplasm with a uridine diphosphate (UDP)–NAM molecule. The amino acids L-alanine, D-glutamic acid, and L-lysine [or diaminopimelic acid (DAP) in Gram-negative organisms] are individually and sequentially added to NAM (**Fig. 27.7**, step 1); and then a dipeptide of D-alanine is attached (step 2). Next, the NAM-pentapeptide is transferred to a membrane-situated, 55-carbon, lipid molecule called bactoprenol (step 3), and uridine monophosphate (UMP) is released. Another sugar molecule, NAG, is then linked to NAM, once again through a UDP intermediate (step 4). All of this takes place on the cytoplasmic side of the membrane. Bactoprenol then "flips," moving NAM-NAG to the outer side of the cytoplasmic membrane (step 5), where transpeptidases and transglycosylases (two so-called **penicillin-binding proteins**, or **PBPs**) bind to the D-Ala-D-Ala part of the pentapeptide. Transglycosylase attaches the new disaccharide unit to an existing peptidoglycan chain (step 6). Transpeptidase then links two peptide side chains with a pentaglycine cross-link (in *Staphylococcus aureus*). The pentaglycine connects L-Lys on one side chain and the penultimate D-Ala on the other side chain (step 7). The terminal D-Ala is removed in the process. Other bacteria do not use a pentaglycine cross-link, but directly form a peptide bond between L-Lys (DAP) and the penultimate D-Ala. Cross-linking, which strengthens the cell wall, can take place within the same strand or between adjacent strands. In the last part of the cycle, one of the phosphates on the now liberated bactoprenol is removed and the lipid moves back to the cytoplasmic side of the membrane, ready to pick up and taxi another unit of peptidoglycan to the growing chain (step 8).

Penicillin and other beta-lactam antibiotics target penicillin-binding proteins. Penicillin is an antibiotic derived from cysteine and valine, which combine to form the beta-lactam ring structure shown in **Figure 27.8A**. Different R groups can be added to the basic ring structure to change the antimicrobial spectrum and stability of the derivative penicillin (**Fig. 27.8B**). Note that the beta-lactam ring of penicillin chemically resembles the D-Ala-D-Ala piece of peptidoglycan (**Fig. 27.8A**). This molecular mimicry allows the drug to bind transpeptidase and transglycosylase (which is why the proteins are called penicillin-binding proteins), preventing their activities and halting synthesis of the chain. In addition, penicillins can bind and activate autolysin proteins in the cell wall that can hydrolyze peptidoglycan. The consequence is a disaster for bacteria that are trying to grow larger and larger. Eventually, the growing cell bursts for lack of cell wall restraint. Penicillin, then, is a bactericidal drug (unless the treated organism is suspended in an isotonic solution). Note that in addition to cell lysis, there is another explanation for why penicillin and other bactericidal drugs kill bacteria that we will discuss later.

Penicillin is more effective against Gram-positive organisms, rather than Gram negatives, because the drug has difficulty passing through the Gram-negative outer membrane. Ampicillin, which was used in the case history, is a modified version of penicillin that more easily penetrates this membrane and is more effective than penicillin against

Gram-negative microbes. Thus, ampicillin has a broader spectrum of activity than penicillin.

As noted earlier, antibiotic resistance is a growing problem throughout the world. Bacteria develop resistance to penicillin in two basic ways. The first is through inheritance of a gene encoding one of the beta-lactamase enzymes, which cleave the critical ring structure of this class of antibiotics. Beta-lactamase is transported out of the cell and into the surrounding medium (for Gram-positives) or the periplasm (for Gram-negatives), where it can destroy penicillin before the drug even gets to the cell. Bacteria that produce beta-lactamase are still susceptible to certain modified penicillins and cephalosporins engineered to be poor substrates for the enzyme. Methicillin, for example, works well against beta-lactamase-producing microbes.

Unfortunately, a new type of beta-lactamase, called New Delhi metallo-beta-lactamase-1 (NDM-1), has emerged that confers resistance to most all beta-lactam antibiotics. Originating in India, NDM-1-containing plasmids are promiscuously transferred, being found in various enterobacterial species, such as *Klebsiella pneumoniae* and *E. coli*, as well as the nonenteric pathogens *Pseudomonas aeruginosa* and *Acinetobacter baumanii*.

Aside from beta-lactamases, the second way a microbe can become resistant to beta-lactam antibiotics is through mutations in a gene encoding a key penicillin-binding protein (PBP). Resistance occurs when the mutated gene produces an altered protein that no longer binds to the antibiotic. Methicillin-resistant *Staphylococcus aureus* (MRSA) uses this strategy. Resistance to methicillin in *S. aureus* is mediated by the *mecA* gene, which is part of a chromosome cassette called staphylococcal cassette chromosome *mec* (SCC*mec*). The *mecA* gene encodes an altered penicillin-binding protein (PBP2a or PBP2′) with low affinity for beta-lactam antibiotics. The low affinity provides resistance to all beta-lactam antibiotics, rendering them useless. Hospitals take special interest in MRSA because very few drugs can kill it.

One of the few remaining antibiotics effective against MRSA is vancomycin. Unfortunately, resistance to this drug is also developing. Vancomycin-resistant *S. aureus* strains are called VRSA. The penicillin-resistant *Streptococcus pneumoniae* in the preceding case history actually had an altered penicillin-binding protein. No beta-lactamase-producing *S. pneumoniae* has yet been found.

Cephalosporins are another type of beta-lactam antibiotic originally discovered in nature but modified in the laboratory to fight microbes that are naturally resistant to penicillins (especially *P. aeruginosa*). Over the years, the basic structure of cephalosporin has undergone a series of modifications to improve its effectiveness against penicillin-resistant pathogens. Each modification is increasingly complex and produces what is referred to as a new "generation" of cephalosporins. There are currently four generations of this semisynthetic antibiotic (**Fig. 27.9**). Unfortunately, the microbial world constantly adapts and eventually becomes resistant to new antibiotics. In the case of the cephalosporins, new beta-lactamases evolve that can attack the sterically buried beta-lactam rings in these molecules. It is also important to note that because the core feature of these drugs is the beta-lactam ring, persons who are sensitive to penicillins may also suffer a hypersensitivity reaction to cephalosporins.

Treatment note: In the preceding case history, the infecting strain of *S. pneumoniae* turned out to be resistant to ampicillin. Had the patient been an adult, a

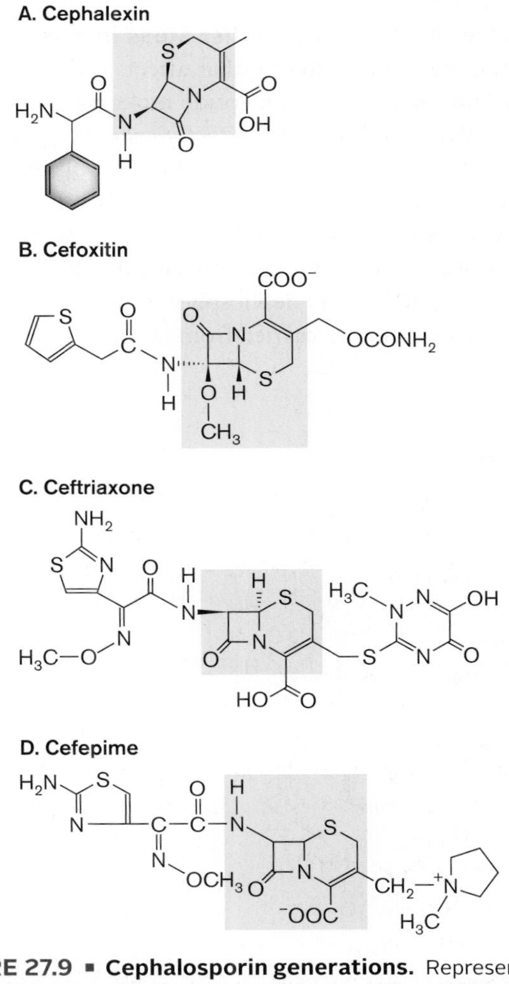
A. Cephalexin
B. Cefoxitin
C. Ceftriaxone
D. Cefepime

FIGURE 27.9 ▪ Cephalosporin generations. Representative examples. **A.** First generation: cephalexin (Keflex). **B.** Second generation: cefoxitin. **C.** Third generation: ceftriaxone. **D.** Fourth generation: cefepime. With each successive generation, the side groups become more complex. Highlighted areas indicate the core structure of each of the cephalosporins, with beta-lactam rings.

fluoroquinolone (see Section 7.2) might have been the best secondary drug of choice because its target, a type II topoisomerase, is unrelated to cell wall synthesis. Quinolones are not recommended for children, as in this case, because of potential side effects. Other beta-lactam antibiotics, such as the third-generation cephalosporins, may still work on penicillin-resistant *S. pneumoniae,* because the modified antibiotic often can still bind the altered PBP. Nevertheless, cephalosporin-resistant strains of *S. pneumoniae* are now appearing, leaving vancomycin or an oxazolidinone as the best last choice.

Note: Archaeal peptidoglycan contains talosaminuronic acid instead of muramic acid and lacks the D-amino acids found in bacterial peptidoglycan. Archaea are thus insensitive to penicillins, which interfere with bacterial transpeptidases. This natural resistance is not a problem, because there are no known archaeal pathogens.

Cell wall antibiotics target other steps in peptidoglycan synthesis. Another antibiotic that affects cell wall synthesis is **bacitracin**, a large polypeptide molecule produced by *Bacillus subtilis* and *Bacillus licheniformis* (**Fig. 27.10A**). The antibiotic inhibits cell wall synthesis by binding to the bactoprenol lipid carrier molecule that normally transports monomeric units of peptidoglycan across the cell membrane and to the growing chain (see **Fig. 27.7**). Bacitracin binds to and inhibits dephosphorylation of the carrier, thereby preventing the carrier from accepting a new unit of UDP-NAM. Resistance to bacitracin can develop if the organism can rapidly recycle the phosphorylated lipid carrier molecule through dephosphorylation or if the organism possesses an efficient drug export system (discussed in Section 27.5). Normally, bacitracin is used only topically because of serious side effects, such as kidney damage, which can occur if bacitracin is ingested.

Cycloserine (made by *Streptomyces garyphalus*) is one of several antimicrobials used to treat tuberculosis (**Fig. 27.10B**). Relative to bacitracin, it acts at an even earlier step in peptidoglycan synthesis. Cycloserine inhibits the two enzymes that make the D-Ala-D-Ala dipeptide. As a result, the complete pentapeptide side chain on *N*-acetylmuramic acid cannot be made (see **Fig. 27.7**). Without these alanines, cross-linking cannot occur and peptidoglycan integrity is compromised.

Vancomycin, a very large and complex glycopeptide produced by the streptomycete *Amycolatopsis orientalis* (**Fig. 27.10C**), binds to the D-Ala-D-Ala terminal end of the disaccharide unit and prevents the action of transglycosylases and transpeptidases (see **Fig. 27.7**). The mechanism of resistance is very different for vancomycin and penicillin, which makes vancomycin particularly useful against penicillin-resistant bacteria. To prevent the development and spread of vancomycin-resistant bacteria, this antibiotic is typically used only as a drug of last resort. Resistance can develop when products from a cluster of *van* genes collaborate to make D-lactate and incorporate it into the

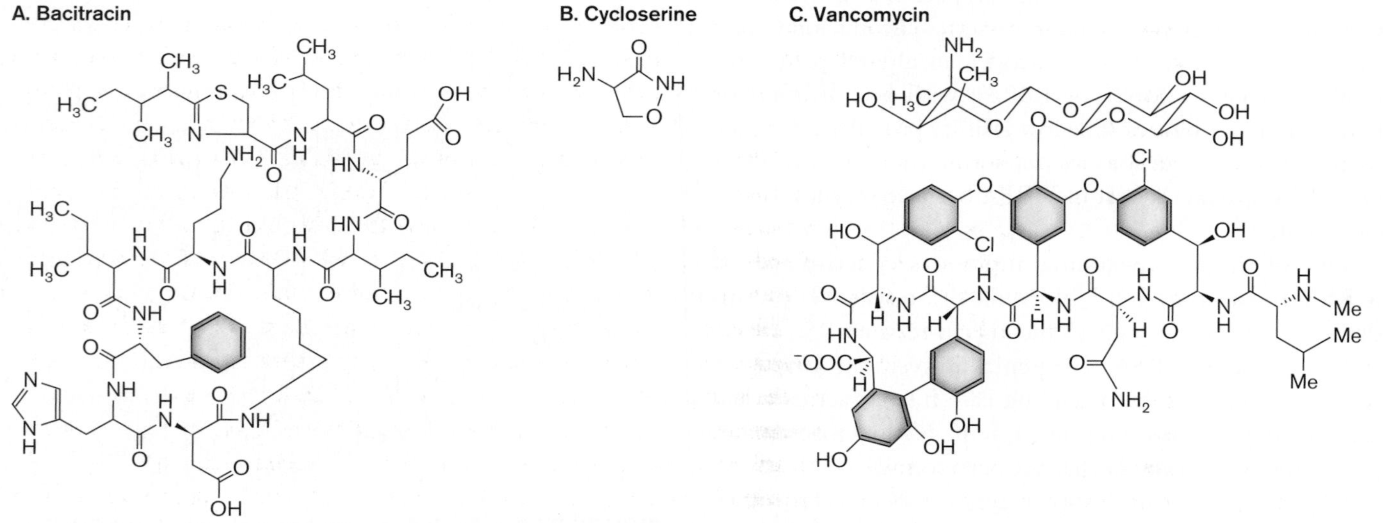
A. Bacitracin
B. Cycloserine
C. Vancomycin

FIGURE 27.10 ▪ Other antibiotics that affect peptidoglycan synthesis. A. Bacitracin is produced by *Bacillus subtilis.* It is generally used only topically to prevent infection. **B.** Cycloserine, an analog of D-alanine, is one of several drugs used to treat tuberculosis. **C.** Vancomycin is a cyclic polypeptide made by *Amycolatopsis orientalis,* previously classified as a streptomycete. These antibiotics, especially bacitracin and vancomycin, are synthesized by exceedingly complex biochemical pathways in the producing organisms. Me = methyl.

ester D-Ala-D-lactate, to which vancomycin cannot bind. Another enzyme in the *van* gene cluster prevents the accumulation of D-Ala-D-Ala; as a result, the D-Ala-D-lactate replaces D-Ala-D-Ala in peptidoglycan. Peptidoglycan containing D-Ala-D-lactate functions just fine, but the organism is resistant to the antibiotic because vancomycin cannot bind the D-lactate form.

Note that antibiotics targeting cell wall biosynthesis generally kill only growing cells. These drugs do not affect static or stationary-phase cells, because in this state the cell has no need for new peptidoglycan.

Thought Question

27.5 When treating a patient for an infection, why would combining a drug such as erythromycin with a penicillin be counterproductive? (Erythromycin is described in Section 8.3.)

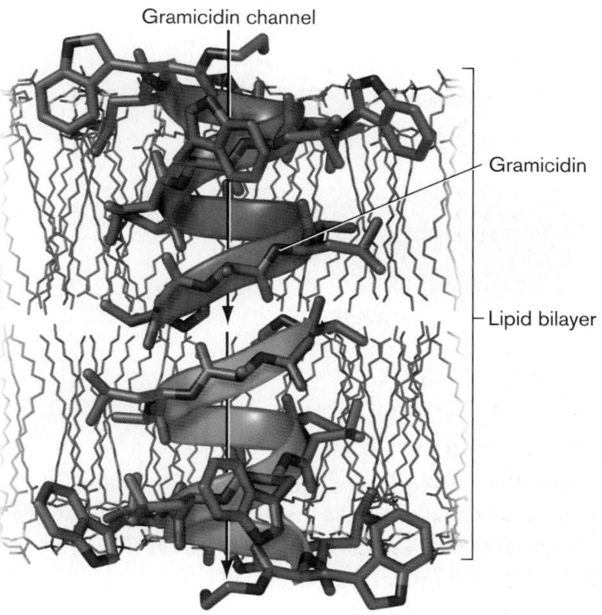
Gramicidin channel
Gramicidin
Lipid bilayer

FIGURE 27.11 ▪ Gramicidin is a peptide antibiotic that affects membrane integrity. As a dimer, gramicidin forms a cation channel across cell membranes through which H^+, Na^+, or K^+ can freely pass. (PDB code: 1GRM)

Drugs That Affect Bacterial Membrane Integrity

Poking holes in a bacterial cytoplasmic membrane is an effective way to kill bacteria. There are a few compounds useful in this regard, among them a group called the peptide antibiotics, of which **gramicidin** is an example. Produced by *Bacillus brevis,* gramicidin is a cyclic peptide composed of 15 alternating D- and L-amino acids. It inserts into the membrane as a dimer, forming a cation channel that disrupts membrane polarity (**Fig. 27.11**). Polymyxin (from *Bacillus polymyxa*), another polypeptide antibiotic, has a positively charged polypeptide ring that binds to the outer (lipid A) and inner membranes of bacteria, both of which are negatively charged. Its major lethal effect seems to be to destroy the inner membrane, much like a detergent. These antibiotics are used only topically to treat or prevent infection. Because they can also form channels across human cell membranes, they should never be ingested. Polymyxin has been fused to some bandage materials used to treat burn patients who are particularly susceptible to Gram-negative infections (for example, *Pseudomonas aeruginosa*).

Drugs That Affect DNA Synthesis and Integrity

Bacteria generally make and maintain their DNA using enzymes that closely resemble those of mammals. Thus, you might think it impossible to selectively target bacterial DNA synthesis, but it is possible, as you will see.

Case History: Pneumonia Due to a Gram-Negative Anaerobe

A 23-year-old woman arrived at the emergency room by ambulance with fever, chills, and severe muscle aches. She developed a nonproductive cough, had difficulty breathing, had pleuritic chest pain, and became hypotensive (low blood pressure). An X-ray showed lower-lobe infiltrate in the lungs, and the clinical laboratory reported the presence of the Gram-negative anaerobe Fusobacterium necrophorum *in blood cultures. The patient was diagnosed with pneumonia and treated with metronidazole, a DNA-damaging agent specific for anaerobes. She fully recovered.*

There are several classes of drugs, including sulfa drugs, quinolones, and metronidazole, that selectively affect the synthesis or integrity of DNA in microorganisms.

Sulfa drugs. The sulfa drugs, originally discovered by Domagk, belong to a group of drugs known as antimetabolites because they interfere with the synthesis of metabolic intermediates. Ultimately, sulfa drugs inhibit the synthesis of nucleic acids. Drugs such as sulfamethoxazole or sulfanilamide work at the metabolic level to prevent the synthesis of tetrahydrofolic acid (THF), an important cofactor in the synthesis of nucleic acid precursors

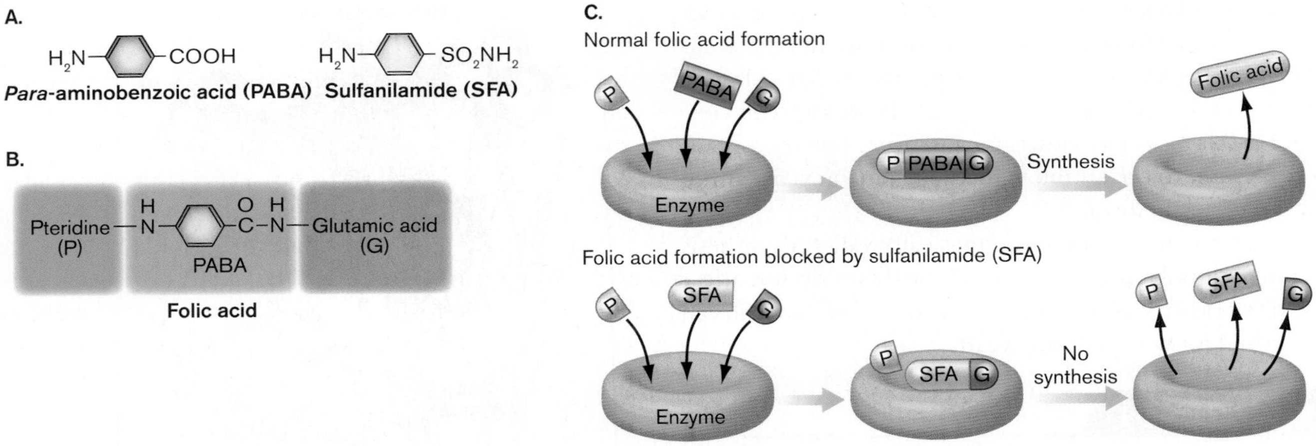
A.
H2N COOH
H2N SO2NH2
Para-aminobenzoic acid (PABA)
Sulfanilamide (SFA)
B.
Pteridine (P)
PABA
Glutamic acid (G)
Folic acid
C.
Normal folic acid formation
P
PABA
G
Enzyme
Synthesis
Folic acid
Folic acid formation blocked by sulfanilamide (SFA)
SFA
No synthesis

FIGURE 27.12 ▪ Mode of action of sulfanilamides. **A.** The structures of PABA and sulfanilamide are very similar. **B.** PABA, pteridine, and glutamic acid combine to make the vitamin folic acid. **C.** Normal synthesis of folic acid requires that all three components engage the active site of the biosynthetic enzyme. The sulfa drugs replace PABA at the active site. The sulfur group, however, will not form a peptide bond with glutamic acid, and the size of sulfanilamide sterically hinders the binding of pteridine, so folic acid cannot be made.

(**Fig. 27.12**). All organisms use THF to synthesize nucleic acids, so why are the sulfa drugs selectively toxic to bacteria? The selectivity occurs because mammalians do not synthesize folic acid, a precursor of THF. Higher mammals generally rely on bacteria and green leafy vegetables as sources of folic acid. Bacteria make folic acid from the combination of PABA, glutamic acid, and pteridine. Sulfanilamide (SFA), a structural analog of PABA, competes for one of the enzymes in the bacterial folic acid pathway and inhibits both folic acid and THF production (**Fig. 27.12C**). Because humans lack that pathway, sulfa drugs are selectively toxic toward bacteria.

Quinolones. Another group of drugs inhibits DNA synthesis by targeting microbial topoisomerases such as DNA gyrase and topoisomerase IV (the mechanism is discussed in Section 7.2). Because these enzymes are structurally distinct from their mammalian counterparts, drugs can be designed to selectively interact with them without interfering with mammalian DNA metabolism. In 1963, one such drug, nalidixic acid, was discovered as a by-product of the synthesis of chloroquine, an antimalarial drug. Nalidixic acid, which targets DNA gyrase, has a very narrow antimicrobial spectrum, covering only a few Gram-negative organisms. However, various chemical modifications, such as adding fluorine and amine groups, have increased its antimicrobial spectrum and its half-life in the bloodstream. The result is the class of drugs known as the **quinolones**. (The mode of action of quinolones and fluoroquinolones was discussed in Section 7.2.)

Metronidazole. Also known as Flagyl, metronidazole is an example of a prodrug—a drug that is harmless until activated. Metronidazole is activated after it receives an electron (is reduced) from the microbial protein cofactors flavodoxin and ferredoxin, found in microaerophilic and anaerobic bacteria such as *Bacteroides* (**Fig. 27.13**). Once activated, the compound begins nicking DNA at random, thus killing the cell. Because the etiological agent in our case history was an anaerobe, metronidazole was an effective therapy. Metronidazole is also effective against protozoa such as *Giardia, Trichomonas,* and *Entamoeba.* Aerobic microbes, although in possession of ferredoxin, are

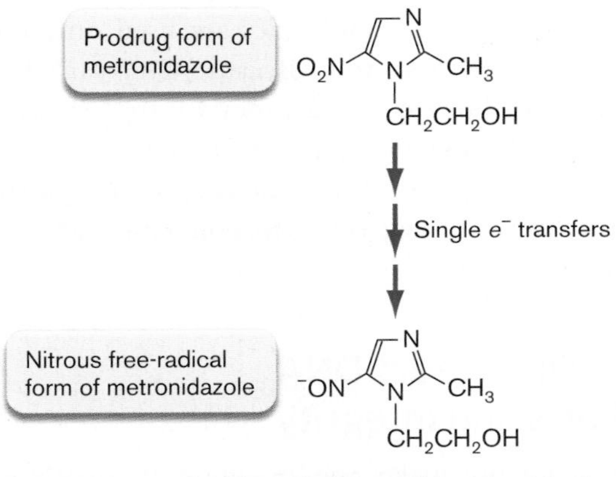
Prodrug form of metronidazole
O2N
N
CH3
CH2CH2OH
Single e− transfers
Nitrous free-radical form of metronidazole
−ON

FIGURE 27.13 ▪ Activation of metronidazole. Single-electron transfers are made by ferredoxin and flavodoxin from anaerobes. Ferredoxin and flavodoxin are reducing agents capable of reducing oxidized molecules such as thioredoxin.

incapable of reducing metronidazole, presumably because oxygen is reduced in preference to metronidazole.

Thought Question

27.6 The enzyme DNA gyrase, a target of the quinolone antibiotics, is an essential protein in DNA replication. The quinolones bind to and inactivate this protein. Research has proved that quinolone-resistant mutants contain mutations in the gene encoding DNA gyrase. If the resistant mutants contain a mutant DNA gyrase and DNA gyrase is essential for growth, then why are these mutations not lethal?

RNA Synthesis Inhibitors

The mode of action of antibiotics that inhibit transcription, such as rifampin and actinomycin D (**Fig. 27.14**), was described in Chapter 8. These drugs are bactericidal and are most active against growing bacteria. The tricyclic ring of actinomycin D binds DNA from any source. As a result, it is not selectively toxic and not used to treat infections. Rifampin (also called rifampicin), on the other hand, does exhibit selective toxicity and is often prescribed for the treatment of tuberculosis or meningococcal meningitis. Curiously, because rifampin is reddish orange, it turns bodily secretions, including breast milk, orange. The astute physician will warn the patient of this highly visible but harmless side effect to avoid unnecessary anxiety when the patient's urine changes color.

Until recently, rifampicin was the only antibiotic known to directly target bacterial RNA polymerase (RNAP). Rifampicin binding to RNAP prevents transcription only after the enzyme has started polymerization (it obstructs the exit tunnel for nascent RNA, as described in Chapter 8). In an exciting development, a new class of antibiotics (pyronins), represented by myxopyronin (produced by *Myxococcus fulvis*), has been discovered, which prevents RNAP from ever starting polymerization. The pyronins bind to RNA polymerase at a site called the hinge region, which is required to separate (melt) DNA strands—a requirement to begin transcription. Rifampicin binds to a completely different region of RNAP. Because the binding sites for these drugs are different, rifampicin-resistant RNAP molecules are still sensitive to the new antibiotics. This is exciting because pathogens such as *Mycobacterium tuberculosis* that are rapidly developing resistance to rifampicin can now be treated with a new drug that targets the same enzyme. It is unlikely (but not impossible) that pathogens will develop resistance to myxopyronin, since any mutation that alters the hinge binding site on RNAP will likely interfere with the enzyme's ability to perform an essential task (that is, melting DNA). Thus, any attempt to become resistant would likely kill the bacterium anyway. Representatives of these new antibiotics are undergoing clinical trials.

Protein Synthesis Inhibitors

The differences between prokaryotic and eukaryotic ribosomes account for the selective toxicity of antibiotics that specifically inhibit bacterial protein synthesis. How various antibiotics inhibit protein synthesis was discussed in

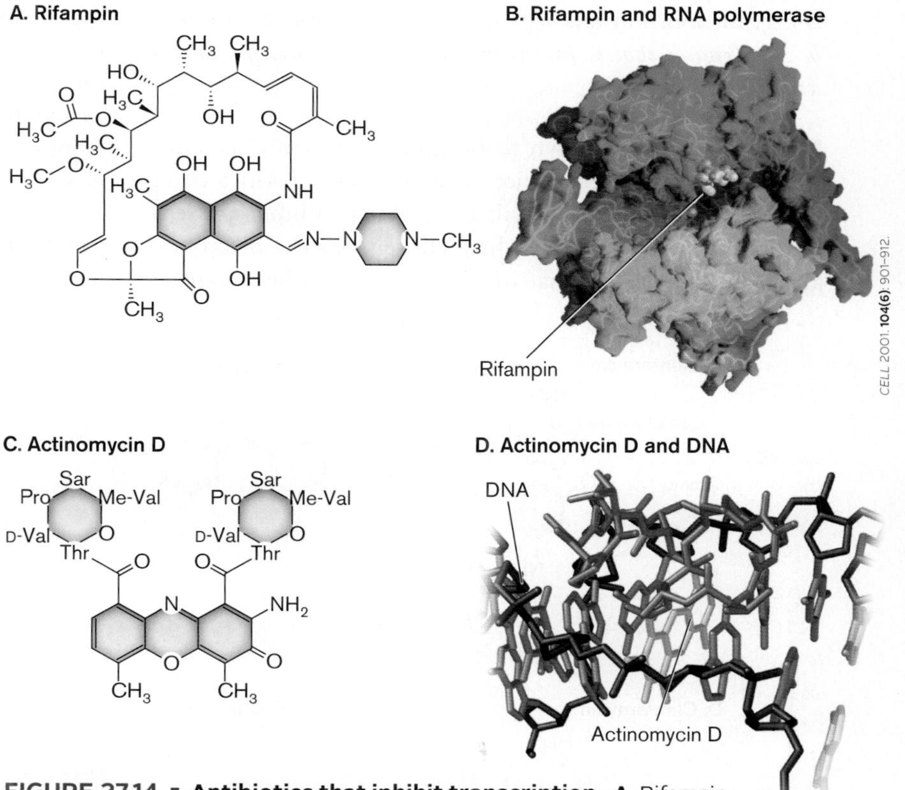
A. Rifampin
B. Rifampin and RNA polymerase
Rifampin
CELL 2001. 104(6) 901-912.
C. Actinomycin D
Sar
Pro
Me-Val
D-Val
Thr
D. Actinomycin D and DNA
DNA
Actinomycin D

FIGURE 27.14 ▪ Antibiotics that inhibit transcription. **A.** Rifampin. **B.** Rifampin-binding site on the RNA polymerase beta subunit. Yellow = alpha subunit; blue = beta subunit; magenta = beta-prime subunit; red and green = DNA double helix; orange = RNA; bright yellow = rifampin. **C.** Actinomycin D. **D.** Actinomycin D (yellow and red) interacting with DNA. Covalent intercalation of actinomycin interferes with DNA synthesis and transcription. (PDB code: 1DSC) *Source:* Part B from A. Tupin et al. 2010. *Int. J. Antimicrob. Agents* **35**:519–523.

Section 8.3. Recall that protein synthesis inhibitors can be classified into several groups based on structure and function (**Fig. 27.15**). Most of these antibiotics work by binding and interfering with the function of bacterial rRNA, which differs from eukaryotic rRNA. Recall, too, that protein synthesis inhibitors are, by and large, bacteriostatic (not bactericidal).

Case History: Erysipelas in a Penicillin-Sensitive Patient

Sixteen-year-old Jamal arrived at the emergency room after 2 days of fever, malaise, chills, and neck stiffness. His most notable symptom was a painful, red, rapidly spreading rash covering the right side of his face. The rash covered his entire cheek, which was swollen, and extended into his scalp. About 7 days earlier, Jamal had had a severe sore throat. Because it subsided in 2 days, however, he was not clinically evaluated. Throat cultures taken on admission revealed group A Streptococcus pyogenes, *suggesting that the rash was a case of erysipelas caused by this organism. Although penicillin would be the drug of choice, Jamal was known to be allergic to this antibiotic.*

When a patient is known to be immunologically sensitive to the usual drug of choice, a structurally distinct drug is best. Often that drug will be one that inhibits protein synthesis; in this case, the drug chosen was the macrolide azithromycin. Drugs that inhibit protein synthesis can be subdivided into different groups based on their structures and on which part of the translation machine is targeted.

Drugs That Affect the 30S Subunit

The classification of antibiotics affecting protein synthesis is initially based on the bacterial ribosomal subunit targeted. Thus, one class of antibiotics interferes with 30S subunit function, and the other scrambles 50S subunit activities.

Aminoglycosides. There is considerable variation in structure among different aminoglycosides, but all contain a cyclohexane ring and amino sugars (**Fig. 27.15A**). The aminoglycosides are unusual among protein synthesis inhibitors in that they are bactericidal rather than bacteriostatic. Most of them bind 16S rRNA and cause translational misreading of mRNA, which is why these drugs are bactericidal. The resulting synthesis of jumbled peptides wreaks havoc with physiology and kills the cell. Streptomycin and gentamicin (**Fig. 27.15A**) are two widely used drugs in this class. Ototoxicity (hearing damage) is a major, but uncommon, side effect of these antibiotics (approximately 0.5%–3% of patients treated with gentamicin suffer from this toxicity). Hearing is generally affected at frequencies above 4,000 Hz.

Tetracyclines. Tetracycline antibiotics are characterized by a structure with four fused cyclic rings—hence the name.

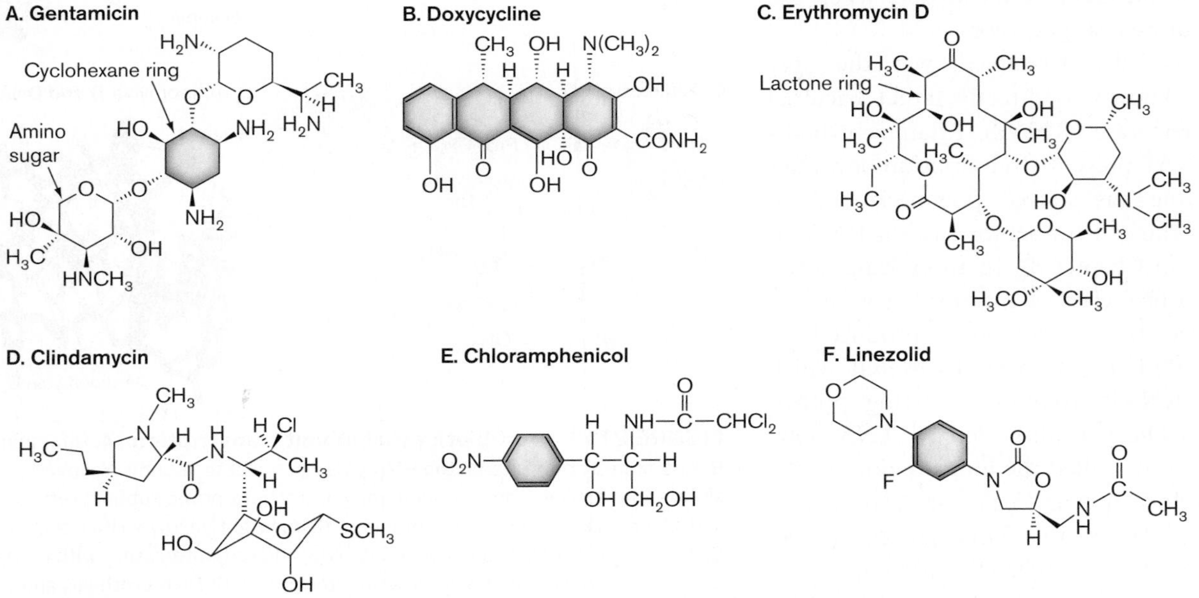
A. Gentamicin
Cyclohexane ring
Amino sugar
B. Doxycycline
C. Erythromycin D
Lactone ring
D. Clindamycin
E. Chloramphenicol
F. Linezolid

FIGURE 27.15 ▪ Protein synthesis inhibitors. A. The aminoglycoside gentamicin. **B.** The tetracycline doxycycline. **C.** The macrolide erythromycin. **D.** The lincosamide clindamycin. **E.** Chloramphenicol. **F.** The oxazolidinone linezolid.

Figure 27.15B shows one frequently used example, called doxycycline. Tetracyclines are bacteriostatic and work by binding to and distorting the ribosomal A site that accepts incoming charged tRNA molecules. Doxycycline is used to treat early stages of Lyme disease (caused by *Borrelia burgdorferi*), acne (*Propionibacterium acnes*), and other infections. An important adverse side effect of tetracyclines is that they can interfere with bone development in a fetus or young child. Tetracycline use by pregnant mothers will also cause yellow discoloration of the infant's teeth. As a result, this drug is not recommended for pregnant women or nursing mothers.

Drugs That Affect the 50S Subunit

Five classes of drugs subvert translation by binding to the 50S ribosomal subunit. Most of these drugs were discussed in Chapter 8 and are recapped here only briefly.

- **Macrolides**, all of which contain a 14- to 16-member lactone ring (**Fig. 27.15C**), inhibit translocation of the growing peptide (bacteriostatic action). Commonly prescribed examples are erythromycin and azithromycin. Azithromycin was the antibiotic used to treat the *Streptococcus pyogenes* infection in our case history, although other drugs could have been used. Because it is structurally dissimilar to any of the beta-lactam antibiotics, such as penicillin, it can be used safely in patients who are penicillin sensitive.
- **Lincosamides** (**Fig. 27.15D**), such as clindamycin, are similar to macrolides in function but have a different structure.
- **Chloramphenicol** (**Fig. 27.15E**) inhibits peptidyltransferase activity (bacteriostatic). Bone marrow depression leading to aplastic anemia is the most common serious side effect and limits its clinical use.
- **Oxazolidinones** (**Fig. 27.15F**) are a recently discovered class of synthetic antibiotics effective against many antibiotic-resistant microbes. In fact, this was the first new class of antibiotics discovered since the "golden age" of antibiotic discovery over 35 years ago. Oxazolidinones such as linezolid bind to the 23S rRNA in the 50S subunit of the prokaryotic ribosome and prevent formation of the protein synthesis 70S initiation complex. This is a novel mode of action; other protein synthesis inhibitors either block polypeptide extension or cause misreading of mRNA. Linezolid binds to the 50S subunit near where chloramphenicol binds, but it does not inhibit peptidyltransferase. Resistance is limited because most bacterial genomes have multiple operons encoding 23S rRNA. Usually more than one of these genes must mutate to confer high-level resistance. The more unmutated 23S rRNA genes there are, the more ribosomes susceptible to the antibiotic will be present. Oxazolidinones are useful primarily against Gram-positive bacteria. Gram-negative bacteria are intrinsically resistant because of multidrug efflux pumps (Section 27.5) and decreased permeability due to the outer membrane.
- **Streptogramins** (**Fig. 27.16**), produced by some *Streptomyces* species, fall into two groups, designated A and B. Streptogramins belonging to group A have a large nonpeptide ring (**Fig. 27.16A**), whereas streptogramin B members are cyclic peptides (**Fig. 27.16B**). The two groups differ in their modes of action, although both inhibit bacterial protein synthesis by binding to the peptidyltransferase site. Group A streptogramins bound to the peptidyltransferase site distort the ribosome to prevent binding of tRNA to the ribosome A site. In contrast, group B streptogramins are thought to narrow the peptide exit channel, preventing exit of the peptide and thereby blocking translocation.

Natural streptogramins are produced as a mixture of A and B, the combination of which is more potent than either individual compound alone (an example of synergy). In

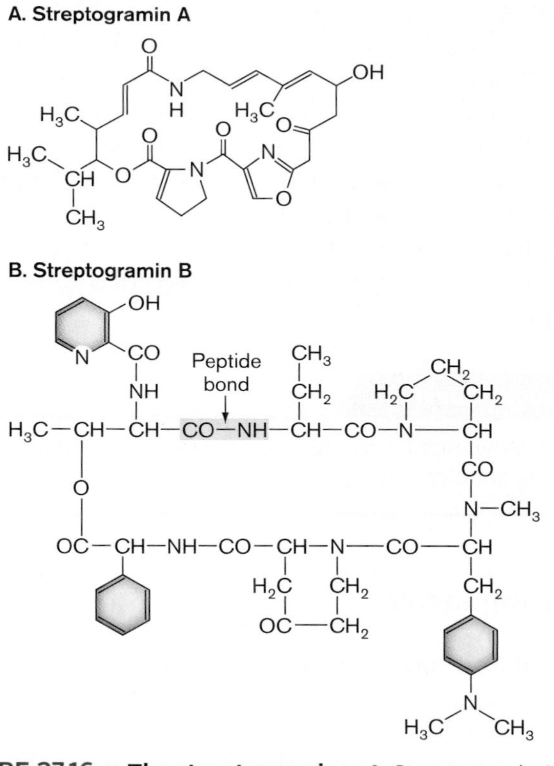
A. Streptogramin A
B. Streptogramin B
Peptide bond

FIGURE 27.16 ■ The streptogramins. A. Streptogramin A is a large nonpeptide ring structure. **B.** Streptogramin B is a cyclic peptide.

tribute to this synergistic action, the drug combination is marketed under the name Synercid. Synergy between the two drugs occurs because the A-type streptogramin alters the binding site for the B-type drug, increasing its affinity. Bacteria can develop resistance through ribosomal modification (the modification in 23S rRNA is the same one that provides resistance to macrolides), via the production of inactivating enzymes, or by active efflux of the antibiotic.

To Kill or Not to Kill: What Makes a Bactericidal Drug?

Bactericidal antibiotics include quinolones that bind to DNA topoisomerases; aminoglycosides that bind to the 30S ribosome subunit; rifampin, which binds to RNA polymerase; and penicillins that inhibit peptidoglycan synthesis. While these antimicrobials do halt critical cell processes, other antibiotics that inhibit some of these same processes are not bactericidal. For example, aminoglycosides and macrolides inhibit protein synthesis, but macrolides do not typically kill bacteria.

A new theory attempts to explain why particular antibiotics are bactericidal. Put simply, a common consequence of the modes of action for bactericidal antibiotics is the generation of highly reactive hydroxyl radicals, which damage DNA, protein, and lipids, leading to cell death. Partial evidence for this model came when James J. Collins and colleagues found that preventing induction of the SOS response that prevents and repairs oxidative damage to DNA (Section 9.5) made cells more sensitive to the bactericidal antibiotics. The antibiotics would still inhibit growth but not kill the cells. A practical consequence of this finding is that determining how to inhibit the SOS response by pathogens could make them more susceptible to antibiotic therapy.

Thought Question

27.7 Why might a combination therapy of an aminoglycoside antibiotic and cephalosporin be synergistic?

To Summarize

- **Antibiotic specificity** for bacteria can be achieved by targeting a process that occurs only in bacteria, not host cells; by targeting small structural differences between components of a process shared by bacteria and hosts; or by exploiting a physiological condition such as anaerobiosis present only in certain bacteria.
- **Antibiotic targets** include cell wall synthesis, cell membrane integrity, DNA synthesis, RNA synthesis, protein synthesis, and metabolism.
- **Antibiotics targeting the cell wall** bind to the transglycosylases, transpeptidases, and lipid carrier proteins involved with peptidoglycan synthesis and cross-linking.
- **Antibiotics interfering with DNA** include the antimetabolite sulfa drugs that inhibit nucleotide synthesis; quinolones that inhibit DNA topoisomerases; and a drug, metronidazole, that, when activated, randomly nicks the phosphodiester backbone.
- **Inhibitors of RNA synthesis** target RNA polymerase (rifampin and pyronins) or bind DNA and inhibit polymerase movement (actinomycin D).
- **Aminoglycosides and tetracyclines** bind the 30S subunit of the prokaryotic ribosome.
- **A variety of antibiotics** bind the 50S ribosomal subunit and inhibit translocation (macrolides, lincosamides), peptidyltransferase (chloramphenicol), formation of the 70S complex (oxazolidinones), or peptide exit through the ribosome exit channel (streptogramins).

27.5 Challenges of Drug Resistance

Why do microbes make antibiotics, and how do they avoid killing themselves in the process? The answers provide insight into the origin of antibiotic-resistant pathogens and our fight to halt their spread throughout the world.

Risky Business: Why Make Antibiotics?

Antibiotics are considered **secondary metabolites** because they often have no apparent primary use in the producing organism. This most likely means that a purpose, either current or ancestral, has yet to be identified. Certainly, antibiotic production today can help one microbe compete favorably with another in nature. Antibiotic production can also forge a mutualistic relationship between a microbe and a colonized host by protecting the host from deadly pathogens (as *Streptomyces* does for leaf-cutter ants; see Section 18.4 and eTopic 17.3). Whatever their use today, growth inhibition by antibiotics may not have been the original purpose of secondary metabolite production. The complexity of the biosynthetic pathways involved in making antibiotics suggests a more immediate purpose—for example, cell-cell signaling—that evolved into cross-species inhibition. Features of antibiotic biosynthetic pathways are discussed in **eTopic 27.2** (penicillin synthesis) and **eTopic 15.3** (vancomycin biosynthesis).

Given that microbes continue to make antibiotics for a reason, how does the producing microorganism avoid committing suicide? Fungi that make penicillin face no consequence for having done so because the organism does not contain peptidoglycan. Actinomycetes that produce compounds such as streptomycin or chloramphenicol, however, could be susceptible to their own secondary metabolite. Ribosomes isolated from *Streptomyces griseus*, for example, are fully sensitive to the streptomycin produced by this organism. *S. griseus* avoids killing itself in two ways. First, the organism synthesizes an inactive precursor of streptomycin, 6-phosphorylstreptomycin, which is secreted from the cell and, once outside the mycelium, becomes activated by a specific phosphatase. In addition, this streptomycete has an enzyme that inactivates any streptomycin that may leak back into the mycelium. Other organisms protect themselves by methylating key residues on their rRNA to prevent drug binding or by setting up permeability barriers that thwart reentry of the antibiotic.

These and other strategies of self-preservation employed by antibiotic-producing microbes are clever. Unfortunately, these mechanisms have been shared via horizontal gene transfers, making many pathogens antibiotic resistant. Horizontal gene transfers are discussed in Sections 9.7 and 25.2.

Case History: Multidrug-Resistant Pneumonia

A 14-year-old boy with fever (39°C; 102.2°F), chills, and left-sided pleuritic chest pain was referred to a hospital emergency department by his general practitioner. A chest X-ray showed left lower-lobe pneumonia. The boy reported that he was allergic to amoxicillin and cephalosporins (as a child he had developed a rash in response to these agents) and had been taking daily doxycycline (tetracycline) for the previous 3 months to treat mild acne. He was admitted to the hospital and treated with intravenous erythromycin because of his reported beta-lactam allergies, but he continued to feel sick. The day after admission, both sputum and blood cultures grew Streptococcus pneumoniae. *After 48 hours, antibiotic susceptibility results indicated that the microbe was resistant to penicillin, erythromycin, and tetracycline. Armed with this information, the clinician immediately changed antibiotic treatment to vancomycin. The boy's fever resolved over the next 12 hours, and he made a slow but full recovery over the next week.*

Unfortunately, the scenario presented in this case is far too common and has become an extremely serious concern. The bar graph in **Figure 27.17** shows the rapid rise of penicillin resistance among *Streptococcus pneumoniae*

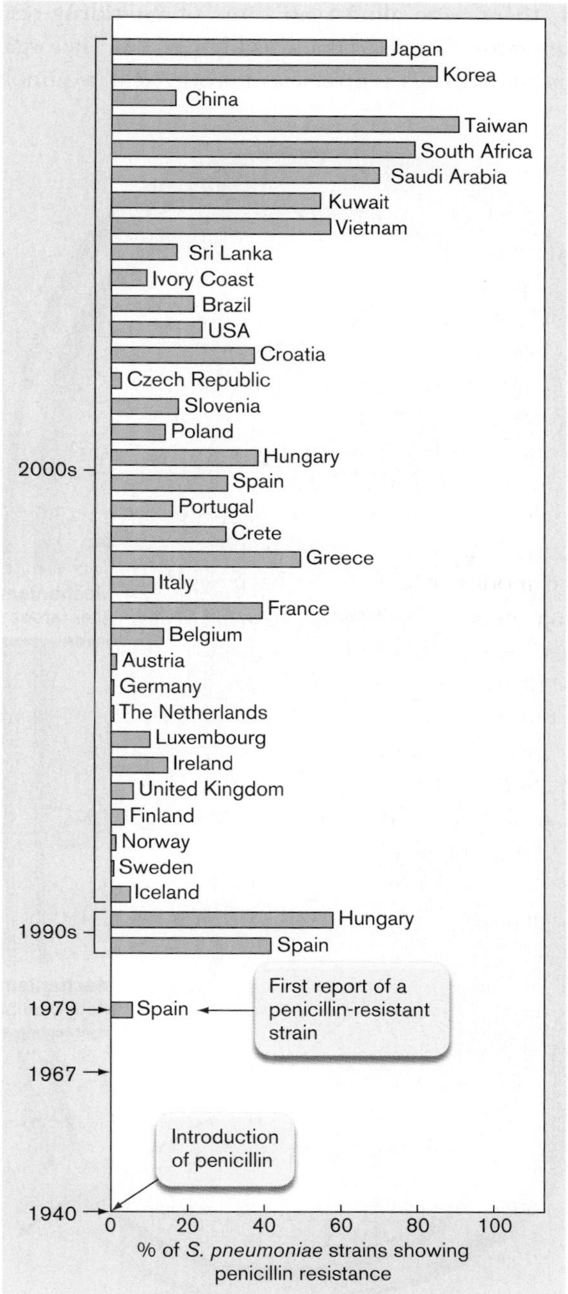
Japan
Korea
China
Taiwan
South Africa
Saudi Arabia
Kuwait
Vietnam
Sri Lanka
Ivory Coast
Brazil
USA
Croatia
Czech Republic
Slovenia
Poland
Hungary
Spain
Portugal
Crete
Greece
Italy
France
Belgium
Austria
Germany
The Netherlands
Luxembourg
Ireland
United Kingdom
Finland
Norway
Sweden
Iceland
Hungary
Spain
Spain
First report of a penicillin-resistant strain
Introduction of penicillin
2000s
1990s
1979
1967
1940
0
20
40
60
80
100
% of S. pneumoniae strains showing penicillin resistance

FIGURE 27.17 ■ The rise of penicillin-resistant *Streptococcus pneumoniae* throughout the world. Numbers reflect the number of penicillin-resistant strains among clinical isolates (strains of disease-causing bacteria isolated from patients from different countries). No resistance among clinical isolates was noted until after 1967.

strains in the world. Another instance of emerging antibiotic resistance is unfolding in Europe and the Far East. The non-Enterobacteriaceae Gram-negative rod *Acinetobacter baumanii* is increasingly seen as a dangerous cause of nosocomial infections. It commonly colonizes hospitalized patients, particularly those in intensive care units. Before 1998, there were almost no cases of multidrug-resistant *A. baumanii*. The rate is now as high as 8%. The organism is resistant to drugs as diverse as ciprofloxacin (a quinolone), amikacin (an aminoglycoside), penicillins, third-generation cephalosporins, tetracycline, and chloramphenicol. Imipenem, one of a relatively new class of beta-lactam drugs, is currently useful, but resistance to it is also likely to develop.

There are four basic forms of antibiotic resistance (**Fig. 27.18**). The resistant organism can:

- **Modify the target so that it no longer binds the antibiotic.** Mutations in key penicillin-binding proteins and ribosomal proteins, for instance, can confer resistance to methicillin and streptomycin, respectively. These mutations occur spontaneously and are not typically transferred between organisms.
- **Destroy the antibiotic before it enters the cell.** For example, the enzyme beta-lactamase (or penicillinase) is made exclusively to destroy penicillins (Section 27.4). The sites of ring cleavage and the structure of the enzyme are illustrated in **Figure 27.19**.
- **Add modifying groups that inactivate the antibiotic.** For instance, there are three classes of enzymes that modify and inactivate aminoglycoside antibiotics. The results of these types of enzyme modifications are illustrated for kanamycin in **Figure 27.20**.
- **Pump the antibiotic out of the cell using specific (for example, tetracycline export) and nonspecific transport proteins.** This strategy works because the pumps bail drugs out of the cell faster than the drugs can enter. Some are single-component pumps present in the cytoplasmic membrane of Gram-negative and Gram-positive bacteria (for example, NorA in *Staphylococcus aureus,* PmrA in *Streptococcus pneumoniae,* and the TetA and B proteins in Gram-negatives). Other drug efflux pumps are multicomponent systems present in Gram-negative bacteria only (discussed shortly). Efflux in either case is usually energized by proton motive force.

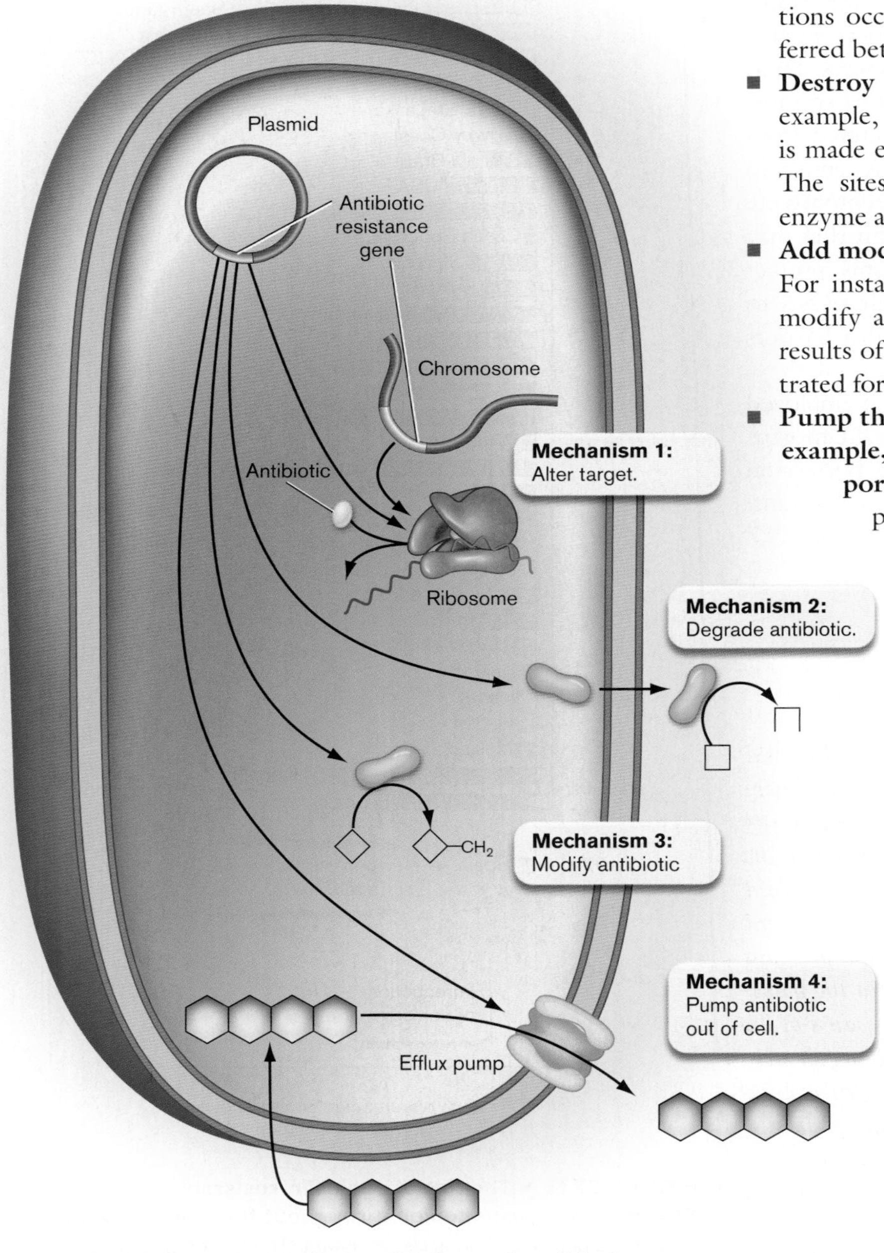

Plasmid
Antibiotic resistance gene
Chromosome
Antibiotic
Mechanism 1:
Alter target.
Ribosome
Mechanism 2:
Degrade antibiotic.
CH_2
Mechanism 3:
Modify antibiotic
Mechanism 4:
Pump antibiotic out of cell.
Efflux pump

FIGURE 27.18 ▪ Alternative mechanisms of antibiotic resistance. Antibiotic resistance genes can be plasmid-borne or they can be part of the chromosome. Each specific antibiotic resistance gene product will use only one of the four mechanisms shown.

Thought Question

27.8 Fusaric acid is a cation chelator that normally does not penetrate the *E. coli* membrane, which means *E. coli* is typically resistant to this compound. Curiously, cells that develop resistance to tetracycline become sensitive to fusaric acid. Resistance to tetracycline is usually the result of an integral membrane efflux pump that pumps tetracycline out of the cell. What might explain the development of fusaric acid sensitivity?

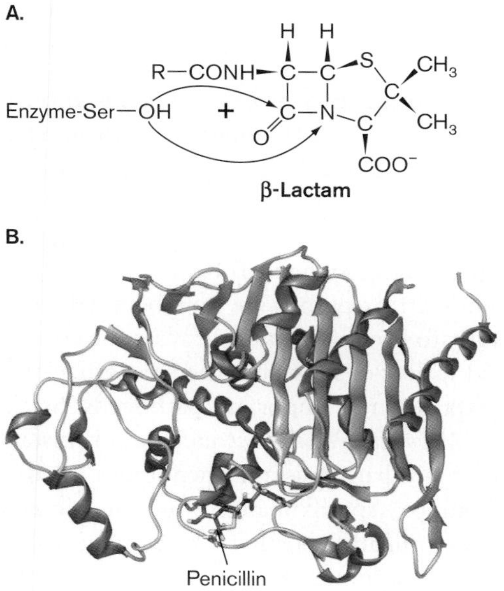
A.
H H
R—CONH—C—C
S
CH3
Enzyme-Ser—OH
+
C—N—C
O
CH3
COO⁻
β-Lactam
B.
Penicillin

FIGURE 27.19 ▪ Destroying penicillin. A. Beta-lactamase (or penicillinase) cleaves the beta-lactam ring of penicillins and cephalosporins. There are two types of penicillinases, based on where the enzyme attacks the ring. In either type, a serine hydroxyl group launches a nucleophilic attack on the ring. **B.** Structure of a beta-lactamase and location of the penicillin-binding site. (PDB code: 1XX2)

Thought Question

27.9 Mutations in the ribosomal protein S12 (encoded by *rpsL*) confer resistance to streptomycin. Given a cell containing both *rpsL*$^{+}$ and *rpsL*^{R} genes, would the cell be streptomycin resistant or sensitive? (Recall that genes encoding ribosome proteins for the small subunit are designated *rps.* "+" indicates the wild-type allele, while "R" indicates a gene whose product is resistant to a certain drug.)

A particularly dangerous type of drug resistance is mediated by what are called **multidrug resistance (MDR) efflux pumps** (**Fig. 27.21**). A single pump in this class can export many different kinds of antibiotics with little regard to structure. MDR pumps of Gram-negative microbes are similar to the ABC export systems described in Section 4.2. They include three proteins: an inner membrane pump protein (fueled by proton motive force—a distinction from true ABC exporters), an outer membrane channel connected to the pump protein, and an accessory protein that

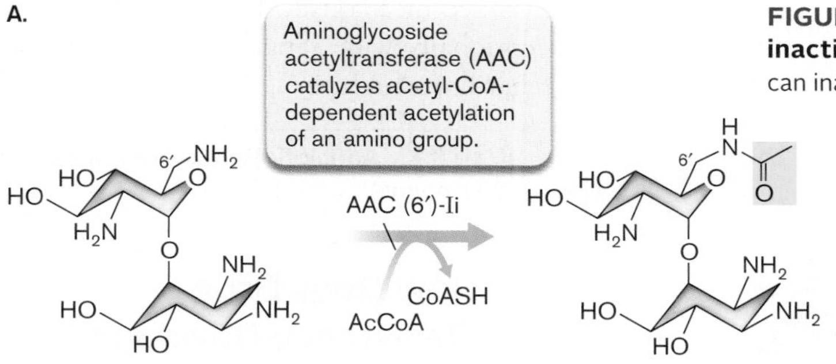
A.
Aminoglycoside acetyltransferase (AAC) catalyzes acetyl-CoA-dependent acetylation of an amino group.
AAC (6′)-Ii
CoASH
AcCoA

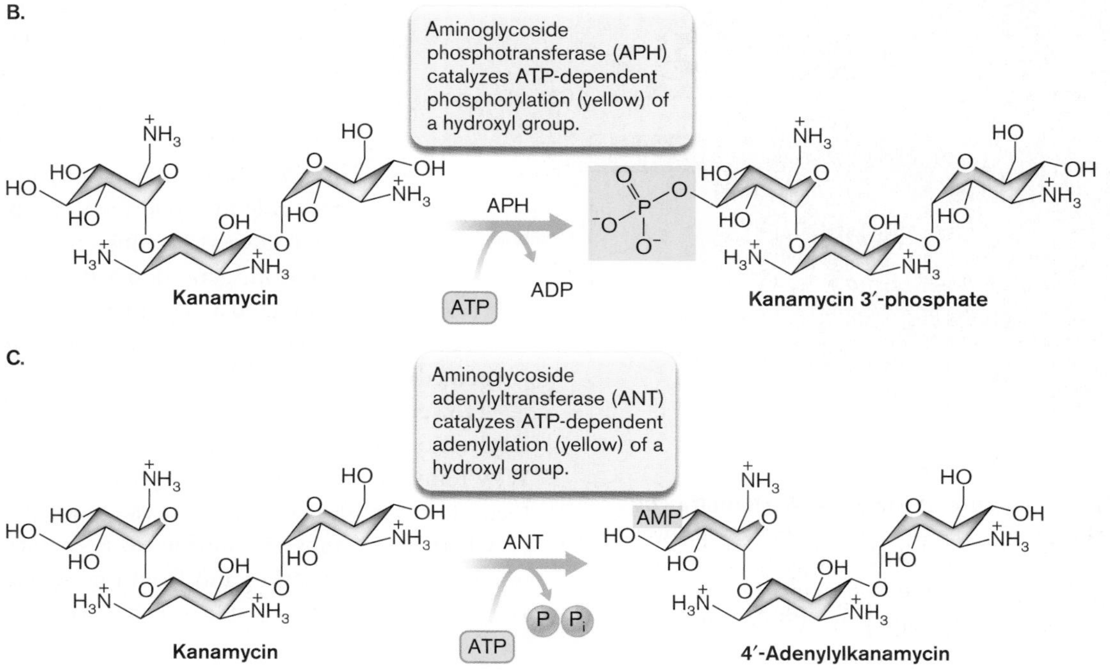
B.
Aminoglycoside phosphotransferase (APH) catalyzes ATP-dependent phosphorylation (yellow) of a hydroxyl group.
APH
ATP
ADP
Kanamycin
Kanamycin 3′-phosphate
C.
Aminoglycoside adenylyltransferase (ANT) catalyzes ATP-dependent adenylylation (yellow) of a hydroxyl group.
ANT
ATP
AMP
Kanamycin
4′-Adenylylkanamycin

FIGURE 27.20 ▪ Aminoglycoside-inactivating enzymes. Different enzymes can inactivate aminoglycoside antibiotics.

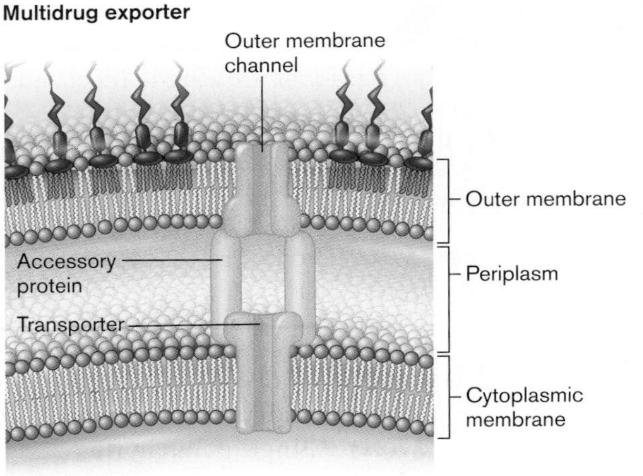
Multidrug exporter
Outer membrane channel
Outer membrane
Accessory protein
Periplasm
Transporter
Cytoplasmic membrane

FIGURE 27.21 ■ Basic structure of a multidrug resistance efflux pump in Gram-negative bacteria. These efflux systems have promiscuous binding sites that can bind and pump a wide range of drugs out of the bacterial cell.

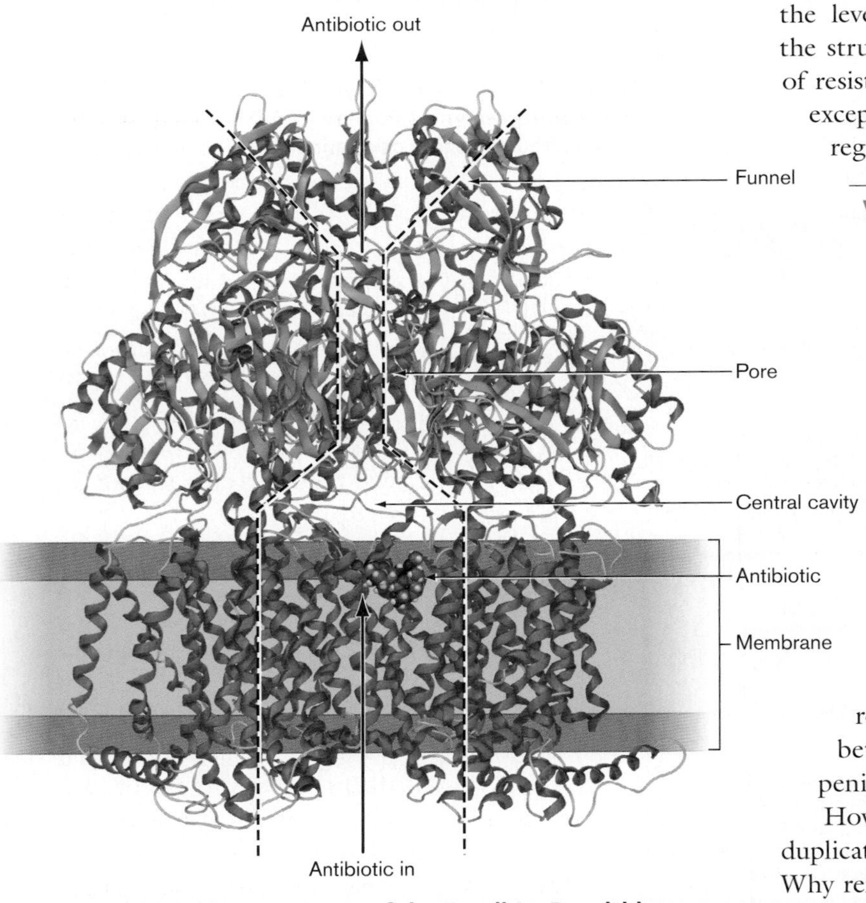
Antibiotic out
Funnel
Pore
Central cavity
Antibiotic
Membrane
Antibiotic in

FIGURE 27.22 ■ Structure of the *E. coli* AcrB multidrug resistance efflux pump. The structure is trimeric and driven by proton motive force. Dotted lines outline the central cavity, pore, and efflux funnel through which the drug is pumped. The connection to TolC in the outer membrane is not shown. (PDB code: 2GIF)

may link the other two proteins. For instance, the AcrB transporter (**Fig. 27.22**) almost indiscriminately binds antibiotics in a large central cavity (a promiscuous binding site) and uses proton motive force to move those compounds through a pore and out a funnel that connects to an outer membrane channel, TolC.

Antibiotic efflux pumps are now believed to contribute significantly to bacterial antibiotic resistance because of the very broad variety of substrates they recognize and because of their expression in important pathogens. Strains of the pathogen *Mycobacterium tuberculosis*, for instance, have developed multidrug-resistant phenotypes in part because of MDR pumps. Approximately 2 million people die from tuberculosis annually, mostly in developing nations. What is even more alarming is that an increasing number of *M. tuberculosis* strains isolated from patients exhibit multidrug resistance. Although most of the antibiotic resistance in the majority of *M. tuberculosis* multidrug-resistant strains is due to the accumulation of independent mutations in several genes, MDR pumps are thought to increase the level of resistance. Chemists typically try to tweak the structure of an antibiotic to overcome a specific type of resistance mechanism, but the MDR pumps act on an exceptionally wide range of antibiotics, almost without regard to structure.

Weblinks Antibiotic Resistance Genes Database (ARDB) (*see ebook*)

How Does Drug Resistance Develop?

As discussed in previous chapters, nature has engineered a certain degree of flexibility in the way genomes are replicated and passed from one generation to the next. DNA repair pathways involving lesion bypass polymerases (for example, UmuDC; see Section 9.5) are thought to play a large role in randomized adaptive and evolutionary processes. For instance, at some point during evolution, gene duplication and mutational reshaping generated a gene product able to cleave the beta-lactam ring, producing an organism resistant to penicillin.

However, de novo antibiotic resistance through gene duplication and/or mutation does not occur in all species. Why reinvent the wheel—or, in this case, drug resistance? Gene transfer mechanisms such as conjugation, described in Chapter 9, can move antibiotic resistance genes from one organism to another and from one species to another. In fact, several drug resistance genes found in pathogenic bacteria actually had their start in the chromosomes of the drug-producing organisms and were passed on through gene swapping. For instance, *Streptomyces clavuligerus* produces a beta-lactamase (encoded by the *bla* gene) that protects this organism from the penicillin it produces. Transfer of a drug resistance gene is particularly evident if the gene has been incorporated into a plasmid and that plasmid is found in a new species.

An interesting case study of antibiotic resistance is provided by the Gram-positive bacterium *Enterococcus faecalis,* a natural inhabitant of the mammalian gastrointestinal tract that can cause life-threatening disease if granted access to other body sites (as in subacute bacterial endocarditis; see Section 26.7). *E. faecalis* is naturally resistant to numerous antibiotics, making disease treatment particularly challenging.

Vancomycin is one of the last lines of defense for treating serious *E. faecalis* infections. Unfortunately, increasing numbers of vancomycin-resistant strains have arisen in recent years. The completed genome sequence of one vancomycin-resistant strain illustrates the reason. The organism has an incredible propensity for incorporating mobile genetic elements that encode drug resistance. About a quarter of the genome consists of mobile or exogenously acquired DNA, including 7 probable phages, 38 insertion elements, numerous transposons, and integrated plasmid genes. One such mobile element encodes vancomycin resistance in the sequenced strain.

Multidrug resistance in various microbes (for example, *Salmonella enterica*) has also been attributed to the presence of integrons. Integrons are gene expression elements that account for rapid transmission of drug resistance because of their mobility and ability to collect resistance gene cassettes (see **eTopic 9.6**).

We should note that the development of antibiotic resistance is not without consequence to the bacterium. Altering one aspect of an organism's physiology for the better may weaken another area. Bacteria may become resistant to a certain antibiotic, but that resistance comes at a price. For example, the altered DNA gyrase that affords resistance to quinolones may not function as well as the "normal" gyrase does. Thus, when both resistant and susceptible organisms cohabit the same environment, the wild-type (sensitive) strain may grow faster and eventually overwhelm the mutant strain—unless fluoroquinolone is present, of course.

How Did We Get into This Mess?

Consider the following case: A young mother brings her 4-year-old child to the physician. The child is screaming because he has an extremely painful sore throat. Simply looking at the throat is not diagnostic. The raw tissue could mean the child is suffering from a bacterial infection, in which case antibiotics are needed. Alternatively, a virus could be the cause—a situation in which antibiotics do nothing but pacify the parent. More often than not, the clinician will prescribe an antibiotic without ever knowing the cause of disease. The problem is this: The more an antibiotic is used, the more opportunities there are to select for an antibiotic-resistant organism. The presence of drug does not <u>cause</u> resistance, but it will kill off or inhibit the growth of competing bacteria that are sensitive, thereby allowing the resistant organism to grow to high numbers. Resistance can form not only in the pathogen itself, but also in a member of the normal microbiota. The danger, then, is that the gene imparting resistance might be horizontally transferred to other bacteria—some of them pathogens.

When <u>should</u> antimicrobials be administered? Certainly, in life-threatening situations where time is of the essence, antibiotics should be administered even before knowing the cause of the infection. On the other hand, the most prudent course to take when a patient has a simple infection is to confirm a bacterial etiology, and then prescribe. An exception may be in an elderly or otherwise immunocompromised individual, who may be more susceptible to secondary bacterial infections that can occur subsequent to viral disease.

Another proposed source of antibiotic resistance is the widespread practice of adding antibiotics to animal feed. No one quite knows why, but giving animals antibiotics in their food makes for larger, and therefore more profitable, animals. Some estimates suggest that 70% of all antibiotics used in the United States are fed to healthy livestock. The consequence is that the animals may serve as incubators for the development of antibiotic resistance. Even if the resistance develops in nonpathogens, the antibiotic genes produced can be transferred to pathogens.

Feeding cattle growth-promoting antibiotics can also stimulate the spread of pathogenicity genes between bacteria. The antibiotics are often added to feed at sub-inhibitory concentrations that do not kill the bacteria, but stress them instead. Stress can trigger SOS responses (Chapter 9) that reactivate prophages embedded into bacterial chromosomes. If those phages carry toxin genes, the new phage can transfer toxin production to new strains or species cohabiting the intestine. For example, the Shiga toxin genes (*stx*) carried by certain pathovars of *E. coli* are associated with prophage genomes. Growth-promoting antibiotics such as carbadox or monensin (not used in humans) have been shown to reactivate these *stx*-carrying prophages and foster distribution of *stx* to other strains. The unusual O104:H4 Shiga toxin–producing *E. coli* that in

2011 caused a European outbreak of hemolytic uremic syndrome may have been generated by this mechanism. Efforts to curb the addition of antibiotics to animal feed have been under way since the 1960s.

Many, or all, of these events conspired to produce an incredibly dangerous bacterium that is resistant to almost every antibiotic known. The organism, a *Klebsiella pneumoniae,* was first isolated in 2008 in Örebro, Sweden, from a Swedish citizen returning from New Delhi, India (**Fig. 27.23**). This organism is resistant to most commonly used antibiotics, such as amino-penicillins, beta-lactam/beta-lactamase inhibitors, aminoglycosides, fluoroquinolones, cephalosporins, tigecycline (structurally similar to tetracycline), and carbapenems. It carries the NDM-1 plasmid noted earlier, and numerous other antibiotic resistance genes. Despite attempted treatment with linezolid, the patient died.

Thought Questions

27.10 **Figure 27.23** shows a distinctive-looking colony morphology. Why are the colonies on this agar plate red and mucoid?

27.11 Could genomics ever predict the drug resistance phenotype of a microbe? If so, how?

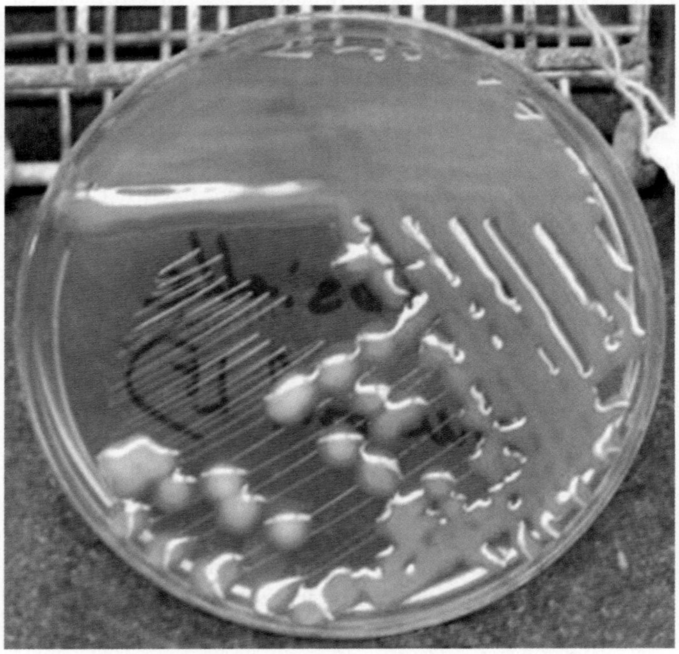

FIGURE 27.23 ■ MacConkey agar plate containing a sputum culture of *Klebsiella pneumoniae*. *K. pneumoniae* carrying the antibiotic resistance gene $bla_{NDM\text{-}1}$ is emerging as a dangerous, drug-resistant pathogen.

Weblinks Video: "Antibiotics: Is a Strong Offense the Best Defense?" (*see ebook*)

Fighting Drug Resistance

Several strategies are being used to stay one step ahead of drug-resistant pathogens. In some instances, dummy target compounds that inactivate resistance enzymes have been developed. Clavulanic acid, for example, is a compound sometimes used in combination with penicillins such as amoxicillin. Clavulanic acid, a beta-lactam compound with no antimicrobial effect, competitively binds to beta-lactamases secreted from penicillin-resistant bacteria. Because the enzyme releases bound clavulanic acid very slowly, the amoxicillin remains free to enter and kill the bacterium. Another strategy is to alter the structure of the antibiotic in a way that sterically hinders the access of modifying enzymes. **Figure 27.24** illustrates how adding a side chain to gentamicin, converting it to amikacin, blocks the activity of various aminoglycoside-modifying enzymes. Of course, we are now seeing resistance to amikacin develop (this resistance is ribosome based, involving mutational alteration of the S12 protein or 16S rRNA).

Linking antibiotics is another strategy currently used to limit resistance. Recent advances have been made in linking a quinolone to an oxazolidinone to form a hybrid antibiotic with dual modes of action (**Fig. 27.25**). Because it has two modes of action, this hybrid antibiotic may limit the development of antibiotic resistance. Here's why: The rate of spontaneous resistance to a given antibiotic is roughly one out of 10^7 cells. For spontaneous resistance to develop to two antibiotics, that probability rises to one out of 10^{14}, making it very unlikely that an organism can become doubly drug resistant. However, multidrug resistance efflux pumps, integron cassettes, and plasmids carrying multiple antibiotic resistance genes can overwhelm that approach.

Biofilms, Persisters, and the Mystery of Antibiotic Tolerance

A well-recognized puzzle of microbiology, unsolved for decades, is the problem of persister cells—cells that neither grow nor die in the presence of bactericidal agents. Joseph Bigger in 1944 noticed that penicillin would lyse a growing culture of *Staphylococcus aureus*, but a small number of persister cells always survived. These persisters were not mutants made permanently resistant through mutation. They acted as though dormant.

Persister cells that tolerate antibiotic treatment can be found in any biofilm or population of late-exponential-

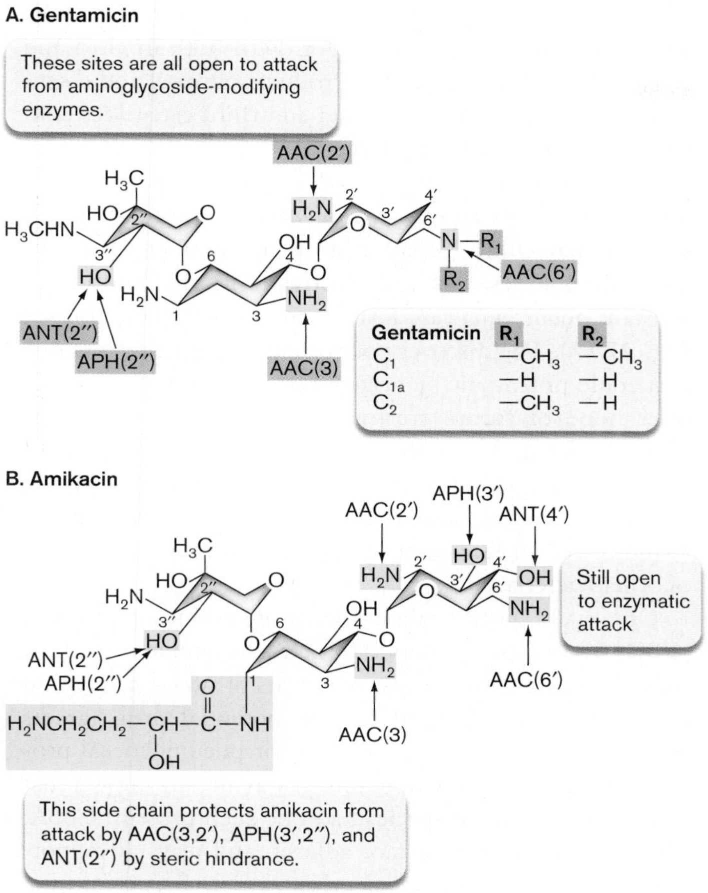
A. Gentamicin
These sites are all open to attack from aminoglycoside-modifying enzymes.
AAC(2′)
AAC(6′)
ANT(2″)
APH(2″)
AAC(3)
Gentamicin R1 R2
C1 −CH3 −CH3
C1a −H −H
C2 −CH3 −H
B. Amikacin
APH(3′)
AAC(2′)
ANT(4′)
Still open to enzymatic attack
ANT(2″)
APH(2″)
AAC(6′)
AAC(3)
This side chain protects amikacin from attack by AAC(3,2′), APH(3′,2″), and ANT(2″) by steric hindrance.

FIGURE 27.24 ▪ Fighting drug resistance. **A.** Sites where gentamicin is vulnerable to enzymatic inactivation. AAC = aminoglycoside acetyltransferase; ANT = aminoglycoside adenylyltransferase; APH = aminoglycoside phosphotransferase. Inset shows the R groups for different gentamicin compounds: C_1, C_{1a}, and C_2. **B.** Gentamicin can be chemically modified at the highlighted sites to prevent loss of activity due to enzyme action. The side groups block access to enzyme active sites by steric hindrance (that is, the added groups prevent the active site from interacting with its target structure), but do not inactivate the antibiotic.

phase cells. It is now thought that persisters are the cause of antibiotic treatment failures and even latent bacterial infections such as recrudescent typhus or latent tuberculosis. Dae-Hyuk Kweon and collaborators in South Korea recently reported finding an antimicrobial agent (called C10) that specifically targets persister cells. Although the precise mechanism is not known, C10 added with a normal antibiotic, such as ampicillin, apparently converts persister cells back into antibiotic sensitive cells. C10, or a similar compound, used in combination with conventional bactericidal antibiotics might in the future prevent patient treatment failures.

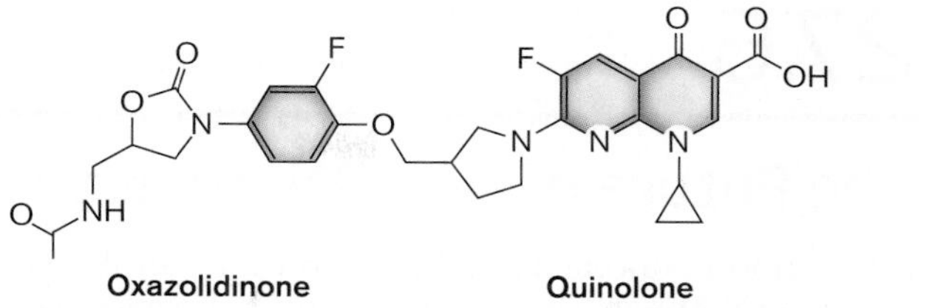
Oxazolidinone
Quinolone

FIGURE 27.25 ▪ Quinolone-oxazolidinone hybrid. Physically combining two different antibiotics may help reduce the emergence of drug-resistant bacteria.

How might dormancy explain antibiotic tolerance? If persisters are dormant and have little or no cell wall synthesis, translation, or topoisomerase activity, then even if antibiotics can bind to their targets, target function cannot be corrupted. Tolerance, then, provides antibiotic resistance at the price of not growing. Thus, an antibiotic can kill all susceptible bacteria in an infection, but the remaining persister cells serve as a source of population regrowth once the antibiotic is removed. You might say persister cells are the microbial equivalent of medieval monks whose simple but critically important role was to keep and protect learned texts. Persister cells, therefore, may be considered monk-like "keepers of the genome." So far, the molecular mechanisms that produce persister cells remain a mystery.

To Summarize

- **Antibiotic resistance** is a growing problem worldwide.
- **Certain microbes make antibiotics** to eliminate competitors in the environment and prevent self-destruction by means of various antibiotic resistance mechanisms. Genes encoding some of these drug resistance mechanisms have been transferred to pathogens.
- **Mechanisms of antibiotic resistance** include modifying the antibiotic, destroying the antibiotic, altering the target to reduce affinity, and pumping the antibiotic out of the cell.
- **Multidrug resistance efflux pumps** use promiscuous binding sites to bind antibiotics of diverse structure.
- **Antibiotic resistance can arise spontaneously** through mutation, can be inherited by gene exchange mechanisms, or can arise de novo through gene duplication and mutational reengineering.
- **Indiscriminate use of antibiotics** has significantly contributed to the rise in antibiotic resistance.
- **Measures to counter antibiotic resistance** include chemically altering the antibiotic, using combination antibiotic therapy, and adding a chemical decoy.

27.6

The Future of Drug Discovery

Is humankind doomed to a future in which antibiotics will no longer work? The general consensus is no. Through prudent use of current antibiotics and innovative approaches for finding new ones, humans should continue to effectively control evolving bacterial pathogens.

How does one screen for new drugs that target specific proteins? Certainly, the classic approach in which microbes, plants, and even animals collected from around the world are screened for their abilities to make new antibiotics is still valid and remains the most fruitful source of new drugs. However, the ability to search for novel bacterial drug targets or validate theoretical targets has been revolutionized by genome sequence analysis and associated genetic techniques. Powerful computational and bioinformatic methods are required in the initial identification and selection of molecular targets, followed by a series of postgenomic approaches to validate and characterize the targets, devise screens for effective inhibitor molecules, and pursue structure-based drug design to handle the inescapable emergence of future resistance. Examples of metagenomic and peptidogenomic approaches to drug discovery are described in **eTopic 15.2** and Special Topic 15.1, respectively.

The modern drug discovery process can be outlined as follows:

1. Identify new targets (such as unique essential enzymes) based on genomics.
2. Find or design compounds that inhibit the target in vitro and show that the inhibitory compound has actual antibacterial activity.
3. Show that the target within the bacterial cell is the same as the in vitro target.
4. Optimize the MICs against susceptible species by altering the compound's structure.
5. Examine the compound's spectrum of activity (Gram-positive, Gram-negative, aerobe, anaerobe, and so on) and the rate at which antibiotic resistance develops in pathogens.
6. Determine the new drug's toxicity to animals and humans, and its pharmacological properties (for instance, how long it stays at therapeutic levels in a patient).

The functions of genes that constitute potential new drug targets fall into three broad categories: those required only for growth of bacteria in laboratory media (in vitro–expressed genes), those required only for bacterial infection in vivo (in vivo–induced genes, discussed in **eTopic 25.2**), and those required for bacterial growth both in vitro and in vivo (housekeeping genes). Products of novel and essential genes falling into the second and third categories constitute new potential drug targets.

Take, for example, the case of *Streptococcus pneumoniae*. Twenty-five years ago, *S. pneumoniae* was uniformly sensitive to penicillin. Today, a growing number of strains have become resistant (as illustrated in the multidrug-resistant pneumonia case history presented earlier; also see **Fig. 27.17**). The discovery of proteins within this microbe that could potentially be targeted by new antibiotics would offer hope of future treatments. Recently, a computer-assisted strategy has helped identify several potential drug targets in this pathogen. The search consisted of scanning the pneumococcal genome for sequence motifs commonly found in cell surface–exposed or virulence-related proteins of other bacteria.

One such motif, called the choline-binding domain (CBD), is used by a variety of bacteria to interact with host cells. These domains consist of repeats of two to ten amino acids. The rationale used to find potential drug targets in the pneumococcus was to look for pneumococcal proteins that contain these domains. It was already known that various streptococcal species had one such protein, CbpA, which functions as a surface adhesin and plays an important role in nasopharyngeal colonization. By performing genome-wide searches with the C-terminal choline-binding region of *cbpA*, Khoosheh Gosink and his collaborators at St. Jude Children's Research Hospital (Memphis, Tennessee) identified six genes predicted to encode CBD-containing proteins ranging from 20 to 80 kDa (CbpD, E, F, G, I, and J). They constructed mutations in each gene and tested the mutants for pathogenicity. Mutants defective in CbpG adhered poorly to nasopharyngeal cells and were less virulent in a mouse infection model. Therefore, CbpG appears to play an important role in invasion and infection of the mucosa and the bloodstream. Its structural similarity to proteases suggests that it may be an excellent candidate for target-based drug development because numerous chemical inhibitors of proteases already exist.

Another advance in drug design utilizes a combination of genomic and proteomic approaches. The basis of this technique is to develop fluorescent molecules that bind only to active proteins of a given family of proteins. The probes can be devised by rational design in which knowledge of the active-site structure allows chemists to engineer chemicals that interact with that site. A true inhibitor of that protein, and thus a potential antibiotic, will prevent binding of the tagged probe—an event that can be viewed after directly blotting cell extracts on a membrane filter, much like a western blot. In the absence of inhibitor, the probe will bind the spot, which will fluoresce. However, the presence of an inhibitor that binds well to the active site will prevent binding of the probe, and the spot will not fluoresce.

Alternatively, if the probe binds to more than one protein, the same test can be performed using gel electrophoresis separation techniques. Proteins are separated on a nondenaturing gel so that the protein in question will retain its shape and its ability to bind probe. Tagged probe, with or without a potential inhibitor, is added to duplicate gels or a blot of those gels. The protein band will fluoresce in the absence of inhibitor or in the presence of a poor inhibitor. The band will not light up if an effective inhibitor is present.

The inhibitors are often made using more or less random combinatorial chemistry techniques, in which random variations of a molecule are made and tested. A library of these related compounds, which could number in the thousands, is robotically screened for inhibitory activity.

Genomic and combinatorial chemistries are certainly the cutting edge of antibiotic discovery. However, the old brute-force screening of natural products was recently used in a novel way to discover a promising new class of antibiotic. Merck scientists screened 250,000 natural product extracts for an ability to specifically inhibit bacterial fatty acid biosynthesis. Fatty acid biosynthesis is an attractive target for antibiotic development because the process and proteins involved are different from those of eukaryotes. The strategy specifically targeted the protein FabF because it is an essential component of fatty acid synthesis and is conserved among key pathogens such as *Staphylococcus aureus.*

Scientists engineered a strain of *S. aureus* to contain a gene expressing an antisense RNA to *fabF* mRNA. When the antisense RNA was induced, it bound to *fabF* mRNA and prevented efficient translation. As a result, the level of FabF protein in the cell decreased. The strain could still grow but would be exquisitely sensitive to any compound that targeted the remaining FabF protein. Fewer molecules of FabF present per cell means that fewer molecules of an active ingredient in a natural product are needed to stop fatty acid synthesis. Thus, zones of inhibition on an agar plate will be wider than they would be if cells produced normal amounts of FabF.

This novel targeted screening method led to the discovery of the antibiotic platensimycin, made by *Streptomyces platensis,* an organism isolated from South African soil. Platensimycin was subsequently shown to bind FabF and exhibit bacteriostatic, broad-spectrum activity, acting on Gram-positive and Gram-negative bacteria. It is only the third entirely new antibiotic developed in the last four decades. The novel chemical structure of platensimycin and its unique mode of action provide a great opportunity to develop a new class of critically needed antibiotics—a class that selectively targets fatty acid biosynthesis. More information on fatty acid biosynthesis and inhibitors can be found in Chapter 15.

Other intriguing ideas may lead to novel antimicrobial therapies. One involves using nanotubes to poke holes in bacterial membranes (**eTopic 27.3**). Another idea involves using photosensitive chemicals that can penetrate the microorganism and generate toxic reactive oxygen species (such as superoxide) when exposed to specific wavelengths of visible light (obviously good only for topical use). A third idea involves generating antimicrobials that "cork" the type III secretion apparatus that delivers virulence proteins into host target cells. Another clever approach is to interfere with quorum-sensing mechanisms of pathogens (**Special Topic 27.1**). Finally, the promise and flexibility of synthetic-biology approaches discussed in Chapter 10 could revolutionize the process of antibiotic discovery and production.

To Summarize

- **Potential targets for rational antimicrobial drug design** include proteins expressed only in vivo or proteins expressed both in vivo and in vitro.
- **Candidate antimicrobial compounds** can be designed to interact at the active site of a known enzyme and inhibit its activity.
- **Combinatorial chemistry** is used to make random combinations of compounds that can be tested for enzyme inhibitory activity and antimicrobial activity.

27.7

Antiviral Agents

A father pleading with a physician to give his child antibiotics when the infant is suffering with a cold is an all-too-common dilemma faced by the general practitioner, but there is nothing of substance the physician can do. The common cold is caused by the rhinovirus, and no antibiotic designed for bacteria can touch it. So why are there so few antiviral agents in the clinician's arsenal? The reason is that applying the principle of selective toxicity is much harder to achieve for viruses than it is for bacteria. Viruses routinely usurp host cell functions to make copies of themselves.

Special Topic 27.1: Anti-Quorum Sensing Drug Blocks Pathogen "Control and Command"

Biofilms enhance the infection process in many ways. From guarding against innate and adaptive immune mechanisms, to secreting toxins and producing antibiotic tolerance, biofilms serve as sturdy, infectious fortresses. **Figure 1** shows, for instance, a biofilm of *Pseudomonas aeruginosa* being built on a suture. This biofilm will constantly seed the wound with bacteria and will be hard to kill with antibiotics.

Since cell-cell signaling (quorum sensing) is critical to biofilm formation and function, jamming these signals may help resolve infections more quickly and prevent chronic infections from developing. As a result, research designed to destroy acyl homoserine lactones (AHLs, the chemical signal molecules) or to interfere with their activities is under way. Some scientists are trying to use paraoxonases, natural enzymes that inactivate quorum-sensing signal molecules. Others are attempting to devise catalytic antibodies that can selectively degrade the signal molecules. Underlying the latter approach is the idea that an antibody raised against a molecule resembling the transition form of an AHL undergoing hydrolysis will actually catalyze AHL hydrolysis.

An intriguing twist to these approaches was published by Vanessa Sperandio (**Fig. 2**) and colleagues. Her group studied the two-component signal transduction system consisting of QseC (a sensor kinase) and QseB (a response regulator). The QseCB system, present in enterohemorrhagic *E. coli* (EHEC), *Salmonella, Francisella,* and other important bacterial pathogens, recognizes an enigmatic bacterial signal molecule (autoinducer 3), but also appears to sense host adrenergic molecules (for example, epinephrine). Upon sensing any of these signals, QseCB activates the transcription of key virulence genes. Sperandio's group screened 150,000 randomly synthesized small organic molecules for any that would interfere with QseC-dependent gene activation—the idea being that such a molecule would prevent induction of virulence genes. One compound was selected for its potency as compared with other molecules, minimal toxicity toward bacterial and human cell lines, and potential for chemical modification: LED209, or *N*-phenyl-4-{[(phenylamino) thioxomethyl] amino}-benzenesulfonamide (**Fig. 3**). When tested for effects on pathogenicity, LED209 interfered with pedestal formation

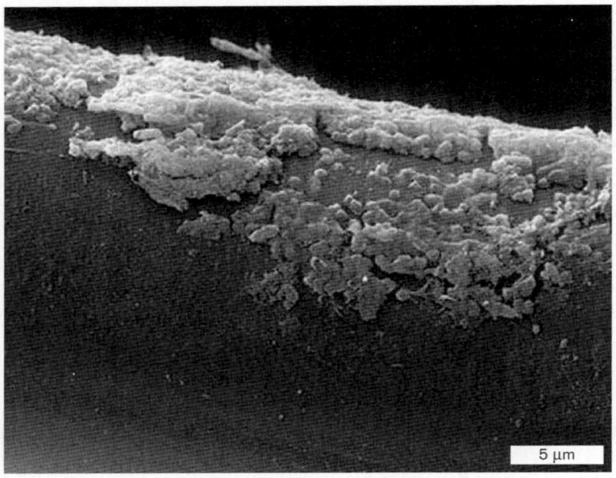
5 µm

COURTESY OF G. A. O'TOOLE, DARTMOUTH MEDICAL SCHOOL.

FIGURE 1 ▪ *Pseudomonas aeruginosa* biofilm forming on a suture.

JOHN W. FOSETER

FIGURE 2 ▪ Interrupting cell-cell communication. Vanessa Sperandio (left) and her student prepare cultures for their studies.

Thus, a drug that hurts the virus is likely to also harm the patient. Nevertheless, there are several useful antiviral agents for which selective viral targets have been found and exploited. Some of these agents are listed in **Table 27.3**. Select examples are discussed here in this section. All of the molecular mechanisms of viruses presented in Chapter 11 are studied as potential drug targets.

Antiviral Agents That Prevent Virus Uncoating or Release

Membrane-coated viruses are vulnerable at two stages: when the virus is invading the host cell and after viral propagation, when the progeny viruses release from the host cell. The flu virus presents a good example of both.

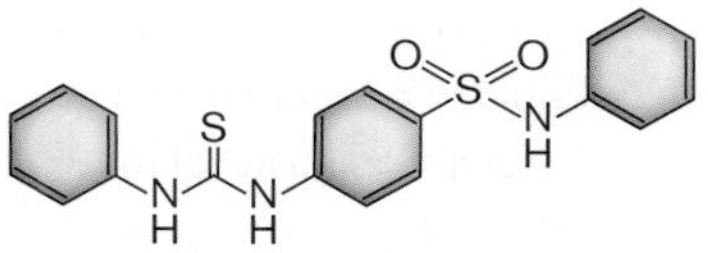

S
N
H
N
H
O
S
O
N
H

FIGURE 3 ▪ **The structure of LED209.** *Source:* From D. A. Rasko et al. 2008. *Science* **321**:1078–1080.

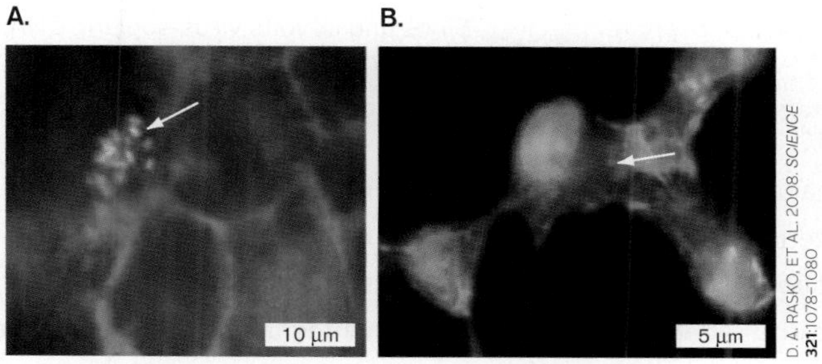

A.
B.
10 µm
5 µm
D. A. RASKO, ET AL. 2008. SCIENCE 321:1078–1080

FIGURE 4 ▪ **LED209 inhibits pedestal formation by EHEC.** Cell nuclei and bacterial cells are stained red (with propidium iodide), and the cytoskeleton is stained green (with fluorescein isothiocyanate-phalloidin). **A.** EHECs are perched atop the green-stained, stemlike pedestals. The arrow points to bacterial cells raised off the surface by short, bright-green pedestals. **B.** Treatment with 5-µM LED209 prevented pedestal formation. The arrow points to a bacterial cell sitting on the host surface without a bright-green pedestal lifting it up.

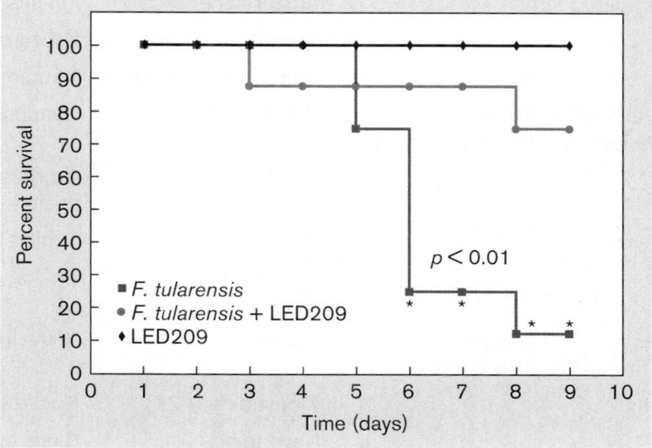

Percent survival
Time (days)
p < 0.01
▪ F. tularensis
• F. tularensis + LED209
♦ LED209
0 10 20 30 40 50 60 70 80 90 100
0 1 2 3 4 5 6 7 8 9 10

FIGURE 5 ▪ **LED209 reduces virulence of *Francisella tularensis*.** Survival plot of mice infected intranasally with *F. tularensis* upon oral treatment with LED209 (20 mg/kg mouse weight). Asterisks indicate results that are significantly different from controls. *Source:* Modified from D. A. Rasko et al. 2008. *Science* **321**:1078–1080.

of EHEC—a key virulence feature of this organism (**Fig. 4**)—and reduced the virulence of *Francisella tularensis* in mice (**Fig. 5**).

While Sperandio's research sounds incredibly promising, we must remain cautious. There is a dearth of knowledge about interactions between bacterial signals and host signals. As recent studies seem to indicate, hosts may gauge their immune response by monitoring the signal molecules produced by pathogens and by their own microbiota. The danger of these therapies lies in developing antagonistic molecules that interfere with host and bacterial receptors or that destroy the signal molecules themselves, not knowing the consequence to the host. The possibility of unintended consequences must be recognized but does not diminish the exciting promise of this research. Further studies designed to unravel quorum-sensing networks that regulate virulence may lead to the discovery of new virulence factors and novel targets for vaccine and drug development.

RESEARCH QUESTION

LED209 does not inhibit the growth of target bacteria. What effect do you think this property should have on the development of drug resistance?

Curtis, Meridith M., and Vanessa Sperandio. 2011. A complex relationship: The interaction among symbiotic microbes, invading pathogens, and their mammalian host. *Mucosal Immunology* **4**:133–138.

Case History: Antiviral Treatment of Infant Influenza

A 9-month-old infant arrived at the Johns Hopkins Hospital with an acute onset of fever, cough, regurgitation from his gastrostomy feeding tube, and dehydration. This illness followed a series of chronic problems, including bronchiolitis (infection and inflammation of the bronchioles) caused by respiratory syncytial virus, and neonatal group B streptococcal sepsis. (Neonatal sepsis is often caused by Lancefield group B Streptococcus agalactiae, *described in Chapter 28.) Physical exam revealed fever, a severe cough resulting in respiratory distress, a rapid heart rate, and moderate dehydration. Nasopharyngeal aspirate was positive for influenza A antigen. The patient was*

TABLE 27.3

Examples of antiviral agents.

Virus	Agent	Mechanism of action	Result
Influenza virus	Amantadine	Inhibits viral M2 protein	Prevents viral uncoating
	Zanamivir	Neuraminidase inhibitor (nasal spray)	Prevents viral release
	Oseltamivir (Tamiflu)	Neuraminidase inhibitor (oral prodrug)	Prevents viral release
Herpes simplex virus and varicella-zoster virus (shingles)	Acyclovir	Guanosine analog	Halts DNA synthesis
	Famciclovir	Prodrug of penciclovir, a guanosine analog	Halts DNA synthesis
Cytomegalovirus	Ganciclovir	Similar to acyclovir	Halts DNA synthesis
	Foscarnet	Analog of inorganic phosphate	Binds and inhibits virus-specific DNA polymerase
Respiratory syncytial virus and chronic hepatitis C virus	Ribavirin	RNA virus mutagen	Causes catastrophic replication errors
HIV	Zidovudine (AZT)	Nucleoside analog; resembles thymine	Inhibits reverse transcriptase
	Nevirapine	Binds to allosteric site	Inhibits reverse transcriptase
	Nelfinavir	Protease inhibitor	Prevents viral maturation
	Raltegravir	Integrase inhibitor	Prevents integration into host genome
	Maraviroc	CCR5 entry inhibitor	Prevents virus entry into host cells

treated with oseltamivir when influenza was diagnosed. He gradually improved and was discharged home 4 days after admission.

In 2003, an unusually severe form of influenza (H3N2) spread across the United States. Most states reported a higher-than-normal number of influenza-related deaths of children and young adults. Several factors, however, helped keep the outbreak from becoming an epidemic of larger proportions. First, the administration of flu vaccine afforded the population what is called herd immunity (discussed in Section 26.9). Herd immunity occurs when only a portion of the population is immunized. The vaccinated individuals will not become infected and so cannot spread the disease to others. At the very least, this slows the progression of infection throughout the population. (It is impossible to immunize all humans against any given disease. However, it is estimated that immunizing about 80% of a population for influenza can halt an epidemic by cutting off transmission. Unfortunately, this level of immunization is rarely achieved.)

The second factor that prevented an influenza pandemic was the availability of antiviral agents that can limit the disease course. As a result of studying the molecular biology of influenza, scientists discovered two selective targets. Influenza virus (200 nm) is encased in a membrane envelope donned when the virion buds from an infected cell. As described in Section 11.3, the envelope contains the viral proteins neuraminidase (NA) and hemagglutinin (HA). Spikes of hemagglutinin bind to sialic acid receptors on the host cell and trigger receptor-mediated endocytosis. The virus ends up inside the resulting endosome. After the endosome is formed, proton pumps in the membrane acidify endosomal contents. The drop in pH changes the structure of hemagglutinin on the viral membrane so that it can now bind to receptors on the endocytic membrane. The result is fusion of the two membranes and release of the virion into the cytoplasm. For this change in hemagglutinin structure to occur, it is essential that the interior of the enveloped virion also be acidified, and this is facilitated by the formation of a channel by the virus-encoded M2 envelope protein. The drug amantadine (**Fig. 27.26A**) is a specific inhibitor of the influenza M2 protein. Inhibiting M2 will prevent viral uncoating. Unfortunately, amantadine-resistant strains of influenza have developed, in part because of the widespread use of amantadine by Chinese poultry farmers. As a result, the drug is no longer recommended as a treatment for influenza.

The second target of the influenza virus is the envelope protein neuraminidase. The newer antiflu drugs, such as zanamivir (Relenza) and oseltamivir (Tamiflu), are **neuraminidase inhibitors** that act against types A and B influenza strains (**Fig. 27.26B** and **C**). Neuraminidase on the

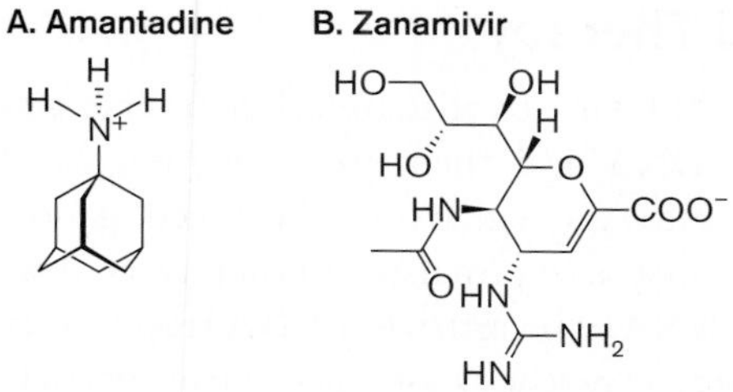
A. Amantadine
B. Zanamivir

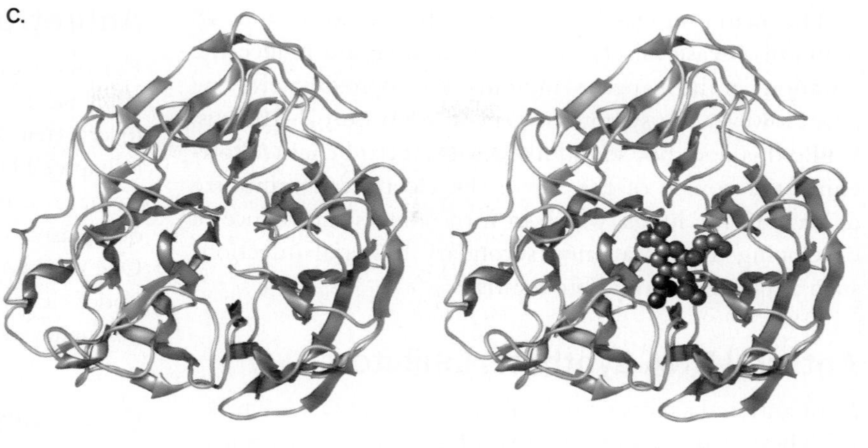
C.

FIGURE 27.26 ▪ Inhibitors of influenza proteins. A. Amantadine inhibits the M2 protein. **B.** Zanamivir inhibits neuraminidase. **C.** Neuraminidase without (left) and with (right) bound inhibitor. (PDB: 2HTQ)

viral envelope allows virus particles to leave the cell in which they were made. The enzyme cleaves any sialic acid on the cell surface that may have bound to the hemagglutinin on the virus. Unencumbered, the virus can leave the cell surface and infect another cell. Neuraminidase inhibitors prevent sialic acid cleavage, causing the virus particles to aggregate at the cell surface. Surface aggregation reduces the number of virus particles released. The contributions of different NA and HA genes to the severity of influenza are discussed in **Special Topic 27.2** and eTopic 27.4.

Special Topic 27.2: Resurrecting the 1918 Pandemic Flu Virus

The influenza virus that swept the globe from 1918 to 1919 took the lives of 20–40 million people, making it a greater killer than the First World War (influenza is described in Chapter 11 and eTopic 26.4). Why was the 1918 flu strain so much more virulent than any strain since? This question may now have been answered.

Teams of scientists working in fortress-like biohazard research facilities used forensic molecular biology techniques to re-create that deadly virus. Recall from Chapter 11 that influenza has a segmented, single-stranded RNA genome. Viral genomic RNA gene pieces were fished from tissues of a 1918 victim frozen in the Alaskan permafrost and from formalin-treated archived tissues. Once sequenced and resynthesized, the eight viral genes were reassembled into a plasmid vector and sent to the laboratory of Terrence Tumpey at the Centers for Disease Control and Prevention. The plasmid was introduced into eukaryotic tissue culture cells that allowed the viral genes to function as a unit and produce infectious 1918 influenza virus.

Scientists then analyzed the ability of the resurrected strain to infect a variety of cell types and cause disease. An important difference noted between seasonal influenza strains and the 1918 version is that replication of typical flu strains is generally limited to lung cells, whereas the 1918 flu seemed able to replicate in many different types of cells. What could explain the more promiscuous replication of the 1918 strain? Previous work by Christopher Basler and colleagues revealed that the specific combination of hemagglutinin (HA) and neuraminidase (NA) from the 1918 strain was critical to virulence (eTopic 27.4). The researchers all knew that proteolytic cleavage of HA is required for the fusing of viral and host cell membranes and that most flu strains infect only those host cells that contain high levels of a particular protease typically found in lung cells. Experiments with the resurrected 1918 strain, however, revealed that the NA produced by the 1918 strain imparts an HA-cleaving ability to proteases found in many different cell types. Consequently, the 1918 flu strain can damage many other tissues beyond lung. It was the combination of HA and NA proteins from the 1918 strain that made the flu disaster possible.

Might today's antiviral drugs prevent a disaster if the 1918 flu strain reemerges? The answer, fortunately, is yes. Research has shown that amantadine and oseltamivir will inhibit replication of reconstructed 1918 flu strains, affording some level of comfort as we await the arrival of the next great flu pandemic.

RESEARCH QUESTION

Many deaths attributed to influenza today, and during the 1918 pandemic, are the result of secondary bacterial lung infections. In fact, patients with influenza are more susceptible to bacterial lung infections than are people with other viral lung infections. What mechanisms of flu infection might cause this heightened susceptibility?

Chertow, Daniel S, and Matthew J. Memoli. 2013. Bacterial coinfection in influenza: A grand rounds review. *JAMA* **309**:275–282.

The neuraminidase inhibitors, when used within 48 hours of disease onset, decrease shedding and reduce the duration of influenza symptoms by approximately one day. However, flu symptoms generally last only 3–10 days. While this does not sound like a substantial benefit, shortening the course of the flu in the elderly can minimize damage to the lungs, which in turn reduces the chance of developing life-threatening secondary bacterial infections such as pneumonia and bronchitis.

Antiviral DNA Synthesis Inhibitors

Most antiviral agents work by inhibiting viral DNA synthesis. These drugs chemically resemble normal DNA nucleosides, molecules containing deoxyribose and analogs of adenine, guanine, cytosine, or thymine. Viral enzymes then add phosphate groups to these nucleoside analogs to form DNA nucleotide analogs. The DNA nucleotide analogs are then inserted into the growing viral DNA strand in place of a normal nucleotide. Once inserted, however, new nucleotides cannot attach to the nucleoside analogs, and DNA synthesis stops. These DNA chain-terminating analogs (**Fig. 27.27**) are selectively toxic because viral polymerases are more prone to incorporate nucleotide analogs into their nucleic acid than are the more selective host cell polymerases. Antiviral DNA synthesis inhibitors work on DNA viruses or retroviruses, but not on viruses such as influenza with its RNA genome.

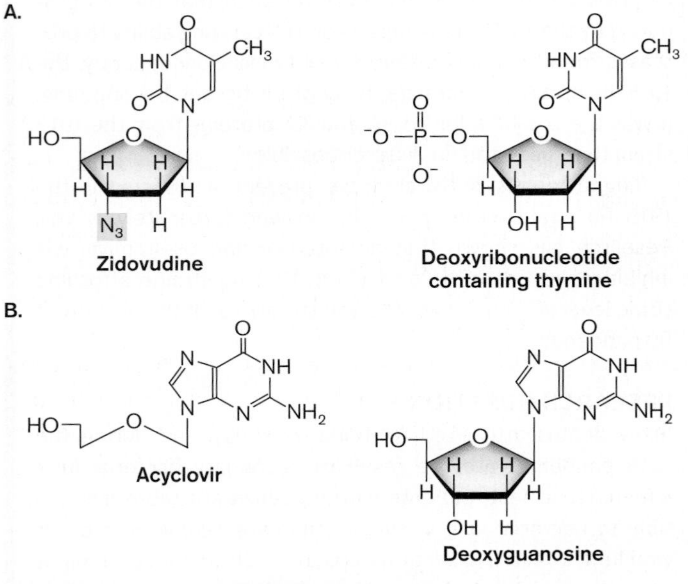
A.
Zidovudine
Deoxyribonucleotide containing thymine
B.
Acyclovir
Deoxyguanosine

FIGURE 27.27 ▪ Antiviral inhibitors that prevent DNA synthesis. Zidovudine (AZT) **(A)** and acyclovir **(B)** are analogs of thymine and guanine nucleotides, respectively. Because the analogs have no 3′ OH to which another nucleotide can add, chain elongation ceases.

Antiretroviral Therapy

Retroviruses are RNA viruses that use viral reverse transcriptase to make DNA and then use viral integrase to insert that DNA into the eukaryotic host cell genome (Chapter 11). The integrated provirus can then be activated to make retroviral RNA. The retroviral RNA travels to the cytoplasm and directs synthesis of more virus particles. One of the most devastating retroviruses is human immunodeficiency virus (HIV), the cause of acquired immunodeficiency syndrome (AIDS; Section 11.4).

Case History: Treatment of HIV

A married couple came to the community clinic for prenatal care. He was 20 years old. She was 19 and reportedly 2 months pregnant with her first child. She denied intravenous (IV) drug use or a history of other sexual partners and had no history of sexually transmitted disease; however, a routine prenatal HIV antibody screen was reported as positive for HIV-1. Careful questioning of the patient and her husband elicited from him a history of IV drug use 5 years earlier. An HIV antibody screen for him was also positive. The laboratory results indicated that the wife might not yet require therapy (she had a low viral load—that is, less than 1,000 copies per milliliter of blood—and a high CD4 T-cell count), but since she was pregnant, a short course of antiretroviral therapy (AZT) would be helpful in preventing transmission of the virus to her child. The husband had an HIV viral load of 10,000 copies per milliliter and was started on combination antiretroviral therapy including a protease inhibitor.

Being diagnosed with HIV is no longer a death sentence. Advances in antiretroviral therapy (ART) over the past 10 years have transformed HIV into a manageable chronic condition. At least half, if not most, of the HIV-positive people in the United States now live long enough to die from diseases of aging, such as heart attacks or strokes. Because it is such a treatable infection, the CDC even recommends that everyone be tested for HIV. The reason for this recommendation is that many people do not know they are infected, but treating an asymptomatic, HIV-positive person reduces the risk of sexually transmitting the virus. Antiretroviral therapy can also prevent transplacental transmission from an HIV-positive, pregnant woman to her fetus, as in our case history. The drugs described next are important components of effective treatment.

Nucleoside, nucleotide, and nonnucleoside reverse transcriptase inhibitors. As an RNA retrovirus, HIV uses a reverse transcriptase to make DNA that then integrates into host nuclear DNA (discussed in Section 11.4). The antiretroviral drug zidovudine (abbreviated ZDV or AZT), which was used in the case history, is a nucleoside analog recognized by reverse transcriptase. Once incorporated into a replicating HIV DNA molecule, the DNA chain-terminating property (lack of a 3′ hydroxyl group) of AZT prevents further DNA synthesis. Nucleoside inhibitors must undergo three successive phosphorylation steps inside the cell to become an active trinucleotide form of the drug. Nucleotide inhibitors are essentially monophosphorylated analogs that require only two phosphorylation steps for activation.

In the case history described, AZT therapy was used to prevent transplacental transmission of the virus from mother to fetus. Because HIV transmission from the mother to the neonate can also occur at delivery or by breast-feeding, treatment of the mother and the child post-delivery are important steps needed to prevent transmission.

In addition to nucleoside inhibitors, there are also non-nucleoside reverse transcriptase inhibitors. For example, the drug delavirdine binds directly to reverse transcriptase and allosterically inactivates the enzyme.

Protease inhibitors. To make optimum use of its limited provirus DNA sequence, HIV makes long, nonfunctional polypeptide chains that are proteolytically cleaved to make the actual proteins and enzymes used to replicate and produce new virions. For example, the *gag* and *pol* genes reside next to each other in the HIV genome and are transcribed as a single mRNA molecule (see Section 11.4). The Gag and Pol open reading frames overlap but are offset by one base. This mRNA produces two polyproteins, called Gag and Gag-Pol, the latter being the result of a shift in reading frames that takes place during translation. Once made, both polyproteins are cleaved by HIV protease. The Gag protein is proteolytically cleaved to make different capsid components (p17, p24, and p15, which is further cleaved to make nucleocapsid protein p7), while Gag-Pol is cleaved to make reverse transcriptase and integrase (**Fig. 27.28A**). Protease inhibitors such as Viracept and Lopinavir belong to a powerful class of drugs that block the HIV protease (**Fig. 27.28B**). When the protease is inactivated, the polyproteins remain uncleaved and the virus cannot mature, even though new virus particles are made. Because immature HIV particles cannot infect other cells, progress of the disease stalls. Note that protease inhibitors do not cure AIDS; they can only decrease the number of infectious copies of HIV.

Antiviral therapy with protease inhibitors is recommended for patients with symptoms of AIDS and for asymptomatic patients with HIV viral loads above 30,000 copies per milliliter of blood. Treatment should be considered even for patients with viral loads as low as 5,000 copies per milliliter, as for the husband in the case history presented earlier.

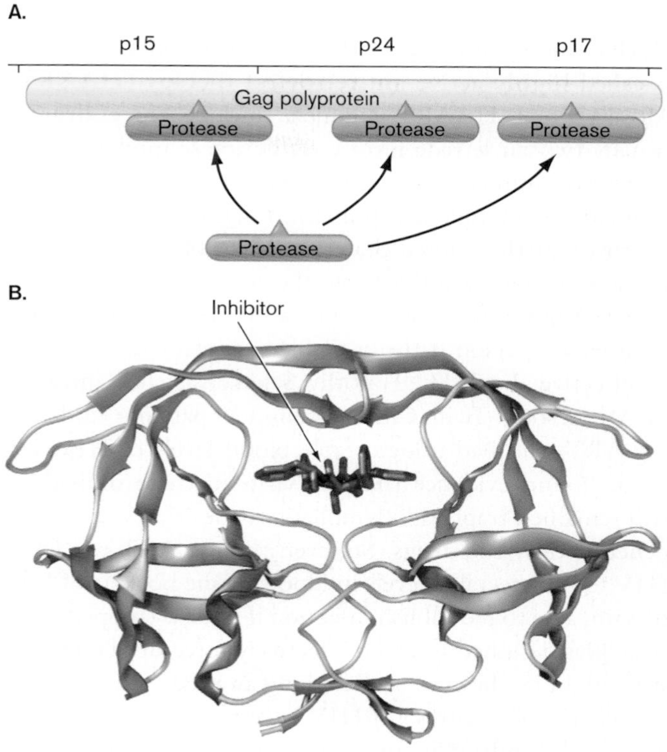
A.
p15
p24
p17
Gag polyprotein
Protease
Protease
Protease
Protease
B.
Inhibitor

FIGURE 27.28 ▪ HIV protease inhibitor. A. Representation of HIV protease cleavage of a single Gag polyprotein into multiple, smaller proteins. **B.** The protease enzyme is shown as a ribbon structure, while the protease inhibitor BEA 369 is shown as a stick model. (PDB code: 1EBY)

Entry inhibitors. Another way to stop HIV is to prevent the virus from infecting cells in the first place. Drugs called entry inhibitors do just that—they stop entry. There are two types of entry inhibitors. CCR5 inhibitors block virus envelope protein gp120 (also known as SU) from binding to host surface protein CCR5, a coreceptor that, together with CD4, is needed for virus binding (see Chapter 11 and Figure 11.24 for details). As a result of CCR5 inhibition, the virus never attaches. Fusion inhibitors, in contrast, do not prevent initial binding, but prevent HIV membranes from fusing with T-cell membranes and thereby halt viral entry. Imagine trying to enter a room through a door. CCR5 inhibitors are like removing the doorknob from the door so there is nothing to grab. Fusion inhibitors, however, are like gluing the door shut. You can grab the knob but still cannot open the door.

Treatment regimens. Because HIV can mutate rapidly and become resistant to single-drug therapies (Section 11.4), treatment today involves administering combinations of three or more antiretroviral drugs. This therapeutic strategy is called **highly active antiretroviral therapy (HAART)**. Most current HAART regimens include three drugs—usually two nucleoside reverse transcriptase inhibitors, plus a protease inhibitor, a nonnucleoside reverse transcriptase inhibitor, or an integrase inhibitor. Integrase inhibitors, first approved in the United States in 2007, block the enzyme needed to insert viral DNA into the host genome.

You might wonder why HIV cannot be eliminated from an infected person if the available antiretroviral drugs are so effective. In 2011, Timothy Schaker at the University of Minnesota, Twin Cities, examined patients undergoing ART who had undetectable blood levels of HIV. His group found evidence that the virus in these individuals still remained trapped in lymphatic tissues that were poorly penetrated by the drugs. So, even though ART can lower HIV to undetectable levels in blood, tissue pockets of HIV remain, able to reestablish infection if ART is stopped. New strategies designed to more effectively force drugs into tissues might be the cure we have long awaited.

Remarkably, "cures" of HIV have recently been accomplished through early and aggressive therapy. In 2013, a newborn baby in rural Mississippi was treated aggressively with antiretroviral drugs starting about 30 hours after birth, something that was not usually done. Then, in that same year, it was revealed that a group of 14 adults treated within 4 to 10 weeks of infection, but whose treatment was inadvertently interrupted for three years, did not relapse while off the drugs. The "cured" patients still had traces of HIV in their blood but these levels were easily kept in check by their immune systems.

HIV treatment as prevention. AIDS can be a devastating disease. Fortunately, we have effective antivirals that can prevent HIV replication and AIDS. Could treating at-risk populations with antivirals before exposure be effective at preventing infection? This strategy is known as preexposure prophylaxis. In fact, the FDA has approved the daily use of an HIV medicine, Truvada (tenofovir/emtricitabine), by healthy, but high-risk, people hoping to lower their risk of infection by a sexual partner. Both drugs are nucleoside analogs of adenosine and cytosine, respectively, and are reverse transcriptase inhibitors. Although this is an approved strategy, it is controversial even among physicians. Some fear the drug would encourage risky behavior.

To Summarize

- **Fewer antiviral agents** than antibacterial agents are available because it is harder to identify viral targets that provide selective toxicity.
- **Preventing viral attachment to, or release from, host cells** is a mechanism of action for antiviral agents, such as amantadine and zanamivir, used to treat influenza virus.
- **Inhibiting DNA synthesis** is the mode of action for most antiviral agents, although they work only for DNA viruses and retroviruses.
- **HIV treatments** include reverse transcriptase inhibitors that prevent the synthesis of DNA, and protease inhibitors that prevent the maturation of viral polyproteins into active forms.

27.8

Antifungal Agents

Fungal infections are much more difficult to treat than bacterial infections, in part because fungal physiology is more similar to that of humans than bacterial physiology is. The other reason is that fungi have an efficient drug detoxification system that modifies and inactivates many antibiotics. Thus, to have a fungistatic effect, repeated applications of antifungal agents are necessary to keep the level of unmodified drug above MIC levels.

Case History: Blastomycosis

A 37-year-old male presented to the emergency department of a Florida hospital with persistent fever, malaise, and a painful right-arm mass. He denied trauma to the arm. White blood cell count was elevated at 27,000/μl, and chest X-ray revealed a left-lung infiltrate. Bronchoscopy revealed granulomatous inflammation containing a single yeastlike mass. Incision and drainage was performed on the arm mass and cultures obtained. Serum cryptococcal antigen tests were negative, as were tests for Bartonella henselae *and* Toxoplasma. *Cultures from the right arm grew out a fungal form similar to that identified from the bronchoscopy specimens. A tentative diagnosis of* Blastomyces dermatitidis *was confirmed using PCR. The patient was placed on amphotericin B, and his fevers and leukocytosis subsequently subsided. His medication was changed to fluconazole for a recommended duration of 6 months.*

Superficial mycoses (fungal infections), such as athlete's foot, and systemic mycoses, such as blastomycosis, require very different treatments. Imidazole-containing drugs (clotrimazole, miconazole) are often used topically in creams for superficial mycoses (**Fig. 27.29A**). Others, such as itraconazole, are administered orally. Superficial mycoses include infections of the skin, hair, and nails, as well as *Candida* infections of moist skin and mucous membranes (for example, vaginal yeast infections). The imidazole-containing drugs appear to disrupt the fungal membrane by inhibiting sterol synthesis. Lamisil (a terbinafine compound) is a different class of agent that selectively inhibits ergosterol synthesis by fungi but not humans. It is currently a popular antifungal agent used to treat superficial mycoses. More chronic dermatophytic infections typically require another antifungal agent, called **griseofulvin**, produced by a *Penicillium* species (**Fig. 27.29B**). Griseofulvin disrupts the mitotic spindle and derails cell division (called metaphase arrest). This does not kill the fungus, but as the hair, skin, or nails grow and are replaced, the fungus is shed.

Vaginal yeast infections caused by *Candida* are often treated with nystatin, a polyene antifungal agent synthesized by *Streptomyces* that forms membrane pores (**Fig. 27.29C**). The name "nystatin" came about because two of the people who discovered it worked for the New York State Public Health Department (now the Wadsworth Center).

The serious, sometimes fatal consequences of systemic mycoses require more aggressive therapy. The drugs used in these instances include **amphotericin B** (produced by *Streptomyces*; see **Fig. 27.29D**) and fluconazole. Amphotericin B binds to the sterols in fungal membranes and destroys membrane integrity. It has a high affinity for ergosterol, which is prevalent in fungal, but not mammalian, membranes. Fluconazole, on the other hand, inhibits the synthesis of ergosterol. Thus, fungal cells grown in the presence of fluconazole make defective membranes. Typically, curing systemic fungal infections requires long-term treatment to prevent disease relapse. **Table 27.4** lists a number of other commonly used antifungal agents.

To Summarize

- **Fungal infections** are difficult to treat because of similarities in human and fungal physiologies.
- **Imidazole-containing** antifungal agents inhibit sterol synthesis.
- **Griseofulvin** inhibits mitotic spindle formation.
- **Nystatin** produces membrane pores.
- **Amphotericin B** binds to membranes and destroys membrane integrity.

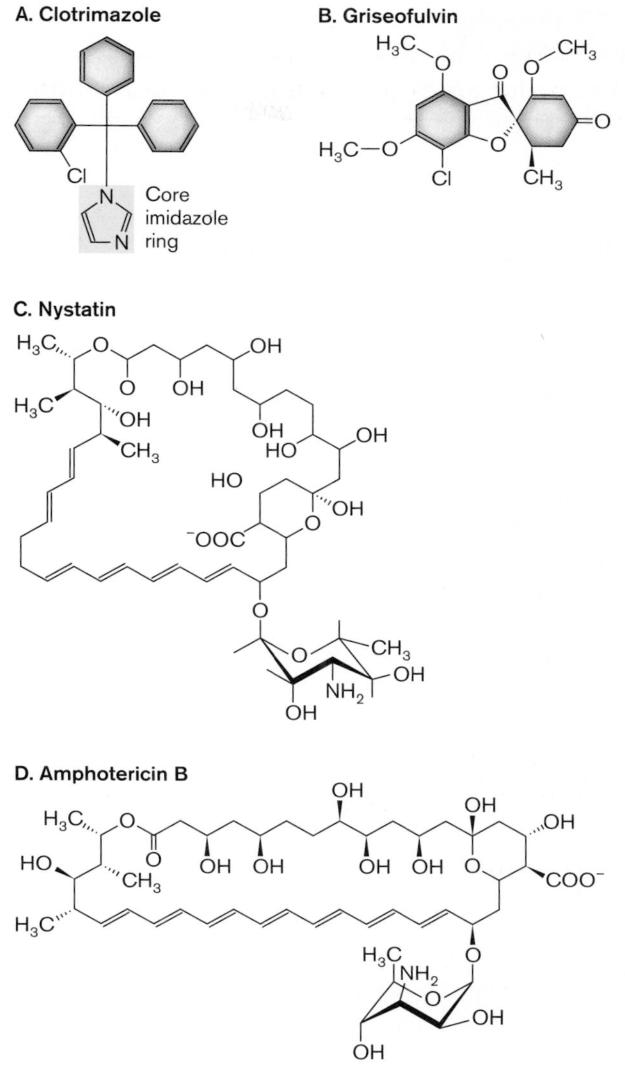
A. Clotrimazole
Core imidazole ring
B. Griseofulvin
C. Nystatin
D. Amphotericin B

FIGURE 27.29 ■ Examples of antifungal agents. A. Clotrimazole belongs to the group of imidazole antifungals, so named because they all contain an imidazole ring. **B.** Griseofulvin is produced by *Penicillium griseofulvum*. **C.** Nystatin is a polyene macrolide produced by *Streptomyces noursei*. **D.** Amphotericin B is a polyene produced by *Streptomyces nodosus*.

Concluding Thoughts

Antibiotics have done much to improve the health and well-being of people and animals around the globe. Although they have successfully kept at bay infectious diseases dreaded for centuries (such as the plague and tuberculosis), a crisis of antibiotic resistance looms because of the irresponsible use of antimicrobial agents. The fact that very few new, clinically useful antimicrobials have been discovered over the last quarter century should give us pause. Redoubling efforts to find new drugs, combined with using existing antibiotics responsibly, is required to maintain our advantage over constantly evolving pathogens.

TABLE 27.4

The major antifungal agents and their common uses.

	Clinical application			
	Systemic mycoses			
Drug	**Coccidioidomycosis**	**Histoplasmosis**	**Blastomycosis**	**Paracoccidioidomycosis**
Polyenes				
Amphotericin B	+	+	+	+
Nystatin	–	–	–	–
Pimaricin	–	–	–	–
Imidazoles				
Clotrimazole	–	–	–	–
Miconazole	–	–	–	–
Ketoconazole	+	+	+	+
Triazoles				
Itraconazole	+	+	+	+
Fluconazole	+	?[b]	?	?
Antimetabolite				
5-Fluorocytosine[c]		–	–	–

[a]mc = mucocutaneous but not systemic candidiasis.

[b]Insufficient data.

[c]Used only in combination with amphotericin B.

CHAPTER REVIEW

Review Questions

1. What is selective toxicity? Provide examples.
2. Explain the difference between antibiotic susceptibility and antibiotic sensitivity.
3. What does the term "spectrum of antibiotic activity" mean?
4. Provide examples of bacteriostatic and bactericidal antibiotics.
5. What is the Kirby-Bauer test? Does it indicate whether a drug is bacteriostatic or bactericidal?
6. Give examples of drugs that target cell wall synthesis; RNA synthesis; protein synthesis; DNA replication. What are their modes of action?
7. What mechanism do producing organisms use to synthesize peptide antibiotics?
8. How do antibiotic-producing microorganisms prevent "suicide"?
9. Why is antibiotic resistance a growing problem?
10. What are the four basic mechanisms of antibiotic resistance?
11. Explain the basic concept of an MDR efflux pump.
12. Discuss the current concepts of the origin of antibiotic resistance.
13. What are some mechanisms used to combat the development of drug resistance?
14. Why are there few antiviral agents available to treat disease?
15. What is herd immunity?
16. How does amantadine inhibit influenza?
17. Discuss the general modes of action of antifungal agents.

Clinical application				
	Opportunistic mycoses			
Aspergillosis	Candidiasis	Cryptococcosis	Dermatophytosis	Other
+	+	+	–	
–	mc[a]	–	–	
–	–	–	–	Mycotic keratitis
–	mc	–	+	
–	mc	–	+	
–	+	–	+	
+	mc	+	+	Sporotrichosis
–	+	+	+	Sporotrichosis
+	+	+	–	Phaeohyphomycosis

Thought Questions

1. Design a summary figure of a bacterium that illustrates the common targets of antimicrobial therapy.
2. A patient presented to the emergency room complaining of a nonproductive cough (no sputum) that had persisted for 6 weeks. The clinician prescribed a 7-day course of cephalosporin. After day 7 the patient returned, no better than before he started the treatment. Laboratory tests later showed that the infection was caused by *Mycoplasma pneumoniae*. Explain why the antibiotic did not work.
3. How would you determine the MIC of an obligate intracellular pathogen such as *Rickettsia prowazekii*, the cause of typhus?

Key Terms

amphotericin B (1143)
antibiotic (1108)
bacitracin (1120)
bactericidal (1112)
bacteriostatic (1112)
chloramphenicol (1125)
cycloserine (1120)
gramicidin (1121)
griseofulvin (1143)
highly active antiretroviral therapy (HAART) (1142)
Kirby-Bauer assay (1113)
lincosamide (1125)
macrolide (1125)
minimal inhibitory concentration (MIC) (1112)
Mueller-Hinton agar (1115)
multidrug resistance (MDR) efflux pump (1129)
neuraminidase inhibitor (1138)
nosocomial (1108)

oxazolidinone (1125)
penicillin (1108)
penicillin-binding protein (PBP) (1118)
quinolone (1122)
secondary metabolite (1126)
selective toxicity (1111)
spectrum of activity (1111)
streptogramin (1125)
sulfa drug (1110)
transglycosylase (1118)
transpeptidase (1118)
vancomycin (1120)
zone of inhibition (1113)

Recommended Reading

Aldridge, Bree B., Marta Fernandez-Suarez, Danielle Heller, Vijay Ambravaneswaran, Daniel Irimia, et al. 2012. Asymmetry and aging of mycobacterial cells lead to variable growth and antibiotic susceptibility. *Science* **335**:100–104.

Bolla, Jean-Michel, Sandrine Alibert-Franco, Jadwiga Handzlik, Jacqueline Chevalier, Abdallah Mahamoud, et al. 2011. Strategies for bypassing the membrane barrier in multidrug resistant Gram-negative bacteria. *FEBS Letters* **585**:1682–1690.

David, Michael Z., and Robert S. Daum. 2010. Community-associated methicillin-resistant *Staphylococcus aureus*: Epidemiology and clinical consequences of an emerging epidemic. *Clinical Microbiological Reviews* **23**:616–687.

Davies, Julian C., and Diana Bilton. 2009. Bugs, biofilms, and resistance in cystic fibrosis. *Respiratory Care* **54**:628–640.

Dawson, Clinton C., Chaidan Intapa, and Mary-Ann Jabra-Rizk. 2011. "Persisters": Survival at the cellular level. *PLoS Pathogens* 7:e1002121.

Dwyer, Daniel J., Diogo M. Camacho, Michael A. Kohanski, Jarrad M. Callura, and James J. Collins. 2012. Antibiotic-induced bacterial cell death exhibits physiological and biochemical hallmarks of apoptosis. *Molecular Cell* **46**:561–572.

Gorski, Andrzej, Ryszard Miedzybrodzki, Jan Borysowski, Beata Weber-Dabrowska, Malgorzata Lobocka, et al. 2009. Bacteriophage therapy for the treatment of infections. *Current Opinion in Investigational Drugs* **10**:766–774.

Kim, Jun-Seob, Paul Heo, Tae-JunYang, Ki-Sing Lee, Da-Hyeong Cho, et al. 2011. Selective killing of bacterial persisters by a single chemical compound without affecting normal antibiotic-sensitive cells. *Antimicrobial Agents and Chemotherapy* **55**:5380–5383.

Kohanski, Michael A., Daniel J. Dwyer, and James J. Collins. 2010. How antibiotics kill bacteria: From targets to networks. *Nature Reviews. Microbiology* **8**:423–435.

Kohanski, Michael A., Daniel J. Dwyer, Boris Hayete, Carolyn A. Lawrence, and James J. Collins. 2007. A common mechanism of cellular death induced by bactericidal antibiotics. *Cell* **130**:797–810.

Marquez, Beatrice. 2005. Bacterial efflux systems and efflux pumps inhibitors. *Biochimie* **87**:1137–1147.

Paulsen, I. T., L. Banerjee, G. S. Myers, K. E. Nelson, R. Seghadri, et al. 2003. Role of mobile DNA in the evolution of vancomycin-resistant *Enterococcus faecalis*. *Science* **299**:2071–2074.

Rogers, Geraint B., Mary P. Carroll, and Kenneth D. Bruce. 2012. Enhancing the utility of existing antibiotics by targeting bacterial behaviour? *British Journal of Pharmacology* **165**:845–857.

Sanyal, Gautam, and Peter Doig. 2012. Bacterial DNA replication enzymes as targets for antibacterial drug discovery. *Expert Opinion in Drug Discovery* 7:327–339.

Sierra-Aragon, Salita, and Hauke Walter. 2012. Targets for inhibition of HIV replication: Entry, enzyme action, release and maturation. *Intervirology* **55**:84–97.

Tumpey, Terrence M., and Jessica A. Belser. 2009. Resurrected pandemic influenza viruses. *Annual Review of Microbiology* **63**:79–98.

Wang, Jun, Stephen M. Soisson, Katherine Young, Wesley Shoop, Srinivas Kodalil, et al. 2006. Platensimycin is a selective FabF inhibitor with potent antibiotic properties. *Nature* **441**:358–361.

Wright, Gerard D., and Hendrik Poinar. 2012. Antibiotic resistance is ancient: Implications for drug discovery. *Trends in Microbiology* **20**:157–159.

Yount, Nannette Y., and Michael R. Yeaman. 2012. Emerging themes and therapeutic prospects for anti-infective peptides. *Annual Review of Pharmacology and Toxicology* **52**:337–360.

CHAPTER 28

Clinical Microbiology and Epidemiology

28.1 Principles of Clinical Microbiology

28.2 Specimen Collection and Processing

28.3 Conventional Approaches to Pathogen Identification

28.4 Rapid Techniques for Pathogen Identification

28.5 Point-of-Care Rapid Diagnostics

28.6 Biosafety Containment Procedures

28.7 Principles of Epidemiology

28.8 Detecting Emerging Microbial Diseases

Throughout history, infectious diseases have killed more people than all of Earth's wars combined. Our present success in controlling the spread of disease is due in large part to worldwide surveillance agencies that are equipped to detect outbreaks quickly, before major epidemics develop. These agencies rely on smaller clinical microbiology laboratories scattered throughout the world that help clinicians diagnose infectious diseases.

Here in Chapter 28 we discuss the principles of clinical microbiology and epidemiology that are used to identify, treat, and contain outbreaks. How do clinical laboratories know what organisms to suspect in a given case? And what tests will unequivocally identify the right pathogen? How are the <u>real</u> pathogens found among the normal microbiota? Finding the <u>source</u> of an outbreak is also critical. How do epidemiologists identify "patient zero" and recognize and contain emerging diseases? Last but not least, we'll consider bioterrorism. Clinical scientists and epidemiologists are trained to detect bioterrorist attacks using the same principles they employ to detect naturally occurring infectious diseases.

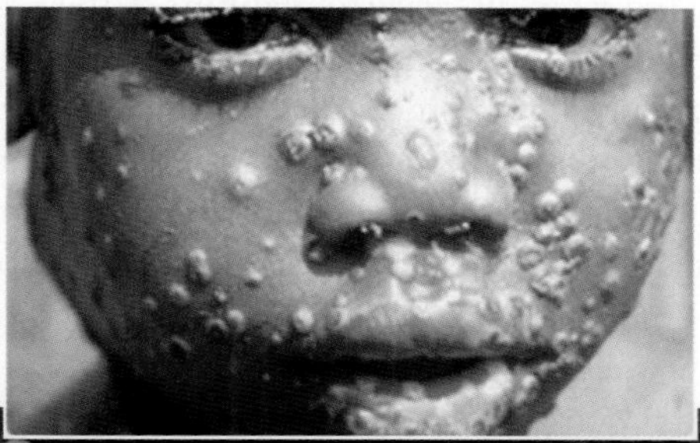

HTTP://WWW.MEDICINENET.COM/MONKEYPOX/PAGE4.HTM

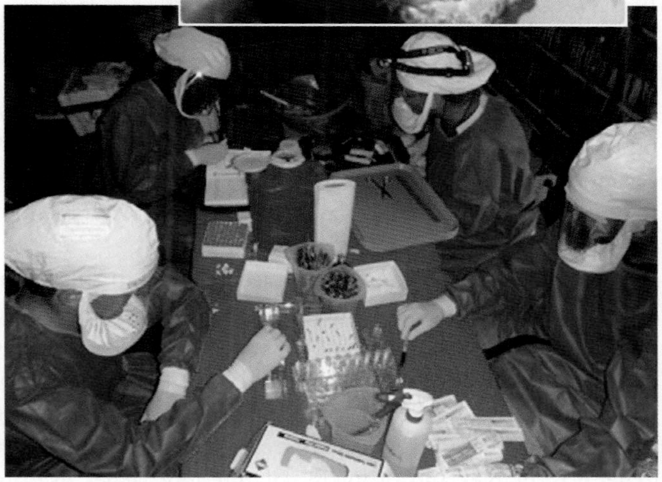

DR. KEVIN KAREM, CDC KDK6@CDC.GOV

CURRENT RESEARCH **highlight**

Modern-day microbe hunters. Monkeypox is a viral disease whose symptoms resemble those of smallpox, complete with vesicular lesions and a death rate of 1%. Monkeypox is endemic to the remote rain forests of central and west Africa, where monkeys and other small mammals serve not only as virus reservoirs, but also as food for people in the area. As a result, monkeypox can be transmitted to humans who hunt these animals (1–2 human cases per 10,000 population). The Centers for Disease Control and Prevention (CDC) periodically flies teams of scientists to these remote areas in order to define the zone of monkeypox distribution and to train the local populace about diagnosis and clinical management. Here, CDC and field museum researchers in 2012 visited the Democratic Republic of Congo to necropsy sampled forest animals. The work site is isolated from the village, and biosafety level 3 practices, including personal protective breathing equipment, were employed for protection.

Wilfrid Saintilus lay immobile on a straw mat when first seen by Dr. Paul Farmer in 1996. The 33-year-old peasant farmer had been sent home without hope by health care workers in a small clinic near his village in Haiti. He was pronounced paralyzed from the waist down; it was said that nothing more could be done. Farmer, however, quickly realized that Wilfrid's problem was not paralysis, but a long-term *Salmonella* infection, very likely contracted from an unclean drinking-water supply—a problem for 80% of Haiti's population. Farmer realized that, while normally a gastrointestinal pathogen, *Salmonella* can occasionally disseminate via the bloodstream and infect bones, causing a painful disease called osteomyelitis. The microbe had infected Wilfrid's hip, causing pain so severe he could not move. His leg muscles had atrophied over time, and he now weighed less than 100 pounds. Contrary to prediction, Wilfrid did not die. Dr. Farmer's diagnostic skills and antibiotics saved his life.

Thousands of stories like this, most with more tragic outcomes, underscore the integral roles that clinical microbiology and epidemiology have in health care. In Wilfrid's case, a simple laboratory test would have revealed a fully treatable *Salmonella* infection and prevented his enormous physical and emotional pain.

As in Chapters 24–27, we use case histories in this chapter to introduce various strategies and methodologies used by modern medicine to diagnose infectious diseases. The approach, by necessity, is selective. Not all techniques, or all diseases, can be described. Our goal is not to catalog the many kinds of infectious diseases, but to demonstrate general principles and problem-solving approaches used in identifying disease-causing microbes. It will become apparent in the process that modern tools of clinical microbiology are indispensable in the diagnosis and treatment of infectious disease.

28.1

Principles of Clinical Microbiology

As in any good detective mystery, the first step in investigating an infectious disease is to identify the most likely suspects. This can be accomplished, in part, from observing the disease symptoms in a patient and knowing which organisms typically produce those symptoms. It also helps if the physician is aware of a similar disease outbreak under way in the community. Beyond these clues, the etiological agent must be identified through biochemical, molecular, serological, or antigen detection strategies.

Why Take the Time to Identify an Infectious Agent?

This is the first question many students ask when contemplating the effort and expense required to identify the genus and species of an organism causing an infection. Why not simply treat the patient with an antibiotic and be done with it? While this approach sounds appealing, there are several compelling reasons for identifying an infectious agent.

Antibiotic resistance. As discussed in Chapter 27, antibiotic resistance is an increasingly serious global problem. Characterizing a microbe's antibiotic resistance profile is part of any microbial identification process. Understanding which antibiotics are effective in treating an infectious agent will help the physician avoid prescribing an inappropriate drug. Furthermore, knowing which drugs are ineffective enables health organizations to track the spread of antibiotic-resistant strains. For example, before 1970, most strains of *Neisseria gonorrhoeae* were susceptible to penicillin. Today, most strains are not only resistant to penicillin but also to many other antibiotics. Remember, too, that viral pathogens are not susceptible to antibiotics, so knowing that a patient has a viral disease can prevent the inappropriate use of antibiotics.

Pathogen-specific disease complications. Many diseases have serious complications that are common to a given organism or strain of organism. For example, children whose sore throats are caused by certain strains of *Streptococcus pyogenes* can develop serious complications affecting the heart and kidney long after the infection has resolved. These complications are the immunological consequence of bacterial and host antigen cross-reactivity; they are called **sequelae** (singular, **sequela**) because they occur after the infection itself is over. Life-threatening sequelae, such as rheumatic fever and acute glomerular nephritis caused by certain strains of *S. pyogenes,* produce severe damage to the heart and kidney, respectively. Knowing early on that *S. pyogenes* has caused a child's sore throat allows the physician to prescribe antibiotics that will quickly eradicate the infection and prevent development of the sequelae.

Note: Penicillin or a penicillin-like antibiotic is the usual treatment for *Streptococcus pyogenes* infection. Contrary to what you might expect, *S. pyogenes* has not developed any penicillin-resistant strains. It is not understood why.

Tracking the spread of a disease. Consider a situation in which ten infants scattered throughout a city develop bloody diarrhea. The clinical laboratory identifies the same strain of *Shigella sonnei,* a Gram-negative bacillus, as the cause in each case. Immunological or nucleic acid amplification tests can be used to type a strain (discussed later). Finding the same strain in all cases suggests they probably originated from the same source. Shigellosis is transferred from person to person by what is called the fecal-oral route. Carriers of *Shigella* will shed this organism in their feces. Inadequate hand washing after defecation will leave bacteria on the hands, which can then transfer the pathogen to foods or utensils or to another person by touching.

Public health officials, armed with the knowledge that all the patients have the same strain of *Shigella,* then question the parents and learn that all of the children attend the same day-care center. By testing the other children and the workers in that center, officials can stop the infection from spreading. This investigative process, called **epidemiology**, is covered more completely in Section 28.7.

To Summarize

Identifying a pathogen enables us to:

- Use appropriate **antibiotics**, if necessary.
- Anticipate possible **sequelae**.
- **Track the spread** of the disease.

28.2

Specimen Collection and Processing

To diagnose an infectious disease, physicians must collect, and the laboratory must process, a wide variety of clinical specimens. The types of specimens range from simple cotton swabs of sore throats to urine and fecal samples. Here in Section 28.2 we describe how these samples should be collected. Note that upon receiving and processing these specimens, the clinical microbiologist must wear latex gloves and use a laminar flow biosafety hood (see Fig. 5.28) as protection against potential infectious agents.

Case History: Abdominal Abscess

A 4-year-old boy was admitted to the hospital for evaluation and treatment of persistent pain in the rectal area. His problem had begun about a week earlier with ill-defined pain in the same area. He had a white blood cell count of 24,900 (normal 4–11,000) with 87% neutrophils (normal 60%). An abdominal computed tomography (CT) scan revealed an abscess adjacent to his rectum. A needle aspiration drained 20 ml of yellowish, foul-smelling fluid from the abscess. Aerobic cultures of this specimen plated on blood and MacConkey agars were negative. Why didn't the infectious agent grow?

The problem in this instance is related to specimen collection and processing. Internal abscesses located near the gastrointestinal tract are often anaerobic infections, in this case caused by the Gram-negative rod *Bacteroides fragilis,* a strict anaerobe (**Fig. 28.1A**). Section 5.6 discusses anaerobes. Intestinal microbes, the majority of which are anaerobic, can sometimes escape the intestine if the organ

A.

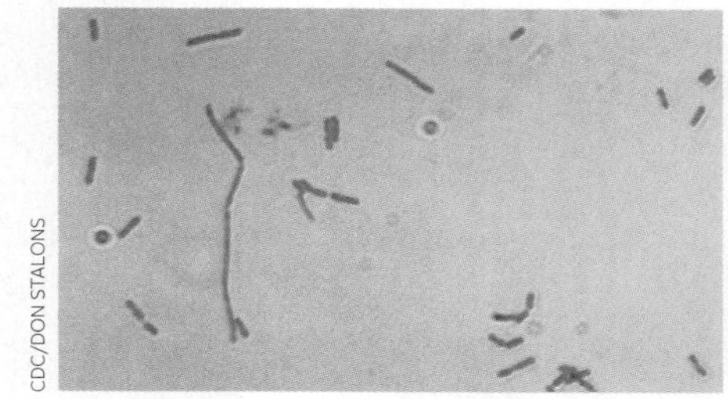
CDC/DON STALONS

B.

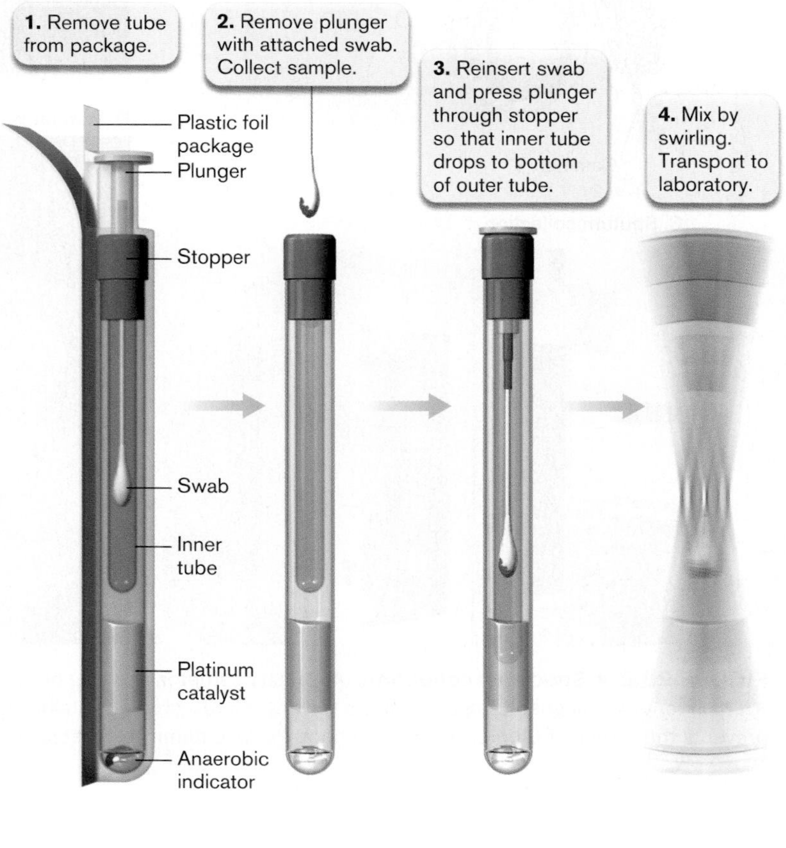
1. Remove tube from package.
2. Remove plunger with attached swab. Collect sample.
3. Reinsert swab and press plunger through stopper so that inner tube drops to bottom of outer tube.
4. Mix by swirling. Transport to laboratory.
Plastic foil package
Plunger
Stopper
Swab
Inner tube
Platinum catalyst
Anaerobic indicator

FIGURE 28.1 ■ Anaerobic infection. A. Gram stain of *Bacteroides fragilis* (1.5–4 μm in length). **B.** Vacutainer anaerobic specimen collector. Plunging the inner tube to the bottom will activate a built-in oxygen elimination system. The anaerobic indicator changes color when anaerobiosis has been achieved.

is damaged in some way. The specimen in this instance should have been collected under anaerobic conditions by aspiration into a nitrogen-filled tube prior to transport to the clinical laboratory. Alternatively, a swab of the abscess material can be inserted into a special transport tube that has a built-in oxygen elimination system (**Fig. 28.1B**). Because the specimen was collected, transported, and handled in air, many anaerobic microbes were probably killed by the oxygen.

Because *B. fragilis* has a stress response system that permits survival of this anaerobe for 1 or 2 days in oxygen, some of the bacteria may have survived transport. The laboratory still had a chance to find the organism, which raises the second problem in the case. After receiving the specimen, the lab cultured it only under aerobic conditions. The laboratory should have also incubated a series of plates anaerobically (see Section 5.6; Fig. 5.21). This case, therefore, illustrates the importance of both proper specimen collection and proper processing.

Some body sites should not contain any microorganisms when collected from a healthy individual. These include blood, cerebrospinal fluid (CSF), and urine from the bladder. Because these sites are sterile, specimens can be plated onto nonselective agar media as well as selective media. Nonselective media, such as blood or chocolate agars, can be used because any organism found in these specimens is considered significant.

Urine collection can be problematic, however. When collected from a catheterized patient, urine should be sterile (**Fig. 28.2A**). Catheterization involves passing thin,

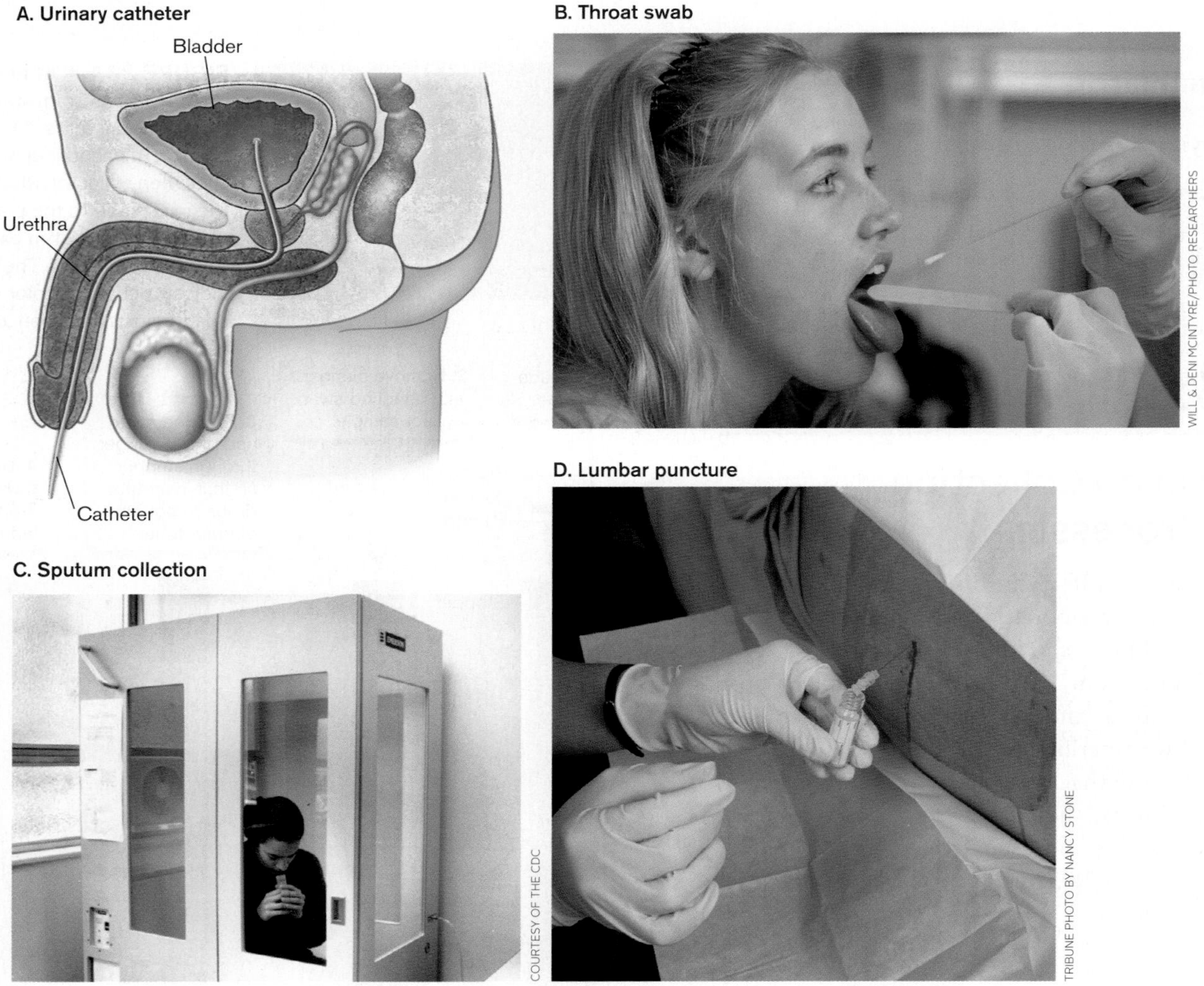
A. Urinary catheter
Bladder
Urethra
Catheter
B. Throat swab
WILL & DENI MCINTYRE/PHOTO RESEARCHERS
C. Sputum collection
COURTESY OF THE CDC
D. Lumbar puncture
TRIBUNE PHOTO BY NANCY STONE

FIGURE 28.2 ■ Specimen collection. A. Urinary catheter, showing placement in the urethra. **B.** Throat swab. **C.** Sputum collection. A TB patient has coughed up sputum and is spitting it into a sterile container. (The patient is sitting in a special sputum collection booth that prevents the spread of tubercle bacilli. The booth is decontaminated between uses.) **D.** Lumbar puncture to obtain cerebrospinal fluid.

sterile tubing through the urethra and directly into the bladder. (Catheterization is used primarily to assist urination by immobilized patients, but it also provides a convenient way to collect urine for bacteriological examination.) Collections made from the tube should be sterile unless an infection is present. Unfortunately, the simple process of inserting the catheter through the nonsterile urethra can sometimes introduce organisms into the bladder and precipitate an infection. In addition, the urine should be collected from the catheter, never from the collection bag. Urine may sit for hours in the collection bag, so organisms initially present at low numbers have time to replicate to high numbers even if the patient does not have a urinary tract infection (UTI).

When a catheter is not in place, urine is most commonly collected by what is called the midstream clean-catch technique, which is performed by the patient. In this procedure, the external genitals are first cleaned with a sterile wipe containing an antiseptic. The patient then partially urinates to wash as many organisms as possible out of the urethra and then collects 5–15 ml of the midstream urine in a sterile cup. This urine sample will usually not be sterile, because of urethral contamination, but the number of bacteria will be low. The clinical laboratory determines how many organisms per milliliter are present in the midstream catch and informs the physician as to whether an infection is present. Finding more than 100,000 organisms per milliliter of urine from a midstream clean catch is considered indicative of an infection even in an asymptomatic patient; 10,000 or fewer is considered normal. If the patient actually has symptoms of a UTI, however, then any count greater than 1,000 colony-forming units (CFUs) per milliliter of a single species is considered significant and should be treated.

Identifying pathogens present at sites that contain normal microbiota is more challenging. A stool or fecal sample, for instance, is normally teeming with microbes. These specimens are typically plated onto selective media—for example, MacConkey, Hektoen, or colistin–naladixic acid (CNA) agar (described in the next section)—to eliminate or decrease the number of normal microbiota that might contaminate the specimen. The following techniques are used to collect specimens from sterile and nonsterile body sites.

- **Collections from sites with normal microbiota:**
  Swabs. Throat swabs (**Fig. 28.2B**), for example, should be placed in specialized liquid nutrient transport medium.
  Sputum. Deep lung secretions expectorated for oral collection (**Fig. 28.2C**).
  Stool samples. Cup or rectal swab; for identifying diarrhea-causing microbes.
  Abscesses. Needle aspirations.
- **Collections from normally sterile sites:**
  Cerebrospinal fluid. Lumbar punctures (spinal taps; **Fig. 28.2D**); testing for meningitis.
  Urine samples. Midstream clean catch (will contain some microflora from urethra) or collected from catheters placed in the bladder (should be sterile); testing for urinary tract infections.
  Blood samples. Generally taken by syringe from two body sites and placed in liquid media for aerobic and anaerobic culture. The same organism isolated from blood samples taken from the two sites is considered the likely etiological agent.

Now that we have collected a sample, how does the clinical microbiologist "divine" the presence and identity of a pathogen? The next sections describe conventional growth-dependent strategies and more rapid, growth-independent, molecular approaches used to catch a pathogen.

Thought Questions

28.1 Two blood cultures, one from each arm, were taken from a patient with high fever. One culture grew *Staphylococcus epidermidis*, but the other blood culture was negative (no organisms grew out). Is the patient suffering from septicemia caused by *S. epidermidis*?

28.2 A 30-year-old woman with abdominal pain went to her physician. After the examination, the physician asked the patient to collect a midstream urine sample that they would send to the lab across town for analysis. The woman complied and handed the collection cup to the nurse. The nurse placed the cup on a table at the nurses' station. Three hours later, the courier service picked up the specimen and transported it to the laboratory. The next day, the report came back: "greater than 200,000 CFUs/ml; multiple colony types; sample unsuitable for analysis." Why was this determination made?

To Summarize

- **Special collection precautions** must be taken when collecting specimens from abscesses of suspected anaerobic etiology.
- **Common specimens** include blood, pus, urine, sputum, throat swabs, stool, and cerebrospinal fluid.
- **Specimens** from sites containing normal microbiota must be handled differently from specimens taken from normally sterile body sites.

28.3

Conventional Approaches to Pathogen Identification

How do we identify pathogens in patient samples, and do it quickly? Pathogens can be identified on the basis of a variety of observations, including the patient's symptoms, the presence of organisms in stained clinical specimens, biochemical clues, and serology (the presence in a patient's serum of antibodies reactive against a specific microbe).

Staining Procedures

Some specimens, such as CSF and sputum, can be directly stained with the Gram stain procedure (described in Section 2.3) or the acid-fast stain. Knowing that an organism is Gram-positive, Gram-negative, or acid-fast guides which additional tests the clinical microbiologist must run. The case that follows illustrates how the acid-fast stain is critical for presumptively identifying mycobacteria.

Case History: Tuberculosis

A 31-year-old male presented to an emergency department (ED) in New York City after experiencing gross hemoptysis (blood in sputum). He had a 2-month history of productive cough, a 25-pound weight loss, night sweats, and fatigue. A chest X-ray revealed bilateral cavitary infiltrates. The initial sputum specimen was negative by Gram stain but positive for acid-fast bacilli (AFB). The specimen was submitted for a nucleic acid amplification assay (NAAT) to detect 16S rRNA, as well as for culture and sensitivity. The patient had a history of heavy alcohol and drug use.

The likely suspect in this case is *Mycobacterium tuberculosis,* although other mycobacterial species are possible causes. *M. tuberculosis* is presumptively diagnosed in the clinical laboratory by the acid-fast stain, a technique first described in 1882 (called the Ziehl-Neelsen stain) that is used to find the tubercle bacillus in a patient's sputum. The acid-fast stain enables a technician to visualize bacteria such as *Mycobacterium* species that are not stained by the Gram stain. Mycobacteria have a very waxy outer coat composed of mycolic acid that resists penetration by most dyes—an obstacle the acid-fast stain was designed to overcome. The original Ziehl-Neelsen acid-fast stain used phenol and heat to drive carbolfuchsin (a red dye) into mycobacterial cells on glass slides. Destaining with an acidic alcohol solution removes the stain from all cell types except mycobacteria. The slide is subsequently counterstained with methylene blue, after which the mycobacteria will be seen as curved, red rods (called acid-fast bacilli), while everything else will appear blue (**Fig. 28.3A**). A more modern version of the acid-fast stain uses the fluorochrome auramine O to stain the mycolic acid. This dye also resists removal by an acidic alcohol wash, so the mycobacteria will fluoresce bright yellow when observed under a fluorescence microscope.

Although the acid-fast stain is very useful, it only detects organisms in the sputum of tuberculosis patients about 60% of the time; and even if they are found, confirmatory tests are needed for a definitive diagnosis. Growth-dependent identification of *M. tuberculosis* typically starts with the sputum sample being inoculated to a blue-green Löwenstein-Jensen medium (selective for mycobacteria) and waiting several weeks for the organism to grow before additional tests can be done (**Fig. 28.3B**). However, while the organism grows, a rapid, nucleic acid amplification test (NAAT) can be used to quickly detect even small amounts of the organism, even in AFB-negative sputum samples. These types of tests are described in Section 28.4.

Weblinks Difco & BBL Manual of Microbiological Culture media (*see ebook*)

A. Acid-fast stain

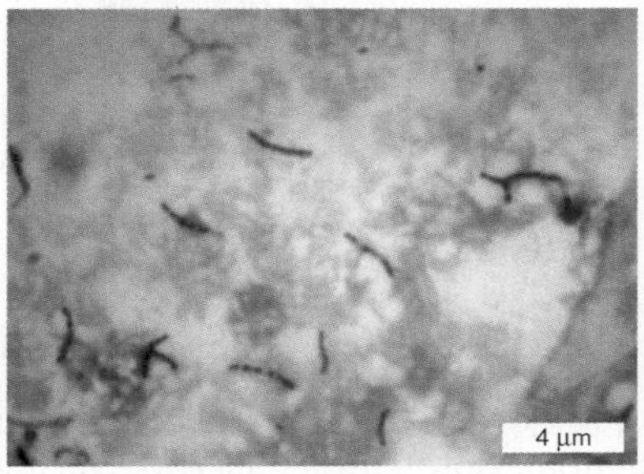
4 µm

CDC

B. *M. tuberculosis* colonies on Löwenstein-Jensen

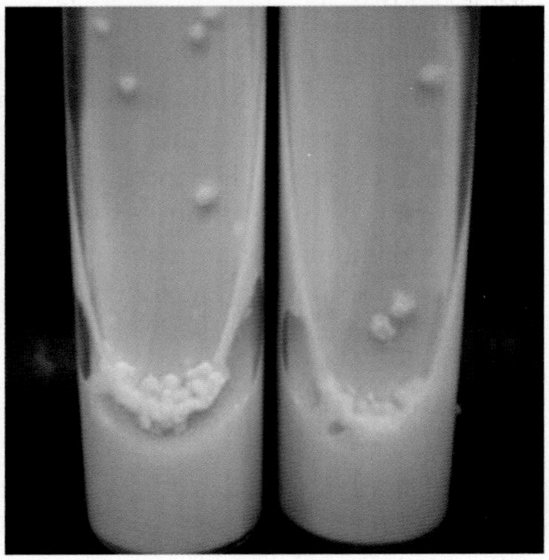

COURTESY OF CLAUDIO BASILICO, M.D./NEW YORK UNIVERSITY

FIGURE 28.3 ■ Acid-fast stain and growth of *Mycobacterium tuberculosis.* **A.** Acid-fast Ziehl-Neelsen stain of *M. tuberculosis.* **B.** Löwenstein-Jensen medium enables growth of mycobacterial species, some of which grow extremely slowly. The colonies have a "bread crumb–like" appearance.

Identifications Based on Growth and Biochemical Testing

Once the Gram reaction of an organism is known, the conventional laboratory approach used to identify bacterial pathogens requires understanding basic microbial physiology and how diverse it is. Thousands of bacterial species are capable of causing disease, but no two species have the same biochemical "signature." The clinical microbiologist can look for reactions, or combinations of reactions, that are unique to a given species, as in the following case history.

Case History: Medical Detective Work

A 38-year-old woman with no significant previous medical history came to the emergency room complaining of a mild sore throat persisting for 3 days. Her symptoms included arthralgia (joint pain), myalgia (muscle pain), and a low-grade fever. The day before coming to the ER she had had a severe headache with neck stiffness, nausea, and vomiting. She was not taking any medications, had no known drug allergies, and did not smoke. She lived with her husband and two children, all of whom were well. Cerebrospinal fluid (CSF) was collected from a spinal tap. The CSF appeared cloudy (it should be clear) and contained 871 white blood cells per microliter (normal is 0–10/μl), the glucose level was 1 mg/dl (normal is 50–80), and the total protein level was 417 mg/dl (normal is less than 45). Gram stain of a CSF smear revealed Gram-negative rods. The CSF sample was sent to the diagnostic laboratory for microbial identification.

As discussed in Thought Question 26.8, low glucose and elevated protein levels in CSF are indicators of bacterial (not viral) infection. The increase in white blood cells revealed that the woman's immune system was trying to fight the disease. The presence of Gram-negative rods in the CSF smear confirmed a diagnosis of bacterial meningitis, since CSF should be sterile. Now it was up to the clinical laboratory to determine the etiological agent.

Problem-Solving Algorithms to Identify Bacteria

Over the years, clinical microbiologists have developed algorithms (step-by-step problem-solving procedures) that expose the most likely cause of a given infectious disease. For instance, there are only a limited number of microbes known to cause meningitis. The microbiologist poses a series of binary yes/no questions about the clinical specimen in the form of biochemical or serological tests. Typical questions in this case might include: Is an organism seen in the CSF of a patient with symptoms of meningitis? Is the organism Gram-positive or Gram-negative? Does it stain acid-fast? Answers to a first round of questions will then dictate the next series of tests to be used.

Because speed is of the essence in deciding how to treat the patient, a series of tests is not always carried out sequentially. To save time, a slew of tests are carried out simultaneously, but the results are interpreted sequentially using the algorithm. In our case history, for instance, consider the most common causes of bacterial meningitis: *Neisseria meningitidis, Streptococcus pneumoniae, Haemophilus influenzae,* and *Escherichia coli.* The CSF sample was Gram-stained by a microbiologist and simultaneously plated onto three media: chocolate agar, blood agar, and Hektoen agar. Chocolate agar is an extremely rich medium that looks brown owing to the presence of heat-lysed red blood cells (**Fig. 28.4A** and **B**). Because it is so nutrient-rich, all four organisms will grow on chocolate agar. However, nutritionally fastidious organisms such as *N. meningitidis* and *H. influenzae* will not grow well, if at all, on ordinary blood agar, because these bacteria cannot lyse red blood cells and release required nutrients. Less fastidious organisms, such as *S. pneumoniae* and *E. coli,* will grow on blood agar, but of these two, only *E. coli* can grow on Hektoen agar (**Fig. 28.4C–E**), which is a selective and differential medium for enteric Gram-negative rods. Hektoen is selective because bile salts and dyes inhibit the growth of Gram positives. It is differential because the medium reveals organisms that ferment lactose or sucrose and produce hydrogen sulfide. Differential and selective media are described in Section 4.3.

In our case history, the Gram stain of the CSF revealed Gram-negative rods, which ruled out *N. meningitidis* (a Gram-negative diplococcus) and *S. pneumoniae* (a Gram-positive diplococcus). The organism in CSF did grow on blood agar, which eliminated *H. influenzae* (a Gram-negative, nonenteric rod) as a candidate. It also grew on Hektoen, where it produced orange lactose-fermenting colonies. Thus, the organism was a Gram-negative, enteric rod, and likely *E. coli.* Additional biochemical tests confirming the identity of the organism had to be carried out, but this simple example shows how simultaneous tests can be interpreted.

Identifying Gram-negative bacteria. The Gram-negative bacterium in this case was subjected to a battery of biochemical tests packaged as 20 separate chambers in a patented analytical profile index (API 20E) strip that can be

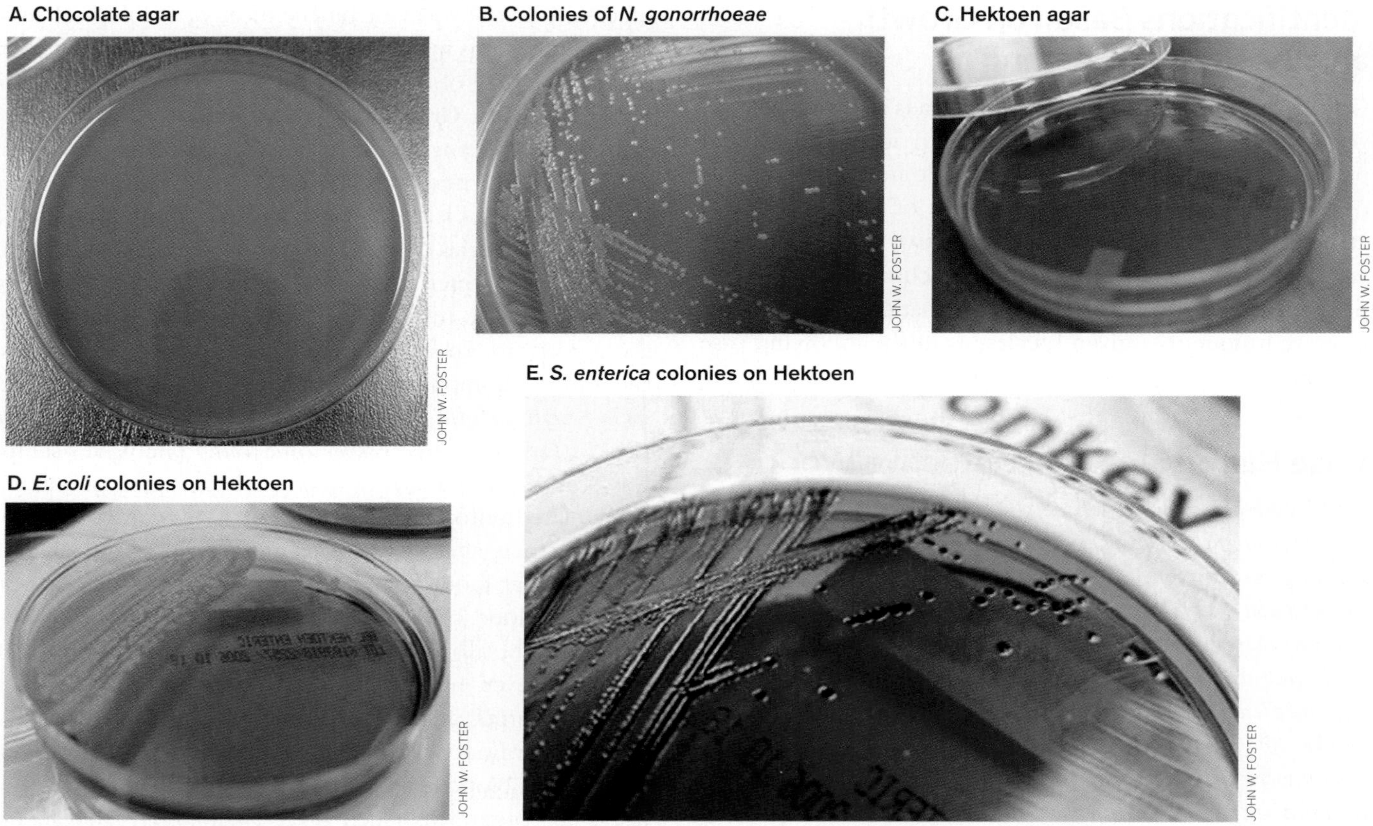
A. Chocolate agar
B. Colonies of *N. gonorrhoeae*
C. Hektoen agar
D. *E. coli* colonies on Hektoen
E. *S. enterica* colonies on Hektoen
JOHN W. FOSTER
JOHN W. FOSTER
JOHN W. FOSTER
JOHN W. FOSTER
JOHN W. FOSTER

FIGURE 28.4 ■ Chocolate agar and Hektoen agar: two widely used clinical media. A. Uninoculated chocolate agar. Its color is due to gently lysed red blood cells that provide a rich source of nutrients for fastidious bacteria. **B.** Chocolate agar inoculated with *Neisseria gonorrhoeae.* This organism will not grow well on typical blood agar, because important nutrients remain locked within intact red blood cells. **C.** Uninoculated Hektoen agar, which contains lactose, peptone, bile salts, thiosulfate, an iron salt, and the pH indicators bromothymol blue and acid fuchsin; the bile salts prevent growth of Gram-positive microbes. **D.** Hektoen agar inoculated with *Escherichia coli.* This organism ferments lactose to produce acidic fermentation products that give the medium an orange color owing to the pH indicators acid fuchsin and bromophenol blue. **E.** Hektoen agar inoculated with *Salmonella enterica.* This organism does not ferment lactose but grows instead on the peptone amino acids. The resulting amines are alkaline and produce a more intense blue color with bromothymol blue. *Salmonella* species also produce hydrogen sulfide gas from the thiosulfate. Hydrogen sulfide reacts with the medium's iron salt to produce an insoluble, black iron sulfide precipitate visible in the center of the colonies.

used for pathogen identification (**Fig. 28.5**). The strip requires overnight incubation and tests whether the organism can ferment a series of carbon sources. It also provides evidence of specific end products formed as a result of fermentation. The results appear as different-colored reactions in each chamber and are scored as positive or negative, depending on the color (**Table 28.1** and eTopic 28.1). For example, in the indole chamber (well 9 in **Fig. 28.5B**), a red reaction at the top of the tube is positive and indicates that the organism can produce indole from tryptophan. A colorless chamber would be a negative result. Similar API strips are available for the identification of Gram-positive bacteria and yeasts.

The results of the API chambers can be used as a dichotomous key, with lab personnel making a stepwise interpretation that begins with a key reaction, such as lactose fermentation (tagged "ONPG" in Table 28.1). The reaction is read as positive or negative depending on the color. Then, following a printed flowchart, the technician goes to the next key reaction—say, indole production—and reads it as positive or negative. If the organism is a lactose fermenter and the indole test is positive, then the choices have been narrowed to *E. coli* or *Klebsiella* species. Another reaction is read to distinguish between the next two choices. The process continues until a single species is identified.

A simplified example is shown in **Figure 28.6** using a limited number of enteric Gram-negative species. In reality, many more reactions than the 11 shown have to be used to make a definitive identification, because of species differences with respect to a single reaction. For example, **Figure 28.6** shows that *Klebsiella pneumoniae* and *Klebsiella oxytoca* exhibit opposite indole reactions, even though they

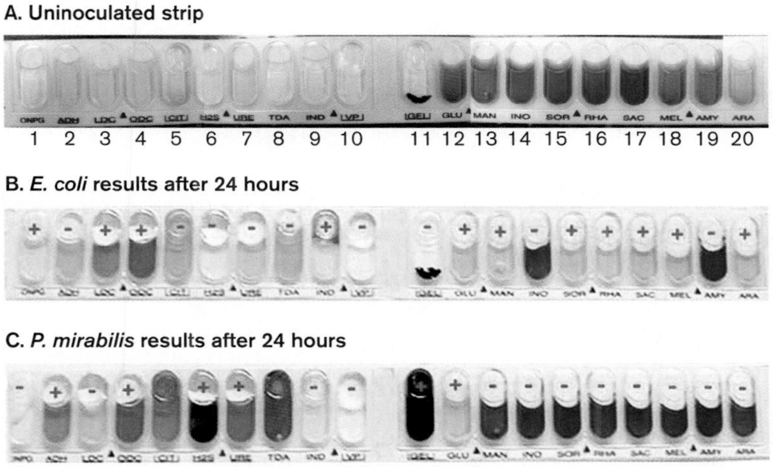
A. Uninoculated strip
1 2 3 4 5 6 7 8 9 10 11 12 13 14 15 16 17 18 19 20
B. E. coli results after 24 hours
C. P. mirabilis results after 24 hours

FIGURE 28.5 ▪ API 20E strip technology for the biochemical identification of Enterobacteriaceae. A. Uninoculated API strip. Each well contains a different medium that tests for a specific biochemical capability. The well numbers correspond to Table 28.1. The color of the media after 24-hour incubation indicates a positive or negative reaction (see Table 28.1). **B.** API results for *E. coli.* Plus (+) and minus (–) indicate positive and negative reactions, respectively. **C.** API results for *Proteus mirabilis.*

are of the same genus. Even within a given species, only a certain percentage of strains might be positive for a given reaction. The inherent danger in using a dichotomous key is that one anomalous result can lead to an incorrect identification.

View **eTopic 28.1** to see how API strips can also be used to generate a seven-digit number that identifies the bacterium. Most clinical laboratories in the United States and Europe now use automated identification systems that are more sophisticated than API. These include bioMérieux's VITEK; Siemens Healthcare Diagnostics' WalkAway; and Becton, Dickinson's (BD's) Phoenix systems (**Fig. 28.7A**). All of them use cards or plates with 30 or more biochemical or enzymatic reactions (**Fig. 28.7B** and **C**). As with the API chambers, the results appear as colored reactions that can be read by computer. The BD Phoenix system, for example, will convert the results into a ten-digit code that is used to identify genus and species.

TABLE 28.1

Reading the API 20E.

Well number	Test	Reaction tested
1	ONPG[a]	Beta-galactosidase
2	ADH	Arginine dihydrolase
3	LDC	Lysine decarboxylase
4	ODC	Ornithine decarboxylase
5	CIT	Citrate utilization
6	H_2S	H_2S production
7	URE	Urea hydrolysis
8	TDA	Tryptophan deaminase
9	IND	Indole production
10	VP	Acetoin production
11	GEL	Gelatinase
12	GLU	Glucose fermentation/oxidation
13	MAN	Mannitol fermentation/oxidation
14	INO	Inositol fermentation/oxidation
15	SOR	Sorbitol fermentation/oxidation
16	RHA	Rhamnose fermentation/oxidation
17	SAC	Sucrose fermentation/oxidation
18	MEL	Melibiose fermentation/oxidation
19	AMY	Amygdalin fermentation/oxidation
20	ARA	Arabinose fermentation/oxidation

[a]ONPG = *ortho*-nitrophenyl-beta-D-galactoside.

Identifying nonenteric Gram-negative bacteria. The procedures we have outlined will accurately identify members of Enterobacteriaceae, but there are pathogenic Gram-negative bacilli found in other, nonenteric, phylogenetic families. For instance, one possibility in the preceding case history is that the Gram-negative bacillus seen in CSF smears would not grow on blood agar or on the other selective media, but would grow as small, glistening colonies on chocolate agar. This result would implicate the Gram-negative rod *Haemophilus influenzae.*

Meningitis caused by *H. influenzae* was a major problem prior to 1988, before the introduction of vaccinations with *H. influenzae* type b capsule material. Growth of *H. influenzae* requires hemin (X factor) and NAD (V factor), so confirming the identity of *H. influenzae* involves growing the organism on agar medium containing hemin and NAD (nicotinamide adenine dinucleotide). This can be done by placing small filter paper disks containing these compounds on a nutrient agar surface (not blood agar) that has been covered with the organism. *H. influenzae* will grow only around a strip containing both X and V factors. Alternatively, X and V factors can be incorporated into Mueller-Hinton agar, as shown in **Figure 28.8A**. The organism grows on chocolate medium because the lysed red blood cells release these factors. Although XV growth phenotype is still used for identification, fluorescent antibody staining is more specific (discussed shortly).

A completely different identification scheme would have been used in the preceding meningitis case if the laboratory had discovered the organism to be a Gram-negative

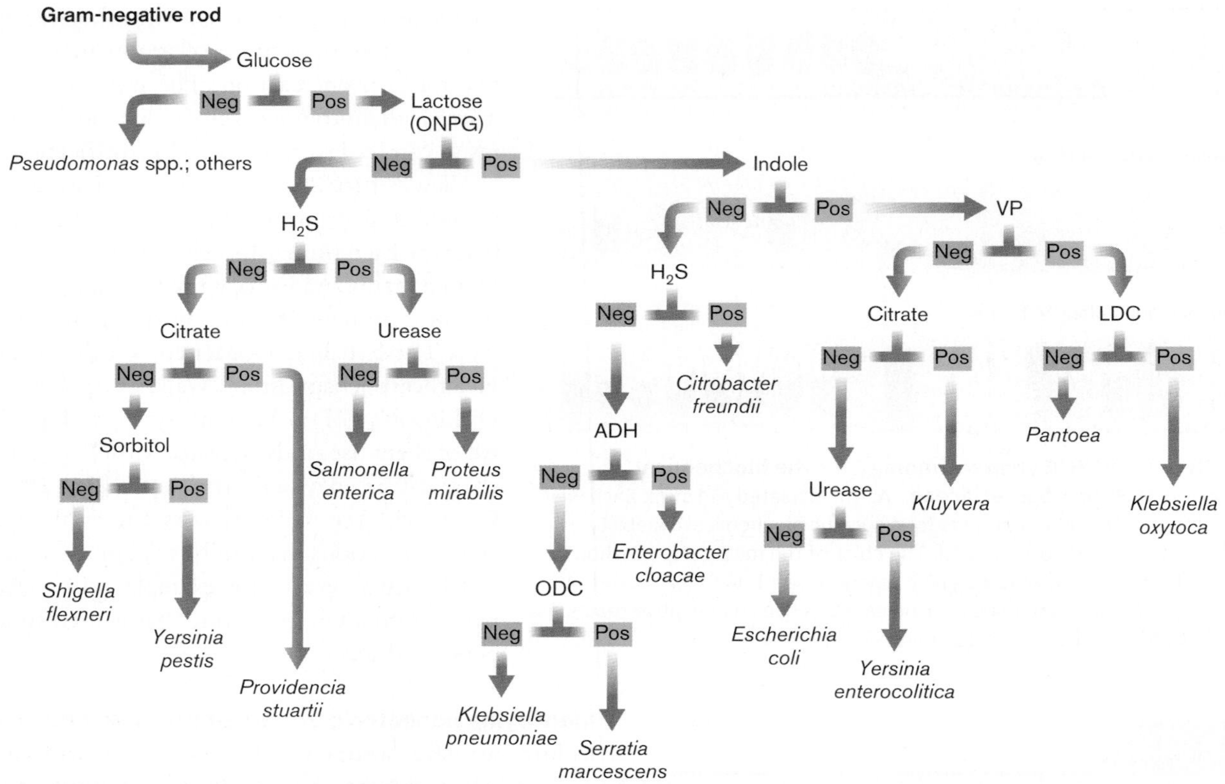
Gram-negative rod
Glucose
Neg
Pos
Lactose (ONPG)
Pseudomonas spp.; others
Neg
Pos
Indole
H_2S
Neg
Pos
VP
Neg
Pos
H_2S
Neg
Pos
Citrate
Urease
Neg
Pos
Citrate
LDC
Neg
Pos
Neg
Pos
Neg
Pos
Neg
Pos
Citrobacter freundii
Sorbitol
ADH
Pantoea
Salmonella enterica
Proteus mirabilis
Neg
Pos
Neg
Pos
Urease
Kluyvera
Klebsiella oxytoca
Neg
Pos
Enterobacter cloacae
Shigella flexneri
ODC
Escherichia coli
Yersinia pestis
Neg
Pos
Yersinia enterocolitica
Providencia stuartii
Klebsiella pneumoniae
Serratia marcescens

FIGURE 28.6 ▪ Simplified biochemical algorithm to identify Gram-negative rods. The diagram presents a dichotomous key using a limited number of biochemical reactions and selected organisms to illustrate how species identifications can be made using biochemistry. Abbreviations and reactions: **ADH** (arginine dihydrolase): stepwise degradation of arginine to citrulline and ornithine; **citrate** (CIT): citrate utilization as a carbon source; **glucose** (GLU): glucose fermentation to produce acid; **H_2S:** production of hydrogen sulfide gas; **indole** (IND): indole production from tryptophan; **lactose** (ONPG): lactose fermentation to produce acid (ortho-nitrophenyl-beta-D-galactoside, cleaved by beta-galactosidase); **LDC** (lysine decarboxylase): cleavage of lysine to produce CO_2 and cadaverine; **ODC** (ornithine decarboxylase): cleavage of ornithine to make CO_2 and putrescine; **sorbitol** (SOR): sorbitol fermentation to produce acid; **urease** (URE): production of CO_2 and ammonia from urea; **VP** (Voges-Proskauer test): production of acetoin or 2,3-butanediol. (Note: "Neg" for Pseudomonas in terms of glucose indicates an inability to ferment glucose. Pseudomonas can still use glucose as a carbon source.)

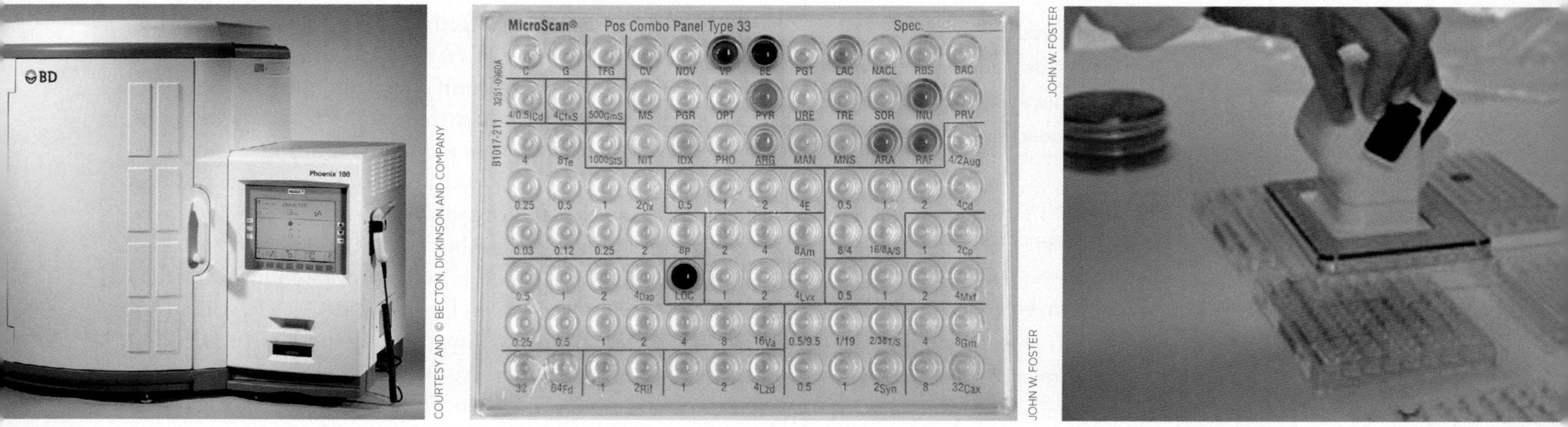
A.
B.
C.
BD
Phoenix 100
MicroScan® Pos Combo Panel Type 33
Spec.
COURTESY AND © BECTON, DICKINSON AND COMPANY
JOHN W. FOSTER
JOHN W. FOSTER

FIGURE 28.7 ▪ Automated microbiology system. A. The BD Phoenix system uses plates with numerous reaction wells and a computerized plate reader to automatically identify pathogenic bacteria. This type of instrument automatically generates and evaluates the numbers. **B.** Microbial identification plate. **C.** Loading a multiwall ID plate with a multichannel pipettor. All wells are simultaneously loaded with the same volume and number of bacteria.

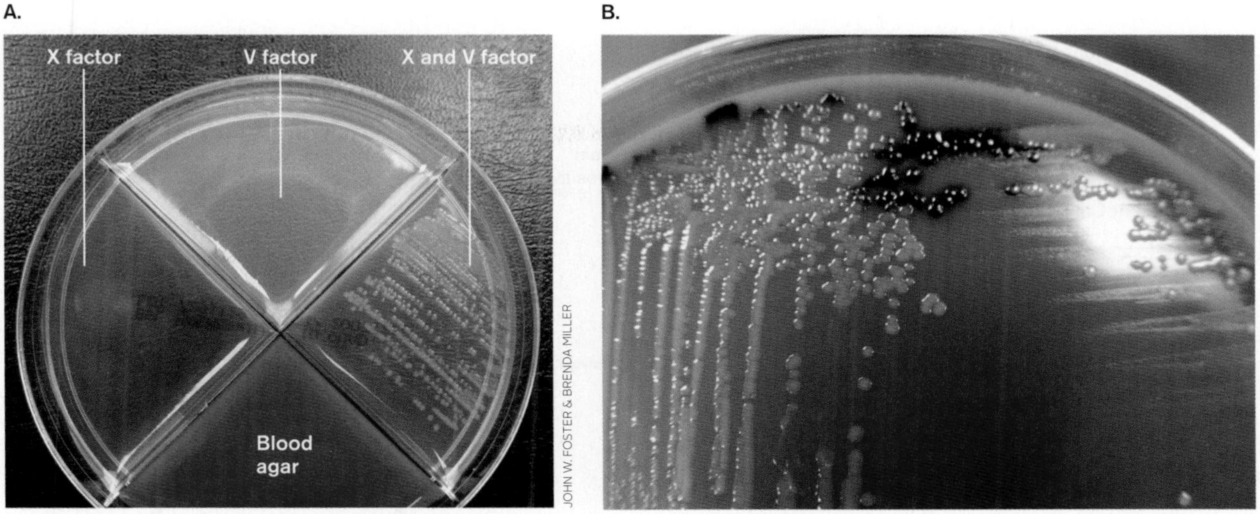
A.
X factor
V factor
X and V factor
Blood agar
JOHN W. FOSTER & BRENDA MILLER
B.
JOHN W. FOSTER

FIGURE 28.8 ■ *Haemophilus influenzae* growth factors and *Neisseria meningitidis* oxidase reaction. **A.** *H. influenzae* will grow on an agar plate (here, Mueller-Hinton agar) only when the medium has been fortified with both X factor (hemin) and V factor (NAD), but not either one alone. Nor will the organism grow on blood agar. **B.** Oxidase-positive reaction for *N. meningitidis.* Oxidase reagent (which is colorless) was dropped onto colonies of *N. meningitidis* grown on chocolate agar. The test is called the cytochrome oxidase test, but it really tests for cytochrome *c*.

diplococcus (rather than a Gram-negative rod). A Gram-negative diplococcus would suggest *Neisseria meningitidis*. *N. gonorrhoeae* is also possible, but it is less likely because the gonococcus lacks the protective capsule that meningococci use to survive in the bloodstream. Without this capsule, *N. gonorrhoeae* is not as resistant to serum complement and cannot disseminate to the meninges.

The first test to determine whether the organism is a species of *Neisseria* is the cytochrome oxidase test (**Fig. 28.8B**). In this test, a few drops of the colorless reagent *N,N,N′,N′*-tetramethyl-*p*-phenylenediamine dihydrochloride are applied to the suspect colonies. The reaction, which takes place only if the organism possesses both cytochrome oxidase and cytochrome *c*, turns the *p*-phenylenediamine reagent (and the colony) a deep purple/black. While many bacteria possess cytochrome oxidase, *Neisseria* is one of only a few genera that also contain cytochrome *c* in their membranes. Oxidase-positive organisms use cytochrome oxidase to oxidize cytochrome *c*, which then oxidizes *p*-phenylenediamine. Other oxidase-positive bacteria include *Pseudomonas, Haemophilus, Bordetella, Brucella,* and *Campylobacter* species—all of them Gram-negative rods. None of the Enterobacteriaceae, however, are oxidase-positive, because they lack cytochrome *c*.

An oxidase-positive, Gram-negative diplococcus is very likely a member of *Neisseria*. Differentiation between species of *Neisseria* is based on their ability to grow on certain carbohydrates, or it can be determined using immunofluorescent antibody staining tests that test for the presence of different capsule antigens in *N. meningitidis*.

Identifying Gram-positive pyogenic cocci. Recall from Section 26.2 the case of the woman with necrotizing fasciitis. How did the laboratory determine that the etiological agent was *Streptococcus pyogenes*? A sample algorithm, or flowchart (**Fig. 28.9**), shows how this is done. The physician sends a cotton swab containing a sample from a lesion to the clinical laboratory. The laboratory technician streaks the material onto several media: (1) blood agar (which will grow both Gram-positive and Gram-negative organisms); (2) blood agar containing the inhibitors colistin and naladixic acid (called a CNA plate, this agar will grow only Gram-positives); and (3) MacConkey agar (which will grow only Gram-negative organisms; see Section 4.3). The suspect organism in this case grows on the CNA and blood plates. Because it grows in the presence of the Gram-negative inhibitory compounds in CNA, one would immediately suspect the organism to be Gram-positive—an assumption borne out by the Gram stain.

Note: Though the skin is normally populated by many different microorganisms, samples from an infected lesion are overwhelmingly populated by the etiological agent. This occurs because the pathogenic microbe outgrows normal microbiota. Using selective media to isolate the infectious agent will further simplify diagnosis by decreasing the growth of any normal microbiota that may still be present.

The algorithm tells the laboratory technician that since the organism is a Gram-positive coccus, the next step is to test for catalase production. Catalase, which converts hydrogen peroxide (H_2O_2) to O_2 and H_2O, clearly

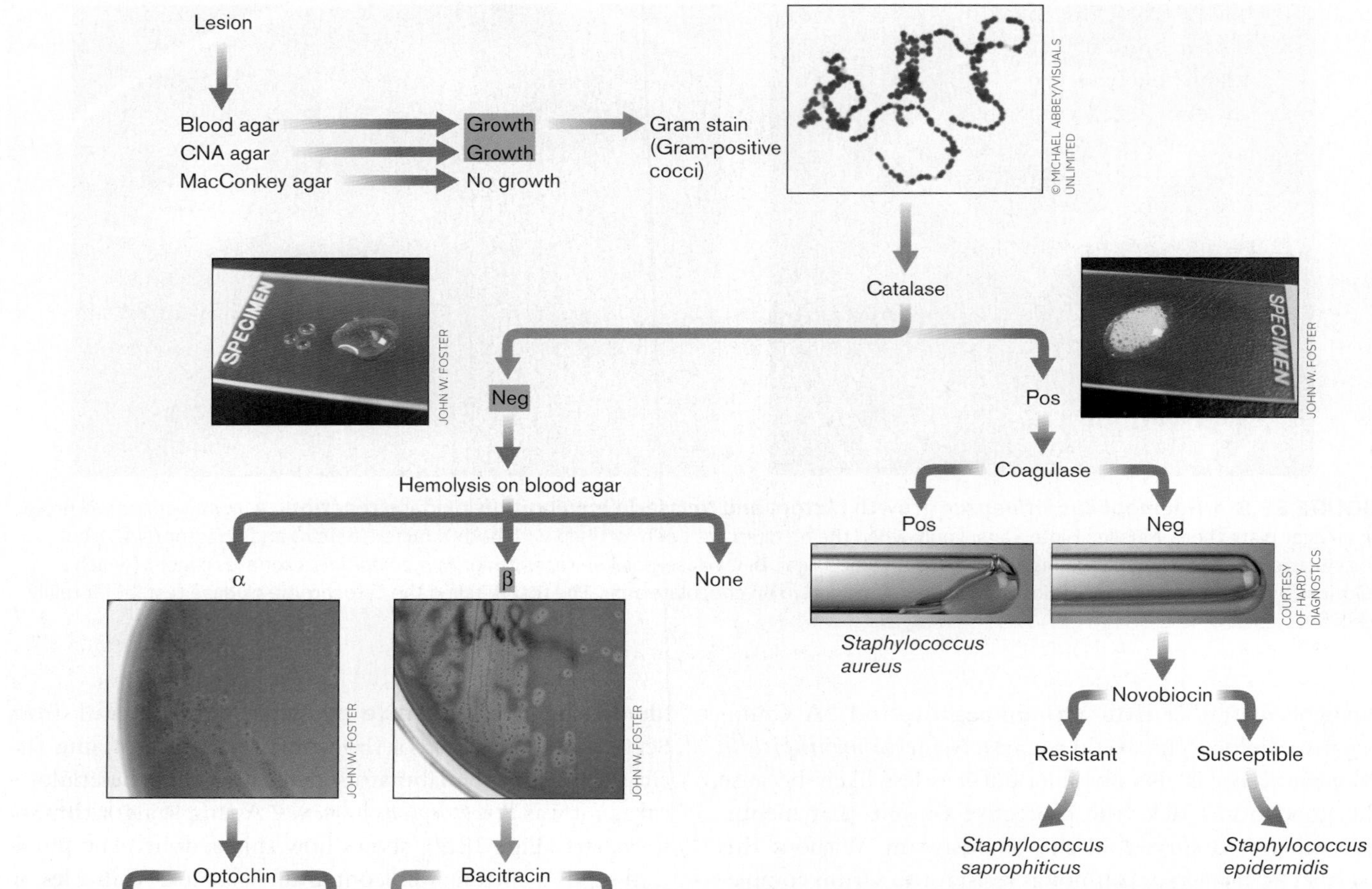
Lesion
Blood agar
CNA agar
MacConkey agar
Growth
Growth
No growth
Gram stain (Gram-positive cocci)
© MICHAEL ABBEY/VISUALS UNLIMITED
Catalase
SPECIMEN
JOHN W. FOSTER
Neg
Pos
SPECIMEN
JOHN W. FOSTER
Hemolysis on blood agar
Coagulase
Pos
Neg
α
β
None
COURTESY OF HARDY DIAGNOSTICS
Staphylococcus aureus
JOHN W. FOSTER
JOHN W. FOSTER
Novobiocin
Resistant
Susceptible
Staphylococcus saprophiticus
Staphylococcus epidermidis

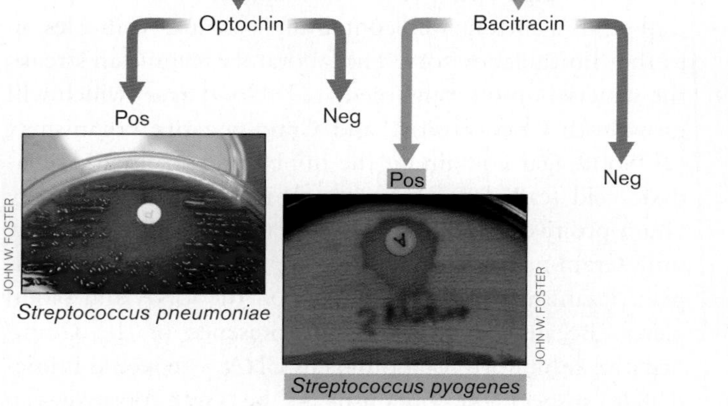
Optochin
Bacitracin
Pos
Neg
Pos
Neg
JOHN W. FOSTER
Streptococcus pneumoniae
JOHN W. FOSTER
Streptococcus pyogenes

FIGURE 28.9 ▪ Algorithm for identifying Gram-positive pathogenic cocci. The red arrows follow the identification of *Streptococcus pyogenes.* The bacitracin and optochin results are designated "positive" if the organism is susceptible and "negative" if the organism is resistant to the agent.

distinguishes staphylococci from streptococci (remember, Gram stain morphology alone is insufficiently reliable to make that distinction). The catalase test is performed by mixing a colony with a drop of H_2O_2 on a glass slide. Effusive bubbling due to the release of oxygen indicates catalase activity (see **Fig. 28.9**). Staphylococci are catalase-positive; streptococci are catalase-negative. Note that many other organisms possess catalase activity, including the Gram-negative rod *E. coli.* According to the algorithm, however, *E. coli* would not be considered, because it does not grow on CNA agar and is not Gram-positive.

Note: When performing a catalase test from colonies grown on blood agar, be sure not to transfer any of the agar, since red blood cells also contain catalase.

Having established that the organism is catalase-negative, the technician examines the blood plate for evidence of hemolysis. Three types of colonies are possible: nonhemolytic, alpha-hemolytic, and beta-hemolytic. Nonhemolytic streptococci do not produce any lytic zone. Alpha-hemolytic strains produce large amounts of hydrogen peroxide that oxidize the heme iron within intact red blood cells to produce a green product. As a result, alpha-hemolytic streptococci produce a green zone around their colonies called alpha hemolysis—even though the red blood cells remain intact. (For example, *Streptococcus mutans,* a cause of dental caries and subacute bacterial endocarditis, is alpha-hemolytic.) Still other streptococci produce a completely clear zone of true hemolysis surrounding their colony. This is called beta hemolysis. Red blood cells are hemolyzed completely by exported enzymes, called hemolysins, that lyse red-cell membranes. The flowchart in **Figure 28.9** indicates that the organism from the case history was beta-hemolytic.

MARINE BIOLOGICAL LABORATORY

FIGURE 28.10 ■ Rebecca Lancefield. In 1918, Dr. Lancefield joined the Rockefeller Institute for Medical Research in New York City, where she studied the hemolytic streptococci, known then as *Streptococcus haemolyticus.* She was the first to use serum precipitation methods to classify *S. haemolyticus* into groups according to differences in cell wall carbohydrate antigens. The basic technique is still used today and is known as the Lancefield classification scheme in her honor.

The final relevant test in this flowchart involves susceptibility to the antibiotic bacitracin, which identifies the most pathogenic group of beta-hemolytic streptococci. The beta-hemolytic streptococci are subdivided into many different groups, known as the Lancefield groups, based on differences in the composition of their carbohydrate peptidoglycans. These cell wall differences are distinguished from each other immunologically and divide the streptococci into Lancefield groups A–U. Rebecca Lancefield (**Fig. 28.10**), for whom the classification scheme is named, was the first to use immunoprecipitation to group the streptococci (immunoprecipitation is described in Section 24.3 and Appendix 3). The vast majority of streptococcal diseases are caused by group A beta-hemolytic streptococci (also called GAS), defined as the species *Streptococcus pyogenes.*

Unfortunately, the Lancefield classification procedure is somewhat time-consuming and not readily amenable as a rapid identification method. However, the group A beta-hemolytic streptococci are uniformly susceptible to the antibiotic bacitracin. Thus, a simple antibiotic disk susceptibility test can be used to indicate GAS (that is, *S. pyogenes*). But beware—many bacteria are bacitracin sensitive, so, like the catalase test, the bacitracin test must be used in conjunction with an algorithm to be useful for identification. The technician must follow the appropriate algorithm before assigning importance to this or any other test result. It is irrelevant, for instance, if an alpha-hemolytic organism is bacitracin susceptible. Some of these may exist, but they are not associated with disease. The organism in our case of necrotizing fasciitis, however, was beta-hemolytic, so bacitracin susceptibility indicated that the organism was *S. pyogenes.*

The other tests named in **Figure 28.9** are equally important for identifying Gram-positive infectious agents. For example, *Streptococcus pneumoniae* is an important cause of pneumonia. Like *S. pyogenes, S. pneumoniae* is a catalase-negative, Gram-positive coccus; but unlike *S. pyogenes,* it is alpha-hemolytic. Optochin susceptibility is a property closely associated with *S. pneumoniae,* while other alpha-hemolytic strains of streptococci are resistant to this compound. Thus, an optochin susceptibility disk test is a useful tool for identifying *S. pneumoniae.*

Coagulase catalyzes a key reaction used to distinguish the pathogen *Staphylococcus aureus,* which causes boils and bone infections, from other staphylococci, such as the normal skin species *Staphylococcus epidermidis.* To conduct a coagulase test, a tube of plasma is inoculated with the suspect organism. If the organism is *S. aureus,* it will secrete the enzyme coagulase. Coagulase will convert fibrinogen to fibrin and produce a clotted, or coagulated, tube of plasma. Coagulase-negative staphylococci can still be medically important, however. *Staphylococcus saprophiticus,* for instance, is an important cause of urinary tract infections. It can be distinguished from *S. epidermidis* by resistance to novobiocin.

Thought Question

28.3 Use **Figure 28.9** to identify the organism from the following case: A sample was taken from a boil located on the arm of a 62-year-old man. Bacteriological examination revealed the presence of Gram-positive cocci that were also catalase-positive, coagulase-positive, and novobiocin resistant.

To Summarize

- **Knowing which patient specimens should be sterile and which should contain normal microbiota** informs the selection of media for processing the specimen. So, too, does understanding whether anaerobes may be expected.
- For some specimens, **direct Gram or acid-fast stains** are appropriate procedures to perform. The results guide the direction of subsequent testing.

- **Growth-dependent pathogen identification** uses numerous, simultaneously run biochemical tests. **An algorithm or a dichotomous key** is applied to the results to identify the species.
- **Different algorithms** (or decision trees) are applied to Gram-positive and Gram-negative bacteria.

28.4

Rapid Techniques for Pathogen Identification

Conventional, petri plate–dependent methods used to identify pathogens take a minimum of 3 days to complete and may take several weeks, depending on the pathogen. *Brucella* species, for instance, can take 14–21 days to grow in blood culture bottles. The wait may be annoying for the physician, but agonizing for the patient awaiting a cure. Today, technologies that offer more rapid identifications, sometimes within minutes, have been introduced into the clinical laboratory.

Nucleic Acid–Based Identifications

Most diagnostic laboratories today use rapid DNA-based methods in addition to conventional petri dish microbiology. The DNA-based methods take mere hours to detect, and type, bacteria and viruses. DNA/RNA detection methods are especially useful for viruses, which otherwise require elaborate electron microscopy to view morphology, or serology to detect an increased presence of antiviral antibodies. The problem with serology is that by the time these antibodies become detectable in blood, the patient is usually already recovering from the disease.

The polymerase chain reaction (PCR). PCR is the most widely used molecular method in the clinical laboratory's diagnostic toolbox (see Section 7.6, Fig. 7.28). DNA primers that bind to unique genes in a pathogen's genome can be used to specifically amplify DNA or RNA present in a clinical or environmental specimen thought to harbor a pathogen. Successful amplification is visualized as an appropriately sized DNA fragment (**amplicon**) in agarose gels following electrophoresis.

Why is PCR needed to detect the presence of these nucleic acids? Without the PCR amplification steps, clinical samples usually provide too little nucleic acid from infecting microorganisms to be detected. For example, sputum samples containing *Mycobacterium tuberculosis*, the cause of tuberculosis, yield minuscule amounts of bacterial DNA. Amplifying *M. tuberculosis* nucleic acid by PCR, however, will turn one copy of DNA into billions of copies.

PCR is very quick. Preparing a clinical specimen for PCR by extracting the DNA or RNA usually takes less than an hour. PCR itself is completed in 1–2 hours. Detection of the PCR-amplified DNA by DNA gel electrophoresis takes an additional 1–2 hours. So, what might take 2–3 days (or sometimes weeks) using biochemical algorithms may take less than a day using PCR.

The table in **eTopic 28.2** lists several instances in which DNA detection tests are useful. A specific example presented in **Figure 28.11** illustrates the use of PCR to type different strains of the anaerobic pathogen *Clostridium botulinum*, the cause of food-borne botulism. These organisms are not typed serologically, as is the case for *Streptococcus pneumoniae*. *C. botulinum* is divided into different types based on the neurotoxin genes they possess. In this example, **multiplex PCR** was used to simultaneously search for these toxin genes. Multiplex PCR uses multiple sets of primers, one pair for each gene, combined in a single tube with a specimen. Care must be taken to be sure that the primers chosen make different-sized products, do not interfere with each other, and do not produce artifactual products that can confuse interpretation. Multiplex PCR

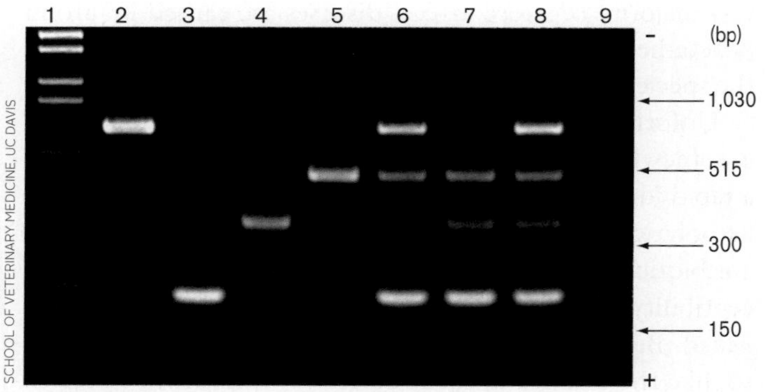
1
2
3
4
5
6
7
8
9
-
(bp)
1,030
515
300
150
+
SCHOOL OF VETERINARY MEDICINE, UC DAVIS

FIGURE 28.11 ■ Multiplex PCR identification of *Clostridium botulinum*. *C. botulinum* cells are typed by which toxin genes a strain possesses, not on the basis of surface antigens. Isolates can be typed in a single PCR reaction that includes primer pairs specific for each of the four major toxin genes: types A, B, E, and F. Because multiple products are sought in a single reaction, this is called multiplex PCR. Each lane in the agarose gel was loaded with multiplex products from different isolates of *C. botulinum* and subjected to electrophoresis. The slower-moving fragments (toward top of gel) are larger than those moving farther down the gel toward the positive pole. Lane 1, DNA size markers; lane 2, type A (*cntA*); lane 3, type B (*cntB*); lane 4, type E (*cntE*); lane 5, type F (*cntF*); lane 6, types A, B, and F; lane 7, types B, E, and F; lane 8, types A, B, E, and F.

can help identify sets of specific genes present in a single species or can screen for the presence of multiple pathogens in a clinical sample. In the latter, primer sets are designed to amplify genes unique to each pathogen.

Reverse transcriptase PCR (RT-PCR). Although you might think PCR detection of pathogens is limited to DNA, the method can also be used to detect microbial RNA in patient samples.

Case History: West Nile Virus

A 55-year-old man was admitted to a local hospital complaining of headache, high fever, and neck stiffness. The man appeared confused and disoriented. He also complained of muscle weakness. History indicated he had received several mosquito bites approximately 2 weeks previously. A blood specimen was sent to the laboratory. The report the following day indicated that the patient was suffering from West Nile virus.

West Nile virus is primarily an infection of birds and culicine mosquitoes (a group of mosquitoes that can transmit human diseases), with humans and horses serving as incidental, dead-end hosts. Replication of virus in this bird-mosquito-bird cycle begins when adult mosquitoes emerge in early spring and continues until fall. Among humans, the incidence of disease peaks in late summer and early fall. Birds provide an efficient means of geographic spread of the virus. As a result, over the past decade the virus has spread throughout much of the United States.

Isolating a disease-causing virus is extremely challenging. Most laboratories are not equipped for the special tissue culture techniques required to grow viruses. Consequently, most viral infections, including human West Nile virus infections, are usually diagnosed by measuring the antibody response of the patient. For instance, the presence of West Nile virus–specific IgM in cerebrospinal fluid is a good indicator of current West Nile virus infection, but it is indirect and inconclusive. Real-time PCR is a molecular test that can quickly reveal the presence of the virus itself.

Real time quantitative reverse transcriptase PCR (qRT-PCR) is used routinely for the high-throughput diagnosis of many viral pathogens, including the West Nile virus. Because West Nile virus (*Flaviviridae* family) contains single-stranded RNA, its RNA must be converted to DNA using reverse transcriptase before PCR can be attempted. The quantitative advantage of qRT-PCR is that you can estimate the number of virus particles present in the sample by the number of viral RNA molecules there.

The basic technique is as follows: RNA is first extracted from the sample. A sub-sequence of viral RNA is then converted to complementary DNA (cDNA) using a single primer and reverse transcriptase (**Fig. 28.12**). The more virus particles there are in the sample (viral load), the more viral RNA will be present and the more cDNA product will be made. The cDNA is then amplified by PCR using two specific primers and a heat-stable DNA polymerase such as Taq polymerase (see Section 12.3; Fig. 12.11). The trick for quantifying is in the method used to detect amplification. In one method, a third, fluorescent oligonucleotide (called the probe) is added to the PCR reaction (see Fig. 12.11A). The probe contains a fluorescent dye at the 3′ end and a chemical dye at the 5′ end that quenches (absorbs) energy emitted from the fluorescent dye. As long as the two chemicals are kept in close proximity by the intact probe, no light is emitted. The probe is designed to anneal to a sequence between the binding sites of the two other primers (modifications on the ends of the probe prevent it from being used as a primer). So, in a successful amplification, Taq polymerase will synthesize DNA from the two outside primers and degrade the probe oligonucleotide as it passes through that area. This cleavage separates the dye from the quencher, and the dye begins to fluoresce. The greater the amount of cDNA there was to begin with, the fewer cycles

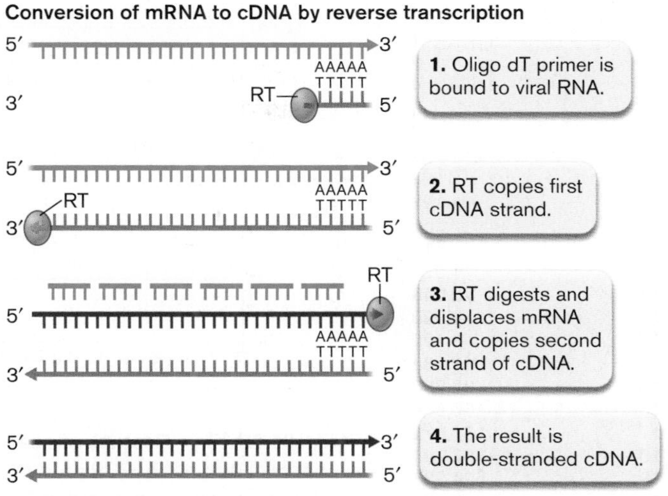
Conversion of mRNA to cDNA by reverse transcription
5′
3′
AAAAA
TTTTT
RT
3′
5′
1. Oligo dT primer is bound to viral RNA.
2. RT copies first cDNA strand.
3. RT digests and displaces mRNA and copies second strand of cDNA.
4. The result is double-stranded cDNA.

FIGURE 28.12 ▪ Identifying viruses by reverse transcriptase PCR. Reverse transcriptase (RT) uses viral RNA as a template to make DNA. The DNA can then be amplified using standard PCR methods. Molecules of eukaryotic and viral RNA usually contain poly-A tails at their 3′ ends. An oligonucleotide poly-T primer added to the reaction tube will anneal to the poly-A tail and allow RT to synthesize the first strand of a complementary DNA (cDNA). A second primer specific to the viral gene is then used to prime RT synthesis of the opposite strand (second-strand synthesis), using the first DNA strand as the template. Subsequent amplification by PCR requires the addition of a thermostable DNA polymerase (Taq) that can withstand the denaturing and DNA synthesis temperatures required for amplifying the cDNA.

A.

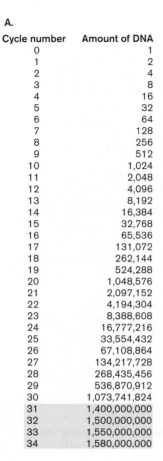

Cycle number	Amount of DNA
0	1
1	2
2	4
3	8
4	16
5	32
6	64
7	128
8	256
9	512
10	1,024
11	2,048
12	4,096
13	8,192
14	16,384
15	32,768
16	65,536
17	131,072
18	262,144
19	524,288
20	1,048,576
21	2,097,152
22	4,194,304
23	8,388,608
24	16,777,216
25	33,554,432
26	67,108,864
27	134,217,728
28	268,435,456
29	536,870,912
30	1,073,741,824
31	1,400,000,000
32	1,500,000,000
33	1,550,000,000
34	1,580,000,000

B.

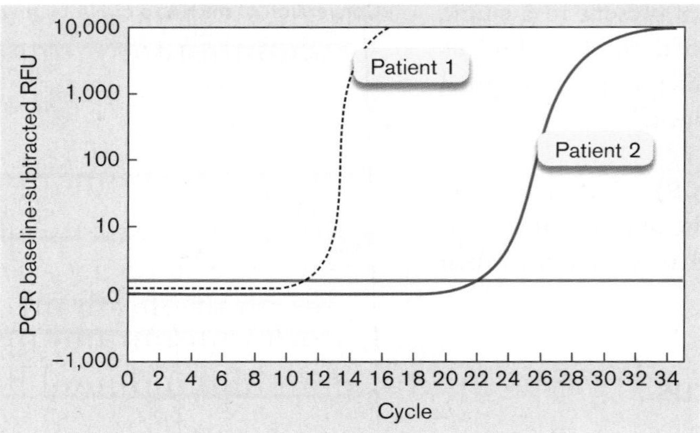

PCR baseline-subtracted RFU
10,000
1,000
100
10
0
−1,000
Patient 1
Patient 2
0 2 4 6 8 10 12 14 16 18 20 22 24 26 28 30 32 34
Cycle

FIGURE 28.13 ▪ Results of real-time PCR. A. The exponential increase in PCR products after each cycle of hybridization and polymerization. The switch to yellow indicates the point where the increase in product plateaus because the primers have been exhausted. **B.** There is an increase in relative fluorescence units (RFUs) as the fluorescent dye is released from the dual-labeled probe during real-time PCR. In patient 1, the dashed curve shows that fluorescent PCR product remained flat for the first 10 cycles because the amount of DNA made, and therefore the level of fluorescent dye released, remained below background level (red line). After cycle 10, fluorescence increased logarithmically. For patient 2, the solid blue curve representing PCR product remained below detection for 22 cycles, and then increased. The more starting DNA there is, the sooner RFU values will increase over background (that is, the fewer cycles will be needed to see the increase over background). The slope eventually decreases because the fluorescent probe has become limiting.

it takes to register a fluorescence increase over background (**Fig. 28.13**).

Weblinks Molecular-beacon PCR (*see ebook*)

Thought Questions

28.4 If the results in **Figure 28.13** were testing for HIV RNA, then which patient would have the higher viral load in their blood?

28.5 Why does finding IgM to West Nile virus indicate current infection? Why wouldn't finding IgG do the same?

Rapid DNA sequencing. The future of nucleic acid–based diagnostic microbiology is a new technology that combines PCR with mass spectrometry (MS). PCR can amplify DNA directly from clinical samples using primers that target conserved bacterial sequences (for instance, 16S rRNA genes). MS then analyzes the molecular weight of the "amplicon" and sequences it. As described for proteins in Appendix 3, sequencing involves electron spray ionization to progressively shorten the PCR product one base at a time, and mass spectrometry to calculate the weights of each new, shortened fragment. The difference in molecular weights between the initial DNA fragment and the progressively shortened ones is used to determine the identity of each base removed and, thus, the sequence of the fragment. Eventually, this technology could replace the time-honored petri dish.

Identifications Based on Serology

In addition to directly identifying a pathogen by isolation or by PCR detection, evidence for a pathogen can be found in a patient's serum, as seen in the following case history.

Case History: Ebola

Between October and November 2000, 62 residents of a small village north of Gulu, a town in Uganda, became ill with high fever, diarrhea, headache, vomiting, and gastrointestinal bleeding from the rectum. As many as 36 patients died. By January, 425 cases had been reported, of whom 224 died. The presumptive diagnosis was Ebola. Laboratory confirmation tests included viral antigen detection and antibody ELISA tests. Laboratory-confirmed Ebola patients were defined as patients who were positive for either Ebola virus antigen or Ebola IgG antibody. Once identified, rigorous quarantine mechanisms were implemented to limit spread of the disease to other villages.

What are these tests?

The ELISA test. Enzyme-linked immunosorbent assay (ELISA) can detect antigens or antibodies present in picogram quantities. One form of ELISA detects serum antibodies. It is carried out in a 96-well microtiter plate, which allows multiple patient serum samples to be tested simultaneously. An antigen from the virus (Ebola in this case) is attached (adsorbed) to the plastic of the wells (**Fig. 28.14**). Albumin or powdered milk is used to block the remaining sites on the plastic that could result in false positives. Patient serum is then added. Ebola-specific antibodies present in the serum

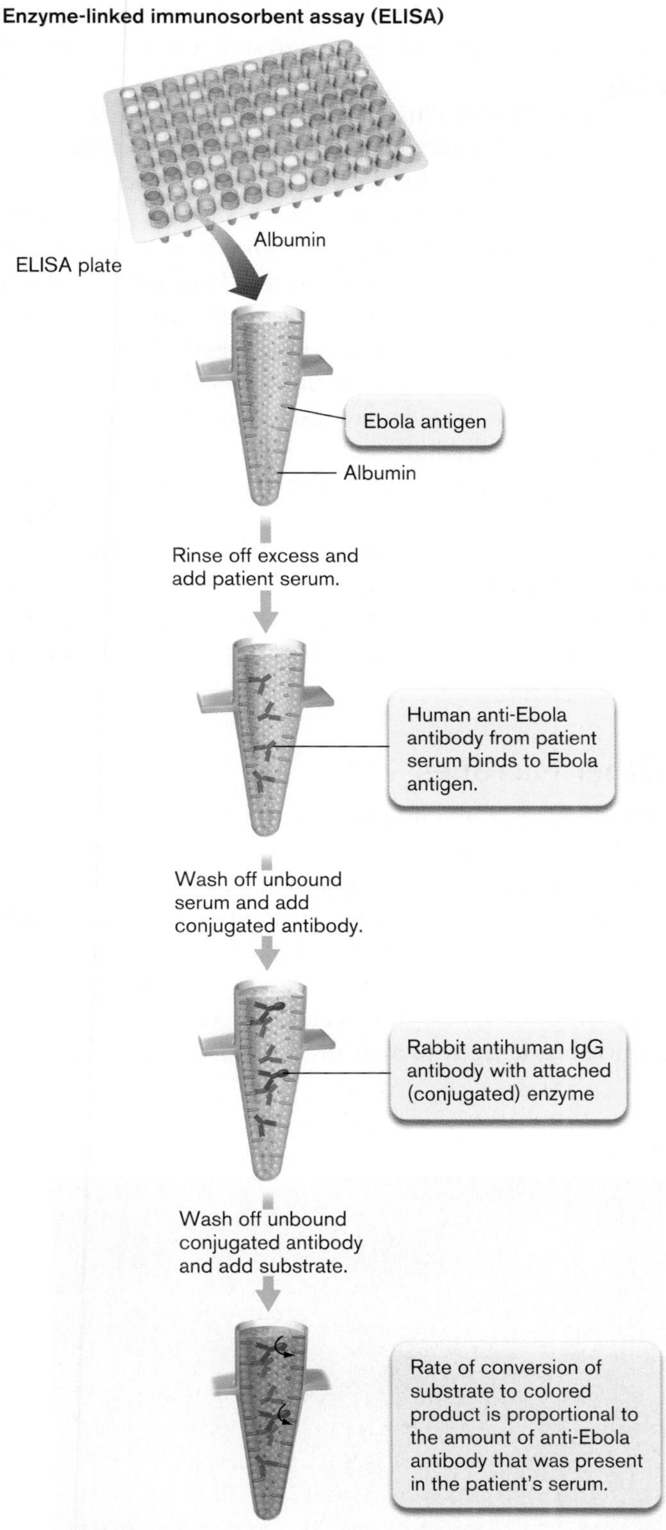

Enzyme-linked immunosorbent assay (ELISA)
ELISA plate
Albumin
Ebola antigen
Albumin
Rinse off excess and add patient serum.
Human anti-Ebola antibody from patient serum binds to Ebola antigen.
Wash off unbound serum and add conjugated antibody.
Rabbit antihuman IgG antibody with attached (conjugated) enzyme
Wash off unbound conjugated antibody and add substrate.
Rate of conversion of substrate to colored product is proportional to the amount of anti-Ebola antibody that was present in the patient's serum.

FIGURE 28.14 ▪ Enzyme-linked immunosorbent assay (ELISA). ELISA to detect anti-Ebola antibodies circulating in patient serum. The 96-well plate can be used to make dilutions of a single patient's serum to more precisely determine the amount of anti-Ebola antibody, or it can be used to test samples from multiple patients.

will react with the antigen attached to the microtiter plate. The antigen-antibody complex is then reacted with goat-antihuman IgG to which an enzyme has been attached, or conjugated (for example, horseradish peroxidase). This forms a chain of viral antigen connected to patient antibody connected to goat antibody-enzyme that links the enzyme to the well. The chromogenic substrate for the enzyme is added next (for example, tetramethylbenzidine). If enzyme-conjugated antibody has bound to human IgG captured by the antigen in the well, the enzyme will convert the substrate to a colored product (blue for tetramethylbenzidine). Enzyme activity can be measured with an ELISA plate reader. The amount of colored product formed, detected as absorbance with a spectrophotometer, will be an indication of the amount of anti-Ebola antibody present in the patient sample.

Thought Question

28.6 Why does adding albumin or powdered milk prevent false positives in ELISA?

Antigen capture is another ELISA technique, but in this instance, anti-Ebola antibody, not viral antigen, is adsorbed to the wells of a microtiter plate (**Fig. 28.15**). Patient serum is then added to the wells. If the serum contains Ebola antigen, the antigen will be captured by the antibody in the well. Then a second, enzyme-conjugated antibody against the Ebola antigen is added. The more antigen there is in the serum, the more enzyme-linked antibody will affix to the well. Addition of the appropriate chromogenic substrate will produce a colored product that can be measured.

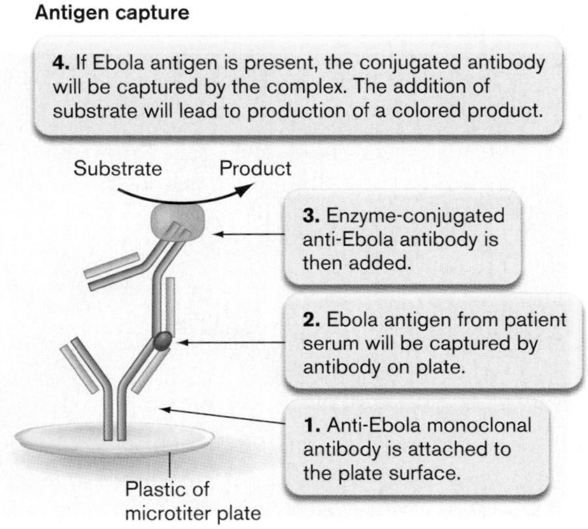

Antigen capture
4. If Ebola antigen is present, the conjugated antibody will be captured by the complex. The addition of substrate will lead to production of a colored product.
Substrate
Product
3. Enzyme-conjugated anti-Ebola antibody is then added.
2. Ebola antigen from patient serum will be captured by antibody on plate.
1. Anti-Ebola monoclonal antibody is attached to the plate surface.
Plastic of microtiter plate

FIGURE 28.15 ▪ Antigen-capture ELISA. This ELISA technique captures Ebola antigens circulating in patient serum.

Antibody against Ebola may be easier to detect than viral antigen because antibodies will be present at higher levels than the virus itself. But because there is a delay between the time when the virus is first present in serum and when the body manages to make antibody, a speedier diagnosis can be made by directly detecting viral antigen.

While bacterial infections are commonly diagnosed by growing the infecting organism on artificial medium in the clinical laboratory, viral diseases are usually diagnosed by immunological means. Viruses are more difficult to grow than bacteria and do not exhibit the biochemical diversity so useful for identifying different bacterial species. In addition to immunological tests, viral diseases can be diagnosed using molecular approaches to identify viral DNA or RNA sequences (discussed earlier).

Thought Question

28.7 Specific antibodies against an infectious agent can persist for years in the bloodstream, long after the infection resolves. How is it possible, then, that antibody titers can be used to diagnose recently acquired diseases such as infectious mononucleosis? Couldn't the antibody be from an old infection?

Fluorescent Antibody Staining

Chapter 26 presents a case history involving an 80-year-old nursing-home resident who contracted pneumonia caused by *Streptococcus pneumoniae*. The laboratory diagnosis was probably made using the biochemical algorithm previously described. However, there are over 80 serological types of *S. pneumoniae*, each one containing a different capsular antigen. How can the lab identify which antigenic type has caused the infection? One way is to stain the organism with antibodies.

Figure 28.16A illustrates the result of staining a smear of the isolated streptococcus with fluorescently tagged antibodies directed against a specific antigenic type of capsule. Viewed under a fluorescence microscope, the organism is "painted" green when the right antibody binds to the capsule. In the case of pneumonia, this knowledge probably will not help in treatment of the individual patient, but its broader value is in determining whether a single type of organism is the cause of an outbreak of pneumonia, which in turn is of epidemiological value for identifying the source of the bacterium.

On the other hand, fluorescent antibody staining techniques are critically important for rapidly identifying organisms that are difficult to grow. Infected tissues can be subjected to direct fluorescent antibody staining. **Figure 28.16B**, for example, shows a direct fluorescent antibody stain of pleural fluid from a patient with Legionnaires' disease.

Other Microbes

Space does not permit a complete listing of the methods used to identify microbial pathogens, but **Table 28.2** presents some additional examples. In general, bacteria that are easily cultured are grown in the laboratory, after which biochemical tests are performed. Bacterial, viral, and fungal species that are difficult to grow are typically identified using immunological techniques that either identify a microbial antigen present in infected tissues or measure

A. B.

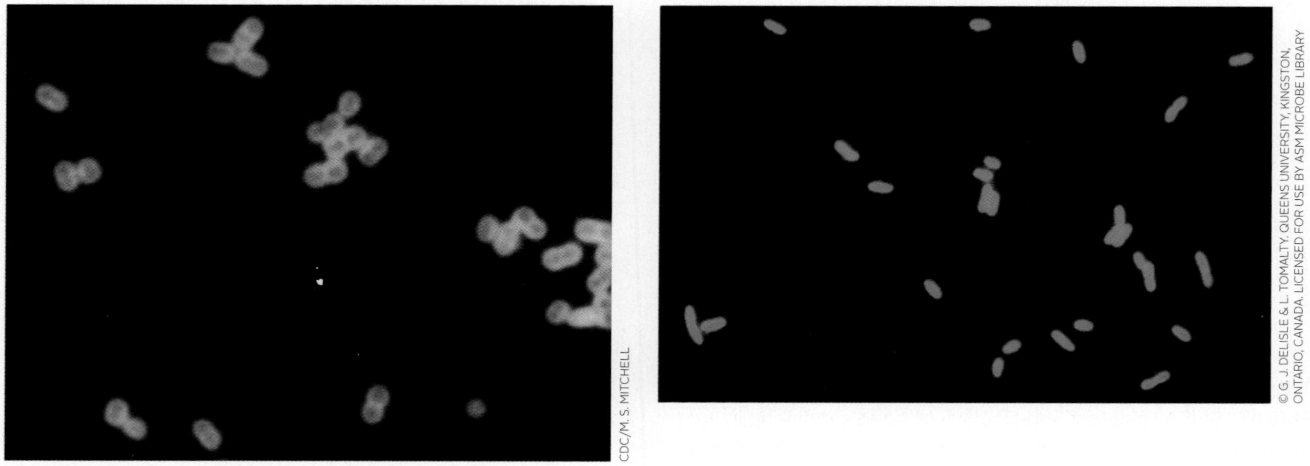
CDC/M. S. MITCHELL
© G. J. DELISLE & L. TOMALTY, QUEENS UNIVERSITY, KINGSTON, ONTARIO, CANADA. LICENSED FOR USE BY ASM MICROBE LIBRARY

FIGURE 28.16 ■ Fluorescent antibody stain. A. *Streptococcus pneumoniae* capsule (cells approx. 0.8 μm, fluorescence microscopy). The capsule is the green halo. The center of the halo is the cell. **B.** *Legionella pneumophila* (approx. 1 μm in length) from a respiratory tract specimen.

TABLE 28.2

Identification procedures for selected diseases.

Agent	Disease	Means of identification
Bacteria		
Corynebacterium diphtheriae	Diphtheria	Material from nose and throat cultured on a special medium; in vivo or in vitro tests for toxin
Bordetella pertussis	Pertussis (whooping cough)	Smears of nasopharyngeal secretions stained with fluorescent antibody; ELISA for toxin in respiratory secretions; culture on special media
Legionella pneumophila	Legionellosis, Legionnaires' disease, Pontiac fever	Culture on special medium is preferred method; antigen detection by ELISA; *Legionella* nucleic acid can be identified in clinical material using PCR and nucleic acid probes
Campylobacter spp.	Campylobacteriosis	Isolation of bacteria on selective media incubated at 42°C in atmosphere of nitrogen containing 5% oxygen and 10% carbon dioxide; then, biochemical testing performed
Leptospira interrogans	Leptospirosis	Serological tests early in the illness and after 2–3 weeks to detect rise in antibody titer
Listeria monocytogenes	Listeriosis	Culture of blood and spinal fluid; selective and enrichment cultures performed on food samples to grow potential pathogens; DNA probe for rapid identification of colonies
Chlamydia trachomatis	Chlamydial genital infections	Identification of *C. trachomatis* antigen in urine or pus using monoclonal antibody; nucleic acid probes
Treponema pallidum	Syphilis	Direct fluorescent antibody staining; serological tests
Francisella tularensis	Tularemia	Cultures using cysteine-containing media; fluorescent antibody stain of pus; detection of rise in antibody titer
Yersinia pestis	Plague	Identification of capsular antigen using fluorescent antibody or ELISA
Viruses		
Rhinovirus	Common cold	Strain identification requires use of specific antibodies
Influenza virus	Flu	Tests comparing influenza antibody levels in blood samples taken during acute and convalescent stages of illness
Hantavirus	Hantavirus pulmonary syndrome	Antigen detection in tissues using electron microscopy or monoclonal antibody; ELISA and western blot tests for IgG and IgM antibodies in victim's blood
Herpes simplex virus	Different strains cause cold sores, ocular lesions, and genital lesions	Identifying the viral antigen in clinical material using fluorescent antibody or DNA probes
Mumps virus	Mumps	Rise in antibody titer, or presence of IgM antibody to mumps virus in patient's blood
Rotavirus	Diarrhea	Electron microscopy or ELISA of diarrheal stool for virus
HIV	AIDS	Detection of antibody to HIV-1 in patient's blood
Fungi		
Coccidioides immitis	Coccidioidomycosis, infection of lung; can disseminate to almost any tissue	Observation of large, thick-walled, round spherules from clinical specimens; PCR identification
Histoplasma capsulatum	Histoplasmosis, intracellular infection of lung; sometimes disseminates	Stained material from pus, sputum, tissue, etc., examined for intracellular *H. capsulatum* yeast phase; blood tests for antibody to the organism

a rise in antibody titer. Nucleic acid–based methodologies are also used to identify organisms that are difficult to grow. Eukaryotic microbial parasites such as *Plasmodium* species (the cause of malaria), *Giardia lamblia* (which causes giardiasis, a diarrheal disease), and *Entamoeba histolytica* (the cause of amebic dysentery) can be identified via their telltale morphologies under the microscope, making biochemical tests unnecessary.

To Summarize

- **Selective media** are used to inhibit growth of one group of organisms while permitting growth of others (such as Gram-positive bacteria versus Gram-negative bacteria). This technique is often used to prevent growth of normal microbiota while permitting growth of pathogens.
- **Differential media** exploit the unique biochemical properties of a pathogen to distinguish it from similar-looking nonpathogens.
- **Bacterial species can be identified** with biochemical analyses, molecular techniques (for example, PCR), and/or immunological methods (for example, ELISA).
- **Viral diseases are often diagnosed using immunological tests**, such as ELISA, that measure the presence of antibody or antigen, or by real-time quantitative PCR.
- **Fluorescent antibody staining** can rapidly identify organisms or antigens present in tissues.

TABLE 28.3

Examples of point-of-care rapid test kits for infectious diseases.[a]

Disease (pathogen)	Type of test	Sample	Indication	Performance[b]	Notes
Bacterial pathogens					
Chlamydia (*Chlamydia trachomatis*)	PCR	Vaginal swab, urine	Screening, suspicion of PID	Sensitivity: 83% Specificity: 99%	
Gonorrhea (*Neisseria gonorrhoeae*)	PCR	Urine	Screening, suspicion of PID	Sensitivity: 98% Specificity: 99%	
Syphilis (*Treponema pallidum*)	ICT	Blood	Screening	Sensitivity: 90%–95% Specificity: 90%–95%	
Strep throat (*Streptococcus pyogenes*)	EIA	Pharyngeal swab	Sore throat	Sensitivity: 53%–99% Specificity: 62%–100%	Confirmation of negative swabs
Legionellosis (*Legionella* spp.)	ICT	Urine	Severe pneumonia; risk factors for legionellosis	Sensitivity: 76% Specificity: 99%	Only serotype 1 reliably detected
Pneumococcal pneumonia (*Streptococcus pneumoniae*)	ICT	Urine; also pleural fluid or CSF	Severe pneumonia; also empyema, meningitis	Sensitivity: 66%–70% Specificity: 90%–100%	Detects capsule antigens
Pseudomembranous enterocolitis (*Clostridium difficile*)	ICT	Stool	Antibiotic-associated diarrhea	Sensitivity: 49%–80% Specificity: 95%–96%	Notably less sensitive than cultures or PCR
Neonatal septicemia (*Streptococcus agalactiae*)	PCR	Vaginal swab	Peripartum detection of colonization	Sensitivity: 92% Specificity: 96%	
Protozoan pathogens					
Malaria (*Plasmodium falciparum*)	ICT	Blood	Fever in returning traveler	Sensitivity: 87%–100% Specificity: 52%–100%	Sensitivity better for *Plasmodium falciparum* (pan-malarial tests)
Trichomoniasis (*Trichomonas* spp.)	ICT	Vaginal swab	Symptoms of vaginitis	Sensitivity: 83% Specificity: 98%	

28.5

Point-of-Care Rapid Diagnostics

Conventional diagnosis of an infection often requires sending a clinical specimen to a faraway laboratory followed by a considerable delay in obtaining results. Inevitably, some patients lose patience and fail to attend follow-up appointments. However, point-of-care (POC) laboratory tests are designed to be used directly at the site of patient care, such as physicians' offices, outpatient clinics, intensive care units, emergency departments, hospital laboratories, and even patients' homes. Most patients are happy to wait 40–50 minutes for a rapid POC test result in order to receive immediate treatment or reassurance.

Commercial POC tests are widely available for the diagnosis of bacterial and viral infections and for parasitic diseases including malaria (**Table 28.3**). However, as convenient as these tests are, sensitivity may be compromised in the quest for a speedy result. Some tests exhibit insufficient sensitivity and should therefore be coupled with confirmatory tests when the results are negative (one test that can produce false negatives is the *Streptococcus pyogenes* rapid antigen detection test). Other POC tests need to be confirmed when positive; for instance, a rapid malaria POC test can produce false positives.

Table 28.3 ***continued***

Disease (pathogen)	Type of test	Sample	Indication	Performance[b]	Notes
Viral pathogens					
Influenza (influenza virus)	ICT	Nasopharyngeal	Flu-like symptoms	Sensitivity: 20%–55% Specificity: 99%	Low sensitivity; probably not helpful during outbreaks; lower in adults
RSV disease (RSV)	ICT	Nasopharyngeal swab		Viral symptoms, especially during winter	Sensitivity: 59%–97% Specificity: 75%–100%
HIV	ICT	Blood; also oral fluid		Screening, prevention of vertical transmission	Sensitivity: 99%–100% Specificity: 99%–100%
Dengue fever (dengue virus)	ICT	Blood	Screening in endemic regions	Sensitivity: 90% Specificity: 100%	
Mononucleosis (Epstein-Barr virus)	ICT	Blood	Screening	Sensitivity: 90% Specificity: 100%	Detects IgM "heterophile antibodies"[c]
Diarrheal disease (rotavirus)	ICT	Stool	Diarrhea	Sensitivity: 88% Specificity: 99%	
Hepatitis B (HBV)	ICT	Blood	Prenatal or transfusion screening; or suspicion of acute or chronic carriage of HBV	Sensitivity: 99% Specificity: 100%	Detects HBs antigen
Rubella (rubella virus)	ICT	Blood	Pregnancy	Sensitivity: 99% Specificity: 99%	

[a]Abbreviations: AIDS, acquired immunodeficiency syndrome; CSF, cerebrospinal fluid; EIA, enzyme immunoassay; HBV, hepatitis B virus; HIV, human immunodeficiency virus; ICT, immunochromatographic test; PCR, polymerase chain reaction; PID, pelvic inflammatory disease; RSV, respiratory syncytial virus.

[b]Sensitivity = proportion of actual positives correctly identified. Specificity = proportion of actual negatives correctly identified.

[c]Epstein-Barr virus randomly infects B cells and causes them to secrete antibodies. Because many thousands of different antibodies are made, they are referred to collectively as "heterophile antibodies."

A.

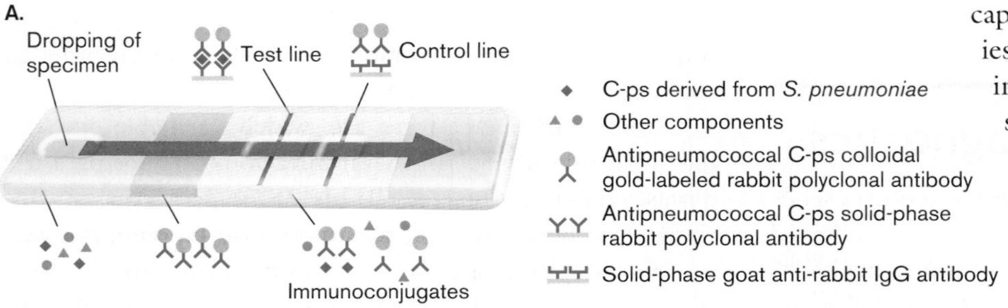
Dropping of specimen
Test line
Control line
Immunoconjugates
C-ps derived from S. pneumoniae
Other components
Antipneumococcal C-ps colloidal gold-labeled rabbit polyclonal antibody
Antipneumococcal C-ps solid-phase rabbit polyclonal antibody
Solid-phase goat anti-rabbit IgG antibody

B.

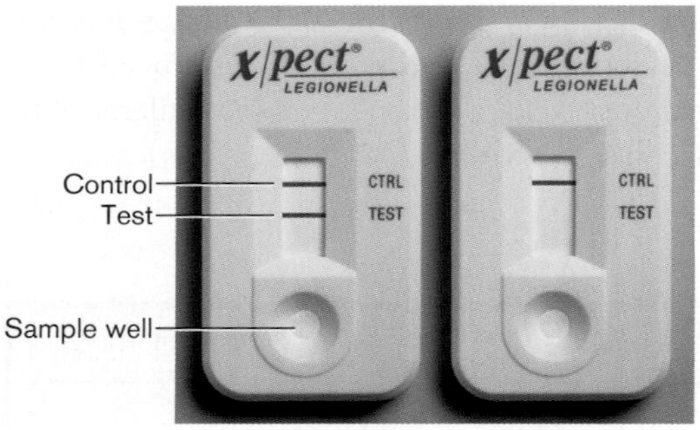
Control
Test
Sample well
x/pect
LEGIONELLA
CTRL
TEST

FIGURE 28.17 ■ Principle of immunochromatographic rapid diagnostic tests. A. The example is a test for the presence of *Streptococcus pneumoniae* in sputum. An extract of sputum is placed onto one end of the strip where *S. pneumoniae* capsular antigen (C-ps) will bind antipneumococcal C-ps polyclonal antibodies. The resulting immunoconjugates move by capillary action to the upper membrane and are captured by the antipneumococcal C-ps solid-phase polyclonal antibodies, thereby forming sandwich conjugates in the sample. A positive test result is indicated by the presence of both a test and a control line, whereas a negative result is indicated by the appearance of only a single control line. **B.** Immunochromatography test for anti–*Treponema pallidum* antibodies. Serum placed on strip 1 contained antibodies to *T. pallidum.*

The typical POC test involves an immunochromatographic assay. The test for *Streptococcus pneumoniae* capsular antigen shown in **Figure 28.17A** is one example. This particular immunochromatographic test, called a "red colloidal gold" test, involves extracting the relevant antigen (for example, capsule antigen) from a clinical specimen and placing a few drops of the extract on a test strip containing rabbit antibodies to the antigen. The antibodies are attached to gold particles in a solution (red colloidal gold). The antigen-antibody complexes, if present, move by capillary action to the upper level of the strip, where they are captured by a line of more anti-antigen antibodies embedded in the strip, forming an antibody-antigen-antibody sandwich. The colloidal gold particles accumulate and eventually produce a red test line, indicating the presence of the antigen. In contrast, rabbit antibodies that are not bound to the antigen pass through the test line but are captured by goat-antirabbit IgG antibodies on a control line, once again forming a red control line that indicates the strip components are working. **Figure 28.17B** shows results of a test for *Treponema pallidum* antibodies.

Weblinks Example of rapid strep test (*see ebook*)

There are several advantages and some disadvantages to POC rapid tests. The advantages are:

- Culturing is not required.
- The clinician can immediately initiate specific antibiotic therapy.
- The consumption of antibiotics is avoided in the case of a viral infection.
- Infection chains among patients with similar symptoms are revealed.
- Compliance with patients who are difficult to reach is improved.

The disadvantages include:

- The tests provide no data on pathogen antibiotic sensitivity.
- There is a higher risk of the technician becoming infected.
- Double or multiple infections are more likely to be overlooked than in culture.
- The levels of false positive or false negative results can vary for different POC tests.

The newer nucleic acid–based tests described earlier exhibit better sensitivity and specificity than most immunochromatographic assays, but they require more expensive instrumentation and training. In the coming years, further evolution of POC tests may lead to new diagnostic approaches, such as panel testing that targets all possible pathogens suspected in a specific clinical setting. The development of next-generation serology-based and/or molecular-based multiplex tests will certainly facilitate quicker diagnosis and improved patient care.

To Summarize

- **Point-of-care (POC) diagnostic tests** can rapidly identify the cause of an infectious disease.
- **Immunochromatography** is the primary platform for POC testing.
- **Therapy** can be initiated rapidly following a positive result.
- **A drawback** is that simultaneous, multiple infections may be missed.

28.6

Biosafety Containment Procedures

Medical and laboratory personnel are exposed to extremely dangerous pathogens on a daily basis. When working with dangerous pathogens, clinical microbiologists must protect themselves from accidental infection and at the same time be certain the pathogen does not escape from the lab.

Case History: Fatal Meningitis

On July 15, an Alabama microbiologist was taken to the emergency room with acute onset of generalized malaise, fever, and diffuse myalgias. She was given a prescription for oral antibiotics and released. On July 16, she became tachycardic and hypotensive and returned to the hospital. She died 3 hours later. Blood cultures were positive for Neisseria meningitidis *serogroup C. Three days before the onset of symptoms, the microbiologist had prepared a Gram stain from the blood culture of a patient subsequently shown to have meningococcal disease; she had also handled agar plates containing cerebrospinal fluid (CSF) cultures from the same patient. Co-workers reported that fluids were aspirated from blood culture bottles at the open laboratory bench. No biosafety cabinets, eye protection, or masks were used for this procedure. Testing at CDC indicated that the isolates from both patients were indistinguishable. The laboratory at the hospital infrequently processed isolates of* N. meningitidis *and had not processed another meningococcal isolate during the previous 4 years.*

The microbiologist in this case did not take appropriate measures to protect herself and ended up with a laboratory-acquired infection leading to meningitis. The CDC has published a series of regulations designed to protect workers at risk of infection by human pathogens. Infectious agents are ranked by the severity of disease and ease of transmission. The more severe the disease or the more easily it is transmitted the higher the risk category. On the basis of this ranking, four levels of biological containment are employed (**Table 28.4**).

Risk group I organisms have little to no pathogenic potential and require the lowest level of containment (biosafety level 1). Standard sterile techniques and laboratory practices are sufficient. Risk group II agents have greater pathogenic potential, but vaccines and/or therapeutic treatments (for example, antibiotics) are readily available. These agents require more rigorous containment procedures, such as limiting laboratory access when experiments are in progress and using biological laminar flow cabinets if aerosolization is possible (biosafety level 2). Risk group III pathogens produce a serious or lethal human disease. Vaccines or therapeutic agents may be available. To safely handle these organisms, level 2 procedures are supplemented with a lab design ensuring that ventilation air flows only into the room and that exhaust air vents directly to the outside, thus creating negative pressure. Negative pressure will keep any organism that may aerosolize from escaping into hallways. In addition, access to the lab is strictly regulated and includes double-door air locks at the entrance (biosafety level 3).

By law, extremely dangerous pathogens such as the Ebola virus, for which there is no treatment or vaccine, may be studied only at a biosafety level 4 containment facility. Practices here dictate that lab personnel wear positive-pressure lab suits connected to a separate air supply (**Fig. 28.18**). The positive pressure ensures that if the suit is penetrated, organisms will be blown away from the

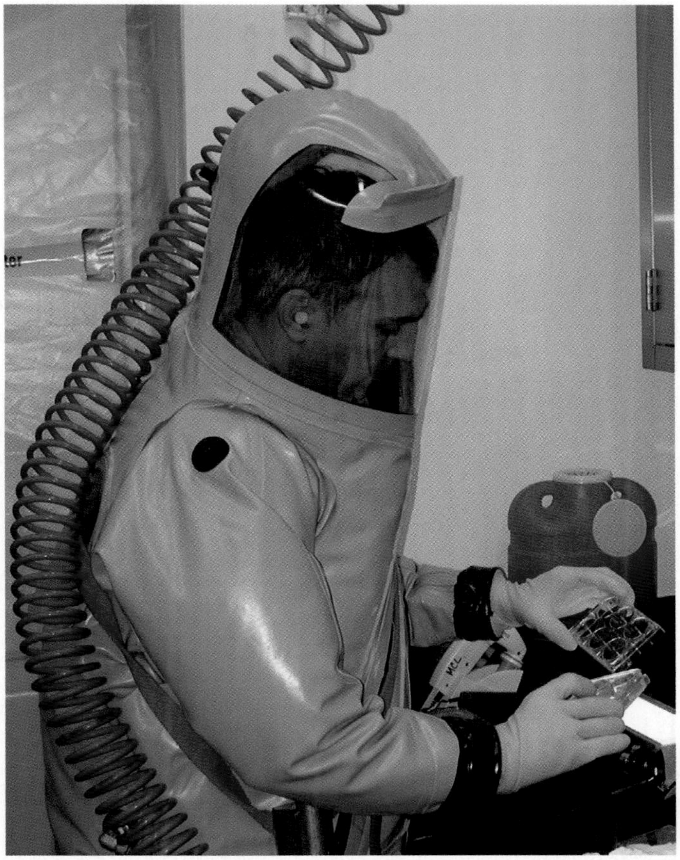

KEVIN KAREM/CDC

FIGURE 28.18 ■ Biosafety level 4 containment. Dr. Kevin Karem at the CDC performs viral plaque assays to determine the neutralization potential of serum from smallpox vaccination trials. He is protected by a positive-pressure suit working in a biosafety level 4 laboratory. The airflow into his suit is so loud he must wear earplugs to protect his hearing.

TABLE 28.4

Biological safety levels and select agents.*

	Containment level			
	Level 1	**Level 2**	**Level 3**	**Level 4**
	Risk group I	**Risk group II**	**Risk group III**	**Risk group IV**
Class of disease agent	Agents not known to cause disease	Agents of moderate potential hazard; also required if personnel may have potential contact with human blood or tissues	Agents may cause disease by inhalation route	Dangerous and exotic pathogens with high risk of aerosol transmission; only 11 labs in the United States handle these
Recommended safety measures	Basic sterile technique; no mouth pipetting	Level 1 procedures plus limited access to lab; biohazard safety cabinets used; hepatitis vaccination recommended	Level 2 procedures plus ventilation providing directional airflow into room, exhaust air directed outdoors; restricted access to lab (no unauthorized persons)	Level 3 procedures plus one-piece positive-pressure suits; lab is completely isolated from other areas present in the same building or is in a separate building
Representative organisms in class	*Bacillus subtilis* *E. coli* K-12 *Saccharomyces* spp.	*Bordetella pertussis* *Campylobacter jejuni* *Chlamydia* spp. *Clostridium* spp. *Corynebacterium diphtheriae* *Cryptococcus neoformans* *Cryptosporidium parvum* Dengue virus Diarrheagenic *E. coli* *Entamoeba histolytica* *Giardia lamblia* *Haemophilus influenzae* *Helicobacter pylori* Hepatitis virus *Legionella pneumophila* *Listeria monocytogenes* *Mycoplasma pneumoniae* *Neisseria* spp. *Salmonella* spp. *Shigella* spp. *Staphylococcus aureus* *Toxoplasma* Pathogenic *Vibrio* spp. *Yersinia enterocolitica*	*Bacillus anthracis* (anthrax) *Brucella* spp. (brucellosis) *Burkholderia mallei* (glanders) California encephalitis virus *Coxiella burnetii* (Q fever) EEE (eastern equine encephalitis) *Francisella tularensis* (tularemia) Japanese encephalitis virus La Crosse encephalitis virus LCM (lymphocytic choriomeningitis) virus *Mycobacterium tuberculosis* Rabies virus *Rickettsia prowazekii* (typhus fever) Rift Valley fever virus SARS (severe acute respiratory syndrome) virus Variola major (smallpox) and other poxviruses VEE (Venezuelan equine encephalitis) virus West Nile virus Yellow fever virus *Yersinia pestis*	Ebola virus Guanarito virus Hantavirus Junin virus Kyasanur Forest disease virus Lassa fever virus Machupo virus Marburg virus Tick-borne encephalitis viruses

*Organisms in blue are on the list of CDC select agents that are considered possible agents of bioterrorism.

breach and not sucked into the suit. As reasonable as these regulations may seem, they were not always in effect. Prior to 1970, scientists had an almost cavalier approach toward handling pathogens. For instance, culture material was routinely transferred from one vessel to another by mouth pipetting (essentially using a glass or plastic pipette as a straw). This practice is now forbidden for obvious reasons. As of this writing there are 11 biosafety level 4 laboratories operating in the United States.

To Summarize

- **Various levels of protective measures** are used in handling potentially infectious biological materials.
- **Risk group I** agents are generally not pathogenic and require the lowest level of containment.
- **Risk group II** agents are pathogenic but not typically transmitted via the respiratory tract. Laminar flow hoods are required.
- **Risk group III** agents are virulent and transmitted by the respiratory route. They require laboratories with special ventilation and air-lock doors.
- **Risk group IV** agents are highly virulent and require the use of positive-pressure suits.

28.7

Principles of Epidemiology

In Section 25.1 we discussed how a new infectious disease can be identified using a version of Koch's postulates. But how do scientists track the spread of a new disease, or find and identify new variants of influenza virus that develop thousands of miles away, and then predict when that virus will arrive on our "doorsteps" many months in advance? We start our discussion of epidemiology with a case in which scientists had to trace a heinous criminal act back to its source.

Case History: Inhalation Anthrax

On October 16, 2001, a 56-year-old African-American U.S. Postal Service worker became ill with a low-grade fever, chills, sore throat, headache, and malaise. These symptoms were followed by minimal dry cough, chest heaviness, shortness of breath, night sweats, nausea, and vomiting. On October 19, the man arrived at a local hospital, where he presented with a normal body temperature and normal blood pressure. He was not in acute distress but had decreased breath sounds and rhonchi (dry sounds in lungs due to congestion). No skin lesions were observed, and he did not smoke. Total white blood cell count was normal, but there was a left shift in the differential—that is, more polymorphonuclear leukocytes (PMNs; see Section 26.3). A chest X-ray showed bilateral pleural effusions (accumulation of fluids in the lung) and a small, right-lower-lobe air space opacity. Within 11 hours, blood cultures taken upon admission grew Bacillus anthracis. *Ciprofloxacin, rifampin, and clindamycin antibiotic treatments were initiated, and the patient recovered. His job at the post office was simply to sort mail.*

From October 4 to November 2, 2001, the Centers for Disease Control and Prevention (CDC) and various state and local public health authorities reported 10 confirmed cases of inhalational anthrax and 12 confirmed or suspected cases of cutaneous anthrax in persons who worked in the District of Columbia, Florida, New Jersey, and New York. Many of them were postal workers. It was clear that a biological attack was in progress.

Painstaking detective work by federal agents and epidemiology scientists proved that the strain of *Bacillus anthracis* used in the 2001 anthrax attack had the same genetic signature as a strain used by Bruce Ivins, a scientist working at the army's Fort Detrick biodefense laboratory. As agents were about to arrest him, Dr. Ivins took his own life.

The word "epidemiology" is derived from the Greek meaning "that which befalls man." In scientific parlance, **epidemiology** examines the distribution and determinants of disease frequency in human populations. Put more simply, epidemiologists determine the source of a disease outbreak and the factors that influence how many individuals will succumb to the disease. Epidemiological principles are also used to determine the effectiveness of therapeutic measures and to identify new syndromes, such as SARS (severe acute respiratory syndrome), MERS (Middle East Respiratory Syndrome), and Lyme disease. Some of the basic concepts of epidemiology were already covered in Chapter 25 when we discussed infection cycles. Now we will explore how those principles are used to track disease.

Epidemiological early-warning systems require an extensive organization that coordinates information from many sources. In the United States, that duty falls to the Centers for Disease Control and Prevention (CDC). On the world stage, it is the World Health Organization (WHO). Any disease considered highly dangerous or infectious is first reported to local public health centers, usually within 48 hours of diagnosis. The local centers forward that information to their state agencies, which then report to the CDC in Atlanta. This is how authorities in 2001 quickly recognized that an outbreak of anthrax was under way.

The terms "endemic" and "epidemic" are often used when referring to disease outbreaks. A disease is **endemic** if it is always present in a population at a low frequency. For example, Lyme disease, caused by the spirochete *Borrelia burgdorferi,* is endemic to the northeastern United States because the organism has found a reservoir in deer and ticks. Recall that a **reservoir** is an animal, bird, or insect that harbors the infectious agent and is indigenous to a geographic area. Humans become infected only when they come in contact with the reservoir. Thus, the disease incidence is low but relatively constant. A disease is **epidemic**, on the other hand, when larger-than-normal numbers of individuals in a population become infected over a short time. Epidemics arise, in part, because of rapid and direct human-to-human transmission.

Figure 28.19A illustrates the difference in the frequency of cases observed between endemic and epidemic disease. An endemic disease can become epidemic if the population of the reservoir increases, allowing for more frequent human contact; or if the infectious agent evolves to spread directly from person to person, bypassing the need for a reservoir. This is the concern with the H5N1 avian flu virus, which is endemic in animals and birds in Asia (see Sections 11.3 and 25.1). A **pandemic** is an epidemic that occurs over a wide geographic area, usually the world. Pandemics may be long-lived, such as the bubonic plague pandemic in the fourteenth century and the AIDS pandemic in the late twentieth and early twenty-first centuries; or they may be short-lived, as with the 1918 flu pandemic.

Finding Patient Zero

When trying to contain the spread of an epidemic, it is vital to track down the first case of the disease (known as the **index case** or patient zero) and then identify everyone who has had contact with that individual so that they can be treated or separated from the general population (**quarantined**). When a new disease arises, the epidemiological search for the index case starts only after a number of patients have been diagnosed and a new disease syndrome declared. This is what happened with AIDS in the 1970s.

Identifying an index case within a specific community is easier if the disease syndrome is already recognized, as was the case with the 2003 SARS outbreak in Singapore. According to the World Health Organization, a suspected case of SARS is defined as an individual who has a fever greater than 38°C (100.4°F), who exhibits lower respiratory tract symptoms, and who has traveled to an area of documented disease or has had contact with a person afflicted with SARS. The index case in Singapore was a 23-year-old woman who had stayed on the ninth floor of a hotel in Hong Kong while on vacation. A physician from southern China who stayed on the same floor of the hotel during this period is believed to have been the source of her infection, as well as that of the index patients who precipitated subsequent outbreaks in Vietnam and Canada.

During the last week of February, the woman, who had returned to Singapore, developed fever, headache, and a dry

A. Endemic versus epidemic

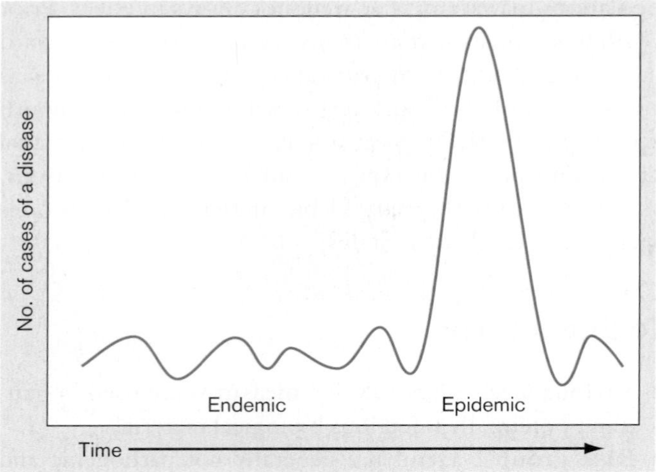
No. of cases of a disease
Endemic
Epidemic
Time

B.

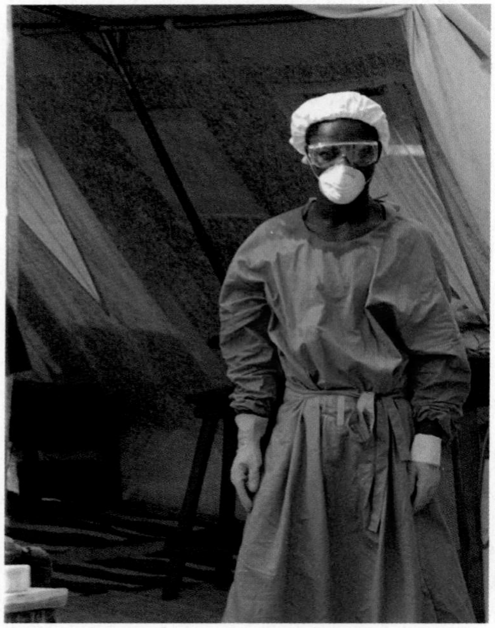
AP PHOTO/JEAN-MARC BOUJU

FIGURE 28.19 ▪ The difference between endemic and epidemic disease. A. An endemic disease is continually present at a low frequency in a population. A sudden rise in disease frequency constitutes an epidemic. **B.** A health care worker stands outside a quarantined area housing Ebola patients in the Ivory Coast. Epidemics can be minimized if infected persons are kept segregated from the general population (quarantined) to avoid spread of the infectious agent.

A.

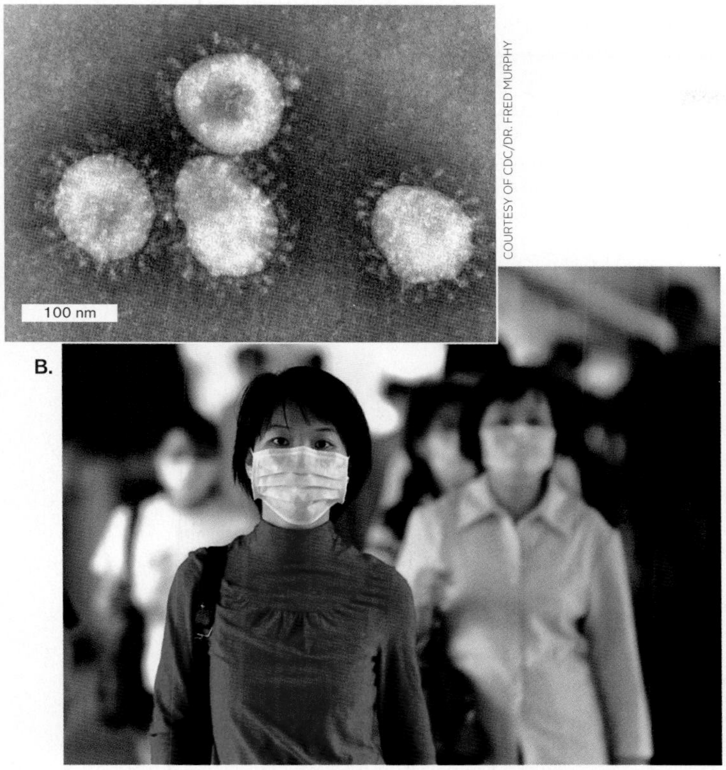
100 nm

COURTESY OF CDC/DR. FRED MURPHY

B.

ANAT GIVON/AP PHOTO

FIGURE 28.20 ■ Severe acute respiratory syndrome (SARS). A. The coronavirus that causes SARS (TEM). **B.** Citizens of China, including the military, donned surgical masks in 2003 to slow the spread of SARS.

cough. She was admitted to Tan Tock Seng Hospital, Singapore, on March 1 with a low white blood cell count and patchy consolidation in the lobes of the right lung. Tests for the usual microbial suspects (*Legionella, Chlamydia, Mycoplasma*) were negative. Electron microscopy of nasopharyngeal aspirations showed virus particles with widely spaced club-like projections (**Fig. 28.20A**). At the time of her admission to the hospital, the clinical features and highly infectious nature of SARS were not known. Thus, for the first 6 days of hospitalization, the patient was in a general ward, without barrier infection control measures. During this period, the index patient infected at least 20 other individuals, including hospital staff, nearby patients, and visitors.

Within weeks, the WHO named the disease in China "SARS" and issued travel alerts (discussed further in Section 28.8). These alerts allowed Singapore health officials to rapidly identify the index patient and her contacts. As a result, they were able to limit spread of the illness. When all was said and done, SARS killed fewer than 1,000 victims worldwide, although thousands more became ill and recovered (**Fig. 28.20B**). Today, SARS remains a threat but one of lesser concern. Methicillin-resistant *Staphylococcus aureus* (MRSA), H5N1 avian flu, and H1N1 swine flu are considered more pressing dangers.

Identifying Disease Trends

How do epidemiologists first recognize that an epidemic is under way, and then identify the agent and its source? Certain diseases, because of their severity and transmissibility, are called reportable, or notifiable, diseases (**Table 28.5**). Physicians are required to report instances of these diseases to a central health organization, such as the CDC in the United States and the WHO. This reporting allows the incidences of certain diseases within a population to be tracked and upsurges noted. An emerging disease not on the list of notifiable diseases can be detected as a cluster of patients with unusual symptoms or combinations of symptoms. Such detection is possible because diseases of unknown etiology are also reported to health authorities. A new disease could manifest with common symptoms (for example, the cough and fever of SARS) that cannot be linked to a known disease agent by clinical tests. An upsurge in cases, of either a reportable disease or an emerging disease, will set off institutional "alarms" that initiate epidemiological efforts to determine the source and cause of the outbreak.

John Snow (1813–1858), the Father of Epidemiology

The first case in which the source of a disease outbreak was methodically investigated took place in the mid-nineteenth century. During a serious outbreak of cholera in London in 1854, John Snow (**Fig. 28.21A**) visited the addresses of all the diarrheal cases he learned about. The source of the infection was unknown, but Snow thought that if the cases clustered geographically, he might gain a clue as to its location. Water in that part of London was pumped from separate wells located in the various neighborhoods. Snow realized there was a close association between the density of cholera cases and a single well located on Broad Street (**Fig. 28.21B**). Simply removing the pump handle of the Broad Street well put an end to the epidemic, proving that the well water was the source of infection (known as the "point source," since all infections originated from that point). This approach succeeded brilliantly, even though the infectious agent that causes cholera, *Vibrio cholerae*, was not recognized until 1905, over 50 years later. Identifying clusters of patients afflicted with a given disease is still used to locate potential sources of infectious disease outbreaks.

TABLE 28.5

Notifiable infectious diseases.

Bacterial		
Anthrax	Hansen's disease (leprosy)	Salmonellosis (non–typhoid fever types)
Botulism	Legionellosis	Shigellosis
Brucellosis	Leptospirosis	Staphylococcal enterotoxin
Campylobacter infection	Listeriosis	Streptococcal invasive disease
Chlamydia infection	Lyme disease	*Streptococcus pneumoniae,* invasive disease
Cholera	Meningitis, infectious	Syphilis
Diphtheria	Pertussis	Tuberculosis
Ehrlichiosis	Plague	Tularemia
E. coli O157:H7 infection	Psittacosis	Typhoid fever
Gonorrhea	Q fever	Vancomycin-resistant *Staphylococcus aureus* (VRSA)
Haemophilus influenzae, invasive disease	Rocky Mountain spotted fever	
Viral		
Dengue fever	Mumps	Smallpox
Hantavirus infection	Poliomyelitis	Varicella (chickenpox), fatal cases only (all types)
Hepatitis, viral	Rabies	Yellow fever
HIV infection	Rubella and congenital rubella syndrome	
Measles		
Fungal		
Coccidioidomycosis		
Parasitic		
Amebiasis	Giardiasis	Trichinosis
Cryptosporidiosis	Malaria	
Cyclosporiasis	Microsporidiosis	

A.

THE JOHN SNOW ARCHIVE AND RESEARCH COMPANION

B.

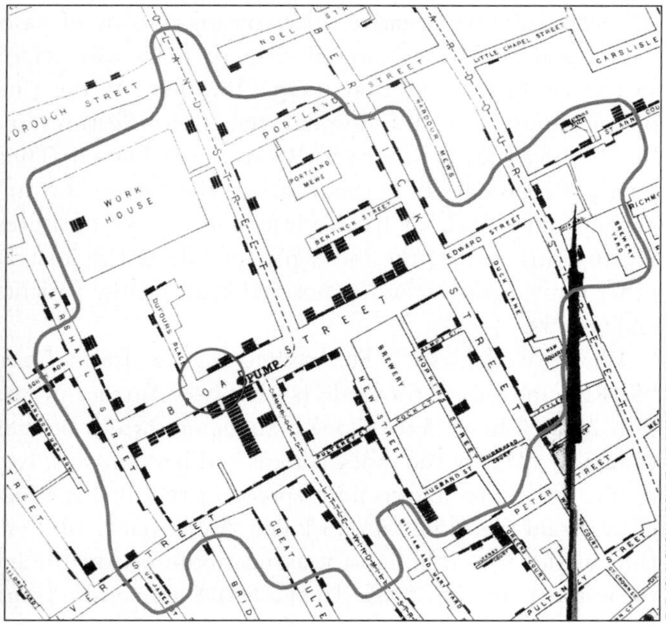
WORK HOUSE
PUMP
PORTLAND MEWS
BENTINCK STREET
EDWARD STREET
BREWERY
LITTLE CHAPEL STREET

THE JOHN SNOW ARCHIVE AND RESEARCH COMPANION

FIGURE 28.21 ▪ Early epidemiology. **A.** John Snow. **B.** A map of London commissioned by Snow in 1855 shows the location of cholera victims of the 1854 outbreak. The map illustrates that each victim within the marked area lived closer to the Broad Street pump than to other nearby pumps. The Broad Street well is found within the red circle. Each black bar represents a death from cholera.

Thought Questions

28.8 Methicillin, a beta-lactam antibiotic, is very useful in treating staphylococcal infections. The emergence of methicillin-resistant strains of *Staphylococcus aureus* (MRSA) is a very serious development because few antibiotics can kill these strains. Imagine a large metropolitan hospital in which there have been eight serious nosocomial infections with MRSA and you are responsible for determining the source of infection so that it can be eliminated. How would you accomplish this task using common bacteriological and molecular techniques?

28.9 What are some reasons why some diseases spread quickly through a population while others take a long time?

Genomic Strategies Help Identify Nonculturable Pathogens

Microbiologists have successfully developed many strategies to identify the causes of infectious diseases. As noted earlier, Robert Koch in the late nineteenth century devised a set of postulates that, when followed, can identify the agent of a new disease (discussed in Sections 1.3 and 25.1). One important tenet of Koch's postulates is that the suspected organism must be grown in pure culture. However, there are bacterial diseases for which the agent cannot be cultured. How might they be identified?

Case History: Whipple's Disease

In April, a 50-year-old man had an abrupt onset of watery diarrhea with a stool frequency of up to 10 times every 24 hours. This was the latest episode in a 6-year history of illness beginning with recurring fevers, flu-like symptoms, profuse night sweats, and painful joint swelling. The current bout of diarrhea was associated with gripping lower abdominal pain, especially after meals. No blood or mucus was found in the stools. Blood and stool cultures tested negative for known infectious agents. Serology was also negative for syphilis, brucellosis, toxoplasmosis, and leptospirosis. The man's weight fell rapidly from 83 to 73 kg within 4 weeks of onset. A flexible sigmoidoscopy showed only diffuse mild erythema in the bowel. However, the appearance of the small bowel was consistent with malabsorption, a disease in which nutrients are poorly absorbed by the intestine. A duodenal biopsy to look for the organisms of Whipple's disease showed large macrophages in the lamina propria (a layer of loose connective tissue beneath the epithelium of an organ). Bacterial rods characteristic of Whipple's disease were seen with electron microscopy. PCR analysis of tissue samples confirmed the diagnosis.

First diagnosed by George Whipple in 1907, Whipple's disease is characterized by malabsorption, weight loss, arthralgia (joint pain), fevers, and abdominal pain. Any organ system can be affected, including the heart, lungs, skin, joints, and central nervous system. The cause of Whipple's disease went undiscovered for 85 years, but the disease was suspected to be of bacterial etiology, even though an organism was never successfully cultured. The agent was finally identified in 1992, not by culturing, but by blindly amplifying 16S rDNA sequences from biopsy tissues.

All bacterial 16S rRNA genes have some sequence regions that are highly homologous across species and other sequences that are unique to a species. Tissues from numerous patients diagnosed with Whipple's disease were subjected to PCR analysis using the common 16S rDNA primers. If a bacterial agent was present, it was predicted that PCR should successfully amplify a DNA fragment corresponding to the agent's 16S rRNA gene. All tissues produced such a fragment, indicating that there were bacteria in the tissues. DNA sequence analysis of these fragments indicated that the organism was similar to actinomycetes but was unlike any of the known species.

The organism is actually a Gram-positive soil-dwelling actinomycete that has been named *Tropheryma whipplei* in honor of the physician who first recognized the disease. Because of its bacterial etiology, Whipple's disease can be treated with antibiotics, usually trimethoprim sulfamethoxazole (Bactrim, Septra, Cotrim). **Figure 28.22** shows an in situ hybridization for *T. whipplei* RNA in a tissue biopsy.

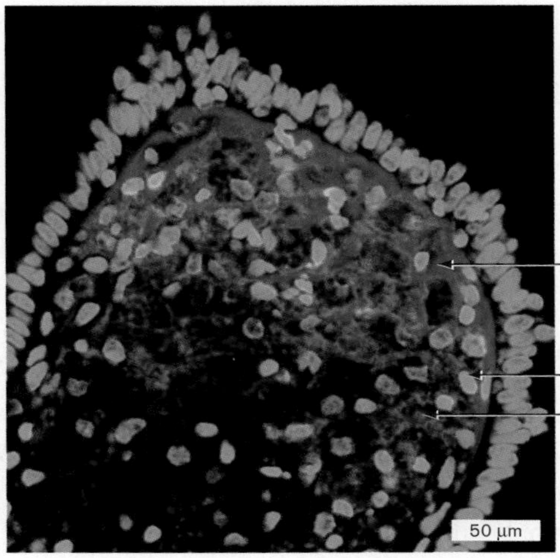
T. whipplei
RNA (blue)
Host nuclei
Host
cytoskeleton
50 µm
D. N. FREDRICKS & D. A. RELMAN. 2001. J INFECT DIS 183:1229

FIGURE 28.22 ▪ Whipple's disease. Fluorescent in situ hybridization of a small intestinal biopsy in a case of Whipple's disease (confocal laser scanning microscopy). In this test, a fluorescently tagged DNA probe that specifically hybridizes to *Tropheryma whipplei* RNA is added to the tissue. Other fluorescent probes are used to visualize host nuclei and cytoskeleton. Blue = *T. whipplei* rRNA; green = nuclei of human cells; red = intracellular cytoskeletal protein vimentin. Magnification approx. 200×.

This story reveals that Koch's postulates (see Figure 1.17) must sometimes be modified when identifying the cause of a new disease. In this instance, the organism could not be cultured in pure form in the laboratory, but it was found, via molecular techniques, in all instances of the disease. More recent studies have successfully cultured *T. whipplei* in vitro. The medium used was formulated on the basis of nutritional requirements deduced from knowing the DNA sequence of the *T. whipplei* genome (discussed in Section 25.7).

Molecular Approaches for Disease Surveillance

A worldwide pandemic of pulmonary tuberculosis currently affects over 2 billion people. Many of the *Mycobacterium tuberculosis* infections are caused by multidrug-resistant strains that are difficult, if not impossible, to kill with existing antibiotics (see Section 26.3). This problem is especially serious among refugee populations attempting to flee war-torn countries. As a result, it is important to screen these refugees as they enter neighboring countries with low incidences of tuberculosis. Although chest X-rays are mandatory in many cases, a positive image will be obtained only if the disease is at a relatively advanced stage. Actively infected individuals who have not developed the characteristic lung tubercles seen on X-ray will not be identified.

Unfortunately, the acid-fast staining of sputum samples (discussed in Section 28.3) also fails to detect individuals at an early stage of infection. Studies have shown, however, that PCR techniques are much more sensitive for detecting these individuals. As time progresses, PCR surveillance strategies will be used more often to track the worldwide ebb and flow of microbial diseases. PCR and restriction fragment length polymorphism (RFLP) strategies are already used for epidemiological purposes to type (that is, determine the relatedness of) different microbial isolates by generating a complex DNA profile that is specific for a particular strain (Section 28.3). For example, DNA profiles were used to link an outbreak of over 2,000 cases of salmonellosis in 2010 to a single strain of *Salmonella enterica* serovar Enteritidis. The results, conducted by a national network of public health agencies (PulseNet), led to the recall of over a half billion eggs.

Bioterrorism

"A wide-scale bioterrorism attack would create mass panic and overwhelm most existing state and local systems within a few days," said Michael T. Osterholm, director of the Center for Infectious Disease Research & Policy at the University of Minnesota, in October 2001. "We know this from simulation exercises."

Less than a month after the September 11 attack on the World Trade Center, a biodefense scientist working at the U.S. Army Medical Research Institute of Infectious Diseases (Fort Detrick, Maryland) allegedly sent weapons-grade anthrax spores through the U.S. mail (see case above on inhalation anthrax). Thankfully, only 5 persons died and a mere 25 became ill. But even though the efficiency of the attack was poor, the impact was enormous. Over 10,000 people took a 2-month course of antibiotics after possible exposure, and mail deliveries throughout the country were affected. The simple act of opening an envelope suddenly became a risky business.

As a result of this attack and other events, the CDC and the National Institutes of Health (NIH) assembled a list of select agents (marked in blue in **Table 28.4**) that could potentially be used as bioweapons. A bioweapon is considered to be any infectious agent or toxin that has high virulence and/or mortality rate. Microorganisms considered bioweapons can be used to conduct biowarfare, with the intent of inflicting massive casualties; or bioterrorism, which may result in only a few casualties but cause widespread psychological trauma.

Although the branding of select agents is recent, biowarfare is not new. In the Middle Ages, victims of the Black Death (plague caused by *Yersinia pestis*) were flung over castle walls using catapults; during the French and Indian War, in the eighteenth century, Jeffrey Amherst distributed smallpox-infected blankets to Native Americans; and during World War II, the Imperial Japanese Army experimented with infectious disease weapons using Chinese prisoners as guinea pigs. Even the United States has participated through the development of weapons-grade anthrax spores. The first documented act of bioterrorism in the United States occurred in 1984, when followers of the cult leader Bhagwan Shree Rajneesh tried to control a local election in The Dalles, Oregon, by infecting salad bars with *Salmonella*. Over 700 people became ill. Rajneesh was given a 10-year suspended sentence, fined $400,000, and deported.

How effective are bioweapons? The method by which a biological agent is dispersed plays a large role in its effectiveness as a weapon. Only a few people became ill during the 2001 anthrax attack, not only because of the epidemiological surveillance, but also because anthrax is inherently difficult to disperse. Thousands of spores must be inhaled to contract disease, which means that effective dispersal of the spores is critical for the use of anthrax as a weapon. Once spores hit the ground, the threat of infection is limited. Weapons-grade spores are very finely ground so that they stay airborne longer. But, as we saw, the letter-borne dispersal system did not effectively generate large numbers of victims. Nevertheless, the potential threat of weapons-grade anthrax on the battlefield led the U.S. military in 2006 to resume vaccinating all soldiers serving in Iraq, Afghanistan, and South Korea.

An effective bioweapon would capitalize on person-to-person transmission. In an easily transmitted disease, one infected person could disseminate disease to scores of others within 1 or 2 days. So, in terms of generating massive numbers of deaths, anthrax was a poor choice. The goal of most terrorists, however, is not to kill large numbers of people, but to terrorize them. In that regard, the anthrax attack succeeded (**Fig. 28.23A**).

The most effective bioweapon in terms of inflicting death (biowarfare) would have a low infectious dose, be easily transmitted between people, and be one to which a large percentage of the population is susceptible. Smallpox fits these criteria and would be the bioweapon of choice (**Fig. 28.23B**). Fortunately, however, smallpox has been eradicated (almost) from the face of the Earth and is not easily obtained. Two laboratories still harbor the virus—one in the United States and one in Russia. It is believed that the virus has been destroyed in all other laboratories. Since smallpox is the perfect biowarfare agent, it is imperative that the last two smallpox repositories remain secure.

While the good news is that smallpox disease has been eradicated, the bad news is that no individual born after 1970 has been vaccinated (actually, some military and a few laboratory personal are vaccinated). As a result, anyone under 40 years of age is susceptible to smallpox. Even those of us who received the smallpox vaccination over 40 years ago are at risk, since our protective antibody titers have diminished. A terrorist attack with smallpox would cause terrible numbers of deaths. A new, safer vaccine does exist, however, and has been stockpiled. Were a smallpox attack to be launched, the vaccine would be rapidly administered to limit the spread of disease. Nevertheless, the economic and psychological impact of a smallpox epidemic would be devastating.

Research with organisms considered to be select agents is tightly regulated. Because *Yersinia pestis,* for instance, is a select agent, laboratory personnel working with it must now possess security clearance with the Department of Justice (even though the organism itself can be handled under biosafety level 2 conditions; see **Table 28.4**). The laboratory must also register with the CDC to legally possess this pathogen, and access to the lab and the organism must be tightly controlled.

Much has improved since Michael Osterholm offered his dire assessment of a wide-scale bioterrorism attack. Education and surveillance procedures have been bolstered, and new detection technologies are being developed (**Special Topic 28.1**; see also eTopic 28.3). We will probably never be fully protected from attack, biological or otherwise, but recent efforts have improved the situation.

Weblinks Bioterrorism preparedness act (*see ebook*)

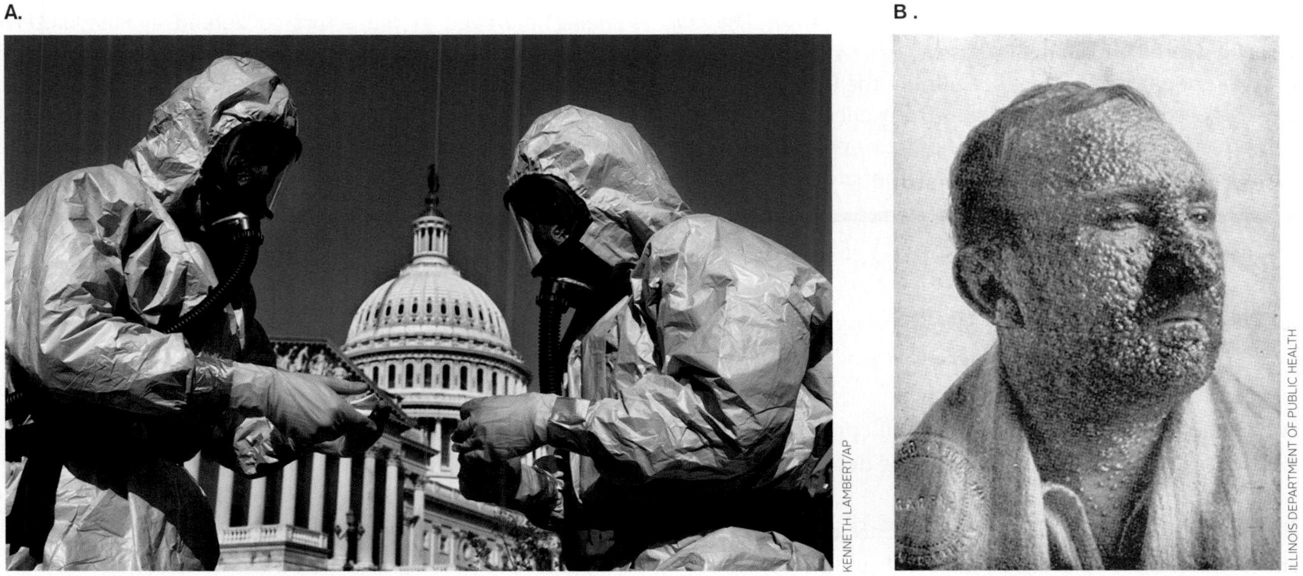
A.
B.
KENNETH LAMBERT/AP
ILLINOIS DEPARTMENT OF PUBLIC HEALTH

FIGURE 28.23 ■ Dealing with bioterrorism. A. Members of a hazardous-materials team near Capitol Hill during the anthrax attacks in 2001. **B.** An Illinois man suffering from smallpox in 1912.

Special Topic 28.1: What's Blowing in the Wind?

In April 1979, workers at a secret Soviet biological weapons facility neglected to replace a filter in a laboratory ventilation system, and a cloud of highly weaponized *Bacillus anthracis* spores quickly spewed into the air outside. The deadly plume drifted downwind, infecting humans and cattle across a wide area. The resulting outbreak ultimately killed more than 64 people. Had a rapid pathogen detection system been available and deployed at the facility, all those lives could have been saved. Unfortunately, 30 years ago such technology was only the stuff of science fiction. Today, that technology exists in small, portable forms.

Several multipathogen molecular detection platforms have been developed. Some involve machines that detect antigen-antibody interactions; others use automated PCR to amplify specific pathogen genes and detect the products through hybridization to immobilized oligonucleotides. The wet chemistries for all of these detection systems are carried out in a roughly 2-inch square called lab-on-a-chip.

An interesting addition to the field of quick pathogen detection is the PANTHER sensor (Pathogen Analyzer for Threatening Environmental Releases), a device developed in 2008 that can detect and identify a pathogen in as little as 3 minutes—significantly faster than traditional methods requiring isolation and growth of the pathogen (which can take 48 hours or longer). The PANTHER sensor (**Fig. 1**) uses a cell-based technology that can detect as few as a dozen particles of a pathogen per liter of air. Currently, it can detect 24 pathogens, including the potential bioterrorism agents of anthrax, plague, smallpox, and tularemia.

REPRINTED WITH PERMISSION OF MIT LINCOLN LABORATORY, LEXINGTON, MA

FIGURE 1 ■ The PANTHER detection system. *Source:* Rider et al. 2003. *Science* **301**:213–215.

The device uses an array of B cells, each displaying antibodies specific to a particular bacterium or virus. The cells are engineered to emit photons of light when they detect their target pathogen. More specifically, the B cells have been bioengineered to express the gene that encodes aequorin, a calcium-sensitive bioluminescent protein. When a pathogen surface antigen cross-links the appropriate B-cell surface antibodies, a signal transduction cascade produces a rapid influx of calcium (**Fig. 2**). Aequorin in the B cell will luminesce within seconds of calcium influx, and the light emitted is detected by a photon detector. The device then displays a list of any pathogens found.

Quick pathogen detection technologies are constantly being improved in the effort to defend against bioterrorism. Small, quick pathogen detection devices could be used in buildings, subways, and other public areas. Eventually, such technology is expected to supplant the classical clinical microbiology practices that require growth of the microbe to achieve identification. There is hope, for example, that such

To Summarize

- **John Snow** founded the discipline of epidemiology.
- **Epidemiology** examines factors that determine the distribution and source of disease.
- **Endemic, epidemic, and pandemic** are terms for different frequencies of disease in different geographic areas.
- **Finding patient zero (the index case)** is important for containing the spread of disease.
- **Molecular approaches using PCR and nucleic acid hybridization** are used to identify nonculturable pathogens and to track disease movements.
- **Bioweapons**, when they have been used, typically kill few people but incite great fear.
- **The CDC** has assembled a list of select agents with bioweapon potential.

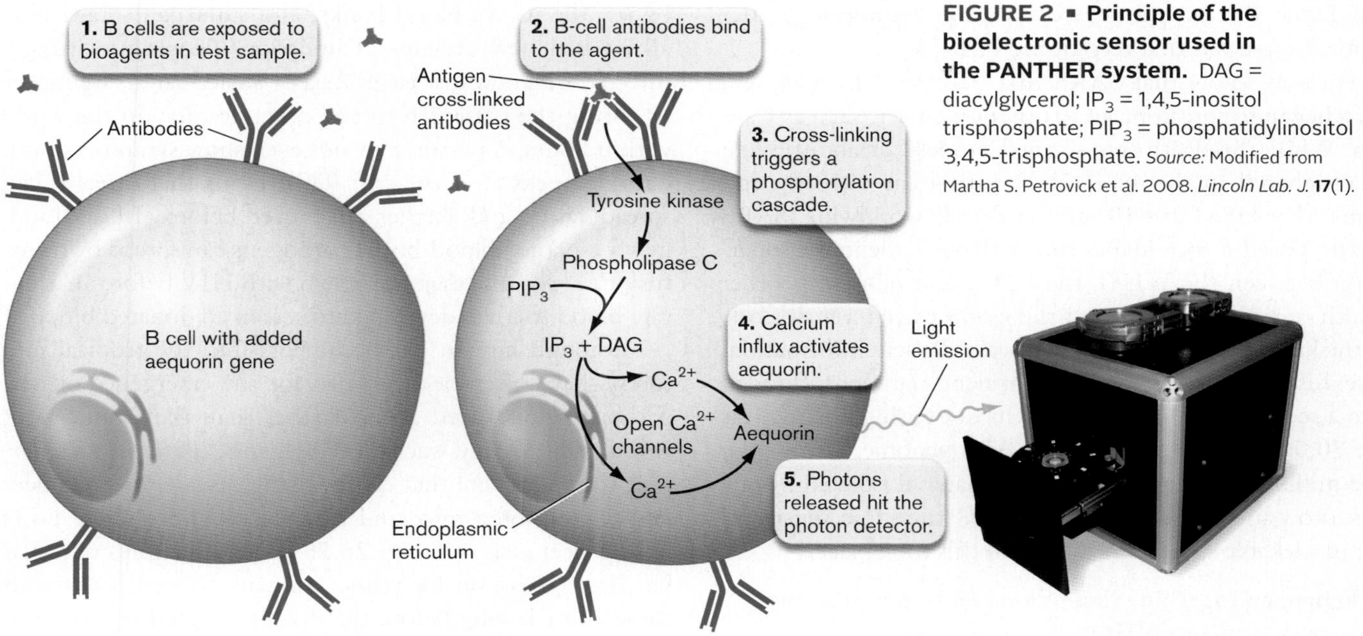

1. B cells are exposed to bioagents in test sample.
2. B-cell antibodies bind to the agent.
Antigen cross-linked antibodies
Antibodies
3. Cross-linking triggers a phosphorylation cascade.
Tyrosine kinase
Phospholipase C
PIP3
IP3 + DAG
B cell with added aequorin gene
4. Calcium influx activates aequorin.
Light emission
Ca2+
Open Ca2+ channels
Aequorin
Ca2+
5. Photons released hit the photon detector.
Endoplasmic reticulum

FIGURE 2 ▪ Principle of the bioelectronic sensor used in the PANTHER system. DAG = diacylglycerol; IP_3 = 1,4,5-inositol trisphosphate; PIP_3 = phosphatidylinositol 3,4,5-trisphosphate. *Source:* Modified from Martha S. Petrovick et al. 2008. *Lincoln Lab. J.* **17**(1).

devices could be used on farms or in food-processing plants to test for contamination by *E. coli, Salmonella,* or other foodborne pathogens. Another potential application is in medical diagnostics, where the technology could be used to test patient samples, giving rapid results without having to send samples to a laboratory. Wherever it is used, successful quick pathogen detection should provide the clinician with better information to quickly prescribe the appropriate course of treatment.

RESEARCH QUESTION

The research article referenced below describes a method of biothreat detection that combines PCR with mass spectrometry (see Section 28.4). Why might this coupled technology be superior for detecting biowarfare agents as compared to the more conventional use of molecular probes to detect PCR amplicons? The authors also published the sequences of DNA primers that their system uses to amplify indicator genes of over 20 potential biowarfare agents. Discuss why this decision to publish was either good or bad.

Sampath, Rangarajan, Niveen Mulholland, Lawrence B. Blyn, Christian Massirel, Chris A. Whitehouse, et al. 2012. Comprehensive biothreat cluster identification by PCR/electrospray-ionization mass spectrometry. *PLoS One* **7**:e36528.

28.8

Detecting Emerging Microbial Diseases

Case History: SARS

A 48-year-old man was hospitalized in a Dutchess County, New York, hospital with a 38.3°C (101°F) fever, headache, and body aches. He was also having difficulty breathing. He had just returned from a business trip to China.

SARS: An Epidemiological Success Story

Within 7 weeks in early 2003, epidemiologists armed with global technologies and rapid DNA sequencing techniques tracked, named, identified, completely sequenced, and contained a newly emerging disease with a scary death rate: severe acute respiratory syndrome (SARS). This was a remarkable feat, especially when we consider that after the first cases of AIDS appeared in 1981, it took over 3 years just to identify the virus; and it took 7 years to track down the Lyme disease spirochete, *Borrelia burgdorferi,* after Lyme disease was first recognized in 1975.

We now know that SARS first developed in Asia, and then began to spread by jet to other countries, such as Canada. While the death rate seemed modest, at about 5%, it was higher than the 4% reached during the 1918 flu epidemic that killed 20–40 million people worldwide. SARS clearly posed a formidable threat. Unprecedented cooperation between the WHO, the CDC, and numerous other health organizations around the world played a major role in tracking and containing the disease. Patient information, case histories, and possible treatment regimens flooded into a secure WHO website. The most exciting posting was the 30,000-bp sequence of the SARS genome, taken from one-millionth of a gram of genetic material isolated from a Toronto patient. The following 2003 time line illustrates the remarkable speed with which all this took place:

February 14: China first reports 305 cases of atypical pneumonia to WHO.

March 15: WHO issues emergency travel alert; names illness "SARS."

March 17: Leading epidemiologists join WHO in conference call; agree to unprecedented cooperation.

April 4: U.S. President George W. Bush authorizes quarantine of SARS patients.

April 12: Canadian scientists in Vancouver post the completed genome on the Internet.

April 16: WHO announces that SARS is caused by a pathogen never before seen in humans—a new coronavirus.

Similar strategies were used in 2012 to identify the cause of Middle East Respiratory Syndrome (MERS). The agent is another coronavirus designated MERS-CoV, whose death rate is estimated at over 50%.

Technology Helps New Infectious Agents Emerge and Spread

Despite all we know of microbes and despite the many ways we have to combat microbial diseases, our species, for all its cleverness, still lives at the mercy of the microbe. Lyme disease, MRSA, SARS, MERS-CoV, Ebola, *E. coli* O157:H7, HIV, "flesh-eating" streptococci, hantavirus, swine flu—all of these and many other new diseases and pathogens have emerged over the last 30 years. Worse yet, forgotten scourges, such as tuberculosis, have reappeared. Yet in the 1970s, medical science was claiming victory over infectious disease. What happened?

Part of the equation has been progress itself. Travel by jet, the use of blood banks, and suburban sprawl have all opened new avenues of infection. People unwittingly infected by a new disease in Asia or Africa can, traveling by jet, bring the pathogen to any other country in the world within hours. A person may not even show symptoms until days or weeks after the trip. This means that diseases can spread faster and farther than ever before. In addition, newly emerged blood-borne pathogens can spread by transfusion. This was a major problem with HIV before an accurate blood test was developed to screen all donated blood.

Although human encroachments into the tropical rain forests have often been blamed for the emergence of new pathogens, one need go no farther than the Connecticut woodlands to find such developments. *Borrelia burgdorferi,* the spirochete that causes Lyme disease, lives on deer and white-footed mice and is passed between these hosts by the deer tick (see Fig. 26.25)—an infection cycle that has been going on for years. Humans crossed paths with these animals long before the disease erupted in our communities. Why have we suddenly become susceptible? The answer appears to be suburban development. In the wild, foxes and bobcats hunt the mice that carry the Lyme agent. These predators disappear as developers clear land and build roads and houses, leaving the infected mice and ticks to proliferate. Humans in these developed areas are more likely to be bitten by an infected tick and contract the disease than in prior decades. Luckily, many diseases that successfully leap from animals to humans find the new host to be a dead end, unable to spread the disease to others.

There are numerous examples in which technology and progress have had the unintended consequence of breeding disease. Here are just a few:

- **Mad cow disease.** Modern farming practices (North America and Europe) of feeding livestock the remains of other animals helped spread transmissible spongiform encephalopathies similar to Creutzfeldt-Jakob disease associated with prions. Because the prion is infectious, the brain matter from one case of mad cow disease could end up infecting hundreds of other cattle, which in turn increases the chance that the disease could spread to humans.
- **Lyme disease.** Suburban development in the northeastern United States destroys predators of the mice that carry *Borrelia burgdorferi.*
- **Hepatitis C.** Transfusions and transplants spread this blood-borne disease.

- **Influenza.** Live poultry markets in Asia serve as breeding grounds for avian flu viruses that can jump to humans.
- **Enterohemorrhagic *E. coli* (for example, *E. coli* O157:H7).** Modern meat-processing plants can accidentally grind trace amounts of these acid-resistant, fecal organisms into beef while making hamburger.

Natural environmental events can also trigger upsurges in the incidence of unusual diseases. For example, an unprecedented outbreak of hantavirus pulmonary disease occurred in 1993 in the Four Corners area of the Southwest, where Arizona, New Mexico, Colorado, and Utah meet, when rain led to greater-than-normal increases in plant and animal numbers. The resulting tenfold increase in deer mice, which carries the virus, made it more likely that infected mice and humans would come in contact. A more recent outbreak of hantavirus occurred among campers at Yosemite National Park in 2012.

Detecting Emerging and Reemerging Pathogens

A world map showing the general locations of emerging and reemerging diseases is shown in **Figure 28.24**. A reemerging disease is one thought to be under control but whose incidence has risen. For example, the incidence rate of tuberculosis dropped sharply in the 1950s, but the number of cases since 1980 has increased just as dramatically. Tuberculosis is a reemerging disease. The trigger for its reemergence was the AIDS pandemic. Immunocompromised AIDS patients are highly susceptible to infection by many organisms, including *Mycobacterium tuberculosis.*

M. tuberculosis is also reemerging among non-AIDS patients, owing to the development of drug-resistant strains. Drug resistance in *M. tuberculosis* developed largely because of noncompliance on the part of patients in completing their full courses of antibiotic treatment. Treatment usually involves three or more antibiotics to reduce the risk that resistance to any one drug will develop. Many patients failed to take all three drugs simultaneously, allowing the organism to develop resistance to one drug at a time, until it became resistant to all of them. These highly drug-resistant strains are almost impossible to kill. The link between noncompliance and the development of drug resistance is the primary reason for the current requirement that tuberculosis patients be in the presence of a medical staff member when taking the multiple antibiotics prescribed.

As described in Section 28.7, highly infectious diseases are identified using aggressive epidemiological surveillance.

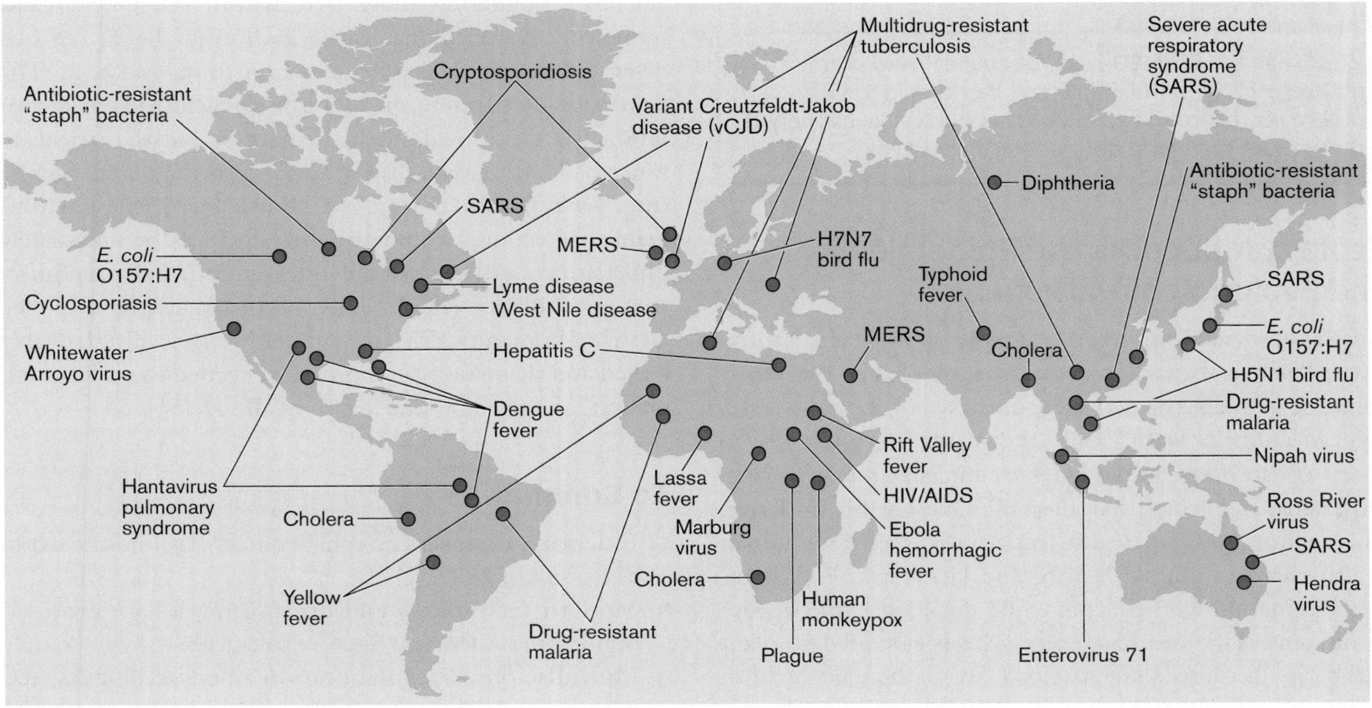
Multidrug-resistant tuberculosis
Severe acute respiratory syndrome (SARS)
Cryptosporidiosis
Antibiotic-resistant "staph" bacteria
Variant Creutzfeldt-Jakob disease (vCJD)
Diphtheria
Antibiotic-resistant "staph" bacteria
SARS
MERS
H7N7 bird flu
E. coli O157:H7
Lyme disease
West Nile disease
Typhoid fever
SARS
Cyclosporiasis
MERS
E. coli O157:H7
Whitewater Arroyo virus
Hepatitis C
Cholera
H5N1 bird flu
Drug-resistant malaria
Dengue fever
Rift Valley fever
Nipah virus
Lassa fever
HIV/AIDS
Hantavirus pulmonary syndrome
Cholera
Ross River virus
Marburg virus
Ebola hemorrhagic fever
SARS
Cholera
Hendra virus
Human monkeypox
Yellow fever
Drug-resistant malaria
Plague
Enterovirus 71

FIGURE 28.24 ▪ Locations of some emerging and reemerging infectious diseases and pathogens. The examples given represent extreme increases in the reported cases over the last 20 years. Many of these diseases, such as HIV/AIDS and cholera, are widespread but show alarming increases in the areas indicated. Cases of H1N1 swine flu have increased around the globe. MERS-CoV emerged in 2012.

Communication among local, national, and world health organizations is critical and helped expose the rise in tuberculosis. To see for yourself how emerging diseases are monitored, search the Internet for a website called ProMED-mail. There you will find daily reports posted from around the world that describe new outbreaks of infectious diseases.

A posting on December 23, 2012, for example, reports nine cases of severe respiratory infection in eastern Mediterranean countries caused by a novel coronavirus related to SARS. Five victims died. This discovery concerns the WHO because of the organization's experience with SARS in 2003. Fortunately this virus does not appear to transmit easily between people, unlike the SARS virus. A January 26, 2013, report from Uganda describes an outbreak of unexplained fever in which 5 people died and 30 were admitted to hospital. Residents describe the disease as causing heat around the chest and itching around the neck; then within a few hours, the patient starts vomiting and bleeding though the nose and mouth. Residents claim the disease is the result of witchcraft, whereas others say it is Ebola. As of this writing, the district health officer is awaiting the results of blood tests but doubts it is Ebola, since Uganda was recently declared Ebola-free. As a fledgling microbiologist, tracking these reports allows you to monitor disease trends almost in real time.

Thought Question

28.10 On the ProMED-mail web page (www.promedmail.org), click on the interactive world map to view outbreaks recorded by WHO. What outbreaks happened throughout the world during the current year?

Ecology Influences the Epidemiology and Evolution of Pathogens

We have provided numerous examples of how the natural environment can harbor pathogens and foster the emergence of new ones. Knowledge that pathogens have reservoirs in animals, arthropods, and plants has given rise to a new collaborative effort among clinicians, scientists, veterinarians, and ecologists called the One Health Initiative. The goal of the One Health Initiative is to control human health through animal health, and vice versa. For example, vaccinating wild rodents could decrease Lyme disease in humans. The idea has recently been extended to plant pathology because some bacteria are pathogens of both plants and humans (for example, *Pantoea agglomerans*, formerly *Enterobacter agglomerans*). In addition, some enteric bacteria can live within a plant vascular system (for example, *E. coli*, *Klebsiella*, and *Salmonella*).

An excellent example of how multidisciplinary collaboration resulted in better understanding of an infectious outbreak occurred in 2006 when approximately 200 people in 26 states were diagnosed with a particularly virulent case of *E. coli* O157:H7. Nearly half of the cases were hospitalized, and many suffered from hemolytic uremic syndrome (HUS; kidney failure described in Section 26.4). The source of the infection was contaminated spinach traced to the Salinas Valley of California. It turned out the organisms were contained within the vascular system of the spinach, so washing the spinach would not remove the pathogen. Had this outbreak been viewed through only the narrow lens of human health, efforts would have focused on morbidity, mortality, outbreak investigation, laboratory diagnosis, and clinical treatment. The origin of the disease would have remained a mystery.

Working together, epidemiologists and veterinarians found a genetically identical organism in cattle close to where the spinach was produced and in wild hogs that ran through the same fields. Ecologists and hydrologists understood that the groundwater and surface water in this region were being mixed because of a drought followed by heavy rains, and that irrigation systems were strained in the effort to keep up with intensified agricultural production. Eventually, the same *E. coli* strain was found in one of the water ditches close to the spinach fields in the area. Scientists pondering these facts within the One Health framework deduced that cattle harboring the *E. coli* had defecated in a field, thereby contaminating wild hogs. The wild hogs, by running through the spinach fields, had contaminated those fields with their feces, and the irrigation water had then swept the pathogen into the plant vasculature. Only by integrating our knowledge of the environment and ecology could this investigation be completely understood and appropriate intervention and prevention strategies implemented. This outbreak exemplifies the fact that human health and animal health are inextricably linked and that a holistic approach is needed to understand, protect, and promote the health of all species.

To Summarize

- **Emerging diseases** can spread quickly around the world as a result of air travel.
- **Modern technology** and urban growth have provided opportunities for new diseases to emerge.
- **Multidisciplinary collaboration** among ecologists, veterinarians, clinicians, and other scientists is necessary for devising appropriate strategies aimed at disease intervention and epidemic prevention.

Concluding Thoughts

Here in Chapter 28 we have examined the basic principles used to collect, detect, and track pathogenic microorganisms. As you can see, the task of controlling the spread of disease is daunting and ever changing because microorganisms continue to evolve. Known pathogens change, eluding eradication efforts by the immune system and antibiotics, while new pathogens keep emerging. The world is depending on the next generation of microbiologists to face the growing threat of these evolving microbes.

CHAPTER REVIEW

Review Questions

1. Why is it important to identify the genus and species of a pathogen?
2. What is an API strip, and what is its use in clinical microbiology?
3. Describe three examples of selective media.
4. If a colony on a nutrient agar plate is catalase-positive, does this mean it is made up of Gram-positive microorganisms? Why or why not?
5. Describe the types of hemolysis visualized on blood agar.
6. What is the clinical significance of a group A, beta-hemolytic streptococcus?
7. How does one distinguish *Staphylococcus aureus* from *Staphylococcus epidermidis*?
8. Why are PCR identification tests preferable to biochemical approaches?
9. How is qRT-PCR performed?
10. Describe an ELISA.
11. Name some sterile and nonsterile body sites.
12. List seven common types of clinical specimens collected for bacteriological examination.
13. Describe key features of the four levels of biological containment.
14. How is a pandemic different from an epidemic?
15. How can genomics help identify nonculturable pathogens?
16. List and briefly describe four emerging diseases.
17. Name four select agents and the diseases they cause.

Thought Questions

The first four thought questions below are based on the following case history: *An infectious disease physician in Florida telephoned the CDC to report two possible cases of botulism. The two male patients presented with drooping eyelids, double vision, difficulty swallowing, and respiratory problems. The physician had drawn sera and collected stool specimens from the men to test for botulinum toxin, but no results were available.*

1. What are the major concerns raised by these two possible cases of botulism?
2. How might you go about swiftly determining if there is a link between the two cases and if there are other cases of botulism?
3. The two patients ate only one food item in common: a cured and fermented fish called "moloha." How would you determine whether the men are indeed suffering from botulism and whether the fermented fish is the source of disease?
4. How could this anaerobic pathogen grow and make toxin in this fish product? Propose rational theories for what has hindered development of a vaccine against botulism.
5. Consider the following hypothetical case: Five small outbreaks of Ebola occurred within days of one another in five different cities in the United States and Canada. Patient histories revealed that one person from each city had been in the Atlanta airport at the same time 2 days prior to becoming seriously ill. None had left the country recently; none flew on the same jet or even crossed paths while in the airport. As an epidemiologist, conjure up a scenario, excluding bioterrorism, that could account for this scattered outbreak. You are aware that an outbreak recently occurred in the Congo.

Key Terms

amplicon (1160)
endemic (1172)
epidemic (1172)
epidemiology (1149, 1171)
index case (1172)
multiplex PCR (1160)
pandemic (1172)
quarantine (1172)
reservoir (1172)
sequela (1148)

Recommended Reading

Bale, James F., Jr. 2012. Emerging viral infections. *Seminars in Pediatric Neurology* **19**:152–157.

Barczak, Amy K., James E. Gomez, Benjamin B. Kaufmann, Ella R. Hinson, Lisa Cosimi, et al. 2012. RNA signatures allow rapid identification of pathogens and antibiotic susceptibilities. *Proceedings of the National Academy of Sciences USA* **109**:6217–6222.

Bengis, R. G., F. A. Leighton, J. R. Fischer, M. Artois, T. Morner, et al. 2004. The role of wildlife in emerging and re-emerging zoonoses. *Reviews in Science and Technology* **23**:497–511.

Espy, Mark, James Uhl, Lynne Sloan, Seanne Buckwalter, Mary Jones, et al. 2006. Real-time PCR in clinical microbiology: Applications for routine laboratory testing. *Clinical Microbiology Reviews* **19**:165–256.

Fraser, Christophe, Christl A. Donnelly, Simon Cauchemez, William P. Hanage, Maria D. Van Kerkhove, et al. 2009. Pandemic potential of a strain of influenza A (H1N1): Early findings. *Science* **324**:1557–1561.

Koser, Claudio U., Matthew J. Ellington, Edward J. Cartwright, Stephen H. Gillespie, Nicholas M. Brown, et al. 2012. Routine use of microbial whole genome sequencing in diagnostic and public health microbiology. *PLoS Pathogens* **8**:e1002824.

Kreft, Rachael, J. William Costerton, and Garth D. Ehrlich. 2013. PCR is changing clinical diagnostics. *Microbe* **8**:15–20.

Madoff, Lawrence C., and John P. Woodall. 2005. The Internet and global monitoring of emerging diseases: Lessons from the first 10 years of ProMED-mail. *Archives of Medical Research* **36**:724–730.

Morse, Stephen S. 2012. Public health surveillance and infectious disease detection. *Biosecurity and Bioterrorism* **10**:6–16.

Pejcic, Bobby, Roland De Marco, and Gordon Parkinson. 2006. The role of biosensors in the detection of emerging infectious diseases. *Analyst* **131**:1079–1090.

Relman, David A., Thomas M. Schmidt, Richard P. MacDermott, and Stanley Falkow. 1992. Identification of the uncultured bacillus of Whipple's disease. *New England Journal of Medicine* **327**:283–301.

Sheffield, Perry E., and Philp J. Landrigan. 2011. Global climate change and children's health: Threats and strategies for prevention. *Environmental Health Perspectives* **119**:291–298.

Shvartzman, Pesach, and Yussuf Nasri. 2004. Urine culture collected from gel-based diapers: Developing a novel experimental laboratory method. *Journal of the American Board of Family Practice* **17**:91–95.

Soto, S. M. 2009. Human migration and infectious diseases. *Clinical Microbiology and Infection* **15**(Suppl.1):26–28.

van Belkum, Alex, Geraldine Durand, Michel Peyret, Sonia Chatellier, Gilles Zambardi, et al. 2013. Rapid clinical bacteriology and its future impact. *Annals of Laboratory Medicine* **33**:14–27.

APPENDIX 1
Biological Molecules

A1.1 Elements, Bonding, and Water

A1.2 Organic Molecules

A1.3 Proteins

A1.4 Carbohydrates

A1.5 Nucleic Acids

A1.6 Lipids

A1.7 Biological Chemistry

Appendix 1 reviews information typically covered in an introductory biology course. We first cover the chemical bonding principles needed to understand biological molecules, with an emphasis on the special properties of water. We then discuss organic molecules, paying particular attention to four important classes of organic biomolecules: proteins, carbohydrates, nucleic acids, and lipids. Finally, we explore common chemical principles, such as concentrations, thermodynamics, equilibrium, pH, and oxidation-reduction reactions.

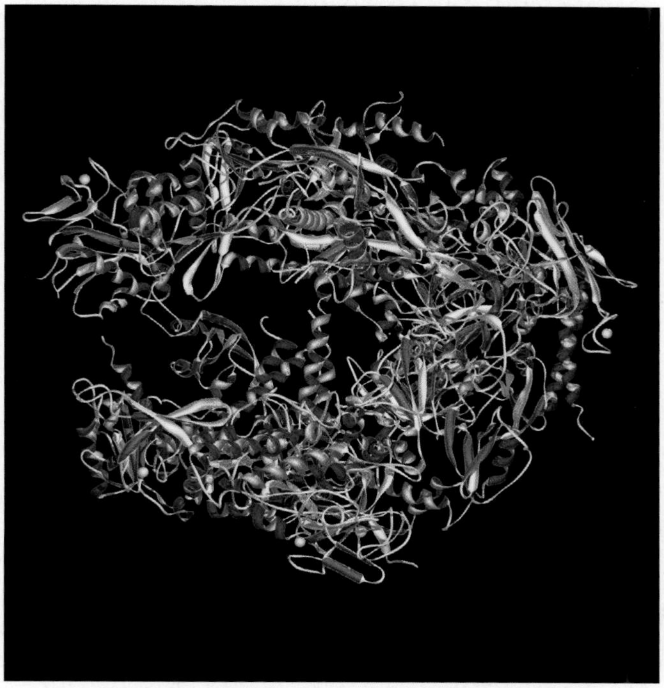

The enzyme RNA polymerase II (computer model). The molecule comprises 12 subunits. This enzyme synthesizes a complementary mRNA strand from a strand of DNA during a process called transcription. It recognizes a start sign on the DNA strand and then moves along the strand, building the mRNA until it reaches a stop sign. Messenger RNA is the intermediary between DNA and its protein product. *Source:* Mark J. Winter/Photo Researchers, Inc.

Living cells are remarkably complex machines, able to integrate and respond to multiple stimuli, to catalyze reactions, and to replicate themselves. Yet despite all the various tasks that cells perform, 98% of the mass of living organisms consists of only six elements: hydrogen (H), oxygen (O), nitrogen (N), phosphorus (P), sulfur (S), and carbon (C); and 90% of the mass is accounted for by just C, H, and O. Of the compounds formed from these elements, the most abundant in cells is water. The remainder of the cell consists, for the most part, of just four different kinds of organic (carbon-based) macromolecules: proteins, nucleic acids, carbohydrates, and lipids. Cells can be thought of as compartments that orchestrate chemical reactions between these organic molecules. It is clear that to understand life, we must understand the properties of water and organic molecules and of their building blocks, the chemical elements.

A1.1

Elements, Bonding, and Water

Cells consist mostly of water and organic molecules. These cellular components are formed from atoms of various elements. An atom consists of a positively charged nucleus that contains protons and neutrons, surrounded by negatively charged electrons. The hydrogen nucleus consists of a single proton. Protons, neutrons, and electrons differ in their mass and charge as summarized in **Table A1.1**.

The elements can be organized into a periodic table as in **Figure A1.1**, indicating each element's **atomic number** (the number of protons) and **atomic mass** (the mass, in grams, of 1 mole of the element). The defining characteristic of an element is the atomic number. For example, all carbon atoms have six protons in their nucleus. To maintain neutrality, atoms have negatively charged electrons equal in number to the positively charged protons.

The atomic mass is the sum of the number of protons and neutrons. The atomic mass is an average that takes into account the relative abundance of each isotope. **Isotopes** are atoms of an element that differ in the number of neutrons. For example, the most abundant isotope of carbon is carbon-12 (with six neutrons), but there are naturally occurring isotopes of carbon-13 (seven neutrons) and carbon-14 (eight neutrons). Because carbon-13 and carbon-14 are rare, the average atomic mass is close to but not exactly 12. While some isotopes are stable, others decay at a known rate and give off radioactivity. Carbon-14 has a **half-life** (the amount of time it takes for half of a sample to decay) of 5,700 years and is used in radiocarbon dating to determine the age of organic material. Carbon-14 and shorter-lived isotopes, such as tritium (hydrogen with a mass of 3—one proton and two neutrons), are used by scientists as tracers to follow specific atoms in metabolic pathways.

TABLE A1.1

The mass and charge of atomic particles.

Particle	Mass (atomic mass unit)	Charge (electronic charge unit)
Proton	1	+1
Neutron	1	0
Electron	0.0005	−1

Bonds between Atoms Form Molecules

Atoms combine by sharing electrons to form molecules. For example, two atoms of oxygen combine to form molecular oxygen, O_2. Molecules may also contain more than one kind of element; an example is water, H_2O. The symbols O_2 and H_2O are examples of **molecular formulas**, a shorthand notation indicating the number and type of atoms present in a molecule.

Each column (group) in the periodic table (**Fig. A1.1**) contains elements of similar reactivity as a result of the similarity in their electronic configurations, particularly of electrons in the outermost shell. The shell closest to the nucleus can hold a maximum of two electrons, and the next shell, a maximum of eight electrons. **Figure A1.2A** shows all the electrons for hydrogen, carbon, nitrogen, and oxygen. Each unpaired electron is capable of participating in a bond. In **Figure A1.2A**, it is clear that H can form one bond; C, four bonds; N, three bonds; and O, two bonds. For example, carbon has four unpaired electrons in its outermost shell; each can form a bond with an unpaired electron from another atom. The two atoms involved in this bond "share" the two electrons. This sharing of electrons is termed a **covalent bond**. Covalent bonds are very strong and difficult to break. In methane (CH_4), carbon forms four covalent bonds with four hydrogen atoms (**Fig. A1.2B**). Methane is stable because both carbon and hydrogen have filled their outer shells. Atoms can also share more than one pair of electrons with

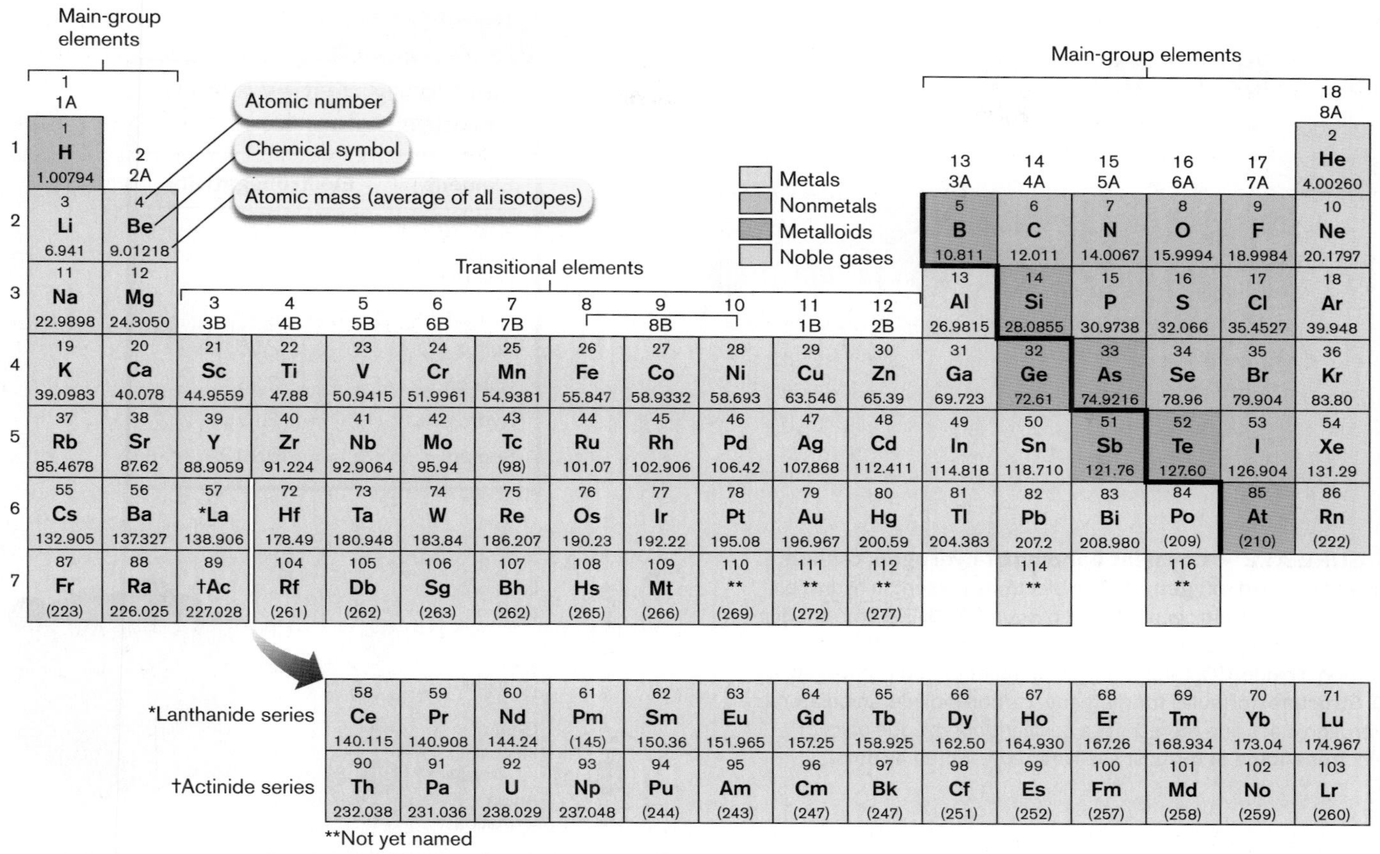
Main-group elements
Main-group elements
Atomic number
Chemical symbol
Atomic mass (average of all isotopes)
Metals
Nonmetals
Metalloids
Noble gases
Transitional elements
1 1A
2 2A
13 3A
14 4A
15 5A
16 6A
17 7A
18 8A
3 3B
4 4B
5 5B
6 6B
7 7B
8 9 10 8B
11 1B
12 2B
1 H 1.00794
2 He 4.00260
3 Li 6.941
4 Be 9.01218
5 B 10.811
6 C 12.011
7 N 14.0067
8 O 15.9994
9 F 18.9984
10 Ne 20.1797
11 Na 22.9898
12 Mg 24.3050
13 Al 26.9815
14 Si 28.0855
15 P 30.9738
16 S 32.066
17 Cl 35.4527
18 Ar 39.948
19 K 39.0983
20 Ca 40.078
21 Sc 44.9559
22 Ti 47.88
23 V 50.9415
24 Cr 51.9961
25 Mn 54.9381
26 Fe 55.847
27 Co 58.9332
28 Ni 58.693
29 Cu 63.546
30 Zn 65.39
31 Ga 69.723
32 Ge 72.61
33 As 74.9216
34 Se 78.96
35 Br 79.904
36 Kr 83.80
37 Rb 85.4678
38 Sr 87.62
39 Y 88.9059
40 Zr 91.224
41 Nb 92.9064
42 Mo 95.94
43 Tc (98)
44 Ru 101.07
45 Rh 102.906
46 Pd 106.42
47 Ag 107.868
48 Cd 112.411
49 In 114.818
50 Sn 118.710
51 Sb 121.76
52 Te 127.60
53 I 126.904
54 Xe 131.29
55 Cs 132.905
56 Ba 137.327
57 *La 138.906
72 Hf 178.49
73 Ta 180.948
74 W 183.84
75 Re 186.207
76 Os 190.23
77 Ir 192.22
78 Pt 195.08
79 Au 196.967
80 Hg 200.59
81 Tl 204.383
82 Pb 207.2
83 Bi 208.980
84 Po (209)
85 At (210)
86 Rn (222)
87 Fr (223)
88 Ra 226.025
89 †Ac 227.028
104 Rf (261)
105 Db (262)
106 Sg (263)
107 Bh (262)
108 Hs (265)
109 Mt (266)
110 ** (269)
111 ** (272)
112 ** (277)
114 **
116 **
*Lanthanide series
58 Ce 140.115
59 Pr 140.908
60 Nd 144.24
61 Pm (145)
62 Sm 150.36
63 Eu 151.965
64 Gd 157.25
65 Tb 158.925
66 Dy 162.50
67 Ho 164.930
68 Er 167.26
69 Tm 168.934
70 Yb 173.04
71 Lu 174.967
†Actinide series
90 Th 232.038
91 Pa 231.036
92 U 238.029
93 Np 237.048
94 Pu (244)
95 Am (243)
96 Cm (247)
97 Bk (247)
98 Cf (251)
99 Es (252)
100 Fm (257)
101 Md (258)
102 No (259)
103 Lr (260)
**Not yet named

Symbol	Name	Symbol	Name	Symbol	Name	Symbol	Name	Symbol	Name
Ac	Actinium	Co	Cobalt	Ir	Iridium	O	Oxygen	Ag	Silver
Al	Aluminum	Cu	Copper	Fe	Iron	Pd	Palladium	Na	Sodium
Am	Americium	Cm	Curium	Kr	Krypton	P	Phosphorus	Sr	Strontium
Sb	Antimony	Db	Dubnium	La	Lanthanum	Pt	Platinum	S	Sulfur
Ar	Argon	Dy	Dysprosium	Lr	Lawrencium	Pu	Plutonium	Ta	Tantalum
As	Arsenic	Es	Einsteinium	Pb	Lead	Po	Polonium	Tc	Technetium
At	Astatine	Er	Erbium	Li	Lithium	K	Potassium	Te	Tellurium
Ba	Barium	Eu	Europium	Lu	Lutetium	Pr	Praseodymium	Tb	Terbium
Bk	Berkelium	Fm	Fermium	Mg	Magnesium	Pm	Promethium	Tl	Thallium
Be	Beryllium	F	Fluorine	Mn	Manganese	Pa	Protactinium	Th	Thorium
Bi	Bismuth	Fr	Francium	Mt	Meitnerium	Ra	Radium	Tm	Thulium
Bh	Bohrium	Gd	Gadolinium	Md	Mendelevium	Rn	Radon	Sn	Tin
B	Boron	Ga	Gallium	Hg	Mercury	Re	Rhenium	Ti	Titanium
Br	Bromine	Ge	Germanium	Mo	Molybdenum	Rh	Rhodium	W	Tungsten
Cd	Cadmium	Au	Gold	Nd	Neodymium	Rb	Rubidium	U	Uranium
Ca	Calcium	Hf	Hafnium	Ne	Neon	Ru	Ruthenium	V	Vanadium
Cf	Californium	Hs	Hassium	Np	Neptunium	Rf	Rutherfordium	Xe	Xenon
C	Carbon	He	Helium	Ni	Nickel	Sm	Samarium	Yb	Ytterbium
Ce	Cerium	Ho	Holmium	Nb	Niobium	Sc	Scandium	Y	Yttrium
Cs	Cesium	H	Hydrogen	N	Nitrogen	Sg	Seaborgium	Zn	Zinc
Cl	Chlorine	In	Indium	No	Nobelium	Se	Selenium	Zr	Zirconium
Cr	Chromium	I	Iodine	Os	Osmium	Si	Silicon		

FIGURE A1.1 ■ **Periodic table of the elements.** The atomic number (number of protons) and atomic mass are shown for each element.

another atom, forming double or triple bonds. In carbon dioxide (CO_2), each oxygen shares two pairs of electrons with carbon; and in diatomic nitrogen (N_2), the nitrogen atoms each share three pairs of electrons (see **Fig. A1.2B**). The bonding in molecules can be represented in a shorthand representation by a **structural formula**, in which covalent bonds are shown as lines between two atoms (**Fig. A1.2C**).

Another way atoms can obtain full outer shells is by gaining or losing electrons. A complete transfer of

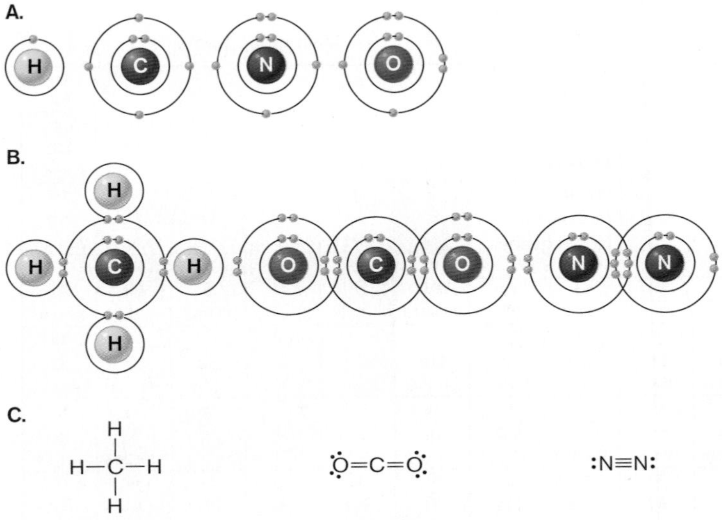
A.
B.
C.

FIGURE A1.2 ▪ Covalent bonding of hydrogen, carbon, nitrogen, and oxygen. A. The electrons present in hydrogen (H), carbon (C), nitrogen (N), and oxygen (O). **B.** Electron sharing in the four single bonds of methane, CH_4; the two double bonds of carbon dioxide, CO_2; and the triple bond of diatomic nitrogen, N_2. **C.** Structural formulas for methane, carbon dioxide, and diatomic nitrogen. Each line represents a covalent bond. Lone pairs of electrons in the outermost shell are represented by dots.

TABLE A1.2

Electronegativities of some common elements.

Element	Electronegativity
Oxygen	3.44
Chlorine	3.16
Nitrogen	3.04
Sulfur	2.58
Carbon	2.55
Hydrogen	2.10
Sodium	0.93

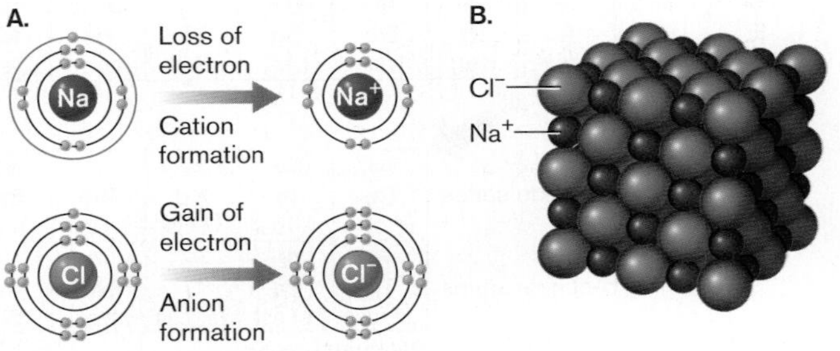
A.
Loss of electron
Cation formation
Gain of electron
Anion formation
B.
Cl⁻
Na⁺

FIGURE A1.3 ▪ Formation of ions and ionic crystals. A. The loss of an electron from an atom of sodium to form the cation Na^+ and the gain of an electron by a chlorine atom to form the chloride anion Cl^-. **B.** Oppositely charged anions and cations—in this case, Cl^- and Na^+—are attracted to one another and form crystals of table salt (sodium chloride).

electrons can occur between two atoms that have a large difference in **electronegativity**, a measure of the affinity of an atom for electrons. A large electronegativity indicates a strong attraction for electrons. Of the elements listed in **Table A1.2**, oxygen has the greatest attraction for electrons, and sodium has the weakest. If two elements with greatly different electronegativities come into close contact, one element can "steal" an electron from the other. For example, sodium (Na) and chlorine (Cl) interact to form table salt, NaCl. The electronegative Cl strips an electron away from Na, and both Cl^- and Na^+ now have full outer shells (**Fig. A1.3A**). Both Cl^- and Na^+ are charged atoms called **ions**, in which the number of electrons and protons are unequal. **Anions** are negatively charged ions, and **cations** are positively charged ions. Anions and cations can form ionic crystals held together by **ionic bonds**, electrostatic attractions between anions and cations (**Fig. A1.3B**). In the absence of water, ionic crystals maintain their integrity, but in the presence of water they are destabilized. To understand why, we need to understand the structure of water.

Water Is the Solvent of Life

Living organisms consist mostly of water, and water has many unique properties that render it particularly suitable for sustaining life. To appreciate these properties, we must understand the forces that hold water together: polar covalent bonds and hydrogen bonds. Molecules such as H_2 and O_2 have even charge distribution because both atoms in the molecule have the same electronegativity. Therefore, the electrons in the covalent bond are shared equally and form **nonpolar covalent bonds**. In contrast, the shared electrons in H_2O spend more of their time around the highly electronegative oxygen than in the vicinity of the less electronegative hydrogen (**Fig. A1.4A**). A bond with unequal electron sharing is a **polar covalent bond**, so called because the molecule has partial positive and negative poles. Polar covalent bonds occur in individual water molecules, forming a partial charge separation within the molecule.

Water molecules experience a strong electrostatic attraction for one another as a result of the charge separation present in individual molecules. This electrostatic attraction, known as a **hydrogen bond**, occurs between a

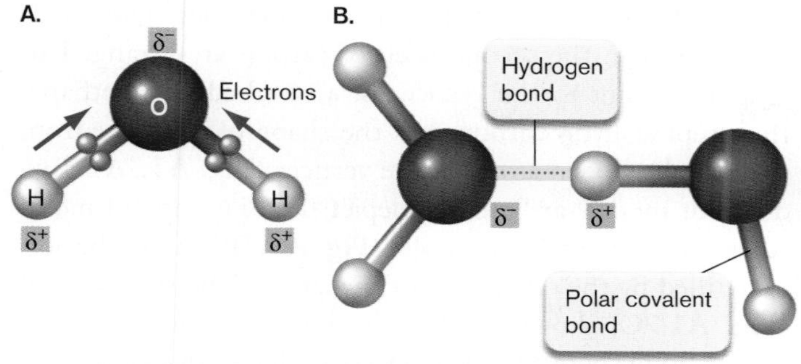
A.
δ⁻
O
Electrons
H
H
δ⁺
δ⁺
B.
Hydrogen bond
δ⁻
δ⁺
Polar covalent bond

FIGURE A1.4 ▪ Polar covalent and hydrogen bonds in water. **A.** The polar covalent bonds in an individual water molecule. Displacement of electrons toward oxygen causes oxygen to have a partial negative charge and the hydrogen atoms to have a partial positive charge. Hence, water is polar. **B.** A hydrogen bond between two water molecules.

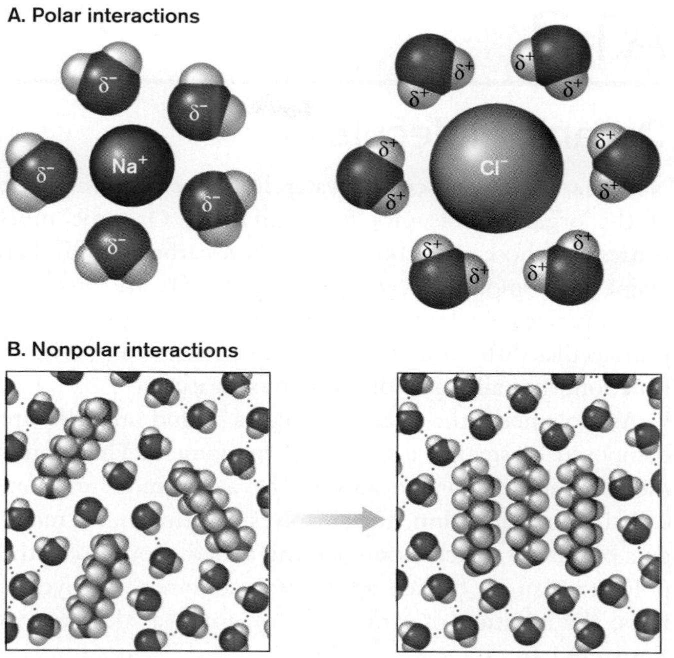
A. Polar interactions
Na⁺
Cl⁻
B. Nonpolar interactions

FIGURE A1.5 ▪ Interactions between water and solutes. **A.** Water surrounds and interacts with individual ions or polar molecules, causing them to dissolve in water. **B.** Nonpolar molecules do not dissolve in water. Instead, to minimize the disruption of hydrogen bonding among water molecules, nonpolar molecules aggregate in water.

hydrogen bonded to an oxygen or nitrogen and a second oxygen or nitrogen, either in the same or a different molecule. Hydrogen bonds are short-lived and constantly break and re-form in liquid water. **Figure A1.4B** shows the polar covalent and hydrogen bonds present in water.

The hydrogen bonds in water contribute the unique properties that enable it to support life. Water is a liquid over a large temperature range because hydrogen bonds cause water molecules to associate, favoring the liquid state over the gas. The hydrogen bonds that keep water a liquid over a wide range of temperatures also endow water with a high specific heat, the amount of energy needed to raise the temperature of 1 gram (g) of a substance by 1°C. The high specific heat of water moderates the temperature of all aqueous environments, oceans and cells alike.

The polar nature of water defines its properties as a solvent. Compounds that are ionic or polar themselves tend to dissolve in water and are termed **hydrophilic**. For example, the ionic bonds in NaCl are very strong in the absence of water, but are easily dissolved in the presence of water. This is because the polar water can surround and interact with the sodium and chloride ions and shield them from each other (**Fig. A1.5A**). Water also dissolves polar compounds. In this case, water does not break the polar covalent bonds; rather, individual, intact polar molecules are surrounded by water owing to electrostatic interactions. In contrast, compounds that are mostly nonpolar do not dissolve in water and are termed **hydrophobic**. Nonpolar molecules have no partial charges to attract water. Because water hydrogen-bonds with other water molecules, it tends to exclude nonpolar compounds, forcing them together (**Fig. A1.5B**). The aggregated nonpolar compounds can be further stabilized by **van der Waals forces**, weak temporary electrostatic attractions between molecules caused by random movements of their electron clouds. **Table A1.3** lists the bonds we have discussed and indicates the strength of the bonds in water.

TABLE A1.3

Bond types and strengths.

Type of bond	Description	Bond strength in water (kJ/mol)[a]
Covalent	Sharing of electrons	210–418
Ionic	Electrostatic attraction between anion and cation	12.5
Hydrogen	Electrostatic attraction between a hydrogen bonded to a nitrogen or oxygen and a second nitrogen or oxygen	4
Van der Waals	Electrostatic attraction between temporary, shifting electron clouds	0.42 (per atom)

[a]Bond strengths are given in water, which is similar to the environment in the aqueous cytoplasm. Anhydrous bond strengths differ from those listed.

A1.2

Organic Molecules

Now that we have discussed water, let's take a look at some of the organic molecules found in cells. **Organic molecules** are those that contain a carbon-carbon bond. The major macromolecular components of cells are proteins, carbohydrates, nucleic acids, and lipids. Although the macromolecules differ from each other in structure and cellular function, they all share some common features.

As polymers, the macromolecules found in cells are composed of smaller units called monomers. The monomers are joined to one another by a common reaction, called a **condensation**, that involves splitting out a molecule of water for each monomer unit added. Conversely, the polymers can be broken apart into monomers by **hydrolysis**, the addition of a molecule of water, as depicted in **Figure A1.6**.

The 3D shape, or structure, of a molecule is critical for proper function. The structure is determined by which atoms are present and how they are bonded together. Structural formulas like those in **Figure A1.2C** represent the sequence of bonds in a molecule; however, bond structure does not adequately convey information about the 3D shape of the molecule. It is particularly important to understand the shape of carbon-containing molecules because carbon is the backbone for most cellular macromolecules. The bonding orbitals in the second shell of carbon are arranged so that they point to the vertices of a tetrahedron; methane, the simplest hydrocarbon, has the shape of a tetrahedron, with the hydrogen atoms at the vertices (**Fig. A1.7A**). Two different models are used to depict the 3D shape of molecules. The **space-filling model** (**Fig. A1.7B**) shows the volume filled by the outer shell of the atoms. The **stick model** (**Fig. A1.7C**) shows the length and orientation of interatomic bonds between the nuclei of each pair of atoms.

An important feature of any organic molecule is the number and type of functional groups present. **Functional groups** are small groups of atoms with characteristic bonding, shape, and reactivity. A few of the more common ones are shown in **Table A1.4**. Knowing which functional groups are present in a molecule allows us to infer something about the structure and reactivity of the molecule. Information about the functional groups present in a molecule is often indicated by a molecule's name; for example, amino acids contain an amino group and a carboxyl (carboxylic acid) group.

We will discuss different classes of molecules individually, but in living cells, molecules of different types are frequently found in combination. For example, sugars decorate some proteins (glycoproteins) and lipids (glycolipids), and the ribosome is a complex of proteins and ribonucleic acids. We now examine the structure of the fundamental biological macromolecules in detail.

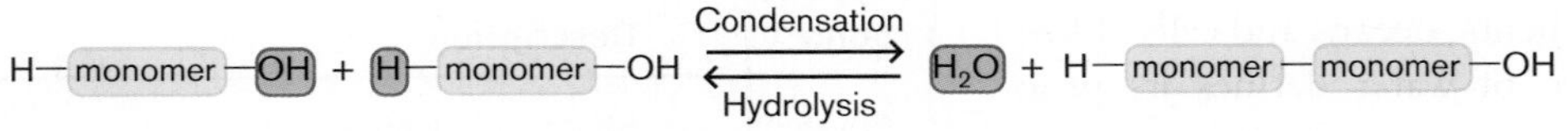

Condensation
H—monomer—OH + H—monomer—OH
Hydrolysis
H_2O + H—monomer—monomer—OH

FIGURE A1.6 ■ Condensation and hydrolysis reactions. Monomers can be covalently bonded to form polymers through a condensation reaction that liberates a water molecule. Polymers can be broken apart into monomers through hydrolysis (water-splitting) reactions.

FIGURE A1.7 ■ Molecular models of methane. A. A molecule of methane, showing the tetrahedral arrangement of the electron orbitals in the second (valence) shell of carbon. **B.** Space-filling model of methane. **C.** Stick model of methane.

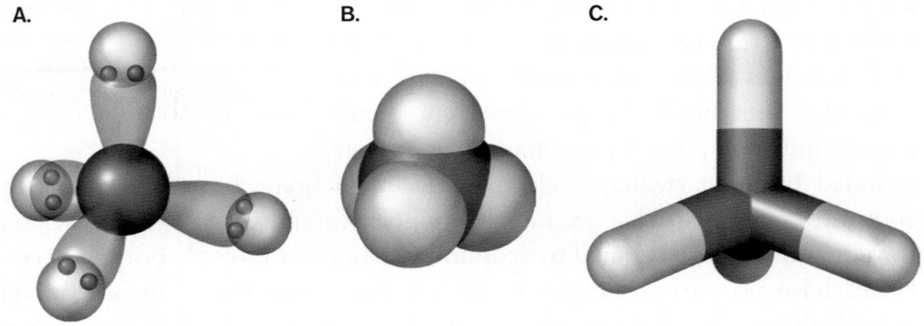

A.
B.
C.

TABLE A1.4

Common functional groups.

Functional group	General structure	Example	Comments
Aldehyde	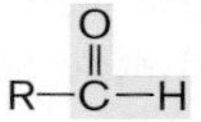	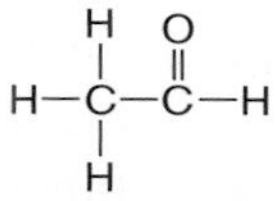 **Acetaldehyde**	Can react with alcohols 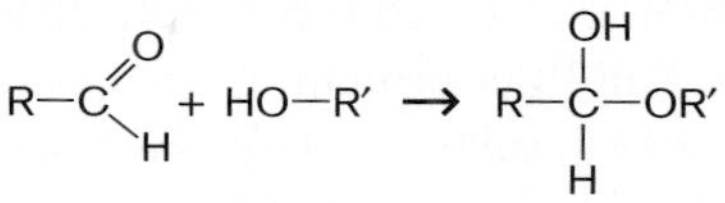
Alkane	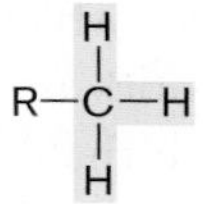	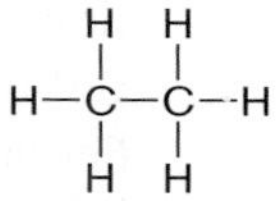 **Ethane**	Nonpolar, tends to make molecules containing it hydrophobic; nonreactive
Amino	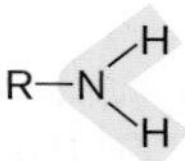	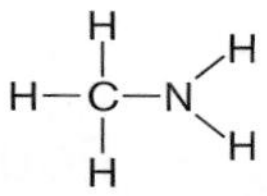 **Methylamine**	Acts as a base by binding a proton; found in amino acids $R{-}NH_2 + H^+ \longrightarrow R{-}NH_3^+$
Carboxyl		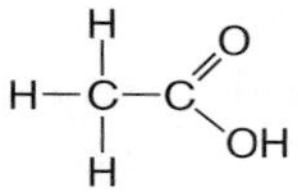 **Acetic acid**	Acts as an acid by releasing a proton; the ionized form name ends in "ate" (e.g., "acetate") 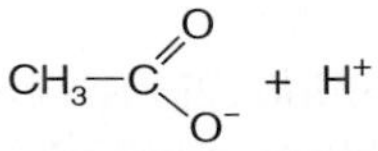 **Acetate** **Proton**
Ester	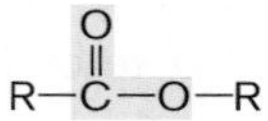	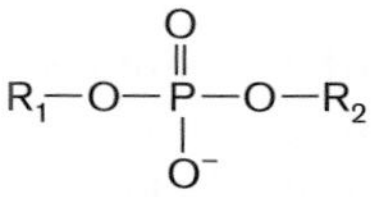 **Phosphodiester**	Common linkage found in lipids
Hydroxyl	R—O—H	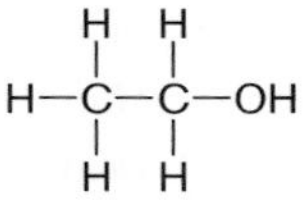 **Ethanol**	Polar, makes compounds more soluble through hydrogen bonding; found in alcohols and sugars
Ketone	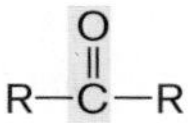	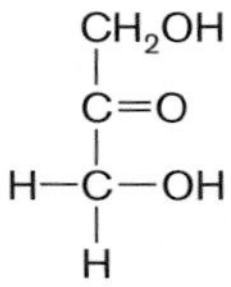 **Dihydroxyacetone**	Found in many intermediates of metabolism
Phosphate	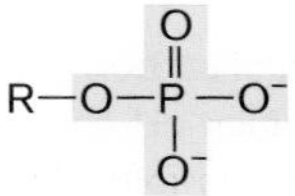	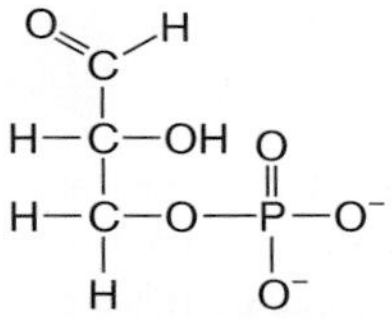 **Glyceraldehyde 3-phosphate**	When two or more phosphoryl groups are linked, a high-energy bond forms because the negative oxygens repel each other

A1.3

Proteins

Cells express thousands of different proteins and may contain more than 2 million protein molecules. Proteins perform many functions, such as catalyzing reactions, serving as receptors and transporters, providing structure, and aiding movement. Proteins can carry out such diverse functions because they can fold into a variety of 3D structures.

Amino Acids Are the Building Blocks of Proteins

Although different proteins can have vastly different structures, all proteins are composed of the same building blocks: amino acids. All 20 of the amino acids commonly occurring in nature have the same general structure. Each **amino acid** contains a central carbon atom (the alpha carbon) covalently bonded to four different moieties: a hydrogen atom, an amino group, a carboxyl group, and a side chain (R, residue) that is unique for each amino acid (**Fig. A1.8A**).

The alpha carbon of an amino acid is a **chiral carbon**, a carbon with four different groups attached to it. Chiral carbons exist in two different forms, called **optical isomers**, or enantiomers. Optical isomers have the same molecular formula and the same order of bonds within the molecules, but a different arrangement of their atoms in space. Optical isomers are actually mirror images of each other, like your left and right hands, and the two forms cannot be superimposed (**Fig. A1.8B**). Each of the amino acid isomers is designated L or D, depending on the configuration at the alpha carbon. The amino acids used to make proteins all have the L configuration at the alpha carbon (**Fig. A1.8**), and all of these amino acids can be derived by substituting the different R groups for one another. Some unusual D-amino acids, with the opposite configuration at the alpha carbon, are found in bacterial cell walls.

The side chains of amino acids can be grouped according to their hydrophobicity, charge, or presence of specific functional groups. **Figure A1.9** depicts the 20 common amino acids, along with their three-letter abbreviations and single-letter codes. All are shown in the L conformation, except for glycine, which does not contain a chiral carbon and therefore does not have optical isomers. Proline has a ring structure that interrupts the regular geometry of the polypeptide chain. The side chain of cysteine is capable of forming inter- and intramolecular disulfide (–S–S–) bonds.

Four Levels of Protein Organization

The first level of organization, the **primary structure**, is simply the linear sequence of amino acids. This is genetically determined; the DNA code specifies the amino acid sequence. During protein synthesis, amino acids are connected together by a condensation reaction to form a covalent **peptide bond**. The peptide bond forms between the carboxyl group of one amino acid and the amino group of a second amino acid (**Fig. A1.10A**). The portion of the amino acid incorporated into the peptide after condensation is called an amino acid residue. Note that the peptide chain has an amino terminus (N terminus) and a carboxyl terminus (C terminus); numbering of the amino acid residues starts at the amino terminus (**Fig. A1.10B**). The bonds on either side of the alpha carbon can rotate, but there is no rotation around the peptide bond, so the backbone of a protein does not rotate freely.

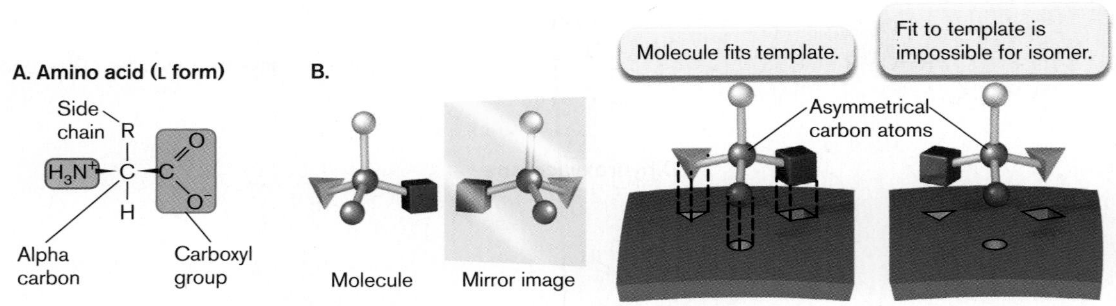
A. Amino acid (L form)
Side chain
R
H_3N^+
C
O
O^-
H
Alpha carbon
Carboxyl group
B.
Molecule
Mirror image
Molecule fits template.
Fit to template is impossible for isomer.
Asymmetrical carbon atoms

FIGURE A1.8 ▪ Amino acid structure and chiral carbons. A. All amino acids contain a central carbon (the alpha carbon) bonded to an amino group, a carboxyl group, a hydrogen, and a variable group or side chain, designated R. At cellular pH, the amino and carboxyl groups are ionized. Because the alpha carbon is bonded to four different molecules, it is a chiral carbon. **B.** Chiral molecules exist in two different forms that are mirror images of each other. The two forms cannot be superimposed, and only the correct isomer can interact with a template (for example, an enzyme).

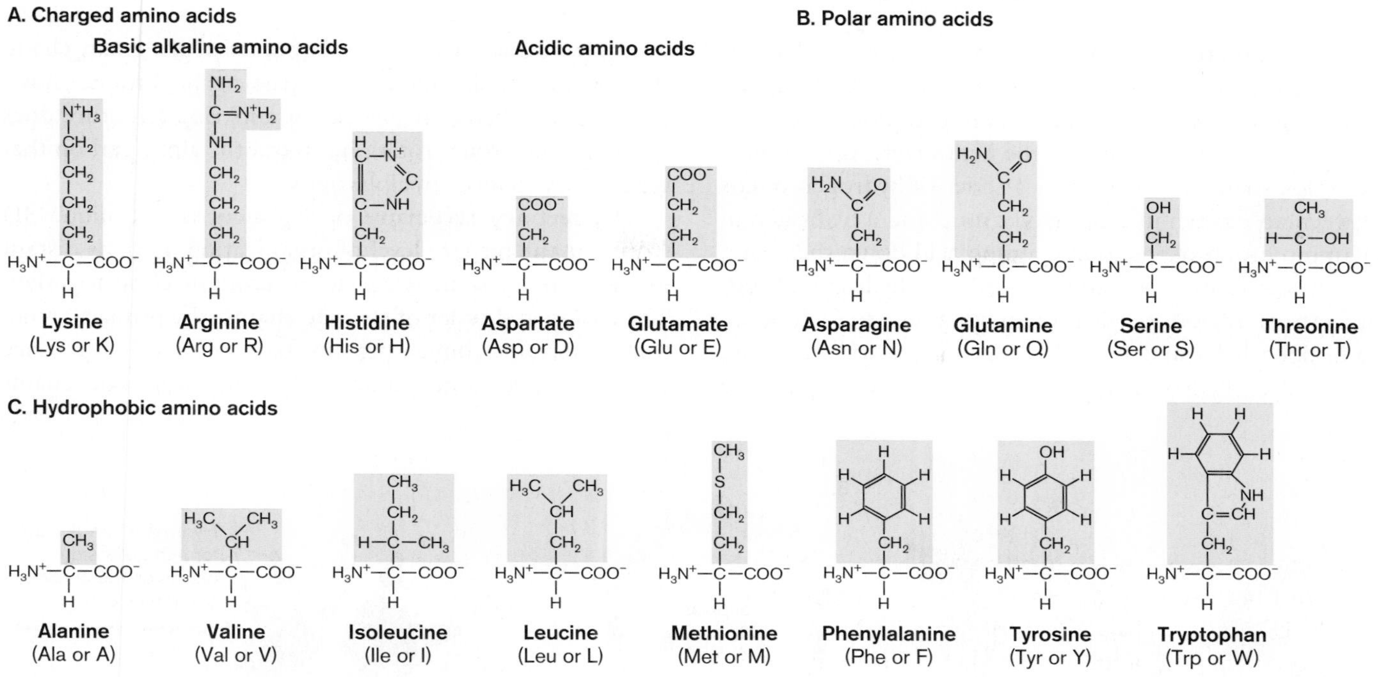
A. Charged amino acids
Basic alkaline amino acids
Acidic amino acids
B. Polar amino acids
Lysine
(Lys or K)
Arginine
(Arg or R)
Histidine
(His or H)
Aspartate
(Asp or D)
Glutamate
(Glu or E)
Asparagine
(Asn or N)
Glutamine
(Gln or Q)
Serine
(Ser or S)
Threonine
(Thr or T)
C. Hydrophobic amino acids
Alanine
(Ala or A)
Valine
(Val or V)
Isoleucine
(Ile or I)
Leucine
(Leu or L)
Methionine
(Met or M)
Phenylalanine
(Phe or F)
Tyrosine
(Tyr or Y)
Tryptophan
(Trp or W)

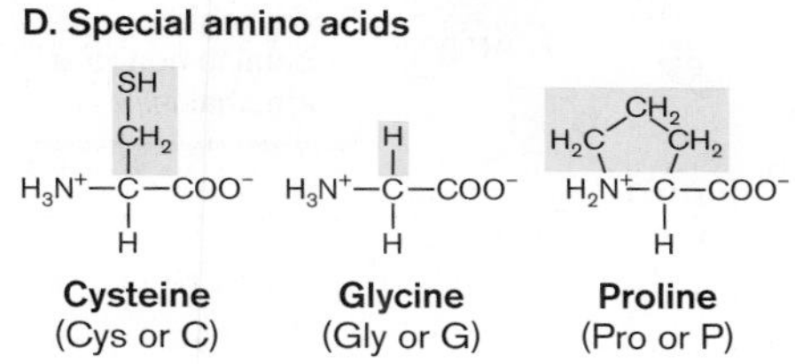
D. Special amino acids
Cysteine
(Cys or C)
Glycine
(Gly or G)
Proline
(Pro or P)

FIGURE A1.9 ■ Twenty common amino acids. The grouping of amino acids is based on their side chains, highlighted in yellow.

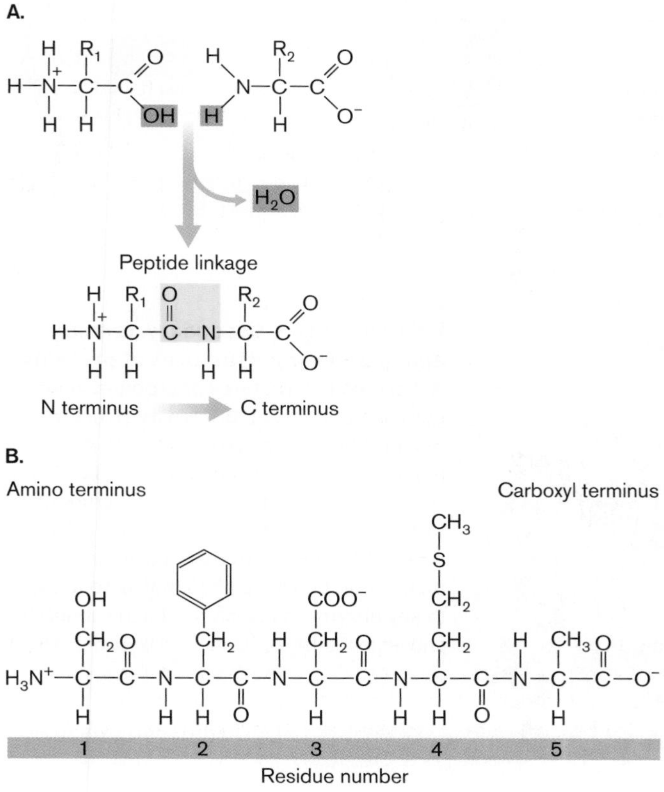
A.
Peptide linkage
N terminus
C terminus
B.
Amino terminus
Carboxyl terminus
Residue number
1
2
3
4
5

FIGURE A1.10 ■ The peptide bond and primary protein structure. A. Formation of a peptide bond. Amino acids are joined by a condensation reaction between the carboxyl group of one amino acid and the amino group of another to form a covalent linkage called a peptide bond. **B.** The primary structure of a pentapeptide, a chain of five amino acids. This pentapeptide has the sequence SFDMA. A protein can be formed of hundreds or thousands of amino acids.

The **secondary structures** of proteins are regular patterns that repeat over short regions of the polypeptide chain. **Figure A1.11A** depicts two common secondary structures: the alpha helix and the beta sheet. Alpha helices are coiled structures formed by a series of hydrogen bonds between an oxygen in a carboxyl group and a hydrogen in an amino group of the fourth amino acid down the chain. Beta sheets form as a result of extensive hydrogen bonding between regions of the protein lying next to each other. Both alpha helix and beta sheet secondary structures are stabilized by hydrogen bonding between the oxygen and hydrogen atoms of the peptide bonds in the main chain; the side chains do not directly participate. Proline, however, acts as a helix breaker because its ring structure does not allow the proper rotation around the alpha carbon that is necessary to form an alpha helix.

The **tertiary structure** of a protein is its unique 3D shape. At the tertiary level of organization, regions distant in the primary structure may be brought close together. The nature and order of the side chains (the primary structure) determines how a protein folds into its final tertiary structure. The large number of amino acid side chains

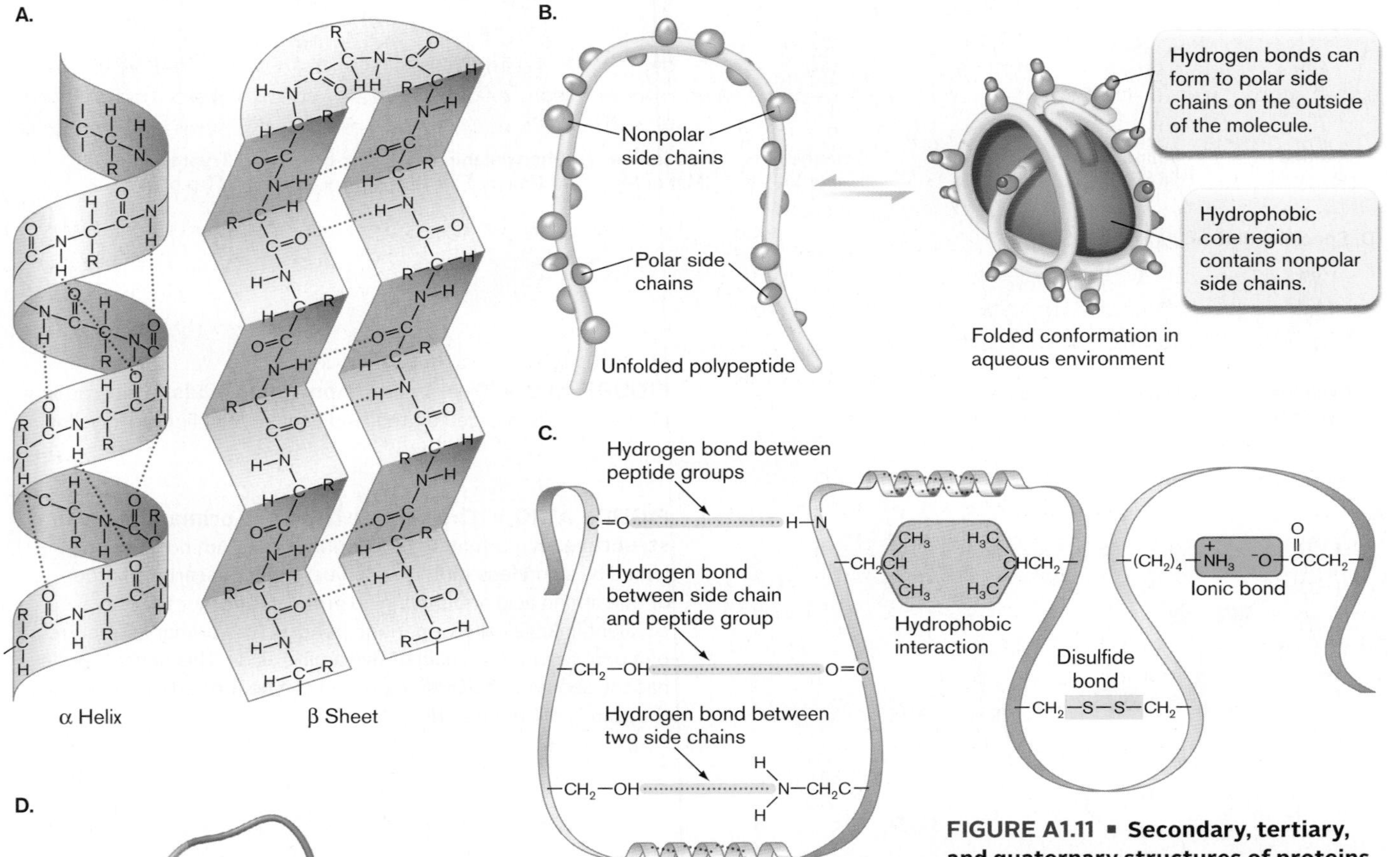
A.
α Helix
β Sheet
B.
Nonpolar side chains
Polar side chains
Unfolded polypeptide
Hydrogen bonds can form to polar side chains on the outside of the molecule.
Hydrophobic core region contains nonpolar side chains.
Folded conformation in aqueous environment
C.
Hydrogen bond between peptide groups
Hydrogen bond between side chain and peptide group
Hydrogen bond between two side chains
Hydrophobic interaction
Disulfide bond
Ionic bond

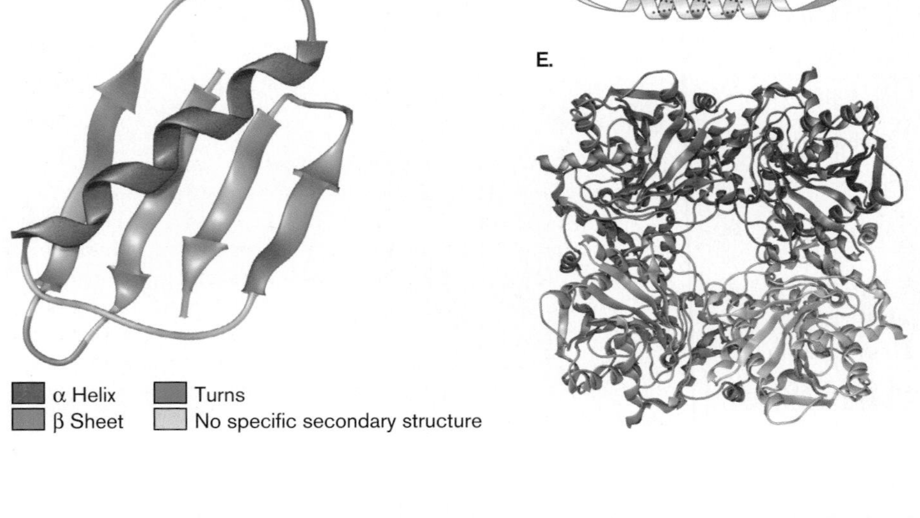
D.
α Helix
β Sheet
Turns
No specific secondary structure
E.

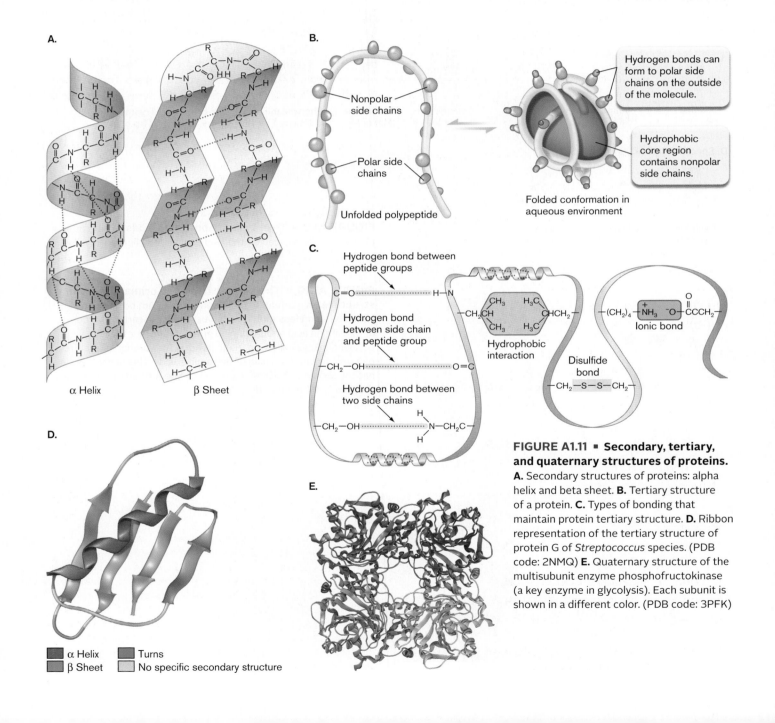

FIGURE A1.11 ■ Secondary, tertiary, and quaternary structures of proteins. **A.** Secondary structures of proteins: alpha helix and beta sheet. **B.** Tertiary structure of a protein. **C.** Types of bonding that maintain protein tertiary structure. **D.** Ribbon representation of the tertiary structure of protein G of *Streptococcus* species. (PDB code: 2NMQ) **E.** Quaternary structure of the multisubunit enzyme phosphofructokinase (a key enzyme in glycolysis). Each subunit is shown in a different color. (PDB code: 3PFK)

gives rise to the diversity of protein structures and, hence, functions. As proteins emerge from the ribosome, they are bound by chaperone proteins that help them fold into their functional shape, known as the **native conformation**. In soluble proteins, amino acids with hydrophobic side chains tend to cluster inside the protein to minimize reactions with water, while polar amino acids are exposed on the surface (**Fig. A1.11B**). Tertiary structure is stabilized by hydrogen bonding between side chains and between side chains and the main chain. Ionic interactions between acidic and basic side chains also contribute to tertiary structure. The one covalent interaction found stabilizing tertiary structure is disulfide bonding between two cysteine residues. The interactions that contribute to protein tertiary structure are shown in **Figure A1.11C**, and a ribbon representation of the 3D structure of a protein is illustrated in **Figure A1.11D**.

Some proteins form stable, functional complexes with other proteins. These multisubunit proteins exhibit the fourth level of protein organization: **quaternary structure** (**Fig. A1.11E**). The forces that hold interacting polypeptide chains together are the same noncovalent interactions and disulfide bonds that determine protein tertiary structure.

Note that proteins are not static structures. The individually weak hydrogen bonds, ionic interactions, and van der Waals forces that contribute to protein structure are constantly being broken and re-formed. Because these noncovalent bonds do not break all at once, the large number of such bonds present act collectively to maintain protein integrity. Some proteins are dynamic entities that may permute through a number of semi-stable states. Such conformational changes may be stochastic, random fluctuations due to the intrinsic thermal (kinetic) energy of the protein. Alternatively, the conformational changes may be influenced by external factors. For many proteins, alterations in conformation are critical for protein function.

A1.4

Carbohydrates

Carbohydrates are so named because their molecular formulas are multiples of CH_2O (hydrated carbon). Polysaccharides are complex carbohydrates, composed of monosaccharide (simple-sugar) monomers. Carbohydrates are a preferred energy source for many cells and also store energy. In addition, they play a structural role, as in cell walls. Moreover, when attached to proteins or lipids, they can serve an informational role.

Monosaccharides

Monosaccharides differ from each other in the number of carbons present, in whether their carbonyl group (C=O) is an aldehyde or a ketone, and in the orientation of the hydroxyl groups. The smallest sugars, glyceraldehyde and dihydroxyacetone, have three carbons (**Fig. A1.12A**). These two sugars have the same molecular formula, $C_3H_6O_3$; but since their atoms are arranged differently, they are **structural isomers** of each other. In addition, because glyceraldehyde contains a chiral carbon at C-2, it also has an optical isomer. In contrast to proteins, where the L-amino acids predominate, it is the D-sugars that are of biological importance.

The six-carbon sugars (hexoses) are particularly important in the cell. Hexose sugars include glucose, mannose, and fructose (**Fig. A1.12B**). Hexoses all have the molecular formula $C_6H_{12}O_6$ and are structural isomers of each other. While fructose is a keto sugar (ketose), mannose and glucose are both aldehydes (aldoses). The difference between mannose and glucose lies in the orientation of the hydroxyl group on one of the chiral carbons, C-2. The carbons in sugars are numbered starting with the carbon of the aldehyde group or the free carbon closest to the ketone group, so the aldehyde group is always C-1.

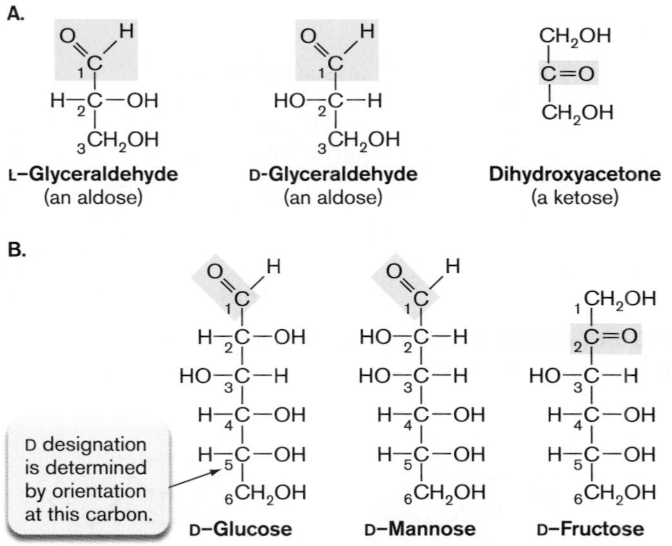
A.
L-Glyceraldehyde
(an aldose)
D-Glyceraldehyde
(an aldose)
Dihydroxyacetone
(a ketose)
B.
D designation is determined by orientation at this carbon.
D-Glucose
D-Mannose
D-Fructose

FIGURE A1.12 ▪ Structural formulas for some monosaccharides. A. The three-carbon sugars glyceraldehyde and dihydroxyacetone. The aldehyde and ketone functional groups are shown in yellow. Glyceraldehyde contains a chiral carbon and has two optical isomers. **B.** The hexose sugars glucose, mannose, and fructose. The numbering of the carbons starts at the carbon of the carbonyl group (highlighted) or at the free carbon closest to the carbonyl. The D designation is determined from the orientation of the hydroxyl group on a chiral carbon farthest from the ketone or aldehyde functional group.

In sugars of five or more carbons, one of the hydroxyl groups can react with the aldehyde or ketone group, forming a ring structure. The ring form predominates because it is more energetically favorable than the straight-chain sugar. The ring forms in two different ways (**Fig. A1.13**). Depending on how the bond between C-1 and C-2 was oriented during formation of the ring, the cyclic form obtained will be alpha-glucose or beta-glucose. The two forms interconvert rapidly in aqueous solution, and both forms are found as components of biological structures.

The hydroxyl groups of sugars can react with other functional groups to form modified sugars. Examples include fructose 1,6-bisphosphate (an intermediate in glycolysis) and *N*-acetylglucosamine (a component of the peptidoglycan cell wall) (**Fig. A1.14**).

Two Monosaccharides Condense to Form a Disaccharide

Disaccharides form when two monosaccharides undergo a condensation reaction to form a covalent glycosidic bond. The glycosidic bond often occurs between the C-1 carbon of one monosaccharide and the C-4 carbon of another monosaccharide (**Fig. A1.15**). The structure of the disaccharide varies, depending on which monosaccharides are involved and whether the C-1 carbon participating in the

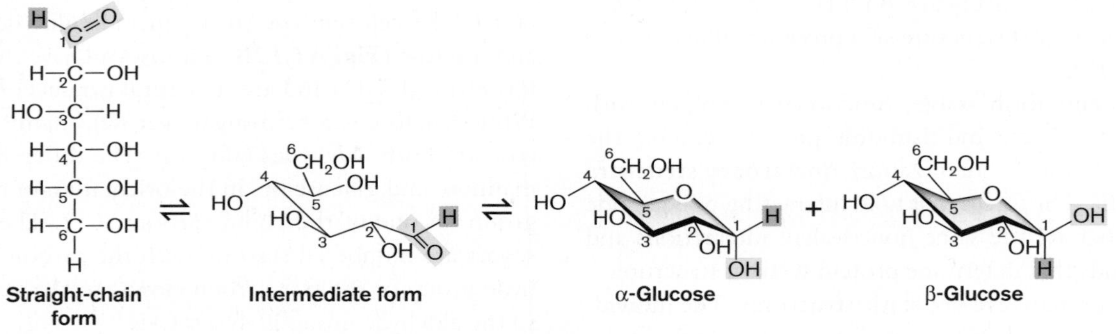
Straight-chain form
Intermediate form
α-Glucose
β-Glucose

FIGURE A1.13 ■ Straight-chain and cyclic forms of glucose. The straight-chain form of glucose can cyclize to form two different ring structures: alpha-glucose and beta-glucose. In this process, a new chiral carbon is formed at C-1, the anomeric carbon.

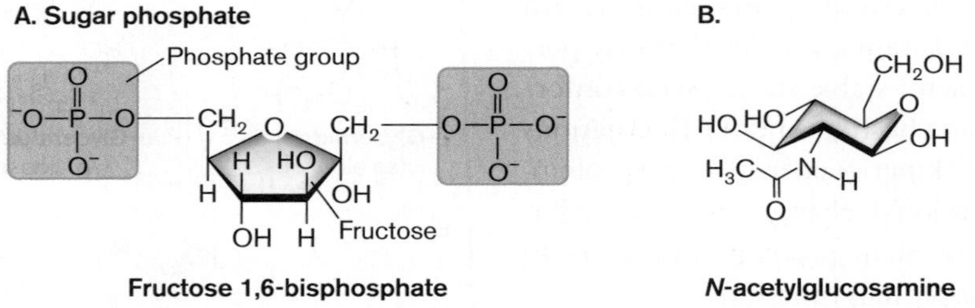
A. Sugar phosphate
Phosphate group
Fructose
Fructose 1,6-bisphosphate
B.
N-acetylglucosamine

FIGURE A1.14 ■ Modified sugars. A. Phosphorylated sugars, such as fructose 1,6-bisphosphate, are intermediates in glycolysis. **B.** *N*-acetylglucosamine is a component of bacterial cell walls.

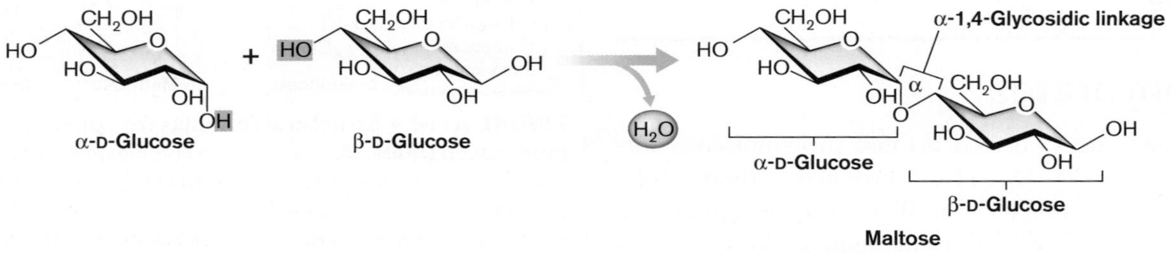
α-D-Glucose
β-D-Glucose
α-1,4-Glycosidic linkage
α-D-Glucose
β-D-Glucose
Maltose

FIGURE A1.15 ■ Formation of the disaccharide maltose. Maltose forms by a condensation reaction between two molecules of D-glucose, so that the C-1 of one molecule of glucose is linked by an oxygen atom to the C-4 of a second molecule of glucose. The hydroxyl groups of the second glucose can be either alpha or beta (the beta form is shown). The covalent glycosidic bond is called an α-1,4-glycosidic linkage because the oxygen on the C-1 carbon of the glycosidic bond is in the alpha position.

bond is in the alpha or beta form. Oligosaccharides consist of a few monosaccharides linked together, while polysaccharides are much longer chains of monosaccharides. Polysaccharides are unbranched if all the linkages are 1,4. In branched polysaccharides, additional sugars are attached to other hydroxyl groups (for example, at C-6).

A1.5

Nucleic Acids

Cells contain two kinds of nucleic acids: deoxyribonucleic acid (DNA) and ribonucleic acid (RNA). DNA is the hereditary material of the cell; it encodes the information to make proteins. Three different types of RNA play key roles in protein synthesis: Messenger RNA (mRNA) is transcribed from DNA and is the template for protein synthesis, transfer RNAs (tRNAs) bind amino acids and deliver them to the ribosome, and ribosomal RNA (rRNA) is a catalytic component of the ribosome.

A Chain of Nucleotides Forms the Primary Structure of Nucleic Acids

Nucleic acids are polymers of nucleotides. **Nucleotides** have three components: a pentose sugar (ribose in RNA, 2-deoxyribose in DNA), a **nucleobase** (also called **nitrogenous base**), and a phosphate group (**Fig. A1.16A**). The nitrogenous base is attached to the C-1 of the sugar. Nucleobases (nitrogenous bases) come in two structural classes: purines (adenine, A; and guanine, G) and pyrimidines (cytosine, C; thymine, T, found only in DNA; and uracil, U, found only in RNA). The phosphate group is esterified to the 5′ hydroxyl group of the sugar. (The prime indicates that the numbering refers to the sugar portion of the nucleotide; the base is numbered without primes.)

A nucleoside is similar to a nucleotide, except that it has only two components—a sugar and a base (that is, it lacks a phosphate group). The names of the nucleosides are listed in **Table A1.5**. Nucleotides are formed when a phosphate group forms a covalent ester bond to the C-5 carbon of the sugar portion of a nucleoside. The nomenclature for nucleotides uses the abbreviation for the nitrogenous base and adds the abbreviation MP, DP, or TP to designate the phosphate group as a monophosphate, diphosphate, or triphosphate, respectively (**Fig. A1.16B**). A lowercase "d" in front of the nucleotide indicates that the sugar is 2-deoxyribose. For example, adenosine diphosphate is ADP, and deoxycytosine triphosphate is dCTP. In addition to being the precursors for nucleic acid synthesis, the ribose nucleotides have cellular functions of their own. For example, adenosine triphosphate (ATP) is an energy carrier in the cell, and guanosine triphosphate (GTP) and cyclic adenosine monophosphate (cAMP) are signaling molecules.

TABLE A1.5

Bases and nucleosides.

Base (abbreviation)[a]	Nucleoside (base plus ribose or deoxyribose)
Adenine (A)	Adenosine
Guanine (G)	Guanosine
Cytosine (C)	Cytidine
Thymine (T)	Thymidine
Uracil (U)	Uridine

[a]A, T, G, C, and U are abbreviations for the nucleobases; however, they are also used when referring to the nucleotides in a DNA or RNA strand to indicate which base is present in the primary structure.

The nucleotide triphosphate monomers (collectively designated NTPs or dNTPs) form nucleic acids through reactions catalyzed by nucleic acid polymerase enzymes. The phosphate at the C-5 position forms a diester linkage with the hydroxyl group at C-3, releasing pyrophosphate. At one end of the nucleic acid is a free phosphate group; at the other end is a free C-3 hydroxyl group. The terminus with a free phosphate is called the 5′ end, and the terminus with the free C-3 hydroxyl is called the 3′ end (**Fig. A1.16C**). Repeating sugar-phosphate linkages form the "backbone" of the molecule, while the information content of nucleic acids resides in the sequence of bases attached to the backbone. This linear sequence of nucleotides, from the 5′ end to the 3′ end, is the primary structure of the nucleic acid.

The DNA Double Helix

Two strands of DNA, held together by hydrogen bonding between a purine and a pyrimidine on opposite strands, twist around each other to form a helix (**Fig. A1.16D** and **E**). Two hydrogen bonds form between adenine and thymine, and three hydrogen bonds form between cytosine and guanine (**Fig. A1.16F**). The sugar-phosphate backbone of DNA is on the outside of the helix. The base pairs are inside the helix, but portions of them are accessible to proteins such as transcription factors at two grooves in the helix: a wide major groove and a narrower minor groove. The double-stranded structure of DNA is critical to its function as the hereditary material. Each strand of DNA encodes the information to make a new double helix based on strict base-pairing rules.

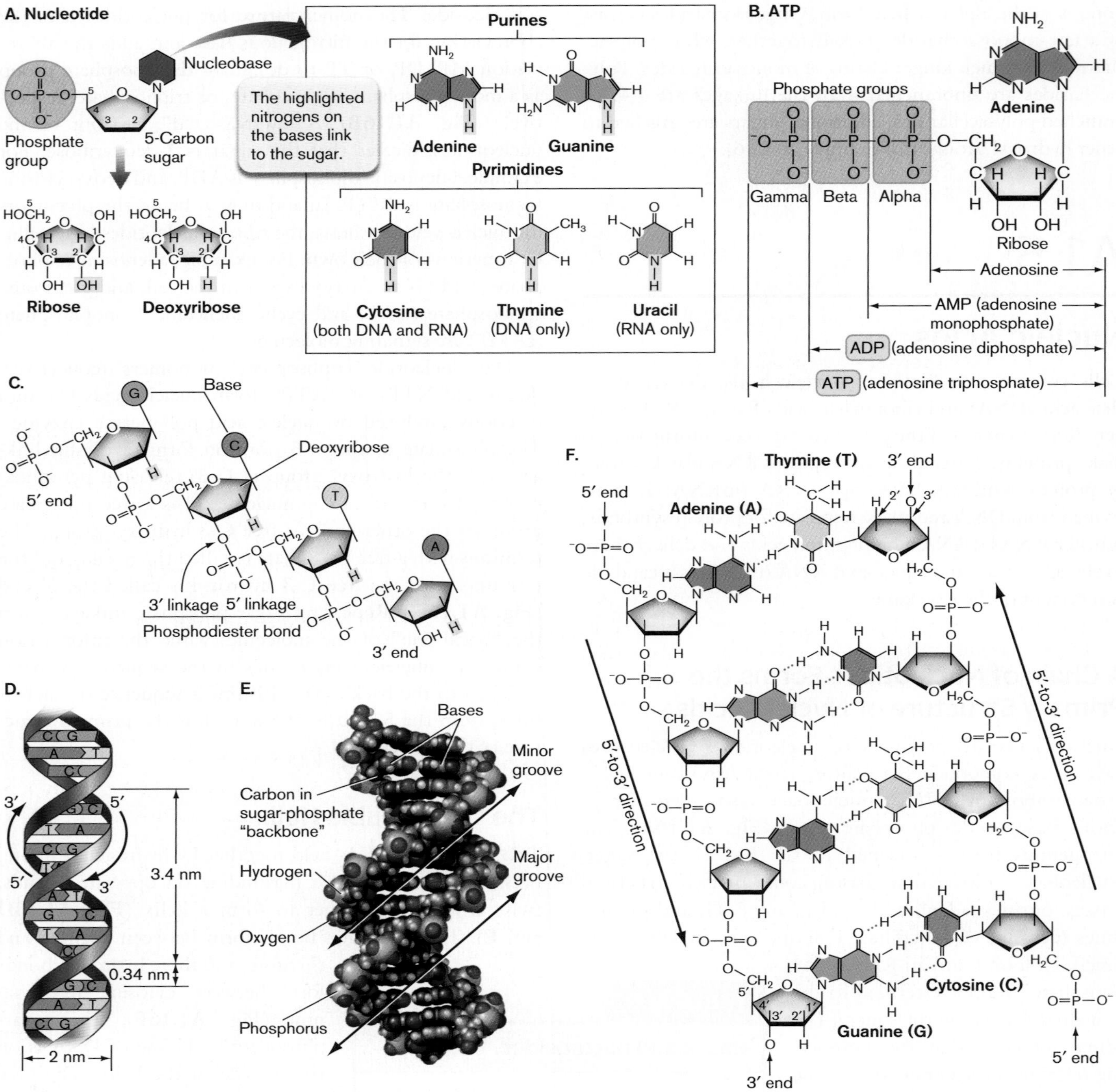
A. Nucleotide
Nucleobase
Phosphate group
5-Carbon sugar
The highlighted nitrogens on the bases link to the sugar.
Ribose
Deoxyribose
Purines
Adenine
Guanine
Pyrimidines
Cytosine (both DNA and RNA)
Thymine (DNA only)
Uracil (RNA only)
B. ATP
Phosphate groups
Gamma
Beta
Alpha
Adenine
Ribose
Adenosine
AMP (adenosine monophosphate)
ADP (adenosine diphosphate)
ATP (adenosine triphosphate)
C.
Base
Deoxyribose
5′ end
3′ end
3′ linkage
5′ linkage
Phosphodiester bond
D.
3.4 nm
0.34 nm
2 nm
E.
Bases
Minor groove
Major groove
Carbon in sugar-phosphate "backbone"
Hydrogen
Oxygen
Phosphorus
F.
5′ end
3′ end
Thymine (T)
Adenine (A)
Cytosine (C)
Guanine (G)
5′-to-3′ direction

FIGURE A1.16 ■ Nucleotide and DNA structure. A. A nucleotide consists of a five-carbon sugar (ribose or deoxyribose), a phosphate group, and a nucleobase. **B.** A nucleotide may be a monophosphate, diphosphate, or triphosphate. The phosphate attached to the sugar is designated alpha, the beta phosphate is in the middle, and the gamma phosphate is distal to the sugar. **C.** The primary structure of DNA. The orientation of the strand is determined by the presence of a free phosphate on the 5′ end and a free hydroxyl on the 3′ end. The sequence of a nucleic acid is read from the 5′ end to the 3′ end, so for this portion of the DNA strand, the primary structure is GCTA. **D.** Ribbon structure of the DNA double helix showing the purine-pyrimidine base pairs. **E.** Space-filling model of DNA. Note that the base pairs stack on top of each other like the rungs of a ladder. **F.** Hydrogen bonding between the base pairs. Two hydrogen bonds form between A and T, three hydrogen bonds between G and C. Note the antiparallel orientation of the two strands.

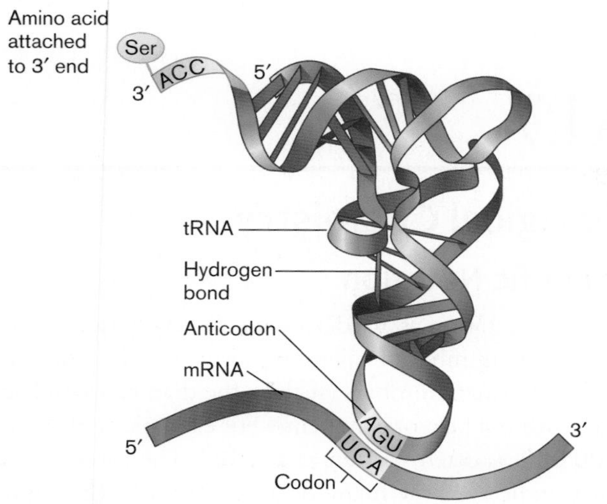
Amino acid attached to 3′ end
Ser
ACC
3′
5′
tRNA
Hydrogen bond
Anticodon
mRNA
AGU
UCA
5′
3′
Codon

FIGURE A1.17 ■ RNA secondary structure. A tRNA folds up into a secondary structure stabilized by the hydrogen bonds shown. The tRNA forms complementary base pairs with an mRNA.

The Secondary Structure of RNA

In contrast to the invariant double-helix structure of DNA, RNA molecules have a variety of secondary structures that allow RNA molecules to perform a number of different functions in the cell. Secondary and tertiary structures are possible because RNA is single-stranded and capable of complementary base pairing. Base pairing in RNA is similar to the base pairing exhibited by DNA, except that in RNA, uracil replaces thymine. Base pairing is used in making mRNA from a DNA template and also in protein synthesis, in which a tRNA charged with an amino acid matches its anticodon with the codon on the mRNA (**Fig. A1.17**). Because RNAs are single-stranded, they can undergo intramolecular as well as intermolecular base pairing. Base pairing within a single RNA molecule can lead to interesting 3D shapes that may allow the RNA to function as a catalyst. The activity of catalytic RNAs (ribozymes) may be enhanced by the reactive hydroxyl group at the C-2 of the ribose sugar, which is lacking in DNA.

A1.6

Lipids

Lipids are a structurally diverse class of molecules that have diverse functions. They are major structural components of cell membranes, they store energy, and they act as cellular signals. The common structural feature of lipid molecules is that they contain a substantial number of nonpolar C–H and C–C covalent bonds. Because lipids are nonpolar, they are hydrophobic and do not dissolve in water.

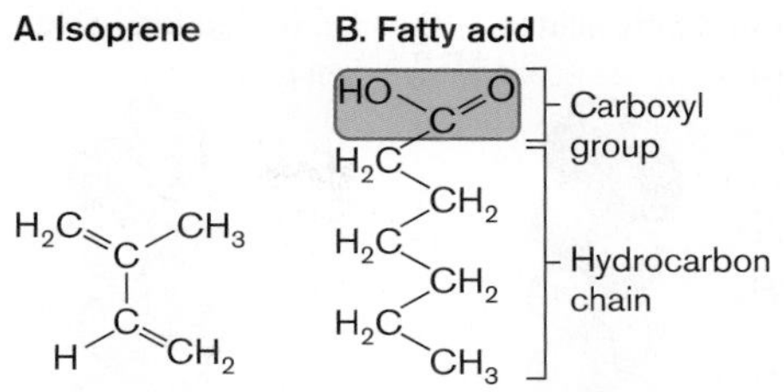
A. Isoprene
B. Fatty acid
Carboxyl group
Hydrocarbon chain

FIGURE A1.18 ■ Lipid building blocks. A. Isoprene is used to make squalene and cholesterol. **B.** Fatty acids are components of triglycerides and phospholipids.

Lipids Are Either Hydrophobic or Amphipathic

An example of a hydrophobic lipid is isoprene (**Fig. A1.18A**), which functions as a building block for more complex cellular lipids, such as squalene. Other lipid building blocks include the fatty acids (**Fig. A1.18B**). In contrast to isoprene, the fatty acids have both a hydrophobic portion (the hydrocarbon tail) and a hydrophilic component (the carboxylic acid). Molecules with both hydrophobic and hydrophilic portions are described as **amphipathic**.

Fatty Acids May Be Saturated or Unsaturated

The cell contains many different fatty acids, which differ in the number of carbons (usually even) and their saturation. In saturated fatty acids, the carbon-carbon bonds are all single bonds, and the carbons are bonded to the maximum number of hydrogen atoms (**Fig. A1.19A**). Unsaturated fatty acids contain one or more double bonds between adjacent carbons in the hydrocarbon tail (**Fig. A1.19B**). Monounsaturated fatty acids have one double bond; polyunsaturated fatty acids have multiple double bonds. Saturated fatty acids can pack together in an arrangement where they can be stabilized by van der Waals forces between adjacent hydrocarbon chains. Thus, saturated fatty acids (including animal fats such as lard) tend to be solids at room temperature. In unsaturated fatty acids, the double bond causes a kink in the chain that prevents tight packing, so unsaturated fatty acids, such as vegetable oils, tend to remain fluid at room temperature.

Glycerol Is a Building Block of Lipids

The three hydroxyl groups on glycerol can undergo a condensation reaction with the carboxylic acid portion of fatty acids to form triesters, known as triglycerides. Triglycerides

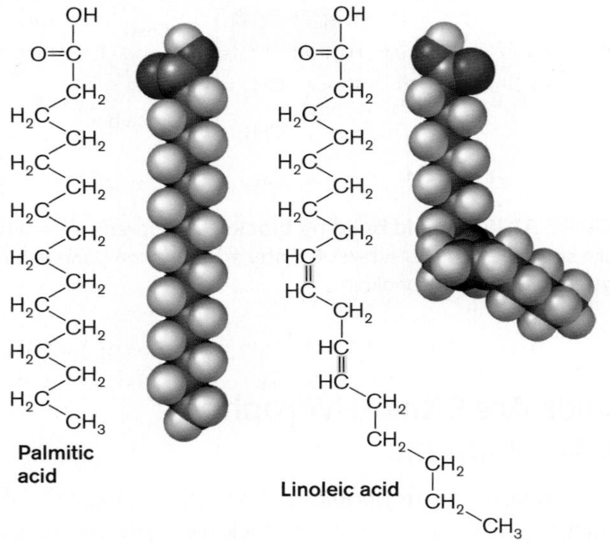
A. Saturated fatty acid
B. Unsaturated fatty acid
Palmitic acid
Linoleic acid

FIGURE A1.19 ▪ Saturated and unsaturated fatty acids. **A.** Palmitic acid is a saturated fatty acid; there are no double bonds between the carbon atoms. **B.** Linoleic acid is an example of a polyunsaturated fatty acid (it has more than one double bond). Note that the double bonds form kinks in the hydrocarbon tail.

are a compact energy source for cells (**Fig. A1.20**). Phospholipids, key components of the cell membrane, contain fatty acids attached to two of the hydroxyl groups of glycerol and a phosphate covalently attached to the third hydroxyl. The structure of phospholipids and how they function in cell membranes is explored further in Appendix 2, Section A2.1.

A1.7

Biological Chemistry

Scientific Notation

Scientific notation is used to express very large numbers (such as the number of microorganisms in a liter of seawater) or very small numbers (such as the diameter of a bacterium) with the help of exponents. For example, a million, or 1,000,000, is often written as 1×10^6. The positive exponent 6 indicates how many decimal places to the right of the number 1 our number is. One-millionth, or 0.000001, is written as 1×10^{-6} to indicate six decimal places to the left of the number 1. Special prefixes can be used to modify the magnitude of units. These are listed in **Table A1.6**. For example, 1×10^{-6} is designated by the symbol μ (which stands for "micro"), so a length of 2 micrometers (μm) is equivalent to 2×10^{-6} meter (m).

When scientists work with a set of numbers that range over many orders of magnitude, they often use a logarithmic scale. For example, the pH scale uses base 10 logarithms. Base 10, or common, logarithms are abbreviated "log," as opposed to natural logarithms, abbreviated "ln." Base 10 logarithms are defined as $\log 10^x = x$. For example, the log of $1 \times 10^{-4} = -4$, the log of 10 = 1, and the log of 1 = 0.

Molarity Is the Unit of Concentration

Scientists often measure concentrations. "Concentration" refers to how much of something is present in a given

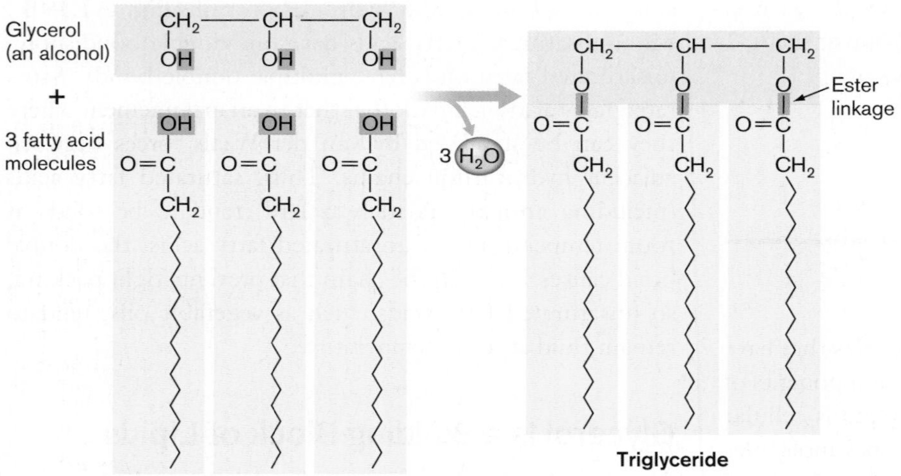
Glycerol
(an alcohol)
+
3 fatty acid
molecules
$3 H_2O$
Ester
linkage
Triglyceride

FIGURE A1.20 ▪ Formation of a triglyceride. Condensation reactions form covalent ester linkages between the carboxyl groups of three fatty acids and the three hydroxyl groups of glycerol.

TABLE A1.6

Common numerical prefixes.

Prefix	Symbol	Factor
kilo	k	10^3
deci	d	10^{-1}
centi	c	10^{-2}
milli	m	10^{-3}
micro	μ	10^{-6}
nano	n	10^{-9}
pico	p	10^{-12}
femto	f	10^{-15}

volume. A frequent way to report concentration is in units of molarity. **Molarity** is defined as the number of moles of substance per liter of solution (usually water). One mole is Avogadro's number (6.02×10^{23}) of molecules. If you wanted to make a 1-molar solution of NaCl, it would be impossible to know when you had added 6.02×10^{23} molecules to a liter of water. It's more convenient to weigh out a mole of NaCl on a balance. The weight of a mole of a substance is equal to Avogadro's number multiplied by the weight of a molecule. For example, NaCl has a mass of 58.5 atomic mass units (amu)—23 from sodium and 35.5 from chlorine (see **Fig. A1.1**). Each amu is equal to 1.66×10^{-24} gram (g), so one molecule of NaCl weighs 9.71×10^{-23} g. A mole of NaCl then weighs 9.7×10^{-23} g multiplied by 6.02×10^{23} (Avogadro's number) = 58.5 g. (Note that this weight is the same as just adding together the atomic masses of Na and Cl and expressing the result in grams.) So, to make a 1-molar solution of NaCl, you would take 58.5 g of NaCl and add water up to a liter.

Change in Free Energy

Reactions in organisms, such as the synthesis of proteins from amino acids by condensation, can occur only if they are energetically favorable. The study of energy and matter changes is called thermodynamics. All systems (including living systems, such as cells) must obey the laws of thermodynamics. The first law of thermodynamics states that energy is neither created nor destroyed. In a closed system (a system that is not exchanging energy with its environment), the total amount of energy remains constant. Although the amount of energy remains constant, energy can be converted from one form to another. The second law of thermodynamics states that in energy transformations, some energy becomes unavailable to do work and is lost as disorder, or **entropy**. In other words, entropy tends to increase. The laws of thermodynamics are summarized in **Figure A1.21**.

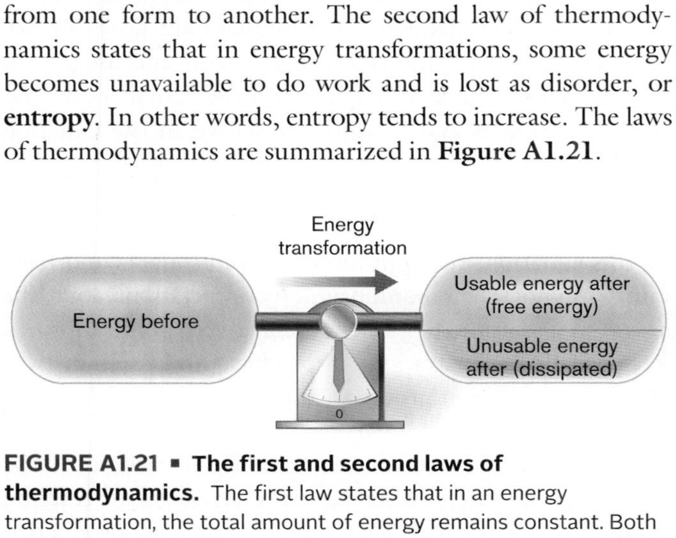
Energy transformation
Energy before
Usable energy after (free energy)
Unusable energy after (dissipated)
0

FIGURE A1.21 ■ The first and second laws of thermodynamics. The first law states that in an energy transformation, the total amount of energy remains constant. Both sides have the same amount of energy (the balance reads zero). The second law states that in an energy transformation, the amount of usable energy, or free energy, decreases and the amount of unusable energy (the product of temperature and entropy) increases.

The total energy in a system, called **enthalpy**, can be expressed in the following equation, where H stands for enthalpy, G stands for free energy, T stands for temperature, and S stands for entropy:

$$H = G + T \times S$$

Because scientists are interested in the amount of free energy—that is, the amount of energy available, or free, to do work—the energy equation is usually written in terms of G:

$$\Delta G = \Delta H - T \times \Delta S$$

The triangle is the Greek letter delta, which signifies "change in." Here, a change in energy is defined as the energy of the products minus the energy of the reactants. Thus, ΔG means $G_{products} - G_{reactants}$, and ΔS indicates $S_{products} - S_{reactants}$.

Systems Tend toward the Most Stable, Lowest Free Energy State

Reactions are favored when the free energy of the products is less than the free energy of the reactants—that is, when ΔG is negative. Reactions with a negative ΔG are termed spontaneous, or **exergonic**, because they are energetically favorable and can occur without an input of energy. Reactions with a positive ΔG are **endergonic**; they are not energetically favorable, and energy needs to be added for them to proceed. Reactions with a ΔG of zero are at equilibrium, and neither products nor reactants are favored.

The sign of ΔG is determined by changes in both enthalpy and entropy. For example, the following reaction—the oxidation, or "burning," of methane to release carbon dioxide and water—is spontaneous and has a negative change in free energy:

$$CH_4 + O_2 \rightleftharpoons CO_2 + H_2O + \text{energy}$$

The bonds in methane and oxygen have more energy than the bonds in CO_2 and H_2O. Because the products CO_2 and H_2O have less energy than the reactants, the change in enthalpy for this reaction, ΔH ($H_{products} - H_{reactants}$), is negative and the reaction is said to be exothermic.

A reaction that absorbs heat from its surroundings (endothermic) can still have a negative value of ΔG, if it has an increase in entropy, ΔS. Consider an ice cube placed at room temperature. The ice cube will spontaneously melt, even though this reaction is endothermic, absorbing heat from the environment. Ice melting is spontaneous because of the increase in entropy in going from a solid to a liquid. Remember that entropy, S, is a measure of the disorder of a system. Gases are more disordered than liquids, and liquids are more disordered than solids. Thus, the change in entropy, ΔS ($S_{products} - S_{reactants}$), from solid water (ice) to liquid water is positive because the product, liquid water, has more entropy than ice has. Positive changes in entropy may result in a negative change in free energy, especially at high temperatures.

Chemical Equilibrium

To compare changes in free energy among different reactions, a standard change in free energy, $\Delta G°$ (the circle is pronounced "naught"), is defined as the change in free energy when the concentrations of all reactants and products are at 1 molar. Each molecule has a value of $\Delta G°$ of formation from its atoms in their standard state. For a given reaction, the overall $\Delta G°$ can be calculated as the sum of the $\Delta G°$ of formation for all the products, minus the sum of $\Delta G°$ of formation for all the reactants (see **Special Topic A1.1**.)

$\Delta G°$ is a constant and depends on the nature of the products and reactants. Another constant, $\Delta G°'$, refers to the standard change in free energy at pH 7. A standard free energy change can be related to an equilibrium constant, K_{eq}, that indicates whether products or reactants will be favored at equilibrium. Reactions between biomolecules can be written as chemical equations. For example, in the following reaction, reactants A and B form products C and D:

$$\mathrm{A} + \mathrm{B} \underset{k_r}{\overset{k_f}{\rightleftharpoons}} \mathrm{C} + \mathrm{D}$$

The double arrow indicates that the reaction is reversible; A and B can be forming C and D at the same time that C and D are reverting to A and B. What determines whether the forward or reverse reaction will predominate? The forward and reverse rates depend on rate constants k_f and k_r, respectively. The rate constants depend on the nature of the interacting molecules and are an indication of how easily they will react. The forward and reverse rates also depend on concentrations of reactants and products:

$$\text{Forward rate} = [\mathrm{A}][\mathrm{B}]k_f$$

$$\text{Reverse rate} = [\mathrm{C}][\mathrm{D}]k_r$$

If the forward and reverse reactions are allowed to proceed, eventually equilibrium will be reached. All chemical reactions have a preferred state called **equilibrium**, where there is no net change in the reaction. Equilibrium is not static; rather, at equilibrium the rate of the forward reaction equals the rate of the reverse reaction:

$$[\mathrm{A}][\mathrm{B}]k_f = [\mathrm{C}][\mathrm{D}]k_r$$

For every reaction, there is an equilibrium constant, K_{eq}, that indicates the relative ratios of products and reactants at equilibrium:

$$K_{eq} = k_f/k_r = [\mathrm{C}][\mathrm{D}]/[\mathrm{A}][\mathrm{B}]$$

A K_{eq} of 1 means that neither products nor reactants are favored; a K_{eq} greater than 1 indicates that products are favored at equilibrium; and a K_{eq} less than 1 means that reactants predominate. K_{eq} and $\Delta G°$ are related as follows, where R is the gas constant [8.315×10^{-3} kJ/(mol · K)] and T is the temperature in kelvins:

$$\Delta G° = -RT \ln K_{eq}$$

A value of 298 K is often used for temperature (this is roughly 25°C). The relationship between K_{eq} and $\Delta G°$ is usually shown with the base 10 logarithm instead of the natural log:

$$\Delta G° = -2.303\, RT \log K_{eq}$$

This equation defines a relationship between $\Delta G°$ and K_{eq} that is discussed in Chapter 13. When K_{eq} is 1 (neither products nor reactants favored), $\Delta G°$ (defined as the change in free energy when concentrations of all substances are at 1 molar) is zero because the reaction is at equilibrium at that point. A K_{eq} of less than 1 (reactants favored) corresponds to a positive $\Delta G°$ because at 1-molar concentrations of reactants and products, the reaction will move to the left. A K_{eq} of more than 1 (products favored) corresponds to a negative $\Delta G°$ because at 1-molar concentrations, the reaction will move to the right.

Free Energy in Cells and the Law of Mass Action

Although the standard changes in free energy are useful for comparing the energetics and equilibria of different reactions, they do not reflect what is happening in the cell, where concentrations of substances are probably not 1 molar. While K_{eq} and $\Delta G°$ are constants and unique for a particular reaction, ΔG depends on the actual concentrations of products and reactants as shown in the equation relating ΔG to $\Delta G°$:

$$\Delta G = \Delta G° + 2.303\, RT \log([\mathrm{C}][\mathrm{D}]/[\mathrm{A}][\mathrm{B}])$$

In this equation, the concentrations of products and reactants are not their equilibrium concentrations, but the actual concentrations present at a given point in time. Because the concentrations of products and reactants will change as a reaction proceeds, the value of ΔG will change over time.

At equilibrium, where the total energy on each side of the reaction is equivalent, ΔG is zero. For an exergonic reaction at equilibrium, the total energy of reactants and products is equivalent because the amount of products is greater than the amount of reactants. If, at equilibrium, reactants (A, B) are added or products (C, D) are removed, ΔG will become negative. A negative ΔG causes the reaction to move to the

Special Topic A1.1: Calculating the Standard Free Energy Change, $\Delta G°$, of Chemical Reactions

For a chemical reaction, the net change in free energy ΔG determines whether the reaction will go forward. The value of ΔG at standard conditions of temperature (25°C), pressure (one atmosphere, P_a), and concentration (each reactant at 1 mole/liter), is designated $\Delta G°$. For a given reaction, how do we find the value of $\Delta G°$? The value is determined by summing the individual values for standard energy of formation ($\Delta G_f°$) for all the products, and then subtracting the sum of the values for all the reactants.

Consider the oxidation of glucose ($C_6H_{12}O_6$) to form water plus carbon dioxide:

$$C_6H_{12}O_6 + 6O_2 \rightarrow 6CO_2 + 6H_2O$$

We must assume that the glucose is present at 1-M concentration, that the O_2 and CO_2 are gases at 1 atmosphere, and that the water is liquid, at 25°C (298 K). The value of $\Delta G°$ for this reaction is given by:

$$\Delta G° = [(6 \times \Delta G_f°\ CO_2) + (6 \times \Delta G_f°\ H_2O)] - [(\Delta G_f°\ C_6H_{12}O_6) + 6 \times (\Delta G_f°\ O_2)]$$

From **Table 1** we can insert the values of $\Delta G_f°$ for each reactant and product. Note that one reactant, O_2, has a value of zero, because oxygen gas is the defined standard state for oxygen:

$$\Delta G° = [(6 \times -394.4\ \text{kJ/mol}) + (6 \times -237.1\ \text{kJ/mol})] - [(-910.4\ \text{kJ/mol}) + 6 \times (0\ \text{kJ/mol})]$$
$$= -2{,}879\ \text{kJ/mol}$$

So the oxidation of glucose has a negative value of ΔG at standard conditions, and it will go forward, releasing energy. The magnitude of the ΔG value suggests that enough energy could be released to form a number of energy carriers, such as ATP; but many additional factors need to be included before we know fully what a cell gains from this reaction.

TABLE 1

Free energies of formation ($\Delta G_f°$), in units of kJ/mol.

Inorganic compounds	$\Delta G_f°$	Inorganic compounds (continued)	$\Delta G_f°$	Organic compounds	$\Delta G_f°$	Organic compounds (continued)	$\Delta G_f°$
CH_4	−50.8	Mn^{2+}	−228.1	Acetaldehyde	−127.6	Guanine	+47.4
CO	−137.2	$MnCl_2$	−490.8	Acetate	−369.7	Lactate	−517.8
CO_2	−394.4	$MnSO_4$	−972.8	Acetone	−152.7	Lactose	−1,567.0
H_2CO_3	−623.2	N_2	0	Arginine	−240.5	Malate	−845.1
HCO_3^-	−586.9	NO	+87.6	Aspartic acid	−730.7	Methanol	−166.6
CO_3^{2-}	−527.9	NO_2	+51.3	Benzene	+124.4	Methionine	−505.8
$HCOO^-$	−351.0	NO_2^-	−32.2	Benzoic acid	−245.3	Methylamine	+35.7
Cu^+	+50.0	NO_3^-	−111.3	1-Butanol	−162.5	Naphthalene	+201.6
Cu^{2+}	+65.5	NH_3	−26.6	Butyrate	+352.6	Oxalate	−697.9
CuS	−53.7	NH_4^+	−79.3	Citrate	−1,236.4	Oxaloacetate	−797.2
Fe^{2+}	−78.9	N_2O	+103.7	Cysteine	+339.8	2-Oxoglutarate	+797.1
Fe^{3+}	−4.7	N_2H_4	+149.3	Ethanol	−174.8	Phenol	−50.4
$FeCO_3$	−666.7	O_2	0	Ethylene	+68.4	Propionate	−361.1
FeS_2	−156.1	OH^-	−157.3	Fructose	−915.4	Pyruvate	+474.6
$FeSO_4$	−823.4	PO_4^{3-}	−1,018.8	Fumarate	−655.6	Ribose	−757.3
H_2	0	S^0	0	Gluconate	−1,128.3	Succinate	−690.2
H^+ (pH 0)	0	SO_3^{2-}	−486.5	Glucose	−910.4	Sucrose	−1,544.7
H^+ (pH 7)	−39.7	$S_2O_3^{2-}$	−522.5	Glutamic acid	−731.3	Toluene	+113.8
HCl	−131.3	H_2S	−27.9	Glutamine	−529.7	Trimethylamine	+93.0
H_2O	−237.1	HS^-	+12.1	Glyceraldehyde	+437.7	Tryptophan	−119.4
H_2O_2	−120.4	S^{2-}	+85.8	Glycerol	−477.0	Tyrosine	−385.7

Sources: James G. Speight. 2005. *Lange's Handbook of Chemistry*, 16th ed., McGraw-Hill, New York; WolframAlpha (http://www.wolframalpha.com); Rudolf K. Thauer. 1977. *Bacteriol. Rev.* **41**:100.

right to reestablish equilibrium. As A and B convert to C and D, ΔG becomes less and less negative until it reaches zero again at equilibrium. If, at equilibrium, products are added or reactants are removed, ΔG will become positive and the reaction will proceed to the left to reestablish equal energy on both sides. This **law of mass action** is the tendency of a reaction to reestablish equilibrium after perturbations in the concentrations of products or reactants.

The law of mass action means that a reaction with a positive $\Delta G°$ can be made spontaneous and driven to the right by keeping the concentration of reactants high or the concentration of products low. The cell employs this strategy in metabolic pathways, where the product of one reaction is constantly removed by a subsequent reaction.

The Rate of a Reaction Depends on the Activation Energy

It is important to realize that although a reaction with a large positive K_{eq} and negative $\Delta G°$ is spontaneous, it may be slow. The value of $\Delta G°$ says nothing about the rate of a reaction. An everyday example of a spontaneous but slow reaction is the rusting of metal. A reaction will be slow, even though it is spontaneous, if it must pass through an unstable, high-energy transition state on the way to forming products. The **activation energy** (E_a) is the energy needed to reach this transition state (**Fig. A1.22**). For reactions with low activation energies, random collisions between reactants may provide enough energy to boost them up and over the E_a. For reactions with high activation energies, collisions between molecules may not provide enough energy for them to reach the transition state.

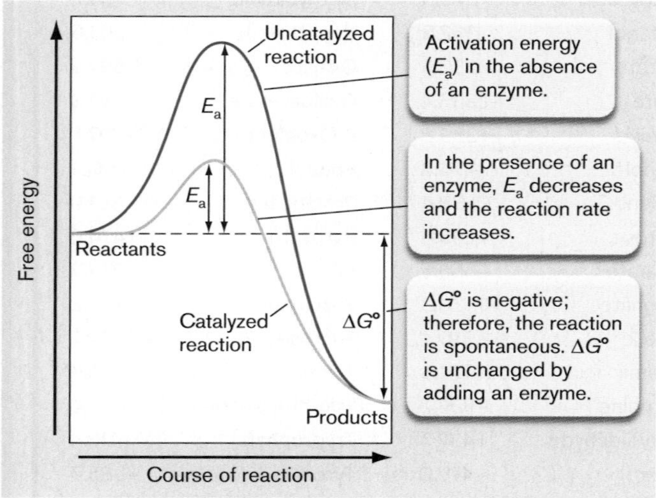
Uncatalyzed reaction
Activation energy (E_a) in the absence of an enzyme.
E_a
E_a
In the presence of an enzyme, E_a decreases and the reaction rate increases.
Reactants
Free energy
Catalyzed reaction
$\Delta G°$
$\Delta G°$ is negative; therefore, the reaction is spontaneous. $\Delta G°$ is unchanged by adding an enzyme.
Products
Course of reaction

FIGURE A1.22 ▪ Enzymes, activation energies, and reaction rates.

Enzymes are biological catalysts that can speed up reaction rates by stabilizing transition states and lowering the activation energy. Most enzymes are proteins that specifically bind reactants and provide an environment that facilitates product formation. Although enzymes increase reaction rates, they do not change the $\Delta G°$ or K_{eq} of a reaction. **Figure A1.22** illustrates how the difference in free energy between products and reactants is unchanged even in the presence of an enzyme that lowers the activation energy.

Biological Processes Depend on pH

Many biological processes occur within only a narrow range of hydrogen ion concentrations. Hydrogen ions (H^+) are also referred to as "protons" because they have lost their single electron and consist of only a proton. Hydrogen ion concentration is reported using a pH (power of hydrogen) scale, where pH is the negative logarithm of the hydrogen ion concentration: $pH = -\log_{10}[H^+]$. In pure water, which is considered neutral, $[H^+]$ is 1×10^{-7} molar (M)—a pH of 7. Because the pH scale is logarithmic, every pH unit corresponds to a tenfold change in hydrogen ion concentration. A solution with a pH of 6 has a hydrogen ion concentration of 1×10^{-6} M, ten times the hydrogen ion concentration at pH 7. Hydrogen ion concentration $[H^+]$ multiplied by hydroxide ion concentration $[OH^-]$ always equals 1×10^{-14}. Hence, the concentrations of H^+ and OH^- are reciprocally related (if one goes up, the other goes down). If the pH is less than 7, the solution is acidic (more H^+, less OH^-). If the pH is greater than 7, the solution is basic, or alkaline (less H^+, more OH^-).

Acids (for example, carboxyl groups) release protons, and bases (for example, amino groups) bind protons (**Fig. A1.23**). At intracellular pH (near 7), the carboxyl and amino groups of amino acids are both ionized (charged), the carboxyl group carrying a negative charge and the amino group, a positive charge. The ionization of these

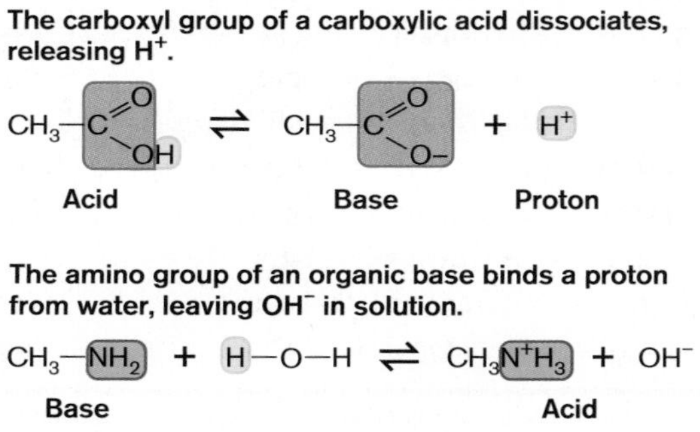
The carboxyl group of a carboxylic acid dissociates, releasing H+.
Acid
Base
Proton
The amino group of an organic base binds a proton from water, leaving OH− in solution.
Base
Acid

FIGURE A1.23 ▪ Organic acids and bases.

groups can change if the pH changes. At lower pH (more acidic solution), protons move back onto the ionized carboxylic acid. For example, the side chain of an acidic amino acid such as glutamate (see **Fig. A1.9**) is ionized at normal cell pH but may regain a proton and become glutamic acid at a lower pH. At higher pH values (lower proton concentrations), amino groups lose protons to become uncharged. Disturbances in the ionization state of carboxyl or amino groups on the side chains of amino acid residues can disrupt ionic bonds between these groups and lead to protein denaturation. This effect of pH on protein tertiary structure is one reason why cells can tolerate only a narrow range of intracellular pH.

Oxidation-Reduction (Redox) Reactions Transfer Energy

Biomolecules may undergo an important class of chemical reactions termed redox reactions. **Redox reactions** involve the transfer of electrons from one molecule to another or from one atom to another. The molecule that gains electrons becomes reduced, and the molecule that loses electrons becomes oxidized (**Fig. A1.24**). (A useful mnemonic device to remember this is "LEO the lion says GER," where LEO stands for "lose electrons, oxidation" and GER stands for "gain electrons, reduction".)

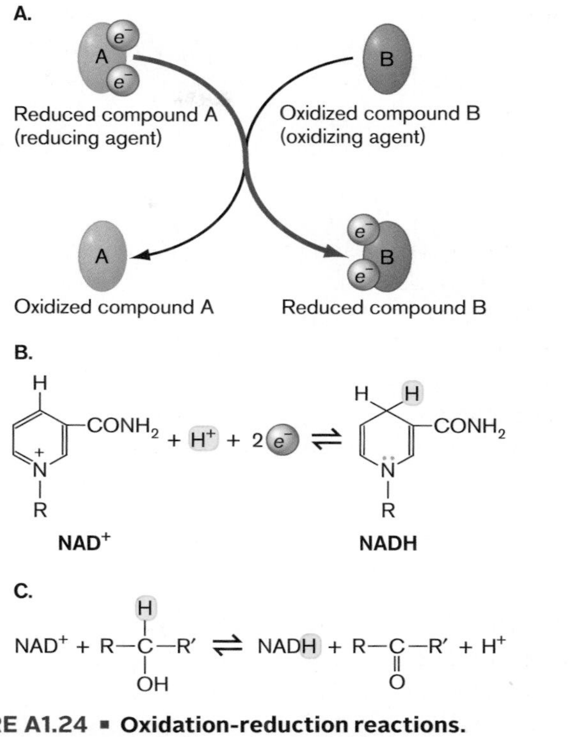
A.
A
B
Reduced compound A (reducing agent)
Oxidized compound B (oxidizing agent)
A
B
Oxidized compound A
Reduced compound B
B.
CONH₂
+ H⁺ + 2e⁻
CONH₂
NAD⁺
NADH
C.
NAD⁺ + R—C—R′ ⇌ NADH + R—C—R′ + H⁺
OH
O

FIGURE A1.24 ▪ Oxidation-reduction reactions. **A.** A reducing agent A donates electrons to reduce compound B. Because A loses electrons, it becomes oxidized itself in the process. B is the oxidizing agent. **B.** NADH is a common biological reducing agent. **C.** NAD^+ is reduced to NADH as an alcohol is oxidized to an aldehyde.

Redox reactions are always coupled. If one atom loses electrons, another atom must gain electrons. Oxygen, with its large electronegativity, usually gains electrons and becomes reduced in redox reactions. Molecules that become reduced are called "oxidizing agents" because they cause something else to become oxidized. In contrast, "reducing agents" can donate electrons, reduce other molecules, and become oxidized themselves in the process. Molecules with reduced carbons contain more energy than their oxidized counterparts. Common reducing agents (electron donors) in the cell are NADH, NADPH, and $FADH_2$. These are all high-energy molecules that can donate electrons.

In addition to achieving a complete transfer of electrons, redox reactions can also occur if electrons are shifted toward or away from an atom. For example, in the burning (oxidation) of methane, electrons move away from the carbon and toward the oxygen (**Fig. A1.25**). Oxygen usually acts as an oxidizing agent (electron acceptor); thus, we expect it to become reduced (gain electrons). By default, then, methane is oxidized to carbon dioxide. Indeed, while the bonding electrons were shared fairly equally between C and H in methane, they are now close to the O in both carbon dioxide and water and farther away from the C and H. Thus, methane has been oxidized and oxygen has been reduced. This reaction releases energy because CO_2 and H_2O are the most stable, lowest-energy forms available when carbon, hydrogen, and oxygen are combined.

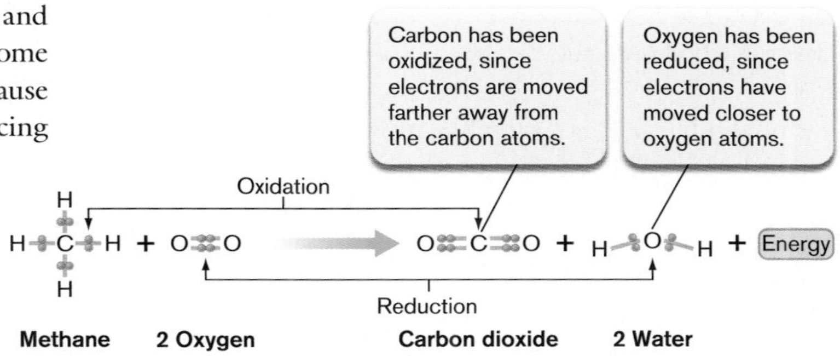
Carbon has been oxidized, since electrons are moved farther away from the carbon atoms.
Oxygen has been reduced, since electrons have moved closer to oxygen atoms.
Oxidation
Energy
Reduction
Methane
2 Oxygen
Carbon dioxide
2 Water

FIGURE A1.25 ▪ The oxidation of methane. Bonding electrons were shared equally by carbon and hydrogen in methane, but they are closer to oxygen in carbon dioxide and water. Carbon and hydrogen have been oxidized; oxygen has been reduced.

APPENDIX 2

Introductory Cell Biology: Eukaryotic Cells

A2.1 The Cell Membrane

A2.2 The Nucleus and Mitosis

A2.3 Problems Faced by Large Cells

A2.4 The Endomembrane System

A2.5 The Cytoskeleton

A2.6 Mitochondria and Chloroplasts

Appendix 2 presents a review of cell biology principles generally covered in an introductory biology class. Cell biology encompasses fundamental principles of cell structure and function. All cells are enclosed by a cell membrane and need to regulate the transport of materials across the membrane. Additional membrane structures are unique to eukaryotic cells.

All eukaryotic cells contain a nucleus, an organelle that houses the DNA. We describe the structure of the nucleus and the processes of chromosome segregation through mitosis and meiosis. We then explore the endomembrane system of organelles that partition the cell into functional compartments, enabling eukaryotic cells to maintain larger sizes than most prokaryotic cells. Eukaryotic subcellular form includes the cytoskeleton, a collection of proteins that provide cell architecture and mediate cell movement. Finally, we discuss mitochondria and chloroplasts, the energy-producing organelles of eukaryotic cells.

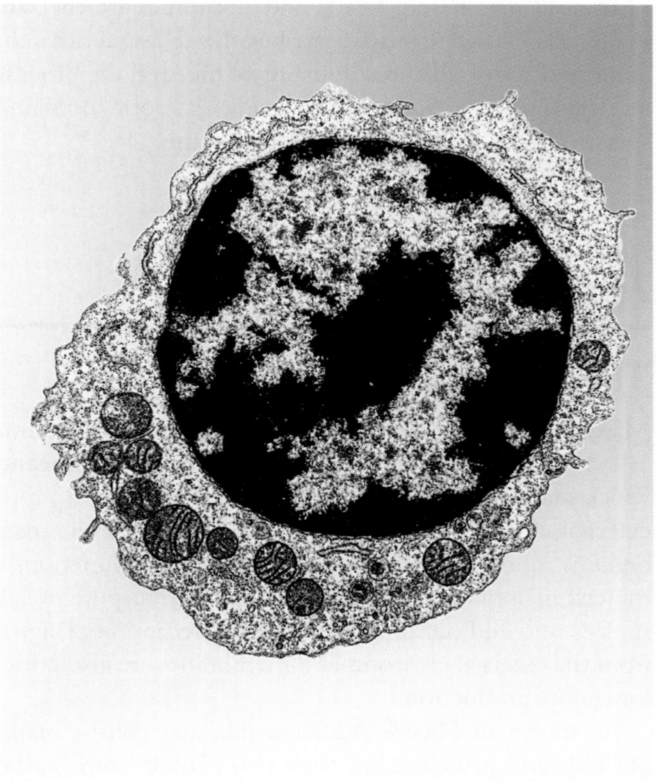

A eukaryotic cell. This eukaryotic cell, a lymphocyte white blood cell, shows numerous major organelles, such as a large nucleus (orange) and multiple mitochondria (blue). (TEM, 20,550×) *Source:* © Gopal Murti/Visuals Unlimited.

The cell is the basic unit of life. All organisms consist of either a single cell or a collection of cells. All cells can potentially perform a common set of tasks, including replication, catalysis, and regulation. The structures that enable these functions are found in all cells and include the cell membrane, DNA, and ribosomes (**Fig. A2.1**). Although the detailed organization of these structures differs in cells from the three domains (**Bacteria**, **Archaea**, and **Eukarya**; see **Table A2.1**), the fundamental structure and function of the cell membrane, DNA, and ribosomes are the same in all cells. This appendix describes the cell structures that are found across all three domains of life, and then focuses on those that are unique to the eukaryotic domain—specifically, organelles and the cytoskeleton.

A2.1

The Cell Membrane

All cells are enclosed by a **cell membrane** (sometimes called the **plasma membrane** or **cytoplasmic membrane**), which maintains an internal environment distinct from the external environment. The aqueous fluid inside the membrane is called **cytoplasm** (or **cytosol**). Major functions of the cell membrane include the regulated transport of substances into and out of the cell and the reception of signals from the external environment. Membranes are also critical for energy production.

As shown in **Figure A2.2**, membranes consist mainly of lipids and proteins, but they also contain some carbohydrates found in hybrid structures such as glycolipids and glycoproteins (sugars joined to lipids or to proteins, respectively). Although the membrane contains a greater number of lipid molecules than protein molecules, proteins are large and contribute about half the mass of the membrane. The number and nature of proteins within a membrane depend on the membrane under consideration. For example, the inner membrane of the mitochondrion contains a rich array of proteins involved in energy production.

Membrane proteins can be classified based on how they interact with membranes (**Fig. A2.2A**). **Transmembrane proteins** (integral proteins) span the bilayer. Transmembrane proteins are **amphipathic**, meaning they have both hydrophobic and hydrophilic portions; the hydrophilic portions face the cytoplasm and the extracellular environment, and the hydrophobic components span the membrane. The transmembrane domains are often alpha helices containing 15–20 amino acid residues (**Fig. A2.2B**). The portions of the protein that are intracellular, in the membrane, and extracellular depend on the protein and can be quite different for different proteins. **Peripheral membrane proteins** are associated with the cell membrane through noncovalent bonds but are not directly inserted into the bilayer.

Membranes Are Composed of Lipids

The predominant lipids in membranes are phospholipids. **Phospholipids** consist of a core of glycerol, to which two fatty acids and a modified phosphate group are attached via ester linkages (**Fig. A2.3A**). Unlike eukaryotes and bacteria, archaea have phospholipids with ether linkages.

Phospholipids are amphipathic: the fatty acid hydrocarbon tails are hydrophobic, and the phosphate head group

TABLE A2.1

Comparison of cell structures in the three domains.

Feature	Bacteria	Archaea	Eukarya
Genome	Usually circular DNA Usually one chromosome Usually lack introns	Usually circular DNA Usually one chromosome A few introns	Linear DNA Multiple chromosomes, in pairs Most genes have introns
Location of DNA	Nucleoid region in cytoplasm	Nucleoid region in cytoplasm	Contained within membrane-enclosed nucleus
Cell membrane	Straight-chain fatty acids ester-linked to glycerol	Branched-chain fatty acids ether-linked to glycerol	Straight-chain fatty acids ester-linked to glycerol
Cell wall	Usually present, composed of peptidoglycan	If present, composed of proteins or pseudopeptidoglycan	If present, composed of cellulose (algae) or chitin (fungi)
Internal membranes	May have energy-transducing lamellae	Uncommon	Extensive membranous organelles

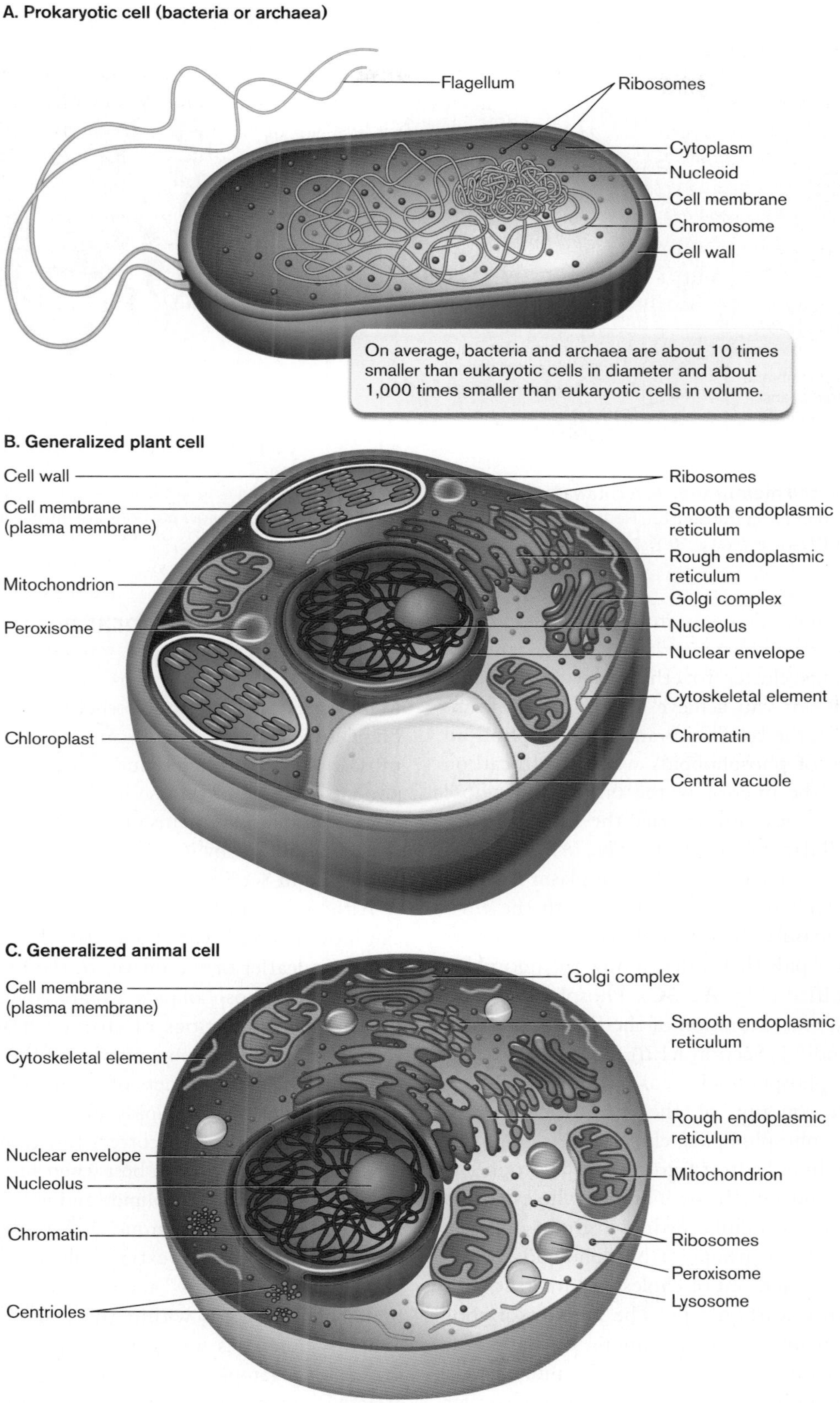
A. Prokaryotic cell (bacteria or archaea)
Flagellum
Ribosomes
Cytoplasm
Nucleoid
Cell membrane
Chromosome
Cell wall
On average, bacteria and archaea are about 10 times smaller than eukaryotic cells in diameter and about 1,000 times smaller than eukaryotic cells in volume.
B. Generalized plant cell
Cell wall
Cell membrane (plasma membrane)
Mitochondrion
Peroxisome
Chloroplast
Ribosomes
Smooth endoplasmic reticulum
Rough endoplasmic reticulum
Golgi complex
Nucleolus
Nuclear envelope
Cytoskeletal element
Chromatin
Central vacuole
C. Generalized animal cell
Cell membrane (plasma membrane)
Cytoskeletal element
Nuclear envelope
Nucleolus
Chromatin
Centrioles
Golgi complex
Smooth endoplasmic reticulum
Rough endoplasmic reticulum
Mitochondrion
Ribosomes
Peroxisome
Lysosome

FIGURE A2.1 ■ The prokaryotic cell and the eukaryotic cell. A. The prokaryotic cell typically contains a single compartment, and its DNA is organized in the nucleoid region. **B, C.** Eukaryotic cells are typically much larger than prokaryotic cells and contain organelles.

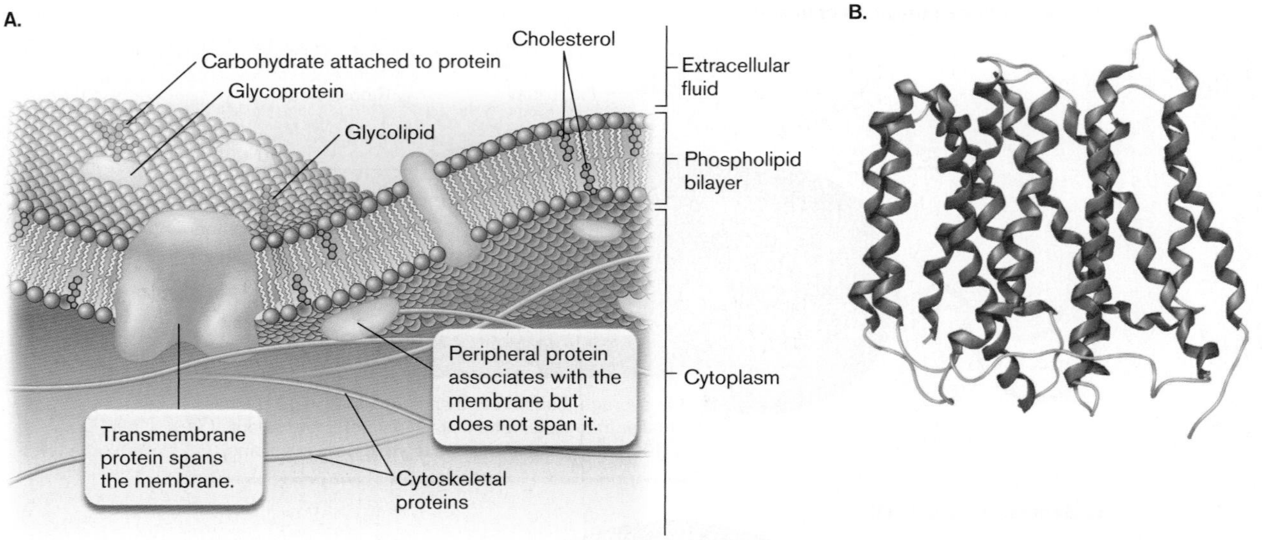
A.
Cholesterol
Carbohydrate attached to protein
Glycoprotein
Glycolipid
Extracellular fluid
Phospholipid bilayer
Cytoplasm
Peripheral protein associates with the membrane but does not span it.
Transmembrane protein spans the membrane.
Cytoskeletal proteins
B.

FIGURE A2.2 ■ The cell membrane. A. A cutaway view of the cell membrane. **B.** Ribbon diagram from X-ray crystallographic data of the *E. coli* EmrD protein, a multidrug transporter. Red portions are transmembrane alpha helices; yellow portions are intracellular and extracellular loops. (PDB code: 2GFP) *Source:* Yin et al. 2006. *Science* **312**:741.

is hydrophilic. Amphipathic lipids are most stable in water when the hydrophilic portions interact with water and the hydrophobic portions cluster together away from water. One way phospholipids can achieve stability is by forming a bilayer. Indeed, the cell membrane is a **phospholipid bilayer**, two layers of phospholipids whose hydrocarbon fatty acid tails face the interior of the bilayer and whose charged phospholipid head groups face the aqueous cytoplasm and extracellular environment (**Fig. A2.3B**). The phospholipid layer in contact with the cytoplasm is called the **inner leaflet**, and the layer in contact with the environment is called the **outer leaflet**. Cells contain a number of different phospholipids that differ in how the phosphate head group is modified (**Fig. A2.3C**). Phospholipids also differ in the length and saturation of the fatty acid chains (discussed in Appendix 1, Section A1.6).

In addition to phospholipids, eukaryotic membranes contain a variable amount of the steroid cholesterol (**Fig. A2.3D**). Like phospholipids, cholesterol is amphipathic, with a hydrophilic hydroxyl group and hydrophobic hydrocarbon rings and tail. In membranes, cholesterol is oriented so that the hydrophilic hydroxyl group interacts with the phosphate head groups of phospholipids, while the hydrophobic rings and tail of cholesterol interact with the phospholipid hydrocarbon tails. The amount of cholesterol present in membranes varies among cells and also among organelles within a cell. Bacterial membranes do not contain cholesterol but do contain molecules of similar form called hopanoids.

Movement of Membrane Lipids and Proteins

Membranes are not static structures; rather, many membrane lipids and proteins can move rapidly. The **fluid mosaic model** of membranes states that membrane components are free to diffuse in the plane of the membrane. Some membrane proteins are restricted to specific regions of the membrane by interactions with cytoskeletal proteins (discussed in Section A2.5).

Although many phospholipids and membrane proteins can move laterally within a leaflet, they do not "flip-flop" from one leaflet of the bilayer to the other (**Fig. A2.3E**). Flip-flop of phospholipids is rare, owing to the highly unfavorable interactions of charged head groups moving through the hydrophobic interior of the membrane. Thus, the inner and outer leaflets of the membrane may be made up of different phospholipids. Such phospholipid asymmetry is important for the correct functioning of membrane proteins, which may work best when surrounded by particular phospholipids. Glycolipids and glycoproteins also contribute to membrane asymmetry, because the carbohydrate moieties always face the extracellular environment.

Membrane "fluidity" refers to the movement of membrane phospholipids within the plane of the membrane, and this fluidity is important for proper membrane function. For example, transport across the membrane is affected by membrane fluidity. Decreased fluidity is associated with decreased transport rates. Because a drop in

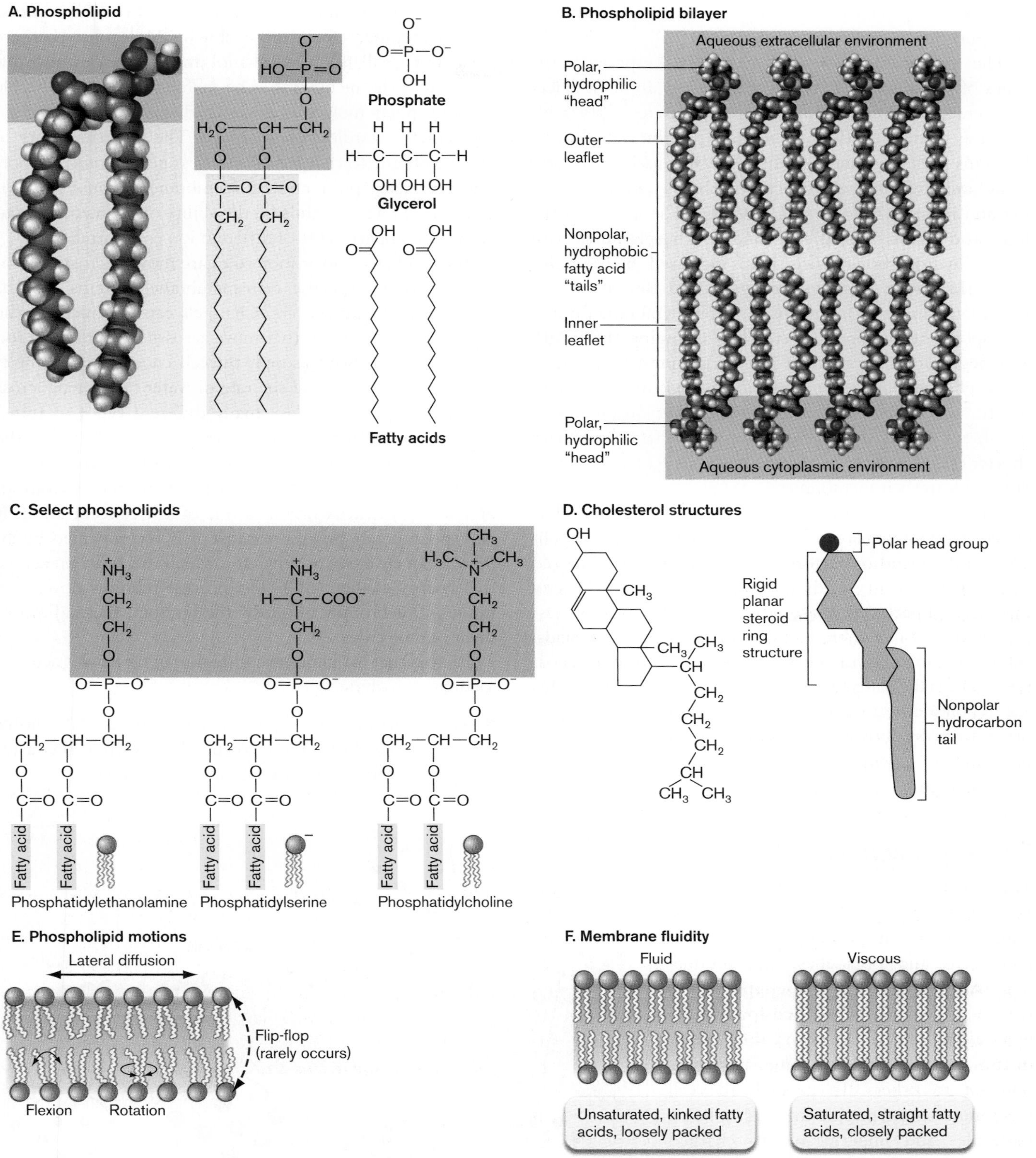
A. Phospholipid
Phosphate
Glycerol
Fatty acids
B. Phospholipid bilayer
Aqueous extracellular environment
Polar, hydrophilic "head"
Outer leaflet
Nonpolar, hydrophobic fatty acid "tails"
Inner leaflet
Polar, hydrophilic "head"
Aqueous cytoplasmic environment
C. Select phospholipids
Fatty acid
Phosphatidylethanolamine
Phosphatidylserine
Phosphatidylcholine
D. Cholesterol structures
Polar head group
Rigid planar steroid ring structure
Nonpolar hydrocarbon tail
E. Phospholipid motions
Lateral diffusion
Flip-flop (rarely occurs)
Flexion
Rotation
F. Membrane fluidity
Fluid
Viscous
Unsaturated, kinked fatty acids, loosely packed
Saturated, straight fatty acids, closely packed

FIGURE A2.3 ■ Phospholipids, cholesterol, the lipid bilayer, and membrane fluidity. **A.** Saturated phospholipid. **B.** Orientation of phospholipids in the bilayer. **C.** Some phospholipids present in cell membranes. **D.** Structural formula and schematic drawing of cholesterol. **E.** Motions of phospholipids in membrane bilayers. **F.** The ratio of saturated and unsaturated fatty acids in the phospholipids affects membrane fluidity.

temperature decreases fluidity, cold temperatures may slow transport processes across the membrane.

The composition of the membrane, especially the types of phospholipids present, can have a dramatic effect on membrane fluidity. For example, saturated fatty acids decrease membrane fluidity because the linear hydrocarbon tails pack together well. In contrast, unsaturated fatty acids have kinks in the hydrocarbon chains that limit packing and increase fluidity (**Fig. A2.3F**). The length of the fatty acid chains also affects fluidity. Phospholipids with longer hydrocarbon chains have increased hydrophobic interactions with neighboring lipids, and thus decreased membrane fluidity. Organisms can alter membrane fluidity in response to temperature stress by changing the length and degree of saturation of fatty acids present in membrane phospholipids. For example, as environmental temperatures drop, both eukaryotes and prokaryotes maintain membrane fluidity by replacing long-chain fatty acids with shorter chains and increasing the percentage of unsaturated fatty acids in their membranes.

Cholesterol also influences membrane fluidity. The effects of cholesterol on membrane fluidity are complicated and depend on factors such as the ratio of saturated to unsaturated fatty acids in the membrane. Cholesterol may prevent packing of saturated fatty acids, thus increasing fluidity. In membranes with unsaturated fatty acids, cholesterol may fill in the spaces between adjacent phospholipids, stabilizing them and decreasing fluidity. In this case, cholesterol can decrease the permeability of the membrane to hydrophobic substances by packing between the hydrocarbon chains and preventing substances from slipping through.

Transport across Membranes

The major functions of membranes (such as containing cytoplasmic components, regulating which substances enter and leave cells and organelles, and producing energy) depend on the semipermeable nature of membranes. **Semipermeable** (also called **selectively permeable**) **membranes** are permeable to some substances but not to others. In general, the cell membrane is permeable to hydrophobic molecules and impermeable to charged molecules (**Fig. A2.4**). Diffusion across the membrane also depends on the size of the molecule. The membrane is freely permeable to small nonpolar molecules such as O_2. Larger nonpolar molecules can also diffuse across the membrane, albeit more slowly. Molecules that are polar but small (such as ethanol and water) can also diffuse across the membrane. The membrane is impermeable to large polar molecules such as glucose and to charged molecules, regardless of their size. The impermeability of the membrane to charged substances such as ions is important for energy production at membranes because the proton motive force depends on the ability of the membrane to separate compartments of different ion concentrations.

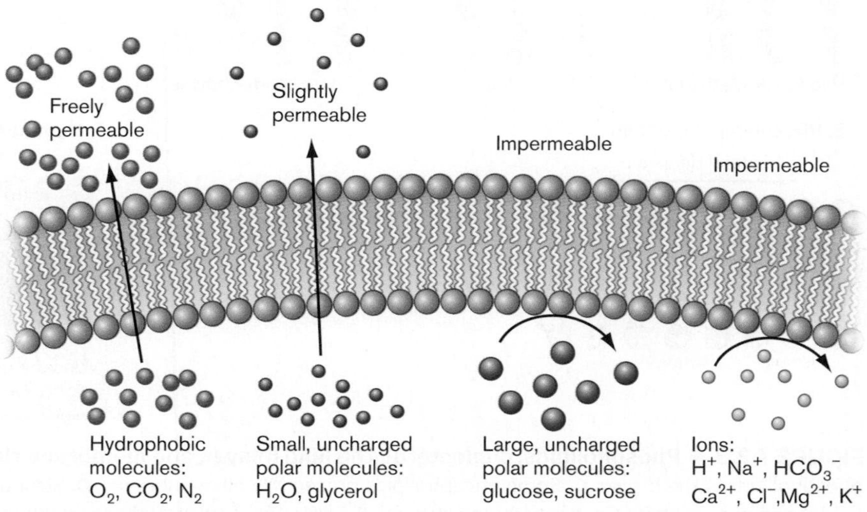

Freely permeable
Slightly permeable
Impermeable
Impermeable
Hydrophobic molecules: O_2, CO_2, N_2
Small, uncharged polar molecules: H_2O, glycerol
Large, uncharged polar molecules: glucose, sucrose
Ions: H^+, Na^+, HCO_3^- Ca^{2+}, Cl^-,Mg^{2+}, K^+

FIGURE A2.4 ■ Selective permeability of cell membranes.

Ions and large polar molecules are moved across a membrane through specific transmembrane proteins such as channels and transporters. Channels can also increase the diffusion of molecules that move across the membrane too slowly on their own to supply the cell's needs. For example, aquaporins can increase the rate of water movement across the membrane. There are many different types of transporters, each differing in its energy requirements and in the types of molecules that it transfers across the membrane.

Diffusion is the net movement of molecules from an area of high concentration to one of low concentration. It is a spontaneous process because it is accompanied by an increase in entropy (positive ΔS), which results in a negative free energy change (ΔG). The process requires no energy input and is brought about by the random, thermal movement of molecules.

Factors that influence the diffusion of molecules across a membrane include:

- **Temperature.** Increased temperatures mean faster motion. The faster the molecules are moving, the faster they will arrive at the membrane and cross it.

- **Solubility of the molecules in the membrane.** To cross the membrane, the molecules must penetrate it. Hydrophobic molecules will dissolve in the membrane and cross it; charged molecules will not.
- **Surface area of the membrane.** To cross the membrane, molecules must first encounter it. The chances of such encounters are increased with an increase in membrane surface area.
- **Concentration gradient of the dissolved molecules.** A larger concentration gradient speeds up diffusion because the more molecules there are, the more will encounter the membrane and cross.
- **Thickness of the membrane.** Diffusion rates are inversely proportional to the square of the distance the solute must travel across the membrane. The thinner the membrane, the faster the molecules can get across.
- **Mass of the molecule.** Friction between a molecule and its medium is a source of resistance that slows down motion. Larger molecules with more mass experience more resistance and cross the membrane more slowly.

These factors can be expressed as follows:

$$\text{Diffusion rate} \propto \frac{\text{temperature} \times \text{surface area} \times \text{concentration gradient}}{\text{mass} \times \text{distance}^2}$$

Conditions for the diffusion of gases and other substances across the cell membrane will be most favorable when the surface area of the membrane is large, the concentration gradient across the membrane is high, and the membrane is thin.

Transport of Water across the Cell Membrane

Osmosis is the diffusion of water across a selectively permeable membrane from regions of high water concentration (low solute) to regions of low water concentration (high solute). Diffusion rates can be determined with an artificial membrane system as depicted in **Figure A2.5A**. Molecules are added to the left side of the beaker, and samples from the right side are analyzed at various time points for the presence of the test molecule.

Cells must maintain osmotic balance with the surrounding environment. The direction of water movement depends on the concentration of dissolved solutes in the cell relative to the cell's environment. When a cell is in an **isotonic** environment (equal concentrations of dissolved solutes inside the cell and out), there is osmotic balance, and water will enter and exit the cell at equal rates (that is, there is no net movement of water). In a **hypertonic** environment (higher

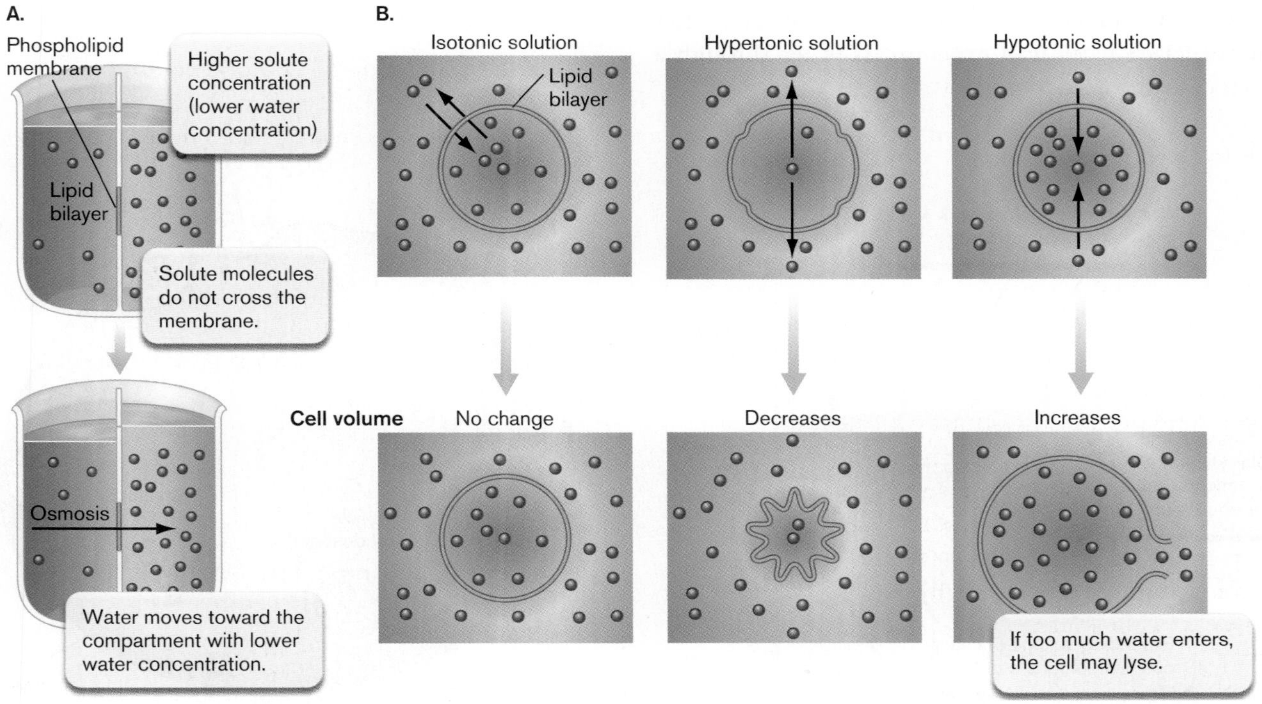
A.
Phospholipid membrane
Higher solute concentration (lower water concentration)
Lipid bilayer
Solute molecules do not cross the membrane.
Osmosis
Water moves toward the compartment with lower water concentration.
B.
Isotonic solution
Lipid bilayer
Hypertonic solution
Hypotonic solution
Cell volume
No change
Decreases
Increases
If too much water enters, the cell may lyse.

FIGURE A2.5 ▪ Osmosis and water balance. A. Osmosis. **B.** Movement of water across the cell membrane, and shrinkage or expansion of the membrane in isotonic, hypertonic, and hypotonic environments. Black arrows indicate net water movement.

concentration of solutes outside the cell), there is a net loss of water from the cell. The cell shrinks, and the concentration of cell contents increases. In a **hypotonic** environment (lower concentration of solutes outside the cell), there is a net uptake of water by the cell; the cell swells, and the cell components are diluted (**Fig. A2.5B**). If enough water enters, the cell is destroyed by **lysis**, a rupturing of the cell membrane and dispersal of cell contents. Both hypertonic and hypotonic environments can cause other problems for cells. Proteins have specific salt requirements, and intracellular environments with salt concentrations that are either higher or lower than normal for a cell can cause denaturation of proteins, with potentially fatal results for the cell. Thus, transport of water across the cell membrane must be tightly controlled, and cells need mechanisms that allow them to live in environments that are not isotonic.

Most cells live in environments that are hypotonic. To deal with this challenge, most bacteria and many eukaryotes have a cell wall external to the cell membrane. As water enters by osmosis and pushes against the cell wall ("turgor pressure"), the wall resists the tension and pushes back with an equal but opposite force known as "wall pressure." The wall pressure is an inward pressure exerted by the cell wall against the cell membrane (**Fig. A2.6**). When turgor pressure and wall pressure are equal in magnitude, the cell is at equilibrium with respect to water movement. Organisms that lack a cell wall employ other strategies to deal with hypotonic environments. For example, some freshwater protists that lack a cell wall expel excess water through a contractile vacuole.

In contrast to freshwater microbes, ocean-dwelling organisms face the problem of water loss. Many of these organisms accumulate solutes known as osmolytes to ensure that they are isotonic to the external environment, thus preventing water loss.

Eukaryotes Transport Molecules by Endocytosis and Exocytosis

All cells use diffusion and transport proteins to move molecules into or out of the cell, but some eukaryotes can, in addition, use endocytosis and exocytosis to achieve this end. In **endocytosis**, parts of the cell membrane bud into the cytoplasm and eventually separate from it to form **endosomes** (**Fig. A2.7**). Endosomes are a type of **vesicle**, a small membranous sphere found within a cell. The interior of these endosomes contains extracellular material. **Phagocytosis** (cell eating) is a form of endocytosis in which large extracellular particles are brought into the cell. **Pinocytosis** (cell drinking) is endocytosis of relatively small volumes of the extracellular fluid. Endocytosis is a controlled, energy-requiring process that relies on many proteins, including cytoskeletal proteins. **Exocytosis** is the reverse of endocytosis. In exocytosis, intracellular vesicles fuse with the cell membrane, and the contents of the vesicles are released to the extracellular environment. Cells can use exocytosis to release wastes.

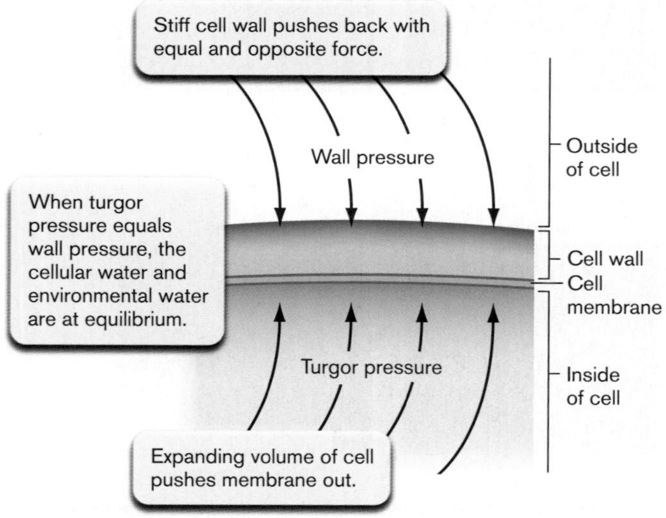

Stiff cell wall pushes back with equal and opposite force.
Wall pressure
When turgor pressure equals wall pressure, the cellular water and environmental water are at equilibrium.
Outside of cell
Cell wall
Cell membrane
Inside of cell
Turgor pressure
Expanding volume of cell pushes membrane out.

FIGURE A2.6 ▪ Turgor pressure and wall pressure.

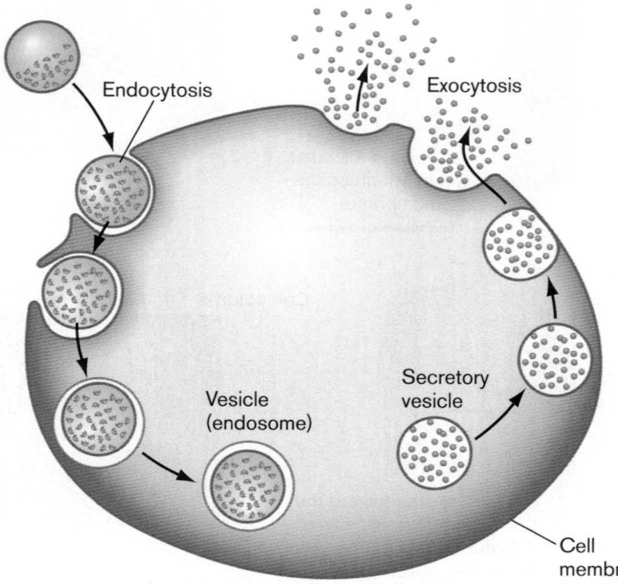

Endocytosis
Exocytosis
Vesicle (endosome)
Secretory vesicle
Cell membrane

FIGURE A2.7 ▪ Endocytosis and exocytosis.

A2.2

The Nucleus and Mitosis

All cells need to synthesize proteins, and all cells contain DNA, the genetic material that encodes the information needed to specify protein primary structure. The central model of molecular biology holds that DNA is transcribed into messenger RNA (mRNA), and mRNA is translated into protein on ribosomes. Although this model holds for all cells, details in the structure of DNA and ribosomes vary among the three domains of life (see **Table A2.1**). For example, ribosomes always function in protein synthesis, but ribosomes from organisms in different domains differ in their size and their sensitivity to various antibiotics. DNA always encodes the information needed for protein synthesis, but whereas bacterial chromosomes are usually circular, eukaryotic chromosomes are linear. Another difference between DNA in prokaryotes and DNA in eukaryotes is its location within the cell. In bacteria and archaea, DNA is found in an area of the cytoplasm known as the nucleoid, while eukaryotic DNA is contained within a nuclear membrane.

Eukaryotic DNA Is Housed in the Membrane-Enclosed Nucleus

Eukaryotes derive their name ("eukaryote" means "true kernel") from the fact that they possess a **nucleus** (plural, **nuclei**), and indeed, the nucleus is often the most prominent feature of eukaryotic cells viewed under a microscope (**Fig. A2.8A**). The nucleus is an **organelle**, an intracellular membrane-enclosed compartment with a specific function. The nucleus contains **chromatin**, a complex of DNA and proteins. The nuclear membrane (envelope) consists of two concentric phospholipid membranes. The outer nuclear membrane is continuous with the membrane of the endoplasmic reticulum (ER), and the space between the two nuclear membranes is continuous with the lumen (inside) of the ER (**Fig. A2.8B**). Nuclei contain a region called the **nucleolus** (plural, **nucleoli**), where ribosome assembly begins. At the nucleolus, multiple rRNA (ribosomal RNA) genes are transcribed, and the resulting rRNA combines with ribosomal proteins imported into the nucleus from the cytoplasm to form the ribosomal subunits. The ribosomal subunits then need to exit the nucleus.

The nuclear membrane contains nuclear pore complexes (NPCs), which allow for transport of material into and out of the nucleus. Metabolites and small proteins can diffuse through the NPCs, but larger proteins and organelles cannot enter by diffusion. Large proteins that need to enter the nucleus are actively transported in through the NPCs. These selectively imported proteins contain a nuclear localization signal, a sequence of amino acids that acts like a zip code to direct them into the nucleus. In addition to their role in protein import, NPCs also function in exporting mRNAs out of the nucleus.

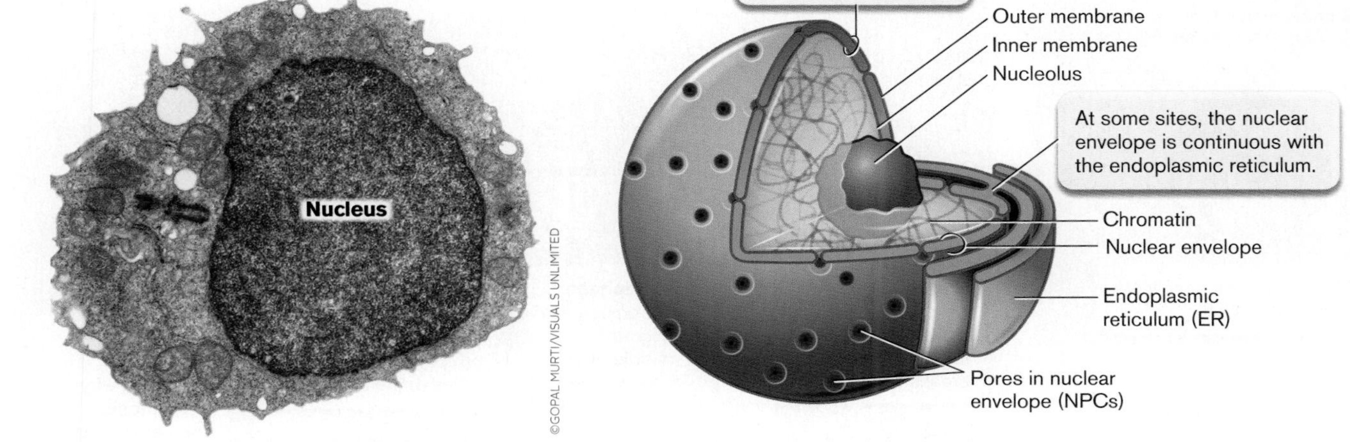
A.
Nucleus
©GOPAL MURTI/VISUALS UNLIMITED
B.
The nuclear envelope is a double membrane.
Outer membrane
Inner membrane
Nucleolus
At some sites, the nuclear envelope is continuous with the endoplasmic reticulum.
Chromatin
Nuclear envelope
Endoplasmic reticulum (ER)
Pores in nuclear envelope (NPCs)

FIGURE A2.8 ▪ The nucleus. A. Electron micrograph of a eukaryotic yeast cell showing the prominent nucleus. **B.** Diagram of a nucleus. NPC = nuclear pore complex.

Eukaryotic Cells Replicate by Mitosis and Meiosis

Cells need to ensure an accurate replication and division of their DNA. Bacteria and archaea replicate by fission, as described in Chapter 3. Eukaryotic cells replicate their nuclear DNA and divide by a completely different process, called "mitosis." For sexual reproduction, sex cells with a halved chromosome number are generated by the related process of "meiosis." Mitosis and meiosis occur only in eukaryotes, never in bacteria and archaea.

Mitosis is a series of steps that segregates duplicated chromosomes and ensures that each daughter cell receives a copy of the genetic material. When the cell is not undergoing mitosis, it is in interphase (**Fig. A2.9A**). During interphase, the individual chromosomes are long and thin and not visible by a light microscope. Interphase can be divided into three phases: G_1, S, and G_2. Cells that are not committed to dividing are in G_1, the first gap phase. If cell division is to occur, then the chromosomes are replicated during S phase (S for "synthesis"). The duplicated chromosomes, called sister chromatids, remain attached to each other at the centromere (**Fig. A2.9B**). Following chromosome replication, a second gap phase, G_2, occurs, after which mitosis can proceed.

Mitosis is divided into four steps (**Fig. A2.9C**):

- **Prophase.** During prophase, the chromosomes condense and become visible by light microscopy. The nuclear membrane may break down. The mitotic spindle, which separates the sister chromatids to opposite poles of the cell, begins to form. The mitotic spindle is a network of microtubules (see Section A2.5) that originate from centrosomes. There are two centrosomes, and these migrate to opposite sides of the cell. Each centrosome contains two centrioles, and each centriole contains nine sets of microtubules in a radial formation (see Section A2.5). The free ends of the microtubules establish connections with the sister chromatids at a structure called the kinetochore.

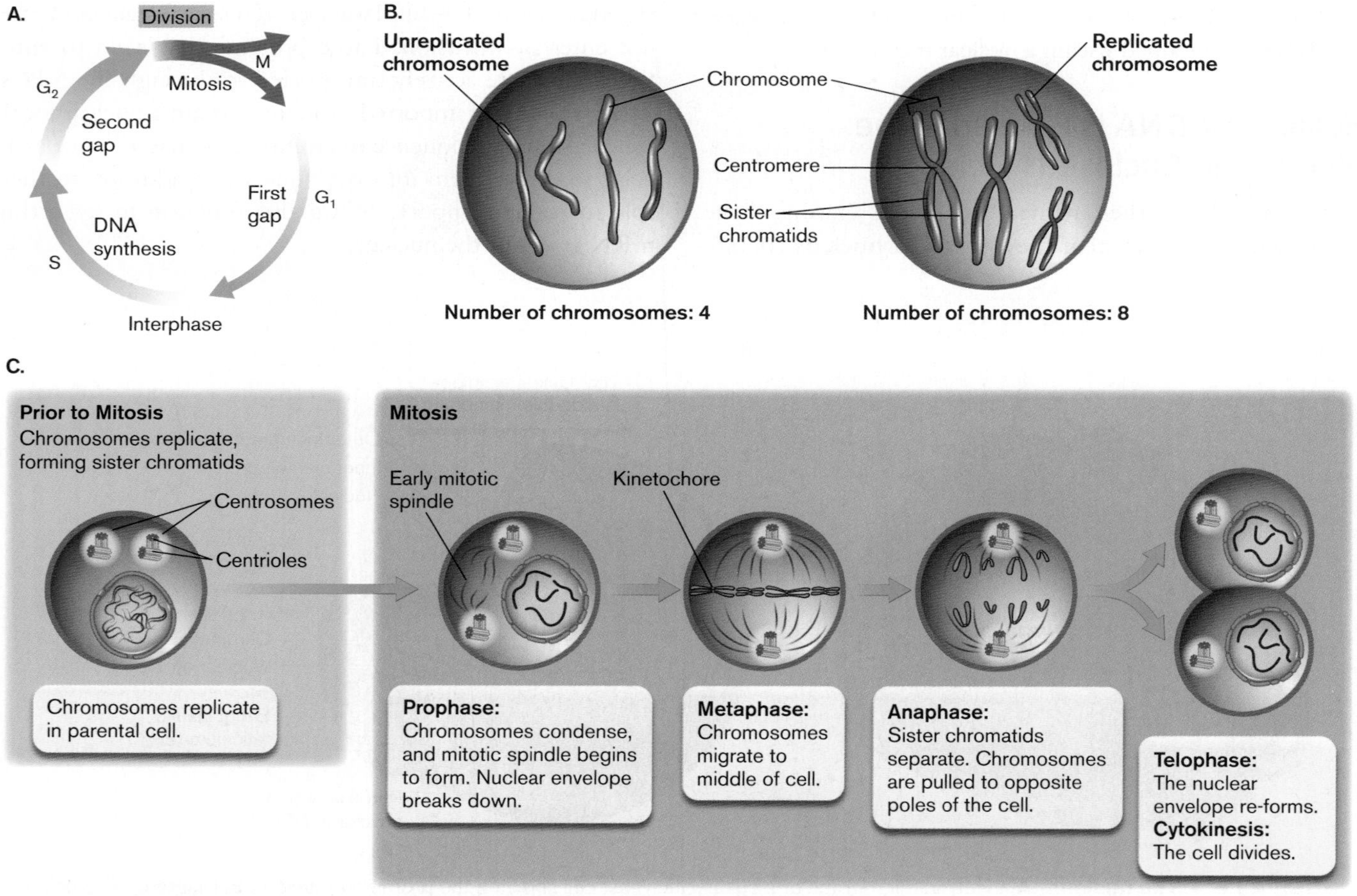

A.
Division
M
Mitosis
G_2
Second gap
First gap
G_1
DNA synthesis
S
Interphase
B.
Unreplicated chromosome
Chromosome
Centromere
Sister chromatids
Replicated chromosome
Number of chromosomes: 4
Number of chromosomes: 8
C.
Prior to Mitosis
Chromosomes replicate, forming sister chromatids
Centrosomes
Centrioles
Chromosomes replicate in parental cell.
Mitosis
Early mitotic spindle
Kinetochore
Prophase: Chromosomes condense, and mitotic spindle begins to form. Nuclear envelope breaks down.
Metaphase: Chromosomes migrate to middle of cell.
Anaphase: Sister chromatids separate. Chromosomes are pulled to opposite poles of the cell.
Telophase: The nuclear envelope re-forms.
Cytokinesis: The cell divides.

FIGURE A2.9 ▪ The cell cycle and mitosis. **A.** Stages of the eukaryotic cell cycle. G_1, S, and G_2 make up interphase (in blue). **B.** Duplication of the chromosomes during S phase. **C.** The phases of mitosis. See text for details.

- **Metaphase.** At metaphase, the spindle apparatus is complete and each sister chromatid is connected to a microtubule. The chromosomes are arranged along an imaginary plane in the middle of the cell.
- **Anaphase.** In anaphase, the microtubules shorten and pull the sister chromatids apart, separating the replicated chromosomes. At the end of anaphase, each set of chromosomes is located on opposite sides of the cell.
- **Telophase.** In this final step of mitosis, the nuclear membrane re-forms around each set of chromosomes and the chromosomes become long and thin again. Cytokinesis also occurs during telophase and partitions the original cell into two daughter cells by the formation of a cell membrane between them.

The reproduction of eukaryotic cells by mitosis is called **asexual reproduction**.

Some variations in the phases of mitosis are seen in microbial eukaryotes. In yeast, for example, the nuclear envelope does not break down. Despite these differences, the end result, the separation of duplicated chromosomes into two daughter cells, is the same.

Unlike asexual reproduction, **sexual reproduction** requires the reassortment of genetic material from different chromosomes. A sexual life cycle alternates between cells that are diploid ($2n$, containing two copies of each chromosome) and sex cells that are haploid ($1n$, containing a single copy of each chromosome). Two of the haploid $1n$ sex cells (called gametes) can join each other by fertilization to regenerate a diploid cell (called a zygote). The diploid thus possesses two **homologs** of each chromosome—that is, two versions of the same chromosome from two different parents.

The process of gamete formation requires a special modification of mitotic cell division called **meiosis** (**Fig. A2.10**). Meiosis includes two cell division series, called meiosis I and meiosis II. Like mitosis, meiosis I must be preceded by replication of all chromosomes during S phase (**Fig. A2.10A**). Thus, a diploid ($2n$) cell becomes temporarily $4n$. Unlike the case in mitosis, in prophase I of meiosis I the replicated chromosomes do not separate; instead, each replicated pair lines up with its homolog, the same chromosome inherited from the other parent. Now the aligned homologs exchange portions of their DNA.

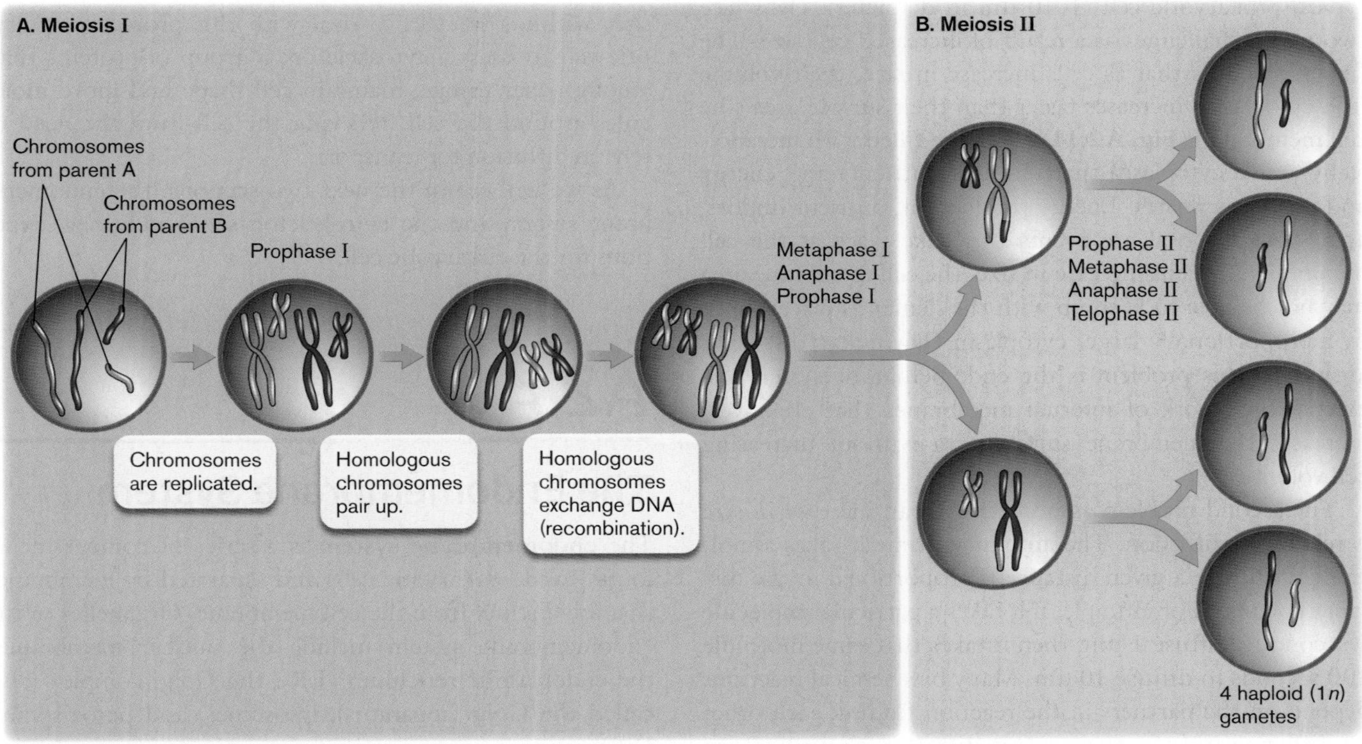
A. Meiosis I
Chromosomes from parent A
Chromosomes from parent B
Prophase I
Chromosomes are replicated.
Homologous chromosomes pair up.
Homologous chromosomes exchange DNA (recombination).
Metaphase I
Anaphase I
Prophase I
B. Meiosis II
Prophase II
Metaphase II
Anaphase II
Telophase II
4 haploid (1n) gametes

FIGURE A2.10 ■ Meiosis. The chromosomes of a diploid ($2n$) cell undergo replication. **A.** In meiosis I, homologous chromosomes exchange DNA, and all the pairs are separated between two daughter cells. **B.** In meiosis II, homologous pairs are separated to produce haploid ($1n$) gametes.

This genetic exchange reassorts the traits and increases genetic diversity of the offspring.

Meiosis II is completed by separation of the paired homologs through metaphase II and anaphase II (**Fig. A2.10B**). A short telophase occurs, in which each daughter cell is now diploid (2*n*). Unlike what happens in the telophase of mitosis, in meiosis I no nuclear membrane forms, and no interphase occurs. Instead, the chromosome pairs immediately separate during meiosis II, which includes prophase II, metaphase II, anaphase II, and telophase II. The result is four haploid (1*n*) gametes, as seen in **Figure A2.10**. Depending on the species, these gametes may develop into specialized forms such as sperm and egg that reunite through fertilization, restoring the diploid form.

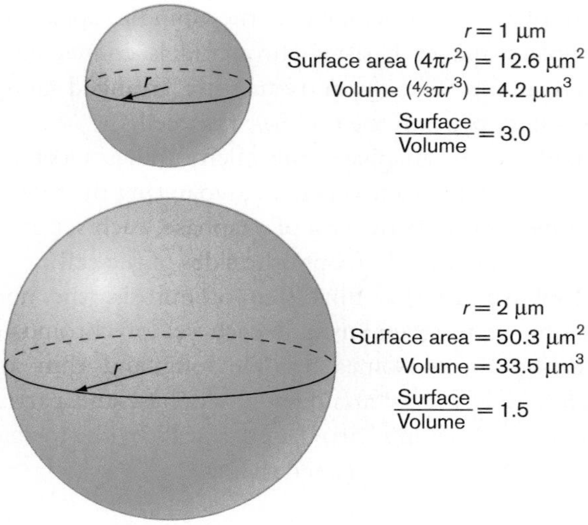
r
$r = 1\ \mu m$
Surface area ($4\pi r^2$) = 12.6 μm^2
Volume ($\frac{4}{3}\pi r^3$) = 4.2 μm^3
$\frac{\text{Surface}}{\text{Volume}} = 3.0$
r
$r = 2\ \mu m$
Surface area = 50.3 μm^2
Volume = 33.5 μm^3
$\frac{\text{Surface}}{\text{Volume}} = 1.5$

FIGURE A2.11 ▪ Cell volume increases faster than surface area. Because the surface area increases by the radius squared and the volume increases by the radius cubed, the volume increases faster than the surface area.

A2.3

Problems Faced by Large Cells

Most eukaryotic cells range in size from 10 to 100 micrometers (μm) in diameter, about ten times as large as the typical prokaryotic cell (1–10 μm in diameter). Cells face two major challenges as a result of increased cell size. The first problem is that as cells increase in size, their volume (the cytoplasm) increases faster than their surface area (the cell membrane) (**Fig. A2.11**). Cells are filled with metabolically active cytoplasm that requires nutrients and energy and produces wastes. Energy production, nutrient import, and waste disposal are events that take place at the cell membrane. As cells increase in size, the cell membrane area may not be able to keep up with the demands placed on it by a proportionally larger cytoplasm. The eukaryotic cell's answer to this problem is the endomembrane system, an extensive network of internal membranes that effectively increases the membrane surface area without increasing cell volume.

The second problem associated with an increase in size is related to diffusion. The amount of time it takes a molecule to diffuse a given distance is proportional to the distance squared. For example, if it takes a particular molecule 1 second to diffuse 1 μm, then it takes that same molecule 100 seconds to diffuse 10 μm. Many biochemical reactions depend on the partners in the reaction finding each other by diffusion (that is, their rate is diffusion controlled), so longer diffusion times result in slower reactions. Furthermore, signals received at the cell membrane need to be communicated throughout the cell. In a large cell, the amount of time it takes for a molecule to traverse the cell by diffusion may be too slow for the cell to rapidly adjust to signals it receives from the environment or from other sites within the cell. To deal with this problem, eukaryotic cells possess a cytoskeleton, a group of proteins that, among other things, maintain cell shape and move molecules around the cell, relieving the cell from the need to rely on diffusion for transport.

As we will see in the next two sections, the endomembrane system and the cytoskeleton serve additional functions for the eukaryotic cell.

A2.4

The Endomembrane System

The **endomembrane system** is a series of compartments found inside eukaryotic cells and separated by membranes that are distinct from the cell membrane. Organelles of the endomembrane system include the nuclear membranes, the endoplasmic reticulum (ER), the Golgi complex (also called the Golgi apparatus), lysosomes, and peroxisomes. Different organelles contain unique subsets of proteins that contribute to their function. To a large extent, the function of the ER and the Golgi complex is to direct to their

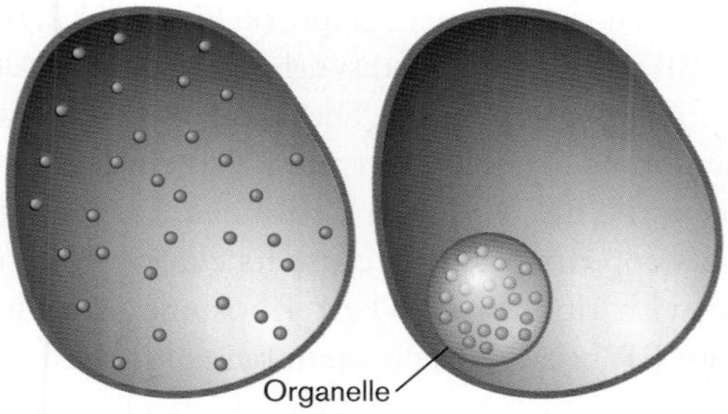
Organelle

FIGURE A2.12 ■ Advantages of organelles. Solutes (green dots) are concentrated in organelles, reactive molecules are separated from the cytoplasm, and membrane surface area is increased without an increase in cell volume.

proper cellular location proteins destined for lysosomes, for the cell membrane, or for secretion from the cell.

Membranous organelles serve many functions for the eukaryotic cell. In general:

- Endomembranes increase the membrane surface area without increasing cell volume.
- Separating cellular contents in small, enclosed compartments increases the concentrations of enzymes and their substrates (**Fig. A2.12**), allowing reactions to proceed faster.
- Organelles can provide different environments that allow disparate reactions to occur simultaneously. For example, proteins are being synthesized by ribosomes in the neutral-pH cytoplasm, while at the same time being hydrolyzed within the acidic organelles known as lysosomes.
- Compartmentalization protects cytoplasmic components from harmful substances. For example, hydrogen peroxide (H_2O_2), a product of cellular oxidation reactions, is produced and converted to water within **peroxisomes**. Localizing the reaction within the peroxisome keeps the toxic peroxide away from other cell components, such as proteins and DNA, that are sensitive to oxidative stress.

Lysosomes Digest Organic Matter

Lysosomes are membrane-enclosed organelles that help eukaryotic cells obtain nourishment from macromolecular nutrients. Lysosomes contain many hydrolytic enzymes (for example, proteases, nucleases, and lipases) and have an acidic pH of about 5. Lysosomes are formed when vesicles containing hydrolytic enzymes and proton pumps bud off from the Golgi complex (**Fig. A2.13A**). Lysosomes then fuse with vesicles containing the material to be digested.

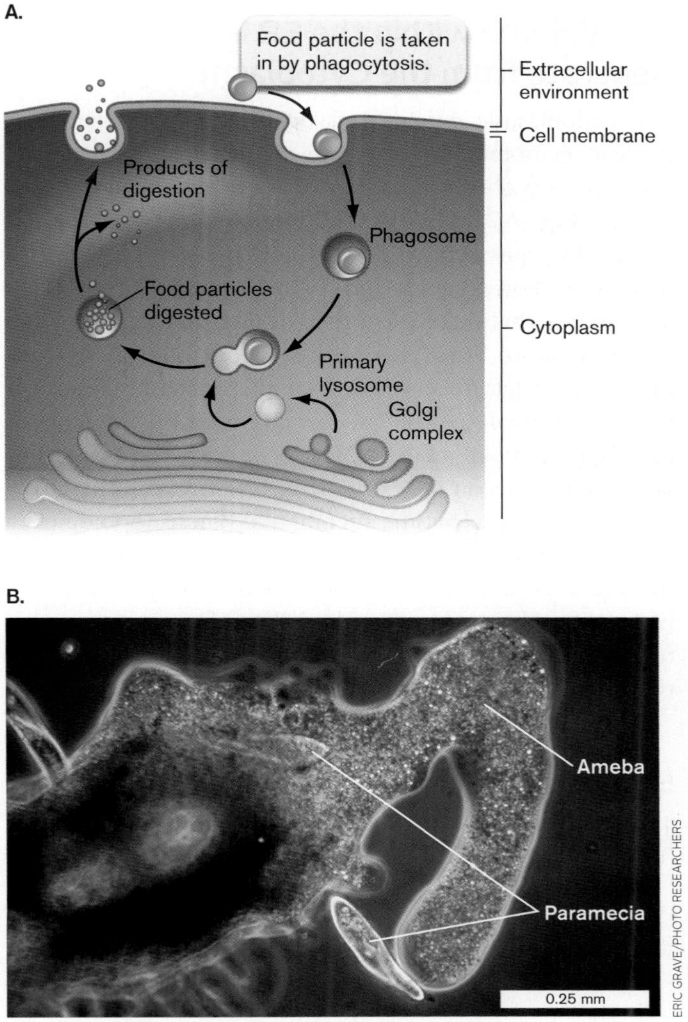
A.
Food particle is taken in by phagocytosis.
Extracellular environment
Cell membrane
Products of digestion
Phagosome
Food particles digested
Cytoplasm
Primary lysosome
Golgi complex
B.
Ameba
Paramecia
0.25 mm
ERIC GRAVE/PHOTO RESEARCHERS

FIGURE A2.13 ■ Lysosomes. A. Lysosomes contain hydrolytic enzymes to digest material brought into the cell by phagocytosis. **B.** Amebas engulfing paramecia (light microscopy). The paramecia will be digested with the aid of lysosomes.

Often this material comes from outside the cell via phagocytosis. Phagocytosis and lysosomal digestion help the eukaryotic cell because they effectively increase the membrane surface area over which nutrients can be absorbed.

Bacteria lack phagocytosis; to obtain nutrition from large molecules in their environment, bacteria must secrete digestive enzymes. The extracellularly digested materials are subsequently transported across the bacterial cell membrane through specific transporters (see Section 4.2). In eukaryotes, by contrast, lysosomes allow for intracellular digestion, and digested material crosses the lysosomal membrane into the cytoplasm. Any waste products left in the lysosome can leave the cell via exocytosis.

The Volume within the ER Is Separate from the Cytoplasm

The endoplasmic reticulum is continuous with the outer nuclear membrane, and the lumen (interior) of the ER is continuous with the space between the two nuclear membranes (**Fig. A2.14A**). In addition, the lumen of the ER is spatially equivalent to the interior spaces of other endomembrane components and to the outside of the cell. This means that material in the ER does not need to cross a membrane to enter these other spaces, and mixing can occur via vesicle fusion. For example, material contained within the lumen of the ER can mix with the contents of the Golgi complex (**Fig. A2.14B**) or with the extracellular milieu by fusion of vesicles from the ER with the Golgi membrane or the plasma membrane. These topologically equivalent areas, indicated by a common color in **Figure A2.14A**, are completely separated from the cytoplasm by endomembranes, so that the ER can be used to sequester substances that must be held at low concentrations in the cytoplasm—for example, calcium ions.

Smooth ER and Rough ER

There are two morphologically and functionally distinct types of endoplasmic reticulum: smooth ER and rough ER, as shown in the micrograph of **Figure A2.14C**. The smooth ER is the site of lipid synthesis and some detoxification of noxious compounds. The rough ER is the site where transmembrane proteins, secreted proteins, and resident proteins of the ER, Golgi, or lysosomes are translated. The rough ER appears rough because its cytoplasmic surface is studded with ribosomes. These ribosomes are located on the rough ER because the protein being synthesized by the ribosome has a hydrophobic signal sequence on its amino terminus (the first part of the protein translated from the mRNA). A **signal sequence** is a specific sequence of amino acids that directs proteins to a specific cellular location, such as a membrane. The signal sequence that directs proteins to the ER recognizes a receptor (the **signal recognition particle**) on the rough ER membranes (**Fig. A2.15A**), so the protein-ribosome complex is directed to the surface of the ER membrane. Note that the ribosomes attached to the rough ER are identical to cytoplasmic ribosomes and attach to the ER only transiently because of the types of proteins they are translating—proteins that contain the correct signal sequence.

After the ribosome docks with the signal recognition particle, the protein is threaded through the ER membrane as it is synthesized. Secreted proteins and proteins destined for the lumen of an organelle are threaded completely through the ER membrane and end up in the lumen

A.

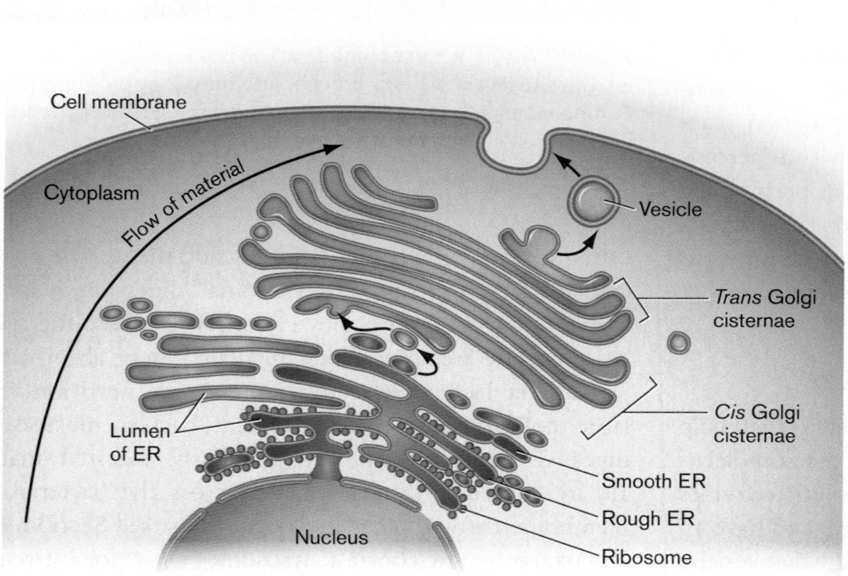
Cell membrane
Cytoplasm
Flow of material
Vesicle
Trans Golgi cisternae
Cis Golgi cisternae
Lumen of ER
Smooth ER
Rough ER
Ribosome
Nucleus

B. Golgi

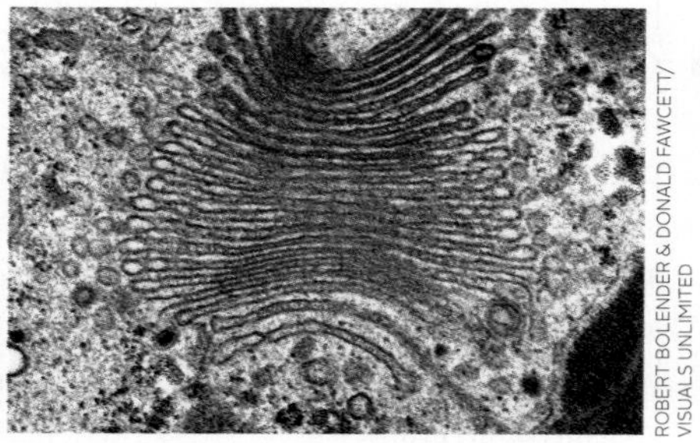
ROBERT BOLENDER & DONALD FAWCETT/ VISUALS UNLIMITED

C. Endoplasmic reticulum

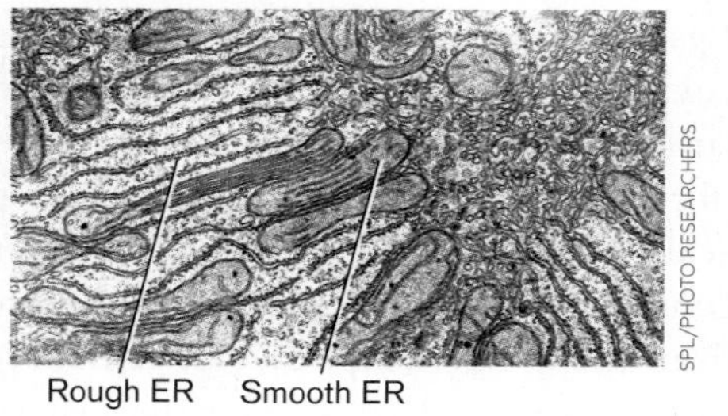
Rough ER
Smooth ER
SPL/PHOTO RESEARCHERS

FIGURE A2.14 ▪ The endoplasmic reticulum and Golgi complex. A. The relationship of the ER to other cellular membranes and the flow of material through vesicles from the rough ER to the cell membrane. **B.** Electron micrograph of Golgi cisternae. **C.** Electron micrograph showing rough and smooth ER.

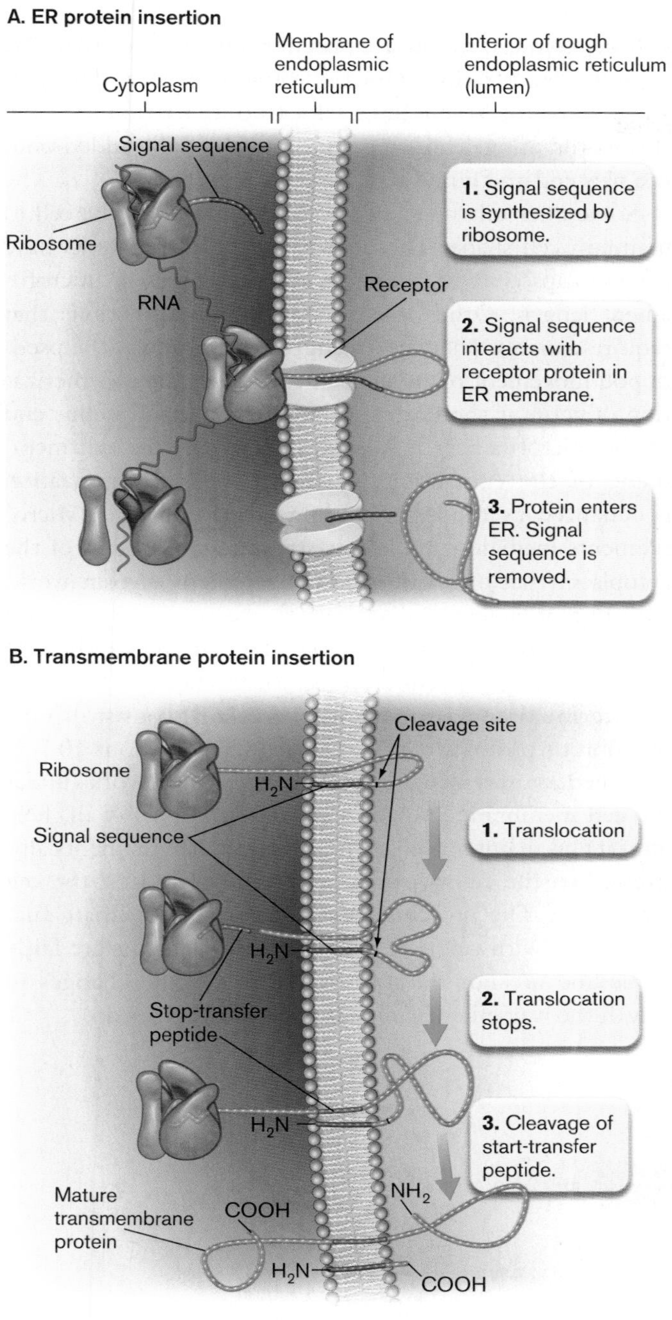
A. ER protein insertion
Cytoplasm
Membrane of endoplasmic reticulum
Interior of rough endoplasmic reticulum (lumen)
Signal sequence
Ribosome
RNA
Receptor
1. Signal sequence is synthesized by ribosome.
2. Signal sequence interacts with receptor protein in ER membrane.
3. Protein enters ER. Signal sequence is removed.
B. Transmembrane protein insertion
Ribosome
Cleavage site
Signal sequence
1. Translocation
Stop-transfer peptide
2. Translocation stops.
3. Cleavage of start-transfer peptide.
Mature transmembrane protein
COOH
NH2
H2N
COOH

FIGURE A2.15 ■ Insertion of proteins into the ER. A. Proteins are targeted to the ER by a signal sequence on the growing polypeptide chain. Secreted proteins and proteins destined for the lumen of an organelle end up in the lumen of the ER. **B.** Transmembrane proteins are not completely inserted through the ER membrane, and a portion of the protein remains within the membrane.

of the ER (see **Fig. A2.15A**). In contrast, transmembrane proteins are not threaded completely through, and part of the protein spans the membrane (**Fig. A2.15B**). The membrane-spanning regions of transmembrane proteins usually contain a continuous stretch of about 20 hydrophobic amino acids that form an alpha helix with the hydrophobic side chains facing out toward the hydrophobic hydrocarbons of the membrane.

As the nascent polypeptide chains are threaded across the ER membrane, the unfolded proteins are bound by chaperone proteins. Chaperones (some of which are heat-shock proteins) prevent partially folded proteins from clumping together and help proteins attain their correct tertiary structure. In the ER, proteins may be modified. The signal sequence that directed the proteins to the ER is usually cleaved off. The environment inside the ER allows disulfide bonds to form between cysteine residues of some proteins. Other ER proteins have oligosaccharide groups covalently attached—a posttranslational modification known as glycosylation. The enzymes that attach the sugars are found only in the ER lumen. In transmembrane proteins of the cell membrane, sugars always face the extracellular environment because the lumen of the ER forms vesicles that open out to the external medium (see **Fig. A2.14A**). Resident ER proteins, those that will stay and function in the ER, are retained inside the ER because they contain a sequence of amino acids that acts as an ER retention signal.

The Golgi Complex Directs Protein Transport

Proteins not retained in the ER move on to the Golgi complex by way of vesicles. The **Golgi complex** consists of separate membrane stacks (cisternae) that each contain unique enzymes. The *cis* face of the Golgi complex is nearest to the ER, and the *trans* face is farthest from the ER, closest to the cell membrane (**Fig. A2.14A** and **C**). As proteins pass through the cisternae, the carbohydrates on them may be trimmed and modified. These modified carbohydrates can serve as address tags for targeting proteins to particular organelles. For example, proteins tagged with mannose 6-phosphate (mannose phosphorylated on its number 6 carbon) are selectively sent to lysosomes; that is, vesicles enriched in proteins containing mannose 6-phosphate bud off from the Golgi and are directed to lysosomes. Proteins not targeted to lysosomes and not marked for retention in the Golgi complex may be sent to the cell membrane. Vesicles leaving the Golgi complex may fuse with the cell membrane, releasing their contents to the extracellular environment (see **Fig. A2.14A**). Transmembrane proteins in these vesicles can then become part of the cell membrane. Regions of transmembrane proteins that were inside vesicles will face the extracellular environment, and cytoplasmic portions will remain cytoplasmic.

A2.5

The Cytoskeleton

Eukaryotic cells contain proteins called intermediate filaments, microfilaments, and microtubules that are collectively termed the **cytoskeleton**. As the name implies, these proteins serve as a cell skeleton and impart specific shapes to eukaryotic cells. However, cytoskeletal proteins are multifunctional and are also involved in whole-cell movements and movements of substances within the cell.

Microfilaments and Intermediate Filaments

Microfilaments, also known as actin filaments, have a diameter of 7 nm (**Fig. A2.16A**). They are formed when individual actin monomers (globular actin, or G-actin) polymerize, in a process fueled by ATP hydrolysis, to form chains of filamentous actin (F-actin). Two F-actin chains twist around each other to form microfilaments that have a plus end and a minus end. Microfilaments are dynamic structures, growing and shrinking in a controlled manner. New monomer units add to the plus end and dissociate from the minus end. Whether actin will polymerize or depolymerize depends on a number of factors, including the concentration of G-actin. The critical concentration is a measure of the ability of actin to polymerize. At G-actin concentrations below the critical concentration, F-actin will depolymerize; and at concentrations greater than the critical concentration, G-actin will polymerize. The plus end of a microfilament has a critical concentration less than that of the minus end, so actin is preferentially added onto the plus end and removed from the minus end.

Some microfilaments play a structural role in the cell to maintain cell shape. These structural microfilaments have protein caps at both ends to prevent changes in microfilament length. Other microfilaments have functions that require dynamic changes in length. For example, the pseudopod movement of an ameba depends on the polymerization of actin at the leading edge of growth. The plus end of the microfilament is located underneath the cell membrane of the extending pseudopod, and polymerization is enhanced by the actin-binding protein profilin. Microfilaments mediate cytoplasmic streaming, a mixing of the cytoplasm that aids diffusion. The protein myosin works with microfilaments to generate the forces needed for cell streaming, pseudopod formation, and cytokinesis, the separation of daughter cells after nuclear division.

Intermediate filaments (**Fig. A2.16B**) consist of various fibrous proteins that have a diameter of about 10 nm. Intermediate filaments often form a meshwork under the cell membrane and, in cells that lack a cell wall, help impart and maintain cell shape. Intermediate filaments also strengthen the cell by resisting tension placed on the cell membrane. The proteins that make up intermediate filaments vary with cell type. Intermediate filaments are fairly stable and are not thought to undergo acute changes in length the way microfilaments and microtubules do.

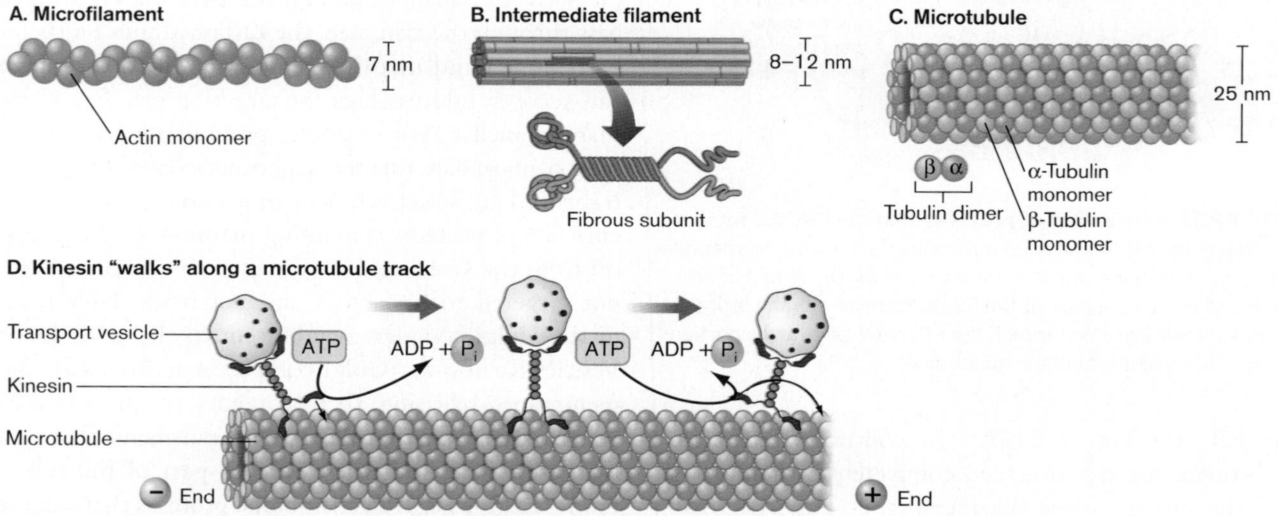
A. Microfilament
7 nm
Actin monomer
B. Intermediate filament
8–12 nm
Fibrous subunit
C. Microtubule
25 nm
β α
Tubulin dimer
α-Tubulin monomer
β-Tubulin monomer
D. Kinesin "walks" along a microtubule track
Transport vesicle
Kinesin
Microtubule
ATP
ADP + Pi
ATP
ADP + Pi
– End
+ End

FIGURE A2.16 ■ **Cytoskeletal proteins. A.** Microfilaments consist of two strands of actin polymers twisted together. **B.** Intermediate filaments are ropelike assemblages of various proteins. **C.** Microtubules are polymers of tubulin dimers. **D.** Fueled by the hydrolysis of ATP, the motor protein kinesin can move vesicles or organelles toward the plus end of microtubules.

Microtubules

Microtubules have a larger diameter (25 nm) than microfilaments and intermediate filaments (**Fig. A2.16C**). The hollow microtubule structure consists of 13 tubulin dimers; one alpha-tubulin protein plus one beta-tubulin protein form one tubulin dimer. Like microfilaments, microtubules have plus (faster-growing) and minus (slower-growing) ends and are dynamic structures that can polymerize and depolymerize. Polymerization is an energy-requiring process, and the necessary energy is obtained by coupling polymerization to GTP (guanosine triphosphate) hydrolysis. Microtubules aid movement of substances within the cell and are also involved in powering whole-cell movement by cilia and flagella.

Traffic of proteins through the endomembrane system (see Section A2.4) relies on the controlled movement of vesicles from one cellular compartment to the next. Microtubules provide tracks that can move vesicles from one organelle to the next in an efficient, directed fashion. Working with microtubules to accomplish this task are motor proteins. Motor proteins such as kinesin and dynein can capture cargo (for example, vesicles or organelles) and walk them along microtubule tracks in an ATP-dependent process (**Fig. A2.16D**). Kinesin moves cargo toward the plus end of the microtubule, while dynein moves cargo toward the minus end. In addition to moving vesicles, microtubules segregate (pull apart) the duplicated chromosomes during mitosis.

Eukaryotic Cilia and Flagella

Cilia and flagella are thin extensions of the cell membrane that can move in a whiplike fashion, driven by interactions between microtubules and the motor protein dynein. **Flagella** (singular, **flagellum**) are relatively long, and cells usually have only one or two of them; **cilia** (singular, **cilium**) are shorter and more numerous. Both flagella and cilia can move cells through space. Cilia may also aid in food capture—for example, by sweeping extracellular fluid into the gullet of a paramecium. Eukaryotic cilia and flagella differ from bacterial cilia and flagella; bacterial flagella depend on the proton motive force to rotate a motor that causes flagellar movement, whereas eukaryotic flagella rely on ATP hydrolysis by dynein to move the flagella in a whiplike fashion.

The study of protists has played a key role in revealing the structure and function of eukaryotic flagella and cilia. Much information about flagellar structure has come from studies of a *Chlamydomonas* species, a unicellular alga that has two long flagella (**Fig. A2.17A**). Dynein was first discovered in the cilia of the unicellular eukaryote *Tetrahymena* (**Fig. A2.17B**). Flagella and cilia have the same

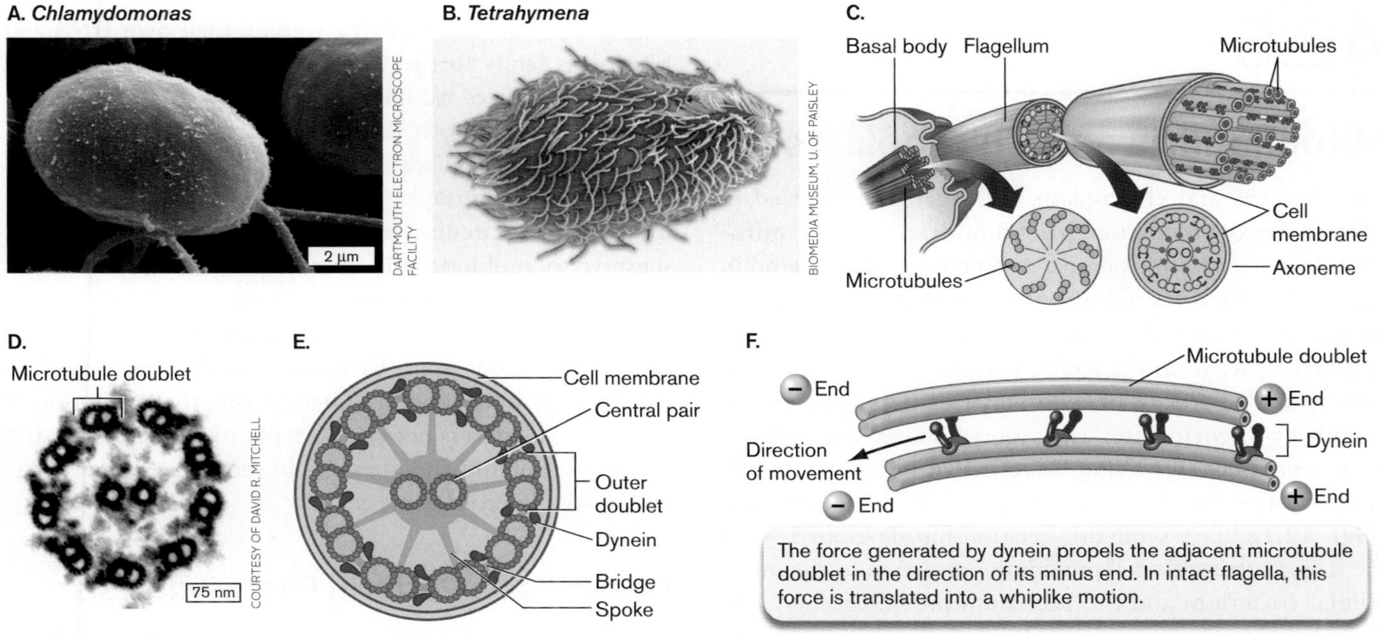
A. Chlamydomonas
2 µm
DARTMOUTH ELECTRON MICROSCOPE FACILITY
B. Tetrahymena
BIOMEDIA MUSEUM, U. OF PAISLEY
C.
Basal body
Flagellum
Microtubules
Cell membrane
Axoneme
Microtubules
D.
Microtubule doublet
75 nm
COURTESY OF DAVID R. MITCHELL
E.
Cell membrane
Central pair
Outer doublet
Dynein
Bridge
Spoke
F.
Microtubule doublet
End
End
Dynein
Direction of movement
End
End
The force generated by dynein propels the adjacent microtubule doublet in the direction of its minus end. In intact flagella, this force is translated into a whiplike motion.

FIGURE A2.17 ■ Flagella and cilia. A. The protist *Chlamydomonas* has two long flagella (fluorescence micrograph). **B.** The protist *Tetrahymena* (size 50–70 µm) has numerous cilia (SEM). **C.** Structure of flagella and cilia. **D.** Cross section of axoneme (TEM). **E.** Cross section of axoneme indicating bridges and spokes formed by cross-linking proteins. **F.** Force generated by dynein on a microtubule doublet.

structure and mechanism of action, so for convenience we will restrict the following discussion to flagella.

A cross section through a flagellum reveals a central bundle of microtubules called the axoneme (**Fig. A2.17C**). The axoneme originates at a microtubule-organizing center called the basal body, which is similar to the centriole (see Section A2.2). The minus ends of the microtubules are at the basal body; the plus ends at the tip of the flagellum are capped to prevent changes in length.

As shown in the electron micrograph in **Figure A2.17D**, the axoneme has a characteristic arrangement of two central microtubules and nine microtubule doublets around the periphery. The microtubule doublets are connected to each other and to the central microtubules through protein cross-links called bridges and spokes (**Fig. A2.17E**). The motor protein dynein connects adjacent microtubule doublets. If the cell membrane is removed from the surface of a flagellum and the protein cross-links (but not dynein) are dissolved, the microtubule doublets are observed to lengthen in the presence of ATP. This lengthening is due to the action of dynein, which slides the microtubule doublets past each other (**Fig. A2.17F**). In intact flagella, the tension imparted as dynein tries to slide microtubules past each other is translated into a bending motion. Controlled cycles of dynein activation and inactivation on opposite sides of the axoneme lead to the whiplike motion of eukaryotic flagella.

A2.6

Mitochondria and Chloroplasts

Mitochondria and chloroplasts are organelles involved in cellular energy production. **Mitochondria** (singular, **mitochondrion**) perform oxidative respiration and are found in nearly all eukaryotes. **Chloroplasts** perform photosynthesis and are found only in photosynthetic eukaryotes, such as green algae. Both organelles became part of eukaryotic cells through a process of endosymbiosis. The **serial endosymbiosis theory** states that mitochondria and chloroplasts were once free-living bacteria that became ingested, but not digested, by a larger (possibly eukaryotic) cell (**Fig. A2.18A**). A symbiotic relationship developed, with the larger eukaryotic cell providing protection to the intracellular bacterium and the bacterium providing energy to the eukaryote. As we will see, the structure of mitochondria and chloroplasts reflects their origin as internalized bacteria.

Mitochondria Produce ATP by Oxidative Respiration

Mitochondria are the powerhouses of the eukaryotic cell. A cell may contain tens or hundreds of mitochondria, depending on its energy needs. Mitochondria have two membranes: an outer membrane and an inner membrane (**Fig. A2.18B**). The inner membrane has numerous infoldings called cristae that increase its surface area. It is thought that as a result of the endocytosis of the bacterium, the inner mitochondrial membrane is derived from the bacterium and the outer membrane is derived from the larger eukaryote. Supporting this idea is the fact that the inner membrane has structural characteristics of a prokaryotic cell membrane, while the outer membrane is similar to the host eukaryotic membrane. For example, 20% of the phospholipids in the inner membrane are cardiolipin, a bacterial phospholipid largely absent from membranes of eukaryotic origin.

Mitochondria contain two distinct compartments: the intermembrane space between the two membranes, and the matrix inside the inner membrane. Different stages of oxidative respiration occur in specific compartments. As predicted by the endosymbiosis theory, the topology of these processes is similar in bacteria and mitochondria (**Table A2.2**). For example, in bacteria and archaea, the tricarboxylic acid cycle (TCA cycle) occurs in the cytoplasm; in mitochondria, the TCA cycle takes place inside the matrix, the metabolic equivalent of the prokaryotic cytoplasm.

The fact that mitochondria contain their own DNA and ribosomes lends further support to endosymbiosis, since these cell features would be a necessary part of any free-living organism. Both the DNA and ribosomes of mitochondria show similarities with the DNA and ribosomes of bacteria. For example, like prokaryotic DNA, mitochondrial DNA is circular, and mitochondrial ribosomes are sensitive to antibiotics that disrupt prokaryotic ribosomes. Although nuclear DNA encodes some mitochondrial proteins, phylogenetic analysis has shown that these nuclear genes originated from bacteria, probably the engulfed ancestors of mitochondria. Furthermore, the generation of new mitochondria is not tied to replication of the cell, and the division of mitochondria within the cell is similar to the fission seen in prokaryotic cells.

Chloroplasts Perform Photosynthesis

Chloroplasts are organelles found only in photosynthetic eukaryotes. In the light reactions of photosynthesis, chloroplasts convert light energy from the Sun to ATP and

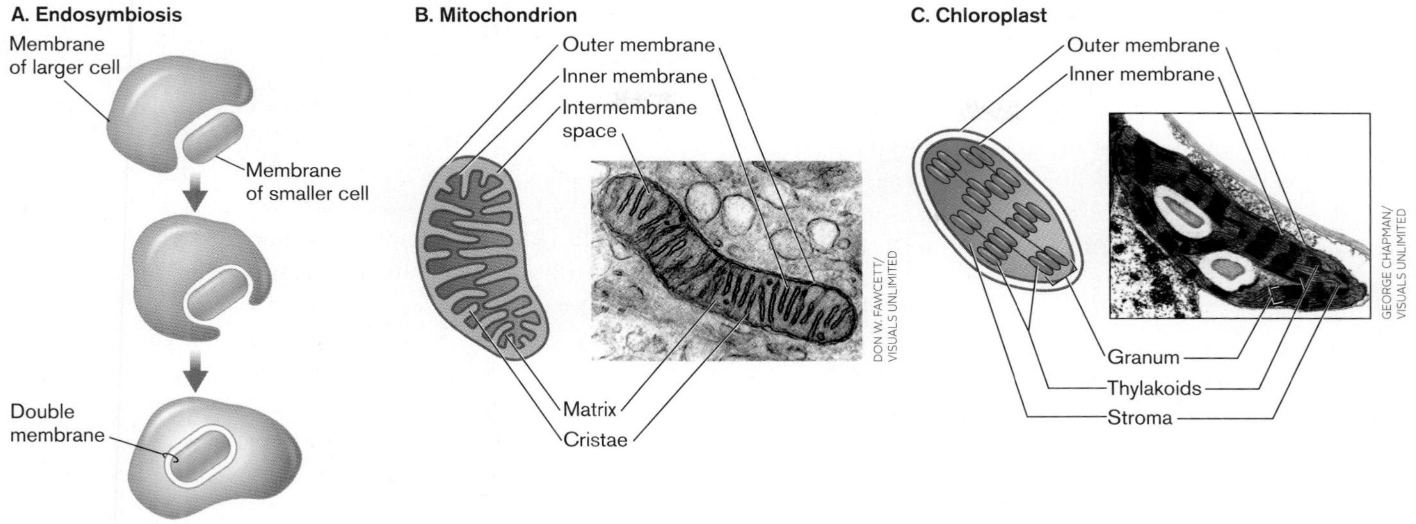
A. Endosymbiosis
Membrane of larger cell
Membrane of smaller cell
Double membrane
B. Mitochondrion
Outer membrane
Inner membrane
Intermembrane space
Matrix
Cristae
DON W. FAWCETT/ VISUALS UNLIMITED
C. Chloroplast
Outer membrane
Inner membrane
Granum
Thylakoids
Stroma
GEORGE CHAPMAN/ VISUALS UNLIMITED

FIGURE A2.18 ▪ Serial endosymbiosis theory, mitochondria, and chloroplasts. **A.** Origin of organelles according to the serial endosymbiosis theory. **B.** Structures in a mitochondrion. **C.** Chloroplast structure.

TABLE A2.2

Location of oxidative respiration components in bacteria and mitochondria.

Feature	Bacteria	Mitochondria
Electron transport system (ETS)	Cell membrane	Inner membrane
ATP synthase	Cell membrane	Inner membrane
TCA cycle	Cytoplasm	Matrix
Proton motive force	Protons diffuse from lower-pH extracellular environment into the cytoplasm across the cell membrane	Protons diffuse from lower-pH intermembrane space into the matrix across the inner membrane

reduced NADPH. In the subsequent light-independent reactions, the ATP and NADPH are used to reduce CO_2 to sugar.

Like mitochondria, chloroplasts originated via endosymbiosis. Bacteria related to cyanobacteria were the prokaryotic partners that gave rise to chloroplasts. Like eukaryotic chloroplasts, cyanobacteria contain chlorophylls *a* and *b* and perform aerobic photosynthesis. In addition, cyanobacteria contain extensive internal membranes called thylakoids that contain chlorophyll and participate in the light reactions of photosynthesis.

Chloroplasts have three membranes whose topology can be understood in light of endosymbiosis (**Fig. A2.18C**). The outer membrane appears to be derived from the host eukaryotic cell, the inner membrane is equivalent to the bacterial cell membrane, and the thylakoid membrane is derived from the bacterial thylakoid membranes. The region inside the inner membrane is called the stroma and is equivalent to the bacterial cytoplasm. The thylakoid membrane is packed with the chlorophyll pigments that give chloroplasts their green color. ATP and NADPH are produced in the stroma and used there in the light-independent reactions of CO_2 fixation. As would be expected as a result of endosymbiosis, chloroplasts, like mitochondria, contain their own circular DNA and their own ribosomes.

APPENDIX 3
Laboratory Methods for Microbiology

A3.1 Isolating Parts of Cells by Using an Ultracentrifuge

A3.2 Agarose Gel Electrophoresis

A3.3 Protein Identification on 2D Gels with Mass Spectrometry

A3.4 RNA and DNA Identification by Northern and Southern Blots

A3.5 Sanger Method of DNA Sequencing

A3.6 Gene Fusions Identify Regulatory Mutants

A3.7 Primer Extension Identifies Transcriptional Start Sites

A3.8 DNA Microarray

A3.9 Multiplex PCR

A3.10 Fluorescence In Situ Hybridization (FISH) and CARD-FISH

A3.11 Immunoprecipitation Techniques

Centrifuge for cell analysis. Centrifugation is a powerful way to isolate cells, such as white blood cells from the blood; or components of cells, such as ribosomes from the cytoplasm. A centrifuge subjects a sample to centrifugal forces associated with rapid rotation. The instrument can be designed to separate components by their mass, or by their density. *Source:* Photo courtesy of Eppendorf AG.

A3.1

Isolating Parts of Cells by Using an Ultracentrifuge

Cellular components, such as ribosomes and flagellar motors, can be isolated readily from cells. The cells must be broken open by techniques that allow subcellular components to remain intact. Here are examples of such techniques:

- **Mild detergent lysis.** Cells can be lysed with a detergent capable of dissolving membranes but not denaturing proteins.
- **Sonication.** Cells can be lysed by intense ultrasonic vibrations above the range of human hearing.
- **Enzymes.** Enzymes such as lysozyme can break the cell wall, allowing the cell to be lysed by mild osmotic shock.
- **Mechanical disruption.** Cells can be broken open by the application of high pressure (with a French press) or through beating with microscopic beads (with a bead beater).

The Ultracentrifuge

Once cells have been broken open, different parts of the cell can be isolated by **cell fractionation**. A key tool of cell fractionation is the ultracentrifuge (**Fig. A3.1A**), a device in which solutions containing cell components are rotated in tubes at high speed. The high rotation rate generates centrifugal forces strong enough to separate subcellular particles. The ultracentrifuge was invented by the Swedish chemist Theodor Svedberg (1884–1971), who won the 1926 Nobel Prize in Chemistry for the use of ultracentrifugation to separate proteins.

Modern ultracentrifuges have titanium rotors that spin in a vacuum to avoid frictional heating, generating forces up to 100,000 times gravity (100,000*g*). For fractionation, cells are first lysed by one of the methods previously described to obtain a cell "lysate," a general term for the contents of a broken cell. The cell lysate is placed in tubes containing a high-density solution, such as sucrose or cesium chloride solution, in which suspended particles sediment slowly. Under a high *g* force, however, particles sediment at different rates, depending on their size and density.

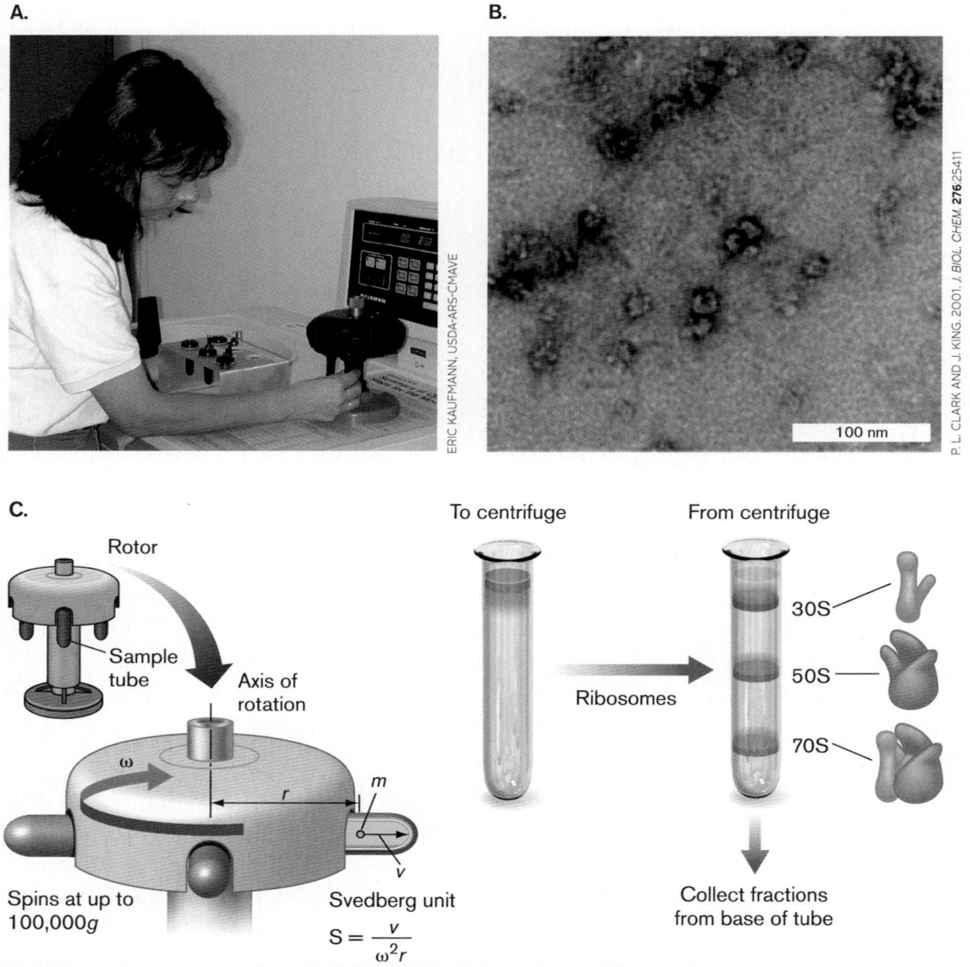
A.
ERIC KAUFMANN, USDA-ARS-CMAVE
B.
100 nm
P. L. CLARK AND J. KING. 2001. J. BIOL. CHEM. 276:25411
C.
Rotor
Sample tube
Axis of rotation
ω
r
m
v
Spins at up to 100,000g
Svedberg unit
$S = \frac{v}{\omega^2 r}$
To centrifuge
From centrifuge
Ribosomes
30S
50S
70S
Collect fractions from base of tube

FIGURE A3.1 ■ Cell fractionation by ultracentrifugation. A. An ultracentrifuge is used to fractionate cell components. **B.** *Salmonella* ribosomes (TEM) isolated from the cytoplasm by ultracentrifugation through a linear sucrose density gradient. A polysome consists of two or more ribosomes attached to mRNA. **C.** High-speed rotation generates high centrifugal forces, measured in units of gravity (*g*). The Svedberg unit (S) offers a measure of particle size based on the particle's rate of travel in a tube subjected to high *g* force. The Svedberg coefficient (number of S units—for example, 30S) is defined in terms of the velocity of the particle in the tube (v), the radius of the rotor (r), and the rotational velocity (ω). The coefficient of S for a given particle depends on its mass (m) and its shape. After centrifugation, fractions are collected from the base of the centrifuge tube. The fractions contain radiolabeled ribosomes. The largest particles (whole ribosomes) sediment near the bottom of the tube, and the smaller particles (separate 50S and 30S subunits) appear in upper fractions.

The **sedimentation rate** is the rate at which particles of a given size and shape travel to the bottom of the tube under centrifugal force. The sedimentation rate is measured by the Svedberg unit, S, which is given by the particle's rate of sedimentation (v), the radius at which the tube rotates (r), and the rotational velocity (ω):

$$S = v/(\omega^2 r)$$

For a given type of particle in suspension, the sedimentation rate also depends on the particle's mass and shape. The contribution of particle mass and shape is defined as its **Svedberg coefficient**; for example, the coefficient of the small subunit of the ribosome is 30, for a sedimentation value of 30S. The value of the Svedberg coefficient increases with the average cross-sectional area of the particle. For bacterial ribosomes, ultracentrifugation yields intact ribosomes (70S), as well as separated ribosomal subunits: the large subunit (50S) and the small subunit (30S). Within cells, ribosomes normally exist as a mixture of joined and separate subunits.

To isolate the ribosomes, a cell lysate is layered onto a tube of sucrose solution, whose density decreases the sedimentation rate and increases the separation of particles of different size (**Fig. A3.1C**). The fractions shown in **Figure A3.1C** were drained sequentially from the base of the tube after centrifugation. The heaviest particle, the 70S ribosome, appears in the fractions nearest the bottom of the tube because it travels fastest under the centrifugal force.

The ribosomes can be seen in collected fractions by electron microscopy (**Fig. A3.1B**). In some cases, two or more ribosomes are connected by a strand of messenger RNA (mRNA). This multiple-ribosome structure is called a **polysome**. Ribosomes isolated by centrifugation can translate messenger RNA in cell-free systems. Experiments in cell-free systems provide the basis of much of our knowledge of protein synthesis (see Chapter 8).

Limitations of Cell Fractionation

Cell fractionation yields clues about internal structure but provides little information about processes that require overall integrity of the cell. For example, the role of the transmembrane electrochemical potential, or proton potential, in ATP synthesis was obscured for many years because biochemists were unable to isolate a cytoplasmic complex that generates it. Transmembrane ion gradients were, in fact, observed within membrane vesicles—spheres of membrane isolated by cell disintegration and centrifugation. But it was harder to demonstrate that the entire cell membrane of an intact cell supports a proton potential.

A3.2

Agarose Gel Electrophoresis

Electrophoresis separates macromolecules as they migrate through a gel under a voltage gradient. The rate of migration is based on the size and charge of the migrating molecules (**Fig. A.3.2**). DNA molecules have a relatively uniform negative charge, one unprotonated phosphoryl group per nucleotide. So within a gel, DNA molecules will separate according to size; the smaller fragments migrate faster.

The gel for DNA separation is formed from agarose, a long-chain polysaccharide that is soluble in water above 50°C but forms a solid gel at room temperature. The agarose gel matrix actually consists of widely separated links, like chicken wire; a water solution can flow through the microscopic holes in the matrix. The concentration of agarose in the gel determines the density of the matrix and the average size of holes through which DNA molecules may penetrate. Agarose concentration determines the size range of DNA that can be separated in a given gel: The higher the agarose concentration, the smaller the sizes of molecules that can be separated.

First a gel must be formed by pouring agarose solution into a tray within an electrophoresis chamber that leads to a voltage source (**Fig. A3.2A** and **B**). The gel tray must include a "comb," a piece of plastic with prongs that form wells within the gel. Once the gel has cooled and solidified, the comb is removed and DNA samples are added. The DNA must be dissolved within a sample buffer that contains a dense substance such as glycerol to settle the sample deep within the well. The sample buffer also contains dyes such as bromophenol blue that can be visualized as they run through the gel under voltage potential.

After the dye has run through the gel, the voltage is turned off and the gel is removed. The gel must be stained to visualize the DNA bands. A typical stain is ethidium bromide, an aromatic molecule that intercalates between DNA nucleotides and fluoresces under UV irradiation. **Figure A3.2C** shows an example of electrophoresis results. Bacterial DNA was obtained from anthrax patients in an outbreak in Sverdlovsk, Russia, in 1979, suspected to be caused by the release of *Bacillus anthracis* from a biowarfare facility of the former USSR. DNA samples (lanes 2–8) were amplified by the polymerase chain reaction (PCR) using primers specific to different strains of *B. anthracis.* The size of the PCR products confirmed the presence of anthrax bacteria.

Note that in electrophoresis, the size range of the fragments is nonlinear; thus the migration distances between

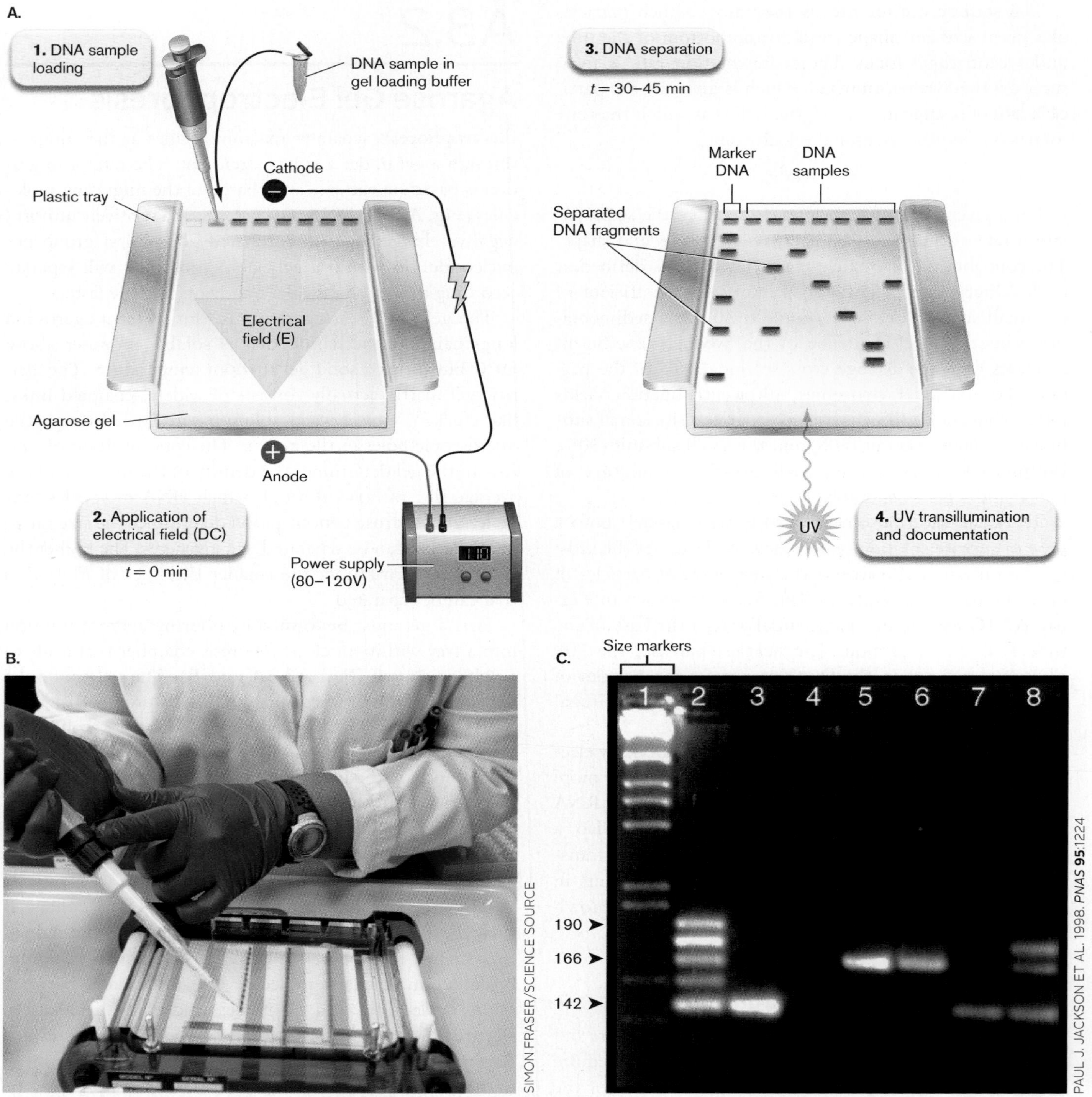
A.
1. DNA sample loading
DNA sample in gel loading buffer
Cathode
Plastic tray
Electrical field (E)
Agarose gel
Anode
2. Application of electrical field (DC)
t = 0 min
Power supply (80–120V)
3. DNA separation
t = 30–45 min
Marker DNA
DNA samples
Separated DNA fragments
UV
4. UV transillumination and documentation
B.
SIMON FRASER/SCIENCE SOURCE
C.
Size markers
1 2 3 4 5 6 7 8
190
166
142
PAUL J. JACKSON ET AL. 1998. PNAS 95:1224

FIGURE A3.2 ■ Agarose gel electrophoresis separation of DNA fragments. A. (1) DNA samples are loaded into slots at the end of an agarose gel. When a voltage is applied (2), the DNA fragments (which have negative charge) are attracted toward the positive electrode (3). The fragments move according to size; the smaller fragments move fastest. When the fragments have run far enough to separate, they can be visualized by fluorescence when exposed to UV light (4). **B.** Apparatus for agarose gel electrophoresis. **C.** DNA fragments in a gel visualized by UV. The fragments are PCR-amplified sequences from *Bacillus anthracis* isolated from victims of anthrax exposure.

large fragments are smaller than the migration distances between smaller fragments. In order to measure DNA sizes accurately, a control sample is run (lane 1) in which DNA fragments of known size appear.

A3.3

Protein Identification on 2D Gels with Mass Spectrometry

Separating and visualizing the proteins in a cell extract is a daunting challenge but can be accomplished by a process known as two-dimensional (2D) gel electrophoresis. The first step in the process is **isoelectric focusing (IEF)**, whereby proteins in a cell extract are sorted according to their individual charges. Every protein is adorned with ionizable side groups (amino and carboxyl) that can impart charge. The degree of protonation (ionization) of each group is affected by pH and by the dissociation constant (p*K*) of the group. Generally, an ionizable group will be protonated at pH values below its p*K*. At pH values below their intrinsic p*K*, amino groups become positively charged ($R\text{-}NH_3^+$), whereas carboxyl groups are neutral (R-COOH). When the pH rises above a group's p*K*, the amino groups will be neutral (NH_2) and carboxyl groups will become negatively charged (COO^-).

Proteins have hundreds of amino and carboxyl groups in different ratios, so for each protein there will be a specific pH, called the **isoelectric point**, where the plus and minus charges cancel out—that is, where the net charge on the protein is zero. In an electrical field, proteins set in a pH gradient gel will migrate through the gel until they reach a gel pH where the charges cancel out. In isoelectric focusing, cell proteins are applied to a gel containing an immobilized pH gradient, called an IPG strip, and are subjected to an electrical field (**Fig. A3.3**). Negatively charged proteins move toward the positive pole until they reach the pH of their isoelectric point, where the proteins no longer have charge. Without a charge, the protein does not move under the influence of the electrical field. Positively charged proteins behave similarly but move toward the negative pole.

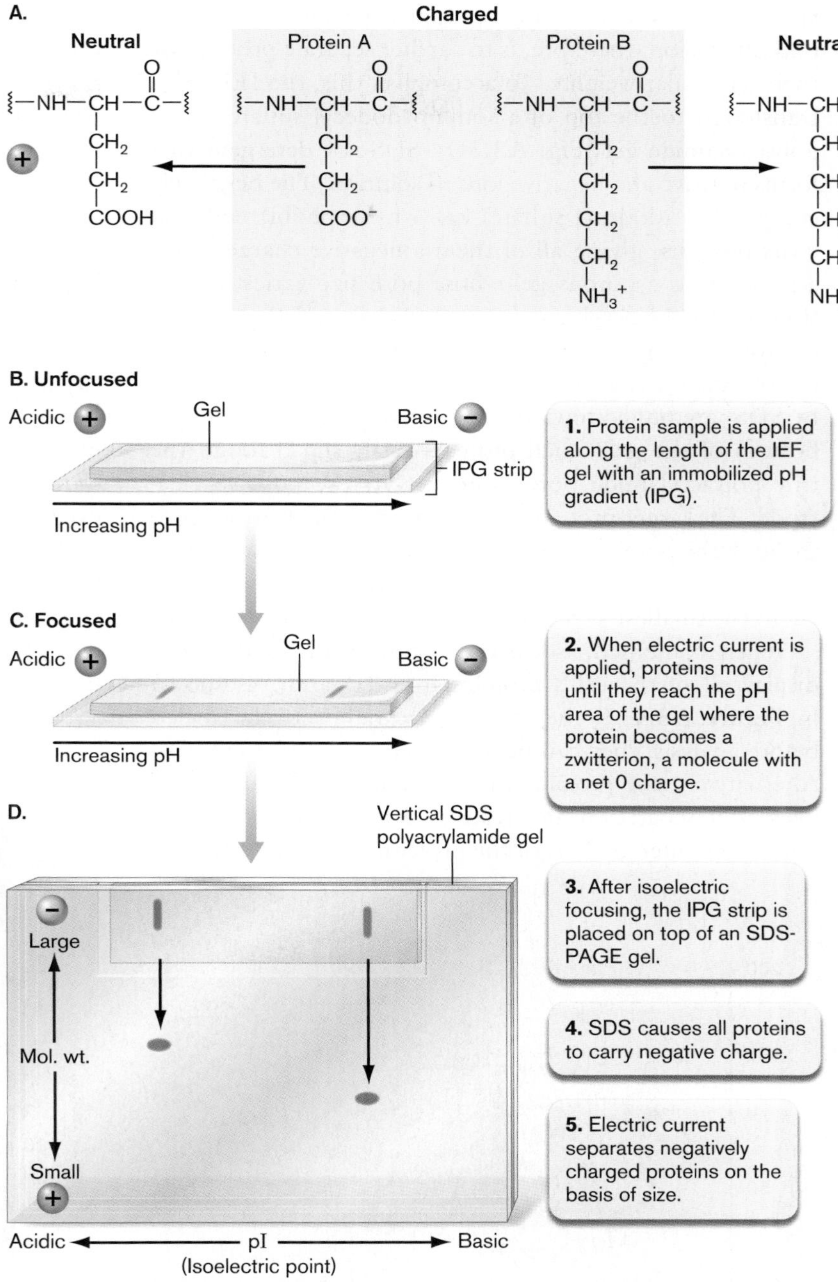
A.
Charged
Neutral
Protein A
Protein B
Neutral
COOH
COO⁻
NH₃⁺
NH₂
B. Unfocused
Acidic
Gel
Basic
IPG strip
Increasing pH
1. Protein sample is applied along the length of the IEF gel with an immobilized pH gradient (IPG).
C. Focused
Acidic
Gel
Basic
Increasing pH
2. When electric current is applied, proteins move until they reach the pH area of the gel where the protein becomes a zwitterion, a molecule with a net 0 charge.
D.
Vertical SDS polyacrylamide gel
Large
Mol. wt.
Small
3. After isoelectric focusing, the IPG strip is placed on top of an SDS-PAGE gel.
4. SDS causes all proteins to carry negative charge.
5. Electric current separates negatively charged proteins on the basis of size.
Acidic
pI
(Isoelectric point)
Basic

FIGURE A3.3 ■ 2D analysis of cell proteins. A. General movement of negatively and positively charged proteins in an electric current applied to an immobilized pH gradient (IPG) strip. Once a protein reaches a position corresponding to its isoelectric point, movement stops and the protein focuses in the area. **B.** Sample applied to an IPG strip. **C.** The IPG strip after focusing. **D.** The focused IPG strip is layered onto a standard SDS polyacrylamide gel and separated by size. The result is separation of the proteins in a 2D pattern. *Source:* Modified from Albert Moat et al. 2002. *Microbial Physiology*, 4th ed.

Proteins with very different molecular weights can share the same isoelectric point. As a result, one band on an IEF gel could contain ten proteins. The goal of the second dimension, therefore, is to further separate proteins by their molecular weights. To accomplish this, the IEF gel is transferred to the top of a sodium dodecyl sulfate (SDS) polyacrylamide gel (**Fig. A3.3D**). SDS is a detergent that forms positive and negative ions in solution. The negatively charged ion (dodecyl sulfate) has a hydrophobic end that coats proteins, giving all of them a negative charge. Polyacrylamide is a porous gel whose pore size varies in relationship to the acrylamide concentration and the amount of cross-linking used to make it a gel. When placed in an electric field, the proteins (which are all negatively charged by SDS) are pulled toward the positive pole, located at the bottom of the gel. Small proteins easily slip through the tiny polyacrylamide pores and race to the positive electrode. The larger proteins find it more difficult to squeeze through the pores and are slow to move. The result is a gradient of proteins ranging from the largest at the top of the gel to the smallest at the bottom.

When combined, isoelectric focusing and SDS-PAGE display all of the cell's proteins in a 2D array, as shown in **Figure A3.4**. If the proteins are radioactively labeled before analysis, they can be visualized by autoradiography. Alternatively, the proteins can be stained with fluorescent dyes after separation and read by a laser scanner. Subsequent computer analysis of the proteome patterns obtained from cells grown under two different conditions will reveal proteins whose levels increase or decrease in response to the changing environment. **Figure A3.5** shows a dual-channel image analysis that compares *Bacillus subtilis* proteomes from cultures grown in a minimal glucose medium to those of cultures grown in the same medium but supplemented with a rich assortment of amino acids. The different colors indicate whether the level of a protein is higher, lower, or the same in the two cultures.

The usefulness of 2D PAGE is starting to be supplanted by a newer technology, mass spectrometry, which can identify proteins by determining their exact molecular weights (see Section 10.9). Protein mass spectrometry linked to 2D gel electrophoresis is a foundational method in the field of proteomics, which looks at the entire set of proteins being expressed in a cell or tissue under specific conditions.

The power of proteomics becomes evident when it is linked to genomics. Knowing the complete sequence of an organism's chromosome makes it easier to identify proteins following growth under any experimental condition. For example, levels of certain proteins increase when

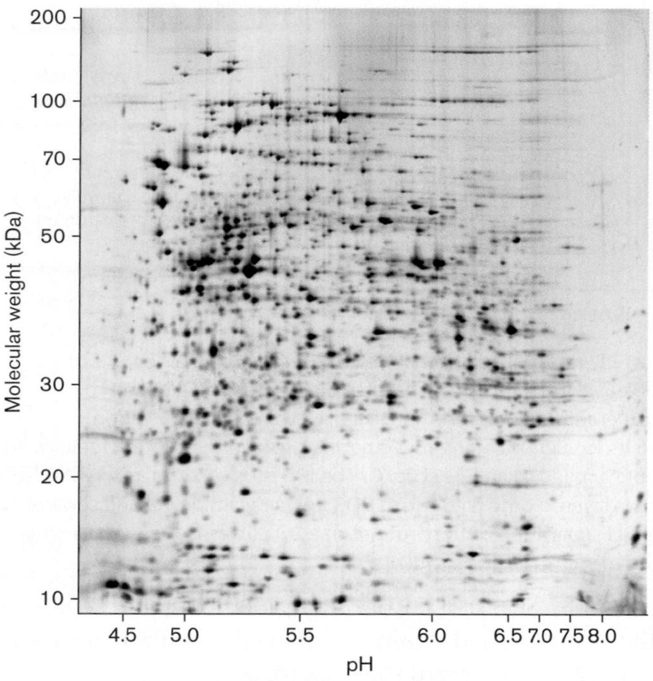
200
100
70
50
30
20
10
Molecular weight (kDa)
4.5
5.0
5.5
6.0
6.5 7.0 7.5 8.0
pH

FIGURE A3.4 ■ Proteome of *Escherichia coli.* Each spot represents a different protein. *Source:* © Swiss Institute of Bioinformatics, Geneva, Switzerland. 2001. *Proteomics* **1**:409.

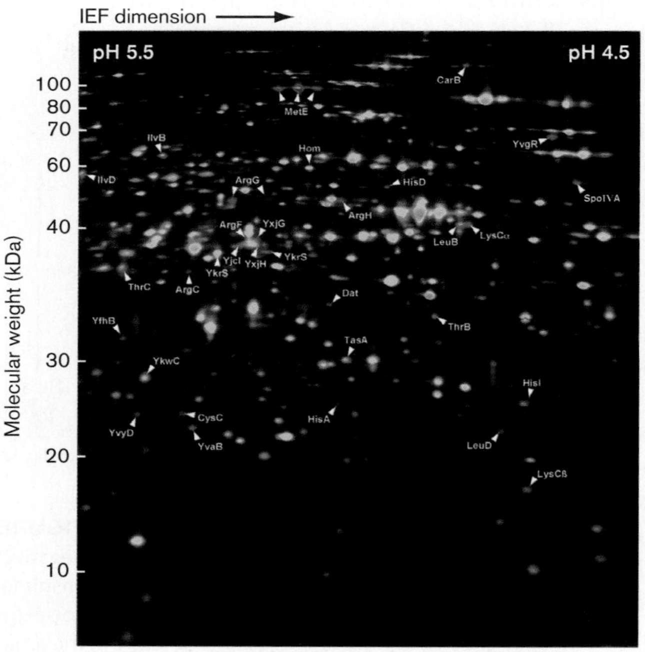
IEF dimension
pH 5.5
pH 4.5
100
80
70
60
40
30
20
10
Molecular weight (kDa)

FIGURE A3.5 ■ Proteomic profile of *Bacillus subtilis* cells grown in minimal media with and without mixed amino acid supplementation (casamino acids). The IEF gradient used in the first dimension was pH 4.5–5.5; this represents only a part of the entire proteome. The figure is the result of dual-channel image analysis of silver-stained gels. A computer assigns the color red to proteins expressed in minimal media and the color green to proteins expressed in amino acid–supplemented media. If the proteins are expressed under both conditions, the red and green colors combine to form yellow or orange. *Source:* Ulrike Mäder et al. 2002. *J. Bacteriol.* **184**:4288.

a bioremediating microbe is grown on benzene instead of glucose. How do we determine which proteins are induced? Those proteins, and the genes encoding them, can be identified from a 2D gel if the organism's genome is sequenced.

The procedure for proteomic identification of proteins in a 2D gel is shown schematically in **Figure A3.6**. The protein spot of interest is excised from the gel and digested into peptide fragments with a protease. The peptide fragment mixture is then analyzed by mass spectrometry, which determines the precise molecular weight of each fragment. In tandem mass spectrometry, each proteolytic fragment is then subfragmented by ionization. Ionization generates progressively smaller secondary fragments missing one or more amino acids. Because the weight of each amino acid is distinct, tandem mass spectrometry analysis will determine the amino acid sequence of the initial proteolytic fragment. Computer programs compare the amino acid sequence of each protein fragment with the predicted sequences of proteolytic fragments from all open reading frames (ORFs) in a genome. Assigning a spot on a 2D gel to a specific protein comes from finding several peptides that match different regions of that protein.

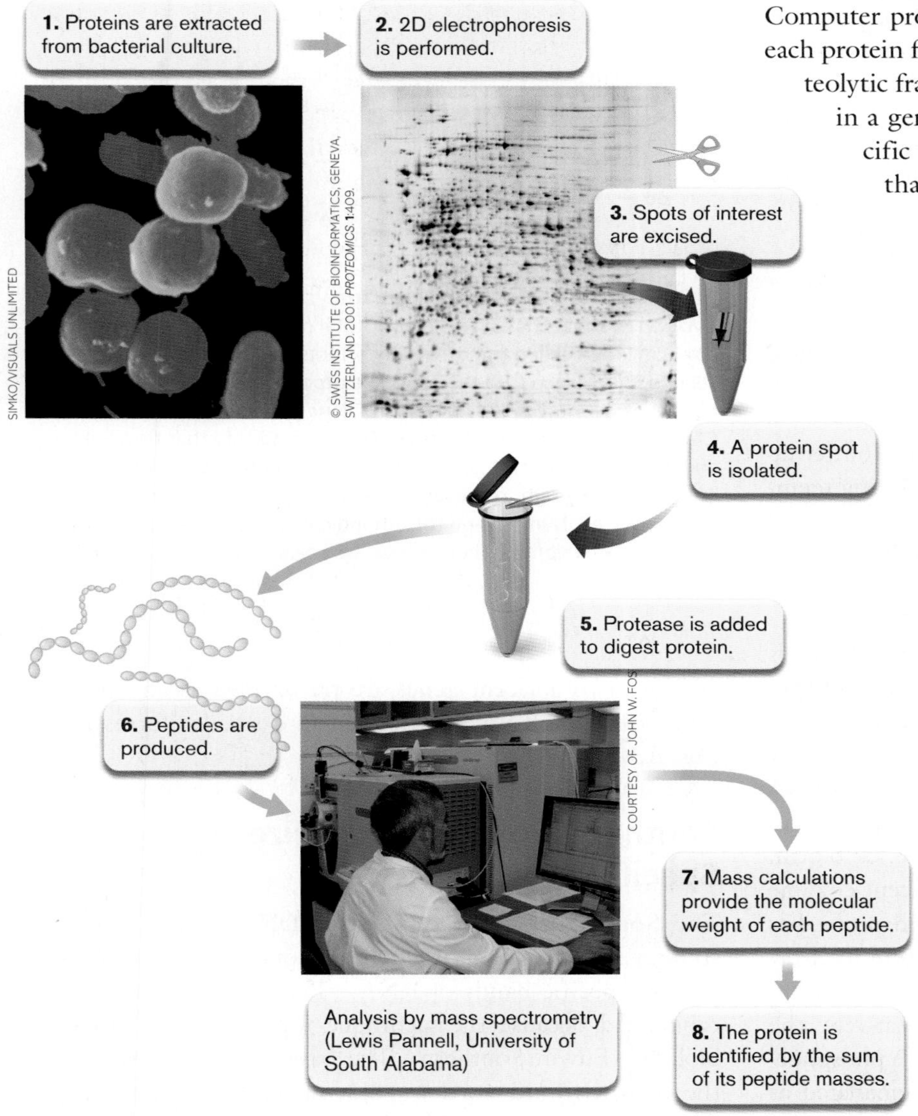
1. Proteins are extracted from bacterial culture.
2. 2D electrophoresis is performed.
SIMKO/VISUALS UNLIMITED
© SWISS INSTITUTE OF BIOINFORMATICS, GENEVA, SWITZERLAND. 2001. *PROTEOMICS* 1:409.
3. Spots of interest are excised.
4. A protein spot is isolated.
5. Protease is added to digest protein.
6. Peptides are produced.
COURTESY OF JOHN W. FOS
7. Mass calculations provide the molecular weight of each peptide.
Analysis by mass spectrometry (Lewis Pannell, University of South Alabama)
8. The protein is identified by the sum of its peptide masses.

FIGURE A3.6 ▪ Proteomic identification of proteins in a 2D gel. Proteins are extracted from a bacterial culture and subjected to 2D electrophoresis. Spots of interest can be cut out of the gel and digested with trypsin. The resulting peptides are analyzed by mass spectrometry. In a process called tandem mass spectrometry (MS-MS), the mass of each peptide is determined and then selected peptides are subjected to additional fragmentation by ionization. Each resulting peptide fragment will differ in size by one or more amino acids. Knowing the mass of each amino acid and the masses of the different peptide fragments allows one to extrapolate the sequence of the original peptide. Then the MS-MS sequences obtained for all of the tryptic peptides are compared by computer to all the predicted open reading frames (ORFs) in a microbial genome. If one ORF contains all the peptides, a match is declared and the protein is identified.

A3.4

RNA and DNA Identification by Northern and Southern Blots

Northern Blots Help Visualize Specific mRNA Messages

In the 1970s, Edwin M. Southern invented a method, now called the Southern blot, for identifying a particular sequence of DNA in a complex mixture of DNA fragments. Scientists later used a similar technique to examine RNA fragments and named their method the **northern blot** in counterpoint to the Southern blot.

Northern blots are used to analyze the presence, size, and processing of a specific RNA molecule in a cell extract. In the northern blot technique, RNA is extracted from the cell. Special precautions must be taken to avoid contaminating these preparations with RNases, which are ubiquitous in the laboratory environment. The RNA is then fractionated by electrophoresis on an agarose-formaldehyde gel (see Section 7.6). Formaldehyde helps keep the RNA denatured and unkinked by preventing base pairing, so the molecules can be separated

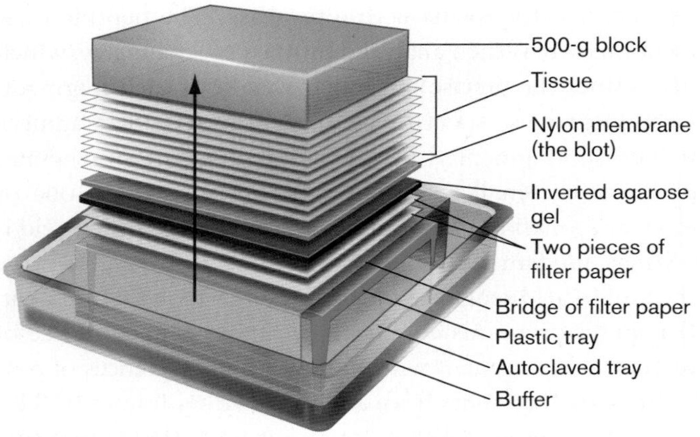
500-g block
Tissue
Nylon membrane (the blot)
Inverted agarose gel
Two pieces of filter paper
Bridge of filter paper
Plastic tray
Autoclaved tray
Buffer

FIGURE A3.7 ▪ Northern blot to view mRNA levels. Northern blots can be used to monitor the quantity and breakdown of RNA. Shown here is the apparatus used to perform the northern blot. Capillary action (arrow) draws buffer through the gel and carries the RNA upward to the membrane, which binds the RNA.

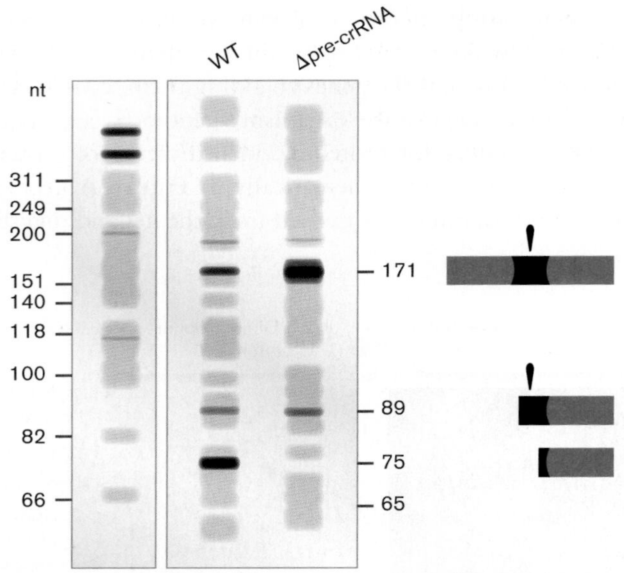
WT
Δpre-crRNA
nt
311
249
200
151
140
118
100
82
66
171
89
75
65

FIGURE A3.8 ▪ Northern blot analysis of *Streptococcus pyogenes* total RNA. This blot of total mRNA was probed for a specific RNA called tracrRNA that affects the maturation of a second RNA, called CRISPR RNA. The processing of tracrRNA into an approx. 75-nt form (wild-type, WT, lane) is abolished in the Δpre-CRISPR RNA (Δpre-crRNA) strain. The data indicate that tracrRNA and pre-crRNA interact and are processed together by an RNase. The bars on the right represent the changing sizes of tracrRNA as it is processed. Numbers on the left indicate sizes (nucleotides, nt) of molecular weight markers. *Source:* Elitza Deltcheva et al. 2011. *Nature* **471**:602–607.

by size. The separated fragments are transferred by simple capillary action onto a nylon or nitrocellulose membrane, forming the blot.

The apparatus pictured in **Figure A3.7** draws buffer up through the agarose gel and then through the facing membrane. RNA (or DNA) travels with the buffer out of the gel and is deposited on the membrane in exactly the same pattern as it was displayed in the gel. Once transferred, the RNA is fixed to the membrane by UV cross-linking.

After the RNA has been transferred, the membrane is hybridized (or "probed") with a small DNA fragment specific for a given mRNA. The probe is first heated to separate its strands. The temperature is then decreased to allow annealing between the DNA probe and matching RNA fragment in a process called **hybridization**. The probe can be labeled with radioactivity, a fluorescent dye, or biotin. The biotin is detected later by a chemiluminescent enzyme assay. RNA fragments to which a radiolabeled probe has bound are visualized by exposing the membrane to X-ray film, followed by photographic development of the film, otherwise known as **autoradiography**. Alternatively, the membrane is subjected to **phosphorimaging**. (A phosphorimager is a machine that records the energy emanated as light or radioactivity from gels or membranes.)

The molecular weights of bands visualized by autoradiography or phosphorimaging can be determined by comparing their locations on the gel with the locations of standard RNA molecular weight markers run in a parallel lane (**Fig. A3.8**). This technique can be used to determine the presence of any given mRNA. By estimating the size of the transcript, which will be larger if multiple genes are transcribed as one, the method can also be used to evaluate whether two adjacent genes are transcribed as an operon or whether an RNA is processed into smaller forms (see **Fig. A3.8**).

Southern Blots Help Visualize Specific DNA Sequences

The **Southern blot** technique for DNA uses a probing strategy similar to that of northern blots to detect the presence of specific bands of RNA. (The term "Southern" is capitalized because the technique was named for British biologist Edwin Southern.) The basic difference between the Southern and northern blot techniques is the starting material. Southern blots begin with restriction endonuclease digestion of genomic DNA to produce discrete DNA fragments that are then separated by electrophoresis in an agarose or acrylamide gel. The DNA fragments are denatured to single strands and then blotted onto a membrane. A DNA probe is generated from one bacterial species by PCR (although other techniques can be used) and labeled with a radioactive or nonradioactive tag. This probe is hybridized to the DNA blot, and the hybridized bands are visualized by autoradiography or phosphorimaging. Standardized double-stranded DNA markers are used to help determine sample fragment sizes. Southern blotting is used, for example, to detect the presence of specific genes among different species or strains of a single species.

A3.5

Sanger Method of DNA Sequencing

Recall from Section 7.6 that the method of DNA sequencing called **Sanger sequencing** mixes a small amount of a dideoxynucleotide with normal deoxynucleotides in a DNA synthesis reaction. Incorporation of dideoxy ATP in a growing DNA chain halts further elongation of that chain because there is no 3′ OH group to which the next base can be linked. But, because only a small amount of the dideoxy ATP terminator is present relative to normal deoxy ATP, many chains can complete their synthesis, while other chains stop at different adenine positions.

Now imagine using four different dideoxynucleotides corresponding to A, T, C, and G, with each dideoxy base tagged with a different-colored fluorescent dye (**Fig. A3.9A**). The result will be a series of different-sized strands of DNA, each tagged at its 3′ end with a color that depends on the base incorporated. Analyzing the results of this reaction by electrophoresis in a single lane of a polyacrylamide gel will separate the various fragments according to size. A laser and detector positioned at the bottom of the gel can read the individual fragments

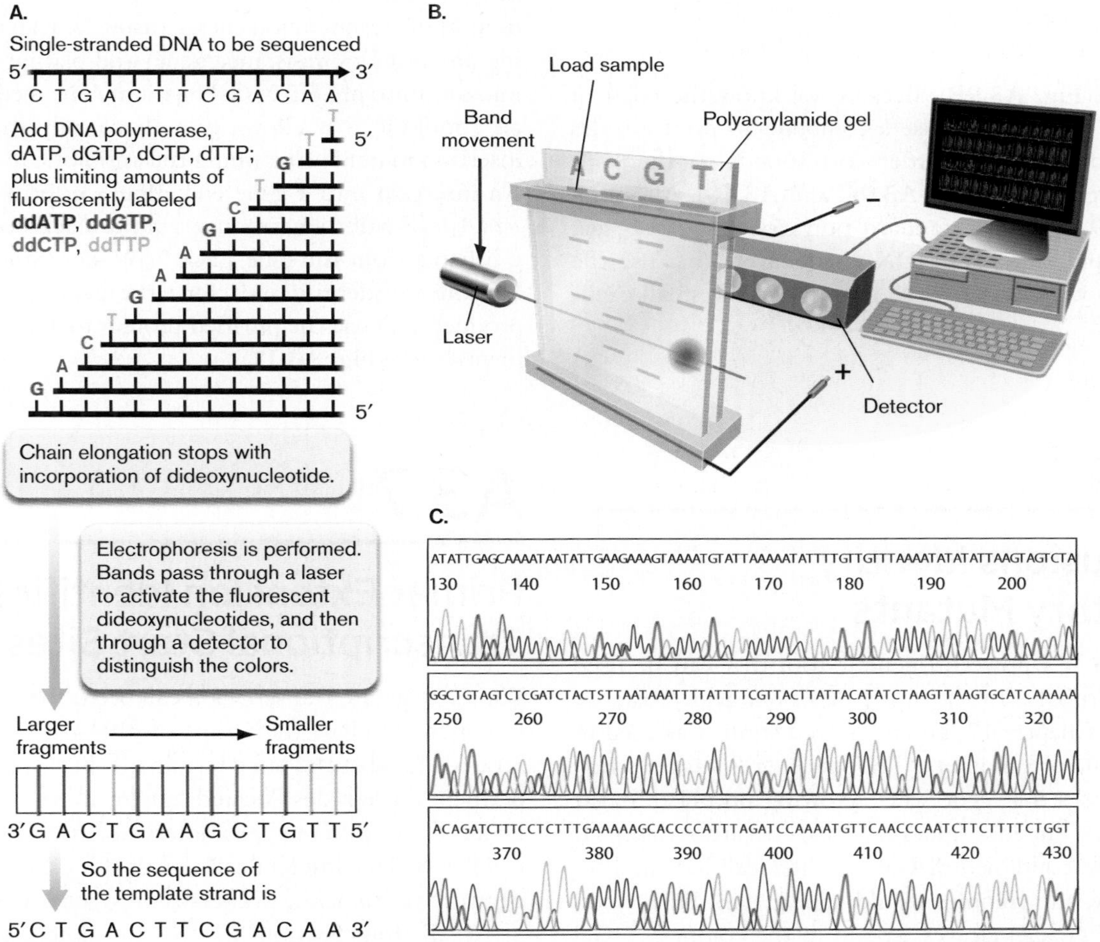
A.
Single-stranded DNA to be sequenced
5′
3′
C T G A C T T C G A C A A
Add DNA polymerase, dATP, dGTP, dCTP, dTTP; plus limiting amounts of fluorescently labeled ddATP, ddGTP, ddCTP, ddTTP
5′
Chain elongation stops with incorporation of dideoxynucleotide.
Electrophoresis is performed. Bands pass through a laser to activate the fluorescent dideoxynucleotides, and then through a detector to distinguish the colors.
Larger fragments
Smaller fragments
3′G A C T G A A G C T G T T 5′
So the sequence of the template strand is
5′C T G A C T T C G A C A A 3′
B.
Load sample
Band movement
Polyacrylamide gel
A C G T
Laser
Detector
C.
ATATTGAGCAAAATAATATTGAAGAAAGTAAAATGTATTAAAAATATTTTGTTGTTTAAAATATATATTAAGTAGTCTA
130 140 150 160 170 180 190 200
GGCTGTAGTCTCGATCTACTSTTAATAAATTTTATTTTCGTTACTTATTACATATCTAAGTTAAGTGCATCAAAAA
250 260 270 280 290 300 310 320
ACAGATCTTTCCTCTTTGAAAAAGCACCCCATTTAGATCCAAAATGTTCAACCCAATCTTCTTTTCTGGT
370 380 390 400 410 420 430

FIGURE A3.9 ■ DNA sequencing using fluorescently tagged dideoxynucleotides to randomly stop chain elongation. **A.** The tagged strands are first synthesized using dideoxynucleotides, and then separated. **B.** The reaction products are separated by size using a polyacrylamide gel apparatus that includes a laser and detector to specifically identify the different tagged fragments. **C.** The bases are then read and printed out as different-colored peaks.

KEITH WELLER, USDA

FIGURE A3.10 ▪ Rapid DNA sequencing using an automated DNA sequencer. Cletus Kurtzman inspects a yeast DNA sequence from a previous run while Larry Tjarks loads samples of new sequencing reactions at the National Center for Agricultural Utilization Research. The lanes on the screen represent the sequences of different DNA molecules.

as they pass (**Fig. A3.9B**). Because we know the color of each tagged base, we can use a computer to print a series of colored peaks whose order corresponds to the template DNA sequence (**Figs. A3.9C** and **A3.10**). Although **Figure A3.9** depicts a standard polyacrylamide slab gel; the more rapid, automated DNA sequencers that use the Sanger method for larger-scale projects utilize small capillary tube gels.

A3.6

Gene Fusions Identify Regulatory Mutants

Gene fusion is a powerful genetic tool that can be used for many purposes. In the study of *E. coli* acid resistance described in Chapter 12, an engineered strain was used to identify regulatory mutants. The *gadA-lacZ* fusion strain highly expresses that gene when growing on pH 5.5 agar media (*gadA* encodes glutamic acid decarboxylase). If that medium contained X-Gal, a chemical indicator of LacZ activity, the colonies would appear dark blue. Beta-galactosidase (LacZ) cleaves a bond in the colorless X-Gal molecule and releases a blue chemical that stays in the colony. One can then search for regulatory genes required to activate *gadA* by subjecting the *gadA-lacZ* strain to random transposon mutagenesis (using a transposon carrying an antibiotic resistance gene) and plating the insertion mutants onto pH 5.5 X-Gal medium. The medium includes an antibiotic that allows growth of only the transposon insertion mutants containing the antibiotic resistance gene. An insertion into a gene encoding a positive regulator of *gadA-lacZ* will then produce a white $LacZ^-$ colony, distinguishable from the blue $LacZ^+$ colonies. Studies can then be done to identify and clone the gene, purify the gene product, and use the purified protein to analyze regulation in vitro. (see **Fig. A3.11**).

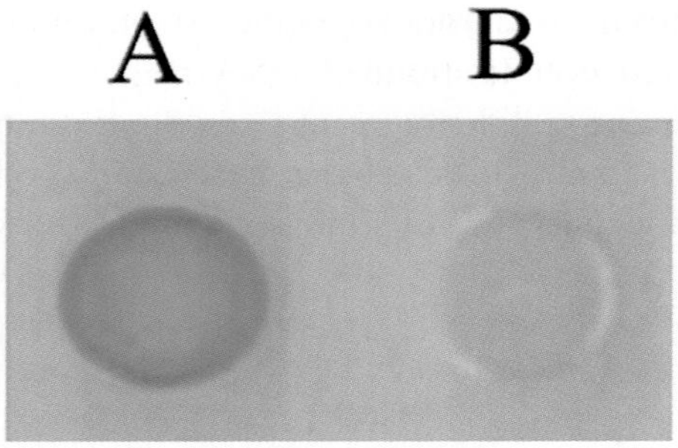
A
B

FIGURE A3.11 ▪ The *Escherichia coli* GadE regulatory protein controls *gadA* expression. Growth of *E. coli* colonies on pH 5.5 agar containing the X-gal indicator. **A.** A *gadA-lacZ* operon fusion strain expresses β-galactorsidase and turns blue. **B.** Introducing a *gadE* mutation into the same strain prevents the expression of the *gadA-lacZ* fusion; no β-galactosidase is made and the colony remains white.

A3.7

Primer Extension Identifies Transcriptional Start Sites

For many genes, transcription can occur from multiple promoters to produce transcripts of different sizes. The transcripts all end at the same terminator but begin at different defined nucleotides located upstream of the structural gene. These different promoters are useful because they can respond to different cell signals (for example, high temperature versus acidic pH versus nitrogen limitation).

Recall from Section 8.1 that promoters are located at –10 and –35 bp from the transcriptional start site. So, if there are different transcriptional start sites upstream of a single ORF (gene), then there must be different –10 and –35 bp sequences (promoters) associated with each transcriptional start. To begin searching for upstream promoter sequences, we need to define where each transcript begins. One method to determine transcript length is called **primer extension**, illustrated for the generic case in **Figure A3.12A**. Knowing the sequence of the gene, the researcher designs a single DNA primer that will anneal to the mRNA of interest near the suspected start site. The primer is used in a reverse transcription reaction (reverse transcriptase makes DNA from RNA) that extends the primer to the 5′ end (the start) of the message. Reverse transcription generates complementary DNA (cDNA) of a precise length. The cDNA is radiolabeled by using radioactive nucleotides in the reaction mixture.

The key to using the primer extension technique is that the same primer is also used in a DNA sequencing reaction with the template DNA (see **Fig. A3.12A**). The cDNA

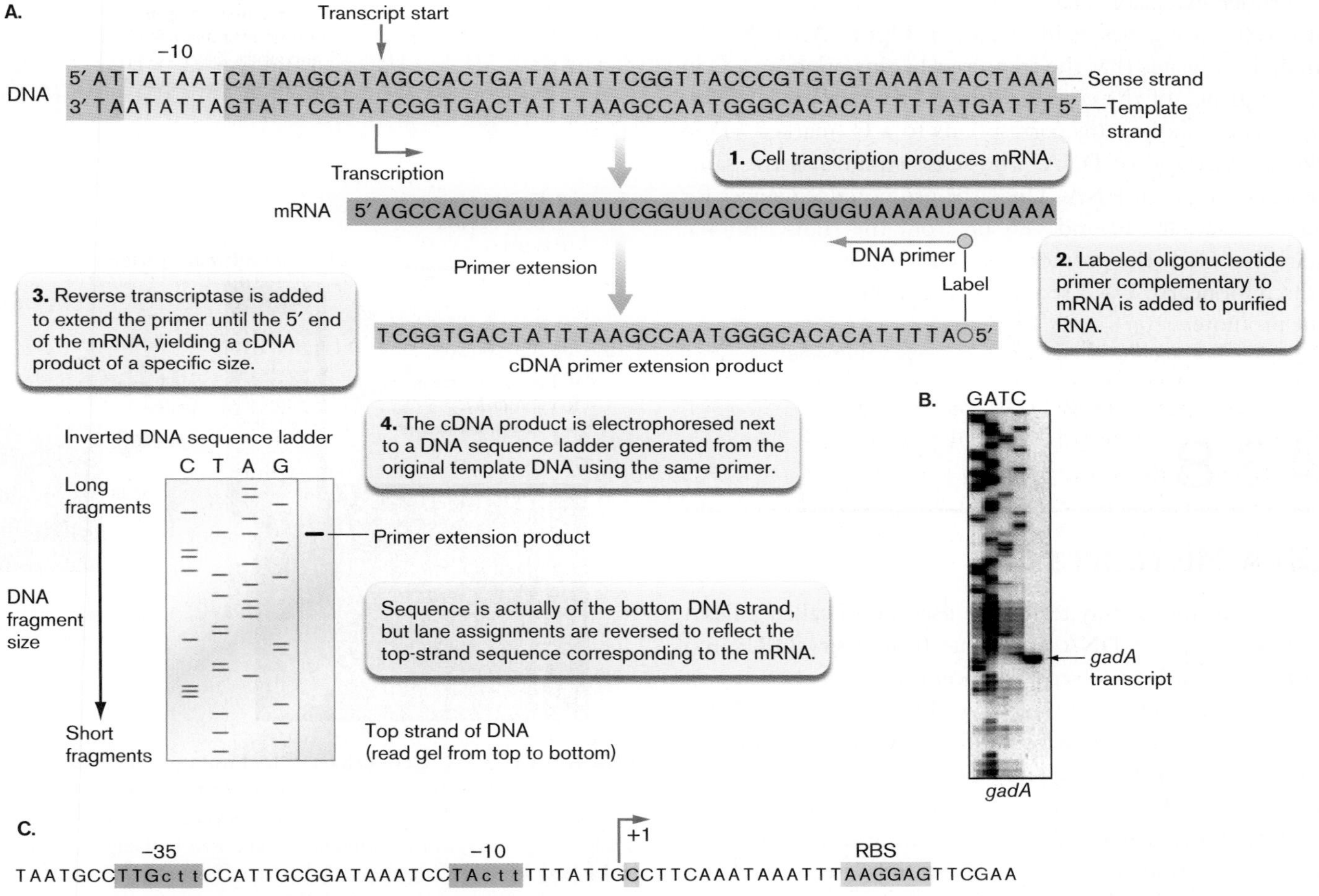
A.
Transcript start
–10
DNA
5′ ATTATAATCATAAGCATAGCCACTGATAAATTCGGTTACCCGTGTGTAAAATACTAAA
3′ TAATATTAGTATTCGTATCGGTGACTATTTAAGCCAATGGGCACACATTTTATGATTT 5′
Sense strand
Template strand
Transcription
1. Cell transcription produces mRNA.
mRNA 5′ AGCCACUGAUAAAUUCGGUUACCCGUGUGUAAAAUACUAAA
Primer extension
DNA primer
Label
2. Labeled oligonucleotide primer complementary to mRNA is added to purified RNA.
3. Reverse transcriptase is added to extend the primer until the 5′ end of the mRNA, yielding a cDNA product of a specific size.
TCGGTGACTATTTAAGCCAATGGGCACACATTTTA 5′
cDNA primer extension product
4. The cDNA product is electrophoresed next to a DNA sequence ladder generated from the original template DNA using the same primer.
Inverted DNA sequence ladder
C T A G
Long fragments
DNA fragment size
Short fragments
Primer extension product
Sequence is actually of the bottom DNA strand, but lane assignments are reversed to reflect the top-strand sequence corresponding to the mRNA.
Top strand of DNA (read gel from top to bottom)
B.
GATC
gadA transcript
gadA
C.
–35
–10
+1
RBS
TAATGCCTTGcttCCATTGCGGATAAATCCTActtTTTATTGCCTTCAAATAAATTTAAGGAGTTCGAA

FIGURE A3.12 ■ Primer extension analysis to determine transcriptional start sites. A. The top sequence represents the DNA sequence of an imaginary promoter. The mRNA product is shown below. After separating the RNA from DNA, a DNA oligonucleotide primer that binds to the mRNA at a defined place is added, followed by a reverse transcriptase reaction. Reverse transcriptase produces a cDNA primer extension product that ends at the beginning (5′ end) of the mRNA. The product is run in a polyacrylamide gel next to a sequencing ladder of the original DNA made with the same primer used in the reverse transcriptase reaction. The primer extension product will migrate to the same position as the DNA fragment ending in the base that represents the start site of transcription. **B.** Primer extension (far-right lane) showing the transcriptional start site of the *gadA* gene (arrow) involved in *E. coli* acid resistance. **C.** The sequence of the promoter is shown with –10, –35, and the transcriptional start site (+1) marked. Lowercase letters indicate bases that differ from consensus –10 and –35 sequences. Note that the first base of the mRNA transcript is marked C (cytosine). This corresponds to a G (guanosine) in the gel sequence ladder (panel B). The C is correct for the transcript because the DNA sequence obtained using the primer is actually that of the strand complementary to the message. *Source:* Daniela De Biase et al. 1999. *Mol. Microbiol.* **32**:1198.

primer extension fragment is then run on a polyacrylamide gel alongside the products of a Sanger DNA sequencing reaction of the gene fragment (see Section 7.6 and Section A3.5). The size of the primer extension cDNA fragment will be identical to the size of one of the "rungs" on the DNA sequence ladder. That rung represents the base in the sequence that correlates to the transcriptional start site. The cDNA from the transcripts of genes that have multiple promoters will produce primer extension fragments of varying sizes that co-migrate with different-sized sequencing fragments.

Primer extension analysis of *gadA,* one of the *E. coli* acid resistance genes, is illustrated in **Figure A3.12B**. The analysis indicates that the transcript begins opposite a G in the sequence of the complementary DNA strand. As **Figure 3.12C** shows, this corresponds to a C (marked +1) in the sense strand of DNA. The sense strand has the same sequence as the mRNA. Thus, the promoter sequences for *gadA* reside at –10 and –35 bp from the transcriptional start site. Once the promoter is identified, we can begin to search for nearby DNA sequences that control expression of the promoter.

A3.8

DNA Microarray

The **DNA microarray** technique uses a tool called a DNA microchip, where DNA fragments from every ORF in a genome are affixed to separate locations on a solid support surface, producing a grid, or array (**Fig. A3.13**). The DNA fragments are generated by PCR or by in vitro chemical synthesis. The key to the technique is that only one strand of a double helix is actually fixed to the slide. Heating the slide breaks the hydrogen bonds holding down the complementary strands. The released strands are washed off, leaving the tethered strands free to anneal with any other complementary DNA that approaches. Think of these strands as a thousand tiny pieces of flypaper, with each piece engineered to catch a different type of fly.

The DNA microchip can be used to analyze RNA extracted from microbes grown under two different conditions. The ultimate goal is to compare the relative expressions of each gene under the two sets of environmental conditions—for example, at a pH of 7 versus a pH of 4.5. The latter is a pH that bacteria such as *Salmonella* might encounter in eukaryotic host cell vesicles. After extraction,

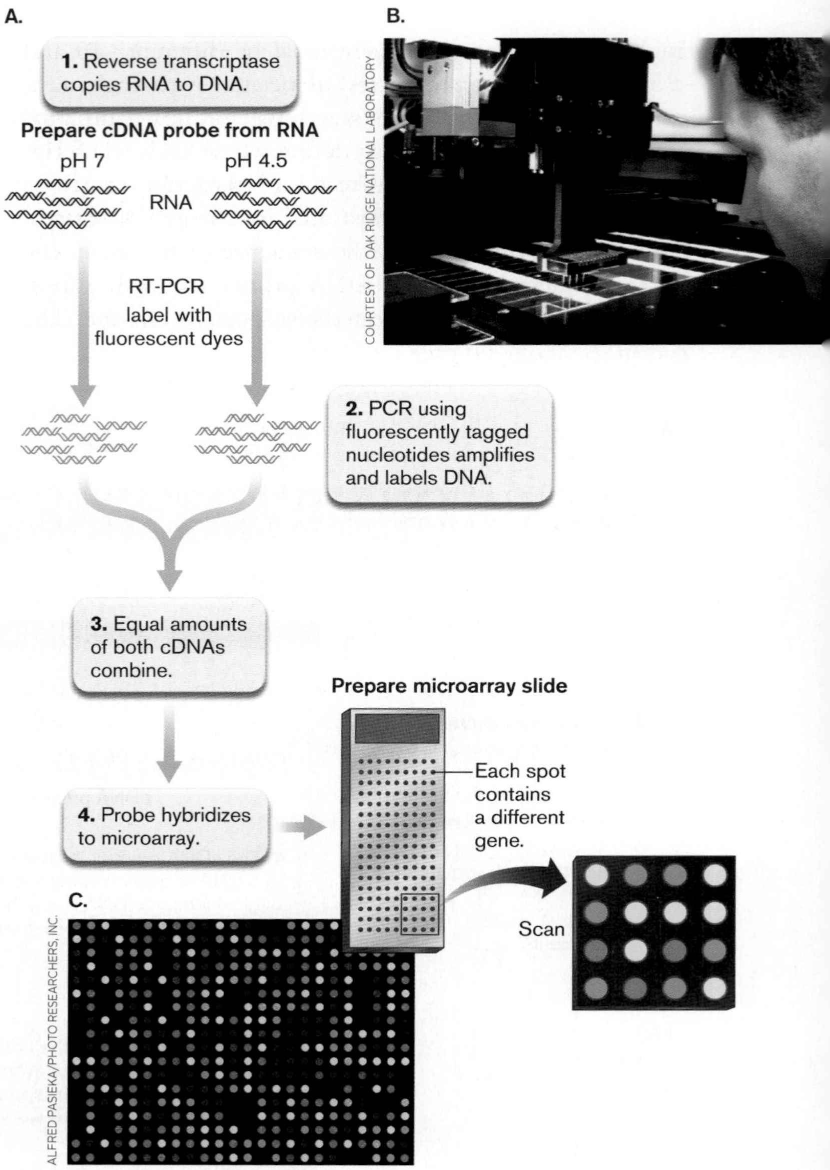
A.
B.
1. Reverse transcriptase copies RNA to DNA.
Prepare cDNA probe from RNA
pH 7
pH 4.5
RNA
RT-PCR
label with fluorescent dyes
COURTESY OF OAK RIDGE NATIONAL LABORATORY
2. PCR using fluorescently tagged nucleotides amplifies and labels DNA.
3. Equal amounts of both cDNAs combine.
Prepare microarray slide
Each spot contains a different gene.
4. Probe hybridizes to microarray.
C.
Scan
ALFRED PASIEKA/PHOTO RESEARCHERS, INC.

FIGURE A3.13 ▪ DNA microarray technology. This procedure is used to determine transcript levels produced from every gene in a cell (transcriptome). It assesses the relative level of each transcript produced by cells grown in different conditions or between mutant and wild-type strains of bacteria. **A.** RNA is extracted from cultures grown under different conditions (pH 7 versus pH 4.5 in this example). The RNA molecules from the two cultures are converted to DNA with reverse transcriptase (RT), and then amplified and quantitatively tagged with different-colored fluorescent tags using PCR techniques. **B.** DNA fragments of each ORF in the bacterial genome are spotted by robotics to a glass slide (DNA microchip). The tagged (fluorescent) RT-PCR mixtures are then applied to the DNA microchip and allowed to anneal. Each tagged RT-PCR fragment will anneal only to the spot corresponding to its gene. **C.** Laser scanning of the annealed chip will reveal red spots if that gene was expressed mostly under pH 4.5 conditions, green spots if it was expressed mostly at pH 7, and yellow or orange if it was equally expressed under both conditions (an equal mix of red tag and green tag).

the two RNA samples are treated separately with the enzyme reverse transcriptase, which makes complementary DNA (cDNA) from RNA templates. Fluorescently tagged nucleotides are used in the reaction so that the product cDNAs also carry a fluorescent tag. How much of one cDNA is made depends on how much mRNA was initially present. Different-colored tags (usually red and green) are used for each sample. In the example shown in **Figure A3.13**, the cDNAs from the pH 7 and pH 4.5 cultures were labeled red and green, respectively.

The two fluorescently tagged batches of cDNA are flooded onto a microchip, where the individual cDNA molecules find and bind their mates on the organized grid. This is the hybridization step. After hybridization, a laser scans the slide and a fluorescence detector reads emissions. Composite images of the two scans, one for red and one for green, are made and analyzed by computer. If a given gene is expressed equally under both test conditions, the corresponding spot on the chip will contain equal amounts of red and green—a result that produces a yellow color. If a gene is expressed more during growth at pH 7, the spot will glow green. If expressed more at pH 4.5, it will fluoresce red. The result is a comprehensive view of what has been called the cell's **transcriptome**—all of a cell's expressed mRNAs (**Fig. A3.13C**). This technology has been used to examine the effect on the transcriptome of several global regulators discussed in the book, such as cAMP receptor protein (CRP), sigma H, NtrC, and many others.

A3.9

Multiplex PCR

Food safety is a major concern for human health. Screening a food product for multiple pathogens one pathogen at a time can be laborious and costly. **Multiplex PCR**, a modification of PCR (Section 7.6), can screen for several organisms simultaneously, saving time and money. Multiplex PCR combines multiple pairs of oligonucleotide probes (that is, DNA primers) that amplify species-specific genes in a single PCR reaction (**Fig. A3.14**). For multiplex PCR to be successful, the oligonucleotide pairs cannot bind to each other or inadvertently amplify some other chromosomal gene. In addition, the amplicon (amplified product) sizes for each pair must be clearly different so that they can be separated on a gel. If these conditions are met, finding a given fragment will unambiguously demonstrate that the corresponding organism was present in the material tested.

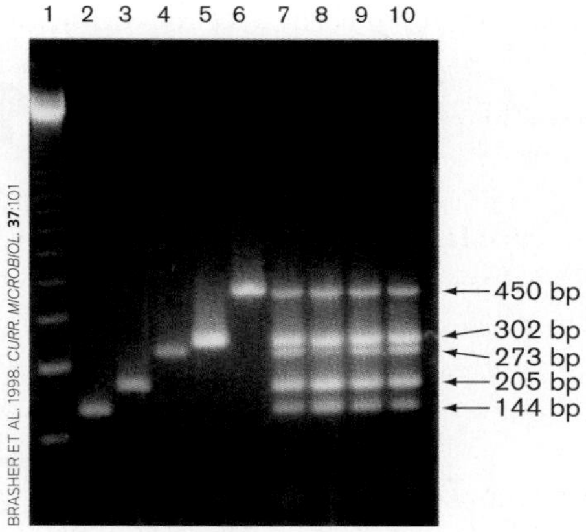
1 2 3 4 5 6 7 8 9 10
450 bp
302 bp
273 bp
205 bp
144 bp
BRASHER ET AL. 1998. CURR. MICROBIOL. 37:101

FIGURE A3.14 ■ Multiplex analysis of pathogenic *E. coli*, *Salmonella*, and *Vibrio* species. PCR products are shown in this agarose gel. Lane 1 shows the 123-bp DNA ladder used to estimate sizes of the amplified fragments. Each of the other lanes represents a multiplex PCR reaction that can detect segments of five different genes representing different bacterial pathogens. Lanes 2–6 used the multiplex oligonucleotide primers to detect single strains. Lane 2 detects the *uidA* (beta-glucuronidase) gene from *E. coli* (144-bp amplicon); lane 3, the *cth* (cytolysin hemolysin) gene from *Vibrio vulnificus* (205 bp); lane 4, the *invA* invasion gene from *Salmonella* Typhimurium (273 bp); lane 5, the *ctx* (cholera toxin) gene from *V. cholerae* (302 bp); and lane 6, the *tl* hemolysin gene from *V. parahaemolyticus* (450 bp). Lanes 7–10 are replicates in which the multiplex reaction was tested on mixtures of all five species. *Source:* Cynthia Brasher et al. 1998. *Curr. Microbiol.* **37**:101.

A3.10

Fluorescence In Situ Hybridization (FISH) and CARD-FISH

A technique to detect individual microbes in an ecological or clinical sample is **fluorescence in situ hybridization (FISH)**. The technique makes use of a fluorophore-labeled oligonucleotide probe (usually a short DNA sequence) that hybridizes to microbial DNA or rRNA. For microbial samples, the probe is most commonly designed to hybridize to rRNA because rRNA is present in approximately 100-fold to 10,000-fold excess over DNA, allowing more sensitive detection. In some cases the technique uses a modified DNA probe such as PNA (peptide nucleic acid) because the binding affinity is even higher than for DNA probes.

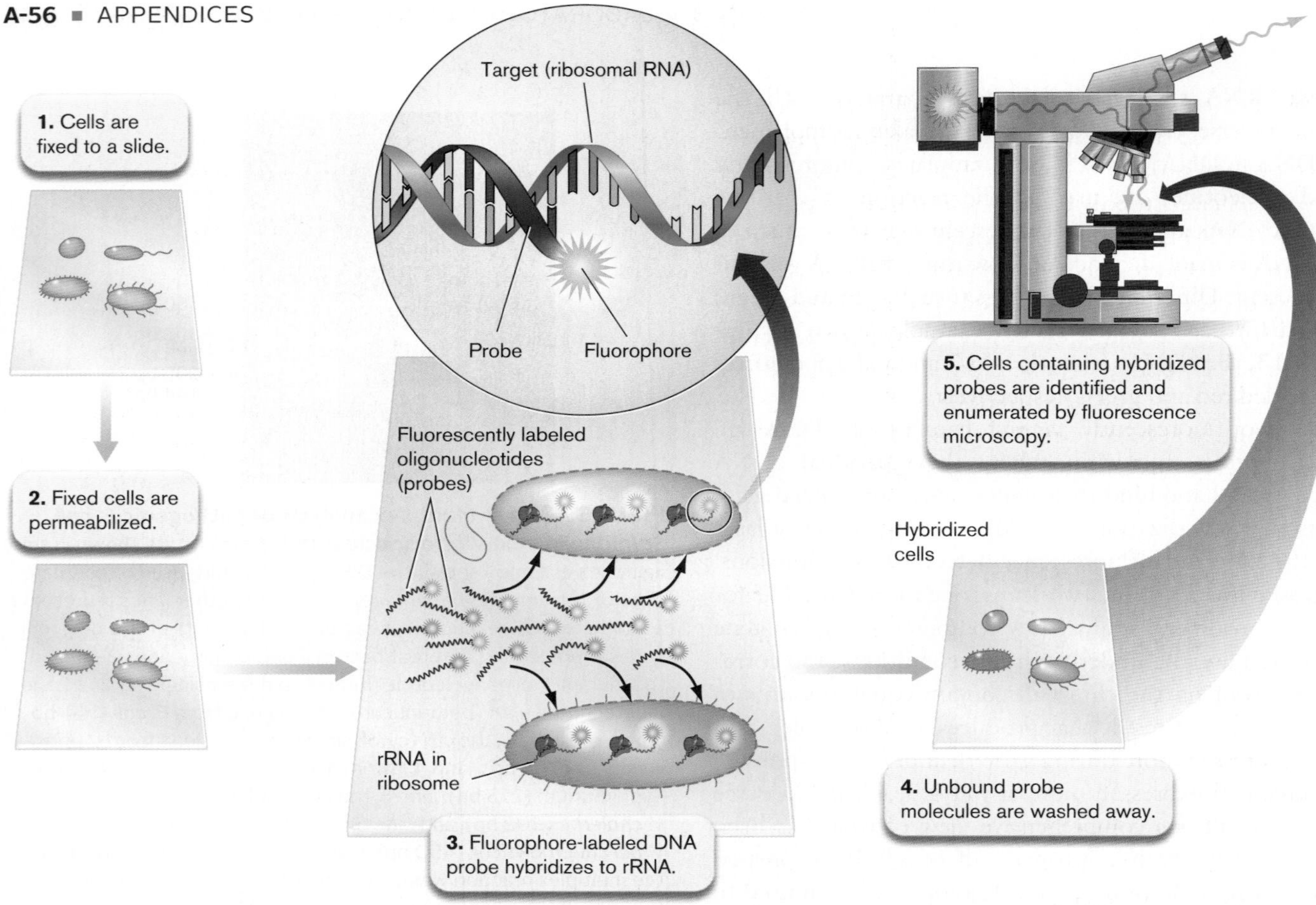
Target (ribosomal RNA)
1. Cells are fixed to a slide.
Probe
Fluorophore
5. Cells containing hybridized probes are identified and enumerated by fluorescence microscopy.
Fluorescently labeled oligonucleotides (probes)
2. Fixed cells are permeabilized.
Hybridized cells
rRNA in ribosome
4. Unbound probe molecules are washed away.
3. Fluorophore-labeled DNA probe hybridizes to rRNA.

FIGURE A3.15 ▪ Fluorescence in situ hybridization. A fluorophore-labeled DNA oligonucleotide hybridizes to a taxon-specific sequence of rRNA molecules within bacteria that are fixed and permeabilized on a microscope slide. *Source:* Modified from Rudolf Amann and Bernhard M. Fuchs. 2008. *Nat. Rev. Microbiol.* **3**:339, fig. 1.

Another variation to increase sensitivity is that of catalyzed reported deposition (CARD) of tyramides, enzyme-activated fluorophores.

The procedure of FISH is outlined in **Figure A3.15**. First the cells of a sample must be fixed to a slide by a treatment that maintains cell integrity while permeabilizing for the fluorophore-labeled DNA probe. Typical chemical treatment involves formaldehyde and ethanol. Next the fixed cells are incubated in a hybridization buffer containing the probe, at a temperature designed to maximize specificity of binding to the sequence of the desired taxa. For example, a probe may have a sequence complementary to a sequence found only in rRNA of anammox bacteria. A probe with more broad specificity might hybridize to all bacterial rRNA, but not to archaeal or eukaryotic rRNAs. Following hybridization, a wash step washes away all unbound probe molecules. Finally, the cells containing hybridized probes are observed by fluorescence microscopy. The labeled cells of various types can be identified and enumerated.

An example of FISH results is shown in **Figure A3.16**. A mixed biofilm of anammox and heterotrophic bacteria

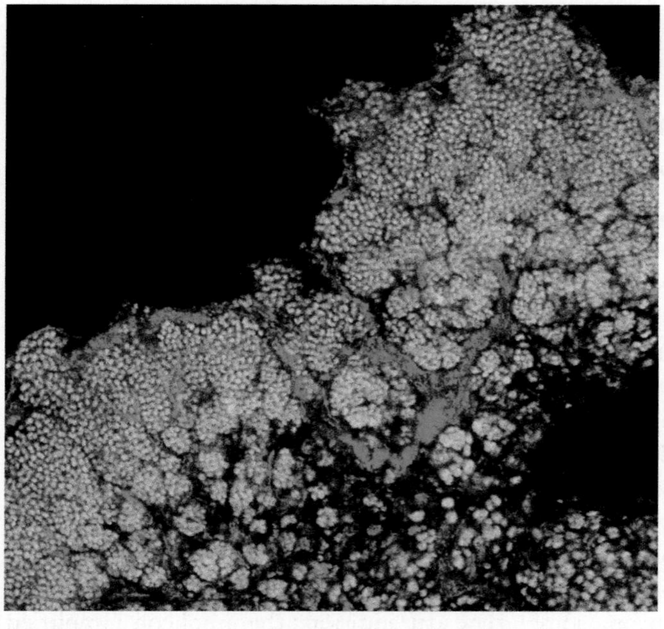

FIGURE A3.16 ▪ Mixed biofilm visualized by FISH. Bacteria are labeled with a bacteria-specific probe (green); anammox bacteria are labeled with the bacterial probe and with an anammox-specific probe (green plus red = yellow). *Source:* Tomonori Kindaichi et al. 2007. *Appl. Environ. Microbiol.* **73**:4931, fig. 5B.

was labeled with a fluorescent probe specific for anammox (red) and a probe that labels all bacteria (green). The coincidence of both probes (yellow) indicates anammox bacteria, whereas heterotrophic bacteria are labeled green.

The CARD-FISH variation on this technique uses a DNA probe bound to the enzyme horseradish peroxidase. The enzyme activates molecules of tyramide, converting the tyramide to a sensitive fluorophore, which then binds to cell protein. A single probe-enzyme can activate and deposit multiple molecules of tyramide; thus, its signal is highly amplified compared to the standard FISH method, in which a probe possesses a single fluorophore.

A3.11

Immunoprecipitation Techniques

Immunoprecipitation, described in Section 24.3, is an important property of antigen-antibody reactions that is also the basis for many techniques used in immunology. But what is needed to precipitate an antigen with an antibody? As seen in Figure 24.8, immunoprecipitation can occur with pure antibody to a single epitope when the target of the antibody has multiple identical epitopes. The cartoon antigen molecules shown in Figure 24.8 are made of two identical epitopes. Immunoprecipitation can also occur when an antigen molecule has different epitopes. Antiserum raised against that antigen will contain antibodies to each epitope. As shown in **Figure A3.17A**, the

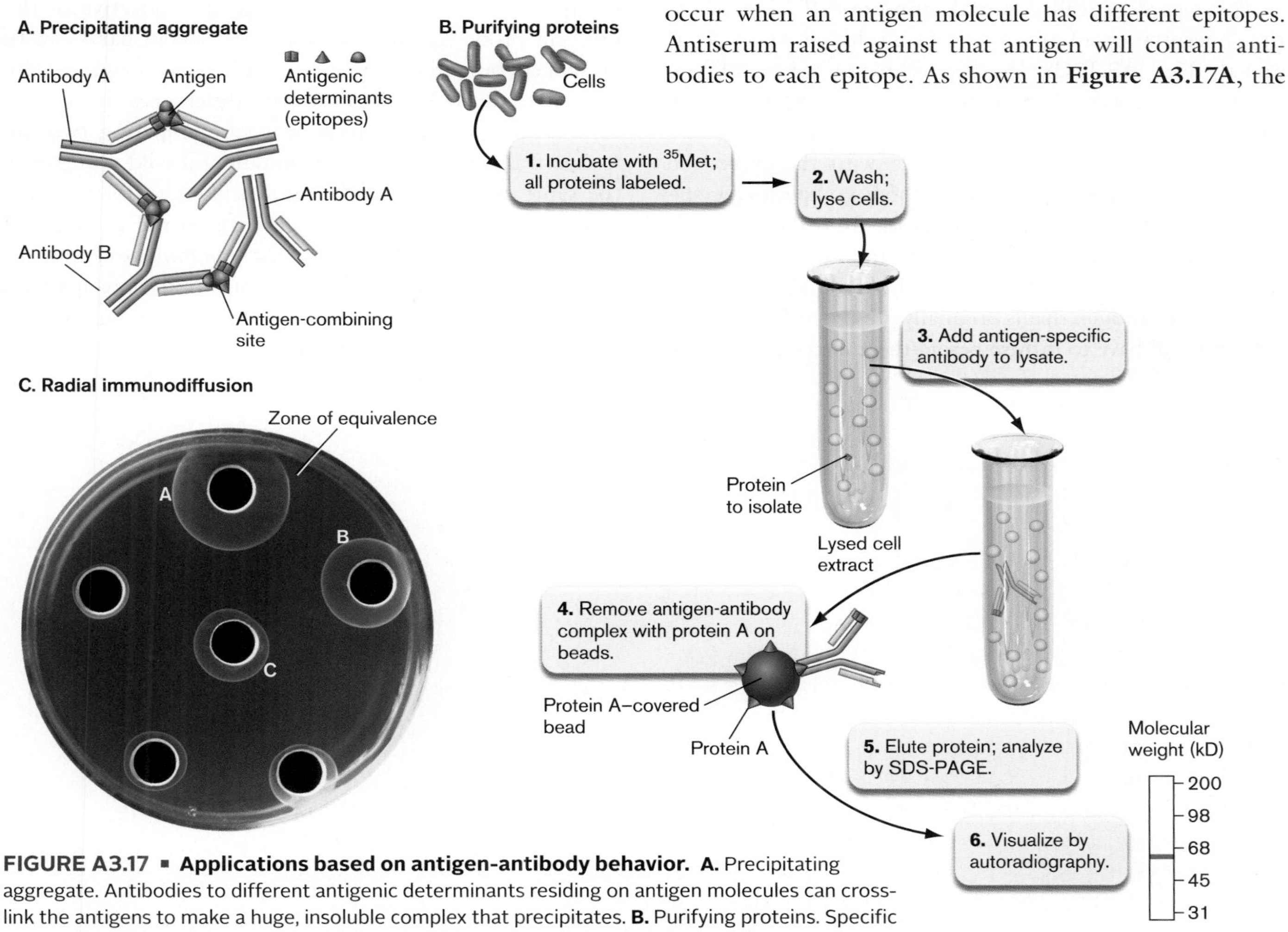
A. Precipitating aggregate
Antibody A
Antigen
Antigenic determinants (epitopes)
Antibody A
Antibody B
Antigen-combining site
B. Purifying proteins
Cells
1. Incubate with 35Met; all proteins labeled.
2. Wash; lyse cells.
3. Add antigen-specific antibody to lysate.
Protein to isolate
Lysed cell extract
4. Remove antigen-antibody complex with protein A on beads.
Protein A–covered bead
Protein A
5. Elute protein; analyze by SDS-PAGE.
6. Visualize by autoradiography.
Molecular weight (kD)
200
98
68
45
31
C. Radial immunodiffusion
Zone of equivalence
A
B
C

FIGURE A3.17 ■ Applications based on antigen-antibody behavior. A. Precipitating aggregate. Antibodies to different antigenic determinants residing on antigen molecules can cross-link the antigens to make a huge, insoluble complex that precipitates. **B.** Purifying proteins. Specific proteins are removed from a complex cell extract using immunoprecipitation. In the experiment, antibody to a specific cellular protein (shown as yellow square) is added to a lysed cell extract. Protein A beads are then added. Protein A will bind to the Fc region of the antigen-antibody complex. Centrifugation then is used to pull down the beads and the protein with them. The protein is eluted from the antibody and analyzed by SDS-PAGE (protein A remains covalently attached to the bead). **C.** Radial immunodiffusion. The agarose plate shown is embedded with antibodies specific for a certain antigen. Different concentrations of the antigen are placed in each well. The antigen diffuses into the agarose. A ring of precipitation occurs when the concentration of the diffusing antigen reaches a zone of equivalence with the antibody in the plate. The more antigen placed in the well, the larger the diameter of the precipitation ring. Well A was loaded with more antigen than wells B or C. *Source:* Part C © 2006 EDVOTEK, Inc.

different antibodies can collaborate to precipitate the antigen by cross-binding identical epitopes on different molecules. **Figure A3.17B** shows how specific proteins can be isolated from a complex cell extract using immunoprecipitation. Antibody is added to a complex mix of different antigens. The antibody, however, can bind only to its specific antigen. Beads coated with a molecule known as protein A (derived from *Staphylococcus aureus*) are added to specifically purify that antigen-antibody complex. Protein A binds to the Fc region of an IgG antibody molecule. Consequently, all the antigen-antibody complexes will become bound to the bead. Because the bead is heavy, centrifugation will remove the desired antigen from the complex mixture.

The technique called **radial immunodiffusion** allows the concentration of an antigen in a solution to be determined (**Fig. A3.17C**). In this technique a ring of precipitation is visualized in an agarose gel impregnated with antibody. Antigen placed within a well will diffuse outward until reaching a zone of equivalence with the embedded antibody. At this point, antigen-antibody complexes precipitate and form a ring a certain distance from the well. The farther the antigen diffuses away from the well, the lower its concentration becomes. Consequently, the higher the concentration of antigen that is originally present in the well, the farther it will have to diffuse before the zone of equivalence is reached and a ring of precipitation forms. The concentration of antigen in an unknown solution is determined by comparing the radius of immunoprecipitation formed to a standard curve in which known concentrations of antigen are plotted against the radius of the ring of precipitation.

Another important research technique involving antibodies is the **western blot**, which is used to detect the presence of a specific protein in cell extracts. As described in Section 12.1, proteins are separated using SDS polyacrylamide gel electrophoresis (SDS-PAGE). The proteins are transferred (blotted) from the gel onto a nitrocellulose or other membrane. The membrane is then probed with an antibody directed against a specific protein. Antibody sticks to the protein in question, and because that antibody is labeled in some way (for example, a radioactive or fluorescent probe has been added), the protein band can be visualized by autoradiography or phosphorimaging. This technique can be used to estimate differences in the concentration of a specific protein when comparing two different strains of cells (such as mutant and wild type), or in the same cell type treated in different ways (for instance, growth in different environments, or in the presence or absence of different chemicals). These techniques and many others that rely on the principles of immunoprecipitation are critical tools in many scientific venues, ranging from basic research to clinical analysis.

APPENDIX 4
Taxonomy

A4.1 Viruses
A4.2 Bacteria
A4.3 Archaea
A4.4 Eukarya

This appendix outlines the major known groups of microbial species. Most of these species are mentioned in Chapter 6 Viruses, and in the diversity chapters (Chapter 18 Bacteria, Chapter 19 Archaea, and Chapter 20 Eukaryotes). The species are grouped according to clades based on genome phylogeny. We indicate their major traits of metabolism, ecology, and relevance to medicine.

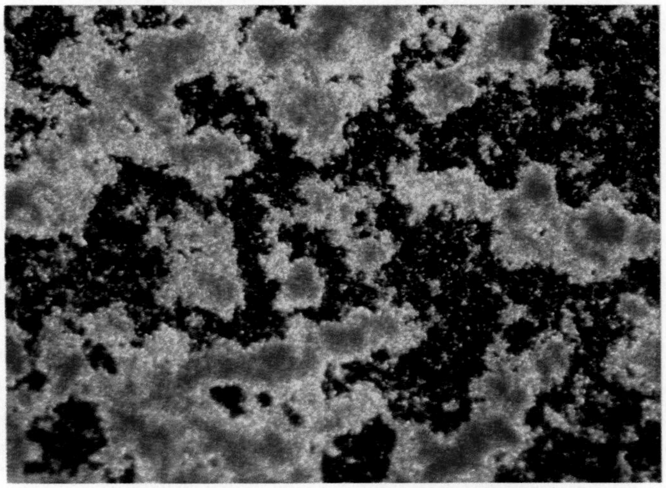

Multispecies biofilm. A multispecies biofilm formed on an agar-coated slide immersed in a slurry of soil bacteria. The biofilm includes Gram-positive bacterial genera such as *Bacillus*, as well as Gram-negative genera such as *Brevundimonas*, *Flavobacterium*, and *Pseudomonas*. *Source:* Mette Burmølle et al. 2007. Microbial Ecology **54**:352.

A4.1

Viruses

TABLE A4.1

Groups of viruses—Baltimore classification. (Expanded version of Table 6.1.)

Group I. Double-stranded DNA viruses

Replicate using host or viral DNA polymerase.

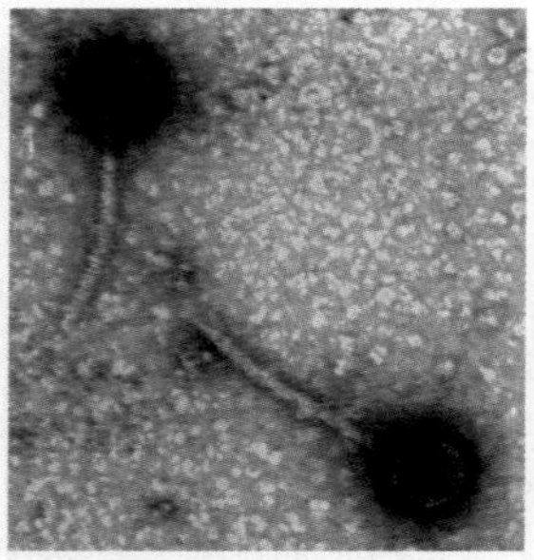

©GOPAL MURTI/VISUALS UNLIMITED

Bacteriophage lambda

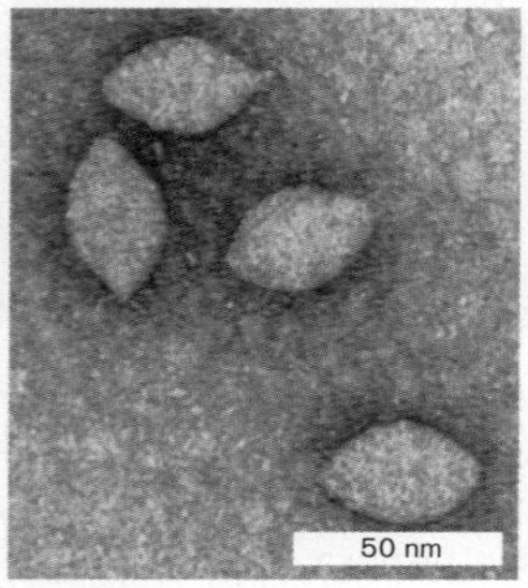
50 nm

ICTVDB

Fusellovirus

Nonenveloped bacteriophages (structure includes head and tail)

Myoviridae. Bacteriophage T4 infects *Escherichia coli.*

Siphoviridae. Bacteriophages lambda and mu infect *E. coli.* Others infect Gram-positive hosts.

Tectiviridae. Infect enteric bacteria.

Nonenveloped viruses of animals and protists

Adenoviridae. Adenovirus generates tumors in humans.

Iridoviridae. Infect insects and amphibians.

Mimiviridae. Large viruses; infect *Acanthamoeba* and human macrophages.

Pandoraviridae. The largest known viruses; infect amebas.

Papillomaviridae. Papillomavirus causes genital warts.

Phycodnaviridae. Infect *Chlorella,* algal symbiont of paramecia and hydras.

Enveloped viruses of animals

Baculoviridae. Baculoviruses infect insects.

Herpesviridae. Herpes simplex causes oral and genital herpes; varicella-zoster virus causes chickenpox.

Poxviridae. Include smallpox and cowpox viruses.

Archaeal viruses

Ampullaviridae. Bottle-shaped viruses; infect *Acidianus.*

Fuselloviridae. Infect thermoacidophiles (*Sulfolobus*) and halophiles (*Haloarcula*).

Guttaviridae. Spindle-shaped viruses; infect *Sulfolobus.*

Haloviridae. Haloviruses infect haloarchea such as *Haloferax, Halobacterium,* and *Haloarcula.*

Rudiviridae, Lipothrixviridae. Infect *Sulfolobus* and *Thermoproteus.*

Group II. Single-stranded DNA viruses

Genome consists of (+) sense DNA; host DNA polymerase used; nonenveloped.

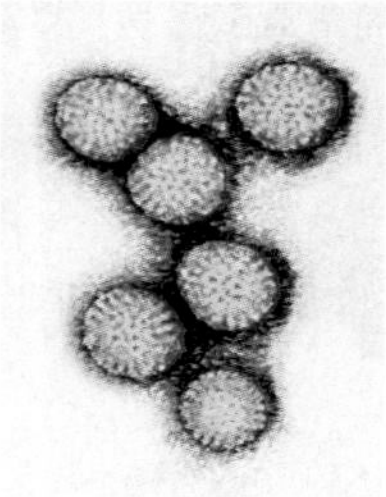

ROBERT G. MILNE, CNR ISTITUTO DE FITOVIROLOGICA APPLICATA, TORINO, ITALY

Geminivirus

Bacteriophages

Inoviridae. Bacteriophage M13 infects *E. coli* and has a slow-release life cycle.

Microviridae. Bacteriophage ϕX174 infects *E. coli.*

Animal viruses

Circoviridae. Infect pigs and birds, causing immunosuppression.

Parvoviridae. Cause various diseases in cats, pigs, and other animals.

Plant viruses

Geminiviridae. Transmitted by aphids to tomato plants and other important crops. Their virions group in "twins," each member of the pair carrying one DNA circle with part of the genome.

Group III. Double-stranded RNA viruses

Require viral RNA-dependent RNA polymerase; usually package the polymerase before exiting host cell.

GARY GAUGLER/VISUALS UNLIMITED

Rotavirus

Nonsegmented, enveloped bacteriophages

Cystoviridae. Infect *Pseudomonas* species of bacteria.

Segmented, nonenveloped viruses of animals and plants

Birnaviridae. Infect marine and aquatic fish.

Reoviridae. Orthoreoviruses and rotaviruses infect humans and other vertebrates. Cypovirus infects insects. Rice dwarf virus (phytoreovirus) devastates rice crops worldwide.

Varicosaviridae. Infect plants.

Group IV. (+) sense single-stranded RNA viruses

Require viral RNA-dependent RNA polymerase to generate (–) template for progeny (+) genome; usually nonsegmented.

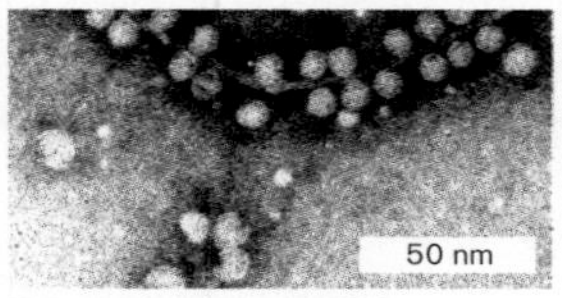
50 nm
Bacteriophage MS2

©HANS ACKERMANN/ VISUALS UNLIMITED

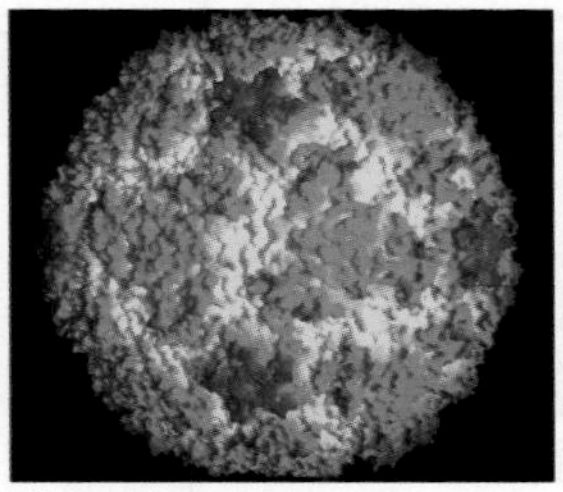
Rhinovirus 14

KENNETH EWARD/PHOTO RESEARCHERS, INC.

Nonenveloped bacteriophages

Leviviridae. Bacteriophages MS2 and Qβ infect *E. coli.*

Nonenveloped animal and plant viruses

Bromoviridae. Infect many kinds of plants and are often carried by beetles.

Picornaviridae. Poliovirus causes poliomyelitis. Rhinovirus causes the common cold. Aphthovirus causes foot-and-mouth disease in cattle and other stock.

Potyviridae. Viruses such as plum pox virus infect fruits, peanuts, potatoes, and other plants.

Tobamoviridae. Tobacco mosaic virus infects plants.

Enveloped animal and plant viruses

Coronaviridae. Coronaviruses include SARS (severe acute respiratory syndrome) and animal viruses.

Flaviviridae. Infect humans; include West Nile virus, yellow fever virus, and hepatitis C virus.

Togaviridae. Include rubella virus and equine encephalitis virus.

Group V. (–) sense single-stranded RNA viruses

Require viral RNA-dependent RNA transcriptase. Genome often segmented. Some viruses include (+) and (–) strand regions.

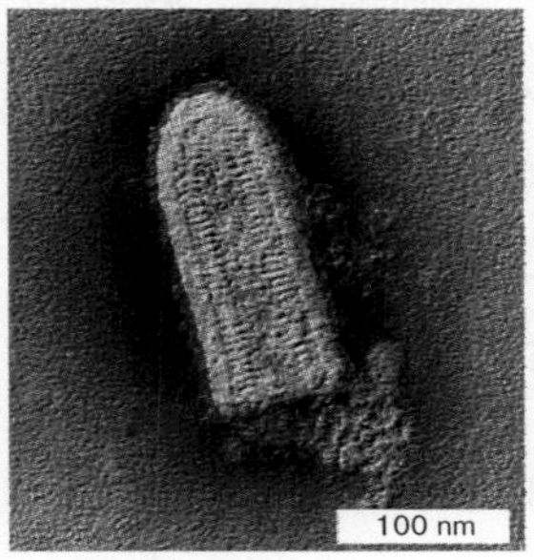
100 nm
Rabies virus

©GOPAL MURTI/PHOTOTAKE/ ALAMY

Segmented, enveloped viruses

Orthomyxoviridae. Influenza virus causes major epidemics among humans and animals.

Nonsegmented, enveloped viruses

Filoviridae. Ebola virus causes outbreaks among humans and chimpanzees.

Paramyxoviridae. Infect humans and cause measles, mumps, and parainfluenza.

Rhabdoviridae. Rabies virus infects mammals (virion length 130–300 nm).

Segmented (+/–) strand, enveloped viruses

Arenaviridae. Spread by rodents; cause hemorrhagic fever and lymphocytic choriomeningitis.

Bunyaviridae. Hantaviruses are spread by rodents and infect humans. Tospoviruses are transmitted by thrips, infecting plants. Rift Valley fever virus infects livestock.

Group VI. Retroviruses (RNA reverse-transcribing viruses)

Viral reverse transcriptase copies RNA to DNA for integration into host chromosome.

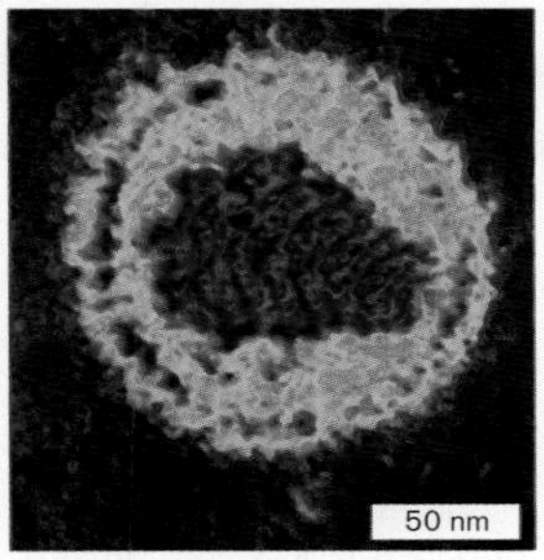
50 nm
HIV-1

©HANS GELDERBLOM/VISUALS UNLIMITED

Retroviridae. Simple retroviruses include oncogenic retroviruses: feline leukemia virus, Rous sarcoma virus, avian leukosis virus. Lentiviruses (complex retroviruses) include human immunodeficiency virus (HIV, the cause of AIDS) and simian immunodeficiency virus (SIV). Lentiviruses are engineered to make lentiviral vectors (lentivectors) for gene therapy.

Group VII. Pararetroviruses (DNA reverse-transcribing viruses)

DNA transcribed to RNA; reverse-transcribed to DNA using host reverse transcriptase or packaged viral reverse transcriptase.

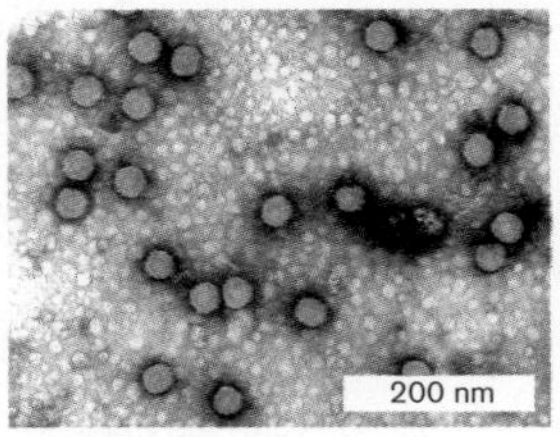
200 nm
Cauliflower mosaic virus

ENCYCLOPEDIA OF VIROLOGY. 1995 ACADEMIC PRESS LTD.

Nonenveloped plant viruses

Badnaviridae. Infect bananas, cocoa plants, citrus, yams, and sugarcane.

Caulimoviridae. Transmitted by aphids; cauliflower mosaic virus and related viruses infect cauliflower, broccoli, groundnuts, soybeans, and cassava. The cauliflower mosaic virus promoter sequence is used to construct vectors to insert genes into transgenic plants.

Enveloped animal viruses

Hepadnaviridae. Hepatitis B virus causes widespread disease of the human liver.

A4.2

Bacteria

TABLE A4.2

Bacterial diversity. (Expanded version of Table 18.1.)

Note: Each trait applies to most members of the taxon described, but exceptions have evolved.

Deep-branching thermophiles

Thermophilic bacteria that diverged early from archaea and eukaryotes. Many genes transferred laterally from archaea.

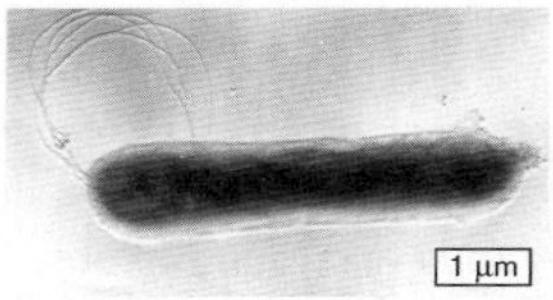
1 μm
Aquifex

K. O. STETTER & R. RACHEL, U. OF REGENSBURG

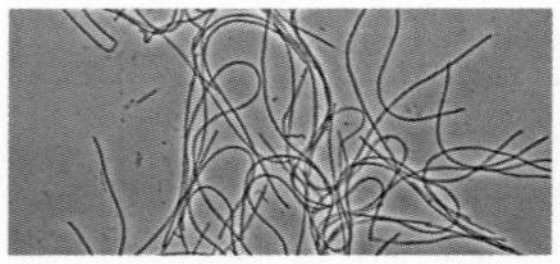
Chloroflexus aurantiacus

DOE-HERTER/VISUALS UNLIMITED

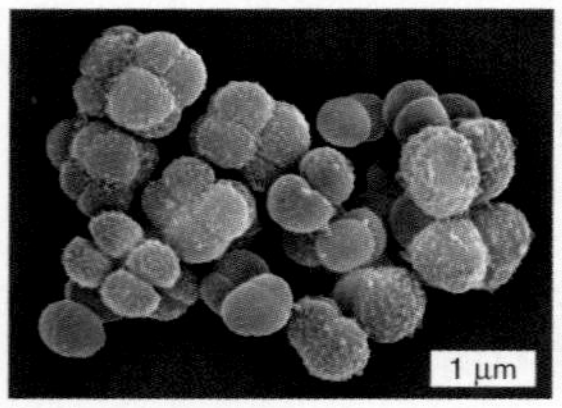
1 μm
Deinococcus

DENNIS KUNKEL/VISUALS UNLIMITED

Aquificae. **Hyperthermophiles (70°C–95°C). Oxidize H_2, H_2S, or thiosulfate.**
- **Aquificales.** Outer membrane; cells may be single with flagella or grow in filaments. *Aquifex, Thermocrinis.*

Chloroflexi. **Filamentous green bacteria.**
- **Chloroflexales.** Filamentous phototrophs; absorb light to oxidize organic compounds or H_2S. Most contain photosystem II in chlorosomes. *Chloroflexus aurantiacus* is thermophilic; forms mats in hot springs.

Deferribacteres. **Anaerobic vent thermophiles.**
- **Deferribacterales.** Single rod or vibrio cells, flagellated. *Deferribacter autotrophicus* reduces Fe^{3+} with H_2; *D. abyssi* respires on small organic molecules.

Deinococcus-Thermus. **Peptidoglycan contains ornithine. Diverse growth temperatures.**
- **Deinococcales.** Thick envelope; stain Gram-positive. Not thermophilic, but extremely resistant to ionizing radiation and to desiccation. *Deinococcus.*
- **Thermales.** Species are filamentous clustered cells or single-celled. Grow at 70°C–75°C. *Thermus aquaticus* is the source of Taq polymerase for PCR.

Thermotogae. **Thermophiles or hyperthermophiles (55°C–100°C).**
- **Thermotogales.** Outer membrane with large periplasm; anaerobic heterotrophs. *Petrotoga, Thermotoga.*

Cyanobacteria

Oxygenic photoautotrophs with thylakoid membranes. Share ancestry with chloroplasts.

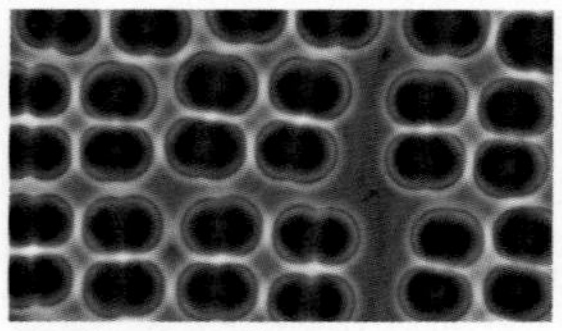
Merismopedia

MICHAEL ABBEY/VISUALS UNLIMITED

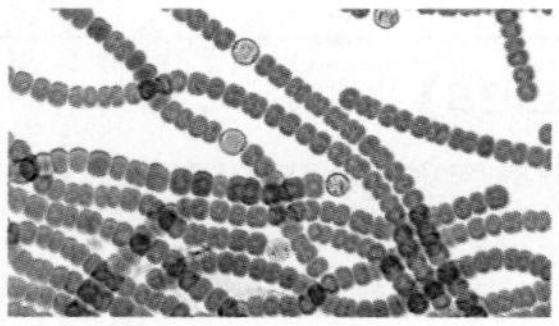
Anabaena

©CAROLINA BIOLOGICAL SUPPLY/VISUALS UNLIMITED

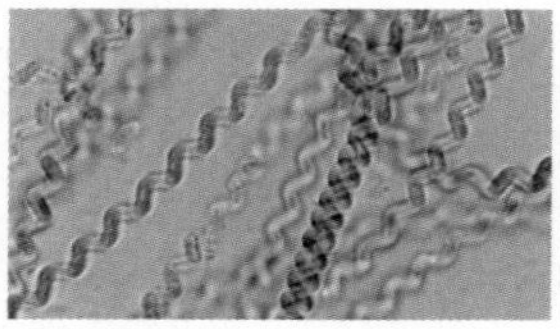
Spirulina

PETER SIVER/VISUALS UNLIMITED

Cyanobacteria. **Oxygenic photoautotrophs with thylakoid membranes. Fix CO_2 using Rubisco. Often mutualists. Share ancestry with chloroplasts.**
- **Chroococcales.** Square colonies based on two division planes.
  - ***Chroococcus.*** Single, double, or quartet of cells (10–20 μm per cell). Grow in pond sediment.
  - ***Merismopedia.*** Platelike colonies of elliptical cells (5 μm).
  - ***Synechococcus.*** Single or double cells. Marine producer.
- **Gloeobacterales.** Lack thylakoids; conduct photosynthesis in cell membrane. *Gloeobacter violaceus.*
- **Nostocales.** Filamentous chains with N_2-fixing heterocysts. Some grow symbiotically with corals or plants.
  - ***Anabaena.*** Aquatic. Grow in association with red water fern (*Azolla*).
  - ***Nostoc.*** Aquatic. Grow independently; or mutualistically with fungi, as lichens; or as endosymbionts of *Gunnera* plant cells.
- **Oscillatoriales.** Filamentous chains with motile hormogonia (short chains) (filament width 50 μm).
  - ***Oscillatoria.*** Aquatic or marine. Grow independently, or as sponge endosymbionts.
  - ***Spirulina.*** Aquatic. Farmed as food supplement (filament width 5 μm).
  - ***Trichodesmium.*** Major marine producers. Form "blooms" of overgrowth.
- **Pleurocapsales.** Globular colonies; reproduce through baeocytes. *Pleurocapsa, Myxosarcina.*
- **Prochlorales.** Tiny single cells, elliptical or spherical (0.5 μm).
  - ***Prochlorococcus.*** Most abundant marine producers. One of the smallest cells of a free-living microbe, with a highly reduced genome.
  - ***Prochloron.*** Tropical marine producers; endosymbionts of sea squirts.

Firmicutes and Actinobacteria (Gram-positive)

Peptidoglycan multiple layers, cross-linked by teichoic acids.

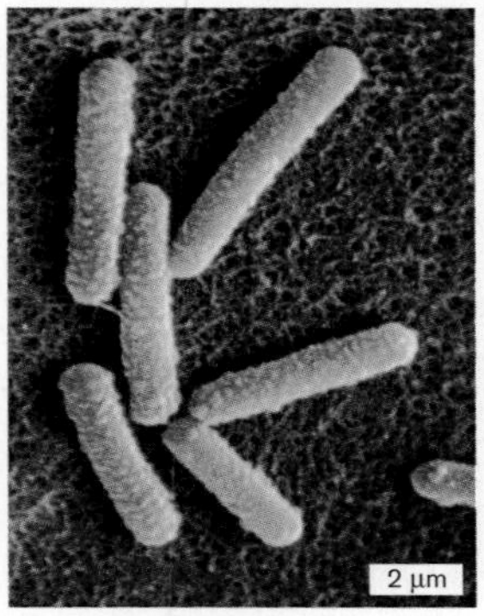
2 μm

Bacillus subtilis

ANDREW SYRED/PHOTO RESEARCHERS, INC.

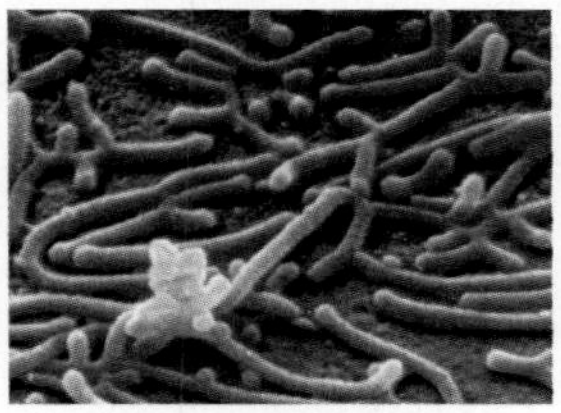

Mycoplasma genitalium

DON W. FAWCETT/PHOTO RESEARCHERS, INC.

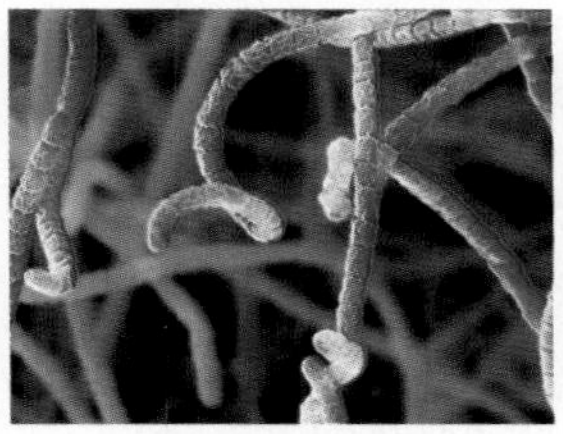

Streptomyces

SCIENCE VU/FREDERICK MERTZ/VISUALS UNLIMITED

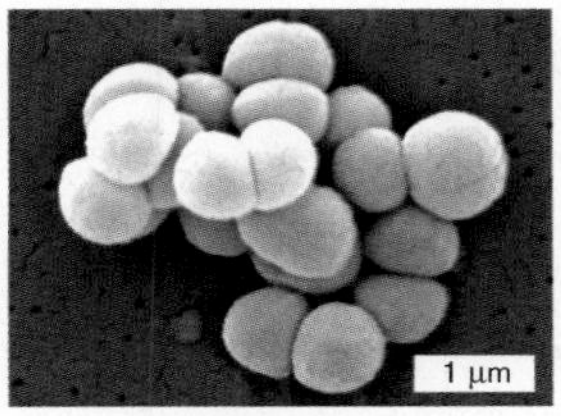
1 μm

Micrococcus luteus

CDC/BETSY CRANE

Firmicutes. **Low-GC, Gram-positive rods and cocci.**

- **Bacillales.** Aerobic or facultative anaerobes (2–10 μm).
  - ***Bacillus.*** Endospore-forming rods. Soil-growing: *B. subtilis, B. cereus, B. anthracis, B. thuringiensis.* Extremophiles: *B. alkalophilus, B. thermophilus, B. halodurans.*
  - ***Listeria.*** Non-spore-forming rods; intracellular pathogens. *L. monocytogenes* causes listeriosis.
  - ***Staphylococcus.*** Non-spore-forming cocci. Skin biota: *S. epidermidis. S. aureus* infects flesh.
- **Clostridiales.** Anaerobic rods.
  - ***Carboxydothermus.*** Oxidize carbon monoxide (CO) to CO_2. Form endospores.
  - ***Clostridium.*** Form endospores. *C. botulinum* and *C. difficile* are pathogens. *C. acetobutylicum* generates butanol.
  - ***Dehalobacter.*** Non-spore-forming rods. Dechlorinate chloroethenes.
  - ***Epulopiscium.*** Exceptionally large cells. Reproduce by "live birth."
  - ***Heliobacterium, Heliophilum.*** Endospore-forming photoheterotrophs.
  - ***Metabacterium.*** Exceptionally large cells; form multiple endospores. *M. polyspora.*
  - ***Ruminococcus.*** Digestive flora of ruminant animals.
- **Lactobacillales.** Non-spore formers. Facultative anaerobes. Ferment, producing lactic acid.
  - ***Enterococcus.*** Enteric cocci. *E. hirae.*
  - ***Lactobacillus.*** *L. acidophilus* is used for dairy culture.
  - ***Lactococcus.*** Used for dairy culture. *L. lactis.*
  - ***Streptococcus.*** Chains of cocci. Group A streptococci cause "strep throat."
- **Tenericutes, class Mollicutes.** Lack cell wall; require animal host. *Mycoplasma genitalium* has one of smallest known genomes. *M. pneumoniae* causes pneumonia.

Actinobacteria. **High-GC, Gram-positive bacteria with moderate salt tolerance.**

- **Actinomycetales.**
  - **Actinomycetaceae.** Filamentous, producing aerial hyphae and spores (width 1 μm).
  - ***Actinomyces.*** *A. israelii* causes actinomycosis.
  - ***Frankia.*** Saprophytes; grow on leaf litter. Fix nitrogen for plants.
  - ***Salinispora.*** Sponge endosymbionts. Produce drug-like secondary products.
  - ***Streptomyces.*** Produce many antibiotics. *S. coelicolor, S. griseus.*

Corynebacteriaceae. Irregularly shaped rods. *Corynebacterium diphtheriae* causes diphtheria.

Micrococcaceae. Nonfilamentous soil bacteria; mostly obligate aerobes. Often airborne.

- ***Arthrobacter.*** Rods. Respire on oxygen or on chlorinated aromatics.
- ***Micrococcus.*** Aerobic cocci such as *M. luteus.* Grow in soil.

Mycobacteriaceae. Exceptionally thick cell envelope holds acid-fast stain.

- ***Mycobacterium.*** *M. tuberculosis* causes tuberculosis; *M. leprae* causes leprosy.

Propionibacteriaceae. Propionic acid fermentation. *Propionibacterium acnes* causes acne; *P. freudenreichii* makes Swiss cheese.

- **Bifidobacteriales.** Ferment without gas. Enteric biota of breast-fed infants. *Bifidobacterium.*

Proteobacteria

Gram-negative. Outer membrane contains LPS. Diverse metabolism. Share ancestry with mitochondria.

Proteobacteria. **Gram-negative. Diverse cell forms and metabolism (1–10 μm).**

Alphaproteobacteria (class). Heterotrophic rods or spirilla.

- **Caulobacterales.** Aquatic oligotrophs; alternate stalk and flagellum. *Caulobacter.*
- **Rhizobiales.** Plant mutualists and pathogens, methyl oxidizers, and animal pathogens.
  - ***Agrobacterium.*** *A. tumefaciens* causes plant tumors; transgenic plant vector.
  - ***Bartonella.*** *B. quintana* causes trench fever; *B. henselae* causes cat scratch fever.
  - ***Brucella.*** Cause brucellosis in horses and sheep. *B. melitensis.*
  - ***Methylobacterium.*** Grow on single-carbon compounds.
  - ***Nitrobacter.*** Nitrite-oxidizing lithotrophs. *N. winogradskyi.*
  - ***Rhizobium* group.** *Bradyrhizobium* and *Sinorhizobium* fix nitrogen within legumes.
  - ***Rhodopseudomonas, Rhodomicrobium.*** Soil photoheterotrophs.

(continued)

TABLE A4.2

Bacterial diversity. *(continued)*

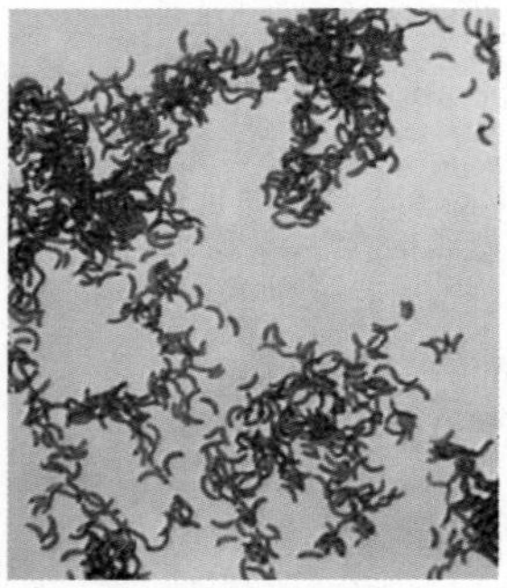

Rhodospirillum rubrum

DAVID M. PHILLIPS/VISUALS UNLIMITED

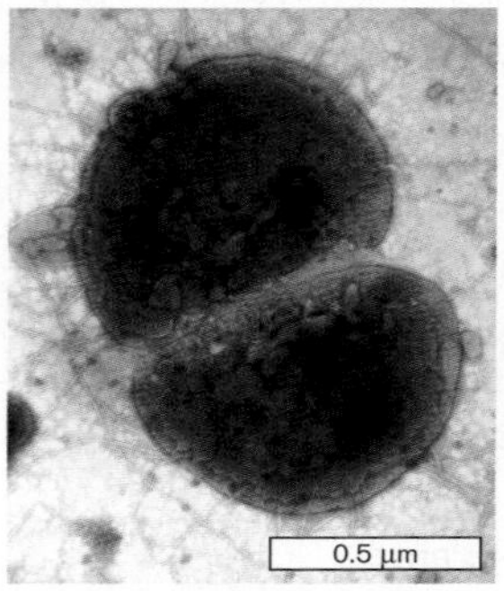
0.5 µm

Neisseria gonorrhoeae

KWANGSHIN KIM/PHOTO RESEARCHERS, INC.

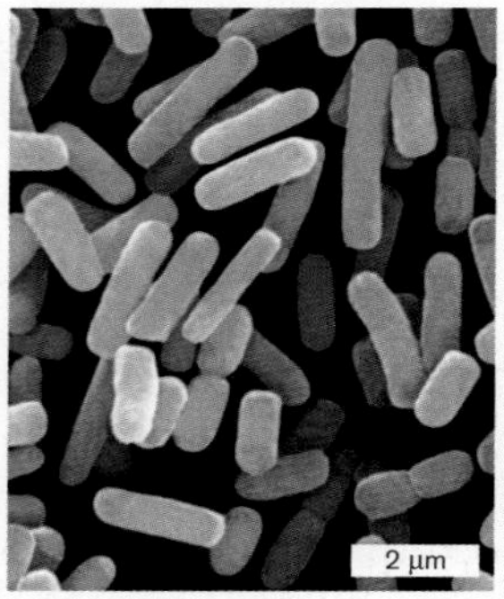
2 µm

Escherichia coli

RICHARD KESSEL & GENE SHIH/VISUALS UNLIMITED

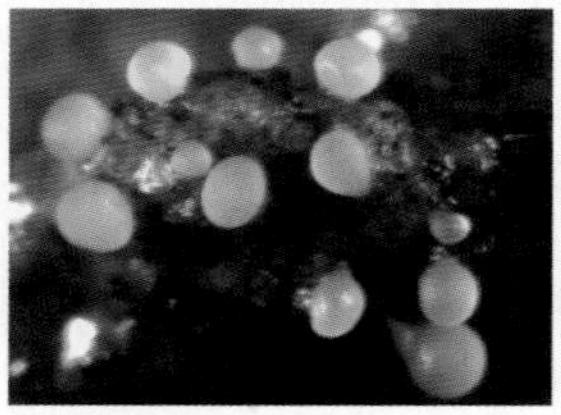

Myxococcus

MICHIEL VOS, U. OF OXFORD

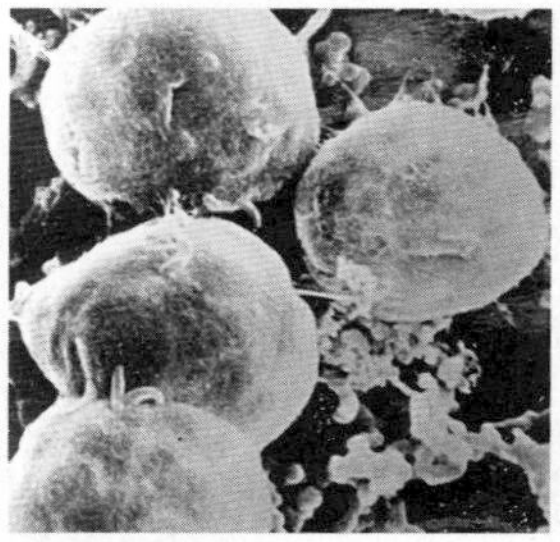

Thiovulum

CARL O. WIRSEN AND HOLGER W. JANNASCH. 1978. *J. BACTERIOL.* **136**:765

- **Rhodobacterales, Rhodospirillales.** Flagellated photoheterotrophs: *Rhodobacter sphaeroides, Rhodospirillum rubrum, Roseobacter* (oxidizes CO). Nonphototrophs: *Acetobacter* makes vinegar.
- **Rickettsiales.** Includes intracellular parasites; related to mitochondria.
  - ***Rickettsia.*** Intracellular parasites. *R. rickettsii* causes Rocky Mountain spotted fever; *R. prowazekii* causes typhus.
  - **SAR11 cluster.** Marine photoheterotrophs use proteorhodopsin. *Pelagibacter.*
- **Sphingomonadales.** Catabolize complex organics; used for bioremediation. Opportunistic pathogen: *Sphingomonas.* Aerobic photoheterotrophs: *Citromicrobium, Erythromicrobium.*

Betaproteobacteria (class). Phototrophs, lithotrophs, and pathogens.

- **Burkholderiales.** *Burkholderia pseudomallei* causes melioidosis in humans and farm animals.
- **Neisseriales.** Mucous-membrane normal flora and pathogens. Aerobic or microaerophilic diplococci. *Neisseria gonorrhoeae* causes gonorrhea; *N. meningitidis* causes meningitis.
- **Nitrosomonadales.** Lithotrophs. *Nitrosomonas europaea* oxidizes ammonia.
- **Rhodocyclales.** Soil heterotrophs and photoheterotrophs. *Azoarcus evansii* catabolizes complex aromatic molecules. *Rhodocyclus* species are purple photoheterotrophs.

Gammaproteobacteria (class). Facultative anaerobes and lithotrophs.

- **Acidithiobacillales.** Lithotrophs. *Acidithiobacillus ferrooxidans* oxidizes iron and sulfur.
- **Aeromonadales.** Aquatic heterotrophs. *Aeromonas hydrophila* infects fish and humans.
- **Alteromonadales.** Aquatic. *Shewanella oneidensis* reduces metals; used in fuel cells.
- **Chromatiales.** Lithotrophs: *Nitrococcus* species oxidize ammonia. Sulfur and iron phototrophs: *Chromatium.* Nitrite phototrophs: *Thiocapsa.*
- **Enterobacteriales.** Enterobacteriaceae. Facultative anaerobes; colonize the human colon.
  - ***Escherichia coli.*** Strains include normal flora and enteric pathogens.
  - ***Salmonella.*** *S. enterica* strains cause gastrointestinal infection and typhoid fever.
  - ***Yersinia.*** *Y. pestis* causes bubonic plague.
- **Legionellales.** *Legionella pneumophila* causes legionellosis pneumonia. *Coxiella burnetti* causes Q fever.
- **Oceanospirillales.** Marine heterotrophs, including degraders of petroleum. *Alcanivorax.*
- **Pseudomonadales.** Rods; aerobic or respire on nitrate; catabolize aromatics. *Pseudomonas aeruginosa* infects lungs in cystic fibrosis patients. *P. fluorescens* suppresses plant diseases.
- **SAR86 cluster.** Marine photoheterotrophs use proteorhodopsin.
- **Thiotrichales.** Lithotrophs and heterotrophs. *Beggiatoa alba* and *Thiomargarita namibiensis* oxidize sulfur. *Cycloclasticus* species catabolize polycyclic aromatic hydrocarbons in petroleum.
- **Vibrionales.** Marine heterotrophs. *Vibrio cholerae* causes cholera.

Deltaproteobacteria (class). Lithotrophs and multicellular communities.

- **Bdellovibrionales.** Periplasmic predators. *Bdellovibrio bacteriovorus* consumes *E. coli.*
- **Desulfobacterales.** Reduce sulfate. *Desulfobacter, Desulfococcus.*
- **Desulfuromonadales.** Lithotrophs; reduce or oxidize sulfur or metals.
  - ***Desulfuromonas.*** Reduce elemental sulfur.
  - ***Geobacter metallireducens.*** Reduces iron oxide.
- **Myxococcales.** Gliding bacteria that form fruiting bodies. *Myxococcus.*

Epsilonproteobacteria (class).

- **Campylobacterales.** Spirillar pathogens.
  - ***Campylobacter.*** *C. jejuni* causes food poisoning.
  - ***Helicobacter.*** *H. pylori* causes gastritis.
  - ***Nautilia,*** *Hydrogenimonas, Sulfurimonas.* Oxidize H_2 with sulfur or nitrate.
  - ***Thiovulum.*** Oxidize sulfides, producing sulfur granules.

Deep-branching Gram-negative phyla

Gram-negative, with outer membrane. Most are anaerobes.

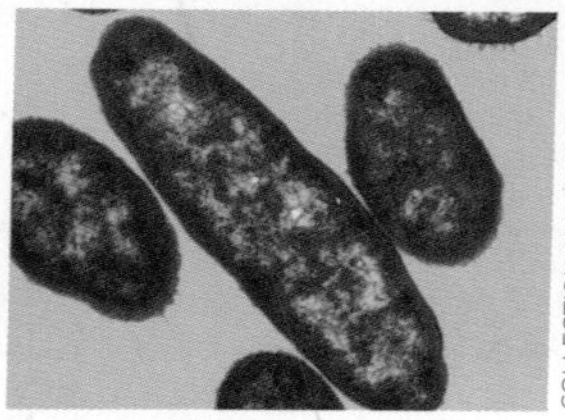
Bacteroides

COLLECTION CNRI/PHOTOTAKE

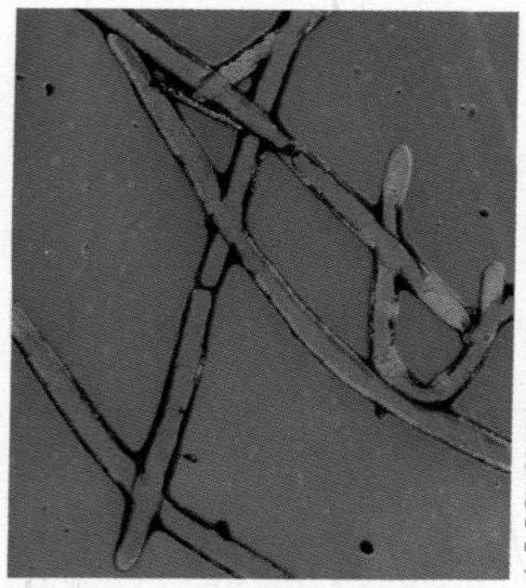
Fusobacterium nucleatum

A. B. DOWSETT/SCIENCE PHOTO LIBRARY

Acidobacteria. **Gram-negative bacteria; include acidophiles, with diverse metabo**

Acidobacterium capsulatum.

Geothrix capsulatum. Oxidizes toluene with Fe III.

Chloracidobacterium thermophilum. Phototrophic thermophile.

Bacteroidetes. **Obligate anaerobes. Heterotrophs that feed on diverse carbon sources.**

- **Bacteroidales.** Obligate anaerobes, soil or human flora.

  Bacteroides. Enteric. *B. thetaiotaomicron* digests complex plant carbohydrates in gut. *Bacteroides* species escaping the colon cause abscesses.

  Porphyromonas. Includes gingival pathogens such as *P. gingivalis.*

- **Flavobacteriales.** Aerobic or facultative heterotrophs in soil and water. *Flavobacterium psychrophilum* infects fish.

Chlorobi. **Green sulfur-oxidizing phototrophs.**

- **Chlorobiales.** Anaerobic H_2S photolysis, absorbing red and infrared. *Chlorobium tepidum.*

Fusobacteria. **Gram-negative anaerobic bacteria found in septicemia and in skin ulcers.**

Fusobacterium nucleatum. Isolated from dental plaque.

Nitrospirae. **Nitrite oxidizers. Aerobic or facultative.**

- **Nitrospirales.** Tight spirilla. In soil and water, oxidize NO_2^- to NO_3^-. *Nitrospira, Leptospirillum.* Includes acidophilic Fe oxidizers.

Spirochetes (Spirochaeta)

Narrow, coiled cell encased by sheath, which encloses polar flagella.

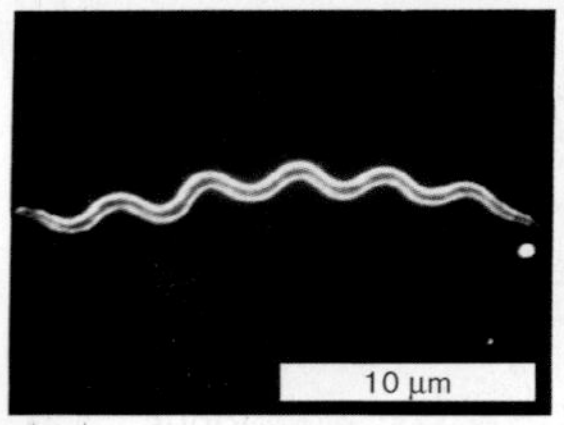
10 µm
Borrelia burgdorferi

CHARLES STRATTON/VISUALS UNLIMITED

Spirochetes. **Narrow, coiled cells with axial filaments, encased by sheath. Polar flagella beneath sheath double back around cell.**

- **Spirochaetales.** Aquatic free-living; endosymbionts and pathogens (width <0.5 µm; length 10–20 µm).

  Borrelia. *B. burgdorferi* causes Lyme disease, transmitted by ticks.

  Hollandina. Termite gut endosymbionts.

  Leptospira. L-shaped animal pathogens; cause leptospirosis.

  Spirochaeta. Aquatic, free-living heterotrophs.

  Treponema. *T. pallidum* causes syphilis.

Chlamydiae, Planctomycetes, and Verrucomicrobia

regular cells lacking peptidoglycan, with subcellular structures analogous to those of eukaryotes.

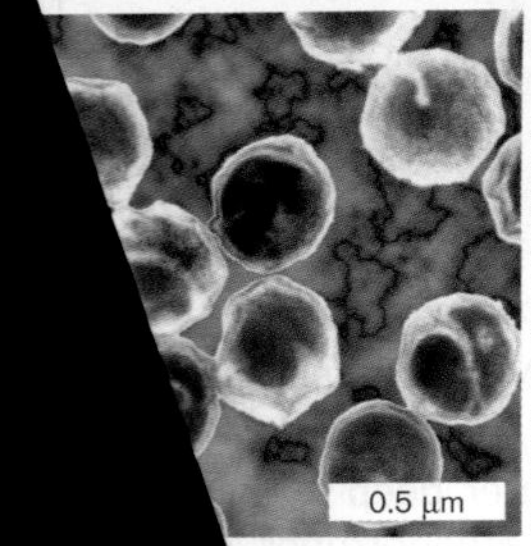
0.5 µm
achomatis

DAVID PHILLIPS/VISUALS UNLIMITED

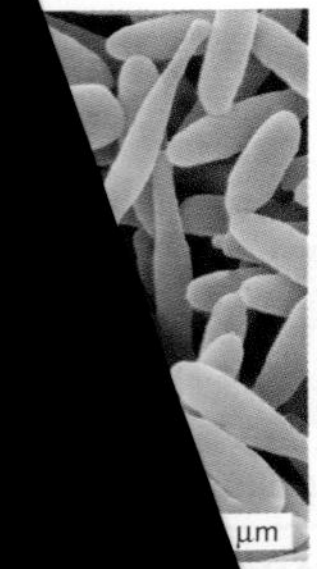
µm
nis

DENNIS KUNKEL/PHOTOTAKE

Chlamydiae. **Intracellular cell wall–less pathogens of animals or protists.**

- **Chlamydiales.** Reticulate bodies form multiple spore-like transfer particles (elementary bodies) that infect the next host.

  Chlamydia. *C. trachomatis* causes sexually transmitted disease and trachoma (eye infection).

  Chlamydophila. *C. pneumoniae* causes pneumonia; *C. abortus* causes spontaneous abortion in animals; *C. psittaci* causes psittacosis in birds and humans.

  Parachlamydia. Chlamydia-like species that infect free-living amebas.

Planctomycetes. **Nucleoid has double membrane analogous to eukaryotic nuclear membrane and other intracytoplasmic subcellular membrane compartments.**

- **Planctomycetales.** Aquatic or marine. Flexible cell shape.

  Brocadia. Anaerobically oxidize ammonium and release N_2, using anammoxosomes.

  Gemmata. Nuclear double membrane, surrounded by additional intracellular membrane.

  Pirellula. Only one intracellular membrane. Has fimbriae and single flagellum. Marine habitat.

  Planctomyces. One intracellular membrane. Halophile.

Scalindua. Conduct anammox in wastewater. Oxidize CO.

Verrucomicrobia. **Stalk-like appendages contain actin filaments. Aquatic oligotrophs.**

- **Verrucomicrobiales.**

  Prosthecobacter. Each cell has a polar cytoplasmic extension called a prostheca.

  Verrucomicrobium. Stellate cells have multiple prosthecae.

A4.3

Archaea

TABLE A4.3

Archaeal diversity. (Expanded version of Table 19.2.)

Note: Each trait applies to most members of the taxon described, but exceptions have evolved.

Crenarchaeota

Temperature: Hyperthermophiles and psychrophiles, as well as mesophiles.

Metabolism: Sulfur and hydrogen oxidizers, aerobic and anaerobic heterotrophs, ammonia oxidizers.

Envelope: Membrane lipids include tetraethers with crenarchaeol. Flexible S-layer surrounds membrane.

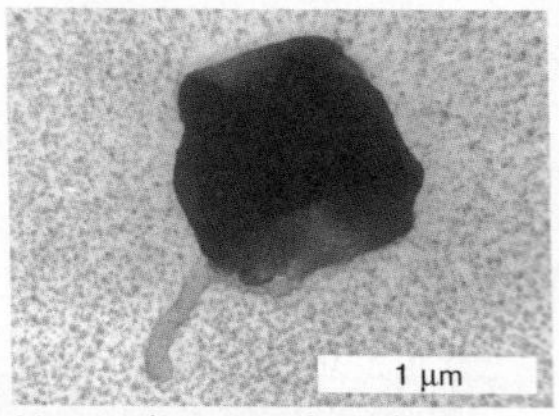
1 µm
Hyperthermus butylicus

ZILLIG ET AL. 1990. *J. BACTERIOL.* 172:3959

1 µm
Sulfolobus sp.

OLIVER MECKES/NICOLE OTTAWA

Marine Benthic Group B (MBGB). **Anoxic marine sediments: deep-sea methane hydrates, hydrothermal vents, coastal regions. Metabolism uncertain.**

Miscellaneous Crenarchaeote Group (MCG). **Widely divergent clades. Terrestrial and marine, cold and hot, water and subsurface environments.**

Thermoprotei. **Many thermophiles in hot springs and marine vents. Also includes marine mesophiles and psychrophiles.**

- **Caldisphaerales.** Thermoacidophilic heterotrophs grow in hot springs. *Caldisphaera.*
- **Desulfurococcales.** Anaerobic sulfur reduction with organic electron donors. Irregularly shaped cells with glycoprotein S-layer; no cell wall.
  - ***Aeropyrum pernix,*** *Desulfurococcus mobilis, Hyperthermus butylicus.*
  - ***Ignicoccus islandicus.*** Unique periplasmic space contains membrane vesicles.
  - ***Pyrodictium abyssi,*** *Pyrodictium occultum, Pyrolobus fumarii.* Grow at marine thermal vents, u to 110°C. Form 3D network of cells and extracellular cannulae.
  - ***Thermosphaera.*** Heterotrophic; inhibited by sulfur.
- **Psychrophilic marine crenarchaeotes (uncharacterized).** Marine water, deep sea, and An Anaerobic heterotrophs, sulfate reducers, nitrite-reducing methanotrophs.
- **Sulfolobales.** Aerobic acidophiles; moderate thermophiles. Oxidize H_2S to H_2SO_4. *Sulf Sulfurisphaera, Acidianus.*
- **Thermoproteales.** *Pyrobaculum, Thermoproteus, Vulcanisaeta.*

Thaumarchaeota

Temperature: Mesophiles, thermophiles, and psychrophiles.

Metabolism: Marine and soil archaea that oxidize NH_3 with O_2. Important marine sources of nitrat of methylphosphonate (CH_3-PO_3^{2-}), which bacteria convert to methane. Includes members of

Envelope: Membrane lipids include tetraethers with crenarchaeol.

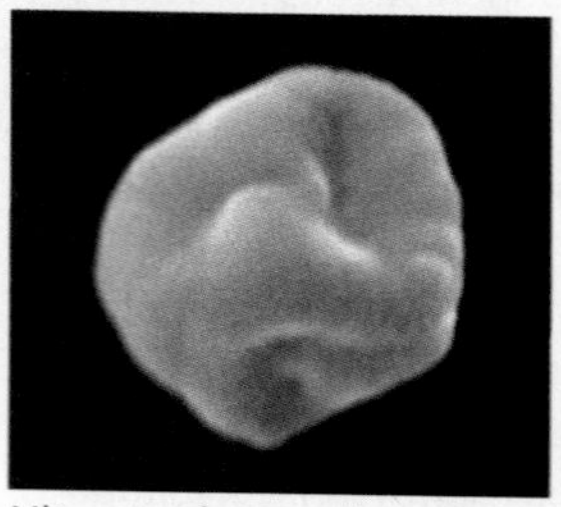
Nitrososphaera viennensis

MICHAEL STIEGLMEIER AND NIKOLAUS LEISCH, DEPARTMENT OF ECOGENOMICS AND SYSTEMS BIOLOGY, UNIVERSITY OF VIENNA

- **Cenarchaeales.** Sponge symbionts; grow at 10°C. *Cenarchae*
- **Nitrosopumilales.** Ammonia-oxidizing archaea (AOA), mar *Nitrososphaera gargensis.*
- **Psychrophilic marine thaumarchaeotes (characte** Antarctica. Anaerobic heterotrophs, sulfate reducer

Deeply branching groups o

Uncultured marine organisms, branching near root of archaeal

Ancient Archaeal Group (AAG). **Vent hy**

Korarchaeota. **Hyperthermophiles in Yellows**

Marine Hydrothermal Vent Group (MHVG). **Vent hy**

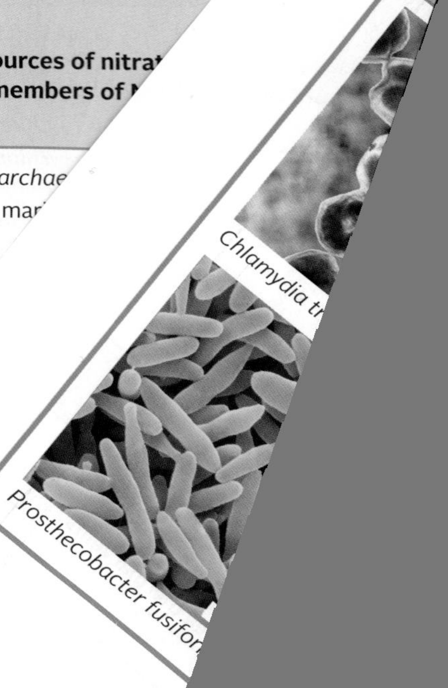
Chlamydia tr
Prosthecobacter fusifor

Euryarchaeota

Temperature: Mesophiles and thermophiles; some psychrophiles.

Metabolism: Methanogens, halophilic photoheterotrophs, and sulfur and hydrogen oxidizers; acidophiles an

Envelope: Methanogens and halophiles have rigid cell walls of glycans, glycoproteins, or pseudopeptidoglycan

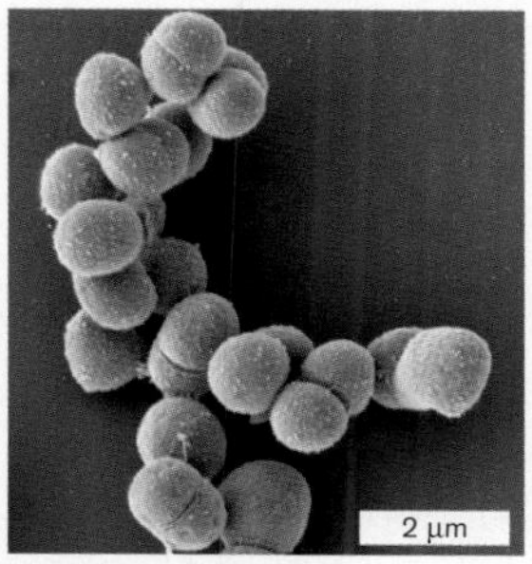
2 µm

Halobacteriales

EYE OF SCIENCE

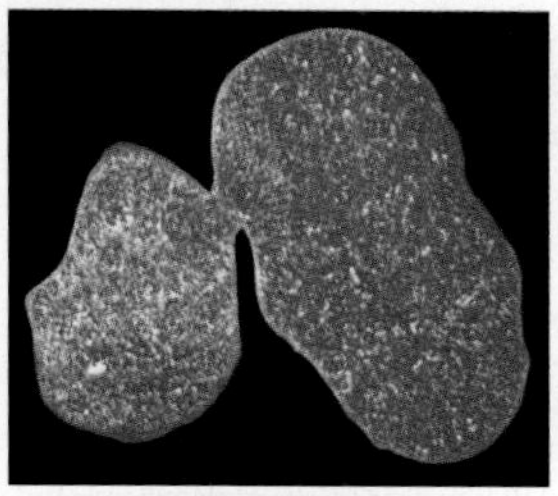

Methanoculleus nigri

T. J. BEVERIDGE/VISUALS UNLIMITED

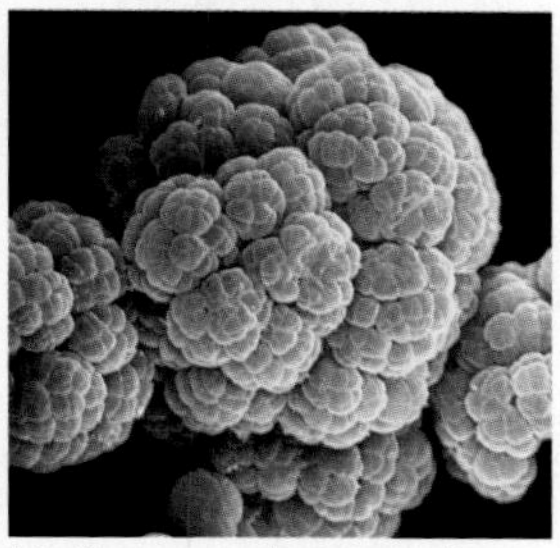

Methanosarcina mazei

©RALPH ROBINSON/VISUALS UNLIMITED

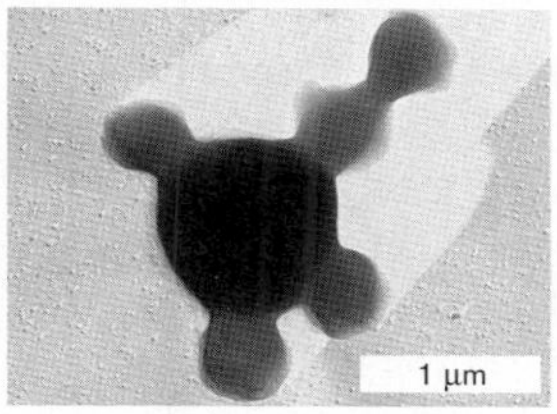
1 µm

Nanoarchaeum equitans (attached to surface of *Ignicoccus* sp.)

HUBER ET AL. 2003. *RESEARCH IN MICROBIOL.* **154**:165

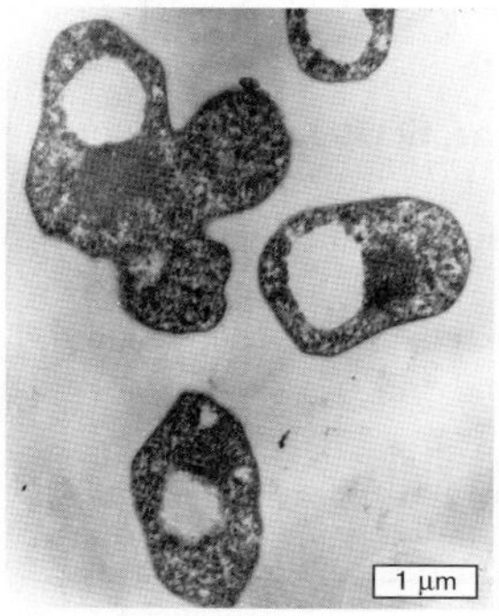
1 µm

Picrophilus oshimae

SCHLEPER ET AL. 1995. *J. BACTERIOL.* **177**:7050

Anaerobic Methane Oxidizers (ANME). **Anoxic marine sediments. Oxidize methane fro methanogens, in syntrophy with sulfate-reducing bacteria. Important control of greenh**

Archaeoglobi. **Hyperthermophiles; metabolize sulfur and use reverse methanogenesis.**

- **Archaeoglobales**. Sulfate oxidation of H_2 or organic hydrogen donors. *Archaeoglobus fulgidu*

Haloarchaea. **Halophiles; grow in brine (concentrated NaCl). Conduct photoheterotrophy, with light-driven H^+ pump and Cl^- pump.**

- **Halobacteriales.**

  Haloarcula, *Halobacterium, Halococcus, Haloferax.* Mesophiles at neutral pH. Grow in salterns and salt lakes. *Haloarcula marismortui, Halobacterium salinarum.*

  Haloquadra (*Haloquadratum*). Square-shaped mesophilic halophiles.

  Halorubrum lacusprofundi. Psychrophile; grows in subglacial Antarctic salt lakes.

  Natronococcus, Natronomonas. Alkaliphiles; grow above pH 9 in soda lakes.

Methanogens (four classes). **Generate methane from CO_2 and H_2, formate, acetate, other small molecules. Strict anaerobes. Must associate with bacteria producing their substrates.**

Cell walls of pseudopeptidoglycan or polysaccharides.

Wide variety of shapes: rods, filaments, cocci, and coccoid clusters.

Wide range of anaerobic habitats: wetlands, landfills, and intestinal tracts of animals. Psychrophiles in deep-ocean floor sediment generate methane hydrates.

- **Methanobacteriales.** Lack cytochromes; reduce CO_2, formate, or methanol with H_2.

  Methanobacterium. Have peptidoglycan cell walls.

  Methanobrevibacter smithii, Methanosphaera stadtmanae. Inhabit human digestive tract. Increase caloric output of human food by reducing methanol, a breakdown product of pectin.

  Methanothermus fervidus. Marine vent thermophile.

- **Methanomicrobiales, Methanococcales, and Methanopyrales.** Lack cytochromes; reduce CO_2, formate, or methanol with H_2.

  Methanocaldococcus jannaschii. Vent thermophile.

  Methanoculleus nigri. Budding cells, found in anaerobic wastewater digester.

  Methanogenium frigidum. Psychrophile.

  Methanopyrus kandleri. Marine vent thermophile.

- **Methanosarcinales.** Possess cytochromes; reduce methylamines and acetate (as well as CO_2 and formate) with H_2.

  Methanococcoides alaskense. Arctic marine habitat.

  Methanohalophilus. Halophile. Inhabits salt lakes, often at high pH.

  Methanosaeta concilii. Filamentous colonies.

  Methanosarcina acetivorans. Globular colonies; cell walls of sulfated polysaccharides.

Nanoarchaeota. **Vent hyperthermophiles; obligate symbionts attached to *Ignicoccus*.**

Nanoarchaeum equitans.

South African Gold Mine Euryarchaeotal Group (SAGMEG). **Terrestrial and marine deep subsurface. Anaerobic heterotrophs.**

Terrestrial Miscellaneous Euryarchaeotal Group (TMEG). **Terrestrial deep subsurface and surface soils, marine and freshwater sediments.**

Thermococci. **Hyperthermophiles (grow above 100°C) and barophiles (up to 200 atm pressure). Anaerobes; reduce sulfur.**

- **Thermococcales.**

  Pyrococcus abyssi, P. furiosus. Use S^0 to oxidize H_2 or organic hydrocarbons. *Thermococcus* species are closely related.

Thermoplasmata. **Extreme acidophiles; oxidize sulfur from pyrite (FeS_2), generating sulfuric acid. Mesophiles or moderate thermophiles.**

- **Thermoplasmatales.**

  Ferroplasma acidiphilum, F. acidarmanus. Grow at 37°C–50°C. Oxidize sulfur from FeS_2, generating ambient pH as low as pH 0. No cell wall.

  Picrophilus torridus. Grows above 60°C. Oxidizes sulfur, generating acid. Possesses cell wall.

  Thermoplasma acidophilum. Grows at 59°C and pH 2.

A4.4

Eukarya

TABLE A4.4

Eukaryotic microbial diversity. (Expanded version of Table 20.1.)

Note: Each trait applies to most members of the taxon described, but exceptions have evolved.

Opisthokonta (fungi and metazoan animals)

Single flagellum on reproductive cells. Includes multicellular animals.

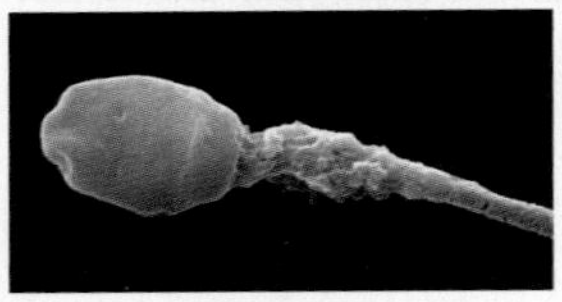

Human spermatozoan

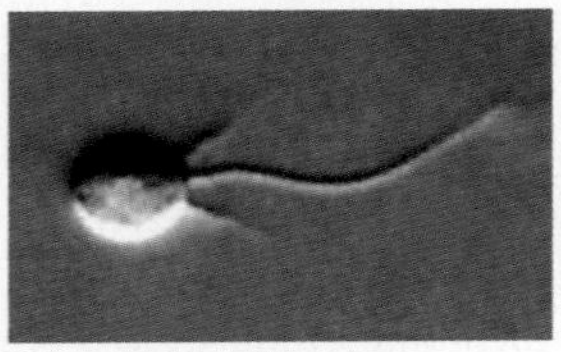

Choanoflagellate

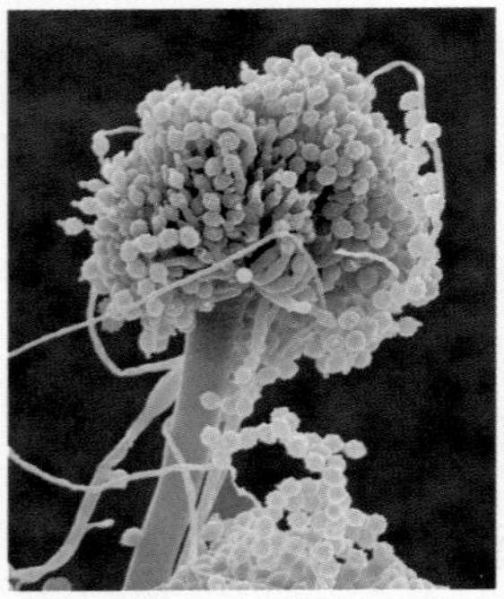

Aspergillus mold

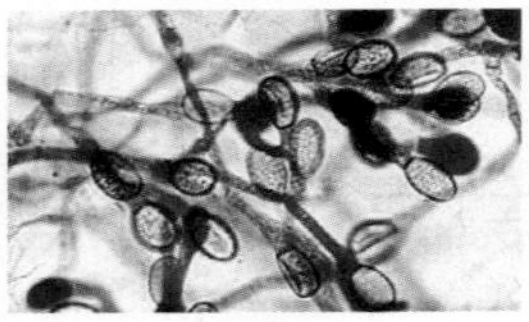

Allomyces zoospores

Metazoa (animals). **Multicellular organisms with motile cells and body parts.**
- **Colonial animals.** Sponges, jellyfish.
- **Invertebrates.** Hydras, mollusks, arthropods, worms.
- **Vertebrates.** Fish, amphibians, reptiles, birds, mammals, including *Homo sapiens.*

Choanoflagellata. **Single flagellum with collar of microvilli. Resemble sponge choanocytes. Possible link to common ancestor of multicellular animals.**

Eumycota (fungi). Cells form hyphae with cell walls of chitin.

Ascomycota. Fruiting bodies form asci containing haploid ascospores.
- **Filamentous species.** *Neurospora, Penicillium.*
    - ***Aspergillus.*** Opportunistic pathogen; produces aflatoxin.
    - ***Geomyces.*** Associated with bat mortality, white nose syndrome.
    - ***Magnaporthe oryzae.*** Causes rice blast, the most serious disease of cultivated rice.
    - ***Microsporum, Trichophyton.*** Cause ringworm skin infection.
    - ***Trichoderma.*** Endophyte within plants; provides nutrients and induces plant defenses.
- **Morels and truffles.** Large fruiting bodies are highly valued foods.
- ***Stachybotrys.*** Known as "black mold"; contaminates homes.
- **Yeasts.** Unicellular species have lost mycelial stages.
    - ***Blastomyces dermatitidis, Candida albicans, Pneumocystis jirovecii.*** Infect immunocompromised patients.
    - ***Saccharomyces cerevisiae.*** Ferments beer and bread dough.

Basidiomycota. Basidiospores form primary and secondary mycelia; generate mushrooms.

Amanita, Lycoperdon, Ustilago maydis. *Amanita* is extremely poisonous; *Lycoperdon* is edible; *U. maydis* causes corn smut (a plant disease).

Cryptococcus neoformans. Yeast-form basidiomycete; an opportunistic pathogen.

Chytridiomycota. **The deepest-branching fungal clade. Zoospores (motile gametes) with a single flagellum resemble the gametes of animals. Saprophytes or anaerobic rumen fungi. *Allomyces.***

Batrachochytrium dendrobatidis. Frog pathogen.

Neocallomastix. Bovine rumen digestive endosymbiont.

Glomeromycota. **Mutualists of plant roots, forming arbuscular mycorrhizae, filamentous networks that share nutrients with and among diverse plants.**

Lichens. **Mutualistic association between an ascomycete and green algae (*Trebouxia*) or cyanobacteria (*Nostoc*).**

Microsporidia. **Single-celled parasites that inject a spore through a tube into a host cell, causing microsporidiosis.**

Encephalitozoon. Commonly infect AIDS patients.

Zygomycota. **Nonmotile gametes grow toward each other and fuse to form the zygote (zygospore). Saprophytes or insect parasites. Some form mycorrhizae.**

Other opisthokonts: *Corallochytrium,* Ichthyosporea, Nucleariidae

Viridiplantae (primary endosymbiotic algae and plants)

Includes algae and multicellular plants. Chloroplasts arose from a primary endosymbiont.

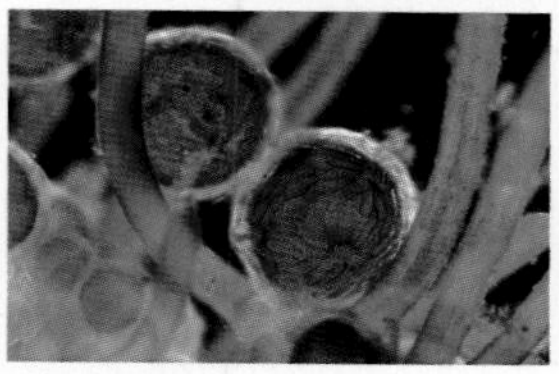

Charophyte

1 cm

Cymopolia barbata

Palmaria palmata

Plants. Adapted to growth on land.

- **Nonvascular plants.** Mosses, ferns.
- **Vascular plants.** Gymnosperms, angiosperms.

Charophyta (stoneworts). **Multicellular algae with rhizoids that adhere to sediment. Form green mats with crust of calcium carbonate, giving the name "stoneworts."**

Chlorophyta (green algae). **Chlorophyll *a* confers green color. Inhabit upper waters.**

- **Unicellular with paired flagella.** *Chlamydomonas* is a unicellular green alga; a model system for research on algae. *Volvox* forms colonies of flagellated cells.
- **Multicellular.** *Ulva* grows in large sheets. *Spirogyra* forms chains of cells. *Cymopolia* forms calcified stalks with filaments.
- **Picoeukaryotes.** *Ostreococcus* and *Micromonas* are unicellular algae <3 μm in diameter.
- **Siphonous algae.** *Caulerpa* species consist of a single cell with multiple nuclei, growing to indefinite size.

Rhodophyta (red algae). **Phycoerythrin obscures chlorophyll, colors the algae red. Absorption of blue-green light enables colonization of deeper waters.**

Porphyra. Form sheets edible by humans.

Sebdenia, *Plocamium.* Form branched fronds. *Palmaria.*

Mesophyllum. Form coralline algae, hardened by calcium carbonate crust; resemble coral.

Amoebozoa (amebas and slime molds)

Lobe-shaped (lobose) pseudopods driven by sol-gel transition of actin filaments.

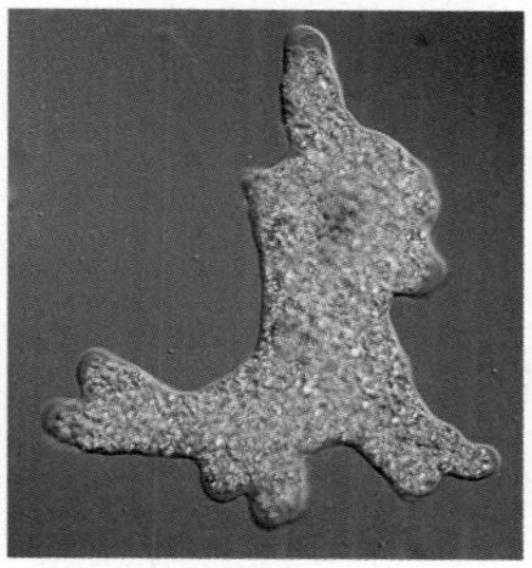

Chaos carolinense (size 1–2 mm)

Amebas. **Unicellular. No microtubules to define shape. Life cycle is primarily asexual. Predators in soil or water.**

Amoeba proteus, Pelomyxa, Chaos chaos. Giant free-living amebas in soil and water; they consume small invertebrates.

Entamoeba histolytica. Intestinal parasite.

Acanthamoeba. Soil predator, an opportunistic parasite causing keratitis and meningitis.

Mycetozoa. **Slime molds. Upon starvation, amebas aggregate to form a fruiting body, which undergoes meiosis and produces spores.**

- **Cellular slime molds.** Multicellular fruiting body alternates with single-celled amebas.

  Dictyostelium discoideum. Provides an important model system for multicellular development.

- **Plasmodial slime molds.** Multinucleate plasmodium alternates with single-celled amebas.

  Physarum polycephalum. In aqueous environment, amebas generate flagella.

Rhizaria (amebas with filament-shaped pseudopods)

Filament-shaped (filose) pseudopods. Some species have a test (shell) of silica or other inorganic materials.

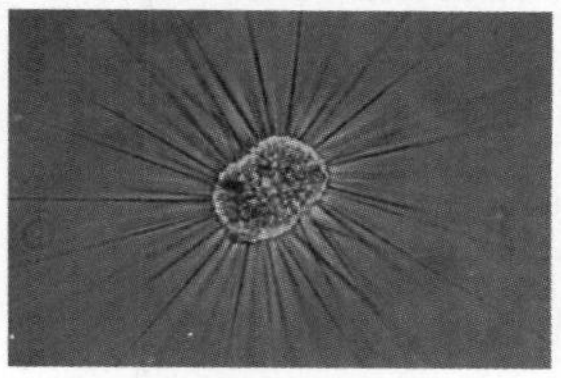

Actinophrys sol (cell size 30–90 μm)

Cercozoa. **Amebas with flagella and/or filamentous pseudopods.**

- **Euglyphida.** Amebas with a test.
- **Chlorarachniophyta.** Have algal endosymbionts.

Foraminifera. **Form spiral tests. Fossil "forams" are an indicator of petroleum deposits.**

Heliozoa. **Form thin pseudopods.** *Actinophrys sol.* Note: Phylogeny is disputed.

Radiolaria. **Form thin pseudopods (filopodia) reinforced by microtubules, radiating starlike from the center. Some possess inorganic skeletons or spicules.**

(continued)

TABLE A4.4

Eukaryotic microbial diversity. *(continued)*

Alveolata (having cortical alveoli)

Cortex contains flattened vesicles called alveoli, reinforced below by lateral microtubules.

Vorticella (length 1 mm)

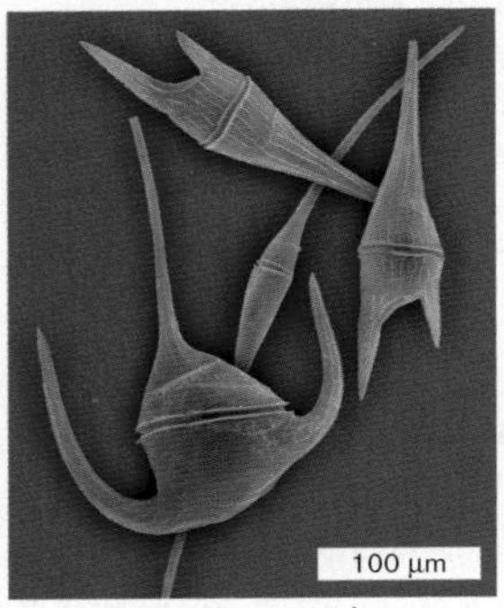
100 µm

Ceratium tripos and *Ceratium furca*

Ciliophora. **Common aquatic predators. Reproduce by conjugation, in which micronuclei are exchanged; then regenerate macronucleus for gene expression. Macronuclear DNA may be cut into thousands of segments.**

Didinium. Has two equatorial rings of flagella.

Paramecium, Spirostomum, Oxytricha. Covered with cilia.

Suctorians such as *Acineta.* Have knobbed tentacles.

Vorticella, Stentor. Stalked, with a mouth ringed by cilia.

Dinoflagellata. **Secondary or tertiary endosymbiont algae, from engulfment of red algae or diatoms. Cortical alveoli contain stiff plates. Pair of flagella, one wrapped around the cell.**

- **Free-living marine and aquatic dinoflagellates** such as *Peridinium* and *Ceratium* supplement their photosynthesis with predation. Some produce bioluminescence. Blooms of marine dinoflagellates generate the "red tide." Cause paralytic shellfish poisoning.
- **Zooxanthellae** such as *Symbiodinium* are endosymbionts of corals, providing essential nutrition through photosynthesis. "Coral bleaching" is a serious condition in which the zooxanthellae are expelled under environmental stress. Zooxanthellae also inhabit clams, flatworms, mollusks, and jellyfish.

Apicomplexa (formerly Sporozoa). **Parasites with complex life cycles. Lack flagella or cilia; possess apical complex for invasion of host cells. Vestigial chloroplasts.**

Cryptosporidium parvum. Waterborne opportunistic parasite.

Plasmodium falciparum. Causes malaria.

Toxoplasma gondii. Causes feline-transmitted toxoplasmosis.

Heterokonta (having pair of dissimilar flagella)

Paired flagella of dissimilar form, one much shorter than the other.

Diatom

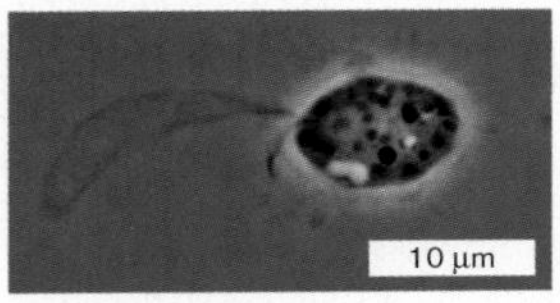
10 µm

Paraphysomonas sp.

Secondary endosymbiont algae. **Free-living in marine and aquatic systems. Mixotrophic; combine photrophy with heterotrophy.**

- **Bacillariophyceae.** Diatoms. Possess intricate silicate shells with radial symmetry (centric) or bilateral symmetry (pennate).
- **Phaeophyceae.** "Brown algae" such as kelps and sargassum weed. Extend for many meters.
- **Chrysophyceae.** "Golden algae," pale-colored flagellates such as *Ochromonas* and *Paraphysomonas.*
- **Prymnesiophyceae.** Coccolithophores. Covered with intricate plates of $CaCO_3$.
- **Xanthophyceae.** "Yellow-green algae" and other less-studied forms of phytoplankton.

Oomycetes. **Water molds.** Superficially resemble fungi; often infect fish or plants. *Phytophthora infestans* destroyed potato crops and caused the Great Irish Famine.

Euglenozoa or Discicristata (having disk-shaped cristae)

Usually possess a deep feeding groove. Disk-shaped cristae of mitochondria.

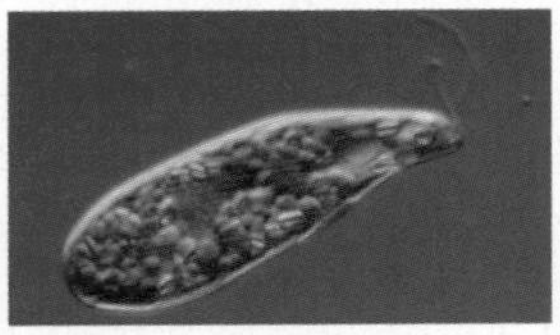

Euglena gracilis

Euglenida. **Free-living flagellates.** Some contain a secondary endosymbiotic chloroplast. *Euglena gracilis* is a common aquatic flagellate.

Jakobida. **Free-living flagellates with lorica (stalk).** *Reclinomonas americana.*

Trypanosomatidae. Parasites. *Trypanosoma brucei* causes sleeping sickness. *T. cruzi* causes Chagas' disease. *Leishmania* species cause leishmaniasis.

Metamonada (vestigial mitochondria)

Parasitic or symbiotic flagellates. Mitochondria and Golgi degenerated through evolution.

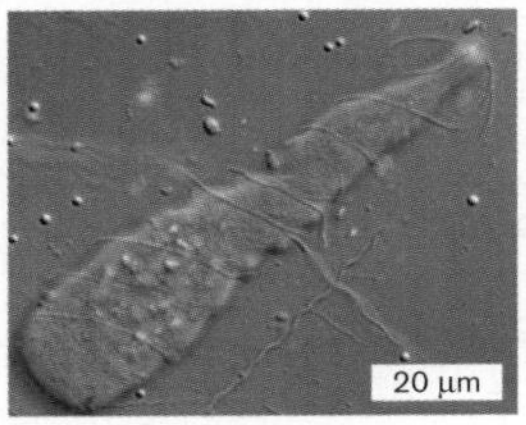
20 µm

Pyrsonympha sp.

Diplomonadida and Retortamonadida. **Human intestinal parasites.** Highly degenerate cells, lacking mitochondria. *Giardia lamblia.* Cause of diarrhea from contaminated drinking water.

Oxymonadida and Parabasalia. **Symbionts of termite gut.** *Pyrsonympha.*

ANSWERS TO THOUGHT QUESTIONS

CHAPTER 1

1.1 The minimum size of known microbial cells is about 0.2 μm. Could even smaller cells be discovered? What factors may determine the minimum size of a cell?

ANSWER: The smallest cells known, about 0.2 μm in length, are cell wall–less bacteria called mycoplasmas—for example, *Mycoplasma pneumoniae*, a causative agent of pneumonia. Bacteria might be discovered that are smaller than 0.2 μm, but it is hard to see how their cell components, such as ribosomes (about a tenth this size), could fit inside such a small cell. The volume required for DNA and the apparatus of transcription and translation probably sets the lower limit on cell size.

1.2 If viruses are not functional cells, are they "alive"?

ANSWER: A traditional definition of a life-form includes the capability for metabolism and homeostasis (maintaining internal conditions of its cytoplasm), as well as reproduction and response to its environment. Viruses reproduce themselves and respond to the environment of the host cell, but they lack metabolism or homeostasis outside their host cell. Nevertheless, viruses such as herpes viruses contain numerous metabolic enzymes that participate in the metabolism of their host. Certain large viruses, such as the mimivirus, appear to have evolved from cells. Some microbiologists argue that viruses should be considered alive if reproduction is the main criterion, and if the viral "environment" is considered the inside of the host cell.

1.3 Why do you think it took so long for humans to connect microbes with infectious disease?

ANSWER: For most of human history, we were unaware that microbes existed. Even after microscopy had revealed their existence, the incredible diversity of the microbial world and the difficulties in isolating and characterizing microbial organisms made it difficult to discern the specific effects of microbes. All healthy people contain microbes, and most disease-causing microbes are indistinguishable from normal microbiota by light microscopy. Not all microbial diseases can be transmitted directly from human to human; they may require complex cycles with intermediate hosts, such as the fleas and rats that carry bubonic plague.

1.4 How could you use Koch's postulates (Fig. 1.18) to demonstrate the causative agent of influenza? What problems not encountered with anthrax would you need to overcome?

ANSWER: Using Koch's postulates to demonstrate the causative agent of influenza would require an animal model host. Secretions from diseased patients could be applied to different animal species, such as monkeys and mice, in order to find an animal showing signs of the disease. To determine the causative agent of disease, the patient's secretions could be filtered in order to separate bacteria and viruses. Only the filtrate would cause disease, because it contains viruses (relevant to Koch's postulates 1 and 3). Viruses, however, are more difficult to isolate in pure culture than are bacteria (postulate 2)—a problem Koch did not address. Furthermore, some viruses, such as HIV (human immunodeficiency virus) have no animal model; they grow only in human cells. Today, viruses are usually isolated in a tissue culture. Once isolated, the virus could be used to inoculate a new host animal (if an animal model exists) or a tissue culture and determine whether infection results (postulates 3 and 4). Another problem Koch did not address was the detection of infectious agents too small to be observed under a microscope. Today, antibody reactions are used to determine whether an individual has been exposed to a putative pathogen. An antibody test could be used to determine whether healthy and diseased individuals have been exposed to the isolated virus.

1.5 Why do you think some pathogens generate immunity readily, whereas others evade the immune system?

ANSWER: Some pathogens (microbes that cause disease) have external coat proteins that strongly stimulate the immune system and induce production of antibodies. Other pathogens have evolved to avoid the immune system by changing the identity of their external proteins. Immunity also varies greatly with the host's status. The very young and very old generally have weaker immune systems than do people in the prime of life. Some pathogens, such as HIV, will directly attack the host's immune system, limiting the immune response to the pathogen.

1.6 How do you think microbes protect themselves from the antibiotics they produce?

ANSWER: Microbes protect themselves from their antibiotics by producing their own resistance factors. As discussed in later chapters, microbes may synthesize pumps to pump the antibiotics out; or they may make altered versions of the target macromolecule, such as the ribosome subunit; or they may make enzymes to cleave the antimicrobial substance.

1.7 Why don't all living organisms fix their own nitrogen?

ANSWER: Nitrogen fixation requires a tremendous amount of energy, about 30 molecules of ATP per dinitrogen molecule converted to ammonia (discussed in Chapter 15). In a community containing adequate nitrogen sources, organisms that lose the nitrogen fixation pathway make more efficient use of their energy reserves than do those that spend energy to fix nitrogen from the atmosphere. Another consideration is that nitrogenase is an oxygen-sensitive enzyme, whereas plants, animals, and fungi are aerobes. In order to fix nitrogen, aerobic organisms need to develop complex mechanisms to exclude oxygen from nitrogenase.

1.8 What arguments support the classification of Archaea as a third domain of life? What arguments support the classification of archaea and bacteria together, as prokaryotes, distinct from eukaryotes?

ANSWER: The sequence of 16S rRNA (small-subunit rRNA) and other fundamental genes differs as much between archaea and bacteria as it does between archaea and eukaryotes. The composition of archaeal cell walls and phospholipids is completely distinct from that of bacteria and eukaryotes. Some aspects of gene expression, such as the RNA polymerase complex, are more similar between archaea and eukaryotes than between archaea and bacteria. On the other hand, archaeal and bacterial cells are prokaryotic; they both lack nuclei and complex membranous organelles. Archaeal metabolism and lifestyles are more similar to those of bacteria than to those of eukaryotes. Some archaea and bacteria sharing the same environment, such as high-temperature springs, have undergone horizontal transfer of genes that encode traits such as heat-stable membrane lipids.

1.9 Do you think engineered strains of bacteria should be patentable? What about sequenced genes or genomes?

ANSWER: Microbes—as well as multicellular organisms such as transgenic mice—have been patented, and the patents have held up in court. DNA sequences per se are not patentable, but specific plans for use of a DNA sequence can include the sequence as part of the patent. The reason for granting these patents is to encourage medical research by companies that need to earn a profit. The disadvantage of patents is that they restrict information flow and undercut competition. Furthermore, religious and philosophical arguments have been made that patenting live organisms cheapens life. Current laws aim for a balance among these concerns.

CHAPTER 2

2.1 (refer to Fig. 2.7) You have discovered a new kind of microbe, never observed before. What kinds of questions about this microbe might be answered by light microscopy? What questions would be better addressed by electron microscopy?

ANSWER: Light microscopy could answer questions such as: What is the overall shape of this cell? Does it form individual cells or chains? Is the organism motile? Only light microscopy can visualize an organism alive. Electron microscopy can answer questions about internal and external subcellular structures. For example, does a bacterial cell possess external filamentous structures, such as flagella or pili? If the dimensions of the unknown microbe are smaller than the lower limits of a light microscope's resolution, EM may be the only way to observe the organism. Viruses are often characterized by shape, and this shape is observed by electron microscopy.

2.2 Explain what happens to the refracted light wave as it emerges from a piece of glass of even thickness. How do its new speed and direction compare with its original (incident) speed and direction?

ANSWER: The part of the wave front that emerges first travels faster than the portion still in the glass, causing the wave front to bend toward the surface of the glass. Ultimately, the wave travels in the same direction and with the same speed as it did before entering the glass. The path of the emerging light ray is parallel to the path of the light ray entering the glass and is shifted over by an amount dependent on the thickness of the glass. This refraction will alter the path of the beam of light and decrease the amount of light reaching the lens of the microscope. Immersion oil has the same refractive index as glass and will limit the amount of light lost in this way.

2.3 Parabolic lenses are generally "biconvex"—that is, curving outward on both sides. What will happen to parallel light rays that pass through a lens that is concave on both sides? Or a lens that is convex on one side and concave with equal curvature on the other?

ANSWER: When light passes through a lens that is concave (curving inward) on both sides, the light rays diverge within the lens material and then diverge again more steeply on their way out. If the lens is concave on one side and convex with equal curvature on the opposite side, the light rays will diverge within the lens but will emerge parallel, although slightly farther apart than when they entered.

2.4 (refer to Fig. 2.14) What angle θ might offer magnification even greater than 100×? What practical problem would you have in designing a lens to generate this light cone?

ANSWER: In theory, an angle theta (θ) of 90° would produce the highest resolution, even greater than 100°. However, a 90° angle of theta generates a cone of 180°, which would require the object to sit in the same position as the objective lens—in other words, to have a focal distance of zero. In practice, the cone of light needs to be somewhat less than 180°, in order to allow room for the object and to avoid substantial aberrations (light-distorting properties) in the lens material.

2.5 Under starvation conditions, bacteria such as *Bacillus thuringiensis*, the biological insecticide, packages its cytoplasm into a spore, leaving behind an empty cell wall. Suppose, under a microscope, you observe what appears to be a hollow cell. How can you tell whether the cell is indeed hollow or is simply out of focus?

ANSWER: You can tell whether the cell is out of focus or actually hollow by rotating the fine-focus knob to move the objective up and down while observing the specimen carefully. If the hollow shape appears to be the sharpest image possible, it is probably a hollow cell. If the hollow shape turns momentarily into a sharp, dark cell, it was probably out of focus before. Alternatively, you could use a confocal microscope to visualize the center of the hollow cell.

2.6 What experiment could you devise to determine the actual order of events in DNA movement toward the pole during formation of an endospore?

ANSWER: One way to track the movement of DNA during sporulation would be to stain the DNA with a dye such as DAPI at various stages of sporulation. Alternatively, green fluorescent protein (GFP) fused to a DNA-binding protein could be used to label a specific sequence of DNA and track its position. Another way to determine the order of events in sporulation could be to observe mutant strains of bacteria that contain defects at different points in the sporulation process. (Sporulation is discussed further in Chapter 4.)

2.7 Some early observers claimed that the rotary motions observed in bacterial flagella could not be distinguished from whiplike patterns, comparable to the motion of eukaryotic flagella. Can you imagine an experiment to distinguish the two and prove that the flagella rotate? *Hint:* Bacterial flagella can get "stuck" to the microscope slide or coverslip.

ANSWER: To prove that flagella rotate, you can "tether" a bacterium to the microscope slide by getting one of its flagella stuck to the slide. A simple way to tether bacteria is by using a slide coated with anti-flagellin antibody. When the flagellum is stuck to the slide, its motor continues to rotate; thus, the entire cell now rotates. The rotation of the cell body can easily be seen by video microscopy. If the flagella moved in a whiplike fashion, the tethered cell would move back and forth, not rotate.

2.8 Compare and contrast fluorescence microscopy with dark-field microscopy. What similar advantage do they provide, and how do they differ?

ANSWER: Both dark-field and fluorescence microscopy enable detection (but not resolution) of objects whose dimensions are smaller than the wavelength of light. Dark-field technique is based on light scattering, which detects all small objects without discrimination. Fluorescence, however, provides a means to label specific parts of cells, such as cell membrane or DNA, or particular species of microbes, using fluorescent antibody tags.

2.9 An electron microscope can be focused at successive powers of magnification, as in a light microscope. At each level, the image rotates at an angle of several degrees. Given the geometry of the electron beam, as shown in Figure 2.36, why do you think the image rotates?

ANSWER: The image rotates because the electron beam is not straight, as for photons, but travels in a spiral through the magnetic field lines. As magnification increases, the spiral expands, and it reaches the image plane at a slightly different angle than before.

2.10 What kinds of research questions could you investigate using SEM? What questions could you answer using TEM?

ANSWER: SEM could be used to examine the surface of cells: Do the cells possess a smooth surface, or does their surface contain protein complexes or bulges that serve special functions? How do pathogens attach to the surface of cells? TEM can be used to determine the intracellular structure of attachment sites, as well as of internal organelles. TEM can also visualize the shapes of macromolecular complexes such as flagellar motors or ribosomes.

2.11 What kinds of experiments could prove or disprove the interpretations of the images of "nanobacteria" in blood plasma?

ANSWER: Attempt to observe the proposed "nanobacteria" in the presence of a general antibacterial agent such as sodium azide. If the particles still appear, despite the sodium azide, they cannot be alive and are probably inorganic in nature. Another approach is to use PCR amplification of 16S ribosomal RNA sequences to establish the existence of a novel isolate. So far, the PCR sequences obtained from "nanobacteria" in blood plasma have been identified as those of an environmental microbe that commonly contaminates PCR samples.

CHAPTER 3

3.1 Which chemicals occur in the greatest number in a bacterial cell? The smallest number? Why does a cell contain 100 times as many lipid molecules as strands of RNA?

ANSWER: Inorganic ions occur in the greatest number in a prokaryotic cell (250 million/cell). They are also the smallest in size. DNA molecules are found in the lowest number (one large molecule, branched during replication). A prokaryotic cell contains a hundred times as many lipid molecules as strands of RNA because lipids are small structural molecules, highly packed. RNA molecules are long macromolecules that either are packed into complexes (such as ribosomal RNA) or are temporary information carriers (messenger RNA), present only as needed to make proteins.

3.2 Amino acids have acidic and basic groups that can dissociate. Why are they not membrane-permeant weak acids or weak bases? Why do they fail to cross the phospholipid bilayer?

ANSWER: At neutral pH, amino acids each have both a positively charged amine and a negatively charged carboxylate; that is, they can act as either a weak acid or a weak base. Charged ions, no matter what their size, will not freely pass through a plasma membrane. In an amino acid, if either charged group becomes neutralized by acid or base, the other group remains charged, so the molecule as a whole will never cross the membrane.

3.3 If the sacculus (cell wall) consists of a single molecule, how do you think it expands as the cell grows?

ANSWER: The sacculus could expand by adding new units of peptidoglycan: (1) at the ends of the cell; (2) in the middle, around the equator where the cell divides; or (3) throughout the length of the expanding cell. Evidence from radiolabel incorporation experiments is consistent with expansion in the middle of the cell for species of cocci such as *Staphylococcus*. Rod-shaped cells such as *Escherichia coli* and *Bacillus subtilis* appear to incorporate new peptidoglycan throughout the expanding cell surface, although *Agrobacterium tumefaciens* extends peptidoglycan solely at a polar end. Wherever the new units are inserted, enzymes must first cleave between glycan strands to provide connection points for cross-bridges. For this reason, growing cells are highly susceptible to antibiotics that target peptidoglycan synthesis; if the wall is cleaved without inserting new units and forming new cross-bridges, it will eventually fall apart.

3.4 Figure 3.15 highlights the similarities and differences between the cell envelopes of Gram-negative and Gram-positive bacteria. What do you think are the advantages and limitations of a cell's having one layer of peptidoglycan (Gram-negative) versus several layers (Gram-positive)?

ANSWER: Having multiple layers of peptidoglycan increases the cell's resistance to osmotic shock, to desiccation stress, and to enzymes that cleave the cell wall. On the other hand, it requires more energy and biomass to build the layers of peptidoglycan. In addition, a thick cell wall could slow the uptake of nutrients. The mycobacteria, which have exceptionally thick cell walls, grow relatively slowly.

3.5 Why would laboratory culture conditions select for evolution of cells lacking an S-layer?

ANSWER: Degeneration of protective traits is a common problem when conducting research on microbes that can produce 30 generations overnight. Their rapid reproductive rate gives ample opportunity for spontaneous mutations to accumulate over an experimental timescale. In the case of the S-layer, in a laboratory test tube free of predators or viruses, mutant bacteria that fail to produce the thick protein layer would save energy compared to S-layer synthesizers, and would therefore grow faster. Such mutants would quickly take over a rapidly growing population.

3.6 Why would proteins be confined to specific cell locations? Why would a protein not be able to function everywhere in the cell?

ANSWER: Proteins have evolved one or more specific functions often optimized for a specific part of the cell. For example, water-conducting porins are found solely in the inner membrane (cell membrane), which is otherwise impermeable to water. The outer membrane, which is water permeable, is the sole location for specific porins that transport small peptides and sugars. The sugars then need to be taken across the inner membrane by transport proteins that have evolved to function best in this location. Similarly, different chaperones (proteins that aid peptide folding) have evolved to function best in the environment of the cytoplasm or periplasm, membrane-enclosed regions that differ substantially in pH and ion concentrations. In a different chemical environment of the cell, a protein may denature and lose its functional structure. A protein may be active only as a part of a complex of proteins. If the protein is placed into a different location within the cell, its protein partners may be absent, rendering the protein nonfunctional.

3.7 What do you think are the advantages and disadvantages of a contractile vacuole, compared with a cell wall?

ANSWER: The disadvantage of a contractile vacuole is that it requires continual input of energy to bail out the water. On the other hand, the contractile vacuole permits the existence of a cell that is flexible enough to engulf other cells as prey. A cell wall does not require energy (other than the initial energy to synthesize it). A cell wall is inflexible and does not allow another cell to be engulfed as prey.

3.8 Figure 3.35 presents data from an experiment that allows the function of the TipN protein of *Caulobacter* to be visualized by microscopy. Can you propose an experiment with mutant strains of *Caulobacter* to test the hypothesis that one of the proteins shown in the next figure (Fig. 3.36) is required for one of the cell changes to occur?

ANSWER: The diagram of **Figure 3.36** proposes that PodJ protein is required for a pole to develop a flagellum. Suppose we construct a mutant strain with a deletion of the gene *podJ*. This *podJ* mutant fails to express PodJ protein. When the *podJ* mutant is supplied with nutrients to grow, the stalked cells should grow and fission; but the progeny from the plain pole should fail to grow a flagellum. The stalked progeny will continue to divide, producing a stalked cell and a cell with plain poles, lacking flagellum or stalk. Other results are possible, but the result described would be consistent with a requirement of PodJ for flagellar development.

3.9 How would a magnetotactic species have to behave if it was in the Southern Hemisphere instead of the Northern Hemisphere?

ANSWER: In the Northern Hemisphere, the field lines for magnetic north point downward; in the Southern Hemisphere, the opposite is true. Thus, if downward direction is the aim of magnetotaxis, bacteria existing in the two hemispheres would have to respond oppositely to the magnetic field; in the Southern Hemisphere, anaerobic magnetobacteria swim toward magnetic south. Near the equator, the proportions of north-seeking and south-seeking bacteria are roughly equal.

3.10 Most laboratory strains of *E. coli* and *Salmonella* commonly used for genetic research lack flagella. Why do you think this is the case? How can a researcher maintain a motile strain?

ANSWER: The motility apparatus requires 50 different genes generating different protein components. Cells that acquire mutations eliminating expression of the motility apparatus gain an energy advantage over cells that continue to invest energy in motors. In a natural environment, the nonmotile cells lose out in competition for nutrients, despite their energetic advantage; but in the laboratory, cells are cultured in isotropic environments such as a shaking test tube, where motility confers no advantage. These culture conditions lead to evolutionary degeneration of motility, as they do for the S-layer (see Thought Question 3.5). In order to maintain a motile strain, bacteria are cultured on a soft agar medium containing an attractant nutrient. As cells consume the attractant, they generate a gradient, and chemotaxis leads them to swim outward. By subculturing only bacteria from the leading edge of swimming cells, one can maintain a motile strain.

CHAPTER 4

4.1 In a mixed ecosystem of autotrophs and heterotrophs, what happens when a heterotroph allows the autotroph to grow and begin to make excess organic carbon?

ANSWER: At first the growth of the heterotroph might outpace the growth of the autotroph, using the carbon sources faster than the autotroph can make them. As the organic carbon sources diminish through consumption, growth of the heterotroph decreases, but the CO_2 formed by the heterotroph will allow the autotroph to grow and make more organic carbon. Ultimately, the ecosystem comes into balance.

4.2 In what situation would antiport and symport be passive rather than active transport?

ANSWER: Antiport and symport are passive when both molecules are moving down their concentration gradients. A symport or an antiport system can do work to move a molecule from a low concentration to a high concentration (against a concentration gradient) as long as the cotransported molecule is moving from high to low concentration. When this happens, it is a form of active transport. If a symport or antiport moves both molecules with the concentration gradient (from high concentration to low), it is passive transport, assuming the molecules are moving no faster than the rate of diffusion.

4.3 What kind of transporter, other than an antiporter, could produce electroneutral coupled transport?

ANSWER: Electroneutral coupled transport can occur by symport if molecules of opposite charge are symported—for example, Na^+ flux together with Cl^-.

4.4 What would be the phenotype (growth characteristic) of a cell that lacks the *trp* genes (genes required for the synthesis of tryptophan)? What would be the phenotype of a cell missing the *lac* genes (genes whose products catabolize the carbohydrate lactose)?

ANSWER: The difference lies in the function of the two pathways. The *trp* operon is a biosynthetic operon. Errors in the biosynthetic pathway will lead to a failure to produce tryptophan. Therefore, a *trp* auxotrophic mutant will grow on defined medium <u>only</u> if tryptophan is added. The lactose operon involves the catabolism of a carbon source, lactose. If any of these genes are damaged, the cells are no longer able to use lactose as a carbon source. A *lac* mutant will not grow on defined medium with lactose as the sole carbon source.

4.5 If lactose was left out of MacConkey medium (Fig. 4.15), would lactose-fermenting *E. coli* bacteria grow, and if so, what color would their colonies be?

ANSWER: Even without lactose in the medium, *E. coli* would grow nonfermentatively on the peptides present. The colonies would appear white because the cells do not make acidic products needed to bring neutral red into the colony.

4.6 The addition of sheep's blood to agar produces a very rich medium called blood agar. Do you think blood agar is a selective medium? A differential medium? *Hint:* Some bacteria can lyse red blood cells.

ANSWER: Blood agar can be considered differential, because different species growing on blood have different abilities to lyse the red blood cells in the agar. Some do not lyse, others completely lyse red blood cells (secreted hemolysin produces complete clearing around a colony), while still others only partially lyse the blood (the secreted hemolysin produces a greening around the colony). It will therefore differentiate between hemolytic and nonhemolytic bacteria. The medium is very rich and supports growth of many species, so it is not considered selective.

4.7 Use the information in Figure 4.16 to determine the concentration (in cells per milliliter) of bacteria shown.

ANSWER: 1.25×10^6 bacteria per ml

SOLUTION:

- Each small square is 0.0025 mm^2 in size, and the depth from coverslip to surface is 0.2 mm.
- 0.0025 $mm^2 \times$ 0.2 mm = 0.0005 mm^3 (1 mm^3 = 1 µl). Each square defines a volume of 0.0005 µl.
- 10 cells observed over 16 squares averages to 0.625 bacteria/0.0005 µl or 0.625 bacteria/square.
- 1,000 ÷ l/ml ÷ 0.0005 µl /square = 2×10^6 squares/ml.
- 2×10^6 squares/ml × 0.625 bacteria/square = 1.25×10^6 bacteria per ml.

4.8 A virus such as influenza virus might produce 800 progeny virus particles from one infected host cell. How would you mathematically represent the exponential growth of the virus? What practical factors might limit such growth?

ANSWER: In theory, the growth rate of the virus would be proportional to 800^n. In practice, however, it is unlikely that the 800 virus particles released from one host cell will find 800 different host cells to infect. Furthermore, it turns out that only a small proportion of the influenza virus progeny are viable (see Chapter 11).

4.9 Suppose you ingest 20 cells of *Salmonella enterica* in a peanut butter cookie. They all survive the stomach and enter the intestine. Suppose further that the sum total of subsequent bacterial replication and death (caused by the host) produces an average generation time of 2 hours and you will feel sick when there are 1,000,000 bacteria. How much time will elapse before you feel sick?

ANSWER: $N = \log_{10}(N_t/N_0)/0.301$; $\log_{10}(1{,}000{,}000/20)/0.301$; 15.61×2 hours = 31 hours.

4.10 Suppose one cell of the nitrogen fixer *Sinorhizobium meliloti* colonizes a plant root. After 5 days (120 hours), there are 10,000 bacteria fixing N_2 within the plant cells. What is the bacterial doubling time?

ANSWER: 9 hours. Note that this generation time is much longer than if these same organisms were grown in a test tube containing suitable medium. In a suitable laboratory medium the generation time is about 1.5 hours.

4.11 It takes 40 minutes for a typical *E. coli* cell to completely replicate its chromosome and about 20 minutes to prepare for another round of replication. Yet the organism enjoys a 20-minute generation time growing at 37°C in complex medium. How is this possible? *Hint:* How might the cell overlap the two processes?

ANSWER: After the DNA is replicated about halfway around the chromosome, each daughter half-chromosome initiates a second round of replication, so the time needed to divide from one cell to two is effectively halved. Most cells in a log-phase culture in rich medium actually have four copies of the DNA origin of replication, each with a separate attachment site on the cell envelope, the future midpoint of a cell two generations ahead (see Chapter 3).

4.12 Figure 4.20C shows growth curves for different population densities of *Acidithiobacillus thiooxidans* when the concentration of sulfur in the medium is constant. Draw the growth curves you would expect to see if the initial population density was constant but the concentration of sulfur varied.

ANSWER:

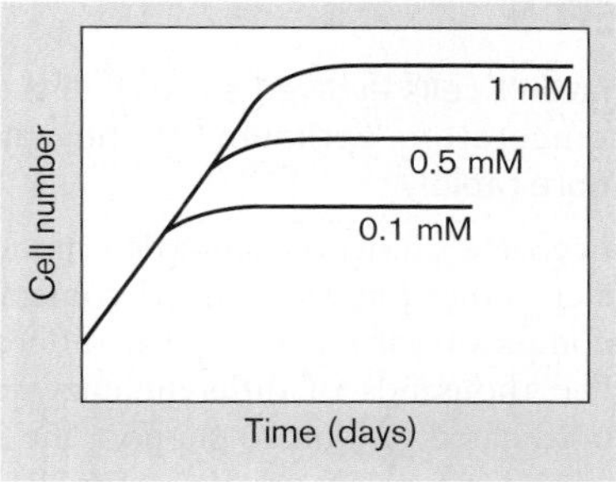
1 mM
0.5 mM
0.1 mM
Cell number
Time (days)

4.13 What can happen to the growth curve when a culture medium contains two carbon sources, if one is a preferred carbon source of growth-limiting concentration and the second is a nonpreferred source?

ANSWER: There are two possibilities. If the enzyme systems needed to utilize both carbon sources are always made, the growth curve will look normal because both will be used simultaneously. Usually the enzyme system for the nonpreferred carbon source is not produced until the preferred source is used up. In this case, a second lag phase will interrupt the exponential phase. This is called a diauxic growth curve and is commonly seen when cells are grown on both glucose and lactose. Lactose is the nonpreferred carbon source and is used second. The second lag phase marks the exhaustion of one nutrient and the gearing up by the cell to use the other (see Chapter 10).

4.14 How would you modify the equations describing microbial growth rate to describe the rate of death?

ANSWER: The death rate applies to a period of declining cell numbers. Therefore, the logarithm of the cell number ratio of N_1 to N_0 will be a negative number, and this factor will need to be preceded by a negative sign to convert it to a positive "halving time," or half-life of the culture.

4.15 Why are cells in log phase larger than cells in stationary phase?

ANSWER: Cells strive to maintain a certain DNA/mass ratio. In so doing, they balance the number of biochemical processes needed to sustain viability. If the cell mass becomes large relative to the number of copies of a given, critical gene, the amount of enzyme produced may not be sufficient to keep the cell alive and growing. In addition, the DNA/mass ratio serves as a signal to trigger cell division. Thus, when a cell divides faster than it replicates its chromosome, it must start a second round of replication before it finishes the first. This type of replication ensures that at least one chromosome duplication will be complete at the time of division. Because fast-growing cells contain more than one chromosome, they will increase in size to maintain the desired DNA/mass ratio. If the ratio were not maintained, cell division would not occur when needed.

4.16 How might *Streptomyces* and *Actinomyces* species avoid committing suicide when they make their antibiotics?

ANSWER: Bacteria that produce antibiotics need to make defenses against the antibiotic within their own cytoplasm. For example, their genes can express an altered form of the target molecule, such as a ribosomal subunit; or they can make pumps to pump the antibiotic out of the cell.

CHAPTER 5

5.1 Why haven't cells evolved so that all their enzymes have the same temperature optimum? If they did, wouldn't they grow even more rapidly?

ANSWER: An enzyme's function is not determined by temperature alone. There are other physicochemical constraints based on the variety and complexity of functions that different enzymes must carry out. The thousands of different enzyme molecules must work in a coordinated fashion to support the basic functions of life. Having some enzymes work below or above their optimum temperatures will alter the rate of the reactions they catalyze. A population's evolution is based on an entire organism's ability to reproduce, not the speed at which each individual chemical reaction is carried out. The primary goal of a microbe is not just to grow fast but also to survive. Growing too fast could deplete food sources and produce toxic by-products too quickly.

5.2 If microbes lack a nervous system, how can they sense a temperature change?

ANSWER: Most bacteria respond to outside stimuli, such as heat, by altering their gene expression. They sense heat by monitoring the concentration of misfolded proteins, a consequence of too high a temperature. The mechanism does not perceive heat per se, but recognizes the deleterious effects of moving outside the optimum growth temperature range so that the cell can launch an emergency response. The same mechanisms can sense other environmental stresses that misfold proteins, such as acid stress. See Chapter 10.

5.3 What could be a relatively simple way to grow barophiles in the laboratory?

ANSWER: High pressures can be maintained using a syringe. Scientists would place medium in a stainless steel syringe and maintain pressure with a calibrated vise.

5.4 How might the concept of water availability be used by the food industry to control spoilage?

ANSWER: Food preservation traditionally includes water exclusion by salt, as seen in hams, back bacon, and salted fish; or by high concentration of sugar, as in canned fruit or jellies. The lower a_w prevents microbial growth. Dehydrating foods will also prevent microbial growth.

5.5 Recall from Section 4.2 that an antiporter couples movement of one ion down its concentration gradient with movement of another molecule uphill, against its gradient. If this is true, how could a Na^+/H^+ antiporter work to bring protons into a haloalkaliphile growing in high salt at pH 10? Since the Na^+ concentration is lower inside the cell than outside and H^+ concentration is higher inside than outside (see Fig. 5.16), both ions are moving against their gradients.

ANSWER: In this situation the cell has to expend some energy to make the antiporter work. The energy involved is rooted in the charge difference between the inside of the cell (negative charge) and the outside of the cell (positive charge, called delta psi, $\Delta\psi$; see Chapter 14). The antiporter in this case must also exchange a different number of Na^+ and H^+, maintaining an electrical charge difference across the membrane; for example, export $2Na^+$ and import $1H^+$.

5.6 If anaerobes cannot live in oxygen gas, how do they incorporate oxygen into their cellular components?

ANSWER: Obligate anaerobes incorporate oxygen from their carbon sources (for example, CO_2 and carbohydrates such as glucose), all of which contain oxygen. This form of oxygen will not damage the cells.

5.7 How can anaerobes grow in the human mouth when there is so much oxygen there?

ANSWER: A synergistic relationship exists between facultatives and anaerobes within a tooth biofilm. The facultative anaerobes consume oxygen within the biofilm microenvironment, which allows the anaerobes to grow underneath them.

5.8 What evidence led people to think about looking for anaerobes? *Hint:* Look up "Spallanzani," "Pasteur," and "spontaneous generation" on the Internet.

ANSWER: The priest Lazzaro Spallanzani (1729–1799), during his quest to disprove spontaneous generation, said, "Every beast on Earth needs air to live, and I am going to show just how animal these little animals are by putting them in a vacuum and watching them die." He dipped a glass tube into a culture, sealed one end, and attached the other to a vacuum. He was astonished to find that the microbes lived for weeks. He then wrote, "How wonderful this is. For we have always believed there is no living being that can live without the advantages air offers it. " Fifty years later, Louis Pasteur observed that air could kill some organisms. After looking at a drop of liquid from a fermentation culture, he wrote, "There is something new here—in the middle of the drop they are lively, going every which way, but on the edge they were stiff as pokers."

5.9 What happens if excess nitrogen is added to the culture illustrated in Figure 5.23? What about excess phosphate and excess nitrogen?

ANSWER: Excess nitrogen would give species A a nutrient advantage, and it might outgrow species B. An excess of both P and N would favor neither strain. Other factors would play more dominant roles, instead, such as relative growth rates, relative abilities to use alternative carbon sources, etc.

5.10 Why would a bacteriostatic antibiotic that only inhibits growth of a pathogenic bacterium be useful for treating an infection? *Hint:* Is the human body a quiet bystander during an infection?

ANSWER: The bacteriostatic agent stops growth of the bacteria and allows the host immune response to kill them.

5.11 If a disinfectant is added to a culture containing 1×10^6 CFUs per milliliter and the D-value of the disinfectant is 2 minutes, how many viable cells are left after 4 minutes of exposure?

ANSWER: 1×10^4 CFUs per ml (90% after 2 min = 1×10^5; 90% after another 2 min = 1×10^4).

5.12 How would you test the killing efficacy of an autoclave?

ANSWER: Construct a death curve by measuring survival of a known quantity of spores (for example, *Bacillus stearothermophilus*) after autoclaving for various lengths of time. Spores should be used because they are more resistant to heat than is any vegetative cell. Typically, autoclaves are regularly checked with spore strips that change color once the endospores are no longer viable.

CHAPTER 6

6.1 Which viruses do you know that have a narrow host range, and which have a broad host range?

ANSWER: Examples of viruses with a narrow host range include poliovirus (poliomyelitis), which infects only humans and chimpanzees; smallpox virus, which infects only humans; and feline T-cell leukemia virus, which infects only cats. Examples of viruses with a broad host range include rabies virus, which infects numerous species of mammals; and influenza strains, which show preference for particular species but can jump between various mammals and birds.

6.2 What will happen if a virus particle remains intact within a host cell and fails to release its genome?

ANSWER: In most cases, a virus particle that fails to release its genome will be unable to reproduce, because DNA polymerases cannot reach its genome for reproduction and RNA polymerase cannot transcribe its genes to make gene products. An exception is that viruses with double-stranded RNA genomes keep their genome partly enclosed in order to protect it from recognition by the host cell immune system.

6.3 For a viral capsid, what is the advantage of an icosahedron, as shown in Figure 6.9, instead of some other polyhedron?

ANSWER: The icosahedron is a polyhedron of 20 triangular faces, the largest number possible. Thus, the icosahedron turns out to be the largest and most economical form to enclose space with a small repeating unit. Natural selection probably favored viruses that could build the largest capsid from the smallest amount of genetic information.

6.4 How could viruses with different kinds of genomes (RNA versus DNA) combine and share genetic content in their progeny?

ANSWER: DNA viruses require messenger RNA intermediates to express their proteins. In a rare event, a DNA virus could mutate and evolve the ability to package its RNA transcript in a capsid, rather than its DNA. Alternatively, RNA retroviruses form DNA intermediates within their host cell; these DNA intermediates might recombine with the DNA genome of another virus.

6.5 What are the relative advantages and disadvantages (to a virus) of the slow-release strategy, compared with the strategy of a temperate phage, which alternates between lysis and lysogeny?

ANSWER: A disadvantage of slow release is that the phages can never reproduce progeny phages as rapidly as in a lytic burst. The drain on resources of the host cell infected by a slow-release virus causes it to grow more slowly compared with uninfected cells; in contrast, a lysogenized cell suffers little or no reproductive deficit compared with uninfected cells. An advantage of reproduction by slow release is the continuous release of phage, while avoiding the possibility of releasing all particles into an environment where no other host cell exists.

6.6 How might a phage evolve resistance to the CRISPR host defense (outlined in Fig. 6.21)?

ANSWER: The phage could have a mutation in its DNA sequence that was cleaved to form the spacer. Now when the Cas-crRNA complex forms, the crRNA will no longer base-pair correctly with the phage DNA and will not cleave the DNA of the infecting phage.

6.7 How could humans undergo natural selection for resistance to rhinovirus infection? Is such evolution likely? Why or why not?

ANSWER: Resistance to all rhinovirus infections might evolve through a mutation in the host gene encoding ICAM-1. The mutation would have to prevent rhinovirus binding without impairing the protein's ability to bind integrin. Such evolution is unlikely because of the importance of integrin binding and because rhinovirus infection is rarely fatal; thus, there is little selection pressure to evolve inherited resistance. Note, however, that the immune system rapidly generates immunity to particular strains of rhinovirus. Over a lifetime, most individuals acquire immunity to many rhinovirus strains but remain susceptible to others.

6.8 From the standpoint of a virus, what are the advantages and disadvantages of replication by the host polymerase, compared with using a polymerase encoded by its own genome?

ANSWER: An advantage of using the host polymerase is energetic: The virus avoids the energetic cost of manufacturing a polymerase to package with each virion. This is an advantage to the virus because its reproductive potential is limited by the energy resources of its host cell. Furthermore, because DNA and RNA polymerases are so central to cell function, the host species is unlikely to evolve a mutant form of the polymerase that resists the virus. On the other hand, the advantage of the virus making its own polymerase is that the viral polymerase can evolve traits that better meet the needs of its own replication, such as high speed and low accuracy to generate frequent variants. One disadvantage of a DNA virus using the host cell DNA polymerase is that the virus must gain access to the host cell nucleus where the polymerase is. In addition, if the host cell is fully differentiated, it has exited the cell cycle and is not replicating. If it is not replicating, a DNA polymerase may not be available unless the virus can force the host cell to start going through the cell cycle. These are two problems that the virus will not have to overcome if it brings in its own DNA polymerase.

6.9 Why does bacteriophage reproduction give a step curve, whereas cellular reproduction generates an exponential growth curve? Could you design an experiment in which viruses generate an exponential curve? Under what conditions does the growth of cellular microbes give rise to a step curve?

ANSWER: Lytic viruses appear to make a step curve because the number of progeny per infected cell is 100 or more, released simultaneously. After two or three generations the cell cycles would fall out of synchrony, and the curve would smooth out, but the later cell cycles are rarely observed in practice because by then the supply of host cells is exhausted. If, however, an extremely low ratio of viruses to host cells is provided, the growth of virus particles will eventually generate an exponential curve. By contrast, the growth of cellular microbes is rarely observed during the first few doublings. By the time we measure the population, the cells are all undergoing different stages of division, and the population growth overall generates a smooth exponential curve. But if we observe the growth of a synchronized population of cells, we see a step curve of cell division too.

6.10 Suppose a certain virus depletes the population of an algal bloom. If some of the algae are selected for resistance, will they grow up again and dominate the producer community?

ANSWER: If some algae are resistant to the virus, they will reproduce and avoid infection. But their population growth will still face competition from other algal species that were never hosts of the virus. Furthermore, if the resistant algae form another bloom, eventually some other species of virus will infect them and cut their population again.

CHAPTER 7

7.1 What do you think happens to two single-stranded DNA molecules isolated from <u>different</u> genes when they are mixed together at very high concentrations of salt? *Hint:* High salt concentrations favor bonding between hydrophobic groups.

ANSWER: In high salt conditions, the stacking of hydrophobic bases is so strongly favored that two single strands of DNA will form a duplex no matter what the sequence of base pairs is.

7.2 How do the kinetics of denaturation and renaturation depend on DNA concentration?

ANSWER: The speed of denaturation does not depend on DNA concentration, but the speed of renaturation does. The higher the concentration of ssDNA, the more likely it is that complementary sequences will find each other and the faster the duplex can re-form.

7.3 DNA gyrase is essential to cell viability. Why, then, are nalidixic acid–resistant cells that contain mutations in *gyrA* still viable?

ANSWER: The *gyrA* mutations alter only the nalidixic acid binding site on GyrA, not its gyrase activity. In other words, active DNA gyrase is still made, but the drug cannot bind to it.

7.4 Bacterial cells contain many enzymes that can degrade linear DNA. How, then, do linear chromosomes in organisms like *Borrelia burgdorferi* (the causative agent in Lyme disease) avoid degradation?

ANSWER: DNA-digesting exonucleases act on free 5′ or 3′ ends. The *Borrelia* linear chromosomes possess covalently closed hairpin ends called telomeres and do not possess free 5′ or 3′ groups.

7.5 Suppose you can label DNA in a bacterium by growing cells in medium containing either ^{14}N nitrogen or the heavier isotope, ^{15}N; you can isolate pure DNA from the organism; and you can subject DNA to centrifugation in a cesium chloride solution, a solution that forms a density gradient when subjected to centrifugal force, thereby separating the light (^{14}N) and heavy (^{15}N) forms of DNA to different locations in the test tube. Given these capabilities, how might you prove that DNA replication is semiconservative?

ANSWER: Grow bacteria in medium containing ^{15}N, so that all the DNA made by the cells will be in the heavy form. Transfer the cells to medium containing only ^{14}N and allow the cells to divide for one generation. Extract the DNA and centrifuge the preparation in cesium chloride. If replication is semiconservative, a hybrid DNA band will be seen at a position located between and equidistant from the point where heavy DNA and light DNA would be located. The hybrid band will be composed of one heavy strand and one light strand, making it of intermediate weight (density). This, in fact, was how Meselson and Stahl proved semiconservative replication in 1958, 5 years after the discovery that DNA is double-stranded (**eTopic 7.5**).

7.6 How fast does *E. coli* DNA polymerase synthesize DNA (in nucleotides per second), if the genome is 4.6 million base pairs, replication is bidirectional, and the chromosome completes a round of replication in 40 minutes?

ANSWER: It takes *E. coli* 40 minutes (2,400 seconds) to complete the replication of its 4,639,221-bp chromosome. Since Pol III works as a dimer to synthesize the two strands simultaneously, we will consider only one strand in the calculation. The rate is 4,639,221 nucleotides (nt) per 2,400 seconds = 1,933 nt per second. But there are two replication forks, so each polymerase dimer synthesizes only one-half of the chromosome, which means that each polymerase still synthesizes a remarkable 800–1,000 nt per second.

7.7 Reexamine Figure 7.16. If you GFP-tagged a protein bound to the *ter* region, where would you expect fluorescence to appear in the cell with two origins?

ANSWER: The *ter* region is the last region of the chromosome to be replicated. It will appear at midcell, between the two origins.

7.8 Individual cells in a population of *E. coli* typically initiate replication at different times (asynchronous replication). However, depriving the population of a required amino acid can synchronize reproduction of the population. Ongoing rounds of DNA synthesis finish, but new rounds do not begin. Replication stops until the amino acid is once again added to the medium—an action that triggers simultaneous initiation in all cells. Why does this synchronization of reproduction happen?

ANSWER: Initiation requires synthesis of the initiator protein DnaA. Depriving the population of an amino acid prevents protein synthesis, which precludes synthesis of DnaA. Because DnaA is not required to complete already-initiated rounds of replication, all rounds already started are completed, but reinitiation cannot occur. Adding the amino acid once again will allow all cells to simultaneously make DnaA so that initiation is triggered in every cell at the same time.

7.9 The antibiotic rifampin inhibits transcription by RNA polymerase, but not by primase (DnaG). What happens to DNA synthesis if rifampin is added to a synchronous culture?

ANSWER: Initiation of DNA synthesis requires primer transcription at the origin by RNA polymerase, an enzyme sensitive to rifampin. Primase (DnaG), which synthesizes RNA primers in the lagging strand throughout DNA synthesis, is resistant to rifampin. So adding rifampin to a synchronized culture will prevent new rounds of DNA replication but will not affect already-initiated rounds.

7.10 Figure 7.26B described how agarose gels separate DNA molecules. If you ran a highly supercoiled plasmid in one lane of the gel and the same plasmid with one phosphodiester bond cut in another lane, where would the two molecules end up relative to each other in this type of gel?

ANSWER: Removing one phosphodiester bond from a supercoiled plasmid will unravel (relax) all supercoils. Even though both plasmids will have the same DNA sequence, the highly supercoiled plasmid will be more compact than the relaxed plasmid and will move more quickly through the pores of the agarose gel. As a result, the supercoiled plasmid will be closer to the bottom of the gel than the relaxed form.

7.11 Given that *E. coli* possesses restriction endonucleases that cleave DNA from other species, how is it possible to clone a gene from one organism to another without the cloned gene sequence being degraded?

ANSWER: The best way to overcome the restriction barrier is to use a mutant strain of *E. coli* in which the restriction system has been inactivated. DNA entering the cell will not be degraded and, if the modification system is still in place, will be modified. Once modified, that DNA can be safely transferred to other *E. coli* strains that still have their restriction endonucleases.

7.12 PCR is a powerful technique, but a sample can be easily contaminated and the wrong DNA amplified—perhaps sending an innocent person to jail. What might you do to minimize this possibility?

ANSWER: Use sterile, filtered micropipette tips. The filters prevent contamination of the micropipette and subsequent reactions. In addition, perform a control PCR reaction without adding a DNA sample. If the water, buffers, or tubes are contaminated with a given DNA, a PCR product will appear even when DNA template has not been added deliberately to the reaction.

CHAPTER 8

8.1 If each sigma factor recognizes a different promoter, how does the cell manage to transcribe genes that respond to multiple stresses, each involving a different sigma factor?

ANSWER: In these situations, a given gene will have multiple promoters. Each promoter will be recognized by a different sigma factor and will begin transcription at different distances from the start codon of the gene.

8.2 Imagine two different sigma factors with different promoter recognition sequences. What would happen to the overall gene expression profile in the cell if one sigma factor were artificially overexpressed? Could there be a detrimental effect on growth?

ANSWER: Since sigma factors compete for the same site on core polymerase, overexpressing one sigma factor could displace the other sigma factors from the RNA polymerase population and compromise expression of those target genes. If those genes were important to survival, the cell could die.

8.3 Why might some genes contain multiple promoters, each one specific for a different sigma factor?

ANSWER: The gene might need to be expressed under multiple conditions at different levels. If a given condition increases expression of an alternate sigma factor, the target gene will need a promoter that the new sigma factor can recognize. As the need disappears and the sigma factor diminishes in concentration, a promoter that uses the housekeeping sigma factor will be needed. For example, the gene for DnaK heat-shock protein has promoters for RpoH (sigma-32) and RpoD (sigma-70), the housekeeping sigma factor. The level of protein needed during normal growth is supplied by sigma-70. Upon encountering heat stress, the RpoH sigma factor level increases and mediates an increase in DnaK production.

8.4 How might the redundancy of the genetic code be used to establish evolutionary relationships between different species? *Hints:* 1. Genomes of different species have different overall GC content. 2. Within a given genome one can find segments of DNA sequence with a GC content distinctly different from the rest of the genome.

ANSWER: The codon preferences of different microorganisms are based in part on their GC content. Thus, an organism with an AT-rich genome will preferentially use codons for a given amino acid that have A's and T's over those with G's and C's. Evolutionarily, finding a long AT-rich region that encodes mRNA with AT codon bias within a chromosome that is otherwise GC-rich suggests that the AT-rich region was inherited by horizontal DNA transfer from another species. (See Chapter 9.)

8.5 The synthesis of ribosomal RNAs and ribosomal proteins represents a major energy drain on the cell. How might the cell regulate the synthesis of these molecules when the growth rate slows because amino acids are limiting? *Hints:* Predict what happens to any translating ribosome when an amino acid is limiting. Look up RelA protein on the Internet.

ANSWER: As energy levels in a cell fall, fewer amino acids are made. The decrease in amino acid production sets off a chain of events starting with fewer charged tRNA molecules being assembled because of a lack of amino acids. Fewer available charged tRNA molecules means ribosomes have to pause during translation until they find the right charged tRNA. This pause causes the synthesis of a signal molecule called guanosine tetraphosphate (ppGpp).

This signal molecule is made from GTP and ATP by a ribosome-associated protein called RelA. Among other effects, ppGpp interacts with RNA polymerase and selectively stops transcription of the rRNA genes. The subsequent decrease in rRNA also stops new ribosome synthesis. How? Fewer rRNAs made means fewer available rRNA binding sites for ribosomal proteins. As a result, ribosomal proteins do not assemble in ribosomes and end up accumulating in the cytoplasm. These free ribosomal proteins bind to sequences in their own mRNA molecules that are similar to the target sequences they would bind to on rRNA. By binding to their own mRNA, these ribosomal proteins can prevent their own translation, so now fewer rRNA molecules, fewer ribosomal proteins, and fewer new ribosomes are made in the cell. (See Chapter 10.) Finally, as cells continue to divide (even though at a slower rate because of amino acid deprivation), the remaining ribosomes will dilute to a lower steady-state level—that is, to a level suitable for sustaining the new growth rate.

8.6 Figure 8.20 illustrates an operon and its relationship to transcripts and protein products. Imagine that a mutation that stops translation (TAA, for example) was substituted for a normal amino acid about midway through the DNA sequence encoding gene A. What happens to the expression of the gene A and gene B proteins?

ANSWER: The part of the gene A protein only up until the stop codon will be made. The complete mRNA transcript, however, will be made and will include gene B (there are some rare exceptions to this). Gene B has its own translation start codon, so the complete gene B protein can be produced.

8.7 While working as a member of a pharmaceutical company's drug discovery team, you find that a soil microbe snatched from the jungles of South America produces an antibiotic that will kill even the most deadly, drug-resistant form of *Enterococcus faecalis,* which is an important cause of heart valve vegetations in bacterial endocarditis. Your experiments indicate that the compound stops protein synthesis. How could you more precisely determine the antibiotic's mode of action? *Hints:* Can you use mutants resistant to the antibiotic?

ANSWER: One way is to take a culture of bacteria susceptible to the antibiotic and isolate resistant mutants (bacteria that are not killed by the antibiotic), purify their ribosomes, and separate the 30S and 50S ribosomal subunits. Cross-mix subunits from sensitive and resistant cells (for example, mix 30S subunits from sensitive cells with 50S subunits from resistant cells). Then measure the ability of the hybrid ribosome to carry out protein synthesis with and without the drug. If resistance is due to an altered ribosomal protein or RNA, the subunit mix containing the altered component will make protein regardless of whether the drug is present. Once identified, the responsible ribosomal subunits from resistant and sensitive cells can be broken down further into their component parts, reconstituted in hybrid form, and again tested for an ability to make protein in the presence of drug. This reductive approach will likely, but not always, uncover the target ribosomal protein or rRNA.

8.8 How might one gene code for two proteins with different amino acid sequences?

ANSWER: One gene can code for two proteins with different amino acid sequences by having two different translation start sites in different reading frames. While this is not a common occurrence, it happens. Hepatitis B virus is one example.

8.9 Why involve RNA in protein synthesis? Why not translate directly from DNA?

ANSWER: Because transcription enables the cell to amplify the gene sequence information into multiple copies of RNA. Amplification means that more ribosomes can be engaged in translating the same protein, causing the concentration of the protein to rise more quickly than if only a single gene were used. The transcriptional process also makes it possible to more easily regulate the production of a protein.

8.10 Codon 45 of a 90-codon gene was changed into a translation stop codon, producing a shortened (truncated) protein. What kind of mutant cell could produce a full-length protein from the gene <u>without</u> removing the stop codon? *Hints:* What molecule recognizes a codon?

ANSWER: If a tRNA gene sequence corresponding to an anticodon is altered by mutation so that the anticodon of the tRNA "sees" the stop codon as an amino acid codon, then the mutant cell could produce a full-length protein from the gene. The mutated tRNA molecule will transfer its amino acid to the peptide chain. The stop codon is still there but now it can direct the addition of an amino acid. The attached amino acid can be used to bridge the gap caused by the stop codon, and a full-length protein is made. These modified tRNAs are called suppressor tRNAs because they suppress the mutant phenotype.

8.11 What would happen if the tyrosine residue UAC in the mRNA-like domain of tmRNA was altered by mutation to UAA?

ANSWER: The peptide stuck on the ribosome would terminate early (UAA is a translation termination codon), but the shortened peptide tag would likely be insufficient to target the defective protein for degradation. The accumulation of this nonfunctional protein could have detrimental effects on cell growth.

8.12 An ORF 1,200 bp in length could encode a protein of what size and molecular weight? *Hints:* Find the molecular weight of an average amino acid.

ANSWER: There are three bases per codon, so the ORF can encode 400 amino acids. The average molecular weight of an amino acid is 110 Da. Therefore, the hypothetical protein will be approximately 44,000 Da (44 kDa).

CHAPTER 9

9.1 Figure 9.1 illustrates the process of transformation in *Streptococcus pneumoniae*. Will a mutant of *Streptococcus* lacking comD be able to transform DNA?

ANSWER: No. The membrane sensor ComD detects the competence factor (CF) and initiates a cascade of events leading to transformasome construction. No ComD, no transformasome, and no DNA transformation.

9.2 Transfer of an F factor from an F^+ cell to an F^- cell converts the recipient to F^+. Why doesn't transfer of an Hfr do the same?

ANSWER: The last piece of an Hfr to transfer is the F factor and *oriT*. Only rarely will an entire chromosome transfer from one cell to another, so most Hfr transfers do not result in transfer of *oriT* and thus cannot initiate conjugation.

9.3 A prophage genome that is embedded in the chromosome of a lysogenic cell will prevent replication of any new phage DNA (from the same phage group) that has infected the cell. By preventing the replication of superinfecting phage DNA, the prophage genome protects itself from destruction by lysis. Is a lysogen also protected from generalized transduction carried out by a similar phage? For example, can a P22 lysogen of *Salmonella* be transduced by P22 grown on a different strain of *Salmonella*? Explain why or why not.

ANSWER: Yes, a P22 lysogen can be transduced by P22 phage grown on a different strain of *Salmonella*. Capsids of generalized transducing phages contain bacterial DNA instead of phage DNA, so the transducing particle has no phage DNA to replicate. Furthermore, transduction does not require replication of the transferred DNA before recombination.

9.4 How do you think phage DNA containing restriction sites evades the restriction-modification screening systems of its host?

ANSWER: Phage DNA will survive because sometimes the modification enzyme reaches the foreign DNA before the restriction enzyme. Once methylated, the phage DNA will be shielded from restriction and the methylated molecule will replicate unchallenged. If, on the other hand, a foreign DNA fragment (not necessarily a phage) has been modified and the conditions are right, it might recombine into the host chromosome and convey a new character to the strain.

9.5 Each type I restriction-modification system includes separate host specificity determinant (Hsd) proteins for restriction (HsdR), modification (HsdM), and sequence recognition (HsdS). Can a plasmid grown in a wild-type strain be used to transform a strain defective in *hsdR*? What about an *hsdS* mutant? Can a plasmid grown in an *hsdM* mutant strain be transformed into an *hsdR* defective strain?

ANSWER: A plasmid grown in a wild-type strain will be methylated; but methylated or not, it can survive in a restrictionless *hsdR* mutant or an *hsdS* mutant that lacks the recognition protein needed by both the restriction and modification subunits. Since an *hsdM* mutant lacks the modification protein, it cannot protect its genes from restriction endonucleases and would commit suicide, so this last experiment cannot be done.

9.6 In a transductional cross between an $A^+B^+C^+$ genotype donor and an $A^-B^-C^-$ genotype recipient, 100 A^+ recombinants were selected. Of those 100, 15% were also B^+, while 75% were C^+. Is gene *B* or gene *C* closer to gene *A*?

ANSWER: Gene *C* is closer to gene *A*, because gene *C* was cotransduced with *A* at the higher frequency.

9.7 You have been asked to calculate the mutation rate for a gene in *Salmonella enterica*, so you dilute an 18-hour broth culture to 1×10^4 cells/ml and let it grow to 1×10^9 cells/ml. At that point you dilute and plate cells on the appropriate agar medium and estimate that there were 10,000,000 mutants in the culture. What you don't know is that the original dilution of 1×10^4 cells/ml already had 10 mutants defective in that gene. Does that affect your estimate of mutation rate? If so, then how?

ANSWER: Mutation rate is an estimate of how many new mutations arise per cell division as a culture grows. The division of preexisting mutants during growth could lead you to estimate a false high mutation rate. By knowing how many mutants existed in the beginning, you can compensate for how many mutants you would have as a result of mutant replication, and subtract that number from the total number of mutants you observe. This adjustment assumes that the growth rates of the mutant and wild-type bacteria are the same. If there were no mutations in the original dilute culture, then the mutation rate would be estimated as follows: (1×10^7 mutants) ÷ approximately 10^9 cell divisions = 10^{-2} mutation rate. This is a very high rate.

However, it takes about 17 generations (doublings) to go from 1×10^4 cells to 1×10^9 cells. If the diluted culture had 10 mutants to begin with, and if those mutants had the same generation time as the wild-type cells, then 10 cells after 17 generations would amount to about 10^6 cells. Subtracting this number from the actual numbers of mutants found gives the following: 1×10^7 (total mutants) – 0.1×10^7 (sibling mutants) = 0.9×10^7 newly derived mutants. So, (0.9×10^7 mutants) ÷ (10^9 cell divisions) = 9×10^{-3} mutation rate, a rate not much different from the original calculation. But if there were 100 mutants in the original 10,000-cell dilution, the situation would change. Now there would be 1×10^7 sibling mutants, which, subtracted from 1×10^7 total mutants, leaves no new mutants. The mutation rate would seem to be $<10^{-9}$, which is a huge difference from the previous calculation.

This exercise illustrates the importance of starting with as few cells as possible when calculating mutation rate, of doing controls to determine the number of preexisting mutants, and of performing replicates.

9.8 During transformation in *Streptococcus pneumoniae,* a single strand of donor DNA is transferred into the recipient. This single strand of DNA can be recombined into the recipient's genome at the homologous area. However, if the segment of recipient DNA had a mutation in it relative to the donor DNA strand, then the result would be a mismatch after recombination. Explain how, in the face of mismatch repair, the recipient cell ever retains the wild-type sequence from the donor.

ANSWER: Because the donor DNA was probably methylated when transferred, the methyl mismatch repair system should not distinguish between donor and recipient strands. Thus, there is equal chance that the system will repair the original recipient sequence (with a mutation) or the donor sequence. However, if the sequence was methylated differently (because there were different numbers of GATC sequences), then the mismatch repair system will preferentially remove the donor strand. The mismatch repair system in *S. pneumoniae* is called Hex and is orthologous to the *E. coli* Mut system.

9.9 It has been reported that hypermutable bacterial strains are overrepresented in clinical isolates. Out of 500 isolates of *Haemophilus influenzae,* for example, 2%–3% were mutator strains having mutation rates 100–1,000 times higher than lab reference strains. Why might mutator strains be beneficial to pathogens?

ANSWER: The mutator strains may speed microbial evolution, which could help the microbe outwit the immune system or escape the effects of administered antibiotics.

9.10 Diagram how a composite transposon like Tn*10* can generate inversions or deletions in target DNA during

THOUGHT QUESTION 9.10

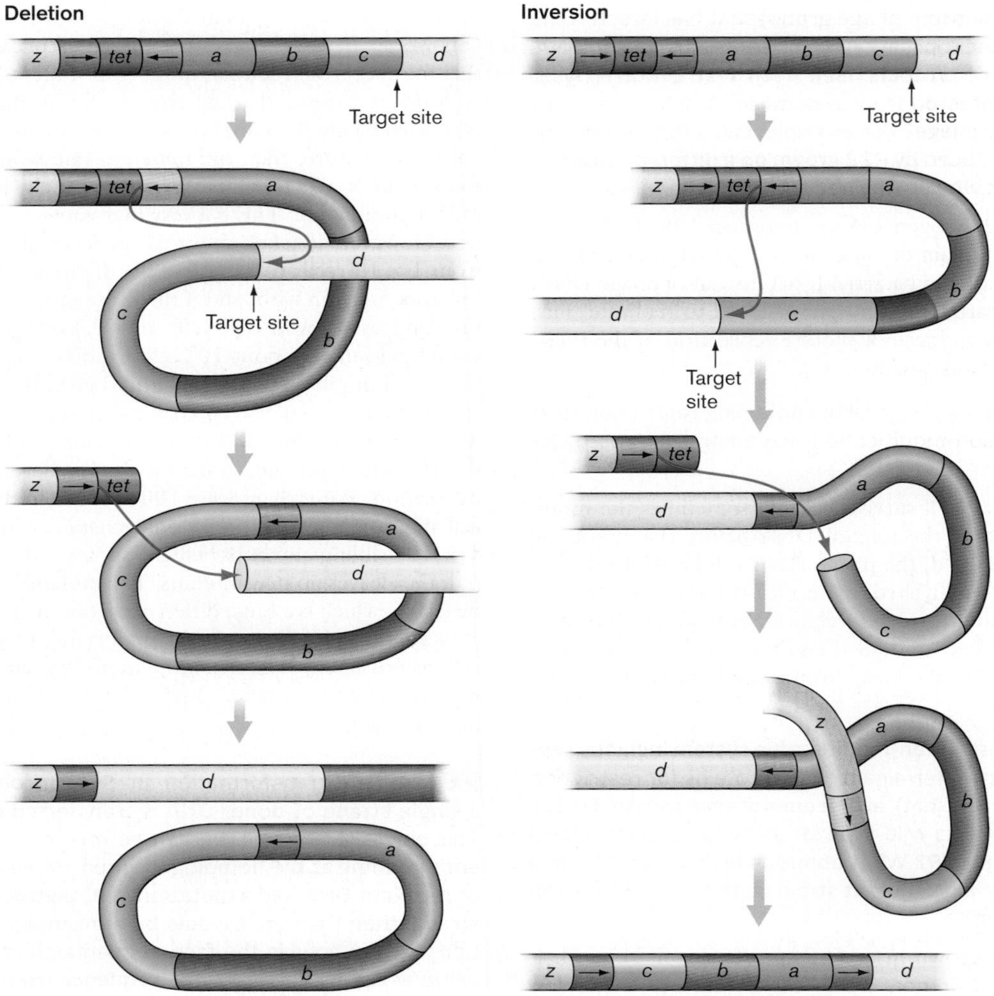
Deletion
z
tet
a
b
c
d
Target site
Inversion
Target site
Target
site

transposition. ***Hint:*** **This happens when transposition within the chromosome occurs using the inverted repeats closest to the tetracycline resistance gene (see Fig. 9.38A). If you draw it right, you will see, in the end, that the tetracycline resistance gene is lost.**

ANSWER: The accompanying diagram shows that inversions or deletions occur when the inner rather than the outer ends of the inverted repeats are the targets of the transposase. The tetracycline resistance gene is lost, and the chromosome will contain a deletion or inversion, depending on the orientation of the chromosome (looped or unlooped) at the time of transposase action.

9.11 Gene homologs of *dnaK* encoding the heat-shock chaperone HSP70 exist in all three domains of life. All bacteria contain HSP70, but only some species of archaea encode a *dnaK* homolog. The archaeal homologs are closely related to those of bacteria. Knowing this information, how do you suppose *dnaK* genes arose in archaea?

ANSWER: The *dnaK* gene is thought to have been moved by some type of horizontal gene transfer mechanism from domain Bacteria to some members of domain Archaea.

CHAPTER 10

10.1 If the gene *lacZ* has a nonsense mutation in its open reading frame, will *lacY* still be translated?

ANSWER: *LacY* mRNA will still be translated, because each gene in this polycistronic mRNA (*lacZYA*) has its own ribosome-binding site and the *lacY* ribosome-binding site is functional. A translational stop codon in an upstream gene does not usually affect transcription of the downstream RNA.

10.2 Null mutations completely eliminate the function of a mutated gene. Predict the effect of the following null mutations on the induction of beta-galactosidase by lactose, and predict whether the *lacZ* gene is expressed at high or low levels in each case: *lacI, lacO, lacP, crp,* and *cya* (the gene encoding adenylyl cyclase). What effect will those mutations have on catabolite repression?

ANSWER: Loss of LacI repressor will lead to constitutive expression of *lacZ* and will partially affect catabolite repression. [Explanation: Because LacI is missing, allolactose inducer is not required, so the glucose effect on the LacY permease is irrelevant. What remains relevant is that the mechanism by which glucose transport reduces cAMP synthesis remains (**eTopic 10.2**). Thus, decreased cAMP levels caused by growth on glucose will cause a decrease in cAMP-CRP-dependent activation of *lac* operon expression.] A *lacO* mutant will not bind LacI repressor, so the phenotype will mimic that of a *lacI* mutation. A *lacP* mutation will prevent expression of *lacZYA* because RNA polymerase will not bind. Mutations in *crp* or *cya* will partially prevent catabolite repression (see preceding explanation), but *lacZYA* induction by lactose will be normal. Without the cAMP-CRP complex, however, expression can never achieve maximal levels.

10.3 Predict what will happen to the expression of *lacZ* under the following partial diploid conditions (see Section 9.2). The genotypes of these strains are presented as chromosomal genes/plasmid gene. (a) *lacI*$^-$ *lacO*$^+$*P*$^+$*Z*$^+$*Y*$^+$*A*$^+$/plasmid lacI$^+$; (b) *lacO*$^-$ *lacI*$^+$*P*$^+$*Z*$^+$*Y*$^+$*A*$^+$/plasmid *lacO*$^+$; (c) *crp*$^-$ *lacI*$^+$*O*$^+$*P*$^+$*Z*$^+$*Y*$^+$*A*$^+$/ plasmid *crp*$^+$.

ANSWER: (a) The *lacI*$^+$ gene on the plasmid will produce LacI repressor protein that can diffuse through the cytoplasm, bind chromosomal *lacO*, and repress the *lacZYA* operon. Because the complementing gene and the mutant gene are on different DNA molecules, the gene is said to work in *trans*. (b) Because the *lacO* gene does not produce a diffusible product (for example, protein or RNA), the plasmid *lacO*$^+$ cannot complement a *lacO* mutation in *trans*, and the strain will not make beta-galactosidase. Thus, the *lacO* gene functions only in *cis*—that is, when it resides next to the gene it regulates. (c) The *crp* gene produces a diffusible protein product, so it can function in *trans* and complement a *crp* mutation. The strain will make beta-galactosidase to the highest level, in the presence of inducer lactose.

10.4 Researchers often use isopropyl-β-D-thiogalactopyranoside (IPTG) rather than lactose to induce the *lacZYA* operon. IPTG resembles lactose, which is why it can interact with the LacI repressor, but it is not degraded by beta-galactosidase. Why do you think the use of IPTG is preferred in these studies?

ANSWER: There are at least two reasons IPTG is used. First, the level of IPTG inducer will not change, but the level of lactose inducer will continually decrease as it is consumed, affecting the kinetics of induction. Second, the act of degrading lactose produces glucose and galactose. Glucose, as the preferred carbon source, will catabolite-repress the *lacZYA* operon, once again affecting the kinetics of induction.

10.5 When the *lacI* gene of *E. coli* is missing because of mutation, the *lacZYA* operon is highly expressed regardless of whether lactose is present in the medium. Based on Figure 10.13 and its illustration of arabinose operon expression, what would happen to *araBAD* expression if AraC were missing? Why?

ANSWER: The operon would be poorly expressed because contact between AraC and RNA polymerase is needed to activate transcription.

10.6 Predict the phenotype of a *spoIIAA* mutant that completely lacks SpoIIA anti-anti-sigma factor (see Fig. 10.18).

ANSWER: The *spoIIAA* mutant will have no way to efficiently remove anti-sigma F factor from sigma F. As a result, sigma F will almost never be active, even in the forespore. The cell can get as far as asymmetrical cell division to make what would be the forespore, but sigma F and anti-sigma F equally distribute in both compartments, so sigma F will not be activated and spores will not form.

10.7 The relationship between the small RNA rhyB, the iron regulatory protein Fur, and succinate dehydrogenase is shown in Figure 10.20. Based on this regulatory circuit, will a *fur* mutant grow on succinate?

ANSWER: No. Without Fur, RhyB is made whether or not iron is present, and RhyB sRNA causes the continuous degradation of the *sucCDAB* mRNA. Succinate dehydrogenase cannot be made, so the cell cannot grow on succinate.

10.8 While viewing Figure 10.24, imagine the phenotype of a cell in which *fljB* has been deleted so that *fljA* is still expressed. Would cells be motile? What type of flagella would be produced? Would the cells undergo phase variation? What would happen if *fliC* was deleted?

ANSWER: A *fljA* mutant lacks the repressor needed to turn off FliC. Thus, a cell will switch back and forth from making H2 flagellin to making both H1 and H2 flagellins. A *fliC* mutant, however, will switch from being motile to nonmotile. In one orientation, the invertible element will allow H2 flagellin to be made; but in the opposite orientation no flagellin will be made, at which point the cell will not be motile.

10.9 Using antibody to flagella, we can tether a cell of *E. coli* to a glass slide via a single flagellum. Looking through a microscope, you will then see the <u>bacterium</u> rotate in opposite directions as the flagellar rotor switches from clockwise to counterclockwise rotation and back again. (See the "tethered *E. coli*" video at the following Internet link.) Which way will the bacillus rotate when an attractant is added? What would the interval between clockwise and counterclockwise switching be if you tethered the mutants *cheY, cheA, cheZ,* and *cheR* to the slide and then added attractant? (See Fig. 10.25.)

ANSWER: When an attractant is added to the slide, the rotor will turn counterclockwise for smooth swimming (CheA becomes less active, so there is less CheY-P and there are less frequent tumbles; thus, motor rotation is biased to counterclockwise). But, because the flagellum is fixed to the slide, the bacillus will rotate in the <u>opposite</u> direction (clockwise). You would expect the following rotating phenotypes from mutants fixed to slides: For mutant *cheY*, the rotor will turn mostly counterclockwise because there is no CheY-P, so there will be longer runs, but fixed cells will turn mostly clockwise. Mutant *cheA* will have the same phenotype as *cheY* (no CheY-P, longer runs). For mutant *cheZ*, the rotor will turn mostly clockwise, because there is more CheY-P (so there will be frequent tumbles and shorter runs), but the fixed cells will turn counterclockwise. For mutant *cheR*, no methylation of MCPs will reactivate CheA kinase after attractant is added, so there will be less CheY-P; therefore, the rotor will more frequently switch to clockwise, but fixed cells will turn counterclockwise.

10.10 Predict the phenotype of a *glnB* mutant. Will it be a glutamine auxotroph? What about a *glnD* or *glnE* mutant?

ANSWER: A *glnB* mutant will not dephosphorylate NtrB or remove AMP from GlnA. The cell will continue to make GlnA and fail to inactivate it. Thus, the cell will overproduce glutamine. A *glnD* mutant, however, will not add UMP to GlnB, so GlnB will not direct GlnE to remove AMP from GlnA (remember: GlnA without AMP is active). As a result, GlnA will remain inactive and the cell will likely require glutamine. The same will occur with a *glnE* mutant.

10.11 Genes encoding luciferase can be used as "reporters." What would happen if the promoter for an SOS response gene was fused to the luciferase open reading frame?

ANSWER: It could serve as a real-time biosensor for an environmental stress response. You could fuse the luciferase gene to the *recA* promoter and examine a real-time increase in fluorescence after ultraviolet irradiation by inserting the whole culture into a spectrofluorometer.

10.12 What would happen if a culture was coinoculated with a *Aliivibrio fischeri luxI* mutant and a *luxA* mutant, neither of which produces light?

ANSWER: The *luxI* mutant would glow, and the *luxA* mutant would still make autoinducer as it grew. This autoinducer would accumulate in the culture medium, and then diffuse and enter the *luxI* mutant cells, where it would trigger induction of the *lux* operon and the production of luciferase.

10.13 Can tandem mass spectrometry (MS-MS) identify where modifications such as phosphorylation or acetylation happen on proteins?

ANSWER: Yes. Recall that the fragmentation of peptides by ionization will release single amino acids. Unambiguous identification of the amino acid lost is based on the mass difference between, for example, a five-amino-acid peptide and the four-amino-acid derivative peptide (the mass of alanine, for instance, is different from the mass of glycine). The presence of a side group like phosphate will change the weight of that amino acid and signal the presence of phosphate.

CHAPTER 11

11.1 The phage T4 chromosome is a linear piece of DNA, yet the genomic map of T4 (as shown in Fig. 11.3) is circular. Why?

ANSWER: Each phage head contains just enough space to package slightly more than the length of a genome. The head-filling mechanism of DNA packaging means that successive genomes on a concatemer each get cut at different positions within the genome sequence. Thus, any map of gene order based on recombination linkages between genes, or based on the sequencing of a population of genome molecules, will generate a circular map.

11.2 Why would phage T4 production require a rate of DNA replication tenfold higher than that of the host cell? Why would the phage substitute all of its cytosine with an unusual base that requires more energy to synthesize?

ANSWER: The phage T4 genome needs to replicate itself and generate progeny as fast as possible, in order to use as much of the host cell's resources as possible before the cell deteriorates. By replacing cytosine with a modified base, the phage can avoid degradation of its chromosome by its own endonucleases (which cleave host DNA) or by those of the host (which are made to cleave phage DNA).

11.3 Which numbered genes from Figure 11.7 might be mutated in each of the defective phage populations presented in Figure 11.8D?

ANSWER: Mutant 1 may have a defect in a gene whose product is needed for tail fiber attachment, such as gene 63. Mutant 2 may be defective for tail fiber assembly (gene 34 or 57). Mutant 3 could be defective for sheath assembly around the internal tube (gene 18).

11.4 If HCV virions develop a quasispecies under strong selection for propagation within a single host, how do isolated progeny virions retain the ability to initiate infection of a new host?

ANSWER: The answer is not fully understood, but the following model is proposed. Over time, rapid mutation and selection lead to a quasispecies population in which host-adapted mutants predominate. But random mutation continues, and the population of virions is so large that in any given population there will always be a few that revert to a form capable of infecting the next host.

11.5 Why does influenza virus have to provide its polymerase ready-made, whereas hepatitis C virus does not?

ANSWER: The influenza viral chromosome is (–) strand RNA. When it enters the host cytoplasm, it cannot be translated by ribosomes, and no host polymerase makes RNA from RNA. Therefore, the virion must provide an RNA-dependent RNA polymerase to generate a (+) strand RNA for translation to proteins. HCV, however, injects (+) strand RNA, whose IRES can immediately initiate translation by host ribosomes to generate RNA-dependent RNA polymerase (protein NS5B).

11.6 Explain why the expression $8!/8^8$ approximates the proportion of infective particles of influenza virus. What assumption might be changed to make the proportion greater or less?

ANSWER: An infective particle can start by packaging any one of the eight chromosomes. Once the first is chosen, seven possibilities remain for the second, six for the third, and so forth. Thus, there are 8! ways to obtain a perfect set out of 8^8 total ways to pick eight segments at random. This expression assumes, however, that exactly eight segments are packaged in every virion. If the average number of segments packaged is less than eight, the proportion of infectives falls off. If the number of segments is usually more than eight, the proportion of virions containing at least one of each segment increases.

11.7 How do attachment and entry of HIV resemble attachment and entry of influenza virus? How do attachment and entry differ between these two viruses?

ANSWER: Attachment of HIV requires the envelope spike proteins to bind receptors in the host plasma membrane, just as the influenza envelope protein hemagglutinin binds the sialic acid protein in the host membrane. However, the entry processes differ between the two viruses. Influenza virus induces formation of an endocytic vesicle, whose acidification triggers membrane fusion and release of the core contents into the host cytoplasm. HIV virions, however, do not induce endocytosis and do not require acidification to induce membrane fusion and release of the core into the cytoplasm. In both cases, receptor binding signals the major rearrangement of a viral envelope protein so that a fusion peptide is inserted into the host membrane: for HIV, it is the plasma membrane; for influenza virus, it is the endocytic membrane.

11.8 Compare and contrast the fate of the HSV chromosome with that of the HIV chromosome.

ANSWER: HSV-1 contains a DNA chromosome, which is transported to the nuclear membrane within an intact capsid. HIV contains two RNA chromosomes, which are released in the cytoplasm upon dissolution of the capsid. The RNA chromosomes of HIV are copied to double-stranded DNA for transport into the nucleus. In HSV-1, the DNA chromosome circularizes and generates concatemeric duplicates by rolling-circle replication. In HIV, the replicated DNA circularizes but immediately integrates into the host chromosome. In both cases, the viral DNA can persist for decades as a latent infection.

CHAPTER 12

12.1 How would you modify the technique illustrated in Figure 12.1 to identify mutants defective in the synthesis of an amino acid, such as glutamic acid? *Hint:* You can use a minimal agar medium.

ANSWER: Cells subjected to transposon mutagenesis are plated onto rich agar medium. Colonies that grow are pressed onto sterile velveteen pads that cover a solid cylinder, as shown in the figure below (use a different pad for each plate). The cylinder is used to transfer the colonies to minimal agar plates with and without glutamic acid. After growth, the plates are visually aligned and any colonies able to grow on the glutamate plate but not on the unsupplemented plate are identified. These mutants require glutamate to grow and are called glutamate auxotrophs.

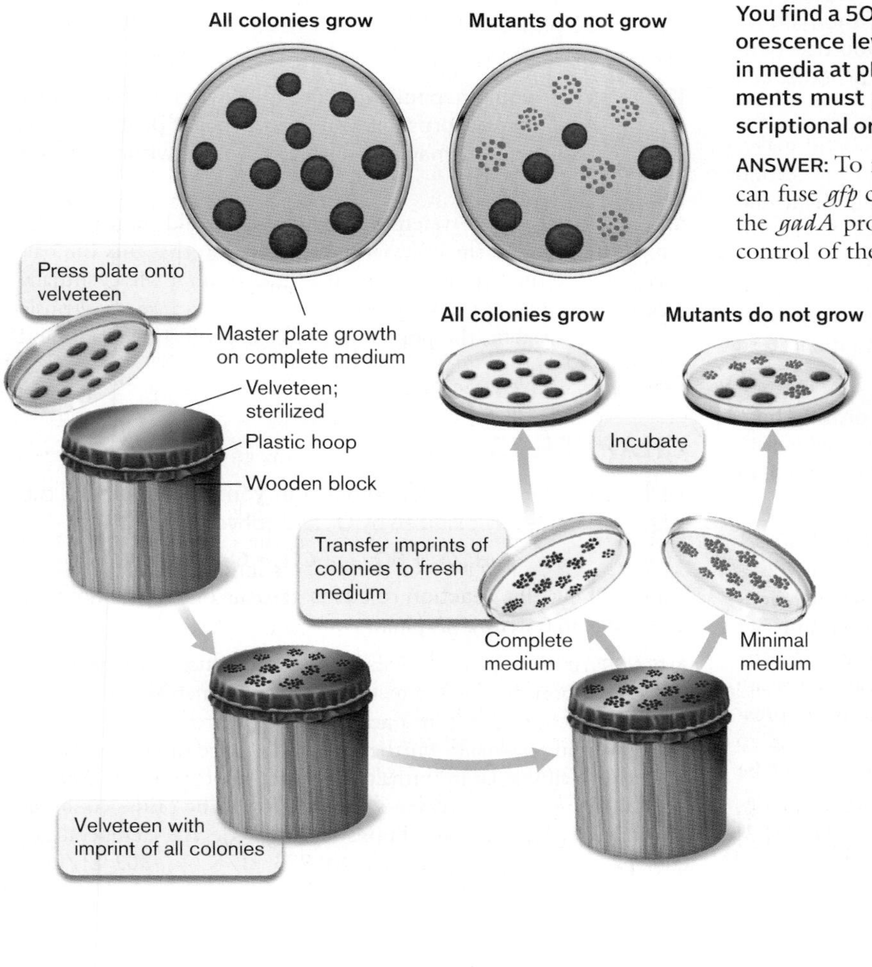
All colonies grow
Mutants do not grow
Press plate onto velveteen
Master plate growth on complete medium
Velveteen; sterilized
Plastic hoop
Wooden block
All colonies grow
Mutants do not grow
Incubate
Transfer imprints of colonies to fresh medium
Complete medium
Minimal medium
Velveteen with imprint of all colonies

12.2 In this postgenomic era, how might you generally design a strategy to construct a strain of *E. coli* that cannot synthesize histidine?

ANSWER: The genomic sequence of *E. coli* is known. We also know which genes encode the steps of histidine biosynthesis. The strategy today would be to engineer a plasmid containing DNA sequences (about 200 bp) that flank the histidine biosynthesis gene you want to delete, but in place of the target gene you would instead insert an antibiotic resistance gene (for example, for kanamycin resistance). The plasmid used would be a "suicide" plasmid that cannot replicate in the strain you wish to mutate, but can replicate in other permissive strains (strains that produce a specific protein the plasmid needs to replicate). The permissive strain allows a lot of the plasmid to be made. The recombinant plasmid would then be moved into the nonpermissive target strain by transformation, and the transformation mixture would be plated on nutrient agar medium containing the antibiotic. Since the plasmid cannot replicate in this strain, the only way the target strain can become resistant to the antibiotic is if the chromosomal sequences flanking the drug marker in the plasmid undergo homologous recombination with the same DNA sequences on the actual chromosome. This double recombination event "surgically" removes the target gene from the chromosome and replaces it with the antibiotic resistance marker. In our example, the newly constructed mutant would be able to grow in the presence of antibiotic, but only if histidine was present.

12.3 You have monitored expression of a gene fusion in which *gfp* is fused to *gadA*, the glutamate decarboxylase gene. You find a 50-fold increase in expression of *gadA* (based on fluorescence level) when the cells carrying this fusion are grown in media at pH 5.5 as compared to pH 8. What additional experiments must you do to determine whether the control is transcriptional or translational?

ANSWER: To measure regulation of the *gadA* promoter only, you can fuse *gfp* containing its native ribosome-binding site (RBS) to the *gadA* promoter. This fusion will expose only transcriptional control of the *gadA* promoter. Translational control of the *gadA* message is absent because the *gadA* RBS (and all other *gadA* sequences) was replaced by the *gfp* RBS. To measure translational control, you must make a new fusion that starts with a constitutive promoter (a constantly expressed promoter), add a section of *gadA* containing the *gadA* ribosome-binding site, and follow that with *gfp* (minus its RBS) fused in-frame to the *gadA* sequence. Any condition that alters the expression of fusion 1 affects transcription of the *gadA* promoter. Any condition that alters the expression of fusion 2 must affect translation of the *gadA* message.

12.4 How could you quickly separate cells in the population that express sigma B–YFP from those that do not express this fusion? And what could you learn by separating these subpopulations? *Hint:* A technique presented in Chapter 4 will help.

ANSWER: The population of cells can be passed single file through a fluorescence activated cell sorter (FACS; Chapter 4). Cells that fluoresce can be collected in one tube, and nonfluorescent cells can be collected in another. The cells can then be analyzed using deep sequencing to determine the transcriptomes, and mass spectrometry can determine what proteins are produced in the different cell populations.

12.5 What would you conclude if the northern blot of the *gad* genes showed no difference in mRNA levels in cells grown at pH 5.5 and pH 7.7, but the western blot showed more protein at pH 5.5 than at pH 7.7?

ANSWER: Regulatory control would likely be at the posttranscriptional level, through increasing mRNA stability, translational efficiency, or protein stability.

12.6 Another regulator of glutamate decarboxylase (Gad) production does not affect the production of *gad* mRNA, but is required to accumulate Gad protein. What two regulatory mechanisms might account for this phenotype?

ANSWER: The two regulatory mechanisms are control of translation and control of protein degradation.

12.7 How would you have to modify standard real-time PCR to quantify the level of a specific mRNA?

ANSWER: You would have to first convert the mRNA into cDNA using the enzyme reverse transcriptase. An oligonucleotide primer hybridizing to the 3′ end of the mRNA is acted on by reverse transcriptase to synthesize DNA (called complementary DNA, or cDNA). Normal real-time PCR techniques can then quantify the cDNA.

12.8 You suspect that proteins A and C can simultaneously bind to protein B, and that this interaction then allows A to interact with C in the complex. How could you use two-hybrid analysis to determine whether these three proteins can interact?

ANSWER: Place three plasmids in the cell. The first plasmid makes protein A fused to the GAL4 DNA-binding domain, the second plasmid makes protein C fused to the GAL4 activator domain, and the third plasmid simply makes protein B. If the yeast cell contains only the first two plasmids, then no interaction will occur and *lacZ* will not be activated. If all three plasmids are in the same yeast cell, then protein B will be the center of a sandwich, linking the A and C fusion proteins. The fusion proteins can then interact and activate *lacZ*.

12.9 Would insect resistance to an insecticidal protein be a concern when developing a transgenic plant? Why or why not? How would you design a transgenic plant to limit the possibility of insects developing resistance?

ANSWER: Insects have, in fact, developed resistance to single insecticidal proteins. The most common resistance mechanism involves a change in the membrane receptors in the midgut to which activated *Bt* toxins bind. Resistance can be due to a reduced number of *Bt* toxin receptors or to a reduced affinity of the receptor for the toxin. Some insects, such as the spruce budworm, can inactivate specific toxins by precipitating them with a protein complex present in the midgut.

While developing a transgenic plant, several steps could be taken to limit the development of resistance in an insect population. The insecticidal gene could be fused to a promoter that is expressed only when the plant is most susceptible to attack. Alternatively, the gene could be fused to a promoter that is expressed only in a tissue of the plant that is most vulnerable to attack. This would limit the time during which the insects can develop resistance. For instance, cotton plants attacked by bollworms could produce toxin only in young boll tissues, the most important part of the plant. In addition to specifically protecting the critical plant tissue, this strategy would affect only one generation of bollworms, avoiding the constant selection pressure that hastens evolution of resistance. Another technique would be to engineer two different insecticidal proteins into the plant genome that will not exhibit cross-resistance. In other words, even if an insect develops resistance to one toxin, it will still remain susceptible to the second.

12.10 You have just employed phage display technology to select for a protein that tightly binds and blocks the eukaryotic cell receptor targeted by anthrax toxin. When this receptor is blocked, the toxin cannot get into the target cell. You suspect that the protein may be a useful treatment for anthrax. How would you recover the gene following phage display and then express and purify the protein product?

ANSWER: The gene can be cut out of the phage genome using restriction endonucleases or amplified by PCR using known phage DNA sequences that flank the gene. The fragment can then be cloned into a His_6 tag expression vector. The sequence of the gene would have to be known so that the DNA encoding the His_6 tag could be placed in-frame with the open reading frame of the receptor-blocking protein. The plasmid containing the His_6-tagged protein gene is then induced to overexpress the protein in *E. coli*. The vector will contain an inducible promoter (for example, *lacP*) to drive expression of the gene. The His_6-tagged protein can then be purified by pouring cell extracts over a nickel column, as described in the text.

12.11 What would happen with a toggle switch if the genetic engineer could set repressor 1 and repressor 2 protein levels to be exactly equal? Imagine that this is done without either inducer present.

ANSWER: In a perfect system, both proteins would remain at the same level, but any small deviation in condition that tilts the balance between the two proteins even a small amount will eventually lock the cell into either GFP on or GFP off. Because of intrinsic noise in the system, the population of cells could be half on and half off.

CHAPTER 13

13.1 Consider glucose catabolism in your blood, where the sugar is completely oxidized by O_2 and converted to CO_2:

$$C_6H_{12}O_6 + 6O_2 \rightarrow 6CO_2 + 6H_2O$$

Do you think this reaction releases greater energy as heat, or by change in entropy? Explain.

ANSWER: At first glance, the breakdown of glucose to six molecules of carbon dioxide seems to incur a large increase in entropy. But the reaction also consumes six molecules of oxygen, so the entropy gain is small. Furthermore, the oxidation reaction is associated with a large enthalpy change, approximately $\Delta H°' = -2{,}540$ kJ/mol. (The degree symbol followed by prime connotes biochemical standard conditions.) At 37°C, the temperature-entropy term $-T\Delta S°' = -(310\ \text{K})(0.973\ \text{kJ/K}) = -302$ kJ/mol.

The overall value of free energy change $\Delta G^{o\prime}$ = −2,540 kJ/mol − 302 kJ/mol = −2,842 kJ/mol. Thus, the entropy term yields some energy, but less than an eighth as much as the enthalpy ($\Delta H^{o\prime}$) term.

13.2 The bacterium *Lactococcus lactis* was voted the official state microbe of Wisconsin, because of its importance for cheese production. During cheese production, *L. lactis* ferments milk sugars to lactic acid:

$$C_6H_{12}O_6 \rightarrow 2C_3H_6O_3 \rightleftharpoons 2C_3H_5O_3^- + 2H^+$$

Large quantities of lactic acid are formed, with relatively small growth of bacterial biomass. Why do you think this occurs? Cheese making usually runs more efficiently at high temperature; why?

ANSWER: The lactic acid fermentation reaction does not involve a strong oxidant, but only breakdown of sugar to smaller molecules, so a larger proportion of the free energy yield is in the entropy term $-T\Delta S$. The overall energy yield is low, so a large amount of sugar must be cycled to lactic acid (lactate) for a relatively small amount of bacterial growth, as compared to bacterial growth with oxygen. Because fermentation depends on entropy change ($-T\Delta S$), the free energy yield can be increased by increasing the temperature.

13.3 The thermophilic bacteria *Thermus* species grow in deep-sea hydrothermal vents at 80°C. It was proposed that they metabolize formate to bicarbonate ion and hydrogen gas:

$$HCOO^- + H_2O \rightarrow HCO_3^- + H_2$$

But the standard $\Delta G^{o\prime}$ is near zero (−2.6 kJ/mol). Under actual conditions, do you think the reaction yields energy? Assume concentrations of 150 mM formate, 20 mM bicarbonate ion, and 10 mM hydrogen gas.

ANSWER: The ability to gain energy from formate will depend on the temperature and the concentrations of reactants and products. Remember that $[H_2O]$ equals 1. Consider the equation:

$$\Delta G = \Delta G^{o\prime} + 2.303\ RT \log\ [\text{products}]/[\text{reactants}]$$

$$= -2.6\ \text{kJ/mol} + 2.303\ [8.315 \times 10^{-3}\ \text{kJ/(mol}\cdot\text{K)}]\ (273\ \text{K} + 80\ \text{K}) \times \log\ [HCO_3^-][H_2]/[HCO_2^-][H_2O]$$

$$= -2.6\ \text{kJ/mol} + 6.760\ \text{kJ} \times \log\ [0.02\ \text{M} \times (0.01\ \text{M}/0.15\ \text{M})]$$

$$= -22\ \text{kJ/mol}$$

The ΔG value is small, but it is enough to drive growth of some species of *Thermus*. Note that this simplified treatment omits the role of gas formation. (Data are based on Yun Jae Kim et al. 2010. *Nature* **467**:352.)

13.4 When ATP phosphorylates glucose to glucose 6-phosphate, what is the net value of $\Delta G^{o\prime}$? What if ATP phosphorylates pyruvate? Can this latter reaction go forward without additional input of energy? (See Table 13.3.)

ANSWER: Using **Table 13.3**, we see that the phosphorylation of glucose by ATP is composed of these two reactions:

Reaction	$\Delta G^{o\prime}$ (kJ/mol)
$ATP + H_2O \rightleftharpoons ADP + P_i + H^+$	−31
$Glucose + P_i \rightleftharpoons glucose\ 6\text{-}P + H_2O$	+14
$ATP + glucose \rightarrow ADP + glucose\ 6\text{-}P$	−17

The net energy lost is −17 kJ/mol, so the phosphorylation can go forward. To phosphorylate pyruvate, however, the $\Delta G^{o\prime}$ of ATP hydrolysis (−31 kJ/mol) must be subtracted from the value of phosphoenolpyruvate formation (+62 kJ/mol), giving a net value of +31 kJ/mol. Since $\Delta G^{o\prime}$ is positive, this reaction cannot go forward without additional energy.

13.5 Linking an amino acid to its cognate tRNA is driven by ATP hydrolysis to AMP (adenosine monophosphate) plus pyrophosphate. Why release PP_i instead of P_i?

ANSWER: The formation of aminoacyl-tRNA must be irreversible until the ribosome is ready to release the tRNA. The pyrophosphate from ATP is immediately cleaved into 2 P_i, preventing the reversal of aminoacyl-tRNA formation.

13.6 In the microbial community of the bovine rumen, the actual ΔG value has been calculated for glucose fermentation to acetate:

$$C_6H_{12}O_6 + 2H_2O \rightarrow 2C_2H_3O_2^- + 2H^+ + 4H_2 + 2CO_2 \qquad \Delta G = -318\ \text{kJ/mol}$$

If the actual ΔG for ATP formation is 44 kJ/mol and each glucose fermentation yields four molecules of ATP, what is the thermodynamic efficiency of energy gain? Where does the lost energy go?

ANSWER: The energy efficiency is (4 × 44 kJ/mol)/(318 kJ/mol) × 100 = 55%. The remaining energy is dissipated as heat.

13.7 What would happen to the cell if pyruvate kinase catalyzed PEP conversion to pyruvate but failed to couple this reaction to ATP production?

ANSWER: If the bacterial cell were to convert PEP to pyruvate without coupling to ATP production, it would lose much of the energy available from glucose and other food substrates converted to glucose. The energy would be lost as heat.

13.8 Some bacteria make an enzyme, dihydroxyacetone kinase, that phosphorylates dihydroxyacetone to dihydroxyacetone phosphate (DHAP). How could this enzyme help the cell yield energy?

ANSWER: Bacteria can obtain dihydroxyacetone from their environment using a transporter protein. The bacterial kinase can then phosphorylate the substrate to DHAP and direct it into glycolysis. Some bacteria can grow on dihydroxyacetone as a sole carbon source.

13.9 Explain why the ED pathway generates only one ATP, whereas the EMP pathway generates two. What is the consequence for cell metabolism?

ANSWER: The EMP pathway primes the six-carbon sugar with two phosphoryl groups. The sugar then splits into two three-carbon units (glyceraldehyde 3-phosphate), each of which generates two ATPs for one of the original ATPs. By contrast, the ED pathway phosphorylates the sugar only once before it splits in two. The phosphorylated end yields glyceraldehyde 3-phosphate, which enters the EMP pathway to generate ATP, ending up as pyruvate. The unphosphorylated three-carbon unit yields pyruvate directly, with no ATP. The consequence for cell metabolism is that the ED pathway needs to cycle more substrate in order for a cell to grow the same amount of biomass as it would with the EMP pathway.

Bacteria growing with the ED pathway may produce greater amounts of a valuable product such as ethanol.

13.10 If a cell respiring on glucose runs out of oxygen and other electron acceptors, what happens to the electrons transferred from the catabolic substrates?

ANSWER: The electrons from the catabolic substrates are transferred to NADH and $FADH_2$ during glycolysis and the TCA cycle. Without a terminal electron acceptor, the cytoplasmic electron carriers cannot use the electron transport system. Instead, they must transfer their electrons back onto pyruvate, acetate, and other products of catabolism in order to complete the reactions of fermentation.

13.11 Compare the reactions catalyzed by pyruvate dehydrogenase and pyruvate formate lyase (see Section 13.5). What conditions favor each reaction, and why?

ANSWER: The pyruvate dehydrogenase complex (PDC) is favored in the presence of oxygen because the electrons transferred to NADH can enter the electron transport chain, eventually combining with oxygen to release energy. In the absence of oxygen, pyruvate formate lyase is favored to yield fermentation products that can be excreted from the cell without reducing more energy carriers. At high pH, formate and acetate production is especially favorable because the extra acid counteracts alkalinity.

13.12 Suppose a cell is pulse-labeled with ^{14}C-acetate (fed the ^{14}C label briefly, then "chased" with unlabeled acetate). Can you predict what will happen to the level of radioactivity observed in isolated TCA intermediates? Plot a curve showing your predicted level of radioactivity as a function of the number of rounds of the cycle.

ANSWER: The amount of radioactivity measured in TCA intermediates will rise steeply as labeled acetate is incorporated, and then will decrease by half with each succeeding cycle, as the order of the carbons is randomized by succinate.

CHAPTER 14

14.1 *Pseudomonas aeruginosa*, a cause of pneumonia in cystic fibrosis patients, oxidizes NADH with nitrate (NO_3^-) at neutral pH. What is the value of $E°'$?

ANSWER: To calculate the reduction potential $E°'$:

$$NADH + H^+ + NO_3^- \rightarrow NAD^+ + NO_2^- + H_2O$$

$$E°' = 320 \text{ mV} + 420 \text{ mV} = 740 \text{ mV}$$

14.2 Could a bacterium obtain energy from succinate as an electron donor with nitrate (NO_3^-) as an electron acceptor?

ANSWER: For nitrate reduction:

$$\text{Succinate} + NO_3^- \rightarrow \text{fumarate} + NO_2^- + H_2O$$

$$E°' = -33 \text{ mV} + 420 \text{ mV} = 387 \text{ mV}$$

Although succinate is a relatively poor electron donor, nitrate is a strong electron acceptor. This reaction should provide energy for bacterial metabolism.

14.3 What do you think happens to $\Delta\psi$ as the cell's external pH increases or decreases? What could happen to the Δp of bacteria that are swallowed and enter the extremely acidic stomach?

ANSWER: As external pH changes, ΔpH increases or decreases, affecting the magnitude of Δp. As enteric bacteria enter the stomach, they encounter pH values (pH 1.5–3.0) below their growth range (pH 5.0–9.0). At first, the cell's transmembrane ΔpH may be very large, as the cell tries to maintain its cytoplasmic pH above 5.0, the limit for viability. But since the cell no longer grows, it loses its energy supply and can no longer spend energy to maintain ΔpH. One way to maintain cytoplasmic pH homeostasis is to reverse the electrical potential $\Delta\psi$ (inside positive) so as to drive out some H^+ and maintain a small ΔpH. Oppositely directed ΔpH and $\Delta\psi$ enable the cell to keep cytoplasmic pH high enough to survive when external pH is extremely low. On the other hand, when the external pH is raised by pancreatic secretions (alkaline), the cell needs to compensate by inverting its ΔpH and maintaining a relatively large $\Delta\psi$.

14.4 Suppose that de-energized cells of *E. coli* ($\Delta p = 0$) with an internal pH 7.6 are placed in a solution at pH 6. What do you predict will happen to the cell's flagella? What does this demonstrate about the function of Δp?

ANSWER: The flagella will rotate, driven solely by the ΔpH component of Δp. This result is consistent with the hypothesis that the transmembrane proton potential Δp drives flagellar rotation.

14.5 In Figure 14.14, what is the advantage of the oxidoreductase transferring electrons to a pool of mobile quinones, which then reduce the terminal reductase (cytochrome complex)? Why does each oxidoreductase not interact directly with a cytochrome complex?

ANSWER: The mobile quinone pool connects diverse electron donors with diverse electron acceptors. If each oxidoreductase had to interact specifically with a different terminal oxidase, the pathways of electron transport would be limited; for example, NADH might donate electrons only to O_2, whereas succinate might donate electrons only to nitrate. Instead, all potential electron donors can be coupled with all potential acceptors.

14.6 In Figure 14.14, why are most electron transport proteins fixed within the cell membrane? What would happen if they "got loose" in aqueous solution?

ANSWER: If the electron transport proteins came away from the membrane into aqueous solution, they could carry their energized electrons back into the cytoplasm or lose them outside the cell. In either case, they could no longer convert the flow of electrons into a proton gradient.

14.7 Would *E. coli* be able to grow in the presence of an uncoupler that eliminates the proton potential supporting ATP synthesis?

ANSWER: Yes. *E. coli* can grow with the proton gradient eliminated, but only with a rich supply of nutrients for substrate phosphorylation to generate ATP (for example, from glycolysis). In addition, the external pH and salt levels must be maintained close to those of the cytoplasm, to minimize the need for ion transport.

14.8 The proposed scheme for uranium removal requires injection of acetate under highly anoxic conditions, with less than 1 part per million (ppm) dissolved oxygen. Why must the acetate be anoxic?

ANSWER: Oxygen is the strongest terminal electron acceptor. If O_2 is present, bacteria will use it preferentially (instead of U^{6+}) to oxidize the acetate to CO_2.

14.9 Hydrogen gas is so light that it rapidly escapes from Earth. Where does all the hydrogen come from to be used for hydrogenotrophy and methanogenesis?

ANSWER: Hydrogen is produced in substantial quantities as a byproduct of fermentation. It may seem surprising that organisms would readily excrete quantities of energy-rich H_2, but in the absence of a good electron acceptor (or the enzymes to utilize electron acceptors), hydrogen may be just another waste product. Hydrogen gas trapped underground supports large communities of methanogens and hydrogenotrophs. The human colonic bacteria generate so much hydrogen that all parts of the body show traces of hydrogen gas.

14.10 Suppose you discover bacteria that require a high concentration of Fe^{2+} for photosynthesis. Can you hypothesize what the role of Fe^{2+} may be? How would you test your hypothesis?

ANSWER: The organism uses reduced iron as an electron donor for its photosystem ($Fe^{2+} \rightarrow Fe^{3+}$). To test this hypothesis, grow the organism on a defined concentration of Fe^{2+}. Measure the amount of iron oxidized and the amount of carbon fixed into biomass; if the Fe^{2+} is an electron donor for the photosystem, the two numbers should show a linear correlation.

CHAPTER 15

15.1 Propose a simple experiment to reveal the key intermediate to receive CO_2.

ANSWER: Add a very short pulse of [^{14}C] CO_2 to the cells. The pulse must contain a level of CO_2 sufficient to begin fixation but insufficient to continue beyond one "turn" of the cycle. In this case, the radiolabel will accumulate in the intermediate that needs to assimilate CO_2 for the next round. This kind of experiment ultimately identified the key intermediate ribulose 1,5-bisphosphate.

15.2 Speculate on why Rubisco catalyzes a competing reaction with oxygen. Why might researchers be unsuccessful in attempting to engineer Rubisco without this reaction?

ANSWER: The oxygenation reaction might have an essential function in regulation of metabolism. For example, it might help prevent excessive reduction of cell components or fixation of too much carbon to be used in biosynthesis. Given the universal existence of the oxygenation reaction in bacterial and chloroplast Rubiscos, it seems unlikely that oxygenation serves no purpose. For this reason, attempts to engineer Rubisco without oxygenation may not succeed.

15.3 Why does ribulose 1,5-bisphosphate have to contain two phosphoryl groups, whereas the other intermediates of the Calvin cycle contain only one?

ANSWER: Only ribulose 1,5-bisphosphate needs to split into two molecules (3-phosphoglycerate). Each of the two products needs to have its own phosphate as a tag for the enzymes to recognize it within the cycle.

15.4 Which catabolic pathway (see Chapter 13) includes some of the same sugar-phosphate intermediates that the Calvin cycle has? What might these intermediates in common suggest about the evolution of the two pathways?

ANSWER: The pentose phosphate pathway includes ribulose 5-phosphate, erythrose 4-phosphate, and sedoheptulose 7-phosphate in a similar series of carbon exchanges. Perhaps the pentose phosphate pathway and the Calvin cycle pathway evolved from a common amphibolic pathway of sugar consumption and biosynthesis. Alternatively, the one pathway evolved earlier and then the sugar intermediates were available for evolution of the second pathway.

15.5 For a given species, uniform thickness of a cell membrane requires uniform chain length of its fatty acids. How do you think chain length may be regulated?

ANSWER: In *E. coli,* the chain length of a growing fatty acid appears to be limited by beta-ketoacyl-ACP synthase, which binds only precursor acyl-ACPs shorter than 18 carbons. Thus, only carbon chains of up to 18 carbons are synthesized.

15.6 Suggest two reasons why transamination is advantageous to cells.

ANSWER: Ammonia is toxic to cells. Transamination enables cells to store amine groups in nontoxic form, readily available for biosynthesis. The availability of multiple enzymes of transamination from different amino acids enables cells to quickly recycle existing resources into the amino acids most needed by the cell in a given environment. For example, if a sudden supply of glutamine appears, cells can immediately distribute its amines into all 20 amino acids.

15.7 Which energy carriers (and how many) are needed to make arginine from 2-oxoglutarate?

ANSWER: Arginine biosynthesis requires three ATP molecules and three NADPH molecules (including two for converting two molecules of 2-oxoglutarate to glutamate). An additional ATP is spent converting acetate to acetyl-CoA.

15.8 Why are purines synthesized onto a sugar-base-phosphate?

ANSWER: Purines are highly hydrophobic, insoluble in the cytoplasm. The ribose phosphate component solubilizes the molecule, enabling synthesis to occur in the cytoplasm, where needed to make RNA and DNA.

15.9 Why are the ribosyl nucleotides synthesized first, and then converted to deoxyribonucleotides as necessary? What does this order suggest about the evolution of nucleic acids?

ANSWER: Ribonucleic acid is believed to be the original chromosomal material of cells. Cells evolved to synthesize RNA first; then later, as DNA was used, pathways evolved to synthesize it by modification of RNA, which the cell already had the ability to make.

CHAPTER 16

16.1 Why do the lipid components of food experience relatively little breakdown during anaerobic fermentation?

ANSWER: Lipids are highly reduced molecules, largely hydrocarbon with relatively low oxidizing potential. Thus, lipids cannot undergo as many intramolecular redox reactions as do sugars, which readily generate energy through anaerobic fermentation.

16.2 Why does oxygen allow excessive breakdown of food, compared with anaerobic processes?

ANSWER: Oxygen functions as the terminal electron acceptor for the complete breakdown of all kinds of organic molecules to water and CO_2.

16.3 In an outbreak of listeriosis from unpasteurized cheese, only the refrigerated cheeses were found to cause disease. Why would this be the case?

ANSWER: In the cheeses kept at room temperature, other naturally occurring bacteria outgrew the pathogenic *Listeria,* whereas in the refrigerator only the *Listeria* could grow. (Note, however, that many other potential pathogens, such as *Salmonella,* are inhibited by refrigeration.)

16.4 Cow's milk contains 4% lipid (butterfat). What happens to the lipid during cheese production?

ANSWER: Lipids undergo little catabolism, because the fermentation conditions are anaerobic. During coagulation, lipid droplets become trapped in the network of denatured protein and are largely retained in the bulk of the cheese. "Low-fat" cheeses are made from skim milk, which eliminates the lipids before fermentation.

16.5 In traditional fermented foods, without pure starter cultures, what determines the kind of fermentation that occurs?

ANSWER: The fermentation can be controlled by introducing a crude starter culture obtained from a previous batch of the food product or from a natural source of a particular microbe; for example, rice straw is a source of *Bacillus natto* for natto production. The fermentation type can be manipulated by the addition of factors, such as brine, that retard growth of all but a few strains. In pidan, for example, the high concentration of sodium hydroxide limits bacterial growth to alkali-tolerant strains of *Bacillus.*

16.6 Compare and contrast the role of fermenting organisms in the production of cheese and bread.

ANSWER: In cheese production, fermentation causes major biochemical changes in the food, such as the buildup of acids and the breakdown of proteins to smaller peptides and amino acids. Minor by-products, such as methanethiols and esters, accumulate to levels that confer flavors. In yeast bread, by contrast, the only significant product of fermentation is the carbon dioxide that leavens the dough. The small amount of ethanol produced evaporates during cooking. A form of bread in which extended fermentation does generate flavor is injera, for which the dough ferments for 3 days.

16.7 Compare and contrast the role of low-concentration by-products in the production of cheese and beer.

ANSWER: In both cheese and beer, minor by-products such as esters contribute flavor. Oxidation of esters can lead to off-flavors. In cheese, however, the exclusion of oxygen usually prevents off-flavors. In beer, the yeast requires a low level of oxygen; thus, significant amounts of acetaldehyde and diacetyl are produced and must be eliminated by a secondary fermentation.

16.8 Why would bacteria convert trimethylamine oxide (TMAO) to trimethylamine? Would this kind of spoilage be prevented by exclusion of oxygen?

ANSWER: TMAO acts as a terminal electron acceptor—that is, an alternative to oxygen for anaerobic respiration, as discussed in Chapter 14. Exclusion of oxygen inhibits only aerobic bacteria; TMAO respirers continue to grow and can spoil the fish.

16.9 Is it possible for physical or chemical preservation methods to completely eliminate microbes from food? Explain.

ANSWER: Preservation methods either slow microbial growth or induce microbial death. Microbial death follows a negative exponential curve, as discussed in Chapter 5. In theory, the exponential curve never reaches zero, so total exclusion of microbes is impossible. In practice, there is a high probability of totally eliminating microbes if the treatment time extends several "half-lives" beyond the time at which microbial concentration declines to less than one per total volume.

16.10 Why would different industrial strains or species be used to express different kinds of cloned products?

ANSWER: Different industrial strains have biochemical systems that favor different products. Some fungi naturally possess the highly complex pathways to generate antibiotics, as well as regulatory timing to turn on these pathways after the culture has grown to high population density. On the other hand, bacteria such as *Bacillus subtilis* are the most genetically tractable and predictable in their growth cycles, and the easiest to manipulate to express recombinant products such as human genes.

16.11 Why would an herbicide resistance gene be desirable in an agricultural plant? What long-term problems might be caused by microbial transfer of herbicide resistance genes into plant genomes?

ANSWER: Introduction of an herbicide resistance gene allows application of higher amounts of herbicide to crops in order to control growth of weeds. But the higher concentrations of herbicide may also have greater side effects on animals and on human consumers of the crop. In the long run, the herbicide resistance gene is likely to escape into weed plants through natural genetic transfer mechanisms. Thus, eventually the weeds may require still higher concentrations of herbicide. While the costs versus benefits of new gene modifications remain poorly understood, it must be recognized that all modern crops today are the product of many generations of genetic manipulation.

CHAPTER 17

17.1 What would have happened to life on Earth if the Sun were a different stellar class, substantially hotter or colder than it is?

ANSWER: If the Sun were hotter, too much ultraviolet and gamma radiation would reach the Earth, breaking chemical bonds of living organisms so rapidly that life could not be sustained. If the Sun were colder, too little radiation with sufficient energy would be available to drive photosynthesis. In either case, life as we know it could not have evolved on Earth.

17.2 Outline the strengths and limitations of each model of the origin of living cells. Which aspects of living cells does each model explain?

ANSWER: The three models are complementary in that each explains aspects of modern cells not addressed by the others. The prebiotic soup model accounts for the major classes of compounds used by cells, such as nucleosides, TCA cycle intermediates, amino acids, and fatty acids. It also suggests the origin of membranes as soap bubble–like micelles. It does not, however, account for the evolution of metabolic pathways and replication of genetic information. The metabolist model accounts for the prevalence of major cellular reactions such as carbon fixation and the TCA cycle. It does not explain the evolution of membranes and genetic material. The RNA world accounts for the central role of RNA in living cells; of all molecular classes, RNA and ribonucleotides probably serve the widest range of functions as information carriers, agents of catalysis, and genetic regulators. Most RNA-world models do not address the origin of membranes.

17.3 Suppose that a NASA rover were to discover living organisms on Mars. How might such a find shed light on the origin and evolution of life on Earth?

ANSWER: If life on Mars showed a completely different basis than that of Earth—for example, it was based on silicon polymers instead of carbon—such a find would support the view that life originated independently on each planet, rather than traveling from one planet to the other; or that both planets were seeded from somewhere else. If life on Mars were based on similar macromolecules, perhaps even showing the same genetic code, this would support the view that life arose on Mars first; or that both planets were seeded from the same source.

17.4 What kinds of DNA sequence changes have no effect on gene function?

ANSWER: Base substitutions that do not change the amino acid specified by the codon have no effect on gene function. For example, CUA → CUG still encodes leucine. In addition, a majority of the amino acids in any given protein can be replaced by an amino acid of similar form (for example, leucine → valine) without significantly affecting function of the gene.

17.5 What are the major sources of error in constructing phylogenetic trees?

ANSWER: Phylogenetic trees are affected by variability in the number of substitutions, or rate of mutation in different strains. The tree is distorted by errors in sequence alignment, and by systematic errors due to failure of the fundamental assumptions of the molecular clock. These assumptions include the constant rate of mutation for all branches, constant generation time, and true orthology of the gene chosen (that is, the encoded product has the same function and hence the same degree of selection pressure in all taxa under consideration).

17.6 What are the limits of evidence for horizontal gene transfer in ancestral genomes? What alternative interpretation might be offered?

ANSWER: Horizontal gene transfer is inferred from the appearance of genes in clade A that are absent from other members of the clade but present in clade B. The degree of similarity between genes in the two clades, however, must be high enough to exclude the possibility that the genes in question were retained from a common ancestor of the two clades but lost from other members of clade A. This possibility is difficult to exclude in the case of deep-branching clades, where all genes have had a long time to diverge. For example, the large number of archaeal genes present in deep-branching thermophilic bacteria such as *Thermotoga* may include some inherited from the last common ancestor, or they may represent archaeal genes that were horizontally transferred to bacteria sharing the high-temperature habitat.

17.7 What would be the identification of a straight, nonsheathed, Gram-negative bacterium that has sulfur granules, is motile, and is 1.0 μm wide? What would happen if you assigned the bacterium a width of 0.9 μm? (See Fig. 17.28.)

ANSWER: The bacterium would key out as *Beggiatoa* sp. If the cell width were measured as 0.9 μm, however, the identification would proceed down a completely different, wrong track, at the first step of the key. This is one disadvantage of the dichotomous key.

17.8 Besides mitochondria and chloroplasts, what other kinds of entities within cells might have evolved from endosymbionts?

ANSWER: Some of the large "megaplasmids" found in bacteria and protists are as large as genomic chromosomes and contain numerous housekeeping genes. These megaplasmids may have originated as endosymbiotic cells that lost all their membranes through reductive evolution. Similarly, some of the giant viruses, such as mimivirus and smallpox virus, as well as phages such as T4, possess a wide spectrum of housekeeping genes. These viruses may have originated as cellular parasites that underwent reductive evolution.

CHAPTER 18

18.1 What taxonomic questions are raised by the apparent high rate of gene transfer between archaea and thermophilic bacteria?

ANSWER: If gene transfer results in a species containing a quarter of its genes from organisms outside its domain, such a mosaic genome raises questions of how to define the species and the domain. How can a species be defined if its genome contains large portions from distantly related sources? Other interesting questions relate to the means of gene transfer. How do such distantly related organisms as bacteria and archaea maintain a compatible mechanism of gene transfer?

18.2 Which taxonomic groups in Table 18.1 stain Gram-positive and which stain Gram-negative? Which group contains both Gram-positive and Gram-negative species? For which groups is the Gram stain undefined, and why?

ANSWER: Most Firmicutes and Actinobacteria stain Gram-positive. These bacteria have relatively thick cell walls that retain the stain. The Proteobacteria, Nitrospirae, and Bacteroidetes/Chlorobi groups stain Gram-negative. The Cyanobacteria have an outer membrane and are considered Gram-negative, although their cell walls are very thick. The Deinococcus-Thermus group includes both Gram-positive- and Gram-negative-staining members. For Chlamydiae and Planctomycetes, the Gram stain is irrelevant because they lack the cell wall that retains the stain. For Spirochetes, many species are too narrow to observe the stain under light microscopy.

18.3 Which groups of bacterial species share common structure and physiology within the group? Which groups show extreme structural and physiological diversity?

ANSWER: Cyanobacteria all carry out oxygenic photosynthesis within thylakoid membranes. Their overall cell structure and organization, however, take diverse forms. All spirochetes are a sheathed flexible spiral with internal flagella; most share anaerobic or facultative heterotrophy. Chlamydiae and Planctomycetes groups of species each share general structural features. Other groups, particularly Firmicutes and Actinobacteria, show considerable diversity of form and physiology. The Proteobacteria display more extreme diversity of metabolism than any other division.

18.4 What are the relative advantages and disadvantages of propagation by hormogonia, as compared with akinetes?

ANSWER: Hormogonia are motile, and thus capable of active chemotaxis toward a more favorable environment. On the other hand, hormogonia have active metabolism that requires nutrition; if the environment lacks nutrients, the hormogonia will die. Akinete cells can persist until environmental conditions improve, but they cannot actively seek out a new location.

18.5 What are the relative advantages and disadvantages of the different strategies for maintaining separation of nitrogen fixation and photosynthesis?

ANSWER: Temporal separation has the advantage that all cells possess the ability to perform both nitrogen fixation and photosynthesis. On the other hand, it eliminates the ability of a chain of cells to conduct both processes simultaneously—the benefit of heterocysts. Heterocysts face the problem of operating in close proximity to photosynthetic cells generating toxic oxygen. This problem may be solved by symbiosis with respiring bacteria. Globular clusters of cells can bury their nitrogen fixers within the cluster; this arrangement effectively excludes oxygen, but it may lack flexibility during environmental change. Endosymbiotic nitrogen fixation within a respiring eukaryote is probably the most effective strategy of all, because the host provides oxygen-removing proteins such as leghemoglobin. Endosymbiosis, however, requires the presence of an appropriate host organism.

18.6 Why would *Streptomyces* produce antibiotics targeting other bacteria?

ANSWER: *Streptomyces* species may produce antibiotics to curb the growth of bacterial competitors with smaller genomes and faster rates of reproduction. The lysed cells release nutrients that feed growing mycelia of *Streptomyces.*

18.7 Why might genes for the proteorhodopsin light-powered proton pump be more likely to transfer horizontally than the bacteriochlorophyll-based photosystems PS I and PS II?

ANSWER: Proteorhodopsin requires only the one gene encoding the pump, plus one or two genes to produce retinal. A relatively small amount of sequence has to be transferred, and the encoded products generate proton potential on their own, without requiring interaction with recipient enzymes. By contrast, PS I and PS II each involves multiple electron carriers that must function together and interact with the recipient electron transport chain.

18.8 Can you hypothesize a mechanism for migration of the daughter nucleoid of *Hyphomicrobium* through the stalk to the daughter cell? For possibilities, see Chapter 3 and consider the various molecular mechanisms of cell division and shape formation.

ANSWER: One mechanism might involve polar localization similar to that seen in *Caulobacter.* A DNA-binding protein might pull the nucleoid through the stalk until it binds a polar localization protein at the end of the daughter cell. Another mechanism might involve formation of a scaffold of cytoskeletal proteins similar to FtsZ or MreB. The cytoskeleton could act as a track for DNA movement through the stalk, perhaps powered by ATP hydrolysis.

18.9 Why do you think it took many years of study to realize that *Escherichia coli* and other Proteobacteria can grow as a biofilm?

ANSWER: *E. coli* and its relatives grow exceptionally well in liquid culture. Liquid culture is attractive because it enables quantitative measurement of defined aliquots of a microbial population. However, repeated subculturing in liquid medium selects for planktonic (nonbiofilm) cells. Eventually, the biofilm-forming property may be lost if non-biofilm mutants evolve to grow faster than the original genotype in liquid medium.

18.10 Compare and contrast the formation of cyanobacterial akinetes, firmicute endospores, actinomycete arthrospores, and myxococcal myxospores.

ANSWER: An endospore forms as the daughter product (forespore) of a single cell. Within the same cell, endospore development is supported by the mother cell, which disintegrates after release of the endospore. Endospores have tough coatings of calcium dipicolinate; they are heat resistant. By contrast, arthrospores and myxospores are less durable and are not heat resistant, although they can persist in the environment for an extended period. Arthrospores form through binary fission of actinomycete filaments. Myxospores are formed by a multicellular fruiting body. In all three cases, spore formation can be induced by depletion of nutrients, and the spore-producing entity is left behind to die.

CHAPTER 19

19.1 If two deeply diverging clades each show a wide range of growth temperature, what does this suggest about the evolution of thermophily or psychrophily?

ANSWER: The two clades diverged before temperature adaptation occurred. Adaptations to high or low temperature must have evolved independently in the two clades.

19.2 What might be the advantages of flagellar motility for a hyperthermophile living in a thermal spring or in a black smoker vent? What would be the advantages of growth in a biofilm?

ANSWER: Flagellar motility enables isolated cells to detect a new nutrient source, or an approximate temperature range, and approach it through chemotaxis. Growth in a biofilm attached to a substrate prevents the microbes from floating away from the nutrient source, or from being carried away in the flow from the vent.

19.3 What problem with cell biochemistry is faced by acidophiles that conduct heterotrophic metabolism?

ANSWER: Heterotrophic metabolism generates fermentation products such as acetate and lactate, which act as permeant acids. Permeant acids become protonated outside the cell, at low pH; the protonated forms then permeate the membrane, returning into the cell. Given the high transmembrane pH difference maintained by *Sulfolobus,* one would expect even small traces of fermentation acids to cross the membrane in the protonated form, and then dissociate and accumulate to toxic levels of organic acids. It is unknown how *Sulfolobus* solves this problem.

19.4 What conclusions might be drawn if viruses of mesophilic archaea are found to have RNA genomes? What if they all have DNA genomes only?

ANSWER: If mesophilic viruses show RNA genomes but thermophiles do not, then it is likely that only double-stranded DNA is sufficiently stable for viruses to persist in the environment of hyperthermophiles. Finding only double-stranded DNA viruses throughout the archaea would suggest that all archaeal viruses evolved from viruses infecting a common ancestral cell that was a thermophile. The latter hypothesis would require supporting evidence from archaeal cell physiology and phylogeny.

19.5 What hypothesis might you propose to explain the reason for the time and depth distribution of marine Crenarchaeota seen in Figure 19.18? How might you test your hypothesis?

ANSWER: The crenarchaeotes appear to be most abundant in the deeper water, below 1,000 meters; and least abundant during the summer months of June and July. Perhaps the crenarchaeotes are adapted to low temperature, or to high pressure. Either hypothesis might be tested by obtaining a sample of seawater and incubating at different combinations of temperature and pressure. The crenarchaeotes and bacteria could then be sampled to see which temperature or pressure levels favor more rapid growth of crenarchaeotes, compared to bacteria.

19.6 What do the multiple metal requirements suggest about how and where the early methanogens evolved?

ANSWER: The requirement for so many different metals may suggest that methanogens evolved in habitats such as geothermal vents where superheated water carries up high concentrations of dissolved metal ions.

19.7 Compare and contrast the metabolic options available for *Pyrococcus* and for the crenarchaeote *Sulfolobus.*

ANSWER: *Sulfolobus* catabolizes sugars and amino acids aerobically, using O_2 as terminal electron acceptor. *Pyrococcus abyssi* catabolizes sugars and amino acids anaerobically, using S^0 as terminal electron acceptor. *P. abyssi* can also reduce sulfur lithotrophically with H_2 to form H_2S. By contrast, *Sulfolobus* uses molecular oxygen (O_2) to oxidize sulfur lithotrophically, from S^{2-} to S^0, SO_3^{2-}, and ultimately SO_4^{2-}.

19.8 Compare and contrast sulfur metabolism in *Pyrococcus* and *Ferroplasma.*

ANSWER: *Pyrococcus* species reduce S^0 with hydrogens from organic substrates, forming HS^- and H_2S. *Ferroplasma* species oxidize sulfur in the form of FeS_2 (using oxidant Fe^{3+}), forming sulfuric acid. The result is extreme acidification of their environment.

CHAPTER 20

20.1 Why would yeasts remain unicellular? What are the relative advantages and limitations of hyphae?

ANSWER Yeasts grow in environments with sufficient dissolved nutrients to absorb from the medium. The advantage of forming hyphae is that they enable penetration of other organisms, and hence provide access to nutrients. On the other hand, hyphae formation limits the rate of dispersal of progeny cells. Since yeasts grow in environments where dissolved nutrients can be absorbed from the medium, they do not need to produce hyphae and can proliferate more rapidly than mycelial fungi.

20.2 Why would some fungi conduct asexual reproduction under most conditions? What are the advantages and limitations of sexual reproduction?

ANSWER The sexual life cycle involves significant genetic and metabolic costs to the organism. Asexual reproduction enables fungi to eliminate an energy drain and produce more offspring using fewer resources. Reductive evolution leading to loss of sexual reproduction might eliminate an energy drain and perhaps enable greater proliferation with fewer resources. On the other hand, the sexual life cycle provides a valuable means of generating diversity through genetic recombination, so that the population may respond to environmental change. Fungi reproducing asexually must rely on mutation and gene transfer by viruses and mobile sequence elements to generate genetic diversity.

20.3 What are the advantages and limitations of motile gametes, as compared to nonmotile spores?

ANSWER Motile gametes have the advantage of rapid dispersal on their own and the potential for chemotaxis toward a food source or toward a gamete of the opposite mating type. On the other hand, motility uses up energy that could alternatively be invested in production of a greater number of nonmotile gametes. Motile gametes are especially useful in a watery habitat but are of little use in a terrestrial habitat, where air currents or animal hosts must be used for dispersal.

20.4 Compare the life cycle of an ascomycete (Fig. 20.14C) with that of a chytridiomycete (Fig. 20.12C). How are they similar, and how do they differ?

ANSWER Both chytridiomycetes and ascomycetes undergo alternation of generations. Each has an alternative route of an asexual cycle of mitotic cell proliferation. In the chytridiomycete asexual cycle, the diploid form develops a mycelium, motile zoospores, and cysts. In the ascomycete, however, the haploid form undergoes mitotic divisions. The sexual cycles of the two fungal groups differ structurally. In the chytridiomycete, the zoosporangium forms motile zoospores that develop and release motile gametes, which fertilize each other to form motile zygotes. In the ascomycete, there are no motile forms. Instead of motile gametes, haploid hyphal antheridia undergo cytoplasmic fusion and form fruiting bodies, or asci, within which ascospores develop. The ascospores are not motile; they are carried by wind or water.

20.5 How are coralline algae able to grow at greater depths than coral?

ANSWER Corals generally grow symbiotically with green algae, which contain chlorophyll but lack accessory photopigments. Corallines are red algae, possessing both chlorophyll and phycoerythrins. The latter absorb blue and green light that is not absorbed by green algae, and thus coralline algae penetrate deeper in the water column.

20.6 What might happen when an ameba phagocytoses algae?

ANSWER If light is available, the algae may be retained as endosymbionts providing energy through photosynthesis. For example,

Chlorarachnion possesses obligate chloroplast-bearing endosymbionts descended from green algae. Alternatively, the ameba may digest all but the algal chloroplast, which persists for some time, providing photosynthetic products.

20.7 What kind of habitat would favor a flagellated ameba?

ANSWER A dilute watery habitat would favor flagella, which allow more rapid propulsion than pseudopods. Pseudopod motility requires a solid substrate, such as debris in the sediment of a pond.

20.8 Compare and contrast the process of conjugation in ciliates and bacteria (see Chapter 9).

ANSWER Conjugation in ciliates is a completely different process from conjugation in bacteria, although the function (gene transfer) is similar. In ciliates, two cells form a bridge allowing cytoplasm to flow directly between them, along with micronuclei containing chromosomes. In bacteria, a donor cell attaches to another (by pili in some cases), and then a protein complex transfers DNA across both cell envelopes, without direct cytoplasmic contact. In bacterial conjugation, DNA is transferred unidirectionally from the donor cell to the recipient, whereas in ciliates there is reciprocal exchange of DNA. A donor bacterium generally transfers only part of its genome, whereas ciliates exchange entire copies of their respective genomes.

20.9 For ciliates, what are the advantages and limitations of conjugation, as compared with gamete production?

ANSWER The process of conjugation avoids the necessity of dissolving the intricate cell structure of the ciliate in order to form gametes that fuse or fertilize each other. On the other hand, conjugation requires two diploid organisms to find each other and make contact for several hours, during which time feeding is suspended and the pair is vulnerable to predation.

CHAPTER 21

21.1 Suppose you are conducting a metagenomic analysis of soil sampled from different parts of a wetland. Your budget for soil analysis covers only ten sample analyses. Would you use one DNA extraction method, or multiple methods?

ANSWER: The advantage of using one DNA extraction method is that it will maximize information about the differences between communities from different parts of the wetland. The advantage of multiple extraction methods applied to a given sample is that they will maximize the detection of diversity within a sample. Your choice of method (or a compromise design) may depend on how different you expect the microhabitats to be within the overall wetland.

21.2 Suppose you plan to sequence a marine metagenome for the purpose of estimating carbon dioxide fixation and release, to improve our model for global climate change. Do you focus your resources on assembling as many complete genomes as possible, or do you focus on identifying all the community's enzymes of carbon metabolism?

ANSWER: To maximize your yield of information about carbon flux, you are likely to learn more about the community as a whole by focusing on those enzymes involved in carbon metabolism. However, you will miss novel genes that were not previously known to participate in carbon flux. If your equipment and computational resources enable genome assembly, it may be informative to learn more about the most abundant species that participate in carbon flux.

21.3 From Chapters 13–15, give examples of microbial metabolism that fit patterns of assimilation and dissimilation.

ANSWER: Microbes can assimilate carbon either by reducing carbon dioxide or by oxidizing methane, and they can dissimilate carbon by fermentation and respiration. Nitrogen is assimilated by N_2 fixation, and by incorporation of NH_4^+ into glutamine and glutamate. Nitrogen is dissimilated by deamination of amino acids, and by lithotrophic oxidation.

21.4 How do viruses select for increased diversity of microbial plankton?

ANSWER: Since viruses tend to infect only a narrow host range, their existence favors the evolution of a large number of different host species with highly dispersed populations. Highly dispersed populations minimize the chance of viral transmission from one host to another. Over generations, mutations that allow host microbes to "escape" viral infection will be selected for, while viral mutations that allow a virus to infect previously unsusceptible hosts will also be selected.

21.5 Design an experiment to test the hypothesis that the presence of mycorrhizae enhances plant growth in nature.

ANSWER: Such an experiment requires a control based on the natural environment, where various unknown factors may be very different from those in the laboratory. One possibility is to compare the growth of seedlings in natural soil versus sterilized natural soil. However, this experiment would not prove that fungi are the cause of enhanced growth in unsterile soil. The sterilization procedure (usually involving heat and pressure) could break down key nutrients in the soil. A follow-up experiment might be to grow the plants in the presence of a fungus inhibitor in sterilized and unsterilized soil.

21.6 How do you think symbiotic rhizobia reproduce? Why do bacteroids develop if they cannot proliferate?

ANSWER: Various answers have been proposed. Not all of the invading bacteria become bacteroids; some continue to undergo cell division, particularly within senescing tissues of the plant. These bacteria benefit from plant growth, which is sustained by the bacteroids whose genes they share. Alternatively, the entire plant-bacteroid system may benefit rhizobia that grow just outside the plant, in the rhizosphere.

21.7 High levels of nitrate or ammonium ion corepress the expression of Nod factors (see Figure 21.35). What is the biological advantage of Nod regulation?

ANSWER: Nitrate and ammonium ion are the main forms of nitrogen assimilated by plants. If they are abundant in the soil, the plant does not need rhizobial symbionts to fix N_2—a process that consumes much energy. The energy required to maintain the symbiosis comes from the plant, in the form of sugars and other nutrients, thus, it is more efficient not to have the symbiosis when fixed nitrogen is already available. Therefore, the presence of alternative nitrogen sources inhibits development of the rhizobia-legume symbiosis.

21.8 Compare and contrast the processes of plant infection by rhizobia and by fungal haustoria.

ANSWER: Both rhizobial bacteria and fungal haustoria penetrate the volume of a plant cell, but they keep the plant cell membrane intact, its invagination always surrounding the invading cell. Rhizobia establish a complex, highly regulated exchange of nutrients with the host, receiving catabolites and oxygen in exchange for ammonium, and cycling the components of amino acids. By contrast, haustoria establish one-way removal of nutrients such as sucrose, while providing no nutrients in return. Fungal pathogens weaken the structure of the host plant and decrease or halt its growth.

21.9 Why does ruminant fermentation provide food molecules that the animal host can use? How is the animal able to obtain nourishment from waste products that the microbes could not use?

ANSWER: The rumen interior is anaerobic. In the absence of oxygen as a terminal electron acceptor, microbes are forced to generate waste products in which the electrons are put back onto the electron donors (fermentation; see Chapter 13). When the short-chain fatty acid wastes enter the animal's bloodstream, the blood is full of oxygen, which enables complete digestion to CO_2 and water.

21.10 How do you think cattle feed might be altered or supplemented to decrease methane production?

ANSWER: Several methods have been proposed to limit methanogenesis. One is to feed cattle an inhibitor of a process that methanogens require, but not bacteria. An example would be inhibitors of sodium transport, which methanogens need to maintain a sodium potential. Another approach is to feed cattle an organic electron acceptor for H_2, such as fumarate, which bacteria use to generate short-chain acids instead of methane. These approaches have been used with only partial success—not surprisingly, given the complexity of the system.

CHAPTER 22

22.1 Why is oxidation state critical for the acquisition, usability, and potential toxicity of cycled compounds? Cite examples based on your study of microbial metabolism.

ANSWER: Many examples can be cited. In the case of carbon, CO_2 can be fixed by many species with substantial input from photosynthesis or hydrogen donors. The reduced form methane, however, can be assimilated only by methanotrophic bacteria, usually with oxygen as electron acceptor. Nitrogen gas can be assimilated only by nitrogen-fixing bacteria and archaea, whereas NH_4^+ can be assimilated by many plants and microbes. But NH_4^+ can also be oxidized by lithotrophs to NO_3^-, a substance potentially toxic to humans.

22.2 What would happen if wastewater treatment lacked microbial predators? Why would the result be harmful?

ANSWER: Without predators, too many planktonic bacteria would remain in the wastewater after sedimentation of the sludge. The bacteria could be killed by chlorination, but the treated water would have significant BOD (biochemical oxygen demand) because the bacterial remains provide an organic carbon source for respirers.

22.3 How many kinds of biomolecules can you recall that contain nitrogen? What are the usual oxidation states for nitrogen?

ANSWER: Amino acids, nucleotide bases, polyamines for DNA stabilization, peptidoglycan (both amino sugar and peptide chains), and the heme derivatives of cytochromes, chlorophyll, and vitamin B_{12} all include nitrogen (as do many other biochemicals). The oxidation states of nitrogen in living organisms are nearly always reduced, either R–NH_2, R=NH, or R–N=R. An exception is the neurotransmitter NO (nitric oxide).

22.4 The nitrogen cycle has to be linked with the carbon cycle, since both contribute to biomass. How might the carbon cycle of an ecosystem be affected by increased input of nitrogen?

ANSWER: One hypothesis is that the injection of nitrogen into an ecosystem accelerates growth of producers (phytoplankton in the ocean, or trees in a forest) and therefore facilitates net removal of CO_2 from the atmosphere. Overall, however, the additional fixed carbon ends up dissipated by consumers and decomposers.

22.5 In the laboratory, which bacterial genus would likely grow on artificial medium including NH_2OH as the energy source: *Nitrosomonas* or *Nitrobacter*?

ANSWER: *Nitrosomonas* is more likely to utilize NH_2OH, since it performs the intermediate oxidation of NH_2OH during nitrification of ammonia.

22.6 Compare and contrast the cycling of nitrogen and sulfur. How are the cycles similar? How are they different?

ANSWER: Cycling of both nitrogen and sulfur involves interconversion between different oxidation states. Most of these interconversion reactions are performed solely by microbes, many of them solely by bacteria. Examples include nitrification of ammonia and denitrification to N_2, as well as sulfide oxidation and photolysis. In both cases, oxidation produces strong acids (HNO_3, H_2SO_4). The major sources and sinks differ; nitrogen is obtained primarily from the atmosphere as N_2, whereas sulfur (in the form of sulfate) is at high levels in the ocean and soil. Sulfur is rarely limiting, whereas nitrogen frequently is. Sulfur participates extensively in phototrophy; nitrogen shows little involvement in phototrophy; phototrophy based on nitrate reduction has been observed.

22.7 Compare and contrast the cycling of nitrogen and phosphorus. How are the cycles similar? How are they different?

ANSWER: Nitrogen and phosphorus are both limiting nutrients in many ecosystems—marine, aquatic, and terrestrial. Addition of either element into an aquatic system may cause algal bloom and eutrophication. On the other hand, the two elements differ in their major sources: the atmosphere for nitrogen, and crustal rock for phosphate. Within biomass, nitrogen exists almost entirely in reduced form, whereas phosphorus is entirely oxidized. Phosphorus cycles through the biosphere mainly as inorganic or organic phosphates, whereas nitrogen cycles through a broad range of oxidation states, from NH_3 to NO_3^-.

CHAPTER 23

23.1 How can an anaerobic microorganism grow on skin or in the mouth, both of which are exposed to air?

ANSWER: Facultative organisms living in proximity to the anaerobes will deplete oxygen in the environment, especially around nooks and crannies (for example, between teeth and gums, in gingival pockets) that would ordinarily prevent anaerobes from growing. These small spaces have limited access to oxygen.

23.2 Why do many Gram-positive microbes that grow on the skin, such as *Staphylococcus epidermidis,* grow poorly or not at all in the gut?

ANSWER: Bile salts present in the intestine (not on the skin) easily gain access to and destroy cytoplasmic membranes of Gram-positive organisms (unless the organism possesses bile salt hydrolases). Gram-negative microbes have extra protection in the form of an outer membrane and so can survive better in the intestine.

23.3 How might normal biota escape the intestine and cause disease at other body sites?

ANSWER: Normal biota can escape through intestinal perforations resulting from gunshot or knife wounds, surgery, or cancer.

23.4 Why can the colon be considered a fermenter?

ANSWER: The contents are continually flowing through the intestinal tube, with food containing substrates for fermentation ingested at one end and waste containing fermentation products removed from the other.

23.5 Figure 23.15 shows how a neutrophil extracellular trap can ensnare a nearby pathogen. What bacterial structure might blunt the microbicidal effect of NETs?

ANSWER: Bacterial capsules can prevent direct contact between the NET and the cell.

23.6 Why do defensins have to be so small? Do defensins kill normal microbiota?

ANSWER: Defensins need to be small so that they can get through the outer membrane of Gram-negative organisms and the thick peptidoglycan maze of Gram-positive organisms. Defensins do kill normal microbiota and in fact are part of what keeps levels of the normal intestinal microbiota in check. Research suggests that a decrease in intestinal defensin production can lead to an imbalance of gastrointestinal microbes in both number and species. This imbalance appears to contribute to conditions such as irritable bowel syndrome (IBS) or inflammatory bowel disease (IBD). Pathogens or normal microbiota that penetrate the intestinal mucosa probably encounter higher concentrations of defensins as they do so.

23.7 As illustrated in Figure 23.25, integrin is important for neutrophil extravasation. Some individuals, however, produce neutrophils that lack integrin. What is the likely consequence of this genetic disorder?

ANSWER: Individuals with neutrophils lacking integrin have leukocyte adhesion deficiency. Their neutrophils are defective in extravasation. These patients are more susceptible to infections because the neutrophils cannot easily get out of the bloodstream.

23.8 What happens to all the neutrophils that enter a site of infection once the infection has resolved?

ANSWER: Once bacteria at the site of infection have been killed, the tissue cells in the area stop making the cytokines and chemokines that attracted neutrophils in the first place, so neutrophils stop coming in, and some wander out. But the majority of neutrophils undergo a self-programmed cell death (called apoptosis) and are cleared by monocytes in the area by phagocytosis. The average life span of a neutrophil is only 5 days.

23.9 If NK cells can attack infected host cells coated with antibody, why won't neutrophils?

ANSWER: Actually, neutrophils (PMNs) <u>can</u> attack infected cells coated with antibody, but the killing mechanism is different from ADCC. Human neutrophils do not make perforin or the other ADCC-related compounds, called granzymes, used by NK cells to kill target cells. In addition to that difference, NK cells possess a type of Fc receptor not found on neutrophils, which means the intracellular signaling pathways are different between NK cells and neutrophils. Neutrophils can be activated, however, when their Fc receptors bind antibody. Activated neutrophils make reactive oxygen products and can release a variety of peptides, including defensins, cathelicidins, and myeloperoxidase, which can all damage target cells.

23.10 Figure 23.31 shows how the complement cascade can destroy a bacterial cell. Factor H (not shown) is a blood protein that regulates complement activity. Factor H binds host cells and inhibits complement from attacking our cells by accelerating degradation of C3b and C3bBb. How could bacteria take advantage of factor H?

ANSWER: Some pathogens have structures on their cell surfaces that can bind factor H. This binding inhibits the alternative pathway from attacking the microbe.

23.11 If increased fever limits bacterial growth, why do bacteria make pyrogenic toxins?

ANSWER: The pyrogenic toxins have other effects that compromise and damage the host. The toxins can induce cytokines that damage local host cells or confuse the immune system. This can provide the pathogen with nutrients and help it hide from the immune system longer. Pyrogenic toxins include lipopolysaccharide and protein toxins such as toxic shock syndrome toxin (see Chapter 25).

CHAPTER 24

24.1 Two different stretches of amino acids in a single protein form a 3D antigenic determinant. Will the specific immune response to that 3D antigen also respond to one of the two amino acid stretches alone?

ANSWER: Most likely no. It is the 3D shape formed by the two stretches that is recognized as an antigen. A denatured protein that contains both amino acid stretches will not possess the 3D shape of the antigen. However, other specific immune responses involving different subsets of lymphocytes can recognize the separate amino acid stretches that together form the 3D antigenic determinant. As an analogy, take a computer image of a friend's face and shuffle the facial features. Turn the nose upside down, exchange the eyes with the mouth, and lower the ears. Since you were programmed to respond to the original facial configuration, you likely would not recognize the rearranged face as a whole. But you might find that the nose looks familiar.

24.2 How does a neutralizing antibody that recognizes a viral coat protein prevent infection by the associated virus?

ANSWER: Neutralizing antibodies usually bind attachment proteins on the virus and sterically prevent them from binding to host cell receptors (see **Fig. 24.5**). Some antibodies to enveloped viruses might trigger the complement cascade (see later in the chapter), thus destroying the membrane.

24.3 The attachment proteins of different rhinovirus strains all bind to ICAM-1. How can all these proteins be immunologically different if they find the same target (ICAM-1)? Why won't antibodies directed against one rhinovirus strain block the attachment of other rhinovirus strains?

ANSWER: The ICAM-1 binding sites on different rhinovirus strains have small, but immunologically significant, differences. ICAM-1 is "promiscuous" in its binding specificity toward the rhinovirus attachment proteins, whereas antibodies are more specific. Thus, ICAM-1 protein is like a master key that can fit dozens of different locks (rhinovirus-binding sites). Each lock (binding site) has a very specific key (antibody) that won't unlock any of the other locks. But the master key (ICAM-1) can turn all locks.

24.4 Can an $F(ab')_2$ antibody fragment prevent the binding of rhinovirus to the ICAM-1 receptor on host cells? And can an $F(ab')_2$ antibody fragment facilitate phagocytosis of a microbe?

ANSWER: An $F(ab')_2$ antibody fragment can prevent binding of rhinovirus to the ICAM-1 receptor on host cell surfaces. The antigen-binding sites will block virus receptor access to ICAM-1. However, an $F(ab')_2$ antibody fragment cannot facilitate phagocytosis of a microbe, because an opsonizing antibody needs its Fc region to bind Fc receptors on phagocytes. Thus, an $F(ab')_2$ antibody can bind to the microbe antigen but cannot link the microbe to a phagocyte's cell surface.

24.5 What types of antibodies will IgG taken from Sherrie raise when injected into John (for example, anti-isotype, anti-allotype, or anti-idiotype)?

ANSWER: Since they are both human, Sherrie's IgG will not elicit anti-isotype antibodies in John. Sherrie's IgG will elicit anti-allotype antibodies and anti-idiotype antibodies in John. Thus, John will develop antibodies that react only to the amino acid differences between John's and Sherrie's IgG antibodies.

24.6 The mother of a newborn was found to be infected with rubella, a viral disease. Infection of the fetus could lead to serious consequences for the newborn. How could you determine whether the newborn was infected while in utero?

ANSWER: Since maternal IgM antibodies cannot cross the placenta, finding IgM antibodies to rubella antigens in the newborn's circulation indicates that the fetus was infected and initiated its own immune response. If the newborn has only IgG antibodies to rubella (no IgM antibodies), then the child was not infected and maternal IgG crossed the placenta.

24.7 B cells in early stages have both IgM and IgD surface antibodies, but the delta region has no switch region. Why does the delta region have no switch region?

ANSWER: If the delta region had a switch region, then the B cell could make IgD only after DNA rearrangement. Then, no single B cell could have IgM and IgD at the same time. Recombination at the DNA level is not involved, because alternative RNA-splicing events after transcription determine whether an IgM or IgD molecule is made. B cells at the early stages have both IgM and IgD surface antibodies.

24.8 Why do individuals with type A blood have anti-B and not anti-A antibodies?

ANSWER: Because the B-cell population that would react to type A antigen was deleted during B-cell maturation.

24.9 Why do immunizations lose their effectiveness over time?

ANSWER: Because memory B cells eventually die. Without some exposure to antigen, those memory cells will not be replaced.

24.10 What would happen to someone lacking CD154 on T_H2 cells because of a genetic mutation?

ANSWER: The person's T-cell and B-cell numbers would remain normal, but the B cells would not undergo heavy-chain class switching. Thus, any plasma cells made would make only IgM, causing serum levels of IgM to rise. No other antibody type would be secreted; the result is called hyper IgM syndrome.

24.11 Transplant rejection is a major consideration when transplanting most tissues, because host T_C cells can recognize allotypic MHC on donor cells. Why, then, are corneas easily transplanted from a donor to just about any other person?

ANSWER: The cornea is not normally vascularized. So even though corneal cells express MHC proteins, circulating host T cells do not have an opportunity to interact with them. The cornea will not be rejected. This is called an immune-privileged site.

24.12 Why does attaching a hapten to a carrier protein allow antihapten antibodies to be produced?

ANSWER: B cells with antihapten surface antibody (as part of the B-cell receptor) can take up hapten but cannot present the hapten to a helper T cell. The same B cell can also take up the hapten bound to a carrier molecule, and because the carrier molecule is larger than the hapten, the B cell will present the carrier epitope to the helper T cell. The helper T cell stimulates the B cell, which was already programmed to make antihapten antibody, to differentiate into plasma and memory B cells.

24.13 Do bone marrow transplants in a patient with severe combined immunodeficiency (SCID) require immunosuppressive chemotherapy?

ANSWER: No. The patient has no T cells to recognize foreign antigens, so the transplant is not rejected.

24.14 How can a stem cell be differentiated from a B cell at the level of DNA?

ANSWER: DNA recombination at switch regions will have taken place in the B cell but not the stem cell. Thus, the B cell will have fewer segments for each of the V, D, and J regions, while the stem cell will have all of them. PCR techniques can be used to view these differences.

CHAPTER 25

25.1 Figure 25.6 presents the association between LD_{50} and virulence. Is a microbe with an LD_{50} of 5×10^4 more or less virulent than a microbe with an LD_{50} of 5×10^7?

ANSWER: Because it takes fewer cells to cause disease, the microbe with the smaller LD_{50} (5×10^4) is more virulent.

25.2 Figure 25.20 illustrates how cholera toxin works to cause diarrhea. To develop a vaccine that generates protective antibodies, which subunit of cholera toxin should be used to best protect a person from the toxin's effects?

ANSWER: Antibodies to the B subunit will be more protective. Inactivating the B subunit will prevent binding of the toxin to cell membranes. The A-subunit active site is typically sequestered in these toxins and inaccessible to antibody. Furthermore, once the A subunit has entered a host cell, antibodies cannot enter and neutralize it.

25.3 How might you experimentally determine whether a pathogen secretes an exotoxin? (*Hint:* You can learn more about how new toxins are found in eTopic 25.7.)

ANSWER: The microbe can be grown in liquid culture and the cells removed by either centrifugation or filtration. If the organism makes an exotoxin, it may well be present in the cell-free supernatant. The presence of a toxin can be determined by injecting the supernatant into an animal model (for example, mice) and examining the result (death or altered function). Alternatively, the supernatant can be administered to a layer of tissue culture cells and the health of the monolayer noted.

25.4 Would patients with iron overload (excess free iron in the blood) be more susceptible to infection?

ANSWER: With a few exceptions, withholding iron from potential pathogens is a host defense strategy, because when iron is plentiful, the microbe does not have to expend energy to get it and so can readily grow. On the other hand, low iron can also be a signal to express various virulence genes, so for some organisms, high iron might hinder infection.

25.5 Use an Internet search engine to determine what other toxins are related to the cholera toxin A subunit. (*Hint:* Start by searching on "protein" at PubMed to find the protein sequence; then use a BLAST program.)

ANSWER: Go to PubMed (http://www.ncbi.nlm.nih.gov/pubmed). At the top of the page, select **Protein** from the Search dropdown list. Then type "cholera enterotoxin A subunit" in the text box and click **Search**. Click on the appropriate result (any designated "258 aa protein"). Highlight and copy the entire amino acid sequence at the bottom of the page. Then, back at the NCBI home page (http://www.ncbi.nlm.nih.gov), click on **BLAST**; select **Protein Blast**; paste the A-subunit sequence into the box; click on **BLAST** at the bottom of the page. After some seconds, all the homology results will be displayed. Items with large negative E values represent the most significant homologies. Scroll down the page to find the actual alignments between the query protein and each homolog. The toxin of *Escherichia coli* possesses an A subunit similar to that of cholera.

25.6 Protein and DNA have very different structures. Why would a protein secretion system be derived from a DNA-pumping system?

ANSWER: Conjugation systems actually move DNA that is attached to a pilot protein at the 5′ end. A pilot protein is made by the conjugation system and then binds to the 5′ end of the DNA to be transferred and "pilots" the DNA through the conjugation pore. So, a modified conjugation system that moves only protein is not as much of a leap as might initially be thought.

25.7 Figure 25.32 shows *Shigella* forming an actin tail at one pole. Why do organisms such as *Shigella* and *Listeria* assemble actin-polymerizing proteins at only one pole?

ANSWER: If actin tails formed at both poles, the organism would spin aimlessly throughout the cell and would likely not be able to protrude into and infect an adjacent host cell. The organisms have mechanisms that localize the key bacterial actin polymerization protein at only one pole (the older pole) of the cell. The molecular basis for this unipolar localization remains unclear but is probably related in some way to polar aging (discussed in Chapter 3).

25.8 How can you determine whether a bacterium is an intracellular pathogen?

ANSWER: Microscopic examination to see whether bacteria are found within cultured mammalian cells is usually not satisfactory. The difficulty lies in determining whether the organism is inside the host cell or just bound to its surface—or, if it is inside, whether it is a live or dead bacterium. One commonly used approach is to add to infected cell monolayers an antibiotic that can kill the microbe but will not penetrate the mammalian cells. The protein synthesis inhibitor gentamicin is typically used. A bacterium that invades a host cell will gain sanctuary from gentamicin and grow intracellularly. Extracellular bacteria and bacteria attached to the outside of the host cell are killed. Counting viable colony-forming units of bacteria released from the mammalian cells by gentle detergent treatments at various times will reveal whether the organism grew intracellularly. This will only work, though, if the microorganism is not an obligate intracellular parasite.

25.9 Why might killing a host be a bad strategy for a pathogen?

ANSWER: The goal of any microbe is to maintain its species. If a microbe did not have an opportunity to easily spread to a new host, killing its host would be tantamount to suicide.

25.10 You have discovered an E3 ligase that specifically modifies surface proteins of *Listeria monocytogenes.* Using what has been presented about toxin entry and protein secretion in Chapter 25, suggest some strategies for delivering this enzyme into host cells infected with this pathogen.

ANSWER: First realize that *Listeria* is an intracellular pathogen that can escape the phagosome and avoid autophagy. Ubiquitylation of the organism's surface by the E3 ligase, however, may mark the organism for autophagy and clearance. There are several ways this new enzyme could be delivered into infected cells using components of microbial toxins. For instance, fuse the E3 ligase to a nontoxic form of the cholera toxin A subunit. Done properly, this fusion protein could interact with cholera toxin B subunits, which would bind to the infected host cell and deliver the E3 ligase. You might also be able to engineer the E3 ligase as part of a T3SS or T6SS effector protein. An avirulent form of another bacterium equipped with one of these secretion systems could then deliver the E3 ligase directly into the cytoplasm of the host cell. Finally, you might be able to engineer an attenuated virus with the E3 ligase gene. Once infected with the modified virus, the host would make the enzyme itself. Keep in mind that these scenarios are entirely hypothetical.

CHAPTER 26

26.1 Does *Staphylococcus aureus* have to disseminate through the circulation to produce the symptoms of scalded skin syndrome (SSS)? Explain why or why not.

ANSWER: Because SSS symptoms are caused by an exotoxin, the infection can be in a single location [a focus of infection, such as the nares (nostrils)], but the toxin can disseminate through the circulatory system.

26.2 Why would treatment of some infections require multiple antibiotics?

ANSWER: Although not typically recommended for simple infections, with some organisms multiple antibiotic therapy is required to effect a cure. The organisms that require multidrug therapy are very hardy in vivo and able to grow in the presence of single antibiotics. Active infection with *Mycobacterium tuberculosis,* for example, is typically treated with isoniazid (thought to inhibit synthesis of mycolic acid) and rifampin (an RNA polymerase inhibitor), and sometimes a third drug, pyrazinamide (which inhibits fatty acid synthesis). Using multiple antibiotics will minimize the chance that the organism will develop resistance to any one antibiotic. Multiple antibiotic therapy is also recommended for *Helicobacter pylori* (ulcers, treated with lansoprazole, amoxicillin, and clarithromycin) and methicillin-resistant *Staphylococcus epidermidis* (MRSE; treated with vancomycin and gentamicin). MRSE can cause serious infections, such as meningitis.

26.3 Explain why patient noncompliance (failure to take drugs as directed) is thought to have led to XDR-TB.

ANSWER: In most cases, an organism causing an infection starts out being susceptible to a given antimicrobial agent. However, the more times an organism divides, the more likely it is that a spontaneous mutation producing drug resistance will arise. The amount of the antibiotic and the duration of its use are calibrated so that all organisms are killed—either by the drug itself or by the immune system. Often the drug is given to stop growth of the bacterium so that the immune system has enough time to actually do the killing. If a patient stops taking an antibiotic before the prescribed end of the treatment, the organism can start to replicate again, renewing the chance that a spontaneous drug-resistant variant will be produced. Even if drug therapy is resumed, the resistant organism will continue to grow and cause disease—and can be transmitted to other individuals.

26.4 Why do you think most urinary tract infections occur in women?

ANSWER: The major reason is anatomy. Most bladder UTIs come from access through the urethra. Since the urethra in men is longer than in women, it usually takes a catheter to introduce bacteria into a male bladder. The trip for the infecting organism in women is much shorter. However, in people older than 50, UTIs become more common in both men and women, with less difference between the sexes. The reason is as yet unclear.

26.5 Urine samples collected from six hospital patients were placed on a table at the nurses' station awaiting pickup from the microbiology lab. Several hours later, a courier retrieved the samples and transported them to the lab. The next day, the lab reported that four of the six patients had UTIs. Would you consider these results reliable? Would you start treatment based on these results?

ANSWER: Because urine is a good growth medium for many bacteria, the delay of several hours in picking up the samples gave the organisms time to replicate and increase their numbers. Consequently, the lab results should be viewed with suspicion.

26.6 Given that *Neisseria gonorrhoeae* is exquisitely sensitive to ceftriaxone, why do you suppose the patient in this case was also treated with tetracycline? And why can one person be infected repeatedly with *N. gonorrhoeae*?

ANSWER: Because STDs often travel in pairs and because the initial symptomology is similar, the clinician will want to "cover" the patient for possible chlamydia infection. Chlamydias are not susceptible to ceftriaxone. The large, single dose of ceftriaxone (as opposed to smaller multiple injections given over days) was given because patients with gonorrhea are typically poorly compliant and fail to return for subsequent injections. A person who experiences gonorrhea is not protected from reinfection because the organism's surface antigens undergo phase variations (see Section 10.6) and may down-regulate the immune system (see **Fig. 26.13C**).

26.7 Like the cause of plague, HIV is a blood-borne disease. Why, then, do fleas and mosquitoes fail to transmit HIV?

ANSWER: When insect vectors take a blood meal, they typically defecate or regurgitate simultaneously. So, theoretically, they could serve as a vector for HIV. Pathogens such as the West Nile virus that are transmitted by insect vectors actually grow in their insect hosts; however, HIV does not. Because it is unusually fragile, HIV will die quickly. There have been no cases of HIV transmitted by insect vectors.

26.8 Normal cerebrospinal fluid is usually low in protein and high in glucose. The protein and glucose content does not change much during a viral meningitis, but bacterial infection leads to greatly elevated protein and lowered glucose levels. What could account for this?

ANSWER: There are several explanations. Bacteria and infiltrating PMNs will consume glucose, and alterations in the blood-brain barrier can lead to decreased transport of glucose into the spinal fluid. As a result, glucose levels plummet. Growth of the bacteria and infiltration of PMNs account for the increase in CSF protein levels. Viruses are very small and consist mostly of nucleic acids, so even at high numbers they will not significantly add to CSF protein content. Since viruses do not grow in CSF directly, they will not consume glucose. In addition, viral meningitis does not cause a great infiltration of PMNs into the CSF—another reason glucose levels remain high and protein levels remain low.

26.9 Knowing the symptoms of tetanus, what kind of therapy would you use to treat the disease?

ANSWER: At the first sign of muscle spasm, antitoxin should be given. If tetany is severe, then muscle relaxants can relieve spasms.

26.10 Figure 26.19 demonstrates that tetanus toxin has a mode of action (spastic paralysis) opposite to that of botulism toxin (flaccid paralysis). Since they have opposing modes of action, can botulism toxin be used to save a patient with tetanus?

ANSWER: When you first think about it, this sounds feasible. Tetanus toxin causes spastic paralysis, while botulinum toxin causes flaccid paralysis. However, the two toxins act at different places in the nervous system. Tetanus toxin produced by *Clostridium tetani* growing in a wound travels to the central nervous system to exert its effect, whereas botulism toxin works at the periphery. If you administer botulinum toxin intravenously to counteract the whole-body effect of tetanus toxin, then it will affect essentially all neuronal junctions, autonomic and voluntary alike. Thus, the dose of botulinum toxin required to counteract the effect of tetanus toxin on voluntary muscle contraction would likely kill the patient by stopping his or her breathing.

26.11 How do the actions of tetanus toxin and botulinum toxin actually help the bacteria colonize or obtain nutrients?

ANSWER: This is a difficult question to answer. Few scientists have speculated. Recall that the toxins are encoded by genes in resident bacteriophages that became part of the clostridial genome through horizontal transfer from some other source. Since the organisms (vegetative, as well as spores) normally reside in soil, the actual function of these toxins may have something to do with survival in that habitat. The toxin's effect on humans may simply be an unfortunate accident. With tetanus toxin, however, there may be benefit in that muscle spasms could limit oxygen delivery to infected tissues, enabling a more anaerobic environment for growth. Cell death may also release iron or other nutrients useful to *Clostridium tetani*.

26.12 If *Clostridium botulinum* is an anaerobe, how might botulism toxin get into foods?

ANSWER: The most common way botulism toxin gets into food today is via home canning. The canning process involves heating food in jars to very high temperatures. The heat destroys microorganisms and drives out oxygen; both processes help to preserve foods. If the jars are not heated to sterilization conditions, spores of *C. botulinum* will survive. When the jars are cooled for storage at room temperature, the spores germinate; the organism then grows in the anaerobic medium and releases the toxin. When the food is eaten, the toxin is eaten too.

26.13 A patient presenting with high fever and in an extremely weakened state is suspected of having a septicemia. Two sets of blood cultures are taken from different arms. One bottle from each set grows *Staphylococcus aureus,* yet the laboratory report states that the results are inconclusive. New blood cultures are ordered. What would make these results inconclusive?

ANSWER: There must be something different about the two strains of staphylococci grown in the separate bottles. For example, they may have different antibiotic susceptibility patterns when tested against a battery of antibiotics. One strain may be susceptible to penicillin, while the other strain is resistant. Since the expectation is that a single strain initiated the infection, both isolates should exhibit the same susceptibility pattern. The laboratory suspects contamination of the bottles from separate sources. These organisms are not the source of infection.

26.14 How might you design/construct a more effective vaccine for an antigen (for instance, the *Yersinia pestis* F1 antigen, or hepatitis A) that would harness the power of a Toll-like receptor (TLR)? *Hint:* TLR5 from Table 23.5.

ANSWER: By genetically fusing DNA encoding a hepatitis antigen to DNA encoding flagellin, a MAMP (microbe-associated molecular pattern) that is recognized by TLR5 on dendritic cells, you would end up with a chimeric protein that would activate innate immune mechanisms, via TLR5, and improve antigen presentation. The result is a better link between innate and adaptive immune mechanisms.

CHAPTER 27

27.1 Figure 27.3 illustrates how MICs are determined. Test your understanding of how MICs are measured in the following example. The drug tobramycin is added to a concentration of 1,000 µg/ml in a tube of broth from which serial twofold dilutions were made. Including the initial tube (tube 1), there are a total of ten tubes. Twenty-four hours after all the tubes are inoculated with *Listeria monocytogenes*, turbidity is observed in tubes 6–10. What is the MIC?

ANSWER: The MIC is 62.5 µg/ml, the concentration in tube 5, the last tube with no growth. (Relative to tube 1, tube 5 has been diluted 2^4—a 16-fold dilution: 1,000 → 500 → 250 → 125 → 62.5.)

27.2 What additional test performed on an MIC series of tubes will tell you whether a drug is bacteriostatic or bactericidal?

ANSWER: Streaking a portion of the broth from the dilution tubes showing no growth onto an agar plate. If the drug is bacteriostatic, colonies will form on the agar plate because during streaking, the bacteria are removed from the presence of the drug. If the antibiotic is bactericidal, no colonies will form, because the organisms are dead before plating. This method determines the minimum bactericidal concentration (MBC) of an antibiotic. MBC would be defined as the lowest dilution that did not yield viable cells.

27.3 A patient with a bacterial lung infection was given the antibiotic shown in Figure 27.6 and was told to take one pill twice a day. The pathogen is susceptible to this drug. Will the prescribed treatment be effective? Explain your answer.

ANSWER: No. **Figure 27.6** shows that the antibiotic remains at effective serum levels for about 8 hours. If the patient takes two pills 12 hours apart, there will be two 4-hour periods (a total of 8 hours) during each 24-hour interval when serum levels fall below an effective MIC concentration. Each 4-hour window of low concentration will allow regrowth and a greater opportunity to select for antibiotic-resistant strains.

27.4 You are testing whether a new antibiotic will be a good treatment choice for a patient with a staph infection. The Kirby-Bauer test using the organism from the patient shows a zone of inhibition of 15 mm around the disk containing this drug. Clearly, the organism is susceptible. But you conclude from other studies that the drug would be ineffective in the patient. What would make you draw this conclusion?

ANSWER: If the average attainable tissue level of the drug is below the MIC, then the drug will be ineffective.

27.5 When treating a patient for an infection, why would combining a drug such as erythromycin with a penicillin be counterproductive? (Erythromycin is described in Section 8.3.)

ANSWER: Erythromycin, a bacteriostatic drug, will stop growth, which indirectly stops cell wall synthesis and renders the microbe insensitive to penicillin.

27.6 The enzyme DNA gyrase, a target of the quinolone antibiotics, is an essential protein in DNA replication. The quinolones bind to and inactivate this protein. Research has proved that quinolone-resistant mutants contain mutations in the gene encoding DNA gyrase. If the resistant mutants contain a mutant DNA gyrase and DNA gyrase is essential for growth, then why are these mutations not lethal?

ANSWER: The mutations cause changes in DNA gyrase that have little to no effect on its function but do prevent the drug from binding. Thus, the mutant organism will continue to twist its DNA and grow well with or without the drug.

27.7 Why might a combination therapy of an aminoglycoside antibiotic and cephalosporin be synergistic?

ANSWER: The two drugs given together could act synergistically because the cephalosporin can weaken the cell wall and allow the aminoglycoside easier access to the cell interior, where it can attack ribosomes. This synergism is especially useful in organisms that have some resistance to both drugs.

27.8 Fusaric acid is a cation chelator that normally does not penetrate the *E. coli* membrane, which means *E. coli* is typically resistant to this compound. Curiously, cells that develop resistance to tetracycline become <u>sensitive</u> to fusaric acid. Resistance to tetracycline is usually the result of an integral membrane efflux pump that pumps tetracycline out of the cell. What might explain the development of fusaric acid sensitivity?

ANSWER: Fusaric acid is imported by the tetracycline efflux pump. This phenomenon can be used to isolate mutants with deletions of transposons encoding tetracycline resistance. Transposons, such as Tn*10*, which carries tetracycline resistance, can spontaneously delete at a frequency of about 10^{-6}, but finding one out of a million tetracycline-susceptible cells is impossible without a positive selection. Fusaric acid provides that positive selection, since the cell with the Tn*10* deletion will be resistant to fusaric acid.

27.9 Mutations in the ribosomal protein S12 (encoded by *rpsL*) confer resistance to streptomycin. Given a cell containing both *rpsL*$^+$ and *rpsL*^{R} genes, would the cell be streptomycin resistant or sensitive? (Recall that genes encoding <u>r</u>ibosome <u>p</u>roteins for the <u>s</u>mall subunit are designated *rps.* "+" indicates the wild-type allele, while "R" indicates a gene whose product is resistant to a certain drug.)

ANSWER: This merodiploid cell would contain two sets of ribosomes: One set, containing normal S12, would be sensitive to streptomycin; another set, containing the resistant S12, would be resistant. Because streptomycin causes mistranslation of mRNA on sensitive ribosomes, inappropriate proteins that can kill the cell would still be synthesized. Thus, the cell would remain sensitive to streptomycin. Note, however, that the recessive nature of antibiotic resistance seen in this case is not the norm. Resistance is a dominant trait in the majority of cases.

27.10 Figure 27.23 shows a distinctive-looking colony morphology. Why are the colonies on this agar plate red and mucoid?

ANSWER: MacConkey medium contains lactose. As described in Chapter 4, an organism that ferments lactose acidifies the medium and takes up neutral red, turning the colonies red. Organisms that produce a large capsule produce slimy, mucoid colonies. *Klebsiella pneumoniae* is one such bacterium that also ferments lactose. A well-trained laboratory technician will immediately suspect *K. pneumoniae* upon seeing these colonies.

27.11 Could genomics ever predict the drug resistance phenotype of a microbe? If so, how?

ANSWER: Yes. If the organism's genome possesses genes whose deduced protein sequences harbor significant similarity to antibiotic resistance proteins from other organisms, then one can predict a similar drug resistance. Definitive proof of drug resistance requires actual in vitro testing.

CHAPTER 28

28.1 Two blood cultures, one from each arm, were taken from a patient with high fever. One culture grew *Staphylococcus epidermidis,* but the other blood culture was negative (no organisms grew out). Is the patient suffering from septicemia caused by *S. epidermidis*?

ANSWER: Probably not. *S. epidermidis* is a common inhabitant of the skin and could easily have contaminated the needle when blood was taken from the patient. The fact that only one of the two cultures grew this organism supports this conclusion. If the patient was really infected with *S. epidermidis,* then both blood cultures would have grown this organism.

28.2 A 30-year-old woman with abdominal pain went to her physician. After the examination, the physician asked the patient to collect a midstream urine sample that they would send to the lab across town for analysis. The woman complied and handed the collection cup to the nurse. The nurse placed the cup on a table at the nurses' station. Three hours later, the courier service picked up the specimen and transported it to the laboratory. The next day, the report came back: "greater than 200,000 CFUs/ml; multiple colony types; sample unsuitable for analysis." Why was this determination made?

ANSWER: Although the CFU number is high enough to consider relevant, UTIs are typically caused by a single organism. The fact that the lab found many different colony types suggests a problem with specimen collection. In this case, not refrigerating the sample allowed the small number of urethral contaminants to overgrow the specimen.

28.3 Use Figure 28.9 to identify the organism from the following case: A sample was taken from a boil located on the arm of a 62-year-old man. Bacteriological examination revealed the presence of Gram-positive cocci that were also catalase-positive, coagulase-positive, and novobiocin resistant.

ANSWER: *Staphylococcus aureus.* The novobiocin test is irrelevant in this situation.

28.4 If the results in Figure 28.13 were testing for HIV RNA, then which patient would have the higher viral load in their blood?

ANSWER: Patient 1. The detection threshold was crossed earlier for patient 1 (cycle 10) than for patient 2 (cycle 22), so patient 1's serum must have a higher number of HIV virus particles in it.

28.5 Why does finding IgM to West Nile virus indicate current infection? Why wouldn't finding IgG do the same?

ANSWER: Upon infection with any organism, IgM antibodies are the first to rise. After a short time, the levels of IgM decline as IgG levels rise. IgG, however, can remain in serum for years, making it a poor prognosticator of current infection.

28.6 Why does adding albumin or powdered milk prevent false positives in ELISA?

ANSWER: Antibodies are proteins. They can stick to plastic just as easily as to the antigen being tested. Without blocking all the possible binding sites on plastic with albumin (a major protein ingredient in milk), any antibody from the patient's serum could stick to the plastic instead of to the antigen and react with the secondary enzyme-conjugated antihuman antibody.

28.7 Specific antibodies against an infectious agent can persist for years in the bloodstream, long after the infection resolves. How is it possible, then, that antibody titers can be used to diagnose recently acquired diseases such as infectious mononucleosis? Couldn't the antibody be from an old infection?

ANSWER: During the course of a disease, the body's immune system increases the amount of antibody made specifically against the infectious agent. Thus, one compares the antibody titer in a blood sample taken from a patient in the active, or acute, phase of disease with the antibody titer several weeks later, when the patient is in the recovery, or convalescent, stage. Seeing a greater-than-fourfold rise in a specific antibody titer (for example, in mononucleosis) indicates that the patient's immune system was responding to the specific agent. Remember, simply finding IgG against an organism or virus in serum indicates only that the patient was exposed to that microbe at some time in the past.

28.8 Methicillin, a beta-lactam antibiotic, is very useful in treating staphylococcal infections. The emergence of methicillin-resistant strains of *Staphylococcus aureus* (MRSA) is a very serious development because few antibiotics can kill these strains. Imagine a large metropolitan hospital in which there have been eight serious nosocomial infections with MRSA and you are responsible for determining the source of infection so that it can be eliminated. How would you accomplish this task using common bacteriological and molecular techniques?

ANSWER: Samples from all the affected patients and from the hospital staff would be screened for the presence of *S. aureus* resistant to methicillin. Each strain would then be rapidly sequenced in part (multilocus sequencing) or in whole (whole-genome sequencing) using the techniques described in Chapter 7. Strains from all the patients will likely have identical restriction patterns or DNA sequences if they came from the same source. The source is then identified by determining which staff member possesses MRSA with the same pattern. The source may also be inanimate, such as surgical equipment or ventilation apparatus. A connection between patients and specific staff members or instruments would also have to be demonstrated.

28.9 What are some reasons why some diseases spread quickly through a population while others take a long time?

ANSWER: There are several factors. One is mode of transmission; airborne diseases can spread more quickly than food-borne diseases, for instance. Sexually transmitted diseases spread more slowly still. Herd immunity is another factor. Herd immunity is based on the number of individuals within a population that are resistant to a disease. Someone immune to the disease cannot pass it on. The more immune people there are in a population (or herd), the slower the epidemic spreads to susceptible people.

28.10 On the ProMED-mail web page (www.promedmail.org), click on the interactive world map to view outbreaks recorded by WHO. What outbreaks happened throughout the world during the current year?

ANSWER: As of this writing (2013), there were outbreaks of respiratory disease by a novel coronavirus in the United Kingdom, polio in Niger, yellow fever in Chad, and avian flu (H5N1) in Cambodia, among many others.

GLOSSARY

2D gels *or* 2D PAGE See **two-dimensional polyacrylamide gel electrophoresis**.

A

A site See **acceptor site**.

ABC transporter An ATP-powered transport system that contains an ATP-binding cassette.

aberration An imperfection in a lens.

abiotic Produced without living organisms; occurring in the absence of life.

absorption In optics, the capacity of a material to absorb light.

acceptor site (A site) The region of the ribosome that binds an incoming charged tRNA.

accessory protein A protein found in the viral capsid or tegument that is needed early in the viral life cycle.

acid-fast stain A diagnostic stain for mycobacteria, which retain the dye fuchsin because of mycolic acids in the cell wall.

Acidobacteria A phylum of Gram-negative bacteria with an outer membrane, related to the Proteobacteria; often found in soil habitats.

acidophile An organism that grows fastest in acid (generally defined as below pH 5).

ACP See **acyl carrier protein**.

acquired immunodeficiency syndrome (AIDS) A disease caused by HIV that leads to the destruction of T cells and the inability to fight off opportunistic infections.

actin filament See **microfilament**.

Actinobacteria A phylum of Gram-positive bacteria with high GC content.

actinomycete A member of the group Actinomycetales, an order of Actinobacteria including branched spore formers such as *Streptomyces*, as well as irregularly shaped corynebacteria.

activation energy The energy needed for reactants to reach the transition state between reactants and products.

activator A regulatory protein that can bind to a specific DNA sequence and increase transcription of genes.

active transport An energy-requiring process that moves molecules across a membrane against their electrochemical gradient.

acyl carrier protein (ACP) A protein that can carry an acetyl group for anabolic pathways such as fatty acid synthesis.

adaptive immunity Immune responses activated by a specific antigen and mediated by B cells and T cells.

ADCC See **antibody-dependent cell-mediated cytotoxicity**.

adenosine triphosphate (ATP) A ribonucleotide with three phosphates and the base adenine. It has many functions in the cell, including precursor for RNA synthesis and energy carrier.

adenylate cyclase An enzyme that converts ATP into cyclic adenosine monophosphate (cAMP).

adherence The ability of an organism to attach to a substrate.

adhesin Any cell-surface factor that promotes attachment of an organism to a substrate.

ADP-ribosyltransferase A bacterial toxin that enzymatically transfers the ADP-ribose group from NAD^+ to target proteins, altering the target protein's structure and function.

aerial mycelium A hypha that extends above the surface and produces spores at its tip.

aerobic respiration The use of oxygen as the terminal electron acceptor in an electron transport system. A proton gradient is generated and used to drive ATP synthesis.

aerotolerant anaerobe An organism that does not use oxygen for metabolism but can grow in the presence of oxygen.

affinity chromatography A chromatographic technique that utilizes the ability of biological molecules to bind to certain ligands specifically and reversibly.

affinity maturation The process by which the antigen-binding site of an antibody gains increased affinity for its target antigen or epitope.

AFM See **atomic force microscopy**.

agar A polymer of galactose that is used as a gelling agent.

AIDS See **acquired immunodeficiency syndrome**.

akinete A specialized spore cell formed by some filamentous cyanobacteria.

alga *pl.* algae A microbial eukaryote that contains chloroplasts.

algal bloom An overgrowth of algae on the water surface, caused by an increase in a limiting nutrient.

alkaline fermentation Bacterial fermentation in conjunction with proteolysis and amino acid catabolism that generates ammonia in amounts that raise pH.

alkaliphile An organism with optimal growth in alkali (generally defined as above pH 9).

allergen An antigen that causes an allergic hypersensitivity reaction.

allosteric site A regulatory site on a biological molecule distinct from the ligand/substrate-binding site.

allotype An amino acid difference in the antibody constant region that distinguishes different individuals within a species.

alternation of generations A life cycle that alternates between populations of haploid cells and diploid cells, which undergo meiosis and fertilization.

alveolar macrophage A type of macrophage, located in the lung alveoli, that phagocytoses foreign material.

alveolate A member of the eukaryotic group Alveolata—ciliated or flagellated protists with complex cortical structure.

ameba *or* amoeba A protist that moves via pseudopods.

amensalism An interaction between species that harms one partner but not the other.

amino acid The monomer unit of proteins. Each amino acid contains a central carbon covalently bonded by a hydrogen, an amino group, a carboxyl group and a side chain. An exception is proline, in which the side chain is cyclized with the central carbon.

aminoacyl-tRNA synthetase An enzyme that attaches a specific amino acid to the correct tRNA, thereby charging the tRNA.

ammonification The generation of ammonia from organic nitrogen.

amphibolic Describing a metabolic pathway that is reversible and can be used for both catabolism and anabolism.

amphipathic Having both hydrophilic and hydrophobic portions.

amphotericin B An antifungal drug that binds the fungus-specific sterol ergosterol and destroys membrane integrity.

amplicon A specific PCR product.

anabolism Also called *biosynthesis.* The building up of complex biomolecules from smaller precursors.

anaerobic methane oxidizers (ANME) A group of archaea that oxidize methane syntrophically with bacteria that reduce sulfate.

anaerobic respiration The use of a molecule other than oxygen as the final electron acceptor of an electron transport chain.

anammox reaction The anaerobic oxidation of ammonium to nitrogen gas, using nitrate as electron acceptor; yields energy.

anaphylatoxin A substance produced as a result of the complement cascade that can trigger degranulation of endothelial cells, mast cells, or phagocytes.

anaphylaxis A severe type I hypersensitivity reaction caused by chemically induced contraction of smooth muscles and dilation of capillaries.

anaplerotic reaction A type of metabolic reaction, occurring in all organisms, that fixes small amounts of CO_2 to regenerate TCA cycle intermediates.

angle of aperture The width of a light cone (theta, θ) that projects from the midline of a lens. Greater angles of aperture increase resolution.

anion A negatively charged ion.

ANME See **anaerobic methane oxidizers**.

annotation Deciphering genome sequences, including identifying genes and predicting gene function.

antenna complex A complex of chlorophylls and accessory pigments in the photosynthetic membrane that collects photons and funnels them to a reaction center.

anthropogenic Caused by humans.

anti-anti-sigma factor A protein that inhibits an anti-sigma factor, allowing the target sigma factor to participate in initiating transcription.

anti-attenuator stem loop An mRNA secondary structure whose formation prevents assembly of a downstream transcriptional termination (attenuator) stem loop. The anti-attenuator stem loop structure permits transcription of the downstream structural genes.

antibiotic A molecule that can kill or inhibit the growth of selected microorganisms.

antibody A host defense protein produced by B cells in response to a specific antigenic determinant. Antibodies, a type of immunoglobulin, bind to their corresponding antigenic determinants.

antibody stain The attachment of a stain to an antibody to visualize cell components recognized by the antibody with high specificity.

antibody-dependent cell-mediated cytotoxicity (ADCC) The process by which natural killer cells destroy viral protein expressing antibody-coated host cells.

anticodon A set of three nucleotides in the middle loop of a tRNA that base-pairs with a codon in mRNA.

antigen A compound, recognized as foreign by the cell, that elicits an adaptive immune response. See also **immunogen**.

antigen-presenting cell (APC) An immune cell that can process antigens into antigenic determinants and display those determinants on the cell surface for recognition by other immune cells.

antigenic determinant Also called *epitope*. A small segment of an antigen that is capable of eliciting an immune response. An antigen can have many different antigenic determinants.

antimicrobial agent A chemical substance that can kill microbes or slow their growth.

antiparallel Oriented such that the two strands are in opposite directions. Commonly refers to a nucleic acid double helix with one strand in the 5′-to-3′ orientation and the other strand in the 3′-to-5′ orientation.

antiport Coupled transport in which the molecules being transported move in opposite directions across the membrane.

antisense RNA (asRNA) Antisense RNA (noncoding) from transcription of a DNA sequence complementary to the template strand of a gene encoding a protein. A *cis*-antisense RNA binds and regulates the coding transcript made from the template strand of the gene. These regulatory RNAs can stop transcription, promote transcript degradation, or prevent translation.

antisepsis The removal of pathogens from living tissues.

antiseptic A chemical that kills microbes.

anti-sigma factor A protein that inhibits a specific sigma factor, preventing transcription initiation.

AP endonuclease An enzyme that cleaves the DNA backbone at regions missing a nitrogenous base.

AP site A position in DNA where there is no base attached to the sugar of the backbone.

APC See **antigen-presenting cell**.

apical complex A conical cell form containing structures that facilitate entry of an apicomplexan parasite into a host cell.

apicomplexan A member of the eukaryotic group Apicomplexa—parasitic alveolates that possess an apical complex used for entry into a host cell.

apurinic site Also called *AP site*. A DNA site missing a purine base because the bond linking the base to the sugar has been hydrolyzed.

arbuscular mycorrhizae Also called *vesicular-arbuscular mycorrhizae (VAM)*. Mutualistic associations between plant roots and certain fungi, involving hyphal penetration of plant root cells.

Archaea One of the three domains of life, consisting of organisms with a last common ancestor not shared with members of Bacteria or Eukarya. Organisms are prokaryotic (lacking nuclei, unlike eukaryotes) and possess ether-linked phospholipid membranes (unlike bacteria).

Archaean eon The second eon (major time period) of Earth's existence, from 4.0 or 3.8 to 2.5 gigayears (Gyr, 10^9 years) before the present. The earliest geological evidence for life dates to this eon.

archaeon *pl.* **archaea** A prokaryotic organism that is a member of the domain Archaea, distinct from bacteria and eukaryotes.

aromatic A planar, unsaturated, ring-shaped organic molecule whose bonding electrons are delocalized equally around the ring.

arthrospore A small vegetative cell, produced by mature mycelia, that gets dispersed.

artifact A structure viewed through a microscope that is incorrectly interpreted.

ascomycete A member of the eukaryotic group Ascomycota—fungi whose mycelia form paired nuclei. Haploid ascospores are produced in pods called asci.

ascospore The spore produced by an ascomycete fungus.

ascus *pl.* **asci** A spore-containing pod produced by ascomycete fungi.

aseptic Free of microbes.

asexual reproduction Reproduction of a cell by fission or by mitosis to form identical daughter cells.

asRNA See **antisense RNA**.

assembly In a virus, the packaging of a viral genome into the capsid to form a complete virion. In metagenomics, the piecing together of DNA sequence reads into contigs, and of contigs into scaffold.
assimilation An organism's acquisition of an element, such as carbon from CO_2, to build into body parts.
assimilatory nitrate reduction The uptake of nitrate and reduction to NH_4^+ by plants and bacteria for use in biosynthetic pathways.
atomic force microscopy (AFM) A technique that maps the 3D topography of a object using van der Waals forces between the object and a probe.
atomic mass The mass (in grams) of one mole of an element.
atomic number The number of protons in an atom; it is unique for each element.
ATP See **adenosine triphosphate**.
ATP synthase A protein complex that synthesizes ATP from ADP and inorganic phosphate using energy derived from the transmembrane proton potential. It is located in the prokaryotic cell membrane and in the mitochondrial inner membrane.
attenuator stem loop Also called *terminator stem loop*. An intramolecular mRNA structure consisting of a base-paired stem connected by a single-stranded loop. The stem loop structure causes transcription to terminate. Its formation requires efficient translation of a leader peptide sequence.
autoclave A device that uses pressurized steam to sterilize materials by raising the temperature above the boiling point of water at standard pressure.
autoimmune response A pathology caused by lymphocytes that can react to self antigens.
autoinducer A secreted molecule that induces quorum-sensing behavior in bacteria.
autophagy Eukaryotic cell function normally used to degrade damaged organelles. Also used to kill intracellular pathogens.
autoradiography The visualization of a radioactive probe by exposing the probed material to X-ray film and then photographically developing the film.
autotroph An organism that can reduce carbon dioxide to produce organic carbon for biosynthesis.
axenic growth The ability of an organism to grow in the absence of any other species, as, for example, in a pure culture.
azidothymidine (AZT) A nucleotide analog that inhibits reverse transcriptase and was the first drug clinically used to fight HIV infections.
AZT See **azidothymidine**.

B

B cell An adaptive immune cell, developed in bone marrow tissue, that can give rise to antibody-producing cells.
B-cell receptor (BCR) A B-cell membrane protein complex containing an antibody in association with the Igα and Igβ immunoglobulins.
B-cell tolerance The exposure of B cells to a high antigen dose, preventing future antibody production against that antigen.
bacillus *pl.* **bacilli** See **rod**.
bacitracin A topical antibiotic that affects cell wall synthesis.
bacteremia A bacterial infection of the blood.
Bacteria One of the three domains of life, consisting of organisms with a last common ancestor not shared with members of Archaea or Eukarya. Organisms are prokaryotic (lacking nuclei, unlike eukaryotes) and possess ester-linked phospholipid membranes (unlike archaea).
bactericidal Having the ability to kill bacterial cells.
bacteriochlorophyll The chlorophyll of anaerobic phototrophic bacteria; it absorbs photons most strongly in the far-red end of the light spectrum.
bacteriophage Also called *phage*. A virus that infects bacteria.
bacteriorhodopsin (BR) An archaeal membrane-embedded protein that contains retinal and acts as a light-driven proton pump; it is homologous to the bacterial proteorhodopsin.
bacteriostatic Having the ability to inhibit the growth of bacterial cells.
bacterium *pl.* **bacteria** A prokaryotic organism that is a member of the domain Bacteria, distinct from archaea and eukaryotes.
bacteroid Cell wall–less, undividing, differentiated rhizobial cell within a plant cell. The bacteroid provides fixed nitrogen for the plant.
Bacteroidetes A phylum of Gram-negative bacteria; nearly all members are obligate anaerobes.
banded iron formation (BIF) A geological formation containing layers of oxidized iron (Fe^{3+}), which indicates formation under oxygen-rich conditions.
barophile Also called *piezophile*. An organism that requires high pressure to grow.
base excision repair (BER) A DNA repair mechanism that cleaves damaged bases off the sugar-phosphate backbone. After endonuclease activity at the AP site, a new, correct DNA strand is synthesized complementary to the undamaged strand.
basidiomycete A member of the eukaryotic group Basidiomycota—fungi that form mushrooms.
basidiospore A haploid spore formed by a basidiomycete through meiosis of a basidium, a reproductive cell of a mushroom.
basophil A white blood cell, stained by basic dyes, that secretes compounds that aid innate immunity.
batch culture The growth of bacteria in a closed system without additional input of nutrients.
BCR See **B-cell receptor**.
benthic organism An organism that lives on the ocean floor or within the sediment.
BER See **base excision repair**.
BIF See **banded iron formation**.
bin A set of sequences composed from metagenomic DNA reads showing a given level of similarity; defines an operational taxonomic unit.
binning The sorting of metagenomic sequences into taxonomic bins.
binary fission The process of replication in which one cell divides to form two genetically equivalent daughter cells of equal size.
bioburden The number of microorganisms found in a pharmaceutical product or in a body part.
biochemical oxygen demand (BOD) Also called *biological oxygen demand*. The amount of oxygen removed from an environment by aerobic respiration.
biofilm A community of microbes growing on a solid surface.
biogeochemical cycle The recycling of elements needed for life (such as carbon or nitrogen) through the biotic and abiotic components of the biosphere.
biogeochemistry Also called *geomicrobiology*. The metabolic interactions of microbial communities with the abiotic (mineral) components of their ecosystems.
bioinformatics A discipline at the intersection of biology and computing that analyzes gene and protein sequence data.
biological oxygen demand See **biochemical oxygen demand**.

biological signature See **biosignature**.
biomass The mass found in the bodies of living organisms.
bioprospecting The search for organisms with potential commercial applications.
bioremediation The use of microbes to detoxify environmental contaminants.
biosignature Also called *biological signature*. A chemical indicator of life.
biosphere The region containing the sum total of all life on Earth.
biosynthesis See **anabolism**.
biotic Caused by living organisms.
black smoker An oceanic thermal vent containing high concentrations of minerals such as iron sulfide.
blood-brain barrier A selectively permeable membrane made up of tightly packed capillaries that supply blood to the brain and spinal cord. Large molecules and most pathogens cannot permeate the narrow spaces. Fat-soluble (lipophilic) molecules and oxygen can dissolve through the capillary cell membranes and are absorbed into the brain.
BOD See **biochemical oxygen demand**.
borreliosis Also called *Lyme disease*. A tick-borne disease caused by *Borrelia burgdorferi*, which may involve skin lesions and arthritis.
botulism A food-borne disease caused by a *Clostridium botulinum* toxin, involving muscle paralysis.
BR See **bacteriorhodopsin**.
bradykinin A cell signaling molecule that promotes extravasation, activates mast cells, and stimulates pain perception.
Braun lipoprotein See **murein lipoprotein**.
bright-field microscopy A type of light microscopy in which the specimen absorbs light and appears dark against a light background.
brown alga See **kelp**.
bubonic plague A disease caused by the bacterium *Yersinia pestis*; characterized by swollen lymph nodes that often turn black.
budding A form of reproduction in which mitosis of the mother cell generates daughter cells of unequal size.
burst size The number of virus particles released from a lysed host cell.

C

C-reactive protein A peptide that stimulates the complement cascade, induced by cytokines in the liver. Elevated levels in the blood may be associated with heart disease.
calorimeter A device used to measure the amount of heat released or absorbed during a reaction.
Calvin cycle Also called *Calvin-Benson cycle*, *Calvin-Benson-Bassham cycle*, or *CBB cycle*. The metabolic pathway of carbon fixation in which the CO_2-condensing step is catalyzed by Rubisco. Found in chloroplasts and in many bacteria.
candidate species A newly described microbial isolate that may become accepted as an official species.
cannula *pl.* **cannulae** A narrow tube.
cannulated cow See **fistulated cow**.
capping The clustering of B-cell receptor molecules on the surface of B cells after binding antigens or epitopes.
capsid The protein shell that surrounds a virion's nucleic acid. Within an enveloped virus, such as HIV, the capsid may be called a *core particle*.
capsule A slippery outer layer composed of polysaccharides that surrounds the cell envelope of some bacteria.
carbon-concentrating mechanism (CCM) The inducible expression of transporters of CO_2 and bicarbonate (HCO_3^-) into carboxysomes to enhance CO_2 levels near Rubisco.
carbon dioxide fixation The enzymatic covalent incorporation of inorganic carbon dioxide (CO_2) into an organic compound.
carbon monoxide reductase pathway The carbon fixation process in methanogenic archaea, so called because the key enzyme can fix both CO and CO_2.
carboxysome Also called *polyhedral body*. A protein-enclosed compartment containing Rubisco to fix CO_2.
cardiolipin Diphosphatidylglycerol, a double phospholipid linked by glycerol.
carotenoid An accessory photosynthetic pigment that absorbs photons in the green end of the spectrum.
catabolism The cellular breakdown of large molecules into smaller molecules, releasing energy.
catabolite repression The inhibition of transcription of an operon encoding catabolic proteins in the presence of a more favorable catabolite, such as glucose.
catalytic RNA Also called *ribozyme*. An RNA molecule that is capable of enzymatic reactions.
catenane Linked rings of DNA found immediately after replication of circular chromosomes.
cation A positively charged ion.
CBB cycle See **Calvin cycle**.
CCM See **carbon-concentrating mechanism**.
CCR See **chemokine receptor**.
cell fractionation A procedure to separate cell components; often includes ultracentrifugation.
cell-mediated immunity A type of adaptive immunity employing mainly T-cell lymphocytes.
cell membrane Also called *plasma membrane* or *cytoplasmic membrane*. The phospholipid bilayer that encloses the cytoplasm.
cell-surface receptor A transmembrane protein that senses a specific extracellular signal and may be the docking site for a specific virus.
cell wall A rigid structure external to the cell membrane. The molecular composition depends on the organism; it is composed of peptidoglycan in bacteria.
cellular slime mold A slime mold in which the individual cells retain their own cell membranes; not a fungus.
cellulitis A spreading infection (inflammation) of connective tissue just below the skin.
CF See **competence factor**.
CFTR See **cystic fibrosis transmembrane conductance regulator**.
chain of infection The serial passage of a pathogenic organism from an infected individual to an uninfected individual, thus transmitting disease.
chancre A painless, hard lesion due to an inflammatory reaction at the site of infection with *Treponema pallidum*, the causative agent of syphilis.
chaperone Also called *chaperonin*. A protein that helps other proteins fold into their correct tertiary structure.
cheddared curd Curd that has been cut and piled in order to remove the liquid whey.
cheese A solid or semisolid food product prepared by coagulating milk proteins. Its production commonly involves microbial fermentation.
chemical imaging microscopy A method of microscopy that maps the distribution of specific elements or chemicals within a sample.

chemiosmotic theory A theory stating that the products of oxidative metabolism store their energy in an electrochemical gradient that can drive cell processes such as ATP synthesis.
chemo- Deriving energy from chemical rearrangement of molecules.
chemoautotroph Also called *chemolithotroph*, *lithotroph*, or *chemolithoautotroph*. An organism that oxidizes inorganic compounds to yield energy and reduce carbon dioxide.
chemoautotrophy Also called *lithotrophy* or *chemolithotrophy*. The metabolic oxidation of inorganic compounds to yield energy used to reduce carbon dioxide.
chemoheterotroph See **chemoorganotroph**.
chemoheterotrophy See **chemoorganotrophy**.
chemokine An attractant for white blood cells that is produced by damaged tissues.
chemokine receptor (CCR) A human T-cell membrane protein that binds chemokine hormones but is also used by HIV for attachment and infection.
chemolithoautotroph See **chemoautotroph**.
chemolithotroph See **chemoautotroph**.
chemolithotrophy See **chemoautotrophy**.
chemoorganotroph Also called *chemoheterotroph* or *organotroph*. An organism that oxidizes organic compounds to yield energy without the use of light.
chemoorganotrophy Also called *chemoheterotrophy* or *organotrophy*. The metabolic oxidation of organic compounds to yield energy without absorption of light.
chemostat A continuous culture system in which the introduced medium contains a limiting nutrient.
chemotaxis The ability of organisms to move toward or away from specific chemicals.
chemotrophy Metabolism that yields energy from oxidation-reduction reactions without using light energy.
ChIP See **chromatin immunoprecipitation**.
ChIP-on-chip analysis A procedure that scans a genome for all DNA sequences capable of binding to a specific DNA-binding protein.
chiral carbon A carbon bonded to four different types of functional groups; it can thus take two different forms exhibiting mirror symmetry.
chlamydia 1. An obligate, intracellular parasitic bacterium of the phylum Chlamydiae. 2. The most frequently reported sexually transmitted disease in the United States. Symptoms range from none, to a burning sensation upon urination, to sterility.
Chlamydiae A phylum of intracellular parasitic bacteria that lose most of their cell envelope during intracellular growth and generate multiple spore-like structures that escape to infect the next host.
chloramphenicol A bacteriostatic antibiotic that acts by inhibiting peptidyltransferase activity of the bacterial ribosome.
Chlorobi A phylum of Gram-negative bacteria. They are obligate anaerobes, "green sulfur" phototrophs that photolyze sulfides or H_2.
chlorophyll A magnesium-containing porphyrin pigment that captures light energy at the start of photosynthesis.
chlorophyte See **green alga**.
chloroplast An organelle of endosymbiotic origin that conducts oxygenic photosynthesis; found in algae and plant cells.
chlorosome A membranous photosynthetic organelle found in some "green" bacteria of the phyla Chloroflexi and Chlorobi.
cholesterol A sterol lipid found in eukaryotic cell membranes.
chromatin Chromosomal DNA complexed with proteins. Usually refers to a eukaryotic chromosome.
chromatin immunoprecipitation (ChIP) An experimental method used to determine the DNA-binding sites on a chromosome to which a DNA protein binds.
chromophore A light-absorbing redox cofactor.
chronic inflammation Inflammation that has persisted over long periods of time, usually months or years.
chrysophyte A member of the eukaryotic group Chrysophyceae (also known as "golden algae")—flagellated heterokont protists possessing chloroplasts as secondary endosymbionts.
cilia *sing.* **cilium** Short, hairlike structures of eukaryotes that beat in waves to propel the cell; structurally similar to flagella.
ciliate An alveolate that has paired cilia.
cis-antisense RNA See **antisense RNA**.
citric acid cycle See **tricarboxylic acid cycle**.
clade Also called *monophyletic group*. A group of organisms that includes an ancestral species and all of its descendants.
class I MHC molecule Membrane surface protein on all nucleated cells of the human body (absent from red blood cells and platelets) that present foreign antigen epitopes to cytotoxic T cells. Cytoplasmic antigens are loaded.
class II MHC molecule Membrane surface protein on antigen-presenting cells (dendritic cells, macrophages) and lymphocytes that present foreign antigen epitopes to helper T cells. Antigens are loaded after phagocytosis.
class switching See **isotype switching**.
classical complement pathway An antibody-mediated pathway for complement activation.
classification The recognition of different forms of life and their placement into different categories.
clonal Giving rise to a population of genetically identical cells, all descendants of a single cell.
clonal selection The rapid proliferation of a subset of B cells during the primary or secondary antibody response.
clone library See **genome library**.
clone pool See **genome library**.
cloning The insertion of DNA into a plasmid where it can be replicated.
cloning vector A small genome that can carry specific genes for cloning.
CoA See **coenzyme A**.
coastal shelf Shallow regions of the ocean, less than 200 meters deep, that are adjacent to land.
coccus *pl.* **cocci** A spherically shaped bacterial or archaeal cell.
codon A set of three nucleotides that encodes a particular amino acid.
coenzyme A (CoA) A nonprotein cellular organic molecule that can carry acetyl groups and participates in metabolism.
coevolution The evolution of two species in response to one another.
cofactor A metallic ion or a coenzyme required by an enzyme to perform normal catalysis.
cointegrate A DNA molecule formed by a single site recombination event joining two participating circular DNA molecules to form a larger single circle.
cold seep A cold region of the seafloor where seeping hydrogen sulfide and methane feed bacteria and their animal symbionts, such as tube worms.
colony A visible cluster of microbes on a plate, all derived from a single founding microbe.

commensal organism An organism that benefits from, but neither helps nor harms, its host. In medical usage, some commensals benefit the host. Commensal organisms are normally found at various nonsterile host body sites.
commensalism An interaction between two different species that benefits only one partner.
community The sum of all populations of organisms interacting within an ecosystem.
competence factor (CF) A species-specific secreted bacterial protein that induces competence for transformation.
competent Able to take up DNA from the environment.
complement Innate immunity proteins in the blood that form holes in bacterial membranes, killing the bacteria.
complex medium Also called *rich medium*. A nutrient-rich growth solution including undefined chemical components such as beef broth.
complex transposon A transposon containing a gene for the transposase enzyme, which is needed for replicative transposition.
composite transposon A transposon containing genes in addition to those for transposition, such as antibiotic resistance or catabolic functions.
compound microscope A microscope with multiple lenses to compensate for lens aberration and increase magnification.
compromised host An animal with a weakened immune system.
condensation In biochemistry, the joining of two molecules to release a water molecule.
condenser In a microscope, a lens that focuses parallel light rays from the light source onto a small area of the specimen to improve the resolution of the objective lens.
conditional lethal mutation A mutation that leads to death under one growth condition but permits growth under a second condition.
confluent Describing a mode of growth that results in a lawn of organisms completely covering a surface.
confocal laser scanning microscopy Also called simply *confocal microscopy*. A type of fluorescence microscopy in which the excitation and emitted light are laser beams focused together, producing high-resolution images.
congenital syphilis Syphilis contracted in utero.
conjugation Horizontal gene transmission involving cell-to-cell contact. In bacteria, pili draw together the donor and recipient cell envelopes, and a protein complex transmits DNA across. In ciliated eukaryotes, a conjugation bridge forms between two cells connecting their cytoplasm, through which micronuclei are exchanged.
conjugative transposon A transposon that can be transferred from one cell to another via conjugation.
consensus sequence A sequence of nucleotides or amino acids with a common function at many nucleic acid or protein positions. Consists of the base pair or amino acid most frequently found at each position in the sequence.
constant region The region of an antibody that defines the class of a heavy chain or a light chain.
consumer An organism that acquires nutrients from producers, either directly or indirectly.
contig A sequence of overlapping fragments of cloned DNA that are contiguous along a chromosome.
continuous culture A culture system in which new medium is continuously added to replace old medium.
contractile vacuole An organelle in eukaryotic microbes that pumps water out of the cell.
contrast Differential absorption or reflection of electromagnetic radiation between an object and background that allows the object to be distinguished from the background.
coral bleaching The death or expulsion of coral algal symbionts. One cause is an increase in temperature.
coralline alga An algal species that calcifies its fronds into hardened shapes similar to corals.
core genome A set of genes shared by a group of related bacterial strains, showing stable inheritance.
core particle A viral capsid that encloses its nucleic acid genome and is surrounded by an envelope.
core polysaccharide A sugar chain that attaches to the glucosamine of lipopolysaccharides and extends outside the cell.
coreceptor A cell-surface receptor needed for viral entry along with a primary receptor.
corepressor A small molecule that must bind to a repressor to allow the repressor to bind operator DNA.
cortical alveolus One of the vesicles that forms a network in the outer covering of an alveolate.
counterstain A secondary stain used to visualize cells that do not retain the first stain.
coupled transport The movement of a substance against its electrochemical gradient (from lower to higher concentration, or from opposite charge to like charge) using the energy provided by the simultaneous movement of a different chemical down its electrochemical gradient.
covalent bond A bond between two atoms that share a pair of electrons.
coverage The proportion of a genome that is included in a metagenomic scaffold or set of scaffolds.
Crenarchaeota One of four major divisions of Archaea, containing sulfur hyperthermophiles, mesophiles, and psychrophiles.
CRISPR Clustered regularly interspaced short palindromic repeats. Consists of short repeated DNA sequences in a bacterial or archaeal genome, derived from previous bacteriophage or viral infection and conferring protection from future infection; considered a prokaryotic "immune system."
cross-bridge An attachment that links parallel molecules, such as the peptide link between glycan chains in peptidoglycan.
cryo-electron microscopy (cryo-EM) Also called *electron cryomicroscopy*. Electron microscopy in which the sample is cooled rapidly in a cryoprotectant medium that prevents freezing. The sample does not need to be stained.
cryo-electron tomography Also called *electron cryotomography*. A method of cryo-electron microscopy in which the electron beam generates multiple views in parallel planes through the specimen.
cryptobiotic soil Also called *cryptogamic crust*. A low-growing desert ground cover composed of cyanobacteria, lichens, and nonlichenous algae, fungi, and mosses.
cryptogamic crust See **cryptobiotic soil**.
curd Coagulated milk proteins produced by the combined action of lactic acid–producing bacteria and stomach enzymes of certain mammals, such as cattle.
Cyanobacteria A phylum of oxygen-producing photoautotrophic bacteria containing chlorophylls *a* and *b*. They are closely related to chloroplasts.
cyclic photophosphorylation A photosynthetic process in which chlorophyll serves as both the initial electron donor and the final electron acceptor. ATP is produced via the proton potential from an electron transport system, but no NADPH is generated.

cycloserine A polypeptide antibiotic that inhibits peptidoglycan synthesis.

cystic fibrosis transmembrane conductance regulator (CFTR) A chloride channel found in respiratory epithelia. Mutations in the CFTR gene lead to cystic fibrosis.

cytochrome A membrane protein that donates and receives electrons.

cytokine A small, secreted host protein that binds to receptors on various endothelial and immune system cells, regulating the cells' responses.

cytoplasm Also called *cytosol*. The aqueous solution contained by the cell membrane in all cells and outside the nucleus (in eukaryotes).

cytoplasmic membrane See **cell membrane**.

cytoskeleton A collection of filamentous proteins that impart structure to and aid movement of cells; in a eukaryote, these include microfilaments, intermediate filaments, and microtubules.

cytosol See **cytoplasm**.

cytotoxic T cell (T_C cell) A T cell that expresses CD8 on its cell surface and can secrete toxic proteins such as perforin and granzymes.

D

D-value See **decimal reduction time**.

dark-field microscopy The detection of microbes too small to be resolved by light rays, by observing the light they scatter.

dead zone Also called *zone of hypoxia*. An anoxic region of an ocean, devoid of most fish and invertebrates.

death phase A period of cell culture, following stationary phase, during which bacteria die faster than they replicate.

death rate The rate at which cells die; it is exponential during the death phase.

decay-accelerating factor A host cell membrane protein that stimulates the decay of complement factors and prevents their deposition at the cell surface.

decimal reduction time (D-value) The length of time it takes for a treatment to kill 90% of a microbial population, and hence a measure of the efficacy of the treatment.

decomposer An organism that consumes dead biomass.

defensin A type of small, positively charged peptide, produced by animal tissues, that destroys the cell membranes of invading microbes.

defined minimal medium A solution of chemically defined compounds for organismal growth that contains only the minimal components required for growth.

degenerative evolution See **reductive evolution**.

degranulate To release antimicrobial granule contents by fusing granule membranes to cytoplasmic or vacuolar membranes.

degron A degradation signal contained within a protein.

dehalorespiration The reduction of halogenated organic molecules by H_2.

delayed-type hypersensitivity (DTH) See **type IV hypersensitivity**.

deletion The loss of nucleotides from a DNA sequence.

denature To lose secondary and tertiary structure in a protein or nucleic acid because of high temperature or chemical treatment.

dendritic cell An antigen-presenting white blood cell that primarily takes up small soluble antigens from its surroundings.

denitrification Also called *dissimilatory nitrate reduction*. Energy-yielding metabolism in which nitrate (NO_3^-) is reduced to nitrite (NO_2^-), diatomic nitrogen (N_2), and in some cases ammonia (NH_3).

depth of field In a microscope, a region of the optical column over which a specimen appears in reasonable focus.

derepression An increase in gene expression caused by the decrease in concentration of a corepressor.

desensitization A clinical treatment to decrease allergic reactions by exposing patients to small doses of an allergen.

Desulfurococcales An order of thermophilic archaea (Crenarchaeota) that metabolize sulfur and organic compounds.

detection The ability to determine the presence of an object.

detritus Discarded biomass that can be consumed by decomposers.

diapedesis See **extravasation**.

diaphragm A device in a microscope to vary the diameter of the light column, changing the amount of light admitted.

diatom A member of the eukaryotic group Bacillariophyceae—protists known for intricate, silica-containing bipartite shells.

diauxic growth A biphasic cell growth curve caused by depletion of the favored carbon source and a metabolic switch to the second carbon source.

DIC See **differential interference contrast microscopy**.

dichotomous key A tool for identifying organisms, in which a series of yes/no decisions successively narrows down the possible categories of species.

differential interference contrast microscopy (DIC) Also called *Nomarski microscopy*. A form of microscopy based on the interference pattern between two light beams split and polarized, one of which passes through a sample. The interference between the two beams generates contrast in a transparent sample.

differential RNA-sequencing (dRNAseq) An RNA sequencing technique that differentiates the true start of an RNA transcript, which begins with a 5′ triphosphate, from an internal cleavage fragment that begins with a 5′ monophosphate.

differential medium A growth medium that can distinguish between various bacteria on the basis of metabolic differences.

differential stain A stain that differentiates among objects by staining only particular types of cells or specific subcellular structures.

diffusion The energy-independent net movement of a substance from a region of high concentration to a region of lower concentration.

dilution streaking A method of spreading bacteria on a plate in order to obtain colonies arising from an individual bacterium.

dinoflagellate A member of the eukaryotic group Dinoflagellata—tertiary endosymbiont algae, alveolates with two flagella, one of which is wrapped distinctively around the cell equator.

diphosphatidylglycerol See **cardiolipin**.

diplococcus The paired cocci of *Neisseria* species.

direct repeat Two identical sequences closely positioned in a DNA molecule, aligned in the same direction (e.g., 5′-ATCGATCGnnnnnnATCGATCG-3′).

directed evolution The application of selective pressure in a laboratory to cause evolution of organisms with desirable properties, such as elevated production of an enzyme.

disinfection The removal of pathogenic organisms from inanimate surfaces.

dissimilation An organism's catabolism or oxidation of nutrients to inorganic minerals that are released into the environment.

dissimilatory denitrification Metabolic reduction of nitrate or nitrite to yield energy; anaerobic respiration of nitrate or nitrite.

dissimilatory metal reduction A type of anaerobic respiration that uses metal cations as terminal electron acceptors.

dissimilatory nitrate reduction See **denitrification**.

DNA control sequence A region of DNA, such as the promoter region, that controls the expression of structural genes but is not itself transcribed to RNA.

DNA ligase An enzyme that cells use to form a covalent bond at a nick in the phosphodiester backbone. Also used in molecular biology laboratories to join pieces of DNA.

DNA microarray A technique, used for measuring the amount of specific mRNA molecules transcribed in cells, in which DNA fragments from every open reading frame in a genome are affixed to separate locations on a solid support surface (a DNA microchip), producing a grid, or array.

DNA replication The biological process of making an identical copy of double-stranded DNA using existing DNA as a template.

DNA reverse-transcribing virus See **pararetrovirus**.

DNA sequencing A technique to determine the order of bases in a DNA sample.

domain (1) In taxonomy, one of three major subdivisions of life, the Archaea, Bacteria, and Eukarya. (2) In protein structure, a portion of a protein that possesses a defined function, such as binding DNA. (3) In membranes, a region of membrane consisting of certain types of phospholipids that are distinct from surrounding lipids.

dormant See **viable but nonculturable**.

doubling time The generation time of bacteria in culture. The amount of time it takes for the population to double.

downstream processing The recovery and purification of a commercial product produced by industrial microbes.

dRNAseq See **differential RNA-sequencing**.

DTH See **type IV hypersensitivity**.

E

E site See **exit site**.

early gene A viral gene expressed early in the infectious cycle.

eclipse period The time after viral genome injection into host cell but before complete virions are formed.

ecosystem A community of species plus their environment (habitat).

ectomycorrhizae Mycorrhizae that colonize the surface of plant roots. Their mycelia do not penetrate the root cells.

ectoparasite A harmful organism that colonizes the surface of a host.

ED pathway See **Entner-Doudoroff pathway**.

edema Tissue swelling due to fluid accumulation.

edema factor (EF) A component of anthrax toxin with adenylate cyclase activity.

EF See **edema factor**.

electrochemical potential A type of potential energy formed by the combined concentration gradient of a molecule and the electrical potential across a membrane.

electrogenic Describing a transport system that results in a net movement of charged molecules across a membrane.

electromagnetic radiation Energy radiating in the form of alternating electrical and magnetic waves, quantized in photons.

electron acceptor An oxidized molecule (e.g., NAD^+) that can accept electrons.

electron bifurcation A biochemical reaction in which an electron transfer that yields energy is coupled to an electron transfer that consumes energy.

electron cryomicroscopy See **cryo-electron microscopy**.

electron cryotomography See **cryo-electron tomography**.

electron donor Also called *reducing agent*. A reduced molecule (e.g., NADH) that can donate electrons.

electron microscope A microscope that obtains high resolution and magnification by focusing electron beams on samples using magnetic lenses.

electron microscopy (EM) A form of microscopy in which a beam of electrons accelerated through a voltage potential is focused by magnetic lenses onto a specimen.

electron transport chain See **electron transport system**.

electron transport system (ETS) Also called *electron transport chain*. A series of membrane-embedded proteins that converts the energy of redox reactions into a proton potential.

electronegativity The affinity of an atom for electrons. The greater the electronegativity, the stronger the attraction for electrons.

electroneutral Describing a transport system that results in no net change in charge across the membrane.

electrophoresis A technique to separate charged proteins and nucleic acids based on how rapidly they migrate in an electrical field through a gel.

electrophoretic mobility shift assay (EMSA) A technique to observe DNA-protein interactions that is based on the ability of a bound protein to slow the voltage-driven migration of DNA through a gel.

electroporation A laboratory technique that temporarily makes the cell membrane more leaky to allow the uptake of DNA.

elementary body The endospore-like form of chlamydias transmitted outside host cells.

EM See **electron microscopy**.

Embden-Meyerhof-Parnas (EMP) pathway See **glycolysis**.

emerging Describing an organism that is newly isolated, defined, or recognized, as in "emerging clade" or "emerging pathogen."

emission wavelength The wavelength of light emitted by a fluorescent molecule. It is of a lower energy and longer wavelength than the excitation wavelength.

EMP pathway See **glycolysis**.

empty magnification Magnification without an increase in resolution.

EMSA See **electrophoretic mobility shift assay**.

enantiomer See **optical isomer**.

endemic Describing a disease that is always present in a population, although the frequency of infection may be low.

endergonic Describing a reaction that requires an input of energy to proceed.

endocarditis An inflammation of the heart's inner lining.

endocytosis The invagination of the cell membrane to form a vesicle that contains extracellular material.

endogenous retrovirus A retroelement (genome sequence descended from a retrovirus) that contains *gag*, *env*, and *pol* genes.

endolith A bacterium that grows within the crystals of solid rock.

endomembrane system A series of membranous organelles that organize uptake, transport, digestion, and expulsion of particles through a eukaryotic cell. Includes endosomes, lysosomes, endoplasmic reticulum, and Golgi complex.

endomycorrhizae Mycorrhizae in which fungal hyphae penetrate plant root cells.

endoparasite A harmful organism that lives inside a host.

endophyte An endosymbiont of vascular plants.

endosome A vesicle formed from the pinching in of the cell membrane.

endospore A durable, inert, heat-resistant spore that can remain viable for thousands of years.

endosymbiont An organism that lives as a symbiont inside another organism.

endosymbiosis An intimate association between different species in which one partner population grows within the body of another organism.

endotoxin A lipopolysaccharide in the outer membrane of Gram-negative bacteria that becomes toxic to the host after the bacterial cell has lysed.

energy The ability to do work.

energy carrier A molecule in the cell, such as ATP or NADH, that serves as energy currency. Energy carriers are produced during catabolic reactions and can be used to drive energy-requiring reactions.

enhancer A noncoding DNA region in eukaryotes that can lead to activation of transcription when bound by an appropriate transcription factor. Its location on the chromosome can be far removed from the regulated gene.

enriched medium A growth solution for fastidious bacteria, consisting of complex medium plus additional components.

enrichment culture The use of selective growth media to allow only certain microbes to grow.

enterotoxin A protein that damages the intestine of the host and causes diarrhea; produced by some Gram-negative pathogens.

enthalpy A measure of the heat energy in a system.

Entner-Doudoroff (ED) pathway A glycolytic pathway in which glucose 6-phosphate is initially oxidized to 6-phosphogluconate, and ultimately yields 1 pyruvate, 1 ATP, 1 NADH and 1 NADPH.

entropy A measure of the disorder in a system.

envelope A structure external to the cell membrane, such as the cell wall or outer membrane of a bacterium. For a virus, the envelope is a membrane enclosing the capsid or core particle.

enzyme A biological catalyst; a protein or RNA that can speed up the progress of a reaction without itself being changed.

eosinophil A white blood cell that stains with the acidic dye eosin and secretes compounds that facilitate innate immunity.

epidemic A disease outbreak in which large numbers of individuals in a population become infected over a short time.

epidemiology The study of factors affecting the health and illness of populations.

epidermis The outer protective cell layer in most multicellular animals.

episome A DNA element that can exist as part of the chromosome or independently, as a plasmid.

epitope See **antigenic determinant**.

EPS See **exopolysaccharide**.

equilibrium *pl.* **equilibria** A dynamic state in which there is no net change in a reaction.

equivalence The antigen/antibody ratio that leads to immunoprecipitation of large, insoluble complexes.

error-prone repair Low-accuracy DNA repair mechanisms that allow mutations.

error-proof repair DNA repair mechanisms that minimize the formation of mutations.

erythema migrans A bull's-eye rash characteristic of borreliosis (Lyme disease).

essential nutrient A compound that an organism cannot synthesize and must acquire from the environment in order to survive.

ethanolic fermentation A fermentation reaction yielding 2 ethanol and 2 CO_2 as products.

ETS See **electron transport system**.

Eukarya One of the three domains of life, consisting of organisms with a last common ancestor not shared with members of Archaea or Bacteria. Cells possess nuclei, unlike cells of bacteria and archaea.

eukaryote An organism whose cells contain a nucleus. All eukaryotes are members of the domain Eukarya.

Eumycota True fungi, a taxonomic group of opisthokont eukaryotes with chitinous cell walls; the group most closely related to animals.

euphotic zone Also called *photic zone*. The region of the ocean that receives sunlight capable of supporting photosynthesis.

Euryarchaeota One of the four major divisions of Archaea, containing methanogens, halophiles, and acidophiles.

eutrophic An aquatic or marine environment in which oversupply of electron donors (usually organic matter) leads to oxygen depletion, and hence loss of animal life.

eutrophic lake A lake in which overgrowth of heterotrophic microbes has eliminated oxygen, leading to a decrease in animal life.

eutrophication A sudden increase of a formerly limiting nutrient in an aquatic environment, leading to overgrowth of algae and grazing bacteria and subsequent oxygen depletion.

excitation wavelength The wavelength of light that must be absorbed by a molecule in order for the molecule to fluoresce. It is of a higher energy and shorter wavelength than the emission wavelength.

exergonic Describing a spontaneous reaction that releases free energy.

exit site (E site) The region of the ribosome that holds the uncharged, exiting tRNA.

exocytosis Fusion of vesicles with the cell membrane to release vesicle contents extracellularly.

exon An expressed (protein-coding) portion of a eukaryotic gene.

exonuclease An enzyme that cleaves DNA from the end.

exopolysaccharide (EPS) A thick extracellular matrix of polysaccharides and entrapped materials that forms around the microbes in a biofilm.

exotoxin A protein toxin, secreted by bacteria, that kills or damages host cells.

experimental evolution Also called *laboratory evolution*. The repeated culturing of a population in a laboratory under defined environmental conditions, leading to evolution of adapted genotypes.

exponential phase Also called *logarithmic (log) phase*. A period of cell culture during which bacteria grow exponentially at their maximal possible rate given the conditions.

extravasation Also called *diapedesis*. The movement of cells of the immune system out of blood vessels and into surrounding infected tissue.

extremophile An organism that grows only in an extreme environment—that is, an environment including one or more conditions that are "extreme" relative to the conditions for human life.
eyepiece See **ocular lens**.

F

F^- cell The DNA recipient cell in conjugation.
F^+ cell The DNA donor cell in conjugation that transmits the fertility factor F^+ to an F^- cell during conjugation.
F factor See **fertility factor**.
F-prime (F′) plasmid Also called *F-prime (F′) factor*. A fertility factor plasmid that contains some chromosomal DNA.
facilitated diffusion A process of passive transport across a membrane that is facilitated by transport proteins.
FACS See **fluorescence-activated cell sorter**.
factor H A normal serum protein that prevents the inadvertent activation of complement in the absence of infection.
facultative Able to grow in the presence or absence of a given environmental factor, such as oxygen.
facultative anaerobe An organism that can grow in either the presence or absence of oxygen.
facultative intracellular pathogen A pathogen that can live either inside host cells or outside host cells.
FAD See **flavin adenine dinucleotide**.
fatty acid synthase complex A collection of all the enzymes and binding proteins necessary for fatty acid synthesis.
Fc region The region of an antibody that binds to specific receptors on host cells in an antigen-independent manner. It is found in the carboxy-terminal "tail" region of the antibody.
feline leukemia virus (FeLV) A retrovirus that is a major cat pathogen.
FeLV See **feline leukemia virus**.
fermentation Also called *fermentative metabolism*. The production of ATP via substrate-level phosphorylation, using organic compounds as both electron donors and electron acceptors. Industrial fermentation is the production of microbial products that are made by microbes grown in fermentation vessels; it may include respiratory metabolism to maximize microbial growth.
fermentation industry Commercial production of microbial products that are made by microbes grown in fermentation vessels; it may include respiratory metabolism to maximize microbial growth.
fermented food Food products that are biochemically modified by microbial growth.
ferredoxin An iron- and sulfur-containing protein that transfers electrons in electron transport systems.
fertility factor (F factor) A specific plasmid (transferred by an F^+ donor cell) that contains the genes needed for pilus formation and DNA export.
filamentous virus A viral structure type consisting of a helical capsid surrounding a single-stranded nucleic acid.
filter In metagenomics, to limit the target community sample on the basis of size or taxonomic group. Filtering can be physical (by pore size or cell sorting) or computational (by DNA sequence analysis).
fimbria *pl.* **fimbriae** See **pilus**.
Firmicutes A phylum of low-GC-content Gram-positive bacteria.
FISH See **fluorescence in situ hybridization**.
fistulated cow Also called *cannulated cow*. A cow in which a hole in the skin has been connected surgically to a hole in the rumen and fitted with a cannula, allowing access for experimental analysis of the rumen.
fixation The adherence of cells to a slide by a chemical or heat treatment.
flagellate A protist that has one or more flagella.
flagellum *pl.* **flagella** A filamentous structure for motility. In prokaryotes, a helical protein filament attached to a rotary motor; in eukaryotes, an undulating cell membrane–enclosed complex of microtubules and ATP-driven motor proteins.
flavin adenine dinucleotide (FAD) An energy carrier in the cell that can donate ($FADH_2$) or accept (FAD^+) electrons.
flavonoid A type of signaling molecule released from legumes to attract nitrogen-fixing rhizobia.
flesh-eating disease See **necrotizing fasciitis**.
floc Particulate matter formed by clumps of microbes during wastewater treatment.
fluid mosaic model A model of the cell membrane in which proteins are free to diffuse laterally within the membrane.
fluorescence The emission of light from a molecule that absorbed light of a shorter, higher-energy wavelength.
fluorescence in situ hybridization (FISH) A technique to detect individual microbes in an ecological or clinical sample, using a fluorophore-labeled oligonucleotide probe (usually a short DNA sequence) that hybridizes to microbial DNA or rRNA.
fluorescence resonance energy transfer (FRET) The detectable transfer of fluorescent energy from one molecule to another. Since the participating molecules must be near each other, FRET can be used to monitor protein-protein interactions in cells and is also used in real-time PCR.
fluorescence-activated cell sorter (FACS) A device that can count cells and sort them according to differences in fluorescence.
fluorescent-focus assay An assay to detect viruses that do not kill host cells, based on intracellular detection of viruses using antiviral antibodies or green fluorescent protein–modified viruses.
fluorophore A fluorescent molecule used to stain specimens for fluorescence microscopy.
focal plane A plane that contains the focal point for a given lens.
focal point The position at which light rays that pass through a lens intersect.
focus *pl.* **foci** 1. The point at which rays of energy converge; in light microscopy, the convergence of light rays maximizes the clarity of the optical image. 2. A group of cells infected by a virus.
folate Folic acid, a heteroaromatic cofactor that is required by some enzymes.
fomite An inanimate object on which pathogens can be transmitted from one host to another.
food poisoning Also called *food contamination*. The presence of human disease-causing microbial pathogens or toxins in food.
food spoilage Microbial changes that render a food unfit or unpalatable for consumption.
food web A network of interactions in which organisms obtain or provide nutrients for each other—for example, by predation or by mutualism.
foraminiferan *or* **foram** An ameba with a calcium carbonate shell and a helical arrangement of chambers.

forespore In sporulation of Gram-positive bacteria, the smaller cell compartment formed through asymmetrical cell division; it develops into the endospore.
fossil fuel Ancient organismal remains that have been converted to hydrocarbons (petroleum and natural gas) or sedimentary rock (coal) through microbial digestion followed by reduction under high pressure underground; fossil fuels are extracted and burned by humans for energy.
frameshift mutation A gene mutation involving the insertion or deletion of nucleotides that cause a shift in the codon reading frame.
free energy change See **Gibbs free energy change**.
freeze-drying Also called *lyophilization*. The removal of water from food, by freezing under vacuum, to limit microbial growth.
FRET See **fluorescence resonance energy transfer**.
fruiting body A multicellular fungal or bacterial reproductive structure.
frustule The silica bipartite shell produced by a diatom.
functional genomics A field of molecular biology that utilizes data from genomic projects to describe gene and protein interactions.
functional group A cluster of covalently bonded atoms that behaves with specific properties and functions as a unit.
fungus *pl.* **fungi** A heterotrophic opisthokont eukaryote with chitinous cell walls. Includes Eumycota, but traditionally may refer to fungus-like protists such as the oomycetes.
fusion peptide A portion of a viral envelope protein that changes shape to facilitate envelope fusion with the host cell membrane.
Fusobacteria A phylum of anaerobic Gram-negative bacteria with an outer membrane, related to the Proteobacteria; includes human pathogens.

G

gain-of-function mutation A mutation that enhances the activity or allows new activity of a gene product.
GALT See **gut-associated lymphoid tissue**.
gamogony The differentiation of parasitic haploid cells into male and female gametes.
gas vesicle An organelle that traps gases to increase buoyancy of aquatic microbes.
GC content The proportion of an organism's genome consisting of guanine-cytosine base pairs.
GDH See **glutamate dehydrogenase**.
gene A distinct series of nucleotides within DNA that has a distinct function (regulatory) or whose product has a distinct function (protein). The functional unit of heredity.
gene duplication The formation of an extra copy of a gene within a genome.
gene fusion Also called *protein fusion* or *translational fusion*. A technique to measure control of gene transcription or translation by inserting a reporter gene into a target gene. The reporter relies on both the promoter and the ribosome-binding site of the target gene.
gene switching Switching between two (out of five) different classes of immunoglobulin genes (e.g., from IgM to IgG) during B-cell development.
gene transfer vector A mobile DNA engineered from a virus or plasmid, designed to insert a genetic sequence into the genome of an organism for experimental study or for medical therapy.
generalized recombination Recombination between two DNA molecules that share long regions of DNA homology.
generalized transduction A phage-mediated gene transfer process in which any donor gene can be transferred to a recipient cell.
generation time The species-specific time period for doubling of a population (e.g., by bacterial cell division) in a given environment, assuming no depletion of resources.
genetic analysis Determination of the function of cell RNAs and proteins based on the phenotype of cells in which the gene encoding the RNA or protein is mutated.
genome The complete genetic content of an organism. The sequence of all the nucleotides in a haploid set of chromosomes.
genome library Also called *clone library* or *clone pool*. A population of host bacteria, each of which carries a DNA molecule cloned from an organism's genome. The set of clones contains overlapping DNA fragments representing the entire genome.
genomic island A region of DNA sequence whose properties indicate that it has been transferred from another genome. Usually comprises a set of genes with shared function, such as pathogenicity or symbiosis support.
genus name The Latin name assigned to the taxonomic rank consisting of closely related species.
geochemical cycling The global interconversion of various inorganic and organic forms of elements.
geomicrobiology See **biogeochemistry**.
geosmin A molecule released from decaying *Streptomyces* cells; it causes the characteristic odor of soil and can affect the taste of drinking water.
germ theory of disease The theory that many diseases are caused by microbes.
germicidal Able to kill cells but not spores.
germination The activation of a dormant spore to generate a vegetative cell.
Gibbs free energy change (ΔG) Also called *free energy change* or *Gibbs energy change*. In a chemical reaction, a measure of how much energy available to do work is released or required as the reaction proceeds.
gliding motility The movement of cells individually or as a collective over surfaces using pili.
gluconeogenesis The biosynthesis of glucose from single-carbon compounds.
glucosamine A glucose modified with an amine group.
glutamate dehydrogenase (GDH) An enzyme that condenses NH_4^+ with 2-oxoglutarate to form glutamate. The condensation requires reduction by NADPH.
glutamate synthase Glutamine:2-oxoglutarate aminotransferase (GOGAT), an enzyme that converts 2-oxoglutarate plus glutamine into two molecules of glutamate.
glutamine synthetase (GS) An enzyme that condenses NH_4^+ with glutamate to form glutamine.
glycan Polysaccharide chain composed of oxygen-linked (O-linked) monosaccharides.
glycolysis Also called *Embden-Meyerhof-Parnas (EMP) pathway*. The catabolic pathway of glucose oxidation to pyruvate, in which glucose 6-phosphate isomerizes to fructose 6-phosphate, ultimately yielding 2 pyruvate, 2 ATP, and 2 NADH.

glyoxylate bypass An alternative to the tricarboxylic acid cycle in which isocitrate is converted to glyoxylate and then malate; induced under low glucose conditions.

gnotobiotic animal An animal that is germ-free or colonized by a known set of microbes.

GOGAT See **glutamate synthase**.

Golgi complex Also called *Golgi apparatus*. A series of membrane stacks that modifies proteins and helps sort them to the correct eukaryotic cell compartment.

Gram stain A differential stain that distinguishes cells that possess a thick cell wall and retain a positively charged stain (Gram-positive) from cells that have a thin cell wall and outer membrane and fail to retain the stain (Gram-negative).

Gram-negative Describing cells that do not retain the Gram stain.

Gram-positive Describing cells that do retain the Gram stain and appear dark purple after staining.

gramicidin A peptide antibiotic that disrupts membrane integrity.

granuloma A thick lesion formed around a site of infection.

granzyme An enzyme, secreted by cytotoxic T cells, that damages target cells.

grazer A first-level consumer, feeding directly on producers.

green alga Also called *chlorophyte*. A member of the eukaryotic group Chlorophyta—microbes that have chloroplasts; primary endosymbionts, closely related to plants (Viridiplantae).

greenhouse effect The trapping of solar radiation heat in the atmosphere by CO_2; a cause of global warming.

griseofulvin An antifungal antibiotic that inhibits cell division.

group translocation A form of active transport in which the transported molecule is modified after it enters the cell, thus keeping a favorable inward concentration gradient for the unmodified extracellular molecule.

growth factor A compound needed for the growth of only certain cells.

growth rate The rate of increase in population number or biomass.

growth rate constant The number of organismal generations per unit time; *k*.

GS See **glutamine synthetase**.

gut-associated lymphoid tissue (GALT) Lymphatic tissues such as tonsils and adenoids that are found in conjunction with the gastrointestinal tract and contain immune cells.

H

HAART See **highly active antiretroviral therapy**.

Haber process Industrial nitrogen fixation, in which dinitrogen is hydrogenated by methane (natural gas) under extreme heat and pressure to form ammonia.

Hadean eon The first eon (major time period) of Earth's existence, from 4.5 to approximately 4.0–3.8 gigayears (Gyr, 10^9 years) before the present.

half-life The amount of time it takes for one-half of a radioactive sample to decay.

Haloarchaea A class of extremely halophilic archaea, inhabiting high-salt environments.

Halobacteriales An order of Euryarchaeota that contains the class Haloarchaea.

halophile An organism that requires a high extracellular sodium chloride concentration for optimal growth.

halorhodopsin (HR) An archaeal light-driven chloride pump.

hapten A small compound that must be conjugated to a larger carrier antigen in order to elicit production of an antibody that binds to it.

haustorium *pl.* **haustoria** A bulbous hyphal extension of a fungal plant pathogen into the host cell.

heat-shock protein (HSP) A chaperone protein whose synthesis is induced by high-temperature stress.

heat-shock response A coordinated response of cells to higher-than-normal temperatures. It includes changes in the membrane and expression of heat-shock genes.

heavy chain The larger of the two protein types comprising an antibody. Each antibody contains two heavy chains and two light chains.

helper T cell (T_H cell) A T cell that expresses CD4 on its cell surface and secretes cytokines that modulate B-cell isotype, or class, switching.

heme An organic molecule containing a ring of conjugated double bonds surrounding an iron atom. It is involved in redox reactions and oxygen binding.

hepatitis An inflammation of the liver, caused by infection or by exposure to a toxic substance.

heterocyst In filamentous cyanobacteria, a specialized nitrogen-fixing cell that maintains a reducing environment and excludes O_2.

heteroduplex A double-stranded nucleic acid in which the two strands come from different sources.

heterokont A member of the eukaryotic group Heterokonta—protists possessing two flagella of unequal length.

heterolactic fermentation A fermentation reaction in which the products are lactic acid, ethanol, and CO_2.

heterotroph An organism that uses external sources of organic carbon compounds for biosynthesis.

heterotrophy The use of external sources of organic carbon compounds for biosynthesis.

Hfr A high-frequency recombination bacterial strain, caused by the presence of a chromosomally integrated F factor.

highly active antiretroviral therapy (HAART) A three-drug antiretroviral cocktail highly effective at inhibiting the replication of HIV in patients.

histone A protein that binds eukaryotic DNA and compacts chromosomes in nucleosomes.

HIV See **human immunodeficiency virus**.

holdfast Adhesion factors secreted by the tip of a stalk to firmly attach an organism to a substrate.

Holliday junction Also called *Holliday structure*. A cross-like configuration of recombining DNA molecules that forms during generalized recombination.

homolog A protein whose amino acid sequence is similar to that of another protein.

homologous Similar, referring to DNA sequences derived from a common ancestral gene.

hopanoid Also called *hopane*. A five-ringed hydrocarbon lipid found in bacterial cell membranes.

horizontal gene transfer Also called *lateral gene transfer* or just *horizontal transfer*. The passage of genes from one genome into another, nonprogeny genome.

horizontal transmission In disease, the transfer of a pathogen from one organism into another, nonprogeny organism.

hormogonium *pl.* **hormogonia** A short motile chain of three to five cells produced by filamentous cyanobacteria to disseminate their cells.

host factor A trait of an individual host that affects susceptibility to disease, in comparison with other individuals.
host range The species that can be infected by a given pathogen.
HR See **halorhodopsin**.
HSP See **heat-shock protein**.
human immunodeficiency virus (HIV) A human-specific retrovirus that is the causative agent of AIDS.
humic material Also called *humus*. Phenolic molecules, derived from lignin, that are resistant to degradation and hence very stable in soil.
humoral immunity A type of adaptive immunity mediated by antibodies.
humus See **humic material**.
hybridization The annealing of a nucleic acid strand with another nucleic acid strand containing a complementary sequence of bases. The binding of one nucleic acid strand with a complementary strand.
hydric soil Soil that undergoes periods of anoxic water saturation.
hydrogen bond An electrostatic attraction between a hydrogen bonded to an oxygen or nitrogen and a second, nearby oxygen or nitrogen.
hydrogenosome Found in some protists, an organelle that ferments carbohydrates in a pathway generating H_2; it may have evolved from mitochondria.
hydrogenotrophy The use of molecular hydrogen (H_2) as an electron donor for a variety of electron acceptors.
hydrologic cycle Also called *water cycle*. The cyclic exchange of water between atmospheric water vapor and Earth's bodies of liquid water.
hydrolysis The cleaving of a bond by the addition of a water molecule.
hydrophilic Soluble in water; either ionic or polar.
hydrophobic Insoluble in water; nonpolar.
hydrothermal vent Also called *thermal vent*. An opening in the seafloor through which superheated water arises, carrying high concentrations of reduced minerals such as sulfides.
3-hydroxypropionate cycle A carbon fixation process in which hydrated CO_2 condenses with acetyl-CoA to form 3-hydroxypropionate.
hyperthermophile An organism adapted for optimal growth at extremely high temperatures, generally above 80°C, and as high as 121°C.
hypertonic Having more solutes than another environment separated by a semipermeable membrane. Water will tend to flow toward the hypertonic solution.
hypha *pl.* **hyphae** Threadlike filament forming the mycelium of a fungus.
hypotonic Having fewer solutes than another environment separated by a semipermeable membrane. Water will tend to flow away from the hypotonic solution.
hypoxia A state of lower-than-normal oxygen concentration.

I

icosahedral capsid For a virus, a crystalline protein shell with 20 identical faces, enclosing the nucleic acid.
ID_{50} See **infectious dose 50%**.
identification The recognition of the species (or higher taxonomic category) of a microbe isolated in pure culture.
idiotype An amino acid difference in the antigen-binding site (N terminus of heavy or light chains) that distinguishes different antibodies within an individual.
IEF See **isoelectric focusing**.
IgA An antibody isotype that contains the alpha heavy chain. It can be secreted and found in tears, saliva, breast milk, and so on.
IgD An antibody isotype that contains the delta heavy chain. It is found on B-cell membranes.
IgE An antibody isotype that contains the epsilon heavy chain. It is involved in degranulation of mast cells.
IgG An antibody isotype that contains the gamma heavy chain. It is found in serum.
IgM The first antibody isotype detected during the early stages of an immune response. It contains the mu heavy chain and is found as a pentamer in serum.
IL-1 See **interleukin-1**.
immediate hypersensitivity See **type I hypersensitivity**.
immersion oil An oil with a refractive index similar to glass that minimizes light ray loss at wide angles, thereby minimizing wavefront interference and maximizing resolution.
immune avoidance The changing of cell-surface proteins by pathogens to prevent antibody detection and prolong infection.
immune system An organism's cellular defense system against pathogens.
immunity Resistance to a specific disease.
immunization The stimulation of an immune response by deliberate inoculation with a weakened pathogen, in hopes of providing immunity to disease caused by the pathogen.
immunogen An antigen that, by itself, can elicit antibody production.
immunogenicity A measure of the effectiveness of an antigen in eliciting an immune response.
immunoglobulin A member of a family of proteins that contain a 110-amino-acid domain with an internal disulfide bond. Members include antibodies and major histocompatibility proteins.
immunological specificity The ability of antibodies produced in response to a particular epitope to bind that epitope almost exclusively. Antibodies made to one epitope bind only weakly, if at all, to other epitopes.
immunomodulin A protein made by normal microbiota that influences the host immune response by modifying the secretion of host proteins, such as a cytokine.
immunopathogenesis The process by which an immune response or the products of an immune response cause disease.
immunoprecipitation The antibody-mediated cross-linking of antigens to form large, insoluble complexes. Immunoprecipitation is used in research labs and is normally seen only in vitro.
index case Also called *patient zero*. The first case of an infectious disease, and an important piece of data for helping contain the spread of disease.
indigenous microbiota (microflora) Microbes found naturally in a particular location, often in association with a food substrate.
inducer A molecule that binds to a repressor and prevents it from binding to the operator sequence.
inducer exclusion The ability of glucose to cause metabolic changes that prevent the cellular uptake of less favorable carbon sources that could cause unnecessary induction.

induction Increased transcription of target genes due to an inducer binding to a repressor and preventing repressor-operator binding.
industrial fermentor The equipment used to grow microbes on an industrial scale.
industrial microbiology The commercial exploitation of microbes.
industrial strain A microbial strain whose characteristics are optimized for industrial use.
infection The growth of a pathogen or parasite in or on a host.
infection cycle The route a pathogen takes as it moves from one host into another.
infection thread A column of rhizobial cells that projects down a tube into a plant root cell during the early stages of the rhizobia-legume symbiosis.
infectious dose 50% (ID_{50}) The number of bacteria or virions required to cause disease symptoms in 50% of an experimental group of hosts.
inflammasome A cytoplasmic multiprotein complex that promotes the maturation of inflammatory cytokines IL-1β and IL-18. Inflammasome assembly is triggered by NLR interactions with MAMPs.
information flow The passing of genetic information from DNA to DNA (replication), or from DNA to RNA (transcription) to protein (translation).
injera A highly fermented Ethiopian flat bread using the grain teff.
innate immunity Also called *nonadaptive immunity*. Nonspecific mechanisms for protecting against pathogens.
inner leaflet The layer of the cell membrane phospholipid bilayer that faces the cytoplasm.
inner membrane In Gram-negative bacteria, the membrane in contact with the cytoplasm, equivalent to the cell membrane.
insertion The addition of nucleotides to the middle of a DNA sequence.
insertion sequence (IS) A simple transposable element consisting of a transposase gene flanked by short, inverted-repeat sequences that are the target of transposase.
integral protein See **transmembrane protein**.
interactome All of the proteins that interact with other proteins in a cell.
interference The interaction of two wavefronts. Interference can be additive (amplitudes in phase, constructive) or subtractive (amplitudes out of phase, destructive).
interferon A host-secreted immunomodulatory protein that inhibits viral replication.
interleukin 1 (IL-1) A cytokine release by macrophages.
intermediate filament A eukaryotic cytoskeletal protein that is composed of various proteins depending on the cell type.
internal ribosome entry site (IRES) A site within an mRNA sequence where a ribosome can bind and initiate translation.
intimin A pathogenic *E. coli* adhesion protein that binds tightly to an *E. coli*–produced receptor injected into host cells.
intracellular pathogen A pathogen that lives within a host cell.
intron In eukaryotic genes, an intervening sequence that does not code for protein and is spliced out of the mRNA prior to translation.
inversion A flipping of a DNA fragment within a chromosome. It may allow or repress the transcription of a particular gene.
inverted repeat A DNA sequence that is found in an identical but inverted form at two sites on the same double helix (e.g., 5′-ATCGATCGnnnnnnCGATCGAT-3′).
ion An atom or molecule containing charge—that is, a number of electrons greater or smaller than the number of protons.
ion gradient A difference in concentration of an ion across a membrane.
ionic bond A bond between ions of positive and negative charge.
IRES See **internal ribosome entry site**.
IS See **insertion sequence**.
isoelectric focusing (IEF) A technique that separates proteins according to their charge—that is, migration of proteins to their isoelectric point in a pH gradient.
isoelectric point The pH at which there is no net charge on an amino acid or a protein.
isolate A microbe that has been obtained from a specific location and grown in pure culture.
isoprenoid A condensed isoprene chain, found in archaeal membrane lipids.
isotonic Being in osmotic balance, having equal concentrations of solutes on both sides of a semipermeable membrane. A cell in an isotonic environment will neither gain nor lose water.
isotope An atom of an element with a specific number of neutrons. For example, carbon-12 (^{12}C), carbon-13 (^{13}C), and carbon-14 (^{14}C) are all isotopes of carbon.
isotope ratio The ratio of amounts of two different isotopes of an element. May serve as a biosignature if the ratio between certain isotopes of a given element is altered by biological activity.
isotype A species-specific antibody class, defined by the structure of the heavy chain. IgG, IgA, and IgE are examples of isotypes.
isotype switching Also called *class switching*. A change in the predominant antibody isotype produced by a cell.

J

joule (J) The standard SI unit for energy.

K

kelp Also known as *brown alga*. A heterokont protist that grows as multicellular sheets at the water's surface.
kimchi A popular Korean food based on brine-fermented cabbage.
Kirby-Bauer assay A method for determining antibiotic susceptibility. Antibiotic-impregnated disks are placed on an agar plate whose surface has been confluently inoculated with a test organism. The antibiotic diffuses away from the disk and inhibits growth of susceptible bacteria. The width of the inhibitory zone is proportional to the susceptibility of the organism.
knockout mutation A mutation that completely eliminates the activity of a gene product.
Koch's postulates Four criteria, developed by Robert Koch, that should be met for a microbe to be designated the causative agent of an infectious disease.
Krebs cycle See **tricarboxylic acid cycle**.

L

labile toxin (LT) An *E. coli* enterotoxin, destroyed by heat, that increases cellular cAMP concentrations.
laboratory evolution See **experimental evolution**.
lactate fermentation Also called *lactic acid fermentation*. A fermentation reaction that generates lactic acid from reduction of pyruvic acid.

lag phase A period of cell culture, occurring right after bacteria are inoculated into new media, during which there is slow growth or no growth.
lamellar pseudopod A pseudopod with a sheetlike morphology.
laminar flow biological safety cabinet An air filtration appliance that removes pathogenic microbes from within the cabinet.
Langerhans cell A specialized, phagocytic dendritic cell that is the predominant cell type in skin-associated lymphatic tissue.
late gene A viral gene expressed late in the infectious cycle.
late-phase anaphylaxis Anaphylaxis caused by leukotrienes released by eosinophils recruited by mast cells.
latent period The time in the viral life cycle when progeny virions have formed but are still within the host cell.
latent state A period of the infection process during which a pathogenic agent is dormant in the host and cannot be cultured.
lateral gene transfer See **horizontal gene transfer**.
law of mass action The tendency of a reaction at equilibrium to return to equilibrium after the concentrations of products or reactants is altered.
LD_{50} See **lethal dose 50%**.
leaching The process of metal dissolution from ores.
leader sequence A short DNA sequence preceding a structural gene. In amino acid operons it contains multiple codons for the amino acid synthesized by the downstream structural genes. The leader sequence and translating ribosome help determine whether the structural genes are transcribed.
leaflet One of the two lipid layers in a phospholipid bilayer. The inner leaflet of the cell membrane faces the cytoplasm.
leaven For bread dough, to cause to rise by generating air spaces, usually through carbon dioxide production by microbial fermentation.
leghemoglobin An iron-bearing plant protein that sequesters oxygen to maintain an anoxic environment for nitrogenase within cells containing bacteroids.
lens An object composed of transparent, refractile material that bends light rays to converge at a focal point (or to diverge from an imaginary point). For electron microscopy, a series of magnets arranged as a magnetic lens bends electron beams to converge or diverge.
lentivirus A member of a family of retroviruses with a long incubation period. An example is HIV.
lethal dose 50% (LD_{50}) A measure of virulence; the number of bacteria or virions required to kill 50% of an experimental group of hosts.
lethal factor (LF) A component of anthrax toxin that cleaves host protein kinases.
LF See **lethal factor**.
lichen A simple multicellular organism formed by a mutualistic relationship between algae and fungi.
light chain The smaller of the two protein types comprising an antibody. Each antibody contains two heavy chains and two light chains.
light microscopy Observation of a microscopic object based on light absorption and transmission.
lignin A complex aromatic organic compound that forms the key structural support for trees and woody stems.
limiting nutrient A nutrient whose depletion generally limits growth in a given ecosystem.
lincosamide Any of a class of bacteriostatic antibiotics (e.g., clindamycin).
lipopolysaccharide (LPS) Structurally unique phospholipids found in the outer leaflet of the outer membrane in Gram-negative bacteria. Many are endotoxins.
lithotroph See **chemoautotroph**.
lithotrophy See **chemoautotrophy**.
littoral zone The upper water layer of aquatic habitats that light can penetrate.
logarithmic (log) phase See **exponential phase**.
long terminal repeat (LTR) A repeated nucleic acid sequence at the 5′ and 3′ ends of a provirus.
loss-of-function mutation A mutation that eliminates or decreases the function of the gene product.
LPS See **lipopolysaccharide**.
LT See **labile toxin**.
LTR See **long terminal repeat**.
lumen The interior of an intracellular membrane-enclosed compartment.
Lyme disease See **borreliosis**.
lymph node A secondary lymphatic organ, formed by the convergence of lymphatic vessels, that traps foreign particles from local tissue and presents them to resident immune cells.
lymphocyte A mononuclear leukocyte (white blood cell) that is a product of lymphoid tissue and participates in immunity (e.g., B cell and T cell).
lyophilization See **freeze-drying**.
lysate The contents of broken cells; may include virus particles.
lysis The rupture of the cell by a break in the cell wall and membrane.
lysogeny A viral life cycle in which the viral genome integrates into and replicates with the host genome, but retains the ability to initiate host cell lysis.
lysosome An acidic eukaryotic organelle that aids digestion of molecules. Not found in plant cells.
lytic cycle A viral life cycle in which the virus produces new virions and lyses the cell, releasing virions.

M

M cell A phagocytic innate immune cell (microfold cell) found between intestinal epithelial cells.
MAC See **membrane attack complex**.
MacConkey medium A differential, selective medium that selects for Gram-negative bacteria and can differentiate between lactose fermenters and nonfermenters.
macrolide Any of a group of antibiotics containing a large lactone ring (e.g., erythromycin).
macronucleus A form of nucleus found in ciliates that is derived from gene amplification and rearrangement of micronuclear DNA; contains actively transcribed genes.
macronutrient A nutrient that an organism needs in large quantity.
macrophage A mononuclear, phagocytic, antigen-presenting cell of the immune system.
magnetosome An organelle containing the mineral magnetite that allows microbes to sense a magnetic field.
magnetotaxis The ability to direct motility along magnetic field lines.
magnification An increase in the apparent size of a viewed object as an optical image.
major histocompatibility complex (MHC) Transmembrane cell proteins important for recognizing self and for presenting foreign antigens to the adaptive immune system.

malaria A disease caused by the apicomplexan *Plasmodium falciparum*, transmitted by mosquitoes.
malolactic fermentation Fermentation of L-malate (a side product of glucose fermentation) by *Oenococcus oeni* bacteria; an important process in winemaking.
MAMP See **microbe-associated molecular pattern**.
marine snow Microbial biofilms on inorganic particles suspended in marine water.
mast cell A white blood cell that secretes proteins that aid innate immunity. Mast cells reside in connective tissues and mucosa and do not circulate in the bloodstream.
matrix protein A protein, found in some viruses, that is located between the capsid and the membrane envelope.
MCP See **methyl-accepting chemotaxis protein**.
MDR efflux pump See **multidrug resistance efflux pump**.
mean generation time The reciprocal of the mean growth rate constant; the mean time period for doubling of a population.
mechanical transmission A nonspecific mode of viral entry into damaged tissue.
meiosis A form of cell division by which a diploid eukaryotic cell generates haploid sex cells that contain recombinant chromosomes.
membrane attack complex (MAC) A cell-destroying pore produced in the membrane of invading bacteria by the host cell complement cascade.
membrane potential Energy stored as an electrical voltage difference across a membrane.
membrane-permeant organic acid See **membrane-permeant weak acid**.
membrane-permeant weak acid An acid that exists in charged and uncharged forms, such as acetic acid. The uncharged form can penetrate the membrane.
membrane-permeant weak base A base that exists in charged and uncharged forms, such as methylamine. The uncharged form can penetrate the membrane.
memory B cell A long-lived type of lymphocyte preprogrammed to produce a specific antibody. After encountering their activating antigen, memory B cells differentiate into antibody-producing plasma cells.
merozoite The form of *Plasmodium falciparum*, the causative agent of malaria, that invades red blood cells.
mesocosm A small, controlled model ecosystem.
mesophile An organism with optimal growth between 20°C and 40°C.
messenger RNA (mRNA) An RNA molecule that encodes a protein.
metabolist model A model of early life in which the central components of intermediary metabolism arose from self-sustaining chemical reactions based on inorganic chemicals.
metagenome The sum of genomes of all members of a community of organisms.
metagenomics The study of community genomes, or metagenomes.
metaproteomics The study of all the proteins synthesized by members of a community, known as a metaproteome.
metastatic lesion A lesion of infection, or of cancerous cells, that develops at secondary sites away from the initial site of infection.
metatranscriptomics The study of all the RNA transcripts expressed by members of a community, known as a metatranscriptome.
methane gas hydrate A crystalline material in which methane molecules are surrounded by a cage of water molecules. This molecular configuration is found in the deep ocean.
methanogen An organism that uses hydrogen to reduce CO_2 and other single-carbon compounds to methane, yielding energy.
methanogenesis An energy-yielding metabolic process that produces methane, commonly from hydrogen gas and oxidized one- or two-carbon compounds. It is unique to archaea.
methanotroph An organism that oxidizes methane to yield energy.
methanotrophy The metabolic oxidation of methane to yield energy.
methyl-accepting chemotaxis protein (MCP) A cell-membrane signal transduction protein that becomes methylated during adaptation to a chemotactic signal.
methyl mismatch repair A DNA repair system that fixes misincorporation of a nucleotide after DNA synthesis. The unmethylated daughter strand is corrected to complement the methylated parental strand.
methylotrophy The metabolic oxidation of single-carbon compounds such as methanol, methylamine, or methane to yield energy.
MHC See **major histocompatibility complex**.
MHC restriction The ability of T cells to recognize only those antigens complexed to self MHC molecules.
MIC See **minimal inhibitory concentration**.
microaerophilic Requiring oxygen at a concentration lower than that of the atmosphere, but unable to grow in high-oxygen environments.
microbe An organism or virus too small to be seen with the unaided human eye.
microbe-associated molecular pattern (MAMP) Formerly called *pathogen-associated molecular pattern (PAMP)*. Molecules associated with groups of microbes, both pathogenic and nonpathogenic, that are recognized by cells of the innate immune system.
microbiome The total community of microbes found within a specified environment.
microbiota The normally occurring microbes of a body part or an environmental habitat.
microcolony A small colony of bacteria visible only with the aid of a microscope.
microfilament Also called *actin filament*. A eukaryotic cytoskeletal protein composed of polymerized actin.
microfossil A microscopic fossil in which calcium carbonate deposits have filled in the form of ancient microbial cells.
micrometer (μm) One-millionth (10^{-6}) of a meter.
micronucleus A form of nucleus found in ciliates; contains a diploid set of chromosomes and undergoes meiosis for sexual exchange by conjugation.
micronutrient A nutrient that an organism needs in small quantity, typically a vitamin or a mineral.
microplankton Plankton consisting of microbes approximately 20–1,000 μm in diameter.
microscope A tool that increases the magnification of specimens to enable viewing at higher resolution.
microtubule A eukaryotic cytoskeletal protein composed of polymerized tubulin.
millimeter (mm) One-thousandth (10^{-3}) of a meter.
minimal inhibitory concentration (MIC) The lowest concentration of a drug that will prevent the growth of an organism.

miso A Japanese condiment, made from ground soy and rice, salted and fermented by the mold *Aspergillus oryzae*.
missense mutation A point mutation that alters the sequence of a single codon, leading to a single amino acid substitution in a protein.
mitochondrion *pl.* **mitochondria** An organelle of endosymbiotic origin that produces ATP through the use of an electron transport chain to generate a proton motive force. O_2 is the final electron acceptor to produce H_2O.
mitosis The orderly replication and segregation of eukaryotic chromosomes, usually prior to cell division.
mitosome An organelle derived from mitochondria, found in certain protists, such as *Giardia*, that lack mitochondria.
mitosporic fungus A species of fungus that generates spores by mitosis and lacks a known sexual cycle.
mixed-acid fermentation A bacterial fermentation process in which pyruvate is converted to several different organic acids, as well as ethanol, CO_2, and H_2O.
mixotroph An organism capable of both photosynthetic and heterotrophic metabolism.
mixotrophic Having the ability to switch among metabolic strategies, such as heterotrophy and phototrophy, depending on the environmental conditions.
modular enzyme A multifunctional enzyme in which several domains or subunits conduct sequential steps to generate a product.
MOI See **multiplicity of infection**.
molarity A unit of concentration measured as the number of moles of solute per liter of solution.
molecular clock The use of DNA or RNA sequence information to measure the time of divergence among different species.
molecular formula A notation indicating the number and type of atoms in a molecule. For example, H_2O is the molecular formula for water.
molecular mimicry A structural similarity between two different molecules.
monocyte A white blood cell with a single nucleus that can differentiate into a macrophage or a dendritic cell.
monophyletic Diverging from a common ancestor.
monophyletic group See **clade**.
monosaccharide The monomer unit of sugars. Monosaccharides have a molecular formula of $(CH_2O)_n$.
mordant A chemical binding agent that causes specimens to retain stains better.
mother cell The larger cell that forms during the asymmetrical cell division leading to spore formation. The mother cell will engulf the forespore.
motility The ability of a microbe to generate self-directed movement.
mRNA See **messenger RNA**.
mucociliary elevator The ciliated mucous lining of the trachea, bronchi, and bronchioles that sweeps foreign particles up and away from the lungs.
mucosal immunity The portion of the innate and adaptive immune systems that protects the mucosa from microbial invasion.
Mueller-Hinton agar A specialized, standardized, *para*-aminobenzoic acid–free medium used for the Kirby-Bauer assay.
multidrug resistance (MDR) efflux pump A transmembrane protein pump that can export many different kinds of antibiotics with diverse structure.
multiplex PCR A polymerase chain reaction that uses multiple pairs of oligonucleotide primers to amplify several different DNA sequences simultaneously.
multiplicity of infection (MOI) The ratio of infecting virions to host cells.
murein See **peptidoglycan**.
murein lipoprotein Also called *Braun lipoprotein*. The major lipoprotein that connects the outer membrane of Gram-negative bacteria to the peptidoglycan cell wall.
mutagen A chemical that damages DNA and increases the rate of mutations.
mutation A heritable change in a DNA sequence.
mutation frequency The fraction of mutant cells (defective in a given gene) within the total cell population.
mutation rate The number of mutations introduced into DNA per generation (cell doubling).
mutator strain A strain of cells with a high mutation rate, usually due to a mutation in a DNA repair enzyme.
mutualism A symbiotic relationship in which both partners benefit.
mycelium *pl.* **mycelia** A fungal hypha that projects into the air (aerial mycelium) or into the growth substrate (surface mycelium).
mycolic acid One of a diverse class of sugar-linked fatty acids found in the cell envelopes of mycobacteria such as *Mycobacterium tuberculosis*.
mycology The study of fungi.
mycorrhizae *sing.* **mycorrhiza** Fungi involved in an intimate mutualism with plant roots, in which nutrients are exchanged.
myxospore A durable spherical cell produced by the fruiting body of myxobacteria.

N

N-terminal rule The tendency of the N-terminal amino acid of a protein to influence protein stability.
NAD See **nicotinamide adenine dinucleotide**.
Nanoarchaeota A deeply branching division of Archaea; includes thermophilic cells of extremely small size that are obligate symbionts of *Ignicoccus*.
nanometer (nm) One-billionth (10^{-9}) of a meter.
nanoplankton Plankton consisting of microbes approximately 2–20 μm in diameter.
nanoscale secondary ion mass spectrometry (NanoSIMS) A technique of chemical imaging in which an ionizing beam breaks off organic ions from a sample, which fly off the sample and are captured for analysis by a mass spectrometer.
nanotube A tube of plasma membrane that connects the cytoplasm of two cells, forming a conduit through which intracellular materials or pathogens may pass.
nasopharynx The passage leading from the nose to the oral cavity.
native conformation The fully folded, functional form of a protein.
natto A soybean product, similar to tempeh, produced by alkaline fermentation.
natural selection The change in frequency of genes in a population under environmental conditions that favor some genes over others.
necrotizing fasciitis Also known as *flesh-eating disease*. A severe skin infection usually caused by the Gram-positive coccus *Streptococcus pyogenes*.

negative selection In immunology, the destruction of T cells bearing T-cell receptors (TCRs) that bind strongly to self MHC proteins displayed on thymus epithelial cells.
negative stain A stain that colors the background and leaves the specimen unstained.
NER See **nucleotide excision repair**.
NET See **neutrophil extracellular trap**.
neuraminidase inhibitor Any of a class of anti-influenza drugs that target neuraminidase on the viral envelope and decrease the number of virus particles produced.
neutralophile An organism with an optimal growth range in environments between pH 5 and pH 8.
neutrophil A white blood cell of the innate immune system that can phagocytose and kill microbes.
neutrophil extracellular trap (NET) A net of chromatin and antimicrobial peptides expelled by dying neutrophils to trap and injure nearby pathogenic bacteria.
NHEJ See **nonhomologous end joining**.
niche An organism's environmental requirements for existence and its relations with other members of the ecosystem.
nicotinamide adenine dinucleotide (NAD) An energy carrier in the cell that can donate (NADH) or accept (NAD^+) electrons.
nitrification The oxidation of reduced nitrogen compounds to nitrite or nitrate.
nitrifier An organism that converts reduced nitrogen compounds to nitrite or nitrate.
nitrogen fixation The ability of some prokaryotes to reduce inorganic diatomic nitrogen gas (N_2) to two molecules of ammonium ion ($2NH_4^+$).
nitrogen-fixing bacterium A bacterium that can reduce diatomic nitrogen gas (N_2) to two molecules of ammonium ion (NH_4^+).
nitrogenase The enzyme that catalyzes nitrogen fixation.
nitrogenous base An organic compound containing nitrogen, that acts as a base. The nucleobases are nitrogenous bases that form the nucleotides of nucleic acids.
Nitrospirae A phylum of Gram-negative bacteria, many of which are lithotrophs, oxidizing nitrite to nitrate.
NLR See **NOD-like receptor**.
node In physiology, a small mass of tissue in the form of a swelling, either normal or pathological. In phylogenetic trees, the most recent common ancestor of branching descendents.
NOD-like receptor (NLR) A eukaryotic cytoplasmic protein that recognizes particular MAMPs present on microorganisms.
noise Variance from a mean in an assay. A system in which individual results vary greatly from a mean that was derived from many results is a noisy system.
Nomarski microscopy See **differential interference contrast microscopy**.
nomenclature The naming of different taxonomic groups of organisms.
nonadaptive immunity See **innate immunity**.
nonhomologous end joining (NHEJ) A pathway that repairs double-strand DNA breaks by direct ligation without the need for large regions of homology.
nonpolar covalent bond A covalent bond in which the electrons in the bond are shared equally by the two atoms.
nonribosomal peptide antibiotic A peptide with antimicrobial activity synthesized by modular enzymes and not by ribosomes.
nonsense mutation A mutation that changes an amino acid codon into a premature stop codon.
nonsense suppressor An altered (mutant) tRNA molecule whose anticodon will bind to a nonsense codon (premature stop codon) in a mutant mRNA and allow translation to proceed.
nori A Japanese food obtained from the red algae *Porphyra* spp.
northern blot A technique to detect specific RNA sequences. Sample RNA is subjected to gel electrophoresis, transferred to a blot, and probed with a labeled cDNA that will hybridize to target RNA sequences.
nosocomial Hospital-acquired; commonly refers to an infectious agent.
NP See **nucleocapsid protein**.
nucleobase A planar, heteroaromatic nitrogenous base that forms a nucleotide of nucleic acids; determines the information content of DNA and RNA.
nucleocapsid protein (NP) A protein that coats a viral genome.
nucleoid The looped coils of a bacterial chromosome.
nucleolus *pl.* **nucleoli** A region inside the nucleus where ribosome assembly begins.
nucleomorph A vestigial nucleus within a eukaryotic cell, evolved by genetic reduction from the nucleus of an endosymbiont.
nucleotide The monomer unit of nucleic acids, consisting of a five-carbon sugar, a phosphate, and a nitrogenous base.
nucleotide excision repair (NER) A DNA repair mechanism that cuts out damaged DNA. New, correctly base-paired DNA is synthesized by DNA polymerase I.
nucleus *pl.* **nuclei** A eukaryotic organelle that contains the DNA.
numerical aperture The product of the refractive index of the medium and sin θ. As numerical aperture increases, the magnification increases.

O

O polysaccharide A sugar chain that connects to the core polysaccharide of lipopolysaccharides.
objective lens In a compound microscope, the lens that is closest to the specimen and generates the initial magnification.
obligate aerobe An organism that requires molecular oxygen to grow.
ocular lens Also called *eyepiece*. In a compound microscope, the lens situated closest to the observer's eye.
Okazaki fragments Short fragments of DNA that are synthesized on the lagging strand during DNA synthesis.
oligotroph An organism that can grow only in environments containing extremely low concentrations of organic nutrients.
oligotrophic lake A lake having low concentration of organic nutrients; the opposite of a eutrophic lake.
OMZ See **oxygen minimum zone**.
oncogene A gene that, through mutation or inappropriate expression, can lead to cancer.
oncogenic virus A virus that causes cancer.
oomycete A member of the eukaryotic group Oomycetes—heterokont protists whose life cycle resembles that of fungi; formerly classified as fungi (Oomycota).
open reading frame (ORF) A DNA sequence predicted to encode a protein.
operational taxonomic unit (OTU) A taxonomic group defined by a designated degree of similarity among members based on DNA sequence.
operon A collection of genes that are in tandem on a chromosome and are transcribed into a single RNA.

operon fusion Also called *transcriptional fusion*. A technique to monitor transcriptional regulation by fusing a reporter gene containing its own ribosome-binding site to the 3′ end of an operon. Unlike the gene fusion technique, only the promoter of the target gene directs expression of the reporter.

opportunistic pathogen A microbe that normally is not pathogenic but can cause infection or disease in an immunocompromised host organism.

opsonin An antibody that renders its target (e.g., bacteria) susceptible to phagocytosis.

opsonization The coating of pathogens with antibodies that aid pathogen phagocytosis by innate immune cells.

opsonize To bind IgG antibodies to microbes in order to enhance microbial phagocytosis by host immune cells.

optical density A measure of how many particles are suspended in a solution, based on light scattering by the suspended particles.

optical isomer Also called *enantiomer*. A molecule that has a mirror image. Molecules that contain a chiral carbon can have optical isomers.

optical tweezers Also called *optical trap*. A laser beam focused to generate an attractive or repulsive force that holds a microscopic object in position.

oral groove A ciliate structure for food uptake.

ORF See **open reading frame**.

organelle A membrane-enclosed compartment within eukaryotic cells that serves a specific function.

organic molecule A molecule that contains a carbon-carbon bond.

organism An individual form of life.

organotroph See **chemoorganotroph**.

organotrophy See **chemoorganotrophy**.

origin (*oriC*) The region of a bacterial chromosome where DNA replication initiates.

origin of replication (*ori*) A DNA sequence at which DNA replication initiates. In a bacterial chromosome this site is also attached to the cell envelope.

oropharynx The area between the soft palate and the upper edge of the epiglottis.

ortholog See **orthologous gene**.

orthologous Describing genes present in more than one species that derived from a common ancestral gene and encode the same function.

orthologous gene Also called *ortholog*. A gene present in more than one species that derived from a common ancestral gene and encodes the same function.

osmolarity A measure of the number of solute molecules in solution.

osmosis The diffusion of water from regions of high water concentration (low solute) to regions of low water concentration (high solute) across a semipermeable membrane.

osmotic pressure Pressure exerted by the osmotic flow of water through a semipermeable membrane.

OTU See **operational taxonomic unit**.

outer leaflet The layer of the cell membrane phospholipid bilayer that faces away from the cytoplasm.

outer membrane In Gram-negative bacteria, a membrane external to the cell wall.

oxazolidinone One of a class of synthetic antibiotics that inhibit protein synthesis.

oxidative burst A large increase in the oxygen consumption of immune cells during phagocytosis of pathogens as the immune cells produce oxygen radicals to kill the pathogen.

oxidative phosphorylation An electron transport chain that uses diatomic oxygen as a final electron acceptor and generates a proton gradient across a membrane for the production of ATP via ATP synthase.

oxidoreductase An electron transport system protein that accepts electrons from one molecule (oxidizing that molecule), and donates electrons to a second molecule, thereby reducing the second molecule.

oxygen minimum zone (OMZ) The region of the marine water column in which oxygen is depleted by respiration; usually at a mid level, between the aerated upper water and the deeper oxygenated water where organic food is scarce.

oxygenic Z pathway An ATP-producing photosynthetic pathway consisting of photosystems I and II. Water serves as the initial electron donor (generating O_2), and $NADP^+$ is the final electron acceptor (generating NADPH).

P

P site See **peptidyl-tRNA site**.

PA See **protective antigen**.

palindrome A DNA sequence in which the top and bottom strands have the same sequence in the 5′-to-3′ direction.

PAMP See **microbe-associated molecular pattern**.

pandemic An epidemic that occurs over a wide geographic area.

pan-genome All the genes possessed by all individual members of a species.

panspermia The hypothesis that life-forms originated elsewhere and "seeded" life on Earth.

paper chromatography A technique to separate compounds on the basis of their differential migration in a solvent that wicks up through paper. Compounds are separated by their relative solubility in the solvent.

paralogous gene Also called *paralog*. A gene that arises by gene duplication within a species and evolves to carry out a different function from that of the original gene.

pararetrovirus Also called DNA *reverse-transcribing virus*. A virus with a double-stranded DNA genome that generates an RNA intermediate and thus requires reverse transcriptase to generate progeny DNA genomes.

parasite Any bacterium, virus, fungus, or protozoan (protist) that colonizes and harms its host; the term commonly refers to protozoa and to invertebrates.

parasitism A symbiotic relationship in which one member benefits and the other is harmed.

parenteral route The introduction of materials into the body via intravenous or intramuscular injection.

parfocal In a microscope with multiple objective lenses, having the objective lenses set at different heights that maintain focus when switching among lenses.

passive transport Net movement of molecules across a membrane without energy expenditure by the cell.

pasteurization The heating of food at a temperature and time combination that will kill spore-like structures of *Coxiella burnetii*.

pathogen A bacterial, viral, or fungal agent of disease.

pathogen-associated molecular pattern (PAMP) See **microbe-associated molecular pattern**.

pathogenesis The processes through which microbes cause disease in a host.

pathogenicity The ability of a microorganism to cause disease.
pathogenicity island A type of genomic island, a stretch of DNA that contains virulence factors and may have been transferred from another genome.
patient zero See **index case**.
PBP See **penicillin-binding protein**.
PCR See **polymerase chain reaction**.
PDC See **pyruvate dehydrogenase complex**.
pelagic zone The water column of the open ocean, away from the shore and the ocean floor.
penicillin An antibiotic, produced by the *Penicillium* mold, that blocks cross-bridge formation during peptidoglycan synthesis.
penicillin-binding protein (PBP) A bacterial protein, involved in cell wall synthesis, that is the target of the penicillin antibiotic.
pentose phosphate cycle See **Calvin cycle**.
pentose phosphate pathway (PPP) Also called *pentose phosphate shunt (PPS)*. An alternate glycolytic pathway in which glucose 6-phosphate is first oxidized and then decarboxylated to ribulose 5-phosphate, ultimately generating 1 ATP and 2 NADPH.
peptide bond The covalent bond that links two amino acid monomers.
peptidoglycan Also called *murein*. A polymer of peptide-linked chains of amino sugars; a major component of the bacterial cell wall.
peptidyl-tRNA site (P site) The ribosomal site that contains the growing protein attached to a tRNA.
peptidyltransferase The rRNA enzymatic ability to form peptide bonds.
perforin A cytotoxic protein, secreted by T cells, that forms pores in target cell membranes.
peripheral membrane protein A protein that is associated with a membrane but does not span the phospholipid bilayer.
permease A substrate-specific carrier protein in the membrane.
permissive temperature A temperature at which a temperature-sensitive mutation in a gene is masked, permitting growth of the organism.
peroxisome A eukaryotic organelle that converts hydrogen peroxide to water.
petechia *pl.* **petechiae** A pinpoint capillary hemorrhage due to the absence of clotting factors; may indicate the presence of endotoxin.
petri dish Also called *petri plate*. A round dish with vertical walls covered by an inverted dish of slightly larger diameter. The smaller dish can be filled with a substrate for growing microbes.
PFU See **plaque-forming unit**.
phage See **bacteriophage**.
phage display A technique in which a phage particle contains recombinant coat proteins expressed by genes encoding a coat protein fused to the protein of interest, such as a vaccine antigen.
phagocytosis A form of endocytosis in which a large extracellular particle is brought into the cell.
phagosome A large intracellular vesicle that forms as a result of phagocytosis.
phase-contrast microscopy Observation of a microscopic object based on the differences in the refractive index between cell components and the surrounding medium. Contrast is generated as the difference between refracted light and transmitted light shifted out of phase.
phase variation A gene regulatory mechanism that changes the amino acid sequence of a protein from one antigenic type to another. One mechanism involves site-specific recombination that flips a DNA sequence in a chromosome.
phenol coefficient test A test of the ability of a disinfectant to kill bacteria; the higher the coefficient, the more effective the disinfectant.
phenol red broth test A clinical test for particular bacterial fermentation pathways, indicating the presence or absence of particular species, based on fermentative acids changing a pH indicator.
phosphatidate A negatively charged phosphate head group of a phospholipid.
phosphatidylethanolamine A type of phospholipid with a positively charged ethanolamine attached to the phosphate group.
phosphatidylglycerol A type of phospholipid with a glycerol attached to the phosphate group.
phosphodiester bond The bond that covalently attaches to adjacent nucleotides in a nucleic acid.
phospholipid The major component of membranes. A typical phospholipid is composed of a core of glycerol to which two fatty acids and a modified phosphate group are attached.
phospholipid bilayer Two layers of phospholipids; the hydrocarbon fatty acid tails face the interior of the bilayer, and the charged phosphate groups face the cytoplasm and extracellular environment. The cell membrane is a phospholipid bilayer.
phosphorimaging The visualization of a radioactive probe by recording the energy emanated as light or radioactivity from gels or membranes.
phosphorylation The enzyme-catalyzed addition of a phosphoryl group onto a molecule.
phosphotransferase system (PTS) A group translocation system that uses phosphoenolpyruvate to transfer phosphoryl groups onto the incoming molecule.
photic zone See **euphotic zone**.
photo- Deriving energy from light absorption.
photoautotroph An organism that performs photosynthesis, using light energy to reduce carbon dioxide.
photoautotrophy The metabolic reduction of carbon dioxide using light as an energy source.
photoexcitation Light absorption that raises an electron to a higher energy state, as in bacteriorhodopsin or in chlorophyll.
photoheterotrophy Metabolism that includes gain of energy from light absorption with biosynthesis from preformed organic compounds. Usually also includes organotrophy, gain of energy from reactions of organic compounds.
photoionization Light absorption that causes electron separation.
photolysis The first energy-yielding phase of photosynthesis; the light-driven separation of an electron from a molecule coupled to an electron transport system.
photoreactivation A light-induced, photolyase-catalyzed repair of pyrimidine dimers.
photosynthesis The metabolic ability to absorb and convert solar energy into chemical energy for biosynthesis. Autotrophic photosynthesis, or photoautotrophy, includes CO_2 fixation.
photosystem I A protein complex that harvests light from a chlorophyll, donates an electron to an electron transport system and receives an electron from a small molecule such as H_2S or H_2O, and stores energy in the form of NADPH.
photosystem II A protein complex that harvests light from a bacteriochlorophyll, donates an electron to an electron transport system, and stores energy in the form of a proton potential.

phototrophy The use of chemical reactions powered by the absorption of light to yield energy.
phylogenetic tree A diagram depicting estimates of the relative amounts of evolutionary divergence among different species.
phylogeny A measurement of genetic relatedness. The classification of organisms based on their genetic relatedness.
phylum *pl.* **phyla** The taxonomic rank one level below domain; a group of organisms sharing a common ancestor that diverged early from other groups.
phytoplankton Phototrophic marine bacteria, algae, and protists, the primary producers in pelagic food webs.
picometer (pm) One-trillionth (10^{-12}) of a meter.
picoplankton Plankton consisting of microbes approximately 0.2–2 μm in diameter.
piezophile See **barophile**.
pilin The protein monomer that polymerizes to form a pilus.
pilus *pl.* **pili** Also called *fimbria*. A straight protein filament composed of a tube of protein monomers that extend from the bacterial cell envelope.
pinocytosis A form of endocytosis in which only extracellular fluid and small molecules are brought into the cell.
Planctomycetes A phylum of free-living bacteria that have stalked cells and reproduce by budding. Their nucleoid is surrounded by a membrane.
plankton Organisms that float in water.
planktonic cell An isolated cell, growing individually in a liquid without connections to other cells.
plaque A cell-free zone on a lawn of bacterial cells caused by viral lysis.
plaque assay An assay to determine the presence of bacteriophages based on their ability to form plaques.
plaque-forming unit (PFU) A measure of the concentration of phage particles in liquid culture.
plasma cell A short-lived antibody-producing cell.
plasma membrane See **cell membrane**.
plasmid An extrachromosomal genetic element that may be present in some cells.
plasmodesma *pl.* **plasmodesmata** A membrane channel in plants that connects adjacent plant cells.
plasmodial slime mold A slime mold in which a fertilized zygote undergoes multiple nuclear divisions, generating a multinucleate single cell (plasmodium).
plasmodium *pl.* **plasmodia** The giant, multinucleate cell formed by a plasmodial slime mold.
platelet A small, anucleate cell fragment found in blood that is involved in clotting.
pneumonic plague A highly virulent and contagious *Yersinia pestis* lung infection.
point mutation A change in a single nucleotide within a nucleic acid sequence.
polar aging The process, during bacterial cell fission, in which septation forms two new poles while the original poles of the dividing cell age. With each new cell division, the preexisting poles age incrementally. The effects of aging vary among bacterial species.
polar covalent bond A covalent bond in which the electrons are distributed unequally between two atoms.
polyamine A molecule containing multiple amine groups, positively charged at neutral pH.
polyhedral body See **carboxysome**.
polymerase chain reaction (PCR) A method to amplify DNA in vitro using many cycles of DNA denaturation, primer annealing, and DNA polymerization with a heat-stable polymerase.
polyphyletic Having multiple evolutionary origins.
polyprotein A long peptide translated from one open reading frame but later cleaved into separate proteins with different functions.
polysaccharide A polymer of sugars. See also **glycan**.
polysome A cell structure consisting of multiple ribosomes performing translation on the same mRNA molecule.
porin A transmembrane protein complex that allows movement of specific molecules across the cell membrane or the outer membrane.
portal of entry Site through which microorganisms enter the body. These can be natural orifices, cuts, or by injection.
positive selection In immunology, the survival of T cells bearing T-cell receptors (TCRs) that don't recognize self MHC proteins displayed on thymus epithelial cells.
pour plate technique A procedure in which organisms are added to melted agar cooled to 45°C–50°C and the mixture is poured into an empty petri plate. Colonies grow in the agar after the agar solidifies.
PPP See **pentose phosphate pathway**.
PPS See **pentose phosphate pathway**.
pre-mRNA See **preliminary mRNA transcript**.
prebiotic soup A model for the origin of life based on the abiotic formation of fundamental biomolecules and cell structures such as membranes out of a "soup" of nutrients present on early Earth.
predator A consumer that feeds on grazers.
preliminary mRNA transcript (pre-mRNA) A eukaryotic messenger RNA prior to intron removal.
primary antibody The first antibody added in an immunologic assay (ELISA or Western blot). This antibody binds to the antigen of interest.
primary antibody response The production of antibodies upon first exposure to a particular antigen. B cells become activated and differentiate into plasma cells and memory B cells.
primary endosymbiont An organism in a lineage derived from a single endosymbiotic event.
primary pathogen A disease-causing microbe that can breach the defenses of a healthy host.
primary producer An organism that produces biomass (reduced carbon) from inorganic carbon sources such as CO_2.
primary recovery The initial isolation of commercial product from industrial microbes.
primary structure The first level of organization of polymers, consisting of the linear sequence of monomers—for example, the sequence of amino acids in a protein or nucleotides in a nucleic acid.
primary syphilis The initial inflammatory reaction (chancre) at the site of infection with *Treponema pallidum*.
primase An RNA polymerase that synthesizes short RNA primers complementary to a DNA template to launch DNA replication.
primer extension A technique to determine the 5′ end of an RNA transcript.
prion An infectious agent that causes propagation of misfolded host proteins; usually consists of a defective version of the host protein.
probabilistic indicator A means of quickly identifying microbes in the clinical setting, based on a battery of biochemical tests performed simultaneously on an isolated strain.

probiotic A food or nutritional supplement that contains live microorganisms and aims to improve health by promoting beneficial bacteria.

prokaryote An organism whose cell or cells lack a nucleus; includes both bacteria and archaea.

promoter A noncoding DNA regulatory region immediately upstream of a structural gene that is needed for transcription initiation.

proofreading An enzymatic activity of some nucleic acid polymerases that attempts to correct mispaired bases.

prophage A phage genome integrated into a host genome.

propionic acid fermentation The fermentation of lactic acid to propionic acid by *Propionibacterium* species; used in the production of Swiss cheese.

protease inhibitor A molecule that inhibits a protease enzyme; some are used as anti-HIV drugs to block the virally encoded protease needed to complete HIV assembly.

protective antigen (PA) The core subunit of anthrax toxin, so called because immunity to this protein protects against disease.

protein A A *Staphylococcus aureus* cell wall protein that binds to the Fc region of antibodies, hiding the *S. aureus* cells from phagocytes.

protein fusion See **gene fusion**.

Proteobacteria A large, metabolically and morphologically diverse phylum of Gram-negative bacteria.

proteome All the proteins expressed in a cell at a given time. The "complete proteome" includes all the proteins the cell can express under any condition. The "expressed proteome" represents the set of proteins made under a given condition.

proteomics The biological field of proteome analysis.

proteorhodopsin A bacterial membrane-embedded protein that contains retinal and acts as a light-driven proton pump; it is homologous to the archaeal protein bacteriorhodopsin.

protist A single-celled eukaryotic microbe, usually motile; not a fungus.

proton motive force See **proton potential**.

proton potential Also called *proton motive force*. The potential energy of the concentration gradient of protons (hydrogen ions, H^+) plus the charge difference across a membrane.

protozoan *pl.* **protozoa** A heterotrophic eukaryotic microbe, usually motile, that is not a fungus.

provirus A viral genome that is integrated into the host cell genome.

pseudogene A gene that is no longer functional.

pseudomurein See **pseudopeptidoglycan**.

pseudopeptidoglycan A peptidoglycan-like molecule composed of sugars and peptides that is found in some archaeal cell walls.

pseudopod A locomotory extension of cytoplasm bounded by the cell membrane.

psychrophile An organism with optimal growth at temperatures below 20°C.

PTS See **phosphotransferase system**.

pure culture A culture containing only a single strain or species of microorganism. A large number of microorganisms that all descended from a single individual cell.

purine A nitrogenous base with fused rings found in nucleotides; examples are adenine and guanine.

putrefaction Food spoilage due to the decomposition of proteins and amino acids.

pyrimidine A single-ring nitrogenous base found in nucleotides; examples are cytosine, thymine, and uracil.

pyrogen Any substance that induces fever.

pyrosequencing A method of DNA sequencing that relies on the detection of pyrophosphate released upon nucleotide incorporation.

pyruvate dehydrogenase complex (PDC) The multisubunit enzyme that couples the oxidative decarboxylation of pyruvate to acetyl-CoA and NADH production.

Q

quarantine The separation of infectious individuals from the general population to limit the spread of infection.

quasispecies A collection of isolates (usually viruses) from a common source of infection that have evolved into many different types within one host.

quaternary structure The highest level of organization of proteins, in which multiple polypeptide chains interact and function together.

quinol A reduced electron carrier that can diffuse laterally within membranes.

quinolone A type of antibiotic drug that inhibits DNA synthesis by targeting bacterial topoisomerases such as DNA gyrase.

quinone An oxidized electron carrier that can diffuse laterally within membranes.

quinone pool The oxidized quinones and reduced quinols that diffuse freely within the phospholipid membrane and are able to transfer electrons between many different redox enzymes.

quorum sensing The ability of bacteria to sense the presence of other bacteria via secreted chemical signals called autoinducers.

R

radial immunodiffusion A technique in which a ring of precipitation is visualized in an agarose gel impregnated with antibody. Antigen placed within a well diffuses outward until reaching a zone of equivalence where antigen-antibody complexes precipitate and form a ring.

radiolarian A member of the eukaryotic group Radiolaria—amebas with a silicate shell penetrated by filamentous pseudopods.

rancidity Food spoilage due to the oxidation of fats; may or may not involve microbial activity.

rarefaction curve A graph representing the number of different operational taxonomic units (species) found in a metagenome, as a function of increasing sample size.

RC See **reaction center**.

reaction center (RC) The complex containing a chlorophyll molecule that donates its excited electron to an electron transport system.

reactivation tuberculosis See **secondary tuberculosis**.

reading frame The position in a nucleic acid sequence from which triplet codons encode amino acids.

real-time PCR A technique using fluorescence to detect the products of PCR amplification as the reaction progresses, in order to quantify the amount of DNA in a sample.

reassortment The packaging of viral chromosome segments from two different viruses into one progeny virion. Refers to separate segments from a segmented genome, without helix recombination.

recombination The process by which two DNA molecules exchange arms by cutting and splicing their helix backbones.

recombination signal sequence (RSS) A DNA region downstream of antibody heavy- and light-chain genes that allows recombination between widely separated gene segments.

recombinational repair A DNA repair mechanism that relies on recombination between an undamaged chromosome and a gap that occurred during replication of damaged DNA.

red alga Also called *rhodophyte*. An alga of the eukaryotic group Rhodophyta, which contain chloroplasts as primary endosymbionts, with red accessory photopigments.

redox couple The oxidized and reduced states of a compound. For example, NAD^+ and NADH form a redox couple.

redox reaction A reaction in which one molecule or functional group becomes reduced and another becomes oxidized.

reducing agent See *electron donor*.

reductive acetyl-CoA pathway A carbon assimilation pathway in which two CO_2 molecules are condensed and reduced by 2 H_2 molecules to form an acetyl group.

reductive evolution Also called *degenerative evolution*. The loss or mutation of DNA encoding unselected traits.

reductive pentose phosphate cycle See **Calvin cycle**.

reductive TCA cycle Also called *reverse TCA cycle*. A CO_2 fixation pathway that generates acetyl-CoA through reversal of TCA cycle reactions. It requires ATP and NADPH.

reflection Deflection of an incident light ray by an object, at an angle equal to the incident angle.

refraction The bending and slowing of light as it passes through a substance.

refractive index The degree to which a substance causes the refraction of light; a ratio of the speed of light in a vacuum to its speed in another medium.

regulatory protein A protein that can bind DNA and modulate transcription in response to a metabolite.

regulatory T cell (Treg) A T cell that regulates the activity of another T cell, usually by suppressing its activity.

regulon A group of genes and operons located at different positions in a genome that are coordinately regulated and share a common biochemical function.

rehydration therapy A medical treatment for dehydration, in which a liquid solution of salts and glucose is delivered orally.

release factor A molecule that enters a ribosome A site containing an mRNA stop codon and initiates protein cleavage from the tRNA.

replication fork During DNA synthesis, the region of the chromosome that is being unwound.

replisome A complex of DNA polymerase and other accessory molecules that performs DNA replication.

reporter gene A gene, such as *lacZ* (beta-galactosidase) or *gfp* (green fluorescent protein), whose protein product can be easily quantified; commonly used in a gene fusion.

repression The down-regulation of gene transcription.

repressor A regulatory protein that can bind to a specific DNA sequence and inhibit transcription of genes.

reservoir 1. The major part of the biosphere that contains a significant amount of an element needed for life. 2. An organism that maintains a virus or bacterial pathogen in an area by serving as a high-titer host.

resolution The smallest distance that two objects can be separated and still be distinguished as separate objects.

resource colimitation A situation in which a population size is limited by the lack of two different resources, such as nitrogen and phosphorus.

respiration The oxidation of reduced organic electron donors through a series of membrane-embedded electron carriers to a final electron acceptor. The energy derived from the redox reactions is stored as an electrochemical gradient across the membrane, which may be harnessed to produce ATP.

response regulator A cytoplasmic protein that is phosphorylated by a sensor kinase and modulates gene transcription depending on its phosphorylation state.

restricted transduction See **specialized transduction**.

restriction endonuclease Also called *restriction enzyme*. A bacterial enzyme that cleaves double-stranded DNA within a specific short sequence, usually a palindrome.

restriction site A DNA sequence recognized and cleaved by a restriction endonuclease.

restrictive temperature A temperature at which a temperature-sensitive mutation in a gene leads to the mutant phenotype, which generally includes failure to grow.

reticulate body The metabolically and reproductively active form of chlamydias.

reticuloendothelial system A collection of cells that can phagocytose and sequester extracellular material.

retinal A vitamin A–related cofactor in opsin proteins; it undergoes a conformational change upon absorbing a photon.

retrotransposon A retroelement that contains only partial retroviral sequences but may encode reverse transcriptase to allow further movement into the host genome.

retrovirus Also called *RNA reverse-transcribing virus*. A single-stranded RNA virus that uses reverse transcriptase to generate a double-stranded DNA.

reverse electron flow An enzyme-catalyzed redox reaction that couples an electron transfer with a positive ΔG (such as NAD^+ reduction to NADH) to a source of energy with a larger negative ΔG (such as a proton potential generated by an electron transport system).

reverse TCA cycle See **reductive TCA cycle**.

reverse transcriptase (RT) An enzyme that produces a double-stranded DNA molecule from a single-stranded RNA template.

reversion A mutation that changes a previous mutation back to its original state.

rhizobium *pl.* **rhizobia** A bacterial species of the order Rhizobiales that forms highly specific mutualistic associations with plants in which the bacteria form intracellular bacteroids that fix nitrogen for the plant.

rhizoplane The surface of plant roots.

rhizosphere The soil environment surrounding plant roots.

Rho-dependent Describing a bacterial transcription termination mechanism that requires Rho protein.

Rho-independent Describing a bacterial transcription termination mechanism that requires a GC-rich region near the transcript terminus.

rhodophyte See **red alga**.

ribosomal RNA (rRNA) An RNA molecule that includes the scaffolding and catalytic components of ribosomes.

ribosome-binding site Also called *Shine-Dalgarno sequence*. In bacteria, a stretch of nucleotides upstream of the start codon in an mRNA that hybridizes to the 16S rRNA of the ribosome, correctly positioning the mRNA for translation.

riboswitch A secondary structure (hairpin) within some mRNA transcripts that can interact with metabolites or antisense RNA molecules, change structure, and affect the production or translation of the mRNA.

ribozyme See **catalytic RNA**.

ribulose 5-phosphate The five-carbon sugar ribulose, phosphorylated at carbon 5.

rich medium See **complex medium**.
ripening The aging of cheese.
rise period During viral culture, the time when cells lyse and viral progeny enter the media.
RNA reverse-transcribing virus See **retrovirus**.
RNA world A model of early life in which RNA performed all the informational and catalytic roles of today's DNA and proteins.
RNA-dependent RNA polymerase An enzyme that produces an RNA complementary to a template RNA strand.
rod 1. Also called *bacillus.* A bacterium with a linear shape. 2. A photoreceptor cell in the retina.
root of a tree The earliest common ancestor of all members of a phylogenetic tree.
rotavirus One of a group of nonenveloped dsRNA viruses that cause severe diarrhea in children.
Rous sarcoma virus (RSV) A retrovirus that was the first virus shown to cause cancer.
rRNA See **ribosomal RNA**.
RSS See **recombination signal sequence**.
RSV See **Rous sarcoma virus**. RSV is also the abbreviation for "respiratory syncytial virus."
RT See **reverse transcriptase**.
Rubisco Ribulose 1,5-bisphosphate carboxylase oxygenase, the enzyme that catalyzes the carbon fixation step in the Calvin cycle.
rumen The first chamber of the digestive tract of ruminant animals such as cattle; the main site for microbial digestion of feed.

S

S-layer A crystalline protein surface layer replacing or external to the cell wall in many species of archaea and bacteria.
sacculus *pl.* **sacculi** The bacterial cell wall, consisting of a single covalent molecule.
SALT See **skin-associated lymphoid tissue**.
Sanger sequencing A method of sequencing DNA based on the incorporation of chain-terminating dideoxynucleotides by a DNA polymerase.
sanitation The safe disposal of wastes hazardous to humans.
saprophyte A fungal decomposer.
sarcina *pl.* **sarcinae** A cubical octad cluster of cells formed by septation at right angles to the previous cell division.
sargassum weed An unrooted kelp that floats in marine water and forms kelp forests.
scaffold The assembly of contigs (regions of contiguous sequence) into a large segment of a draft genome.
scanning electron microscopy (SEM) Electron microscopy in which the electron beams scan across the specimen's surface to reveal the 3D topology of the specimen.
scattering Interaction of light with an object resulting in propagation of spherical light waves at relatively low intensity.
schizogony Mitotic reproduction of parasitic cells to achieve a large population within a host tissue.
second messenger A regulatory molecule such as cAMP that is produced in response to a primary signal. Second messengers typically affect the expression of numerous genes.
secondary antibody The second antibody added in an immunologic assay (ELISA or Western blot). A secondary antibody carries a fluorescent or enzymatic tag and binds to the primary antibody in the assay.
secondary antibody response A memory B cell–mediated rapid increase in the production of antibodies in response to a repeat exposure to a particular antigen.
secondary endosymbiont An organism evolved through engulfment of a primary endosymbiont.
secondary metabolite Also called *secondary product.* A biosynthetic product that is not an essential nutrient but enhances nutrient uptake or inhibits competing species (e.g., an antibiotic).
secondary structure The second level of organization of polymers, consisting of regular patterns that repeat, such as the double helix in DNA or the beta sheet in proteins.
secondary syphilis A rash that may appear at some point after the primary latent stage of syphilis.
secondary tuberculosis Also called *reactivation tuberculosis.* A new round of serious disease that is caused by *Mycobacterium tuberculosis* in patients with latent tuberculosis who have become immunocompromised. Symptoms include severe cough, blood sputum, night sweats, and weight loss.
sedimentation rate The rate at which particles of a given size and shape travel to the bottom of a tube under centrifugal force. The rate depends on the particle's mass and cross-sectional area.
segmented genome A viral genome that consists of more than one nucleic acid molecule.
selectin One of a family of cell adhesion molecules.
selective medium A medium that allows the growth of certain species or strains of organisms but not others.
selective toxicity The ability of a drug, at a given dose, to harm the pathogen and not the host.
selectively permeable membrane See **semipermeable membrane**.
SEM See **scanning electron microscopy**.
semiconservative Describing the mode of DNA replication whereby each new double helix contains one old, parental strand and one newly synthesized daughter strand.
semipermeable membrane Also called *selectively permeable membrane.* A membrane that is permeable to some substances but impermeable to other substances.
sensor kinase A transmembrane protein that phosphorylates itself in response to an extracellular signal, and transfers the phosphoryl group to a receiver protein.
septation The formation of a septum, a new section of cell wall and envelope to separate two prokaryotic daughter cells.
septicemia An infection of the bloodstream.
septicemic plague Infection of the bloodstream by *Yersinia pestis.*
septum *pl.* **septa** A plate of cell wall and envelope that forms to separate two daughter cells.
sequela *pl.* **sequelae** A serious, harmful immunological consequence of bacterial and host antigen cross-reactivity that occurs after the infection itself is over. An example is rheumatic fever.
sequencing by synthesis Determining the sequence of a large DNA molecule through the automated, stepwise synthesis of fluorescently tagged DNA using millions of small, overlapping, single-stranded DNA templates derived from that larger sequence.
serial endosymbiosis theory The theory that mitochondria and chloroplasts were originally free-living prokaryotes that formed an internal symbiosis with early eukaryotes.
serum The noncell, liquid component of the blood.

sex pilus A pilus specialized to attach a DNA-donating bacterium to a DNA recipient bacterium, enabling DNA transfer through a conjugation apparatus.

sexual reproduction Reproduction involving the joining of gametes generated by meiosis.

Shine-Dalgarno sequence See **ribosome-binding site**.

shuttle vector A plasmid with origins of replication recognized by both bacteria and eukaryotes.

siderophore A high-affinity iron-binding protein used to scavenge iron from the environment and deliver it to a siderophore-producing organism.

sigma factor A protein needed to bind RNA polymerase for the initiation of transcription in bacteria.

signal recognition particle (SRP) A receptor that recognizes the signal sequence of peptides undergoing translation. The complex attaches to the cell membrane of prokaryotes (or the rough endoplasmic reticulum of eukaryotes), where it docks the protein-ribosome complex to the membrane for protein membrane insertion or secretion.

signal sequence A specific amino acid sequence on the amino terminus of proteins that directs them to the endoplasmic reticulum (of a eukaryote) or the cell membrane (of a prokaryote).

silencer A noncoding DNA region in eukaryotes that can lead to decreased transcription when bound by an appropriate transcription factor.

silent mutation A mutation that does not change the amino acid sequence encoded by an open reading frame. The changed codon encodes the same amino acid as the original codon.

simple stain A stain that makes an object more opaque, increasing its contrast with the external medium or surrounding tissue.

single-celled protein An edible microbe of high food value, such as *Spirulina* or some yeasts.

sink A part of the biosphere that can receive or assimilate significant quantities of an element; may be biotic (as in plants fixing carbon) or abiotic (as in the ocean absorbing carbon dioxide).

site-specific recombination Recombination between DNA molecules that do not share long regions of homology but do contain short regions of homology specifically recognized by the recombination enzyme.

skin-associated lymphoid tissue (SALT) Immune cells, such as dendritic cells, located under the skin that help eliminate bacteria that have breached the skin surface.

sliding clamp A protein that keeps DNA polymerase affixed to DNA during replication.

slime mold An organism in which unicellular amebas can aggregate into a fruiting body; not a fungus.

sludge The solid products of wastewater treatment.

small RNA (sRNA) A non-protein-coding regulatory RNA molecule that modulates translation or mRNA stability.

small-subunit (SSU) rRNA In bacteria and archaea, 16S rRNA; in eukaryotes, 18S rRNA. A ribosomal RNA found in the small subunit of the ribosome. Its gene is often sequenced for phylogenetic comparisons.

soil A complex mixture of decaying organic and mineral matter that covers the terrestrial portions of the planet.

solute Any dissolved molecule.

SOS response A coordinated cellular response to extensive DNA damage. It includes error-prone repair.

source A part of the biosphere that stores a significant quantity of a given element; may be biotic (as in tree biomass, a source of carbon) or abiotic (as in carbonate rock).

sourdough An undefined yeast population, derived from a previous batch of dough, that is used in bread production.

Southern blot A technique (named for its inventor, Edward M. Southern) to detect specific DNA sequences. Sample DNA segments are separated by gel electrophoresis, transferred to a blot, and probed with a labeled DNA that will hybridize to complementary DNA sequences.

space-filling model A molecular model that represents the volume of the electron orbitals of the atoms, usually to the limit of the van der Waals radii.

specialized transduction Also called *restricted transduction*. Transduction in which the phage can transfer only a specific, limited number of donor genes to the recipient cell.

species A single, specific type of organism, designated by a genus and species name.

species name The scientific name of a specific type of organism; presented following the genus name, as in *Escherichia* (genus) *coli* (species).

spectrum of activity The range of pathogens for which an antimicrobial agent is effective.

spike protein A viral glycoprotein that connects the membrane to the capsid or the matrix and may be involved in viral binding to host cell receptors.

spirochete A bacterium with a tight, flexible spiral shape; a species of the phylum Spirochetes (Spirochaeta).

Spirochetes Also called *Spirochaeta*. A phylum of bacteria with a unique morphology: a flexible, extended spiral that twists via intracellular flagella.

spongiform encephalopathy A brain-wasting disease caused by a prion.

spontaneous generation The theory, much debated in the nineteenth century, that under current Earth conditions life can arise spontaneously from nonliving matter.

sporangiospore A haploid spore that can germinate to form a haploid mycelium.

sporangium *pl.* **sporangia** A fungal organ that releases nonmotile spores.

spore stain A type of differential stain that is specific for the endospore coat of various bacteria, typically a firmicute species.

spread plate A method to grow separate bacterial colonies by plating serial dilutions of a liquid culture.

sRNA See **small RNA**.

SRP See **signal recognition particle**.

SSU rRNA See **small-subunit rRNA**.

staining The process of treating microscopic specimens with a stain to enhance their detection or to visualize specific cell components.

stalk An extension of the cytoplasm and envelope that attaches a microbe to a substrate.

standard reduction potential The reduction potential (tendency of a chemical to gain electrons and thereby become reduced) under standard conditions of 1-M concentration, 25°C temperature, and 1-atm pressure.

stalked ciliate A ciliate that adheres to a substrate and uses its cilia to obtain prey.

staphylococcus A hexagonal arrangement of cells formed by septation in random orientations.

start codon A codon (usually AUG) that signals the first amino acid of a protein.

starter culture A mixture of fermenting microbes added to a food substrate to generate a fermented product.

stationary phase A period of cell culture, following exponential phase, during which there is no net increase in replication.

sterilization The destruction of all cells, spores, and viruses on an object.

stick model A molecular model in which stick lengths represent the distances between bonded pairs of atomic nuclei.

Stickland reaction An energy-yielding reaction between two amino acids in which one oxidizes the other. The reaction typically produces short organic acids plus $2NH_4^+$; may also produce CO_2 and H_2.

stop codon One of three codons (UAA, UAG, UGA) that do not encode an amino acid, and thus trigger the end of translation.

streptogramin An antibiotic that binds 23S rRNA and blocks elongation of protein synthesis in bacteria.

strict aerobe An organism that performs aerobic respiration and can grow only in the presence of oxygen.

strict anaerobe An organism that cannot grow in the presence of oxygen.

stringent response A cellular response to idle ribosomes (often indicating low carbon and energy stores) that includes a decrease in rRNA and tRNA production.

stroma The compartment contained by the inner chloroplast membrane where the light-independent reactions (CO_2 fixation) of photosynthesis occur.

stromatolite A mass of sedimentary layers of limestone produced by a marine microbial community over many years.

structural formula A representation of molecular structure in which covalent bonds are shown as a line between atoms.

structural gene A string of nucleotides that encodes a functional RNA molecule.

structural isomer A molecule with the same molecular formula as a different molecule but a different arrangement of atoms.

substrate-binding protein An extracytoplasmic protein that binds specific substrates and delivers them to their cognate uptake ABC transporters.

substrate-level phosphorylation The formation of ATP by the enzymatic transfer of phosphate from a substrate molecule onto ADP.

sulfa drug An antibiotic that inhibits folic acid synthesis and, thus, nucleotide synthesis.

Sulfolobales An order of thermophilic archaea (Crenarchaeota) that includes sulfur oxidizers.

superantigen A molecule that directly stimulates T cells without undergoing antigen-presenting cell processing and surface presentation.

supercoil Also called *superhelical turn*. An extra twist or turn found in DNA, either positive (increases DNA winding) or negative (decreases DNA winding).

supernova An exploding star that has used up most of the nuclei available for fusion reactions.

Svedberg coefficient A measure of particle size based on the particle's sedimentation rate in a tube subjected to a high *g* force.

swarming A behavior in which some microbial cells differentiate into large swarmer cells and swim together as a unit.

switch region A repeating DNA sequence interspersed between antibody constant-region genes that serves as a recombination site during isotype, or class, switching.

symbiogenesis An evolutionary process by which two or more species become intimately associated.

symbiont An organism that lives in a close association with another organism.

symbiosis *pl.* **symbioses** The intimate association of two different species.

symbiosome A bacteroid enclosed by a plant membrane that mediates the exchange of nutrients between the bacteroid and the host cell.

symport Coupled transport in which the molecules being transported move in the same direction across the membrane.

synergism Cooperation between species in which both species benefit but can grow independently. The cooperation is less intimate than symbiosis.

synthetic biology The genetic construction of novel organisms with useful functions.

synthetic medium A bacterial growth solution that contains defined, known components.

syntrophy Metabolic cooperation between two different species; usually one member releases a product whose removal by the second species enables the pair to metabolize with a negative value of ΔG.

T

T cell An adaptive immune cell, developed in the thymus, that can give rise to antigen-specific helper cells and cytotoxic T cells.

tailed phage A phage such as T4 that contains a genome delivery device called the tail.

tandem repeat A stretch of directly repeating DNA sequence (direct repeats) without any intervening DNA.

TaqMan A real-time PCR technique in which Taq polymerase, in the process of synthesizing DNA along a template, degrades a downstream fluorescent oligonucleotide probe. The increase in fluorescence indicates the production of an amplified DNA product.

target community A community whose genomes are sequenced for metagenomic analysis.

taxon *pl.* **taxa** A category of organisms with a shared genetic ancestor.

taxonomy The description of distinct life-forms and their organization into different categories.

T_C cell See **cytotoxic T cell**.

TCA cycle See **tricarboxylic acid cycle**.

tegument The contents of a virion between the capsid and the envelope.

teichoic acid A chain of phosphodiester-linked glycerol or ribitol that threads through and reinforces the cell wall in Gram-positive bacteria.

TEM See **transmission electron microscopy**.

tempeh A mold-fermented soy product, popular as a food in parts of Asia.

temperate phage A phage capable of lysogeny.

template strand A DNA strand (or an RNA strand in some viruses) that is used as a template for the synthesis of mRNA.

Tenericutes (Mollicutes) A clade of bacteria lacking a cell wall; closely related to Gram-positive bacteria.

terminal electron acceptor The final electron acceptor at the end of an electron transport system.

termination (*ter*) site A sequence of DNA that halts replication of DNA by DNA polymerase.

terpenoid A branched lipid derived from isoprene that is found in hydrocarbon chains of archaeal membranes.
terraforming The idea of transforming the environment of another planet to make it suitable for life from Earth.
tertiary structure The third level of organization of polymers; the unique 3D shape of a polymer.
tertiary syphilis A final stage of syphilis, manifested by cardiovascular and nervous system symptoms.
tetanospasmin The tetanus-causing potent exotoxin produced by *Clostridium tetani*.
tetraether A molecule containing four ether links. An example is found in archaeal membranes, when two lipid side chains form ether linkages with a pair of side chains from the other side of the bilayer.
tetrapyrrole An essential primary product that contains four pyrroles (five-membered rings containing nitrogen), each with one or two double bonds. A precursor for many important cell cofactors, such as chlorophylls and vitamin B_{12}.
T_H cell See **helper T cell**.
Thaumarchaeota One of four major divisions of Archaea, containing ammonia oxidizers, marine invertebrate symbionts, and others.
thermal vent See **hydrothermal vent**.
thermocline A region of the ocean where temperature decreases steeply with depth, and water density increases.
thermophile An organism adapted for optimal growth at high temperatures, usually 55°C or higher.
threshold dose The concentration of antigen needed to elicit adequate antibody production.
thylakoid An intracellular chlorophyll-containing membrane folded within a phototrophic bacterium or a chloroplast.
tight junction A type of junction between the membranes of two adjacent vertebrate cells that form an impermeable barrier.
Ti plasmid A plasmid found in tumorigenic strains of *Agrobacterium tumefaciens* that can be used as a vector to introduce DNA into plant cells.
TLR See **Toll-like receptor**.
tmRNA A molecule resembling both tRNA and mRNA that rescues ribosomes stalled on damaged mRNAs lacking a stop codon.
TNF See **tumor necrosis factor**.
Toll-like receptor (TLR) A member of a eukaryotic transmembrane glycoprotein family that recognizes a particular microbe-associated molecular pattern (MAMP) present on pathogenic microorganisms.
tomography The acquisition of projected images of a transparent specimen from different angles that are digitally combined to visualize the entire specimen.
topoisomerase An enzyme that can change the supercoiling of DNA.
total magnification The magnification of the ocular lens multiplied by the magnification of the objective lens.
transamination The transfer of an ammonium ion between two metabolites.
transcript An RNA copy of a DNA template.
transcription The synthesis of RNA complementary to a DNA template.
transcriptional attenuation A transcriptional regulatory mechanism in which translation of a leader peptide affects transcription of downstream structural genes.
transcriptional fusion See **operon fusion**.
transcriptome The set of transcribed genes in a cell at given time. The "complete transcriptome" includes all the possible RNA transcription products from a given genome. The "expressed transcriptome" is the set of RNAs present during a given condition.
transcytosis The movement of a cell or substance from one side of a polarized cell to the other side, using an intracellular route.
transduction The transfer of host genes between bacterial cells via a phage head coat.
transertion The membrane insertion of a nascent polypeptide chain during translation of a messenger RNA undergoing transcription from DNA. Overall, the growing peptide-mRNA complex connects DNA to the membrane.
transfection Deliberate transfer of DNA (usually viral) into cells.
transfer RNA (tRNA) An RNA that carries an amino acid to the ribosome. The anticodon on the tRNA base-pairs with the codon on the mRNA.
transform To cause bacteria to take up exogenous DNA. In eukaryotes, to convert cultured cells into cancer cells.
transformasome A bacterial cell membrane protein complex that imports external DNA during transformation.
transformation The internalization of free DNA from the environment into bacterial cells.
transformed-focus assay The detection of oncogenic viruses based on their ability to transform cells, generating foci of unrestricted cell growth.
transgene A gene that has been transferred by genetic engineering techniques from one organism to another.
transglycosylase An enzyme that links *N*-acetylglucosamine and *N*-acetylmuramic acid into chains during bacterial cell wall synthesis.
transition A point mutation in which a purine is replaced by a different purine or a pyrimidine is replaced by a different pyrimidine.
translation The ribosomal synthesis of proteins based on triplet codons present in mRNA.
translational control A regulatory mechanism that modulates protein production by influencing the translation of mRNA.
translational fusion See **gene fusion**.
translocation The energy-dependent movement of the ribosome to the next triplet codon along an mRNA.
transmembrane protein Also called *integral protein*. A protein with a membrane-spanning region.
transmission electron microscopy (TEM) Electron microscopy in which electron beams are transmitted through a thin specimen to reveal internal structure.
transovarial transmission The transfer of a pathogen from parent to offspring by infection of the egg cell. Typically seen in insects.
transpeptidase An enzyme that cross-links the side chains from adjacent peptidoglycan strands during bacterial cell wall synthesis.
transport protein Also called *transporter*. A membrane protein that moves specific molecules across a membrane.
transposable element A segment of DNA that can move from one DNA region to another.
transposase A transposable element–encoded enzyme that catalyzes the transfer of the transposable element from one DNA region to another.
transposition The process of moving a transposable element from one DNA region to another.

transposon A transposable DNA element that contains genes in addition to those required for transposition.
transversion A point mutation in which a purine is replaced by a pyrimidine or vice versa.
Treg See **regulatory T cell**.
tricarboxylic acid (TCA) cycle Also called *citric acid cycle* or *Krebs cycle*. A metabolic cycle that catabolizes the acetyl group from acetyl-CoA to $2CO_2$ with the concomitant production of NADH, $FADH_2$, and ATP.
tRNA See **transfer RNA**.
trophic level A level of the food web representing the consumption of biomass of organisms from another level, usually closer to producers.
tropism The ability of a virus to infect a particular tissue type.
trypanosome A parasitic excavate protist that has a cortical skeleton of microtubules culminating in a long flagellum.
tumor necrosis factor (TNF) A cytokine released by several cell types (e.g., macrophages) in response to cell damage.
tumor necrosis factor alpha (TNF-α) A cytokine involved in systemic inflammation.
twitching motility A type of bacterial movement on solid surfaces where a specific pilus extends and retracts.
two-component signal transduction system A message relay system composed of a sensor kinase protein and a response regulator protein that regulates gene expression in response to a signal (usually an extracellular signal).
two-dimensional polyacrylamide gel electrophoresis (2D PAGE, 2D gels) A technique to separate proteins based on differences in charge and molecular weight.
type I hypersensitivity Also called *immediate hypersensitivity*. An IgE-mediated allergic reaction that causes degranulation of mast cells within minutes of exposure to the antigen. The severe reaction known as anaphylaxis is triggered by type I hypersensitivity.
type I pilus A pilus that adheres to mannose residues on host cell surfaces.
type II hypersensitivity An immune response in which antibodies bind to the patient's own cell-surface antigens or to foreign antigens adsorbed onto the patients cells. Antibody binding triggers cell-mediated cytotoxicity or activation of the complement cascade.
type II secretion system A bacterial protein secretion system that uses a type IV pilus-like extraction/retraction mechanism to push proteins out of the cell.
type III hypersensitivity An immune reaction triggered when IgG antibody binds to an excess of soluble foreign antigen in the blood. The immune complexes deposit in small blood vessels, where they interact with complement to initiate an inflammatory response.
type III pilus A bacterial pilus that does not bind to mannose but binds to tannic acid.
type III secretion system (T3SS) A bacterial protein secretion system that uses a molecular syringe to inject bacterial proteins into the host cytoplasm.
type IV hypersensitivity Also called *delayed-type hypersensitivity (DTH)*. An immune response that develops 24–72 hours after exposure to an antigen that the immune system recognizes as foreign. The response is triggered by antigen-specific T cells. It is delayed because the T cells need time to proliferate after being activated by the allergen.
type IV pilus A dynamic pilus that can repeatedly assemble and disassemble; it mediates twitching motility.

U

ultracentrifuge A machine that subjects samples to high centrifugal forces and can be used to separate subcellular components.
uncoating The release of a viral genome from its capsid, following entry of the virion into a host cell.
uncoupler A molecule that makes a membrane permeable to protons, dissipating the proton motive force and uncoupling electron transport from ATP synthesis.
unculturable See **uncultured**.
uncultured Describing an organism whose requirements for culture remain unknown.
upstream processing The culturing of industrial microbes to produce large quantities of a desired product.

V

vaccination Exposure of an individual to a weakened version of a microbe or a microbial antigen to provoke immunity and prevent development of disease upon reexposure.
VAM See **arbuscular mycorrhizae**.
van der Waals force Weak, temporary electrostatic attraction between molecules caused by shifting electron clouds.
vancomycin A glycopeptide antibiotic that inhibits bacterial cell wall synthesis in a mechanism distinct from penicillin inhibition.
variable region The amino-terminal portions of antibody light and heavy chains that confer specificity to antigen binding and define the antibody idiotype.
vasoactive factor A cell signaling molecule that increases capillary permeability.
VBNC See **viable but nonculturable**.
vector 1. An organism (e.g., insect) that can carry infectious agents from one animal to another. 2. In molecular biology, a molecule of DNA into which exogenous DNA can be inserted to be cloned.
vegetative cell A metabolically active, replicating bacterial cell.
vegetative mycelium A branched filament produced by vegetative cells that expands into the substrate.
Verrucomicrobia A phylum of free-living aquatic bacteria with wart-like, protruding structures containing actin.
vertical gene transfer Gene transfer from parent to offspring through reproduction.
vertical transmission In genetics, the passage of genes from parent to offspring. In disease, the transfer of a pathogen from parent to offspring. See also **transovarial transmission**.
vesicle A small, membrane-enclosed sphere found within a cell.
vesicular-arbuscular mycorrhizae (VAM) See **arbuscular mycorrhizae**.
viable Capable of replicating—for instance, by forming a colony on an agar plate.
viable but nonculturable (VBNC) Also called *dormant*. Metabolically active but unable to replicate to form a colony on a plate by current means of culture.
viremia The presence of large numbers of virions in the bloodstream.
virion A virus particle.
viroid An infectious naked nucleic acid.
virome The genomes of all the viruses that inhabit a particular organism or environment.
virulence A measure of the severity of a disease caused by a pathogenic agent.
virulence factor A trait of a pathogen that enhances the pathogen's disease-producing capability.
virus A noncellular particle containing a genome that can replicate only inside a cell.

W

wastewater treatment A series of wastewater transformations designed to lower biological oxygen demand and eliminate human pathogens before water is returned to local rivers.

water activity A measure of the water that is not bound to solutes and is available for use by organisms.

water cycle See **hydrologic cycle**.

water table The layer of soil that is permanently saturated with water.

WBC differential See **white blood cell differential**.

western blot A technique to detect specific proteins. Proteins are subjected to gel electrophoresis, transferred to a blot, and probed with enzyme-linked or fluorescently tagged antibodies that specifically bind the protein of interest.

wet mount A technique to view living microbes with a microscope by placing the microbes in water on a slide under a coverslip.

wetland A region of land that undergoes seasonal fluctuations in water level and aeration.

whey The liquid portion of milk after proteins have precipitated out of solution, usually during cheese production.

white blood cell (WBC) differential A laboratory test that counts the different types of white blood cells in a patient's blood.

Winogradsky column A column containing a stratified environment that causes specific microbes to grow at particular levels; a type of enrichment culture for the growth of microbes from wetland environments.

X

X-ray crystallography Also called *X-ray diffraction analysis.* A technique to determine the positions of atoms (atomic coordinates) within a molecule or molecular complex, based on the diffraction of X-rays by the molecule.

Y

yeast A unicellular fungus.

yeast two-hybrid system An in vivo technique to determine protein-protein interactions in which DNA sequences encoding proteins of interest are fused separately to the DNA-binding and activation domains of a yeast transcription factor. The recombinant yeast is then tested for expression of a reporter gene.

yogurt A semisolid food produced through acidification of milk by lactic acid–producing bacteria.

Z

zone of hypoxia See **dead zone**.

zone of inhibition A region of no bacterial growth on an agar plate due to the diffusion of a test antibiotic. Correlates to the minimal inhibitory concentration.

zoonotic disease An infection that normally affects animals but can be transmitted to humans.

zoospore A flagellated reproductive cell produced by chytridiomycete fungi.

zooxanthella *pl.* **zooxanthellae** A phototrophic coral endosymbiont, most commonly a dinoflagellate of the genus *Symbiodinium*.

Z pathway See **oxygenic Z pathway**.

zygomycete A member of the eukaryotic group Zygomycota—fungi forming nonmotile haploid gametes that grow toward each other, fusing to form the zygospore.

zygospore In zygomycetes, the diploid structure formed by the fusion of two gamete-bearing hyphae.

FIGURE CREDITS

Special Topic Figure 10.2: Gyanendra P. Dubey and Sigal Ben-Yehuda. Figure 4b from "Intercellular Nanotubes Mediate Bacterial Communication." *Cell,* 144:590–600, February 18, 2011. Copyright © 2011 Elsevier Inc. All rights reserved.

Figure 11.3: Christopher K. Matthews et al.: Figure from *Bacteriophage T4,* 1983, ASM Press, Washington, DC. Reprinted with permission from ASM Press.

Figure 11.11A: Darius Moradpour, François Penin and Charles M. Rice. Figure 2 from "Replication of the Hepatitis C Virus," *Nature Reviews Microbiology,* 5:453, June 2007. Copyright © 2007, Nature Publishing Group, a division of Macmillan Publisher Ltd. All rights reserved.

Figure 11.11B: Thomas Pietschmann. Figure 1 from "Virology: Final Entry Key for Hepatitis C," *Nature* 4578:797, February 2009. Copyright © 2011, Nature Publishing Group, a division of Macmillan Publisher Ltd. All rights reserved.

Figure 11.14: Patrizia Farci. Figure B from "Early changes in hepatitis C viral quasispecies during interferon therapy predict the therapeutic outcome," *Proceedings of the National Academy of Sciences* 99:3081. March 2002. Reprinted by permission of the National Academy of Sciences.

Figure 11.17B: Gavin Smith et al. 2009. Figure 1 from "Origins and evolutionary genomics of the 2009 swine-origin H1N1 influenza A epidemic," *Nature* 459:1122. June 2009. Copyright © 2009, Nature Publishing Group, a division of Macmillan Publisher Ltd. All rights reserved.

Figure 11.23A: Swenson. Figure from "Estimating dates using HIV evolution patterns," *Nothing in Biology Makes Sense* Blog, December 6, 2011.

Figure 11.23B: Andrew Rambaut, David Posada, Keith A. Crandall, and Edward C. Holmes. Figure 5 from "The causes and consequences of HIV evolution," *Nature Reviews Genetics* 5, 52–61, January 2004. Copyright © 2004, Nature Publishing Group, a division of Macmillan Publisher Ltd. All rights reserved.

Special Topic Figure 11.1: Min Hua Lo et al. Figure 1 from "Human Cytomegalovirus Infection Causes Premature and Abnormal Differentiation of Human Neural Progenitor Cells" *Journal of Virology,* 84: 3528, April 2010. Reprinted by permission of the American Society for Microbiology.

Figure 12.20: J. Stricker et al. Figure 1 from "A fast, robust and tunable synthetic gene oscillator," *Nature* 456:516–519. November 2008. Copyright © 2008, Nature Publishing Group, a division of Macmillan Publisher Ltd. All rights reserved.

Figure 12.22: Jarred M. Callura. Figure 1 from "Genetic switchboard for synthetic biology applications," *Proceedings of the National Academy of Sciences,* Vol. 109, No. 15. April 2012. Reprinted by permission of the National Academy of Sciences.

Figure 12.23: Jarred M. Callura. Figure 3 from "Genetic switchboard for synthetic biology applications," *Proceedings of the National Academy of Sciences,* Vol. 109, No. 15. April 2012. Reprinted by permission of the National Academy of Sciences.

Figure 12.24: Jarred M. Callura. Figure 3A from "Tracking, tuning, and terminating microbial physiology using synthetic riboregulators," *Proceedings of the National Academy of Sciences* Vol. 107, No. 36. September 2010. Reprinted by permission of the National Academy of Sciences.

Figure 12.32: Georg Fuchs et al. Figure from "Microbial degradation of aromatic compounds—from one strategy to four," *Nature Reviews Microbiology* 9, 803–816 (November 2011). Copyright © 2011, Nature Publishing Group, a division of Macmillan Publisher Ltd. All rights reserved.

Figure 12.33: Hongzhi Tang et al. Figure 1 from "Genomic analysis of *Pseudomonas putida*: genes in a genome island are crucial for nicotine degradation," *Nature Scientific Reports* 2:377. April 2012. © 2012, Macmillan Publishers Limited. All rights reserved.

Figure 15.24: Daniel Smith and Matthew Chapman. Figure 2 from "Economical evolution: microbes reduce the synthetic cost of extracellular proteins." *mBio* 1(3):e00131–10. (August 2010). © 2010, American Society for Microbiology. Reprinted with permission.

Special Topic 17.1 Figure 2A: Scott T. Kelley. Figure 2 from "Molecular Analysis of Shower Curtain Biofilm Microbes," *Applied and Environmental Microbiology,* 2004 July; 70(7): 4187–4192. © 2004, American Society for Microbiology. Reprinted with permission.

Figure 17.22: W. Ford Doolittle: Figure from "Phylogenetic Classification and the Universal Tree", *Science* 284:2124 (25 June 1999). Copyright © 1999, The American Association for the Advancement of Science. Reprinted with permission from AAAS.

Figure 17.25: Adapted from Zachary Blount et al. Figure 1 from "Historical contingency and the evolution of a key innovation in an experimental population of *Escherichia coli,*" *Proceedings of the National Academy of Sciences* 105:7899. June 2008. Reprinted by permission of the National Academy of Sciences.

Figure 18.1: The Human Microbiome Project Consortium, Figure 4 from "Structure, function and diversity of the healthy human microbiome" *Nature* 486, 207–214 (14 June 2012). Copyright © 2012, Nature Publishing Group, a division of Macmillan Publisher Ltd. All rights reserved.

Special Topic Figure 19.1: Matsutani, Naoki, Tatsunori Nakagawa, Kyoko Nakamura, Reiji Takahashi, Kiyoshi Yoshihara et al. 2011. Figure from "Enrichment of a novel marine ammonia-oxidizing archaeon obtained from sand of an eelgrass zone." *Microbes and Environments* 26:23–29. © The Japanese Society of Microbial Ecology/The Japanese Society of Soil Microbiology.

Figure 19.19: Markus B. Karner, Edward F. DeLong, and David M. Karl: Figure reprinted by permission from Macmillan Publishers Ltd. from "Archael Dominance in the Mesopelagic Zone of the Pacific Ocean", *Nature,* Vol. 409, No. 507. Copyright © 2001, Nature Publishing Group.

Special Topic 21.1, Figure 2: Molly C. Redmond and David L. Valentine. Figure 1 from "Natural gas and temperature structured a microbial community response to the *Deepwater Horizon* oil spill," *Proceedings of the National Academy of Sciences* 109: 20292, December 2012. Reprinted by permission of the National Academy of Sciences.

Figure 21.2: Raina Maier et al.: Figure from *Environmental Microbiology,* Copyright © 2000 by Academic Press. Reprinted by permission of Elsevier, Ltd.

Figure 21.6A: Susannah Tringe et al. 2008. Figure A from "The Airborne Metagenome in an Indoor Urban Environment," PLoS ONE 3:e1862. © 2008 Tringe et al.

Figure 21.6B: Susannah Tringe et al. 2008. Figure C from "The Airborne Metagenome in an Indoor Urban Environment," PLoS ONE 3:e1862. © 2008 Tringe et al.

Figure 21.7: Noah Fierer, Figure 3 from "Cross-biome metagenomic analyses of soil microbial communities and their functional attributes," *Proceedings of the National Academy of Sciences* 105:7899. December 2012. Reprinted by permission of the National Academy of Sciences.

Figure 21.11A: Lynn Margulis: Figure from *Symbiosis in Cell Evolution* by Lynn Margulis. Copyright © 1981, 1993 by W.H. Freeman and Company. Used with permission.

Figure 22.22: "A schematic diagram of a SAR plant" from insituarsenic.org. Reprinted with permission.

Figure 26.14A: "HIV infections and AIDS deaths worldwide," Figure from *The Millennium Development Goals Report,* 2010. Page 40. Copyright © 2010 United Nations. Reprinted with permission.

INDEX

Page numbers followed by f or t denote figures or tables, respectively. Page numbers set in **bold** type refer to glossary terms.

A
ABC (ATP-binding cassette) transporters, 88, 89, 130–32, 309f, 310–11, 361
Aberdeen Proving Ground, 902
Aberrations, **51**
ABI 3730 sequencing method, 845t
Abiotic materials and reactions, **677**, **891**
  and origins of life, 671, 672, 677, 678
  oxidation, 641–42
Abiotrophia, 121t
ABO blood group system, 968, eTopic 24.2
Abscesses, 749, 750
Abshire, Kelly, 177
Absorption, **47–48**
Acanthamoeba, 825, 1072, A-69t
Acanthocorbis unguiculata, 803f
Accelerator cells, 108–9
Acceptor site (A site), **292**, 294, 295, 296, 297f
Accessory proteins, 435f, **436**, 437t
Accidental transmission, 1008f
Acenaphthene, 513
Acetate, 527f, 528, 529f, 530, 539, 561, 783
Acetic acid, 88, 123f
Acetic acid bacteria, 635–36
Acetobacter, 635
Acetone, 525
Acetylcholine, 1087
Acetyl-CoA
  3-hydroxypropionate cycle and, 600f, 601
  activation in fatty acid biosynthesis, 602
  in aromatic catabolism, 533, 534f
  in carbon monoxide reductase pathway, 761
  conversion to acetate, 527f, 528
  entry into TCA cycle, 527–28, 588f
  fermentation intermediate, 523
  in glyoxylate bypass, 531
  from pyruvate decarboxylation, 527–28
  reductive acetyl-CoA pathway, 590t, 599, 601
  reverse TCA cycle and, 598f, 599
  structure, 523, 525f
Acetyl-CoA carboxylase, 602, 603
N-Acetylglucosamine, 92, 93f, 515
N-Acetylmuramic acid, 92, 93f
Acetyl phosphate, 527f, 528
Acid-fast bacilli (AFB), 1152
Acid-fast stain, 57, 59f, **732**, 1152
Acidianus brierleyi, 601
Acidithiobacillales, A-64t
Acidithiobacillus, 742, 743f, 910
Acidithiobacillus ferrooxidans, 564, 589f, 590, 742, 743f, 911, 912
Acidithiobacillus thiooxidans, 122t, 142f, 143
Acidobacteria, **717**, 748–49, A-65t
Acidobacterium capsulatum, A-65t
Acidophiles, 159t, 168f, **169–70**, 566–67, 765, 769–71, 852
Acidovorax citrulli (watermelon fruit blotch), 133f
Acid resistance of *E. coli,* 1025, 1070
  acid fitness island, 360
  acid-sensitive mutants, 459, 460f
  GadE (YhiE) regulatory protein, 465, 468, 469f
  GadX regulatory protein, 468, 469f
  gene fusion as a tool for studying, A-51–A-52
  gene regulation, 460–61, 463–65, 468, 469f
  genetic analysis, 458–60
  glutamate decarboxylase (GadA and GadB), 460, 461f, 463, 464f, 465, 466
  glutamate/GABA antiporter (GadC), 460
  overview, 458
  primer extension analysis, 466–67, A-52–A-53
  proposed model for, 460, 461f
Acid stress, 168–69, 172
Acids used for food preservation, 649
Acid tolerance, 172
Acineta, 830f
Acinetobacter baumanii, 1119, 1128
Acinetobacter calcoaceticus, 186, 324
Acne, 510, 928, 929f
Acquired immunodeficiency syndrome (AIDS), **432**. *See also* Human immunodeficiency virus (HIV)
  AIDS memorial quilt, 11f
  AIDS-related complex (ARC), 1082
  deaths, 1081f
  decline in $CD4^+$ T cells, 991, 1082
  definition (CDC), 1082
  diagnosis, 1082
  history, 14, 193, 433–34
  incidence in United States, 1081
  Kaposi's sarcoma, 1081f, 1082
  locations of recent increases, 1181f
  opportunistic infections, 1006, 1063, 1082
  Pneumocystis jirovecii infections, 808–9, 817, 1006, 1081f, 1082
  President's Emergency Plan for AIDS Relief (PEPFAR), 433
  and risk for tuberculosis, 652, 1064, 1082, 1181
  secondary infections by *Candida albicans,* 1081f, 1082
  treatments, 433, 434, 1082
Acrylamide, 654f, 655
Acrylaway, 655
Actin
  actin filaments (microfilaments), A-38
  actin rearrangement in *Salmonella* pathogenesis, 1032f
  actin tails, 59f, 60, 727, 728f, 741–42, 1038, 1040f, 1043
  F-actin, A-38
  G-actin, A-38
  polymerization or depolymerization by toxins, 1019
  pseudopod motility, 825, A-38
Actinobacteria, 711, 712f, **716–17**, 723, 730–34, 839, A-63t
Actinomyces, 731, 928, A-63t
Actinomyces bovis, 949
Actinomycetaceae, A-63t
Actinomycetales, A-63t
Actinomycetes, **717**, 730–34
  association with animals and plants, 731
  cell envelopes, 96
  decomposition of plant material, 872
  hierarchy of classification, 700
  hyphae, 152, 153f
  irregularly shaped actinomycetes, 733–34
  Micrococcaceae, 734
  mycelia, 152, 153f
  nonmycelial actinobacteria, 732–33
  nonribosomal peptide antibiotics, 618
  polyketides, 602, 603
  production of antibiotics that effect transcription, 283
  sponge-associated actinomycetes, 883f
  unipolar extension, 107
Actinomycin D, 283, 284f, 1123
Actinomycosis, 731
Actinophrys sol, A-70f
Actinorhodin, 153f, 585
Actin tails, 59f, 60, 104f, 727, 728f, 741–42, 1038
Activation energy, **A-20**
Activation energy (E_a), 508, 509f
Activators, **366**, 369
Active transport, **89**, 128–32
Acute glomerular nephritis, 1148
Acute viruses, 228
Acyclovir, 444, 1140f
Acyl carrier protein (ACP), **602**, 604
Acyl homoserine lactone (AHL), 399, 400, 478–79, 1136
Acyltransferase, 604

Adaptive evolution, 692–96
experimental evolution in the laboratory, 694–96
genome analysis, 692–93
kinds of evidence for study of, 692
strongly selective environments, 693
Adaptive immunity, **936**, **962**. *See also* Antibodies; Immune system
cell-mediated immunity, 963, 964–65
complement as part of, 992–94
definition, 220–21, 936
humoral immunity, 963–64, 965
immunogenicity, 965–68
microbial evasion of, 991
overview, 962–65
Adenine (A), 239, 243, 501, 679
Adenoids, 942, 944, 1018
Adenosine diphosphate (ADP), 504
Adenosine monophosphate (AMP), 617
Adenosine triphosphate (ATP), **501**, **504–5**
ABC transporters, 88, 89, 130–31
complex with $Mg2^+$, 504
as energy carrier, 501, 504–5
hydrolysis, 89, 131, 498f, 499
synthesis, 34, 125, 130, 548–49
S-adenosylmethionine, 388
Adenoviridae, A-60t
Adenoviruses, 1092
as DNA virus, 443, 444f
as gene therapy vectors, 450, 453
uncoating and endocytosis, 214f, 215
Adenylate cyclase, **996**, 1023, 1024, 1026, eTopic 25.5
Adherence, **113**, 115
Adhesins, **1013**, 1014f, 1014t, 1016–17, 1046. *See also* Pili
ADP-ribosyltransferase, **1022**, 1023
Adrenaline, 996
Aedes aegypti, 192, 1008f
Aedes albopictus, 192
Aedes mosquito, 1009
Aequorea victoria, 61
Aequorin, 1178, 1179f
Aeras Global TB Vaccine Foundation, 652
Aerial mycelia (or hyphae), 152, 153f, 154f, **730**, 807
Aerobes
aerobes *vs.* anaerobes, 174
aerobic and anaerobic metabolism, 851t
classification, 159t, 173t
strict aerobes, 174
Aerobic aromatic catabolism, 533
Aerobic respiration, 124, **173**, 175, 531, 541, 542
Aeromonadales, A-64t
Aeromonas hydrophila, 1035
Aeropyrum, 690
Aeropyrum pernix, 767, A-66t
Aerotolerant anaerobes, **175**
Affinity chromatography, **466**, 466–67f, 473
Affinity maturation, **981**
Aflatoxin, 814
African sleeping sickness, 44, 314–15, 390, 797, 805, 833–34
Agar, **21**, 133–34, 137, 821, 1022
Agaricus bisporus, 625–26
Agarose gel electrophoresis, 263–64, 821, A-45–A-46
Agriculture
cauliflower mosaic virus effect on, 193, 206, 220
genetic resistance to disease, 220
virus-host mutualism, 227, 229
virus-to-bacteria ratio in soil, 230
Agrobacterium, 688, 741, 747, A-63t
Agrobacterium tumefaciens
crown gall formation, 329, 741, 881
endosymbiosis, 741
gene transfer to plants, 329
genome of, 239, 240t
Ti plasmid, 329, 657, 1033
type IV secretion, 1033
as vector for plant engineering, 657, 659, 741
AhpC (antioxidant stress protein), 85f
AIDS. *See* Acquired immunodeficiency syndrome (AIDS)
AiiA enzyme, 478–79
AIRE gene activator, 986
Air filtration, 181, 182f
Airy, George, 51
Airy disks, 50f, 51
Akinetes, **720–22**
Alanine, 92, 93, 613, 614
Albuterol, 997
Alcalase, 654
Alcaligenes, 126
Alcaligenes eutrophus, 593f
Alcaligenes faecalis, 526
Alcaligenes xylosoxidans, 68, 69f
Alcohol
as disinfectant, 184f
as fermentation waste product, 627
Alcohol dehydrogenase, 639
Aldehydes, 184f, 185
Aldoses, A-11
Aleuria, 815f
Algae (singular, alga), **800**, **802–4**, 818–36
biofuels produced from, 857
brown algae (kelp), 623, 624, 799t, 800, 801f, 805, 824
cell walls, 99, 819
coralline algae, 821
cryptophyte algae, 803f
diversity, 797
edible algae, 626
endosymbiotic algae, 882–83
filamentous algae, 798f
green algae (Chlorophyta), 803, 805, 818–21
marine algae, 228, 626, 722–23
microfossils, 672f
mixotrophic, 127
photoautotropy, 127
phytoplankton, 818
primary endosymbionts, 803, 818–22
protist–algae secondary endosymbionts, 706, 707f, 797, 803–4, 822
red algae (Rhodophyta), 626, 672f, 803, 818, 821–22
secondary endosymbionts, 706, 707f, 803, 818, 822–24, 830, A-70t
siphonous algae, 821
symbiotic partnerships, 822
tertiary endosymbionts, 830–31
virus infection, 228
Algal blooms, **867**
blooms controlled by viruses, 228, 229f
coccolithophores (coccoliths), 797, 823
dinoflagellate blooms, 830f, 831
from eutrophication, 178, 867, 868
iron-induced algal bloom, 913, 914f
LOHAFEX experiment, 914, 916
and nutrient availability, 898, 898f, 912
red tide, 830f, 831
satellite sensing, 823, 913, 914f
Alginate, 149, 623–24, 945
Aliivibrio fischeri, 366, 397–98. *See also Vibrio fischeri*
Alkaline fermentation, **629**
Alkaline stress, 172
Alkaliphiles, 159t, **170–71**, 725, 765, 852
Alkaloids, 877
Alkene double bonds (kinks), 603
Allergens, **995–98**
Allergic hypersensitivity. *See* Hypersensitivity
Allograft rejection, 998
Allolactose, 371
Allomyces, A-68f
Allosteric regulation, 518
Allosteric sites, **509**
Allotypes, **971**
Alnus glutinosa (alder tree), 731f
Alpha-amanitin, 625, 815
Alpha helix structures, 73, 75, 76f
Alpha hemolysis, 1158
Alpha herpes viruses, 206f, 207
Alphaproteobacteria, 736t, 740–42, A-64t. *See also* Rhizobia
aquatic and soil oligotrophs, 737
CO oxidation, 739
endosymbionts, 740–42
in marine habitats, 861
methanotrophy, 737, 740
methylotrophy, 737, 740
photoheterolithotrophs, 735
photoheterotrophs, 735, 736–37
photosystem II, 577, 578–79
phylogeny, 713t, 715t
Alternation of generations, 808f, **809**, 819f, 820
Alternative complement pathway, 956–57
Alternator cells, 108–9
Alteromonadales, A-64t
Altman, Sidney, 19, 679
Alu sequence, 442
Alveolar macrophages, **944**
Alveolar plates, 830f, 831
Alveolates (Alveolata), **804–5**, 827–33, A-70t
apicomplexans, 831
ciliates, 804, 827–30
dinoflagellates, 830–31
phylogeny, 799t, 800, 801f, 803–4, 804–5

Alvin (submersible vessel), 860, 865
Alviniconcha gastropods, 663f
Alzheimer's disease, 197, 810–11
Amanita, 625, 815
Amanita, A-68t
Amanita phalloides, 815f
Amantadine, 430, 1138, 1139f
Amebas, **804**, A-69t. *See also* Slime molds
amebas engulfing paramecia, A-35f
amebic dysentery, 1066, 1072
amebic meningoencephalitis, 1072
definition, 804
filamentous and shelled amebas, 827
genomes, 825
Legionella colonization of, 745, 824
meningoencephalitis, 1072
mimivirus infection of, 195, 196f
phylogeny, 799t, 800, 801f, 804
pseudopod motility, 825
pseudopods, 44, 745f, 824–25, 827
reproduction, 825
Amensalism, **856**
American Society for Microbiology (ASM), 923
Ames, Bruce, 347
Ames test for mutagenesis, 347–48
Amherst, Jeffrey, 1176
Amikacin, 298, 1132
Amino acid decarboxylases, 172
Amino acids, **A-8**
amino acid catabolism and flavor generation, 633f
aminoacyl-tRNA synthetases, 289, 290t
arginine biosynthesis, 613, 615–16
aromatic amino acid biosynthesis, 616
biosynthesis, 532, 601, 612–16
computer programs to analyze sequences, 314f, 316
conservative replacements, 342
expense of biosynthesis, 613, 614f
genetic code, 286–89
identity, 315
in meteorites, 613, 677–78
multiple-sequence alignment, 314f, 315
NH_4^+ assimilation, 394–96, 613–14
nonconservative replacements, 342
similarity, 315
structure, A-8, A-9f
Aminoacyl-AMP, 290f
Aminoacyl-tRNA synthetases, 196, **289**, 290t
Aminoglycoside-inactivating enzymes, 1128, 1129f
Aminoglycoside phosphotransferase, 298
Aminoglycosides, 130, 1124, 1126, 1128
5-aminolevulinicacid (ALA), 619
Ammonia. *See also* Nitrogen cycle; Nitrogen fixation
ammonium assimilation into amino acids, 394–96, 613–14
ammonium oxidation by lithotrophs, 123, 608, 908
as fermentation waste product, 627
nitrate reduction to ammonia, 605, 907–8
nitrite reduction to ammonia, 605
oxidation to nitrate, 26, 126
production by urease, 172
Ammonia oxidizing archaea (AOA), 765, 774–75, 776, 777f, 906, 907, 911
Ammonification, **906**
Amoeba proteus, 44f, 824f, A-69t. *See also* Amebas
Amoebozoa, 799t, 800, 801f, 802, 804, 825, A-69t
Amoxicillin, 1127, 1132
Amphibolic pathways, **518–19**, 532
Amphipathic molecules, **A-15**, **A-24**
Amphotericin B, 1063, 1142, **1143**
Ampicillin, 98, 99f, 187f, 1111, 1117, 1118–19
Amplicon, definition, **1160**
Ampullaviridae, A-60t
Amycolatopsis mediterranei, 283
Amycolatopsis orientalis, 1120
Anabaena, 45f, 152, 240t, 608, 718, 720–22, 868f, A-62t
Anabaena oscillarioides, 75
Anabaena spiroides, 608f
Anabolism, **493**, **586**. *See also* Biosynthesis
Anaerobe jar, 175
Anaerobes
aerobes *vs.* anaerobes, 174
aerobic and anaerobic metabolism, 851t
aerotolerant anaerobes, 175
anaerobic dechlorinators, 728, 729f
classification, 159t, 173t
culturing in the laboratory, 175–76
facultative anaerobes, 175
facultative microbes *vs.* anaerobes, 174–75
magnetotactic organisms, 113
strict anaerobes, 174
Anaerobic aromatic catabolism, 534
Anaerobic gas gangrene, 935f
Anaerobic glove box, 175f, 176
Anaerobic metabolism, 500, 502–3, 851f
Anaerobic methane oxidizers (ANME), **781**, 895, 896
Anaerobic Methane Oxidizers (ANME), A-67t
Anaerobic photosynthesis, 534
Anaerobic respiration, **174**, 175, **515**, 534, 542, **559–63**
bacterial electric power, 562–63
dissimilatory metal reduction, 561–62
electron acceptors and donors, 559–61
in lake water, 561f
Analytical profile index (API 20E) strips, 1153–55, eTopic 28.1
Anammoxosome membranes, 566f
Anammox reaction, **565–66**, 753, 908
Anaphase, A-32f, A-33
Anaphylatoxins, **957**
Anaphylaxis, 973, **996**
Anaplerotic reactions, **597**
Anchor bacteria, 854, 855f
Ancient Archaeal Group (AAG), 763, 764t, 765, 793, A-68t
Anderson, Mark, 329
Andromeda Strain, The (Crichton), 66
Androstenol, 625
And the Band Played On (Shilts), 433
Anfinsen, Christian, 303
Angert, Esther, 726
Angle of aperture, **52**
Animals
animal cell structure, A-25f
animal microbial communities, 882–86
gnotobiotic animals, 935, 994–95
host defense against viruses, 220–21
viral limitations on population density, 228
Animal viruses, 213–18
binding host receptors, 213
DNA virus replication, 215–16
genome entry and uncoating, 213–15
host defenses, 220–21
oncogenic viruses, 216, 218
one-step growth curve, 222f
plaque isolation and assay of, 224–25
RNA retroviruses, 217–18
RNA virus replication, 216–17
tissue culture, 222–23, 224
tissue tropism, 213
Anions, **A-4**
Annotation, **312–13**
Annular ring, 65, 66f
Anopheles mosquito, 1092
Antarctica
lichens in, 7f
microbes in, 6
psychrophiles, 161–62, 655, 765
Antenna complexes, **575**, 576f, 578
Antenna system, **576**
Anthocyanin, 640
Anthony, Katey, 889
Anthracnose fungus, 881
Anthrax. *See also Bacillus anthracis*
as bioterror agent, 20, 150, 181, 1025, 1171, 1176–77
cutaneous anthrax, 1171
edema factor (EF), 1025–26
inhalation anthrax, 1171
Koch's discovery of cause of, 1004
lethal factor (LF), 75, 76f, 1025–26
protective antigen (PA), 1025–26
treatment for, 248
vaccines, 1177
Anti-anti-sigma factors, **382**, 383–84, 385f
Anti-attenuator stem loops, **379–80**
Antibiotic resistance, 1126–33. *See also* Drug efflux pumps
alternative mechanism of, 1128f
ampicillin resistance, 357f
antibiotic tolerance, 1132–33
beta-lactamase cleavage of penicillin, 93, 1119, 1128
biofilms, 149, 1018, eTopic 4.3
chloramphenicol resistance, 299, 402–3
development of drug resistance, 1130–31
DNA cassettes, 341, 358
forms of antibiotic resistance, 1128
horizontal gene transfer and, 1127, 1130–31
integrons, 1131
kanamycin resistance, 341
lincomycin resistance, 402–3

Antibiotic resistance (*continued*)
methicillin resistance, 1128
methicillin-resistant *S. aureus* (MRSA), 25, 387f, 693, 696
mobile genetic elements, 1131
multidrug resistance efflux pumps, 130, 1129, 1130f
multidrug-resistant *A. baumanii,* 1128
multidrug-resistant MRSA, 693
multidrug-resistant *Mycobacterium tuberculosis,* 25, 108–9, 1107, 1130, 1181
multidrug-resistant pneumonia, 1127
nanotubes exchanges of, 402
Neisseria gonorrhoeae, 1080–81
from overuse of antibiotics, 24–25, 93, 1131–32
penicillin-resistant *S. pneumoniae,* 1127–28
persister cells, 1132–33
plasmids and, 259, 260, 328, 1131
prevention, 1132, 1133f
RNA synthesis research, 283–84
Streptococcus pneumoniae, 1062
streptomycin resistance, 298, 1128
tetracycline resistance, 299, 355, 358
transposable elements and, 355, 356, 357f, 358
Antibiotics, **24**, 152, 153f, **186**, **1108**. *See also* Antibiotic resistance; Control of microbes; ***individual antibiotics***
actinomycetes production of, 152, 153f
bactericidal antibiotics, 1112, 1126
bacteriostatic antibiotics, 1112
biosynthesis, 603–5, 618, eTopic 15.3, eTopic 27.2
cell wall antibiotics, 87, 105, 186, 1117–21
definition, 24
designing for biofilms, 5
drugs that affect cell membrane integrity, 1121
drugs that affect DNA synthesis and integrity, 1121–23
drugs that affect the 30S subunit, 1124–25
drugs that affect the 50S subunit, 1125–26
drugs that affect transcription, 280f, 282–84, 1123
drugs that affect translation, 103, 298–300, 1123–26
effect on host-microbiota interactions, 1111–12
effect on intestinal microbiota, 933, 1111
effect on vaginal microbiota, 934
fatty acid biosynthesis as target for, 601–2, 1135
future of drug discovery, 1134–35
history, 24–25, 1108–10
hybrid antibiotics, 1132, 1133f
importance of pathogen identification, 1148
Kirby-Bauer assay, 1113–15
and lipid biosynthesis, 86
measuring drug susceptibility, 1112–15
mechanisms of action, 1116–26
minimal inhibitory concentration (MIC), 1112–13
modular enzyme biosynthesis of, 605, 606–7, eTopic 15.3
molecular mimicry, 299–300, 1118
nonribosomal peptide antibiotics, 605, 606–7, 618
polyketide antibiotics, 601, 602, 603–5
S. coelicolor production of, 585
search for new antibiotics, 187, 606–7, 653
selective toxicity, 1111
spectrum of activity, 1111–12, eTopic 27.1
targets of common antibiotics, 1117t
topoisomerase as target for, 248
toxicity, 1111
zone of inhibition, 1113
Antibodies, **963**. *See also* Humoral immunity; Immunoglobulins
allotypes, 971
antibody sandwich, 465, 1168
antigen/antibody equivalence, 969f, 970
as antigens, 971
antinuclear antibodies, 940
cell-cell interactions involved in production, 963–64
constant regions, 970–71, 980–81
definition, 963
diversity from combinatorial joining, 978t
DNA rearrangements and hypermutation, 978–80
enzyme-linked immunosorbent assay (ELISA), 1072
F(ab) regions, 969f, 970f
Fc region, 951, 952f, 954, 970f, 971, 972
fluorescent labeling, 61
gene switching, 978, 979f
genetics of antibody production, overview, 977–80
heavy chains, 969, 970–71, 978, 979f, 980
hypermutation, 980
idiotypes, 971
IgA, 970, 971, 972
IgD, 970, 971, 972–73, 975
IgE, 939, 970, 971, 972, 973, 996, 997–98
IgG, 970, 971, 972, 974, 1037
IgM, 970, 971, 972, 973, 974, 975, 976
and immune avoidance, 389
immunochromatographic assays, 1168
immunoprecipitation, 969–70
isotypes, 971, 972–73
isotype switching (class switching), 974, 980–81, 987
light chains, 969, 970–71, 978, 981
opsonization of microbes, 951, 952f, 972
primary antibody response, 974–75, 980, 984–87
production elicited by haptens, 967–68
properties of human immunoglobulins, 973t
and recombination signal sequences (RSS), 978, 979f
secondary antibody response, 974–75
steps in antibody formation, 977f
structure, 968–73
switch regions in genes, 981
variable regions, 970f, 971
western blot analysis, 464–65
Antibody-dependent cell-mediated cytotoxicity (ADCC), **954**
Antibody-secreting plasma cells, 941
Antibody stains, 58, **58**
Anticodons, **288**
Antifreeze proteins, 162
Antifungal agents, 1142–43
amphotericin B, 1063, 1142, 1143
clotrimazole, 1143f
fluconazole, 1142, 1143
griseofulvin, 1143
imidazole-containing drugs, 1142–43
inhibitors of chitin synthesis, 806
inhibitors of ergosterol synthesis, 806, 1143
itraconazole, 1143
Lamisil (terbinafine), 1143
major antifungal agents and their common uses, 1144t–1145t
nikkomycins, 806
nystatin, 806, 807, 1143
polyoxins, 806
triazoles, 806
Antigenic determinants, **963**
Antigenic drift, 428
Antigenic shift, 426, 1092
Antigen-presenting cells (APCs), **939**, 961, **963–64**, 965, **982–85**, 986. *See also* Dendritic cells; Macrophages; Mast cells
Antigens, **936**, **963**
ABO blood group system, 968, eTopic 24.2
antigen/antibody equivalence, 969f, 970
on antigen-presenting cells (APCs), 963
binding to major histocompatibility complex (MHC), 983f
definition, 936, 963
epitopes, 963, 971, 984, 999
haptens, 967–68
immunogenicity, 965, eTopic 24.1
immunogens, 963, 967–68
processing and presentation by antigen-presenting cells (APCs), 984–85
superantigens, 989
Antigen-sampling microfold (M) cells, 1003
Antigen-specific helper cells, 941
Antigermination therapies, 726
Antihistamines, 996–97
Antimetabolites, 1121
Antimicrobial agents, 178–80, 181, 185, **641**
Antimicrobial peptides, 940, 941f, 945–46, 951, eTopic 23.2
Antimicrobial touch surfaces, 185
Antinuclear antibodies, 940
Antiparallel arrangement, **243**

Antiport transport systems (antiporters), **128–30**, 170, 171f, 172, 551
Antiretroviral therapy (ART), 194, 1082, 1140–42
Antisense RNAs, 369, 385, 387–88
Antisepsis, **179**
Antiseptics, **24**
Anti-sigma factors, **382**, 383–84, 385f
Antiviral agents, 1135–42
  acyclovir, 444, 1140f
  amantadine, 430, 1138, 1139f
  antiretroviral therapy (ART), 194, 1082, 1140–42
  azidothymidine (AZT), 194, 432, 438, 439f, 1140
  CCR5 inhibitors, 1141
  delaviridine, 1141
  DNA synthesis inhibitors, 1140
  entry inhibitors, 1141
  examples of antiviral agents, 1138t
  fusion inhibitors, 1141
  highly active antiretroviral therapy (HAART), 1142
  integrase inhibitors, 1142
  Lopinavir, 1141
  prevention of viral uncoating or release, 1136–40
  protease inhibitors, 194, 420, 1141, 1142
  Relenza (zanamivir), 1138, 1139f
  reverse transcriptase inhibitors, 1140–41, 1142
  Tamiflu (oseltamivir), 426, 431, 1138
  Truvada (tenofovir/emtricitabine), 1142
  Viracept, 1141
  zidovudine (ZDV or AZT), 1140f, 1141
AP endonucleases, **351**
Aphotic zone, 857f, **858**
API 20E (analytical profile index) strips, 1153–55
Apical complex, **831–32**
Apicomplexa (formerly Sporozoa), A-70t
Apicomplexans, 804, **805**, 831–33
Apicoplast, 831, 833
Apis mellifera, 268
Aplastic anemia, 1111
Apoptosis, 448–49, 981, 988, 991, 1037, 1042, 1046, 1047
Aporepressors, 378, 379f
Appendix, 942, 1067
AP sites, **351**
Apurinic sites, **344**, 345f
Aquaporins, 87, 127–28, 166, A-28
Aquifex, A-62f
Aquifex aeolicus, 714
Aquifex pyrophilus, 714
Aquificae, 714, A-62t
Aquificales, 691, A-62t
Arabidopsis thaliana, 711, 839
Arabinans, 96
Arabinogalactans, 96
Arabinose, 481, 484
Arabinose catabolism (*ara*) operon, 376–78, 391
araC gene, 481
AraC protein, 376–78, 391, 481
AraC/Xyls family of transcriptional regulators, 376
Arbuscular mycorrhizae, 816, 874, 875f
Archaea (domain), **9**, 81, 685. *See also* Archaeal diversity; Methanogens
  ammonia oxidizing archaea (AOA), 765, 774–75, 906, 907, 911
  archaeal signatures, 758
  archaeal traits, 758–62
  archaeosine in tRNA, 762
  biofilms, 150, 758
  cell structure and metabolism, 758–61
  comparison of archaea, bacteria, and eukaryotes, 31, 32f, 91, 688, 689f, A-24t, A-25f, A-31
  cooperation with bacteria and eukaryotes, 758
  cyclic diether membranes, 758
  discovery, 685
  DNA supercoiling, 761
  in extreme environments, 7f, 19, 28, 164, 247, 271, 763, 765
  gene regulation, 762
  genomes, 241, 322, 761–62, 763
  glucose catabolism, 760–61
  halophilic archaea (*See* Haloarchaea)
  histones, 762
  initiation factors for translation, 294
  introns and intron splicing, 688, 762
  isoprenoid membranes, 758–59, 760f
  membrane lipids, 90–91, 758–59, 760f
  methanogenic archaea (*See* Methanogens)
  naming of, 669
  nucleoid, 101–2
  penicillin resistance, 1120
  phototrophy, 761
  phylogeny, 762–65, A-66t–A-68t
  proteasomes, 305
  pseudomurein in cell walls, 760
  pseudopeptidoglycan in cell walls, 713, 760
  secretion systems, 309
  S-layer and cell wall, 760
  sulfur-oxidizing archaea, 170, 566–67, 763, 765
  sulfur-reducing archaea, 598
  unique metabolic pathways, 760–61
Archaeal diversity, 757–94. *See also specific types*
  deeply branching groups of uncultured Archaea, 763, 764t, 765, 793
  overview, 757–58
  phylogenetic tree, 763f
  phylogeny, 762–65
  representative groups of archaea, 764t, A-66t–A-68t
Archaean eon, 669, 670f, 671–72, 673, 676f
Archaeoglobales, 791
Archaeoglobi, A-67t
Archaeoglobus, 765
Archaeoglobus fulgidus, 781, 791, 792f
Archaeosine, 762
Arginine biosynthesis, 613, 615–16
Aromatic amino acid synthesis, 616
Aromatic catabolism, 512–13, 532–35, eTopic 13.5
  aerobic benzene catabolism, 533
  anaerobic benzene catabolism, 534
  benzoate catabolism, 502–3, 532, 533f, 534
  halogenated aromatic catabolism, 513
  lignin catabolism, 512–13
  polycyclic aromatic hydrocarbons (PAHs), 513
  toluene catabolism, 533
Aromatic molecules, **506**, 511f, 512–13, 532
  halogenated aromatics, 513
  polycyclic aromatic hydrocarbons (PAHs), 513
Arrhenius equation, 160, eTopic 5.1
Arsenic, 478–79, 737, 877, 914, 915f, 915t
Artemisinin, 478
Arthritis with Lyme disease, 1096–97
Arthrobacter, 733–34, A-63t
Arthrobacter globiformis, 733f
Arthrospores, 152–53, 721f, 723, **730**
Artifacts, **69**
Ascomycetes (Ascomycota), **812**, 813f, 814, 816, 817, 874
Ascomycota, A-68t
Ascospores, 813f, **814**
Ascus (plural, asci), **809**, 813f, 814
Aseptic environments, **24**
Asexual reproduction, **A-33**
Asimov, Isaac, 916
Asparaginase, 654f, 655
Asparagine, 654f, 655
Aspartate, 612, 613, 616, 618
Aspergillus, 642, 655, 809, 816–17, A-68t
Aspergillus nidulans, 814f
Aspergillus niger, 656
Aspergillus oryzae, 633–34
Aspirin (acetylsalicylic acid), 88
Assembly (of DNA fragments), **844–45**
Assembly (viruses), **414–15**
Assimilation, **849**
Assimilatory nitrate reduction, **906**
Asthma, 996, 997
Astrobiology, 916–19. *See also* Extraterrestrial life; Mars
  comets or meteorites as source of life, 19, 613, 680
  definition, 916
  extrasolar planets, 919
  panspermia, 680
  possible ocean on Europa, 918–19
  terraforming, 918
Atacama Desert, Chile, 6
Ataxia, 706
Athalassic lakes, 786
Atherosclerosis, 1090, eTopic 26.8
Athlete's foot, 1005
Atomic force microscopy (AFM), **45**, 46f, **72**
Atomic mass, **A-2**
Atomic number, **A-2**
Atopic disease, 996
ATP. *See* Adenosine triphosphate (ATP)
ATP-binding cassette (ABC) transporters, 88, 89, 130–32, 309f, 310–11, 361

ATP citrate lyase, 598
ATP sulfurylase, 268
ATP synthase, 99, **125**
in anammoxosome membrane, 566
antibiotic targeting of, 557
ATP synthesis at high pH, 559, eTopic 14.4
F_1F_o ATP synthase, 125, 547f, 551f, 557–59, 572, 575, 579f
location on cell membrane, 80–81f, 86
Na^+-dependent ATP synthase, 550
proton motive force, 547, 548–49, 557–59
Attenuated pathogens in vaccines, 23
Attenuation (transcriptional), 369, 378–80, 616
Attenuator stem loops, **379–80**
Attractant signals, 115, 116f
Ausmees, Nora, 235
Australopithecus africanus, 1004, 1005f
Autoantibodies, 940, 941f
Autoclaves, **17**
Autoimmune diseases, 949, 950, 999–1000
Autoimmune response, **999–1000**
Autoimmunity, 949, 999–1000
Autoinducers, **398**, 1037
Autophagosomes, 952
Autophagy, **952**, 954, **1042–43**
Autoradiography, **465**, **A-50**
Autotrophs, **123**, **124**, 736, 743, 750
Auxospores, 822f, 823
Auxotrophs, 343
Avatar, 806
Average nucleotide identity (ANI), 697
Avery, Oswald, 34, 322, 923
Avian influenza strain H5N1, 213, 409, 426, 427–28, 1172, 1173
Avian influenza strain H7N9, 6, 228
Avian influenza virus, 1007
Avian leukosis virus, 195, 205t
Avogadro's number, A-17
Axenic growth, **121**
Axial filaments, 71
Axinella mexicana, 774
Axoneme, A-39f, A-40
Azam, Farooq, 862
Azidothymidine (AZT), 194, 432, **438**, 439f, 1140
Azithromycin, 1062, 1063, 1124, 1125
Azoarcus evansii, 534
Azotobacter, 608, 610, 611
Azotobacter vinelandii, 559

B

Babesiosis, 1093
Bachman, Herwig, 665
Bacillales, 724–25, A-63t
Bacillariophyceae, A-70t. *See also* Diatoms (Bacillariophyceae)
Bacillary dysentery, 1006
Bacille Calmette-Guérin (BCG) vaccine, 652
Bacilli (singular, bacillus), **44**, 45f. *See also* Rods
Bacillus, 57, 107, 126, A-63t
enzymes developed as commercial products, 654, 655
fermentation by, 628t, 629, 636
genome similarity among species, 696
natural transformation, 323
nitrite reduction to ammonia, 560
nonhomologous end joining (NHEJ), 354
skin microbiota, 928
spore formation, 382, 723, 724–25
thermophilic species, 248
xylanase genes from, 698
Bacillus alkalophilus, 725
Bacillus anthracis. See also Anthrax
accidental release of, 1178, A-45
anthrax lethal factor, 75
and anthrax toxin, 1026f
fatty acid profiles, 90
genome, 696, 697
separation of DNA fragments, A-45–A-46
septicemia, 1098
spore stain, 57
transmission electron microscopy, 68, 69f
Bacillus brevis, 1121
Bacillus cereus, 697
Bacillus halodurans, 725
Bacillus licheniformis, 1120
Bacillus megaterium, 44f
Bacillus natto, 636
Bacillus polymyxa, 1121
Bacillus subtilis, A-63f
bacitracin production, 1120
biofilms, 119
cell envelope, 94f
chemotaxis, 394
culture media, 135
DNA replication, 104, 105f, 237
engineered to detect meat spoilage, 485f
filtration sterilization, 182f
glutamine synthase gene (*gltB*), 315
as industrial strain, 656
intrachain transfer of transposons, 321
nanotubes, 114, 402–3
proteomic profile, 404f, A-47, A-48f
S. coelicolor antibiotic production against, 585
sigma B expression, 462, 463f
sporulation, 62, 150, 384, 724
stress response, 724
temperature and growth, 161
Bacillus thermophilus, 725
Bacillus thuringiensis (*Bt*)
aiiA gene, 478
B. thuringiensis subsp. *kurstaki,* 475f
as biological control agent, 724–25
genome, 696
insecticidal toxin genes, 329, 475
insecticide, 474–75, 1007
parasporal body, 474, 475f
parasporal crystals, 474, 475f
spores, 475f
spore stain, 57
sporulation, 724–25
transgenic plants, 475, 725
Bacitracin, 1116t, **1120**, 1159
Bacteremia, **930**, 1062, **1091**
Bacteria (domain), **9**, 685. *See also* Bacterial diversity
biochemical composition of bacteria, 83–85
cell components, 81f, 82
cell division, 104–7
cell membrane and transport, 86–91
cell model, 80–81f, 82
cell overview, 81–83
cell wall, 91–93
comparison of archaea, bacteria, and eukaryotes, 31, 32f, 91, 688, 689f, A-24t, A-25f, A-31
cytoskeleton, 92, 100–101, 234, eTopic 3.4
and DNA revolution, 35
edible bacteria, 626
genomes, 9–10, 80, 102, 239, 241, 359t
intermediate filaments, 235
nucleoid, 101–2
specialized structures, 112–16
Bacterial computers, 457
Bacterial diversity, 711–54. *See also* Phylogeny; Taxonomy; *specific types*
bacterial phylogenetic tree, 713–14
common traits, 713–14
overview, 711–18
phototrophic bacteria, 719t
representative groups of bacteria, 715t, A-62t–A-65t
Bacterial electric power, 562–63
Bacterial lawn, 223, 224f
Bacterial "lobster traps," 365
Bacterial spot disease of tomato, 746f
Bactericidal agents, definition, **179**, **1112**
Bacteriochlorophylls, **574**, **575**
bacteriochlorophyll P680, 580
bacteriochlorophyll P840, 577–78
bacteriochlorophyll P870, 578–79
in Chlorobi, 750
in *Chloroflexus,* 714
in proteobacteria, 723, 735, 737
Bacteriocins, 885
Bacteriophage CTXϕ, 200, 210
Bacteriophage DMS3, 337f
Bacteriophage fd, 476
Bacteriophage lambda, A-60f
on bacteriophage proteomic tree, 207f
binding to receptors, 208, 210
CI repressor protein, 367, 368f
classification, 205t
and DNA revolution, 32
early and late gene expression, 210
lysis-lysogeny decision, 210, 397, eTopic 10.5
lysogeny, 411, 413
plaques, 224f
reproduction, 209f, 210
site-specific recombination, 341
specialized transduction, 332f
Bacteriophage M13
classification, 205t
filamentous structure, 200
phage display technology, 476
on proteomic tree, 207f
slow release, 224
vaccines and nanowires, eTopic 11.2

Bacteriophage MS2, 211, A-61f
Bacteriophage P22 (*Salmonella*), 331, 355
Bacteriophage phiX174, 193, 207f
Bacteriophage Qβ, 211
Bacteriophages (phages), **194**. *See also* Viruses
attachment to host cells, 208, 209f
bacterial host defenses, 211–12, 333–37
batch culture, 221
burst size, 210
classification, A-60t–A-61t
as cloning vectors, 194, 210
diphtheria toxin gene carried within, 1004
DNA injection into cell, 209–10
and DNA revolution, 35
early study of, 192–93, 208–9
filamentous bacteriophages, 200
gene expression, 413–15
lysis, 209f, 210
lysis-lysogeny decision, 210, eTopic 10.5
lysogeny, 209f, 210
lytic cycle, 209f, 210
as medical tools, 411
and nanotechnology, 411
one-step growth curve, 221f
phage display technology, 476–77
phage therapy as medical treatment, 188
plaque isolation and assay of, 223–24
plaques, 194
prophages, 210, 333, 353–54
proteomic tree, 207
replication, 208–13, eTopic 7.3
reproduction within host cells, 208–11
RNA phages, 211
site-specific recombination, 210
slow release, 211
tailed phages, 410–11, 412
temperate phages, 210
transducing particles, 331, 332f, 333
transduction of host genes, 210–11, 331–33
virulence genes in genomes of, 1011, 1012
Bacteriophage T2, 193, 194
Bacteriophage T4, 410–17
adsorption to host and DNA injection, 412–13
assembly, 414–16
on bacteriophage proteomic tree, 207f
capsid structure, 410f, 411
as classic molecular model, 410–17
conditional lethal mutations, 416
developmental mutants, 415–16
dihydrofolate reductase, 415
DNA injection into cell, 209
early genes, 413f, 414
genome, 195, 411–12
ghost capsid, 209
infection of bacteria, 194
injector, 412–13
late genes, 413f, 414–15
lysis, 415
lytic cycle, 210
methylated cytosine in DNA, 413f, 414
nanomotor, 413f, 415
P18 protein, 412
rolling circle replication, 414
sheath, 412
structure, 201–2, 411–12
virulent replication, 413–15
Bacteriorhodopsin, **571–73**, **575**, 761, 787. *See also* Phototrophy; Retinal
haloarchaea, 571, 676–77, 761
light-driven cycle, 572
photoheterotrophy, 572
proteorhodopsin, 571–72
proton pump, 549, 572
purple membrane, 572, 573f
structure, 572f
Bacterioruberin, 784
Bacteriostatic agents, definition, **179**, **1112**
Bacteroidales, A-65t
Bacteroides, A-65t
activation of metronidazole, 1122
anaerobic gas gangrene, 935f
biofilms, 28
interaction with digestive system, 515–16, 717, 749, 856, 885
intestinal microbiota, 932
Bacteroides fragilis, 175, 749, 934, 935, 1098, 1149–50
Bacteroides plebeius, 514–15f, 516
Bacteroides thetaiotaomicron
decay-accelerating factor upregulation, 993
induction of mucus production, 520
as normal gut flora, 701, 749
polysaccharide catabolism, 514f, 515, 781, 931
Bacteroidetes, 711, 712f, **717**, 749, 865, 932, A-65t
Bacteroids, **740**, 741f, **877–78**, 879f, 880f
Baculoviridae, A-60t
Baculovirus, 1007
Badnaviridae, A-61t
Baeocytes, 722
Bait proteins, 473
Baldauf, Sandra, 801
Baltimore, David, 203, 204f
Baltimore virus classification, 203–4, 205t, 206, A-60t–A-61t
Group I, double-stranded DNA viruses, 203, 204f, 205t, A-60t
Group II, single-stranded DNA viruses, 203, 204f, 205t, A-60t
Group III, double-stranded RNA viruses, 203–4, 205t, A-60t
Group IV, (+) sense single-stranded RNA viruses, 204–6, A-61t
Group V, (–) sense single-stranded RNA viruses, 205t, 206, A-61t
Group VI, retroviruses, or RNA reverse-transcribing viruses, 205t, 206, A-61t
Group VII, pararetroviruses, or DNA reverse transcribing viruses, 205t, 206, A-61t
Banded iron formations (BIFs), 671t, **674–75**
Banfield, Jillian, 10, 842, 843f, 845
Bangia, 672f
Barley, 639
Barns, Susan, 683–84, 793
Barophiles, **164–65**, 763, 765, **766**, 767–69, **865**
Barosensitive microbes, 165
Barotolerant microbes, 159t, 165
Barré-Sinoussi, Françoise, 434
Barry, Clifton, 653
Bartlett, Douglass, 865
Bartonella, A-63t
Bartonella henselae, 737
Barzantny, Helena, 405
Basal bodies, 115f, A-39f, A-40
Basal epithelial cells, infection by HPV, 215–16
Base excision repair (BER), 348t, **350**
Basidiomycetes (Basidiomycota), **815–16**, 874
Basidiospores, 815f, **816**
Basler, Christopher, 1139
Basophils, **938**, 939, 973
Bassham, James, 589
Bassi de Lodi, Agostino, 16, 20
Bassler, Bonnie, 400
Batch culture, **141**, 142f, **221–22**
of bacteriophages and viruses, 221–22
stages of growth in, 141–43
Bates, Paul, 1100
Batrachochytrium dendrobatidis, 810, 812f, A-68t
Battista, John, 183
B-cell receptor (BCR), 972, **975–76**, 980, 987
B cells, 937f, **941–42**
activation, 964, 975, 976, 986–88
affinity maturation, 981
antibodies bound to surface of, 972–73, 975
B-cell receptor, 972, 975–76, 980, 987
capping, 976, 987
clonal selection, 975–76
differentiation into plasma cells and memory B cells, 964, 974, 975, 976, 977f
DNA rearrangements and hypermutation, 978–80
humoral immunity, 963–64
immune system memory, 961
isotype switching (class switching), 974, 976, 977f, 980–81
memory B cells, 964, 974, 975, 981, 987–88
primary antibody response, 974, 980, 986–87
in SCID, 962
secondary antibody response, 974, 987–88
T cell–independent activation route, 976
T_H2 cell activation route, 986–88
B-cell tolerance, **965**
Bdellovibrio, 746–47
Bdellovibrio bacteriovorus, 747f
Bdellovibrionales, A-64t
BD Phoenix system, 1155, 1156f
Bead beater, 82
Beadle, George, 812, 814

Becher, Dörte, 405
Beclin 1, 1042
Bedaquiline, 557, 1107
Beef extract, 134–35
Beer production, 120, 627, 628t, 639, 640f, eTopic 16.1
Beggiatoa, 26, 608, 743–44, 856, 910
Beijerinck, Martinus, 25, 28, 192, 608
Bej, Asim, 161f
Béjà, Oded, 571–72
Bell's palsy, 1096
Benson, Andrew, 589
Benson, David, 401, 404
Benthic organisms, **858**
Benthos (ocean floor), 857f, **858**, 865, 868, 869f
Ben-Yehuda, Sigal, 402–3
Benzalconium chloride, 184f
Benzene as hapten, 968
Benzene catabolism, 492, 512, 532–34, eTopic 13.5
Benzoate catabolism, 502, 503f, 532, 533f, 534
Benzoic acid, 649
Benzothiazinones, 96
Benzoyl-CoA, 532, 534
Berg, Howard, 64, 114, 392f
Berg, Paul, 9, 35–36
Bergey's Manual of Determinative Bacteriology, 701
Berkmen, Melanie, 105f
Bernal, John, 74, 76f
Beta-amyloid toxicity, 810–11
Beta barrels, 98
Betadine, 184
Beta hemolysis, 1158–59
Beta herpes viruses, 206f, 207
Beta-lactam antibiotics, 1118–20
Beta-lactamase (β-lactamase), 93, 1119, 1128, 1129f, 1131, 1132
Betapropiolactone, 184f, 185
Betaproteobacteria, 742–43
  lithotrophy, 736t, 742
  in metagaenomes of honeybee gut communities, 840
  pathogens, 742–43
  photoheterotrophy, 735, 742
  phylogeny, 713f, 715t, 717, 739f, A-64t
Beta sheets, 73, 75, 76f
Beutler, Bruce, 954
Biased random walk, 115, 116f
Bicarbonate, 596
Bickhart, Derek, 401, 404
Bifidobacteriales, A-63t
Bifidobacterium, 188, 885, 933
Bifidobacterium infantis, 994–95
Bigger, Joseph, 1132
Bile salts, 135
Binary fission, **139**, 140
Binning, **846**
Bins (genome bins), 845f, **846**
BioBricks, 484, 485f
Bioburden, **926**
Biochemical oxygen demand (BOD), **859**, 867, **897**, 898, 899, 906
Biocomplexity initiative, 28
Biocrystallization, 102
Biofilms, **145**, 145–50
  Alcaligenes xylosoxidans, 68, 69f
  antibiotic development for, 5
  antibiotic resistance, 149, 1018, eTopic 4.3
  antibiotic tolerance, 1132–33
  archaea in, 758
  on artificial heart valves, 1018
  Bacillus subtilis, 119, 149
  bacterial factors important to formation, 1018
  in bacterial fuel cells, 562
  Bacteroides, 28
  Beggiatoa, 744
  catheter biofilms, 68, 69f, 145, 146–47, 744, 1018
  cdiGMP and, 396–97
  Chloroflexus, 601
  confocal microscopy, 5, 62, 63f
  conjugation, 324–25
  and CRISPR, 337
  development, 148–49
  disinfectant resistance, 185–86
  dissolution and dispersal, 149–50
  electricity generation, 539
  Enterobacteriaceae, 744
  Escherichia coli, 28, 41
  exopolysaccharides (EPS), 148f, **149–50**
  in extreme environments, 150
  Ferroplasma acidarmanus, 567, 791, 793
  flagella and, 115
  floating, 149, 150
  flocs, 900
  Geobacter sulfurreducens, 539
  green fluorescent protein, 61, 474, eTopic 12.2
  on implanted medical devices, 1018
  and infections, 1017–18
  intestinal, 29f
  intracellular biofilms, 1076, eTopic 26.5
  iron-oxidizing bacteria, 113–14, 750
  Legionella pneumophila, 686
  Leptospirillum, 750
  marine snow, 150, 229
  matrix of biofilms, 488–89
  on medical instrumentation, 68, 69f, 145, 146, 148
  mixed-species biofilms, 150, 186, 325, 402–3, 758, A-56f, A-59f
  oral biofilms, 148f, 501, 542, 750
  persister cells, 1132–33
  phagocytosis resistance, 149
  pili and biofilm formation, 149, 324, eTopic 9.1
  Pseudomonas aeruginosa, 62, 63f, 148–50, 186f, 337, 488–89, 745, 945f, 1018, 1136
  quorum sensing, 149
  sessile, 149–50
  shower curtain biofilm, 684
  Staphylococcus epidermidis, 146
  of sulfate-reducing bacteria, 913
  Synechococcus, 148f
  transformation, 323–24
  twitching motility, 116
  Vibrio anguillarum, 400
  Vibrio cholerae, 5
  viral infection, 229
  Yersinia pestis, 1095–96
Biofuel-producing microbes, 159, 540–41, 857
Biogeochemical cycles, **890–93**
  chemical and spectroscopic analysis, 893
  measuring flux of elements in biosphere, 893
  microbial cycling of essential elements, 605, 891
  oxidation states of cycled compounds, 892t
  radioisotope incorporation, 893
  sources and sinks of essential elements, 891, 893, 894
  stable isotope ratio measurements, 893
Biogeochemistry, **891**
Bioinformatics, 312–18, **313**
  annotation of DNA sequences, 312–13
  bioinformatic analysis, 158–59, 207, 276
  BLAST (Basic Local Alignment Search Tool), 316
  ClustalW2, 314f
  *coli*Base, 316
  computer analysis and Web science, 313–15
  ExPASy (Expert Protein Analysis System), 316
  functional genomics, 316
  genomic predictions of transport and metabolic pathways, 316–17f
  homologs, orthologs, and paralogs, 315–16
  Joint Genome Institute, 316
  KEGG (Kyoto Encyclopedia of Genes and Genomes), 316
  minimum number of genes required for life, 316, eTopic 8.4
  Motif Search, 316
  Multiple Sequence Alignment, 316
  nicotine catabolism research, 534–35
  sequence alignment, 314–15
Biological chemistry, overview, A-16–A-21
Biological control of microbes, 187–88
Biological oxygen demand, **859**
Bioluminescence, 28, 366, 397–98, 400
Biomass, **850**
  biomass building through autotrophy or heterotrophy, 123
  definition, 850
  determination in plankton, 862
  and Gibbs free energy change (ΔG), 495–96
  recycling, 851
  storage of solar energy in, 495
Biomolecular motor, 125
Biopanning, 477
Biophotonic imaging, 994, 995f
Bioprospecting, **654–56**
Bioremediation, **183**, 477, 491, **890**
  dechlorinization of toxic pollutants, 569, 728, 729f, 902
  dehalorespiration, 569, 902
  of halogenated aromatics, 513

of hexavalent chromium, 734
of metal contamination, 737, 865, 877, 914, eTopic 22.2
of nicotine, 534–35
Sphingomonas, 686
subterranean arsenic removal, 914, 915f
of uranium-contaminated water, 561
wastewater treatment, 899–902
Biosafety containment procedures, 1169–71
Biosignatures (biological signatures), **672–73**, **917–18**, 919
Biosphere
Earth's crust and, 668
energy and entropy, 494, **668**
microbial biosphere, 6, 26
oxygen in, 674
rare biosphere, 848
Biosynthesis, 510, **585–621**. *See also* Carbon dioxide fixation; Nitrogen fixation
amino acids, 532, 601, 612–16
antibiotics, 603–5, 618, eTopic 15.3
definition, 586
energy costs of biosynthesis, 588–89
fatty acids, 601–3
medical and industrial applications, 586, 603
overview, 586–89
polyesters, 586, 601, 603
polyketides, 601, 602, 603–5, eTopic 15.2
purines, 617
pyrimidines, 617, 618
substrates for biosynthesis, 587–88
tetrapyrroles, 618–20
Biotechnology. *See also* Molecular analysis; Synthetic biology
applied biotechnology, 458, 474–77
basic tools, 458, 459f
ChIP-on-chip analysis, 471–72
DNA-inversion based data storage system, 457
genetic analysis, 83, 458–63
history, 457–58
mapping protein-protein interactions, 472–73, eTopic 12.1
tracking cells with light, 473–74
Bioterrorism, 1176–79
agents that could be used as bioweapons, 1170t
anthrax as bioterror agent, 20, 150, 181, 248, 1171, 1176–77
avian influenza virus as potential weapon, 409
Clostridium botulinum, 175
history, 1176
PANTHER detection system, 1178–79
Salmonella, 1176
smallpox as bioweapon, 1177
Biotic entities, **891**
Biowarfare, 1094, 1176
Birnaviruses, 191, 205t
Black Death, 1030. *See also* Bubonic plague
Black mold, 814f
Black smokers, 6, **767–68**, 769, 865
BLAST (Basic Local Alignment Search Tool), 316
Blastochloris, 868f
Blastochloris viridis, 735
Blastomyces dermatitidis, 809, 1063, 1142, A-68t
Blastomycosis, 809, 1063, 1142
Blood agar, 745, 1022, 1153, 1157, 1158f
Blood-brain barrier, **1083**, 1084
Blood in culture media, 135
Blount, Zack, 3
Blue baby syndrome, 608, 906
Boceprevir, 420
BOD (biochemical oxygen demand) analyzers, 897
Bohr, Niels, 923
Boils, 1054, 1056
Bordetella, 121t, 1016, 1157
Bordetella bronchiseptica, 360
Bordetella pertussis, 360, 1017f, 1024, 1033, 1058, 1102, eTopic 26.3
Borrelia, 752, A-65t
Borrelia burgdorferi, 45f, 1097f, A-65f. *See also* Lyme disease (borreliosis)
as agent of Lyme disease, 21, 1180
doxycycline use for, 1125
essential nutrients, 121
genome, 240t
linear chromosomes, 239, 752
Lyme disease lesion, 751f
mutipartite genome, 752
outer membrane proteins isolated for vaccines, 99, eTopic 3.3
plasmids, 752
reservoir in deer, mice, and ticks, 1097f
Borrelia recurrentis, 751f, 752
Borreliosis, 21, **1096**. *See also* Lyme disease (borreliosis)
Botox, 1087
Botstein, David, 355
Botulism, 646, 648, **1084–87**. *See also* *Clostridium botulinum*
botulinum toxin, 646, 648, 725, 1026, 1085, 1086–87
case history, 1084–85
flaccid paralysis, 1085
home canning and, 180, 1085
infant botulism, 1086
Bovine serum albumin (BSA), 967–68
Bovine spongiform encephalopathy (BSE, mad cow disease), 1090, 1180
Boyd, Phillip, 913
Boyer, Herb, 264–65
Boyle, Robert, 14
Bradykinin, **948**, 949f
Bradyrhizobium, 126, 360, 738, 740, 877
Bradyrhizobium japonicum, 608
Brain coral, 882f
Brasier, Martin, 672
Braun lipoprotein, 97
Bread making, 627, 628t, 629
history, 637–38
injera (extended fermentation), 628t, 638–39
leavening, 637–39
sourdough, 629, 638
Breakbone fever, 191–92
Bregoff, Herta, 608
Brennan, Anthony, 146
Bridge amplification, 270
Bright-field microscopy
compound microscope, 52–54
fixation and staining, 55–58
focusing the object, 54–55
magnification, 51–52
preparing a specimen, 54
Brine pools beneath ocean, 786
Brocadia, A-65t
Brochothrix thermosphacta, 642
Brock, Thomas, 162, 768
Bromoviridae, A-61t
Bronchodilators, 997
Brown algae (kelp, Phaeophyceae), 623, 624, 799t, 800, 801f, 805, 824
Brucella (brucellosis)
in *Australopithecus africanus,* 1004, 1005f
changes in host gene expression, 1045–46
cytochrome oxidase test, 1157
denitrifier, 560
slow growth of, 1160
soil pathogen, 737
Brucella melitensis, 1046–47
Brugia malayi, 705
Bubonic plague, **1094–96**. *See also* *Yersinia pestis*
biowarfare, 1094, 1176
buboes, 1094f, 1095
case history, 1094
history, 11f, 20, 1094, 1095, 1096, 1176
infective cycles of the plague bacillus, 1094–95
pneumonic plague, 1095
septicemic plague, 1095
Buchnera, 684–85
Budding, 139, **808**
Buffer gates, 480
Bulgarelli, Davide, 711
Bunyaviridae, A-61t
Burkholderia, 1034, 1038
Burkholderia cepacia, 240t, 743
Burkholderiales, A-64t
Burst size, **210**, 221f, **222**
Bush, George W., 380f, 433
Butanol, 525
Bütikofer, Peter, 315
Butyric acid (butyrate), 525
Byrd, Matthew, 488

C

Cadaverine, 512, 642
Caffeine catabolism, 491, 492, 534
CagA effector protein, 1033, 1045f, 1046, 1072
Calcium in spore coats, 151
Caldisphaera, 771
Caldisphaerales, 771, A-66t
Calmodulin, 1024, 1026
Calorimeter, **497**
Calvin, Melvin, 589, 591f, 818

Calvin-Benson-Bassham cycle. *See* Calvin cycle
Calvin-Benson cycle. *See* Calvin cycle
Calvin cycle, **589–97**
carbon-concentrating mechanism (CCM), 594, 595–96
carbon isotope labeling, 590–91
carboxylation and splitting, 592
by chloroplasts, 590
in detail, 595f
discovery, 589, 591f
by facultatively anaerobic purple bacteria, 590
intermediates, 590–92
lithotrophs, 590
overview, 592, 593f
by oxygenic phototrophic bacteria, 590
paper chromatography, 591–92
regulation, 594, 596–97
ribulose 1,5-bisphosphate, 592
ribulose 1,5-bisphosphate and CO_2 fixation, 592, 593f, 594
ribulose 1,5-bisphosphate reduction, 592
ribulose 1,5-bisphosphate regeneration, 592, 594
cAMP. *See* Cyclic adenosine monophosphate (cyclic AMP or cAMP)
Campi, Carlo, 775
cAMP receptor protein (CRP)
cAMP-CRP activation of *araBAD* operon, 378, 391
cAMP-CRP activation of *lac* operon, 371–73, 391
interactions with RNA polymerase, 372f, 373, eTopic 10.1
Campylobacter, 647t, 747, 1066, 1157, A-64t
Campylobacterales, A-64t
Campylobacter jejuni, 175
Candida, 635, 638, 928, 1143
Candida albicans, 286, 626, 808, 809, 932, 934, A-68t
Candidate species, **701**
Candidatus Chloracidobacterium thermophilum, 748–49
Candidatus Korarchaeum cryptofilum, 793
Candidatus Nitrosopumilus maritimus, 701
Candida utilis, 656
Canine parvovirus, 203
Cannibalism, 119, 152
Canning of foods, 180, 649
Cannulae (singular, cannula), **768–69**
Cannulated cows, 843f, **884**
Capone, Douglas, 75
Capping, **976**
Capsase 1, 1042
Capsids, 191, **192**, 194, 198–202, 209, **412**
Cap snatching, 430
Capsules, 82, 94f, **95**, 96f, 1037
Carbadox, 1131
Carbamoyl phosphate, 617, 618
Carbohydrate catabolism, 510, 511f, 512f. *See also* Glucose catabolism
Carbohydrates, A-11–A-13. *See also* Polysaccharides; *specific types*
disaccharides, 510, A-12–A-13
general formula, 123
monosaccharides, 510, A-11–A-12
oligosaccharides, 510, 515, A-13
polysaccharides, definition, A-13
Carbolfuchsin, 57, 96, 1152
Carbolic acid, 24
Carbon
as essential nutrient, 120, 123, 891
global reservoirs of carbon, 892t, 894
microbial cycling of carbon, 605, 891
oxidation states, 891, 892t
recycling by marine viruses, 230
uptake in cyanobacteria, 75
Carbon-12 (^{12}C), mass, 590
Carbon-13 (^{13}C)
isotope ratios as biosignature, 672–73
labeling of acenaphthene for NMR, 513
labeling of carbon sources for *M. tuberculosis,* 532
mass of, 590
Carbon-14 (^{14}C)
carbon isotope labeling, overview, 590–91
CO_2 flux measurement by radioisotope incorporation, 893
discovery, 529, 591, eTopic 15.1
half-life, A-2
labeling of TCA cycle, 529f, 530
radioactive decay, 590–91
Carbon-concentrating mechanism (CCM), 594, 595–96
Carbon cycle, 123–24, 894–96. *See also* Biogeochemical cycles; Carbon dioxide fixation
aerobic carbon cycling, 894, 895f
anaerobic carbon cycling, 571, 894–95
autotrophs, 123
carbon assimilation, 849, 850
carbon dissimilation, 849–50, 850–51
carbon flux in ocean, 896
carbon flux in wetlands, 896, eTopic 22.1
chemical and spectroscopic analysis, 893
chemoautotrophs, chemoautotrophy, 123, 124
chemoheterotrophs, chemoheterotrophy, 124
chemolithotrophs, 123
chemotrophy, 124
CO_2 flux measurement by radioisotope incorporation, 893
CO_2 from fossil fuel combustion, 889, 890f, 894, 896
coccoliths, 823–24, 894
cyanobacteria as keystone species, 75
forests as carbon sinks, 894, 896
global carbon balance, 896
global reservoirs of carbon, 892t, 894
heterotrophs, 123
lithotrophs, 123, 124
marine viruses and global carbon balance, 230
measuring flux of CO_2 in biosphere, 893
overview, 123f, 895f
oxidation states of cycled compounds, 891, 892t
photoautotrophs, photoautotrophy, 123, 124
phototrophy, 124
Carbon dioxide
covalent bonding and structural formula, A-3, A-4f
food storage under increased CO_2, 647
greenhouse effect, 669
reduction to methane, 561, 569–70
Carbon dioxide fixation, 75, **589–601**. *See also* Biosynthesis; Photosynthesis
alternative CO_2 fixation systems, 597
autotrophs, 123, 124
Beggiatoa, 26
Calvin cycle, 589–97
carboxysomes, 112
chloroplasts, 590
cyanobacteria, 75, 589–90
3-hydroxypropionate cycle, 590t, 600f, 601
and metabolist models of early life, 678
model suggested by paper chromatography, 591f
pathways, 590t
reductive acetyl-CoA pathway, 590t, 599, 601
reductive (reverse) TCA cycle, 590t, 597–99
Carbon flux in ocean, 230
Carbonic anhydrase, 596
Carbon isotope labeling, 590–91
Carbon monoxide (CO), 599, 736, 737, 738–39
Carbon monoxide dehydrogenase, 599
Carbon monoxide reductase pathway, **761**
Carbon-nitrogen-oxygen (CNO) cycle, 667, 668f
Carboxydothermus, 723, A-63t
Carboxylate, 533, 534
Carboxylic acids, 497, 512, 516, 627
Carboxysomes, **112**, **596–97**, 720, 750
Carbuncles, 1054, 1056
CARD-FISH technique, A-55
Cardiolipin, **89**, 315, A-40
Cardiolipin synthases, 314–15
Cardiovascular disease, 957
Cardiovascular system infections. *See also* Endocarditis
atherosclerosis, 1090, eTopic 26.8
malaria, 831–33, 1092–93
myocarditis, 1092
pericarditis, 1090
septicemia, 1090, 1091
Carezyme, 654–55
Carnivores in marine food chain, 230
Carotenoids, **574**, **575**, 831
Carrageenan, 821
Cas-crRNA complex, 212, 336–37
Caseins, 630, 633f
Catabolism, **492**. *See also* Fermentation; Glucose catabolism; Lactose catabolism; Photoheterotrophs, photoheterotrophy; Respiration
amino acid catabolism, 525
aromatic catabolism, 502–3, 512–13, 532–35

lipid catabolism, 510–11, 512f, 531–32
products of, 513–15
of sewage components, 502–3
of soil contaminants, 491, 492, 534–35
starch utilization system, 515–16
substrates for, 510–13
Catabolite repression, **374**, **510**
Catalase, 174, 175, 693, 861, 1157–58
Catalase test, 1157–58
Catalytic RNA, 19f, 196–97, **285**. *See also* Ribozymes
Catalyzed reporter deposition fluorescence in situ hybridization (CARD-FISH), 711, A-55
Catechol dioxygenase, 533
Catechols, 533
Catenanes, **256**, 257f
Caterpillar viruses produce commercial products, eTopic 16.2
Cathelicidins, 945t, 946, eTopic 23.3
Cathepsins, 1100
Catheter biofilms, 68, 69f, 145, 146–47, 744, 1018
Cations, 120
Cat scratch fever, 737
Caulerpa taxifola, 8, 821
Cauliflower mosaic virus (caulimovirus), A-61f
classification, 205t, 206, A-61t
effect on agriculture, 193, 206, 220
genome, 195
life cycle, 219–20
promoter as gene transfer vector, 220
Caulimoviridae, A-61t
Caulobacter
localization of signaling proteins, 234
as model system, 234–35
peptidoglycan cell wall, 235
and study of pathogen cell cycles, 235
xerD mutant, 257
Caulobacterales, A-64t
Caulobacter crescentus
asymmetrical cell division, 107–10, 139, 235f
cell cycle, 108–10, 111f
cell differentiation, 150
DNA array and proteomics, 405
flagella, 107–10, 111f, 114
Hyphomicrobium compared to, 740
as model system, 234–35
oligotrophy, 737
shape determining proteins, 100f
stalks, 107, 110, 111f, 113
Cavanaugh, Colleen, 662
CBB cycle. *See* Calvin cycle
CC10 (lung development protein), 653
CCD (Charge-coupled device camera), 60f
CcdB (DNA damaging toxin), 484
CD3 complex, 985
CD4 cell surface protein, 982, 983, 989–91
CD8 cell surface protein, 982, 983, 988
CD14 protein, 954
CD47 protein, 951
CD59 cell surface protein, 993
CDC. *See* Centers for Disease Control and Prevention (CDC)
CdiGMP (cyclic dimeric guanosine monophosphate), 150
Cech, Thomas, 19, 679
Cefepime, 1119f
Cefotaxime, 1084
Cefoxitin, 1119f
Ceftriaxone, 1080, 1081, 1119f
Cell biology, 32–34, 62, A-23–A-41
Cell-cell adhesion protein, 5f
Cell density
in continuous culture, 144–45
and quorum sensing, 323–24, 365–66, 397–98
Cell differentiation, 150–54
C. crescentus, 107–8, 150
endospores, 150–52
eukaryotic microbes, 150
filamentous structures, 152–53, 154f
fruiting bodies, 152
nitrogen-fixing heterocysts, 152
Cell division, 104–7
asymmetrical cell division, 107–10, 111f, 139, 150, 383–84
Pyrodictium, 768–69
symmetrical cell division, 139
xerD mutation, 257
Cell envelope, 94–97, 104–6, 114, 760
Cell fission, 16, 104–10, 111f
Cell fractionation, **82–83**, 99, **A-44–A-45**
Cell-free systems, 83, 276, 283–84, A-45
Cell fusion, 442
Cell-mediated immunity, **963**, 964–65, 987f. *See also* Adaptive immunity; T cells
Cell membrane, 33, 80–81f, **82**, **A-24**, **A-26**. *See also* Membrane lipids
cytochromes, 545–46
Gram-positive *vs.* Gram-negative bacteria, 94f
hyperthermophilic Archaea, 164
isoprenoid membranes in archaea, 758–59, 760f
lipid composition, A-24, A-26
membrane fluidity, 160, 165, A-26–A28
movement of membrane lipids and proteins, A-26–A-28
prokaryotic protein synthesis and, 104
proteins, 86–87
selective permeability of cell membranes, A-28f
thermophiles compared to mesophiles, 164
transport across, 87–89, A-28–A-29
transport of water across, A-29–A-30
Cell shape, 92, 100–101, eTopic 3.4
Cell structure, 33
cell fractionation, 82–83
model of bacterial cell, 80–81f, 82
overview, 79–82
specialized structures, 112–16
Cell-surface receptors, **208**
Cell surface sugar chains, fluorescent labeling of, 5f
Cellular slime molds, **826**
Cellulitis, **1057**
Cellulose, 99, 510, 511f, 515
Cell walls, **82**. *See also* Sacculus
in alkaliphiles, 170
archaea, 760
bacterial cell model, 80–81f
biosynthesis, 87, 105, 1118
eukaryotes, 99
Gram-positive *vs.* Gram-negative bacteria, 94f
mycobacteria, 96
peptide cross-bridges, 92–93
peptidoglycan, 79, 85, 92–93, 1116f, 1117–18
during replication, 104–5
during septation, 105–6
structure in bacteria, 91–93
Cenarchaeales, A-66t
Cenarchaeum symbiosum, 774
Centers for Disease Control and Prevention (CDC), 14, 418, 433, 1171, 1173, 1177, 1180
Central nervous system infections, 1083–90. *See also specific diseases*
Centrioles, A-32
Centrosomes, A-32
Century egg, 636
Cephalexin (Keflex), 1119f
Cephalosporins, 1117, 1119–20, 1127
Ceratium, 830
Ceratium furca, A-70f
Ceratium tripos, A-70f
Cercozoa, 804, A-70t
Cerebrospinal fluid (CSF), 1084–85, 1151
Cetylpyridinium chloride, 184f
CFP (cyan fluorescent protein), 104, 105f
CFU (colony-forming unit), 134
Chagas' disease, 805, 835
Chain, Ernst, 24, 1109
Chain of infection, **20**
Chalfie, Martin, 61
Chan, Russel, 355
Chancres, **1078**
Chang, Annie, 265
Chaos carolinense, A-69f
Chaos chaos, A-69t
Chaperones, 80–81f, 82, **303–4**, A-37
DnaJ (HSP40), 304, 306f
DnaK (HSP70), 304, 306f
DnaK-DnaJ-GrpE chaperone system, 382, 383f
GroEL (HSP60), 275, 304, 306f
GroES, 304
and heat-shock response, 164
periplasmic chaperones, 307, 308, 309
protein folding, 303–4
refolding damaged proteins, 305–6, 382–83
trigger factor, 304, 306f, 307, 308
Charge-coupled device camera (CCD), 60f
Charophyta (stoneworts), A-69t
Charophyte, A-69f
Chase, Martha, 193, 208
Cheddared, **631**, 632f
Cheese, 627, 628t, 629
blue cheese, 631
Brie, 631
brined cheese, 631–32

Cheese (*continued*)
Camembert, 631, 632
cheddar cheese, 630
cottage cheese, 630, 631f
curd formation, 630
Emmentaler cheese, 630f, 631f
feta, 631
flavor generation in cheese, 632, 633f
Gouda, 631–32
lactic acid fermentation, 630–32, 727
mold-ripened cheese, 629, 631
Muenster, 630
Parmesan, 630
production flowchart, 631f
ricotta, 630
ripening, 630–31
Romano, 630
Roquefort cheese, 11f, 629, 630, 631
spoilage, 642
starter culture, 631, 632f
Swiss cheese, 525, 629, 630, 632, eTopic 13.2
varieties, 630–31
Chemical bonding, A-2–A-5
Chemical elements
electron configuration, A-2
elemental composition of Gram-negative cells, 891t
global reservoirs of carbon, nitrogen, and sulfur, 892t
macronutrients, 120
microbial cycling of essential elements, 605, 891
micronutrients, 120
periodic table, A-2, A-3f
reservoirs, 891
sources and sinks of essential elements, 891, 893
Chemical energy, 124–25
Chemical equilibrium, A-18
Chemical imaging microscopy, 41, **46**, 72, 74–75
Chemiosmotic theory, 34, 547, 548–49
Chemoautotrophs, chemoautotrophy, **123**, **124**, 126, 493f
Chemoheterotrophs, chemoheterotrophy, 123f, **124**
Chemokine receptors (CCRs), 437f, **438**
Chemokines, **948–49**, 982
Chemolithoautotrophs, **542**, 589f, 590
Chemolithotrophs, **26**, **123**, 124. *See also* Lithotrophs
Chemolithotrophy, **492**, 493f, **542**, **563–69**. *See also* Lithotrophy (chemolithotrophy)
Chemoorganotrophy, **492**, 493f, **542**
Chemostats, **144**, 145
Chemotactic receptors, 115, 116f
Chemotaxis, **115–16**, **391–94**
CheA kinase, 392–94
chemoreceptor array in *E. coli,* 393f
CheY protein, 391–94
and flagellar rotation, 115–16, 391–92
methyl-accepting chemotaxis proteins (MCPs), 392–94
signaling pathway in *E. coli,* 391–93
suppression of tumble frequency, 391–93
Chemotrophy, **124**
Chert, 668, 671, 674f
Chickenpox, 206, 443, 444, 1055t. *See also* Varicella$zoster virus (VZV)
ChIP-on-chip analysis, **471–72**, 922
Chiral carbon, **A-8**
Chirality, 16–17
Chisholm, Sallie, 861–62
Chitin in fungal cell walls, 99, 806
Chiu, Wah, 70
Chlamydia, 185, 704, **1079**
Chlamydia, 752, 753f, 1038, 1041, 1075, 1079, A-65t
Chlamydiae, 713f, 715t, **717**, 752, A-65t
Chlamydiales, A-65t
Chlamydia trachomatis, 752, 753f, 1079, A-65f
Chlamydomonas, A-39
Chlamydomonas reinhardtii, 818–20
alternation of generations, 819f
cell cycle regulation, 818
cell structure, 819
flagella, 818
Chlamydophila, 752, A-65t
Chlamydophila pneumoniae, 752, 1079, 1093, eTopic 26.8
Chloracidobacterium thermophilum, A-65t
Chloramine, 184t
Chloramphenicol, 299, 300f, 1111, 1124f, 1125
Chlorella, 591, 704, 705, 818
Chloride transport, 788
Chlorinated pollutants
bioremediation, 728, 902
dehalorespiration, 569, 902
microbial catabolism of, 492, 511f, 513
Chlorination, 900
Chlorine, as disinfectant, 184
Chlorobenzene, 569
2-chlorobenzoate catabolism, 533
Chlorobi, **717**, 719t, 749–50, A-65t
Chlorobia, 577–78
Chlorobiales, A-65t
Chlorobium
electron microscopy, 33f
in eutrophic lakes, 868
photosystem I, 576–78, 749
Chlorobium tepidum, 598, 750f
Chloroflexales, A-62t
Chloroflexi, 577, 714, 715t
Chloroflexi, A-62t
Chloroflexus, 714, 716f
Chloroflexus aggregans, 123f
Chloroflexus aurantiacus, 600f, 601, A-62f
Chlorofluorocarbons, 918
Chlorophylls, 573, **574–75**
absorbance spectra, 575f, 868f
antenna complexes/systems, 574, 575
bacteriochlorophylls, 574, 575f
bacteriorhodopsin compared to, 676–77
biosynthesis of, 620f
chlorophyll *a,* 574, 575f, 716
chlorophyll *b,* 574, 575f, 716
chromophores, 574
reaction centers, 575, 576–77
structure, 574f, 619
Chlorophyta. *See* Green algae (Chlorophyta)
Chloroplasts, **802**, **A-40–A-41**. *See also* Oxygenic Z pathway; Photosynthesis
in algae, 802–3, 818f, 819, 820
Calvin cycle, 590
chemiosmotic theory demonstration, 548, 549f
endosymbiosis theory of origin, 30–31, 127, 705–7, A-41
genome, 706, 707f
photosynthetic membranes, 575, 577f
serial endosymbiosis theory, 30–31
stroma, 577f, A-41
structure, A-41f
thylakoids, 577f, A-41
Chloroquine, 1093
Chlorosomes, 33, 578, 598f, **714**, 749, 750
Choanocytes, 802, 803f
Choanoflagellates (Choanoflagellata), 802, 803f, A-68f, A-68t
Chocolate, 627, 628t, 634–36
Chocolate agar, 1080f, 1153, 1154f
Cholera, 5, 20. *See also Vibrio cholerae*
cholera toxin (CT), 86, 200, 210, 365, 1019, 1022, 1023–24
diagnosis, 1053
fowl cholera, 23
locations of recent increases, 1181f
London outbreak in 1854, 1173, 1174f
pathogenesis of, 1023f
resistance to, 931
water decontamination and, 856
Cholesterol, **90**, 91, A-26, A-27f, A-28
Choline-binding domain (CBD), 1134
Chorismate, 616
Christner, Brent, 163
Chromatiales, A-64t
Chromatin, **A-31**
Chromatin immunoprecipitation (ChIP), **471**
Chromatium, 743, 750
Chromium, 915t
Chromium-6, 914
Chromophores, **574**
Chromosomes. *See also* DNA replication
eukaryotic chromosomes, 260–62
histones, 261
in microbial nucleoid, 80–81f, 82, 101–2
organization in nucleoid, 101–2, 245–47
origin (*oriC*), 249f, 250, 252, 252f, 253f
shape and size in prokaryotes, 239–41
during sporulation, 151
telomeres, 261
tracking replication of, 237
Chronic granulomatous disease, 737
Chronic inflammation, **949**, 950f
Chroococcales, A-62t
Chroococcus, 722, A-62t
Chrysophyceae, A-70t
Chrysophyceae (golden algae), 823
Chrysophytes, **803**
Chu, Steven, 5
Chymosin, 630
Chymotrypsin, 632
Chytridiomycetes, 884
Chytridiomycota, 809–11, 812f, A-68t

Chytrids (Chytridiomycota), 799f, 802, 809–11, 812f
Cilia (singular, cilium), 7f, **804**, 827–28, 930, 1058, **A-39**, eTopic 26.3
Ciliates, **804**, 827–30
  cell structure, 828
  genetics and reproduction, 828–29
  phylogeny, 799t, 800, 801f, 804
  stalked ciliates, 829–30, 900f, 901
Ciliophora, A-70t. *See also* Ciliates
Cinnamon, 650
Ciprofloxacin, 248, 1084, 1111, 1171
Circoviridae, A-60t
Circular display map of genome, 312f
Circular polarization, 919
cis-antisense RNA (asRNA), **385**, 387–88
cis-repressed mRNA (crRNA), 482–83, 484
Citrate, 527, 528, 529f
Citric acid, 635
Citromicrobium, 737
Clades, **681**, 682
Clarassance, 653
Class I MHC molecules, 954, **983**, 984–85, 988
Class II MHC molecules, **983**, 984–85, 986, 987, 989, 1042
Classical complement pathway, 955–56, **992**, 993f
Classification, **697**
  challenges of classification, 29
  classification systems, 30
  environmental, 159t
  taxa (singular, taxon), 697, 700
  taxomic hierarchy, 700t
  of viruses, 203–6
Claudin-1, 420, 421f
Clavulanic acid, 1132
Clemmensen, Karina, 894
Climate change. *See also* Global warming
  human causes of, 608
  methanogenesis, 680
  and pathogen epidemics, 228, eTopic 6.2
Clindamycin, 928, 1057, 1066, 1124f, 1125, 1171
Clinical microbiology, 1147–83. *See also* Culturing bacteria; Pathogen identification
  biosafety containment procedures, 1169–71
  overview, 1147–49
  specimen collection, 1149–51
Cloaked mutagens, 347
Clonal, definition, **963**
Clonal selection of B cells, **975–76**
Cloning, 264–66, **266**, 652–54, 656
Cloning vectors, **194**
Clostridiales, 724, 725–26, 727f, 845f, 846, A-63t
Clostridium, 523, 723, 725–26, 905, 932, A-63t
Clostridium acetobutylicum, 388, 525, 725
Clostridium botulinum. See also Botulism
  bioterrorism potential, 175, 1087
  botulinum toxin, 646, 648, 725, 1026, 1085, 1086–87
  and canning of food, 180, 649, 1085
  heat-resistant spores, 150
  multiplex PCR identification, 1160–61
  prophage-encoded toxin, 210
  spores in honey, 1086
Clostridium difficile, 93, 151f, 188, 618, 725–26, 923, 933, 1066–67
Clostridium pasteurianum, 609
Clostridium tetani, 59f, 150, 725, 1085f, 1087, 1102. *See also* Tetanus
Clostridium thermoaceticum, 599
Cloves, 650
Clp proteases, 301, 302f, 305, 384
ClustalW2 (computer program), 314f
CmpABCD (high-affinity HCO_3^- transporter), 596–97
CNA agar (colistin and naladixic acid), 1157, 1158f
Coagulase, 1158f, 1159
Coastal shelf, **857**
Cobalt, 120
Cocci (singular, coccus), **44**, 45f, 105–6
Coccidioides, 1063
Coccolithophores (coccoliths, Prymnesiophyceae), 229f, 797, 823–24, 914, 916
Coccolithus pelagicus, 797
Cockayne syndrome, 354
Cocoa fermentation, 634–36
Cocoa mass, 635–36
Codons, **286**. *See also* Genetic code
  anticodons, 288
  codon–anticodon pairing, 288, 295
  codon synonyms, 286–87, 289
  start codons, 287, 292–93f, 293, 313
  stop codons, 287, 296, 297f, 301, 302f, 313
Coenzyme A (CoA), **523**, 525f, 527–28, 531, 602
Coenzyme B_{12}. *See* Vitamin B_{12}
Coenzyme F_{420}, 570, 782
Coenzyme M–SH, 570
Coevolution, **704**
Cofactors, **120**, 551–53, 570, 599
Cohen, Stanley, 264–65
Cohen-Bazire, Germaine, 573
Cohesive ends of DNA, 263, 264, 265f
Cointegrate molecules, **338**, 356, 357f
Colanic acid, 149
Cold seeps, **865**, eTopic 21.2
Cold sores, 1006
*coli*Base, 316
Collagenase, 1045, 1046f
Collins, James, 483–84, 1126
Colon, 950f
Colonies, **134**
Colonoscopy, 523
Colony collapse disorder (CCD), 268
Colony-forming unit (CFU), 134
Colwell, Rita, 20, 28, 856, 858, 860, eTopic 1.1
Colwellia, 858–59
Combinatorial library of mutant genes, 699
ComD sensory protein, 323, 324f
ComE regulatory protein, 323, 324f
Commensalism, **856**
Commensal organisms, **926**, 934–35, 994–95. *See also* Human microbiota (microbiome)
Common cold, 227, 930. *See also* Rhinoviruses
Communities, **840**. *See also* Microbial communities
Community-acquired infection, 1056
Compatible solutes, 167
Competence factor (CF), **323**, 324f
Competent cells, **323**
Complement, **936**, 955–58, 992–94. *See also* Innate immunity
  activation pathways, overview, 955–57
  and acute-phase reactants, 957
  alternative complement pathway, 956–57, 992
  anaphylatoxins, 957
  and B cell activation, 986
  classical complement pathway, 955–56, 992, 993f
  complement factors and cleavage fragments, 956, 957, 992, 993, 1037
  and C-reactive protein, 957
  decay-accelerating factor, 993
  factor H control of activation, 993
  and heart disease, 957
  lectin activation pathway, 956, 992
  membrane attack complex (MAC), 956, 992, 993
  and *N. gonorrhoeae,* 936, 993
  opsonin, 957, 993
  regulation of activation, 993
Complementary DNA (cDNA), 466–67
Complex media, **134–35**, 142, 143
Complex transposons, **356–57**
Composite transposons, **356**, 357f
Composts and composting, 162, 492, 494, 497
Compound microscopes, 52–54
Compromised host, **934**
Computational filtration, 843
Concentration gradients, 127–28, 131, 500–501
Condensation, **A-6**
Condensers (microscopes), **53**, 54
Conditional lethal mutations, **416**
Confluent growth, **134**
Confocal laser scanning microscopy, **62**, 63f, eTopic 2.1
Congenital rubella syndrome, 1058
Congenital syphilis, **1078**
Conidia, 1005f
Conidiophores, 814
Conjugated double bonds, 552
Conjugation, **239**, **324**, **829**
  from bacteria to eukaryotes, 329
  between ciliates, 829
  conjugation bridge, 329, 819, 820f, 829
  conjugation used for genetic engineering, 329
  conjugative transposons, 321
  F^- cell, 325, 326f
  F^+ cell, 325, 326f
  fertility factor (F factor), 325–28

Conjugation (*continued*)
F-prime (F′) factor, 328
Hfr (high-frequency recombination) strains, 327
mapping gene position by conjugation, 327, eTopic 9.2
plasmid transmission, 260, 325–29
relaxase, 325, 326f
relaxosome complex, 325, 326f
rolling circle replication, 325, 326f
sex pili, 113, 324–25, 326f
in *Spirogyra,* 819, 820f
Streptococcus agalactiae evolution through, eTopic 25.3
and type IV secretion, 1033
Conjugative transposons, 321, **358**
Consensus sequences, **278**
Constant regions, **970–71**
Constructive interference, 49, 50f
Consumers, 495, **850**, 851
Contact dermatitis, 998–99
Contigs, **271**, **845**
Continuous culture, **144–45**
Contractile vacuoles, **99**, 100f, 819, **828**
Contrast, **47**, 51
Control of microbes, 178–88. *See also* Antibiotics
antibiotics, 186–87
antimicrobials, overview, 178–79
antisepsis, definition, 179
bactericidal agents, definition, 179
bacteriostatic agents, definition, 179
biological control, 187–88
chemical agents, 183–86
death rates, 179–80
decimal reduction time (D-value), 179
disinfectants, 178–79, 184–86
disinfection, definition, 178–79
germicidal agents, definition, 179
physical agents that kill microbes, 180–83
sanitation, definition, 179
sterilization, definition, 178
Convergent evolution, 798
Conway, Tyrrell, 520
Copeland, Herbert, 30, 800
Copepods, 856
Copiotrophs, 848
Copper, 120, 185
Copper mining, 568f, 743f
Coprotease activity, 352–54
Coral bleaching, **882–83**
Coralline algae, **821**
Corallochytrium, A-69t
Corals
brain coral (*Diploria* sp.), 882f
coral bleaching, 882–83
coralline algae as foundation for coral reefs, 821
zooxanthella endosymbionts, 831, 882–83
Coreceptors, **438**
Core envelope of poxviruses, 202
Core gene pool, 359
Core genes, 845–46
Core genome, **697**
Core particle, **434–35**
Core polysaccharide, 97f, **98**
Corepressors, **367**
Coronary artery disease, 1093, eTopic 26.8
Coronaviridae, A-61t
Coronaviruses, 205t, 206, 225f, 227, 417t, 1173f
Corrin ring structure, 76f
Cortical alveoli, **804**
Corynebacteria, 717
Corynebacteriaceae, A-63t
Corynebacterium, 730, 732, 733
Corynebacterium diphtheriae, 107, 210, 733, 1004, 1011, 1025
Corynebacterium jeikeium strain K411, 405
Corynebacterium striatum, 1004
Coulter counters, 137
Counterstains, **57**
Counting bacteria, 136–39
biochemical assays, 137–38
direct microscopic counts of cells, 136–37
optical density measurements of cell growth, 138–39
viable counts, 137
Coupled transport, 89, **128–30**. *See also* Transport across membranes
Covalent bonds, A-2, **A-2**, A-3f
Coverage, 845f, **846**
Cowman, Alan, 1092–93
Cowpox (vaccinia virus), 23, 228, 966
Coxiella, 1038f–1039f
Coxiella burnetii, 121, 181, 746, 1038f, 1040, 1053–54
Crabb, Brendan, 1092–93
C-reactive protein, **957**
Crenarchaeol, 773
Crenarchaeota, **763**, 766–73, A-66t
effect of depth and season, 772
hyperthermophiles, 763, 765, 766–71
hyperthermophilic species, 767t
mesophiles, 771–72
methanotrophs, 773
phylogeny, 763, 764t, 765
psychrophiles, 772–73
psychrophilic marine crenarchaeotes, A-66t
symbiosis, 771–72
thermophiles, 760, 763, 765, 766
Creosote, 513
CreS (crescentin), 100
Cresols, 183, 184t
Creutzfeldt-Jakob disease (CJD), 197, 1090, 1180
Crichton, Michael, 66
Crick, Francis, 34–35
Crinipellis perniciosa, 644
CRISPR (clustered regularly interspaced short palindromic repeats), **212**, **335–37**
anatomy, 335, 336f
biofilms and, 337
Cas-crRNA complex, 212, 336–37
cas genes, 335, 336f, 337
function, 212, 335–37
RAMP (repeat-associated mysterious proteins) module, 335–37
Cristae of mitochondrial inner membrane, 545, 805
Crohn's disease, 949, 950f
Cro repressor protein, 244
Cross-bridges, **92**, 93f
Crown gall disease, 329, 741f, 881f, 1033. *See also Agrobacterium tumefaciens*
CRP (cAMP receptor protein). *See* cAMP receptor protein (CRP)
Cryocrystallography, 76
Cryo-electron microscopy (cryo-EM) (electron cryomicroscopy), **68**, **69–70**, 191, 199
Cryo-electron tomography (electron cryotomography), **70–71**
Cryoprotectants, 162
Cryptobiotic soil, **854**
Cryptococcosis, 1063
Cryptococcus, 1063
Cryptococcus neoformans, 809, A-68t
Cryptogamic crust, **822**, **854**
Cryptophyte algae, 803f
Cryptosporidiosis, 1072
Cryptosporidium parvum, 1072, A-70t
Crystal violet, 55f, 57f, 58f, 135
Culex mosquito, 1007
Culture media, 121, 122t, 133
agar, 133–34
complex *vs.* synthetic, 134–35
selective and differential, 135–36
Culturing bacteria, 133–36
anaerobes, 175–76
culture media, 121, 122t, 133, 134–36
dilution streaking, 133–34
spread plate technique, 134
Culturing viruses, 221–25
batch culture, 221–22
burst size, 221f, 222
challenges, 192, 221
eclipse period, 221, 222
latent period, 221–22
lysates, 222
multiplicity of infection (MOI), 221
one-step growth curve for animal virus, 222f
one-step growth curve for bacteriophage, 221f
plaque assay of animal viruses, 224–25
plaque assay of bacteriophages, 223–24
rise period, 221f, 222
in tissue culture, 222–23
in whole animals, 222
Curd, **630**, 632f
Curiosity rover, 6f
Curvularia protuberata, 226–27
Curvularia thermal tolerance virus (CThTV), 226–27
Cyan fluorescent protein (CFP), 104, 105f
Cyanobacteria, **716**, 718–23
biofilms, 150
carbon-concentrating mechanism, 596f
carboxysomes, 720
cell structure, 720
colonial cyanobacteria, 720–22
cyanobacterial communities, 722–23
diversity, 714

Entophysalis, 672f
filamentous cyanobacteria, 7f, 720–22
gas vesicles, 720
growth and cell division, 140
heterocyst formation, 608, 611, 720, 721f, 852
marine food chain, 124
microbial mats, 722–23
microfossils, 18, 671, 672f
motility, 116
nitrogen fixation, 75, 152, 608, 611, 720, 905
as origin of chloroplasts, 705, A-41
oxygenic phototrophy, 718, 719t
photosynthesis, 26–27, 573, 577
phylogeny, 713f, 715t, 716, A-62t
single-celled cyanobacteria, 720, 722f
Spirulina, 626, 627f
stromatolites, 666f, 669
thermophiles, 766
thylakoids, 112, 720
toxic blooms, 722
Cyclic adenosine monophosphate (cyclic AMP or cAMP). *See also* cAMP receptor protein (CRP)
during hypersensitivity reaction, 996
and lacZYA operon, 371, 373f
microbial exotoxins and elevation of, 1019, 1023, 1024, 1026
during starvation, 176, 826
Cyclic dimeric guanosine monophosphate (cdiGMP), 150, 396–97
Cyclic guanosine monophosphate (cGMP), 1019
Cyclic photophosphorylation, 540, 541f, **579**
Cycloclasticus, 858, 859f
Cyclodextrin glucanotransferase, 171, eTopic 5.2
Cyclooxygenase (COX), 948
Cyclopentane rings, 90–91, 759, 760f, 770
Cyclophilin A, 434, 435f
Cyclopropane fatty acid, 90
Cycloserine, 109, 1116t, **1120**
Cyclosporia, 831
Cyclosporine, 606
Cymopolia, 821
Cymopolia barbata, A-69f
Cysteine disulfide bonds, 308
Cystic fibrosis, 398, 488–89, 743, 745, 944–45, 1018
Cystic fibrosis transmembrane conductance regulator (CFTR), **944–45**, 1023–24
Cystoviridae, A-60t
Cytidine monophosphate (CMP), 618
Cytochrome oxidase test, 1157
Cytochromes, 172–73, **545–46**
cofactors, 551–53
cytochrome *b,* 554–55
cytochrome ba_3, 565f
cytochrome *bc,* 579f
cytochrome bc_1, 565f
cytochrome *bd* oxidase, 555
cytochrome *bf,* 580, 581f
cytochrome *bo* quinol oxidase, 554f, 555, 556
cytochrome *c,* 552f, 564, 578f, 579, 1157
cytochrome complexes, 554–55, 557
cytochrome oxidase test, 1157
light absorbance spectrum, 545f, 546
and metal reduction, 561, 562, 564
proton pumping, 125, 556, 557
terminal oxidases, 554–55, 556
Cytokines, **935**, **980**. *See also* Interferons
anti-inflammatory cytokines, 991
cytokine cascades, 1009
cytokine receptor, 962
definition, 935
downregulation by *Bifidobacterium,* 885
endotoxin and, 1027
initiation of extravasation, 947–49
interleukin 1 (IL-1), 948, 957
interleukin 2 (IL-2), 988
interleukin 4 (IL-4), 986, 987, 990
interleukin 6 (IL-6), 987–88, 991
interleukin 10 (IL-10), 990, 991
lymphokines, 990
and mucosal immunity, 982
pathogen alteration of production, 1041
pyrogens and release of, 958
select cytokines that modulate the immune response, 990t
synthesis by cytotoxic T cells, 964–65
Toll-like receptors and release of, 954
tumor necrosis factor (TNF), 989
tumor necrosis factor alpha (TNF-α), 948, 956
Cytomegalovirus (CMV), 448–49
birth defects, 448–49
effect on neural progenitor cells (NPCs), 448–49
evasion of adaptive immunity, 991
phylogeny, 206f, 207
prevalence of latent infection in humans, 446, 448
promoter use in viral vectors, 451–52
Cytoplasm
Escherichia coli, 80–81f
fluorescent labeling of, 5f
Gram-negative cells, 99
nanotubes, 114
Cytoplasm (cytosol), 33, **A-24**
Cytoplasmic bridges, 819, 820f, 821, 829
Cytoplasmic membrane, **A-24**. *See also* Cell membrane
Cytoproct, 828, 829
Cytosine (C), 239, 243, 618
Cytoskeleton, 92, 100–101, 104–5, 234, **A-38-A-40**, eTopic 3.4
Cytotoxic T cells (T_C cells), 940, 964–65, **982**, 983, 988

D

Dalgarno, Lynn, 293
Dalsgaard, Tage, 908
Daly, Michael, 183
DAPI (4′,6-diamidino-2-phenylindole), 59f, 61, 89f, 686, 862, 863f
Daptomycin, 606
Dark-field condenser system, 64f
Dark-field microscopy, **63–64**
Darwin, Erasmus, 666
DasSarma, Priya, 786–87
DasSarma, Shiladitya, 167f, 758, 785, 786
Data storage in bacteria, 457
Dawadawa, 628t, 636
Dead zones, **898–99**, 906, 907
Deaminases, 172
Deamination, 512
Death curve, 179f
Death phase, **143**
Death rate, **143**, 179–80
cis-2-decenoic acid, 150
Decarboxylation, 512
Decay-accelerating factor, **993**
Dechlorinating bacteria, 902
Decimal reduction time (D-value), 179
Decomposers, 495, **850–51**, 872–73
Deep Aquarium, 768, 769f
Deep-branching Gram-negative bacteria, 715t, 717, 748–50, A-65t
Deep-branching thermophiles, 713f, 714, 715t, 716f, A-62t
Deep-branching uncultured Archaea, A-68t
Deepwater Horizon oil spill, 6, 532, 858–59, 865, 898, 899f
Defensins, **945–46**, eTopic 23.2
Deferribacterales, A-62t
Deferribacteres, A-62t
Defined minimal medium, **121**, 122t, 142
Degenerative evolution, **681**. *See also* Reductive evolution
Degranulate, **946**
Degrons, **304**
Dehalobacter, 902, A-63t
Dehalobacter restrictus, 728, 729f
Dehalococcoides, 902
Dehalococcoides strain CBDB1, 569f
Dehalorespiration, **569**, 902
Dehydratase, 603, 604
Dehydrogenases, 553–54, 560
Deinococcales, A-62t
Deinococcus, A-62f
Deinococcus radiodurans, 158, 183, 234, 340, 716, eTopic 9.3
Deinococcus-Thermus, 716, A-62t
Delaviridine, 1141
Deletions, **341**, 342, 801
DeLong, Edward, 772, 861–62
Deltaproteobacteria, 713f, 715t, 717, 736t, 746–47, A-64t
DeMontigny, Bree, 780, 781f
Denaturation of DNA, **244**, 247, 248
Dendritic cells, **939**, 940, 941f, 963, 964f, 998
Dengue hemorrhagic fever, 191–92, 1009
Dengue virus, 191–92, 417, 1009
Denitrification, **126**, 560, 608, 905f, **907**
Denitrifiers, 126, 893, 901, 907
Dental caries, 929
Deoxyadenosine methylase (Dam), 250–51, 349
Deoxyribonucleic acid. *See* DNA (deoxyribonucleic acid)
Depolarization, 130
Depth of field, **54**
Derepression, **367**, 378
Desensitization, **997**

Desmids, 798
Destructive interference, 49, 50f
Desulfobacter, 913
Desulfobacterales, A-64t
Desulfobacterium autotrophicum, 599
Desulfovibrio, 502–3, 913
Desulfovibrio desulfuricans, 913f
Desulfovibrio vulgaris, 503
Desulfuration, 910
Desulfurococcales, **766–67**, A-66t
Desulfurococcus, 781
Desulfurococcus fermentans, 684, 766, 767f
Desulfuromonadales, A-64t
Desulfuromonas, A-64t
Detection, **43**
Detergents as disinfectants, 184–85
Detritus, **851**
Deubiquitylases, 1043–44
"The Devil in the Dark" (Star Trek episode), 566, 567f
d'Herrelle, Félix, 188, 193, 223
Diabetes, 554
Diagnosing microbial diseases, 1053–54
4′,6-diamidino-2-phenylindole (DAPI), 59f, 61, 89f, 686, 862, 863f
M-Diaminopimelic acid, 92, 93f, 97
Diapedesis. *See* Extravasation
Diaphragms (microscopes), **53**, 54
Diarrhea, 135–36, 188, 1024, 1025, 1030, 1066, 1072–73. *See also* Gastrointestinal tract infections
Diatoms (Bacillariophyceae), **805**, A-70f
- cell walls, 99
- definition, 805, 823
- diatomaceous earth, 823
- life cycle, 822f, 823
- microbial mats, 723
- phylogeny, 799t, 800, 801f

Diauxic growth, **374**
Dibenzofuran, 686
DIC (differential interference contrast microscopy), **65**
Dichanthelium lanuginosum, 226–27
Dichotomous keys, **702**, 703f, 1154–55
Dichroic mirror, 60f
Dictyostelium discoideum, 150, 826, A-69t
Dideoxynucleotides, 267
Didinium, 828, A-70t
Differential interference contrast microscopy (DIC), **65**
Differential media, **135–36**
Differential RNA-sequencing (dRNAseq), **467**, 468f
Differential stains, **56–58**
Diffusion, 500–501, **A-28–A-29**, A-34
Digestive proteins, 127
Digestive vacuoles, 828
Digital dermatitis, 752
Digital gene expression (DGE), 1046–47
Diguanylate cyclases (DCGs), 396–97
Dihydrofolate reductase, 415
Dihydrouridine, 289
Dihydroxyacetone phosphate (DHAP), 518, 519f
Dilution rate, 144–45
Dilution streaking, **133–34**
Dimethyl sulfide, 643, 909
2,4-dinitrophenol (DNP), 550–51
Dinoflagellata, A-70t
Dinoflagellates (Dinophyceae), **803–4**, 830–31
- cell structure, 830f, 831
- endosymbiosis in other organisms, 882
- flagella, 830
- hornlike extensions, 830
- phylogeny, 799t, 800, 801f, 803–4
- red tide, 830f, 831
- zooxanthellae, 831

Dioxin, 511f, 531
Dioxygenases, 533
Diphosphatidylglycerol, **89**
Diphtheria, 1004. *See also Corynebacterium diphtheriae*
Diphtheria toxin, 1019, 1022, eTopic 25.6
Diphtheroids, 928, 929
Diphytanylglycerol diether, 759, 760f
Dipicolinic acid, 151
Diplococci, **742–43**
Diplomonadida, A-71t
Diploria, 882f
Directed evolution, 477, **696**, 698–99, eTopic 12.5
Disaccharides, 510, A-12–A-13
Discosphaera tubifera, 823
Disinfectant resistance, 185–86
Disinfectants, 178–79, 184–86
Disinfection, **178–79**
Dissimilation, **849–50**
Dissimilatory denitrification, **560**
Dissimilatory metal reduction, **561–62**
Dissimilatory nitrate reduction, **907**
Dissimilatory nitrate reduction to ammonia (DNRA), 907–8
Disulfide bond catalyst (DsbA), 308
Disulfide reductases, 308
DivJ protein, 110, 111f
DNA (deoxyribonucleic acid). *See also* DNA replication
- agarose gel separation of fragments, 263–64
- antiparallel arrangement, 243
- background and overview, 238–39
- in bacterial cell, 80–81f, 82
- in bacterial transformation, 34
- cohesive ends, 263, 264, 265f
- complementary DNA (cDNA), 466–67
- complementary pairing of bases, 34–35, 243
- denaturing of hydrogen bonds, 244, 247, 248
- direct repeats, 389
- domains, 102
- double helix, 34–35, 102, 104, 243f, A-13, A-14f
- as genetic material, 34
- genome-wide repeats, 262f
- hydrogen bonding, 243
- introns, 261, 262f
- inverted repeats, 374–75, 389
- isolation and purification, 263
- major and minor groove, 244
- models, 244f
- noncoding DNA, 241, 261–62
- nontemplate (sense) strand, 315
- nucleotides, 239, 243
- palindromes, 263
- phosphodiester backbone, 243
- protein-binding sites, 374–75
- proteins interacting with, 244
- purines, 243
- pyrimidines, 243
- radiation damage, 182, 183, 210, 343, 345, 350
- separation of fragments, A-45–A-46
- single-strand nicks in, 245
- structure, 242f, 243–44, A-13, A-14f
- supercoils, 102, 245–48, 256
- tandem repeats, 389
- template strand, 277, 315

DNA-based detection tests, 1160–62, eTopic 28.2
DNA-binding proteins
- Dps, 102
- histone-like proteins, 244, 245
- H-NS, 102, 245
- HU, 102, 245
- in nucleoid, 80, 81f, 102, 103f
- protection from denaturation, 244
- repressor proteins, 367
- single-stranded DNA-binding proteins (SSBs), 254f, 255
- in thermophiles, 164

DNA cassettes, 341, 389, 390f
DNA control sequences, **239**
DNA fingerprinting, 644, 645f
DNA gyrase, 247, 248, 252f, 256, 1122, 1131
DNA helicase (DnaB), 252, 251f, 253, 254f, 256
DNA helicase (UvrD), 349
DNA helicase loader (DnaC), 252, 251f
DNA hybridization. *See also* Hybridization
- fluorescent labeling, 61
- and working definition of microbial species, 696–97

DNA ligase, **255–56**, 259f, 264, 265f, 349, 351
DNA methylases, 250–51, 334, 349
DNA methylation
- initiation of replication, 250–51
- methyl mismatch repair, 349
- protection against endonucleases, 263, 333f, 334

DNA microarrays, 404, **A-53–A-54**
DNA polymerases
- in bacterial cells, 102, 104, 237
- DinB (Pol IV), 352
- direction of replication, 250, 253
- discovery of enzymes at Yellowstone, 35, 162
- DNA polymerase I (Pol I), 252f, 253, 255–56, 259f, 349, 350, 351, 352
- DNA polymerase III (Pol III), 252, 251f, 253–55, 259f, 325, 326f, 349, 352, 353
- *Escherichia coli,* 80–81f
- housekeeping DNA polymerases, 353
- proofreading, 253

RNA primers, 252, 251f, 253– 256
Taq polymerase and PCR reaction, 35, 162, 266, 470, 790
UmuDC (Pol V), 352–53
vent polymerases, 790
virus replication, 203–4, 216
DNA primase (DnaG), 252, 251f, 253–55
DNA protection analysis, 467–68, 469f
DnaQ (epsilon subunit of Pol III), 253
DNA rearrangements, 369, 389–90
DNA regulatory sequences, 374–75
DNA repair, 348–54
AP endonucleases, 351
AP sites, 351
base excision repair (BER), 350
basic types of DNA repair, 348t
error-prone repair pathways, 349, 352–54
error-proof repair pathways, 348–52
errors in, 345
human homologs of bacterial repair genes, 354
methyl mismatch repair, 349, 354
mutator strains, 349
nonhomologous end joining (NHEJ), 354
nucleotide excision repair (NER), 350
photoreactivation, 350
RecA (synaptase), 340, 349, 352–54
recombinational repair, 338, 352
SOS response, 352–54
transcription-coupled repair, 350
DNA replication, **238**, 248–58. *See also* DNA polymerases; Origin of replication (*ori*)
B. subtilis sporulation, 62
in bacterial cell, 80–81f, 103, 104–5, 249–57
bidirectional replication in prokaryotes, 104–5, 248f, 249f, 250
catenanes, 256, 257f
clamp loader, 251f, 254–55
deoxyadenosine methylase (Dam), 250–52
DnaA-ATP complex, 252, 251f
DnaA replication initiator protein, 250, 252
DnaB (helicase), 252, 251f, 253, 254f, 255
DnaC (helicase loader), 252, 251f
DNA ligase, 255–56, 259f
DNA primase, 252, 251f, 253, 254f
duplication during replication, 361f
elongation, 250, 253–56
in eukaryotes, 103
initiation, 250–53
Okazaki fragments, 249f, 250, 255
origin (*oriC*), 249f, 250, 252, 251f, 253f
overlapping rounds in *E. coli,* 250
plasmids, 258–59
process of, 249f, 250
replication bubble, 249f, 250
replication forks, 249, 250
replisomes, 104, 105f, 253, 254f, 255
rolling-circle replication, 258–59, 325, 326f, 414
semiconservative replication, 249
SeqA replication inhibitor protein, 250–51
single-stranded DNA-binding proteins (SSBs), 254f, 255, 259
sliding clamp, 252, 251f, 254–55, eTopic 7.2
sporulation, 151
termination, 250, 256–57
termination (*ter*) sites, 249f, 250, 256, 257f
20 components, eTopic 7.1
DNA restriction and modification systems, 333–35
DNA reverse-transcribing viruses, **206**. *See also* Retroviruses
DNA revolution, 34–36, 264
DNaseI footprinting technique, 467–68, 469f
DNA sequence analysis. *See also* DNA polymerases; RNA deep sequencing
annotation of DNA sequences, 312–13
bridge amplification, 270
cloning, 264–66
computer analysis and Web science, 313–15
computer programs for, 316
dideoxy chain termination (Sanger) method, 9, 267, 270, A-50–A-51
DNA isolation and purification, 263, 843
DNA sequencing, definition, 35
of entire genomes, 267–70
fluorogram, 35f
genome sequencing, 9–10, 267–68, 798
high-throughput sequencing methods, 845t
homologous genes, 315–16
homologous sequences, 314–15
metagenomics, 270–72, 844–46
pyrosequencing, 267, 268–69, 271, 401
rapid DNA sequencing using an automated sequencer, A-51f
restriction endonuclease digestion, 263–64
reversible fluorescent termination, 270, 271f
sequence alignment for molecular clock, 682–83
sequencing by synthesis, 267, 270, 270–71f, 271, 401
of transposon insertion sites, 460
of unculturable microorganisms, 121–23
and Whipple's disease, 1046
whole-genome shotgun (WGS) sequencing methods, 271
DNA sequencing, definition, **35**
DNA shuffling for in vitro evolution, 477, eTopic 12.6
DNA-specific fluorescence, 58, 59f, 61
DNA synthesis. *See also* DNA replication
drugs that affect, 1121–23
DNA vaccines, 652, eTopic 12.3
Doherty, Peter, 986
Domagk, Gerhard, 1109–10, 1121
Domains (DNA), 102
Domains (lipids), 89, 102
Domains (proteins), 102
Domains of life (taxonomy), 9, 31, 32f, 79, 91, **685**, 688–89. *See also* Archaea (domain); Bacteria (domain); Eukarya (domain)
comparison of three domains, 688, 689t
divergence of rRNA genes, 31, 32f
Doolittle, Ford, 690, 691
Door into Ocean, A (Slonczewski), 28–29
Dorrestein, Pieter, 606–7
Doubling time, **140**
Downstream processing, **657**, 658f
Doxycycline, 1096, 1097, 1124f, 1125, 1127
Dps (DNA binding protein), 102
Drosophila, 270, 330, 415, 704
Drosophila ananassae, 330
Drug discovery. *See also* Antibiotics
combination of genomic and proteomic approaches, 1134–35
drug discovery process, overview, 1134
fatty acid biosynthesis as target, 1135
future of, 1134–35
golden age of antibiotic discovery, 1108–10
interference with quorum-sensing mechanism, 1136–37
nanotubes poking holes in bacterial membranes, 1135, eTopc 27.3
search for potential drug targets in genome, 1134
tagged probes, 1135
Drug efflux pumps. *See also* Antibiotic resistance
contribution to antibiotic resistance, 1130
multidrug resistance (MDR) efflux pumps, 88, 130, 1129, 1130f
proton-driven efflux pumps, 547, 549, 551, 1128
tetracyclines transport by efflux proteins, 88, 299, 1128
Drug sensitivity, 1111
Drug susceptibility, 1111, 1112–16
Dual-channel image analysis, A-47, A-48f
Dubey, Gyanendra P., 402–3
Dubilier, Nicole, 662–63, 865
Duchesne, Ernest, 1108, 1109
Dulbecco, Renato, 203, 204f, 224
DuPont Industrial Biosciences, 624–25
DuPont Laboratory, 609
Dust mite feces, 997f
Dutch elm disease, 881
DWH Oceanospirillales, 858–59
Dynein, A-39–A-40
Dysentery, 1006, 1066, 1072

E

E3 ligases, 1043–44
Early genes, 413f, **414**
Earth. *See also* Biosphere
crust, 668
early atmosphere, 668, 677
elemental composition, 668
geological composition, 669f
reduced compounds in environment of early Earth, 18–19
temperature, 668–69

Eastern equine encephalitis (EEE), 1007, 1009, 1087, 1089
Ebola virus, 1100–1101
  animal reservoirs, 1101
  broad tissue tropism, 213
  case history, 1162
  Ebola-Reston, 1101
  fatality rate, 1006, 1100
  filaments, 200
  quarantine, 1172f
  rapid replication and release from cells, 1100
  size, 1007f
  symptoms, 1007f, 1100
  tetherin disabled, 1100, 1101f
Eclipse period, **221**, 222
Ecological niche (ecotype), 696, 701
EcoRI restriction site, 334f
Ecosystems, **840–41**. *See also* Food webs and food chains; Microbial communities; Microbial ecology
  boreal forest ecosystems, 854
  cold-seep ecosystems, 165, 850, 865, 866f
  definition, 840
  human influence on microbial ecosystems, 177–78
  mesocosms, 893
  microbial niches in ecosystems, 159, 168, 841
  microbial support of natural ecosystems, 26–27
  role of microbes in, 841, 849–52
  thermal vent ecosystems, 165, 850, 865, 866f, 910
  viral roles in ecosystems, 851
  wetland ecosystems, 875–76
  Winogradsky column, 26–27
Ectomycorrhizae, **874**
Ectoparasites, **1005**
Ectosymbionts, 754
Edema, **996**
Edema factor (EF), **1025–26**
Edgar, Robert, 416
Edwards, Katrina, 567
Edwards, Rob, 207
EF-1α (elongation factor 1α), 801, 802f
Effector proteins, 1028, 1030–33, 1035, 1045f, 1046
EF-G (elongation factor G), 295–96
EF-Tu (elongation factor Tu), 85f, 295–96
Ehrlichia equi, 1096
Elanco, 188
Electrical potential (Δψ; charge difference), 547, 549f, 550, eTopic 14.2
Electric current in living organisms, 539–40
Electrochemical potential, **125**, 130, 541
Electrogenic transport, **128**
Electromagnetic radiation, **46–47**, 183
Electron acceptors, **501**, **541–44**
  alternative electron acceptors and donors, 559–60
  definition, 501
  in electron transport system (ETS), 541–44
  lithotrophy, 564t
  terminal electron acceptor, 172–73, 507
Electron bifurcation, **782**
Electron configuration, A-2, A-4f
Electron donors, **501**, **541–44**
  alternative electron acceptors and donors, 559–60, 561
  definition, 501
  in electron transport system (ETS), 541–44
  initial electron donor, 542
  lithotrophy, 564t
Electronegativity, **A-4**
Electroneutral coupled transport, **128–29**
Electron microscope, **32–33**, 66–68
Electron microscopy (EM), 41, **45**, **66**
  cryo-electron microscopy and tomography, 68, 69–71
  H. pylori in stomach lining, 42
  limitations of, 82
  magnetic lenses, 33, 66–67
  sample preparation, 68–69
  scanning electron microscopy (SEM), 45, 46f, 67–68, 69f
  transmission electron microscopy (TEM), 45, 46f, 67, 68, 69f
  ultracentrifugation combined with, 34
Electron spray ionization, 1162
Electron tower of standard reduction potentials, 543t
Electron transport chain, 130, **172–73**, 530, **541**. *See also* Electron transport system (ETS)
Electron transport system (ETS), **34**, **506–8**, 540, 541–46. *See also* Respiratory electron transport system
  in aerobic benzene catabolism, 533
  in aerobic respiration, 172–73
  in anaerobic respiration, 174
  anammox ETS within planctomycete, 566f
  chemiosmotic theory, 34
  common ancestral ETS, 541
  and complete oxidation of glucose, 530
  coupled O_2 reduction and NADH oxidation, 507
  cytochromes, 172–73, 545–46
  electric current generation, 539, 540
  electron donors and acceptors, 541–44
  electron tower of standard reduction potentials, 543t
  electron transport chain, definition, 172–73
  energy storage, 542–44
  initial electron donor, 542
  iron oxidation ETS, 565f
  location and function within membranes, 544–46
  and methanogenesis, 782
  NADH transfer of electrons to, 506
  photolytic electron transport system, 574, 575–76, 577
  proton pumping, 546–47, 548
  standard reduction potentials, 507t
  terminal electron acceptor, 172–73, 507
Electrophoresis, **84–85**, 644, 645f, **A-45–A-46**
Electrophoretic mobility shift assay (EMSA), **465**
Electroporation, **323**
Elementary bodies, **752**, 753f
Elements (chemical). *See* Chemical elements
Elephantiasis (filariasis), 704–5, 1005, 1006f
Eloe, Emiley, 865
Elongation factor 1α (EF-1α), 801, 802f
Elongation factor 2 (EF2), 1025
Elongation factor G (EF-G), 295–96
Elongation factor Tu (EF-Tu), 85f, 295–96
Elongation of replicating DNA, 250, 253–56
Elongation of RNA transcripts, 277, 281–82, 283
Elowitz, Michael, 462, 463f
Embden–Meyerhof–Parnas (EMP) pathway, **516**, **517–19**, 760–61, 790–91. *See also* Glycolysis
Emerging diseases. *See also* Ebola virus; Escherichia coli O157:H7; Human immunodeficiency virus (HIV); *Legionella pneumophila*; Lyme disease; Methicillin-resistant *S. aureus* (MRSA)
  and climate change, 228
  Clostridium difficile, 93, 151f, 188, 618, 725–26, 933, 1066–67
  detecting emerging and reemerging pathogens, 1181–82
  emergence of viral pathogens, overview, 227–28
  emerging fungal pathogens, 816–17
  Encephalitozoon intestinalis, 817
  hantavirus pulmonary disease, 1181
  locations of emerging and reemerging diseases, 1181f
  mad cow disease (bovine spongiform encephalopathy, BSE), 197, 1090, 1180
  Middle Eastern Respiratory Syndrome (MERS), 1004, 1180, 1181f
  monitoring recent developments, 1182
  multidrug-resistant *Mycobacterium tuberculosis,* 25, 108–9, 1107, 1130, 1181
  role of technology, 1180–81
  SARS, 1004, 1065, 1172–73, 1179–80
  swine flu (influenza strain H1N1), 410, 426, 1173, 1181f
  toxic shock syndrome, 1012, 1052, 1056
  West Nile virus, 194, 206, 227, 417, 1007, 1161, eTopic 6.2
Emerging organisms, **701**, **717**, 1047
Emiliana, 824
Emiliana huxleyi, 228, 229f, 823
Emission wavelength, **60**
Empty magnification, **49**
Encephalitis
  amebic meningoencephalitis, 1072
  Cryptococcus meningoencephalitis, 1063
  eastern equine encephalitis (EEE), 1007, 1009, 1087, 1089

measles, 1057
microbes that cause meningitis/encephalitis, 1088t
Encephalitozoon, A-69t
Encephalitozoon intestinalis, 817
Encephalopathy, 885
Endemic diseases, **1172**
Endergonic reactions, **A-17**
Enders, John F., 222
Endocarditis, **1090**, 1091–92
acute endocarditis, 1091
bacterial endocarditis, 1091–92
dental work and, 1091, 1092
in intravenous drug users, 1091
and mitral valve prolapse, 930, 1091, 1092
prophylactic antibiotic treatment, 1092
subacute bacterial endocarditis (SBE), 930, 1091
vegetations, 146, 930, 1091f, 1092
viral endocarditis, 1091, 1092
Endocytosis, 99, 132, **215**, **A-30**, eTopic 4.2
Endogenous retroviruses, **442**, 443f
Endoliths, 840f, **871**
Endomembrane system, **A-34–A-37**
Endomycorrhizae, **874**, 875f
Endonucleases. *See also* Restriction endonucleases
AP endonucleases, 351
TraI endonuclease (relaxase), 325, 326f
UvrABC, 350, 352
Endoparasites, **1005**, 1006f
Endophytes, **646**, **731**, 839, **877**
Endoplasmic reticulum (ER), 217f, 218, A-34–A-35, A-36–A-37
Endosomes, **A-30**
Endospores, **717**, 721t
asymmetrical cell division, 107, 150
Bacillus subtilis, 62, 150, 724
contamination with, 17
Firmicutes, 717, 724–26
forespores, 151, 724, 726
germination, 151
mother cells, 151, 724, 726
resistance to sporulation, 152
septation, 151
sporangium, 151
sporulation stages, 150–51
variant sporulation and "live birth," 726, 727f
Endosymbionts, **28–29**, **704**, 706, 707f
Endosymbiosis, 28–29, **704**, **800**
cyanobacteria in sponges, 723
evolution of, 703–5
and evolution of eukaryotes, 30–31, 705–7, 798
metabolic fluxes within termite hindgut, 854–55
microbial endosymbiosis with plants and animals, 28, 854–55
primary endosymbiosis, 800
rhizobia in legume roots, 28, 704
secondary endosymbiosis, 706, 707f, 797, 800, 801f
tertiary endosymbiosis, 706
Endotoxin, **98**, **1018**, 1027–28
Endy, Drew, 457
Energy, **494–501**. *See also* Gibbs free energy change (ΔG)
in biochemical reactions, 498–501
capture of, 494–95
definition, 494
dissipation as heat, 495
reactions yielding energy in bacteria and archaea, 492, 493f
solar energy, 495f, 540f
sources, 124, 127
storage, 124–25
Energy carriers, **501**, **504–6**, 507–8
adenosine diphosphate (ADP), 504, 508
adenosine triphosphate (ATP), 501, 504–5, 508
flavin adenine dinucleotide (FAD, $FADH_2$), 507, 508
nicotinamide adenine dinucleotide (NADH, NAD^+), 501, 505–6, 508, 578
nicotinamide adenine dinucleotide phosphate (NADPH), 507, 508, 578
observing energy carriers in living cells, 508, eTopic 13.1
photolytic electron transport system, 574, 575–77
reasons for different energy carriers, 508
Energy stress, 462, 463f
Enhancers, **241**, 262, 394–95, 452
Enoyl reductase, 86, 601, 602f, 604f
Enriched media, **135**
Enrichment culture, **26**, **841**
Entamoeba, 1122
Entamoeba hartmanni, 932
Entamoeba histolytica, 825, 1066, 1072, 1166, A-69t
Enterobacter, 744, 1075
Enterobacter agglomerans, 1182
Enterobacteriaceae, 744, 932, 934, 1155
Enterobacteriales, A-64t
Enterochelin, 386f
Enterococcus, 358, 728, 934, 1075, 1098, A-63t
Enterococcus faecalis, 559, 728, 1131
Enteromorpha, 400
Enterotoxins, **935**, 1023–24
Enteroviruses, 204, 417, 1092
Enthalpy (ΔH), **496**, 497, 518, **A-17**
Entner–Doudoroff (ED) pathway, 512f, **516–17**, **520**, 760–61, 770, 788, 932
Entophysalis, 672f
Entropy (ΔS), **494**, **496**, 497, 518, **A-17–A-18**
Envelope (bacterial cell), 80–81f, **82**, 200
Envelope (viral), 192, 198, **199–200**
Environmental classification of microbes, 159t
Environmental Sample Processor, 768, 769f
Enzyme-linked immunosorbent assay (ELISA), 1072, 1162–64
Enzyme prosthetic groups, 135
Enzymes, **508–10**
alkaliphilic enzymes, 171, eTopic 5.2
as commercial products, 654–55
contained by viruses, 202
directed evolution of hyperthermophilic xylanase, 698–99
effect on reaction rates and activation energy, A-20
embedded in filters, 181, 181f
essential nutrients and, 120
thermophilic enzymes, 162–64, 758
Eosinophils, 937f, **938**, 997
Epidemic diseases, **1172**
Epidemiology, **1149**, **1171–79**. *See also* Bioterrorism; Pandemics
Centers for Disease Control and Prevention (CDC), 1171, 1173, 1177, 1180
definition, 1171
detecting emerging microbial diseases, 1179–82
ecological influences on pathogens, 1182
egg-related salmonellosis outbreak in 2010, 1176
endemic *vs.* epidemic diseases, 1172
finding index case (patient zero), 1172–73
genomic detection of unculturable microbes, 121, 123, 1175–76
history, 1173, 1174f
molecular approaches for disease surveillance, 1176
notifiable infectious diseases, 1174t
physician notification of health organizations, 1173
of *Shigella* in a day care center, 1149
World Health Organization (WHO), 1171, 1173, 1180
Epidermis, **928**. *See also* Skin
Epigenetic silencing, 1093
Epilimnion, 866–67
Epimerization, 618
Epinephrine, 996, 997
Episomes, **327**
Epitopes, **963**, 971, 984, 999
Epsilonproteobacteria, 598, 713f, 715t, 717, 747–48, A-64t
Epstein, Richard, 416
Epstein-Barr virus, 193, 206f, 207, 218
death of bubble boy by, 962
DNA replication, 443
as DNA virus, 444t
persistence in B-cell lymphocytes, 448
prevalence of latent infection in humans, 446
Epulopiscium, A-63t
Epulopiscium fishelsoni, 688, 726, 727f
Equilibrium, **A-18**
Equilibrium density gradient centrifugation, eTopic 7.5
Equine herpes virus, 206f, 207
Equivalence, **970**
Ergosterol, 806
Error-prone repair, **349**, 352–54
Error-proof repair, **348–52**
Erwinia, 163, 644, 744, 881

Erwinia amylovora, 158
Erwinia carotovora, 744
Erysipelas, 728, 1124
Erythema migrans, **1096**
Erythromicrobium ramosum, 737
Erythromycin. *See also* Antibiotics; Polyketides
antibiotic resistance, 1127
binding to ribosomal subunits, 103, 299, 300f, 1125
biosynthesis, 603, 604f
endocarditis prevention, 1092
structure, 300f, 1124f
Erythrose 4-phosphate, 520, 522f, 587, 594
Escherichia, 684
Escherichia coli, A-64t. *See also* Acid resistance of *E. coli*
2D gel, 85f
ABC transporters, 130
acetyl-CoA conversion to acetate, 528
AcrB multidrug resistance efflux pump, 1130f
afimbriate adhesins, 1016
alternative electron donors and electron acceptors, 560f
ampicillin effect on growth, 187
API 20E strip results, 1155f
arginine biosynthesis, 615–16
arsenic sensing by, 478–79
as bacterial computer, 457
bacteriophage lambda receptor protein, 208
bacteriophage M13 infection, 200
bacteriophage T2 infection, 194
bacteriophage T4 infection, 201f, 412–13
Bdellovibrio infection, 747
biofilms, 28, 41, 146, 186
cell division timing, 250
cell model, 80–81f, 82
cell structure changes, 150
chemotaxis proteins and signaling pathway, 391–93
chromosome length, 238, 245
circular chromosomes, 239, 240t
conjugation, 239, 325
consumption of intestinal mucus, 520
CRISPR locus, 336f
culture media, 135
disease categories, 701
DNA replication, 250–57
dysentery, 1066
electron donor-acceptor systems in colon, 544
EmrD protein, A-26f
enteroaggregative *E. coli* (EAEC), 1067, 1068
enterohemorrhagic *E. coli* (EHEC), 1067, 1070, 1136–37, 1181
enteroinvasive *E. coli* (EIEC), 1067
enteropathogenic *E. coli* (EPEC), 1031, 1053, 1067
enterotoxigenic *E. coli* (ETEC), 1067
essential nutrients, 120–21
ETS for aerobic NADH oxidation, 555f, 556
evolution of citrate-utilizing strain, 2f, 3, 694–95
exopolysaccharides (EPSs), 149
experimental evolution, 2–3, 694–96
eye microbiota, 928
fitness islands, 360
flagella, 64f, 114
as food-borne pathogen, 1009
fts mutants, 106–7
genome, 261, 262f, 322, 359
genome evolution, 358
glutamine synthase gene (*gltB*), 315
glycerol transport, 127–28
GroEL chaperone, 275
growth, 65f, 120
growth rate and temperature, 160f, 161
heat-shock response, 164, 383f
on Hektoen agar, 1154f
horizontal gene transfer, 690
in human intestines, 932, 933
industrial strains, 586
interactome, 473, eTopic 12.1
internal pH (pH_{int}), 168, 169, 172f
internal structures, 46f
interspecies mating with *Salmonella,* 329
irradiation, 181, 182
labile toxin (LT), 1023–24
lactose transport, 128
lacZ gene and recombination, 340
lipids, 89, 90
MAM7 expression in, 1017
as model system, 158, 744
molecular composition of *E. coli* cells, 84t
multiplex analysis, A-55f
NDMS1 plasmids, 1119
nucleoid, 101
oscillator switches constructed in, 478–79, 481
pathogenicity island, 360f
phase variation in expression of type I pili, 341
and phenol red broth test, 526
phosphorylation by succinyl-CoA synthetase in, 528
phosphotransferase system (PTS), 132f
pH range, 168
in plant vascular system, 1182
plasmid vectors, 264–66
polar aging, 110, 111f
protein catalog, 84, 85f
proteome, 85, A-48f
sacculus (cell wall), 91–92
septicemia, 1098
shape determining proteins, 100f
Shiga toxin-producing *E. coli* (STEC), 1068, 1131–32
sigma S, 382
size, 8t, 44f
sRNA genes, 385f
stable toxin (ST), 1019
and starch bloat, 884
stationary phase, 143
in stomata of a lettuce leaf cell, 7f
stress survival responses, 176, 177f
synchronous clock engineered into, 478–79
synthetic biology applications, 478
transcriptomics of acid resistance genes, 401f
tryptophan biosynthetic pathway, 379f
type IV pili in pathogenic strains, 1015–16
type VI secretion, 1034
uropathogenic *E. coli* (UPEC), 1013, 1015, 1067, 1075–76
virulence genes, 1011
Escherichia coli K-12, 224f, 240t, 358, 359, 1012
Escherichia coli O104:H4, 1012, 1068–69, 1131–32
Escherichia coli O111, 946f
Escherichia coli O157:H7
acid resistance, 458, 1070
case history, 1067
effect of pH on survival, 649
as emergent pathogen, 646, 647t, 744, 1070
endophytic growth in crop plants, 877
epidemiology, 1070
GC content of genome, 359
horizontal gene transfer, 358–59, 691, 1012, eTopic 17.2
intestinal crypt cells covered with, 646f
irradiation, 181, 1070
multispecies biofilms, 186
PCR detection, 266–67
phage therapy, 188
within plant vascular system, 1182
rumen pH and *E. coli,* 884
Shiga toxins, 210, 646, 1025, 1069f, 1070
sorbitol fermentation, 526
type III secretion, 1070–71
Espirito Santo virus, 191
Essential nutrients, **120**
Esters of organic acids, 650
Etest, 1113
Ethambutol, 96, 653, 1065
Ethanol, 87
alcohol dehydrogenase, 639
carbon cycle, 123f
as disinfectant, 184
toxic effects of, 639
Ethanolic fermentation, 513, 514t, **521**, **629**, 637–41
beer production, 120, 627, 628t, 639, 640f, eTopic 16.1
bread dough leavening, 629, 637–39
and entropy, 497
history, 17, 639
NADH reduction of pyruvate, 506
wine production, 624, 627, 628t, 629, 639–40
Ethidium bromide, A-45
Ethylene oxide, 184f, 185
Eugenol, 650
Euglena, 127, 833
Euglena gracilis, A-71f
Euglenas, 798, 805
Euglenida, 833, A-71t
Euglenozoa, 805, A-71t
Eukarya (domain), **9**, 685. *See also* Eukaryotes; Eukaryotic diversity

Eukaryotes, **9**, 81
cell cycle and mitosis, 139, A-32–A-33
cell membrane, A-24–A-30
cell structure, A-23, A-25
cilia, A-39
classification, 30
comparison of archaea, bacteria, and eukaryotes, 31, 32f, 91, 688, 689f, A-24t, A-25f, A-31
consumers and producers, 126–27
endomembrane system, A-34–A-37
endosymbiosis theory of evolution, 30–31, 126, 127, 705–7, 742, 798, A-40–A-41
flagella, 801–2, 804, 805, A-39–A-40
genomes, 261–62
historical overview, 798, 800
meiosis, A-33–A-34
membrane lipids, 90, 91, A-24, A-26
microbial cell size, 44
mitosis, A-32–A-33
nanoeukaryotes, 805
newly discovered eukaryotes, 805
noncoding DNA sequences in, 313–14
nucleus, 101, 261, A-31
osmotic shock protection, 99
picoeukaryotes, 805
polysaccharides in cell walls, 713
problems faced by large cells, A-34
proteasomes, 305
reductive evolution, 798, 802
secretion systems, 309
serial endosymbiosis theory, 30–31, A-40
stationary phase, 143
Eukaryotic diversity, 797–836. *See also* Phylogeny; Taxonomy; *specific types*
emerging eukaryotes, 805
historical overview, 798, 800
overview, 797–98
phylogeny, 798–805, A-68t–A-71t
phylogeny, based on DNA sequence data, 800, 801f
representative groups of eukaryotes, 799t
Eukaryotic initiation factor 2 (eIF2), 953
Eumycota, A-68t
Eumycota (true fungi), **801–2**
Euphotic (photic) zone, **857**
Euprymna scolopes, 366, 397–98
Europa (moon), 918–19
Euryarchaeota, **763**, **765**. *See also* Haloarchaea; Methanogens
acidophiles, 791–93
halophiles, 784–90
methanogenic euryarchaeotes, 775–83
Nanoarchaeota, 793
phylogeny, 763f, 764t, 765, A-67t
thermophiles, 790–93
Eutrophication, 177f, **178**, 867–68, 906
Eutrophic lakes, **867–68**
Evolution. *See also* Reductive evolution
adaptive evolution, 692–96
coevolution, 704
directed evolution of hyperthermophilic xylanase, 696, 698–99
directed evolution through phage display, 477, eTopic 12.5
DNA shuffling for in vitro evolution, 477, eTopic 12.6
endosymbiosis theory of evolution, 30–31, 126, 127, 705–7, 742, 798, A-40–A-41
experimental evolution, 2–3, 694–96
historical contingency model, 694
Lactococcus lactis, 665
molecular mechanisms, 357–61
site-directed mutagenesis, 477, eTopic 12.7
time line, 675f
Excitation wavelength, **60**
Exergonic reactions, **A-17**
Exfoliative toxin, 1056
Exit site (E site), **292**, 295–96
Exocytosis, **A-30**, A-35
Exons, 313f, **314**
Exopolysaccharides (EPS), 148f, **149–50**, 945f
Exoskeletons, 99, 823, 894
Exotoxins, **1018–26**. *See also* Shiga toxins
AB toxins, 1022, 1023–24
ADP-ribosyltransferase toxins, 1022, 1023
alpha (α) toxin, 1019, 1022
anthrax toxin, 1025–26
botulinum toxin, 646, 648, 725, 1026, 1085
characteristics and classes, 1019, 1021t–1022t
cholera toxin (CT), 86, 200, 210, 365, 1019, 1022, 1023–24
diphtheria toxin, 1019, 1022, eTopic 25.6
E. coli labile toxin (LT), 1023–24
E. coli stable toxin (ST), 1019
edema factor (EF) in anthrax toxin, 1025–26
identifying new microbial toxins, eTopic 25.7
lethal factor (LF) in anthrax toxin, 75, 76f, 1025–26
protective antigen (PA) in anthrax toxin, 1025–26
tetanus toxin, 1019, 1026, 1085
ExPASy (Expert Protein Analysis System), 316
Experimental evolution, 2–3, **694–96**
Exponential death, 143, 179–80
Exponential growth, 140–41
Exponential phase, **142–43**, 144
Exserohilum rostratum, 817, 1051
Extraterrestrial life. *See also* Astrobiology; Mars; Origins of life
comets or meteorites as source of life, 19, 613, 680
comparison to extremophiles, 158
extrasolar planets, 919
panspermia, 680
possible ocean on Europa, 918–19
Extravasation, **947–49**
Extremophiles, **28**, **158**, 851–52. *See also* Acidophiles; Halophiles; Hyperthermophiles; Thermophiles
alkaliphiles, 159t, 170–71, 725, 765, 852
Archaea, 7f, 19, 28, 164, 247, 559
barophiles, 159t, 164–65
bioinformatic analysis, 158–59
bioprospecting for, 655
and biotechnology, 758
classes and environments, 852t
comparison to extraterrestrial life, 158
psychrophiles, 159t, 160f, 161–62, 164, 642, 655
radiation resistance, 183, 234, 340
Extrusomes, 830f, 831
Exxon Valdez oil spill, 898
Eye microbiota, 928

F

F_1F_o ATP synthase, 125, 547f, 551f, 557–59, 572, 575, 579f
F(ab) regions, 969f, 970f
Facilitated diffusion, **127–28**
Factor H, **993**
Facultative anaerobes, **175**, 590, 642, 744
Facultative intracellular pathogens, **1038**
Facultative microbes, 159t, 173t, **175**
anaerobes *vs.* facultative microbes, 174–75
FAD (flavin adenine dinucleotide), 174
Falkow, Stanley, 264, 1010
Fang, Ferric, 922–23
Faraday constant, 544, 547
Farmer, Paul, 1148
Fatty acid degradation (FAD) pathway, 510–11
Fatty acids
biosynthesis, 601–3, 1135
catabolism, 510–11, 512f, 531
in cell membranes, 86, 89–90, 165
cyclopropane fatty acids, 90
micelle formation from fatty acid derivatives, 678
saturated fatty acids, A-15, A-16f, A-27f, A-28
unsaturated fatty acids, A-15, A-16f, A-27f, A-28
Fatty acid synthase complex, **601–2**
F^- cells, 325, 326f
F^+ cells, **325**, 326f
Fc receptors, 954, 972
Fc region, 951, 952f, 954, 970f, **971**, 972
Fecal-oral route, defined, 1149
Fecal transplants, 934
Feces, human, 931, 932
Feline leukemia virus (FeLV), 205t, 206, 218, **432**, 433f. *See also* Retroviruses
FeMO cluster of nitrogenase, 609–10
Femtoplankton, 862
Fenton reaction, 174f
Fenugreek sprouts, 1069f
Fermentation, **17**, **174**, 493f, **513**, 521–27. *See also* Ethanolic fermentation
acidic fermentation of dairy products, 628t, 630–32
acidic fermentation of vegetables, 628t, 632–34
alkaline fermentation, 628t, 636
anaerobic fermentative metabolism, 174, 175
carbon cycle, 123f

Fermentation (*continued*)
cheese production, 525, 630–32, eTopic 13.2
classes of fermentation reactions, 629f
cocoa bean fermentation, 634–36
diagnostic applications, 526
electron flow, 541
entropy and, 497
fermentation industry history, 624–25
fermentation pathways, 524f
food and industrial applications, overview, 525–26
foregut fermentation, 885
heterolactic fermentation, 521
hindgut fermentation, 885
history, 17, 624
lactate fermentation, 513, 514t, 521, 525
mixed-acid fermentation, 523–25, 563, 780
permeant weak acid production, 188
vanilla bean fermentation, 872
in vertebrate gut, 883–85
waste products, 523
Fermentation industry, **624–25**
Fermentative metabolism, **174**
Fermented foods, **627–29**
acid-fermented foods, 628t, 629–36
alcoholic beverages, 639–40
alkali-fermented foods, 628t, 636
cabbage and other vegetables, 634, 636
chemical conversions in, 629f
chocolate, 634–36
commercially produced fermented foods, 627, 629
dairy products, 630–32
leavened bread dough, 628t, 637–39
overview, 627–29
soy products, 632–34, 636
traditional fermented foods, 627, 629
Ferredoxin, **578**, 598f, 599, 1122–23
Ferredoxin-NAD$^+$ reductase (FNR), 578
Ferroplasma, 91, 765, 791
Ferroplasma acidarmanus, 567, 763, 791, 792f, 793, 845, A-67t
Ferroplasma acidiphilum, 792f, A-67t
Fertility factor (F factor), **325–28**
Fescue, 877
Fever, 958
Fibrin, 930, 939
Fibrobacter flavefaciens, 884
Fibronectin, 1016, 1017
Fibrosis, 949
Filament 021N, 703f
Filamentous algae, 672f
Filamentous fungi, A-68t
Filamentous hemagglutinin (FHA), 1017f
Filamentous mutants, 83
Filamentous prokaryotes, 672f, 702, 703f, 720–22, 900
Filamentous structures, differentiation into, 152–53
Filamentous viruses, **200–201**
Filarial nematodes, 704–5
Filariasis (elephantiasis), 704–5, 1005, 1006f
Filopodia, 825
Filoviridae, 1100, A-61t
Filoviruses, 205t
Filtering samples, **843**, 844f
Filtration sterilization, 181, 182f
Fimbriae (singular, fimbria), **113**, **1013**
Finalyse, 188
Finlay, Brett, 1031
Fire blight, 158
Firmicutes, **716–17**, 723–30
anaerobic dechlorinators, 728, 729f
CG content, 717, 723
endospore-forming rods, 723–26, 727f
Gram-positive cells, 94, 716–17, 723
human microbiota, 712f
intestinal biota, 932
"live birth," 726, 727f
non-spore-forming Firmicutes, 726–30
photoheterotrophy in Heliobacteriaceae, 719t, 726
phylogeny, 713f, 715t, 716–17, A-63t
thermophiles, 766
Fish and marine food chain, 124, 127
Fish tank granuloma, 950f
Fistulated cows, 842, 843f, **884**
Fitness islands, 360
Fixation, **55**, 56f, 57f
Flagella (singular, flagellum), 44, **64**, 65f, **82**, **114**, **A-39–A-40**
anti-sigma factor (FigM) and flagella assembly, 382
bacterial contrasted with eukaryotic, 116
biofilm development, 148f, 149, 396
biosynthetic expense of components, 613, 614f
C. crescentus, 107–10, 111f, 114
eukaryotic flagella, 801–2, 804, 805, 818, 820–21, 830, 836, A-39–A-40
flagellar genes and type III secretion, 1030
flagellar motility, 503
flagellar motors, 68, 69f, 80f, 114–15, 125
flagellin, 389, 390f
H antigen, 389
Planctomyces, 753f
rotary flagella, 28f, 54, 114–16, 753
Spirochetes, 717, 751
unpaired flagellum, 801–2
Flagellates, **804**
Flagellin, 389, 390f
Flammulina velutipes (enoki mushrooms), 626
Flash pasteurization, 181, 631
Flavin adenine dinucleotide (FAD, $FADH_2$), 174, 506f, **507**
Flavin mononucleotide (FMN), 552, 553
Flaviviridae, 1099
Flaviviruses, 192, 205t, 206, 215, 417t, 1007, 1008f
Flavobacteriales, A-65t
Flavobacterium, 161f, 749, 900
Flavobacterium capsulatum, 57, 59f
Flavodoxin, 1122
Flavonoids, **878**
Fleming, Alexander, 24, 186, 457, 1108–9
Flesh-eating disease, 1056–57
Flexibacter, 702
Flexible gene pool, 359
Flocs, **900**
Florey, Howard, 24, 1109
Florida Everglades, 901
Flow cell, 54, 55f
Fluconazole, 1142, 1143
Fluid mosaic model of membranes, **A-26–A-28**
Fluorescence, **48**, 60. *See also* Fluorescence microscopy
immunofluorescence, 61
Fluorescence-activated cell sorter (FACS), **137**, 138f
Fluorescence in situ hybridization (FISH), 739, 862, 1175f, **A-55–A-56**
Fluorescence microscopy, 41, 58–63, 60f
cardiolipin, 89
detection of planktonic microbes, 862, 863f
DNA replication, 104, 105f
fluorophores, 58, 59f, 60, 61–63, 89
Listeria monocytogenes, 59f, 60
living and dead cells, 136–37
marine microbes, 58, 59f
rotary flagella, 28f, 54
Trypanosoma brucei, 44f
Fluorescence resonance energy transfer (FRET), **470**
Fluorescent-focus assay, **224–25**
Fluorochrome auramine O, 1152
Fluorophores, **58**, 59f, 60, 61–63. *See also* Green fluorescent protein (GFP)
cardiolipiin-specific, 89
catalyzed reported deposition (CARD) of tyramides, A-55
cyan fluorescent protein, 104, 105f
DAPI, 59f, 61, 89f
fluorescent-focus assay, 224–25
FM4-64, 61, 62, 100f, 105f
yellow fluorescent protein (YFP), 104, 105f
Fluoroquinolones, 248, 1119–20
FLUXNET study (NASA), 893, 894
The Fly (movie), 330
FM4-64, 61, 62, 100f, 105f
fMet peptides, 303, 948
Focal plane, 49f, **54**
Focal point, **48–49**
Focus (viral infection), **224**
Folate, **599**
Folic acid, 1110, 1122
Follicular-associated epithelial (FAE) cells, 1003
Fomites, **1007**, 1008f, 1009
Food and Drug Administration (FDA), 182, 644, 657
Food contamination
Clostridium botulinum, 646, 648
definition, 642
Escherichia coli O157:H7, 646
food-borne pathogens in the United States, 647t
food poisoning, definition, 642
Listeria monocytogenes, 645–46

noroviruses (Norwalk-type viruses), 646
Salmonella enterica serovar Typhimurium in peanut butter, 644–45
Salmonella pathogenicity island, 646
Food irradiation, 181–83, 649
Food microbiology. *See also* Fermentation; Fermented foods
acid stress, 168–69
dairy industry, 656, 727
edible algae, 626
edible bacteria and yeasts, 626, 627f
edible fungi, 625–26
food contamination by pathogens, 642, 644–48
food preservation, 623, 627, 648–50
food spoilage, 642–44
overview, 623–25
single-celled protein, 626, 627f
Food poisoning, **642**
Food preservation, 648–50
chemical methods, 649–50
overview, 623, 627
physical methods, 165, 648–49
Food Safety Modernization Act (2011), 644
Food spoilage, **642–44**
dairy products, 642
meat and poultry, 642
plant foods, 644
seafood, 642–43
types of spoilage, 643t
Food webs and food chains, **849–52**. *See also* Ecosystems; Microbial ecology
abiotic factors affecting, 851–52
carbon assimilation, 849, 850
carbon dissimilation, 849–50, 850–51
consumers, 850
decomposers, 850–51
detritus, 851
food web, definition, 850
grazers, 850
lithotrophs, 850
marine compared to terrestrial, 851
marine food chain, 124, 127, 229–30, 798, 850f
microbes within food webs, 850f
phototrophs, 850
planktonic food webs, 863–65
predators, 850
primary producers, 849, 850
trophic levels, 850
Foraminifera, A-70t
Foraminiferans (Foraminifera), 804, 825, **827**
Forespores, **151**, 383–84, **724**, 726
Formaldehyde, 184f, 184t, 185
Formate, 523f, 599
N-formylmethionine (fMet) peptides, 303, 948
N-formylmethionyl-tRNA (fMet-tRNA), 293, 294
Forterre, Patrick, 790
Fortunato, Lee, 448–49
Fortune, Sarah, 108–9
Fossil fuels, **896**
carbon reservoir, 892t, 894
CO_2 from combustion, 889, 890f, 894, 896
formation, 895
isotope ratios, 673
Fouchier, Ron, 409
F-prime (F′) factor, **328**
F-prime (F′) plasmids, **328**
Fracastoro, Girolamo, 20
Fragilariopsis kerguelensis, 913
Frameshift mutations, **342**
Francisella, 121t, 1098, 1136
Francisella tularensis, 1041, 1053, 1136–37
Frank, A. B., 874
Frankenstein (Shelley), 540
Frankia, 404, 730, 731, A-63t
Franklin, Rosalind, 25, 34–35, 75, 193
Fraser-Liggett, Claire, 9f, 10
Freeze-drying, 181, **648**
French press, 82
Frequency (ν), 47
Frère, Jean-Marie, 234
Frog skin infection with chytrids, 810, 812f
Fructose, A-11
Fructose 1,6-bisphosphate, 509, 518, 519f
Fructose 6-phosphate, 516, 518, 519f, 521, 522f, 594
Fructose bisphosphatase, 518–19
Fruiting bodies, **625**, **746**, 802
Aspergillus, 814
basidiomycetes, 815–16
morels (*Morchella hortensis*), 813f, 814
Myxococcus, 153f, 746, 747f
Penicillium, 814
slime molds, 802, 804, 826
starvation, 152
truffles (*Tuber aestivum*), 625, 814
Frustule, 822f, **823**
FtsY protein, 307
FtsZ protein (Z ring subunit), 100, 106–7, 353
Fucose, 515
Fumarate
in arginine biosynthesis, 616
as electron acceptor, 539, 542, 544, 560
as TCA cycle intermediate, 508, 528, 529f
Fumarate reductase, 598
Functional gene categories, 847, 848f
Functional genomics, **316**
Functional groups, **A-6**, A-7f
Fungi, **801–2**, 806–18
absorptive nutrition, 806
antifungal agents, 806, 807, 1142–43, 1144t–1145t
ascomycetes, 812, 813f, 814
basidiomycetes, 815–16
Candida infections, 934, 1143
cellular basis of hyphal extension, 807f
cell walls, 99, 806
chitin in cell walls, 806
chytrids, 799f, 802, 809–11, 812f
dimorphic fungi, 809
diversity, 797, 802, 806
edible fungi, 625–26
emerging fungal pathogens, 816–17
ergosterol, 806
fairy rings, 798, 800f, 816
fruiting bodies, 625
fungal lung infections, 1062–64
Glomeromycota, 816
heterotrophy, 126–27
hyphae, 802, 806–7
meningitis caused by, 1051
mitosporic fungi, 809
mushrooms, 800f, 802
mycelia, 806, 807
mycorrhizae, 806, 815f, 816, 872, 873–75
organisms reclassified as, 817
as plant pathogens, 881–82
ruminal fungi, 884
shared traits of fungi, 806
superficial mycoses (fungal infections), 1142–43
symbiotic partnerships, 822
three-way mutualism, 226–27
traditional views of, 798, 800
unicellular fungi, 802, 807–9
unpaired flagelllum, 801
upper temperature limit, 161
urinary tract infections, 1075
yeasts (single-celled fungi), 626, 801, 802, 807–9
zygomycetes, 812, 813f
Fur box DNA sequence, 375f, 386f
Fur (ferric uptake regulator) protein, 385–86
Fuselloviridae, A-60t
Fuselloviruses, 770, 771f, A-60f
Fusidic acid, 300
Fusion peptides, **426**
Fusobacteria, **717**, 750, A-65t
Fusobacterium, 928, 929f, 930
Fusobacterium necrophorum, 1121
Fusobacterium nucleatum, 750, A-65t

G

GadA and GadB (glutamate decarboxylase), 460, 461f, 463, 464f, 465, 466
GadC (glutamate/GABA antiporter), 460, 461f, 463, 464f
GadE (YhiE) regulatory protein, 465, 468, 469f
GadX regulatory protein, 468, 469f
Gain-of-function mutations, **342**
GAL1 gene promoter region, 472
GAL4 transcription factor, 472
Galactans, 96
beta-Galactosidase (LacZ)
expression in riboswitchboard, 483f
expression in yeast two-hybrid system, 472–73
in gene fusion experiments, A-51–A-52
induction, 370–71
lactose catabolism, 340, 370, 372f
repression, 374f
Galen, 1080
Galileo space probe, 918
Gallionella ferruginea, 113–14
Gallo, Robert, 432, 433–34, 436
Galvani, Luigi, 540
Gametes, A-33, A-34
Gamma-aminobutyric acid (GABA), 1019, 1087

Gamma herpes viruses, 206f, 207
Gammaproteobacteria, 736t, 742–46, A-64t
aerobic rods, 745–46
ammonia oxidation, 742
CO oxidation, 739f
enteric fermenters and respirers, 744
in honeybee guts, 840
methylotrophy, 737
photoheterotrophy, 735
phylogeny, 684f, 713f, 715t, 717
plant pathogens, 746
sulfur and iron phototrophs, 743
sulfur lithotrophs, 743–44
Gamogony, **832**
Ganglioside GM1, 1023
Gangrene, 1057f
Gannon, Victor, 1068
Gardasil vaccine, 198, 216, 1102
Gas chromatography, 893
Gas discharge plasma sterilization, 185
Gases as disinfectants, 184f, 185
Gastric ulcers and *Helicobacter pylori,* 22, 42, 172, 747–48, 931, 1045–46, 1071–72
Gastrointestinal tract infections, 644–45, 1066–74. *See also* Human intestines, microbiota; Stomach microbiota; *specific diseases*
amebic dysentery, 1066
antibiotic treatment, 1066–67
diarrhea, overview, 1066
diarrhea and dysentery, case history, 1067
gastroenteritis, overview, 1066
microbes that cause gastrointestinal diseases, 1074t
protozoa (protists) that cause diarrheal diseases, 1072–73
Gas vesicles, **112**, **720**, **785**, 787
Gattaca (Niccol), 267
GC content, **717**
difference between Firmicutes and Actinobacteria, 717, 723
and DNA denaturation, 244
of genomic islands, 1011, 1012f
in haloarchaea, 784
and horizontal gene transfer, 359, 690
in *Mycoplasma mycoides genome,* 312f
and pausing of RNA polymerase, 301
in *Salmonella* Typhimurium LT2 genome, 241f
Geminiviridae, A-60t
Geminiviruses, 205t, 218, A-60f
Gemmata, A-65t
Gemmata obscuriglobus, 753f
Gene array technology, 1045–46
Gene assembly, 698–99
Gene-calling in metagenomics, 846–47
Gene cloning, 264–66
Gene duplication, 361, **692**, 695
Gene expression. *See also* Gene regulation; Transcription; Translation
in bacteriophages, 413–15
carbon-concentrating mechanism, 596–97
herpes simplex virus (HSV), 446, 447f, 448
during nutrient stress, 176
in prokaryotes *vs.* eukaryotes, 103
proteomics, 404–5
regulation of (*See* Gene regulation)
in response to changing environments, 159
transcriptomics, 401, 404
Gene fusions (translational fusions), **461**, 462f, **A-51–A-52**
Gene inversion, 389, 390f
General Electric & *Science* Prize, 234
Generalized recombination, **338–40**
Generalized transduction, **331**
Generation time, **140–41**, 144–45
Gene regulation, 35, 366–69, 603. *See also* Gene expression; Lactose (*lac*) operon; Regulatory proteins
acid resistance in *E. coli,* 463–65, 468, 469f
aporepressors, 378, 379f
in bacteriophage lambda, eTopic 10.5
corepressors, 367, 378
derepression, 367, 378
DNA sequence controls, 369
holorepressors, 379f
inducers, 366–67, 371
induction, 366–67, 367f
integrated control circuits, 369, 391–97
leader sequences, 285, 292f, 293, 378–79
mRNA stability controls, 369
posttranslational controls, 369
protein-DNA binding experiments, 374–75
regulatory sequences, 366, 367f, 375f
reporter fusions, 460–63
repression, definition, 366–67
repression of anabolic (biosynthetic pathways), 378
repression of catabolic pathways, 374, 375f
sigma factor regulation, 382–84
stringent response, 381, 603
trailer sequences, 293
transcriptional attenuation, 369, 378–80
transcriptional controls, 366–67, 369
translational controls, 369
Genes, **239**. *See also* Operons
alignment of gene with mRNA transcript, 292–93f, 293
annotation of DNA sequences, 312–13
duplication of, 361
exons, 313f, 314
functional units, 242
gene organization, 242f, 262f
homologous genes (homologs), 315
informational genes, 691, 696
introns, 261, 262f, 313–14
nomenclature, 243
oncogenes, 216, 218
open reading frames (ORFs), 282
operational genes, 691
orthologous genes (orthologs), 206, 315, 361, 696, 697
paralogous genes (paralogs), 315–16, 361, 688, 692
promoters, 241, 242f, 262, 277–78
pseudogenes, 261, 262f, 360, 733
regulatory sequences, 242
regulons, 242, 382
reporter genes, 83, 224–25, 348
structural genes, 239
tRNA genes, 359, 360f
Gene site saturation mutagenesis (GSSM), 698f, 699
Gene switching, **978**
Gene therapy with viruses, 450–53, eTopic 12.4
adenoviruses as vectors, 450, 453
adrenoleukodystrophy (ALD) treatment, 450
cancer treatment, 453
gene transfer vectors, 450
helper plasmids, 450, 451–52
HIV as vector, 194, 218, 434, 436, 443, 450, 451–52
human embryonic kidney 293T cells, 452
lentiviral gene therapy, 450, 451–53
leukemia treatment, 450
nerve growth factor (NGF) transfer, 452–53
pseudotyping, 451
Reolysin, 453
safety of viral vectors, 450–51
for SCID, 962
transfection, 452
transgenes, 452
vector integration into neurons, 452, 453f
vision restoration, 450
Genetically modified organisms
crops, 475
kill switches for, 484
Genetic analysis, **83**, 458–63
of acid resistance in *E. coli,* 458–60, 461f
identification of marked genes, 460
marking target genes with transposons, 458–59, 460f
reporter fusions, 460–63
Genetic code, 35, 276, 286–89, eTopic 1.2. *See also* Codons
degeneracy or redundancy, 287, 289
origins on early Earth, 678–79
transfer RNA, 287–89
wobble, 289, 617
Genetic drift, 681
Genetic engineering, 32, 35, 329. *See also* Biotechnology; Synthetic biology
Genetics, 34–36
Gene transfer, 321–37. *See also* Conjugation; Horizontal gene transfer
between bacteria and eukaryotes, 329–30
competence factor (CF), 323, 324f
competent cells, 323
conjugation, 324–29
electroporation, 323
fertility factor (F factor), 325
Gram-negative organisms, 324
Gram-positive organisms, 323–24
horizontal gene transfer, 323

natural transformation, 323–24
quorum sensing, 149, 323–24
transduction, 331–33
transformasomes, 323–24
transformation, 264, 322–24
Gene transfer vectors, **450**
Genital herpes virus (HSVS2), 206f, 207, 444, 445f
Genital warts, 215f. *See also* Human papillomavirus (HPV)
Genitourinary tract
microbiota, 934, 935
sexually transmitted diseases, 1077–83
urinary tract infections (UTIs), 744, 934, 1018, 1075–76
vaginal yeast infections, 934, 1143
Genome bins. *See* Bins (genome bins)
Genome evolution, 358–61
Genome libraries, **265**, 271
Genome mining, 606–7, 655–56
Genome reduction, 360
Genomes, **9**, **239**
archaeal genomes, 262
background and overview, 237–38
bacterial genomes, 9–10, 80, 102, 239, 359t
circular display map, 312f
core gene pool, 359
eukaryotic compared to prokaryotic, 261–62
evolution of, 358–61
flexible gene pool, 359
GC content, 359
genomic islands, 271, 359–60
mosaic nature, 322
pathogenicity islands, 359, 360f
of representative bacteria and archaea, 240t
sequencing, 9–10
sizes of genomes, 240–41
viral genomes, 193, 195–96, 198, 203–6, 409
Genome transplants, 485
Genomic islands, 271, 359–60, 535, **691**, **1011**, 1012
Genotype, definition, 343
Gentamicin, 298, 1057, 1091, 1096, 1124, 1132, 1133f
Genus name, **700**
Geobacter, 545, 552, 561
Geobacter metallireducens, 534, 561, 746, A-64t
Geobacter sulfurreducens, 539
Geochemical cycling, **27**. *See also* Biogeochemical cycles; Carbon cycle; Nitrogen cycle
global reservoirs of carbon, nitrogen, and sulfur, 892t
hydrologic cycle (water cycle), 896, 897f
microbial metabolism of metals, 914, 915t
oxidation states of cycled compounds, 892t
sources and sinks of essential elements, 891, 893
Geogemma, 28, 28f
Geological evidence for early life, 669–75. *See also* Origins of life
Archaean eon, 669, 670f, 671–72, 673, 676f
banded iron formations (BIFs), 671t, 674–75
biosignatures (biological signatures), 672–73
cyanobacterial hopanoids, 673
Hadean eon, 669, 670f
isotope ratios, 671t, 672–73
stromatolites, 666, 669, 670f, 671t, 672
time line, 675f
Geomicrobiology, **891**
Geomyces, A-68t
Geosmin, **730**
Geothrix capsulatum, A-65t
German measles (rubella virus), 1057–58
Germicidal agents, definition, **179**
Germination, **151**
Germ theory of disease, **20**
Germ tubes, 152, 154f
Geysers, 766
GFP (green fluorescent protein). *See* Green fluorescent protein (GFP)
GGDEF/EAL proteins, 396–97
Ghon complexes, 1064
Giant clams, 165
Giant kelp, 623, 624
Giant tube worms, 662
Giardia, 800, 1122
Giardia intestinalis, 65, 805, 835–36
Giardia lamblia, 65, 805, 835–36, 1073, 1166
Giardiasis, 1073
Gibbs free energy change (ΔG), **495–96**, A-17–A-18
in biochemical reactions, 498–501
in cells, A-18, A-20
and concentration gradients, 500–501
and concentration of reactants and products, 499–500
and direction of reactions, 495, 496
effect of concentration ratio, 499t
endergonic reactions, A-17
exergonic reactions, A-17
low-ΔG metabolic pathways, 500, 503
relationship with ΔH, ΔS, and T, 496, 497, 498, A-17–A-18
standard free energy change, 498–99, A-18, A-19
Gilbert, Walter, 9
Glaucocystophyta, 705
Gliding motility, 116, **152**, 746, 1015, 1029
GlnB regulator protein, 394–96
GlnD protein, 395
GlnE protein, 395–96
Global Ocean Sampling Expedition (GOS) metagenome, 895
Global warming. *See also* Climate change; Greenhouse effect
and Calvin cycle, 589
correlation with industrial age, 890
and ocean carbon flux, 230
permafrost melting and, 889, 896
Globigerinella aequilateralis, 827
Gloeobacterales, A-62t
Gloeocapsa, 722
Glomeromycota, 816, 874, A-69t
Glomus, 816, 874
GlpF and glycerol transport, 127–28
Gluconate, 520, 932
Gluconeogenesis, 500, 531, **598**
Glucosamine, 97f, **98**
Glucose
biosynthesis of, 500, 531
carbon cycle, 123f
fermentation to lactic acid, 629
glucose transport, 374, eTopic 10.2
pathways to pyruvate, 516–17
phosphorylation, 498f, 499
repression of *lac* operon, 374
straight-chain and cyclic forms, A-12f
structural formula, A-11f
transport by phosphotransferase system (PTS), 374, 375f
Glucose 6-phosphate
in Entner-Doudoroff pathway, 516, 520, 521f
isomerization to fructose 6-phosphate, 516, 518, 519f
oxidation to 6-phosphogluconate, 516, 520, 521f, 522f
from phosphorylation of glucose, 498f, 499, 519f
transport through phosphotransferase system, 132, 375f
Glucose catabolism, 516–27. *See also* Entner–Doudoroff (ED) pathway; Glycolysis; Pentose phosphate pathway (PPP)
in archaea, 760–61
ATP production, 505, 530–31
electron transfer, 501, 530–31
substrates for biosynthesis from, 587–88
Glucose respiration, 531
Glutamate, 613–14, 616
Glutamate decarboxylase (GadA and GadB), 460, 461f, 463, 464f, 465, 466
Glutamate dehydrogenase (GDH), 613–14
Glutamate/GABA antiporter (GadC), 460, 461f, 463, 464f
Glutamate/glutamine balance, 394–96
Glutamate pathway of tetrapyrrole biosynthesis, 618–19, 620f
Glutamate synthase (GOGAT) (glutamine:2-oxoglutarate aminotransferase), 613–14
Glutamic acid, 634
D-Glutamic acid, 92, 93f
Glutamine, 613–14, 616, 617
Glutamine synthase gene (*gltB*), 315
Glutamine synthetase (GS; GlnA), 303, 394–96, 613–14
Glutaraldehyde, 184f
Gluten, 638
Glycan, **92**, 93f
Glyceraldehyde 3-phosphate (G3P)
Calvin cycle, 591, 592
in Entner-Doudoroff pathway, 520, 521f
in glycolysis (EMP pathway), 517, 518, 519f

Glycerol
as building block of lipids, A-15–A-16
catabolism, 510–11, 512f
cold storage of bacteria with, 181
D-glycerol, 758
as fermentation product, 525
L-glycerol, 758
in membrane lipids, 86, 87f, 89, 90f, 164
nonitol, 758
Glycerol transport, 127–28
Glycine, 163, 613
Glycine-succinate pathway of tetrapyrrole biosynthesis, 618–19, 620f
Glycocalyx, 929, 930, 1092
Glycogen, 112, 176
Glycolysis, 512f, **516**, **517–19**
ATP generation, 518, 519f
irreversible steps, 518, 519f
phosphorylation and splitting of glucose, 518, 519f
regulation, 518–19
Glycoproteins and biofilm development, 149
Glycosylases, 351
Glynn House, Cornwall, England, 548
Glyoxylate, 600f, 601
Glyoxylate bypass, **531–32**
Gnotobiotic animals, **935**, 994–95
Goiters, 1000
Golgi complex (Golgi apparatus), A-34–A-35, A-36f, **A-37**
Gonorrhea, 185, 743, 936, 1079–81. *See also Neisseria gonorrhoeae*
Gophna, Uri, 757
Gordon, Jeffrey, 780–81
Gosink, Khoosheh, 1134
Gottesman, Susan, 386, 387f
Gout, 617, 626
G phase, 103
Gram, Hans Christian, 56
Gramicidins, **1121**
Gram-negative cells, **57**, 82
cell envelope, 94, 97
conjugation, 325
culture media, 122t, 135
electron transport in, 544
endotoxin, 98, 1018, 1027–28
Gram-negative bacteria, elemental composition, 891t
Gram-negative bacteria, phylogeny, 713f, 715t, 717
iron transport, 131
outer membrane, 97–99
protein secretion, 306–7
substrate-binding proteins, 130–31
transformation, 324
urinary tract infections, 1075–76
Gram-positive cells, **57**
cell envelope, 94–96
cell wall, 95, 716–17
culture media, 122t, 135
Gram-positive bacteria, phylogeny, 713f, 715t, 716–17
S-layer, 95–96, 717
spore formation, 150–51
substrate-binding proteins, 130–31
transformation, 323–24
Gram stain, **56–57**, 58f, 94, 95
Granulibacter bethesdensis, 737
Granulocytes. *See* Polymorphonuclear leukocytes (PMNs)
Granulomas, **949**, 950f
Granzymes, **988**
Grapes, 639–40, 641f
Graves' disease, 999–1000
Gray (Gy), 182
Grazers, 230, **850**
Green algae (Chlorophyta), **803**, 805, 818–21, A-69t
Greenberg, Peter, 400
Green chemistry, 654
Green fluorescent protein (GFP), 61–62
discovery, 61
expression in riboswitchboard, 483f
fluorescence-activated cell sorting, 137, 138f
genetic toggle switches, 480–81
GFP–TBGp3 fusion protein, 474
observing cell pH, 41
oscillating flashing in *E. coli,* 478–79, 481
QS-dependent expression of, 365
reporter fusions, 61, 83, 100–101, 224–25, 461, 462–63
tracking cell movements in biofilms, 474, eTopic 12.2
tracking movement of proteins, 473–74
transfer through interspecies nanotubes, 402, 403f
Greenhouse effect, 601, **669**, 680, **894**
Greenhouse gases
increases in atmospheric CO_2 levels, 890, 894
increases in atmospheric methane levels, 890
increases in atmospheric nitrous oxide (N_2O) levels, 890, 907
Kyoto Protocol, 890
Green nonsulfur bacteria, 714
Green sulfur bacteria, 27f, 577–78, 598, 717, 749–50
Griffith, Frederick, 34, 238–39, 322
Griseofulvin, **1143**
Grossman, Alan, 105f, 321
Ground beef, bacterial growth in, 648f
Group A beta-hemolytic streptococci (GAS). *See Streptococcus pyogenes*
Group translocation, 131–32
Growth curves, 142f, 143
one-step curve for animal virus, 222f
one-step curve for bacteriophage, 221f
Growth cycle, 139–45
exponential growth, 140
generation time, 140–41
overview, 139
Growth factors, **121**
Growth rate, **140**, 160–61, 164–65
Growth rate constant, **140–41**, 160
Guanine (G), 239, 243
Guanosine monophosphate (GMP), 617
Guanosine tetraphosphate (ppGpp), 176, 381, 396, 603, 693
Guanosine triphosphate (GTP), 505
Guillardia theta, 706, 707f
Gulf of Mexico dead zone, 898, 899f
Gut-associated lymphoid tissue (GALT), 942, **944**, 982, 994
Gutman, Antoinette, 209
Guttaviridae, A-60t
Gymnodinium, 830
GyrA and GyrB proteins, 247
Gyrase, 102

H

Haber, Fritz, 904
Haber process, **608**, 904
Hadean eon, **669**, 670f
Haeckel, Ernst, 30
Haemophilus, 121t, 323, 324, 1157
Haemophilus influenzae, 9–10
genome, 10f
identification, 1155
meningitis, 1084, 1153, 1155
pinkeye, 928
vaccines, 1155
X and V factors, 1155, 1157f
Hajela, Neerja, 3
Half-life, definition, **A-2**
Haloarchaea, 7f, **784–90**, A-67t
acidic proteins, 784
alkaliphilic haloarchaea, 170, 785, 787
applications in research and industry, 790
bacteriorhodopsin, 571, 676–77, 761, 787–88
bacteriorhodopsin photoswitch, 790
bacterioruberin, 784
cell form and physiology, 784–85
classroom study, 786–87
diversity, 765
gas vesicles, 785, 787, 790
halocins (antibiotics), 790
halorhodopsin, 787–88
high GC content of DNA, 784
hypersaline habitats, 786–87, 851–52
hyperthermophilic halophiles, 786
interspecies hybrids, 757
metabolic map (*Halobacterium* NRC-1), 789f
phototaxis, 788
representative species and properties, 785t
retinal-based photoheterotrophy, 784, 787–90
sensory rhodopsins I and II, 788
sodium ion pumps, 167, 559
square haloarchaea, 785f
vaccine delivery, 758, 787, 790
Haloarcula, 786, A-67t
Haloarcula marismortui, 240t
Halobacteriales, **784**, A-67t
Halobacterium, 167f, 549, 760–61, 785, 786, 787
Halobacterium halobium, 761
Halobacterium NRC-1, 758f, 765, 785, 786–87, 788, 789f
Halobacterium salinarum, 170, 571, 572
Halochromatium roseum, 112f
Halocins, 790

Haloferax dombrowskii, 785
Haloferax mediterranei, 757
Haloferax volcanii, 545f, 546, 757
Halogenated aromatics, 513
Halogenation, 618
Halophiles, 159t, **167**. *See also* Haloarchaea
Halophilic salt flats, 167f
Haloquadra, A-67t
Haloquadratum walsbyi, 697, 785
Halorhodopsin (HR), **787**
Halorubrum lacusprofundi, 756, A-67t
Halothiobacillus neapolitanus, 596f
Haloviridae, A-60t
Han, Yejun, 159f
Handelsman, Jo, 841, 847–48
Hanseniaspora, 640
Hantavirus, 1181
Haptens, **967–68**, 999
Harper, Kristin, 1078
Hartwell, Leland, 809
Harwood, Caroline, 540–41, 735, eTopic 14.1
Hasty, Jeff, 478–79, 481
Haustoria (singular, haustorium), **881–82**
Hawaiian squid (bobtailed squid, *Euprymna scolopes*), 366, 397–98
Hawkins, Aaron, 159f
Hay fever (allergic rhinitis), 973, 996–97
Heat-shock proteins (H-Ps), 161, 382–83
Heat-shock response, **164**, 172, 383f
Heavy chains, **969**, 970–71, 978, 979f
Heinzen, Robert, 121
Hektoen agar, 1153, 1154f
Helicases
  DNA helicase (DnaB), 252, 251f, 253, 254f, 256
  DNA helicase (UvrD), 349
  DNA helicase loader (DnaC), 252, 251f
  RNA-DNA helicase activity of Rho, 282
  RNA helicase, 286
Helicobacter, 1033, 1045–46, A-64t
Helicobacter hepaticus, 935
Helicobacter pylori
  ammonia production by urease, 172, 748, 931, 1072
  average nucleotide identity of orthologs, 697
  and cancer, 931f
  culturing in the laboratory, 175–76
  discovery, 22–23, 747
  enzyme-linked immunosorbent assay (ELISA), 1072
  ETS in inner membrane, 545f
  gastric cancer, 1072
  in gastric crypt cells, 42, 1071f
  gastritis and stomach ulcers, 22, 42, 172, 747–48, 931, 1045–46, 1071–72
  genetic variation, 361, 696, 697
  genomic predictions of transport and metabolic pathways, 316, 317f
  growth in mucous lining of stomach, 931
  hydrogen gas oxidation, 523
  metabolism, 748
  oxidation of hydrogen gas in colon, 563
  pathogenesis, 1045–46
  steps in colonization, 1046f
  T cell apoptosis, 991
  urease, 172, 316–17f, 747, 1072
  VacA toxin, 1019
Heliobacteriaceae, 719t, 726
Heliobacterium, A-63t
Heliophilum, A-63t
Heliozoa, A-70t
Helix-turn-helix motif, 368f
Helling, Robert, 265
Helper T cells (T_H cells), **980**, 982
  B-cell activation, 982
  CD4 helper T cells, 982, 983, 989–91
  cytokine regulation of T_H1 and T_H2 cells, 990
  precursor T_H0 cells, 982, 986
  T_H1 cells and activation of cytotoxic T cells, 988, 990, 994
  T_H1 cells and activation of macrophages, 991–92
  T_H2 cells and B cell differentiation, 982, 990
Hemagglutinin, 410, 425f, 426, 427, 428, 429f, 431, 1138–39
Hemes, 135, **552–53**, 555, 619, 620
Hemocytometers, 136
Hemolysins, 309f, 310, 311, 478, 1022, 1038
Hemolytic uremic syndrome (HUS), 1025, 1067, 1068–69, 1070, 1182
Hemoptysis, 1152
Henson, Jim, 989
Hepadnaviridae, 1099, A-61t
Hepadnaviruses, 205t, 206, 219
HEPA filters, 181, 182f
Hepatitis, **1099–1100**
Hepatitis A virus (HAV), 1099, 1100
Hepatitis B virus (HBV), 206, 1099, 1100
Hepatitis C virus (HCV), 417–24, 1099–1100
  attachment and host cell entry, 420, 421f
  background and history, 193, 417–18
  cell-to-cell transmission, 420, 422f
  effects of technology and progress on spread, 1180
  genome uncoating, 214f, 215
  host receptors and tight junction proteins, 420, 421f
  hypervariable regions, 422–23
  internal ribosome entry site (IRES), 419f, 420
  liver cell culture, 418, 423f
  membranous replication complexes, 421–22, 423f
  mouse model for HCV infection, 420, 421f
  mutation and quasispecies formation, 422–23
  nonstructural proteins (NS), 419, 420, 422
  as oncogenic virus, 218
  overview, 1099–1100
  polyproteins, 420, 422f
  recombination and template switching, 423
  replication cycle, 420, 422
  (+) strand RNA, 206, 419–20
  structural proteins (C, E1, E2), 419, 422
  treatments, 418, 420, 422, 423
  virion structure, 418f, 419
Herd immunity, 1102, 1138
Herpes simplex virus (HSV), 443–50
  attachment and entry to host cell, 446
  envelope and tegument surrounding herpes capsid, 199f
  fetal and infant infections, 444
  gene expression, 446, 447f, 448
  genital herpes virus (HSV-2), 206f, 207, 445f
  genome, 199, 445f, 446
  icosahedral capsid, 198–99, 444, 445f
  infection of neurons, 444, 448
  infection of oral and genital mucosa, 444
  latent infections (latent state), 444, 446
  LAT proteins, 446, 447f, 448
  lytic infection, 444, 446
  oral herpes virus (HSV-1), 206f, 207
  persistent infections, 446, 448–49
  phylogeny, 206f, 207
  prevalence of infection, 193
  replicative cycle, 446, 447f
  rolling-circle replication, 446, 447f
  spike proteins, 199, 446
  structure, 198–99, 444–46
  tegument, 199, 444–46
Herpes viruses. *See also* Epstein-Barr virus
  cryo-EM, 70
  cytomegalovirus (CMV), 206f, 207, 446, 448–49, 451–52
  evasion of immune responses, 449
  evolution, 196, 448
  gamma herpes virus in mice, 446, 448
  genomes and classification, 25, 195, 203, 205t
  human herpes viruses 6 and 7, 446
  latent infections (latent state), 1006, 1042
  microbicides, 185
  mutualism, 448
  phylogeny of genomes, 206–7
  structure, 198
  varicella-zoster virus (VZV), 206, 207, 444
Hershey, Alfred, 193, 208
Hess, Matthias, 845–46
Hesse, Angelina, 21
Hesse, Walther, 21
Heteroaromatic molecules, 506, 552
Heterochromatin, 448
Heterocysts, 75, 152, **608**, 611, **720**, 721f, 852
Heteroduplex, **340**
Heterokonts (Heterokonta), **805**, 823–24, A-70t
  coccolithophores, 823–24
  definition, 805
  diatoms, 805, 822f, 823
  kelps, 805, 824
  oomycetes, 802, 805
  phylogeny, 799t, 800, 801f, 802, 805
Heterolactic fermentation, **521**, **629**, 634
Heterorhabditis bacteriophora, 872
Heterotrophs, heterotrophy, **123**, **124**, 126–27, **492**, 495

Hexachlorocyclohexane, 686
Hexachlorophene, 184f
Hexokinase, 499, 519f
Hexoses, A-11
Hfr (high-frequency recombination) strains, **327**
The Highest Frontier (novel), 625, 857
Highly active antiretroviral therapy (HAART), 433, 1142, **1142**
High-performance liquid chromatography (HPLC), 592, 600f
Hill, Russell, 883
Hin recombinase, 389, 390f
His_6 peptide tag, 466, 466–67f
Histamine, 948, 973, 996
Histidine, 616
Histidine kinases, 788
Histocompatibility, 983
Histone-like proteins, 244, 245
Histones, **261**, **762**
Histoplamosis, 1063
Histoplasma, 1007
Histoplasma capsulatum, 816, 817f, 1082
Historical contingency model, 694
HIV. *See* Human immunodeficiency virus (HIV)
Hmf histone protein, 244
H-NS (DNA binding protein), 81f, 102, 245, 922, 923f
Hodgkin, Dorothy Crowfoot, 34, 75, 76f
Hoffman, Jules, 954
Holdfasts, **113**, 119
Hollandina, A-65t
Holley, Robert, 276
Holliday, Robin, 340
Holliday junction, 339f, **340**
Holorepressors, 379f
Homer Price (McCloskey), 625
Homochirality, 919
Homologous genes or proteins (homologs), **315–16**
Homologous sequences, 314–15
Homologs, **A-33**
Honeybee (*Apis mellifera*), 268, 840
Hooke, Robert, 15
Hopanoids (hopanes), 86f, **90**, 91, 673, A-26
Hops, 639
Horizontal gene transfer, 238–39, **323**, 681, **690–91**
  and antibiotic resistance, 1127
  in archaea, 763
  from archaea to bacteria, 714
  conjugation, 239
  and digestion of seaweed, 516
  effect on phylogeny, 690–91
  Escherichia coli O157:H7, 358–59, 691, eTopic 17.2
  from eukaryotes to bacteria, 754
  evidence for horizontal transfer, 359–60
  and genome evolution, 323, 358–60
  and genomic islands, 271, 1011
  between hyperthermophiles, 714, 791
  and nicotine catabolism, 34, 239, 535
  obscuring of phylogeny, 690f, 691f
  pathogen evolution by, 1012
  of proteorhodopsin genes, 571
  transformation, 34, 239
  and xenogenic silencing, 922
Horizontal transmission (genetics), **238**
Horizontal transmission (infection), **1007**, 1008f
Hormogonia (singular, hormogonium), **720**
Horseradish peroxidase, 465
Hospital-acquired infections. *See* Nosocomial infections
Host cell membrane ruffling, 1031f, 1032
Host cell reprogramming, 1011
Host factors, **426–28**
Host-pathogen interactions, 1004–9
Host range, **194**
Hot springs, 31f, 35, 89, 162, 655, 766
Houston, Clifford W., eTopic 1.3
Howard Hughes Medical Institute, 36
HTST (high-temperature/short-time) process for pasteurization, 181
Htz histone protein, 244
HU (DNA binding protein), 81f, 102, 245
Human aging, 828, eTopic 20.2
Human body odor, 405
Human breast milk, 515, 932
Human cytomegalovirus, 991
Human embryonic kidney 293T cells, 452
Human feces, 931, 932
Human immunodeficiency virus (HIV), **432**, 432–43. *See also* Acquired immunodeficiency syndrome (AIDS); Retroviruses
  absence of animal host, 22
  accessory proteins, 435f, 436, 437t, 440, 442
  antiretroviral therapy (ART), 194, 1082, 1140–42
  apoptosis prevention, 449
  attachment and host cell entry, 436–38
  azidothymidine (AZT), 194, 432, 438, 439f, 1140
  binding to CCR5 cell surface protein, 437f, 438
  case history, 1140
  causative role in AIDS, 432–33
  CD4 cell surface protein, 436, 437f, 438, 442, 991
  cell-to-cell transmission, 442
  and chemokine receptors (CCRs), 437f, 438
  chronic asymptomatic HIV, 1082
  commercial disinfectants, 184, 185
  coreceptors, 220
  core particle, 434–35
  "cures," 1142
  cyclophilin A, 434, 435f
  detection by polymerase chain reaction (PCR), 21–22
  drug resistance, 436f
  emergence as pathogen, 227
  env sequence, 436, 451
  fusion peptides, 437f, 438
  gag sequence, 436, 451, 452
  genetic resistance to, 220, 438, eTopic 11.3
  genome, 25, 434–36
  highly active antiretroviral therapy (HAART), 433
  history, 14, 193, 433–34
  HIV-1, 432–33, 434, 435, A-61f
  HIV-2, 433, 435
  host range, 194
  incubation period, 223
  integrase, 434, 435f, 440
  integrase and Tn*5* transposase, 358, eTopic 9.5
  life cycle, 217–18
  long terminal repeat (LTR), 439f, 440
  maraviroc therapy, 437f, 438, eTopic 11.3
  matrix, 435
  mutation of envelope protein, 410
  nucleocapsid proteins, 434, 435f
  origin and evolution, 196, 435–36
  pathogenesis, eTopic 26.6
  pol sequence, 436, 451
  polypurine tract (PPT), 440, 452
  preexposure prophylaxis, 1142
  prevention of autophagosome formation, 1042
  protease, 434, 435f, 442
  protease inhibitors, 442
  provirus in host chromosome, 439f, 440
  quasispecies, 435–36, 438
  raltegravir, 439f, 440
  replication cycle, 440–42
  resistance by CCR5 cell surface protein, 1017
  as retrovirus, 206
  reverse-transcribed DNA insertion into human genome, 439f, 440
  reverse transcriptase, 434, 435f, 438–40
  rev gene, 451
  spike proteins, 434, 435, 436–38
  structure, 70, 434–35
  tethering by host proteins, 442
  treatments, 433, 434, 436, 438, 440, 442
  trimer complex of SU and TM, 435f, 436–38
  as tuberculosis risk factor, 652, 1064, 1082, 1181
  vaccine development challenges, 434
  as vector for gene therapy, 194, 218, 434, 436, 443, 450, 451–52
Human intestines, microbiota, 931–34. *See also* Gastrointestinal tract infections
  composition and human disease, 272
  diet and, 932
  and digestion, 514f, 515–16, 885, 931–32
  and immune system, 885
  inhibition of pathogen growth, 187–88
  metagenomics, 841–42, 932
  mucus farming, 520, 932
  organisms inhabiting intestines, 932
  probiotics, 933–34
Human Microbiome Project, 271–72, 712

Human Microbiome Project Consortium survey, 712
Human microbiota (microbiome), 926–35. *See also* Human intestines, microbiota
  bacterial ecosystems of humans, 928t
  body sites colonized by bacteria, 927f
  commensal organisms, definition, 926
  eye, 928
  genitourinary tract, 934
  intestines, 931–34
  oral and nasal cavities, 928–30
  respiratory tract, 930
  risks and benefits of, 934–35
  skin, 187, 928
  stomach, 930–31
  virome, 928
Human papillomavirus (HPV), 198, 205t, 215–16, 218, eTopic 26.2
Human spermatozoan, A-68f
Human T-cell leukemia virus (HTLV). *See* Primate T-lymphotrophic virus (PTLV-1)
Humic material (humus), 561, **872–73**
Humoral immunity, **963–64**, 965, 987f. *See also* Adaptive immunity; Antibodies; B cells
Hungate, Robert, 884
Huq, Anwar, 856
Hurricane Katrina, 814
Hybridization, **244**, **A-50**, A-55–A-56
Hydrazine, 566
Hydrazine synthase complex, 566
Hydric soil, **875**
Hydrogen
  as electron donor, 542, 568–69
  as essential nutrient, 120
  hydrogen gas production by microbes, 540–41
  hydrogen reduction of carbon dioxide, 570
  oxidation by sulfur or nitrate, 492, 748
  syntrophy, 500, 502–3
  ΔG of oxidation to form water, 496
Hydrogen bonds, **A-4–A-5**
Hydrogenimonas, 748
Hydrogenosomes, **810**
Hydrogenotrophy, **568–69**, 676, 748, 902, 910
Hydrogen peroxide, 174, 184t, 353–54, 952
Hydrogen sulfide, 643
Hydrologic cycle (water cycle), **896**, 897f
Hydrolysis, **504–5**, **A-6**
Hydronium ion, 167
Hydrophilic compounds, **A-5**
Hydrophobic compounds, **A-5**, A-15
Hydrothermal (thermal) vents, **865**
  acidophilic hyperthermophiles, 771
  barophilic hyperthermophiles, 766, 767–69
  biofilms, 769, 770f
  black smokers, 6, 767–68, 769
  discovery of Ancient Arachaeal Group in, 765
  microbial communities, 850, 865
  and origins of life, 676f, 680
  temperature near vents on ocean floor, 161
  thermal vent ecosystems, 165, 850, 865, 866f, 910
Hydroxide ions, 952
Hydroxyl radical (*OH), 174, 952
3-hydroxypropionate cycle, 590t, 600f, 601
Hypersensitivity, 994–99
  anaphylaxis, 973, 996, 997
  case studies, 996, 998, eTopic 24.4
  desensitization, 997
  late-phase anaphylaxis, 997
  to penicillin, 967
  summary of hypersensitivity reactions, 996t
  type I hypersensitivity (immediate), 996–98
  type II hypersensitivity, 999
  type III hypersensitivity, 999
  type IV hypersensitivity (DTH) (delayed-type), 998–99
Hyperthermophiles
  acidophilic hyperthermophiles, 769–71
  barophilic hyperthermophiles, 766, 767–69
  biofilms, 769, 770f
  classification, 159t
  conjugation, 324–25
  Crenarchaeota, 763, 765, 766–71
  horizontal gene transfer in, 714, 791
  Korarchaeota, 793
  reverse gyrase, 248, 761
  sulfur reduction, 851
  temperature and growth rate, 160f, 851
  thermal vent communities, 865
Hyperthermus butylicus, A-66f
Hyperthyroidism, 999–1000
Hypertonic environment, **A-29–A-30**
Hyphae (singular, hyha), **152**, 153f, 154f, **802**
  absorption zone, 807
  aerial hyphae, 153f
  in filamentous bacteria, 152, 153f
  in fungi, 806–7
  hyphal extension, 806–7
  senescence zone, 807
Hyphomicrobium, 139, 740
Hyphomicrobium sulfonivorans, 740
Hypochlorhydria, 931
Hypochlorous acid, 952
Hypolimnion, 867
Hypothalamus, 958
Hypotonic environment, A-29f, **A-30**
Hypoxanthine, 351
Hypoxia, **898**, 899f

I

ICAM-1 (intracellular adhesion molecule 1)
  C5a upregulation of, 957
  rhinovirus attachment, 213, 214f, 216, 966–67
  and selectins, 948, 949f
Ice-nucleating bacteria, 162, 163
Ichthyosporea, A-69t
Icosahedral capsids, **198–99**
Identification, **697**
Idiotypes, **971**
IgA, 970, 971, **972**
IgD, 970, 971, **972–73**, 975
IgE, 939, 970, 971, 972, **973**, 996, 997–98
IgG, 970, 971, **972**, 974, 1037
IgM, 970, 971, **972**, 973, 974, 975, 976
Ignicoccus, 765, 793
Ignicoccus hospitalis, 793f
Ignicoccus islandicus, 766–67, A-66t
Illumina, Inc., 270
Imipenem, 1128
Immersion oil, **52**, 53f
Immobilized pH gradient (IPG) strip, A-46, A-47f
Immune avoidance, **389**, 1037–41
Immune system, **23**, 220–21, **935–42**. *See also* Innate host defenses
  adaptive immunity, 936, 937, 940
  antigens, definition, 936
  autoimmune response, 999–1000
  cells of the immune system, 937–40
  complement, 936
  gut microbiota and development of, 925
  hypersensitivity, 995–99
  infection *vs.* disease, 936–37
  innate immunity, 936, 937
  lymphoid organs, 940–42
  memory, 961
  microbial strategies to misdirect, 1041–44
  monocytes, 937f, 939
  mucosal immunity, 981–82
  overview, 935–42
Immunity, definition, **23**
Immunization, **23**, 1102–3. *See also* Vaccines
  booster doses, 1102
  definition, 23
  diseases with available immunizations, 967t
  herd immunity, 1102, 1138
  history, 22–23
  recommended immunization schedule, 1102, 1103t
  secondary antibody response and, 975, 1102
Immunochromatographic assays, 1168
Immunofluorescence microscopy, 61, 835
Immunogenicity, **965–68**. *See also* Antigens
  contributing factors, 965, eTopic 24.1
  definition, 965
  immunological specificity, 965–67
  threshold dose, 965
Immunogens, **963**, **967–68**. *See also* Antigens
Immunoglobulins, **968**, 973t. *See also* Antibodies
  IgA, 970, 971, 972
  IgD, 970, 971, 972–73
  IgE, 939, 970, 971, 972, 973, 996, 997–98
  IgG, 970, 971, 972, 974, 1037
  IgM, 970, 971, 972, 973, 974, 975, 976
  immunoglobulin superfamily, 968
Immunological specificity, **965–67**, 968
Immunomodulation, 933–34

Immunomodulins, **935**
Immunopathogenesis, **1009**
Immunoprecipitation, **969–70**, **A-56–A-58**
  radial immunodiffusion, 970, A-57–A-58
  western blotting, 970, A-58
Inactivated polio vaccine (IPV), 981
Index case, **1172–73**
Indigenous microbiota, **627**, **629**, 636
Inducer exclusion, **374**
Inducers, **366–67**
Induction, **366–67**
Industrial effluents, bioremediation of, 901–2
Industrial fermentors, **656–57**
Industrial microbiology, **650–60**, 657, 658f. *See also* Bioremediation; Food microbiology; Wastewater treatment
  bioprospecting, 654–56
  commercial products from microbes, 651–56
  directed evolution of xylanase, 696, 698–99
  downstream processing, 657, 658f
  fermentation systems, 656–57
  goal of commercial success, 650–51
  industrial strains, attributes and engineering, 656
  laboratory evolution, 696
  overview, 623–25
  primary recovery, 657, 658f
  production in animal or plant systems, 657, 659, eTopic 16.2
  upstream processing, 657, 658f
Industrial strains, **656**
Infection, **1005–6**
  biofilms and infections, 1017–18
  chemical barriers to, 945–46
  definition, 1005
  infection *vs.* disease, 936–37
  physical barriers to, 943–45
Infection and Immunity (journal), 923
Infection cycles, **1007–9**
Infection thread, 740, 741f, **878**, 879f
Infectious dose (ID_{50}), 1006
Infectious mononucleosis, 443, 444t. *See also* Epstein-Barr virus
Inflammasomes, **954–55**, 1041, 1042
Inflammation
  acute inflammatory response, 946–49
  chronic inflammation, 949, 950f, 1018
Inflammatory bowel disease (IBS), 188, 934
Influenza virus, 424–32
  acid activation of virion, 428–30
  antigenic drift, 428
  antigenic shift, 426
  antiviral treatment, 1137–38
  attachment and entry to host cell, 426–30
  avian influenza strain H5N1, 213, 409, 426, 427–28, 1172
  avian influenza strain H7N9, 6, 228, 426
  avian influenza virus, 1007
  background and history, 424
  classification, 206
  common cold *vs.* influenza, 1065, eTopic 26.4
  effects on technology and progress on spread, 1181
  emergence of new strains, 227–28, 424
  envelope synthesis and assembly, 431
  epidemic of 1918, 193
  epidemic of 2009, 221
  ferret model system, 409, 428
  fusion peptides, 426
  genome packaging, 201
  hemagglutinin (HA), 410, 425f, 426, 427, 428, 429f, 431, 1138–39
  Hong Kong strain, 426
  host factors, 426–28
  influenza strain H3N2, 1138
  M2 protein, 429f, 430, 1042, 1138
  matrix protein (M1), 425, 431
  neuraminidase (NA), 410, 425f, 426, 427f, 431, 1138–39
  neuraminidase inhibitors, 1138–40
  nucleocapsid proteins (NPs), 425, 430, 431
  origin, 196
  pandemic of 1918, 424, 426, 427f, 1139, 1180, eTopic 27.4
  prevention of autophagosome formation, 1042
  reassortment of genome, 424, 426, 427f
  replication cycle, 430–31
  RNA-dependent RNA polymerase, 425, 430, 431
  RNA genome, 25
  segmented genome, 424, 426
  sialic acid in receptor proteins, 427–28, 429f
  (–) strand RNA, 424, 425–26, 430
  (+) strand RNA, 425f, 430–31
  swine as mixing bowls, 428
  swine flu (influenza strain H1N1), 410, 426, 1173, 1181f
  Tamiflu (oseltamivir), 426, 431, 1138
  vaccines, 1102, 1138
  virion structure, 425
Information flow, **238**
Infrared absorption spectroscopy, 893
Initiation of DNA replication, 250–53
Initiation of transcription, 277–78, 280–81
Initiation of translation, 294
Injera, 628t, **638–39**
Innate host defenses, 943–58. *See also* Interferons
  chemical barriers to infection, 945–46
  fever, 958
  interferons, 953
  lungs, 944–45
  mucous membranes, 943–44
  natural killer (NK) cells, 937f, 940, 953–54
  NOD-like receptors (NLRs), 954–55
  overview of innate (nonspecific) defenses, 943–46
  physical barriers to infection, 943–45
  skin, 943
  Toll-like receptors (TLRs), 952, 953, 954, 955t
Innate immunity, **936**, 937, 939, 940. *See also* Complement; Innate host defenses; Phagocytosis
  acute inflammatory response, 946–49
  chronic inflammation, 949–50
  cooperation with adaptive immunity, 951, 954
Inner leaflet, **A-26**
Inner membrane, 71, 79, 80–81f, 82, 94f, **97**, 99, 544
Inorganic ions
  biochemical composition of bacteria, 84
  transport across membranes, 88
Inosine, 288, 289
Inosine monophosphate, 617
Inositol hexaphosphate, 632
Inositol triphosphate, 996
Inoviridae, A-60t
Insects
  bacterial endosymbionts, 330
  crop protection against, 474–75
  mutualism with viruses, 229
  as pathogen vectors, 1007, 1008f
  transmission of fungal infections, 881
  transmission of protozoan parasites, 314–15, 831, 832, 833–35, 1092
  transmission of viruses, 191–92, 218, 226, 228, 417
Insertions, **341**, 342, 801, 802f
Insertion sequences (ISs), **355–56**. *See also* Transposable elements
Insulin, 74
Integral membrane proteins, 307
Integrase inhibitors, 1142
Integrated control circuits, 369, 391–97
  biofilms and second messengers, 396–97
  chemotaxis, 391–94
  nitrogen regulation, 394–96
Integrins, 213, 214f, 948, 949f, 982, 1016
Integrons, 358, 1131, eTopic 9.6
Interactome, **472–73**, eTopic 12.1
Interference, **49–51**, 65, 73–74
Interferons, 220, **953**
  HCV treatment with, 418, 420, 423
  interferon-alpha, 940, 941f
  interferon-gamma, 990, 991
  interferon signal cascade, 1043–44
  type II interferons, 953
  type I interferons, 940, 953
Interleukin 1 (IL-1), **948**, 957
Interleukin 2 (IL-2), 988
Interleukin 4 (IL-4), 986, 987, 990
Interleukin 6 (IL-6), 987–88
Interleukin 10 (IL-10), 986, 987, 990
Intermediate filaments, 235, **A-38**
Internal pH (pH_{int}), 168, 169, 170, 172
Internal ribosome entry site (IRES), 419f, **420**
International Committee on Systematics of Prokaryotes (ICSP), 701
International Committee on Taxonomy of Viruses (ICTV), 203
International Genetically Engineered Machine (iGEM) competition, 484, 485f

International Journal of Systematic and Evolutionary Microbiology, 701
Intestinal mucus consumption by bacteria, 520, 932
Intestines, microbiota. *See* Human intestines, microbiota
Intimin, 1016, **1031–32**
Intracellular pathogens, 952, **1037–41**
Introns, **261**, 262f, **313–14**, 688, 762
Invadosome-like protrusions (ILPs), 961
Invasin, 478, 1031f
Inversion mutations, **341**, 342–43
Iodine, as disinfectant, 184, 185
Iodophors, 184
Ion gradients, 87, **88**
Ionic bonds, **A-4**
Ion potentials, 541
Ion pumps, 167, 551, 676–77
IPTG, 481
Iridoviridae, A-60
Irish potato famine, 644, 817
Iron
  banded iron formations (BIFs), 671f, 674–75
  corrosion of iron by sulfur-reducing bacteria, 567, 568f, 913
  as electron donor, 542
  elemental composition of Earth, 668, 669f
  generation of reactive oxygen species, 174
  iron cycle, 912–14
  iron fertilization in the ocean, 913–14
  iron-induced phytoplankton bloom, 913, 914f
  iron phototrophy, 674–75
  iron precipitate due to lithotrophy, 912f, 913
  iron-sulfur clusters, 552, 553, 574
  iron transport, 131, 913
  marine iron, 913–14
  reduction by *Geogemma,* 28
  reduction through anaerobic respiration, 561
  regulation of uptake by Fur, 385–86
  Shiga toxin and, 1025
  as transition metal, 782
Iron cycle, 912–14
Iron Mountain, 567
Iron ore, 674
Iron oxidation, 564–65, 671f, 674–75, 678, 913
Iron-oxidizing bacteria, 113–14, 499, 564, 742, 743f
Iron pyrite ore mining, 791, 792f, 793
Iron sulfide, 767, 768f
Iron-sulfur clusters, 552, 553, 574
Iron transport, 131
Irradiation, 181–83, 649, 1070
Irritable bowel syndrome, 272
Ishiwata, Shigetane, 724
Isocitrate, 528, 529f, 599
Isocitrate dehydrogenase, 303
Isoelectric focusing (IEF), **A-46**, A-47
Isoelectric point, **A-46**
Isoflavonoids, 905
Isolates, **701**
Isoleucine, 613
Isoniazid, 1065, 1110, 1111
Isoprene, 90, 91f, 164, A-15f
Isoprenoid chains, **759**, 760f
Isopropanol, 184f
Isotonic environment, **A-29**
Isotope ratios, **672–73**
  evidence of early life, 671t, 672–73, 917
  measurement of element flux in biosphere, 893
Isotopes, definition, **A-2**
Isotypes, **971**, 972–73
Isotype switching (class switching), **974**, 976, 977f, 980–81, 987
Israeli acute paralysis virus of bees, 268
Ivanovsky, Dmitri, 25, 192
Ivins, Bruce, 1171
Ivors, Kelly, 625f
Ixodes (deer tick), 1097f
Ixodes pacificus, 1096
Ixodes scapularis, 1096

J

Jacob, François, 369–70
Jacobs-Wagner, Christine, 108–10, 234–35
Jagendorf, André, 548
Jakana, Joanita, 70f
Jakobida, A-71t
Jannasch, Holger, 860, 865
J chain, 972
Jenner, Edward, 23, 966
Jensen, Grant, 71, 1034–35
Jiang, 634
Joint, Ian, 400
Joint Genome Institute, 316
Joule (J), **497**
Joyce, Gerald, 679
Jupiter, moons of, 918–19

K

K22 isolate of acidobacteria, 748, 749f
Kaback, H. Ronald, 128, 129f
Kamen, Martin, 529, 591, 608, 818, eTopic 15.1
Kanamycin, 298, 1129f
Kangaroos, 885
Kaposi's sarcoma, 1081f, 1082
Karem, Kevin, 1169f
Kashefi, Kazem, 28
Kasugamycin, 299–300
Keasling, Jay, 478
KEGG (Kyoto Encyclopedia of Genes and Genomes), 316
KEGG Pathway Database, 527
Keilin, David, 545, 546, 547
Kelley, Scott, 686
Kelly, Robert, 159
Kelp forests, 824
Kelps, **805**. *See also* Brown algae (kelp, Phaeophyceae)
Keratin, 943
Keratinocytes, 215f, 216, 943
Keratitis, 825
Kerns, Steve, 901
Kessler, John, 858f
2-ketoisovalerate, 614
Ketoreductase, 604
Ketoses, A-11
Ketosynthase, 604
Khorana, Har Gobind, 276, 286
Kidney stones, 626
Kimchi, 627, 628t, **634**
Kinesin, A-38f, A-39
Kinetochore, A-32
Kinetoplast, 833
Kirby-Bauer assay, 1113–15
Klebsiella, 684f, 744, 905, 928, 1075, 1182
Klebsiella oxytoca, 1154–55
Klebsiella pneumoniae, 610, 611, 938f, 1119, 1132, 1154–55
Kleckner, Nancy, 355
Kleptoplasty, 804
Kloeckera, 635, 640
Knockout mutations, **342**
Koch, Robert, 20–22, 26, 55, 841, 859, 1004, 1175
Koch's postulates, **21–22**, 1004, 1010, 1175, 1176
Korarchaeota, 765, 793, A-68t
Kornberg, Arthur, 252, 253
Kornberg, Sylvy, 253f
Koudelka, Gerald, 1025
Kovach, Janet, 787f
Krebs, Hans, 33, 523, 527
Krebs cycle. *See* Tricarboxylic acid (TCA) cycle
Krulwich, Terry, 559
Kuenenia, 908
Ku protein, 354
Kurtzman, Cletus, A-51f
Kuru, 197
Kustu, Sydney, 611
Kuypers, Marcel, 908
Kweon, Dae-Hyuk, 1133
Kyoto Protocol, 890

L

L1 sequences, 329
Labile toxin (LT), **1023**
Lab-on-a-chip, 1178
Laboratory evolution, **694–96**
LacA gene, 243, 370, 371f. *See also* Thiogalactoside transacetylase (LacA)
LacI (lactose catabolism regulator), 370–71, 372f, 373, 376, 385, 391, 481
lacI gene, 370–71, 481
LacO$_I$ operator sequence, 371
LacO operator sequence, 370–71
Lac operon. *See* Lactose (*lac*) operon
Lactate fermentation, 513, 514t, **521**, 525. *See also* Fermentation; Lactic acid fermentation
Lactic acid, 168, 629, 630
Lactic acid bacteria, 635, 642, 727, 840
Lactic acid fermentation, **629**
  acidic fermentation of dairy products, 628t, 630–32
  cheese, 630–32
  curd formation, 630
  yogurt, 630
Lactobacillales, A-63t

Lactobacillus, 723, 727, A-63t
kimchi preparation, 634
lactic acid fermentation, 628t, 629, 630, 632
oral microbiota, 928
as probiotic, 188, 933
in yogurt, 168, 188
Lactobacillus acidophilus, 45f, 188, 727, 933f, 934
Lactobacillus helveticus, 525, 630f
Lactobacillus johnsonii, 840
Lactobacillus lactis, 45f
Lactobacillus plantarum, 635
Lactococcus, 130, 193, 727, A-63t
Lactococcus lactis, 665
Lactococcus lactis subsp. *cremoris,* 665
Lactoferrin, 951
Lactoperoxidase, 944
Lactose catabolism, 340, 370, 372f, 391, 515
Lactose fermentation, 135–36, 346, 642
Lactose (*lac*) operon, 369–75. *See also* Operons
activation by cAMP and CRP, 371–73, 391
catabolite repression, 374
inducer exclusion, 374
as inducible system, 366
induction by lactose, 370–71, 510
lac promoter, 278f, 370, 371f
and lactose catabolism, 243, 370
lacZ reporter gene, 348
mutation, 346
naming conventions, 243
repression by glucose, 374, 510
Lactose permease (LacY), 128, 370, 371, 372f, 374, 375f
LacY gene, 243, 370, 371f. *See also* Lactose permease (LacY)
LacZ gene, 243, 340, 348, 370, 371f, 461, 462. *See also* beta-Galactosidase (LacZ)
LacZ reporter fusions, 461, 462, 472
LacZYA operon. *See* Lactose (*lac*) operon
Ladderanes, 566
Lag phase, **141–42**
Lainhart, William, 1025
Lake, James, 691
Lake Magadi, Kenya, 170, 171f
Lamellae in respiratory membranes, 545
Lamellar pseudopods, **825**
Laminar flow biological safety cabinets, **181**, 182f
Laminin receptor 1, 931f
Lancefield, Rebecca, 1159
Lancefield classification scheme, 1159
Landmark protein, 109, 110f
Landsteiner, Karl, 968, eTopic 24.2
Langerhans cells, 436, **943**, 998
Larson, Ann, 923
Laser confocal microscopy, 5f
Late genes, 413f, **414–15**
Latent period, **221–22**
Latent state, **1006**
Late-phase anaphylaxis, **997**
Law of mass action, **A-20**
Leaching, **568**, 742, 743f
Leader sequences, 285, 292f, 293, **378–79**
Leaf-cutter ants with partner fungi and bacteria, 704, 731, 856, eTopic 17.3
Leaflets, **86**, 97, 98
Leaven, leavening, **637**
Lebeis, Sarah, 839
Leber congenital amaurosis, 450
Lectin complement activation pathway, 956, 992
Lectins, 632, 636
LED209, 1136–37
Leeuwenhoek, Antonie van, 15–16, 42
Leghemoglobin, 611, **740**, 878, **879**, 880f
Legionella, 121t, 1038
Legionellales, A-64t
Legionella pneumophila
in amebas, 745–46, 824
biofilms, 686
fluorescent antibody staining, 1164f
in natural environment, 1007
phagosome–lysosome fusion inhibition, 1038–39
polyester granule synthesis, 603
survival inside macrophages, 1038–39
threonine catabolism, 511
Legionellosis, 824
Legionnaires' disease, 1039. *See also* *Legionella pneumophila*
Legumes
bacteroids in, 740, 741f, 877–78, 879f, 880f
flavonoids, 878
nitrogen fixation in nodules, 28, 878–81
symbiosis with rhizobia, 28, 740, 741f, 877–81
symbiosomes, 878, 880f
Leishmania, 1041, 1042
Leishmania major, 833
Leishmaniasis, 833f
Lenses, **48–49**
aberrations, 51, 66–67
magnetic lenses, 33, 66–67
objective lens, 51–52, 54
ocular lens, 53, 54
Lenski, Richard, 2–3, 667, 694–95
Lentinula edodes (black forest mushroom, or shiitake), 626
Lentivector, **450**, 451–53
Lentiviral vector, **450**, 451–53, 811
Lentiviruses, 205t, **223**, **432**, 433t, 443
Leprosy, 283, 732–34
Leptospira, 752, A-65t
Leptospira biflexa, 79
Leptospira interrogans, 45f
Leptospirillum, 842, 843f, 845
Leptospirosis, 45, 79
Leptothrix, 671, 672f
Lethal dose 50% (LD_{50}), **1006**, 1007f
Lethal factor (LF), **1025–26**
Leucine, 613, 616
Leuckart, Rudolf, 800f
Leucocin, 824
Leuconostoc, 629
Leuconostoc mesenteroides, 634
Leukocytes, 825. *See also* White blood cells
Leukotrienes, 947, 997
Levin, Bruce, 2
Leviviruses, 211
Lewis, Kim, 861
LexA repressor, 352–53
LH-1 central antenna complex, 576f
LH-2 accessory antenna complex, 576f
Li, Changsheng, 780
Lian, Hong, 159f
Lichens, **822**, **853–54**, A-69t
in boreal forest ecosystems, 854
in cryptogamic crust, 822, 854
ground cover, 854f
mutualism, 7f, 853–54
parasitism, 856
on stone or rock, 7f, 853f
Life, extraterrestrial. *See* Extraterrestrial life
Life, origins of. *See* Origins of life
Ligands, 366–67
LigD protein, 354
Light. *See also* Light microscopy; Magnification; Resolution
absorption, 47–48
contrast, 47, 51
electromagnetic spectrum, 47f
fluorescence, 48
frequency (ν), 47
information carried, 47
interactions with matter, 47–48
interference, 49–51
magnification and resolution, 49–51
magnification by a lens, 48–49
reflection, 48
refraction, 48–49, 65, 66f
scattering of, 48, 63–64
speed of light (*c*), 47
wavelength (λ), 47, 51, 60–61
Light chains, **969**, 970–71, 978, 981
Light microscopy, **42**, **45**. *See also* Microscopy
bright-field microscopy, 45, 51–58
dark-field microscopy, 63–64
fluorescence microscopy, 41, 58–63, 60f
phase-contrast microscopy, 64–65, 66f
size scales, 45–46
Lightning triggered by volcanic eruptions, 677
Lignin, 126–27, **513–14**, 561, 745, 806, 870, **872–73**
Limiting nutrients, **868**
Limnology, 865
Lincoln, Tracey, 679
Lincosamides, 1125
Lindahl, Björn, 894
Lindquist, Susan, 810–11
Linezolid, 693, 1056, 1124f, 1125
Linné, Carl von (Carolus Linnaeus), 29, 841
Linoleic acid, A-16f
Lipid autooxidation, 642
Lipid catabolism, 510–11, 512f, 531–32
Lipid monolayers compared to bilayers, 164
Lipids, A-15–A-16. *See also* Fatty acids; Membrane lipids
amphipathic lipids, A-15
biosynthesis, 86
cyclopentane rings, 90–91, 759, 760f
diether lipids, 170, 759, 760f
diversity and environmental stress, 89

domains, 89, 102
ether-linked, 90, 91f, 759, 760f
hydrophobic lipids, A-15
isoprenoid chains, 759, 760f
phospholipids, 86, 87f, 89–90, 96f, A-16, A-24, A-26
side chains, 90f
terpene-derived, 90–91
tetraether lipids, 90, 95f, 759, 760f
triglyceride structure, 511f, A-16
Lipmann, Fritz, 523
Lipolase, 654
Lipopeptides, 606–7
Lipopolysaccharides (LPS), **82**, 97f, **98**
biofilm development, 149
endotoxin, 1027
and macrophage activation, 991
O antigens, 1027
and phage T4 adsorption to *E. coli,* 412
Proteobacteria outer membrane, 717, 734, 737
Lipoproteins, 97
Liposomes, 549
Lipothrixviridae, A-60t
Liquid nitrogen for freezing cells, 181
Lister, Joseph, 24, 727
Listeria, 631, 648, 727, 951, 1038f–1039f, 1041, A-63t
Listeria monocytogenes
actin tails, 59f, 60, 727, 728f
cold resistance, 162, 181
escape from phagosomes, 1038
as food-borne pathogen, 645–46, 647t
hemolysin gene, 478
horizontal gene transfer, 1012
resistance in herpes-infected mice, 446, 448
transmission of, 645f
Listeriosis, 60, 727
List of Prokaryotic Names with Standing in Nomenclature (LPSN), 29
Lithoautotrophs. *See* Chemolithoautotrophs
Lithotrophs (chemolithotrophs), **26**, **123**, **124**
bacterial leaching in mining, 11, 568, 743
betaproteobacteria, 742
deltaproteobacteria, 746
facultative lithotrophs, 542
gammaproteobacteria, 743–44
microbial corrosion of statues, 11f, 567
Nitrospirae, 750
obligate lithotrophs, 563–64
production of acidity, 852
used for bioremediation, 914, 915f
Lithotrophy (chemolithotrophy), **492**, **540**, **542**, 563–69
alternative oxidoreductases, 563
biomass production, 894
carbon assimilation, 850
carbon cycle, 123f
electron donors and acceptors, 540–542, 563–64
energy acquisition in bacteria and archaea, 492, 493t
iron oxidation, 544, 564–65
nitrogen oxidation, 565–66, 905–6
proteobacteria, 735–36, 742, 743f
reverse electron flow, 565
sulfur and metal oxidation, 566–68
Littoral zone, **867**
Liu, Qianhong, 1046–47
"Live birth" in *Epulopiscium,* 726, 727f
Liver granulomas, 949, 950f
LmrP transporter, 130
Lobaria pulminaria, 853f, 854
Loder, Andrew, 159f
Logarithmic (log) phase, **142–43**
Logic gates, 479–81
LOHAFEX experiment (iron fertilization), 914, 916
Long interspersed nuclear elements (LINEs), 329, 442–43
Long terminal repeat (LTR), 439f, **440**
Lon protein, 305, 353
Lophotrichous cells, 114
Lorah, Michelle, 902
Losick, Richard, 384
Loss-of-function mutations, **342**
Lotus corniculatus, 322
Low-density granulocytes, 940, 941f
Low-density lipoprotein receptor (LDLR), 420, 421f
Löwenstein-Jensen medium, 1152
LPS (lipopolysaccharides), **82**, 97f, **98**
LPSN (List of Prokaryotic Names with Standing in Nomenclature), 29
LTLT (low-temperature/long-time) process for pasteurization, 181
Lu, Haiping, 488
Luciferase, 268, 398, 399, 483f
Lumen, **575**, 577f
Lundberg, Derek, 839
Lungs, 944–45, 1062–64. *See also* Respiratory tract
Lupus. *See* Systemic lupus erythematosus
Luria Bertani medium, 122t
LuxL synthase, 478–79
LuxO response regulator, 399f, 400
LuxR regulatory molecule (*A. fischeri*), 398, 478–79
LuxR regulatory molecule (*V. harveyi*), 399f, 400
Lwoff, André, 209, 370
Lycoperdon, A-68t
Lyme disease (borreliosis), 21, 44, 751f, **1096–97**. *See also Borrelia burgdorferi*
arthritis, 1097
deer tick (*Ixodes scapularis*), 1096
detection of cause, 1180
disease stages, 1096–97
effects of suburban development, 1180
erythema migrans, 1096, 1097f
reservoir in deer, mice, and ticks, 1097f, 1180
treatment, 1097
vaccination of wild rodents, 1182
Lymph nodes, 941, **942**, 963–64
Lymphocytes, 937f, 939, **940–42**, 949, A-23. *See also* B cells; T cells
Lymphoid organs, 940–42, 963
Lymphoid stem cells, 937f
Lymphokines, 990
Lyngbya, 722
Lyophilization, **181**, **648**
Lysates, **222**
Lysis, **82**, **210**, **415**, **A-30**
Lysogeny, **210**, 411, eTopic 10.5
Lysol, 183, 184t
Lysosomes, **A-34–A-35**
Lysozyme, 82, 928, 944, 951, 957
Lytic cycle, **210**, **444**

M

M9 medium, 122t
Ma, Luyan, 488
MacConkey medium, **135–36**, 1157, 1158f
MacElroy, R., 158
MacLeod, Colin, 322
Macnab, Robert, 64, 115
Macrocystis pyrifera (giant kelp), 624
Macrolides, 130, 299, 1125, 1126
Macromolecules
biochemical composition of bacteria, 84–85
Macronuclei, **828–29**
Macronutrients, **120**
Macrophages, **939**
activation by T cells, 991–92
alveolar macrophages, 944
as antigen-presenting cells, 941–42, 963
definition, 939
inflammation, 947, 949
opsonization, 952f
S. enterica invasion of, 177
Mad cow disease (bovine spongiform encephalopathy, BSE), 197, 1090, 1180
Magnaporthe oryzae, 814, A-68t
Magnesium, 84, 504
Magnetic lenses, 33, 66–67
Magnetic resonance imaging (MRI), 508
Magnetite, 28, 71, 113
Magnetosomes, 71, 113
Magnetospirillum gryphiswaldense, 113
Magnetospirillum magneticum, 71f
Magnetotactic bacteria, 113
Magnetotaxis, **113**
Magnification, **43**
in bright-field microscopy, 51–52
empty magnification, 49
and refraction, 48–49
and resolution, 47, 49–51
total magnification, 54
Mahendran, Kozhinjampara R., 99f
Maillard reaction, 655
Majcher, Emily, 902
Major histocompatibility complex (MHC), **953**, **965**, **982–83**
antigen binding, 983f
class I MHC molecules, 954, 983, 984–85, 988
class II MHC molecules, 983, 984–85, 986, 987, 989, 1042
MHC I downregulation by viruses, 991
MHC restriction, 986
transplanted organ rejection and MHC type, eTopic 24.3
ubiquitylation, 1043, 1044f
Malaria, 20, 797, **831–33**, 1092–93

Malate, 528, 529f, 531–32, 640
MALDI-TOF (matrix-associated laser desorption/ionization time-of-flight) mass spectrometry, 606–7
MalE (maltose-binding protein), 85f
Malolactic fermentation, **640**
Malonyl-ACP, 602, 603–4
Malonyl-CoA, 600f, 602
Maltose, A-12f
Maltose-binding protein (MBP), 308
Maltose porin, 208, 210
Malyl-CoA, 601
Manganese, 120, 183, 561, 915t, 917–18
Mannose
  group translocation, 132f
  mannose-resistant attachment, 1013
  mannose-sensitive attachment, 1013
  structural formula, A-11f
Maraviroc, 437f, 438, eTopic 11.3
Marburg virus, 1006, 1007f, 1100
March of Dimes, 36, 193
Margulis, Lynn, 30, 705, 706f, 800
Mariana Trench, 165f
Marine and aquatic microbiology, 857–69. *See also* Marine habitats
  algal blooms, 178, 228, 229f, 867, 868
  aquatic microbial metabolism in lakes, 867–69
  benthic organisms, definition, 858
  biochemical oxygen demand (BOD), 859
  cold-seep ecosystems, 865, eTopic 21.2
  epilimnion, 866–67
  eutrophic lakes, 867–68
  freshwater microbial communities, 865–69
  genes expressed in a marine microbial community, 861f
  hypolimnion, 867
  littoral zone, 867
  marine metagenomes, 861–62
  oligotrophic lakes, 866–67
  thermal vent ecosystems, 165, 850, 865, 866f
  uncultured organisms, 861
  virus control of algal blooms, 229, 229f
  virus-to-bacteria ratio, 230
Marine Benthic Group B (MBGB), A-66t
Marine food chain, 124, 127, 229–30, 798
Marine habitats, 857–59. *See also* Marine and aquatic microbiology
  aphotic zone, 857f, 858
  benthos (ocean floor), 857f, 858, 865, 866f, 868, 869f
  euphotic (photic) zone, 857
  marine snow, 150, 229, 862
  neuston, 857
  pelagic zone, 857–58
  regions of marine habitat, 857f
  thermocline, 858
Marine Hydrothermal Vent Group (MHVG), 764t, A-68t
Marine snow, 150, 229, **862**
Mars, 916–18. *See also* Astrobiology
  atmosphere, 668
  Curiosity rover, 6, 6f, 917
  evidence for water, 6, 917
  Mars Reconnaissance Orbiter, 917
  models for life on, 6, 6f
  possible source of Earth life, 680
  search for microbial life, 6, 6f, 668, 917–18
  terraforming, 918
  Viking Lander, 918
Marshall, Barry, 22, 42, 747, 1071–72
Mars Science Laboratory, 6, 6f
Massion, Gene, 769f
Mass spectroscopy/spectrometry
  identification of purified proteins, 473
  MALDI-TOF, 606–7
  measurement of environmental elements, 893
  nanoDESI, 585
  nanoscale secondary ion mass spectrometry (NanoSIMS), 72, 74–75, 663
  PCR combined with, 1162
  proteomics, 404–5, A-47–A-48
Mast cells, **938–39**, 948, 952, 963, 973, 996, 997
Mather, Cotton, 22
Matrix proteins (M1), **425**
Matthaei, Heinrich, 276
Matthews, Keith, 835
McCarty, Maclyn, 322
McClintock, Barbara, 355
McCloskey, Robert, 625
McDougall, Andrew, 889
M cells, **944**, 1003, 1032, 1090
MCherry fluorescent protein, 186f, 483f
McInerney, Michael, 500, 502–3
Mcl-1 protein, 981
Mean generation time, **141**
Measles (rubeola), 194, 1057
Measles, mumps, and rubella (MMR) vaccine, 1057
Measles virus (Paramyxovirus), 194, 213–15, 1057
Mechanical transmission, **218–19**
Mechnikov, Ilya, 188
Medical microbiology, 20–26
Mefloquine, 1093
Megasphaera elsdenii, 884
Meiosis, **A-33–A-34**
Mekalanos, John, 1034–35
Membrane attack complex (MAC), **956**
Membrane filtration apparatus, 181f
Membrane fluidity, 160, 165, A-26–A28
Membrane lipids, 86, 87f, A-24, A-26. *See also* Cell membrane
  in acidophilic microbes, 170
  in alkaliphilic microbes, 170
  in barophilic microbes, 165
  cardiolipin, 89
  cholesterol, 90, 91, A-26, A-27f, A-28
  diphosphatidylglycerol, 89
  fatty acid side chains, 87f, 89–90
  hopanoids (hopanes), 86f, 90, 91, A-26
  movement, A-26–A-28
  phosphatidates, 86
  phosphatidylcholine, A-27f
  phosphatidylethanolamine, 86, 87f, A-27f
  phosphatidylglycerol, 86, 87f
  phosphatidylserine, A-27f
  phospholipids, 86, 87f, 89–90, A-16, A-24, A-26, A-27f
  phosphoryl head groups, 86, 87f, 89
  terpene derivatives, 90–91
  in thermophiles compared to mesophiles, 164
Membrane-permeant organic acids, **168–69**
Membrane-permeant weak acids, **87–88**, 188, 550–51
Membrane-permeant weak bases, **87–88**
Membrane potential, **125**
Membrane proteins, 86–87
Memory B cells, 964, **974**, 975, 976, 977f, 981, 987–88
Menaquinone, 554
Mendelian rules of inheritance, 34
Meningitis
  bacterial meningitis, treatments, 283
  case histories, 1083–84, 1117, 1153
  caused by amebas, 804
  definition, 1084
  Escherichia coli, 1153
  Exserohilum rostratum, 1051
  Haemophilus influenzae, 1084, 1153
  Lyme disease, 1096
  microbes that cause meningitis/encephalitis, 1088t
  Neisseria meningitidis, 283, 560, 696, 743, 1015, 1027, 1083–84, 1153, 1169
  rifampin treatment of meningococcal meningitis, 1123
  Streptococcus pneumoniae, 1062, 1083f, 1084, 1117, 1153
  vaccines, 1084, 1088t
  viral meningitis, 1084
Meningoencephalitis, 1063, 1072
Mercury, 915t
Mercury chloride, 184t
Mercury removal, 477, eTopic 22.2
Merismopedia, 573f, 722, A-62t
Meropenem, 109
Merozoite, **831–32**
Merozoites, 1092
MERS-CoV, 1180, 1181f
Meselson-Stahl experiment, eTopic 7.5
Mesocosm, **893**
Mesophiles, 159t, 160f, **161**, 162, 164
Mesophyllum, 821, A-69t
Mesorhizobium loti, 322, 360
Messenger RNA (mRNA), 46f, 54, 76, **284–85**, A-1
  alignment of gene with mRNA transcript, 292–93f
  in bacterial cell, 80–81f, 82, 102–3
  cis-repressed mRNA, 482–83, 484
  in eukaryotes, 103
  leader sequences, 285, 292f, 293
  mRNA stability controls of gene regulation, 369
  northern blot analysis, 463–64, A-48–A-50
  polycistronic RNA, 293, 300
  preliminary mRNA transcript (pre-mRNA), 314

production by viruses, 203
ribosome-binding sites, 292–93f, 293–94, eTopic 8.2
ribosomes stuck on mRNA, 301, 302f
Shine-Dalgarno sequence, 293
splicing, 313f, 314
stability and half-life, 285–86
trailer sequences, 293, 293f
translation of, 294–97
Metabacterium, A-63t
Metabacterium polyspora, 726
Metabolic fluxes, 854–55
Metabolic islands, 360f
Metabolic switchboards, 483–84
Metabolism, genomic analysis, eTopic 13.3
Metabolist models, **677**, 678–79
Metagenomes, **10**, **271**, 841–42, **926**
of *Arabidopsis* endophytes, 839
definition, 10, 271
GOS metagenome, 895
of honeybee gut communities, 840
of human gut microbiome, 841–42
human metagenome, 515, 926
Metagenomics, 268–69, **270–72**, **701–2**, **840**, **841–49**
assembling genomes, 844–46
DNA sequencing, 844, 845t
examples of metagenomic analysis, 842t
filtering samples, 843
functional annotation, 846–47, 848f
of human intestinal microbiota, 932
isolating DNA, 843
limitations of, 847–48
of marine microbial communities, 861–62
sampling the target community, 842–43
species diversity, 846, 847f
steps in sequencing a metagenome, 844f
target communities, 842
Metal and sulfur oxidation, 566–68
Metal-reducing bacteria, 545, 561–63
Metamonada, 799t, 800, 801f, 805, 835–36, A-71t
Metaphase, A-32f, A-33
Metaproteomics, **848**
Metastatic lesions, **1064**
Metatranscriptomics, **848**, 861–62
Metazoa, 124, A-68t
Metcalf, William M., 895
Meteorites, 19, 613, 669, 677–78, 680, 916
Methane. *See also* Methanogenesis; Methanogens
covalent bonding and structural formula, A-3, A-4f
from freshwater lakes, 868
as greenhouse gas, 601, 680, 737, 780
methylphosphonate as source, 895, 911–12
molecular models, A-6f
oxidation of methane, 781, A-21
oxidation of methane coupled with sulfate reduction, 781
permafrost melting and production of, 889
production in rumen, 780, 781f
recycling by methanotrophs, 737, 740, 773
wastewater treatment and production of, 901
Methane gas hydrates, 571, **781**, 895, 896
Methanethiol, 632, 633f
Methanobacteriales, 777, 778t, A-67t
Methanobacterium, 778, 782, 783f, A-67t
Methanobacterium thermoautotrophicum, 777, 778, 778t, 779f
Methanobrevibacter ruminantium, 777, 778t
Methanobrevibacter smithii, 780–81, 885, A-67t
Methanocaldococcus jannaschii, A-67t
cell form, 162f, 599f, 778
CO_2 fixation, 599f
genome, 240t, 322
initiation of transcription, 294
taxonomic hierarchy of classification, 700t
as thermophilic methanogen, 162f, 700t
Methanococcales, A-67t
Methanococcoides alaskense, A-67t
Methanococcus, 778, 782
Methanococcus maripaludis, 503
Methanoculleus nigri, A-67t
Methanofuran (MFR), 570, 782
Methanogenesis, **569–70**, **761**
from acetate, 783
biochemistry of methanogenesis, 780–81
carbon monoxide reductase pathway, 761
cattle, 601, 780
from CO_2 and H_2, 541, 782–83
cofactors for methanogenesis, 781–82
on early Earth, 677, 680
energy acquisition, 493t
in human digestion, 780–81
in landfills, 780
methane gas hydrates on the seafloor, 571, 781, 895
pathways, 775–77
from retreating glaciers, 780
reverse methanogenesis by *Archaeoglobus,* 781, 791, 792f
in soil, 780, 876
termites, 780
as trait unique to archaea, 758
in wetlands, 876
Methanogenium frigidum, A-67t
Methanogens, **570**, **775**
anaerobic habitats, 780–81
in bovine rumen, 780, 781f, 884
cell forms, 778–79
communities of methanogens, 570
digestive methanogenic symbionts, 780
diversity, 765, 777–78
ecological role, 561, 775
filamentous methanogens, 901
highly divergent genomes, 677
and human digestion, 780–81, 885
in human intestines, 932
hyperthermophilic methanogens, 777
methanogenic archaea, 775–83
phylogeny, 777, 778t, A-67t
pseudopeptidoglycan in cell walls, 760, 778
sewage waste treatment, 778–79
sodium requirement, 782
in thermal vent ecosystem, 865, 866f
Methanohalophilus, A-67t
Methanol, 571, 686, 737, 780, 783
Methanomicrobiales, A-67t
Methanomicrobium, 778
Methanopterin (MPT), 599, 781, 791
Methanopyrales, A-67t
Methanopyrus, 777
Methanopyrus kandleri, A-67t
Methanosaeta, 778–79
Methanosaeta concilii, A-67t
Methanosarcina, 778, 782, 783f
Methanosarcina acetivorans, A-67t
Methanosarcina barkeri, 497
Methanosarcinales, A-67t
Methanosarcina mazei, 778, 779f, A-67f
Methanosarcina thermophila, 305
Methanosphaera stadtmanae, 780, A-67t
Methanospirillum, 502
Methanospirillum hungatei, 778
Methanothermus, 760
Methanothermus fervidus, 762, 777, 778, 779f, A-67t
Methanotrophs, **737**, **740**, 865, 866f, 873, 876
Methanotrophy, 571
Methemoglobinemia, 906
Methicillin, 1119
Methicillin-resistant *S. aureus* (MRSA), 387f
alternative drugs, 696, 1056
evolution of antibiotic resistance, 693
Kirby-Bauer disk susceptibility test, 1114f
as nosocomial infection, 727–28, 1056, 1108
resistance strategy, 1119
as significant current threat, 25, 1056, 1173
Methionine, 388
Methionine deformylase, 303
Methyl-accepting chemotaxis proteins (MCPs), **392–94**
3-methyladenine, 351
Methylamine, 571, 686, 737, 783
Methylene blue, 55–56, 1152
2-Methylhopane, 673
Methylmalonyl-CoA, 600f, 601
Methyl mismatch repair, 348t, **349**
Methylobacterium, 686, 687f, 737, A-63t
Methylobacterium extorquens, 686
Methylobacterium zatmanii, 686
Methylococcaceae, 859
Methylotrophy, **571**, 736t, **737**, **740**, 859
Methylphosphonate, 895, 911–12
Metronidazole (Flagyl), 1057, 1073, 1082, 1122–23
MexCD-OprJ efflux system, 185
MHC restriction, **986**
Micelles, 678
Michaud, Alex, 163
Microaerophilic microorganisms, 159t, 173t, 174f, **175**, 175–76

Microarray analysis, 835
Microbe-associated molecular patterns (MAMPs), **944**, 954, 986, 994, 1027, 1043, 1044f
Microbes, **8–9**
  and human history, 12–13
  sizes of, 44
Microbial communities, 8, 839–41. *See also* Biofilms; Microbial ecology
  animal microbial communities, 882–86
  bovine rumen microbial community, 883–84
  communities of methanogens, 570
  communities within rock, 840
  digestive communities, 883–85
  endophytes, 839
  freshwater microbial communities, 865–69
  honeybee gut communities, 840
  hydrothermal vent communities, 850, 865
  identification of members, 841
  kelp forests, 824
  metagenomics of marine communities, 861–62
  planktonic communities, 862–63
  plant microbial communities, 839, 876–82
  recycling of organic material, 840
  sponge microbial communities, 883
Microbial diseases. *See also* Emerging diseases; Infection; Pathogenesis; *specific diseases*
  cardiovascular system infections, 1090–93
  central nervous system infections, 1083–90
  characterizing and diagnosing microbial diseases, 1052–54
  detecting emerging microbial diseases, 1179–82
  gastrointestinal tract infections, 1066–74
  genitourinary tract infections, 1075–83
  history of effects of disease, 11, 14
  infection *vs.* disease, 936–37
  locations of emerging and reemerging infectious diseases, 1181f
  organism approach to diseases, 1052–53
  organ system approach to diseases, 1052, 1053
  portal-of-entry approach to diseases, 1052–53
  respiratory tract infections, 1058–65
  skin and soft tissue infections, 1054–58
  systemic infections, 1094–1102
Microbial ecology, 26–29, 839–86. *See also* Ecosystems; Food webs and food chains; Metagenomics; Microbial communities; Symbiosis; Viral ecology
  abiotic factors, 851–52
  biomass building through autotrophy or heterotrophy, 123
  energy from phototrophy or chemotrophy, 124
  eutrophication, 177f, 178, 867–68
  history, 26–27
  human influence on microbial ecosystems, 177–78
  methanogens, ecological role, 561, 775
  natural ecosystems support, 26–27
  niches, 159, 168
  van Niel hypotheses, 841, 849
Microbial ecosystems
  maintaining diversity, 177–78
  oligotrophy, eTopic 5.4
Microbial fuel cells, 562–63
Microbial genetics. *See* Genetics
Microbial mats
  cyanobacteria in, 722, 723f
  fossil stromatolites, 666
  oxidation of methane hydrates by, 895
  reactive mats, 902
  of white, sulfur-oxidizing bacteria, 909f, 910
  in Yellowstone National Park, 148f, 714, 716f
Microbial toxins. *See* Endotoxin; Exotoxins
Microbiology careers and education, 36–37, eTopic 1.3
Microbiomes, **841–42, 926**
Microbiota, **926**. *See also* Human microbiota
Micrococcaceae, 734, A-63t
Micrococcus, 734, A-63t
Micrococcus luteus, 734, A-63f
Micrococcus tetragenus, 106
Microcolonies, **134**, 148f, 149
Microfilaments, **A-38**
Microfluidic chamber, 108
Microfluidic chemostat, 145
Microfossils, 18, **671–72**, 673f, 917, 918
Micrographia (Hooke), 15
Micronuclei, **828–29**
Micronutrients, **120**, **891**
Microplankton, **862**
MicroRNAs (miRNAs), 1042
Microscilla, 514–15f, 516
Microscopes, 14–16, **42**
Microscopy. *See also* Electron microscopy; Light microscopy
  background and overview, 41–46
  counting bacteria with, 136–37
  detection *vs.* resolution, 43
  emerging methods, 71–73
  optics and properties of light, 46–51
  resolution of the eye, 42–43, 46f
  size scales, 45–46
Microsporidia, 268, 817, A-69t
Microsporum, 814, A-68t
Microthrombi, 1067
Microtubules, **A-39**
  in cilia and flagella, 116, 804, 828f, A-40
  in contractile vacuoles, 100f
  in cytoskeleton, A-38f, A-39
  mitotic spindle, A-32
  and pseudopod motility, 825
  use by herpes simplex, 446
Microviridae, A-60t
MIC strip test, 1113
Middle Eastern Respiratory Syndrome (MERS), 1004, 1180, 1181f
Miller, Stanley, 18, 19, 677
Mimiviridae, A-60t
Mimivirus, 8, 195–96, 198, 205t
Mine leaching, 477
Mineral oxidation, 123, 124, 590
Minimal inhibitory concentration (MIC), **1112–15**
Minnesota soil data set, 271
Miracle of Bolsena, 644
Mirte charcoal stove, 638f
Miscellaneous Crenarchaeote Group (MCG), A-66t
Miso, 627, 628t, **633**
Missense mutations, **341–42**
Mitchell, Peter, 34, 547, 548, 549
Mitochondria (singular, mitochondrion), **A-40**
  cristae, 805, A-40
  endosymbiosis theory of origin, 30–31, 126, 127, 705–7, 742, 798, A-40–A-41
  genetic code, 286
  genome, 706, 707f
  oxidative respiration, A-40, A-41f
  proton efflux demonstration, 548
  reductive evolution, 706
  respiratory membrane, 545
  serial endosymbiosis theory, 30–31, A-40
  structure, A-41f
  TCA cycle in, A-40
Mitochondrial respiration, 556–57f, 557
Mitomycin C, 1068, 1069f
Mitosis, 106–7, 139, 261, **A-32–A-33**
Mitosomes, **805**, 836
Mitosporic fungi, **809**
Mitotic spindle, A-32
Mitral valve prolapse, 930
Mixed-acid fermentation, **523–25**, 563, 780
Mixotricha paradoxa, 854–55
Mixotrophs, **127**, 804, **863**, 864f, 913
Mixotrophy, 580, 818
Mobile genetic elements, 261, 299, 355–58, 1131
Mobile symbiosis islands, 360
Models of early life, 677–79. *See also* Origins of life
  abiotic creation of macromolecules, 677, 678
  early-Earth simulation experiments, 18, 675, 677
  metabolist models, 677, 678–79
  prebiotic soup, 677–78
  RNA world, 679, eTopic 17.1
Modified sugars, A-12f
Modular enzymes, **603–5**, 606–7, 618
Molarity, **A-16–A-17**
Molecular analysis, 463–74. *See also* Biotechnology
  affinity chromatography, 466, 466–67f, 473
  DNA protection analysis, 467–68
  electrophoretic mobility shift assay (EMSA), 465
  mapping transcriptional start sites, 466–67, 468f
  northern blot analysis, 226, 463–64

real-time PCR, 468–70
Southern blot analysis, 463
two-hybrid analysis of protein–protein interactions, 472–73
western blot analysis, 464–65
Molecular clocks, 31, 32f, 292–93, 361, **681–83**
Molecular formulas, **A-2**
Molecular Koch's postulates, 1010
Molecular mimicry, **295**, 299–300, 1041, 1118, eTopic 25.8
Molecular structure, visualizing, 73–76
Mollicutes, 728–30, A-63t
Molybdenum, 120, 610, 783
Monensin, 1131
Monera, 30
Monkeypox, 1147
Monocistronic RNA, 242
Monocytes, 937f, **939**
Monod, Jacques, 369–70, 573
Monophyletic ancestry, **30**, 32f, 690f, 691f
Monophyletic groups, **681**
Monosaccharides, 510, **A-11–A-12**
Monosodium glutamate (MSG), 634
Monotrichous cells, 114
Montagnier, Luc, 432, 433–34
Montagu, Mary, 22, 23f
Moraxella catarrhalis, 929
Morchella hortensis (morels), 813f, 814
Mordants, **57**
Morowitz, Harold, 678–79
Mother cells, **151**, 383–84, **724**, 726
Motif Search (computer program), 316
Motility, 64, **114–16**
Motor neuron disease, 706
Motor proteins, A-38f, A-39
Moyle, Jennifer, 34, 548
MreB protein, 100
MRNA. *See* Messenger RNA (mRNA)
MRSA (methicillin-resistant *S. aureus*). *See* Methicillin-resistant *S. aureus* (MRSA)
Mucilage, 720, 721f
Mucinase, 1045, 1046f
Mucociliary elevator, **930**, 944, 1058–59, eTopic 26.3
Mucor, 812, 813f
Mucosa-associated lymphoid tissue (MALT), 981–82
Mucosal immunity, **981–82**, 994
Mucous membranes, 943–44
Mueller-Hinton agar, 1114f, **1115**, 1155, 1157f
Multidrug resistance (MDR) efflux pumps, 88, 185, **1129**, 1130f
Multiple fission, 140
Multiple Sequence Alignment, 314f, 316
Multiple symbionts, 854–55
Multiplex PCR, **1160–61**, **A-55**
Multiplicity of infection (MOI), **221**
Multivalent adhesion molecule 7 (MAM7), 1016–17
Murchison meteorite, 613f
Murein, **92**. *See also* Peptidoglycan
Murein lipoproteins, **97**, 105
Murray, Alison, 772–73
Mushrooms
Agaricus bisporus, 625–26
basidiomycetes, 815–16
commercial production, 625–26
edible fungi, 625–26
fairy rings, 798, 800f, 816
Flammulina velutipes (enoki mushrooms), 626
Lentinula edodes (black forest mushroom, or shiitake), 626
morels (*Morchella hortensis*), 813f, 814, A-68t
Pleurotus (oyster mushrooms), 626
poisonous mushrooms, 625, 815
truffles (*Tuber aestivum*), 625, 814, 874, A-68t
Mutagens, **343**, 345, 347–48, 351
Mutation frequency, **346**
Mutation rates, **346**, eTopic 9.4
Mutations, **341**. *See also* DNA repair
apurinic sites, 344, 345f
deletions, 341, 342, 801
frameshift mutations, 342
gain-of-function mutations, 342
insertions, 341, 342, 801, 802f
inversions, 341, 342–43
knockout mutations, 342
lacZ reporter gene for mutagenicity testing, 348
loss-of-function mutations, 342
methylation, 344–45
missense mutations, 341–42
mutagens, 343, 345, 347–48
nonsense mutations, 342
point mutations, 341, 342
potentiating mutations, 695
pyrimidine dimers, 345, 350, 352
reactive oxygen species, 344, 345f
reversions, 341
silent mutations, 341
spontaneous mutations, 343–45
tautomeric shifts, 344
transitions, 341, 344f
transversions, 341
Mutator strains, 346f, **349**
Mutualism, **852–55**. *See also* Lichens; Symbiosis
cyanobacteria and sponges, 723
definition, 703–4
digestive mutualism, 854–55
nitrogen fixation, 126, 704, 731
Streptomyces and leaf-cutter ants, 731
virus-host mutualism, 226–27, 229, 448
Mycelium (plural, mycelia), **152**, **806**
aerial mycelia (or hyphae), 152, 153f, 154f, 730, 807
in filamentous bacteria, 152, 153f, 154f, 730
in fungi, 806, 807, 815f, 816
primary mycelium, 815f, 816
secondary mycelium, 815f, 816
vegetative mycelia, 730
Mycena, 872
Mycetozoa, A-69t
Mycobacteria, 96, 107, 108–9, 949, 991
Mycobacteriaceae, A-63t
Mycobacterium, 121t, 730, 732, 951, 991, 1038, A-63t
Mycobacterium avium-intracellulare, 1082
Mycobacterium bovis, 991f
Mycobacterium bovis BCG strain, 998
Mycobacterium leprae (leprosy), 57, 96, 360, 717, 732–33
Mycobacterium marinum, 949, 950f
Mycobacterium smegmatis, 108–9, 733
Mycobacterium tuberculosis. See also Tuberculosis (TB)
acid-fast stain, 57, 59f, 732, 1064f, 1152, 1176
as actinomycetes relative, 717
antibiotic development, 653, 1123
antibiotic resistance, 25, 108–9, 1107, 1123, 1130, 1181
bedaquiline inhibition of ATP synthase, 557, 1107
cell envelope, 96
and chronic inflammation, 949, 950f
CO oxidation, 739
crinkled colonies, 732f
fatty acids, 89, 90
flipping cytokine profiles, 1041
genome, 240t, 733
glyoxylate bypass, 531f, 532
growth in pure culture, 20–21
history, 14
infecting a lung macrophage, 1107f
as intracellular pathogen, 952
isoniazid, 1065, 1110, 1111
manipulation of host cell suicide pathways, 1042
mycolic acids, 601
nonhomologous end joining (NHEJ), 354
PCR identification, 1176
prevention of pyroptotic cell death, 1042
proteasomes, 305
as reemerging pathogen, 1181
vaccine development, 652
Mycolic acids, 96, 601, 717, **732**
Mycology, **798**, **800**
Mycolyl-arabinogalactan-peptidoglycan complex, 732
Mycoplasma, 728–30
Mycoplasma capricolum, 485
Mycoplasma genitalium, 240t, 241, A-63f
Mycoplasma mobile, 729f, 730
Mycoplasma mycoides, 312f, 485
Mycoplasma pneumoniae, 261, 701
Mycoplasmas, 91, 96, 286, 713, 728–30
Mycorrhizae (singular, mycorrhiza), **816**, **873–75**, 894
Avatar, 806
as carbon sink, 894
definition, 815f, 816, 872, 874
ectomycorrhizae, 874
endomycorrhizae, 874, 875f
vesicular-arbuscular mycorrhizae (VAM), 874
Mycotoxins, 814
Myeloid stem cells, 937f
Myeloperoxidase, 952
Myers, Charles, 562

Myocarditis, 1092
Myoviridae, A-60t
Myrothecium, 187
Myxobacteria, 116, 746, 747f
Myxococcales, A-64t
Myxococcus, 1029
Myxococcus fulvis, 1123
Myxococcus xanthus, 152, 153f, 746, 747f, 1015
Myxopyronin, 1123
Myxosarcina, 722
Myxospores, **746**, 747f
Myxovirus, 228

N

Nacy, Carol, 652–53
NADH, NAD^+. *See* Nicotinamide adenine dinucleotide (NADH, NAD^+)
NADH dehydrogenase (NDH-1), 553–54, 556
NADH dehydrogenase 2 (NDH-2), 554
NADPH (nicotinamide adenine dinucleotide phosphate), 507
NADPH oxidase, 952
Naegleria, 1072
Na^+/H^+ antiporters, 128–29, 130, 170, 171f, 172
Nakagawa, Tatsunori, 776–77
Nalidixic acid, 1121
Nanoaerobes, 749
Nanoarchaeota, **765**, 793, A-67t
Nanoarchaeum equitans, 793, A-67t
NanoDESI, 585
Nanoeukaryotes, 805
Nanoplankton, **862**
Nanoscale secondary ion mass spectrometry (NanoSIMS), **72**, **74–75**, 663
Nanotechnology, 200, 411, 790, eTopic 11.2
Nanotubes, 114, 402–3
  Bacillus subtilis, 114, 402–3
  cell-to-cell transmission of HIV, 442
  electric current through nanotubes, 539
  interspecies nanotubes, 402–3
  in syntrophy, 503
  transient nonhereditary phenotype transfer, 402–3
Nanowires, 539, 540, 561, 562
Naphthalene catabolism, 534
Naqvi, Syed Wajih, 907
Narcissus, 985
NASA, 6, 668, 768
Nasopharynx, **929**
National Center for Biotechnology Information (NCBI), 10f, 11
National Center for Biotechnology Information (NCBI) Taxonomy Database, 701, **712**, **758**
National Institutes of Health (NIH), 36, 386
National Medal of Science, 347, 616
National Science Foundation (NSF), 28, 36
Native conformation of proteins, **A-11**
Natronobacterium gregoryi, 170, 171f
Natronococcus, 765, A-67t
Natronomonas, A-67t
Natto, 628t, 629, **636**
Natural killer (NK) cells, 937f, 940, 953–54, 990, 991, 999
Natural selection, **692**
Nautilia, 748, A-64t
Navarre, Will, 922
NdhF3 (high-affinity CO_2 transporter), 596–97
NdhF4 (low-affinity CO_2 transporter), 596–97
Nealson, Kenneth, 562–63
Nebula RCW-49, 919
Necrosis, 1042
Necrotizing fasciitis, **1056–57**
Nef accessory protein, 435f, 437t, 1042
Negative selection (T cells), **986**
Negative stain, 57–58, 59f
Neidhardt, Fred, 84, 85f, 177
Neisseria
  horizontal gene transfer, 690
  natural transformation, 323, 324
  oral microbiota, 928
  oxidase (cytochrome oxidase) test, 1157
  pathogenicity, 742–43
  pili, 324
  twitching motility, 1029
Neisseria gonorrhoeae, A-64f
  antibiotic resistance, 1080–81, 1148
  binding to $CD4^+$ T cells, 1080
  on chocolate agar, 1154f
  clinical test for nitrate reduction, 560
  complement sensitivity, 1080
  dissemination in bloodstream, 1053
  gene transfer, 329
  gonorrhea, 185, 743, 936, 1079–81
  immune response, 936
  iron complex binding, 131
  lack of capsule, 1157
  as microaerophile, 743
  Opa membrane proteins, 1016
  phase variation, 1080
  slipped-strand mispairing, 390, eTopic 10.3
  transformation, 323
  type IV pili, 1015
Neisseria lactamica, 696
Neisseriales, A-64t
Neisseria meningitidis, 1028f, 1083f
  capsules, 1037, 1084
  complement resistance, 1084
  denitrification, 560
  genome compared to other *Neisseria* species, 329, 696
  identification, 1157
  laboratory exposure to, 1169
  meningitis, 283, 560, 696, 743, 1015, 1027, 1153, 1169
  petechial rash, 1027, 1028f
  purpuric rash, 1083f
  rifampin, 283
  septicemic phase, 1027
  type IV pili, 1015, 1016f, 1084
Neisseria sicca, 743
Nelmes, Sarah, 23f
Neocallomastix, 809–10, 812f, A-68t
Neonatal group B streptococcal sepsis, 1137
Neotyphodium coenophialum, 877
Nephritis, 752
NETosis, 938, 940, 941f
Neural progenitor cells (NPCs), 448–49
Neuraminidase, 410, 425f, 426, 427f, 431, 1138–39
Neuraminidase inhibitors, **1138–40**
Neurospora, 797, 809, 812, 814
Neurotransmitters, 885, 1019
Neuston, **857**
Neutralophiles, 159t, **169**, 170f
Neutral red, 135
Neutrophil-activating protein (NAP), 1046
Neutrophil extracellular traps (NETs), **938**, 940, 941f
Neutrophils, **937–38**, 940, 941f, 946, 947, 948, 949f, 952
New Delhi metallo-beta-lactamase-1 (NDM-1), 1119, 1132
Newman, Dianne, 674–75
Newton, Isaac, 14
NF-kappaB transcription factor, 994–95, 1043, 1044f
Niccol, Andrew, 267
Niches, 159, 168, **841**
Nickel, 120, 782
Nicotinamide adenine dinucleotide (NADH, NAD^+), **505–6**
  DNA ligase use of, 255–56
  as energy carrier, 124, 173f, 501, 505–6, 507
  small energy transitions, 552
Nicotinamide adenine dinucleotide phosphate (NADPH), 507
Nicotine catabolism, 534–35
Nightingale, Florence, 14
NIH (National Institutes of Health), 36
Nikkomycins, 806
Nirenberg, Marshall, 276, 286
Nitrate
  ammonia oxidation to nitrate, 26, 126
  contamination of drinking water, 906
  expression of nitrate reductase, 560
  nitrate influx and algal blooms, 906
  nitrate reduction to ammonia, 605, 907–8
  nitrate reduction to nitrogen gas, 560, 561, 608, 907
  as terminal electron acceptor, 174, 542, 560
Nitrate reductase, 560
Nitric oxide (NO), 605, 952
Nitric oxide synthetase, 952
Nitrification, **126**, 608, **905–6**
Nitrifiers, **565**, **742**, **905**, 906
Nitrifying bacteria, 26, 126, 561
Nitrite
  nitrite-induced blue baby syndrome, 906
  nitrite oxidation to nitrate, 717, 750
  nitrite reduction to ammonium ion, 605
  nitrite reduction to nitrogen gas, 608
Nitrite oxidizing bacteria, 717, 750
Nitrite reductase, 560
Nitrites used for food preservation, 650
Nitrobacter, 906, A-63t
Nitrobacter winogradskyi, 905f
Nitrobenzene catabolism, 533

Nitrogen
diatomic nitrogen covalent bonding and structural formula, A-3, A-4f
as essential nutrient, 120, 605, 891
global reservoirs of nitrogen, 892t
microbial cycling of nitrogen, 605, 891
oxidation states, 891, 892t
oxidized forms, 560–61, 678
pollution of biosphere, 608
removal from wastewater, 899–900
runoff and algal blooms, 868
uptake in cyanobacteria, 75
Nitrogenase, **609**. *See also* Nitrogen fixation
effect of oxygen, 611–12
four cycles of reduction, 610–11
genes, 9
molybdenum, 609–10
nitrifying bacteria, 126
oxygen sensitivity, 152, 905
reaction mechanism, 609–10
structure, 609f
Nitrogen cycle, 27, 125–26, 903–8. *See also* Nitrogen fixation
ammonia oxidizers (AOA), 765, 774–75
anammox reaction, 565–66, 753, 908
chemical and spectroscopic analysis, 893
cyanobacteria as keystone species, 75
denitrification, 126, 560, 608, 905f, 907–8
dependence on prokaryotes, 125–26, 903
global reservoirs of nitrogen, 892t
nitrification, 126, 608, 905–6
"nitrogen triangle," 904–8
overview, 903f
oxidation states of cycled compounds, 891, 892t, 903
sources of nitrogen, 892t, 904
stable isotope ratio measurements, 893
Nitrogen fixation, 6, **27**, 605, 608–12, **904–5**. *See also* Nitrogen cycle
anaerobiosis and, 611
Azotobacter, 610, 611
cyanobacteria, 75, 152, 608, 611
discovery, 608
energy and oxygen regulation, 878–79, 880f
Haber process, 608, 904
Klebsiella pneumoniae, 610
leghemoglobin, 611, **740**, 878, **879**, 880f
in legume nodules, 740, 741f, 878, 880f
loss of H_2, 610–11
mechanism, 609–11
molecular regulation, 611–12
nif genes, 611–12
nitrogen assimilation, 605, 608
photosynthesis and, 152, 611
rhizobia, 28, 126, 608, 610–11, 740, 878–81
Rhodopseudomonas palustris, 540, 541f
Rhodospirillum rubrum, 608
Nitrogen-fixing bacteria, **126**, 608. *See also* Rhizobia
Nitrogen gas, 125–26
Nitrogen metabolism regulation, 394–97
Nitrogenous base, 243. *See also* Nucleobase
Nitrogen oxidation by lithotrophs, 565–66
Nitrosamines, 906
Nitrosococcus, 742
Nitrosolobus, 742
Nitrosomonadales, A-64t
Nitrosomonas, 742, 850, 906
Nitrosomonas europaea, 542, 742
Nitrosopumilales, 774–75, A-66t
Nitrosopumilus, 775, 776–77, 895, 906, 907, 911
Nitrosopumilus maritimus, 775, 776–77
Nitrosovibrio, 742
Nitrospira, 545, 574, 750, 906
Nitrospirae, **717**, 736t, 742, 750, A-65t
Nitrospirales, A-65t
Nitrous oxide gas (N_2O), 890, 907
Nobel Prize, 9
Altman, Sidney, 19
Anfinsen, Christian, 303
Baltimore, David, 203, 204f
Barré-Sinoussi, Françoise, 434
Berg, Paul, 9
Beutler, Bruce, 954
Calvin, Melvin, 589, 591f
Cech, Thomas, 19, 679
Chain, Ernst, 24
Crick, Francis, 35
Doherty, Peter, 986
Dulbecco, Renato, 203, 204f
Fleming, Alexander, 1108–9
Florey, Howard, 24
Gilbert, Walter, 9
Haber, Fritz, 904
Hartwell, Leland, 809
Hodgkin, Dorothy Crowfoot, 34, 75, 76f
Hoffman, Jules, 954
Holley, Robert, 276
Jacob, François, 369–70
Khorana, Har Gobind, 276
Koch, Robert, 20–22, 26, 55, 841, 859, 1175
Kornberg, Arthur, 252, 253
Krebs, Hans, 523, 527
Lipmann, Fritz, 523, 527
Lwoff, André, 209, 370
Marshall, Barry, 22, 747, 1071–72
Mechnikov, Ilya, 188
Mitchell, Peter, 549
Monod, Jacques, 369–70
Montagnier, Luc, 432, 433–34
Nirenberg, Marshall, 276
Nurse, Paul, 809
Rous, Peyton, 193
Sanger, Fred, 9, 193
Stanley, Wendell, 25
Steinman, Ralph, 963, 964f
Svedberg, Theodor, A-44
Temin, Howard, 203, 204f
Tonegawa, Susumu, 978
Waksman, Selman, 1110
Warren, Robin, 22, 747, 1071–72
Watson, James, 35
Wilkins, Maurice, 35
Zernike, Frits, 65
Zinkernagel, Rolf, 986
Nocardia, 733, 900
Nodes (phylogeny), 683f, **684**
Nod factors, 87, 878
NOD-like receptors (NLRs), 954–55, 985, 991
Noise, **481–82**
Nomenclature, **697**
Nonadaptive immunity, **936**
Noncoding DNA, 241
Nonhomologous end joining (NHEJ), **354**
Nonpolar covalent bonds, **A-4**
Nonribosomal peptide antibiotics, 605, 606–7, **618**
Nonsense mutations, **342**, 416
Nonsense suppressors, **416**
Nori, **626**, 821
Normak, Staffan, 234
Normanski microscopy, 65
Noroviruses (Norwalk-type viruses), 417, 646, 647t
Northern blots, 226, **463–64**, **A-48–A-50**
Nosocomial infections, 185, 1056, 1076, 1091, **1108**, 1128
Nostoc, 718, 720, 721f, A-62t
Nostocales, A-62t
NOT gates, 480, 481
Novobiocin sensitivity test, 1158f, 1159
Novozymes, 625, 654–55
NSF (National Science Foundation), 28
N-terminal rule, **304**
NtrB sensor kinase, 394–95
NtrC response regulator, 394–95, 611–12, 614
Nucleariidae, A-69t
Nuclear magnetic resonance (NMR), 508, 513, eTopic 13.1
Nuclear pore complexes (NPCs), A-31
Nucleases, 252
Nucleic acid amplification assay (NAAT), 1152
Nucleic acids, 85, A-13–A-15. *See also* DNA (deoxyribonucleic acid); RNA (ribonucleic acid)
Nucleobase, 243. *See also* Nitrogenous base
Nucleocapsid proteins (NPs), **425**, 430, 431, 434, 435f
Nucleoid, **82**, **245**
DNA organization in nucleoid, 101–2
electron microscopy, 33
Escherichia coli, 80–81f, 100
packing of DNA in, 245
Nucleolus (plural, nucleoli), **A-31**
Nucleomorph, **803**
Nucleosides, A-13
Nucleosomes, 261
Nucleotide excision repair (NER), 348t, **350**
Nucleotides, 239, 243, 267f, 286, 617, **A-13**, A-14f
Nucleus (plural, nuclei), 30, 261, **A-31**
Numerical aperture, **52**
Nurse, Paul, 809
NusA protein, 282, 301
NusG protein, 301
Nüsslein-Volhard, Christiane, 954
Nutrient-binding proteins, 127

Nutrient deprivation and starvation, 152, 176–78, 185
Nutrient pollution, 178
Nutrient shifts, 142–43
Nutrient uptake, 127–32
  ABC transporters, 130–31
  active transport, 128–30
  facilitated diffusion, 127–28
  transporters, 88, 98, 128–31, 176
Nutrition, 119–27. *See also* Nutrient uptake
  carbon cycle, 75, 123
  energy sources and storage, 124–25
  essential nutrients, 120
  growth factors, 121
  macronutrients, 120
  micronutrients, 120
  nitrogen cycle, 27, 75, 125–27
  nutrient supplies and growth, 120–23
Nystatin, 806, 807

O

Obama, Barack, 433, 644
Objective lenses, **51**, 52f, 54, 60f, 64f
Obligate aerobes, **559**
Obligate autotrophs, 714
Obligate lithotrophs, 563–64
Occludin, 420, 421f
Oceanic Microbial Observatory, 58
Oceanospirillales, 858–59, A-64t
Ocular lenses, **53**, 54, 60f
Oenococcus oeni, 640, 641f
Ogiri, 628t, 636
Oil spills
  bacterial remediation, 139, 162, 898, 899
  dead zones, 898–99
  Deepwater Horizon oil spill, 6, 532, 858–59, 865, 898, 899f
  Exxon Valdez oil spill, 898
Okazaki, Reiji and Tsuneko, 250
Okazaki fragments, 249f, **250**, 255, 790
Oleic acid, 90f
Oligosaccharides
  definition, 510, A-13
  in human milk, 515
Oligotrophic lakes, **866–67**
Oligotrophs, 700, 737, 752, 754, 785, eTopic 5.4
Omalizumab (Xolair), 997
OMP (outer membrane porin), 98
OmpC (porin), 412
OmpF (porin), 98, 99f, eTopic 3.2
OmpX (periplasmic transporter), 85f
Oncogenes, 216, **218**, 450
Oncogenic viruses, 216, **218**, 225
One gene-one protein theory, 814
One Health Initiative, 1182
Onesimus, 22
Oomycetes, 799t, 800, 801f, 802, 805, **817**, A-70t, eTopic 20.1
Oparin, Aleksandr, 677
Open reading frames (ORFs), **282**, 313
Operational taxonomic units (OTUs), **846**, 847f
Operon fusions (transcriptional fusions), **461**, 462f
Operons, **242**, **370**. *See also* Genes; Lactose (*lac*) operon
  accBC, 603
  alignment with mRNA transcript, 292–93f
  ara, 376–78, 391
  aro, 616
  control of transcription, 369
  definition, 242
  sense and template strands, 292–93f, 293
  thrABC, 261, 262f
  trp, 367, 616
  ubiG-mccB-mccA, 388f
  vir, 657
  Ophiostoma novo-ulmi, 881
Ophthalmia neonatorum, 1080
Opines, 329, 657
Opisthokonta, A-68t–A69t
Opisthokonts, 801–2. *See also* Animals; Fungi
O polysaccharide, 97f, **98**
Opportunistic pathogens, **934**, **1006**
Opsonins, **957**, 993
Opsonize (opsonization), **951**, 952f, **972**, 1037
Optical density, **138–39**, 141, 142f
Optical isomers, **A-8**
Optical trap, **54**
Optical tweezers, **54**, 298–99, 415
Optics. *See also* Light
  bright-field microscopy, 51–54
  dark-field optics, 63–64
  differential interference contrast microscopy, 65
  phase-contrast optics, 65, 66f
  and properties of light, 46–51
Optochin sensitivity test, 1158f, 1159
Oral and nasal cavities, microbiota, 928–30
Oral candidiasis (thrush), 1081f
Oral groove, **828**, 829
Oral herpes virus (HSV-1), 206f, 207
Oral polio vaccine (IPV), 981
ORF Finder, 313
Organelles, 30–31, 79, 112, **A-31**, A-34–A-35. *See also* Cell structure
Organic molecules, **A-6**, A-7t
  biochemical composition of bacteria, 84
  transport across membranes, 88
Organic phosphonates, 911
Organic pollutants, 868
Organic respiration, 493f
Organism approach to classification, **1052–53**
Organotrophy, organotrophs, **124**, **492**, 540, 542
Organ transplant rejection, eTopic 24.3
OR gates, 480
Origin (*oriC*), 249f, **250**, 252, 252f, 253f
Origin of replication (*ori*)
  bacterial chromosome, **102**, 103f, 104, 105f
  plasmids, 258f, 259f, 265–66
Origins of life, 6, 18–19, 665–81. *See also* Geological evidence for early life; Models of early life
  clues from extremophiles, 18–19
  comets or meteorites as source, 19, 613, 680
  controversy surrounding, 666–67
  elements of life, 667–69
  metabolism of first cells, 676–78
  methanogenesis, 677, 680
  microfossils, 18, 671–72, 673f
  panspermia, 680
  photosynthetic oxidation, 674–75
  reduced compounds in environment of early Earth, 18–19
  stellar origin of atomic nuclei, 667–68
  thermophilic *vs.* psychrophilic origin, 680
  time line, 675f
Ornithine, 616
Oró, Juan, 18, 19, 677
Oropharynx, **929**
Orth, Kim, 1017
Orthologous genes (orthologs), **206**, **315**, 361, 696, 697
Orthomyxoviridae, A-61t
Orthomyxoviruses, 205t
Orthophenylphenol, 183
Oscillator constructed in *E. coli,* 481
Oscillatoria, 589f, 720, 721f, A-62t
Oscillatoriales, A-62t
Oseltamivir, 426, 431, 1138
Osmolarity, 159t, **166-167**
Osmolytes, A-30
Osmosis, 87, **87**, **A-29**
Osmotic pressure, **87**
Osmotic shock, 82, 85, 96, 99, 100f
Osmotic stress, 166–67
Osteomyelitis, 1018, 1148
Osterholm, Michael T., 1176, 1177
Ostreococcus tauri, 805
Otitis media, 1018
Ototoxicity, 1124
Outer leaflet, **A-26**
Outer membrane, **82**
  beta barrel assembly machine (BAM), 309
  E. coli proteins, 85f
  lipoproteins and lipopolysaccharides, 97–98
  in model of bacterial cell, 80f–81f, 82
  proteins, 98, 99, 309, eTopic 3.3
  Pseudomonas aeruginosa, 94f
  viewed with cryo-EM, 71, 79
Outer membrane of poxvirus, 202
Outer membrane porin (OMP), 98
Oxaloacetate, 527, 528–30, 531–32, 588, 598, 612, 614
Oxazolidinones, 1120, 1125, 1132, 1133f
Oxidative phosphorylation, 530f, **531**
Oxidation–reduction (redox) reactions, 676, A-21
Oxidative burst, **952**
Oxidative decarboxylation, 527–28, 529f
Oxidative phosphorylation, 557
Oxidoreductase protein complexes, 553–55
Oxidoreductases, **546**, **553**, 559–60
N-(2-oxocyclohexyl)-3-oxododecanamide, 1037
2-oxoglutarate (alpha-ketoglutarate)
  amino acid biosynthesis, 613–14, 616
  reverse TCA cycle, 598f, 599
  TCA cycle, 528, 529f, 532, 588

2-oxoglutarate:ferredoxin oxidoreductase, 598
Oxyclines, 898
Oxygen
aerobes *vs.* anaerobes, 174
aerobic respiration, 173
anaerobes *vs.* facultative microbes, 174–75
classification, 159t
in early biosphere, 674
electron transport chain, 172–73
as essential nutrient, 120
oxygen-related growth zones in a standing test tube, 174f
role in electrochemical gradient, 173
as terminal electron acceptor, 172–73, 507, 542
Oxygenic Z pathway, **577**, 580, 581f. *See also* Photolysis; Photosystem I; Photosystem II
Oxygen minimum zones (OMZ), **898**, 906
Oxymonadida, A-71t
Oxyrase, 175
Oxytetracycline, 158
Oxytricha, A-70t
Oxytricha nova, 19f
Ozone layer, 907

P

P18 protein in phage T4, 412
Pace, Norman, 683–84, 686–87, 793, 841
Padan, Etana, 129
Paenibacillus dendritiformis, 157, 177f
Palindromes, **263**
Palmaria palmata, A-69f
Palmitic acid, 90f, A-16f
Palsson, Bernhard, 696
Pandemics, **1172**. *See also* Epidemiology
AIDS pandemic, 193, 453, 1081, 1172
antigenic shifts and influenza pandemics, 426
bubonic plague pandemic, 1172
concern about avian influenza strain H5N1, 1172, 1173
definition, 1172
influenza pandemic of 1918, 424, 427f, 1139, 1172, 1180, eTopic 27.4
pulmonary tuberculosis, 1176
swine flu pandemic of 2008, 426
Pandoravirus, **196**
Paneth cells, 946
Pan-genomes, **697**
Panspermia, **680**
PANTHER detection system, 1178–79
Pantoea agglomerans, 1182
Paper chromatography, **591–92**
Papillomaviridae, A-60t
Papillomaviruses, 7f, 193, 213, 443, 444t. *See also* Human papillomavirus (HPV)
Para-aminobenzoic acid (PABA), 1110, 1115, 1122
Parabasalia, A-71t
Parabens, 650
Parachlamydia, A-65t
Paracoccus denitrificans, 126
Paralogous genes (paralogs), **315–16**, 361, 688, **692**
Paramecium, 44f, 46f, 100f, 804, 828, 829f, A-35, A-70t
Paramecium bursaria, 704, 705
Paramecium tetraurelia, 804f
Paramyxoviridae, A-61t
Paramyxoviruses, 205t, 213–15
Paraphysomonas, A-70f
Pararetroviruses, 204f, 205t, **206**, 219–20, A-61t
Parasites, **1005**
Parasitism, **704**, **856**
Parasomal sacs, 828
Parasporal bodies, 474, 474f
Parenteral route of infection, 1009
Parfocal systems, **53**
Parkinson's disease, 554, 706
Partial diploids, 328, 332f, 333
Parvoviridae, A-60t
Parvoviruses, 203, 205t
Passive diffusion, 87
Passive transport, 89, **89**, 127–28
Pasteur, Louis
discovery of chirality, 16–17
fermentation research, 17, 513, 625, 637
fowl cholera vaccination, 23
pasteurization, 181, 648–49
photograph, 16f
rabies vaccination, 23, 24f
swan-necked flask experiment, 16f, 17
Pasteurella multocida, 1019
Pasteur Institute, 23, 36, 188
Pasteurization, **181**, 648–49
Pathogenesis, **1003–47**. *See also* Endotoxin; Exotoxins; Infection; Microbial diseases; Secretion
attachment to host, 1013–18
experimental tools, 1045–47
host susceptibility, 1017
immunopathogenesis, 1009
infection, definition, 1004–5
latent state, definition, 1006
pathogenesis, definition, 1004
portals of entry, 1009, 1052–53
Pathogenicity, **1006**. *See also* Pathogenicity islands
infectious dose, definition, 1006
LD_{50} (lethal dose 50%), definition, 1006
pathogenicity, definition, 1006
virulence, definition, 222, 1006
Pathogenicity islands, **359**, **646**, **1010–12**. *See also* Genomic islands; Pathogenicity; Virulence factors
E. coli, 360f, 1076
examples of, 1011t
model pathogenicity island, 1011f
Salmonella, 359, 646
Staphylococcus aureus, 360f, 1012
type III secretion systems, 1030–32
uropathogenic *E. coli,* 1076
Pathogen identification
acid-fast stain, 57, 59f, 732, 1152
alpha hemolysis, 1158
for antibiotic selection, 1148
for anticipating sequelae, 1148
antigen capture ELISA, 1163–64
API 20E strip for identifying Gram-negative rods, 1153–55, eTopic 28.1
automated identification systems, 1155, 1156f
bacitracin susceptibility, 1158f, 1159
beta hemolysis, 1158–59
catalase test, 1157–58
coagulase, 1158f, 1159
DNA-based detection tests, 1160–62, eTopic 28.2
electronic detection of pathogens, 1178–79, eTopic 28.3
enzyme-linked immunosorbent assay (ELISA), 1072, 1162–64
fluorescent antibody staining, 1164
Gram-negative diplococci, 1155, 1157
Gram-negative enteric bacteria, 1153–55, 1156f
Gram-negative nonenteric rods, 1155, 1157f
Gram-positive pyogenic cocci, 1157–59
immunochromatographic assays, 1168
Lancefield classification scheme, 1159
multiplex PCR identification, 1160–61
novobiocin sensitivity test, 1158f, 1159
optochin sensitivity test, 1158f, 1159
oxidase (cytochrome oxidase) test, 1157
point-of-care (POC) rapid diagnostics, 1166t, 1167–68
polymerase chain reaction (PCR), 1160–62, 1176
procedures for selected diseases, 1165t
rapid DNA sequencing, 1162
real time quantitative reverse transcriptase PCR (qRT-PCR), 1161–62
reasons for identifying pathogens, 1148–49
restriction fragment length polymorphisms (RFLPs), 1176
selective media for stool or fecal samples, 1151
simplified biochemical algorithm to identify Gram-negative rods, 1154–55, 1156f
specimen collection and processing, 1149–51
for tracking spread of disease, 1148–49
virus identification, 1161–64
Pathogens, **1005**
definition, 1005
emerging pathogens, 1047, 1179–82
evolution by horizontal gene transfer, 1012
extracellular immune avoidance, 1037
facultative intracellular pathogens, 1038
food-borne pathogens, 644–48, 1009, 1052–53
host–pathogen interactions, 1004–9
human mortality from, 6
infection cycles, overview, 1007–9
intracellular pathogens, 952, 1037–41
location sensing systems, 1036–37
opportunistic pathogens, definition, 1006

Pathogens (*continued*)
pathogens whose genomes have been sequenced, eTopic 26.1
primary pathogens, definition, 1006
quorum sensing, 398, 1037
researching, 7f
strategies to misdirect immune system, 1041–44
virulence signals, eTopic 5.3
Patient histories, 1053–54
Pavlova vivescens, 228
PDB (Protein Data Bank), 75
Pearson, Ann, 775
Pectins, 510, 511f
Pedestal formation, 1031f, 1032
Pelagibacter, 735, 861, 895
Pelagibacter ubique, 700t, 742
Pelagic zone, **857–58**
Pellicles, 99, 149
Pelomyxa, 8, A-69t
Pelvic inflammatory disease, 1079, 1080
Penicillin, 24, 34, 74, **1108–9**
as bactericidal antibiotic, 1126
destruction by beta-lactamase (penicillinase), 93, 1119, 1128, 1129f
discovery, 1108–9
formation by *Penicillium,* 814, 1109
hypersensitivity to, 967, 1111, 1124
membrane-permeant weak acid, 88
penicillin-binding proteins as target of, 1118–20
penicillin G, 1091, 1109f
peptidoglycan and cell wall synthesis, 79, 87, 93, 186–87, 713, 1111, 1116f
prophylactic treatment with, 1092
resistance to, 93, 1119–20, 1127–28
S. pyogenes lack of resistant strains, 1148
spectrum of activity, 1111
structure, 1118f
synthesis, eTopic 27.2
Penicillin-binding proteins (PBPs), 1056, **1118**, 1119
Penicillium, 632, 642, 797, 809, 1108–9, 1143
Penicillium griseofulvum, 1143f
Penicillium notatum, 24, 186–87
Penicillium roqueforti, 631
Pentadecylcatechols, 999f
Pentaglycine, 713
Pentose phosphate pathway (PPP), **517**, **520–21**, 522f
Pepsin, 630
Peptic nucleic acid (PNA), A-55
Peptide bonds, 95, 97, 102, **A-8**, A-9f
Peptide chains, 102, 103f
Peptides, 476–77, 511–12
Peptide tags, 466, 473
Peptidogenomics, 606–7
Peptidoglycan, **85**
bacterial cell walls, 71, 82, 85, 713, 716
biochemical composition of bacteria, 85
biofilm dissolution, 149
digestion upon phage DNA injection, 413
Gram-negative bacteria, 94, 97
Gram-positive bacteria, 94, 96
and Gram stain, 57, 58f
sporulation, 151
structure, 92, 93f
synthesis as antibiotic target, 79, 93, 98, 109, 186, 1111, 1116f, 1117–21
Peptidyltransferase, **292**, 296, 297f
Peptidyl-tRNA site (P site), **292**, 294, 295, 299
Peptones, 135
Peptostreptococcus, 932
Peptostreptococcus anaerobius, 884
Perchlorate, 917
Perchloroethene, 569
Perforin, 953f, **954**, **988**
Pericarditis, 1090
Periodic table, A-3f
Periodontal disease, 929f, 1018
Peripheral membrane proteins, **A-24**
Periplasm, 82, 97, 98, 99
cell envelope, 94f
chaperones, 307, 308, 309
nanotubes, 114
protein export to, 308–9
Peritrichous cells, 114, 116f
Permafrost melting and global warming, 889, 896
Permeases, **127**, 128
Permissive temperature, **106**
Peroxidase, 174, 175
Peroxisomes, **A-35**
Persistent viruses, 228–29
Persister cells, 1132–33
Pertactin, 1016, 1017f
Pertussis toxin, 1024, 1033
Petechiae, **1027**, 1028f, 1067, 1083f
Petri, Julius Richard, 21
Petri dish, **21**
Petroff-Hausser counting chamber, 136
Petroleum, 513, 533f
Peyer's patches, 942, 944, 1003, 1090
pH, 159, 168–72
acidophiles, 159t, 168f, 169–70
alkaliphiles, 159t, 170–71
classification of organisms by optimum growth pH, 159t, 168f
of common substances, 169f
and fluorescence microscopy, 41
internal pH (pH_{int}), 168, 169, 170, 172f
membrane-permeant organic acids and pH, 168–69
neutralophiles, 159t, 169, 170f
pH difference (ΔpH), 169, 546–47, 549f, 550, 564, eTopic 14.2
pH effect on biological processes, A-20–A-21
pH homeostasis and acid tolerance, 172
pH homeostasis and proton circulation, 172f
pH optima, minima, and maxima, 168–69, 170
PHA and PHB polymers, 112
Phaeophyceae, A-70t. *See also* Brown algae (kelp, Phaeophyceae)
Phage display, **476–77**
Phages, **194**
Phage therapy as medical treatment, 188
Phagocytosis, **824**, 950–52, **A-30**
amebas, 824
autophagy, 952
definition, 937–38
lysosomes and, A-35
monocytes and macrophages, 939
neutrophils, 937–38
opsonization, 951, 952f, 972, 993
oxidative burst, 952
oxygen-dependent killing pathways, 951, 952
oxygen-independent killing pathways, 951
pathogen resistance to, 1037
pellicle, 99
phagosome–lysosome fusion, 938f, 951, 952
quorum sensing and resistance to, 149
recognition of foreign particles, 951
white blood cells attacking bacteria, 951f
Phagosome–lysosome fusion, 938f, 951, 952, 1038–40
Phagosomes, **938**, 951, 1038, 1038f–1039f
Phanizomenon, 152f
Pharmaceutical drugs, 88
Phase-contrast microscopy, 64–65, 66f
Phase plate, 65, 66f
Phase variation, **341**, 369, **389–90**
DNA cassette inversion, 341, 389, 390f
flagellar proteins in *Salmonella enterica,* 341, 389–90
and immune avoidance, 1037
site-specific recombination, 341
slipped-strand mispairing, 389–90
pH difference (ΔpH), 169, 546–47, 549f, 550, 564, eTopic 14.2
Phenol and phenolics, 183–84
Phenol coefficient test, **184**
Phenolic flavor compounds, 640
Phenolic glycolipids, 96f, 732
Phenol red broth test, **526**
Phenotype, definition, 343
Phenylalanine, 616
Phipps, James, 23f
Phosphate cycle, 911–12
Phosphate fertilizers and algal blooms, 912
Phosphates and algal blooms, 868
Phosphatidates, **86**
Phosphatidylethanolamine, **86**, 87f
Phosphatidylglycerol, **86**, 87f
Phosphine, 911
Phosphodiesterases (PDEs), 396–97
Phosphodiester bonds, **243**, 505
Phosphoenolpyruvate (PEP), 132, 374, 375f, 508, 510, 519f
Phosphofructokinase, 518–19
6-phosphogluconate, 520, 521f, 522f
Phosphoglucose isomerase, 518 519f
3-phosphoglycerate (PGA), 591, 592, 596
Phosphoimaging, 465, **A-50**
Phospholipase A2, 958
Phospholipid bilayer, 86, 88, **A-26**, A-27f

Phospholipids, **86**, **A-24**, **A-26**
bilayer formation, a-26
diversity, 89-90
in mycobacterial capsule, 96f
structure, 86, 87f, 511f, A-16, A-24, A-27f
5-phosphoribosyl-1-pyrophosphate (PRPP), 617, 618
Phosphorus
as essential nutrient, 120
global cycling, 605
phosphate cycle, 911–12
removal from wastewater, 899–900
uptake in cyanobacteria, 75
Phosphorylation, **504**, 505, 518
Phosphorylation cascade, 323, 1026, eTopic 4.1
Phosphoryl groups, hydrolysis of, 505t
Phosphotransferase system (PTS), **132**, 343, 374, **505**, eTopic 4.1
Photoautotrophs, photoautotrophy, **123**, **124**
Photobacterium profundum, 684
Photodegradation, 739
Photoexcitation, 572, 574, 576–77, 578
Photoferrotrophy, 675, 743
Photoheterotrophs, photoheterotrophy, **124**, **513**, **572**. *See also* Heterotrophs, heterotrophy; Phototrophy
acidobacteria, 749
Calvin cycle, 590
carbon cycle, 123f
energy acquisition in bacteria and archaea, 124, 492, 493t, 513, 542
Haloarchaea, 765
proteobacteria, 579, 735, 736–37, 742
Photoionization, 573
Photolithoheterotrophs, 540–41, 735
Photolithotrophs, 749–50
Photolyase, 350
Photolysis, 542, 574–77
chlorophylls, 574
overview, 574
oxygenic photolysis, 577, 580, 581f
photolytic electron transport system, 574, 575–76, 577
photosystems I and II, 576–79
reaction centers, 575, 578f, 579f, 581f
Photons, 47
Photopigment classes, 575
Photoreactivation, 348t, **350**
Photorhabdus luminescens, 684f, 872
Photosynthesis, **26–27**, 112, **492**. *See also* Chlorophylls; Chloroplasts; Phototrophy
algae, 127
anaerobic photosynthesis, 534, 577–79
coupled to Calvin cycle, 590
cyanobacteria, 26–27, 573, 577, 580
cyclic photophosphorylation, 540, 541f, 579
and nitrogen fixation, 152, 611
oxygenic Z pathway, 577, 580, 581f
photopigment classes, 575
photosynthetic membranes, 575, 577f, 737, 750
R. palustris, 540, 541f
solar energy capture, 494–95
Photosystem I, **576**
in acidobacteria, 749
in chlorobia, 577–78
electron separation from sulfides and organic molecules, 577, 578f
ferredoxin, 578
reaction center, 578f
Photosystem II, **576**
in alphaproteobacteria, 578–79
cyclic photophosphorylation, 579
electron separation from bacteriochlorophyll, 578–79
reaction center, 579f
Phototrophy, **124**, **492**, **540**, **542**, 571–82. *See also* Bacteriorhodopsin; Photosynthesis
carbon assimilation, 850
energy acquisition in bacteria and archaea, 492, 493t
facultative phototrophy, 580
iron phototrophy, 674–75
organelles of, 112–13
oxygenic Z pathway, 577, 580
photoexcitation, 572
photoionization, 573
photolysis, 542, 574–77
photosystems I and II, 576–79
phototrophic bacteria, 719t
retinal-based proton pumps, 571–73
Phycocyanin, 718
Phycodnaviridae, A-60t
Phycoerythrin, 718, 818, 821
Phylloquinone/phylloquinol (PQ), 578
Phylogenetic trees, **683–85**
of bacterial symbionts of *Alviniconcha,* 663f
calibration, 685
of coastal CO-oxidizing bacteria, 739f
DNA sequence divergence, 682f, 683, 684
"fans" or "bushy branches," 685
of Gammaproteobacteria, 684f
for herpes viruses, 206–7
node, 683f, 684
radial phylogenetic tree (unrooted), 683f, 684
rectangular phylogenetic tree (rooted), 683f, 684
root of a tree, 683f, 684, 688
for thaumarchaeote NM25, 777f
of thermophiles isolated from Yellowstone National Park, 683–84
tree of life, 685, 688, 690f
Phylogeny, **681**. *See also* Evolution; Phylogenetic trees
archaeal phylogeny, 762–65
bacterial phylogeny, 713–14
clade, definition, 681
definition, 681
divergence of three domains of life, 685, 688–89
divergence through natural selection, 681
divergence through random mutation, 681
eukaryotic phylogeny, 798–805
gene flow model, 691f
horizontal gene transfer, 690–91
molecular clocks, 31, 32f, 292–93, 361, 681–83
monophyletic group, definition, 681
reductive evolution, 527, 532, 679, 681
of shower curtain biofilm, 686–87
and species definitions, 696
Phylum, defined, **714**
Physarum polycephalum, A-69t
Phytates, 632, 636
Phytoestrogens, 878
Phytophthora cinnamomi, 187
Phytophthora infestans, 458, 817
Phytophthora ramosum, 817
Phytoplankton, **818**, **863**
and carbon balance, 230, 862
definition, 804, 818, 863
diversity, 228
iron-induced phytoplankton bloom, 913, 914f
mapping marine phytoplanktons, eTopic 21.1
oxygen production, 862
phytoplankton infection by marine viruses, 228, 230f, 863
Pickles, 634
Picoeukaryotes, 805, A-69t
Picoplankton, **862**, 863f
Picornaviridae, 1099, A-61t
Picornaviruses, 216–17, 417t
Picrophilus oshimae, A-67f
Picrophilus torridus, A-67t
Pidan, 628t, **636**
Piezophiles, **164–65**
Pili (singular, pilus), **113**, 116, **1013**
attachment pili, 1013, 1014f
biofilm development, 149, 324, eTopic 9.1
curli pili, 396
E. coli, 200, 341
encoding by pathogenicity genes, 1011
pili tip protein, 1013, 1014f, 1015, eTopic 25.4
pilus assembly, 1013, 1015–16, 1029
pyelonephritis-associated pili (Pap), 1013, 1015
sex pili, 324–25, 326f
type I pili, 341, 1013, 1014f, 1015
type III pili, 1013
type IV pili, 1013, 1015–16, 1029, 1084
uropathogenic *E. coli,* 1076
Pilin, **113**
Pinkeye, 928
Pink slime, 843f
Pinocytosis, **A-30**
Piptoporus, 815
Pirellula, A-65t
Planctomyces, 753–54, 767, A-65t
Planctomyces bekefii, 753–54

Planctomycetales, A-65t
Planctomycetes, **717**, 752–54
anammox reaction, 566, 736t, 908
internal membrane, 752–53
in metagenome of benthos, 865
phylogeny, 713t, 715t, 717, A-65t
Plankton, 58, 59f, **862**
biomass determination, 862
enumeration by fluorescence microscopy, 862, 863f
femtoplankton, 862
mapping marine phytoplanktons, eTopic 21.1
measurement of incorporation of radiolabeled substrates, 862–63
measuring planktonic populations, 862–63
microplankton, 862
nanoplankton, 862
phytoplankton, definition, 804, 818, 863
phytoplankton infection by marine viruses, 228, 230f, 863
picoplankton, 862, 863f
planktonic food webs, 863–65
zooplankton, 804
Planktonic cells, **139**, 148f, 149, 396
Plant biotechnology, 220
Plant pathogens, 218–20, 644, 881–82
Plants, A-69t
Crenarchaeotes on surface of roots, 772f
gene transfer from bacteria to plants, 329
host defense against viruses, 220–21
plant cell structure, A-25f
plant microbial communities, 839, 876–82
thermotolerant plants, 226–27
three-way mutualism, 226–27
Plant viruses, 193, 194, 196, 218–21, 881
Plaque assay, **223**
of animal viruses, 224–25
of bacteriophages, 223–24
Plaque-forming units (PFUs), **223**
Plaques, **194**, **223**
Plasma cells, from B cell differentiation, 964, **974**, 975, 976, 977f
Plasmacytoid dendritic cells, 940, 941f
Plasma membrane, **A-24**. *See also* Cell membrane
Plasmids, 240t, **258**
addiction modules, 260, eTopic 7.4
and antibiotic resistance, 259, 260, 328
bidirectional replication, 258, 259
as cloning vectors, 264–66
cointegration with transposable elements, 356, 357f
DNA map, 258f
DNA supercoiling, 258
episomes, 327
fertility factor (F factor), 325–28
F-prime (F′) plasmids, 328
high-copy-number plasmids, 259
host survival genes in plasmids, 260
inheritance, 259–60
low-copy-number plasmids, 259
mobilizable, 328
NDM-1 plasmids, 1119, 1132
origin of replication (*ori*), 258f, 259f, 265–66
pBR322, 258f, 260
plasmid R1, 259
replication initiator RepA, 259
R factors, 328
rolling-circle replication, 258–59, 325, 326f
segregation, 259, 260f
shuttle vectors, 265–66
size, 258f
Ti plasmid, 329, 657, 1033
transferable, 325, 328
transmission between cells, 260, 325–27
virulence genes on, 1011
Plasmodesmata, **219**
Plasmodial slime molds, **826**
Plasmodium, 1166
Plasmodium (plural, plasmodia), **826**
Plasmodium falciparum, 831–33, 1092–93, A-70t
antigenic shift, 1092, 1093
apicoplast, 831, 833
drug resistance, 1093
epigenetic silencing, 1093
erythrocytic cycle, 1092
gamogony, 832
genome, 833
hemoglobin consumption, 832, 833
life cycle, 832, 1092
in liver, 832
malaria, 797, 805, 831–33, 1092–93
merozoite form, 831–32
mitotic reproduction, 832
in red blood cells, 831, 832
schizogony, 832
schizonts, 831, 832, 1092, 1093f
Plasmodium malariae, 1092
Plasmodium ovale, 1092
Plasmodium vivax, 1092
Plastocyanin, 580, 581f
Platelet-activating factor, 947
Platelets, 937, **939–40**
Platensimycin, 1135
Pleurocapsa, 722
Pleurocapsales, A-62t
Pleurotus (oyster mushrooms), 626
Plocamium, 821
Ploss, Alexander, 420
Plum pox virus, 218–19
Pneumococcus, 690
Pneumocystis jirovecii, A-68t
Pneumocystis jirovecii (formerly *P. carinii*), 808–9, 817, 1006
Pneumonia
case histories, 1059, 1121, 1127
diagnosis, 1059, 1062
multidrug-resistant pneumonia, 1127
relative incidence caused by various microorganisms, 1059f
Streptococcus pneumoniae, 728, 930, 1059, 1062
Pneumonic plague, 1094f, **1095**
PodJ protein, 110, 111f
Poggio, Sebastian, 235
Point mutations, **341**, 342
Point-of-care (POC) rapid diagnostics, 1166t, 1167–68
Poison ivy, 998, 999f
Polar aging, **107–10**
Polar area chart, 14f
Polar covalent bonds, **A-4**, **A-5f**
Polar tube, 817
Poliomyelitis (polio), 36, 193, 222
Poliovirus. *See also* Picornaviruses
classification, 204, 205t
DNA genome, 25
molecular biology, eTopic 11.1
PVR receptor protein, 213, 216
replication cycle, eTopic 11.1
Sabin (attenuated, oral) vaccine, 981
Salk (killed, IPV) vaccine, 981
(+) strand RNA, 417
structure, 202
tissue culture, 222, 223f
vaccine development, 36, 193
Poly-3-hydroxyalkanoate (PHA), 112
Poly-3-hydroxybutyrate, 586
Poly-4-hydroxybutyrate, 586f
Polyacrylamide gel electrophoresis, 84–85, 404, 464, 465, A-47, A-50–A-51, A-58
Polyamines, **84**
Polychlorinated aromatic compounds, 513
Polycistronic RNA, 242, 293, 300
Polycyclic aromatic hydrocarbons (PAHs), 513, 898
Polydnaviruses, 229
Polyesters, 586, 601, 603
Polyglutamate, 636
Polyhydroxyalkanoates, 112, 586
Polyhydroxybutyrate (PHB), 112, 586, 603, 744
Polyketides, 601, 602, 603–5, eTopic 15.2
Polyketide synthase, 604
Polymerase chain reaction (PCR), **6**, **266–67**
bioinformatic analysis, 158
detection of *C. difficile,* 9223
detection of HIV, 21–22
multiplex PCR, 1160–61, A-55
for pathogen identification, 1160–62
real-time PCR, 121–23, 468–70
reverse-transcriptase PCR (RT-PCR), 596–97, 1161
SSU rRNA used for creating primers, 682, 686
Thermus aquaticus, 35, 162
unculturable microorganisms, 121–23
Polymixin, 946f, 1121
Polymorphonuclear leukocytes (PMNs), 937, 957
Polyomaviruses, 218
Polyoxins, 806
Polyphenols, 635
Polyphyletic ancestry, **30**
Polyproteins, **420**
Polysaccharides
alginate, 624
biofilm development, 149
carbon cycle, 123f
catabolism of, 510, 511f, 512f, 514–15f, 515–16

Polysomes, 83, **103**, A-44f, **A-45**
Polyvinyl chloride (PVC), 569
Polyvinylidene fluoride (PVDF) membrane, 464
Popa, Radu, 75
Populations, **840**
Porins, **98**, 99
  aquaporins, 87, 127–28, 166
  maltose porin, 208, 210
  OmpC, 412
  OmpF, 98, 99f, eTopic 3.2
  outer membrane porin (OMP), 98
  Proteobacteria outer membrane, 717
  sucrose porin, 98
Porphobilinogen, 619
Porphyra (red alga), 514–15f, 516, 626, 821, A-69t
Porphyromonas, 930, A-65t
Porphyromonas gingivalis, 523
Porphyromonas gingivalis (periodontal disease), 113
Portal-of-entry approach to classification, **1052–53**
Positive-pressure lab suits, 1169, 1171
Positive selection (T cells), **986**
Potassium transport, 167, 172, 549, 806–7
Potato blight disease, 458
Potato spindle tuber viroid, 196, 197f
Potato virus X (potexvirus), 474–75
Potyviridae, A-61t
Potyvirus (plum pox virus), 218–19
Pour plate technique, **137**
Poxviridae, A-60t
Poxviruses, 202
ppGpp. *See* Guanosine tetraphosphate (ppGpp)
Prebiotic soup, 677–78
Predators, **850**
Preliminary mRNA transcript (pre-mRNA), **314**
President's Emergency Plan for AIDS Relief (PEPFAR), 433
Pressure
  and microbial growth, 159t, 164–65
  steam sterilization, 180
Prestegaard, Karen L., 114f
Prevotella, 884, 928, 929f, 930, 932
Prieur, Daniel, 790
Primaquine, 1093
Primary antibody, **464–65**
Primary antibody response, **974–75**, 980, 986–87
Primary endosymbionts, **800**, 803
Primary pathogens, **1006**
Primary producers, **849**, 850, 851
Primary recovery, **657**, 658f
Primary structure of proteins, **A-8**, A-9f
Primary syphilis, **1078**
Primases, **252**
Primate T-lymphotrophic virus (PTLV-1), 432
Primer extension, **466–67**, **A-52–A-53**
Prindle, Arthur, 478–79
Prion diseases, 1089–90, 1180
Prions, 25, 182, **197**, **1089–90**
Probabilistic indicators, **702**
Probiotics, **188**, 727, **933–34**
Prochlorales, A-62t
Prochlorococcus, A-62t
  carbon flux, 896
  culturing of, 21, 861
  genomic islands, 271
  reductive evolution, 693
  size, 8
  substitution of nonphosphorus lipids for phospholipids, 911
Prochlorococcus marinus
  open pan-genome, 697
  photosynthesis, 573
  size, 44f
  thylakoids and carboxysomes, 112f, 720
Prochloron, A-62t
Prodigiosin, 585
Prodrugs, 1122
Producer microbes, engineering of, 477
Progenesis Technologies, 623, 624
Prokaryotes, **8**
  cell structure and shapes, 44, 45f
  classification, 30
  microbial cell size, 44
  nucleoid, 101–2
  thermal range, 161
Proline, 613, 616
Promoters, **241**, 242f, 262, **277–78**
  binding by RNA polymerases, 277–78, 280
  binding by sigma (σ) factors, 277–78, 279
  E. coli promoter sequences, 278f
  GAL1 gene promoter region, 472
  primer extension analysis, 466–67
  and regulatory proteins, 366–67
Prontosil, 1109–10
Proofreading, **253**
Prophages, **210**, 333, 353–54, 1057, 1131
Prophase, A-32
Propionibacteriaceae, A-63t
Propionibacterium, 629, 632
Propionibacterium acnes, 510, 928, 1125
Propionibacterium freudenreichii, 200, 525, 618
Propionic acid, 649
Propionic acid fermentation, **629**
Propionigenium modestum, 523
Prostaglandins, 947, 948, 949f, 958
Prosthecobacter, A-65t
Prosthecobacter dejongeii, 754
Prosthecobacter fusiformis, A-65f
Protamex, 654
Protease inhibitors, 194, 420, **442**, 632, 1141, 1142
Proteases, 202, 301, 302f, 305–6
Proteasomes, 305
Protective antigen (PA), **1025–26**
Protein A, **1037**, A-57
Protein channels or pores, 127
Protein Data Bank (PDB), 75
Protein-DNA binding, 374–75, 467–68, 469f, 471–72
"Protein Jive Sutra," 36
Protein-protein interactions, 472–73
Proteins, A-8–A-11. *See also* Transport proteins (transporters)
  affinity tagging, 466, 466–67f, 473
  biochemical composition of bacteria, 84–85
  degradation, 304–6
  degrons, 304
  denaturation of, 963f
  DNA-binding proteins, 80–81f, 102
  domains, 102
  folding of proteins, 275, 303–4
  fold-or-destroy triage system, 305–6
  Golgi complex direction of transport, A-37
  identification with 2D gels and mass spectrometry, 404–5, A-46–A-48
  immunogenicity, 965, eTopic 24.1
  insertion into ER, A-36–A-37
  integral membrane proteins, 307
  modification after translation, 303
  native conformation of proteins, A-11
  N-terminal rule, 304
  peptide bonds, A-8, A-9f
  peripheral membrane proteins, A-24
  primary structure of proteins, A-8, A-9f
  quaternary structure of proteins, A-10f, A-11
  radiation damage, 183
  ribosomal proteins, 290, 291f, 303
  secondary structure of proteins, A-10
  secretion (*See* Secretion)
  sequence analysis, 314–16
  superfamilies, 361
  tertiary structure of proteins, A-10–A-11
  transition from RNA world to proteins, 679
  transmembrane proteins, A-24
Protein synthesis. *See* Translation
Protein Synthesis: An Epic on the Cellular Level, 35–36
Protein tagging, 301, 302f
Proteobacteria, **717**, 734–48. *See also* Alphaproteobacteria; Gammaproteobacteria
  anaerobic photosystem II, 576-77, 735
  Betaproteobacteria, 717, 735, 736t, 739t, 742–43
  CO oxidation, 739
  Deltaproteobacteria, 717, 736t
  as endophytes, 839
  Epsilonproteobacteria, 598, 717, 747–48
  Gram-negative cells, 94, 717, 734
  lithotrophy, 735–36
  in marine habitats, 861, 865
  metabolism, 734–36
  microbial mats, 723, 744
  photoheterotrophy, 719t, 735
  phylogeny, 713t, 715t, 717, A-63t–A-64t
  proteorhodopsin, 571–72, 735
  species relative abundance, 846, 847f
Proteome, **84**, 85, **207**, **404–5**
Proteomics, **207**, 404–5, 848, A-47–A-48, A-49f
Proteorhodopsin, **571–72**, **735**, 847, 861
Proteus, 928, 1075
Proteus mirabilis, 56, 744, 745f, 1155f

Proteus vulgaris, 744
Protists, **800**
algae, 127, 804
definition, 800
diversity, 797, 804–5
ectosymbionts of, 754
heterotrophy, 126, 804
marine food chain, 124, 127, 798
organisms reclassified as, 802
upper temperature limit, 161
Proton motive force, **125**, **546–51**
active transport, 130
ATP synthesis, 83, 541-42, 547, 548–49, 557–59
bacteriorhodopsin generation of, 572
and chemiosmotic theory, 34, 547, 548–49
dissipation, 550–51
drug efflux pumps, 547, 551
electrical potential ($\Delta\psi$; charge difference), 547, 549f, 550, eTopic 14.2
and electron transport systems, 530, 539, 541
flagellar motors, 172f, 551
in iron-oxidizing bacteria, 564
methanogenesis, 570
oxygen's role in electrochemical gradient, 173
and pH difference (ΔpH), 169, 546–47, 549f, 550, eTopic 14.2
photosystem I, 578
proton potential (Δp), definition, 125, 546–47
proton pumping by ETS, 546–47
uptake of nutrients, 551
Proton potential, definition, **125**. *See also* Proton motive force
Protozoa (singular, protozoan), **800**
definition, 800
parasitic, 314–15, 949
Provirus, 439f, **440**
ProX (periplasmic transporter), 85f
Prozac (fluoxetine), 88
Prymnesiophyceae, A-70t. *See also* Coccolithophores (coccoliths, Prymnesiophyceae)
Pseudogenes, **261**, 262f, 360, 733
Pseudomembranous enterocolitis, 933
Pseudomonadacea (pseudomonads), 745, 849
Pseudomonadales, A-64t
Pseudomonas, 46f, 126
Bdellovibrio infection, 747
benzoate catabolism, 532
biofilms, 745
catabolism of soil contaminants, 491, 492
as floc microbes, 900
GGDEF/EAL proteins, 396–97
meat spoilage, 642
nitrogen fixation, 905
oxidase-positive, 1157
phosphorylation by succinyl-CoA synthetase in, 528
and study of emerging pathogens, 489
twitching motility, 1029
type VI secretion, 1034
urinary tract infections, 1075
Pseudomonas aeruginosa, 62, 63f, 94f, 116
alginate production, 623, 624f
biofilms, 62, 63f, 148–50, 186f, 337, 488–89, 745, 945f, 1018, 1136
CO oxidation, 739
and cystic fibrosis patients, 398, 488–89, 745, 944–45, 1018
cytochrome *c,* 551–52
disinfectant resistance, 185
exopolysaccharides (EPS), 149
glyoxylate bypass, 531
LPS membrane, 1027f
multivalent vaccine development, 489
NDM-1 plasmids, 1119
prevalence of infection in hospitalized patients, 488
Psl in biofilm matrix, 488–89
quorum sensing and pathogenesis, 398, 399t
susceptibility in cystic fibrosis, 944–45
type IV pili, 1015
type VI secretion system, 1035
versatility, 489
virulence factors, 489
Pseudomonas fluorescens, 745
Pseudomonas putida, 491
Pseudomonas putida S16, 534–35
Pseudomonas sp. strain A2279, 513f
Pseudomonas syringae, 163, 1031
Pseudomurein, **760**
Pseudopeptidoglycan, 713, **760**, 778
Pseudopods, 44, 745f, **804**, 824, 825, 827, 831
Pseudouridine (Ψ), 289
Psychotolerant bacteria, 161f, 162, 163
Psychotrophic organisms, 645–46
Psychrophiles, **161–62**, **865**
classification, 159t
cold-seep ecosystems, 865
collection in Antarctica, 161–62, 655, 765
ice nucleation by bacteria, 162, 163
membranes, 162
psychrophilic barophiles, 165
psychrotrophic bacteria in food, 162, 642
temperature adaptations, 160f, 161–62
Puerperal fever, 23–34
Pulque, or "cactus beer," 520
PulseNet, 1176
Pure culture, **21**, **133**, 134
Purified protein derivative (PPD), 1062
Purines, **243**, 617, 626
Puromycin, 300
Purple bacteria
absorption spectra, 574, 575f
Calvin cycle, 590
carotenoids, 574, 575f
glycine–succinate pathway, 618–19
phototrophy, 574–75, 577, 578–79, 735
Purple nonsulfur bacteria, 719t, 735
Purple sulfur bacteria, 27, 719t, 723f, 735
Purpuras, 1067, 1083f
Putrefaction, **642**
Putrescine, 512, 642
PVDF (polyvinylidene fluoride) membrane, 464
PVR receptor protein, 213, 216
Pyelonephritis-associated pili (Pap), 1013, 1015
Pyrazinamide, 1065
Pyrenoids, 819
Pyrimidine dimers, 345, 350, 352
Pyrimidines, **243**, 617, 618
Pyrobaculum, 598
Pyrobaculum aerophilum, 684
Pyrococcus, 690, 790–91
Pyrococcus abyssi, 790, A-67t
Pyrococcus furiosus, 31f, 336f, 761, 790, A-67t
Pyrococcus horikoshii, 791f
Pyrococcus woesei, 790
Pyrodictium, 768–69, 910
Pyrodictium abyssi, 765, 768–69, A-66t
Pyrodictium brockii, 568-69, 768
Pyrodictium occultum, 492, 684, 768, 865
Pyrogens, **958**, 1019
Pyronins, 1123
Pyrophosphate (PP_i; PRPP), 268, 281, 617, 618
Pyroptosis, 1042
Pyrosequencing, 267, **268–69**, 271
Pyrsonympha, A-71f
Pyruvate
3-hydroxypropionate cycle and, 601
amino acid biosynthesis from, 613, 614
carbon cycle, 123f
glucose conversion to pyruvate, 499, 512f, 516–19
in mixed-acid fermentation, 523
from phosphoenolpyruvate, 505t, 508
to phosphoenolpyruvate, 510
pyruvate conversion to acetyl-CoA, 527–28
reduction by NADH, 506
reverse TCA cycle and, 599
structure, 508
Pyruvate dehydrogenase complex (PDC), **527–28**, eTopic 13.4
Pyruvate formate lyase, 523
Pyruvate kinase, 81f, 508–10, 519f, 523

Q

Q fever, 121, 181, 746, 1040, 1053–54. *See also Coxiella burnetii*
QseC sensor kinase, 1136
Quarantined, **1172**
Quasispecies, **423**, 435–36, 438
Quaternary ammonium compounds, 184f
Quaternary structure of proteins, A-10f, **A-11**
Queuosine, 762
Quinine, 832
Quinolone-oxazolidinone hybrid, 1132, 1133f
Quinolones, 102, 130, **248**, 1120, **1122**, 1126
Quinone pool, **554**, 560f, 581f
Quinones and quinols, **552**, 553, 554

Quorum sensing, **149**, 397–400, **398**
Aliivibrio fischeri, 366, 397–98
anti-quorum sensing drug, 1136–37
autoinducers, 398, 399–400
bacterial lobster traps, 365
biofilms, 149, 398
and cell competence, 323–24
drawbacks to, 402
examples of microbial quorum-sensing systems, 399t
interspecies communication, 399–400
and pathogenesis, 398, 1037
Pseudomonas aeruginosa, 398, 399t
synthetic biology applications, 478–79
Vibrio harveyi, 399–400

R

Rabies vaccination, 23, 24f
Rabies virus, 227, 451, A-61f
Racker, Efrem, 549
Radial immunodiffusion, **A-57–A-58**
Radioactive decay, 590–91
Radioisotope tracers, 529–30, eTopic 15.1
Radiolaria, 825, A-70t
Radiolarians, **827**
Ragsdale, Steven, 780, 781f
Rainfall and psychotolerant bacteria, 163
Rajneesh, Bhagwan Shree, 1176
Ralstonia, 496
Ralstonia eutropha, 586, 603
Raltegravir, 439f, 440
RAMP (repeat-associated mysterious proteins) module, 335–37
Rancidity, **642**
Raoult, Didier, 196, 1046
Rare biosphere, 848
Rarefaction curve, **846**, 847f
Rational drug design, 187
Reaction centers (RC), **575**, 578f, 579f, 581f, 722, 723f
Reactive mats, 902
Reactive oxygen species (ROS), 174, 175, 344, 345f, 405
Reading frames, **195**
Real-time PCR, **468–70**, 1161–62. *See also* Polymerase chain reaction (PCR)
Real time quantitative reverse transcriptase PCR (qRT-PCR), 1161–62
Reassortment, **424**, 426, 427f
RecA (synaptase), 338f, 339–40, 352–54
Recombinant DNA, 32, 35, 223, 263, 264–66
Recombination, **239**, 338–41
branch migration, 339f, 340
Chi (crossover hot-spot instigator), 339f, 340
cointegrate molecules, 338
functions, 338
and gene duplication, 361
generalized recombination, 338–40
heteroduplex region, 340
Holliday junction, 339f, 340
phase variation, 341
RecA (synaptase), 338f, 339–40, 352–54
RecBCD, 339f, 340
RuvA, B, and C proteins, 339f, 340
site-specific recombination, 338, 341
Recombinational repair, 338, **352**
Recombination signal sequences (RSSs), **978**
Red algae (Rhodophyta), 626, 672f, **803**, 818, 821–22, A-69t
Red blood cells, 936f, 937
Red colloidal gold, 1168
Reddy, Shravanthi T., 146, 147f
Redi, Francesco, 16
Redman, Regina S., 226
Red Mars (Robinson), 918
Redmond, Molly, 858–59
Redox couples, **542–44**, 560–61
Redox reactions, **A-21**
Red tide, 830f, 831
Reduction potential (*E*), 507, 542–44
Reductive acetyl-CoA pathway, 590t, **599**, **601**
Reductive evolution, **681**, **692–93**. *See also* Evolution
definition, 681
diminished RNA components in cells, 679
endosymbiosis, 704–5
in eukaryotes, 798, 802, 831
loss of TCA cycle, 527, 532
mitochondria, 706
Mycobacterium leprae, 733
Reductive pentose phosphate cycle, **589**. *See also* Calvin cycle
Reductive (reverse) TCA cycle, 590t, **597–99**
Reed, Walter, 1008f
Reeve, John, 762
Reflection, **48**
Refraction, **48–49**, 65, 66f
Refractive index, **48–49**, 52, 64–65
Refrigeration, 631, 642, 643, 646, 648
Regulatory proteins, **366–67**. *See also* Gene regulation
activator proteins, 366–67, 369
AraC regulator protein, 376–78, 391
AraC/XylS family of regulators, 376
CRP (cAMP receptor protein), 371–73, 378, 391
degradation, 304
derepression, 367
GadE (YhiE), 465, 468, 469f
GadX, 468, 469f
GAL4 transcription factor, 472
helix-turn-helix motif, 368f
induction, 366
LacI (lactose catabolism regulator), 370–71, 372f, 373, 376, 385, 391
and ligands, 366–67
and regulons, 242
repression, 366–67
repressor proteins, 366–67, 368f, 369
response regulators, 368
sensing extracellular environment, 368
sensor kinases, 368
transcription control, general, 366–67
TrpR (tryptophan synthesis regulator) protein, 378, 379f, 385
two-component signal transduction systems, 368
XylS activator, 376
Regulatory RNAs, 369, 385–88
cis-antisense RNA molecules, 385, 387–88
small RNA (sRNA) molecules, 385–87
Regulatory T cells (Tregs), **986**, **994–95**
Regulons, **242**, **382**, 693
Rehydration therapy, **1066**
Relaxase, 325, 326f
Relaxosome complex, 325, 326f
Release factors (RF1, Rf2), **296–97**
Relenza (zanamivir), 1138, 1139f
Remotely operated vehicles (ROVs), 663
Rennet, 630, 632f
Reolysin, 204, 453
Reoviruses, 204, 205t, 453
Repellent signals, 115
Replication forks, 104, **249**, 250
Replisomes, **104**, 105f, **253**, 254f, 255
Reporter genes, 83, **461**
green fluorescent protein (GFP), 83, 224–25, 461, 462–63
lacZ reporter gene, 348, 461, 472
nerve growth factor (NGF), 452–53
reporter fusions, 460–63
yellow fluorescent protein (YFP), 462, 463f
Repression, definition, **366**
Repressors, **366–67**, 368f, 369
Reservoir (of infection), **1007**, **1009**
Reservoirs, **891**, 892t, 894
Resistance islands, 360f
Resolution, **42–43**
in bright-field microscopy, 51–52
and magnification, 47, 49–51
Resolvase, 356–57
Resource colimitation, **914**, **916**
Respiration, **513**, **515**, 531, 541–42
Respiratory electron transport system, 551–59. *See also* Electron transport system (ETS)
alternative oxidoreductases, 556–57
cofactors for small energy transitions, 551–53
E. coli ETS for aerobic NADH oxidation, 555f, 556
environmental modulation of the ETS, 556–57, eTopic 14.3
initial substrate oxidoreductases, 553–54
mitochondrial respiration, 556–57f, 557
pathways for organotrophy, 555–57
proton fates, 556
quinone pool, 554
terminal oxidases, 554–55, 556
Respiratory syncytial virus (RSV), 451, 452f, 1065, 1137
Respiratory tract
fungal diseases of the lung, 1060t–1061t, 1062–64
microbes that cause respiratory tract infections, 1060t–1061t
microbiota, 930

Respiratory tract (*continued*)
mucociliary elevator, 930, 944, 1058–59, eTopic 26.3
respiratory tract infections, 1058–65
tuberculosis as a reemerging disease, 1064–65
viral diseases of the lung, 1060t–1061t, 1065
Response regulators, **368**
Restriction endonucleases, **263–64**, **333–34**
bacterial host defense, 211–12, 333–34
and cloning, 264–65
digestion of DNA, 263–64, A-50
EcoRI, 264f, 265f, 334
types of restriction–modification systems, 334, 335t
Restriction fragment length polymorphisms (RFLPs), 1176
Restriction sites, 263, **266**, 334
Restrictive temperature, **106**
Reticulate bodies, **752**, 753f
Reticulitermes flavipes, 855f
Reticulitermes santonensis, 855f
Reticuloendothelial system, **939**
Retinal, **572**
Retinal-based proton pumps, 571–73, 761
Retortamonadida, A-71t
Retrotransposons, **442**, 443f
Retroviridae, A-61t
Retroviruses, 204f, 205t, **206**, **432**, A-61t. *See also* Human immunodeficiency virus (HIV)
Avian leukosis virus, 195, 205t
definition, 1140
endogenous retroviruses, 442, 443f
evasion of immune responses, 449
feline leukemia virus (FeLV), 205t, 206, 218, 432, 433t
gene transfer to human genome, 211
lentiviruses, 205t, 223, 432, 433t
life cycle, 217–18
primate T-lymphotrophic virus (PTLV-1), 432, 433t
retroviruses of animals, 433t
Rous sarcoma virus (RSV), 193, 205t, 432
simian immunodeficiency virus (SIV), 22, 194, 205t, 433
simple retroviruses, 432, 433t
Rev accessory protein, 435f, 437t
Reverse electron flow, **565**, **579**
Reverse gyrase in hyperthermophiles, 248, 761
Reverse transcriptase (RT), **206**, **438–40**
APOBEC3G inhibitor protein, 440
and azidothymidine (AZT), 438
and evolution, 261
high error rate of replication, 438
HIV, 202, 432, 438–40
initiation of reverse transcription, 440
origin, 196
pararetroviruses, 206, 220
retroviruses, 206, 217–18
reverse transcriptase inhibitors, 1140–41, 1142
telomerase, 261
Reverse transcriptase PCR (RT-PCR), 596–97, 1161–62
Reversible fluorescent termination, 270, 271f
Reversions, **341**
Reversion tests, 347–48
Rhabdoviridae, A-61t
Rhabdovirus, 205
Rheumatic fever, 999, 1148
Rheumatoid arthritis, 949
Rhinoviruses, 193, 930. *See also* Picornaviruses
attachment to glycoprotein ICAM-1, 213, 214f, 216, 966–67
common cold *vs.* influenza, 1065, eTopic 26.4
different antigenic determinants between strains, 966–67
evolution, 227
infection cycle, 1007, 1008f, 1009
low virulence, 1006
neutralizing antibodies, 966–67
rhinovirus 14, A-61f
Rhizaria, 799t, 800, 801f, 804, 825, A-70t
Rhizobia, **877**
bacteroids, 740, 741f, 877–78, 879f, 880f
endosymbiosis with legumes, 28, 740, 741f, 877–81
increased crop yields, 905, 906f
infection thread, 740, 741f, 878, 879f
Mesorhizobium loti, 322
mobile symbiosis islands, 360
nitrogen fixation, 28, 126, 608, 610–11, 740, 878–81
Nod factors, 87, 878
plasmids, 260
Rhizobiales, A-64t
Rhizobium, A-63t
bacteroids, 880f
Bdellovibrio infection, 747
nitrogen fixation, 126, 608, 740, 877, 878f
plasmid-borne genes, 260
root hair curing around *Rhizobium, 879f*
symbiosis with legumes, 126, 608, 740, 877, 878f
Rhizoplane, **873**, 874
Rhizopus, 812, 813f
Rhizopus oligosporus, 632–33
Rhizosphere, 329, 848, **872**, **873**, 876
Rhl quorum-sensing system, 398, 399t
Rhodanobacter, 126
Rho-dependent termination, **282**
Rhodobacter, 574, 575, 576, 590
Rhodobacterales, A-64t
Rhodobacter sphaeroides, 589f, 590, 597, 736, 737f
Rhodococcus, 470f, 532, 733
Rhodocyclales, A-64t
Rhodocyclus, 742
Rhodomicrobium, 737, A-63t
Rhodomicrobium vannielii, 737f
Rhodophyta. *See* Red algae (Rhodophyta)
Rhodopseudomonas, 776, 868, 910, A-63t
Rhodopseudomonas palustris, 534, 540–41, 578, 675, 735, 849
Rhodopseudomonas viridis, 576f
Rhodospirillum, 574, 575, 576, 590
Rhodospirillum rubrum, A-64f
carbon and energy acquisition, 124
in eutrophic lakes, 868
flagella, 114
light absorption, 578
microscopic observation, 43, 54f, 55
nitrogen fixation, 608
photoheterotrophy, 124, 736–37
photolysis, 574
Rubisco, 594
tetrapyrrole biosynthesis, 619–20
Rho-independent termination, **282**
Rhopaloeides odorabile, 883
Rhoptries, 832
RhyB sRNA, 385–86, 387t
RhyB sRNA, 385–86, 387t
Ribavirin, 418, 420, 422
Ribonucleic acid. *See* RNA (ribonucleic acid)
Ribonucleoproteins, 941f
Ribonucleoside monophosphate (rNMP), 281
Ribonucleoside triphosphates (rNTPs), 280–81
Ribose 5-phosphate, 520, 522f, 617
Ribosomal RNA (rRNA), **285**
5S rRNA in 50S ribosomal subunit, 290
5S rRNA structure, 291f
16S rRNA and antibiotics, 298–99
16S rRNA and metagenomics, 270–71
16S rRNA as molecular clock, 31
16S rRNA in 30S ribosomal subunit, 290, 293–94
23S rRNA in 50S ribosomal subunit, 290
23S rRNA structure, 291f
in bacterial cell, 82, 83, 102
catalytic role in protein synthesis, 292
highly conserved base sequence, 121, 292, 315
modified bases, 285
as molecular clock, 292–93, 361, 681
Ribosome-binding sites, 292–93f, **293–94**, 482–83, 484, eTopic 8.2
Ribosome recycling factors (RRF), 296, 297f
Ribosomes, 46f, 54, 289–92
5S rRNA in 50S ribosomal subunit, 290
16S rRNA in 30S ribosomal subunit, 290, 293–94
23S rRNA in 50S ribosomal subunit, 290
30S (small) subunit structure, 289–90
50S (large) subunit structure, 289–90
70S ribosome complex, 290, 292
acceptor site (A site), 292, 294, 295, 296, 297f
and antibiotics, 103, 1111
in bacterial cell, 80–81f, 82, 83, 102–3
in barophilic organisms, 165
binding of tRNA, 292f
complexed with RNA, 76
in eukaryotes, 103
exit site (E site), 292, 295–96

finding reading frame, 293–94
isolation and analysis, A-44f, A-45, eTopic 3.1
and mRNA translation, 294–97
peptidyl-tRNA site (P site), 292, 294, 295, 299
ribosomal proteins, 290, 291f, 303
ribosomal protein S12, 298
ribosome assembly, 290, 291f, eTopic 8.1
ribosome disassembly, 296–97
ribosome recycling factors (RRF), 296, 297f
ribosomes stuck on mRNA, 301, 302f
ribosome trap experiment, 298–99
stringent response, and decreased synthesis, 381
Riboswitchboards, 483–84
Riboswitches, 381, 388, **482–84**, 620, eTopic 15.4
Ribotyping, 932
Ribozymes, 19, 197, **292**, **679**
peptidyltransferase, 292, 296, 297f
Ribulose 1,5-bisphosphate, 592
Ribulose 1,5-bisphosphate carbon dioxide reductase/oxidase. *See* Rubisco (ribulose 1,5-bisphosphate carbon dioxide reductase/oxidase)
Ribulose 5-phosphate, **520**, 594
Rice blast, 814
Rice dwarf virus, 70f, 193
Rickettsia, 235, A-64t
Rickettsiales, 700, A-64t
Rickettsia prowazekii (typhus fever), 121, 122f, 316, 1006, 1041
Rickettsia rickettsii (Rocky Mountain spotted fever), 741, 1040f, 1096
Rickettsias, 545, 741–42, 861, 1038
Rifampin (rifampicin), 103f, 109, 283, 1065, 1084, 1123, 1126, 1171
Rifamycin (rifamycin B), 280f, 283, 284
Riftia, 865, 866f
Ringworm, 814
Ripening (cheese), **630–31**
Rise period, 221f, **222**
Rivers, Thomas, 1004
R-malonyl groups, 603–4
RNA (ribonucleic acid)
antisense RNAs, 369, 385, 387–88
catalytic RNA, 19f, 196–97, 285, 292, 679, A-15
cis-repressed mRNA, 482–83, 484
classes of, 284–85
CRISPR RNA, 212, 336–37
differences from DNA, 244
hairpins, 244–45, 298–99
messenger RNA (*See* Messenger RNA (mRNA))
microRNAs (miRNAs), 1042
regulatory RNA molecules, 369, 385–88
ribosomal RNA (*See* Ribosomal RNA (rRNA))
secondary structure, A-15
small RNA (*See* Small RNA (sRNA))
stability, 285–86
structure, 35, 242f, A-15
tmRNA, 285, 301, 302f
trans-activating RNA (taRNA), 482–83, 484
transfer RNA (*See* Transfer RNA (tRNA))
RNA deep sequencing, 386, 387–88, 401f, 404, 467, 468f, 1047
RNA degradosome, 286
RNA-dependent RNA polymerase, **203–4**, 216, 217f, 425, 430, 431
RNAIII sRNA, 386, 387f
RNA interference (RNAi), 221
RNA polymerases
and activator proteins, 367
alpha (α) subunits, 277
antibiotics targeting RNA polymerase, 1123
archaeal, 762
in bacterial cell, 80–81f, 82, 102–3
beta-prime (β′) subunit, 277, 280, 281f
beta (β) subunit, 277, 281f, 283, 284
closed complex, 280
collisions caused by asRNA transcription, 388
core polymerase, 277–78
DNA-dependent RNA polymerase, 277
DnaG (DNA primase), 252, 251f
holoenzyme, 277
housekeeping RNA polymerase, 252
inhibitors, 625
initiation of DNA replication, 252
initiation of transcription, 280–81
and magnesium, 84, 277, 281
open complex, 280–81
pausing of, 300–301
plasmid replication, 259f
promoter binding, 277–78, 280
RNA-dependent RNA polymerase, 203–4, 216, 217f, 425, 430, 431
RNA polymerase II structure, A-1
rudder complex, 280, 281f
sigma factors, 277–80
similarity in archaea and eukaryotes, 762
subunit structure, 277
Taq RNA polymerase, 279f, 281f
as target of antibiotics, 103f, 109
three-dimensional structure, 277f, 279f, 281f
viroids and, 196, 197f
virus replication, 203–4, 206, 216–17
RNA reverse-transcribing viruses, **206**. *See also* Retroviruses
RNaseH, 255, 256f
RNases, 285–86, 301
RNA synthesis. *See also* Transcription
drugs that inhibit, 1123
RNA world, **17**, 19, 285, **677**, 679, 680, eTopic 17.1
Robbins, Frederick, 222
Robinson, Kim Stanley, 918
Roche 454 sequencing method, 845t
Rocky Mountain spotted fever, 545
Rodriguez, Russell, 226
Rods, **44**, 45f, 106–7
Rohwer, Forest, 207
Ronson, Clive, 322, 360
Roosevelt, Franklin Delano, 193
Roossinck, Marilyn, 226–27
Root of a tree (phylogeny), 683f, **684**, 688
ROS (reactive oxygen species), 174, 175
Roseobacter, 739
Rosing, Minik, 673
Rossman, Michael, 412
Rotary flagella, 28f, 54, 114–16
Rotavirus, 204, 1009, 1043, 1066, **1072**, A-60f
Rough endoplasmic reticulum, A-36
Rous, Peyton, 193
Rous sarcoma virus (RSV), 193, 205t, **432**
RpoH gene, 382, 383f
rRNA (ribosomal RNA). *See* Ribosomal RNA (rRNA)
Rubella virus, 1057–58
Rubisco (ribulose 1,5-bisphosphate carbon dioxide reductase/oxidase), **592**. *See also* Calvin cycle
archaeal Rubisco, 592
in carboxysomes, 112f, 596–97
isotope ratios in carbon fixed, 672–73
mechanism, 593f, 594
structure, 592, 593f, 594
Ruby, Edward, 397f
Rumen, **883**
bovine rumen, 780, 781f, 883–84
chytrid inhabitants of, 809–10, 812f
metabolism, 884, 885f
methane production in, 780, 781f, 884
microbial community, 883–84
microbiology research, 780, 781f
nitrate reduction to ammonia, 908
partial genomes assembled from microbiome, 845–46
pH and *E. coli* growth, 884
sampling of target communities, 842, 843f
Ruminococcus, 515, A-63t
Ruminococcus albus, 843f, 884
Ruska, Ernst, 33
Rusticyanin, 564, 565f

S

Saccharomyces, 286, 626, 635
Saccharomyces cerevisiae, A-68t
as baker's or brewer's yeast, 637, 640, 807–8
budding, 808
as edible yeast, 626
genome, 261, 809
life cycle, 808f, 809
as model system, 808f, 809, 810–11
secretion system, 309
shmoos, 808f, 809
Sacculus, **91–93**, 105, 760. *See also* Cell walls
Sadoff, Jerald, 652
Safranin, 57
Saintilus, Wilfrid, 1148
Salinispora, 730, A-63t
Salk, Jonas, 193
Salmonella, 83f, 97f, 114, 135, A-64t
anti-sigma factor (FigM) and flagella assembly, 382
bacteriophage P22, 331, 355

Salmonella (*continued*)
biofilm formation, 396f
deubiquitylase, 1044f
diarrhea, 1030, 1032
facultative intracellular pathogen, 1038
flipping cytokine profiles, 1041
as food-borne pathogen, 647t, 1009, 1032
GGDEF/EAL proteins, 396–97
interspecies communication with *V. harveyi,* 400
interspecies mating with *E. coli,* 329
irradiation, 181, 182
as model system, 922
nitrogen fixation, 905
osteomyelitis, 1148
pathogenesis, schematic overview, 1032f
pathogenicity islands, 646, 1032–33
phagosome–lysosome fusion inhibition, 1033, 1038–40
on phylogenetic tree of Gammaproteobacteia, 684f
in plant vascular system, 1182
plasmid R1, 259
probiotics and, 188
rapid growth, 922
replication efficiency, 248
ribosome isolation, A-44f, A-45
Salmonella pathogenicity island 1 (SPI-1), 1032
Salmonella pathogenicity island 2 (SPI-2), 1032–33
sensing of change in external pH, 172, eTopic 5.3
transduction in, 331
Salmonella enterica, 69f, 98f, 115f, 923f
conversion of host cells, 1003
culture media, 133f, 135
deubiquitylase, 1043
diarrhea, 1066
endophytic growth in crop plants, 877
on Hektoen agar, 1154f
hisG mutant, 347
horizontal gene transfer, 690
internal pH (pH_{int}), 168, 169
intracellular pathogen, 1067
outbreak (2008), 644–45
phase variation, 341, 389, 390f, 1037
phenol coefficients, 184t
sigma S, 382
stress survival responses, 177
superoxide dismutase (SOD), 308
tetrathionate as terminal electron acceptor, 551
type III secretion system, 1032–33
variation in gene distribution, 241
Salmonella enterica serovar Enteritidis (S. Enteritidis), 1053, 1176
Salmonella enterica serovar Typhi (S. Typhi)
antibiotic therapy, 1066
DNA microarray, 404
prevention of phagosome-lysosome fusion, 1039–40
septicemia, 1052–53, 1098
Salmonella enterica serovar Typhimurium (S. Typhimurium)
Bifidobacterium and decreased inflammation, 994–95
as frequent cause of diarrhea, 1066
genome, 241f
multiplex analysis, A-55f
needle complex of type III secretion system, 1030f
pathogenicity islands, 359
suicide bomber cells, 1032
Salmonella senftenberg, 360f
Salt, adaptations to high concentrations, 167, 170–71
Salterns, 784, 786
Salt pockets (underground), 787
Salvarsan, 1111
Samuel, Buck, 780–81
Sands, David, 163
Sanger, Fred, 9, 193, 267
Sanger sequencing, 9, 267, 270, **A-50**
Sanitation, **179**
Saprophytes, **873**
SAR11 cluster of marine species, 700, A-64t
SAR86 cluster of marine species, A-64t
Sarcinae (singular, sarcina), 106, **734**
Sargasso Sea metagenome, 271, 847
Sargassum natans, 824f
Sargassum weeds, 805, **824**
SARS (severe acute respiratory syndrome), 227
Sauerkraut, 627, 628t, 629, 634
Sausage, 627, 628t, 629
SbtA (high-affinity HCO_3^- transporter), 596–97
Scaffolds, 845f, **846**
Scalindua, 908, A-65t
Scanning electron microscopy (SEM), **45**, 46f, **67–68**, 69f
Scanning probe microscopy (SPM), 72
Scarlet fever, 728
Scattering of light, **48**
Schaker, Timothy, 1142
Schirmacher Oasis, 161f
Schizogony, **832**
Schizonts, 831, 832, 1092, 1093f
Schopf, William, 672
Schwalm, Christina K., 787f
Schwan, Rosane, 634–35, 636
Scientific notation, A-16
Scopes, John, 666
Scopes trial, 666, 667f
Scrapie, 197
SDS polyacrylamide electrophoresis (SDS-PAGE), 404, 464, 466f, A-47, A-48f, A-58
Sebaceous glands, 928, 929f, 943
Sebdenia, A-69t
Sebum, 943
Sec-dependent general secretion pathway, 308–9, 1015, 1029, 1033
Secondary antibody, **465**
Secondary antibody response, **974–75**, 987–88
Secondary endosymbionts, 706, 707f, **800**, 801f, 803, **822–23**
Secondary metabolites, **603**, 618, **1126–27**
Secondary products, **603**
Secondary structure of proteins, **A-10**
Secondary syphilis, **1078**
Secondary (reactivation) tuberculosis, **1065**
Second messengers, **396–97**
Secretion, 306–11. *See also* Exotoxins; Pathogenesis
ABC (ATP-binding cassette) transporters, 309f, 310–11
beta barrel channels, 309, 311f
HylABC transporter, 309f, 310, 311
LepB, 308
outer membrane protein secretion mechanisms comparison, 310t
protein export out of cytoplasm, 307
protein export through outer membrane, 309–10
protein export to cell membrane, 307
protein export to periplasm, 308–9
protein secretion overview, 306–7
SecA, 308
SecB piloting protein, 308
Sec-dependent general secretion pathway, 308–9
secretion systems for bacterial toxins, 1029t
SecYEG translocon, 307, 308
signal recognition particles (SRPs), 307
signal sequences, 307
TolC channel protein, 309f, 311
twin arginine translocase (TAT), 309
type I secretion, 309f, 310–11
type II secretion, 1029–30
type III secretion, 360, 1030–33, 1070–71, 1096
type IV secretion, 360, 1033
type VI secretion, 1033, 1034–35, 1036f
Secretory IgA (sIGA), 972, 994
Sectored colonies, 346f
Sedimentation rate, **A-45**
Segmented filamentous bacilli (SFBs), 925, 932
Segmented genome, **203**, 206, **424**, 426
Selectins, 947, **948**, 949f
Selective media, **135–36**
Selective toxicity, **1111**
Selenium, 737, 915t
Seliberia stellata, 737
Semiconservative replication, **249**
Semipermeable (selectively permeable) membranes, **A-28**
Semmelweis, Ignaz, 23–24
Sengupta, Bhaskar, 914
Senior cells, 108–9
Sensor kinases, **368**
Sensory rhodopsins I and II, 788
Sepsis, 24
Septation, **105–7**, 110f, 151
Septicemia, 146, 993, 1027, 1053, 1090, **1091**. *See also* Systemic infections
Septicemic plague, **1095**

Septin cages, 1043
Septum, 62f, **105**, 106–7, 109, 110f, 256–57
SeqA replication inhibitor protein, 250–51
Sequelae (singular, sequela), **1148**
Sequella, 652–53
Sequencing by synthesis, **267**, **270**, 270–71f, 271
Serial endosymbiosis theory, 30–31, **A-40–A-41**
Serovars, 241
Serratia marcescens, 644
Serricchio, Mauro, 315
Serum, **974**
Setlow, Peter, 151
Severe acute respiratory syndrome (SARS), 1004, 1065, 1172–73, 1179–80
Severe combined immunodeficiency (SCID), 962
Sewage effluents and algal blooms, 868
Sewage treatment. *See* Wastewater treatment
Sewer pipe corrosion, 910f, 911
Sex pili, **113**, 324–25, 326f
Sexually transmitted diseases, 1077–83. *See also* Human immunodeficiency virus (HIV); Syphilis
  chlamydia, 185, 704, 752, 1079
  genital herpes virus (HSV-2), 206f, 207, 444, 445f
  gonorrhea, 185, 743, 936, 1079–81
  human papilloma virus (genital warts), 215
  microbes that cause sexually transmitted diseases, 1077t
  trichomoniasis, 1082–83
Sexual reproduction, **A-33–A-34**
Shape-shifting, 157
Sharkskin and Sharklet, 146, 147f
Shelley, Mary, 540
Shewanella, 545, 684
Shewanella oneidensis, 744
Shewanella oneidensis MRS1, 562
Shewanella violacea, 165f
Shiga toxins, 1067, 1070. *See also* Exotoxins
  E. coli O104:H4, 1012, 1068, 1069f
  E. coli O157:H7, 210, 646, 1025, 1067, 1069f
  genes on prophage genomes, 1131
  hemolytic uremic syndrome, 107, 1067, 1068–69
  protein synthesis disruption, 1019f, 1067, 1070
  Shigella, 210, 1025, 1067
  thromobotic thrombocytopenic purpura, 1067
Shigella
  actin motility, 1038, 1040f
  culture media, 135, 136
  diarrhea, 1030, 1148–49
  dysentery, 1066
  escape from phagosomes, 1038f–1039f
  facultative intracellular pathogen, 1038, 1040f
  as food-borne pathogen, 647t, 1009
  genome reduction, 360
  growth factors and natural habitats, 121t
  infection cycle, 1007
  molecular mimicry, eTopic 25.8
  on phylogenetic tree of Gammaproteobacteia, 684f
  septin cages, 1043
  Shiga toxin, 210, 1025, 1067, 1070
  tracking spread of, 1148–49
Shigella dysenteriae, 1025, 1038, 1066, 1067
Shigella flexneri, 1006, 1012, 1031, 1040f
Shigella sonnei, 1148–49
Shilts, Randy, 433
Shimomura, Osamu, 61
Shine, John, 293
Shine-Dalgarno sequence, **293**
Shingles (varicella-zoster virus), 444
Short interspersed nuclear elements (SINEs), 442–43
Shuttle vectors, **265–66**
Sialic acid, 427–28, 515
Sialidase, 948
Siderophores, 131, 861, **913**
Sieber, Jessica, 502f, 503
Siefert, Hans, 329
Sigma factors, **277**. *See also* RNA polymerases
  anti-anti-sigma factors, 382, 383–84, 385f
  anti-sigma factors, 382, 383–84, 385f
  consensus sequences, 278
  conserved regions, 280
  dissociation from transcribing complex, 280f, 281
  in elongation of transcripts, 281
  promoter binding, 277–78
  and proteolytic degradation, 382–83
  regulation, 382–84
  regulation of physiological responses, 278–80
  RpoE, 923
  SigH, 323, 324f
  sigma-70 housekeeping sigma factor, 278f, 279
  sigma B, 462, 463f
  sigma F (RpoF), 382, 383–84, 385f
  sigma H (RpoH), 382–83, 923
  sigma S (RpoS), 382, 923
  and sporulation, 383–84
  in transcription initiation, 280
Signal recognition particles (SRPs), 103f, **307**, **A-36**, A-37f
Signal sequences, **307**, **A-36**, A-37f
Signature-tagged mutagenesis, 1045-1046, eTopic 25.1
Silent mutations, **341**
Simian immunodeficiency virus (SIV), 22, 194, 205t, 433, 435–36
Simian virus 40 (SV40), 443, 444t
Simple stains, **55–56**
Singer, Maxine, 35, 276
Single-celled protein, **626**, 627f
Single-stranded DNA–binding proteins (SSBs), 254f, 255, 259
Singulair, 997
Sinks, **891**
Sinorhizobium, 82, 126, 610–11, 740, 877
Sinorhizobium meliloti, 237, 559, 741f
Siphonous algae, 821
Siphonous algae, A-69t
Siphoviridae, A-60t
Site-directed mutagenesis, 477, eTopic 12.7
Site-specific recombination, **210**, **338**, 341
SIV (simian immunodeficiency virus), 22, 194, 205t
Skatole, 642
Skin
  acne, 510, 928, 929f
  common infectious diseases of the skin, 1055t
  granulomas, 949, 950f
  Langerhans cells, 436, 943
  microbiota, 187, 928
  physical barrier to infection, 943
  sebaceous glands, 928, 929f, 943
  skin and soft tissue infections, 1054–58
  viral diseases causing skin rashes, 1055t, 1057–58
Skin-associated lymphoid tissue (SALT), **943**
S-layer, 91, 94f, **95–96**, 717, 760
Sliding clamp (beta clamp), **252**, 251f, 254–55, eTopic 7.2
Slime molds, **826**
  cellular slime molds, 826
  fruiting bodies, 802, 804, 826
  phylogeny, 799t, 800, 801f
  plasmodial slime molds, 826
  plasmodium, 826
  reclassification as protists, 802
  reproduction, 826
Slime molds, A-69t
Slipped-strand mispairing, 389–90, 1037, eTopic 10.3
Slonczewski, Joan, 28–29
Slow-release replication cycle, 211
Slow viruses, 223
Sludge, **899**, 900, 901
Small acid-soluble proteins (SASPs), 151
Smallpox, 1055t, 1177f
  attenuated virus, 22, 23
  as bioweapon, 1177
  DNA genome, 443, 444t
  DNA polymerase, 203
  history, 14, 22–23, 965–66
  and immunological specificity, 965–66
  inoculation, 22, 23
  origin, 196
  RNA polymerase, 203
  vaccination, 23, 966, 1177
  virus structure, 202, 966f
Small RNA (sRNA), **285**, 369, 385–87. *See also* RNA (ribonucleic acid)
  in eukaryotes, 386–87
  examples, 387t
  finding in genome, 386
  RhyB expansion of Fur, 385–86, 387t
  RNAIII function in *Staphylococcus aureus,* 386, 387f, 387t

Small-subunit (SSU) rRNA, **682**
  highly conserved genes, 271, 682
  for phylogeny and molecular clocks, 32f, 682, 683–84
  as standard for identifying environmental taxa, 841, 846
  used for creating PCR primers, 682, 686
  and working definition of microbial species, 697
Smith, Hamilton, 10
Smooth endoplasmic reticulum, A-36
Sneezing, 944
Snow, John, 1173, 1174f
Snyder, Tonya, 1051
Soda lakes, 170, 171f, 787, 852
Sodium dodecyl sulfate (SDS), A-47. *See also* SDS polyacrylamide electrophoresis (SDS-PAGE)
Sodium exchange
  in alkaliphiles, 170–71
  cation transport, 167, 172
  in halophiles, 167
  Na^+/H^+ antiporters, 128–29, 130, 170, 171f, 172
Sodium motive force (ΔNa^+), 170, 550, 559, 570, 782
Soil, **869–70**. *See also* Mycorrhizae
  endoliths, 840f, 870
  humic material (humus), 872–73
  hydric soil, 875
  methanogenesis in, 780, 876
  microbes associated with roots, 873–74
  microbes in soil and rock, 870f
  remediation of contamination, 491, 492, 877
  rhizoplane, 873, 874
  rhizosphere, 329, 848, 872, 873, 876
  soil food web, 871–72
  soil microbiology, 869–76
  soil profile and structure, 870–71
  virus-to-bacteria ratio, 230
  water table, 870f, 871
  wetland soils, 873, 875–76
Solar energy, 495f, 540f
Solar spectrum, 495f
Solexa, 267, 270
Solexa Illumina sequencing method, 845t
Sol-gel transition, 825
Solutes, **87**
Sonication, 82
Sorangium cellulosum, 746
Sorbic acid, 649
Sorbitol MacConkey agar, 526
Sortases, 95
SOS response, **352–54**, 1025, 1126, 1131
Sources, **891**
Sourdough, 629, **638**
South African Gold Mine Euryarchaeotal Group (SAGMEG), A-67t
Southern, Edwin M., A-48
Southern blots, **463**, A-48, **A-50**
Soy fermentation, 632–34, 636
Soylent Green (science fiction film), 626, 627f
Soy sauce, 627, 628t, 634
Space-filling molecular models, **A-6**
Space science and medicine, 36, eTopic 1.3
Spallanzani, Lazzaro, 16, 17
Sparks, William, 919
Specialized transduction, **331**, 332f, 333
Species, **696–703**. *See also* Taxonomy
  candidate species, 701
  classification and nomenclature, 697, 700–701
  definition, for eukaryotes, 696
  ecological niche (ecotype), 696
  emerging organisms, 701
  genus name, 700
  identification, 697, 702
  pan-genomes, 697
  phylogenetic relatedness, 696
  rules for defining and naming, 701–2
  species name, 700
  working definition of microbial species, 696–97
Species name, **700**
Specimen collection
  abscesses, 1149–50, 1151
  anaerobic infections, 1149–50
  blood culture samples from two sites, 1151
  lumbar puncture, 1150f, 1151
  midstream urine collection, 1075, 1151
  from normally sterile sites, 1151
  selective media for stool or fecal samples, 1151
  sputum collection, 1150f, 1151
  throat swabs, 1150f, 1151
  urinary catheters, 1150–51
Spectrum of activity, **1111–12**, eTopic 27.1
Speed of light (c), 47
Sperandio, Vanessa, 1136–37
Spermatozoan, 801
Sphaerotilus natans, 702
S phase, 103
Sphingomonadales, A-64t
Sphingomonas, 686, 687f
Sphingomonas paucimobilis, 686, 687f
Spider light stops, 64
Spike proteins, **199**, **434**, 435, 436–38
Spirilla, 44, 748
Spirochaeta, 751, A-65t
Spirochaetales, A-65t
Spirochetes, **44**, **717**, 750–52
  in association with *Mixotricha,* 854, 855f
  cell structure, 44, 45f, 751
  dark-field observation, 63f
  mystery organelles, 79
  phylogeny, 713f, 714, 715t, 717, A-65t
  spirochete diversity, 751–52
Spirogyra, 819, 820f
Spiroplasma, 730, 1090
Spirostomum, A-70t
Spirulina, 170, 171f, 626, 627f, 718, A-62t
Spitzer Space Telescope, 919
Spleen, 941–42
Spliceosome, 386
SPM (scanning probe microscopy), 72
SpoIIAA protein, 384, 385f
SpoIIAB protein, 384, 385f
Sponges
  choanocyte cells, 802, 803f
  endosymbiosis, 723, 731
  sponge microbial communities, 883
Spongiform encephalopathies, **1090**, eTopic 26.7
Spontaneous generation, 16–17, 666
Sporangiospores, **812**, 813f
Sporangium (plural, sporangia), **812**, 813f
Spores. *See also* Endospores
  akinetes, 720–22
  arthrospores, 723
  ascospores, 813f, 814
  auxospores, 822f, 823
  basidiospores, 815f, 816
  elementary bodies, 752, 753f
  forespores, 151, 383–84, 724
  myxospores, 746, 747f
  of slime molds, 826
  sporangiospores, 812, 813f
  spore types in bacteria, 721t
  zoospores, 801–2, 809–11, 812f
  zygospores, 812, 813f
Spore stain, 57, 59f
Sporozoites, 1092
Sporulation, 150–52, 152–53, 383–84
Spread plate technique, **134**
Spyrogyra, 8t
SQ109 antibiotic, 653
Squalene, 91f, A-15
SRNA. *See* Small RNA (sRNA)
SRPs (signal recognition particles), 103f
SspB protein, 301, 302f
SsrA (tmRNA in *E. coli*), 302f
Stachybotrys, 814, A-68t
Stahl, David, 775
Staining, **55**, 56f, 57f
Stains, 55–58
  acid-fast stain, 57, 59f, 732, 1152
  antibody stains, 58
  carbolfuchsin, 57, 96, 1152
  counterstains, 57
  crystal violet, 55f, 57f, 58f
  DAPI (4′,6-diamidino-2-phenylindole), 59f, 61, 89f, 686, 862, 863f
  differential stains, 56–58
  fluorophores, 58, 59f, 60, 61–63
  Gram stain, 56–57, 58f
  heavy-atom stains, 68–69
  live-dead stain, 136–37
  methylene blue, 55–56, 1152
  and mordants, 57
  negative stain, 57–58, 59f
  propidium iodide, 136–37
  safranin, 57
  simple stains, 55–56
  spore stain, 57, 59f
  Syto-9, 137, 148f
  Wright stain, 1093
  Ziehl-Neelsen, 57
Stainzyme, 655
Stalk, 107–10, 111f, **113**, 139
Stalked ciliates, **829–30**
Standard reduction potential (E°), **542–44**
Stanier, Roger, 573

Stanley, Wendell, 25
Staphylococcal cassette chromosome *mec* (SCC*mec*), 1119
Staphylococcal toxic shock syndrome toxin (TSST), 989
Staphylococci (Greek, "grape clusters"), **106**, 727–28, 1158
Staphylococcus, 24, 93, 100f, 121t, 729f, A-63t
  biological control on human skin, 187
  catalase-positive, 1158
  coagulase production and differentiation of species, 927f
  colony forming units, 134
  salt tolerance, 728
  urinary tract infections, 1076
Staphylococcus aureus. See also Methicillin-resistant *S. aureus* (MRSA)
  alpha (α) toxin, 1019, 1022
  antibiotic susceptibility data, 1114t
  biofilms, 146, 147f
  boils, 1054, 1056
  cellulitis, 1057
  coagulase, 927f, 1159
  exfoliative toxin, 1019
  as food-borne pathogen, 647t, 1066
  horizontal gene transfer, 239, 1012
  inflammation caused by, 946, 947f
  nosocomial infections, 727–28
  in oral and nasal cavities, 929
  pathogenicity islands (SaPIs), 360f, 1012
  penicillin inhibition of growth, 186f, 1109
  peptidoglycan structure, 92, 713
  persister cells, 1132
  phenol coefficients, 184t
  prophage activation, 353–54
  protein A, 1037
  pyrogenic toxins, 1019
  resistance island, 360f
  RNAIII (sRNA) function, 386, 387f, 387t
  scalded skin syndrome, 1054f, 1056
  septation, 105–6
  septicemia, 1098
  size, 44f
  skin infections, 1054
  stress response to nitric oxide, 405
  toxic shock syndrome, 1019, 1052, 1056
  urinary tract infections, 1075
  vancomycin-resistant *S. aureus* (VRSA), 1119
Staphylococcus epidermidis
  biofilms, 146
  eye microbiota, 928
  novobiocin susceptibility, 1158f, 1159
  in oral and nasal cavities, 929
  skin microbiota, 727, 928
  in urethra, 934
Staphylococcus saprophiticus, 1075, 1158f, 1159
Staphylococcus scalded skin syndrome, 1054f, 1056
Starches, 127, 510, 511f, 515
Starch utilization system (SUS), 514–15f, 515–16
Starfish stinkhorn, 815
STAR-FISH technique, 739
Stars and origins of life, 667–68
Start codons, **287**, 292–93f, 293, 313
Starter cultures, **631**, 632f
Starvation, 152, 176–78, 185, 603
Stationary phase, **143**
Steam autoclaves, 180
Stedman, Ken, 769–70
Steinman, Ralph, 963, 964f
Steitz, Joan, 293–94
Stem cells, 450, 937, 962
Stendomycin I-VI, 606–7
Stenotrophomonas, 877
Stentor, 7f, 829, 830f, A-70t
Stephenson, Marjorie, 548
Sterilization, **178**, 180, 649
Steroids, 997
Stetter, Karl, 714, eTopic 18.1
Stickland reaction, **523**
Stick molecular models, **A-6**
Stock, Ann, 392f
Stolfa, Gino, 1025
Stomach microbiota, 930–31
Stop codons, **287**, 296, 297f, 301, 302f, 313
Storage granules, 112–13, 586, 603
Storz, Gisela, 386, 387f
Streptococcus, 106, 723, 728, 729f, A-63t
  catalase-negative, 1158
  colony forming units, 134, 137
  group A streptococci (GAS), 728
  kimchi preparation, 634
  lactic acid fermentation, 630, 728
  M protein, 1016
  natural transformation, 323, 324f
  oral microbiota, 928
  Prontosil used to treat, 1109–10
  protein F, 1016
  protein G, A-10f
  tooth decay, 728
Streptococcus agalactiae, 1012, eTopic 25.3
Streptococcus haemolyticus, 1159
Streptococcus mutans, 929, 1091, 1158
Streptococcus oralis, 929
Streptococcus pneumoniae
  antibiotic resistance, 1062, 1117, 1119–20, 1127–28
  bacteremia, 1062
  capsule, 1037, 1164
  competence factor, 323
  competitor displacement, 353–54
  fluorescent antibody staining, 1164
  Gram stain, 56
  human microbiota, 927f
  humoral immunity and, 965
  identification, 1158f, 1159, 1168
  immunochromatographic assay, 1168
  Kirby-Bauer disk susceptibility test for, 1114f
  meningitis, 1062, 1083f, 1084, 1117, 1153
  open pan-genome, 697
  optochin sensitivity test, 1159
  phagocytosis of, 951
  pinkeye, 928
  pneumonia, 728, 930, 1059, 1062, 1063–64, 1127
  transformation, 238–39, 322, 323
Streptococcus pyogenes
  bacitracin sensitivity, 1159
  cellulitis, 1057
  erysipelas, 1124
  growth factors, 121
  identification, 1157–59
  M protein, 999, 1017f
  necrotizing fasciitis, 1056–57
  northern blot analysis of total RNA, A-50f
  pyrogenic toxins, 1019
  rapid antigen test, 1167
  rheumatic fever, 999
  sequelae, 1148
  strep throat, 94, 728, 999
  superantigen, 989
  treatment of infections, 1125
  virulence factors, 1057
Streptococcus salivarius, 525, 929
Streptococcus thermophilus, 336f, 630f
Streptogramins, 130, 1125–26
Streptomyces, 730–31, A-63t
  aerial mycelia, 730
  alternative metabolic pathways, 776
  antibiotic production, 152, 298–300, 731, 1125
  arthrospores, 152–53, 723, 730
  association with leaf-cutter ants, 731
  cell differentiation, 152–53
  chromosomes, 730–31
  decomposition of plant material, 872
  filaments, 730–31
  geosmin, 730
  hyphae and mycelia, 152, 153f
  laboratory evolution, 696
  life cycle, 152–53, 154f
  polyketide production, 603, eTopic 15.2
  production of antifungal agents, 1143
  vegetative mycelia, 730
Streptomyces clavuligerus, 1131
Streptomyces coelicolor, 88, 153f, 154f, 585, 589, 700, 730–31
Streptomyces erythraeus, 299
Streptomyces garyphalus, 1120
Streptomyces griseus, 187, 872f, 1110, 1127
Streptomyces hygroscopicus, 606–7
Streptomyces lavendulae, 153f
Streptomyces nodosus, 1143f
Streptomyces noursei, 1143f
Streptomyces platensis, 1135
Streptomyces scabies, 731
Streptomycin, 158, 187, 298, 300f, 589, 1110, 1124
Streptomycin 6-kinase, 187
Stress survival responses, 157, 176, 462, 463f
Strict aerobes, **174**
Strict anaerobes, **174**, 176
Strigolactones, 816

Stringent response, **381**, 603
Stroma, **575**, 577f, A-41
Stromatolites, **666**, 669, 670f, 671t, 672
Structural formulas, **A-3**
Structural genes, **239**
Structural isomers, **A-11**
Subacute bacterial endocarditis, 930
Subacute bacterial endocarditis (SBE), 930
Substrate adhesion protein, 5f
Substrate-binding proteins (SBP), **130–31**
Substrate-level phosphorylation, **518**
Substrate mycelia, 152, 153f, 154f, 807
Substrate-tracking microautoradiography (STAR), 739
Subterranean arsenic removal (SAR), 914, 915f
Succinate, 508, 529f, 530, 531, 544, 560, 613
Succinate dehydrogenase, 386, 557
Succinyl-CoA, 528, 529f, 530, 533, 587, 598f, 599
Succinyl-CoA synthetase, 528, 529f
Sucrose, 515
Sucrose density gradient, 83f, A-44f, A-45
Sucrose porin, 98
Suctorians, 829–30, A-70t
Sudden infant death syndrome, 725
Sudden oak death, 817
SulA protein, 353
Sulfa drugs, **1110**, 1121–22
Sulfamethoxazole, 1121
Sulfanilamide, 1110, 1121–22
Sulfated polygalactans, 821
Sulfate-reducing archaea, 791, 792f
Sulfate-reducing bacteria, 27, 561, 599, 676, 781, 865, 866f, 895, 910, 913
Sulfites used for food preservation, 650
Sulfolobales, **769–71**, A-66t
Sulfolobus, A-66f
  acidic environment, 170f, 566–67
  cell membrane, 770
  conjugation, 324–25
  as double extremophile, 770
  glucose catabolism, 760
  horizontal gene transfer to *T. acidophilum,* 791
  metabolism, 770
  multiple origins of replication, 770
  RNA polymerase, 762
  S-layer instead of cell wall, 770, 771f
  sulfuric acid production, 770
  thermophilic sulfur metabolism, 499, 759–70, 766
  virus infection of, 770–71
Sulfolobus acidocaldarius, 170, 250, 567f, 684, 770
Sulfolobus metallicus, 601
Sulfolobus solfataricus, 769, 770–71
Sulfolobus turreted icosahedralvirus (STIV), 770–71
Sulfonamides, 1111
Sulfur
  anaerobic reactions between sulfur and iron, 567
  decomposition and generation of sulfur compounds, 909, 910
  as essential nutrient, 120
  global reservoirs of sulfur, 892t
  hydrogen oxidation, 492, 568
  iron-sulfur clusters, 552, 553, 574
  microbes in sulfur springs, 31, 170f, 567
  microbial cycling of sulfur, 605, 891
  oxidation states, 891, 892t, 909–10
  oxidized forms, 561, 564, 678
  storage granules or globules, 112–13, 598, 743, 744, 750, 910
Sulfur and metal oxidation, 566–68
Sulfur Cauldron, 170f
Sulfur cycle, 891, 892t, 908–11, 913
Sulfuric acid production, 566–67, 766, 770, 791, 910f, 911, 912
Sulfur oxidizers medium, 122t
Sulfur-oxidizing archaea, 566–67, 763, 765, 766, 769–71, 791
Sulfur-oxidizing bacteria, 26, 143, 742, 748, 749–50, 865, 866f, 909f, 910–11
Sulfur-reducing archaea, 598, 763, 766–67, 768
Sulfur-reducing bacteria, 112–13, 567, 748, 851, 910
Sulfur triangle, 909
Sullivan, John, 322, 360
Summers, Ryan, 491
Sun, 667–68
Superantigens, **989**, 1012, 1019, 1052, 1056
Supercoils, **102**, 245–48
  chromosome compaction, 102, 245–47
  DNA gyrase, 102, 247–48, 256
  domains, 245
  generation during DNA replication, 256
  negative supercoils, 246–47
  positive supercoils, 247, 248, 256, 281–82
  topoisomerases, 246f, 247–48
Superfamilies of proteins, 361
Supernovas, **668**
Superoxide dismutase (SOD), 174, 175, 308, 952
Superoxide ion, 952
Superoxide radicals ($*O_2^-$), 174
Super-size microbial cells, 8
Surface mycelia, 807
Surfactants, 184
Surgeonfish, 726
Sushi, 626
Sutherland, John, 679
Sutovec, Martin, 1069f
Svedberg, Theodor, 33–34, A-44
Svedberg coefficient, A-44f, **A-45**
Svedberg units, 289–90, A-44f, A-45
Swan-necked flask, 16f, 17
Swarmer cells, 107f, 110, 111f, 114, 744, 745f
Swarming, **744**
Swine flu (influenza strain H1N1), 410, 426, 1173, 1181f
Switch regions, **981**
SYBR green dye, 470
Sylvestre, Diana, 418
Symbiodinium, 882–83, 882f
Symbiogenesis, **706**
Symbionts, **126**, 662–63, 854–55
Symbiosis (plural, symbioses), **703–5**, **852–57**. *See also* Endosymbiosis; Mutualism
  amensalism, 856
  coevolution, 852
  commensalism, 856
  ectosymbionts, 754
  leaf-cutter ants with partner fungi and bacteria, 704, 731, 856, eTopic 17.3
  between marine animals and bacteria, 662–63
  microbial mutualism with plants and animals, 28–29, 854–55
  multiple symbionts, 854–55
  parasitism, definition, 704, 856
  symbiogenesis, 706
  synergism, 856
  transfer of genes encoding symbiosis, 322
  types of symbiotic associations, 853t
Symbiosomes, **878**, 880f
Symbiotic worms, 165, 662–63
Symport transport systems (symporters), **128–30**, 551
Synaptosomal-associated protein (SNAP-25), 1087
Synchronous clock in *E. coli,* 478–79
Syncytia, 442
Synechococcus, 148, 718, 720, 841, 911, A-62t
Synechocystis, 596f, 597f
Synercid, 1125–26
Synergism, **856**
Syntaxin, 1087
Synthetic biology, **37**, 457, **477**, 477–85, 625
  BioBricks, 484, 485f
  DIY synthetic biology, 484
  electrical engineering principles, 479–80
  genome transplants, 485
  industrial microbiology, 625
  kill switches, 484
  logic gates, 479–80
  oscillator switches, 478–79, 481
  riboswitches and switchboards, 482–84
  synchronous clock in *E. coli,* 478–79
  system noise, 481–82
  toggle switches, 480–81
Synthetic media, **134–35**
Syntrophomonas wolfei, 502
Syntrophus aciditrophicus, 502–3
Syntrophy, **500**, 502–3, **776**, 781, 854–55
Syphilis, 1077–79. *See also Treponema pallidum*
  case history, 1077
  chancres, 1078
  congenital syphilis, 1078
  New World theory of syphilis, 1078
  primary syphilis, 1078
  secondary syphilis, 1078
  spirochetes, 44, 751
  tertiary syphilis, 1078
  treatment with Salvarsan, 1111
  Tuskegee experiment, 1079

Systemic infections, 1094–1102. *See also* Bubonic plague; Septicemia
Ebola, 200, 213, 1006, 1007f, 1100–1101
hepatitis, 1099–1100
Lyme disease, 21, 44, 751f, 1096–97
microbes that cause systemic disease, 1098t
Systemic lupus erythematosus, 938, 940–41
Szostak, Jack, 678

T

T3SS needle complexes, 1030–31
Tailed phages, 410–11, **412**
Takacs, Constantin (Nick), 235
Tamiflu (oseltamivir), 426, 431, 1138
Tandem mass spectrometry (MS-MS), 405
Tang, Hongzhi, 534–35
TaqMan, **470**
Taq polymerase and PCR reaction, 35, 162, 266, 470, 790, 1161
Target communities, **842**
TasA protein, 149
TATA-binding protein (TBP), 762
Tat accessory protein, 436, 437t, 449
Tatum, Edward, 812, 814
Tau particle of Alzheimer's disease, 197
Tautomeric shifts, 344
Taxa, **697**, **700**
Taxonomy, **696–703**, A-59–A71t. *See also* Bacterial diversity; Species
classification systems, 30, 697, 700
definition, 697
dichotomous keys, 702, 703f
domains, 9, 31, 32f, 79, 91, 685, 688–89
endosymbiosis theory of evolution, 30–31
formal and informal categories, 701, 718
hierarchy of classification, 700t
identification, 697, 702
International Committee on Systematics of Prokaryotes (ICSP), 701
List of Prokaryotic Names with Standing in Nomenclature (LPSN), 29, 700
microbial species and taxonomy, 696–97, A-59–A-71t
nomenclature, 697, 700–701
nongenetic systems of categorization, 701
probabilistic indicator, 702
species name, 700
of viruses, 203
Taylor, Alison, 797
TBGp3 protein, 474
TCA cycle. *See* Tricarboxylic acid (TCA) cycle
T-cell help, 987
T cells, 937f, 939, **941–42**. *See also* Cytotoxic T cells (T_C cells); Helper T cells (T_H cells)
activation, 964, 982–83, 986, 987f
activation by superantigens, 989
antigen recognition, 961, 965, 983, 986
CCR5 surface protein, 1017
CD3 complex, 985
$CD4^+$ T cells, 995, 1009, 1080
cell-mediated immunity, 963, 964–65
classes of T cells, 982, 983t
education and deletion, 985–86
immune system memory, 961
lack in SCID, 962
link between humoral and cell-mediated immunity, 982
major histocompatibility complex (MHC), 965, 982–83, 985, 986, 988
MHC restriction, 986
negative selection, 986
positive selection, 986
processing in thymus, 937f, 982, 986
regulatory T cells (Tregs), 986, 994–95
T-cell activation of B cells, 964, 975, 976, 977f, 986–88
T-cell receptors (TCRs), 985, 986, 988
T-DNA, 657, 659
Tectiviridae, A-60t
Teff, 638
Tegument, 199, **444–46**
Teichoic acids, 94, **95**, 717, 723
Telaprevir, 420
Tellurium, 737
Telomerase, 196, 261
Telomeres, 730–31, 828, eTopic 20.2
Telophase, A-32f, A-33
Temin, Howard, 203, 204f
Tempeh, 627, 628t, 629, **632–33**
Temperate phages, **210**
Temperature and fatty acid composition, 603
Temperature and microbial growth, 159, 160–64
classification, 159t, 161–64
enthalpy and entropy, 497
evolutionary relationships to, 164
growth rate, 160–61
heat-shock response, 164
low temperature, 181
pasteurization, 181
steam sterilization, 180
Temperature-sensitive mutants, 416
Template strand, **277**, 315
Template switching, 423
Tenericutes, **728–30**, A-63t
TephaFLEX surgical sutures, 586f
Tequila, 627, 628t
Terminal electron acceptor, 172–73, **507**. *See also* Electron transport system (ETS)
Terminal organelle, 730
Terminal oxidases, 554–55, 556
Terminal reductases, 559–60, 561
Terminal redundancy, 414
Termination of replication, 250, 256–57
Termination of transcription, 277, 282
Termination (*ter*) sites, 249f, **250**, 256, 257f
Terminator exonuclease (TEX), 467, 468f
Termites, 751, 854–55
Terpenoids, **90–91**
Terraforming, **918**
Terrestrial Miscellaneous Euryarchaeotal Group (TMEG), A-67t
Tertiary endosymbiosis, 706, 803, 830–31
Tertiary structure of proteins, **A-10–A-11**
Tertiary syphilis, **1078**
Testerman, Traci, 923
Tests (inorganic shells), 804, 825
Tetanospasmin (tetanus toxin), **1085**, 1087
Tetanus. *See also Clostridium tetani*
spastic paralysis, 1085
tetanus toxin, 1019, 1026, 1085, 1086–87
vaccine, 1087, 1102
Tetherin, 1100, 1101f
Tetrachloroethene, 569f, 729f, 7287
Tetracyclines
acne treatment, 928, 1125, 1127
antibiotic resistance, 130, 299
binding to ribosomal subunits, 103, 298-299, 300f, 1111
biosynthesis, 601
bubonic plague treatment, 1096
doxycycline, 1096, 1097, 1124f, 1125, 1127
gonorrhea treatment, 1080
Lyme disease treatment (doxycycline), 1096, 1097, 1125
membrane-permeant weak base, 88
side effects, 1125
structure, 88f, 300f
transport by efflux proteins, 130, 299
used with kill switches, 484
Tetrads of cells, 106
Tetraethers, 90, 95f, **759**, 760f, 770
Tetrahydrofolate (THF), 599, 617f
Tetrahydromethanopterin, 570
Tetrahymena, 679, A-39
Tetrahymena thermophila, 1025
Tetrapyrroles, **618–20**
TetR protein, 484
Thalassic lakes, 786
Thalassiosira, 822f, 911
Thaumarchaeota, **763**, **765**, 773–75, 906
ammonia oxidizers, 774–75, 776, 777f
crenarchaeol, 773
marine sponge symbionts, 773–74
phylogeny, 763f, 764t, 765, 777f, A-66t
psychrophilic marine thaumarchaeotes, A-66t
reassignment of some organisms from Crenarchaeota, 772, 773
strain NM25, 776–77
The Institute for Genomic Research (TIGR), 9f, 10
Theisen, Ulrike, 686
Theobroma cacao, 636
Thermales, A-62t
Thermal vents, 161, 162, 165, 568, 578, **865**. *See also* Hydrothermal (thermal) vents
Thermoanaerobacter sulfurigignens, 113f
Thermocline, **858**, 866, 867f, eTopic 21.1
Thermococcales, 765, 790–91, A-67t
Thermococci, A-67t
Thermococcus, 790
Thermococcus litoralis, 790
Thermocrinis ruber, 714
Thermodynamic extremophiles, 503

Thermodynamics, A-17. *See also* Energy; Enthalpy (ΔH); Entropy (ΔS); Gibbs free energy change (ΔG)
first law of thermodynamics, A-17
and limits on cell growth, 161, 500
second law of thermodynamics, A-17
standard reaction conditions, 498–99
Thermomyces lanuginosus, 654
Thermophiles, **161**, **865**. *See also* Hyperthermophiles
acidobacteria, 748–49
Aquificales group of thermophilic bacteria, 691
biofuel-producing, 159f
chaperone proteins, 164
classification, 159t, 161
Crenarchaeota, 760, 763, 765, 766
deep-branching thermophiles, 713f, 714, 715t
DNA-binding proteins, 164
DNA replication, 248
Euryarchaeota, 765
membrane lipids, 90–91, 164
sulfur-oxidizing archaea, 170, 566–67
temperature and growth rate, 160f, 162
Thaumarchaeota, 775
thermal vent communities, 865
thermophilic enzymes, 162–64, 266
thermostability of proteins, 725
X-ray diffraction analysis, 76, 161
Thermoplasma, 685, 760
Thermoplasma acidophilum, 305, 791, A-67t
Thermoplasmata, A-67t
Thermoplasmatales, 791, 793, A-67t
Thermoproteales, 771, A-66t
Thermoprotei, 771, A-66t
Thermoproteus, 598, 771
Thermoproteus tenax, 95f
Thermosphaera, 766, A-66t
Thermosphaera aggregans, 769, 770f
Thermotogae, 714, A-62t
Thermotogales, A-62t
Thermotoga maritima, 690, 692, 714
Thermotolerance, 226–27
Thermozymes, 162–63
Thermus, 716
Thermus aquaticus
ribosome, 292f
Taq DNA polymerase and PCR, 35, 162, 266, 470, 790
Taq RNA polymerase, 279f, 281f
Thiobacillus, 574
Thiobacillus ferrooxidans, 567–68
Thiocapsa, 743, 906
Thioesterase (TE), 604
Thiogalactoside transacetylase (LacA), 370, 371f
Thioglycolate, 175
Thiomargarita namibiensis, 8, 44
Thiomicrospira, 865
Thioploca, 744
Thiosemicarbazones, 1110
Thiothrix, 703f
Thiotrichales, A-64t
Thiovulum, 748, A-64f, A-64t
Thrasher, Adrian J., 962
Threonine, 616
Threonine catabolism, 511
Threshold dose, **965**
Thrombocytopenia, 1067
Thrombotic thrombocytopenic purpura (TPP), 1067
Thylakoids, **112**, **575**, 577f
Thymidine monophosphate (CMP), 618
Thymine (T), 239, 243, 618, 679
Thyroid-stimulating hormone (TSH), 1000
Thyroxine (T4), 1000
Tight junctions, 420, 421f, **943**, 948, 1015–16
TIGR (The Institute for Genomic Research), 9f, 10
Tinoco, Ignacio, 298–99
Ti plasmid, 329, **657**, 1033
TipN protein, 108–10, 111f
Tissue culture of animal viruses, 222–23, 224
Tjarks, Larry, A-51f
TLR4 (Toll-like receptor), 1027
TmRNA, **285**
Tobacco mosaic disease, 25, 194
Tobacco mosaic virus (TMV), 25, 35, 193, 194, 200, 200f, 205t
Togaviridae, 1089, A-61t
Toggle switches (genetic), 480–81
Tolli, John D., 739
Toll/interleukin 1 receptor domain (TIR domain), 954
Toll-like receptors (TLRs), 952, 953, **954**, 955t
Toluene catabolism, 533
Tomography, **70**
cryo-electron tomography, 70–71
Tonegawa, Susumu, 978
Tonsils, 942, 944, 1018
Tooth decay, 929
Topoisomerases, 246f, **247–48**
antibiotic targeting of, 248, 1121
DNA gyrase, 247, 248, 252f, 256, 1122
Topo IV, 256, 257f
type II topoisomerases, 247–48, 256, 1120
type I topoisomerases, 246f, 247
TorA protein, 309
Total magnification, **54**
Toxic shock gene, 239
Toxic shock syndrome, 1012, 1019, 1052, 1056
Toxic shock syndrome toxin (TSST), 1052, 1056
Toxin-antitoxin (TA) DNA modules, eTopic 10.4
Toxoplasma gondii, 647t, 831, A-70t
ToxR membrane protein, 86
Trachoma, 752
TraI endonuclease (relaxase), 325, 326f
Trans-activating RNA (taRNA), 482–83, 484
Transamination, **613–14**, 615
Transcription, **238**, **277**. *See also* RNA polymerases
antibiotics that affect transcription, 280f, 282–84, 1123
in bacterial cell, 80–81f, 82, 102–3, 277
cap snatching, 430
coupled transcription and translation, 103, 300–301
elongation of RNA transcripts, 277, 281–82, 283
in eukaryotes, 103
initiation, 277–78, 280–81
modulation of speed, 300–301
NusA protein, 282
pause sites, 282
Rho-dependent transcription termination, 282
Rho-independent transcription termination, 282
ribonucleoside triphosphates (rNTPs), 280–81
RNA stem loop, 282
template strand, 277
termination of transcription, 277, 282
transcriptional control of gene regulation, 366–67, 369
transcripts, 277, 285–86
Transcriptional attenuation, 369, **378–80**
Transcriptional fusions (operon fusions), **461**, 462f
Transcription factor B (TFIIB), 762
Transcriptomes, 386–87, **401**, 404, 835, 1047, **A-54**
Transcriptomics, 401, 404, 848
Transcripts, **277**
Transcytosis, **1084**
Transdermal patch test for TB, 653
Transduction, **210–11**, **331–33**
double recombination, 339f, 340
generalized transduction, 331
plasmid transmission, 260
S. aureus, 1012
specialized transduction, 331, 332f, 333
transducing particles, 331, 332f, 333
Transertion, **103–4**
Transfection, **452**
Transfer RNA (tRNA), 76, **285**
acceptor end, 288f
aminoacyl-tRNA synthetases, 289, 290f
anticodon loop, 288f, 289
in archaea, 762
in bacterial cell, 82, 102–3
binding to ribosome, 292f
cloverleaf structure, 288–89
codon–anticodon pairing, 288, 295
DHU loop, 288f, 289
during elongation of peptide, 295–96
and genetic code, 286–89
modified bases, 285, 288, 289f
N-formylmethionyl-tRNA (fMet-tRNA), 293, 294
nonsense suppressors, 416
orientation in ribosome, 297f
pathogenicity/genomic islands near genes for, 359, 360f, 1011

secondary structure, A-15
three-dimensional structure, 288f
TPC (or TΨC) loop, 288f, 289
Transform (oncogenesis), 215f, 216, **218**
Transformasomes, **323–24**
Transformation, 264, **322–24**
Transformed-focus assay, **225**
Transgenes, **452**
Transglycosylases, **1118**
Transitions, **341**, 344f
Translation, **238**, **286**
aminoacyl-tRNA synthetases, 289
antibiotics that affect translation, 298–300, 1123–24
in bacterial cell, 80–81f, 82, 83, 102–3, 294–97
catalytic RNA, 679
coupled transcription and translation, 103, 300–301
elongation factors, 85f, 295–96
elongation of peptide, 295–96, 298–99
in eukaryotes, 103, 300
genetic code and tRNA molecules, 286–89
initiation factors (IF1, IF2, IF3), 294, 297
initiation of translation, 294
Protein Synthesis: An Epic on the Cellular Level, 35–36
release factors (RF1, RF2, RF3), 296–97
termination of translation, 296–97
tmRNA and protein tagging, 301, 302f
translational control of gene regulation, 369, 381
translocation, 295, 298–99
Translational control, 369, **381**
Translational fusions (gene fusions), **461**, 462f
Translocation (movement of protein between cell compartments), 308
Translocation (movement of ribosome along mRNA), **295**, 298–99
Transmembrane ion gradients, 84
Transmembrane proteins, **A-24**, A-37f
Transmission electron microscopy (TEM), **45**, 46f, **67**, 68, 69f, 199
Transovarial transmission, **1007**
Transpeptidases, 93, **1118**
Transplacental transfer, 1081
Transport across membranes, 87–89, A-28–A-29. *See also* Coupled transport; Transport proteins (transporters)
active transport, 89, 128–32
carbon-concentrating mechanism and, 596–97
against concentration gradient, 500f, 501
diffusion, A-28–A-29
facilitated diffusion, 127–28
membrane-permeant weak acids and bases, 87–88
osmosis, 87
passive diffusion, 87
passive transport, 89, 127–28
Transport proteins (transporters), 86f, 87, **88**, 89f, A-28
ABC transporters, 88, 89, 130–32, 309f, 310–11, 361
antiport transport systems (antiporters), 128–30, 170, 171f, 172, 551
CO_2 and HCO_3^- transporters, 596–97
porins, 98
symport transport systems (symporters), 128–30, 551
Transposable elements, **355–58**
and antibiotic resistance, 355, 356, 357f, 358
and gene duplication, 361
insertion sequences (ISs), 355–56
insertion symbol (::), 355
integrons, 358, eTopic 9.6
inverted repeats, 355f, 356, 357f
nonreplicative transposition, 356, 357f
replicative transposition, 356–57
resolvase, 356–57
target site duplication, 357f
tetracycline (Tc) resistance gene in P22 phage, 355
transposase, 355, 356, 357f, 358
transposition, definition, 356
transposition-mediated cointegrate formation, 356, 357f
transposons (*See* Transposons)
transpososome complexes, 356
Transposase, **355**, 356, 357f, 358
Transposition, definition, **356**
Transposons, **356–58**. *See also* Transposable elements
complex transposons, 356, 357f
composite transposons, 356, 357f
conjugative transposons, 321, 358
as genetic tools, 458–60, A-52
random transposon mutagenesis, A-52
retrotransposons, 442–43
Tn*3,* 356–57
Tn*5,* 358
Tn*10,* 356, 357f
Tn*916,* 358
Transversions, **341**
Travisano, Mike, 3
Trehalose, 162
Treponema, 751, 752, A-65t
Treponema pallidum, eTopic 13.3. *See also* Syphilis
genomic analysis of metabolism, eTopic 13.3
immunochromatographic assay, 1168f
loss of TCA cycle, 527, 532, 693
microscopy, 63f, 751f
Salvarsan, 1111
syphilis, 1078, 1111
T. pallidum subspecies *pertenue,* 1078
Triazoles, 806
Tricarboxylic acid (TCA) cycle, **527–32**, 529f
acetyl-CoA entry into TCA cycle, 527f, 528, 529f, 533
carbon cycle, 123f
^{14}C labeling, 529f, 530
catabolism pathways and, 511, 512, 513
citrate, 527, 528, 529f
discovery, 33, 527
fumarate, 508, 528, 529f
glyoxylate bypass, 531–32
isocitrate, 528, 529f, 531
Krebs pathway, 527
malate, 528, 529f, 531–32
and metabolite models of early life, 678–79
observing TCA cycle intermediates, 528–30
oxaloacetate, 527, 528, 529f, 530, 531–32, 588
and oxidative phosphorylation, 530–31
2-oxoglutarate (alpha-ketoglutarate), 528, 529f, 532, 588
substrates for biosynthesis from, 532, 587–88
succinate, 508, 529f, 530, 531
succinyl-CoA, 528, 529f, 530, 533, 587
Trichoderma, 698, 814
Trichodesmium, 722, 911, A-62t
Trichomonas, 1122
Trichomonas hominis, 932
Trichomonas vaginalis, 1082–83
Trichophyton, 814, A-68t
Trichophyton rubrum, 1005
Triclosan, 86, 185, 601, 833
Trigger factor, 304, 306f, 307, 308
Triglycerides, 511f, A-15–A-16
Triiodothyronine (T3), 1000
Trimethylamine, 560, 643
Trinitrotoluene (TNT) catabolism, 492, 511f
Tropheryma whipplei, 1046, 1175–76
Trophic levels, **850**, 863
Trophozoites, 836, 1073, 1082, 1092
Tropism, **213**
trpE gene, 378–79
TrpR (tryptophan synthesis regulator) protein, 378, 379f, 385
Trypanasoma brucei, 44, 314–15, 390, 833–35
Trypanasoma cruzi, 835, 1041, 1042
Trypanosomatidae, A-71t
Trypanosomes, 797, 805, **833–35**
differentiation among shapes, 834–35
kinetoplast, 833
life cycle, 834
PAD1 protein, 835
PAD2 protein, 835
parasitism, 833
surface protein variation, 834
Trypanosomiasis. *See* African sleeping sickness
Trypsin, 632
Tryptophan, 616
Tryptophan biosynthetic pathway, 135, 379f
Tryptophan (*trp*) operon, 367, 378–80
Tsetse fly, 314–15
Tsien, Roger, 61
Tuber aestivum (truffles), 625, 814

Tuberculosis (TB). *See also Mycobacterium tuberculosis*
antibiotic development, 653
antibiotic resistance, 108–9
biotechnology companies addressing TB, 652–53
case history, 1152
delayed-type hypersensitivity reaction, 1064
extensively drug-resistant tuberculosis (XDR-TB) strains, 1065
extrapulmonary TB, 1064
Ghon complexes, 1064
history, 14, 20–21
latent tuberculosis infection (LTBI), 1064–65, 1107
multidrug resistant (MDR) strains, 1064, 1065, 1107
as reemerging disease, 1064–65, 1181
secondary or reactivation TB, 1065
transdermal patch test, 652–53
treatments, 283, 557, 732, 1107, 1120, 1123
tuberculin skin test, 652–53, 998, 1062, 1064
vaccines, 652
Tube worms, 865, 866f
Tubulin, 106, 754
Tularemia, diagnosis, 1053
Tulip streaking from viral infection, 881
Tumor necrosis factor (TNF), **989**
Tumor necrosis factor alpha (TNF-α), **948**, 957
Tumpey, Terrence, 427f, 1139
Tungsten, 783, 791
Tungsten arc lamp, 60f
Turgor pressure, 85, 92, 806–7, A-30
Turner, Caroline, 3
Tus (terminus utilization substance), 256
Tuskegee experiment, 1079
Twitching motility, 116, **149**, 1015, 1029
Two-component signal transduction systems, **368**, 391
Two-dimensional polyacrylamide gel electrophoresis (2D PAGE or 2D gels), **84–85**, 404, A-46–A-48
Twort, Frederick William, 193
Tye, Bik-Kwoon, 355
Tyndall, John, 17
Type I hypersensitivity (immediate), **996–98**
Type II hypersensitivity, **999**, eTopic 24.4
Type III hypersensitivity, **999**, eTopic 24.4
Type IV hypersensitivity (DTH) (delayed-type), **998–99**
Type I pili, 341, **1013**, 1014f, 1015
Type III pili, **1013**
Type IV pili, **1013**, 1015–16, 1029, 1084
Type II secretion system, **1029**
Type III secretion system (T3SS), **1030–33**, 1070–71, 1096
Type IV secretion system, 1033
Type VI secretion systems (T6SSs), 1033, 1034–35, 1036f
Typhoid fever, 1066. *See also Salmonella enterica* serovar Typhi (S. Typhi)
Typhoid vaccine, 758
Typhus fever, 1006, 1041. *See also Rickettsia prowazekii*
Tyramides, A-55
Tyrosine, 616
Tyson, Gene, 845

U

UbiG-mccB-mccA operon, 388f
Ubiquinone, 552, 553, 554
Ubiquitin, 305, 1043–44
Ubiquitylation (ubiquitination), 1043–44, eTopic 8.3
Ugba, 636
Ulcerative colitis, 885, 934
Ultracentrifugation, A-44–A-45
Ultracentrifuge, **33–34**, **82–83**
Ultraviolet (UV) radiation, 182, 210, 345, 350, 676, 678
Ulva (sea lettuce), 819–20
Uncoating, 214f, **215**
Uncouplers, 550f, **551**
Unculturable microorganisms, 121–23
Uncultured microbes, **860–61**
Unsaturation, 603
Upstream processing, **657**, 658f
Uracil (U), 244, 351, 618, 679
Uranium, 561, 748, 915t
Ureaplasma urealyticum, 730
Urease, 172, 316–17f, 1045, 1046f
Urethritis, 64
Urey, Harold C., 18, 677
Uric acid, 626
Urinary tract infections (UTIs), 744, 934, 1018, 1075–76, 1151
Uroporphyrinogen III, 619
U.S. Centers for Disease Control (CDC), 644
U.S. Department of Energy, 561
U.S. Public Health Service, 646, 647t
Ustilago maydis, A-68t

V

Vaccination, **23**, **966**. *See also* Immunization
primary *vs.* secondary response, 974–75
recommended immunization schedule, 1103t
smallpox vaccination with vaccinia, 23, 966
Vaccines. *See also* Immunization
anthrax vaccine, 1177
attenuated pathogens in vaccines, 23, 966, 981
delivery by haloarchaea, 758, 787, 790
DNA vaccines, 652, eTopic 12.3
DTaP (diphtheria, tetanus, acellular pertussis), 1087
edible vaccines (fruits and vegetables), 475
Gardasil vaccine, 198, 216, 1102
hepatitis A (HepA), 1099
hepatitis B (HepB), 1099
HIV vaccine development challenges, 434
influenza vaccines, 1102, 1138
killed-organism vaccines, 981
measles, mumps, and rubella (MMR) vaccine, 1057
meningitis vaccines, 1084, 1088t
microbes as vaccine delivery systems, eTopic 23.1
multivalent vaccines, 1062
pneumococcal polysaccharide vaccine (PPSV), 1062
polio vaccine, 36, 193, 981
side effects of vaccines, 1102
tuberculosis vaccines, 652
vaccine antigens produced in plants, 475–76
vaccines against viral and bacterial pathogens, 967f
Vaccinia virus (cowpox), 202, 443, 444t, 966
Vacuolating cytotoxin (VacA), 1046
Vacutainer anaerobic specimen container, 1149f
Vaginal microbiota, 934, 935
Valentine, David, 858–59
Valine, 613, 614, 616
Valinomycin, 549
Vampirella, 872
VAMP protein, 1087
Vanadium, 610, 915t
Vancini, Ricardo, 191
Vancomycin, **1120–21**
biosynthesis, 618, eTopic 15.3
peptidoglycan as target of, 79, 93f, 713
resistant strains, 1119, 1120–21, 1131
targeting of cell wall biosynthesis, 1116f, 1120–21
use against *Clostridium difficile,* 618
use against methicillin-resistant *S. aureus* (MRSA), 1056, 1108
use against *Streptococcus pneumoniae,* 1127
Vancomycin-resistant *S. aureus* (VRSA), 1119
Van der Waals forces, 72, **A-5**
Vanillin, 872
van Niel, Cornelius B., 841
van Niel hypotheses, 841, 849
Variable regions, 970f, **971**
Variant Creutzfeldt-Jakob disease (vCJD), 1090
Variant surface glycoprotein (VSG), 834
Varicella-zoster virus (VZV), 206, 207, 444. *See also* Chickenpox
Varicella-zoster virus 1, 8t
Varicosaviridae, A-60t
Varicosaviruses, 205t
Variola major virus, 966f
Varroa destructor, 268
Vasoactive factors, **947–48**, 949f
Vasoconstriction, 958
Vasodilation, 947–48, 958
Vasopressors, 1057f
VCAM-1 (vascular cell adhesion molecule 1), 948, 949f
Vectorial metabolism, 548
Vectors, **1007**, 1008f
Vegetations, definition, 930

Vegetative cells, 75, **724**
Vegetative mycelia, **730**
Venter, Craig, 10, 847, 861
Vent polymerases, 790
Verenium Corporation, 698–99
Verrucomicrobia, 713f, 715t, **717–18**, 754, A-65t
Verrucomicrobiales, A-65t
Verrucomicrobium, A-65t
Vertical gene transfer, **690–91**
Vertical transmission (genetics), **238**
Vertical transmission (infection), **1007**, 1008f, 1009
Very Large Array (VLA) dishes, 576f
Vesicles, 113f, 807, **A-30**
Vesicular-arbuscular mycorrhizae (VAM), **874**
Vesicular stomatitis virus (VSV), 450, 451f, 452f
Vetter, David Philip ("bubble boy"), 962
Viable bacteria, **134**, 137
Viable but nonculturable (VBNC), **134, 860**
Vibrio, 1017
Vibrio anguillarum, 400
Vibrio cholerae. See also Cholera
  bacteriophage CTXϕ, 200, 210
  biofilm, 5, 860f
  glutamine synthase gene (*gltB*), 315
  multiplex analysis, A-55f
  RTX toxin, 1019
  salt-tolerance, 857
  sari cloth filtration, 856
  sodium pump, 550, 559
  stomach acidity and susceptibility to, 931
  symbiosis with copepods, 856
  T6SS mechanism, 1034–35, 1036f
  toxin production, 365
  ToxR membrane protein, 86
  type IV pili, 1015
  uncultured cells, 860f
Vibrio fischeri, 28, 478. *See also Aliivibrio fischeri*
Vibrio harveyi, 399–400
Vibrionales, A-64t
Vibrio parahaemolyticus, A-55f
Vibrio vulnificus, 647t, 926, A-55f
Vif accessory protein, 437t, 440
Viral ecology, 192, 226–30
  background and overview, 226
  emergence of viral pathogens, 227–28
  viral roles in ecosystems, 228–30
Viral genomes and classification, 193, 195–96, 198, 203–6
  Baltimore virus classification, 203–6
  diversity, 409
  DNA viruses, 443, 444t
  International Committee on Taxonomy of Viruses (ICTV), 203
  retroviruses, 432
  segmented genome, 203, 206, 424, 426
  (-) strand RNA virus species, 424
  (+) strand RNA virus species, 417t
Viremia, 1057, **1091**
Virgin, Herbert, 446, 448
Viridans streptococci, 1091
Viridiplantae, 800, 801f, 802–3, A-69t
Virions, **192**, 194, 198
Viroids, 25, 191, **196–97**
Virome, **928**
Vir operon, 657, 659f
Virulence, **222**, **1006**, 1007f
Virulence factors, **1010**
Virulence genes, 359, 1010–12, eTopic 25.1, eTopic 25.2
Virulence signals, eTopic 5.3
Virulent phages, 210
Viruses, **8**, **191–230**. *See also* Animal viruses; Bacteriophages (phages); Retroviruses; Virus structure; *specific viruses*
  acute viruses, 228
  and animal disease, 194
  antiviral drugs, 194
  archaeal viruses, 770–71
  background and overview, 191–92
  caterpillar viruses produce commercial products, eTopic 16.2
  cell-to-cell transmission, 420, 442
  classification, 203–4, 205t, 206–7, A-60t–A-61t
  computer viruses, 195
  culturing in laboratory, 192, 221–25
  definition, 192
  discovery, 25, 192
  DNA viruses, 25, 193, 195, 203, 204f, 205t, A-60t
  filtration and, 181, 192
  gene therapy with, 194, 218, 450–53
  genetic resistance to, 211, 220
  genome differences, 415
  genomes, 193, 195–96, 198, 203–6
  highly virulent viruses, 1006, 1007f
  host gene pickup, 415
  host range, 194, 203, 227–28
  and human disease, 191–92, 193, 194, 200
  insect transmission, 191–92, 218, 226, 228
  irradiation resistance, 182
  marine food chain infection, 229–30
  mutualism with fungi and plants, 226–27
  mutualism with insects, 229
  oncogenic viruses, 216, 218, 225
  oncolytic viruses, 453
  origins of, 196, eTopic 6.1
  pararetroviruses, 204f, 205t, 206, 219–20
  persistent viral infections, 228–29, 446, 448–49
  phylogeny, 206–7
  plant viruses, 193, 194, 196, 218–21
  plaques, 194
  propagation, 195
  ribozymes, 196–97
  RNA viruses, 25, 35, 192, 193, 195, 203–6, A-60t–A-61t
  size, 8, 25, 203
  slow viruses, 206
  structure, 70, 198–202, 203
  as tools and model systems for molecular biology, 192
  tumor-causing, 193
  viral diseases causing skin rashes, 1055t, 1057–58
  viral diseases of the lung, 1060t–1061t, 1065
  viral gene transfer, 193
  viremia, 1057
  virions, 192, 194, 196–97, 198
  viroids, 25, 191, 196–97
  virulence, 222
  virus infections, 192, 193–94
  virus-to-bacteria ratio, 229–30
Virus-host mutualism, 226–27, 229
Virus structure, 70, 198–202, 203
  accessory proteins, 202
  capsids, 25f, 191, 192, 194, 198–202, 209
  complex and asymmetrical virus particles, 201–2
  core envelope, 202
  envelopes, 192, 198, 199–200
  filamentous virus, 200–201
  helical capsids, 200–201
  icosahedral capsids, 198–99
  outer membrane, 202f
  spike proteins, 199
  symmetrical virus particles, 198–201
  tegument, 199
Virus-to-bacteria ratio, 230
Vitamin B_{12}
  biosynthesis, 6, 587, 618, 620f
  cobalt, 120
  E. coli production of, 933
  in fermented foods, 627, 632
  transport across membrane, 88, 89f, 131
  yeasts as source of, 626
Vitamin B_{12}-like cofactor TB, 599
Vitamin B_{12} transporter, 88, 89f, 131
Vitis vinifera, 639
Voges-Proskauer test, 744
Vollmer, Waldemar, 235
Volta, Alessandro, 775
Voltage potential, 27
Volvox, 820–21
von Tappeiner, Hermann, 775
Vorticella, 829, A-70t
Vpr accessory protein, 437t, 440
Vpu accessory protein, 437t, 442

W

Wächterhäuser, Günter, 678
Waksman, Selman, 1110
Wall pressure, A-30
Warren, Robin, 22, 747, 1071–72
Washout, 145
Wastewater treatment, **899–901**
  anammox reaction (anaerobic ammonium oxidation), 565–66, 908
  filamentous bacteria, 702, 829, 900
  filamentous methanogens, 778–79, 901
  flocs of biofilm, 900
  magnetotactic bacteria, 113
  metal contamination and bioremediation, eTopic 22.2

Wastewater treatment (*continued*)
- methane production from acetate, 783, 901
- microbes used, 900–901
- microbial dissimilation, 849–50
- nitrifier bacteria, 742
- predators of bacteria, 901
- preliminary treatment, 899, 900f
- primary treatment, 899, 900f
- secondary treatment, 899–900
- sludge, 899
- syntrophy, 502–3
- tertiary treatment, 900
- weapons waste bioremediation, 902
- wetland filtration, 901

Water, A-4–A-5
- biochemical composition of bacteria, 84
- passage across membranes, 87
- transport across cell membrane, A-29–A-30

Water activity (a_w), **166**, 167
Water molds, 798, 802, 805, 817, A-70t. *See also* Oomycetes
Water table, **871**
Watrous, Jeramie, 585
Watson, James, 34–35
Wave interference. *See* Interference
Wavelength (λ), 47, 51, 52, 60–61
Weak acids, 88
Weak bases, 88
Web biology, 316
Weber, Peter, 75
Weiss, Gabriel, 36
Weiss, Jackie Bennington, 36
Weitzmann, Chaim, 525
Wellcome Trust, 36
Weller, Thomas J., 222
Wescodyne, 184
Western blots, **464–65**, 970, **A-58**
West Nile virus, 194, 206, 227, 417, 1007, 1161, eTopic 6.2
Wetlands, 780, **875–76**, 901
Wet mount preparation, **54**
Wheat, 634, 638, 731
Whey, **630**, 632f
Whipple, George, 1175
Whipple's disease, 1046, 1175–76
White blood cell (WBC) differential, **940**, 942t
White blood cells, 937–39. *See also* Immune system
- basophils, 938, 939
- dendritic cells, 939, 940, 941f
- development of white blood cells, 937f
- eosinophils, 937f, 938
- lymphocytes, 937f, 939, 940–42
- macrophages, 177, 939, 941–42, 944
- mast cells, 938–39
- monocytes, 937f, 939
- neutrophils, 937–38, 940, 941f, 946
- phagocytes, 937–38, 947f, 949, 951–53
- polymorphonuclear leukocytes (PMNs), 937
- pseudopods, 825, 937
- size and morphology, 936f
- types of white blood cells, 937f

Whitehead, Emily, 450, 451f
White oak leaf spotted with anthracnose fungus, 881f
White rot fungi, 872, 873f
White sulfur bacteria, 909f, 910
Whittaker, Robert, 30, 800
Whole-genome DNA-binding analysis (ChIP-on-chip), 471–72
Whole-genome shotgun (WGS) sequencing methods, 271
Whooping cough, 1024, 1033, 1058, 1102. *See also Bordetella pertussis*
Wilkins, Maurice, 34–35
Wilson, Kent, 36f
Wine production, 624, 627, 628t, 629, 639–40, 641f
Winogradsky, Sergei, 26–27, 841
Winogradsky column, **26–27**
Woese, Carl, 31, 32f, 669, 681, 684f
Wolbachia, 270, 705
Wolbachia pipientis, 330, 704
Wong-Staal, Flossie, 435f, 436
Woodchuck hepatitis virus (WHV), 452
World Health Organization (WHO), 1171, 1173, 1180
Wozniak, Dan, 488–89
Wright stain, 1093
Wuchereria bancrofti, 1005, 1006f
Wybutosine (yW), 288, 289f

X

Xanthomonas, 746, 881
Xanthomonas campestris, 467
Xanthophyceae, A-70t
Xanthophyceae (yellow-green algae), 823
Xenogeneic silencing, 992
XerC and XerD, 257
Xeroderma pigmentosum, 249, 354
X-ray crystallography, 25, 34, **45**, 46f, **73**, 76f, 83
X-ray diffraction analysis, **73–76**
Xylanase, 696, 698–99
Xylobolus frustulatus, 873f
Xylulose 5-phosphate, 594

Y

Yamamori, Tetsuo, 164
Yanofsky, Charles, 378, 380f, 616
Yaws, 1078
YbtA regulator protein, 376
Yeast extract, 134–35
Yeasts, **807–9**, A-68t. *See also Saccharomyces cerevisiae*
- alcoholic beverages, 120, 628t, 629, 640, 641f
- alternation of generations, 808f, 809
- bread leavening, 627, 628t, 629, 637–39
- cocoa fermentation, 635
- edible yeasts, 626
- as model system, 808f, 809, 810–11
- oral microbiota, 928
- reproductive cycles, 809

Yeast two-hybrid system, **472–73**
Yellow fever, 1007, 1008f, 1009
Yellow fluorescent protein (YFP), 104, 105f, 462, 463f
Yellowstone National Park
- acidobacteria, 749
- archaea in hot springs, 31f, 766, 770, 771
- bioprospecting, 654–55
- boiling sulfur springs, 31
- *Candidatus* Korarchaeum cryptofilum, 793
- hot springs, 31f, 35, 89, 162, 655, 766
- microbes in acidic springs, 170f
- microbes in sulfur springs, 31, 170f
- microbial mats, 148f, 714, 716f
- phylogenetic tree of Obsidian Pool thermophiles, 683
- and polymerase chain reaction (PCR), 35, 162
- *Sulfolobus,* 170f
- Sulfur Caldron, 170f
- thermotolerant plants, 226–27
- *Thermus aquaticus,* 35, 162

Yersinia, 1017, 1030, 1041, A-64t
Yersinia enterocolitica, 181, 991, 1042, 1053
Yersinia pestis. See also Bubonic plague
- bipolar staining, 1094f
- history, 11, 29
- on phylogenetic tree of Gammaproteobacteia, 684f
- as potential bioweapon, 1177
- resistance in herpes-infected mice, 448
- sodium pump in, 559
- YbtA regulator protein in, 376

Yersinia pseudotuberculosis, 478, 1042
YFP (yellow fluorescent protein), 104, 105f
Yogurt, 627, 628t, 629, **630**, 727
Yura, Takashi, 164

Z

Zanamivir (Relenza), 1138, 1139f
Zernike, Frits, 65
Ziehl-Neelsen stain, 1152. *See also* Acid-fast stain
Zinc, 120
Zinder, Norton, 331
Zinkernagel, Rolf, 986
Zinser, Erik, 861
Zobellia, 514–15f, 516
Zone of inhibition, **1113**
Zones of hypoxia, **898**, 907
Zoogloea, 900
Zoonotic disease, 1046, **1053–54**
Zooplankton, 804
Zoosporangia, 810, 811, 812f
Zoospores, **801–2**, 809–11, 812f, 819f, 820
Zooxanthellae, A-70t
Zooxanthellae (singular, zooxanthella), **831**, **882–83**
Z-ring, 100, 106–7
Zygomycetes (Zygomycota), **812**, 813f
Zygomycota, A-69t
Zygospores, **812**, 813f
Zymomonas, 520

Replication of influenza virus.

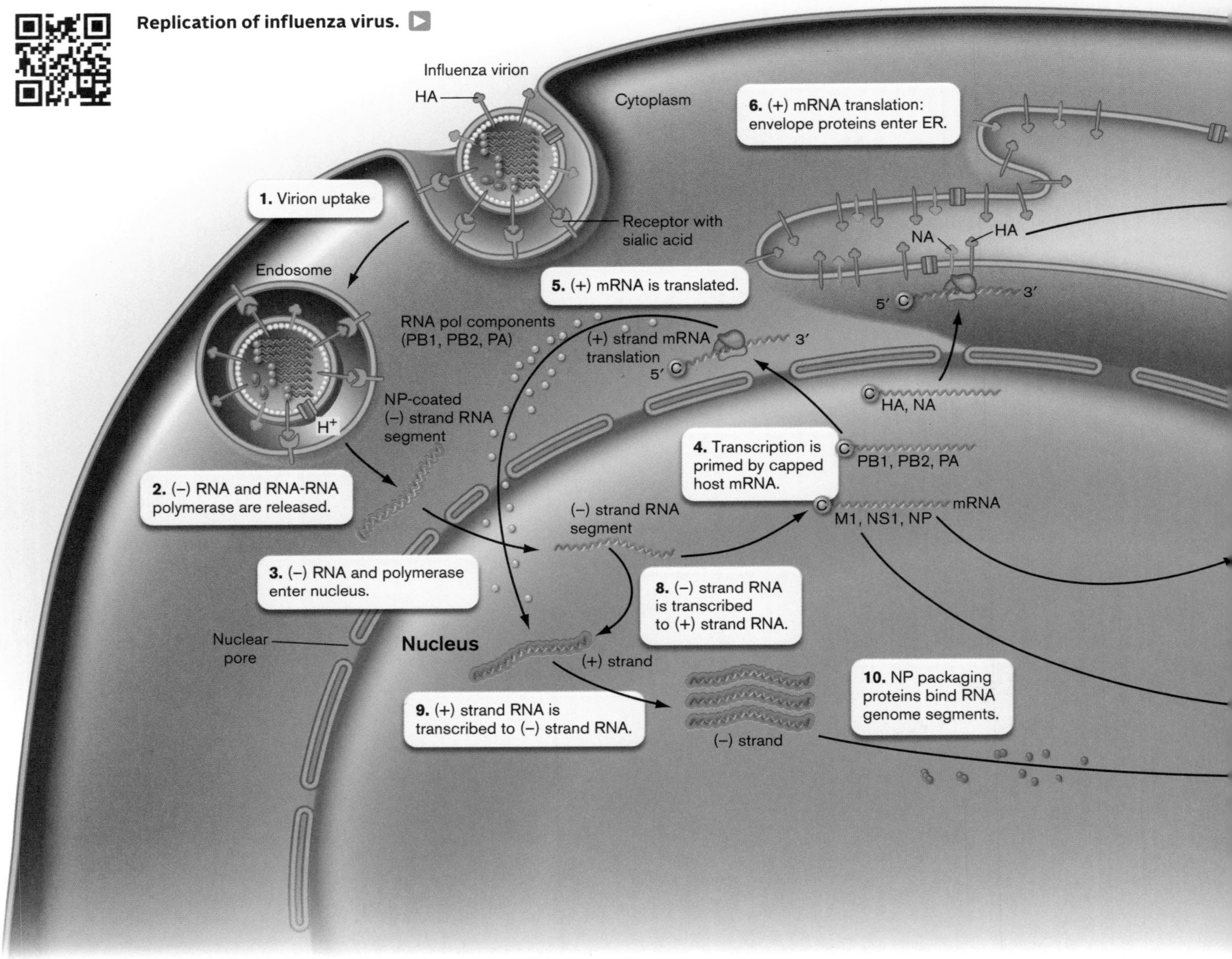
Influenza virion
HA
Cytoplasm
6. (+) mRNA translation: envelope proteins enter ER.
1. Virion uptake
Receptor with sialic acid
NA
HA
Endosome
5. (+) mRNA is translated.
5′
3′
RNA pol components (PB1, PB2, PA)
(+) strand mRNA translation
5′
3′
HA, NA
NP-coated (–) strand RNA segment
H+
4. Transcription is primed by capped host mRNA.
PB1, PB2, PA
2. (–) RNA and RNA-RNA polymerase are released.
(–) strand RNA segment
mRNA
M1, NS1, NP
3. (–) RNA and polymerase enter nucleus.
8. (–) strand RNA is transcribed to (+) strand RNA.
Nuclear pore
Nucleus
(+) strand
10. NP packaging proteins bind RNA genome segments.
9. (+) strand RNA is transcribed to (–) strand RNA.
(–) strand